D1087274

THE

COMPUTER SCIENCE AND ENGINEERING

HANDBOOK

T H E

COMPUTER SCIENCE AND ENGINEERING

H A N D B O O K

EDITOR-IN-CHIEF

ALLEN B. TUCKER, JR.

 **CRC PRESS**

A CRC Handbook Published in Cooperation with ACM, The Association for Computing

Cover image: A fractal terrain model ©1992 by F. Kenton Musgrave. Used by permission.

Acquiring Editors: *Jerry Papke and Bob Stern*
Cover Designer: *Denise Craig*
Direct Marketing Manager: *Becky McEldowney*
Marketing Manager, Marketing Management: *Susie Carlisle*
Senior Project Editor: *Carol Whitehead*
Manufacturing Assistant: *Sheri Schwartz*

Library of Congress Cataloging-in-Publication Data

The computer science and engineering handbook / edited by Allen B. Tucker, Jr.
    p.   cm.
  Includes bibliographical references and index.
  ISBN 0-8493-2909-4 (alk. paper)
    1. Computer science--Handbooks, manuals, etc.   2. Engineering--Handbooks, manuals, etc.
  I. Tucker, Allen B.
  QA76.C57315   1996
  004--dc20   96-20442
                                                     CIP

Preface

Purpose

The purpose of *The Computer Science and Engineering Handbook* is to provide a single substantive and comprehensive reference for practicing computer scientists and engineers, as well as other professionals with strong computing interests. Its goal is to provide the most current information in each of the following 10 major subfields of computer science and engineering (CS&E): algorithms and data structures, architecture, artificial intelligence and robotics, computational science, database and information retrieval, graphics, human–computer interaction, operating systems and networks, programming languages, and software engineering.

In addition, the appendices provide useful information about the major professional organizations for computer science and engineering, standards, programming languages, and other contemporary topics. Different points of access to this rich collection of theory and practice are provided through the table of contents, two introductory chapters, and four different indexes: a comprehensive subject index; an index of key algorithms and equations; an index of figures and tables; and an index of cited authors in the field.

A more complete overview of the Handbook's contents can be found in Chapter 1, which summarizes both the rapid evolution of CS&E throughout the last 50 years and the salient features of the 10 subfields just listed (and hence the key topics in each of the 10 major sections of this Handbook).

Acknowledgments

A work of this magnitude cannot be completed without the efforts of many individuals working toward a common goal. Throughout the last two years, it has been my pleasure to know and work with 10 very distinguished, talented, and dedicated editorial advisors, whose collective insights and hard work have combined to create this Handbook. I thank each of them for providing substantial and varied contributions throughout the Handbook's two-year development period.

Sincerest thanks are also due to the more than 150 authors, all experts in their subfields of CS&E, who have developed and refined the individual chapters that make up this Handbook. Some are close friends and colleagues, others are new acquaintances, and still others are former advisors and students whose work and mine have crossed paths over several years. The quality of this Handbook is a direct reflection of the quality of their technical insights and expositions. The names of all the editorial advisors and authors are listed separately in the next few pages.

Bowdoin College deserves credit for providing the institutional, academic, and computing support for this work. My personal thanks are due to Matthew Jacobson-Carroll, Alice Morrow, and Aaron Olmstead for the editorial and technical assistance they provided at different stages of this project.

CRC Press deserves recognition for its support and guidance throughout the many phases of this Handbook's development. Thanks are due to Bob Stern for not only conceiving this project but also providing the editorial guidance and personal support for turning a good idea into a real product. Thanks are also due to Jerry Papke and Ron Powers for seeing it through the production process, and Nora Konopka and Carol Whitehead for their day-to-day care and feeding of all of the details that need attention throughout the development life of this project.

Finally, as always, I owe a great deal of gratitude to my wife Meg. I'll not forget all the weekends we spent together in our office at home, listening to "Car Talk" while editing chapter manuscripts for this Handbook. Supporting a spouse in a project like this is a true act of love and generosity, and hers is abundant beyond measure.

Allen B. Tucker, Jr.
Brunswick, Maine

Editor-in-Chief

Allen B. Tucker, Jr. is Professor of Computer Science at Bowdoin College. He has held similar positions at Colgate University and Georgetown University. He served 16 years as department chair in these institutions and 2 years as Associate Dean of the Faculty. He held the John D. and Catherine T. MacArthur Chair in Computer Science while at Colgate from 1983 to 1988.

Professor Tucker earned a B.A. in mathematics from Wesleyan University in 1963 and an M.S. and Ph.D. in computer science from Northwestern University in 1970. He is the author or coauthor of numerous books and articles in the areas of programming languages, natural language processing, and computer science education. He has given many invited talks, panel and workshop presentations, and has served as a reviewer for various journals, NSF grant programs, and curriculum projects. He has also served as a consultant to various colleges and universities, as well as private and public institutions, in several areas of computer science curriculum, software design, programming languages, and natural language processing applications.

A Fellow of the ACM, Professor Tucker co-chaired the ACM/IEEE-CS Joint Curriculum Task Force that developed the report Computing Curricula 1991, for which he received the ACM's Outstanding Contribution Award and shared the IEEE's Meritorious Service Award in 1991. He is a member of the ACM, the IEEE Computer Society, CPSR, and the Liberal Arts Computer Science Consortium.

Editor-in-Chief

Advisory Board

Contributors

James L. Alty
Loughborough University
Leicester, United Kingdom

Thomas E. Anderson
University of California
Berkeley, California

John K. Antonio
Texas Tech University
Lubbock, Texas

Anthony A. Apodaca
Pixar
Richmond, California

M. Pauline Baker
National Center for
   Supercomputing Applications
University of Illionis
Urbana, Illinois

François Bancilhon
O_2 Technology
Versailles, France

Steven Bellovin
AT&T Bell Laboratories
Murray Hill, New Jersey

Andrew P. Bernat
University of Texas at
   El Paso
El Paso, Texas

Brian N. Bershad
University of Washington
Seattle, Washington

Christopher M. Bishop
Aston University
Birmingham, United Kingdom

Guy E. Blelloch
Carnegie Mellon University
Pittsburgh, Pennsylvania

Jonathan P. Bowen
The University of Reading
Reading, England

Ruven Brooks
Schlumberger Austin Product
   Center
Austin, Texas

Kim Bruce
Williams College
Williamstown, Massachusetts

Steve Bryson
MRJ, Inc./NASA Ames Research
   Center
Moffett Field, California

Douglas Burger
University of Wisconsin–Madison
Madison, Wisconsin

Colleen Bushell
National Center for
   Supercomputing Applications
University of Illinois
Urbana, Illinois

William L. Bynum
College of William and Mary
Williamsburg, Virginia

Bryan Cantrill
Brown University
Providence, Rhode Island

Luca Cardelli
Digital Equipment Corporation
   Systems Research Center
Palo Alto, California

David A. Caughey
Sibley School of Mechanical and
   Aerospace Engineering
Cornell University
Ithaca, New York

Stefano Ceri
Politecnico di Milano
Milan, Italy

Vijay Chandru
Indian Institute of Science
Bangalore, India

Steve J. Chapin
University of Virginia
Charlottesville, Virginia

Pasquale Cinnella
Mississippi State University
Mississippi State, Mississippi

Jacques Cohen
Brandeis University
Waltham, Massachusetts

Carey F. Cox
Mississippi State University
Mississippi State, Mississippi

Alan B. Craig
National Center for
  Supercomputing Applications
University of Illinois
Urbana, Illinois

Maxime Crochemore
University of Marne-la-Vallée
Noisy Le Grand, France

Robert D. Cupper
Allegheny College
Meadville, Pennsylvania

Thomas Dean
Brown University
Providence, Rhode Island

Gerald DeJong
University of Illinois at
  Urbana–Champaign
Urbana, Illinois

Steven A. Demurjian, Sr.
The University of Connecticut
Storrs, Connecticut

Peter J. Denning
George Mason University
Fairfax, Virginia

Henry G. Dietz
Purdue University
West Lafayette, Indiana

T. W. Doeppner, Jr.
Brown University
Providence, Rhode Island

Wolfgang Dzida
GMD—German National
  Research Center for Information
  Technology
Sankt Augustin, Germany

Angel Díaz
Rensselaer Polytechnic Institute
Troy, New York

David S. Ebert
University of Maryland–Baltimore
  County
Baltimore, Maryland

Raimund Ege
Florida International University
Miami, Florida

James Feldman
Northeastern University
Boston, Massachusetts

David Ferbrache
Defence Research Agency
Malvern, Worcester
United Kingdom

Raphael Finkel
University of Kentucky
Lexington, Kentucky

M. Fitzgerald
University of Texas at Arlington
Fort Worth, Texas

Michael J. Flynn
Stanford University
Stanford, California

Kenneth D. Forbus
Northwestern University
Evanston, Illinois

Stephanie Forrest
University of New Mexico
Albuquerque, New Mexico

Michael J. Franklin
University of Maryland
College Park, Maryland

John D. Gannon
University of Maryland
College Park, Maryland

Carlo Ghezzi
Politecnico di Milano
Milan, Italy

Benjamin Goldberg
New York University
New York, New York

James R. Goodman
University of Wisconsin–Madison
Madison, Wisconsin

Jonathan Grudin
University of California
Irvine, California

Michael G. Hinchey
New Jersey Institute of
  Technology
Newark, New Jersey and
University of Limerick
Limerick, Ireland

Stuart Hirshfield
Hamilton College
Clinton, New York

Lee Hollaar
University of Utah
Salt Lake City, Utah

Donald H. House
Texas A&M University
College Station, Texas

Windsor W. Hsu
University of California
Berkeley, California

Daniel Huttenlocher
Cornell University and Xerox Palo
  Alto Research Center
Ithaca, New York and
Palo Alto, California

Yannis E. Ioannidis
University of Wisconsin–Madison
Madison, Wisconsin

Robert J. K. Jacob
Tufts University
Medford, Massachusetts

Sushil Jajodia
George Mason University
Fairfax, Virginia

Mehdi Jazayeri
Technische Universität Wien
Vienna, Austria

Tao Jiang
McMaster University
Hamilton, Ontario
Canada

Michael J. Jipping
Hope College
Holland, Michigan

Deborah G. Johnson
Rensselaer Polytechnic Institute
Troy, New York

Johndan Johnson-Eilola
Purdue University
West Lafayette, Indiana

Michael I. Jordan
Massachusetts Institute of
  Technology
Cambridge, Massachusetts

David R. Kaeli
Northeastern University
Boston, Massachusetts

Erich Kaltofen
North Carolina State University
Raleigh, North Carolina

Subbarao Kambhampati
Arizona State University
Tempe, Arizona

Lakshmi Kantha
Naval Research Laboratory
Stennis Space Center
Mississippi and
University of Colorado
Boulder, Colorado

Arie E. Kaufman
State University of New York
Stony Brook, New York

Samir Khuller
University of Maryland
College Park, Maryland

David Kieras
University of Michigan
Ann Arbor, Michigan

David T. Kingsbury
Johns Hopkins University School
of Medicine
Baltimore, Maryland

Danny Kopec
Richard Stockton College
Pomona, New Jersey

Henry F. Korth
Bell Laboratories
Lucent Technologies, Inc.
Murray Hill, New Jersey

Carl E. Landwehr
Naval Research Laboratory
Washington, D.C.

Andrea S. LaPaugh
Princeton University
Princeton, New Jersey

Edward D. Lazowska
University of Washington
Seattle, Washington

Thierry Lecroq
University of Rouen
Mount-Saint-Aignan, France

D. T. Lee
Northwestern University
Evanston, Illinois

Henry M. Levy
University of Washington
Seattle, Washington

F. L. Lewis
University of Texas at Arlington
Fort Worth, Texas

Ming Li
University of Waterloo
Waterloo, Ontario
Canada

Kai Liu
University of Texas at Arlington
Fort Worth, Texas

Kenneth C. Louden
San Jose State University
San Jose, California

Michael C. Loui
University of Illinois at
    Urbana–Champaign
Urbana, Illinois

James J. Lu
Bucknell University
Lewisburg, Pennsylvania

Bruce M. Maggs
Carnegie Mellon University
Pittsburgh, Pennsylvania

Kavi Mahesh
New Mexico State University
Las Cruces, New Mexico

Dino Mandrioli
Politecnico di Milano
Milan, Italy

M. Lynne Markus
The Claremont Graduate School
Claremont, California

T. A. Marsland
University of Alberta
Edmonton, Alberta
Canada

Marshall Kirk McKusick
Marshall K. McKusick
    Consultancy
Berkeley, California

Brad Mehlenbacher
North Carolina State University
Raleigh, North Carolina

Jim Melton
Sybase, Inc.
Sandy, Utah

Keith Miller
University of Illinois at Springfield
Springfield, Illinois

Stuart Mort
Defence Research Agency
Malvern, Worcester
United Kingdom

Rajeev Motwani
Stanford University
Stanford, California

Sape J. Mullender
University of Twente
Enschede, The Netherlands

Brad A. Myers
Carnegie Mellon University
Pittsburgh, Pennsylvania

Peter G. Neumann
SRI International
Menlo Park, California

Jakob Nielsen
Sun Microsystems
Mountain View, California

Sergei Nirenburg
New Mexico State University
Las Cruces, New Mexico

Robert E. Noonan
College of William and Mary
Williamsburg, Virginia

Ahmed K. Noor
University of Virginia
Charlottesville, Virginia and
NASA Langley Research Center
Hampton, Virginia

Patrick O'Neil
University of
    Massachusetts–Boston
Boston, Massachusetts

M. Tamer Özsu
University of Alberta
Edmonton, Alberta
Canada

Victor Pan
Lehman College
City University of New York
Bronx, New York

Judea Pearl
University of California
Los Angeles, California

Jih-Kwon Peir
University of Florida
Gainesville, Florida

Radia Perlman
Novell, Inc.
Acton, Massachusetts

Steve Piacsek
Naval Research Laboratory
Stennis Space Center
Mississippi

Benjamin C. Pierce
Indiana University
Bloomington, Indiana

Roger S. Pressman
R. S. Pressman & Associates Inc.
Orange, Connecticut

J. R. Quinlan
University of Sydney
Sydney, Australia

Balaji Raghavachari
University of Texas at Dallas
Richardson, Texas

Prabhakar Raghavan
IBM Almaden Research Center
San Jose, California

Raghu Ramakrishnan
University of Wisconsin–Madison
Madison, Wisconsin

M. R. Rao
Indian Institute of Management
Bangalore, India

Bala Ravikumar
University of Rhode Island
Kingston, Rhode Island

Brian Reid
Digital Equipment Corporation
Palo Alto, California

Edward M. Reingold
University of Illinois at
Urbana–Champaign
Urbana, Illinois

Steven P. Reiss
Brown University
Providence, Rhode Island

Ellen Riloff
University of Utah
Salt Lake City, Utah

Alyn P. Rockwood
Arizona State University
Tempe, Arizona

Erik Rosenthal
University of New Haven
West Haven, Connecticut

Paul W. Ross
Millersville University
Millersville, Pennsylvania

Mary Beth Rosson
Virginia Polytechnic Institute and
State University
Blacksburg, Virginia

Kevin Rudd
Stanford University
Stanford, California

Betty Salzberg
Northeastern University
Boston, Massachusetts

Pierangela Samarati
Universitá degli Studi di Milano
Milan, Italy

Ravi S. Sandhu
George Mason University
Fairfax, Virginia

Stephen R. Schach
Vanderbilt University
Nashville, Tennessee

David A. Schmidt
Kansas State University
Manhattan, Kansas

Stuart A. Selber
Texas Tech University
Lubbock, Texas

Stephanie Seneff
Massachusetts Institute of
Technology
Cambridge, Massachusetts

J. S. Shang
Wright Laboratory
Wright-Patterson Air Force Base
Ohio

Dennis Shasha
New York University and
AT&T Bell Laboratories
New York, New York

William R. Sherman
National Center for
Supercomputing Applications
University of Illinois
Urbana, Illinois

Howard Jay Siegel
Purdue University
West Lafayette, Indiana

Avi Silberschatz
Bell Laboratories
Lucent Technologies, Inc.
Murray Hill, New Jersey

Gurindar S. Sohi
University of Wisconsin–Madison
Madison, Wisconsin

Ian Sommerville
Lancaster University
Lancaster, United Kingdom

Bharat K. Soni
Mississippi State University
Mississippi State, Mississippi

William Stallings
Consultant and Writer
Brewster, Massachusetts

John A. Stankovic
University of Massachusetts
Amherst, Massachusetts

Stephen Stuart
Digital Equipment Corporation
Palo Alto, California

S. Sudarshan
Indian Institute of
Technology–Bombay
Bombay, India

Earl E. Swartzlander, Jr.
University of Texas at Austin
Austin, Texas

Roberto Tamassia
Brown University
Providence, Rhode Island

Daniel Thalmann
Swiss Federal Institute of
Technology (EPFL)
Lausanne, Switzerland

Nadia Magnenat Thalmann
University of Geneva
Geneva, Switzerland

Allen B. Tucker, Jr.
Bowdoin College
Brunswick, Maine

Patrick Valduriez
INRIA, Rocquencourt
Chesnay, France

Colin Ware
University of New Brunswick
Fredericton, New Brunswick
Canada

Alan Watt
University of Sheffield
Sheffield, United Kingdom

Nigel P. Weatherill
Mississippi State University
Mississippi State, Mississippi

Peter Wegner
Brown University
Providence, Rhode Island

Craig E. Wills
Worcester Polytechnic Institute
Worcester, Massachusetts

George Wolberg
City College of New York
New York, New York

Michael Wolfe
The Portland Group
Wilsonville, Oregon

Jürgen Ziegler
Fraunhofer Institute IAO
Stuttgart, Germany

Victor Zue
Massachusetts Institute of
    Technology
Cambridge, Massachusetts

Contents

SECTION I   Algorithms and Data Structures

SECTION II Architecture

Keith Barker, Section Adviser

SECTION III   Artificial Intelligence and Robotics

Harold Abelson, Section Adviser

SECTION IV   Computational Science

SECTION V Database and Information Retrieval

Raghu Ramakrishnan, Section Adviser

SECTION VI   Graphics

Donald H. House, Section Adviser

SECTION VII Human–Computer Interaction

John M. Carroll, Section Adviser

SECTION VIII Operating Systems and Networks

SECTION IX   Programming Languages

SECTION X   Software Engineering

Steven A. Demurjian, Sr., Section Adviser

Appendixes

Indexes

This Handbook is dedicated to the memory of Paris Kanellakis, who had been working on a chapter for the Database and Information Retrieval section of this Handbook before his untimely death in December 1995. Paris' significant contributions to several subfields of computer science and engineering are well known by his colleagues throughout academia and industry. A more complete memorial summary of Paris' abundant scholarly contributions to the field appears in the 50th anniversary issue of ACM Computing Surveys.

THE
COMPUTER SCIENCE AND ENGINEERING
HANDBOOK

1

Computer Science and Engineering: The Discipline and Its Impact

Allen B. Tucker, Jr.
Bowdoin College

Peter Wegner
Brown University

1.1 Introduction

The discipline and profession of computer science and engineering (CS&E) has undergone dramatic changes in its short 50-year life. As the field has matured, new areas of research and development have emerged and joined with older areas to revitalize the discipline. In the 1930s, fundamental mathematical concepts of computing were developed by Turing and Church. Early computers implemented by von Neumann, Eckert, Atanasoff, and others in the 1940s led to the birth of commercial computing in the 1950s and to numerical programming languages like Fortran, commercial languages like COBOL, and artificial-intelligence languages like LISP. In the 1960s the rapid development and consolidation of the subjects of algorithms, data structures, databases, and operating systems formed the core of what we now call traditional computer science; the 1970s saw the emergence of software engineering, structured programming, and object-oriented programming. The emergence of personal computing and networks in the 1980s set the stage for dramatic advances in computer graphics, software technology, and parallelism. This Handbook aims to characterize computing in the 1990s, incorporating the explosive growth of networks like the World Wide Web and the increasing importance of areas like human–computer interaction, computational science, and other subfields that would not have appeared in such an encyclopedia even ten years ago.

This introductory chapter reviews the evolution of CS&E during the last two decades. It introduces those fundamental contemporary themes that form the nucleus of the subject matter and methodologies of the discipline, identifying the social context and scientific challenges that will continue to stimulate rapid growth and evolution of CS&E into the next century. Finally, it provides an overview of the discipline of CS&E, serving as a conceptual introduction to the ten major sections and 122 chapters and appendices

0-8493-2909-4/97/$0.00+$.50

that constitute the entire Handbook. These ten sections, corresponding to ten major subject areas, reflect a useful classification of the subject matter in CS&E:

- Algorithms and Data Structure
- Architecture
- Artificial Intelligence and Robotics
- Computational Science
- Database and Information Retrieval
- Graphics
- Human–Computer Interaction
- Operating Systems and Networks
- Programming Languages
- Software Engineering

Section 1.2 of this chapter presents a brief history of the computing industry and the parallel development of the computing curriculum. Section 1.3 frames the practice of CS&E in terms of four major conceptual paradigms: theory, abstraction, design, and the social context. Section 1.4 identifies the "grand challenges" that promise to extend the field's vitality and reshape its definition for the next generation and beyond, and section 1.5 summarizes the contents of the ten sections of the Handbook.

This Handbook is designed as a professional reference for researchers and practitioners in the field. Readers interested in exploring specific subject topics may prefer to move directly to the appropriate section of the Handbook. To facilitate rapid inquiry, the Handbook contains a Table of Contents and three indexes (Subject, Who's Who, and Key Algorithms and Formulas) for immediate access to specific topics at various levels of detail.

1.2 Growth of the Industry and the Profession

The computer industry has experienced tremendous growth and change over the last several decades, and most recently some retrenchment. The transition that began in the 1980s, from centralized mainframes to a decentralized networked microcomputer–server technology, was accompanied by the rise and fall of major corporations. The old monopolistic, vertically integrated industry epitomized by IBM's comprehensive client services gave way to a highly competitive industry in which the major players changed almost overnight. In 1992 alone, emergent companies like Dell and Microsoft had spectacular profit gains of 77% and 53%. In contrast, traditional companies like IBM and DEC suffered combined record losses of $7.1 billion in the same year [Economist 1993]. The exponential decrease in computer cost and increase in power by a factor of two every eighteen months, known as Moore's law, shows no signs of abating, though underlying physical limits must eventually be reached.

Overall, the rapid 18% annual growth rate that the computer industry had enjoyed in earlier decades gave way in the early 1990s to a 6% growth rate, caused in part by a saturation of the personal computer market. Another reason for this slowing of growth is that the performance of computers (speed, storage capacity) has improved at a rate of 30% per year in relation to their cost. Today, it is not unusual for a desktop computer to run at hundreds of times the speed and capacity of a typical mainframe computer of the 1980s, and at a fraction of the cost. However, it is not clear whether this slowdown in growth represents a temporary plateau or whether a new round of fundamental technical innovations in areas such as networking and human–computer interaction might again propel the computer industry to more spectacular rates of growth.

Curriculum Development

The computer industry's evolution has been strongly affected by the evolution of both theory and practice in the last several years. Changes in theory and practice are intertwined with the parallel evolution of the field's undergraduate and graduate curricula during the last three decades, and those curricula have, in turn, defined the conceptual and methodological framework for understanding the discipline itself.

The first coherent and widely cited curriculum for CS&E was developed in 1968 by the ACM Curriculum Committee on Computer Science [ACM 1968] in response to widespread demand for systematic under-graduate and graduate programs [Rosser 1966]. "Curriculum 68" defined computer science as comprising three main areas: information structures and processes, information processing systems, and methodologies. The first area included programming languages, data structures, and formal models of computation; the second computer architecture, compilers, and operating systems; the third numerical mathematics, file management, text processing, graphics, simulation, information retrieval, artificial intelligence, process control, and instructional systems. Curriculum 68 used this taxonomy to define computer science as a discipline and to provide concrete recommendations and guidance to colleges and universities in developing undergraduate, master's, and Ph.D. programs to respond to the widespread demand for computer scientists in research, education, and industry. Curriculum 68 stood as a robust and exemplary model for degree programs at all levels for a decade or more.

In 1978, a new ACM Curriculum Committee on Computer Science developed a revised and updated undergraduate curriculum [ACM 1978]. The "Curriculum 78" report responded to the rapid evolution of the discipline and practice of computing and to a demand for a more detailed elaboration of the computer science (as distinguished from the mathematical) elements of the courses that would comprise the core curriculum. Around the same time, the IEEE Computer Society developed a model curriculum for engineering-oriented undergraduate programs in CS&E [IEEE-CS 1976]. Updated and published in 1983 by the Computer Society as a "Model Program in Computer Science and Engineering" [IEEE-CS 1983], this curriculum was designed not only to define a course of study for computer science programs in engineering schools but also to meet a more extensive set of engineering accreditation criteria.

In 1988, the ACM Task Force on the Core of Computer Science and the IEEE Computer Society [ACM 1988] cooperated in developing a fundamental redefinition of the discipline of CS&E. Called "Computing as a Discipline," this report aimed to provide a contemporary foundation for undergraduate curriculum design by responding to the changes in computing research, development, and industrial applications in the previous decade. This report also acknowledged some fundamental methodological changes in the field. No longer could the "computer science = programming" model hope to encompass the richness of the field. Instead, three perspectives—*theory, abstraction,* and *design*—were used to characterize how various kinds of computer professionals and researchers did their work. These three points of view, those of the theoretical mathematician or scientist (theory), the experimental or applied scientist (abstraction, or modeling), and the engineer (design), were essential components of research and development throughout all the nine major subject areas (similar to the ten Handbook areas) into which the field was divided.

"Computing as a Discipline" led directly to the formation of a joint ACM/IEEE-CS Curriculum Task Force, which developed a comprehensive model for undergraduate curriculum design in the 1990s called "Curricula 91" [ACM/IEEE 1991]. Acknowledging that undergraduate computer science programs could be effectively supported in colleges of engineering, arts and sciences, and liberal arts, Curricula 91 proposed a core curriculum of common knowledge that undergraduate majors in any of these programs should cover. This core curriculum also contained sufficient theory, abstraction, and design content that students would become familiar with the fundamentally different but complementary ways of "doing" CS&E. It also ensured that students would gain a broad exposure to the nine major subject areas, and their social and ethical context. A significant laboratory component ensured that undergraduates gained significant abstraction (experimentation) and design experience.

Growth of Academic Programs

Fueling the rapid evolution of curricula in CS&E during the last three decades was an enormous growth in demand, by industry and academia, for computer professionals, researchers, and scientists at all levels. In response, the number of CS&E Ph.D.-granting programs in the U.S. grew from 12 in 1964 to 132 in 1994. During the period 1966–1993, the annual number of bachelor's degrees awarded grew from 89 to 24,580,

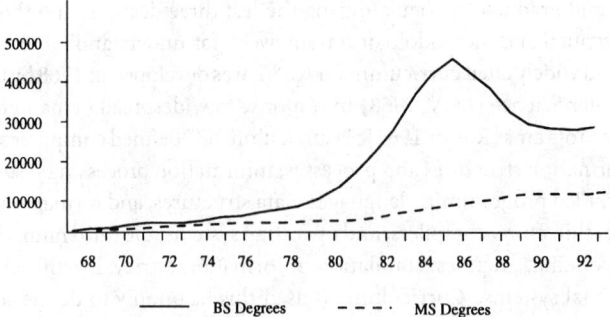

FIGURE 1.1 U.S. bachelor's and master's degrees in CS&E.

master's degrees grew from 238 to 10,349, and Ph.D. degrees grew from 19 to 969 [ACM 1968, Andrews 1995].

Figure 1.1 shows the number of bachelor's and master's degrees awarded by U.S. colleges and universities in CS&E from 1966 to 1993. The number of bachelor's degrees peaked at 42,195 in 1986, and then declined, leveling off to a fairly steady 25,000 by 1993. In contrast, master's degree production in computer science has grown steadily throughout the same period and shows no signs of leveling off. The rapid falloff in bachelor's degree production in 1986 may be attributed to the saturation of industry demand for programmers, while the steady growth of master's degrees in recent years may reflect a recognition by industry that an undergraduate computer science degree by itself, while providing good preparation for some positions, is not adequate for many of the newly emerging positions in the technology industry.

Figure 1.2 shows the number of U.S. Ph.D. degrees in computer science and computer engineering during the same 1966–1993 period [Andrews 1995]. The annual number of Ph.D.'s in computer science in the U.S. grew from 19 in 1966 to 878 in 1993, while the overall number of Ph.D.'s in computer science and computer engineering peaked at 1113 in 1992 and leveled off at 969 in 1993 and 1005 in 1994 [Andrews 1995].

Production of M.S. and Ph.D. degrees in computer science and engineering continued to grow into the 1990s, fueled by continuing demand from industry for graduate-level talent and continuing stong demand in academia to staff growing undergraduate and graduate research programs in CS&E. However, there is also a widely held belief that the period of growth in Ph.D. production in CS&E has now leveled off and reached a steady state with respect to demand from industry and academia.

Academic R&D and Growth of Industry Positions

University and industrial research and development (R&D) investments in CS&E grew rapidly in the period 1986–1993. Figure 1.3 shows that academic research and development in computer science nearly

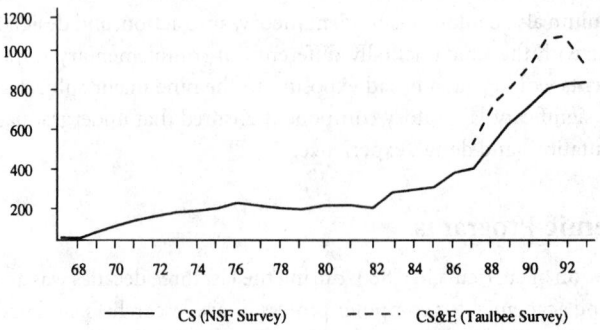

FIGURE 1.2 U.S. Ph.D. degrees in computer science and CS&CE.

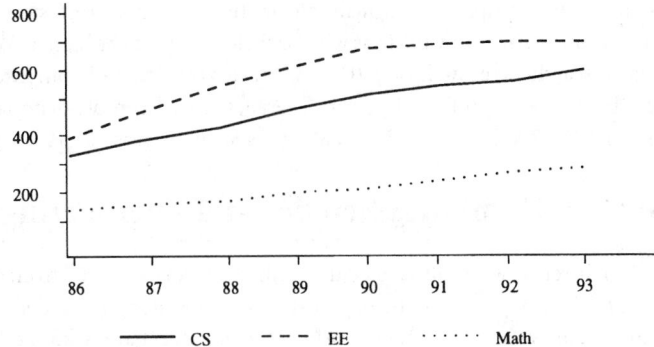

FIGURE 1.3   Academic R&D in computer science and related fields (millions of dollars).

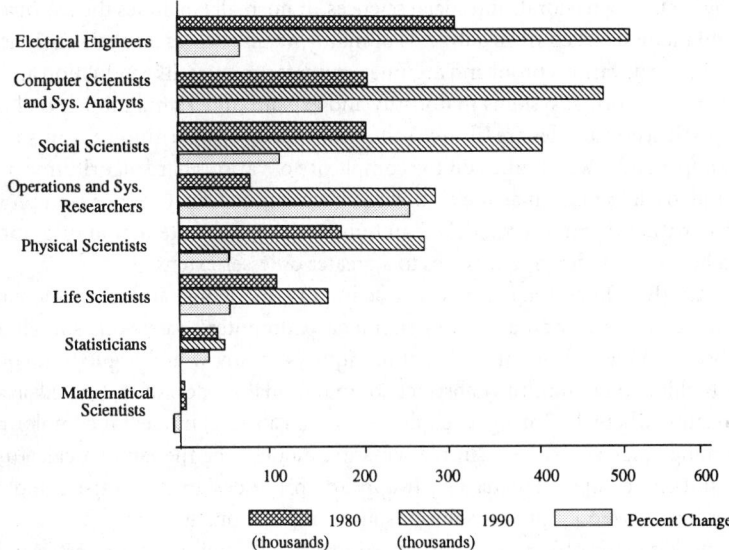

FIGURE 1.4   Nonacademic computer scientists and engineers: 1980 and 1990.

doubled, from $321 million to $597 million, during this time period, a growth rate somewhat higher than that of academic R&D in the related fields of electrical engineering and mathematics. During this same period, the overall growth of academic R&D in engineering and the social sciences also nearly doubled, while that in the physical sciences grew by only about 65%. In 1993, about 68% of the total support for academic R&D came from federal and state sources, while 7% came from industry and 18% came from the institutions themselves [NSF 1995a].

Figure 1.4 shows the growth between 1980 and 1990 in the number of persons with at least a bachelor's degree who were employed in nonacademic (industry and government) computer science positions. Overall, the total number of computer scientists and systems analysts grew by 150%, from 194,100 in 1980 to 485,200 in 1990. In 1990, there were 9400 Ph.D.'s in nonacademic computer science and systems analyst positions (of course, many of these Ph.D.'s were not in computer science) [NSF 1995b]. An informal survey conducted by the Computing Research Association (CRA) suggests that slightly more than half of the domestically employed new Ph.D.'s accepted positions in industry or government in 1994, and the rest accepted academic positions in colleges and universities. This survey also suggests that nearly a third of the total number of 1994 CS&E Ph.D.'s accepted positions abroad [Andrews 1995].

Figure 1.4 also provides some comparative growth information for other professions, again using 1980 and 1990 census data and considering only persons with bachelor's degrees or higher. We see that only the operations and systems researchers' growth of 250% was greater than that of computer scientists, while most professions' growth rates were significantly lower. Overall, the total number of nonacademic scientists and engineers grew from 2,136,200 in 1980 to 3,512,800 in 1990, an increase of 64.4% [NSF 1995b].

1.3 Perspectives in Computer Science and Engineering

Computer science and engineering is a multifaceted discipline that can be viewed from at least four different perspectives. Three of the perspectives—theory, abstraction, and design—underscore the point that computer scientists and engineers approach their work and their subject areas from different intellectual viewpoints. A fourth perspective—the social and professional context—acknowledges that computing directly affects the quality of people's lives, and that computing professionals must be prepared to understand and confront the social issues that arise from their work.

The theory of CS&E draws from principles of mathematics and logic as well as from the formal methods of the physical, biological, behavioral, and social sciences. It normally includes the use of advanced mathematical ideas and methods taken from subfields of mathematics such as algebra, analysis, and statistics. Theory includes the use of various proof and argumentation techniques, like induction and contradiction, to establish properties of formal systems that justify and explain underlying models and paradigms supporting computer science; examples are Church's thesis, the study of algorithmically unsolvable problems, and the study of upper and lower bounds on the complexity of various hard algorithmic problems. Fields like algorithms and to a lesser extent artificial intelligence, computational science, and programming languages have more mature theoretical models than human–computer interaction or graphics, but all ten areas considered here have underlying theories to a greater or lesser extent.

Abstraction in CS&E includes the use of scientific inquiry, modeling, and experimentation to test the validity of hypotheses about computational phenomena. Computer professionals in all ten areas of the discipline use abstraction as a fundamental tool of inquiry—many would argue that computer science is the science of building and examining abstract computational models of reality. Abstraction arises in computer architecture, where the Turing machine serves as an abstract model for complex real computers, and in programming languages, where simple semantic models like the lambda calculus are used as a framework for studying complex languages. It appears in the design of heuristic and approximation algorithms for problems whose optimal solutions are computationally intractable. It is surely used in graphics, where models of 3D objects are constructed mathematically, given properties of lighting, color, and surface texture, and projected in a realistic way on a two-dimensional video screen.

Design is a process used to describe the essential structure of complex systems as a prelude to their implementation. It also encompasses the use of traditional engineering methods, including the classical life-cycle model, to implement efficient and effective computational systems in hardware and software. It includes the use of tools like cost/benefit analysis of alternatives, risk analysis, and fault tolerance that ensure that computing applications are brought to market effectively. Design is a central preoccupation of architects and software engineers developing hardware systems and software applications. Like abstraction, it is an important activity in computational science, database and information retrieval, human–computer interaction, operating systems and networks, and the other areas considered here.

The social and professional context includes many issues that arise at the computer–human interface, such as liability for hardware and software errors, security and privacy of databases and networks, intellectual property issues (patent and copyright), and equity issues (universal access to the technology and the profession). Computing professionals in all subject areas must consider the ethical context in which their work occurs and the special responsibilities that attend their work. The next preliminary chapter discusses these issues, and several other chapters address topics in which specific social and professional issues come into play. For example, security and privacy issues in databases, operating systems, and networks are discussed in Chapters 49 and 89. Risks in software are discussed in several chapters of section X of the Handbook.

1.4 Broader Horizons: HPCC and Grand Challenge Applications

The 1992 report "Computing the Future" (CTF) [CSNRCTB 1992], written by a group of leading computer professionals in response to a request by the Computer Science and Technology Board (CSTB), identifies the need for CS&E to broaden its research agenda and its educational horizons. The view that the research agenda should be broadened initially caused concerns among researchers that funding and other incentives might overemphasize short-term at the expense of long-term goals. This Handbook reflects the broader view of the discipline in its inclusion of computational science, graphics, and computer–human interaction among the major subfields of computer science.

CTF aimed to bridge the gap between suppliers of research in CS&E and consumers of research such as industry, the Federal government, and funding agencies like NSF, DARPA, and DOE. It addresses fundamental challenges to the field and suggests responses that encourage greater interaction between research and computing practice. Its overall recommendations focus on three priorities:

1. To sustain the core effort that creates the theoretical and experimental science base on which applications build.
2. To broaden the field to reflect the centrality of computing in science and society.
3. To improve education at both the undergraduate and graduate levels.

CTF includes recommendations to federal policy makers and universities regarding research and education:

Recommendations to federal policy makers regarding research:
- The High-Performance Computing and Communication (HPCC) program passed by Congress in 1989 [OST 1989] should be fully supported.
- Application-oriented computer science and engineering research should be strongly encouraged through special funding programs.

Recommendations to universities regarding research:
- Academic research should broaden its horizons, embracing application-oriented and technology-transfer research as well as core applications.
- Laboratory research with experimental as well as theoretical content should be supported.

Recommendation to federal policy makers regarding education:
- Basic and human resources research of HPCC and other areas should be expanded to address educational needs.

Recommendations to universities regarding education:
- Broaden graduate education to include requirements and incentives to study application areas.
- Reach out to women and minorities to broaden the talent pool.

Though this report was motivated by the desire to provide a rationale for the HPCC program, its message that computer science must be responsive to the needs of society is much broader. The years since publication of CTF have seen a swing away from pure research towards application-oriented research that is reflected in this Handbook. However, it is important to maintain a balance between short-term applications and long-term research in core disciplines.

The HPCC program encourages universities, research programs, and industry to develop specific capabilities to address the "grand challenges" of the future. Realizing these grand challenges requires both fundamental and applied research, including the development of high-performance computing systems whose speed is two to three orders of magnitude greater than that of current systems, advanced software technology and algorithms that enable scientists and mathematicians to effectively address these grand challenges, networking to support R&D for a gigabit National Research and Educational Network (NREN), and human resources that expand basic research in all areas relevant to high-performance computing.

The grand challenges themselves were identified in HPCC as those fundamental problems in science and engineering with potentially broad economic, political, or scientific impact that can be advanced by applying high-performance computing technology and that can be solved only by high-level collaboration among computer professionals, scientists, and engineers. A list of grand challenges developed by agencies like NSF, DOD, DOE, and NASA in 1989 includes:

- Prediction of weather, climate, and global change.
- Challenges in materials sciences.
- Semiconductor design.
- Superconductivity.
- Structural biology.
- Design of drugs.
- Human genome.
- Quantum chromodynamics.
- Astronomy.
- Transportation.
- Vehicle dynamics and signature.
- Turbulence.
- Nuclear fusion.
- Combustion systems.
- Oil and gas recovery.
- Ocean science.
- Speech.
- Vision.
- Undersea surveillance for antisubmarine warfare.

More recently, an HPCC budget request identifies the following items as the "National Challenges" that face American CS&E and related professions:

- Digital libraries.
- Crisis and emergency management.
- Educational and lifelong learning.
- Electronic commerce.
- Energy management.
- Environmental monitoring and waste management.
- Health care.
- Manufacturing processes and products.
- Public access to government information.

As an outcome of HPCC and CTF, two new subject areas, "computational science" [Stevenson 1994] and "organizational informatics" [Kling 1993] emerged to influence the structure of the original nine subject areas identified in the report "Computing as a Discipline." In this Handbook, we view computational science as an extension of the area of numerical and symbolic computation. This area includes as a central element the fundamental interaction between computation and scientific research. For instance, fields like computational astrophysics, computational fluid dynamics, and computational chemistry all emphasize applications of computing in science and engineering, algorithms, and special considerations for computer architecture. Much of the research and early accomplishments of the emerging computational science field is reported in section IV of this Handbook.

Organizational informatics, on the other hand, emphasizes applications of computing in business and management, information systems and networks, as well as their implementation, risks, and human factors. Some of these intersect in major ways with human–computer interaction, while others fall more directly within the realm of management information systems (MIS), which is usually treated as a separate discipline from computer science and engineering. Thus, in this Handbook, we do not attempt to cover all of organizational informatics as a separate subject area within CS&E. Rather, we include many of its concerns within section VII on Human–Computer Interaction.

In addition, the growth of computer graphics and networks in the last few years provides strong arguments for their inclusion as major subject areas in the discipline. This Handbook distinguishes *graphics* from *human–computer interaction*, of which it had been a subarea in "Computing as a Discipline." Finally, the area of *operating systems and networks* in this Handbook has evolved out of what had been called *operating systems* in "Computing as a Discipline," in recognition of the rapid growth of distributed computing in the last few years.

1.5 Organization and Content

In the 1940s computing was identified with number crunching, and numerical analysis was considered a central tool. Hardware, logical design, and information theory emerged as important subfields in the early 1950s. Software and programming emerged as important subfields in the mid 1950s and soon dominated hardware as topics of study in computer science. In the 1960s computer science could be comfortably classified into theory, systems (including hardware and software), and applications. Software engineering emerged as an important subdiscipline in the late 1960s. The 1980 Computer Science and Engineering Research Study [Arden 1980] classified the discipline into nine subfields:

- Numerical computation.
- Theory of computation.
- Hardware systems.
- Artificial intelligence.
- Programming languages.
- Operating systems.
- Database management systems.
- Software methodology.
- Applications.

This Handbook's classification into ten subfields is quite similar to that of the COSERS study, suggesting that the discipline of CS&E is stabilizing:

- Algorithms and data structures.
- Architecture.
- Artificial intelligence.
- Computational science.
- Database and information retrieval.
- Graphics.
- Human–computer interaction.
- Operating systems and networks.
- Programming languages.
- Software engineering.

This Handbook's classification has discarded numerical analysis and added the new areas of human–computer interaction and graphics. The other eight areas appear in both classifications with some name changes (theory of computation has become algorithms and data structures, applications has become computational science, hardware systems has become architecture, operating systems has added networks, and database has added information retrieval as important new directions).

Though the high-level classification has remained stable, the content of each area has evolved and matured. We examine below the scope of each area and the topics within each area treated in the Handbook.

Algorithms and Data Structures

The subfield of algorithms and data structures is interpreted broadly to include core topics in the theory of computation as well as data structures and practical algorithm techniques. Its thirteen chapters provide a comprehensive overview that spans both theoretical and applied topics in the analysis of algorithms. Chapter 3 introduces fundamental concepts such as computability and undecidability and formal models such as Turing machines and Chomsky grammars, while Chapter 4 reviews techniques of algorithm design like divide and conquer, dynamic programming, recurrence relations, and greedy heuristics.

Chapter 5 covers data structures both descriptively and in terms of their space–time complexity, while Chapter 6 reviews methods and techniques of computational geometry, and Chapter 7 presents the rich area of randomized objects. Pattern matching and text compression algorithms are examined in Chapter 8, graph and network algorithms in Chapter 9, and algebraic algorithms in Chapter 10. Chapter 11 examines topics in complexity like P vs. NP, NP-completeness, and circuit complexity, while Chapter 12 examines parallel algorithms, and Chapter 13 considers combinatorial optimization. Chapter 14 concludes section I with a case study in VLSI layout that makes use of partitioning, divide and conquer, and other algorithm techniques common in VLSI design.

Architecture

Computer architecture is the design of efficient and effective computer hardware at all levels, from the most fundamental concerns of logic and circuit design to the broadest concerns of parallelism and high-performance computing. The chapters in section II span all these levels, providing a sampling of the principles, accomplishments, and applications of modern computer architectures.

Chapters 15 and 16 introduce the fundamentals of logic design components, including elementary circuits, Karnaugh maps, programmable array logic, circuit complexity and minimization issues, arithmetic processes, and speedup techniques. The architecture of buses is covered in Chapter 17, while the principles of memory architecture are addressed in Chapter 18. Topics there include associative memories, cache design, interleaving, and memories for pipelined and vector processors.

Chapter 19 concerns the design of effective and efficient computer arithmetic units. Chapter 20 extends the design horizon by considering the various models of parallel architectures, including the performance of contemporary machines that fall into the SIMD, MISD, and MIMD categories.

Artificial Intelligence and Robotics

Artificial intelligence (AI) is the study of the computations that make it possible to simulate human perception, reasoning, and action. Current efforts are aimed at constructing computational mechanisms that process visual data, understand speech and written language, control robot motion, and model physical and cognitive processes. Robotics is a complex field, drawing heavily from AI as well as other areas of science and engineering.

AI includes techniques for automated learning, planning, and representing knowledge. Chapter 21 opens this section with a discussion of deductive learning. The use of decision trees and neural networks in learning and other areas is the subject of Chapters 22 and 23, while Chapter 24 introduces genetic algorithms.

Chapter 25 focuses on the area of computer vision. Chapter 26 addresses issues related to the mechanical understanding of spoken language. Chapter 27 presents the rationale and uses of planning and scheduling models in AI research. Chapter 28 describes the principles of knowledge representation and their applications in natural-language processing (NLP).

Artificial-intelligence work requires a number of distinct tools and models. These include the use of fuzzy, temporal, and other logics, as described in Chapter 29. The use of a variety of specialized search techniques to address the combinatorial explosion of alternatives in many AI problems is the subject of Chapter 30. Many AI applications must handle the notion of uncertainty. Chapter 31 discusses the modeling of decision making under uncertainty, while the related idea of qualitative modeling is discussed in Chapter 32.

Chapter 33 concludes section III with a thorough discussion of the principles and major results in the field of robotics: the design of effective devices that simulate mechanical, sensory, and intellectual functions of humans in specific task domains such as factory production lines.

Computational Science

The emerging area of computational science unites computational simulation, experimental investigations, and theoretical pursuits as three fundamental modes of scientific discovery. It uses scientific visualization, made possible by computational simulation, as a window into the analysis of physical phenomena and processes, providing a virtual microscope/telescope for inquiry and investigation at an unprecedented level of detail.

Section IV focuses on the challenges and opportunities offered by computers in aiding scientific analysis and engineering design. Chapter 34 introduces the section by presenting the fundamental subjects of computational geometry and grid generation. The design of graphical models for scientific visualization of complex physical and biological phenomena is the subject of Chapter 35.

Each of the remaining chapters in this section covers the computational science challenges and discoveries in a particular scientific or engineering field. Chapter 36 presents the computational aspects of structural mechanics, while Chapter 37 does the same for fluid dynamics. Computational reacting flow is the subject of Chapter 38, while Chapter 39 summarizes the progress in the area of computational electromagnetics. Chapter 40 addresses the grand challenge of computational ocean modeling. This section closes with a discussion of computational biological modeling in Chapter 41.

Database and Information Retrieval

The subject area of database and information retrieval addresses the general problem of storing large amounts of data in such a way that they are reliable, up to date, and efficiently retrieved. This problem is prominent in a wide range of applications in industry, government, and academic research. Availability of such data on the Internet and in forms other than text (e.g., audio and video) makes this problem increasingly complex.

At the foundation are the fundamental data models (relational, hierarchical, entity–relationship, network, and object-oriented) discussed in Chapter 42. The conceptual, logical, and physical levels of designing a database for high performance in a particular application domain are discussed in Chapter 43.

A number of basic issues surround the design of database models and systems. These include choosing from alternative access methods (Chapter 44), optimizing database queries (Chapter 45), controlling concurrency (Chapter 46), and benchmarking database workloads and performance (Chapter 47).

The design of heterogeneous databases and interoperability is discussed in Chapter 48. The issue of database security and privacy protection, in stand-alone and networked environments, is the subject of Chapter 49. The special considerations involved in storing and retrieving information from text databases are covered in Chapter 50.

A special topic in database research is the study of deductive (rule-based) databases, addressed in Chapter 51. Chapter 53 closes section V with a case study on SQL, a widely used database query language standard.

Graphics

Computer graphics is the study and realization of complex processes for representing physical and conceptual objects visually on a computer screen. These processes include the internal modeling of objects, rendering, hidden-surface elimination, color, shading, projection, and representing motion. An overview of these processes and their interaction is presented in Chapter 54.

Fundamental to all graphics applications are the processes of modeling and rendering. Modeling is the design of an effective and efficient internal representation for geometric objects (points, lines, polygons, solids, fractals, and their transformations), which is the subject of Chapters 55 and 56. Rendering, the process of representing the objects in a three-dimensional scene on a two-dimensional screen, is discussed in Chapter 57. Among its special challenges are the elimination of hidden surfaces, color, illumination, and shading.

The reconstruction of scanned images is another important area of computer graphics. Sampling, filtering, reconstruction, and antialiasing are the focus of Chapter 58. The representation and control of motion, or animation, is another complex and important area of computer graphics. Its special challenges are presented in Chapter 59.

Chapter 60 discusses volume data sets, and Chapter 61 looks at the emerging field of virtual reality and its particular challenges for computer graphics. Chapter 62 concludes section VI with a discussion of Renderman as a case study of a particularly effective application of the principles of computer graphics in the real world.

Human–Computer Interaction

This area, the study of how humans and computers interact, has the goal of improving the quality of the interaction and the effectiveness of those who use technology. This includes the conception, design, implementation, risk analysis, and effects of user interfaces and tools on those who use them in their work.

Chapter 63 opens section VII with a discussion of methods of overall system modeling, including users and modes of use. Modeling the organizational environments in which technology users work is the subject of Chapter 64. Usability engineering is the focus of Chapter 65, while user interface design methods are discussed in Chapter 66. The impact of international standards for user interfaces on the design process is the main concern of Chapter 67.

Specific devices, tools, and techniques for effective user-interface design form the basis for the next few chapters in this section. Chapters 68 and 69 discuss, respectively, the characteristics of input devices like the mouse and keyboard and output devices like computer screens and multimedia audio devices. Chapter 70 focuses on design techniques for effective interaction with users through these devices. The special concerns for integrating multimedia with user interaction are presented in Chapter 71. Lower-level concerns for the design of interface software technology are addressed in Chapter 72. The programming and software development process for user-interface implementation is discussed in Chapter 73. The effective presentation of documentation, training, and help facilities for users is a perennial concern for software designers, and its current status is reviewed in Chapter 74.

Operating Systems and Networks

Operating systems form the software interface between the computer and its applications. Section VIII covers their analysis and design, their performance, and their special challenges in a networked computing environment. Chapter 75 briefly traces the historical development of operating systems and introduces the fundamental terminology, including process scheduling, memory management, synchronization, I/O management, and distributed systems.

The process is a key unit of abstraction in operating-system design. Chapter 76 discusses the dynamics of processes and threads. Strategies for process and device scheduling are presented in Chapter 77. The special requirements for operating systems in real-time and embedded system environments are the subject of Chapter 78. Algorithms and techniques for process synchronization and interprocess communication are the subject of Chapter 79.

Memory and input/output device management is also a central concern of operating systems. Chapter 80 discusses the concept of virtual memory, from its early incarnations to its uses in present-day systems and networks. The different types and access methods for secondary storage and file systems are covered in Chapter 81.

Extending operating system functionality across a networked environment adds another level of complexity to the design process. Chapter 82 presents an overview of network organization and topologies, while Chapter 83 describes network routing protocols. The topology and functionality of internetworking, with the Internet as prime example, are presented in Chapter 84.

The influence of networked environments on the design of distributed operating systems is considered in Chapter 85. Distributed file and memory systems are discussed in Chapter 86, while distributed and multiprocessor scheduling are the focus of attention in Chapter 87. Finally, the forward-looking notion of dynamically partitioning a computing task across a network of heterogeneous computers is the topic of Chapter 88.

Operating systems and networks, especially the Internet, must make provisions for ensuring system integrity in the event of inappropriate access, unexpected malfunction and breakdown, and violations of security or privacy principles. Chapter 89 introduces some of the security and privacy issues that arise in a networked environment. Models for system security and protection are the subject of Chapter 90, while Chapter 91 discusses authentication, access control, and intrusion detection. Chapter 92 focuses on security issues that arise in networks, while a case discussion of some noteworthy malicious software and hacking events appears in Chapter 93.

Programming Languages

In section IX the design space of programming languages is partitioned into paradigms, mechanisms for compiling, and run-time management, and the theoretical areas of foundational models, type systems, and semantics are examined. Overall, this section provides a good balance between considerations of language paradigms, implementation issues, and theoretical models.

Chapter 94 considers traditional language and implementation questions for imperative programming languages like Fortran, C, Pascal, and Ada 83. Chapter 95 considers topics in functional programming like lazy and eager evaluation, and Chapter 96 examines object-oriented concepts like classes, inheritance, encapsulation, and polymorphism. Chapter 97 considers declarative programming in the logic/constraint programming paradigm, while Chapter 98 considers issues in concurrent/distributed programming as well as parallel models of computation. Compilers and interpreters for sequential languages are considered in Chapter 99, while compilers for parallel architectures and dataflow languages are considered in Chapter 100. The issues surrounding run-time environments and memory management for compilers and interpreters are addressed in Chapter 101.

Chapters 102, 103, and 104 deal with foundations and theoretical models. Chapter 102 deals with foundational calculi like the lambda calculus and the pi calculus and with the influence of input/output automata, Petri nets, and other models of computation on language design. Chapter 103 examines issues of type theory in programming, including static versus dynamic type checking, type safety, and polymorphism. Chapter 104 examines models of programming-language semantics, including denotational, operational, and axiomatic models.

Software Engineering

Section X on software engineering examines formal specification, design, verification and testing, project management, and other aspects of the software life cycle. Chapter 105 considers models of the software life cycle such as the waterfall and spiral models as well as specific phases of the life cycle. Chapter 106 examines software qualities like maintainability, portability, and reuse that are needed for high-quality software systems, while Chapter 107 considers formal models, specification languages, and the specification process.

Chapter 108 deals with the traditional and object-oriented design processes, featuring a case study in top-down functional design. Chapter 109 on verification and validation deals with the use of systematic techniques like verification and testing for quality assurance, while Chapter 110 examines testing models as well as risk and reliability issues.

Chapter 111 considers methods of project design such as chief programmer teams and rapid prototyping, as well as project scheduling and evaluation. Chapter 112 considers software tools like compilers, editors, and CASE tools and surveys graphical environments. Chapter 113 on interoperability considers architectures for communicating among heterogeneous software components such as OMG's Common Object Request Broker Architecture (CORBA) and Microsoft's Common Object Model (COM).

1.6 Conclusion

In 1997, the ACM celebrates its 50th anniversary. The first 50 years of CS&E are characterized by dramatic growth and evolution. While it is safe to affirm today that the field has reached a certain level of maturity, it would be foolish to assume that it will remain unchanged in the future. Already, conferences are calling for new visions that will enable the discipline to continue its rapid evolution into the twenty-first century. This Handbook is designed to convey the modern spirit, accomplishments, and direction of CS&E as we see it in 1996. It interweaves theory with practice, highlighting "best practices" in the field as well as current research directions. It provides today's answers to well-formed questions posed by professionals and researchers across the ten major subject areas. Finally, it identifies key professional and social issues that lie at the intersection of the technical aspects of CS&E and its impact in the world.

The future holds great promise for the next generations of computer scientists and engineers. These people will solve problems that have only recently been conceived, such as those suggested by the HPCC as "grand challenges." To address these problems, and to extend these solutions in a way that benefits the lives of significant numbers of the world's population, will require substantial energy, commitment, and real investment on the part of institutions and professionals throughout the world. The challenges are complex, and the solutions are not likely to be obvious.

References

ACM Curriculum Committee on Computer Science 1968. Curriculum 68: recommendations for the undergraduate program in computer science. *Commun. ACM* 11(3):151–97, Mar.

ACM Curriculum Committee on Computer Science 1978. Curriculum 78: recommendations for the undergraduate program in computer science. *Commun. ACM* 22(3):147–166, Mar.

ACM Task Force on the Core of Computer Science: Denning, P., Comer, D., Gries, D., Mulder, M., Tucker, A., and Young, P., 1988. *Computing as a Discipline*. Abridged version, *Commun. ACM*, Jan. 1989.

ACM/IEEE Joint Curriculum Task Force 1991. Computing Curricula 1991. ACM Press. Abridged version, *Commun. ACM*, June 1991, and *IEEE Comput.* Nov. 1991.

Andrews, G. R., 1995. CRA Taulbee Survey: PhDs Holding Steady. Computing Research Assoc.

Arden, B., ed. 1980. *What Can be Automated?* Computer Science and Engineering Research (COSERS) Study. MIT Press, Boston, MA.

CSNRCTB 1992. Computer Science and National Research Council Telecommunications Board. *Computing the Future: A Broader Agenda for Computer Science and Engineering*. National Academy Press, Washington, DC.

Economist 1993. The computer industry: reboot system and start again. *Economist*, Feb. 27.

IEEE-CS 1976. Education Committee of the IEEE Computer Society. *A Curriculum in Computer Science and Engineering*. IEEE Pub. EH0119-8, Jan. 1977.

IEEE-CS 1983. Educational Activities Board. *The 1983 Model Program in Computer Science and Engineering. Tech. Rep. 932.* Computer Society of the IEEE, Dec.

Kling, R. 1993. Organizational analysis in computer science. *Inform. Soc.* Mar.–June.

NSF 1995a. National Science Foundation. *Survey of Scientific and Engineering Expenditures at Universities and Colleges.* NSF, Arlington, VA.

NSF 1995b. National Science Foundation. *Nonacademic Scientists and Engineers: Trends from the 1980 and 1990 Censuses.* NSF 95-306. Arlington, VA.

OST 1989. Office of Science and Technology. *The Federal High Performance Computing and Communication Program.* Executive Office of the President, Washington, DC.

Rosser, J. B. et al. 1966. *Digital Computer Needs in Universities and Colleges.* Publ. 1233, National Academy of Sciences, National Research Council, Washington, DC.

Stevenson, D. E. 1994. Science, computational science, and computer science. *Commun. ACM,* Dec.

2

Ethical Issues for Computer Scientists and Engineers

Deborah G. Johnson
Rensselaer Polytechnic Institute

Keith Miller
University of Illinois at Springfield

2.1 Introduction: Why a Chapter on Ethical Issues?

Computers have had a powerful impact on our world and are destined to shape our future. This observation, now commonplace, is the starting place for any discussion of professionalism and ethics in computing. The work of computer scientists and engineers is part of the social, political, economic, and cultural world in which we live, and affects many aspects of that world. Professionals who work with computers have special knowledge and that knowledge, when combined with computers, has significant power to change people's lives.

In this chapter of the Handbook we provide a perspective on the role of computer and engineering professionals and we examine the relationships and responsibilities that go with having and using computing expertise. In addition to the topic of professional ethics, we briefly discuss several of the social-ethical issues created or exacerbated by increasing use of computers: privacy, property, risk and reliability, and global communication.

Computers, digital data, and telecommunications have changed work, travel, education, business, entertainment, government, and manufacturing. For example, work now increasingly involves sitting in front of a computer screen and using a keyboard to make things happen in a manufacturing process or to keep track of records where in the past these same tasks would have involved physically lifting, pushing, and twisting or using pens, paper, and file cabinets. Changes such as these—in the way we do things—have, in turn, fundamentally changed who we are as individuals, communities, and nations. Some would argue, for example, that new kinds of communities (e.g., cyberspace on the Internet) are forming, individuals are developing new types of personal identity, and new forms of authority and control are taking hold as a result of this evolving technology.

Computer technology is shaped by social-cultural concepts, laws, the economy, and politics. These same concepts, laws, and institutions have been pressured, challenged, and modified by computer technology. Technological advances can antiquate laws, concepts, and traditions, compelling us to reinterpret and create new laws, concepts, and moral notions. Our attitudes about work and play, our values, and our laws and customs are deeply involved in technological change.

When it comes to the social-ethical issues surrounding computers, some have argued that the issues are not unique. All of the ethical issues raised by computer technology can, it is said, be classified and worked out using traditional moral concepts, distinctions, and theories. There is nothing new here in the sense that all of the issues have to do with traditional moral concepts such as privacy, property, responsibility, and traditional moral ends such as maximizing individual freedom or holding individuals accountable. These concepts and values predate computers; hence, it would seem there is nothing unique about *computer ethics*.

On the other hand, those who argue for the uniqueness of the topic point to the fundamental ways that computers have changed so many human activities, such as manufacturing, record keeping, banking, and communicating. The change is so radical, it is claimed, that traditional moral concepts, distinctions, and theories, if not abandoned, must be significantly reinterpreted and extended. For example, they must be extended to computer-mediated relationships, computer software, computer art, electronic bulletin boards, and so on. The uniqueness of the ethical issues surrounding computers can be argued for in a number of ways. Computer technology makes possible a scale of activities not possible before. This includes a larger scale of record keeping of personal information, as well as larger scale calculations which, in turn, allow us to build and do things not possible before, such as undertaking space travel and operating a global communication system. In addition to scale, computer technology has involved the creation of new kinds of entities for which no rules initially existed: entities such as computer files, computer programs, and user interfaces. The uniqueness argument can also be made in terms of the power and pervasiveness of computer technology. It seems to be bringing about a magnitude of change comparable to that which took place during the Industrial Revolution, transforming our social, economic, and political institutions, our understanding of what it means to be human, and the distribution of power in the world. Hence, it would seem the issues are special if not unique.

A synthesis of these two views of computer ethics seems necessary since analysis of a computer ethical issue generally involves both working on something new and drawing on something old. Issues in computer ethics are new species of older ethical problems [Johnson 1994]. Most of the issues can be understood using traditional moral concepts such as autonomy, privacy, property, and responsibility. Most arise in contexts in which there are already social, ethical, and legal norms; that is, the issues arise in the context of the workplace, government, business, role relationships, and so on. In this respect, the issues are not new or unique. Nevertheless, when a computer is involved, the situation may have special features which have not been addressed by prevailing norms, and these features make a moral difference. For example, although property rights and even intellectual property rights have been worked out in the past, software has raised new property issues: Should the arrangement of icons appearing on the screen of a user interface be ownable? Is there anything intrinsically wrong with copying software? Software has features which make the distinction between idea and expression (a distinction at the core of copyright law) almost incoherent. As well, it has features which make standard intellectual property laws difficult to enforce. Hence, questions about what should be owned when it comes to software and how to evaluate violations of software ownership rights are not new in the sense that they are property rights issues, but they are new in the sense that nothing with the characteristics of software has been addressed before. We have, then, a new species of traditional property rights.

Similarly, although our understanding of rights and responsibilities in the employer–employee relationship have been evolving for centuries, never before have employers had the capacity to monitor their workers electronically, keeping track of every keystroke, and recording and reviewing all work done by an employee (covertly or with prior consent). When we evaluate this new monitoring capability and ask whether employers should use it, we are working on an issue that has never arisen before, though many other issues involving employer–employee rights have. We address a new species of the tension between employer–employee rights and interests.

The social-ethical issues posed by computer technology are significant in their own right, but they are of special interest here because computer and engineering professionals bear responsibility for this technology. It is of critical importance that they understand the social change brought about by their work and the difficult social-ethical issues posed. Just as some have argued that the social-ethical issues posed by computer technology are not unique, some have argued that the issues of professional ethics surrounding computers are not unique. We propose, in parallel with our previous genes-species argument, that the professional ethics issues arising for computer scientists and engineers are species of generic issues of professional ethics. All professionals have responsibilities to their employers, clients, coprofessionals, and to the public. Managing these types of responsibilities poses a challenge in all professions. Moreover, all professionals bear some responsibility for the impact of their work. In this sense, the professional ethics issues arising for computer scientists and engineers are generally similar to those in other professions. Nevertheless, it is also true to say that the issues arise in unique ways for computer scientists and engineers because of the special features of computer technology.

In what follows, we discuss ethics in general, professional ethics, and finally, the ethical issues surrounding computer technology.

2.2 Ethics in General

There is a lively history of ethical theories. Ethicists explore theories to: (1) explain and justify prevailing moral notions, (2) critique ordinary moral beliefs, and (3) assist in rational, ethical decision making. It is not our purpose here to propose, defend, or attack any particular ethical theory. Rather, we offer brief descriptions of three theories to illustrate the nature of ethical analysis. We also include a decision making method that combines elements of each theory.

Ethical analysis involves giving reasons for moral claims and commitments. It is not just a matter of articulating intuitions. When the reasons given for a claim are developed into a moral theory, the theory can be incorporated into techniques for improved technical decision making. Three traditions in ethical analysis and problem solving are described. This is by no means an exhaustive account, nor is our description of any of the three any more than a brief introduction. The three traditions are utilitarianism, deontology, and social contract theory.

Utilitarianism

Utilitarianism has greatly influenced 20th-century thinking. According to this theory, we should make decisions about what to do by focusing on the consequences of our actions. Ethical rules are derived from their usefulness (their utility) in bringing about happiness. Utilitarianism offers one seemingly simple moral principle which everyone should use to determine what to do in a given situation: everyone ought to act so as to bring about the greatest amount of happiness for the greatest number of people.

According to utilitarianism, happiness is the only value that can serve as the fundamental base for ethics. Since happiness is the ultimate good, morality must be based on creating as much of this good as possible. The utilitarian principle provides a decision procedure. When you want to decide what to do, the alternative that produces the most overall net happiness (good minus bad) is the right action. The right action may be one that brings about some unhappiness but that is justified if the action also brings about enough happiness to counterbalance the unhappiness, or if the action brings about the least unhappiness of all possible alternatives.

Be careful not to confuse utilitarianism with egoism. Egoism is a theory claiming that one should act so as to bring about the most good consequences for one's self. Utilitarianism does not say that you should maximize your own good. Rather, total happiness in the world is what is at issue; when you evaluate your alternatives you have to ask about their effects on the happiness of everyone. It may turn out to be right for you to do something that will diminish your own happiness because it will bring about an increase in overall happiness.

The emphasis on consequences found in utilitarianism is very much a part of decision making in our society, in particular as a framework for law and public policy. Cost-benefit and risk-benefit forms of analysis are, for example, consequentialist in character.

Utilitarians do not all agree on the details of utilitarianism; there are different kinds of utilitarians. One issue is whether the focus should be on *rules* of behavior or individual *acts*. Utilitarians have recognized that it would be counter to overall happiness if each one of us had to calculate at every moment what the consequences of every one of our actions would be. Sometimes we must act quickly, and often the consequences are difficult or impossible to foresee. Thus, there is a need for general rules to guide our actions in ordinary situations. Hence, *rule utilitarians* argue that we ought to adopt rules which, if followed by everyone, would, in general and in the long run, maximize happiness. *Act utilitarians*, on the other hand, put the emphasis on judging individual actions rather than creating rules.

Both rule utilitarians and act utilitarians share an emphasis on consequences; deontological theories do not share this emphasis.

Deontological Theories

Deontological theories can be understood as a response to the criticisms of utilitarian theories. A traditional criticism of utilitarianism is that it sometimes leads to conclusions that are incompatible with our most strongly held moral intuitions. Utilitarianism seems, for example, open to the possibility of justifying enormous burdens on some individuals for the sake of others. To be sure, every person counts equally; no one person's happiness or unhappiness is more important than any other person's. However, since utilitarians are concerned with the total amount of happiness, we can imagine situations where great overall happiness might result from sacrificing the happiness of a few. Suppose, for example, that having a small number of slaves would create great happiness for large numbers of people; or suppose we kill one healthy person and use the resulting body parts to save ten people in need of transplants.

Critics of utilitarianism say that if utilitarianism justifies such practices, then the theory must be wrong. Utilitarians have a defense, arguing that such practices could not be justified in utilitarianism because of the long-term consequences. Such practices would produce so much fear that the happiness temporarily created would never counterbalance the unhappiness of everyone living in fear that they might be taken for sacrifice.

We need not debate utilitarianism here. The point is that deontologists find utilitarianism problematic because it puts the emphasis on the consequences of an act rather than on the act itself. Deontological theories claim that the internal character of the act is what is important. The rightness or wrongness of an action depends on the principles inherent in the action. If an action is done from a sense of duty, and if the principle of the action can be universalized, then the action is right. For example, if I tell the truth because it is convenient for me to do so, or because I fear the consequences of getting caught in a lie, my action is not worthy. A worthy action is an action that is done from duty, which involves respecting other people, recognizing them as ends in themselves, not as means to some good effect.

According to deontologists, utilitarianism is wrong because it treats individuals as means to an end (maximum happiness). For deontologists, what grounds morality is not happiness, but human beings as rational agents. Human beings are capable of reasoning about what they want to do. The laws of nature determine most activities: plants grow towards the sun, water boils at a certain temperature, and objects fall at a constant rate in a vacuum. Human action is different in that it is self-determining; humans initiate action after thinking, reasoning, and deciding. The human capacity for rational decisions makes morality possible, and it grounds deontological theory. Because each human being has this capacity, each human being must be treated accordingly: with respect. No one else can make our moral choices for us, and each of us must recognize this capacity in others.

Although deontological theories can be formulated in a number of ways, one formulation is particularly important: Immanuel Kant's categorical imperative [Kant 1785]. There are three versions of it, and the second version goes as follows: *Never treat another human being merely as a means but always as an end.* It is important to note the *merely* in the categorical imperative. Deontologists do not insist that we never use another person; only that we never *merely* use them. For example, if I own a company and hire employees to work in my company, I might be thought of as using those employees as a means to my end (i.e., the success of my business). This, however, is not wrong if the employees agree to work for me and if I pay them a fair wage. I thereby respect their ability to choose for themselves and I respect the value of their labor. What would be wrong would be to take them as slaves and make them work for me, or to pay them so little that they must borrow from me and must remain always in my debt. This would show disregard for the value of each person as a freely choosing, rationally valuing, specially efficacious person.

Social Contract Theories

A third tradition in ethics thinks of ethics on the model of a social contract. There are many different social contract theories, and some, at least, are based on a deontological principle. Individuals are rational free agents; hence, it is immoral to exert undue power over them, to coerce them. Government and society are problematic insofar as they seem to force individuals to obey rules, apparently treating individuals as means to social good. Social contract theories get around this problem by claiming that morality (and government policy) are, in effect, the outcome of rational agents agreeing to social rules. In agreeing to live by certain rules, we make a contract. Morality and government are not, then, systems imposed on individuals; they do not exactly involve coercion. Rather, they are systems created by freely choosing individuals (or they are institutions that rational individuals would choose if given the opportunity).

Philosophers such as Rousseau, Locke, Hobbes, and more recently Rawls [1971] are generally considered social contract theorists. They differ in how they get to the social contract and what it implies. For our purposes, however, the key idea is that principles and rules guiding behavior may be derived from identifying what it is that rational (even self-interested) individuals would agree to in making a social contract. Such principles and rules are the basis of a shared morality. For example, it would be rational for me to agree to live by rules that forbid killing and lying. Even though such rules constrain me, they also give me some degree of protection: if they are followed, I will not be killed or lied to.

The social contract theory cannot be used simply by asking what rules you would agree to now. Most theorists recognize that what you would agree to now is influenced by your present position in society. Most individuals would opt for rules that would benefit their particular situation and characteristics. Hence, most social contract theorists insist that the principles or rules of the contract must be derived by assuming certain things about human nature or the human condition. Rawls, for example, insists that we imagine ourselves behind a *veil of ignorance*. We are not allowed to know important features about ourselves, e.g., what talents we have, what race, gender we will be, for if we know these things we will not agree to just rules, but only rules that will maximize our self-interest. Justice consists of the rules we would agree to when we do not know who we are, for we would want rules that would give us a fair situation no matter where we ended up in the society.

A Paramedic Method for Computer Ethics

Drawing on elements of the three theories described, Collins and Miller [1992] have proposed a decision assisting method, called the paramedic method for computer ethics. This is not an algorithm for solving ethical problems; it is not nearly detailed or objective enough for that designation. It is merely a guideline for an organized approach to ethical problem solving.

Assume that a computer professional is faced with a decision that involves human values in a significant way. There may already be some obvious alternatives, and there also may be creative solutions not already discovered. The paramedic method is designed to help the professional analyze alternative actions and to encourage the development of creative solutions. The method proceeds as follows:

1. Identify alternative actions; list the few alternatives that seem most promising. If an action requires a long description, summarize it as a title with just a few words. Call the actions $A1, A2, \ldots, Aa$. No more than five actions should be analyzed at a time.

2. Identify people, groups of people, or organizations that will be affected by the decision that must be made. Again, hold down the number of entities to the five or six that are affected most. Label the people $P1, P2, \ldots, P_p$.

3. Make a table with the horizontal rows labeled by the identified people and the vertical columns labeled with the identified actions. We call such a table a $P \times A$ table. Make two copies of the $P \times A$ table, and label one as the *opportunities* table and the other as the *vulnerabilities* table. In the opportunities table, list in each interior cell of the table at entry $[x, y]$ the possible good that is likely to happen to person x if action y is taken. Similarly, in the vulnerability table, at position $[x, y]$ list all of the things that are likely to happen badly for x if the action y is taken. These two graphs represent benefit/cost calculations for a consequentialist, utilitarian analysis.

4. Make a new table with the set of persons marking both the columns and the rows (a $P \times P$ table). In each cell $[x, y]$ name any responsibilities or duties that x owes y in this situation. (The cells on the diagonal $[x, x]$ are important; they list things one owes oneself.) Now make copies of this table, labeling one copy for each of the alternative actions being considered. Work through each cell $[x, y]$ of each table and place a + next to a duty if the action for that table is likely to fulfill the duty x owes y; mark the duty with a − if the action is unlikely to fulfill that duty; mark the duty with a +/− if the action partially fulfills it and partially does not; and mark the duty with a ? if the action is irrelevant to the duty or if it is impossible to predict whether or not the duty will be fulfilled. (Few cells generally fall into this last category.)

5. Review the tables from steps 3 and 4. Envision a meeting of all of the parties (or one representative from each of the groups) in which no one knows which role they will take or when they will leave the negotiation. Which alternative do you think such a group would adopt, if any? Do you think such a group could discover a new alternative, perhaps combining the best elements of the previously listed actions? If this thought experiment produces a new alternative, expand the $P \times A$ tables from step 3 to include the new alternative action, and make a new copy of the $P \times P$ table in step 4 and do the + and − marking for the new table.

6. If any one of the alternatives seems to be clearly preferred (i.e., it has high opportunity and low vulnerability for all parties, and tends to fulfill all the duties in the $P \times P$ table), then that becomes the recommended decision. If no one alternative action stands out, the professionals can examine tradeoffs using the charts, or can iteratively attempt step 5 (perhaps with outside consultations) until an acceptable alternative is generated.

Using the paramedic method can be time consuming, and it does not eliminate the need for judgement. But it can help organize and focus analysis as an individual or group works through the details of a case situation to arrive at a decision.

Easy and Hard Ethical Decision Making

Sometimes ethical decision making is easy; for example, when it is clear that an action will prevent a serious harm and has no drawbacks, then that action is the ethical thing to do. Sometimes, however, ethical decision making is more complicated and challenging. Take the following case: your job is to make decisions about which parts to buy for a computer manufacturing company; a person who sells parts to the company offers you tickets to an expensive Broadway show; should you accept the tickets? In this case the right thing to do is more complicated because you may be able to accept the tickets and not have this affect your decision about parts. You owe your employer a decision on parts that is in the best interests of the company, but will accepting the tickets influence future decisions? Other times you know what the right thing to do is but doing it will have such great personal costs, that you can not bring yourself to do it; for example, you might be considering blowing the whistle on your employer who has been extremely kind and generous with you, but who now has asked you to cheat on the testing results on a life-critical software system designed for a client.

To make good decisions, professionals must be aware of potential issues and must have a fairly clear sense of their responsibilities in various kinds of situations. This often requires sorting out complex relationships and obligations, anticipating the effects of various actions, and balancing responsibilities to multiple parties. This activity is part of professional ethics.

2.3 Professional Ethics

Ethics is not just a matter for individuals as individuals. We all occupy a variety of social roles which carry with them special responsibilities and privileges. As parents, we have special responsibilities for children. As citizens, members of churches, officials in clubs, and so on, we have special rights and duties and so it is with professional roles. Being a professional is often distinguished from merely having an occupation because a professional makes a different sort of commitment. Being a professional means more than just having a job. The difference is commitment to doing the right thing because you are a member of a group that has taken on responsibility for a domain of social activity—a social function. The group is accountable to society for this domain, and for this reason, professionals must behave in ways that are worthy of public trust.

Some theorists explain this commitment in terms of a social contract between a profession and the society in which it functions. Society grants special rights and privileges to the professional group, such as control of admission to the group, access to educational institutions, and confidentiality in professional–client relationships. Society, in turn, may even grant the group a monopoly over a domain of activity (e.g., only licensed engineers can sign off on construction designs, only doctors can prescribe drugs). In exchange, the professional group promises to self-regulate and practice its profession in ways that are beneficial to society, i.e., to promote safety, health, welfare. The social contract idea is a way of illustrating the importance of the trust that clients and the public put in professionals; it shows the importance of professionals acting so as to be worthy of that trust.

The special responsibilities of professionals have been accounted for in other theoretical frameworks as well. Davis [1995], for example, argues that members of professions implicitly, if not explicitly, agree among themselves to adhere to certain standards because this elevates the level of activity. If all computer scientists and engineers, for example, agreed never to release software that has not met certain testing standards, this would prevent market pressures from driving down the quality of software being produced. Davis's point is that the special responsibilities of professionals are grounded in what members of a professional group owe to one another; they owe it to one another to live up to agreed-upon rules and standards. Yet other theorists have tried to ground the special responsibilities of professionals in ordinary morality. Alpern [1991] argues, for example, that the engineer's responsibility for safety derives from the ordinary moral edict, *do no harm*. Since engineers are in a position to do greater harm than others, engineers have a special responsibility in their work to take greater care.

In the case of the role of computing professionals, responsibilities are not always well articulated because of several factors. Computing is a relatively new field. Moreover, many computer scientists and engineers are both employees of companies and simultaneously members of a profession. This can create a tension blurring a professional's responsibilities. Being a professional means having the independence to make decisions on the basis of special expertise, but being an employee of a company often means acting for the best interests of the company, being loyal, and so on. The demands of a business (expectations of one's employer) can conflict with the demands of professional responsibility.

Another difficulty in defining and maintaining professional ethics for computing professionals is the diversity of the field. Computing professionals are employed in a wide variety of contexts, have a wide variety of expertise, and come from diverse educational backgrounds. There is no single unifying organization, no uniform admission standards, and no single identifiable professional role. To be sure, there are signs of the field moving in the direction of more professionalization, but as yet, computing is still a loose cluster of overlapping fields composed of individuals following diverse educational and career paths, and engaged in a wide variety of job activities.

Despite the lack of well-articulated unifying professional standards and ideals, there are expectations for professional practice. It is these expectations that form the basis of an emerging professional ethic that may, in the future, be refined to the point where there will be a strongly differentiated role for computer professionals.

These expectations, in particular their evolving character, can be seen in the growing sophistication of ethical codes in the field of computing. Professional codes play an important role in articulating a collective sense of what is both the ideal of the profession as well as the minimum standards required. Codes of conduct state the consensus views of members as well as shaping behavior.

A number of professional organizations have codes of ethics that are of interest here. The most well known include the Association for Computing Machinery (ACM) Code of Ethics and Professional Conduct (see Appendix B), the Institute of Electrical and Electronic Engineers (IEEE) Code of Ethics, the Data Processing Managers Association (DPMA) Code of Ethics and Standards of Conduct, the Institute for Certification of Computer Professionals (ICCP) Code of Ethics, the Canadian Information Processing Society Code of Ethics, and the British Computer Society Code of Conduct. Each of these codes has different emphases and goals. Each in its own way, however, deals with issues that arise in the context in which computer scientists and engineers typically practice.

The codes are relatively consistent in identifying computer professionals as having responsibilities to be faithful to their employers, to clients, and to protect public safety and welfare. The most salient ethical issues that arise in professional practice have to do with balancing these responsibilities together with personal (or nonprofessional) responsibilities. Two common areas of tension are worth mentioning here, albeit briefly.

As previously mentioned, computer engineers and scientists may find themselves in situations where their responsibility as professionals to protect the public comes in conflict with loyalty to their employer. Such situations sometimes escalate to the point where the computer professional has to decide whether to blow the whistle. Such a situation might arise, for example, when the computer professional believes that a piece of software has not been tested enough but her employer wants to deliver the software on time and within the allocated budget (which means immediate release and no more resources being spent on the project). The decision to blow the whistle or not to blow the whistle is one of the most difficult computer engineers and scientists may have to face. Whistle blowing has received a good deal of attention in the popular press and in the literature on professional ethics because this tension seems to be built into the role of engineers and scientists. Ideally, corporations and professional societies will, in the future, develop mechanisms to help avoid the need to blow the whistle. For example, if corporations had ombudspersons to whom engineers and scientists could report their concerns (anonymously) or if professional societies maintained hotlines that professionals could call for advice on how to get their concerns addressed, these would lessen the need to blow the whistle.

Another important professional ethics issue that often arises is directly tied to the importance of being worthy of client (and indirectly public) trust. Professionals can find themselves in situations in which they have (or are likely to have) a conflict of interest. A conflict of interest situation is one in which the professional is hired to perform work for a client and the professional has some personal or professional interest that may (or may appear to) interfere with their judgement on behalf of the client. For example, suppose a computer professional is hired by a company to evaluate their needs and recommend hardware and software that will best suit the company's needs. The computer professional does precisely what is requested, but fails to mention being a silent partner in a company that manufactures the hardware and software that has been recommended. In other words, the professional has a personal interest—financial benefit—in the company buying certain equipment. If the company were to find this out later on, it might rightly be thought that there had been deception. The professional was hired to evaluate the needs of the company and to determine how best to meet them, and in so doing to have the best interests of the company fully in mind. Now it is suspected that the professional's judgment may have been biased. The professional had an interest that might have biased his or her judgement.

There are a number of strategies that professions use to avoid these situations. A code of conduct may, for example, specify that professionals reveal all relevant interests to their clients before they accept a job. Or, the code might specify that members never work in a situation where there is even the appearance of a conflict of interest.

This brings us to the special character of computer technology and the effects that the work of computer professionals can have on the shape of our world. Some may argue that computer professionals have very little say in what technologies get designed and built. This seems to be mistaken on at least two counts. First, we can distinguish computer professionals as individuals and computer professionals as a group. Even if individuals have little power in the jobs they hold, they can exert power collectively. Second, individuals can have an effect if they think of themselves as professionals and consider it their responsibility to think about the impact of their work.

2.4 Ethical Issues that Arise from Computer Technology

The effects of a new technology on society can draw attention to an old issue, and can change our understanding of that issue. The issues listed in this section—privacy, property rights, risk and reliability, and global communication—were of concern, even problematic, before computers were an important technology. But computing and, more generally, electronic telecommunications, have added new twists and new intensity to each of these issues. Although computer professionals cannot be expected to be experts on all of these impacts, it is important for them to understand how computer technology is shaping the world. And, it is important for them to keep these impacts in mind as they work with computer technology. Those who are aware of privacy issues are more likely to take this into account when they design database management systems, those who are aware of risk and reliability issues are more likely to articulate these to clients and attend to them in design and documentation, and so on.

Privacy

Privacy is a central topic in computer ethics. Some have even suggested that privacy is a notion that has been antiquated by technology and that it should be replaced by a new openness. Others think that computers must be harnessed to help restore as much privacy as possible to our society. Although they may not like it, computer professionals are at the center of this controversy. Some are designers of the systems that facilitate information gathering and manipulation; others maintain and protect the information. As the saying goes, *information is power*, but power can be used and/or abused.

Computer technology creates wide ranging possibilities for tracking and monitoring of human behavior. Consider just two ways in which personal privacy may be affected by computer technology. First, because of the capacity of computers, massive amounts of information can be gathered by record keeping organizations such as banks, insurance companies, government agencies, educational institutions. The information gathered can be kept and used indefinitely, and shared with other organizations, rapidly and frequently. A second way in which computers have enhanced the possibilities for monitoring and tracking of individuals is by making possible new kinds of information. When activities are done using a computer, transactional information is created. When individuals use automated bank teller machines, records are created; when certain software is operating, keystrokes on a computer keyboard are recorded; the content and destination of electronic mail can be tracked, and so on. With the assistance of newer technologies, much more of this transactional information is likely to be created. For example, television advertisers may be able to monitor television watchers with scanning devices that record who is sitting in a room facing the television, and new highway systems may allow drivers to pass through toll booths without stopping as a beam reading a bar code on the automobile will automatically charge the toll to a credit card, creating a record of individual travel patterns. All of this information (transactional and otherwise) can be brought together to create a detailed portrait of a person's life, a portrait the individual may never see, though it is used by others to make decisions about the individual.

This picture of computer technology suggests that computer technology poses a serious threat to personal privacy. However, one can counter this picture in a number of ways. Is it computer technology per se that poses the threat or is it just the way the technology has been used (and is likely to be used in the future)? Only those who understand the technology are in a position to design or change computer technology so that it does not eradicate privacy.

At the same time we think about changing technology, we also have to ask deeper questions about privacy itself and what it is that individuals need, want, or are entitled to when they express concerns about the loss of privacy. In this sense, computers and privacy issues are ethical issues. They compel us to ask deep questions about what makes for a good and just society. Should individuals have more choice about who has what information about them? What is the proper relationship between citizens and government, between individuals and private corporations? As previously suggested, the questions are not completely new; but some of the possibilities created by computers are new, and these possibilities do not readily fit the concepts and frameworks used in the past.

Property Rights and Computing

The protection of intellectual property rights has become an active legal and ethical debate, involving national and international players. Should software be copyrighted, patented, or free? Is computer software a process, a creative work, a mathematical formalism, an idea, or some of all of these? What is society's stake in protecting software rights? What is society's stake in widely disseminating software? How do corporations and other institutions protect their rights to ideas developed by individuals, and what are the individuals' rights? These kinds of questions must be answered publicly through legislation, through corporate policies, and with the advice of computing professionals. Some of the answers will involve technical details, and all should be informed by ethical analysis and debate.

Perhaps the issue that has received the most legal and public attention is that concerning the ownership of software. In the course of history, software is a relatively new entity. Whereas western legal systems have developed property laws that encourage invention by granting certain rights to inventors, there are provisions against ownership of things that might interfere with the development of the technological arts and sciences. For this reason, copyrights protect only the expression of ideas, not the ideas themselves, and we do not grant patents on laws of nature, mathematical formulas, and abstract ideas. The problem with computer software is that it has not been clear that we could grant ownership of it without, in effect, granting ownership of numerical sequences or mental steps. Software can be copyrighted, because a copyright gives the holder ownership of the *expression* of the idea (not the idea itself), but this does not give software inventors as much protection as they claim to need to *fairly* compete. Competitors may see the software, grasp the idea, and write a somewhat different program to do the same thing. The competitor can sell the software at less cost because the cost of developing the first software does not have to be paid. Patenting would provide stronger protection, but until quite recently the courts have been reluctant to grant this protection because of the problem previously mentioned: patents on software would appear to give the holder control of the building blocks of the technology, an ownership comparable to owning ideas themselves.

Like the questions surrounding privacy, property rights in computer software also lead back to broader ethical and philosophical questions about what constitutes a just society. In computing, as in other areas of technology, we want a system of property rights that promotes invention (creativity, progress), but at the same time, we want a system that is fair in the sense that it rewards those who make significant contributions but does not give anyone so much control that they prevent others from creating. Policies with regard to property rights in computer software cannot be made without an understanding of the technology, and this is why it is so important for computer professionals to be involved in public discussion and policy setting on this topic.

Risk, Reliability, and Accountability

As computer technology becomes more important to the way we live, its risks become more worrisome. System errors can lead to physical danger, sometimes catastrophic in scale. There are security risks due to hackers and crackers. Unreliable data as well as intentional misinformation are risks that are increased because of the technical and economic characteristics of digital data. Furthermore, the use of computer programs is, in a practical sense, inherently unreliable.

Each of these issues (and many more) requires computer professionals to face the linked problems of risk, reliability, and accountability. Professionals must be candid about the risks of a particular application or system. Computing professionals should take the lead in educating customers and the public about what predictions we can and cannot make about software and hardware reliability. Computer professionals should make realistic assessments about costs and benefits, and be willing to take on both for projects they are involved with.

There are also issues of sharing risks as well as resources. Should liability fall to the individual who buys software or to the corporation that developed it? Should society acknowledge the inherent risks in using software in life-critical situations and shoulder some of the responsibility when something goes wrong? Or should software providers (both individuals and institutions) be exclusively responsible for software safety? All of these issues require us to look at the interaction of technical decisions, human consequences, rights, and responsibilities. They call not just for technical solutions but for solutions which recognize the kind of society we want to have and the values we want to preserve.

Rapidly Evolving Globally Networked Telecommunications

The system of computers and connections known as the Internet is forming a new kind of community or sets of communities—electronic communities. Questions of individual accountability and social control, as well as matters of etiquette that arise in all societies are taking shape in a new way, in the electronic medium. It is as if we have society (societies) forming in a new physical environment. A new way of living together is evolving as we watch. What will the Internet be like in five years? Who will and won't have access? How much freedom will we trade for security? How will commercial interests and citizens opposed to commercialization coexist? What will electronic communications mean to our worlds of work and play? Will the Internet begin to change who we are or who we are to each other?

Speculating about the Internet is now a popular pastime. But some researchers in computer ethics think that more serious thought, and perhaps action, should be applied to shaping the society of network users. Commercial, governmental, and recreational groups are already changing what the Internet used to be, often making unilateral statements or actions. Instead of asking What will happen to the Internet? we should perhaps be asking What *should* happen to the Internet? Questions of *should* are exactly the questions that ethics addresses.

2.5 Final Thoughts

Computer technology will, no doubt, continue to evolve and will continue to affect the character of the world we live in. Computer scientists and engineers will play an important role in shaping the technology. The technologies we use shape how we live and who we are. They make every difference in the moral environment in which we live. Hence, it seems of utmost importance that computer scientists and engineers understand just how their work affects humans and human values.

References

Alpern, K. D. 1991. Moral responsibility for engineers. In *Ethical Issues in Engineering*, D. G. Johnson, ed., pp. 187–195. Prentice–Hall, Englewood Cliffs, NJ.

Collins, W. R. and Miller, K. 1992. A paramedic method for computing professionals. *J. Syst. Software.* 17(1):47–84.

Davis, M. 1995. Thinking like an engineer: The place of a code of ethics in the practice of a profession. In *Computers, Ethics, and Social Values*, D. G. Johnson and H. Nissenbaum, eds., pp. 586–597. Prentice–Hall, Englewood Cliffs, NJ.

Johnson, D. G. 1994. *Computer Ethics*, 2nd ed. Prentice–Hall, Englewood Cliffs, NJ.

Kant, I. 1785. Foundations of the Metaphysics of Morals. L. Beck, Trans., 1959. Library of Liberal Arts, 1959.

Rawls, J. 1971. *A Theory of Justice*. Harvard Univ. Press, Cambridge, MA.

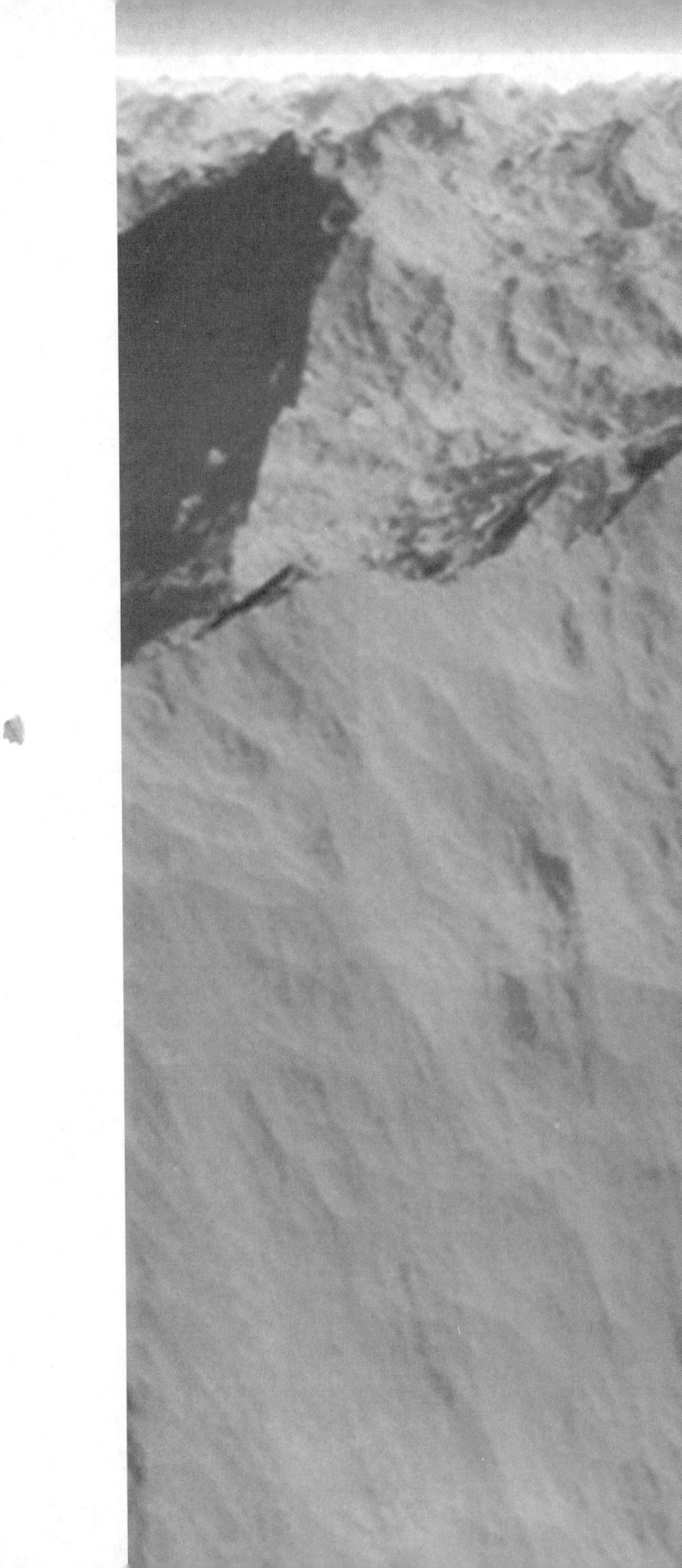

I

Algorithms and Data Structures

Mikhail J. Atallah, Section Adviser
Purdue University

THIS SECTION ADDRESSES THE CHALLENGE of solving hard problems algorithmically and efficiently. It covers basic methodologies (divide and conquer), complexity classes P and NP, space and time measures, approximation methods, parallel algorithms, hard problems, unsolvable problems, and an application in VLSI design. Other application areas covered include pattern matching and text compression algorithms, computational geometry, and algebraic algorithms.

3

Formal Models and Computability

Tao Jiang
McMaster University

Ming Li
University of Waterloo

Bala Ravikumar
University of Rhode Island

3.1 Introduction

The concept of **algorithms** is perhaps almost as old as human civilization. The famous Euclid's algorithm is more than 2000 years old. Angle trisection, solving diophantine equations, and finding polynomial roots in terms of radicals of coefficients are some well-known examples of algorithmic questions. However, until the 1930s the notion of algorithms was used informally (or rigorously but in a limited context). It was a major triumph of logicians and mathematicians of this century to offer a rigorous definition of this fundamental concept. The revolution that resulted in this triumph was a collective achievement of many mathematicians, notably Church, Gödel, Kleene, Post, and Turing. Of particular interest is a machine model proposed by Turing in 1936, which has come to be known as a **Turing machine** [Turing 1936].

This particular achievement had numerous significant consequences. It led to the concept of a general-purpose computer or universal computation, a revolutionary idea originally anticipated by Babbage in the 1800s. It is widely acknowledged that the development of a universal Turing machine was prophetic of the modern all-purpose digital computer and played a key role in the thinking of pioneers in the development of modern computers such as von Neumann [Davis 1980]. From a mathematical point of view, however, a more interesting consequence was that it was now possible to show the *nonexistence* of algorithms, hitherto impossible due to their elusive nature. In addition, many apparently different definitions of an algorithm proposed by different researchers in different continents turned out to be equivalent (in a precise technical sense, explained later). This equivalence led to the widely held hypothesis known as the *Church–Turing thesis* that mechanical solvability is the same as solvability on a Turing machine.

Formal languages are closely related to algorithms. They were introduced as a way to convey mathematical proofs without errors. Although the concept of a formal language dates back at least to the time of Leibniz, a systematic study of them did not begin until the beginning of this century. It became a vigorous field of study when Chomsky formulated simple grammatical rules to describe the syntax of a language

[Chomsky 1956]. **Grammars** and **formal languages** entered into computability theory when Chomsky and others found ways to use them to classify algorithms.

The main theme of this chapter is about formal models, which include Turing machines (and their variants) as well as grammars. In fact, the two concepts are intimately related. Formal computational models are aimed at providing a framework for computational problem solving, much as electromagnetic theory provides a framework for problems in electrical engineering. Thus, formal models guide the way to build computers and the way to program them. At the same time, new models are motivated by advances in the technology of computing machines. In this chapter, we will discuss only the most basic computational models and use these models to classify problems into some fundamental classes. In doing so, we hope to provide the reader with a conceptual basis with which to read other chapters in this Handbook.

3.2 Computability and a Universal Algorithm

Turing's notion of mechanical computation was based on identifying the basic steps of such computations. He reasoned that an operation such as multiplication is not primitive because it can be divided into more basic steps such as digit-by-digit multiplication, shifting, and adding. Addition itself can be expressed in terms of more basic steps such as add the lowest digits, compute, carry, and move to the next digit, etc. Turing thus reasoned that the most basic features of mechanical computation are the abilities to read and write on a storage medium (which he chose to be a linear tape divided into cells or squares) and to make some simple logical decisions. He also restricted each tape cell to hold only one among a finite number of symbols (which we call the *tape alphabet*).[1] The decision step enables the computer to control the sequence of actions. To make things simple, Turing restricted the next action to be performed on a cell neighboring the one on which the current action occurred. He also introduced an instruction that told the computer to stop. In summary, Turing proposed a model to characterize mechanical computation as being carried out as a sequence of instructions of the form: write a symbol (such as 0 or 1) on the tape cell, move to the next cell, observe the symbol currently scanned and choose the next step accordingly, or stop.

These operations define a language we call the GOTO language.[2] Its instructions are

> PRINT i (i is a tape symbol)
> GO RIGHT
> GO LEFT
> GO TO STEP j IF i IS SCANNED
> STOP

A **program** in this language is a sequence of instructions (written one per line) numbered 1–k. To run a program written in this language, we should provide the *input*. We will assume that the input is a string of symbols from a finite input alphabet (which is a subset of the tape alphabet), which is stored on the tape before the computation begins. How much memory should we allow the computer to use? Although we do not want to place any bounds on it, allowing an infinite tape is not realistic. This problem is circumvented by allowing *expandable memory*. In the beginning, the tape containing the input defines its boundary. When the machine moves beyond the current boundary, a new memory cell will be attached with a special symbol B (blank) written on it. Finally, we define the result of computation as the contents of the tape when the computer reaches the STOP instruction.

We will present an example program written in the GOTO language. This program accomplishes the simple task of doubling the number of 1s (Fig. 3.1). More precisely, on the input containing k 1s, the program produces $2k$ 1s. Informally, the program achieves its goal as follows. When it reads a 1, it changes

---

[1] This bold step of using a discrete model was perhaps the harbinger of the digital revolution that was soon to follow.

[2] Turing's original formulation is closer to our presentation in section 3.5. But the GOTO language presents an equivalent model.

the 1 to 0, moves left looking for a new cell, writes a 1 in the cell, returns to the starting cell and rewrites as 1, and repeats this step for each 1. Note the way the GOTO instructions are used for repetition. This feature is the most important aspect of programming and can be found in all of the imperative style programming languages.

1	PRINT 0
2	GO LEFT
3	GO TO STEP 2 IF 1 IS SCANNED
4	PRINT 1
5	GO RIGHT
6	GO TO STEP 5 IF 1 IS SCANNED
7	PRINT 1
8	GO RIGHT
9	GO TO STEP 1 IF 1 IS SCANNED
10	STOP

FIGURE 3.1   The doubling program in the GOTO language.

The simplicity of the GOTO language is rather deceptive. There is strong reason to believe that it is powerful enough that any mechanical computation can be expressed by a suitable program in the GOTO language. Note also that the programs written in the GOTO language may not always halt, that is, on certain inputs, the program may never reach the STOP instruction. In this case, we say that the output is undefined.

We can now give a precise definition of what an algorithm is. An algorithm is any program written in the GOTO language with the additional property that it halts on all inputs. Such programs will be called *halting programs.* Throughout this chapter, we will be interested mainly in computational problems of a special kind called *decision problems* that have a yes/no answer. We will modify our language slightly when dealing with decision problems. We will augment our instruction set to include ACCEPT and REJECT (and omit STOP). When the ACCEPT (REJECT) instruction is reached, the machine will output yes or 1 (no or 0) and halt.

Some Computational Problems

We will temporarily shift our focus from the tool for problem solving (the computer) to the problems themselves. Throughout this chapter, a computational problem refers to an input/output relationship. For example, consider the problem of squaring an integer input. This problem assigns to each integer (such as 22) its square (in this case 484). In technical terms, this input/output relationship defines a function. Therefore, solving a computational problem is the same as computing the function defined by the problem. When we say that an algorithm (or a program) solves a problem, what we mean is that, for all inputs, the program halts and produces the correct output. We will allow inputs of arbitrary size and place no restrictions. A reader with primary interest in software applications is apt to question the validity (or even the meaningfulness) of allowing inputs of arbitrary size because it makes the set of all *possible* inputs infinite, and thus unrealistic, in real-world programming. But there are no really good alternatives. Any finite bound is artificial and is likely to become obsolete as the technology and our requirements change. Also, in practice, we do not know how to take advantage of restrictions on the size of the inputs. (See the discussion about nonuniform models in section 3.5.) Problems (functions) that can be solved by an algorithm (or a halting GOTO program) are called *computable.*

As already remarked, we are interested mainly in decision problems. A decision problem is said to be decidable if there is a halting GOTO program that solves it correctly on all inputs. An important class of problems called **partially decidable problems** can be defined by relaxing our requirement a little bit; a decision problem is partially decidable if there is a GOTO program that halts and outputs 1 on all inputs for which the output should be 1 and either halts and outputs 0 or loops forever on the other inputs. This means that the program may never give a wrong answer but is not required to halt on negative inputs (i.e., inputs with 0 as output).

We now list some problems that are fundamental either because of their inherent importance or because of their historical roles in the development of computation theory:

> Problem 1 (**halting problem**). The input to this problem is a program P in the GOTO language and a binary string x. The expected output is 1 (or yes) if the program P halts when run on the input x, 0 (or no) otherwise.

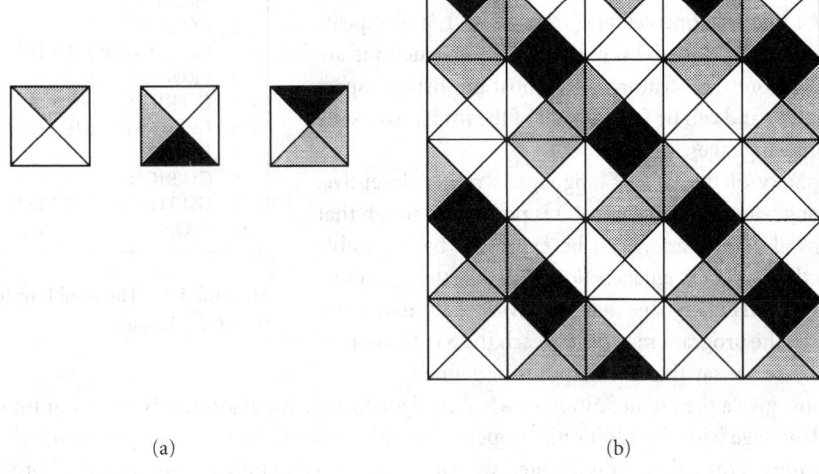

(a)   (b)

FIGURE 3.2   An example of tiling.

Problem 2 (universal computation problem). A related problem takes as input a program P and an
input x and produces as output what (if any) P would produce on input x. (Note that this is
a decision problem if P is restricted to a yes/no program.)

Problem 3 (string compression). For a string x, we want to find the shortest program in the GOTO
language that when started with the empty tape (i.e., tape containing one B symbol) halts
and prints x. Here shortest means the total number of symbols in the program is as small as
possible.

Problem 4 (tiling). A tile[3] is a square card of unit size (i.e., 1×1) divided into four quarters by two
diagonals, each quarter colored with some color (selected from a finite set of colors). The tiles
have fixed orientation and cannot be rotated. Given some finite set T of such tiles as input,
the program is to determine if finite rectangular areas of all sizes (i.e., $k \times m$ for all positive
integers k and m) can be tiled using only the given tiles such that the colors on any two touching
edges are the same. It is assumed that an unlimited number of cards of each type is available.
Figure 3.2(b) shows how the base set of tiles given in Fig. 3.2(a) can be used to tile a 5×5
square area.

Problem 5 (linear programming). Given a system of linear inequalities (called constraints), such as
$3x - 4y \leq 13$ with integer coefficients, the goal is to find if the system has a solution satisfying
all of the constraints.

Some remarks must be made about the preceding problems. The problems in our list include nonnu-
merical problems and *meta problems*, which are problems about other problems. The first two problems
are motivated by a quest for reliable program design. An algorithm for problem 1 (if it exists) can be used
to test if a program contains an infinite loop. Problem 2 is motivated by an attempt to design a **universal
algorithm**, which can simulate any other. This problem was first attempted by Babbage, whose analytical
engine had many ingredients of a modern electronic computer (although it was based on mechanical
devices). Problem 3 is an important problem in information theory and arises in the following setting.
Physical theories are aimed at creating simple laws to explain large volumes of experimental data. A famous
example is Kepler's laws, which explained Tycho Brahe's huge and meticulous observational data. Problem
3 asks if this compression process can be automated. When we allow the inference rules to be sufficiently
strong, this problem becomes **undecidable**. We will not discuss this problem further in this section but
will refer the reader to some related formal systems discussed in Li and Vitányi [1993]. The tiling problem

---

[3]More precisely, a Wang tile, after Hao Wang, who wrote the first research paper on it.

is not merely an interesting puzzle. It is an art form of great interest to architects and painters. Tiling has recently found applications in crystallography. Linear programming is a problem of central importance in economics, game theory, and operations research.

In the remainder of the section, we will present some basic algorithm design techniques and sketch how these techniques can be used to solve some of the problems listed (or their special cases). The main purpose of this discussion is to present techniques for showing the decidability (or partial decidability) of these problems. The reader can learn more advanced techniques of algorithm design in some later sections of this chapter as well as in many later chapters of this volume.

Table Lookup

The basic idea is to create a table for a function f, which needs to be computed by tabulating in one column an input x and the corresponding $f(x)$ in a second column. Then the table itself can be used as an algorithm. This method cannot be used directly because the set of all inputs is infinite. Therefore, it is not very useful, although it can be made to work in conjunction with the technique described subsequently.

Bounding the Search Domain

The difficulty of establishing the decidability of a problem is usually caused by the fact that the object we are searching for may have no known upper limit. Thus, if we can place such an upper bound (based on the structure of the problem), then we can reduce the search to a finite domain. Then table lookup can be used to complete the search (although there may be better methods in practice). For example, consider the following special case of the tiling problem: Let k be a fixed integer, say 1000. Given a set of tiles, we want to determine whether all rectangular rooms of shape $k \times n$ can be tiled for all n. (Note the difference between this special case and the general problem. The general one allows k and n both to have unbounded value. But here we allow only n to be unbounded.) It can be shown (see section 3.5 for details) that there are two bounds n_0 and n_1 (they depend on k) such that if there is at least one tile of size $k \times t$ that can be tiled for some $n_0 \leq t \leq n_1$ then every tile of size $k \times n$ can be tiled. If no $k \times t$ tile can be tiled for any t between n_0 and n_1, then obviously the answer is no. Thus, we have reduced an infinite search domain to a finite one.

As another example, consider the linear programming problem. The set of possible solutions to this problem is infinite, and thus a table search cannot be used. But it is possible to reduce the search domain to a finite set using the geometric properties of the set of solutions of the linear programming problem. The fact that the set of solutions is convex makes the search especially easy.

Use of Subroutines

This is more of a program design tool than a tool for algorithm design. A central concept of programming is repetitive (or iterative) computation. We already observed how GOTO statements can be used to perform a sequence of steps repetitively. The idea of a subroutine is another central concept of programming. The idea is to make use of a program P itself as a single step in another program Q. Building programs from simpler programs is a natural way to deal with the complexity of programming tasks. We will illustrate the idea with a simple example. Consider the problem of multiplying two positive integers i and j. The input to the problem will be the form $11 \ldots 1011 \ldots 1$ (i 1s followed by a 0, followed by j 1s) and the output will be $i * j$ 1s (with possibly some 0s on either end). We will use the notation $1^i 0 1^j$ to denote the starting configuration of the tape. This just means that the tape contains i 1s followed by a 0 followed by j 1s. The basic idea behind a GOTO program for this problem is simple; add j 1s on the right end of tape exactly $i - 1$ times and then erase the original sequence of i 1s on the left. A little thought reveals that the subroutine we need here is to duplicate a string of 1s so that if we start with $x 0 2^k 1^j$ a call to the subroutine will produce $x 0 2^{k+j} 1^j$. Here x is just any sequence of symbols. Note the role played by the symbol 2. As new 1s are created on the right, the old 1s change to 2s. This will ensure that there are exactly j 1s on the right end of the tape all of the time. This duplication subroutine is very similar to the doubling program, and the reader should have very little difficulty writing this program. Finally, the multiplication program can be done using the copy subroutine ($i - 1$) times.

A Universal Algorithm

We will now present in some detail a (partial) solution to problem 2 by arguing that there is a program U written in the GOTO language, which takes as input a program P (also written using the GOTO language) and an input x and produces as output $P(x)$, the output of P on input x. For convenience, we will assume that all programs written in the GOTO language use a fixed alphabet containing just 0, 1, and B. Because we have assumed this for all programs in

TABLE 3.1 Coding the GOTO Instructions

Instruction	Code
PRINT i	0001^{i+1}
GO LEFT	001
GO RIGHT	010
GO TO j IF i IS SCANNED	$0111^j 01^{i+1}$
STOP	100

the GOTO language, we should first address the issue of how an input to program U will look. We cannot directly place a program P on the tape because the alphabet used to write the program P uses letters G, O, T, O, etc. This minor problem can be easily circumvented by coding. The idea is to represent each instruction using only 0 and 1. One such coding scheme is shown in Table 3.1.

To encode an entire program, we simply write down in order (without the line numbers) the code for each instruction as given in the table. For example, here is the code for the doubling program shown in Fig. 3.1:

$$000100101111011000110100111111011000110100111011100$$

Note that the encoded string contains all of the information about the program so that the encoding is completely reversible. From now on, if P is a program in the GOTO language, then code(P) will denote its binary code as just described. When there is no confusion, we will identify P and code(P). Before proceeding further, the reader may want to test his/her understanding of the encoding/decoding process by decoding the following string: 010011101100.

The basic idea behind the construction of a universal algorithm is simple, although the details involved in actually constructing one are enormous. We will present the central ideas and leave out the actual construction. Such a construction was carried out in complete detail by Turing himself and was simplified by others.[4] U has as its input code(P) followed by the string x. U simulates the computational steps of P on input x. It divides the input tape into three segments, one containing the program P, the second one essentially containing the contents of the tape of P as it changes with successive moves, and the third one containing the line number in program P of the instruction being currently simulated (similar to a *program counter* in an actual computer).

We now describe a *cycle* of computation by U, which is similar to a central processing unit (CPU) cycle in a real computer. A single instruction of P is implemented by U in one cycle. First, U should know which location on the tape that P is currently reading. A simple artifact can handle this as follows: U uses in its tape alphabet two special symbols $0'$ and $1'$. U stores the tape of P in the tape segment alluded to in the previous paragraph exactly as it would appear when the program P is run on the input x with one minor modification. The symbol currently being read by program P is stored as the *primed version* ($0'$ is the primed version of 0, etc.). As an example, suppose after completing 12 instructions, P is reading the fourth symbol (from left) on its tape containing 01001001. Then the tape region of U after 12 cycles looks like $0100'1001$. At the beginning of a new cycle, U uses a subroutine to move to the region of the tape that contains the ith instruction of program P where i is the value of the program counter. It then decodes the ith instruction. Based on what type it is, U proceeds as follows: If it is a PRINT i instruction, then U scans the tape until the unique primed symbol in the *tape region* is reached and rewrites it as instructed. If it is a GO LEFT or GO RIGHT symbol, U locates the primed symbol, unprimes it, and primes its left or right neighbor, as instructed. In both cases, U returns to the program counter and increments it. If the instruction is GO TO i IF j IS SCANNED, U reads the primed symbol, and if it is j', U changes the program counter to i. This completes a cycle. Note that the three regions may grow and contract while U

---

[4]A particularly simple exposition can be found in Robinson [1991].

executes the cycles of computation just described. This may result in one of them running into another. U must then shift one of them to the left or right and make room as needed.

It is not too difficult to see that all of the steps described can be done using the instructions of the GOTO language. The main point to remember is that these actions will have to be coded as a single program, which has nothing whatsoever to do with program P. In fact, the program U is totally independent of P. If we replace P with some other program Q, it should simulate Q as well. The preceding argument shows that problem 2 is partially decidable. But it does not show that this problem is decidable. Why? It is because U may not halt on all inputs; specifically, consider an input consisting of a program P and a string x such that P does not halt on x. Then U will also keep executing cycle after cycle the moves of P and will never halt. In fact, in section 3.3, we will show that problem 2 is not decidable.

3.3 Undecidability

Recall the definition of an undecidable problem. In this section, we will establish the undecidability of problem 2, section 3.2. The simplest way to establish the existence of undecidable problems is as follows: There are more problems than there are programs, the former set being uncountable, whereas the latter is countably infinite.[5] But this argument is purely existential and does not identify any specific problem as undecidable. In what follows, we will show that problem 2 introduced in section 3.2 is one such problem.

Diagonalization and Self-Reference

Undecidability is inextricably tied to the concept of self-reference, and so we begin by looking at this rather perplexing and sometimes paradoxical concept. The idea of self-reference seems to be many centuries old and may have originated with a barber in ancient Greece who had a sign board that read: "I shave all those who do not shave themselves." When the statement is applied to the barber himself, we get a self-contradictory statement. Does he shave himself? If the answer is yes, then he is one of those who shaves himself, and so the barber should not shave him. The contrary answer no is equally untenable. So neither yes nor no seems to be the correct answer to the question; this is the essence of the paradox. The barber's paradox has made entry into modern mathematics in various forms. We will present some of them in the next few paragraphs.[6]

The first version, called Berry's paradox, concerns English descriptions of natural numbers. For example, the number 7 can be described by many different phrases: seven, six plus one, the fourth smallest prime, etc. We are interested in the *shortest* of such descriptions, namely, the one with the fewest letters in it. Clearly there are (infinitely) many positive integers whose shortest descriptions exceed 100 letters. (A simple counting argument can be used to show this. The set of positive integers is infinite, but the set of positive integers with English descriptions in fewer than or equal to 100 letters is finite.) Let D denote the set of positive integers that do not have English descriptions with fewer than 100 letters. Thus, D is not empty. It is a well-known fact in set theory that any nonempty subset of positive integers has a smallest integer. Let x be the smallest integer in D. Does x have an English description with fewer than or equal to 100 letters? By the definition of the set D and x, we have: x is "the smallest positive integer that cannot be described in English in fewer than 100 letters." This is clearly absurd because part of the last sentence in quotes is a description of x and it contains fewer than 100 letters in it. A similar paradox was found by the British mathematician Bertrand Russell when he considered the set of all sets that do

---

[5]The reader who does not know what countable and uncountable infinities are can safely ignore this statement; the rest of the section does not depend on it.

[6]The most enchanting discussions of self-reference are due to the great puzzlist and mathematician R. Smullyan who brings out the breadth and depth of this concept in such delightful books as *What is the name of this book?* published by Prentice–Hall in 1978 and *Satan, Cantor, and Infinity* published by Alfred A. Knopf in 1992. We heartily recommend them to anyone who wants to be amused, entertained, and, more importantly, educated on the intricacies of mathematical logic and computability.

not include themselves as elements, that is, $S = \{x \mid x \notin x\}$. The question "Is $S \in S$?" leads to a similar paradox.

As a last example, we will consider a charming self-referential paradox due to mathematician William Zwicker. Consider the collection of all two-person games (such as chess, tic-tac-toe, etc.) in which players make alternate moves until one of them loses. Call such a game *normal* if it has to end in a finite number of moves, no matter what strategies the two players use. For example, tic-tac-toe must end in at most nine moves and so it is normal. Chess is also normal because the 50-move rule ensures that the game cannot go forever. Now here is *hypergame*. In the first move of the hypergame, the first player calls out a normal game, and then the two players go on to play the game, with the second player making the first move. The question is: "Is hypergame normal?" Suppose it is normal. Imagine two players playing hypergame. The first player can call out hypergame (since it is a normal game). This makes the second player call out the name of a normal game, hypergame can be called out again and they can keep saying hypergame without end, and this contradicts the definition of a normal game. On the other hand, suppose it is not a normal game. But now in the first move, player 1 cannot call out hypergame and would call a normal game instead, and so the infinite move sequence just given is not possible, and so hypergame is normal after all!

In the rest of the section, we will show how these paradoxes can be modified to give nonparadoxical but surprising conclusions about the decidability of certain problems. Recall the encoding we presented in section 3.2 that encodes any program written in the GOTO language as a binary string. Clearly this encoding is reversible in the sense that if we start with a program and encode it, it is possible to decode it back to the program. However, not every binary string corresponds to a program because there are many strings that cannot be decoded in a meaningful way, for example, 11010011000110. For the purposes of this section, however, it would be convenient if we can treat *every* binary string as a program. Thus, we will simply stipulate that any undecodable string be decoded to the program containing the single statement

1.   REJECT

In the following discussion, we will identify a string x with a GOTO program to which it decodes. Now define a function f_D as follows: $f_D(x) = 1$ if x, decoded into a GOTO program, does not halt when started with x itself as the input. Note the self-reference in this definition. Although the definition of f_D seems artificial, its importance will become clear in the next section when we use it to show the undecidability of problem 2. First we will prove that f_D is not computable. Actually, we will prove a stronger statement, namely, that f_D is not even partially decidable. [Recall that a function is partially decidable if there is a GOTO program (not necessarily halting) that computes it. An important distinction between computable and semicomputable functions is that a GOTO program for the latter need not halt on inputs with output $= 0$.]

Theorem 3.1.   *Function f_D is not partially decidable.*

The proof is by contradiction. Suppose a GOTO program P' computes the function f_D. We will modify P' into another program P in the GOTO language such that P computes the same function as P' but has the additional property that it will never terminate its computation by ending up in a REJECT statement.[7] Thus, P is a program with the property that it computes f_D and halts on an input y if and only if $f_D(y) = 1$. We will complete the proof by showing that there is at least one input in which the program produces a wrong output, that is, there is an x such that $f_D(x) \neq P(x)$.

Let x be the encoding of program P. Now consider the question: Does P halt when given x as input? Suppose the answer is yes. Then, by the way we constructed P, here $P(x) = 1$. On the other hand, the

---

[7]The modification needed to produce P from P' is straightforward. If P' did not have any REJECT statements at all, then no modification would be needed. If it had, then we would have to replace each one by a looping statement, which keeps repeating the same instruction forever.

definition of f_D implies that $f_D(x) = 0$. (This is the punch line in this proof. We urge the reader to take a few moments and read the definition of f_D a few times and make sure that he or she is convinced about this fact!) Similarly, if we start with the assumption that $P(x) = 0$, we are led to the conclusion that $f_D(x) = 1$. *In both cases,* $f_D(x) \neq P(x)$ and thus P is not the correct program for f_D. Therefore, P' is not the correct program for f_D either because P and P' compute the same function. This contradicts the hypothesis that such a program exists, and the proof is complete.

Note the crucial difference between the paradoxes we presented earlier and the proof of this theorem. Here we do not have a paradox because our conclusion is of the form $f_D(x) = 0$ if and only if $P(x) = 1$ and not $f_D(x) = 1$ if and only if $f_D(x) = 0$. But in some sense, the function f_D was motivated by Russell's paradox. We can similarly create another function f_Z (based on Zwicker's paradox of hypergame). Let f be any function that maps binary strings to $\{0, 1\}$. We will describe a method to generate successive functions f_1, f_2, etc., as follows: Suppose $f(x) = 0$ for all x. Then we cannot create any more functions, and the sequence stops with f. On the other hand, if $f(x) = 1$ for some x, then choose one such x and decode it as a GOTO program. This defines another function; call it f_1 and repeat the same process with f_1 in the place of f. We call f a normal function if no matter how x is selected at each step, the process terminates after a finite number of steps. A simple example of a nonnormal function is as follows: Suppose $P(Q) = 1$ for some program P and input Q and at the same time $Q(P) = 1$ (note that we are using a program and its code interchangeably), then it is easy to see that the functions defined by both P and Q are not normal. Finally, define $f_Z(X) = 1$ if X is a normal program, 0 if it is not. We leave it as an instructive exercise to the reader to show that f_Z is not semicomputable. A perceptive reader will note the connection between Berry's paradox and problem 3 in our list (string compression problem) just as f_Z is related to Zwicker's paradox. Such a reader should be able to show the undecidability of problem 3 by imitating Berry's paradox.

Reductions and More Undecidable Problems

Theory of computation deals not only with the behavior of individual problems but also with relations among them. A **reduction** is a simple way to relate two problems so that we can deduce the (un)decidability of one from the (un)decidability of the other. Reduction is similar to using a subroutine. Consider two problems A and B. We say that problem A can be reduced to problem B if there is an algorithm for B provided that A has one. To define the reduction (also called a *Turing reduction*) precisely, it is convenient to augment the instruction set of the GOTO programming language to include a new instruction CALL X, i, j where X is a (different) GOTO program, and i and j are line numbers. In detail, the execution of such augmented programs is carried out as follows: When the computer reaches the instruction CALL X, i, j, the program will simply start executing the instructions of the program from line 1, treating whatever is on the tape currently as the input to the program X. When (if at all) X finishes the computation by reaching the ACCEPT statement, the execution of the original program continues at line number i and, if it finishes with REJECT, the original program continues from line number j.

We can now give a more precise definition of a reduction between two problems. Let A and B be two computational problems. We say that A is reducible to B if there is a halting program Y in the GOTO language for problem A in which calls can be made to a halting program X for problem B. The algorithm for problem A described in the preceding reduction does not assume the availability of program X and cannot use the details behind the design of this algorithm. The right way to think about a reduction is as follows: Algorithm Y, from time to time, needs to know the solutions to different instances of problem B. It can query an algorithm for problem B (as a black box) and use the answer to the query for making further decisions. An important point to be noted is that the program Y actually can be implemented even if program X was never built as long as someone can correctly answer some questions asked by program Y about the output of problem B for certain inputs. Programs with such calls are sometimes called *oracle programs*. Reduction is rather difficult to assimilate at the first attempt, and so we will try to explain it using a puzzle. How do you play two chess games, one each with Kasparov and Anand (perhaps currently the world's two best players) and ensure that you get at least one point? (You earn one point for a win,

0 for a loss, and 1/2 for a draw.) Because you are a novice and are pitted against two Goliaths, you are allowed a concession. You can choose to play white or black on either board. The well-known answer is the following: Take white against one player, say, Anand, and black against the other, namely, Kasparov. Watch the first move of Kasparov (as he plays white) and make the same move against Anand, get his reply and play it back to Kasparov and keep playing back and forth like this. It takes only a moment's thought that you are guaranteed to win (exactly) 1 point. The point is that your game involves taking the position of one game, applying the algorithm of one player, getting the result and applying it to the other board, etc., and you do not even have to know the rules of chess to do this. This is exactly how algorithm Y is required to use algorithm X.

We will use reductions to show the undecidability as follows: Suppose A can be reduced to B as in the preceding definition. If there is an algorithm for problem B, it can be used to design a program for A by essentially imitating the execution of the augmented program for A (with calls to the oracle for B) as just described. But we will turn it into a negative argument as follows: If A is undecidable, then so is B. Thus, a reduction from a problem known to be undecidable to problem B will prove B's undecidability.

First we define a new problem, problem $2'$, which is a special case of problem 2. Recall that in problem 2 the input is (the code of) a program P in GOTO language and a string x. The output required is $P(x)$. In problem $2'$, the input is (only) the code of a program P and the output required is $P(P)$, that is, instead of requiring P to run on a given input, this problem requires that it be run on its own code. This is clearly a special case of problem 2. The reader may readily see the self-reference in problem $2'$ and suspect that it may be undecidable; therefore, the more general problem 2 may be undecidable as well. We will establish these claims more rigorously as follows.

We first observe a general statement about the decidability of a function f (or problem) and its *complement*. The complement function is defined to take value 1 on all inputs for which the original function value is 0 and vice versa. The statement is that a function f is decidable if and only if the complement $\bar{f}$ is decidable. This can be easily proved as follows. Consider a program P that computes f. Change P into $\bar{P}$ by interchanging all of the ACCEPT and REJECT statements. It is easy to see that $\bar{P}$ actually computes $\bar{f}$. The converse also is easily seen to hold. It readily follows that the function defined by problem $2'$ is undecidable because it is, in fact, the complement of f_D.

Finally, we will show that problem 2 is uncomputable. The idea is to use a reduction from problem $2'$ to problem 2. (Note the direction of reduction. This always confuses a beginner.) Suppose there is an algorithm for problem 2. Let X be the GOTO language program that implements this algorithm. X takes as input code(P) (for any program P) followed by x, produces the result $P(x)$, and halts. We want to design a program Y that takes as input code(P) and produce the output $P(P)$ using calls to program X. It is clear what needs to be done. We just create the input in proper form code(P) followed by code(P) and call X. This requires first duplicating the input, but this is a simple programming task similar to the one we demonstrated in our first program in section 3.2. Then a call to X completes the task. This shows that problem $2'$ reduces to problem 2, and thus the latter is undecidable as well.

By a more elaborate reduction (from f_D), it can be shown that tiling is not partially decidable. We will not do it here and refer the interested reader to Harel [1992]. But we would like to point out how the undecidability result can be used to infer a result about tiling. This deduction is of interest because the result is an important one and is hard to derive directly. We need the following definition before we can state the result. A different way to pose the tiling problem is whether a given set of tiles can tile *an entire plane* in such a way that all of the adjacent tiles have the same color on the meeting quarter. (Note that this question is different from the way we originally posed it: Can a given set of tiles tile any *finite* rectangular region? Interestingly, the two problems are identical in the sense that the answer to one version is yes if and only if it is yes for the other version.) Call a tiling of the plane periodic if one can identify a $k \times k$ square such that the entire tiling is made by repeating this $k \times k$ square tile. Otherwise, call it *aperiodic*. Consider the question: Is there a (finite) set of unit tiles that can tile the plane, but only aperiodically? The answer is yes and it can be shown from the total undecidability of the tiling problem. Suppose the answer is no. Then, for any given set of tiles, the entire plane can be tiled if and only if the plane can be tiled periodically. But a periodic tiling can be found, if one exists, by trying to tile a $k \times k$ region for successively increasing values of k. This process will eventually succeed (in a finite number of steps) if the tiling exists. This will make the tiling problem partially decidable, which contradicts the total undecidability of the problem. This means that the assumption that the entire plane can be tiled if and only if some $k \times k$ region can be tiled is wrong. Thus, there exists a (finite) set of tiles that can tile the entire plane, but only aperiodically.

3.4 Formal Languages and Grammars

The universe of strings is probably the most general medium for the representation of information. This section is concerned with sets of strings called *languages* and certain systems generating these languages such as *grammars*. Every programming language including Pascal, C, or Fortran can be precisely described by a grammar. Moreover, the grammar allows us to write a computer program (called the lexical analyzer in a compiler) to determine if a piece of code is syntactically correct in the programming language. Would not it be nice to also have such a grammar for English and a corresponding computer program which can tell us what English sentences are grammatically correct?[8] The focus of this brief exposition is the formalism and mathematical properties of various languages and grammars. Many of the concepts have applications in domains including natural language and computer language processing, string matching, etc. We begin with some standard definitions about languages.

Definition 3.1. An *alphabet* is a finite nonempty set of *symbols*, which are assumed to be *indivisible*.

For example, the alphabet for English consists of 26 uppercase letters $A, B, \ldots, Z$ and 26 lowercase letters $a, b, \ldots, z$. We usually use the symbol Σ to denote an alphabet.

Definition 3.2. A *string* over an alphabet Σ is a finite sequence of symbols of Σ.

The number of symbols in a string x is called its *length*, denoted $|x|$. It is convenient to introduce an empty string, denoted ϵ, which contains no symbols at all. The length of ϵ is 0.

Definition 3.3. Let $x = a_1 a_2 \cdots a_n$ and $y = b_1 b_2 \cdots b_m$ be two strings. The *concatenation* of x and y, denoted xy, is the string $a_1 a_2 \cdots a_n b_1 b_2 \cdots b_m$.

Thus, for any string x, $\epsilon x = x \epsilon = x$. For any string x and integer $n \geq 0$, we use x^n to denote the string formed by sequentially concatenating n copies of x.

Definition 3.4. The set of all strings over an alphabet Σ is denoted Σ^* and the set of all nonempty strings over Σ is denoted Σ^+. The empty set of strings is denoted $\emptyset$.

Definition 3.5. For any alphabet Σ, a *language* over Σ is a set of strings over Σ. The members of a language are also called the *words* of the language.

Example 3.1. The sets $L_1 = \{01, 11, 0110\}$ and $L_2 = \{0^n 1^n \mid n \geq 0\}$ are two languages over the binary alphabet $\{0, 1\}$. The string 01 is in both languages, whereas 11 is in L_1 but not in L_2.

Because languages are just sets, standard set operations such as union, intersection, and complementation apply to languages. It is useful to introduce two more operations for languages: *concatenation* and *Kleene closure*.

---

[8] Actually, English and the other natural languages have grammars; but these grammars are not precise enough to tell apart the correct and incorrect sentences with 100% accuracy. The main problem is that *there is no universal agreement* on what are grammatically correct English sentences.

Definition 3.6. Let L_1 and L_2 be two languages over Σ. The concatenation of L_1 and L_2, denoted $L_1 L_2$, is the language $\{xy \mid x \in L_1, y \in L_2\}$.

Definition 3.7. Let L be a language over Σ. Define $L^0 = \{\epsilon\}$ and $L^i = LL^{i-1}$ for $i \geq 1$. The Kleene closure of L, denoted L^*, is the language

$$L^* = \bigcup_{i \geq 0} L^i$$

and the *positive closure* of L, denoted L^+, is the language

$$L^+ = \bigcup_{i \geq 1} L^i$$

In other words, the Kleene closure of language L consists of all strings that can be formed by concatenating some words from L. For example, if $L = \{0, 01\}$, then $LL = \{00, 001, 010, 0101\}$ and L^* includes all binary strings in which every 1 is preceded by a 0. L^+ is the same as L^* except it excludes ϵ in this case. Note that, for any language L, L^* always contains ϵ and L^+ contains ϵ if and only if L does. Also note that Σ^* is in fact the Kleene closure of the alphabet Σ when viewed as a language of words of length 1, and Σ^+ is just the positive closure of Σ.

Representation of Languages

In general, a language over an alphabet Σ is a subset of Σ^*. How can we describe a language rigorously so that we know if a given string belongs to the language or not? As shown in the preceding paragraphs, a finite language such as L_1 in Example 3.1 can be explicitly defined by enumerating its elements, and a simple infinite language such as L_2 in the same example can be described using a rule characterizing all members of L_2. It is possible to define some more systematic methods to represent a wide class of languages. In the following, we will introduce three such methods: regular expressions, pattern systems, and grammars. The languages that can be described by this kind of system are often referred to as *formal languages*.

Definition 3.8. Let Σ be an alphabet. The *regular expressions* over Σ and the languages they represent are defined inductively as follows.

1. The symbol $\emptyset$ is a regular expression, denoting the empty set.
2. The symbol ϵ is a regular expression, denoting the set $\{\epsilon\}$.
3. For each $a \in \Sigma$, a is a regular expression, denoting the set $\{a\}$.
4. If r and s are regular expressions denoting the languages R and S, then $(r + s)$, (rs), and (r^*) are regular expressions that denote the sets $R \cup S$, RS, and R^*, respectively.

For example, $((0(0 + 1)^*) + ((0 + 1)^*0))$ is a regular expression over $\{0, 1\}$, and it represents the language consisting of all binary strings that begin or end with a 0. Because the set operations union and concatenation are both associative, many parentheses can be omitted from regular expressions if we assume that Kleene closure has higher precedence than concatenation and concatenation has higher precedence than union. For example, the preceding regular expression can be abbreviated as $0(0 + 1)^* + (0 + 1)^*0$. We will also abbreviate the expression rr^* as r^+. Let us look at a few more examples of regular expressions and the languages they represent.

Example 3.2. The expression $0(0 + 1)^*1$ represents the set of all strings that begin with a 0 and end with a 1.

Example 3.3. The expression $0 + 1 + 0(0 + 1)^*0 + 1(0 + 1)^*1$ represents the set of all nonempty binary strings that begin and end with the same bit.

Example 3.4. The expressions 0^*, 0^*10^*, and $0^*10^*10^*$ represent the languages consisting of strings that contain no 1, exactly one 1, and exactly two 1s, respectively.

Example 3.5. The expressions $(0 + 1)^*1(0 + 1)^*1(0 + 1)^*$, $(0 + 1)^*10^*1(0 + 1)^*$, $0^*10^*1(0 + 1)^*$, and $(0 + 1)^*10^*10^*$ all represent the same set of strings that contain at least two 1s.

For any regular expression r, the language represented by r is denoted as $L(r)$. Two regular expressions representing the same language are called *equivalent*. It is possible to introduce some identities to algebraically manipulate regular expressions to construct equivalent expressions, by tailoring the set identities for the operations union, concatenation, and Kleene closure to regular expressions. For more details, see Salomaa [1966]. For example, it is easy to prove that the expressions $r(s + t)$ and $rs + rt$ are equivalent and $(r^*)^*$ is equivalent to r^*.

Example 3.6. Let us construct a regular expression for the set of all strings that contain no consecutive 0s. A string in this set may begin and end with a sequence of 1s. Because there are no consecutive 0s, every 0 that is not the last symbol of the string must be followed by at least a 1. This gives us the expression $1^*(01^+)^*1^*(\epsilon + 0)$. It is not hard to see that the second 1^* is redundant, and thus the expression can in fact be simplified to $1^*(01^+)^*(\epsilon + 0)$.

Regular expressions were first introduced in Kleene [1956] for studying the properties of neural nets. The preceding examples illustrate that regular expressions often give very clear and concise representations of languages. Unfortunately, not every language can be represented by regular expressions. For example, it will become clear that there is no regular expression for the language $\{0^n1^n \mid n \geq 1\}$. The languages represented by regular expressions are called the **regular languages**. Later, we will see that regular languages are exactly the class of languages generated by the so-called **right-linear grammars**. This connection allows one to prove some interesting mathematical properties about regular languages as well as to design an efficient algorithm to determine whether a given string belongs to the language represented by a given **regular expression**.

Another way of representing languages is to use *pattern systems* [Angluin 1980, Jiang et al. 1995].

Definition 3.9. A *pattern system* is a triple (Σ, V, p), where Σ is the alphabet, V is the set of *variables* with $\Sigma \cap V = \emptyset$, and p is a string over $\Sigma \cup V$ called the *pattern*.

An example pattern system is $(\{0, 1\}, \{v_1, v_2\}, v_1v_10v_2)$.

Definition 3.10. The language generated by a pattern system (Σ, V, p) consists of all strings over Σ that can be obtained from p by replacing each variable in p with a string over Σ.

For example, the language generated by $(\{0, 1\}, \{v_1, v_2\}, v_1v_10v_2)$ contains words 0, 00, 01, 000, 001, 010, 011, 110, etc., but does not contain strings, 1, 10, 11, 100, 101, etc. The pattern system $(\{0, 1\}, \{v_1\}, v_1v_1)$ generates the set of all strings, which is the concatenation of two equal substrings, that is, the set $\{xx \mid x \in \{0, 1\}^*\}$. The languages generated by pattern systems are called the *pattern languages*.

Regular languages and pattern languages are really different. One can prove that the pattern language $\{xx \mid x \in \{0, 1\}^*\}$ is not a regular language and the set represented by the regular expression 0^*1^* is not a pattern language. Although it is easy to write an algorithm to decide if a string is in the language generated by a given pattern system, such an algorithm most likely would have to be very inefficient [Angluin 1980].

Perhaps the most useful and general system for representing languages is based on grammars, which are extensions of the pattern systems.

Definition 3.11. A grammar is a quadruple (Σ, N, S, P), where:

1. Σ is a finite nonempty set called the alphabet. The elements of Σ are called the *terminals*.
2. N is a finite nonempty set disjoint from Σ. The elements of N are called the *nonterminals* or *variables*.
3. $S \in N$ is a distinguished nonterminal called the *start symbol*.
4. P is a finite set of *productions* (or *rules*) of the form

$$\alpha \rightarrow \beta$$

where $\alpha \in (\Sigma \cup N)^* N (\Sigma \cup N)^*$ and $\beta \in (\Sigma \cup N)^*$, that is, α is a string of terminals and nonterminals containing at least one nonterminal and β is a string of terminals and nonterminals.

Example 3.7. Let $G_1 = (\{0, 1\}, \{S, T, O, I\}, S, P)$, where P contains the following productions:

$$S \rightarrow OT$$
$$S \rightarrow OI$$
$$T \rightarrow SI$$
$$O \rightarrow 0$$
$$I \rightarrow 1$$

As we shall see, the grammar G_1 can be used to describe the set $\{0^n 1^n \mid n \geq 1\}$.

Example 3.8. Let $G_2 = (\{0, 1, 2\}, \{S, A\}, S, P)$, where P contains the following productions.

$$S \rightarrow 0SA2$$
$$S \rightarrow \epsilon$$
$$2A \rightarrow A2$$
$$0A \rightarrow 01$$
$$1A \rightarrow 11$$

This grammar G_2 can be used to describe the set $\{0^n 1^n 2^n \geq n \geq 0\}$.

Example 3.9. To construct a grammar G_3 to describe English sentences, the alphabet Σ contains all words in English. N would contain nonterminals, which correspond to the structural components in an English sentence, for example, ⟨sentence⟩, ⟨subject⟩, ⟨predicate⟩, ⟨noun⟩, ⟨verb⟩, ⟨article⟩, etc. The start symbol would be ⟨sentence⟩. Some typical productions are

$$\langle\text{sentence}\rangle \rightarrow \langle\text{subject}\rangle\langle\text{predicate}\rangle$$

$$\langle\text{subject}\rangle \rightarrow \langle\text{noun}\rangle$$

$$\langle\text{predicate}\rangle \rightarrow \langle\text{verb}\rangle\langle\text{article}\rangle\langle\text{noun}\rangle$$

$$\langle\text{noun}\rangle \rightarrow \text{mary}$$

$$\langle\text{noun}\rangle \rightarrow \text{algorithm}$$

$$\langle\text{verb}\rangle \rightarrow \text{wrote}$$

$$\langle\text{article}\rangle \rightarrow \text{an}$$

The rule ⟨sentence⟩ → ⟨subject⟩⟨predicate⟩ follows from the fact that a sentence consists of a subject phrase and a predicate phrase. The rules ⟨noun⟩ → mary and ⟨noun⟩ → algorithm mean that both mary and algorithms are possible nouns.

To explain how a grammar represents a language, we need the following concepts.

Definition 3.12. Let (Σ, N, S, P) be a grammar. A *sentential form* of G is any string of terminals and nonterminals, that is, a string over $\Sigma \cup N$.

Definition 3.13. Let (Σ, N, S, P) be a grammar and γ_1 and γ_2 two sentential forms of G. We say that γ_1 *directly derives* γ_2, denoted $\gamma_1 \Rightarrow \gamma_2$, if $\gamma_1 = \sigma \alpha \tau$, $\gamma_2 = \sigma \beta \tau$, and $\alpha \to \beta$ is a production in P.

For example, the sentential form $00S11$ directly derives the sentential form $00OT11$ in grammar G_1, and $A2A2$ directly derives $AA22$ in grammar G_2.

Definition 3.14. Let γ_1 and γ_2 be two sentential forms of a grammar G. We say that γ_1 *derives* γ_2, denoted $\gamma_1 \Rightarrow^* \gamma_2$, if there exists a sequence of (zero or more) sentential forms $\sigma_1, \ldots, \sigma_n$ such that

$$\gamma_1 \Rightarrow \sigma_1 \Rightarrow \cdots \Rightarrow \sigma_n \Rightarrow \gamma_2$$

The sequence $\gamma_1 \Rightarrow \sigma_1 \Rightarrow \cdots \Rightarrow \sigma_n \Rightarrow \gamma_2$ is called a derivation from γ_1 to γ_2

For example, in grammar G_1, $S \Rightarrow^* 0011$ because

$$S \Rightarrow \underline{OT} \Rightarrow 0\underline{T} \Rightarrow 0S\underline{I} \Rightarrow 0\underline{S}1 \Rightarrow 0\underline{O}I1 \Rightarrow 00\underline{I}1 \Rightarrow 0011$$

and in grammar G_2, $S \Rightarrow^* 001122$ because

$$S \Rightarrow 0\underline{S}A2 \Rightarrow 00\underline{S}A2A2 \Rightarrow 00\underline{A}2A2 \Rightarrow 0012\underline{A}2 \Rightarrow 0011\underline{A}22 \Rightarrow 001122$$

Here the left-hand side of the relevant production in each derivation step is underlined for clarity.

Definition 3.15. Let (Σ, N, S, P) be a grammar. The language generated by G, denoted $L(G)$, is defined as

$$L(G) = \{x \mid x \in \Sigma^*, S \Rightarrow^* x\}$$

The words in $L(G)$ are also called the *sentences* of $L(G)$.

Clearly, $L(G_1)$ contains all strings of the form $0^n 1^n$, $n \geq 1$, and $L(G_2)$ contains all strings of the form $0^n 1^n 2^n$, $n \geq 0$. Although only a partial definition of G_3 is given, we known that $L(G_3)$ contains sentences such as "mary wrote an algorithm" and "algorithm wrote an algorithm" but does not contain sentences such as "an wrote algorithm".

The introduction of formal grammars dates back to the 1940s [Post 1943], although the study of rigorous description of languages by grammars did not begin until the 1950s [Chomsky 1956]. In the next subsection, we consider various restrictions on the form of productions in a grammar and see how these restrictions can affect the power of a grammar in representing languages. In particular, we will know that regular languages and pattern languages can all be generated by grammars under different restrictions.

Hierarchy of Grammars

Grammars can be divided into four classes by gradually increasing the restrictions on the form of the productions. Such a classification is due to Chomsky [1956, 1963] and is called the *Chomsky hierarchy*.

Definition 3.16. Let $G = (\Sigma, N, S, P)$ be a grammar.

1. G is also called a *type-0 grammar* or an *unrestricted grammar*.

2. G is *type*-1 or **context sensitive** if each production $\alpha \rightarrow \beta$ in P either has the form $S \rightarrow \epsilon$ or satisfies $|\alpha| \leq |\beta|$.

3. G is *type*-2 or **context free** if each production $\alpha \rightarrow \beta$ in P satisfies $|\alpha| = 1$, that is, α is a nonterminal.

4. G is *type*-3 or right linear or regular if each production has one of the following three forms:

$$A \rightarrow aB, \qquad A \rightarrow a, \qquad A \rightarrow \epsilon$$

where A and B are nonterminals and a is a terminal.

The language generated by a type-i is called a type-i language, $i = 0, 1, 2, 3$. A type-1 language is also called a **context-sensitive language** and a type-2 language is also called a **context-free language**. It turns out that every type-3 language is in fact a regular language, that is, it is represented by some regular expression, and vice versa. See the next section for the proof of the equivalence of type-3 (right-linear) grammars and regular expressions.

The grammars G_1 and G_3 given in the last subsection are context free and the grammar G_2 is context sensitive. Now we give some examples of unrestricted and right-linear grammars.

Example 3.10. Let $G_4 = (\{0, 1\}, \{S, A, O, I, T\}, S, P)$, where P contains

$$S \rightarrow AT$$
$$A \rightarrow 0AO \qquad A \rightarrow 1AI$$
$$00 \rightarrow 00 \qquad 01 \rightarrow 10$$
$$10 \rightarrow 0I \qquad I1 \rightarrow 1I$$
$$OT \rightarrow 0T \qquad IT \rightarrow 1T$$
$$A \rightarrow \epsilon \qquad T \rightarrow \epsilon$$

Then G_4 generates the set $\{xx \mid x \in \{0, 1\}^*\}$. For example, we can derive the word 0101 from S as follows:

$$S \Rightarrow \underline{A}T \Rightarrow 0\underline{A}OT \Rightarrow 01\underline{A}IOT \Rightarrow 01I\underline{OT} \Rightarrow 01\underline{I0}T \Rightarrow 010\underline{IT} \Rightarrow 0101\underline{T} \Rightarrow 0101$$

Example 3.11. We give a right-linear grammar G_5 to generate the language represented by the regular expression in Example 3.3, that is, the set of all nonempty binary strings beginning and ending with the same bit. Let $G_5 = (\{0, 1\}, \{S, O, I\}, S, P)$, where P contains

$$S \rightarrow 0O \qquad S \rightarrow 1I$$
$$S \rightarrow 0 \qquad S \rightarrow 1$$
$$O \rightarrow 0O \qquad O \rightarrow 1O$$
$$I \rightarrow 0I \qquad I \rightarrow 1I$$
$$O \rightarrow 0 \qquad I \rightarrow 1$$

The following theorem is due to Chomsky [1956, 1963].

Theorem 3.2. *For each $i = 0, 1, 2$, the class of type-i languages properly contains the class of type-$(i + 1)$ languages.*

For example, one can prove by using a technique called *pumping* that the set $\{0^n 1^n \mid n \geq 1\}$ is context free but not regular, and the sets $\{0^n 1^n 2^n \mid n \geq 0\}$ and $\{xx \mid x \in \{0, 1\}^*\}$ are context sensitive but not context free [Hopcroft and Ullman 1979]. It is, however, a bit involved to construct a language that is of type-0 but not context sensitive. See, for example, Hopcroft and Ullman [1979] for such a language.

The four classes of languages in the Chomsky hierarchy also have been completely characterized in terms of Turing machines and their restricted versions. We have already defined a Turing machine in section 3.2. Many restricted versions of it will be defined in the next section. It is known that type-0 languages are exactly those recognized by Turing machines, context-sensitive languages are those recognized by Turing machines running in linear space, context-free languages are those recognized by Turing machines whose worktapes operate as pushdown stacks [called **pushdown automata** (PDA)], and regular languages are those recognized by Turing machines without any worktapes (called **finite-state machine** or **finite automata**) [Hopcroft and Ullman 1979].

Remark 3.1. Recall our definition of a Turing machine and the function it computes from section 3.2. In the preceding paragraph, we refer to *a language recognized* by a Turing machine. These are two seemingly different ideas, but they are essentially the same. The reason is that the function f, which maps the set of strings over a finite alphabet to $\{0, 1\}$, corresponds in a natural way to the language L_f over Σ defined as: $L_f = \{x \mid f(x) = 1\}$. Instead of saying that a Turing machine computes the function f, we say equivalently that it recognizes L_f.

Because $\{xx \mid x \in \{0, 1\}^*\}$ is a pattern language, the preceding discussion implies that the class of pattern languages is not contained in the class of context-free languages. The next theorem shows that the class of pattern languages is contained in the class of context-sensitive languages.

Theorem 3.3. *Every pattern language is context sensitive.*

The theorem follows from the fact that every pattern language is recognized by a Turing machine in linear space [Angluin 1980] and linear space-bounded Turing machines recognize exactly context-sensitive languages. To show the basic idea involved, let us construct a context-sensitive grammar for the pattern language $\{xx \mid x \in \{0, 1\}^*\}$. The grammar G_4 given in Example 3.10 for this language is almost context-sensitive. We just have to get rid of the two ϵ-productions: $A \to \epsilon$ and $T \to \epsilon$. A careful modification of G_4 results in the following grammar $G_6 = (\{0, 1\}, \{S, A_0, A_1, O, I, T_0, T_1\}, S, P)$, where P contains

$$
\begin{aligned}
S &\to \epsilon \\
S &\to A_0 T_0 & S &\to A_1 T_1 \\
A_0 &\to 0 A_0 O & A_0 &\to 1 A_0 I \\
A_1 &\to 0 A_1 O & A_1 &\to 1 A_1 I \\
A_0 &\to 0 & A_1 &\to 1 \\
O0 &\to 0O & O1 &\to 1O \\
I0 &\to 0I & I1 &\to 1I \\
OT_0 &\to 0T_0 & IT_0 &\to 1T_0 \\
OT_1 &\to 0T_1 & IT_1 &\to 1T_1 \\
T_0 &\to O & T_1 &\to 1,
\end{aligned}
$$

which is context sensitive and generates $\{xx \mid x \in \{0, 1\}^*\}$. For example, we can derive 011011 as

$$\Rightarrow \underline{A_1} T_1 \Rightarrow 0 \underline{A_1} O T_1 \Rightarrow 01 \underline{A_1} I O T_1$$

$$\Rightarrow 011 \underline{I\,O} T_1 \Rightarrow 011 \underline{I0} T_1 \Rightarrow 0110 \underline{I T_1} \Rightarrow 01101 \underline{T_1} \Rightarrow 011011$$

For a class of languages, we are often interested in the so-called *closure properties* of the class.

Definition 3.17.   A class of languages (e.g., regular languages) is said to be *closed* under a particular operation (e.g., union, intersection, complementation, concatenation, Kleene closure) if each application of the operation on language(s) of the class results in a language of the class.

These properties are often useful in constructing new languages from existing languages as well as proving many theoretical properties of languages and grammars. The closure properties of the four types of languages in the Chomsky hierarchy are now summarized [Harrison 1978, Hopcroft and Ullman 1979, Gurari 1989].

Theorem 3.4.

1. *The class of type-0 languages is closed under union, intersection, concatenation, and Kleene closure but not under complementation.*
2. *The class of context-free languages is closed under union, concatenation, and Kleene closure but not under intersection or complementation.*
3. *The classes of context-sensitive and regular languages are closed under all five of the operations.*

For example, let $L_1 = \{0^m 1^n 2^p \mid m = n \text{ or } n = p\}$, $L_2 = \{0^m 1^n 2^p \mid m = n\}$, and $L_3 = \{0^m 1^n 2^p \mid n = p\}$. It is easy to see that all three are context-free languages. (In fact, $L_1 = L_2 \cup L_3$.) However, intersecting L_2 with L_3 gives the set $\{0^m 1^n 2^p \mid m = n = p\}$, which is not context free.

We will look at context-free grammars more closely in the next subsection and introduce the concept of **parsing** and ambiguity.

Context-Free Grammars and Parsing

From a practical point of view, for each grammar $G = (\Sigma, N, S, P)$ representing some language, the following two problems are important:

1. (Membership) Given a string over Σ, does it belong to $L(G)$?
2. (Parsing) Given a string in $L(G)$, how can it be derived from S?

The importance of the membership problem is quite obvious: given an English sentence or computer program we wish to know if it is grammatically correct or has the right format. Parsing is important because a derivation usually allows us to interpret the meaning of the string. For example, in the case of a Pascal program, a derivation of the program in Pascal grammar tells the compiler how the program should be executed. The following theorem illustrates the decidability of the membership problem for the four classes of grammars in the Chomsky hierarchy. The proofs can be found in Chomsky [1963], Harrison [1978], and Hopcroft and Ullman [1979].

Theorem 3.5.   *The membership problem for type-0 grammars is undecidable in general and is decidable for any context-sensitive grammar (and thus for any context-free or right-linear grammars).*

Because context-free grammars play a very important role in describing computer programming languages, we discuss the membership and parsing problems for context-free grammars in more detail. First, let us look at another example of context-free grammar. For convenience, let us abbreviate a set of productions with the same left-hand side nonterminal

$$A \to \alpha_1, \ldots, A \to \alpha_n$$

as

$$A \rightarrow \alpha_1 \mid \cdots \mid \alpha_n$$

Example 3.12. We construct a context-free grammar for the set of all valid Pascal real values. In general, a real constant in Pascal has one of the following forms:

$$m.n, \qquad m\mathbf{e}q, \qquad m.n\mathbf{e}q,$$

where m and q are signed or unsigned integers and n is an unsigned integer. Let $\Sigma = \{0, 1, 2, 3, 4, 5, 6, 7, 8, 9, \mathbf{e}, +, -, .\}$, $N = \{S, M, N, D\}$, and the set P of the productions contain

$$S \rightarrow M.N \mid M\mathbf{e}M \mid M.N\mathbf{e}M$$
$$M \rightarrow N \mid + N \mid - N$$
$$N \rightarrow DN \mid D$$
$$D \rightarrow 0|1|2|3|4|5|7|8|9$$

Then the grammar generates all valid Pascal real values (including some absurd ones like 001.200e000). The value 12.3**e** − 4 can be derived as

$$S \Rightarrow \underline{M}.N\mathbf{e}M \Rightarrow \underline{N}.N\mathbf{e}M \Rightarrow \underline{D}N.N\mathbf{e}M \Rightarrow 1\underline{N}.N\mathbf{e}M \Rightarrow 1\underline{D}.N\mathbf{e}M$$

$$\Rightarrow 12.\underline{N}\mathbf{e}M \Rightarrow 12.\underline{D}\mathbf{e}M \Rightarrow 12.3\mathbf{e}\underline{M} \Rightarrow 12.3\mathbf{e} - \underline{N} \Rightarrow 12.3\mathbf{e} - \underline{D} \Rightarrow 12.3\mathbf{e} - 4$$

Perhaps the most natural representation of derivations for a context-free grammar is *a derivation tree* or *a parse tree*. Each *internal node* of such a tree corresponds to a nonterminal and each *leaf* corresponds to a terminal. If A is an internal node with children $B_1, \ldots, B_n$ ordered from left to right, then $A \rightarrow B_1 \cdots B_n$ must be a production. The concatenation of all leaves from left to right yields the string being derived. For example, the derivation tree corresponding to the preceding derivation of 12.3**e** − 4 is given in Fig. 3.3. Such a tree also makes possible the extraction of the parts 12, 3, and −4, which are useful in the storage of the real value in a computer memory.

Definition 3.18. A context-free grammar G is **ambiguous** if there is a string $x \in L(G)$, which has two distinct derivation trees. Otherwise G is *unambiguous*.

Unambiguity is a very desirable property to have as it allows a unique interpretation of each sentence in the language. It is not hard to see that the preceding grammar for Pascal real values and the grammar G_1 defined in Example 3.7 are all unambiguous. The following example shows an ambiguous grammar.

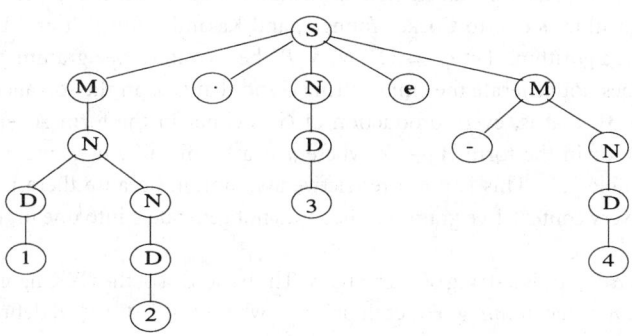

FIGURE 3.3 The derivation tree for 12.3**e** − 4.

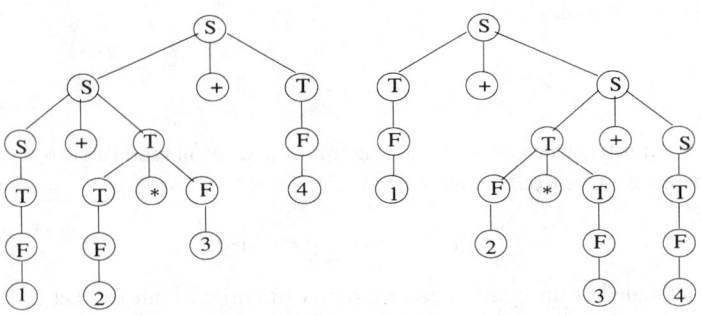

FIGURE 3.4   Different derivation trees for the expression $1 + 2 * 3 + 4$.

Example 3.13.   Consider a grammar G_7 for all valid arithmetic expressions that are composed of unsigned positive integers and symbols $+$, $*$, $($, $)$. For convenience, let us use the symbol n to denote any unsigned positive integer. This grammar has the productions

$$S \to T + S \mid S + T \mid T$$

$$T \to F * T \mid T * F \mid F$$

$$F \to n \mid (S)$$

Two possible different derivation trees for the expression $1 + 2 * 3 + 4$ are shown in Fig. 3.4. Thus, G_7 is ambiguous. The left tree means that the first addition should be done before the second addition and the right tree says the opposite.

Although in the preceding example different derivations/interpretations of any expression always result in the same value because the operations addition and multiplication are associative, there are situations where the difference in the derivation can affect the final outcome. Actually, the grammar G_7 can be made unambiguous by removing some (redundant) productions, for example, $S \to T + S$ and $T \to F * T$. This corresponds to the convention that a sequence of consecutive additions (or multiplications) is always evaluated from left to right and will not change the language generated by G_7. It is worth noting that there are context-free languages which cannot be generated by any unambiguous context-free grammar [Hopcroft and Ullman 1979]. Such languages are said to be *inherently ambiguous*. An example of inherently ambiguous languages is the set

$$\{0^m 1^m 2^n 3^n \mid m, n > 0\} \cup \{0^m 1^n 2^m 3^n \mid m, n > 0\}$$

We end this section by presenting an efficient algorithm for the membership problem for context-free grammars. The algorithm is due to Cocke, Younger, and Kasami [Hopcroft and Ullman 1979] and is often called the CYK algorithm. Let $G = (\Sigma, N, S, P)$ be a context-free grammar. For simplicity, let us assume that G does not generate the empty string ϵ and that G is in the so-called **Chomsky normal form** [Chomsky 1963], that is, every production of G is either in the form $A \to BC$ where B and C are nonterminals, or in the form $A \to a$ where a is a terminal. An example of such a grammar is G_1 given in Example 3.7. This is not a restrictive assumption because there is a simple algorithm which can convert every context-free grammar that does not generate ϵ into one in the Chomsky normal form.

Suppose that $x = a_1 \cdots a_n$ is a string of n terminals. The basic idea of the CYK algorithm, which decides if $x \in L(G)$, is *dynamic programming*. For each pair i, j, where $1 \le i \le j \le n$, define a set $X_{i,j} \subseteq N$ as

$$X_{i,j} = \{A \mid A \Rightarrow^* a_i \cdots a_j\}$$

Thus, $x \in L(G)$ if and only if $S \in X_{1,n}$. The sets $X_{i,j}$ can be computed inductively in the ascending order of $j - i$. It is easy to figure out $X_{i,i}$ for each i because $X_{i,i} = \{A \mid A \to a_i \in P\}$. Suppose that we have computed all $X_{i,j}$ where $j - i < d$ for some $d > 0$. To compute a set $X_{i,j}$, where $j - i = d$, we just have to find all the nonterminals A such that there exist some nonterminals B and C satisfying $A \to BC \in P$ and for some k, $i \leq k < j$, $B \in X_{i,k}$, and $C \in X_{k+1,j}$. A rigorous description of the algorithm in a Pascal style pseudocode is given as follows.

TABLE 3.2 An Example Execution of the CYK Algorithm

		0	0	0	1	1	1
		\multicolumn{6}{c}{$j \to$}					
		1	2	3	4	5	6
	1	O					S
	2		O			S	T
i	3			O	S	T	
$\downarrow$	4				I		
	5					I	
	6						I

Algorithm CYK($x = a_1 \cdots a_n$):

1. for $i \leftarrow 1$ to n do
2. $\quad$ $X_{i,i} \leftarrow \{A \mid A \to a_i \in P\}$
3. for $d \leftarrow 1$ to $n - 1$ do
4. $\quad$ for $i \leftarrow 1$ to $n - d$ do
5. $\quad\quad$ $X_{i,i+d} \leftarrow \emptyset$
6. $\quad\quad$ for $t \leftarrow 0$ to $d - 1$ do
7. $\quad\quad\quad$ $X_{i,i+d} \leftarrow X_{i,i+d} \cup \{A \mid A \to BC \in P$ for some $B \in X_{i,i+t}$ and $C \in X_{i+t+1,i+d}\}$

Table 3.2 shows the sets $X_{i,j}$ for the grammar G_1 and the string $x = 000111$. It just so happens that every $X_{i,j}$ is either empty or a singleton. The computation proceeds from the main diagonal toward the upper-right corner.

3.5 Computational Models

In this section, we will present many restricted versions of Turing machines and address the question of what kinds of problems they can solve. Such a classification is a central goal of computation theory. We have already classified problems broadly into (totally) decidable, partially decidable, and totally undecidable. Because the decidable problems are the ones of most practical interest, we can consider further classification of decidable problems by placing two types of restrictions on a Turing machine. The first one is to restrict its structure. This way we obtain many machines of which a finite automaton and a pushdown automaton are the most important. The other way to restrict a Turing machine is to bound the amount of resources it uses, such as the number of time steps or the number of tape cells it can use. The resulting machines form the basis for *complexity theory*.

Finite Automata

The finite automaton (in its deterministic version) was first introduced by McCulloch and Pitts [1943] as a logical model for the behavior of neural systems. Rabin and Scott [1959] introduced the nondeterministic version of the finite automaton and showed the equivalence of the nondeterministic and deterministic versions. Chomsky and Miller [1958] proved that the set of languages that can be recognized by a finite automaton is precisely the regular languages introduced in section 3.4. Kleene [1956] showed that the languages accepted by finite automata are characterized by regular expressions as defined in section 3.4.

In addition to their original role in the study of neural nets, finite automata have enjoyed great success in many fields such as sequential circuit analysis in circuit design [Kohavi 1978], asynchronous circuits [Brzozowski and Seger 1994], lexical analysis in text processing [Lesk 1975], and compiler design. They also led to the design of more efficient algorithms. One excellent example is the development of linear-time string-matching algorithms, as described in Knuth et al. [1977]. Other applications of finite automata can be found in computational biology [Searls 1993], natural language processing, and distributed computing.

A finite automaton, as in Fig. 3.5, consists of an input tape which contains a (finite) sequence of input symbols such as *aabab*···, as shown in the figure, and a finite-state control. The tape is read by the one-way *read-only* input head from left to right, one symbol at a time. Each time the input head reads an input symbol, the finite control changes its state according to the symbol and the current state of the machine. When the

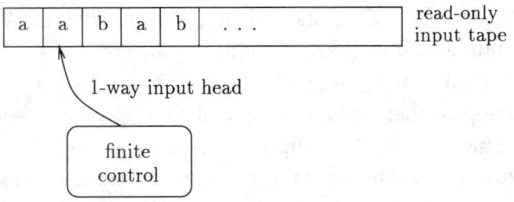

FIGURE 3.5   A finite automaton.

input head reaches the right end of the input tape, if the machine is in a final state, we say that the input is accepted; if the machine is not in a final state, we say that the input is rejected. The following is the formal definition.

Definition 3.19.   A *nondeterministic finite automaton* (NFA) is a quintuple $(Q, \Sigma, \delta, q_0, F)$, where:

- Q is a finite set of *states*.
- Σ is a finite set of *input symbols*.
- δ, the *state transition function*, is a mapping from $Q \times \Sigma$ to subsets of Q.
- $q_0 \in Q$ is the *initial state* of the NFA.
- $F \subseteq Q$ is the set of *final states*.

If δ maps $|Q| \times \Sigma$ to singleton subsets of Q, then we call such a machine a *deterministic finite automaton* (DFA).

When an automaton, M, is nondeterministic, then from the current state and input symbol, it may go to one of several different states. One may imagine that the device goes to all such states in parallel. The DFA is just a special case of the NFA; it always follows a single deterministic path. The device M *accepts* an input string x if, starting with q_0 and the read head at the first symbol of x, one of these parallel paths reaches an accepting state when the read head reaches the end of x. Otherwise, we say M *rejects* x. A language, L, is accepted by M if M accepts all of the strings in L and nothing else, and we write $L = L(M)$. We will also allow the machine to make ϵ-*transitions*, that is, changing state without advancing the read head. This allows transition functions such as $\delta(s, \epsilon) = \{s'\}$. It is easy to show that such a generalization does not add more power.

Remark 3.2.   The concept of a nondeterministic automaton is rather confusing for a beginner. But there is a simple way to relate it to a concept which must be familiar to all of the readers. It is that of a solitaire game. Imagine a game like *Klondike*. The game starts with a certain arrangement of cards (the input) and there is a well-defined final position that results in success; there are also dead ends where a further move is not possible; you lose if you reach any of them. At each step, the precise rules of the game dictate how a new arrangement of cards can be reached from the current one. But the most important point is that there are many possible moves at each step. (Otherwise, the game would be no fun!) Now consider the following question: What starting positions are *winnable*? These are the starting positions for which *there is a winning move sequence*; of course, in a typical play a player may not achieve it. But that is beside the point in the definition of what starting postions are winnable. The connection between such games and a nondeterministic automaton should be clear. The multiple choices at each step are what make it *nondeterministic*. Our definition of winnable positions is similar to the concept of acceptance of a string by a nondeterministic automaton. Thus, an NFA may be viewed as a formal model to define solitaire games.

Example 3.14.   We design a DFA to accept the language represented by the regular expression $0(0 + 1)^*1$ as in Example 3.2, that is, the set of all strings in $\{0, 1\}$ which begin with a 0 and end with a 1. It is usually convenient to draw our solution as in Fig. 3.6. As a convention, each circle represents a state; the state a, pointed at by the initial arrow, is the initial state. The darker circle represents the final states (state c). The

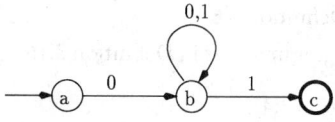

FIGURE 3.6 An NFA accepting $0(0+1)^*1$.

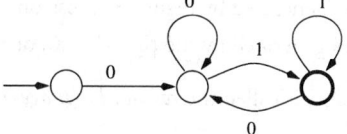

FIGURE 3.7 A DFA accepting $0(0+1)^*1$.

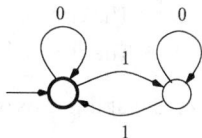

FIGURE 3.8 A DFA accepting $(0^*10^*1)^*0^*$.

transition function δ is represented by the labeled edges. For example, $\delta(a, 0) = \{b\}$. When a transition is missing, for example on input 1 from a and on inputs 0 and 1 from c, it is assumed that all of these lead to an implicit nonaccepting trap state, which has transitions to itself on all inputs.

The machine in Fig. 3.6 is nondeterministic because from b on input 1 the machine has two choices: stay at b or go to c.

Figure 3.7 gives an equivalent DFA, accepting the same language.

Example 3.15. The DFA in Fig. 3.8 accepts the set of all strings in $\{0, 1\}^*$ with an even number of 1s. The corresponding regular expression is $(0^*10^*1)^*0^*$.

Example 3.16. As a final example, consider the special case of the tiling problem that we discussed in section 3.2. This version of the problem is as follows: Let k be a fixed positive integer. Given a set of unit tiles, we want to know if they can tile any $k \times n$ area for all n. We show how to deal with the case $k = 1$ and leave it as an exercise to generalize our method for larger values of k. Number the quarters of each tile as in Fig. 3.9. The given set of tiles will tile the area if we can find a sequence of the given tiles $T_1, T_2, \ldots, T_m$ such that (1) the third quarter of T_1 has the same color as the first quarter of T_2, and the third quarter of T_2 has the same color as the first quarter of T_3, etc., and (2) the third quarter of T_m has the same color as T_1. These conditions can be easily understood as follows. The first condition states that the tiles T_1, T_2, etc., can be placed adjacent to each other along a row in that order. The second condition implies that the whole sequence $T_1 T_2 \cdots T_m$ can be replicated any number of times. And a little thought reveals that this is all we need to answer yes on the input. But if we cannot find such a sequence, then the answer must be no. Also note that in the sequence no tile needs to be repeated and so the value of m is bounded by the number of tiles in the input. Thus, we have reduced the problem to searching a finite number of possibilities and we are done.

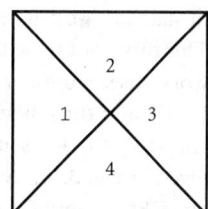

FIGURE 3.9 Numbering the quarters of a tile.

How is the preceding discussion related to finite automata? To see the connection, define an alphabet consisting of the unit tiles and define a language $L = \{T_1 T_2 \cdots T_m \mid T_1 T_2 \cdots T_m$ is a valid tiling, $m \geq 0\}$. We will now construct an NFA for the language L. It consists of states corresponding to *distinct* colors contained in the tiles plus two states, one of them the start state and another state called the dead state. The NFA makes transitions as follows: From the start state there is an ϵ-transition to each color state, and all states except the dead state are accepting states. When in the state corresponding to color i, suppose it receives input tile T. If the first quarter of this tile has color i, then it moves to the color of the third quarter of T; otherwise, it enters the dead state. The basic idea is to remember the only relevant piece of information after processing some input. In this case, it is the third quarter color of the last tile seen. Having constructed this NFA, the question we are asking is if the language accepted by this NFA is infinite. There is a simple algorithm for this problem [Hopcroft and Ullman 1979].

The next three theorems show a satisfying result that all the following language classes are identical:

- The class of languages accepted by DFAs
- The class of languages accepted by NFAs

- The class of languages generated by regular expressions, as in Definition 3.8
- The class of languages generated by the right-linear, or type-3, grammars, as in Definition 3.16

Recall that this class of languages is called the *regular languages* (see section 3.4).

Theorem 3.6. *For each NFA, there is an equivalent DFA.*

PROOF An NFA might look more powerful because it can carry out its computation in parallel with its nondeterministic branches. But because we are working with a *finite number* of states, we can simulate an NFA $M = (Q, \Sigma, \delta, q_0, F)$ by a DFA $M' = (Q', \Sigma, \delta', q_0', F')$, where

- $Q' = \{[S] : S \subseteq Q\}$.
- $q_0' = [\{q_0\}]$.
- $\delta'([S], a) = [S'] = [\cup_{q_l \in S} \delta(q_l, a)]$.
- F' is the set of all subsets of Q containing a state in F.

It can now be verified that $L(M) = L(M')$.   □

Example 3.17. Example 3.1 contains an NFA and an equivalent DFA accepting the same language. In fact, the proof provides an effective procedure for converting an NFA to a DFA. Although each NFA can be converted to an equivalent DFA, the resulting DFA might be exponentially large in terms of the number of states, as we can see from the previous procedure. This turns out to be the best thing one can do in the worst case. Consider the language: $L_k = \{x : x \in \{0, 1\}^*$ and the kth letter from the right of x is a $1\}$. An NFA of $k + 1$ states (for $k = 3$) accepting L_k is given in Fig. 3.10. A counting argument shows that any DFA accepting L_k must have at least 2^k states.

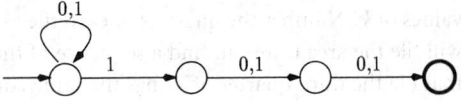

FIGURE 3.10   An NFA accepting L_3.

Theorem 3.7. *L is generated by a right-linear grammar iff it is accepted by an NFA.*

PROOF Let L be accepted by a right-linear grammar $G = (\Sigma, N, S, P)$. We design an NFA $M = (Q, \Sigma, \delta, q_0, F)$ where $Q = N \cup \{f\}, q_0 = S, F = \{f\}$. To define the δ function, we have $C \in \delta(A, b)$ iff $A \to bC$. For rules $A \to b, \delta(A, b) = \{f\}$. Obviously, $L(M) = L(G)$.

Conversely, if L is accepted by an NFA $M = (Q, \Sigma, \delta, q_0, F)$, we define an equivalent right-linear grammar $G = (\Sigma, N, S, P)$, where $N = Q, S = q_0, q_i \to aq_j \in N$ if $q_j \in \delta(q_i, a)$, and $q_j \to \epsilon$ iff $q_j \in F$. Again it is easily seen that $L(M) = L(G)$.   □

Theorem 3.8. *L is generated by a regular expression iff it is accepted by an NFA.*

PROOF Idea   **Part 1.** We inductively convert a regular expression to an NFA which accepts the language generated by the regular expression as follows.

- Regular expression ϵ converts to $(\{q\}, \Sigma, \emptyset, q, \{q\})$.
- Regular expression $\emptyset$ converts to $(\{q\}, \Sigma, \emptyset, q, \emptyset)$.
- Regular expression a, for each $a \in \Sigma$ converts to $(\{q, f\}, \Sigma, \delta(q, a) = \{f\}, q, \{f\})$.
- If α and β are regular expressions, converting to NFAs M_α and M_β, respectively, then the regular expression $\alpha \cup \beta$ converts to an NFA M, which connects M_α and M_β in parallel: M has an initial state q_0 and all of the states and transitions of M_α and M_β; by ϵ-transitions, M goes from q_0 to the initial states of M_α and M_β.

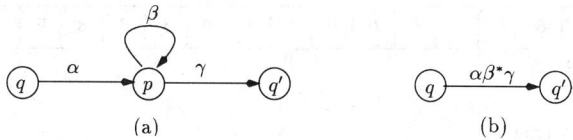

FIGURE 3.11 Converting an NFA to a regular expression.

- If α and β are regular expressions, converting to NFAs M_α and M_β, respectively, then the regular expression $\alpha\beta$ converts to NFA M, which connects M_α and M_β sequentially: M has all of the states and transitions of M_α and M_β, with M_α's initial state as M's initial state, ϵ-transition from the final states of M_α to the initial state of M_β, and M_β's final states as M's final states.
- If α is a regular expression, converting to NFA M_α, then connecting all of the final states of M_α to its initial state with ϵ-transitions gives α^+. Union of this with the NFA for ϵ gives the NFA for α^*.

Part 2. We now show how to convert an NFA to an equivalent regular expression. The idea used here is based on Brzozowski and McCluskey [1963]; see also Brzozowski and Seger [1994] and Wood [1987].

Given an NFA M, expand it to M' by adding two extra states i, the initial state of M', and t, the only final state of M', with ϵ transitions from i to the initial state of M and from all final states of M to t. Clearly, $L(M) = L(M')$. In M', remove states other than i and t one by one as follows. To remove state p, for each triple of states q, p, q' as shown in Fig. 3.11(a), add the transition as shown in Fig. 3.11(b).

If p does not have a transition leading back to itself, then $\beta = \epsilon$. After we have considered all such triples, delete state p and transitions related to p. Finally, we obtain Fig. 3.12 and $L(\alpha) = L(M)$.

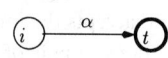

FIGURE 3.12
The reduced NFA.

Apparently, DFAs cannot serve as our model for a modern computer. Many extremely simple languages cannot be accepted by DFAs. For example, $L = \{xx : x \in \{0, 1\}^*\}$ cannot be accepted by a DFA. One can prove this by counting, or using the so-called pumping lemmas; one can also prove this by arguing that x contains more information than a *finite* state machine can *remember*. We refer the interested readers to textbooks such as Hopcroft and Ullmann [1979], Gurari [1989], Wood [1987], and Floyd and Beigel [1994] for traditional approaches and to Li and Vitányi [1993] for a nontraditional approach. One can try to generalize the DFA to allow the input head to be *two way* but still read only. But such machines are not more powerful, they can be simulated by normal DFAs. The next step is apparently to add *storage* space such that our machines can *write* information in.

Turing Machines

In this section we will provide an alternative definition of a Turing machine to make it compatible with our definitions of a DFA, PDA, etc. This also makes it easier to define a nondeterministic Turing machine. But this formulation (at least the deterministic version) is essentially the same as the one presented in section 3.2.

A Turing machine (TM), as in Fig. 3.13, consists of a *finite control*, an infinite *tape* divided into cells, and a read/write *head* on the tape. We refer to the two directions on the tape as *left* and *right*. The finite control can be in any one of a finite set Q of states, and each tape cell can contain a 0, a 1, or a *blank* B. Time is discrete and the time instants are ordered $0, 1, 2, \ldots$ with 0 the time at which the machine starts its computation. At any time, the head is positioned over a particular cell, which it is said to *scan*. At time 0 the head is situated on a distinguished cell on the tape called the *start cell*, and the finite control is in the initial state q_0. At time 0 all cells contain Bs, except a contiguous finite sequence of cells, extending from the start cell to the right, which contain 0s and 1s. This binary sequence is called the *input*.

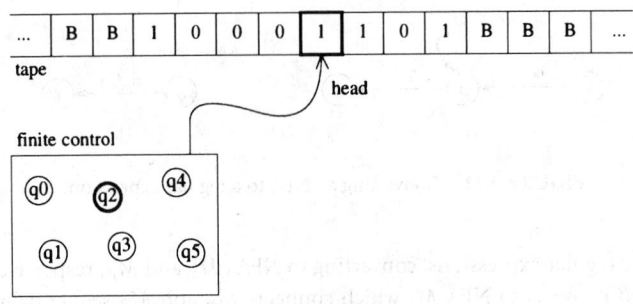

FIGURE 3.13   A Turing machine.

The device can perform the following basic operations:

1. It can write an element from the tape alphabet $\Sigma = \{0, 1, B\}$ in the cell it scans.
2. It can shift the head one cell left or right.

Also, the device executes these operations at the rate of one operation per time unit (a *step*). At the conclusion of each step, the finite control takes on a state in Q. The device operates according to a finite set P of *rules*.

The rules have format (p, s, a, q) with the meaning that if the device is in state p and s is the symbol under scan then write a if $a \in \{0, 1, B\}$ or move the head according to a if $a \in \{L, R\}$ and the finite control changes to state q. At some point, if the device gets into a special *final* state q_f, the device stops and accepts the input.

If every pair of distinct quadruples differs in the first two elements, then the device is *deterministic*. Otherwise, the device is *nondeterministic*. Not every possible combination of the first two elements has to be in the set; in this way we permit the device to perform *no* operation. In this case, we say the device *halts*. In this case, if the machine is not in a final state, we say that the machine *rejects* the input.

Definition 3.20.   A Turing machine is a quintuple $M = (Q, \Sigma, P, q_0, q_f)$ where each of the components has been described previously.

Given an input, a deterministic Turing machine carries out a uniquely determined succession of operations, which may or may not terminate in a finite number of steps. If it terminates, then the nonblank symbols left on the tape are the output. Given an input, a **nondeterministic Turing machine** behaves much like an NFA. One may imagine that it carries out its computation in parallel. Such a computation may be viewed as a (possibly infinite) tree. The root of the tree is the starting configuration of the machine. The children of each node are all possible configurations one step away from this node. If any of the branches terminates in the final state q_f, we say the machine accepts the input. The reader may want to test understanding this new formulation of a Turing machine by redoing the doubling program on a Turing machine with states and transitions (rather than a GOTO program).

A Turing machine *accepts* a language L if $L = \{w : M$ accepts $w\}$. Furthermore, if M halts on all inputs, then we say that L is *Turing decidable*, or *recursive*. The connection between a recursive language and a decidable problem (function) should be clear. It is that function f is decidable if and only if L_f is recursive. (Readers who may have forgotten the connection between function f and the associated language L_f should review Remark 3.1.)

Theorem 3.9.   *All of the following generalizations of Turing machines can be simulated by a one-tape deterministic Turing machine defined in Definition 3.20.*

- *Larger tape alphabet Σ*
- *More work tapes*

- *More access points, read/write heads, on each tape*
- *Two- or more dimensional tapes*
- *Nondeterminism*

Although these generalizations do not make a Turing machine compute more, they do make a Turing machine more efficient and easier to program. Many more variants of Turing machines are studied and used in the literature. Of all simulations in Theorem 3.9, the last one needs some comments. A nondeterministic computation branches like a tree. When simulating such a computation for n steps, the obvious thing for a deterministic Turing machine to do is to try all possibilities; thus, this requires up to c^n steps, where c is the maximum number of nondeterministic choices at each step.

Example 3.18. A DFA is an extremely simple Turing machine. It just reads the input symbols from left to right. Turing machines naturally accept more languages than DFAs can. For example, a Turing machine can accept $L = \{xx : x \in \{0, 1\}^*\}$ as follows:

- Find the middle point first: it is trivial by using two heads; with one head, one can mark one symbol at the left and then mark another on the right, and go back and forth to eventually find the middle point.
- Match the two parts: with two heads, this is again trivial; with one head, one can again use the marking method matching a pair of symbols each round; if the two parts match, accept the input by entering q_f.

There are types of storage media other than a tape:

- A *pushdown store* is a semi-infinite work tape with one head such that each time the head moves to the left, it erases the symbol scanned previously; this is a last-in first-out storage.
- A *queue* is a semi-infinite work tape with two heads that move only to the right, the leading head is write-only and the trailing head is read-only; this is a first-in first-out storage.
- A *counter* is a pushdown store with a single-letter alphabet (except its one end, which holds a special marker symbol). Thus, a counter can store a nonnegative integer and can perform three operations.

A queue machine can simulate a normal Turing machine, but the other two types of machines are not powerful enough to simulate a Turing machine.

Example 3.19. When the Turing machine tape is replaced by a pushdown store, the machine is called a *pushdown automaton*. Pushdown automata have been thoroughly studied because they accept the class of context-free languages defined in section 3.4. More precisely, it can be shown that if L is a context-free language, then it is accepted by a PDA, and if L is accepted by a PDA, then there is a CFG generating L. Various types of PDAs have fundamental applications in compiler design.

The PDA is more restricted than a Turing machine. For example, $L = \{xx : x \in \{0, 1\}^*\}$ cannot be accepted by a PDA, but it can be accepted by a Turing machine as in Example 3.18. But a PDA is more powerful than a DFA. For example, a PDA can accept the language $L' = \{0^k 1^k : k \geq 0\}$ easily. It can read the 0s and push them into the pushdown store; then, after it finishes the 0s, each time the PDA reads a 1, it removes a 0 from the pushdown store; at the end, it accepts if the pushdown store is empty (the number of 0s matches that of 1s). But a DFA cannot accept L', because after it has read all of the 0s, it cannot remember k when k has higher information content than the DFA's finite control.

Two pushdown stores can be used to simulate a tape easily. For comparisons of powers of pushdown stores, queues, counters, and tapes, see van Emde Boas [1990] and Li and Vitányi [1993].

The idea of the universal algorithm was introduced in section 3.2. Formally, a *universal Turing machine*, U, takes an encoding of a pair of parameters (M, x) as input and simulates M on input x. U accepts (M, x)

iff M accepts x. The universal Turing machines have many applications. For example, the definition of Kolmogorov complexity [Li and Vitányi 1993] fundamentally relies on them.

Example 3.20. Let $L_u = \{\langle M, w \rangle : M \text{ accepts } w\}$. Then L_u can be accepted by a Turing machine, but it is not Turing decidable. The proof is omitted.

If a language is Turing acceptable but not Turing decidable, we call such a language *recursively enumerable* (r.e.). Thus, L_u is r.e. but not recursive. It is easily seen that if both a language and its complement are r.e., then both of them are recursive. Thus, $\bar{L}_u$ is not r.e.

Time and Space Complexity

With Turing machines, we can now formally define what we mean by **time and space complexities**. Such a formal investigation by Hartmanis and Stearns [1965] marked the beginning of the field of *computational complexity*. We refer the readers to Hartmanis' Turing Award lecture [Hartmanis 1994] for an interesting account of the history and the future of this field.

To define the space complexity properly (in the sublinear case), we need to slightly modify the Turing machine of Fig. 3.13. We will replace the tape containing the input by a read-only input tape and give the Turing machine some extra work tapes.

Definition 3.21. Let M be a Turing machine. If for each n, for each input of length n, and for each sequence of choices of moves when M is nondeterministic, M makes at most $T(n)$ moves we say that M is of *time complexity* $T(n)$; similarly, if M uses at most $S(n)$ tape cells of the work tape, we say that M is of space complexity $S(n)$.

Theorem 3.10. *Any Turing machine using $s(n)$ space can be simulated by a Turing machine, with just one work tape, using $s(n)$ space. If a language is accepted by a k-tape Turing machine running in time $t(n)$ [space $s(n)$], then it also can be accepted by another k-tape Turing machine running in time $ct(n)$ [space $cs(n)$], for any constant $c > 0$.*

To avoid writing the constant c everywhere, we use the standard big-O notation: we say $f(n)$ is $O(g(n))$ if there is a constant c such that $f(n) \leq cg(n)$ for all but finitely many n. The preceding theorem is called the linear speedup theorem; it can be proved easily by using a larger tape alphabet to encode several cells into one and hence compress several steps into one. It leads to the following definitions.

Definition 3.22.

DTIME[$t(n)$] is the set of languages accepted by multitape deterministic TMs in time $O(t(n))$.
NTIME[$t(n)$] is the set of languages accepted by multitape nondeterministic TMs in time $O(t(n))$.
DSPACE[$s(n)$] is the set of languages accepted by multitape deterministic TMs in space $O(s(n))$.
NSPACE[$s(n)$] is the set of languages accepted by multitape nondeterministic TMs in space $O(s(n))$.
P is the complexity class $\bigcup_{c \in \mathcal{N}} \text{DTIME}[n^c]$.
NP is the complexity class $\bigcup_{c \in \mathcal{N}} \text{NTIME}[n^c]$.
PSPACE is the complexity class $\bigcup_{c \in \mathcal{N}} \text{DSPACE}[n^c]$.

Example 3.21. We mentioned in Example 3.18 that $L = \{xx : x \in \{0, 1\}^*\}$ can be accepted by a Turing machine. The procedure we have presented in Example 3.18 for a one-head one-tape Turing machine takes $O(n^2)$ time because the single head must go back and forth marking and matching. With two heads, or two tapes, L can be easily accepted in $O(n)$ time.

It should be clear that any language that can be accepted by a DFA, an NFA, or a PDA can be accepted by a Turing machine in $O(n)$ time. The type-1 grammar in Definition 3.16 can be accepted by a Turing machine in $O(n)$ space. Languages in P, that is, languages acceptable by Turing machines in *polynomial* time, are considered as *feasibly* computable. It is important to point out that all generalizations of the Turing machine, except the nondeterministic version, can all be simulated by the basic one-tape deterministic Turing machine with at most polynomial slowdown. The class NP represents the class of languages accepted in polynomial time by a nondeterministic Turing machine. The nondeterministic version of PSPACE turns out to be identical to PSPACE [Savitch 1970]. The following relationships are true:

$$P \subseteq NP \subseteq PSPACE$$

Whether or not either of the inclusions is proper is one of the most fundamental open questions in computer science and mathematics. Research in computational complexity theory centers around these questions. To solve these problems, one can identify the hardest problems in NP or PSPACE. These topics will be discussed in Chapter 11. We refer the interested reader to Gurari [1989], Hopcroft and Ullman [1979], Wood [1987], and Floyd and Beigel [1994].

Other Computing Models

Over the years, many alternative computing models have been proposed. With reasonable complexity measures, they can all be simulated by Turing machines with at most a polynomial slowdown. The reference van Emde Boas [1990] provides a nice survey of various computing models other than Turing machines. Because of limited space, we will discuss a few such alternatives very briefly and refer our readers to van Emde Boas [1990] for details and references.

Random Access Machines. The *random access machine* (RAM) [Cook and Reckhow 1973] consists of a finite control where a program is stored, with several arithmetic registers and an infinite collection of memory registers $R[1], R[2], \ldots$. All registers have an unbounded word length. The basic instructions for the program are LOAD, ADD, MULT, STORE, GOTO, ACCEPT, REJECT, etc. Indirect addressing is also used. Apparently, compared to Turing machines, this is a closer but more complicated approximation of modern computers. There are two standard ways for measuring time complexity of the model:

- The *unit-cost RAM*: in this case, each instruction takes one unit of time, no matter how big the operands are. This measure is convenient for analyzing some algorithms such as sorting. But it is unrealistic or even meaningless for analyzing some other algorithms, such as integer multiplication.
- The *log-cost RAM*: each instruction is charged for the sum of the lengths of all data manipulated implicitly or explicitly by the instruction. This is a more realistic model but sometimes less convenient to use.

Log-cost RAMs and Turing machines can simulate each other with polynomial overheads. The unit-cost RAM might be exponentially (but unrealistically) faster when, for example, it uses its power of multiplying two large numbers in one step.

Pointer Machines. The pointer machines were introduced by Kolmogorov and Uspenskii [1958] (also known as the Kolmogorov–Uspenskii machine) and by Schönhage in 1980 (also known as the storage modification machine, see Schönhage [1980]). We informally describe the pointer machine here. A pointer machine is similar to a RAM but differs in its memory structure. A pointer machine operates on a storage structure called a Δ structure, where Δ is a finite alphabet of size greater than one. A Δ-structure S is a finite directed graph (the Kolmogorov–Uspenskii version is an undirected graph) in which each node has $k = |\Delta|$ outgoing edges, which are labeled by the k symbols in Δ. S has a distinguished node called the *center*, which acts as a starting point for addressing, with words over Δ, other nodes in the structure. The pointer machine has various instructions to redirect the pointers or edges and thus modify the storage structure. It should be clear that Turing machines and pointer machines can simulate each other with at most polynomial delay if we use the log-cost model as with the RAMs. There are many interesting studies on the efficiency of the preceding simulations. We refer the reader to van Emde Boas [1990] for more pointers on the pointer machines.

Circuits and Nonuniform Models. A *Boolean circuit* is a finite, labeled, directed acyclic graph. Input nodes are nodes without ancestors; they are labeled with input variables $x_1, \ldots, x_n$. The internal nodes are labeled with functions from a finite set of Boolean operations, for example, {and, or, not} or {$\oplus$}. The number of ancestors of an internal node is precisely the number of arguments of the Boolean function that the node is labeled with. A node without successors is an output node. The circuit is naturally evaluated from input to output: at each node the function labeling the node is evaluated using the results of its ancestors as arguments. Two cost measures for the circuit model are:

- *Depth:* the length of a longest path from an input node to an output node
- *Size:* the number of nodes in the circuit

These measures are applied to a family of circuits $\{C_n : n \geq 1\}$ for a particular problem, where C_n solves the problem of size n. If C_n can be computed from n (in polynomial time), then this is a *uniform measure*. Such circuit families are equivalent to Turing machines. If Cn cannot be computed from n, then such measures are *nonuniform* measures, and such classes of circuits are more powerful than Turing machines because they simply can compute any function by encoding the solutions of all inputs for each n. See van Emde Boas [1990] for more details and pointers to the literature.

Acknowledgment

We would like to thank John Tromp and the reviewers for reading the initial drafts and helping us to improve the presentation.

Defining Terms

Algorithm: A finite sequence of instructions that is supposed to solve a particular problem.

Ambiguous context-free grammar: For some string of terminals the grammar has two distinct derivation trees.

Chomsky normal form: Every rule of the context-free grammar has the form $A \rightarrow BC$ or $A \rightarrow a$, where A, B, and C are nonterminals and a is a terminal.

Computable or decidable function/problem: A function/problem that can be solved by an algorithm (or equivalently, a Turing machine).

Context-free grammar: A grammar whose rules have the form $A \rightarrow \beta$, where A is a nonterminal and β is a string of nonterminals and terminals.

Context-free language: A language that can be described by some context-free grammar.

Context-sensitive grammar: A grammar whose rules have the form $\alpha \rightarrow \beta$, where α and β are strings of nonterminals and terminals and $|\alpha| \leq |\beta|$.

Context-sensitive language: A language that can be described by some context-sensitive grammar.

Derivation or parsing: An illustration of how a string of terminals is obtained from the start symbol by successively applying the rules of the grammar.

Finite automaton or finite-state machine: A restricted Turing machine where the head is read only and shifts only from left to right.

(Formal) grammar: A description of some language typically consisting of a set of terminals, a set of nonterminals with a distinguished one called the start symbol, and a set of rules (or productions) of the form $\alpha \rightarrow \beta$, depicting what string α of terminals and nonterminals can be rewritten as another string β of terminals and nonterminals.

(Formal) language: A set of strings over some fixed alphabet.

Halting problem: The problem of deciding if a given program (or Turing machine) halts on a given input.

Nondeterministic Turing machine: A Turing machine that can make any one of a prescribed set of moves on a given state and symbol read on the tape.

Partially decidable decision problem: There exists a program that always halts and outputs 1 for every input expecting a positive answer and either halts and outputs 0 or loops forever for every input expecting a negative answer.

Program: A sequence of instructions that is not required to terminate on every input.

Pushdown automaton: A restricted Turing machine where the tape acts as a pushdown store (or a stack).

Reduction: A computable transformation of one problem into another.

Regular expression: A description of some language using operators union, concatenation, and Kleene closure.

Regular language: A language that can be described by some right-linear/regular grammar (or equivalently by some regular expression).

Right-linear or regular grammar: A grammar whose rules have the form $A \to aB$ or $A \to a$, where A, B are nonterminals and a is either a terminal or the null string.

Time/space complexity: A function describing the maximum time/space required by the machine on any input of length n.

Turing machine: A simplest formal model of computation consisting of a finite-state control and a semi-infinite sequential tape with a read–write head. Depending on the current state and symbol read on the tape, the machine can change its state and move the head to the left or right.

Uncomputable or undecidable function/problem: A function/problem that cannot be solved by any algorithm (or equivalently, any Turing machine).

Universal algorithm: An algorithm that is capable of simulating any other algorithms if properly encoded.

References

Angluin, D. 1980. Finding patterns common to a set of strings. *J. Comput. Syst. Sci.* 21:46–62.

Brzozowski, J. and McCluskey, E., Jr. 1963. Signal flow graph techniques for sequential circuit state diagram. *IEEE Trans. Electron. Comput.* EC-12(2):67–76.

Brzozowski, J. A. and Seger, C.-J. H. 1994. *Asynchronous Circuits.* Springer–Verlag, New York.

Chomsky, N. 1956. Three models for the description of language. *IRE Trans. Inf. Theory* 2(2):113–124.

Chomsky, N. 1963. Formal properties of grammars. In *Handbook of Mathematical Psychology*, Vol. 2, pp. 323–418. John Wiley and Sons, New York.

Chomsky, N. and Miller, G. 1958. Finite-state languages. *Information and Control* 1:91–112.

Cook, S. and Reckhow, R. 1973. Time bounded random access machines. *J. Comput. Syst. Sci.* 7:354–375.

Davis, M. 1980. What is computation? In *Mathematics Today–Twelve Informal Essays.* L. Steen, ed., pp. 241–259. Vintage Books, New York.

Floyd, R. W. and Beigel, R. 1994. *The Language of Machines: An Introduction to Computability and Formal Languages.* Computer Science Press, New York.

Gurari, E. 1989. *An Introduction to the Theory of Computation.* Computer Science Press, Rockville, MD.

Harel, D. 1992. *Algorithmics: The Spirit of Computing.* Addison–Wesley, Reading, MA.

Harrison, M. 1978. *Introduction to Formal Language Theory.* Addison–Wesley, Reading, MA.

Hartmanis, J. 1994. On computational complexity and the nature of computer science. *Commun. ACM* 37(10):37–43.

Hartmanis, J. and Stearns, R. 1965. On the computational complexity of algorithms. *Trans. Amer. Math. Soc.* 117:285–306.

Hopcroft, J. and Ullman, J. 1979. *Introduction to Automata Theory, Languages and Computation.* Addison–Wesley, Reading, MA.

Jiang, T., Salomaa, A., Salomaa, K., and Yu, S. 1995. Decision problems for patterns. *J. Comput. Syst. Sci.* 50(1):53–63.

Kleene, S. 1956. Representation of events in nerve nets and finite automata. In *Automata Studies*, pp. 3–41. Princeton University Press, Princeton, NJ.

Knuth, D., Morris, J., and Pratt, V. 1977. Fast pattern matching in strings. *SIAM J. Comput.* 6:323–350.

Kohavi, Z. 1978. *Switching and Finite Automata Theory.* McGraw–Hill, New York.

Kolmogorov, A. and Uspenskii, V. 1958. On the definition of an algorithm. *Usp. Mat. Nauk.* 13:3–28.

Lesk, M. 1975. LEX–a lexical analyzer generator. *Tech. Rep. 39. Bell Labs.* Murray Hill, NJ.

Li, M. and Vitányi, P. 1993. *An Introduction to Kolmogorov Complexity and Its Applications.* Springer–Verlag, Berlin.

McCulloch, W. and Pitts, W. 1943. A logical calculus of ideas immanent in nervous activity. *Bull. Math. Biophys.* 5:115–133.

Post, E. 1943. Formal reductions of the general combinatorial decision problems. *Am. J. Math.* 65:197–215.

Rabin, M. and Scott, D. 1959. Finite automata and their decision problems. *IBM J. Res. Dev.* 3:114–125.

Robinson, R. 1991. Minsky's small universal Turing machine. *Int. J. Math.* 2(5):551–562.

Salomaa, A. 1966. Two complete axiom systems for the algebra of regular events. *J. ACM* 13(1):158–169.

Savitch, J. 1970. Relationships between nondeterministic and deterministic tape complexities. *J. Comput. Syst. Sci.* 4(2)177–192.

Schönhage, A. 1980. Storage modification machines. *SIAM J. Comput.* 9:490–508.

Searls, D. 1993. The computational linguistics of biological sequences. In *Artificial Intelligence and Molecular Biology.* L. Hunter, ed., pp. 47–120. MIT Press, Cambridge, MA.

Turing, A. 1936. On computable numbers with an application to the Entscheidungsproblem. *Proc. London Math. Soc., Ser. 2* 42:230–265.

van Emde Boas, P. 1990. Machine models and simulations. In *Handbook of Theoretical Computer Science.* J. van Leeuwen, ed., pp. 1–66. Elsevier/MIT Press.

Wood, D. 1987. *Theory of Computation.* Harper and Row.

Further Information

The fundamentals of the theory of computation, automata theory, and formal languages can be found in many text books including Floyd and Beigel [1994], Gurari [1989], Harel [1992], Harrison [1978], Hopcroft and Ullman [1979], and Wood [1987]. The central focus of research in this area is to understand the relationships between the different resource complexity classes. This work is motivated in part by some major open questions about the relationships between resources (such as time and space) and the role of control mechanisms (nondeterminism/randomness). At the same time, new computational models are being introduced and studied. One such recent model that has led to the resolution of a number of interesting problems is the interactive proof systems. They exploit the power of randomness and interaction. Among their applications are new ways to encrypt information as well as some unexpected results about the difficulty of solving some difficult problems even approximately. Another new model is the quantum computational model that incorporates quantum-mechanical effects into the basic move of a Turing machine. There are also attempts to use molecular or cell-level interactions as the basic operations of a computer. Yet another research direction motivated in part by the advances in hardware technology is the study of neural networks, which model (albeit in a simplistic manner) the brain structure of mammals. The following chapters of this volume will present state-of-the-art information about many of these developments. The following annual conferences present the leading research work in computation theory: Association of Computer Machinery (ACM) Annual Symposium on Theory of Computing; Institute of Electrical and Electronics Engineers (IEEE) Symposium on the Foundations of Computer Science; IEEE Conference on Structure in Complexity Theory; International Colloquium on Automata, Languages and Programming; Symposium on Theoretical Aspects of Computer Science; Mathematical

Foundations of Computer Science; and Fundamentals of Computation Theory. There are many related conferences such as Computational Learning Theory, ACM Symposium on Principles of Distributed Computing, etc., where specialized computational models are studied for a specific application area. Concrete algorithms is another closely related area in which the focus is to develop algorithms for specific problems. A number of annual conferences are devoted to this field. We conclude with a list of major journals whose primary focus is in theory of computation: *The Journal of the Association of Computer Machinery, SIAM Journal on Computing, Journal of Computer and System Sciences, Information and Computation, Mathematical Systems Theory, Theoretical Computer Science, Computational Complexity, Journal of Complexity, Information Processing Letters, International Journal of Foundations of Computer Science,* and *ACTA Informatica.*

4

Basic Techniques for Design and Analysis of Algorithms

Edward M. Reingold*
*University of Illinois at
Urbana-Champaign*

We outline the basic methods of algorithm design and analysis that have found application in the manipulation of discrete objects such as lists, arrays, sets, graphs, and geometric objects such as points, lines, and polygons. We begin by discussing recurrence relations and their use in the analysis of algorithms. Then we discuss some specific examples in algorithm analysis, sorting, and priority queues. In the final three sections, we explore three important techniques of algorithm design: divide-and-conquer, dynamic programming, and greedy heuristics.

4.1 Analyzing Algorithms

It is convenient to classify algorithms based on the relative amount of time they require: how fast does the time required grow as the size of the problem increases? For example, in the case of arrays, the size of the problem is ordinarily the number of elements in the array. If the size of the problem is measured by a variable n, we can express the time required as a function of n, $T(n)$. When this function $T(n)$ grows rapidly, the algorithm becomes unusable for large n; conversely, when $T(n)$ grows slowly, the algorithm remains useful even when n becomes large.

We say an algorithm is $\Theta(n^2)$ if the time it takes quadruples (asymptotically) when n doubles; an algorithm is $\Theta(n)$ if the time it takes doubles when n doubles; an algorithm is $\Theta(\log n)$ if the time it takes increases by a constant, independent of n, when n doubles; an algorithm is $\Theta(1)$ if its time does not increase at all when n increases. In general, an algorithm is $\Theta(T(n))$ if the time it requires on problems of size n grows proportionally to $T(n)$ as n increases. Table 4.1 summarizes the common growth rates encountered in the analysis of algorithms.

The analysis of an algorithm is often accomplished by finding and solving a recurrence relation that describes the time required by the algorithm. The most commonly occurring families of recurrences in

*Supported in part by the National Science Foundation, Grants CCR-93-20577 and CCR-95-30297.

TABLE 4.1 Common Growth Rates of Times of Algorithms

Rate of Growth	Comment	Examples
$\Theta(1)$	Time required is constant, independent of problem size	Expected time for hash searching
$\Theta(\log\log n)$	Very slow growth of time required	Expected time of interpolation search
$\Theta(\log n)$	Logarithmic growth of time required: doubling the problem size increases the time by only a constant amount	Computing x^n, binary search of an array
$\Theta(n)$	Time grows linearly with problem size: doubling the problem size doubles the time required	Adding/subtracting n-digit numbers, linear search of an n-element array
$\Theta(n\log n)$	Time grows worse than linearly, but not much worse: doubling the problem size more than doubles the time required	Merge sort, heapsort, lower bound on comparison-based sorting
$\Theta(n^2)$	Time grows quadratically: doubling the problem size quadruples the time required	Simple-minded sorting algorithms
$\Theta(n^3)$	Time grows cubically: doubling the problem size results in an 8-fold increase in the time required	Ordinary matrix multiplication
$\Theta(c^n)$	Time grows exponentially: increasing the problem size by 1 results in a c-fold increase in the time required; doubling the problem size *squares* the time required	Traveling salesman problem

the analysis of algorithms are linear recurrences and divide-and-conquer recurrences. In the following subsection we describe the *method of operators* for solving linear recurrences; in the next subsection we describe how to transform divide-and-conquer recurrences into linear recurrences by substitution to obtain an asymptotic solution.

Linear Recurrences

A *linear recurrence with constant coefficients* has the form

$$c_0 a_n + c_1 a_{n-1} + c_2 a_{n-2} + \cdots + c_k a_{n-k} = f(n) \tag{4.1}$$

for some constant k, where each c_i is constant. To solve such a recurrence for a broad class of functions f (that is, to express a_n in closed form as a function of n) by the *method of operators*, we consider two basic operators on sequences: $\mathcal{S}$, which shifts the sequence left,

$$\mathcal{S}\langle a_0, a_1, a_2, \ldots \rangle = \langle a_1, a_2, a_3, \ldots \rangle$$

and C, which, for any constant C, multiplies each term of the sequence by C

$$C\langle a_0, a_1, a_2, \ldots \rangle = \langle Ca_0, Ca_1, Ca_2, \ldots \rangle$$

Then, given operators A and B, we define the sum and product

$$(A + B)\langle a_0, a_1, a_2, \ldots \rangle = A\langle a_0, a_1, a_2, \ldots \rangle + B\langle a_0, a_1, a_2, \ldots \rangle$$

$$(AB)\langle a_0, a_1, a_2, \ldots \rangle = A(B\langle a_0, a_1, a_2, \ldots \rangle)$$

Thus, for example,

$$(\mathcal{S}^2 - 4)\langle a_0, a_1, a_2, \ldots \rangle = \langle a_2 - 4a_0, a_3 - 4a_1, a_4 - 4a_2, \ldots \rangle$$

which we write more briefly as

$$(S^2 - 4)\langle a_i \rangle = \langle a_{i+2} - 4a_i \rangle$$

With the operator notation, we can rewrite Eq. (4.1) as

$$P(S)\langle a_i \rangle = \langle f(i) \rangle$$

where

$$P(S) = c_0 S^k + c_1 S^{k-1} + c_2 S^{k-2} + \cdots + c_k$$

is a polynomial in S.

Given a sequence $\langle a_i \rangle$, we say that the operator $P(S)$ *annihilates* $\langle a_i \rangle$ if $P(S)\langle a_i \rangle = \langle 0 \rangle$. For example, $S^2 - 4$ annihilates any sequence of the form $\langle u2^i + v(-2)^i \rangle$, with constants u and v. In general,

> The operator $S^{k+1} - c$ annihilates $\langle c^i \times$ a polynomial in i of degree $k \rangle$.

The *product* of two annihilators annihilates the *sum* of the sequences annihilated by each of the operators; that is, if A annihilates $\langle a_i \rangle$ and B annihilates $\langle b_i \rangle$, then AB annihilates $\langle a_i + b_i \rangle$. Thus determining the annihilator of a sequence is tantamount to determining the sequence; moreover, it is straightforward to determine the annihilator from a recurrence relation.

For example, consider the Fibonacci recurrence

$$F_0 = 0$$

$$F_1 = 1$$

$$F_{i+2} = F_{i+1} + F_i$$

The last line of this definition can be rewritten as $F_{i+2} - F_{i+1} - F_i = 0$, which tells us that $\langle F_i \rangle$ is annihilated by the operator

$$S^2 - S - 1 = (S - \phi)(S + 1/\phi)$$

where $\phi = (1 + \sqrt{5})/2$. Thus, we conclude that

$$F_i = u\phi^i + v(-\phi)^{-i}$$

for some constants u and v. We can now use the initial conditions $F_0 = 0$ and $F_1 = 1$ to determine u and v. These initial conditions mean that

$$u\phi^0 + v(-\phi)^{-0} = 0$$

$$u\phi^1 + v(-\phi)^{-1} = 1$$

and these linear equations have the solution

$$u = v = 1/\sqrt{5}$$

and hence,

$$F_i = \phi^i/\sqrt{5} + (-\phi)^{-i}/\sqrt{5}$$

In the case of the similar recurrence,

$$G_0 = 0$$

$$G_1 = 1$$

$$G_{i+2} = G_{i+1} + G_i + i$$

the last equation tells us that

$$(S^2 - S - 1)\langle G_i \rangle = \langle i \rangle$$

so the annihilator for $\langle G_i \rangle$ is $(S^2 - S - 1)(S - 1)^2$ since $(S - 1)^2$ annihilates $\langle i \rangle$ (a polynomial of degree 1 in i) and hence the solution is

$$G_i = u\phi^i + v(-\phi)^{-i} + \text{(a polynomial of degree 1 in } i)$$

that is,

$$G_i = u\phi^i + v(-\phi)^{-i} + wi + z$$

Again, we use the initial conditions to determine the constants u, v, w, and x.

In general, then, to solve the recurrence (4.1), we factor the annihilator

$$P(S) = c_0 S^k + c_1 S^{k-1} + c_2 S^{k-2} + \cdots + c_k$$

multiply it by the annihilator for $\langle f(i) \rangle$, write down the form of the solution from this product (which is the annihilator for the sequence $\langle a_i \rangle$), and then use the initial conditions for the recurrence to determine the coefficients in the solution.

Divide-and-Conquer Recurrences

The divide-and-conquer paradigm of algorithm construction that we will discuss in section 4.3 leads naturally to divide-and-conquer recurrences of the type

TABLE 4.2 Rate of Growth of the Solution to the Recurrence $T(n) = g(n) + uT(n/v)$, the Divide-and-Conquer Recurrence Relations

$g(n)$	u, v^a	Growth Rate of $T(n)$
$\Theta(1)$	$u = 1$	$\Theta(\log n)$
	$u \neq 1$	$\Theta(n^{\log_v u})$
$\Theta(\log n)$	$u = 1$	$\Theta[(\log n)^2]$
	$u \neq 1$	$\Theta(n^{\log_v u})$
$\Theta(n)$	$u < v$	$\Theta(n)$
	$u = v$	$\Theta(n \log n)$
	$u > v$	$\Theta(n^{\log_v u})$
$\Theta(n^2)$	$u < v^2$	$\Theta(n^2)$
	$u = v^2$	$\Theta(n^2 \log n)$
	$u > v^2$	$\Theta(n^{\log_v u})$

[a]Positive constants, independent of n, and $v > 1$.

$$T(n) = g(n) + uT(n/v)$$

for constants u and v, $v > 1$, and sufficient initial values to define the sequence $\langle T(0), T(1), T(2), \ldots \rangle$. The growth rates of $T(n)$ for various values of u and v are given in Table 4.2. The growth rates in this table are derived by transforming the divide-and-conquer recurrence into a linear recurrence for a subsequence of $\langle T(0), T(1), T(2), \ldots \rangle$.

To illustrate this method, we derive the penultimate line in Table 4.2. We want to solve

$$T(n) = n^2 + v^2 T(n/v)$$

Thus, we want to find a subsequence of $\langle T(0), T(1), T(2), \ldots \rangle$ that will be easy to handle. Let $n_k = v^k$; then

$$T(n_k) = n_k^2 + v^2 T(n_k/v)$$

or

$$T(v^k) = v^{2k} + v^2 T(v^{k-1})$$

Defining $t_k = T(v^k)$,

$$t_k = v^{2k} + v^2 t_{k-1}$$

The annihilator for t_k is then $(\mathcal{S} - v^2)^2$ and thus

$$t_k = v^{2k}(ak + b)$$

for constants a and b. Expressing this in terms of $T(n)$,

$$T(n) \approx t_{\log_v n} = v^{2\log_v n}(a \log_v n + b) = an^2 \log_v n + bn^2$$

or

$$T(n) = \Theta(n^2 \log n)$$

4.2 Some Examples of the Analysis of Algorithms

In this section we introduce the basic ideas of analyzing algorithms by looking at some data structure problems that occur commonly in practice, problems relating to maintaining a collection of n objects and retrieving objects based on their relative size. For example, how can we determine the smallest of the elements? Or, more generally, how can we determine the kth largest of the elements? What is the running time of such algorithms in the worst case—or, on the average, if all $n!$ permutations of the input are equally likely? What if the set of items is dynamic—that is, the set changes through insertions and deletions—how efficiently can we keep track of, say, the largest element?

Sorting

The most demanding request that we can make of an array of n values $\mathbf{x[1]}, \mathbf{x[2]}, \ldots, \mathbf{x[n]}$ is that they be kept in perfect order so that $\mathbf{x[1]} \leq \mathbf{x[2]} \leq \cdots \leq \mathbf{x[n]}$. The simplest way to put the values in order is to mimic what we might do by hand: take item after item and insert each one into the proper place among those items already inserted:

```
1  void insert (float x[], int i, float a) {
2      // Insert a into x[1] ... x[i]
3      // x[1] ... x[i-1] are sorted;  x[i] is unoccupied
4      if (i == 1 || x[i-1] <= a)
5          x[i] = a;
6      else {
7          x[i] = x[i-1];
```

```
 8            insert(x, i-1, a);
 9      }
10   }
11
12   void insertionSort (int n, float x[]) {
13      // Sort x[1] ... x[n]
14      if (n > 1) {
15         insertionSort(n-1, x);
16         insert(x, n, x[n]);
17      }
18   }
```

To determine the time required in the worst case to sort n elements with **insertionSort,** we let t_n be the time to sort n elements and derive and solve a recurrence relation for t_n. We have

$$t_n = \begin{cases} \Theta(1) & \text{if } n = 1, \\ t_{n-1} + s_{n-1} + \Theta(1) & \text{otherwise} \end{cases}$$

where s_m is the time required to insert an element in place among m elements using **insert.** The value of s_m is also given by a recurrence relation:

$$s_m = \begin{cases} \Theta(1) & \text{if } m = 1 \\ s_{m-1} + \Theta(1) & \text{otherwise} \end{cases}$$

The annihilator for $\langle s_i \rangle$ is $(S - 1)^2$, and so $S_m = \Theta(m)$. Thus the annihilator for $\langle t_i \rangle$ is $(S - 1)^3$, and so $t_n = \Theta(n^2)$. The analysis of the average behavior is nearly identical; only the constants hidden in the Θ notation change.

We can design better sorting methods using the divide-and-conquer idea of the next section. These algorithms avoid $\Theta(n^2)$ worst-case behavior, working in time $\Theta(n \log n)$. We can also achieve time $\Theta(n \log n)$ by using a clever way of viewing the array of elements to be sorted as a tree: consider **x[1]** as the root of the tree and, in general, **x[2*i]** is the root of the left subtree of **x[i]** and **x[2*i+1]** is the root of the right subtree of **x[i].** If we further insist that parents be greater than or equal to children, we have a *heap*; Fig. 4.1 shows a small example.

A heap can be used for sorting by observing that the largest element is at the root, that is, **x[1]:** thus to put the largest element in place, we swap **x[1]** and **x[n].** To continue, we must restore the heap property, which may now be violated at the root. Such restoration is accomplished by swapping **x[1]** with its larger child, if that child is larger than **x[1],** and then continuing to swap it downward until it reaches either the bottom or a spot where it is greater than or equal to its children. Since the tree-cum-array has height $\Theta(\log n)$, this restoration process takes time $\Theta(\log n)$. Now, with the heap in **x[1]** to **x[n-1]** and **x[n]** the largest value in the array, we can put the second largest element in place by swapping **x[1]** and **x[n-1];** then we restore the heap property in **x[1]** to **x[n-2]** by propagating **x[1]** downward— this takes time $\Theta(\log(n - 1))$. Continuing in this fashion, we find we can sort the entire array in time

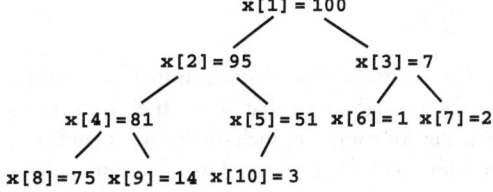

FIGURE 4.1   A heap, that is, an array, interpreted as a binary tree.

$$\Theta(\log n + \log(n - 1) + \cdots + \log 1) = \Theta(n \log n)$$

The initial creation of the heap from an unordered array is done by applying the restoration process successively to **x[n/2], x[n/2-1],..., x[1]**, which takes time $\Theta(n)$.

Hence we have the following $\Theta(n \log n)$ sorting algorithm:

```
1  void heapify (int n, float x[], int i) {
2      // Repair heap property below x[i] in x[1] ... x[n]
3      int largest = i; // largest of x[i], x[2 * i], x[2 * i+1]
4      if (2 * i <= n && x[2*i] > x[i])
5          largest = 2*i;
6      if (2*i+1 <= n && x[2 * i+1] > x[largest])
7          largest = 2*i+1;
8      if (largest != i) {
9          // swap x[i] with larger child and repair heap below
10         float t = x[largest]; x[largest] = x[i]; x[i] = t;
11         heapify(n, x, largest);
12     }
13 }
14
15 void makeheap (int n, float x[]) {
16     // Make x[1] ... x[n] into a heap
17     for (int i=n/2; i>0; i--)
18         heapify(n, x, i);
19 }
20
21 void heapsort (int n, float x[]) {
22     // Sort x[1] ... x[n]
23     float t;
24     makeheap(n, x);
25     for (int i=n; i>1; i--) {
26         // put x[1] in place and repair heap
27         t = x[1]; x[1] = x[i]; x[i] = t;
28         heapify(i-1, x, 1);
29     }
30 }
```

Can we find sorting algorithms that take time less than $\Theta(n \log n)$? The answer is no if we are restricted to sorting algorithms that derive their information from comparisons between the values of elements. The flow of control in such sorting algorithms can be viewed as binary trees in which there are $n!$ leaves, one for every possible sorted output arrangement. Because a binary tree with height h can have at most 2^h leaves, it follows that the height of a tree with $n!$ leaves must be at least $\log_2 n! = \Theta(n \log n)$. Since the height of this tree corresponds to the longest sequence of element comparisons possible in the flow of control, any such sorting algorithm must, in its worst case, use time proportional to $n \log n$; this is a *lower bound* on sorting—no algorithm for sorting can be guaranteed always to use time less than $\Theta(n \log n)$.

Priority Queues

Aside from its application to sorting, the heap is an interesting data structure in its own right. In particular, heaps provide a simple way to implement a *priority queue*; a priority queue is an abstract data structure that keeps track of a dynamically changing set of values allowing the operations

> **create:** Create an empty priority queue.
> **insert:** Insert a new element into a priority queue.
> **decrease:** Decrease an element in a priority queue.
> **minimum:** Report the minimum element in a priority queue.
> **deleteMinimum:** Delete the minimum element in a priority queue.
> **delete:** Delete an element in a priority queue.
> **merge:** Merge two priority queues.

A heap can implement a priority queue by altering the heap property to insist that parents are less than or equal to their children, so that that smallest value in the heap is at the root, that is, in the first array position. Creation of an empty heap requires just the allocation of an array, an $\Theta(1)$ operation; we assume that once created, the array containing the heap can be extended arbitrarily at the right end. Inserting a new element means putting that element in the $(n + 1)$st location and *bubbling it* up by swapping it with its parent until it reaches either the root or a parent with a smaller value. Since a heap has logarithmic height, insertion to a heap of n elements thus requires worst-case time $O(\log n)$. Decreasing a value in a heap requires only a similar $O(\log n)$ bubbling up. The minimum element of such a heap is always at the root, and so reporting it takes $\Theta(1)$ time. Deleting the minimum is done by swapping the first and last array positions, bubbling the new root value downward until it reaches its proper location, and truncating the array to eliminate the last position. Delete is handled by decreasing the value so that it is the least in the heap and then applying the **deleteMinimum** operation; this takes a total of $O(\log n)$ time.

The merge operation, unfortunately, is not so economically accomplished; there is little choice but to create a new heap out of the two heaps in a manner similar to the **makeheap** function in heap sort. If there are a total of n elements in the two heaps to be merged, this recreation will require time $O(n)$.

There are better data structures than a heap for implementing priority queues, however. In particular, the *Fibonacci heap* provides an implementation of priority queues in which the **delete** and **deleteMinimum** operations take $O(\log n)$ time and the remaining operations take $\Theta(1)$ time, provided we consider the times required for a sequence of priority queue operations, rather than individual times. That is, we must consider the cost of the individual operations *amortized over the sequence of operations*. Given a sequence of n priority queue operations, we will compute the total time $T(n)$ for all n operations. In doing this computation, however, we do not simply add the costs of the individual operations; rather, we subdivide the cost of each operation into two parts, the *immediate cost* of doing the operation and the *long-term savings* that result from doing the operation. The long-term savings represent costs *not* incurred by later operations as a result of the present operation. The immediate cost minus the long-term savings gives the amortized cost of the operation.

It is easy to calculate the immediate cost (time required) of an operation, but how can we measure the long-term savings that result? We imagine that the data structure has associated with it a bank account; at any given moment the bank account must have a nonnegative balance. When we do an operation that will save future effort, we are making a deposit to the savings account and when, later on, we derive the benefits of that earlier operation we are making a withdrawal from the savings account. Let $\mathcal{B}(i)$ denote the balance in the account after the ith operation, $\mathcal{B}(0) = 0$. We define the amortized cost of the ith operation to be

$$\text{amortized cost of } i\text{th operation} = (\text{immediate cost of } i\text{th operation}) + (\text{change in bank account})$$
$$= (\text{immediate cost of } i\text{th operation}) + (\mathcal{B}(i) - \mathcal{B}(i - 1))$$

Since the bank account $\mathcal{B}$ can go up or down as a result of the ith operation, the amortized cost may be less than or more than the immediate cost. By summing the previous equation, we get

$$\sum_{i=1}^{n}(\text{amortized cost of } i\text{th operation}) = \sum_{i=1}^{n}(\text{immediate cost of } i\text{th operation}) + (\mathcal{B}(n) - \mathcal{B}(0))$$

$$= (\text{total cost of all } n \text{ operations}) + \mathcal{B}(n)$$

$$\geq \text{total cost of all } n \text{ operations}$$

$$= T(n)$$

because $\mathcal{B}(i)$ is nonnegative. Thus defined, the sum of the amortized costs of the operations gives us an upper bound on the total time $T(n)$ for all n operations.

It is important to note that the function $\mathcal{B}(i)$ is not part of the data structure but is just our way to measure how much time is used by the sequence of operations. As such, we can choose *any rules* for $\mathcal{B}$, provided $\mathcal{B}(0) = 0$ and $\mathcal{B}(i) \geq 0$ for $i \geq 1$. Then the sum of the amortized costs defined by

$$\text{amortized cost of } i\text{th operation} = (\text{immediate cost of } i\text{th operation}) + (\mathcal{B}(i) - \mathcal{B}(i-1))$$

bounds the overall cost of the operation of the data structure.

Now, we apply this method to priority queues. A Fibonacci heap is a list of heap-ordered trees (not necessarily binary); since the trees are heap-ordered, the minimum element must be one of the roots and we keep track of which root is the overall minimum. Some of the tree nodes are *marked*. We define

$$\mathcal{B}(i) = (\text{number of trees after the } i\text{th operation})$$

$$+ 2 \times (\text{number of marked nodes after the } i\text{th operation})$$

The clever rules by which nodes are marked and unmarked, and the intricate algorithms that manipulate the set of trees, are too complex to present here in their complete form, and so we just briefly describe the simpler operations and show the calculation of their amortized costs:

create: To create an empty Fibonacci heap we create an empty list of heap-ordered trees. The immediate cost is $\Theta(1)$; since the numbers of trees and marked nodes are zero before and after this operation, $\mathcal{B}(i) - \mathcal{B}(i-1)$ is zero and the amortized time is $\Theta(1)$.

insert: To insert a new element into a Fibonacci heap we add a new one-element tree to the list of trees constituting the heap and update the record of what root is the overall minimum. The immediate cost is $\Theta(1)$. $\mathcal{B}(i) - \mathcal{B}(i-1)$ is also 1 since the number of trees has increased by 1, while the number of marked nodes is unchanged. The amortized time is thus $\Theta(1)$.

decrease: Decreasing an element in a Fibonacci heap is done by cutting the link to its parent, if any, adding the item as a root in the list of trees, and decreasing its value. Furthermore, the marked parent of a cut element is itself cut, propagating upward in the tree. Cut nodes become unmarked, and the unmarked parent of a cut element becomes marked. The immediate cost of this operation is $\Theta(c)$, where c is the number of cut nodes. If there were t trees and m marked elements before this operation, the value of $\mathcal{B}$ before the operation was $t + 2m$. After the operation, the value of $\mathcal{B}$ is $(t + c) + 2(m - c + 2)$ so $\mathcal{B}(i) - \mathcal{B}(i-1) = 4 - c$. The amortized time is thus $\Theta(c) + 4 - c = \Theta(1)$ by changing the definition of $\mathcal{B}$ by a multiplicative constant large enough to dominate the constant hidden in $\Theta(c)$.

minimum: Reporting the minimum element in a Fibonacci heap takes time $\Theta(1)$ and does not change the numbers of trees and marked nodes; the amortized time is thus $\Theta(1)$.

deleteMinimum: Deleting the minimum element in a Fibonacci heap is done by deleting that tree root, making its children roots in the list of trees. Then, the list of tree roots is consolidated in a complicated $O(\log n)$ operation that we do not describe. The result takes amortized time $O(\log n)$.

delete: Deleting an element in a Fibonacci heap is done by decreasing its value to $-\infty$ and then doing a **deleteMinimum**. The amortized cost is the sum of the amortized cost of the two operations, $O(\log n)$.

merge: Merging two Fibonacci heaps is done by concatenating their lists of trees and updating the record of which root is the minimum. The amortized time is thus $\Theta(1)$.

Notice that the amortized cost of each operation is $\Theta(1)$ except **deleteMinimum** and **delete**, both of which are $O(\log n)$.

4.3 Divide-and-Conquer Algorithms

One approach to the design of algorithms is to decompose a problem into subproblems that resemble the original problem, but on a reduced scale. Suppose, for example, that we want to compute x^n. We reason that the value we want can be computed from $x^{\lfloor n/2 \rfloor}$ because

$$x^n = \begin{cases} 1 & \text{if } n = 0 \\ (x^{\lfloor n/2 \rfloor})^2 & \text{if } n \text{ is even} \\ x \times (x^{\lfloor n/2 \rfloor})^2 & \text{if } n \text{ is odd} \end{cases}$$

This recursive definition can be translated directly into

```
1  int power (float x, int n) {
2      // Compute the n-th power of x
3      if (n == 0)
4          return 1;
5      else {
6          int t = power(x, floor(n/2));
7          if ((n % 2) == 0)
8              return t*t;
9          else
10             return x*t*t;
11     }
12 }
```

To analyze the time required by this algorithm, we notice that the time will be proportional to the number of multiplication operations performed in lines 8 and 10, and so the divide-and-conquer recurrence

$$T(n) = 2 + T(\lfloor n/2 \rfloor)$$

with $T(0) = 0$ describes the rate of growth of the time required by this algorithm. By considering the subsequence $n_k = 2^k$, we find, using the methods of the previous section, that $T(n) = \Theta(\log n)$. Thus, the preceding algorithm is considerably more efficient than the more obvious

```
1  int power (int k, int n) {
2      // Compute the n-th power of k
3          int product = 1;
```

```
4        for (int i = 1; i <= n; i++)
5            // at this point power is k*k*k*...*k (i times)
6            product = product * k;
7        return product;
8  }
```

which requires time $\Theta(n)$.

An extremely well-known instance of a divide-and-conquer algorithm is *binary search* of an ordered array of n elements for a given element—we probe the middle element of the array, continuing in either the lower or upper segment of the array, depending on the outcome of the probe:

```
1  int binarySearch (int x, int w[], int low, int high) {
2  //Search for x among sorted array w[low..high]. The integer returned
3  //is either the location of x in w, or the location where x belongs.
4      if (low > high) // Not found
5          return low;
6      else {
7          int middle := (low+high)/2;
8          if (w[middle] < x)
9              return binarySearch(x, w, middle+1, high);
10         else if (w[middle] == x)
11             return middle;
12         else
13             return binarySearch(x, w, low, middle-1);
14     }
15 }
```

The analysis of binary search in an array of n elements is based on counting the number of probes used in the search, since all remaining work is proportional to the number of probes. But, the number of probes needed is described by the divide-and-conquer recurrence

$$T(n) = 1 + T(n/2)$$

with $T(0) = 0$, $T(1) = 1$. We find from Table 4.2 (top line) that $T(n) = \Theta(\log n)$. Hence binary search is much more efficient than a simple linear scan of the array.

To multiply two very large integers x and y, assume that x has exactly $l \geq 2$ digits and y has at most l digits. Let $x_0, x_1, x_2, \ldots, x_{l-1}$ be the digits of x and $y_0, y_1, \ldots, y_{l-1}$ be the digits of y (some of the significant digits at the end of y may be zeros, if y is shorter than x), so that

$$x = x_0 + 10x_1 + 10^2 x_2 + \cdots + 10^{l-1} x_{l-1}$$

and

$$y = y_0 + 10y_1 + 10^2 y_2 + \cdots + 10^{l-1} y_{l-1}$$

We apply the divide-and-conquer idea to multiplication by chopping x into two pieces, the leftmost n digits and the remaining digits

$$x = x_{\text{left}} + 10^n x_{\text{right}}$$

where $n = l/2$. Similarly, chop y into two corresponding pieces

$$y = y_{\text{left}} + 10^n y_{\text{right}}$$

Because y has at most the number of digits that x does, y_{right} might be 0. The product $x \times y$ can be now written

$$
\begin{aligned}
x \times y &= (x_{\text{left}} + 10^n x_{\text{right}}) \times (y_{\text{left}} + 10^n y_{\text{right}}) \\
&= x_{\text{left}} \times y_{\text{left}} + 10^n (x_{\text{right}} \times y_{\text{left}} + x_{\text{left}} \times y_{\text{right}}) \\
&\quad + 10^{2n} x_{\text{right}} \times y_{\text{right}}
\end{aligned}
$$

If $T(n)$ is the time to multiply two n-digit numbers with this method, then

$$T(n) = kn + 4T(n/2)$$

The kn part is the time to chop up x and y and to do the needed additions and shifts; each of these tasks involves n-digit numbers and hence $\Theta(n)$ time. The $4T(n/2)$ part is the time to form the four needed subproducts, each of which is a product of about $n/2$ digits.

The line for $g(n) = \Theta(n)$, $u = 4 > v = 2$ in Table 4.2 tells us that $T(n) = \Theta(n^{\log_2 4}) = \Theta(n^2)$, so the divide-and-conquer algorithm is no more efficient than the elementary-school method of multiplication. However, we can be more economical in our formation of subproducts

$$
\begin{aligned}
x \times y &= (x_{\text{left}} + 10^n x_{\text{right}}) \times (y_{\text{left}} + 10^n y_{\text{right}}) \\
&= B + 10^n C + 10^{2n} A
\end{aligned}
$$

where

$$A = x_{\text{right}} \times y_{\text{right}}$$

$$B = x_{\text{left}} \times y_{\text{left}}$$

$$C = (x_{\text{left}} + x_{\text{right}}) \times (y_{\text{left}} + y_{\text{right}}) - A - B$$

The recurrence for the time required changes to

$$T(n) = kn + 3T(n/2)$$

The kn part is the time to do the two additions that form $x \times y$ from A, B, and C and the two additions and the two subtractions in the formula for C; each of these six additions/subtractions involves n-digit numbers. The $3T(n/2)$ part is the time to (recursively) form the three needed products, each of which is a product of about $n/2$ digits. The line for $g(n) = \Theta(n)$, $u = 3 > v = 2$ in Table 4.2 now tells us that

$$T(n) = \Theta(n^{\log_2 3})$$

Now

$$\log_2 3 = \frac{\log_{10} 3}{\log_{10} 2} \approx 1.5849625 \cdots$$

which means that this divide-and-conquer multiplication technique will be faster than the straightforward $\Theta(n^2)$ method for large numbers of digits.

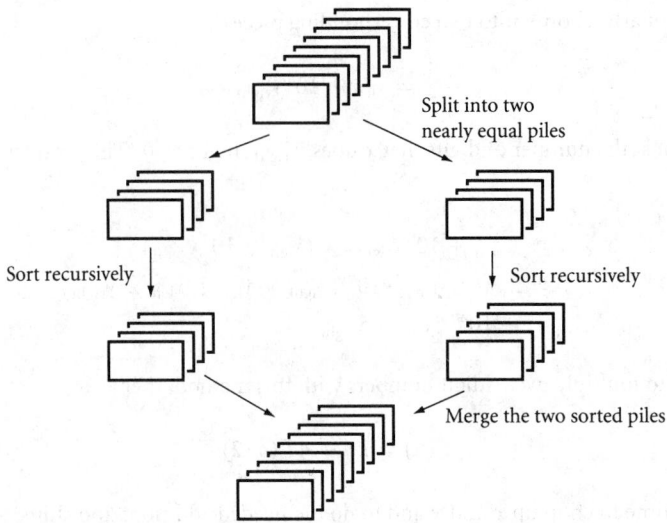

FIGURE 4.2   Schematic description of merge sort.

Sorting a sequence of n values efficiently can be done using the divide-and-conquer idea. Split the n values arbitrarily into two piles of $n/2$ values each, sort each of the piles separately, and then merge the two piles into a single sorted pile. This sorting technique, pictured in Fig. 4.2, is called *merge sort*. Let $T(n)$ be the time required by merge sort for sorting n values. The time needed to do the merging is proportional to the number of elements being merged, so

$$T(n) = cn + 2T(n/2)$$

because we must sort the two halves [time $T(n/2)$ each] and then merge (time proportional to n). We see by Table 4.2 that the growth rate of $T(n)$ is $\Theta(n \log n)$, since $u = v = 2$ and $g(n) = \Theta(n)$.

4.4 Dynamic Programming

In the design of algorithms to solve optimization problems, we need to make the optimal (lowest cost, highest value, shortest distance, and so on) choice from among a large number of alternative solutions; *dynamic programming* is an organized way to find an optimal solution by systematically exploring all possibilities without unnecessary repetition. Often, dynamic programming leads to efficient, polynomial-time algorithms for problems that appear to require searching through exponentially many possibilities.

Like the divide-and-conquer method, dynamic programming is based on the observation that many optimization problems can be solved by solving similar subproblems and then composing the solutions of those subproblems into a solution for the original problem. In addition, the problem is viewed as a sequence of decisions, with each decision leading to different subproblems; if a wrong decision is made, a suboptimal solution results, and so all possible decisions need to be accounted for.

As an example of dynamic programming, consider the problem of constructing an optimal search pattern for probing an ordered sequence of elements. The problem is similar to searching an array—in the previous section we described binary search in which an interval in an array is repeatedly bisected until the search ends. Now, however, suppose we know the frequencies with which the search will seek various elements (both in the sequence and missing from it). For example, if we know that the last few elements in the sequence are frequently sought—binary search does not make use of this information—it might be more efficient to begin the search at the right end of the array, not in the middle. Specifically, we are given an ordered sequence $x_1 < x_2 < \cdots < x_n$ and associated frequencies of access $\beta_1, \beta_2, \ldots, \beta_n$, respectively;

furthermore, we are given $\alpha_0, \alpha_1, \ldots, \alpha_n$ where α_i is the frequency with which the search will fail because the object sought, z, was missing from the sequence, $x_i < z < x_{i+1}$ (with the obvious meaning when $i = 0$ or $i = n$). What is the optimal order to search for an unknown element z? In fact, how should we describe the optimal search order?

We express a search order as a *binary search tree*, a diagram showing the sequence of probes made in every possible search. We place at the root of the tree the sequence element at which the first probe is made, say, x_i; the left subtree of x_i is constructed recursively for the probes made when $z < x_i$ and the right subtree of x_i is constructed recursively for the probes made when $z > x_i$. We label each item in the tree with the frequency that the search ends at that item. Figure 4.3 shows a simple example. The search of sequence $x_1 < x_2 < x_3 < x_4 < x_5$ according the tree of Fig. 4.3 is done by comparing the unknown element z with x_4 (the root); if $z = x_4$, the search ends. If $z < x_2$, then z is compared with x_2 (the root of the left subtree); if $z = x_2$, then the search ends. Otherwise, if $z < x_2$, then z is compared with x_1 (the root of the left subtree of x_2); if $z = x_1$, then the search ends. Otherwise, if $z < x_1$, then the search ends unsuccessfully at the leaf labeled α_0. Other results of comparisons lead along other paths in the tree from the root downward. By its nature, a binary search tree is

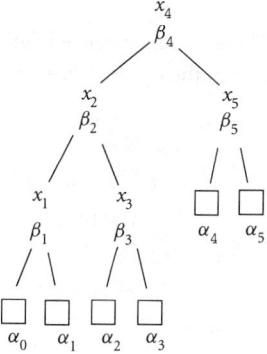

FIGURE 4.3 A binary search tree.

lexicographic in that for all nodes in the tree, the elements in the left subtree of the node are smaller and the elements in the right subtree of the node are larger than the node.

Because we are to find an optimal search pattern (tree), we want the cost of searching to be minimized. The cost of searching is measured by the *weighted path length* of the tree,

$$\sum_{i=1}^{n} \beta_i \times [1 + \text{level}(\beta_i)] + \sum_{i=0}^{n} \alpha_i \times \text{level}(\alpha_i)$$

defined formally as

$$W(\square) = 0$$

$$W\left(T = \bigwedge_{T_l \, T_r}\right) = W(T_l) = W(T_r) + \sum \alpha_i + \sum \beta_i$$

where the summations $\sum \alpha_i$ and $\sum \beta_i$ are over all α_i and β_i in T. Since there are exponentially many possible binary trees, finding the one with minimum-weighted path length could, if done naïvely, take exponentially long.

The key observation we make is that a *principle of optimality* holds for the cost of binary search trees: subtrees of an optimal search tree must themselves be optimal. This observation means, for example, if the tree shown in Fig. 4.3 is optimal, then its left subtree must be the optimal tree for the problem of searching the sequence $x_1 < x_2 < x_3$ with frequencies $\beta_1, \beta_2, \beta_3$ and $\alpha_0, \alpha_1, \alpha_2, \alpha_3$. (If a subtree in Fig. 4.3 were *not* optimal, we could replace it with a better one, reducing the weighted path length of the entire tree because of the recursive definition of weighted path length.) In general terms, the principle of optimality states that subsolutions of an optimal solution must themselves be optimal.

The optimality principle, together with the recursive definition of weighted path length, means that we can express the construction of an optimal tree recursively. Let $C_{i,j}$, $0 \le i \le j \le n$, be the cost of an optimal tree over $x_{i+1} < x_{i+2} < \cdots < x_j$ with the associated frequencies $\beta_{i+1}, \beta_{i+2}, \ldots, \beta_j$ and $\alpha_i, \alpha_{i+1}, \ldots, \alpha_j$. Then,

$$C_{i,i} = 0$$

$$C_{i,j} = \min_{i < k \le j} (C_{i,k-1} + C_{k,j}) + W_{i,j}$$

where

$$W_{i,i} = \alpha_i$$

$$W_{i,j} = W_{i,j-1} + \beta_j + \alpha_j$$

These two recurrence relations can be implemented directly as recursive functions to compute $C_{0,n}$, the cost of the optimal tree, leading to the following two functions

```
1  int W (int i, int j) {
2    if (i == j)
3      return alpha[j];
4    else
5      return W(i,j-1) + beta[j] + alpha[j];
6  }
7
8  int C (int i, int j) {
9    if (i == j)
10     return 0;
11   else {
12     int minCost = MAXINT;
13     int cost;
14     for (int k = i+1; k <= j; k++) {
15         cost = C(i,k-1) + C(k,j) + W(i,j);
16       if (cost < minCost)
17           minCost = cost;
18     }
19     return minCost;
20   }
21 }
```

These two functions correctly compute the cost of an optimal tree; the tree itself can be obtained by storing the values of **k** when **cost < minCost** in line 16.

However, the preceding functions are unnecessarily time consuming (requiring exponential time) because the same subproblems are solved repeatedly. For example, each call **W(i,j)** uses time $\Theta(j - i)$ and such calls are made repeatedly for the same values of **i** and **j**. We can make the process more efficient by caching the values of **W(i,j)** in an array as they are computed and using the cached values when possible:

```
1  int w[n][n];
2  for (int i = 0; i < n; i++)
3    for (int j = 0; j < n; j++)
4      W[i][j] = MAXINT;
5
6  int W (int i, int j) {
7    if (W[i][j] = MAXINT)
8      if (i == j)
```

```
 9              W[i][j] = alpha[j];
10          else
11              W[i][j] = W(i, j-1) + beta[j] + alpha[j];
12      return W[i][j];
13  }
```

In the same way, we should cache the values of **C(i,j)** in an array as they are computed:

```
 1  int C[n][n];
 2  for (int i = 0; i < n; i++)
 3      for (int j = 0; j < n; j++)
 4          C[i][j] = MAXINT;
 5
 6  int C (int i, int j) {
 7      if (C[i][j] == MAXINT)
 8          if (i == j)
 9              C[i][j] = 0;
10          else {
11              int minCost = MAXINT;
12              int cost;
13              for (int k = i+1; k <= j; k++) {
14                  cost = C(i, k-1) + C(k, j) + W(i, j);
15                  if (cost < minCost)
16                      minCost = cost;
17              }
18              C[i][j] = minCost;
19          }
20      return C[i][j];
21  }
```

The idea of caching the solutions to subproblems is crucial to making the algorithm efficient. In this case, the resulting computation requires time $\Theta(n^3)$; this is surprisingly efficient, considering that an optimal tree is being found from among exponentially many possible trees.

By studying the pattern in which the arrays **C** and **W** are filled in, we see that the main diagonal **C[i][i]** is filled in first, then the first upper super diagonal **C[i][i+1]**, then the second upper super diagonal **C[i][i+2]**, and so on until the upper right corner of the array is reached. Rewriting the code to do this directly, and adding an array **R[][]** to keep track of the roots of subtrees, we obtain

```
 1  int W[n][n];
 2  int R[n][n];
 3  int C[n][n];
 4
 5  // Fill in main diagonal
 6  for (int i = 0; i < n; i++) {
 7      W[i][i] = alpha[i];
```

```
 8     R[i][i] = 0;
 9     C[i][i] = 0;
10  }
11
12  int minCost, cost;
13  for (int d = 1; d < n; d++)
14      // Fill in d-th upper super-diagonal
15      for (i = 0; i < n-d; i++) {
16          W[i][i+d] = W[i][i+d-1] + beta[i+d] + alpha[i+d];
17          R[i][i+d] = i+1;
18          C[i][i+d] = c[i][i] + c[i+1][i+d] + W[i][i+d];;
19          for (int k = i+2; k <= i+d; k++) {
20              cost = C[i][k-1] + C[k][i+d] + W[i][i+d];
21              if (cost < C[i ][i+d]) {
22                  R[i][i+d] = k;
23                  C[i][i+d] = cost;
24              }
25          }
26  }
```

which more clearly shows the $\Theta(n^3)$ behavior.

As a second example of dynamic programming, consider the *traveling salesman problem* in which a salesman must visit n cities before returning to his starting point and is required to minimize the cost of the trip. The cost of going from city i to city j is $C_{i,j}$. To use dynamic programming we must specify an optimal tour in a recursive framework, with subproblems resembling the overall problem. Thus we define

$$T(i; j_1, j_2, \ldots, j_k) = \begin{cases} \text{cost of an optimal tour from city } i \text{ to city 1} \\ \text{that goes through each of the cities } j_1, j_2, \ldots, \\ j_k \text{ exactly once, in any order, and through no} \\ \text{other cities.} \end{cases}$$

The principle of optimality tells us that

$$T(i; j_1, j_2, \ldots, j_k) = \min_{1 \le m \le k} \{C_{i, j_m} + T(j_m; j_1, j_2, \ldots, j_{m-1}, j_{m+1}, \ldots, j_k)\}$$

where, by definition,

$$T(i; j) = C_{i,j} + C_{j,1}$$

We can write a function **T** that directly implements the preceding recursive definition, but as in the optimal search tree problem, many subproblems would be solved repeatedly, leading to an algorithm requiring time $\Theta(n!)$. By caching the values $T(i; j_1, j_2, \ldots, j_k)$, we reduce the time required to $\Theta(n^2 2^n)$, still exponential, but considerably less than without caching.

4.5 Greedy Heuristics

Optimization problems always have an objective function to be minimized or maximized, but it is not often clear what steps to take to reach the optimum value. For example, in the optimum binary search tree problem of the previous section, we used dynamic programming to examine systematically all possible trees; but perhaps there is a simple rule that leads directly to the best tree—say, by choosing the largest β_i to be the root and then continuing recursively. Such an approach would be less time consuming than the $\Theta(n^3)$ algorithm we gave, but it does not necessarily give an optimum tree (if we follow the rule of choosing the largest β_i to be the root, we get trees that are no better, on the average, than a randomly chosen tree). The problem with such an approach is that it makes decisions that are *locally optimum*, though perhaps not *globally optimum*. But such a *greedy* sequence of locally optimum choices does lead to a globally optimum solution in some circumstances.

Suppose, for example, $\beta_i = 0$ for $1 \le i \le n$, and we remove the lexicographic requirement of the tree; the resulting problem is the determination of an optimal prefix code for $n + 1$ letters with frequencies $\alpha_0, \alpha_1, \ldots, \alpha_n$. Because we have removed the lexicographic restriction, the dynamic programming solution of the previous section no longer works, but the following simple greedy strategy yields an optimum tree. Repeatedly combine the two lowest frequency items as the left and right subtrees of a newly created item whose frequency is the sum of the two frequencies combined. Here is an example of this construction; we start with five leaves with weights

$$\square \qquad \square \qquad \square \qquad \square \qquad \square \qquad \square$$
$$\alpha_0 = 25 \quad \alpha_1 = 34 \quad \alpha_2 = 38 \quad \alpha_3 = 58 \quad \alpha_4 = 95 \quad \alpha_5 = 21$$

First, combine leaves $\alpha_0 = 25$ and $\alpha_5 = 21$ into a subtree of frequency $25 + 21 = 45$

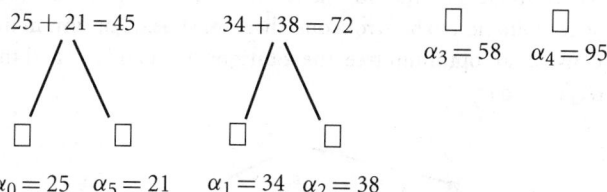

Then, combine leaves $\alpha_1 = 34$ and $\alpha_2 = 38$ into a subtree of frequency $34 + 38 = 72$

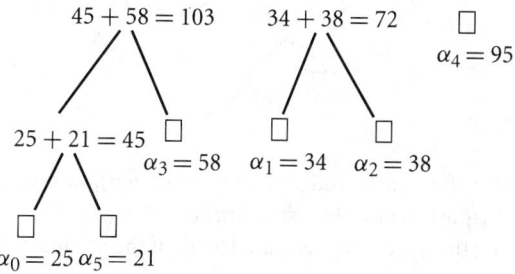

Next, combine the subtree of frequency $\alpha_0 + \alpha_5 = 45$ with $\alpha_3 = 58$

Then, combine the subtree of frequency $\alpha_1 + \alpha_2 = 72$ with $\alpha_4 = 95$

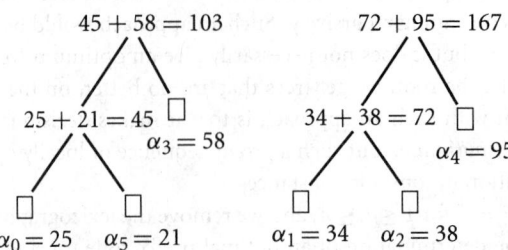

Finally, combine the only two remaining subtrees

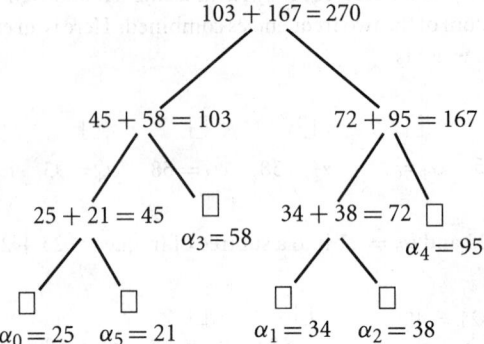

How do we know that the outlined process leads to an optimum tree? The key to proving that the tree is optimum is to assume, by way of contradiction, that it is not optimum. In this case, the greedy strategy must have erred in one of its choices, so let us look at the *first* error this strategy made. Since all previous greedy choices were not errors, and hence lead to an optimum tree, we can assume that we have a sequence of frequencies $\alpha_0, \alpha_1, \ldots, \alpha_n$ such that the first greedy choice is erroneous—without loss of generality assume that α_0 and α_1 are the two smallest frequencies, those combined erroneously by the greedy strategy. For this combination to be erroneous, there must be no optimum tree in which these two leaves are siblings, so consider an optimum tree, the locations of α_0 and α_1, and the location of the two deepest leaves in the tree, α_i and α_j

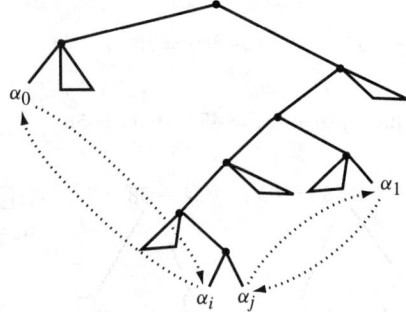

By interchanging the positions of α_0 and α_i and α_1 and α_j (as shown), we obtain a tree in which α_0 and α_1 are siblings. Because α_0 and α_1 are the two lowest frequencies (because they were the greedy algorithm's choice), $\alpha_0 \leq \alpha_i$ and $\alpha_1 \leq \alpha_j$; thus, the weighted path length of the modified tree is no larger than before the modification since $\text{level}(\alpha_0) \geq \text{level}(\alpha_i)$, $\text{level}(\alpha_1) \geq \text{level}(\alpha_j)$, and hence

$$\text{level}(\alpha_i) \times \alpha_0 + \text{level}(\alpha_j) \times \alpha_1 \leq \text{level}(\alpha_0) \times \alpha_0 + \text{level}(\alpha_1) \times \alpha_1$$

In other words, the first so-called mistake of the greedy algorithm was in fact not a mistake since there is an optimum tree in which α_0 and α_1 are siblings. Thus, we conclude that the greedy algorithm never makes a first mistake; that is, it never makes a mistake at all!

The greedy algorithm just described is called *Huffman's algorithm*. If the subtrees are kept on a priority queue by cumulative frequency, the algorithm needs to insert the $n + 1$ leaf frequencies onto the queue, repeatedly remove the two least elements on the queue, unite those to elements into a single subtree, and put that subtree back on the queue. This process continues until the queue contains a single item, the optimum tree. Reasonable implementations of priority queues will yield $O(n \log n)$ implementations of Huffman's greedy algorithm.

The idea of making greedy choices, facilitated with a priority queue, works to find optimum solutions to other problems too. For example, a *spanning tree* of a weighted, connected, undirected graph $G = (V, E)$ is a subset of $|V| - 1$ edges from E connecting all of the vertices in G; a spanning tree is minimum if the sum of the weights of its edges is as small as possible. *Prim's algorithm* uses a sequence of greedy choices to determine a minimum spanning tree. Start with an arbitrary vertex $v \in V$ as the spanning tree-to-be. Then, repeatedly add the cheapest edge connecting the spanning tree-to-be to a vertex not yet in it. If the vertices not yet in the tree are stored in a priority queue implemented by a Fibonacci heap, then the total time required by Prim's algorithm will be $O(|E| + |V| \log |V|)$. But why does the sequence of greedy choices lead to a minimum spanning tree?

Suppose Prim's algorithm does *not* result in a minimum spanning tree. As we did with Huffman's algorithm, we ask what the state of affairs must be when Prim's algorithm makes its first mistake; we will see that the assumption of a first mistake leads to a contradiction, proving the correctness of Prim's algorithm. Let the edges added to the spanning tree be, in the order added, $e_1, e_2, e_3, \ldots$, and let e_i be the first mistake. In other words, there is a minimum spanning tree $T_{\min}$ containing $e_1, e_2, \ldots, e_{i-1}$, but no minimum spanning tree contains $e_1, e_2, \ldots, e_i$. Imagine what happens if we add the edge e_i to $T_{\min}$; since $T_{\min}$ is a spanning tree, the addition of e_i causes a cycle containing e_i. Let $e_{\max}$ be the highest cost edge on that cycle. Because Prim's algorithm makes a greedy choice—that is, chooses the lowest cost available edge—the cost of $e_{\max}$ is at least that of e_i, so the cost of the spanning $T_{\min} - \{e_{\max}\} \cup \{e_i\}$ is at most that of $T_{\min}$; in other words, $T_{\min} - \{e_{\max}\} \cup \{e_i\}$ is also a minimum spanning tree, contradicting our assumption that the choice of e_i is the first mistake. Therefore, the spanning tree constructed by Prim's algorithm must be a minimum spanning tree.

We can apply the greedy heuristic to many optimization problems, and even if the results are not optimal, they are often quite good. For example, in the n-city traveling salesman problem, we can get near-optimal tours in time $O(n^2)$ when the intercity costs are symmetric ($C_{i,j} = C_{j,i}$ for all i and j) and satisfy the triangle inequality ($C_{i,j} \leq C_{i,k} + C_{k,j}$ for all i, j, and k). The *closest insertion algorithm* starts with a tour consisting of a single, arbitrarily chosen city and successively inserts the remaining cities into the tour, making a greedy choice about which city to insert next and where to insert it. The city chosen for insertion is the city not on the tour but closest to a city on the tour; the chosen city is inserted adjacent to the city on the tour to which it is closest.

Given an $n \times n$ symmetric distance matrix C that satisfies the triangle inequality, let I_n be the tour of length $|I_n|$ produced by the closest insertion heuristic and let O_n be an optimal tour of length $|O_n|$. Then

$$\frac{|I_n|}{|O_n|} < 2$$

This bound is proved by an incremental form of the optimality proofs for greedy heuristics we have seen earlier; we ask not where the first error is, but by how much we are in error at each greedy insertion to the tour—we establish a correspondence between edges of the optimal tour and cities inserted on the closest insertion tour. We show that at each insertion of a new city to the closest insertion tour, the cost of that insertion is at most twice the cost of the corresponding edge of the optimal tour. To establish the correspondence, imagine the closest insertion algorithm keeping track not only of the current tour but also of a spiderlike configuration including the edges of the current tour (the body of the spider) and pieces of the optimal tour (the legs of the spider). We show the current tour in solid lines and the pieces of the optimal tour as dotted lines:

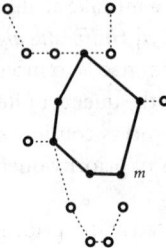

Initially, the spider consists of the arbitrarily chosen city with which the closest insertion tour begins and the legs of the spider consist of all of the edges of the optimal tour *except* for one edge eliminated arbitrarily. As each city is inserted into the closest insertion tour, the algorithm will delete from the spiderlike configuration one of the dotted edges from the optimal tour. When city k is inserted between cities l and m, the edge deleted is the one attaching the spider to the leg containing the city inserted (from city x to city y), shown here in bold:

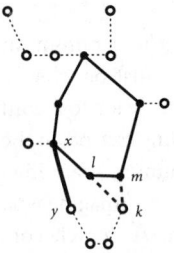

Now,

$$C_{k,m} \leq C_{x,y}$$

because the of greedy choice to add city k to the tour and not city y. By the triangle inequality,

$$C_{l,k} \leq C_{l,m} + C_{m,k}$$

and by symmetry we can combine these two inequalities to get

$$C_{l,k} \leq C_{l,m} + C_{x,y}$$

Adding this last inequality to the first one just given

$$C_{l,k} + C_{k,m} \leq C_{l,m} + 2C_{x,y}$$

that is,

$$C_{l,k} + C_{k,m} - C_{l,m} \leq 2C_{x,y}$$

Thus, adding city k between cities l and m adds no more to I_n than $2C_{x,y}$. Summing these incremental amounts over the cost of the entire algorithm tells us

$$I_n \leq 2O_n$$

as we claimed.

Acknowledgment

The comments of Ken Urban are gratefully acknowledged.

Defining Terms

Amortized cost: The cost of an operation considered to be spread over a sequence of many operations.

Average-case cost: The sum of costs over all possible inputs divided by the number of possible inputs.

Divide-and-conquer: A paradigm of algorithm design in which a problem is solved by reducing it to subproblems of the same structure.

Dynamic programming: A paradigm of algorithm design in which an optimization problem is solved by a combination of caching subproblem solutions and appealing to the "principle of optimality."

Greedy heuristic: A paradigm of algorithm design in which an optimization problem is solved by a by making locally optimum decisions.

Lower bound: A function (or growth rate) below which solving a problem is impossible.

Principle of optimality: The observation, in some optimization problems, that components of a globally optimum solution must the themselves be globally optimal.

Recurrence relation: The specification of a sequence of values in terms of earlier values in the sequence.

Worst-case cost: The cost of an algorithm in the most pessimistic input possibility.

References

Cormen, T. H., Leiserson, C. E., and Rivest, R. L. 1990. *Introduction to Algorithms.* McGraw–Hill, New York.

Fredman, M. L. and Tarjan, R. E. 1987. Fibonacci heaps and their use in improved network optimization problems. *J. ACM* 34:596–615.

Greene, D. H. and Knuth, D. E. 1990. *Mathematics for the Analysis of Algorithms,* 3rd ed. Birkhäuser, Boston MA.

Knuth, D. E. 1973. *The Art of Computer Programming, Volume 2: Seminumerical Algorithms,* 2nd ed. Addison–Wesley, Reading, MA.

Knuth, D. E. 1973. *The Art of Computer Programming, Volume 3: Sorting and Searching.* Addison–Wesley, Reading, MA.

Lueker, G. S. 1980. Some techniques for solving recurrences. *Comput. Surv.* 12:419–436.

Reingold, E. M. and Hansen, W. J. 1986. *Data Structures in Pascal.* Little, Brown and Co., Boston, MA.

Reingold, E. M., Nievergelt, J., and Deo, N. 1977. *Combinatorial Algorithms: Theory and Practice.* Prentice–Hall, Englewood Cliffs, NJ.

Rosencrantz, D. J., Stearns, R. E., and Lewis, P. M. 1977. An analysis of several heuristics for the traveling salesman problem. *SIAM J. Comp.* 6:563–581.

Further Information

General discussions of the analysis of algorithms and data structures can be found in Cormen [1990], Knuth [1973] and Reingold [1977]; Reingold [1986] has a more elementary treatment. Both Greene [1990] and Lueker [1980] contain detailed treatments of recurrences, especially in regard to the analysis of algorithms. A derivation of Stirling's approximation can be found in Greene [1990] or in Knuth [1973]. Knuth [1973] discusses algorithms for problems such as computing powers, evaluating polynomials, and multiplying large numbers. Our discussion of Fibonacci heaps is from Fredman [1987]; our discussion of the heuristics for the traveling salesman problem is from Rosencrantz [1977].

5

Data Structures

Roberto Tamassia
Brown University

Bryan Cantrill
Brown University

5.1 Introduction

The study of data structures, that is, methods for organizing data that are suitable for computer processing, is one of the classic topics of computer science. At the hardware level, a computer views storage devices such as internal memory and disk as holders of elementary data units (bytes), each accessible through its address (an integer). When writing programs, instead of manipulating the data at the byte level, it is convenient to organize them into higher level entities, called *data structures*.

Containers, Elements, and Locators

Most data structures can be viewed as **containers** that store a collection of objects of a given type, called the *elements* of the container. Often a total order is defined among the elements (e.g., alphabetically ordered names, points in the plane ordered by x-coordinate). We assume that the elements of a container can be accessed by means of variables called **locators**. When an object is inserted into the container, a locator is returned, which can be later used to access or delete the object. A locator is typically implemented with a pointer or an index into an array.

A data structure has an associated repertory of operations, classified into *queries*, which retrieve information on the data structure (e.g., return the number of elements, or test the presence of a given element), and *updates*, which modify the data structure (e.g., insertion and deletion of elements). The performance of a data structure is characterized by the space requirement and the time complexity of the operations in its repertory. The *amortized* time complexity of an operation is the average time over a suitably defined sequence of operations.

However, efficiency is not the only quality measure of a data structure. Simplicity and ease of implementation should be taken into account when choosing a data structure for solving a practical problem.

0-8493-2909-4/97/$0.00+$.50

Abstract Data Types

Data structures are concrete implementations of **abstract data types** (ADTs). A *data type* is a collection of objects. A data type can be mathematically specified (e.g., real number, directed graph) or concretely specified within a programming language (e.g., `int` in C, `set` in Pascal). An ADT is a mathematically specified data type equipped with operations that can be performed on the objects. Object-oriented programming languages, such as C++, provide support for expressing ADTs by means of *classes*. ADTs specify the data stored and the operations to be performed on them.

Main Issues in the Study of Data Structures

The following issues are of foremost importance in the study of data structures.

Static vs Dynamic. A *static* data structure supports only queries, whereas a dynamic data structure also supports updates. A *dynamic* data structure is often more complicated than its static counterpart supporting the same repertory of queries. A *persistent* data structure (see, e.g., Driscoll et al. [1989]) is a dynamic data structure that supports operations on past versions. There are many problems for which no efficient dynamic data structures are known. It has been observed that there are strong similarities among the classes of problems that are difficult to parallelize and those that are difficult to dynamize (see, e.g., Reif [1987]). Further investigations are needed to study the relationship between parallel and incremental complexities [Miltersen et al. 1994].

Implicit vs Explicit. Two fundamental data organization mechanisms are used in data structures. In an *explicit* data structure, pointers (i.e., memory addresses) are used to link the elements and access them (e.g., a singly linked list, where each element has a pointer to the next one). In an *implicit* data structure, mathematical relationships support the retrieval of elements (e.g., array representation of a heap, see section 5.3). Explicit data structures must use additional space to store pointers. However, they are more flexible for complex problems. Most programming languages support pointers and basic implicit data structures, such as arrays.

Internal vs External Memory. In a typical computer, there are two levels of memory: internal memory [random access memory (RAM)] and external memory (disk). The internal memory is much faster than external memory but has much smaller capacity. Data structures designed to work for data that fit into internal memory may not perform well for large amounts of data that need to be stored in external memory. For large-scale problems, data structures need to be designed that take into account the two levels of memory [Aggarwal and Vitter 1988]. For example, two-level indices such as B-trees [Comer 1979] have been designed to efficiently search in large databases.

Space vs Time. Data structures often exhibit a tradeoff between space and time complexity. For example, suppose we want to represent a set of integers in the range $[0, N]$ (e.g., for a set of social security numbers $N = 10^{10} - 1$) such that we can efficiently query whether a given element is in the set, insert an element, or delete an element. Two possible data structures for this problem are an N-element bit array (where the bit in position i indicates the presence of integer i in the set), and a balanced search tree (such as a 2–3 tree or a red–black tree). The bit array has optimal time complexity, since it supports queries, insertions, and deletions in constant time. However, it uses space proportional to the size N of the range, irrespectively of the number of elements actually stored. The balanced **search tree** supports queries, insertions, and deletions in logarithmic time but uses optimal space proportional to the current number of elements stored.

Theory vs Practice. A large and ever growing body of theoretical research on data structures is available, where the performance is measured in asymptotic terms (big-Oh notation). Although asymptotic complexity analysis is an important mathematical subject, it does not completely capture the notion of efficiency of data structures in practical scenarios, where constant factors cannot be disregarded and the difficulty of implementation substantially affects design and maintenance costs. Experimental studies comparing the practical efficiency of data structures for specific classes of problems should be encouraged to bridge the gap between the theory and practice of data structures.

Fundamental Data Structures

The following four data structures are ubiquitously used in the description of discrete algorithms, and serve as basic building blocks for realizing more complex data structures. They are covered in detail in the textbooks listed in the Further Information section and in the additional references provided.

Sequence. A **sequence** is a container that stores elements in a certain linear order, which is imposed by the operations performed. The basic operations supported are retrieving, inserting, and removing an element given its position. Special types of sequences include stacks and queues, where insertions and deletions can be done only at the head or tail of the sequence. The basic realization of sequences are by means of arrays and linked lists. Concatenable queues (see, e.g., Hoffman et al. [1986]) support additional operations such as splitting and splicing, and determining the sequence containing a given element. In external memory, a sequence is typically associated with a file.

Priority Queue. A **priority queue** is a container of elements from a totally ordered universe that supports the basic operations of inserting an element and retrieving/removing the largest element. A key application of priority queues is sorting algorithms. A **heap** is an efficient realization of a priority queue that embeds the elements into the ancestor/descendant partial order of a **binary tree**. A heap also admits an implicit realization where the nodes of the tree are mapped into the elements of an array (see section 5.3). Sophisticated variations of priority queues include min–max heaps, pagodas, deaps, binomial heaps, and Fibonacci heaps. The buffer tree is an efficient external-memory realization of a priority queue.

Dictionary. A **dictionary** is a container of elements from a totally ordered universe that supports the basic operations of inserting/deleting elements and searching for a given element. **Hash tables** provide an efficient implicit realization of a dictionary. Efficient explicit implementations include skip lists [Pugh 1990], tries, and balanced search trees (e.g., **AVL-trees**, red–black trees, 2–3 trees, 2–3–4 trees, weight-balanced trees, biased search trees, splay trees). The technique of fractional cascading [Chazelle and Guibas 1986] speeds up searching for the same element in a collection of dictionaries. In external memory, dictionaries are typically implemented as B-trees and their variations.

Union-Find. A union-find data structure represents a collection of disjoint sets and supports the two fundamental operations of merging two sets and finding the set containing a given element. There is a simple and optimal union-find data structure (rooted tree with path compression) whose time complexity analysis is very difficult to analyze. See, for example, Galil and Italiano [1991].

Fundamental Data Structures. Examples of fundamental data structures used in three major application domains are as follows:

1. *Graphs and Networks:* adjacency matrix, adjacency lists, link-cut tree [Sleator and Tarjan 1983], dynamic expression tree [Cohen and Tamassia 1995], topology tree [Frederickson 1993], SPQR-tree [Di Battista and Tamassia 1990], sparsification tree [Eppstein et al. 1992]. See also, for example, Even [1979], Mehlhorn [1984], and Tarjan [1983].
2. *Text Processing:* string, suffix tree, Patricia tree. See, for example, Gonnet and Baeza-Yates [1991].
3. *Geometry and Graphics:* binary space partition tree, chain tree, trapezoid tree, range tree, segment tree, interval tree, priority search tree, hull tree, quad tree, R-tree, grid file, metablock tree. For example, see Chiang and Tamassia [1992], Edelsbrunner [1987], Foley et al. [1990], Mehlhorn [1984], Nievergelt and Hinrichs [1993], O'Rourke [1994], and Preparata and Shamos [1985].

Organization of the Chapter

The rest of this chapter focuses on three fundamental abstract data types: sequences, priority queues, and dictionaries. Examples of efficient data structures and algorithms for implementing them are presented in detail in sections 5.2, 5.3, and 5.4, respectively. Namely, we cover arrays, singly and doubly linked lists, heaps, search trees, **(a, b)-trees**, AVL-trees, **bucket arrays**, and hash tables.

5.2 Sequence

Introduction

A *sequence* is a container that stores elements in a certain order, which is imposed by the operations performed. The basic operations supported are:

- INSERT RANK: insert an element in a given position.
- REMOVE: remove an element.

Sequences are a basic form of data organization, and are typically used to realize and implement other data types and data structures.

Operations

Using locators (see first subsection of section 5.1), we can define a more complete repertory of operations for a sequence S:

SIZE(N) return the number of elements N of S.

HEAD(c) assign to c a locator to the first element of S; if S is empty, then c is a null locator.

TAIL(c) assign to c a locator to the last element of S; if S is empty, then a null locator is returned.

LOCATERANK(r, c) assign to c a locator to the rth element of S; if $r < 1$ or $r > N$, where N is the size of S, then c is a null locator.

PREV(c', c'') assign to c'' a locator to the element of S preceding the element with locator c'; if c' is the locator of the first element of S, then c'' is a null locator.

NEXT(c', c'') assign to c'' a locator to the element of S following the element with locator c'; if c' is the locator of the last element of S, then c'' is a null locator.

INSERTAFTER(e, c', c'') insert element e into S after the element with locator c', and return a locator c'' to e.

INSERTBEFORE(e, c', c'') insert element e into S before the element with locator c', and return a locator c'' to e.

INSERTHEAD(e, c) insert element e at the beginning of S, and return a locator c to e.

INSERTTAIL(e, c) insert element e at the end of S, and return a locator c to e.

INSERTRANK(e, r, c) insert element e in the rth position of S; if $r < 1$ or $r > N + 1$, where N is the current size of S, then c is a null locator.

REMOVE(c, e) remove from S and return element e with locator c.

MODIFY(c, e) replace with e the element with locator c.

Some of the preceding operations can be easily expressed by means of other operations of the repertory. For example, operations HEAD and TAIL can be easily expressed by means of LOCATERANK and SIZE.

Implementation with an Array

The simplest way to implement a sequence is to use a (one-dimensional) array, where the ith element of the array stores the ith element of the list, and to keep a variable that stores the size N of the sequence. With this implementation, accessing elements takes $O(1)$ time, whereas insertions and deletions take $O(N)$ time.

TABLE 5.1 Performance of a Sequence Implemented with an Array

Operation	Time
SIZE	$O(1)$
HEAD	$O(1)$
TAIL	$O(1)$
LOCATERANK	$O(1)$
PREV	$O(1)$
NEXT	$O(1)$
INSERTAFTER	$O(N)$
INSERTBEFORE	$O(N)$
INSERTHEAD	$O(N)$
INSERTTAIL	$O(1)$
INSERTRANK	$O(N)$
REMOVE	$O(N)$
MODIFY	$O(1)$

TABLE 5.2 Performance of a Sequence Implemented with a Singly Linked List

Operation	Time
SIZE	$O(1)$
HEAD	$O(1)$
TAIL	$O(1)$
LOCATERANK	$O(N)$
PREV	$O(N)$
NEXT	$O(1)$
INSERTAFTER	$O(1)$
INSERTBEFORE	$O(N)$
INSERTHEAD	$O(1)$
INSERTTAIL	$O(1)$
INSERTRANK	$O(N)$
REMOVE	$O(N)$
MODIFY	$O(1)$

TABLE 5.3 Performance of a Sequence Implemented with a Doubly Linked List

Operation	Time
SIZE	$O(1)$
HEAD	$O(1)$
TAIL	$O(1)$
LOCATERANK	$O(N)$
PREV	$O(1)$
NEXT	$O(1)$
INSERTAFTER	$O(1)$
INSERTBEFORE	$O(1)$
INSERTHEAD	$O(1)$
INSERTTAIL	$O(1)$
INSERTRANK	$O(N)$
REMOVE	$O(1)$
MODIFY	$O(1)$

Table 5.1 shows the time complexity of the implementation of a sequence by means of an array. In the table we denote with N the number of elements in the sequence at the time the operation is performed. The space complexity is $O(N)$.

Implementation with a Singly Linked List

A sequence can also be implemented with a singly linked list, where each element has a pointer to the next one. We also store the size of the sequence and pointers to the first and last element of the sequence.

With this implementation, accessing elements takes $O(N)$ time, since we need to traverse the list, whereas some insertions and deletions take $O(1)$ time.

Table 5.2 shows the time complexity of the implementation of sequence by means of singly linked list. In the table we denote with N the number of elements in the sequence at the time the operation is performed. The space complexity is $O(N)$.

Implementation with a Doubly Linked List

Better performance can be achieved, at the expense of using additional space, by implementing a sequence with a doubly linked list, where each element has pointers to the next and previous elements. We also store the size of the sequence and pointers to the first and last element of the sequence.

Table 5.3 shows the time complexity of the implementation of sequence by means of a doubly linked list. In the table we denote with N the number of elements in the sequence at the time the operation is performed. The space complexity is $O(N)$.

5.3 Priority Queue

Introduction

A priority queue is a container of elements from a totally ordered universe that supports the following two basic operations:

- INSERT: insert an element into the priority queue.
- REMOVEMAX: remove the largest element from the priority queue.

Here are some simple applications of a priority queue:

Scheduling. A scheduling system can store the tasks to be performed into a priority queue, and select the task with highest priority to be executed next.

Sorting. To sort a set of N elements, we can insert them one at a time into a priority queue by means of N INSERT operations, and then retrieve them in decreasing order by means of N REMOVEMAX operations. This two-phase method is the paradigm of several popular sorting algorithms, including *selection sort, insertion sort,* and *heap-sort.*

Operations

Using locators, we can define a more complete repertory of operations for a priority queue Q:

SIZE(N) return the current number of elements N in Q.

MAX(c) return a locator c to the maximum element of Q.

INSERT(e, c) insert element e into Q and return a locator c to e.

REMOVE(c, e) remove from Q and return element e with locator c.

REMOVEMAX(e) remove from Q and return the maximum element e from Q.

MODIFY(c, e) replace with e the element with locator c.

Note that operation REMOVEMAX(e) is equivalent to MAX(c) followed by REMOVE(c, e).

Realization with a Sequence

We can realize a priority queue by reusing and extending the sequence abstract data type (see section 5.2). Operations SIZE, MODIFY, and REMOVE correspond to the homonymous sequence operations.

Unsorted Sequence

We can realize INSERT by an INSERTHEAD or an INSERTTAIL, which means that the sequence is not kept sorted. Operation MAX can be performed by scanning the sequence with an iteration of NEXT operations, keeping track of the maximum element encountered. Finally, as observed earlier, operation REMOVEMAX is a combination of MAX and REMOVE. Table 5.4 shows the time complexity of this realization, assuming that the sequence is implemented with a doubly linked list. In the table we denote with N the number of elements in the priority queue at the time the operation is performed. The space complexity is $O(N)$.

TABLE 5.4 Performance of a Priority Queue Realized by an Unsorted Sequence, Implemented with a Doubly Linked List

Operation	Time
SIZE	$O(1)$
MAX	$O(N)$
INSERT	$O(1)$
REMOVE	$O(1)$
REMOVEMAX	$O(N)$
MODIFY	$O(1)$

Sorted Sequence

An alternative implementation uses a sequence that is kept sorted. In this case, operation MAX corresponds to simply accessing the last element of the sequence. However, operation INSERT now requires scanning the sequence to find the appropriate position to insert the new element. Table 5.5 shows the time complexity of this realization, assuming that the sequence is implemented with a doubly linked list. In the table we denote with N the number of elements in the priority queue at the time the operation is performed. The space complexity is $O(N)$.

Realizing a priority queue with a sequence, sorted or unsorted, has the drawback that some operations require linear time in the worst case. Hence, this realization is not suitable in many applications where fast running times are sought for all of the priority queue operations.

TABLE 5.5 Performance of a Priority Queue Realized by a Sorted Sequence, Implemented with a Doubly Linked List

Operation	Time
SIZE	$O(1)$
MAX	$O(1)$
INSERT	$O(N)$
REMOVE	$O(1)$
REMOVEMAX	$O(1)$
MODIFY	$O(N)$

Sorting

For example, consider the sorting application (see the first introduction to this section). We have a collection of N elements from a totally ordered universe, and we want to sort them using a priority queue Q. We assume that each element uses $O(1)$ space, and any two elements can be compared in $O(1)$ time. If we realize Q with an unsorted sequence, then the first phase (inserting the N elements into Q) takes $O(N)$ time. However the second phase (removing N times the maximum element) takes time

$$O\left(\sum_{i=1}^{N} i\right) = O(N^2)$$

Hence, the overall time complexity is $O(N^2)$. This sorting method is known as *selection sort.*

However, if we realize the priority queue with a sorted sequence, then the first phase takes time

$$O\left(\sum_{i=1}^{N} i\right) = O(N^2)$$

while the second phase takes time $O(N)$. Again, the overall time complexity is $O(N^2)$. This sorting method is known as *insertion sort.*

Realization with a Heap

A more sophisticated realization of a priority queue uses a data structure called a heap. A heap is a binary tree T whose internal nodes each store one element from a totally ordered universe, with the following properties (see Fig. 5.1):

> *Level property.* All of the levels of T are full, except possibly for the bottommost level, which is left filled.
> *Partial order property.* Let μ be a node of T distinct from the root, and let ν be the parent of μ; then the element stored at μ is less than or equal to the element stored at ν.

The leaves of a heap do not store data and serve only as placeholders. The level property implies that heap T is a minimum-height binary tree. More precisely, if T stores N elements and has height h, then each level i with $0 \leq i \leq h - 2$ stores exactly 2^i elements, whereas level $h - 1$ stores between 1 and 2^{h-1} elements. Note that level h contains only leaves. We have

$$2^{h-1} = 1 + \sum_{i=0}^{h-2} 2^i \leq N \leq \sum_{i=0}^{h-1} 2^i = 2^h - 1$$

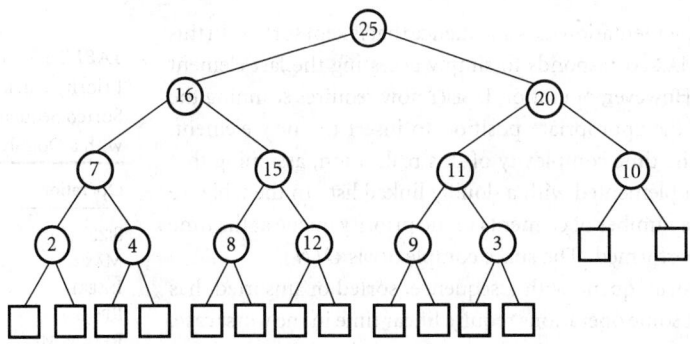

FIGURE 5.1   Example of a heap storing 13 elements.

from which we obtain

$$\log_2(N + 1) \leq h \leq 1 + \log_2 N$$

Now, we show how to perform the various priority queue operations by means of a heap T. We denote with $x(\mu)$ the element stored at an internal node μ of T. We denote with ρ the root of T. We call the *last node* of T the rightmost internal node of the bottommost internal level of T.

By storing a counter that keeps track of the current number of elements, SIZE consists of simply returning the value of the counter. By the partial order property, the maximum element is stored at the root, and hence operation MAX can be performed by accessing node ρ.

Operation INSERT

To insert an element e into T, we add a new internal node μ to T such that μ becomes the new last node of T, and set $x(\mu) = e$. This action ensures that the level property is satisfied, but may violate the partial-order property. Hence, if $\mu \neq \rho$, we compare $x(\mu)$ with $x(\nu)$, where ν is the parent of μ. If $x(\mu) > x(\nu)$, then we need to restore the partial order property, which can be locally achieved by exchanging the elements stored at μ and ν. This causes the new element e to move up one level. Again, the partial order property may be violated, and we may have to continue moving up the new element e until no violation occurs. In the worst case, the new element e moves up to the root ρ of T by means of $O(\log N)$ exchanges. The upward movement of element e by means of exchanges is conventionally called *upheap*.

An example of a sequence of insertions into a heap is shown in Fig. 5.2.

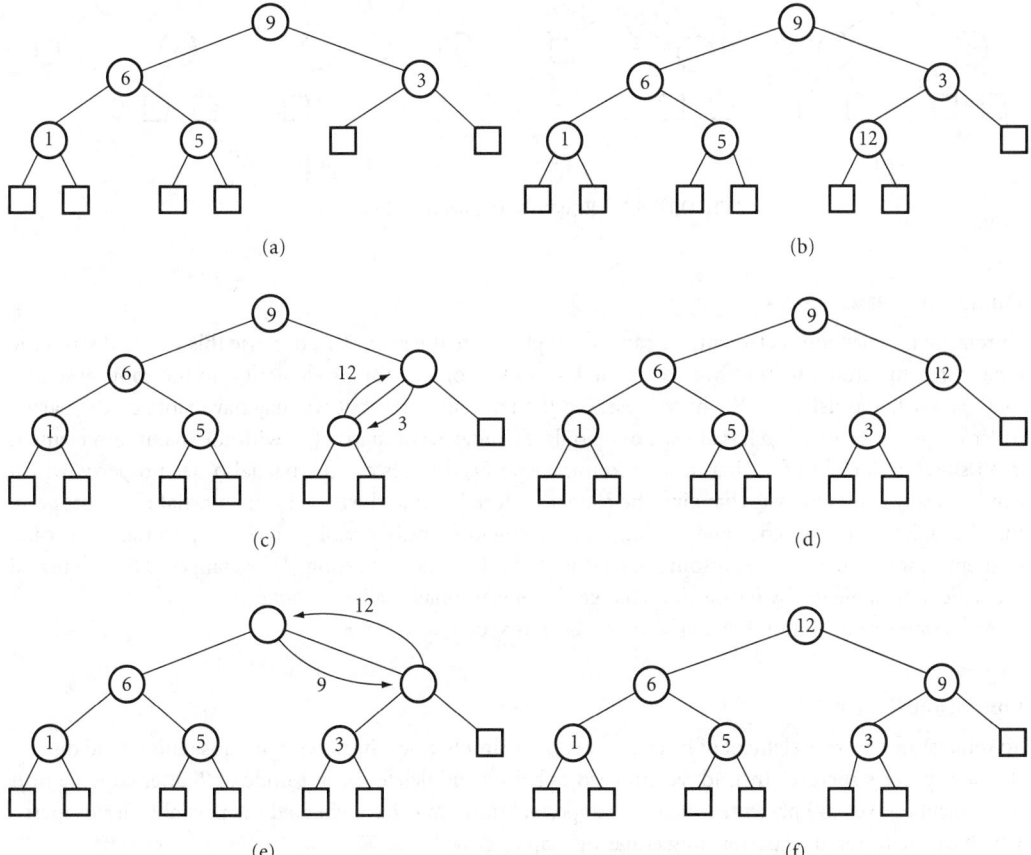

FIGURE 5.2 Example of insertion into a heap.

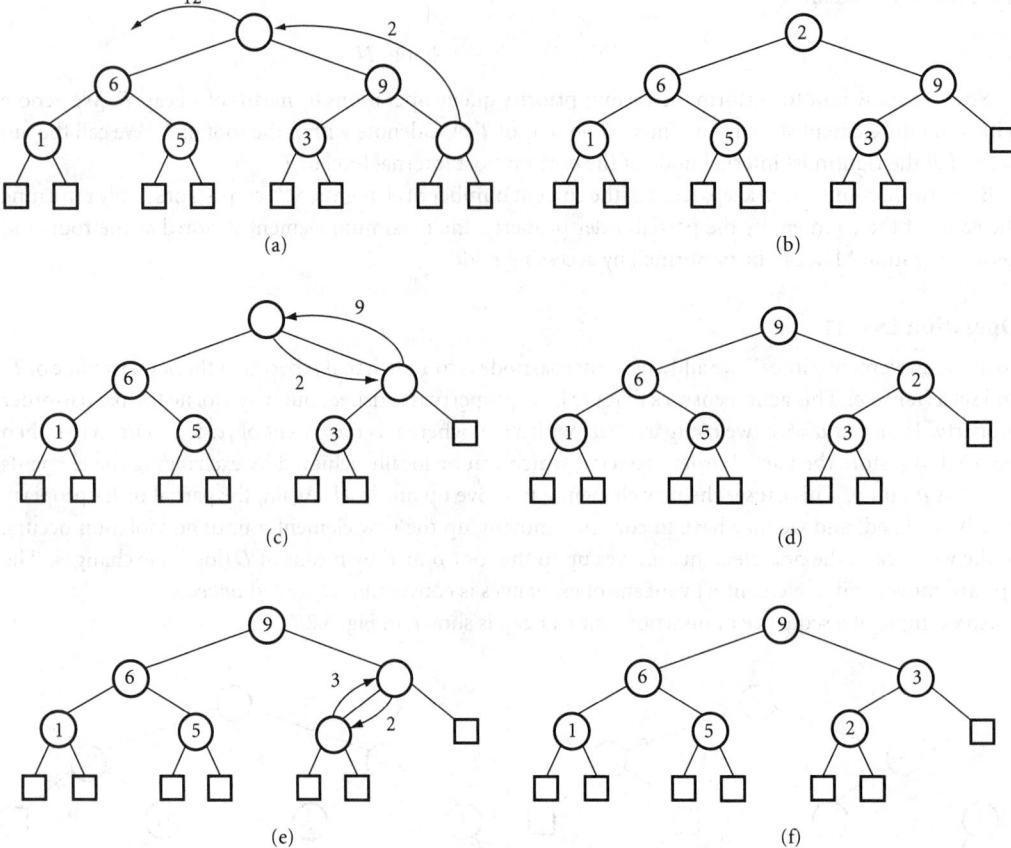

FIGURE 5.3   REMOVEMAX operation in a heap.

Operation REMOVEMAX

To remove the maximum element, we cannot simply delete the root of T, because this would disrupt the binary tree structure. Instead, we access the last node λ of T, copy its element e to the root by setting $x(\rho) = x(\lambda)$, and delete λ. We have preserved the level property, but we may have violated the partial order property. Hence, if ρ has at least one nonleaf child, we compare $x(\rho)$ with the maximum element $x(\sigma)$ stored at a child of ρ. If $x(\rho) < x(\sigma)$, then we need to restore the partial order property, which can be locally achieved by exchanging the elements stored at ρ and σ. Again, the partial order property may be violated, and we continue moving down element e until no violation occurs. In the worst case, element e moves down to the bottom internal level of T by means of $O(\log N)$ exchanges. The downward movement of element e by means of exchanges is conventionally called *downheap*.

An example of operation REMOVEMAX in a heap is shown in Fig. 5.3.

Operation REMOVE

To remove an arbitrary element of heap T, we cannot simply delete its node μ, because this would disrupt the binary tree structure. Instead, we proceed as before and delete the last node of T after copying to μ its element e. We have preserved the level property, but we may have violated the partial order property, which can be restored by performing either upheap or downheap.

Finally, after modifying an element of heap T, if the partial order property is violated, we just need to perform either upheap or downheap.

Time Complexity

Table 5.6 shows the time complexity of the realization of a priority queue by means of a heap. In the table we denote with N the number of elements in the priority queue at the time the operation is performed. The space complexity is $O(N)$. We assume that the heap is itself realized by a data structure for binary trees that supports $O(1)$-time access to the children and parent of a node. For instance, we can implement the heap explicitly with a linked structure (with pointers from a node to its parents and children), or implicitly with an array (where node i has children $2i$ and $2i + 1$).

Let N be the number of elements in a priority queue Q realized with a heap T at the time an operation is performed. The time bounds of Table 5.6 are based on the following facts:

TABLE 5.6 Performance of a Priority Queue Realized by a Heap, Implemented with a Suitable Binary Tree Data Structure

Operation	Time
SIZE	$O(1)$
MAX	$O(1)$
INSERT	$O(\log N)$
REMOVE	$O(\log N)$
REMOVEMAX	$O(\log N)$
MODIFY	$O(\log N)$

- In the worst case, the time complexity of upheap and downheap is proportional to the height of T.
- If we keep a pointer to the last node of T, we can update this pointer in time proportional to the height of T in operations INSERT, REMOVE, and REMOVEMAX, as illustrated in Fig. 5.4.
- The height of heap T is $O(\log N)$,

The $O(N)$ space complexity bound for the heap is based on the following facts:

- The heap has $2N + 1$ nodes (N internal nodes and $N + 1$ leaves).
- Every node uses $O(1)$ space.
- In the array implementation, because of the level property the array elements used to store heap nodes are in the contiguous locations 1 through $2N - 1$.

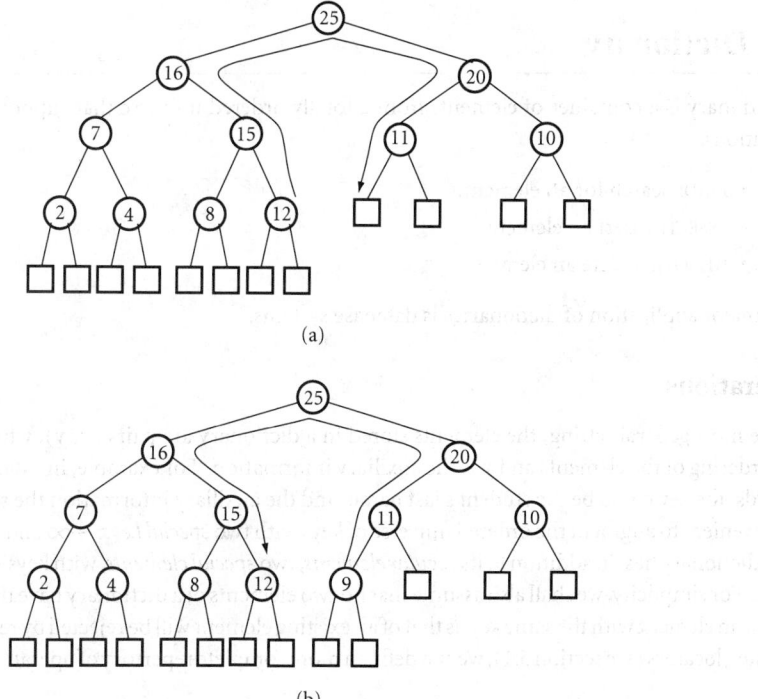

(a)

(b)

FIGURE 5.4 Update of the pointer to the last node: (a) INSERT and (b) REMOVE or REMOVEMAX.

Note that we can reduce the space requirement by a constant factor implementing the leaves of the heap with null objects, such that only the internal nodes have space associated with them.

Sorting

Realizing a priority queue with a heap has the advantage that all of the operations take $O(\log N)$ time, where N is the number of elements in the priority queue at the time the operation is performed. For example, in the sorting application (see first subsection of this section), both the first phase (inserting the N elements) and the second phase (removing N times the maximum element) take time

$$O\left(\sum_{i=1}^{N} \log i\right) = O(N \log N)$$

Hence, sorting with a priority queue realized with a heap takes $O(N \log N)$ time. This sorting method is known as *heap sort*, and its performance is considerably better than that of selection sort and insertion sort (see the section on sorting section), where the priority queue is realized as a sequence.

Realization with a Dictionary

A priority queue can be easily realized with a dictionary (see section 5.4). Indeed, all of the operations in the priority queue repertory are supported by a dictionary. To achieve $O(1)$ time for operation MAX, we can store the locator of the maximum element in a variable, and recompute it after an update operation. This realization of a priority queue with a dictionary has the same asymptotic complexity bounds as the realization with a heap, provided the dictionary is suitably implemented, for example, with an (a, b)-tree (see section "Realization with an (a, b)-tree") or an AVL-tree (see section "Realization with an AVL-tree"). However, a heap is simpler to program than an (a, b)-tree or an AVL-tree.

5.4 Dictionary

A dictionary is a container of elements from a totally ordered universe that supports the following basic operations:

- FIND: search for an element.
- INSERT: insert an element.
- REMOVE: delete an element.

A major application of dictionaries is database systems.

Operations

In the most general setting, the elements stored in a dictionary are pairs (x, y), where x is the *key* giving the ordering of the elements and y is the auxiliary information. For example, in a database storing student records, the key could be the student's last name, and the auxiliary information the student's transcript. It is convenient to augment the ordered universe of keys with two *special keys*: $+\infty$ and $-\infty$, and assume that each dictionary has, in addition to its *regular elements*, two *special elements*, with keys $+\infty$ and $-\infty$, respectively. For simplicity, we shall also assume that no two elements of a dictionary have the same key. An insertion of an element with the same key as that of an existing element will be rejected by returning a null locator.

Using locators (see section 5.1), we can define a more complete repertory of operations for a dictionary D:

SIZE(N) return the number of regular elements N of D.

FIND(x, c) if D contains an element with key x, assign to c a locator to such an element, otherwise set c equal to a null locator.

LOCATEPREV(x, c) assign to c a locator to the element of D with the largest key less than or equal to x; if x is smaller than all of the keys of the regular elements, then c is a locator to the special element with key $-\infty$; if $x = -\infty$, then c is a null locator.

LOCATENEXT(x, c) assign to c a locator to the element of D with the smallest key greater than or equal to x; if x is larger than all of the keys of the regular elements, then c is a locator to the special element with key $+\infty$; then if $x = +\infty$, then c is a null locator.

LOCATERANK(r, c) assign to c a locator to the rth element of D; if $r < 1$, then c is a locator to the special element with key $-\infty$; if $r > N$, where N is the size of D, then c is a locator to the special element with key $+\infty$.

PREV(c', c'') assign to c'' a locator to the element of D with the largest key less than that of the element with locator c'; if the key of the element with locator c' is smaller than all of the keys of the regular elements, then this operation returns a locator to the special element with key $-\infty$.

NEXT(c', c'') assign to c'' a locator to the element of D with the smallest key larger than that of the element with locator c'; if the key of the element with locator c' is larger than all of the keys of the regular elements, then this operation returns a locator to the special element with key $+\infty$.

MIN(c) assign to c a locator to the regular element of D with minimum key; if D has no regular elements, then c is a null locator.

MAX(c) assign to c a locator to the regular element of D with maximum key; if D has no regular elements, then c is a null locator.

INSERT(e, c) insert element e into D, and return a locator c to e; if there is already an element with the same key as e, then this operation returns a null locator.

REMOVE(c, e) remove from D and return element e with locator c.

MODIFY(c, e) replace with e the element with locator c.

Some of these operations can be easily expressed by means of other operations of the repertory. For example, operation FIND is a simple variation of LOCATEPREV or LOCATENEXT; MIN and MAX are special cases of LOCATERANK, or can be expressed by means of PREV and NEXT.

Realization with a Sequence

We can realize a dictionary by reusing and extending the sequence abstract data type (see section 5.2). Operations SIZE, INSERT, and REMOVE correspond to the homonymous sequence operations.

Unsorted Sequence

We can realize INSERT by an INSERTHEAD or an INSERTTAIL, which means that the sequence is not kept sorted. Operation FIND(x,c) can be performed by scanning the sequence with an iteration of NEXT operations, until we either find an element with key x, or we reach the end of the sequence. Table 5.7 shows the time complexity of this realization, assuming that the sequence is implemented with a doubly linked list. In the table we denote with N the number of elements in the dictionary at the time the operation is performed. The space complexity is $O(N)$.

Sorted Sequence

We can also use a sorted sequence to realize a dictionary. Operation INSERT now requires scanning the sequence to find the appropriate position to insert the new element. However, in a FIND operation, we can stop scanning the sequence as soon as we find an element with a key larger than the search key. Table 5.8 shows the time complexity of this realization by a sorted sequence, assuming that the sequence is implemented with a doubly linked list. In the table we denote with N the number of elements in the dictionary at the time the operation is performed. The space complexity is $O(N)$.

TABLE 5.7 Performance of a Dictionary Realized by an Unsorted Sequence, Implemented with a Doubly Linked List

Operation	Time
SIZE	$O(1)$
FIND	$O(N)$
LOCATEPREV	$O(N)$
LOCATENEXT	$O(N)$
LOCATERANK	$O(N)$
NEXT	$O(N)$
PREV	$O(N)$
MIN	$O(N)$
MAX	$O(N)$
INSERT	$O(1)$
REMOVE	$O(1)$
MODIFY	$O(1)$

TABLE 5.8 Performance of a Dictionary Realized by a Sorted Sequence, Implemented with a Doubly Linked List

Operation	Time
SIZE	$O(1)$
FIND	$O(N)$
LOCATEPREV	$O(N)$
LOCATENEXT	$O(N)$
LOCATERANK	$O(N)$
NEXT	$O(1)$
PREV	$O(1)$
MIN	$O(1)$
MAX	$O(1)$
INSERT	$O(N)$
REMOVE	$O(1)$
MODIFY	$O(N)$

TABLE 5.9 Performance of a Dictionary Realized by a Sorted Sequence, Implemented with an Array

Operation	Time
SIZE	$O(1)$
FIND	$O(\log N)$
LOCATEPREV	$O(\log N)$
LOCATENEXT	$O(\log N)$
LOCATERANK	$O(1)$
NEXT	$O(1)$
PREV	$O(1)$
MIN	$O(1)$
MAX	$O(1)$
INSERT	$O(N)$
REMOVE	$O(N)$
MODIFY	$O(N)$

Sorted Array

We can obtain a different performance tradeoff by implementing the sorted sequence by means of an array, which allows constant-time access to any element of the sequence given its position. Indeed, with this realization we can speed up operation FIND(x, c) using the *binary search* strategy, as follows. If the dictionary is empty, we are done. Otherwise, let N be the current number of elements in the dictionary. We compare the search key k with the key x_m of the middle element of the sequence, that is, the element at position $\lfloor N/2 \rfloor$. If $x = x_m$, we have found the element. Else, we recursively search in the subsequence of the elements preceding the middle element if $x < x_m$, or following the middle element if $x > x_m$. At each recursive call, the number of elements of the subsequence being searched halves. Hence, the number of sequence elements accessed and the number of comparisons performed by binary search is $O(\log N)$. While searching takes $O(\log N)$ time, inserting or deleting elements now takes $O(N)$ time.

Table 5.9 shows the performance of a dictionary realized with a sorted sequence, implemented with an array. In the table we denote with N the number of elements in the dictionary at the time the operation is performed. The space complexity is $O(N)$.

Realization with a Search Tree

A *search tree* for elements of the type (x, y), where x is a key from a totally ordered universe, is a rooted ordered tree T such that:

- Each internal node of T has at least two children and stores a nonempty set of elements.
- A node μ of T with d children $\mu_1, \ldots, \mu_d$ stores $d - 1$ elements $(x_1, y_1) \cdots (x_{d-1}, y_{d-1})$, where $x_1 \leq \cdots \leq x_{d-1}$.
- For each element (x, y) stored at a node in the subtree of T rooted at μ_i, we have $x_{i-1} \leq x \leq x_i$, where $x_0 = -\infty$ and $x_d = +\infty$.

In a search tree, each internal node stores a nonempty collection of keys, whereas the leaves do not store any key and serve only as placeholders. An example of search tree is shown in Fig. 5.5(a). A special type of search tree is a *binary search tree*, where each internal node stores one key and has two children.

We will recursively describe the realization of a dictionary D by means of a search tree T, since we will use dictionaries to implement the nodes of T. Namely, an internal node μ of T with children $\mu_1, \ldots, \mu_d$ and elements $(x_1, y_1) \cdots (x_{d-1}, y_{d-1})$ is equipped with a dictionary $D(\mu)$ whose regular elements are the pairs $(x_i, (y_i, \mu_i)), i = 1, \ldots, d - 1$ and whose special element with key $+\infty$ is $(+\infty, (\cdot, \mu_d))$. A regular

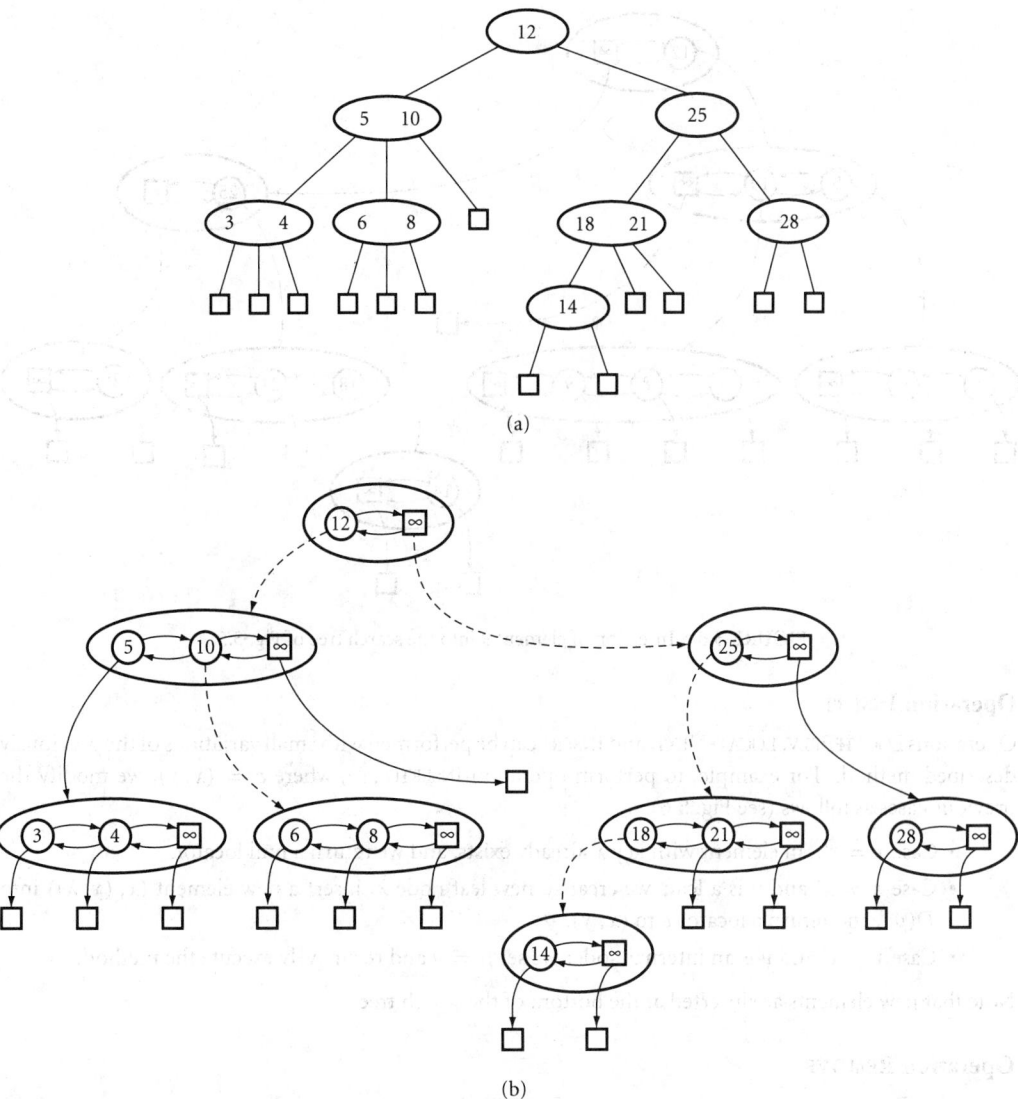

FIGURE 5.5 Realization of a dictionary by means of a search tree: (a) a search tree T, (b) realization of the dictionaries at the nodes of T by means of sorted sequences. The search paths for elements 9 (unsuccessful search) and 14 (successful search) are shown with dashed lines.

element (x, y) stored in D is associated with a regular element $(x, (y, v))$ stored in a dictionary $D(\mu)$, for some node μ of T. See the example in Fig. 5.5(b).

Operation FIND

Operation FIND(x, c) on dictionary D is performed by means of the following recursive method for a node μ of T, where μ is initially the root of T [see Fig. 5.5(b)]. We execute LOCATENEXT(x, c') on dictionary $D(\mu)$ and let $(x', (y', v))$ be the element pointed by the returned locator c'. We have three cases:

- Case $x = x'$: we have found x and return locator c to (x', y').
- Case $x \neq x'$ and v is a leaf: we have determined that x is not in D and return a null locator c.
- Case $x \neq x'$ and v is an internal node: we set $\mu = v$ and recursively execute the method.

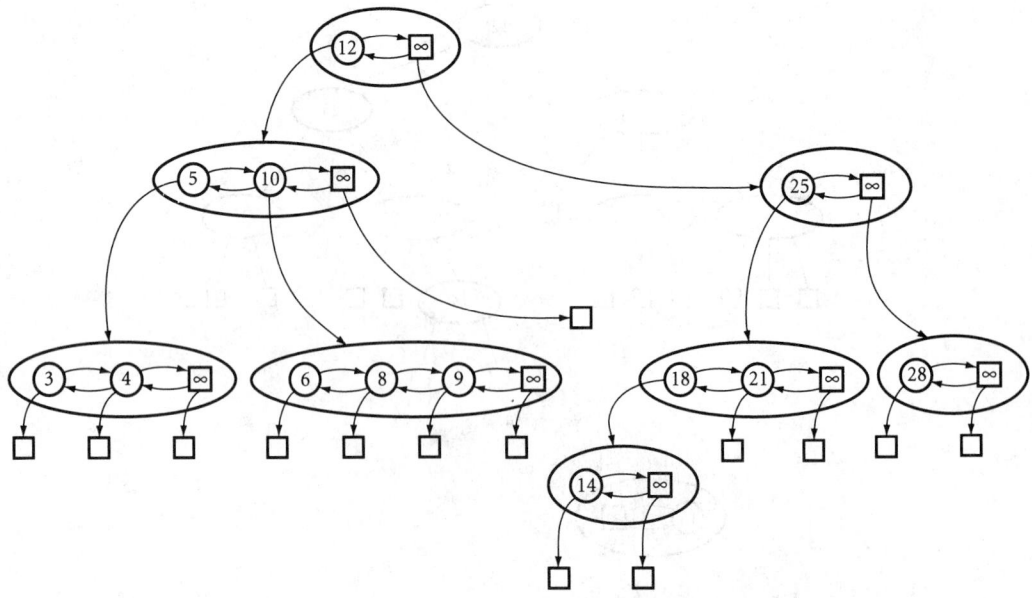

FIGURE 5.6   Insertion of element 9 into the search tree of Fig. 5.5.

Operation INSERT

Operations LOCATEPREV, LOCATENEXT, and INSERT can be performed with small variations of the previously described method. For example, to perform operation INSERT(e, c), where $e = (x, y)$, we modify the previous cases as follows (see Fig. 5.6):

- Case $x = x'$: an element with key x already exists, and we return a null locator.
- Case $x \neq x'$ and v is a leaf: we create a new leaf node λ, insert a new element $(x, (y, \lambda))$ into $D(\mu)$, and return a locator c to (x, y).
- Case $x \neq x'$ and v is an internal node: we set $\mu = v$ and recursively execute the method.

Note that new elements are inserted at the bottom of the search tree.

Operation REMOVE

Operation REMOVE(e, c) is more complex (see Fig. 5.7). Let the associated element of $e = (x, y)$ in T be $(x, (y, v))$, stored in dictionary $D(\mu)$ of node μ:

- If node v is a leaf, we simply delete element $(x, (y, v))$ from $D(\mu)$.
- Else (v is an internal node), we find the successor element $(x', (y', v'))$ of $(x, (y, v))$ in $D(\mu)$ with a NEXT operation in $D(\mu)$. (1) If v' is a leaf, we replace v' with v, that is, change element $(x', (y', v'))$ to $(x', (y', v))$, and delete element $(x, (y, v))$ from $D(\mu)$. (2) Else (v' is an internal node), while the leftmost child v'' of v' is not a leaf, we set $v' = v''$. Let $(x'', (y'', v''))$ be the first element of $D(v')$ (node v'' is a leaf). We replace $(x, (y, v))$ with $(x'', (y'', v))$ in $D(\mu)$ and delete $(x'', (y'', v''))$ from $D(v')$.

The listed actions may cause dictionary $D(\mu)$ or $D(v')$ to become empty. If this happens, say for $D(\mu)$ and μ is not the root of T, we need to remove node μ. Let $(+\infty, (\cdot, \kappa))$ be the special element of $D(\mu)$ with key $+\infty$, and let $(z, (w, \mu))$ be the element pointing to μ in the parent node π of μ. We delete node μ and replace $(z, (w, \mu))$ with $(z, (w, \kappa))$ in $D(\pi)$.

Note that, if we start with an initially empty dictionary, a sequence of insertions and deletions performed with the described methods yields a search tree with a single node. In the next sections, we show how to avoid this behavior by imposing additional conditions on the structure of a search tree.

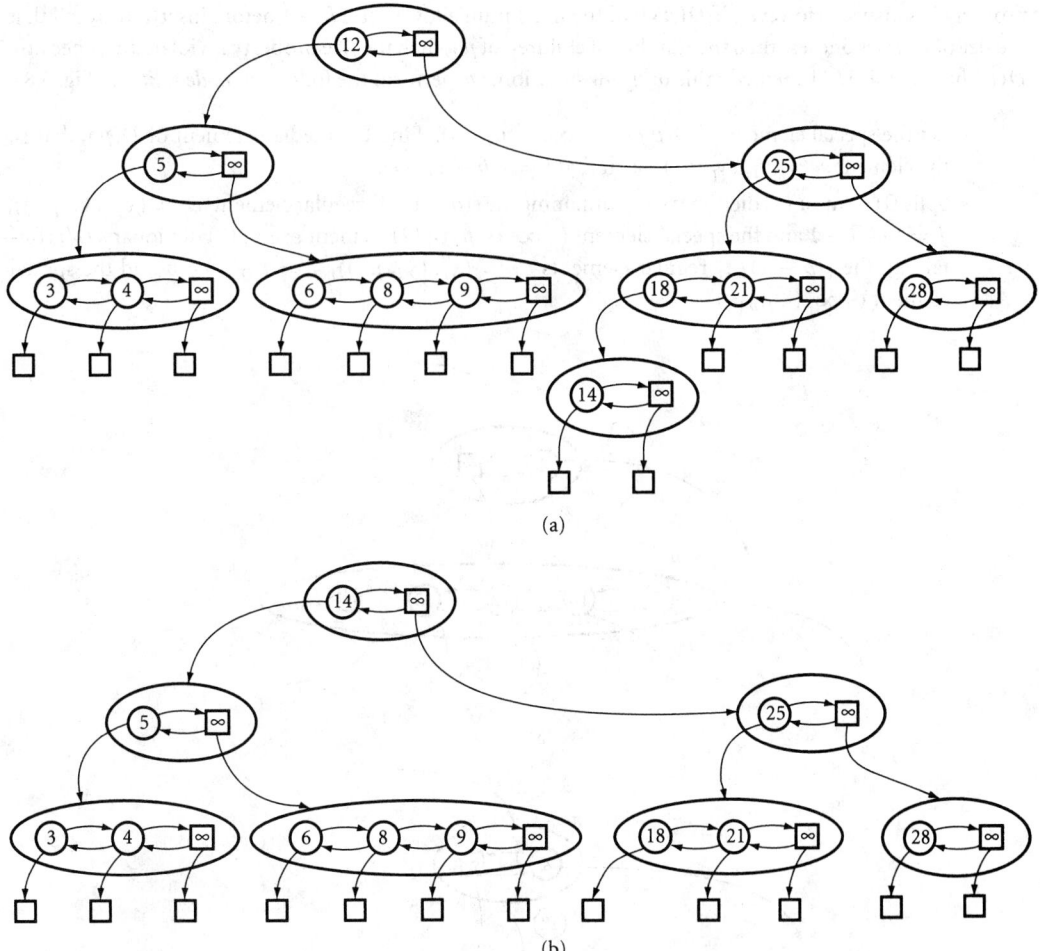

(a)

(b)

FIGURE 5.7 (a) Deletion of element 10 from the search tree of Fig. 5.6. (b) Deletion of element 12 from the search tree of part a.

Realization with an (a, b)-Tree

An (a, b)-*tree*, where a and b are integer constants such that $2 \leq a \leq (b + 1)/2$, is a search tree T with the following additional restrictions:

> *Level property.* All of the levels of T are full, that is, all of the leaves are at the same depth.
> *Size property.* Let μ be an internal node of T, and d be the number of children of μ; if μ is the root of T, then $d \geq 2$, else $a \leq d \leq b$.

The height of an (a, b)-tree storing N elements is $O(\log_a N) = O(\log N)$. Indeed, in the worst case, the root has two children, and all of the other internal nodes have a children.

The realization of a dictionary with an (a, b)-tree extends that with a search tree. Namely, the implementation of operations INSERT and REMOVE need to be modified in order to preserve the level and size properties. Also, we maintain the current size of the dictionary, and pointers to the minimum and maximum regular elements of the dictionary.

Insertion

The implementation of operation INSERT for search trees given earlier in this section adds a new element to the dictionary $D(\mu)$ of an existing node μ of T. Since the structure of the tree is not changed, the level property is satisfied. However, if $D(\mu)$ had the maximum allowed size $b - 1$ before insertion (recall that the size of $D(\mu)$ is one less than the number of children of μ), then the size property is violated at μ because $D(\mu)$ has now size b. To remedy this *overflow* situation, we perform the following *node split* (see Fig. 5.8):

- Let the special element of $D(\mu)$ be $(+\infty, (\cdot, \mu_{b+1}))$. Find the median element of $D(\mu)$, that is, the element $e_i = (x_i, (y_i, \mu_i))$ such that $i = \lceil (b+1)/2 \rceil$.
- Split $D(\mu)$ into: (1) dictionary D' containing the $\lceil (b-1)/2 \rceil$ regular elements $e_j = (x_j, (y_j, \mu_j))$, $j = 1 \cdots i - 1$ and the special element $(+\infty, (\cdot, \mu_i))$; (2) element e; and (3) dictionary D'', containing the $\lfloor (b-1)/2 \rfloor$ regular elements $e_j = (x_j, (y_j, \mu_j))$, $j = i + 1 \cdots b$ and the special element $(+\infty, (\cdot, \mu_{b+1}))$.

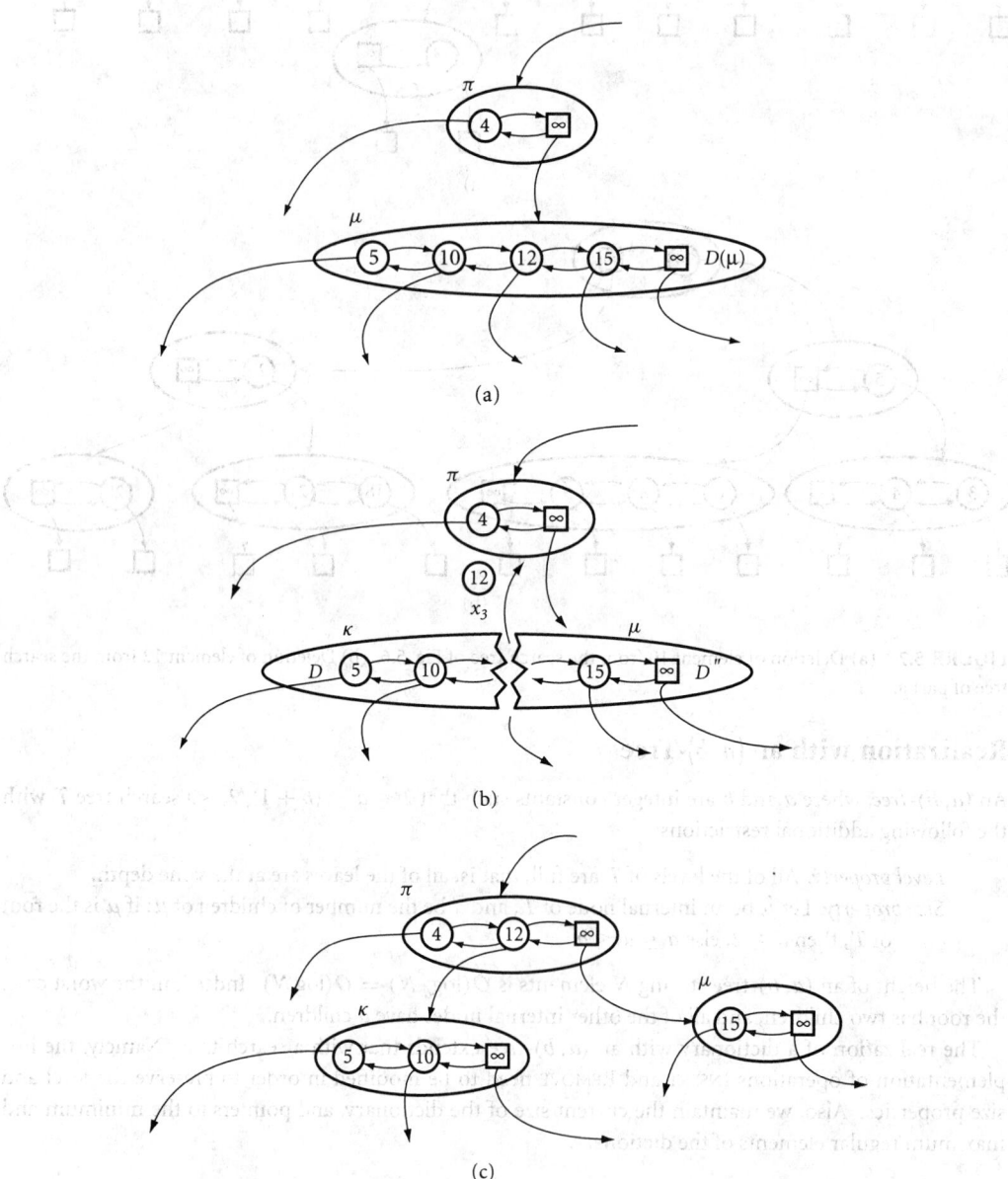

FIGURE 5.8   Example of node split in a 2–4 tree: (a) initial configuration with an overflow at node μ, (b) split of the node μ into μ' and μ'' and insertion of the median element into the parent node π, and (c) final configuration.

- Create a new tree node κ, and set $D(\kappa) = D'$. Hence, node κ has children $\mu_1 \cdots \mu_i$.
- Set $D(\mu) = D''$. Hence, node μ has children $\mu_{i+1} \cdots \mu_{b+1}$.
- If μ is the root of T, create a new node π with an empty dictionary $D(\pi)$. Else, let π be the parent of μ.
- Insert element $(x_i, (y_i, \kappa))$ into dictionary $D(\pi)$.

After a node split, the level property is still verified. Also, the size property is verified for all of the nodes of T, except possibly for node π. If π has $b + 1$ children, we repeat the node split for $\mu = \pi$. Each time we perform a node split, the possible violation of the size property appears at a higher level in the tree. This guarantees the termination of the algorithm for the INSERT operation. We omit the description of the simple method for updating the pointers to the minimum and maximum regular elements.

Deletion

The implementation of operation REMOVE for search trees given earlier in this section removes an element from the dictionary $D(\mu)$ of an existing node μ of T. Since the structure of the tree is not changed, the level property is satisfied. However, if μ is not the root, and $D(\mu)$ had the minimum allowed size $a - 1$ before deletion (recall that the size of the dictionary is one less than the number of children of the node), then the size property is violated at μ because $D(\mu)$ has now size $a - 2$. To remedy this *underflow* situation, we perform the following *node merge* (see Figs. 5.9 and 5.10):

- If μ has a right sibling, then let μ'' be the right sibling of μ and $\mu' = \mu$; else, let μ' be the left sibling of μ and $\mu'' = \mu$. Let $(+\infty, (\cdot, v))$ be the special element of $D(\mu')$.
- Let π be the parent of μ' and μ''. Remove from $D(\pi)$ the regular element $(x, (y, \mu'))$ associated with μ'.
- Create a new dictionary D containing the regular elements of $D(\mu')$ and $D(\mu'')$, regular element $(x, (y, v))$, and the special element of $D(\mu'')$.
- Set $D(\mu'') = D$, and destroy node μ'.
- If μ'' has more than b children, perform a node split at μ''.

After a node merge, the level property is still verified. Also, the size property is verified for all of the nodes of T, except possibly for node π. If π is the root and has one child (and thus an empty dictionary), we remove node π. If π is not the root and has fewer than $a - 1$ children, we repeat the node merge for $\mu = \pi$. Each time we perform a node merge, the possible violation of the size property appears at a higher level in the tree. This guarantees the termination of the algorithm for the REMOVE operation. We omit the description of the simple method for updating the pointers to the minimum and maximum regular elements.

Complexity

Let T be an (a, b)-tree storing N elements. The height of T is $O(\log_a N) = O(\log N)$. Each dictionary operation affects only the nodes along a root-to-leaf path. We assume that the dictionaries at the nodes of T are realized with sequences. Hence, processing a node takes $O(b) = O(1)$ time. We conclude that each operation takes $O(\log N)$ time.

Table 5.10 shows the performance of a dictionary realized with an (a, b)-tree. In the table we denote with N the number of elements in the dictionary at the time the operation is performed. The space complexity is $O(N)$.

TABLE 5.10 Performance of a Dictionary Realized by an (a, b)-Tree

Operation	Time
SIZE	$O(1)$
FIND	$O(\log N)$
LOCATEPREV	$O(\log N)$
LOCATENEXT	$O(\log N)$
LOCATERANK	$O(\log N)$
NEXT	$O(\log N)$
PREV	$O(\log N)$
MIN	$O(1)$
MAX	$O(1)$
INSERT	$O(\log N)$
REMOVE	$O(\log N)$
MODIFY	$O(\log N)$

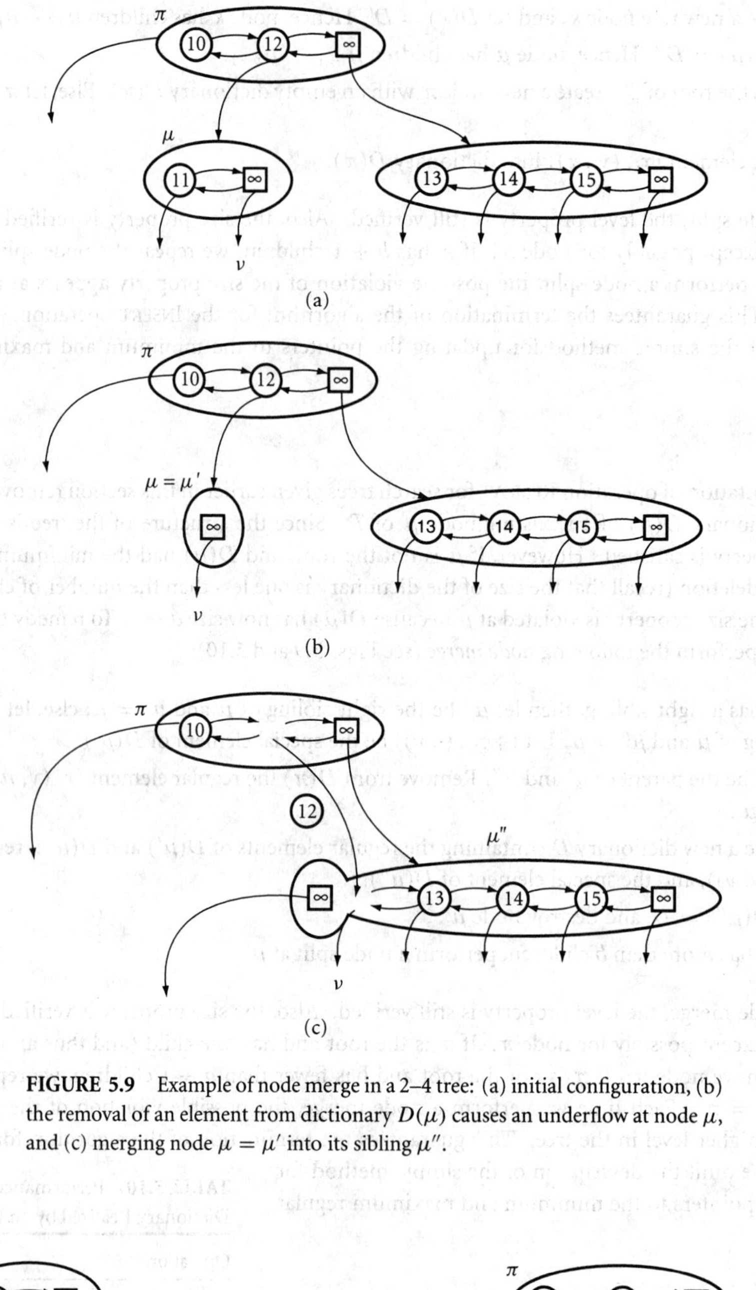

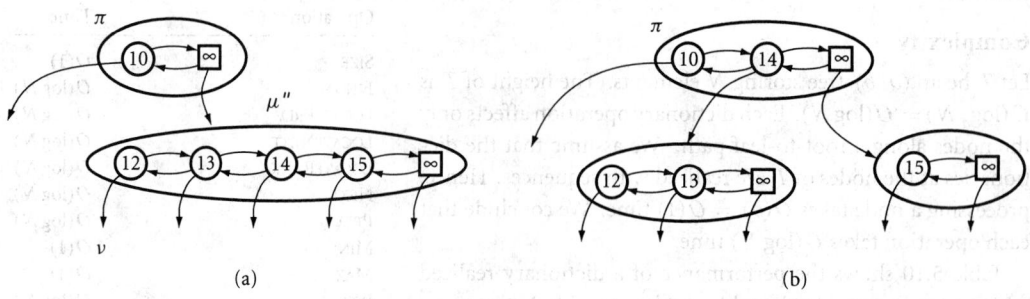

FIGURE 5.9 Example of node merge in a 2–4 tree: (a) initial configuration, (b) the removal of an element from dictionary $D(\mu)$ causes an underflow at node μ, and (c) merging node $\mu = \mu'$ into its sibling μ''.

FIGURE 5.10 Example of subsequent node merge in a 2–4 tree: (a) overflow at node μ'' and (b) final configuration after splitting node μ''.

Realization with an AVL-Tree

An *AVL-tree* is a search tree T with the following additional restrictions:

> *Binary property.* T is a binary tree, that is, every internal node has two children (left and right child), and stores one key.
>
> *Height-balance property.* For every internal node μ, the heights of the subtrees rooted at the children of μ differ at most by one.

An example of AVL-tree is shown in Fig. 5.11. The height of an AVL-tree storing N elements is $O(\log N)$. This can be shown as follows. Let N_h be the minimum number of elements stored in an AVL-tree of height h. We have $N_0 = 0$, $N_1 = 1$, and

$$N_h = 1 + N_{h-1} + N_{h-2}, \quad \text{for } h \geq 2$$

The preceding recurrence relation defines the well-known Fibonacci numbers. Hence, $N_h = \Omega(\phi^N)$, where $1 < \phi < 2$.

The realization of a dictionary with an AVL-tree extends that with a search tree. Namely, the implementation of operations INSERT and REMOVE need to be modified in order to preserve the binary and height-balance properties after an insertion or deletion.

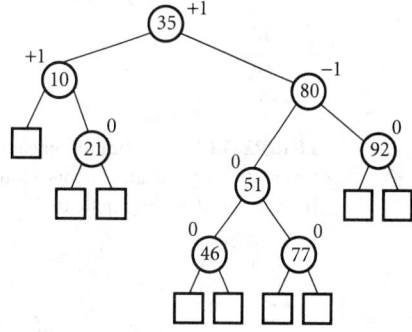

FIGURE 5.11 Example of AVL-tree storing 9 elements. The keys are shown inside the nodes, and the balance factors (see subsequent section on rebalancing) are shown next to the nodes.

Insertion

The implementation of INSERT for search trees given earlier in this section adds the new element to an existing node. This violates the binary property, and hence cannot be done in an AVL-tree. Hence, we modify the three cases of the INSERT algorithm for search trees as follows:

- Case $x = x'$: an element with key x already exists, and we return a null locator c.
- Case $x \neq x'$ and v is a leaf: we replace v with a new internal node κ with two leaf children, store element (x, y) in κ, and return a locator c to (x, y).
- Case $x \neq x'$ and v is an internal node: we set $\mu = v$ and recursively execute the method.

We have preserved the binary property. However, we may have violated the height-balance property, since the heights of some subtrees of T have increased by one. We say that a node is balanced if the difference between the heights of its subtrees is -1, 0, or 1, and is unbalanced otherwise. The unbalanced nodes form a (possibly empty) subpath of the path from the new internal node κ to the root of T. See the example of Fig. 5.12.

Rebalancing

To restore the height-balance property, we *rebalance* the lowest node μ that is unbalanced, as follows:

- Let μ' be the child of μ whose subtree has maximum height, and μ'' be the child of μ' whose subtree has maximum height.
- Let (μ_1, μ_2, μ_3) be the left-to-right ordering of nodes $\{\mu, \mu', \mu''\}$, and (T_0, T_1, T_2, T_3) be the left-to-right ordering of the four subtrees of $\{\mu, \mu', \mu''\}$ not rooted at a node in $\{\mu, \mu', \mu''\}$.
- Replace the subtree rooted at μ with a new subtree rooted at μ_2, where μ_1 is the left child of μ_2 and has subtrees T_0 and T_1, and μ_3 is the right child of μ_2 and has subtrees T_2 and T_3.

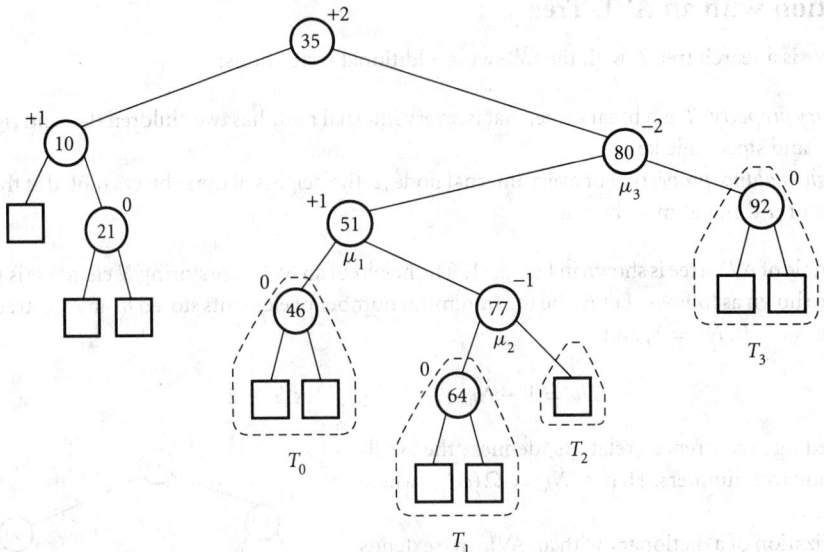

FIGURE 5.12   Insertion of an element with key 64 into the AVL-tree of Fig. 5.11. Note that two nodes (with balance factors +2 and −2) have become unbalanced. The dashed lines identify the subtrees that participate in the rebalancing, as illustrated in Fig. 5.14.

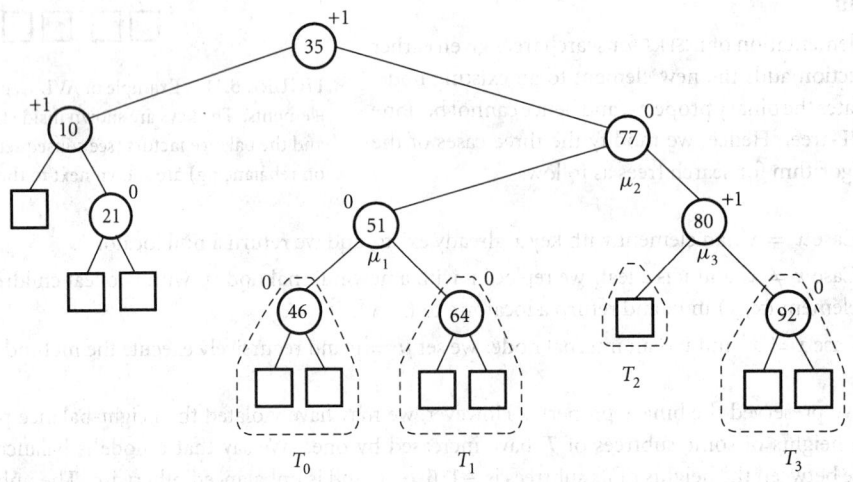

FIGURE 5.13   AVL-tree obtained by rebalancing the lowest unbalanced node in the tree of Fig. 5.11. Note that all of the nodes are now balanced. The dashed lines identify the subtrees that participate in the rebalancing, as illustrated in Fig. 5.14.

Two examples of rebalancing are schematically shown in Fig. 5.14. Other symmetric configurations are possible. In Fig. 5.13, we show the rebalancing for the tree of Fig. 5.12.

Note that the rebalancing causes all of the nodes in the subtree of μ_2 to become balanced. Also, the subtree rooted at μ_2 now has the same height as the subtree rooted at node μ before insertion. This causes all of the previously unbalanced nodes to become balanced. To keep track of the nodes that become unbalanced, we can store at each node a *balance factor*, which is the difference of the heights of the left and right subtrees. A node becomes unbalanced when its balance factor becomes +2 or −2. It is easy to modify the algorithm for operation INSERT such that it maintains the balance factors of the nodes.

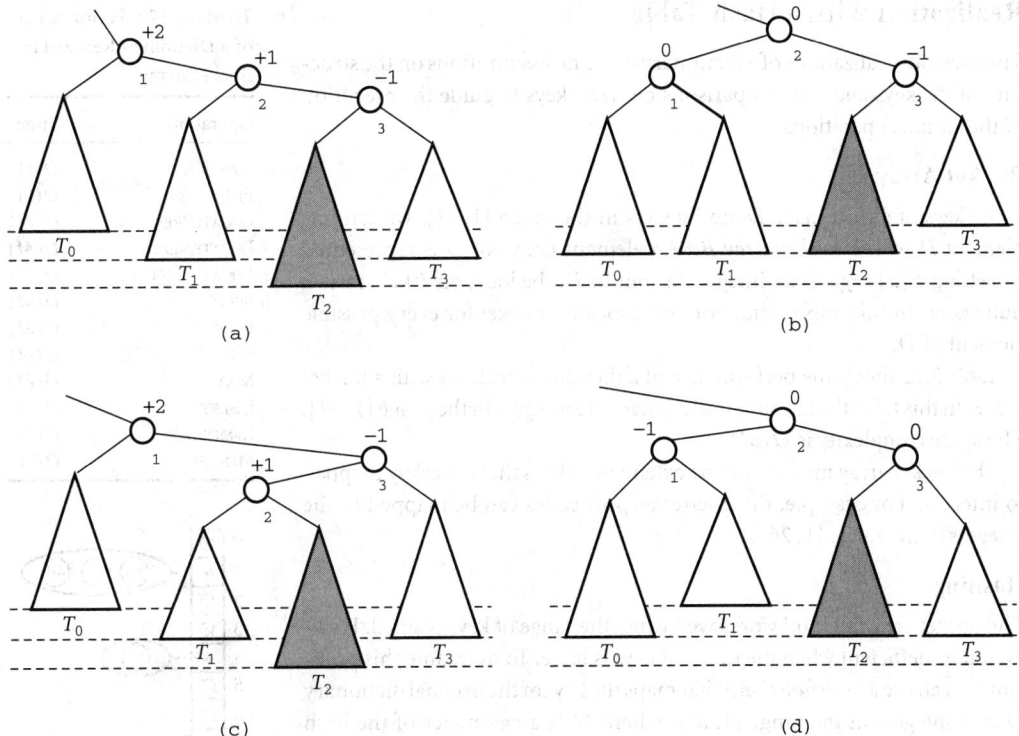

FIGURE 5.14 Schematic illustration of rebalancing a node in the INSERT algorithm for AVL-trees. The shaded subtree is the one where the new element was inserted. (a) and (b) Rebalancing by means of a single rotation. (c) and (d) Rebalancing by means of a double rotation.

Deletion

The implementation of REMOVE for search trees given earlier in this section preserves the binary property, but may cause the height-balance property to be violated. After deleting a node, there can be only one unbalanced node, on the path from the deleted node to the root of T.

To restore the height-balance property, we *rebalance* the unbalanced node using the previous algorithm. Notice, however, that the choice of μ'' may not be unique, since the subtrees of μ' may have the same height. In this case, the height of the subtree rooted at μ_2 is the same as the height of the subtree rooted at μ before rebalancing, and we are done. If, instead, the subtrees of μ' do not have the same height, then the height of the subtree rooted at μ_2 is one less than the height of the subtree rooted at μ before rebalancing. This may cause an ancestor of μ_2 to become unbalanced, and we repeat the rebalancing step. Balance factors are used to keep track of the nodes that become unbalanced, and can be easily maintained by the REMOVE algorithm.

Complexity

Let T be an AVL-tree storing N elements. The height of T is $O(\log N)$. Each dictionary operation affects only the nodes along a root-to-leaf path. Rebalancing a node takes $O(1)$ time. We conclude that each operation takes $O(\log N)$ time.

Table 5.11 shows the performance of a dictionary realized with an AVL-tree. In this table we denote with N the number of elements in the dictionary at the time the operation is performed. The space complexity is $O(N)$.

TABLE 5.11
Performance of a Dictionary Realized by an AVL-Tree

Operation	Time
SIZE	$O(1)$
FIND	$O(\log N)$
LOCATEPREV	$O(\log N)$
LOCATENEXT	$O(\log N)$
LOCATERANK	$O(\log N)$
NEXT	$O(\log N)$
PREV	$O(\log N)$
MIN	$O(1)$
MAX	$O(1)$
INSERT	$O(\log N)$
REMOVE	$O(\log N)$
MODIFY	$O(\log N)$

Realization with a Hash Table

The previous realizations of a dictionary make no assumptions on the structure of the keys and use comparisons between keys to guide the execution of the various operations.

Bucket Array

If the keys of a dictionary D are integers in the range $[1, M]$, we can implement D with a *bucket array* B. An element (x, y) of D is represented by setting $B[x] = y$. If an integer x is not in D, the location $B[x]$ stores a null value. In this implementation, we allocate a bucket for every possible element of D.

Table 5.12 shows the performance of a dictionary realized with a bucket array. In this table the keys in the dictionary are integers in the range $[1, M]$. The space complexity is $O(M)$.

The bucket array method can be extended to keys that are easily mapped to integers. For example, three-letter airport codes can be mapped to the integers in the range $[1, 26^3]$.

Hashing

The bucket array method works well when the range of keys is small. However, it is inefficient when the range of keys is large. To overcome this problem, we can use a *hash function* h that maps the keys of the original dictionary D into integers in the range $[1, M]$, where M is a parameter of the hash function. Now, we can apply the bucket array method using the *hashed value* $h(x)$ of the keys. In general, a *collision* may happen, where two distinct keys x_1 and x_2 have the same hashed value, that is, $x_1 \neq x_2$ and $h(x_1) = h(x_2)$. Hence, each bucket must be able to accommodate a collection of elements.

A hash table of size M for a function $h(x)$ is a bucket array B of size M (primary structure) whose entries are dictionaries (secondary structures), such that element (x, y) is stored in the dictionary $B[h(x)]$. For simplicity of programming, the dictionaries used as secondary structures are typically realized with sequences. An example of a hash table is shown in Fig. 5.15.

If all of the elements in the dictionary D collide, they are all stored in the same dictionary of the bucket array, and the performance of the hash table is the same as that of the kind of dictionary used for the secondary structures. At the other end of the spectrum, if no two elements of the dictionary D collide, they are stored in distinct one-element dictionaries of the bucket array, and the performance of the hash table is the same as that of a bucket array.

A typical hash function for integer keys is $h(x) = x \bmod M$ (here the range is $[0, M - 1]$). The size M of the hash table is usually chosen as a prime number. An example of a hash table is shown in Fig. 5.15. It is interesting to analyze the performance of a hash table from a probabilistic viewpoint. If we assume that the hashed values of the keys are uniformly distributed in the range $[1, M]$, then each bucket holds on average N/M keys, where N is the size of the dictionary. Hence, when $N = O(M)$, the average size of the secondary data structures is $O(1)$.

Table 5.13 shows the performance of a dictionary realized with a hash table. Both the worst-case and average time complexity in the preceding probabilistic model are indicated. In this table we denote with N the number of elements in the

TABLE 5.12　Performance of a Dictionary Realized by Bucket Array

Operation	Time
SIZE	$O(1)$
FIND	$O(1)$
LOCATEPREV	$O(M)$
LOCATENEXT	$O(M)$
LOCATERANK	$O(M)$
NEXT	$O(M)$
PREV	$O(M)$
MIN	$O(M)$
MAX	$O(M)$
INSERT	$O(1)$
REMOVE	$O(1)$
MODIFY	$O(1)$

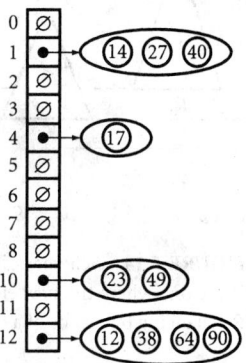

FIGURE 5.15　Example of a hash table of size 13 storing 10 elements. The hash function is $h(x) = x \bmod 13$.

TABLE 5.13　Performance of a Dictionary Realized by a Hash Table of Size M

Operation	Time	
	Worst Case	Average
SIZE	$O(1)$	$O(1)$
FIND	$O(N)$	$O(N/M)$
LOCATEPREV	$O(N + M)$	$O(N + M)$
LOCATENEXT	$O(N + M)$	$O(N + M)$
LOCATERANK	$O(N + M)$	$O(N + M)$
NEXT	$O(N + M)$	$O(N + M)$
PREV	$O(N + M)$	$O(N + M)$
MIN	$O(N + M)$	$O(N + M)$
MAX	$O(N + M)$	$O(N + M)$
INSERT	$O(1)$	$O(1)$
REMOVE	$O(1)$	$O(1)$
MODIFY	$O(1)$	$O(1)$

dictionary at the time the operation is performed. The space complexity is $O(N + M)$. The average time complexity refers to a probabilistic model where the hashed values of the keys are uniformly distributed in the range $[1, M]$.

Defining Terms

(a, b)-**Tree:** Search tree with additional properties (each node has between a and b children, and all the levels are full).

Abstract data type: Mathematically specified data type equipped with operations that can be performed on the objects.

AVL-tree: Binary search tree such that the subtrees of each node have heights that differ by at most one.

Binary search tree: Search tree such that each internal node has two children.

Bucket array: Implementation of a dictionary by means of an array indexed by the keys of the dictionary elements.

Container: Abstract data type storing a collection of objects (elements).

Dictionary: Container storing elements from a sorted universe supporting searches, insertions, and deletions.

Hash table: Implementation of a dictionary by means of a bucket array storing secondary dictionaries.

Heap: Binary tree with additional properties storing the elements of a priority queue.

Locator: Variable that allows access to an object stored in a container.

Priority queue: Container storing elements from a sorted universe that supports finding the maximum element, insertions, and deletions.

Search tree: Rooted ordered tree with additional properties storing the elements of a dictionary.

Sequence: Container storing objects in a certain order, supporting insertions (in a given position) and deletions.

References

Aggarwal, A. and Vitter, J. S. 1988. The input/output complexity of sorting and related problems. *Commun. ACM* 31:1116–1127.

Aho, A. V., Hopcroft, J. E., and Ullman, J. D. 1983. *Data Structures and Algorithms.* Addison–Wesley, Reading, MA.

Chazelle, B. and Guibas, L. J. 1986. Fractional cascading: I. a data structuring technique. *Algorithmica* 1:133–162.

Chiang, Y.-J. and Tamassia, R. 1992. Dynamic algorithms in computational geometry. *Proc. IEEE* 80(9):1412–1434.

Cohen, R. F. and Tamassia, R. 1995. Dynamic expression trees. *Algorithmica* 13:245–265.

Comer, D. 1979. The ubiquitous B-tree. *ACM Comput. Surv.* 11:121–137.

Cormen, T. H., Leiserson, C. E., and Rivest, R. L. 1990. *Introduction to Algorithms.* MIT Press, Cambridge, MA.

Di Battista, G. and Tamassia, R. 1990. On-line graph algorithms with SPQR-trees. In *Automata, Languages and programming (Proc. 17th ICALP)*, Vol. 442, pp. 598–611. *Lecture Notes in Computer Science.*

Driscoll, J. R., Sarnak, N., Sleator, D. D., and Tarjan, R. E. 1989. Making data structures persistent. *J. Comput. Syst. Sci.* 38:86–124.

Edelsbrunner, H. 1987. *Algorithms in Combinatorial Geometry,* Vol. 10, *EATCS Monographs on Theoretical Computer Science.* Springer–Verlag, Heidelberg, Germany.

Eppstein, D., Galil, Z., Italiano, G. F., and Nissenzweig, A. 1992. Sparsification: a technique for speeding up dynamic graph algorithms. In *Proc. 33rd Annu. IEEE Symp. Found. Comput. Sci.*, pp. 60–69.

Even, S. 1979. *Graph Algorithms.* Computer Science Press, Potomac, MD.

Foley, J. D., van Dam, A., Feiner, S. K., and Hughes, J. F. 1990. *Computer Graphics: Principles and Practice.* Addison–Wesley, Reading, MA.

Frederickson, G. N. 1993. A data structure for dynamically maintaining rooted trees. In *Proc. 4th ACM-SIAM Symp. Discrete Algorithms*, pp. 175–184.

Galil, Z. and Italiano, G. F. 1991. Data structures and algorithms for disjoint set union problems. *ACM Comput. Surv.* 23(3):319–344.

Gonnet, G. H. and Baeza-Yates, R. 1991. *Handbook of Algorithms and Data Structures*. Addison–Wesley, Reading, MA.

Hoffmann, K., Mehlhorn, K., Rosenstiehl, P., and Tarjan, R. E. 1986. Sorting Jordan sequences in linear time using level-linked search trees. *Inf. Control* 68:170–184.

Horowitz, E., Sahni, S., and Metha, D. 1995. *Fundamentals of Data Structures in C++*. Computer Science Press.

Knuth, D. E. 1968. *Fundamental Algorithms*. Vol. I. In *The Art of Computer Programming*. Addison–Wesley, Reading, MA.

Knuth, D. E. 1973. *Sorting and Searching*, Vol. 3. In *The Art of Computer Programming*. Addison–Wesley, Reading, MA.

Lewis, H. R. and Denenberg, L. 1991. *Data Structures and Their Algorithms*. Harper Collins.

Mehlhorn, K. 1984. *Data Structures and Algorithms*. Vol. 1–3. Springer–Verlag.

Mehlhorn, K. and Naher, S. 1995. LEDA: a platform for combinatorial and geometric computing. *CACM*, 38:96–102; http://www.mpi-sb.mpg.de/guide/staff/uhrig/leda.html.

Mehlhorn, K. and Tsakalidis, A. 1990. Data structures. In *Algorithms and Complexity*. J. van Leeuwen, ed. Vol. A, *Handbook of Theoretical Computer Science*. Elsevier, Amsterdam.

Miltersen, P. B., Sairam, S., Vitter, J. S., and Tamassia, R. 1994. Complexity models for incremental computation. *Theoret. Comput. Sci.* 130:203–236.

Nievergelt, J. and Hinrichs, K. H. 1993. *Algorithms and Data Structures: With Applications to Graphics and Geometry*. Prentice–Hall, Englewood Cliffs, NJ.

O'Rourke, J. 1994. *Computational Geometry in C*. Cambridge Univ. Press.

Overmars, M. H. 1983. *The Design of Dynamic Data Structures*, Vol. 156, *Lecture Notes in Computer Science*. Springer–Verlag.

Preparata, F. P. and Shamos, M. I. 1985. *Computational Geometry: An Introduction*. Springer–Verlag, New York.

Pugh, W. 1990. Skip lists: a probabilistic alternative to balanced trees. *Commun. ACM* 35:668–676.

Reif, J. H. 1987. A topological approach to dynamic graph connectivity. *Inform. Process. Lett.* 25:65–70.

Sedgewick, R. 1992. *Algorithms in C++*. Addison–Wesley, Reading, MA.

Sleator, D. D. and Tarjan, R. E. 1993. A data structure for dynamic trees. *J. Comput. Syst. Sci.* 26(3):362–381.

Tarjan, R. E. 1983. *Data Structures and Network Algorithms, Vol. 44, CBMS-NSF Regional Conference Series in Applied Mathematics*. Society for Industrial Applied Mathematics.

Vitter, J. S. and Flajolet, P. 1990. Average-case analysis of algorithms and data structures. In *Algorithms and Complexity*, J. van Leeuwen, ed., Vol. A, *Handbook of Theoretical Computer Science*, pp. 431–524. Elsevier, Amsterdam.

Wood, D. 1993. *Data Structures, Algorithms, and Performance*. Addison–Wesley, Reading, MA.

Further Information

Many textbooks and monographs have been written on data structures, for example, Aho et al. [1983], Cormen et al. [1990], Gonnet and Baeza-Yates [1990], Horowitz et al. [1995], Knuth [1968, 1973], Lewis and Denenberg [1991], Mehlhorn [1984], Nievergelt and Hinrichs [1993], Overmars [1983], Preparata and Shamos [1995], Sedgewick [1992], Tarjan [1983], and Wood [1993].

Recent papers surveying the state-of-the art in data structures include Chiang and Tamassia [1992], Galil and Italiano [1991], Mehlhorn and Tsakalidis [1990], and Vitter and Flajolet [1990].

The LEDA project [Mehlhorn and Näher 1995] aims at developing a C++ library of efficient and reliable implementations of advanced data structures.

6

Computational Geometry

D. T. Lee*
Northwestern University

6.1 Introduction

Computational geometry evolves from the classical discipline of design and analysis of algorithms, and has received a great deal of attention in the past two decades since its identification in 1975 by Shamos. It is concerned with the computational complexity of geometric problems that arise in various disciplines such as pattern recognition, computer graphics, computer vision, robotics, very large-scale integrated (VLSI) layout, operations research, statistics, etc. In contrast with the classical approach to proving mathematical theorems about geometry-related problems, this discipline emphasizes the computational aspect of these problems and attempts to exploit the underlying geometric properties possible, e.g., the metric space, to derive efficient algorithmic solutions.

The classical theorem, for instance, that a set S is convex if and only if for any $0 \leq \alpha \leq 1$ the convex combination $\alpha p + (1 - \alpha)q = r$ is in S for any pair of elements $p, q \in S$, is very fundamental in establishing convexity of a set. In geometric terms, a body S in the Euclidean space is convex if and only if the line segment joining any two points in S lies totally in S. But this theorem per se is not suitable for computational purposes as there are infinitely many possible pairs of points to be considered. However, other properties of convexity can be utilized to yield an algorithm. Consider the following problem. Given a simple closed Jordan polygonal curve, determine if the interior region enclosed by the curve is convex. This problem can be readily solved by observing that if the line segments defined by all pairs of vertices of the polygonal curve, $\overline{v_i, v_j}, i \neq j, 1 \leq i, j \leq n$, where n denotes the total number of vertices, lie totally inside the region, then the region is convex. This would yield a straightforward algorithm with time complexity $O(n^3)$, as there are $O(n^2)$ line segments, and to test if each line segment lies totally in the region takes $O(n)$ time by comparing it against every polygonal segment. As we shall show, this problem can be solved in $O(n)$ time by utilizing other geometric properties.

*This material is based on work supported in part by the National Science Foundation under Grant CCR-9309743 and by the Office of Naval Research under Grants N00014-93-1-0272 and N00014-95-1-1007.

At this point, an astute reader might have come up with an $O(n)$ algorithm by making the observation: Because the interior angle of each vertex must be strictly less than π in order for the region to be convex, we just have to check for every consecutive three vertices v_{i-1}, v_i, v_{i+1} that the angle at vertex v_i is less than π. (A vertex whose internal angle has a measure less than π is said to be *convex*; otherwise, it is said to be *reflex*.) One may just be content with this solution. Mathematically speaking, this solution is fine and indeed runs in $O(n)$ time. The problem is that the algorithm implemented in this straightforward manner without care may produce an incorrect answer when the input polygonal curve is ill formed. That is, if the input polygonal curve is not simple, i.e., it self-intersects, then the *enclosed* region by this closed curve is not well defined. The algorithm, without checking this simplicity condition, may produce a wrong answer. Note that the preceding observation that all of the vertices must be convex in order to have a convex region is only a necessary condition. Only when the input polygonal curve is *verified* to be simple will the algorithm produce a correct answer. But to verify whether the input polygonal curve self-intersects or not is no longer as straightforward. The fact that we are dealing with computer solutions to geometric problems may make the task of designing an algorithm and proving its correctness nontrivial.

An objective of this discipline in the theoretical context is to prove lower bounds of the complexity of geometric problems and to devise algorithms (giving upper bounds) whose complexity *matches* the lower bounds. That is, we are interested in the *intrinsic* difficulty of geometric computational problems under a certain computation model and at the same time are concerned with the algorithmic solutions that are provably optimal in the worst or average case. In this regard, the asymptotic time (or space) complexity of an algorithm is of interest. Because of its applications to various science and engineering related disciplines, researchers in this field have begun to address the efficacy of the algorithms, the issues concerning robustness and numerical stability [Fortune 1993], and the actual running times of their implementations.

In this chapter, we concentrate mostly on the theoretical development of this field in the context of sequential computation. Parallel computation geometry is beyond the scope of this chapter. We will adopt the *real* random access machine (RAM) model of computation in which all arithmetic operations, comparisons, kth-root, exponential or logarithmic functions take unit time. For more details refer to Edelsbrunner [1987], Mulmuley [1994], and Preparata and Shamos [1985]. We begin with a summary of problem solving techniques that have been developed [Lee and Preparata 1982, O'Rourke 1994, Yao 1994] and then discuss a number of topics that are central to this field, along with additional references for further reading about these topics.

6.2 Problem Solving Techniques

We give an example for each of the eight major problem-solving paradigms that are prevalent in this field. In subsequent sections we make reference to these techniques whenever appropriate.

Incremental Construction

This is the simplest and most intuitive method, also known as *iterative method*. That is, we compute the solution in an iterative manner by considering the input incrementally.

Consider the problem of computing the line arrangements in the plane. Given is a set $\mathcal{L}$ of n straight lines in the plane, and we want to compute the partition of the plane induced by $\mathcal{L}$. One obvious approach is to compute the partition iteratively by considering one line at a time [Chazelle et al. 1985]. As shown in Fig. 6.1, when line i is inserted, we need to traverse the regions that are intersected by the line and construct the new partition at the same time. One can show that the traversal and repartitioning of the intersected regions can be done in $O(n)$ time per insertion, resulting in a total of $O(n^2)$

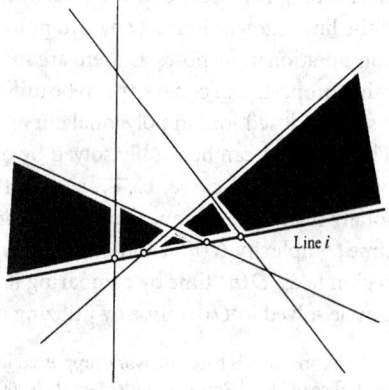

FIGURE 6.1 Incremental construction of line arrangement: phase i.

time. This algorithm is asymptotically optimal because the running time is proportional to the amount of space required to represent the partition. This incremental approach also generalizes to higher dimensions. We conclude with the theorem [Edelsbrunner et al. 1986].

Theorem 6.1. *The problem of computing the arrangement $\mathcal{A}(H)$ of a set H of n hyperplanes in $\Re^k$ can be solved iteratively in $O(n^k)$ time and space, which is optimal.*

Plane Sweep

This approach works most effectively for two-dimensional problems for which the solution can be computed incrementally as the entire input is scanned in a certain order. The concept can be easily generalized to higher dimensions [Bieri and Nef 1982]. This is also known as the *scan-line* method in computer graphics and is used for a variety of applications such as shading and polygon filling, among others.

Consider the problem of computing the *measure* of the union of n isothetic rectangles, i.e., whose sides are parallel to the coordinate axes. We would proceed with a *vertical* sweep line, sweeping across the plane from left to right. As we sweep the plane, we need to keep track of the rectangles that intersect the current sweep line and those that are yet to be visited. In the meantime we compute the area covered by the union of the rectangles seen so far. More formally, associated with this approach there are two basic data structures containing all *relevant* information that should be maintained.

1. *Event schedule* defines a sequence of *event points* that the sweep-line status will change. In this example, the sweep-line status will change only at the left and right boundary edges of each rectangle.
2. *Sweep-line status* records the information of the geometric structure that is being swept. In this example the sweep-line status keeps track of the set of rectangles intersecting the current sweep line.

The event schedule is normally represented by a *priority queue*, and the list of events may change dynamically. In this case, the events are static; they are the x-coordinates of the left and right boundary edges of each rectangle. The sweep-line status is represented by a suitable data structure that supports insertions, deletions, and computation of the partial solution at each event point. In this example a *segment tree* attributed to Bentley is sufficient [Preparata and Shamos 1985]. Because we are computing the area of the rectangles, we need to be able to know the *new* area covered by the current sweep line between two adjacent event points. Suppose at event point x_{i-1} we maintain a partial solution $\mathcal{A}_{i-1}$. In Fig. 6.2 the shaded area S needs to be added to the partial solution, that is, $\mathcal{A}_i = \mathcal{A}_{i-1} + S$. The shaded area is equal to the total measure, denoted sum_ℓ, of the union of vertical line segments representing the intersection of the rectangles and the current sweep line times the distance between the two event points x_i and x_{i-1}. If the next event corresponds to the left boundary of a rectangle, the corresponding vertical segment, $\overline{p,q}$ in Fig. 6.2, needs to be inserted to the segment tree. If the next event corresponds to a right boundary edge, the segment, $\overline{u,v}$ needs to be deleted from the segment tree. In either case, the total measure sum_ℓ should be updated accordingly. The correctness of this algorithm can be established by observing that the partial solution obtained for the rectangles to

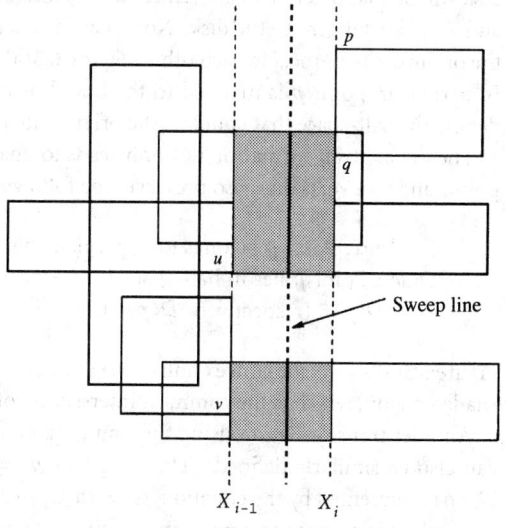

FIGURE 6.2 The plane-sweep approach to the measure problem in two-dimensions.

the *left* of the sweep line is maintained correctly. In fact, this property is typical of any algorithm based on the plane-sweep technique.

Because the segment tree structure supports segment insertions and deletions and the update (of sum_ℓ) operation in $O(\log n)$ time per event point, the total amount of time needed is $O(n \log n)$.

The measure of the union of rectangles in higher dimensions also can be solved by the plane-sweep technique with quad trees, a generalization of segment trees.

Theorem 6.2. *The problem of computing the measure of n isothetic rectangles in k dimensions can be solved in $O(n \log n)$ time, for $k \le 2$ and in $O(n^{k-1})$ time for $k \ge 3$.*

The time bound is asymptotically optimal. Even in one dimension, i.e., computing the total length of the union of n intervals requires $\Omega(n \log n)$ time (see Preparata and Shamos [1985]).

We remark that the sweep line used in this approach is not necessarily a straight line. It can be a topological line as long as the objects stored in the sweep line status are ordered, and the method is called *topological sweep* [Asano et al. 1994, Edelsbrunner and Guibas 1989]. Note that the measure of isothetic rectangles can also be solved using the divide-and-conquer paradigm to be discussed.

Geometric Duality

This is a geometric transformation that maps a given problem into its equivalent form, preserving certain geometric properties so as to manipulate the objects in a more convenient manner. We will see its usefulness for a number of problems to be discussed. Here let us describe a transformation in k-dimensions, known as *polarity* or *duality*, denoted $\mathcal{D}$, that maps d-dimensional varieties to $(k-1-d)$-dimensional varieties, $0 \le d < k$.

Consider any point $p = (\pi_1, \pi_2, \ldots, \pi_k) \in \Re^k$ other than the origin. The dual of p, denoted $\mathcal{D}(p)$, is the hyperplane $\pi_1 x_1 + \pi_2 x_2 + \cdots + \pi_k x_k = 1$. Similarly, a hyperplane that does not contain the origin is mapped to a point such that $\mathcal{D}(\mathcal{D}(p)) = p$. Geometrically speaking, point p is mapped to a hyperplane whose normal is the vector determined by p and the origin and whose distance to the origin is the reciprocal of that between p and the origin. Let S denote the unit sphere $S : x_1^2 + x_2^2 + \cdots + x_k^2 = 1$. If point p is external to S, then it is mapped to a hyperplane $\mathcal{D}(p)$ that intersects S at those points q that admit supporting hyperplanes h such that $h \cap S = q$ and $p \in h$. In two dimensions a point p outside of the unit disk will be mapped to a line intersecting the disk at two points, q_1 and q_2, such that line segments $\overline{p, q_1}$ and $\overline{p, q_2}$ are tangent to the disk. Note that the distances from p to the origin and from the line $\mathcal{D}(p)$ to the origin are reciprocal to each other. Figure 6.3(a) shows the duality transformation in two dimensions. In particular, point p is mapped to the line shown in boldface. For each hyperplane $\mathcal{D}(p)$, let $\mathcal{D}(p)^+$ denote the half-space that contains the origin and let $\mathcal{D}(p)^-$ denote the other half-space.

The duality transformation not only leads to dual arrangements of hyperplanes and configurations of points and vice versa, but also preserves the following properties.

> *Incidence:* Point p belongs to hyperplane h if and only if point $\mathcal{D}(h)$ belongs to hyperplane $\mathcal{D}(p)$.
> *Order:* Point p lies in half-space h^+ (respectively, h^-) if and only if point $\mathcal{D}(h)$ lies in half-space $\mathcal{D}(p)^+$ (respectively, $\mathcal{D}(p)^-$).

Figure 6.3(a) shows the convex hull of a set of points that are mapped by the duality transformation to the shaded region, which is the common intersection of the half-planes $\mathcal{D}(p)^+$ for all points p.

Another transformation using the unit paraboloid U, represented as $U : x_k = x_1^2 + x_2^2 + \cdots + x_{k-1}^2$, can also be similarly defined. That is, point $p = (\pi_1, \pi_2, \ldots, \pi_k) \in R^k$ is mapped to a hyperplane $\mathcal{D}_u(p)$ represented by the equation $x_k = 2\pi_1 x_1 + 2\pi_2 x_2 + \cdots + 2\pi_{k-1} x_{k-1} - \pi_k$. And each nonvertical hyperplane is mapped to a point in a similar manner such that $\mathcal{D}_u(\mathcal{D}_u(p)) = p$. Figure 6.3(b) illustrates the two-dimensional case, in which point p is mapped to a line shown in boldface. For more details see, e.g., Edelsbrunner [1987] and Preparata and Shamos [1985].

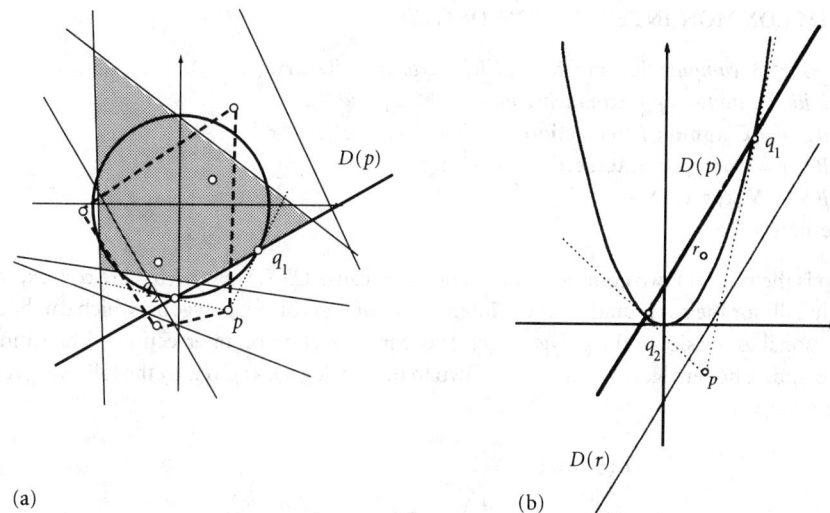

(a)   (b)

FIGURE 6.3   Geometric duality transformation in two dimensions.

Locus

This approach is often used as a preprocessing step for a geometric *searching* problem to achieve faster query-answering response time. For instance, given a *fixed* database consisting of geographical locations of post offices, each represented by a point in the plane, one would like to be able to efficiently answer queries of the form: "what is the nearest post office to location q?" for some query point q. The locus approach to this problem is to partition the plane into n regions, each of which consists of the locus of *query* points for which the *answer* is the same. The partition of the plane is the so-called *Voronoi* diagram discussed subsequently. In Fig. 6.7, the post office closest to query point q is site s_i. Once the Voronoi diagram is available, the query problem reduces to that of locating the region that contains the query, an instance of the point-location problem discussed in section 6.3.

Divide-and-Conquer

This is a classic problem-solving technique and has proven to be very powerful for geometric problems as well. This technique normally involves partition-ing of the given problem into several subproblems, recursively solving each subproblem, and then com-bining the solutions to each of the subproblems to obtain the final solution to the original problem. We illustrate this paradigm by considering the problem of computing the common intersection of n half-planes in the plane. Given is a set S of n half-planes, h_i, represented by $a_i x + b_i y \le c_i, i = 1, 2, \ldots, n$. It is well known that the common intersection of half-planes, denoted $CI(S) = \bigcap_{i=1}^{n} h_i$, is a con-vex set, which may or may not be bounded. If it is bounded, it is a convex polygon. See Fig. 6.4, in which the shaded area is the common intersection.

The divide-and-conquer paradigm consists of the following steps.

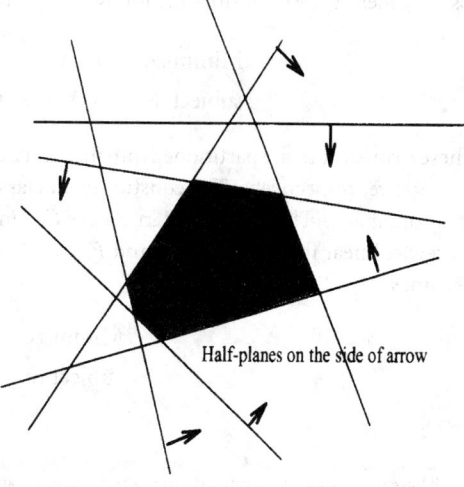

Half-planes on the side of arrow

FIGURE 6.4   The common intersection of half-planes.

ALGORITHM COMMON_INTERSECTION_D&C (S).

1. If $|S| \leq 3$, *compute the intersection CI(S) explicitly.* **Return** $(CI(S))$.
2. *Divide S into two approximately equal subsets S_1 and S_2.*
3. $CI(S_1) = $ **Common_Intersection_D&C**(S_1).
4. $CI(S_2) = $ **Common_Intersection_D&C**(S_2).
5. $CI(S) = $ **Merge**$(CI(S_1), CI(S_2))$.
6. **Return** $(CI(S))$.

The key step is the *merge* of two common intersections. Because $CI(S_1)$ and $CI(S_2)$ are convex, the merge step basically calls for the computation of the intersection of two convex polygons, which can be solved in time proportional to the size of the polygons (cf. subsequent section on intersection). The running time of the divide-and-conquer algorithm is easily shown to be $O(n \log n)$, as given by the following recurrence formula, where $n = |S|$:

$$T(3) = O(1)$$

$$T(n) = 2T\left(\frac{n}{2}\right) + O(n) + M\left(\frac{n}{2}, \frac{n}{2}\right)$$

where $M(n/2, n/2) = O(n)$ denotes the merge time (step 5).

Theorem 6.3. *The common intersection of n half-planes can be solved in $O(n \log n)$ time by the divide-and-conquer method.*

The time complexity of the algorithm is asymptotically optimal, as the problem of sorting can be reduced to it [Preparata and Shamos 1985].

Prune-and-Search

This approach, developed by Dyer [1986] and Megiddo [1983a, 1983b, 1984], is a very powerful method for solving a number of geometric optimization problems, one of which is the well-known linear programming problem. Using this approach, they obtained an algorithm whose running time is linear in the number of constraints. For more development of linear programming problems, see Megiddo [1983c, 1986]. The main idea is to prune away a fraction of *redundant* input constraints in each iteration while searching for the solution. We use a two-dimensional linear programming problem to illustrate this approach. Without loss of generality, we consider the following linear programming problem:

$$\begin{aligned} \text{Minimize} \quad & Y \\ \text{subject to} \quad & \alpha_i X + \beta_i Y + \gamma_i \leq 0, \quad i = 1, 2, \ldots, n \end{aligned}$$

These n constraints are partitioned into three classes, C_0, C_+, C_-, depending on whether β_i is zero, positive, or negative, respectively. The constraints in class C_0 define an X-interval $[x_1, x_2]$, which constrains the solution, if any. The constraints in classes C_+ and C_- define, however, upward- and downward-convex piecewise linear functions $F_+(X)$ and $F_-(X)$ delimiting the feasible region[1] (Fig. 6.5). The problem now becomes

$$\begin{aligned} \text{Minimize} \quad & F_-(X) \\ \text{subject to} \quad & F_-(X) \leq F_+(X) \\ & x_1 \leq X \leq x_2 \end{aligned}$$

---

[1] These upward- and downward-convex functions are also known as the upper and lower *envelopes* of the line arrangements for lines belonging to classes C_- and C_+, respectively.

Let λ^* denote the optimal solution, if it exists. The values of $F_-(\lambda)$ and $F_+(\lambda)$ for any λ can be computed in $O(n)$ time, based on the slopes $-\alpha_i/\beta_i$. Thus, in $O(n)$ time one can determine for any $\lambda' \in [x_1, x_2]$ if (1) λ' is infeasible, and there is no solution, (2) λ' is infeasible, and we know a feasible solution is less or greater than λ', (3) $\lambda' = \lambda^*$, or (4) λ' is feasible, and whether λ^* is less or greater than λ'.

To choose λ' we partition constraints in classes C_- and C_+ into pairs and find the abscissa $\lambda_{i,j}$ of their intersection. If $\lambda_{i,j} \notin [x_1, x_2]$ then one of the constraints can be eliminated as redundant. For those $\lambda_{i,j}$ that are in $[x_1, x_2]$ we find in $O(n)$ time [Dobkin and Munro 1981] the median $\lambda'_{i,j}$ and com-

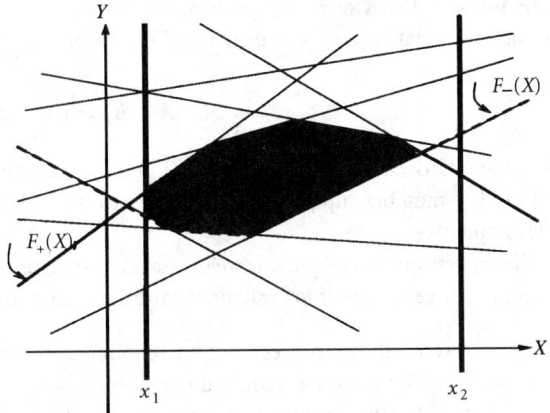

FIGURE 6.5 Feasible region defined by upward- and downward-convex piecewise linear functions.

pute $F_-(\lambda'_{i,j})$ and $F_+(\lambda'_{i,j})$. By the preceding arguments that we can determine where λ^* should lie, we know one-half of the $\lambda_{i,j}$ do not lie in the region containing λ^*. Therefore, one constraint of the corresponding pair can be eliminated. The process iterates. In other words, in each iteration at least a fixed fraction $\delta = 1/4$ of the current constraints can be eliminated. Because each iteration takes $O(n)$ time, the total time spent is $Cn + C\delta n + \cdots = O(n)$. In higher dimensions, we have the following result due to Dyer [1986] and Clarkson [1986].

Theorem 6.4. *A linear program in k-dimensions with n constraints can be solved in $O(3^{k^2}n)$ time.*

We note here some of the new recent developments for linear programming. There are several randomized algorithms for this problem, of which the best expected complexity, $O(k^2 n + k^{k/2+O(1)} \log n)$ is due to Clarkson [1988], which is later improved by Matoušek et al. [1992] to run in $O(k^2 n + e^{O(\sqrt{k \ln k})} \log n)$. Clarkson's [1988] algorithm is applicable to work in a general framework, which includes various other geometric optimization problems, such as *smallest enclosing ellipsoid*. The best known deterministic algorithm for linear programming is due to Chazelle and Matoušek [1993], which runs in $O(k^{7k+o(k)}n)$ time.

Dynamization

Techniques have been developed for query-answering problems, classified as *geometric searching* problems, in which the underlying database is changing over (discrete) time. A typical geometric searching problem is the *membership* problem, i.e., given a set $\mathcal{D}$ of objects, determine if x is a member of $\mathcal{D}$, or the *nearest neighbor searching* problem, i.e., given a set $\mathcal{D}$ of objects, find an object that is closest to x according to some distance measure. In the database area, these two problems are referred to as the *exact match* and *best match* queries. The idea is to make use of good data structures for a static database and enhance them with dynamization mechanisms so that updates of the database can be accommodated on line and yet queries to the database can be answered efficiently.

A general query Q contains a variable of type $T1$ and is asked of a set of objects of type $T2$. The answer to the query is of type $T3$. More formally, Q can be considered as a mapping from $T1$ and subsets of $T2$ to $T3$, that is, $Q: T1 \times 2^{T2} \rightarrow T3$. The class of geometric searching problems to which the dynamization techniques are applicable is the class of *decomposable searching problems* [Bentley and Saxe 1980].

Definition 6.1. A searching problem with query Q is decomposable if there exists an efficiently computable associative, and communtative binary operator @ satisfying the condition

$$Q(x, A \cup B) = @(Q(x, A), Q(x, B))$$

In other words, the answer to a query Q in $\mathcal{D}$ can be computed by the answers to two subsets $\mathcal{D}_1$ and $\mathcal{D}_2$ of $\mathcal{D}$. The membership problem and the nearest-neighbor searching problem previously mentioned are decomposable.

To answer queries efficiently, we have a data structure to support various update operations. There are typically three measures to evaluate a static data structure $\mathcal{A}$. They are:

1. $P_{\mathcal{A}}(N)$, the preprocessing time required to build $\mathcal{A}$
2. $S_{\mathcal{A}}(N)$, the storage required to represent $\mathcal{A}$
3. $Q_{\mathcal{A}}(N)$, the query response time required to search in $\mathcal{A}$

where N denotes the number of elements represented in $\mathcal{A}$. One would add another measure $U_{\mathcal{A}}(N)$ to represent the *update* time.

Consider the nearest-neighbor searching problem in the Euclidean plane. Given a set of n points in the plane, we want to find the nearest neighbor of a query point x. One can use the Voronoi diagram data structure $\mathcal{A}$ (cf. subsequent section on Voronoi diagrams) and point location scheme (cf. subsequent section on point location) to achieve the following: $P_{\mathcal{A}}(n) = O(n \log n)$, $S_{\mathcal{A}}(n) = O(n)$, and $Q_{\mathcal{A}}(n) = O(\log n)$. We now convert the static data structure $\mathcal{A}$ to a dynamic one, denoted $\mathcal{D}$, to support insertions and deletions as well. There are a number of dynamization techniques, but we describe the technique developed by van Leeuwan and Wood [1980] that provides the general flavor of the approach.

The general principle is to decompose $\mathcal{A}$ into a collection of separate data structures so that each update can be confined to one or a small, fixed number of them; however, to avoid degrading the query response time we cannot afford to have excessive fragmentation because queries involve the entire collection.

Let $\{x_k\}_{k \geq 1}$ be a sequence of increasing integers, called *switch points*, where x_k is divisible by k and $x_{k+1}/(k + 1) > x_k/k$. Let $x_0 = 0$, $y_k = x_k/k$, and n denote the current size of the point set. For a given *level* k, $\mathcal{D}$ consists of $(k + 1)$ static structures of the same type, one of which, called *dump* is designated to allow for insertions. Each substructure $\mathcal{B}$ has size $y_k \leq s(\mathcal{B}) \leq y_{k+1}$, and the dump has size $0 \leq s(dump) < y_{k+1}$. A block $\mathcal{B}$ is called *low* or *full* depending on whether $s(\mathcal{B}) = y_k$ or $s(\mathcal{B}) = y_{k+1}$, respectively, and is called *partial* otherwise. When an insertion to the dump makes its size equal to y_{k+1}, it becomes a full block and any nonfull block can be used as the dump. If all blocks are full, we switch to the next level. Note that at this point the total size is $y_{k+1} * (k + 1) = x_{k+1}$. That is, at the beginning of level $k + 1$, we have $k + 1$ low blocks and we create a new dump, which has size 0. When a deletion from a low block occurs, we need to borrow an element either from the dump, if it is not empty, or from a partial block. When all blocks are low and $s(dump) = 0$, we switch to level $k - 1$, making the low block from which the latest deletion occurs the *dump*. The level switching can be performed in $O(1)$ time. We have the following:

Theorem 6.5. *Any static data structure $\mathcal{A}$ used for a decomposable searching problem can be transformed into a dynamic data structure $\mathcal{D}$ for the same problem with the following performance. For $x_k \leq n < x_{k+1}$, $Q_{\mathcal{D}}(n) = O(k Q_{\mathcal{A}}(y_{k+1}))$, $U_{\mathcal{D}}(n) = O(C(n) + U_{\mathcal{A}}(y_{k+1}))$, and $S_{\mathcal{D}}(n) = O(k S_{\mathcal{A}}(y_{k+1}))$, where $C(n)$ denotes the time needed to look up the block which contains the data when a deletion occurs.*

If we choose, for example, x_k to be the first multiple of k greater than or equal to 2^k, that is, $k = \log_2 n$, then y_k is about $n/\log_2 n$. Because we know there exists an $\mathcal{A}$ with $Q_{\mathcal{A}}(n) = O(\log n)$ and $U_{\mathcal{A}}(n) = P_{\mathcal{A}}(n) = O(n \log n)$, we have the following corollary.

Corollary 6.1. *The nearest-neighbor searching problem in the plane can be solved in $O(\log^2 n)$ query time and $O(n)$ update time. [Note that $C(n)$ in this case is $O(\log n)$.]*

There are other dynamization schemes that exhibit various query-time/space and query-time/update-time tradeoffs. The interested reader is referred to Chiang and Tamassia [1992], Edelsbrunner [1987], Mehlhorn [1984], Overmars [1983], and Preparata and Shamos [1985] for more information.

Random Sampling

Randomized algorithms have received a great deal of attention recently because of their potential applications. See Chapter 4 for more information. For a variety of geometric problems, randomization techniques help in building geometric subdivisions and data structures to quickly answer queries about such subdivisions. The resulting randomized algorithms are simpler to implement and/or asymptotically faster than those previously known. It is important to note that the focus of randomization is *not* on random input, such as a collection of points randomly chosen uniformly and independently from a region. We are concerned with algorithms that use a source of random numbers and analyze their performance for an arbitrary input. Unlike *Monte Carlo* algorithms, whose output may be incorrect (with very low probability), the randomized algorithms, known as *Las Vegas* algorithms, considered here are guaranteed to produce a correct output.

There are a good deal of newly developed randomized algorithms for geometric problems. See Du and Hwang [1992] for more details. Randomization gives a general way to divide and conquer geometric problems and can be used for both parallel and serial computation. We will use a familiar example to illustrate this approach.

Let us consider the problem of nearest-neighbor searching discussed in the preceding subsection. Let $\mathcal{D}$ be a set of n points in the plane and q be the query point. A simple approach to this problem is:

ALGORITHM S

- *Compute the distance to q for each point $p \in \mathcal{D}$.*
- *Return the point p whose distance is the smallest.*

It is clear that Algorithm S, requiring $O(n)$ time, is not suitable if we need to answer many queries of this type. To obtain faster query response time one can use the technique discussed in the preceding subsection. An alternative is to use the *random sampling* technique as follows. We pick a random sample, a subset $\mathcal{R} \subset \mathcal{D}$ of size r. Let point $p \in \mathcal{R}$ be the nearest neighbor of q in $\mathcal{R}$. The open disk $K_{\mathcal{R}}(q)$ centered at q and passing through p does not contain any other point in $\mathcal{R}$. The answer to the query is either p or some point of $\mathcal{D}$ that lies in $K_{\mathcal{R}}(q)$.

We now extend the above observation to a finite region G in the plane. Let $K_{\mathcal{R}}(G)$ be the union of disks $K_{\mathcal{R}}(r)$ for all $r \in G$. If a query q lies in G, the nearest neighbor of q must be in $K_{\mathcal{R}}(G)$ or in $\mathcal{R}$. Let us consider the Voronoi diagram, $\mathcal{V}(\mathcal{R})$ of $\mathcal{R}$ and a triangulation, $\Delta(\mathcal{V}(\mathcal{R}))$. For each triangle T with vertices a, b, c of $\Delta(\mathcal{V}(\mathcal{R}))$ we have $K_{\mathcal{R}}(T) = K_{\mathcal{R}}(a) \cup K_{\mathcal{R}}(b) \cup K_{\mathcal{R}}(c)$, shown as the shaded area in Fig. 6.6. A probability lemma [Clarkson 1988] shows that with probability at least $1 - O(1/n^2)$

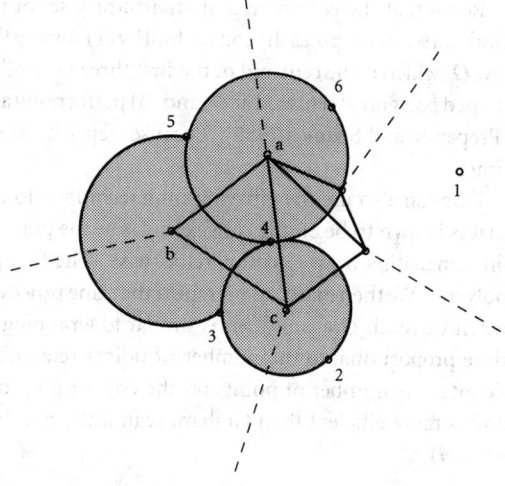

FIGURE 6.6 A triangulation of the Voronoi diagram of six sites and $K_{\mathcal{R}}(T)$, $T = \Delta(a, b, c)$.

the candidate set $\mathcal{D} \cap K_{\mathcal{R}}(T)$ for all $T \in \Delta(\mathcal{V}(\mathcal{R}))$ contains $O(\log n)n/r$ points. More precisely, if $r > 5$ then with probability at least $1 - e^{-C/2+3\ell_{n}r}$ each open disk $K_{\mathcal{R}}(r)$ for $r \in \mathcal{R}$ contains no more than Cn/r points of $\mathcal{D}$. If we choose r to be $\sqrt{n}$, the query time becomes $O(\sqrt{n}\log n)$, a speedup from Algorithm S. If we apply this scheme recursively to the candidate sets of $\Delta(\mathcal{V}(\mathcal{R}))$, we can get a query time $O(\log n)$ [Clarkson 1988].

There are many applications of these random sampling techniques. Derandomized algorithms were also developed. See, e.g., Chazelle and Friedman [1990] for a deterministic view of random sampling and its use in geometry.

6.3 Classes of Problems

In this section we aim to touch upon classes of problems that are fundamental in this field and describe solutions to them, some of which may be nontrivial. The reader who needs further information about these problems is strongly encouraged to refer to the original articles cited in the references.

Convex Hull

The convex hull of a set of points in $\mathfrak{R}^k$ is the most fundamental problem in computational geometry. Given is a set of points, and we are interested in computing its convex hull, which is defined to be the smallest convex body containing these points. Of course, the first question one has to answer is how to represent the convex hull. An implicit representation is just to list all of the extreme points,[2] whereas an explicit representation is to list all of the extreme d-faces of dimensions $d = 0, 1, \ldots, k - 1$. Thus, the complexity of any convex hull algorithm would have two parts, computation part and the output part. An algorithm is said to be *output sensitive* if its complexity depends on the size of output.

Definition 6.2. The convex hull of a set S of points in $\mathfrak{R}^k$ is the smallest convex set containing S. In two dimensions, the convex hull is a convex polygon containing S; in three dimensions it is a convex polyhedron.

Convex Hulls in Two and Three Dimensions

For an arbitrary set of n points in two and three dimensions, we can compute the convex hull using the *Graham scan*, *gift-wrapping*, or *divide-and-conquer* paradigm, which are briefly described next.

Recall that the convex hull of an arbitrary set of points in two dimensions is a convex polygon. The Graham scan computes the convex hull by (1) sorting the input set of points with respect to an interior point, say, O, which is the centroid of the first three noncollinear points, (2) connecting these points into a star-shaped polygon P centered at O, and (3) performing a linear scan to compute the convex hull of the polygon [Preparata and Shamos 1985]. Because step 1 is the dominating step, the Graham scan, takes $O(n \log n)$ time.

One can also use the gift-wrapping technique to compute the convex polygon. Starting with a vertex that is known to be on the convex hull, say, the point O, with the smallest y-coordinate, we sweep a half-line emanating from O counterclockwise. The first point v_1 we hit will be the next point on the convex polygon. We then march to v_1, repeat the same process, and find the next vertex v_2. This process terminates when we reach O again. This is similar to wrapping an object with a *rope*. Finding the next vertex takes time proportional to the number of points remaining. Thus, the total time spent is $O(n\mathcal{H})$, where $\mathcal{H}$ denotes the number of points on the convex polygon. The gift-wrapping algorithm is output sensitive and is more efficient than Graham scan if the number of points on the convex polygon is small, that is, $o(\log n)$.

---

[2]A point in S is an extreme point if it cannot be expressed as a convex combination of other points in S. In other words, the convex hull of S would change when an extreme point is removed from S.

One can also use the divide-and-conquer paradigm. As mentioned previously, the key step is the merge of two convex hulls, each of which is the solution to a subproblem derived from the recursive step. In the division step, we can recursively separate the set into two subsets by a vertical line L. Then the merge step basically calls for computation of two common tangents of these two convex polygons. The computation of the common tangents, also known as *bridges* over line L, begins with a segment connecting the rightmost point l of the left convex polygon to the leftmost point r of the right convex polygon. Advancing the endpoints of this segment in a *zigzag* manner we can reach the top (or the bottom) common tangent such that the entire set of points lies on one side of the line containing the tangent. The running time of the divide-and-conquer algorithm is easily shown to be $O(n \log n)$.

A more sophisticated output-sensitive and optimal algorithm, which runs in $O(n \log \mathcal{H})$ time, has been developed by Kirkpatrick and Seidel [1986]. It is based on a variation of the divide-and-conquer paradigm. The main idea in achieving the optimal result is that of eliminating *redundant* computations. Observe that in the divide-and-conquer approach after the common tangents are obtained, some vertices that used to belong to the left and right convex polygons must be deleted. Had we known these vertices were not on the final convex hull, we could have saved time by not computing them. Kirkpatrick and Seidel capitalized on this concept and introduced the *marriage-before-conquest* principle. They construct the convex hull by computing the upper and lower hulls of the set; the computations of these two hulls are symmetric. It performs the *divide* step as usual that decomposes the problem into two subproblems of approximately equal size. Instead of computing the upper hulls recursively for each subproblem, it finds the common tangent segment of the two yet-to-be-computed upper hulls and proceeds recursively. One thing that is worth noting is that the points known not to be on the (convex) upper hull are discarded before the algorithm is invoked recursively. This is the key to obtaining a time bound that is both output sensitive and asymptotically optimal.

The divide-and-conquer scheme can be easily generalized to three dimensions. The merge step in this case calls for computing common supporting faces that *wrap* two recursively computed convex polyhedra. It is observed by Preparata and Hong that the common supporting faces are computed from connecting two *cyclic* sequences of edges, one on each polyhedron [Preparata and Shamos 1985]. The computation of these supporting faces can be accomplished in linear time, giving rise to an $O(n \log n)$ time algorithm. By applying the marriage-before-conquest principle Edelsbrunner and Shi [1991] obtained an $O(n \log^2 \mathcal{H})$ algorithm.

The gift-wrapping approach for computing the convex hull in three dimensions would mimic the process of wrapping a gift with a piece of paper and has a running time of $O(n\mathcal{H})$.

Convex Hulls in k-Dimensions, $k > 3$

For convex hulls of higher dimensions, a recent result by Chazelle [1993] showed that the convex hull can be computed in time $O(n \log n + n^{\lfloor k/2 \rfloor})$, which is optimal in all dimensions $k \geq 2$ in the worst case. But this result is insensitive to the output size. The gift-wrapping approach generalizes to higher dimensions and yields an output-sensitive solution with running time $O(n\mathcal{H})$, where $\mathcal{H}$ is the total number of i-faces, $i = 0, 1, \ldots, k - 1$, and $\mathcal{H} = O(n^{\lfloor k/2 \rfloor})$ [Edelsbrunner 1987]. One can also use the *beneath-beyond* method of adding points one at a time in ascending order along one of the coordinate axes.[3] We compute the convex hull $CH(S_{i-1})$ for points $S_{i-1} = \{p_1, p_2, \ldots, p_{i-1}\}$. For each added point p_i, we update $CH(S_{i-1})$ to get $CH(S_i)$, for $i = 2, 3, \ldots, n$, by deleting those t-faces, $t = 0, 1, \ldots, k - 1$, that are internal to $CH(S_{i-1} \cup \{p_i\})$. It is shown by Seidel (see Edelsbrunner [1987])that $O(n^2 + \mathcal{H} \log n)$ time is sufficient. Most recently Chan [1995] obtained an algorithm based on the gift-wrapping method that runs in $O(n \log \mathcal{H} + (n\mathcal{H})^{1-1/(\lfloor k/2 \rfloor+1)} \log^{O(1)} n)$ time. Note that the algorithm is optimal when $k = 2, 3$. In particular, it is optimal when $\mathcal{H} = o(n^{1-\epsilon})$ for some $0 < \epsilon < 1$.

We conclude this subsection with the following theorem [Chan 1995].

---

[3]If the points of S are not given a priori, the algorithm can be made *on line* by adding an extra step of checking if the newly added point is internal or external to the current convex hull. If internal, just discard it.

Theorem 6.6. *The convex hull of a set S of n points in $\Re^k$ can be computed in $O(n \log \mathcal{H})$ time for $k = 2$ or $k = 3$, and in $O(n \log \mathcal{H} + (n\mathcal{H})^{1-1/(\lfloor k/2 \rfloor+1)} \log^{O(1)} n)$ time for $k > 3$, where $\mathcal{H}$ is the number of i-faces, $i = 0, 1, \ldots, k - 1$.*

Proximity

In this subsection we address proximity related problems.

Closest Pair

Consider a set S of n points in $\Re^k$. The closest pair problem is to find in S a pair of points whose distance is the minimum, i.e., find p_i and p_j, such that $d(p_i, p_j) = \min_{k \neq l}\{d(p_k, p_l)$, for all points $p_k, p_l \in S\}$, where $d(a, b)$ denotes the Euclidean distance between a and b. (The subsequent result holds for any distance metric in Minkowski's norm.) The brute force method takes $O(d \cdot n^2)$ time by computing all $O(n^2)$ interpoint distances and taking the minimum; the pair that gives the minimum distance is the closest pair. In one dimension, the problem can be solved by sorting these points and then scanning them in order, as the two closest points must occur consecutively. And this problem has a lower bound of $\Omega(n \log n)$ even in one dimension following from a linear time transformation from the *element uniqueness problem*. See Preparata and Shamos [1985].

But sorting is not applicable for dimension $k > 1$. Indeed this problem can be solved in optimal time $O(n \log n)$ by using the divide-and-conquer approach as follows. Let us first consider the case when $k = 2$. Consider a vertical cutting line λ that divides S into S_1 and S_2 such that $|S_1| = |S_2| = n/2$. Let δ_i be the minimum distance defined by points in S_i, $i = 1, 2$. Observe that the minimum distance defined by points in S can be either δ_1, δ_2, or defined by two points, one in each set. In the former case, we are done. In the latter, these two points must lie in the vertical strip of width $\delta = \min\{\delta_1, \delta_2\}$ on each side of the cutting line λ. The problem now reduces to that of finding the closest pair between points in S_1 and S_2 that lie inside the strip of width 2δ. This subproblem has a special property, known as the *sparsity* condition, i.e., the number of points in a box[4] of length 2δ is bounded by a constant $c = 4 \cdot 3^{k-1}$, because in each set S_i, there exists no point that lies in the interior of the δ-ball centered at each point in S_i, $i = 1, 2$ [Preparata and Shamos 1985]. It is this sparsity condition that enables us to solve the bichromatic closest pair problem (cf. the following subsection for more information) in $O(n)$ time. Let $\overline{S}_i \subseteq S_i$ denote the set of points that lies in the vertical strip. In two dimensions, the sparsity condition ensures that for each point in $\overline{S}_1$ the number of candidate points in $\overline{S}_2$ for the closest pair is at most 6. We therefore can scan these points $\overline{S}_1 \cup \overline{S}_2$ in order along the cutting line λ and compute the distance between each point scanned and its six candidate points. The pair that gives the minimum distance δ_3 is the bichromatic closest pair. The minimum distance of all pairs of points in S is then equal to $\delta_S = \min\{\delta_1, \delta_2, \delta_3\}$.

Since the merge step takes linear time, the entire algorithm takes $O(n \log n)$ time. This idea generalizes to higher dimensions, except that to ensure the sparsity condition the cutting hyperplane should be appropriately chosen to obtain an $O(n \log n)$ algorithm [Preparata and Shamos 1985].

Bichromatic Closest Pair

Given two sets of *red* and *blue* points, denoted R and B, respectively, find two points, one in R and the other in B, that are closest among all such mutual pairs.

The special case when the two sets satisfy the sparsity condition defined previously can be solved in $O(n \log n)$ time, where $n = |R| + |B|$. In fact a more general problem, known as *fixed radius all nearest-neighbor problem in a sparse set* [Bentley 1980, Preparata and Shamos 1985], i.e., given a set M of points in $\Re^k$ that satisfies the sparsity condition, find all pairs of points whose distance is less than a given parameter δ, can be solved in $O(|M| \log |M|)$ time [Preparata and Shamos 1985]. The bichromatic closest pair problem in general, however, seems quite difficult. Agarwal et al. [1991] gave an $O(n^{2(1-1/(\lceil k/2 \rceil+1))+\epsilon})$

---

[4]A box is also known as a hypercube.

time algorithm and a randomized algorithm with an expected running time of $O(n^{4/3} \log^c n)$ for some constant c. Chazelle et al. [1993] gave an $O(n^{2(1-1/(\lfloor k/2 \rfloor +1))+\epsilon})$ time algorithm for the bichromatic farthest pair problem, which can be used to find the diameter of a set S of points by setting $R = B = S$.

A lower bound of $\Omega(n \log n)$ for the bichromatic closest pair problem can be established. (See e.g., Preparata and Shamos [1985].) However, when the two sets are given as two simple polygons, the bichromatic closest pair problem can be solved relatively easily. Two problems can be defined. One is the *closest visible vertex pair* problem, and the other is the *separation problem*. In the former, one looks for a red–blue pair of vertices that are visible to each other and are the closest; in the latter, one looks for two boundary points that have the shortest distance. Both the closest visible vertex pair problem and the separation problem can be solved in linear time [Amato 1994, 1995]. But if both polygons are convex, the separation problem can be solved in $O(\log n)$ time [Chazelle and Dobkin 1987, Edelsbrunner 1985].

Additional references about different variations of closest pair problems can be found in Bespamyatnikh [1995], Callahan and Kosaraju [1995], Kapoor and Smid [1996], Schwartz et al. [1994], and Smid [1992].

Voronoi Diagrams

The Voronoi diagram $\mathcal{V}(S)$ of a set S of points, called *sites*, $S = \{s_1, s_2, \dots, s_n\}$ in $\mathfrak{R}^k$ is a partition of $\mathfrak{R}^k$ into Voronoi cells $V(s_i)$, $i = 1, 2, \dots, n$, such that each cell contains points that are closer to site s_i than to any other site s_j, $j \neq i$, i.e.,

$$V(s_i) = \{x \in \mathfrak{R}^k \mid d(x, s_i) \leq d(x, s_j) \forall s_j \in \mathfrak{R}^k, j \neq i\}$$

Figure 6.7(a) shows the Voronoi diagram of 16 point sites in two dimensions. Figure 6.7(b) shows the straight-line dual graph of the Voronoi diagram, which is called the Delaunay triangulation.

In two dimensions, $\mathcal{V}(S)$ is a planar graph and is of size linear in $|S|$. In dimensions $k \geq 2$, the total number of d-faces of dimensions $d = 0, 1, \dots, k-1$, in $\mathcal{V}(S)$ is $O(n^{\lceil d/2 \rceil})$.

Construction of Voronoi Diagram in Two Dimensions. The Voronoi diagram possesses many properties that are proximity related. For instance, the closest pair problem for S can be solved in linear time after the Voronoi diagram has been computed. Because this pair of points must be adjacent in the Delaunay

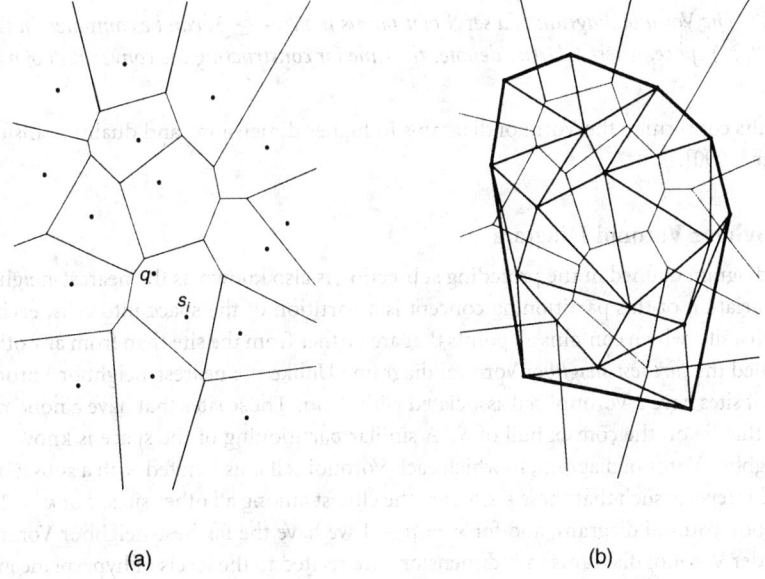

(a)   (b)

FIGURE 6.7 The Voronoi diagram of a set of 16 points in the plane.

triangulation, all one has to do is examine all adjacent pairs of points and report the pair with the smallest distance. A divide-and-conquer algorithm to compute the Voronoi diagram of a set of points in the Euclidean plane was first given by Shamos and Hoey and generalized by Lee to L_p-metric for all $1 \leq p \leq \infty$ [Preparata and Shamos 1985]. A *plane-sweep* technique for constructing the diagram is proposed by Fortune [1987] that runs in $O(n \log n)$ time. There is a rich body of literature concerning the Voronoi diagram. The interested reader is referred to a recent survey by Fortune in Du and Hwang [1992, pp. 192–234].

Although $\Omega(n \log n)$ is the lower bound for computing the Voronoi diagram for an arbitrary set of n sites, this lower bound does not apply to special cases, e.g., when the sites are on the vertices of a convex polygon. In fact the Voronoi diagram of a convex polygon can be computed in linear time [Aggarwal et al. 1989]. This demonstrates further that an additional property of the input is to help reduce the complexity of the problem.

Construction of Voronoi Diagrams in Higher Dimensions. The Voronoi diagrams in $\Re^k$ are related to the convex hulls $\Re^{k+1}$ via a geometric transformation similar to duality discussed earlier in the subsection on geometric duality. Consider a set of n sites in $\Re^k$, which is the hyperplane $\mathcal{H}^0$ in $\Re^{k+1}$ such that $x_{k+1} = 0$, and a paraboloid $\mathcal{P}$ in $\Re^{k+1}$ represented as $x_{k+1} = x_1^2 + x_2^2 + \cdots + x_k^2$. Each site $s_i = (\mu_1, \mu_2, \ldots, \mu_k)$ is transformed into a hyperplane $\mathcal{H}(s_i)$ in $\Re^{k+1}$ denoted as

$$x_{k+1} = 2 \sum_{j=1}^{k} \mu_j x_j - \left(\sum_{j=1}^{k} \mu_j^2 \right)$$

That is, $\mathcal{H}(s_i)$ is tangent to the paraboloid $\mathcal{P}$ at a point $\mathcal{P}(s_i) = (\mu_1, \mu_2, \ldots, \mu_k, \mu_1^2 + \mu_2^2 + \cdots + \mu_k^2)$, which is just the vertical projection of site s_i onto the paraboloid $\mathcal{P}$. The half-space defined by $\mathcal{H}(s_i)$ and containing the paraboloid $\mathcal{P}$ is denoted as $\mathcal{H}^+(s_i)$. The intersection of all half-spaces, $\bigcap_{i=1}^{n} \mathcal{H}^+(s_i)$ is a convex body, and the boundary of the convex body is denoted $CH(\mathcal{H}(S))$. Any point $p \in \Re^k$ lies in the Voronoi cell $V(s_i)$ if the vertical projection of p onto $CH(\mathcal{H}(S))$ is contained in $\mathcal{H}(s_i)$. In other words, every κ-face of $CH(\mathcal{H}(S))$ has a vertical projection on the hyperplane $\mathcal{H}^0$ equal to the κ-face of the Voronoi diagram of S in $\mathcal{H}^0$.

We thus obtain the result which follows from Theorem 6.6 [Edelsbrunner 1987].

Theorem 6.7. *The Voronoi diagram of a set S of n points in $\Re^k$, $k \geq 3$, can be computed in $O(CH_{RH}(n))$ time and $O(n^{\lceil k/2 \rceil})$ space, where $CH_\ell(n)$ denotes the time for constructing the convex hull of n points in $\Re^\ell$.*

For more results concerning the Voronoi diagrams in higher dimensions and duality transformation see Aurenhammer [1990].

Farthest-Neighbor Voronoi Diagram

The Voronoi diagram defined in the preceding subsection is also known as the nearest-neighbor Voronoi diagram. A variation of this partitioning concept is a partition of the space into cells, each of which is associated with a site, which contains all points that are farther from the site than from any other site. This diagram is called the *farthest-neighbor* Voronoi diagram. Unlike the nearest-neighbor Voronoi diagram, only a subset of sites have a Voronoi cell associated with them. Those sites that have a nonempty Voronoi cell are those that lie on the convex hull of S. A similar partitioning of the space is known as the order κ-nearest-neighbor Voronoi diagram, in which each Voronoi cell is associated with a subset of κ sites in S for some fixed integer κ such that these κ sites are the closest among all other sites. For $\kappa = 1$ we have the nearest-neighbor Voronoi diagram, and for $\kappa = n - 1$ we have the farthest-neighbor Voronoi diagram. The higher order Voronoi diagrams in k-dimensions are related to the levels of hyperplane arrangements in $k + 1$ dimensions using the paraboloid transformation [Edelsbrunner 1987].

Because the farthest-neighbor Voronoi diagram is related to the convex hull of the set of sites, one can use the marriage-before-conquest paradigm of Kirkpatrick and Seidel [1986] to compute the farthest-neighbor Voronoi diagram of S in two dimensions in time $O(n \log \mathcal{H})$, where $\mathcal{H}$ is the number of sites on the convex hull.

Weighted Voronoi Diagrams

When the sites are associated with weights such that the distance function from a point to the sites is weighted, the structure of the Voronoi diagram can be drastically different than the unweighted case.

Power Diagrams. Suppose each site s in $\mathfrak{R}^k$ is associated with a nonnegative weight, w_s. For an arbitrary point p in $\mathfrak{R}^k$ the weighted distance from p to s is defined as

$$\delta(s, p) = d(s, p)^2 - w_s^2$$

If w_s is positive, and if $d(s, p) \geq w_s$, then $\sqrt{\delta(s, p)}$ is the length of the tangent of p to the ball $b(s)$ of radius w_s and centered at s. Here $\delta(s, p)$ is also called the *power of p with respect to the ball $b(s)$*. The locus of points p equidistant from two sites $s \neq t$ of equal weight will be a hyperplane called the *chordale* of s and t. See Fig. 6.8. Point q is equidistant to sites a and b, and the distance is the length of the tangent line $\overline{q, c} = \overline{q, d}$.

The power diagram of two dimensions can be used to compute the contour of the union of n disks and the connected components of n disks in $O(n \log n)$ time, and in higher dimensions it can be used to compute the union or intersection of n axis-parallel cones in $\mathfrak{R}^k$ with apices in a common hyperplane in time $O(CH_{k+1}(n))$, the multiplicative

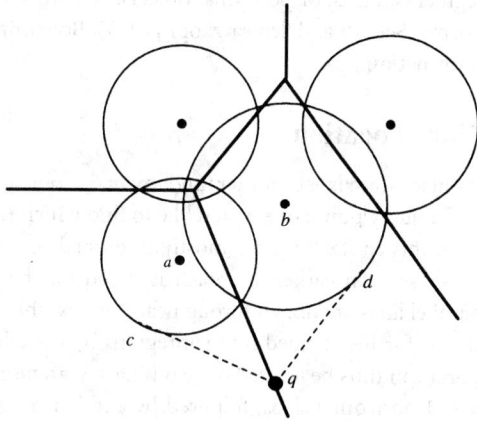

FIGURE 6.8 The power diagram in two dimensions; solid lines are equidistant to two sites.

weighted nearest-neighbor Voronoi diagram (defined subsequently) for n points in $\mathfrak{R}^k$ in time $O(CH_{k+2}(n))$, and the Voronoi diagrams for n spheres in $\mathfrak{R}^k$ in time $O(CH_{k+2}(n))$, where $CH_\ell(n)$ denotes the time for constructing the convex hull of n points in $\mathfrak{R}^\ell$ [Aurenhammer 1987]. For the best time bound for $CH_\ell(n)$ consult the subsection on convex hulls.

Multiplicative-Weighted Voronoi Diagrams. Each site $s \in \mathfrak{R}^k$ has a positive weight w_s, and the distance from a point p to s is defined as

$$\delta_{\text{multi}-w}(s, p) = d(p, s)/w_s$$

In two dimensions, the locus of points equidistant to two sites $s \neq t$ is a circle, if $w_s \neq w_t$, and a perpendicular bisector of line segment $\overline{s, t}$, if $w_s = w_t$. Each cell associated with a site s consists of all points closer to s than to any other site and may be disconnected. In the worst case the nearest-neighbor Voronoi diagram of a set S of n points in two dimensions can have an $O(n^2)$ regions and can be found in $O(n^2)$ time. In one dimension, the diagram can be computed optimally in $O(n \log n)$ time. However, the farthest-neighbor multiplicative-weighted Voronoi diagram has a very different characteristic. Each Voronoi cell associated with a site remains connected, and the size of the diagram is still linear in the number of sites. An $O(n \log^2 n)$ time algorithm for constructing such a diagram is given in Lee and Wu [1993]. See Schaudt and Drysdale [1991] for more applications of the diagram.

Additive-Weighted Voronoi Diagrams. The distance of a point p to a site s of a weight w_s is defined as

$$\delta_{\text{add}-w}(s, p) = d(p, s) - w_s$$

In two dimensions, the locus of points equidistant to two sites $s \neq t$ is a branch of a hyperbola, if $w_s \neq w_t$, and a perpendicular bisector of line segment $\overline{s, t}$ if $w_s = w_t$. The Voronoi diagram has properties similar to the ordinary unweighted diagram. For example, each cell is still connected and the size of the diagram is linear. If the weights are positive, the diagram is the same as the Voronoi diagram of a set of spheres centered at site s and of radius w_s, in two dimensions this diagram for n disks can be computed in $O(n \log^2 n)$ time [Lee and Drysdale 1981, Sharir 1985], and in $k \geq 3$ one can use the notion of power diagram to compute the diagram [Aurenhammer 1987].

Other Generalizations

The sites mentioned so far are point sites. They can be of different shapes. For instance, they can be line segments, disks, or polygonal objects. The metric used can also be a convex distance function or other norms. See Alt and Schwarzkopf [1995], Boissonnat et al. [1995], Klein [1989], and Yap [1987] for more information.

Point Location

Point location is yet another fundamental problem in computational geometry. Given a planar subdivision and a query point, one would like to find which region contains the point in question.

In this context, we are mostly interested in fast response time to answer repeated queries to a fixed database. An earlier approach is based on the *slab method* [Preparata and Shamos 1985], in which parallel lines are drawn through each vertex, thus partitioning the plane into parallel slabs. Each parallel slab is further divided into subregions by the edges of the subdivision that can be ordered. Any given point can thus be located by two binary searches: one to locate the slab containing the point among the $n + 1$ horizontal slabs, followed by another to locate the region defined by a pair of consecutive edges that are ordered from left to right. This requires preprocessing of the planar subdivision, and setting up suitable search tree structures for the slabs and the edges crossing each slab. We use a three-tuple, $(P(n), S(n), Q(n)) = $ (preprocessing time, space requirement, query time) to denote the performance of the search strategy (cf. section on dynamization). The slab method gives an $(O(n^2), O(n^2), O(\log n))$ algorithm. Because preprocessing time is only performed once, the time requirement is not as critical as the space requirement. The primary goal of the query processing problems is to minimize the query time and the space required.

Lee and Preparata first proposed a *chain decomposition* method to decompose a monotone planar subdivision with n points into a collection of $m \leq n$ monotone chains organized in a complete binary tree [Preparata and Shamos 1985]. Each node in the binary tree is associated with a monotone chain of at most n edges, ordered in the y-coordinate. Between two adjacent chains, there are a number of disjoint regions. Each query point is compared with the node, hence the associated chain, to decide on which side of the chain the query point lies. Each chain comparison takes $O(\log n)$ time, and the total number of nodes visited is $O(\log m)$. The search on the binary tree will lead to two adjacent chains and hence identify a region that contains the point. Thus, the query time is $O(\log m \log n) = O(\log^2 n)$. Unlike the slab method in which each edge may be stored as many as $O(n)$ times, resulting in $O(n^2)$ space, it can be shown that each edge in the planar subdivision, with an appropriate chain assignment scheme, is stored only once. Thus, the space requirement is $O(n)$. The chain decomposition scheme gives rise to an $(O(n \log n), O(n), O(\log^2 n))$ algorithm. The binary search on the chains is not efficient enough. Recall that after each *chain comparison*, we will move down the binary search tree to perform the next chain comparison and start over another binary search on the y-coordinate to find an edge of the chain, against which a comparison is made to decide if the point lies to the left or right of the chain. A more efficient scheme is to perform a binary search of the y-coordinate at the root node and to spend only

$O(1)$ time per node as we go down the chain tree, shaving off an $O(\log n)$ factor from the query time [Edelsbrunner et al. 1986]. This scheme is similar to the ones adopted by Chazelle and Guibas [1986] in a fractional cascading search paradigm and by Willard [1985] in his range tree search method. With the linear time algorithm for triangulating a simple polygon due to Chazelle [1991] (cf. subsequent subsection on triangulation) we conclude with the following optimal search structure for planar point location.

Theorem 6.8. *Given a planar subdivision of n vertices, one can preprocess the subdivision in linear time and space such that each point location query can be answered in $O(\log n)$ time.*

The point location problem in arrangements of hyperplanes is also of significant interest. See, e.g., Chazelle and Friedman [1990]. Dynamic versions of the point location problem have also been investigated. See Chiang and Tamassia [1992] for a survey of dynamic computational geometry.

Motion Planning: Path Finding Problems

The problem is mostly cast in the following setting. Given are a set of obstacles O, an object, called *robot*, and an initial and final position, called source and destination, respectively. We wish to find a path for the robot to move from the source to the destination, avoiding all of the obstacles. This problem arises in several contexts. For instance, in robotics this is referred to as the *piano movers' problem* [Yap 1987] or *collision avoidance* problem, and in VLSI routing this is the *wiring* problem for 2-terminal nets. In most applications we are searching for a collision avoidance path that has a shortest length, where the distance measure is based on the Euclidean or L_1-metric. For more information regarding motion planning see, e.g., Alt and Yap [1990] and Yap [1987].

Path Finding in Two Dimensions

In two dimensions, the Euclidean shortest path problem in which the robot is a point and the obstacles are simple polygons, is well studied. A most fundamental approach is by using the notion of *visibility graph*. Because the shortest path must make turns at polygonal vertices, it is sufficient to construct a graph whose vertices are the vertices of the polygonal obstacles and the source and destination and whose edges are determined by vertices that are mutually *visible*, i.e., the segment connecting the two vertices does not intersect the interior of any obstacle. Once the visibility graph is constructed with edge weight equal to the Euclidean distance between the two vertices, one can then apply Dijkstra's shortest path algorithms [Preparata and Shamos 1985] to find a shortest path between the source and destination. The Euclidean shortest path between two points is referred to as the *geodesic path* and the distance as the *geodesic distance*. The computation of the visibility graph is the dominating factor for the complexity of any visibility graph-based shortest path algorithm. Research results aiming at more efficient algorithms for computing the visibility graph and for computing the geodesic path in time proportional to the size of the graph have been obtained. Ghosh and Mount [1991] gave an output-sensitive algorithm that runs in $O(E + n \log n)$ time for computing the visibility graph, where E denotes the number of edges in the graph.

Mitchell [1993] used the so-called *continuous Dijkstra* wave front approach to the problem for the general polygonal domain of n obstacle vertices and obtained an $O(n^{5/3+\epsilon})$ time algorithm. He constructed a *shortest path map* that partitions the plane into regions such that all points q that lie in the same region have the same vertex sequence in the shortest path from the given source to q. The shortest path map takes $O(n)$ space and enables us to perform shortest path queries, i.e., find a shortest path from the given source to any query points, in $O(\log n)$ time. Hershberger and Suri [1993] on the other hand, used a plane subdivision approach and presented an $O(n \log^2 n)$-time and $O(n \log n)$-space algorithm to compute the shortest path map of a given source point. They later improved the time bound to $O(n \log n)$. If the source-destination path is confined in a simple polygon with n vertices, the shortest path can be found in $O(n)$ time [Preparata and Shamos 1985].

In the context of VLSI routing one is mostly interested in rectilinear paths (L_1-metric) whose edges are either horizontal or vertical. As the paths are restricted to be rectilinear, the shortest path problem can be solved more easily. Lee et al. [1996] gave a survey on this topic.

In a two-layer VLSI routing model, the number of segments in a rectilinear path reflects the number of *vias*, where the wire segments change layers, which is a factor that governs the fabrication cost. In robotics, a straight-line motion is not as costly as making turns. Thus, the number of segments (or *turns*) has also become an objective function. This motivates the study of the problem of finding a path with the smallest number of segments, called the *minimum link path problem* [Mitchell et al. 1992, Suri 1990].

These two cost measures, length and number of links, are in conflict with each other. That is, a shortest path may have far too many links, whereas a minimum link path may be arbitrarily long compared with a shortest path. Instead of optimizing both measures *simultaneously*, one can seek a path that either optimizes a linear function of both length and the number of links or optimizes them in a lexicographical order. For example, we optimize the length first, and then the number of links, i.e., among those paths that have the same shortest length, find one whose number of links is the smallest, and vice versa.

A generalization of the collision-avoidance problem is to allow collision with a cost. Suppose each obstacle has a weight, which represents the cost if the obstacle is *penetrated*. Mitchell and Papadimitriou [1991] first studied the weighted region shortest path problem. Lee et al. [1991] studied a similar problem in the rectilinear case. Another generalization is to include in the set of obstacles some subset $F \subset O$ of obstacles, whose vertices are *forbidden* for the solution path to make turns. Of course, when the weight of obstacles is set to be ∞, or the forbidden set $F = \emptyset$, these generalizations reduce to the ordinary collision-avoidance problem.

Path Finding in Three Dimensions

The Euclidean shortest path problem between two points in a three-dimensional polyhedral environment turns out to be much harder than its two-dimensional counterpart. Consider a convex polyhedron P with n vertices in three dimensions and two points s, d on the surface of P. A shortest path from s to d on the surface will cross a sequence of edges, denoted $\xi(s, d)$. Here $\xi(s, d)$ is called the *shortest path edge sequence* induced by s and d and consists of distinct edges. If the edge sequence is known, the shortest path between s and d can be computed by a planar unfolding procedure so that these faces crossed by the path lie in a common plane and the path becomes a straight-line segment.

Mitchell et al. [1987] gave an $O(n^2 \log n)$ algorithm for finding a shortest path between s and d even if the polyhedron may not be convex. If s and d lie on the surface of two different polyhedra, Sharir [1987] gave an $O(N^{O(k)})$ algorithm, where N denotes the total number of vertices of k obstacles. In general, the problem of determining the shortest path edge sequence of a path between two points among k polyhedra is NP-hard [Canny and Reif 1987].

Motion Planning of Objects

In the previous sections, we discussed path planning for moving a point from the source to a destination in the presence of polygonal or polyhedral obstacles. We now briefly describe the problem of moving a polygonal or polyhedral object from an initial position to a final position subject to translational and/or rotational motions.

Consider a set of k convex polyhedral obstacles, $O_1, O_2, \ldots, O_k$, and a convex polyhedral robot, R in three dimensions. The motion planning problem is often solved by using the so-called *configuration space*, denoted $\mathcal{C}$, which is the space of parametric representations of possible robot placements [Lozano-Pérez 1983]. The free placement (FP) is the subspace of $\mathcal{C}$ of points at which the robot does not intersect the interior of any obstacle. For instance, if only translations of R are allowed, the free configuration space will be the union of the Minkowski sums $M_i = O_i \oplus (-R) = \{a - b \mid a \in O_i, b \in R\}$ for $i = 1, 2, \ldots, k$. A *feasible* path exists if the initial placement of R and final placement belong to the same connected component of FP. The problem is to find a continuous curve connecting the initial and final positions in FP. The combinatorial complexity, i.e., the number of vertices, edges, and faces on the boundary of FP, largely influences the efficiency of any $\mathcal{C}$-based algorithm. For translational motion planning, Aronov and Sharir [1994] showed that the combinatorial complexity of FP is $O(nk \log^2 k)$, where k is the number of obstacles defined above and n is the total complexity of the Minkowski sums M_i, $1 \leq i \leq k$.

Moving a ladder (represented as a line segment) among a set of polygonal obstacles of size n can be done in $O(K \log n)$ time, where K denotes the number of pairs of obstacle vertices whose distance is less than the length of the ladder and is $O(n^2)$ in general [Sifrony and Sharir 1987]. If the moving robot is also a polygonal object, Avnaim et al. [1988] showed that $O(n^3 \log n)$ time suffices. When the obstacles are *fat*[5] Van der Stappen and Overmars [1994] showed that the two preceding two-dimensional motion planning problems can be solved in $O(n \log n)$ time, and in three dimensions the problem can be solved in $O(n^2 \log n)$ time, if the obstacles are ℓ-fat for some positive constant ℓ.

Geometric Optimization

The geometric optimization problems arise in operations research, pattern recognition, and other engineering disciplines. We list some representative problems.

Minimum Cost Spanning Trees

The minimum (cost) spanning tree MST of an undirected, weighted graph $G(V, E)$, in which each edge has a nonnegative weight, is a well-studied problem in graph theory and can be solved in $O(|E| \log |V|)$ time [Preparata and Shamos 1985]. When cast in the Euclidean or other L_p-metric plane in which the input consists of a set S of n points, the complexity of this problem becomes different. Instead of constructing a *complete* graph whose edge weight is defined by the distance between its two endpoints, from which to extract an MST, a sparse graph, known as the *Delaunay triangulation* of the point set, is computed. It can be shown that the MST of S is a subgraph of the Delaunay triangulation. Because the MST of a planar graph can be found in linear time [Preparata and Shamos 1985], the problem can be solved in $O(n \log n)$ time. In fact, this is asymptotically optimal, as the closest pair of the set of points must define an edge in the MST, and the closest pair problem is known to have an $\Omega(n \log n)$ lower bound, as mentioned previously.

This problem in three or more dimensions can be solved in subquadratic time. For instance, in three dimensions $O((n \log n)^{1.5})$ time is sufficient [Chazelle 1985] and in $k \geq 3$ dimensions $O(n^{2(1-1/(\lceil k/2 \rceil+1))+\epsilon})$ time suffices [Agarwal et al. 1991].

Minimum Diameter Spanning Tree

The minimum *diameter* spanning tree (MDST) of an undirected, weighted graph $G(V, E)$ is a spanning tree such that the total weight of the longest path in the tree is minimum. This arises in applications to communication networks where a tree is sought such that the maximum delay, instead of the total cost, is to be minimized. A graph-theoretic approach yields a solution in $O(|E||V| \log |V|)$ time [Handler and Mirchandani 1979]. Ho et al. [1991] showed that by the triangle inequality there exists an MDST such that the longest path in the tree consists of no more than *three* segments. Based on this an $O(n^3)$ time algorithm was obtained.

Theorem 6.9. *Given a set S of n points, the minimum diameter spanning tree for S can be found in $\theta(n^3)$ time and $O(n)$ space.*

We remark that the problem of finding a spanning tree whose total cost and the diameter are both bounded is NP-complete [Ho et al. 1991]. A similar problem that arises in VLSI clock tree routing is to find a tree from a source to multiple sinks such that every source-to-sink path is the shortest and the total wire length is to be minimized. This problem still is not known to be solvable in polynomial time or

---

[5]An object $O \subseteq R^k$ is said to be ℓ-*fat* if for all hyperspheres S centered inside O and not fully containing O we have $\ell \cdot$ *volume* $(O \cap S) \geq$ *volume*(S).

NP-hard. Recently, we have shown that the problem of finding a minimum spanning tree such that the longest source-to-sink path is bounded by a given parameter is NP-complete [Seo and Lee 1995].

Minimum Enclosing Circle Problem

Given a set S of points, the problem is to find the smallest disk enclosing the set. This problem is also known as the (unweighted) one-center problem. That is, find a center such that the maximum distance from the center to the points in S is minimized. More formally, we need to find the center $c \in \Re^2$ such that $\max_{p_j \in S} d(c, p_j)$ is minimized. The weighted one-center problem, in which the distance function $d(c, p_j)$ is multiplied by the weight w_j, is a well-known minimax problem, also known as the *emergency center problem* in operations research. In two dimensions, the one-center problem can be solved in $O(n)$ time [Dyer 1986, Megiddo 1983b]. The minimum enclosing ball problem in higher dimensions is also solved by using a linear programming technique [Megiddo 1983b, 1984].

Largest Empty Circle Problem

This problem, in contrast to the minimum enclosing circle problem, is to find a circle centered in the interior of the convex hull of the set S of points that does not contain any given point and the radius of the circle is to be maximized. This is mathematically formalized as a maximin problem; the minimum distance from the center to the set is maximized. The weighted version is also known as the *obnoxious center* problem in facility location. An $O(n \log n)$ time solution for the unweighted version can be found in [Preparata and Shamos 1985].

Minimum Annulus Covering Problem

The *minimum annulus covering problem* is defined as follows. Given a set of S of n points find an annulus (defined by two concentric circles) whose center lies internal to the convex hull of S such that the *width* of the annulus is minimized. The problem arises in mechanical part design. To measure whether a circular part is *round*, an American National Standards Institute (ANSI) standard is to use the width of an annulus covering the set of points obtained from a number of measurements. This is known as the *roundness* problem [Le and Lee 1991]. It can be shown that the center of the annulus is either at a vertex of the nearest-neighbor Voronoi diagram, a vertex of the farthest-neighbor Voronoi diagram, or at the intersection of these two diagrams [Le and Lee 1991]. If the input is defined by a simple polygon P with n vertices, and the problem is to find a minimum-width annulus that contains the boundary of P, the problem can be solved in $O(n \log n + k)$, where k denotes the number of intersection points of the *medial axis* of the simple polygon and the boundary of P [Le and Lee 1991]. When the polygon is known to be convex, a linear time is sufficient [Swanson et al. 1995]. If the center of the smallest annulus of a point set can be arbitrarily placed, the center may lie at infinity and the annulus degenerates to a pair of parallel lines enclosing the set of points. This problem is different from the problem of finding the width of a set, which is to find a pair of parallel lines enclosing the set such that the distance between them is minimized. The width of a set of n points can be found in $O(n \log n)$ time, which is optimal [Lee and Wu 1986]. In three dimensions the *width* of a set is also used as a measure for flatness of a *plate*—*flatness* problem. Houle and Toussaint [1988] gave an $O(n^2)$ time algorithm, and Chazelle et al. [1993] improved it to $O(n^{8/5+\epsilon})$.

Decomposition

Polygon decomposition arises in pattern recognition in which recognition of a shape is facilitated by first decomposing it into simpler parts, called *primitives*, and comparing them to templates previously stored in a library via some similarity measure. The primitives are often convex, with the simplest being the shape of a triangle.

We consider two types of decomposition, *partition* and *covering*. In the former type, the components are pairwise disjoint except they may have some boundary edges in common. In the latter type, the components may overlap. A minimum decomposition is one such that the number of components is minimized. Sometimes additional points, called *Steiner points*, may be introduced to obtain a minimum decomposition. Unless otherwise specified, we assume that no Steiner points are used.

Triangulation

Triangulating a simple polygon or, in general, triangulating a planar straight-line graph, is a process of introducing noncrossing edges so that each face is a triangle. It is also a fundamental problem in computer graphics, geographical information systems, and finite-element methods.

Let us begin with the problem of triangulating a simple polygon with n vertices. It is obvious that for a simple polygon with n edges, one needs to introduce at most $n - 3$ diagonals to triangulate the interior into $n - 2$ triangles. This problem has been studied very extensively. A pioneering work is due to Garey et al., which gave an $O(n \log n)$ algorithm and a linear algorithm if the polygon is monotone [O'Rourke 1994, Preparata and Shamos 1985]. A recent breakthrough linear time triangulation result of Chazelle [1991] settled the long-standing open problem. As a result of this linear triangulation algorithm, a number of problems can be solved in linear time, for example, the simplicity test, defined subsequently, and many other shortest path problems inside a simple polygon [Guibas and Hershberger 1989]. Note that if the polygons have holes, the problem of triangulating the interior requires $\Omega(n \log n)$ time [Asano et al. 1986].

Sometimes we want to look for *quality* triangulation instead of just an arbitrary one. For instance, triangles with large or small angles are not desirable. It is well known that the Delaunay triangulation of points in general position is unique, and it will maximize the minimum angle. In fact, the characteristic angle vector[6] of the Delaunay triangulation of a set of points is *lexicographically maximum* [Lee 1978]. The notion of Delaunay triangulation of a set of points can be generalized to a planar straight-line graph $G(V, E)$. That is, we would like to have G as a subgraph of a triangulation $G'(V, E')$, $E \subseteq E'$, such that each triangle satisfies the *empty circumcircle* property; no vertex visible from the vertices of a triangle is contained in the interior of the circle. This *generalized* Delaunay triangulation was first introduced by Lee [1978] and an $O(n^2)$ (respectively, $O(n \log n)$) algorithm for constructing the generalized triangulation of a planar graph (respectively, a simple polygon) with n vertices was given in Lee and Lin [1986b]. Chew [1989] later improved the result and gave an $O(n \log n)$ time algorithm using divide-and-conquer. Triangulations that minimize the maximum angle or maximum edge length were also studied. But if constraints on the measure of the triangles, for instance, each triangle in the triangulation must be nonobtuse, then Steiner points must be introduced. See Bern and Eppstein (in Du and Hwang [1992, pp. 23–90]) for a survey of different criteria of triangulations and discussions of triangulations in two and three dimensions.

The problem of triangulating a set P of points in $\Re^k$, $k \geq 3$, is less studied. In this case, the convex hull of P is to be partitioned into $\mathcal{F}$ nonoverlapping simplices, the vertices of which are points in P. A simplex in k-dimensions consists of exactly $k + 1$ points, all of which are extreme points. Avis and ElGindy [1987] gave an $O(k^4 n \log_{1+1/k} n)$ time algorithm for triangulating a simplicial set of n points in $\Re^k$. In $\Re^3$ an $O(n \log n + \mathcal{F})$ time algorithm was presented and $\mathcal{F}$ is shown to be linear if no three points are collinear and at most $O(n^2)$ otherwise. See Du and Hwang [1992] for more references on three-dimensional triangulations and Delaunay triangulations in higher dimensions.

Other Decompositions

Partitioning a simple polygon into shapes such as convex polygons, star-shaped polygons, spiral polygons, monotone polygons, etc., has also been investigated [Toussaint 1985]. A linear time algorithm for partitioning a polygon into star-shaped polygons was given by Avis and Toussaint [1981] after the polygon has been triangulated. This algorithm provided a very simple proof of the traditional art gallery problem originally posed by Klee, i.e., $\lfloor n/3 \rfloor$ vertex guards are always sufficient to see the entire region of a simple polygon with n vertices. But if a minimum partition is desired, Keil [1985] gave an $O(n^5 N^2 \log n)$ time,

---

[6]The characteristic angle vector of a triangulation is a vector of minimum angles of each triangle arranged in nondescending order. For a given point set, the number of triangles is the same for all triangulations, and therefore each of them is associated with a characteristic angle vector.

where N denotes the number of reflex vertices. However, the problem of *covering* a simple polygon with a minimum number of star-shaped parts is *NP*-hard [Lee and Lin 1986a]. The problem of partitioning a polygon into a minimum number of convex parts can be solved in $O(N^2 n \log n)$ time [Keil 1985]. The minimum covering problem by star-shaped polygons for rectilinear polygons is still open. For variations and results of art gallery problems the reader is referred to O'Rourke [1987] and Shermer [1992]. Polynomial time algorithms for computing the minimum partition of a simple polygon into simpler parts while allowing Steiner points can be found in Asano et al. [1986] and Toussaint [1985].

The minimum partition or covering problem for simple polygons becomes NP-hard when the polygons are allowed to have *holes* [Keil 1985, O'Rourke and Supowit 1983]. Asano et al. [1986] showed that the problem of partitioning a simple polygon with h holes into a minimum number of trapezoids with two horizontal sides can be solved in $O(n^{h+2})$ time and that the problem is NP-complete if h is part of the input. An $O(n \log n)$ time 3-approximation algorithm was presented. Imai and Asano [1986] gave an $O(n^{3/2} \log n)$ time and $O(n \log n)$ space algorithm for partitioning a rectilinear polygon with holes into a minimum number of rectangles (allowing Steiner points). The problem of covering a rectilinear polygon (without holes) with a minimum number of rectangles, however, is also NP-hard [Culberson and Reckhow 1988].

The problem of minimum partition into convex parts and the problem of determining if a nonconvex polyhedron can be partitioned into tetrahedra without introducing Steiner points are NP-hard [O'Rourke and Supowit 1983, and Ruppert and Seidel 1992].

Intersection

This class of problems arises in architectural design, computer graphics [Dorward 1994], etc., and encompasses two types of problems, *intersection detection* and *intersection computation*.

Intersection Detection Problems

The intersection detection problem is of the form: Given a set of objects, do any two intersect? The intersection detection problem has a lower bound of $\Omega(n \log n)$ [Preparata and Shamos 1985]. The pairwise intersection detection problem is a precursor to the general intersection detection problem.

In two dimensions the problem of detecting if two polygons of r and b vertices intersect was easily solved in $O(n \log n)$ time, where $n = r + b$ using the red–blue segment intersection algorithm [Mairson and Stolfi 1988]. However, this problem can be reduced in linear time to the problem of detecting the self-intersection of a polygonal curve. The latter problem is known as the *simplicity* test and can be solved optimally in linear time by Chazelle's [1991] linear time triangulation algorithm. If the two polygons are convex, then $O(\log n)$ suffices [Chazelle and Dobkin 1987, Edelsbrunner 1985]. We remark here that, although detecting whether two convex polygons intersect can be done in logarithmic time, detecting whether the boundary of the two convex polygons intersects requires $\Omega(n)$ time [Chazelle and Dobkin 1987].

In three dimensions, detecting if two convex polyhedra intersect can be solved in linear time by using a hierarchical representation of the convex polyhedron, or by formulating it as a linear programming problem in three variables [Chazelle and Dobkin 1987, Dobkin and Kirkpatrick 1985, Dyer 1984, Megiddo 1983b].

For some applications, we would not only detect intersection but also *report* all such intersecting pairs of objects or *count* the number of intersections, which is discussed next.

Intersection Reporting/Counting Problems

One of the simplest of such intersecting reporting problems is that of *reporting* all intersecting pairs of line segments in the plane. Using the plane sweep technique, one can obtain an $O((n + \mathcal{F}) \log n)$ time, where $\mathcal{F}$ is the output size. It is not difficult to see that the lower bound for this problem is $\Omega(n \log n + \mathcal{F})$; thus the preceding algorithm is $O(\log n)$ factor from the optimal. Recently, this segment intersection reporting problem was solved optimally by Chazelle and Edelsbrunner [1992], who used several important algorithm design and data structuring techniques as well as some crucial combinatorial analysis. In contrast to this asymptotically optimal *deterministic* algorithm, a simpler randomized algorithm for this problem that takes $O(n \log n + \mathcal{F})$ time but requires only $O(n)$ space (instead of $O(n + \mathcal{F})$) was obtained [Du and Hwang 1992]. Balaban [1995] most recently reported a deterministic algorithm that solves this problem optimally both in time and space.

On a separate front, the problem of finding intersecting pairs of segments from different sets was considered. This is called the *bichromatic line segment* intersection problem. Nievergelt and Preparata [1982] considered the problem of merging two planar convex subdivisions of total size n and showed that the resulting subdivision can be computed in $O(n \log n + \mathcal{F})$ time. This result [Nievergelt and Preparata 1982] was extended in two ways. Mairson and Stolfi [1988] showed that the bichromatic line segment intersection reporting problem can be solved in $O(n \log n + \mathcal{F})$ time. Guibas and Seidel [1987] showed that merging two convex subdivisions can actually be solved in $O(n + \mathcal{F})$ time using topological plane sweep.

Most recently, Chazelle et al. [1994] used *hereditary segment trees* structure and *fractional cascading* [Chazelle and Guibas 1986] and solved both segment intersection reporting and counting problems optimally in $O(n \log n)$ time and $O(n)$ space. (The term $\mathcal{F}$ should be included for reporting.)

The *rectangle intersection reporting* problem arises in the design of VLSI circuitry, in which each rectangle is used to model a certain circuitry component. This is a well-studied classic problem and optimal algorithms ($O(n \log n + \mathcal{F})$ time) have been reported (see Lee and Preparata [1984] for references). The k-dimensional hyperrectangle intersection reporting (respectively, counting) problem can be solved in $O(n^{k-2} \log n + \mathcal{F})$ time and $O(n)$ space [respectively, in time $O(n^{k-1} \log n)$ and space $O(n^{k-2} \log n)$].

Intersection Computation

Computing the actual intersection is a basic problem, whose efficient solutions often lead to better algorithms for many other problems.

Consider the problem of computing the common intersection of half-planes discussed previously. Efficient computation of the intersection of two convex polygons is required. The intersection of two convex polygons can be solved very efficiently by plane sweep in linear time, taking advantage of the fact that the edges of the input polygons are ordered. Observe that in each vertical strip defined by two consecutive sweep lines, we only need to compute the intersection of two trapezoids, one derived from each polygon [Preparata and Shamos 1985].

The problem of intersecting two convex polyhedra was first studied by Muller and Preparata [Preparata and Shamos 1985], who gave an $O(n \log n)$ algorithm by reducing the problem to the problems of intersection detection and convex hull computation. From this one can easily derive an $O(n \log^2 n)$ algorithm for computing the common intersection of n half-spaces in three dimensions by the divide-and-conquer method. However, using geometric duality and the concept of separating plane, Preparata and Muller [Preparata and Shamos 1985] obtained an $O(n \log n)$ algorithm for this problem, which is asymptotically optimal. There appears to be a difference in the approach to solving the common intersection problem of half-spaces in two and three dimensions. In the latter, we resorted to geometric duality instead of divide-and-conquer. This *inconsistency* was later resolved. Chazelle [1992] combined the hierarchical representation of convex polyhedra, geometric duality, and other ingenious techniques to obtain a linear time algorithm for computing the intersection of two convex polyhedra. From this result several problems can be solved optimally: (1) the common intersection of half-spaces in three dimensions can now be solved by divide-and-conquer optimally, (2) the merging of two Voronoi diagrams in the plane can be done in linear time by observing the relationship between the Voronoi diagram in two dimensions and the convex hull in three dimensions (cf. subsection on Voronoi diagrams), and (3) the medial axis of a simple polygon or the Voronoi diagram of vertices of a convex polygon can be solved in linear time.

Geometric Searching

This class of problems is cast in the form of query answering as discussed in the subsection on dynamization. Given a collection of objects, with preprocessing allowed, one is to find objects that satisfy the queries. The problem can be static or dynamic, depending on whether the database is allowed to change over the course of query-answering sessions, and it is studied mostly in modes, *count-mode* and *report-mode*. In the former case only the number of objects satisfying the query is to be answered, whereas in the latter the actual identity of the objects is to be reported. In the report mode the query time of the algorithm consists of two components, *search time* and *output*, and expressed as $Q_A(n) = O(f(n) + \mathcal{F})$, where n denotes the size of the database, $f(n)$ a function of n, and $\mathcal{F}$ the size of output. It is obvious that algorithms that handle the report-mode queries can also handle the count-mode queries ($\mathcal{F}$ is the answer). It seems natural to expect that the algorithms for count-mode queries would be more efficient (in terms of the order of magnitude of the space required and query time), as they need not search for the objects. However, it was argued that in the report-mode range searching, one could take advantage of the fact that since reporting takes time, the more there is to report, the *sloppier* the search can be. For example, if we were to know that the ratio $n/\mathcal{F}$ is $O(1)$, we could use a sequential search on a linear list. Chazelle in his seminal paper on filtering search capitalizes on this observation and improves the time complexity for searching for several problems [Chazelle 1986]. As indicated subsequently, the count-mode range searching problem is harder than the report-mode counterpart.

Range Searching Problems

This is a fundamental problem in database applications. We will discuss this problem and the algorithm in two dimensions. The generalization to higher dimensions is straightforward using a known technique [Bentley 1980]. Given is a set of n points in the plane, and the ranges are specified by a product $(l_1, u_1) \times (l_2, u_2)$. We would like to find points $p = (x, y)$ such that $l_1 \leq x \leq u_1$ and $l_2 \leq y \leq u_2$. Intuitively we want to find those points that lie inside a query rectangle specified by the range. This is called *orthogonal range searching*, as opposed to other kinds of range searching problems discussed subsequently. Unless otherwise specified, a range refers to an orthogonal range. We discuss the static case; as this belongs to the class of decomposable searching problems, the dynamization transformation techniques can be applied. We note that the range tree structure mentioned later can be made dynamic by using a weight-balanced tree, called a $BB(\alpha)$ tree [Mehlhorn 1984, Willard and Luecker 1985].

For count-mode queries this problem can be solved by using the locus method as follows. Divide the plane into $O(n^2)$ cells by drawing horizontal and vertical lines through each point. The answer to the query q, i.e., find the number of points dominated by q (those points whose x- and y-coordinates are both no greater than those of q) can be found by locating the cell containing q. Let it be denoted by $Dom(q)$. Thus, the answer to the count-mode range queries can be obtained by some simple arithmetic operations of $Dom(q_i)$ for the four corners of the query rectangle. We have $Q(k, n) = O(k \log n)$, $S(k, n) = P(k, n) = O(n^k)$. To reduce the space requirement at the expense of query time has been a goal of further research on this topic. Bentley [1980] introduced a data structure, called *range trees*. Using this structure the following results were obtained: for $k \geq 2$, $Q(k, n) = O(\log^{k-1} n)$, $S(k, n) = P(k, n) = O(n \log^{k-1} n)$. (See Lee and Preparata [1984] and Willard [1985] for more references.)

For report-mode queries, Chazelle [1986] showed that by using a filtering search technique the space requirement can be further reduced by a $\log \log n$ factor. In essence we use less space to allow for more objects than necessary to be found by the search mechanism, followed by a filtering process leaving out unwanted objects for output. If the range satisfies additional conditions, e.g., *grounded* in one of the coordinates, say, $l_1 = 0$, or the aspect ratio of the intervals specifying the range is fixed, then less space is needed. For instance, in two dimensions, the space required is linear (a saving of $\log n / \log \log n$ factor) for these two cases. By using the so-called functional approach to data structures Chazelle [1988] developed a *compression* scheme to encode the *downpointers* used by Willard [1985] to reduce further the space requirement. Thus in k-dimensions, $k \geq 2$, for the count-mode range queries we have $Q(k, n) = O(\log^{k-1} n)$ and $S(k, n) = O(n \log^{k-2} n)$ and for report-mode range queries $Q(k, n) = O(\log^{k-1} n + \mathcal{F})$, and $S(k, n) = O(n \log^{k-2+\epsilon} n)$ for some $0 < \epsilon < 1$.

Other Range Searching Problems

There are other range searching problems, called the simplex range searching problem and the half-space range searching problem that have been well studied. A simplex range in $\mathfrak{R}^k$ is a range whose boundary is specifed by $k + 1$ hyperplanes. In two dimensions it is a triangle.

The report-mode half-space range searching problem in the plane is optimally solved by Chazelle et al. [1985] in $Q(n) = O(\log n + \mathcal{F})$ time and $S(n) = O(n)$ space, using geometric duality transform. But this method does not generalize to higher dimensions. For $k = 3$, Chazelle and Preparata [1986] obtained an optimal $O(\log n + \mathcal{F})$ time algorithm using $O(n \log n)$ space. Agarwal and Matoušek [1995] obtained a more general result for this problem: for $n \le m \le n^{\lfloor k/2 \rfloor}$, with $O(m^{1+\epsilon})$ space and preprocessing, $Q(k, n) = O((n/m^{1/\lfloor k/2 \rfloor}) \log n + \mathcal{F})$. As the half-space range searching problem is also decomposable (cf. earlier subsection on dynamization) standard dynamization techniques can be applied.

A general method for simplex range searching is to use the notion of the *partition tree*. The search space is partitioned in a hierarchical manner using cutting hyperplanes, and a search structure is built in a tree structure. Willard [1982] gave a sublinear time algorithm for count-mode half-space query in $O(n^\alpha)$ time using linear space, where $\alpha \approx 0.774$, for $k = 2$. Using Chazelle's cutting theorem Matoušek showed that for k-dimensions there is a linear space search structure for the simplex range searching problem with query time $O(n^{1-1/k})$, which is optimal in two dimensions and within $O(\log n)$ factor of being optimal for $k > 2$. For more detailed information regarding geometric range searching see Matoušek [1994].

The preceding discussion is restricted to the case in which the database is a collection of points. One may consider other kinds of objects, such as line segments, rectangles, triangles, etc., depending on the needs of the application. The inverse of the orthogonal range searching problem is that of the *point enclosure searching problem*. Consider a collection of isothetic rectangles. The point enclosure searching problem is to find all rectangles that contain the given query point q. We can cast these problems as the *intersection searching* problems, i.e., given a set S of objects and a query object q, find a subset $\mathcal{F}$ of S such that for any $f \in \mathcal{F}$, $f \cap q \ne \emptyset$. We then have the rectangle enclosure searching problem, rectangle containment problem, segment intersection searching problem, etc. We list only a few references about these problems [Bistiolas et al. 1993, Imai and Asano 1987, Lee and Preparata 1982]. Janardan and Lopez [1993] generalized intersection searching in the following manner. The database is a collection of *groups* of objects, and the problem is to find all groups of objects intersecting a query object. A group is considered to be intersecting the query object if any object in the group intersects the query object. When each group has only one object, this reduces to the ordinary searching problems.

6.4 Conclusion

We have covered in this chapter a wide spectrum of topics in computational geometry, including several major problem solving paradigms developed to date and a variety of geometric problems. These paradigms include incremental construction, plane sweep, geometric duality, locus, divide-and-conquer, prune-and-search, dynamization, and random sampling. The topics included here, i.e., convex hull, proximity, point location, motion planning, optimization, decomposition, intersection, and searching, are not meant to be exhaustive. Some of the results presented are classic, and some of them represent the state of the art of this field. But they may also become classic in months to come. The reader is encouraged to look up the literature in major computational geometry journals and conference proceedings given in the references. We have not discussed parallel computational geometry, which has an enormous amount of research findings. Atallah [1992] gave a survey on this topic.

We hope that this treatment will provide sufficient background information about this field and that researchers in other science and engineering disciplines may find it helpful and apply some of the results to their own problem domains.

References

Agarwal, P., Edelsbrunner, H., Schwarzkopf, O., and Welzl, E. 1991. Euclidean minimum spanning trees and bichromatic closest pairs. *Discrete Comput. Geom.* 6(5):407–422.

Agarwal, P. and Matoušek, J. 1995. Dynamic half-space range reporting and its applications. *Algorithmica* 13(4):325–345.

Aggarwal, A., Guibas, L. J., Saxe, J., and Shor, P. W. 1989. A linear-time algorithm for computing the Voronoi diagram of a convex polygon. *Discrete Comput. Geom.* 4(6):591–604.

Alt, H. and Schwarzkopf, O. 1995. The Voronoi diagram of curved objects, pp. 89–97. In *Proc. 11th Ann. ACM Symp. Comput. Geom.*, June.

Alt, H. and Yap, C. K. 1990. Algorithmic aspect of motion planning: a tutorial, part 1 & 2. *Algorithms Rev.* 1(1, 2):43–77.

Amato, N. 1994. Determining the separation of simple polygons. *Int. J. Comput. Geom. Appl.* 4(4):457–474.

Amato, N. 1995. Finding a closest visible vertex pair between two polygons. *Algorithmica* 14(2):183–201.

Aronov, B. and Sharir, M. 1994. On translational motion planning in 3-space, pp. 21–30. In *Proc. 10th Ann. ACM Comput. Geom.*, June.

Asano, T., Asano, T., and Imai, H. 1986. Partitioning a polygonal region into trapezoids. *J. ACM* 33(2):290–312.

Asano, T., Asano, T., and Pinter, R. Y. 1986. Polygon triangulation: efficiency and minimality. *J. Algorithms* 7:221–231.

Asano, T., Guibas, L. J., and Tokuyama, T. 1994. Walking on an arrangement topologically. *Int. J. Comput. Geom. Appl.* 4(2):123–151.

Atallah, M. J. 1992. Parallel techniques for computational geometry. *Proc. of IEEE* 80(9):1435–1448.

Aurenhammer, F. 1987. Power diagrams: properties, algorithms and applications. *SIAM J. Comput.* 16(1):78–96.

Aurenhammer, F. 1990. A new duality result concerning Voronoi diagrams. *Discrete Comput. Geom.* 5(3):243–254.

Avis, D. and ElGindy, H. 1987. Triangulating point sets in space. *Discrete Comput. Geom.* 2(2):99–111.

Avis, D. and Toussaint, G. T. 1981. An efficient algorithm for decomposing a polygon into star-shaped polygons. *Pattern Recog.* 13:395–398.

Avnaim, F., Boissonnat, J. D., and Faverjon, B. 1988. A practical exact motion planning algorithm for polygonal objects amidst polygonal obstacles, pp. 67–86. In *Proc. Geom. Robotics Workshop*. J. D. Boissonnat and J. P. Laumond, eds. LNCS Vol. 391.

Balaban, I. J. 1995. An optimal algorithm for finding segments intersections, pp. 211–219. In *Proc. 11th Ann. Symp. Comput. Geom.*, June.

Bentley, J. L.1980. Multidimensional divide-and-conquer. *Comm. ACM* 23(4):214–229.

Bentley, J. L. and Saxe, J. B. 1980. Decomposable searching problems I: static-to-dynamic transformation. *J. Algorithms* 1:301–358.

Bespamyatnikh, S. N. 1995. An optimal algorithm for closest pair maintenance, pp. 152–166. In *Proc. 11th Ann. Symp. Comput. Geom.*, June.

Bieri, H. and Nef, W. 1982. A recursive plane-sweep algorithm, determining all cells of a finite division of R^d. *Computing* 28:189–198.

Bistiolas, V., Sofotassios, D., and Tsakalidis, A. 1993. Computing rectangle enclosures. *Comput. Geom.: Theory Appl.* 2(6):303–308.

Boissonnat, J.-D., Sharir, M., Tagansky, B., and Yvinec, M. 1995. Voronoi diagrams in higher dimensions under certain polyhedra distance functions, pp. 79–88. In *Proc. 11th Annu. ACM Symp. Comput. Geom.*, June.

Callahan, P. and Kosaraju, S. R. 1995. Algorithms for dynamic closests pair and *n*-body potential fields, pp. 263–272. In *Proc. 6th ACM–SIAM Symp. Discrete Algorithms*.

Canny, J. and Reif, J. R. 1987. New lower bound techniques for robot motion planning problems, pp. 49–60. In *Proc. 28th Annual Symp. Found. Comput. Sci.* Oct.

Chan, T. M. 1995. Output-sensitive results on convex hulls, extreme points, and related problems, pp. 10–19. In *Proc. 11th ACM Ann. Symp. Comput. Geom.*, June.

Chazelle, B. 1985. How to search in history. *Inf. Control* 64:77–99.

Chazelle, B. 1986. Filtering search: a new approach to query-answering, *SIAM J. Comput.* 15(3):703–724.

Chazelle, B. 1988. A functional approach to data structures and its use in multidimensional searching. *SIAM J. Comput.* 17(3):427–462.

Chazelle, B. 1991. Triangulating a simple polygon in linear time. *Discrete Comput. Geom.* 6:485–524.

Chazelle, B. 1992. An optimal algorithm for intersecting three-dimensional convex polyhedra. *SIAM J. Comput.* 21(4):671–696.

Chazelle, B. 1993. An optimal convex hull algorithm for point sets in any fixed dimension. *Discrete Comput. Geom.* 8(2):145–158.

Chazelle, B. and Dobkin, D. P. 1987. Intersection of convex objects in two and three dimensions. *J. ACM* 34(1):1–27.

Chazelle, B. and Edelsbrunner, H. 1992. An optimal algorithm for intersecting line segments in the plane. *J. ACM* 39(1):1–54.

Chazelle, B., Edelsbrunner, H., Guibas, L. J., and Sharir, M. 1993. Diameter, width, closest line pair, and parametric searching. *Discrete Comput. Geom.* 8(2):183–196.

Chazelle, B., Edelsbrunner, H., Guibas, L. J., and Sharir, M. 1994. Algorithms for bichromatic line-segment problems and polyhedral terrains. *Algorithmica* 11(2):116–132.

Chazelle, B. and Friedman, J. 1990. A deterministic view of random sampling and its use in geometry. *Combinatorica* 10(3):229–249.

Chazelle, B. and Friedman, J. 1994. Point location among hyperplanes and unidirectional ray-shooting. *Comput. Geom. Theory Appl.* 4(2):53–62.

Chazelle, B. and Guibas, L. J. 1986. Fractional cascading: I. a data structuring technique. *Algorithmica* 1(2):133–186.

Chazelle, B., Guibas, L. J., and Lee, D. T. 1985. The power of geometric duality. *BIT* 25:76–90.

Chazelle, B. and Matoušek, J. 1993. On linear-time deterministic algorithms for optimization problems in fixed dimension, pp. 281–290. In *Proc. 4th ACM–SIAM Symp. Discrete Algorithms.*

Chazelle, B. and Preparata, F. P. 1986. Halfspace range search: an algorithmic application of *k*-sets. *Discrete Comput. Geom.* 1(1):83–93.

Chew, L. P. 1989. Constrained Delaunay triangulations. *Algorithmica* 4(1):97–108.

Chiang, Y.-J. and Tamassia, R. 1992. Dynamic algorithms in computational geometry. *Proc. IEEE* 80(9):1412–1434.

Clarkson, K. L. 1986. Linear programming in $O(n3^{d^2})$ time. *Inf. Proc. Lett.* 22:21–24.

Clarkson, K. L. 1988. A randomized algorithm for closest-point queries. *SIAM J. Comput.* 17(4): 830–847.

Culberson, J. C. and Reckhow, R. A. 1988. Covering polygons is hard, pp. 601–611. In *Proc. 29th Annu. IEEE Symp. Found. Comput. Sci.*

Dobkin, D. P. and Kirkpatrick, D. G. 1985. A linear algorithm for determining the separation of convex polyhedra. *J. Algorithms* 6:381–392.

Dobkin, D. P. and Munro, J. I. 1981. Optimal time minimal space selection algorithms. *J. ACM* 28(3):454–461.

Dorward, S. E. 1994. A survey of object-space hidden surface removal. *Int. J. Comput. Geom. Appl.* 4(3):325–362.

Du, D. Z. and Hwang, F. K., eds. 1992. *Computing in Euclidean Geometry.* World Scientific, Singapore.

Dyer, M. E. 1984. Linear programs for two and three variables. *SIAM J. Comput.* 13(1):31–45.

Dyer, M. E. 1986. On a multidimensional search technique and its applications to the Euclidean one-center problem. *SIAM J. Comput.* 15(3):725–738.

Edelsbrunner, H. 1985. Computing the extreme distances between two convex polygons. *J. Algorithms* 6:213–224.

Edelsbrunner, H. 1987. *Algorithms in Combinatorial Geometry.* Springer–Verlag.

Edelsbrunner, H. and Guibas, L. J. 1989. Topologically sweeping an arrangement. *J. Comput. Syst. Sci.* 38:165–194; (1991). *Corrigendum* 42:249–251.

Edelsbrunner, H., Guibas, L. J., and Stolfi, J. 1986. Optimal point location in a monotone subdivision. *SIAM J. Comput.* 15(2):317–340.

Edelsbrunner, H., O'Rourke, J., and Seidel, R. 1986. Constructing arrangements of lines and hyperplanes with applications. *SIAM J. Comput.* 15(2):341–363.

Edelsbrunner, H. and Shi, W. 1991. An $O(n \log^2 h)$ time algorithm for the three-dimensional convex hull problem. *SIAM J. Comput.* 20(2):259–269.

Fortune, S. 1987. A sweepline algorithm for Voronoi diagrams. *Algorithmica* 2(2):153–174.

Fortune, S. 1993. Progress in computational geometry. In *Directions in Geom. Comput.*, pp. 81–128. R. Martin, ed. Information Geometers Ltd.

Ghosh, S. K. and Mount, D. M. 1991. An output-sensitive algorithm for computing visibility graphs. *SIAM J. Comput.* 20(5):888–910.

Guibas, L. J. and Hershberger, J. 1989. Optimal shortest path queries in a simple polygon. *J. Comput. Syst. Sci.* 39:126–152.

Guibas, L. J. and Seidel, R. 1987. Computing convolutions by reciprocal search. *Discrete Comput. Geom.* 2(2):175–193.

Handler, G. Y. and Mirchandani, P. B. 1979. *Location on Networks: Theory and Algorithm.* MIT Press, Cambridge, MA.

Hershberger, J. and Suri, S. 1993. Efficient computation of Euclidean shortest paths in the plane, pp. 508–517. In *Proc. 34th Annu. IEEE Symp. Found. Comput. Sci.*

Ho, J. M., Chang, C. H., Lee, D. T., and Wong, C. K. 1991. Minimum diameter spanning tree and related problems. *SIAM J. Comput.* 20(5):987–997.

Houle, M. E. and Toussaint, G. T. 1988. Computing the width of a set. *IEEE Trans. Pattern Anal. Machine Intelligence* PAMI-10(5):761–765.

Imai, H. and Asano, T. 1986. Efficient algorithms for geometric graph search problems. *SIAM J. Comput.* 15(2):478–494.

Imai, H. and Asano, T. 1987. Dynamic orthogonal segment intersection search. *J. Algorithms* 8(1):1–18.

Janardan, R. and Lopez, M. 1993. Generalized intersection searching problems. *Int. J. Comput. Geom. Appl.* 3(1):39–69.

Kapoor, S. and Smid, M. 1996. New techniques for exact and approximate dynamic closest-point problems. *SIAM J. Comput.* 25(4):775–796.

Keil, J. M. 1985. Decomposing a polygon into simpler components. *SIAM J. Comput.* 14(4):799–817.

Kirkpatrick, D. G. and Seidel, R. 1986. The ultimate planar convex hull algorithm? *SIAM J. Comput.*, 15(1):287–299.

Klein, R. 1989. *Concrete and Abstract Voronoi Diagrams.* LNCS Vol. 400, Springer–Verlag.

Le, V. B. and Lee, D. T. 1991. Out-of-roundness problem revisited. *IEEE Trans. Pattern Anal. Machine Intelligence* 13(3):217–223.

Lee, D. T. 1978. *Proximity and Reachability in the Plan.* Ph.D. Thesis, *Tech. Rep.* R-831, Coordinated Science Lab., University of Illinois, Urbana.

Lee, D. T. and Drysdale, R. L., III 1981. Generalization of Voronoi diagrams in the plane. *SIAM J. Comput.* 10(1):73–87.

Lee, D. T. and Lin, A. K. 1986a. Computational complexity of art gallery problems. *IEEE Trans. Inf. Theory* 32(2):276–282.

Lee, D. T. and Lin, A. K. 1986b. Generalized Delaunay triangulation for planar graphs. *Discrete Comput. Geom.* 1(3):201–217.

Lee, D. T. and Preparata, F. P. 1982. An improved algorithm for the rectangle enclosure problem. *J. Algorithms* 3(3):218–224.

Lee, D. T. and Preparata, F. P. 1984. Computational geometry: a survey. *IEEE Trans. Comput.* C-33(12):1072–1101.

Lee, D. T. and Wu, V. B. 1993. Multiplicative weighted farthest neighbor Voronoi diagrams in the plane, pp. 154–168. In *Proc. Int. Workshop Discrete Math. and Algorithms.* Hong Kong, Dec.

Lee, D. T. and Wu, Y. F. 1986. Geometric complexity of some location problems. *Algorithmica* 1(2):193–211.

Lee, D. T., Yang, C. D., and Chen, T. H. 1991. Shortest rectilinear paths among weighted obstacles. *Int. J. Comput. Geom. Appl.* 1(2):109–124.

Lee, D. T., Yang, C. D., and Wong, C. K. 1996. Rectilinear paths among rectilinear obstacles. In *Perspectives in Discrete Applied Math.* K. Bogart, ed.

Lozano-Pérez, T. 1983. Spatial planning: a configuration space approach. *IEEE Trans. Comput.* C-32(2):108–120.

Mairson, H. G. and Stolfi, J. 1988. Reporting and counting intersections between two sets of line segments, pp. 307–325. In *Proc. Theor. Found. Comput. Graphics CAD.* Vol. F40, Springer–Verlag.

Matoušek, J. 1994. Geometric range searching. *ACM Computing Sur.* 26:421–461.

Matoušek, J., Sharir. M., and Welzl, E. 1992. A subexponential bound for linear programming, pp. 1–8. In *Proc. 8th Ann. ACM Symp. Comput. Geom.*

Megiddo, N. 1983a. Applying parallel computation algorithms in the design of serial algorithms. *J. ACM* 30(4):852–865.

Megiddo, N. 1983b. Linear time algorithm for linear programming in R^3 and related problems. *SIAM J. Comput.* 12(4):759–776.

Megiddo, N. 1983c. Towards a genuinely polynomial algorithm for linear programming. *SIAM J. Comput.* 12(2):347–353.

Megiddo, N. 1984. Linear programming in linear time when the dimension is fixed. *J. ACM* 31(1):114–127.

Megiddo, N. 1986. New approaches to linear programming. *Algorithmica* 1(4):387–394.

Mehlhorn, K. 1984. *Data Structures and Algorithms*, Vol. 3, Multi-dimensional searching and computational geometry. Springer–Verlag.

Mitchell, J. S. B. 1993. Shortest paths among obstacles in the plane, pp. 308–317. In *Proc. 9th ACM Symp. Comput. Geom.* May.

Mitchell, J. S. B., Mount, D. M., and Papadimitriou, C. H. 1987. The discrete geodesic problem. *SIAM J. Comput.* 16(4):647–668.

Mitchell, J. S. B. and Papadimitriou, C. H. 1991. The weighted region problem: finding shortest paths through a weighted planar subdivision. *J. ACM* 38(1):18–73.

Mitchell, J. S. B., Rote, G., and Wöginger, G. 1992. Minimum link path among obstacles in the planes. *Algorithmica* 8(5/6):431–459.

Mulmuley, K. 1994. *Computational Geometry: An Introduction through Randomized Algorithms.* Prentice–Hall, Englewood Cliffs, NJ.

Nievergelt, J. and Preparata, F. P. 1982. Plane-sweep algorithms for intersecting geometric figures. *Commun. ACM* 25(10):739–747.

O'Rourke, J. 1987. *Art Gallery Theorems and Algorithms.* Oxford University Press, New York.

O'Rourke, J. 1994. *Computational Geometry in C.* Cambridge University Press, New York.

O'Rourke, J. and Supowit, K. J. 1983. Some NP-hard polygon decomposition problems. *IEEE Trans. Inform. Theory* IT-30(2):181–190,

Overmars, M. H. 1983. *The Design of Dynamic Data Structures.* LNCS Vol. 156, Springer–Verlag.

Preparata, F. P. and Shamos, M. I. 1985. In *Computational Geometry: An Introduction.* Springer–Verlag.

Ruppert, J. and Seidel, R. 1992. On the difficulty of triangulating three-dimensional non-convex polyhedra. *Discrete Comput. Geom.* 7(3):227–253.

Schaudt, B. F. and Drysdale, R. L. 1991. Multiplicatively weighted crystal growth Voronoi diagrams, pp. 214–223. In *Proc. 7th Annu. ACM Symp. Comput. Geom.*

Schwartz, C., Smid, M., and Snoeyink, J. 1994. An optimal algorithm for the on-line closest-pair problem. *Algorithmica* 12(1):18–29.

Seo, D. Y. and Lee, D. T. 1995. On the complexity of bicriteria spanning tree problems for a set of points in the plane. *Tech. Rep. Dept.* EE/CS, Northwestern University, June.

Sharir, M. 1985. Intersection and closest-pair problems for a set of planar discs. *SIAM J. Comput.* 14(2):448–468.

Sharir, M. 1987. On shortest paths amidst convex polyhedra. *SIAM J. Comput.* 16(3):561–572.

Shermer, T. C. 1992. Recent results in art galleries. *Proc. IEEE* 80(9):1384–1399.

Sifrony, S. and Sharir, M. 1987. A new efficient motion planning algorithm for a rod in two-dimensional polygonal space. *Algorithmica* 2(4):367–402.

Smid, M. 1992. Maintaining the minimal distance of a point set in polylogarithmic time. *Discrete Comput. Geom.* 7:415–431.

Suri, S. 1990. On some link distance problems in a simple polygon. *IEEE Trans. Robotics Automation* 6(1):108–113.

Swanson, K., Lee, D. T., and Wu, V. L. 1995. An optimal algorithm for roundness determination on convex polygons. *Comput. Geom. Theory Appl.* 5(4):225–235.

Toussaint, G. T., ed. 1985. *Computational Geometry*. North–Holland.

Van der Stappen, A. F. and Overmars, M. H. 1994. Motion planning amidst fat obstacle, pp. 31–40. In *Proc. 10th Ann. ACM Comput. Geom.*, June.

van Leeuwen, J. and Wood, D. 1980. Dynamization of decomposable searching problems. *Inf. Proc. Lett.* 10:51–56.

Willard, D. E. 1982. Polygon retrieval. *SIAM J. Comput.* 11(1):149–165.

Willard, D. E. 1985. New data structures for orthogonal range queries. *SIAM J. Comput.* 14(1):232–253.

Willard, D. E. and Lueker, G. S. 1985. Adding range restriction capability to dynamic data structures. *J. ACM* 32(3):597–617.

Yao, F. F. 1994. Computational geometry. In *Handbook of Theoretical Computer Science*, Vol. A: Algorithms and Complexity, J. van Leeuwen, ed., pp. 343–389.

Yap, C. K. 1987. An $O(n \log n)$ algorithm for the Voronoi diagram of a set of simple curve segments. *Discrete Comput. Geom.* 2(4):365–393.

Yap, C. K. 1987. Algorithmic motion planning. In *Advances in Robotics, Vol I: Algorithmic and Geometric Aspects of Robotics*. J. T. Schwartz and C. K. Yap, eds., pp. 95–143. Lawrence Erlbaum, London.

Further Information

We remark that there are new efforts being made in the applied side of algorithm development. A library of geometric software including visualization tools and applications programs is under development at the Geometry Center, University of Minnesota, and a concerted effort is being put together by researchers in Europe and in the United States to organize a system library containing primitive geometric abstract data types useful for geometric algorithm developers and practitioners.

Those who are interested in the implementations or would like to have more information about available software may consult the Proceedings of the Annual ACM Symposium on Computational Geometry, which has a video session, or the WWW page on *Geometry in Action* by David Eppstein (http://www.ics.uci.edu/~eppstein/geom.html).

7

Randomized Algorithms

Rajeev Motwani*
Stanford University

Prabhakar Raghavan
*IBM Almaden Research
Center*

7.1 Introduction

A **randomized algorithm** is one that makes random choices during its execution. The behavior of such an algorithm may thus be random even on a fixed input. The design and analysis of a randomized algorithm focus on establishing that it is likely to behave well on *every* input; the likelihood in such a statement depends only on the probabilistic choices made by the algorithm during execution and not on any assumptions about the input. It is especially important to distinguish a randomized algorithm from the *average-case analysis* of algorithms, where one analyzes an algorithm assuming that its input is drawn from a fixed probability distribution. With a randomized algorithm, in contrast, no assumption is made about the input.

Two benefits of randomized algorithms have made them popular: simplicity and efficiency. For many applications, a randomized algorithm is the simplest algorithm available, or the fastest, or both. In the following, we make these notions concrete through a number of illustrative examples. We assume that the reader has had undergraduate courses in algorithms and complexity, and in probability theory. A comprehensive source for randomized algorithms is the book by the authors [Motwani and Raghavan 1995]. The articles by Karp [1991], Maffioli et al. [1985], and Welsh [1983] are good surveys of randomized algorithms. The book by Mulmuley [1993] focuses on randomized geometric algorithms.

Throughout this chapter, we assume the random access memory (RAM) model of computation, in which we have a machine that can perform the following operations involving registers and main memory: input–output operations, memory-register transfers, indirect addressing, branching, and arithmetic operations. Each register or memory location may hold an integer which can be accessed as a unit, but an algorithm

*Supported by an Alfred P. Sloan Research Fellowship, an IBM Faculty Partnership Award, an ARO MURI Grant DAAH04-96-1-0007, and NSF Young Investigator Award CCR-9357849, with matching funds from IBM, Schlumberger Foundation, Shell Foundation, and Xerox Corporation.

has no access to the representation of the number. The arithmetic instructions permitted are $+, -, \times, /$. In addition, an algorithm can compare two numbers and evaluate the square root of a positive number. In this chapter, $\mathbf{E}[X]$ will denote the expectation of random variable X, and $\mathbf{Pr}[A]$ will denote the probability of event A.

7.2 Sorting and Selection by Random Sampling

Some of the earliest randomized algorithms included algorithms for sorting the set S of numbers and the related problem of finding the kth smallest element in S. The main idea behind these algorithms is the use of *random sampling*: a randomly chosen member of S is unlikely to be one of its largest or smallest elements; rather, it is likely to be near the middle. Extending this intuition suggests that a random sample of elements from S is likely to be spread roughly uniformly in S. We now describe randomized algorithms for sorting and selection based on these ideas.

ALGORITHM RQS.

Input: A set of numbers, S.
Output: The elements of S sorted in increasing order.

1. *Choose element y uniformly at random from S: every element in S has equal probability of being chosen.*
2. *By comparing each element of S with y, determine the set S_1 of elements smaller than y and the set S_2 of elements larger than y.*
3. *Recursively sort S_1 and S_2. Output the sorted version of S_1, followed by y, and then the sorted version of S_2.*

Algorithm RQS is an example of a *randomized algorithm*: an algorithm that makes random choices during execution. It is inspired by the Quicksort algorithm due to Hoare [1962], and described in Motwani and Raghavan [1995]. We assume that the random choice in step 1 can be made in unit time. What can we prove about the running time of RQS?

We now analyze the *expected* number of comparisons in an execution of RQS. Comparisons are performed in step 2, in which we compare a randomly chosen element to the remaining elements. For $1 \leq i \leq n$, let $S_{(i)}$ denote the element of *rank i* (the ith smallest element) in the set S. Define the random variable X_{ij} to assume the value 1 if $S_{(i)}$ and $S_{(j)}$ are compared in an execution and the value 0 otherwise. Thus, the total number of comparisons is $\sum_{i=1}^{n} \sum_{j>i} X_{ij}$. By linearity of expectation, the expected number of comparisons is

$$\mathbf{E}\left[\sum_{i=1}^{n} \sum_{j>i} X_{ij}\right] = \sum_{i=1}^{n} \sum_{j>i} \mathbf{E}[X_{ij}] \qquad (7.1)$$

Let p_{ij} denote the probability that $S_{(i)}$ and $S_{(j)}$ are compared during an execution. Then

$$\mathbf{E}[X_{ij}] = p_{ij} \times 1 + (1 - p_{ij}) \times 0 = p_{ij} \qquad (7.2)$$

To compute p_{ij}, we view the execution of RQS as binary tree T, each node of which is labeled with a distinct element of S. The root of the tree is labeled with the element y chosen in step 1; the left subtree of y contains the elements in S_1 and the right subtree of y contains the elements in S_2. The structures of the two subtrees are determined recursively by the executions of RQS on S_1 and S_2. The root y is compared to the elements in the two subtrees, but no comparison is performed between an element of the left subtree and an element of the right subtree. Thus, there is a comparison between $S_{(i)}$ and $S_{(j)}$ if and only if one of these elements is an ancestor of the other.

Consider the permutation π obtained by visiting the nodes of T in increasing order of the level numbers and in a left-to-right order within each level; recall that the ith level of the tree is the set of all nodes at a distance exactly i from the root. The following two observations lead to the determination of p_{ij}:

1. There is a comparison between $S_{(i)}$ and $S_{(j)}$ if and only if $S_{(i)}$ or $S_{(j)}$ occurs earlier in the permutation π than any element $S_{(\ell)}$ such that $i < \ell < j$. To see this, let $S_{(k)}$ be the earliest in π from among all elements of rank between i and j. If $k \notin \{i, j\}$, then $S_{(i)}$ will belong to the left subtree of $S_{(k)}$ and $S_{(j)}$ will belong to the right subtree of $S_{(k)}$, implying that there is no comparison between $S_{(i)}$ and $S_{(j)}$. Conversely, when $k \in \{i, j\}$, there is an ancestor–descendant relationship between $S_{(i)}$ and $S_{(j)}$, implying that the two elements are compared by RQS.
2. Any of the elements $S_{(i)}, S_{(i+1)}, \ldots, S_{(j)}$ is equally likely to be the first of these elements to be chosen as a partitioning element and hence to appear first in π. Thus, the probability that this first element is either $S_{(i)}$ or $S_{(j)}$ is exactly $2/(j - i + 1)$.

It follows that $p_{ij} = 2/(j - i + 1)$. By Eqs. (7.1) and (7.2), the expected number of comparisons is given by

$$\sum_{i=1}^{n} \sum_{j>i} p_{ij} = \sum_{i=1}^{n} \sum_{j>i} \frac{2}{j - i + 1}$$

$$\leq \sum_{i=1}^{n-1} \sum_{k=1}^{n-i} \frac{2}{k + 1}$$

$$\leq 2 \sum_{i=1}^{n} \sum_{k=1}^{n} \frac{1}{k}.$$

It follows that the expected number of comparisons is bounded above by $2nH_n$, where H_n is the nth *harmonic number*, defined by $H_n = \sum_{k=1}^{n} 1/k$.

Theorem 7.1. *The expected number of comparisons in an execution of RQS is at most $2nH_n$.*

Now $H_n = \ell_n n + \Theta(1)$, so that the expected running time of RQS is $O(n \log n)$. Note that this expected running time *holds for every input*. It is an expectation that depends only on the random choices made by the algorithm and *not* on any assumptions about the distribution of the input.

Randomized Selection

We now consider the use of random sampling for the problem of selecting the kth smallest element in set S of n elements drawn from a totally ordered universe. We assume that the elements of S are all distinct, although it is not very hard to modify the following analysis to allow for multisets. Let $r_S(t)$ denote the rank of element t (the kth smallest element has rank k) and recall that $S_{(i)}$ denotes the ith smallest element of S. Thus, we seek to identify $S_{(k)}$. We extend the use of this notation to subsets of S as well. The following algorithm is adapted from one due to Floyd and Rivest [1975].

ALGORITHM LAZYSELECT.

Input: A set, S, of n elements from a totally ordered universe and an integer, k, in $[1, n]$.
Output: The kth smallest element of S, $S_{(k)}$.

1. *Pick $n^{3/4}$ elements from S, chosen independently and uniformly at random with replacement; call this multiset of elements R.*
2. *Sort R in $O(n^{3/4} \log n)$ steps using any optimal sorting algorithm.*

3. *Let* $x = kn^{-1/4}$. *For* $\ell = \max\{\lfloor x - \sqrt{n} \rfloor, 1\}$ *and* $h = \min\{\lceil x + \sqrt{n} \rceil, n^{3/4}\}$, *let* $a = R_{(\ell)}$ *and* $b = R_{(h)}$. *By comparing* a *and* b *to every element of* S, *determine* $r_S(a)$ *and* $r_S(b)$.

4. *if* $k < n^{1/4}$, *let* $P = \{y \in S \mid y \le b\}$ *and* $r = k$;
   else if $k > n - n^{1/4}$, *let* $P = \{y \in S \mid y \ge a\}$ *and* $r = k - r_S(a) + 1$;
   else if $k \in [n^{1/4}, n - n^{1/4}]$, *let* $P = \{y \in S \mid a \le y \le b\}$ *and* $r = k - r_S(a) + 1$;
   Check whether $S_{(k)} \in P$ *and* $|P| \le 4n^{3/4} + 2$. *If not, repeat steps 1–3 until such a set, P, is found.*

5. *By sorting* P *in* $O(|P| \log |P|)$ *steps, identify* P_r, *which is* $S_{(k)}$.

Figure 7.1 illustrates step 3, where small elements are at the left end of the picture and large ones are to the right. Determining (in step 4) whether $S_{(k)} \in P$ is easy because we know the ranks $r_S(a)$ and $r_S(b)$ and we compare either or both of these to k, depending

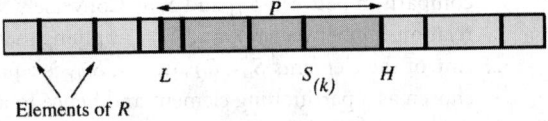

FIGURE 7.1   The LazySelect algorithm.

on which of the three *if* statements in step 4 we execute. The sorting in step 5 can be performed in $O(n^{3/4} \log n)$ steps.

Thus, the idea of the algorithm is to identify two elements a and b in S such that both of the following statements hold with high probability:

1. The element $S_{(k)}$ that we seek is in P, the set of elements between a and b.
2. The set P of elements is not very large, so that we can sort P inexpensively in step 5.

As in the analysis of RQS, we measure the running time of LazySelect in terms of the number of comparisons performed by it. The following theorem is established using the *Chebyshev bound* from elementary probability theory; a full proof may be found in Motwani and Raghavan [1995].

Theorem 7.2. *With probability* $1 - O(n^{-1/4})$, *LazySelect finds* $S_{(k)}$ *on the first pass through steps 1–5 and thus performs only* $2n + o(n)$ *comparisons.*

This adds to the significance of LazySelect: the best known deterministic selection algorithms use $3n$ comparisons in the worst case and are quite complicated to implement.

7.3 A Simple Min-Cut Algorithm

Two events $\mathcal{E}_1$ and $\mathcal{E}_2$ are said to be *independent* if the probability that they both occur is given by

$$\Pr[\mathcal{E}_1 \cap \mathcal{E}_2] = \Pr[\mathcal{E}_1] \times \Pr[\mathcal{E}_2] \tag{7.3}$$

More generally when $\mathcal{E}_1$ and $\mathcal{E}_2$ are not necessarily independent,

$$\Pr[\mathcal{E}_1 \cap \mathcal{E}_2] = \Pr[\mathcal{E}_1 \mid \mathcal{E}_2] \times \Pr[\mathcal{E}_2] = \Pr[\mathcal{E}_2 \mid \mathcal{E}_1] \times \Pr[\mathcal{E}_1] \tag{7.4}$$

where $\Pr[\mathcal{E}_1 \mid \mathcal{E}_2]$ denotes the *conditional probability* of $\mathcal{E}_1$ given $\mathcal{E}_2$. When a collection of events is not independent, the probability of their intersection is given by the following generalization of Eq. (7.4):

$$\Pr\left[\bigcap_{i=1}^{k} \mathcal{E}_i \right] = \Pr[\mathcal{E}_1] \times \Pr[\mathcal{E}_2 \mid \mathcal{E}_1] \times \Pr[\mathcal{E}_3 \mid \mathcal{E}_1 \cap \mathcal{E}_2] \cdots \Pr\left[\mathcal{E}_k \,\middle|\, \bigcap_{i=1}^{k-1} \mathcal{E}_i \right] \tag{7.5}$$

Let G be a connected, undirected multigraph with n vertices. A *multigraph* may contain multiple edges between any pair of vertices. A *cut* in G is a set of edges whose removal results in G being broken into two or more components. A *min-cut* is a cut of minimum cardinality. We now study a simple algorithm due to Karger [1993] for finding a min-cut of a graph.

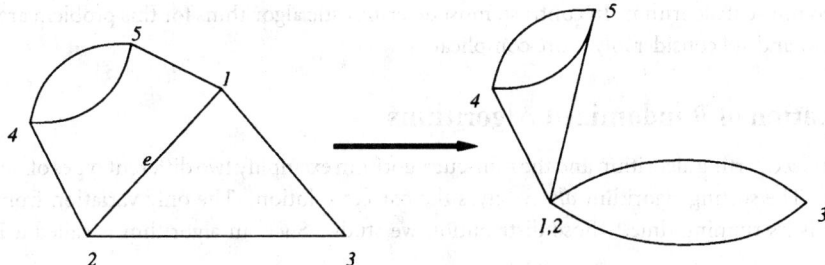

FIGURE 7.2 A step in the min-cut algorithm; the effect of contracting edge $e = (1, 2)$ is shown.

We repeat the following step: pick an edge uniformly at random and merge the two vertices at its end points. If as a result there are several edges between some pairs of (newly formed) vertices, retain them all. Remove edges between vertices that are merged, so that there are never any self-loops. This process of merging the two endpoints of an edge into a single vertex is called the *contraction* of that edge. See Fig. 7.2. With each contraction, the number of vertices of G decreases by one. Note that as long as at least two vertices remain, an edge contraction does not reduce the min-cut size in G. The algorithm continues the contraction process until only two vertices remain; at this point, the set of edges between these two vertices is a cut in G and is output as a candidate min-cut. What is the probability that this algorithm finds a min-cut?

Definition 7.1. For any vertex v in the multigraph G, the *neighborhood* of G, denoted $\Gamma(v)$, is the set of vertices of G that are adjacent to v. The *degree* of v, denoted $d(v)$, is the number of edges incident on v. For the set S of vertices of G, the neighborhood of S, denoted $\Gamma(S)$, is the union of the neighborhoods of the constituent vertices.

Note that $d(v)$ is the same as the cardinality of $\Gamma(v)$ when there are no self-loops or multiple edges between v and any of its neighbors.

Let k be the min-cut size and let C be a particular min-cut with k edges. Clearly, G has at least $kn/2$ edges (otherwise there would be a vertex of degree less than k, and its incident edges would be a min-cut of size less than k). We bound from below the probability that no edge of C is ever contracted during an execution of the algorithm, so that the edges surviving until the end are exactly the edges in C.

For $1 \leq i \leq n - 2$, let $\mathcal{E}_i$ denote the event of *not* picking an edge of C at the ith step. The probability that the edge randomly chosen in the first step is in C is at most $k/(nk/2) = 2/n$, so that $\Pr[\mathcal{E}_1] \geq 1 - 2/n$. Conditioned on the occurrence of $\mathcal{E}_1$, there are at least $k(n - 1)/2$ edges during the second step so that $\Pr[\mathcal{E}_2 \mid \mathcal{E}_1] \geq 1 - 2/(n - 1)$. Extending this calculation, $\Pr[\mathcal{E}_i \mid \cap_{j=1}^{i-1}\mathcal{E}_j] \geq 1 - 2/(n - i + 1)$. We now invoke Eq. (7.5) to obtain

$$\Pr\left[\bigcap_{i=1}^{n-2} \mathcal{E}_i\right] \geq \prod_{i=1}^{n-2}\left(1 - \frac{2}{n - i + 1}\right) = \frac{2}{n(n - 1)}$$

Our algorithm may err in declaring the cut it outputs to be a min-cut. But the probability of discovering a particular min-cut (which may in fact be the unique min-cut in G) is larger than $2/n^2$, so that the probability of error is at most $1 - 2/n^2$. Repeating the preceding algorithm $n^2/2$ times making independent random choices each time, the probability that a min-cut is not found in any of the $n^2/2$ attempts is [by Eq. (7.3)] at most

$$\left(1 - \frac{2}{n^2}\right)^{n^2/2} < \frac{1}{e}$$

By this process of repetition, we have managed to reduce the probability of failure from $1 - 2/n^2$ to less than $1/e$. Further executions of the algorithm will make the failure probability arbitrarily small (the only consideration being that repetitions increase the running time). Note the extreme simplicity of this randomized min-cut algorithm. In contrast, most deterministic algorithms for this problem are based on network flow and are considerably more complicated.

Classification of Randomized Algorithms

The randomized sorting algorithm and the min-cut algorithm exemplify two different types of randomized algorithms. The sorting algorithm *always* gives the correct solution. The only variation from one run to another is its running time, whose distribution we study. Such an algorithm is called a **Las Vegas algorithm**.

In contrast, the min-cut algorithm may sometimes produce a solution that is incorrect. However, we prove that the probability of such an error is bounded. Such an algorithm is called a **Monte Carlo algorithm**. In section 7.3, we observed a useful property of a Monte Carlo algorithm: if the algorithm is run repeatedly with independent random choices each time, the failure probability can be made arbitrarily small, at the expense of running time. In some randomized algorithms both the running time and the quality of the solution are random variables; sometimes these are also referred to as Monte Carlo algorithms. The reader is referred to Motwani and Raghavan [1995] for a detailed discussion of these issues.

7.4 Foiling an Adversary

A common paradigm in the design of randomized algorithms is that of *foiling an adversary*. Whereas an adversary might succeed in defeating a **deterministic algorithm** with a carefully constructed *bad* input, it is difficult for an adversary to defeat a randomized algorithm in this fashion. Due to the random choices made by the randomized algorithm the adversary cannot, while constructing the input, predict the precise behavior of the algorithm. An alternative view of this process is to think of the randomized algorithm as first picking a series of random numbers, which it then uses in the course of execution as needed. In this view, we may think of the random numbers chosen at the start as *selecting* one of a family of deterministic algorithms. In other words, a randomized algorithm can be thought of as a probability distribution on deterministic algorithms. We illustrate these ideas in the setting of AND–OR *tree evaluation*; the following algorithm is due to Snir [1985].

For our purposes, an AND–OR tree is a rooted complete binary tree in which internal nodes at even distance from the root are labeled AND and internal nodes at odd distance are labeled OR. Associated with each leaf is a Boolean *value*. The *evaluation* of the game tree is the following process. Each leaf *returns* the value associated with it. Each OR node returns the Boolean OR of the values returned by its children, and each AND node returns the Boolean AND of the values returned by its children. At each step an evaluation algorithm chooses a leaf and reads its value. We do not charge the algorithm for any other computation. We study the number of such steps taken by an algorithm for evaluating an AND–OR tree, the worst case being taken over all assignments of Boolean values of the leaves.

Let T_k denote an AND-OR tree in which every leaf is at distance $2k$ from the root. Thus, any root-to-leaf path passes through k AND nodes (including the root itself) and k OR nodes, and there are 2^{2k} leaves. An algorithm begins by specifying a leaf whose value is to be read at the first step. Thereafter, it specifies such a leaf at each step based on the values it has read on previous steps. In a deterministic algorithm, the choice of the next leaf to be read is a deterministic function of the values at the leaves read so far. For a randomized algorithm, this choice may be randomized. It is not hard to show that for any deterministic evaluation algorithm, there is an instance of T_k that forces the algorithm to read the values on all 2^{2k} leaves.

We now give a simple randomized algorithm and study the expected number of leaves it reads on any instance of T_k. The algorithm is motivated by the following simple observation. Consider a single AND node with two leaves. If the node were to return 0, at least one of the leaves must contain 0. A deterministic algorithm inspects the leaves in a fixed order, and an adversary can therefore always *hide* the 0 at the second of the two leaves inspected by the algorithm. Reading the leaves in a random order foils this strategy. With probability $1/2$, the algorithm chooses the hidden 0 on the first step, so that its expected number of steps is 3/2, which is better than the worst case for any deterministic algorithm. Similarly, in the case of an OR node, if it were to return a 1, then a randomized order of examining the leaves will reduce the expected number of steps to 3/2. We now extend this intuition and specify the complete algorithm.

To evaluate an AND node, v, the algorithm chooses one of its children (a subtree rooted at an OR node) at random and evaluates it by recursively invoking the algorithm. If 1 is returned by the subtree, the algorithm proceeds to evaluate the other child (again by recursive application). If 0 is returned, the algorithm returns 0 for v. To evaluate an OR node, the procedure is the same with the roles of 0 and 1 interchanged. We establish by induction on k that the expected cost of evaluating any instance of T_k is at most 3^k.

The basis ($k = 0$) is trivial. Assume now that the expected cost of evaluating any instance of T_{k-1} is at most 3^{k-1}. Consider first tree T whose root is an OR node, each of whose children is the root of a copy of T_{k-1}. If the root of T were to evaluate to 1, at least one of its children returns 1. With probability 1/2 this child is chosen first, incurring (by the inductive hypothesis) an expected cost of at most 3^{k-1} in evaluating T. With probability 1/2 both subtrees are evaluated, incurring a net cost of at most $2 \times 3^{k-1}$. Thus, the expected cost of determining the value of T is

$$\leq \frac{1}{2} \times 3^{k-1} + \frac{1}{2} \times 2 \times 3^{k-1} = \frac{3}{2} \times 3^{k-1} \tag{7.6}$$

If, on the other hand, the OR were to evaluate to 0 both children must be evaluated, incurring a cost of at most $2 \times 3^{k-1}$.

Consider next the root of the tree T_k, an AND node. If it evaluates to 1, then both its subtrees rooted at OR nodes return 1. By the discussion in the previous paragraph and by linearity of expectation, the expected cost of evaluating T_k to 1 is at most $2 \times (3/2) \times 3^{k-1} = 3^k$. On the other hand, if the instance of T_k evaluates to 0, at least one of its subtrees rooted at OR nodes returns 0. With probability 1/2 it is chosen first, and so the expected cost of evaluating T_k is at most

$$2 \times 3^{k-1} + \frac{1}{2} \times \frac{3}{2} \times 3^{k-1} \leq 3^k$$

Theorem 7.3. *Given any instance of T_k, the expected number of steps for the preceding randomized algorithm is at most 3^k.*

Because $n = 4^k$ the expected running time of our randomized algorithm is $n^{\log_4 3}$, which we bound by $n^{0.793}$. Thus, the expected number of steps is smaller than the worst case for any deterministic algorithm. Note that this is a Las Vegas algorithm and always produces the correct answer.

7.5 The Minimax Principle and Lower Bounds

The randomized algorithm of the preceding section has an expected running time of $n^{0.793}$ on any uniform binary AND–OR tree with n leaves. Can we establish that *no randomized algorithm* can have a lower expected running time? We first introduce a standard technique due to Yao [1977] for proving such lower bounds. This technique applies only to algorithms that terminate in finite time on all inputs and sequences of random choices.

The crux of the technique is to relate the running times of randomized algorithms for a problem to the running times of deterministic algorithms for the problem *when faced with randomly chosen inputs*. Consider a problem where the number of distinct inputs of a fixed size is finite, as is the number of distinct (deterministic, terminating, and always correct) algorithms for solving that problem. Let us define the **distributional complexity** of the problem at hand as the expected running time of the best deterministic algorithm for the worst distribution on the inputs. Thus, we envision an adversary choosing a probability distribution on the set of possible inputs and study the best deterministic algorithm for this distribution. Let p denote a probability distribution on the set $\mathcal{I}$ of inputs. Let the random variable $C(I_p, A)$ denote the running time of deterministic algorithm $A \in \mathcal{A}$ on an input chosen according to p. Viewing a randomized algorithm as a probability distribution q on the set $\mathcal{A}$ of deterministic algorithms, we let the random variable $C(I, A_q)$ denote the running time of this randomized algorithm on the worst-case input.

Proposition 7.1 (Yao's Minimax Principle). *For all distributions p over $\mathcal{I}$ and q over $\mathcal{A}$,*

$$\min_{A \in \mathcal{A}} \mathrm{E}[C(I_p, A)] \leq \max_{I \in \mathcal{I}} \mathrm{E}[C(I, A_q)]$$

In other words, the expected running time of the optimal deterministic algorithm for an arbitrarily chosen input distribution p is a lower bound on the expected running time of the optimal (Las Vegas) randomized algorithm for Π. Thus, to prove a lower bound on the randomized complexity, it suffices to choose any distribution p on the input and prove a lower bound on the expected running time of deterministic algorithms for that distribution. The power of this technique lies in the flexibility in the choice of p and, more importantly, the reduction to a lower bound on deterministic algorithms. It is important to remember that the deterministic algorithm "knows" the chosen distribution p.

The preceding discussion dealt only with lower bounds on the performance of Las Vegas algorithms. We briefly discuss Monte Carlo algorithms with error probability $\epsilon \in [0, 1/2]$. Let us define the distributional complexity with error ϵ, denoted $\min_{A \in \mathcal{A}} \mathrm{E}[C_\epsilon(I_p, A)]$, to be the minimum expected running time of any deterministic algorithm that errs with probability at most ϵ under the input distribution p. Similarly, we denote by $\max_{I \in \mathcal{I}} \mathrm{E}[C_\epsilon(I, A_q)]$ the expected running time (under the worst input) of any randomized algorithm that errs with probability at most ϵ (again, the randomized algorithm is viewed as probability distribution q on deterministic algorithms). Analogous to Proposition 7.1, we then have:

Proposition 7.2. *For all distributions p over $\mathcal{I}$ and q over $\mathcal{A}$ and any $\epsilon \in [0, 1/2]$.*

$$\frac{1}{2} \left(\min_{A \in \mathcal{A}} \mathrm{E}[C_{2\epsilon}(I_p, A)] \right) \leq \max_{I \in \mathcal{I}} \mathrm{E}[C_\epsilon(I, A_q)]$$

Lower Bound for Game Tree Evaluation

We now apply Yao's minimax principle to the AND-OR tree evaluation problem. A randomized algorithm for AND-OR tree evaluation can be viewed as a probability distribution over deterministic algorithms, because the length of the computation as well as the number of choices at each step are both finite. We may imagine that all of these coins are tossed before the beginning of the execution.

The tree T_k is equivalent to a balanced binary tree, all of whose leaves are at distance $2k$ from the root and all of whose internal nodes compute the NOR function: a node returns the value 1 if both inputs are 0, and 0 otherwise. We proceed with the analysis of this tree of NORs of depth $2k$.

Let $p = (3 - \sqrt{5})/2$; each leaf of the tree is independently set to 1 with probability p. If each input to a NOR node is independently 1 with probability p, its output is 1 with probability

$$\left(\frac{\sqrt{5} - 1}{2} \right)^2 = \frac{3 - \sqrt{5}}{2} = p$$

Thus, the value of every node of NOR tree is 1 with probability p, and the value of a node is independent of the values of all of the other nodes on the same level. Consider a deterministic algorithm that is evaluating a tree furnished with such random inputs, and let v be a node of the tree whose value the algorithm is trying to determine. Intuitively, the algorithm should determine the value of one child of v before inspecting any leaf of the other subtree. An alternative view of this process is that the deterministic algorithm should inspect leaves visited in a depth-first search of the tree, except of course that it ceases to visit subtrees of node v when the value of v has been determined. Let us call such an algorithm a *depth-first pruning* algorithm, referring to the order of traversal and the fact that subtrees that supply no additional information are pruned away without being inspected. The following result is due to Tarsi [1983]:

Proposition 7.3. *Let T be a NOR tree each of whose leaves is independently set to 1 with probability q for a fixed value $q \in [0, 1]$. Let $W(T)$ denote the minimum, over all deterministic algorithms, of the expected* *number of steps to evaluate T. Then, there is a depth-first pruning algorithm whose expected number of steps to evaluate T is W(T).*

Proposition 7.3 tells us that for the purposes of our lower bound, we may restrict our attention to depth-first pruning algorithms. Let $W(h)$ be the expected number of leaves inspected by a depth-first pruning algorithm in determining the value of a node at distance h from the leaves, when each leaf is independently set to 1 with probability $(3 - \sqrt{5})/2$. Clearly,

$$W(h) = W(h-1) + (1-p) \times W(h-1)$$

where the first term represents the work done in evaluating one of the subtrees of the node, and the second term represents the work done in evaluating the other subtree (which will be necessary if the first subtree returns the value 0, an event occurring with probability $1 - p$). Letting h be $\log_2 n$ and solving, we get $W(h) \geq n^{0.694}$.

Theorem 7.4. *The expected running time of any randomized algorithm that always evaluates an instance of T_k correctly is at least $n^{0.694}$, where $n = 2^{2^k}$ is the number of leaves.*

Why is our lower bound of $n^{0.694}$ less than the upper bound of $n^{0.793}$ that follows from Theorem 7.3? The reason is that we have not chosen the best possible probability distribution for the values of the leaves. Indeed, in the NOR tree if both inputs to a node are 1, no reasonable algorithm will read leaves of both subtrees of that node. Thus, to prove the best lower bound we have to choose a distribution on the inputs that precludes the event that both inputs to a node will be 1; in other words, the values of the inputs are chosen at random but not independently. This stronger (and considerably harder) analysis can in fact be used to show that the algorithm of section 7.4 is optimal; the reader is referred to the paper of Saks and Wigderson [1986] for details.

7.6 Randomized Data Structures

Recent research into data structures has strongly emphasized the use of randomized techniques to achieve increased efficiency without sacrificing simplicity of implementation. An illustrative example is the randomized data structure for dynamic dictionaries called *skip list* that is due to Pugh [1990].

The dynamic dictionary problem is that of maintaining the set of keys X drawn from a totally ordered universe so as to provide efficient support of the following operations: find(q, X)—decide whether the query key q belongs to X and return the information associated with this key if it does indeed belong to X; insert(q, X)—insert the key q into the set X, unless it is already present in X; delete(q, X)—delete the key q from X, unless it is absent from X. The standard approach for solving this problem involves the use of a binary search tree and gives worst-case time per operation that is $O(\log n)$, where n is the size of X at the time the operation is performed. Unfortunately, achieving this time bound requires the use of complex rebalancing strategies to ensure that the search tree remains balanced, i.e., has depth $O(\log n)$. Not only does rebalancing require more effort in terms of implementation, it also leads to significant overheads in the running time (at least in terms of the constant factors subsumed by the big-O notation). The skip list data structure is a rather pleasant alternative that overcomes both of these shortcomings.

Before getting into the details of randomized skip lists, we will develop some of the key ideas without the use of randomization. Suppose we have a totally ordered data set, $X = \{x_1 < x_2 < \cdots < x_n\}$. A *gradation of X* is a sequence of nested subsets (called *levels*)

$$X_r \subseteq X_{r-1} \subseteq \cdots \subseteq X_2 \subseteq X_1$$

such that $X_r = \emptyset$ and $X_1 = X$. Given an ordered set, X, and a gradation for it, the level of any element $x \in X$ is defined as

$$L(x) = \max\{i \mid x \in X_i\}$$

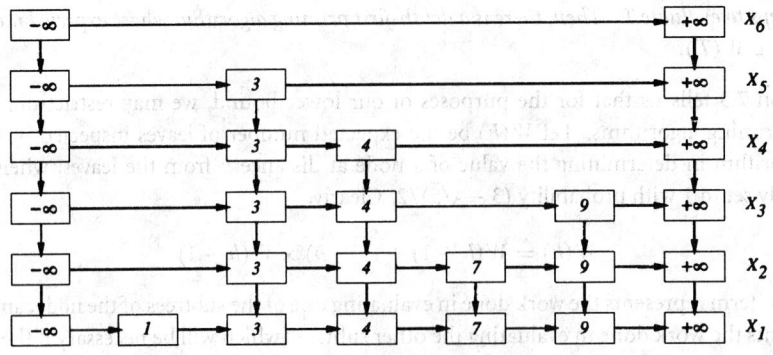

FIGURE 7.3   A skip list.

that is, $L(x)$ is the largest index i such that x belongs to the ith level of the gradation. In what follows, we will assume that two special elements $-\infty$ and $+\infty$ belong to each of the levels, where $-\infty$ is smaller than all elements in X and $+\infty$ is larger than all elements in X.

We now define an ordered list data structure with respect to a gradation of the set X. The first level, X_1, is represented as an ordered linked list, and each node x in this list has a stack of $L(x) - 1$ additional nodes directly above it. Finally, we obtain the skip list with respect to the gradation of X by introducing horizontal and vertical pointers between these nodes as illustrated in Fig. 7.3. The skip list in Fig. 7.3 corresponds to a gradation of the data set $X = \{1, 3, 4, 7, 9\}$ consisting of the following 6 levels:

$$X_6 = \emptyset$$

$$X_5 = \{3\}$$

$$X_4 = \{3, 4\}$$

$$X_3 = \{3, 4, 9\}$$

$$X_2 = \{3, 4, 7, 9\}$$

$$X_1 = \{1, 3, 4, 7, 9\}$$

Observe that starting at the ith node from the bottom in the leftmost column of nodes and traversing the horizontal pointers in order yields a set of nodes corresponding to the elements of the ith level X_i.

Additionally, we will view each level i as defining a set of *intervals*, each of which is defined as the set of elements of X spanned by a horizontal pointer at level i. The sequence of levels X_i can be viewed as successively coarser partitions of X. In Fig. 7.3, the levels determine the following partitions of X into intervals:

$$X_6 = [-\infty, +\infty]$$

$$X_5 = [-\infty, 3] \cup [3, +\infty]$$

$$X_4 = [-\infty, 3] \cup [3, 4] \cup [4, +\infty]$$

$$X_3 = [-\infty, 3] \cup [3, 4] \cup [4, 9] \cup [9, +\infty]$$

$$X_2 = [-\infty, 3] \cup [3, 4] \cup [4, 7] \cup [7, 9] \cup [9, +\infty]$$

$$X_1 = [-\infty, 1] \cup [1, 3] \cup [3, 4] \cup [4, 7] \cup [7, 9] \cup [9, +\infty]$$

An alternative view of the skip list is in terms of a tree defined by the interval partition structure, as illustrated in Fig. 7.4 for the preceding example. In this tree, each node corresponds to an interval, and the intervals at a given level are represented by nodes at the corresponding level of the tree. When the interval

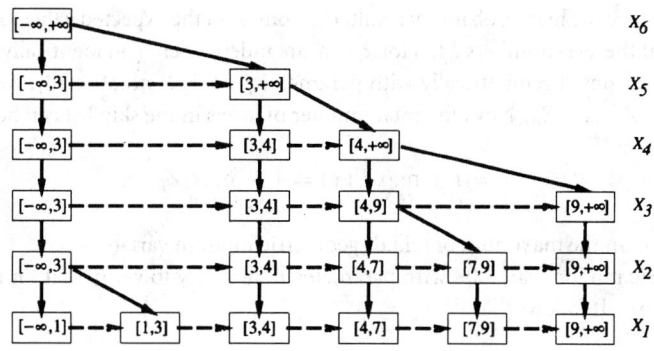

FIGURE 7.4   Tree representation of a skip list.

J at level $i + 1$ is a superset of the interval I at level i, then the corresponding node J has the node I as a child in this tree. Let $C(I)$ denote the number of children in the tree of a node corresponding to the interval I, i.e., it is the number of intervals from the previous level that are subintervals of I. Note that the tree is not necessarily binary because the value of $C(I)$ is arbitrary. We can view the skip list as a threaded version of this tree, where each thread is a sequence of (horizontal) pointers linking together the nodes at a level into an ordered list. In Fig. 7.4, the broken lines indicate the threads, and the full lines are the actual tree pointers.

Finally, we need some notation concerning the membership of element x in the intervals already defined, where x is not necessarily a member of X. For each possible x, let $I_j(x)$ be the interval at level j containing x. In the degenerate case where x lies on the boundary between two intervals, we assign it to the leftmost such interval. Observe that the nested sequence of intervals containing y,

$$I_r(y) \subseteq I_{r-1}(y) \subseteq \cdots \subseteq I_1(y)$$

corresponds to a root-leaf path in the tree corresponding to the skip list.

It remains to specify the choice of the gradation that determines the structure of a skip list. This is precisely where we introduce randomization into the structure of a skip list. The idea is to define a random gradation. Our analysis will show that, with high probability, the search tree corresponding to a random skip list is balanced, and then the dictionary operations can be efficiently implemented.

We define the *random gradation* for X as follows: given level X_i, the next level X_{i+1} is determined by independently choosing to retain each element $x \in X_i$ with probability $1/2$. The random selection process begins with $X_1 = X$ and terminates when for the first time the resulting level is empty. Alternatively, we may view the choice of the gradation as follows: for each $x \in X$, choose the level $L(x)$ independently from the geometric distribution with parameter $p = 1/2$ and place x in the levels $X_1, \ldots, X_{L(x)}$. We define r to be one more than the maximum of these level numbers. Such a random level is chosen for every element of X upon its insertion and remains fixed unit until its deletion.

We omit the proof of the following theorem bounding the space complexity of a randomized skip list. The proof is a simple exercise, and it is recommended that the reader verify this to gain some insight into the behavior of this data structure.

Theorem 7.5.   *A random skip list for a set, X, of size n has expected space requirement $O(n)$.*

We will go into more detail about the time complexity of this data structure. The following lemma underlies the running time analysis.

Lemma 7.1.   *The number of levels r in a random gradation of a set, X, of size n has expected value $\mathrm{E}[r] = O(\log n)$. Further, $r = O(\log n)$ with high probability.*

PROOF. We will prove the high probability result; the bound on the expected value follows immediately from this. Recall that the level numbers $L(x)$ for $x \in X$ are independent and identically distributed (i.i.d.) random variables distributed geometrically with parameter $p = 1/2$; notationally, we will denote these random variables by $Z_1, \ldots, Z_n$. Now, the total number of levels in the skip list can be determined as

$$r = 1 + \max_{x \in X} L(x) = 1 + \max_{1 \le i \le n} Z_i$$

that is, as one more than the maximum of n i.i.d. geometric random variables.

For such geometric random variables with parameter p, it is easy to verify that for any positive real t, $\Pr[Z_i > t] \le (1-p)^t$. It follows that

$$\Pr\left[\max_i Z_i > t\right] \le n(1-p)^t = \frac{n}{2^t}$$

because $p = 1/2$ in this case. For any $\alpha > 1$, setting $t = \alpha \log n$, we obtain that

$$\Pr[r > \alpha \log n] \le \frac{1}{n^{\alpha-1}}. \qquad \square$$

We can now infer that the tree representing the skip list has height $O(\log n)$ with high probability. To show that the overall search time in a skip list is similarly bounded, we must first specify an efficient implementation of the find operation. We present the implementation of the dictionary operations in terms of the tree representation; it is fairly easy to translate this back into the skip list representation.

To implement find (y, X), we must walk down the path

$$I_r(y) \subseteq I_{r-1}(y) \subseteq \cdots \subseteq I_1(y)$$

For this, at level j, starting at the node $I_j(y)$, we use the vertical pointer to descend to the leftmost child of the current interval; then, via the horizontal pointers, we move rightward until the node $I_j(y)$ is reached. Note that it is easily determined whether y belongs to a given interval or to an interval to its right. Further, in the skip list, the vertical pointers allow access only to the leftmost child of an interval, and therefore we must use the horizontal pointers to scan its children.

To determine the expected cost of find(y, X) operation, we must take into account both the number of levels and the number of intervals/nodes scanned at each level. Clearly, at level j, the number of nodes visited is no more than the number of children of $I_{j+1}(y)$. It follows that the cost of find can be bounded by

$$O\left(\sum_{j=1}^{r}(1 + C(I_j(y)))\right)$$

The following lemma shows that this quantity has expectation bounded by $O(\log n)$.

Lemma 7.2. *For any y, let $I_r(y), \ldots, I_1(y)$ be the search path followed by find(y, X) in a random skip list for a set, X, of size n. Then,*

$$\mathrm{E}\left[\sum_{j=1}^{r}(1 + C(I_j(y)))\right] = O(\log n)$$

PROOF. We begin by showing that for any interval I in a random skip list, $\mathrm{E}[C(I)] = O(1)$. By Lemma 7.1, we are guaranteed that $r = O(\log n)$ with his probability, and so we will obtain the desired bound. It is important to note that we really do need the high-probability bound on Lemma 7.1 because it is incorrect to multiply the expectation of r with that of $1 + C(I)$ (the two random variables need not be independent). However, in the approach we will use, because $r > \alpha \log n$ with probability at most $1/n^{\alpha-1}$ and $\sum_j (1 + C(I_j(y))) = O(n)$, it can be argued that the case $r > \alpha \log n$ does not contribute significantly to the expectation of $\sum_j C(I_j(y))$.

To show that the expected number of children of interval J at level i is bounded by a constant, we will show that the expected number of siblings of J (children of its parent) is bounded by a constant; in fact, we will bound only the number of right siblings because the argument for the number of left siblings is identical. Let the intervals to the right of J be the following:

$$J_1 = [x_1, x_2]; \; J_2 = [x_2, x_3]; \; \ldots; \; J_k = [x_k, +\infty]$$

Because these intervals exist at level i, each of the elements $x_1, \ldots, x_k$ belongs to X_i. If J has s right siblings, then it must be the case that $x_1, \ldots, x_s \notin X_{i+1}$, and $x_{s+1} \in X_{i+1}$. The latter event occurs with probability $1/2^{s+1}$ because each element of X_i is independently chosen to be in X_{i+1} with probability $1/2$. Clearly, the number of right siblings of J can be viewed as a random variable that is geometrically distributed with parameter $1/2$. It follows that the expected number of right siblings of J is at most 2. $\square$

Consider now the implementation of the insert and delete operations. In implementing the operation insert(y, X), we assume that a random level, $L(y)$, is chosen for y as described earlier. If $L(y) > r$, then we start by creating new levels from $r + 1$ to $L(y)$ and then redefine r to be $L(y)$. This requires $O(1)$ time per level because the new levels are all empty prior to the insertion of y. Next we perform find(y, X) and determine the search path $I_r(y), \ldots, I_1(y)$, where r is updated to its new value if necessary. Given this search path, the insertion can be accomplished in time $O(L(y))$ by splitting around y the intervals $I_1(y), \ldots, I_{L(y)}(y)$ and updating the pointers as appropriate. The delete operation is the converse of the insert operation; it involves performing find(y, X) followed by collapsing the intervals that have y as an endpoint. Both operations incur costs that are the cost of a find operation and additional cost proportional to $L(y)$. By Lemmas 7.1 and 7.2, we obtain the following theorem.

Theorem 7.6. *In a random skip list for a set, X, of size n, the operations find, insert, and delete can be performed in expected time $O(\log n)$.*

7.7 Random Reordering and Linear Programming

The *linear programming problem* is a particularly notable example of the two main benefits of randomization: simplicity and speed. We now describe a simple algorithm for linear programming based on a paradigm for randomized algorithms known as *random reordering*. For many problems, it is possible to design natural algorithms based on the following idea. Suppose that the input consists of n elements. Given any subset of these n elements, there is a solution to the partial problem defined by these elements. If we start with the empty set and add the n elements of the input one at a time, maintaining a partial solution after each addition, we will obtain a solution to the entire problem when all of the elements have been added. The usual difficulty with this approach is that the running time of the algorithm depends heavily on the order in which the input elements are added; for any fixed ordering, it is generally possible to force this algorithm to behave badly. The key idea behind random reordering is to *add the elements in a random order*. This simple device often avoids the pathological behavior that results from using a fixed order.

The linear programming problem is to find the extremum of a linear objective function of d real variables subject to set H of n constraints that are linear functions of these variables. The intersection of the n half-spaces defined by the constraints is a polyhedron in d-dimensional space (which may be empty, or possibly unbounded). We refer to this polyhedron as the *feasible region*. Without loss of generality [Schrijver 1986] we assume that the feasible region is nonempty and bounded. (Note that we are not assuming that we can *test* an arbitrary polyhedron for nonemptiness or boundedness; this is known to be equivalent to solving a linear program.) For a set of constraints, S, let $\mathcal{B}(S)$ denote the optimum of the linear program defined by S; we seek $\mathcal{B}(S)$.

Consider the following algorithm due to Seidel [1991]: add the n constraints in random order, one at a time. After adding each constraint, determine the optimum subject to the constraints added so far. This algorithm also may be viewed in the following backwards manner, which will prove useful in the sequel.

ALGORITHM SLP.

Input: A set of constraints H, and the dimension d.
Output: The optimum $\mathcal{B}(H)$.

    0. *If there are only d constraints, output $\mathcal{B}(H) = H$.*
    1. *Pick a random constraint $h \in H$;*
        Recursively find $\mathcal{B}(H \backslash \{h\})$.
  2.1. *if $\mathcal{B}(H \backslash \{h\})$ does not violate h, output $\mathcal{B}(H \backslash \{h\})$ to be the optimum $\mathcal{B}(H)$.*
  2.2. *else project all of the constraints of $H \backslash \{h\}$) onto h and recursively solve this new linear programming problem of one lower dimension.*

The idea of the algorithm is simple. Either h (the constraint chosen randomly in step 1) is redundant (in which case we execute step 2.1), or it is not. In the latter case, we know that the vertex formed by $\mathcal{B}(H)$ must lie on the hyperplane bounding h. In this case, we project all of the constraints of $H \backslash \{h\}$ onto h and solve this new linear programming problem (which has dimension $d - 1$).

The optimum $\mathcal{B}(H)$ is defined by d constraints. At the top level of recursion, the probability that random constraint h violates $\mathcal{B}(H \backslash \{h\})$ is at most d/n. Let $T(n, d)$ denote an upper bound on the expected running time of the algorithm for any problem with n constraints in d dimensions. Then, we may write

$$T(n, d) \leq T(n - 1, d) + O(d) + \frac{d}{n}[O(dn) + T(n - 1, d - 1)] \qquad (7.7)$$

In Eq. (7.7), the first term on the right denotes the cost of recursively solving the linear program defined by the constraints in $H \backslash \{h\}$. The second accounts for the cost of checking whether h violates $\mathcal{B}(H \backslash \{h\})$. With probability d/n it does, and this is captured by the bracketed expression, whose first term counts the cost of projecting all of the constraints onto h. The second counts the cost of (recursively) solving the projected problem, which has one fewer constraint and dimension. The following theorem may be verified by substitution and proved by induction.

Theorem 7.7. *There is a constant b such that the recurrence (7.7) satisfies the solution $T(n, d) \leq bnd!$.*

In contrast, if the choice in step 1 of SLP were not random, the recurrence (7.7) would be

$$T(n, d) \leq T(n - 1, d) + O(d) + O(dn) + T(n - 1, d - 1) \qquad (7.8)$$

whose solution contains a term that grows quadratically in n.

7.8 Algebraic Methods and Randomized Fingerprints

Some of the most notable randomized results in theoretical computer science, particularly in complexity theory, have involved a nontrivial combination of randomization and algebraic methods. In this section, we describe a fundamental randomization technique based on algebraic ideas. This is the randomized fingerprinting technique, originally due to Freivalds [1977], for the verification of identities involving matrices, polynomials, and integers. We also describe how this generalizes to the so-called Schwartz–Zippel technique for identities involving multivariate polynomials (independently due to Schwartz [1987] and Zippel [1979]; see also DeMillo and Lipton [1978]. Finally, following Lovász [1979], we apply the technique to the problem of detecting the existence of perfect matchings in graphs.

The *fingerprinting* technique has the following general form. Suppose we wish to decide the equality of two elements x and y drawn from some large universe U. Assuming any reasonable model of computation, this problem has a deterministic complexity $\Omega(\log|U|)$. Allowing randomization, an alternative approach is to choose a random function from U into a smaller space V such that with high probability x and y are identical if and only if their images in V are identical. These images of x and y are said to be their *fingerprints*, and the equality of fingerprints can be verified in time $O(\log|V|)$. Of course, for any fingerprint function the average number of elements of U mapped to an element of V is $|U|/|V|$; thus, it would appear impossible to find good fingerprint functions that work for arbitrary or worst-case choices of x and y. However, as we will show subsequently, when the identity checking is required to be correct only for x and y chosen from the small subspace S of U, particularly a subspace with some algebraic structure, it is possible to choose good fingerprint functions without any a priori knowledge of the subspace, provided the size of V is chosen to be comparable to the size of S.

Throughout this section, we will be working over some unspecified field $\mathcal{F}$. Because the randomization will involve uniform sampling from a finite subset of the field, we do not even need to specify whether the field is finite. The reader may find it helpful in the infinite case to assume that $\mathcal{F}$ is the field $\mathcal{Q}$ of rational numbers and in the finite case to assume that $\mathcal{F}$ is $\mathcal{Z}_p$, the field of integers modulo some prime number p.

Freivalds' Technique and Matrix Product Verification

We begin by describing a fingerprinting technique for verifying matrix product identities. Currently, the fastest algorithm for matrix multiplication (due to Coppersmith and Winograd [1990]) has running time $O(n^{2.376})$, improving significantly on the obvious $O(n^3)$ time algorithm; however, the fast matrix multiplication algorithm has the disadvantage of being extremely complicated. Suppose we have an implementation of the fast matrix multiplication algorithm and, given its complex nature, are unsure of its correctness. Because program verification appears to be an intractable problem, we consider the more reasonable goal of verifying the correctness of the output produced by executing the algorithm on specific inputs. (This notion of verifying programs on specific inputs is the basic tenet of the theory of *program checking* recently formulated by Blum and Kannan [1989].) More concretely, suppose we are given three $n \times n$ matrices X, Y, and Z over field $\mathcal{F}$, and would like to verify that $XY = Z$. Clearly, it does not make sense to use a simpler but slower matrix multiplication algorithm for the verification, as that would defeat the whole purpose of using the fast algorithm in the first place. Observe that, in fact, there is no need to recompute Z; rather, we are merely required to verify that the product of X and Y is indeed equal to Z. Freivalds' technique gives an elegant solution that leads to an $O(n^2)$ time randomized algorithm with bounded error probability.

The idea is to first pick the random vector $r \in \{0, 1\}^n$, that is, each component of r is chosen independently and uniformly at random from the set $\{0, 1\}$ consisting of the additive and multiplicative identities of the field $\mathcal{F}$. Then, in $O(n^2)$ time, we can compute $y = Yr$, $x = Xy = XYr$, and $z = Zr$. We would like to claim that the identity $XY = Z$ can be verified merely by checking that $x = z$. Quite clearly, if $XY = Z$, then $x = z$; unfortunately, the converse is not true in general. However, given the random choice of r, we can show that for $XY \neq Z$, the probability that $x \neq z$ is at least $1/2$. Observe that the fingerprinting algorithm errs only if $XY \neq Z$ but x and z turn out to be equal, and this has a bounded probability.

Theorem 7.8. *Let X, Y, and Z be $n \times n$ matrices over some field $\mathcal{F}$ such that $XY \neq Z$; further, let r be chosen uniformly at random from $\{0, 1\}^n$ and define $x = XYr$ and $z = Zr$. Then,*

$$\Pr[x = z] \leq 1/2$$

PROOF. Define $W = XY - Z$ and observe that W is not the all-zeroes matrix. Because $Wr = XYr - Zr = x - z$, the event $x = z$ is equivalent to the event that $Wr = 0$. Assume, without loss of generality, that the first row of W has a nonzero entry and that the nonzero entries in that row precede all of the zero entries. Define the vector w as the first row of W, and assume that the first $k > 0$ entries in w are nonzero. Because the first component of Wr is $w^T r$, giving an upper bound on the probability that the inner product of w and r is zero will give an upper bound on the probability that $x = z$.

Observe that $w^T r = 0$ if and only if

$$r_1 = \frac{-\sum_{i=2}^{k} w_i r_i}{w_1} \tag{7.9}$$

Suppose that while choosing the random vector r, we choose $r_2, \ldots, r_n$ before choosing r_1. After the values for $r_2, \ldots, r_n$ have been chosen, the right-hand side of Eq. (7.9) is fixed at some value $v \in \mathcal{F}$. If $v \notin \{0, 1\}$, then r_1 will never equal v; conversely, if $v \in \{0, 1\}$, then the probability that $r_1 = v$ is 1/2. Thus, the probability that $w^T r = 0$ is at most 1/2, implying the desired result.   □

We have reduced the matrix multiplication verification problem to that of verifying the equality of two vectors. The reduction itself can be performed in $O(n^2)$ time and the vector equality can be checked in $O(n)$ time, giving an overall running time of $O(n^2)$ for this Monte Carlo procedure. The error probability can be reduced to $1/2^k$ via k independent iterations of the Monte Carlo algorithm. Note that there was nothing magical about choosing the components of the random vector r from $\{0, 1\}$, because any two distinct elements of $\mathcal{F}$ would have done equally well. This suggests an alternative approach toward reducing the error probability, as follows: each component of r is chosen independently and uniformly at random from some subset S of the field $\mathcal{F}$; then, it is easily verified that the error probability is no more than $1/|S|$.

Finally, note that Freivalds' technique can be applied to the verification of any matrix identity $A = B$. Of course, given A and B, just comparing their entries takes only $O(n^2)$ time. But there are many situations where, just as in the case of matrix product verification, computing A explicitly is either too expensive or possibly even impossible, whereas computing Ar is easy. The random fingerprint technique is an elegant solution in such settings.

Extension to Identities of Polynomials

The fingerprinting technique due to Freivalds is fairly general and can be applied to many different versions of the identity verification problem. We now show that it can be easily extended to identity verification for symbolic polynomials, where two polynomials $P_1(x)$ and $P_2(x)$ are deemed identical if they have identical coefficients for corresponding powers of x. Verifying integer or string equality is a special case because we can represent any string of length n as a polynomial of degree n by using the kth element in the string to determine the coefficient of the kth power of a symbolic variable.

Consider first the polynomial product verification problem: given three polynomials $P_1(x), P_2(x), P_3(x) \in \mathcal{F}[x]$, we are required to verify that $P_1(x) \times P_2(x) = P_3(x)$. We will assume that $P_1(x)$ and $P_2(x)$ are of degree at most n, implying that $P_3(x)$ has degree at most $2n$. Note that degree n polynomials can be multiplied in $O(n \log n)$ time via fast Fourier transforms and that the evaluation of a polynomial can be done in $O(n)$ time.

The randomized algorithm we present for polynomial product verification is similar to the algorithm for matrix product verification. It first fixes set $S \subseteq \mathcal{F}$ of size at least $2n + 1$ and chooses $r \in S$ uniformly at random. Then, after evaluating $P_1(r), P_2(r)$, and $P_3(r)$ in $O(n)$ time, the algorithm declares the identity $P_1(x)P_2(x) = P_3(x)$ to be correct if and only if $P_1(r)P_2(r) = P_3(r)$. The algorithm makes an error only in the case where the polynomial identity is false but the value of the three polynomials at r indicates otherwise. We will show that the error event has a bounded probability.

Consider the degree $2n$ polynomial $Q(x) = P_1(x)P_2(x) - P_3(x)$. The polynomial $Q(x)$ is said to be *identically zero*, denoted by $Q(x) \equiv 0$, if each of its coefficients equals zero. Clearly, the polynomial identity $P_1(x)P_2(x) = P_3(x)$ holds if and only if $Q(x) \equiv 0$. We need to establish that if $Q(x) \not\equiv 0$, then with high probability $Q(r) = P_1(r)P_2(r) - P_3(r) \neq 0$. By elementary algebra we know that $Q(x)$ has at most $2n$ distinct roots. It follows that unless $Q(x) \equiv 0$, not more that $2n$ different choices of $r \in S$ will cause $Q(r)$ to evaluate to 0. Therefore, the error probability is at most $2n/|S|$. The probability of error can be reduced either by using independent iterations of this algorithm or by choosing a larger set S. Of course, when $\mathcal{F}$ is an infinite field (e.g., the reals), the error probability can be made 0 by choosing r uniformly from the entire field $\mathcal{F}$; however, that requires an infinite number of random bits!

Note that we could also use a deterministic version of this algorithm where each choice of $r \in S$ is tried once. But this involves $2n + 1$ different evaluations of each polynomial, and the best known algorithm for multiple evaluations needs $\Theta(n \log^2 n)$ time, which is more than the $O(n \log n)$ time requirement for actually performing a multiplication of the polynomials $P_1(x)$ and $P_2(x)$.

This verification technique is easily extended to a generic procedure for testing any polynomial identity of the form $P_1(x) = P_2(x)$ by converting it into the identity $Q(x) = P_1(x) - P_2(x) \equiv 0$. Of course, when P_1 and P_2 are explicitly provided, the identity can be deterministically verified in $O(n)$ time by comparing corresponding coefficients. Our randomized technique will take just as long to merely evaluate $P_1(x)$ and $P_2(x)$ at a random value. However, as in the case of verifying matrix identities, the randomized algorithm is quite useful in situations where the polynomials are implicitly specified, e.g., when we have only a *black box* for computing the polynomials with no information about their coefficients, or when they are provided in a form where computing the actual coefficients is expensive. An example of the latter situation is provided by the following problem concerning the determinant of a symbolic matrix. In fact, the determinant problem will require a technique for the verification of polynomial identities of *multivariate* polynomials that we will discuss shortly.

Consider the $n \times n$ matrix M. Recall that the determinant of the matrix M is defined as follows:

$$\det(M) = \sum_{\pi \in S_n} \text{sgn}(\pi) \prod_{i=1}^{n} M_{i,\pi(i)} \tag{7.10}$$

where S_n is the symmetric group of permutations of order n, and $\text{sgn}(\pi)$ is the sign of a permutation π. [The sign function is defined to be $\text{sgn}(\pi) = (-1)^t$, where t is the number of pairwise exchanges required to convert the identity permutation into π.] Although the determinant is defined as a summation with $n!$ terms, it is easily evaluated in polynomial time provided that the matrix entries M_{ij} are explicitly specified. Consider the Vandermonde matrix $M(x_1, \ldots, x_n)$, which is defined in terms of the indeterminates $x_1, \ldots, x_n$ such that $M_{ij} = x_i^{j-1}$, that is,

$$M = \begin{pmatrix} 1 & x_1 & x_1^2 & \cdots & x_1^{n-1} \\ 1 & x_2 & x_2^2 & \cdots & x_2^{n-1} \\ & & \cdot & & \\ & & \cdot & & \\ & & \cdot & & \\ 1 & x_n & x_n^2 & \cdots & x_n^{n-1} \end{pmatrix}$$

It is known that for the Vandermonde matrix, $\det(M) = \prod_{i<j}(x_i - x_j)$. Consider the problem of verifying this identity without actually devising a formal proof. Computing the determinant of a symbolic matrix is infeasible as it requires dealing with a summation over $n!$ terms. However, we can formulate the identity verification problem as the problem of verifying that the polynomial $Q(x_1, \ldots, x_n) = \det(M) - \prod_{i<j}(x_i - x_j)$ is identically zero. Based on our discussion of Freivalds' technique, it is natural to consider the substitution of random values for each x_i. Because the determinant can be computed in polynomial time for any specific assignment of values to the symbolic variables $x_1, \ldots, x_n$, it is easy to evaluate the polynomial Q for random values of the variables. The only issue is that of bounding the error probability for this randomized test.

We now extend the analysis of Freivalds' technique for univariate polynomials to the multivariate case. But first, note that in a multivariate polynomial $Q(x_1, \ldots, x_n)$, the degree of a term is the sum of the exponents of the variable powers that define it, and the total degree of Q is the maximum over all terms of the degrees of the terms.

Theorem 7.9. Let $Q(x_1, \ldots, x_n) \in \mathcal{F}[x_1, \ldots, x_n]$ be a multivariate polynomial of total degree m. Let S be a finite subset of the field $\mathcal{F}$, and let $r_1, \ldots, r_n$ be chosen uniformly and independently from S. Then

$$\Pr[Q(r_1 \ldots, r_n) = 0 \mid Q(x_1, \ldots, x_n) \not\equiv 0] \leq \frac{m}{|S|}$$

PROOF. We will proceed by induction on the number of variables n. The basis of the induction is the case $n = 1$, which reduces to verifying the theorem for a univariate polynomial $Q(x_1)$ of degree m. But we have already seen for $Q(x_1) \not\equiv 0$ the probability that $Q(r_1) = 0$ is at most $m/|\mathcal{S}|$, taking care of the basis.

We now assume that the induction hypothesis holds for multivariate polynomials with at most $n - 1$ variables, where $n > 1$. In the polynomial $Q(x_1, \ldots, x_n)$ we can factor out the variable x_1 and thereby express Q as

$$Q(x_1, \ldots, x_n) = \sum_{i=0}^{k} x_1^i P_i(x_2, \ldots, x_n)$$

where $k \leq m$ is the largest exponent of x_1 in Q. Given our choice of k, the coefficient $P_k(x_2, \ldots, x_n)$ of x_1^k cannot be identically zero. Note that the total degree of P_k is at most $m - k$. Thus, by the induction hypothesis, we conclude that the probability that $P_k(r_2, \ldots, r_n) = 0$ is at most $(m - k)/|\mathcal{S}|$.

Consider now the case where $P_k(r_2, \ldots, r_n)$ is indeed not equal to 0. We define the following univariate polynomial over x_1 by substituting the random values for the other variables in Q:

$$q(x_1) = Q(x_1, r_2, r_3, \ldots, r_n) = \sum_{i=0}^{k} x_1^i P_i(r_2, \ldots, r_n)$$

Quite clearly, the resulting polynomial $q(x_1)$ has degree k and is not identically zero (because the coefficient of x_i^k is assumed to be nonzero). As in the basis case, we conclude that the probability that $q(r_1) = Q(r_1, r_2, \ldots, r_n)$ evaluates to 0 is bounded by $k/|\mathcal{S}|$.

By the preceding arguments, we have established the following two inequalities:

$$\Pr[P_k(r_2, \ldots, r_n) = 0] \leq \frac{m - k}{|\mathcal{S}|}$$

$$\Pr[Q(r_1, r_2, \ldots, r_n) = 0 \mid P_k(r_2, \ldots, r_n) \neq 0] \leq \frac{k}{|\mathcal{S}|}$$

Using the elementary observation that for any two events $\mathcal{E}_1$ and $\mathcal{E}_2$, $\Pr[\mathcal{E}_1] \leq \Pr[\mathcal{E}_1 \mid \bar{\mathcal{E}}_2] + \Pr[\mathcal{E}_2]$, we obtain that the probability that $Q(r_1, r_2, \ldots, r_n) = 0$ is no more than the sum of the two probabilities on the right-hand side of the two obtained inequalities, which is $m/|\mathcal{S}|$. This implies the desired results. □

This randomized verification procedure has one serious drawback: when working over large (or possibly infinite) fields, the evaluation of the polynomials could involve large intermediate values, leading to inefficient implementation. One approach to dealing with this problem in the case of integers is to perform all computations modulo some small random prime number; it can be shown that this does not have any adverse effect on the error probability.

Detecting Perfect Matchings in Graphs

We close by giving a surprising application of the techniques from the preceding section. Let $G(U, V, E)$ be a bipartite graph with two independent sets of vertices $U = \{u_1, \ldots, u_n\}$ and $V = \{v_1, \ldots, v_n\}$ and edges E that have one endpoint in each of U and V. We define a matching in G as a collection of edges $M \subseteq E$ such that each vertex is an endpoint of at most one edge in M; further, a perfect matching is defined to be a matching of size n, that is, where each vertex occurs as an endpoint of exactly one edge in M. Any perfect matching M may be put into a one-to-one correspondence with the permutations in $\mathcal{S}_n$, where the matching corresponding to a permutation $\pi \in \mathcal{S}_n$ is given by the collection of edges $\{(u_i, v_{\pi(i)}) \mid 1 \leq i \leq n\}$. We now relate the matchings of the graph to the determinant of a matrix obtained from the graph.

Theorem 7.10. *For any bipartite graph $G(U, V, E)$, define a corresponding $n \times n$ matrix A as follows:*

$$A_{ij} = \begin{cases} x_{ij} & (u_i, v_j) \in E \\ 0 & (u_i, v_j) \notin E \end{cases}$$

Let the multivariate polynomial $Q(x_{11}, x_{12}, \ldots, x_{nn})$ denote the determinant $\det(A)$. Then G has a perfect matching if and only if $Q \neq 0$.

PROOF. We may express the determinant of A as follows:

$$\det(A) = \sum_{\pi \in S_n} \text{sgn}(\pi) A_{1,\pi(1)} A_{2,\pi(2)} \cdots A_{n,\pi(n)}$$

Note that there cannot be any cancellation of the terms in the summation because each indeterminate x_{ij} occurs at most once in A. Thus, the determinant is not identically zero if and only if there exists some permutation π for which the corresponding term in the summation is nonzero. Clearly, the term corresponding to a permutation π is nonzero if and only if $A_{i,\pi(i)} \neq 0$ for each i, $1 \leq i \leq n$; this is equivalent to the presence in G of the perfect matching corresponding to π. □

The matrix of indeterminates is sometimes referred to as the *Edmonds matrix* of a bipartite graph. The preceding result can be extended to the case of nonbipartite graphs, and the corresponding matrix of indeterminates is called the Tutte matrix. Tutte [1947] first pointed out the close connection between matchings in graphs and matrix determinants; the simpler relation between bipartite matchings and matrix determinants was given by Edmonds [1967].

We can turn the preceding result into a simple randomized procedure for testing the existence of perfect matchings in a bipartite graph (due to Lovász [1979]): using the algorithm from the preceding subsection, determine whether the determinant is identically zero. The running time of this procedure is dominated by the cost of computing a determinant, which is essentially the same as the time required to multiply two matrices. Of course, there are algorithms for *constructing* a maximum matching in a graph with m edges and n vertices in time $O(m\sqrt{n})$ (see Hopcroft and Karp [1973], Micali and Vazirani [1980], Vazirani [1994], and Feder and Motwani [1991]). Unfortunately, the time required to compute the determinant exceeds $m\sqrt{n}$ for small m, and so the benefit in using this randomized *decision* procedure appears marginal at best. However, this technique was extended by Rabin and Vazirani [1984, 1989] to obtain simple algorithms for the actual *construction* of maximum matchings; although their randomized algorithms for matchings are simple and elegant, they are still slower than the deterministic $O(m\sqrt{n})$ time algorithms known earlier. Perhaps more significantly, this randomized decision procedure proved to be an essential ingredient in devising fast *parallel* algorithms for computing maximum matchings [Karp et al. 1988, Mulmuley et al. 1987].

Defining Terms

Deterministic algorithm: An algorithm whose execution is completely determined by its input.

Distributional complexity: The expected running time of the best possible deterministic algorithm over the worst possible probability distribution of the inputs.

Las Vegas algorithm: A randomized algorithm that always produces correct results, with the only variation from one run to another being in its running time.

Monte Carlo algorithm: A randomized algorithm that may produce incorrect results but with bounded error probability.

Randomized algorithm: An algorithm that makes random choices during the course of its execution.

Randomized complexity: The expected running time of the best possible randomized algorithm over the worst input.

References

Aleliunas, R., Karp, R. M., Lipton, R. J., Lovász, L., and Rackoff, C. 1979. Random walks, universal traversal sequences, and the complexity of maze problems. In *Proc. 20th Ann. Symp. Found. Comput. Sci.*, pp. 218–223. San Juan, Puerto Rico, Oct.

Aragon, C. R. and Seidel, R. G. 1989. Randomized search trees. In *Proc. 30th Ann. IEEE Symp. Found. Comput. Sci.*, pp. 540–545.

Ben-David, S., Borodin, A., Karp, R. M., Tardos, G., and Wigderson, A. 1994. On the power of randomization in on-line algorithms. *Algorithmica* 11(1):2–14.

Blum, M. and Kannan, S. 1989. Designing programs that check their work. In *Proc. 21st Ann. ACM Symp. Theory Comput.*, pp. 86–97. ACM.

Coppersmith, D. and Winograd, S. 1990. Matrix multiplication via arithmetic progressions. *J. Symbolic Comput.* 9:251–280.

DeMillo, R. A. and Lipton, R. J. 1978. A probabilistic remark on algebraic program testing. *Inf. Process. Lett.* 7:193–195.

Edmonds, J. 1967. Systems of distinct representatives and linear algebra. *J. Res. Nat. Bur. Stand.* 71B, 4:241–245.

Feder, T. and Motwani, R. 1991. Clique partitions, graph compression and speeding-up algorithms. In *Proc. 25th Ann. ACM Symp. Theory Comput.*, pp. 123–133.

Floyd, R. W. and Rivest, R. L. 1975. Expected time bounds for selection. *Commun. ACM* 18:165–172.

Freivalds, R. 1977. Probabilistic machines can use less running time. In *Inf. Process. 77, Proc. IFIP Congress 77*, B. Gilchrist, ed., pp. 839–842, North-Holland, Amsterdam, Aug.

Goemans, M. X. and Williamson, D. P. 1994. 0.878-approximation algorithms for MAX-CUT and MAX-2SAT. In *Proc. 26th Annu. ACM Symp. Theory Comput.*, pp. 422–431.

Hoare, C. A. R. 1962. Quicksort. *Comput. J.* 5:10–15.

Hopcroft, J. E. and Karp, R. M. 1973. An $n^{5/2}$ algorithm for maximum matching in bipartite graphs. *SIAM J. Comput.* 2:225–231.

Karger, D. R. 1993. Global min-cuts in $\mathcal{RNC}$, and other ramifications of a simple min-cut algorithm. In *Proc. 4th Annu. ACM–SIAM Symp. Discrete Algorithms.*

Karger, D. R., Klein, P. N., and Tarjan, R. E. 1995. A randomized linear-time algorithm for finding minimum spanning trees. *J. ACM* 42:321–328.

Karger, D., Motwani, R., and Sudan, M. 1994. Approximate graph coloring by semidefinite programming. In *Proc. 35th Annu. IEEE Symp. Found. Comput. Sci.*, pp. 2–13.

Karp, R. M. 1991. An introduction to randomized algorithms. *Discrete Appl. Math.* 34:165–201.

Karp, R. M., Upfal, E., and Wigderson, A. 1986. Constructing a perfect matching is in random $\mathcal{NC}$. *Combinatorica* 6:35–48.

Karp, R. M., Upfal, E., and Wigderson, A. 1988. The complexity of parallel search. *J. Comput. Sys. Sci.* 36:225–253.

Lovász, L. 1979. On determinants, matchings and random algorithms. In *Fundamentals of Computing Theory*. L. Budach, ed. Akademia-Verlag, Berlin.

Maffioli, F., Speranza, M. G., and Vercellis, C. 1985. Randomized algorithms. In *Combinatorial Optimization: Annotated Bibliographies*, M. O'Eigertaigh, J. K. Lenstra, and A. H. G. Rinooy Kan, eds., pp. 89–105. Wiley, New York.

Micali, S. and Vazirani, V. V. 1980. An $O(\sqrt{|V|}|e|)$ algorithm for finding maximum matching in general graphs. In *Proc. 21st Ann. IEEE Symp. Found. Comput. Sci.*, pp. 17–27.

Motwani, R., Naor, J., and Raghavan, P. 1996. Randomization in approximation algorithms. In *Approximation Algorithms*, D. Hochbaum, ed. PWS.

Motwani, R. and Raghavan, P. 1995. *Randomized Algorithms*. Cambridge Univ. Press, New York.

Mulmuley, K. 1993. *Computational Geometry: An Introduction Through Randomized Algorithms*. Prentice–Hall, New York.

Mulmuley, K., Vazirani, U. V., and Vazirani, V. V. 1987. Matching is as easy as matrix inversion. *Combinatorica* 7:105–113.

Pugh, W. 1990. Skip lists: a probabilistic alternative to balanced trees. *Commun. ACM* 33(6):668–676.

Rabin, M. O. 1980. Probabilistic algorithm for testing primality. *J. Number Theory* 12:128–138.

Rabin, M. O. 1983. Randomized Byzantine generals. In *Proc. 24th Annu. Symp. Found. Comput. Sci.*, pp. 403–409.

Rabin, M. O. and Vazirani, V. V. 1984. Maximum matchings in general graphs through randomization. *Aiken Computation Lab. Tech. Rep.* TR-15-84, Harvard University, Cambridge, MA.

Rabin, M. O. and Vazirani, V. V. 1989. Maximum matchings in general graphs through randomization. *J. Algorithms* 10:557–567.

Raghavan, P. and Snir, M. 1994. Memory versus randomization in on-line algorithms. *IBM J. Res. Dev.* 38:683–707.

Saks, M. and Wigderson, A. 1986. Probabilistic Boolean decision trees and the complexity of evaluating game trees. In *Proc. 27th Annu. IEEE Symp. Found. Comput. Sci.*, pp. 29–38. Toronto, Ontario.

Schrijver, A. 1986. *Theory of Linear and Integer Programming.* Wiley, New York.

Schwartz, J. T. 1987. Fast probabilistic algorithms for verification of polynomial identities. *J. ACM* 27(4):701–717.

Seidel, R. G. 1991. Small-dimensional linear programming and convex hulls made easy. *Discrete Comput. Geom.* 6:423–434.

Sinclair, A. 1992. *Algorithms for Random Generation and Counting: A Markov Chain Approach, Progress in Theoretical Computer Science.* Birkhauser, Boston, MA.

Snir, M. 1985. Lower bounds on probabilistic linear decision trees. *Theor. Comput. Sci.* 38:69–82.

Solovay, R. and Strassen, V. 1977. A fast Monte-Carlo test for primality. *SIAM J. Comput.* 6(1):84–85. See also 1978. *SIAM J. Comput.* 7(Feb.):118.

Tarsi, M. 1983. Optimal search on some game trees. *J. ACM* 30:389–396.

Tutte, W. T. 1947. The factorization of linear graphs. *J. London Math. Soc.* 22:107–111.

Valiant, L. G. 1982. A scheme for fast parallel communication. *SIAM J. Comput.* 11:350–361.

Vazirani, V. V. 1994. A theory of alternating paths and blossoms for proving correctness of $O(\sqrt{V}E)$ graph maximum matching algorithms. *Combinatorica* 14(1):71–109.

Welsh, D. J. A. 1983. Randomised algorithms. *Discrete Appl. Math.* 5:133–145.

Yao, A. C.-C. 1977. Probabilistic computations: towards a unified measure of complexity. In *Proc. 17th Annu. Symp. Found. Comput. Sci.*, pp. 222–227.

Zippel, R. E. 1979. Probabilistic algorithms for sparse polynomials. In *Proc. EUROSAM 79*, Vol. 72, Lecture Notes in Computer Science., pp. 216–226. Marseille, France.

Further Information

In this section we give pointers to a plethora of randomized algorithms not covered here. The reader should also note that the examples in the text are but a (random!) sample of the many randomized algorithms for each of the problems considered. These algorithms have been chosen to illustrate the main ideas behind randomized algorithms rather than to represent the state of the art for these problems. The reader interested in other algorithms for these problems is referred to Motwani and Raghavan [1995].

Randomized algorithms also find application in a number of other areas: in load balancing [Valiant 1982], approximation algorithms and combinatorial optimization [Goemans and Williamson 1994, Karger et al. 1994, Motwani et al. 1996], graph algorithms [Aleliunas et al. 1979, Karger et al. 1995], data structures [Aragon and Seidel 1989], counting and enumeration [Sinclair 1992], parallel algorithms [Karp et al. 1986, 1988], distributed algorithms [Rabin 1983], geometric algorithms [Mulmuley 1993], online algorithms [Ben-David et al. 1994, Raghavan and Snir 1994], and number-theoretic algorithms [Rabin 1983, Solovay and Strassen 1977]. The reader interested in these applications may consult these articles or Motwani and Raghavan [1995].

8

Pattern Matching and Text Compression Algorithms

Maxime Crochemore
University of Marne-la-Vallée

Thierry Lecroq
University of Rouen

8.1 Processing Texts Efficiently

The present chapter describes a few standard algorithms used for processing texts. They apply, for example, to the manipulation of texts (text editors), to the storage of textual data (text compression), and to data retrieval systems. The algorithms of this chapter are interesting in different respects. First, they are basic components used in the implementations of practical software. Second, they introduce programming methods that serve as paradigms in other fields of computer science (system or software design). Third, they play an important role in theoretical computer science by providing challenging problems.

Although data are stored in various ways, text remains the main form of exchanging information. This is particularly evident in literature or linguistics where data are composed of huge corpora and dictionaries. This applies as well to computer science where a large amount of data are stored in linear files. And this is also the case in molecular biology where biological molecules can often be approximated as sequences of nucleotides or amino acids. Moreover, the quantity of available data in these fields tends to double every 18 months. This is the reason why algorithms should be efficient even if the speed of computers increases more slowly.

0-8493-2909-4/97/$0.00+$.50
© 1997 by CRC Press, Inc.

Pattern matching is the problem of locating a specific pattern inside raw data. The pattern is usually a collection of strings described in some formal language. Two kinds of textual patterns are presented: single strings and approximated strings. We also present two algorithms for matching patterns in images that are extensions of string-matching algorithms.

In several applications, texts need to be structured before being searched. Even if no further information is known about their syntactic structure, it is possible and indeed extremely efficient to build a data structure that supports searches. From among several existing data structures equivalent to indices, we present the suffix tree, along with its construction.

The comparison of strings is implicit in the approximate pattern searching problem. Since it is sometimes required to compare just two strings (files, or molecular sequences) we introduce the basic method based on longest common **subsequences**.

Finally, the chapter contains two classical text compression algorithms. Variants of these algorithms are implemented in practical compression software, in which they are often combined together or with other elementary methods.

The efficiency of algorithms is evaluated by their running times, and sometimes by the amount of memory space they require at run time as well.

8.2 String-Matching Algorithms

String matching consists of finding one, or more generally, all of the **occurrences** of a pattern in a text. The pattern and the text are both strings built over a finite alphabet (a finite set of symbols). Each algorithm of this section outputs all occurrences of the pattern in the text. The pattern is denoted by $x = x[0 \ldots m-1]$; its length is equal to m. The text is denoted by $y = y[0 \ldots n-1]$; its length is equal to n. The alphabet is denoted by Σ and its size is equal to σ.

String-matching algorithms of the present section work as follows: they first align the left ends of the pattern and the text, then compare the aligned symbols of the text and the pattern—this specific work is called an attempt or a scan—and after a whole match of the pattern or after a mismatch they shift the pattern to the right. They repeat the same procedure again until the right end of the pattern goes beyond the right end of the text. This is called the scan and shift mechanism. We associate each attempt with the position i in the text when the pattern is aligned with $y[i \ldots i+m-1]$.

The brute force algorithm consists of checking, at all positions in the text between 0 and $n-m$, whether an occurrence of the pattern starts there or not. Then, after each attempt, it shifts the pattern exactly one position to the right. This is the simplest algorithm, which is described in Fig. 8.1.

The time complexity of the brute force algorithm is $O(mn)$ in the worst case but its behavior in practice is often linear on specific data.

```
void BF(char *y, char *x, int n, int m) {
    int i, j;

    /* Searching */
    for (i=0; i <= n-m; i++) {
        j=0;
        while (j < m && y[i+j] == x[j]) j++;
        if (j >= m) OUTPUT(i);
    }
}
```

FIGURE 8.1   The brute force string-matching algorithm.

Karp–Rabin Algorithm

Hashing provides a simple method for avoiding a quadratic number of symbol comparisons in most practical situations. Instead of checking at each position of the text whether the pattern occurs, it seems to be more efficient to check only if the portion of the text aligned with the pattern "looks like" the pattern. In order to check the resemblance between these portions a hashing function is used. To be helpful for the string-matching problem the hashing function should have the following properties:

- efficiently computable
- highly discriminating for strings
- $hash(y[i + 1 \ldots i + m])$ must be easily computable from $hash(y[i \ldots i + m - 1])$:
  $hash(y[i + 1 \ldots i + m]) = rehash(y[i], y[i + m], hash(y[i \ldots i + m - 1]))$.

For a word w of length k, its symbols can be considered as digits, and we define $hash(w)$ by

$$hash(w[0 \ldots k - 1]) = (w[0] * 2^{k-1} + w[1] * 2^{k-2} + \cdots + w[k - 1]) \bmod q$$

where q is a large number. Then, *rehash* has a simple expression

$$rehash(a, b, h) = ((h - a * d) * 2 + b) \bmod q$$

where $d = 2^{k-1}$.

During the search for the pattern x, it is enough to compare $hash(x)$ with $hash(y[i \ldots i + m - 1])$ for $0 \le i \le n - m$. If an equality is found, it is still necessary to check the equality $x = y[i \ldots i + m - 1]$ symbol by symbol.

In the algorithm of Fig. 8.2 all of the multiplications by 2 are implemented by shifts. Furthermore, the computation of the modulus function is avoided by using the implicit modular arithmetic given by the hardware that forgets carries in integer operations. Thus, q is chosen as the maximum value of an integer.

The worst-case time complexity of the Karp–Rabin algorithm is quadratic (as it is for the brute force algorithm) but its expected running time is $O(m + n)$.

Example 8.1. Let $x = $ **ing**. Then $hash(x) = 105 * 2^2 + 110 * 2 + 103 = 743$ (symbols are assimilated with their ASCII codes).

$y =$	s	t	r	i	n	g		m	a	t	c	h	i	n	g
$hash =$			806	797	776	743	678	585	443	746	719	766	709	736	743

Knuth–Morris–Pratt Algorithm

This section presents the first discovered linear-time string-matching algorithm. Its design follows a tight analysis of the brute force algorithm, and especially in the way this latter algorithm wastes the information gathered during the scan of the text.

Let us look more closely at the brute force algorithm. It is possible to improve the length of shifts and simultaneously remember some portions of the text that match the pattern. This saves comparisons between characters of the text and of the pattern, and consequently increases the speed of the search.

Consider an attempt at position i, that is, when the pattern $x[0 \ldots m - 1]$ is aligned with the window $y[i \ldots i + m - 1]$ in the text. Assume that the first mismatch occurs between symbols $y[i + j]$ and $x[j]$ for $0 \le j < m$. Then, $y[i \ldots i + j - 1] = x[0 \ldots j - 1] = u$ and $a = y[i + j] \ne x[j] = b$. When shifting, it is reasonable to expect that a **prefix** v of the pattern matches some **suffix** of the portion u of the text. Moreover, if we want to avoid another immediate mismatch, the letter following the prefix v in the pattern must be different from b. The longest such prefix v is called the **border** of u (it occurs at both ends of u). This introduces the notation: let $next[j]$ be the length of the longest (proper) border

```
#define REHASH (a, b, h) ((h-1*d)<<1)+b)

void KR(char *y, char *x, int n, int m) {
  int hy, hx, d, i;

  /* Preprocessing */
  /* computes d = 2^(m-1) with the left-shift operator */
  d=1;
  for (i=1; i < m; i++) d<<=1;

  hy=hx=0;
  for (i=0; i < m; i++) {
    hx=((hx<<1)+x[i]);
    hy=((hy<<1)+y[i]);
  }

  /* Searching */
  for (i=m; i <= n; i++) {
    if (hy == hx && strncmp (y+i-m, x, m) == 0) OUTPUT(i-m);
    hy=REHASH(y[i-m], y[i], hy);
  }
}
```

FIGURE 8.2   The Karp–Rabin string-matching algorithm.

of $x[0 \ldots j-1]$ followed by a character c different from $x[j]$. Then, after a shift, the comparisons can resume between characters $y[i+j]$ and $x[next[j]]$ without missing any occurrence of x in y, and avoid a backtrack on the text (see Fig. 8.3).

Example 8.2.   Here

$$
\begin{array}{lllllllllllllllll}
y = & . & . & . & \mathbf{a} & \mathbf{b} & \mathbf{a} & \mathbf{b} & \mathbf{a} & \mathbf{a} & . & . & . & . & . & . & . \\
x = & & & & \underline{\mathbf{a}} & \underline{\mathbf{b}} & \underline{\mathbf{a}} & \underline{\mathbf{b}} & \underline{\mathbf{a}} & \underline{\mathbf{b}} & \mathbf{a} \\
x = & & & & & & & & \underline{\mathbf{a}} & \underline{\mathbf{b}} & \mathbf{a} & \mathbf{b} & \mathbf{a} & \mathbf{b} & \mathbf{a}
\end{array}
$$

Compared symbols are underlined. Note that the empty string is the suitable border of **ababa**. Other borders of **ababa** are **aba** and **a**.

The Knuth–Morris–Pratt algorithm is displayed in Fig. 8.4. The table *next* it uses is computed in $O(m)$ time before the search phase, applying the same searching algorithm to the pattern itself, as if $y = x$ (see

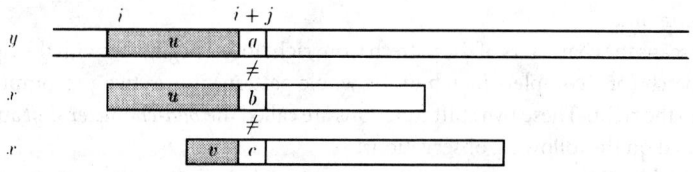

FIGURE 8.3   Shift in the Knuth–Morris–Pratt algorithm (v suffix of u).

```
void KMP (char *y, char *x, int n, int m) {
  /* XSIZE is the maximum size of a pattern */
  int i, j, next[XSIZE];

  /* Preprocessing */
  PRE_KMP(x, m, next);

  /* Searching */
  i=j=0;
  while (i < n) {
    while (j > -1 && x[j] != y[i] j=next[j];
    i++, j++;
    if (j >= m) { OUTPUT(i-j); j=next[m]; }
  }
}
```

FIGURE 8.4   The Knuth–Morris–Pratt string-matching algorithm.

```
void PRE_KMP(char *x, int m, int next[]) {
  int i, j;

  i=0; j=next[0]=-1;
  while (i < m) {
    while (j > -1 && x[i] != x[j]) j=next[j];
    i++, j++;
    if (x[i] == x[j] next[i]=next[j];
    else next[i]=j;
  }
}
```

FIGURE 8.5   Preprocessing phase of the Knuth–Morris–Pratt algorithm: computing **next**.

Fig. 8.5). The worst-case running time of the algorithm is $O(m + n)$ and it requires $O(m)$ extra space. These quantities are independent of the size of the underlying alphabet.

Boyer–Moore Algorithm

The Boyer–Moore algorithm is considered the most efficient string-matching algorithm in usual applications. A simplified version of it, or the entire algorithm, is often implemented in text editors for the search and substitute commands.

The algorithm scans the characters of the pattern from right to left beginning with the rightmost symbol. In case of a mismatch (or a complete match of the whole pattern) it uses two precomputed functions to shift the pattern to the right. These two shift functions are called the *bad-character shift* and the *good-suffix shift*. They are based on the following observations.

Assume that a mismatch occurs between the character $x[j] = b$ of the pattern and the character $y[i + j] = a$ of the text during an attempt at position i. Then, $y[i + j + 1 \ldots i + m - 1] = x[j +$

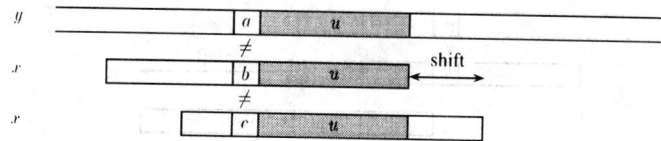

FIGURE 8.6 The good-suffix shift, u reappears preceded by a character different from b.

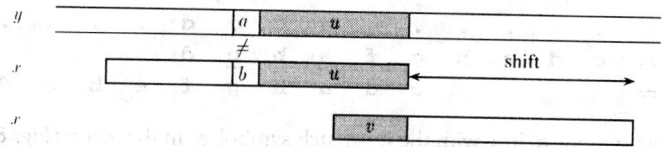

FIGURE 8.7 The good-suffix shift, only a suffix of u reappears as a prefix of x.

$1 \ldots m - 1] = u$ and $y[i + j] \neq x[j]$. The good-suffix shift consists of aligning the **segment** $y[i + j + 1 \ldots i + m - 1] = x[j + 1 \ldots m - 1]$ with its rightmost occurrence in x that is preceded by a character different from $x[j]$. (See Fig. 8.6.) If there exists no such segment, the shift consists of aligning the longest suffix v of $y[i + j + 1 \ldots i + m - 1]$ with a matching prefix of x. (see Fig. 8.7.)

Example 8.3. Here

```
y =  .  .  .  a  b  b  a  a  b  b  a  b  b  a  .  .  .
x =  a  b  b  a  a  b  b  a  b  b  a
x =           a  b  b  a  a  b  b  a  b  b  a
```

The shift is driven by the suffix **abba** of x found in the text. After the shift, the segment **abba** in the middle of y matches a segment of x as in Fig. 8.6. The same mismatch does not recur.

Example 8.4. Here

```
y =  .  .  .  a  b  b  a  a  b  b  a  b  b  a  b  b  a  .  .
x =        b  b  a  b  b  a  b  b  a
x =                 b  b  a  b  b  a  b  b  a
```

The segment **abba** found in y partially matches a prefix of x after the shift, as in Fig. 8.7.

The bad-character shift consists of aligning the text character $y[i + j]$ with its rightmost occurrence in $x[0 \ldots m - 2]$. (See Fig. 8.8.) If $y[i + j]$ does not appear in the pattern x, no occurrence of x in y can overlap the symbol $y[i + j]$, then the left end of the pattern is aligned with the character at position $i + j + 1$. (See Fig. 8.9.)

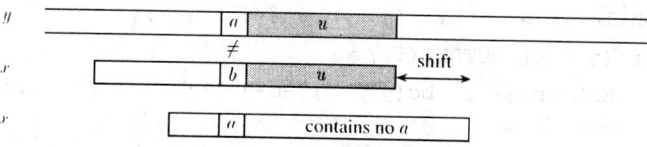

FIGURE 8.8 The bad-character shift, a appears in x.

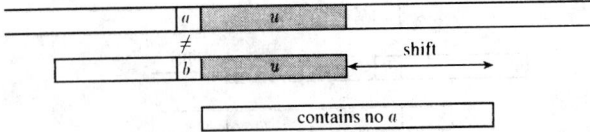

FIGURE 8.9   The bad-character shift, *a* does not appear in *x*.

Example 8.5.   Here

$$
\begin{array}{llllllllllll}
y = & . & . & . & . & . & . & \mathbf{a} & \mathbf{b} & \mathbf{c} & \mathbf{d} & . & . & . & . \\
x = & c & d & a & h & g & f & \underline{e} & \underline{b} & \underline{c} & \underline{d} \\
x = & & & & & c & d & a & h & g & f & e & b & c & \underline{d}
\end{array}
$$

The shift aligns the symbol **a** in *x* with the mismatch symbol **a** in the text *y* (Fig. 8.8).

Example 8.6.   Here

$$
\begin{array}{lllllllllll}
y = & . & . & . & . & . & \mathbf{a} & \mathbf{b} & \mathbf{c} & \mathbf{d} & . & . & . & . & . \\
x = & c & d & h & g & f & \underline{e} & \underline{b} & \underline{c} & \underline{d} \\
x = & & & & c & d & h & g & f & e & b & c & \underline{d}
\end{array}
$$

The shift positions the left end of *x* right after the symbol **a** of *y* (Fig. 8.9).

The Boyer–Moore algorithm is shown in Fig. 8.10. For shifting the pattern, it applies the maximum between the bad-character shift and the good-suffix shift. More formally, the two shift functions are defined as follows. The bad-character shift is stored in a table *bc* of size σ and the good-suffix shift is stored in a table *gs* of size $m + 1$. For $a \in \Sigma$

$$
bc[a] = \begin{cases} \min\{j / 1 \le j < m \text{ and } x[m-1-j] = a\} & \text{if } a \text{ appears in } x \\ m & \text{otherwise} \end{cases}
$$

```
void BM (char *y, char *x, int n, int m) {
    /* XSIZE is the maximum size of a pattern */
    /* ASIZE is the size of the alphabet      */
    int i, j, gs[XSIZE], bc[ASIZE];

    /* Preprocessing */
    PRE_GS(x, m, gs);
    PRE_BC(x, m, bc);

    /* Searching */
    i=0;
    while (i <= n-m) {
      j=m-1;
      while (j >= 0 && x[j] == y[i+j]) j--;
      if (j < 0) OUTPUT(i);
      i+=MAX(gs[j+1], bc[y[i+j]]-m+j+1);        /* shift */
    }
}
```

FIGURE 8.10   The Boyer–Moore string-matching algorithm.

Let us define two conditions,

$$cond_1(j, s): \quad \text{for each } k \text{ such that } j < k < m, s \geq k \text{ or } x[k - s] = x[k]$$

$$cond_2(j, s): \quad \text{if } s < j \text{ then } x[j - s] \neq x[j]$$

Then, for $0 \leq j < m$,

$$gs[j + 1] = \min\{s > 0 / cond_1(j, s) \text{ and } cond_2(j, s) \text{ hold}\}$$

and we define $gs[0]$ as the length of the smallest period of x.

Tables bc and gs can be precomputed in time $O(m + \sigma)$ before the search phase and require an extra space in $O(m + \sigma)$. (See Fig. 8.11 and Fig. 8.12.) The worst-case running time of the algorithm is quadratic. However, on large alphabets (relative to the length of the pattern) the algorithm is extremely fast. Slight modifications of the strategy yield linear-time algorithms (see the bibliographic notes). When searching for $a^{m-1}b$ in a^n the algorithm makes only $O(n/m)$ comparisons, which is the absolute minimum for any string-matching algorithm in the model where the pattern only is preprocessed.

```
void PRE_BC (char *x, int m, int bc[]) {
   /* ASIZE is the size of the alphabet */
   int j;

   for (j=0; j < ASIZE; j++) bc[j]=m;
   for (j=0; j < m-1; j++) bc[x[j]]=m-j-1;
}
```

FIGURE 8.11 Computation of the bad-character shift.

```
void PRE_GS (char *x, int m, int gs[]) {
   /* XSIZE is the maximum size of a pattern */
   int i, j, p, f[XSIZE];

   for (i=0; i <= m; i++) gs[i]=0;
   f[m]=j=m+1;
   for (i=m; i > 0; i--) {
      while (j <= m && x[i-1] != x[j-1]) {
         if (!gs [j]) gs[j]=j-i;
         j=f[j];
      }
      f[i-1]=--j;
   }
   p=f[0];
   for (j=0; j <= m; j++) {
      if (!gs[j]) gs[j]=p;
      if (j == p) p=f[p];
   }
}
```

FIGURE 8.12 Computation of the good-suffix shift.

Quick Search Algorithm

The bad-character shift used in the Boyer–Moore algorithm is not very efficient for small alphabets, but when the alphabet is large compared with the length of the pattern, as is often the case with the ASCII table and ordinary searches made under a text editor, it becomes very useful. Using it only produces a very efficient algorithm in practice that is described now.

After an attempt where x is aligned with $y[i \ldots i + m - 1]$, the length of the shift is at least equal to one. Thus, the character $y[i + m]$ is necessarily involved in the next attempt, and thus can be used for the bad-character shift of the current attempt. In the present algorithm, the bad-character shift is slightly modified to take into account the observation as follows ($a \in \Sigma$):

$$bc[a] = \begin{cases} \min\{j/0 \le j < m \text{ and } x[m-1-j] = a\} & \text{if } a \text{ appears in } x \\ m & \text{otherwise} \end{cases}$$

Indeed, the comparisons between text and pattern characters during each attempt can be done in any order. The algorithm of Fig. 8.13 performs the comparisons from left to right. It is called Quick Search after its inventor and has a quadratic worst-case time complexity but good practical behavior.

Example 8.7. Here

```
y =  s  t  r  i  n  g  -  m  a  t  c  h  i  n  g
x =  i  n  g
x =           i  n  g
x =                    i  n  g
x =                                   i  n  g
x =                                      i  n  g
```

The Quick Search algorithm makes nine comparisons to find the two occurrences of **ing** inside the text of length 15.

```
void QS(char *y, char *x, int n, int m) {
  /* ASIZE is the size of the alphabet */
  int i, j, bc[ASIZE];

  /* Preprocessing */
  for (j=0; j < ASIZE; j++) bc[j]=m;
  for (j=0; j < m; j++) bc[x[j]]=m-j-1;

  /* Searching */
  i=0;
  while (i <= n-m) {
    j=0;
    while (j < m && x[j] == y[i+j]) j++;
    if (j >= m) OUTPUT(i);
    i+=bc [y[i+m]]+1;                        /* shift */
  }
}
```

FIGURE 8.13 The Quick Search string-matching algorithm.

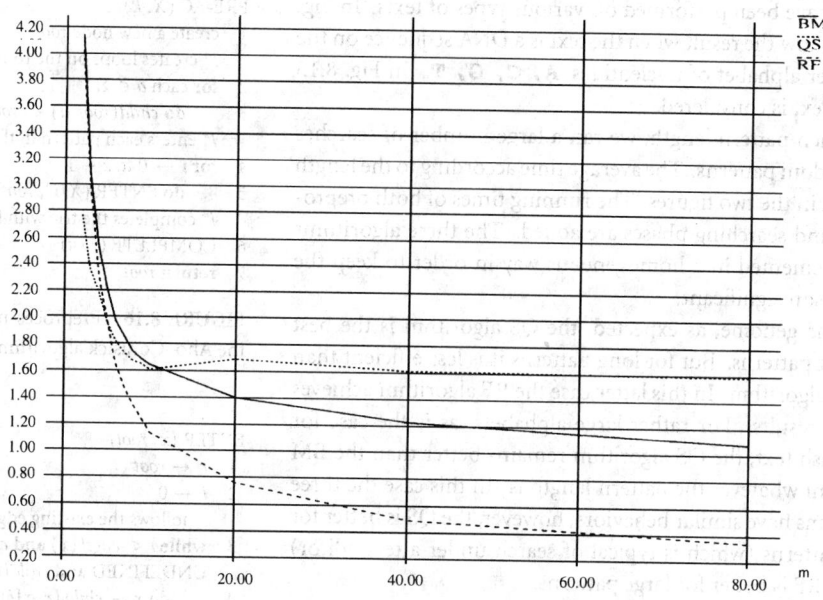

FIGURE 8.14   Running times for a DNA sequence.

Experimental Results

In Fig. 8.14 and Fig. 8.15 we present the running times of three string-matching algorithms: the Boyer–Moore algorithm (BM), the Quick Search algorithm (QS), and the Reverse-Factor algorithm (RF). The Reverse-Factor algorithm can be viewed as a variation of the Boyer–Moore algorithm where factors (segments) rather than suffixes of the pattern are recognized. The RF algorithm uses a data structure to store all of the factors of the reversed pattern: a suffix automaton or a **suffix tree** (see section 8.4).

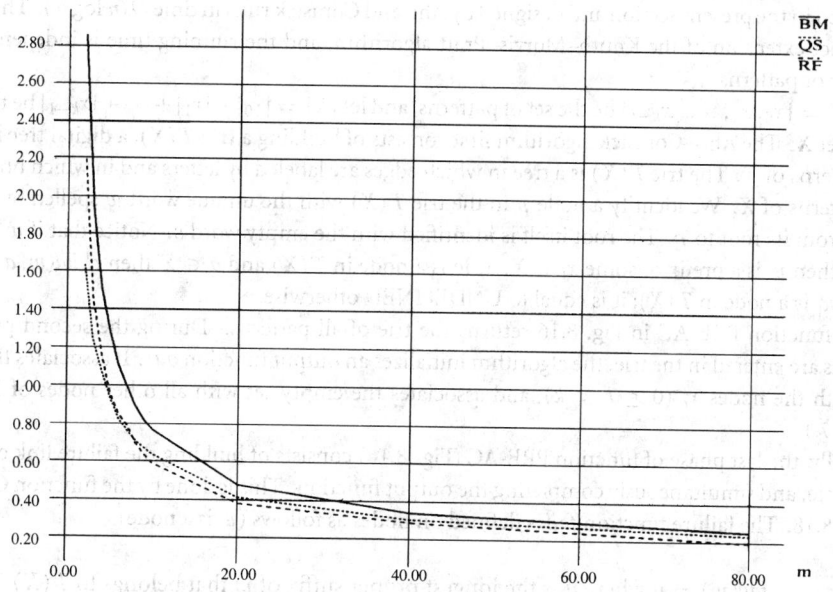

FIGURE 8.15   Running times for an English text.

Tests have been performed on various types of texts. In Fig. 8.14 we show the result when the text is a DNA sequence on the four-letter alphabet of nucleotides **A, C, G, T**. In Fig. 8.15 English text is considered.

For each pattern length, we ran a large number of searches with random patterns. The average time according to the length is shown in the two figures. The running times of both preprocessing and searching phases are added. The three algorithms are implemented in a homogeneous way in order to keep the comparison significant.

For the genome, as expected, the QS algorithm is the best for short patterns. But for long patterns it is less efficient than the BM algorithm. In this latter case the RF algorithm achieves the best results. For rather large alphabets, as is the case for an English text, the QS algorithm remains better than the BM algorithm whatever the pattern length is. In this case the three algorithms have similar behaviors; however, the QS is better for short patterns (which is typical of search under a text editor) and the RF is better for large patterns.

Aho–Corasick Algorithm

The UNIX operating system provides standard text (or file) facilities. Among them is the series of **grep** commands that locate patterns in files. We describe in this section the algorithm underlying the **fgrep** command of UNIX. It searches files for a finite set of strings, and can, for instance, output lines containing at least one of the strings.

If we are interested in searching for all occurrences of all patterns taken from a finite set of patterns, a first solution consists of repeating some string-matching algorithm for each pattern. If the set contains k patterns, this search runs in time $O(kn)$. The solution described in the present section and designed by Aho and Corasick runs in time $O(n \log \sigma)$. The algorithm is a direct extension of the Knuth–Morris–Pratt algorithm, and the running time is independent of the number of patterns.

Let $X = \{x_0, x_1, \ldots, x_{k-1}\}$ be the set of patterns, and let $|X| = |x_0| + |x_1| + \cdots + |x_{k-1}|$ be the total size of the set X. The Aho–Corasick algorithm first consists of building a **trie** $T(X)$, a digital tree recognizing the patterns of X. The trie $T(X)$ is a tree in which edges are labeled by letters and in which branches spell the patterns of X. We identify a node p in the trie $T(X)$ with the unique word w spelled by the path of $T(X)$ from its root to p. The root itself is identified with the empty word ε. Notice that if w is a node in $T(X)$; then w is a prefix of some $x_i \in X$. If w is a node in $T(X)$ and $a \in \Sigma$ then $child(w, a)$ is equal to wa if wa is a node in $T(X)$; it is equal to UNDEFINED otherwise.

The function PRE-AC in Fig. 8.16 returns the trie of all patterns. During the second phase, where patterns are entered in the trie, the algorithm initializes an output function out. It associates the singleton $\{x_i\}$ with the nodes x_i $(0 \leq i < k)$, and associates the empty set with all other nodes of $T(X)$. (See Fig. 8.17.)

Finally, the last phase of function PRE-AC (Fig. 8.16) consists of building the failure link of each node of the trie, and simultaneously completing the output function. This is done by the function COMPLETE in Fig. 8.18. The failure function *fail* is defined on nodes as follows (w is a node):

$$fail(w) = u \text{ where } u \text{ is the longest proper suffix of } w \text{ that belongs to } T(X)$$

PRE-AC (X, k)
1   create a new node *root*
    /* creates loops on the root of the trie */
2   **for** each $a \in \Sigma$
3   **do** $child(root, a) \leftarrow root$
    /* enters each pattern in the trie */
4   **for** $i \rightarrow 0$ to $k - 1$
5   **do** ENTER $(X[i], root)$
    /* completes the trie with failure links */
6   COMPLETE $(root)$
7   **return** *root*

FIGURE 8.16 Preprocessing phase of the Aho–Corasick algorithm.

ENTER $(x, root)$
1   $r \leftarrow root$
2   $i \leftarrow 0$
    /* follows the existing edges */
3   **while** $i < length(x)$ **and** $child \neq (r, x[i]$
    UNDEFINED **and** $child(r, x[i]) \neq root$
4   **do** $r \leftarrow child(r, x[i])$
5           $i \leftarrow i + 1$
    /*creates new edges */
6   **while** $i < length(x)$
7   **do** create a new node s
8           $child(r, x[i]) \leftarrow s$
9           $r \leftarrow s$
10          $i \leftarrow i + 1$
11 $out(r) \leftarrow x$

FIGURE 8.17 Construction of the trie.

```
COMPLETE (root)
 1   q ← empty queue
 2   l ← list of the edges (root, a, p) for any character a ∈ Σ and
       any node p ≠ root
 3   while the list l is not empty
 4     do (r, a, p) ← FIRST(l)
 5        l ← NEXT(l)
 6        ENQUEUE(q, p)
 7        fail(p) ← root
 8   while the queue q is not empty
 9     do r ← DEQUEUE(q)
10        l ← list of the edges (r, a, p) for any character a ∈ Σ and any node p
11        while the list l is not empty
12          do (r, a, p) ← FIRST(l)
13             l ← NEXT (l)
14             ENQUEUE(q, p)
15             s ← fail(r)
16             while child(s, a) = UNDEFINED
17               do s ← fail(s)
18             fail(p) ← child(s, a)
19             out(p) ← out(p) ∪ out(child(s, a))
```

FIGURE 8.18 Completion of the output function and construction of failure links.

Computation of failure links is done during a breadth-first traversal of $T(X)$. Completion of the output function is done while computing the failure function *fail* using the following rule:

$$\text{if } fail(w) = u \text{ then } out(w) = out(w) \cup out(u)$$

Example 8.8. Here $X = \{\texttt{search, ear, arch, chart}\}$

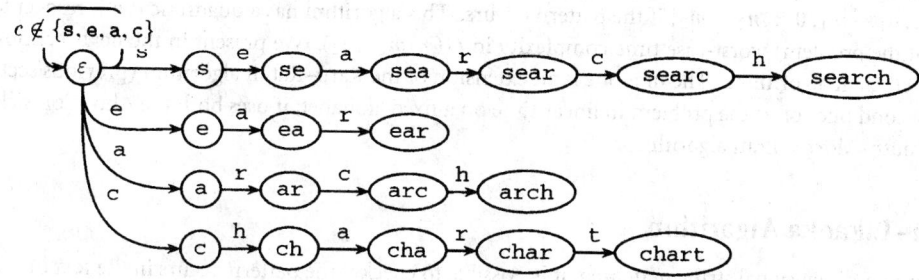

nodes	ε	s	se	sea	sear	searc	search	e	ea	ear
fail	ε	ε	e	ea	ear	arc	arch	ε	a	ar

nodes	a	ar	arc	arch	c	ch	cha	char	chart	
fail	ε	ε	c	ch	ε	ε	a	ar	ε	

nodes	sear	search	ear	arch	chart
out	{ear}	{search , arch}	{ear}	{arch}	{chart}

To stop going back with failure links during the computation of the failure links, and also to pass text characters for which no transition is defined from the root, a loop is added on the root of the trie for these symbols. This is done at the first phase of function PRE-AC.

After the preprocessing phase is completed, the searching phase consists of parsing all of the characters of the text y with $T(X)$. This starts at the root of $T(X)$ and uses failure links whenever a character in y does not match any label of outgoing edges of the current node. Each time a node with a nonempty output is encountered, this means that the patterns of the output have been discovered in the text, ending at the current position. Then, the position is output.

An implementation of the Aho–Corasick algorithm from the previous discussion is shown in Fig. 8.19. Note that the algorithm processes the text in an on-line way, so that the buffer on the text can be limited to only one symbol. Also note that the instruction $r \leftarrow fail(r)$ in Fig. 8.19 is the exact analog of instruction `j=next[j]` in Fig. 8.4. A unified view of both algorithms exists but is beyond the scope of the chapter.

The entire algorithm runs in time $O(|X| + n)$ if the *child* function is implemented to run in constant time. This is the case for any fixed alphabet. Otherwise a $\log \sigma$ multiplicative factor comes from access to the children of the nodes.

```
AC(y, n, X, k)
      /* Preprocessing */
1     r ← PRE-AC (X, k);
      /* Searching */
2     for i ← 0 to n − 1
3         do while child (r, y[i]) = UNDEFINED
4             do  r ← fail(r);
5             r ← child (r, y[i]);
6             if out(r) ≠ ∅
7                 then  OUTPUT (out(r), i);
```

FIGURE 8.19 The complete Aho–Corasick algorithm.

8.3 Two-Dimensional Pattern Matching Algorithms

In this section we consider only two-dimensional arrays. Arrays may be thought of as bit map representations of images, where each cell of arrays contains the codeword of a pixel. The string-matching problem finds an equivalent formulation in two dimensions (and even in any number of dimensions), and algorithms of section 8.2 can be extended to operate on arrays.

The problem now is to locate all occurrences of a two-dimensional pattern $x = x[0 \ldots m_1 - 1, 0 \ldots m_2 - 1]$ of size $m_1 \times m_2$ inside a two-dimensional text $y = [0 \ldots n_1 - 1, 0 \ldots n_2 - 1]$ of size $n_1 \times n_2$. The brute force algorithm for this problem is given in Fig. 8.20. It consists of checking at all positions of $y[0 \ldots n_1 - m_1, 0 \ldots n_2 - m_2]$ if the pattern occurs. This algorithm has a quadratic (with respect to the size of the problem) worst-case time complexity in $O(m_1 m_2 n_1 n_2)$. We present in the next sections two more efficient algorithms. The first one is an extension of the Karp–Rabin algorithm (previous section). The second one solves the problem in linear time on a fixed alphabet; it uses both the Aho–Corasick and the Knuth–Morris–Pratt algorithms.

Zhu–Takaoka Algorithm

As for one-dimensional string matching, it is possible to check if the pattern occurs in the text only if the *aligned* portion of the text looks like the pattern. To do that, the idea is to use vertically the hash function method proposed by Karp and Rabin. To initialize the process, the two-dimensional arrays x and y are translated into one-dimensional arrays of numbers x' and y'. The translation from x to x' is done as follows $(0 \leq i < m_2)$:

$$x'[i] = hash(x[0, i]x[1, i] \ldots x[m_1 - 1, i])$$

and the translation from y to y' is done by $(0 \leq i < m_2)$:

$$y'[i] = hash(y[0, i]y[1, i] \ldots y[m_1 - 1, i]).$$

```
/* YSIZE is the maximum size for a BIG    IMAGE */
/* XSIZE is the maximum size for a SMALL IMAGE */
typedef char BIG_IMAGE[YSIZE][YSIZE];
typedef char SMALL_IMAGE[XSIZE][XSIZE];


void BF_2D(BIG_IMAGE y, SMALL_IMAGE x, int n1, int n2, int m1, int m2) {
  int i, j, k

  /* Searching */
  for (i=0; i <= n1-m1; i++)
    for (j=0; j <= n2-m2; j++) {
      k=0;
      while (k < m1 && strncmp(&y[i+k][j], x[k], m2) == 0) k++;
      if (k >= m1) OUTPUT(i, j);
  }

}
```

FIGURE 8.20   The brute force two-dimensional pattern matching algorithm.

The fingerprint y' helps to find occurrences of x starting at row $j = 0$ in y. It is then updated for each new row in the following way ($0 \le i < m_2$):

$$hash(y[j+1, i]y[j+2, i] \ldots y[j+m_1, i])$$
$$= rehash(y[j, i], y[j+m_1, i], hash(y[j, i]y[j+1, i] \ldots y[j+m_1 - 1, i]))$$

(functions *hash* and *rehash* were described in the section on the Karp–Rabin algorithm).

Example 8.9.

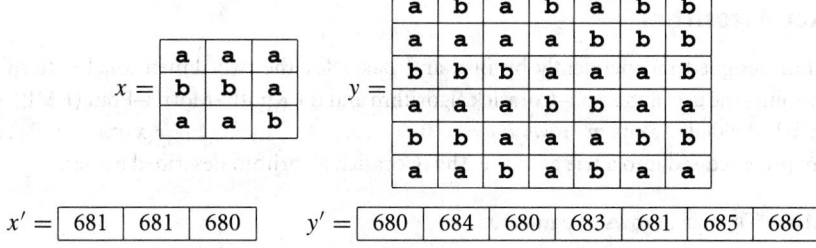

Since the alphabet of x' and y' is large, searching for x' in y' must be done by a string-matching algorithm for which the running time is independent of the size of the alphabet: the Knuth–Morris–Pratt suits this application perfectly. Its adaptation is shown in Fig. 8.21.

When an occurrence of x' is found in y', then we still have to check if an occurrence of x starts in y at the corresponding position. This is done naively by the procedure of Fig. 8.22.

The Zhu–Takaoka algorithm as explained is displayed in Fig. 8.23. The search for the pattern is performed row by row starting at row 0 and ending at row $n_1 - m_1$.

```
void KMP_IN_LINE(BIG_IMAGE Y, SMALL_IMAGE X, int YB[], int XB[],
                 int n2, int m1, int m2, int next[], int row) {
  int i, j;

  i=j=0;
  while (j < n2) {
    while (i > -1 && XB[i] != YB[j]) i=next[i];
    i++; j++;
    if (i >= m2) {
      DIRECT_COMPARE (Y, X, m1, m2, row, j-1);
      i=next[m2];
    }
  }
}
```

FIGURE 8.21   Search for x' in y' using KMP algorithm.

```
void DIRECT_COMPARE(BIG_IMAGE Y, SMALL_IMAGE X, int m1, int m2,
                    int row, int column) {
  int i, j, i0, j0;

  i0=row-m1+1;
  j0=column-m2+1;
  for (i=0; i < m1; i++)
    for (j=0; j < m2; j++)
      if (X[i][j] != Y[i0+i][j0+j]) return;
  OUTPUT(i0, j0);
}
```

FIGURE 8.22   Naive check of an occurrence of x in y at position (*row, column*).

Bird/Baker Algorithm

The algorithm designed independently by Bird and Baker for the two-dimensional pattern matching problem combines the use of the Aho–Corasick algorithm and the Knuth–Morris–Pratt (KMP) algorithm. The pattern x is divided into its m_1 rows $R_0 = x[0, 0 \ldots m_2 - 1]$ to $R_{m_1-1} = x[m_1 - 1, 0 \ldots m_2 - 1]$. The rows are preprocessed into a trie as in the Aho–Corasick algorithm described earlier.

Example 8.10.   The trie of rows of pattern x:

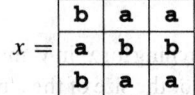

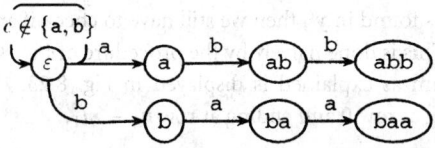

The search proceeds as follows. The text is read from the upper left corner to the bottom right corner, row by row. When reading the character $y[i, j]$ the algorithm checks whether the portion $y[i, j - m_2 + 1 \ldots j] = R$ matches any of $R_0, \ldots R_{m_1-1}$ using the Aho–Corasick machine. An additional one-dimensional array a of size n_1 is used as follows: $a[j] = k$ means that the $k - 1$ first rows $R_0, \ldots, R_{k-2}$ of the pattern match, respectively, the portions of the text: $y[i - k + 1, j - m_2 + 1 \ldots j], \ldots, y[i - 1, j - m_2 + 1 \ldots j]$. Then, if $R = R_{k-1}$, $a[j]$ is incremented to $k + 1$. If not, $a[j]$ is set to $s + 1$ where s is the maximum i such that

$$R_0 \ldots R_i = R_{k-s+1} \ldots R_{k-2} R$$

```
#define REHASH(a, b, h) (((h-a*d)<<1)+b)


void ZT (BIG_IMAGE Y, SMALL_IMAGE X, int n1, int n2, int m1, int m2) {
    int YB [YSIZE], XB[XSIZE], next [XSIZE], j, i, row, d;


    /* Preprocessing */
    /* Computes the first value of y' */
    for (j=0; j < n2; j++) {
      YB[j]=0;
      for (i=0; i < m1; i++) YB[j]=(YB[j]=(YB[j]<<1)+Y[i][j];
    }


    /* Computes x' */
    for (j=0; j < m2; j++) {
      XB [j]=0;
      for (i=0; i < m1; i++) XB[j]=(XB[j]<<1)+X[i][j];
    }


    row=m1-1;
    /* computes d=2^(m1-1) using the left shift operator */
    d=1;
    for (j=1; j < m1; j++) d<<=1;


    PRE_KMP(XB, m2, next);


    /* Searching */
    while (row < n1) {
      KMP_IN_LINE (Y, X, YB, XB, n2, m1, m2, next, row);
      if (row < n1-1)
         for (j=0; j < n2; j++)
            YB[j]=REHASH(Y[row-m1+1][j], Y[row+1][j], YB[j]);
      row++;
    }

}
```

FIGURE 8.23 The Zhu–Takaoka two-dimensional pattern matching algorithm.

```
PRE-KMP-FOR-B (X, m₁, next)
1    i ← 0
2    next[0] ← −1
3    j ← −1
4    while i < m₁
5        do while j > −1 and X[i, 0...m₂ − 1] ≠ X[j, 0...m₂ − 1]
6            do j ← next[j]
7        i ← i + 1
8        j ← j + 1
9        if X[i, 0...m₂−1] = X[j, 0...m₂− 1]
10           then   next[i] ← next[j]
11           else   next[i] ← j
```

FIGURE 8.24   Computes the function *next* for rows of *X*.

```
B (Y, n₁, n₂, X, m₁, m₂)
     /* Preprocessing */
1    for i ← 0 to n₂ − 1
2        do a[i] ← 0
3    root ← PRE-AC (set of lines of X, m₁)
4    PRE-KMP-FOR-B (X, m₁, next)
     /* Searching */
5    for row ← 0 to n₁ − 1
6        do r ← root
7            for column 0 to n₂ − 1
8                do while child (r, Y[row, column]) = UNDEFINED
9                    do r ← fail(r)
10                   r ← child (r, Y[row, column])
11                   if out(r) ≠ ∅
12                       then   k ← a[column]
13                           while k > 0 and X[k, 0...m₂ − 1] = out(r)
14                               do k ← next[k]
15                           a[column] ← k + 1
16                           if a[column] = m₁
17                               then   OUTPUT (row − m₁ + 1, column − m₂ + 1)
18                       else a[column] ← 0
```

FIGURE 8.25   The Bird/Baker two-dimensional pattern matching algorithm.

The value s is computed using the KMP algorithm vertically (in columns). If there exists no such s, $a[j]$ is set to 0. Finally, if at some point $a[j] = m_1$ an occurrence of the pattern appears at position $(i − m_1 + 1, j − m_2 + 1)$ in the text.

The Bird/Baker algorithm is presented in Figs. 8.24 and 8.25. It runs in time $O((n_1 n_2 + m_1 m_2) \log \sigma)$.

8.4 Suffix Trees

The suffix tree $S(y)$ of a string y is a trie (described earlier) containing all of the suffixes of the string, and having the properties described subsequently. This data structure serves as an index on the string: it provides a direct access to all segments of the string, and gives the positions of all their occurrences in the string.

Once the suffix tree of a text y is built, searching for x in y remains to spell x along a branch of the tree. If this walk is successful the positions of the pattern can be output. Otherwise, x does not occur in y.

Any kind of trie that represents the suffixes of a string can be used to search it. But the suffix tree has additional features which imply that its size is linear. The suffix tree of y is defined by the following properties:

- All branches of $S(y)$ are labeled by all suffixes of y.
- Edges of $S(y)$ are labeled by strings.
- Internal nodes of $S(y)$ have at least two children (when y is not empty).
- Edges outgoing an internal node are labeled by segments starting with different letters.
- The preceding segments are represented by their starting positions and their lengths in y.

Moreover, it is assumed that y ends with a symbol occurring nowhere else in it (the dollar sign is used in examples). This avoids marking nodes, and implies that $S(y)$ has exactly n leaves (number of nonempty suffixes). The other properties then imply that the total size of $S(y)$ is $O(n)$, which makes it possible to design a linear-time construction of the trie. The algorithm described in the present section has this time complexity provided the alphabet is fixed, or with an additional multiplicative factor $\log \sigma$ otherwise.

SUFFIX-TREE (y, n)
1 $T_{-1} \leftarrow$ one-node tree
2 **for** $i \leftarrow 0$ **to** $n - 1$
3   **do** $T_i \leftarrow$ INSERT $(T_{i-1}, y[i \ldots n - 1])$
4 **return** T_{n-1}

FIGURE 8.26 Construction of a suffix tree for y.

The algorithm inserts all nonempty suffixes of y in the data structure from the longest to the shortest suffix, as shown in Fig. 8.26. We introduce two definitions to explain how the algorithm works:

- $head_i$ is the longest prefix of $y[i \ldots n - 1]$ which is also a prefix of $y[j \ldots n - 1]$ for some $j < i$
- $tail_i$ is the word such that $y[i \ldots n - 1] = head_i \, tail_i$.

The strategy to insert the ith suffix in the tree is based on these definitions and described in Fig. 8.27.

The second step of the insertion (Fig. 8.27) is clearly performed in constant time. Thus, finding the node h is critical for the overall performance of the algorithm. A brute-force method to find it consists of spelling the current suffix $y[i \ldots n - 1]$ from the root of the tree, giving an $O(|head_i|)$ time complexity for the insertion at step i, and an $O(n^2)$ running time to build $S(y)$. Adding short-cut links leads to an overall $O(n)$ time complexity, although there is no guarantee that insertion at step i is realized in constant time.

Example 8.11. The different tries during the construction of the suffix tree of $y = $ **CAGATAGAG$**. Leaves are black and labeled by the position of the suffix they represent. Plain arrows are labeled by pairs: the pair (i, l) stands for the segment $y[i \ldots i + l - 1]$. Dashed arrows represent the nontrivial suffix links.

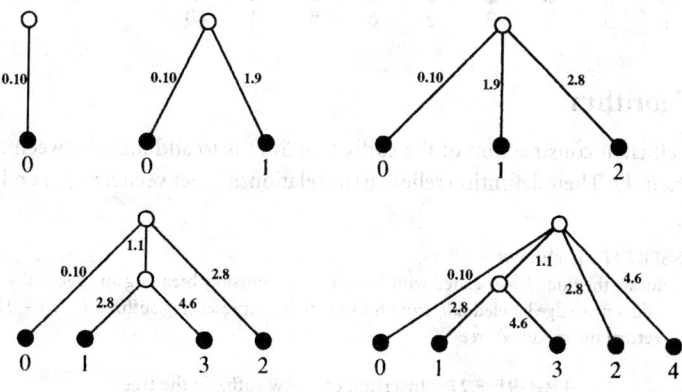

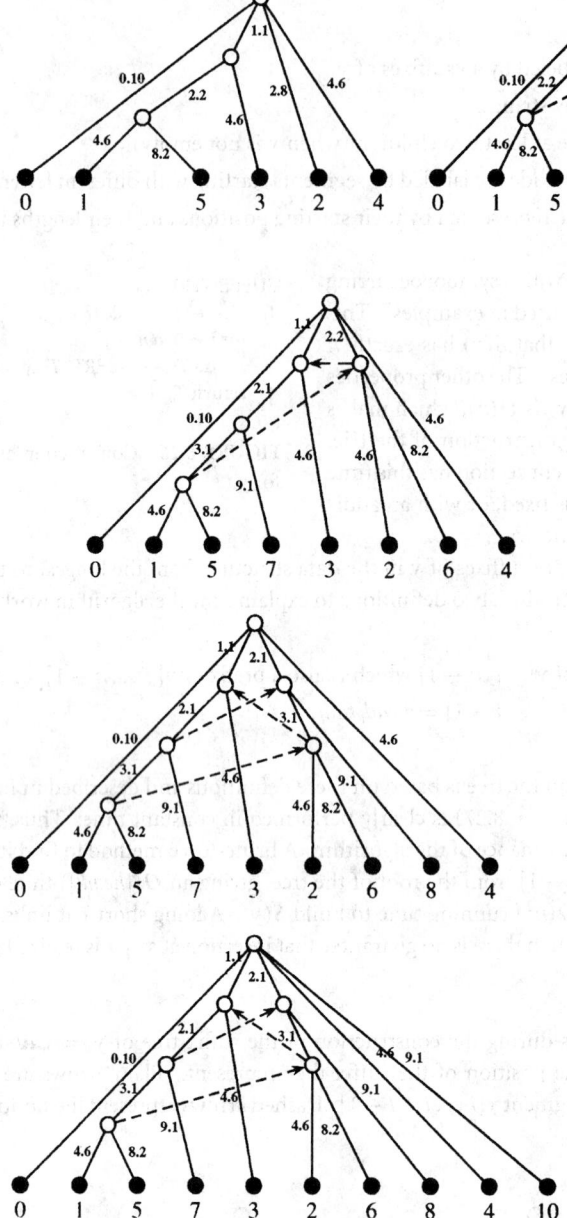

McCreight Algorithm

The key to get an efficient construction of the suffix tree $S(y)$ is to add links between nodes of the tree: they are called *suffix links*. Their definition relies on the relationship between $head_{i-1}$ and $head_i$: if $head_{i-1}$

INSERT $(T_{i-1}, y[i \ldots n-1])$
1   locate the node h associated with $head_i$ in T_{i-1}, possibly breaking an edge
2   add a new edge labeled $tail_i$ from h to a new left representing suffix $y[i \ldots n-1]$
3   **return** the modified tree

FIGURE 8.27   Insertion of a new suffix in the tree.

is of the form az ($a \in \Sigma, z \in \Sigma^*$), then z is a prefix of $head_i$. In the suffix tree the node associated with z is linked to the node associated with az. The suffix link creates a shortcut in the tree that helps in finding the next head efficiently. The insertion of the next suffix, namely, $head_i tail_i$, in the tree reduces to the insertion of $tail_i$ from the node associated with $head_i$.

The following property is an invariant of the construction: in T_i, only the node h associated with $head_i$ can fail to have a valid suffix link. This effectively happens when h has just been created at step i. The procedure to find the next head at step i is composed of two main phases:

A Rescanning: Assume that $head_{i-1} = az$ ($a \in \Sigma, z \in \Sigma^*$) and let d' be the associated node. If the suffix link on d' is defined, it leads to a node d from which the second step starts. Otherwise, the suffix link on d' is found by rescanning as follows. Let c' be the parent of d', and let (i, l) be the label of edge (c', d'). For the ease of the description, assume that $az = av(y[i \ldots i+l-1])$ (it may happen that $az = y[i \ldots i+l-1]$). There is a suffix link defined on c' and going to some node c associated with v. The crucial observation here is that $y[i \ldots i+l-1]$ is the prefix of the label of some branch starting at node c. Then, the algorithm rescans $y[i \ldots i+l-1]$ in the tree: let e be the child of c along that branch, and let (j, m) be the label of edge (c, e). If $m < l$, then a recursive rescan of $q = y[i+m \ldots i+l-1]$ starts from node e. If $m > l$, the edge (c, e) is broken to insert a new node d; labels are updated correspondingly. If $m = l$, d is simply set to e. If the suffix link of d' is currently undefined, it is set to d.

B Scanning: A downward search starts from d to find the node h associated with $head_i$. The search is dictated by the characters of $tail_{i-1}$ one at a time from left to right. If necessary a new internal node is created at the end of the scanning.

After the two phases A and B are executed, the node associated with the new head is known, and the tail of the current suffix can be inserted in the tree.

```
M(y, n)
 1   root ← INIT (y, n)
 2   head ← root
 3   tail ← child (root, y[0])
 4   n ← n − 1
 5   while n > 0
 6       do /* Phase A (rescanning) */
 7           if head = root
 8               then   d ← root
 9                      (i, l) ← label (tail)
10                      γ ← (i + 1, l − 1)
11               else   γ ← label (tail)
12                      if link(head) ≠ UNDEFINED
13                          then d ← link (head)
14                          else (i, l) ← label (head)
15                              if parent (head) = root
16                                  then   d ← RESCAN (root, i + 1, l − 1)
17                                  else   d ← RESCAN (link (parent (head)), i, l)
18                              link (head)  ← d
             /* Phase B (scanning) */
19           (head, γ) ← SCAN (d, γ)
20           create a new node tail
21           parent (tail) ← head
22           label (tail) ← γ
23           (i, l) ← γ
24           child (head, y[i] ← tail
25           n ← n − 1
26   return root
```

FIGURE 8.28 Suffix tree construction.

INIT (y, n)
1 create a new node *root*
2 create a new node *c*
3 *parent*(*root*) ← UNDEFINED
4 *parent*(*c*) ← *root*
5 *child*(*root*, *y*[0]) ← *c*
6 *label*(*root*) ← UNDEFINED
7 *label*(*c*) ← (0, *n*)
8 **return** *root*

FIGURE 8.29 Initialization procedure.

RESCAN (c, i, l)
1 (j, m) ← *label*(*child*(*c*, *y*[*i*]))
2 **while** $l > 0$ **and** $l \geq m$
3   **do** c ← *child*(*c*, *y*[*i*])
4        $l \leftarrow l - m$
5        $i \leftarrow i + m$
6   (j, m) ← *label*(*child*(*c*, *y*[*i*]))
7 **if** $l > 0$
8   **then**   **return** BREAK-EDGE $(child(c, y[i]), l)$
9   **else**   **return** c

FIGURE 8.30 The crucial rescan operation.

BREAK-EDGE (c, k)
1 create a new node *g*
2 *parent*(*g*) ← *parent*(*c*)
3 (i, l) ← *label*(*c*)
4 *child*(*parent*(*c*), *y*[*i*]) ← *g*
5 *label*(*g*) ← (*i*, *k*)
6 *parent*(*c*) ← *g*;
7 *label*(*c*) ← (*i* + *k*, *l* − *k*)
8 *child*(*g*, *y*[*i* + *k*]) ← *c*
9 *link*(*g*) ← UNDEFINED
10 **return** *g*

FIGURE 8.31 Breaking an edge.

SCAN (d, γ)
1 (i, l) ← γ
2 **while** *child*(*d*, *y*[*i*]) ≠ UNDEFINED
3   **do** g ← *child*(*d*, *y*[*i*])
4        $k \leftarrow 1$
5   (s, lg) ← *label*(*g*)
6        $s \leftarrow s + 1$
7        $l \leftarrow l - 1$
8        $i \leftarrow i + 1$
9   **while** $k < lg$ **and** $y[i] = y[s]$
10   **do** $i \leftarrow i + 1$
11             $s \leftarrow s + 1$
12             $k \leftarrow k + 1$
13             $l \leftarrow l - 1$
14   **if** $k < lg$
15   **then**   **return** (BREAK-EDGE$(g, k), (i, l)$)
16       $d \leftarrow g$
17 **return** $(d, (i, l))$

FIGURE 8.32 The scan operation.

To analyze the time complexity of the entire algorithm we mainly have to evaluate the total time of all scannings, and the total time of all rescannings. We assume that the alphabet is fixed, so that branching from a node to one of its children can be implemented to take constant time. Thus, the time spent for all scannings is linear because each letter of *y* is scanned only once. The same holds true for rescannings because each step downward (through node *e*) increases strictly the position of the segment of *y* considered there, and this position never decreases.

An implementation of McCreight's algorithm is shown in Fig. 8.28. The next figures (Figs. 8.29–8.32) give the procedures used by the algorithm, especially procedures RESCAN and SCAN.

We use the following notation:

- *parent(c)* is the parent node of the node *c*.
- *label(c)* is a pair (i, l) if node *c* is associated with the factor $y[i \ldots i + l − 1]$.
- *child(c, a)* is the only node that can be reached from the node *c* with the character *a*.
- *link(c)* is the suffix node of the node *c*.

8.5 Longest Common Subsequence of Two Strings

The notion of a longest common subsequence of two strings is widely used to compare files. The **diff** command of UNIX operating system implements an algorithm based on this notion, in which lines of the files are treated as symbols. The output of a comparison made by **diff** gives the minimum number of operations (insert a symbol, or delete a symbol) to transform one file into the other, which introduces what is known as the **edit distance** between the strings. (See section 8.6.) The comparison of molecular sequences is basically done with a closed concept: the alignment of strings, which consists of aligning their symbols on vertical lines. This is related to an edit distance with the additional operation of substitution.

A subsequence of a word x is obtained by deleting zero or more characters from x. More formally $w[0 \ldots i-1]$ is a subsequence of $x[0 \ldots m-1]$ if there exists an increasing sequence of integers $(k_j/j = 0, \ldots, i-1)$ such that for $0 \leq j \leq i-1, w[j] = x[k_j]$. We say that a word is an lcs(x, y) if it is a longest common subsequence of the two words x and y. Note that two strings can have several lcs(x, y). Their common length is denoted by llcs(x, y).

A brute-force method to compute an lcs(x, y) would consist of computing all of the subsequences of x, checking if they are subsequences of y, and keeping the longest one. The word x of length m has 2^m subsequences, and so this method could take $O(2^m)$ time, which is impractical even for fairly small values of m.

Dynamic Programming

The commonly used algorithm to compute an lcs(x, y) is a typical application of the dynamic programming method. Decomposing the problem into subproblems produces wide overlaps between them. So memorization of intermediate values is necessary to avoid recomputing them many times. Using dynamic programming it is possible to compute an lcs(x, y) in $O(mn)$ time and space. The method naturally leads to computing lcss for longer and longer prefixes of the two words. To do so, we consider the two-dimensional table L defined by

$$L[i, 0] = L[0, j] = 0, \quad \text{for } 0 \leq i \leq m \quad \text{and} \quad 0 \leq j \leq n$$

$$L[i+1, j+1] = \text{llcs}(x[0 \ldots i], y[0 \ldots j]), \quad \text{for } 0 \leq i \leq m-1 \quad \text{and} \quad 0 \leq j \leq n-1$$

Computing llcs$(x, y) = L[m, n]$ relies on a basic observation that yields the simple recurrence relation $(0 \leq i < m, 0 \leq j < n)$,

$$L[i+1, j+1] = \begin{cases} L[i, j] + 1 & \text{if } x[i] = y[j] \\ \max(L[i, j+1], L[i+1, j]) & \text{otherwise} \end{cases}$$

The relation is used by the algorithm of Fig. 8.33 to compute all the values from $L[0, 0]$ to $L[m, n]$. The computation takes $O(mn)$ time and space. It is possible afterward to trace back a path from $L[m, n]$ to exhibit an lcs(x, y). (See Fig. 8.34.)

```
int LCS(char *x, char *y, int m, int n, int L[YSIZE] [YSIZE]) {
    int i, j;

    for (i=0; i <= m; i++) L[i] [0]=0;
    for (j=0; j <= n; j++) L[0] [j]=0;

    for (i=0; i < m; i++)
        for (j=0; j < n; j++)
            if (x[i] == y[j]) L[i+1][j+1]=L[i] [j]+1;
            else L[i+1][j+1]=MAX (L(i+1][j], L[i] [j+1]);
    return L[m] [n];
}
```

FIGURE 8.33 Dynamic programming algorithm to compute llcs$(x, y) = L[m, n]$.

```
char *TRACE (char *x, char *y, int m, int n, int L[YSIZE] [YSIZE]) {
  int i, j, l;
  char z[YSIZE];


  i=m; j=n; l=L[m][n];
  z[l--]='\0';
  while (i>0 && j>0) {
    if (L[i][j] == L[i-1][j-1]+1 && x[i-1] == y[j-1] {
      z [l--]=x[i-1];
      i--; j--;
    }
    else if (L[i-1][j]>L[i][j-1])i--;
    else j--;
  }
  return(z);
}
```

FIGURE 8.34 Production of an lcs(x, y).

Example 8.12.   The value $L[5, 9] = 4$ is llcs(x, y) for $x =$ **AGCGA** and $y =$ **CAGATAGAG**. String **AGGA** is an lcs of x and y.

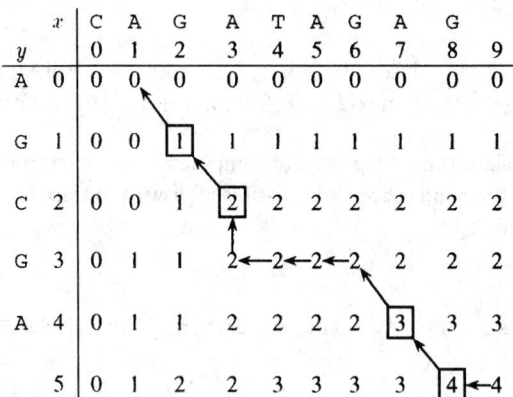

Reducing the Space: Hirschberg Algorithm

If only the length of an lcs(x, y) is required, it is easy to see that only one row (or one column) of the table L needs to be stored during the computation. The space complexity becomes $O(\min(m, n))$ as can be checked on the algorithm of Fig. 8.35. Indeed, the Hirschberg algorithm computes an lcs(x, y) in linear space and not only the value llcs(x, y). The computation uses the algorithm of Fig. 8.35.
    Let us define

$$L^*[i, n] = L^*[m, j] = 0, \quad \text{for } 0 \le i \le m \quad \text{and} \quad 0 \le j \le n$$

$$L^*[m - i, n - j] = \text{llcs}((x[i \dots m - 1])^R, (y[j \dots n - 1])^R)$$

$$\text{for } 0 \le i \le m - 1 \quad \text{and} \quad 0 \le j \le n - 1$$

```
int LLCS (char *x, char *y, int m, int n, int *L) {
  int i, j, last;

    for (j=0; j <= n; j++) L[j]=0;
    for (i=0; i < m; i++) {
      last=0;
      for (j=0; j < n; j++)
        if (last > L[j+1]) L[j+1]=last;
        else if (last < L[j+1]) last=L(j+1);
        else if (x[i] == y[j] {
          L(j+1]++;
          last++;
        }
    }
    return L[n];
}
```

FIGURE 8.35 $O(\min(m, n))$-space algorithm to compute llcs(x, y).

and

$$M(i) = \max_{0 \le j < n} \{L[i, j] + L^*[m - i, n - j]\}$$

where the word w^R is the reverse (or mirror image) of the word w. The algorithm of Fig. 8.36 computes the table L^*. The following property is the key observation to compute an lcs(x, y) in linear space:

$$M(i) = L[m, n], \quad \text{for } 0 \le i < m$$

In the algorithm shown in Fig. 8.37 the integer i is chosen as $m/2$. After $L[i, j]$ and $L^*[m - i, n - j]$ ($0 \le j < m$) are computed, the algorithm finds an integer k such that $L[i, k] + L^*[m - i, n - k] = L[m, n]$.

```
void LLCS_REVERSE (char *x, char *y, int a, int m, int n, int *Lstar) {
  int i, j, last;

  for (j=0; j <= n; j++) Lstar[j]=0;
  for (i=m-1; i >= a; i--) {
    last=0;
    for (j=n-1; j >= 0; j--)
      if (last > Lstar[n-j]) Lstar[n-j]=last;
      else if (last < Lstar[n-j]) last=Lstar [n-j];
      else if (x[i] == y[j]) {
        Lstar [n-j]++;
        last++;
      }
  }
}
```

FIGURE 8.36 Computation of L^*.

```
char *HIRSCHBERG (char *x, char *y, int m, int n) {
  int i, j, k, M;
  static char z[YSIZE];
  static int L[YSIZE], Lstar[YSIZE];
  static int count=0;

  if (m == 0) z[count]='\0';
  else if (m == 1) {
    for (i=0; i < n; i++)
      if (x[0] == y[i] {
        z[count++]=x[0];
        z[count]='\0';
        return (z);
      }
    z[count]='\0';
  }
  else{
    i=m/2;
    LLCS(x, y, i, n, L);
    LLCS_REVERSE(x, y, i, m, n, Lstar);
    k=n;
    M=L[n]+Lstar[0];
    for (j=n-1; j >= 0; j--)
      if (L[j]+Lstar[n-j] >= M) {
        M=L[j]+Lstar[n-j];
        k=j;
      }
    HIRSCHBERG (x, y, i, k);
    HIRSCHBERG (x+i, y+k, m-i, n-k);
    z[count]='\0';
  }
  return(z);
}
```

FIGURE 8.37 $O(\min(m, n))$-space computation of $\mathrm{lcs}(x, y)$.

Then, recursively, it computes an $\mathrm{lcs}(x[0 \ldots i], y[0 \ldots k])$ and an $\mathrm{lcs}(x[i + 1 \ldots m - 1], y[k + 1 \ldots n - 1])$, and concatenates them to get an $\mathrm{lcs}(x, y)$.

The running time of the Hirschberg algorithm is still $O(mn)$ but the amount of space required for the computation becomes $O(\min(m, n))$ instead of being quadratic as described in the section on dynamic programming.

8.6 Approximate String Matching

Approximate string matching is the problem of finding all approximate occurrences of a pattern x of length m in a text y of length n. Approximate occurrences of x are segments of y that are close to x according to a specific distance: the distance between segments and x must be not greater than a given integer k. We consider two distances in this section, the **Hamming distance** and the **Levenshtein distance**.

With the Hamming distance, the problem is also known as approximate string matching with k mismatches. With the Levenshtein distance (or edit distance), the problem is known as approximate string matching with k differences.

The Hamming distance between two words w_1 and w_2 of the same length counts the number of positions with different characters. The Levenshtein distance between two words w_1 and w_2 (not necessarily of the same length) is the minimal number of differences between the two words. A difference is one of the following operations:

- A substitution: a character of w_1 corresponds to a different character in w_2.
- An insertion: a character of w_1 corresponds to no character in w_2.
- A deletion: a character of w_2 corresponds to no character in w_1.

The *Shift-Or algorithm* of the next section is a method that is both very fast in practice and very easy to implement. It solves the Hamming distance and the Levenshtein distance problems. We initially describe the method for the exact string-matching problem and then we show how it can handle the cases of k mismatches and of k insertions, deletions, or substitutions. The method is flexible enough to be adapted to a wide range of similar approximate matching problems.

Shift-Or Algorithm

We first present an algorithm to solve the exact string-matching problem using a technique different from those developed in section 8.2, but which extends to the approximate string-matching problem.

Let $\mathbf{R}^0$ be a bit array of size m. Vector $\mathbf{R}_i^0$ is the value of the entire array $\mathbf{R}^0$ after text character $y[i]$ has been processed. (See Fig. 8.38.) It contains information about all matches of prefixes of x that end at position i in the text ($0 \leq j \leq m - 1$).

$$\mathbf{R}_i^0[j] = \begin{cases} 0 & \text{if } x[0 \ldots j] = y[i - j \ldots i] \\ 1 & \text{otherwise} \end{cases}$$

Therefore, $\mathbf{R}_i^0[m - 1] = 0$ is equivalent to saying that an (exact) occurrence of the pattern x ends at position i in y.

The vector $\mathbf{R}_i^0$ can be computed after $\mathbf{R}_{i-1}^0$ by the following recurrence relation:

$$\mathbf{R}_i^0[j] = \begin{cases} 0 & \text{if } \mathbf{R}_{i-1}^0[j - 1] = 0 \quad \text{and} \quad x[j] = y[i] \\ 1 & \text{otherwise} \end{cases}$$

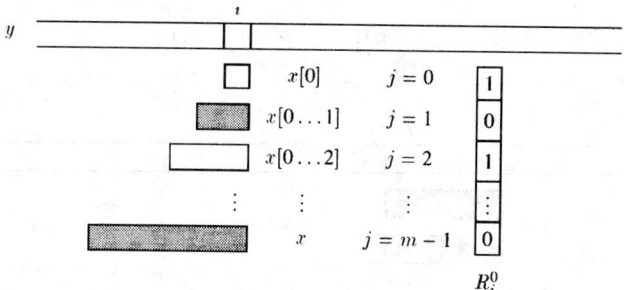

FIGURE 8.38 Meaning of vector $\mathbf{R}_i^0$.

and

$$\mathbf{R}_i^0[0] = \begin{cases} 0 & \text{if } x[0] = y[i] \\ 1 & \text{otherwise} \end{cases}$$

The transition from $\mathbf{R}_{i-1}^0$ to $\mathbf{R}_i^0$ can be computed very fast as follows. For each $a \in \Sigma$, let S_a be a bit array of size m defined by

$$S_a[j] = 0 \quad \text{iff} \quad x[j] = a, \quad \text{for } 0 \le j \le m - 1$$

The array S_a denotes the positions of the character a in the pattern x. Each S_a can be preprocessed before the search. And the computation of $\mathbf{R}_i^0$ reduces to two operations, _SHIFT_ and _OR_:

$$\mathbf{R}_i^0 = SHIFT\,(\mathbf{R}_{i-1}^0) \; OR \; S_{y[i]}$$

Example 8.13. String $x = $ **GATAA** occurs at position 2 in $y = $ **CAGATAAGAGAA**

$S_\mathbf{A}$	$S_\mathbf{C}$	$S_\mathbf{G}$	$S_\mathbf{T}$
1	1	0	1
0	1	1	1
1	1	1	0
0	1	1	1
0	1	1	1

	C	A	G	A	T	A	A	G	A	G	A	A
G	1	1	0	1	1	1	1	0	1	0	1	1
A	1	1	1	0	1	1	1	1	0	1	0	1
T	1	1	1	1	0	1	1	1	1	1	1	1
A	1	1	1	1	1	0	1	1	1	1	1	1
A	1	1	1	1	1	1	0	1	1	1	1	1

String Matching with k Mismatches

The Shift-Or algorithm easily adapts to support approximate string matching with k mismatches. To simplify the description, we shall present the case where at most one substitution is allowed.

We use arrays $\mathbf{R}^0$ and S as before, and an additional bit array $\mathbf{R}^1$ of size m. Vector $\mathbf{R}_{i-1}^1$ indicates all matches with at most one substitution up to the text character $y[i - 1]$. The recurrence on which the computation is based splits into two cases:

- There is an exact match on the first j characters of x up to $y[i - 1]$ (i.e., $\mathbf{R}_{i-1}^0[j - 1] = 0$). Then, substituting $y[i]$ to $x[j]$ creates a match with one substitution. (See Fig. 8.39.) Thus,

$$\mathbf{R}_i^1[j] = \mathbf{R}_{i-1}^0[j - 1]$$

FIGURE 8.39 If $\mathbf{R}_{i-1}^0[j - 1] = 0$ then $\mathbf{R}_i^1[j] = 0$.

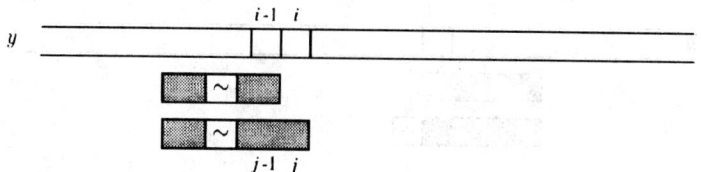

FIGURE 8.40 $\mathbf{R}_i^1[j] = \mathbf{R}_{i-1}^1[j-1]$ if $y[i] = x[j]$.

- There is a match with one substitution on the first j characters of x up to $y[i-1]$ and $y[i] = x[j]$. Then, there is a match with one substitution of the first $j + 1$ characters of x up to $y[i]$. (See Fig. 8.40.) Thus,

$$\mathbf{R}_i^1[j] = \begin{cases} \mathbf{R}_{i-1}^1[j-1] & \text{if } y[i] = x[j] \\ 1 & \text{otherwise} \end{cases}$$

This implies that $\mathbf{R}_i^1$ can be updated from $\mathbf{R}_{i-1}^1$ by the relation:

$$\mathbf{R}_i^1 = \left(SHIFT(\mathbf{R}_{i-1}^1) \; OR \; S_{y[i]}\right) \quad AND \quad SHIFT(\mathbf{R}_{i-1}^0)$$

Example 8.14. String $x = $ **GATAA** occurs at positions 2 and 7 in $y = $ **CAGATAAGAGAA** with no more than one mismatch.

	C	A	G	A	T	A	A	G	A	G	A	A
G	0	0	0	0	0	0	0	0	0	0	0	0
A	1	0	1	0	1	0	0	1	0	1	0	0
T	1	1	1	1	0	1	1	1	1	0	1	0
A	1	1	1	1	1	0	1	1	1	1	0	1
A	1	1	1	1	1	1	0	1	1	1	1	0

String Matching with k Differences

We show in this section how to adapt the Shift-Or algorithm to the case of only one insertion, and then to the case of only one deletion. The method is based on the following elements.

One insertion is allowed: Here, vector $\mathbf{R}_{i-1}^1$ indicates all matches with at most one insertion up to text character $y[i-1]$. $\mathbf{R}_{i-1}^1[j-1] = 0$ if the first j characters of $x(x[0 \ldots j-1])$ match j symbols of the last $j + 1$ text characters up to $y[i-1]$. Array $\mathbf{R}^0$ is maintained as before, and we show how to maintain array $\mathbf{R}^1$. Two cases can arise:

- There is an exact match on the first j characters of $x(x[0 \ldots j-1])$ up to $y[i-1]$. Then inserting $y[i]$ creates a match with one insertion up to $y[i]$. (See Fig. 8.41.) Thus,

$$\mathbf{R}_i^1[j] = \mathbf{R}_{j-1}^0[j-1]$$

- There is a match with one insertion on the j first characters of x up to $y[i-1]$. Then if $y[i] = x[j]$ there is a match with one insertion on the first $j + 1$ characters of x up to $y[j]$. (See Fig. 8.42.) Thus,

$$\mathbf{R}_i^1[j] = \begin{cases} \mathbf{R}_{i-1}^1[j-1] & \text{if } y[i] = x[j] \\ 1 & \text{otherwise} \end{cases}$$

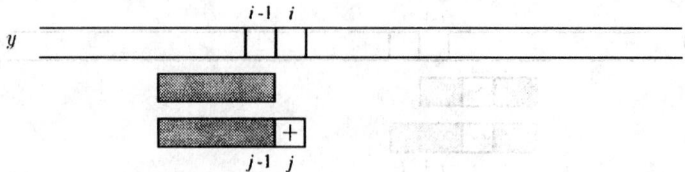

FIGURE 8.41   If $R_{i-1}^0[j-1] = 0$ then $R_i^1[j] = 0$.

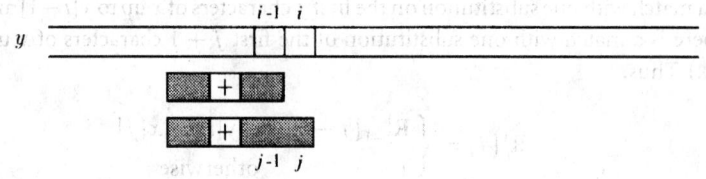

FIGURE 8.42 $R_i^1[j] = R_{i-1}^1[j-1]$ if $y[i] = x[j]$.

This shows that R_i^1 can be updated from R_{i-1}^1 with the formula

$$R_i^1 = \left(SHIFT(R_{i-1}^1) \text{ OR } S_{y[i]}\right) AND R_{i-1}^0$$

Example 8.15.   **GATAAG** is an occurrence of $x =$ **GATAA** with one insertion in $y =$ **CAGATAAGAGAA**

	C	A	G	A	T	A	A	G	A	G	A	A
G	1	1	1	0	1	1	1	1	0	1	0	1
A	1	1	1	1	0	1	1	1	1	0	1	0
T	1	1	1	1	1	0	1	1	1	1	1	1
A	1	1	1	1	1	1	0	1	1	1	1	1
A	1	1	1	1	1	1	1	0	1	1	1	1

One deletion is allowed: We assume here that R_{i-1}^1 indicates all possible matches with at most one deletion up to $y[i-1]$. As in previous problems, two cases arise:

- There is an exact match on the first $j+1$ characters of $x(x[0 \ldots j])$ up to $y[i]$ (i.e., $R_i^0[j] = 0$). Then, deleting $x[j]$ creates a match with one deletion. (See Fig. 8.43.) Thus,

$$R_i^1[j] = R_i^0[j]$$

- There is a match with one deletion on the first j characters of x up to $y[i-1]$ and $y[i] = x[j]$. Then, there is a match with one deletion on the first $j+1$ characters of x up to $y[i]$.

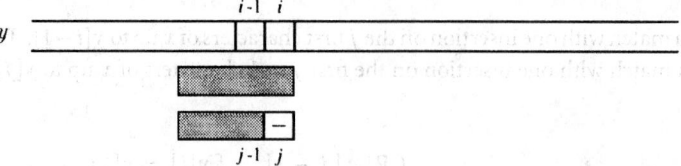

FIGURE 8.43   If $R_i^0[j] = 0$ then $R_i^1[j] = 0$.

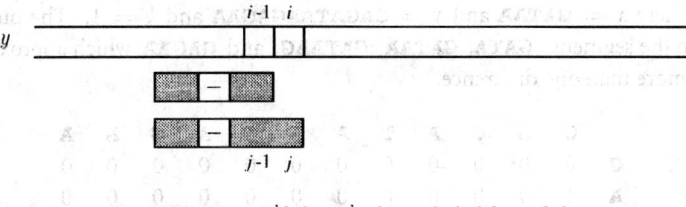

FIGURE 8.44 $R_i^1[j] = R_{i-1}^1[j-1]$ if $y[i] = x[j]$.

(See Fig. 8.44.) Thus,

$$R_i^1[j] = \begin{cases} R_{i-1}^1[j-1] & \text{if } y[i] = x[j] \\ 1 & \text{otherwise} \end{cases}$$

The discussion provides the following formula used to update R_i^1 from R_{i-1}^1:

$$R_i^1 = \left(SHIFT\left(R_{i-1}^1\right) OR \ S_{y[i]}\right) \quad AND \quad SHIFT\left(R_i^0\right)$$

Example 8.16. **GATA** and **ATAA** are two occurrences with one deletion of $x =$ **GATAA** in $y =$ **CAGA TAAGAGAA**

	C	A	G	A	T	A	A	G	A	G	A	A
G	0	0	0	0	0	0	0	0	0	0	0	0
A	1	0	0	0	1	0	0	0	0	0	0	0
T	1	1	1	0	0	1	1	1	0	1	0	1
A	1	1	1	1	0	0	1	1	1	1	1	0
A	1	1	1	1	1	0	0	1	1	1	1	1

Wu–Manber Algorithm

We present in this section a general solution for the approximate string-matching problem with at most k differences of the types: insertion, deletion, and substitution. It is an extension of the problems presented previously. The algorithm maintains $k + 1$ bit arrays $R^0, R^1, \ldots, R^k$ that are described now. The vector R^0 is maintained similarly as in the exact matching case (section "Shift-Or Algorithm"). The other vectors are computed with the formula ($1 \leq j \leq k$)

$$R_i^j = \left(SHIFT(R_{i-1}^j) \ OR \ S_{y[i]}\right)$$
$$AND \ SHIFT\left(R_i^{j-1}\right)$$
$$AND \ SHIFT\left(R_{i-1}^{j-1}\right)$$
$$AND \ R_{i-1}^{j-1}$$

which can be rewritten into

$$R_i^j = \left(SHIFT\left(R_{i-1}^j\right) \ OR \ S_{y[i]}\right)$$
$$AND \ SHIFT\left(R_i^{j-1} \ AND \ R_{i-1}^{j-1}\right)$$
$$AND \ R_{i-1}^{j-1}.$$

Example 8.17. Here $x =$ **GATAA** and $y =$ **CAGATAAGAGAA** and $k = 1$. The output 5, 6, 7, and 11 corresponds to the segments **GATA**, **GATAA**, **GATAAG**, and **GAGAA** which approximate the pattern **GATAA** with no more than one difference.

	C	A	G	A	T	A	A	G	A	G	A	A
G	0	0	0	0	0	0	0	0	0	0	0	0
A	1	0	0	0	0	0	0	0	0	0	0	0
T	1	1	1	0	0	0	1	1	0	0	0	0
A	1	1	1	1	0	0	0	1	1	1	0	0
A	1	1	1	1	1	0	0	0	1	1	1	0

The method, called the Wu–Manber algorithm, is implemented in Fig. 8.45. It assumes that the length of the pattern is no more than the size of the memory word of the machine, which is often the case in applications.

The preprocessing phase of the algorithm takes $O(\sigma m + km)$ memory space, and runs in time $O(\sigma m + k)$. The time complexity of its searching phase is $O(kn)$.

```
void WM (char *y, char *x, int n, int m, int k) {
    unsigned int j, last1, last2, lim, mask, S[ASIZE], R[KSIZE];
    int i;


    /* Preprocessing */
    for (i=0, i < ASIZE; i++) S[i]=~0;
    lim=0;
    for (i=0, j=1, i<m; i++, j <<= 1) {
      S[x[i]]&=~j;
      lim|=j;
    }
    lim = ~(lim >>1);
    R[0]=~0;
    for (j=1; j <= k; j++) R[j]=R[j-1] >> 1;

    /* Search */
    for (i=0; i<n; i++) {
      last1=R[0];
      masks=S[y[i]];
      R[0]=(r[0]<<1) | mask;
      for (j=1; j <= k; j++) {
        last2=R[j];
        R[j]=(R[j]<<1)| mask) & ((last1&R[j-1]) <<1)&last1;
        last1=last 2;
    }
    if (R[k] < lim)OUTPUT (i);
    }
}
```

FIGURE 8.45   Wu–Manber approximate string-matching algorithm.

8.7 Text Compression

In this section we are interested in algorithms that compress texts. Compression serves both to save storage space and to save transmission time. We shall assume that the uncompressed text is stored in a file. The aim of compression algorithms is to produce another file containing the compressed version of the same text. Methods in this section work with no loss of information, so that decompressing the compressed text restores exactly the original text.

We apply two strategies to design the algorithms. The first strategy is a statistical method that takes into account the frequencies of symbols to build a uniquely decipherable code optimal with respect to the compression. The code contains new codewords for the symbols occurring in the text. In this method fixed-length blocks of bits are encoded by different codewords. *A contrario* the second strategy encodes variable-length segments of the text. To put it simply, the algorithm, while scanning the text, replaces some already read segments just by a pointer to their first occurrences.

Huffman Coding

The Huffman method is an optimal statistical coding. It transforms the original code used for characters of the text (ASCII code on 8 b, for instance). Coding the text is just replacing each symbol (more exactly each occurrence of it) by its new codeword. The method works for any length of blocks (not only 8 b), but the running time grows exponentially with the length. In the following, we assume that symbols are originally encoded on 8 b to simplify the description.

The Huffman algorithm uses the notion of **prefix code**. A prefix code is a set of words containing no word that is a prefix of another word of the set. The advantage of such a code is that decoding is immediate. Moreover, it can be proved that this type of code does not weaken the compression.

A prefix code on the binary alphabet {0, 1} can be represented by a trie (see section on the Aho–Corasick algorithm) that is a binary tree. In the present method codes are complete: they correspond to complete tries (internal nodes have exactly two children). The leaves are labeled by the original characters, edges are labeled by 0 or 1, and labels of branches are the words of the code. The condition on the code implies that codewords are identified with leaves only. We adopt the convention that, from an internal node, the edge to its left child is labeled by 0, and the edge to its right child is labeled by 1.

In the model where characters of the text are given new codewords, the Huffman algorithm builds a code that is optimal in the sense that the compression is the best possible (the length of the compressed text is minimum). The code depends on the text, and more precisely on the frequencies of each character in the uncompressed text. The more frequent characters are given short codewords, whereas the less frequent symbols have longer codewords.

Encoding

The coding algorithm is composed of three steps: count of character frequencies, construction of the prefix code, and encoding of the text.

The first step consists of counting the number of occurrences of each character in the original text. (See Fig. 8.46.) We use a special end marker (denoted by **END**), which (virtually) appears only once at the end of the

COUNT (*fin*)
1   **for** each character $a \in \Sigma$
2   **do** *freq*(*a*) ← 0
3   **while** not end of file *fin* and *a* is the next symbol
4   **do** *freq*(a) ← *freq*(*a*) + 1
5   *freq* (**END**) ← 1

FIGURE 8.46   Counts the character frequencies.

text. It is possible to skip this first step if fixed statistics on the alphabet are used. In this case the method is optimal according to the statistics, but not necessarily for the specific text.

The second step of the algorithm builds the tree of a prefix code using the character frequency *freq* (*a*) of each character *a* in the following way:

- create a one-node tree *t* for each character *a*, setting *weight*(*t*) = *freq* (*a*) and *label*(*t*) = *a*;
- repeat (1), extract the two least weighted trees t_1 and t_2, and (2) create a new tree t_3 having left subtree t_1, right subtree t_2, and weight *weight*(t_3) = *weight* (t_1)+ *weight*(t_2);
- until only one tree remains.

```
BUILD-TREE
1    for each a ∈ Σ ∪ { END }
2        do if freq (a) ≠ 0
3            then   create a new node t
4                   weight (t) ← freq(a)
5                   label (t) ← a
6    lleaves ← list of all the nodes in increasing order of weight
7    ltrees ← empty list
8    while LENGTH (lleaves) + LENGTH (ltrees) > 1
9        do (l, r) ← extract the two nodes of smallest weight (among the two nodes at the beginning
                    of lleaves and the two nodes at the beginning of ltress)
10           create a new node t
11           weight (t) ← weight (l) + weight (r)
12           left(t) ← l
13           right(t) ← r
14           insert t at the end of ltrees
15   return t
```

FIGURE 8.47   Builds the coding tree.

The tree is constructed by the algorithm BUILD-TREE in Fig. 8.47. The implementation uses two linear lists. The first list contains the leaves of the future tree, each associated with a symbol. The list is sorted in the increasing order of the weight of the leaves (frequency of symbols). The second list contains the newly created trees. Extracting the two least weighted trees consists of extracting the two least weighted trees among the two first trees of the list of leaves and the two first trees of the list of created trees. Each new tree is inserted at the end of the list of the trees. The only tree remaining at the end of the procedure is the coding tree.

After the coding tree is built, it is possible to recover the codewords associated with characters by a simple depth-first search of the tree (see Fig. 8.48); *codeword(a)* is then the binary code associated with the character *a*.

In the third step, the original text is encoded. Since the code depends on the original text, in order to be able to decode the compressed text, the coding tree and the original codewords of symbols must be stored with the compressed text. This information is placed in a header of the compressed file, to be read at decoding time just before the compressed text. The header is made via a depth-first traversal of the tree. Each time an internal node is encountered a 0 is produced. When a leaf is encountered a 1 is produced followed by the original code of the corresponding character on 9 b (so that the end marker can be equal to 256 if all of the characters appear in the original text). This part of the encoding algorithm is shown in Fig. 8.49.

```
BUILD-CODE (t, length)
1   if t is not a leaf
2      then    temp [length] ← 0
3              BUILD-CODE (left(t), length + 1)
4              temp[length] ← 1
5              BUILD-CODE (right(t), length + 1)
6      else    codeword (label(t)) ← temp[0 ... length − 1]
```

FIGURE 8.48   Builds the character codes from the coding tree.

After the header of the compressed file is computed, the encoding of the original text is realized by the algorithm of Fig. 8.50.

```
CODE-TREE (fout, t)
1   if t is not a leaf
2      then   write a 0 in the file fout
3             CODE-TREE (fout, left(t))
4             CODE-TREE (fout, right(t))
5      else   write a 1 in the file fout
6             write the original code of label(t) in the file fout
```

FIGURE 8.49   Memorizes the coding tree in the compressed file.

```
CODE-TEXT (fin, fout)
1   while not end of file fin and a is the next symbol
2      do write codeword (a) in the file fout
3   write codeword (END) in the file fout
```

FIGURE 8.50   Encodes the characters in the compressed file.

A complete implementation of the Huffman algorithm, composed of the three steps just described, is given in Fig. 8.51.

Example 8.18. Here $y = $ **CAGATAAGAGAA**. The length of $y = 12 \times 8 = 96$ b (assuming an 8-b code). The character frequencies are

A	C	G	T	END
7	1	3	1	1

The different steps during the construction of the coding tree are

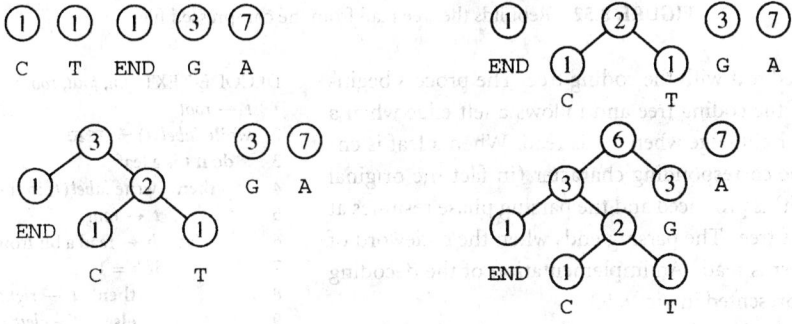

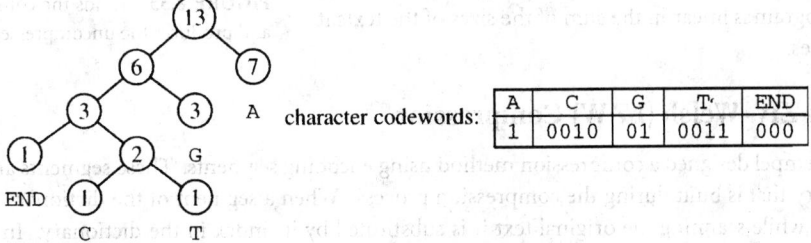

character codewords:

A	C	G	T'	END
1	0010	01	0011	000

The encoded tree is **0001** binary (**END**, 9)**01**binary (**C**, 9)**1**binary(**T**, 9) **1**binary (**G**, 9)**1**binary (**A**, 9), which produces a header of length 54 b,

$$\textbf{0001}\ 100000000\ 01\ 001000011\ 1\ 001010100\ 1\ 001000111\ 1\ 001000001$$

The encoded text

$$0010\ 1\ 01\ 1\ 0011\ 1\ 1\ 01\ 1\ 01\ 1\ 1\ 000$$

is of length 24 b. The total length of the compressed file is 78 b.

The construction of the tree takes $O(\sigma \log \sigma)$ time if the sorting of the list of the leaves is implemented efficiently. The rest of the encoding process runs in linear time in the sum of the sizes of the original and compressed texts.

Decoding

Decoding a file containing a text compressed by the Huffman algorithm is a mere programming exercise. First, the coding tree is rebuilt by the algorithm of Fig. 8.52. Then, the uncompressed text is recovered by parsing

CODING (*fin, fout*)
1 COUNT (*fin*)
2 $t \leftarrow$ BUILD-TREE (*Q, size*)
3 BUILD-CODE (*t*, 0)
4 CODE-TREE (*fout, t*)
5 CODE-TEXT (*fin, fout*)

FIGURE 8.51 Complete function for Huffman coding.

REBUILD-TREE (*fin, t*)
```
 1    b ← read a bit from the file fin
 2    if b = 1 /* leaf */
 3      then   left(t) ← NIL
 4             right(t) ← NIL
 5             label(t) ← symbol corresponding to the 9 next bits in the file fin
 6      else   create a new node l
 7             left(t) ← l
 8             REBUILD-TREE (fin, l)
 9             create a new node r
10             right(t) ← r
11             REBUILD-TREE (fin, r)
```

FIGURE 8.52   Rebuilds the tree read from the compressed file.

the compressed text with the coding tree. The process begins at the root of the coding tree and follows a left edge when a 0 is read or a right edge when a 1 is read. When a leaf is encountered, the corresponding character (in fact the original codeword of it) is produced and the parsing phase resumes at the root of the tree. The parsing ends when the codeword of the end marker is read. An implementation of the decoding of the text is presented in Fig. 8.53.

The complete decoding program is given in Fig. 8.54. It calls the preceding functions. The running time of the decoding program is linear in the sum of the sizes of the texts it manipulates.

DECODE-TEXT (*fin, fout, root*)
```
1    t ← root
2    while label(t) ≠ END
3      do if t is a leaf
4          then   write label(t) in the file fout
5                 t ← root
6          else   b ← read a bit from the file fin
7                 if b = 1
8                    then   t ← right(t)
9                    else   t ← left(t)
```

FIGURE 8.53   Reads the compressed text and produces the uncompressed text.

Lempel–Ziv–Welsh (LZW) Compression

Ziv and Lempel designed a compression method using encoding segments. These segments are stored in a dictionary that is built during the compression process. When a segment of the dictionary is encountered later while scanning the original text it is substituted by its index in the dictionary. In the model where portions of the text are replaced by pointers on previous occurrences, the Ziv–Lempel compression scheme can be proved to be asymptotically optimal (on large enough texts satisfying good conditions on the probability distribution of symbols).

The dictionary is the central point of the algorithm. It has the property of being prefix closed (every prefix of a word of the dictionary is in the dictionary), so that it can be implemented as a tree. Furthermore, a hashing technique makes its implementation efficient. The version described in this section is called the Lempel–Ziv–Welsh method after several improvements introduced by Welsh. The algorithm is implemented by the **compress** command existing under the UNIX operating system.

DECODING (*fin, fout*)
```
1    create a new node root
2    REBUILD-TREE (fin, root)
3    DECODE-TEXT (fin, fout, root)
```

FIGURE 8.54   Complete function for Huffman decoding.

Compression Method

We describe the scheme of the compression method. The dictionary is initialized with all of the characters of the alphabet. The current situation is when we have just read a segment w in the text. Let a be the next symbol (just following w). Then we proceed as follows:

- If wa is not in the dictionary, we write the index of w to the output file, and add wa to the dictionary. We then reset w to a and process the next symbol (following a).
- If wa is in the dictionary, we process the next symbol, with segment wa instead of w.

Initially, the segment w is set to the first symbol of the source text.

Example 8.19. Here $y = \text{CAGTAAGAGAA}$

C	A	G	T	A	A	G	A	G	A	A		w	written	added
↑												C	67	**CA**, 257
	↑											A	65	**AG**, 258
		↑										G	71	**GT**, 259
			↑									T	84	**TA**, 260
				↑								A	65	**AA**, 261
					↑							A		
						↑						AG	258	**AGA**, 262
							↑					A		
								↑				AG		
									↑			AGA	262	**AGAA**, 262
										↑		A		
													65	
													256	

Decompression Method

The decompression method is symmetrical to the compression algorithm. The dictionary is recovered while the decompression process runs. It is basically done in this way:

- Read a code c in the compressed file.
- Write in the output file the segment w which has index c in the dictionary.
- Add to the dictionary the word wa where a is the first letter of the next segment.

In this scheme, a problem occurs if the next segment is the word which is being built. This arises only if the text contains a segment $azazax$ for which az belongs to the dictionary but aza does not. During the compression process the index of az is written into the compressed file, and aza is added to the dictionary. Next, aza is read and its index is written into the file. During the decompression process the index of aza is read while the word az has not been completed yet: the segment aza is not already in the dictionary. However, since this is the unique case where the situation arises, the segment aza is recovered taking the last segment az added to the dictionary concatenated with its first letter a.

Example 8.20. Here the decoding is 67, 65, 71, 84, 65, 258, 262, 65, 256

read	written	added
67	C	
65	A	**CA**, 257
71	G	**AG**, 258
84	T	**GT**, 259
65	A	**TA**, 260
258	AG	**AA**, 261
262	AGA	**AGA**, 262
65	A	**AGAA**, 263
256		

Implementation

For the compression algorithm shown in Fig. 8.55, the dictionary is stored in a table D. The dictionary is implemented as a tree; each node z of the tree has the three following components:

- $parent(z)$ is a link to the parent node of z.
- $label(z)$ is a character.
- $code(z)$ is the code associated with z.

```
COMPRESS (fin, fout)
1    count ← −1
2    for each character a ∈ Σ
3        do count ← count + 1
4            HASH-INSERT (D, (−1, a, count))
5    count ← count + 1)
6    HASH-INSERT (D, (−1, END, count))
7    p ← −1
8    while not end of file fin
9        do a ← next character of fin
10           q ← HASH-SEARCH (D, (p, a))
11           if q = NIL
12               then   write code(p) on 1 + log(count) bits in fout
13                      count ← count + 1
14                      HASH-INSERT (D, (p, a, count))
15                      p ← HASH-SEARCH (D, (−1, a))
16               else p ← q
17    write p on 1+ log(count) bits in fout
18    write code (HASH-SEARCH (D, (−1, END)) in 1+ log(count) bits in fout
```

FIGURE 8.55   LZW compression algorithm.

The tree is stored in a table that is accessed with a hashing function. This provides fast access to the children of a node. The procedure HASH-INSERT $(D, (p, a, c))$ inserts a new node z in the dictionary D with $parent(z) = p$, $label(z) = a$, and $code(z) = c$. The function HASH-SEARCH $(D, (p, a))$ returns the node z such that $parent(z) = p$ and $label(z) = a$.

For the decompression algorithm, no hashing technique is necessary. Having the index of the next segment, a bottom-up walk in the trie implementing the dictionary produces the mirror image of the segment. A stack is used to reverse it. We assume that the function $string(c)$ performs this specific work for a code c. The bottom-up walk follows the parent links of the data structure. The function $first(w)$ gives the first character of the word w. These features are part of the decompression algorithm displayed in Fig. 8.56.

The Ziv–Lempel compression and decompression algorithms run both in linear time in the sizes of the files provided a good hashing technique is chosen. Indeed, it is very fast in practice. Its main advantage compared to Huffman coding is that it captures long repeated segments in the source file.

```
UNCOMPRESS (fin, fout)
1    count ← −1
2    for each character a ∈ Σ
3        do count ← count + 1
4            HASH-INSERT (D, (−1, a, count))
5    count ← count + 1
6    HASH-INSERT (D, (−1, END, count))
7    c ← first code on 1 + log(count) bits in fin
8    write string(c) in fout
9    a ← first (string(c))
10   repeat
11       d ← next code on 1 + log(count) in the fin
12       if d > count
13           then   count ← count + 1
14                  parent (count) ← c
15                  label (count) ← a
16                  write string(c) a in fout
17                  c ← d
18           else   a ← first (string(d))
19               if a ≠ END
20                   then   count ← count + 1
21                          parent (count) ← c
22                          label (count) ← a
23                          write string(d) in fout
17                          c ← d
24                   else   exit
25   forever
```

FIGURE 8.56   LZW decompression algorithm.

Experimental Results

Table 8.1 contains a sample of experimental results showing the behavior of compression algorithms on different types of texts. The table is extracted from Zipstein [1992].

The source files are: French text, C sources, Alphabet, and Random. Alphabet is a file containing a repetition of the line **abc...zABC...Z**. Random is a file where the symbols have been generated randomly, all with the same probability and independently of each other.

TABLE 8.1   Sizes of Texts Compressed with Three Algorithms

Source Texts	French	C Sources	Alphabet	Random
Sizes in Bytes	62816	684497	530000	70000
Huffman, %	53.27	62.10	72.65	**55.58**
Ziv–Lempel, %	**41.46**	34.16	2.13	63.60
Factor, %	47.43	**31.86**	**0.09**	73.74

Source: Zipstein, M. 1992. Data compression with factor automata. *Theor. Comput. Sci.* 92(1):213–221.

The compression algorithms reported in the table are: the Huffman algorithm of the "Huffman Coding" section, the Ziv–Lempel algorithm of the "Lempel–Ziv–Welsh Compression" section, and a third algorithm called Factor. This latter algorithm encodes segments of the source text as the Ziv–Lempel algorithm does. But the segments are taken among all segments already encountered in the text before the current position. The method usually gives a better compression ratio but is more difficult to implement.

Table 8.1 gives in percentage the sizes of compressed files. Results obtained by the Ziv–Lempel and Factor algorithms are similar. Huffman coding gives the best result for the Random file. Finally, experience shows that exact compression methods often reduce the size of data to 30–50% of their original size.

8.8 Research Issues and Summary

The algorithm for string searching by hashing was introduced by Harrison in 1971, and later fully analyzed by Karp and Rabin [1987].

The linear-time string-matching algorithm of Knuth, Morris, and Pratt is from 1976. It can be proved that, during the search, a character of the text is compared to a character of the pattern no more than $\log_{\Phi}(|x| + 1)$ (where Φ is the golden ratio $(1 + \sqrt{5})/2$). Simon [1993] gives an algorithm similar to the previous one but with a delay bounded by the size of the alphabet (of the pattern x). Hancart [1993] proves that the delay of Simon's algorithm is, indeed, no more than $1 + \log_2 |x|$. He also proves that this is optimal among algorithms searching the text through a window of size 1.

Galil [1981] gives a general criterion to transform searching algorithms of that type into real-time algorithms.

The Boyer–Moore algorithm was designed by Boyer and Moore [1977]. The first proof on the linearity of the algorithm when restricted to the search of the first occurrence of the pattern is in Knuth et al. [1977]. Cole [1995] proves that the maximum number of symbol comparisons is bounded by $3n$, and that this bound is tight.

Knuth et al. [1977] consider a variant of the Boyer–Moore algorithm in which all previous matches inside the current window are memorized. Each window configuration becomes the state of what is called the Boyer–Moore automaton. It is still unknown whether the maximum number of states of the automaton is polynomial or not.

Several variants of the Boyer–Moore algorithm avoid the quadratic behavior when searching for all occurrences of the pattern. Among the more efficient in terms of the number of symbol comparisons are: the algorithm of Apostolico and Giancarlo (1986), Turbo-BM algorithm by Crochemore et al. (1992) (the two algorithms are analyzed in Lecroq [1995]), and the algorithm of Colussi [1994].

The general bound on the expected time complexity of string matching is $O(|y| \log |x|/|x|)$. The probabilistic analysis of a simplified version of the Boyer–Moore algorithm, similar to the Quick Search algorithm of Sunday [1990] described in the chapter, was studied by several authors.

String searching can be solved by a linear-time algorithm requiring only a constant amount of memory in addition to the pattern and the (window on the) text. This can be proved by different techniques presented in Crochemore and Rytter [1994].

It is known that any string searching algorithm, working with symbol comparisons, makes at least $n + [9/4m](n - m)$ comparisons in the worst case (see Cole et al. [1995]). Some string searching algorithms make less than $2n$ comparisons at search phase. The presently known upper bound on the problem is $n + [8/3(m+1)](n-m)$, but with a quadratic-time preprocessing step [Cole et al. 1995]. With a linear-time preprocessing step, the current upper bound is $n + [(4 \log m + 2)/m](n - m)$ by Breslauer and Galil [1993]. Except in a few cases (patterns of length 3, for example), lower and upper bounds do not meet. Thus, the problem of the exact complexity of string searching is open.

The Aho–Corasick algorithm is from Aho and Corasick [1975]. It is implemented by the **fgrep** command under the UNIX operating system. Commentz–Walter (1979) has designed an extension of the Boyer–Moore algorithm to several patterns. It is fully described in Aho [1990].

On general alphabets the two-dimensional pattern matching can be solved in linear time, whereas the running time of the Bird/Baker algorithm has an additional $\log \sigma$ factor. It is still unknown whether the problem can be solved by an algorithm working simultaneously in linear time and using only a constant amount of memory space [see Crochemore and Rytter 1994].

The suffix tree construction of section 8.4 is by McCreight [1976]. Other data structures to represent indices on text files are: direct acyclic word graph [Blumer et al. 1985], suffix automata [Crochemore 1986], and suffix arrays [Manber and Myers 1993]. All of these techniques are presented in Crochemore and Rytter [1994]. The data structures implement full indices with standard operations, whereas applications sometimes need only incomplete indices. The design of compact indices is still unsolved.

Hirchsberg [1975] presents the computation of the lcs in linear space. This is an important result because the algorithm is classically run on large sequences. The quadratic time complexity of the algorithm to compute the Levenshtein distance is a bottleneck in practical string comparison for the same reason.

Approximate string searching is a lively domain of research. It includes, for instance, the notion of regular expressions to represent sets of strings. Algorithms based on regular expression are commonly found in books related to compiling techniques. The algorithms of section 8.6 are by Baeza-Yates and Gonnet [1992] and Wu and Manber [1992].

The statistical compression algorithm of Huffman (1951) has a dynamic version where symbol counting is done at coding time. The current coding tree is used to encode the next character and then updated. At decoding time a symmetrical process reconstructs the same tree, so the tree does not need to be stored with the compressed text. The command **compact** of UNIX implements this version.

Several variants of the Ziv and Lempel algorithm exist. The reader can refer to Bell et al. [1990] for a discussion on them. Nelson [1992] presents practical implementations of various compression algorithms.

Defining Terms

Border: A word $u \in \Sigma^*$ is a border of a word $w \in \Sigma^*$ if u is both a prefix and a suffix of w (there exist two words $v, z \in \Sigma^*$ such that $w = vu = uz$). The common length of v and z is a period of w.

Edit distance: The metric distance between two strings that counts the minimum number of insertions and deletions of symbols to transform one string into the other.

Hamming distance: The metric distance between two strings of same length that counts the number of mismatches.

Levenshtein distance: The metric distance between two strings that counts the minimum number of insertions, deletions, and substitutions of symbols to transform one string into the other.

Occurrence: An occurrence of a word $u \in \Sigma^*$, of length m, appears in a word $w \in \Sigma^*$, of length n, at position i if for $0 \le k \le m - 1, u[k] = w[i + k]$.

Prefix: A word $u \in \Sigma^*$ is a prefix of a word $w \in \Sigma^*$ if $w = uz$ for some $z \in \Sigma^*$.

Prefix code: Set of words such that no word of the set is a prefix of another word contained in the set. A prefix code is represented by a coding tree.

Segment: A word $u \in \Sigma^*$ is a segment of a word $w \in \Sigma^*$ if u occurs in w (see occurrence), that is, $w = vuz$ for two words $v, z \in \Sigma^*$ (u is also referred to as a factor or a subword of w).

Subsequence: A word $u \in \Sigma^*$ is a subsequence of a word $w \in \Sigma^*$ if it is obtained from w by deleting zero or more symbols that need not be consecutive (u is sometimes referred to as a subword of w, with a possible confusion with the notion of segment).

Suffix: A word $u \in \Sigma^*$ is a suffix of a word $w \in \Sigma^*$ if $w = vu$ for some $v \in \Sigma^*$.

Suffix tree: Trie containing all of the suffixes of a word.

Trie: Tree in which edges are labeled by letters or words.

References

Aho, A. V. 1990. Algorithms for finding patterns in strings. In *Handbook of Theoretical Computer Science,* vol. A. *Algorithms and Complexity,* J. van Leeuwen, ed., pp. 255–300. Elsevier, Amsterdam.

Aho, A. V. and Corasick, M. J. 1975. Efficient string matching: an aid to bibliographic search. *Comm. ACM* 18(6):333–340.

Baeza-Yates, R. A. and Gonnet, G. H. 1992. A new approach to text searching. *Comm. ACM* 35(10):74–82.

Baker, T. P. 1978. A technique for extending rapid exact-match string matching to arrays of more than one dimension. *SIAM J. Comput.* 7(4):533–541.

Bell, T. C., Cleary, J. G., and Witten, I. H. 1990. *Text Compression.* Prentice–Hall, Englewood Cliffs, NJ.

Bird, R. S. 1977. Two-dimensional pattern matching. *Inf. Process. Lett.* 6(5):168–170.

Blumer, A., Blumer, J., Ehrenfeucht, A., Haussler, D., Chen, M. T., and Seiferas, J. 1985. The smallest automaton recognizing the subwords of a text. *Theor. Comput. Sci.* 40:31–55.

Boyer, R. S. and Moore, J. S. 1977. A fast string searching algorithm. *Comm. ACM* 20(10):762–772.

Breslauer, D. and Galil, Z. 1993. Efficient comparison based string matching. *J. Complexity* 9(3): 339–365.

Breslauer, D., Colussi, L., and Toniolo, L. 1993. Tight comparison bounds for the string prefix matching problem. *Inf. Process. Lett.* 47(1):51–57.

Cole, R. 1994. Tight bounds on the complexity of the Boyer–Moore pattern matching algorithm. *SIAM J. Comput.* 23(5):1075–1091.

Cole, R., Hariharan, R., Zwick, U., and Paterson, M. S. 1995. Tighter lower bounds on the exact complexity of string matching. *SIAM J. Comput.* 24(1):30–45.

Colussi, L. 1994. Fastest pattern matching in strings. *J. Algorithms* 16(2):163–189.

Crochemore, M. 1986. Transducers and repetitions. *Theor. Comput. Sci.* 45(1):63–86.

Crochemore, M. and Rytter, W. 1994. *Text Algorithms.* Oxford University Press.

Galil, Z. 1981. String matching in real time. *J. ACM* 28(1):134–149.

Hancart, C. 1993. On Simon's string searching algorithm. *Inf. Process. Lett.* 47(2):95–99.

Hirchsberg, D. S. 1975. A linear space algorithm for computing maximal common subsequences. *Comm. ACM* 18(6):341–343.

Hume, A. and Sunday, D. M. 1991. Fast string searching. *Software–Practice Exp.* 21(11):1221–1248.

Karp, R. M. and Rabin, M. O. 1987. Efficient randomized pattern-matching algorithms. *IBM J. Res. Dev.* 31(2):249–260.

Knuth, D. E., Morris, J. H., Jr, and Pratt, V. R. 1977. Fast pattern matching in strings. *SIAM J. Comput.* 6(1):323–350.

Lecroq, T. 1995. Experimental results on string-matching algorithms. *Software–Practice Exp.* 25(7):727–765.

McCreight, E. M. 1976. A space-economical suffix tree construction algorithm. *J. Algorithms* 23(2):262–272.

Manber, U. and Myers, G. 1993. Suffix arrays: a new method for on-line string searches. *SIAM J. Comput.* 22(5):935–948.

Nelson, M. 1992. *The Data Compression Book.* M&T Books.

Simon, I. 1993. String matching algorithms and automata. In *First American Workshop on String Processing,* Baeza-Yates and Ziviani, eds., pp. 151–157. Universidade Federal de Minas Gerais.

Stephen, G. A. 1994. *String Searching Algorithms.* World Scientific Press.

Sunday, D. M. 1990. A very fast substring search algorithm. *Comm. ACM* 33(8):132–142.

Welch, T. 1984. A technique for high-performance data compression. *IEEE Comput.* 17(6):8–19.

Wu, S. and Manber, U. 1992. Fast text searching allowing errors. *Comm. ACM* 35(10):83–91.

Zhu, R. F. and Takaoka, T. 1989. A technique for two-dimensional pattern matching. *Comm. ACM* 32(9):1110–1120.

Zipstein, M. 1992. Data compression with factor automata. *Theor. Comput. Sci.* 92(1):213–221.

Further Information

Problems and algorithms presented in the chapter are just a sample of questions related to pattern matching. They share the formal methods used to design solutions and efficient algorithms. A wider panorama of algorithms on texts may be found in a few books such as:

Bell, T. C., Cleary, J. G., and Witten, I. H. 1990. *Text Compression.* Prentice–Hall, Englewood Cliffs, NJ.

Crochemore, M. and Rytter, W. 1994. *Text Algorithms.* Oxford University Press.

Nelson, M. 1992. *The Data Compression Book.* M&T Books.

Stephen, G. A. 1994. *String Searching Algorithms.* World Scientific Press.

Research papers in pattern matching are disseminated in a few journals, among which are: *Communications of the ACM, Journal of the ACM, Theoretical Computer Science, Algorithmica, Journal of Algorithms, SIAM Journal on Computing.*

Finally, two main annual conferences present the latest advances of this field of research:

Combinatorial Pattern Matching, which started in 1990 and was held in Paris (France), London (England), Tucson (AZ), Padova (Italy), Asilomar (CA), and Helsinki (Finland).

Data Compression Conference, which is regularly held at Snowbird.

General conferences in computer science often have sessions devoted to pattern matching algorithms.

Several books on the design and analysis of general algorithms contain chapters devoted to algorithms on texts. Here is a sample of these books:

Cormen, T. H., Leiserson, C. E., and Rivest, R. L. 1990. *Introduction to Algorithms.* MIT Press.

Gonnet, G. H. and Baeza-Yates, R. A. 1991. *Handbook of Algorithms and Data Structures.* Addison–Wesley.

9

Graph and Network Algorithms

Samir Khuller*
University of Maryland

Balaji Raghavachari†
University of Texas at Dallas

9.1 Introduction

Graphs are useful in modeling many problems from different scientific disciplines, since they capture the basic concept of objects (vertices) and relationships between objects (edges). Indeed, many optimization problems can be formulated in graph theoretic terms. Hence algorithms on graphs have been widely studied. In this chapter, a few fundamental graph algorithms are described. For a more detailed treatment of graph algorithms the reader is referred to textbooks on graph algorithms [Cormen et al. 1989, Even 1979, Gibbons 1985, Tarjan 1983].

An undirected *graph* $G = (V, E)$ is defined as a set V of *vertices* and a set E of *edges*. An edge $e = (u, v)$ is an unordered pair of vertices. A *directed graph* is defined similarly, except that its edges are ordered pairs of vertices, i.e., for a directed graph, $E \subseteq V \times V$. The terms *nodes* and vertices are used interchangeably. In this chapter, it is assumed that the graph has neither self-loops, edges of the form (v, v), nor multiple

*Research supported by National Science Foundation (NSF) Research Initiation Award CCR-9307462 and NSF CAREER Award CCR-9501355.

†Research supported by NSF Research Initiation Award CCR-9409625.

edges connecting two given vertices. The number of vertices of a graph, $|V|$, is often denoted by n. A graph is a **sparse graph** if $|E| \ll |V|^2$.

Bipartite graphs form a subclass of graphs and are defined as follows. A graph $G = (V, E)$ is bipartite if the vertex set V can be partitioned into two sets X and Y such that $E \subseteq X \times Y$. In other words, each edge of G connects a vertex in X with a vertex in Y. Such a graph is denoted by $G = (X, Y, E)$. Since bipartite graphs occur commonly in practice, algorithms are often specially designed for them.

A vertex w is *adjacent* to another vertex v if $(v, w) \in E$. An edge (v, w) is said to be *incident* to vertices v and w. The *neighbors* of a vertex v are all vertices $w \in V$ such that $(v, w) \in E$. The number of edges incident to a vertex v is called the **degree** of vertex v. For a directed graph, if (v, w) is an edge, then we say that the edge goes from v to w. The *out-degree* of a vertex v is the number of edges from v to other vertices. The *in-degree* of v is the number of edges from other vertices to v.

A *path* $p = [v_0, v_1, v_2, \ldots, v_k]$ from v_0 to v_k is a sequence of vertices such that (v_i, v_{i+1}) is an edge in the graph for $0 \leq i < k$. Any edge may be used only once in a path. A **cycle** is a path whose end vertices are the same, that is, $v_0 = v_k$. A path is *simple* if all its internal vertices are distinct. A **walk** $w = [v_0, v_1, \ldots, v_k]$ from v_0 to v_k is a sequence of vertices such that (v_i, v_{i+1}) is an edge in the graph for $0 \leq i < k$. A *closed walk* is one in which $v_0 = v_k$. A graph is said to be **connected** if there is a path between every pair of vertices. A directed graph is said to be **strongly connected** if there is a path between every pair of vertices in each direction. An acyclic, undirected graph is a **forest**, and a **tree** is a connected forest. A directed graph that does not have any cycles is known as a **directed acyclic graph** (DAG). Consider a binary relation C between the vertices of an undirected graph G such that for any two vertices u and v, uCv if and only if there is a path in G between u and v. It can be shown that C is an equivalence relation, and it partitions the vertices of G into equivalence classes, known as the connected components of G.

There are two convenient ways of representing graphs on computers. We first discuss the *adjacency list* representation. Each vertex has a linked list: there is one entry in the list for each of its adjacent vertices. The graph is thus represented as an array of linked lists, one list for each vertex. This representation uses $O(|V| + |E|)$ storage, which is good for sparse graphs. Such a storage scheme allows one to scan all vertices adjacent to a given vertex in time proportional to the degree of a vertex. The second representation, the *adjacency matrix*, is as follows. In this scheme, an $n \times n$ array is used to represent the graph. The $[i, j]$ entry of this array is 1 if the graph has an edge between vertices i and j, and 0 otherwise. This representation permits one to test if there is an edge between any pair of vertices in constant time. Both these representation schemes can be used in a natural way to represent directed graphs. For all algorithms in this chapter, it is assumed that the given graph is represented by an adjacency list.

Section 9.2 discusses various types of tree traversal algorithms. Sections 9.3 and 9.4 discuss depth-first and breadth-first search techniques. Section 9.5 discusses the single source shortest path problem. Section 9.6 discusses minimum spanning trees. Section 9.7 discusses the bipartite matching problem and the single commodity maximum flow problem. Section 9.8 discusses some traversal problems in graphs, and the Further Information section concludes with some pointers to current research on graph algorithms.

9.2 Tree Traversals

A tree is *rooted* if one of its vertices is designated as the root vertex and all edges of the tree are oriented (directed) to point away from the root. In a rooted tree, there is a directed path from the root to any vertex in the tree. For any directed edge (u, v) in a rooted tree, u is v's *parent* and v is u's *child*. The *descendants* of a vertex w are all vertices in the tree (including w) that are reachable by directed paths starting at w. The *ancestors* of a vertex w are those vertices for which w is a descendant. Vertices that have no children are called **leaves**. A *binary tree* is a special case of a rooted tree in which each node has at most two children, namely, the left child, and the right child. The trees rooted at the two children of a node are called the *left subtree* and *right subtree*.

In this section we study techniques for processing the vertices of a given binary tree in various orders. We assume that each vertex of the binary tree is represented by a record that contains fields to hold attributes of that vertex and two special fields *left* and *right* that point to its left and right subtree, respectively.

The three major tree traversal techniques are *preorder*, *inorder* and *postorder*. These techniques are used as procedures in many tree algorithms where the vertices of the tree have to be processed in a specific order. In a preorder traversal, the root of any subtree has to be processed *before* any of its descendants. In a postorder traversal, the root of any subtree has to be processed *after* all of its descendants. In an inorder traversal, the root of a subtree is processed after all vertices in its left subtree have been processed, but before any of the vertices in its right subtree are processed. Preorder and postorder traversals generalize to arbitrary rooted trees. In the example to follow, we show how postorder can be used to count the number of descendants of each node and store the value in that node. The algorithm runs in linear time in the size of the tree:

POSTORDER ALGORITHM. *PostOrder* (T):

```
1  if T ≠ nil then
2    lc ← PostOrder (T ↑. left).
3    rc ← PostOrder (T ↑. right).
4    T ↑. desc ← lc + rc + 1.
5    return T ↑. desc.
6  else
7    return 0.
8  end-if
end-proc
```

9.3 Depth-First Search

Depth-first search (DFS) is a fundamental graph searching technique [Tarjan 1972, Hopcroft and Tarjan 1973]. Similar graph searching techniques were given earlier by Tremaux (see Fraenkel [1970] and Lucas [1882]). However, the structure of DFS enables efficient algorithms for many other graph problems such as biconnectivity, triconnectivity, and planarity [Even 1979].

The algorithm first initializes all vertices of the graph as being unvisited. Processing of the graph starts from an arbitrary vertex, known as the root vertex. Each vertex is processed when it is first discovered (also referred to as *visiting* a vertex). It is first marked as visited, and its adjacency list is then scanned for unvisited vertices. Each time an unvisited vertex is discovered, it is processed recursively by DFS. After a node's entire adjacency list has been explored, that invocation of the DFS procedure returns. This procedure eventually visits all vertices that are in the same connected component of the root vertex. Once DFS terminates, if there are still any unvisited vertices left in the graph, one of them is chosen as the root and the same procedure is repeated.

The set of edges such that each one led to the discovery of a new vertex form a maximal forest of the graph, known as the **DFS forest**; a *maximal forest* of a graph G is an acyclic subgraph of G such that the addition of any other edge of G to the subgraph introduces a cycle. The algorithm keeps track of this forest using parent pointers. In each connected component, only the root vertex has a *nil* parent in the DFS tree.

The Depth-First Search Algorithm

DFS is illustrated using an algorithm that labels vertices with numbers 1, 2, ... in such a way that vertices in the same component receive the same label. This labeling scheme is a useful preprocessing step in many problems. Each time the algorithm processes a new component, it numbers its vertices with a new label.

DEPTH-FIRST SEARCH ALGORITHM. *DFS-Connected-Component (G):*

```
 1  c ← 0.
 2  for all vertices v in G do
 3      visited[v] ← false.
 4      finished[v] ← false.
 5      p[v] ← nil.
 6  end-for
 7  for all vertices v in G do
 8      if not visited [v] then
 9          c ← c + 1.
10          DFS (v, c).
11      end-if
12  end-for
end-proc
```

DFS (v, c):

```
 1  visited[v] ← true.
 2  component[v] ← c.
 3  for all vertices w in adj[v] do
 4      if not visited[w] then
 5          p[w] ← v.
 6          DFS (w, c).
 7      end-if
 8  end-for
 9  finished[v] ← true.
end-proc
```

Sample Execution

Figure 9.1 shows a graph having two connected components. DFS was started at vertex a, and the DFS forest is shown on the right. DFS visits the vertices b, d, c, e, and f, in that order. DFS then continues with vertices g, h, and i. In each case, the recursive call returns when the vertex has no more unvisited

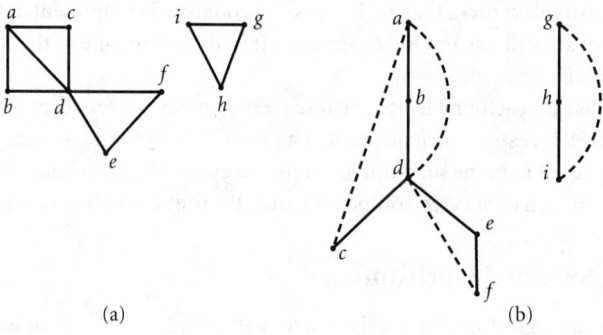

(a)   (b)

FIGURE 9.1 Sample execution of DFS on a graph having two connected components: (a) graph, (b) DFS forest.

neighbors. Edges (d, a), (c, a), (f, d), and (i, g) are called *back edges* (these do not belong to the DFS forest).

Analysis

A vertex v is processed as soon as it is encountered, and therefore at the start of DFS (v), *visited*[v] is *false*. Since *visited*[v] is set to true as soon as DFS starts execution, each vertex is visited exactly once. Depth-first search processes each edge of the graph exactly twice, once from each of its incident vertices. Since the algorithm spends constant time processing each edge of G, it runs in $O(|V| + |E|)$ time.

Remark 9.1. In the following discussion, there is no loss of generality in assuming that the input graph is connected. For a rooted DFS tree, vertices u and v are said to be *related*, if either u is an ancestor of v, or vice versa.

DFS is useful due to the special nature by which the edges of the graph may be classified with respect to a DFS tree. Notice that the DFS tree is not unique, and which edges are added to the tree depends on the order in which edges are explored while executing DFS. Edges of the DFS tree are known as tree edges. All other edges of the graph are known as *back* edges, and it can be shown that for any edge (u, v), u and v must be related. The graph does not have any *cross* edges, edges that connect two vertices that are unrelated. This property is utilized by a DFS-based algorithm that classifies the edges of a graph into **biconnected** components, maximal subgraphs that cannot be disconnected by the removal of any single vertex [Even 1979].

Directed Depth-First Search

The DFS algorithm extends naturally to directed graphs. Each vertex stores an adjacency list of its outgoing edges. During the processing of a vertex, first mark it as visited, and then scan its adjacency list for unvisited neighbors. Each time an unvisited vertex is discovered, it is processed recursively. Apart from tree edges and back edges (from vertices to their ancestors in the tree), directed graphs may also have *forward* edges (from vertices to their descendants) and *cross* edges (between unrelated vertices). There may be a cross edge (u, v) in the graph only if u is visited after the procedure call DFS (v) has completed execution.

Sample Execution

A sample execution of the directed DFS algorithm is shown in Fig. 9.2. DFS was started at vertex a, and the DFS forest is shown on the right. DFS visits vertices b, d, f, and c in that order. DFS then returns and continues with e, and then g. From g, vertices h and i are visited in that order. Observe that (d, a) and (i, g) are back edges. Edges (c, d), (e, d), and (e, f) are cross edges. There is a single forward edge (g, i).

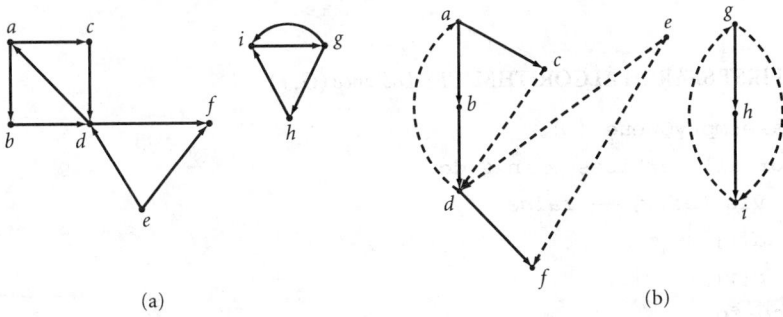

(a)   (b)

FIGURE 9.2 Sample execution of DFS on a directed graph: (a) graph, (b) DFS forest.

Applications of Depth-First Search

Directed DFS may be used to design a linear-time algorithm that classifies the edges of a given directed graph into *strongly connected* components: maximal subgraphs that have directed paths connecting any pair of vertices in them. The algorithm itself involves running DFS twice, once on the original graph, and then a second time on G^R, which is the graph obtained by reversing the direction of all edges in G. During the second DFS, we are able to obtain all of the strongly connected components. The proof of this algorithm is somewhat subtle, and the reader is referred to Cormen et al. [1989] for details.

Checking if a graph has a cycle can be done in linear time using DFS. A graph has a cycle if and only if there exists a back edge relative to any of its depth-first search trees. A directed graph that does not have any cycles is known as a directed acyclic graph. DAGs are useful in modeling precedence constraints in scheduling problems, where nodes denote jobs/tasks, and a directed edge from u to v denotes the constraint that job u must be completed before job v can begin execution. Many problems on DAGs can be solved efficiently using dynamic programming.

A useful concept in DAGs is that of a **topological order**: a linear ordering of the vertices that is consistent with the partial order defined by the edges of the DAG. In other words, the vertices can be labeled with distinct integers in the range $[1 \ldots |V|]$ such that if there is a directed edge from a vertex labeled i to a vertex labeled j, then $i < j$. The vertices of a given DAG can be ordered topologically in linear time by a suitable modification of the DFS algorithm. We keep a counter whose initial value is $|V|$. As each vertex is marked finished, we assign the counter value as its topological number, and decrement the counter. Observe that there will be no back edges; and that for all edges (u, v), v will be marked finished before u. Thus the topological number of v will be higher than that of u. Topological sort has applications in diverse areas such as project management, scheduling, and circuit evaluation.

9.4 Breadth-First Search

Breadth-first search (BFS) is another natural way of searching a graph. The search starts at a root vertex r. Vertices are added to a queue as they are discovered, and processed in (first-in–first-out) (FIFO) order.

Initially, all vertices are marked as unvisited, and the queue consists of only the root vertex. The algorithm repeatedly removes the vertex at the front of the queue, and scans its neighbors in the graph. Any neighbor not visited is added to the end of the queue. This process is repeated until the queue is empty. All vertices in the same connected component as the root are scanned and the algorithm outputs a spanning tree of this component. This tree, known as a breadth-first tree, is made up of the edges that led to the discovery of new vertices. The algorithm labels each vertex v by $d[v]$, the distance (length of a shortest path) of v from the root vertex, and stores the BFS tree in the array p, using parent pointers. Vertices can be partitioned into levels based on their distance from the root. Observe that edges not in the BFS tree always go either between vertices in the same level, or between vertices in adjacent levels. This property is often useful.

BREADTH-FIRST SEARCH ALGORITHM. *BFS-Distance* (G, r):

```
1  MakeEmptyQueue (Q).
2  for all vertices v in G do
3      visited[v] ← false.
4      d[v] ← ∞.
5      p[v] ← nil.
6  end-for
7  visited[r] ← true.
```

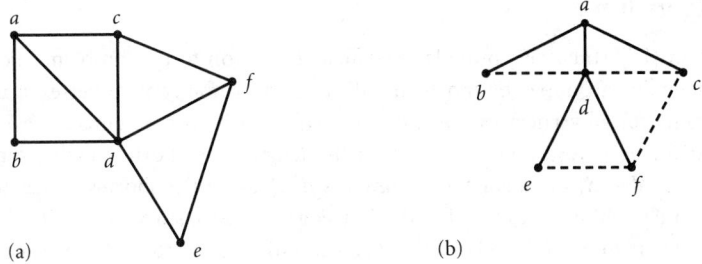

FIGURE 9.3   Sample execution of BFS on a graph: (a) graph, (b) BFS tree.

```
 8  d[r] ← 0.
 9  Enque (Q, r).
10  while not Empty (Q) do
11    v ← Deque (Q).
12    for all vertices w in adj[v] do
13      if not visited [w] then
14        visited[w] ← true.
15        p[w] ← v.
16        d[w] ← d[v] + 1.
17        Enque (w, Q).
18      end-if
19    end-for
20  end-while
end-proc
```

Sample Execution

Figure 9.3 shows a connected graph on which BFS was run with vertex a as the root. When a is processed, vertices b, d, and c are added to the queue. When b is processed nothing is done since all its neighbors have been visited. When d is processed, e and f are added to the queue. Finally c, e, and f are processed.

Analysis

There is no loss of generality in assuming that the graph G is connected, since the algorithm can be repeated in each connected component, similar to the DFS algorithm. The algorithm processes each vertex exactly once, and each edge exactly twice. It spends a constant amount of time in processing each. Hence the algorithm runs in $O(|V| + |E|)$ time.

9.5 Single-Source Shortest Paths

A natural problem that often arises in practice is to compute the shortest paths from a specified node to all other nodes in a graph. BFS solves this problem if all edges in the graph have the same length. Consider the more general case when each edge is given an arbitrary, non-negative length, and one needs to calculate a shortest length path from the root vertex to all other nodes of the graph, where the length of a path is defined to be the sum of the lengths of its edges. The distance between two nodes is the length of a shortest path between them.

Dijkstra's Algorithm

Dijkstra's algorithm [Dijkstra 1959] provides an efficient solution to this problem. For each vertex v, the algorithm maintains an upper bound to the distance from the root to vertex v in $d[v]$; initially $d[v]$ is set to infinity for all vertices except the root. The algorithm maintains a set S of vertices with the property that for each vertex $v \in S$, $d[v]$ is the length of a shortest path from the root to v. For each vertex u in $V - S$, the algorithm maintains $d[u]$ to be the shortest distance from the root to u that goes entirely within S, except for the last edge. It selects a vertex u in $V - S$ that minimizes $d[u]$ and adds it to S, and updates the distance estimates to the other vertices in $V - S$. In this update step it checks to see if there is a shorter path to any vertex in $V - S$ from the root that goes through u. Only the distance estimates to vertices that are adjacent to u are updated in this step. Since the primary operation is the selection of a vertex with minimum distance estimate, a priority queue is used to maintain the d-values of vertices. The priority queue should be able to handle the DE-CREASEKEY operation to update the d-value in each iteration. The next algorithm implements Dijkstra's algorithm.

DIJKSTRA'S ALGORITHM. *Dijkstra-Shortest Paths* (G, r):

```
 1 for all vertices v in G do
 2     visited[v] ← false.
 3     d[v] ← ∞.
 4     p[v] ← nil.
 5 end-for
 6 d[r] ← 0.
 7 BuildPQ (H, d).
 8 while not Empty (H) do
 9     u ← DELETEMIN (H).
10     visited[u] ← true.
11     for all vertices v in adj[u] do
12         Relax (u, v).
13     end-for
14 end-while
end-proc


Relax (u, v)
 1 if not visited[v] and d[v] > d[u] + w(u, v) then
 2     d[v] ← d[u] + w(u, v).
 3     p[v] ← u.
 4     DecreaseKey (H, v, d[v]).
 5 end-if
end-proc
```

Sample Execution

Figure 9.4 shows a sample execution of the algorithm. The column titled Iter specifies the number of iterations that the algorithm has executed through the while loop in step 8. In iteration 0 the initial values of the distance estimates are ∞. In each subsequent line of the table, the column marked u shows the vertex that was chosen in step 9 of the algorithm, and the change to the distance estimates at the end of that iteration of the while loop. In the first iteration, vertex r was chosen, after that a was chosen as it had

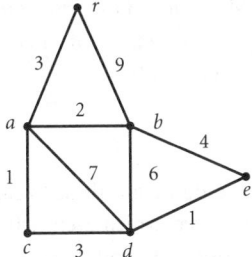

Iter	u	$d[a]$	$d[b]$	$d[c]$	$d[d]$	$d[e]$
0	—	∞	∞	∞	∞	∞
1	r	3	9	∞	∞	∞
2	a	3	5	4	10	∞
3	c	3	5	4	7	∞
4	b	3	5	4	7	9
5	d	3	5	4	7	8
6	e	3	5	4	7	8

FIGURE 9.4 Dijkstra's shortest path algorithm.

the minimum distance label among the unvisited vertices, and so on. The distance labels of the unvisited neighbors of the visited vertex are updated in each iteration.

Analysis

The running time of the algorithm depends on the data structure that is used to implement the priority queue H. The algorithm performs $|V|$ DELETEMIN operations and at most $|E|$ DECREASEKEY operations. If a binary heap is used to find the records of any given vertex, each of these operations run in $O(\log |V|)$ time. There is no loss of generality in assuming that the graph is connected. Hence the algorithm runs in $O(|E| \log |V|)$. If a Fibonacci heap is used to implement the priority queue, the running time of the algorithm is $O(|E| + |V| \log |V|)$. Even though the Fibonacci heap gives the best asymptotic running time, the binary heap implementation is likely to give better running times for most practical instances.

Bellman–Ford Algorithm

The shortest path algorithm previously described directly generalizes to directed graphs, but it does not work correctly if the graph has edges of negative length. For graphs that have edges of negative length, but no cycles of negative length, there is a different algorithm due to Bellman [1958] and Ford and Fulkerson [1962] that solves the single source shortest paths problem in $O(|V||E|)$ time.

The key to understanding this algorithm is the RELAX operation applied to an edge. In a single scan of the edges, we execute the RELAX operation on each edge. We then repeat the step $|V| - 1$ times. No special data structures are required to implement this algorithm, and the proof relies on the fact that a shortest path is simple and contains at most $|V| - 1$ edges (see Cormen et al. [1989] for a proof).

This problem also finds applications in finding a feasible solution to a system of linear equations, where each equation specifies a bound on the difference of two variables. Each constraint is modeled by an edge in a suitably defined directed graph. Such systems of equations arise in real-time applications.

9.6 Minimum Spanning Trees

The following fundamental problem arises in network design. A set of sites needs to be connected by a network. This problem has a natural formulation in graph-theoretic terms. Each site is represented by a vertex. Edges between vertices represent a potential link connecting the corresponding nodes. Each edge is given a nonnegative cost corresponding to the cost of constructing that link. A tree is a minimal network that connects a set of nodes. The cost of a tree is the sum of the costs of its edges. A minimum cost tree connecting the nodes of a given graph is called a minimum-cost spanning tree, or simply a **minimum spanning tree**.

The problem of computing a minimum spanning tree (MST) arises in many areas, and as a subproblem in combinatorial and geometric problems. MSTs can be computed efficiently using algorithms that are greedy in nature, and there are several different algorithms for finding an MST. One of the first algorithms was due to Boruvka [1926]. The two algorithms that are popularly known as Prim's algorithm and Kruskal's algorithm are described here. (Prim's algorithm was first discovered by Jarnik [1930].)

Prim's Algorithm

Prim's [1957] algorithm for finding an MST of a given graph is one of the oldest algorithms to solve the problem. The basic idea is to start from a single vertex and gradually grow a tree, which eventually spans the entire graph. At each step, the algorithm has a tree that covers a set S of vertices, and looks for a *good* edge that may be used to extend the tree to include a vertex that is currently not in the tree. All edges that go from a vertex in S to a vertex in $V - S$ are candidate edges. The algorithm selects a minimum-cost edge from these candidate edges and adds it to the current tree, thereby adding another vertex to S.

As in the case of Dijkstra's algorithm, each vertex $u \in V - S$ can attach itself to only one vertex in the tree (so that cycles are not generated in the solution). Since the algorithm always chooses a minimum-cost edge, it needs to maintain a minimum-cost edge that connects u to some vertex in S as the candidate edge for including u in the tree. A priority queue of vertices is used to select a vertex in $V - S$ that is incident to a minimum-cost candidate edge.

PRIM'S ALGORITHM. *Prim-MST* (G, r):

```
 1  for all vertices v in G do
 2      visited[v] ← false.
 3      d[v] ← ∞.
 4      p[v] ← nil.
 5  end-for
 6  d[r] ← 0.
 7  BuildPQ (H, d).
 8  while not Empty (H) do
 9      u ← DeleteMin (H).
10      visited[u] ← true.
11      for all vertices v in adj[u]do
12          if not visited[v] and d[v] > w(u,v) then
13              d[v] ← w(u,v).
14              p[v] ← u.
15              DecreaseKey (H, v, d[v]).
16          end-if
17      end-for
18  end-while
end-proc
```

Analysis

First observe the similarity between Prim's and Dijkstra's algorithms. Both algorithms start building the tree from a single vertex and grow it by adding one vertex at a time. The only difference is the rule for deciding when the current label is updated for vertices outside the tree. Both algorithms have the same structure and therefore have similar running times. Prim's algorithm runs in $O(|E| \log |V|)$ time if the priority queue is implemented using binary heaps, and it runs in $O(|E| + |V| \log |V|)$ if the priority queue is implemented using Fibonacci heaps.

Kruskal's Algorithm

Kruskal's [1956] algorithm for finding an MST of a given graph is another classical algorithm for the problem, and is also greedy in nature. Unlike Prim's algorithm which grows a single tree, Kruskal's algorithm grows a forest. First the edges of the graph are sorted in nondecreasing order of their costs. The algorithm starts with the empty spanning forest (no edges). The edges of the graph are scanned in sorted order, and if the addition of the current edge does not generate a cycle in the current forest, it is added to the forest. The main test at each step is: does the current edge connect two vertices in the same connected component? Eventually the algorithm adds $n - 1$ edges to make a spanning tree in the graph.

The main data structure needed to implement the algorithm is for the maintenance of connected components, to ensure that the algorithm does not add an edge between two nodes in the same connected component. An abstract version of this problem is known as the Union-Find problem for a collection of disjoint sets. Efficient algorithms are known for this problem, where an arbitrary sequence of UNION and FIND operations can be implemented to run in almost linear time [Cormen et al. 1989, Tarjan 1983].

KRUSKAL'S ALGORITHM. *Kruskal-MST (G)*:

```
1  T ← φ.
2  for all vertices v in G do
3    p[v] ← v.
4  Sort the edges of G by nondecreasing order of costs.
5  for all edges e = (u, v) in G in sorted order do
6    if Find (u) ≠ Find (v) then
7      T ← T ∪ (u, v).
8      Union (u, v).
9  end-proc
```

Analysis

The running time of the algorithm is dominated by step 4 of the algorithm in which the edges of the graph are sorted by nondecreasing order of their costs. This takes $O(|E| \log |E|)$ [which is also $O(|E| \log |V|)$] time using an efficient sorting algorithm such as heap sort. Kruskal's algorithm runs faster in the following special cases: if the edges are presorted, if the edge costs are within a small range, or if the number of different edge costs is bounded. In all of these cases, the edges can be sorted in linear time, and the algorithm runs in near-linear time, $O(|E|\alpha(|E|, |V|))$, where $\alpha(m, n)$ is the inverse Ackermann function [Tarjan 1983].

Remark 9.2. The MST problem can be generalized to directed graphs. The equivalent of trees in directed graphs are called arborescences or **branchings**, and since edges have directions, there are incoming branchings with its edges directed toward the root, and outgoing branchings with its edges directed away from the root. The input is a directed graph with arbitrary costs on the edges, and a root vertex r. The output is a minimum-cost branching rooted at r. The algorithms discussed in this section for finding minimum spanning trees do not directly extend to the problem of finding optimal branchings. There are efficient algorithms that run in $O(|E|+|V| \log |V|)$ time using Fibonacci heaps for finding minimum-cost branchings [Gibbons 1985, Gabow et al. 1986]. These algorithms are based on techniques for weighted matroid intersection [Lawler 1976]. Almost linear-time deterministic algorithms for the MST problem in undirected graphs are also known [Fredman and Tarjan 1987].

9.7 Matchings and Network Flows

Networks are important both for electronic communication and for transporting goods. The problem of efficiently moving entities (such as bits, people, or products) from one place to another in an underlying network is modeled by the **network flow** problem. The problem plays a central role in the fields of operations research and computer science, and much emphasis has been placed on the design of efficient algorithms for solving it. Many of the basic algorithms studied earlier in this chapter play an important role in developing various implementations for network flow algorithms.

First the **matching** problem, which is a special case of the flow problem, is introduced. Then the **assignment problem**, which is a generalization of the matching problem to the weighted case, is studied. Finally the network flow problem is introduced and algorithms for solving it are outlined.

The maximum matching problem is studied here in detail only for bipartite graphs. Even though this restricts the class of graphs, the same principles are used to design polynomial time algorithms for graphs that are not necessarily bipartite. The algorithms for general graphs are complex due to the presence of structures called *blossoms* and the reader is referred to Papadimitriou and Steiglitz [1982, Ch. 10], or Tarjan [1983, Ch. 9] for a detailed treatment of how blossoms are handled. Edmonds (see Even [1979]) gave the first algorithm to solve the matching problem in polynomial time. Micali and Vazirani [1980] obtained an $O(\sqrt{|V|}|E|)$ algorithm for nonbipartite matching, by extending the algorithm by Hopcroft and Karp [1973] for the bipartite case.

Matching Problem Definitions

Given a graph $G = (V, E)$, a matching M is a subset of the edges such that no two edges in M share a common vertex. In other words, the problem is that of finding a set of independent edges that have no incident vertices in common. The cardinality of M is usually referred to as its *size*.

The following terms are defined with respect to a matching M. The edges in M are called *matched edges* and edges not in M are called *free edges*. Likewise, a vertex is a *matched vertex* if it is incident to a matched edge. A *free vertex* is one that is not matched. The *mate* of a matched vertex v is its neighbor w that is at the other end of the matched edge incident to v. A matching is called *perfect* if all vertices of the graph are matched in it. The objective of the maximum matching problem is to maximize $|M|$, the size of the matching. If the edges of the graph have weights, then the *weight* of a matching is defined to be the sum of the weights of the edges in the matching. A path $p = [v_1, v_2, \ldots, v_k]$ is called an *alternating path* if the edges (v_{2j-1}, v_{2j}), $j = 1, 2, \ldots$, are free and the edges (v_{2j}, v_{2j+1}), $j = 1, 2, \ldots$, are matched. An **augmenting path** $p = [v_1, v_2, \ldots, v_k]$ is an alternating path in which both v_1 and v_k are free vertices. Observe that an augmenting path is defined with respect to a specific matching. The symmetric difference of a matching M and an augmenting path P, $M \oplus P$, is defined to be $(M - P) \cup (P - M)$. The operation can be generalized to the case when P is any subset of the edges.

Applications of Matching

Matchings are the underlying basis for many optimization problems. Problems of assigning workers to jobs can be naturally modeled as a bipartite matching problem. Other applications include assigning a collection of jobs with precedence constraints to two processors, such that the total execution time is minimized [Lawler 1976]. Other applications arise in chemistry, in determining structure of chemical bonds, matching moving objects based on a sequence of photographs, and localization of objects in space after obtaining information from multiple sensors [Ahuja et al. 1993].

Matchings and Augmenting Paths

The following theorem gives necessary and sufficient conditions for the existence of a perfect matching in a bipartite graph.

Theorem 9.1 (Hall's Theorem.). *A bipartite graph $G = (X, Y, E)$ with $|X| = |Y|$ has a perfect matching if and only if $\forall S \subseteq X$, $|N(S)| \geq |S|$, where $N(S) \subseteq Y$ is the set of vertices that are neighbors of some vertex in S.*

Although Theorem 9.1 captures exactly the conditions under which a given bipartite graph has a perfect matching, it does not lead to an algorithm for finding maximum matchings directly. The following lemma shows how an augmenting path with respect to a given matching can be used to increase the size of a matching. An efficient algorithm will be described later that uses augmenting paths to construct a maximum matching incrementally.

Lemma 9.1. *Let P be the edges on an augmenting path $p = [v_1, \ldots, v_k]$ with respect to a matching M. Then $M' = M \oplus P$ is a matching of cardinality $|M| + 1$.*

PROOF. Since P is an augmenting path, both v_1 and v_k are free vertices in M. The number of free edges in P is one more than the number of matched edges. The symmetric difference operator replaces the matched edges of M in P by the free edges in P. Hence the size of the resulting matching, $|M'|$, is one more than $|M|$. □

The following theorem provides a necessary and sufficient condition for a given matching M to be a maximum matching.

Theorem 9.2. *A matching M in a graph G is a maximum matching if and only if there is no augmenting path in G with respect to M.*

PROOF. If there is an augmenting path with respect to M, then M cannot be a maximum matching, since by Lemma 9.1 there is a matching whose size is larger than that of M. To prove the converse we show that if there is no augmenting path with respect to M, then M is a maximum matching. Suppose that there is a matching M' such that $|M'| > |M|$. Consider the set of edges $M \oplus M'$. These edges form a subgraph in G. Each vertex in this subgraph has degree at most two, since each node has at most one edge from each matching incident to it. Hence, each connected component of this subgraph is either a path or a simple cycle. For each cycle, the number of edges of M is the same as the number of edges of M'. Since $|M'| > |M|$, one of the paths must have more edges from M' than from M. This path is an augmenting path in G with respect to the matching M, contradicting the assumption that there were no augmenting paths with respect to M. □

Bipartite Matching Algorithm

High-Level Description

The algorithm starts with the empty matching $M = \emptyset$, and augments the matching in phases. In each phase, an augmenting path with respect to the current matching M is found, and it is used to increase the size of the matching. An augmenting path, if one exists, can be found in $O(|E|)$ time, using a procedure similar to breadth-first search that was described in section 9.4.

The search for an augmenting path proceeds from the free vertices. At each step when a vertex in X is processed, all its unvisited neighbors are also searched. When a matched vertex in Y is considered, only its matched neighbor is searched. This search proceeds along a subgraph referred to as the *Hungarian tree*.

Initially, all free vertices in X are placed in a queue that holds vertices that are yet to be processed. The vertices are removed one by one from the queue and processed as follows. In turn, when vertex v is removed from the queue, the edges incident to it are scanned. If it has a neighbor in the vertex set Y that is free, then the search for an augmenting path is successful; procedure AUGMENT is called to update the matching, and the algorithm proceeds to its next phase. Otherwise, add the mates of all of the matched neighbors of v to the queue if they have never been added to the queue, and continue the search for an augmenting path. If the algorithm empties the queue without finding an augmenting path, its current matching is a maximum matching and it terminates.

The main data structure that the algorithm uses are the arrays *mate* and *free*. The array *mate* is used to represent the current matching. For a matched vertex $v \in G$, *mate*[v] denotes the matched neighbor of vertex v. For $v \in X$, *free*[v] is a vertex in Y that is adjacent to v and is free. If no such vertex exists then *free*[v] = 0.

BIPARTITE MATCHING ALGORITHM. *Bipartite Matching* ($G = (X, Y, E)$):

```
 1 for all vertices v in G do
 2     mate[v] ← 0.
 3 end-for
 4 found ← false.
 5 while not found do
 6       Initialize.
 7       MakeEmptyQueue (Q).
 8       for all vertices x ∈ X do
 9         if mate[x] = 0 then
10            Push (Q,x).
11            label[x] ← 0.
12         endif
13       end-for
14       done ← false.
15       while not done and not Empty (Q) do
16          x ← Pop (Q).
17          if free[x] ≠ 0 then
18             Augment(x).
19             done ← true.
20          else
21             for all edges (x,x') ∈ A do
22                if label[x'] = 0 then
23                   label[x'] ← x.
24                   Push (Q,x').
25                end-if
26             end-for
27          end-if
28          if Empty (Q) then
29             found ← true.
30          end-if
31       end-while
32 end-while
end-proc
```

Initialize :

```
1 for all vertices x ∈ X do
2     free[x] ← 0.
```

```
3   end-for
4   for all edges (x, y) ∈ E do
5       if mate[y] = 0 then free[x] ← y
6       else if mate[y] ≠ x then A ← A ∪ (x, mate[y]).
7       end-if
8   end-for
end-proc
```

Augment(x):

```
1   if label[x] = 0 then
2       mate[x] ← free[x].
3       mate[free[x]] ← x
4   else
5       free[label[x]] ← mate[x]
6       mate[x] ← free[x]
7       mate[free[x]] ← x
8       Augment label[x])
9   end-if
end-proc
```

Sample Execution

Figure 9.5 shows a sample execution of the matching algorithm. We start with a partial matching and show the structure of the resulting Hungarian tree. An augmenting path from vertex b to vertex u is found by the algorithm.

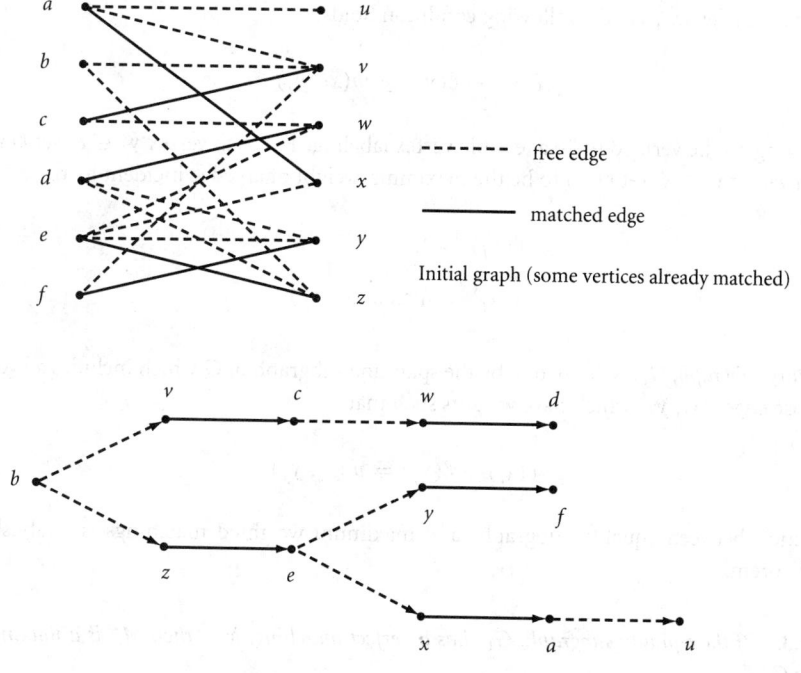

FIGURE 9.5 Sample execution of matching algorithm.

Analysis

If there are augmenting paths with respect to the current matching, the algorithm will find at least one of them. Hence, when the algorithm terminates, the graph has no augmenting paths with respect to the current matching and the current matching is optimal. Each iteration of the main while loop of the algorithm runs in $O(|E|)$ time. The construction of the auxiliary graph A and computation of the array *free* also take $O(|E|)$ time. In each iteration, the size of the matching increases by one and thus there are at most $\min(|X|, |Y|)$ iterations of the while loop. Therefore, the algorithm solves the matching problem for bipartite graphs in time $O(\min(|X|, |Y|)|E|)$. Hopcroft and Karp [1973] showed how to improve the running time by finding a maximal set of disjoint augmenting paths in a single phase in $O(|E|)$ time. They also proved that the algorithm runs in only $O(\sqrt{|V|})$ phases.

Assignment Problem

We now introduce the assignment problem, which is that of finding a maximum-weight matching in a given bipartite graph in which edges are given nonnegative weights. There is no loss of generality in assuming that the graph is complete, since zero-weight edges may be added between pairs of vertices that are nonadjacent in the original graph without affecting the weight of a maximum-weight matching. The minimum-weight perfect matching can be reduced to the maximum-weight matching problem as follows: choose a constant M that is larger than the weight of any edge. Assign each edge a new weight of $w'(e) = M - w(e)$. Observe that maximum-weight matchings with the new weight function are minimum-weight perfect matchings with the original weights. We restrict our attention to the study of the maximum-weight matching problem for bipartite graphs. Similar techniques have been used to solve the maximum-weight matching problem in arbitrary graphs (see Lawler [1976] and Papadimitriou and Steiglitz [1982]).

The input is a complete bipartite graph $G = (X, Y, X \times Y)$ and each edge e has a nonnegative weight of $w(e)$. The following algorithm, known as the Hungarian method, was first given by Kuhn [1955]. The method can be viewed as a primal-dual algorithm in the linear programming framework [Papadimitriou and Steiglitz 1982]. No knowledge of linear programming is assumed here.

A *feasible vertex-labeling* ℓ is defined to be a mapping from the set of vertices in G to the real numbers such that for each edge (x_i, y_j) the following condition holds:

$$\ell(x_i) + \ell(y_j) \geq w(x_i, y_j)$$

The following can be verified to be a feasible vertex labeling. For each vertex $y_j \in Y$, set $\ell(y_j)$ to be 0, and for each vertex $x_i \in X$, set $\ell(x_i)$ to be the maximum weight of an edge incident to x_i,

$$\ell(y_j) = 0,$$

$$\ell(x_i) = \max_j w(x_i, y_j)$$

The *equality subgraph*, G_ℓ, is defined to be the spanning subgraph of G which includes all vertices of G but only those edges (x_i, y_j) which have weights such that

$$\ell(x_i) + \ell(y_j) = w(x_i, y_j)$$

The connection between equality subgraphs and maximum-weighted matchings is established by the following theorem.

Theorem 9.3. *If the equality subgraph, G_ℓ, has a perfect matching, M^*, then M^* is a maximum-weight matching in G.*

PROOF. Let M^* be a perfect matching in G_ℓ. By definition,

$$w(M^*) = \sum_{e \in M^*} w(e) = \sum_{v \in X \cup Y} \ell(v)$$

Let M be any perfect matching in G. Then

$$w(M) = \sum_{e \in M} w(e) \leq \sum_{v \in X \cup Y} \ell(v) = w(M^*)$$

Hence M^* is a maximum-weight perfect matching. □

High-Level Description

The Theorem 9.3 is the basis of the algorithm for finding a maximum-weight matching in a complete bipartite graph. The algorithm starts with a feasible labeling, then computes the equality subgraph and a maximum cardinality matching in this subgraph. If the matching found is perfect, by Theorem 9.3 the matching must be a maximum-weight matching and the algorithm returns it as its output. Otherwise the matching is not perfect, and more edges need to be added to the equality subgraph by *revising* the vertex labels. The revision should ensure that edges from the current matching do not leave the equality subgraph. After more edges are added to the equality subgraph, the algorithm grows the Hungarian trees further. Either the size of the matching increases because an augmenting path is found, or a new vertex is added to the Hungarian tree. In the former case, the current phase terminates and the algorithm starts a new phase since the matching size has increased. In the latter case, new nodes are added to the Hungarian tree. In n phases, the tree includes all of the nodes, and therefore there are at most n phases before the size of the matching increases.

It is now described in more detail how the labels are updated and which edges are added to the equality subgraph. Suppose M is a maximum matching found by the algorithm. Hungarian trees are grown from all of the free vertices in X. Vertices of X (including the free vertices) that are encountered in the search are added to a set S and vertices of Y that are encountered in the search are added to a set T. Let $\overline{S} = X - S$ and $\overline{T} = Y - T$. Figure 9.6 illustrates the structure of the sets S and T. Matched edges are shown in bold; the other edges are the edges in G_ℓ. Observe that there are no edges in the equality subgraph from S to $\overline{T}$, even though there may be edges from T to $\overline{S}$. The algorithm now revises the labels as follows. Decrease all of the labels of vertices in S by a quantity δ (to be determined later) and increase the labels of the vertices in T by δ. This ensures that edges in the matching continue to stay in the equality subgraph. Edges in G (not in G_ℓ) that go from vertices in S to vertices in $\overline{T}$ are candidate edges to enter the equality subgraph, since one label is decreasing and the other is unchanged. The algorithm chooses δ to be the smallest value such that some edge of $G - G_\ell$ enters the equality subgraph. Suppose this edge goes from $x \in S$ to $y \in \overline{T}$. If y is free then an augmenting path has been found. On the other hand if y is matched, the Hungarian tree is grown by moving y to T and its matched neighbor to S and the process of revising labels continues.

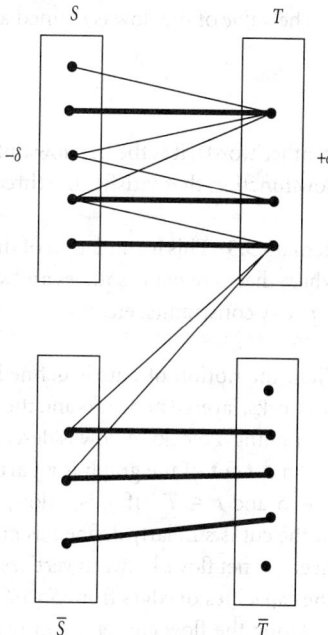

FIGURE 9.6 Sets S and T as maintained by the algorithm. Only edges in G_l are shown.

Network Flows

A number of polynomial time flow algorithms have been developed over the last two decades. The reader is referred to Ahuja et al. [1993], for a detailed account of the historical development of the various flow methods. Cormen et al. [1989] review the preflow push method in detail, and to complement their coverage an implementation of the blocking flow technique of Malhotra et al. [1978] is discussed here.

Network Flow Problem Definitions

First the network flow problem is defined and the basic terminology is defined.

Flow network: A flow network $G = (V, E)$ is a directed graph, with two specially marked nodes, namely, the source s, and the sink t. There is a *capacity* function $c : E \mapsto R^+$ that maps edges to positive real numbers.

Max-flow problem: A flow function $f : E \mapsto R$ maps edges to real numbers. For an edge $e = (u, v)$, $f(u, v)$ refers to the flow on edge e, which is also called the net flow from vertex u to vertex v. This notation is extended to sets of vertices as follows: If X and Y are sets of vertices then $f(X, Y)$ is defined to be $\sum_{x \in X} \sum_{y \in Y} f(x, y)$. A flow function is required to satisfy the following constraints:

- Capacity constraint. For all edges e, $f(e) \leq c(e)$.
- Skew symmetry constraint. For an edge $e = (u, v)$, $f(u, v) = -f(v, u)$.
- Flow conservation. For all vertices $u \in V - \{s, t\}$, $\sum_{v \in V} f(u, v) = 0$.

The capacity constraint says that the total flow on an edge does not exceed its capacity. The skew symmetry condition says that the flow on an edge is the negative of the flow in the reverse direction. The flow conservation constraint says that the total net flow out of any vertex other than the source and sink is zero.

The *value* of the flow is defined as

$$|f| = \sum_{v \in V} f(s, v)$$

In other words, it is the net flow out of the source. In the *maximum-flow problem* the objective is to find a flow function that satisfies the three constraints, and also maximizes the total flow value $|f|$.

Remark 9.3 This formulation of the network flow problem is powerful enough to capture generalizations where there are many sources and sinks (single commodity flow), and where both vertices and edges have capacity constraints, etc.

First, the notion of cuts is defined, and the max-flow min-cut theorem is introduced. Then, residual networks, layered networks and the concept of blocking flows are introduced. Finally, an efficient algorithm for finding a blocking flow is described.

An *s–t cut* of the graph is a partitioning of the vertex set V into two sets S and $T = V - S$ such that $s \in S$ and $t \in T$. If f is a flow, then the net flow across the cut is defined as $f(S, T)$. The capacity of the cut is similarly defined as $c(S, T) = \sum_{x \in X} \sum_{y \in Y} c(x, y)$. The net flow across a cut may include negative net flows between vertices, but the capacity of the cut includes only nonnegative values, i.e., only the capacities of edges from S to T.

Using the flow conservation principle, it can be shown that the net flow across an $s–t$ cut is exactly the flow value $|f|$. By the capacity constraint, the flow across the cut cannot exceed the capacity of the cut. Thus the value of the maximum flow is no greater than the capacity of a minimum s-t cut. The well-known *max-flow min-cut theorem* [Elias et al. 1956, Ford and Fulkerson 1962] proves that the two numbers are actually equal. In other words, if f^* is a maximum flow, then there is some cut $(X, \overline{X})$ such that $|f^*| = c(X, \overline{X})$. The reader is referred to Cormen et al. [1989] and Tarjan [1983] for further details.

The *residual capacity* of a flow f is defined to be a function on vertex pairs given by $c'(v, w) = c(v, w) - f(v, w)$. The residual capacity of an edge (v, w), $c'(v, w)$, is the number of additional units of flow that can be pushed from v to w without violating the capacity constraints. An edge e is *saturated* if $c(e) = f(e)$, that is, if its residual capacity, $c'(e)$, is zero. The residual graph $G_R(f)$ for a flow f is the graph with vertex set V, source and sink s and t, respectively, and those edges (v, w) for which $c'(v, w) > 0$.

An augmenting path for f is a path P from s to t in $G_R(f)$. The residual capacity of P, denoted by $c'(P)$, is the minimum value of $c'(v, w)$ over all edges (v, w) in the path P. The flow can be increased by $c'(P)$, by increasing the flow on each edge of P by this amount. Whenever $f(v, w)$ is changed, $f(w, v)$ is also correspondingly changed to maintain skew symmetry.

Most flow algorithms are based on the concept of augmenting paths pioneered by Ford and Fulkerson [1956]. They start with an initial zero flow and augment the flow in stages. In each stage, a residual graph $G_R(f)$ with respect to the current flow function f is constructed and an augmenting path in $G_R(f)$ is found to increase the value of the flow. Flow is increased along this path until an edge in this path is saturated. The algorithms iteratively keep increasing the flow until there are no more augmenting paths in $G_R(f)$, and return the final flow f as their output.

The following lemma is fundamental in understanding the basic strategy behind these algorithms.

Lemma 9.2. *Let f be any flow and f^* a maximum flow in G, and let $G_R(f)$ be the residual graph for f. The value of a maximum flow in $G_R(f)$ is $|f^*| - |f|$.*

PROOF. Let f' be any flow in $G_R(f)$. Define $f + f'$ to be the flow defined by the flow function $f(v, w) + f'(v, w)$ for each edge (v, w). Observe that $f + f'$ is a feasible flow in G of value $|f| + |f'|$. Since f^* is the maximum flow possible in G, $|f'| \le |f^*| - |f|$. Similarly define $f^* - f$ to be a flow in $G_R(f)$ defined by $f^*(v, w) - f(v, w)$ in each edge (v, w), and this is a feasible flow in $G_R(f)$ of value $|f^*| - |f|$, and it is a maximum flow in $G_R(f)$. □

Blocking flow: A flow f is a **blocking flow** if every path in G from s to t contains a saturated edge. Blocking flows are also known as maximal flows. It is important to note that a blocking flow is not necessarily a maximum flow. There may be augmenting paths that increase the flow on some edges and decrease the flow on other edges (by increasing the flow in the reverse direction).

Layered networks: Let $G_R(f)$ be the residual graph with respect to a flow f. The level of a vertex v is the length of a shortest path from s to v in $G_R(f)$. The level graph L for f is the subgraph of $G_R(f)$ containing vertices reachable from s and only the edges (v, w) such that $dist(s, w) = 1 + dist(s, v)$. L contains all shortest length augmenting paths and can be constructed in $O(|E|)$ time.

The maximum flow algorithm proposed by Dinitz [1970] starts with the zero flow, and iteratively increases the flow by augmenting it with a blocking flow in $G_R(f)$ until t is not reachable from s in $G_R(f)$. At each step the current flow is replaced by the sum of the current flow and the blocking flow. This algorithm terminates in $|V| - 1$ iterations since in each iteration the shortest distance from s to t in the residual graph increases. The shortest path from s to t is at most $|V| - 1$, and this gives an upper bound on the number of iterations of the algorithm.

An algorithm to find a blocking flow that runs in $O(|V|^2)$ time is described here, and this yields an $O(|V|^3)$ max-flow algorithm. There are a number of $O(|V|^2)$ blocking flow algorithms available [Karzanov 1974, Malhotra et al. 1978, Shiloach and Vishkin 1982, Tarjan 1983], some of which are described in detail in Tarjan [1983].

Blocking Flows

Dinitz's algorithm to find a blocking flow runs in $O(|V||E|)$ time [Dinitz 1970]. The main step is to find paths from the source to the sink and saturate them by pushing as much flow as possible on these paths.

Every time the flow is increased by pushing more flow along an augmenting path, one of the edges on this path becomes saturated. It takes $O(|V|)$ time to compute the amount of flow that can be pushed on the path. Since there are $|E|$ edges, this yields an upper bound of $O(|V||E|)$ steps on the running time of the algorithm.

Malhotra–Kumar–Maheshwari BLOCKING FLOW ALGORITHM. *The algorithm has a current flow function f and its corresponding residual graph $G_R(f)$. Define for each node $v \in G_R(f)$, a quantity $tp[v]$ that specifies its maximum throughput, i.e., either the sum of the capacities of the incoming arcs or the sum of the capacities of the outgoing arcs, which ever is smaller. $tp[v]$ represents the maximum flow that could pass through v in any feasible blocking flow in the residual graph. Vertices for which the throughput is zero are deleted from $G_R(f)$.*

The algorithm selects a vertex u for which its throughput is a minimum among all vertices with nonzero throughput. It then greedily pushes a flow of $tp[u]$ from u toward t, level by level in the layered residual graph. This can be done by creating a queue, which initially contains u and which is assigned the task of pushing $tp[u]$ out of it. In each step, the vertex v at the front of the queue is removed, and the arcs going out of v are scanned one at a time, and as much flow as possible is pushed out of them until v's allocated flow has been pushed out. For each arc (v, w) that the algorithm pushed flow through, it updates the residual capacity of the arc (v, w) and places w on a queue (if it is not already there) and increments the net incoming flow into w. Also, $tp[v]$ is reduced by the amount of flow that was sent through it now. The flow finally reaches t, and the algorithm never comes across a vertex that has incoming flow that exceeds its outgoing capacity since u was chosen as a vertex with the smallest throughput. The preceding idea is again repeated to pull a flow of $tp[u]$ from the source s to u. Combining the two steps yields a flow of $tp[u]$ from s to t in the residual network that goes through u. The flow f is augmented by this amount. Vertex u is deleted from the residual graph, along with any other vertices that have zero throughput.

This procedure is repeated until all vertices are deleted from the residual graph. The algorithm has a blocking flow at this stage since at least one vertex is saturated in every path from s to t. In the algorithm, whenever an edge is saturated it may be deleted from the residual graph. Since the algorithm uses a greedy strategy to pump flows, at most $O(|E|)$ time is spent when an edge is saturated. When finding flow paths to push $tp[u]$, there are at most n times, one each per vertex, when the algorithm pushes a flow that does not saturate the corresponding edge. After this step, u is deleted from the residual graph. Hence in $O(|E| + |V|^2) = O(|V|^2)$ steps, the algorithm to compute blocking flows terminates.

Goldberg and Tarjan [1988] proposed a preflow push method that runs in $O(|V||E|\log|V|^2/|E|)$ time without explicitly finding a blocking flow at each step.

Applications of Network Flow

There are numerous applications of the maximum flow algorithm in scheduling problems of various kinds. See Ahuja et al. [1993] for further details.

9.8 Tour and Traversal Problems

There are many applications for finding certain kinds of paths and tours in graphs. We briefly discuss some of the basic problems.

The **traveling salesman problem (TSP)** is that of finding a shortest tour that visits all of the vertices in a given graph with weights on the edges. It has received considerable attention in the literature [Lawler et al. 1985]. The problem is known to be computationally intractable (NP-hard). Several heuristics are known to solve practical instances. Considerable progress has also been made for finding optimal solutions for graphs with a few thousand vertices.

One of the first graph-theoretic problems to be studied, the **Euler tour problem** asks for the existence of a closed walk in a given connected graph that traverses each edge exactly once. Euler proved that such a closed walk exists if and only if each vertex has even degree [Gibbons 1985]. Such a graph is known as an **Eulerian graph**. Given an Eulerian graph, an Euler tour in it can be computed using DFS in linear time.

Given an edge-weighted graph, the **Chinese postman problem** is that of finding a shortest closed walk that traverses each edge at least once. Although the problem sounds very similar to the TSP problem, it can be solved optimally in polynomial time by reducing it to the matching problem [Ahuja et al. 1993].

Defining Terms

Assignment problem: That of finding a perfect matching of maximum (or minimum) total weight.

Augmenting path: An alternating path that can be used to augment (increase) the size of a matching.

Biconnected graph: A graph that cannot be disconnected by the removal of any single vertex.

Bipartite graph: A graph in which the vertex set can be partitioned into two sets X and Y, such that each edge connects a node in X with a node in Y.

Blocking flow: A flow function in which any directed path from s to t contains a saturated edge.

Branching: A spanning tree in a rooted graph, such that the root has a path to each vertex.

Chinese postman problem: Asks for a minimum length tour that traverses each edge at least once.

Connected: A graph in which there is a path between each pair of vertices.

Cycle: A path in which the start and end vertices of the path are identical.

Degree: The number of edges incident to a vertex in a graph.

DFS forest: A rooted forest formed by depth-first search.

Directed acyclic graph: A directed graph with no cycles.

Eulerian graph: A graph that has an Euler tour.

Euler tour problem: Asks for a traversal of the edges that visits each edge exactly once.

Forest: An acyclic graph.

Leaves: Vertices of degree one in a tree.

Matching: A subset of edges that do not share a common vertex.

Minimum spanning tree: A spanning tree of minimum total weight.

Network flow: An assignment of flow values to the edges of a graph that satisfies flow conservation, skew symmetry, and capacity constraints.

Path: An ordered list of edges such that any two consecutive edges are incident to a common vertex.

Sparse graph: A graph in which $|E| \ll |V|^2$.

s–t cut: A partitioning of the vertex set into S and T such that $s \in S$ and $t \in T$.

Strongly connected: A directed graph in which there is a directed path in each direction between each pair of vertices.

Topological order: A linear ordering of the edges of a DAG such that every edge in the graph goes from left to right.

Traveling salesman problem: Asks for a minimum length tour of a graph that visits all of the vertices exactly once.

Tree: An acyclic graph with $|V| - 1$ edges.

Walk: An ordered sequence of edges (in which edges could repeat) such that any two consecutive edges are incident to a common vertex.

References

Ahuja, R. K., Magnanti, T., and Orlin, J. 1993. *Network Flows*. Prentice–Hall.

Bellman, R. 1958. On a routing problem. *Q. App. Math.* 16(1):87–90.

Boruvka, O. 1926. O jistem problemu minimalnim. *Praca Moravske Prirodovedecke Spolecnosti* 3:37–58 (in Czech).

Cormen, T. H.; Leiserson, C. E., and Rivest, R. L. 1989. *Introduction to Algorithms*. The MIT Press.

DiBattista, G., Eades, P., Tamassia, R., and Tollis, I. 1994. Annotated bibliography on graph drawing algorithms. *Comput. Geom.: Theory Applic.* 4:235–282.

Dijkstra, E. W. 1959. A note on two problems in connexion with graphs. *Numerische Mathematik* 1:269–271.

Dinitz, E. A. 1970. Algorithm for solution of a problem of maximum flow in a network with power estimation. *Soviet Math. Dokl.* 11:1277–1280.

Elias, P., Feinstein, A., and Shannon, C. E. 1956. Note on maximum flow through a network. *IRE Trans. Inf. Theory* IT-2:117–119.

Even, S. 1979. *Graph Algorithm.* Computer Science Press, Potomac.

Ford, L. R., Jr. and Fulkerson, D. R. 1956. Maximal flow through a network. *Can. J. Math.* 8:399–404.

Ford, L. R., Jr. and Fulkerson, D. R. 1962. *Flows in Networks.* Princeton University Press.

Fraenkel, A. S. 1970. Economic traversal of labyrinths. *Math. Mag.* 43:125–130.

Fredman, M. and Tarjan, R. E. 1987. Fibonacci heaps and their uses in improved network optimization algorithms. *J. ACM* 34(3):596–615.

Gabow, H. N., Galil, Z., Spencer, T., and Tarjan, R. E. 1986. Efficient algorithms for finding minimum spanning trees in undirected and directed graphs. *Combinatorica* 6(2):109–122.

Gibbons, A. M. 1985. *Algorithmic Graph Theory.* Cambridge University Press, New York.

Goldberg, A. V. and Tarjan, R. E. 1988. A new approach to the maximum-flow problem. *J. ACM* 35:921–940.

Hochbaum, D. S., ed. 1996. *Approximation Algorithms for NP-Hand Problems.* PWS Publishing.

Hopcroft, J. E. and Karp, R. M. 1973. An $n^{2.5}$ algorithm for maximum matching in bipartite graphs. *SIAM J. Comput.* 2(4):225–231.

Hopcroft, J. E. and Tarjan, R. E. 1973. Efficient algorithms for graph manipulation. *Commun. ACM* 16:372–378.

Jarnik, V. 1930. O jistem problemu minimalnim. *Praca Moravske Prirodovedecke Spolecnosti* 6:57–63 (in Czech).

Karzanov, A. V. 1974. Determining the maximal flow in a network by the method of preflows. *Soviet Math. Dokl.* 15:434–437.

Kruskal, J. B. 1956. On the shortest spanning subtree of a graph and the traveling salesman problem. *Proc. Am. Math. Soc.* 7:48–50.

Kuhn, H. W. 1955. The Hungarian method for the assignment problem. *Nav. Res. Logistics Q.* 2:83–98.

Lawler, E. L. 1976. *Combinatorial Optimization: Networks and Matroids.* Holt, Rinehart and Winston.

Lawler, E. L., Lenstra, J. K., Rinnooy Kan, A. H. G., and Shmoys, D. B. 1985. *The Traveling Salesman Problem: A Guided Tour of Combinatorial Optimization.* Wiley.

Lucas, E. 1882. *Recreations Mathematiques.* Paris.

Malhotra, V. M., Kumar, M. P., and Maheshwari, S. N. 1978. An $O(|V|^3)$ algorithm for finding maximum flows in networks. *Inf. Process. Lett.* 7:277–278.

Micali, S. and Vazirani, V. V. 1980. An $O(\sqrt{|V|}|E|)$ algorithm for finding maximum matching in general graphs, pp. 17–27. In *Proc. 21st Annu. Symp. Found. Comput. Sci.*

Papadimitriou, C. H. and Steiglitz, K. 1982. *Combinatorial Optimization: Algorithms and Complexity.* Prentice Hall.

Prim, R. C. 1957. Shortest connection networks and some generalizations. *Bell Sys. Tech. J.* 36:1389–1401.

Shiloach, Y. and Vishkin, U. 1982. An $O(n^2 \log n)$ parallel max-flow algorithm. *J. Algorithms* 3:128–146.

Tarjan, R. E. 1972. Depth first search and linear graph algorithms. *SIAM J. Comput.* 1:146–160.

Tarjan, R. E. 1983. *Data Structures and Network Algorithms.* SIAM.

Further Information

The area of graph algorithms continues to be a very active field of research. There are several journals and conferences that discuss advances in the field. Here we name a partial list of some of the important meetings: ACM Symposium on Theory of Computing, IEEE Conference on Foundations of Computer Science, ACM–SIAM Symposium on Discrete Algorithms, the International Colloquium on Automata,

Languages and Programming, and the European Symposium on Algorithms. There are many other regional algorithms/theory conferences that carry research papers on graph algorithms. The journals that carry articles on current research in graph algorithms are *Journal of the ACM, SIAM Journal on Computing, SIAM Journal on Discrete Mathematics, Journal of Algorithms, Algorithmica, Journal of Computer and System Sciences, Information and Computation, Information Processing Letters,* and *Theoretical Computer Science.*

To find more details about some of the graph algorithms described in this chapter we refer the reader to the books by Cormen et al. [1989], Even [1979], and Tarjan [1983]. For network flows and matching, a more detailed survey regarding various approaches can be found in Tarjan [1983]. Papadimitriou and Steiglitz [1982] discuss the solution of many combinatorial optimization problems using a primal–dual framework.

Current research on graph algorithms focuses on approximation algorithms [Hochbaum 1996], dynamic algorithms, and in the area of graph layout and drawing [DiBattista et al. 1994].

10

Algebraic Algorithms*

Angel Díaz
Rensselaer Polytechnic Institute

Erich Kaltofen
North Carolina State University

Victor Pan
Lehman College, City University of New York

10.1 Introduction

The title's subject is the algorithmic approach to algebra: arithmetic with numbers, polynomials, matrices, differential polynomials, such as $y'' + (1/2 + x^4/4)y$, truncated series, and algebraic sets, i.e., quantified expressions such as $\exists x \in \mathbb{R}: x^4 + p \cdot x + q = 0$, which describes a subset of the two-dimensional space with coordinates p and q for which the given quartic equation has a real root. Algorithms that manipulate such objects are the backbone of modern symbolic mathematics software such as the Maple and Mathematica systems, to name but two among many useful systems. This chapter restricts itself to algorithms in four areas: linear matrix algebra, root finding of univariate polynomials, solution of systems of nonlinear algebraic equations, and polynomial factorization.

10.2 Matrix Computations and Approximation of Polynomial Zeros

This section covers several major algebraic and numerical problems of scientific and engineering computing that are usually solved numerically, with rounding off or chopping the input and computed values to a

*This material is based on work supported in part by the National Science Foundation under Grants CCR-9319776 (first and second author) and CCR-9020690 (third author), by GTE under a Graduate Computer Science Fellowship (first author), and by PSC CUNY Awards 665301 and 666327 (third author). Part of this work was done while the second author was at the Department of Computer Science at Rensselaer Polytechnic Insititute in Troy, New York.

fixed number of bits that fit the computer precision (sections 10.2 and 10.3 are devoted to some fundamental infinite precision symbolic computations, and within section 10.2 we comment on the infinite precision techniques for some matrix computations). We also study approximation of polynomial zeros, which is an important, fundamental, as well as very popular subject. In our presentation, we will very briefly list the major subtopics of our huge subject and will give some pointers to the references. We will include brief coverage of the topics of the algorithm design and analysis, regarding the complexity of matrix computation and of approximating polynomial zeros. The reader may find further material on these subjects in the survey articles by Pan [1984a, 1991, 1992a, 1995b] and in the books by Bini and Pan [1994, 1996].

Products of Vectors and Matrices, Convolution of Vectors

An $m \times n$ matrix $A = (a_{i,j}, \ i = 0, 1, \ldots, m - 1; \ j = 0, 1, \ldots, n - 1)$ is a two-dimensional array, whose (i, j) entry is $(A)_{i,j} = a_{i,j}$. A is a column vector of dimension m if $n = 1$ and is a row vector of dimension n if $m = 1$. Transposition, hereafter, indicated by the superscript T, transforms a row vector $v^T = [v_0, \ldots, v_{n-1}]$ into a column vector $v = [v_0, \ldots, v_{n-1}]^T$.

For two vectors, $u^T = (u_0, \ldots, u_{m-1})$ and $v^T = (v_0, \ldots, v_{n-1})^T$, their *outer product* is an $m \times n$ matrix,

$$W = uv^T = [w_{i,j}, \ i = 0, \ldots, m - 1; \ j = 0, \ldots, n - 1]$$

where $w_{i,j} = u_i v_j$, for all i and j, and their *convolution* vector is said to equal

$$w = u \circ v = (w_0, \ldots, w_{m+n-2})^T, \qquad w_k = \sum_{i=0}^{k} u_i v_{k-i}$$

where $u_i = v_j = 0$, for $i \geq m, j \geq n$; in fact, w is the coefficient vector of the product of two polynomials,

$$u(x) = \sum_{i=0}^{m-1} u_i x^i \qquad \text{and} \qquad v(x) = \sum_{i=0}^{n-1} v_i x^i$$

having coefficient vectors u and v, respectively.

If $m = n$, the scalar value

$$v^T u = u^T v = u_0 v_0 + u_1 v_1 + \cdots + u_{n-1} v_{n-1} = \sum_{i=0}^{n-1} u_i v_i$$

is called the *inner (dot, or scalar) product* of u and v.

The straightforward algorithms compute the inner and outer products of u and v and their convolution vector by using $2n - 1$, mn, and $mn + (m - 1)(n - 1) = 2mn - m - n + 1$ arithmetic operations (hereafter, referred to as **ops**), respectively.

These upper bounds on the numbers of ops for computing the inner and outer products are sharp, that is, cannot be decreased, for the general pair of the input vectors u and v, whereas (see, e.g. Bini and Pan [1994]) one may apply the *fast fourier transform* (FFT) in order to compute the convolution vector $u \circ v$ much faster, for larger m and n; namely, it suffices to use $4.5K \log K + 2K$ ops, for $K = 2^k$, $k = \lceil \log(m + n + 1) \rceil$. (Here and hereafter, all logarithms are binary unless specified otherwise.)

If $A = (a_{i,j})$ and $B = (b_{j,k})$ are $m \times n$ and $n \times p$ matrices, respectively, and $v = (v_k)$ is a p-dimensional vector, then the straightforward algorithms compute the vector

$$w = Bv = (w_0, \ldots, w_{n-1})^T, \qquad w_i = \sum_{j=0}^{p-1} b_{i,j} v_j, \quad i = 0, \ldots, n - 1$$

by using $(2p - 1)n$ ops (sharp bound), and compute the *matrix product*

$$AB = (w_{i,k}, \ i = 0, \ldots, m - 1; \ k = 0, \ldots, p - 1)$$

by using $2mnp - mp$ ops, which is $2n^3 - n^2$ if $m = n = p$. The latter upper bound is not sharp: the subroutines for $n \times n$ matrix multiplication on some modern computers, such as CRAY and Connection Machines, rely on algorithms using $O(n^{2.81})$ ops, and some nonpractical algorithms involve $O(n^{2.376})$ ops [Bini and Pan 1994, Golub and Van Loan 1989].

In the special case, where all of the input entries and components are bounded integers having short binary representation, each of the preceding operations with vectors and matrices can be reduced to a single multiplication of 2 longer integers, by means of the techniques of *binary segmentation* (cf. Pan [1984b, section 40], Pan [1991], Pan [1992b], or Bini and Pan [1994, Examples 36.1–36.3]).

For an $n \times n$ matrix B and an n-dimensional vector v, one may compute the vectors $B^i v$, $i = 1, 2, \ldots, k - 1$, which define *Krylov sequence* or *Krylov matrix*

$$[B^i v, \ i = 0, 1, \ldots, k - 1]$$

used as a basis of several computations. The straightforward algorithm takes on $(2n - 1)nk$ ops, which is order n^3 if k is of order n. An alternative algorithm first computes the matrix powers

$$B^2, B^4, B^8, \ldots, B^{2^s}, \quad s = \lceil \log k \rceil - 1$$

and then the products of $n \times n$ matrices B^{2^i} by $n \times 2^i$ matrices, for $i = 0, 1, \ldots, s$,

$$
\begin{aligned}
&B \quad\ v \\
&B^2 \quad (v, \ Bv) = (B^2 v, \ B^3 v) \\
&B^4 \quad (v, \ Bv, \ B^2 v, \ B^3 v) = (B^4 v, \ B^5 v, \ B^6 v, \ B^7 v) \\
&\ \vdots
\end{aligned}
$$

The last step completes the evaluation of the Krylov sequence, which amounts to $2s$ matrix multiplications, for $k = n$, and, therefore, can be performed (in theory) in $O(n^{2.376} \log k)$ ops.

Some Computations Related to Matrix Multiplication

Several fundamental matrix computations can be ultimately reduced to relatively few [that is, to a constant number, or, say, to $O(\log n)$] $n \times n$ matrix multiplications. These computations include the evaluation of det A, the **determinant** of an $n \times n$ matrix A; of its *inverse* A^{-1} (where A is nonsingular, that is, where det $A \neq 0$); of the coefficients of its **characteristic polynomial**, $c_A(x) = \det(xI - A)$, x denoting a scalar variable and I being the $n \times n$ identity matrix, which has ones on its diagonal and zeros elsewhere; of its *minimal polynomial, $m_A(x)$*; of its *rank*, rank A; of the solution vector $x = A^{-1} v$ to a nonsingular *linear system of equations, $Ax = v$*; of various *orthogonal* and *triangular factorizations* of A; and of a submatrix of A having the maximal rank, as well as some fundamental computations with singular matrices. Consequently, all of these operations can be performed by using (theoretically) $O(n^{2.376})$ ops (cf. Bini and Pan [1994, Chap. 2]). The idea is to represent the input matrix A as a block matrix and, operating with its blocks (rather than with its entries), to apply fast matrix multiplication algorithms. In practice, due to various other considerations (accounting, in particular, for the overhead constants hidden in the O notation, for the memory space requirements, and particularly, for numerical stability problems), these computations are based either on the straightforward algorithm for matrix multiplication or on other methods allowing order n^3 arithmetic operations (cf. Golub and Van Loan [1989]). Many block matrix algorithms supporting the (nonpractical) estimate $O(n^{2.376})$, however, become practically important for parallel computations (see later section).

In the next six sections, we will more closely consider the solution of a linear system of equations, $Av = b$, which is the most frequent operation in practice of scientific and engineering computing and is highly important theoretically. We will partition the known solution methods depending on whether the coefficient matrix A is *dense and unstructured*, **sparse**, or *dense and* **structured**.

Gaussian Elimination Algorithm

The solution of a nonsingular linear system $Ax = v$ uses only about n^2 ops if the system is lower (or upper) triangular, that is, if all subdiagonal (or superdiagonal) entries of A vanish. For example (cf. Pan [1992b]), let $n = 3$,

$$x_1 + 2x_2 - x_3 = 3$$
$$-2x_2 - 2x_3 = -10$$
$$-6x_3 = -18$$

Compute $x_3 = 3$ from the last equation, substitute into the previous ones, and arrive at a triangular system of $n - 1 = 2$ equations. In $n - 1$ (in our case, 2) such recursive substitution steps, we compute the solution.

The triangular case is itself important; furthermore, every nonsingular linear system is reduced to two triangular ones by means of *forward elimination* of the variables, which essentially amounts to computing the PLU factorization of the input matrix A, that is, to computing two lower triangular matrices L and U^T (where L has unit values on its diagonal) and a permutation matrix P such that $A = PLU$. [A permutation matrix P is filled with zeros and ones and has exactly one nonzero entry in each row and in each column; in particular, this implies that $P^T = P^{-1}$. Pu has the same components as u but written in a distinct (fixed) order, for any vector u]. As soon as the latter factorization is available, we may compute $x = A^{-1} v$ by solving two triangular systems, that is, at first, $Ly = P^T v$, in y, and then $Ux = y$, in x. Computing the factorization (elimination stage) is more costly than the subsequent *back substitution stage*, the latter involving about $2n^2$ ops. The Gaussian classical algorithm for elimination requires about $2n^3/3$ ops, not counting some comparisons, generally required in order to ensure appropriate *pivoting*, also called *elimination ordering*. Pivoting enables us to avoid divisions by small values, which could have caused numerical stability problems. Theoretically, one may employ fast matrix multiplication and compute the matrices P, L, and U in $O(n^{2.376})$ ops [Aho et al. 1974] [and then compute the vectors y and x in $O(n^2)$ ops]. Pivoting can be dropped for some important classes of linear systems, notably, for *positive definite* and for *diagonally dominant* systems (Golub and Van Loan 1989, Pan 1991, 1992b, Bini and Pan 1994).

We refer the reader to Golub and Van Loan [1987, pp. 82–83], or Pan [1992b, p. 794], on sensitivity of the solution to the input and roundoff errors in numerical computing. The output errors grow with the **condition number** of A, represented by $\|A\|\|A^{-1}\|$ for an appropriate matrix norm or by the ratio of maximum and minimum singular values of A. Except for ill-conditioned linear systems $Ax = v$, for which the condition number of A is very large, a rough initial approximation to the solution can be rapidly refined (cf. Golub and Van Loan [1989]) via the *iterative improvement algorithm*, as soon as we know P and rough approximations to the matrices L and U of the PLU factorization of A. Then b correct bits of each output value can be computed in $(b + n)n^2$ ops as $b \to \infty$.

Singular Linear Systems of Equations

If the matrix A is **singular** (in particular, if A is rectangular), then the linear system $Ax = v$ is either overdetermined, that is, has no solution, or underdetermined, that is, has infinitely many solution vectors. All of them can be represented as $\{x_0 + y\}$, where x_0 is a fixed solution vector and y is a vector from the *null space* of A, $\{y : Ay = 0\}$, that is, y is a solution of the homogeneous linear system $Ay = 0$. (The null space of an $n \times n$ matrix A is a linear space of the dimension n–rank A.) A vector x_0 and a basis for the null-space of A can be computed by using $O(n^{2.376})$ ops if A is an $n \times n$ matrix or by using $O(mn^{1.736})$ ops if A is an $m \times n$ or $n \times m$ matrix and if $m \geq n$ (cf. Bini and Pan [1994]).

For an overdetermined linear system $Ax = v$, having no solution, one may compute a vector x minimizing the norm of the residual vector, $\|v - Ax\|$. It is most customary to minimize the Euclidean norm,

$$\|\boldsymbol{u}\| = \left(\sum_i |u_i|^2 \right)^{1/2}, \quad \boldsymbol{u} = \boldsymbol{v} - A\boldsymbol{x} = (u_i)$$

This defines a least-squares solution, which is relatively easy to compute both practically and theoretically ($O(n^{2.376})$ ops suffice in theory) (cf. Bini and Pan [1994] and Golub and Van Loan [1989]).

Sparse Linear Systems (Including Banded Systems), Direct and Iterative Solution Algorithms

A matrix is sparse if it is filled mostly with zeros, say, if its all nonzero entries lie on 3 or 5 of its diagonals. In many important applications, in particular, solving partial and ordinary differential equations (PDEs and ODEs), one has to solve linear systems whose matrix is sparse and where, moreover, the disposition of its nonzero entries has a certain structure. Then, memory space and computation time can be dramatically decreased (say, from order n^2 to order $n \log n$ words of memory and from n^3 to $n^{3/2}$ or $n \log n$ ops) by using some special data structures and special solution methods. The methods are either direct, that is, are modifications of Gaussian elimination with some special policies of elimination ordering that preserve sparsity during the computation (notably, *Markowitz rule* and *nested dissection* [George and Liu 1981, Gilbert and Tarjan 1987, Lipton et al. 1979, Pan 1993]), or various iterative algorithms. The latter algorithms rely either on computing Krylov sequences [Saad 1995] or on multilevel or multigrid techniques [McCormick 1987, Pan and Reif 1992], specialized for solving linear systems that arise from discretization of PDEs. An important particular class of sparse linear systems is formed by *banded linear systems* with $n \times n$ coefficient matrices $A = (a_{i,j})$ where $a_{i,j} = 0$ if $i - j > g$ or $j - i > h$, for $g + h$ being much less than n. For banded systems, the nested dissection methods are known under the name of *block cyclic reduction* methods and are highly effective, but Pan et al. [1995] give some alternative algorithms, too. Some special techniques for computation of Krylov sequences for sparse and other special matrices A can be found in Pan [1995a]; according to these techniques, Krylov sequence is recovered from the solution of the associated linear system $(I - A)\boldsymbol{x} = \boldsymbol{v}$, which is solved fast in the case of a special matrix A.

Dense and Structured Matrices and Linear Systems

Many dense $n \times n$ matrices are defined by $O(n)$, say, by less than $2n$, parameters and can be multiplied by a vector by using $O(n \log n)$ or $O(n \log^2 n)$ ops. Such matrices arise in numerous applications (to signal and image processing, coding, algebraic computation, PDEs, integral equations, particle simulation, Markov chains, and many others). An important example is given by $n \times n$ *Toeplitz matrices* $T = (t_{i,j})$, $t_{i,j} = t_{i+1,j+1}$ for $i, j = 0, 1, \ldots, n-1$. Such a matrix can be represented by $2n - 1$ entries of its first row and first column or by $2n - 1$ entries of its first and last columns. The product $T\boldsymbol{v}$ is defined by vector convolution, and its computation uses $O(n \log n)$ ops. Other major examples are given by *Hankel matrices* (obtained by reflecting the row or column sets of Toeplitz matrices), *circulant* (which are a subclass of Toeplitz matrices), and *Bezout, Sylvester, Vandermonde,* and *Cauchy* matrices. The known solution algorithms for linear systems with such dense structured coefficient matrices use from order $n \log n$ to order $n \log^2 n$ ops. These properties and algorithms are extended via associating some linear operators of displacement and scaling to some more general classes of matrices and linear systems. We refer the reader to Bini and Pan [1994] for many details and further bibliography.

Parallel Matrix Computations

Algorithms for matrix multiplication are particularly suitable for parallel implementation; one may exploit natural association of processors to rows and/or columns of matrices or to their blocks, particularly, in the implementation of matrix multiplication on loosely coupled multiprocessors (cf. Golub and Van Loan

[1989] and Quinn [1994]). This motivated particular attention to and rapid progress in devising effective parallel algorithms for block matrix computations. The complexity of parallel computations is usually represented by the computational and communication time and the number of processors involved; decreasing all of these parameters, we face a tradeoff; the product of time and processor bounds (called potential work of parallel algorithms) cannot usually be made substantially smaller than the sequential time bound for the solution. This follows because, according to a variant of *Brent's scheduling principle*, a single processor can simulate the work of s processors in time $O(s)$. The usual goal of designing a parallel algorithm is in decreasing its parallel time bound (ideally, to a constant, logarithmic or polylogarithmic level, relative to n) and keeping its work bound at the level of the record sequential time bound for the same computational problem (within constant, logarithmic, or at worst polylog factors). This goal has been easily achieved for matrix and vector multiplications, but turned out to be nontrivial for linear system solving, inversion, and some other related computational problems. The recent solution for general matrices [Kaltofen and Pan 1991, 1992] relies on computation of a Krylov sequence and the coefficients of the minimum polynomial of a matrix, by using randomization and auxiliary computations with structured matrices (see the details in Bini and Pan [1994]).

Rational Matrix Computations, Computations in Finite Fields and Semirings

Rational algebraic computations with matrices are performed for a rational input given with no errors, and the computations are also performed with no errors. The precision of computing can be bounded by reducing the computations modulo one or several fixed primes or prime powers. At the end, the exact output values $z = p/q$ are recovered from $z \bmod M$ (if M is sufficiently large relative to p and q) by using the continued fraction approximation algorithm, which is the Euclidean algorithm applied to integers (cf. Pan [1991, 1992a], and Bini and Pan [1994, section 3 of Chap. 3]). If the output z is known to be an integer lying between $-m$ and m and if $M > 2m$, then z is recovered from $z \bmod M$ as follows:

$$
z = \begin{cases} z \bmod M & \text{if } z \bmod M < m \\ -M + z \bmod M & \text{otherwise} \end{cases}
$$

The reduction modulo a prime p may turn a nonsingular matrix A and a nonsingular linear system $Ax = v$ into singular ones, but this is proved to occur only with a low probability for a random choice of the prime p in a fixed sufficiently large interval (see Bini and Pan [1994, section 3 of Chap. 4]). To compute the output values z modulo M for a large M, one may first compute them modulo several relatively prime integers $m_1, m_2, \ldots, m_k$ having no common divisors and such that $m_1, m_2, \ldots, m_k > M$ and then easily recover $z \bmod M$ by means of the Chinese remainder algorithm. For matrix and polynomial computations, there is an effective alternative technique of *p-adic* (*Newton–Hensel*) *lifting* (cf. Bini and Pan [1994, section 3 of Chap. 3]), which is particularly powerful for computations with dense structured matrices, since it preserves the structure of a matrix. We refer the reader to Bareiss[1968] and Geddes et al. [1992] for some special techniques, which enable one to control the growth of all intermediate values computed in the process of performing rational Gaussian elimination, with no roundoff and no reduction modulo an integer.

Gondran and Minoux [1984] and Pan [1993] describe some applications of matrix computations on semirings (with no divisions and subtractions allowed) to graph and combinatorial computations.

Matrix Eigenvalues and Singular Values Problems

The matrix eigenvalue problem is one of the major problems of matrix computation: given an $n \times n$ matrix A, one seeks a $k \times k$ diagonal matrix Λ and an $n \times k$ matrix V of full rank k such that

$$
AV = \Lambda V \tag{10.1}
$$

The diagonal entries of Λ are called the *eigenvalues* of A; the entry (i, i) of Λ is associated with the ith column of V, called an *eigenvector* of A. The eigenvalues of an $n \times n$ matrix A coincide with the zeros of the characteristic polynomial

$$c_A(x) = \det(xI - A)$$

If this polynomial has n distinct zeros, then $k = n$, and V of Eq. (10.1) is a nonsingular $n \times n$ matrix. The matrix $A = I + Z$, where $Z = (z_{i,j})$, $z_{i,j} = 0$ unless $j = i + 1$, $z_{i,i+1} = 1$, is an example of a matrix for which $k = 1$, so that the matrix V degenerates to a vector.

In principle, one may compute the coefficients of $c_A(x)$, the characteristic polynomial of A, and then approximate its zeros (see the next section) in order to approximate the eigenvalues of A. Given the eigenvalues, the corresponding eigenvectors can be recovered by means of the inverse power iteration [Golub and Van Loan 1989, Wilkinson 1965]. Practically, the computation of the eigenvalues via the computation of the coefficients of $c_A(x)$ is not recommended, due to arising numerical stability problems [Wilkinson 1965], and most frequently, the eigenvalues and eigenvectors of a general (unsymmetric) matrix are approximated by means of the QR *algorithm* [Wilkinson 1965, Watkins 1982, Golub and Van Loan 1989]. Before application of this algorithm, the matrix A is simplified by transforming it into the more special (*Hessenberg*) *form* H, by a *similarity transformation*,

$$H = UAU^H \tag{10.2}$$

where $U = (u_{i,j})$ is a unitary matrix, where $U^H U = I$, where $U^H = (\overline{u}_{j,i})$ is the Hermitian transpose of U, with $\overline{z}$ denoting the complex conjugate of z; $U^H = U^T$ if U is a real matrix [Golub and Van Loan 1989]. Similarity transformation into Hessenberg form is one of examples of *rational transformations* of a matrix into special *canonical forms*, of which transformations into *Smith* and *Hermite forms* are two other most important representatives [Kaltofen et al. 1990, Geddes et al. 1992, Giesbrecht 1995].

In practice, the eigenvalue problem is very frequently symmetric, that is, arises for a real symmetric matrix A, for which

$$A^T = (a_{j,i}) = A = (a_{i,j})$$

or for complex Hermitian matrices A, for which

$$A^H = (\overline{a}_{j,i}) = A = (a_{i,j})$$

For real symmetric or Hermitian matrices A, the eigenvalue problem (called symmetric) is treated much more easily than in the unsymmetric case. In particular, in the symmetric case, we have $k = n$, that is, the matrix V of Eq. (10.1) is a nonsingular $n \times n$ matrix, and moreover, all of the eigenvalues of A are real and little sensitive to small input perturbations of A (according to the Courant–Fisher minimization criterion [Parlett 1980, Golub and Van Loan 1989]).

Furthermore, similarity transformation of A to the Hessenberg form gives much stronger results in the symmetric case: the original problem is reduced to one for a symmetric tridiagonal matrix H of Eq. (10.2) (this can be achieved via the Lanczos algorithm, cf. Golub and Van Loan [1989] or Bini and Pan [1994, Section 3 of Chap. 2]). For such a matrix H, application of the QR algorithm is dramatically simplified; moreover, two competitive algorithms are also widely used, that is, the *bisection* [Parlett 1980] (a slightly slower but very robust algorithm) and the *divide-and-conquer* method [Cuppen 1981, Golub and Van Loan 1989]. The latter method has a modification [Bini and Pan 1991] that only uses $O(n \log^2 n (\log n + \log^2 b))$ arithmetic operations in order to compute all of the eigenvalues of an $n \times n$ symmetric tridiagonal matrix A within the output error bound $2^{-b} \|A\|$, where $\|A\| \le n \max |a_{i,j}|$.

The eigenvalue problem has a generalization, where generalized eigenvalues and eigenvectors for a pair A, B of matrices are sought, such that

$$AV = B\Lambda V$$

(the solution algorithm should proceed without computing the matrix $B^{-1}A$, so as to avoid numerical stability problems).

In another highly important extension of the symmetric eigenvalue problem, one seeks a singular value decomposition (SVD) of a (generally unsymmetric and, possibly, rectangular) matrix A: $A = U\Sigma V^T$, where U and V are unitary matrices, $U^H U = V^H V = I$, and Σ is a diagonal (generally rectangular) matrix, filled with zeros, except for its diagonal, filled with (positive) singular values of A and possibly, with zeros. The SVD is widely used in the study of numerical stability of matrix computations and in numerical treatment of singular and ill-conditioned (close to singular) matrices. An alternative tool is orthogonal (QR) factorization of a matrix, which is not as refined as SVD but is a little easier to compute [Golub and Van Loan 1989]. The squares of the singular values of A equal the eigenvalues of the Hermitian (or real symmetric) matrix $A^H A$, and the SVD of A can be also easily recovered from the eigenvalue decomposition of the Hermitian matrix

$$\begin{bmatrix} 0 & A^H \\ A & 0 \end{bmatrix}$$

but more popular are some effective direct methods for the computation of the SVD [Golub and Van Loan 1989].

Approximating Polynomial Zeros

Solution of an nth degree polynomial equation,

$$p(x) = \sum_{i=0}^{n} p_i x^i = 0, \quad p_n \neq 0$$

(where one may assume that $p_{n-1} = 0$; this can be ensured via shifting the variable x) is a classical problem that has greatly influenced the development of mathematics throughout the centuries [Pan 1995b]. The problem remains highly important for the theory and practice of present day computing, and dozens of new algorithms for its approximate solution appear every year. Among the existent implementations of such algorithms, the practical heuristic champions in efficiency (in terms of computer time and memory space used, according to the results of many experiments) are various modifications of *Newton's iteration*, $z(i+1) = z(i) - a(i)p(z(i))/p'(z(i))$, $a(i)$ being the step-size parameter [Madsen 1973], *Laguerre's method* [Hansen et al. 1977, Foster 1981], and the randomized *Jenkins–Traub algorithm* [1970] [all three for approximating a single zero z of $p(x)$], which can be extended to approximating other zeros by means of deflation of the input polynomial via its numerical division by $x - z$. For simultaneous approximation of all of the zeros of $p(x)$ one may apply the Durand–Kerner algorithm, which is defined by the following recurrence:

$$z_j(i+1) = \frac{z_j(i) - p((z_j(i)))}{z_j(i) - z_k(i)}, \quad j = 1, \ldots, n, \quad i = 1, 2, \ldots \tag{10.3}$$

Here, the customary choice for the n initial approximations $z_j(0)$ to the n zeros of

$$p(x) = p_n \prod_{j=1}^{n}(x - z_j)$$

is given by $z_j(0) = Z \exp(2\pi\sqrt{-1}/n)$, $j = 1, \ldots, n$, with Z exceeding (by some fixed factor $t > 1$) $\max_j |z_j|$; for instance, one may set

$$Z = 2t \max_{i<n}(p_i/p_n) \tag{10.4}$$

For a fixed i and for all j, the computation according to Eq. (10.3) is simple, only involving order n^2 ops, and according to the results of many experiments, the iteration Eq. (10.3) rapidly converges to the solution, though no theory confirms or explains these results. Similar is the situation with various modifications of this algorithm, which are now even more popular than the original algorithms and many of which are listed in [Pan 1992a, 1992b] (also cf. Bini and Pan [1996] and McNamee [1993]).

On the other hand, there are two groups of algorithms that, when implemented, promise to be competitive or even substantially superior to Newton's and Laguerre's iteration, the algorithm by Jenkins and Traub, and all of the algorithms of the Durand–Kerner type. One such group is given by the modern modifications and improvements (due to Pan [1987, 1994a, 1994b] and Renegar [1989]) of *Weyl's quadtree construction* of 1924. In this approach, an initial square S, containing all the zeros of $p(x)$ [say, $S = \{x, |Im\ x| < Z, |Re\ x| < Z\}$ for Z of Eq. (10.4)], is recursively partitioned into four congruent subsquares. In the center of each of them, a proximity test is applied that estimates the distance from this center to the closest zero of $p(x)$. If such a distance exceeds one-half of the diagonal length, then the subsquare contains no zeros of $p(x)$ and is discarded. When this process ensures a strong isolation from each other for the components formed by the remaining squares, then certain extensions of Newton's iteration [Renegar 1989, Pan 1994a, 1994b], or some iterative techniques based on numerical integration [Pan 1987] are applied and very rapidly converge to the desired approximations to the zeros of $p(x)$, within the error bound $2^{-b}Z$ for Z of Eq. (10.4). As a result, the algorithms of [Pan 1987, 1994a, 1994b] solve the entire problem of approximating (within $2^{-b}Z$) all of the zeros of $p(x)$ at the overall cost of performing $O((n^2 \log n) \log(bn))$ ops (cf. Bini and Pan [1996]), versus order n^2 operations at each iteration of Durand–Kerner type.

The second group is given by the divide-and-conquer algorithms. They first compute a sufficiently wide annulus A, which is free of the zeros of $p(x)$ and contains comparable numbers of such zeros (that is, the same numbers up to a fixed constant factor) in its exterior and its interior. Then the two factors of $p(x)$ are numerically computed, that is, $F(x)$ having all its zeros in the interior of the annulus, and $G(x) = p(x)/F(x)$ having no zeros there. The same process is recursively repeated for $F(x)$ and $G(x)$ until factorization of $p(x)$ into the product of linear factors is computed numerically. From this factorization, approximations to all of the zeros of $p(x)$ are obtained. The algorithms of Pan [1995a, 1996] based on this approach only require $O(n \log(bn) (\log n)^2)$ ops in order to approximate all of the n zeros of $p(x)$ within $2^{-b}Z$ for Z of Eq. (10.4). (Note that this is a quite sharp bound: at least n ops are necessary in order to output n distinct values.)

The computations for the polynomial zero problem are ill conditioned, that is, they generally require a high precision for the worst-case input polynomials in order to ensure a required output precision, no matter which algorithm is applied for the solution. Consider, for instance, the polynomial $(x - \frac{6}{7})^n$ and perturb its x-free coefficient by 2^{-bn}. Observe the resulting jumps of the zero $x = 6/7$ by 2^{-b}, and observe similar jumps if the coefficients p_i are perturbed by $2^{(i-n)b}$ for $i = 1, 2, \ldots, n-1$. Therefore, to ensure the output precision of b bits, we need an input precision of at least $(n-i)b$ bits for each coefficient p_i, $i = 0, 1, \ldots, n-1$. Consequently, for the worst-case input polynomial $p(x)$, any solution algorithm needs at least about a factor n increase of the precision of the input and of computing vs. the output precision.

Numerically unstable algorithms may require even a higher input and computation precision, but inspection shows that this is not the case for the algorithms of Pan [1987, 1994a, 1994b, 1995a, 1996] and Renegar [1989] (cf. Bini and Pan [1996]).

Fast Fourier Transform and Fast Polynomial Arithmetic

To yield the record complexity bounds for approximating polynomial zeros, one should exploit fast algorithms for basic operations with polynomials (their multiplication, division, and transformation under the shift of the variable), as well as FFT, both directly and for supporting the fast polynomial arithmetic. The FFT and fast basic polynomial algorithms (including those for multipoint polynomial evaluation and interpolation) are the basis for many other fast polynomial computations, performed both numerically and symbolically (compare the next sections). These basic algorithms, their impact on the field of algebraic computation, and their complexity estimates have been extensively studied in Aho et al. [1974], Borodin and Munro [1975], and Bini and Pan [1994].

10.3 Systems of Nonlinear Equations and Other Applications

Given a system $\{p_1(x_1, \ldots, x_n), p_2(x_1, \ldots, x_n), \ldots, p_r(x_1, \ldots, x_n)\}$ of nonlinear polynomials with rational coefficients [each $p_i(x_1, \ldots, x_n)$ is said to be an element of $\mathbb{Q}[x_1, \ldots, x_n]$, the ring of polynomials in $x_1, \ldots, x_n$ over the field $\mathbb{Q}$ of rational numbers], the n-tuple of complex numbers $(a_1, \ldots, a_n)$ is a common solution of the system, if $f_i(a_1, \ldots, a_n) = 0$ for each i with $1 \leq i \leq r$. In this section, we explore the problem of exactly solving a system of nonlinear equations over the field $\mathbb{Q}$. We provide an overview and cite references to different symbolic techniques used for solving systems of algebraic (polynomial) equations. In particular, we describe methods involving *resultant* and *Gröbner basis* computations.

The *Sylvester resultant method* is the technique most frequently utilized for determining a common zero of two polynomial equations in one variable [Knuth 1981]. However, using the Sylvester method successively to solve a system of multivariate polynomials proves to be inefficient. Successive resultant techniques, in general, lack efficiency as a result of their sensitivity to the ordering of the variables [Kapur and Lakshman 1992]. It is more efficient to eliminate all variables together from a set of polynomials, thus leading to the notion of the *multivariate resultant*. The three most commonly used multivariate resultant formulations are the *Dixon* [Dixon 1908, Kapur and Saxena 1995], *Macaulay* [Macaulay 1916, Canny 1990, Kaltofen and Lakshman 1988], and *sparse resultant formulations* [Canny and Emiris 1993a, Sturmfels 1991].

The theory of Gröbner bases provides powerful tools for performing computations in multivariate polynomial rings. Formulating the problem of solving systems of polynomial equations in terms of polynomial ideals, we will see that a Gröbner basis can be computed from the input polynomial set, thus allowing for a form of back substitution (cf. section 10.2) in order to compute the common roots.

Although not discussed, it should be noted that the *characteristic set algorithm* can be utilized for polynomial system solving. Ritt [1950] introduced the concept of a characteristic set as a tool for studying solutions of algebraic differential equations. Wu [1984, 1986], in search of an effective method for automatic theorem proving, converted Ritt's method to ordinary polynomial rings. Given the before mentioned system P, the characteristic set algorithm transforms P into a triangular form, such that the set of common zeros of P is equivalent to the set of roots of the triangular system [Kapur and Lakshman 1992].

Throughout this exposition we will also see that these techniques used to solve nonlinear equations can be applied to other problems as well, such as computer-aided design and automatic geometric theorem proving.

Resultant Methods

The question of whether two polynomials $f(x), g(x) \in \mathbb{Q}[x]$,

$$f(x) = f_n x^n + f_{n-1} x^{n-1} + \cdots + f_1 x + f_0$$

$$g(x) = g_m x^m + g_{m-1} x^{m-1} + \cdots + g_1 x + g_0$$

have a common root leads to a condition that has to be satisfied by the coefficients of both f and g. Using a derivation of this condition due to Euler, the *Sylvester matrix* of f and g (which is of order $m + n$) can be formulated. The vanishing of the determinant of the Sylvester matrix, known as the *Sylvester resultant*, is a necessary and sufficient condition for f and g to have common roots [Knuth 1981].

As a running example let us consider the following system in two variables provided by Lazard [1981]:

$$f = x^2 + xy + 2x + y - 1 = 0$$

$$g = x^2 + 3x - y^2 + 2y - 1 = 0$$

The Sylvester resultant can be used as a tool for eliminating several variables from a set of equations [Kapur and Lakshman 1992]. Without loss of generality, the roots of the Sylvester resultant of f and g treated as polynomials in y, whose coefficients are polynomials in x, are the x-coordinates of the common zeros of f and g. More specifically, the Sylvester resultant of the Lazard system with respect to y is given by the following determinant:

$$\det\left(\begin{bmatrix} x+1 & x^2+2x-1 & 0 \\ 0 & x+1 & x^2+2x-1 \\ -1 & 2 & x^2+3x-1 \end{bmatrix}\right) = -x^3 - 2x^2 + 3x$$

The roots of the Sylvester resultant of f and g are $\{-3, 0, 1\}$. For each x value, one can substitute the x value back into the original polynomials yielding the solutions $(-3, 1)$, $(0, 1)$, $(1, -1)$.

The method just outlined can be extended recursively, using *polynomial GCD computations*, to a larger set of multivariate polynomials in $\mathbb{Q}[x_1, \ldots, x_n]$. This technique, however, is impractical for eliminating many variables, due to an explosive growth of the degrees of the polynomials generated in each elimination step.

The Sylvester formulations have led to a *subresultant theory*, developed simultaneously by G. E. Collins and W. S. Brown and J. Traub. The subresultant theory produced an efficient algorithm for computing polynomial GCDs and their resultants, while controlling intermediate expression swell [Brown 1971, Brown and Traub 1971, Collins 1967, 1971, Knuth 1981].

It should be noted that by adopting an implicit representation for symbolic objects, the intermediate expression swell introduced in many symbolic computations can be palliated. Recently, polynomial GCD algorithms have been developed that use implicit representations and thus avoid the computationally costly content and primitive part computations needed in those GCD algorithms for polynomials in explicit representation [Diaz and Kaltofen 1995, Kaltofen 1988, Kaltofen and Trager 1990].

The solvability of a set of nonlinear multivariate polynomials over the field $\mathbb{Q}$ can be determined by the vanishing of a generalization of the Sylvester resultant of two polynomials in a single variable.

Due to the special structure of the Sylvester matrix, Bézout developed a method for computing the resultant as a determinant of order $\max(m, n)$ during the 18th century. Cayley [1865] reformulated Bézout's method leading to Dixon's [1908] extension to the bivariate case. Dixon's method can be generalized to a set

$$\{p_1(x_1, \ldots, x_n), p_2(x_1, \ldots, x_n), \ldots, p_{n+1}(x_1, \ldots, x_n)\}$$

of $n+1$ generic n-degree polynomials in n variables [Kapur et al. 1994]. The vanishing of the Dixon resultant is a necessary and sufficient condition for the polynomials to have a nontrivial projective common zero, and also a necessary condition for the existence of an affine common zero. The Dixon formulation gives the resultant up to a multiple, and hence in the affine case it may happen that the vanishing of the Dixon resultant does not necessarily indicate that the equations in question have a common root. A nontrivial multiple, known as the *projection operator*, can be extracted via a method based on so-called *rank subdeterminant computation* (RSC) [Kapur et al. 1994]. It should be noted that the RSC method can also be applied to the Macaulay and sparse resultant formulations as is detailed here.

In 1916, Macaulay constructed a resultant for n homogeneous polynomials in n variables, which simultaneously generalizes the Sylvester resultant and the determinant of a system of linear equations [Canny et al. 1989, Kapur and Lakshman 1992]. Like the Dixon formulation, the Macaulay resultant is a multiple of the resultant (except in the case of generic homogeneous polynomials, where it produces the exact resultant). For the Macaulay formulation, Canny [1990] has invented a general method that perturbs any polynomial system and extracts a nontrivial projection operator.

Using recent results pertaining to sparse polynomial systems [Gelfand et al. 1994, Sturmfels 1991, Sturmfels and Zelevinsky 1992], the mixed sparse resultant of a system of $n + 1$ sparse polynomials in n variables in its matrix form was given by Canny and Emiris [1993a] and consequently improved in Canny and Emiris [1993b, 1994]. Here, sparsity denotes that only certain monomials in each of the $n + 1$ polynomials have nonzero coefficients. The determinant of the sparse resultant matrix, such as the Macaulay and Dixon matrices, only yields a projection operation, not the exact resultant.

Suppose we are asked to find the common zeros of a set of n polynomials in n variables $\{p_1(x_1, \ldots, x_n), p_2(x_1, \ldots, x_n), \ldots, p_n(x_1, \ldots, x_n)\}$. By augmenting the polynomial set by a generic linear form [Canny 1990, Canny and Manocha 1991, Kapur and Lakshman 1992], one can construct the *u-resultant* of a given system of polynomials. The u-resultant factors into linear factors over the complex numbers, providing the common zeros of the given polynomials equations. The u-resultant method takes advantage of the properties of the multivariate resultant, and hence can be constructed using either Dixon's, Macaulay's, or sparse formulations.

Consider the previous example augmented by a generic linear form

$$f_1 = x^2 + xy + 2x + y - 1 = 0$$
$$f_2 = x^2 + 3x - y^2 + 2y - 1 = 0$$
$$f_1 = ux + vy + w = 0$$

As described in Canny et al. [1989], the following matrix M corresponds to the Macaulay u-resultant of the preceding system of polynomials, with z being the homogenizing variable:

$$M = \begin{bmatrix}
1 & 0 & 0 & 1 & 0 & 0 & 0 & 0 & 0 & 0 \\
1 & 1 & 0 & 0 & 1 & 0 & u & 0 & 0 & 0 \\
2 & 0 & 1 & 3 & 0 & 1 & 0 & u & 0 & 0 \\
0 & 1 & 0 & -1 & 0 & 0 & v & 0 & 0 & 0 \\
1 & 2 & 1 & 2 & 3 & 0 & w & v & u & 0 \\
-1 & 0 & 2 & -1 & 0 & 3 & 0 & w & 0 & u \\
0 & 0 & 0 & 0 & -1 & 0 & 0 & 0 & 0 & 0 \\
0 & 1 & 0 & 0 & 2 & -1 & 0 & 0 & v & 0 \\
0 & -1 & 1 & 0 & -1 & 2 & 0 & 0 & w & v \\
0 & 0 & -1 & 0 & 0 & -1 & 0 & 0 & 0 & w
\end{bmatrix}$$

It should be noted that

$$\det(M) = (u - v + w)(-3u + v + w)(v + w)(u - v)$$

corresponds to the affine solutions $(1, -1)$, $(-3, 1)$, $(0, 1)$, and one solution at infinity. An empirical comparison of the detailed resultant formulations can be found in Kapur and Saxena [1995]. Recently, the multivariate resultant formulations are being used for other applications such as *algebraic and geometric reasoning* [Kapur et al. 1994], *computer-aided design* [Stederberg and Goldman 1986], and for *implicitization and finding base points* [Chionh 1990].

Gröbner Bases

Solving systems of nonlinear equations can be formulated in terms of polynomial ideals [Becker and Weispfenning 1993, Geddes et al. 1992, Winkler 1996]. Let us first establish some terminology.

The ideal generated by a system of polynomial equations $p_1, \ldots, p_r$ over $\mathbb{Q}[x_1, \ldots, x_n]$ is the set of all linear combinations

$$(p_1, \ldots, p_r) = \{h_1 p_1 + \cdots + h_r p_r \mid h_1, \ldots, h_r \in \mathbb{Q}[x_1, \ldots, x_n]\}$$

The algebraic variety of $p_1, \ldots, p_r \in \mathbb{Q}[x_1, \ldots, x_n]$ is the set of their common zeros,

$$V(p_1, \ldots, p_r) = \{(a_1, \ldots, a_n) \in \mathbb{C}^n \mid f_1(a_1, \ldots, a_n) = \cdots = f_r(a_1, \ldots, a_n) = 0\}.$$

A version of the *Hilbert Nullstellensatz* states that

$$V(p_1, \ldots, p_r) = \text{the empty set } \emptyset \iff 1 \in (p_1, \ldots, p_r) \text{ over } \mathbb{Q}[x_1, \ldots, x_n]$$

which relates the solvability of polynomial systems to the ideal membership problem.

A term $t = x_1^{e_1} x_2^{e_2} \ldots x_n^{e_n}$ of a polynomial is a product of powers with $\deg(t) = e_1 + e_2 + \cdots + e_n$. In order to add needed structure to the polynomial ring we will require that the terms in a polynomial be ordered in an admissible fashion [Geddes et al. 1992, Kapur and Lakshman 1992]. Two of the most common admissible orderings are the **lexicographic order** ($\prec_l$), where terms are ordered as in a dictionary, and the **degree order** ($\prec_d$), where terms are first compared by their degrees with equal degree terms compared lexicographically. A variation to the lexicographic order is the *reverse lexicographic order*, where the lexicographic order is reversed [Davenport et al. 1988, p. 96].

It is this previously mentioned structure that permits a type of simplification known as polynomial reduction. Much like a polynomial remainder process, the process of polynomial reduction involves subtracting a multiple of one polynomial from another to obtain a smaller degree result [Becker and Weispfenning 1993, Geddes et al. 1992, Kapur and Lakshman 1992, Winkler 1996].

A polynomial g is said to be reducible with respect to a set $P = \{p_1, \ldots, p_r\}$ of polynomials if it can be reduced by one or more polynomials in P. When g is no longer reducible by the polynomials in P, we say that g is *reduced* or is *a normal form* with respect to P.

For an arbitrary set of basis polynomials, it is possible that different reduction sequences applied to a given polynomial g could reduce to different normal forms. A basis $G \subseteq \mathbb{Q}[x_1, \ldots, x_n]$ is a *Gröbner basis* if and only if every polynomial in $\mathbb{Q}[x_1, \ldots, x_n]$ has a unique normal form with respect to G. Buchberger [1965, 1976, 1983, 1985] showed that every basis for an ideal $(p_1, \ldots, p_r)$ in $\mathbb{Q}[x_1, \ldots, x_n]$ can be converted into a Gröbner basis $\{p_1^*, \ldots, p_s^*\} = GB(p_1, \ldots, p_r)$, concomitantly designing an algorithm that transforms an arbitrary ideal basis into a Gröbner basis. Another characteristic of Gröbner bases is that by using the previously mentioned reduction process we have

$$g \in (p_1, \ldots, p_r) \iff (g \bmod p_1^*, \ldots, p_s^*) = 0$$

Further, by using the Nullstellensatz it can be shown that $p_1, \ldots, p_r$ viewed as a system of algebraic equations is solvable if and only if $1 \notin GB(p_1, \ldots, p_r)$.

Depending on which admissible term ordering is used in the Gröbner bases construction, an ideal can have different Gröbner bases. However, an ideal cannot have different (reduced) Gröbner bases for the same term ordering.

Any system of polynomial equations can be solved using a lexicographic Gröbner basis for the ideal generated by the given polynomials. It has been observed, however, that Gröbner bases, more specifically lexicographic Gröbner bases, are hard to compute [Becker and Weispfenning 1993, Geddes et al. 1992, Lakshman 1990, and Winkler 1996]. In the case of zero-dimensional ideals, those whose varieties have only isolated points, Faugère, et al. [1993] outlined a change of basis algorithm which can be utilized for solving zero-dimensional systems of equations. In the zero-dimensional case, one computes a Gröbner basis for the ideal generated by a system of polynomials under a degree ordering. The so-called *change of basis algorithm* can then be applied to the degree ordered Gröbner basis to obtain a Gröbner basis under a lexicographic ordering.

Turning to Lazard's example in the form of a polynomial basis,

$$f_1 = x^2 + xy + 2x + y - 1$$
$$f_2 = x^2 + 3x - y^2 + 2y - 1$$

one obtains (under lexicographical ordering with $x \prec_l y$) a Gröbner basis in which the variables are triangularized such that the finitely many solutions can be computed via back substitution:

$$f_1^* = x^2 + 3x + 2y - 2$$

$$f_2^* = xy - x - y + 1$$

$$f_3^* = y^2 - 1$$

It should be noted that the final univariate polynomial is of minimal degree and the polynomials used in the back substitution will have degree no larger than the number of roots.

As an example of the process of polynomial reduction with respect to a Gröbner basis, the following demonstrates two possible reduction sequences to the same normal form. The polynomial $x^2 y^2$ is reduced with respect to the previously computed Gröbner basis $\{f_1^*, f_2^*, f_3^*\} = GB(f_1, f_2)$ along the following two distinct reduction paths, both yielding $-3x - 2y + 2$ as the normal form.

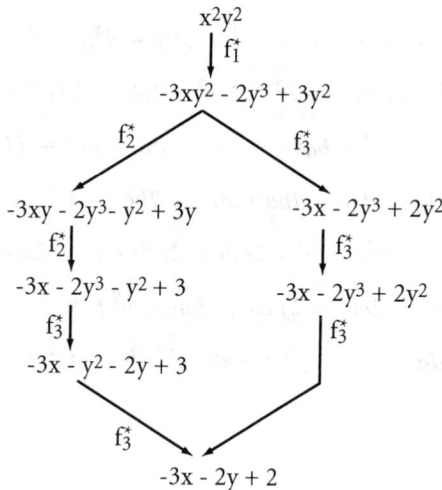

There is a strong connection between lexicographic Gröbner bases and the previously mentioned resultant techniques. For some types of input polynomials, the computation of a reduced system via resultants might be much faster than the computation of a lexicographic Gröbner basis. A good comparison between the Gröbner computations and the different resultant formulations can be found in Kapur and Saxena [1995].

In a survey article, Buchberger [1985] detailed how Gröbner bases can be used as a tool for many polynomial ideal theoretic operations. Other applications of Gröbner basis computations include automatic geometric theorem proving [Kapur 1986, Wu 1984, 1986], multivariate polynomial factorization and GCD computations [Gianni and Trager 1985], and polynomial interpolation [Lakshman and Saunders 1994, 1995].

10.4 Polynomial Factorization

The problem of factoring polynomials is a fundamental task in symbolic algebra. An example in one's early mathematical education is the factorization $x^2 - y^2 = (x + y) \cdot (x - y)$, which in algebraic terms is a factorization of a polynomial in two variables with integer coefficients. Technology has advanced to a state where most polynomial factorization problems are doable on a computer, in particular, with any of the popular mathematical software, such as the Mathematica or Maple systems. For instance, the factorization of the determinant of a 6×6 symmetric Toeplitz matrix over the integers is computed in Maple as

```
> readlib(showtime):
> showtime():
O1 := T := linalg[toeplitz]([a,b,c,d,e,f]);
```

$$T := \begin{bmatrix} a & b & c & d & e & f \\ b & a & b & c & d & e \\ c & b & a & b & c & d \\ d & c & b & a & b & c \\ e & d & c & b & a & b \\ f & e & d & c & b & a \end{bmatrix}$$

```
time 0.03 words 7701
O2 := factor(linalg[det](T));
```

$$-(2dca - 2bce + 2c^2a - a^3 - da^2 + 2d^2c + d^2a + b^3 + 2abc - 2c^2b$$

$$+ d^3 + 2ab^2 - 2dcb - 2cb^2 - 2ec^2 + 2eb^2 + 2fcb + 2bae$$

$$+ b^2f + c^2f + be^2 - ba^2 - fdb - fda - fa^2 - fba + e^2a - 2db^2$$

$$+ dc^2 - 2deb - 2dec - dba)(2dca - 2bce - 2c^2a + a^3$$

$$- da^2 - 2d^2c - d^2a + b^3 + 2abc - 2c^2b + d^3 - 2ab^2 + 2dcb$$

$$+ 2cb^2 + 2ec^2 - 2eb^2 - 2fcb + 2bae + b^2f + c^2f + be^2 - ba^2$$

$$- fdb + fda - fa^2 + fba - e^2a - 2db^2 + dc^2 + 2deb - 2dec$$

$$+ dba)$$

```
time 27.30 words 857700
```

Clearly, the Toeplitz determinant factorization requires more than tricks from high school algebra. Indeed, the development of modern algorithms for the polynomial factorization problem is one of the great successes of the discipline of symbolic mathematical computation. Kaltofen [1982, 1990, 1992] has surveyed the algorithms until 1992, mostly from a computer science perspective. In this chapter we shall focus on the applications of the known fast methods to problems in science and engineering. For a more extensive set of references, please refer to Kaltofen's survey articles.

Polynomials in a Single Variable over a Finite Field

At first glance, the problem of factoring an integer polynomial modulo a prime number appears to be very similar to the problem of factoring an integer represented in a prime radix. That is simply not so. The factorization of the polynomial $x^{511} - 1$ can be done modulo 2 on a computer in a matter of milliseconds, whereas the factorization of the integer $2^{511} - 1$ into its integer factors is a computational challenge. For those interested: the largest prime factors of $2^{511} - 1$ have 57 and 67 decimals digits, respectively, which makes a tough but not undoable 123 digit product for the number field sieve factorizer [Leyland 1995]. Irreducible factors of polynomials modulo 2 are needed to construct finite fields. For example, the factor $x^9 + x^4 + 1$ of $x^{511} - 1$ leads to a model of the finite field with 2^9 elements, GF(2^9), by simply computing with the polynomial remainders modulo $x^9 + x^4 + 1$ as the elements. Such irreducible polynomials are used for setting up error-correcting codes, such as the BCH codes [MacWilliams and Sloan 1977]. Berlekamp's [1967, 1970] pioneering work on factoring polynomials over a finite field by linear algebra is done with this motivation. The linear algebra tools that Berlekamp used seem to have been introduced to the subject as early as in 1937 by Petr (cf. Št. Schwarz [1956]).

Today, factoring algorithms for univariate polynomials over finite fields form the innermost subalgorithm to lifting-based algorithms for factoring polynomials in one [Zassenhaus 1969] and many [Musser 1975] variables over the integers. When Maple computed the factorization of the previous Toeplitz determinant, it began with factoring a univariate polynomial modulo a prime integer. The case when the prime integer is very large has led to a significant development in computer science itself. As it turns out, by selecting random residues the expected performance of the algorithms can be speeded up exponentially [Berlekamp 1970, Rabin 1980]. Randomization is now an important tool for designing efficient algorithms and has proliferated to many fields of computer science. Paradoxically, the random elements are produced by a congruential random number generator, and the actual computer implementations are quite deterministic, which leads some computer scientists to believe that random bits can be eliminated in general at no exponential slow down. Nonetheless, for the polynomial factoring problem modulo a large prime, no fast methods are known to date that would work without this *probabilistic* approach.

One can measure the computing time of selected algorithms in terms of n, the degree of the input polynomial, and p, the cardinality of the field. When counting arithmetic operations modulo p (including reciprocals), the best known algorithms are quite recent. Berlekamp's 1970 method performs $O(n^\omega + n^{1+o(1)} \log p)$ residue operations. Here and subsequently, ω denotes the exponent implied by the used linear system solver, i.e., $\omega = 3$ when classical methods are used, and $\omega = 2.376$ when asymptotically fast (though impractical) matrix multiplication is assumed. The correction term $o(1)$ accounts for the $\log n$ factors derived from the FFT-based fast polynomial multiplication and remaindering algorithms. An approach in the spirit of Berlekamp's but possibly more practical for $p = 2$ has recently been discovered by Niederreiter [1994]. A very different technique by Cantor and Zassenhaus [1981] first separates factors of different degrees and then splits the resulting polynomials of equal degree factors. It has $O(n^{2+o(1)} \log p)$ complexity and is the basis for the following two methods. Algorithms by von zur Gathen and Shoup [1992] have running time $O(n^{2+o(1)} + n^{1+o(1)} \log p)$ and those by Kaltofen and Shoup [1995] have running time $O(n^{1.815} \log p)$, the latter with fast matrix multiplication.

For n and p simultaneously large, a variant of the method by Kaltofen and Shoup [1995] that uses classical linear algebra and runs in $O(n^{2.5} + n^{1+o(1)} \log p)$ residue operations is the current champion among the practical algorithms. With it Shoup[1996], using his own fast polynomial arithmetic package, has factored a randomlike polynomial of degree 2048 modulo a 2048-bit prime number in about 12 days on a Sparc-10 computer using 68 megabyte of main memory. For even larger n, but smaller p, parallelization helps, and Kaltofen and Lobo [1994] could factor a polynomial of degree $n = 15\,001$ modulo $p = 127$ in about 6 days on 8 computers that are rated at 86.1 MIPS. At the time of this writing, the largest polynomial factored modulo 2 is $X^{216\,091} + X + 1$; this was accomplished by Peter Montgomery in 1991 by using Cantor's fast polynomial multiplication algorithm based on additive transforms [Cantor 1989].

Polynomials in a Single Variable over Fields of Characteristic Zero

As mentioned before, generally usable methods for factoring univariate polynomials over the rational numbers begin with the Hensel lifting techniques introduced by Zassenhaus [1969]. The input polynomial is first factored modulo a suitable prime integer p, and then the factorization is lifted to one modulo p^k for an exponent k of sufficient size to accommodate all possible integer coefficients that any factors of the polynomial might have. The lifting approach is fast in practice, but there are hard-to-factor polynomials on which it runs an exponential time in the degree of the input. This slowdown is due to so-called parasitic modular factors. The polynomial $x^4 + 1$, for example, factors modulo all prime integers but is irreducible over the integers: it is the cyclotomic equation for eighth roots of unity. The products of all subsets of modular factors are candidates for integer factors, and irreducible integer polynomials with exponentially many such subsets exist [Kaltofen et al. 1983].

The elimination of the exponential bottleneck by giving a polynomial-time solution to the integer polynomial factoring problem, due to Lenstra et al. [1982] is considered a major result in computer science algorithm design. The key ingredient to their solution is the construction of integer relations to real or complex numbers. For the simple demonstration of this idea, consider the polynomial

$$x^4 + 2x^3 - 6x^2 - 4x + 8$$

A root of this polynomial is $\alpha \approx 1.236067977$, and $\alpha^2 \approx 1.527864045$. We note that $2\alpha + \alpha^2 \approx 4.000000000$, hence $x^2 + 2x - 4$ is a factor. The main difficulty is to efficiently compute the integer linear relation with relatively small coefficients for the high-precision big-float approximations of the powers of a root. Lenstra et al. [1982] solve this diophantine optimization problem by means of their now famous lattice reduction procedure, which is somewhat reminiscent of the ellipsoid method for linear programming.

The determination of linear integer relations among a set of real or complex numbers is a useful task in science in general. Very recently, some stunning identities could be produced by this method, including the following formula for π [Finch 1995]:

$$\pi = \sum_{n=0}^{\infty} \frac{1}{16^n} \left(\frac{4}{8n+1} - \frac{2}{8n+4} - \frac{1}{8n+5} - \frac{1}{8n+6} \right)$$

Even more surprising, the lattice reduction algorithm can prove that no linear integer relation with integers smaller than a chosen parameter exists among the real or complex numbers. There is an efficient alternative to the lattice reduction algorithm, originally due to Ferguson and Forcade [1982] and recently improved by Ferguson and Bailey.

The complexity of factoring an integer polynomial of degree n with coefficients of no more than l bits is thus a polynomial in n and l. From a theoretical point of view, an algorithm with a low estimate is by Miller [1992] and has a running time of $O(n^{5+o(1)}l^{1+o(1)} + n^{4+o(1)}l^{2+o(1)})$ bit operations. It is expected that the relation-finding methods will become usable in practice on hard-to-factor polynomials in the near future. If the hard-to-factor input polynomial is irreducible, an alternate approach can be used to prove its irreducibility. One finds an integer evaluation point at which the integral value of the polynomial has a large prime factor, and the irreducibility follows by mathematical theorems. Monagan [1992] has proven large hard-to-factor polynomials irreducible in this way, which would be hopeless by the lifting algorithm.

Coefficient fields other than finite fields and the rational numbers are of interest. Computing the factorizations of univariate polynomials over the complex numbers is the root finding problem described in the earlier section Approximating Polynomial Zeros. When the coefficient field has an extra variable, such as the field of fractions of polynomials (rational functions) the problem reduces, by an old theorem of Gauss, to factoring multivariate polynomials, which we discuss subsequently. When the coefficient field is the field of Laurent series in t with a finite segment of negative powers,

$$\frac{c_{-k}}{t^k} + \frac{c_{-k+1}}{t^{k-1}} + \cdots + \frac{c_{-1}}{t} + c_0 + c_1 t + c_2 t^2 + \cdots, \quad \text{where } k \geq 0$$

fast methods appeal to the theory of Puiseux series, which constitute the domain of algebraic functions [Walsh 1993].

Polynomials in Two Variables

Factoring bivariate polynomials by reduction to univariate factorization via homomorphic projection and subsequent lifting can be done similarly to the univariate algorithm [Musser 1975]. The second variable y takes the role of the prime integer p and $f(x, y) \bmod y = f(x, 0)$. Lifting is possible only if $f(x, 0)$ had no multiple root. Provided that $f(x, y)$ has no multiple factor, which can be ensured by a simple GCD computation, the *squarefreeness* of $f(x, 0)$ can be obtained by variable translation $\hat{y} = y + a$, where a is an easy-to find constant in the coefficient field. For certain domains, such as the rational numbers, any irreducible multivariate polynomial $h(x, y)$ can be mapped to an irreducible univariate polynomial $h(x, b)$ for some constant b. This is the important *Hilbert irreducibility theorem*, whose consequence is that the combinatorial explosion observed in the univariate lifting algorithm is, in practice, unlikely. However, the magnitude and probabilistic distribution of good points b is not completely analyzed.

For so-called non-Hilbertian coefficient fields good reduction is not possible. An important such field is the complex number. Clearly, all $f(x, b)$ completely split into linear factors, while $f(x, y)$ may be irreducible over the complex numbers. An example of an irreducible polynomial is $f(x, y) = x^2 - y^3$. Polynomials that remain irreducible over the complex numbers are called absolutely irreducible. An additional problem is the determination of the algebraic extension of the ground field in which the absolutely irreducible factors can be expressed. In the example

$$x^6 - 2x^3 y^2 + y^4 - 2x^3 = (x^3 - \sqrt{2}x - y^2) \cdot (x^3 + \sqrt{2}x - y^2)$$

the needed extension field is $\mathbb{Q}(\sqrt{2})$. The relation-finding approach proves successful for this problem. The root is computed as a Taylor series in y, and the integrality of the linear relation for the powers of the series means that the multipliers are polynomials in y of bounded degree. Several algorithms of polynomial-time complexity and pointers to the literature are found in Kaltofen [1995].

Bivariate polynomials constitute implicit representations of algebraic curves. It is an important operation in geometric modeling to convert from implicit to parametric representation. For example, the circle

$$x^2 + y^2 - 1 = 0$$

has the rational parameterization

$$x = \frac{2t}{1 + t^2}, \qquad y = \frac{1 - t^2}{1 + t^2}, \qquad \text{where } -\infty \le t \le \infty.$$

Algorithms are known that can find such rational parameterizations provided that they exist [Sendra and Winkler 1991]. It is crucial that the inputs to these algorithms are absolutely irreducible polynomials.

Polynomials in Many Variables

Polynomials in many variables, such as the symmetric Toeplitz determinant previously exhibited, are rarely given explicitly, due to the fact that the number of possible terms grows exponentially in the number of variables: there can be as many as $\binom{n+v}{n} \ge 2^{\min\{n, v\}}$ terms in a polynomial of degree n with v variables. Even the factors may be dense in canonical representation, but could be sparse in another basis: for instance, the polynomial

$$(x_1 - 1)(x_2 - 2) \cdots (x_v - v) + 1$$

has only two terms in the shifted basis, whereas it has 2^v terms in the power basis, i.e., in expanded format.

Randomized algorithms are available that can efficiently compute a factor of an implicitly given polynomial, say, a matrix determinant, and even can find a shifted basis with respect to which a factor would be sparse, provided, of course, that such a shift exists. The approach is by manipulating polynomials in so-called black box representations [Kaltofen and Trager 1990]: a black box is an object that takes as input a value for each variable, and then produces the value of the polynomial it represents at the specified point. In the Toeplitz example the representation of the determinant could be the Gaussian elimination program which computes it. We note that the size of the polynomial in this case would be nearly constant, only the variable names and the dimension need to be stored. The factorization algorithm then outputs procedures which will evaluate all irreducible factors at an arbitrary point (supplied as the input). These procedures make calls to the black box given as input to the factorization algorithm in order to evaluate them at certain points, which are derived from the point at which the procedures computing the values of the factors are probed. It is, of course, assumed that subsequent calls evaluate one and the same factor and not associates that are scalar multiples of one another. The algorithm by Kaltofen and Trager [1990] finds procedures that with a controllably high probability evaluate the factors correctly. Randomization is needed to avoid parasitic factorizations of homomorphic images which provide some static data for the factor boxes and cannot be avoided without mathematical conjecture. The procedures that evaluate the individual factors are deterministic.

Factors constructed as black box programs are much more space efficient than those represented in other formats, for example, the straight-line program format [Kaltofen 1989]. More importantly, once the black box representation for the factors is found, sparse representations can be rapidly computed by any of the new sparse interpolation algorithms. See Grigoriev and Lakshman [1995] for the latest method allowing shifted bases and pointers to the literature of other methods, including those for the standard power bases.

The black box representation of polynomials is normally not supported by commercial computer algebra systems such as Axiom, Maple, or Mathematica. Díaz is currently developing the FOXBOX system in C++ that makes black box methodology available to users of such systems. It is anticipated that factorizations as those of large symmetric Toeplitz determinants will be possible on computers. Earlier implementations based on the straight-line program model [Freeman et al. 1988] could factor 16×16 group determinants, which represent polynomials of over 300 million terms.

Defining Terms

Characteristic polynomial: A polynomial associated with a square matrix, the determinant of the matrix when a single variable is subtracted to its diagonal entries. The roots of the characteristic polynomial are the eigenvalues of the matrix.

Condition number: A scalar derived from a matrix that measures its relative nearness to a singular matrix. Very close to singular means a large condition number, in which case numeric inversion becomes an unstable process.

Degree order: An order of the terms in a multivariate polynomial; for two variables x and y with $x \prec y$ the ascending chain of terms is $1 \prec x \prec y \prec x^2 \prec xy \prec y^2 \cdots$.

Determinant: A polynomial in the entries of a square matrix with the property that its value is nonzero if and only if the matrix is invertible.

Lexicographic order: An order of the terms in a multivariate polynomial; for two variables x and y with $x \prec y$ the ascending chain of terms is $1 \prec x \prec x^2 \prec \cdots \prec y \prec xy \prec x^2 y \cdots \prec y^2 \prec xy^2 \cdots$.

Ops: Arithmetic operations, i.e., additions, subtractions, multiplications, or divisions; as in floating point operations (*flops*).

Singularity: A square matrix is singular if there is a nonzero second matrix such the the product of the two is the zero matrix. Singular matrices do not have inverses.

Sparse matrix: A matrix where many of the entries are zero.

Structured matrix: A matrix where each entry can be derived by a formula depending on few parameters. For instance, the Hilbert matrix has $1/(i + j - 1)$ as the entry in row i and column j.

References

Anderson, E. et al. 1992. *LAPACK Users' Guide.* SIAM Pub. Philadelphia, PA.

Aho, A., Hopcroft, J., and Ullman, J. 1974. *The Design and Analysis of Algorithms.* Addison–Wesley, Reading, MA.

Bareiss, E. H. 1968. Sylvester's identity and multistep integers preserving Gaussian elimination. *Math. Comp.* 22:565–578.

Becker, T. and Weispfenning, V. 1993. *Gröbner Bases: A Computational Approach to Commutative Algebra.* Springer–Verlag, New York.

Berlekamp, E. R. 1967. Factoring polynomials over finite fields. *Bell Systems Tech. J.* 46:1853–1859; rev. 1968. *Algebraic Coding Theory.* Chap. 6, McGraw–Hill, New York.

Berlekamp, E. R. 1970. Factoring polynomials over large finite fields. *Math. Comp.* 24:713–735.

Bini, D. and Pan, V. Y. 1991. Parallel complexity of tridiagonal symmetric eigenvalue problem. In *Proc. 2nd Annu. ACM-SIAM Symp. on Discrete Algorithms*, pp. 384–393. ACM Press, New York, SIAM Pub., 1994. Philadelphia, PA.

Bini, D. and Pan, V. Y. 1994. *Polynomial and Matrix Computations Vol. 1, Fundamental Algorithms*. Birkhäuser, Boston, MA.

Bini, D. and Pan, V. Y. 1996. *Polynomial and Matrix Computations, Vol. 2*. Birkhäuser, Boston, MA.

Borodin, A. and Munro, I. 1975. *Computational Complexity of Algebraic and Numeric Problems*. American Elsevier, New York.

Brown, W. S. 1971. On Euclid's algorithm and the computation of polynomial greatest common divisors. *J. ACM* 18:478–504.

Brown, W. S. and Traub, J. F. 1971. On Euclid's algorithm and the theory of subresultants. *J. ACM* 18:505–514.

Buchberger, B. 1965. *Ein Algorithmus zum Auffinden der Basiselemente des Restklassenringes nach einem nulldimensionalen Polynomideal*. Ph.D. dissertation. University of Innsbruck, Austria.

Buchberger, B. 1976. A theoretical basis for the reduction of polynomials to canonical form. *ACM SIGSAM Bull.* 10(3):19–29.

Buchberger, B. 1983. A note on the complexity of constructing Gröbner-bases. In *Proc. EUROCAL '83*, J. A. van Hulzen, ed. *Lecture Notes in Computer Science*, pp. 137–145. Springer.

Buchberger, B. 1985. Gröbner bases: an algorithmic method in polynomial ideal theory. In *Recent Trends in Multidimensional Systems Theory*, N. K. Bose, ed., pp. 184–232. D. Reidel, Dordrecht, Holland.

Cantor, D. G. 1989. On arithmetical algorithms over finite fields. *J. Combinatorial Theory, Serol. A* 50:285–300.

Canny, J. 1990. Generalized characteristic polynomials. *J. Symbolic Comput.* 9(3):241–250.

Canny, J. and Emiris, I. 1993a. An efficient algorithm for the sparse mixed resultant. In *Proc. AAECC-10*, G. Cohen, T. Mora, and O. Moreno, ed. Vol. 673, *Lecture Notes in Computer Science*, pp. 89–104. Springer.

Canny, J. and Emiris, I. 1993. A practical method for the sparse resultant. In *ISSAC '93, Proc. Internat. Symp. Symbolic Algebraic Comput.*, M. Bronstein, ed., pp. 183–192. ACM Press, New York.

Canny, J. and Emiris, I. 1994. Efficient incremental algorithms for the sparse resultant and the mixed volume. Tech. Rep., Univ. California-Berkeley, CA.

Canny, J., Kaltofen, E., and Lakshman, Y. 1989. Solving systems of non-linear polynomial equations faster. In *Proc. ACM-SIGSAM Internat. Symp. Symbolic Algebraic Comput.*, pp. 121–128.

Canny, J. and Manocha, D. 1991. Efficient techniques for multipolynomial resultant algorithms. In *ISSAC '91, Proc. Internat. Symp. Symbolic Algebraic Comput.*, S. M. Watt, ed., pp. 85–95, ACM Press, New York.

Cantor, D. G. and Zassenhaus, H. 1981. A new algorithm for factoring polynomials over finite fields. *Math. Comp.* 36:587–592.

Cayley, A. 1865. On the theory of eliminaton. *Cambridge and Dublin Math. J.* 3:210–270.

Chionh, E. 1990. *Base Points, Resultants and Implicit Representation of Rational Surfaces*. Ph.D. dissertation. Department of Computer Science, University of Waterloo, Waterloo, Canada.

Collins, G. E. 1967. Subresultants and reduced polynomial remainder sequences. *J. ACM* 14:128–142.

Collins, G. E. 1971. The calculation of multivariate polynomial resultants. *J. ACM* 18:515–532.

Cuppen, J. J. M. 1981. A divide and conquer method for the symmetric tridiagonal eigenproblem. *Numer. Math.* 36:177–195.

Davenport, J. H., Tournier, E., and Siret, Y. 1988. *Computer Algebra Systems and Algorithms for Algebraic Computation*. Academic Press, London.

Díaz, A. and Kaltofen, E. 1995. On computing greatest common divisors with polynomials given by black boxes for their evaluation. In *ISSAC '95 Proc. 1995 Internat. Symp. Symbolic Algebraic Comput.*, A. H. M. Levelt, ed., pp. 232–239, ACM Press, New York.

Dixon, A. L. 1908. The elimination of three quantics in two independent variables. In *Proc. London Math. Soc.* Vol. 6, pp. 468–478.

Dongarra, J. et al. 1978. *LAPACK Users' Guide*. SIAM Pub., Philadelphia, PA.

Faugère, J. C., Gianni, P., Lazard, D., and Mora, T. 1993. Efficient computation of zero-dimensional Gröbner bases by change of ordering. *J. Symbolic Comput.* 16(4):329–344.

Ferguson, H. R. P. and Forcade, R. W. 1982. Multidimensional Euclidean algorithms. *J. Reine Angew. Math.* 334:171–181.

Finch, S. 1995. The miraculous Bailey–Borwein–Plouffe pi algorithm. Internet document, Mathsoft Inc., http://www.mathsoft.com/asolve/plouffe/plouffe.html, Oct.

Foster, L. V. 1981. Generalizations of Laguerre's method: higher order methods. *SIAM J. Numer. Anal.* 18:1004–1018.

Freeman, T. S., Imirzian, G., Kaltofen, E., and Lakshman, Y. 1988. Dagwood: a system for manipulating polynomials given by straight-line programs. *ACM Trans. Math. Software* 14(3):218–240.

Garbow, B. S. et al. 1972. *Matrix Eigensystem Routines: EISPACK Guide Extension.* Springer, New York.

Geddes, K. O., Czapor, S. R., and Labahn, G. 1992. *Algorithms for Computer Algebra.* Kluwer Academic.

Gelfand, I. M., Kapranov, M. M., and Zelevinsky, A. V. 1994. *Discriminants, Resultants and Multidimensional Determinants.* Birkhäuser Verlag, Boston, MA.

George, A. and Liu, J. W.-H. 1981. *Computer Solution of Large Sparse Positive Definite Linear Systems.* Prentice–Hall, Englewood Cliffs, NJ.

Gianni, P. and Trager, B. 1985. GCD's and factoring polynomials using Gröbner bases. *Proc. EUROCAL '85*, Vol. 2, *Lecture Notes in Computer Science*, 204, pp. 409–410.

Giesbrecht, M. 1995. Nearly optimal algorithms for canonical matrix forms. *SIAM J. Comput.* 24(5):948–969.

Gilbert, J. R. and Tarjan, R. E. 1987. The analysis of a nested dissection algorithm. *Numer. Math.* 50:377–404.

Golub, G. H. and Van Loan, C. F. 1989. *Matrix Computations.* Johns Hopkins Univ. Press, Baltimore, MD.

Gondran, M. and Minoux, M. 1984. *Graphs and Algorithms.* Wiley–Interscience, New York.

Grigoriev, D. Y. and Lakshman, Y. N. 1995. Algorithms for computing sparse shifts for multivariate polynomials. In *ISSAC '95 Proc. 1995 Internat. Symp. Symbolic Algebraic Comput.*, A. H. M. Levelt, ed., pp. 96–103, ACM Press, New York.

Hansen, E., Patrick, M., and Rusnack, J. 1977. Some modifications of Laguerre's method. *BIT* 17:409–417.

Heath, M. T., Ng, E., and Peyton, B. W. 1991. Parallel algorithms for sparse linear systems. *SIAM Rev.* 33:420–460.

Jenkins, M. A., and Traub, J. F. 1970. A three-stage variable-shift iteration for polynomial zeros and its relation to generalized Rayleigh iteration. *Numer. Math.* 14:252–263.

Kaltofen, E. 1982. Polynomial factorization. In 2nd ed. *Computer Algebra*, B. Buchberger, G. Collins, and R. Loos, eds., pp. 95–113. Springer–Verlag, Vienna.

Kaltofen, E. 1988. Greatest common divisors of polynomials given by straight-line programs. *J. ACM* 35(1):231–264.

Kaltofen, E. 1989. Factorization of polynomials given by straight-line programs. In *Randomness and Computation*, S. Micali, ed. Vol. 5 of Advances in computing research, pp. 375–412. JAI Press, Greenwhich, CT.

Kaltofen, E. 1990. Polynomial factorization 1982–1986. 1990. In *Computers in Mathematics*, D. V. Chudnovsky and R. D. Jenks, eds. Vol. 125, *Lecture Notes in Pure and Applied Mathematics*, pp. 285–309. Marcel Dekker, New York.

Kaltofen, E. 1992. Polynomial factorization 1987–1991. In *Proc. LATIN '92*, I. Simon, ed. Vol. 583, *Lecture Notes in Computer Science*, pp. 294–313.

Kaltofen, E. 1995. Effective Noether irreducibility forms and applications. *J. Comput. Syst. Sci.* 50(2):274–295.

Kaltofen, E., Krishnamoorthy, M. S., and Saunders, B. D. 1990. Parallel algorithms for matrix normal forms. *Linear Algebra Appl.* 136:189–208.

Kaltofen, E. and Lakshman, Y. 1988. Improved sparse multivariate polynomial interpolation algorithms. *Proc. ISSAC '88*, Vol. 358, *Lecture Notes in Computer Science*, pp. 467–474.

Kaltofen, E. and Lobo, A. 1994. Factoring high-degree polynomials by the black box Berlekamp algorithm. In *ISSAC '94, Proc. Internat. Symp. Symbolic Algebraic Comput.*, J. von zur Gathen and M. Giesbrecht, eds., pp. 90–98, ACM Press, New York.

Kaltofen, E., Musser, D. R., and Saunders, B. D. 1983. A generalized class of polynomials that are hard to factor. *SIAM J. Comp.* 12(3):473–485.

Kaltofen, E. and Pan, V. 1991. Processor efficient parallel solution of linear systems over an abstract field. In *Proc. 3rd Ann. ACM Symp. Parallel Algor. Architecture*, pp. 180–191, ACM Press, New York.

Kaltofen, E. and Pan, V. 1992. Processor-efficient parallel solution of linear systems II: the positive characteristic and singular cases. In *Proc. 33rd Annual Symp. Foundations of Comp. Sci.*, pp. 714–723, Los Alamitos, CA. IEEE Computer Society Press.

Kaltofen, E. and Shoup, V. 1995. Subquadratic-time factoring of polynomials over finite fields. In *Proc. 27th Annual ACM Symp. Theory Comp.*, pp. 398–406, ACM Press, New York.

Kaltofen, E. and Trager, B. 1990. Computing with polynomials given by black boxes for their evaluations: greatest common divisors, factorization, separation of numerators and denominators. *J. Symbolic Comput.* 9(3):301–320.

Kapur, D. 1986. Geometry theorem proving using Hilbert's nullstellensatz. *J. Symbolic Comp.* 2:399–408.

Kapur, D. and Lakshman, Y. N. 1992. Elimination methods: an introduction. In *Symbolic and Numerical Computation for Artificial Intelligence*. B. Donald, D. Kapur, and J. Mundy, eds. Academic Press.

Kapur, D. and Saxena, T. 1995. Comparison of various multivariate resultant formulations. In *Proc. Internat. Symp. Symbolic Algebraic Comput. ISSAC '95*, A. H. M. Levelt, ed., pp. 187–195, ACM Press, New York.

Kapur, D., Saxena, T., and Yang, L. 1994. Algebraic and geometric reasoning using Dixon resultants. In *ISSAC '94, Proc. Internat. Symp. Symbolic Algebraic Comput.* J. von zur Gathen and M. Giesbrecht, ed., pp. 99–107, ACM Press, New York.

Knuth, D. E. 1981. *The Art of Computer Programming, Vol. 2, Seminumerical Algorithms*, 2nd ed. Addison–Wesley, Reading, MA.

Lakshman, Y. N. 1990. *On the complexity of computing Gröbner bases for zero dimensional polynomia*. Ph.D. thesis, Dept. Comput. Sci., Rensselaer Polytechnic Inst. Troy, NY, Dec.

Lakshman, Y. N. and Saunders, B. D. 1994. On computing sparse shifts for univariate polynomials. In *ISSAC '94, Proc. Internat. Symp. Symbolic Algebraic Comput.*, J. von zur Gathen and M. Giesbrecht, eds., pp. 108–113, ACM Press, New York.

Lakshman, Y. N. and Saunders, B. D. 1995. Sparse polynomial interpolation in non-standard bases. *SIAM J. Comput.* 24(2):387–397.

Lazard, D. 1981. Résolution des systèmes d'équation algébriques. *Theoretical Comput. Sci.* 15:77–110. (In French).

Lenstra, A. K., Lenstra, H. W., and Lovász, L. 1982. Factoring polynomials with rational coefficients. *Math. Ann.* 261:515–534.

Leyland, P. 1995. Cunningham project data. Internet document, Oxford Univ., ftp://sable.ox.ac.uk/pub/math/cunningham/, November.

Lipton, R. J., Rose, D., and Tarjan, R. E. 1979. Generalized nested dissection. *SIAM J. on Numer. Analysis* 16(2):346–358.

Macaulay, F. S. 1916. Algebraic theory of modular systems. *Cambridge Tracts* 19, Cambridge.

MacWilliams, F. J. and Sloan, N. J. A. 1977. *The Theory of Error-Correcting Codes*. North–Holland, New York.

Madsen, K. 1973. A root-finding algorithm based on Newton's method. *BIT* 13:71–75.

McCormick, S., ed. 1987. *Multigrid Methods*. SIAM Pub., Philadelphia, PA.

McNamee, J. M. 1993. A bibliography on roots of polynomials. *J. Comput. Appl. Math.* 47(3):391–394.

Miller, V. 1992. Factoring polynomials via relation-finding. In *Proc. ISTCS '92*, D. Dolev, Z. Galil, and M. Rodeh, eds. Vol. 601, *Lecture Notes in Computer Science*, pp. 115–121.

Monagan, M. B. 1992. A heuristic irreducibility test for univariate polynomials. *J. Symbolic Comput.* 13(1):47–57.

Musser, D. R. 1975. Multivariate polynomial factorization. *J. ACM* 22:291–308.

Niederreiter, H. 1994. New deterministic factorization algorithms for polynomials over finite fields. In *Finite Fields: Theory, Applications and Algorithms*, L. Mullen and P. J.-S. Shiue, eds. Vol. 168, Contemporary mathematics, pp. 251–268, Amer. Math. Soc., Providence, RI.

Ortega, J. M., and Voight, R. G. 1985. Solution of partial differential equations on vector and parallel computers. *SIAM Rev.* 27(2):149–240.

Pan, V. Y. 1984a. How can we speed up matrix multiplication? *SIAM Rev.* 26(3):393–415.

Pan, V. Y. 1984b. How to multiply matrices faster. *Lecture Notes in Computer Science,* 179.

Pan, V. Y. 1987. Sequential and parallel complexity of approximate evaluation of polynomial zeros. *Comput. Math. (with Appls.),* 14(8):591–622.

Pan, V. Y. 1991. Complexity of algorithms for linear systems of equations. In *Computer Algorithms for Solving Linear Algebraic Equations (State of the Art),* E. Spedicato, ed. Vol. 77 of *NATO ASI Series,* Series F: computer and systems sciences, pp. 27–56, Springer–Verlag, Berlin.

Pan, V. Y. 1992a. Complexity of computations with matrices and polynomials. *SIAM Rev.* 34(2):225–262.

Pan, V. Y. 1992b. Linear systems of algebraic equations. In *Encyclopedia of Physical Sciences and Technology* 2nd ed. Marvin Yelles, ed. Vol. 8, pp. 779–804, 1987. 1st ed. Vol. 7, pp. 304–329.

Pan, V. Y. 1993. Parallel solution of sparse linear and path systems. In *Synthesis of Parallel Algorithms,* J. H. Reif, ed. Ch. 14, pp. 621–678. Morgan Kaufmann, San Mateo, CA.

Pan, V. Y. 1994a. Improved parallel solution of a triangular linear system. *Comput. Math. (with Appl.),* 27(11):41–43.

Pan, V. Y. 1994b. *On approximating polynomial zeros: modified quadtree construction and improved Newton's iteration.* Manuscript, Lehman College, CUNY, Bronx, New York.

Pan, V. Y. 1995a. Parallel computation of a Krylov matrix for a sparse and structured input. *Math. Comput. Modelling* 21(11):97–99.

Pan, V. Y. 1995b. *Solving a polynomial equation: some history and recent progress.* Manuscript, Lehman College, CUNY, Bronx, New York.

Pan, V. Y. 1996. Optimal and nearly optimal algorithms for approximating polynomial zeros. *Comput. Math. (with Appl.).*

Pan, V. Y. and Preparata, F. P. 1995. Work-preserving speed-up of parallel matrix computations. *SIAM J. Comput.* 24(4):811–821.

Pan, V. Y. and Reif, J. H. 1992. Compact multigrid. *SIAM J. Sci. Stat. Comput.* 13(1):119–127.

Pan, V. Y. and Reif, J. H. 1993. Fast and efficient parallel solution of sparse linear systems. *SIAM J. Comp.,* 22(6):1227–1250.

Pan, V. Y., Sobze, I. and Atinkpahoun, A. 1995. On parallel computations with band matrices. *Inf. and Comput.* 120(2):227–250.

Parlett, B. 1980. *Symmetric Eigenvalue Problem.* Prentice–Hall, Englewood Cliffs, NJ.

Quinn, M. J. 1994. *Parallel Computing: Theory and Practice.* McGraw–Hill, New York.

Rabin, M. O. 1980. Probabilistic algorithms in finite fields. *SIAM J. Comp.* 9:273–280.

Renegar, J. 1989. On the worst case arithmetic complexity of approximating zeros of systems of polynomials. *SIAM J. Comput.* 18(2):350–370.

Ritt, J. F. 1950. *Differential Algebra.* AMS, New York.

Saad, Y. 1992. *Numerical Methods for Large Eigenvalue Problems: Theory and Algorithms.* Manchester Univ. Press, U.K., Wiley, New York. 1992.

Saad, Y. 1995. *Iterative Methods for Sparse Linear Systems.* PWS Kent, Boston, MA.

Sendra, J. R. and Winkler, F. 1991. Symbolic parameterization of curves. *J. Symbolic Comput.* 12(6): 607–631.

Shoup, V. 1996. A new polynomial factorization algorithm and its implementation. *J. Symbolic Comput.*

Smith, B. T. et al. 1970. *Matrix Eigensystem Routines: EISPACK Guide,* 2nd ed. Springer, New York.

St. Schwarz, 1956. On the reducibility of polynomials over a finite field. *Quart. J. Math. Oxford Ser. (2),* 7:110–124.

Stederberg, T. and Goldman, R. 1986. Algebraic geometry for computer-aided design. *IEEE Comput. Graphics Appl.* 6(6):52–59.

Sturmfels, B. 1991. Sparse elimination theory. In *Proc. Computat. Algebraic Geom. and Commut. Algebra,* D. Eisenbud and L. Robbiano, eds. Cortona, Italy, June.

Sturmfels, B. and Zelevinsky, A. 1992. Multigraded resultants of the Sylvester type. *J. Algebra.*

von zur Gathen, J. and Shoup, V. 1992. Computing Frobenius maps and factoring polynomials. *Comput. Complexity* 2:187–224.

Walsh, P. G. 1993. *The computation of Puiseux expansions and a quantitative version of Runge's theorem on diophantine equations.* Ph.D. dissertation. University of Waterloo, Waterloo, Canada.

Watkins, D. S. 1982. Understanding the QR algorithm. *SIAM Rev.* 24:427–440.

Watkins, D. S. 1991. Some perspectives on the eigenvalue problem. *SIAM Rev.* 35(3):430–471.

Wilkinson, J. H. 1965. *The Algebraic Eigenvalue Problem.* Clarendon Press, Oxford, England.

Winkler, F. 1996. *Introduction to Computer Algebra.* Springer–Verlag, Heidelberg, Germany.

Wu, W. 1984. Basis principles of mechanical theorem proving in elementary geometries. *J. Syst. Sci. Math Sci.* 4(3):207–235.

Wu, W. 1986. Basis principles of mechanical theorem proving in elementary geometries. *J. Automated Reasoning* 2:219–252.

Zassenhaus, H. 1969. On Hensel factorization I. *J. Number Theory* 1:291–311.

Zippel, R. 1993. *Effective Polynomial Computations*, p. 384. Kluwer Academic, Boston, MA.

Further Information

The books by Knuth [1981], Davenport et al. [1988], Geddes et al. [1992], and Zippel [1993] provide a much broader introduction to the general subject. There are well-known libraries and packages of subroutines for the most popular numerical matrix computations, in particular, Dongarra et al. [1978] for solving linear systems of equations, Smith et al. [1970] and Garbow et al. [1972] approximating matrix eigenvalues, and Anderson et al. [1992] for both of the two latter computational problems. There is a comprehensive treatment of numerical matrix computations [Golub and Van Loan 1989], with extensive bibliography, and there are several more specialized books on them [George and Liu 1981, Wilkinson 1965, Parlett 1980, Saad 1992, 1995], as well as many survey articles [Heath et al. 1991, Watkins 1991, Ortega and Voight 1985, Pan 1992b] and thousands of research articles.

Special (more efficient) parallel algorithms have been devised for special classes of matrices, such as sparse [Pan and Reif 1993, Pan 1993], banded [Pan et al. 1995], and dense structured [Bini and Pan (cf. [1994])]. We also refer to Pan and Preparata [1995] on a simple but surprisingly effective extension of Brent's principle for improving the processor and work efficiency of parallel matrix algorithms and to Golub and Van Loan [1989], Ortega and Voight [1985], and Heath et al. [1991] on practical parallel algorithms for matrix computations.

11

Complexity Theory

Michael C. Loui*
University of Illinois
at Urbana-Champaign

11.1 Introduction

Computational complexity is the study of the difficulty of solving computational problems, in terms of the required computational resources, such as time and space (memory). The formal theoretical study of computational complexity began with the paper of Hartmanis and Stearns [1965], who introduced the basic concepts and proved the first results. For their achievement, Hartmanis and Stearns received the 1993 Turing Award of the Association for Computing Machinery (ACM).

Whereas the analysis of algorithms focuses on the time or space required by an algorithm to solve a specific computational problem, such as sorting, complexity theory focuses on the **complexity class**, which consists of all problems solvable within the same amount of time or space. Two important complexity classes are P, the set of problems that can be solved in polynomial time, and NP, the set of problems whose solutions can be verified in polynomial time. Complexity theorists have discovered that most common computational problems fall into a small number of complexity classes.

*Supported by the National Science Foundation under Grant CCR-9315696.

0-8493-2909-4/97/$0.00+$.50
© 1997 by CRC Press, Inc.

By quantifying the resources required to solve a problem, complexity theory has profoundly affected our thinking about computation. Computability theory establishes the existence of undecidable problems, which cannot be solved in principle, regardless of the amount of time invested. In contrast, complexity theory establishes the existence of decidable problems that, although solvable in principle, cannot be solved in practice, because the time and space required would be larger than the age and size of the known universe [Stockmeyer and Chandra 1979]. Thus, complexity theory characterizes the computationally feasible problems.

The quest for the boundaries of the set of feasible problems has led to the most important unsolved question in all of computer science: Is P different from NP? Hundreds of fundamental problems, including many ubiquitous optimization problems of operations research, are **NP-complete**; they are the hardest problems in NP. If someone could find a polynomial-time algorithm for any one NP-complete problem, then there would be polynomial-time algorithms for all of them. Despite the concerted efforts of many scientists over several decades, no polynomial-time algorithm has been found for any NP-complete problem. Although we do not yet know whether P is different from NP, showing that a problem is NP-complete provides strong evidence that the problem is computationally infeasible and justifies the use of heuristics for solving the problem.

In this chapter, we define P, NP, and related complexity classes. We illustrate the use of **diagonalization** and **padding** techniques to prove relationships between classes. Next, we define NP-completeness, and we show how to prove that a problem is NP-complete. Finally, we define complexity classes for probabilistic and interactive computations.

Throughout this chapter, all numeric functions take integer arguments and produce integer values. All logarithms are taken to base 2. In particular, $\log n$ means $\lceil \log_2 n \rceil$.

11.2 Models of Computation

To develop a theory of the difficulty of computational problems, we need to specify precisely what a problem is, what an algorithm is, and what a measure of difficulty is. For simplicity, complexity theorists have chosen to represent problems as languages, to model algorithms by off-line multitape **Turing machines**, and to measure computational difficulty by the time and space required by a Turing machine. To justify these choices, some theorems of complexity theory show how to translate statements about, say, the time complexity of language recognition by Turing machines into statements about computational problems on more realistic models of computation. These theorems imply that the principles of complexity theory are not artifacts of Turing machines, but intrinsic properties of computation.

This section defines different kinds of Turing machines. The deterministic Turing machine models actual computers. The nondeterministic Turing machine is not a realistic model, but it helps classify the complexity of important computational problems. The alternating Turing machine models a form of parallel computation, and it helps elucidate the relationship between time and space.

Computational Problems and Languages

Computer scientists have invented many elegant formalisms for representing data and control structures. Fundamentally, all representations are patterns of symbols. Therefore, we represent an instance of a computational problem as a sequence of symbols. (See Chap. 4 on formal models and computability.)

Let Σ be a finite set, called the *alphabet*. A *word* over Σ is a finite sequence of symbols from Σ. Sometimes a word is called a *string*. Let Σ^* denote the set of all words over Σ. For example, if $\Sigma = \{0,1\}$, then

$$\Sigma^* = \{ \ \lambda, \ 0, \ 1, \ 00, \ 01, \ 10, \ 11, \ 000, \ \ldots \}$$

is the set of all binary words, including the empty word λ. (Binary representations of numbers and data, such as ASCII, are pervasive in computing.) The *length* of a word w, denoted $|w|$, is the number of symbols in w. A *language* over Σ is a subset of Σ^*.

A *decision problem* is a computational problem whose answer is simply yes or no. For example: Is the input graph connected? or Is the input a sorted list of integers? A decision problem can be expressed as a membership problem for a language L: for an input x, does x belong to L? For a language L that represents connected graphs, the input word x might represent an input graph G, and $x \in L$ if and only if G is connected.

For every decision problem, the representation should allow for easy parsing, to determine whether a word represents a legitimate instance of the problem. Furthermore, the representation should be concise. In particular, it would be unfair to encode the answer to the problem into the representation of an instance

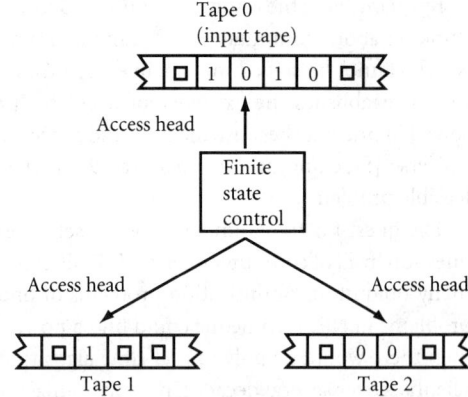

FIGURE 11.1   A 2-tape Turing machine.

of the problem; for example, for the problem of deciding whether an input graph is connected, the representation should not have an extra bit that tells whether the graph is connected. A set of integers $S = \{x_1, \ldots, x_m\}$ is represented by listing the binary representation of each x_i, with the representations of consecutive integers in S separated by a nonbinary symbol. A graph is naturally represented by giving either its adjacency matrix or a set of adjacency lists, where the list for each vertex v specifies the vertices adjacent to v.

Whereas the solution to a decision problem is yes or no, the solution to an optimization problem is more complicated; for example, determine the shortest path from vertex u to vertex v in an input graph G. Nevertheless, for every optimization (minimization) problem, with objective function g, there is a corresponding decision problem that asks whether there exists a feasible solution z such that $g(z) \le k$, where k is a given target value. Clearly, if there is an algorithm that solves an optimization problem, then that algorithm can be used to solve the corresponding decision problem. Conversely, if algorithm A solves the decision problem, then with a binary search on the range of values of g, we can determine the optimal value. A fortiori, for many optimization problems, with multiple calls to the decision algorithm A, we can even construct an optimal solution. Therefore, there is little loss of generality in considering only decision problems, represented as language membership problems.

Turing Machines

This subsection and the next three give precise, formal definitions of Turing machines and their variants. These subsections are intended for reference. For the rest of this chapter, from section 11.3 on, the reader need not understand these definitions in detail, but may generally substitute program or computer for each reference to Turing machine.

A k-tape **Turing machine** M consists of the following:

- A finite set of states Q, with special states q_0 (initial state), q_A (accept state), and q_R (reject state).
- A finite alphabet Σ, and a special blank symbol $\square \notin \Sigma$.
- The $k + 1$ linear tapes, each divided into cells. Tape 0 is the input tape, and tapes $1, \ldots, k$ are the worktapes. Each tape is infinite to the left and to the right. Each cell holds a single symbol from $\Sigma \cup \{\square\}$. By convention, the input tape is read only. Each tape has an access head, and at every instant, each access head scans one cell. See Fig. 11.1
- A finite transition table δ, which comprises tuples of the form

$$(q, s_0, s_1, \ldots, s_k, q', s'_1, \ldots, s'_k, d_0, d_1, \ldots, d_k)$$

where $q, q' \in Q$, each $s_i, s'_i \in \Sigma \cup \{\square\}$, and each $d_i \in \{-1, 0, +1\}$.

A tuple specifies a step of M: if the current state is q, and $s_0, s_1, \ldots, s_k$ are the symbols in the cells scanned by the access heads, then M replaces s_i by s_i' for $i = 1, \ldots, k$ simultaneously, changes state to q', and moves the head on tape i one cell to the left ($d_i = -1$) or right ($d_i = +1$) or not at all ($d_i = 0$) for $i = 0, \ldots, k$. Note that M cannot write on tape 0, that is, M may write only on the worktapes, not on the input tape.

- In a tuple, no s_i' can be the blank symbol $\square$. Since M may not write a blank, the worktape cells that its access heads previously visited are nonblank.

- No tuple contains q_A or q_R as its first component. Thus, once M enters state q_A or state q_R, it stops.

- Initially, M is in state q_0, an input word in Σ^* is inscribed on contiguous cells of the input tape, the access head on the input tape is on the leftmost symbol of the input word, and all other cells of all tapes contain the blank symbol $\square$.

The Turing machine M that we have defined is *nondeterministic*: δ may have several tuples with the same combination of state q and symbols $s_0, s_1, \ldots, s_k$ as the first $k + 2$ components, so that M may have several possible next steps. A machine M is *deterministic* if for every combination of state q and symbols $s_0, s_1, \ldots, s_k$, at most one tuple in δ contains the combination as its first $k + 2$ components. A deterministic machine always has at most one possible next step.

A *configuration* of a Turing machine M specifies a state, contents of all tapes, and positions of all access heads.

A *computation path* is a sequence of configurations $C_0, C_1, \ldots, C_t, \ldots$, where C_0 is the initial configuration of M, and each C_{j+1} follows from C_j in one step according to the changes specified by a tuple in δ. If no tuple is applicable to C_t, then C_t is *terminal*, and the computation path is *halting*. If M has no infinite computation paths, then M *always halts*. Without loss of generality, we may assume that in every terminal configuration, the state is q_A or q_R.

A halting computation path is *accepting* if the state in the last configuration C_t is q_A; it is *rejecting* if the state in C_t is q_R.

M *accepts* an input word x if there exists an accepting computation path that starts from the initial configuration in which x is on the input tape. M *rejects* x if the only halting computation paths are rejecting; a machine may reject x by failing to halt. If M is deterministic, then there is at most one halting computation path, hence at most one accepting path.

The *language accepted by* M, written $\mathcal{L}(M)$, is the set of words accepted by M. If $L = \mathcal{L}(M)$, and M always halts, then M *decides* L.

Generally, in this chapter, we consider only Turing machines that always halt. When we define complexity classes, we consider only time- and space-constructible functions (see Constructibility in section 11.3) as bounds, and Turing machines that accept within those time or space bounds can be converted into machines that always halt.

In addition to deciding languages, deterministic Turing machines can compute functions. Designate tape 1 to be the *output tape*. If M halts on input word x, then the nonblank word on tape 1 in the final configuration is the output of M. A function f is *total recursive* if there exists a deterministic Turing machine M that always halts such that for each input word x, the output of M is the value of $f(x)$.

Almost all results in complexity theory are insensitive to minor variations in the underlying computational models. For example, we could have chosen Turing machines whose tapes are restricted to be only one-way infinite or whose alphabet is restricted to $\{0,1\}$. It is straightforward to simulate a Turing machine as defined by one of these restricted Turing machines, one step at a time: each step of the original machine can be simulated by $O(1)$ steps of the restricted machine. The simulation of one Turing machine by a Turing machine with a different structure is analogous to the implementation of virtual machines in multilayer computer systems. (See section VIII of this Handbook, on operating systems and networks.)

Universal Turing Machines

A k-tape *universal Turing machine* is a Turing machine U with two input tapes and k worktapes that interprets the word i on one input tape as an encoding of a k-tape Turing machine M_i and the word x on the other input tape as the input to M_i. Machine U simulates the operation of M_i on x, consulting the transition table encoded in i to determine each next step. (See Chap. 4, on formal models and computability, for a universal GOTO program.)

Each symbol in the alphabet of M_i is encoded by $O(1)$ symbols of the alphabet of U. Thus, to simulate the writing of a cell on a worktape of M_i, machine U writes on $O(1)$ of its cells on the corresponding worktape. Overall, provided that the encoding i is reasonable, U takes $O(|i|)$ steps to simulate each step of M_i.

We can think of U with a fixed i as a machine U_i and define $\mathcal{L}(U_i) = \{x : U \text{ accepts } \langle i, x \rangle\}$. Then $\mathcal{L}(U_i) = \mathcal{L}(M_i)$.

Alternating Turing Machines

By definition, a nondeterministic Turing machine M accepts its input word x if there exists an accepting computation path, starting from the initial configuration with x on the input tape. Let us call a *configuration* C accepting if there is a computation path of M that starts in C and ends in a configuration whose state is q_A. Equivalently, a configuration C is accepting if either the state in C is q_A or there exists an accepting configuration C' reachable from C by one step of M. Then M accepts x if the initial configuration with input word x is accepting.

The *alternating Turing machine* generalizes this notion of acceptance. In an alternating Turing machine M, each state is labeled either existential or universal. (Do not confuse the universal state in an alternating Turing machine with the universal Turing machine.) A nonterminal configuration C is existential (respectively, universal) if the state in C is labeled existential (universal). A terminal configuration is accepting if its state is q_A. A nonterminal existential configuration C is accepting if there exists an accepting configuration C' reachable from C by one step of M. A nonterminal universal configuration C is accepting if for every configuration C' reachable from C by one step of M, the configuration C' is accepting. Finally, M accepts x if the initial configuration with input word x is an accepting configuration.

A nondeterministic Turing machine is a special case of an alternating Turing machine in which every state is existential.

The computation of an alternating Turing machine M alternates between existential states and universal states. Intuitively, from an existential configuration, M guesses a step that leads toward acceptance; from a universal configuration, M checks whether each possible next step leads toward acceptance—in a sense, M checks all possible choices in parallel. An alternating computation captures the essence of a two-player game: player 1 has a winning strategy if there exists a move for player 1 such that for every move by player 2, there exists a subsequent move by player 1, etc., such that player 1 eventually wins.

Oracle Turing Machines

Some computational problems remain difficult even when solutions to instances of a particular, different decision problem are available for free. When we study the complexity of a problem *relative* to a language L, we assume that answers about membership in L have been precomputed and stored in a (possibly infinite) table and that there is no cost to obtain an answer to a membership query: Is w in L? The language L is called an **oracle**. Conceptually, an algorithm queries the oracle whether a word w is in L, and it receives the correct answer.

An *oracle Turing machine* is a Turing machine M with a special *oracle tape* and special states QUERY, YES, and NO. The computation of the oracle Turing machine M^L, with oracle language L, is the same as that of an ordinary Turing machine, except that when M enters the QUERY state with a word w on the oracle tape, in one step, M enters either the YES state if $w \in L$ or the NO state if $w \notin L$. Furthermore, during this step, the oracle tape is erased, so that the time for setting up each query is accounted separately.

11.3 Resources and Complexity Classes

In this section, we define the measures of difficulty of solving computational problems: time and space. We introduce complexity classes, which enable us to classify problems according to the difficulty of their solution.

Time and Space

We measure the difficulty of a computational problem by the running time and the space (memory) requirements of an algorithm that solves the problem. Clearly, in general, a finite algorithm cannot have a table of all answers to infinitely many instances of the problem, although an algorithm could look up precomputed answers to a finite number of instances; in terms of Turing machines, the finite answer table is built into the set of states and the transition table. For these instances, the running time is negligible—just the time needed to read the input word. Consequently, our complexity measure should consider a whole problem, not only specific instances.

We express the complexity of a problem, in terms of the growth of the required time or space, as a function of the length n of the input word that encodes a problem instance. As in Chap. 5, on algorithm analysis, we consider the worst-case complexity, that is, for each n, the maximum time or space required among all inputs of length n.

The *time* taken by a Turing machine M on input word x, denoted $\text{Time}_M(x)$, is defined as follows:

- If M accepts x, then $\text{Time}_M(x)$ is the number of steps in the shortest accepting computation path for x.
- If M rejects x, then $\text{Time}_M(x)$ is the number of steps in the longest computation path for x or $+\infty$ if M has an infinite computation path for x.

For a deterministic machine M, for every input x, there is at most one halting computation path, and $\text{Time}_M(x)$ is unambiguous. For a nondeterministic machine M, if $x \in \mathcal{L}(M)$, then M can guess the correct steps to take toward an accepting configuration, and $\text{Time}_M(x)$ measures the length of the path on which M always makes the best guess.

The *space* used by a Turing machine M on input x, denoted $\text{Space}_M(x)$, is defined as follows. The space used by a halting computation path is the number of nonblank worktape cells in the last configuration; this is the number of different cells ever written by the worktape heads of M during the computation path, since M never writes the blank symbol. Because the space occupied by the input word is not counted, a machine can use a sublinear ($o(n)$) amount of space.

- If M accepts x, then $\text{Space}_M(x)$ is the minimum space used among all accepting computation paths for x.
- If M rejects x, then $\text{Space}_M(x)$ is the maximum space used among all computation paths for x, possibly $+\infty$.

The **time complexity** of a machine M is the function

$$t(n) = \max\{\text{Time}_M(x) : |x| = n\}$$

We assume that M reads all of its input word, and the blank symbol after the right end of the input word, so $t(n) \geq n + 1$. The **space complexity** of M is the function

$$s(n) = \max\{\text{Space}_M(x) : |x| = n\}$$

Because few interesting languages can be decided by machines of sublogarithmic space complexity, we henceforth assume that $s(n) \geq \log n$.

A function $f(x)$ is computable in polynomial time if there exists a deterministic Turing machine M of polynomial time complexity such that for each input word x, the output of M is $f(x)$.

Constructibility

A function $t(n)$ is **time constructible** if there exists a deterministic Turing machine that halts after exactly $t(n)$ steps for every input of length n. A function $s(n)$ is **space constructible** if there exists a 1-tape deterministic Turing machine that uses exactly $s(n)$ worktape cells for every input of length n.

For example, $t(n) = n + 1$ is time constructible. Furthermore, if $t_1(n)$ and $t_2(n)$ are time constructible, then so are the functions $t_1 + t_2$, $t_1 t_2$, $t_1^{t_2}$, and c^{t_1} for every integer $c > 1$. Consequently, if $p(n)$ is a polynomial, then $p(n) = \Theta(t(n))$ for some time-constructible polynomial function $t(n)$. Similarly, $s(n) = \log n$ is space constructible, and if $s_1(n)$ and $s_2(n)$ are space constructible, then so are the functions $s_1 + s_2$, $s_1 s_2$, $s_1^{s_2}$, and c^{s_1} for every integer $c > 1$. Many common functions are space constructible: for example, $n \log n$, n^3, 2^n, $n!$.

Suppose $t(n)$ is time constructible by a deterministic Turing machine M_t. For every Turing machine M that may have infinite computation paths, we can convert M into a machine M' of time complexity $O(t(n))$ that always halts, as follows: first, make a copy of the input word on a separate worktape; then concurrently run M with M_t, and halt as soon as M_t halts. The total time taken is $2n + t(n) + O(1)$, which is $O(t(n))$, because $t(n) \geq n + 1$. In particular, for a deterministic (respectively, nondeterministic) Turing machine M that accepts each word in $\mathcal{L}(M)$ in $t(n)$ time, but may reject words by not halting, this construction produces a deterministic (nondeterministic) Turing machine M' of time complexity $O(t(n))$ that accepts the same language—i.e., $\mathcal{L}(M') = \mathcal{L}(M)$—but M' always halts.

Analogously, though by a different argument, if $s(n)$ is space constructible and $s(n) \geq \log n$, then for every deterministic (respectively, nondeterministic) Turing machine M that accepts each word in $\mathcal{L}(M)$ in $s(n)$ space, there is a deterministic (nondeterministic) Turing machine M' of space complexity $O(s(n))$ that accepts the same language, but M' always halts. On input word x of length n, machine M' simulates one step of M at a time, counting the number of steps and the number of worktape cells used by M. If M accepts x, then M has an accepting computation path whose number of steps is at most the total number of distinct configurations of M that use at most $s(n)$ space, which is $O(nc^{s(n)})$ for some constant c. Thus, if the number of steps taken by M exceeds this bound, then this computation path is not accepting—machine M might be in an infinite loop—and M' halts in its rejecting state. If the number of worktape cells used by M exceeds $s(n)$, then M' rejects and halts. The space used by M' is $s(n)$ to simulate M, plus

$$\log(nc^{s(n)}) = O(\log n + s(n)) = O(s(n))$$

tape cells for its step counter, and $\log s(n)$ cells for its cell counter.

Complexity Classes

Having defined the time complexity and space complexity of individual Turing machines, we now define classes of languages with particular complexity bounds. These definitions will lead to definitions of P and NP.

Let $t(n)$ be a time-constructible function, and let $s(n)$ be a space-constructible function. Define the following classes of languages:

- DTIME$(t(n))$ is the class of languages decided by deterministic Turing machines of time complexity $O(t(n))$.
- NTIME$(t(n))$ is the class of languages decided by nondeterministic Turing machines of time complexity $O(t(n))$.
- DSPACE$(s(n))$ is the class of languages decided by deterministic Turing machines of space complexity $O(s(n))$.
- NSPACE$(s(n))$ is the class of languages decided by nondeterministic Turing machines of space complexity $O(s(n))$.

Defining each class asymptotically (with the big-O) has several advantages. First, the definition of each complexity class is insensitive to the alphabet used by the Turing machines; without loss of generality, we may assume that every Turing machine uses alphabet $\Sigma_0 = \{\mathbf{0}, \mathbf{1}\}$, because every step of a Turing machine with a larger alphabet can be simulated by $O(1)$ steps of a Turing machine whose alphabet is Σ_0. Second, the big-O allows us to concentrate on the high-order term of the complexity function, ignoring smaller order terms, such as terms that arise from the time-constructibility maneuver. Third, because the big-O bound holds for all n sufficiently large, languages L and L' that differ by a finite number of words belong to the same class; a machine M that decides L can be converted into a machine that decides L' by adding to M a table of answers for inputs up to a certain length.

The following are the *canonical complexity classes*:

- L $=$ DSPACE($\log n$) (deterministic log space)
- NL $=$ NSPACE($\log n$) (nondeterministic log space)
- P $=$ DTIME($n^{O(1)}$) $= \bigcup_{k \geq 1}$ DTIME(n^k) (polynomial time)
- NP $=$ NTIME($n^{O(1)}$) $= \bigcup_{k \geq 1}$ NTIME(n^k) (nondeterministic polynomial time)
- PSPACE $=$ DSPACE($n^{O(1)}$) $= \bigcup_{k \geq 1}$ DSPACE(n^k) (polynomial space)
- E $=$ DTIME($O(1)^n$) $= \bigcup_{k \geq 1}$ DTIME(k^n)
- NE $=$ NTIME($O(1)^n$) $= \bigcup_{k \geq 1}$ NTIME(k^n)
- EXP $=$ DTIME($2^{n^{O(1)}}$) $= \bigcup_{k \geq 1}$ DTIME(2^{n^k}) (deterministic exponential time)
- NEXP $=$ NTIME($2^{n^{O(1)}}$) $= \bigcup_{k \geq 1}$ NTIME(2^{n^k}) (nondeterministic exponential time)
- EXPSPACE $=$ DSPACE($2^{n^{O(1)}}$) $= \bigcup_{k \geq 1}$ DSPACE(2^{n^k}) (exponential space)

The space classes PSPACE and EXPSPACE are defined in terms of DSPACE; by Savitch's theorem (see Theorem 11.1 in first subsection of section 11.4), PSPACE and EXPSPACE could also be defined by NSPACE classes.

Each of these classes contains important computational problems, some of which are listed here and in section 11.5.

The class P contains many familiar problems that can be solved efficiently, such as finding shortest paths in networks, parsing for context-free languages, sorting, matrix multiplication, and linear programming. Consequently, P has become accepted as representing the set of computationally feasible problems. Although one could legitimately argue that a problem whose best algorithm has time complexity $\Theta(n^{99})$ is really infeasible, in practice, the time complexities of the vast majority of known polynomial-time algorithms have low degrees: they run in $O(n^4)$ time or less. Moreover, P is a robust class: though defined by Turing machines, P remains the same when defined by other models of sequential computation. For example, random access machines (RAMs) (a more realistic model of computation defined in Chap. 4, on models and computability) can be used to define P, because Turing machines and RAMs can simulate each other with polynomial-time overhead.

The class NP also enjoys alternative characterizations. In one characterization, NP comprises the problems whose solutions can be verified quickly, by a deterministic Turing machine in polynomial time; equivalently, NP comprises the languages whose membership proofs can be checked quickly. For example, one language in NP is the set of composite numbers, written in binary. A proof that a number z is composite would consist of two factors $z_1 \geq 2$ and $z_2 \geq 2$ whose product $z_1 z_2 = z$. A nondeterministic Turing machine for composite numbers takes a computation path on which it guesses z_1 and z_2 and then deterministically computes the product to check whether $z_1 z_2 = z$. The machine accepts z if there exists a computation path on which $z_1 z_2 = z$. Another important language in NP is the set of satisfiable Boolean formulas, SAT. A Boolean formula ϕ is satisfiable if there exists a truth assignment of **true** or **false** to each of its variables such that under this truth assignment, the value of ϕ is true. For example, $x \wedge (\overline{x} \vee y)$ is satisfiable, but $x \wedge \overline{y} \wedge (\overline{x} \vee y)$ is not satisfiable. A nondeterministic Turing machine takes a computation path that guesses a truth assignment and then deterministically evaluates ϕ for this truth assignment to check whether this truth assignment proves that ϕ is satisfiable. The machine accepts ϕ if and only if there exists a computation path that finds a satisfying truth assignment.

The characterization of NP as the set of problems with easily verified solutions is formalized as follows: $L \in$ NP if and only if there exist a language $L' \in$ P and a polynomial p such that for every $x, x \in L$ if and only if there exists a y such that $|y| \le p(|x|)$ and $(x, y) \in L'$. Here, y is interpreted as the solution to the problem represented by x or, equivalently, as a proof that x belongs to L.

The difference between P and NP is the difference between finding a proof of a mathematical theorem and checking that a proof is correct. In essence, P comprises the theorems that can be proved quickly from scratch (in polynomial time), and NP comprises the theorems whose proofs are short (of polynomial length).

11.4 Relationships Between Complexity Classes

The P vs NP question asks about the relationship between these complexity classes: Is P a proper subset of NP, or does P $=$ NP? Much of complexity theory focuses on the relationships between complexity classes, because these relationships have implications for the difficulty of solving computational problems. In this section, we summarize important known relationships. We demonstrate two techniques for proving relationships between classes: diagonalization and padding.

Basic Relationships

Clearly, for every $t(n)$ and $s(n)$, DTIME$(t) \subseteq$ NTIME(t) and DSPACE$(s) \subseteq$ NSPACE(s), because a deterministic machine is a special case of a nondeterministic machine. Furthermore, DTIME$(t) \subseteq$ DSPACE(t) and NTIME$(t) \subseteq$ NSPACE(t), because at each step, a k-tape Turing machine can write on at most $k = O(1)$ previously unwritten cells. The next theorem presents additional important relationships between classes.

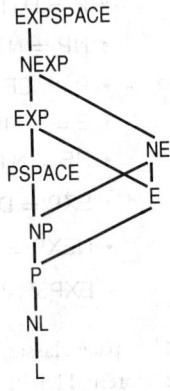

FIGURE 11.2 Inclusion relationship between the canonical classes.

Theorem 11.1. *Let $t(n)$ be a time-constructible function, and let $s(n)$ be a space-constructible function, $s(n) \ge \log n$.*

1. NTIME$(t) \subseteq$ DTIME$(O(1)^t)$.
2. NSPACE$(s) \subseteq$ DTIME$(n + O(1)^s)$.
3. DTIME$(t) \subseteq$ DSPACE$(t/\log t)$ [*Hopcroft et al.* 1977].
4. NTIME$(t) \subseteq$ DSPACE(t).
5. (*Savitch's theorem*) NSPACE$(s) \subseteq$ DSPACE(s^2).

As a consequence of the first part of this theorem, NP $\subseteq$ EXP. No better general upper bound on deterministic time is known for languages in NP, however. See Fig. 11.2 for other known inclusion relationships between the canonical complexity classes.

Although we do not know whether allowing nondeterminism strictly increases the class of languages decided in polynomial time, Savitch's theorem says that for space classes, nondeterminism does not help by more than a polynomial amount. Savitch's theorem implies that both PSPACE and EXPSPACE can be defined by either deterministic or nondeterministic Turing machines.

Complementation

For a language L over an alphabet Σ, define $\overline{L}$ to be the complement of L in the set of words over Σ: $\overline{L} = \Sigma^* - L$. For a class of languages $\mathcal{C}$, define co-$\mathcal{C} = \{\overline{L} : L \in \mathcal{C}\}$. If $\mathcal{C} =$ co-$\mathcal{C}$, then $\mathcal{C}$ is closed under complementation.

In particular, co-NP is the class of languages that are complements of languages in NP. For the language SAT of satisfiable Boolean formulas, $\overline{\text{SAT}}$ is the set of unsatisfiable formulas, whose value is **false** for every truth assignment, together with the syntactically incorrect formulas. A closely related language in co-NP is the set of Boolean tautologies, formulas whose value is **true** for every truth assignment. Most complexity theorists believe that NP $\neq$ co-NP; it is unclear how to check that a formula on m variables is a tautology without checking all 2^m possible truth assignments, which would take exponential time.

Questions about complementation bear directly on the P vs. NP question. It is easy to show that P is closed under complementation (see the next theorem). Consequently, if NP $\neq$ co-NP, then P $\neq$ NP.

Theorem 11.2 (Complementation Theorems). *Let $t(n)$ be a time-constructible function, and let $s(n)$ be a space-constructible function, $s(n) \geq \log n$.*

1. *DTIME(t) is closed under complementation.*
2. *DSPACE(s) is closed under complementation.*
3. *NSPACE(s) is closed under complementation.*

The complementation theorems are used to prove the hierarchy theorems in the next section.

Hierarchy Theorems and Diagonalization

Intuitively, with more time (or more space), we should be able to solve more problems. More precisely, the class of languages decided by Turing machines with a large time complexity should strictly include the class decided by machines with a small time complexity. The next theorem confirms this intuition. In the following, $\subset$ denotes *strict* inclusion between complexity classes.

Theorem 11.3 (Hierarchy Theorems). *Let $t_1(n)$ and $t_2(n)$ be time-constructible functions, and let $s_1(n)$ and $s_2(n)$ be space-constructible functions, with $s_1(n), s_2(n) \geq \log n$.*

1. *If $t_1(n) \log t_1(n) = o(t_2(n))$, then DTIME($t_1$) $\subset$ DTIME(t_2).*
2. *If $t_1(n+1) = o(t_2(n))$, then NTIME(t_1) $\subset$ NTIME(t_2).*
3. *If $s_1(n) = o(s_2(n))$, then DSPACE(s_1) $\subset$ DSPACE(s_2).*
4. *If $s_1(n) = o(s_2(n))$, then NSPACE(s_1) $\subset$ NSPACE(s_2).*

As a corollary of the hierarchy theorem for DTIME,

$$\text{P} \subseteq \text{DTIME}(n^{\log n}) \subset \text{DTIME}(2^n) \subseteq \text{E}$$

Hence, we have the strict inclusion P $\subset$ E. Although we do not know whether P $\subset$ NP, there exists a problem in E that cannot be solved in polynomial time. Other consequences of the hierarchy theorems are NE $\subset$ NEXP and L $\subset$ PSPACE.

In combination with the relationship between DTIME and DSPACE in Theorem 11.1, the hierarchy theorem for DSPACE implies

$$\text{DTIME}(t) \subseteq \text{DSPACE}(t/\log t) \subset \text{DSPACE}(t)$$

In other words, space is more valuable than time. Intuitively, Turing machines can decide more languages with t units of space than with t units of time, because tape cells can be reused.

In the hierarchy theorem for DTIME, the hypothesis on t_1 and t_2 is $t_1(n) \log t_1(n) = o(t_2(n))$, instead of $t_1(n) = o(t_2(n))$, for technical reasons related to the simulation of machines with multiple worktapes by a single universal machine with a fixed number of worktapes. Other computational models, such as random access machines, enjoy tighter time hierarchy theorems.

The gap theorem shows that the constructibility hypotheses are necessary. The gap theorem implies, for example, that there exists a function $t(n)$ such that DTIME(t) = DTIME(2^t); if $t(n)$ were time constructible, then this would be a contradiction of the hierarchy theorem for DTIME.

The proofs of the hierarchy theorems use the **diagonalization** technique. The proof for DTIME constructs a Turing machine M of time complexity t_2 that considers all machines $M_1, M_2, \ldots$, whose time complexity is t_1; for each i, the proof finds a word x_i that is accepted by M if and only if $x_i \notin \mathcal{L}(M_i)$. Consequently, $\mathcal{L}(M)$ differs from each $\mathcal{L}(M_i)$, hence $\mathcal{L}(M) \notin$ DTIME(t_1). The diagonalization technique resembles the classic method used to prove that the real numbers are uncountable by constructing a number whose jth digit differs from the jth digit of the jth number on the list. To illustrate the diagonalization technique, we outline a proof of the Hierarchy Theorem for DSPACE.

PROOF. We construct a deterministic Turing machine M that decides a language L such that $L \in$ DSPACE(s_2) $-$ DSPACE(s_1). Let U be a deterministic universal Turing machine. On input x of length n, machine M performs the following:

1. Lay out $s_2(n)$ cells on a worktape.
2. On another worktape, copy the first $\min\{n, s_2(n)\}$ symbols of x, and let i be this word.
3. Simulate U on input $\langle i, x \rangle$. Accept x if U tries to use more than s_2 worktape cells. (We omit some technical details, such as interleaving multiple worktapes onto the fixed number of worktapes of M and then using the constructibility of s_2 to ensure that this process halts.)
4. If U_i accepts x, then reject; if U_i rejects x, then accept.

Clearly, M always halts and uses space $O(s_2(n))$. Let $L = \mathcal{L}(M)$.

Suppose $L \in$ DSPACE($s_1(n)$). By carefully constructing U, we can ensure that there exists a word y such that U_y decides L, and for every word z, U_{yz} also decides L in space $cs_1(n)$, where the constant c depends only on L; in essence, y is a complete description of a machine that decides L, and U_{yz} ignores symbols beyond the end of y.

Since $s_1(n) = o(s_2(n))$, there is an n_0 such that $cs_1(n) \leq s_2(n)$ for all $n \geq n_0$. Choose z so that $|y| \leq s_2(|yz|)$ and $|yz| \geq n_0$, so that $cs_1(|yz|) \leq s_2(|yz|)$. On input yz, machine M has enough space to simulate U_y on input yz. By construction, M accepts yz if and only if U_y rejects yz. Contradiction! $\square$

Although the diagonalization technique successfully separates some pairs of complexity classes, diagonalization does not seem strong enough to separate P from NP. (See Theorem 11.8 in section 11.6.)

Padding Arguments

A useful technique for establishing relationships between complexity classes is the *padding argument*. Let L be a language over alphabet Σ, and let # be a symbol not in Σ. Let f be a numeric function. The f-*padded version* of L is the language

$$L' = \{x\#^{f(n)} : x \in L \text{ and } n = |x|\}$$

That is, each word of L' is a word in L concatenated with $f(n)$ consecutive # symbols. The padded version L' has the same information content as L, but, because each word is longer, the computational complexity of L' is smaller!

The proof of the next theorem illustrates the use of a padding argument.

Theorem 11.4. *If* $P = NP$, *then* $E = NE$.

PROOF. Since E $\subseteq$ NE, we prove that NE $\subseteq$ E.

Let $L \in$ NE be decided by a nondeterministic Turing machine M in at most $t(n) = k^n$ time for some constant integer k. Let L' be the $t(n)$-padded version of L. From M, we construct a nondeterministic Turing machine M' that decides L' in linear time. M' checks that its input has the correct format, using the time-constructibility of t; then M' runs M on the prefix of the input preceding the first # symbol. Thus, $L' \in$ NP.

If P = NP, then there is a deterministic Turing machine D' that decides L' in at most $p'(n)$ time for some polynomial p'. From D', we construct a deterministic Turing machine D that decides L, as follows. On input x of length n, since $t(n)$ is time constructible, machine D constructs $x\#^{t(n)}$, whose length is $n + t(n)$, in $O(t(n))$ time. Then D runs D' on this input word. The time complexity of D is at most $O(t(n)) + p'(n + t(n)) = O(1)^n$. Therefore, NE $\subseteq$ E. □

11.5 Reducibility and Completeness

In this section, we discuss relationships between problems: informally, if one problem reduces to another problem, then in a sense, the second problem is harder than the first. The hardest problems in NP are the NP-complete problems. We define NP-completeness precisely, and we show how to prove that a problem is NP-complete. Finally, we list some problems that are complete for NP and other complexity classes.

Resource-Bounded Reducibilities

In mathematics, as in everyday life, a typical way to solve a new problem is to reduce it to a previously solved problem. Frequently, an instance of the new problem is expressed completely in terms of an instance of the previous problem, and the solution is then interpreted in the terms of the new problem. For example, the maximum weighted matching problem for bipartite graphs reduces to the network flow problem. (See Chap. 10, on graph and network problems.) This kind of reduction is called *many–one reducibility*, and is defined subsequently.

A different way to solve the new problem is to use a subroutine that solves the previous problem. For example, we can solve an optimization problem whose solution is feasible and maximizes the value of an objective function g by repeatedly calling a subroutine that solves the corresponding decision problem of whether there exists a feasible solution z whose value $g(z)$ satisfies $g(z) \geq k$. This kind of **reduction** is called *Turing reducibility* and is defined subsequently.

Let L_1 and L_2 be languages. L_1 is many–one reducible to L_2, written $L_1 \leq_m L_2$, if there exists a total recursive function f such that for all x, $x \in L_1$ if and only if $f(x) \in L_2$. The function f is called the *transformation function*. L_1 is Turing reducible to L_2, written $L_1 \leq_T L_2$, if L_1 can be decided by a deterministic oracle Turing machine M using L_2 as its oracle, that is, $L_1 = \mathcal{L}(M^{L_2})$.

A reduction between problems may not be helpful if it takes too much time. To study complexity classes defined by bounds on time and space resources, it is natural to consider resource-bounded reducibilities.

Let L_1 and L_2 be languages. L_1 is *Karp reducible* to L_2, written $L_1 \leq_m^P L_2$, if L_1 is many–one reducible to L_2 via a transformation function that is computable deterministically in polynomial time. Karp reducibility is also called *polynomial-time reducibility*.

L_1 is *log-space reducible* to L_2, written $L_1 \leq_m^{\log} L_2$, if L_1 is many–one reducible to L_2 via a transformation function that is computable by a deterministic Turing machine in $O(\log n)$ space.

L_1 is *Cook reducible* to L_2, written $L_1 \leq_T^P L_2$, if L_1 is Turing reducible to L_2 via a deterministic oracle Turing machine of polynomial time complexity.

A reduction from a language L_1 to a language L_2, together with a method for deciding membership in L_2, yields a method for deciding L_1. Suppose L_1 is Karp reducible to L_2 via the transformation f. If machine M_2 decides L_2, and machine M_f computes f, then to decide whether an input word x is in L_1, first use M_f to compute $f(x)$, and then run M_2 on input $f(x)$. A fortiori, if the time complexity of M_2 is a polynomial t_2, and the time complexity of M_f is a polynomial t_f, then on inputs x of length $|x| = n$, the time taken by this method for deciding membership in L_1 is at most $t_f(n) + t_2(t_f(n))$, which is also a polynomial in n. In summary, if L_2 is feasible, and there is an efficient reduction from L_1 to L_2, then L_1 is feasible. We formally state this property of Karp reducibility for P in Theorem 11.7 after stating other properties of these reducibilities.

Log-space reducibility is useful for complexity classes within P, such as NL, for which Karp reducibility allows too many reductions. By definition, for every nontrivial language L_0, (i.e., $L_0 \neq \emptyset$ and $L_0 \neq \Sigma^*$)

and for every L in P, necessarily $L \leq_m^P L_0$ via a transformation that simply runs a deterministic Turing machine that decides L in polynomial time. It is not known whether log-space reducibility is different from Karp reducibility, however, all transformations for known Karp reductions can be computed in $O(\log n)$ space. Even for decision problems, L is not known to be a proper subset of P.

Theorem 11.5. *Log-space reducibility implies Karp reducibility, which implies Cook reducibility:*

1. *If $L_1 \leq_m^{\log} L_2$, then $L_1 \leq_m^P L_2$.*
2. *If $L_1 \leq_m^P L_2$, then $L_1 \leq_T^P L_2$.*

Theorem 11.6. *Log-space reducibility, Karp reducibility, and Cook reducibility are transitive:*

1. *If $L_1 \leq_m^{\log} L_2$ and $L_2 \leq_m^{\log} L_3$, then $L_1 \leq_m^{\log} L_3$.*
2. *If $L_1 \leq_m^P L_2$ and $L_2 \leq_m^P L_3$, then $L_1 \leq_m^P L_3$.*
3. *If $L_1 \leq_T^P L_2$ and $L_2 \leq_T^P L_3$, then $L_1 \leq_T^P L_3$.*

A class of languages $\mathcal{C}$ is *closed under a reducibility* $\leq$ if for all languages L_1 and L_2, whenever $L_1 \leq L_2$ and $L_2 \in \mathcal{C}$, necessarily $L_1 \in \mathcal{C}$.

Theorem 11.7.

1. *P is closed under log-space reducibility, Karp reducibility, and Cook reducibility.*
2. *NP is closed under log-space reducibility and Karp reducibility.*
3. *L and NL are closed under log-space reducibility.*

We shall see the importance of closure under a reducibility in conjunction with the concept of completeness, which we define in the next section.

Complete Languages

Let $\mathcal{C}$ be a class of languages that represent computational problems. A language L_0 is $\mathcal{C}$-*hard* under a reducibility $\leq$ if for all L in $\mathcal{C}$, $L \leq L_0$. A language L_0 is $\mathcal{C}$-*complete* under $\leq$ if L_0 is $\mathcal{C}$-hard, and $L_0 \in \mathcal{C}$. Informally, if L_0 is $\mathcal{C}$-hard, then L_0 represents a problem that is at least as difficult to solve as any problem in $\mathcal{C}$. If L_0 is $\mathcal{C}$-complete, then in a sense, L_0 is one of the most difficult problems in $\mathcal{C}$.

Unless stated otherwise, Karp reducibility is generally assumed. Thus, a language L_0 is *NP-hard* if L_0 is NP-hard under Karp reducibility. L_0 is *NP-complete* if L_0 is NP-complete under Karp reducibility.

Let L_0 be NP-complete. If there exists a deterministic Turing machine that decides L_0 in polynomial time—that is, if $L_0 \in$ P—then because P is closed under Karp reducibility (Theorem 11.7 in the preceding subsection), it would follow that NP $\subseteq$ P, hence P = NP. In essence, the question of whether P is the same as NP reduces to whether any particular NP-complete language is in P.

A common misconception is that this property of NP-complete languages is actually their definition: that is, if $L \in$ NP, and $L \in$ P implies P = NP, then L is NP-complete. This definition is wrong. It is known that P $\neq$ NP if and only if there exists a language L^* in NP $-$ P such that L^* is not NP-complete. Thus, if P $\neq$ NP, then L^* is a counterexample to the definition.

We have noted that an NP-complete language L_0 is unlikely to belong to P. It is also unlikely to belong to co-NP, because, by an elementary argument, if $L_0 \in$ co-NP, then NP = co-NP.

Proving NP-Completeness

After one language has been proved complete for a class, others can be proved complete by constructing transformations. For NP, if L_0 is NP-complete, then to prove that another language L_1 is NP-complete, it suffices to prove that $L_1 \in$ NP, and to construct a polynomial-time transformation that establishes

$L_0 \leq_m^P L_1$. Since L_0 is NP-complete, for every language L in NP, $L \leq_m^P L_0$, hence, by transitivity (Theorem 11.6 previously presented), $L \leq_m^P L_1$.

Cook [1971] defined NP-completeness and proved that SAT, the language of satisfiable Boolean formulas defined in section 11.3, is NP-complete. Consequently, if deciding SAT is easy (in polynomial time), then factoring integers is easy—a surprising connection between ostensibly unrelated problems.

Beginning with Cook [1971] and Karp [1972], hundreds of computational problems in many fields of science and engineering have been proved to be NP-complete, almost always by reduction from a problem that was previously known to be NP-complete. The following NP-complete decision problems are frequently used in these reductions (the language corresponding to each problem is the set of instances whose answers are yes):

- 3-Satisfiability (3SAT)
  Instance: A Boolean expression ϕ in conjunctive normal form with three literals per clause [e.g., $(w \vee x \vee \overline{y}) \wedge (\overline{x} \vee y \vee z)$]
  Question: Is ϕ satisfiable?

- Vertex Cover
  Instance: A graph G and an integer k
  Question: Does G have a set W of k vertices such that every edge in G is incident on a vertex of W?

- Clique
  Instance: A graph G and an integer k
  Question: Does G have a set K of k vertices such that every two vertices in K are adjacent in G?

- Hamiltonian Circuit
  Instance: A graph G
  Question: Does G have a circuit that includes every vertex exactly once?

- Three-Dimensional Matching
  Instance: Sets W, X, Y with $|W| = |X| = |Y| = q$ and a subset $S \subseteq W \times X \times Y$
  Question: Is there a subset $S' \subseteq S$ of size q such that no two triples in S' agree in any coordinate?

- Partition
  Instance: A set S of positive integers
  Question: Is there a subset $S' \subseteq S$ such that the sum of the elements of S' equals the sum of the elements of $S - S'$?

Here is an example of an NP-completeness proof, for the following decision problem:

- Traveling Salesman Problem (TSP)
  Instance: A set of m cities $C_1, \ldots, C_m$, with an integer distance $d(i, j)$ between every pair of cities C_i and C_j, and an integer D.
  Question: Is there a tour of the cities whose total length is at most D? That is, a permutation $c_1, \ldots, c_m$ of $\{1, \ldots, m\}$, such that

$$d(c_1, c_2) + \cdots + d(c_{m-1}, c_m) + d(c_m, c_1) \leq D$$

First, it is easy to see that TSP is in NP: a nondeterministic Turing machine simply guesses a tour and checks that the total length is at most D.

Next, we construct a reduction from Hamiltonian Circuit to TSP. (The reduction goes from the known NP-complete problem, Hamiltonian Circuit, to the new problem, TSP, not vice versa!)

From a graph G on m vertices $v_1, \ldots, v_m$, define the distance function d as follows:

$$d(i, j) = \begin{cases} 1 & \text{if } (v_i, v_j) \text{ is an edge in } G \\ m + 1 & \text{otherwise} \end{cases}$$

Set $D = m$. Clearly, d and D can be computed in polynomial time from G. Each vertex of G corresponds to a city in the constructed instance of TSP.

If G has a Hamiltonian circuit, then the length of the tour that corresponds to this circuit is exactly m. Conversely, if there is a tour whose length is at most m, then each step of the tour must have distance 1, not $m + 1$. Thus, each step corresponds to an edge of G, and the corresponding sequence of vertices in G is a Hamiltonian circuit.

Complete Problems for Other Classes

Besides NP, the following canonical complexity classes have natural complete problems. The three problems now listed are complete for their respective classes under log-space reducibility.

- NL: Graph Accessibility Problem
  Instance: A directed graph G with nodes $1, \ldots, N$
  Question: Does G have a directed path from node 1 to node N?
- P: Circuit Value Problem
  Instance: A Boolean circuit (see section 11.9) with output node u, and an assignment I of $\{0,1\}$ to each input node
  Question: Is 1 the value of u under I?
- PSPACE: Quantified Boolean Formulas
  Instance: A Boolean expression with all variables quantified with either $\forall$ or $\exists$ [e.g., $\forall x \forall y \exists z (x \wedge (\overline{y} \vee z))$].
  Question: Is the expression **true**?

The theory of P-completeness, analogous to the theory of NP-completeness, is explained in Chap. 14, on parallel algorithms.

Stockmeyer and Meyer [1973] defined a natural decision problem that they proved to be complete for NE. If this problem were in P, then by closure under Karp reducibility (Theorem 11.7 earlier in this section), we would have NE $\subseteq$ P, a contradiction of the hierarchy theorems (Theorem 11.3 in section 11.4). Therefore, this decision problem is infeasible: it has no polynomial-time algorithm. In contrast, decision problems in NE $-$ P constructed by diagonalization are unnatural.

11.6 Relativization of the P vs NP Problem

Let L be a language. Define P^L (respectively, NP^L) to be the class of languages decided in polynomial time by deterministic (nondeterministic) oracle Turing machines with oracle L.

Theorem 11.8. *There exist languages A and B such that $P^A = NP^A$, and $P^B \neq NP^B$.*

This theorem suggests that resolving the P vs NP question demands techniques that do not relativize, i.e., that do not apply to oracle Turing machines too. Proofs that use the diagonalization technique on Turing machines without oracles generally relativize to oracle Turing machines; thus, diagonalization is unlikely to succeed in separating P from NP. The only major nonrelativizing proof technique in complexity theory appears to be the technique used to prove that IP=PSPACE (see section 11.1).

11.7 The Polynomial Hierarchy

The oracle B in Theorem 11.8 is an ad hoc language. Let us explore what classes we can define with oracle languages from known complexity classes.

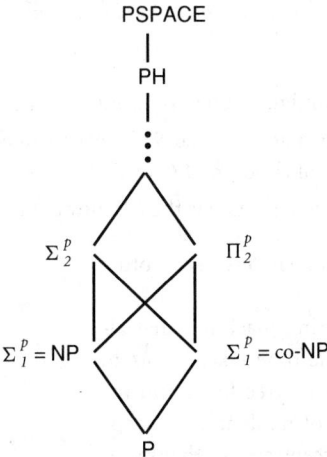

FIGURE 11.3 The polynomial hierarchy.

Let $\mathcal{C}$ be a class of languages. Define

$$NP^{\mathcal{C}} = \bigcup_{L \in \mathcal{C}} NP^{L}$$

Define

$$\Sigma_0^P = \Pi_0^P = P$$

For $k \geq 0$, define

$$\Sigma_{k+1}^P = NP^{\Sigma_k^P}$$

$$\Pi_{k+1}^P = \text{co-}\Sigma_{k+1}^P$$

Observe that $\Sigma_1^P = NP^P = NP$, because each of polynomially many queries to an oracle language in P can be answered directly by a (nondeterministic) Turing machine in polynomial time. Consequently, $\Pi_1^P = \text{co-NP}$. For each k, $\Sigma_k^P \subseteq \Sigma_{k+1}^P$, and $\Pi_k^P \subseteq \Sigma_{k+1}^P$, but these inclusions are not known to be strict. See Fig. 11.3.

The classes Σ_k^P and Π_k^P constitute the *polynomial hierarchy*. Define

$$PH = \bigcup_{k \geq 0} \Sigma_k^P$$

It is straightforward to prove that $PH \subseteq PSPACE$, but it is not known whether the inclusion is strict. In fact, if $PH = PSPACE$, then the polynomial hierarchy collapses to some level, i.e., $PH = \Sigma_m^P$ for some m. In the next section, we define the polynomial hierarchy in two other ways, one of which is in terms of alternating Turing machines.

11.8 Alternating Complexity Classes

The possible computations of an alternating Turing machine M on an input word x can be represented by a tree T_x in which the root is the initial configuration, and the children of a nonterminal node C are the configurations reachable from C by one step of M.

For a word x in $\mathcal{L}(M)$, define an *accepting subtree S* of T_x as follows:

- S is finite.
- The root of S is the initial configuration with input word x.
- If S has an existential configuration C, then S has exactly one child of C in T_x; if S has a universal configuration C, then S has all children of C in T_x.
- Every leaf is a configuration whose state is the accepting state q_A.

See Fig. 11.4. Observe that each node in S is an accepting configuration (see section 11.2).

We consider only alternating Turing machines that always halt. For $x \in \mathcal{L}(M)$, define the time taken by M to be the height of the shortest accepting tree for x and the space to be the maximum number of nonblank worktape cells among configurations in the accepting tree that minimizes this number. For $x \notin \mathcal{L}(M)$, define the time to be the height of T_x, and the space to be the maximum number of nonblank worktape cells among configurations in T_x.

Let $t(n)$ be a time-constructible function, and let $s(n)$ be a space-constructible function. Define the following complexity classes:

- ATIME($t(n)$) is the class of languages decided by alternating Turing machines of time complexity $O(t(n))$.
- ASPACE($s(n)$) is the class of languages decided by alternating Turing machines of space complexity $O(s(n))$.

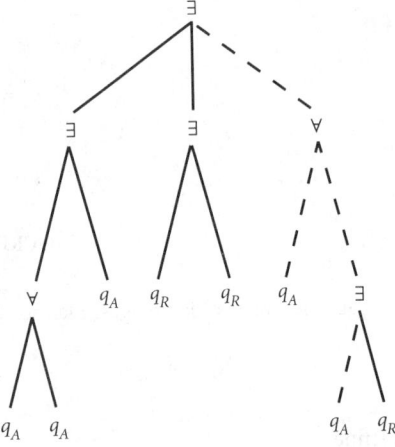

FIGURE 11.4   A computation tree of an alternating machine. Each $\exists$ marks an existential configuration; each $\forall$ marks a universal configuration. The edges of one accepting subtree are drawn with dashed lines.

Because a nondeterministic Turing machine is a special case of an alternating Turing machine, for every $t(n)$ and $s(n)$, NTIME(t) $\subseteq$ ATIME(t) and NSPACE(s) $\subseteq$ ASPACE(s). The next theorem states further relationships between computational resources used by alternating Turing machines and resources used by deterministic and nondeterministic Turing machines.

Theorem 11.9.   *Let $t(n)$ be a time-constructible function, and let $s(n)$ be a space-constructible function, $s(n) \geq \log n$.*

1. *NSPACE(s) $\subseteq$ ATIME($n + s^2$).*
2. *ATIME(t) $\subseteq$ DSPACE(t).*
3. *ASPACE(s) $\subseteq$ DTIME($n + O(1)^s$).*
4. *DTIME(t) $\subseteq$ ASPACE($\log t$).*

In other words, space on deterministic and nondeterministic Turing machines is polynomially related to time on alternating Turing machines. Space on alternating Turing machines is exponentially related to time on deterministic Turing machines. In particular, logarithmic space on alternating Turing machines corresponds to P. Polynomial time on alternating Turing machines corresponds to PSPACE. Polynomial space on alternating Turing machines corresponds to EXP.

In section 11.7, we defined the classes of the polynomial hierarchy in terms of oracles, but we can also define them in terms of alternating Turing machines with restrictions on the number of alternations between existential and universal states. Define a *k-alternating Turing machine* to be a machine such that on every computation path, the number of changes from an existential state to universal state, or from a universal state to an existential state, is at most $k - 1$. Thus, a nondeterministic Turing machine, which stays in existential states, is a 1-alternating Turing machine.

Theorem 11.10.   *The following are equivalent:*

1. $L \in \Sigma_k^P$.
2. *L is decided in polynomial time by a k-alternating Turing machine that starts in an existential state.*
3. *There exists a language L' in P and a polynomial p such that $x \in L$ if and only if*

$$(\exists y_1 : |y_1| \leq p(|x|))(\forall y_2 : |y_2| \leq p(|x|)) \cdots (Q y_k : |y_k| \leq p(|x|))[(x, y_1, \ldots, y_k) \in L']$$

where the quantifier Q is $\exists$ if k is odd, $\forall$ if k is even.

Alternating Turing machines are closely related to Boolean circuits, which are defined in the next section.

11.9 Circuit Complexity

The hardware of electronic digital computers is based on digital logic gates, connected into combinational networks. (See Chap. 17, on architecture components.) Here, we specify a model of computation that formalizes the (bounded fan-in) combinational network.

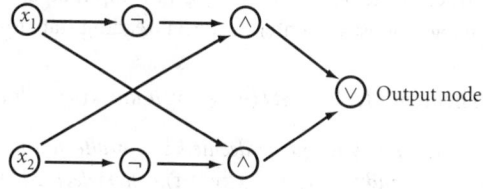

FIGURE 11.5   A Boolean circuit.

A *Boolean circuit* on n input variables $x_1, \ldots, x_n$ is a directed acyclic graph with exactly n input nodes of indegree 0 labeled $x_1, \ldots, x_n$, and other nodes of indegree 1 or 2, called *gates*, labeled with the Boolean operators in $\{\wedge, \vee, \neg\}$. One node is designated as the output of the circuit. See Fig. 11.5. Without loss of generality, we assume that there are no extraneous nodes; there is a directed path from each node to the output node.

An input assignment I is a function that maps each variable x_i to either 0 or 1. The value of each gate g under I is obtained by applying the Boolean operation that labels g to the values of the immediate predecessors of g. The function computed by the circuit is the value of the output node for each input assignment.

A Boolean circuit computes a finite function: a function of only n binary input variables. To decide membership in a language, we need a circuit for each input length n.

A *circuit family* is an infinite set of circuits $C = \{c_1, c_2, \ldots\}$ in which each c_n is a Boolean circuit on n inputs. C *decides* a language $L \subseteq \{\mathbf{0}, \mathbf{1}\}^*$ if for every n and every assignment $a_1, \ldots, a_n$ of $\{\mathbf{0}, \mathbf{1}\}$ to the n inputs, the value of the output node of c_n is $\mathbf{1}$ if and only if the word $a_1 \cdots a_n \in L$. The *size complexity* of C is the function $z(n)$ that specifies the number of nodes in each c_n. The *depth complexity* of C is the function $d(n)$ that specifies the length of the longest directed path in c_n. Clearly, since the fan-in of each gate is at most 2, $d(n) \geq \log z(n) \geq \log n$.

With a different circuit for each input length, a circuit family could solve an undecidable problem such as the halting problem (see Chap. 4, on models and computability)! For each input length, a table of all answers for machine descriptions of that length could be encoded into the circuit. Thus, we need to restrict our circuit families. The most natural restriction is that all circuits in a family should have a concise, uniform description, to disallow a different answer table for each input length. Several uniformity conditions have been studied, and the following is the most convenient.

A circuit family $\{c_1, c_2, \ldots\}$ of size complexity $z(n)$ is *log-space uniform* if there exists a deterministic Turing machine M such that on each input of length n, machine M produces a description of c_n, using space $O(\log z(n))$.

Now we define complexity classes for uniform circuit families and relate these classes to previously defined classes. Define the following complexity classes:

- SIZE($z(n)$) is the class of languages decided by log-space uniform circuit families of size complexity $O(z(n))$.
- DEPTH($d(n)$) is the class of languages decided by log-space uniform circuit families of depth complexity $O(d(n))$.

Theorem 11.11.

1. *If $t(n)$ is a time-constructible function, then DTIME(t) $\subseteq$ SIZE($t \log t$).*
2. *SIZE(z) $\subseteq$ DTIME($z^{O(1)}$).*
3. *If $s(n)$ is a space-constructible function and $s(n) \geq \log n$, then NSPACE(s) $\subseteq$ DEPTH(s^2).*
3. *DEPTH(d) $\subseteq$ DSPACE(d).*

The next theorem shows that size and depth on Boolean circuits are closely related to space and time on alternating Turing machines, provided that we permit sublinear running times for alternating Turing machines, as follows. We augment alternating Turing machines with a random-access input capability. To access the cell at position j on the input tape, M writes the binary representation of j on a special tape, in $\log j$ steps, and enters a special reading state to obtain the symbol in cell j.

Theorem 11.12.   *Let $t(n) \geq \log n$ and $s(n) \geq \log n$.*

1. *Every language decided by an alternating Turing machine of simultaneous space complexity $s(n)$ and time complexity $t(n)$ can be decided by a uniform circuit family of simultaneous size complexity $O(1)^{s(n)}$ and depth complexity $O(t(n))$.*
2. *If $d(n) \geq (\log z(n))^2$, then every language decided by a uniform circuit family of simultaneous size complexity $z(n)$ and depth complexity $d(n)$ can be decided by an alternating Turing machine of simultaneous space complexity $O(\log z(n))$ and time complexity $O(d(n))$.*

In a sense, the Boolean circuit family is a model of parallel computation, because all gates compute independently, in parallel. A fortiori, Boolean circuits (or, equivalently, alternating Turing machines) can be used to define the parallel complexity classes NC^k. (See Chap. 12, on parallel algorithms.)

11.10 Probabilistic Complexity Classes

Since the 1970s, with the development of randomized algorithms for computational problems (see Chap. 8 on randomized algorithms), complexity theorists have placed randomized algorithms on a firm intellectual foundation. Several definitions of randomness have been compared and various kinds of errors distinguished. In this section, we outline some basic concepts in this area.

A *probabilistic Turing machine M* is a nondeterministic Turing machine with exactly two choices at each step. During a computation, M chooses each possible next step with independent probability $1/2$. Intuitively, at each step, M flips a fair coin to decide what to do next. The probability of a computation path of t steps is $1/2^t$. The probability that M accepts an input word x, denoted $p_M(x)$, is the sum of the probabilities of the accepting computation paths.

Throughout this section, we consider only machines whose time complexity $t(n)$ is time constructible. Without loss of generality, we may assume that every halting computation path of the machine has exactly t steps and terminates in either the accepting state q_A or the rejecting state q_R.

Let L be a language. A probabilistic Turing machine M decides L with errors as given in Table 11.1.

For example, the Solovay–Strassen primality testing algorithm of Chap. 7 (on randomized algorithms) makes one-sided errors: when the input x is a prime number, the algorithm always says prime; when

TABLE 11.1   Types of Errors for Probabilistic Turing Machines

		For all $x \in L$	For all $x \notin L$
two-sided error	if	$p_M(x) > 1/2$	$p_M(x) \le 1/2$
bounded two-sided error	if	$p_M(x) > 1/2 + \epsilon$	$p_M(x) < 1/2 - \epsilon$
		for some constant ϵ	
one-sided error	if	$p_M(x) > 1/2$	$p_M(x) = 0$

x is composite, the algorithm usually says composite, but may occasionally say prime. (Actually, the Solovay–Strassen algorithm is a compositeness testing algorithm, because it errs only for inputs that are composite numbers.) Define the following complexity classes:

- PP is the class of languages decided by probabilistic Turing machines of polynomial time complexity with two-sided error.
- BPP is the class of languages decided by probabilistic Turing machines of polynomial time complexity with bounded two-sided error.
- RP is the class of languages decided by probabilistic Turing machines of polynomial time complexity with one-sided error.

In the literature, RP is also called R.

A probabilistic Turing machine M is a PP-machine (respectively, a BPP-machine, an RP-machine) if M has polynomial time complexity, and M decides with two-sided error (bounded two-sided error, one-sided error).

Through repeated Bernoulli trials, we can make the error probabilities of BPP-machines and RP-machines arbitrarily small.

Theorem 11.13.   *If $L \in BPP$, then for every polynomial $q(n)$, there exists a BPP-machine M such that $p_M(x) > 1 - 1/2^{q(n)}$ for every $x \in L$, and $p_M(x) < 1/2^{q(n)}$ for every $x \notin L$.*

If $L \in RP$, then for every polynomial $q(n)$, there exists an RP-machine M such that $p_M(x) > 1 - 1/2^{q(n)}$ for every x in L.

Next, we define a class of problems that have probabilistic algorithms that make no errors. Define

$$ZPP = RP \cap co\text{-}RP$$

The letter Z in ZPP is for zero probability of error, as we now demonstrate. Suppose $L \in ZPP$. Here is an algorithm that checks membership in L. Let M be an RP-machine that decides L, and let M' be an RP-machine that decides $\overline{L}$. For an input word x, alternately run M and M' on x, repeatedly, until a computation path of one machine accepts x. If M accepts x, then accept x; if M' accepts x, then reject x. This algorithm works correctly because when an RP-machine accepts its input, it does not make a mistake. This algorithm might not terminate, but with high probability, the algorithm terminates after a few iterations.

The next theorem expresses some known relationships between probabilistic complexity classes and other complexity classes, such as classes in the polynomial hierarchy (see section 11.7).

Theorem 11.14.

$P \subseteq ZPP \subseteq RP \subseteq BPP \subseteq PP \subseteq PSPACE.$
$RP \subseteq NP \subseteq PP.$
$BPP \subseteq \Sigma_2^P \cap \Pi_2^P.$
$PH \subseteq P^{PP}$ [*Toda 1991*].

See Fig. 11.6.

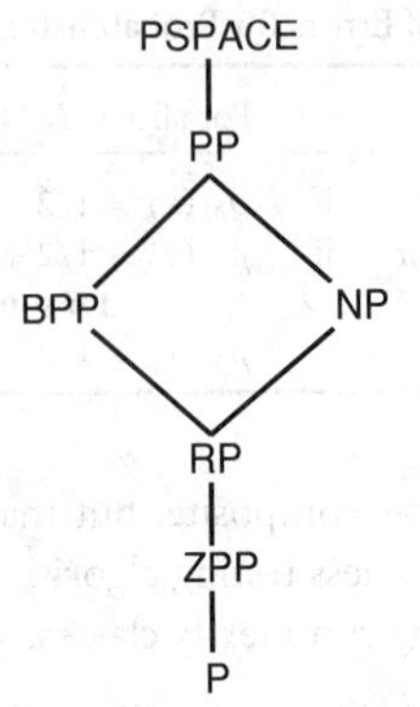

FIGURE 11.6   Probabilistic complexity classes.

11.11 Interactive Models and Complexity Classes

Interactive Proofs

In section 11.3, we characterized NP as the set of languages whose membership proofs can be checked quickly by a deterministic Turing machine M of polynomial time complexity. A different notion of proof involves interaction between two parties, a prover P and a verifier V, who exchange messages. In an *interactive proof system* [Goldwasser et al. 1989], the prover is an all-powerful machine, with unlimited computational resources, analogous to a teacher. The verifier is a computationally limited machine, analogous to a student. One of the original papers on interactive proof systems [Babai and Moran 1988] called them Arthur-Merlin games: the wizard Merlin corresponds to P, and the dim-witted Arthur corresponds to V.

Formally, an interactive proof system comprises the following:

- A read-only input tape on which an input word x is written
- A *prover* P, whose behavior is not restricted.
- A *verifier* V, which is a probabilistic Turing machine augmented with the capability to send and receive messages. The running time of V is bounded by a polynomial in $|x|$.
- A tape on which V writes messages to send to P, and a tape on which P writes messages to send to V. The length of every message is bounded by a polynomial in $|x|$.

A computation of an interactive proof system (P, V) proceeds in rounds, as follows. For $j = 1, 2, \ldots$. in round j, V performs some steps, writes a message m_j, and temporarily stops. Then P reads m_j and responds with a message m'_j, which V reads in round $j + 1$. An interactive proof system (P, V) accepts an input word x if the probability of acceptance by V satisfies $p_V(x) > 1/2$.

In an interactive proof system, a prover can convince the verifier about the truth of a statement without exhibiting an entire proof. For example, consider the graph isomorphism problem: the input consists of two graphs G and H, and the decision is yes if and only if G is isomorphic to H. Here is an interactive proof system (P, V) for this problem. On the first round, V asks P whether the input graphs are isomorphic, but V does not immediately believe the response. If P says yes on the first round, then V repeatedly presents queries to P to try to construct an isomorphism. If P says no on the first round, then V challenges P by repeatedly asking further queries, as follows. In each round, V randomly chooses either G or H with equal probability; if V chooses G, then V computes a random permutation G' of G, presents G' to P, and asks P whether G' came from G or from H (and similarly if V chooses H). If P gave an erroneous answer on the first round, and G is isomorphic to H, then after k subsequent rounds, the probability that P answers all the subsequent queries correctly is $1/2^k$. Thus, in polynomial time, with high probability, V can check whether the original answer of P was correct.

The complexity class IP comprises the languages L for which there exists a verifier V and an ϵ such that:

- There exists a prover $\hat{P}$ such that for all x in L, the interactive proof system $(\hat{P}, V)$ accepts x with probability greater than $1/2 + \epsilon$.
- For every prover P and every $x \notin L$, the interactive proof system (P, V) rejects x with probability greater than $1/2 + \epsilon$.

By substituting random choices for existential choices in the proof that $\text{ATIME}(t) \subseteq \text{DSPACE}(t)$ (Theorem 11.9 in section 11.8), it is straightforward to show that $\text{IP} \subseteq \text{PSPACE}$. As evidence of strict inclusion, Fortnow and Sipser [1988] constructed an oracle language A for which co-$\text{NP}^A - \text{IP}^A \neq \emptyset$, and hence IP^A is strictly included in PSPACE^A. Using a proof technique that does not relativize, however, Shamir [1992] proved that, in fact, IP and PSPACE are the same class.

Theorem 11.15. *IP = PSPACE.*

If NP is a proper subset of PSPACE, as is widely believed, then Theorem 11.15 says that interactive proof systems can decide a larger class of languages than NP.

Probabilistically Checkable Proofs

In an interactive proof system, the verifier does not need a complete conventional proof to become convinced about the membership of a word in a language but uses random choices to query parts of a proof that the prover may know. A new notion of proof explicitly quantifies how much of the proof needs to be inspected.

A language L has a *probabilistically checkable proof* if there exists an oracle BPP-machine M such that:

- For all $x \in L$, there exists an oracle language B_x such that M^{B_x} accepts x.
- For all $x \notin L$, and for every language B, machine M^B rejects x.

Intuitively, the oracle language B_x represents a proof of membership of x in L. Notice that B_x can be finite since the length of each possible query during a computation of M^{B_x} on x is bounded by the running time of M. The oracle language takes the role of the prover in an interactive proof system. The next theorem makes this role precise.

Theorem 11.16. *L has a probabilistically checkable proof if and only if L is decided by an interactive proof system with multiple provers.*

Let $\text{PCP}(r(n), q(n))$ denote the class of languages with probabilistically checkable proofs in which the probabilistic oracle Turing machine M makes $O(r(n))$ random binary choices, and queries its oracle $O(q(n))$ times. (For this definition, we assume that M has either one or two choices for each step.) From the definitions, it follows that $\text{BPP} = \text{PCP}(n^{O(1)}, 0)$, and $\text{NP} = \text{PCP}(0, n^{O(1)})$.

Theorem 11.17. *NP = PCP(log n, 1) [Arora et al. 1992].*

Theorem 11.17 asserts that for every language L in NP, a proof that $x \in L$ can be encoded so that the verifier can be convinced of the correctness of the proof (or detect an incorrect proof) by using only $O(\log n)$ random choices and inspecting only a *constant* number of bits of the proof!

Computational Learning Theory

When we defined interactive proof systems, we compared the prover to a teacher, and the verifier to a student. In this section, we define formal models of learning, with a reliable teacher, and a computationally limited student (the learner).

Let X be a set, called the *domain*; an element of X is an *example*. A *concept class* C is a collection of subsets of X, that is, $C \subseteq 2^X$. For a concept c in C, an example x is *positive* if $x \in c$, *negative* if $x \notin c$. The learner's task is to learn an initially unknown concept c in C.

For instance, the learner may be required to learn a Boolean formula ϕ on n variables from a class of Boolean formulas Φ. In this case, the domain is the set of binary n-tuples, $\{\mathbf{0,1}\}^n$, and a concept is the set of n-tuples $(x_1, \ldots, x_n)$ on which $\phi(x_1, \ldots, x_n) = 1$. Each n-tuple example represents a truth assignment, with $\mathbf{0 = false}$ and $\mathbf{1 = true.}$ An example $(x_1, \ldots, x_n)$ is positive if $\phi(x_1, \ldots, x_n) = 1$, negative if $\phi(x_1, \ldots, x_n) = 0$. The task of the learner is to output a Boolean formula in Φ that is equivalent to ϕ and hence specifies the same concept. Each Boolean formula is a particular representation of an underlying concept, which formally is a subset of $\{0, 1\}^n$; learnability results often depend on the *representation* of concept classes.

In computational learning theory, there are two basic models. In the exact learning model, the learner (student) presents queries to an oracle (teacher), which provides the correct answers. In the probably approximately correct (PAC) model, the learner receives positive and negative examples but has no choice about the examples.

Exact Learning Model. In each round, the learning algorithm proposes a hypothesis c'. If $c' = c$, then the oracle responds yes. If not, then the oracle provides a counterexample: an example on which c and c' differ. A concept class C is *learnable* if there is a learning algorithm that eventually outputs a correct hypothesis for each possible c in C, such that the total running time of the algorithm is a polynomial in the size $|c|$ of (the representation of) c.

Probably Approximately Correct (PAC) Model. The learning algorithm receives a sequence of examples, generated according to an arbitrary probability distribution D on the domain. Each example is labeled correctly as positive or negative. A concept class C is learnable if for every c in C, for all ϵ and δ such that $0 < \epsilon, \delta < 1$, and for all probability distributions D, there is a learning algorithm with the following behavior: the algorithm runs in polynomial time $t(|c|, 1/\epsilon, 1/\delta)$ with probability at least $1 - \delta$, and the algorithm outputs (the representation of) a concept c' such that the probability measure of the examples on which c and c' differ, $D(c \oplus c')$, satisfies $D(c \oplus c') \leq \epsilon$.

The two models differ primarily in their criteria for success. The exact learning model requires logical equivalence, whereas the PAC model requires only approximate distribution-weighted equivalence.

Theorem 11.18. *Every concept class learnable under the exact learning model is also learnable under the PAC Model.*

In both models, to learn a concept, a learning algorithm finds a hypothesis that is consistent with the examples that it has received. The Occam's razor principle asserts that if a sufficiently short hypothesis (not necessarily the shortest) consistent with sufficiently many examples can be computed in polynomial time, then the concept is PAC learnable.

Many results on learnability and nonlearnability (under complexity theoretic assumptions) have been proved for classes of Boolean formulas. Here are two examples.

Theorem 11.19. *The following classes of Boolean formulas are PAC learnable:*

1. *Monomials: conjunctions of literals [e.g., $(w \wedge \bar{x} \wedge \bar{y} \wedge z)$].*
2. *k-DNF: formulas in disjunctive normal form (DNF) with at most k literals per (monomial) term*
3. *k-CNF: formulas in conjunctive normal form (CNF) with at most k literals per clause (e.g., 3SAT).*

Theorem 11.20. *If RP $\neq$ NP, then the following classes of Boolean formulas are not PAC learnable for $k \geq 2$:*

1. *k-term-DNF: formulas in disjunctive normal form with at most k terms*
2. *k-clause-CNF: formulas in conjunctive normal form with at most k clauses.*

Results such as Theorem 11.20 indicate that only simple kinds of concepts are learnable in the two basic models. Thus, to expand the classes of learnable concepts, learning theorists have devised numerous enhancements to the models. Both models can be augmented by allowing the learning algorithm to ask additional queries, such as requests for positive or negative examples. For a *membership query*, the learning algorithm presents an example x to the oracle, which tells whether x is a positive or negative example.

See section III of this Handbook, on artificial intelligence, for other work on models and algorithms for machine learning and concept acquisition.

11.12 Kolmogorov Complexity

Until now, we have considered only dynamic complexity measures, namely, the time and space used by running Turing machines. *Kolmogorov complexity* is a static complexity measure that captures the difficulty of describing a word. For example, the word consisting of three million zeroes can be described with fewer than three million symbols (as in this sentence). In contrast, for a word consisting of three million randomly generated bits, there is probably no shorter description than the word itself.

Let U be a universal Turing machine. Let λ denote the empty word. The Kolmogorov complexity of a binary word y with respect to U, denoted $K_U(y)$, is the length of the shortest binary word i such that on input $\langle i, \lambda \rangle$, machine U outputs y. In essence, i is a description of y, for it tells U how to generate y.

The next theorem states that different choices for the universal Turing machine affect the definition of Kolmogorov complexity in only a small way.

Theorem 11.21 (Invariance Theorem). *There exists a universal Turing machine U such that for every universal Turing machine U', there is a constant c such that for all y*

$$K_U(y) \leq K_{U'}(y) + c$$

Henceforth, let K be defined by the universal Turing machine of Theorem 11.21. For every integer n and every binary word y of length n, because y can be described by giving itself explicitly, $K(y) \leq n + c'$ for a constant c'. Call y *incompressible* if $K(y) \geq n$. Since there are 2^n binary words of length n, and only $2^n - 1$ possible shorter descriptions, there exists an incompressible word for every length n.

Kolmogorov complexity has been used to prove many lower bounds on computational complexity. For example, Maass et al. [1987] constructed a language L_{SMT} (sparse matrix transposition) that can be decided by a 2-tape deterministic Turing machine of time complexity $O(n)$, but every 1-tape Turing machine that decides L_{SMT} requires $\Omega(n \log n / \log \log n)$ time.

11.13 Research Issues and Summary

The core research questions in complexity theory are expressed in terms of separating complexity classes:

- Is L different from NL?
- Is P different from RP or BPP?
- Is P different from NP?
- Is NP different from PSPACE?

Motivated by these questions, much current research is devoted to efforts to understand the power of nondeterminism, randomization, and interaction. In these studies, researchers have gone well beyond the theory presented in this chapter:

- Beyond Turing machines and Boolean circuits, to restricted and specialized models in which nontrivial lower bounds on complexity can be proved
- Beyond Karp reducibility and Cook reducibility, to other kinds of reducibilities

- Beyond worst-case complexity, to average-case complexity
- Beyond decision problems, to enumeration problems and optimization problems

Recent research in complexity theory has had direct applications to other areas of computer science and mathematics. Results on the existence of probabilistically checkable proofs imply that obtaining approximate solutions to NP-complete problems can be as difficult as solving them exactly. Complexity theory provides new tools for studying questions in finite model theory, a branch of mathematical logic. Some questions about logical expressibility are equivalent to open questions about relationships between complexity classes. Fundamental questions in complexity theory are intimately linked to practical questions about the use of cryptography for computer security, such as the existence of one-way functions and the strength of public key cryptosystems.

With precisely defined models and mathematically rigorous proofs, research in complexity theory will continue to provide sound insights into the difficulty of solving real computational problems.

Acknowledgments

Eric Allender, Donna Brown, Bevan Das, Raymond Greenlaw, Lane Hemaspaandra, John Jozwiak, Leonard Pitt, Kenneth Regan, and Martin Tompa kindly read earlier versions of this chapter and suggested numerous helpful improvements. Karen Walny drew the figures and checked the references.

Defining Terms

Complexity class: A set of languages that are decided within a particular resource bound. For example, $\text{NTIME}(n^2 \log n)$ is the set of languages decided by nondeterministic Turing machines within $O(n^2 \log n)$ time.

Constructibility: A function $f(n)$ is time (respectively, space) constructible if there exists a deterministic Turing machine that halts after exactly $f(n)$ steps [after using exactly $f(n)$ worktape cells] for every input of length n.

Diagonalization: A technique for constructing a language L that differs from every $\mathcal{L}(M_i)$ for a list of machines $M_1, M_2, \ldots$.

NP-complete: A language L_0 is NP-complete if $L_0 \in$ NP and $L \leq_m^P L_0$ for every L in NP; that is, for every L in NP, there exists a function f computable in polynomial time such that for every x, $x \in L$ if and only if $f(x) \in L_0$.

Oracle: An oracle is a language L to which a machine presents queries of the form "Is w in L" and receives the correct answers at no cost.

Padding: A technique for establishing relationships between complexity classes that uses padded versions of languages, in which each word is padded out with multiple occurrences of a new symbol—the word x is replaced by the word $x \# f^{(n)}$—in order to artificially reduce the complexity of the language.

Reduction: A language L_1 reduces to a language L_2 if a machine that decides L_2 can be used to decide L_1 efficiently.

Time and space complexity: The time (respectively, space) complexity of a deterministic Turing machine M is the maximum number of steps taken (nonblank cells used) by M among all input words of length n.

Turing machine: A Turing machine M is a model of computation with a read-only input tape and multiple worktapes. At each step, M reads the tape cells on which its access heads are located, and depending on its current state and the symbols in those cells, M changes state, writes new symbols on the worktape cells, and moves each access head one cell left or right or not at all.

References

Angluin, D. 1992. Computational learning theory: survey and selected bibliography. In *Proc. 24th Annu. ACM Symp. Theory Comput.*, pp. 351–369. Victoria, B.C., Canada.

Arora, S., Lund, C., Motwani, R., Sudan, M., and Szegedy, M. 1992. Proof verification and hardness of approximation problems. In *Proc. 33rd Annu. Symp. Found. Comput. Sci.*, pp. 14–23. Pittsburgh, PA, IEEE.

Babai, L. and Moran, S. 1988. Arthur-Merlin games: a randomized proof system, and a hierarchy of complexity classes. *J. Comput. Sys. Sci.* 36(2):254–276.

Balcázar, J. L., Díaz, J., and Gabarró, J. 1988. *Structural Complexity I.* Springer–Verlag, Berlin.

Balcázar, J. L., Díaz, J., and Gabarró, J. 1990. *Structural Complexity II.* Springer–Verlag, Berlin.

Bovet, D. P. and Crescenzi, P. 1994. *Introduction to the Theory of Complexity.* Prentice–Hall International Ltd, Hertfordshire, U.K.

Cook, S. A. 1971. The complexity of theorem-proving procedures. In *Proc. 3rd Annu. ACM Symp. Theory Comput.*, pp. 151–158. Shaker Heights, OH.

Fortnow, L. and Sipser, M. 1988. Are there interactive protocols for co-NP languages? *Inform. Process. Lett.* 28(5):249–251.

Garey, M. R. and Johnson, D. S. 1979. *Computers and Intractability: A Guide to the Theory of NP-Completeness.* W. H. Freeman, San Francisco.

Goldwasser, S., Micali, S., and Rackoff, C. 1989. The knowledge complexity of interactive proof systems. *SIAM J. Comput.* 18(1):186–208.

Hartmanis, J., ed. 1989. *Computational Complexity Theory.* American Mathematical Society, Providence, RI.

Hartmanis, J. 1994. On computational complexity and the nature of computer science. *Commun. ACM* 37(10):37–43.

Hartmanis, J. and Stearns, R. E. 1965. On the computational complexity of algorithms. *Trans. Amer. Math. Soc.* 117:285–306.

Hopcroft, J., Paul, W., and Valiant, L. 1977. On time versus space. *J. ACM* 24(3):332–337.

Karp, R. M. 1972. Reducibility among combinatorial problems. In *Complexity of Computer Computations.* R. E. Miller and J. W. Thatcher, eds., pp. 85–103. Plenum Press, New York.

Kearns, M. J. and Vazirani, U. V. 1994. *Introduction to Computational Learning Theory*, M.I.T. Press, Cambridge, MA.

Li, M. and Vitányi, P. M. B. 1993. *An Introduction to Kolmogorov Complexity and Its Applications.* Springer–Verlag, New York.

Maass, W., Schnitger, G., and Szemeredi, E. 1987. Two tapes are better than one for off-line Turing machines. In *Proc. of the 19th Ann. ACM Symp. on Theory of Comput.*, pp. 94–100. New York.

Papadimitriou, C. H. 1994. *Computational Complexity.* Addison–Wesley, Reading, MA.

Shamir, A. 1992. IP = PSPACE. *J. ACM* 39(4):869–877.

Sipser, M. 1992. The history and status of the P versus NP question. In *Proc. 24th Annu. ACM Symp. Theory Comput.*, pp. 603–618. Victoria, B.C., Canada.

Stearns, R. E. 1990. Juris Hartmanis: the beginnings of computational complexity. In *Complexity Theory Retrospective.* A. L. Selman, ed., pp. 5–18, Springer–Verlag, New York.

Stockmeyer, L. J. and Chandra, A. K. 1979. Intrinsically difficult problems. *Sci. Am.* 240(5):140–159.

Stockmeyer, L. J. and Meyer, A. R. 1973. Word problems requiring exponential time: preliminary report. In *Proc. 5th Annu. ACM Symp. Theory Comput.*, pp. 1–9. Austin, TX.

Toda, S. 1991. PP is as hard as the polynomial-time hierarchy. *SIAM J. Comput.* 20(5):865–877.

van Leeuwen, J. 1990. *Handbook of Theoretical Computer Science, Volume A: Algorithms and Complexity.* Elsevier Science, Amsterdam, and M.I.T. Press, Cambridge, MA.

Wagner, K. and Wechsung, G. 1986. *Computational Complexity.* D. Reidel, Dordrecht, The Netherlands.

Further Information

Three contemporary textbooks on complexity theory are by Balcázar et al. [1988, 1990], by Bovet and Crescenzi [1994], and by Papadimitriou [1994]. The exhaustive survey of complexity theory by Wagner and Wechsung [1986] covers work published before 1986.

A good general reference is the *Handbook of Theoretical Computer Science* [van Leeuwen 1990]. The following chapters in the van Leeuwen *Handbook* are particularly relevant: Machine models and simulations, by P. van Emde Boas, pp. 1–66; A catalog of complexity classes, by D. S. Johnson, pp. 67–161; Machine-independent complexity theory, by J. I. Seiferas, pp. 163–186; Kolmogorov complexity and its applications, by M. Li and P. M. B. Vitányi, pp. 187–254; and The complexity of finite functions, by R. B. Boppana and M. Sipser, pp. 757–804, which covers circuit complexity.

A collection of articles edited by Hartmanis [1989] includes an overview of complexity theory and chapters on sparse complete languages, on relativizations, on interactive proof systems, and on applications of complexity theory to cryptography.

For specific topics in complexity theory, the following references are helpful. Garey and Johnson [1979] explain NP-completeness thoroughly, with examples of NP-completeness proofs, and a collection of hundreds of NP-complete problems. Li and Vitányi [1993] provide a comprehensive scholarly treatment of Kolmogorov complexity, with many applications. Angluin [1992] and Kearns and Vazirani [1994] give up-to-date introductions to computational learning theory.

For historical perspectives on complexity theory, see Hartmanis [1994], Sipser [1992], and Stearns [1990].

Research papers on complexity theory are presented at several annual conferences, including the annual ACM Symposium on Theory of Computing; the annual International Colloquium on Automata, Languages, and Programming, sponsored by the European Association for Theoretical Computer Science (EATCS); and the annual Symposium on Foundations of Computer Science, sponsored by the IEEE. The annual Conference on Computational Complexity (formerly Structure in Complexity Theory), also sponsored by the IEEE, is entirely devoted to complexity theory. Research articles on complexity theory regularly appear in the following journals, among others: *Computational Complexity, Information and Computation, Journal of the ACM, Journal of Computer and System Sciences, Mathematical Systems Theory, SIAM Journal on Computing,* and *Theoretical Computer Science.* Each issue of *ACM SIGACT News* and *Bulletin of the EATCS* contains a column on complexity theory.

12

Parallel Algorithms

Guy E. Blelloch
Carnegie Mellon University

Bruce M. Maggs
Carnegie Mellon University

12.1 Introduction

The subject of this chapter is the design and analysis of parallel algorithms. Most of today's computer algorithms are sequential, that is, they specify a sequence of steps in which each step consists of a single operation. As it has become more difficult to improve the performance of sequential computers, however, researchers have sought performance improvements in another place: parallelism. In contrast to a sequential algorithm, a parallel algorithm may perform multiple operations in a single step. For example, consider the problem of computing the sum of a sequence, A, of n numbers. The standard sequential algorithm computes the sum by making a single pass through the sequence, keeping a running sum of the numbers seen so far. It is not difficult, however, to devise an algorithm for computing the sum that performs many operations in parallel. For example, suppose that, in parallel, each element of A with an even index is paired and summed with the next element of A, which has an odd index, i.e., $A[0]$ is paired with $A[1]$, $A[2]$ with $A[3]$, and so on. The result is a new sequence of $\lceil n/2 \rceil$ numbers whose sum is identical to the sum that we wish to compute. This pairing and summing step can be repeated, and after $\lceil \log_2 n \rceil$ steps, only the final sum remains.

The parallelism in an algorithm can yield improved performance on many different kinds of computers. For example, on a parallel computer, the operations in a parallel algorithm can be performed simultaneously by different processors. Furthermore, even on a single-processor computer it is possible to exploit the

0-8493-2909-4/97/$0.00+$.50
© 1997 by CRC Press, Inc.

parallelism in an algorithm by using multiple functional units, pipelined functional units, or pipelined memory systems. As these examples show, it is important to make a distinction between the parallelism in an algorithm and the ability of any particular computer to perform multiple operations in parallel. Typically, a parallel algorithm will run efficiently on a computer if the algorithm contains at least as much parallelism as the computer. Thus, good parallel algorithms generally can be expected to run efficiently on sequential computers as well as on parallel computers.

The remainder of this chapter consists of eight sections. Section 12.2 begins with a discussion of how to model parallel computers. Next, in section 12.3 we cover some general techniques that have proven useful in the design of parallel algorithms. Sections 12.4–8 present algorithms for solving problems from different domains. We conclude in section 12.9 with a brief discussion of parallel complexity theory. Throughout this chapter, we assume that the reader has some familiarity with sequential algorithms and asymptotic analysis.

12.2 Modeling Parallel Computations

To analyze parallel algorithms it is necessary to have a formal model in which to account for costs. The designer of a sequential algorithm typically formulates the algorithm using an abstract model of computation called a *random-access machine* (RAM) [Aho et al. 1974, ch. 1]. In this model, the machine consists of a single processor connected to a memory system. Each basic central processing unit (CPU) operation, including arithmetic operations, logical operations, and memory accesses, requires one time step. The designer's goal is to develop an algorithm with modest time and memory requirements. The random-access machine model allows the algorithm designer to ignore many of the details of the computer on which the algorithm ultimately will be executed, but it captures enough detail that the designer can predict with reasonable accuracy how the algorithm will perform.

Modeling parallel computations is more complicated than modeling sequential computations because in practice parallel computers tend to vary more in their organizations than do sequential computers. As a consequence, a large proportion of the research on parallel algorithms has gone into the question of modeling, and many debates have raged over what the *right* model is, or about how practical various models are. Although there has been no consensus on the right model, this research has yielded a better understanding of the relationships among the models. Any discussion of parallel algorithms requires some understanding of the various models and the relationships among them.

Parallel models can be broken into two main classes: **multiprocessor models** and **work-depth models**. In this section we discuss each and then discuss how they are related.

Multiprocessor Models

A multiprocessor model is a generalization of the sequential RAM model in which there is more than one processor. Multiprocessor models can be classified into three basic types: local memory machines, modular memory machines, and **parallel random-access machines (PRAMs)**. Figure 12.1 illustrates the structures of these machines. A local memory machine consists of a set of n processors, each with its own local memory. These processors are attached to a common communication network. A modular memory machine consists of m memory modules and n processors all attached to a common network. A PRAM consists of a set of n processors all connected to a common shared memory [Fortune and Wyllie 1978, Goldshlager 1978, Savitch and Stimson 1979].

The three types of multiprocessors differ in the way memory can be accessed. In a local memory machine, each processor can access its own local memory directly, but it can access the memory in another processor only by sending a memory request through the network. As in the RAM model, all local operations, including local memory accesses, take unit time. The time taken to access the memory in another processor, however, will depend on both the capabilities of the communication network and the pattern of memory accesses made by other processors, since these other accesses could congest the network. In a modular memory machine, a processor accesses the memory in a memory module by sending a memory request through the network. Typically, the processors and memory modules are arranged so that the time for any processor to access any memory module is roughly uniform. As in a local memory machine, the exact amount of time depends on the communication network and the memory access pattern. In a PRAM, in a single step each processor can simultaneously access any word of the memory by issuing a memory request directly to the shared memory.

The PRAM model is controversial because no real machine lives up to its ideal of unit-time access to shared memory. It is worth noting, however, that the ultimate purpose of an abstract model is not to directly model a real machine but to help the algorithm designer produce efficient algorithms. Thus, if an algorithm designed for a PRAM (or any other model) can be translated to an algorithm that runs efficiently on a real computer, then the model has succeeded. Later in this section, we show how algorithms designed for one parallel machine model can be translated so that they execute efficiently on another model.

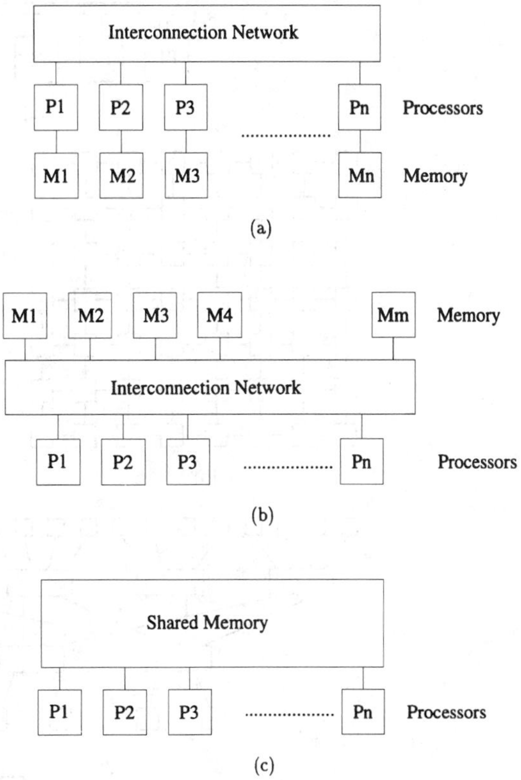

The three types of multiprocessor models that we have defined are very broad, and these

FIGURE 12.1 The three classes of multiprocessor machine models: (a) a local memory machine, (b) a modular memory machine, and (c) a parallel random-access machine (PRAM).

models further differ in network topology, network functionality, control, synchronization, and cache coherence. Many of these issues are discussed elsewhere in this volume. Here we will briefly discuss some of them.

Network Topology

A network is a collection of switches connected by communication channels. A processor or memory module has one or more communication ports that are connected to these switches by communication channels. The pattern of interconnection of the switches is called the network topology. The topology of a network has a large influence on the performance and also on the cost and difficulty of constructing the network. Figure 12.2 illustrates several different topologies.

The simplest network topology is a bus. This network can be used in both local memory machines and modular memory machines. In either case, all processors and memory modules are typically connected to a single bus. In each step, at most one piece of data can be written onto the bus. This datum might be a request from a processor to read or write a memory value, or it might be the response from the processor or memory module that holds the value. In practice, the advantages of using buses are that they are simple to build, and, because all processors and memory modules can observe the traffic on a bus, it is relatively easy to develop protocols that allow processors to cache memory values locally. The disadvantage of using a bus is that the processors have to take turns accessing the bus. Hence, as more processors are added to a bus, the average time to perform a memory access grows proportionately.

A two-dimensional *mesh* is a network that can be laid out in a rectangular fashion. Each switch in a mesh has a distinct label (x, y) where $0 \leq x \leq X - 1$ and $0 \leq y \leq Y - 1$. The values X and Y determine the length of the sides of the mesh. The number of switches in a mesh is thus $X \cdot Y$. Every switch, except those on the sides of the mesh, is connected to four neighbors: one to the north, one to the south, one to

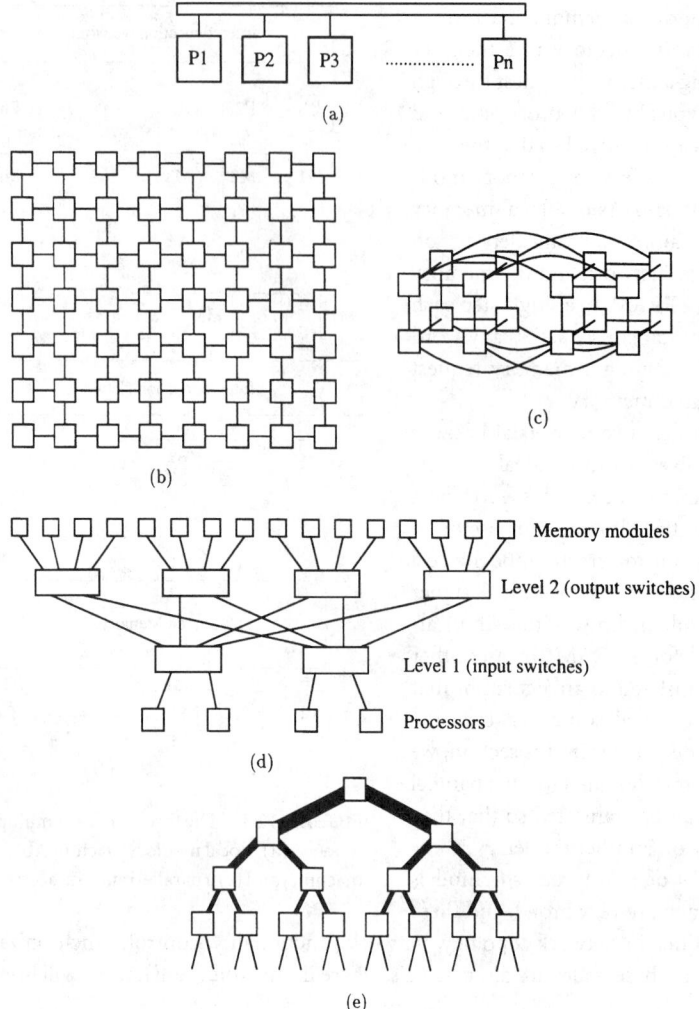

FIGURE 12.2 Various network topologies: (a) bus, (b) two-dimensional mesh, (c) hypercube, (d) two-level multistage network, and (e) fat-tree.

the east, and one to the west. Thus, a switch labeled (x, y), where $0 < x < X - 1$ and $0 < y < Y - 1$ is connected to switches $(x, y + 1)$, $(x, y - 1)$, $(x + 1, y)$, and $(x - 1, y)$. This network typically appears in a local memory machine, i.e., a processor along with its local memory is connected to each switch, and remote memory accesses are made by routing messages through the mesh. Figure 12.2(b) shows an example of an 8×8 mesh.

Several variations on meshes are also popular, including three-dimensional meshes, toruses, and hypercubes. A *torus* is a mesh in which the switches on the sides have connections to the switches on the opposite sides. Thus, every switch (x, y) is connected to four other switches: $(x, y + 1 \bmod Y)$, $(x, y - 1 \bmod Y)$, $(x + 1 \bmod X, y)$, and $(x - 1 \bmod X, y)$. A hypercube is a network with 2^n switches in which each switch has a distinct n-bit label. Two switches are connected by a communication channel in a hypercube if their labels differ in precisely one-bit position.

A *multistage network* is used to connect one set of switches called the *input switches* to another set called the *output switches* through a sequence of stages of switches. Such networks were originally designed for telephone networks [Beneš 1965]. The stages of a multistage network are numbered 1 through L, where L is the **depth** of the network. The input switches form stage 1 and the output switches form stage L. In most multistage networks, it is possible to send a message from any input switch to any output switch along a path that traverses the stages of the network in order from 1 to L. Multistage networks are frequently used in modular memory computers; typically, processors are attached to input switches, and memory modules to output switches. There are many different multistage network topologies. Figure 12.2(d), for example, shows a 2-stage network that connects 4 processors to 16 memory modules. Each switch in this network has two channels at the bottom and four channels at the top. The ratio of processors to memory modules in this example is chosen to reflect the fact that, in practice, a processor is capable of generating memory access requests faster than a memory module is capable of servicing them.

A *fat-tree* is a network whose overall structure is that of a tree [Leiserson 1985]. Each edge of the tree, however, may represent many communication channels, and each node may represent many network switches (hence the name fat). Figure 12.2(e) shows a fat-tree whose overall structure is that of a binary tree. Typically the capacities of the edges near the root of the tree are much larger than the capacities near the leaves. For example, in this tree the two edges incident on the root represent 8 channels each, whereas the edges incident on the leaves represent only 1 channel each. One way to construct a local memory machine is to connect a processor along with its local memory to each leaf of the fat-tree. In this scheme, a message from one processor to another first travels up the tree to the least common ancestor of the two processors and then down the tree.

Many algorithms have been designed to run efficiently on particular network topologies such as the mesh or the hypercube. For an extensive treatment such algorithms, see Leighton [1992]. Although this approach can lead to very fine-tuned algorithms, it has some disadvantages. First, algorithms designed for one network may not perform well on other networks. Hence, in order to solve a problem on a new machine, it may be necessary to design a new algorithm from scratch. Second, algorithms that take advantage of a particular network tend to be more complicated than algorithms designed for more abstract models such as the PRAM because they must incorporate some of the details of the network. Nevertheless, there are some operations that are performed so frequently by a parallel machine that it makes sense to design a fine-tuned network-specific algorithm. For example, the algorithm that routes messages or memory access requests through the network should exploit the network topology. Other examples include algorithms for broadcasting a message from one processor to many other processors, for collecting the results computed in many processors in a single processor, and for synchronizing processors.

An alternative to modeling the topology of a network is to summarize its routing capabilities in terms of two parameters, its latency and bandwidth. The latency L of a network is the time it takes for a message to traverse the network. In actual networks this will depend on the topology of the network, which particular ports the message is passing between, and the congestion of messages in the network. The latency, however, often can be usefully modeled by considering the worst-case time assuming that the network is not heavily congested. The bandwidth at each port of the network is the rate at which a processor can inject data into the network. In actual networks this will depend on the topology of the network, the bandwidths of the network's individual communication channels, and, again, the congestion of messages in the network. The bandwidth often can be usefully modeled as the maximum rate at which processors can inject messages into the network without causing it to become heavily congested, assuming a uniform distribution of message destinations. In this case, the bandwidth can be expressed as the minimum *gap g* between successive injections of messages into the network.

Three models that characterize a network in terms of its latency and bandwidth are the postal model [Bar-Noy and Kipnis 1992], the bulk-synchronous parallel (BSP) model [Valiant 1990], and the LogP model [Culler et al. 1993]. In the postal model, a network is described by a single parameter, L, its latency. The bulk-synchronous parallel model adds a second parameter, g, the minimum ratio of computation steps to communication steps, i.e., the gap. The LogP model includes both of these parameters and adds a third parameter, o, the overhead, or wasted time, incurred by a processor upon sending or receiving a message.

Primitive Operations

As well as specifying the general form of a machine and the network topology, we need to define what operations the machine supports. We assume that all processors can perform the same instructions as a typical processor in a sequential machine. In addition, processors may have special instructions for issuing nonlocal memory requests, for sending messages to other processors, and for executing various global operations, such as synchronization. There can also be restrictions on when processors can simultaneously issue instructions involving nonlocal operations. For example a machine might not allow two processors to write to the same memory location at the same time. The particular set of instructions that the processors can execute may have a large impact on the performance of a machine on any given algorithm. It is therefore important to understand what instructions are supported before one can design or analyze a parallel algorithm. In this section we consider three classes of nonlocal instructions: (1) how global memory requests interact, (2) synchronization, and (3) global operations on data.

When multiple processors simultaneously make a request to read or write to the same resource—such as a processor, memory module, or memory location—there are several possible outcomes. Some machine models simply forbid such operations, declaring that it is an error if more than one processor tries to access a resource simultaneously. In this case we say that the machine allows only *exclusive* access to the resource. For example, a PRAM might allow only exclusive read or write access to each memory location. A PRAM of this type is called an **exclusive-read exclusive-write (EREW)** PRAM. Other machine models may allow unlimited access to a shared resource. In this case we say that the machine allows *concurrent* access to the resource. For example, a **concurrent-read concurrent-write (CRCW)** PRAM allows both concurrent read and write access to memory locations, and a **CREW** PRAM allows **concurrent reads but only exclusive writes**. When making a concurrent write to a resource such as a memory location there are many ways to resolve the conflict. Some possibilities are to choose an arbitrary value from those written, to choose the value from the processor with the lowest index, or to take the *logical or* of the values written. A final choice is to allow for *queued* access, in which case concurrent access is permitted but the time for a step is proportional to the maximum number of accesses to any resource. A **queue-read queue-write (QRQW)** PRAM allows for such accesses [Gibbons et al. 1994].

In addition to reads and writes to nonlocal memory or other processors, there are other important primitives that a machine may supply. One class of such primitives supports synchronization. There are a variety of different types of synchronization operations and their costs vary from model to model. In the PRAM model, for example, it is assumed that all processors operate in lock step, which provides implicit synchronization. In a local-memory machine the cost of synchronization may be a function of the particular network topology. Some machine models supply more powerful primitives that combine arithmetic operations with communication. Such operations include the prefix and **multiprefix** operations, which are defined in the subsections on scans and multiprefix and fetch-and-add.

Work-Depth Models

Because there are so many different ways to organize parallel computers, and hence to model them, it is difficult to select one multiprocessor model that is appropriate for all machines. The alternative to focusing on the machine is to focus on the algorithm. In this section we present a class of models called work-depth models. In a work-depth model, the cost of an algorithm is determined by examining the total number of operations that it performs and the dependencies among those operations. An algorithm's **work** W is the total number of operations that it performs; its *depth* D is the longest chain of dependencies among its operations. We call the ratio $\mathcal{P} = W/D$ the *parallelism* of the algorithm. We say that a parallel algorithm is work-efficient relative to a sequential algorithm if it does at most a constant factor more work.

The work-depth models are more abstract than the multiprocessor models. As we shall see, however, algorithms that are efficient in the work-depth models often can be translated to algorithms that are efficient in the multiprocessor models and from there to real parallel computers. The advantage of a work-depth model is that there are no machine-dependent details to complicate the design and analysis of algorithms. Here we consider three classes of work-depth models: circuit models, vector machine models, and language-based models. We will be using a language-based model in this chapter, and so we will return to these models later in this section.

The most abstract work-depth model is the *circuit model*. In this model, an algorithm is modeled as a family of directed acyclic circuits. There is a circuit for each possible size of the input. A circuit consists of nodes and arcs. A node represents a basic operation, such as adding two values. For each input to an operation (i.e., node), there is an incoming arc from another node or from an input to the circuit. Similarly, there are one or more outgoing arcs from each node representing the result of the operation. The work of a circuit is the total number of nodes. (The work is also called the *size*.) The depth of a circuit is the length of the longest directed path between any pair of nodes. Figure 12.3 shows a circuit in which the inputs are at the top, each + is an adder circuit, and each of the arcs carries the result of an adder circuit. The final sum is returned at the bottom. Circuit models have been used for many years to study various theoretical aspects of parallelism, for example, to prove that certain problems are hard to solve in parallel (see Karp and Ramachandran [1990] for an overview).

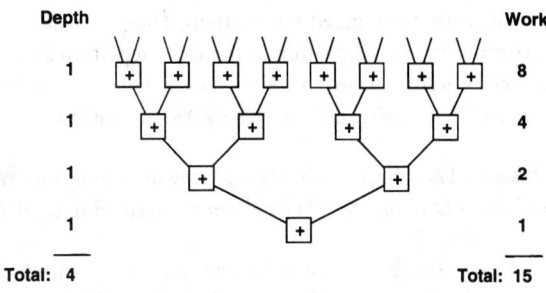

FIGURE 12.3 Summing 16 numbers on a tree. The total depth (longest chain of dependencies) is 4 and the total work (number of operations) is 15.

In a *vector model*, an algorithm is expressed as a sequence of steps, each of which performs an operation on a vector (i.e., sequence) of input values, and produces a vector result [Pratt and Stockmeyer 1976, Blelloch 1990]. The work of each step is equal to the length of its input (or output) vector. The work of an algorithm is the sum of the work of its steps. The depth of an algorithm is the number of vector steps.

In a *language* model, a work-depth cost is associated with each programming language construct [Blelloch and Greiner 1995, Blelloch 1996]. For example, the work for calling two functions in parallel is equal to the sum of the work of the two calls. The depth, in this case, is equal to the maximum of the depth of the two calls.

Assigning Costs to Algorithms

In the work-depth models, the cost of an algorithm is determined by its work and by its depth. The notions of work and depth also can be defined for the multiprocessor models. The work W performed by an algorithm is equal to the number of processors times the time required for the algorithm to complete execution. The depth D is equal to the total time required to execute the algorithm.

The depth of an algorithm is important because there are some applications for which the time to perform a computation is crucial. For example, the results of a weather-forecasting program are useful only if the program completes execution before the weather does!

Generally, however, the most important measure of the cost of an algorithm is the work. This can be justified as follows. The cost of a computer is roughly proportional to the number of processors in the computer. The cost for purchasing time on a computer is proportional to the cost of the computer times the amount of time used. The total cost of performing a computation, therefore, is roughly proportional to the number of processors in the computer times the amount of time used, i.e., the work.

In many instances, the cost of running a computation on a parallel computer may be slightly larger than the cost of running the same computation on a sequential computer. If the time to completion is sufficiently improved, however, this extra cost often can be justified. As we shall see, in general there is a tradeoff between work and time to completion. It is rarely the case, however, that a user is willing to give up any more than a small constant factor in cost for an improvement in time.

Emulations Among Models

Although it may appear that a different algorithm must be designed for each of the many parallel models, there are often automatic and efficient techniques for translating algorithms designed for one model into algorithms designed for another. These translations are *work preserving* in the sense that the work performed by both algorithms is the same, to within a constant factor. For example, the following theorem, known as Brent's Theorem [1974], shows that an algorithm designed for the circuit model can be translated in a work-preserving fashion to a PRAM algorithm.

Theorem 12.1 (Brent's Theorem). *Any algorithm that can be expressed as a circuit of size (i.e., work) W and depth D in the circuit model can be executed in $O(W/P + D)$ steps in the PRAM model.*

PROOF. The basic idea is to have the PRAM emulate the computation specified by the circuit in a level-by-level fashion. The level of a node is defined as follows. A node is on level 1 if all of its inputs are also inputs to the circuit. Inductively, the level of any other node is one greater than the maximum of the level of the nodes with arcs into it. Let l_i denote the number of nodes on level i. Then, by assigning $\lceil l_i/P \rceil$ operations to each of the P processors in the PRAM, the operations for level i can be performed in $O(\lceil l_i/P \rceil)$ steps. Summing the time over all D levels, we have

$$T_{\text{PRAM}}(W, D, P) = O\left(\sum_{i=1}^{D} \left\lceil \frac{l_i}{P} \right\rceil\right)$$

$$= O\left(\sum_{i=1}^{D} \left(\frac{l_i}{P} + 1\right)\right)$$

$$= O\left(\frac{1}{P}\left(\sum_{i=1}^{D} l_i\right) + D\right)$$

$$= O(W/P + D) \qquad\qquad \square$$

The total work performed by the PRAM, i.e., the processor-time product, is $O(W + PD)$. This emulation is work preserving to within a constant factor when the parallelism ($\mathcal{P} = W/D$) is at least as large as the number of processors P, in this case the work is $O(W)$. The requirement that the parallelism exceed the number of processors is typical of work-preserving emulations.

Brent's theorem shows that an algorithm designed for one of the work-depth models can be translated in a work-preserving fashion on to a multiprocessor model. Another important class of work-preserving translations is those that translate between different multiprocessor models. The translation we consider here is the work-preserving translation of algorithms written for the PRAM model to algorithms for a more realistic machine model. In particular, we consider a *butterfly machine* in which P processors are attached through a butterfly network of depth $\log P$ to P memory banks. We assume that, in constant time, a processor can hash a virtual memory address to a physical memory bank and an address within that bank using a sufficiently powerful hash function. This scheme was first proposed by Karlin and Upfal [1988] for the EREW PRAM model. Ranade [1991] later presented a more general approach that allowed the butterfly to efficiently emulate CRCW algorithms.

Theorem 12.2. *Any algorithm that takes time T on a P-processor PRAM can be translated into an algorithm that takes time $O(T(P/P' + \log P'))$, with high probability, on a P'-processor butterfly machine.*

SKETCH OF PROOF. Each of the P' processors in the butterfly machine emulates a set of P/P' PRAM processors. The butterfly machine emulates the PRAM in a step-by-step fashion. First, each butterfly processor emulates one step of each of its P/P' PRAM processors. Some of the PRAM processors may wish to perform memory accesses. For each memory access, the butterfly processor hashes the memory address to a physical memory bank and an address within the bank and then routes a message through the network to that bank. These messages are pipelined so that a processor can have multiple outstanding requests. Ranade proved that if each processor in a P-processor butterfly machine sends at most P/P' messages whose destinations are determined by a sufficiently powerful hash function, then the network can deliver all of the messages, along with responses, in $O(P/P' + \log P')$ time. The $\log P'$ term accounts for the latency of the network and for the fact that there will be some congestion at memory banks, even if each processor sends only a single message. □

This theorem implies that, as long as $P \geq P' \log P'$, i.e., if the number of processors employed by the PRAM algorithm exceeds the number of processors in the butterfly machine by a factor of at least $\log P'$, then the emulation is work preserving. When translating algorithms from a guest multiprocessor model (e.g., the PRAM) to a host multiprocessor model (e.g., the butterfly machine), it is not uncommon to require that the number of guest processors exceed the number of host processors by a factor proportional to the latency of the host. Indeed, the latency of the host often can be hidden by giving it a larger guest to emulate. If the bandwidth of the host is smaller than the bandwidth of a comparably sized guest, however, it usually is much more difficult for the host to perform a work-preserving emulation of the guest.

For more information on PRAM emulations, the reader is referred to Harris [1994] and Valiant [1990].

Model Used in This Chapter

Because there are so many work-preserving translations between different parallel models of computation, we have the luxury of choosing the model that we feel most clearly illustrates the basic ideas behind the algorithms, a work-depth language model. Here we define the model we will use in this chapter in terms of a set of language constructs and a set of rules for assigning costs to the constructs. The description we give here is somewhat informal, but it should suffice for the purpose of this chapter. The language and costs can be properly formalized using a profiling semantics [Blelloch and Greiner 1995].

Most of the syntax that we use should be familiar to readers who have programmed in Algol-like languages, such as Pascal and C. The constructs for expressing parallelism, however, may be unfamiliar. We will be using two parallel constructs—a parallel *apply-to-each* construct and a *parallel-do* construct—and a small set of parallel primitives on sequences (one-dimensional arrays). Our language constructs, syntax, and cost rules are based on the NESL language [Blelloch 1996].

The apply-to-each construct is used to apply an expression over a sequence of values in parallel. It uses a setlike notation. For example, the expression

$$\{a * a : a \in [3, -4, -9, 5]\}$$

squares each element of the sequence $[3, -4, -9, 5]$ returning the sequence $[9, 16, 81, 25]$. This can be read: "in parallel, for each a in the sequence $[3, -4, -9, 5]$, square a." The apply-to-each construct also provides the ability to subselect elements of a sequence based on a filter. For example,

$$\{a * a : a \in [3, -4, -9, 5] \mid a > 0\}$$

can be read: "in parallel, for each a in the sequence $[3, -4, -9, 5]$ such that a is greater than 0, square a." It returns the sequence $[9, 25]$. The elements that remain maintain their relative order.

The parallel-do construct is used to evaluate multiple statements in parallel. It is expressed by listing the set of statements after an **in parallel do**. For example, the following fragment of code calls FUN1(X) and assigns the result to A and in parallel calls FUN2(Y) and assigns the result to B:

in parallel do

$A := \text{FUN1}(X)$

$B := \text{FUN2}(Y)$

The parallel-do completes when all the parallel subcalls complete.

Work and depth are assigned to our language constructs as follows. The work and depth of a scalar primitive operation is one. For example, the work and depth for evaluating an expression such as $3 + 4$ is one. The work for applying a function to every element in a sequence is equal to the sum of the work for each of the individual applications of the function. For example, the work for evaluating the expression

$$\{a * a : a \in [0..n)\}$$

which creates an n-element sequence consisting of the squares of 0 through $n - 1$, is n. The depth for applying a function to every element in a sequence is equal to the maximum of the depths of the individual applications of the function. Hence, the depth of the previous example is one. The work for a parallel-do construct is equal to the sum of the work for each of its statements. The depth is equal to the maximum depth of its statements. In all other cases, the work and depth for a sequence of operations is the sum of the work and depth for the individual operations.

In addition to the parallelism supplied by apply-to-each, we will use four built-in functions on sequences, *dist*, **++** (append), *flatten*, and ← (write), each of which can be implemented in parallel. The function *dist* creates a sequence of identical elements. For example, the expression *dist* $(3, 5)$ creates the sequence

$$[3, 3, 3, 3, 3]$$

The **++** function appends two sequences. For example, $[2, 1]$**++**$[5, 0, 3]$ create the sequence $[2, 1, 5, 0, 3]$. The *flatten* function converts a nested sequence (a sequence for which each element is itself a sequence) into a flat sequence. For example,

$$flatten([[3, 5], [3, 2], [1, 5], [4, 6]])$$

creates the sequence

$$[3, 5, 3, 2, 1, 5, 4, 6]$$

The ← function is used to write multiple elements into a sequence in parallel. It takes two arguments. The first argument is the sequence to modify and the second is a sequence of integer-value pairs that specify what to modify. For each pair (i, v), the value v is inserted into position i of the destination sequence. For example,

$$[0, 0, 0, 0, 0, 0, 0, 0] \leftarrow [(4, -2), (2, 5), (5, 9)]$$

inserts the -2, 5, and 9 into the sequence at locations 4, 2, and 5, respectively, returning

$$[0, 0, 5, 0, -2, 9, 0, 0]$$

As in the PRAM model, the issue of concurrent writes arises if an index is repeated. Rather than choosing a single policy for resolving concurrent writes, we will explain the policy used for the individual algorithms. All of these functions have depth one and work n, where n is the size of the sequence(s) involved. In the case of the ←, the work is proportional to the length of the sequence of integer-value pairs, not the modified sequence, which might be much longer. In the case of **++**, the work is proportional to the length of the second sequence.

We will use a few shorthand notations for specifying sequences. The expression $[-2..1]$ specifies the same sequence as the expression $[-2, -1, 0, 1]$. Changing the left or right brackets surrounding a sequence omits the first or last elements, i.e., $[-2..1)$ denotes the sequence $[-2, -1, 0]$. The notation $A[i..j]$ denotes the subsequence consisting of elements $A[i]$ through $A[j]$. Similarly, $A[i, j)$ denotes the subsequence $A[i]$ through $A[j - 1]$. We will assume that sequence indices are zero based, i.e., $A[0]$ extracts the first element of the sequence A.

Throughout this chapter, our algorithms make use of random numbers. These numbers are generated using the functions *rand_bit()*, which returns a random bit, and *rand_int(h)*, which returns a random integer in the range $[0, h - 1]$.

12.3 Parallel Algorithmic Techniques

As with sequential algorithms, in parallel algorithm design there are many general techniques that can be used across a variety of problem areas. Some of these are variants of standard sequential techniques, whereas others are new to parallel algorithms. In this section we introduce some of these techniques, including parallel divide-and-conquer, randomization, and parallel pointer manipulation. In later sections on algorithms we will make use of them.

Divide-and-Conquer

A divide-and-conquer algorithm first splits the problem to be solved into subproblems that are easier to solve than the original problem either because they are smaller instances of the original problem, or because they are different but easier problems. Next, the algorithm solves the subproblems, possibly recursively. Typically, the subproblems can be solved independently. Finally, the algorithm merges the solutions to the subproblems to construct a solution to the original problem.

The divide-and-conquer paradigm improves program modularity and often leads to simple and efficient algorithms. It has, therefore, proven to be a powerful tool for sequential algorithm designers. Divide-and-conquer plays an even more prominent role in parallel algorithm design. Because the subproblems created in the first step are typically independent, they can be solved in parallel. Often the subproblems are solved recursively and thus the next divide step yields even more subproblems to be solved in parallel. As a consequence, even divide-and-conquer algorithms that were designed for sequential machines typically have some inherent parallelism. Note, however, that in order for divide-and-conquer to yield a highly parallel algorithm, it often is necessary to parallelize the divide step and the merge step. It is also common in parallel algorithms to divide the original problem into as many subproblems as possible, so that they all can be solved in parallel.

As an example of parallel divide-and-conquer, consider the sequential mergesort algorithm. Mergesort takes a set of n keys as input and returns the keys in sorted order. It works by splitting the keys into two sets of $n/2$ keys, recursively sorting each set, and then merging the two sorted sequences of $n/2$ keys into a sorted sequence of n keys. To analyze the sequential running time of mergesort we note that two sorted sequences of $n/2$ keys can be merged in $O(n)$ time. Hence, the running time can be specified by the recurrence

$$T(n) = \begin{cases} 2T(n/2) + O(n) & n > 1 \\ O(1) & n = 1 \end{cases}$$

which has the solution $T(n) = O(n \log n)$. Although not designed as a parallel algorithm, mergesort has some inherent parallelism since the two recursive calls can be made in parallel. This can be expressed as:

ALGORITHM: MERGESORT(A).

1 **if** $(|A| = 1)$ **then return** A
2 **else**
3   **in parallel do**
4        $L := $ MERGESORT($A[0..|A|/2]$)
5        $R := $ MERGESORT($A[|A|/2..|A|]$)
6   **return** MERGE(L, R)

Recall that in our work-depth model we can analyze the depth of an algorithm that makes parallel calls by taking the maximum depth of the two calls, and the work by taking the sum. We assume that the merging remains sequential so that the work and depth to merge two sorted sequences of $n/2$ keys is $O(n)$.

Thus, for mergesort the work and depth are given by the recurrences:

$$W(n) = 2W(n/2) + O(n)$$
$$D(n) = \max(D(n/2), D(n/2)) + O(n)$$
$$= D(n/2) + O(n)$$

As expected, the solution for the work is $W(n) = O(n \log n)$, i.e., the same as the time for the sequential algorithm. For the depth, however, the solution is $D(n) = O(n)$, which is smaller than the work. Recall that we defined the parallelism of an algorithm as the ratio of the work to the depth. Hence, the parallelism of this algorithm is $O(\log n)$ (not very much). The problem here is that the merge step remains sequential, and this is the bottleneck.

As mentioned earlier, the parallelism in a divide-and-conquer algorithm often can be enhanced by parallelizing the divide step and/or the merge step. Using a parallel merge [Shiloach and Vishkin 1982], two sorted sequences of $n/2$ keys can be merged with work $O(n)$ and depth $O(\log n)$. Using this merge algorithm, the recurrence for the depth of mergesort becomes

$$D(n) = D(n/2) + O(\log n)$$

which has solution $D(n) = O(\log^2 n)$. Using a technique called **pipelined divide-and-conquer**, the depth of mergesort can be further reduced to $O(\log n)$ [Cole 1988]. The idea is to start the merge at the top level before the recursive calls complete.

Divide-and-conquer has proven to be one of the most powerful techniques for solving problems in parallel. In this chapter we will use it to solve problems from computational geometry, sorting, and performing fast Fourier transforms. Other applications range from linear systems to factoring large numbers to n-body simulations.

Randomization

The use of random numbers is ubiquitous in parallel algorithms. Intuitively, randomness is helpful because it allows processors to make local decisions which, with high probability, add up to good global decisions. Here we consider three uses of randomness.

Sampling. One use of randomness is to select a representative sample from a set of elements. Often, a problem can be solved by selecting a sample, solving the problem on that sample, and then using the solution for the sample to guide the solution for the original set. For example, suppose we want to sort a collection of integer keys. This can be accomplished by partitioning the keys into buckets and then sorting within each bucket. For this to work well, the buckets must represent nonoverlapping intervals of integer values and contain approximately the same number of keys. **Random sampling** is used to determine the boundaries of the intervals. First, each processor selects a random sample of its keys. Next, all of the selected keys are sorted together. Finally, these keys are used as the boundaries. Such random sampling also is used in many parallel computational geometry, graph, and string matching algorithms.

Symmetry Breaking. Another use of randomness is in **symmetry breaking**. For example, consider the problem of selecting a large independent set of vertices in a graph in parallel. (A set of vertices is *independent* if no two are neighbors.) Imagine that each vertex must decide, in parallel with all other vertices, whether to join the set or not. Hence, if one vertex chooses to join the set, then all of its neighbors must choose not to join the set. The choice is difficult to make simultaneously by each vertex if the local structure at each vertex is the same, for example, if each vertex has the same number of neighbors. As it turns out, the impasse can be resolved by using randomness to break the symmetry between the vertices [Luby 1985].

Load Balancing.   A third use is load balancing. One way to quickly partition a large number of data items into a collection of approximately evenly sized subsets is to randomly assign each element to a subset. This technique works best when the average size of a subset is at least logarithmic in the size of the original set.

Parallel Pointer Techniques

Many of the traditional sequential techniques for manipulating lists, trees, and graphs do not translate easily into parallel techniques. For example, techniques such as traversing the elements of a linked list, visiting the nodes of a tree in postorder, or performing a depth-first traversal of a graph appear to be inherently sequential. Fortunately, these techniques often can be replaced by parallel techniques with roughly the same power.

Pointer Jumping.   One of the earliest parallel pointer techniques is **pointer jumping** [Wyllie 1979]. This technique can be applied to either lists or trees. In each pointer jumping step, each node in parallel replaces its pointer with that of its successor (or parent). For example, one way to label each node of an n-node list (or tree) with the label of the last node (or root) is to use pointer jumping. After at most $\lceil \log n \rceil$ steps, every node points to the same node, the end of the list (or root of the tree). This is described in more detail in the subsection on pointer jumping.

Euler Tour.   An Euler tour of a directed graph is a path through the graph in which every edge is traversed exactly once. In an undirected graph each edge is typically replaced with two oppositely directed edges. The Euler tour of an undirected tree follows the perimeter of the tree visiting each edge twice, once on the way down and once on the way up. By keeping a linked structure that represents the Euler tour of a tree, it is possible to compute many functions on the tree, such as the size of each subtree [Tarjan and Vishkin 1985]. This technique uses linear work and parallel depth that is independent of the depth of the tree. The Euler tour often can be used to replace standard traversals of a tree, such as a depth-first traversal.

Graph Contraction.   **Graph contraction** is an operation in which a graph is reduced in size while maintaining some of its original structure. Typically, after performing a graph contraction operation, the problem is solved recursively on the contracted graph. The solution to the problem on the contracted graph is then used to form the final solution. For example, one way to partition a graph into its connected components is to first contract the graph by merging some of the vertices into their neighbors, then find the connected components of the contracted graph, and finally undo the contraction operation. Many problems can be solved by contracting trees [Miller and Reif 1989, 1991], in which case the technique is called **tree contraction**. More examples of graph contraction can be found in section 12.5.

Ear Decomposition.   An ear decomposition of a graph is a partition of its edges into an ordered collection of paths. The first path is a cycle, and the others are called ears. The endpoints of each ear are anchored on previous paths. Once an ear decomposition of a graph is found, it is not difficult to determine if two edges lie on a common cycle. This information can be used in algorithms for determining biconnectivity, triconnectivity, 4-connectivity, and planarity [Maon et al. 1986, Miller and Ramachandran 1992]. An ear decomposition can be found in parallel using linear work and logarithmic depth, independent of the structure of the graph. Hence, this technique can be used to replace the standard sequential technique for solving these problems, depth-first search.

Other Techniques

Many other techniques have proven to be useful in the design of parallel algorithms. Finding small graph separators is useful for partitioning data among processors to reduce communication [Reif 1993, ch. 14]. Hashing is useful for load balancing and mapping addresses to memory [Vishkin 1984, Karlin and Upfal 1988]. Iterative techniques are useful as a replacement for direct methods for solving linear systems [Bertsekas and Tsitsiklis 1989].

12.4 Basic Operations on Sequences, Lists, and Trees

We begin our presentation of parallel algorithms with a collection of algorithms for performing basic operations on sequences, lists, and trees. These operations will be used as subroutines in the algorithms that follow in later sections.

Sums

As explained at the opening of this chapter, there is a simple recursive algorithm for computing the sum of the elements in an array:

ALGORITHM: SUM(A).

1   **if** $|A| = 1$ **then return** $A[0]$
2   **else return** SUM($\{A[2i] + A[2i + 1] : i \in [0..|A|/2)\}$)

The work and depth for this algorithm are given by the recurrences

$$W(n) = W(n/2) + O(n) = O(n)$$

$$D(n) = D(n/2) + O(1) = O(\log n)$$

which have solutions $W(n) = O(n)$ and $D(n) = O(\log n)$. This algorithm also can be expressed without recursion (using a **while** loop), but the recursive version forshadows the recursive algorithm for implementing the **scan** function.

As written, the algorithm works only on sequences that have lengths equal to powers of 2. Removing this restriction is not difficult by checking if the sequence is of odd length and separately adding the last element in if it is. This algorithm also can easily be modified to compute the sum relative to any associative operator in place of $+$. For example, the use of max would return the maximum value of a sequence.

Scans

The *plus-scan* operation (also called **all-prefix-sums**) takes a sequence of values and returns a sequence of equal length for which each element is the sum of all previous elements in the original sequence. For example, executing a plus-scan on the sequence $[3, 5, 3, 1, 6]$ returns $[0, 3, 8, 11, 12]$. The scan operation can be implemented by the following algorithm [Stone 1975]:

ALGORITHM: SCAN(A).

1   **if** $|A| = 1$ **then return** $[0]$
2   **else**
3       $S = $ SCAN($\{A[2i] + A[2i + 1] : i \in [0..|A|/2)\}$)
4       $R = \{$**if** $(i \bmod 2) = 0$ **then** $S[i/2]$ **else** $S[(i - 1)/2] + A[i - 1] : i \in [0..|A|)\}$
5   **return** R

The algorithm works by elementwise adding the even indexed elements of A to the odd indexed elements of A and then recursively solving the problem on the resulting sequence (line 3). The result S of the recursive call gives the plus-scan values for the even positions in the output sequence R. The value for each of the odd positions in R is simply the value for the preceding even position in R plus the value of the preceding position from A.

The asymptotic work and depth costs of this algorithm are the same as for the SUM operation, $W(n) = O(n)$ and $D(n) = O(\log n)$. Also, as with the SUM operation, any associative function can be used in place of the $+$. In fact, the algorithm described can be used more generally to solve various recurrences, such as the first-order linear recurrences $x_i = (x_{i-1} \otimes a_i) \oplus b_i, 0 \leq i \leq n$, where $\otimes$ and $\oplus$ are both associative [Kogge and Stone 1973].

Scans have proven so useful in the implementation of parallel algorithms that some parallel machines provide support for scan operations in hardware.

Multiprefix and Fetch-and-Add

The multiprefix operation is a generalization of the scan operation in which multiple independent scans are performed. The input to the multiprefix operation is a sequence A of n pairs (k, a), where k specifies a key and a specifies an integer data value. For each key value, the multiprefix operation performs an independent scan. The output is a sequence B of n integers containing the results of each of the scans such that if $A[i] = (k, a)$ then

$$B[i] = \text{sum}(\{b : (t, b) \in A[0..i) | t = k\})$$

In other words, each position receives the sum of all previous elements that have the same key. As an example,

$$\text{MULTIPREFIX}([(1, 5), (0, 2), (0, 3), (1, 4), (0, 1), (2, 2)])$$

returns the sequence

$$[0, 0, 2, 5, 5, 0]$$

The *fetch-and-add* operation is a weaker version of the multiprefix operation, in which the order of the input elements for each scan is not necessarily the same as their order in the input sequence A. In this chapter we omit the implementation of the multiprefix operation, but it can be solved by a function that requires work $O(n)$ and depth $O(\log n)$ using concurrent writes [Matias and Vishkin 1991].

Pointer Jumping

Pointer jumping is a technique that can be applied to both linked lists and trees [Wyllie 1979]. The basic pointer jumping operation is simple. Each node i replaces its pointer $P[i]$ with the pointer of the node that it points to, $P[P[i]]$. By repeating this operation, it is possible to compute, for each node in a list or tree, a pointer to the end of the list or root of the tree. Given set P of pointers that represent a tree (i.e., pointers from children to their parents), the following code will generate a pointer from each node to the root of the tree. We assume that the root points to itself.

ALGORITHM: POINT_TO_ROOT(P).

```
1   for j from 1 to ⌈log |P|⌉
2       P := {P[P[i]] : i ∈ [0..|P|)}
```

The idea behind this algorithm is that in each loop iteration the distance spanned by each pointer, with respect to the original tree, will double, until it points to the root. Since a tree constructed from $n = |P|$ pointers has depth at most $n - 1$, after $\lceil \log n \rceil$ iterations each pointer will point to the root. Because each iteration has constant depth and performs $\Theta(n)$ work, the algorithm has depth $\Theta(\log n)$ and work $\Theta(n \log n)$.

List Ranking

The problem of computing the distance from each node to the end of a linked list is called *list ranking*. Algorithm POINT_TO_ROOT can be easily modified to compute these distances, as follows.

ALGORITHM: LIST_RANK(P).

```
1   V = {if P[i] = i then 0 else 1 : i ∈ [0..|P|)}
2   for j from 1 to ⌈log |P|⌉
3       V := {V[i] + V[P[i]] : i ∈ [0..|P|)}
4       P := {P[P[i]] : i ∈ [0..|P|)}
5   return V
```

In this function, $V[i]$ can be thought of as the distance spanned by pointer $P[i]$ with respect to the original list. Line 1 initializes V by setting $V[i]$ to 0 if i is the last node (i.e., points to itself), and 1 otherwise. In each iteration, line 3 calculates the new length of $P[i]$. The function has depth $\Theta(\log n)$ and work $\Theta(n \log n)$.

It is worth noting that there is a simple sequential solution to the list-ranking problem that performs only $O(n)$ work: you just walk down the list, incrementing a counter at each step. The preceding parallel algorithm, which performs $\Theta(n \log n)$ work, is not **work efficient**. There are, however, a variety of work-efficient parallel solutions to this problem.

The following parallel algorithm uses the technique of random sampling to construct a pointer from each node to the end of a list of n nodes in a work-efficient fashion [Reid-Miller 1994]. The algorithm is easily generalized to solve the list-ranking problem:

1. Pick m list nodes at random and call them the *start* nodes.
2. From each start node u, follow the list until reaching the next start node v. Call the list nodes between u and v the *sublist* of u.
3. Form a shorter list consisting only of the start nodes and the final node on the list by making each start node point to the next start node on the list.
4. Using pointer jumping on the shorter list, for each start node create a pointer to the last node in the list.
5. For each start node u, distribute the pointer to the end of the list to all of the nodes in the sublist of u.

The key to analyzing the work and depth of this algorithm is to bound the length of the longest sublist. Using elementary probability theory, it is not difficult to prove that the expected length of the longest sublist is at most $O((n \log m)/m)$. The work and depth for each step of the algorithm are thus computed as follows:

1. $W(n, m) = O(m)$ and $D(n, m) = O(1)$.
2. $W(n, m) = O(n)$ and $D(n, m) = O((n \log m)/m)$.
3. $W(n, m) = O(m)$ and $D(n, m) = O(1)$.
4. $W(n, m) = O(m \log m)$ and $D(n, m) = O(\log m)$.
5. $W(n, m) = O(n)$ and $D(n, m) = O((n \log m)/m)$.

Thus, the work for the entire algorithm is $W(m, n) = O(n + m \log m)$, and the depth is $O((n \log m)/m)$. If we set $m = n / \log n$, these reduce to $W(n) = O(n)$ and $D(n) = O(\log^2 n)$.

Using a technique called **contraction**, it is possible to design a list ranking algorithm that runs in $O(n)$ work and $O(\log n)$ depth [Anderson and Miller 1988, 1990]. This technique also can be applied to trees [Miller and Reif 1989, 1991].

Removing Duplicates

Given a sequence of items, the remove-duplicates algorithm removes all duplicates, returning the resulting sequence. The order of the resulting sequence does not matter.

Approach 1: Using an Array of Flags

If the items are all nonnegative integers drawn from a small range, we can use a technique similar to bucket sort to remove the duplicates. We begin by creating an array equal in size to the range and initializing all of its elements to 0. Next, using concurrent writes we set a flag in the array for each number that appears in the input list. Finally, we extract those numbers whose flags are set. This algorithm is expressed as follows.

ALGORITHM: REM_DUPLICATES (V).

1 RANGE := $1 + \text{MAX}(V)$
2 FLAGS := $dist(0, \text{RANGE}) \leftarrow \{(i, 1) : i \in V\}$
3 **return** $\{j : j \in [0..\text{RANGE}) \mid \text{FLAGS}[j] = 1\}$

This algorithm has depth $O(1)$ and performs work $O(\text{MAX}(V))$. Its obvious disadvantage is that it explodes when given a large range of numbers, both in memory and in work.

Approach 2: Hashing

A more general approach is to use a hash table. The algorithm has the following outline. First, we create a hash table whose size is prime and approximately two times as large as the number of items in the set V. A prime size is best, because it makes designing a good hash function easier. The size also must be large enough that the chances of collisions in the hash table are not too great. Let m denote the size of the hash

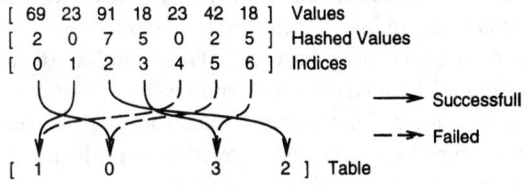

FIGURE 12.4 Each key attempts to write its index into a hash table entry.

table. Next, we compute a hash value, $hash(V[j], m)$, for each item $V[j] \in V$ and attempt to write the index j into the hash table entry $hash(V[j], m)$. For example, Fig. 12.4 describes a particular hash function applied to the sequence [69, 23, 91, 18, 23, 42, 18]. We assume that if multiple values are simultaneously written into the same memory location, one of the values will be correctly written. We call the values $V[j]$ whose indices j are successfully written into the hash table *winners*. In our example, the winners are $V[0]$, $V[1]$, $V[2]$, and $V[3]$, that is, 69, 23, 91, and 18. The winners are added to the duplicate-free set that we are constructing, and then set aside. Among the losers, we must distinguish between two types of items: those that were defeated by an item with the same value, and those that were defeated by an item with a different value. In our example, $V[5]$ and $V[6]$ (23 and 18) were defeated by items with the same value, and $V[4]$ (42) was defeated by an item with a different value. Items of the first type are set aside because they are duplicates. Items of the second type are retained, and we repeat the entire process on them using a different hash function. In general, it may take several iterations before all of the items have been set aside, and in each iteration we must use a different hash function.

Removing duplicates using hashing can be implemented as follows:

ALGORITHM: REMOVE_DUPLICATES (V).

1   $m := \text{NEXT\_PRIME} (2 * |V|)$
2   TABLE := $dist(-1, m)$
3   $i := 0$
4   $R := \{\}$
5   **while** $|V| > 0$
6   TABLE := TABLE $\leftarrow \{(hash(V[j], m, i), j) : j \in [0..|V|)\}$
7       $W := \{V[j] : j \in [0..|V|) \mid \text{TABLE}[hash(V[j], m, i)] = j\}$
8       $R := R \,\text{++}\, W$
9   TABLE := TABLE $\leftarrow \{(hash(k, m, i), k) : k \in W\}$
10      $V := \{k \in V \mid \text{TABLE}[hash(k, m, i)] \neq k\}$
11      $i := i + 1$
12 **return** R

The first four lines of function REMOVE_DUPLICATES initialize several variables. Line 1 finds the first prime number larger than $2 * |V|$ using the built-in function NEXT_PRIME. Line 2 creates the hash table and initializes its entries with an arbitrary value (-1). Line 3 initializes i, a variable that simply counts iterations of the **while** loop. Line 4 initializes the sequence R, the result, to be empty. Ultimately, R will contain a single copy of each distinct item in the sequence V.

The bulk of the work in function REMOVE_DUPLICATES is performed by the **while** loop. Although there are items remaining to be processed, we perform the following steps. In line 6, each item $V[j]$ attempts to write its index j into the table entry given by the hash function $hash(V[j], m, i)$. Note that the hash function takes the iteration i as an argument, so that a different hash function is used in each iteration. Concurrent writes are used so that if several items attempt to write to the same entry, precisely one will win. Line 7 determines which items successfully wrote their indices in line 6 and stores their values in an array called W (for *winners*). The winners are added to the result array R in line 8. The purpose of lines 9 and 10 is to remove all of the items that are either winners or duplicates of winners. These lines reuse the hash table. In line 9, each winner writes its value, rather than its index, into the hash table. In this step there are no concurrent writes. Finally, in line 10, an item is retained only if it is not a winner, and the item that defeated it has a different value.

It is not difficult to prove that, with high probability, each iteration reduces the number of items remaining by some constant fraction until the number of items remaining is small. As a consequence, $D(n) = O(\log n)$ and $W(n) = O(n)$.

The remove-duplicates algorithm is frequently used for set operations; for instance, there is a trivial implementation of the set union operation given the code for REMOVE_DUPLICATES.

12.5 Graphs

Graphs present some of the most challenging problems to parallelize since many standard sequential graph techniques, such as depth-first or priority-first search, do not parallelize well. For some problems, such as minimum spanning tree and biconnected components, new techniques have been developed to generate efficient parallel algorithms. For other problems, such as single-source shortest paths, there are no known efficient parallel algorithms, at least not for the general case.

We have already outlined some of the parallel graph techniques in section 12.3. In this section we describe algorithms for breadth-first search, connected components, and minimum spanning trees. These algorithms use some of the general techniques. In particular, randomization and graph contraction will play an important role in the algorithms. In this chapter we will limit ourselves to algorithms on sparse undirected graphs. We suggest the following sources for further information on parallel graph algorithms Reif [1993, chs. 2–8], JáJá [1992, ch. 5], and Gibbons and Ritter [1990, ch. 2].

Graphs and Their Representation

A *graph* $G = (V, E)$ consists of a set of *vertices* V and a set of *edges* E in which each edge connects two vertices. In a *directed graph* each edge is directed from one vertex to another, whereas in an *undirected graph* each edge is symmetric, i.e., goes in both directions. A *weighted graph* is a graph in which each edge $e \in E$ has a weight $w(e)$ associated with it. In this chapter we will use the convention that $n = |V|$ and $m = |E|$. Qualitatively, a graph is considered sparse if $m \ll n^2$ and dense otherwise. The *diameter* of a graph, denoted $D(G)$, is the maximum, over all pairs of vertices (u, v), of the minimum number of edges that must be traversed to get from u to v.

There are three standard representations of graphs used in sequential algorithms: edge lists, adjacency lists, and adjacency matrices. An *edge list* consists of a list of edges, each of which is a pair of vertices. The list directly represents the set E. An *adjacency list* is an array of lists. Each array element corresponds to one vertex and contains a linked list of the neighboring vertices, i.e., the linked list for a vertex v would contain pointers to the vertices $\{u \mid (v, u) \in E\}$. An *adjacency matrix* is an $n \times n$ array A such that A_{ij} is 1 if $(i, j) \in E$ and 0 otherwise. The adjacency matrix representation is typically used only when the graph is dense since it requires $\Theta(n^2)$ space, as opposed to $\Theta(m)$ space for the other two representations. Each of these representations can be used to represent either directed or undirected graphs.

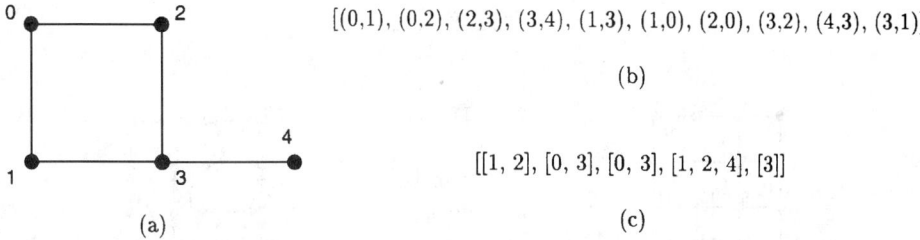

$$[(0,1), (0,2), (2,3), (3,4), (1,3), (1,0), (2,0), (3,2), (4,3), (3,1)]$$

(b)

$$[[1, 2], [0, 3], [0, 3], [1, 2, 4], [3]]$$

(a)   (c)

FIGURE 12.5 Representations of an undirected graph: (a) a graph, G, with 5 vertices and 5 edges, (b) the edge-list representation of G, and (c) the adjacency-list representation of G. Values between square brackets are elements of an array, and values between parentheses are elements of a pair.

For parallel algorithms we use similar representations for graphs. The main change we make is to replace the linked lists with arrays. In particular, the edge list is represented as an array of edges and the adjacency list is represented as an array of arrays. Using arrays instead of lists makes it easier to process the graph in parallel. In particular, they make it easy to grab a set of elements in parallel, rather than having to follow a list. Figure 12.5 shows an example of our representations for an undirected graph. Note that for the edge-list representation of the undirected graph each edge appears twice, once in each direction. We assume these double edges for the algorithms we describe in this chapter.[1] To represent a directed graph we simply store the edge only once in the desired direction. In the text we will refer to the left element of an edge pair as the *source vertex* and the right element as the *destination vertex*.

In algorithms it is sometimes more efficient to use the edge list and sometimes more efficient to use an adjacency list. It is, therefore, important to be able to convert between the two representations. To convert from an adjacency list to an edge list (representation c to representation b in Fig. 12.5) is straightforward. The following code will do it with linear work and constant depth:

$$flatten(\{\{(i, j) : j \in G[i]\} : i \in [0 \cdots |G|]\})$$

where G is the graph in the adjacency list representation. For each vertex i this code pairs up each of i's neighbors with i and then flattens the results.

To convert from an edge list to an adjacency list is somewhat more involved but still requires only linear work. The basic idea is to sort the edges based on the source vertex. This places edges from a particular vertex in consecutive positions in the resulting array. This array can then be partitioned into blocks based on the source vertices. It turns out that since the sorting is on integers in the range $[0 \ldots |V|]$, a radix sort can be used (see radix sort subsection in section 12.6), which can be implemented in linear work. The depth of the radix sort depends on the depth of the multiprefix operation. (See previous subsection on multiprefix.)

Breadth-First Search

The first algorithm we consider is parallel breadth-first search (BFS). BFS can be used to solve various problems such as finding if a graph is connected or generating a spanning tree of a graph. Parallel BFS is similar to the sequential version, which starts with a source vertex s and visits levels of the graph one after the other using a queue. The main difference is that each level is going to be visited in parallel and no queue is required. As with the sequential algorithm, each vertex will be visited only once and each edge, at most twice, once in each direction. The work is therefore linear in the size of the graph $O(n + m)$. For a graph with diameter D, the number of levels processed by the algorithm will be at least $D/2$ and at most D, depending on where the search is initiated. We will show that each level can be processed in constant depth assuming a concurrent-write model, so that the total depth of parallel BFS is $O(D)$.

---

[1] If space is of serious concern, the algorithms can be easily modified to work with edges stored in just one direction.

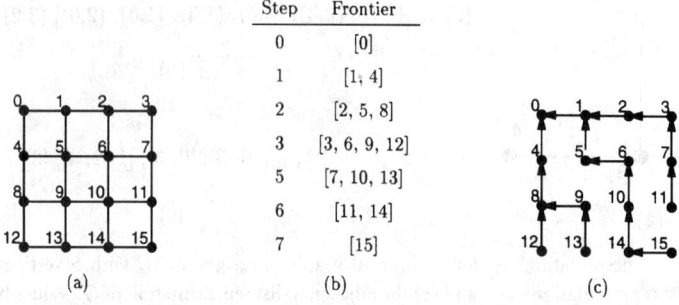

Step	Frontier
0	[0]
1	[1, 4]
2	[2, 5, 8]
3	[3, 6, 9, 12]
5	[7, 10, 13]
6	[11, 14]
7	[15]

(a)   (b)   (c)

FIGURE 12.6 Example of parallel breadth-first search: (a) a graph, G, (b) the frontier at each step of the BFS of G with $s = 0$, and (c) a BFS tree.

The main idea of parallel BFS is to maintain a set of frontier vertices, which represent the current level being visited, and to produce a new frontier on each step. The set of frontier vertices is initialized with the singleton s (the source vertex) and during the execution of the algorithm each vertex will be visited only once. A new frontier is generated by collecting all of the neighbors of the current frontier vertices in parallel and removing any that have already been visited. This is not sufficient on its own, however, since multiple vertices might collect the same unvisited vertex. For example, consider the graph in Fig. 12.6. On step 2 vertices 5 and 8 will both collect vertex 9. The vertex will therefore appear twice in the new frontier. If the duplicate vertices are not removed, the algorithm can generate an exponential number of vertices in the frontier. This problem does not occur in the sequential BFS because vertices are visited one at a time. The parallel version therefore requires an extra step to remove duplicates.

The following algorithm implements the parallel BFS. It takes as input a source vertex s and a graph G represented as an adjacency array and returns as its result a breadth-first search tree of G. In a BFS tree each vertex processed at level i points to one of its neighbors processed at level $i - 1$ [see Fig. 12.6(c)]. The source s is the root of the tree.

ALGORITHM: BFS (s, G).

```
1  Fr := [s]
2  Tr := dist(−1, |G|)
3  Tr[s] := s
4  while (|Fr| ≠ 0)
5      E := flatten({{(u, v) : u ∈ G[v]} : v ∈ Fr})
6      E' := {(u, v) ∈ E | Tr[u] = −1}
7      Tr := Tr ← E'
8      Fr := {u : (u, v) ∈ E' | v = Tr[u]}
9  return Tr
```

In this code Fr is the set of frontier vertices, and Tr is the current BFS tree, represented as an array of indices (pointers). The pointers in Tr are all initialized to -1, except for the source s, which is initialized to point to itself. The algorithm assumes the arbitrary concurrent-write model.

We now consider each iteration of the algorithm. The iterations terminate when there are no more vertices in the frontier (line 4). The new frontier is generated by first collecting together the set of edges from the current frontier vertices to their neighbors into an edge array (line 5). An edge from v to u is represented as the pair (u, v). We then remove any edges whose destination has already been visited (line 6). Now each edge writes its source index into the destination vertex (line 7). In the case that more than one edge has the same destination, one of the source vertices will be written arbitrarily; this is the only place the algorithm will require a concurrent write. These indices will act as the back pointers for the BFS tree, and they also will be used to remove the duplicates for the next frontier set. In particular, each edge checks whether it succeeded by reading back from the destination, and if it succeeded, then the destination is included in the new frontier (line 8). Since only one edge that points to a given destination vertex will succeed, no duplicates will appear in the new frontier.

The algorithm requires only constant depth per iteration of the while loop. Since each vertex and its associated edges are visited only once, the total work is $O(m + n)$. An interesting aspect of this parallel BFS is that it can generate BFS trees that cannot be generated by a sequential BFS, even allowing for any order of visiting neighbors in the sequential BFS. We leave the generation of an example as an exercise. We note, however, that if the algorithm used a priority concurrent write (see previous subsection describing the model used in this chapter) on line 7, then it would generate the same tree as a sequential BFS.

Connected Components

We now consider the problem of labeling the connected components of an undirected graph. The problem is to label all of the vertices in a graph G such that two vertices u and v have the same label if and only if there is a path between the two vertices. Sequentially, the connected components of a graph can easily be labeled using either depth-first or breadth-first search. We have seen how to implement breadth-first search, but the technique requires a depth proportional to the diameter of a graph. This is fine for graphs with a small diameter, but it does not work well in the general case. Unfortunately, in terms of work, even the most efficient polylogarithmic depth parallel algorithms for depth-first search and breadth-first search are very inefficient. Hence, the efficient algorithms for solving the connected components problem use different techniques.

The two algorithms we consider are based on graph contraction. Graph contraction proceeds by contracting the vertices of a connected subgraph into a single vertex to form a new smaller graph. The techniques we use allow the algorithms to make many such contractions in parallel across the graph. The algorithms, therefore, proceed in a sequence of steps, each of which contracts a set of subgraphs, and forms a smaller graph in which each subgraph has been converted into a vertex. If each such step of the algorithm contracts the size of the graph by a constant fraction, then each component will contract down to a single vertex in $O(\log n)$ steps. By running the contraction in reverse, the algorithms can label all of the vertices in the components. The two algorithms we consider differ in how they select subgraphs for contraction. The first uses randomization and the second is deterministic. Neither algorithm is work efficient because they require $O((n + m) \log n)$ work for worst-case graphs, but we briefly discuss how they can be made to be work efficient in the subsequent improved version subsection. Both algorithms require the concurrent-write model.

Random Mate Graph Contraction

The random mate technique for graph contraction is based on forming a set of star subgraphs and contracting the stars. A *star* is a tree of depth one; it consists of a root and an arbitrary number of children. The random mate algorithm finds a set of nonoverlapping stars in a graph and then contracts each star into a single vertex by merging the children into their parents. The technique used to form the stars uses randomization. It works by having each vertex flip a coin and then identify itself as either a parent or a child based on the outcome. We assume the coin is unbiased so that every vertex has a 50% probability of being a parent. Now every vertex that has come up a child looks at its neighbors to see if any are parents. If at least one is a parent, then the child picks one of the neighboring parents as its parent. This process has selected a set of stars, which can be contracted. When contracting, we relabel all of the edges that were incident on a contracting child to its parent's label. Figure 12.7 illustrates a full contraction step. This contraction step is repeated until all components are of size 1.

To analyze the costs of the algorithm we need to know how many vertices are expected to be removed on each contraction step. First, we note that the step is going to remove only children and only if they have a neighboring parent. The probability that a vertex will be deleted is therefore the probability that it is a child multiplied by the probability that at least one of its neighbors is a parent. The probability that it is a child is 1/2 and the probability that at least one neighbor is a parent is at least 1/2 (every vertex has

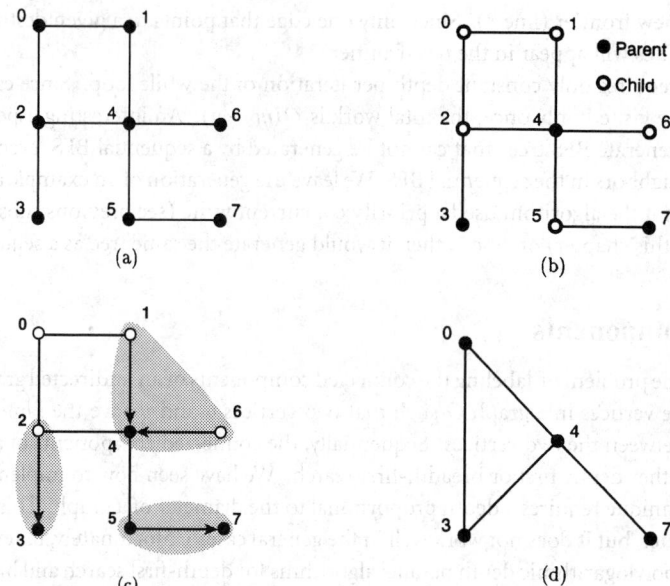

FIGURE 12.7 Example of one step of random mate graph contraction:
(a) the original graph G, (b) G after selecting the parents randomly, (c) contracting the children into the parents (the shaded regions show the subgraphs),
and (d) the contracted graph G'.

one or more neighbors, otherwise it would be completed). We, therefore, expect to remove at least 1/4 of
the remaining vertices at each step and expect the algorithm to complete in no more than $\log_{4/3} n$ steps.
The full probabilistic analysis is somewhat more involved since we could have a streak of bad flips, but it
is not too hard to show that the algorithm is very unlikely to require more than $O(\log n)$ steps.

The following algorithm implements the random mate technique. The input is a graph G in the edge
list representation (note that this is a different representation than used in BFS), along with the labels L of
the vertices. We assume the labels are initialized to the index of the vertex. The output of the algorithm is
a label for each vertex, such that all vertices in a component will be labeled with one of the original labels
of a vertex in the component.

ALGORITHM: CC_RANDOM_MATE (L, E).

1   **if** $(|E| = 0)$ **then return** L
2   **else**
3   CHILD := $\{rand\_bit() : v \in [1..n]\}$
4        $H := \{(u, v) \in E \mid \text{CHILD}[u] \wedge \neg\text{CHILD}[v]\}$
5        $L := L \leftarrow H$
6        $E' := \{(L[u], L[v]) : (u, v) \in E \mid L[u] \neq L[v]\}$
7        $L' := \text{CC\_RANDOM\_MATE}(L, E')$
8        $L' := L' \leftarrow \{(u, L'[v]) : (u, v) \in H\}$
9   **return** L'

The algorithm works recursively by contracting the graph, labeling the components of the contracted
graph, and then passing the labels to the children of the original graph. The termination condition is when
there are no more edges (line 1). To make a contraction step the algorithm first flips a coin on each vertex
(line 3). Now the algorithm subselects the edges-with a child on the left and a parent on the right (line
4). These are called the *hook edges*. Each of the hook edges-writes the parent index into the child's label
(line 5). If a child has multiple neighboring parents, then one of the parents will be written arbitrarily; we

are assuming an arbitrary concurrent write. At this point each child is labeled with one of its neighboring parents, if it has one. Now all edges update themselves to point to the parents by reading from their two endpoints and using these as their new endpoints (line 6). In the same step the edges can check if their two endpoints are within the same contracted vertex (self-edges) and remove themselves if they are. This gives a new sequence of edges E^1. The algorithm has now completed the contraction step and is called recursively on the contracted graph (line 7). The resulting labeling L' of the recursive call is used to update the labels of the children (line 8).

Two things should be noted about this algorithm. First, the algorithm flips coins on all of the vertices on each step even though many have already been contracted (there are no more edges that point to them). It turns out that this will not affect our worst-case asymptotic work or depth bounds, but in practice it is not hard to flip coins only on active vertices by keeping track of them: just keep an array of the labels of the active vertices. Second, if there are cycles in the graph, then the algorithm will create redundant edges in the contracted subgraphs. Again, keeping these edges is not a problem for the correctness or cost bounds, but they could be removed using hashing as previously discussed in the section on removing duplicates.

To analyze the full work and depth of the algorithm we note that each step requires only constant depth and $O(n+m)$ work. Since the number of steps is $O(\log n)$ with high probability, as mentioned earlier, the total depth is $O(\log n)$ and the work is $O((n + m)\log n)$, both with high probability. One might expect that the work would be linear since the algorithm reduces the number of vertices on each step by a constant fraction. We have no guarantee, however, that the number of edges also is going to contract geometrically, and in fact for certain graphs they will not. Subsequently, in this section we will discuss how this can be improved to lead to a work-efficient algorithm.

Deterministic Graph Contraction

Our second algorithm for graph contraction is deterministic [Greiner 1994]. It is based on forming trees as subgraphs and contracting these trees into a single vertex using pointer jumping. To understand the algorithm, consider the graph in Fig. 12.8(a). The overall goal is to contract all of the vertices of the graph into a single vertex. If we had a spanning tree that was imposed on the graph, we could contract the graph by contracting the tree using pointer jumping as discussed previously. Unfortunately, finding a spanning tree turns out to be as hard as finding the connected components of the graph. Instead, we will settle for finding a number of trees that cover the graph, contract each of these as our subgraphs using pointer jumping, and then recurse on the smaller graph. To generate the trees, the algorithm hooks each vertex into a neighbor with a smaller label. This guarantees that there are no cycles since we are only generating pointers from larger to smaller numbered vertices. This hooking will impose a set of disjoint trees on the graph. Figure 12.8(b) shows an example of such a hooking step. Since a vertex can have more than one neighbor with a smaller label, there can be many possible hookings for a given graph. For example, in Fig. 12.8, vertex 2 could have hooked into vertex 1.

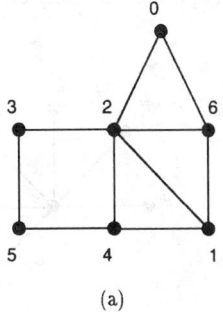
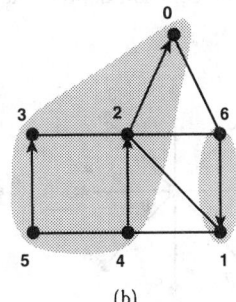

(a) (b)

FIGURE 12.8 Tree-based graph contraction: (a) a graph, G, and (b) the hook edges induced by hooking larger to smaller vertices and the subgraphs induced by the trees.

The following algorithm implements the tree-based graph contraction. We assume that the labels L are initialized to the index of the vertex.

ALGORITHM: CC_TREE_CONTRACT(L, E).

```
1   if(|E| = 0)
2   then return L
3   else
4       H := {(u, v) ∈ E | u < v}
5       L := L ← H
6       L := POINT_TO_ROOT(L)
7       E' := {(L[u], L[v]) : (u, v) ∈ E | L[u] ≠ L[v]}
8       return CC_TREE_CONTRACT(L, E')
```

The structure of the algorithm is similar to the random mate graph contraction algorithm. The main differences are inhow the hooks are selected (line 4), the pointer jumping step to contract the trees (line 6), and the fact that no relabeling is required when returning from the recursive call. The hooking step simply selects edges that point from smaller numbered vertices to larger numbered vertices. This is called a *conditional hook*. The pointer jumping step uses the algorithm given earlier in section 12.4. This labels every vertex in the tree with the root of the tree. The edge relabeling is the same as in a random mate algorithm. The reason we do not need to relabel the vertices after the recursive call is that the pointer jumping will do the relabeling.

Although the basic algorithm we have described so far works well in practice, in the worst case it can take $n - 1$ steps. Consider the graph in Fig. 12.9(a). After hooking and contracting, only one vertex has been removed. This could be repeated up to $n - 1$ times. This worst-case behavior can be avoided by trying to hook in both directions (from larger to smaller and from smaller to larger) and picking the hooking that hooks more vertices. We will make use of the following lemma.

Lemma 12.1. *Let $G = (V, E)$ be an undirected graph in which each vertex has at least one neighbor, then either $|\{u \mid (u, v) \in E, u < v\}| \geq |V|/2$ or $|\{u|(u, v) \in E, u > v\}| > |V|/2$.*

PROOF. Every vertex must have either a neighbor with a lesser index or a neighbor with a greater index. This means that if we consider the set of vertices with a lesser neighbor and the set of vertices with a greater neighbor, then one of those sets must consist of at least one-half the vertices.   □

This lemma will guarantee that if we try hooking in both directions and pick the better one we will remove at least one-half of the vertices on each step, so that the number of steps is bounded by $\log n$.

We now consider the total cost of the algorithm. The hooking and relabeling of edges on each step takes $O(m)$ work and constant depth. The tree contraction using pointer jumping on each step requires

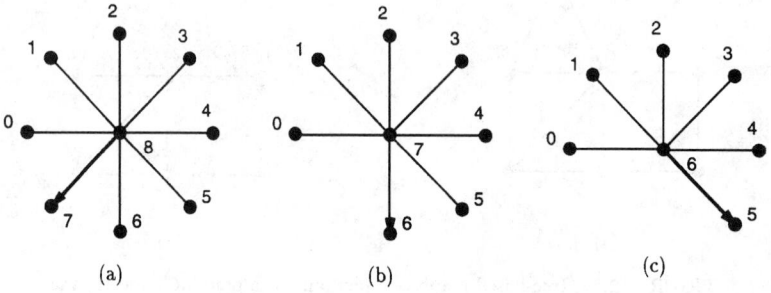

FIGURE 12.9   A worst-case graph: (a) a star graph, G, with the maximum index at the root of the star, (b) G after one step of contraction, and (c) G after two steps of contraction.

$O(n \log n)$ work and $O(\log n)$ depth, in the worst case. Since there are $O(\log n)$ steps, in the worst case, the total work is $O((m + n \log n) \log n)$ and depth $O(\log^2 n)$. However, if we keep track of the active vertices (the roots) and only pointer jump on active vertices, then the work is reduced to $O((m + n) \log n)$ since the number of vertices geometrically decreases. This requires that the algorithm relabels on the way back up the recursion as done for the random mate algorithm. The total work with this modification is the same work as the randomized technique, although the depth has increased.

Improved Versions of Connected Components

There are many improvements to the two basic connected component algorithms we described. Here we mention some of them.

The deterministic algorithm can be improved to run in $O(\log n)$ depth with the same work bounds [Awerbuch and Shiloach 1983, Shiloach and Vishkin 1982]. The basic idea is to interleave the hooking steps with the **shortcutting** steps. The one tricky aspect is that we must always hook in the same direction (i.e., from smaller to larger), so as not to create cycles. Our previous technique to solve the star-graph problem, therefore, does not work. Instead, each vertex checks if it belongs to any tree after hooking. If it does not, then it can hook to any neighbor, even if it has a larger index. This is called an *unconditional hook*.

The randomized algorithm can be improved to run in optimal work $O(n + m)$ [Gazit 1991]. The basic idea is to not use all of the edges for hooking on each step and instead use a sample of the edges. This basic technique developed for parallel algorithms has since been used to improve some sequential algorithms, such as deriving the first linear work algorithm for minimum spanning trees [Klein and Tarjan 1994].

Another improvement is to use the EREW model instead of requiring concurrent reads and writes [Halperin and Zwick 1994]. However, this comes at the cost of greatly complicating the algorithm. The basic idea is to keep circular linked lists of the neighbors of each vertex and then to splice these lists when merging vertices.

Extensions to Spanning Trees and Minimum Spanning Trees

The connected component algorithms can be extended to finding a spanning tree of a graph or minimum spanning tree of a weighted graph. In both cases we assume the graphs are undirected.

A *spanning tree* of a connected graph $G = (V, E)$ is a connected graph $T = (V, E')$ such that $E' \subseteq E$ and $|E'| = |V| - 1$. Because of the bound on the number of edges, the graph T cannot have any cycles and therefore forms a tree. Any given graph can have many different spanning trees.

It is not hard to extend the connectivity algorithms to return the spanning tree. In particular, whenever two components are hooked together the algorithm can keep track of which edges were used for hooking. Since each edge will hook together two components that are not connected yet, and only one edge will succeed in hooking the components, the collection of these edges across all steps will form a spanning tree (they will connect all vertices and have no cycles). To determine which edges were used for contraction, each edge checks if it successfully hooked after the attempted hook.

A minimum spanning tree of a connected weighted graph $G = (V, E)$ with weights $w(e)$ for $e \in E$ is a spanning tree $T = (V, E')$ of G such that

$$w(T) = \sum_{e \in E'} w(e)$$

is minimized. The connected component algorithms also can be extended to determine the minimum spanning tree. Here we will briefly consider an extension of the random mate technique. The algorithm will take advantage of the property that, given any $W \subset V$, the minimum edge from W to $V - W$ must be in some minimum spanning tree. This implies that the minimum edge incident on a vertex will be on a minimum spanning tree. This will be true even after we contract subgraphs into vertices since each subgraph is a subset of V.

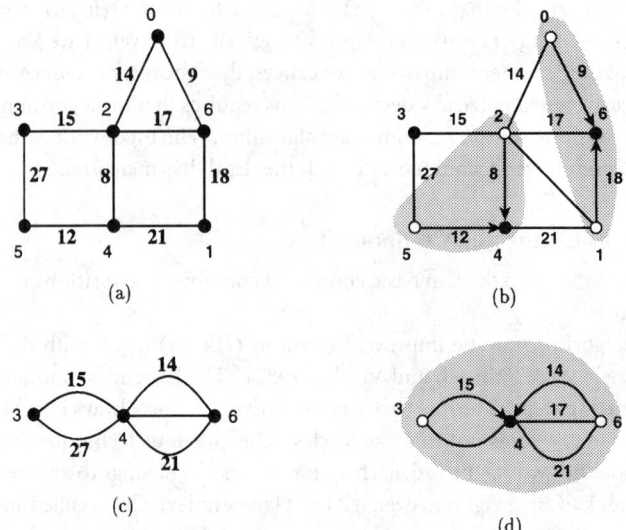

FIGURE 12.10 Example of the minimum spanning tree algorithm. (a) The original weighted graph G. (b) Each child (light) hooks across its minimum weighted edge to a parent (dark), if the edge is incident on a parent. (c) The graph after one step of contraction. (d) The second step in which children hook across minimum weighted edges to parents.

To implement the minimum spanning tree algorithm we therefore modify the random mate technique so that each child u, instead of picking an arbitrary parent to hook into, finds the incident edge (u, v) with minimum weight and hooks into v if it is a parent. If v is not a parent, then the child u does nothing (it is left as an orphan). Figure 12.10 illustrates the algorithm. As with the spanning tree algorithm, we keep track of the edges we use for hooks and add them to a set E'. This new rule will still remove $1/4$ of the vertices on each step on average since a vertex has $1/2$ probability of being a child, and there is $1/2$ probability that the vertex at the other end of the minimum edge is a parent. The one complication in this minimum spanning tree algorithm is finding for each child the incident edge with minimum weight. Since we are keeping an edge list, this is not trivial to compute. If we had an adjacency list, then it would be easy, but since we are updating the endpoints of the edges, it is not easy to maintain the adjacency list. One way to solve this problem is to use a priority concurrent write. In such a write, if multiple values are written to the same location, the one coming from the leftmost position will be written. With such a scheme the minimum edge can be found by presorting the edges by their weight so that the lowest weighted edge will always win when executing a concurrent write. Assuming a priority write, this minimum spanning tree algorithm has the same work and depth as the random mate connected components algorithm.

12.6 Sorting

Sorting is a problem that admits a variety of parallel solutions. In this section we limit our discussion to two parallel sorting algorithms, QuickSort and radix sort. Both of these algorithms are easy to program, and both work well in practice. Many more sorting algorithms can be found in the literature. The interested reader is referred to Akl [1985], JáJá [1992], and Leighton [1992] for more complete coverage.

QuickSort

We begin our discussion of sorting with a parallel version of QuickSort. This algorithm is one of the simplest to code.

ALGORITHM: QUICKSORT(A).

```
1  if |A| = 1 then return A
2  i := rand_int(|A|)
3  p := A[i]
4  in parallel do
5      L := QUICKSORT({a : a ∈ A | a < p})
6      E := {a : a ∈ A | a = p}
7      G := QUICKSORT({a : a ∈ A | a > p})
8  return L ++ E ++ G
```

We can make an optimistic estimate of the work and depth of this algorithm by assuming that each time a partition element, p, is selected, it divides the set A so that neither L nor H has more than half of the elements. In this case, the work and depth are given by the recurrences

$$W(n) = 2W(n/2) + O(n)$$
$$D(n) = D(n/2) + 1$$

whose solutions are $W(n) = O(n \log n)$ and $D(n) = O(\log n)$. A more sophisticated analysis [Knuth 1973] shows that the expected work and depth are indeed $W(n) = O(n \log n)$ and $D(n) = O(\log n)$, independent of the values in the input sequence A.

In practice, the performance of parallel QuickSort can be improved by selecting more than one partition element. In particular, on a machine with P processors, choosing $P - 1$ partition elements divides the keys into P sets, each of which can be sorted by a different processor using a fast sequential sorting algorithm. Since the algorithm does not finish until the last processor finishes, it is important to assign approximately the same number of keys to each processor. Simply choosing $p - 1$ partition elements at random is unlikely to yield a good partition. The partition can be improved, however, by choosing a larger number, sp, of candidate partition elements at random, sorting the candidates (perhaps using some other sorting algorithm), and then choosing the candidates with ranks $s, 2s, \ldots, (p - 1)s$ to be the partition elements. The ratio s of candidates to partition elements is called the *oversampling ratio*. As s increases, the quality of the partition increases, but so does the time to sort the sp candidates. Hence, there is an optimum value of s, typically larger than one, which minimizes the total time. The sorting algorithm that selects partition elements in this fashion is called *sample sort* [Blelloch et al. 1991, Huang and Chow 1983, Reif and Valiant 1983].

Radix Sort

Our next sorting algorithm is radix sort, an algorithm that performs well in practice. Unlike QuickSort, radix sort is not a *comparison sort*, meaning that it does not compare keys directly in order to determine the relative ordering of keys. Instead, it relies on the representation of keys as b-bit integers.

The basic radix sort algorithm (whether serial or parallel) examines the keys to be sorted one *digit* at a time, starting with the least significant digit in each key. Of fundamental importance is that this intermediate sort on digits be *stable*: the output ordering must preserve the input order of any two keys whose bits are the same.

The most common implementation of the intermediate sort is as a counting sort. A counting sort first counts to determine the *rank* of each key—its position in the output order—and then we permute the keys to their respective locations. The following algorithm implements radix sort assuming one-bit digits.

ALGORITHM: RADIX_SORT(A, b)

```
1   for i from 0 to b − 1
2       B := {(a ≫ i) mod 2 : a ∈ A}
3       NB := {1 − b : b ∈ B}
4       R₀ := SCAN(NB)
5       s₀ := SUM(NB)
6       R₁ := SCAN(B)
7       R := {if B[j] = 0 then R₀[j] else R₁[j] + s₀ : j ∈ [0..|A|)}
8       A := A ← {(R[j], A[j]) : j ∈ [0..|A|)}
9   return A
```

For keys with b bits, the algorithm consists of b sequential iterations of a **for** loop, each iteration sorting according to one of the bits. Lines 2 and 3 compute the value and inverse value of the bit in the current position for each key. The notation $a \gg i$ denotes the operation of shifting a i bit positions to the right. Line 4 computes the rank of each key whose bit value is 0. Computing the ranks of the keys with bit value 1 is a little more complicated, since these keys follow the keys with bit value 0. Line 5 computes the number of keys with bit value 0, which serves as the rank of the first key whose bit value is 1. Line 6 computes the relative order of the keys with bit value 1. Line 7 merges the ranks of the even keys with those of the odd keys. Finally, line 8 permutes the keys according to their ranks.

The work and depth of RADIX_SORT are computed as follows. There are b iterations of the **for** loop. In each iteration, the depths of lines 2, 3, 7, 8, and 9 are constant, and the depths of lines 4, 5, and 6 are $O(\log n)$. Hence, the depth of the algorithm is $O(b \log n)$. The work performed by each of lines 2–9 is $O(n)$. Hence, the work of the algorithm is $O(bn)$.

The radix sort algorithm can be generalized so that each b-bit key is viewed as b/r blocks of r bits each, rather than as b individual bits. In the generalized algorithm, there are b/r iterations of the **for** loop, each of which invokes the SCAN function 2^r times. When r is large, a multiprefix operation can be used for generating the ranks instead of executing a SCAN for each possible value [Blelloch et al. 1991]. In this case, and assuming the multiprefix runs in linear work, it is not hard to show that as long as $b = O(\log n)$, the total work for the radix sort is $O(n)$, and the depth is the same order as the depth of the multiprefix.

Floating-point numbers also can be sorted using radix sort. With a few simple bit manipulations, floating-point keys can be converted to integer keys with the same ordering and key size. For example, IEEE double-precision floating-point numbers can be sorted by inverting the mantissa and exponent bits if the sign bit is 1 and then inverting the sign bit. The keys are then sorted as if they were integers.

12.7 Computational Geometry

Problems in computational geometry involve determining various properties about sets of objects in a k-dimensional space. Some standard problems include finding the closest distance between a pair of points (closest pair), finding the smallest convex region that encloses a set of points (convex hull), and finding line or polygon intersections. Efficient parallel algorithms have been developed for most standard problems in computational geometry. Many of the sequential algorithms are based on divide-and-conquer and lead in a relatively straightforward manner to efficient parallel algorithms. Some others are based on a technique called plane sweeping, which does not parallelize well, but for which an analogous parallel technique, the *plane sweep tree* has been developed [Aggarwal et al. 1988, Atallah et al. 1989]. In this section we describe parallel algorithms for two problems in two dimensions—closest pair and convex hull. For the convex hull we describe two algorithms. These algorithms are good examples of how sequential algorithms can be parallelized in a straightforward manner.

We suggest the following sources for further information on parallel algorithms for computational geometry: Reif [1993, chs. 9, 11], JáJá [1992, ch. 6], and Goodrich [1996].

Closest Pair

The *closest pair problem* takes a set of points in k dimensions and returns the two points that are closest to each other. The distance is usually defined as Euclidean distance. Here we describe a closest pair algorithm for two-dimensional space, also called the planar closest pair problem. The algorithm is a parallel version of a standard sequential algorithm [Bentley and Shamos 1976], and, for n points, it requires the same work as the sequential versions $O(n \log n)$ and has depth $O(\log^2 n)$. The work is optimal.

The algorithm uses divide-and-conquer based on splitting the points along lines parallel to the y axis and is implemented as follows.

ALGORITHM: CLOSEST_PAIR(P).

```
1   if (|P| < 2) then return (P, ∞)
2   x_m := MEDIAN ({x : (x, y) ∈ P})
3   L := {(x, y) ∈ P | x < x_m}
4   R := {(x, y) ∈ P | x ≥ x_m}
5   in parallel do
6       (L', δ_L) := CLOSEST_PAIR(L)
7       (R', δ_R) := CLOSEST_PAIR(R)
8   P' := MERGE_BY_Y(L', R')
9   δ_P := BOUNDARY_MERGE(P', δ_L, δ_R, x_m)
10  return (P', δ_P)
```

This function takes a set of points P in the plane and returns both the original points sorted along the y axis and the distance between the closest two points. The sorted points are needed to help merge the results from recursive calls and can be thrown away at the end. It would be easy to modify the routine to return the closest pair of points in addition to the distance between them. The function works by dividing the points in half based on the median x value, recursively solving the problem on each half, and then merging the results. The MERGE_BY_Y function merges L' and R' along the y axis and can use a standard parallel merge routine. The interesting aspect of the code is the BOUNDARY_MERGE routine, which works on the same principle as described by Bentley and Shamos [1976] and can be computed with $O(\log n)$ depth and $O(n)$ work. We first review the principle and then show how it is implemented in parallel.

The inputs to BOUNDARY_MERGE are the original points P sorted along the y axis, the closest distance within L and R, and the median point x_m. The closest distance in P must be either the distance δ_L, the distance δ_R, or the distance between a point in L and a point in R. For this distance to be less than δ_L or δ_R, the two points must lie within $\delta = \min(\delta_L, \delta_R)$ of the line $x = x_m$. Thus, the two vertical lines at $x_r = x_m + \delta$ and $x_l = x_m - \delta$ define the borders of a region M in which the points must lie (see Fig. 12.11). If we could find the closest distance in M, call it δ_M, then the closest overall distance is $\delta_P = \min(\delta_L, \delta_R, \delta_M)$.

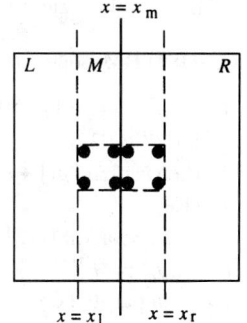

To find δ_M, we take advantage of the fact that not many points can be packed closely together within M since all points within L or R must be separated by at least δ. Figure 12.11 shows the tightest possible packing of points in a $2\delta \times \delta$ rectangle within M. This packing implies that if the points in M are sorted along the y axis, each point can determine the minimum distance to another point in M by looking at a fixed number of neighbors in the sorted order, at most seven in each direction. To see this, consider one of the points along the top of the $2\delta \times \delta$ rectangle. To find if there are any points below it that are closer than δ, it needs only to consider the points within the rectangle (points below the rectangle must be farther than δ away). As the figure illustrates, there can be at most seven other points within the rectangle. Given this property, the following function implements the border merge.

FIGURE 12.11 Merging two rectangles to determine the closest pair. Only 8 points can fit in the $2\delta \times \delta$ dashed rectangle.

ALGORITHM: BOUNDARY_MERGE($P, \delta_L, \delta_R, x_m$).

1 $\delta := \min(\delta_L, \delta_R)$
2 $M := \{(x, y) \in P \mid (x \geq x_m - \delta) \wedge (x \leq x_m + \delta)\}$
3 $\delta_M := \min(\{= \min(\{distance(M[i], M[i + j]) : j \in [1..7]\})$
4 $\qquad\qquad : i \in [0..|P - 7)\}$
5 **return** $\min(\delta, \delta_M)$

In this function each point in M looks at seven points in front of it in the sorted order and determines the distance to each of these points. The minimum over all distances is taken. Since the distance relationship is symmetric, there is no need for each point to consider points behind it in the sorted order.

The work of BOUNDARY_MERGE is $O(n)$ and the depth is dominated by taking the minimum, which has $O(\log n)$ depth.[2] The work of the merge and median steps in CLOSEST_PAIR is also $O(n)$, and the depth of both is bounded by $O(\log n)$. The total work and depth of the algorithm therefore can be solved with the recurrences

$$W(n) = 2W(n/2) + O(n) \quad = O(n \log n)$$
$$D(n) = D(n/2) + O(\log n) = O(\log^2 n)$$

Planar Convex Hull

The convex hull problem takes a set of points in k dimensions and returns the smallest convex region that contains all of the points. In two dimensions, the problem is called the planar convex hull problem and it returns the set of points that form the corners of the region. These points are a subset of the original points. We will describe two parallel algorithms for the planar convex hull problem. They are both based on divide-and-conquer, but one does most of the work before the divide step, and the other does most of the work after.

QuickHull

The parallel *QuickHull* algorithm [Blelloch and Little 1994] is based on the sequential version [Preparata and Shamos 1985], so named because of its similarity to the QuickSort algorithm. As with QuickSort, the strategy is to pick a *pivot* element, split the data based on the pivot, and recurse on each of the split sets. Also as with QuickSort, the pivot element is not guaranteed to split the data into equally sized sets, and in the worst case the algorithm requires $O(n^2)$ work; however, in practice the algorithm is often very efficient, probably the most practical of the convex hull algorithms. At the end of the section we briefly describe how the splits can be made precisely so the work is bounded by $O(n \log n)$.

The QuickHull algorithm is based on the recursive function SUBHULL, which is implemented as follows.

ALGORITHM: SUBHULL(P, p_1, p_2).

1   $P' := \{p \in P \mid$ RIGHT_OF ?$(p, (p_1, p_2))\}$
2   **if** $(|P'| < 2)$
3   **then return** $[p_1]$ ++ P'
4   **else**
5      $i :=$ MAX_INDEX($\{$DISTANCE$(p, (p_1, p_2)) : p \in P'\}$)
6      $p_m := P'[i]$
7   **in parallel do**
8         $H_l :=$ SUBHULL(P', p_1, p_m)
9         $H_r :=$ SUBHULL(P', p_m, p_2)
10   **return** H_l ++ H_r

---

[2] The depth of finding the minimum or maximum of a set of numbers actually can be improved to $O(\log \log n)$ with concurrent reads [Shiloach and Vishkin 1981].

This function takes a set of points P in the plane and two points p_1 and p_2 that are known to lie on the convex hull and returns all of the points that lie on the hull clockwise from p_1 to p_2, inclusive of p_1, but not of p_2. For example, in Fig. 12.12 SUBHULL($[A, B, C, \ldots, P]$, A, P) would return the sequence $[A, B, J, O]$.

The function SUBHULL works as follows. Line 1 removes all of the elements that cannot be on the hull because they lie to the right of the line from p_1 to p_2. This can easily be calculated using a cross product. If the remaining set P' is either empty or has just one element, the algorithm is done. Otherwise, the algorithm finds the point p_m farthest from the line (p_1, p_2). The point p_m must be on the hull since as a line at infinity parallel to (p_1, p_2) moves toward (p_1, p_2), it must first hit p_m. In line 5, the function MAX_INDEX returns the index of the maximum value of a sequence, using $O(n)$ work $O(\log n)$ depth, which is then used to extract the point p_m. Once p_m is found, SUBHULL is called twice recursively to find the hulls from p_1 to p_m and from p_m to p_2. When the recursive calls return, the results are appended.

The algorithm function uses SUBHULL to find the full convex hull.

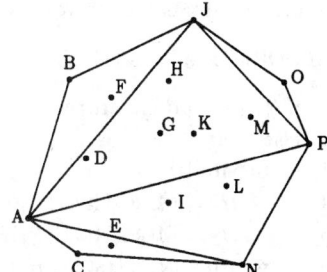

$[A\ B\ C\ D\ E\ F\ G\ H\ I\ J\ K\ L\ M\ N\ O\ P]$

$A\ [B\ D\ F\ G\ H\ J\ K\ M\ O]\ P\ [C\ E\ I\ L\ N]$

$A\ [B\ F]\ J\ [O]\ P\ N\ [C\ E]$

$A\ B\ J\ O\ P\ N\ C$

FIGURE 12.12 An example of the QuickHull algorithm.

ALGORITHM: QUICK_HULL(P).

1   $X := \{x : (x, y) \in P\}$
2   $x_{\min} := P[\text{min\_index}(X)]$
3   $x_{\max} := P[\text{max\_index}(X)]$
4   **return** SUBHULL($P, x_{\min}, x_{\max}$) **++** SUBHULL($P, x_{\max}, x_{\min}$)

We now consider the costs of the parallel QuickHull. The cost of everything other than the recursive calls is $O(n)$ work and $O(\log n)$ depth. If the recursive calls are balanced so that neither recursive call gets much more than half the data, then the number of levels of recursion will be $O(\log n)$. This will lead to the algorithm running in $O(\log^2 n)$ depth. Since the sum of the sizes of the recursive calls can be less than n (e.g., the points within the triangle AJP will be thrown out when making the recursive calls to find the hulls between A and J and between J and P), the work can be as little as $O(n)$ and often is in practice. As with QuickSort, however, when the recursive calls are badly partitioned, the number of levels of recursion can be as bad as $O(n)$ with work $O(n^2)$. For example, consider the case when all of the points lie on a circle and have the following unlikely distribution: $x_{\min}$ and $x_{\max}$ appear on opposite sides of the circle. There is one point that appears halfway between $x_{\min}$ and $x_{\max}$ on the sphere and this point becomes the new $x_{\max}$. The remaining points are defined recursively. That is, the points become arbitrarily close to $x_{\min}$ (see Fig. 12.13).

Kirkpatrick and Seidel [1986] have shown that it is possible to modify QuickHull so that it makes provably good partitions. Although the technique is shown for a sequential algorithm, it is easy to parallelize. A simplification of the technique is given by Chan et al. [1995]. This parallelizes even better and leads to an $O(\log^2 n)$ depth algorithm with $O(n \log h)$ work where h is the number of points on the convex hull.

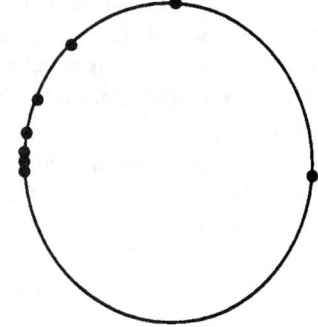

FIGURE 12.13 Contrived set of points for worst-case QuickHull.

MergeHull

The MergeHull algorithm [Overmars and Van Leeuwen 1981] is another divide-and-conquer algorithm for solving the planar convex hull problem. Unlike QuickHull, however, it does most of its work after returning from the recursive calls. The algorithm is implemented as follows.

ALGORITHM: MERGEHULL(P).

1 **if** ($|P| < 3$) **then return** P
2 **else**
3   **in parallel do**
4      $H_1 = $ MERGEHULL ($P[0..|P|/2)$)
5      $H_2 = $ MERGEHULL ($P[|P|/2..|P|)$)
6   **return** JOIN_HULLS(H_1, H_2)

This function assumes the input P is presorted according to the x coordinates of the points. Since the points are presorted, H_1 is a convex hull on the left and H_2 is a convex hull on the right. The JOIN_HULLS routine is the interesting part of the algorithm. It takes the two hulls and merges them into one. To do this, it needs to find upper and lower points u_1 and l_1 on H_1 and u_2 and l_2 on H_2 such that u_1, u_2 and l_1, l_2 are successive points on H (see Fig. 12.14). The lines b_1 and b_2 joining these upper and lower points are called the upper and lower bridges, respectively. All of the points between u_1

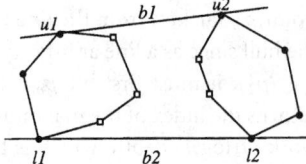

FIGURE 12.14 Merging two convex hulls.

and l_1 and between u_2 and l_2 on the *outer* sides of H_1 and H_2 are on the final convex hull, whereas the points on the *inner* sides are not on the convex hull. Without loss of generality we consider only how to find the upper bridge b_1. Finding the lower bridge b_2 is analogous.

To find the upper bridge, one might consider taking the points with the maximum y. However, this does not work in general; u_1 can lie as far down as the point with the minimum x or maximum x value (see Fig. 12.15). Instead, there is a nice solution based on binary search. Assume that the points on the convex hulls are given in order (e.g., clockwise). At each step the search algorithm will eliminate half the remaining points from consideration in either H_1 or H_2 or both. After at most $\log|H_1| + \log|H_2|$ steps the search will be left with only one point in each hull, and these will be the desired points u_1 and u_2. Figure 12.16 illustrates the rules for eliminating part of H_1 or H_2 on each step.

We now consider the cost of the algorithm. Each step of the binary search requires only constant work and depth since we only need to consider the middle two points M_1 and M_2, which can be found in constant time if the hull is kept sorted. The cost of the full binary search to find the upper bridge is therefore bounded by $D(n) = W(n) = O(\log n)$. Once we have found the upper and lower bridges, we need to remove the points on H_1 and H_2 that are not on H and append the remaining convex hull points. This requires linear work and constant depth. The overall costs of MERGEHULL are, therefore,

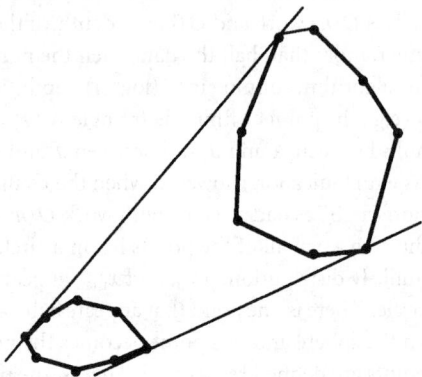

FIGURE 12.15 A bridge that is far from the top of the convex hull.

$$D(n) = D(n/2) + \log n = O(\log^2 n)$$

$$W(n) = 2W(n/2) + \log n + n = O(n \log n)$$

This algorithm can be improved to run in $O(\log n)$ depth using one of two techniques. The first involves implementing the search for the bridge points such that it runs in constant depth with linear work [Atallah and Goodrich 1988]. This involves sampling every $\sqrt{n}$th point on each hull and comparing all pairs of these two samples to narrow the search region down to regions of size $\sqrt{n}$ in constant depth. The patches then can be finished in constant depth by comparing all pairs between the two patches. The second

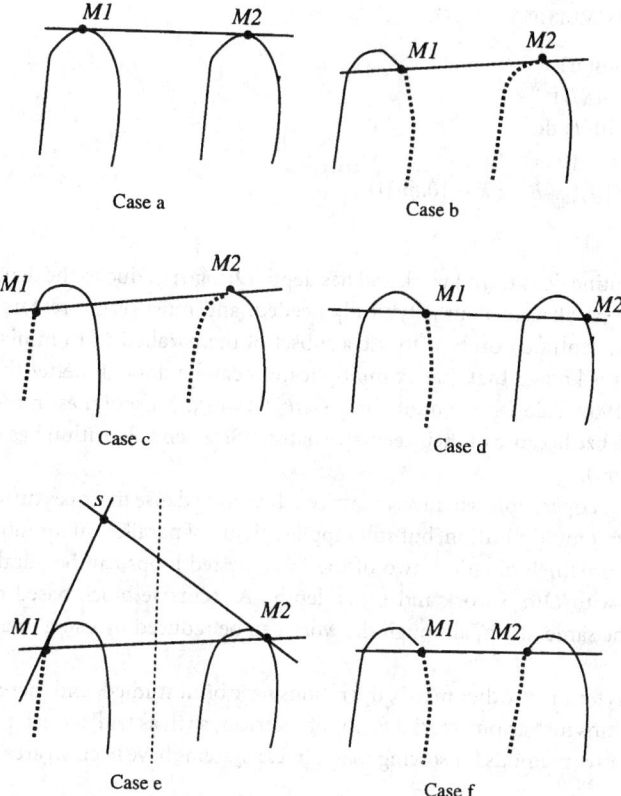

FIGURE 12.16 Cases used in the binary search for finding the upper bridge for the MergeHull. The points $M1$ and $M2$ mark the middle of the remaining hulls. The dotted lines represent the part of the hull that can be eliminated from consideration. The mirror images of cases b–e are also used. In case e, the region to eliminate depends on which side of the separating line the intersection of the tangents appears.

technique [Aggarwal et al. 1988, Atallah and Goodrich 1986] uses a divide-and-conquer to separate the point set into $\sqrt{n}$ regions, solves the convex hull on each region recursively, and then merges all pairs of these regions using the binary search method. Since there are $\sqrt{n}$ regions and each of the searches takes $O(\log n)$ work, the total work for merging is $O((\sqrt{n})^2 \log n) = O(n \log n)$ and the depth is $O(\log n)$. This leads to an overall algorithm that runs in $O(n \log n)$ work and $O(\log n)$ depth.

12.8 Numerical Algorithms

There has been an immense amount of work on parallel algorithms for numerical problems. Here we briefly discuss some of the problems and results. We suggest the following sources for further information on parallel numerical algorithms: Reif [1993, chs. 12–14], JáJá [1992, ch. 8], Kumar et al. [1994, chs. 5, 10, 11], and Bertsekas and Tsitsiklis [1989].

Matrix Operations

Matrix operations form the core of many numerical algorithms and led to some of the earliest work on parallel algorithms. The most basic matrix operation is matrix multiply. The standard triply nested loop for multiplying two dense matrices is highly parallel since each of the loops can be parallelized:

ALGORITHM: MATRIX_MULTIPLY (A, B).

```
1   (l, m) := dimensions(A)
2   (m, n) := dimensions(B)
3   in parallel for i ∈ [0..l) do
4       in parallel for j ∈ [0..n) do
5           R_ij := sum({A_ik * B_kj : k ∈ [0..m)})
6   return R
```

If $l = m = n$, this routine does $O(n^3)$ work and has depth $O(\log(n))$, due to the depth of the summation. This has much more parallelism than is typically needed, and most of the research on parallel matrix multiplication has concentrated on how to use a subset of the parallelism to minimize communication costs. Sequentially, it is known that matrix multiplication can be done in better than $O(n^3)$ work. For example, Strassen's [1969] algorithm requires only $O(n^{2.81})$ work. Most of these more efficient algorithms are also easy to parallelize because of their recursive nature (Strassen's algorithm has $O(\log n)$ depth using a simple parallelization).

Another basic matrix operation is to invert matrices. Inverting dense matrices turns out to be somewhat less parallel than matrix multiplication, but still supplies plenty of parallelism for most practical purposes. When using Gauss–Jordan elimination, two of the three nested loops can be parallelized leading to an algorithm that runs with $O(n^3)$ work and $O(n)$ depth. A recursive block-based method using matrix multiplies leads to the same depth, although the work can be reduced by using one of the more efficient matrix multiplies.

Parallel algorithms for many other matrix operations have been studied, and there has also been significant work on algorithms for various special forms of matrices, such as tridiagonal, triangular, and general sparse matrices. Iterative methods for solving sparse linear systems have been an area of significant activity.

Fourier Transform

Another problem for which there has been a long history of parallel algorithms is the discrete Fourier transform (DFT). The fast Fourier transform (FFT) algorithm for solving the DFT is quite easy to parallelize and, as with matrix multiplication, much of the research has gone into reducing communication costs. In fact, the butterfly network topology is sometimes called the FFT network since the FFT has the same communication pattern as the network [Leighton 1992, section 3.7]. A parallel FFT over complex numbers can be expressed as follows.

ALGORITHM: FFT(A).

```
1   n := |A|
2   if (n = 1) then return A
3   else
4       in parallel do
5           E := FFT({A[2i] : i ∈ [0..n/2)})
6           O := FFT({A[2i + 1] : i ∈ [0..n/2)})
7       return {E[j] + O[j]e^{2πij/n} : j ∈ [0..n/2)} ++ {E[j] − O[j]e^{2πij/n} : j ∈ [0..n/2)}
```

It simply calls itself recursively on the odd and even elements and then puts the results together. This algorithm does $O(n \log n)$ work, as does the sequential version, and has a depth of $O(\log n)$.

12.9 Parallel Complexity Theory

Researchers have developed a complexity theory for parallel computation that is in some ways analogous to the theory of NP-completeness. A problem is said to belong to the class NC (Nick's class) if it can be solved in depth polylogarithmic in the size of the problem using work that is polynomial in the size of the problem [Cook 1981, Pippenger 1979]. The class NC in parallel complexity theory plays the role of P in sequential complexity, i.e., the problems in NC are thought to be tractable in parallel. Examples of problems in NC include sorting, finding minimum cost spanning trees, and finding convex hulls. A problem is said to be P-complete if it can be solved in polynomial time and if its inclusion in NC would imply that $NC = P$. Hence, the notion of P-completeness plays the role of NP-completeness in sequential complexity. (And few believe that $NC = P$.)

Although much early work in parallel algorithms aimed at showing that certain problems belong to the class NC (without considering the issue of efficiency), this work tapered off as the importance of work efficiency became evident. Also, even if a problem is P-complete, there may be efficient (but not necessarily polylogarithmic time) parallel algorithms for solving it. For example, several efficient and highly parallel algorithms are known for solving the maximum flow problem, which is P-complete.

We conclude with a short list of P-complete problems. Full definitions of these problems and proofs that they are P-complete can be found in textbooks and surveys such as Gibbons and Rytter [1990], JáJá [1992], and Karp and Ramachandran [1990]. P-complete problems are:

1. **Lexicographically first maximal independent set and clique.** Given a graph G with vertices $V = 1, 2, \ldots, n$, and a subset $S \subseteq V$, determine if S is the lexicographically first maximal independent set (or maximal clique) of G.
2. **Ordered depth-first search.** Given a graph $G = (V, E)$, an ordering of the edges at each vertex, and a subset $T \subset E$, determine if T is the depth-first search tree that the sequential depth-first algorithm would construct using this ordering of the edges.
3. **Maximum flow.**
4. **Linear programming.**
5. **The circuit value problem.** Given a Boolean circuit, and a set of inputs to the circuit, determine if the output value of the circuit is one.
6. **The binary operator generability problem.** Given a set S, an element e not in S, and a binary operator·, determine if e can be generated from S using·.
7. **The context-free grammar emptiness problem.** Given a context-free grammar, determine if it can generate the empty string.

Defining Terms

CRCW: This refers to a shared memory model that allows for concurrent reads (CR) and concurrent writes (CW) to the memory.

CREW: This refers to a shared memory model that allows for concurrent reads (CR) but only exclusive writes (EW) to the memory.

Depth: The longest chain of sequential dependences in a computation.

EREW: This refers to a shared memory model that allows for only exclusive reads (ER) and exclusive writes (EW) to the memory.

Graph contraction: Contracting a graph by removing a subset of the vertices.

List contraction: Contracting a list by removing a subset of the nodes.

Multiprefix: A generalization of the scan (prefix sums) operation in which the partial sums are grouped by keys.

Multiprocessor model: A model of parallel computation based on a set of communicating sequential processors.

Pipelined divide-and-conquer: A divide-and-conquer paradigm in which partial results from recursive calls can be used before the calls complete. The technique is often useful for reducing the depth of various algorithms.

Pointer jumping: In a linked structure replacing a pointer with the pointer it points to. Used for various algorithms on lists and trees. Also called **recursive doubling.**

PRAM model: A multiprocessor model in which all of the processors can access a shared memory for reading or writing with uniform cost.

Prefix sums: A parallel operation in which each element in an array or linked list receives the sum of all of the previous elements.

Random sampling: Using a randomly selected sample of the data to help solve a problem on the whole data.

Recursive doubling: Same as pointer jumping.

Scan: A parallel operation in which each element in an array receives the sum of all of the previous elements.

Shortcutting: Same as pointer jumping.

Symmetry breaking: A technique to break the symmetry in a structure such as a graph which can locally look the same to all of the vertices. Usually implemented with randomization.

Tree contraction: Contracting a tree by removing a subset of the nodes.

Work: The total number of operations taken by a computation.

Work-depth model: A model of parallel computation in which one keeps track of the total work and depth of a computation without worrying about how it maps onto a machine.

Work efficient: When an algorithm does no more work than some other algorithm or model. Often used when relating a parallel algorithm to the best known sequential algorithm but also used when discussing emulations of one model on another.

References

Aggarwal, A., Chazelle, B., Guibas, L., Ò'Dùnlaing, C., and Yap, C. 1988. Parallel computational geometry. *Algorithmica* 3(3):293–327.

Aho, A. V., Hopcroft, J. E., and Ullman, J. D. 1974. *The Design and Analysis of Computer Algorithms.* Addison–Wesley, Reading, MA.

Akl, S. G. 1985. *Parallel Sorting Algorithms.* Academic Press, Toronto, Canada.

Anderson, R. J. and Miller, G. L. 1988. Deterministic parallel list ranking. In *Aegean Workshop on Computing: VLSI Algorithms and Architectures.* J. Reif, ed. Vol. 319, Lecture notes in computer science, pp. 81–90. Springer–Verlag, New York.

Anderson, G. L. and Miller, G. L. 1990. A simple randomized parallel algorithm for list-ranking. *Inf. Process. Lett.* 33(5):269–273.

Atallah, M. J., Cole, R., and Goodrich, M. T. 1989. Cascading divide-and-conquer: a technique for designing parallel algorithms. *SIAM J. Comput.* 18(3):499–532.

Atallah, M. J. and Goodrich, M. T. 1986. Efficient parallel solutions to some geometric problems. *J. Parallel Distrib. Comput.* 3(4):492–507.

Atallah, M. J. and Goodrich, M. T. 1988. Parallel algorithms for some functions of two convex polygons. *Algorithmica* 3(4):535–548.

Awerbuch, B. and Shiloach, Y. 1987. New connectivity and MSF algorithms for shuffle-exchange network and PRAM. *IEEE Trans. Comput.* C-36(10):1258–1263.

Bar-Noy, A. and Kipnis, S. 1992. Designing broadcasting algorithms in the postal model for message-passing systems, pp. 13–22. In *Proc. 4th Annu. ACM Symp. Parallel Algorithms Architectures.* ACM Press, New York.

Beneš, V. E. 1965. *Mathematical Theory of Connecting Networks and Telephone Traffic.* Academic Press, New York.

Bentley, J. L. and Shamos, M. 1976. Divide-and-conquer in multidimensional space, pp. 220–230. In *Proc. ACM Symp. Theory Comput.* ACM Press, New York.

Bertsekas, D. P. and Tsitsiklis, J. N. 1989. *Parallel and Distributed Computation: Numerical Methods.* Prentice–Hall, Englewood Cliffs, NJ.

Blelloch, G. E. 1990. *Vector Models for Data-Parallel Computing.* MIT Press, Cambridge, MA.

Blelloch, G. E. 1996. Programming parallel algorithms. *Commun. ACM* 39(3):85–97.

Blelloch, G. E., Chandy, K. M., and Jagannathan, S., eds. 1994. *Specification of Parallel Algorithms.* Vol. 18, DIMACS series in discrete mathematics and theoretical computer science. American Math. Soc. Providence, RI.

Blelloch, G. E. and Greiner, J. 1995. Parallelism in sequential functional languages, pp. 226–237. In *Proc. ACM Symp. Functional Programming Comput. Architecture.* ACM Press, New York.

Blelloch, G. E., Leiserson, C. E., Maggs, B. M., Plaxton, C. G., Smith, S. J., and Zagha, M. 1991. A comparison of sorting algorithms for the connection machine CM-2, pp. 3–16. In *Proc. ACM Symp. Parallel Algorithms Architectures.* Hilton Head, SC, July. ACM Press, New York.

Blelloch, G. E. and Little, J. J. 1994. Parallel solutions to geometric problems in the scan model of computation. *J. Comput. Syst. Sci.* 48(1):90–115.

Brent, R. P. 1974. The parallel evaluation of general arithmetic expressions. *J. Assoc. Comput. Mach.* 21(2):201–206.

Chan, T. M. Y., Snoeyink, J., and Yap, C. K. 1995. Output-sensitive construction of polytopes in four dimensions and clipped Voronoi diagrams in three, pp. 282–291. In *Proc. 6th Annu. ACM–SIAM Symp. Discrete Algorithms.* ACM–SIAM, ACM Press, New York.

Cole, R. 1988. Parallel merge sort. *SIAM J. Comput.* 17(4):770–785.

Cook, S. A. 1981. Towards a complexity theory of synchronous parallel computation. *Enseignement Mathematique* 27:99–124.

Culler, D., Karp, R., Patterson, D., Sahay, A., Schauser, K. E., Santos, E., Subramonian, R., and von Eicken, T. 1993. LogP: towards a realistic model of parallel computation, pp. 1–12. In *Proc. 4th ACM SIGPLAN Symp. Principles Pract. Parallel Programming.* ACM Press, New York.

Cypher, R. and Sanz, J. L. C. 1994. *The SIMD Model of Parallel Computation.* Springer–Verlag, New York.

Fortune, S. and Wyllie, J. 1978. Parallelism in random access machines, pp. 114–118. In *Proc. 10th Annu. ACM Symp. Theory Comput.* ACM Press, New York.

Gazit, H. 1991. An optimal randomized parallel algorithm for finding connected components in a graph. *SIAM J. Comput.* 20(6):1046–1067.

Gibbons, P. B., Matias, Y., and Ramachandran, V. 1994. The QRQW PRAM: accounting for contention in parallel algorithms, pp. 638–648. In *Proc. 5th Annu. ACM–SIAM Symp. Discrete Algorithms.* Jan. ACM Press, New York.

Gibbons, A. and Rytter, W. 1990. *Efficient Parallel Algorithms.* Cambridge University Press, Cambridge, England.

Goldshlager, L. M. 1978. A unified approach to models of synchronous parallel machines, pp. 89–94. In *Proc. 10th Annu. ACM Symp. Theory Comput.* ACM Press, New York.

Goodrich, M. T. 1996. Parallel algorithms in geometry. In *CRC Handbook of Discrete and Computational Geometry.* CRC Press, Boca Raton, FL.

Greiner, J. 1994. A comparison of data-parallel algorithms for connected components, pp. 16–25. In *Proc. 6th Annu. ACM Symp. Parallel Algorithms Architectures.* June. ACM Press, New York.

Halperin, S. and Zwick, U. 1994. An optimal randomized logarithmic time connectivity algorithm for the EREW PRAM, pp. 1–10. In *Proc. ACM Symp. Parallel Algorithms Architectures.* June. ACM Press, New York.

Harris, T. J. 1994. A survey of pram simulation techniques. *ACM Comput. Surv.* 26(2):187–206.

Huang, J. S. and Chow, Y. C. 1983. Parallel sorting and data partitioning by sampling, pp. 627–631. In *Proc. IEEE Comput. Soc. 7th Int. Comput. Software Appl. Conf.* Nov.

JáJá, J. 1992. *An Introduction to Parallel Algorithms.* Addison–Wesley, Reading, MA.

Karlin, A. R. and Upfal, E. 1988. Parallel hashing: an efficient implementation of shared memory. *J. Assoc. Comput. Mach.* 35:876–892.

Karp, R. M. and Ramachandran, V. 1990. Parallel algorithms for shared memory machines. In *Handbook of Theoretical Computer Science—Volume A: Algorithms and Complexity*. J. Van Leeuwen, ed., MIT Press, Cambridge, MA.

Kirkpatrick, D. G. and Seidel, R. 1986. The ultimate planar convex hull algorithm? *SIAM J. Comput.* 15:287–299.

Klein, P. N. and Tarjan, R. E. 1994. A randomized linear-time algorithm for finding minimum spanning trees. In *Proc. ACM Symp. Theory Comput.* May. ACM Press, New York.

Knuth, D. E. 1973. *Sorting and Searching*. Vol. 3. The Art of Computer Programming. Addison–Wesley, Reading, MA.

Kogge, P. M. and Stone, H. S. 1973. A parallel algorithm for the efficient solution of a general class of recurrence equations. *IEEE Trans. Comput.* C-22(8):786–793.

Kumar, V., Grama, A., Gupta, A., and Karypis, G. 1994. *Introduction to Parallel Computing: Design and Analysis of Algorithms*. Benjamin Cummings, Redwood City, CA.

Leighton, F. T. 1992. *Introduction to Parallel Algorithms and Architectures: Arrays, Trees, and Hypercubes*. Morgan Kaufmann, San Mateo, CA.

Leiserson, C. E. 1985. Fat-trees: universal networks for hardware-efficient supercomputing. *IEEE Trans. Comput.* C-34(10):892–901.

Luby, M. 1985. A simple parallel algorithm for the maximal independent set problem, pp. 1–10. In *Proc. ACM Symp. Theory Comput.* May. ACM Press, New York.

Maon, Y., Schieber, B., and Vishkin, U. 1986. Parallel ear decomposition search (eds) and st-numbering in graphs. *Theor. Comput. Sci.* 47:277–298.

Matias, Y. and Vishkin, U. 1991. On parallel hashing and integer sorting. *J. Algorithms* 12(4):573–606.

Miller, G. L. and Ramachandran, V. 1992. A new graph triconnectivity algorithm and its parallelization. *Combinatorica* 12(1):53–76.

Miller, G. and Reif, J. 1989. Parallel tree contraction part 1: fundamentals. In *Randomness and Computation. Vol. 5. Advances in computing research*, pp. 47–72. JAI Press, Greenwich, CT.

Miller, G. L. and Reif, J. H. 1991. Parallel tree contraction part 2: further applications. *SIAM J. Comput.* 20(6):1128–1147.

Overmars, M. H. and Van Leeuwen, J. 1981. Maintenance of configurations in the plane. *J. Comput. Syst. Sci.* 23:166–204.

Padua, D., Gelernter, D., and Nicolau, A., eds. 1990. *Languages and Compilers for Parallel Computing Research Monographs in Parallel and Distributed*. MIT Press, Cambridge, MA.

Pippenger, N. 1979. On simultaneous resource bounds, pp. 307–311. In *Proc. 20th Annu. Symp. Found. Comput. Sci.*

Pratt, V. R. and Stockmeyer, L. J. 1976. A characterization of the power of vector machines. *J. Comput. Syst. Sci.* 12:198–221.

Preparata, F. P. and Shamos, M. I. 1985. *Computational Geometry—An Introduction*. Springer–Verlag, New York.

Ranade, A. G. 1991. How to emulate shared memory. *J. Comput. Syst. Sci.* 42(3):307–326.

Reid-Miller, M. 1994. List ranking and list scan on the Cray C-90, pp. 104–113. In *Proc. 6th Annu. ACM Symp. Parallel Algorithms Architectures*. June. ACM Press, New York.

Reif, J. H., ed. 1993. *Synthesis of Parallel Algorithms*. Morgan Kaufmann, San Mateo, CA.

Reif, J. H. and Valiant, L. G. 1983. A logarithmic time sort for linear size networks, pp. 10–16. In *Proc. 15th Annu. ACM Symp. Theory Comput.* April. ACM Press, New York.

Savitch, W. J. and Stimson, M. 1979. Time bounded random access machines with parallel processing. *J. Assoc. Comput. Mach.* 26:103–118.

Shiloach, Y. and Vishkin, U. 1981. Finding the maximum, merging and sorting in a parallel computation model. *J. Algorithms* 2(1):88–102.

Shiloach, Y. and Vishkin, U. 1982. An $O(\log n)$ parallel connectivity algorithm. *J. Algorithms* 3:57–67.

Stone, H. S. 1975. Parallel tridiagonal equation solvers. *ACM Trans. Math. Software* 1(4):289–307.

Strassen, V. 1969. Gaussian elimination is not optimal. *Numerische Mathematik* 14(3):354–356.

Tarjan, R. E. and Vishkin, U. 1985. An efficient parallel biconnectivity algorithm. *SIAM J. Comput.* 14(4):862–874.

Valiant, L. G. 1990. A bridging model for parallel computation. *Commun. ACM* 33(8):103–111.

Valiant, L. G. 1990. General purpose parallel architectures, pp. 943–971. In *Handbook of Theoretical Computer Science*. J. van Leeuwen, ed. Elsevier Science, B. V., Amsterdam, The Netherlands.

Vishkin, U. 1984. Parallel-design distributed-implementation (PDDI) general purpose computer. *Theor. Comp. Sci.* 32:157–172.

Wyllie, J. C. 1979. The Complexity of Parallel Computations. *Department of Computer Science, Tech. Rep.* TR-79-387, Cornell University, Ithaca, NY. Aug.

13

Combinatorial Optimization

Vijay Chandru
Indian Institute of Science

M. R. Rao
Indian Institute
of Management

13.1 Introduction

Bin packing, routing, scheduling, layout, and network design are generic examples of combinatorial optimization problems that often arise in computer engineering and decision support. Unfortunately, almost all interesting generic classes of combinatorial optimization problems are $\mathcal{N}P$-hard. The scale at which these problems arise in applications and the explosive exponential complexity of the search spaces preclude the use of simplistic enumeration and search techniques. Despite the worst-case intractability of combinatorial optimization, in practice we are able to solve many large problems and often with off-the-shelf software. Effective software for combinatorial optimization is usually problem specific and based on sophisticated algorithms that combine approximation methods with search schemes and that exploit mathematical (and not just syntactic) structure in the problem at hand.

Multidisciplinary interests in combinatorial optimization have led to several fairly distinct paradigms in the development of this subject. Each paradigm may be thought of as a particular combination of a *representation scheme* and a *methodology* (see Table 13.1). The most established of these, the **integer programming** paradigm, uses implicit algebraic forms (linear constraints) to represent combinatorial optimization and **linear programming** and its extensions as the workhorses in the design of the solution algorithms. It is this paradigm that forms the central theme of this chapter.

0-8493-2909-4/97/$0.00+$.50
© 1997 by CRC Press, Inc.

TABLE 13.1 Paradigms in Combinatorial Optimization

Paradigm	Representation	Methodology
`Integer programming`	Linear constraints, Linear objective, Integer variables	Linear programming and extensions
`Search`	State space, Discrete control	Dynamic programming, $\mathcal{A}^*$
`Local improvement`	Neighborhoods Fitness functions	Hill climbing, Simulated annealing, Tabu search, Genetic algorithms
`Constraint logic programming`	Horn rules	Resolution, constraint solvers

Other well known paradigms in combinatorial optimization are **search**, **local improvement**, and **constraint logic programming**. Search uses state-space representations and partial enumeration techniques such as $\mathcal{A}^*$ and dynamic programming. Local improvement requires only a representation of neighborhood in the solution space, and methodologies vary from simple hill climbing to the more sophisticated techniques of simulated annealing, tabu search, and genetic algorithms. Constraint logic programming uses the syntax of Horn rules to represent combinatorial optimization problems and uses resolution to orchestrate the solution of these problems with the use of domain-specific constraint solvers. Whereas integer programming was developed and nurtured by the mathematical programming community, these other paradigms have been popularized by the artificial intelligence community.

An abstract formulation of combinatorial optimization is

$$(\text{CO}) \quad \min\{f(I) : I \in \mathcal{I}\}$$

where $\mathcal{I}$ is a collection of subsets of a finite ground set $E = \{e_1, e_2, \ldots, e_n\}$ and f is a criterion (objective) function that maps 2^E (the power set of E) to the reals. A **mixed integer linear program** (MILP) is of the form

$$(\text{MILP}) \quad \min_{x \in \Re^n}\{\mathbf{cx} : A\mathbf{x} \geq \mathbf{b}, \; \mathbf{x}_j \text{ integer } \forall \, j \in J\}$$

which seeks to minimize a linear function of the decision vector $\mathbf{x}$ subject to linear inequality constraints and the requirement that a subset of the decision variables is integer valued. This model captures many variants. If $J = \{1, 2, \ldots, n\}$, we say that the integer program is *pure*, and *mixed* otherwise. Linear equations and bounds on the variables can be easily accommodated in the inequality constraints. Notice that by adding in inequalities of the form $0 \leq \mathbf{x}_j \leq 1$ for a $j \in J$ we have forced $\mathbf{x}_j$ to take value 0 or 1. It is such Boolean variables that help capture combinatorial optimization problems as special cases of MILP.

Pure integer programming with variables that take arbitrary integer values is a class which has strong connections to number theory and particularly the geometry of numbers and Presburgher arithmetic. Although this is a fascinating subject with important applications in cryptography, in the interests of brevity we shall largely restrict our attention to MILP where the integer variables are Boolean.

The fact that mixed integer linear programs subsume combinatorial optimization problems follows from two simple observations. The first is that a collection $\mathcal{I}$ of subsets of a finite ground set E can always be represented by a corresponding collection of incidence vectors, which are $\{0, 1\}$-vectors in $\Re^E$. Further, arbitrary nonlinear functions can be represented via piecewise linear approximations by using linear constraints and mixed variables (continuous and Boolean).

The next section contains a primer on linear inequalities, polyhedra, and linear programming. These are the tools we will need to analyze and solve integer programs. Section 13.4, is a testimony to the earlier cryptic comments on how integer programs model combinatorial optimization problems. In addition to working a number of examples of such integer programming formulations, we shall also review a formal representation theory of (Boolean) mixed integer linear programs.

With any mixed integer program we associate a **linear programming relaxation** obtained by simply ignoring the integrality restrictions on the variables. The point being, of course, that we have polynomial-time (and practical) algorithms for solving linear programs. Thus, the linear programming relaxation of (MILP) is given by

$$\text{(LP)} \quad \min_{x \in \Re^n}\{\mathbf{cx} : A\mathbf{x} \geq \mathbf{b}\}$$

The thesis underlying the integer linear programming approach to combinatorial optimization is that this linear programming relaxation retains enough of the structure of the combinatorial optimization problem to be a useful weak representation. In section 13.5 we shall take a closer look at this thesis in that we shall encounter special structures for which this relaxation is *tight*. For general integer programs, there are several alternative schemes for generating linear programming relaxations with varying qualities of approximation. A general principle is that we often need to disaggregate integer formulations to obtain higher quality linear programming relaxations. To solve such huge linear programs we need specialized techniques of large-scale linear programming. These aspects will be the content of section 13.3.

The reader should note that the focus in this chapter is on solving hard combinatorial optimization problems. We catalog the special structures in integer programs that lead to tight linear programming relaxations (section 13.5) and hence to polynomial-time algorithms. These include structures such as network flows, matching, and matroid optimization problems. Many hard problems actually have pieces of these nice structures embedded in them. Practitioners of combinatorial optimization have always used insights from special structures to devise strategies for hard problems.

The computational art of integer programming rests on useful interplays between search methodologies and linear programming relaxations. The paradigms of branch and bound and branch and cut are the two enormously effective partial enumeration schemes that have evolved at this interface. These will be discussed in section 13.6. It may be noted that all general purpose integer programming software available today uses one or both of these paradigms.

The inherent complexity of integer linear programming has led to a long-standing research program in approximation methods for these problems. Linear programming relaxation and Lagrangian relaxation are two general approximation schemes that have been the real workhorses of computational practice. Primal–dual strategies and semidefinite relaxations are two recent entrants that appear to be very promising. Section 13.7 of this chapter reviews these developments in the approximation of combinatorial optimization problems.

We conclude the chapter with brief comments on future prospects in combinatorial optimization from the algebraic modeling perspective.

13.2 A Primer on Linear Programming

Polyhedral combinatorics is the study of embeddings of combinatorial structures in Euclidean space and their algebraic representations. We will make extensive use of some standard terminology from polyhedral theory. Definitions of terms not given in the brief review below can be found in Nemhauser and Wolsey [1988].

A (convex) **polyhedron** in $\Re^n$ can be algebraically defined in two ways. The first and more straightforward definition is the *implicit* representation of a polyhedron in $\Re^n$ as the solution set to a finite system of linear inequalities in n variables. A single linear inequality $\mathbf{ax} \leq a_0$; $\mathbf{a} \neq \mathbf{0}$ defines a *half-space* of $\Re^n$. Therefore, geometrically a polyhedron is the intersection set of a finite number of half-spaces.

A *polytope* is a bounded polyhedron. Every polytope is the convex closure of a finite set of points. Given a set of points whose convex combinations generate a polytope, we have an explicit or *parametric* algebraic representation of it. A *polyhedral cone* is the solution set of a system of homogeneous linear inequalities. Every (polyhedral) cone is the conical or positive closure of a finite set of vectors. These generators of the cone provide a parametric representation of the cone. And finally, a polyhedron can be alternatively defined as the Minkowski sum of a polytope and a cone. Moving from one representation of any of these polyhedral objects to another defines the essence of the computational burden of polyhedral combinatorics. This is particularly true if we are interested in *minimal* representations.

A set of points $\mathbf{x}^1, \ldots, \mathbf{x}^m$ is *affinely independent* if the unique solution of $\sum_{i=1}^m \lambda_i \mathbf{x}^i = 0$, $\sum_{i=1}^m \lambda_i = 0$ is $\lambda_i = 0$ for $i = 1, \ldots, m$. Note that the maximum number of affinely independent points in $\Re^n$ is $n + 1$. A polyhedron P is of *dimension k*, dim $P = k$, if the maximum number of affinely independent points in P is $k + 1$. A polyhedron $P \subseteq \Re^n$ of dimension n is called *full dimensional*. An inequality $\mathbf{a}\mathbf{x} \leq a_0$ is called *valid* for a polyhedron P if it is satisfied by all $\mathbf{x}$ in P. It is called *supporting* if in addition there is an $\bar{\mathbf{x}}$ in P that satisfies $\mathbf{a}\bar{\mathbf{x}} = a_0$. A *face* of the polyhedron is the set of all $\mathbf{x}$ in P that also satisfies a valid inequality as an equality. In general, many valid inequalities might represent the same face. Faces other than P itself are called *proper*. A *facet* of P is a maximal nonempty and proper face. A facet is then a face of P with a dimension of dim $P - 1$. A face of dimension zero, i.e., a point v in P that is a face by itself, is called an **extreme point** of P. The extreme points are the elements of P that cannot be expressed as a strict convex combination of two distinct points in P. For a full-dimensional polyhedron, the valid inequality representing a facet is unique up to multiplication by a positive scalar, and facet-inducing inequalities give a minimal implicit representation of the polyhedron. Extreme points, on the other hand, give rise to minimal parametric representations of polytopes.

The two fundamental problems of linear programming (which are polynomially equivalent) follow:

- *Solvability.* This is the problem of checking if a system of linear constraints on real (rational) variables is solvable or not. Geometrically, we have to check if a polyhedron, defined by such constraints, is nonempty.

- *Optimization.* This is the problem (LP) of optimizing a linear objective function over a polyhedron described by a system of linear constraints.

Building on polarity in cones and polyhedra, duality in linear programming is a fundamental concept which is related to both the complexity of linear programming and to the design of algorithms for solvability and optimization. We will encounter the solvability version of duality (called Farkas' Lemma) while discussing the Fourier elimination technique subsequently. Here we will state the main duality results for optimization. If we take the *primal* linear program to be

$$(P) \quad \min_{x \in \Re^n}\{\mathbf{c}\mathbf{x} : A\mathbf{x} \geq \mathbf{b}\}$$

there is an associated *dual* linear program

$$(D) \quad \max_{y \in \Re^m}\{\mathbf{b}^T \mathbf{y} : A^T \mathbf{y} = \mathbf{c}^T, \; \mathbf{y} \geq \mathbf{0}\}$$

and the two problems satisfy the following:

1. For any $\hat{\mathbf{x}}$ and $\hat{\mathbf{y}}$ feasible in (P) and (D) (i.e., they satisfy the respective constraints), we have $\mathbf{c}\hat{\mathbf{x}} \geq \mathbf{b}^T\hat{\mathbf{y}}$ (**weak duality**). Consequently, (P) has a finite optimal solution if and only if (D) does.
2. The pair $\mathbf{x}^*$ and $\mathbf{y}^*$ are optimal solutions for (P) and (D), respectively, if and only if $\mathbf{x}^*$ and $\mathbf{y}^*$ are feasible in (P) and (D) (i.e., they satisfy the respective constraints) and $\mathbf{c}\mathbf{x}^* = \mathbf{b}^T\mathbf{y}^*$ (**strong duality**).
3. The pair $\mathbf{x}^*$ and $\mathbf{y}^*$ are optimal solutions for (P) and (D), respectively, if and only if $\mathbf{x}^*$ and $\mathbf{y}^*$ are feasible in (P) and (D) (i.e., they satisfy the respective constraints) and $(A\mathbf{x}^* - \mathbf{b})^T\mathbf{y}^* = 0$ (**complementary slackness**).

The strong duality condition gives us a good stopping criterion for optimization algorithms. The complementary slackness condition, on the other hand, gives us a constructive tool for moving from dual to primal solutions and vice versa. The weak duality condition gives us a technique for obtaining lower bounds for minimization problems and upper bounds for maximization problems.

Note that the properties just given have been stated for linear programs in a particular form. The reader should be able to check that if, for example, the primal is of the form

$$(P') \quad \min_{\mathbf{x} \in \Re^n}\{\mathbf{cx} : A\mathbf{x} = \mathbf{b}, \ \mathbf{x} \geq \mathbf{0}\}$$

then the corresponding dual will have the form

$$(D') \quad \max_{\mathbf{y} \in \Re^m}\{\mathbf{b}^T\mathbf{y} : A^T\mathbf{y} \leq \mathbf{c}^T\}$$

The tricks needed for seeing this are that any equation can be written as two inequalities, an unrestricted variable can be substituted by the difference of two nonnegatively constrained variables, and an inequality can be treated as an equality by adding a nonnegatively constrained variable to the lesser side. Using these tricks, the reader could also check that duality in linear programming is involutory (i.e., the dual of the dual is the primal).

Algorithms for Linear Programming

We will now take a quick tour of some algorithms for linear programming. We start with the classical technique of Fourier, which is interesting because of its really simple syntactic specification. It leads to simple proofs of the duality principle of linear programming (solvability) that has been alluded to. We will then review the simplex method of linear programming, a method that has been finely honed over almost five decades. We will spend some time with the ellipsoid method and, in particular, with the polynomial equivalence of solvability (optimization) and separation problems, for this aspect of the ellipsoid method has had a major impact on the identification of many tractable classes of combinatorial optimization problems. We conclude the primer with a description of Karmarkar's [1984] breakthrough, which was an important landmark in the brief history of linear programming. A noteworthy role of interior point methods has been to make practical the theoretical demonstrations of tractability of various aspects of linear programming, including solvability and optimization, that were provided via the ellipsoid method.

Fourier's Scheme for Linear Inequalities

Constraint systems of linear *inequalities* of the form $A\mathbf{x} \leq \mathbf{b}$, where A is an $m \times n$ matrix of real numbers, are widely used in mathematical models. Testing the solvability of such a system is equivalent to linear programming.

Suppose we wish to eliminate the first variable $\mathbf{x}_1$ from the system $A\mathbf{x} \leq \mathbf{b}$. Let us denote

$$I^+ = \{i : A_{i1} > 0\} \qquad I^- = \{i : A_{i1} < 0\} \qquad I^0 = \{i : A_{i1} = 0\}$$

Our goal is to create an equivalent system of linear inequalities $\tilde{A}\tilde{\mathbf{x}} \leq \tilde{\mathbf{b}}$ defined on the variables $\tilde{\mathbf{x}} = (\mathbf{x}_2, \mathbf{x}_3, \ldots, \mathbf{x}_n)$:

- If I^+ is empty then we can simply delete all the inequalities with indices in I^- since they can be trivially satisfied by choosing a large enough value for $\mathbf{x}_1$. Similarly, if I^- is empty we can discard all inequalities in I^+.
- For each $k \in I^+$, $l \in I^-$ we add $-A_{l1}$ times the inequality $A_k\mathbf{x} \leq \mathbf{b}_k$ to A_{k1} times the inequality $A_l\mathbf{x} \leq \mathbf{b}_l$. In these new inequalities the coefficient of $\mathbf{x}_1$ is wiped out, that is, $\mathbf{x}_1$ is eliminated. Add these new inequalities to those already in I^0.

- The inequalities $\{\tilde{A}_{i1}\tilde{\mathbf{x}} \leq \tilde{\mathbf{b}}_i\}$ for all $i \in I^0$ represent the equivalent system on the variables $\tilde{\mathbf{x}} = (\mathbf{x}_2, \mathbf{x}_3, \ldots, \mathbf{x}_n)$.

Repeat this construction with $\tilde{A}\tilde{\mathbf{x}} \leq \tilde{\mathbf{b}}$ to eliminate $\mathbf{x}_2$ and so on until all variables are eliminated. If the resulting $\tilde{\mathbf{b}}$ (after eliminating $\mathbf{x}_n$) is nonnegative, we declare the original (and intermediate) inequality systems as being consistent. Otherwise,[1] $\tilde{\mathbf{b}} \not\geq 0$ and we declare the system inconsistent.

As an illustration of the power of elimination as a tool for theorem proving, we show now that Farkas Lemma is a simple consequence of the correctness of Fourier elimination. The lemma gives a direct proof that solvability of linear inequalities is in $\mathcal{NP} \bigcap co\mathcal{NP}$.

Farkas Lemma 13.1 (Duality in Linear Programming : Solvability). *Exactly one of the alternatives*

$$I. \quad \exists \, \mathbf{x} \in \Re^n : A\mathbf{x} \leq \mathbf{b}$$

$$II. \quad \exists \, \mathbf{y} \in \Re^m_+ : \mathbf{y}^t A = \mathbf{0}, \, \mathbf{y}^t \mathbf{b} < 0$$

is true for any given real matrices A, $\mathbf{b}$.

PROOF. Let us analyze the case when Fourier elimination provides a proof of the inconsistency of a given linear inequality system $A\mathbf{x} \leq \mathbf{b}$. The method clearly converts the given system into $RA\mathbf{x} \leq R\mathbf{b}$ where RA is zero and $R\mathbf{b}$ has at least one negative component. Therefore, there is some row of R, say, $\mathbf{r}$, such that $\mathbf{r}A = \mathbf{0}$ and $\mathbf{r}\mathbf{b} < 0$. Thus $\neg I$ implies II. It is easy to see that I and II cannot both be true for fixed A, $\mathbf{b}$. $\qquad\square$

In general, the Fourier elimination method is quite inefficient. Let k be any positive integer and n the number of variables be $2^k + k + 2$. If the input inequalities have left-hand sides of the form $\pm\mathbf{x}_r \pm \mathbf{x}_s \pm \mathbf{x}_t$ for all possible $1 \leq r < s < t \leq n$, it is easy to prove by induction that after k variables are eliminated, by Fourier's method, we would have at least $2^{n/2}$ inequalities. The method is therefore exponential in the worst case, and the explosion in the number of inequalities has been noted, in practice as well, on a wide variety of problems. We will discuss the central idea of minimal generators of the projection cone that results in a much improved elimination method.

First, let us identify the set of variables to be eliminated. Let the input system be of the form

$$P = \{(\mathbf{x}, \mathbf{u}) \in \Re^{n_1+n_2} \mid A\mathbf{x} + B\mathbf{u} \leq \mathbf{b}\}$$

where $\mathbf{u}$ is the set to be eliminated. The projection of P onto $\mathbf{x}$ or equivalently the effect of eliminating the $\mathbf{u}$ variables is

$$P_{\mathbf{x}} = \{\mathbf{x} \in \Re^{n_1} \mid \exists \mathbf{u} \in \Re^{n_2} \text{ such that } A\mathbf{x} + B\mathbf{u} \leq \mathbf{b}\}$$

Now W, the *projection cone* of P, is given by

$$W = \{\mathbf{w} \in \Re^m \mid \mathbf{w}B = \mathbf{0}, \, \mathbf{w} \geq \mathbf{0}\}$$

A simple application of Farkas Lemma yields a description of $P_{\mathbf{x}}$ in terms of W.

Projection Lemma 13.2. *Let G be any set of generators (e.g., the set of extreme rays) of the cone W. Then $P_{\mathbf{x}} = \{\mathbf{x} \in \Re^{n_1} \mid (\mathbf{g}A)\mathbf{x} \leq \mathbf{g}\mathbf{b} \, \forall \, \mathbf{g} \in G\}$.*

The lemma, sometimes attributed to Černikov [1961], reduces the computation of $P_{\mathbf{x}}$ to enumerating the extreme rays of the cone W or equivalently the extreme points of the polytope $W \cap \{\mathbf{w} \in \Re^m \mid \sum_{i=1}^m \mathbf{w}_i = 1\}$.

---

[1]Note that the final $\tilde{\mathbf{b}}$ may not be defined if all of the inequalities are deleted by the monotone sign condition of the first step of the construction described. In such a situation, we declare the system $A\mathbf{x} \leq \mathbf{b}$ *strongly consistent* since it is consistent for any choice of $\mathbf{b}$ in $\Re^m$. To avoid making repeated references to this exceptional situation, let us simply assume that it does not occur. The reader is urged to verify that this assumption is indeed benign.

Simplex Method

Consider a polyhedron $\mathcal{K} = \{\mathbf{x} \in \mathfrak{R}^n : A\mathbf{x} = \mathbf{b}, \mathbf{x} \geq \mathbf{0}\}$. Now $\mathcal{K}$ cannot contain an infinite (in both directions) line since it is lying within the nonnegative orthant of $\mathfrak{R}^n$. Such a polyhedron is called a *pointed* polyhedron. Given a pointed polyhedron $\mathcal{K}$ we observe the following:

- If $\mathcal{K} \neq \emptyset$, then $\mathcal{K}$ has at least one extreme point.
- If $\min\{\mathbf{cx} : A\mathbf{x} = \mathbf{b}, \mathbf{x} \geq \mathbf{0}\}$ has an optimal solution, then it has an optimal extreme point solution.

These observations together are sometimes called the fundamental theorem of linear programming since they suggest simple finite tests for both solvability and optimization. To generate all extreme points of $\mathcal{K}$, in order to find an optimal solution, is an impractical idea. However, we may try to run a partial search of the space of extreme points for an optimal solution. A simple local improvement search strategy of moving from extreme point to adjacent extreme point until we get to a local optimum is nothing but the simplex method of linear programming. The local optimum also turns out to be a global optimum because of the convexity of the polyhedron $\mathcal{K}$ and the linearity of the objective function $\mathbf{cx}$.

The simplex method walks along edge paths on the combinatorial graph structure defined by the boundary of convex polyhedra. Since these graphs are quite dense (Balinski's theorem states that the graph of d-dimensional polyhedron must be d-connected [Ziegler 1995]) and possibly large (the Lower Bound Theorem states that the number of vertices can be exponential in the dimension [Ziegler 1995]), it is indeed somewhat of a miracle that it manages to get to an optimal extreme point as quickly as it does. Empirical and probabilistic analyses indicate that the number of iterations of the simplex method is just slightly more than linear in the dimension of the primal polyhedron. However, there is no known variant of the simplex method with a worst-case polynomial guarantee on the number of iterations. Even a polynomial bound on the diameter of polyhedral graphs is not known.

Procedure 13.1. Primal Simplex $(\mathcal{K}, c)$:

0. **Initialize:**

   $\mathbf{x}_0 :=$ an extreme point of $\mathcal{K}$
   $k := 0$

1. **Iterative step:**

   do
   If for all edge directions $\mathcal{D}_k$ at $\mathbf{x}_k$, the objective function is nondecreasing, i.e.

   $$\mathbf{cd} \geq 0 \quad \forall \, \mathbf{d} \in \mathcal{D}_k$$

   then exit and return optimal $\mathbf{x}_k$.
   Else pick some $\mathbf{d}_k$ in $\mathcal{D}_k$ such that $\mathbf{cd}_k < 0$.
   If $\mathbf{d}_k \geq 0$ **then** declare the linear program unbounded in objective value and exit.
   Else $\mathbf{x}_{k+1} := \mathbf{x}_k + \theta_k * \mathbf{d}_k$, where

   $$\theta_k = \max\{\theta : \mathbf{x}_k + \theta * \mathbf{d}_k \geq 0\}$$

   $k := k + 1$
   od

2. **End**

Remark 13.1. In the initialization step, we assumed that an extreme point x_0 of the polyhedron K is available. This also assumes that the solvability of the constraints defining K has been established. These assumptions are reasonable since we can formulate the solvability problem as an optimization problem, with a self-evident extreme point, whose optimal solution either establishes unsolvability of $Ax = b$, $x \geq 0$ or provides an extreme point of K. Such an optimization problem is usually called a phase I model. The point being, of course, that the simplex method, as just described, can be invoked on the phase I model and, if successful, can be invoked once again to carry out the intended minimization of cx. There are several different formulations of the phase I model that have been advocated. Here is one:

$$\min\{v_0 \; : \; Ax + bv_0 = b, \; x \geq 0, \; v_0 \geq 0\}$$

The solution $(x, \; v_0)^T = (0, \ldots, 0, 1)$ is a self-evident extreme point and $v_0 = 0$ at an optimal solution of this model is a necessary and sufficient condition for the solvability of $Ax = b$, $x \geq 0$.

Remark 13.2. The scheme for generating improving edge directions uses an algebraic representation of the extreme points as certain bases, called feasible bases, of the vector space generated by the columns of the matrix A. It is possible to have linear programs for which an extreme point is geometrically overdetermined (degenerate), i.e., there are more than d facets of K that contain the extreme point, where d is the dimension of K. In such a situation, there would be several feasible bases corresponding to the same extreme point. When this happens, the linear program is said to be *primal degenerate*.

Remark 13.3. There are two sources of nondeterminism in the primal simplex procedure. The first involves the choice of edge direction d_k made in step 1. At a typical iteration there may be many edge directions that are improving in the sense that $cd_k < 0$. Dantzig's rule, the maximum improvement rule, and steepest descent rule are some of the many rules that have been used to make the choice of edge direction in the simplex method. There is, unfortunately, no clearly dominant rule and successful codes exploit the empirical and analytic insights that have been gained over the years to resolve the edge selection nondeterminism in simplex methods. The second source of nondeterminism arises from degeneracy. When there are multiple feasible bases corresponding to an extreme point, the simplex method has to pivot from basis to adjacent basis by picking an entering basic variable (a pseudoEdge direction) and by dropping one of the old ones. A wrong choice of the leaving variables may lead to cycling in the sequence of feasible bases generated at this extreme point. Cycling is a serious problem when linear programs are highly degenerate as in the case of linear relaxations of many combinatorial optimization problems. The lexicographic rule (perturbation rule) for the choice of leaving variables in the simplex method is a provably finite method (i.e., all cycles are broken). A clever method proposed by Bland (cf. Schrijver [1986]) preorders the rows and columns of the matrix A. In the case of nondeterminism in either entering or leaving variable choices, Bland's rule just picks the lowest index candidate. All cycles are avoided by this rule also.

The simplex method has been the veritable workhorse of linear programming for four decades now. However, as already noted, we do not know of a simplex method that has worst-case bounds that are polynomial. In fact, Klee and Minty exploited the sensitivity of the original simplex method of Dantzig, to projective scaling of the data, and constructed exponential examples for it. The ellipsoid method of Shor [1970] was devised to overcome poor scaling in convex programming problems and, therefore, turned out to be the natural choice of an algorithm to first establish polynomial-time solvability of linear programming. Later Karmarkar [1984] took care of both projection and scaling simultaneously and arrived at a superior algorithm.

The Ellipsoid Algorithm

The ellipsoid algorithm of Shor [1970] gained prominence in the late 1970s when Hačijan [1979] (pronounced Khachyan) showed that this convex programming method specializes to a polynomial-time algorithm for linear programming problems. This theoretical breakthrough naturally led to intense study of this method and its properties. The survey paper by Bland et al. [1981] and the monograph by Akgül [1984] attest to this fact. The direct theoretical consequences for combinatorial optimization problems was independently documented by Padberg and Rao [1981], Karp and Papadimitriou [1982], and Grötschel et al. [1988]. The ability of this method to implicitly handle linear programs with an exponential list of constraints and maintain polynomial-time convergence is a characteristic that is the key to its applications in combinatorial optimization. For an elegant treatment of the many deep theoretical consequences of the ellipsoid algorithm, the reader is directed to the monograph by Lovász [1986] and the book by Grötschel et al. [1988].

Computational experience with the ellipsoid algorithm, however, showed a disappointing gap between the theoretical promise and practical efficiency of this method in the solution of linear programming problems. Dense matrix computations as well as the slow average-case convergence properties are the reasons most often cited for this behavior of the ellipsoid algorithm. On the positive side though, it has been noted (cf. Ecker and Kupferschmid [1983]) that the ellipsoid method is competitive with the best known algorithms for (nonlinear) convex programming problems.

Let us consider the problem of testing if a polyhedron $Q \in \Re^d$, defined by linear inequalities, is nonempty. For technical reasons let us assume that Q is rational, i.e., all extreme points and rays of Q are rational vectors or, equivalently, that all inequalities in some description of Q involve only rational coefficients. The ellipsoid method does not require the linear inequalities describing Q to be explicitly specified. It suffices to have an oracle representation of Q. Several different types of oracles can be used in conjunction with the ellipsoid method (Karp and Papadimitriou [1982], Padberg and Rao [1981], Grötschel et al. [1988]). We will use the *strong separation oracle*:

Oracle: **Strong Separation**$(Q, \mathbf{y})$

> Given a vector $\mathbf{y} \in \Re^d$, decide whether $\mathbf{y} \in Q$, and if not find a hyperplane that separates $\mathbf{y}$ from Q; more precisely, find a vector $\mathbf{c} \in \Re^d$ such that $\mathbf{c}^T \mathbf{y} < \min\{\mathbf{c}^T \mathbf{x} \mid \mathbf{x} \in Q\}$.

The ellipsoid algorithm initially chooses an ellipsoid large enough to contain a part of the polyhedron Q if it is nonempty. This is easily accomplished because we know that if Q is nonempty then it has a rational solution whose (binary encoding) length is bounded by a polynomial function of the length of the largest coefficient in the linear program and the dimension of the space.

The center of the ellipsoid is a feasible point if the separation oracle tells us so. In this case, the algorithm terminates with the coordinates of the center as a solution. Otherwise, the separation oracle outputs an inequality that separates the center point of the ellipsoid from the polyhedron Q. We translate the hyperplane defined by this inequality to the center point. The hyperplane slices the ellipsoid into two halves, one of which can be discarded. The algorithm now creates a new ellipsoid that is the minimum volume ellipsoid containing the remaining half of the old one. The algorithm questions if the new center is feasible and so on. The key is that the new ellipsoid has substantially smaller volume than the previous one. When the volume of the current ellipsoid shrinks to a sufficiently small value, we are able to conclude that Q is empty. This fact is used to show the polynomial-time convergence of the algorithm.

The crux of the complexity analysis of the algorithm is on the a priori determination of the iteration bound. This in turn depends on three factors. The volume of the initial ellipsoid E_0, the rate of volume shrinkage ($vol(E_{k+1})/vol(E_k) < e^{-\frac{1}{(2d)}}$), and the volume threshold at which we can safely conclude that Q must be empty. The assumption of Q being a rational polyhedron is used to argue that Q can be modified into a full-dimensional polytope without affecting the decision question: "Is Q non-empty ?" After careful accounting for all of these technical details and some others (e.g., compensating for the roundoff errors caused by the square root computation in the algorithm), it is possible to establish the following fundamental result.

Theorem 13.1. *There exists a polynomial $g(d, \phi)$ such that the* **ellipsoid method** *runs in time bounded by $T g(d, \phi)$ where ϕ is an upper bound on the size of linear inequalities in some description of Q*

and T *is the maximum time required by the oracle* **Strong Separation** $(Q, \mathbf{y})$ *on inputs* $\mathbf{y}$ *of size at most* $g(d, \phi)$.

The size of a linear inequality is just the length of the encoding of all of the coefficients needed to describe the inequality. A direct implication of the theorem is that solvability of linear inequalities can be checked in polynomial time if strong separation can be solved in polynomial time. This implies that the standard linear programming solvability question has a polynomial-time algorithm (since separation can be effected by simply checking all of the constraints). Happily, this approach provides polynomial-time algorithms for much more than just the standard case of linear programming solvability. The theorem can be extended to show that the optimization of a linear objective function over Q also reduces to a polynomial number of calls to the strong separation oracle on Q. A converse to this theorem also holds, namely, separation can be solved by a polynomial number of calls to a solvability/optimization oracle (Grötschel et al. [1982]). Thus, optimization and separation are polynomially equivalent. This provides a very powerful technique for identifying tractable classes of optimization problems. Semidefinite programming and submodular function minimization are two important classes of optimization problems that can be solved in polynomial time using this property of the ellipsoid method.

Semidefinite Programming

The following optimization problem defined on symmetric $(n \times n)$ real matrices

$$\text{(SDP)} \quad \min_{X \in \Re^{n \times n}} \left\{ \sum_{ij} C \bullet X : A \bullet X = B, \; X \succeq 0 \right\}$$

is called a semidefinite program. Note that $X \succeq 0$ denotes the requirement that X is a positive semidefinite matrix, and $F \bullet G$ for $n \times n$ matrices F and G denotes the product matrix $(F_{ij} * G_{ij})$. From the definition of positive semidefinite matrices, $X \succeq 0$ is equivalent to

$$\mathbf{q}^T X \mathbf{q} \geq 0 \quad \text{for every } \mathbf{q} \in \Re^n$$

Thus semidefinite programming (SDP) is really a linear program on $O(n^2)$ variables with an (uncountably) infinite number of linear inequality constraints. Fortunately, the strong separation oracle is easily realized for these constraints. For a given symmetric X we use Cholesky factorization to identify the minimum eigenvalue λ_{min}. If λ_{min} is nonnegative then $X \succeq 0$ and if, on the other hand, λ_{min} is negative we have a separating inequality

$$\gamma_{min}^T X \gamma_{min} \geq 0$$

where γ_{min} is the eigenvector corresponding to λ_{min}. Since the Cholesky factorization can be computed by an $O(n^3)$ algorithm, we have a polynomial-time separation oracle and an efficient algorithm for SDP via the ellipsoid method. Alizadeh [1995] has shown that interior point methods can also be adapted to solving SDP to within an additive error ϵ in time polynomial in the size of the input and $\log 1/\epsilon$.

This result has been used to construct efficient approximation algorithms for maximum stable sets and cuts of graphs, Shannon capacity of graphs, and minimum colorings of graphs. It has been used to define hierarchies of relaxations for integer linear programs that strictly improve on known exponential-size linear programming relaxations. We shall encounter the use of SDP in the approximation of a maximum weight cut of a given vertex-weighted graph in section 13.7.

Minimizing Submodular Set Functions

The minimization of submodular set functions is another important class of optimization problems for which ellipsoidal and projective scaling algorithms provide polynomial-time solution methods.

Definition 13.1. Let N be a finite set. A real valued set function f defined on the subsets of N is submodular if $f(X \cup Y) + f(X \cap Y) \leq f(X) + f(Y)$ for $X, Y \subseteq N$.

Example 13.1. Let $G = (V, E)$ be an undirected graph with V as the node set and E as the edge set. Let $c_{ij} \geq 0$ be the weight or capacity associated with edge $(ij) \in E$. For $S \subseteq V$, define the cut function $c(S) = \sum_{i \in S, j \in V \setminus S} c_{ij}$. The cut function defined on the subsets of V is submodular since $c(X) + c(Y) - c(X \cup Y) - c(X \cap Y) = \sum_{i \in X \setminus Y, j \in Y \setminus X} 2c_{ij} \geq 0$.

The optimization problem of interest is

$$\min\{f(X) : X \subseteq N\}$$

The following remarkable construction that connects submodular function minimization with convex function minimization is due to Lovász (see Grötschel et al. [1988]).

Definition 13.2. The Lovász extension $\hat{f}(.)$ of a submodular function $f(.)$ satisfies

- $\hat{f} : [0, 1]^N \to \Re$.
- $\hat{f}(\mathbf{x}) = \sum_{I \in \mathcal{I}} \lambda_I f(\mathbf{x}_I)$ where $\mathbf{x} = \sum_{I \in \mathcal{I}} \lambda_I \mathbf{x}_I$, $\mathbf{x} \in [0, 1]^N$, $\mathbf{x}_I$ is the incidence vector of I for each $I \in \mathcal{I}$, $\lambda_I > 0$ for each I in $\mathcal{I}$, and $\mathcal{I} = \{I_1, I_2, \ldots, I_k\}$ with $\emptyset \neq I_1 \subset I_2 \subset \cdots \subset I_k \subseteq N$. Note that the representation $\mathbf{x} = \sum_{I \in \mathcal{I}} \lambda_I \mathbf{x}_I$ is unique given that the $\lambda_I > 0$ and that the sets in $\mathcal{I}$ are nested.

It is easy to check that $\hat{f}(.)$ is a convex function. Lovász also showed that the minimization of the submodular function $f(.)$ is a special case of convex programming by proving

$$\min\{f(X) : X \subseteq N\} = \min\{\hat{f}(\mathbf{x}) : \mathbf{x} \in [0, 1]^N\}$$

Further, if $\mathbf{x}^*$ is an optimal solution to the convex program and

$$\mathbf{x}^* = \sum_{I \in \mathcal{I}} \lambda_I \mathbf{x}_I$$

then for each $\lambda_I > 0$, it can be shown that $I \in \mathcal{I}$ minimizes f. The ellipsoid method can be used to solve this convex program (and hence submodular minimization) using a polynomial number of calls to an oracle for f [this oracle returns the value of $f(X)$ when input X].

Interior Point Methods

The announcement of the polynomial solvability of linear programming followed by the probabilistic analyses of the simplex method in the early 1980s left researchers in linear programming with a dilemma. We had one method that was good in a theoretical sense but poor in practice and another that was good in practice (and on average) but poor in a theoretical worst-case sense. This left the door wide open for a method that was good in both senses. Narendra Karmarkar closed this gap with a breathtaking new projective scaling algorithm. In retrospect, the new algorithm has been identified with a class of nonlinear programming methods known as logarithmic barrier methods. Implementations of a primal–dual variant of the logarithmic barrier method have proven to be the best approach at present. It is this variant that we describe.

It is well known that moving through the interior of the feasible region of a linear program using the negative of the gradient of the objective function, as the movement direction, runs into trouble because of getting *jammed* into corners (in high dimensions, corners make up most of the interior of a polyhedron). This jamming can be overcome if the negative gradient is balanced with a *centering* direction. The centering direction in Karmarkar's algorithm is based on the *analytic center* $\mathbf{y}_c$ of a full-dimensional polyhedron $\mathcal{D} = \{\mathbf{y} : A^T \mathbf{y} \leq c\}$ which is the unique optimal solution to

$$\max\left\{\sum_{j=1}^{n} \ln(\mathbf{z}_j) : A^T \mathbf{y} + \mathbf{z} = c\right\}$$

Recall the primal and dual forms of a linear program may be taken as

$$(P) \quad \min\{\mathbf{cx} : A\mathbf{x} = \mathbf{b}, \ \mathbf{x} \geq \mathbf{0}\}$$

$$(D) \quad \max\{\mathbf{b}^T\mathbf{y} : A^T\mathbf{y} \leq \mathbf{c}\}$$

The logarithmic barrier formulation of the dual (D) is

$$(D_\mu) \quad \max\left\{\mathbf{b}^T\mathbf{y} + \mu \sum_{j=1}^{n} \ell n(\mathbf{z}_j) : \ A^T\mathbf{y} + \mathbf{z} = \mathbf{c}\right\}$$

Notice that (D_μ) is equivalent to (D) as $\mu \to 0^+$. The optimality (Karush–Kuhn–Tucker) conditions for (D_μ) are given by

$$D_\mathbf{x} D_\mathbf{z} \mathbf{e} = \mu \mathbf{e}$$

$$A\mathbf{x} = \mathbf{b}$$

$$A^T\mathbf{y} + \mathbf{z} = \mathbf{c}$$

where $D_\mathbf{x}$ and $D_\mathbf{z}$ denote $n \times n$ diagonal matrices whose diagonals are $\mathbf{x}$ and $\mathbf{z}$, respectively. Notice that if we set μ to 0, the above conditions are precisely the primal–dual optimality conditions: complementary slackness, primal and dual feasibility of a pair of optimal (P) and (D) solutions. The problem has been reduced to solving the equations in $\mathbf{x}, \mathbf{y}, \mathbf{z}$. The classical technique for solving equations is Newton's method, which prescribes the directions,

$$\Delta\mathbf{y} = -\left(AD_\mathbf{x} D_\mathbf{z}^{-1} A^T\right)^{-1} AD_\mathbf{z}^{-1}(\mu\mathbf{e} - D_\mathbf{x} D_\mathbf{z}\mathbf{e}) \Delta\mathbf{z} = -A^T \Delta\mathbf{y}\Delta\mathbf{x}$$

$$= D_\mathbf{z}^{-1}(\mu\mathbf{e} - D_\mathbf{x} D_\mathbf{z}\mathbf{e}) - D_\mathbf{x} D_\mathbf{z}^{-1}\Delta\mathbf{z} \tag{13.1}$$

The strategy is to take one Newton step, reduce μ, and iterate until the optimization is complete. The criterion for stopping can be determined by checking for feasibility $(\mathbf{x}, \mathbf{z} \geq \mathbf{0})$ and if the duality gap $(\mathbf{x}^t\mathbf{z})$ is close enough to 0. We are now ready to describe the algorithm.

Procedure 13.2. Primal-Dual Interior:

   0. **Initialize:**

        $\mathbf{x}_0 > 0, \mathbf{y}_0 \in \Re^m, \mathbf{z}_0 > 0, \mu_0 > 0, \epsilon > 0, \rho > 0$
        $k := 0$

   1. **Iterative step:**

        do
        Stop if $A\mathbf{x}_k = \mathbf{b}, A^T\mathbf{y}_k + \mathbf{z}_k = \mathbf{c}$ and $\mathbf{x}_k^T \mathbf{z}_k \leq \epsilon$.
        $\mathbf{x}_{k+1} \leftarrow \mathbf{x}_k + \alpha_k^P \Delta\mathbf{x}_k$
        $\mathbf{y}_{k+1} \leftarrow \mathbf{y}_k + \alpha_k^D \Delta\mathbf{y}_k$
        $\mathbf{z}_{k+1} \leftarrow \mathbf{z}_k + \alpha_k^D \Delta\mathbf{z}_k$
        `/*` $\Delta\mathbf{x}_k, \Delta\mathbf{y}_k, \Delta\mathbf{z}_k$ `are the Newton directions from (1) */`
        $\mu_{k+1} \leftarrow \rho\mu_k$
        $k := k + 1$
        od

   2. **End**

Remark 13.4. The step sizes α_k^P and α_k^D are chosen to keep $\mathbf{x}_{k+1}$ and $\mathbf{z}_{k+1}$ strictly positive. The ability in the primal–dual scheme to choose separate step sizes for the primal and dual variables is a major advantage that this method has over the pure primal or dual methods. Empirically this advantage translates to a significant reduction in the number of iterations.

Remark 13.5. The stopping condition essentially checks for primal and dual feasibility and near complementary slackness. Exact complementary slackness is not possible with interior solutions. It is possible to maintain primal and dual feasibility through the algorithm, but this would require a phase I construction via artificial variables. Empirically, this feasible variant has not been found to be worthwhile. In any case, when the algorithm terminates with an interior solution, a post-processing step is usually invoked to obtain optimal extreme point solutions for the primal and dual. This is usually called the *purification* of solutions and is based on a clever scheme described by Megiddo [1991].

Remark 13.6. Instead of using Newton steps to drive the solutions to satisfy the optimality conditions of (D_μ), Mehrotra [1992] suggested a predictor–corrector approach based on power series approximations. This approach has the added advantage of providing a rational scheme for reducing the value of μ. It is the predictor–corrector based primal–dual interior method that is considered the current winner in interior point methods. The OB1 code of Lustig et al. [1994] is based on this scheme.

Remark 13.7. CPLEX 4.0 [1993], a general purpose linear (and integer) programming solver, contains implementations of interior point methods. A very recent[2] computational study of parallel implementations of simplex and interior point methods on the SGI Power Challenge (SGI R8000) platform indicates that on all but a few small linear programs in the NETLIB linear programming benchmark problem set, interior point methods dominate the simplex method in run times. New advances in handling Cholesky factorizations in parallel are apparently the reason for this exceptional performance of interior point methods.

Remark 13.8. Karmarkar [1990] has proposed an interior-point approach for integer programming problems. The main idea is to reformulate an integer program as the minimization of a quadratic energy function over linear constraints on continuous variables. Interior-point methods are applied to this formulation to find local optima.

13.3 Large-Scale Linear Programming in Combinatorial Optimization

Linear programming problems with thousands of rows and columns are routinely solved either by variants of the simplex method or by interior point methods. However, for several linear programs that arise in combinatorial optimization, the number of columns (or rows in the dual) are too numerous to be enumerated explicitly. The columns, however, often have a structure which is exploited to generate the columns as and when required in the simplex method. Such an approach, which is referred to as **column generation**, is illustrated next on the *cutting stock problem* (Gilmore and Gomory [1963]), which is also known as the *bin packing problem* in the computer science literature.

Cutting Stock Problem

Rolls of sheet metal of standard length L are used to cut required lengths $l_i, i = 1, 2, \ldots, m$. The jth cutting pattern should be such that a_{ij}, the number of sheets of length l_i cut from one roll of standard length L, must satisfy $\sum_{i=1}^{m} a_{ij} l_i \leq L$. Suppose $n_i, i = 1, 2, \ldots, m$ sheets of length l_i are required. The problem is to find cutting patterns so as to minimize the number of rolls of standard length L that are used to meet the requirements. A linear programming formulation of the problem is as follows.

---

[2]This comment is based on presentations by R. L. Bixby and I. J. Lustig at the INFORMS Computer Science Technical Section Meeting held in Dallas, Texas, in Jan. 1996.

Let $\mathbf{x}_j, j = 1, 2, \ldots, n$, denote the number of times the jth cutting pattern is used. In general, $\mathbf{x}_j, j = 1, 2, \ldots, n$ should be an integer but in the next formulation the variables are permitted to be fractional.

$$(\text{P1}) \quad \text{Min} \quad \sum_{j=1}^{n} \mathbf{x}_j$$

$$\text{Subject to} \quad \sum_{j=1}^{n} a_{ij}\mathbf{x}_j \geq n_i \quad i = 1, 2, \ldots, m$$

$$\mathbf{x}_j \geq 0 \quad j = 1, 2, \ldots, n$$

$$\text{where} \quad \sum_{i=1}^{m} l_i a_{ij} \leq L \quad j = 1, 2, \ldots, n$$

The formulation can easily be extended to allow for the possibility of p standard lengths $L_k, k = 1, 2, \ldots, p$, from which the n_i units of length $l_i, i = 1, 2, \ldots, m$, are to be cut.

The cutting stock problem can also be viewed as a bin packing problem. Several bins, each of standard capacity L, are to be packed with n_i units of item i, each of which uses up capacity of l_i in a bin. The problem is to minimize the number of bins used.

Column Generation

In general, the number of columns in (P1) is too large to enumerate all of the columns explicitly. The simplex method, however, does not require all of the columns to be explicitly written down. Given a basic feasible solution and the corresponding simplex multipliers $\mathbf{w}_i, i = 1, 2, \ldots, m$, the column to enter the basis is determined by applying dynamic programming to solve the following knapsack problem:

$$(\text{P2}) \quad z = \text{Max} \quad \sum_{i=1}^{m} \mathbf{w}_i a_i$$

$$\text{Subject to} \quad \sum_{i=1}^{m} l_i a_i \leq L$$

$$a_i \geq 0 \text{ and integer,} \quad \text{for } i = 1, 2, \ldots, m$$

Let $a_i^*, i = 1, 2, \ldots, m$, denote an optimal solution to (P2). If $z > 1$, the kth column to enter the basis has coefficients $a_{ik} = a_i^*, i = 1, 2, \ldots, m$.

Using the identified columns, a new improved (in terms of the objective function value) basis is obtained, and the column generation procedure is repeated. A major iteration is one in which (P2) is solved to identify, if there is one, a column to enter the basis. Between two major iterations, several minor iterations may be performed to optimize the linear program using only the available (generated) columns.

If $z \leq 1$, the current basic feasible solution is optimal to (P1). From a computational point of view, alternative strategies are possible. For instance, instead of solving (P2) to optimality, a column to enter the basis can be indentified as soon as a feasible solution to (P2) with an objective function value greater than 1 has been found. Such an approach would reduce the time required to solve (P2) but may increase the number of iterations required to solve (P1).

A column once generated may be retained, even if it comes out of the basis at a subsequent iteration, so as to avoid generating the same column again later on. However, at a particular iteration some columns, which appear unattractive in terms of their reduced costs, may be discarded in order to avoid having to store a large number of columns. Such columns can always be generated again subsequently, if necessary. The rationale for this approach is that such unattractive columns will rarely be required subsequently.

The dual of (P1) has a large number of rows. Hence column generation may be viewed as row generation in the dual. In other words, in the dual we start with only a few constraints explicitly written down. Given an optimal solution $\mathbf{w}$ to the current dual problem (i.e., with only a few constraints which have been explicitly written down) find a constraint that is violated by $\mathbf{w}$ or conclude that no such constraint exists.

The problem to be solved for identifying a violated constraint, if any, is exactly the separation problem that we encountered in the section on algorithms for linear programming.

Decomposition and Compact Representations

Large-scale linear programs sometimes have a block diagonal structure with a few additional constraints linking the different blocks. The linking constraints are referred to as the master constraints and the various blocks of constraints are referred to as subproblem constraints. Using the representation theorem of polyhedra (see, for instance, Nemhauser and Wolsey [1988]), the decomposition approach of Dantzig and Wolfe [1961] is to convert the original problem to an equivalent linear program with a small number of constraints but with a large number of columns or variables. In the cutting stock problem described in the preceding section, the columns are generated, as and when required, by solving a knapsack problem via dynamic programming. In the Dantzig–Wolfe decomposition scheme, the columns are generated, as and when required, by solving appropriate linear programs on the subproblem constraints.

It is interesting to note that the reverse of decomposition is also possible. In other words, suppose we start with a statement of a problem and an associated linear programming formulation with a large number of columns (or rows in the dual). If the column generation (or row generation in the dual) can be accomplished by solving a linear program, then a *compact* formulation of the original problem can be obtained. Here compact refers to the number of rows and columns being bounded by a polynomial function of the input length of the original problem. This result due to Martin [1991] enables one to solve the problem in the polynomial time by solving the compact formulation using interior point methods.

13.4 Integer Linear Programs

Integer linear programming problems (ILPs) are linear programs in which all of the variables are restricted to be integers. If only some but not all variables are restricted to be integers, the problem is referred to as a mixed integer program. Many combinatorial problems can be formulated as integer linear programs in which all of the variables are restricted to be 0 or 1. We will first discuss several examples of combinatorial optimization problems and their formulation as integer programs. Then we will review a general representation theory for integer programs that gives a formal measure of the expressiveness of this algebraic approach. We conclude this section with a representation theorem due to Benders [1962], which has been very useful in solving certain large-scale combinatorial optimization problems in practice.

Example Formulations

Covering and Packing Problems

A wide variety of location and scheduling problems can be formulated as set covering or set packing or set partitioning problems. The three different types of **covering and packing** problems can be succinctly stated as follows: Given (1) a finite set of elements $\mathcal{M} = \{1, 2, \ldots, m\}$, and (2) a family F of subsets of $\mathcal{M}$ with each member F_j, $j = 1, 2, \ldots, n$ having a profit (or cost) c_j associated with it, find a collection, S, of the members of F that maximizes the profit (or minimizes the cost) while ensuring that every element of $\mathcal{M}$ is in one of the following:

(P3): at most one member of S (set packing problem)
(P4): at least one member of S (set covering problem)
(P5): exactly one member of S (set partitioning problem)

The three problems (P3), (P4), and (P5) can be formulated as ILPs as follows:

Let A denote the $m \times n$ matrix where

$$A_{ij} = \begin{cases} 1 & \text{if element } i \in F_j \\ 0 & \text{otherwise} \end{cases}$$

The decision variables are x_j, $j = 1, 2, \ldots, n$ where

$$x_{ij} = \begin{cases} 1 & \text{if } F_j \text{ is chosen} \\ 0 & \text{otherwise} \end{cases}$$

The set packing problem is

$$(P3) \quad \text{Max } \mathbf{cx}$$

$$\text{Subject to} \quad A\mathbf{x} \le \mathbf{e}_m$$

$$x_j = 0 \quad \text{or} \quad 1, \quad j = 1, 2, \ldots, n$$

where $\mathbf{e}_m$ is an m-dimensional column vector of ones.

The set covering problem (P4) is (P3) with less than or equal to constraints replaced by greater than or equal to constraints and the objective is to minimize rather than maximize. The set partitioning problem (P5) is (P3) with the constraints written as equalities. The set partitioning problem can be converted to a set packing problem or set covering problem (see Padberg [1995]) using standard transformations. If the right-hand side vector $\mathbf{e}_m$ is replaced by a nonnegative integer vector $\mathbf{b}$, (P3) is referred to as the generalized set packing problem.

The airline crew scheduling problem is a classic example of the set partitioning or the set covering problem. Each element of $\mathcal{M}$ corresponds to a flight segment. Each subset F_j corresponds to an acceptable set of flight segments of a crew. The problem is to cover, at minimum cost, each flight segment exactly once. This is a set partitioning problem. If *dead heading* of crew is permitted, we have the set covering problem.

Packing and Covering Problems in a Graph

Suppose A is the node-edge incidence matrix of a graph. Now, (P3) is a weighted matching problem. If in addition, the right-hand side vector $\mathbf{e}_m$ is replaced by a nonnegative integer vector $\mathbf{b}$, (P3) is referred to as a weighted $\mathbf{b}$-matching problem. In this case, each variable x_j which is restricted to be an integer may have a positive upper bound of u_j. Problem (P4) is now referred to as the weighted edge covering problem. Note that by substituting for $x_j = 1 - y_j$, where $y_j = 0$ or 1, the weighted edge covering problem is transformed to a weighted $\mathbf{b}$-matching problem in which the variables are restricted to be 0 or 1.

Suppose A is the edge-node incidence matrix of a graph. Now, (P3) is referred to as the weighted vertex packing problem and (P4) is referred to as the weighted vertex covering problem. The *set packing* problem can be transformed to a weighted vertex packing problem in a graph G as follows:

> G contains a node for each x_j and an edge between nodes j and k exists if and only if the columns $A_{.j}$ and $A_{.k}$ are not orthogonal. G is called the *intersection graph* of A. The set packing problem is equivalent to the weighted vertex packing problem on G. Given G, the complement graph $\overline{G}$ has the same node set as G and there is an edge between nodes j and k in $\overline{G}$ if and only if there is no such corresponding edge in G. A clique in a graph is a subset, k, of nodes of G such that the subgraph induced by k is complete. Clearly, the weighted vertex packing problem in G is equivalent to finding a maximum weighted clique in $\overline{G}$.

Plant Location Problems

Given a set of customer locations $N = \{1, 2, \ldots, n\}$ and a set of potential sites for plants $M = \{1, 2, \ldots, m\}$, the plant location problem is to identify the sites where the plants are to be located so that the customers are served at a minimum cost. There is a fixed cost $\mathbf{f}_i$ of locating the plant at site i and the cost of serving customer j from site i is $\mathbf{c}_{ij}$. The decision variables are: y_i is set to 1 if a plant is located at site i and to 0 otherwise; x_{ij} is set to 1 if site i serves customer j and to 0 otherwise.

A formulation of the problem is

$$(P6) \quad \text{Min} \sum_{i=1}^{m} \sum_{j=1}^{n} \mathbf{c}_{ij} \mathbf{x}_{ij} + \sum_{i=1}^{m} \mathbf{f}_i \mathbf{y}_i$$

$$\text{subject to} \quad \sum_{i=1}^{m} \mathbf{x}_{ij} = 1 \qquad\qquad j = 1, 2, \ldots, n$$

$$\mathbf{x}_{ij} - \mathbf{y}_i \leq 0 \qquad\qquad i = 1, 2, \ldots, m; \quad j = 1, 2, \ldots, n$$

$$\mathbf{y}_i = 0 \quad \text{or} \quad 1 \qquad i = 1, 2, \ldots, m$$

$$\mathbf{x}_{ij} = 0 \quad \text{or} \quad 1 \qquad i = 1, 2, \ldots, m; \quad j = 1, 2, \ldots, n$$

Note that the constraints $\mathbf{x}_{ij} - \mathbf{y}_i \leq 0$ are required to ensure that customer j may be served from site i only if a plant is located at site i. Note that the constraints $\mathbf{y}_i = 0$ or 1 force an optimal solution in which $\mathbf{x}_{ij} = 0$ or 1. Consequently, the $\mathbf{x}_{ij} = 0$ or 1 constraints may be replaced by nonnegativity constraints $\mathbf{x}_{ij} \geq 0$.

The linear programming relaxation associated with (P6) is obtained by replacing constraints $\mathbf{y}_i = 0$ or 1 and $\mathbf{x}_{ij} = 0$ or 1 by nonnegativity contraints on $\mathbf{x}_{ij}$ and $\mathbf{y}_i$. The upper bound constraints on $\mathbf{y}_i$ are not required provided $\mathbf{f}_i \geq 0, i = 1, 2, \ldots, m$. The upper bound constraints on $\mathbf{x}_{ij}$ are not required in view of constraints $\sum_{i=1}^{m} \mathbf{x}_{ij} = 1$.

Remark 13.9. It is frequently possible to formulate the same combinatorial problem as two or more different ILPs. Suppose we have two ILP formulations (F1) and (F2) of the given combinatorial problem with both (F1) and (F2) being minimizing problems. Formulation (F1) is said to be stronger than (F2) if (LP1), the the linear programming relaxation of (F1), always has an optimal objective function value which is greater than or equal to the optimal objective function value of (LP2), which is the linear programming relaxation of (F2).

It is possible to reduce the number of constraints in (P6) by replacing the constraints $\mathbf{x}_{ij} - \mathbf{y}_i \leq 0$ by an aggregate:

$$\sum_{j=1}^{n} \mathbf{x}_{ij} - n\mathbf{y}_i \leq 0 \quad i = 1, 2, \ldots, m$$

However, the disaggregated (P6) is a stronger formulation than the formulation obtained by aggregrating the constraints as previously. By using standard transformations, (P6) can also be converted into a set packing problem.

Satisfiability and Inference Problems:

In propositional logic, a truth assignment is an assignment of true or false to each atomic proposition $\mathbf{x}_1, \mathbf{x}_2, \ldots \mathbf{x}_n$. A literal is an atomic proposition $\mathbf{x}_j$ or its negation $\neg \mathbf{x}_j$. For propositions in conjunctive normal form, a clause is a disjunction of literals and the proposition is a conjunction of clauses. A clause is obviously satisfied by a given truth assignment if at least one of its literals is true. The satisfiability problem consists of determining whether there exists a truth assignment to atomic propositions such that a set S of clauses is satisfied.

Let T_i denote the set of atomic propositions such that if any one of them is assigned true, the clause $i \in S$ is satisfied. Similarly, let F_i denote the set of atomic propositions such that if any one of them is assigned false, the clause $i \in S$ is satisfied.

The decision variables are

$$\mathbf{x}_j = \begin{cases} 1 & \text{if atomic proposition } j \text{ is assigned true} \\ 0 & \text{if atomic proposition } j \text{ is assigned false} \end{cases}$$

The satisfiability problem is to find a feasible solution to

$$(\text{P7}) \quad \sum_{j \in T_i} \mathbf{x}_j - \sum_{j \in F_i} \mathbf{x}_j \geq 1 - \mid F_i \mid \quad i \in S$$

$$\mathbf{x}_j = 0 \quad \text{or} \quad 1 \quad \text{for } j = 1, 2, \ldots, n$$

By substituting $\mathbf{x}_j = 1 - \mathbf{y}_j$, where $\mathbf{y}_j = 0$ or 1, for $j \in F_i$, (P7) is equivalent to the set covering problem

$$(\text{P8}) \quad \text{Min} \sum_{j=1}^{n} (\mathbf{x}_j + \mathbf{y}_j) \tag{13.2}$$

$$\text{subject to} \quad \sum_{j \in T_i} \mathbf{x}_j + \sum_{j \in F_i} \mathbf{y}_j \geq 1 \quad i \in S \tag{13.3}$$

$$\mathbf{x}_j + \mathbf{y}_j \geq 1 \quad j = 1, 2, \ldots, n \tag{13.4}$$

$$\mathbf{x}_j, \mathbf{y}_j = 0 \quad \text{or} \quad 1 \quad j = 1, 2, \ldots, n \tag{13.5}$$

Clearly (P7) is feasible if and only if (P8) has an optimal objective function value equal to n.

Given a set S of clauses and an additional clause $k \notin S$, the logical inference problem is to find out whether every truth assignment that satisfies all of the clauses in S also satisfies the clause k. The logical inference problem is

$$(\text{P9}) \quad \text{Min} \sum_{j \in T_k} \mathbf{x}_j - \sum_{j \in F_k} \mathbf{x}_j$$

$$\text{subject to} \quad \sum_{j \in T_i} \mathbf{x}_j - \sum_{j \in F_i} \mathbf{x}_j \geq 1 - |F_i| \quad i \in S$$

$$\mathbf{x}_j = 0 \quad \text{or} \quad 1 \quad j = 1, 2, \ldots, n$$

The clause k is implied by the set of clauses S, if and only if (P9) has an optimal objective function value greater than $-|F_k|$. It is also straightforward to express the MAX-SAT problem (i.e., find a truth assignment that maximizes the number of satisfied clauses in a given set S) as an integer linear program.

Multiprocessor Scheduling

Given n jobs and m processors, the problem is to allocate each job to one and only one of the processors so as to minimize the make span time, i.e., minimize the completion time of all of the jobs. The processors may not be identical and, hence, job j if allocated to processor i requires p_{ij} units of time. The multiprocessor scheduling problem is

$$(\text{P10}) \quad \text{Min} \ T$$

$$\text{subject to} \quad \sum_{i=1}^{m} \mathbf{x}_{ij} = 1 \quad j = 1, 2, \ldots, n$$

$$\sum_{j=1}^{n} \mathbf{p}_{ij} \mathbf{x}_{ij} - T \leq 0 \quad i = 1, 2, \ldots, m$$

$$\mathbf{x}_{ij} = 0 \quad \text{or} \quad 1$$

Note that if all $\mathbf{p}_{ij}$ are integers, the optimal solution will be such that T is an integer.

Jeroslow's Representability Theorem

Jeroslow [1989], building on joint work with Lowe [1984], characterized subsets of n-space that can be represented as the feasible region of a mixed integer (Boolean) program. They proved that a set is the feasible region of some mixed integer/linear programming problem (MILP) if and only if it is the union of finitely many polyhedra having the same recession cone (defined subsequently). Although this result is not widely known, it might well be regarded as the fundamental theorem of mixed integer modeling.

The basic idea of Jeroslow's results is that any set that can be represented in a mixed integer model can be represented in a disjunctive programming problem (i.e., a problem with either/or constraints). A *recession direction* for a set S in n-space is a vector $\mathbf{x}$ such that $s + \alpha\mathbf{x} \in S$ for all $s \in S$ and all $\alpha \geq 0$. The set of recession directions is denoted $rec(S)$. Consider the general mixed integer constraint set

$$\mathbf{f}(\mathbf{x}, \mathbf{y}, \boldsymbol{\lambda}) \leq \mathbf{b}$$

$$\mathbf{x} \in \Re^n, \qquad \mathbf{y} \in \Re^p \tag{13.6}$$

$$\boldsymbol{\lambda} = (\lambda_1, \ldots, \lambda_k), \quad \text{with} \quad \lambda_j \in \{0, 1\} \quad \text{for } j = 1, \ldots, k$$

Here $\mathbf{f}$ is a vector-valued function, so that $\mathbf{f}(\mathbf{x}, \mathbf{y}, \boldsymbol{\lambda}) \leq \mathbf{b}$ represents a set of constraints. We say that a set $S \subset \Re^n$ is *represented* by Eq. (13.6) if,

$$\mathbf{x} \in S \quad \text{if and only if } (\mathbf{x}, \mathbf{y}, \boldsymbol{\lambda}) \text{ satisfies Eq. (13.6) for some } y, \lambda.$$

If $\mathbf{f}$ is a linear transformation, so that Eq. (13.6) is a MILP constraint set, we will say that S is *MILP representable*. The main result can now be stated.

Theorem 13.2 [Jeroslow and Lowe 1984, Jeroslow 1989]. *A set in n-space is MILP representable if and only if it is the union of finitely many polyhedra having the same set of recession directions.*

Benders's Representation

Any mixed integer linear program can be reformulated so that there is only one continuous variable. This reformulation, due to Benders [1962], will in general have an exponential number of constraints. Analogous to column generation, discussed earlier, these rows (constraints) can be generated as and when required.

Consider the (MILP)

$$\max \{\mathbf{cx} + \mathbf{dy} : A\mathbf{x} + G\mathbf{y} \leq \mathbf{b}, \ \mathbf{x} \geq \mathbf{0}, \ \mathbf{y} \geq \mathbf{0} \text{ and integer}\}$$

Suppose the integer variables $\mathbf{y}$ are fixed at some values, then the associated linear program is

$$(\text{LP}) \quad \max \{\mathbf{cx} : \mathbf{x} \in \mathcal{P} = \{\mathbf{x} : A\mathbf{x} \leq \mathbf{b} - G\mathbf{y}, \ \mathbf{x} \geq \mathbf{0}\}\}$$

and its dual is

$$(\text{DLP}) \quad \min \{\mathbf{w}(\mathbf{b} - G\mathbf{y}) : \mathbf{w} \in Q = \{\mathbf{w} : \mathbf{w}A \geq \mathbf{c}, \ \mathbf{w} \geq \mathbf{0}\}\}$$

Let $\{\mathbf{w}^k\}$, $k = 1, 2, \ldots, K$ be the extreme points of Q and $\{\mathbf{u}^j\}$, $j = 1, 2, \ldots, J$ be the extreme rays of the recession cone of Q, $C_Q = \{\mathbf{u} : \mathbf{u}A \geq \mathbf{0}, \ \mathbf{u} \geq \mathbf{0}\}$. Note that if Q is nonempty, the $\{\mathbf{u}^j\}$ are all of the extreme rays of Q.

From linear programming duality, we know that if Q is empty and $\mathbf{u}^j(\mathbf{b} - G\mathbf{y}) \geq 0$, $j = 1, 2, \ldots, J$ for some $\mathbf{y} \geq \mathbf{0}$ and integer then (LP) and consequently (MILP) have an unbounded solution. If Q is nonempty and $\mathbf{u}^j(\mathbf{b} - G\mathbf{y}) \geq 0$, $j = 1, 2, \ldots, J$ for some $\mathbf{y} \geq \mathbf{0}$ and integer then (LP) has a finite optimum given by

$$\min_k \{\mathbf{w}^k(\mathbf{b} - G\mathbf{y})\}$$

Hence an equivalent formulation of (MILP) is

$$\text{Max } \alpha$$

$$\alpha \leq \mathbf{d}\mathbf{y} + \mathbf{w}^k(\mathbf{b} - G\mathbf{y}), \quad k = 1, 2, \ldots, K$$

$$\mathbf{u}^j(\mathbf{b} - G\mathbf{y}) \geq 0, \quad j = 1, 2, \ldots, J$$

$$\mathbf{y} \geq \mathbf{0} \text{ and integer}$$

$$\alpha \quad \text{unrestricted}$$

which has only one continuous variable α as promised.

13.5 Polyhedral Combinatorics

One of the main purposes of writing down an algebraic formulation of a combinatorial optimization problem as an integer program is to then examine the linear programming relaxation and understand how well it represents the discrete integer program. There are somewhat special but rich classes of such formulations for which the linear programming relaxation is sharp or tight. These correspond to linear programs that have integer valued extreme points. Such polyhedra are called **integral polyhedra**.

Special Structures and Integral Polyhedra

A natural question of interest is whether the LP associated with an ILP has only integral extreme points. For instance, the linear programs associated with matching and edge covering polytopes in a bipartite graph have only integral vertices. Clearly, in such a situation, the ILP can be solved as LP. A polyhedron or a polytope is referred to as being integral if it is either empty or has only integral vertices.

Definition 13.3. A 0, ± 1 matrix is totally unimodular if the determinant of every square submatrix is 0 or ± 1.

Theorem 13.3 [Hoffman and Kruskal 1956]. *Let*

$$A = \begin{pmatrix} A_1 \\ A_2 \\ A_3 \end{pmatrix}$$

be a 0, ± 1 *matrix and*

$$\mathbf{b} = \begin{pmatrix} \mathbf{b}_1 \\ \mathbf{b}_2 \\ \mathbf{b}_3 \end{pmatrix}$$

be a vector of appropriate dimensions. Then A is totally unimodular if and only if the polyhedron

$$P(A, \mathbf{b}) = \{\mathbf{x} : A_1\mathbf{x} \leq \mathbf{b}_1; A_2\mathbf{x} \geq \mathbf{b}_2; A_3\mathbf{x} = \mathbf{b}_3; \mathbf{x} \geq \mathbf{0}\}$$

is integral for all integral vectors $\mathbf{b}$.

The constraint matrix associated with a network flow problem (see, for instance, Ahuja et al.[1993]) is totally unimodular. Note that for a given integral $\mathbf{b}$, $P(A, \mathbf{b})$ may be integral even if A is not totally unimodular.

Definition 13.4. A polyhedron defined by a system of linear constraints is totally dual integral (TDI) if for each objective function with integral coefficient the dual linear program has an integral optimal solution whenever an optimal solution exists.

Theorem 13.4 [Edmonds and Giles 1977]. *If $P(A) = \{x : Ax \leq b\}$ is TDI and b is integral, then $P(A)$ is integral.*

Hoffman and Kruskal [1956] have, in fact, shown that the polyhedron $P(A, b)$ defined in Theorem 13.3 is TDI. This follows from Theorem 13.3 and the fact that A is totally unimodular if and only if A^T is totally unimodular.

Balanced matrices, first introduced by Berge [1972] have important implications for packing and covering problems (see also Berge and Las Vergnas [1970]).

Definition 13.5. A $0, 1$ matrix is balanced if it does not contain a square submatrix of odd order with two ones per row and column.

Theorem 13.5 [Berge 1972, Fulkerson et al. 1974]. *Let A be a balanced $0, 1$ matrix. Then the set packing, set covering, and set partitioning polytopes associated with A are integral, i.e., the polytopes*

$$P(A) = \{x : x \geq 0; Ax \leq 1\}$$

$$Q(A) = \{x : 0 \leq x \leq 1; Ax \geq 1\}$$

$$R(A) = \{x : x \geq 0; Ax = 1\}$$

are integral.

Let

$$A = \begin{pmatrix} A_1 \\ A_2 \\ A_3 \end{pmatrix}$$

be a balanced $0, 1$ matrix. Fulkerson et al. [1974] have shown that the polytope $P(A) = \{x : A_1 x \leq 1; A_2 x \geq 1; A_3 x = 1; x \geq 0\}$ is TDI and by the theorem of Edmonds and Giles [1977] it follows that $P(A)$ is integral.

Truemper [1992] has extended the definition of balanced matrices to include $0, \pm 1$ matrices.

Definition 13.6. A $0, \pm 1$ matrix is balanced if for every square submatrix with exactly two nonzero entries in each row and each column, the sum of the entries is a multiple of 4.

Theorem 13.6 [Conforti and Cornuejols 1992b]. *Suppose A is a balanced $0, \pm 1$ matrix. Let $n(A)$ denote the column vector whose ith component is the number of -1s in the ith row of A. Then the polytopes*

$$P(A) = \{x : Ax \leq 1 - n(A); 0 \leq x \leq 1\}$$

$$Q(A) = \{x : Ax \geq 1 - n(A); 0 \leq x \leq 1\}$$

$$R(A) = \{x : Ax = 1 - n(A); 0 \leq x \leq 1\}$$

are integral.

Note that a $0, \pm 1$ matrix A is balanced if and only if A^T is balanced. Moreover, A is balanced (totally unimodular) if and only if every submatrix of A is balanced (totally unimodular). Thus, if A is balanced (totally unimodular) it follows that Theorem 13.6 (Theorem 13.3) holds for every submatrix of A.

Totally unimodular matrices constitute a subclass of balanced matrices, i.e., a totally unimodular $0, \pm 1$ matrix is always balanced. This follows from a theorem of Camion [1965], which states that a $0, \pm 1$ is totally unimodular if and only if for every square submatrix with an even number of nonzero entries in each row and in each column, the sum of the entries equals a multiple of 4. The 4×4 matrix in

Fig. 13.1 illustrates the fact that a balanced matrix is not necessarily totally unimodular. Balanced $0, \pm 1$ matrices have implications for solving the satisfiability problem. If the given set of clauses defines a balanced $0, \pm 1$ matrix, then as shown by Conforti and Cornuejols [1992b], the satisfiability problem is trivial to solve and the associated MAXSAT problem is solvable in polynomial time by linear programming. A survey of balanced matrices is in Conforti et al. [1994].

$$A = \begin{bmatrix} 1 & 1 & 0 & 0 \\ 1 & 1 & 1 & 1 \\ 1 & 0 & 1 & 0 \\ 1 & 0 & 0 & 1 \end{bmatrix} \qquad A = \begin{bmatrix} 1 & 1 & 0 \\ 0 & 1 & 1 \\ 1 & 0 & 1 \\ 1 & 1 & 1 \end{bmatrix}$$

FIGURE 13.1 A balanced matrix and a perfect matrix. (*Source:* Chandru, V. and Rao, M. R. Combinatorial optimization: an integer programming perspective. *ACM Comput. Surveys*, spec. iss. (to appear).)

Definition 13.7. A 0, 1 matrix A is perfect if the set packing polytope $P(A) = \{\mathbf{x} : A\mathbf{x} \le \mathbf{1}; \mathbf{x} \ge \mathbf{0}\}$ is integral.

The chromatic number of a graph is the minimum number of colors required to color the vertices of the graph so that no two vertices with the same color have an edge incident between them. A graph G is perfect if for every node induced subgraph H, the chromatic number of H equals the number of nodes in the maximum clique of H. The connections between the integrality of the set packing polytope and the notion of a perfect graph, as defined by Berge [1961, 1970], are given in Fulkerson [1970], Lovasz [1972], Padberg [1974], and Chvátal [1975].

Theorem 13.7 [Fulkerson 1970, Lovasz 1972, Chvátal 1975] *Let A be 0, 1 matrix whose columns correspond to the nodes of a graph G and whose rows are the incidence vectors of the maximal cliques of G. The graph G is perfect if and only if A is perfect.*

Let G_A denote the intersection graph associated with a given 0, 1 matrix A (see section 13.4). Clearly, a row of A is the incidence vector of a clique in G_A. In order for A to be perfect, every maximal clique of G_A must be represented as a row of A because inequalities defined by maximal cliques are facet defining. Thus, by Theorem 13.7, it follows that a 0, 1 matrix A is perfect if and only if the undominated (a row of A is dominated if its support is contained in the support of another row of A) rows of A form the clique-node incidence matrix of a perfect graph.

Balanced matrices with 0, 1 entries, constitute a subclass of 0, 1 perfect matrices, i.e., if a 0, 1 matrix A is balanced, then A is perfect. The 4×3 matrix in Fig. 13.1 is an example of a matrix that is perfect but not balanced.

Definition 13.8. A 0, 1 matrix A is ideal if the set covering polytope

$$Q(A) = \{\mathbf{x} : A\mathbf{x} \ge \mathbf{1}; 0 \le \mathbf{x} \le \mathbf{1}\}$$

is integral.

Properties of ideal matrices are described by Lehman [1979], Padberg [1993], and Cornuejols and Novick [1994]. The notion of a 0, 1 perfect (ideal) matrix has a natural extension to a 0, ± 1 perfect (ideal) matrix. Some results pertaining to 0, ± 1 ideal matrices are contained in Hooker [1992], whereas some results pertaining to 0, ± 1 perfect matrices are given in Conforti et al. [1993].

An interesting combinatorial problem is to check whether a given 0, ± 1 matrix is totally unimodular, balanced, or perfect. Seymour's [1980] characterization of totally unimodular matrices provides a polynomial-time algorithm to test whether a given matrix 0, 1 matrix is totally unimodular. Conforti et al. [1991] give a polynomial-time algorithm to check whether a 0, 1 matrix is balanced. This has been extended by Conforti et al. [1994] to check in polynomial time whether a 0, ± 1 matrix is balanced. An open problem is that of checking in polynomial time whether a 0, 1 matrix is perfect. For linear matrices (a matrix is linear if it does not contain a 2×2 submatrix of all ones), this problem has been solved by Fonlupt and Zemirline [1981] and Conforti and Rao [1993].

Matroids

Matroids and submodular functions have been studied extensively, especially from the point of view of combinatorial optimization (see, for instance, Nemhauser and Wolsey [1988]). Matroids have nice properties that lead to efficient algorithms for the associated optimization problems. One of the interesting examples of a matroid is the problem of finding a maximum or minimum weight spanning tree in a graph. Two different but equivalent definitions of a matroid are given first. A greedy algorithm to solve a linear optimization problem over a matroid is presented. The matroid intersection problem is then discussed briefly.

Definition 13.9. Let $N = \{1, 2, \cdot, n\}$ be a finite set and let $\mathcal{F}$ be a set of subsets of N. Then $I = (N, \mathcal{F})$ is an independence system if $S_1 \in \mathcal{F}$ implies that $S_2 \in \mathcal{F}$ for all $S_2 \subseteq S_1$. Elements of $\mathcal{F}$ are called independent sets. A set $S \in \mathcal{F}$ is a maximal independent set if $S \cup \{j\} \notin \mathcal{F}$ for all $j \in N \backslash S$. A maximal independent set T is a maximum if $|T| \geq |S|$ for all $S \in \mathcal{F}$.

The rank $r(Y)$ of a subset $Y \subseteq N$ is the cardinality of the maximum independent subset $X \subseteq Y$. Note that $r(\phi) = 0$, $r(X) \leq |X|$ for $X \subseteq N$ and the rank function is nondecreasing, i.e., $r(X) \leq r(Y)$ for $X \subseteq Y \subseteq N$.

A matroid $M = (N, \mathcal{F})$ is an independence system in which every maximal independent set is a maximum.

Example 13.2. Let $G = (V, E)$ be an undirected connected graph with V as the node set and E as the edge set.

1. Let $I = (E, \mathcal{F})$ where $F \in \mathcal{F}$ if $F \subseteq E$ is such that at most one edge in F is incident to each node of V, that is, $F \in \mathcal{F}$ if F is a matching in G. Then $I = (E, \mathcal{F})$ is an independence system but not a matroid.
2. Let $M = (E, \mathcal{F})$ where $F \in \mathcal{F}$ if $F \subseteq E$ is such that $G_F = (V, F)$ is a forest, that is, G_F contains no cycles. Then $M = (E, \mathcal{F})$ is a matroid and maximal independent sets of M are spanning trees.

An alternative but equivalent definition of matroids is in terms of submodular functions.

Definition 13.10. A nondecreasing integer valued submodular function r defined on the subsets of N is called a matroid rank function if $r(\phi) = 0$ and $r(\{j\}) \leq 1$ for $j \in N$. The pair (N, r) is called a matroid.

A nondecreasing, integer-valued, submodular function f, defined on the subsets of N is called a polymatroid function if $f(\phi) = 0$. The pair (N, r) is called a polymatroid.

Matroid Optimization

To decide whether an optimization problem over a matroid is polynomially solvable or not, we need to first address the issue of representation of a matroid. If the matroid is given either by listing the independent sets or by its rank function, many of the associated linear optimization problems are trivial to solve. However, matroids associated with graphs are completely described by the graph and the condition for independence. For instance, the matroid in which the maximal independent sets are spanning forests, the graph $G = (V, E)$ and the independence condition of no cycles describes the matroid.

Most of the algorithms for matroid optimization problems require a test to determine whether a specified subset is independent. We assume the existence of an oracle or subroutine to do this checking in running time, which is a polynomial function of $|N| = n$.

Maximum Weight Independent Set. Given a matroid $M = (N, \mathcal{F})$ and weights w_j for $j \in N$, the problem of finding a maximum weight independent set is $\max_{F \in \mathcal{F}} \left\{ \sum_{j \in F} w_j \right\}$. The greedy algorithm to solve this problem is as follows:

Procedure 13.3. Greedy:

    0. **Initialize:** Order the elements of N so that $w_i \geq w_{i+1}$, $i = 1, 2, \ldots, n - 1$. Let $T = \phi, i = 1$.

1. **If** $w_i \leq 0$ or $i > n$, **stop** T is optimal, i.e., $x_j = 1$ for $j \in T$ and $x_j = 0$ for $j \notin T$. If $w_i > 0$ and $T \cup \{i\} \in \mathcal{F}$, add element i to T.
2. **Increment** i by 1 and return to step 1.

Edmonds [1970, 1971] derived a complete description of the *matroid polytope*, the convex hull of the characteristic vectors of independent sets of a matroid. While this description has a large (exponential) number of constraints, it permits the treatment of linear optimization problems on independent sets of matroids as linear programs. Cunningham [1984] describes a polynomial algorithm to solve the separation problem for the matroid polytope. The matroid polytope and the associated greedy algorithm have been extended to polymatroids (Edmonds [1970], McDiarmid [1975]).

The separation problem for a polymatroid is equivalent to the problem of minimizing a submodular function defined over the subsets of N (see Nemhauser and Wolsey [1988]). A class of submodular functions that have some additional properties can be minimized in polynomial time by solving a maximum flow problem [Rhys 1970, Picard and Ratliff 1975]. The general submodular function can be minimized in polynomial time by the ellipsoid algorithm [Grötschel et al. 1988].

The uncapacitated plant location problem formulated in section 13.4 can be reduced to maximizing a submodular function. Hence, it follows that maximizing a submodular function is $\mathcal{NP}$-hard.

Matroid Intersection

A matroid intersection problem involves finding an independent set contained in two or more matroids defined on the same set of elements.

Let $G = (V_1, V_2, E)$ be a bipartite graph. Let $M_i = (E, \mathcal{F}_i), i = 1, 2$, where $F \in \mathcal{F}_i$ if $F \subseteq E$ is such that no more than one edge of F is incident to each node in V_i. The set of matchings in G constitutes the intersection of the two matroids $M_i, i = 1, 2$. The problem of finding a maximum weight independent set in the intersection of two matroids can be solved in polynomial time [Lawler 1975, Edmonds 1970, 1979, Frank 1981]. The two (poly) matroid intersection polytope has been studied by Edmonds [1979].

The problem of testing whether a graph contains a Hamiltonian path is $\mathcal{NP}$-complete. Since this problem can be reduced to the problem of finding a maximum cordinality independent set in the intersection of three matroids, it follows that the matroid intersection problem involving three or more matroids is $\mathcal{NP}$-hard.

Valid Inequalities, Facets, and Cutting Plane Methods

Earlier in this section, we were concerned with conditions under which the packing and covering polytopes are integral. But, in general, these polytopes are not integral, and additional inequalities are required to have a complete linear description of the convex hull of integer solutions. The existence of finitely many such linear inequalities is guaranteed by Weyl's [1935] Theorem.

Consider the feasible region of an ILP given by

$$P_I = \{\mathbf{x} \;:\; A\mathbf{x} \leq \mathbf{b}; \mathbf{x} \geq \mathbf{0} \text{ and integer}\} \tag{13.7}$$

Recall that an inequality $\mathbf{f}\mathbf{x} \leq f_0$ is referred to as a valid inequality for P_I if $\mathbf{f}\mathbf{x}^* \leq f_0$ for all $\mathbf{x}^* \in P_I$. A valid linear inequality for $P_I(A, \mathbf{b})$ is said to be facet defining if it intersects $P_I(A, \mathbf{b})$ in a face of dimension one less than the dimension of $P_I(A, \mathbf{b})$. In the example shown in Fig. 13.2, the inequality $\mathbf{x}_2 + \mathbf{x}_3 \leq 1$ is a facet defining inequality of the integer hull.

Let $\mathbf{u} \geq \mathbf{0}$ be a row vector of appropriate size. Clearly $\mathbf{u}A\mathbf{x} \leq \mathbf{u}\mathbf{b}$ holds for every $\mathbf{x}$ in P_I. Let $(\mathbf{u}A)_j$ denote the jth component of the row vector $\mathbf{u}A$ and $\lfloor (\mathbf{u}A)_j \rfloor$ denote the largest integer less than or equal to $(\mathbf{u}A)_j$. Now, since $\mathbf{x} \in P_I$ is a vector of nonnegative integers, it follows that $\sum_j \lfloor (\mathbf{u}A)_j \rfloor \mathbf{x}_j \leq \lfloor \mathbf{u}\mathbf{b} \rfloor$ is a valid inequality for P_I. This scheme can be used to generate many valid inequalities by using different $\mathbf{u} \geq \mathbf{0}$. Any set of generated valid inequalities may be added to the constraints in Eq. (13.7) and the process of generating them may be repeated with the enhanced set of inequalities. This iterative procedure of generating valid inequalities is called Gomory–Chvátal (GC) rounding. It is remarkable that this simple

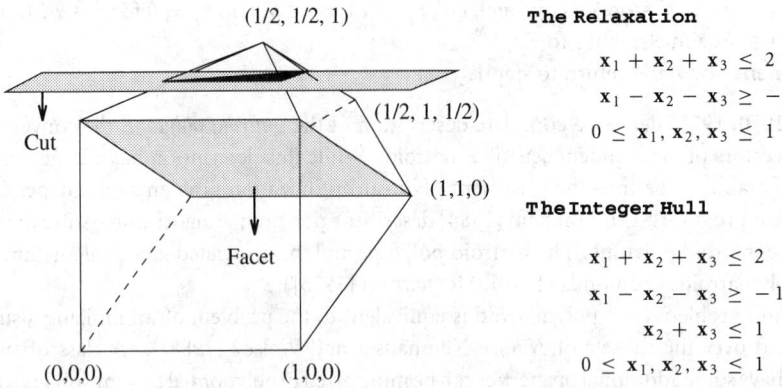

FIGURE 13.2   Relaxation, cuts, and facets (*Source:* Chandru, V. and Rao, M. R. Combinatorial optimization: an integer programming perspective. *ACM Comput. Surveys*, spec. iss. (to appear).)

scheme is complete, i.e., every valid inequality of P_I can be generated by finite application of GC rounding (Chvátal [1973], Schrijver [1986]).

The number of inequalities needed to describe the convex hull of P_I is usually exponential in the size of A. But to solve an optimization problem on P_I, one is only interested in obtaining a partial description of P_I that facilitates the identification of an integer solution and prove its optimality. This is the underlying basis of any cutting plane approach to combinatorial problems.

The Cutting Plane Method

Consider the optimization problem

$$\max\{\mathbf{cx} : \mathbf{x} \in P_I = \{\mathbf{x} : A\mathbf{x} \le \mathbf{b}; \ \mathbf{x} \ge \mathbf{0} \text{ and integer}\}\}$$

The generic **cutting plane** method as applied to this formulation is given as follows.

Procedure 13.4.   Cutting Plane:

1. Initialize $A' \leftarrow A$ and $\mathbf{b}' \leftarrow \mathbf{b}$.
2. Find an optimal solution $\bar{\mathbf{x}}$ to the linear program

$$\max\{\mathbf{cx} : A'\mathbf{x} \le \mathbf{b}'; \ \mathbf{x} \ge \mathbf{0}\}$$

   If $\bar{\mathbf{x}} \in P_I$, stop and return $\bar{\mathbf{x}}$.
3. Generate a valid inequality $\mathbf{fx} \le f_0$ for P_I such that $\mathbf{f\bar{x}} > f_0$ (the inequality "cuts" $\bar{\mathbf{x}}$).
4. Add the inequality to the constraint system, update

$$A' \leftarrow \begin{pmatrix} A' \\ \mathbf{f} \end{pmatrix}, \qquad \mathbf{b}' \leftarrow \begin{pmatrix} \mathbf{b}' \\ f_0 \end{pmatrix}$$

   Go to step 2.

In step 3 of the cutting plane method, we require a suitable application of the GC rounding scheme (or some alternative method of identifying a cutting plane). Notice that while the GC rounding scheme will generate valid inequalities, the identification of one that cuts off the current solution to the linear programming relaxation is all that is needed. Gomory [1958] provided just such a specialization of the rounding scheme that generates a cutting plane. Although this met the theoretical challenge of designing a sound and complete cutting plane method for integer linear programming, it turned out to be a weak method in practice. Successful cutting plane methods, in use today, use considerable additional insights into the structure of facet-defining cutting planes. Using facet cuts makes a huge difference in the speed of convergence of these methods. Also, the idea of combining cutting plane methods with search methods has been found to have a lot of merit. These branch and cut methods will be discussed in the next section.

The *b*-Matching Problem

Consider the **b**-matching problem:

$$\max\{\mathbf{cx} \ : \ A\mathbf{x} \leq \mathbf{b}, \ \mathbf{x} \geq \mathbf{0} \text{ and integer}\} \tag{13.8}$$

where A is the node-edge incidence matrix of an undirected graph and **b** is a vector of positive integers. Let G be the undirected graph whose node-edge incidence matrix is given by A and let $W \subseteq V$ be any subset of nodes of G (i.e., subset of rows of A) such that

$$\mathbf{b}(W) = \sum_{i \in W} \mathbf{b}_i$$

is odd. Then the inequality

$$\mathbf{x}(W) = \sum_{e \in E(W)} \mathbf{x}_e \leq \frac{1}{2}(\mathbf{b}(W) - 1) \tag{13.9}$$

is a valid inequality for integer solutions to Eq. (13.8) where $E(W) \subseteq E$ is the set of edges of G having both ends in W. Edmonds [1965] has shown that the inequalities Eq. (13.8) and Eq. (13.9) define the integral **b**-matching polytope. Note that the number of inequalities Eq. (13.9) is exponential in the number of nodes of G. An instance of the successful application of the idea of using only a partial description of P_I is in the blossom algorithm for the matching problem, due to Edmonds [1965].

As we saw, an implication of the ellipsoid method for linear programming is that the linear program over P_I can be solved in polynomial time if and only if the associated separation problem (also referred to as the constraint identification problem, see section 13.2) can be solved in polynomial time, see Grötschel et al. [1982], Karp and Papadimitriou [1982], and Padberg and Rao [1981]. The separation problem for the **b**-matching problem with or without upper bounds was shown by Padberg and Rao [1982], to be solvable in polynomial time. The procedure involves a minor modification of the algorithm of Gomory and Hu [1961] for multiterminal networks. However, no polynomial (in the number of nodes of the graph) linear programming formulation of this separation problem is known. A related unresolved issue is whether there exists a polynomial size (compact) formulation for the **b**-matching problem. Yannakakis [1988] has shown that, under a symmetry assumption, such a formulation is impossible.

Other Combinatorial Problems

Besides the matching problem, several other combinatorial problems and their associated polytopes have been well studied and some families of facet defining inequalities have been identified. For instance, the set packing, graph partitioning, plant location, max cut, traveling salesman, and Steiner tree problems have been extensively studied from a polyhedral point of view (see, for instance, Nemhauser and Wolsey [1988]).

These combinatorial problems belong to the class of $\mathcal{NP}$-complete problems. In terms of a worst-case analysis, no polynomial-time algorithms are known for these problems. Nevertheless, using a cutting plane approach with branch and bound or branch and cut (see section 13.6), large instances of these problems have been successfully solved, see Crowder et al. [1983], for general $0-1$ problems, Barahona et al. [1989] for the max cut problem, Padberg and Rinaldi [1991] for the traveling salesman problem, and Chopra et al. [1992] for the Steiner tree problem.

13.6 Partial Enumeration Methods

In many instances, to find an optimal solution to integer linear programing problems (ILP), the structure of the problem is exploited together with some sort of partial enumeration. In this section, we review the branch and bound (B-and-B) and branch and cut (B-and-C) methods for solving an ILP.

Branch and Bound

The branch bound (B-and-B) method is a systematic scheme for implicitly enumerating the finitely many feasible solutions to an ILP. Although, theoretically the size of the enumeration tree is exponential in the problem parameters, in most cases, the method eliminates a large number of feasible solutions. The key features of branch and bound method are:

1. **Selection/removal** of one or more problems from a candidate list of problems
2. **Relaxation** of the selected problem so as to obtain a lower bound (on a minimization problem) on the optimal objective function value for the selected problem
3. **Fathoming**, if possible, of the selected problem
4. **Branching** strategy is needed if the selected problem is not fathomed. Branching creates subproblems, which are added to the candidate list of problems

The four steps are repeated until the candidate list is empty. The B-and-B method sequentially examines problems that are added and removed from a candidate list of problems.

Initialization.   Initially, the candidate list contains only the original ILP, which is denoted as

$$(P) \quad \min\{\mathbf{cx} : A\mathbf{x} \leq \mathbf{b}, \ \mathbf{x} \geq \mathbf{0} \text{ and integer}\}$$

Let $F(P)$ denote the feasible region of (P) and $z(P)$ denote the optimal objective function value of (P). For any $\bar{\mathbf{x}}$ in $F(P)$, let $z_P(\bar{\mathbf{x}}) = \mathbf{c}\bar{\mathbf{x}}$.

Frequently, heuristic procedures are first applied to get a good feasible solution to (P). The best solution known for (P) is referred to as the current incumbent solution. The corresponding objective function value is denoted as z_I. In most instances, the initial heuristic solution is neither optimal nor at least immediately certified to be optimal. Thus, further analysis is required to ensure that an optimal solution to (P) is obtained. If no feasible solution to (P) is known, z_I is set to ∞.

Selection/Removal.   In each iterative step of B-and-B, a problem is selected and removed from the candidate list for further analysis. The selected problem is henceforth referred to as the candidate problem (CP). The algorithm terminates if there is no problem to select from the candidate list. Initially, there is no issue of selection since the candidate list contains only the problem (P). However, as the algorithm proceeds, there would be many problems on the candidate list and a selection rule is required. Appropriate selection rules, also referred to as branching strategies, are discussed later. Conceptually, several problems may be simultaneously selected and removed from the candidate list. However, most sequential implementations of B-and-B select only one problem from the candidate list and this is assumed henceforth. Parallel aspects of B-and-B on $0 - 1$ integer linear programs are discussed in Cannon and Hoffman [1990] and for the case of traveling salesman problems in Applegate et al. [1994].

The computational time required for the B-and-B algorithm depends crucially on the order in which the problems in the candidate list are examined. A number of clever heuristic rules may be employed in devising such strategies. Two general purpose selection strategies that are commonly used are as follows:

1. Choose the problem that was added last to the candidate list. This last-in–first-out rule (LIFO) is also called depth first search (DFS) since the selected candidate problem increases the depth of the active enumeration tree.

2. Choose the problem on the candidate list that has the least lower bound. Ties may be broken by choosing the problem that was added last to the candidate list. This rule would require that a lower bound be obtained for each of the problems on the candidate list. In other words, when a problem is added to the candidate list, an associated lower bound should also be stored. This may be accomplished by using ad hoc rules or by solving a relaxation of each problem before it is added to the candidate list.

Rule 1 is known to empirically dominate rule 2 when storage requirements for candidate list and computation time to solve (P) are taken into account. However, some analysis indicates that rule 2 can be shown to be superior if minimizing the number of candidate problems to be solved is the criterion (see Parker and Rardin [1988]).

Relaxation. In order to analyze the selected candidate problem (CP), a **relaxation** (CP_R) of (CP) is solved to obtain a lower bound $z(CP_R) \leq z(CP)$. (CP_R) is a relaxation of (CP) if:

1. $F(CP) \subseteq F(CP_R)$
2. For $\bar{x} \in F(CP)$, $z_{CP_R}(\bar{x}) \leq z_{CP}(\bar{x})$
3. For $\bar{x}, \hat{x} \in F(CP)$, $z_{CP_R}(\bar{x}) \leq z_{CP_R}(\hat{x})$ implies that $z_{CP}(\bar{x}) \leq z_{CP}(\hat{x})$

Relaxations are needed because the candidate problems are typically hard to solve. The relaxations used most often are either linear programming or Lagrangian relaxations of (CP), see section 13.7 for details. Sometimes, instead of solving a relaxation of (CP), a lower bound is obtained by using some ad hoc rules such as penalty functions.

Fathoming. A candidate problem is fathomed if:

(FC1) analysis of (CP_R) reveals that (CP) is infeasible. For instance, if $F(CP_R) = \phi$, then $F(CP) = \phi$.

(FC2) analysis of (CP_R) reveals that (CP) has no feasible solution better than the current incumbent solution. For instance, if $z(CP_R) \geq z_I$, then $z(CP) \geq z(CP_R) \geq z_I$.

(FC3) analysis of (CP_R) reveals an optimal solution of (CP). For instance, if the optimal solution, x_R, to (CP_R) is feasible in (CP), then (x_R) is an optimal solution to (CP) and $z(CP) = cx_R$.

(FC4) analysis of (CP_R) reveals that (CP) is dominated by some other problem, say, CP^*, in the candidate list. For instance, if it can shown that $z(CP^*) \leq z(CP)$, then there is no need to analyze (CP) further.

If a candidate problem (CP) is fathomed using any of the preceding criteria, then further examination of (CP) or its descendants (subproblems) obtained by separation is not required. If (FC3) holds, and $z(CP) < z_I$, the incumbent is updated as x_R and z_I is updated as $z(CP)$.

Separation/Branching. If the candidate problem (CP) is not fathomed, then CP is separated into several problems, say, $(CP_1), (CP_2), \ldots, (CP_q)$, where $\bigcup_{t=1}^q F(CP_t) = F(CP)$ and, typically,

$$F(CP_i) \cap F(CP_j) = \phi \; \forall i \neq j$$

For instance, a separation of (CP) into (CP_i), $i = 1, 2, \ldots, q$, is obtained by fixing a single variable, say, x_j, to one of the q possible values of x_j in an optimal solution to (CP). The choice of the variable to fix depends on the separation strategy, which is also part of the branching strategy. After separation, the subproblems are added to the candidate list. Each subproblem (CP_t) is a restriction of (CP) since $F(CP_t) \subseteq F(CP)$. Consequently, $z(CP) \leq z(CP_t)$ and $z(CP) = \min_t z(CP_t)$.

The various steps in the B-and-B algorithm are outlined as follows.

Procedure 13.5. B-and-B:

0. **Initialize:** Given the problem (P), the incumbent value z_I is obtained by applying some heuristic (if a feasible solution to (P) is not available, set $z_I = +\infty$). Initialize the candidate list $C \leftarrow \{(P)\}$.

1. **Optimality:** If $C = \emptyset$ and $z_I = +\infty$, then (P) is infeasible, stop. Stop also if $C = \emptyset$ and $z_I < +\infty$, the incumbent is an optimal solution to (P).

2. **Selection:** Using some candidate selection rule, select and remove a problem $(CP) \in C$.

3. **Bound:** Obtain a lower bound for (CP) by either solving a relaxation (CP_R) of (CP) or by applying some ad-hoc rules. If (CP_R) is infeasible, return to Step 1. Else, let x_R be an optimal solution of (CP_R).

4. **Fathom:** If $z(CP_R) \geq z_I$, return to step 1. Else if x_R is feasible in (CP) and $z(CP) < z_I$, set $z_I \leftarrow z(CP)$, update the incumbent as x_R and return to step 1. Finally, if x_R is feasible in (CP) but $z(CP) \geq z_I$, return to step 1.

5. **Separation:** Using some separation or branching rule, separate (CP) into (CP_i), $i = 1, 2, \ldots, q$ and set $C \leftarrow C \cup \{CP_1), (CP_2), \ldots, (CP_q)\}$ and return to step 1.

6. **End Procedure.**

Although the B-and-B method is easy to understand, the implementation of this scheme for a particular ILP is a nontrivial task requiring the following:

1. A relaxation strategy with efficient procedures for solving these relaxations
2. Efficient data-structures for handling the rather complicated bookkeeping of the candidate list
3. Clever strategies for selecting promising candidate problems
4. Separation or branching strategies that could effectively prune the enumeration tree

A key problem is that of devising a relaxation strategy, that is, to find *good relaxations*, which are significantly easier to solve than the original problems and tend to give sharp lower bounds. Since these two are conflicting, one has to find a reasonable tradeoff.

Branch and Cut

In the past few years, the branch and cut (B-and-C) method has become popular for solving NP-complete combinatorial optimization problems. As the name suggests, the B-and-C method incorporates the features of both the branch and bound method just presented and the cutting plane method presented previously. The main difference between the B-and-C method and the general B-and-B scheme is in the bound step (step 3).

A distinguishing feature of the B-and-C method is that the relaxation (CP_R) of the candidate problem (CP) is a linear programming problem, and, instead of merely solving (CP_R), an attempt is made to solve (CP) by using cutting planes to tighten the relaxation. If (CP_R) contains inequalities that are valid for (CP) but not for the given ILP, then the GC rounding procedure may generate inequalities that are valid for (CP) but not for the ILP. In the B-and-C method, the inequalities that are generated are always valid for the ILP and hence can be used globally in the enumeration tree.

Another feature of the B-and-C method is that often heuristic methods are used to convert some of the fractional solutions, encountered during the cutting plane phase, into feasible solutions of the (CP) or more generally of the given ILP. Such feasible solutions naturally provide upper bounds for the ILP. Some of these upper bounds may be better than the previously identified best upper bound and, if so, the current incumbent is updated accordingly.

We thus obtain the B-and-C method by replacing the bound step (step 3) of the B-and-B method by steps 3(a) and 3(b) and also by replacing the fathom step (step 4) by steps 4(a) and 4(b) given subsequently.

3(a) **Bound:** Let (CP_R) be the LP relaxation of (CP). Attempt to solve (CP) by a cutting plane method which generates valid inequalities for (P). Update the constraint System of (P) and the incumbent as appropriate.

Let $F\mathbf{x} \leq \mathbf{f}$ denote all of the valid inequalities generated during this phase. Update the constraint system of (P) to include all of the generated inequalities, i.e., set $A^T \leftarrow (A^T, F^T)$ and $\mathbf{b}^T \leftarrow (\mathbf{b}^T, \mathbf{f}^T)$. The constraints for all of the problems in the candidate list are also to be updated.

During the cutting plane phase, apply heuristic methods to convert some of the identified fractional solutions into feasible solutions to (P). If a feasible solution, $\bar{\mathbf{x}}$, to (P), is obtained such that $\mathbf{c}\bar{\mathbf{x}} < z_I$, update the incumbent to $\bar{\mathbf{x}}$ and z_I to $\mathbf{c}\bar{\mathbf{x}}$. Hence, the remaining changes to B-and-B are as follows:

3(b) **If** (CP) is solved go to step 4(a). **Else**, let $\hat{\mathbf{x}}$ be the solution obtained when the cutting plane phase is terminated, (we are unable to identify a valid inequality of (P) that is violated by $\hat{\mathbf{x}}$). Go to step 4(b).

4(a) **Fathom by Optimality:** Let $\mathbf{x}^*$ be an optimal solution to (CP). If $z(CP) < z_I$, set $x_I \leftarrow z(CP)$ and update the incumbent as $\mathbf{x}^*$. Return to step 1.

4(b) **Fathom by Bound:** If $\mathbf{c}\hat{\mathbf{x}} \geq z_I$, return to Step 1.
Else go to step 5.

The incorporation of a cutting plane phase into the B-and-B scheme involves several technicalities which require careful design and implementation of the B-and-C algorithm. Details of the state of the art in cutting plane algorithms including the B-and-C algorithm are reviewed in Jünger et al. [1995].

13.7 Approximation in Combinatorial Optimization

The inherent complexity of integer linear programming has led to a long-standing research program in approximation methods for these problems. Linear programming relaxation and Lagrangian relaxation are two general approximation schemes that have been the real workhorses of computational practice. Semidefinite relaxation is a recent entrant that appears to be very promising. In this section, we present a brief review of these developments in the approximation of combinatorial optimization problems.

In the past few years, there has been significant progress in our understanding of performance guarantees for approximation of $\mathcal{NP}$-hard combinatorial optimization problems. A ρ-**approximate** algorithm for an optimization problem is an approximation algorithm that delivers a feasible solution with objective value within a factor of ρ of optimal (think of minimization problems and $\rho \geq 1$). For some combinatorial optimization problems, it is possible to *efficiently* find solutions that are arbitrarily close to optimal even though finding the true optimal is hard. If this were true of most of the problems of interest, we would be in good shape. However, the recent results of Arora et al. [1992] indicate exactly the opposite conclusion.

A polynomial-time approximation scheme (PTAS) for an optimization problem is a family of algorithms, A_ρ, such that for each $\rho > 1$, A_ρ is a polynomial-time ρ-approximate algorithm. Despite concentrated effort spanning about two decades, the situation in the early 1990s was that for many combinatorial optimization problems, we had no PTAS and no evidence to suggest the nonexistence of such schemes either. This led Papadimitriou and Yannakakis [1991] to define a new complexity class (using reductions that preserve approximate solutions) called MAXSNP, and they identified several complete languages in this class. The work of Arora et al. [1992] completed this agenda by showing that, assuming $\mathcal{P} \neq \mathcal{NP}$, there is no PTAS for a MAXSNP-complete problem.

An implication of these theoretical developments is that for most combinatorial optimization problems, we have to be quite satisfied with performance guarantee factors ρ that are of some small fixed value. (There are problems, like the general traveling salesman problem, for which there are no ρ-approximate algorithms for any finite value of ρ, assuming of course that $\mathcal{P} \neq \mathcal{NP}$.) Thus, one avenue of research is to go problem by problem and knock ρ down to its smallest possible value. A different approach would be to look for other notions of good approximations based on probabilistic guarantees or empirical validation. Let us see how the polyhedral combinatorics perspective helps in each of these directions.

LP Relaxation and Randomized Rounding

Consider the well-known problem of finding the *smallest weight vertex cover* in a graph. We are given a graph $G(V, E)$ and a nonnegative weight $\mathbf{w}(v)$ for each vertex $v \in V$. We want to find the smallest total weight subset of vertices S such that each edge of G has at least one end in S. (This problem is known to be MAXSNP-hard.) An integer programming formulation of this problem is given by

$$\min \left\{ \sum_{v \in V} \mathbf{w}(v)\mathbf{x}(v) : \mathbf{x}(u) + \mathbf{x}(v) \geq 1, \ \forall (u, v) \in E, \ \mathbf{x}(v) \in \{0, 1\} \ \forall v \in V \right\}$$

To obtain the linear programming relaxation we substitute the $\mathbf{x}(v) \in \{0, 1\}$ constraint with $\mathbf{x}(v) \geq 0$ for each $v \in V$. Let $\mathbf{x}^*$ denote an optimal solution to this relaxation. Now let us round the fractional parts of $\mathbf{x}^*$ in the usual way, that is, values of 0.5 and up are rounded to 1 and smaller values down to 0. Let $\hat{\mathbf{x}}$ be the 0–1 solution obtained. First note that $\hat{\mathbf{x}}(v) \leq 2\mathbf{x}^*(v)$ for each $v \in V$. Also, for each $(u, v) \in E$, since $\mathbf{x}^*(u) + \mathbf{x}^*(v) \geq 1$, at least one of $\hat{\mathbf{x}}(u)$ and $\hat{\mathbf{x}}(v)$ must be set to 1. Hence $\hat{\mathbf{x}}$ is the incidence vector of a vertex cover of G whose total weight is within twice the total weight of the linear programming relaxation (which is a lower bound on the weight of the optimal vertex cover). Thus, we have a 2-approximate algorithm for this problem, which solves a linear programming relaxation and uses rounding to obtain a feasible solution.

The deterministic rounding of the fractional solution worked quite well for the vertex cover problem. One gets a lot more power from this approach by adding in randomization to the rounding step. Raghavan and Thompson [1987] proposed the following obvious randomized rounding scheme. Given a 0 – 1 integer program, solve its linear programming relaxation to obtain an optimal $\mathbf{x}^*$. Treat the $\mathbf{x}_j{}^* \in [0, 1]$ as probabilities, i.e., let probability $\{\mathbf{x}_j = 1\} = \mathbf{x}_j{}^*$, to randomly round the fractional solution to a 0 – 1 solution. Using Chernoff bounds on the tails of the binomial distribution, Raghavan and Thompson [1987] were able to show, for specific problems, that with high probability, this scheme produces integer solutions which are close to optimal. In certain problems, this rounding method may not always produce a feasible solution. In such cases, the expected values have to be computed as conditioned on feasible solutions produced by rounding. More complex (nonlinear) randomized rounding schemes have been recently studied and have been found to be extremely effective. We will see an example of nonlinear rounding in the context of semidefinite relaxations of the max-cut problem in the following.

Primal–Dual Approximation

The linear programming relaxation of the vertex cover problem, as we saw previously, is given by

$$(P_{VC}) \quad \min \left\{ \sum_{v \in V} \mathbf{w}(v)\mathbf{x}(v) : \mathbf{x}(u) + \mathbf{x}(v) \geq 1, \ \forall (u, v) \in E, \ \mathbf{x}(v) \geq 0 \ \forall v \in V \right\}$$

and its dual is

$$(D_{VC}) \quad \max \left\{ \sum_{(u,v) \in E} \mathbf{y}(u, v) : \sum_{(u,v) \in E} \mathbf{y}(u, v) \leq w(v), \ \forall v \in V, \ \mathbf{y}(u, v) \geq 0 \ \forall (u, v) \in E \right\}$$

The primal–dual approximation approach would first obtain an optimal solution $\mathbf{y}^*$ to the dual problem (D_{VC}). Let $\hat{V} \subseteq V$ denote the set of vertices for which the dual constraints are tight, i.e.,

$$\hat{V} = \left\{ v \in V : \sum_{(u,v) \in E} \mathbf{y}^*(u, v) = \mathbf{w}(v) \right\}$$

The approximate vertex cover is taken to be $\hat{V}$. It follows from complementary slackness that $\hat{V}$ is a vertex cover. Using the fact that each edge (u, v) is in the star of at most two vertices (u and v), it also follows that $\hat{V}$ is a 2-approximate solution to the minimum weight vertex cover problem.

In general, the primal–dual approximation strategy is to use a dual solution to the linear programming relaxation, along with complementary slackness conditions as a heuristic to generate an integer (primal) feasible solution, which for many problems turns out to be a good approximation of the optimal solution to the original integer program.

Semidefinite Relaxation and Rounding

The idea of using semidefinite programming to solve combinatorial optimization problems appears to have originated in the work of Lovász [1979] on the Shannon capacity of graphs. Grötschel et al. [1988] later used the same technique to compute a maximum stable set of vertices in perfect graphs via the ellipsoid method. Recently, Lovasz and Schrijver [1991] resurrected the technique to present a fascinating theory of semidefinite relaxations for general 0–1 integer linear programs. We will not present the full-blown theory here but instead will present a lovely application of this methodology to the problem of finding the maximum weight cut of a graph. This application of semidefinite relaxation for approximating MAXCUT is due to Goemans and Williamson [1994].

We begin with a quadratic Boolean formulation of MAXCUT

$$\max \left\{ \frac{1}{2} \sum_{(u,v) \in E} \mathbf{w}(u,v)(1 - \mathbf{x}(u)\mathbf{x}(v)) : \mathbf{x}(v) \in \{-1, 1\} \; \forall \, v \in V \right\}$$

where $G(V, E)$ is the graph and $\mathbf{w}(u, v)$ is the nonnegative weight on edge (u, v). Any $\{-1, 1\}$ vector of $\mathbf{x}$ values provides a bipartition of the vertex set of G. The expression $(1 - \mathbf{x}(u)\mathbf{x}(v))$ evaluates to 0 if u and v are on the same side of the bipartition and to 2 otherwise. Thus, the optimization problem does indeed represent exactly the MAXCUT problem.

Next we reformulate the problem in the following way:

- We square the number of variables by substituting each $\mathbf{x}(v)$ with $\chi(v)$ an n-vector of variables (where n is the number of vertices of the graph).
- The quadratic term $\mathbf{x}(u)\mathbf{x}(v)$ is replaced by $\chi(u) \cdot \chi(v)$, which is the inner product of the vectors.
- Instead of the $\{-1, 1\}$ restriction on the $\mathbf{x}(v)$, we use the Euclidean normalization $\|\chi(v)\| = 1$ on the $\chi(v)$.

Thus, we now have a problem

$$\max \left\{ \frac{1}{2} \sum_{(u,v) \in E} \mathbf{w}(u,v)(1 - \chi(u) \cdot \chi(v)) : \|\chi(v)\| = 1 \; \forall \, v \in V \right\}$$

which is a relaxation of the MAXCUT problem (note that if we force only the first component of the $\chi(v)$ to have nonzero value, we would just have the old formulation as a special case).

The final step is in noting that this reformulation is nothing but a semidefinite program. To see this we introduce $n \times n$ Gram matrix Y of the unit vectors $\chi(v)$. So $Y = X^T X$ where $X = (\chi(v) : v \in V)$. Thus, the relaxation of MAXCUT can now be stated as a semidefinite program,

$$\max \left\{ \frac{1}{2} \sum_{(u,v) \in E} \mathbf{w}(u,v)(1 - Y_{(u,v)}) : Y \succeq 0, \; Y_{(u,v)} = 1 \; \forall \, v \in V \right\}$$

Recall from section 13.2, that we are able to solve such semidefinite programs to an additive error ϵ in time polynomial in the input length and $\log 1/\epsilon$ by using either the ellipsoid method or interior point methods.

Let χ^* denote the near optimal solution to the semidefinite programming relaxation of MAXCUT (convince yourself that χ^* can be reconstructed from an optimal Y^* solution). Now we encounter the final trick of Goemans and Williamson. The approximate maximum weight cut is extracted from χ^* by randomized rounding. We simply pick a random hyperplane H passing through the origin. All of the $v \in V$ lying to one side of H get assigned to one side of the cut and the rest to the other. Goemans and Williamson observed the following inequality.

Lemma 13.3. *For χ_1 and χ_2, two random n-vectors of unit norm, let $\mathbf{x}(1)$ and $\mathbf{x}(2)$ be ± 1 values with opposing signs if H separates the two vectors and with same signs otherwise. Then $\tilde{E}(1 - \chi_1 \cdot \chi_2) \leq 1.1393 \cdot \tilde{E}(1 - \mathbf{x}(1)\mathbf{x}(2))$ where $\tilde{E}$ denotes the expected value.*

By linearity of expectation, the lemma implies that the expected value of the cut produced by the rounding is at least 0.878 times the expected value of the semidefinite program. Using standard conditional probability techniques for derandomizing, Goemans and Williamson show that a deterministic polynomial-time approximation algorithm with the same margin of approximation can be realized. Hence we have a cut with value at least 0.878 of the maximum cut value.

Lagrangian Relaxation

We end our discussion of approximation methods for combinatorial optimization with the description of Lagrangian relaxation. This approach has been widely used for about two decades now in many practical applications. Lagrangian relaxation, like linear programming relaxation, provides bounds on the combinatorial optimization problem being relaxed (i.e., lower bounds for minimization problems).

Lagrangian relaxation has been so successful because of a couple of distinctive features. As was noted earlier, in many hard combinatorial optimization problems, we usually have embedded some nice tractable subproblems which have efficient algorithms. Lagrangian relaxation gives us a framework to *jerry-rig* an approximation scheme that uses these efficient algorithms for the subproblems as subroutines. A second observation is that it has been empirically observed that well-chosen Lagrangian relaxation strategies usually provide very tight bounds on the optimal objective value of integer programs. This is often used to great advantage within partial enumeration schemes to get very effective pruning tests for the search trees.

Practitioners also have found considerable success with designing heuristics for combinatorial optimization by starting with solutions from Lagrangian relaxations and constructing good feasible solutions via so-called *dual ascent* strategies. This may be thought of as the analogue of rounding strategies for linear programming relaxations (but with no performance guarantees, other than empirical ones).

Consider a representation of our combinatorial optimization problem in the form

$$(P) \quad z = \min\{\mathbf{cx} : A\mathbf{x} \geq \mathbf{b}, \ \mathbf{x} \in X \subseteq \Re^n\}$$

Implicit in this representation is the assumption that the explicit constraints ($A\mathbf{x} \geq \mathbf{b}$) are *small* in number. For convenience, let us also assume that that X can be replaced by a finite list $\{\mathbf{x}^1, \mathbf{x}^2, \ldots, \mathbf{x}^T\}$.

The following definitions are with respect to (P):

- Lagrangian. $L(\mathbf{u}, \mathbf{x}) = \mathbf{u}(A\mathbf{x} - \mathbf{b}) + \mathbf{cx}$ where $\mathbf{u}$ are the Lagrange multipliers.
- Lagrangian-dual function. $\mathcal{L}(\mathbf{u}) = \min_{\mathbf{x} \in X}\{L(\mathbf{u}, \mathbf{x})\}$.
- Lagrangian-dual problem. $(D) \quad d = \max_{\mathbf{u} \geq 0}\{\mathcal{L}(\mathbf{u})\}$.

It is easily shown that (D) satisfies a weak duality relationship with respect to (P), i.e., $z \geq d$. The discreteness of X also implies that $\mathcal{L}(\mathbf{u})$ is a piecewise linear and concave function (see Shapiro [1979]). In practice, the constraints X are chosen such that the evaluation of the Lagrangian dual function $\mathcal{L}(\mathbf{u})$ is easily made (i.e., the *Lagrangian subproblem* $\min_{\mathbf{x} \in X}\{L(\mathbf{u}, \mathbf{x})\}$ is easily solved for a fixed value of $\mathbf{u}$).

Example 13.3.   Traveling salesman problem (TSP). For an undirected graph G, with costs on each edge, the TSP is to find a minimum cost set H of edges of G such that it forms a Hamiltonian cycle of the graph. H is a Hamiltonian cycle of G if it is a simple cycle that spans all the vertices of G. Alternatively, H must satisfy: (1) exactly two edges of H are adjacent to each node, and (2) H forms a connected, spanning subgraph of G.

Held and Karp [1970] used these observations to formulate a Lagrangian relaxation approach for TSP that relaxes the degree constraints (1). Notice that the resulting subproblems are minimum spanning tree problems which can be easily solved.

The most commonly used general method of finding the optimal multipliers in Lagrangian relaxation is subgradient optimization (cf. Held et al. [1974]). Subgradient optimization is the non-differentiable counterpart of steepest descent methods. Given a dual vector $\mathbf{u}^k$, the iterative rule for creating a sequence of solutions is given by:

$$\mathbf{u}^{k+1} = \mathbf{u}^k + t_k \gamma(\mathbf{u}^k)$$

where t_k is an appropriately chosen step size, and $\gamma(\mathbf{u}^k)$ is a subgradient of the dual function $\mathcal{L}$ at $\mathbf{u}^k$. Such a subgradient is easily generated by

$$\gamma(\mathbf{u}^k) = A\mathbf{x}^k - \mathbf{b}$$

where $\mathbf{x}^k$ is a maximizer of $\min_{\mathbf{x} \in X}\{L(\mathbf{u}^k, \mathbf{x})\}$.

Subgradient optimization has proven effective in practice for a variety of problems. It is possible to choose the step sizes $\{t_k\}$ to guarantee convergence to the optimal solution. Unfortunately, the method is not finite, in that the optimal solution is attained only in the limit. Further, it is not a pure descent method. In practice, the method is heuristically terminated and the best solution in the generated sequence is recorded. In the context of nondifferentiable optimization, the ellipsoid algorithm was devised by Shor [1970] to overcome precisely some of these difficulties with the subgradient method.

The ellipsoid algorithm may be viewed as a scaled subgradient method in much the same way as variable metric methods may be viewed as scaled steepest descent methods (cf. Akgul [1984]). And if we use the ellipsoid method to solve the Lagrangian dual problem, we obtain the following as a consequence of the polynomial-time equivalence of optimization and separation.

Theorem 13.8. *The Lagrangian dual problem is polynomial-time solvable if and only if the Lagrangian subproblem is. Consequently, the Lagrangian dual problem is $\mathcal{NP}$-hard if and only if the Lagrangian subproblem is.*

The theorem suggests that, in practice, if we set up the Lagrangian relaxation so that the subproblem is tractable, then the search for optimal Lagrangian multipliers is also tractable.

13.8 Prospects in Integer Programming

The current emphasis in software design for integer programming is in the development of shells (for example, CPLEX [1993], MINTO (Savelsbergh et al. [1994]), and OSL [1991]) wherein a general purpose solver like branch and cut is the driving engine. Problem-specific codes for generation of cuts and facets can be easily interfaced with the engine. We believe that this trend will eventually lead to the creation of general purpose problem solving languages for combinatorial optimization akin to AMPL (Fourer et al. [1993]) for linear and nonlinear programming.

A promising line of research is the development of an empirical science of algorithms for combinatorial optimization (Hooker [1993]). Computational testing has always been an important aspect of research on the efficiency of algorithms for integer programming. However, the standards of test designs and empirical analysis have not been uniformly applied. We believe that there will be important strides in this aspect of integer programming and more generally of algorithms. J. N. Hooker argues that it may be useful to stop looking at algorithmics as purely a deductive science and start looking for advances through repeated application of "hypothesize and test" paradigms, i.e., through empirical science. Hooker and Vinay [1995] develop a science of selection rules for the Davis–Putnam–Loveland scheme of theorem proving in propositional logic by applying the empirical approach.

The integration of logic-based methodologies and mathematical programming approaches is evidenced in the recent emergence of constraint logic programming (CLP) systems (Saraswat and Van Hentenryck [1995], Borning [1994]) and logico-mathematical programming (Jeroslow [1989], Chandru and Hooker [1991]). In CLP, we see a structure of Prolog-like programming language in which some of the predicates are constraint predicates whose truth values are determined by the solvability of constraints in a wide range of algebraic and combinatorial settings. The solution scheme is simply a clever orchestration of constraint solvers in these various domains and the role of conductor is played by resolution. The clean semantics of logic programming is preserved in CLP. A bonus is that the output language is symbolic and expressive. An orthogonal approach to CLP is to use constraint methods to solve inference problems in logic. Imbeddings of logics in mixed integer programming sets were proposed by Williams [1987] and Jeroslow [1989]. Efficient algorithms have been developed for inference algorithms in many types and fragments of logic, ranging from Boolean to predicate to belief logics (Chandru and Hooker [1996]).

A persistent theme in the integer programming approach to combinatorial optimization, as we have seen, is that the representation (formulation) of the problem deeply affects the efficacy of the solution methodology. A proper choice of formulation can therefore make the difference between a successful solution of an optimization problem and the more common perception that the problem is insoluble and one must be satisfied with the best that heuristics can provide. Formulation of integer programs has been treated more as an art form than a science by the mathematical programming community. (See Jeroslow [1989] for a refreshingly different perspective on representation theories for mixed integer programming.) We believe that progress in representation theory can have an important influence on the future of integer programming as a broad-based problem solving methodology in combinatorial optimization.

Defining Terms

Column generation: A scheme for solving linear programs with a huge number of columns.

Cutting plane: A valid inequality for an integer polyhedron that separates the polyhedron from a given point outside it.

Extreme point: A corner point of a polyhedron.

Fathoming: Pruning a search tree.

Integer polyhedron: A polyhedron, all of whose extreme points are integer valued.

Linear program: Optimization of a linear function subject to linear equality and inequality constraints.

Mixed integer linear program: A linear program with the added constraint that some of the decision variables are integer valued.

Packing and covering: Given a finite collection of subsets of a finite ground set, to find an optimal subcollection that is pairwise disjoint (packing) or whose union covers the ground set (covering).

Polyhedron: The set of solutions to a finite system of linear inequalities on real-valued variables. Equivalently, the intersection of a finite number of linear half-spaces in $\Re^n$.

Relaxation: An enlargement of the feasible region of an optimization problem. Typically, the relaxation is considerably easier to solve than the original optimization problem.

ρ-Approximation: An approximation method that delivers a feasible solution with an objective value within a factor ρ of the optimal value of a combinatorial optimization problem.

References

Ahuja, R. K., Magnati, T. L., and Orlin, J. B. 1993. *Network Flows: Theory, Algorithms and Applications.* Prentice–Hall, Englewood Cliffs, NJ.

Akgul, M. 1984. *Topics in Relaxation and Ellipsoidal Methods, Research Notes in Mathematics,* Pitman.

Alizadeh, F. 1995. Interior point methods in semidefinite programming with applications to combinatorial optimization. *SIAM J. Optimization* 5(1):13–51.

Applegate, D., Bixby, R. E., Chvátal, V., and Cook, W. 1994. Finding cuts in large TSP's. *Tech. Rep.,* Aug.

Arora, S., Lund, C., Motwani, R., Sudan, M., and Szegedy, M. 1992. Proof verification and hardness of approximation problems. In *Proc. 33rd IEEE Symp. Found. Comput. Sci.,* pp. 14–23.

Barahona, F., Jünger, M., and Reinelt, G. 1989. Experiments in quadratic $0 - 1$ programming. *Math. Programming* 44:127–137.

Benders, J. F. 1962. Partitioning procedures for solving mixed-variables programming problems. *Numerische Mathematik* 4:238–252.

Berge, C. 1961. Farbung von Graphen deren samtliche bzw. deren ungerade Kreise starr sind (Zusammen-fassung). Wissenschaftliche Zeitschrift, Martin Luther Universitat Halle-Wittenberg, Mathematisch-Naturwiseenschaftliche Reihe, pp. 114–115.

Berge, C. 1970. Sur certains hypergraphes generalisant les graphes bipartites. In *Combinatorial Theory and its Applications I*. P. Erdos, A. Renyi, and V. Sos, eds., Colloq. Math. Soc. Janos Bolyai, 4, pp. 119–133. North-Holland, Amsterdam.

Berge, C. 1972. Balanced matrices. *Math. Programming* 2:19–31.

Berge, C. and Las Vergnas, M. 1970. Sur un theoreme du type Konig pour hypergraphes, pp. 31–40. In *Int. Conf. Combinatorial Math.*, Ann. New York Acad. Sci. 175.

Bixby, R. E. 1994. Progress in linear programming. *ORSA J. Comput.* 6(1):15–22.

Bland, R., Goldfarb, D., and Todd, M. J. 1981. The ellipsoid method: a survey. *Operations Res.* 29:1039–1091.

Borning, A., ed. 1994. *Principles and Practice of Constraint Programming*, LNCS Vol. 874, Springer–Verlag.

Camion, P. 1965. Characterization of totally unimodular matrices. *Proc. Am. Math. Soc.* 16:1068–1073.

Cannon, T. L. and Hoffman, K. L. 1990. Large-scale zero-one linear programming on distributed work-stations. *Ann. Operations Res.* 22:181–217.

Černikov, R. N. 1961. The solution of linear programming problems by elimination of unknowns. *Doklady Akademii Nauk* 139:1314–1317 (translation in 1961. *Soviet Mathematics Doklady* 2:1099–1103).

Chandru, V. and Hooker, J. N. 1991. Extended Horn sets in propositional logic. *JACM* 38:205–221.

Chandru, V. and Hooker, J. N. 1996. *Optimization Methods for Logical Inference*, manuscript to be published by Wiley Interscience.

Chopra, S., Gorres, E. R., and Rao, M. R. 1992. Solving Steiner tree problems by branch and cut. *ORSA J. Comput.* 3:149–156.

Chvátal, V. 1973. Edmonds polytopes and a hierarchy of combinatorial problems. *Discrete Math.* 4:305–337.

Chvátal, V. 1975. On certain polytopes associated with graphs. *J. Combinatorial Theory B* 18:138–154.

Conforti, M. and Cornuejols, G. 1992a. Balanced 0, ±1 matrices, bicoloring and total dual integrality. preprint, Carnegie Mellon University.

Conforti, M. and Cornuejols, G. 1992b. A class of logical inference problems solvable by linear programming. *FOCS* 33:670–675.

Conforti, M., Cornuejols, G., and De Francesco, C. 1993. Perfect 0, ±1 matrices. Preprint, Carnegie Mellon University.

Conforti, M., Cornuejols, G., Kapoor, A., and Vuskovic, K. 1994. Balanced 0, ±1 matrices. Pts. I–II, preprints, Carnegie Mellon University.

Conforti, M., Cornuejols, G., Kapoor, A. Vuskovic, K., and Rao, M. R. 1994. Balanced matrices. In *Mathematical Programming, State of the Art 1994*. J. R. Birge and K. G. Murty, eds., University of Michigan.

Conforti, M., Cornuejols, G., and Rao, M. R. 1991. Decomposition of balanced 0, 1 matrices. Pts. I–VII, preprints, Carnegie Mellon University.

Conforti, M. and Rao, M. R. 1993. Testing balancedness and perfection of linear matrices. *Math. Programming* 61:1–18.

Cook, W., Lovász, L., and Seymour, P., eds. 1995. *Combinatorial Optimization: Papers from the DIMACS Special Year*. Series in discrete mathematics and theoretical computer science, Vol. 20, AMS.

Cornuejols, G. and Novick, B. 1994. Ideal 0, 1 matrices. *J. Combinatorial Theory* 60:145–157.

CPLEX. 1993. Using the CPLEX Callable Library and CPLEX Mixed Integer Library, CPLEX Optimization, Inc.

Crowder, H., Johnson, E. L., and Padberg, M. W. 1983. Solving large scale 0–1 linear programming problems. *Operations Res.* 31:803–832.

Cunningham, W. H. 1984. Testing membership in matroid polyhedra. *J. Combinatorial Theory* 36B:161–188.

Dantzig, G. B. and Wolfe, P. 1961. The decomposition algorithm for linear programming. *Econometrica* 29:767–778.

Ecker, J. G. and Kupferschmid, M. 1983. An ellipsoid algorithm for nonlinear programming. *Math. Programming* 27.

Edmonds, J. 1965. Maximum matching and a polyhedron with 0–1 vertices. *J. Res. Nat. Bur. Stand.* 69B:125–130.

Edmonds, J. 1970. Submodular functions, matroids and certain polyhedra. In *Combinatorial Structures and their Applications*, R. Guy, ed., pp. 69–87. Gordon Breach, New York.

Edmonds, J. 1971. Matroids and the greedy algorithm. *Math. Programming* 127–136.

Edmonds, J. 1979. Matroid intersection. *Ann. Discrete Math.* 4:39–49.

Edmonds, J. and Giles, R. 1977. A min-max relation for submodular functions on graphs. *Ann. Discrete Math.* 1:185–204.

Edmonds, J. and Johnson, E. L. 1970. Matching well solved class of integer linear polygons. In *Combinatorial Structure and Their Applications*. R. Guy, ed., Gordon Breach, New York.

Fonlupt, J. and Zemirline, A. 1981. A polynomial recognition algorithm for $K_4 \backslash e$-free perfect graphs. *Res. Rep.*, University of Grenoble.

Fourer, R., Gay, D. M., and Kernighian, B. W. 1993. *AMPL: A Modeling Language for Mathematical Programming*, Scientific Press.

Fourier, L. B. J. 1827. In: Analyse des travaux de l'Academie Royale des Sciences, pendant l'annee 1824, Partie mathematique, *Histoire de l'Academie Royale des Sciences de l'Institut de France 7* (1827) xlvii–lv. (Partial English translation Kohler, D. A. 1973. *Translation of a Report by Fourier on his Work on Linear Inequalities. Opsearch* 10:38–42.)

Frank, A. 1981. A weighted matroid intersection theorem. *J. Algorithms* 2:328–336.

Fulkerson, D. R. 1970. The perfect graph conjecture and the pluperfect graph theorem, pp. 171–175. In *Proc. 2nd Chapel Hill Conf. Combinatorial Math. Appl.* R. C. Bose et al., eds.

Fulkerson, D. R., Hoffman, A., and Oppenheim, R. 1974. On balanced matrices. *Math. Programming Study* 1:120–132.

Gilmore, P. and Gomory, R. E. 1963. A linear programming approach to the cutting stock problem. Pt. I. *Operations Res.* 9:849–854; Pt. II. *Operations Res.* 11:863–887.

Goemans, M. X. and Williamson, D. P. 1994. .878 approximation algorithms MAX CUT and MAX 2SAT. pp. 422–431. In *Proc. ACM STOC*.

Gomory, R. E. 1958. Outline of an algorithm for integer solutions to linear programs. *Bull. Am. Math. Soc.* 64:275–278.

Gomory, R. E. and Hu, T. C. 1961. Multi-terminal network flows. *SIAM J. Appl. Math.* 9:551–556.

Grötschel, M., Lovasz, L., and Schrijver, A. 1982. The ellipsoid method and its consequences in combinatorial optimization. *Combinatorica* 1:169–197.

Grötschel, M., Lovász, L., and Schrijver, A. 1988. *Geometric Algorithms and Combinatorial Optimization.* Springer–Verlag.

Hacijan, L. G. 1979. A polynomial algorithm in linear programming. *Soviet Math. Dokl.* 20:191–194.

Held, M. and Karp, R. M. 1970. The travelling-salesman problem and minimum spanning trees. *Operations Res.* 18:1138–1162, Pt. II. 1971. *Math. Programming* 1:6–25.

Held, M., Wolfe, P., and Crowder, H. P. 1974. Validation of subgradient optimization. *Math. Programming* 6:62–88.

Hoffman, A. J. and Kruskal, J. K. 1956. Integral boundary points of convex polyhedra. In *Linear Inequalities and Related Systems*, H. W. Kuhn and A. W. Tucker, eds., pp. 223–246. Princeton University Press, Princeton, NJ.

Hooker, J. N. 1988. Resolution vs cutting plane solution of inference problems: some computational experience. *Operations Res. Lett.* 7:1–7.

Hooker, J. N. 1992. Resolution and the integrality of satisfiability polytopes. preprint, GSIA, Carnegie Mellon University.

Hooker, J. N. 1993. Towards and empirical science of algorithms. *Operations Res.* 42:201–212.

Hooker, J. N. and Vinay, V. 1995. Branching rules for satisfiability. In *Automated Reasoning* 15:359–383.

Huynh, T., Lassez C., and Lassez, J.-L. 1992. Practical issues on the projection of polyhedral sets. *Ann. Math. Artif. Intell.* 6:295–316.

IBM. 1991. *Optimization Subroutine Library—Guide and Reference (Release 2)*, 3rd ed.

Jeroslow, R. E. 1987. Representability in mixed integer programming, I: characterization results. *Discrete Appl. Math.* 17:223–243.

Jeroslow, R. E. and Lowe, J. K. 1984. Modeling with integer variables. *Math. Programming Stud.* 22:167–184.

Jeroslow, R. G. 1989. *Logic-Based Decision Support: Mixed Integer Model Formulation.* Ann. discrete mathematics, Vol. 40, North-Holland.

Jünger, M., Reinelt, G., and Thienel, S. 1995. Practical problem solving with cutting plane algorithms. In *Combinatorial Optimization: Papers from the DIMACS Special Year.* Series in discrete mathematics and theoritical computer science, Vol. 20, pp. 111–152. AMS.

Karmarkar, N. K. 1984. A new polynomial-time algorithm for linear programming. *Combinatorica* 4:373–395.

Karmarkar, N. K. 1990. An interior-point approach to NP-complete problems—Part I. In *Contemporary Mathematics*, Vol. 114, pp. 297–308.

Karp, R. M. and Papadimitriou, C. H. 1982. On linear characterizations of combinatorial optimization problems. *SIAM J. Comput.* 11:620–632.

Lawler, E. L. 1979. Matroid intersection algorithms. *Math. Programming* 9:31–56.

Lehman, A. 1979. On the width-length inequality, mimeographic notes (1965). *Math. Programming* 17:403–417.

Lovasz, L. 1972. Normal hypergraphs and the perfect graph conjecture. *Discrete Math.* 2:253–267.

Lovasz, L. 1979. On the Shannon capacity of a graph. *IEEE Trans. Inf. Theory* 25:1–7.

Lovasz, L. 1986. *An Algorithmic Theory of Numbers, Graphs and Convexity*, SIAM Press.

Lovasz, L. and Schrijver, A. 1991. Cones of matrices and setfunctions. *SIAM J. Optimization* 1:166–190.

Lustig, I. J., Marsten, R. E., and Shanno, D. F. 1994. Interior point methods for linear programming: computational state of the art. *ORSA J. Comput.* 6(1):1–14.

Martin, R. K. 1991. Using separation algorithms to generate mixed integer model reformulations. *Operations Res. Lett.* 10:119–128.

McDiarmid, C. J. H. 1975. Rado's theorem for polymatroids. *Proc. Cambridge Philos. Soc.* 78:263–281.

Megiddo, N. 1991. On finding primal- and dual-optimal bases. *ORSA J. Comput.* 3:63–65.

Mehrotra, S. 1992. On the implementation of a primal-dual interior point method. *SIAM J. Optimization* 2(4):575–601.

Nemhauser, G. L. and Wolsey, L. A. 1988. *Integer and Combinatorial Optimization.* Wiley.

Padberg, M. W. 1973. On the facial structure of set packing polyhedra. *Math. Programming* 5:199–215.

Padberg, M. W. 1974. Perfect zero-one matrices. *Math. Programming* 6:180–196.

Padberg, M. W. 1993. Lehman's forbidden minor characterization of ideal 0, 1 matrices. *Discrete Math.* 111:409–420.

Padberg, M. W. 1995. *Linear Optimization and Extensions.* Springer–Verlag.

Padberg, M. W. and Rao, M. R. 1981. The Russian method for linear inequalities. Part III, bounded integer programming. Preprint, New York University, New York.

Padberg, M. W. and Rao, M. R. 1982. Odd minimum cut-sets and b-matching. *Math. Operations Res.* 7:67–80.

Padberg, M. W. and Rinaldi, G. 1991. A branch and cut algorithm for the resolution of large scale symmetric travelling salesman problems. *SIAM Rev.* 33:60–100.

Papadimitriou, C. H. and Yannakakis, M. 1991. Optimization, approximation, and complexity classes. *J. Comput. Syst. Sci.* 43:425–440.

Parker, G. and Rardin, R. L. 1988. *Discrete Optimization.* Wiley.

Picard, J. C. and Ratliff, H. D. 1975. Minimum cuts and related problems. *Networks* 5:357–370.

Pulleyblank, W. R. 1989. Polyhedral combinatorics. In *Handbooks in Operations Research and Management Science.* Vol. 1, Optimization, G. L. Nemhauser, A. H. G. Rinooy Kan, and M. J. Todd, eds., pp. 371–446. North–Holland.

Raghavan, P. and Thompson, C. D. 1987. Randomized rounding: a technique for provably good algorithms and algorithmic proofs. *Combinatorica* 7:365–374.

Rhys, J. M. W. 1970. A selection problem of shared fixed costs and network flows. *Manage. Sci.* 17:200–207.

Saraswat, V. and Van Hentenryck, P., eds. 1995. *Principles and Practice of Constraint Programming,* MIT Press, Cambridge, MA.

Savelsbergh, M. W. P., Sigosmondi, G. S., and Nemhauser, G. L. 1994. MINTO, a mixed integer optimizer. *Operations Res. Lett.* 15:47–58.

Schrijver, A. 1986. *Theory of Linear and Integer Programming.* Wiley.

Seymour, P. 1980. Decompositions of regular matroids. *J. Combinatorial Theory* B 28:305–359.

Shapiro, J. F. 1979. A survey of lagrangian techniques for discrete optimization. *Ann. Discrete Math.* 5:113–138.

Shmoys, D. B. 1995. Computing near-optimal solutions to combinatorial optimization problems. In *Combinatorial Optimization: Papers from the DI'ACS special year.* Series in discrete mathematics and theoretical computer science, Vol. 20, pp. 355–398. AMS.

Shor, N. Z. 1970. Convergence rate of the gradient descent method with dilation of the space. *Cybernetics* 6.

Truemper, K. 1992. Alpha-balanced graphs and matrices and GF(3)-representability of matroids. *J. Combinatorial Theory* B 55:302–335.

Weyl, H. 1935. Elemetere Theorie der konvexen polyerer. *Comm. Math. Helv.* Vol. pp. 3–18 (English translation 1950. *Ann. Math. Stud.* 24, Princeton).

Williams, H. P. 1987. Linear and integer programming applied to the propositional calculus. *Int. J. Syst. Res. Inf. Sci.* 2:81–100.

Yannakakis, M. 1988. Expressing combinatorial optimization problems by linear programs, pp. 223–228. In *Proc. ACM Symp. Theory Comput.*

Ziegler, M. 1995. *Convex Polytopes.* Springer–Verlag.

14

A Case Study in Algorithms: Very Large-Scale Integrated (VLSI) Circuit Layout

Andrea S. LaPaugh*
Princeton University

14.1 Introduction

One of the many application areas which has made effective use of algorithm design and analysis is computer-aided design (CAD) of digital circuits. Many aspects of circuit design yield to combinatorial models and the algorithmic techniques discussed in the preceding chapters. In this chapter we focus on one area within the field of CAD: layout of very large-scale integrated (VLSI) circuits, which is a particularly good example of the effective use of algorithm design and analysis. We will discuss specific problems in VLSI layout and how algorithmic techniques have been successfully applied. This chapter will not provide a broad survey of CAD techniques for layout, but will highlight a few problems that are important and have particularly nice algorithmic solutions. The reader may find more complete discussions of the field in the references discussed in the Further Information section at the end of this chapter.

Integrated circuits are made by arranging active elements (usually transistors) on a planar substrate and interconnecting these elements with conducting wires that are also patterned on the planar substrate [Wolf 1994]. There may be several layers that can be used for wires, but there are restrictions on how wires in different layers can connect as well as requirements of separations between wires and elements. Therefore, the layout of integrated circuits usually is modeled as a planar embedding problem with several layers in the plane.

*Supported in part by the National Science Foundation, FAW award MIP-9023542.

VLSI circuits contain hundreds of thousands to millions of transistors. Therefore, it is not feasible to consider the positioning of each transistor separately. Transistors are organized into subcircuits called components; this may be done hierarchically, resulting in several levels of component definition between the individual transistors and the complete VLSI circuit. The layout problem for VLSI circuits becomes one of positioning components and their interconnecting wires on a plane, following the design rules, and optimizing some measure such as area or wire length. Within this basic problem structure are a multiple of variations arising from changes in design rules and flexibilities within components as to size, shape, and regions where wires may connect to components. Graph models are used heavily, both to model the components and interconnections themselves and to capture constraints between objects. Geometric aspects of the layout problem also must be modeled. Most layout problems are optimization problems and most are NP-complete. Therefore, heuristics are also employed heavily. In the following, we present several of the best known and best understood problems in VLSI layout.

14.2 Background

We will consider a design style known as general cell. In **general cell layout**, components vary in size and degree of functional complexity. Some components may be from a predesigned component library and have rigidly defined layouts (e.g., a register bank) and others may be full custom designs, in which the building blocks are individual transistors and wires. Components may be defined hierarchically, so that the degree of flexibility in the layout of each component is quite variable. Other design styles, such as standard cell, gate array, and sea-of-gates, are more constrained but share many of the same layout techniques.

The layout problem for a VLSI chip often is decomposed into two stages: placement of components and routing of wires. For this decomposition, the circuit is described as a set of components and a set of interconnections among those components. The components are first placed on the plane based on their size, shape, and interconnectivity. Paths for wires are then found to interconnect specified positions on the components. Thus, a placement problem is to position a set of components in a planar region; either the region is bounded, or a measure such as total area of the region used is optimized. The area needed for the yet undetermined routing must be taken into account. A routing problem is, given a collection of sets of points in the plane, to interconnect each sets of points (called a **net**) using paths from an allowable set of paths. The allowable set of paths captures all of the constraints on wire routes. In routing problems, the width of wires is abstracted away by representing only the midline of each wire and ensuring enough room for the actual wires through the definition of the set of allowable paths.

The preceding description of the decomposition of a layout problem into placement and routing is meant to be very general. To discuss specific problems and algorithms, we will use a more constrained model. In our model, components will be rectangles. Each component will contain a set of points along its boundary, the **terminals**. Sets of these terminals are the nets, which must be interconnected. A layout consists of a placement of the components in the plane and a set of paths in the plane that do not intersect the components except at terminals and interconnect the terminals as specified by the nets. The paths are composed of segments in various layers of the plane. Further constraints on the set of allowable paths define the routing style and will be discussed for each style individually. The area of a layout will be the area of the minimum-area rectangle that contains the components and wire paths. (See Fig. 14.1.)

Although we still have a fairly general model, we have now restricted our component shapes to be rectangular, our terminals to be single points on component boundaries, and our routing paths to avoid components. (Components are thus assumed to be densely populated with circuitry.) Although these assumptions are common and allow us to illustrate important algorithmic results, there is quite a bit of work on nonrectangular components (e.g., Dai et al. [1987]), more flexible terminals (see Sherwani [1995]), and over-the-cell routing (see Sherwani [1995]). Often, layouts are further constrained to be **rectilinear**. In rectilinear layouts, there is an underlying pair of orthogonal axes defining *horizontal* and *vertical* and the sides of the components are oriented parallel to these axes. The paths of wires are composed of horizontal and vertical segments. In our discussion to follow, we, too, often will assume rectilinear layouts.

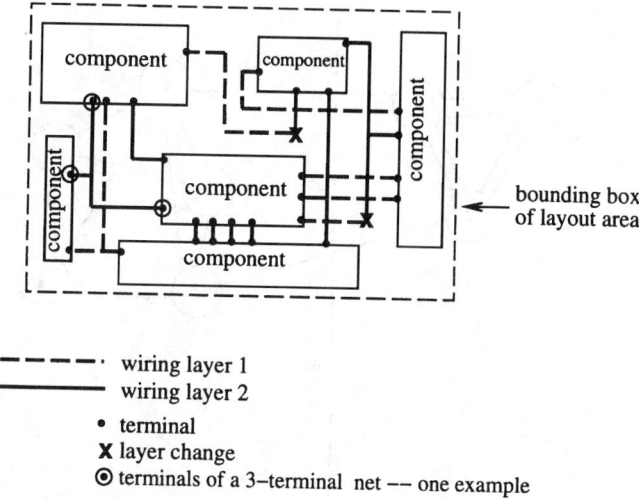

FIGURE 14.1 Example of a layout. This layout is rectilinear and has two layers of wiring.

If a VLSI system is too large to fit on one chip, then it is first partitioned into chip-sized pieces. During partitioning, the goal is to create the fewest chips with the fewest connections between chips. Estimates are used for the amount of space needed by wires to interconnect the components on one chip. The underlying graph problem for this task is **graph partitioning**, which is discussed subsequently.

Closely related to the placement problem is the **floor planning** problem. Floor planning occurs before the designs of components in a general cell design have been completed. The resulting approximate layout is called a **floor plan**. Estimates are used for the size of each component, based on either the functionality of the component or a hierarchical decomposition of the component. Rough positions are determined for the components. These positions can influence the shape and terminal placement within each component as its layout is refined. For hierarchically defined components, one can work bottom up to get rough estimates of size, then top down to get rough estimates of position, and then bottom up again to refine positions and sizes.

Once a layout is obtained for a VLSI circuit, either through the use of tools or by hand with a layout editor, there still may be room for improvement. **Compaction** refers to the process of modifying a given layout to remove extra space between features of the layout, space not required by design rules. Humans may introduce such space by virtue of the complexity of the layout task. Tools may place artificial restrictions on layout in order to have tractable models for the main problems of placement and routing. Compaction becomes a postprocessing step to make improvements too difficult to do during placement and routing.

14.3 Placement Techniques

Placement algorithms can be divided into two types: constructive initial placement algorithms and iterative improvement algorithms. A constructive initial placement algorithm has as input a set of components and a set of nets. The algorithm constructs a legal placement with the goal of optimizing some cost function for the layout. Common cost functions measure component area, estimated routing area, estimated total wire length, estimated maximum wire length of a net, or a combination of these. An iterative improvement algorithm has as input the set of components, set of nets, and an initial placement; it modifies the placement, usually repeatedly, to improve a cost function. The initial placement may be a random placement or may be the output of a constructive initial placement algorithm.

Iterative improvement of placements can proceed in a variety of ways. A set of allowable moves, that is, ways in which a placement can be modified, must be identified. These moves should be simple to carry

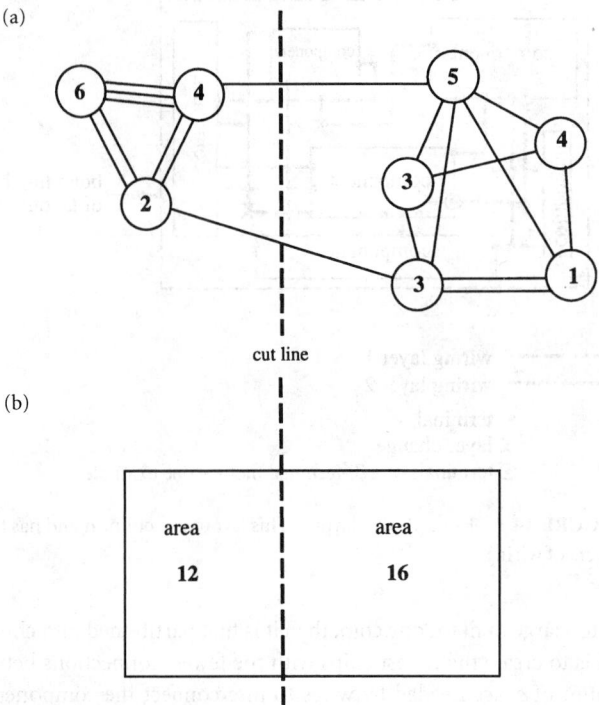

FIGURE 14.2 Partitioning used in placement construction. (a) Partitioning the circuit: Each vertex represents a component; the area of the component is the number inside the vertex. Connections between components are represented by edges. Multiple edges between vertices indicate multiple nets connecting components. (b) Partitioning the layout rectangle proportionally to the partition of component area.

out, including re-evaluating the cost function. Typically, a component is rotated or a pair of components are exchanged. An iterative improvement algorithm repeatedly modifies the placement, evaluates the new placement, and decides whether the move should be kept as a starting point for further moves or as a tentative final placement. In the simplest of iterative improvement algorithms, a move is kept only if the placement cost function is improved by it. One more sophisticated paradigm for iterative improvement is simulated annealing. (See Chapter 13.) It has been applied successfully to the placement problem, e.g., by Sechen [1988].

For general cell placement, one of the most widely used initial placement algorithms is a recursive partitioning method, first proposed by Lauther [1980]. In this method, a rectangular area for the layout is estimated based on component sizes and connectivity. The algorithm partitions the set of components into two roughly equal-sized subsets such that the number of interconnections between the two subsets is minimized and simultaneously partitions the layout area into two subrectangles of sizes equal to the sizes of the subsets. (See Fig. 14.2.) This partitioning proceeds recursively on the two subsets and subrectangles until each component is in its own subset and the rectangle contains a region for each component.

The fundamental problem underlying placement by partitioning is graph partitioning.[1] Given a graph $G = (V, E)$, a vertex weight function $w : V \to N$, an edge cost function $c : E \to N$, and a balance factor

---

[1]Our definitions follow those of Lengauer [1990].

$\beta \epsilon [1/2, 1]$, the graph partitioning problem is to partition V into two subsets V_1 and V_2 such that

$$\sum_{v \in V_1} w(v) \leq \beta \sum_{v \in V} w(v) \tag{14.1}$$

$$\sum_{v \in V_2} w(v) \leq \beta \sum_{v \in V} w(v) \tag{14.2}$$

and the cost of the partition,

$$\sum_{e \in E \cap (V_1 \times V_2)} c(e) \tag{14.3}$$

is minimized. This problem is NP-complete. (See Garey and Johnson [1979, pp. 209, 210].) Graph partitioning is a well-studied problem. The version we have defined is actually *bipartitioning*. Heuristics for this problem form the core of heuristic algorithms for more general versions of the partition problem where one partitions into more than two vertex sets. The hypergraph version of partitioning, in which each edge is a set of two or more vertices rather than simply a pair of vertices, is a more accurate version for placement, since nets may contain many terminals on many components. But heuristics for the hypergraph version again are based on techniques used for the graph bipartitioning problem.

Among many techniques for graph partitioning, two—the Kernighan–Lin [1970] algorithm and simulated annealing—are best known. Both are techniques for iteratively improving a partition. The Kernighan–Lin approach involves exchanging pairs of vertices across the partition. It was improved in the context of layout problems by Fiduccia and Mattheyses [1982], who move a single vertex at a time. As applied to graph partitioning, simulated annealing also considers the exchange of vertices across the partition or the movement of a vertex from one side of the partition to the other. The methods of deciding which partitions are altered, which moves are tried, and when to stop the iteration process differentiate the two techniques. The reader is referred to Sherwani [1995] for a more complete discussion of graph partitioning heuristics.

A second group of algorithms for constructing placements is based on clustering. In this approach, components are selected one at a time and placed in the layout area according to their connectivity to components previously placed. Components are clustered so that highly connected sets of components are close to each other.

When the cost function for a layout involves estimating wire length, several methods can be used. The goal is to define a measure for each net that estimates the length of that net after it is routed. These estimated lengths can then be summed over all nets to get the estimate on the total wire length, or the maximum can be taken over all nets to get maximum net length. Two estimates are commonly used: (1) the half-perimeter of the smallest rectangle containing all terminals of a net; (2) the minimum Euclidean spanning tree of the net. Given a placement, the Euclidean spanning tree of a net is the spanning tree of a graph whose vertices are the terminals of the net, whose edges are all edges between vertices, and whose edge costs are the Euclidean distances in the given placement between the pair of terminals that are endpoints of the edges. Another often used estimate is the minimum rectilinear spanning tree. This is because rectilinear layouts are common. For a rectilinear layout, the length of the shortest path between a pair of points, (x_a, y_a) and (x_b, y_b), is $|x_a - x_b| + |y_a - y_b|$ (assuming no obstacles). This distance, rather than the Euclidean distance, is used as the distance between terminals. (This is also called the L_1 or Manhattan metric.) A more accurate estimate of the wire length of a net would be the minimum Euclidean (or rectilinear) **Steiner tree** for the net. A Steiner tree for a set of points in the plane is a spanning tree for a superset of the points, i.e., additional points may be introduced to decrease the length of the tree. Finding minimum Steiner trees is NP-hard (see Garey and Johnson [1979]), whereas finding minimum spanning trees can be done in $O(|E| + |V| \log |V|)$ time for general graphs and in $O(|V| \log |V|)$ time for Euclidean or rectilinear spanning trees. (See Chapters 6 and 9.) The cost of a minimum spanning tree is an upper bound on the cost of a minimum Steiner tree. For rectilinear Steiner trees, the half-perimeter measure and two-thirds the cost of a minimum rectilinear spanning tree are lower bounds on the cost of a Steiner tree [Hwang 1976].

The minimum spanning tree also is useful for estimating routing area. The minimum spanning tree for each net is used as an approximation of the set of paths for the wires of the net. Congested areas of the layout then can be identified and space for routing allocated accordingly.

14.4 Compaction and the Single-Source Shortest Path Problem

Compaction can be done at various levels of design: an entire layout can be compacted at the level of transistors and wires; the layouts of individual components can be compacted; a layout of components can be compacted without changing the layout within components. To simplify our discussion, we will assume that layouts are rectilinear. For compaction, we model a layout as composed entirely of rectangles. These rectangles may represent the most basic geometric building blocks of the circuit: pieces of transistors and segments of wires, or they may represent more complex objects such as complex components. We refer to these rectangles as the *features* of the layout. We distinguish two types of features: those that are of fixed size and shape and those that can be stretched or shrunk in one or both dimensions. For example, a wire segment may be able to stretch or shrink in one dimension, representing a lengthening or shortening of the wire, but be fixed in the other dimension, representing a wire of fixed width. We refer to the horizontal dimension of a feature as its *width* and the vertical dimension as its *height*.

Compaction is fundamentally a two-dimensional problem. However, two-dimensional compaction is very difficult. Algorithms based on branch and bound techniques for integer linear programming (see Chapter 13) have been developed, but none is efficient enough to use in practice. (See Lengauer [1990] and Wolf and Dunlop [1988].) Therefore, the problem is commonly simplified by compacting in each dimension separately: first all features of a layout are pushed together horizontally as much as the design rules will allow, keeping their vertical positions fixed; then all features of a layout are pushed together vertically as much as the design rules will allow, keeping their horizontal positions fixed. The vertical compaction may, in fact, make possible more horizontal compaction, and so the process may be iterated. This method is not guaranteed to find a minimum area compaction, but for each dimension the compaction problem can be solved optimally. We are assuming that we start with a legal layout and that one-dimensional compaction cannot change left-to-right or top-to-bottom relationships. That is, if two features are intersected by the same horizontal (vertical) line and one feature is to the left of (above) the other, then horizontal (vertical) compaction will maintain this property. The algorithm we present is based on the single-source shortest path algorithm. (See Chapter 9.) It is an excellent example of a widely used application of this graph algorithm.

The compaction approach we are presenting is called *constraint-based* compaction because it models constraints on and between features explicitly. We shall discuss the algorithm in terms of horizontal compaction, with vertical compaction being analogous. We use a graph model in which vertices represent the horizontal positions of features; edges represent constraints between the positions of features. Constraints capture the layout design rules, relationships between features such as connectivity, and possibly other desirable constraints such as performance-related constraints. Design rules are of two types: feature-size rules and separation rules. Feature-size rules give exact sizes or minimum dimensions of features. For example, each type of wire has a minimum width; each transistor in a layout is of a fixed size. Separation rules require that certain features of a layout be at least a minimum distance apart to avoid electrical interaction or problems during fabrication. Connectivity constraints occur when a wire segment is allowed to connect to a component (or another wire segment) anywhere in a given interval along the component boundary. Performance requirements may dictate that certain elements are not too far apart. A detailed discussion of the issues in the extraction of constraints from a layout can be found in Wolf and Dunlop [1988].

In the simplest case, we start with a legal layout and consider only feature-size rules and separation rules. We assume all wire segments connect at fixed positions on component boundaries and there are no performance constraints. Furthermore, we assume all features that have a variable width attach at their left and right edges to features with fixed width, e.g., a wire segment stretched between two components.

Then, we need represent only features with a fixed width; we can use one variable for each feature. In this case, any constraints on the width of a variable-width feature are translated into constraints on the positions of the fixed-width features attached to either end. (See Fig. 14.3.) We are left with only separation constraints, which are of the form

$$x_B \geq x_A + d_{\min} \tag{14.4}$$

or, equivalently,

$$x_B - x_A \geq d_{\min} \tag{14.5}$$

where B is a feature to the right of A, x_A is the horizontal position of A, x_B is the horizontal position of B, and the minimum separation between x_A and x_B is $d_{\min}$. In our graph model, there is an edge from the vertex for feature A to the vertex for feature B with length $d_{\min}$. We add a single extra source vertex to the graph and a 0-length edge from this source vertex to every other vertex in the graph. This source vertex represents the left edge of the layout. Then finding the longest path from this source vertex to every vertex in the graph will give the left-most legal position of each feature—as if we had pushed each feature as far to the left as possible. Finding the longest path is converted to a single-source shortest path problem by negating all of the lengths on edges. This is equivalent to rewriting the constraint as

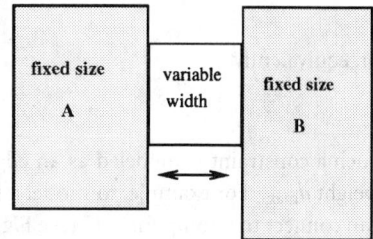

FIGURE 14.3 Generating separation constraints. The constraint on the separation between features A and B is the larger of the minimum separation between them and the minimum width of the flexible feature connecting them.

$$x_A - x_B \leq -d_{\min} \tag{14.6}$$

From now on, we will write constraints in this form. Note that this graph is acyclic. Therefore, as explained subsequently, the single-source shortest path problem can be solved in time $O(n + |E|)$ by **topological sort**, where n is the number of features and E is the set of edges in the constraint graph.

Topological sort is an algorithm for visiting the vertices of a directed acyclic graph (DAG). The edges of a DAG induce a partial order on the vertices: $v < u$ if there is a (directed) path from v to u in the graph. A topological order is any total ordering of the vertices that is consistent with this partial order. Topological sort visits the vertices in some topological order. For a graph $G = (V, E)$ it can be expressed as follows.

TOPOLOGICAL SORT ALGORITHM (INPUT G).

```
1   S ← all vertices with no incoming edges (sources)
2   U ← V
3   while S is not empty
4       do choose any vertex v from S
5           VISIT v
6           for each vertex u such that (v, u) ∈ E
7               do E ← E − {(v, u)}
8                   if u is now a source
9                       then S ← S ∪ {u}
10          U ← U − {v}
11          S ← S − {v}
12  if U is not empty
13      then error ▷ G is not acyclic
```

In our single-source shortest path problem, we start with only one source s, the vertex representing the left edge of the layout. We compute the length of the shortest path from s to each vertex v, denoted $\ell(v)$.

We initialize $\ell(v)$ before line 3 to be 0 for $\ell(s)$ and ∞ for all other vertices. Then for each vertex v we select at line 4, and each edge (v, u) we delete at line 7, we update for all shortest paths that go through v by $\ell(u) \leftarrow \min\{\ell(u), \ell(v) + length(v, u)\}$. When the topological sort has completed, $\ell(v)$ will contain the length of the shortest path from s to v (unless G was not acyclic to begin with). The algorithm takes $O(|V| + |E|)$ time.

In our simplest case, all our constraints were minimum separation constraints. In the general case, we may have maximum separation constraints as well. These occur when connectivity constraints are used and also when performance constraints that limit the length of wires are used. Then we have constraints of the form

$$x_B \leq x_A + d_{max} \tag{14.7}$$

or, equivalently,

$$x_B - x_A \leq d_{max} \tag{14.8}$$

Such a constraint is modeled as an edge from the vertex for feature B to the vertex for feature A with weight d_{max}. For example, to model a horizontal wire W that has an interval from l to r along which it can connect to a component C (see Fig. 14.4), we use the pair of constraints

$$x_C - x_W \leq -l \tag{14.9}$$

and

$$x_W - x_C \leq r \tag{14.10}$$

Once we allow both minimum and maximum separation constraints, we have a much more general linear constraint system. All constraints are still of the form

$$x - y \leq d \tag{14.11}$$

but the resulting constraint graph need not be acyclic. (Note that equality constraints $y - x = d$ may be expressed as $y - x \leq d$ and $x - y \leq -d$. Equality constraints also may be handled in a preprocessing step that merges vertices that are related by equality constraints.) To solve the single-source shortest path algorithm, we now need the $O(|V||E|)$-time Bellman–Ford algorithm (see Cormen et al. [1990]). This algorithm works only if the graph contains no negative cycle. If the constraint graph is derived from a layout that satisfies all of the constraints, this will be true, since a negative cycle represents an infeasible set of constraints. However, if the initial layout does not satisfy all of the constraints, for example, the designer adds constraints for performance to a rough layout, then the graph may be cyclic. The Bellman–Ford algorithm can detect this condition, but since the set of constraints is infeasible, no layout can be produced.

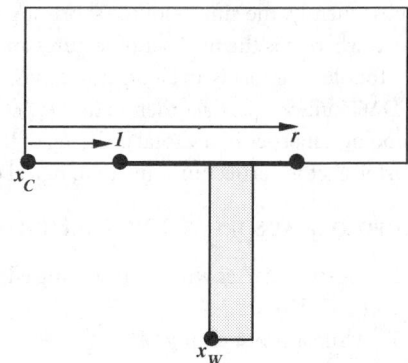

FIGURE 14.4 Modeling a connection that can be made along an interval of a component boundary.

If the constraint graph is derived from a layout that satisfies all of the constraints, an observation by Maley [1987a] allows us to use Dijkstra's algorithm to compute the shortest paths more efficiently. To use Dijkstra's algorithm, the weights on all of the edges must be positive (see Chapter 9). Maley observed that when an initial layout exists, the initial positions of the features can be used to convert all lengths to positive lengths as follows. Let p_A and p_B be initial positions of features A and B. The constraint graph is modified so that the length of an edge (v_A, v_B) from the vertex for A to the vertex for B becomes $length(v_A, v_B) + p_B - p_A$. Since the initial layout satisfies the constraint $x_A - x_B \leq length(v_A, v_B)$, we have $p_B - p_A \geq -length(v_A, v_B)$ and $p_B - p_A + length(v_A, v_B) \geq 0$. Maley shows that this transformation of the edge lengths preserves the shortest paths. Since all edge weights have been converted to positive lengths, Dijkstra's algorithm can be used, giving a running time of $O(|V| \log |V| + |E|)$ or $O(|E| \log |V|)$, depending on the implementation used.[2]

Even when the constraint graph is not acyclic and an initial feasible layout is not available, restrictions on the type or structure of constraints can be used to get faster algorithms. For example, Lengauer and Mehlhorn [1986] give an $O(|V| + |E|)$-time algorithm when the constraint graph has a special structure called a chain DAG that is found when the only constraints other than minimum separation constraints are those coming from flexible connections such as those modeled by Eqs. (14.9) and (14.10). Liao and Wong [1983] and Mata [1984][3] present $O(|E_x| \times |E|)$-time algorithms, where E_x is the set of edges derived from constraints other than the minimum-separation constraints. These algorithms are based on the observation that $E - E_x$ is a directed acyclic graph (as in our simple case). Topological sort is used as a subroutine to solve the single-source shortest path problem with edges $E - E_x$. The solution to this problem may violate constraints represented by E_x. Therefore, after finding the shortest paths for $E - E_x$, positions are modified in an attempt to satisfy the other constraints (represented by E_x), and the single-source shortest path algorithm for $E - E_x$ is run again. This technique is iterated until it converges to a solution for the entire set of constraints or the set of constraints is shown to be infeasible, which is proven to be within $|E_x|$ iterations. If $|E_x|$ is small, this algorithm is more efficient than using Bellman–Ford.

The single-dimensional compaction that we have discussed ignores many practical issues. One major issue is the introduction of bends in wires. The fact that a straight wire segment connects two components may be an artifact of the layout but it puts the components in lock-step during compaction. Adding a bend to the wire would allow the components to move independently, stretching the bend accordingly. Although the bend may require extra area, the overall area might improve through compaction with the components moving separately.

Another issue is allowing components to change their order from left to right or top to bottom. This change might allow for a smaller layout, but the compaction problem becomes much more difficult. In fact, a definition of one-dimensional compaction that allows for such exchanges in NP-complete [Doenhardt and Lengauer 1987]. Practically speaking, such exchanges may cause problems for wire routing. The compaction problem we have presented requires that the topological relationships between the layout features remain unchanged while space is compressed.

14.5 Floor Plan Sizing and Classic Divide-and-Conquer

The problem we will now consider, called **floor plan sizing**, is one encountered during certain styles of placement or floor planning. With some reasonable assumptions about the form of the layout, the problem can be solved optimally by a polynomial-time algorithm that is an example of classic divide-and-conquer.

Floor plan sizing occurs when a floor plan is initially specified as a partitioning of a rectangular layout area, representing the chip, into subrectangles, representing components. [See Fig. 14.5(a).] Each subrectangle corresponds to some component. We assume that the rectangular partitioning is rectilinear. For this discussion, given a set of components C, by "a floor plan for C", we shall mean such a rectangular partition of a rectangle into $|C|$ subrectangles and a one-to-one mapping from C to the subrectangles. This partition indicates the relative position of components, but the subrectangles are constrained only to have approximately the same area as the components (possibly with some bloating to account for routing), not to have the same aspect ratio as the components. When the actual components are placed in the locations, the layout will change. [See Fig. 14.5(b).] Furthermore, it may be possible to orient each component so that the the longer dimension may be either horizontal or vertical, affecting the ultimate size of the

---

[2] The $O(|V| \log |V| + |E|)$ running time depends on using Fibonacci heaps for a priority queue. If the simpler binary heap is used, the running time is $O(|E| \log |V|)$. This comment also holds for finding minimum spanning trees. See Chapter 9 for a discussion of the running times of Dijkstra's algorithm and Prim's algorithm for finding a minimum spanning tree.

[3] The technique used by Mata is the same as that used by Liao and Wong, but Mata has a different stopping condition for his search, which can lead to more efficient execution.

(a)

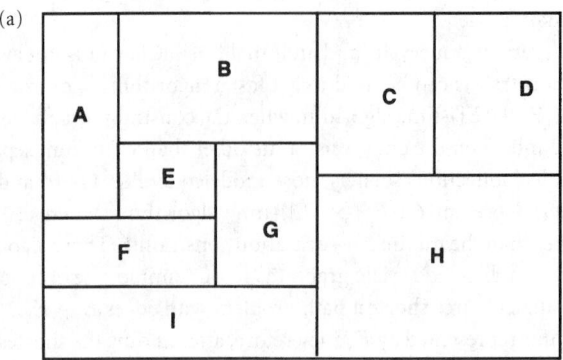

(b)

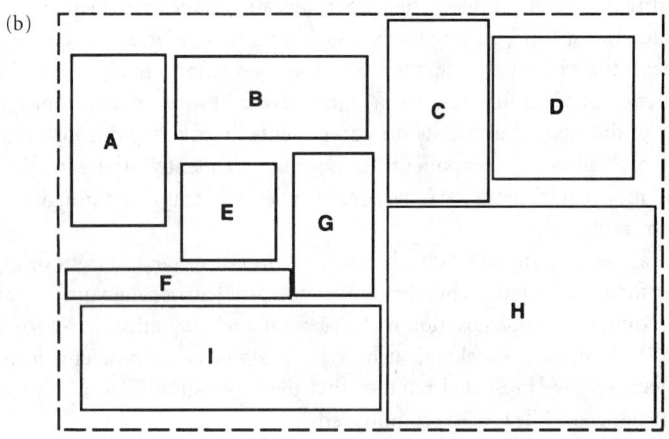

(c)

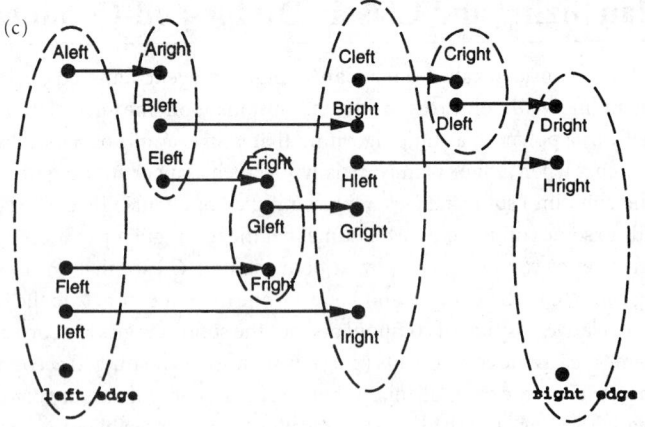

FIGURE 14.5 A floor plan and the derived layout. (a) A partition of a rectangle representing a floor plan. (b) A layout with actual components corresponding to the floor plan. (c) The horizontal constraint graph for the layout. Vertices in the same oval represent edges with the same horizontal position. They are taken to be a single vertex.

layout. In fact, if the component layouts have not been completed, the components may be able to assume many shapes while satisfying an area bound that is represented by the corresponding subrectangle. We will formalize this through the use of a **shape function**.

Definition 14.1. A shape function for a component is a mapping $s : [w_{min}, \infty] \rightarrow [h_{min}, \infty]$ such that s is monotonically decreasing, where $[w_{min}, \infty]$ and $[h_{min}, \infty]$ are subsets of $\Re^+$.

The interpretation of $s(w)$ is that it is the minimum height (vertical dimension) of any rectangle of width w that contains a layout of the component. Here w_{min} is the minimum width of any rectangle that contains a layout and h_{min} is the minimum height. The monotonicity requirement represents the fact that if there is a layout for a component that fits in a $w \times s(w)$ rectangle, it certainly must fit in a $(w + d) \times s(w)$ rectangle for any $d \geq 0$; therefore, $s(w + d) \leq s(w)$. In this discussion we will restrict ourselves to *piecewise linear* shape functions.

Given an actual shape (width and height) for each component, determining the minimum width and minimum height of the rectangular layout area becomes a simple compaction problem as previously discussed. Each dimension is done separately, and two constraint graphs are built. We will discuss the horizontal constraint graph; the vertical constraint graph is analogous. The reader should refer to Fig. 14.5(c). The horizontal constraint graph has a vertex for each vertical side of the layout rectangle and each vertical side of a subrectangle (representing a component) in the rectangular partition. There is a directed edge from each vertex representing the left side of a subrectangle to each vertex representing the right side of a subrectangle; this edge has a length which is the width of the corresponding component. There also are two directed edges (one in each direction) between the vertices representing any two overlapping sides of the layout rectangle or subrectangles; the length of these edges is 0. Thus, the vertex representing the left side of the layout rectangle has 0-length edges between it and the vertices representing the left sides of the leftmost subrectangles. Note that these constraints force two subrectangles that do not touch but are related through a chain of 0-length edges between one's left side and the other's right side to lie on opposite sides of a vertical line in the layout [e.g., components B and H in Fig. 14.5(a)]. This is an added restriction to the layout but an important one for the correctness of the algorithm for floor plan sizing of slicing floor plans presented subsequently.

Given an actual width for each component and having constructed the horizontal constraint graph as described in the preceding paragraph, to determine the minimum width of the rectangular layout area one simply finds the longest path from the vertex representing the left side of the layout rectangle to the vertex representing the right side of the layout rectangle. To simplify the problem to one in an acyclic graph, vertices connected by pairs of 0-length edges can be collapsed into a single vertex; only the longest edge between each pair of (collapsed) vertices is needed. Then topological sort can be used to find the longest path between the left side and the right side of the floor plan, as discussed in the preceding section of this chapter.

We now have the machinery to state the problem of interest.

PROBLEM 16.1. *Floor plan sizing.* Given a set C of components, a piecewise linear shape function for each component, and a floor plan for C, find an assignment of specific shapes to the components so that the area of the layout is minimized.

Stockmeyer [1983] showed that for general floor plans, the Floor Plan Sizing Problem is NP-complete. This holds even if the components are of fixed shape but can be rotated 90°. In this case, the shape function of each component is a step function with at most two steps: $s(x) = d_2$ for $d_1 \leq x < d_2$ and $s(x) = d_1$ for $d_2 \leq x$, where the dimensions of the component are d_1 and d_2 with $d_1 \leq d_2$. However, for floor plans of a special form, called **slicing floor plans**, Stockmeyer gave a polynomial-time algorithm for the Floor Plan Sizing Problem when components are of fixed shape but can rotate. Otten [1983] generalized this result to any piecewise-linear shape function. A slicing floor plan is one in which the partition of the layout rectangle can be constructed by a recursive cutting of a rectangular region into two subregions using either a vertical or a horizontal line segment. (See Fig. 14.6.) The rectangular regions that are not further partitioned are the subrectangles corresponding to components. The recursive partitioning method of constructing a placement previously discussed produces a slicing floor plan.

A slicing floor plan can be represented by a binary tree. The root of the tree represents the entire layout rectangle and is labeled with the direction of the first cut. Other interior vertices represent rectangular subregions that are to be further subdivided, and the label on any such vertex is the direction of the cut used to subdivide it. The two children of any vertex are the rectangular regions resulting from cutting the rectangle represented by the vertex. The leaves of the binary tree represent the rectangles corresponding to components.

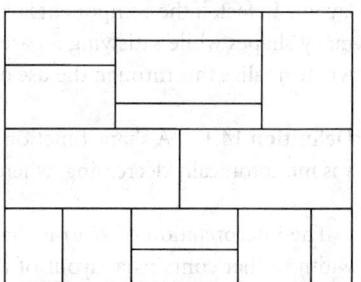

FIGURE 14.6   A slicing floor plan.

The algorithm for the sizing of a slicing floor plan uses the binary tree representation in a fundamental way. The key observation is that one needs only the shape functions of the two subregions represented by the children of a vertex to determine the shape function of a vertex. If the shape functions can be represented succinctly and combined efficiently for each vertex, the shape function of the root can be determined efficiently. We will present the combining step for shape functions that are step functions (i.e., piecewise constant) following the description in Lengauer [1990] since it illustrates the technique but is a straightforward calculation. Otten [1983] shows how to combine piecewise linear slicing functions, but Lengauer [1990] comments that arithmetic precision can become an issue in this case.

We shall represent a shape function that is a step function by a list of pairs $(w_i, s(w_i))$ for $0 \leq i \leq b_s$ and $w_0 = w_{\min}$. The interpretation is that for all x, $w_i \leq x < w_{i+1}$ (with $w_{b_s+1} = \infty$), $s(x) = s(w_i)$. Parameter b_s is the number of *breaks* in the step function. The representation of the function is linear in the number of breaks. (This represents a step function whose constant intervals are left closed and right open and is the most logical form of step function for shape functions. However, other forms of step functions also can be represented in size linear in the number of breaks.)

Given step functions s_l and s_r for the shapes of two children of a vertex, the shape function s for the vertex will also be a step function. When the direction of the cut is horizontal, the shape functions can simply be added, that is, $s(x) = s_l(x) + s_r(x)$. The $w_{\min}$ for s is the maximum of $w_{\min,l}$ for s_l and $w_{\min,r}$ for s_r. Each subsequent break point for s_l or s_r is a break point for s, so that $b_s \leq b_{s_l} + b_{s_r}$. Combining the shape functions takes time $O(b_{s_l} + b_{s_r})$. When the direction is vertical, the step functions must first be inverted, the inverted functions combined, and then the combined function inverted back. The inversion of a step function s is straightforward and can be done in $O(b_s)$ time.

To compute the shape function for the root of a slicing floor plan, one simply does a postorder traversal of the binary tree (see Chapter 9), computing the shape function for each vertex from the shape functions for the children of the vertices. The number of breaks in the shape functions for any vertex is no more than the number of breaks in the shape functions for the leaves of the subtree rooted at that vertex. Let b be the maximum number of breaks in any shape function of a component. Then the running time of this algorithm for a slicing floor plan with n components is

$$T(n) \leq T(n_l) + T(n_r) + bn \qquad (14.12)$$

$$\leq dbn \qquad (14.13)$$

where d is the depth of the tree (the length of the longest path from the root to a leaf). We have the following.

> Given an instance of the Floor Plan Sizing Problem that has a slicing floor plan and step functions as shape functions for the components there is an $O(dbn)$-time algorithm to compute the shape function of the layout rectangle.

Given the shape function for the layout rectangle, the minimum area shape can be found by computing the area at each break in time linear in the number of breaks, which is at most $O(bn)$.

14.6 Routing Problems

We shall only discuss the most common routing model for general cell placement: the rectilinear channel routing model. In this model, the layout is rectilinear. The regions of the layout that are not covered by components are partitioned into nonoverlapping rectangles, called **channels**. The allowed partitions are restricted so that each channel has components touching only its horizontal edges (a horizontal channel) or its vertical edges (a vertical channel). These edges will be referred to as the top and bottom of the channel, regardless of whether the channel is horizontal or vertical. The orthogonal edges, referred to as the left edge and right edge of the channel, can touch only another channel. These channels compose the area for the wire routing. There are several strategies for partitioning the routing area into channels, i.e., *defining* the channels, but most use maximal rectangles where possible (i.e., no two channels can be merged to form a larger channel). The layout area becomes a rectangle that is partitioned into subrectangles of two types: components and channels. (See Fig. 14.7.)

Given a layout with channels defined, the routing problem can be decomposed into two subproblems: global routing and local or detailed routing. **Global routing** is the problem of choosing which channels will be used to make the interconnections for each net. Actual paths are not produced. By doing global routing first, one can determine the actual paths for wires in each channel separately. The problem of detailed routing is to determine these actual paths and is more commonly referred to as **channel routing**. Of course, the segments of a wire path in each channel must join at the edges of the channel to produce an actual interconnection of terminals. To understand the approaches for handling this interfacing, we must have a more detailed definition of the channel routing problem, which we give next.

The channel routing problem is defined so that there are initial positional constraints in only one dimension. Recall that we define channels to components on only two parallel sides, the top and bottom. This is so that the routes in the channel will be constrained by terminal positions on only two sides. The standard channel routing problem has the following input: a rectangle (the channel) containing points (terminals) at fixed positions along its top and bottom edges, a set of nets that must be routed in the channel, and an assignment of each of the terminals on the channel to one of these nets. Also, two

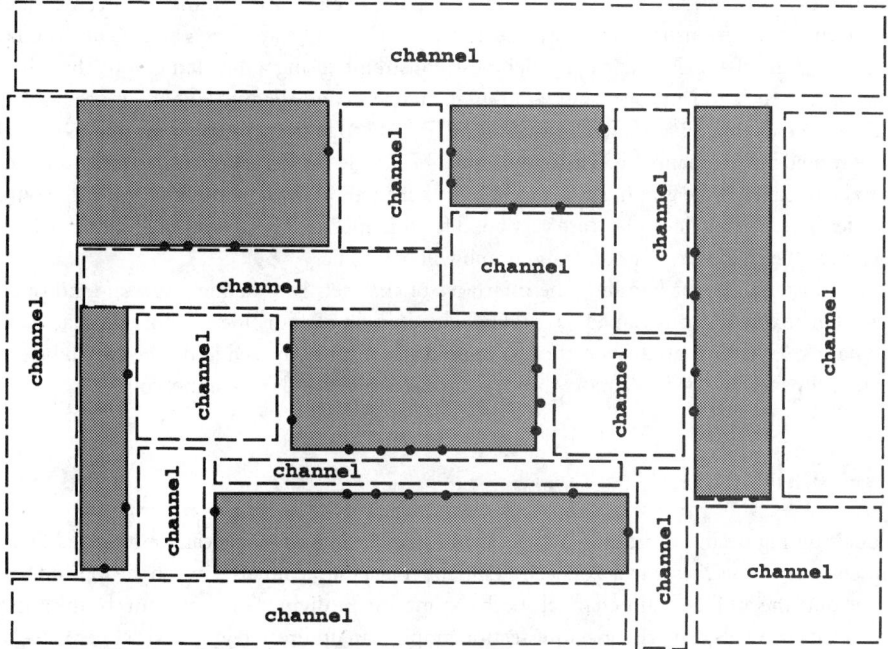

FIGURE 14.7 The decomposition of a layout into routing channels and components.

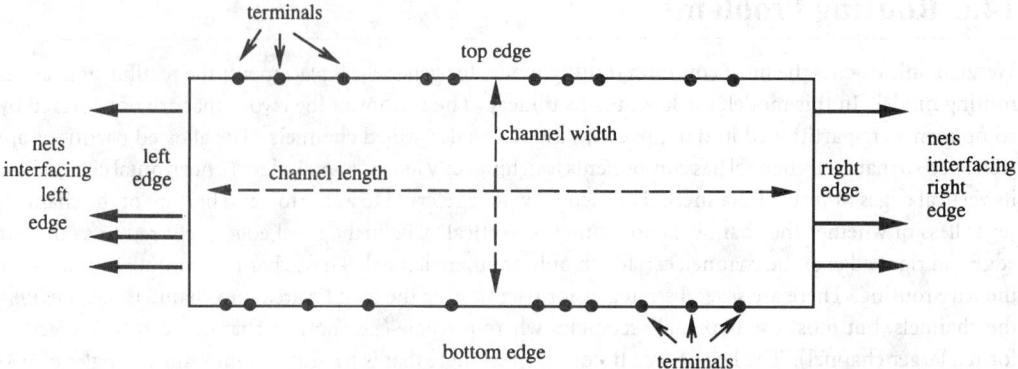

FIGURE 14.8   A channel. Nets interfacing at the left and right edges are not in a given order.

(possibly empty) subsets of nets are specified: one containing nets whose routing must interface with the left edge of the channel and one containing nets whose routing must interface with the right edge of the channel. The positions at which wires must intersect the left and right edges of the channel are not specified. The dimension of the channel from the left edge to the right edge is called the *length* of the channel; the dimension from the top edge to the bottom edge is the *width* of the channel. (See Fig. 14.8.) Since there are no terminals on the left and right edges, the width often is taken to be variable, and the goal is to find routing paths achieving the connections specified by the nets and minimizing the width of the channel. In this case, the space needed for the wires determines the width of the channel. The length of the channel is more often viewed as fixed, although there are channel routing models in which routes are allowed to cross outside the left and right edges of the channel.

We now can discuss the problem of interfacing routes at channel edges. There are two main approaches to handling the interfacing of channels. One is to define the channels so that all adjacent channels form Ts (no +s). Then if the channel routing is done for the base of the T first, the positions of paths leaving the base of the T and entering the crosspiece of the T are fixed by the routing of the base and can be treated as terminal positions for the crosspiece of the T. Using this approach places constraints on the routing order of the channels. We can model these constraints using a directed graph: there is a vertex v_C for each channel C, and an edge from v_A to v_B if channel A and channel B abut with channel A as the base of the T and channel B as the crosspiece of the T. The edge from v_A to v_B represents that channel A must be routed before channel B. This graph must be acyclic for the set of constraints on the order of routing to be feasible. If the graph is not acyclic, another method must be used to deal with some of the channel interfaces. Slicing floor plans are very good for this approach because if each slice is defined to be a channel, then the channel order constraint graph will be acyclic.

The second alternative for handling the interfaces of channels is to define a special kind of channel, called a **switch box** that is constrained by terminal locations on all four sides. In this alternative, some or all of the standard channels abut switch boxes. Special algorithms are used to do the switch box routing, since, practically speaking, this is a more difficult problem than standard channel routing.

14.7 Global Routing

Since global routing need not produce actual paths for wires, the channels can be modeled by a graph, called the *channel intersection graph*. For each channel, project the terminals lying on the top and bottom of the channel onto the midline of the channel. Each channel can be divided into segments by the positions of these projected terminals. The channel intersection graph is an undirected graph with one vertex for each projected terminal and one vertex for each intersection of channels. There is one edge for each segment of a channel, which connects the pair of vertices representing the terminals and/or channel intersections bounding the segment. A length and a capacity can be assigned to each edge, representing, respectively, the length between the ends of the channel segment and the width of the channel. Different versions of the global routing problem use one or both of the length and capacity parameters.

Given the channel intersection graph, the problem of finding a global routing for a set of nets becomes the problem of finding a Steiner tree in the channel intersection graph for the terminals of each net. Earlier in this chapter, we defined Steiner trees for points in the plane. For a general graph $G = (V, E)$, a Steiner tree for a subset of vertices $U \subset V$ is a set of edges of G that form a tree in G and whose set of endpoints contains the set U. Various versions of the global routing problem are produced by the constraints and optimization criteria used. For example, one can simply ask for the minimum length Steiner tree for each net, or one can ask for a set of Steiner trees that does not violate capacity constraints and has total minimum length. For each edge, the capacity used by a set of Steiner trees is the number of Steiner trees containing that edge; this must be no greater than the capacity of the edge. Another choice is not to have constraints on the capacity of edges but to minimize the maximum capacity used for any edge. In general, any combination of constraints and cost functions using length and capacity can be used. However, regardless of the criteria, the global routing problem is invariably NP-complete. A more detailed discussion of variations of the problem and their complexity can be found in Lengauer [1990]. There, a number of sophisticated algorithms for Steiner tree problems also are discussed. Here we will discuss only two techniques based on basic graph algorithms: breadth-first search and Dijkstra's single-source shortest path algorithm.

The minimum Steiner tree problem is itself NP-complete (see Garey and Johnson [1979, pp. 208–209]). Therefore, one approach to global routing is to avoid finding Steiner trees by breaking up each net into a collection of point-to-point connections. One way to do this is to find the minimum Euclidean or rectilinear spanning tree for the terminals belonging to each net (ignoring the channel structure) and use the edges of this tree to define the point-to-point connections. Then one can use Dijkstra's single-source shortest path algorithm on the channel intersection graph to find a shortest path for each point-to-point connection. Paths for connections of the same net that share edges can then be merged, yielding a Steiner tree. If there are no capacity constraints on edges, the quality of this solution is limited only by the quality of the approximation of a minimum Steiner tree by the chosen collection of point-to-point paths. If there are capacity constraints, then after solving each shortest path problem, one must remove from the channel intersection graph the edges whose used capacity already equals the edge capacity. In this case, the order in which nets and terminals within a net are routed is significant. Heuristics are used to choose this order. One can better approximate Steiner trees for the nets by choosing, at each iteration for connections within one net, a terminal not yet connected to any other terminals in the net as the source of Dijkstra's algorithm. Since this algorithm computes the shortest path to every other vertex in the graph from the source, the shortest path which connects to any other vertex in the channel intersection graph that is already on a path connecting terminals of the net can be used. Of course, there are variations on this idea.

For any graph, breadth-first search from a vertex v will find a shortest path from v to every other vertex in the graph when the length of each edge is 1. Breadth-first search takes time $O(|V| + |E|)$ compared to the best worst-case running time known for Dijkstra's algorithm: $O(|V| \log |V| + |E|)$ time. It is also very straightforward to implement. It is easy to incorporate heuristics that take into account the capacity of an edge already used and bias the search toward edges with little capacity used. Furthermore, breadth-first search can be started from several vertices simultaneously, so that all terminals of a net could be starting points of the search simultaneously. If it is adequate to view all channel segments as being of equal length, then the edge lengths can all be assigned value 1 and breadth-first search can be used. This might occur when the terminals are uniformly distributed and so divide channels into approximately equal-length segments. Alternatively, one can add new vertices to the channel intersection graph to further decompose the channel segments into unit-length segments. This can substantially increase $|V|$ and $|E|$ so that they are proportional to the dimensions of the underlying grid defining the unit of length rather than the number of channels and terminals in the problem. However, this allows one to use breadth-first search to compute shortest paths while modeling the actual length of channels. In fact, breadth-first search was developed by Lee [1961][4] for routing of circuit boards in exactly this manner. He modeled the entire board by a grid graph and modeled obstacles to routing paths as grid vertices that were missing from the graph. Each wire route was found by doing a breadth-first search in the grid graph.

14.8 Channel Routing

Channel routing is not one single problem but rather a family of problems based on the allowable paths for wires in the channel. We will limit our discussion to grid-based routing. Although both grid-free rectilinear and nonrectilinear routing techniques exist, the most basic techniques are grid based. We assume that there is a grid underlying the channel, the sides of the channel lie on grid edges, terminals lie on grid points, and all routing paths must follow edges in the grid. For ease of discussion, we shall refer to channel directions as though the channel were oriented with its length running horizontally. The vertical segments of the grid that run from the top to the bottom of the channel are referred to as *columns*. The horizontal segments of the grid that run from the left edge to the right edge of the channel are referred to as *tracks*. We will consider channel routing problems that allow the width of the channel to vary. Therefore, the number of columns determining the channel length will be fixed, but the number of tracks determining the channel width will be variable. The goal is to minimize the number of tracks used to route the channel.

The next distinction is based on how many routing layers are presumed. If there are ℓ routing layers, then there are ℓ overlaid copies of the grid, one for each layer. Routes that use the same edge on different layers do not interact and are considered disjoint. Routes change layer at grid points. The exact rules for how routes can change layers vary, but the most common is to view a route that goes from layer i to layer $j (j > i)$ at a grid point as using layers $i + 1, \ldots, j - 1$ as well at the grid point. One can separate the channel routing problem into two subproblems: finding paths in the grid that achieve the interconnections and finding a **layer assignment** for each edge in each path so that the resulting set of routes is legal. One channel routing model for which this is done is knock-knee routing. (See Lengauer [1990, section 9.5].) However, in our discussion, we will consider only two routing models: single-layer routing and two-layer **Manhattan routing**. In single-layer routing there is no issue of layer assignment; in Manhattan routing, the model is such that the layer assignment is automatically derived from the paths. Routing models that allow more than two layers, called multilayer models, are becoming more popular as technology is able to provide more layers for wires. However, many of the multilayer routing techniques are derived from single-layer or two-layer Manhattan techniques (e.g., Greenberg et al. [1988]), so we focus this restricted discussion on those models. A more detailed review of channel routing algorithms can be found in LaPaugh and Pinter [1990].

Manhattan Routing

Manhattan routing is the dominant 2-layer routing model. It dates back to printed circuit boards [Hashimoto and Stevens 1971]. It dominates because it finesses the issue of layer assignment by defining all vertical wire segments to be on one layer and all horizontal wire segments to be on the other layer. Therefore, a horizontal routing path and a vertical routing path can cross without interacting, but any path that bends at a grid point is on both layers at that point and no path for a disjoint net can use the same point. Thus, under the Manhattan model, the problem of routing can be stated completely in terms of finding a set of paths such that paths for distinct nets may cross but do not share edges or bend points.

Although Manhattan routing provides a simple model of legal routes, the resulting channel routing problem is NP-complete [Szymanski 1985]. An important lower bound on the width of a channel is the **channel density**. The density at any vertical line cutting the channel (not necessarily a column) is the number of nets that have terminals both to the left and right of the vertical line. The interpretation is

---

[4]The first published description of breadth-first search was by E. F. Moore for finding a path in a maze. Lee developed the algorithm for routing in grid graphs under a variety of path costs. See the discussion on page 394 of Lengauer [1990].

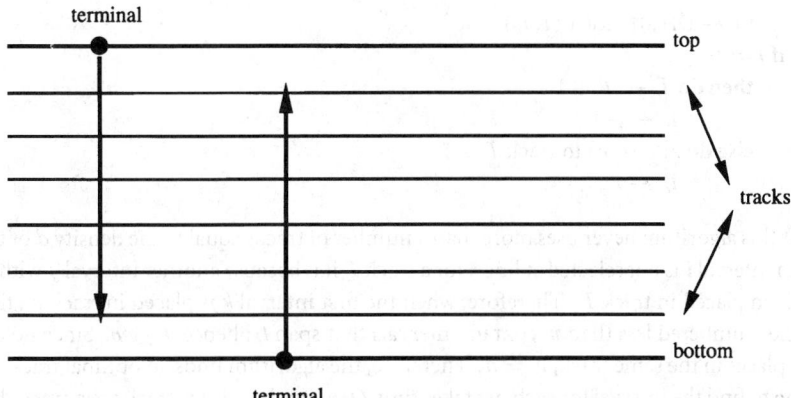

FIGURE 14.9 Connecting to tracks without conflicts. Each of the nets can reach any of
the tracks with a vertical segment in the column containing the terminal.

that each net must have at least one wire crossing the vertical line, and thus a number of tracks equal to
the density at the vertical line is necessary. For columns, nets that contain a terminal on the column are
counted in the density unless the net contains exactly two terminals and these are at the top and bottom of
the column. (Such a net can be routed with a vertical path the extent of the column.) The channel density
is the maximum density of any vertical cut of the channel. In practice, the channel routing problem is
solved with heuristic algorithms that find routes giving a channel width within one or two of the density,
although such performance is not provably achievable.

If a channel routing instance has no top terminal and bottom terminal on the same column, then the
number of tracks equal to the channel density suffices, and a route achieving this density can be solved
in $O(m + n \log n)$ time, where n is the number of nets and m is the number of terminals. Under this
restriction, the channel routing problem becomes equivalent to the problem of **interval graph coloring**.
The equivalence is based on the observation that if no column contains terminals at its top and bottom,
then any terminal can connect to any track by a vertical segment that does not conflict with any other
path (see Fig. 14.9). In this case, density can be achieved by a greedy algorithm. Each net will use one
horizontal segment that spans all of the columns containing its terminals. The issue is to pack these
horizontal segments into the minimum number of tracks so that no segments intersect. Equivalently, the
goal is to color the segments (or intervals) with the minimum number of colors so that no two segments
that intersect are the same color. Hence we have the relationship to interval graphs, which have a set of
vertices that represent intervals on a line and edges between vertices of intersecting intervals.

A classic greedy algorithm can do the track assignment for a set I of intervals. The assignment is made
track-by-track (the actual position of each track does not matter). Intervals are assigned to each track from
left to right in a greedy fashion: at any point in the algorithm, the first interval that fits is put in a track. The
set of interval endpoints is first sorted so that given the right endpoint of the last interval put in a track, the
next interval to be put in the track can be found in constant time. Function DELETERIGHT (I, t_r) deletes
from I and returns the interval (l_i, r_i) whose left endpoint l_i is the leftmost endpoint to the right of t_r,
that is, $l_i > t_r$ and for no $(l_j, r_j) \in I$ is $t_r < l_j < l_i$. If no such interval exists, DELETERIGHT (I, t_r) returns
$(0, 0)$. As each track is filled, variable t_r will hold the value of the right endpoint of the most recently placed
interval.

TRACK-BY-TRACK ASSIGNMENT ALGORITHM (INPUT I).

1   Sort the endpoints of intervals in I, maintaining left endpoint to right endpoint correspondence
2   $T \leftarrow 1$
3   $t_r \leftarrow -1$
4   **while** I is not empty

```
5        do (l, r) ← DELETERIGHT (I, t_r)
6          if r = 0
7            then do T ← T + 1
8                  t_r ← -1
9            else do ADD (l, r) to track T
10                 t_r ← r
```

To see that this algorithm never uses more than a number of tracks equal to the density d of the channel, note that if an interval i is not selected at line 5 for a track T it is because another interval j with $l_j \leq l_i$ and $r_j \geq l_i$ has been placed in track T. Therefore, when the first interval k is placed in track w, the last track used, all tracks numbered less than w contain intervals that span l_k; hence $d \geq w$. Since no overlapping intervals are places in the same track, $w = d$. Therefore, the algorithm finds an optimal track assignment. Preprocessing to find the interval for each net takes time $O(m)$ and track-by-track assignment has running time $O(|I| \log |I|) = O(n \log n)$, due to the initial sorting of I.

Once one allows terminals at both the top and bottom of a column (except when all such pairs of terminals belong to the same net), one introduces a new set of constraints called *vertical constraints*. These constraints capture the fact that if net i has a terminal at the top of column c and net j has a terminal at the bottom of column c, then to connect these terminals to horizontal segments using vertical segments at column c, the horizontal segment for i must be above the horizontal segment for j. One can construct a vertical constraint graph that has a vertex v_i for each net i and a directed edge between v_i and v_j if there is a column that contains a terminal in i at the top and a terminal in j at the bottom. If one considers only routes that use at most one horizontal segment per net, then the constraint graph represents order constraints on the tracks used by the horizontal segments. If the vertical constraint graph is cyclic, then the routing cannot be done with one horizontal segment per net. If the vertical constraint graph is acyclic, LaPaugh [1980] has shown that it is NP-complete to determine if the routing can be achieved in a given number of tracks. Furthermore, even if an optimal or good routing using one horizontal segment per net is found, the number of tracks required often is substantially larger than what could be obtained using more horizontal segments. For these reasons, practical channel routing algorithms allow the route for each net to traverse portions of several tracks. Each time a route changes from one track to another, it uses a section of a column; this is called a **jog** (see Fig. 14.10).

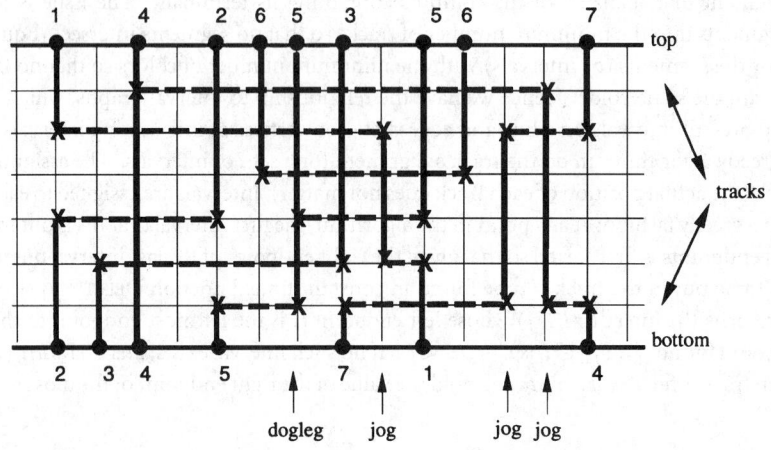

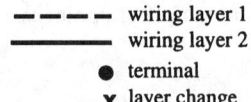

FIGURE 14.10   A channel routing in the 2-layer Manhattan model showing jogs.

The NP-completeness result of Szymanski [1985] applies to Manhattan channel routing with unrestricted jogs. Therefore, the practical routing algorithms for this problem use heuristics. The earliest of these, by Deutsch [1976], was based on the previously described track-by-track algorithm. Deutsch allowed the route for a net to jog only in a column that contained a terminal of the net; he called these jogs *doglegs* (see Fig. 14.10). This approach effectively breaks up each net into two-point subnets, and one can then define a vertical constraint graph in which each vertex represents a subnet. Deutsch's basic algorithm does track-by-track assignment but does not assign an interval for a subnet to a track if the assignment would violate a vertical constraint. Embellishments on the algorithm try to improve its performance and minimize the number of doglegs. Others have also modified the approach. (See the discussion in Lengauer [1990, section 9.6.1.4].) The class of algorithms based on track-by-track assignment is also known as *left edge algorithms.*

Manhattan channel routing is arguably the most widely used detailed routing model and many algorithms have been developed. The reader is referred to Sherwani [1995] or LaPaugh and Pinter [1990] for a survey of algorithms. In this chapter, we will discuss only one more algorithm, an algorithm that proceeds column-by-column in contrast to the previously described track-by-track algorithm. The column-by-column algorithm was originally proposed by Rivest and Fiduccia [1982] and was called by them a greedy router. This algorithm routes all nets simultaneously. It starts at the leftmost column of the channel and proceeds to the right, considering each column in order. As it proceeds left to right it creates, destroys, and continues horizontal segments for nets in the channel. Using this approach, it is easy to introduce a jog for a net in any column. At each column, the algorithm connects terminals at the top and bottom to horizontal segments for their nets, starting new segments when necessary, and ending segments when justified. At each column, for each continuing net with terminals to the right, it also may create a jog to bring a horizontal segment of the route closer to the channel edge containing the next terminal to the right. Thus, the algorithm employs some lookahead in determining what track to use for each net at each column. The algorithm is actually a framework with many parameters that can be adjusted. It may create, for one net, multiple horizontal segments that cross the same column and may extend routes beyond the left and right edges of the channel. It is a very flexible framework that allows many competing criteria to be considered and allows the interaction of nets more directly than strategies that route one net at a time. Many routing tools have adopted this approach.

Single-Layer Routing

Although single-layer channel routing plays a role in the routing of multilayer channels (e.g., Greenberg et al. [1988]), its greatest significance comes from the fact that even in its most general form, it can be solved optimally in polynomial time. There is a rich theory of single-layer detailed routing that has been developed not only for channel routing but for routing in more general regions (see Maley [1987b]). The first algorithmic results for single-layer channel routing were by Tompa [1981], who considered **river routing** problems. A river routing problem is a single-layer channel routing problem in which each net contains exactly two terminals, one at the top edge of the channel and one at the bottom edge of the channel. The nets have terminals in the same order along the top and bottom, a requirement if the problem is to be routable in one layer. Tompa considered unconstrained (vs. rectilinear) wire paths and gave a $O(n^2)$-time algorithm for n nets to test routability and find the route that minimizes both the individual wire lengths and the total wire length when the width of the channel is fixed. This algorithm can be used as a subroutine within binary search to find the minimum-width channel in $O(n^2 \log n)$ time. Tompa also suggested how to modify his algorithm for the rectilinear case. Dolev et al. [1981] built upon Tompa's theory for the rectilinear case and presented an $O(n)$-time algorithm to compute the minimum width of the channel and an $O(n^2)$-time algorithm to actually produce the rectilinear routing. The difference in running times comes from the fact that the routing may actually have n segments per net and thus would take $O(n^2)$-time to generate (see Fig. 14.11). In contrast, the testing for routability can be done by examining a set of constraints for the channel. The results were generalized to multiterminal nets by Greenberg and Maley [1992], where the time to calculate the minimum width remains linear in the number of terminals.

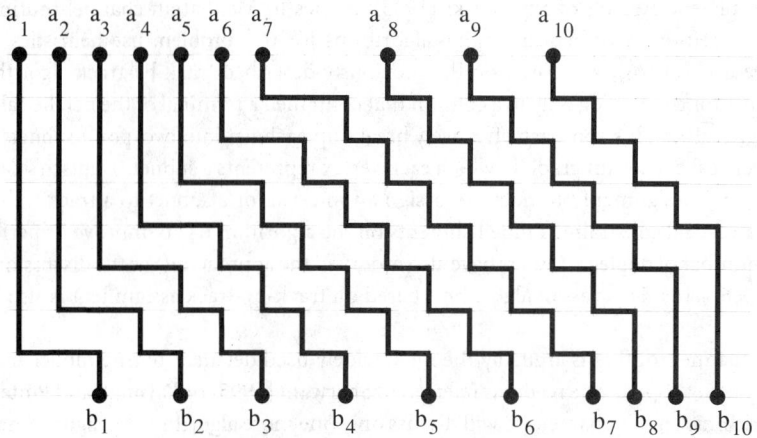

FIGURE 14.11   A river routing. The route for a single net may bend $O(n)$ times.

We now present the theory that allows the width of a river routing channel to be computed in linear time. Our presentation is an amalgam of the presentations in Dolev et al. [1981] and Leiserson and Pinter [1983]. The heart of the theory is the observation that for river routing, cut lines other than the vertical lines that define channel density also contribute to a lower bound for the width of the channel. This lower bound is then shown to be an upper bound as well. Indexing from left to right, let the ith net of a routing channel, $1 \leq i \leq n$, be denoted by the pair (a_i, b_i) where a_i is the horizontal position of its terminal on the top of the channel and b_i is the horizontal position of its terminal on the bottom of the channel. Consider a line from b_i to a_{i+k}, $k > 0$, cutting the channel. There are $k + 1$ nets that must cross this line. Measuring slopes from $0°$ to $180°$, if the line has slope $\geq 90°$, then $k + 1$ nets must cross the vertical ($90°$) line at b_i, and there must be $k + 1$ tracks. If the line has slope $< 90°$ and $> 45°$, then each vertical grid segment that crosses the line can be paired with a horizontal grid segment that also must cross the line and that cannot be used by a different net. Therefore, the line must cross $k + 1$ tracks. Finally, if the line has slope $\leq 45°$, then each horizontal grid segment that crosses the line can be paired with a vertical grid segment that also must cross the line and that cannot be used by a different net. Therefore, there must be $k + 1$ columns crossing the line, that is, $a_{i+k} - b_i \geq k$. Similarly, by considering a line from b_{i+k} to a_i, we conclude $k + 1$ tracks are necessary unless $b_{k+i} - a_i \geq k$. In the case of $k = 0$, a_i must equal b_i for no horizontal track to be required. Based on this observation, it can be proved that the minimum number of tracks t required by an instance of river routing is the least t such that for all $1 \leq i \leq n - t$,

$$b_{i+t} - a_i \geq t \tag{14.14}$$

and

$$a_{i+t} - b_i \geq t \tag{14.15}$$

To find the minimum such t in linear time, observe that $a_{i+k+1} \geq a_{i+k} + 1$ and $b_{i+k+1} \geq b_{i+k} + 1$. Therefore,

$$\text{if } b_{i+k} - a_i \geq k \quad \text{then} \quad b_{i+k+1} - a_i \geq b_{i+k} + 1 - a_i \geq k + 1 \tag{14.16}$$

and

$$\text{if } a_{i+k} - b_i \geq k \quad \text{then} \quad a_{i+k+1} - b_i \geq a_{i+k} + 1 - b_i \geq k + 1 \tag{14.17}$$

Therefore, we can start with $t = 0$ and search for violated constraints from $i = 1$ to $n - t$; each time a constraint of the form of Eq. (14.14) or (14.15) is violated, we increase t by one and continue the search; t can be no larger than n. Let N denote the set of nets in a river routing channel, $|N| = n$. The following algorithm calculates the minimum number of tracks needed to route this channel.

RIVER ROUTING WIDTH ALGORITHM (INPUT N).

```
1   i ← 1
2   t ← 0
4   while i ≤ |N| − t
5       do if b_{i+t} − a_i ≥ t and a_{i+t} − b_i ≥ t
6           then do i ← i + 1
6           else do t ← t + 1
8   return t
```

The actual routing for a river routing channel can be produced in greedy fashion by routing one net at a time from left to right and routing each net beginning with its left terminal. The route of any net travels vertically whenever it is not blocked by a previously routed net and travels horizontally right until it can travel vertically again or until it reaches the horizontal position of the right terminal of the net. This routing takes worst-case time $O(n^2)$, as the example in Fig. 14.11 illustrates.

14.9 Summary

This chapter has given an overview of the design problems arising from the computer-aided layout of VLSI circuits and some of the algorithmic approaches used. The algorithms presented in this chapter draw on the theoretical foundations discussed in preceding chapters and also on the optimization paradigm presented in Chapter 13. Graph models are predominant and frequently are used to capture constraints. Since many of the problems are NP-complete, heuristics are used. Research continues in this field, both to find better methods of solving these difficult problems—both in efficiency and quality of solution—and to model and solve new layout problems arising from the ever-changing technology of VLSI fabrication and packaging. The reader is referred to the references given in Further Information for broader descriptions of the field and descriptions of current research.

Defining Terms

Channel: A rectangular region for routing wires, with terminals lying on two opposite edges, called the top and bottom. The other two edges contain no terminals, but wires may cross these edges for nets that enter the channel from other channels. The routing area of a layout is decomposed into several channels.

Channel density: Orient a channel so that the top and bottom are horizontal edges. Then the density at any vertical line cutting the channel is the number of nets that have terminals both to the left and right of the vertical line. Nets with a terminal on the vertical line contribute to the density unless all of the terminals of the net lie on the vertical line. The channel density is the maximum density of any vertical cut of the channel.

Channel routing: The problem of determining the routes, i.e., paths and layers, for wires in a routing channel.

Compaction: The process of modifying a given layout to remove extra space between features of the layout.

Floor plan: An approximate layout of a circuit that is made before the layouts of the components composing the circuit have been completely determined.

Floor plan sizing: Given a floor plan and a set of components, each with a shape function, finding an assignment of specific shapes to the components so that the area of the layout is minimized.

Floor planning: Designing a floor plan.

General cell layout: A style of layout in which the components may be of arbitrary height and width and functional complexity.

Global routing: When a layout area is decomposed into channels, global routing is the problem of choosing which channels will be used to make the interconnections for each net.

Graph partitioning: Given a graph with weights on its vertices and costs on its edges, the problem of partitioning the vertices into some given number k of approximately equal-weight subsets such that the cost of the edges that connect vertices in different subsets is minimized.

Interval graph coloring: Given a finite set of n intervals $\{[l_i, r_i], 1 \le i \le n\}$, for integer l_i and r_i, color the intervals with a minimum number of colors so that no two intervals that intersect are the same color. The graph representation is direct: each vertex represents an interval, and there is an edge between two vertices if the corresponding intervals intersect. Then coloring the interval graph corresponds to coloring the intervals.

Jog: In a rectilinear routing model, a vertical segment in a path that is generally running horizontally, or vice versa.

Layer assignment: Given a set of paths in the plane interconnecting terminals, an assignment of a routing layer to each segment of each path so that the resulting wiring layout achieves the interconnections and satisfies the design rules.

Manhattan routing: A popular rectilinear channel routing model in which paths for disjoint nets can cross (a vertical segment crosses a horizontal segment) but cannot contain segments that overlap in the same direction at even a point.

Net: A set of terminals to be connected together.

Rectilinear: With respect to layouts, describes a layout for which there is an underlying pair of orthogonal axes defining horizontal and vertical; the features of the layout, such as the sides of the components and segments of the paths of wires, are horizontal and vertical line segments.

River routing: A single-layer channel routing problem in which each net contains exactly two terminals, one at the top edge of the channel and one at the bottom edge of the channel. The nets have terminals in the same order along the top and bottom: a requirement if the problem is to be routable in one layer.

Shape function: A function that gives the possible dimensions of the layout of a component with a flexible (or not yet completely determined) layout. For a shape function $s: [w_{min}, \infty] \to [h_{min}, \infty]$ with $[w_{min}, \infty]$ and $[h_{min}, \infty]$ subsets of $\Re^+$, $s(w)$ is the minimum height of any rectangle of width w that contains a layout of the component.

Slicing floor plan: A floor plan that can be obtained by the recursive bipartitioning of a rectangular layout area using vertical and horizontal line segments.

Steiner tree: Given a graph $G = (V, E)$ a Steiner tree for a subset of vertices U of V is a subset of edges of G that form a tree and contain among their endpoints all of the vertices of U. The tree may contain vertices other than those in U. For a Euclidean Steiner tree, U is a set of points in the Euclidean plane, and the tree interconnecting U can contain arbitrary points and line segments in the plane.

Switch box: A rectangular routing region containing terminals to be connected on all four sides of the rectangle boundary and for which the entire interior of the rectangle can be used by wires (contains no obstacles).

Terminal: A position within a component where a wire attaches. Usually a terminal is a single point on the boundary of a component, but a terminal can be on the interior of a component and may consist of a set of points, any of which may be used for the connection. A typical set of points is an interval along the component boundary.

Topological sort: Given a directed, acyclic graph, a topological sort of the vertices of the graph is a total ordering of the vertices such that if vertex u comes before vertex v in the ordering, there is no directed path from v to u.

References

Cormen, T. H., Leiserson, C. E., and Rivest, R. L. 1990. *Introduction to Algorithms*. MIT Press, Cambridge, MA.

Dai, W.-M., Sato, M., and Kuh, E. S. 1987. A dynamic and efficient representation of building-block layout, pp. 376–384. In *Proc. 24th ACM/IEEE Design Automation Conf*. IEEE.

Deutsch, D. N. 1976. A dogleg channel router, pp. 425–433.In *Proc. 13th ACM/IEEE Design Automation Conf.* IEEE.

Doenhardt, J. and Lengauer, T. 1987. Algorithmic aspects of one-dimensional layout compaction. In *IEEE Trans. Comput.-Aided Design* CAD-6(5):863–879.

Dolev, D., Karplus, K., Siegel, A., Strong, A., and Ullman, J. D. 1981. Optimal wiring between rectangles, pp. 312–317. In *Proc. 13th Annu. ACM Symp. Theory Comput.* ACM.

Fiduccia, C. M. and Mattheyses, R. M. 1982. A linear-time heuristic for improving network partitions, pp. 175–181. In *Proc. 19th ACM/IEEE Design Automation Conf.* IEEE.

Garey, M. R. and Johnson, D. S. 1979. *Computers and Intractability: A Guide to the Theory of NP-Completeness.* W. H. Freeman, San Francisco, CA.

Greenberg, R. I., Ishii, A. T., and Sangiovanni-Vincentelli, A. L. 1988. MulCh: a multi-layer channel router using one, two and three layer partitions, pp. 88–91. In *IEEE Int. Conf. Comput.-Aided Design,* IEEE.

Greenberg, R. I. and Maley, F. M. 1992. Minimum separation for single-layer channel routing. *Inf. Process. Lett.* 43:201–205.

Hashimoto, A and Stevens, J. 1971. Wire routing by optimizing channel assignment within large apertures, pp. 155–169. In *Proc. 8th IEEE Design Automation Workshop.* IEEE.

Hwang, F. K. 1976. On Steiner minimal trees with rectilinear distance. *SIAM J. Appl. Math.* 30(1):104–114.

Joy, D. and Ciesielski, M. 1992. Layer assignment for printed circuit boards and integrated circuits. *Proc. IEEE* 80(2):311–331.

Kernighan, W. and Lin, S. 1970. An efficient heuristic procedure for partitioning graphs. *Bell Syst. Tech. J.* 49:291–307.

Kuh, E. S. and Ohtsuki, T. 1990. Recent advances in VLSI layout. *Proc. IEEE* 78(2):237–263.

LaPaugh, A. S. 1980. *Algorithms for Integrated Circuit Layout: An Analytic Approach.* Ph.D. Thesis, Department of Electrical Engineering and Computer Science, Massachusetts Institute of Technology, Cambridge.

LaPaugh, A. S. and Pinter, R. Y. 1990. Channel routing for integrated circuits. In *Annual Review of Computer Science,* vol. 4, J. Traub, ed., pp. 307–363. Annual Reviews, Palo Alto, CA.

Lauther, U. 1980. A min-cut based algorithm for general cell assemblies based on a graph representation. *J. Digital Syst.* 4(1):21–34.

Lee, C. Y. 1961. An algorithm for path connection and its applications. *IRE Trans. Electron. Comput.* EC-10(3):346–365.

Leiserson, C. E. and Pinter, R. Y. 1983. Optimal placement for river routing. *SIAM J. Comput.* 12(3):447–462.

Lengauer, T. 1990. *Combinatorial Algorithms for Integrated Circuit Layout.* Wiley, West Sussex, England.

Lengauer, T. and Mehlhorn, K. 1986. VLSI complexity, efficient VLSI algorithms, and the HILL design system. In *Algorithmics for VLSI,* C. Trullemans, ed., pp. 33–89. Academic Press, New York.

Liao, Y. Z. and Wong, C. K. 1983. An algorithm to compact a VLSI symbolic layout with mixed constraints. *IEEE Trans. Comput.-Aided Design* CAD-2(2):62–69.

Maley, F. M. 1987a. An observation concerning constraint-based compaction. *Inf. Process. Lett.* 25(2):119–122.

Maley, F. M. 1987b. *Single-Layer Wire Routing.* Ph.D. Thesis, Department of Electrical Engineering and Computer Science, Massachusetts Institute of Technology, Cambridge.

Mata, J. M. 1984. Solving systems of linear equalities and inequalities efficiently. In *15th Southeastern Conf. Combinatorics, Graph Theory Comput.*

Otten, R. H. J. M. 1983. Efficient floorplan optimization, pp. 499–502. In *Proc. Int. Conf. Comput. Design: VLSI in Comput.* IEEE.

Preas, B. T. and Karger, P. G. 1986. Automatic placement: a review of current techniques, pp. 622–629. In *Proc. 23rd ACM/IEEE Design Automation Conf.* IEEE.

Preas, B. T. and Lorenzetti, M. J., eds. 1988. *Physical Design Automation of VLSI Systems.* Benjamin Cummings, Menlo Park, CA.

Rivest, R. L. and Fiduccia, C. M. 1982. A "greedy" channel router, pp. 418–424. In *Proc. 19th ACM/IEEE Design Automation Conf.* IEEE.

Sechen, C. 1988. Chip-planning, placement, and global routing of macro/custom cell integrated circuits using simulated annealing, pp. 73–80. In *Proc. 25th ACM/IEEE Design Automation Conf.* IEEE.

Sherwani, N. 1995. *Algorithms for VLSI Physical Design Automation*, 2nd ed. Kluwer Academic, Norwell, MA.

Stockmeyer, L. 1983. Optimal orientations of cells in slicing floorplan designs. *Inf. Control* 57:91–101.

Szymanski, T. G. 1985. Dogleg channel routing is NP-complete. In *IEEE Trans. Comput.-Aided Design*. CAD-4(1):31–41.

Tompa, M. 1981. An optimal solution to a wire-routing problem. *J. Comput. Syst. Sci.* 23(2):127–150.

Wolf, W. H. 1994. *Modern VLSI Design: A Systems Approach*. Prentice–Hall, Englewood Cliffs, NJ.

Wolf, W. H. and Dunlop, A. E. 1988. Symbolic layout and compaction, pp. 211–281. In *Physical Design Automation of VLSI Systems*. B. T. Preas and M. J. Lorenzetti, eds. Benjamin Cummings, Menlo Park, CA.

Further Information

This chapter has given several examples of the successful application of the theory of combinatorial algorithms to problems in VLSI layout. It is by no means a survey of all of the important problems and algorithms in the area. Several text books have been written on algorithms for VLSI layout, such as Preas and Lorenzetti [1988], Lengauer [1990], and Sherwani [1995], and the reader is referred to these for more complete coverage of the area. Other review articles of interest are Preas and Karger [1986] on placement, Kuh and Ohtsuki [1990] on layout, LaPaugh and Pinter [1990] on channel routing, and Joy and Ciesielski [1992] on layer assignment. There are many conferences and workshops on computer-aided design that contain papers presenting algorithms for layout. *The IEEE/ACM Design Automation Conference* and the *IEEE International Conference on Computer-Aided Design* are the richest sources of algorithms among these conferences. The premier journal on the topic is *IEEE Transactions on Computer-Aided Design of Integrated Circuits and Systems*. Other journals include *Integration, the VLSI Journal* published by North-Holland Publishing, and *IEEE Transaction on VLSI Systems*. The ACM has just begun a new journal *Transactions on Design Automation of Electronic Systems* that has the potential to become another good source of papers on layout algorithms. Layout algorithms also appear in journals focusing on general algorithm development such as *SIAM Journal on Computing*, published by the Society of Industrial and Applied Mathematics, *Journal of Algorithms*, published by Academic Press, and *Algorithmica*, published by Springer–Verlag.

Architecture

Keith Barker, Section Adviser
The University of Connecticut

COMPUTER ARCHITECTURE is the design of efficient and effective computer hardware at all levels, from the most fundamental concerns of logic and circuit design to the broadest concerns of RISC, parallel, and high-performance computing. The chapters in this section cover classical architectural models, optimization, control, arithmetic, input/output, memory, addressing, and various architectures for high-performance computing.

15

Digital Logic

James Feldman
Northeastern University

15.1 Introduction

This chapter explores combinational and sequential Boolean logic and some of the rules and tools that are used to develop efficient, fast switching circuits. Some of the most common devices used in computers and general logic circuits are developed in depth. Sections 15.2 through 15.4 introduce the fundamental concepts of logic circuits and in particular the rules and theorems upon which *combinational logic* (logic with no internal memory) is based. Section 15.5 develops some tools for minimizing the number of gates and the time necessary to compute a given combinational-logic result. Section 15.6 introduces the subject of *sequential logic* (logic in which feedback and thus internal memory exist.) Section 15.7 pursues two of the most important elements of sequential logic design: the *data flip-flop* and the *register*. The final section of the chapter examines the span of field-programmable logic arrays which now provide fast, economical solutions to large and diverse logic problems.

15.2 Overview of Logic

Logic has been a favorite academic subject, certainly since the Middle Ages and arguably since the days of the greatness of Athens. That use of *logic* connoted the pursuit of orderly methods for defining theorems and proving their consistency with certain accepted propositions. In the middle of the nineteenth century, George Boole put the whole subject on a sound mathematical basis and spread "logic" from the Philosophy Department into Engineering and Mathematics. Specifically, what Boole did was to create an algebra of

0-8493-2909-4/97/$0.00+$.50
© 1997 by CRC Press, Inc.

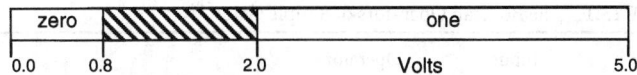

FIGURE 15.1   The states zero and one as defined in TTL logic.

two-valued (*binary*) variables. Initially designed as *true* or *false*, these two values can represent any parameter which has two clearly defined states. Boolean algebras of more than two values have been explored, but the original binary variable of Boole dominates the design of circuitry for reasons that we will explore. This chapter presents some of the rules and methods of binary Boolean algebra and shows how it is used to design digital hardware to meet specific engineering applications.

One of the first things that must strike a reader who sees *true* or *false* proposed as the two identifiable, discrete states is that we live in a world with many half-truths, with hung juries which end somewhere between *guilty* and *not guilty*, and with "not bad" being a response which does not necessarily mean "good." The answer to the question: "Does a two-state variable really describe anything?" is properly: "Yes and no." This apparent conflict between the continuum that appears to represent life and the underlying reality of atomic physics, which is inherently and absolutely discrete, never quite goes away at any level. We use the words "quantum leap" to describe a transition between two states with no apparent state between them. Yet we know that the leaper spends some time between the two states.

A system that is well adapted to digital (discrete) representation is one that spends little time in a state of ambiguity. All digital systems spend some time in indeterminate states. One very common definition of the two states (called **transistor–transistor logic** or TTL) is made for systems operating between 5 V and ground. It is shown in Fig. 15.1. One state, usually called *one*, is defined as any voltage greater than 2 V. The other state, usually called *zero*, is defined as any voltage less than 0.8 V.

The hatched area in the middle is *ambiguous*. When an input signal is between 0.8 and 2.0 V in a TTL digital circuit, you cannot predict the output value. Most of the rest of what you will read in this chapter will assume that input variables may be clearly assigned to the state *one* or the state *zero*. In real designs, there are always moments when the inputs are ambiguous. A good design is one in which the system never makes decisions based on ambiguous data. That requirement limits the speed of response of real systems; they must wait for the ambiguities to settle out.

15.3 Concept and Realization of a Digital Gate

A *gate* is the basic building block of digital circuits. A gate is a circuit with one or more inputs and a single output. From a logical perspective in a binary system, any input or output can take on only the values *one* and *zero*. From an analog perspective (a perspective which will vanish for most of this chapter), the gates make transitions through the ambiguous region with great rapidity and quickly achieve an unambiguous state of *oneness* or *zeroness*.

Since the gates are the fundamental building blocks, a most proper question is: How many different gates do we need? The answer is one. We normally admit three or four to our algebra, but one is enough. If we pick the right gate, we can build all the others. A logical place to start is with the building blocks that we want. Later we can explore the minimal set.

In Boolean algebra, a good place to begin is with three operations: AND, OR, and NOT. These are a sufficient set to build all other logical functions, and they have the added appeal that they do in algebra what they do in plain speech. The usual modern algebraic operators for the three operations are shown in Table 15.1 along with some others we will want to have.

Since NAND and NOR are the electronic building blocks, AND and OR are really the negations of NAND and NOR, respectively. XOR is a synonym for *exclusive OR*. With two inputs, *A* and *B*, XOR is TRUE if only one of *A* and *B* is TRUE. XNOR is the negation of XOR. With two inputs, *A* and *B*, XNOR is TRUE if both *A* and *B* are 1 or if both *A* and *B* are 0.

TABLE 15.1   The Boolean Operators of Simple Gates

Operation	Input Variables	Operator Symbol	High If[a]
NOT	A	A'	$A = 0$
AND	$A, B, \ldots$	$A \bullet B \bullet \cdots$	All of the set $[A, B, \ldots]$ are 1.
OR	$A, B, \ldots$	$A + B + \cdots$	Any of the set $[A, B, \ldots]$ are 1.
NAND	$A, B, \ldots$	$(A \bullet B \bullet \cdots)'$	Any of the set $[A, B, \ldots]$ are 0.
NOR	$A, B, \ldots$	$(A + B + \cdots)'$	All of the set $[A, B, \ldots]$ are 0.
XOR	$A, B, \ldots$	$A \oplus B \oplus \cdots$	The set $[A, B, \ldots]$ contains an odd number of 1's.
XNOR	$A, B, \ldots$	$A \otimes B \otimes \cdots$	The set $[A, B, \ldots]$ contains an even number of 1's.

[a]That is, output is 1 if ....

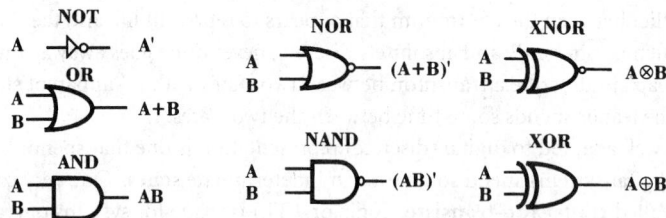

FIGURE 15.2   The commonly used graphical symbols for the seven gates defined in Table 15.1.

There are also widely used graphical symbols for these same operations. These are presented in Fig. 15.2. The symbol for NOT includes both a buffer (the triangle) and the actual inversion operation (the open circle). Often, the inversion operation alone is used, as seen in the outputs of NAND, NOR, and XNOR. The AND operation is often implied, just as multiplication is, by adjacency. Thus, $A \cdot B \Rightarrow AB$.

To illustrate the use of these symbols and operators and to see how well these definitions fit common speech, Fig. 15.3 shows two constructs made from the gates of Fig. 15.2. These two examples show how to build the expression $AB + CD$ and how to construct XOR from the gates in column 1 of Fig. 15.2.

The first construct of Fig. 15.3 would fit the logic of the sentence: "I will be content if my federal and state taxes are lowered (A and B, respectively), or if the money that I send is spent on reasonable things and spent effectively (C and D, respectively)." You would certainly expect the speaker to be content if either pair is TRUE and most definitely content if both are TRUE. The output on the right side of the construct is TRUE if either or both of the inputs to the OR is TRUE. The outputs of the AND gates are TRUE when both of their inputs are TRUE. In other words, both state and federal taxes must be reduced to make the top AND's output TRUE.

The right construct in Fig. 15.3 gives an example of how one can build one of the elemental gates from several of the others. The three gates that are used in this construct are sufficient to build all the others; that

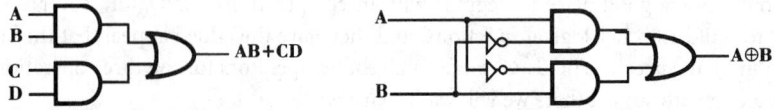

FIGURE 15.3   Two constructs built from the gates in column 1 of Fig. 15.2. The first is a common construct in which if either of two paired propositions is TRUE, the output is TRUE. The second is XOR constructed from the more primitive gates, AND, OR, and NOT.

is, they form a complete and primitive set. But as we shall show, a single gate—either NAND or NOR—is also complete and, in fact, more primitive.

For the moment, let us consider only the relationship of this construct to common speech. The sentence: "With the time remaining, we should eat dinner or go to a movie." The implication is that one cannot do both. The circuit on the right of Fig. 15.3 would indicate an acceptable decision (TRUE if acceptable) if either movie or dinner were selected (*asserted* or made TRUE) but an unacceptable decision if both or neither were asserted.

This may seem like much effort to assert the obvious, but what makes logic gates so very useful is their speed and remarkably low cost. On-chip logic gates today can respond in less than a nanosecond and can cost less than 0.01 cent each. Furthermore, a rather sophisticated decision-making apparatus can be designed by combining many simple-minded binary decisions. The fact that it takes many gates to build a useful apparatus leads us back directly to one of the reasons why binary logic is so popular.

Reasons for Binary Logic

A modern microcomputer chip contains considerably more than one million gates. If all of those gates were generating lots of heat at all times, the chip would melt. Keeping them cool is one of the most critical issues in computer design. Good thermal designs were significant parts of the success of Cray, IBM, and DEC. One of the principal reasons for embracing a binary logic world is that it can be made to be much more energy-conserving.

If you have only ON and OFF, you can have an ideal switch which has either a current and no voltage or a voltage and no current. In this way, you control the energy sent to the load without dissipating any power except during the brief transition between ON and OFF. If you have more than two states in this simple system, it means that the circuit must have a state with some voltage and some current. That state must be dissipative.

Gates are classified as **saturated logic** or **active logic** depending on whether they control the current continuously or simply switch it on or off. In active logic, the gate has a considerable voltage across it and conducts current in all of its states; in saturated logic, the TRUE–FALSE dichotomy has the gate striving to be perfectly connected to the power bus when the output voltage is high and perfectly connected to the ground bus when the voltage is low. These are zero-dissipation ideals that are not achieved in real gates, but the closer one gets to the ideal, the better the gate. When you start with more than one million gates per chip, small reductions in dissipation make the difference between usable and unusable chips.

Saturated logic is *saturated* because it is driven hard enough to ensure that it is in a minimum-dissipation state. Because it takes some effort to bring such logic out of saturation, it is a little slower than active logic. Active logic, on the other hand, is always dissipative. It is very fast, but it is always getting hot. Although it has often been the choice for the most active circuits in the fastest computers, active logic has never been a major player, and it owns a diminishing role in today's designs. Binary gates include many families of saturated logic and at least one family of active logic. (Chapter 3 in [Wakerly 1994] gives more details.)

When one goes to multivalued logic, inevitably one ends up with active logic. The disincentives for using active logic are just too strong to make up for the somewhat higher density of multivalued logic. Multivalued logic is there, and there are IEEE committees devoted to it, but it has not happened in any commercial sense. That being the case, the chapter focuses on today's dominant family of binary, saturated logic: CMOS.

CMOS Switching Model for NOT, NAND, and NOR

The metal–oxide–semiconductor (MOS) transistor is the oldest transistor in concept and still the best in one particular aspect: its control electrode—also called a *gate* but in a different meaning of that word—is a purely capacitive load. Holding it at constant voltage takes no energy whatsoever. These MOS transistors,

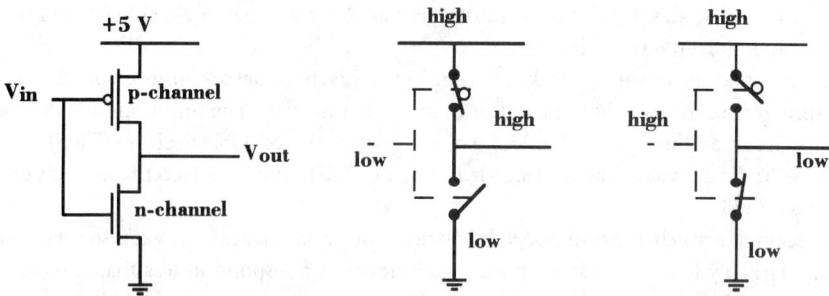

FIGURE 15.4 A CMOS inverter shown as a pair of transistors with voltage and ground and also as pairs of switches with logic levels. The open circle indicates logical negation (NOT).

like most transistors, come in two types. One turns on with a positive voltage; the other turns off with a positive voltage. This pairing allows one to build *complementary* gates, which have the property that they dissipate no energy except when switching. With what we have said about the number of gates and the criticality of energy dissipation, zero dissipation in the static state is enormously compelling. It is small wonder that the complementary metal–oxide–semiconductor gate (CMOS) gate dominates current digital technology. Consider how we can construct a set of primitive gates in this family.

The basic element is a pair of switches in series, the NOT gate. This basic building block is shown in Fig. 15.4.

The switching operation is shown in the two drawings to the right. If the input is low, the upper switch is closed and the lower one is open—complementary operation. This connects the output to the high side. Apart from voltage drops across the switch itself, the output voltage becomes the voltage of the high bus. If the input now goes high, both switches flip and the output is connected, through the resistance of the switch, to the ground bus. High–in, low–out, and vice versa. We have an inverter. Only while the switches are switching is there significant current flowing from one bus to the other. Furthermore, if the loads are other CMOS switches, only while the gates are charging is any current flowing from bus to load. Thus, in the static state, these devices dissipate almost no power at all.

The transistor operation is understandable if you know how a *p–n* junction diode works. Figure 15.5 presents a simple but essentially correct description of this fundamental element of digital electronics.

Once one has the CMOS switch concept, it is easy to show how to build NAND and NOR gates with multiple inputs.

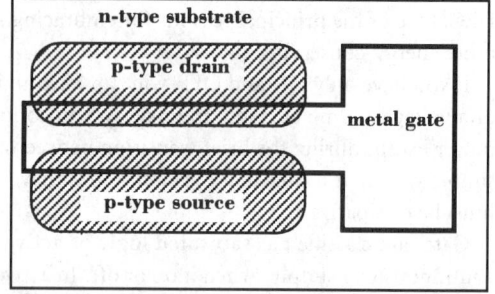

FIGURE 15.5 Top view of a *p*-channel planar MOS transistor. The two *p*-type regions are totally surrounded by *n*-type material forming *p–n* junctions. The gate floats on top of a very thin layer of quartz (the native silicon oxide). Connections are made to substrate, source, drain, and gate. The junctions to the substrate are back-biased to block current flow. To do that, the bias of the substrate is positive. If the gate is negatively biased with respect to the substrate, it will drive electrons away from the surface and attract holes. (In crystals, the composite behavior of the valence electrons can appear as "positively charged electrons.") With sufficient bias, the density of holes exceeds the density of electrons and the *n*-type material underneath the gate turns into *p*-type material. The thin layer of material transformed to *p*-type forms a channel between the two *p*-type regions. With enough gate bias, the conductivity between source and drain can be quite substantial. Negative bias (with respect to the substrate) turns this device on; more positive bias turns it off. Reverse all of the *p* and *n* regions, and the same statement works with the opposite bias. When both types of devices are formed on the same semiconductor chip, the resulting pair is said to form a complementary switch pair; hence CMOS.

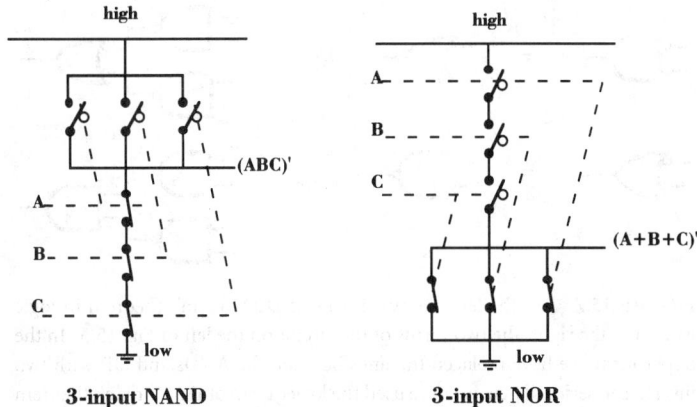

FIGURE 15.6 Three pairs of CMOS switches arranged on the left to execute the three-input NAND function and on the right the three-input NOR. The switches are shown with all the inputs high, putting the output in the low state.

Multiple Inputs and Our Basic Primitives

To prove that we can build everything out of NAND gates, all that must be done is to show how to build AND, OR, and NOT. That will be done shortly. First, let us look at the switching structure of a 3-input NAND and 3-input NOR, just to show how multiple-input gates are created.

The basic NOT gate of Fig. 15.4 is our paradigm; if the lower switch is closed, the upper one is open, and vice versa. To go from NOT to an N-input NAND, make the single lower switch in the NOT a series of N switches, so only one of these need be open to open the circuit. Then change the upper complementary switch in the NOT into N parallel switches. With these, only one switch need be closed to connect the circuit. Such an arrangement with $N = 3$ is shown on the left in Fig. 15.6. On the left, if any input is low, the output is high. On the right is the construction for NOR. All three inputs must be low to drive the output high.

An interesting question at this point is: How many inputs could such a circuit support? The answer is called the *fan-in* of the circuit. The fan-in depends mostly on the resistance of each switch in the series string. That series of switches must be able to *sink* a certain amount of current to ground and still hold the output voltage at 0.8 V or less over the entire temperature range specified for the particular class of gate. While in most cases, 6 or 7 inputs would be considered a reasonable limit, the standard TTL family includes a 13-input NAND (LS133). The analogous question at the output is: How many gates can this one gate drive? This is the *fan-out* of the gate. It too is determined by the requirement that it sink a certain amount of current through the series string. This minimum sink current represents a central design parameter, although it is not absolutely limiting. To a reasonable degree, the designer can make the switches wider to carry more current at a given voltage.

Doing It All with NAND

First, let us have a reason for doing our design with NAND. After all, AND, OR, and NOT are much more natural to normal thinking. The reason is simple. To build an AND or an OR, you take a NAND or NOR and add an inverter. The NAND and NOR are simpler. A design with a million gates might need something close to one million extra NOTs to accomplish what one can do quite directly without them using NAND. The more primitive nature of NAND and NOR comes about because transistor switches are inherently inverting. Thus, a single-stage gate will be NAND or NOR. AND and OR require an extra stage. Extra stages mean longer propagation delays and more heat. Eliminating delay and unnecessary heat are two of the most important objectives of logic design.

Now consider the declaration: "Fred and Jack will come over this afternoon." This is quite equivalent to saying in modern kid-speak: "Fred will stay away or Jack will stay away, NOT". This strange construct in

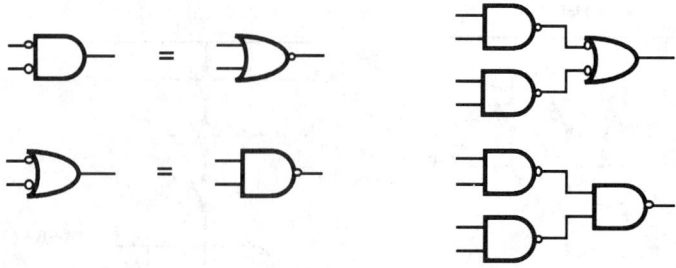

FIGURE 15.7 On the left, the two forms of De Morgan's theorem in logic gates. On the right, the two forms of the circuit on the left of Fig. 15.3. In the upper form, we have replaced the lines between the ANDs and OR with two inverters in series. Then, we have used the lower form of De Morgan's theorem to replace the OR and its two inverters with a NAND. The resulting circuit is all-NAND and both faster and cooler than the construction from AND and OR in Fig. 15.3.

English is an exact formulation of the second of two relationships known in logic as De Morgan's theorems. More formally:

$$(AB)' = A' + B'$$

$$(A + B)' = A'B'$$

These two statements can be represented at the gate level by Fig. 15.7.

Since De Morgan's theorems show that a NAND can be converted into a NOR if we have inverters, to show that NAND could "do it all," it is necessary only to show that we can construct a NOT gate from a NAND. Tying the inputs of the NAND together produces the logic function $(AA)'$, but since A AND A is TRUE only if A is TRUE ($AA = A$), we have our inverter. If we actually wanted an inverter, we would not use a two-input gate where a one-input gate would do. But we could. That was all that was required.

15.4 Rules and Objectives in Combinational Design

Once the concept of a logic gate is established, the next natural question is: What useful devices can you build with them? We will look at a few of the most common of these useful devices. Those that we will examine are basic building blocks rather than finished systems. On the other hand, it will not be too difficult to extrapolate conceptually from the devices we do examine to more complex structures.

The components of digital circuits can be divided into two classes. The first class of circuits has outputs that are simply some logical combination of their inputs. Such circuits are called **combinational**. Examples include the gates we have just looked at and those that we will examine in this section and in section 15.5. The other class of circuits, constructed from combinational gates but with the addition of feedback from output to input, have the property of *memory*. Thus, their output is a function not only of their inputs but also of their previous state(s). Since such circuits go through a sequence of states, they are called **sequential**. These will be discussed in section 15.6.

The two principal objectives in digital design are functionality and minimum cost. Functionality requires not only that the circuit generate the correct outputs for any possible inputs but also that those outputs be available quickly enough to serve the application. Minimum cost must include both the design effort and the cost of production and operation. For very small production runs (<10, 000), one wants to "program" off-the-shelf devices. For very large runs, costs focus mostly on manufacture and operation. The *operation* costs are dominated by cooling or battery drain, where these necessary peripherals add weight and complexity to the finished product. To fit in off-the-shelf devices, to reduce delays between input and output, and to reduce the gate count and thus the dissipation for a given functionality, designs must be realized with a reasonably minimum number of gates. Many design tools have been developed for achieving designs with minimum gate count. In this section and the next, we will develop the basis for such minimization in a way which assures that the design achieves logical functionality. We begin, however, with several simple devices which will provide both some flavor and some functionalities which are basic and widely used.

Elementary Digital Devices That We need: MUX, DEC, ENC, ADD/SUB

MUX

Many systems have multiple inputs which are handled one at a time. *Call waiting* is an example. You are talking on one connection when a clicking noise signals that someone else is calling. You switch to the other, talk briefly, and then switch back. You can toggle between calls as often as you like. Since you are using one phone to talk on two different circuits, you need a **multiplexer** or MUX. There is also an inverse MUX gate called either a DEMUX or a DECODER. Again, a telephone example is the selection of an available line among, say, eight lines between two exchanges. That is, you have one line in and eight possible out, but only one output line is connected at any time. An algorithm based on which lines are currently free determines the choice. Let us design these two devices, beginning with a two-to-one MUX.

What we want in a two-input MUX is a circuit with one output. The value of that output should be the same as the input that we select—if we select an input. That last condition is normally stated as having an *enable* control line. Note that we now have a new dichotomy of inputs. We will call some of them *inputs* and the others *controls*. They are not inherently different, but from the human perspective, we would like to separate them. In logic circuit drawings, *inputs* come in from the left and outputs go out to the right. *Controls* are set off by being brought in from top or bottom.

Our two-input MUX will have, as the name suggests, two inputs. It also will have two controls, one to select which input to transfer to the output and one to indicate whether the transfer is to take place (*enable*). A simple realization of this circuit is shown in Fig. 15.8.

Another standard rule of design is that any input or control should represent no more than one *standard load* (one input to NAND or NOR). Although this rule can be set aside in some circumstances, it normally is followed so that the designer has a simple rule for combining many gates without overloading the output stage of any gate. This rule requires that many inputs be *buffered* with an inverter. For example, the enable (/*e*) and select (*S*) inputs each drive two loads. Accordingly, they must be buffered to isolate their input from the internal loads. If the input were asserted high, that would mean that it would be low on the inside—unless it were double buffered. To eliminate the need for extra buffering, it is customary to design the enable so that the control is *asserted low*. This polarity is indicated in the circuit by having a slash "/" in front of the letter symbol. Thus, /*e* means that the signal is asserted low. (In logic circuit diagrams, the other common "asserted low" symbols are an asterisk, as in *e**; an explicit NOT on the chip in the form of an open circle; and a bar over a symbol, as in *ē*. We will always use the leading slash.)

By De Morgan's theorem, the circuit of Fig. 15.8 can (and would) be executed by replacing both the AND and OR with NANDs. This simple circuit is quite useful and is found packaged four to a chip as a small-scale integrated circuit (SSI). The SSI chip has eight inputs (two sets of four inputs), four outputs, two control lines, and a power and a ground line. The reason for four pairs rather than six or eight is that 16-pin chips are a standard and the cost of such circuits is directly related to the number of pins. Such a chip allows the simultaneous selection of one or another group of four lines. If you need to switch between

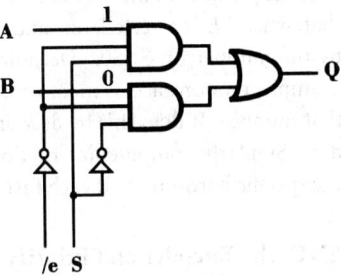

FIGURE 15.8 A two-to-one MUX with enable. If the enable is asserted, this simple but very useful circuit delivers at its output, *Q*, the value of *A* or the value of *B*, depending on the value of *S*. In this sense, the output is "connected" to one of the input lines. If the enable is not asserted, the output *Q* is low.

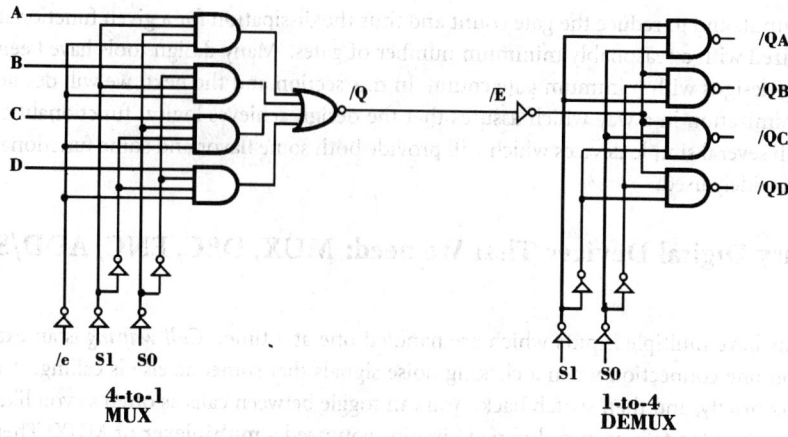

FIGURE 15.9   A four-to-one inverting MUX feeding a one-to-four DEMUX. The value on MUX select lines S1:S0 determines the input connected to $/Q$. E, in turn, is connected to the output of choice by S1:S0 on the DEMUX.

two groups of eight lines, you simply connect two chips. Note that this doubles the load on the control lines. Buffering those two lines keeps that load from escalating too rapidly.

DEMUX/DECODER

If you mentally turn the circuit of Fig. 15.8 around, you get the concept of a DEMUX. That is, a DEMUX has one line in and two lines out. Instead of keeping it quite so small, let us increase our pool of examples by constructing a four-to-one MUX and a one-to-four DEMUX. To choose one of four, we must use two select lines. This leads immediately to the circuit of Fig. 15.9. The most noticeable change here is the structure of the selector. Thinking of the inputs on the select lines as the binary numbers $0, \ldots, 3$, we understand the selection process as enabling the top (A) AND when the input is 00 and then progressively lower ANDs as the numbers become 01, 10, and 11.

The MUX in Fig. 15.9 is *inverting*. That is, if the selected input is high, $/Q$ is asserted low. This fits nicely with the DEMUX, whose buffered input line is asserted low. The overall transfer from MUX input to DEMUX output is inverting. These two devices are available in a variety of logic families, each two to a chip, as the $x352$ and $x139$ (where x is the logic family code, such as 74LS or 74 ALS).

If the $/E$ line on the DEMUX is treated as an *enable*, the DEMUX becomes a DECODER, in the sense that, when $/E$ is asserted, one and only one of the four outputs is asserted, that being the output selected by the number on S1:S0. Decoding is an essential function in many places in computer design. For example, random-access memory (RAM) is fed an address—a number—and must return data based on that number. It does this by decoding the address to assert lines which enable the output of the selected data. Similarly, computer instructions are numbers which must be decoded to assert the lines which enable the specific hardware that each instruction requires.

ENC, the Encoder and Priority Encoder

Just as the DECODER asserts the one line corresponding to the input number, a very useful function is a circuit which puts out the number of the input line that is asserted. Such a circuit would **encode** the asserted line. An encoder which could deal with more than one assertion would be even more useful, but how would we define the output if more than one line were asserted? One simple choice is to have the ENCODER deliver the value of the highest-ranking line that is asserted. In that sense, it is a **priority encoder**. The chips available for this function convert the assertion of eight lines to 3 bits or 10 lines to 4 bits (that is, *binary coded decimal* or BCD). They are both modestly tangled circuits, so to see the principle without the tangle, let us design a four-to-two line priority encoder.

The logic is easy to follow if you look at the output of this circuit as a binary number. If input lines 2 and 1 are both asserted, the output should say 2 **(10)** and not 1 **(01)**. Once you accept that viewpoint, it follows that the low bit must be asserted if the number is odd (1 or 3) and the upper bit must be asserted if the number is 2 or 3. But how can we tell the difference between line 0 being asserted and nothing being asserted? And what about an enable? To answer the first question, we will need an output that says: "Something is asserted." By enabling that output, we have an enable. We will assume that line assertion is low, giving us Fig. 15.10.

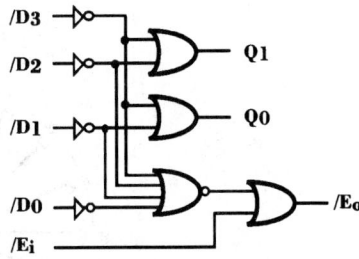

FIGURE 15.10 The four-to-two-line priority encoder.

The line encoding is just what we said—a pair of simple ORs. The *enable out* (E_o) is more interesting. The NOR gate is asserted low if at least one of the $/D$ lines is asserted. $/E_o$ is then asserted if both the NOR and $/E_i$ are low (asserted).

Adder/Subtracter

As our final example of informal design, consider the basic arithmetic building block, the one-bit adder/subtracter. N of these circuits will be connected in a parallel–series arrangement to accomplish an N-bit binary add or subtract. As an introduction to this design, a brief review of complement arithmetic is in order.

The basic presumption in complement arithmetic is that the number of digits in a number is fixed. Most modern computers allow integers of 8, 16, or 32 binary digits. A few of the most powerful modern chips routinely do arithmetic with 64-bit integers. With a fixed register size, a limited set of integers can be represented. For example, in decimal, if you have three digits, there are only 1000 integers. Add 1 to 999, and the result is 000, not 1000. Such an algebra is more efficiently represented in complement form than the more familiar signed-magnitude notation.

The fact that the numbers go from 0 to all 9's and then back to 0 again suggests representing the full set of numbers as a circle. The circle diagram for a four-digit binary system, represented by single hexadecimal digits, is shown in Fig. 15.11. The number systems of particular interest to binary logic are the unsigned and the complement representations described in that figure caption.

The base-complement number system gets its name from the fact that positive–negative pairs sum to 0. Thus, they form a "full complement." Consider ±4 from Fig. 15.11, 0100 and 1100, respectively. Add them as directed displacements on the circle—no carry-out. They sum to 0000. To form the complement of a number, subtract it from 0: that is, perform an oppositely directed rotation with no need for a borrow: $0000 - 1100 = 0100$. Another method to form the complement is to flip all the bits and add 1. Thus, flip 1100 to get 0011 (or 3), and add 1 to get 0100. This method makes it easy to do subtraction by inverting the bits of the subtrahend and adding them and 1 to the minuend. Doing the sum requires an adder. That is our next design objective.

Just as we do addition one digit at a time, the adder circuit handles two input bits, A_j and B_j, plus a carry-in Ci_j. We can arrange as many of these circuits in parallel as we have bits. The jth circuit gets the jth bits of two operands plus the carry-out of the previous stage. It forms both the sum and difference of these three bits and the carry-out borrow of the operation. The arrangement of the circuits is shown in Fig. 15.12.

The adder/subtracter can be broken into three separate problems:

1. choosing B_i or the complement of B_i,
2. forming the sum of the three input bits, $S_j = A_j + B_j + Ci_j$,
3. forming the carry-out of the three bits, $Co_j = f(A_j, B_j, Ci_j)$.

The first task is easy. Feeding the SB signal and B_j into an XOR makes the desired choice. If SB is 0, the output of the XOR is B_j. If SB is 1, the output of the XOR is the complement (also called the *logical*

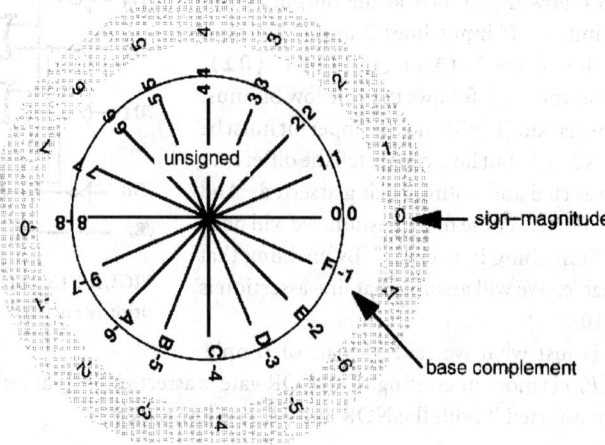

FIGURE 15.11 Three ways to represent numbers in a four-bit system. The inner ring, inside the circle, is the set of *unsigned numbers* 0000 to 1111 (0 to F in hexadecimal or 0 to 15 in decimal). If the leading bit is used as a sign bit, we get the outer ring (*sign–magnitude* number system). That familiar notation is really quite awkward for finite register applications, with −7 being the apparent predecessor to 0 and −0 the successor to 7. Number-system graphs are best when continuous, so the better choices are the inner one (unsigned) and the middle one (the *base complement* number system). In the latter, represented by the numbers just outside the circle, 0 is at the center of the graph, nicely surrounded by ±1. There is only a single discontinuity of the complement number system, just as there is in the unsigned one, but for the complement system the discontinuity is between 0111 (7) and 1000 (−8).

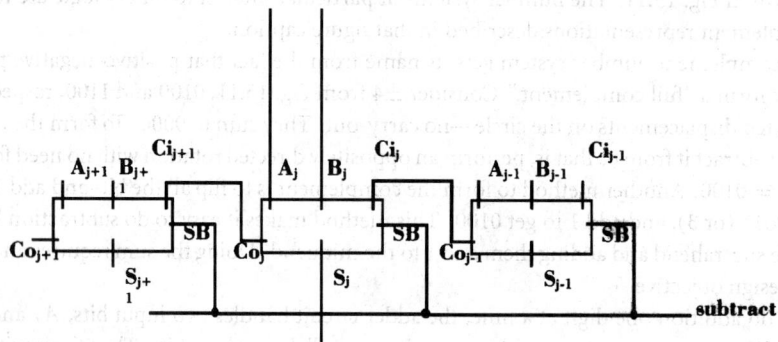

FIGURE 15.12 Connection of adder/subtracters to form an N-bit ripple-carry adder. At the rightmost adder/subtracter, the subtract line is connected to both SB and Ci_0.

complement) of B_j. The second problem is solved by thinking of summing $0, \ldots, 3$ ones. If there are an odd number of ones, the output is 1. Thus

$$S_j = A_j B'_j Ci'_j + A'_j B_j Ci'_j + A'_j B'_j Ci_j + A_j B_j Ci_j$$

The right-hand side has the four combinations that are odd. Finally, for the third problem, there is a carry-out if at least two bits are set. Listing all such cases, we obtain

$$\text{Co}_j = A_j B_j \text{Ci}'_j + A'_j B_j \text{Ci}_j + A_j B'_j \text{Ci}_j + A_j B_j \text{Ci}_j$$

But is that the best we can do? This question has not arisen before except in our brief discussion of objectives. By "best we can do," we mean the fewest gates. Let us consider how we might get a smaller expression for Co_j.

An Example of Algebraic Reduction

George Boole created an algebra based on a set of postulates. Most of the theorems in this *switching* algebra will seem familiar, but a few of them, while logical enough, are unexpected. Let us be formal here and state the postulates:

1. Variables are binary. This means that every variable in the algebra can take on one of two values and these two values are not the same. Usually, we will choose to call the two values 1 and 0, but other binary pairs such as TRUE and FALSE, HIGH and LOW, and ASSERTED and DEASSERTED are widely used and often more descriptive. Two binary operators, AND ($\cdot$) and OR ($+$), and one unary operator, NOT ($'$), can transform variables into other variables. These operators are defined in Table 15.1.
2. Closure: The AND or OR of any two variables is also a binary variable.
3. Commutativity: $A \cdot B = B \cdot A$ and $A + B = B + A$.
4. Associativity: $(A \cdot B) \cdot C = A \cdot (B \cdot C)$ and $(A + B) + C = A + (B + C)$.
5. Identity elements: $A \cdot 1 = 1 \cdot A = A$ and $A + 0 = 0 + A = A$.
6. Distributivity: $A \cdot (B + C) = A \cdot B + A \cdot C$ and $A + (B \cdot C) = (A + B) \cdot (A + C)$. (The usual rules of algebraic hierarchy are used here where $\cdot$ is done before $+$.)
7. Complement pairs: $A \cdot A' = 0$ and $A + A' = 1$.

These are the postulates or axioms of this algebra. As long as they are not inconsistent, we need not prove them. The first two pairs are consistent with what we have defined as desired binary variables. The fact that all the algebraic relationships have *duals* proves quite useful. To get the dual of a theorem (or postulate), one simply interchanges AND and OR as well as 0 and 1.

In general, one may prove a Boolean theorem by exhaustion—that is, by listing all of the possible cases—although more abstract algebraic reasoning may be more efficient. Since our initial objective here is only to show how we may reduce the expression for Co_j in the previous section, we will need only one theorem beyond our set of postulates:

Theorem 15.1 (Idempotency). $A \cdot A = A$ *and* $A + A = A$.

PROOF. The definition of AND in Table 15.1 can be used with exhaustion to complete the proof for the first form. □

$$A \text{ is } 1: \quad 1 \cdot 1 = 1 = A$$
$$A \text{ is } 0: \quad 0 \cdot 0 = 0 = A$$

The second form follows as the dual of the first.

Now let us reconsider the expression for

$$\text{Co}_j = A_j B_j \text{Ci}'_j + A'_j B_j \text{Ci}_j + A_j B'_j \text{Ci}_j + A_j B_j \text{Ci}_j$$

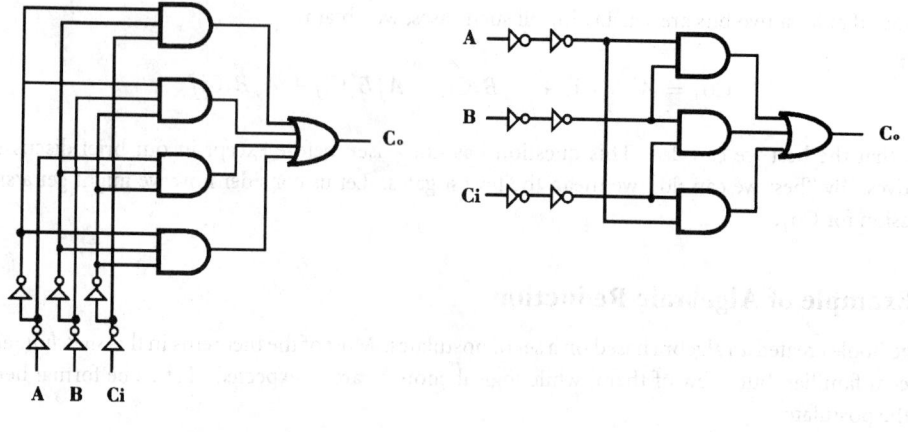

Direct

Reduced

FIGURE 15.13 The direct and reduced circuits for computing the carry-out from the three inputs to the full adder.

Now we apply idempotency twice to the last term on the right and put the duplicates after the first and second terms:

$$Co_j = A_j B_j Ci'_j + A_j B_j Ci_j + A'_j B_j Ci_j + A_j B_j Ci_j + A_j B'_j Ci_j + A_j B_j Ci_j$$

Now we apply postulates 2, 3, and 6 to obtain

$$Co_j = A_j B_j (Ci'_j + Ci_j) + (A'_j + A_j) B_j Ci_j + A_j Ci_j (B'_j + B_j)$$

And finally, we take advantage of postulate 7 to obtain

$$Co_j = A_j B_j + B_j Ci_j + A_j Ci_j$$

Although the reduced equation certainly looks simpler, consider the gate representation of the two equations. This is shown in Fig. 15.13. From four 3-input ANDs to three 2-input ANDs and from a 4-input OR to a 3-input OR is a major saving in a basically simple circuit.

The reduction is clear. The savings in a chip containing more than a million gates should build some enthusiasm for gate simplification. What is probably not so clear is how you could know that the key to all of this saving was knowing to make two extra copies of the fourth term in the direct expression. It turns out that there is a fairly direct way to see what you have to do, one that takes advantage of the eye's remarkable ability to see a pattern. This tool, the **Karnaugh map**, is the next topic.

15.5 Design, Gate-Count Reduction, and SOP/POS Conversions

The *truth table* is one of the earliest tools of logic. This table enumerates the outputs that are wanted for every possible input. The name assumes that you are listing the outputs as TRUE or FALSE. A truth table for the functions $S(A, B, Ci)$ and $Co(A, B, Ci)$ is Table 15.2. All possible combinations of the three input bits appear in the column $ABCi$ with the appropriate output bits for S and Co in their respective columns. Looking at the bits in $ABCi$ as a number gives the input values listed to the left of the table. Although the order seems a little peculiar, if you look at the upper and lower groups of four, note that adjacent values differ by only one bit. There is method in this strange order.

Consider the last two lines in Table 15.2. Both have Co $= 1$. On the input side, the pair is represented as $A_j B_j \text{Ci}'_j + A_j B_j \text{Ci}_j = A_j B_j (\text{Ci}_j + \text{Ci}'_j)$. The algebraic reduction operation shows up as adjacency in the table. In the same way, the 7:5 pair can be reduced. The two are adjacent and both Co outputs are 1. It is less obvious in the truth table, but notice that 3:7 also forms just such a pair. In other words, all of the steps proposed in algebra are "visible" in the truth table. To make adjacency even clearer, we arrange the groups of four one above the other in a table called a Karnaugh map after its inventor, M. Karnaugh [1953]. In this map, each possible input gets its own box. The adjacent boxes all have numerical values exactly one bit different from their neighbors on any side. It is customary to mark the asserted outputs (the 1's) but to leave the unasserted cells blank (for improved readability). The tables for S and Co are shown in Fig. 15.14. The two rows are just the first and second group of four from the truth table with the output values of the appropriate column. First convince yourself that each and every cell differs from any of its neighbors (no diagonals) by precisely one bit. The neighbors of an outside cell include the opposite outside cell. That is, they wrap. Thus, 2 and 0 or 4 and 6 are neighbors. The Karnaugh map (or K-map) simply shows the relationships of the outputs of conjugate pairs.

The item that most people find difficult about K-maps is the meaning and arrangement of the input variables around the map. If you think of these input variables as the bits in a binary number, the arrangement is more logical. The difference between the first four rows of the truth table and the second four is that A is set in the second set and reset in the first. In the map, this is shown by having A indicated as asserted in the second row. In other words, where the input parameter is placed, it is asserted. Where it isn't placed, it is unasserted. Accordingly, the middle two columns are those cells which have Ci asserted. The right two columns have B asserted. Column 3 has both B and Ci asserted.

Let us look at how the CARRY-OUT map implies gate reduction while SUM's K-map shows that no reduction is possible. Since we are looking for conjugate pairs of asserted cells, we simply look for adjacent pairs of 1's. CARRY-OUT has three such pairs; SUM has none. We take pairs, pairs of pairs, or pairs of pairs of pairs—any rectangular grouping of 2^n cells with all 1's. With CARRY-OUT, this gives us the groupings shown in Fig. 15.15.

The three groupings do the three things that we must always achieve:

1. The groups must cover all of the 1's (and none of the 0's).
2. Each group must include at least one cell not included in any other group.
3. Each group must be as large a rectangular box of 2^n cells as can be drawn.

The last rule says that in Fig. 15.15 none of these groups can cover only one cell. Once we fulfill these three rules, we are assured of a minimal set. Although there is no ambiguity in the application of these rules in this example, it is not always so obvious, but for the problems solvable with K-maps (functions of up to six input variables) it is usually reasonably easy to spot correct solutions.

TABLE 15.2 Truth Table for the Two Functions SUM and CARRY-OUT

Input	$AB\text{Ci}$	S	Co
0	000	0	0
1	001	1	0
3	011	0	1
2	010	1	0
4	100	1	0
5	101	0	1
7	111	1	1
6	110	0	1

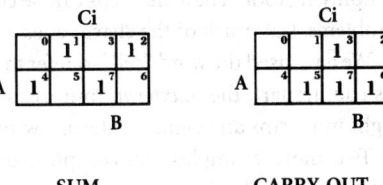

FIGURE 15.14 Karnaugh maps for SUM and CARRY-OUT. The numbers in the cell corners give the bit patterns of $AB\text{Ci}$. The cells whose outputs are 1 are marked; those whose outputs are 0 are left blank.

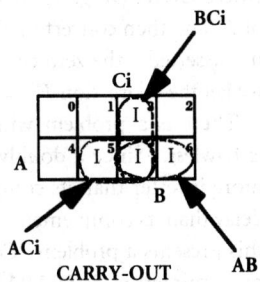

FIGURE 15.15 The groupings of conjugate pairs in CARRY-OUT.

Writing down the solution once you have done the groupings is done by reading the specification of the groups. The vertical pair is BCi. In other words, that pair of cells is uniquely defined as having B and Ci both 1. The other two groups are indicated in Fig. 15.15. The sum of those three (where "+" is OR) is the very function we derived algebraically in the last section.

Notice how you could know to twice replicate cell 7. It occurs in three different groups. It is important to keep in mind that the Karnaugh map simply represents the algebraic steps in a highly visual way. It is not magical or intrinsically different from the algebra. For the limited range of inputs for which such visual mapping is possible (two to six inputs), it is a marvelously simplifying tool. The same steps can be computerized for larger problems, but much of the charm goes out of it.

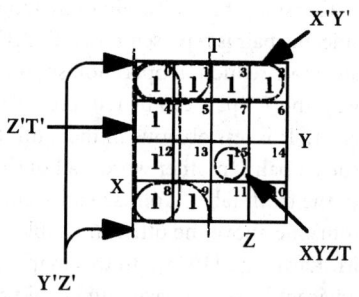

FIGURE 15.16 The K-map for $F(X, Y, Z, T) = \sum(0, 1, 2, 3, 4, 8, 9, 12, 15)$ with the minterm groupings shown.

We have used the word "cell" to refer to a single box in the K-map. The formal name for a cell is *minterm*. Its counterpart, the *maxterm*, comprises all the cells which are not the minterm. In Fig. 15.15, there are eight minterms and eight maxterms, with each maxterm containing seven cells.

Two more examples will complete our coverage of K-maps. One way to specify a function is to list the asserted minterms in the form of a summation. Consider the arbitrary four-input function $F(X, Y, Z, T) = \sum(0, 1, 2, 3, 4, 8, 9, 12, 15)$. With four input variables, there are sixteen possible input states, and every minterm must contact four neighbors. That can be accomplished in a 4×4 array of cells as shown in Fig. 15.16. Convince yourself that each cell is properly adjacent to its neighbors. For example, 11 (1011) is adjacent to 9 (1001), 10 (1010), and 3 (0011) with each neighbor differing by one bit. Now consider the groupings. Minterm 15 has no asserted neighbors. It must stand alone, represented by the AND of all four inputs. The top row and first columns can each be grouped as a pair of pairs. It takes only two variables to specify such a group. For example, the top row includes all terms of the form $00xx$, and the first column includes all the terms of the form $xx00$. This leaves us but one uncovered cell, 9. You might be tempted to group it with its neighbor, 8, but rule 3 demands that we make as large a covering as possible. We can make a group of four by including the neighbors 0 and 1 on top. Had we not done that, the bottom pair would be $XY'Z'$, but by increasing the coverage, we get that down to $Y'Z'$, a 2-input AND versus a 3-input AND. The final expression is

$$F(X, Y, Z, T) = X'Y' + Y'Z' + Z'T' + XYZT$$

Such an expression is a set of "products" (ANDs) which are "summed" (ORed) together. It is called a sum-of-products form or SOP. It is just as easy to generate the function with a *product of sums* (POS) where several OR gates are joined by a single AND. To get to that expression, we find F', the complement of F, and then convert to F using De Morgan's theorem. F' is obtained by grouping the cells where F is not asserted—the zero cells. This is shown in Fig. 15.17, where we get the expression $F' = X'YZ + XZT' + XY'Z + YZ'T$.

There is a problem with this expression, though the logic is flawless. Since a doubly buffered input passes through one more inverter than its complement, it arrives with a little more delay than its complement. The two bars in the map show where this presents a problem. Consider the horizontal one that connects minterms 5 **(0101)** with 7 **(0111)**. In going from 5 to 7, Z changes from 0 to 1. Unfortunately, because of the delay between Z' and Z, Z' goes to 0 before Z goes to 1. Accordingly, the AND asserting 5 ($YZ'T$) goes to 0 earlier than the AND asserting 7 goes to 1. During that brief instant, F' would go to 0. Such

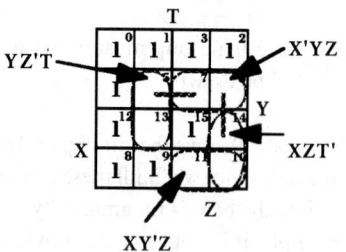

FIGURE 15.17 The K-map for the complement of F from Fig. 15.16.

a brief, false signal is called a **glitch**. In some circuits, you might not care; in others, you definitely would care. Glitches occur wherever two groupings touch but don't overlap. (Again, diagonals do not count.) To avoid a glitch, it is necessary to add a term which is indifferent to the transition being made. These would be the cell pairs indicated by the two bars: $X'YT$ and YZT'. They do not do anything for the static logic, but they prevent glitches.

Let us ignore the glitches in the interest of simplicity and convert from F' to F using De Morgan's theorem to get the POS form:

$$F = (X'YZ + XZT' + XY'Z + YZ'T)'$$
$$= (X + Y' + Z')(X' + Z' + T)(X' + Y + Z')(Y' + Z + T)$$

Why would one want to do this? Economy of gates. Sometimes the SOP form has fewer gates, sometimes the POS form does. In this example, the SOP form is somewhat more economical.

Don't-Cares

There are always a finite number of possible inputs to a digital circuit, but it is often true that not all of the possible states occur. These unused "words" in the circuit's vocabulary are a useful resource in minimizing the gate count. An example might be the classic seven-segment numerical display that is common in watches and digital meters. These are driven by binary-coded-decimal (BCD) numbers, a 4-bit representation

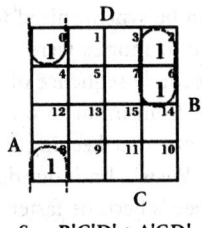

FIGURE 15.18 Segment e of the seven-segment display whose decoder we are going to minimize.

with 16 possible input combinations, but only the 10 numbers 0, ..., 9 ever occur. The states 10, ..., 15 are called **don't-cares**. One can assign them to achieve minimum gate count. Consider the entire number set that one can display using seven line segments. We will consider the one line segment indicated by the arrows in Fig. 15.18. It is generally referred to as "segment e," and it is asserted only for the numbers 0, 2, 6, and 8.

Now we will minimize Se(A, B, C, D) with and without the use of the don't-cares. We put an "x" wherever the don't-cares may lie in the K-map and then treat each one as either 0 or 1 in such a way as to minimize the gate count. This is shown in Fig. 15.19.

We are not doing something intrinsically different on the right and left. On the left, all of the don't-cares are assigned to 0. In other words, if someone enters a 14 into this 0 : 9 decoder, it won't light up segment e. But since this is a don't-care event, we examine

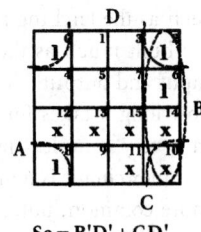

Se = B'C'D' + A'CD' Se = B'D' + CD'

FIGURE 15.19 Minimization of Se without and with deliberate assignment of the don't-cares.

the map to see if letting it light up on 14 will help. The grouping with the choice of don't-care values is decidedly better. We choose to assert e only for *don't-cares* 10 and 14, but those assignments reduce the gates required from two 3-input ANDs to two 2-input ANDs. For this little circuit, that is a substantial reduction.

15.6 Sequential Circuits

Concept of a Sequential Device

One of the oldest and most familiar sequential devices is a clock. In its mechanical implementation, ticks from a mechanical oscillator—pendulum or spring and balance wheel—are tallied in a complex base-12 counter. Typically, the counter recycles every 12 hours. All the states are specifiable, and they form an orderly sequence. Apart from transitions from one state to its successor, the clock is always in a discrete state. To be in a discrete state requires some form of memory. These properties are quite different from the combinational circuits we have considered so far. In those, the inputs uniquely determined the output. History had nothing to do with it.

Memory has many manifestations. One of the earliest forms of computer memory was a tube of mercury with piezoelectric transducers at each end. A set of sound pulses launched from one end traveled to the far end with a delay of the order of a millisecond. As the sequence emerged, it could be restored and fed back into the source end to retain the value indefinitely. The maximum length of a sequence was determined by the length of the tube.

Yet another early form of computer memory is a latching relay. The relay has two positions, one sustained by a spring and the other by a latch. It can be driven from the unlatched to the latched positions by a pulse to the relay coil. It can be unlatched by a pulse to the reset coil. The polarity of the pulse determines which action happens. The state of the relay can be read from contacts which make or break in the two positions.

Finally, one of the most ubiquitous and essential memory elements in our electrically driven world is the toggle switch. It snaps from one position to the other and retains memory of its current position even without power. It is a slow but rather ideal memory element, consuming no power in retaining its memory and dissipating only a tiny fraction of the power it controls.

Like clocks, computers are *finite-state machines*. Although not so dull and regular of state as a clock, all of the states of a computer can be denumerated. Saying this does not in any way restrict what you can compute any more than saying you can completely describe the states of a clock limits the life of the universe. We depend on the absolute predictability of a computer and always look a bit horrified when we find some behavior that we failed to predict.

Accordingly, we can say that memory elements are essential and that by linking such elements together, we can build predictable sequential machines that do very important and interesting tasks. Only the electronic "latch" and some very simple state machines are included in this short chapter, but it is from these very elements that complex machines can be built. Right from the beginning, there were two kinds of sequential circuits, called *clocked* and *asynchronous*. The clocked circuits are built from components such as the latching relay, which can be synchronized to a common clock. The asynchronous circuits are built much as the early acoustic memories were. Their "memory" is the intrinsic delay between input and output. To maintain an orderly sequence of events, they depend on knowing quite precisely how long it takes for a signal to get from input to output. Although that sounds difficult to manage in a very complex device, it turns out that keeping a common clock synchronized over a large and complex circuit is nontrivial as well. We will limit our discussion to clocked sequential circuits. They are more common, but, as computer speeds become faster, the asynchronous approach is receiving greater attention.

15.7 The Data Flip-Flop and the Register

The *SR* Building Block. Set, Reset, Hold, and Muddle

In an electronic circuit, positive feedback can be used to force the circuit into a self-limiting latched state. Since saturated logic goes into such states quite normally, it is a very small step to generate an electronic latching relay from a pair of NAND or NOR gates. The simplest such circuit is shown in Fig. 15.20.

Analyzing Fig. 15.20 requires only walking through the circuit. Recall that parameters specified with a leading "/" are asserted low. Thus, both Q and $/Q$ will be asserted at the same time, but the asserted state of the first will be high and the second low. Start with both $/S$ and $/R$ deasserted and Q asserted. The inputs to B will be high, so $/Q$ will be low. Accordingly, you could toggle $/S$ and no change would take place. Now, with $/S$ high, assert $/R$. First, $/Q$ goes high. This makes both of the inputs to A high, so Q goes low. Now the upper input to B is low, so deasserting $/R$ will have no effect. Thus, asserting $/R$ has reset the latch. At this point, asserting $/S$ and holding it sufficiently long will set the latch.

That phrase "holding it sufficiently long" has an ominous ring. What happens if you do not hold it that long? The latch may—in fact, probably will—break into oscillation as the switching pulse chases its tail through the delays of the two NANDs. The only way to stop the oscillation once started is to assert either $/S$ or $/R$ and hold it long enough to force a single stable state.

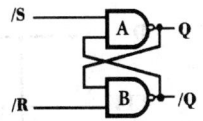

FIGURE 15.20 The basic set/reset (SR) latch. If $/S$ is asserted, Q and $/Q$ are asserted (set). If $/R$ is asserted, Q and $/Q$ are deasserted (reset). If neither $/S$ nor $/R$ is asserted (both high), the latch retains its current state. If both are asserted, the latch goes into a *muddle* state where Q is asserted and $/Q$ is deasserted (both high), but upon simultaneous release of the inputs, the next state is unpredictable.

The *muddle* state is a similar "illogical" problem. If both $/S$ and $/R$ are asserted at the same time—for example, by driving them from a common source—the initial result is to have both Q and $/Q$ go high simultaneously. That is illogical enough, but it gets worse. Now, deassert both inputs simultaneously. What happens? You cannot tell. It may go into either the set or the reset state, or it may oscillate or simply hang there for a remarkably long time in a metastable state.

There is another problem with this circuit. To hold its value, both $/S$ and $/R$ must be deasserted. Glitches and other noise in a circuit might cause the state to flip when it should not. Instead, with a little extra logic, we can improve upon this basic latch to build circuits less likely to oscillate, hang up, or switch inadvertently. These better designs also eliminate the muddle state.

The Transparent *D*-Latch

A simple way to avoid having someone press two buttons at once is to provide them with a toggle switch. You can push it only one way at one time. We can also provide a single line to enable the latch. By custom and frequent application, this enable control is usually called the **clock**. With the addition of two NANDs and two inverters, we can accomplish both purposes, as shown in Fig. 15.21.

Tying the data line, *D*, to the two buffer NANDs assures us that only one of the two NAND outputs can be low at one time. The CLK signal allows us to open the latch (let data through) or latch the data at will. The inverter on the CLK signal is necessary to buffer the clock line. This device is useful enough to be available as a commercial chip with four or eight of them in one package. On those, the clock line is buffered at the chip input and again at each latch, so the latch is transparent on CLK high.

Has this device solved all of the problems we described for the *SR*-latch? Not at all. We have a well-behaved device as long as we don't fool the logic by moving too quickly. Consider what might happen if *D* went from low to high just as the clock went from low to high. For the brief period before the change has propagated through the *D*-inverter, both NANDs see both inputs high. Thus, at least briefly, both $/R$ and $/S$ are asserted. This is the very situation we wanted to avoid. This muddle situation would last only for the propagation time of the inverter, but then the CLK signal arrives and drives both $/S$ and $/R$ high. The latch might oscillate or flip either way. In any case, it will be unpredictable.

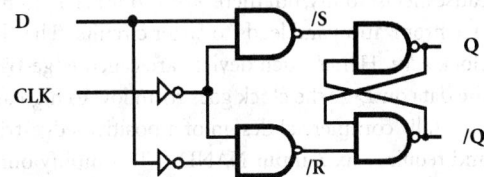

FIGURE 15.21 The transparent D-latch. The circuit is transparent when CLK is low (that is, the current value D appears at Q) and latched when CLK is high (the value of D when the clock went low is held at Q.)

There is another problem with this circuit. It is indeed transparent during the low clock signal. This means that Q will mirror D while CLK is low. If D thrashes around, so will Q. Sometimes you care; sometimes you don't. Sometimes you may want transparency. However, if you do not want transparency, you really want a different circuit. The device you want is a flip-flop (FF).

Master–Slave DFF to Eliminate Transparency

The problem with transparent gates is not a new one. A solution which first appeared in King Solomon's time (ninth century B.C.E.) will work here as well. The Solomonic gate was a series pair of two quite ordinary city gates. They were arranged so that both were never open at the same time. You entered

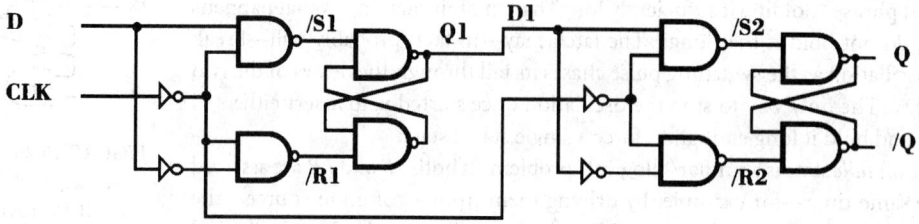

FIGURE 15.22   The master–slave data flip-flop constructed of two *D*-latches in series.

the first and it was shut behind you. While you were stuck between the two gates, a well-armed, suspicious soldier asked your business. Only if you satisfied him was the second gate opened. The solution of putting out-of-phase transparent latches between input and output is certainly one obvious solution to generating a **data flip-flop** (DFF). Such an arrangement of two *D*-latches is shown in Fig. 15.22.

The latch on the left is called the **master**; that on the right is called the *slave*. This master–slave (MS) DFF solves the transparency problem but does nothing to ameliorate the timing problems. While timing problems are not entirely solvable in any FF, accommodating the number of delays in this circuit tends to make the MSFF a slow device and thus a less attractive solution. Why should it be slow? The issue is that to be sure that you do not put either of these devices into a metastable or oscillatory state, you must hold *D* constant for a relatively long setup time and continue it past the clock transition for a sufficient hold time. This accommodation limits the speed with which the whole system can switch.

Can we do better? Yes, not perfect, but better. The device of choice is the edge-triggered FF. This will be the last new device we will consider. Using it, we will build a few machines to do quite useful work.

Edge Triggering to Get Better State Timing

Carried to the limit, the double-gate perspective on what a FF should do becomes a tiny slice in time in which the new datum is read and passed to the output. There still will be a setup time, and violating it may cause metastability, but there is only one brief moment when any critical changes take place. That improves synchronization and leads to faster circuits. The "brief moment" occurs during the clock transition—the clock *edge*. Hence, such devices are called **edge-triggered flip-flops**. The one that we will look at latches the data on *D* as the clock goes from low to high and then moves it to the output.

A fully commercial design of a positive-edge-triggered DFF, such as the 74LS74, includes *set* and *clear* and requires six 3-input NANDs. To simplify our analysis, we will exclude those two peripheral inputs and limit ourselves to two inputs, CLK and *D*. That still leaves us with three latches but only one 3-input NAND. The circuit is shown in Fig. 15.23.

Feedback circuits are difficult to analyze because propagation delays in the several gates mean that the circuit goes through several intermediate states to get from the initial state to the final state. Where multiple loops occur, as in the DFF, it is not always clear which signal "gets there" first. When the different possibilities lead to different final states, you have what is called a *critical race condition*. So how can we analyze a circuit with all these feedback loops?

There are two approaches. The first is to keep to our logic states but impose quite artificial control of the internal flow of signals. In this approach, what we do is make cuts in each feedback path as shown in the figure. This turns the circuit into a simple, feedforward combinational circuit, but at the cuts, we now have three new internal input variables. These are what we propose to control, choosing to hold their old values or connect them to their appropriate internal output. Thus, in Fig. 15.23, there would now be five input variables: D, CLK, and a, b, and c. There are four outputs: $/Q, a^*, b^*$, and c^*. Here Q and c^* are the same, so we have not included Q. We can write the four static equations describing the

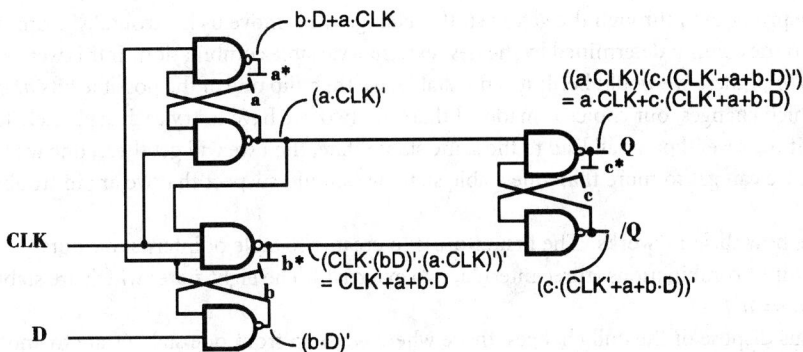

FIGURE 15.23 The essential part of a positive-edge-triggered DFF. The breaks in the feedback loops of each latch are inserted for purposes of open-loop analysis of the circuit. Equations for the outputs of each NAND are shown in the figure. Liberal use of De Morgan's theorem reduces the equations as shown.

outputs:

$$a^* = b \cdot D + a \cdot \text{CLK}$$

$$b^* = \text{CLK}' + a + b \cdot D$$

$$c^* = a \cdot \text{CLK} + c \cdot \text{CLK}' + a \cdot c + b \cdot c \cdot D$$

$$/Q = c' + \text{CLK} \cdot a' \cdot b' + \text{CLK} \cdot a' \cdot D'$$

These are no great joy to behold, but the point is that we can specify the four output variables for all possible variations of the five input variables. Think of yourself as having three switches, each of which passes a "∗" variable such as a^* to the associated internal input variable such as a.

The critical presumption in this approach is to allow only one variable to change at a time. If things really did proceed in this deliberate a fashion, we would then be plotting all of the possible ways to get from one state to another. We have a stable state when closing any of the switches produces no change. This is a fairly easy method and sometimes is quite revealing. The method is particularly well developed in [Breeding 1992]. Unfortunately, though most useful in design, it is only a crude approximation of reality.

Reality requires the flip-flops to pass through active states to get back and forth between the saturated states that we were considering in the previous method. To analyze what is happening in the active state, we must treat these gates for what they are—analog devices. That requires a completely different approach, one that is completely outside this chapter. For interested readers, there exist multiple versions of SPICE, an analog simulation program which handles quite realistic nonlinear-device models and does a very fine if somewhat slow job of analyzing such circuits.

Our first task is to build a table of all possible states. With five binary variables, that is 32 states. This sounds a bit more ominous than it really is. Two of the variables are *external*. We list the four possibilities for them across the top of Table 15.3. The first column lists the internal variables. There are three internal variables, so we have eight possible combinations. We will assume that we are in one stable state when one (and only one) of these external variables changes. Since this change in input

TABLE 15.3 The Values of the Internal Outputs $a^*b^*c^*$ and External Outputs ($Q \cdot /Q$) for the Various Input Combinations of abc and CLK and D[a]

| abc | CLK D | | | |
	00	01	11	10
000	010 (01)	010 (01)	000 (01)	000 (01)
001	011 (10)	011 (10)	000 (01)	000 (01)
011	011 (10)	111 (10)	111 (10)	000 (01)
010	010 (01)	110 (01)	110 (01)	000 (01)
100	010 (01)	010 (01)	111 (10)	111 (10)
101	011 (10)	011 (10)	111 (10)	111 (10)
111	011 (10)	111 (10)	111 (10)	111 (10)
110	010 (01)	110 (01)	111 (10)	111 (10)

[a]The stable states are indicated in bold outlines.

cannot have propagated through the gates yet, this change will move us horizontally from the column we were in to the column determined by the new external-variable combination. If this were a stable state ($abc = a^*b^*c^*$), there we would be. If it is not stable, we then flip one of the possible bits abc. If there is only one which changes, our choice is made. If there are two, we have to try each separately to see where it leads us. If the several paths all lead to the same stable state, then we will get there, one way or another. If, however, we can get to more than one stable state, we should suspect that we are in trouble with this circuit.

Let us see how this one works. The first effort is to create the table of internal outputs for each of the 32 possible input combinations (three internal, two external.) The eight states which are stable are those in which $abc = a^*b^*c^*$.

First, let us dispose of the dull changes, those where we snap from one stable state to another. These lie in the 111 and 000 rows. Don't yet ask how we got there, but if we are already in the $abc = 000$, CLK $D = 11$ state, flipping D down or up makes no difference to the internal state. Similarly, in the 111 row, only CLK $D = 00$ gets us out to unstable territory. That sounds interesting. Let us see what happens next.

First note that to those of us on the outside, the output in all the states in row 111 is the same—10. That is, the flip-flop is set. Now, let D be 0 and the clock go low. That is, we move from the 10 column in row 111 to the 00 column in 111. The only internal output bit that changes in this move is a^*. It goes low. As soon as we propagate that signal (set $a = 0$), we jump to row 011. Not only is that stable, but it has the same output that we already had. This means that we get an *internal* change but no external change. Good. This is a *positive*-edge-triggered device, and we have just analyzed a negative clock transition.

Now we should consider two sorts of transitions: $D \Rightarrow 1$ or CLK $\Rightarrow 1$. In other words, in the first case, we "take our datum back"; in the second, we log it through. Switching D up takes us right one column into $a^*b^*c^* = 111$. That puts us back in the stable state where we began. If the clock goes up now, we simply move one column to the right into another stable state. So far, no output the public can see has changed.

All right, let us hold D low and move out of the stable state 011,00 to the unstable one 011,10. That has $a^*b^*c^* = 000$ and $Q = 0$. We then move to the state 000,01 and find that it is stable. Thus, as predicted, the flip-flop has logged through the datum that was valid as the clock swung up. If you follow through for the transition from reset to set, you will see that it works just fine as well.

Does it all really work this well? As long as you do not change D too close to the time that the clock swings high, it does. If you do make D change just before CLK goes high, SPICE or experiment tells you that the circuit does unpredictable things. For classic small-scale integrated circuits like the 74LS74, D must be stable a full 20 ns before the clock goes high. With more modern, on-chip circuits, the setup time is much less, as befits circuits operating at up to 500 MHz and above. However, the setup time is never 0.

From DFF to a Functional Register and a Few More Applications

The first device that we would like to consider is perhaps the most essential and obvious application of a DFF—a register. This is the device where a processor stores its internal data. It has two actions that it can take on the appropriate clock edge: *read new data* or *read its old data*. In the latter mode, it stores data; in the former, it logs in new data. An n-bit register with *three-state output* is shown in Fig. 15.24. The top two layers are devices we have already discussed. The bottom layer—the three-state output buffer—is really, its name not withstanding, a two-state device. It is simply a buffer with the additional property of being able to be turned off so that it presents an open circuit. To design such a device out of CMOS switches, you simply must have logic which allows the /EN signal to shut off simultaneously both the upper and lower switches in an inverter. This then allows the three-state to pass the inverted input signal or present an open circuit to the outside. The "three-state" name derives from the fact that although any old inverter can provide outputs of 0 or 1, the three-state can deliver 0, 1, or OPEN.

The presence of a three-state output buffer allows one to hook multiple registers to a common bus and then select one and only one of this set of registers to the common bus. Selection of the register is done with a decoder.

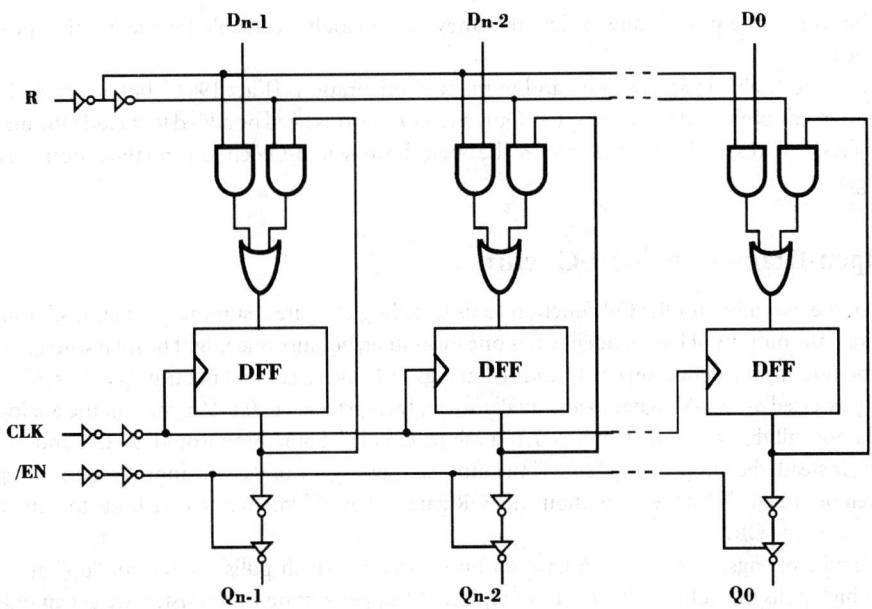

FIGURE 15.24 An *n*-bit register with three-state output. The upper layer is a set of *n* 2-input MUXs. The middle layer is a set of *n* positive-edge-triggered DFFs. The bottom layer comprises an inverter–three-state pair.

To bring in new data, one delivers the data to the *D*-lines and asserts the *R* (read) line. On the next positive edge of the clock, the data are logged into the DFFs, displacing whatever data were previously held. If *R* is not asserted, on each positive edge, the old data are read in.

15.8 ASICs—Faster, Cheaper, More Reliable Logic

No one who has watched the astonishing decline in the cost of digital electronics brought about by *very large-scale integrated circuits* (VLSI) would expect to find engineers generating circuits by hooking up vast arrays of 14-pin DIP packages with four 2-input NANDs per package. In a world where powerful computer chips roll off the line with 10 million transistors all properly connected and functioning at clock speeds in excess of 100 MHz, why would we be manually hooking up hundreds of these small packages with 16 transistors in a chip that takes 15 to 20 ns to get a signal through a single NAND? Today, several pages of random logic circuit diagrams can be implemented in a single chip, programmed on site, and able to perform many layers of logic in the time it takes a signal to propagate through one of the old NAND chips. These devices are called **field-programmable gate arrays** (FPGAs) and are available from several different manufacturers. En route from the simple gates of small-scale integration (SSI) to VLSI and FPGAs, a series of ever more complex and field-programmable logic devices have evolved. Two of these, **programmable array logic** (PAL) and **generic array logic** (GAL), will be described below.

One major advantage of an FPGA is that all of the steps which take the initial design (a symbolic or tabular description of the circuit) to finished chip can be automated. These steps fold into a few computer applications that reduce the design to programming data for the FPGA and then download the data to the chip. For the circuit contained in the FPGA, no circuit boards, no solder, and, best of all, almost no delay from design to working hardware. One can make a change in a complex design and have a working realization of the design in less than an hour. By tightening the design cycle, such rapid prototyping has dramatically reduced the cost of designing and producing complex circuits for specific applications. Such **application-specific integrated circuits** (ASICs) used to be justifiable only for applications with huge production runs. The price is now so low that they are generally justifiable for the limiting production run of one unit.

FPGAs are realizable in several ways and in many configurations [Katz 1994], but all are derivative of principles already presented in this chapter. Only a few extensions need be added to include the underlying circuits of PALs, GALs, and FPGAs. These are the open-drain–wired-OR circuit and the concept of a quartz isolated gate.

The Open-Drain–Wired-OR Circuit

The use of the + symbol for the OR function leads to calling OR gates "summing junctions." You cannot simply hook the outputs of independent gates one to another, because one might be high when the other is low. Remembering the definition of "1" and "0" in Fig. 15.1, such a conflict in outputs represents hooking a strongly asserted 3- to 5-V signal to an equally strongly asserted 0- to 0.8-V signal. In the analog world, we would not call that a "conflict" but rather a "short circuit." Something would get hot and eventually burn out. Instead, the OR properly "sums" the inputs, asserting a 1 or 0 as the inputs require. A question might well be raised: "Can we do without the OR gate?" The affirmative answer leads to a most useful concept, the wired OR.

The circuits of Figs. 15.4 and 15.6 have an upper section which pulls the output "up" and a lower section which pulls the output "down." If we replace the upper section by a resistor, we get an older form of circuit called *resistor–transistor logic* or RTL. In that case, the resistor is the "pullup" and the transistor(s) is the "pulldown." What makes this old and out-of-date circuit of interest to us is its ready application to local reconfigurability. Consider Fig. 15.25, which shows how one could build a 4-input OR from such components. The "out-of-date" issue can be dealt with by replacing R with a transistor, but what is most interesting about this circuit is its collection of fuses.

The gate can be "programmed," one input at a time, by connecting a moderately high voltage to the summing line and asserting the input line of choice for long enough to blow the fuse. The fuse is indicated in the logic diagram to the right in the same figure by the ×. Note, however, that in blowing any or all of the fuses, you simply prevent that set of inputs from pulling down the inverter's input. You do not leave a floating input in the OR gate. Floating inputs give indeterminate outputs.

The Oxide-Isolated Gate

The fuse-blowing programming scheme is good—at least as long as you can assure that the detritus of the blown fuse does not short out other circuit elements (a problem with the earliest versions). It was the

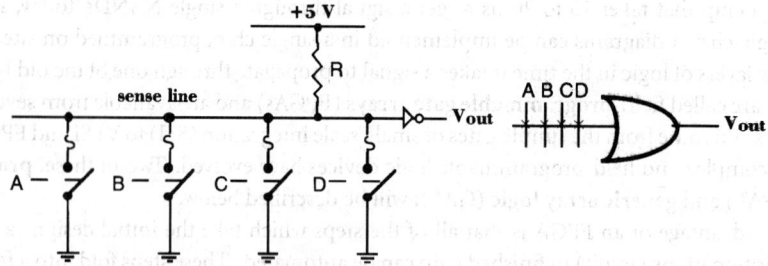

FIGURE 15.25 Four MOS switches arranged as independent pulldowns on the sense line. A high voltage on an input closes the switch. The pullup resistor (or transistor) pulls the sense line up if all the pulldowns are open. If any of $A, \ldots, D$ are asserted, the sense line will be low and V_{out} will be high. The circuit thus forms a 4-input OR. The fuses can be blown to remove any of the inputs, thus creating a 3-, 2-, 1-, or even 0-input OR. The logic circuit on the right is the equivalent.

essential element in a variety of programmable logic elements developed in the late seventies and still in use to this day. The principal problem with such circuits is that what is blown cannot be unblown. They are programmable but not reprogrammable. Reprogrammability can mean not only that you can reuse a part for further prototyping or other use, but also that you might even dynamically reprogram a circuit in use. Some reprogramming schemes require off-line deprogramming. These make the chip reusable but not dynamically reprogrammable. Reprogrammability has been carried so far as to require a chip to download its current configuration upon power-up. The secret behind all forms of reprogrammability is the oxide-isolated gate. The ultimate in reprogrammability is to have the ability to reconfigure the chip on the fly, just as a program in a computer can reprogram itself by executing self-modifying code. Some FPGAs offer this dynamic reconfigurability capability.

All MOS devices, such as that shown in Fig. 15.5, isolate their control gate by a thin oxide layer separating the gate from the channel being controlled. If the gate itself can be charged and then isolated from the outside world, it will retain its state essentially indefinitely. This is done in most cases by using a floating gate of polysilicon totally surrounded on all sides by quartz (SiO_2). The charging is accomplished by creating a very large electric field using an auxiliary electrode. This injects carriers into the quartz, allowing the charging of the floating gate. When the high field is removed, the accumulated charge is trapped on the gate. The gate can be deprogrammed by shining ultraviolet light on the chip. That too makes the quartz weakly conducting, allowing the accumulated charge on the floating gate to leak off. This UV deprogramming is a lengthy affair (of the order of half an hour) and is useful only for reuse.

When rapid reprogrammability is needed, the gates become memory elements just as in static RAM chips. In that case, the circuit loses its program every time it is depowered. This means that you can reprogram it every time you use it; in fact, you must reprogram it every time you use it. To hold a given configuration, the circuit must remain powered, though not much power is required in standby mode.

PALs

The most successful early concept, *Programmable Array Logic* or PAL®, was developed by Monolithic Memories (now part of AMD) in the late 1970s. At the simplest, PALs use the circuit of Fig. 15.25 to provide an array of two-layer AND–OR circuits much like Fig. 15.3. What makes the devices so useful is that they provide a remarkably large number of input and output combinations in a small, programmable, fast package. A typical unit, such as the 16L8 or 16H8 will have as many as 16 inputs with 2 to 8 AND–OR outputs, with each OR summing 7 inputs. Consider Fig. 15.26, which shows schematically the logic of a typical PAL.

Eight 32-input ANDs feed an 8-input OR. The output of that OR feeds a three-state inverter, which is controlled in turn by another 32-input AND. The three-state inverter selects whether the inverted output or an external source connected to that same line is fed back to the 32 input lines. The 32 input lines are 16 inputs and their complements. The PAL allows for 10 up to 16 inputs with 8 down to 2 outputs, respectively.

The output of unwanted ANDs can be set to 0 by leaving any pair of inputs connected (since $AA' = 0$). In practice, you would leave all pairs connected to prevent glitches. Make all of the inputs to the OR 0, and the input of the 3-state inverter is 0. Alternatively, if you want a permanent 1 into the inverter, simply open all the connections to one of the ANDs.

The thought of a 32-input gate is somewhat overwhelming, but the actual circuit which realizes Fig. 15.26 is much closer to Fig. 15.25. If you simply reverse the polarity labels on the vertical lines, you have, in effect, put inverters at the input to an OR. By De Morgan's theorem (Fig. 15.7), that makes it a NAND. Follow that with a 7-input NAND, and you have the hard parts of Fig. 15.26.

PALs have been available for almost 20 years. They are very useful for board-level logic designs, but they are more limited than Fig. 15.26 at first might suggest. As soon as you start packing logic diagrams into one of these chips, you find that your appetite for more gates grows rapidly. Programs exist, particularly ABEL (Advanced Boolean Equation Language from Data I/O Corp.), for making optimum use of any particular PAL, but one is often frustrated by annoying limitations such as having to use many more gates

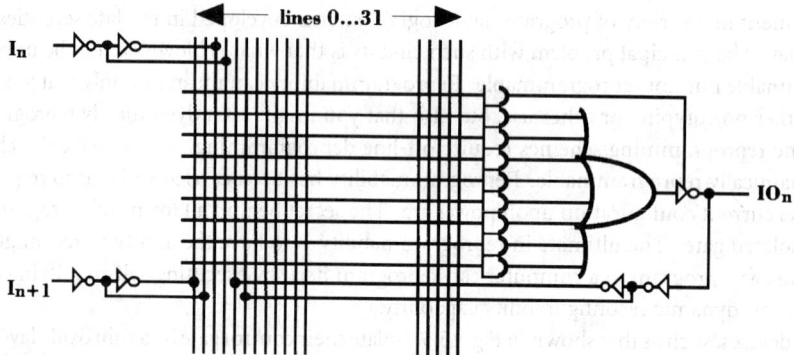

FIGURE 15.26   One of eight AND–OR–INVERT units in the 16L8 PAL. Six of the units are identical to this one. Two others do not connect back into the vertical lines. A connection can be programmed at any of the points where a vertical line intersects a horizontal line to one of the ANDs. Where all the connections are broken, the line is forever high.

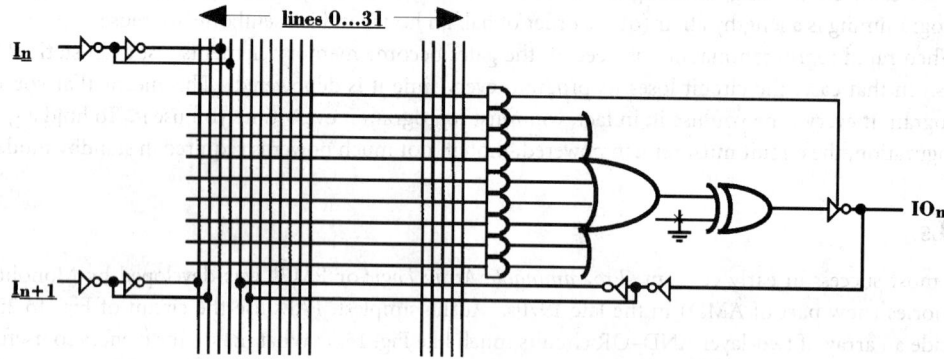

FIGURE 15.27   The basic circuit unit in the GAL. The programmable XOR allows the designer to obtain either the inverse or direct function, programming the other XOR lead to be high or low, respectively.

to achieve a particular function because all the outputs are inverted (or not inverted). Although both output polarities are available, only one polarity is available on a given chip. Other frustrations include not having edge-triggered flip-flops or other basic building blocks.

Control of I/O polarity was an obvious shortcoming of the PAL. Lattice Semiconductor introduced the GAL (Generic Array Logic), which solved that and several other problems for PAL designers. The inverter problem was solved by inserting a programmable XOR into the output circuit as shown in Fig. 15.27.

ABEL can be applied to GALs too, so the same software package can be used to program either device. Because the inverse and direct logic expressions will generally need different numbers of AND gates, being able to choose the output polarity means that GALs will yield greater circuit density (fewer chips). AMD responded to the GAL circuits, so PAL equivalents are available.

GALs and PALs are also available with D-flops on chip. With this addition, quite useful sequential circuits can be programmed into these standard and very inexpensive chips. For greater versatility, the more recent GALs and PALs include logic to configure the outputs to have either D-flop outputs or direct access to the programmable inverters. These modern versions can be configured to be anything you can put down on a logic diagram—as long as the diagram is not too big. But what about those "pages of logic diagrams" that were mentioned at the beginning of this section? How does VLSI, which gives us gigabit memory chips and 64-bit processors on a single chip, influence ASIC realization?

Higher Levels of Complexity

Three "truths" in digital circuit design suggest that the more logic that you can pack into a single chip, the better that product will be for most customers. These truths comprise:

1. The cost of a product rises quickly with numbers of chips.
2. The reliability of a product declines with increasing chip count.
3. The speed of a logic circuit is strongly dependent on how many times a signal must be propagated from one chip to another.

Essentially, if a PAL or GAL is limited and we want to do larger designs while keeping our chip count low, why not put more than one GAL or PAL on a single chip? The objective is to move in one step with no human intervention from a design that is tested in emulation to a design that is realized in a single chip. The problems that this raises include interconnectability (essentially arranging for and programming the interconnection of the many logic blocks), vastly increased complexity of programming, testability of the finished design, and all the implications of having a much more expensive single chip. The last item gives high value to the requirement that such devices, when used for prototyping, be reprogrammable. Most but not all FPGA families are reprogrammable, and one even requires that the chip be programmed every time it is turned on.

The complexity issue is sufficiently imposing that the solutions have not yet converged to families of basically similar devices (as PALs and GALs have). Instead, the FPGAs from each manufacturer tend to be unique products. Even the software to convert designs from diagram, truth description, or algebraic expression to the efficient programming of the chip itself is unique to each manufacturer and an essential tool for each user of these products. Some of the principal products that are available today include the Actel programmable gate arrays, which provide particularly versatile solutions to the interconnection problem but are not reprogrammable, and the Xilinx FPGAs, which store their configuration (program) in separate, standard ROMs and download the configuration on startup. These chips have lots of D-flops, many independent combinational logic blocks, and great versatility in interconnection. For example, in the Xilinx 3000 family, you can have up to 320 independent logic blocks and 144 versatile I/O blocks. The equivalently loaded circuit board would be immense and vastly more expensive and less reliable.

To complete this chapter, let us look briefly at the Xilinx FPGAs. Our objective is not to learn how to program them—a task normally accomplished by software—but rather to show the relationship of these sophisticated chips to the logic we have already developed.

The Xilinx chip is organized with its configurable logic blocks (CLBs) in the middle and its I/O blocks (IOBs) on the periphery and lots of distributed configuration memory to control not only the CLBs and IOBs but also the interconnections to the wiring channels, which determine the versatility of the chips to adapt to any design and to utilize a substantial fraction of their component count. The overview is presented in Fig. 15.28. Its resemblance to a printed circuit board with chips installed is not accidental. Each CLB is statically programmed and does one or two logic tasks. Data enter or exit the chip through the moderately sophisticated IOBs. The IOBs are dynamically reconfigurable for data direction and data hold, with three-state outputs and latchable input and output data.

The circuit internals are presented in Figs. 15.29 and 15.30. In each of these two figures, small, unconnected inputs are indicators of statically programmed control inputs. The IOB layout of Fig. 15.29 is simple but complete. It allows for a latched or direct datum in both directions. On output, not only is the datum three-stated and invertible, but one of the output buffers has controllable slew rate to limit ringing and glitches in the external circuits. (*Slew rate* is the speed with which the circuit traverses the active region between saturated states. *Ringing* is oscillation around the stable state.) The timing of data latching in either direction is dynamically programmable. The input D-flop can be statically configured to be either a transparent latch or an edge-triggered D-flop.

The CLB in Fig. 15.30 does not look much more complicated than the IOB, but it is considerably more programmable. The most reconfigurable element in the CLB is the *combinational function generator* (CFG) with its five external and two internal data inputs. It can be operated in one of three modes:

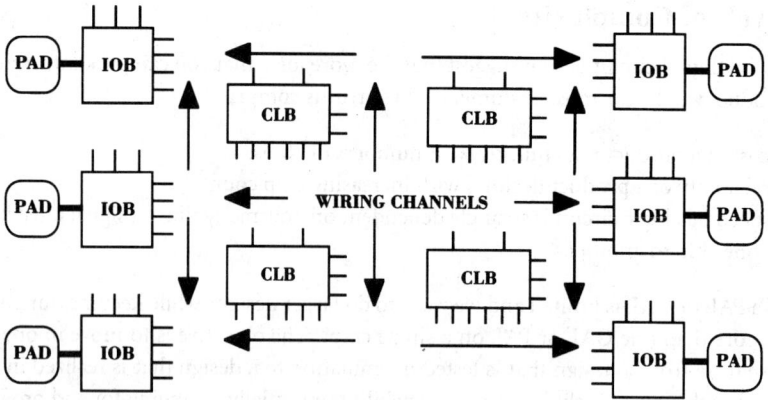

FIGURE 15.28 Overview of the Xilinx FPGA. Pads are connected to the chip-carrier pins. The wiring inside between IOBs and CLBs is determined entirely by programming data which are read in on bootup (normally automatically from a ROM) and shifted serially into a remarkably long serial memory.

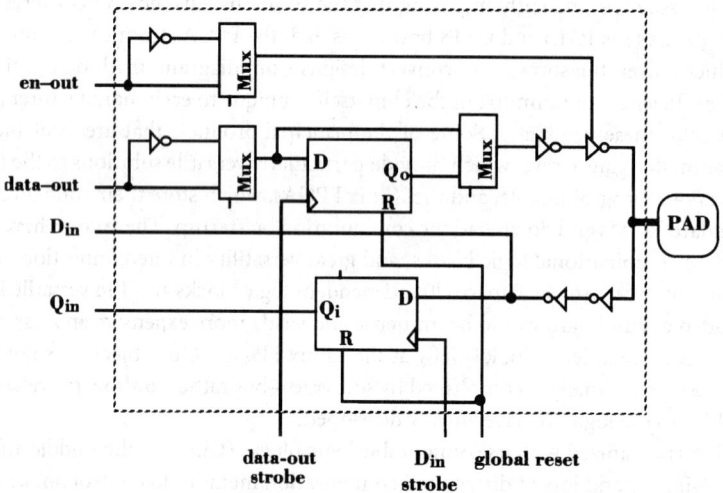

FIGURE 15.29 The internal logic of the IOB. Circuits with small unconnected control leads indicate static configurability determined by the program read in at bootup. The global reset line is fed to all IOBs and resets all their *D*-flops simultaneously.

1. Outputs *F* and *G* are the same and can be any combinational function of five of the seven input variables. The choice of five includes *A*, *D*, *E*, and one each from $B/Q_0/Q_1$ and $C/Q_0/Q_1$.
2. Outputs *F* and *G* are independent, and each can be any combination of four variables. In this case, the choice of inputs is *A* and one each from $B/Q_0/Q_1$, $C/Q_0/Q_1$, and *D/E*.
3. Outputs *F* and *G* are the same, but variable *E* selects between two different functions of four variables from the set *A*, *D*, $B/Q_0/Q_1$, and $C/Q_0/Q_1$. In this mode, the CLB is dynamically reconfigurable, since the data input *E* selects between two functions to be computed.

Output data can be latched or direct. The latched data can come from *F*, *G*, or D_{in}. The *D*-flops are edge-triggered with the triggering edge (rising or falling) being statically programmed. The *en* lead permits dynamic programming to determine whether data in the two *D*-flops get updated on the chosen clock edge.

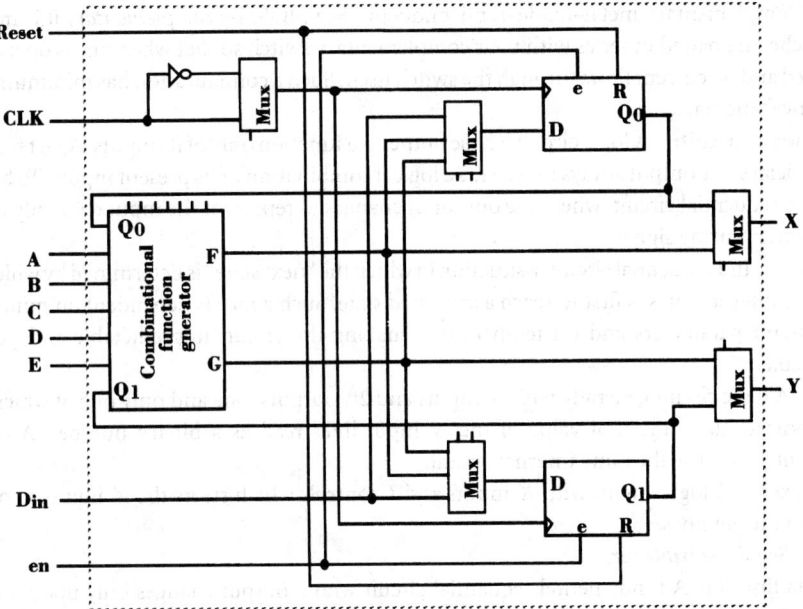

FIGURE 15.30   The internal logic of the CLB. Circuits with small unconnected control leads indicate static configurability determined by the program read in at bootup.

Having described the several parts of this circuit board on a chip, what can one do with it? We used one to realize a Booth's algorithm 8-bit multiplier that was a classroom project and downloaded it into a Xilinx FPGA. The equivalent circuit-board product, which is what was designed, included the following list of 18 distinct SSI and MSI chips plus a few extra gates which would have brought the chip count up above 20:

2	LS244	8-bit 3-state buffers
1	LS164	8-bit shift register
3	LS373	octal transparent latch (modified)
1	LS04	hex inverter
3	LS86	quad 2-bit XOR
4	LS157	quad 2 to 1 MUX
2	LS83	4-bit adder
1	LS139	dual 2-bit decoder
1	LS74	dual D-flop

Defining Terms

Active logic: Digital logic which operates all of the time in the active, dissipative region of the electronic amplifiers from which it is constructed. The output of such a gate is determined primarily by the gate and not by the load. Such logic is generally faster than saturated logic, but it dissipates much more energy. See **saturated logic**.

ASIC: Application-specific integrated circuit. In the broadest sense, integrated circuits that are designed for a specific application. The term is used to describe VLSI circuits of standard form which can be configured either in manufacture or on site to meet the specific needs of the application. The configuration process is generally so inexpensive that such circuits are economical to produce in extremely small runs.

CLB: Control logic block in a Xylinx FPGA.

Clock: The input which provides the timing signal for a circuit. In general, the oscillator circuit which generates a synchronization signal.

CMOS: Complementary metal–oxide–semiconductor. See MOS. By *complementary* it is meant that all switches are paired in series with their complementary switch so that when one is open, the other is closed and no current flows through the switch itself. Such a configuration has minimum dissipation in any static state.

Combinational circuit: A logic circuit whose output is a function only of its inputs. Apart from propagation delays, the output always represents a logical combination of its present inputs. To be contrasted with a sequential circuit, where the output(s) changes to represent the input data only upon a clock or synchronizing signal.

Critical race: In a sequential circuit, a situation in which the "next state" is determined by which of two (or more) internal gates is first to reach a saturated state. Such a race is dependent on minor variations in circuit parameters and on temperature, making the circuit unpredictable and possibly even unstable.

Decoder: A logic circuit generally with N inputs and 2^N outputs, one and only one of which is asserted to indicate the numerical value of the N input lines read as a binary number. A decoder and demultiplexer use the same internal circuit.

Demultiplexer: A logic circuit with K inputs and I controls which steers the K inputs to one set of 2^I sets of output lines.

DEMUX: See *demultiplexer*.

DFF: Data flip-flop. A fundamental sequential circuit whose output changes only upon a clock signal and whose output represents the data on its input at the time of the last clock.

Don't-care: In a truth table or Karnaugh-map, a state which is irrelevant to the functioning of the circuit (e.g., because it never occurs in the intended application). Thus, the designer "doesn't care" whether that state is asserted, and he or she may choose the output that best minimizes the number of gates.

Edge-triggered FF: A flip-flop which changes state on a clock transition from low to high or high to low rather than responding to the level of the clock signal. Contrast to master–slave FF.

Encoder: A logic circuit with 2^N inputs and N outputs, the outputs indicating the number of the one input line that is asserted. See also *priority encoder*.

Flip-flop: Any of several related bistable circuits which form the memory elements in clocked, sequential circuits. They are analogous to toggle switches or latching relays.

FPGA: Field-programmable gate array. VLSI chips with a large number of reconfigurable gates that can be "programmed" to function as complex logic circuits. The programming can be done on site and may be statically (out of circuit) or dynamically (in circuit) reprogrammable.

GAL: Generic Array Logic. PLAs with more flexible output control than the original PALS. GAL is a trade name of Lattice Semiconductor.

Glitch: A transient transition between logic states caused by different delays through parallel paths in a logic circuit. They are unintentional transitions, so they do not correctly represent the logic of the intended design.

Karnaugh map: A mapping of a truth table into a rectangular array of cells in which the nearest neighbors of any cell differ from that cell by exactly one binary input variable.

Master-slave FF: A flip-flop which changes state when the clock voltage reaches a threshold level. Contrast to edge-triggered FF.

MOS: Metal–oxide–semiconductor. A layered capacitor with the added property that by applying a voltage of the proper polarity, the underlying semiconductor can be switched from n- to p-type or vice versa. Such structures form the active part of an MOS transistor.

Multiplexer: A circuit with N control lines to connect one of 2^N input lines to the single output line.

MUX: See **Multiplexer**.

PAL: Programmable array logic, a brand name of Monolithic Memories (now AMD), for programmable logic arrays (PLAs) generally characterized as having a programmable AND layer and a fixed (unprogrammable) summing (OR) layer.

PLA: Programmable logic array. A general term used to cover PALs, GALs and, more particularly, programmable arrays with both AND and OR layers programmable.

Priority encoder: An encoder (see definition) with the additional property that if several inputs are asserted simultaneously, the output number indicates the numerically highest input that is asserted. For example, if lines 1 and 3 were both asserted, the output value would be 3.

Saturated logic: Logic gates whose output is fully on or fully off. The output voltage from such a switch or the current through such a switch is determined principally by the external circuit. The opposite of *saturated logic* is **active logic.**

Sequential circuit: A circuit which goes through a sequence of stable states, transitioning between such states at times determined by a clock signal. Complex examples would include an electronic counter and a digital wristwatch.

Transparent latch: Essentially, a flip-flop which continuously passes the input to the output (thus *transparent*) when the clock is high (low) but holds the last output during any interval when the clock is low (high). The circuit is said to have *latched* when it is holding its output constant regardless of the value of the input.

TTL: Transistor–transistor logic. One of several general families of logic gates. TTL gets its name from the fact that it uses transistors for both the pullup and pulldown functions. That is, the output of the gate is amplified in going from 0 to 1 as well as from 1 to 0. (CMOS families do this as well, but the earlier logic families did not.) Using active pullups leads to faster gates with the ability to drive more loads.

References

Breeding, K. J. 1992. *Digital Design Fundamentals,* 2nd ed. Prentice–Hall.

Hsu, Y.-C. 1995. *VHDL Modeling for Digital Design Systems.* Kluwer Academic.

Karnaugh, M. 1953. A map method for synthesis of combinational circuits. *Trans. AIEE. Comm. and Electron.* 72(1):593–599.

Katz, R. H. 1994. *Contemporary Logic Design.* Benjamin Cummings.

Navabi, Z. 1993. *VHDL Analysis and Modeling for Digital Systems.* McGraw–Hill.

Nelson, V. P., Nagle, H. T., Carroll, B. D., and Irwin, J. D. 1995. *Digital Logic Circuit Analysis & Design.* Prentice–Hall.

Pick, J. 1996. *VHDL Techniques, Experiments and Caveats.* McGraw–Hill.

Wakerly, J. F. 1994. *Digital Design Principles and Practice,* 2nd ed. Prentice–Hall.

Further Information

This is a very quick pass through digital circuit design. A more thorough grounding, with many examples and excellent development of field-programmable logic arrays and computer methods for gate minimization, can be found in [Nelson et al. 1995], [Wakerly 1994], or [Katz 1994]. What has been covered herein provides a good overview of the principles as well as specific designs for all of the logic needed to read the chapter on computer architecture in this volume.

Much of the current research and development in digital design is concerned with automation of the design process. Just as high-level computer languages have greatly reduced the cost of generating computer code, there is a belief that it should be possible to automate a very large fraction of the work which lies between a functional description of a complex digital circuit and the generation of production masks for the integrated circuit which realizes that circuit. The first step in building such an integrated toolset is to define a standard *hardware description language* (HDL) through which to describe the objectives of a design. Then, one builds compilers which convert the description successively through all of the stages which lead to production masks. A widely accepted example of such a language is VHDL. Readers interested in pursuing this topic would find a good place to begin in [Pick 1996], [Hsu 1995], or [Navabi 1993].

16

Digital Computer Architecture

David R. Kaeli
Northeastern University

16.1 Introduction

A computer architecture is a specification which defines the interface between the hardware and software. This specification is a contract, describing the features of the hardware which the software writer can depend upon and identifying design implementation issues which need to be supported in the hardware. This very basic definition of architecture serves as the starting point for this chapter. We will begin with a programming example and attempt to identify the interface between programming and the underlying architecture. From there we will expand upon features of a computer architecture, describing how one might tune the performance of a computer architecture.

Tasks are carried out on a computer by software specific to each program task. A C program to compute the difference of two integers might look something like the text shown in Fig. 16.1. Even within a simple programming example we are able to identify some of the necessary elements in a computer architecture (e.g., arithmetic and assignment operations, integers).

The program is then compiled, producing a machine-language representation of the task to be performed. An assembly-code version of the subtraction machine code is shown in Fig. 16.2 (assembly code is an intermediate format between high-level language and machine code which closely resembles the machine code).

The machine code of the computer system comprises a set of primitive operations which are performed with great rapidity. We refer to these operations as instructions. The set of instructions provided is defined in the specification of the architecture. An instruction set is just one aspect of defining an architecture.

0-8493-2909-4/97/$0.00+$.50
© 1997 by CRC Press, Inc.

```
x = 5;          /* Initialize x to 5 */
y = 3;          /* Initialize y to 3 */
z = x - y;      /* Compute the difference */
```

FIGURE 16.1   High-level language program.

Instructions are executed on the execution processor (commonly called the CPU or processor), which may comprise a number of units, including (1) an *arithmetic logical unit* (ALU), (2) a *floating-point unit* (FPU), (3) local memory, and (4) external bus control. A particular computer architecture can be realized in hardware in a wide variety of hardware organizations. What remains constant across these different implementations of the architecture is a common software interface that programmers can depend upon. The IBM 360 and 370 architectures are two good examples of a well-defined computer architecture.

```
x:      .int 5
y:      .int 3
z:      .

load r1, x
load r2, y
sub r3, r1, r2
store r3, z
```

FIGURE 16.2   Assembly-language version of HLL program.

In this chapter various aspects of a computer architecture are presented. The design of a digital computer will also include the supporting memory system and busing necessary to interconnect these units. We will begin our discussion by describing the components of an instruction set.

16.2 The Instruction Set

An instruction set includes the machine-language primitives which can be directly executed on a processor. Instruction classes present in all instruction sets include:

1. ALU instructions (e.g., integer add/subtract, shift left/right, logical and/or/xor/inversion).
2. Memory accessing instructions (e.g., load and store).
3. Control transfer instructions (e.g., conditional/unconditional branch, call/return, and interrupt).

Two prevalent implementation paradigms are *reduced instruction set computers* (RISC) [Patterson 1985] and *complex instruction set computers* (CISC) [IBM 1981]. RISC microprocessors are designed around the concept that by keeping instructions simple, the compiler will have greater freedom to produce optimized code (i.e., code which will execute faster). CISC architectures use a very different principle, attempting to perform operations specified in the high-level language in a single instruction (commonly referred to as reducing the semantic gap). While RISC architectures have become the paradigm of the future, the Intel *x*86 and Motorola 680*x*0 architectures continue to employ a CISC architecture.

While these two models are based on very different principles, they both contain instructions from the three classes above. Nearly all of today's instruction sets fall into one of these two categories. The underlying principles are that RISC instruction sets are simple (or reduced) and that CISC instruction sets are complicated (or complex).

Most architectures include floating-point instructions. Those implementations of the architecture which contain a floating-point unit (FPU) are able to execute floating-point operations at speeds approaching integer operations. If a floating-point unit is not provided, then floating-point instructions are emulated by the integer processor (using a software program). When the hardware encounters a floating-point instruction, and if no FPU is present, a message (i.e., a software interrupt) is presented to the operating system. In response to this message, the floating-point instruction is executed using a number of integer instructions (i.e., it is emulated). The performance of emulated floating-point instructions is typically 3–4 orders of magnitude slower than if an FPU were present. Specific instructions may also be provided to manipulate decimal or string data formats. Most modern architectures include instructions for graphics and video.

ALU Instructions

An ALU, which builds on the adder/subtracter presented in the chapter on digital logic, is used to perform simple operations upon data values. The ALU performs a variety of operations on input data fed from its two input registers and stores the result in a third register. Figure 16.3 shows an example of an ALU which might be found in the CPU.

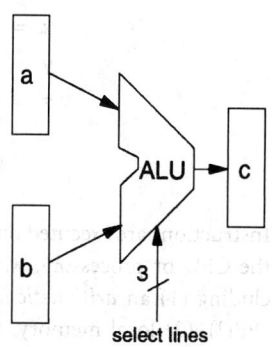

The ALU is supplied data from a pair of registers a and b (registers are typically designed using D-flip-flops). The resulting answer is placed into register c. The function performed is determined based on the value of the select lines, which is directly derived by the instruction currently being executed. Table 16.1 shows the possible values for the three control lines shown in Fig. 16.3.

FIGURE 16.3   ALU with input and output registers.

The select lines are decoded (e.g., with a 3-to-8 decoder) to specify the desired operation. For example, the machine code format for the subtract instruction's operation code would contain (or be decoded into) the bit values 001. The assignment of these values is defined by the architecture and generally appears in a programmer's reference manual for the particular instruction set.

TABLE 16.1   Sample ALU Operations

Operation Selected	Select-Line Value
$c = a + b$	000
$c = a - b$	001
$c = a$ shifted left b bits	010
$c = a$ shifted right b bits	011
$c = a$ OR b	100
$c = a$ AND b	101
$c = a$ XOR b	110
$c = $ NOT a	111

Memory Referencing Instructions

When a program begins execution, it is loaded into memory by the operating system. Both the instructions and the data which will be used in the program are initially stored in memory. For example, the values of the variables **x** and **y** in our high-level language program need to be initially stored in memory. The compiler will reserve memory locations to store these values and will provide instructions which will retrieve these values to initialize **x** and **y**. Then **x** and **y** are supplied to the input of the ALU via instructions that load them from registers **r1** and **r2**. The subtract instruction will tell the ALU to produce the difference **(r1 − r2)** in register **r3**.

A computer architecture defines how data are stored in memory and how they are retrieved. To retrieve the values of **x** and **y** from memory, a load instruction is used. A load retrieves data values from memory and loads them into registers (e.g., data values **x** and **y**, and registers **r1** and **r2**). A store instruction takes the contents of a register (e.g., register **r3**) and stores it to the specified memory location (e.g., the memory location which the compiler assigned to **z**). These instructions are necessary for obtaining the data upon which the CPU will operate and to store away results produced by the CPU.

The CPU needs a way to differentiate between different data. Just as we are assigned Social Security numbers to differentiate one taxpayer from another, memory locations are assigned unique numbers called memory addresses. Using a memory address, the CPU can retrieve the desired datum. We will discuss addressing a little later and will focus on the aspects of addressing which are defined by the computer architecture.

RISC CPUs provide individual instructions for loading from and storing to memory. Pure RISC architectures provide only two memory-referencing instructions, load and store. All ALU instructions can have only input and output operands specified as registers or integer values. In contrast, CISC architectures provide a variety of instructions which both access memory and perform operations in a single instruction. A good discussion comparing the implications of these capabilities can be found in [Colwell et al. 1985].

Control Transfer Instructions

Control transfer instructions cause a break in the sequential flow of program execution. The transfer can be performed unconditionally or conditionally. When an unconditional control transfer occurs (commonly referred to as a jump), the execution processor will begin processing instructions from a different portion of the program. Jumps include unconditional jump instructions, call instructions, and return instructions.

Interrupts are another form of jump. They break the sequential instruction flow unconditionally and transfer control not to another part of the current application but instead to an interrupt service routine. Interrupts can be programmed (inserted by the programmer or compiler) or they can occur due to hardware or software events (e.g., input/output, timers). Interrupts transfer control over to the operating system. The event which caused the interruption is handled, and then control is later passed back to the program.

A conditional control transfer (commonly referred to as a conditional branch) is dependent on the execution processor state determined by the execution of past instructions. Conditional branches are used for making decisions in control logic, checking for errors or overflows, or a variety of other situations. For instance, we may want to check if the result of adding our two integer numbers produced an overflow (a result which cannot be represented in the range of values provided for in the result).

Conditional branch instructions can perform both the decision making and the control transfer in a single branch instruction, or a separate comparison instruction (i.e., an ALU operation) can perform the comparison upon which the branch decision is dependent. Next, we explore the memory elements provided in the support of the execution processor.

16.3 Memory

The interface between the memory system and CPU is also defined by the architecture. This includes defining the amount of space addressable, the addressing format, and the organization of memory in various levels of the memory hierarchy.

In traditional computer systems, memory is used for storing both the instructions and the data for the program to be executed. Instructions are fetched from memory by the CPU, and once decoded (instructions are typically in an encoded format, similar to that found in Table 16.1 for ALU instructions), the data operands to be operated upon by the instruction are retrieved from, or stored to, memory. Memory is typically organized in a hierarchy. The closer the memory is to the CPU, the faster (and more expensive) it is. The memory hierarchy shown in Fig. 16.4 includes the following levels:

1. Register file.
2. Cache memory.
3. Main memory.
4. Secondary memory (disk, tapes, etc.).

Above each level in Fig. 16.4 is a measure of its typical size and access speed in contemporary technology.

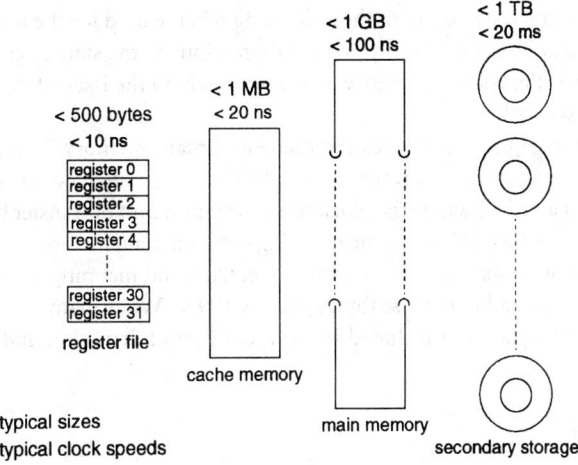

FIGURE 16.4 The memory hierarchy.

Register File

The register file is an integral part of the CPU and, as such, is clearly defined in the architecture. The register file can provide operands directly to the ALU. Memory referencing instructions either load to or store from the registers contained in the register file. The register file can contain both general-purpose registers (GPRs) and floating-point registers (FPRs). Additional registers for managing the CPUs state and addressing information are generally provided.

The register file represents the lowest level in the memory hierarchy, since it is closest to the processor. The registers are constructed out of fast flip-flops. The typical size of a GPR in current designs is 32 or 64 bits, as defined by the architecture. Registers are typically accessible on either a bit, byte, halfword, or fullword granularity (a word refers to either 32 or 64 bits).

Main Memory and Cache

Main memory is physical (versus virtual) memory which typically resides off the CPU chip. The memory is usually organized in banks and supplies instructions and data to the processor. Main memory is typically byte-addressable (meaning that the smallest addressable quantity is 8 bits of instruction or data). Main memory is generally implemented in dynamic random-access memory (DRAM) to take advantage of DRAM's low cost, low power drain, and high storage density. The costs of using this memory technology include reduced storage response time and increased design complexity (DRAM needs to be periodically refreshed, since it is basically a tiny capacitor).

Main memory is typically organized to provide efficient access to sequential memory addresses. This technique of accessing many memory locations in parallel is called interleaving. Interleaved memory allows memory references to be multiplexed between different banks of memory. Since memory references tend to be sequential in nature, allowing the processor to obtain multiple addresses in a single access cycle can be advantageous. Multiplexing can also provide a substantial benefit when using cache memories.

Cache memory is used to hold a small subset of the main memory. The cache is typically developed in static random access memory (SRAM), which is faster than DRAM, but is more expensive, more power-hungry, and less dense. SRAM technology does not need to be refreshed. The cache contains the most recently accessed code and data. The cache memory is used to provide instructions and data to the processor faster than would be possible if only main memory were used. In current CPUs, separate caches for instructions and data are provided on the CPU chip.

The amount of the memory system defined by the architectural specification varies greatly. The design and layout of the cache and main memory are typically not defined by the architecture. The architecture specifies the bit and byte arrangement of data and the addressing formats within instructions. The interface (e.g., address lines) to external (main) memory can also be specified in the architecture. The virtual memory systems may also place some restrictions on the addressing scheme used for the main memory system. In some architectures, instructions have been provided to manipulate the state of cache memory to ensure the memory coherency of the system (memory coherency refers to the issue of having only a single valid copy of a datum in the system).

Since the performance gap between accessing cache and main memory is so great, maximizing the probability of finding a memory request resident in cache when requested is of great interest. Some of the tradeoffs in the design of a cache include the block size (minimum unit of transfer between main memory and the cache), associativity (used to define the mapping between main memory and the cache), handling of cache writes (for data and mixed caches), number of entries, and mapping strategies. Handy provides a thorough discussion on a number of these topics [Handy 1993]. While many of these design parameters are important, they typically are not included in the architectural definition and are left as part of the design not specified.

Secondary Storage

On the initial run of your program, the program code will reside in secondary storage (disk storage). When the program is run, it will be transferred to main memory, then to cache, and then to the execution processor (generally these last two transfers are performed simultaneously). Disk storage is commonly used for secondary storage, since it is nonvolatile storage (the contents are maintained even when power is turned off).

The disk is designed using magnetic media, and data are stored in the form of a magnetic polarization. The disk comprises one or more platters which are always spinning. When a request is made for instructions or data, the disk must rotate to the proper location in order for the read heads to be able to access the information at that location. Because disk rotation is a mechanical operation, disk accesses are many orders of magnitude slower than the access time of the DRAM used for main memory.

After a program has been run once, it will reside for a period of time in either the cache or main memory. Access to the program will be faster upon subsequent runs if the execution processor does not have to wait for the program to be reloaded from disk.

The architecture of a system does not typically impose any limitations on the organization of secondary storage besides defining the smallest addressable unit (typically called a block) and the total amount of addressable storage.

16.4 Addressing

All instructions and data in memory are accessed by memory address. Next we look at various aspects of memory addressing: addressing format, physical addressing, virtual addressing, and byte ordering.

Addressing Format

Defined within all architectures are the various addressing formats which are permissible with particular instructions. Standard addressing formats provided with most architectures include:

1. *Direct:* The full address is provided in the instruction. No address calculation is necessary. The address field can be used directly (thus, direct addressing). Direct addressing is commonly used for accessing static data.
2. *Register indirect:* The number of the GPR which contains the memory address is specified. Register addressing is commonly used for accessing via a pointer value.
3. *Memory indirect:* The number of the GPR is specified which contains the address where the address of the desired data is located. In this case, two memory references are performed to obtain the desired data. Indirect addressing is commonly used to dereference pointers.
4. *Base displacement:* The number of the GPR which contains the base address of some data structure is specified. The base value is added to a displacement field to obtain the final memory address. Base-displacement addressing is commonly used for addressing sequential data patterns (e.g., arrays, structures).
5. *Indexed:* A GPR is added to a base register (GPR) to obtain the memory address. Some architectures provide separate index registers for this purpose. Other architectures add an index to a base register and possibly even include a displacement field. Indexed addressing is commonly used for traversing complex data structures such as link lists of structures.

Physical and Virtual Memory

Virtual memory refers to the ability of a computer system to address a larger address range than the amount of physically installed memory (DRAM). This is a very cost-effective approach to computing. We need to have only a small amount of DRAM memory installed on our system, compared to the actual addressable space. The range of the virtual address space places a limit on the amount of addressable secondary storage (including disks, printers, devices, etc.). Prior to the introduction of the concept of virtual memory, the programmer had to insert explicit commands to load programs in from disk storage. Since one segment supplanted another in memory, this is called *overlay*. With the introduction of virtual

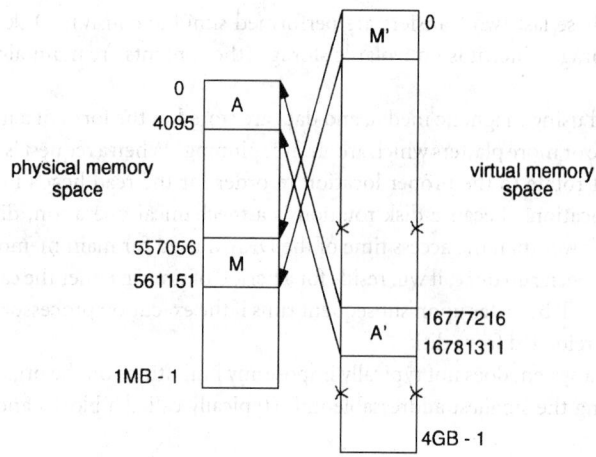

FIGURE 16.5   Mapping of physical to virtual memory addresses.

memory, this was no longer necessary. Some motivating factors behind using virtual memory include:
(1) providing efficient usage of physical memory without the need for explicit overlays, and (2) allowing
full flexibility of placement of instructions and data.

A virtual memory address generally refers to any memory location in the available address range as
defined by the architecture. A physical memory address refers to a memory location in main memory.
The physical-memory-address range is defined by the amount of installed main memory and is bounded
by the number of bits provided for in the address scheme of the architecture. The virtual-address range is
defined by the largest permissible address value. Figure 16.5 provides an example of how a 32-bit virtual
address can address over 4 billion different memory locations. Since the virtual-address range is generally
2 or more orders of magnitude greater than the installed physical memory, a mapping from virtual to
physical memory is performed. See Fig. 16.5 for an example of how different regions (i.e., pages) in the
virtual address space map to the physical address space.

The operating system performs the mapping between the virtual and physical address spaces. The
mapping is performed on either a fixed or a variable memory block size (pages versus segments, respectively). In Fig. 16.5 we see that a fixed-size page (4096 bytes) is being mapped to the physical memory
address space. This mapping is stored in a table in memory (called a page table). The page table will hold
virtual-to-physical mapping for all pages present in the physical memory. In Fig. 16.5 we see that virtual
page A' is mapped to physical memory location A and that virtual page M' is mapped to physical page M.
Thus, sequentiality is limited to a page and generally does not extend past the page boundary, which is to
say, two sequential pages in the virtual address space need not be mapped to two sequential physical pages.

Pages are brought from secondary storage into physical (main) memory when they are requested by
the execution processor. This is referred to as *demand paging*. The size of a page is typically fixed (e.g.,
4096 bytes in our example). Segmented virtual memory systems provide a variable-sized unit of transfer
between the virtual and physical memory spaces. A good discussion of the different types of virtual
memory systems can be found in [Feldman and Retter 1995] as well as in Chapter 80.

Since every instruction execution in the CPU would require that at least two memory accesses be
performed to obtain instructions (one access to obtain the physical address stored in the page table, and
a second access to obtain the instruction stored at that physical address), a hardware feature called a
translation lookaside buffer (TLB) is commonly used. The TLB caches the recently accessed portions of the
page table and quickly provides a virtual-to-physical translation of the address. The TLB is generally located
on the CPU to provide fast translation capability. Further discussion on TLBs can be found in [Teller 1991].

Now that we know a little bit more about instructions, addressing, and memory, we can begin to
understand how instructions are processed by the CPU.

16.5 Instruction Execution

To this point we have focused on "architected" features of a digital computer system. These will be used to document the interface to which compiler and application developers must program and the features which the hardware designer must provide. Next we will describe some "nonarchitected" features of a digital computer system which are typically left to the implementer to decide upon. The first feature looks at how we organize elements of the CPU in order to execute an instruction efficiently.

Instructions are requested from the memory system. This involves sending an address out to the memory and waiting until the memory system produces the requested address. Once retrieved, the instruction enters the processor. The execution of an instruction involves a number of steps. Figure 16.6 provides the buses and logic to be used to describe the steps for nonpipelined instruction execution.

In a nonpipelined system, a single instruction at a time enters the processor and waits the necessary amount of time for all of the steps associated with an instruction to be completed. For instance, in Fig. 16.6 the subtract instruction is loaded into the instruction register and decoded by the control logic. Then the input registers to the ALU are enabled and the ALU is programmed to perform a subtract. After a sufficient delay to allow all of these operations to be performed and for the output of the ALU to reach a steady-state value, the output register is latched. This all happens during a single processor clock cycle (the CPU clock is typically used to synchronously capture the new state of memory devices.

If instructions were allowed to flow through the execution processor only one at a time, the throughput of the execution processor (the rate at which instructions exit the processor) would be low. This is because a majority of the elements of the CPU would remain idle while different phases of the instruction execution progressed.

Instead, consider the execution of the same instruction broken up into a number of stages, as is shown in Fig. 16.7. A single clock is used to latch results at the end of each stage (in practice, different clock edges may be used to latch particular elements in a stage). The key ideas here are that all stages work synchronously and that multiple instructions are being processed simultaneously (similar in concept to Henry Ford's assembly line). This is called *pipelining*. Given a pipeline of n stages, n instructions can be in process simultaneously (each at different stages of the pipeline). This can dramatically increase instruction throughput.

To complete the execution of an instruction, a series of tasks are performed by a number of functional units. To begin with, the instruction must be present in the processor. We will label this stage the instruction fetch stage of the pipeline, since it is fetching the next instruction to be executed.

The second stage of the pipeline is instruction decode. During this stage, the machine language is decoded and the operation to be performed is discovered. The decoding process generates the control bit

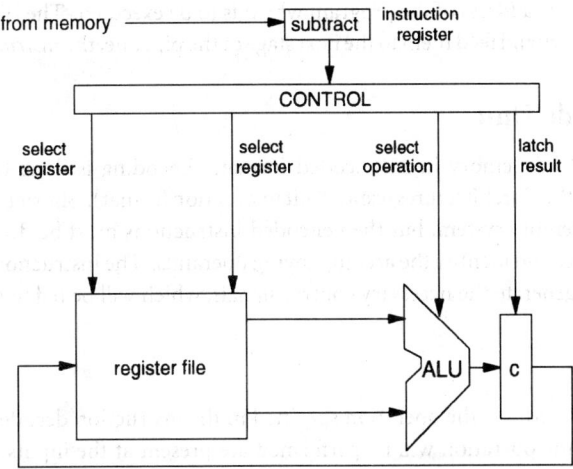

FIGURE 16.6 Nonpipelined instruction execution.

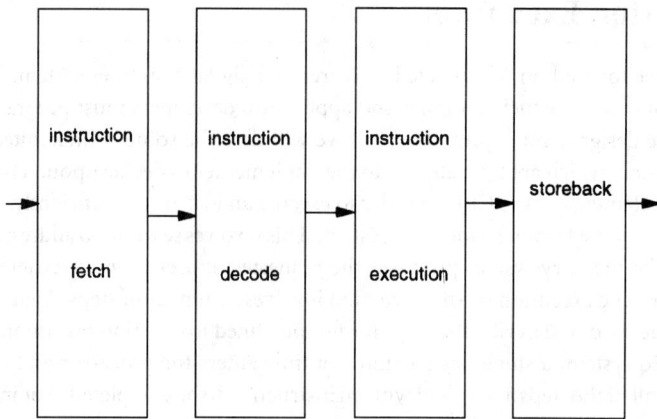

FIGURE 16.7 Pipelined execution of the subtraction example.

values which will enable buses, latch registers, and program ALUs. These are represented in Fig. 16.6 by the vertical lines produced by the control logic. Also during this stage, the necessary operands (e.g., **x** and **y** for the subtract) may also be discovered.

The third stage of the pipeline is the instruction execution stage. During this stage, the operation specified in the instruction operation code is actually performed (e.g., in our example program, the subtract will take place during this stage and the result will be latched into the ALU result register **c**).

The final stage is the storeback stage. During this stage, the results of the execution stage are stored back to the specified register or storage location. While the discussion here has been for ALU-type instructions, it can be generalized for memory referencing and control-transfer instructions. Kogge describes a number of pipeline organizations in [Kogge 1981].

Next we discuss the individual contents of each of these stages.

Instruction Fetch Unit

An executable computer program is a contiguous block of instructions which is stored in memory (copies may reside simultaneously in secondary, main, and cache memory). The *instruction fetch unit* (IFU) is responsible for retrieving the instructions which are next to be executed. To start (or restart) a program, the dispatch portion of the operating system will point the execution processor's *program counter* (PC) to the beginning (or current) address of the program which is to be executed. The IFU will begin to retrieve instructions from memory and feed them to the next stage of the pipeline, the *instruction decode unit* (IDU).

Instruction Decode Unit

Instructions are stored in memory in an encoded format. Encoding is done to reduce the length of instructions (common RISC architectures use a 32-bit instruction format). Shorter instructions will reduce the demands on the memory system, but then encoded instructions must be decoded to determine the desired control bit values and identify the accompanying operands. The instruction decode unit performs this decoding and will generate the necessary control signals, which will be fed to the *execution unit*.

Execution Unit

The execution unit will perform the operation specified in the instruction decoded operation code. The operands upon which the operation will be performed are present at the inputs of the ALU. If this is a memory referencing instruction (we will assume we are using a RISC processor for this discussion, so ALU operations are performed only on either immediate or register operands), address calculations will

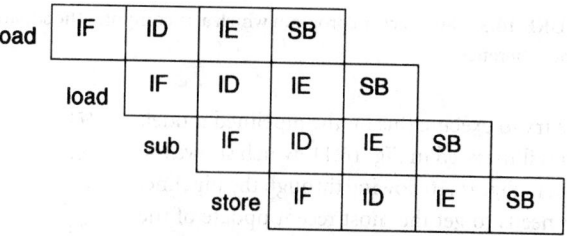

pipelined execution

FIGURE 16.8 Comparison of nonpipelined and pipelined execution.

be performed during this stage. The execution unit will also perform any comparisons needed to execute conditional branch instructions. The result of the execution unit is then fed to the *storeback unit*.

Storeback Unit

Once the result of the requested operation is available, it must be stored away so that the next instruction can utilize the execution unit. The storeback unit is used to store the results of ALU operation to the register file (again, we are considering only RISC processors in this discussion), to update a register with a new value from memory (for LOAD instructions), and to update memory with a register value (for STORE instructions). The storeback unit is also used to update the program counter on branch instructions.

Figure 16.8 shows our subtraction code flowing through both a nonpipelined and a pipelined execution processor. The width of the boxes in the figures is meant to depict the length of a processor clock cycle. The nonpipelined clock cycle is longer (slower), since all of the work accomplished in the four separate stages of the instruction execution are completed in a single clock tick. The pipelined execution clock cycle is dominated by the time to stabilize and latch the result at the end of each stage. This is why a single pipelined instruction execution will typically take longer to execute than a nonpipelined instruction. The advantages of pipelining are reaped only when instructions are overlapped. As we can see, the time to execute the subtraction program is significantly smaller for the pipelined example (but not nearly four times smaller). Also, other executions can be overlapped with this example code (i.e., instructions in the pipeline prior to the first load and after the store instruction). This is not the case for nonpipelined execution.

Note that in our examples in Fig. 16.8 we are assuming that all nonpipelined instructions and pipelined stages take a single clock cycle. This is one of the underlying principles in RISC architectures. If instructions are kept simple enough, they can be executed in a single cycle.

Pipelining can provide an advantage only if the pipeline is supplied with instructions which can be issued without any delay or uncertainty. The benefits of pipelining can be greatly reduced if stalls occur due to different types of hazards. We will discuss this topic next.

16.6 Execution Hazards

Consider attempting to process the high-level language example in Fig. 16.9, which builds upon our subtraction program. If we look at what the compiler would do with this code, it might look something like the instruction sequence shown in Fig. 16.10.

```
x = 5;          /* Initialize x to 5 */
y = 3;          /* Initialize y to 3 */
z = x - y;      /* Compute the difference */
w = z * z;      /* compute the square */
```

FIGURE 16.9 Subtraction program which also computes the square of the difference.

A problem occurs if we try to execute this on the pipelined model. The cause of the problem is illustrated in Fig. 16.11, which shows the instruction sequence given in Fig. 16.10 flowing through the pipeline. The multiply instruction needs to get the most recent update of the variable **z** (which will reside in **r3**). Given the pipeline model described above, the multiply instruction will be attempting to direct the contents of **r3** to the inputs of the ALU during its instruction decode stage. In the same cycle the subtract instruction will be trying to store the results of the subtraction to **r3** during the storeback stage. This is just one example of a *hazard*, called a data hazard.

There are three classes of hazards:

1. *data hazards*, which include read-after-write (RAW), write-after-read (WAR), and write-after-write hazards;
2. *control hazards*, which include any instructions or interruptions which break sequential instruction execution; and
3. *structural hazards*, which occur when multiple instructions vie for a single functional element (e.g., an ALU).

```
w:      .
x:      .int 5
y:      .int 3
z:      .

        load r1, x
        load r2, y
        sub r3, r1, r2
        store r3, z
        mult r4, r3, r3
```

FIGURE 16.10   Assembly-language version of Fig. 16.9.

Data Hazards

Data hazards occur when there exist dependencies between instructions. For the hazard to occur, the instructions need to be close enough in the execution stream to coexist in the pipeline. The example provided in Fig. 16.10 is just one type of interinstruction dependence, called a *read-after-write* hazard. The multiply is attempting to read **r3** as the subtract is writing **r3**. This is the most commonly encountered type of data hazard.

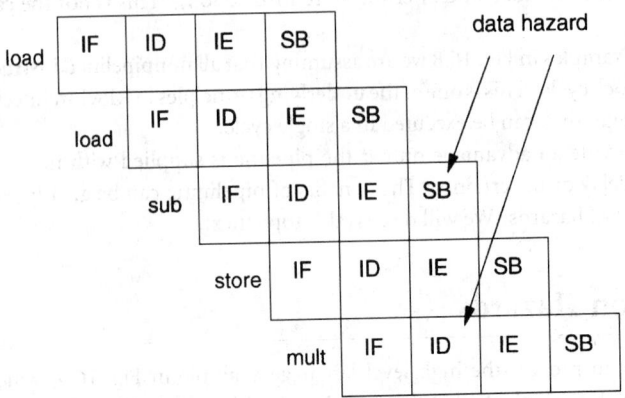

FIGURE 16.11   Example of a RAW hazard.

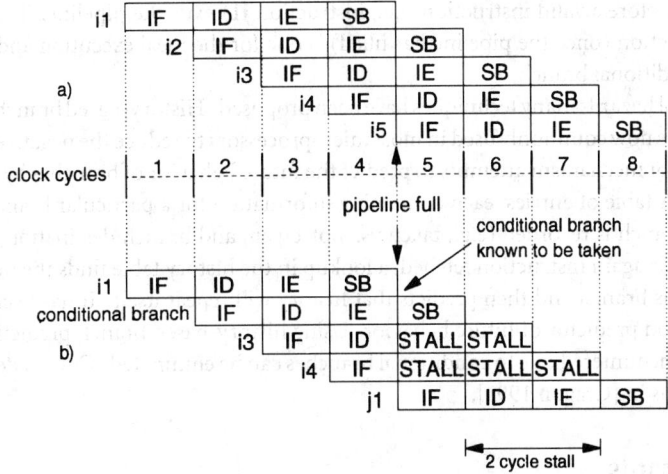

FIGURE 16.12   a) Ideal execution. b) A taken conditional branch.

A second type of data hazard occurs when a subsequent instruction modifies an operand of the current instruction before it is read by the current instruction. This is called a *write-after-read* hazard. This will occur only if writes are allowed to be performed early in the pipeline and reads are performed late in the pipeline.

A third type of data hazard occurs when a subsequent instruction modifies an operand of the current instruction before it is written by the current instruction. This is called a *write-after-write* hazard. This will occur only if writes are allowed to be performed both early and late in the pipeline.

A number of solutions to detecting data hazards have been proposed. Hardware techniques-of-forwarding data values can overcome a large number of these hazards which frequently occur. This technique is integrated into the decoding logic and keeps track of the source and destination of all instructions currently active in the pipeline. Hazard detection then becomes part of the instruction decoding stage. Compilers can also be tuned to eliminate the possibility of data hazards occurring.

Control Hazards

Another type of hazard which can affect pipeline throughput is a control hazard. This class of hazard includes any instructions which break the sequential pipeline flow. These include branches (both conditional and unconditional), jumps, calls and returns, and interrupts. Since the instruction fetch stage of the pipeline assumes that the sequential instruction path will be followed, when a change in the path takes place, the instructions which have entered the pipeline need to be flushed and the program counter needs to be adjusted to point to the new instruction stream. Once the instruction fetch unit redirects instruction fetching, the pipeline will be refilled. The result is that the benefits of pipelining will be dramatically reduced.

Figure 16.12(a) shows an ideal execution which does not contain any breaks in sequential execution. After the pipeline fills (i.e., after cycle 4) an instruction exits the pipeline each cycle. Figure 16.12(b) shows how a taken conditional branch will degrade pipeline performance. After clock cycle 4, the pipeline is full of valid instructions and the instruction **i2** is a taken conditional branch instruction. The pipeline will know that this instruction is a branch after it has passed through the ID stage (instruction decode). The pipeline forges ahead, hoping that the branch will not be taken. The pipeline detects that the conditional branch is taken when it completes the IE stage (at the end of clock cycle 4). Instructions **i3** and **i4** have already entered the pipeline and will need to be flushed. This introduces a 2-cycle stall into the pipeline. The branch is taken to instruction **j1**, and the pipeline is refilled. There is a 2-cycle delay in this example, since 2 cycles pass before a valid instruction (i.e., instruction **j1**) exits the pipeline. The average number of cycles per instruction (once the pipeline has filled) is one for the ideal execution and two for the code containing the conditional branch.

To handle control hazards many techniques have been proposed. History-based branch prediction is one mechanism which is now commonly used in most microprocessors to reduce the negative effects of control hazards. This type of mechanism attempts to predict the future behavior of branches by recording history of past execution. A table of entries, each containing information for a particular branch, is maintained. History for each branch is recorded (e.g., taken vs. not-taken, and branch destination). Then when the branch instruction is again instruction fetched, a lookup in the history table finds the outcome of the last execution(s) for this branch and then predicts that history will repeat itself. It has been found that past history is a very good predictor of future behavior. Using history-based branch prediction, a majority of the pipeline stalls encountered due to conditional branches can be eliminated. Cragon describes a number of these mechanisms in [Cragon 1992].

Structural Hazards

Structural hazards are the third class of pipeline delays which can occur. They occur when multiple instructions active in the pipeline vie for shared resources. Some examples of structural hazards include: two instructions trying to compute a memory address in the same cycle when a single address-generation ALU is provided or two instructions both attempting to access the data cache in the same cycle when only a single data-cache access port is provided.

Two approaches can be taken to alleviate the delays introduced by structural hazards. The first approach further exploits the principle of pipelining by employing pipelined stages within each pipeline unit. This technique is called *superpipelining*. This will allow multiple instructions which are active in the pipeline to coexist in a single pipeline stage.

Second, we can provide multiple functional units (e.g., two cache ports or multiple copies of the register file). This approach is commonly used when high performance is critical or when we want to be able to issue multiple instructions in a single clock cycle. Multiple issue processors, also called *superscalar* processors, are discussed next.

16.7 Superscalar Design

If we can solve all the problems associated with hazards, an instruction should be exiting the pipeline every processor clock cycle. While this level of performance is seldom achieved (mainly due to latencies in the memory system and the limitations of effectively handling control hazards), we would like to be able to see multiple instructions exit the pipeline in a single clock cycle if possible. This approach has been labeled *superscalar* design. The idea is that if the compiler can produce groups of instructions which can be issued in parallel (which do not contain any data or control dependencies), then we can attain our goal of having multiple instructions exit the pipeline in a single cycle.

Some of the initial ideas which have motivated this direction date back to the 1960s and were initially implemented in early IBM [Anderson et al. 1967] and CDC machines [Thornton 1964]. The problem with this approach is finding a large number of instructions which are independent of one another. The compiler cannot exploit the scheduling to perfection because some conflicts are data-dependent. We can instead design complex hazard detection logic in our execution processor. This has been the approach taken by most superscalar designers.

Two issues occur in superscalar execution. First, can we issue nonsequential instructions in parallel? This is referred to as *out-of-order issue*. A second question is whether we can allow instructions to exit the pipeline in nonsequential order. This is referred to as *out-of-order completion*. A thorough discussion of the trade-offs associated with superscalar execution and issue/completion design can be found in [Johnson 1990].

16.8 Summary

This chapter has introduced many of the features provided in a digital computer architecture. The instruction set, memory hierarchy, and memory addressing elements, which are central to the definition of a computer architecture, were covered. Then some optimization techniques, which attempt to improve the efficiency of instruction execution, were presented. The hope is that this introductory material provides enough background for the nonspecialist to gain an appreciation for the pipelining and superscalar techniques that are currently used in today's CPU designs.

Defining Terms

Branch prediction: A mechanism used to predict the outcome of branches prior to their execution.
Cache memory: Fast memory, located between the CPU and main storage, that stores the most recently accessed portions of memory for future use.
Control hazards: Breaks in sequential instruction execution flow.
Data hazards: Dependencies between instructions that coexist in the pipeline.
Memory coherency: Ensuring that there is only one valid copy of any memory address at any time.
Pipelining: Splitting the CPU into a number of stages, which allows multiple instructions to be executed concurrently.
Structural hazards: A situation where shared resources are simultaneously accessed by multiple instructions.
Superpipelining: Dividing each pipeline stage into substages, providing for further overlap of multiple instruction execution.
Superscalar: Having the ability to simultaneously issue multiple instructions to separate functional units in a CPU.

References

Anderson, D. W., Sparacio, F. J., and Tomasulo, R. M. 1967. The IBM 360 Model 91: processor philosophy and instruction handling. *IBM J. Res. Dev.* 11(1):8–24, Jan.

Colwell, R. P., Hitchcock, C. Y., Jensen, E. D., Sprunt, H. M. B., and Kollar, C. P. 1985. Computers, complexity, and controversy. *IEEE Comput. Mag.* Sept., pp. 8–19.

Cragon, H. C. 1992. *Branch Strategy Taxonomy and Performance Models.* IEEE Computer Society Press, Los Alamitos, CA.

Feldman, J. and Retter, C. 1995. *Computer Architecture: A Designer's Guide to a Generic RISC.* Prentice–Hall, Reading, MA.

Handy, J. 1993. *The Cache Memory Book.* Academic Press, Boston.

IBM Corp. 1981. *IBM System/370 Principles of Operation.* Document No. GA22-7000-7, IBM Corp. New York, Mar.

Johnson, M. 1990. *Superscalar Microprocessor Design.* Prentice–Hall, Englewood Cliffs, NJ.

Kogge, P. M. 1981. *The Architecture of Pipelined Computers.* McGraw–Hill, New York.

Patterson, D. 1985. Reduced instruction set computers. *Commun. ACM* 28(1):8–21, Jan.

Teller, P. 1991. *Translation Lookaside Buffer Consistency in Highly-Parallel Shared Memory Multiprocessors,* Ph.D. dissertation. New York University, May. Also available as an *IBM Research Report,* RC 16858, #74685, May 14.

Thornton, J. E. 1964. Parallel operation in the Control Data 6600. In *AFIPS Proc. Fall Joint Computer Conf.* No. 27.

Further Information

To learn more about the recent advances in computer architecture, you will find articles on a variety of related subjects in the following list of IEEE and ACM publications:

IEEE Transactions on Computers.

ACM SIGARCH Newsletter.

IEEE TCCA Newsletter.

Proceedings of the International Symposium on Computer Architecture, IEEE Computer Society Press.

Proceedings of the International Conference on High-Performance Computer Architecture, IEEE Computer Society Press.

Proceedings of the Conference on Architectural Support for Programming Languages and Operating Systems, ACM.

17

Buses

Windsor W. Hsu
University of California

Jih-Kwon Peir
University of Florida

17.1 Introduction

The *bus* is the most popular communication pathway among the various components of a computer system. The distinguishing feature of the bus is that it consists of a single set of shared communication links to which many components can be attached. The bus is not only a very cost-effective means of connecting various components together, but also is very versatile in that new components can be added easily. Furthermore, the bus has a broadcasting capability which can be extremely useful. The downside of the shared communication links is that they allow only one communication to occur at a time and the bandwidth does not scale with the number of components attached. Nevertheless, the bus is very popular because there are many situations where several components need to be connected together but they need not all transmit at the same time. This kind of requirement maps naturally onto a bus, allowing a very cost-effective solution. However, there are cases where the bus does become a communication bottleneck. In such cases, very aggressive bus designs have been attempted, but there comes a point where the fundamental characteristic of the bus cannot be overcome and more expensive solutions such as point-to-point links have to be used.

Buses are used at every level in the computer system. For instance, within the processor itself, the bus is often the means of communication between the register file and the various execution units. At a higher level, the processor is connected to the memory subsystem through the *system bus*. Today's computers typically have a fast peripheral bus called a *local bus* which directly interfaces onto the system bus to provide a high bandwidth for demanding devices such as the graphics adaptor. Other less demanding peripheral devices are attached to the *I/O bus*. In the old days, the processor, memory subsystem, and I/O devices were all plugged onto the *backplane bus,* which is so called because the bus runs physically along the backplane of the computer chassis. The various buses are each optimized for a particular set of performance requirements and cost constraints and may thus seem very different from one another. However, their underlying issues are fundamentally the same.

A major requirement for designing a bus or simply comprehending a bus design is a proper understanding of the electrical and mechanical behavior of the bus. As buses are pushed to provide higher data rates, physical phenomena such as signal reflection, crosstalk, skew, etc. are becoming more significant and have to be handled carefully. Because the communication medium of a bus is shared by multiple devices, at most one transmission can be initiated at any time by any device. The other devices can act only as receivers or **bus slaves** for the transmission. A device that is capable of initiating and controlling a communication is called a **bus master**. In order to ensure that only one bus master is talking at any one time, the bus masters have to go through a **bus arbitration** process before they can gain control of the bus. Once a bus master has been granted control of the bus, a **bus protocol** has to be followed by the master and the slave in order for them to understand one another. The specifics of the protocol can vary widely, depending on the functional and performance requirements. For instance, the protocol used in the system bus of a uniprocessor is dramatically different from that used in the system bus of a shared-memory **symmetric multiprocessor** (SMP).

In this chapter, an overview of the basic underlying physics of computer buses is presented first. This is followed by a discussion of important issues in bus designs, including bus arbitration and various communication protocols. Because of the increasing prevalence of SMP systems and the special challenges they pose for buses, a separate section is devoted to discussing special bus issues in such machines. The discussion is wrapped up with a case study of a modern SMP system bus design. Finally, a historical perspective of computer buses and the related research issues are given in the last section.

17.2 Bus Physics

Computer buses are becoming wider and are being run at higher frequencies in order to keep up with the phenomenal improvement in CPU performance. As the physical and temporal margins in bus designs are reduced, it is imperative that electrical phenomena such as signal reflections, skew, crosstalk, etc. be understood and properly handled. In this section, we introduce the basic ideas behind these phenomena. A more detailed discussion can be found in [Giacomo 1990].

Transmission-Line Concepts

Electrical signals propagate with a finite speed that depends on the propagation medium. For instance, it takes approximately 5 ns for an electrical signal to travel 1 m along a typical copper wire. However, when analyzing electrical circuits, we often ignore their spatial properties. For example, we are seldom concerned with where each element of the circuit is located. We can do this because the circuits we usually encounter are lumped. In other words, their physical dimensions are small enough that for their particular applications, electromagnetic waves propagate across the circuits virtually instantaneously. However, this is not the case with high-speed buses. In this subsection, we introduce the basic concepts of the transmission-line model which are needed to understand the electrical behavior of today's buses.

In general, a connection is considered a transmission line if the propagation time of an electrical signal through it is a significant part of the rise time, fall time, or width of the signal. A good rule of thumb is to consider anything above $\frac{1}{4}$ as a "significant part." A transmission line has a characteristic impedance Z_0. If the line impedance Z_L changes as a signal propagates down the line, part of the power travels backwards as a reflected signal. The reflection coefficient is given by $\Gamma = (Z_L - Z_0)/(Z_L + Z_0)$. In other words, the strength of the reflected signal increases with the magnitude of the impedance mismatch.

Consider the circuit in Fig. 17.1. It contains a voltage source V_s with internal resistance R_s connected by a pair of transmission lines to a load of resistance Z_L. We assume that it takes T units of time for an electrical signal to propagate one-way across the circuit. At time $t = 0$, the switch is closed and a voltage pulse is sent down the transmission lines toward the load. Figure 17.2 contains a short chronicle of the voltage

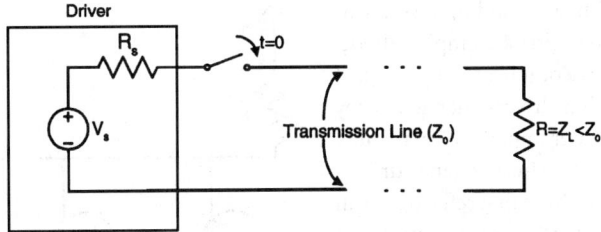

FIGURE 17.1 Circuit for transmission-line example.

waveform seen on the lines. For time $t < T$, the signal sees only Z_0, the impedance of the line. Thus, the voltage of the incident wave $V_{inc} = V_s Z_0 / (R_s + Z_0)$. When the incident wave hits the load at time $t = T$, it sees an additional impedance Z_L which reflects the incident wave with a coefficient of $\Gamma = (Z_L - Z_0)/(Z_L + Z_0)$. In Fig. 17.2, we assume that $Z_L < Z_0$ so that the reflected wave is negative with voltage $V_{re} = \Gamma V_{inc}$. If the transmission lines are not properly terminated at the driver, i.e., $R_s \neq Z_0$, there will be additional reflections before the system settles down. The composite signal at any point along the line is the instantaneous sum of all the incident and reflected signals. Note that all the signals are subject to line losses, especially in their high-frequency components. Thus, pulses eventually lose their shape.

When the circuit is in steady state, the above analysis should agree with the lumped circuit model. For simplicity, let us assume that the transmission lines are properly terminated at the driver, i.e., $R_s = Z_0$. In this case, there is only one reflection and the circuit settles down at time $t = 2T$. Thus, we would expect the steady-state voltage to be the sum of V_{inc} and V_{re}. This works out to $V_s Z_L / (Z_L + Z_0)$, a result which agrees with that predicted by the lumped-circuit model. An intuitive way to reason about transmission lines is to think in terms of feedback. In the beginning, the signal has no clear idea of how much impedance it will en-

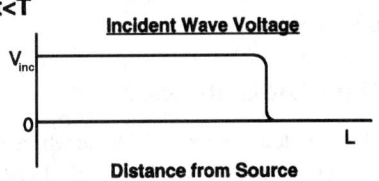

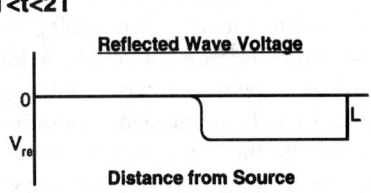

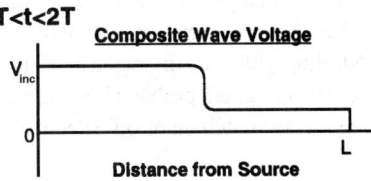

FIGURE 17.2 Transmission-line voltage waveform.

counter in the circuit. Thus, it makes a guess based on the impedance it has already seen. As the signal propagates, it learns more about the circuit, and this information is fed back in the form of reflections, so that eventually Kirchhoff's circuit laws are satisfied.

Signal Reflections

The consequence of having signal reflections in the system is that glitches and extra pulses may appear on the bus. This may cause some unexpected and obscure problems. Some of the more common symptoms of reflection problems include the following:

- A board that stops working after another board is plugged into the system.
- A board that works only in a particular slot on the bus.
- A system that works only when the boards are arranged in a specific order.

Reflections are typically most significant at the various sources and loads on the bus. We can reduce the magnitude of these reflections by matching the impedance of the sources and loads to that of the lines.

This can be accomplished by adding a series or parallel resistance or by using a clamping diode. Impedance matching is complicated by the fact that the properties of bus drivers change as they switch on or off. For better impedance matching, small voltage swings and low-capacitance drivers and receivers are helpful. Note that reflections can also occur at other impedance discontinuities such as interboard connections, board layer changes, etc. To accurately model all these effects, computer simulations using tools such as SPICE are often needed.

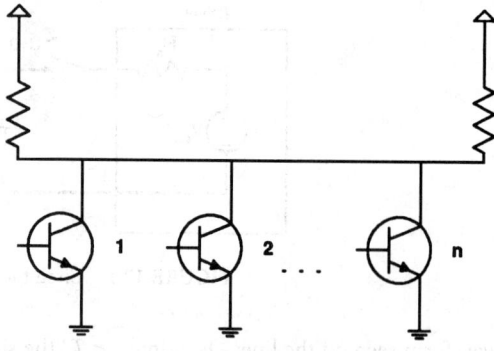

FIGURE 17.3   Wire-OR circuit.

Wire-OR Glitches

Wire-OR logic is a kind of logic where the outputs of several open-collector gates are connected together in such a way as to realize a logical *OR* function. A sample circuit is shown in Fig. 17.3. Notice that the voltage on the line is low as long as any one of the transistors is turned on. Thus this circuit implements the logical *NOR* function or the *OR* function with the output asserted low. Wire-OR is very useful in bus arbitration. For instance, it enables the system to determine whether any bus master wishes to use the bus. Most buses use wire-OR logic for at least a few lines.

However, wire-OR lines are subject to a fundamental phenomenon known as *wire-OR glitch*. During an active to high-impedance transition, a glitch of up to one round-trip delay in width may appear on a wire-OR line. This phenomenon is a result of the finite propagation speed of electrical signals on a transmission line. Consider the case where only transistors 1 and *n* are initially turned on. Suppose that transistor *n* is now turned off. The current that it was previously sinking continues to flow, thus creating a signal which propagates along the line.

A more detailed explanation of the wire-OR glitch is given in [Gustavson and Theus 1983]. Various ways of dealing with it are discussed in [Gustavson and Theus 1983; Taub 1983a, 1983b]. In the IEEE Futurebus, the wire-OR glitch problem is mitigated by the fact that the bus specification imposes constraints on when devices can switch on or off, effectively setting a limit on the maximum glitch duration [Taub 1984].

Signal Skew

Another important electrical phenomenon in buses is *signal skew*. Because of differences in transmission lines, loading, etc., slight differences in the propagation delay of different bits in a word are inevitable. These differences are known as signal skew.

In a transmission, the receiver must be able to somehow determine when all the bits of a word have arrived and can be sampled. The effect of skew is to reduce the window during which the receiver can assume that the data are valid. This effectively limits the data rate of the bus. A wide bus consists of more parallel lines and is thus subject to more skew. In general, skew can be reduced by paying meticulous attention to the impedance of the bus lines. Synchronous buses have to deal with the additional problem of clock to data skew. An approach that has been taken to minimize this skew is to loop back the clock line at the end of the bus. When doing a data transfer, the clock signal that propagates in the same direction as the data transfer is used. Rambus uses this technique to minimize skew with respect to its aggressive 250-MHz clock [Rambus 1992].

Cross-Coupling Effects

A signal-carrying line sets up electrostatic and magnetic fields around it. In a bus, the lines run parallel and close to one another. Thus the fields from nearby lines intersect, causing a signal on one line to affect the signal on another. This is called crosstalk or coupling noise.

A simple way to reduce this effect is to spatially separate the bus lines so that the fields do not interfere with one another. However there is clearly a limit to how far we can carry this. Another way to reduce both the mutual capacitance and inductance of the lines is to introduce ground planes or wires near the bus lines. But this has undesirable side effects such as increasing the self-capacitance of the lines. An approach commonly taken to reduce coupling effects is to separate the lines with an insulator that has a low dielectric constant. Typically, combinations of these techniques are used in a bus design.

17.3 Bus Arbitration

Buses are ubiquitous in computer systems because they are a cost-effective and versatile means of connecting several devices together. The cost-effectiveness and versatility of buses stems from the fact that a bus has only one communication medium, which is shared by all the devices. In other words, at most one communication can occur on a bus at any one time. This implies that there must be some mechanism to decide which bus master has control of the bus at a given time. The process of arbitrating between requests for bus control is called **bus arbitration**. Bus arbitration can be handled in different ways depending on the performance requirements and cost constraints.

Centralized Arbitration

In *centralized arbitration*, there is a special device, the central arbiter, which is in charge of granting bus control to one of the bus masters. The fact that there is only one arbiter means that the centralized scheme has a single point of failure. In general, centralized arbitration can be further divided into two schemes, depending on how the bus masters are connected to the arbiter.

In the first scheme, the central arbiter is connected to each of the bus masters through private two-way connections. Because the bus requests can be made independently and in parallel by the bus masters, this is sometimes known as *centralized independent requests arbitration* or *centralized parallel arbitration*. Notice that the connections in this system form star networks emanating from the central arbiter. One such network carries the bus request signals from the bus masters to the arbiter. Another carries the bus grant signal from the arbiter back to one of the bus masters. Various arbitration policies can be implemented in the arbiter, making this a very flexible scheme. This scheme is fast because there are direct connections between any bus master and the arbiter. However, all these direct connections lead to a high implementation cost. Furthermore, the use of direct connections means that the arbitration signals do not appear on the bus. This makes bus monitoring for debugging and diagnostic purposes difficult.

The second centralized arbitration scheme is known as *centralized serial priority arbitration*. In this scheme, there is a single bus grant signal line, which is routed through each of the bus masters as shown in Fig. 17.4. This form of connection is known as a daisy chain. Hence, centralized serial priority arbitration is more commonly referred to as *daisy-chain arbitration*. In this scheme, there is a common wire-OR bus request line. A bus master may take control of the bus only if it has made a request and its incoming grant line is asserted. A bus master that does not wish to use the bus is required to forward the bus grant signal along the daisy chain. Notice that this implies an implicit priority assignment—the nearer a bus master is to the arbiter, the higher is its priority. The main advantage of daisy-chain arbitration is that

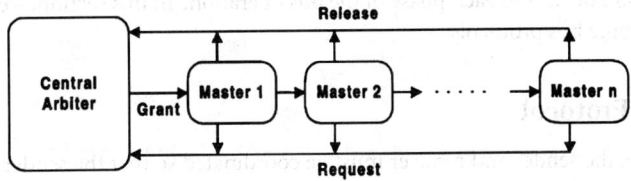

FIGURE 17.4 Daisy-chain bus arbitration.

it requires very few interconnections and the interface logic is simple. However, bus allocation in this scheme is slow because the grant signal has to travel along the daisy chain. Furthermore, the implicit priority scheduling may cause low-priority requests to be locked out indefinitely. Finally, as in centralized parallel arbitration, daisy-chain arbitration does not facilitate debugging and diagnosis. The VMEbus uses a variation of this scheme with four daisy-chained grant lines, which enable it to implement a variety of scheduling algorithms [Giacomo 1990].

Decentralized Arbitration

In *decentralized arbitration*, each bus master has its own arbitration and allocation logic. The responsibility of deciding who has control of the bus is distributed among the bus masters. Thus, this scheme is also known as *distributed arbitration*.

Typically, the bus contains n arbitration wire-OR lines and each bus master is assigned a unique n-bit arbitration number according to some priority scheme. During arbitration, if a master wishes to use the bus, it drives the arbitration lines with its arbitration number. If a master detects that the arbitration lines are carrying a higher priority number, it stops driving the less significant bits of its number. When the system settles down, the arbitration lines will indicate the winner, which is the participating master with the highest priority. The time for the system to settle down can be specified as a fixed time in the bus protocol. An alternative is to use an additional wire-OR line to indicate whether competing bus masters have completed arbitration. As with any priority scheme, the possibility of access starvation of the low-priority masters has to be considered. In the IEEE Futurebus, bus masters are divided into priority and fairness modules depending on whether they have any particularly urgent needs such as having to meet real-time constraints [Taub 1984]. A priority module can issue bus requests whenever it needs to. A fairness module, after winning an arbitration, will have to wait for all pending requests to be serviced before it can issue another request [Taub 1984]. Note that this does not guarantee but only helps to ensure that every module will be able to get a portion of the bus bandwidth.

The major disadvantage of the distributed scheme is that it requires relatively complex arbitration logic in each bus master and several arbitration lines on the bus. However, this scheme allows very fast bus allocation and a flexible assignment of priorities to the bus masters. Distributed arbitration is also more fault-tolerant than the centralized schemes in that the failure of a single bus master does not necessarily affect the operation of the bus. Variations of this scheme are widely used in buses such as the IEEE Futurebus, Nubus, Multibus II, Fastbus [Borrill 1985], and the Powerpath-2 System Bus of the SGI Challenge [Galles and Williams 1994].

The scheme described above is sometimes known as *distributed arbitration by self-selection* because each master decides whether it has won the race, effectively letting the winner select itself. Some computer networks, such as Ethernet, use another form of distributed arbitration which relies on collision detection [Metcalfe and Boggs 1976].

17.4 Bus Protocol

The bus is a shared resource. In order for it to function properly, all the devices on it must cooperate and adhere to a protocol or set of rules. This protocol defines precisely the bus signals that have to be asserted by the master and slave devices in each phase of the bus operation. In this section, we discuss some of the key options in designing bus protocols.

Asynchronous Protocol

In a communication, the sender and receiver must be coordinated so that the sender knows when to talk and the receiver knows when to listen. There are two basic ways to achieve proper coordination. This subsection discusses the asynchronous protocol. The next describes the synchronous design.

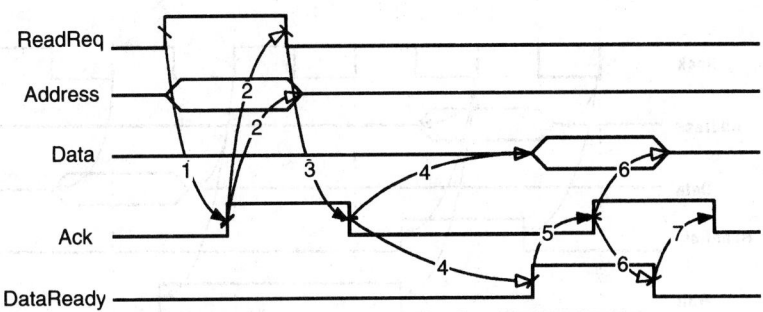

FIGURE 17.5 A basic handshaking protocol.

An asynchronous system does not have an explicit clock signal to coordinate the sender and receiver. Instead a *handshaking protocol* is used. In a handshaking protocol, the sender and receiver can proceed to the next step in the bus operation only if both of them are ready. In other words, both parties have to shake hands and agree to proceed before they can do so.

Figure 17.5 shows a basic handshaking protocol. Assume that *ReadReq* is used to request a read from memory, *DataReady* is used to indicate that data are ready on the data lines, and *Ack* is used to acknowledge the *ReadReq* or *DataReady* signals of the other party. Suppose a device wishes to initiate a memory read. When memory sees the *ReadReq*, it reads the address on the address bus and raises the *Ack* in acknowledgment of the request (step 1). When the device sees the *Ack*, it deasserts *ReadReq* and stops driving the address bus (step 2). Once memory sees that *ReadReq* has been deasserted, it drops the *Ack* line to acknowledge that it has seen the *ReadReq* signal (step 3). A similar exchange is carried out when memory is ready with the data that has been requested (steps 5–7). Notice that the two parties involved in the protocol take turns to respond to one another. They proceed in lockstep. This is how the handshaking protocol is able to coordinate the two devices.

The handshaking protocol is relatively insensitive to noise because of its self-timing nature. This self-timing nature also allows data to be transferred between devices of any speed, giving asynchronous designs the flexibility to handle new and faster devices as they appear. Thus, asynchronous designs are better able to scale with technology improvements. Furthermore, because clock skew is not an issue, asynchronous buses can be physically longer than their synchronous counterparts. The disadvantage of asynchronous designs lies in the fact that the handshaking protocol adds significant overhead to each data transfer and is thus slower than a synchronous protocol. As in any communication between parties with different clocks, there is an additional problem of synchronization failure when an asynchronous signal is sampled. Asynchronous designs are typically used when there is a need to accommodate many devices with a wide performance range and when the ability to incrementally upgrade the system is important. Thus, many of the asynchronous buses such as VMEbus, Futurebus, MCA, and IPI are backplane or I/O buses [Giacomo 1990].

Synchronous Protocol

In synchronous buses, the coordination of devices on the bus is achieved by distributing an explicit clock signal throughout the system. This clock signal is used as a reference to determine when the various bus signals can be assumed to be valid. Figure 17.6 shows a basic synchronous protocol coordinating a memory read transaction. Notice that all the signal changes happen with respect to the clock. In this particular example, the system is negative-edge triggered, which means that the signals are sampled on the falling edge of the clock.

An important design decision in synchronous buses is the choice of the clock frequency. Once a frequency is selected, it becomes locked into the protocol. The clock frequency must be chosen to allow sufficient time for the signals to propagate and settle throughout the system. Allowances must also be

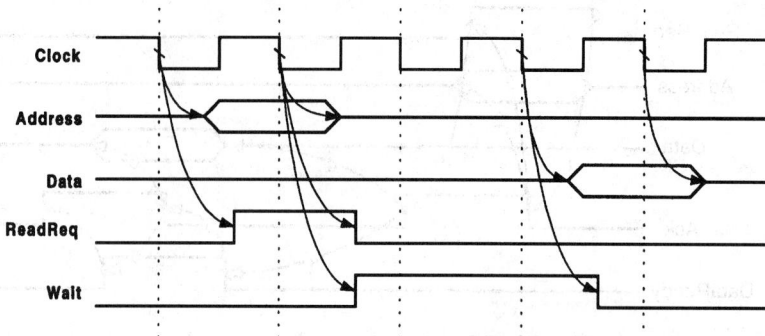

FIGURE 17.6 A basic synchronous protocol.

made for clock skew. Thus, the clock frequency is limited by the length of the bus and the speed of the interface logic. All things being equal, shorter buses can be designed to run at higher speeds.

The main advantage of the synchronous protocol is that it is fast. It also requires relatively few bus lines and simple interface logic, making it easy to implement and test. However, the synchronous protocol is less flexible than the asynchronous protocol in that it requires all the devices to support the same clock rate. Furthermore, this clock rate is fixed and cannot be raised compatibly to take advantage of technological advances. In addition, the length of synchronous buses is limited by the difficulty of distributing the clock signal to all the devices at the same time. Synchronous buses are typically used where there is a need to connect a small number of very tightly coupled devices and where speed is of paramount importance. Thus, synchronous buses are often used to connect the processor and the memory subsystem.

Split-Transaction Protocol

In order to increase the effective bandwidth of a bus, a **split-transaction protocol** can be used. The basic observation behind this protocol is that the bus is not being used to transmit information throughout the entire duration of a transaction, but only at the start and toward the end of the transaction. The idea is thus to split a bus transaction into a request transaction and a reply transaction so as to allow the bus to be released for other uses in between the request and reply stages. Figure 17.7 illustrates the idea. Clearly, this protocol only makes sense when the system has more than one bus master and the memory system is sophisticated enough to handle multiple overlapping transactions. This protocol is sometimes also known as a *connect/disconnect protocol*, *pipelined protocol*, or *packet-switched protocol*.

Although the split-transaction protocol allows more efficient utilization of bus bandwidth than a protocol that holds on to the bus for the whole transaction, it usually has a higher latency because the bus has to be acquired twice—once for the request and once for the reply. Furthermore, the split-transaction protocol is expensive to implement because it requires that the bus transactions be tagged and tracked by each device. Split-transaction protocols are widely used in the system buses of shared-memory SMPs, because bus bandwidth is a big issue in these machines.

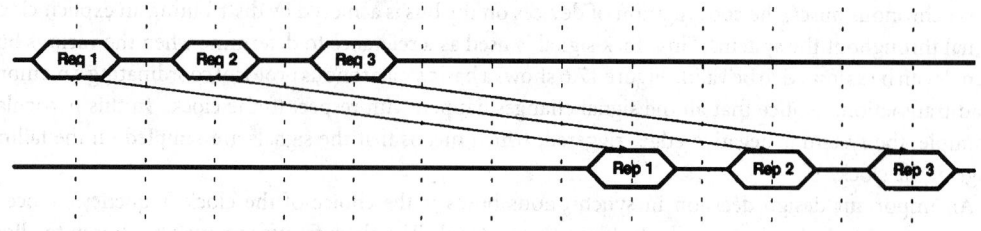

FIGURE 17.7 A split-transaction protocol.

Notice from Fig. 17.7 that even with split transactions, the bus bandwidth is not totally utilized. This is because some bus cycles are needed to acquire the bus and to set up the transfer. Furthermore, some buses require a cycle of turnaround time between different masters driving the bus. Another way to increase effective bus bandwidth is thus to amortize this fixed cost over several words by allowing the bus to transfer multiple contiguous words back to back in one transaction. This is known as a *burst protocol* or *block transfer protocol* because a contiguous block of several words is transferred in each transaction.

17.5 Issues in SMP System Buses

Microprocessor-based **symmetric multiprocessors** (SMPs) with shared snooping buses have become an industry standard for building mid-range departmental servers [Peir et al. 1993, Galles and Williams 1994]. In this section, we devote special attention to the bus issues that arise in SMP designs.

A typical SMP bus architecture (usually referred to as a *SMP system bus*, or simply *system bus*) is illustrated in Fig. 17.8. SMP system buses require high bandwidth and low latency to connect multiple processors, memory modules, and I/O bridges. A system bus is composed of independent signal lines for transmitting control, command/address, and data information. These lines can be grouped into what is commonly called the *control bus*, the *command/address bus*, and the *data bus*, respectively. Each bus can be acquired and used independently.

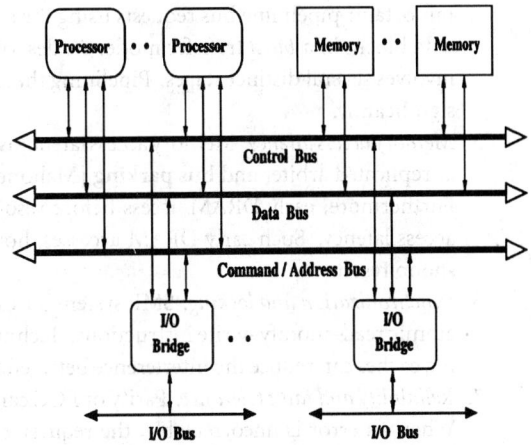

FIGURE 17.8 A typical multiprocessor system bus.

Some of the control lines, such as the bus request signal to the arbiter, are point-to-point connections; therefore, they can be used before the bus is acquired. Also, the wire-OR control lines need not be acquired before they are used. Note also that the signal lines for sending the command and the address are considered as the same bus because they are always acquired together.

In general, each system bus request traverses through a number of stages, each of which may take one or more bus cycles.

- *Bus arbitration:* The requesting processor needs to go through an arbitration process in order to gain access to the shared system bus.
- *Command issuing:* After winning the bus arbitration, the processor issues the command along with an address on the command/address bus. Certain requests, e.g., a cache line writeback, also require access to the data bus in this stage.
- *Cache snooping:* Once a valid command is received, all the system bus masters (processors, I/O bridges) search their own cache directory and initiate proper cache coherence activities to maintain data coherence among multiple caches. This is a unique requirement for SMP system buses. A description of **cache coherence protocols** is given in Cache Coherence Protocols subsection of this section.
- *Acknowledgment:* The snoop results are driven onto the bus. The issuing processor has to update its cache directory based on the results.
- *Data transfer:* When a bus request incurs a data transfer, such as a line-fill request issued upon a cache miss, the transfer of data is carried out at this stage through the data bus.
- *Completion:* The bus transaction is completed.

Several important and unique issues in SMP system bus design are outlined as follows.

1. *Bus physics:* Due to the high bandwidth requirement, SMP system buses face the challenge of a high speed and wide data bus, and have to overcome heavy signal loading to accommodate a reasonable number of ports for connecting multiple system devices.

2. *Cache coherence:* Cache memory is a critical component in SMP systems. In order to maintain data coherence among multiple caches, a cache coherence protocol must be incorporated into the bus protocol.

3. *Bus arbitration:* Each of the processors should be given an equal opportunity to gain access to the bus. Logic to prevent, detect, and resolve starvation and deadlock is typically needed.

4. *Bus bandwidth:* Besides faster and wider data buses, two other bandwidth-increasing features are important: pipelining bus requests using the **split-transaction bus** design, and transferring large data blocks in a *burst* transfer mode. As described before, the processing of a system bus request involves several distinct stages. Pipelining these requests can increase the effective bus bandwidth significantly.

5. *Memory access latency:* Memory access latency is a classic performance bottleneck. Techniques such as replicated arbiter and **bus parking** [Mahoney 1990] can reduce the latency in bus arbitration. Furthermore, early DRAM access before resolving cache coherence issues can reduce memory access latency. Such early DRAM accesses, however, may have to be canceled depending on the snoop results.

6. *Synchronization and locking:* SMP system buses need to provide an efficient way of implementing atomic read–modify–write instructions. Techniques such as address-based locking and nonblocking caches can reduce the interference between locking requests and other normal bus operations.

7. *Reliability and fault tolerance:* Parity or ECC can be used to detect and correct transmission errors. When an error is uncorrectable, the requester will receive a negative acknowledgment and will retry the request. The timeout scheme is commonly included to detect the loss of a command.

In the following, these important issues will be discussed in detail. Because the techniques for handling bus physics issues for the SMP system bus are basically the same as those in other buses, we will omit further discussion.

Cache Coherence Protocols

Cache memory is a critical component in SMP systems. It helps the processors to execute near their full speed by substantially reducing the average memory access time. This is achieved through fast cache hits for the majority of memory accesses and through reduced memory contention in the entire system. However, in designing a shared-memory SMP system where each processor is equipped with a cache memory, it is necessary to maintain coherence among the caches such that any memory access is guaranteed to return the latest version of the data in the system [Censier and Feautrier 1978]. Cache coherence can be enforced through a shared **snooping bus** [Goodman 1983, Sweazey and Smith 1986]. The basic idea is to rely on the broadcast nature of the bus to keep all the cache controllers informed of each other's activities so that they can perform the necessary operations to maintain coherency.

A number of snooping cache coherence protocols have been proposed [Archibald and Baer 1986, Sweazey and Smith 1986]. They can be broadly classified into the write-invalidate scheme and the write-broadcast scheme. In both schemes, read requests are carried out locally if a valid copy exists in the local cache. For write requests, these two schemes work differently. When a processor updates a cache line, all other copies of the same cache line must be invalidated according to the write-invalidate scheme to prevent other processors from accessing the stale data. Under the write-broadcast scheme, the new data of a write request will be broadcast to all the other caches to enable them to update any old copies. These two cache coherence schemes normally operate in conjunction with the *writeback* policy, because the *writethrough* policy generates memory traffic on every write request and is thus not suitable for a bus-based multiprocessor system.

TABLE 17.1   Four States in MESI Coherence Protocol

State	Description
Modified (M)	The M-state indicates that the corresponding line is valid and is exclusive to the local cache. It also indicates that the line has been modified by the local processor. Therefore, the local cache has the latest copy of the line.
Exclusive (E)	The E-state indicates that the corresponding line is valid and is exclusive to the local cache. No modification has been made to the line. A write to an E-state line can be performed locally without producing any snooping bus traffic.
Shared (S)	The S-state indicates that the corresponding line is valid but may also exist in other caches in a multiprocessor system. Writing to an S-state line updates the local cache and generates a request to invalidate other shared copies.
Invalid (I)	The I-state indicates that the corresponding line is not available in the local cache. A cache miss occurs in accessing an I-state line. Typically, a line-fill request is issued to the memory to bring in the valid copy of the requested cache line.

The MESI (modified-exclusive-shared-invalid) write-invalidate cache coherence protocol with the writeback policy is considered in the following discussion. With minor variations, this protocol has been implemented in several commercial systems [Intel 1994, Greenley et al. 1995, Levitan et al. 1995]. In this protocol, each cache line has an associated MESI state recorded in the cache *directory* (also called cache tag array). The definitions of the four states are given in Table 17.1. When a memory request from either the processor or the snooping bus arrives at a cache controller, the cache directory is searched to determine cache hit/miss and the coherence action to be taken. The state transition diagram of the MESI protocol is illustrated in Fig. 17.9, in which solid arrows represent state transitions due to requests issued by the local processor, and dashed arrows indicate state transitions due to requests from the snooping bus.

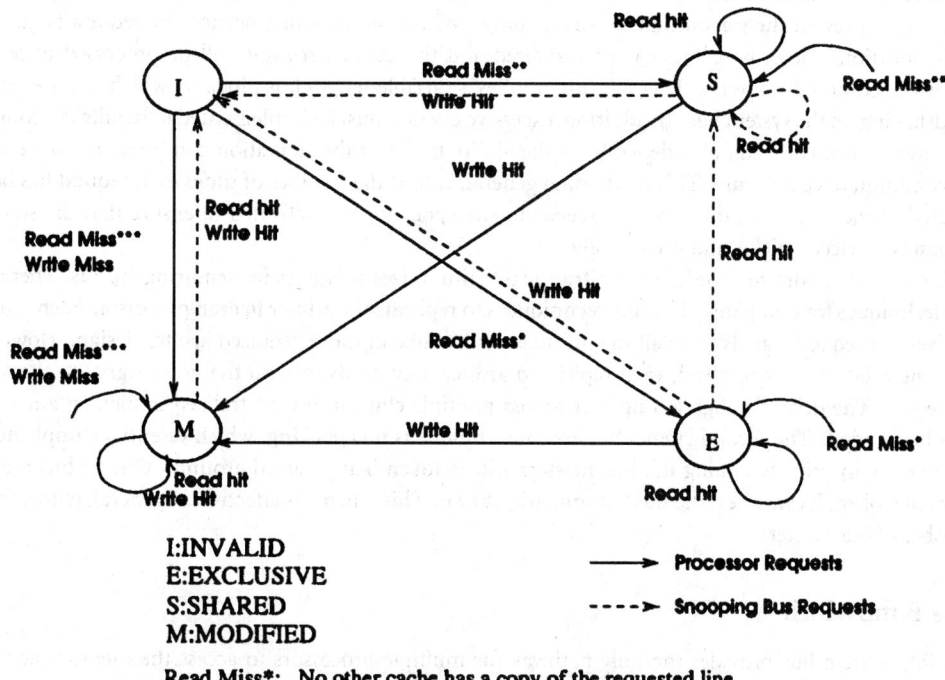

FIGURE 17.9   State transition diagram of MESI coherence protocol.

In general, when a read request from the processor hits a line in the local cache, the state of the cache line will not be altered. A write hit, on the other hand, will change the line state to M. In the meantime, if the original line state is S upon the write hit, an invalidation request will be issued to the snooping bus to invalidate the same line if it is present in any of the other caches. When a read miss occurs, the requested line will be brought into the cache in different states depending on whether the line is present in other caches and on its state in these caches. When a write miss occurs, the target line is fetched into the local cache; the new state becomes M and any copy of the cache line in any other cache is invalidated. For the requests from the snooping bus, a write hit always causes invalidation. A read hit to an M-state line will result in a writeback of the modified line and a transfer of ownership of the target line to the requesting processor. A read hit to an E- or S-state line for a snooping read request will cause a state transition to S.

Bus Arbitration

There are two issues in bus arbitration that are especially important for SMP system buses. The first is to ensure that the system is fair, deadlock-free, and starvation-free. In general, the fairness issue can be handled by using a first-come-first-served (FCFS) priority scheme, which will guarantee that requests are serviced in order of arrival. In fact, a simple random priority scheme may be adequate to provide a fair arbitration scheme for all the bus masters. The deadlock problem typically arises in a split-transaction bus where it is possible for a cycle of dependencies to form among the requests and replies. This can be handled by distinguishing between requests and replies and always ensuring that replies are able to make forward progress. At first sight, it might seem that a fair arbitration policy should be starvation-free. However, this is not the case, because when overlapping bus requests are allowed, a request that has been granted by the bus may have to be later rejected due to interference with other requests. On a subsequent retry, the request may again encounter another conflict. In the pathological case, such a request may retry indefinitely. This situation is worse if there are lock requests and other resource conflicts on the system bus.

There are two general solutions to this starvation problem. The first solution eliminates any request rejection to prevent the starvation from happening. Whenever a conflict occurs, the request is queued at the location where the conflict is encountered and the queued request will be processed once the conflict condition disappears. This solution implies a variable-length bus pipeline, which requires extra handshaking on the system bus. In addition, excessive queues must be implemented to handle the conflict situation. The second solution depends on the ability to detect the starvation condition and to resolve the condition once it occurs. This method, in general, counts the number of times each request has been rejected. When a certain threshold is exceeded, emergency logic is activated to ensure that the starved request is serviced quickly and successfully.

The second important issue in bus arbitration is to minimize the latency for acquiring the bus. There are two techniques for doing this. The first technique is to replicate the arbiter in each processor. Each arbiter receives the request signals from all the bus masters just like in the centralized arbiter design. However, after the arbitration is resolved, each replicated arbiter only needs to send the grant signal to the local processor. The delay through a long wire across multiple chips in the centralized implementation can thus be avoided. The second latency-reduction technique is **bus parking**, which essentially implements a round-robin priority among the bus masters with a token being passed around. Once a bus master owns the token, it can access the bus without arbitration. This scheme is effective, in general, with a small number of bus masters.

Bus Bandwidth

An SMP system bus provides the only pathway for multiple processors to access the memory and the I/O devices. Therefore, the bandwidth of the system bus effectively determines the number of processors that can be supported. The simplest system bus design is to allow only one request at a time. A second command cannot be issued before the current bus request completes. This so-called *simple bus* design has very limited bus performance. A **split-transaction bus**, on the other hand, allows the overlapping of multiple bus requests using the pipelining technique. When a bus transaction is split, it only occupies the bus when the bus is really needed. For example, when a request is at the cache snooping stage or at the stage of accessing memory, the bus is available for other bus transactions. Even when the request is at the data transfer stage, the command/address bus is free to accept another request which does not require the data bus in the same cycle. The split-transaction approach utilizes the critical system bus much better. Most of the modern SMP system buses employ split-transaction designs to maximize the bus bandwidth [Peir et al. 1993, Galles and Williams 1994].

Even with the aggressive split-transaction design, the system bus can support a very limited number of today's high-performance microprocessors. For instance, consider a 50-MHz system bus connecting multiple 100 MIPS processors together. Assume that each processor has on-chip instruction/data caches as well as an external second-level cache; together, the instruction-per-miss rate is 50. In this case, each processor will generate bus traffic for handling two million cache misses every second. As a result, 25 processors will produce enough traffic to use up 100% of the bus bandwidth. This calculation does not allow for the fact that about 30% of the cache lines being replaced are typically modified and have to be written back to memory. In addition, this calculation does not take into account bus traffic due to I/O instructions, cache coherence transactions, etc. All these factors will further limit the number of processors that can be effectively supported by the system bus.

In the above discussion, two ideal conditions are assumed. The first assumption is that the bus utilization can reach 100%. In reality, the rule-of-thumb is that heavy queuing will occur when the bus utilization reaches above 60%. The second assumption is that each bus request only occupies a single bus cycle. This is very difficult to achieve for a split-transaction bus running at 50 MHz. Typically, both the arbitration stage and the cache snooping stage may require more than one cycle. Even though these two stages do not occupy the bus directly, they need to access other related critical components such as the arbiter and the cache directory, respectively. Realistically, it takes a minimum of two cycles to initiate a new command on the system bus. This two-cycle-per-request design is already very aggressive, because the subsequent command is issued before the current command is acknowledged. In order to avoid potential interference between the active commands, protection hardware must be implemented. In addition, the data bus must be able to transfer a cache line in two cycles. For instance, a 128-bit data bus is required when the Intel Pentium processor is used in the above example, because the Pentium has 32-byte cache lines. Furthermore, any *idle* cycles during the data transfer must be eliminated. This typically requires a high-performance memory system with multiple modules each with independent memory banks.

Current projections suggest that processor performance will continue to improve at over 50% a year [Hennessy and Patterson 1996]. It is unlikely that improvements in the system bus will be able to keep up with the processor curve. Thus, the number of processors that the snooping bus can support is expected to decrease even further. There has been some work to extend the single bus architecture to multiple interleaved buses [Hopper et al. 1989] or hierarchical buses [Wilson 1987]. There have also been proposals to abandon the snooping bus approach and to use a directory method to enforce cache coherence [Lenoski et al. 1992, Gustavson 1992]. The detailed descriptions of these proposals are beyond the scope of this chapter.

Memory Access Latency

When a requested data item is not present in a processor's caches, the item must be fetched from memory. The delay in obtaining the data from memory may stall the requesting processor even when advanced techniques, such as out-of-order execution, nonblocking cache, relaxed consistency model, etc., are used to overlap processor execution with cache misses [Johnson 1990, Farkas and Jouppi 1992, Gharachorloo et al. 1990]. Therefore, it is important to design the system bus to minimize the number of cycles needed to return data upon a cache miss. Techniques such as replicated arbiter and bus parking to reduce arbitration cycles have been described in the subsection on Bus Arbitration above.

Another way to reduce the memory latency is to trigger DRAM access once the command arrives at the memory controller. This early DRAM access may have to be canceled if the requested line turns out to be present in a modified state in another processor's cache. Such a condition will only be known after the acknowledgment. Early DRAM access is very useful because the chance of hitting a modified line in another cache is not very high. In the unlikely case that the requested line is in fact modified in another cache, a cache-to-cache data and ownership transfer can be implemented to send the requested data directly to the requesting processor. The cache-to-cache transfer normally has higher priority and may bypass other requests in the write buffer of the processor that owns the modified line. In some implementations, the memory is also updated with the modified data during the cache-to-cache transfer.

Synchronization and Locking

The basic hardware primitive for implementing synchronization and locking is an atomic read–modify–write instruction such as Test&Set and Compare&Swap. These instructions must be guaranteed to read the contents of a memory location, test and modify the data, and write the result back to the same memory location in an indivisible sequence of operations. In a bus-based SMP system, the Test&Set instruction can be executed in two separate steps: a read-lock operation followed by a write-unlock operation. The read-lock operation reads the memory word and at the same time sets up certain hardware lock signals or registers so that no other requests for the same memory word will be permitted. After the memory word has been modified, the write-unlock operation returns the new result back to the memory location and releases the hardware lock.

There are two ways of implementing the hardware lock on the system bus. The first is to lock the bus completely during the period from the read-lock operation to the write-unlock operation; no other request is allowed in between. This approach has poor performance but may be suitable for a simple bus design which allows only one request at a time anyway. The second approach is to implement **address locking** on the system bus. When a read-lock operation is issued, an invalidation request is sent across the system bus to knock out any copy of the cache line from the other caches. In the meantime, the address of the cache line is recorded in a *lock register* to prevent any snooping on the same cache line until the lock is released by the write-unlock operation. Depending on the data alignment and the size of the synchronization variable, a read-lock operation may have to lock two cache lines at a time.

The address locking method minimizes the interference of a Test&Set instruction with other system bus requests, because only those requests that access the same cache line as the Test&Set will be rejected. Multiple lock requests, each from a different processor, are permitted as long as they target different cache lines. However, the lock request is still relatively expensive because the read-lock operation has to be broadcast to all the other processors and the issuing processor cannot proceed until confirmation is received from each processor.

17.6 Putting It All Together—CCL-XMP System Bus

CCL-XMP [Peir et al. 1993], an Intel Pentium-based SMP system, was designed and developed at the Computer and Communication Laboratories (CCL) of the Industrial Technology Research Institute (ITRI) of Taiwan under a collaborative effort with Acer, ICL, and Intel. The initial target system consists of eight 66-MHz Pentium processors, each with a 256-Kbyte second-level cache. The system bus, operating at 33 MHz, has independent control, command/address, and data buses. The MESI protocol is incorporated to enforce coherence among multiple caches. The system bus can arbitrate and accept one request every two cycles. A dual 64-bit data bus sustains a data transfer rate of over 400 Mbyte/s. Based on odd–even line interleaving, each data bus is connected to one memory module. In order to maximize the bus bandwidth, each memory module is divided into four independent banks. Furthermore, each bank is designed as a two-way interleaved DRAM array to provide zero-wait-state operation in the burst-mode data transfer at 33 MHz. CCL-XMP requires a powerful I/O subsystem. Both VESA local bus and PCI bus can be connected to the system bus through high-performance I/O bridges. The I/O bridges act as both a bus master and slave to transfer data between the system bus and the I/O buses.

The CCL-XMP system bus architecture is very similar to the general SMP bus architecture shown in Fig. 17.8. Each request traverses through the same number of stages as described in section 17.5. In this section, some of the relevant features of the CCL-XMP system bus design are described.

The CCL-XMP system bus is a synchronous bus. All the signals are driven at the rising edge of the clock and latched and sampled at the next rising edge. TTL voltage levels are used, and most signals are tristate with the exception of a few wire-OR control signals. The first version of the motherboard measures 32 cm × 32 cm. The system bus is 15 cm in length and supports eight bus slots. A single 64-bit data bus is included in the first prototype to support up to six Pentium processors. The clock tree is carefully laid out on the motherboard to limit the clock skew to within 2 ns. The total delay between the sender and the receiver through the system bus is less than 28 ns. This includes the delay of latches (flip-flops), the output TTL pad (with 120-pF bus loading), the input pad, boundary scan, clock skews, and other combinational logic. This is tolerable under the 33-MHz clock rate. In designing CCL-XMP, it is more challenging to manage the delay between the system bus interface chip and the second-level cache controller, which is operating at 66 MHz.

Figure 17.10 shows the detailed operation of a line-fill request on the CCL-XMP system bus. Arbitration of the command/address bus takes two cycles. Once the bus is granted, the memory read command and address are issued to the bus in the third cycle. It takes two cycles for all the bus masters (including I/O bridges) to search their cache directories and to update the respective line states upon a hit. At the same time, the memory controller initiates the DRAM access on the arrival of the read command. In the sixth cycle, an acknowledgment is sent back to the requester from each device based on the snooping result. The requester then updates its cache directory according to the acknowledgment and the cache coherence protocol. The memory controller begins arbitrating for the data bus two cycles before the data are fetched out of the DRAM array. A read request is given priority for using the data bus over a write request. After the data bus has been granted, the target 64-bit data is driven onto the bus in the ninth cycle. This is followed by three consecutive cycles to transfer the entire cache line to the requester. Finally, the command completes in the thirteenth cycle.

The CCL-XMP system bus uses replicated arbiters to achieve a two-cycle arbitration. Each arbiter receives *n* request signals, one from each bus master; but the grant signal is only sent to the local processor to avoid chip-crossing and long wiring delays. Basically, the arbiter latches the requests from all the bus masters at the end of the first cycle. In the second cycle, a resolution of the bus priority is carried out, and the winner is notified by the local arbiter. In addition, the CCL-XMP system bus uses the concept of *arbitration group* to achieve fairness without implementing FIFO queues. An arbitration group consists of the bus masters that are simultaneously requesting access to the system bus. The arbiter will serve all the members in the group according to some priority scheme. Any other master outside the current group is not allowed to join the group until all the requesters in the group have been granted the bus. After that, a new arbitration group can be formed.

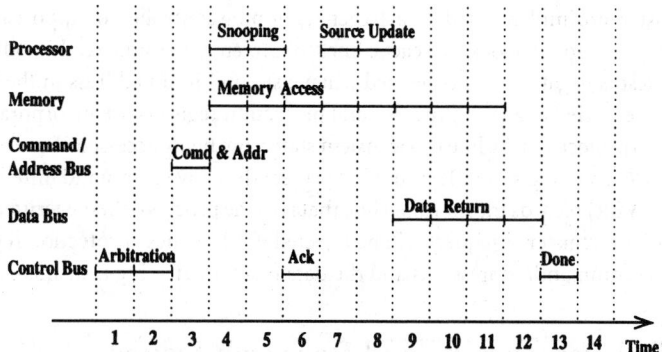

FIGURE 17.10 The pipeline cycles of a line-fill request on the CCL-XMP system bus.

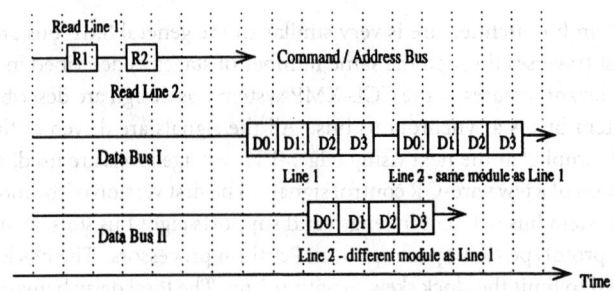

FIGURE 17.11   Split transactions on the CCL-XMP system bus.

Emergency logic is included in the arbiter design to resolve the starvation conditions. Each bus master has an associated *urgent counter*. The counter is incremented when a command receives a negative acknowledgment on the bus. A command can be rejected for several reasons, including resource busy, address locking, or some erroneous conditions. When the counter exceeds a certain threshold, the processor will raise its urgent request control line to the arbiter. The arbiter will then give the highest priority to the urgent requester and activate the emergency logic for resolving any blocking condition for the urgent request until the command is executed successfully and the urgent line is dropped.

The CCL-XMP system bus is a split-transaction bus; it allows the pipelining of multiple requests to sustain maximum bus bandwidth. Figure 17.11 illustrates two consecutive cache line-fill instructions on the command/address bus and the data transfer cycles for the two cache lines on the data buses. These two requests must have originated from different processors, because the Pentium processor only allows one outstanding memory request at a time. When the two requested lines are located in different memory modules, concurrent data transfers are possible. However, the data transfers have to be serialized when both accesses hit the same module. Note that there is a dead cycle on the data bus between the two cache line transfers on the same data bus. This is required to prevent a potential signal conflict. After the read command is issued, it takes six cycles including ECC check for the memory controller to return the target 64-bit data. Because the command bus can accept one command every two cycles, a dual 64-bit bus is designed to balance the performance. The dual data bus can transfer two 32-bytes of data every five bus cycles (including the dead cycle). This provides a sustained bus bandwidth of over 400 Mbyte/s.

CCL-XMP supports cache-to-cache data transfer with memory reflection when a request hits a modified line located in another cache. The memory copy is updated during the data transfer to the requester. The cache-to-cache transfer request has higher priority and may bypass other requests in the write buffer or the replacement writeback queues. This design reduces the penalty in accessing a modified line located in another processor's cache. The update of the memory copy provides the flexibility to transfer the ownership of the requested line to the requester and to switch the requested line to a shared state when there are subsequent read requests for the same line from other processors.

A pair of lock registers are implemented in each cache snooping controller to support address locking. At any given time, each processor can lock two cache lines to implement atomic read–modify–write memory instructions. A request is negative-acknowledged when a snoop hits an address in the lock register. The rejected request will be retried at a later time. Several protection registers are incorporated into the cache snooping controller to protect a cache line in a transient state from being accessed. For example, in addition to the lock registers, a current register (CR) is used to protect the active command until it completes. Also, a writeback register (WBR) records the modified line that is in the process of being written back to memory. As in the case of the lock register, a request will be rejected if it hits these protection registers. Parity bits are added to both the command/address bus and the data bus for detecting transmission errors.

17.7 Historical Perspective and Research Issues

Computer systems based on a shared backplane bus are very popular because they are both cost-effective and highly flexible. The early to mid 1980s saw the establishment of a number of standard buses such as the VME bus [Pri-Tal 1986], Fastbus [Gustavson 1986], Nubus [Taylor 1989], Multibus II [Mahoney 1990], and Futurebus [Borrill 1986]. All these bus architectures have been standardized by the IEEE to allow different vendors to design various boards that can be attached to each bus. Although the option of connecting multiple processors was considered in all these bus designs, only the Futurebus fully supports cache coherence protocols for both writethrough and writeback caches. Other important aspects, such as synchronous/asynchronous communication protocols, bus arbitration, bus bandwidth, locking mechanisms, TTL/BTL/ECL bus interfaces, connectors, pins, etc., are also significantly different among these open bus architectures. A comprehensive comparison can be found in [Borrill 1985].

The continuing search for higher-performance system interconnection has led to three more recent IEEE standards: the IEEE 896.x Futurebus+ [Aichinger 1992], the P1394 High-Performance Serial Bus [Teener 1992], and the P1596 Scalable Coherence Interface (SCI) [Gustavson 1992]. Futurebus+ is an expanded version of the original Futurebus (IEEE 896.1-1987). The work was initiated by the navy in 1988 and later gained widespread support from the working groups of the VMEbus and Multibus as well as many of the major computer companies. The Futurebus+ was designed to be a truly open standard. The standard defines in detail the architectural, electrical, and mechanical specifications. The data path of Futurebus+ varies from 32 to 256 bits wide to provide better performance scalability. The centralized arbitration mechanism is used for its simplicity and high performance. **Live insertion** is adopted to answer the needs of the market for high-availability and fault-tolerant computers. In addition, Futurebus+ can be reconfigured as an I/O bus. Using the standard "Profile B" bridge, a Futurebus+ can be connected to other open-system buses.

Parallel backplane buses need large physical connectors. Such connectors are usually costly and are the primary source of failure. In addition, a point-to-point, unidirectional link can provide much faster transmission speed. The serial buses were introduced based on these advantages. The P1394 serial bus was first presented at IEEE CompCon 1992. It was designed for low cost, yet it provides the data transmission speed and low latency needed for a peripheral bus or as a possible backup of the backplane bus. The transmission protocol has three layers: transaction, link, and physical. Read, write, and lock are the three transactions supported, and each transaction can be divided into four stages: request, indication, response, and confirmation. The link layer provides a half-duplex data-packet delivery service. Each packet delivery needs to go through three stages: an arbitration stage to gain the access of the physical bus, a packet transmission stage to deliver the packet to the physical layer, and an acknowledgment stage to confirm the transmission with the receiver. The last stage is only required for synchronous package delivery. The split-transaction protocol and a Test&Set instruction are implemented in the P1394 serial bus to enable higher transaction rates and to properly handle locking activities.

As microprocessor performance increased at the rate of over 50% per year [Hennessy and Patterson 1996], it soon became apparent that the use of shared buses as the fundamental interconnection in multiprocessor systems creates a performance bottleneck. In July 1988, the P1596 SCI working group was formed with the aim of defining a scalable interconnect architecture for future shared-memory cache-coherent multiprocessor systems. There are a number of fundamental issues that must be solved in order to achieve this goal. First, signal speed has to be made independent of the size of the system. Second, multiple signal paths (links) have to be used so that multiple independent transfers can take place concurrently. Third, multiple cache coherence activities must be allowed to occur in parallel to overcome the bottleneck in the snooping-bus approach. To resolve the signal-speed problem, SCI uses point-to-point, unidirectional buses with low-voltage differential signals. *Ring* is the most common structure to connect multiple processing nodes (processors) through the fast SCI bus. *Distributed directory* is chosen to maintain cache coherence. The main memory and the directory which records the status of the associated lines in cache are distributed among all the nodes. Instead of maintaining full presence bits to indicate the caches where each line is located, a double link list is constructed to link all the copies of each cache line. Certain cache coherence actions, e.g., invalidation of shared copies of a cache line, need to go through the link list in a sequential fashion.

While all these standardization efforts were under way, a different approach was being taken by Rambus Inc. Rambus Inc. was set up in 1990 to develop technology that will enable systems to keep up with the increasing bandwidth requirements of processors [Rambus 1992]. The core of the Rambus technology is a proprietary chip-to-chip bus, the Rambus Channel. The Rambus Channel consists of a small number of very high-speed lines clocked at an aggressive 250 MHz. Data are transferred on both edges of the clock, allowing nine data lines to achieve a peak transfer rate of 500 Mbyte/s. A block-oriented protocol is used to allow effective utilization of this peak bandwidth. The key to achieving the high data rate lies in paying special attention to the physics of the bus. First, the Rambus interface consists of only 32 carefully terminated transmission lines. The small number of lines helps to reduce bus noise and power consumption, which is significant given the high clock rate. Second, clock-to-data skew is minimized by having a clock-to-master signal and a clock-from-master signal. The appropriate clock is chosen depending on the direction of the data transfer. Third, the channel is designed to operate with low-voltage swings, which further reduces power consumption. Finally, the maximum bus length is limited to about 10 cm and vertical surface-mount packages are used to enable devices to be densely packed onto the bus.

Although flexible and cost-effective, the bus is fundamentally not scalable in performance. This becomes increasingly apparent as processors become faster and SMP designs start pushing the performance limits of buses with fewer and fewer processors. Widening the data bus will not help, because of the bottleneck in the command/address bus. Using multiple buses [Hopper et al. 1989] to achieve more than one transfer at a time results in a complex bus interface design to maintain cache coherence with bus snooping mechanisms. The Scalable Coherence Interface uses a distributed-directory approach to build coherent shared-memory multiprocessors with high-performance SCI rings. Although the SCI approach allows concurrent coherence activities, the cost of the large directory and the latency of coherence transaction remain serious problems. The Stanford DASH project [Lenoski et al. 1992] advocates the same distributed directory method using mesh-connected networks. Again, the latency in accessing remote memory may cause severe performance degradation [Kuskin et al. 1994]. The "right" way to build a scalable cache-coherent shared-memory multiprocessor remains an active research area.

An issue intimately related to bus performance is the performance of the memory subsystem. The standard RAS/CAS DRAM interface is designed for low cost and high density. In order to achieve sufficient memory bandwidth for today's demanding processors, an interleaved memory subsystem is often required. For SMPs, the memory interleaving is typically very aggressive. Recently, a couple of improvements to the existing DRAM interface have been announced. These include the extended-data-out (EDO) mode and the pipelined burst mode (PBM) [Kumanoya et al. 1995]. In addition, several new DRAM interfaces promising higher bandwidth and lower latency have been proposed. These include the synchronous DRAM, cached DRAM, Rambus DRAM, and others designed specifically for graphics application. The interested reader is referred to [Przybylski 1993, Kumanoya et al. 1995] for more information on these novel DRAM interfaces.

Defining Terms

Address locking: A mechanism to protect a specific memory address so that it can be accessed exclusively by a single processor.

Bus arbitration: The process of determining which competing bus master should be granted control of the bus.

Bus master: A bus device that is capable of initiating and controlling a communication on the bus.

Bus parking: A priority scheme which allows a bus master to gain control of the bus without arbitration.

Bus protocol: The set of rules which define precisely the bus signals that have to be asserted by the master and slave devices in each phase of a bus operation.

Bus slave: A bus device that can only act as a receiver.

Cache coherence protocol: A mechanism to maintain data coherence among multiple caches so that every data access will always return the latest version of that datum in the system.

Cache line: A block of data associated with a cache tag.

Live insertion: The process of inserting devices into a system or removing them from the system while the system is up and running.

Snooping bus: A multiprocessor bus that is continually monitored by the cache controllers to maintain cache coherence.

Split-transaction bus: A bus that overlaps multiple bus transactions, in contrast to the *simple bus* that services one bus request at a time.

Symmetric multiprocessor (SMP): A multiprocessor system where all the processors, memories, and I/O devices are equally accessible without a master–slave relationship.

References

Aichinger, B. P. 1992. Futurebus+ as an I/O bus: profile B, pp. 300–307. In *Proc. 19th Int. Symp. on Computer Architecture.*

Archibald, J. and Baer, J. L. 1986. Cache-coherence protocols: evaluation using a multiprocessor simulation model. *ACM Trans. Comput. Systems* 4(4):273–298.

Agarwal, A., Simoni, R., Hennessy, J., and Horowitz, M. 1988. An evaluation of directory schemes for cache coherence, pp. 280–289. In *Proc. 15th Int. Symp. on Computer Architecture.*

Borrill, P. 1985. MicroStandards special feature: a comparison of 32-bit buses. *IEEE Micro* 5(6):71–79.

Borrill, P. 1986. Futurebus: the ultimate in advanced system buses, pp. 210–216. In *Proc. Buscon '86 West.*

Censier, L. and Feautrier, P. 1978. A new solution to coherence problems in multicache systems. *IEEE Trans. Comput.* C-27(12):1112–1118.

Chaiken, D., Fields, C., Kurihara, K., and Agarwal, A. 1990. Directory-based cache coherence in large-scale multiprocessors. *IEEE Comput.* 23(6):49–59.

Farkas, K. and Jouppi, N. 1992. Complexity/performance tradeoffs with non-blocking loads, pp. 211–222. In *Proc. 21st Int. Symp. on Computer Architecture.*

Galles, M. and Williams, E. 1994. Performance optimizations, implementation, and verification of the SGI challenge multiprocessor, pp. 134–143. In *Proc. 1994 Hawaii Int. Conf. on System Science, Architecture Track.*

Gharachorloo, K. et al. 1990. Memory consistency and event ordering in scalable shared-memory multiprocessors, pp. 15–26. In *Proc. 17th Int. Symp. on Computer Architecture.*

Giacomo, J. D. 1990. *Digital Bus Handbook.* McGraw–Hill, New York.

Goodman, J. 1983. Using cache memory to reduce processor-memory traffic, pp. 124–131. In *Proc. 10th Int. Symp. on Computer Architecture.*

Greenley, D. et al. 1995. Ultrasparc: the next generation superscalar 64-bit SPARC, pp. 442–451. In *Proc. COMPCON'95.*

Gustavson, D. B. and Theus, J. 1983. Wire-OR logic on transmission lines. *IEEE Micro* 3(3):51–55.

Gustavson, D. B. 1984. Computer buses—a tutorial. *IEEE Micro* (4):7–22.

Gustavson, D. B. 1986. Introduction to the Fastbus. *Microprocessors and Microsystems* 10(2):77–85.

Gustavson, D. B. 1992. The scalable coherent interface and related standards projects. *IEEE Micro* 12(1):10–22.

Hennessy, J. and Patterson, D. 1996. *Computer Architecture, a Quantitative Approach*, 2nd ed. Morgan Kaufmann, San Francisco.

Hopper, A., Jones, A., and Lioupis, D. 1989. Multiple vs wide shared bus multiprocessors, pp. 300–306. In *Proc. 16th Int. Symp. on Computer Architecture.*

IBM Corp. 1982. IBM 3081 Functional Characteristics, GA22-7076. IBM Corp., Poughkeepsie, NY.

Intel Corp. 1994. *Pentium Processor User's Manual*, Vols. 1, 2. Intel Corp. Order nos. 241428, 241429.

Johnson, M. 1990. *Superscalar Microprocessor Design.* Prentice–Hall, Englewood Cliffs, NJ.

Kumanoya, M., Ogawa, T., and Inoue, K. 1995. Advances in DRAM interfaces. *IEEE Micro* 15(6):30–36.

Kuskin, J. et al. The Stanford FLASH multiprocessor, pp. 302–313. In *Proc. 21st Int. Symp. on Computer Architecture.*

Lenoski, D. et al. 1992. The Stanford Dash multiprocessor. *IEEE Comput.* 25(3):63–79.

Levitan, D., Thomas, T., and Tu, P. 1995. The PowerPC 620 microprocessor: a high performance superscalar RISC microprocessor, pp. 285–291. In *Proc. COMPCON '95.*

Mahoney, J. 1990. Overview of Multibus II architecture. *SuperMicro J.* No. 4, pp. 58–67.

Metcalfe, R. M., and Boggs, D. R. 1976. Ethernet: distributed packet switching for local computer networks. *Commun. ACM* 19(7):395–404.

Peir, J. K. et al. 1993. CCL-XMP: a Pentium-based symmetric multiprocessor system, pp. 545–550. In *Proc. 1993 Int. Conf. on Parallel and Distributed Systems.*

Pri-Tal, S. 1986. The VME subsystem bus. *IEEE Micro* 6(2):66–71.

Przybylski, S. 1993. DRAMs for new memory systems. parts 1, 2, 3. *Microprocessor Rep.*, Mar. 8, pp. 18–21.

Rambus. 1992. *Rambus Architectural Overview.* Rambus, Inc., Mountain View, CA.

Stenstorm, P. 1990. A survey of cache coherence schemes for multiprocessors. *IEEE Comput.* 23(6):12–25.

Sweazey, P. and Smith, A. J. 1986. A class of compatible cache consistency protocols and their support by IEEE Futurebus, pp. 414–423. In *Proc. 13th Int. Symp. on Computer Architecture.*

Taub, D. M. 1983. Overcoming the effects of spurious pulses on wired-OR lines in computer bus systems. *Electron. Lett.* 19(9):340–341.

Taub, D. M. 1983. Limitations of looped-line scheme for overcoming wired-OR glitch effects. *Electron. Lett.* 19(15):579–580.

Taub, D. M. 1984. Arbitration and control acquisition in the proposed IEEE 896 FutureBus. *IEEE Micro* 4(4):28–41.

Taylor, B. G. 1989. Developing for the Macintosh NuBus, pp. 143–175. In *Proc. Eurobus/UK Conference.*

Teener, M. 1992. A bus on a diet—the serial bus alternative: an introduction to the P1394 high performance serial bus, pp. 316–321. In *Compcon '92.*

Wilson, A. 1987. Hierarchical cache/bus architecture for shared memory multiprocessors, pp. 244–252. In *Proc. 14th Int. Symp. on Computer Architecture.*

Zalewski, J. 1995. *Advanced Multimicroprocessor Bus Architecture.* IEEE Computer Society Press.

Further Information

Advanced Multimicroprocessor Bus Architectures by J. Zalewski [Zalewski 1995] contains a comprehensive collection of papers covering bus basics, physics, arbitration and protocols, board and interface designs, cache coherence, various standard bus architectures, and their performance evaluations. The bibliography section at the end provides a complete list of references for each of the topics discussed in the book.

Digital Bus Handbook by J. D. Giacomo [Giacomo 1990] is a good source of information for the details of the various standard bus architectures. In addition, this book has several chapters devoted to the electrical and mechanical issues in bus design.

The bimonthly journal *IEEE Micro* and the *Proceedings of International Symposium on Computer Architecture* are good sources of the latest papers in computer architecture area including computer buses.

18

Memory Systems

Douglas Burger
*University of
Wisconsin—Madison*

James R. Goodman
*University of
Wisconsin—Madison*

Gurindar S. Sohi
*University of
Wisconsin—Madison*

18.1 Introduction

The *memory system* serves as the repository of information (data) in a computer system. The processor (also called the central processing unit, or CPU) accesses (reads or loads) data from the memory system, performs computations on them, and stores (writes) them back to memory. The memory system is a collection of storage locations. Each storage location, or *memory word*, has a numerical *address*. A collection of storage locations form an *address space*. Figure 18.1 shows the essentials of how a processor is connected to a memory system via address, data, and control lines.

When a processor attempts to load the contents of a memory location, the request is very urgent. In virtually all computers, the work soon comes to a halt (in other words, the processor *stalls*) if the memory request does not return quickly. Modern computers are generally able to continue briefly by overlapping memory requests, but even the most sophisticated computers will frequently exhaust their ability to process data and stall momentarily in the face of long memory delays. Thus, a key performance parameter in the design of any computer, fast or slow, is the effective speed of its memory.

Ideally, the memory system must be both infinitely large, so that it can contain an arbitrarily large amount of information and infinitely fast, so that it does not limit the processing unit. Practically, however, this is not possible. There are three properties of memory that are inherently in conflict: speed, capacity, and cost. In general, technology tradeoffs can be employed to optimize any two of the three factors at the expense of the third. Thus it is possible to have memories that are (1) large and cheap, but not fast; (2) cheap and fast, but small; or (3) large and fast, but expensive. The last of the three is further limited by physical constraints. A large-capacity memory that is very fast is also physically large, and speed-of-light delays place a limit on the speed of such a memory system.

The **latency** (L) of the memory is the delay from when the processor first requests a word from memory until that word arrives and is available for use by the processor. The latency of a memory system is one attribute of performance. The other is **bandwidth** (BW), which is the rate at which information can be transferred from the memory system. The bandwidth and the latency are closely related. If R is the

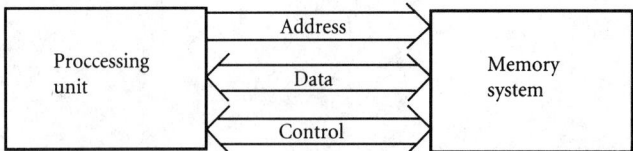

FIGURE 18.1   The memory interface. (*Source:* Dorf, R. C. 1992. *The Electrical Engineering Handbook*, 1st ed., p. 1928. CRC Press, Inc., Boca Raton, FL.)

number of requests that the memory can service simultaneously, then

$$\text{BW} = \frac{R}{L} \tag{18.1}$$

From Eq. (18.1) we see that a decrease in the latency will result in an increase in bandwidth, and vice versa, if R is unchanged. We can also see that the bandwidth can be increased by increasing R, if L does not increase proportionately. For example, we can build a memory system that takes 20 ns to service the access of a single 32-bit word. Its latency is 20 ns per 32-bit word, and its bandwidth is

$$\frac{32}{20 \times 10^{-9}} \text{ bits/s}$$

or 200 Mbyte/s. If the memory system is modified to accept a new (still 20-ns) request for a 32-bit word every 5 ns by overlapping requests, then its bandwidth is

$$\frac{32}{5 \times 10^{-9}} \text{ bits/s}$$

or 800 Mbyte/s. This memory system must be able to handle four requests at a given time.

Building an ideal memory system (infinite capacity, zero latency, and infinite bandwidth, with affordable cost) is not feasible. The challenge is, given the cost and technology constraints, to engineer a memory system whose abilities match the abilities that the processor demands of it. That is, engineering a memory system that performs as close to an ideal memory system (for the given processing unit) as is possible. For a processor that stalls when it makes a memory request (some current microprocessors are in this category), it is important to engineer a memory system with the lowest possible latency. For those processors that can handle multiple outstanding memory requests (vector processors and high-end CPUs), it is important not only to reduce latency but also to increase bandwidth (over what is possible by latency reduction alone) by designing a memory system that is capable of servicing multiple requests simultaneously.

Memory hierarchies provide decreased average latency and reduced bandwidth requirements, whereas parallel or **interleaved** memories provide higher bandwidth.

18.2 Memory Hierarchies

Technology does not permit memories that are cheap, large, and fast. By recognizing the nonrandom nature of memory requests, and emphasizing the *average* rather than worst-case latency, it is possible to implement a hierarchical memory system that performs well. A small amount of very fast memory, placed in front of a large, slow memory, can be designed to satisfy most requests at the speed of the small memory. This, in fact, is the primary motivation for the use of registers in the CPU: in this case, the programmer makes sure that the most commonly accessed variables are allocated to registers.

A variety of techniques, using hardware, software, or a combination of the two, can be employed to assure that most memory references are satisfied by the faster memory. The foremost of these techniques is the exploitation of the *locality of reference* principle. This principle captures the fact that some memory

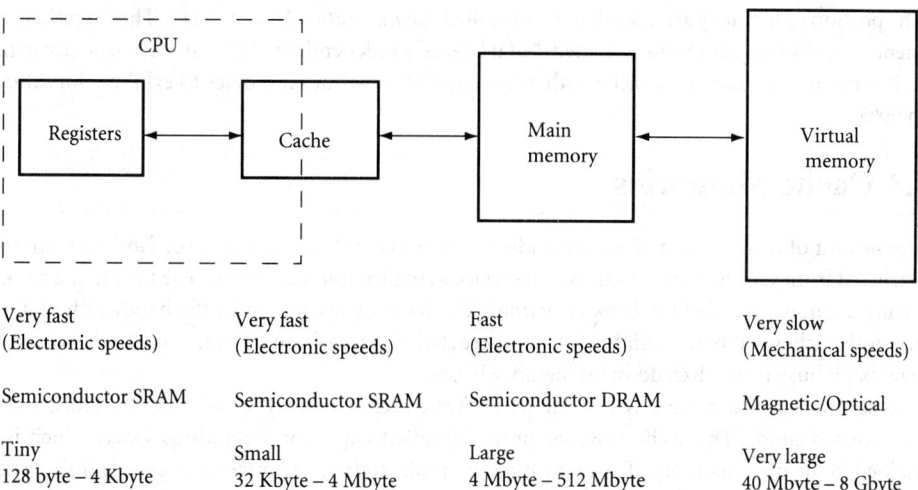

FIGURE 18.2   A memory hierarchy. (*Source:* Dorf, R. C. 1992. *The Electrical Engineering Handbook*, 1st ed., p. 1932. CRC Press, Inc., Boca Raton, FL.)

locations are referenced much more frequently than others. *Spatial locality* is the property that an access to a given memory location greatly increases the probability that neighboring locations will be accessed immediately. This is largely, but not exclusively, a result of the tendency to access memory locations sequentially. *Temporal locality* is the property that an access to a given memory location greatly increases the probability that the same location will be accessed again soon. This is largely, but not exclusively, a result of the high frequency of programs' looping behavior. Particularly for temporal locality, a good predictor of the future is the past; the longer a variable has gone unreferenced, the less likely it is to be accessed soon.

Figure 18.2 depicts a common construction of a memory hierarchy. At the top of the hierarchy are the CPU registers, which are small and extremely fast. The next level down in the hierarchy is a special, high-speed semiconductor memory, known as a **cache memory**. The cache can actually be divided into multiple distinct levels; most current systems have between one and three levels of cache. Some of the levels of cache may be on the CPU chip itself, they may be on the same module as the CPU, or they may all be entirely distinct. Below the cache is the conventional memory, referred to as *main memory*, or *backing storage*. Like a cache, main memory is semiconductor memory, but it is slower, cheaper, and denser than a cache. Below the main memory is the virtual memory, which is generally stored on magnetic or optical disks. Accessing the virtual memory can be tens of thousands of times slower than accessing the main memory, since it involves moving, mechanical parts.

As requests go deeper into the memory hierarchy, they encounter levels that are larger (in terms of capacity) and slower than the higher levels (moving left to right in Fig. 18.2). In addition to size and speed, the bandwidth between adjacent levels in the memory hierarchy is smaller for the lower levels. The bandwidth between the registers and top cache level, for example, is higher than that between cache and main memory or between main memory and virtual memory. Since each level presumably intercepts a fraction of the requests, the bandwidth to the level below need not be as great as that to the intercepting level.

A useful performance parameter is the *effective latency*. If the needed word is found in a level of the hierarchy, it is a *hit*; if a request must be sent to the next lower level, the request is said to *miss*. If the latency L_{HIT} is known in the case of a hit and the latency in the case of a miss is L_{MISS}, the effective latency for that level in the hierarchy can be determined from the *hit ratio* (H), the fraction of memory accesses that are hits:

$$L_{\mathrm{average}} = L_{\mathrm{HIT}}H + L_{\mathrm{MISS}}(1 - H) \qquad (18.2)$$

The portion of memory accesses that miss is called the *miss ratio* ($M = 1 - H$). The hit ratio is strongly influenced by the program being executed, but it is largely independent of the ratio of cache size to memory size. It is not uncommon for a cache with a capacity of a few thousand bytes to exhibit a hit ratio greater than 90%.

18.3 Cache Memories

The basic unit of construction of a semiconductor memory system is a *module* or *bank*. A memory bank, constructed from several memory chips, can service a single request at a time. The time that a bank is busy servicing a request is called the *bank busy time*. The bank busy time limits the bandwidth of a memory bank. Both caches and main memories are constructed in this fashion, although caches have significantly shorter bank busy times than do main memory banks.

The hardware can dynamically allocate parts of the cache memory for addresses deemed most likely to be accessed soon. The cache contains only redundant copies of the address space, which is wholly contained in the main memory. The cache memory is *associative*, or *content-addressable*. In an associative memory, the address of a memory location is stored, along with its content. Rather than reading data directly from a memory location, the cache is given an address and responds by providing data which may or may not be the data requested. When a cache miss occurs, the memory access is then performed with respect to the backing storage, and the cache is updated to include the new data.

The cache is intended to hold the most active portions of the memory, and the hardware dynamically selects portions of main memory to store in the cache. When the cache is full, bringing in new data must be matched by deleting old data. Thus a strategy for cache management is necessary. Cache management strategies exploit the principle of locality. Spatial locality is exploited by the choice of what is brought into the cache. Temporal locality is exploited by the choice of which block is removed. When a cache miss occurs, hardware copies a large, contiguous block of memory into the cache, which includes the requested word. This fixed-size region of memory, known as a cache *line* or *block*, may be as small as a single word, or up to several hundred bytes. A block is a set of contiguous memory locations, the number of which is usually a power of two. A block is said to be *aligned* if the lowest address in the block is exactly divisible by the block size. That is to say, for a block of size B beginning at location A, the block is aligned if

$$A \bmod B = 0 \qquad (18.3)$$

Conventional caches require that all blocks be aligned.

When a block is brought into the cache, it is likely that another block must be evicted. The selection of the evicted block is based on an attempt to capture temporal locality. Since prescience is difficult to achieve, other methods are generally used to predict future memory accesses. A least-recently-used (LRU) policy is often the basis for the replacement choice. Other replacement policies are sometimes used, particularly because true LRU replacement requires extensive logic and hardware bookkeeping.

The cache often comprises two conventional memories: the data memory and the tag memory, shown in Fig. 18.3. The address of each cache line contained in the data memory is stored in the tag memory, as well as other information (*state* information), particularly the fact that a valid cache line is present. The state also keeps track of which cache lines the processor has modified. Each line contained in the data memory is allocated a corresponding entry in the tag memory to indicate the full address of the cache line.

The requirement that the cache memory be associative (content-addressable) complicates the design. Addressing data by content is inherently more complicated than by its address. All the tags must be compared concurrently, of course, because the whole point of the cache is to achieve low latency. The cache can be made simpler, however, by introducing a mapping of memory locations to cache cells. This mapping limits the number of possible cells in which a particular line may reside. The extreme case is known as *direct mapping*, in which each memory location is mapped to a single location in the cache. Direct mapping makes many aspects of the design simpler, since there is no choice of where the line

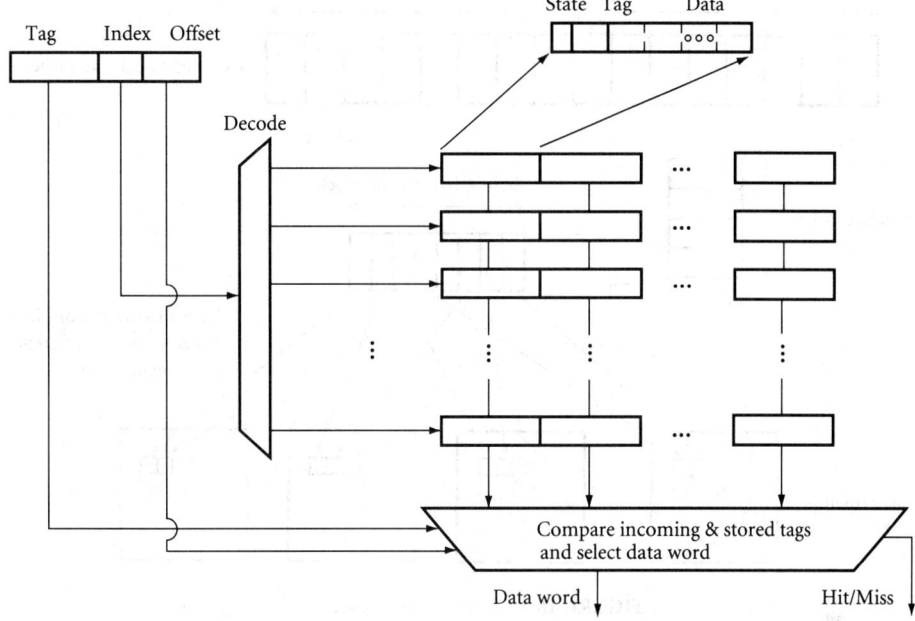

FIGURE 18.3 Components of a cache memory. (*Source:* Hill, M. D. 1988. A case for direct-mapped caches. *IEEE Comput.* 21(12):27. IEEE Computer Society, New York. With permission.)

might reside and no choice as to which line must be replaced. Direct mapping, however, can result in poor utilization of the cache when two memory locations are alternately accessed and must share a single cache cell.

A hashing algorithm is used to determine the cache address from the memory address. The conventional mapping algorithm consists of a function with the form

$$A_{\text{cache}} = \frac{A_{\text{memory}} \bmod \text{cache\_size}}{\text{cache\_line\_size}} \tag{18.4}$$

where A_{cache} is the address within the cache for main memory location A_{memory}, cache_size is the capacity of the cache in addressable units (usually bytes), and cache_line_size is the size of the cache line in addressable units. Since the hashing function is simple bit selection, the tag memory need contain only the part of the address not implied by the result of the hashing function. That is,

$$A_{\text{tag}} = A_{\text{memory}} \text{ div size\_of\_cache} \tag{18.5}$$

where A_{tag} is stored in the tag memory and *div* is the integer divide operation. In testing for a match, the complete address of a line stored in the cache can be inferred from the tag and its storage location within the cache.

A *two-way set-associative* cache maps each memory location into either of two locations in the cache, and can be constructed essentially as two identical direct-mapped caches. However, both caches must be searched at each memory access and the appropriate data selected and multiplexed on a tag match (hit). On a miss, a choice must be made between the two possible cache lines as to which is to be replaced. A single LRU bit can be saved for each such pair of lines to remember which line has been accessed more recently. This bit must be toggled to the current state each time either of the cache lines is accessed.

In the same way, an *M-way associative* cache maps each memory location into any of *M* memory locations in the cache and can be constructed from *M* identical direct-mapped caches. The problem of

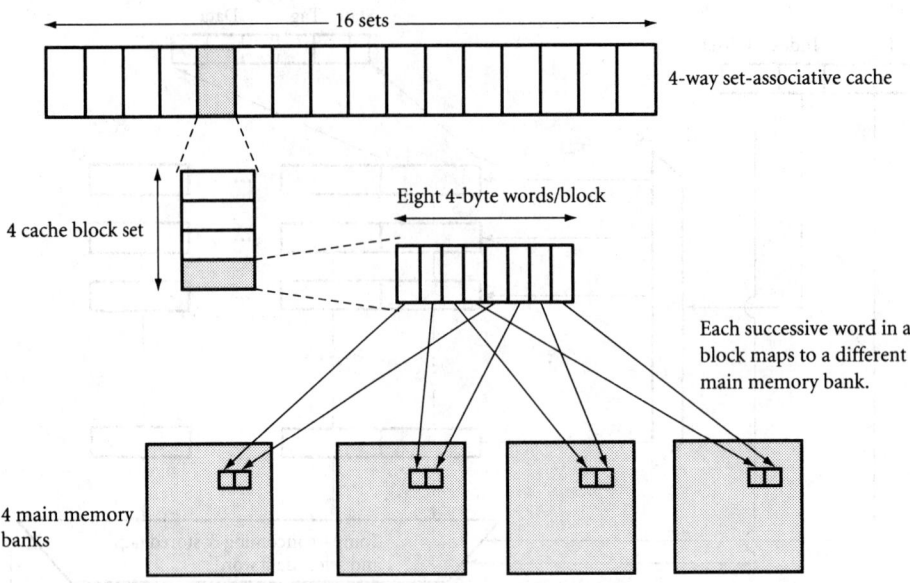

FIGURE 18.4 Organization of a cache.

maintaining the LRU ordering of M cache lines quickly becomes hard, however, since there are $M!$ possible orderings, so it takes at least

$$\lceil \log_2(M!) \rceil \qquad\qquad (18.6)$$

bits to store the ordering. In practice, this requirement limits true LRU replacement to three- or four-way set associativity.

Figure 18.4 shows how a cache is organized into sets, blocks, and words. The cache shown is a 2-Kbyte, four-way set-associative cache, with 16 sets. Each set consists of four blocks. The cache block size in this example is 32 bytes, so each block contains eight 4-byte words. Also depicted at the bottom of Fig. 18.4 is a four-way interleaved main memory system (see the next section for details). Each successive word in the cache block maps into a different main memory bank. Because of the cache's mapping restrictions, each cache block obtained from main memory will be loaded into its corresponding set, but it may appear anywhere within that set.

Write operations require special handling in the cache. If the main memory copy is updated with each write operation—a technique known as *write-through* or *store-through*—the writes may force operations to stall while the write operations are completing. This can happen after a series of write operations even if the processor is allowed to proceed before the write to the memory has completed. If the main memory copy is not updated with each write operation—a technique known as *write-back* or *copy-back* or *deferred writes*—the main memory locations become stale, that is, memory no longer contains the correct values and must not be relied upon to provide data. This is generally permissible, but care must be exercised to make sure that it is always updated before the line is purged from the cache and that the cache is never bypassed. Such a bypass could occur with *direct memory access* (DMA), in which the I/O system writes directly into main memory without the involvement of the processor.

Even for a system that implements a write-through policy, care must be exercised if memory requests may bypass the cache. While the main memory is never stale, a write that bypasses the cache, such as from I/O, could have the effect of making the cached copy stale. A later access by the CPU could then provide an incorrect value. This can be avoided only by making sure that cached entries are invalidated even if the cache is bypassed. The problem is relatively easy to solve for a single processor with I/O, but it becomes very difficult to solve for multiple processors, particularly so if multiple caches are involved as well. This is known in general as the cache *coherence* or *consistency* problem.

The cache exploits spatial locality by loading an entire cache line after a miss. This tends to result in bursty traffic to the main memory, since most accesses are filtered out by the cache. After a miss, however, the memory system must provide an entire line at once. Cache memory nicely complements an interleaved, high-bandwidth main memory (described in the next section), since a cache line can be interleaved across many banks in a regular manner—thus avoiding memory conflicts and being loaded rapidly into the cache. The example of main memory shown in Fig. 18.4 can provide the entire cache line with two parallel memory accesses.

Conventional caches traditionally could not accept requests while they were servicing a miss request. In other words, they *locked-up* or *blocked* when servicing a miss. The growing penalty for cache misses has made it necessary for high-end commodity memory systems to continue to accept (and service) requests from the processor while a miss is being serviced. Some systems are able to service multiple miss requests simultaneously. To allow this mode of operation, the cache design is *lockup-free* or *nonblocking* [Kroft 1981]. Lockup-free caches have one structure for each simultaneous outstanding miss that they can service. This structure holds the information necessary to correctly return the loaded data to the processor, even if the misses come back in a different order than that in which they were sent.

Two factors drive the existence of multiple levels of cache memory in the memory hierarchy: access times and a limited number of transistors on the CPU chip. Larger banks with greater capacity are slower than smaller banks. If the time needed to access the cache limits the clock frequency of the CPU, then the first-level cache size may need to be constrained. Much of the benefit of a large cache may be obtained by placing a small first-level cache above a larger second-level cache; the first is accessed quickly, and the second holds more data close to the processor. Since many modern CPUs have caches on the CPU chip itself, the size of the cache is limited by the CPU silicon real estate. Some CPU designers have assumed that system designers will add large off-chip caches to the one or two levels of caches on the processor chip. The complexity of this part of the memory hierarchy may continue to grow as main memory access penalties continue to increase.

Caches that appear on the CPU chip are manufactured by the CPU vendor. Off-chip caches, however, are a commodity part sold in large volume. An incomplete list of major cache manufacturers includes Hitachi, IBM Micro, Micron, Motorola, NEC, Samsung, SGS-Thomson, Sony, and Toshiba. Although most personal computers and all major workstations now contain caches, very high-end machines (such as multimillion-dollar supercomputers) do not usually have caches. These ultraexpensive computers can afford to implement their main memory in a comparatively fast semiconductor technology such as static RAM (SRAM) and can afford so many banks that cacheless bandwidth out of the main memory system is sufficient. Massively parallel processors (MPPs), however, are often constructed out of workstation-like nodes to reduce cost. MPPs therefore contain cache hierarchies similar to those found in the workstations on which the nodes of the MPPs are based.

Cache sizes have been steadily increasing on personal computers and workstations. Intel Pentium-based personal computers come with 8 Kbyte each of instruction and data caches. Two of the Pentium chip sets, manufactured by Intel and OPTi, allow level-two caches ranging from 256 to 512 Kbyte and 64 Kbyte to 2 Mbyte, respectively. The newer Pentium Pro systems also have 8-Kbyte first-level instruction and data caches, but they also have either a 256-Kbyte or a 512-Kbyte second-level cache on the same module as the processor chip. Higher-end workstations—such as DEC Alpha 21164-based systems—are configured with substantially more cache. The 21164 also has 8-Kbyte first-level instruction and data caches. Its second-level cache is entirely on chip and is 96 Kbyte. The third-level cache is off chip, and can have a size ranging from 1 to 64 Mbyte.

For all desktop machines, cache sizes are likely to continue to grow—although the rate of growth compared to processor speed increases and main memory size increases is unclear.

18.4 Parallel and Interleaved Main Memories

Main memories are composed of a series of semiconductor memory chips. A number of these chips, like caches, form a *bank*. Multiple memory banks can be connected together to form an **interleaved** (or parallel) memory system. Since each bank can service a request, an interleaved memory system with K banks can service K requests simultaneously, increasing the peak bandwidth of the memory system to K times the bandwidth of a single bank. In most interleaved memory systems, the number of banks is a power of two, that is, $K = 2^k$. An n-bit memory word address is broken into two parts: a k-bit bank number and an m-bit address of a word within a bank. Though the k bits used to select a bank number could be any k bits of the n-bit word address, typical interleaved memory systems use the low-order k address bits to select the bank number; the higher order $m = n - k$ bits of the word address are used to access a word in the selected bank. The reason for using the low-order k bits will be discussed shortly. An interleaved memory system which uses the low-order k bits to select the bank is referred to as a *low-order* or a *standard* interleaved memory.

There are two ways of connecting multiple memory banks: *simple interleaving* and *complex interleaving*. Sometimes simple interleaving is also referred to as *interleaving*, and complex interleaving is referred to as *banking*.

Figure 18.5 shows the structure of a simple interleaved memory system. m address bits are simultaneously supplied to every memory bank. All banks are also connected to the same read/write control line (not shown in Fig. 18.5). For a read operation, the banks start the read operation and deposit the data in their latches. Data can then be read from the latches, one by one, by appropriately setting the switch. Meanwhile, the banks could be accessed again to carry out another read or write operation. For a write operation, the latches are loaded one by one. When all the latches have been written, their contents can be

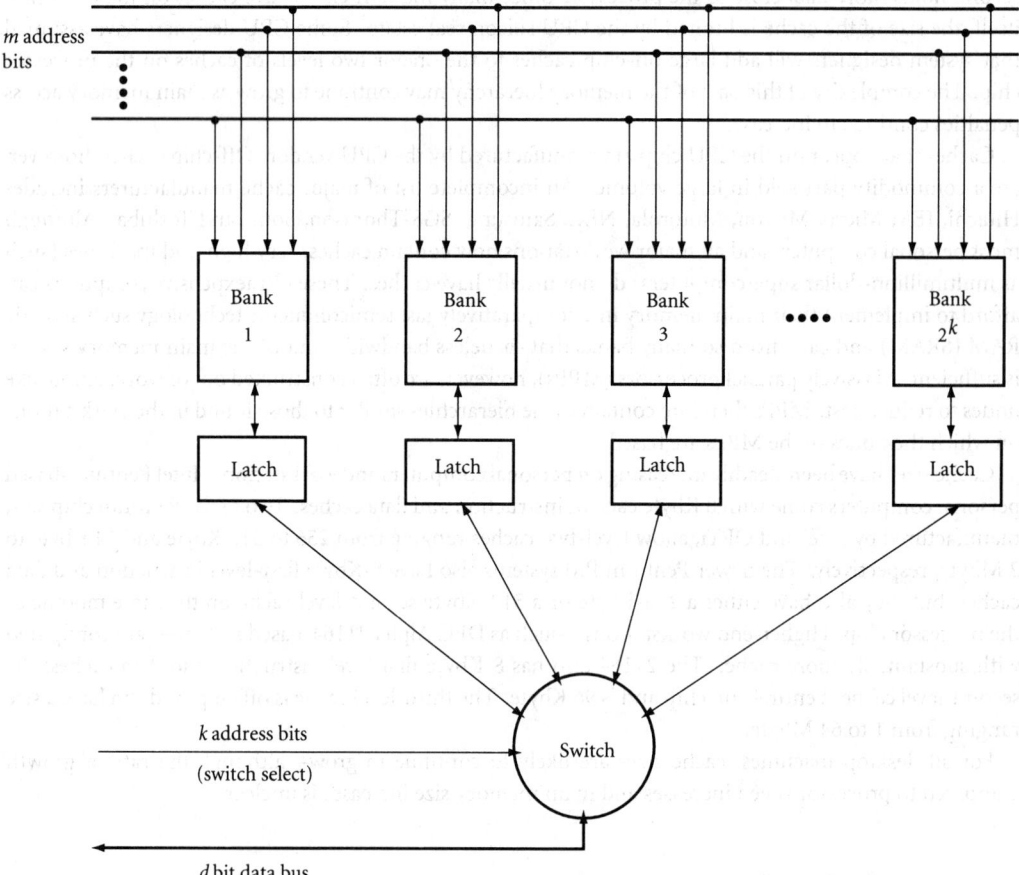

FIGURE 18.5 A simple interleaved memory system. (*Source:* Adapted from Kogge, P. M. 1981. *The Architecture of Pipelined Computers,* 1st ed., p. 41. McGraw–Hill, New York. With permission.)

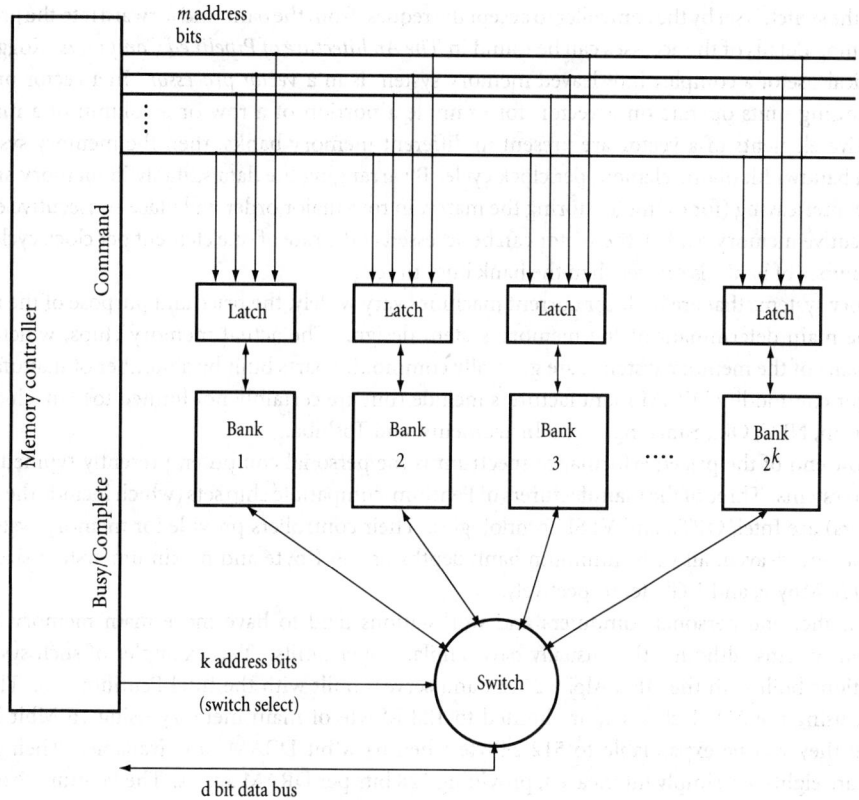

FIGURE 18.6   A complex interleaved memory system. (*Source:* Adapted from Kogge, P. M. 1981. *The Architecture of Pipelined Computers*, 1st ed., p. 42. McGraw–Hill, New York. With permission.)

written into the memory banks by supplying m bits of address (they will be written into the same word in each of the different banks). In a simple interleaved memory, all banks are cycled at the same time; each bank starts and completes its individual operations at the same time as every other bank; a new memory cycle can start (for all banks) once the previous cycle is complete. Timing details of the accesses can be found in *The Architecture of Pipelined Computers* [Kogge 1981].

One use of a simple interleaved memory system is to back up a cache memory. To do so, the memory must be able to read blocks of contiguous words (a cache block) and supply them to the cache. If the low-order k bits of the address are used to select the bank number, then consecutive words of the block reside in different banks; they can all be read in parallel and supplied to the cache one by one. If some other address bits are used for bank selection, then multiple words from the block might fall in the same memory bank, requiring multiple accesses to the same bank to fetch the block.

Figure 18.6 shows the structure of a complex interleaved memory system. In such a system, each bank is set up to operate on its own, independent of the other banks' operation. In this example, bank 1 could carry out a read operation on a particular memory address while bank 2 carries out a write operation on a completely unrelated memory address. (Contrast this with the operation in a simple interleaved memory where all banks are carrying out the same operation, read or write, and the locations accessed within each bank represent a contiguous block of memory.) Complex interleaving is accomplished by providing an address latch and a read/write command line for each bank. The *memory controller* handles the overall operation of the interleaved memory. The processing unit submits the memory request to the memory controller, which determines the bank that needs to be accessed. The controller then determines if the bank is busy (by monitoring a busy line for each bank). The controller holds the request if the bank is busy, submitting it later when the bank is available to accept the request. When the bank responds to a read request, the switch is set by the controller to accept the request from the bank and forward it to the processing unit. Timing details of the accesses can be found in *The Architecture of Pipelined Computers* [Kogge 1981].

A typical use of a complex interleaved memory system is in a *vector processor*. In a vector processor, the processing units operate on a vector, for example a portion of a row or a column of a matrix. If consecutive elements of a vector are present in different memory banks, then the memory system can sustain a bandwidth of one element per clock cycle. By arranging the data suitably in memory and using standard interleaving (for example, storing the matrix in row-major order will place consecutive elements in consecutive memory banks), the vector can be accessed at the rate of one element per clock cycle as long as the number of banks is greater than the bank busy time.

Memory systems that are built for current machines vary widely, the price and purpose of the machine being the main determinant of the memory system design. The actual memory chips, which are the components of the memory systems, are generally commodity parts built by a number of manufacturers. The major commodity DRAM manufacturers include (but are certainly not limited to) Hitachi, Fujitsu, LG Semicon, NEC, Oki, Samsung, Texas Instruments, and Toshiba.

The low end of the price/performance spectrum is the personal computer, presently typified by Intel Pentium systems. Three of the manufacturers of Pentium-compatible chip sets (which include the memory controllers) are Intel, OPTi, and VLSI Technologies. Their controllers provide for memory systems that are simply interleaved, all with minimum bank depths of 256 Kbyte and maximum system sizes of 192 Mbyte, 128 Mbyte, and 1 Gbyte, respectively.

Both higher-end personal computers and workstations tend to have more main memory than the lower-end systems, although they usually have similar upper limits. Two examples of such systems are workstations built with the DEC Alpha 21164 and servers built with the Intel Pentium Pro. The Alpha systems, using the 21171 chip set, are limited to 128 Mbyte of main memory using 16 Mbit DRAMs, although they will be expandable to 512 Mbyte when 64-Mbit DRAMs are available. Their memory systems are eight-way simply interleaved, providing 128 bits per DRAM access. The Pentium Pro systems support slightly different features. The 82450KX and 82450GX chip sets include memory controllers that allow reads to bypass writes (performing writes when the memory banks are idle). These controllers can also buffer eight outstanding requests simultaneously. The 82450KX controller permits one- or two-way interleaving and up to 256 Mbyte of memory when 16-Mbit DRAMs are used. The 82450GX chip set is more aggressive, allowing up to four separate (complex-interleaved) memory controllers, each of which can be up to four-way interleaved and have up to 1 Gbyte of memory (again with 16-Mbit DRAMs).

Interleaved memory systems found in high-end *vector supercomputers* are slight variants on the basic complex interleaved memory system of Fig. 18.6. Such memory systems may have hundreds of banks, with multiple memory controllers that allow multiple independent memory requests to be made every clock cycle. Two examples of modern vector supercomputers are the Cray T-90 series and the NEC SX series. The Cray T-90 models come with varying numbers of processors—up to 32 in the largest configuration. Each of these processors is coupled with 256 Mbyte of memory, split into 16 banks of 16 Mbyte each. The T-90 has complex interleaving among banks. The largest configuration (the T-932) has 32 processors, for a total of 512 banks and 8 Gbyte of main memory. The T-932 can provide a peak of 800-GByte/s bandwidth out of its memory system. NEC's SX-4 product line, their most recent vector supercomputer series, has numerous models. Their largest single-node model (with one processor per node) contains 32 processors, with a maximum of 8 Gbyte of memory, and a peak bandwidth of 512 Gbyte/s out of main memory. Although the sizes of the memory systems are vastly different between workstations and vector machines, the techniques that both use to increase total bandwidth and minimize bank conflicts are similar.

18.5 Virtual Memory

Cache memory contains portions of the main memory in dynamically allocated cache lines. Since the data portion of the cache memory is itself a conventional memory, each line present in the cache has two addresses associated with it: its main memory address and its cache address. Thus, the main memory address of a word can be divorced from a particular storage location and abstractly thought of as an element in the address space. The use of a two-level hierarchy—consisting of main memory and a slower, larger disk storage device—evolved by making a clear distinction between the address space and the locations in memory. An address generated during the execution of a program is known as a *virtual address*, which must be translated to a *physical address* before it can be accessed in main memory. The total address space is only an abstraction.

A **virtual memory** address is mapped to a physical address, which indicates the location in main memory where the data actually reside [Denning 1970]. The mapping is maintained through a structure called the *page table*, which is maintained in software by the operating system. Like the tag memory of a cache memory, the page table is accessed through a virtual address to determine the physical (main memory) address of the entry. Unlike the tag memory, however, the table is usually sorted by virtual addresses, making the translation process a simple matter of an extra memory access to determine the physical address of the desired item. A system maintaining the page table in a way analogous to a cache tag memory is said to have *inverted page tables*. In addition to the physical address mapped to a virtual page, and an indication of whether the page is present at all, a page table entry often contains other information. For example, the page table may contain the location on the disk where each block of data is stored when not present in main memory.

The virtual memory can be thought of as a collection of blocks. These blocks are often aligned and of fixed size, in which case they are known as *pages*. Pages are the unit of transfer between the disk and main memory and are generally larger than a cache line—usually thousands of bytes. A typical page size for machines in 1995 is 4 Kbyte. A page's virtual address can be broken into two parts: a virtual page number and an offset. The page number specifies which page is to be accessed, and the page offset indicates the distance from the beginning of the page to the indicated address.

A physical address can also be broken into two parts: a physical page number (also called a *page frame number*) and an offset. This mapping is done at the level of pages, so the page table can be indexed by means of the virtual page number. The page frame number is contained in the page table and is read out during the translation along with other information about the page. In most implementations the page offset is the same for a virtual address and the physical address to which it is mapped.

The virtual memory hierarchy is different than the cache/main memory hierarchy in a number of respects, resulting primarily from the fact that there is a much greater difference in latency between accesses to the disk and to main memory. While a typical latency ratio for cache and main memory is one order of magnitude (main memory has a latency ten times larger than the cache), the latency ratio between disk and main memory is often four orders of magnitude or more. This large ratio exists because the disk is a mechanical device—with a latency partially determined by velocity and inertia—whereas main memory is limited only by electronic and energy constraints. Because of the much larger penalty for a page miss, many design decisions are affected by the need to minimize the frequency of misses. When a miss does occur, the processor could be idle for a period during which it could execute tens of thousands of instructions. Rather than stall during this time, as may occur upon a cache miss, the processor invokes the operating system and may switch to a different task. Because the operating system is being invoked anyway, it is convenient to rely on it to set up and maintain the page table, unlike cache memory, where it is done entirely in hardware. The fact that this accounting occurs in the operating system enables the system to use virtual memory to enforce protection on the memory. This ensures that no program can corrupt the data in memory that belong to any other program.

Hardware support provided for a virtual memory system generally includes the ability to translate the virtual addresses provided by the processor into the physical addresses needed to access main memory. Thus, only upon a virtual-address miss is the operating system invoked. An important aspect of a computer that implements virtual memory, however, is the necessity of freezing the processor at the point at which a miss occurs, servicing the page table fault, and later returning to continue the execution as if no page fault had occurred. This requirement means either that it must be possible to halt execution at any point—including possibly in the middle of a complex instruction—or that it must be possible to guarantee that all memory accesses will be to pages resident in main memory.

As described above, virtual memory requires two memory accesses to fetch a single entry from memory: one into the page table to map the virtual address into the physical address, and the second to fetch the actual data. This process can be sped up in a variety of ways. First, a special-purpose cache memory to store the active portion of the page table can be used to speed up the first access. This special-purpose cache is usually called a *translation lookaside buffer* (TLB). Second, if the system also employs a cache memory, it may be possible to overlap the access of the cache memory with the access to the TLB, ideally allowing the requested item to be accessed in a single cache access time. The two accesses can be fully overlapped if the virtual address supplies sufficient information to fetch the data from the cache before the virtual-to-physical address translation has been accomplished. This is true for an M-way set associative cache of capacity C if the following relationship holds:

$$\text{Page\_size} \geq \frac{C}{M} \tag{18.7}$$

For such a cache, the index into the cache can be determined strictly from the page offset. Since the virtual page offset is identical to the physical page offset, no translation is necessary, and the cache can be accessed concurrently with the TLB. The physical address must be obtained before the tag can be compared, of course.

An alternative method applicable to a system containing both virtual memory and a cache is to store the virtual address in the tag memory instead of the physical address. This technique introduces consistency problems in virtual memory systems that either permit more than a single address space, or allow a single physical page to be mapped to more than one single virtual page. This problem is known as the *aliasing* problem.

Chapter 80 is devoted to virtual memory and contains significantly more material on this topic for the interested reader.

18.6 Research Issues

Research is occurring on all levels of the memory hierarchy. At the register level, researchers are exploring techniques to provide more registers than are architecturally visible to the compiler. A large volume of work exists (and is occurring) for cache optimizations and alternative cache organizations. For instance, modern processors now commonly split the top level of the cache into separate physical caches, one for instructions (code) and one for program data. Due to the increasing cost of cache misses (in terms of processor cycles), some research trades off increasing the complexity of the cache for reducing the miss rate. Two examples of cache research from opposite ends of the hardware/software spectrum are *blocking* [Lam et al. 1991] and *skewed-associative caches* [Seznec 1993]. Blocking is a software technique in which the programmer or compiler reorganizes algorithms to work on subsets of data that are smaller than the cache instead of streaming entire large data structures repeatedly through the cache. This reorganization greatly improves temporal locality. The skewed-associative cache is one example of a host of hardware techniques that map blocks into the cache differently, with the goal of reducing misses from set conflicts. In skewed-associative caches, either one of two hashing functions may determine where a block should be placed in the cache, as opposed to just the one hashing function (low-order index bits) that traditional caches use. An important cache-related research topic is *prefetching* [Mowry et al. 1992], in which the processor issues requests for data well before the data are actually needed. Speculative prefetching is also a current research topic. In speculative prefetching, prefetches are issued based on guesses as to which data will be needed soon. Other cache-related research examines placing special structures in parallel with the cache, trying to optimize for workloads that do not lend themselves well to caches. Stream buffers [Jouppi 1990] are one such example. A stream buffer automatically detects when a linear access through a data structure occurs. The stream buffer issues multiple sequential prefetches upon detection of a linear array access.

Much of the ongoing research on main memory involves improving the bandwidth from the memory system without greatly increasing the number of banks. Multiple banks are expensive, particularly with the large and growing capacity of modern DRAM chips. Rambus [Rambus 1992] and Ramlink [IEEE Computer Society 1993] are two such examples.

Research issues associated with improving the performance of the virtual memory system fall into the domain of operating system research. One proposed strategy for reducing page faults allows each running program to specify its own page replacement algorithm, enabling each program to optimize the choice of page replacements based on its reference pattern [Engler et al. 1995]. Other recent research focuses on improving the performance of the TLB. Two techniques for doing this are the use of a two-level TLB (the motivation is similar to that for a two-level cache) and the use of superpages [Talluri and Hill 1994]. With superpages, each TLB entry may represent a mapping for more than one consecutive page, thus increasing the total address range that a fixed number of TLB entries may cover.

18.7 Summary

A computer's memory system is the repository for all the information that the CPU uses and produces. A perfect memory system is one that can immediately supply any datum that the CPU requests. This ideal memory is not implementable, however, as the three factors of memory—capacity, speed, and cost—are directly in opposition.

By staging smaller, faster memories in front of larger, slower, and cheaper memories, the performance of the memory system may approach that of a perfect memory system—at a reasonable cost. The memory hierarchies of modern general-purpose computers generally contain registers at the top, followed by one or more levels of cache memory, main memory, and virtual memory on a magnetic or optical disk.

Performance of a memory system is measured in terms of latency and bandwidth. The latency of a memory request is how long it takes the memory system to produce the result of the request. The bandwidth of a memory system is the rate at which the memory system can accept requests and produce results. The memory hierarchy improves average latency by quickly returning results that are found in the higher levels of the hierarchy. The memory hierarchy generally reduces bandwidth requirements by intercepting a fraction of the memory requests at higher levels of the hierarchy. Some machines—such as high-performance vector machines—may have fewer levels in the hierarchy, increasing memory cost for better predictability and performance. Some of these machines contain no caches at all, relying on large arrays of main memory banks to supply very high bandwidth, with pipelined accesses of operands that mitigate the adverse performance impact of long latencies.

Cache memories are a general solution to improving the performance of a memory system. Although caches are smaller than typical main memory sizes, they ideally contain the most frequently accessed portions of main memory. By keeping the most heavily used data near the CPU, caches can service a large fraction of the requests without accessing main memory (the fraction serviced is called the hit rate). Caches assume locality of reference to work well transparently—they assume that accessed memory words will be accessed again quickly (temporal locality) and that memory words adjacent to an accessed word will be accessed soon after the access in question (spatial locality). When the CPU issues a request for a datum not in the cache (a cache miss), the cache loads that datum and some number of adjacent data (a cache block) into itself from main memory.

To reduce cache misses, some caches are associative—a cache may place a given block in one of several places, collectively called a set. This set is content-addressable; a block may or may not be accessed based on an address tag, one of which is coupled with each block. When a new block is brought into a set and the set is full, the cache's replacement policy dictates which of the old blocks should be removed from the cache to make room for the new block. Most caches use an approximation of least-recently-used (LRU) replacement, in which the block last accessed farthest in the past is the one that the cache replaces.

Main memory, or backing store, consists of banks of dense semiconductor memory. Since each memory chip has a small off-chip bandwidth, rows of these chips are placed together to form a bank, and multiple banks are used to increase the total bandwidth out of main memory. When a bank is accessed, it remains busy for a period of time, during which the processor may make no other accesses to that bank. By increasing the number of interleaved (parallel) banks, the chance that the processor issues two conflicting requests to the same bank is reduced.

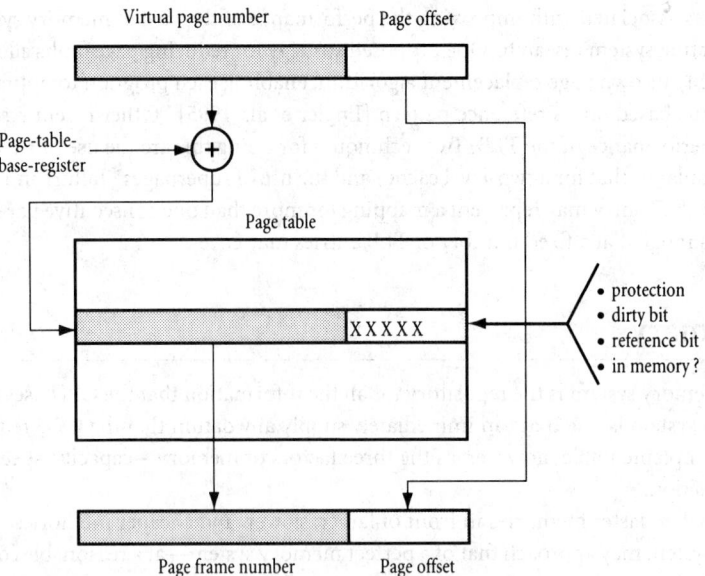

Virtual page number   Page offset

Page-table-
base-register

Page table

X X X X X

- protection
- dirty bit
- reference bit
- in memory ?

Page frame number   Page offset

FIGURE 18.7   Virtual-to-physical address translation. (*Source:* Dorf, R. C. 1992. *The Electrical Engineering Handbook*, 1st ed., p. 1935. CRC Press, Inc., Boca Raton, FL.)

Systems generally require a greater number of memory locations than are available in the main memory (i.e., a larger address space). The entire address space that the CPU uses is kept on large magnetic or optical disks; this is called the virtual address space, or virtual memory. The most frequently used sections of the virtual memory are kept in main memory (physical memory) and are moved back and forth in units called pages. The place at which a virtual address lies in main memory is called its physical address. Since a much larger address space (virtual memory) is mapped onto a much smaller one (physical memory), the CPU must translate the memory addresses issued by a program (virtual addresses) into their corresponding locations in physical memory (physical addresses) (see Fig. 18.7). This mapping is maintained in a memory structure called the page table. When the CPU attempts to access a virtual address that does not have a corresponding entry in physical memory, a page fault occurs. Since a page fault requires an access to a slow mechanical storage device (such as a disk), the CPU usually switches to a different task while the needed page is read from the disk.

Every memory request issued by the CPU requires an address translation, which in turn requires an access to the page table stored in memory. A translation lookaside buffer (TLB) is used to reduce the number of page table lookups. The most frequent virtual-to-physical mappings are kept in the TLB, which is a small associative memory tightly coupled with the CPU. If the needed mapping is found in the TLB, the translation is performed quickly and no access to the page table need be made. Virtual memory allows systems to run larger or more programs than are able to fit in main memory, enhancing the capabilities of the system.

Defining Terms

Bandwidth: The rate at which the memory system can service requests.

Cache memory: A small, fast, redundant memory used to store the most frequently accessed parts of the main memory.

Interleaving: Technique for connecting multiple memory modules together in order to improve the bandwidth of the memory system.

Latency: The time between the initiation of a memory request and its completion.

Memory hierarchy: Successive levels of different types of memory, which attempt to approximate a single large, fast, and cheap memory structure.

Virtual memory: A memory space implemented by storing the more frequently accessed data in main memory and less frequently accessed data on disk.

References

Denning, P. J. 1970. Virtual memory. *Comput. Surveys* 2(3):153–170.

Dorf, R. C. 1992. *The Electrical Engineering Handbook,* 1st ed. CRC Press, Inc., Boca Raton, FL.

Engler, D. R., Kaashoek, M. F., and O'Toole, J., Jr. 1995. Exokernel: an operating system architecture for application-level resource management, pp. 251–266. In *Proc. 15th Symp. on Operating Systems Principles.*

Hennessy, J. L. and Patterson, D. A. 1990. *Computer Architecture: A Quantitative Approach,* 1st ed. Morgan Kaufmann, San Mateo, CA.

Hill, M. D. 1988. A case for direct-mapped caches. *IEEE Comput.* 21(12):25–40.

IEEE Computer Society 1993. *IEEE Standard for High-Bandwidth Memory Interface Based on SCI Signaling Technology (RamLink).* Draft 1.00 IEEE P1596.4-199X.

Jouppi, N. 1990. Improving direct-mapped cache performance by the addition of a small fully-associative cache and prefetch buffers, pp. 364–373. In *Proc. 17th Annual Int. Symp. on Computer Architecture.*

Kogge, P. M. 1981. *The Architecture of Pipelined Computers,* 1st ed. McGraw–Hill, New York.

Kroft, D. 1981. Lockup-free instruction fetch/prefetch cache organization, pp. 81–87. In *Proc. 8th Annual Int. Symp. on Computer Architecture.*

Lam, M. S., Rothberg, E. E., and Wolf, M. E. 1991. The cache performance and optimizations of blocked algorithms, pp. 63–74. In *Proc. 4th Annual Symp. on Architectural Support for Programming Languages and Operating Systems.*

Mowry, T. C., Lam, M. S., and Gupta, A. 1992. Design and evaluation of a compiler algorithm for prefetching, pp. 62–73. In *Proc. 5th Annual Symp. on Architectural Support for Programming Languages and Operating Systems.*

Rambus 1992. *Rambus Architectural Overview.* Rambus Inc., Mountain View, CA.

Seznec, A. 1993. A case for two-way skewed-associative caches, pp. 169–178. In *Proc. 20th International Symposium on Computer Architecture.*

Smith, A. J. 1986. Bibliography and readings on CPU cache memories and related topics. *ACM SIGARCH Comput. Architecture News* 14(1):22–42.

Smith, A. J. 1991. Second bibliography on cache memories. *ACM SIGARCH Comput. Architecture News* 19(4):154–182.

Talluri, M. and Hill, M. D. 1994. Surpassing the TLB performance of superpages with less operating system support, pp. 171–182. In *Proc. 6th Int. Symp. on Architectural Support for Programming Languages and Operating Systems.*

Further Information

Some general information on the design of memory systems is available in *High-Speed Memory Systems* by A. V. Pohm and O. P. Agarwal. 1983. Reston Publishing, Reston, VA.

Computer Architecture: A Quantitative Approach by John Hennessy and David Patterson [Hennessy and Patterson 1990] contains a detailed discussion on the interaction between memory systems and computer architecture.

For information on memory system research, the recent proceedings of the *International Symposium on Computer Architecture* contain annual research papers in computer architecture, many of which focus on the memory system. To obtain copies, contact the IEEE Computer Society Press, at 10662 Los Vaqueros Circle, P.O. Box 3014, Los Alamitos, CA 90720-1264.

19

High-Speed Computer Arithmetic

Earl E. Swartzlander, Jr.*
University of Texas at Austin

19.1 Introduction

The speeds of memory and arithmetic units are the primary determinants of the speed of a computer. Whereas the speed of both units depends directly on the implementation technology, arithmetic unit speed also depends strongly on the logic design. Even for an integer adder, speed can easily vary by an order of magnitude, whereas the complexity varies by less than 50%.

This chapter begins with a discussion of binary fixed point number systems in section 19.2. Section 19.3 provides examples of fixed point implementations of the four basic arithmetic operations (i.e., add, subtract, multiply, and divide). Finally, section 19.4 describes algorithms that implement floating point arithmetic.

Regarding notation, capital letters represent digital numbers (i.e., words), whereas subscripted lower case letters represent bits of the corresponding word. The subscripts range from 0 to $n - 1$ to indicate the bit position within the word (x_0 is the least significant bit of X, x_{n-1} is the most significant bit of X, etc.). The logic designs presented in this chapter are based on positive logic with AND, OR, and INVERT operations. Depending on the technology used for implementation, different operations (such as NAND and NOR) may be used, but the basic concepts are not likely to change significantly.

19.2 Fixed Point Number Systems

Most arithmetic is performed with fixed point numbers which have constant scaling (i.e., the position of the binary point is fixed). The numbers can be interpreted as fractions, integers, or mixed numbers depending on the application. Pairs of fixed point numbers are used to create floating point numbers, as discussed in section 19.4.

*Revision of chapter originally presented in Swartzlander, E. E., Jr. 1992. Computer arithmetic. In *Computer Engineering Handbook*. C. H. Chen, ed., Ch. 4, pp. 4-1–4-20. McGraw–Hill, New York. With permission.

0-8493-2909-4/97/$0.00+$.50
© 1997 by CRC Press, Inc.

At the present time, fixed point binary numbers are generally represented using the two's complement number system. This choice has prevailed over the sign magnitude and one's complement number systems, because the frequently performed operations of addition and subtraction are easiest to perform on two's complement numbers. Sign magnitude numbers are more efficient for multiplication but the lower frequency of multiplication and the development of Booth's efficient two's complement multiplication algorithm have resulted in the nearly universal selection of the two's complement number system for most applications. The algorithms presented in this chapter assume the use of two's complement numbers.

Fixed point number systems represent numbers, say, A, by n bits: a sign bit and $n - 1$ data bits. By convention, the most significant bit a_{n-1} is the sign bit, which is a 1 for negative numbers and a 0 for positive numbers. The $n - 1$ data bits are $a_{n-2}, a_{n-3}, \ldots, a_1, a_0$. In the material that follows, fixed point fractions will be described for each of the three systems.

Two's Complement

In the two's complement fractional number system, the value of a number is the sum of $n - 1$ positive binary fractional bits and a sign bit, which has a weight of -1

$$A = -a_{n-1} + \sum_{i=0}^{n-2} a_i 2^{i-n+1} \tag{19.1}$$

Two's complement numbers are negated by complementing all bits and adding a 1 to the least significant bit (lsb) position. For example, to form $-3/8$,

$$
\begin{array}{lll}
+3/8 & = & 0011 \\
\text{invert all bits} & = & 1100 \\
\text{add 1 lsb} & & 0001 \\
& & 1101 & = & -3/8
\end{array}
$$

Check:

$$
\begin{array}{lll}
\text{invert all bits} & = & 0010 \\
\text{add 1 lsb} & & 0001 \\
& & 0011 & = & +3/8
\end{array}
$$

Sign Magnitude

Sign magnitude numbers consist of a sign bit and $n - 1$ bits that express the magnitude of the number,

$$A = (1 - 2a_{n-1}) \sum_{i=0}^{n-2} a_i 2^{i-n+1} \tag{19.2}$$

Sign magnitude numbers are negated by complementing the sign bit. For example, to form $-3/8$,

$$
\begin{array}{lll}
+3/8 & = & 0011 \\
\text{invert sign bit} & = & 1011 & = & -3/8
\end{array}
$$

Check:

$$
\begin{array}{lll}
\text{invert sign bit} & = & 0011 & = & +3/8
\end{array}
$$

One's Complement

One's complement numbers are negated by complementing all of the bits of the corresponding positive number,

$$A = \sum_{i=0}^{n-2} (a_i - a_{n-1}) 2^{i-n+1} \tag{19.3}$$

In this equation, the subtraction $(a_i - a_{n-1})$ is an arithmetic operation (not a logical operation) that produces values of 1 or 0 (if $a_{n-1} = 0$) or values of 0 or -1 (if $a_{n-1} = 1$).

The negative of a one's complement number is formed by inverting all bits. For example, to form $-3/8$,

$$\begin{aligned} +3/8 \quad &= \quad 0011 \\ \text{invert all bits} \quad &= \quad 1100 \quad = \quad -3/8 \end{aligned}$$

Check:

$$\text{invert all bits} \quad = \quad 0011 \quad = \quad +3/8$$

Table 19.1 compares 4-b fractional fixed point numbers in the three number systems. Note that both the sign magnitude and one's complement number systems have two zeros and that only two's complement is capable of representing -1. For positive numbers all three number systems use identical representations.

A significant difference between the three number systems is their behavior under truncation. Figure 19.1 shows the effect of truncating high-precision fixed point fractions X, to form three bit fractions $T(X)$. Truncation of two's complement numbers never increases the value of the number (i.e., the truncated numbers have values that are unchanged or shift toward negative infinity), as can be seen from Eq. (19.1) where any truncated bits have positive weight. This bias can cause an accumulation of errors for computations that involve summing many truncated numbers (which may occur in scientific, matrix, and signal processing applications). In both the sign magnitude and one's complement number systems, truncated numbers are unchanged or shifted toward zero, so that if approximately half of the numbers are positive and half are negative, the errors will tend to cancel.

TABLE 19.1 Example of 4-b Fractional Fixed Point Numbers

Number	Two's Complement	Sign Magnitude	One's Complement
+7/8	0111	0111	0111
+3/4	0110	0110	0110
+5/8	0101	0101	0101
+1/2	0100	0100	0100
+3/8	0011	0011	0011
+1/4	0010	0010	0010
+1/8	0001	0001	0001
+0	0000	0000	0000
−0	N/A	1000	1111
−1/8	1111	1001	1110
−1/4	1110	1010	1101
−3/8	1101	1011	1100
−1/2	1100	1100	1011
−5/8	1011	1101	1010
−3/4	1010	1110	1001
−7/8	1001	1111	1000
−1	1000	N/A	N/A

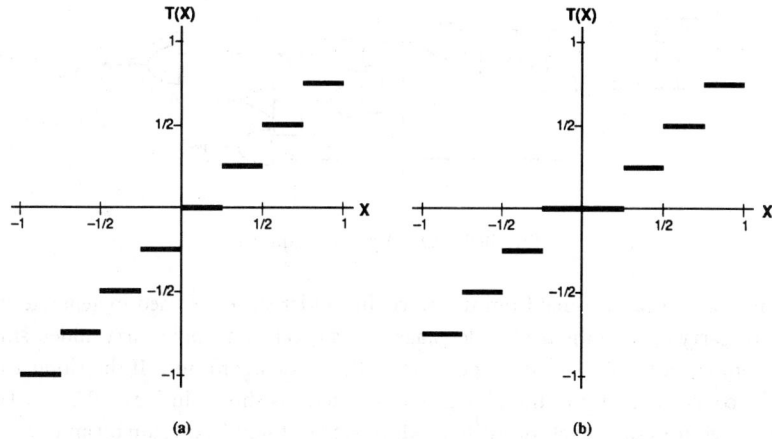

FIGURE 19.1 Behavior of fixed point fractions under truncation: (a) two's complement and (b) sign magnitude and one's complement.

19.3 Fixed Point Arithmetic Algorithms

This section presents a reasonable assortment of typical fixed point algorithms for addition, subtraction, multiplication, and division.

Fixed Point Addition

Addition is performed by summing the corresponding bits of the two n-bit numbers, including the sign bit. Subtraction is performed by summing the corresponding bits of the minuend and the two's complement of the subtrahend. Overflow is detected in a two's complement adder by comparing the carry signals into and out of the most significant adder stage (i.e., the stage which computes the sign bit). If the carries differ, the arithmetic has overflowed and the result is invalid.

Full Adder

The full adder is the fundamental building block of most arithmetic circuits. Its operation is defined by the truth table shown in Table 19.2. The sum and carry outputs are described by the following equations:

$$s_k = a_k \bar{b}_k \bar{c}_k + \bar{a}_k b_k \bar{c}_k + \bar{a}_k \bar{b}_k c_k + a_k b_k c_k \qquad (19.4)$$

$$c_{k+1} = \bar{a}_k b_k c_k + a_k \bar{b}_k c_k + a_k b_k \bar{c}_k + a_k b_k c_k = a_k b_k + a_k c_k + b_k c_k \qquad (19.5)$$

Where a_k, b_k, and c_k are the inputs to the kth full adder stage, and s_k and c_{k+1} are the sum and carry outputs, respectively.

In evaluating the relative complexity of implementations it is often convenient to assume a nine gate realization of the full adder, as shown in Fig. 19.2. For this implementation, the delay from either a_k or b_k to s_k is six gate delays and the delay from c_k to c_{k+1} is two gate delays. Some technologies, such as CMOS, form inverting gates (e.g., NAND and NOR gates) more efficiently than the noninverting gates that are assumed in this chapter. Circuits with equivalent speed and complexity can be constructed with inverting gates.

Ripple Carry Adder

A ripple carry adder for n-bit numbers is implemented by concatenating n full adders as shown in Fig. 19.3. At the kth-bit position, the kth bits of

TABLE 19.2 Full Adder Truth Table

Inputs			Outputs	
a_k	b_k	c_k	s_k	c_{k+1}
0	0	0	0	0
0	0	1	1	0
0	1	0	1	0
0	1	1	0	1
1	0	0	1	0
1	0	1	0	1
1	1	0	0	1
1	1	1	1	1

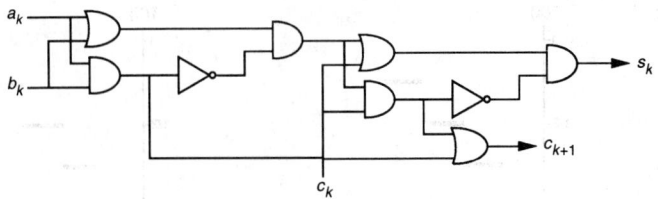

FIGURE 19.2   9-gate full adder.

operands A and B and a carry signal from the preceding adder stage are used to generate the kth bit of the sum s_k and a carry c_{k+1} to the next adder stage. This is called a ripple carry adder, since the carry signals *ripple* from the least significant bit position to the most significant. If the ripple carry adder is implemented by concatenating n of the nine gate full adders, as shown in Figs. 19.2 and 19.3, an n-bit ripple carry adder requires $2n + 4$ gate delays to produce the most significant sum bit and $2n + 3$ gate delays to produce the carry output. A total of $9n$ logic gates are required to implement the n-bit ripple carry adder. In comparing adders, the delay from data input to most significant sum output and the complexity (i.e., the gate count) will be used. These will be denoted by DELAY and GATES (subscripted by RCA to indicate ripple carry adder), respectively. These simple metrics are suitable for first-order comparisons. More accurate comparisons require consideration of both the number and the types of gates (since gates with fewer inputs are generally faster and smaller than gates with more inputs).

$$\text{DELAY}_{\text{RCA}} = 2n + 4 \qquad (19.6)$$

$$\text{GATES}_{\text{RCA}} = 9n \qquad (19.7)$$

Carry Lookahead Adder

Another approach is the carry lookahead adder [Weinberger and Smith 1958, MacSorley 1961]. Here specialized carry logic computes the carries in parallel. The carry lookahead adder uses modified full adders (modified in the sense that a carry output is not formed) for each bit position and lookahead modules. Each lookahead module forms individual carry outputs and blocks carry generate and propagate outputs that indicate that a carry is generated within the module or that an incoming carry would propagate across the module. This is seen by rewriting Eq. (19.5) with $g_k = a_k b_k$ and $p_k = a_k + b_k$,

$$c_{k+1} = g_k + p_k c_k \qquad (19.8)$$

This helps to explain the concept of carry generation and propagation: At a given stage a carry is generated if g_k is true (i.e., both a_k and b_k are 1), and a stage propagates an input carry to its output if p_k is true (i.e., either a_k or b_k is a 1). The nine gate full adder shown in Fig. 19.2 has AND and OR gates that produce g_k and p_k with no additional complexity. In fact, since the carry out is produced in the lookahead logic,

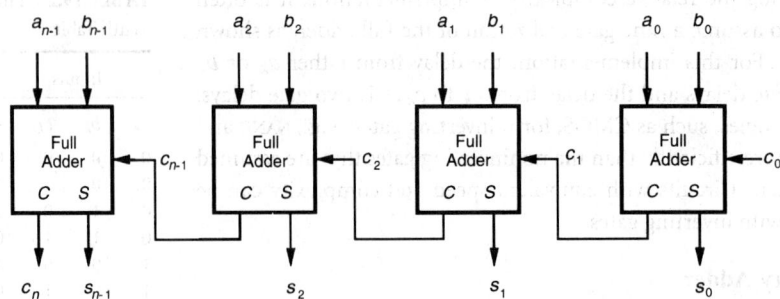

FIGURE 19.3   Ripple carry adder.

the OR gate that produces the c_{k+1} can be eliminated. The result is an eight gate adder module. Extending Eq. (19.8) to a second stage,

$$c_{k+2} = g_{k+1} + p_{k+1}c_{k+1}$$
$$= g_{k+1} + p_{k+1}(g_k + p_kc_k)$$
$$= g_{k+1} + p_{k+1}g_k + p_{k+1}p_kc_k \tag{19.9}$$

This equation results from evaluating Eq. (19.8) for the $k+1$st stage and substituting c_{k+1} from Eq. (19.8). Carry c_{k+2} exits from stage $k+1$ if: (1) a carry is generated there, or (2) a carry is generated in stage k and propagates across stage $k+1$, or (3) a carry enters stage k and propagates across both stages k and $k+1$, etc. Extending to a third stage,

$$c_{k+3} = g_{k+2} + p_{k+2}c_{k+2}$$
$$= g_{k+2} + p_{k+2}(g_{k+1} + p_{k+1}g_k + p_{k+1}p_kc_k)$$
$$= g_{k+2} + p_{k+2}g_{k+1} + p_{k+2}p_{k+1}g_k + p_{k+2}p_{k+1}p_kc_k \tag{19.10}$$

Although it would be possible to continue this process indefinitely, each additional stage increases the size (i.e., the number of inputs) of the logic gates. Four inputs [as required to implement Eq. (19.10)] is frequently the maximum number of inputs per gate for current technologies. To continue the process, we define generate and propagate signals over four bit blocks (stages k to $k+3$), $g_{k+3:k}$ and $p_{k+3:k}$, respectively,

$$g_{k+3:k} = g_{k+3} + p_{k+3}g_{k+2} + p_{k+3}p_{k+2}g_{k+1} + p_{k+3}p_{k+2}p_{k+1}g_k \tag{19.11}$$

and

$$p_{k+3:k} = p_{k+3}p_{k+2}p_{k+1}p_k \tag{19.12}$$

Equation (19.8) can be expressed in terms of the 4-b block generate and propagate signals,

$$c_{k+4} = g_{k+3:k} + p_{k+3:k}c_k \tag{19.13}$$

Thus, the carry out from a 4-b-wide block can be computed in only four gate delays [the first to compute p_i and g_i for $i = k$–$k+3$, the second to evaluate $p_{k+3:k}$, the second and third to evaluate $g_{k+3:k}$, and the third and fourth to evaluate c_{k+4} using Eq. (19.13)].

An n-bit carry lookahead adder requires $\lceil (n-1)/(r-1) \rceil$ lookahead blocks, where r is the width of the block. A 4-b lookahead block is a direct implementation of Eq. (19.8–19.12) with 14 logic gates. In general, an r-bit lookahead block requires $\frac{1}{2}(3r + r^2)$ logic gates. The Manchester carry chain [Kilburn et al. 1960] is an alternative switch-based technique for the implementation of the lookahead block.

Figure 19.4 shows the interconnection of 16 adders and 5 lookahead logic blocks to realize a 16-b carry lookahead adder. The sequence of events which occur during an add operation is as follows: (1) apply A, B, and carry in signals; (2) each adder computes P and G; (3) first-level lookahead logic computes the 4-b propagate and generate signals; (4) second-level lookahead logic computes c_4, c_8, and c_{12}; (5) first-level lookahead logic computes the individual carries; and (6) each adder computes the sum outputs. This process may be extended to larger adders by subdividing the large adder into 16-b blocks and using additional levels of carry lookahead (e.g., a 64-b adder requires three levels).

The delay of carry lookahead adders is evaluated by recognizing that an adder with a single level of carry lookahead (for 4-b words) has six gate delays and that each additional level of lookahead increases

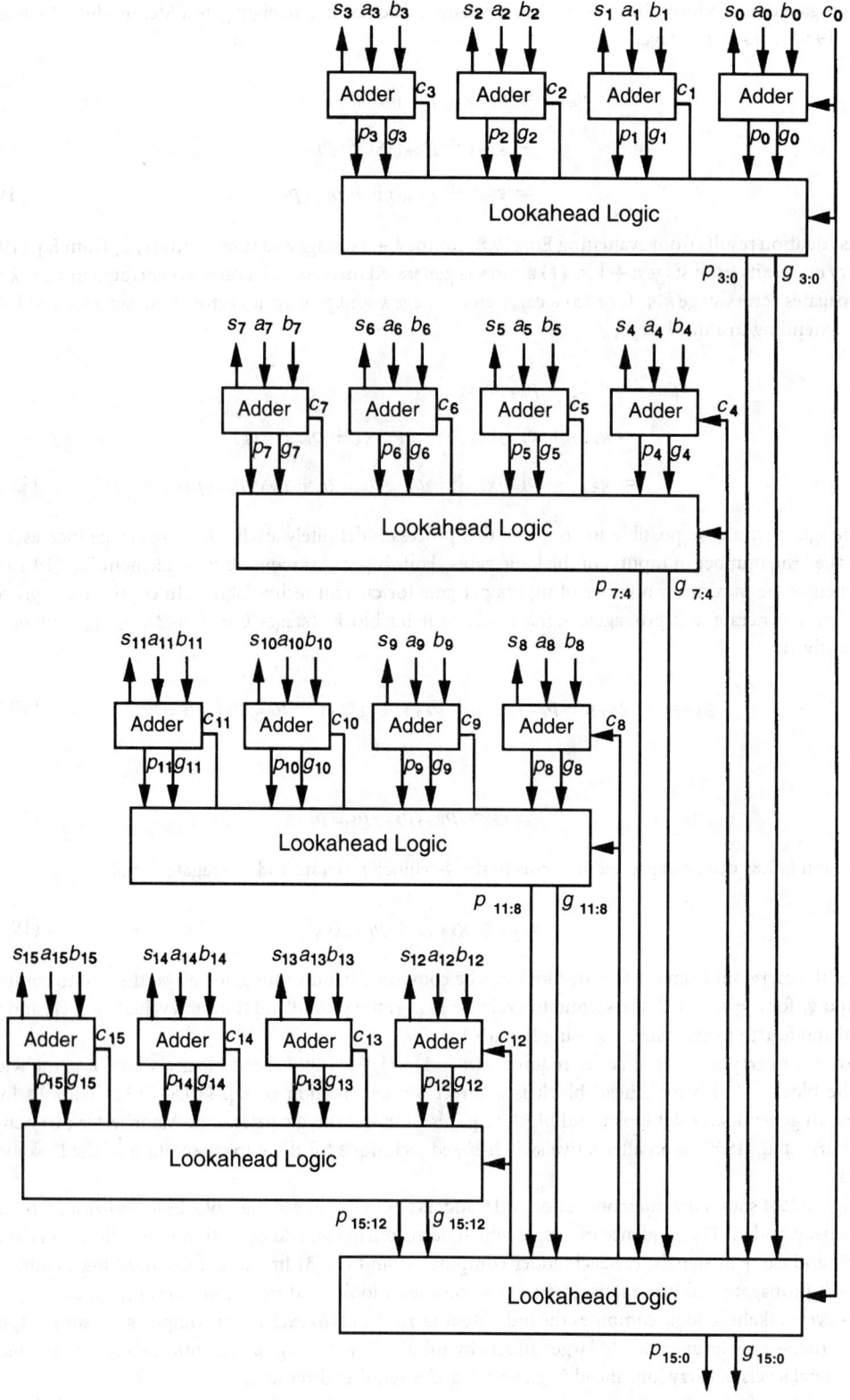

FIGURE 19.4   16-bit two level carry lookahead adder.

the maximum word size by a factor of four and adds four gate delays. More generally [Waser and Flynn 1982, pp. 83–88], the number of lookahead levels for an n-bit adder is $\lceil \log_r n \rceil$ where r is the maximum number of inputs per gate. Since an r-bit carry lookahead adder has six gate delays and there are four additional gate delays per carry lookahead level after the first,

$$\text{DELAY}_{\text{CLA}} = 2 + 4\lceil \log_r n \rceil \tag{19.14}$$

The complexity of an n-bit carry lookahead adder implemented with r-bit lookahead blocks is n modified full adders (each of which requires 8 gates) and $\lceil (n-1)/(r-1) \rceil$ lookahead logic blocks [each of which requires $\frac{1}{2}(3r + r^2)$ gates].

$$\text{GATES}_{\text{CLA}} = 8n + \frac{1}{2}(3r + r^2)\left\lceil \frac{n-1}{r-1} \right\rceil \tag{19.15}$$

For the currently common case of $r = 4$,

$$\text{GATES}_{\text{CLA}} \approx 12\frac{2}{3}n - 4\frac{2}{3} \tag{19.16}$$

The carry lookahead approach reduces the delay of adders from increasing linearly with the word size (as is the case for ripple carry adders) to increasing as the logarithm of the word size. As with ripple carry adders, the carry lookahead adder complexity grows linearly with the word size (for $r = 4$, this occurs at a 40% faster rate than the ripple carry adders).

Carry Skip Adder

The carry skip adder divides the words to be added into blocks (like the carry lookahead adder). The basic structure of a 16-b carry skip adder implemented with five 3-b blocks and one 1-b-wide block is shown in Fig. 19.5. Within each block, ripple carry is used to produce the sum bits and the carry (which is used as a block generate). In addition, an AND gate is used to form the block propagate signal. These signals are combined using Eq. (19.13) to produce a fast carry signal.

For example, with $k = 4$, Eq. (19.13) yields $c_8 = g_{7:4} + p_{7:4}c_4$. The carry out of the second ripple carry adder is a block generate signal if it is evaluated when carries generated by the data inputs (i.e., $a_{7:4}$ and $b_{7:4}$ in Fig. 19.5) are valid but before the carry that results from c_4. Normally, these two types of carries coincide in time, but in the carry skip adder the c_4 signal is produced by a 4-b ripple carry adder, so the carry output is a block generate from 11 gate delays after application of A and B until it becomes c_8 at 19 gate delays after application of A and B.

In the carry skip adder, the first and last blocks are simple ripple carry adders, whereas the $\lceil n/k \rceil - 2$ intermediate blocks are ripple carry adders augmented with three gates. The delay of a carry skip adder is the sum of $2k + 3$ gate delays to produce the carry in the first block, 2 gate delays through each of the intermediate blocks, and $2k + 1$ gate delays to produce the most significant sum bit in the last block. To simplify the analysis, the ceiling function in the count of intermediate blocks is ignored. If the block width is k,

$$\text{DELAY}_{\text{SKIP}} = 2k + 3 + 2\left(\frac{n}{k} - 2\right) + 2k + 1$$

$$= 4k + 2\frac{n}{k} \tag{19.17}$$

where $\text{DELAY}_{\text{SKIP}}$ is the total delay of the carry skip adder with a single level of k-bit-wide blocks. The optimum block size is determined by taking the derivative of $\text{DELAY}_{\text{SKIP}}$ with respect to k, setting it to

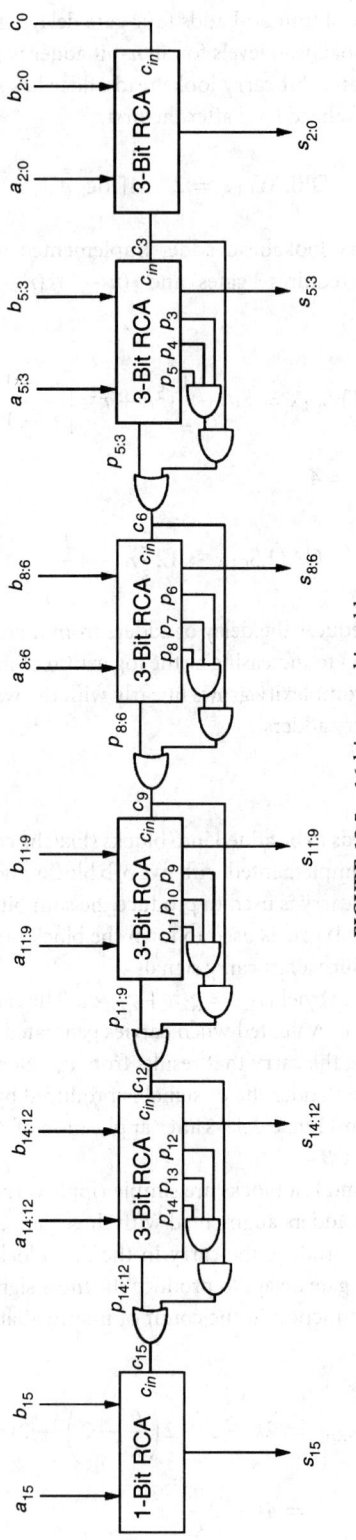

FIGURE 19.5   16-bit carry skip adder.

zero, and solving for k. The resulting optimum values for k and $\text{DELAY}_{\text{SKIP}}$ are

$$k = \sqrt{n/2} \qquad (19.18)$$

$$\text{DELAY}_{\text{SKIP}} = 4\sqrt{2n} \qquad (19.19)$$

Better results can be obtained by varying the block width so that the first and last blocks are smaller and the intermediate blocks are larger and by using multiple levels of carry skip [Chen and Schlag 1990, Turrini 1989].

The complexity of the carry skip adder is only slightly greater than that of a ripple carry adder because the first and last blocks are ripple carry adders and the intermediate blocks are ripple carry adders with three gates added for carry skipping,

$$\text{GATES}_{\text{SKIP}} = 9n + 3\left(\left\lceil \frac{n}{k} \right\rceil - 2\right) \qquad (19.20)$$

Carry Select Adder

The carry select adder divides the words to be added into blocks and forms two sums for each block in parallel (one with a carry in of 0 and the other with a carry in of 1). As shown for a 16-b carry select adder in Fig. 19.6, the carry out from the previous block controls a multiplexer that selects the appropriate sum. The carry out is computed using Eq. (19.13), since the group propagate signal is the carry out of an adder with a carry input of 1 and the group generate signal is the carry out of an adder with a carry input of 0.

If a constant block width of k is used, there will be $\lceil n/k \rceil$ blocks and the delay to generate the sum is $2k + 3$ gate delays to form the carry out of the first block, 2 gate delays for each of the $\lceil n/k \rceil - 2$ intermediate blocks, and 3 gate delays (for the multiplexer) in the final block. To simplify the analysis, the ceiling function in the count of intermediate blocks is ignored. Thus, the total delay is

$$\text{DELAY}_{\text{C-SEL}} = 2k + 2\frac{n}{k} + 2 \qquad (19.21)$$

The optimum block size is determined by taking the derivative of $\text{DELAY}_{\text{C-SEL}}$ with respect to k, setting it to zero, and solving for k. The result is

$$k = \sqrt{n} \qquad (19.22)$$

$$\text{DELAY}_{\text{C-SEL}} = 2 + 4\sqrt{n} \qquad (19.23)$$

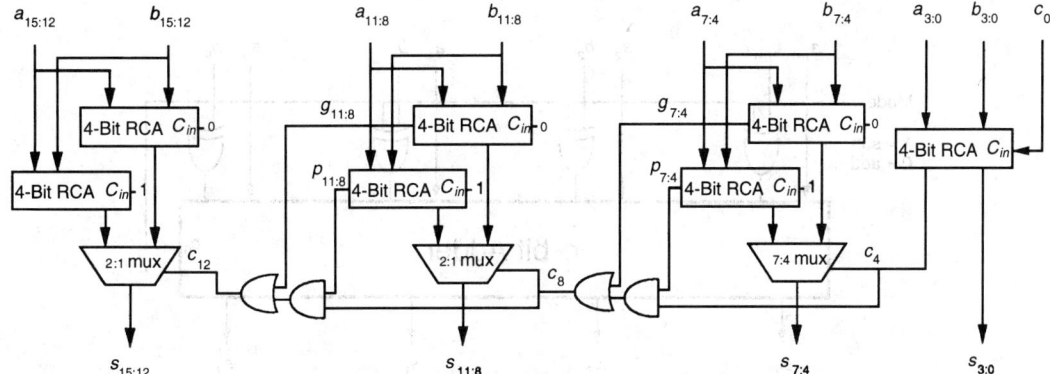

FIGURE 19.6 16-bit carry select adder.

As for the carry skip adder, better results can be obtained by varying the width of the blocks. In this case, the optimum is to make the two least significant blocks the same size and each successively more significant block 1-b larger. For this configuration, the delay for each block's most significant sum bit will equal the delay to the multiplexer control signal [Goldberg 1990, p. A-38].

The complexity of the carry select adder is $2n - k$ ripple carry adder stages, the intermediate carry logic and ($\lceil n/k \rceil - 1$) k-b-wide 2:1 multiplexers.

$$\text{GATES}_{\text{C-SEL}} = 9(2n - k) + 2\left(\left\lceil \frac{n}{k} \right\rceil - 2\right) + 3(n - k) + \left\lceil \frac{n}{k} \right\rceil - 1 \tag{19.24}$$

$$= 21n - 12k + 3\left\lceil \frac{n}{k} \right\rceil - 5$$

This is somewhat more than twice the complexity of a ripple carry adder.

Fixed Point Subtraction

To produce an adder/subtracter, the adder is modified as shown in Fig. 19.7 by including EXCLUSIVE-OR gates to complement operand B when performing subtraction. It forms either $A + B$ or $A - B$ with two's complement arithmetic. In the case of $A + B$, the mode selector is set to logic 0, which causes the EXCLUSIVE-OR gates to pass operand B directly to the ripple carry adder. The carry into the least significant adder stage is ZERO, so standard addition occurs. Subtraction is implemented by setting the mode selector to logic 1, which causes the EXCLUSIVE-OR gates to complement the bits of B; formation of the two's complement is completed by setting the carry into the least significant adder stage to a 1.

Fixed Point Multiplication

Multiplication is generally implemented either via a sequence of addition, subtraction, and shift operations or with direct logic implementations.

Sequential Booth Multiplier

The Booth [1951] algorithm is widely used for two's complement multiplication, since it is easy to implement. Earlier two's complement multipliers (e.g., Shaw [1950]) require data-dependent correction cycles if either operand is negative. To multiply AB, the product P is initially set to 0. Then, the bits of the multiplier A are examined in pairs of adjacent bits starting with the least significant bit (i.e., $a_0 \ a_{-1}$) and assuming $a_{-1} = 0$:

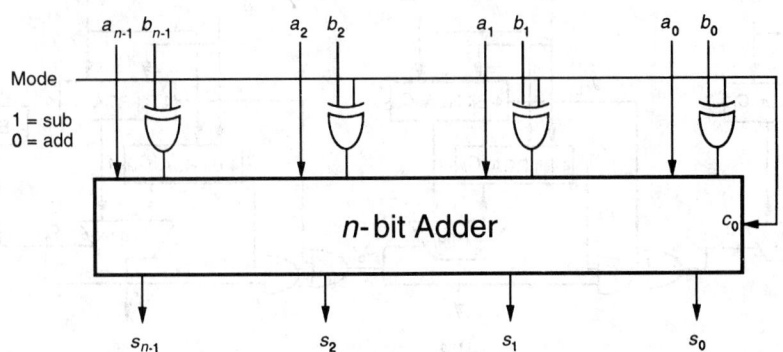

FIGURE 19.7   Adder/subtracter.

Positive Times Positive $A = \frac{5}{8} = 0.101$ $B = \frac{3}{4} = 0.110$

i	a_i	a_{i-1}	Operation	Result
0	1	0	$P = (P - B)/2$	1.1010
1	0	1	$P = (P + B)/2$	0.00110
2	1	0	$P = (P - B)/2$	1.101110
3	0	1	$P = P + B$	0.011110

Thus: $P = 0.011110 = \frac{15}{32}$

Negative Times Positive $A = -\frac{5}{8} = 1.011$ $B = \frac{3}{4} = 0.110$

i	a_i	a_{i-1}	Operation	Result
0	1	0	$P = (P - B)/2$	1.1010
1	1	1	$P = P/2$	1.11010
2	0	1	$P = (P + B)/2$	0.010010
3	1	0	$P = P - B$	1.100010

Thus: $P = 1.100010 = -\frac{15}{32}$

Positive Times Negative $A = \frac{5}{8} = 0.101$ $B = -\frac{3}{4} = 1.010$

i	a_i	a_{i-1}	Operation	Result
0	1	0	$P = (P - B)/2$	0.0110
1	0	1	$P = (P + B)/2$	1.1010
2	1	0	$P = (P - B)/2$	0.010010
3	0	1	$P = P + B$	1.100010

Thus: $P = 1.100010 = -\frac{15}{32}$

Negative Times Negative $A = -\frac{5}{8} = 1.011$ $B = -\frac{3}{4} = 1.010$

i	a_i	a_{i-1}	Operation	Result
0	1	0	$P = (P - B)/2$	0.0110
1	1	1	$P = P/2$	0.00110
2	0	1	$P = (P + B)/2$	1.101110
3	1	0	$P = P - B$	0.011110

Thus: $P = 0.011110 = \frac{15}{32}$

FIGURE 19.8 Example of sequential Booth multiplication.

- If $a_i = a_{i-1}$, $P = P/2$.
- If $a_i = 0$ and $a_{i-1} = 1$, $P = (P + B)/2$.
- If $a_i = 1$ and $a_{i-1} = 0$, $P = (P - B)/2$.

The division by 2 is not performed on the last stage (i.e., when $i = n - 1$). All of the divide by 2 operations are simple arithmetic right shifts (i.e., the word is shifted right one position and the old sign bit is repeated for the new sign bit), and overflows in the addition process are ignored. The algorithm is illustrated in Fig. 19.8, in which products for all combinations of $\pm 5/8$ times $\pm 3/4$ are computed for 4-b operands. The sequential Booth multiplier requires n cycles to form the product of a pair of n-bit numbers, where each cycle consists of an n-bit addition and a shift, an n-bit subtraction and a shift, or a shift without any other arithmetic operation.

Sequential Modified Booth Multiplier

The radix-4 modified Booth multiplier described by MacSorley [1961] uses $n/2$ cycles where each cycle examines three adjacent bits; adds or subtracts 0, B, or $2B$; and shifts 2 b to the right. Table 19.3 shows the operations as a function of the three bits: a_{i+1}, a_i, and a_{i-1}. The radix-4 modified Booth multiplier takes half the number of cycles as the standard (radix-2) Booth multiplier although the operations performed during a cycle are slightly more complex (since it is necessary to select one of five possible addends instead of one of three). Extensions to higher radices that examine

TABLE 19.3 Operations for the Radix-4 Modified Booth Algorithm

a_{i+1}	a_i	a_{i-1}	Operation
0	0	0	$P = P/4$
0	0	1	$P = (P + B)/4$
0	1	0	$P = (P + B)/4$
0	1	1	$P = (P + 2B)/4$
1	0	0	$P = (P - 2B)/4$
1	0	1	$P = (P - B)/4$
1	1	0	$P = (P - B)/4$
1	1	1	$P = P/4$

more than 3 b are possible but generally are not attractive because the addition/subtraction operations involve nonpower of two multiples (such as 3, 5, etc.) of B, which raises the complexity.

Array Multipliers

An alternative approach to multiplication involves the combinational generation of all bit products and their summation with an array of full adders. The block diagram of an 8 by 8 array multiplier is shown in Fig. 19.9. It uses an 8 by 8 array of AND gates to form the bit products (some of the AND gates are denoted G in Fig. 19.9 and the remainder are in the FA and HA blocks) and an 8 by 7 array of adders [48 full adders (denoted FA in Fig. 19.9) and 8 half-adders (denoted HA in Fig. 19.9)] to sum the 64-b products. Summing the bits requires 14 adder delays.

Modification of the array multiplier to multiply two's complement numbers requires inverting some of the signals in the most significant row and column while forming the bit product matrix and adding a few correction terms [Baugh and Wooley 1973, Blankenship 1974]. Array multipliers are easily laid out in a cellular fashion, making them suitable for very large-sized integrated (VLSI) implementation, where minimizing the design effort may be more important than maximizing the speed.

Wallace Tree/Dadda Fast Multiplier

A method for fast multiplication was developed by Wallace [1964] and refined by Dadda [1965]. With this method, a three-step process is used to multiply two numbers: (1) the bit products are formed, (2)

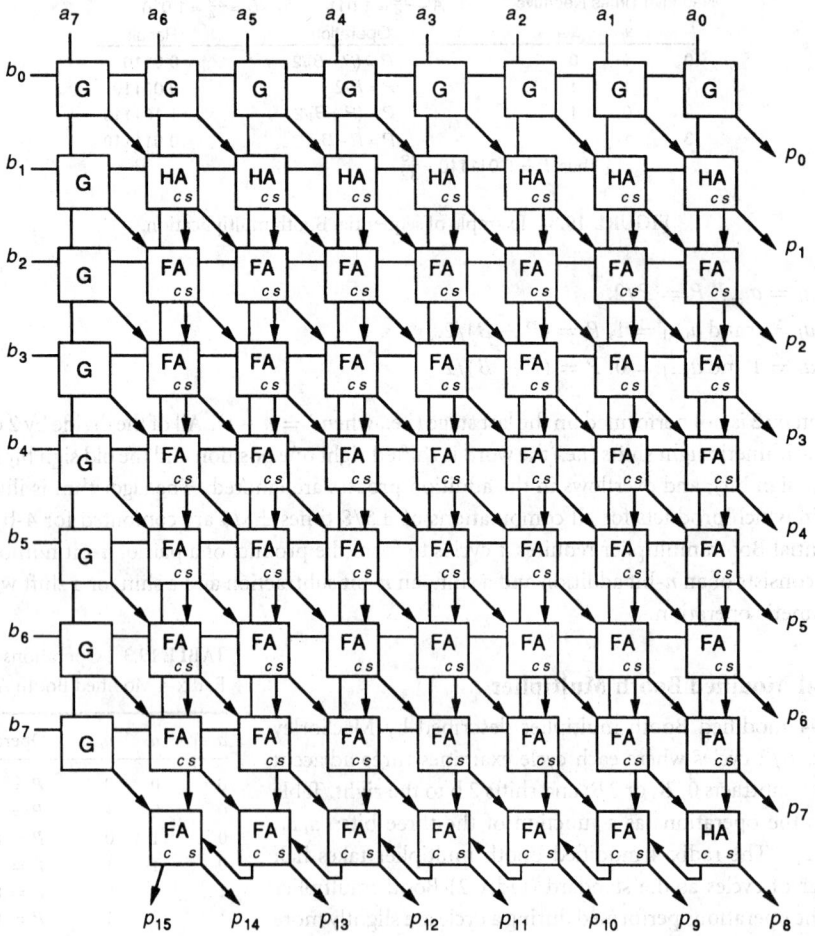

FIGURE 19.9 Unsigned 8 by 8 array multiplier.

the bit product matrix is reduced to a two row matrix where the sum of the rows equals the sum of the bit products, and (3) the two numbers are summed with a fast adder to produce the product. Although this may seem to be a complex process, it yields multipliers with delay proportional to the logarithm of the operand word size, which is faster than the Booth multiplier, the modified Booth multiplier, or array multipliers, which all have delays proportional to the word size.

The second step in the fast multiplication process is shown for an 8 by 8 Dadda multiplier in Fig. 19.10. An input 8 by 8 matrix of dots (each dot represents a bit product) is shown as matrix 0. Columns having more than six dots (or that will grow to more than six dots due to carries) are reduced by the use of half-adders (each half-adder takes in two dots and outputs one in the same column and one in the next more significant column) and full adders (each full adder takes in three dots and outputs one in the same column and one in the next more significant column) so that no column in matrix 1 will have more than six dots. Half-adders are

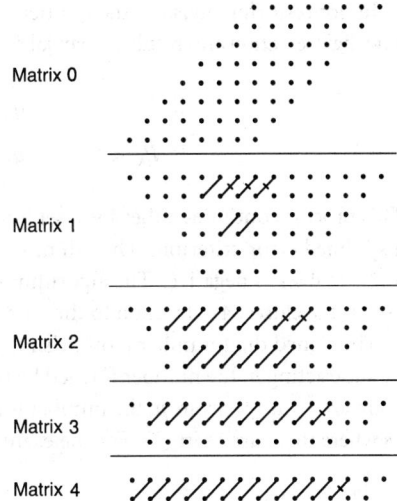

FIGURE 19.10 Unsigned 8 by 8 Dadda multiplier.

shown by a crossed line in the succeeding matrix and full adders are shown by a line in the succeeding matrix. In each case, the rightmost dot of the pair that are connected by a line is in the column from which the inputs were taken for the adder. In the succeeding steps reduction to matrix 2 with no more than four dots per column, matrix 3 with no more than three dots per column, and finally matrix 4 with no more than two dots per column is performed. The height of the matrices is determined by working back from the final (two row) matrix and limiting the height of each matrix to the largest integer that is no more than 1.5 times the height of its successor. Each matrix is produced from its predecessor in one adder delay. Since the number of matrices is logarithmically related to the number of bits in the words to be multiplied, the delay of the matrix reduction process is proportional to log n. Since the adder that reduces the final two row matrix can be implemented as a carry lookahead adder (which also has logarithmic delay), the total delay for this multiplier is proportional to the logarithm of the word size.

Fixed Point Division

Division is traditionally implemented as a digit recurrence requiring a sequence of shift, subtract, and compare operations in contrast to the shift and add approach employed for multiplication. The comparison operation is significant: It results in a serial process, which is not amenable to parallel implementation. There are several digit recurrence-based division schemes including binary restoring, nonperforming, nonrestoring, and Sweeney, Robertson, and Tocher (SRT) division algorithms for a variety of radices [Ercegovac and Lang 1994].

Nonrestoring Divider

Traditional nonrestoring division is based on selecting digits of the quotient Q (where $Q = N/D$) to satisfy the following equation:

$$P_{k+1} = rP_k - q_{n-k-1}D \qquad \text{for } k = 1, 2, \ldots, n-1 \qquad (19.25)$$

where P_k is the partial remainder after the selection of the kth quotient digit, $P_0 = N$ (subject to the constraint $|P_0| < |D|$), r is the radix ($=2$ for binary nonrestoring division), q_{n-k-1} is the kth quotient digit to the right of the binary point, and D is the divisor. In this section, it is assumed that both N and D are positive, see Ercegovac and Lang [1994] for details on handling the general case.

In nonrestoring division, the quotient digits are constrained to be ± 1 (i.e., q_k is selected from $\{1, \bar{1}\}$). The digit selection and resulting partial remainder are given for the kth iteration by the following relations:

$$\text{If } P_k \geq 0, \qquad q_{n-k-1} = 1 \qquad \text{and} \qquad P_{k+1} = rP_k - D \qquad (19.26)$$

$$\text{If } P_k < 0, \qquad q_{n-k-1} = \bar{1} \qquad \text{and} \qquad P_{k+1} = rP_k + D \qquad (19.27)$$

This process continues either for a set number of iterations or until the partial remainder is smaller than a specified error criterion. The kth most significant bit of the quotient is a 1 if P_k is 0 or positive and is a $\bar{1}(-1)$ if P_k is negative. The algorithm is illustrated in Fig. 19.11 where $\frac{5}{16}$ is divided by $\frac{3}{8}$. The result ($\frac{13}{16}$) is the closest 4-b fraction to the correct result of $\frac{5}{6}$.

The signed digit number (comprising ± 1 digits) can be converted into a conventional binary number by subtracting n, the number formed by the negative digits (with 0s where there are 1s in Q and 1s where there are $\bar{1}$s in Q), from p, the number formed by the positive digits (with 0s where there are $\bar{1}$s in Q and 1s where there are 1s in Q). For the example of Fig. 19.11:

$$Q = 0.111\bar{1}$$

$$n = 0.0001$$

$$p = 0.1110$$

$$p - n = 0.1110 - 0.0001$$

$$= 0.1110 + 1.1111$$

$$= 0.1101$$

Other digit recurrence division algorithms include the restoring algorithm and the SRT algorithms [Robertson 1958]. The restoring algorithm is similar to the nonrestoring algorithm presented here except that either subtract and shift or shift are permitted for each iteration instead of subtract and shift

Nonrestoring division

$N = \frac{5}{16} = 0.0101$

$D = \frac{3}{8} = 0.0110$

$P_{(0)} = N$

Since $P_{(0)} \geq 0$, $q_3 = 1$ and $P_{(1)} = 2P_{(0)} - D$

$2P_{(0)} =$	0.1010
$-D$	1.1010
$P_{(1)} =$	0.0100

Since $P_{(1)} \geq 0$, $q_2 = 1$ and $P_{(2)} = 2P_{(1)} - D$

$2P_{(1)} =$	0.1000
$-D$	1.1010
$P_{(2)} =$	0.0010

Since $P_{(2)} \geq 0$, $q_1 = 1$ and $P_{(3)} = 2P_{(2)} - D$

$2P_{(2)} =$	0.0100
$-D$	1.1010
$P_{(3)} =$	1.1110

Since $P_{(3)} < 0$, $q_0 = \bar{1}$ and $P_{(4)} = 2P_{(3)} + D$

$2P_{(3)} =$	1.1100
$+D$	0.0110
$P_{(4)} =$	0.0010

$Q = 0.111\bar{1} = 0.1101 = \frac{13}{16}$

FIGURE 19.11   Example of nonrestoring division.

or add and shift as in Eqs. (19.26) and (19.27). Restoring division forms the quotient from the digits 0 and 1 so that no signed digit to binary conversion is required (as it is for nonrestoring division). For all of the digit recurrence algorithms the number of steps is proportional to the number of quotient digits.

The more advanced SRT division process for radix 2 and higher radices is similar to nonrestoring division in that the recurrence of Eq. (19.25) is used, but the set of allowable quotient digits is increased. The process of quotient digit selection is sufficiently complex that a research monograph has been devoted to SRT division [Ercegovac and Lang 1994].

Newton–Raphson Divider

A second division technique uses a form of Newton–Raphson iteration to derive a quadratically convergent approximation to the reciprocal of the divisor, which is then multiplied by the dividend to produce the quotient. In systems which include a fast multiplier, this process is often faster than conventional division [Ferrari 1967].

The Newton–Raphson division algorithm to compute $Q = N/D$ consists of three basic steps:

1. Calculating a starting estimate of the reciprocal of the divisor $R_{(0)}$. If the divisor D is normalized (i.e., $\frac{1}{2} \leq D < 1$), then $R_{(0)} = 3 - 2D$ exactly computes $1/D$ at $D = 0.5$ and $D = 1$ and exhibits maximum error (of approximately 0.17) at $D = \sqrt{\frac{1}{2}}$. Adjusting $R_{(0)}$ downward to by half the maximum error gives

$$R_{(0)} = 2.915 - 2D \qquad (19.28)$$

   This produces an initial estimate that is within about 0.087 of the correct value for all points in the interval $\frac{1}{2} \leq D < 1$.

2. Computing successively more accurate estimates of the reciprocal by the following iterative procedure:

$$R_{(i+1)} = R_{(i)}(2 - DR_{(i)}) \qquad \text{for } i = 0, 1, \ldots, k \qquad (19.29)$$

3. Computing the quotient by multiplying the dividend times the reciprocal of the divisor,

$$Q = NR(k) \qquad (19.30)$$

   Where i is the iteration count and N is the numerator. Figure 19.12 illustrates the operation of the Newton–Raphson algorithm. For this example, three iterations (one shift, four subtractions, and seven multiplications) produce an answer accurate to nine decimal digits (approximately 30 b).

With this algorithm, the error decreases quadratically, so that the number of correct bits in each approximation is roughly twice the number of correct bits on the previous iteration. Thus, from a $3\frac{1}{2}$-b initial approximation, two iterations produce a reciprocal estimate accurate to 14 b, four iterations produce a reciprocal estimate accurate to 56 b, etc. The number of iterations is proportional to the logarithm of the number of accurate quotient digits.

The efficiency of this process is dependent on the availability of a fast multiplier, since each iteration of Eq. (19.29) requires two multiplications and a subtraction. The complete process for the initial estimate, three iterations, and the final quotient determination requires a shift, four subtractions and seven multiplications to produce a 16-b quotient. This is faster than a conventional nonrestoring divider if multiplication is roughly as fast as addition, a condition which is usually satisfied for systems which include hardware multipliers.

$$N = 0.625$$

$$D = 0.75$$

$R_{(0)}$	$= 2.915 - 2 \cdot D$	1 Shift, 1 Subtract
	$= 2.915 - 2 \cdot .75$	
$R_{(0)}$	$= 1.415$	

$R_{(1)}$	$= R_{(0)} (2 - D \cdot R_{(0)})$	2 Multiplys, 1 Subtract
	$= 1.415 (2 - .75 \cdot 1.415)$	
	$= 1.415 \cdot .93875$	
$R_{(1)}$	$= 1.32833125$	

$R_{(2)}$	$= R_{(1)} (2 - D \cdot R_{(1)})$	2 Multiplys, 1 Subtract
	$= 1.32833125 (2 - .75 \cdot 1.32833125)$	
	$= 1.32833125 \cdot 1.00375156$	
$R_{(2)}$	$= 1.3333145677$	

$R_{(3)}$	$= R_{(2)} (2 - D \cdot R_{(2)})$	2 Multiplys, 1 Subtract
	$= 1.3333145677 (2 - .75 \cdot 1.3333145677)$	
	$= 1.3333145677 \cdot 1.00001407$	
$R_{(3)}$	$= 1.3333333331$	

Q	$= N \cdot R_{(3)}$	1 Multiply
	$= 0.625 \cdot 1.3333333331$	
Q	$= 0.83333333319$	

FIGURE 19.12   Example of Newton–Raphson division.

19.4 Floating Point Arithmetic

Recent advances in VLSI have increased the feasibility of hardware implementations of floating point arithmetic units. The main advantage of floating point arithmetic is that its wide dynamic range virtually eliminates overflow for most applications.

Floating Point Number Systems

A floating point number A consists of a significand (or mantissa) S_a and an exponent E_a. The value of a number A is given by the equation

$$A = S_a r^{E_a} \tag{19.31}$$

where r is the radix (or base) of the number system. The significand is generally normalized by requiring that the most significant digit be nonzero. Use of the binary radix (i.e., $r = 2$) gives maximum accuracy but may require more frequent normalization than higher radices.

The IEEE Standard 754 single precision (32-b) floating point format, which is widely implemented, has an 8-b biased integer exponent, which ranges between 1 and 254 [IEEE 1985]. The exponent is expressed in excess 127 code so that its effective value is determined by subtracting 127 from the stored value. Thus, the range of effective values of the exponent is -126 to 127, corresponding to stored values of 1–254, respectively. A stored exponent value of 0 (E_{min}) serves as a flag for 0 (if the significand is 0) and for denormalized numbers. A stored exponent value of 255 (E_{max}) serves as a flag for infinity (if the significand is 0) and for "not a number." The significand is a 25-b sign magnitude mixed number (the binary point is to the right of the most significant bit, which is always a 1 except for denormalized numbers). More details on floating point formats and on the various considerations that arise in the implementation of floating point arithmetic units are given in Goldberg [1990], Gosling [1980], and Koren [1993].

Floating Point Addition

A flow chart for floating point addition is shown in Fig. 19.13. For this flow chart (and those that follow for multiplication and division), the significands are assumed to have magnitudes in the range $[1, 2)$.

On the flow chart, the operands are (E_a, S_a) and (E_b, S_b), and the result is (E_s, S_s). In step 1 the operand exponents are compared; if they are unequal, the significand of the number with the smaller exponent is shifted right in step 3 or 4 by the difference in the exponents to properly align the significands. For example, to add 0.867×10^5 and 0.512×10^4, the latter would be shifted right and 0.867 added to 0.0512 to give a sum of 0.9182×10^5. The addition of the significands is performed in step 5. Steps 6–8 test for overflow and correct if necessary by shifting the significand one position to the right and incrementing the exponent. Step 9 tests for a zero significand. The loop of steps 10 and 11 scales unnormalized (but non-0) significands upward to normalize the significand. Step 12 tests for underflow.

Floating point subtraction is implemented with a similar algorithm. Many refinements are possible to improve the speed of the addition and subtraction algorithms, but floating point addition will, in general, be much slower than fixed point addition as a result of the need for preaddition alignment and postaddition normalization.

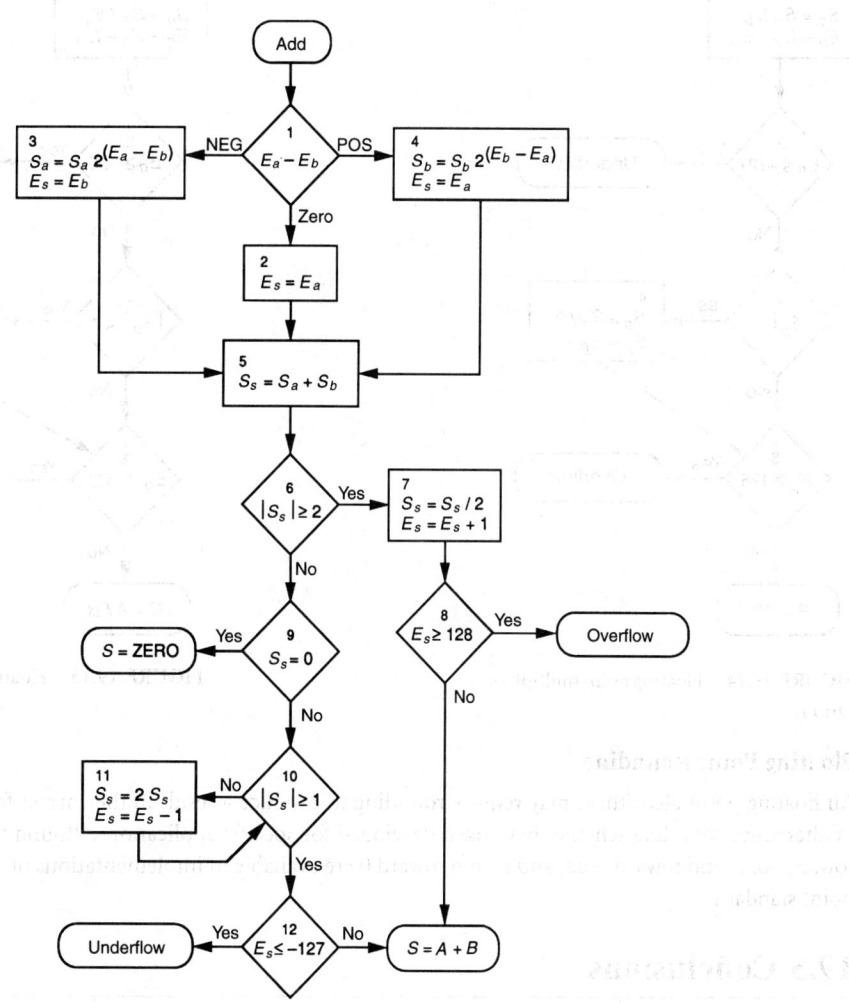

FIGURE 19.13 Floating point addition.

Floating Point Multiplication

The algorithm for floating point multiplication is shown in Fig. 19.14. In step 1, the product of the operand significands and the sum of the operand exponents are computed. Step 2 tests for underflow. Steps 3 and 4 normalize the significand if necessary. For radix 2 floating point numbers, if the operands are normalized, at most a single shift is required to normalize the product. Finally, step 5 tests the exponent for overflow.

Floating Point Division

The basic steps of a floating point division algorithm are shown in Fig. 19.15. The quotient of the significands and the difference of the exponents are computed in step 1. The second step tests for overflow. The quotient is normalized (if necessary) in steps 3 and 4 by shifting the quotient significand while the quotient exponent is adjusted appropriately. For radix 2, if the operands are normalized only a single shift is required to normalize the quotient. Finally, the computed exponent is tested for underflow in step 5.

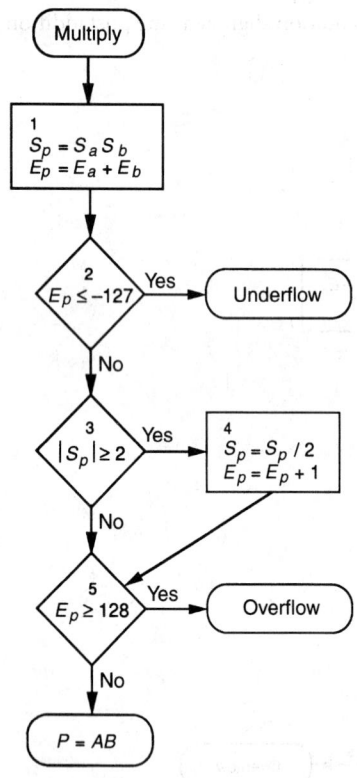

FIGURE 19.14   Floating point multiplication.

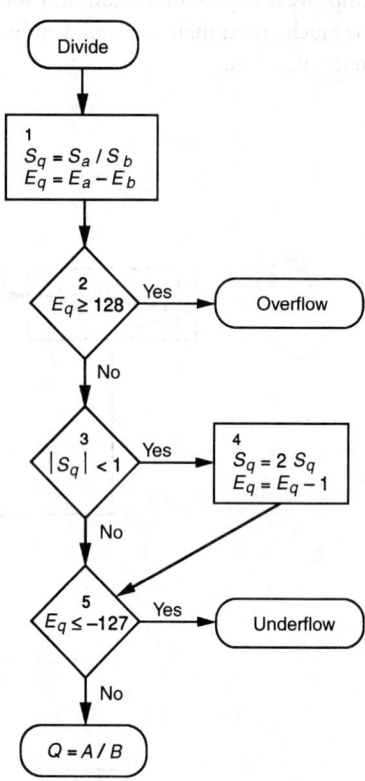

FIGURE 19.15   Floating point division.

Floating Point Rounding

All floating point algorithms may require rounding to produce a result in the correct format. A variety of alternative rounding schemes have been developed for specific applications. Round to nearest, round toward ∞, round toward $-\infty$, and round toward 0 are available in implementations of the IEEE floating point standard.

19.5 Conclusions

This chapter has presented an overview of binary number systems, algorithms for the basic integer arithmetic operations, and a brief introduction to floating point operations. When implementing arithmetic units there is often an opportunity to optimize the performance and the area to the requirements of the specific application. In general, faster algorithms require more area or more complex control: it is often useful to use the fastest algorithm that will fit the available area.

References

Baugh, C. R. and Wooley, B. A. 1973. A two's complement parallel array multiplication algorithm. *IEEE Trans. Comput.* C-22:1045–1047.

Blankenship, P. E. 1974. Comments on "A two's complement parallel array multiplication algorithm." *IEEE Trans. Comput.* C-23:1327.

Booth, A. D. 1951. A signed binary multiplication technique. *Q. J. Mech. Appl. Math.* 4(Pt. 2):236–240.

Chen, P. K. and Schlag, M. D. F. 1990. Analysis and design of CMOS Manchester adders with variable carry skip. *IEEE Trans. Comput.* 39:983–992.

Dadda, L. 1965. Some schemes for parallel multipliers. *Alta Frequenza* 34:349–356.

Ercegovac, M. D. and Lang, T. 1994. *Division and Square Root: Digit-Recurrence Algorithms and Their Implementations*. Kluwer Academic, Boston, MA.

Ferrari, D. 1967. A division method using a parallel multiplier. *IEEE Trans. Electron. Comput.* EC-16:224–226.

Goldberg, D. 1990. Computer arithmetic (App. A). In *Computer Architecture: A Quantitative Approach*, D. A. Patterson and J. L. Hennessy. App. A, Morgan Kauffmann, San Mateo, CA.

Gosling, J. B. 1980. *Design of Arithmetic Units for Digital Computers*. Macmillan, New York.

IEEE. 1985. *IEEE Standard for Binary Floating-Point Arithmetic*. IEEE Std 754-1985, Reaffirmed 1990. IEEE Press, New York.

Kilburn, T. , Edwards, D. B. G., and Aspinall, D. 1960. A parallel arithmetic unit using a saturated transistor fast-carry circuit. *Proc. IEE.* 107(Pt. B):573–584.

Koren, I. 1993. *Computer Arithmetic Algorithms*. Prentice–Hall, Englewood Cliffs, NJ.

MacSorley, O. L. 1961. High-speed arithmetic in binary computers. *Proc. IRE.* 49:67–91.

Robertson, J. E. 1958. A new class of digital division methods. *IRE Trans. Electron. Comput.* EC-7:218–222.

Shaw, R. F. 1950. Arithmetic operations in a binary computer. *Rev. Sci. Instrum.* 21:687–693.

Turrini, S. 1989. Optimal group distribution in carry-skip adders. In *Proc. 9th Symp. Computer Arithmetic*, pp. 96–103. IEEE Computer Society Press, Los Alamitos, CA.

Wallace, C. S. 1964. A suggestion for a fast multiplier. *IEEE Trans. Electron. Comput.* EC-13:14–17.

Waser, S. and Flynn, M. J. 1982. *Introduction to Arithmetic for Digital Systems Designers*. Holt, Rinehart and Winston, New York.

Weinberger, A. and Smith, J. L. 1958. A logic for high-speed addition. *Nat. Bur. Stand. Circular 591*, pp. 3–12. National Bureau of Standards, Washington, DC.

20

Parallel Architectures

Michael J. Flynn
Stanford University

Kevin W. Rudd
Stanford University

20.1 Introduction

Parallel or concurrent operation has many different forms within a computer system. Multiple computers can be executing pieces of the same program in parallel, or a single computer can be executing multiple instructions in parallel, or some combination of the two. Parallelism can arise at a number of levels: task level, instruction level, or some lower machine level. The parallelism may be exhibited in space with multiple independently functioning units, or in time, where a single function unit is many times faster than several instruction-issuing units. This chapter attempts to remove some of the complexity regarding parallel architectures (unfortunately, there is no hope of removing the complexity of programming some of these architectures, but that is another matter).

With all the possible kinds of parallelism, a framework is needed to describe particular instances of parallel architectures. One of the oldest and simplest such structures is the stream approach [Flynn 1966] that is used here as a basis for describing developments in parallel architecture. Using the stream model, different architectures will be described and the defining characteristics for each architecture will be presented. These characteristics provide a qualitative feel for the architecture for high-level comparisons between different processors—they do not attempt to characterize subtle or quantitative differences that, while important, do not provide a significant benefit in a larger view of an architecture.

20.2 The Stream Model

A parallel architecture has, or at least appears to have, multiple interconnected **processor elements** (PE in Fig. 20.1) that operate concurrently, solving a single overall problem. Initially, the various parallel architectures can be described using the *stream* concept. A stream is simply a sequence of objects or actions. Since there are both instruction streams and data streams (I and D in Fig. 20.1), there are four combinations that describe most familiar parallel architectures:

0-8493-2909-4/97/$0.00+$.50
© 1997 by CRC Press, Inc.

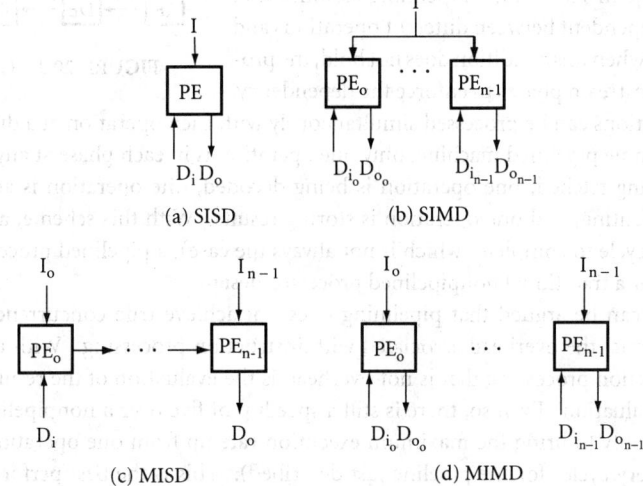

FIGURE 20.1 The stream model.

1. SISD—single instruction, single data stream. This is the traditional uniprocessor [Fig. 20.1(a)].
2. SIMD—single instruction, multiple data stream. This includes vector processors as well as massively parallel processors [Fig. 20.1(b)].
3. MISD—multiple instruction, single data stream. These are typically systolic arrays [Fig. 20.1(c)].
4. MIMD—multiple instruction, multiple data stream. This includes traditional multiprocessors as well as newer work in the area of networks of workstations [Fig. 20.1(d)].

The stream description of architectures uses as its reference point the programmer's view of the machine. If the processor architecture allows for parallel processing of one sort or another, then this information must be visible to the programmer at some level for this reference point to be useful.

An additional limitation of the stream categorization is that, while it serves as a useful shorthand, it ignores many subtleties of an architecture or an implementation. Even an SISD processor can be highly parallel in its execution of operations. This parallelism is typically not visible to the programmer even at the assembly language level, but it becomes visible at execution time with improved performance.

There are many factors that determine the overall effectiveness of a parallel processor organization, and hence its eventual speedup when implemented. Some of these, including networks of interconnected streams, will be touched upon in the remainder of this chapter. The characterizations of both processors and networks are complementary to the stream model and, when coupled with the stream model, enhance the qualitative understanding of a given processor configuration.

20.3 SISD

The SISD class of processor architectures is the most familiar class—it can be found in video games, home computers, engineering workstations, and mainframe computers. From the reference point of the assembly language programmer there is no parallelism in an SISD organization, yet a good deal of concurrency can be present. **Pipelining** is an early technique that is used in almost all current processor implementations. Other techniques aggressively exploit parallelism in executing code whether it is declared statically or determined dynamically from an analysis of the code stream.

Pipelining is a straightforward approach to exploiting parallelism that is based on concurrently performing different phases of processing an instruction. These phases often include fetching an **instruction** from memory, decoding an instruction to determine its **operation** and **operands**, accessing its operands, performing the computation specified by the operation, and storing the computed value (IF, DE, RF, EX,

and WB, respectively, in Fig. 20.2). Pipelining assumes that
these phases are independent between different operations and
can be overlapped—when this condition does not hold, the processor stalls the downstream phases to enforce the dependency.

FIGURE 20.2   Canonical pipeline.

Thus multiple operations can be processed simultaneously with each operation at a different phase of its
processing. For a simple pipelined machine, only one operation is in each phase at any given time—thus
one operation is being fetched, one operation is being decoded, one operation is accessing operands,
one operation is executing, and one operation is storing results. With this scheme, assuming that each
phase takes a single cycle to complete (which is not always the case), a pipelined processor could achieve
a speedup of five over a traditional nonpipelined processor design.

Unfortunately, it can be argued that pipelining does not achieve true concurrency but that it only
eliminates (in the limit) the overhead associated with instruction processing. With this viewpoint, the
only phase of instruction processing that is not overhead is the evaluation of the result—everything else
just supports this evaluation. Even so, there is still a speedup of five over a nonpipelined processor, but
this speedup serves only to bring the maximum execution rate up from one operation every five cycles
to one operation every cycle (for the pipeline just described). This is the best performance that can be
achieved by a processor without true parallel execution.

While pipelining does not necessarily lead to achieving true concurrency, there are other techniques
that do. These techniques use some combination of static scheduling and dynamic analysis to perform
concurrently the actual evaluation phase of several different operations—potentially yielding an execution rate of greater than one operation every cycle. This kind of parallelism exploits concurrency at the
computation level (in contrast to pipelining, which exploits concurrency at the overhead level). Since
historically most instructions consist of only a single operation (and thus a single computation), this kind
of parallelism has been named *instruction-level parallelism* (ILP).

Two architectures that exploit ILP—**superscalar** and **VLIW** (very long instruction word)—use radically
different techniques to achieve more than one operation per cycle. A superscalar processor dynamically
examines the instruction stream to determine which operations are independent and can be executed
concurrently. Figure 20.3(a) shows the issue of ready instructions from a window of available instructions
and the routing of these instructions to the appropriate function units (FU). A VLIW processor depends
on the compiler to analyze the available operations (OP) and to schedule independent operations into
wide instruction words; it then executes these operations in parallel with no further analysis. Figure
20.3(b) shows the issue of operations from a wide instruction word to all function units in parallel. In the
superscalar processor, even if two operations have been determined to be independent by the compiler
and are scheduled properly for the current processor, at execution time the dependency analysis must still
be performed to ensure that the proper ordering is maintained. In the VLIW processor, since the compiler is depended upon to ensure the proper scheduling of operations, any operations that are improperly
scheduled result in indeterminate (and probably bad!) results.

Considering the programmer's reference point, the kind of parallelism that superscalar processors
exploit is invisible, while the kind of parallelism that VLIW processors exploit is visible only at the assembly level where the explicit packing of mul-
tiple operations into instructions is visible—
the high-level language programmer in both
cases is isolated from the machine-exploited
ILP. Actually, these statements are true only in
the general sense; for the superscalar processor, the assembly language programmer may
be aware of the organization and characteristics of the machine and be able to schedule instructions so that they can be executed in parallel by the processor (this scheduling is usually
performed by the compiler, although there

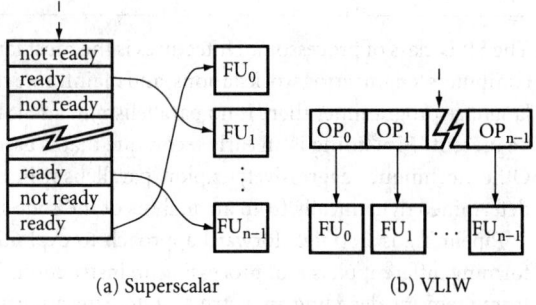

(a) Superscalar   (b) VLIW

FIGURE 20.3   Instruction issue for superscalar processors.

TABLE 20.1 Typical Scalar Processors

Processor	Year of Introduction	Number of Function Units	Issue Width	Scheduling	Issue/Complete Order
Intel x86	1978	2	1	Dynamic	In-order/In-order
Stanford MIPS-X	1981	1	1	Dynamic	In-order/In-order
Berkeley RISC-I	1981	1	1	Dynamic	In-order/In-order
Sun SPARC	1987	2	1	Dynamic	In-order/In-order
MIPS R3000	1988	2	1	Dynamic	In-order/In-order
MIPS R4000	1992	2	1	Dynamic	In-order/In-order
HP PA-RISC 1.1	1992	1	1	Dynamic	In-order/In-order

are assemblers which perform minor scheduling transformations). For the VLIW processor, the assembly language programmer must be aware of the specific characteristics of the machine to ensure the proper scheduling of operations (the assembler could perform the analysis and scheduling although this is typically not desired by the programmer). For both processors, even at the high-level language level there are ways of writing programs that make it easy for the compiler to find the latent parallelism.

Both superscalar and VLIW use the same compiler techniques to achieve their super-scalar performance. However, a superscalar processor is able to execute code scheduled for any instruction-compatible processor while a VLIW processor can only execute code that was specifically scheduled for execution on that particular processor.[1] This flexibility is not for free. While a superscalar processor does execute inappropriately scheduled code, the achieved performance can be significantly worse than if it were appropriately scheduled. Nevertheless, the flexibility is an important feature in a marketplace with a significant investment in software where binary compatibility is more important than raw performance.

A SISD processor has four defining characteristics. The first characteristic is whether or not the processor is capable of executing multiple operations concurrently. The second characteristic is the mechanism by which operations are scheduled for execution—statically at compile time, dynamically at execution, or possibly both. The third characteristic is the order in which operations are issued and retired relative to the original program order—these can be in order or out of order. The fourth characteristic is the manner in which exceptions are handled by the processor—precise, imprecise, or a combination. This last characteristic is not of immediate concern to the applications programmer, although it is certainly important to the compiler writer or operating system programmer, who must be able to properly handle exceptional conditions. Most processors implement precise exceptions, although a few high-performance architectures allow imprecise floating-point exceptions (with the ability to select between precise and imprecise exceptions).

Tables 20.1, 20.2, and 20.3 present some representative (pipelined) scalar and super-scalar (both superscalar and VLIW) processor families. As Table 20.1 and Table 20.2 show, the trend has been from a scalar to a compatible superscalar processor (except for the DEC Alpha and the IBM/Motorola/Apple PowerPC processors, which were designed from the ground up to be capable of super-scalar performance). There have been very few VLIW processors to date, although advances in compiler technology may cause this to change. Philips has explored VLIW processors internally for years, and the TM-1 is the first of a planned series of processors. After the demise of both Multiflow and Cydrome, HP acquired both the technology and some of the staff of these companies and has continued research in VLIW processors—at the time of writing, a joint HP–Intel venture is rumored to be focused on developing a new product line based on VLIW technology.

---

[1]This does not have to be true—although past VLIW processors have had this restriction, this has been due to engineering decisions in the implementation and not to anything inherent in the specification. For two variant approaches to providing support for dynamic scheduling in VLIW processors see [Rau 1993] and [Rudd 1994].

TABLE 20.2 Typical Superscalar Processors

Processor	Year of Introduction	Number of Function Units	Issue Width	Scheduling	Issue/Complete Order
DEC 21064	1992	4	2	Dynamic	In-order/In-order
Sun UltraSPARC	1992	9	4	Dynamic	In-order/Out-of-order
MIPS R8000	1994	6	4	Dynamic	In-order/Out-of-order
DEC 21164	1994	4	2	Dynamic	In-order/In-order
Motorola PowerPC 620	1995	6	4	Dynamic	In-order/Out-of-order
HP PA-RISC 8000	1995	10	4	Dynamic	Out-of-order/Out-of-order
MIPS R10000	1995	5	5	Dynamic	Out-of-order/Out-of-order
Intel Pentium Pro	1996	5	3	Dynamic	In-order x86 Out-of-order uops/Out-of-order
AMD K5	1996	6	4	Dynamic	In-order x86 Out-of-order ROPs/Out-of-order

TABLE 20.3 Typical VLIW Processors

Processor	Year of Introduction	Number of Function Units	Issue Width	Scheduling	Issue/Complete Order
Multiflow Trace 7/200	1987	7	7	Static	In-order/In-order
Multiflow Trace 28/200	1987	28	28	Static	In-order/In-order
Cydrome Cydra 5	1987	7	7	Static	In-order/In-order
Philips TM-1	1996	27	5	Static	In-order/In-order

20.4 SIMD

The SIMD class of processor architectures includes both array and vector processors. The SIMD processor is a natural response to the use of certain regular data structures, such as vectors and matrices. From the reference point of an assembly language programmer, programming an SIMD architecture appears to be very similar to programming a simple SISD processor except that some operations perform computations on aggregate data. Since these regular structures are widely used in scientific programming, the SIMD processor has been very successful in these environments.

Two types of SIMD processor will be considered: the **array processor** and the **vector processor**. They differ both in their implementations and in their data organizations. An array processor consists of many interconnected processor elements that each have their own local memory space. A vector processor consists of a single processor that references a single global memory space and has special function units that operate specifically on vectors.

Array Processors

The array processor is a set of parallel processor elements (typically hundreds to tens of thousands) connected via one or more networks (possibly including local and global interelement data and control communications). Processor elements operate in lockstep in response to a single broadcast instruction from a control processor. Each processor element has its own private memory, and data are distributed across the elements in a regular fashion that is dependent on both the actual structure of the data and also the computations to be performed on the data. Direct access to global memory or another processor element's local memory is expensive (although scalar values can be broadcast along with the instruction), so intermediate values are propagated through the array through local interelement connections. This requires that the data are distributed carefully so that the routing required to propagate these values is simple and regular. It is sometimes easier to duplicate data values and computations than it is to effect a complex or irregular routing of data between processor elements.

TABLE 20.4 Typical Array Processors

Processor	Year of Introduction	Memory Model	Processor Element	Number of Processors
Burroughs BSP	1979	Shared	General purpose	16
Thinking Machines CM-1	1985	Distributed	Bit-serial	Up to 65,536
Thinking Machines CM-2	1987	Distributed	Bit-serial	4,096–65,536
MasPar MP-1	1990	Distributed	Bit-serial	1,024–16,384

Since instructions are broadcast, there is no means local to a processor element of altering the flow of the instruction stream; however, individual processor elements can conditionally disable instructions based on local status information—these processor elements are idle when this condition occurs. The actual instruction stream consists of more than a fixed stream of operations—an array processor is typically coupled to a general-purpose control processor that provides both scalar operations (that operate locally within the control processor) as well as array operations (that are broadcast to all processor elements in the array). The control processor performs the scalar sections of the application, interfaces with the outside world, and controls the flow of execution; the array processor performs the array sections of the application as directed by the control processor.

A suitable application for use on an array processor has several key characteristics: a significant amount of data which has a regular structure; computations on the data which are uniformly applied to many or all elements of the data set; simple and regular patterns relating the computations and the data. An example of an application that has these characteristics is the solution of the Navier–Stokes equations, although any application that has significant matrix computations is likely to benefit from the concurrent capabilities of an array processor.

The programmer's reference point for an array processor is typically the high-level language level—the programmer is concerned with describing the relationships between the data and the computations, but is not directly concerned with the details of scalar and array instruction scheduling or the details of the interprocessor distribution of data within the processor. In fact, in many cases the programmer is not even concerned with size of the array processor. In general, the programmer specifies the size and any specific distribution information for the data, and the compiler maps the implied virtual processor array onto the physical processor elements that are available and generates code to perform the required computations. Thus, while the size of the processor is an important factor in determining the performance that the array processor can achieve, it is not a defining characteristic of an array processor.

The primary defining characteristic of a SIMD processor is whether the memory model is shared or distributed. In this chapter, only processors using a distributed memory model are described, since this is the configuration used by SIMD processors today and the cost of scaling a shared-memory SIMD processor to a large number of processor elements would be prohibitive. Processor element and network characteristics are also important in characterizing a SIMD processor, and these are described in sections 20.2 and 20.6.

There have not been a significant number of SIMD architectures developed, due to a limited application base and market requirement. Table 20.4 shows several representative architectures.

Vector Processors

A vector processor is a single processor that resembles a traditional SISD processor except that some of the function units (and registers) operate on vectors—sequences of data values that are seemingly operated on as a single entity. These function units are deeply pipelined and have a high clock rate; while the vector pipelines have as long or longer latency than a normal scalar function unit, their high clock rate and the rapid delivery of the input vector data elements results in a significant throughput that cannot be matched by scalar function units.

Early vector processors processed vectors directly from memory. The primary advantage of this approach was that the vectors could be of arbitrary lengths and were not limited by processor resources;

however, the high startup cost, limited memory system bandwidth, and memory system contention proved to be significant limitations. Modern vector processors require that vectors be explicitly loaded into special vector registers and stored back into memory—the same course that modern scalar processors have taken for similar reasons. However, since vector registers can rapidly produce values for or collect results from the vector function units and have low startup costs, modern register-based vector processors achieve significantly higher performance than the earlier memory-based vector processors for the same implementation technology.

Modern processors have several features that enable them to achieve high performance. One feature is the ability to concurrently load and store values between the vector register file and main memory while performing computations on values in the vector register file. This is an important feature, since the limited length of vector registers requires that vectors that are longer than the registers be processed in segments—a technique called strip mining. Not being able to overlap memory accesses and computations would pose a significant performance bottleneck.

Just like their SISD cousins, vector processors support a form of result bypassing—in this case called chaining—which allows a follow-on computation to commence as soon as the first value is available from the preceding computation. Thus, instead of waiting for the entire vector to be processed, the follow-on computation can be significantly overlapped with the preceding computation that it is dependent on. Sequential computations can be efficiently compounded and behave as if they were a single operation with a total latency equal to the latency of the first operation with the pipeline and chaining latencies of the remaining operations but none of the startup overhead that would be incurred without chaining. For example, division could be synthesized by chaining a reciprocal with a multiply operation. Chaining typically works for the results of load operations as well as normal computations. Most vector processors implement some form of chaining.

A typical vector processor configuration might have a vector register file, one vector addition unit, one vector multiplication unit, and one vector reciprocal unit (used in conjunction with the vector multiplication unit to perform division); the vector register file contains multiple vector registers (eight registers with 64 double-precision floating-point values is typical). In addition to the vector registers there are also a number of auxiliary and control registers, the most important of which is the vector length register. The vector length register contains the length of the vector (or the loaded subvector if the full vector length is longer than the vector register itself) and is used to control the number of elements processed by vector operations—there is no reason to perform computations on nondata that are useless or could cause an exception.

As with the array processor, the programmer's reference point for a vector machine is the high-level language. In most cases, the programmer sees a traditional SISD machine; however, since vector machines excel on vectorizable loops, the programmer can often improve the performance of the application by carefully coding the application—in some cases explicitly writing the code to perform strip mining—and by providing hints to the compiler that help to locate the vectorizable sections of the code. This situation is purely an artifact of the fact that the programming languages are scalar-oriented and do not support the treatment of vectors as an aggregate data type but only as a collection of individual values. As languages are defined (such as Fortran 90 or High Performance Fortran) that make vectors a fundamental data type, then the programmer is exposed less to the details of the machine and to its SIMD nature.

The vector processor has one primary characteristic. This characteristic is the location of the vectors—vectors can be memory- or register-based. There are many features that vector processors have that are not included here due to their number and many variations. These include variations on chaining, masked vector operations based on a Boolean mask vector, indirectly addressed vector operations (scatter/gather), compressed/expanded vector operations, reconfigurable register files, multiprocessor support, etc.

Vector processors have developed dramatically from simple memory-based vector processors to modern multiple-processor vector processors that exploit both SIMD vector and MIMD-style processing. Table 20.5 shows some representative vector processors.

TABLE 20.5 Typical Vector Processors

Processor	Year of Introduction	Memory- or Register-Based	Number of Processor Units	Maximum Vector Length	Number of Vector Units
Cray 1	1976	Register	1	64	7
CDC Cyber 205	1981	Memory	1	65,535	3–5
Cray X-MP	1982	Register	1–4	64	8
Hitachi HITAC S-810	1984	Register	1		4–8
Fujitsu FACOM VP-100/200[a]	1985	Register	3	32–1024	4
Cray 2	1985	Register	5	64	4
ETA ETA	1987	Memory	2–8	65,535	3–5
Cray C90		Register		64	
NEC SX-3	1990	Register	1–4		4–16
NEC SX-4	1994	Register	1–512		4–8
Cray T90	1995		1–32		2

[a] Sold as the Amdahl VP1100/VP1200 in the United States.

20.5 MISD

While it is easy to both envision and design MISD processors, there has been little interest in this type of parallel architecture. The reason, so far anyway, is that there are no ready programming constructs that easily map programs into the MISD organization.

Abstractly, the MISD can be represented as multiple independently executing function units operating on a single stream of data, forwarding results from one function unit to the next. On the microarchitecture level, this is exactly what the vector processor does. However, in the vector pipeline the operations are simply fragments of an assembly-level operation, as distinct from being a complete operation in themselves. Surprisingly, some of the earliest attempts at computers in the 1940s could be seen as the MISD concept. They used plugboards for programs, where data in the form of a punched card were introduced into the first stage of a multistage processor. A sequential series of actions were taken where the intermediate results were forwarded from stage to stage until at the final stage a result was punched into a new card.

There are, however, more interesting uses of the MISD organization. Nakamura [1995] has pointed out the value of an MISD machine called the SHIFT machine. In the SHIFT machine, all data memory is decomposed into shift registers. Various function units are associated with each shift column. Data are initially introduced into the first column and are shifted across the shift-register memory. In the SHIFT-machine concept, data are regularly shifted from memory region to memory region (column to column) for processing by various function units. The purpose behind the SHIFT machine is to reduce memory latency. In a traditional organization, any function unit can access any region of memory and the worst-case delay path for accessing memory must be taken into account. In the SHIFT machine, we must allow for access time only to the worst element in a data column. The memory latency in modern machines is becoming a major problem—the SHIFT machine has a natural appeal for its ability to tolerate this latency.

20.6 MIMD

The MIMD class of parallel architecture brings together multiple processors with some form of interconnection. In this configuration, each processor executes completely independently, although most applications require some form of synchronization during execution to pass information and data between processors. While there is no requirement that all processor elements be identical, most MIMD configurations are homogeneous with all processor elements being identical. There have been heterogeneous MIMD configurations that use different kinds of processor elements to perform different kinds of tasks, but (with the possible exception of recent work aimed at using networked workstations as a loosely coupled MIMD configuration) these configurations have not lent themselves to general-purpose applications. We limit ourselves to homogeneous MIMD organizations in the remainder of this section.

Up to this point, the MIMD processor with its multiple processor elements interconnected by a network appears to be very similar to a SIMD array processor. This similarity is deceptive, since there is a significant difference between these two configurations of processor elements—in the array processor the instruction stream delivered to each processor element is the same, while in the MIMD processor the instruction stream delivered to each processor element is independent and specific to each processor element. Recall that in the array processor, the instruction stream for each processor element is generated by the control processor and that the processor elements operate in lockstep. In the MIMD processor, the instruction stream for each processor element is generated independently by that processor element as it executes its program. While it is often the case that each processor element is running pieces of the same program, there is no reason that different processor elements could not run different programs.

The interconnection network in both the array processor and the MIMD processor passes data between processor elements; however, in the MIMD processor it is also used to synchronize the independent execution streams between processor elements. When the memory of the processor is distributed across all processors and only the local processor element has access to it, all data sharing is performed explicitly using messages and all synchronization is handled within the message system. When the memory of the processor is shared across all processor elements, synchronization is more of a problem—certainly messages can be used through the memory system to pass data and information between processor elements, but this is not necessarily the most effective use of the system.

When communications between processor elements is performed through a shared memory address space—either global or distributed between processor elements (called distributed shared memory to distinguish it from distributed memory)—there are two significant problems that arise. The first is maintaining **memory consistency**—the programmer-visible ordering effects of memory references both within a processor element and between different processor elements. The second is **cache coherency**— the programmer-invisible mechanism to ensure that all processor elements see the same value for a given memory location. Neither of these problems is significant in SISD or SIMD array processors. In a SISD processor, there is only one instruction stream and the amount of reordering is limited, so the hardware can easily guarantee the effects of perfect memory reference ordering and thus there is no consistency problem; since a SISD processor has only one processor element, cache coherency is not applicable. In a SIMD array processor (assuming distributed memory), there is still only one instruction stream and typically no instruction reordering; since all interprocessor element communication is via messages, there is neither a consistency problem nor a coherency problem.

The memory consistency problem is usually solved through a combination of hardware and software techniques. At the processor element level, the appearance of perfect memory consistency is usually guaranteed for local memory references only—this is usually a feature of the processor element itself. At the MIMD processor level, memory consistency is often guaranteed only through explicit synchronization between processors. In this case, all nonlocal references are only ordered relative to these synchronization points (such as fences or acquire/release points). While the programmer must be aware of the limitations imposed by the ordering scheme, the added performance achieved using nonsequential ordering can be significant. Table 20.6 shows the common memory consistency schemes and a brief description of their basic characteristics ($\sqrt{}$ indicates that the given feature exists, and $\sim \sqrt{}$ indicates that a restricted form of the feature exists).

The cache coherency problem is usually solved exclusively through hardware techniques. This problem is significant because of the possibility that multiple processor elements will have copies of data in their local caches and these copies could have differing values. There are two primary techniques to maintain cache coherency. The first technique is to ensure that all processor elements are informed of any change to the shared memory state—these changes are broadcast throughout the MIMD processor, and each processor element monitors these changes (commonly referred to as "snooping"). The second technique is to keep track of all users of a memory address or block in a directory structure and to specifically inform

TABLE 20.6 Simple Categorization of Consistency Models

Model[a]	Relaxation:[b] $W \to R$ Order	$W \to W$ Order	$W \to RW$ Order	Read Other's Write Early	Read Own Write Early	Explicit Synchronization[c]
SC					$\checkmark$	
TSO, PC	$\checkmark$			$\sim\checkmark$	$\checkmark$	RMW
PSO	$\checkmark$	$\checkmark$			$\checkmark$	RMW, fence
WO	$\checkmark$	$\checkmark$	$\checkmark$		$\checkmark$	Fence
RC	$\checkmark$	$\checkmark$	$\checkmark$	$\sim\checkmark$	$\checkmark$	Release, acquire, nsync, RMW

[a] SC = sequential consistency, TSO = total store order, PC = processor consistency, PSO = partial store order, WO = weak order, RC = release consistency.
[b] $x \to y$ represents the relaxation of the logical ordering between the reference x and a following reference y; R = read reference, W = write reference, RW = read or write reference.
[c] RMW is an atomic read–modify–write operation. Fetch-and-add is a common example of the general Fetch-and-Φ operation. (*Source:* Based on information in Adve and Gharachorloo [1995] and used with their permission.)

each user when there is a change made to the shared memory state. In either case the result of a change can be one of two things—either the new value is provided and the local value is updated, or all other copies of the value are invalidated.

As the number of processor elements in a system increases, a directory-based system becomes significantly better, since the amount of communications required to maintain coherency is limited to only those processors holding copies of the data. Snooping is frequently used within a small cluster of processor elements to track local changes—here the local interconnection can support the extra traffic used to maintain coherency since each cluster has only a few (typically two to eight) processor elements in it. Table 20.7 shows the common cache coherency schemes and a brief description of their basic characteristics ($\checkmark$ indicates that the given state exists).

The primary characteristic of a MIMD processor is the nature of the memory address space—it is either separate or shared for all processor elements. The interconnection network is also important in characterizing a MIMD processor and is described in the next section. With a separate address space (distributed memory), the only means of communications between processor elements is through messages, and thus these processors force the programmer to use a message-passing paradigm. With a shared address space (shared memory), communication between processor elements is through the memory system—depending on the application needs or programmer preference, either a shared-memory or a message-passing paradigm can be used.

The implementation of a distributed-memory machine is far easier than the implementation of a shared-memory machine when memory consistency and cache coherency are taken into account. However, programming a distributed-memory processor can be much more difficult, since the applications must be written to exploit and not be limited by the use of message passing as the only form of communications between processor elements. On the other hand, despite the problems associated with maintaining consistency and coherency, programming a shared-memory processor can take advantage of whatever communications paradigm is appropriate for a given communications requirement and can be much

TABLE 20.7 Cache Coherency Summary

Coherency Model	Protocol Type	Modification Policy	Exclusive Use State	Exclusive Write State
Write once	Invalidate	Copyback on first write		
Synapse $N + 1$	Invalidate	Copyback	$\checkmark$	
Berkeley	Invalidate	Copyback		$\checkmark$
Illinois	Invalidate	Copyback	$\checkmark$	
Firefly	Broadcast	Copyback private, writethrough shared	$\checkmark$	

TABLE 20.8   Typical MIMD Systems

System	Year of Introduction	Processor Element	Number of Processors	Memory Distribution	Programming Paradigm	Interconnection Type
Alliant FX/2800	1990	Intel i860	4–28	Central	Shared memory	Bus + crossbar
Stanford DASH	1992	MIPS R3000	4–64	Distributed	Shared memory	Bus + mesh
CRAY T3D	1993	DEC Alpha 61064	128–2048	Distributed	Shared memory	3D torus
MIT Alewife	1994	Sparcle	1–512	Distributed	Message passing	Mesh
Convex C4/XA	1994	Custom	1–4	Global	Shared memory	Crossbar
Thinking Machines CM-500	1995	SuperSPARC	16–2048	Distributed	Message passing	Fat tree
Tera Computers MTA	1995	Custom	16–256	Distributed	Shared memory	3D torus
SGI Power Challenge XL	1995	MIPS R8000	2–18	Global	Shared memory	Bus
Convex SPP1200/XA	1995	HP PA-RISC 7200	8–128	Global	Shared memory	Crossbar + ring
CRAY T3E	1996	DEC Alpha 61164	16–2048	Distributed	Shared memory	3D torus
Network of Workstations	1990s	Various	Any	Distributed	Message passing	Ethernet

easier to program. Both distributed and shared-memory processors can be extremely scalable and neither approach is significantly more difficult to scale than the other. Some typical MIMD systems are described in Table 20.8.

20.7 Network Interconnections

Both SIMD array processors and MIMD processors rely on networks for the transfer of data between processor elements or processors. A bus is a simple kind of network—it serves to interconnect all devices that are plugged into it—but is not commonly referred to as a network. We discuss here only the aspects of networks that are of interest in characterizing a processor—particularly the SIMD array processors and MIMD processors—and present some network characteristics that provide a qualitative sense that is useful for understanding the basic nature of a multiprocessor interconnect.

There are three primary characteristics of networks. The first is the method used to transfer the information through the network—either using packet routing or circuit switching. The second characteristic is the mechanism that connects source and destination nodes—either the connections are static and fixed or they are dynamic and reconfigurable. The third characteristic is whether the network is a single-level or a multiple-level network. While these characteristics leave out a significant amount of detail about the actual network, they qualitatively describe the network connections and how information is routed between processor elements.

Packet routing is efficient for small random packets, but it has the drawback that neither the latency nor the bandwidth is necessarily deterministic and thus packets may not be delivered in the same order that they were sent; circuit switching achieves high bandwidth for a given connection between processor nodes and guarantees uniform latency and proper receipt ordering, but it has the drawback that the latency for small packets becomes the latency for setting up and breaking down the connection.

Dynamic networks allow network reconfiguration so that there are essentially direct connections between nodes across the network, producing high bandwidth and low latency but limiting the scalability of the system; static networks improve the scalability, since connections are node to node and any two nodes can be connected either directly or through intermediate nodes, resulting in longer latency and lower-bandwidth connections. Use of multilevel networks, which use clusters of processor elements at each network node, increases the complexity of the system but reduces congestion on the global interconnect and leads to a more scalable system—intracluster communications are performed on a local interconnect that is much faster and does not leave the cluster. Single-level networks are more general but less scalable, since all communications must use the global interconnect, and traffic can be much higher for the same number of processor elements.

20.8 Afterword

In this article we have reviewed a number of different parallel architectures organized by the stream model. We have described some general characteristics that offer some insight into the qualitative differences between different parallel architectures but, in the general case, provide little quantitative information about the architectures themselves—this would be a much more significant task, although there is no reason why a significantly increased level of detail could not be described. Just as the stream model is incomplete and overlapping (consider that a vector processor can be considered to be a SIMD, MISD, or SISD processor depending on the particular categorization desired), so the characteristics for each class of architecture are also incomplete and overlapping. However, the general insight gained from considering these general characteristics leads to an understanding of the qualitative differences between gross classes of computer architectures, so the characteristics that we have described provide similar benefits and liabilities.

In a sense, the characterizations that we have provided for each architectural class of processor can be considered as specializations on the stream model. Thus a superscalar processor could be described as a SISD processor which supports concurrent execution of multiple operations that are scheduled dynamically, performs issue and retire out of order, and provides precise interrupts. A similar superscalar processor that does not provide precise interrupts would be described almost identically, but the description would provide an insight into one significant difference between these two processors—the first superscalar processor would be more complicated to design and might run more slowly than the second superscalar processor, but it would support more efficient exception recovery. While this comparison does not, in all likelihood, provide sufficient information to make a design choice in many cases, it does provide a basis for processor comparison.

For a MIMD or a SIMD processor, although the primary characteristic is the characterization of the memory address space, the system can be more completely described by including the description of both the processor element (most likely a SISD processor) along with a description of any networks that are included in the processor. This results in a description of the processor that provides a more complete understanding of the system as a whole but now including much more information about the remainder of the system.

This is not meant to imply that the aggregate of the stream model along with the relevant characteristics is a complete and formal extension to the original taxonomy—far from it. There are still a wide range of processors that are problematic to describe well in this (and likely in any) framework. The example was given earlier concerning the appropriate placement of a vector processor. Another example is the proper placement of an architecture that is designed to support multiple threads on a single processor element. These processor elements could be considered to be just SISD processors which have a specialized operating system that provides this support (albeit requiring hardware support as well), but there is some reason to believe that multiple threads are a significant feature, especially in the case of the Tera MTA architecture [Alverson et al. 1990], where the threads are interleaved through the execution units on a cycle-by-cycle basis—clearly a distinct difference beyond simply performing efficient task switches.

Whatever the problems with classifying and characterizing a given architecture, processor architectures, particularly multiprocessor architectures, are developing rapidly. Much of this growth is the result of significant improvements in compiler technology that allow the unique capabilities of an architecture to be efficiently exploited. In many cases, the design of a system is based on the ability of a compiler to produce code for it. It may be that a feature is unable to be utilized if a compiler cannot exploit it and thus the feature is wasted (although perhaps the inclusion of such a feature would spur compiler development). It may also be that an architectural feature is added specifically to support a capability that a compiler readily supports and thus performance is improved. Compiler development is clearly an integral part of system design and architectural effectiveness is no longer limited only to concerns for the processor itself.

Defining Terms

Array processor: An array of processor elements operating in lockstep in response to a single instruction and performing computations on data that are distributed accross the processor elements.

Cache coherency: The programmer-invisible mechanism that ensures that all caches within a computer system have the same value for the same shared-memory address.

Instruction: Specification of a collection of operations that may be treated as an atomic entity with a guarantee of no dependencies between these operations. A typical processor uses an instruction containing one operation.

Memory consistency: The programmer-visible mechanism that guarantees that multiple processor elements in a computer system receive the same value on a request to the same shared-memory address.

Operand: Specification of a storage location—typically either a register of a memory location—that provides data to or receives data from the results of an operation.

Operation: Specification of one or a set of computations on the specified source operands placing the results in the specified destination operands.

Pipelining: The technique used to overlap stages of instruction execution in a processor so that processor resources are more efficiently used.

Processor element: The element of a computer system that is able to process a data stream (sequence) based on the content of an instruction stream (sequence). A processor element may or may not be capable of operating as a stand-alone processor.

Superscalar processor: A popular term to describe a processor that dynamically analyzes the instruction stream and attempts to execute multiple ready operations independently of their ordering within the instruction stream.

Vector processor: A computer architecture with specialized function units designed to operate very efficiently on vectors represented as streams of data.

VLIW processor: A popular term to describe a processor that performs no dynamic analysis on the instruction stream and executes operations precisely as ordered in the instruction stream.

References

Adve, V. and Gharachorloo, K. 1995. Shared memory consistency models: a tutorial. Research Report 95/7. Digital Equipment Corp. Western Research Lab.

Alverson, R., Callahan, D., Cummings, D., Koblenz, B., Porterfield, A., and Smith, B. 1990. The Tera computer system. In *Int. Conf. on Supercomputing*, ACM, New York, June.

Flynn, M. J. 1966. Very high-speed computing systems. *Proc. IEEE* 54(12):1901–1909.

Flynn, M. J. 1995. *Computer Architecture: Pipelined and Parallel Processor Design*. Jones and Bartlett, Boston.

Hennessy, J. L. and Patterson, D. A. 1996. *Computer Architecture A Quantitative Approach*, 2nd ed. Morgan Kaufmann, San Francisco.

Hockney, R. W. and Jesshope, C. R. 1988. *Parallel Computers: Architecture, Programming and Algorithms*. Adam Hilger, Bristol.

Hwang, K. 1993. *Advanced Computer Architecture: Parallelism, Scalability, Programmability*. McGraw–Hill, New York.

Ibbett, R. N. and Topham, N. P. 1989a. *Architecture of High Performance Computers. Vol. I. Uniprocessors and Vector Processors*. Springer–Verlag, New York.

Ibbett, R. N. and Topham, N. P. 1989b. *Architecture of High Performance Computers, Vol. II. Array Processors and Multiprocessor Systems*. Springer–Verlag, New York.

Lilja, D. J. 1991. *Architectural Alternatives for Exploiting Parallelism*. IEEE Computer Society Press, Los Alamitos, CA.

Nakamura, T. 1995. The SHIFT machine. Personal correspondence.

Rau, B. R. 1993. Dynamically scheduled VLIW processors, pp. 80–92. In *26th Annual International Symp. on Microarchitecture*.

Rudd, K. W. 1994. Instruction level parallel processors—a new architectural model for simulation and analysis. Technical Report CSL-TR-94-657. Stanford Univ.

Trew, A. and Wilson, G., eds. 1991. *Past, Present, Parallel: A Survey of Available Parallel Computing Systems.* Springer–Verlag, London.

Further Information

There are many good sources of information on different aspects of parallel architectures. The references for this chapter provide a selection of texts that cover a wide range of issues in this field. There are many professional journals that cover different aspects of this area either specifically or as part of a wider coverage of related areas. Some of these are:

IEEE Transactions on Computers, Transactions on Parallel and Distributed Systems, Computer, Micro.

ACM Transactions on Computer Systems, Computer Surveys.

Journal of Supercomputing.

Journal of Parallel and Distributed Computing.

There are also a number of conferences that deal with various aspects of parallel processing. The proceedings from these conferences provide a current view of research on the topic. Some of these are:

International Symposium on Computer Architecture (ISCA).

Supercomputing.

International Symposium on Microarchitecture (MICRO).

International Conference on Parallel Processing (ICPP).

International Symposium on High Performance Computer Architecture (HPCA).

Symposium on the Frontiers of Massively Parallel Computation.

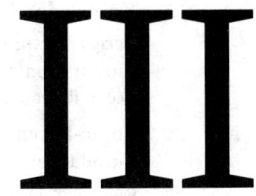

Artificial Intelligence and Robotics

Harold Abelson, Section Adviser
MIT

A RTIFICIAL INTELLIGENCE (AI) is the study of the computations that make it possible to simulate human perception, reasoning, and action. Current efforts are aimed at constructing computational mechanisms that process visual data, understand speech and written language, control robot motion, and model physical and cognitive processes. Robotics is a complex field, drawing heavily from AI as well as other areas of science and engineering.

21

Explanation-Based Learning

Gerald DeJong
University of Illinois at Urbana-Champaign

21.1 Introduction

The ability to learn is one of the most important facets of intelligence. Acquiring concepts from experience is at the heart of learning. Experiences provide us with information on how the world behaves in specific circumstances. From these examples we must form a general characterization, which can be employed to predict the future, to diagnose troubles, and to influence the world in desirable ways. Thus, concept acquisition consists of deriving a general truth about the world from positive and perhaps negative exemplars. The acquired general truth is called the **concept** or **concept descriptor** and the collection of exemplars is called the **training set**.

Suppose that we wish to construct a concept descriptor for where John would like to live. Acceptable cities include San Francisco, Paris, Boston, and Toronto, but he would not like to live in Peoria, Boise, Topeka, or El Paso. What principle would accurately predict John's attitude toward cities in general? We might guess that John wants to live in a large city. Or perhaps he only likes to live in cities that have a prestigious university near by. A plethora of general descriptors are consistent with this training set. As we discover John's opinion of other cities, the set of possible concept descriptors narrows.

But we wish to investigate a more specific phenomenon. Many candidate concepts such as these seem reasonable given this training set. Interestingly, others just as accurate on the training set seem far less reasonable. The positive examples all have poorer air quality than the negative ones. Perhaps John wants to live in a place with unhealthy air pollution levels. Perhaps John wishes to avoid cities with the letter "e" in their names. We as humans have a natural predisposition toward certain descriptors and away from others. Indeed, this *common sense* ability to avoid silly hypotheses is central to the computational tractability of human concept acquisition. Human common sense is intuitively compelling but is difficult to define or implement. Why do these latter concept descriptors seem worse than the former ones? Whereas they are as accurate on the available training data, we are quite correctly skeptical of their accuracy on future untested cities.

The reason for this preference is, of course, our prior knowledge of the world. Existing beliefs greatly influence our interpretation of the training examples. We believe John is a human like us, and we know a great deal about human desires and priorities. An automated learning system must incorporate similar prior world knowledge along with the training data as evidence for and against concept descriptors. Explanation-based learning (EBL) offers a systematic way to incorporate such prior world knowledge into an automated inductive concept learning system.

21.2 Background

Induction can be defined as reasoning from the specific to the general. In the example, we conjectured general characterizations of John's preference from his opinions of specific cities. Importantly, inductive conclusions may be incorrect (unlike the deductive conclusions of conventional logic). Only rarely or in artificial circumstances can we be certain that we have arrived at the correct generalization. Instead, the labeled examples of the training set count as evidence for or against the various generalizations. Forming a generalization can be characterized as an empirical or statistical conclusion. The large literature that exists on statistical inference is directly relevant here.

To formalize induction we define two spaces: an instance space and a concept space. The **instance space** is the set of all items of interest in the world. In the example, this is the set of all possible cities. Note that the instance space is a mathematical construct and we need not explicitly represent all of its elements. It is sufficient to specify set membership rules and to be able to assert properties about some small number of explicitly represented elements. What city properties need to be represented? For this example, we might include the city's population, whether or not it has a prestigious university, number of parks, and so on. The instance space must represent enough properties to support the distinctions required by the concept descriptor.

Most often each instance is represented as a vector of conjoined feature values. The values themselves are ground predicate logic expressions. Thus, we might represent one city, CITY381, as

name=Toronto ∧ population=697,000 ∧ area=221.77 ∧ . . .

An equal sign separates each feature name from its value. The symbol ∧ indicates the logical AND connective. We interpret the expression as saying that CITY381 is equivalent to something whose name is Toronto, and whose population is 697,000, and . . . . Thus, CITY381 is now defined by its features. As in other logical formalisms the actual symbol used to denote this city is arbitrary. If CITY381 were replaced everywhere by CITY382, nothing would change. One can imagine including many additional features such as the city's crime rate or the average housing price.

Toronto is a city in which John would be happy. Thus, its classification is positive. For John, city classification is binary (each city is preferred by John or it is not). In other applications (e.g., identifying aircraft types from visual cues or diagnosing tree diseases from symptoms) many different classes may exist.

Figure 21.1 is a graphical depiction of the instance space for John's city preference. San Francisco, Paris, Boston, and Toronto (the four positive examples in the training set) are represented by the symbol +; Peoria, Boise, Topeka, and El Paso are marked as −. For pedagogical purposes we will simply use two dimensions.

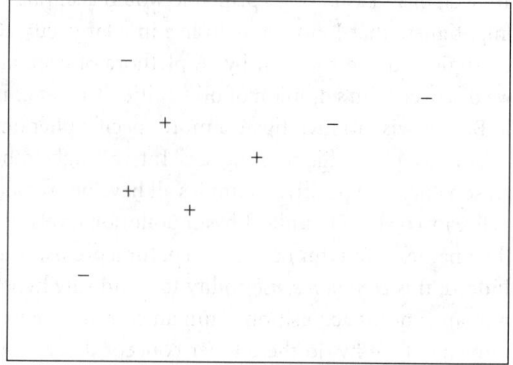

FIGURE 21.1   The training set.

Importantly, distinct cities correspond to distinct points in the space.

A concept descriptor is any function that partitions the instance space. Figure 21.2 depicts three sample concept descriptors for John's city preference. Each concept classifies some instances as positive (those within its boundary) and the others as negative. Concept C1 misclassifies four cities; one undesirable city is included and three desirable cities are excluded. Concepts C2 and C3 both correctly classify all eight cities of the training set but embody quite different partitions of the instance space. Thus, they

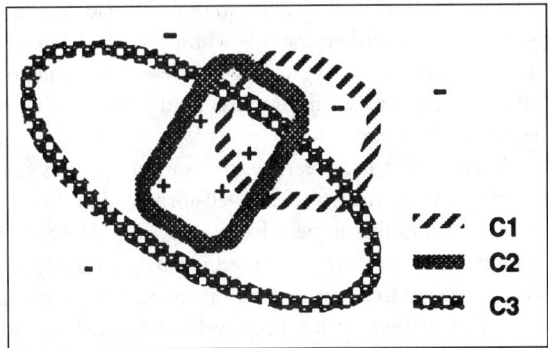

FIGURE 21.2   Three concept descriptors.

venture different labelings for unseen cities. C2 might classify all large cities as desirable whereas C3 might classify any city without the letter e in its name as desirable.

A concept descriptor can be represented using the same features employed to represent instances. Constrained variables augment ground expressions as feature values. Thus, a concept descriptor for large safe cities might be

$$\text{name}=?x \wedge \text{population}=?y \wedge \text{area}=?z \wedge \text{crime-index}=?w$$

with constraints $?y > 500,000$ and $?w < .03$ while variables $?x$ and $?z$ are left unconstrained

This concept descriptor is a conjunction of two features. It denotes all cities with a large population and a low crime rate. We might choose to allow disjunctions or other connectives. We might also allow more complex relations to serve as constraints among descriptor variables. By increasing the expressive power of the concept descriptor vocabulary we allow more varied partitioning of the instance space. The set of all expressible concept descriptors is the **concept space**. Just as looking for the same needle is more challenging in a larger haystack, increasing the size of the concept space by allowing a more expressive concept vocabulary can make the learning problem more difficult.

A **target concept** is the unknown but objectively correct partitioning of the instance space for the task at hand. In this example the target concept is embodied by John himself who is the unimpeachable final authority on where he would like to live. Sometimes, as in this example, the target concept is fundamentally unknowable. Most likely we cannot fully capture John's city preferences. He might, for example, have an unnatural fear of suspension bridges, a fondness for large groves of maple trees, or other subtle or peculiar personality quirks requiring distinctions not possible with our concept vocabulary. But there may still be good approximations to the target concept within the concept space.

In some applications we might be certain that within the concept space lurks the true target concept. Even then we may prefer an approximation. The target concept may be very complex, difficult to learn, or expensive to use, whereas a simpler approximation might be almost as accurate and much more efficient. Additionally, in most applications there is the possibility of noise. Some elements of the training set may be misclassified, or the values of some features may be incorrect. Requiring perfect classification accuracy over the training set in the presence of noise can corrupt the learning process perhaps precluding concept acquisition altogether.

Thus, the concept acquisition task is to find an element of the concept space that is a good approximation to the target concept. We have reduced the slippery problem of inventing a new concept to the problem of finding a descriptor from a pre-existing space of possible answers. Have we trivialized the concept formation task? No, this restriction is quite mild. In fact, we have only committed to a representational vocabulary. This is nothing more than a syntactic well-formedness criterion and need not be overly constraining. It is a bit like a publisher commissioning a novel with the stipulation that the manuscript be in English or requiring that it be typed with roman characters. Such restrictions are severe in the sense that they rule out far more items in our universe than they permit, but the author's creative latitude is not significantly diminished. So it is with properly defined concept spaces. The commitment to a particular representation vocabulary for concept descriptors rules out far more things than it permits. But a properly defined concept space still supports a sufficiently rich variety of concept descriptors so as not to trivialize the learning process.

A learning algorithm searches the concept space as guided by the training data. It entertains alternative hypotheses from the set of all well-formed descriptors until sufficient training set evidence confirms one or indicates that none is likely to be found. The search is not uniquely determined by the training set elements. The concept formed is partly a function of the learning algorithm's characteristics. Such concept preferences that go beyond the training data are collectively termed the **inductive bias** of the learning algorithm. It has been well established that inductive bias is inescapable in the formation of general concepts. Incidentally, among the important implications of this result is the impossibility of any form of Lockean tabula rasa development.

Discipline is necessary in formulating the representational inductive bias. A concept vocabulary that is overly expressive can dilute the concept space and greatly increase the complexity of concept acquisition. On the other hand, a vocabulary that is not expressive enough may preclude finding an adequate concept or (more often) may trivialize the search so that the algorithm is condemned to find a desired concept without relying on the training data as it should.

21.3 Explanation-Based and Similarity-Based Learning

Explanation-based learning is best viewed as a principled method for extracting the maximum information from the training examples. It works by constructing and generalizing explanations for the training examples. The explanations interpret the examples, bringing into focus those aspects which are important for classification. The explanations also augment the examples, filling in by inference aspects relevant to their classification.

The conventional approach to inductive concept formation is sometimes called similarity-based learning (SBL) to distinguish it from explanation-based learning. We now explore a brief and intuitive example illustrating the difference between the explanation-based and similarity-based approaches.

Suppose we are lost in the jungle with our pet gerbil. We have only enough food to keep ourselves alive and decide that bugs, which are plentiful and easy to catch, will have to suffice for the gerbil. Unfortunately, a significant number of insectlike jungle creatures are poisonous. To save our pet we must quickly acquire a descriptor that identifies nonpoisonous bugs.

Again, we represent examples as feature/value pairs. Features might include a bug's number of legs, the number of body parts, the average size, the bug's coloring, how many wings it has, what it seems to like to eat, what seems to like to eat it, where it lives, a measure of how social it is, etc. One insect we see might be represented as

$$\text{legs}=6 \land \text{body-parts}=3 \land \text{size}=2\text{cm} \land \text{color}=\text{bright-purple}$$

$$\land \text{wings}=4 \land \text{wing-type}=\text{folding} \land \dots$$

Let us call this bug I7 for the 7th instance of a bug that we catch. The bug representation vocabulary can also serve as the concept descriptor vocabulary so long as we allow constrained variables to be feature values. A concept descriptor for nonpoisonous bugs might be something like

$$\text{legs}=?\text{x1} \land \text{body-parts}=?\text{x1}+2 \land \text{size}>1.5\text{cm} \land \text{color}=\text{purple}$$

which says that anything that has twice as many legs as body parts, is over 1.5 cm, and is purple will be considered to be nonpoisonous. This descriptor includes insects like I7 and long dark purple centipedes but excludes spiders (because they do not have enough body parts for their 8 legs) and yellow butterflies (because they are the wrong color).

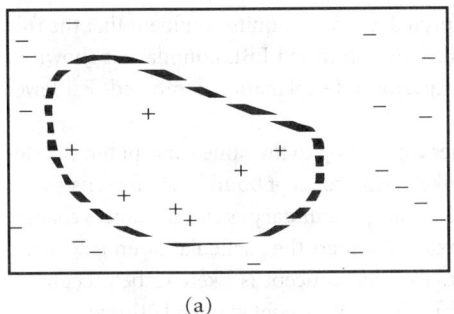
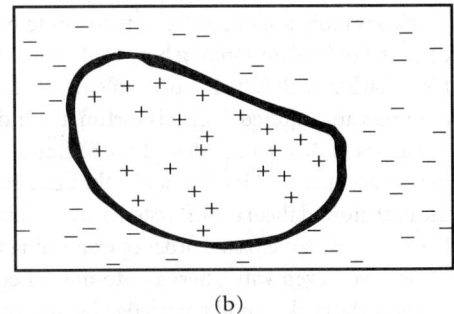

(a)   (b)

FIGURE 21.3 Training examples for an SBL concept: (a) hypothesized and (b) confirmed.

Similarity-based learning uncovers emergent patterns in the instance space by viewing the training data as representative of all instances of interest. Patterns found within the training set are likely to be found in the world provided the training set is a statistically adequate sample. In this example, we feed sample bugs to our gerbil to view the effect. An SBL system searches for an adequate concept descriptor by sampling broadly over the available bugs, revising estimates of the likelihood of various descriptors as evidence mounts. We may observe that the gerbil becomes nauseous and lethargic after eating I7 so we label it as poisonous. Next, I8, a reddish brown many-legged worm which we find hidden on the jungle floor, is consumed with no apparent ill effects. It is labeled as nonpoisonous. This process continues for several dozen bugs. A pattern begins to emerge. This is illustrated in Fig. 21.3(a). We can only be confident of the descriptor after testing it on a statistically significant sample of bugs. This number can be quite large [see Fig. 21.3(b)]. Perhaps after sampling several hundred we adopt the pattern that bugs with more than 10 legs whose size is less than 3 cm are mostly nonpoisonous.

legs>10 ∧ size<3cm

By contrast, the EBL approach searches for an explanation of how gerbil poisoning/nonpoisoning may be connected to the observed bug's features by using our background knowledge of the world. After observing that I7 makes our gerbil ill, we might recall the fact that poisonousness is a characteristic evolved to protect individuals of a species from being eaten. Furthermore, it is an effective deterrent only if the poisonous species itself is easily identifiable by its would-be predator. Thus, the bug should have a unique appearance, sharp patterns, bright colors, great regularity among individuals, etc. This explains the poisonousness of I7: the bright purple coloring is relevant, the number of legs, the fact that it feeds on hemlock blossoms, etc., are irrelevant. The explanation also indicates that bugs that are plain-looking, camouflaged, or change their color to blend into their surroundings, are probably not poisonous. Such an explanation provides evidence for and against the nonpoisonousness of many untested bugs. The palatability of I8 is justified by its plainness. The explanation applies to many yet unseen instances. Figure 21.4(a) shows the negative example (I7) and the positive example (I8) along with the explanation boundary of each, the subspace beyond which their justifications do not hold.

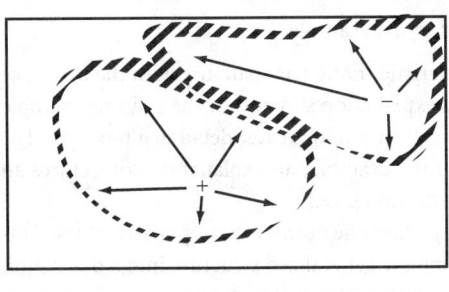
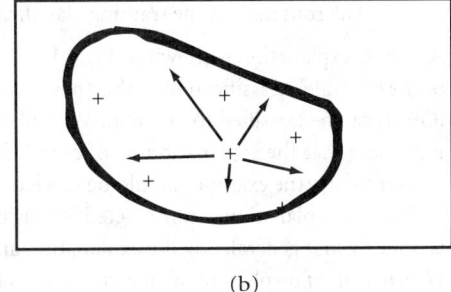

(a)   (b)

FIGURE 21.4 Training examples for an EBL concept: (a) hypothesized and (b) confirmed.

If a few more test examples all turn out to be edible as expected, we can be quite confident that the rule of eating only plain-looking bugs will suffice for our gerbil. The confirmed EBL boundary is shown in Fig. 21.4(b). With EBL, the pattern emerges from the training data and explanation combined. Far fewer examples are required, which is fortunate for our gerbil.

Figures 21.3(b) and 21.4(b) show EBL and SBL approaches converging to the same concept boundaries in instance space. This need not be the case. Indeed it is unlikely. EBL concept boundaries are sensitive to the background theory. SBL systems also exhibit variations in concept boundary as circumstances change. For some SBL systems the order of examining the training examples alters the particular boundary so on.

However, even with alternate **domain theories** the EBL-acquired concept is likely to be acceptable. Suppose the background knowledge includes additional information. We might explain I7 this way: Bugs are very simple organisms; they are more likely to collect and concentrate poison from elsewhere than to manufacture it within their bodies. If we also know a good deal about which plants are poisonous we might choose only those bugs which, like I8, are seen to prefer feeding on nonpoisonous plants. Additional background beliefs may yield further explanations. We might know that gerbils are rodents similar to jungle mice. We notice that jungle mice feed almost exclusively on grubs of the mound-building beetles which construct vast colonies in jungle clearings. We might reason that mound-building beetle grubs are, therefore, likely to be edible for our gerbil. After a few sample grubs are successfully devoured we might acquire a concept descriptor that fits only these grubs. Although quite different boundaries result, each concept descriptor is adequate for the task of keeping our gerbil alive and, thus, satisfies the purpose of the concept acquisition task.

21.4 Explanation-Based Learning

As we have seen, by getting the most out of each training example EBL can acquire a concept descriptor using relatively few training examples. The price for such example efficiency is the requirement of a domain theory. EBL also requires that a functional goal specification be provided in addition to the training set. Acquiring a concept descriptor with EBL consists of three steps:

1. Constructing one or more explanations from training examples
2. Generalizing the explanation(s) to construct a set of hypothesized concept descriptors
3. Selecting one or more of the descriptors to apply to the task at hand

The first step, constructing explanations, employs an inference mechanism over the domain theory. For our purposes, a domain theory is any set of prior beliefs about the world and an inference mechanism is any procedure that draws conclusions by combining beliefs. The inference mechanism will generally be some logical procedure such as first-order resolution. However, any method of combining beliefs to yield new beliefs is quite acceptable; inference by analogy and other unsound procedures is perfectly consistent with explanation-based learning.

An **explanation** for a training example is any tree-structured graph with the following properties:

- Each leaf node is either a prior belief or a property of the example being explained.
- Each nonleaf node is the result of applying the inference procedure to prior nodes.
- The root node is the training classification assigned to the example.

A sample explanation is shown in Fig. 21.7. It has a characteristic triangular structure with the inference of the example's classification at the apex (root node). This explanation shows how the training example (OBJ1) can be classified to be a cup. We will examine this explanation in greater detail in a moment. For now, note that the explanation justifies the classification of the example; an explanation conjectures an answer to *why* the example should be labeled with the classification given.

Once an explanation is constructed, it may be used to conjecture one or more concept descriptors. This is done by first generalizing the explanation and transforming the generalized structure into a descriptor. *Generalization* involves removing constraints from the explanation. An explanation by its very nature is narrowly focused. It applies to the training example but to little else. To yield a useful descriptor the range of applicability must be broadened. This is done by removing constraints while maintaining the veracity of the conclusion. For instance, we might be given a particular circuit to illustrate a two-stage amplifier. Suppose that the coupling capacitor between the two amplifier stages is $2\,\mu F$. Building in a requirement that coupling capacitors must always be $2\,\mu F$ is overly constraining and will probably not result in a useful concept. Our background knowledge about capacitor characteristics dictates that the signal strength at the output by the capacitor is a function of the signal strength input, the signal frequency, etc. Generalizing the explanation involves removing constraints that tie the explanation to the training example. What remains are the functional requirements and mutual constraints for a broader class of two-stage amplifiers. The *transformation* of a general explanation into a concept descriptor involves collecting the remaining constraints among the leaves of the generalized explanation. Any object that satisfies these constraints is to be given the same classification. Among the many constraints that form the concept descriptor for a two-stage amplifier is the one requiring a sufficient coupling between the two amplifier stages just discussed. For a 20-MHz frequency application, the concept descriptor may dictate, via the remaining constraints among its variables, that the coupling capacitor must be in the range of 10–20 pF.

The *selection* of an hypothesized concept descriptor requires that it improve the performance of the application system. Even though a concept is accurate, it may not be useful. That is, its inclusion may degrade rather than improve system performance. Note that this need not be due to deficiencies of the domain theory. The theory of linear amplification is relatively straightforward. Provided we have represented it accurately and provided we have employed a sound inference mechanism, we can be confident that constraints such as the one concerning the coupling capacitor are correct. However, the application system to which this concept is given may perform its job better if it does not make the distinction between two-stage amplifier circuits and others. Perhaps there is already a concept for multistage amplifiers which provides all of the benefit that this concept would offer. Or perhaps there is insufficient knowledge about individual amplifier stages to warrant decomposing circuits in this fashion. Ensuring performance improvement is a deep issue in its own right.

21.5 Constructing Explanations for Learning

We now examine these three steps in greater detail by way of an example. We consider learning a concept descriptor for a simplified drinking cup. A suitable domain theory for this task is given in Fig. 21.5. The domain knowledge is represented as a collection of first-order Horn clauses. Horn theories are a popular formalism for knowledge representation in artificial intelligence (AI). They embody an effective compromise between expressiveness and computational tractability. But there is nothing particularly special about first-order Horn theories as far as EBL is concerned. The important feature is only that the knowledge representation language support the construction of an explanation.

EBL also requires a functional specification of the desired concept. This specification often has the same form as the background knowledge:

$$R1:\ \forall x\ \text{drinkable-from}(x) \Rightarrow \text{cup}(x)$$

The functional specification has been termed a nonoperational goal definition. However, it should not be viewed as giving the learning system a true definition of the target concept. Rather it is better to

R2: $\forall x$ (liftable(x) $\wedge$ open(x) $\wedge$ stable(x) $\wedge$ liquid-container(x)] $\Rightarrow$ drinkable-from(x)

R3: $\forall x\,\forall y$ (weight(x, LIGHT) $\wedge$ has-part(x, y) $\wedge$ isa(y, HANDLE) $\Rightarrow$ liftable(x)

R4: $\forall x\,\forall y$ (has-part(x,y) $\wedge$ isa(y,CONCAVITY)] $\Rightarrow$ OPEN(x)

R5: $\forall x\,\forall y$ (has-part(x,y) $\wedge$ isa(y,CONCAVITY)] $\wedge$ orientation(x, UPWARD)] $\Rightarrow$ liquid-container(x)

R6: $\forall x\,\forall y$ (has-part(x,y) $\wedge$ isa(y,FLAT-BOTTOM)] $\Rightarrow$ stable(x)

FIGURE 21.5 The CUP domain theory.

think of the functional specification as an effective procedure with which to recognize that the object in question possesses the desired functionality. For example, we may specify the goal of designing a *Star Trek* transporter mechanism. We may have no idea how to build one nor even an opinion on whether it is possible to do so. Even so, its functional requirements are easy to state: it makes people disappear from one location and appear somewhere else. Such a specification provides little guidance in designing a transporter, but it provides a ready success criterion. In our cup example, a cup is anything one can drink from. This statement also includes no information on how a cup might achieve its goal.

Because of this specification, EBL-acquired concepts are individuated functionally. Any object with the specified functionality is an instance of the desired concept. This feature of EBL enforces a homogeneity among instances of a concept. Most other forms of machine learning permit a concept to be *any* subset of objects in the world regardless of their functional similarity. Many partitionings of the example space that could be acquired by induction using a similarity-based method (e.g., neural net training methods, ID3, CART, AQ11, PLS, etc.) cannot be acquired using EBL. Instances of an EBL-acquired concept must share a functionality. The additional limitation on EBL-acquired concepts is seldom a problem in real-world applications. Concepts are usually not random collections of instances; they are acquired for some purpose. SBL allows this; EBL requires it.

In our example, we provide the single training object, OBJ1, which is a known positive instance of a cup. It is important to understand that there is nothing special about the name. OBJ1 simply names a particular collection of properties. These are shown in Figs. 21.6(a) and 21.6(b). Two different representation schemes are presented. The first is a predicate calculus representation. It specifies a separate logical sentence for each of the relevant relations. The second is a semantic net representation. Here each arrow or directed arc points from an object to a feature value that the object possesses. The arrow is labeled with the name of the feature. Both representations specify that OJB1 has three parts called CONC12, HAN31, and BOT7. OBJ1 is colored RED, is owned by HERMAN, and so on.

Once given OBJ1 as a positive example, an EBL system attempts to construct an explanation for why OBJ1 is indeed a cup, why it satisfies the cup functional specification. Figure 21.7 shows such an explanation.

owner(OBJ1, HERMAN)	weight(OBJ1, LIGHT)
has-part(OBJ!, HAN31)	has-part(OBJ1, BOT7)
color(OBJ1, RED)	has-part(OBJ1, CONC12)
isa(CONC12, CONCAVITY)	orientation(CONC12, UPWARD)
shape(CONC12, CYLINDER)	isa(BOT7, FLAT-BOTTOM)
isa(HAN31, HANDLE)	shape(HAN31, CIRCLE)

FIGURE 21.6(a)   OBJ1, a cup, in predicate notation.

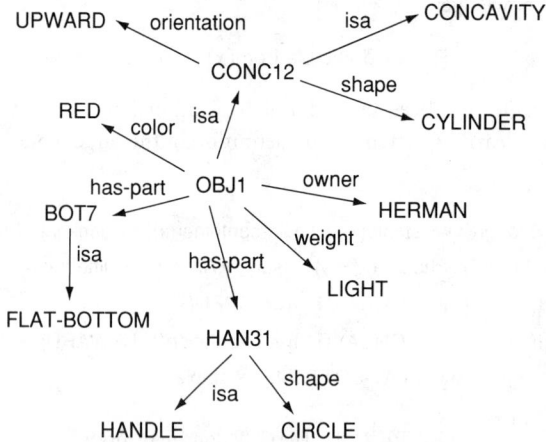

FIGURE 21.6(b)   OBJ1 in semantic net notation.

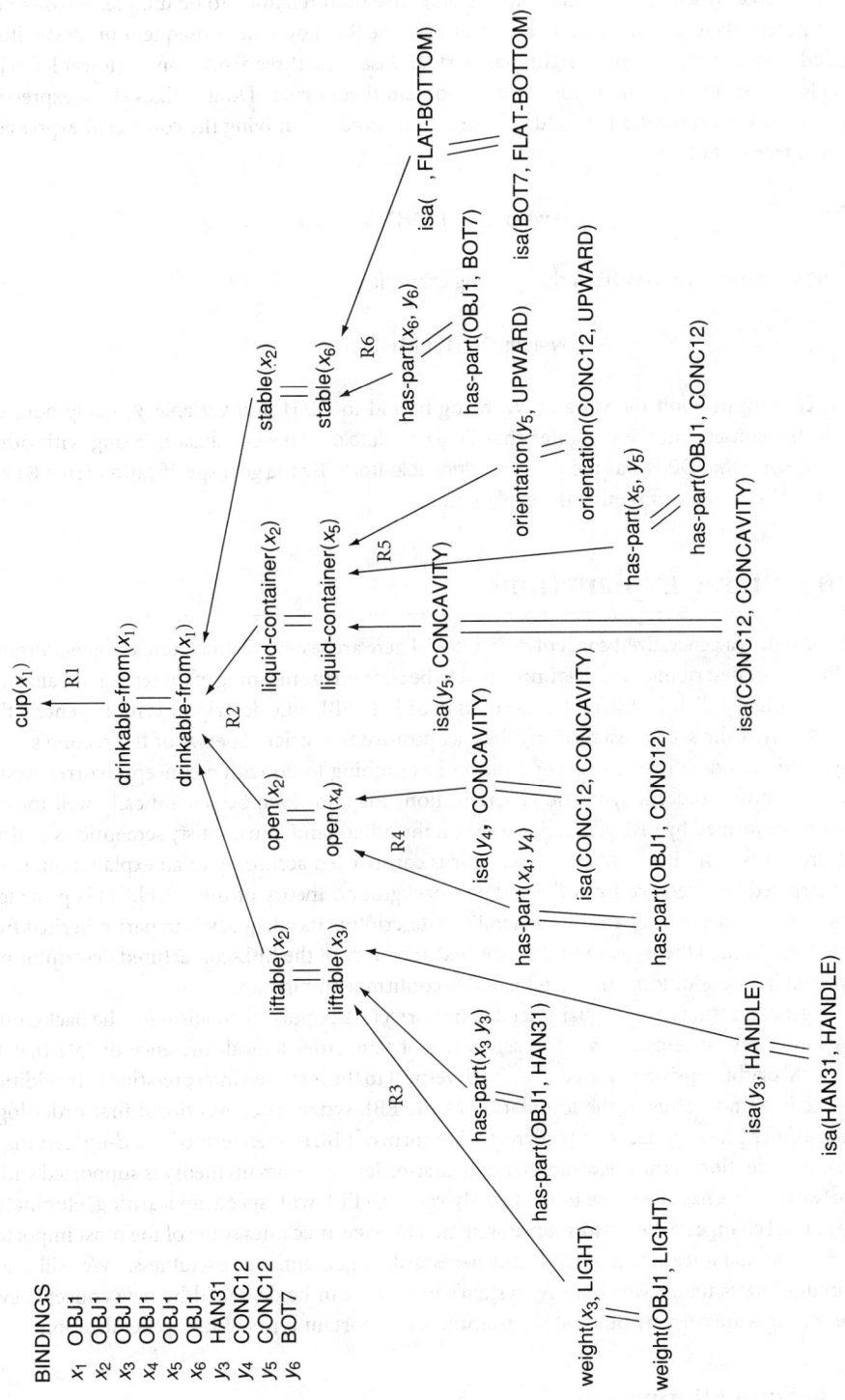

FIGURE 21.7 Explanation for the training example OBJ1.

Arrows denote the contribution of each background knowledge rule to the explanation. For clarity, the quantification variables (each rule's formal parameters) have been renamed to be unique. Arrows point from a rule's antecedents to its consequent. For example, rule R3 allows the consequent property liftable to be concluded from the antecedents weight, has-part, and isa. The three arrows on the lower left which are labeled as R3 constitute an instantiation of this domain theory rule. Double lines show expressions that must match for the explanation to hold. These are enforced by unifying the connected expressions. Thus, the first antecedent of R3,

$$\text{weight}(x_3, \text{LIGHT})$$

is unified with the known property from the training example

$$\text{weight}(\text{OBJ1}, \text{LIGHT})$$

This match is contingent upon the variable x_3, being bound to OBJ1 and variable y_3 being bound to HAN31. From the consequent we can infer that OBJ1 is liftable. This conclusion, along with others, supports the inference that OBJ1 has the property drinkable-from. By the goal specification (rule R1), we can infer that OBJ1 is a cup, completing the explanation.

21.6 Generalizing Explanations

The next step produces a generalized concept descriptor. There are several distinct kinds of generalization performed. But before describing them it is important to be clear on the meaning of the term generalization, which is used in EBL in a slightly different sense than in SBL. In SBL, one descriptor is more general than a second if and only if the set of instances it labels as positive is a strict superset of the second's. This is a purely syntactic notion. Syntactic generalization has nothing to do with a concept's correctness or adequacy for a purpose. Indeed, syntactic generalizations may not even be semantically well formed. Concept descriptors formed by EBL generalizations, on the other hand, must satisfy semantic as well as a syntactic requirements. An EBL-produced descriptor is constructed according to an explanation, which is in turn constructed by inference from the system's background theory of the world. This provides a priori evidence for the adequacy of an EBL-generalized descriptor. Its adequacy is in part inherited from the adequacy of the domain theory, and so the expected adequacy of the EBL-conjectured descriptor may be quite high even before additional training instances confirm it empirically.

Indeed, EBL generalizations can be guaranteed to be correct. A popular formalism for the background theory is first-order predicate calculus. The semantics of first-order logical inference dictate that the sentences of the theory be consistent and correct (with respect to the intended interpretation). In addition, logical inference is sound. Thus, if the foundation of the EBL system is conventional first-order logic, EBL-generalized descriptors are necessarily correct. This form of EBL is often termed *speed-up* learning. It is a common misconception in the literature that only first-order logic domain theory is supported within the EBL framework. This has given rise to mistakenly *equating* EBL with speed-up learning, eliminating any **knowledge level change** in the performance system. This view precludes some of the most important strengths of EBL. A guarantee of accuracy is not necessarily a guarantee of usefulness. We will see in a moment circumstances under which an AI system's behavior can be degraded by new concepts even though those concepts are correct. But first we examine six important types of EBL generalization.

Irrelevant Feature Elimination

The explanation provides sufficient grounds for believing that the training example (in our case OBJ1) satisfies the goal specification (cupness). Clearly, any feature of OJB1 not mentioned in the explanation could have a different value without affecting the explanation's veracity. The owner could have been George instead of Herman; the cup could be blue instead of red, etc. In general, most of an object's properties will not participate in the explanation. This is particularly true for more realistic representations. These might include many additional properties: its current location in the room, distances to other objects, what it is resting upon, how full it is, what it contains, the temperature of the contained liquid, whether it is clean or used and who drank from it last, where it was purchased, how valuable it is, etc.

SBL systems can become overwhelmed in such situations. Coincidences abound when objects have many features, training data is limited, or the space of well-formed concepts is large. This is often the case in the real world. Coincidences, by their very nature, are not predictive of future examples. SBL techniques are vulnerable to such coincidences. The phenomenon of **overfitting** is related to this issue. Recall our first informal example of John's city preference. It is most likely a mere coincidence that testing for the letter e correctly classifies the training data. An SBL approach would have no legitimate mechanism to choose among descriptors that are equally complex and behave similarly on the training data. An EBL system, on the other hand, is unlikely to be sidetracked by such coincidences.

By irrelevant feature elimination, the example explanation of Fig. 21.7 gives rise to the concept descriptor C1:

$$[\text{weight}(\text{OBJ1, LIGHT}) \wedge \text{has-part}(\text{OBJ!, HAN31}) \wedge \text{isa}(\text{HAN31, HANDLE})$$

$$\wedge \text{has-part}(\text{OBJ1, CONC12}) \wedge \text{isa}(\text{CONC12, CONCAVITY})$$

$$\wedge \text{orientation}(\text{CONC12,UPWARD}) \wedge \text{has-part}(\text{OBJ1, BOT7})$$

$$\wedge \text{isa}(\text{BOT7, FLAT-BOTTOM})] \Rightarrow \text{cup}(\text{OBJ1})$$

This descriptor is not very general. Importantly, however, it is in the same form as the original background knowledge. That is, it provides an alternative method to infer the cupness of an object, and it states that weight, orientation, and the various isa and has-part relations are sufficient for this inference. No other features of the object are needed if these are present.

Identity Elimination

The second generalization type removes unnecessary dependence on the particular objects referred to in the explanation. According to the explanation, OBJ1 is liftable by virtue of its having handle HAN31. Without a handle, the explanation's veracity is compromised. But the training instance need not have had this particular handle. If it had some other handle (say, HAN32 or HAN4597) the conclusion of cupness would still be justified. Thus, HAN31 can be generalized into a predicate calculus variable. Likewise, OBJ1 can be generalized to a different predicate calculus variable. But we must ensure that required relations hold among these variables. The variable that replaces HAN31 must still be a handle; it must be a part of the variable that replaces OBJ1, and so on.

We can find the minimal general conditions required to conclude cupness by pruning the lowest portions of the explanation. Pruning involves breaking unification links within the explanation and keeping only the central structure (the subgraph that includes the goal justification). After this generalization step we can obtain the descriptor C2:

$$\forall x_1, y_3, y_4, y_6 \ [\text{weight}(x_1, \text{LIGHT}) \wedge \text{has-part}(x_1, y_3) \wedge \text{isa}(y_3, \text{HANDLE})$$

$$\wedge \text{has-part}(x_1, y_4) \wedge \text{isa}(y_4, \text{CONCAVITY}) \wedge \text{orientation}(y_4,\text{UPWARD})$$

$$\wedge \text{has-part}(x_1, y_6) \wedge \text{isa}(y_6, \text{FLAT-BOTTOM})] \Rightarrow \text{cup}(x_1)$$

This says that anything which has a handle, an upward pointing concavity, a flat bottom, and is light is a cup. This rule applies to many objects in addition to OBJ1 and can be used as a concept descriptor for cup classification. We can be confident of this descriptor's correctness even though only one example has been seen. Identity elimination works because of generalities already built into the domain theory. These pre-existing domain theory generalities are essential to EBL. If the domain theory were changed to include an alternate rule R3:

$$\text{R3a: } \forall x \, [\text{weight}(x,\text{LIGHT}) \wedge \text{has-part}(x,\text{HAN31}) \wedge \text{isa}(\text{HAN31},\text{HANDLE})] \Rightarrow \text{liftable}(x)$$

Then EBL generalization of HAN31 would not be possible although the conclusion of OBJ1's cupness would still be supported. For the sake of EBL we would prefer to avoid domain rules such as R3a. Ideally, the role that an object plays in the domain theory is entirely determined by its properties, never by its identity. Philosophically this has some interesting ramifications, but it is uncontroversial so far as EBL is concerned. This has been termed the principle of no *function in form*. It is often adhered to in AI and usually results in a theory with fewer rules. This property is also important in the next generalization type.

Operationality Pruning

Easily reconstructable subexplanations of the original explanation may also be eliminated. For example, we could prune the substructure added by rule R6 by breaking the unification at the stable predicate. This results in a slightly different concept, C3:

$$\forall \, x_1, y_3, y_4, y_6 \, [\text{weight}(x_1, \text{LIGHT}) \wedge \text{has-part}(x_1, y_3) \wedge \text{isa}(y_3, \text{HANDLE}) \wedge \text{has-part}(x_1, y_4)$$

$$\wedge \, \text{isa}(y_4, \text{CONCAVITY}) \wedge \text{orientation}(y_4,\text{UPWARD}) \wedge \text{stable}(x_1)] \Rightarrow \text{cup}(x_1)$$

This descriptor is syntactically simpler than the previous one. It has one fewer antecedent conjuncts. However, to determine that a new object is a cup, that object's stability must now be justified at the time the concept is applied. The previous descriptor only consults properties expressed directly in the definition of the test object. In that rule the test object's stability is never an issue. Sufficient ancillary properties are tested to justify that the object is stable. In particular, all objects are required to have flat bottoms. In point of fact, many objects are stable even though they do not have flat bottoms. Indeed many cups do not possess flat bottoms. The zarf (Fig. 21.8) is often used in Middle-Eastern countries to support hot coffee cups which have rounded bottoms. The zarf is a cylindrical chalicelike holder into which the rounded cup bottom is nestled. There are other common and not so common ways to achieve stability. C3 latter one is a more general concept descriptor. The price of this generality is that an inferencer must conclude stability when the concept descriptor is evaluated on an object.

FIGURE 21.8   The zarf.

Thus, it is more expensive to use. We say it is less *operational*. Constructing a subexplanation on demand is typically a harder task than the straightforward lookup of several object properties.

Likewise we could entertain concept descriptors that sever other of the explanation's unifications. Cups could be allowed to be liftable by other means than having a handle, which include Styrofoam cups and the like. Liquid containment could be other than by an open concavity, giving rise to covered travel mugs. The higher in the explanation that unifications are broken, the more general is the resulting concept descriptor but the more expensive it is to evaluate.

Thus, there is a tradeoff between concept descriptor generality and operationality. The minimal generalization is the result of applying identity elimination. The maximal generalization is an uninformative repetition of the goal specification:

$$\forall x_1 \, \text{drinkable-from}(x_1) \Rightarrow \text{cup}(x_1)$$

In between there are many alternative choices requiring progressively harder subexplanations to be constructed at the time of concept use.

Figure 21.9(a) shows a schematic explanation tree. An explanation's **operationality boundary** is any coherent choice of unification pruning traversing the explanation structure. Figure 21.9(b) shows several schematic choices for the operationality boundary.

In the following section on selecting a concept descriptor we discuss different approaches that have been advanced for deciding how to choose the operationality boundary. For now it is sufficient to realize that even once a particular explanation has been constructed, there are many potential concept descriptors to choose among.

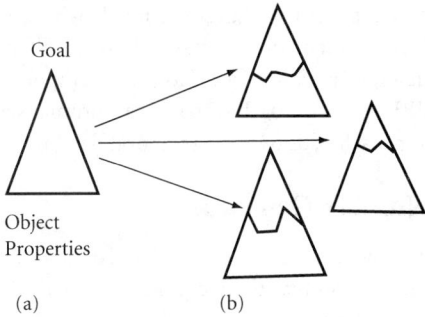

(a)   (b)

FIGURE 21.9 Generality–Operationality trade-off: (a) explanation and (b) operationality boundaries.

Disjunctive Augmentation

Pruning requires a subexplanation to be constructed from scratch at the time the concept is used. This can be expensive. Disjunctive augmentation supports similar flexibility at a greatly reduced cost. Instead of simply pruning a subexplanation we add one or more alternative subexplanations, any one of which may be used to support the requisite property. At the time the concept is used the system need only choose among several pregeneralized subexplanations.

Suppose we give the learning system a second positive example of a cup. This cup's stability is achieved by a zarf as in Fig. 21.8. We also suppose that the system has sufficient background theory to conclude that this cup is indeed stable. The explanation of this example's cupness will be similar to the explanation of Fig. 21.7 except that stability is achieved by the zarf instead of the conventional flat bottom. Generalizing the combined explanations results in a disjunct supporting the stable predicate. Note that this is quite different from operationality pruning. No operationality boundary severs the stable predicate from its support. In this case two distinct methods of achieving stability are determined. Satisfying either qualifies the object as a cup.

Disjunctive augmentation opens the door to potential inefficiencies. It is well established that disjunction is a primary source of computational complexity in automated inference. It is possible, indeed likely, that in any interesting explanation there are many augmentations possible which make subtle distinctions and result in only marginal improvements in a descriptor's generality. Discovering alternative and generalizing alternative subexplanations is expensive. Similarly, use of the concept descriptor can be made more complex. These issues will be addressed in the section on the utility problem. For now we will keep in mind that the EBL generalization defines a concept space and does not adopt any particular concept from that space.

Temporal Generalization

Many concepts have a temporal component. Allowing temporal variations for such concept descriptors yields an important form of generalization. This is a particularly important issue in AI planning: automatically finding a sequence of actions that change the world so as to achieve some specified goal. Here a training example demonstrates how a goal is achieved by a particular sequence of operators. It is possible that a different sequence of actions would work as well. The example's explanation explicitly specifies ordering requirements: the timing of some operators may be arbitrary, others may require a particular partial ordering, some may allow other subsequences to be interleaved, and so on. The most general problem solving concept allows maximal flexibility for variations of operator orderings.

In general, determining minimal constraints on operator ordering is a difficult problem. Algorithms to perform temporal generalization for explanations of STRIPS-type operator sequences can be involved and expensive. Things get much worse when considering a more general specification of operators. Noninstantaneous processes allow simultaneous and overlapping changes in the world (as is common in qualitative reasoning). However, temporal generalization has received significant research attention in EBL. One approach discovers a minimal set of constraints through theorem proving. A more tractable approach adds only the temporal flexibility that can be determined cheaply.

Number Generalization

Number generalization results in a concept descriptor with replicatable subparts. These result from the recognition that particular subexplanations might have been repeated a different number of times than in the training example. Consider an EBL system learning to stack blocks into a tower. Suppose its background knowledge includes information about immediate support and stability. A training example is given in which three red blocks are stacked. With the generalization types described so far, the resulting concept will be limited to building three-block towers. The system will generalize the particular blocks used in the example, their color, etc. It will require that the lower blocks be flat on top, that they be rigid, and so on. However, towers of four, five or six blocks will fall outside the descriptor's abilities. A separate tower concept descriptor will be acquired for each number of blocks. Number generalization remedies this.

Number generalization opportunities can be computationally difficult to recognize because the parameter being generalized (in this example, the number of blocks) is not explicitly represented anywhere in the explanation. Rather the *threeness* of the tower is implicitly coded in the topology of the explanation. There are three subexplanations proving the resulting stability after a block is grasped and moved. Nor are the three subexplanations identical: the blocks are different as are their initial and final locations, etc. Number generalization crucially involves a representation transformation of the explanation into a form in which *sets* or *loops* are included in the theory's ontology.

It is interesting to note that not all cases in which number generalization is theoretically supportable should result in number-generalized concepts. Consider rotating the tires on an automobile. Even though the procedure readily generalizes to automobiles with 5, 6, or 7 tires (and such automobiles are logically possible), there is no particular advantage in complicating a Rotate-Tire problem-solving concept to include them. This is an aspect of the utility problem to which we will turn next. Before we do, we will note that the last three generalization types, disjunctive augmentation, temporal generalization, and number generalization, alter the internal structure of the explanation. Of these, only disjunctive augmentation is easily illustrated in our on-going drinking cup illustration.

21.7 Selecting a Concept Descriptor: The Utility Problem

Concepts are typically acquired for some purpose. There is no point in acquiring a new concept unless an AI application program (a scheduling system, diagnosis system, database retrieval agent, etc.) works better with it. It is important to appreciate that performance improvement in an application system is the ultimate goal of any machine learning system. Ensuring that acquired concepts help rather than hurt performance is the **utility problem**.

Concept accuracy or concept truth is no guarantee of concept utility. In the performance system there is some cost associated with possessing concepts. A newly acquired concept may be accurate so that it indeed saves work by simplifying the task. The benefit of applying that concept may not outweigh the overhead of including it in the system. To illustrate, consider a commuter who drives home every workday. The standard well-traveled route, Avenue A, takes 15 min. Suppose a learning system discovers two additional routes home. One follows Avenue B and takes only 10 min, whereas the other follows Avenue C and takes 11 min. Since they are poorer roads, they are often closed for repair, but since there is but one road crew they are never simultaneously closed. The newly learned concept chooses between these routes, both of which are quicker than Avenue A. The concept must include an antecedent to ensure that the selected road be open since each route only works when there is no construction. The acquired concept might say something like "if there is no maintenance on Avenue B then take it and save 5 min, else take Avenue

C and save 4 min." The overhead cost of a piece of knowledge in the performance system includes three components: 1) the cost of locating the knowledge in memory, 2) the cost of evaluating precondition antecedents in the situation, and 3) the slow down of locating other knowledge for other purposes caused by the presence of this knowledge. Consider cost 2 in the context of our example. Suppose the only reliable way to determine whether or not there is maintenance is by placing a call to the Public Works Department. If it happens that such calls are completed within, say, 2 min, then the newly acquired concept is a useful one: with the concept, the performance system saves 2–3 min over solving the problem without the concept. However, if the public works office is typically busy or inefficient so that the average call takes 6 min, then the performance system incurs a penalty of 1–2 min with each application. In either case, the concept is accurate; it adequately captures the relevant world knowledge, shortens driving time, and never causes a solvable problem to become unsolvable. However, its inclusion may result in poorer system performance.

The decision on what concepts to adopt can be quite subtle. A concept's utility can be strongly influenced by attributes far removed from it. Here the crucial feature in determining positive or negative concept utility (the efficiency of the Public Works telephone operator) would at first seem quite distant from process of driving home. A concept's utility can also depend on the particular distribution of problems to which the performance system will be applied. It can be influenced by other concepts that have been previously adopted. Thus, a utility judgment may be difficult to make reliably.

The fact that correct knowledge can harm the performance system is particularly troublesome for EBL. Since EBL is more example efficient, a particular training set may yield many more concepts using an EBL approach than an SBL approach. The performance system can quickly become swamped even if the average concept utility is only slightly negative. Each concept's utility must be evaluated before it can be given to the performance system. Only concepts with positive utilities should be added.

There are two broad approaches to judging a concept's utility for a performance system. In the first or *empirical* approach a hypothesized concept is temporarily added to the performance system to measure its effect, which is then taken as an estimate of its future utility. In the second or *analytic* approach the decision to retain a new concept is inferred directly from the concept's definition.

Figure 21.10 illustrates the empirical approach. Newly hypothesized concepts from the explanation generalizer are represented by the oval labeled Trial Concepts. The performance system, composed of the Native Performance System together with the Concept Library, processes problems from the world. The Empirical Utility Evaluation monitors the efficiency of the performance system augmented with individual trial concepts. As statistics are gathered and compared to baseline performance, evidence mounts for or against the utility of the currently conjectured concept. When there is sufficient evidence to warrant a confident judgment about the concept's utility, it is permanently added to the system's concept library or discarded.

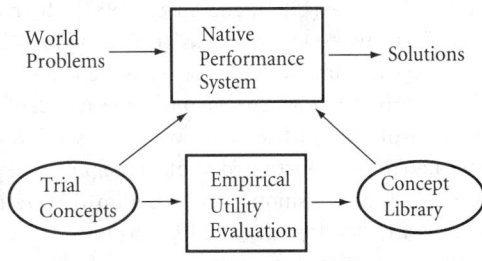

FIGURE 21.10 Empirical utility analysis.

The analytic approach is illustrated in Fig. 21.11. The native performance system is not directly involved in the utility judgment. Rather, information on what constitutes efficient processing is represented in a theory of utility. Trial concepts are evaluated with respect to this a priori utility theory. Concepts that are judged likely to improve the performance system are added to the system's concept library; others are discarded.

In an empirical procedure, the definitions of the concept are not directly examined. The performance system simply continues its normal

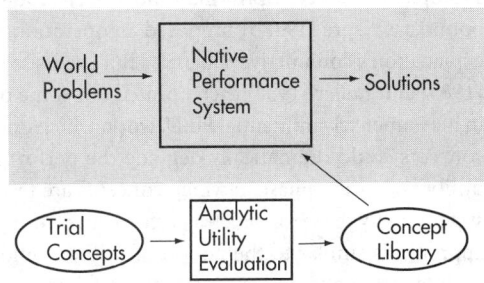

FIGURE 21.11 Analytic utility analysis.

processing although it can now draw upon the conjectured concept. The empirical method quite naturally incorporates information implicit in the distribution of problems. The expected utility of a concept can vary even though the concept itself is static. The knowledge of how to control a high-speed car with a flat tire is of little use to a Bedouin who travels only by camel but is crucial to a race car driver. Such phenomena do not preclude an analytic approach. The theory of utility may itself include information on estimated problem distributions, or the effect of problem distribution on utility may be insignificant in the face of other considerations. Hybrid approaches have also been examined.

Finally, we return to the issue of incorrect knowledge. One advantage of EBL is that the approach is quite forgiving of overly general knowledge representations. The functional specification of our cup (R1) is such a case. It allows a drinking glass to be a functional cup. However, concepts C1–C3 do not apply to drinking glasses. The acquired concepts insist that the item in question have a handle. This robustness is due to the interplay between the background theory and the teacher's positive training examples. If the teacher had labeled a glass as a cup, the EBL system might quite happily have conjectured a rather different set of descriptors. However, provided the teacher does not lie to the system in the training set in a way that admits explanation, the EBL system is often safe from such overgeneralizing. By the same token, it is sometimes useful to endow the EBL system with an overly powerful inferential ability, one that is able to generate or conjecture explanations that need not be true under the semantics of conventional logic. The domain theory need not be complete or correct. This is in striking contrast with conventional inference or theorem proving systems in which the presence of any contradiction in the theory can have disastrous effects. The use of *approximate, overly general,* or *plausible* background theories has been studied by several researchers and is the subject of ongoing research.

21.8 Annotated Summary of EBL Issues

Two early AI works served as inspirations for the development of the EBL approach. These were the notion of goal regression [Waldinger 1977] and the MACROPS learning of the STRIPS system [Fikes et al. 1972]. The Ph.D. work of Soloway [1977] and the analytic learning mode of Waterman's automated poker player [Waterman 1970] also foreshadowed the EBL approach. The modern approach to EBL began with DeJong [1981], Mitchell [1983], and Silver [1983]. It enjoyed an explosion of research interest throughout the 1980s. The two standard works on EBL are by Mitchell et al. [1986] and DeJong and Mooney [1986]. A more recent framework is by Segre and Elkan [1994].

The early research focus was on the mechanics of *how* to construct a new concept using explanations. The question of *when* such learning would be useful was largely ignored. Many researchers incorrectly assumed that any concept so acquired would be helpful. They were very much mistaken. An important development in EBL was the transition from the notion of operationality to one of concept utility. Operationality, central to the early development of EBL, was introduced by Mostow [1981] to aid in re-expressing correct but ineffective advice from an expert into a form more directly useable by an AI system. The same issue of usability arises in EBL and the use of operationality was a natural one. Explanations should terminate in predicates that are easily evaluated by the system [Mitchell et al. 1986]. Operationality became an important subtopic of research. Braverman and Russell [1988] described improvements in selecting an operationality boundary. Segre [1988b] proposed an operationality criterion based on linearizing the interaction of an explanation's conjunctive subgoals. But the issue of operationality was always somewhat murky. Minton [1985] and Keller [1988] each pointed out some of its difficulties. Minton [1988] expanded on this issue in his immensely influential Ph.D. work, which documented that the acquisition of apparently operational concepts could dramatically degrade the performance of a classical planning system. He argued for a utility evaluation phase in which concepts are tested in the context of a normally executing performance system. Only those for which a performance improvement is demonstrated are actually adopted. Minton's approach is similar to the notion of *utility* employed in decision theory. Etzioni [1990] soon espoused an interesting alternative. Instead of trial evaluations, he proposed the nonrecursive hypothesis which judges concept utility directly from its intrinsic properties and interactions with the background theory.

Soon researchers demonstrated situations in which the nonrecursive hypothesis also breaks down. An improved utility evaluation phase was proposed as an alternative [Gratch and DeJong 1992, Greiner and Jurisica 1992]. We can now understand these tensions as a contrast between empirical utility evaluation (including Minton, Gratch, and Greiner) and analytic utility estimation (from Mitchell's operationality to Etzioni's nonrecursive hypothesis). The latter can be viewed as committing to an a priori theory of concept utility whereas the former now consists of statistically sound concept adoption decisions.

Number and temporal generalization require the structure of the explanation itself to be modified. Two influential approaches to number invariant are Shavlik and DeJong [1987] and Cohen [1988]. In the worst case, generalizing number can be exceedingly difficult. However, many important number generalization tasks are comparatively simple. The same can be said for temporal generalization. Mooney [1988] discusses one approach in which an action sequence (representing a specific linear plan) is temporally generalized. Another interesting approach is Kambhampati and Kedar [1994], in which explanation-based generalization is extended to partially ordered and partially instantiated plans.

Combining explanation-based with similarity-based methods has received a good deal of research attention. Lebowitz [1986] points out that EBL's ability to generalize based on few examples can also be a weakness. A poor concept can result if the system happens to be given a few nonrepresentative examples. To remedy this, his system UNIMEM employed SBL to focus attention of the EBL component. Only after detecting a pattern is an explanation attempted. Conversely, EBL can be performed first to focus the SBL component on likely relevant features [Danyluk 1987] resulting in more efficient induction. Cohen [1992a] has explored abduction to bridge EBL and SBL. Furthermore, the approach ameliorates the SBL problem of overfitting training data [Breiman et al. 1984, Quinlan 1987]. This and other methods for combining EBL with SBL can automatically extend the system's domain theory, a research area known as *discovery* or *theory revision*. A number of important theory revision systems have included some kind of EBL component [Doyle 1986, Sims 1987, Mooney and Ourston 1989, Pazzani 1989, Ourston and Mooney 1990]. A more purely EBL approach can be found in Rajamoney and DeJong [1987].

A hybrid EBL/SBL approach can also be applied to more conventional concept acquisition. Flann and Dietterich [1989] advocate a method for combining EBL and SBL in which EBL is first applied to several positive examples. The intersection of these multiple explanations is constructed. The intersection represents a generalization since any disparate constraints are dropped. The result is then specialized inductively. Porter and Kibler [1986] have proposed another method that they call experimental goal regression. Integrating an explanation-based approach with other methods is also notable in the work from the machine learning group at the University of Turin (e.g., Bergadano and Giordana [1988], Saitta et al. [1991], Botta et al. [1992]). Another important hybridization is the DISCIPLE system [Kodratoff and Tecuci 1989] which integrates a number of learning approaches including EBL. These latter systems blend into what has become known as *multistrategy learning* [Michalski and Tecuci 1994].

Several researchers have investigated the process of analogical reasoning in ways relevant to explanation-based learning. Of particular note are Kedar-Cabelli [1987], Russell [1989], and Velosa and Carbonell [1993], although the latter two produce logically entailed inferences and therefore differ from the standard view of analogy [DeJong 1988].

Another related body of work involves memory-based reasoning. Case-based reasoning (CBR) [Schank and Leake 1989, Kolodner 1993] is a memory-based approach. Rather than generalizations, specific cases are stored. A flexible indexing mechanism retrieves past cases that are similar and relevant to the system's needs. The retrieved cases are then adjusted to satisfy the current demands. In EBL, generalization is performed prior to storing the new concept. In CBR generalization is built into the mechanisms that peform indexing and adjustment of cases. The process of acquiring knowledge is simply the storage of new cases. The price for ease of knowledge acquisition is the difficulty of constructing (or automatically learning) an effective indexing and adjusting technique. The related instance-based methods [Aha et al. 1991] employ a numeric similarity function. Classification is based on nearness of the current instance in this metric space to other known instances. Here generalization is built into the similarity function; a good similarity function is one in which instances of the same classification are near to each other and far from other instances. The K-nearest neighbor algorithms are examples of this approach.

The area of automatic planning and design has been proved a fertile ground for investigating EBL [Mitchell 1983, Mitchell et al. 1985, Rosenbloom and Laird 1986, Silver 1986, Minton 1988, Segre 1988a, Chien 1990, Katukam and Kambhampati 1994]. These are natural domains since the operator definitions provide a ready-made background theory of the world with which to drive the EBL component. The problem of scheduling, which can be cast in the same constraint satisfaction framework, has also received attention [Gervasio and DeJong 1992, Minton et al. 1992, Zweben et al. 1992]. Another important application of EBL is in learning apprentice systems [Mitchell et al. 1985, Wilkins 1988]. This research addresses the knowledge acquisition bottleneck in expert systems. The specific behaviors of a human expert are observed. An explanation is constructed to justify how the specific problem was solved. The generalization of this explanation postulates new knowledge to add to the expert system. A major advantage of the approach is that only noninvasive observation of the human expert is required.

The psychological interest in human concept formation has, in recent years, shifted away from sterile artificial concepts to the study of more natural and real-world concepts [Smith and Medin 1981, Murphy and Medin 1985]. A natural question is "Can EBL serve as a computational model for this kind of human concept formation?" The answer appears to be in the affirmative. The explanation-based approach plays an important role in the cognitive architecture system SOAR [Laird et al. 1987], which is motivated primarily by psychological considerations. Furthermore, psychological experiments indicate that humans are capable of explanation-based concept formation and will behave according to its principles over standard SBL ones when appropriate [Ahn et al. 1987, Anderson 1987]. Given the example efficiency enjoyed by EBL over SBL it is perhaps not surprising that adult learning manifests explanation-based behavior. Adults possess a good deal of knowledge about how the world works. It would be surprising if this background knowledge did not aid in the acquisition of later concepts. Interestingly, however, there is also evidence of EBL-like behavior from infant cognitive development studies. Baillargeon has documented that infants can build upon prior understanding in explanation-based ways [Kotovsky and Baillargeon 1994]. The appearance of EBL behavior in infants less than a year old may indicate that the approach offers greater computational advantage than first thought, at least when applied to real-world concepts. More research is needed in this area.

One of the most promising areas of current EBL research concerns the problem of imperfect prior knowledge. Traditionally, EBL's background theory has been viewed (in the conventional terms of standard logic) as constraints on how the world can behave. The qualification problem [McCarthy 1977] virtually guarantees that these constraints cannot be perfectly represented. Several approaches view prior knowledge differently, as a bias on the concept space [Russell and Grosof 1987, Mooney and Ourston 1989, Cohen 1992b, Mahadevan and Tadepalli 1994]. Importantly, this bias is both semantic and declarative (unlike the typical bias of a conventional induction system). Of particular note in this regard are systems that integrate explanation-based learning with neural net learning tuning [Shavlik and Towell 1990, Thrun and Mitchell 1993] or function fitting [DeJong 1994]. In these systems, the result of the explanation processed is tuned in response to training observations. Thus, the explanation-based component determines a subspace of hypotheses, which is then tuned inductively. The role of the explanation-based processing is no longer to produce an accurate concept. Rather its job is to suggest a good starting place for the concept tuning component. This is a hypothesis from which the tuning phase can formulate an accurate concept. The explanation-based hypothesis itself can be arbitrarily bad. Consider backpropagation which efficiently performs multidimensional hill climbing. The summit of the highest peak corresponds to the best concept. The set of initial hypotheses that backpropagation transforms into this best concept include the deep slopes of the valleys surrounding the peak. Indeed, there is little difference to backpropagation in starting near the summit of the peak or near the bottom of the deepest valley adjacent to the peak; if the explanation-based component can produce an hypothesis anywhere within this area backpropagation will quickly find the desired peak. Other tuning methods based on hill climbing behave similarly. With hybrid systems such as these, the explanation-based component is relieved of any requirement of concept accuracy. This is fortunate from the point of view of the qualification problem. However, it raises troublesome questions: What now constitutes a good domain theory if it is not accuracy with respect to the world? What might be a suitable semantic interpretation of the background knowledge statements and symbols since it is no longer appropriate to appeal to conventional logical or probabilistic meanings as constraints on the world?

Defining Terms

Concept descriptor: The declarative representation of a partitioning of the instance space to reflect instance classification.

Concept space: The set of all well-formed concept descriptors.

Domain theory: Any information supplied to the concept acquisition system about the application task, the application system, the nature of expected inputs, etc. This is also known as background knowledge.

Explanation: A tree-structured graph such that each leaf node is either a prior belief or a property of the example being explained, each nonleaf node is the result of applying an inference procedure to prior nodes, the root node is the training classification assigned to the example.

Inductive bias: Any preference not due to the training set which is exhibited by a concept acquisition algorithm for one concept descriptor over another.

Instance space: The set of all classifiable objects.

Knowledge level change: Any change to an AI system's representation that goes beyond the inferential closure of its previously represented knowledge.

Operationality boundary: In a generalized explanation, any division between the root subtree and the peripheral subtrees such that the root subtree yields a useful concept.

Overfitting: The selection of a concept descriptor which captures a pattern exhibited by the training set but not exhibited by the target concept. In section 21.1 the concept that avoids cities with the letter "e" overfits the training data.

Target concept: The correct or desired partitioning of the instance space.

Training set: A collection of examples whose classification is known. This is an input to a concept acquisition system from which a concept descriptor is induced.

Utility problem: The difficulty in ensuring that an acquired concept enhances the performance of the application system.

References

Aha, D., Kibler, D., and Albert, M. 1991. Instance-based learning algorithms. *Machine Learning* 6(1):37–66.

Ahn, W., Mooney, R. J., Brewer, W. F., and DeJong, G. F. 1987. Schema acquisition from one example: psychological evidence for explanation-based learning. In *9th Ann. Conf. Cognitive Sci. Soc.*, pp. 50–57. Lawrence Erlbaum Associates.

Anderson, J. R. 1987. Causal analysis and inductive learning. In *4th Int. Workshop Machine Learning*, pp. 288–299. Morgan Kaufmann.

Bergadano, F. and Giordana, A. 1988. A knowledge intensive approach to concept induction. In *5th Int. Conf. Machine Learning*, pp. 305–317. Morgan Kaufmann.

Botta, M., Ravotto, S., and Saitta, L. 1992. Use of causal models and abduction in learning diagnostic knowledge. *Int. J. Man-Machine Stud.* 36:289–307.

Braverman, M. and Russell, S. 1988. IMEX: overcoming intractability in explanation based learning. In *7th Nat. Conf. Artif. Intelligence*, pp. 575–579. AAAI Press.

Breiman, L., Friedman, J., Olshen, R., and Stone, C. 1984. *Classification and Regression Trees*. Wadsworth.

Chien, S. 1990. *An Explanation-Based Learning Approach to Incremental Planning*. Ph.D. thesis. Computer Science Department, University of Illinois, Urbana.

Cohen, W. 1988. Generalizing number and learning from multiple examples in explanation based learning. In *5th Int. Conf. Machine Learning*, pp. 256–269. Morgan Kaufmann.

Cohen, W. 1992a. Abductive explanation-based learning: a solution to the multiple inconsistent explanation problem. *Machine Learning* 8(2):167–219.

Cohen, W. 1992b. Compiling prior knowledge into an explicit bias. In *9th Int. Conf. Machine Learning*, pp. 102–110.

Danyluk, A. 1987. The use of explanations for similarity-based learning. In *10th Int. J. Conf. Artif. Intelligence*, pp. 274–276. Morgan Kaufmann.

DeJong, G. 1981. Generalizations based on explanations. In *7th Int. J. Conf. Artif. Intelligence*, pp. 67–70. IJCAI.

DeJong, G. 1988. The role of explanation in analogy or the curse of an alluring name. In *Similarity and Analogical Reasoning*. S. Vosniadou and A. Ortony, eds., pp. 346–365. Cambridge University Press.

DeJong, G. 1994. Learning to plan in continuous domains. *Artif. Intelligence* 64(1):71–141.

DeJong, G. and Mooney, R. 1986. Explanation-based learning: an alternative view. *Machine Learning* 1(2):145–176.

Doyle, R. 1986. Constructing and refining causal explanations from an inconsistent domain theory. In *5th Nat. Conf. Artif. Intelligence*, pp. 538–544. AAAI Press.

Etzioni, O. 1990. Why Prodigy/EBL works. In *8th Nat. Conf. Artif. Intelligence*, pp. 916–922. MIT Press.

Fikes, R., Hart, P., and Nilsson, N. 1972. Learning and executing generalized robot plans. *Artif. Intelligence* 3(4):251–288.

Flann, N. S. and Dietterich, T. G. 1989. A study of explanation-based learning methods for inductive learning. *Machine Learning* 4(2):187–226.

Gervasio, M. and DeJong, G. 1992. A completable approach to integrating planning and scheduling. In *1st Int. Conf. Artif. Intelligence Plann. Syst.*, pp. 275–276. Morgan Kaufmann.

Gratch, J. and DeJong, G. 1992. COMPOSER: a probabilistic solution to the utility problem in speed-up learning. In *10th Nat. Conf. Artif. Intelligence*, pp. 235–240. MIT Press.

Greiner, R. and Jurisica, I. 1992. A statistical approach to solving the EBL utility problem. In *10th Nat. Conf. Artif. Intelligence*, pp. 241–248. MIT Press.

Kambhampati, S. and Kedar, S. 1994. A unified framework for explanation-based generalization of partially ordered and partially instantiated plans. *Artif. Intelligence* 67(1):29–70.

Katukam, S. and Kambhampati, S. 1994. Learning explanation based search control rules for partial order planning. In *12th Nat. Conf. Artif. Intelligence*, pp. 582–587. MIT Press.

Kedar-Cabelli, S. 1987. Formulating concepts according to purpose. In *6th Nat. Conf. Artif. Intelligence*, pp. 477–481. Morgan Kaufmann.

Keller, R. 1988. Operationality and generality in explanation-based learning: separate dimensions or opposite endpoints? In *AAAI Symp. Explanation-Based Learning*, pp. 153–157. AAAI Press.

Kodratoff, Y. and Tecuci, G. 1989. The central role of explanations in disciple. In *Knowledge Representation and Organization in Machine Learning*. K. Morik, ed., pp. 135–147. Springer–Verlag.

Kolodner, J. 1993. *Case-Based Reasoning*, Morgan Kaufmann.

Kotovsky, L. and Baillargeon, R. 1994. Calibration-based reasoning about collision events in 11-month-old infants. *Cognition* 51:107–129.

Laird, J., Newell, A., and Rosenbloom, P. 1987. SOAR: an architecture for general intelligence. *Artif. Intelligence* 33(1):1–64.

Lebowitz, M. 1986. Integrated learning: controlling explanation. *Cognitive Sci.* 10(2):219–240.

Mahadevan, S. and Tadepalli, P. 1994. Quantifying prior determinations knowledge using the PAC learning model. *Machine Learning* 17(1):69–105.

McCarthy, J. 1977. Epistemological problems of artificial intelligence. In *5th Int. Conf. Artif. Intelligence*, pp. 1038–1044. IJCAI.

Michalski, R. and Tecuci, G., eds. 1994. *Machine Learning: A Multistrategy Approach*. Morgan Kaufmann.

Minton, S. 1985. Selectively generalizing plans for problem-solving. In *9th Int. J. Conf. Artif. Intelligence*, pp. 596–599. Morgan Kaufmann.

Minton, S. 1988. *Learning Search Control Knowledge: An Explanation-Based Approach*. Kluwer Academic.

Minton, S., Johnston, M., Philips, A., and Laird, P. 1992. Minimizing conflicts: a heuristic repair method for constraint satisfaction and scheduling problems. *Art. Intelligence* 58(1–3):161–206.

Mitchell, T. 1983. Learning and problem solving. In *8th Int. J. Conf. Artif. Intelligence*, pp. 1139–1151. Morgan Kaufmann.

Mitchell, T., Keller, R., and Kedar-Cabelli, S. 1986. Explanation-based generalization: a unifying view. *Machine Learning* 1(1):47–80.

Mitchell, T., Mahadevan, S., and Steinberg, L. 1985. LEAP: a learning apprentice for VLSI design. In *9th Int. J. Conf. Artif. Intelligence*, pp. 573–580. Morgan Kaufmann.

Mooney, R. 1988. Generalizing the order of operators in macro-operators. In *5th Int. Conf. Machine Learning*, pp. 270–283. Morgan Kaufmann.

Mooney, R. and Ourston, D. 1989. Induction over the unexplained: integrated learning of concepts with both explainable and conventional aspects. In *6th Int. Workshop Machine Learning*, pp. 5–7. Morgan Kaufmann.

Mostow, J. 1981. *Mechanical Transformation of Task Heuristics into Operational Procedures*. Ph.D. thesis. Computer Science Department, Carnegie Mellon University, Pittsburgh.

Murphy, G. and Medin, D. 1985. The role of theories in conceptual coherence. *Psychological Rev.* 92:289–316.

Ourston, D. and Mooney, R. 1990. Changing the rules: a comprehensive approach to theory refinement. In *8th Nat. Conf. Artif. Intelligence*, pp. 815–820. MIT Press.

Pazzani, M. 1989. Creating high level knowledge structures from simple events. In *Knowledge Representation and Organization in Machine Learning*. K. Morik, ed., pp. 258–287. Springer–Verlag.

Porter, B. W. and Kibler, D. F. 1986. Experimental goal regression: a method for learning problem-solving. *Machine Learning* 1(3):249–286.

Quinlan, J. R. 1987. Simplifying decision trees. *Int. J. Man-Machine Stud.* 12:221–234.

Rajamoney, S. and DeJong, G. 1987. The classification, detection and handling of imperfect theory problems. In *10th Int. J. Conf. Artif. Intelligence*, pp. 205–207. Morgan Kaufmann.

Rosenbloom, P. and Laird, J. 1986. Mapping explanation-based generalization onto SOAR. In *5th Nat. Conf. Artif. Intelligence*, pp. 561–567. AAAI Press.

Russell, S. 1989. *The Use of Knowledge in Analogy and Induction*. Morgan Kaufmann.

Russell, S. and Grosof, B. 1987. A declarative approach to bias in concept learning. In *Nat. Conf. Artif. Intelligence*, pp. 505–510. AAAI Press.

Saitta, L., Botta, M., Ravotta, S. and Sperotto, S. 1991. Improving learning by using deep models. In *1st Int. Workshop Multistrategy Learning*, pp. 131–143. George Mason University Press.

Schank, R. and Leake, D. 1989. Creativity and learning in a case-based explainer. *Artif. Intelligence* 40(1–3):353–385.

Segre, A. 1988a. *Machine Learning of Robot Assembly Plans*, Kluwer Academic.

Segre, A. 1988b. Operationality and real-world plans. In *AAAI Symp. Explanation-Based Learning*, pp. 158–163. AAAI Press.

Segre, A. and Elkan, C. 1994. A high-performance explanation-based learning algorithm. *Artif. Intelligence* 69(1–2):1–50.

Shavlik, J. and Towell, G. 1990. An approach to combining explanation-based and neural learning algorithms. In *Readings in Machine Learning*. J. Shavlik and T. Dietterich, eds., pp. 828–839. Morgan Kaufmann.

Shavlik, J. W. and DeJong, G. F. 1987. An explanation-based approach to generalizing number. In *10th Int. J. Conf. Artif. Intelligence*, pp. 236–238. IJCAI.

Silver, B. 1983. Learning equation solving methods from worked examples. In *Proc. 1983 Int. Workshop Machine Learning*, pp. 99–104. CS Dept., University of Illinois.

Silver, B. 1986. Precondition analysis: learning control information. In *Machine Learning: An AI Approach*. R. Michalski, J. Carbonell, and T. Mitchell, eds., pp. 647–670. Morgan Kaufmann.

Sims, M. H. 1987. Empirical and analytic discovery in IL. In *4th Int. Conf. Machine Learning*, pp. 274–280. Morgan Kaufmann.

Smith, E. E. and Medin, D. L. 1981. *Categories and Concepts*. Harvard University Press.

Soloway, E. 1977. *Knowledge Directed Learning Using Multiple Levels of Description.* Ph.D. thesis. University of Massachusetts, Amherst.

Thrun, S. and Mitchell, T. 1993. Integrating inductive nerual network learning and explanation-based learning. In *13th Int. J. Conf. Artif. Intelligence,* pp. 930–936. Morgan Kaufmann.

Velosa, M. and Carbonell, J. 1993. Derivational analogy in PRODIGY: automating case acquisition, storage, and utilization. *Machine Learning* 10(3):249–278.

Waldinger, R. 1977. Achieving several goals simultaneously. In *Machine Intelligence.* E. Elcock and D. Michie, eds., pp. 94–136. Ellis Horwood.

Waterman, D. A. 1970. Generalization learning techniques for automating the learning of heuristics. *Artif. Intelligence* 1(2):121–170.

Wilkins, D. C. 1988. Knowledge based refinement using apprenticeship learning techniques. In *7th Nat. Conf. Artif. Intelligence,* pp. 646–651. Morgan Kaufmann.

Zweben, M., Davis, E., Daun, B., Drascher, E., Deale, M., and Eskey, M. 1992. Learning to improve constraint-based scheduling. *Art. Intelligence* 58(1–3):271–296.

22

Decision Trees and Instance-Based Classifiers

J. R. Quinlan
University of Sydney

22.1 Introduction

This chapter looks at two of the common learning paradigms used in artificial intelligence (AI), both of which are also well known in statistics. These methods share an approach to learning that is based on exploiting regularities among observations, so that predictions are made on the basis of similar previously encountered situations. The methods differ, however, in the way that similarity is expressed; trees make important shared properties explicit, whereas instance-based approaches equate (dis)similarity with some measure of distance.

Attribute-Value Representation

Decision tree and instance-based methods both represent each **instance** using a collection $\{A_1, A_2, \ldots, A_x\}$ of properties or **attributes**. Attributes are grouped into two broad types: *continuous* attributes have real or integer values, whereas *discrete* attributes have unordered nominal values drawn from a (usually small) set of possibilities defined for that attribute. Each instance also belongs to one of a fixed set of mutually exclusive **classes** $c_1, c_2, \ldots, c_k$. Both families of methods use a **training set** of classified instances to develop a mapping from attribute values to classes; this mapping can then be used to predict the class of a new instance from its attribute values.

Figure 22.1 shows a small collection of instances described in terms of four attributes. Attributes Outlook and Windy are discrete, with possible values {sunny, overcast,

Outlook	Temp, °F	Humidity, %	Windy	Class
rain	70	96	false	yes
sunny	80	90	true	no
overcast	64	65	true	yes
sunny	75	70	true	yes
sunny	85	85	false	no
sunny	72	95	false	no
rain	75	80	false	yes
sunny	69	70	false	yes
overcast	83	78	false	yes
rain	65	70	true	no
overcast	72	90	true	yes
overcast	81	75	false	yes
rain	68	80	false	yes
rain	71	80	true	no

FIGURE 22.1 An illustrative training set of instances.

rain} and {true, false} respectively, whereas the other two attributes have numeric values. Each instance belongs to one of the classes yes or no.

The x attributes define an x-dimensional **description space** in which each instance becomes a point. From this geometrical perspective, both instance-based and decision tree approaches divide the description space into regions, each associated with one of the classes.

22.2 Decision Trees

Methods for generating decision trees were pioneered by Hunt and his co-workers in the 1960s, although their popularity in statistics stems from the independent work of Breiman et al. [1984]. The techniques are embodied in software packages such as CART [Breiman et al. 1984] and C 4.5 [Quinlan 1993].

Decision tree learning systems have been used in numerous industrial applications, particularly diagnosis and control. In one early success, Leech [1986] learned comprehensible trees from data logged from a complex and imperfectly understood uranium sintering process. The trees pointed the way to improved control of the process with substantial gains in throughput and quality. Evans and Fisher [1994] describe the use of decision trees to prevent banding, a problem in high-speed rotogravure printing. The trees are used to predict situations in which banding is likely to occur so that preventive action can be taken, leading to a dramatic reduction in print delays. Several other tree-based applications are discussed in Langley and Simon [1995].

Method for Constructing Decision Trees

Decision trees are constructed by a recursive *divide-and-conquer* algorithm that generates a partition of the data. The tree for set D of instances is formed as follows:

- If D satisfies a specified **stopping criterion**, the tree for D is a **leaf** that identifies the most frequent class among the instances. The most common stopping criterion is that all instances of D belong to the same class, but some systems also stop when D contains very few instances.
- Otherwise, select some **test** T with mutually exclusive outcomes $T_1, T_2, \ldots, T_n$ and let D_i be the subset of D containing those instances with outcome T_i, $1 \leq i \leq n$. The decision tree for D then has T as its root with a subtree for each outcome T_i of T. If D_i is empty, the subtree corresponding to outcome T_i is a leaf that nominates the majority class in D; otherwise, the subtree for T_i is obtained by applying the same procedure to subset D_i of D.

In the example of Fig. 22.1, the test chosen for the root of the tree might be Outlook $=$? with possible outcomes sunny, overcast, and rain. The subset of instances with outcome sunny might then be further subdivided by a test Humidity ≤ 75 with outcomes true and false. All instances with outlook overcast belong to the same class, so no further subdivision would be necessary. The instances with outlook rain might be further divided by a test Windy $=$? with outcomes true and false. The resulting decision tree appears in Fig. 22.2 and the corresponding partition of the training instances is in Fig. 22.3.

The tree provides a mechanism for classifying any instance. Starting at the root, the outcome of the test for that instance is determined and the process continues with the corresponding subtree. When a leaf is encountered, the instance is predicted to belong to the class identified by the leaf. In the preceding

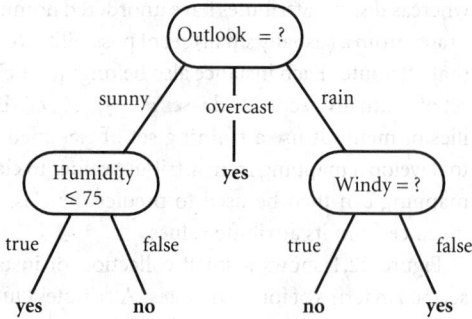

FIGURE 22.2   Decision tree for training instances of Fig. 22.1.

example, a new instance

$$\text{Outlook} = \text{sunny, Temp} = 82, \text{Humidity} = 85, \text{Windy} = \text{true}$$

would follow the outcome sunny, then the outcome false before reaching a leaf labeled no.

Choosing Tests

The example does not explain how tests like Outlook $=$? or Humidity ≤ 75 came to be used. Provided that the chosen test T always produces a nontrivial partition of the instances so that no subset D_i contains all of them, the process will terminate. Nevertheless, the choice of T determines the structure of the final tree and so can affect the class predicted for a new instance. Most decision tree systems are biased toward producing compact trees since, if two trees account equally well for the training instances, the simpler tree seems likely to have higher predictive accuracy.

The first step in selecting a test is to delineate the possibilities. Many systems consider only tests that involve a single attribute as follows:

Outlook	Temp, °F	Humidity, %	Windy	Class
sunny	75	70	true	yes
sunny	69	70	false	yes
sunny	80	90	true	no
sunny	85	85	false	no
sunny	72	95	false	no
overcast	72	90	true	yes
overcast	83	78	false	yes
overcast	64	65	true	yes
overcast	81	75	false	yes
rain	71	80	true	no
rain	65	70	true	no
rain	75	80	false	yes
rain	68	80	false	yes
rain	70	96	false	yes

FIGURE 22.3   Partition of the instances of Fig. 22.1.

- For a discrete attribute A_i with possible values $v_1, v_2, \ldots, v_m$, a single test $A_i =$? with m outcomes could be considered. Additional possibilities are the m binary tests $A_i = v_j$, each with outcomes true and false.
- A continuous attribute A_i usually appears in a thresholded test such as $A_i \leq t$ (with outcomes true and false) for some constant t. Although there are infinitely many possible thresholds t, the number of distinct values of A_i that appear in a set D of instances is at most $|D|$. If these values are sorted into an ascending sequence, say, $n_1 < n_2 < \cdots < n_l$, any value of t in the interval $[n_i, n_{i+1})$ will give the same partition of D, so only one threshold in each interval need be considered.

Most systems carry out an exhaustive comparison of simple tests such as those just described, although more complex tests (see "Extensions" subsections) may be examined heuristically.

Tests are evaluated with respect to some **splitting criterion** that allows the desirability of different tests to be assessed and compared. Such criteria are often based on the class distributions in the set D and subsets $\{D_i\}$ induced by a test. Two examples should illustrate the idea.

Gini Index and Impurity Reduction

Breiman et al. [1984] determine the *impurity* of a set of instances from its class distribution. If the relative frequency of instances belonging to class c_j in D is denoted by r_j, $1 \leq j \leq k$, then

$$Gini(D) = 1 - \sum_{j=1}^{k} (r_j)^2$$

The Gini index of a set of instances assumes its minimum value of zero when all instances belong to a single class.

Suppose now that test T partitions D into subsets $D_1, D_2, \ldots, D_n$ as before. The expected reduction in impurity associated with this test is given by

$$Gini(D) - \sum_{i=1}^{n} \frac{|D_i|}{|D|} \times Gini(D_i)$$

whose value is always greater than or equal to zero.

Gain Ratio

Criteria such as impurity reduction tend to improve with the number of outcomes n of a test. If possible tests have very different numbers of outcomes, such metrics do not provide a fair basis for comparison.

The gain ratio criterion [Quinlan 1993] is an information-based measure that attempts to allow for different numbers (and different probabilities) of outcomes. The residual uncertainty about the class to which an instance in D belongs can be expressed in a form similar to the preceding Gini index as

$$Info(D) = -\sum_{j=1}^{k} r_j \times \log_2(r_j)$$

and the corresponding information gained by a test T as

$$Info(D) - \sum_{i=1}^{n} \frac{|D_i|}{|D|} \times Info(D_i)$$

Like reduction in impurity, information gain focuses on class distributions. On the other hand, the potential information obtained by partitioning a set of instances is based on knowing the subset D_i into which an instance falls; this *split information* is given by

$$-\sum_{i=1}^{n} \frac{|D_i|}{|D|} \times \log_2\left(\frac{|D_i|}{|D|}\right)$$

and tends to increase with the number of outcomes of a test. The gain ratio criterion uses the ratio of the information gain of a test T to its split information as the measure of its usefulness.

There have been numerous studies of the behavior of different splitting criteria, e.g., Liu and White [1994]. Some authors, including Breiman et al. [1984], see little operational difference among a broadly defined class of metrics.

Overfitting

Most data collected in practical applications involve some degree of *noise*. Values of continuous attributes are subject to measurement errors, discrete attributes such as color depend on subjective interpretation, instances are misclassified, and mistakes are made in recording.

When the divide-and-conquer algorithm is applied to such data, it often results in very large trees that *fit* the noise in addition to the meaningful structure in the task. The resulting over-elaborate trees are more difficult to understand and generally exhibit degraded predictive accuracy when classifying unseen instances.

Overfitting can be prevented either by restricting the growth of the tree, usually by means of significance tests of one form or another, or by **pruning** back the full tree to an appropriate size. The latter is generally preferred since it allows interactions of tests to be explored before deciding how much structure is justifiable; on the downside, though, growing and then pruning a tree requires more computation. Three common pruning strategies illustrate the idea.

Cost-Complexity Pruning

Breiman et al. [1984] describe a two-stage process in which a sequence of trees $Z_0, Z_1, \ldots, Z_z$ is generated, one of which is then selected as the final pruned tree. Consider a decision tree Z used to classify each of the $|D|$ instances in the training set from which it was constructed, and let e of them be misclassified. If $L(Z)$ is the number of leaves in Z, the *cost complexity* of Z is defined as the sum

$$\frac{e}{|D|} + \alpha \times L(Z)$$

for some value of the parameter α. Now, suppose we were to replace a subtree S of Z by a leaf identifying the most frequent class among the instances from which S was constructed. In general, the new tree would misclassify Δe more of the instances in the training set but would contain $L(S) - 1$ fewer leaves. This new tree would have the same cost complexity as Z if

$$\alpha = \frac{\Delta e}{|D| \times (L(S) - 1)}$$

The sequence of trees starts with Z_0 as the original tree. To produce Z_{i+1} from Z_i, each nonleaf subtree of Z_i is examined to find the minimum value of α. All subtrees with that value of α are then replaced by their respective best leaves, and the process continues until the final tree Z_z consists of a single leaf.

If a *pruning set* of instances separate from the training set D is available, a final tree can be selected simply by evaluating each Z_i on this set and picking the most accurate tree. If no pruning set is available, Breiman et al. [1984] employ a strategy based on **cross validation** that allows the **true error rate** of each Z_i to be estimated.

Reduced Error Pruning

The previous method considers only some subtrees of the original tree as candidates for the final pruned tree. Reduced error pruning [Quinlan 1987] presumes the existence of a separate pruning set and identifies among all subtrees of the original tree the one with the lowest error on the pruning set. This can be accomplished efficiently as follows.

Every instance in the pruning set is classified by the tree. The method records the number of errors at each leaf and also notes, for each internal node, the number of errors that would be made if that node were to be changed to a leaf. (As with a leaf, the class associated with an internal node is the most frequent class among the instances from which that subtree was constructed.) When all of these error counts have been determined, each internal node is investigated starting from the bottom levels of the tree. The number of errors made by the subtree rooted at that node is compared with the number of errors that would result from changing the node to a leaf and, if the latter is not greater than the former, the change is effected. Since the total number of errors made by a tree is the sum of the errors at its leaves, it is clear that the final subtree minimizes the number of errors on the pruning set.

Minimum Description Length Pruning

Rissanen's minimum description length (MDL) principle and Wallace and Boulton's similar minimum message length principle provide a rationale for offsetting fit on the training data against the complexity of the tree. The idea is to encode, as a single message, a theory (such as a tree) derived from training data together with the data given the theory. A complex theory that explains the data well might be expensive to encode, but the second part of the message should then be short. Conversely, a simple theory can be encoded cheaply but will not account for the data as well as a more complex theory, so that the second part of the message will require more bits. These principles advocate choosing a theory to minimize the length of the complete message; under certain measures of error or *loss functions*, this policy can be shown to maximize the probability of the theory given the data.

In this context, the alternative theories are pruned variants of the original tree. The scheme does not require a separate pruning set and is computationally simple, but its performance is sensitive to the encoding schemes used: the method for encoding a tree, for instance, implies different prior probabilities for trees of various shapes and sizes. The details would take us too far afield here, but Quinlan and Rivest [1989] and Wallace and Patrick [1993] discuss coding schemes and present comparative results.

Missing Attribute Values

Another problem often encountered with real-world datasets is that they are rarely complete; some instances do not have a recorded value for every attribute. This can impact decision tree methods at three stages:

- When comparing tests on attributes with different numbers of missing values
- When partitioning a set D on the outcomes of the chosen test, since the outcomes for some instances may not be known
- When classifying an unseen instance whose outcome for a test is again undetermined.

These problems are usually handled in one of three ways:

- Filling in missing values. For example, if the value of a discrete attribute is not known, it can be assumed to be that attribute's most frequent value, and a missing numeric value can be replaced by the mean of the known values.
- Estimating test outcomes by some other means. Breiman et al. [1984] define the notion of a *surrogate split* for test T, viz., a test on a different attribute that produces a similar partition of D. When the value of a tested attribute is not known, the best surrogate split whose outcome is known is used to predict the outcome of the original test.
- Treating test outcomes probabilistically. Rather than determining a single outcome, this approach uses the probabilities of the outcomes as determined by their relative frequencies in the training data. In the task of Fig. 22.1, for instance, the probabilities of the outcomes sunny, overcast, and rain for the test Outlook $=$? are 5/14, 4/14, and 5/14, respectively. If the tree of Fig. 22.2 is used to classify an instance whose value of Outlook is missing, all three outcomes are explored. The predicted classes associated with each outcome are then combined with the corresponding relative frequencies to give a probability distribution over the classes; this is straightforward, since the outcomes are mutually exclusive. Finally, the class with highest probability is chosen as the predicted class.

The approaches are discussed in more detail in Quinlan [1989] together with comparative trials of different combinations of methods.

Extensions

The previous sections sketch what might be called the fundamentals of constructing and using decision trees. We now look at extensions in various directions aimed at producing trees with higher predictive accuracies on new instances and/or reducing the computation required for learning.

More Complex Tests

Many authors have considered ways of enlarging the repertoire of possible tests beyond those set out in the section on choosing tests. More flexible tests allow greater freedom in dividing the description space into regions and so increase the number of classification functions that can be represented as decision trees.

Subset Tests. If an attribute A_i has numerous discrete values $v_1, v_2, \ldots, v_m$, a test $A_i =$? with one branch for every outcome will divide D into many small subsets. The ability to find meaningful structure in data depends on having sufficient instances to distinguish random and systematic association between attribute values and classes, so this *data fragmentation* generally makes learning more difficult.

One alternative to tests of this form is to group the values of A_i into a small number of subsets $S_1, S_2, \ldots, S_q$ ($q \ll m$), giving a test with outcomes $A_i \in S_j$, $1 \leq j \leq q$. Since there are $\sum_{q=2}^{m-1} q^{m-1}$ possible groupings of values, it is generally impossible to evaluate all of them.

In two-class learning tasks where the values are to be grouped into two subsets, Breiman et al. [1984] give the following algorithm for finding the subsets that optimize convex splitting criteria such as impurity reduction:

- For each value v_j, determine the proportion of instances with this value that belong to one of the classes (the majority class, say).
- Order the values on this proportion, giving $v'_1, v'_2, \ldots, v'_m$.
- The optimal subsets are then $\{v'_1, v'_2, \ldots, v'_l\}$ and $\{v'_{l+1}, v'_{l+2}, \ldots, v'_m\}$ for some value of l in the range 1 to $m - 1$.

This reduces the number of candidate subsets from 2^{m-1} to $m - 1$ and makes it feasible to find the true optimal grouping. This procedure depends on there being only two classes, but CART extends the idea to multiclass tasks by first assembling the classes themselves into two superclasses and then finding the optimal subsets with respect to the superclasses.

Another approach is to grow the subsets heuristically. C4.5 [Quinlan 1993] starts with each subset containing a single value and iteratively merges subsets. At each stage, the subsets to be merged are chosen so that the gain ratio of the new test is maximized; the process stops when merging any pair of subsets would lead to a lower value. Since this algorithm is based on greedy search, optimality of the final subsets cannot be guaranteed.

Linear Multiattribute Tests. If tests all involve a single attribute, the resulting regions in the description space are bounded by hyperplanes that are orthogonal to one of the axes. When the real boundaries are not so simple, the divide-and-conquer algorithm will tend to produce complex trees that approximate general boundaries by successions of small axis-orthogonal segments.

One generalization allows tests that involve a linear combination of attribute values, such as

$$w_0 + \sum_{i=1}^{x} w_i \times A_i \leq 0$$

with outcomes true and false. This clearly makes sense only when each attribute A_i has a numeric value. However, a discrete attribute with m values can be replaced by m binary-valued attributes, each having the value 1 when A_i has the particular value and 0 otherwise; when this is done, the linear test can also include multivalued discrete attributes. Systems such as LMDT [Utgoff and Brodley 1991] and OC1 [Murthy et al. 1994] that implement linear tests of this kind have been found to produce smaller trees, often with higher predictive accuracy on unseen instances. Brodley and Utgoff [1995] provide a summary of methods used to find the coefficients $w_0, w_1, \ldots, w_x$ and compare their performance empirically on several real-world datasets.

Symbolic Multiattribute Tests. One disadvantage of tests that compute a linear combination of attributes is that the tree can become more difficult to understand (although the complexity of the tests must be offset against the smaller overall tree size). Other multiattribute tests that do not use weights suffer less in this respect.

FRINGE [Pagallo and Haussler 1990] uses conjunctions of single-attribute tests. Consider the situation in which an instance belongs to a class if p Boolean conditions are satisfied. A conventional tree using p single-attribute tests would have to partition the training data into $p + 1$ subsets to represent this rule, risking the same problems with data fragmentation mentioned earlier. If a single test consisting of the conjunction of the p tests were used instead, the data would be split into only two subsets. FRINGE finds such conjunctions iteratively, starting with pairs of tests near the leaves of the tree and adding the conjunctions as new attributes for subsequent tree-building stages. In contrast, LFC [Ragavan and Rendell

1993] constructs conjunctive tests directly, in a manner reminiscent of the lookahead employed by the pioneering CLS system [Hunt et al. 1966].

Another form of combination is seen in the *m*-*of*-*n* test whose outcome is true if at least *m* of *n* single-attribute tests are satisfied and false otherwise. Tests of this kind are commonly used in biomedical domains but are extremely cumbersome to represent as trees with single-attribute tests. ID2-of-3 [Murphy and Pazzani 1991] constructs *m*-*of*-*n* tests at each node using a greedy search and often produces smaller trees as a result. Zheng [1995] has recently generalized the idea to an *x*-*of*-*n* test that has one outcome for each possible number of conditions that can be satisfied, rather than just two outcomes based on a specified threshold number *m* of conditions being satisfied. Zheng shows that *x*-*of*-*n* tests are easier to construct than their *m*-*of*-*n* counterparts and have greater representational power.

Multiclass Problems

The effectiveness of decision tree methods is most easily seen in two-class learning tasks when each test contributes to discriminating one class from the other. When there are more than two classes, and especially when classes are very numerous, the goal of a test becomes less clear: should it try to separate one class from all of the others or one group of classes from other groups?

A task with k classes can also be viewed as k two-class tasks, each focusing on distinguishing a single class from all other classes. A separate decision tree can be grown for each class, and a new instance can be classified by looking at the predictions from all k trees. This poses a problem if two or more of the class trees claim the instance, or if none do; the procedure has to be augmented with conflict resolution and default strategies.

A similar idea motivates the *error-correcting output codes* of Dietterich and Bakiri [1995]. With each class is associated a pattern of d binary digits chosen so that the minimum Hamming distance h between any two patterns (i.e., the number of bits in which the patterns differ) is as large as possible. A separate tree is then learned to predict each bit of the class patterns. When a new instance is classified by the d trees, the d output bits may not correspond to the pattern for any class. However, if there are at most $(h-1)/2$ errors in the output bits, the nearest class pattern will indicate the correct class. Case studies presented by Dietterich and Bakiri demonstrate that this technique can result in a large improvement in classification accuracy for domains with numerous classes.

Growing Multiple Trees

An interesting feature of the divide-and-conquer procedure is its sensitivity to the training data. Often two or more attributes will have nearly equal values of the splitting criterion and removing even a single instance from the training set will cause the selected attribute and the associated subtrees to change. (This can be seen as follows: Suppose that omitting one instance typically makes no difference to the learned tree. The correctness or otherwise with which the tree classifies this instance would then be the same whether or not the instance is included in the training set. That is, the **resubstitution error rate** of the tree on the training instances would be the same as the error rate determined by leave-one-out cross validation. However, the resubstitution error rate is known to be an extremely biased measure, substantially underestimating the error rate on new instances, whereas the error rate obtained from leave-one-out cross validation is unbiased.)

The greedy search employed by the divide-and-conquer algorithm will generate only one tree, but a more careful exploration of the space of possible trees will generally uncover many equally appealing trees, compounding the problem of selecting one of them. Buntine [1990] suggests a way around this difficulty that avoids choice altogether! Conceptually, Buntine's idea is to retain all trees that are strong candidates and, when an unseen instance is to be classified, to determine a consensus result by averaging over all predictions. Since it is computationally intractable to explore the complete space of candidate trees, Buntine approximates the process by the use of limited lookahead and by incorporating a small random component in the splitting criterion so that several trees can be constructed from the same set of instances D. Even this constrained search involves substantial increases in computation but, in domains for which classification accuracy is paramount, the additional effort seems justifiable. In a comparison of four methods over 10 learning task domains, Buntine found that the averaging approach using two-ply lookahead gave consistently more accurate predictions on unseen instances. Similarly, Breiman [1996] uses bootstrap samples from the original training set to generate different trees, leading to dramatic improvements on several real-world datasets.

Efficiency Issues

Divide-and-conquer is an efficient algorithm, and current decision tree systems require only seconds on a workstation to generate trees from thousands of instances. In some tasks, however, very large numbers of training instances or the need to constantly regrow trees makes even this computational requirement too demanding.

Determining Thresholds for Continuous Attributes. The most time-consuming operation in growing a tree is finding the possible thresholds for tests on continuous attributes since the values of the attribute that are present in the current set of instances must be sorted, a process of complexity $\Omega(|D| \times \log(|D|))$. This could be avoided, of course, if continuous attributes were thresholded once and for all before growing the tree, thus converting all continuous attributes to discrete attributes. Although this is an active research area, it is clear that continuous attributes will have to be divided into more than just two intervals; recent papers by Fayyad and Irani [1993] and Van de Merckt [1993] suggest algorithms for finding multiple thresholds.

Peepholing. Catlett [1991] investigates efficiency of induction from very large datasets. He first demonstrates that speedup cannot be achieved trivially by learning from samples of the data; in the several domains studied, the accuracy of the final classifier is always reduced when a significant fraction of the data is ignored. In Catlett's approach, a small subset of the training data is studied to determine which continuous-valued attributes can be eliminated from contention for the next test and, for the remainder, the interval in which a good threshold might lie. For the small overhead cost of processing the sample, this method allows the learning algorithm to avoid sorting on some attributes altogether and to sort only those values of the candidate attributes that lie within the indicated limits. As a result, the growth of learning time with the number of training instances is very much closer to linear.

Incremental Tree Construction. In some applications the data available for learning grow continually as new information comes to hand. The divide-and-conquer method is a batch-type process that uses all of the training instances to decide questions such as the choice of the next test. When the training set is enlarged, the previous tree must be discarded and the whole process repeated from scratch to generate a new tree. In contrast, Utgoff [1994] has developed incremental tree-growing algorithms that allow the existing tree to be modified as new training data arrive. Two key ideas are the retention of sufficient counting information at each node to determine whether the test at that node must be changed and a method of pulling up a test from somewhere in a subtree to its root. Utgoff's approach carries an interesting guarantee: the revised tree is identical to the tree that would be produced by divide-and-conquer using the enlarged training set.

22.3 Instance-Based Approaches

Although these approaches (usually under the name of *nearest neighbor* methods) have long interested researchers in pattern recognition, their use in the machine learning community has largely dated from Aha's influential work [Aha et al. 1991]. A useful summary of key developments from the perspective of someone outside AI is provided by the introductory chapter of Dasarthy [1991].

Outline of the Method

Recall that, in the geometrical view, attributes define a description space in which each instance is represented by a point. The fundamental assumption that underlies instance-based classification is that nearby instances in the description space will tend to belong to the same class, i.e., that closeness implies similarity. This does not suggest the converse (similarity implies closeness); there is no implicit assumption that instances belonging to a single class will form one cluster in the description space.

Unlike decision tree methods, instance-based approaches do not rely on a symbolic theory formed from the training instances to predict the class of an unseen instance. Instead, some or all of the training instances are remembered and a new instance is classified by finding instances that lie close to it in the description space and taking the most frequent class among them as the predicted class of the new instance. The central questions in this process are as follows:

- How should closeness in the description space be measured?
- Which training instances should be retained?
- How many neighbors should be used when making a prediction?

These are addressed in the following subsections.

Similarity Metric, or Measuring Closeness

Continuous Attributes

If all attributes are continuous, as was generally the case in early pattern recognition work [Nilsson 1965], the description space is effectively Euclidean. The square of the distance between two instances P and Q, described by their values for the x attributes ($P = \langle p_1, p_2, \ldots, p_x \rangle$ and $Q = \langle q_1, q_2, \ldots, q_x \rangle$) is

$$d^2(P, Q) = \sum_{i=1}^{x} (p_i - q_i)^2$$

and closeness can be equated with small distance. Alternatively, the attributes can be ascribed weights that reflect their relative magnitudes or importances, giving

$$d_w^2(P, Q) = \sum_{i=1}^{x} w_i^2 \times (p_i - q_i)^2$$

Common choices for weights to normalize magnitudes are as follows:

- $w_i = 1/\text{range}_i$. Here range_i is the difference between the largest and smallest values of attribute A_i observed in the training set.
- $w_i = 1/\text{sd}_i$. Here sd_i is the standard deviation of the values of A_i.

The former has the advantage that differences in values of an individual attribute range from 0 to 1, whereas the latter is particularly useful when attribute A_i is known to have a normal distribution.

Discrete Attributes

The difference between unordered values of a discrete attribute is more problematic. The obvious approach is to map the difference $p_i - q_i$ between two values of a discrete attribute A_i to 0 if p_i equals q_i and to 1 otherwise.

Stanfill and Waltz [1986] describe a significant improvement to this two-valued difference that takes account of the similarity of values with respect to the classes. Their *value difference metric* (VDM) first computes a weight for each discrete value of an instance and for each pair of discrete values. Let $n_i(v, c_j)$ denote the number of training instances that have value v for attribute A_i and also belong to class c_j, and let $n_i(v, \cdot)$ denote the sum of these over all classes. An attribute value is important to the extent that it differentiates among the classes. The weight associated with attribute A_i and instance P is taken as

$$w_i(P) = \sqrt{\sum_{j=1}^{k} \left(\frac{n_i(p_i, c_j)}{n_i(p_i, \cdot)} \right)^2}$$

The value difference between p_i and q_i is given by an analogous expression

$$\Delta v_i^2(P, Q) = \sum_{j=1}^{k} \left(\frac{n_i(p_i, c_j)}{n_i(p_i, \cdot)} - \frac{n_i(q_i, c_j)}{n_i(q_i, \cdot)} \right)^2$$

Combining these, the distance between instances P and Q becomes

$$d_{\mathrm{VDM}}(P, Q) = \sum_{i=1}^{x} w_i(P) \times \Delta v_i^2(P, Q)$$

In the task of learning how to pronounce English words, Stanfill and Waltz [1986] found that VDM gave substantially improved performance over simple use of a 0–1 value difference.

Cost and Salzberg [1993] point out that VDM is not symmetric; $d_{\mathrm{VDM}}(P, Q)$ is not generally equal to $d_{\mathrm{VDM}}(Q, P)$ since only the first instance is used to determine the attribute weights. Their *modified value difference metric* (MVDM) drops the attribute weights in favor of an instance weight. They also prefer computing the value difference as the sum of the absolute values of the differences for each class rather that using the square of these differences. In summary,

$$d_{\mathrm{MVDM}}(P, Q) = w(P) \times w(Q) \times \sum_{i=1}^{x} |\Delta v|_i(P, Q)$$

where

$$|\Delta v|_i(P, Q) = \sum_{j=1}^{k} \left| \frac{n_i(p_i, c_j)}{n_i(p_i, \cdot)} - \frac{n_i(q_i, c_j)}{n_i(q_i, \cdot)} \right|$$

The instance weights $w(P)$ and $w(Q)$ depend on their relative success in previous classification trials. If an instance P has been found to be closest to a test instance in t trials, in e of which the test instance belongs to a class different from P, the weight of P is

$$w(P) = \frac{t + 1}{t - e + 1}$$

This means that instances with a poor track record of classification will have a high weight and so appear to be more distant from (and thus less similar to) an unseen instance.

Mixed Continuous and Discrete Attributes

In learning tasks that involve attributes of both types, one strategy to measure distance would be simply to sum the different components as shown earlier, using the weighted square of distance (say) for continuous attributes and the MVDM difference for discrete attributes. Ting [1995] has found that instance-based learners employing nonuniform metrics of this kind have relatively poor performance. His experimental results suggest that it is preferable to convert continuous attributes to discrete attributes using thresholding (as discussed by Fayyad and Irani [1993] or Van de Merckt [1993]) and then to employ a uniform MVDM scheme throughout.

Choosing Instances to Remember

The performance of instance-based methods degrades in the presence of noisy training data. Dasarthy [1991, p. 4] states:

> [Nearest neighbor] classifiers perform best when the training data set is essentially noise free, unlike the other parametric and non-parametric classifiers that perform best when trained in an environment paralleling the operational environment in its noise characteristics.

Performance should improve, then, if noisy training instances are discarded or **edited**. Two approaches to selecting the instances to retain give a flavor of the methods.

IB3 [Aha et al. 1991] starts with training instances arranged in an arbitrary sequence. Each in turn is classified with reference to the (initially empty) pool of retained instances. Those that are classified correctly by the current pool are discarded, whereas misclassified instances are held as potential additions to the pool. Performance statistics for these potential instances are kept and an instance is pooled when a significance test indicates that it would lead to improved classification.

Cameron-Jones [1992] uses an MDL-based approach (see the section on minimum description length pruning). A subset of training instances is chosen heuristically, the goal being to minimize the number of bits in a message specifying the retained instances and the exceptions to the classes that they predict for the training data. This approach usually retains remarkably few instances and yet leads to excellent predictive accuracy.

How Many Neighbors?

Most instance-based approaches use a fixed number of neighbors when classifying a new instance. The size of the neighborhood is important for good classification performance: if it is too small, predictions will be unduly sensitive to the presence of misclassified training instances, whereas too large a value will cause regions of the description space containing fewer exemplars to be merged with surrounding regions. The number of neighbors is usually odd so as to minimize problems with tied class frequencies. Popular choices are one (e.g., Cost and Salzberg [1993]), three, five, and even more (Stanfill and Waltz [1986]).

It is also possible to determine an appropriate number of neighbors from the training instances themselves. A leave-one-out cross validation is performed: each instance in turn is classified using the remaining instances with various neighborhood sizes. The number of neighbors that gives the least number of errors over all instances is then chosen.

Irrelevant Attributes

Instance-based approaches are *parallel* classifiers that use the values of all attributes for each prediction, in contrast with *sequential* classifiers like decision trees that use only a subset of the attributes in each prediction [Quinlan 1994]. When some of the attributes are irrelevant, a random element is introduced to the measurement of distance between instances. Consequently, the performance of instance-based methods can degrade sharply in tasks that have many irrelevant attributes, whereas decision trees are more robust in this respect.

Techniques like MVDM go a long way toward relieving this problem. If a discrete attribute A_i is not related to the instances's classes, the ratio $n_i(v, c_j)/n_i(v, \cdot)$ should not change much for different attribute values v, so that $|\Delta v|_i$ should be close to zero. As a result, the contribution of A_i to the distance calculation should be slight, so that irrelevant attributes are effectively ignored.

Irrelevant attributes can also be excluded more directly by finding the subset of attributes that gives the highest accuracy on a leave-one-out cross validation. There are, of course, $2^x - 1$ nonempty subsets of x attributes, a number that can be too large to investigate if x is greater than 20 or so. Moore and Lee [1994] describe techniques called *racing* and *schemata search* that increase the efficiency of exploring large combinatorial spaces like this. The essence of racing is that competitive subsets are investigated in parallel and a subset is eliminated as soon as it becomes unlikely to win. Schemata search allows subsets of attributes to be described stochastically, using values 0, 1, and * to indicate whether each attribute is definitely excluded, definitely included, or included with probability 0.5. As it becomes clear that subsets including (or excluding) an attribute are performing better, the asterisks for this attribute are resolved in remaining schemata to 1 or 0, respectively.

22.4 Composite Classifiers

This short discussion of decision trees and instance-based methods should not leave the impression that they are solved problems; both are the subject of considerable research. One of the more interesting areas concerns the use of multiple approaches in classifier design. This is motivated by the observation that, even within a single task, there are likely to be regions of the description space in which one or another type of classifier has the edge. For example, instance-based approaches support more general region boundaries than the axis-orthogonal hyperplanes constructed by decision trees. In regions whose true boundaries are complex, the former should provide better models and so lead to more accurate predictions. Conversely, in regions where some attributes are irrelevant, decision trees are likely to prove more robust.

The most general scheme for combining classifiers is *stacking* [Wolpert 1992]. Suppose that y different learning methods are available. For each training instance in turn, y classifiers can be constructed from the remaining instances and used to predict the class of the training instance. This instance thus gives rise to a *first-level* instance with $x + y$ attributes, namely, all of the original attributes plus the y predictions. One of the learning methods can be used with this new dataset; its predictions may employ (selectively) the predictions made by other methods. The process can be repeated to form second-level data with $x + 2y$ attributes, and so on.

In contrast, Brodley [1993] uses hand-crafted rules to decide when a particular classification method is appropriate. One such rule relates to the use of single attribute versus multiattribute tests and can be paraphrased as: If the number of instances is less than the number of attributes, use a single-attribute test, otherwise prefer multiattribute tests.

Finally, Jordan [1994] generalizes the idea of a decision tree to one in which the outcomes of all tests are inherently fuzzy or probabilistic. Constructing a tree then involves not only determining its structure, but also learning a model for estimating the outcome probabilities at each node. Since the latter can involve techniques such as hidden Markov models, the resulting structure is a flexible hybrid.

Acknowledgments

I am most grateful for comments and suggestions from Nitin Indurkhya, Kai Ming Ting, Will Uther, and Zijian Zheng.

Defining Terms

Attribute: A property or feature of all instances. May have *discrete* (nominal) or *continuous* (numeric) values. In statistical terms, an independent variable.

Class: The nominal category to which an instance belongs. The goal of learning is to be able to predict an instance's class from its attribute values. In statistical terms, a dependent variable.

Cross validation: A method for estimating the true error rate of a theory learned from a set of instances. The data are divided into N (e.g., 10) equal-sized groups and, for each group in turn, a theory is learned from the remaining groups and tested on the hold-out group. The estimated true error rate is the total number of test misclassifications divided by the number of instances.

Description space: A conceptual space with one dimension for each attribute. An instance is represented by a point in this space.

Editing: A process of discarding instances from the training set.

Instance: A single observation or datum described by its values of the attributes.

Leaf: A terminal node of a decision tree; has a class label.

Pruning: A process of simplifying a decision tree; each subtree that is judged to add little to the tree's predictive accuracy is replaced by a leaf.

Resubstitution error rate: The misclassification rate of a learned theory on the data from which it was constructed.

Similarity metric: The method used to measure the closeness of two instances in instance-based learning.

Splitting criterion: The basis for selecting one of a set of possible tests.

Stopping criterion: The conditions under which a set of instances is not further subdivided.

Test: An internal node of a decision tree that computes an *outcome* as some function of the attribute values of an instance. A test node is linked to subtrees, one for every possible outcome.

Training set: The collection of instances with known classes that is given to a learning system.

True error rate: The misclassification rate of a theory on unseen instances.

References

Aha, D. W., Kibler, D., and Albert, M. K. 1991. Instance-based learning algorithms. *Machine Learning* 6(1):37–66.

Breiman, L. 1996. Bagging predictors. *Machine Learning* (to appear).

Breiman, L., Friedman, J. H., Olshen, R. A., and Stone, C. J. 1984. *Classification and Regression Trees.* Wadsworth, Belmont, CA.

Brodley, C. E. 1993. Addressing the selective superiority problem: automatic algorithm/model class selection. In *Proc. 10th Int. Conf. Machine Learning,* pp. 17–24. Morgan Kaufmann, San Francisco.

Brodley, C. E. and Utgoff, P. E. 1995. Multivariate decision trees. *Machine Learning* 19(1):45–77.

Buntine, W. L. 1990. *A Theory of Learning Classification Rules.* Ph.D. Thesis. School of Computing Sciences, University of Technology, Sydney, Australia.

Cameron-Jones, R. M. 1992. Minimum description length instance-based learning. In *Proc. 5th Australian J. Conf. Artif. Intelligence,* pp. 368–373. World Scientific, Singapore.

Catlett, J. 1991. *Megainduction.* Ph.D. Thesis. Basser Department of Computer Science, University of Sydney, Australia.

Cost, S. and Salzberg, S. 1993. A weighted nearest-neighbor algorithm for learning with symbolic features. *Machine Learning* 10(1):57–78.

Dasarthy, B. V., ed. 1991. *Nearest Neighbor Norms: NN Pattern Classification Techniques.* IEEE Computer Society Press, Los Alamitos, CA.

Dietterich, T. G. and Bakiri, G. 1995. Solving multiclass learning problems via error correcting output codes. *J. Artif. Intelligence Res.* 2:263–286.

Evans, R. and Fisher, D. 1994. Overcoming process delays with decision tree induction. *IEEE Expert* 9(1):60–66.

Fayyad, U. M. and Irani, K. B. 1993. Multi-interval discretization of continuous-valued attributes for classification learning. In *Proc. 13th Int. J. Conf. Artif. Intelligence,* pp. 1022–1027. Morgan Kaufmann, San Francisco.

Hunt, E. B., Marin, J., and Stone, P. J. 1966. *Experiments in Induction.* Academic Press, New York.

Jordan, M. I. 1994. A statistical approach to decision tree modeling. In *Proc. 11th Int. Conf. on Machine Learning,* pp. 363–370. Morgan Kaufmann, San Francisco.

Langley, P. and Simon, H. A. 1995. Applications of machine learning and rule induction. *Commun. ACM* 38(11):55–64.

Leech, W. J. 1986. A rule based process control method with feedback. In *Proc. Instrument Soc. of Am. Conf.,* pp. 169–175, Houston, TX.

Liu, W. Z. and White, A. P. 1994. The importance of attribute selection measures in decision tree induction. *Machine Learning* 15(1):25–41.

Moore, A. W. and Lee, M. S. 1994. Efficient algorithms for minimizing cross validation error. In *Proc. 11th Int. Conf. Machine Learning*, pp. 190–198. Morgan Kaufmann, San Francisco.

Murphy, P. M. and Pazzani, M. J. 1991. ID2-of-3: constructive induction of M-of-N concepts for discriminators in decision trees. In *Proc. 8th Int. Workshop Machine Learning*, pp. 183–187. Morgan Kaufmann, San Francisco.

Murthy, S. K., Kasif, S., and Salzberg, S. 1994. A system for induction of oblique decision trees. *J. Artif. Intelligence Res.* 2:1–32.

Nilsson, N. J. 1965. *Learning Machines*. McGraw–Hill, New York; 1990. Republished as *The Mathematical Foundations of Learning Machines*, Morgan Kaufmann, San Francisco.

Pagallo, G. and Haussler, D. 1990. Boolean feature discovery in empirical learning. *Machine Learning* 5(1):71–100.

Quinlan, J. R. 1987. Simplifying decision trees. *Int. J. Man-Machine Studies* 27:221–234.

Quinlan, J. R. 1989. Unknown attribute values in induction. In *Proc. 6th Int. Machine Learning Workshop*, pp. 164–168. Morgan Kaufmann, San Francisco.

Quinlan, J. R. 1993. *C4.5: Programs for Machine Learning*. Morgan Kaufmann, San Francisco.

Quinlan, J. R. 1994. Comparing connectionist and symbolic learning methods. In *Computational Learning Theory and Natural Learning Systems*. Vol. 1. S. J. Hanson, G. A. Drastal, and R. L. Rivest, eds., pp. 445–456. MIT Press, Cambridge, MA.

Quinlan, J. R. and Rivest, R. L. 1989. Inferring decision trees using the minimum description length principle. *Inf. Comput.* 80(3):227–248.

Ragavan, H. and Rendell, L. 1993. Lookahead feature construction for learning hard concepts. In *Proc. 11th Int. Conf. Machine Learning*, pp. 252–259. Morgan Kaufmann, San Francisco.

Stanfill, C. and Waltz, D. 1986. Toward memory-based reasoning. *Commun. ACM* 29(12):1213–1228.

Ting, K. M. 1995. *Common Issues in Instance-Based and Naive Bayesian Classifiers*. Ph.D. Thesis. Basser Department of Computer Science, University of Sydney, Australia.

Utgoff, P. E. 1994. An improved algorithm for incremental induction of decision trees. In *Proc. 11th Int. Conf. Machine Learning*, pp. 318–325. Morgan Kaufmann, San Francisco.

Utgoff, P. E. and Brodley, C. E. 1991. Linear machine decision trees. University of Massachusetts, Amherst. COINS Tech. Rep. 91-10.

Van de Merckt, T. 1993. Decision trees in numerical attribute spaces. In *Proc. 13th Int. J. Conf. Artif. Intelligence*, pp. 1016–1021. Morgan Kaufmann, San Francisco.

Wallace, C. S. and Patrick, J. D. 1993. Coding decision trees. *Machine Learning* 11(1):7–22.

Wolpert, D. H. 1992. Stacked generalization. *Neural Networks* 5:241–259.

Zheng, Z. 1995. Constructing nominal X-of-N attributes. In *Proc. 14th Int. J. Conf. Artif. Intelligence*, pp. 1064–1070. Morgan Kaufmann, San Francisco.

Further Information

The principal computer science journals that report advances in learning techniques are *Machine Learning* (Kluwer), *Artificial Intelligence* (Elsevier), and *Journal of Artificial Intelligence Research*. The latter is an electronic journal; details are available at http://www.cs.washington.edu/research/jair/home.html or from jair-ed@ptolemy.arc.nasa.gov.

Papers on learning techniques are presented at the International Conferences in Machine Learning, the International Joint Conferences on Artificial Intelligence, the AAAI National Conferences on Artificial Intelligence, and the European Conferences on Machine Learning. Applications are not as easy to follow, although the Workshops and Conferences on Knowledge Discovery in Databases have relevant papers.

There are two moderated electronic newsletters that often contain relevant material: the *Machine Learning List* (http://www.ics.uci.edu/AI/ML/home.html) and *KDD Nuggets* (http://info.gte.com/~kdd).

23

Neural Networks

Michael I. Jordan
Massachusetts Institute of Technology

Christopher M. Bishop
Aston University

23.1 Introduction

Within the broad scope of the study of artificial intelligence (AI), research in neural networks is characterized by a particular focus on pattern recognition and pattern generation. Many neural network methods can be viewed as generalizations of classical pattern-oriented techniques in statistics and the engineering areas of signal processing, system identification, and control theory. As in these parent disciplines, the notion of "pattern" in neural network research is essentially probabilistic and numerical. Neural network methods have had their greatest impact in problems where statistical issues dominate and where data are easily obtained.

A neural network is first and foremost a graph, with patterns represented in terms of numerical values attached to the nodes of the graph and transformations between patterns achieved via simple message-passing algorithms. Many neural network architectures, however, are also statistical processors, characterized by making particular probabilistic assumptions about data. As we will see, this conjunction of graphical algorithms and probability theory is not unique to neural networks but characterizes a wider family of probabilistic systems in the form of chains, trees, and networks that are currently being studied throughout AI [Spiegelhalter et al. 1993].

Neural networks have found a wide range of applications, the majority of which are associated with problems in pattern recognition and control theory. In this context, neural networks can best be viewed as a class of algorithms for statistical modeling and prediction. Based on a source of *training data*, the aim is to produce a statistical model of the process from which the data are generated, so as to allow the best predictions to be made for new data. We shall find it convenient to distinguish three broad types of statistical modeling problem, which we shall call **density estimation**, **classification**, and **regression**.

For density estimation problems (also referred to as *unsupervised learning* problems), the goal is to model the unconditional distribution of data described by some vector x. A practical example of the application of density estimation involves the interpretation of X-ray images (mammograms) used for breast cancer screening [Tarassenko 1995]. In this case, the training vectors x form a sample taken from

0-8493-2909-4/97/$0.00+$.50
© 1997 by CRC Press, Inc.

normal (noncancerous) images, and a network model is used to build a representation of the density $p(x)$. When a new input vector x' is presented to the system, a high value for $p(x')$ indicates a normal image, whereas a low value indicates a novel input which might be characteristic of an abnormality. This is used to label regions of images that are unusual, for further examination by an experienced clinician.

For classification and regression problems (often referred to as *supervised learning* problems), we need to distinguish between *input* variables, which we again denote by x, and *target* variables, which we denote by the vector t. Classification problems require that each input vector x be assigned to one of C classes $\mathcal{C}_1, \ldots, \mathcal{C}_C$, in which case the target variables represent class labels. As an example, consider the problem of recognizing handwritten digits [LeCun et al. 1989]. In this case, the input vector would be some (preprocessed) image of the digit, and the network would have 10 outputs, one for each digit, which can be used to assign input vectors to the appropriate class (as discussed in section 23.2).

Regression problems involve estimating the values of continuous variables. For example, neural networks have been used as part of the control system for adaptive optics telescopes [Sandler et al. 1991]. The network input x consists of one in-focus and one defocused image of a star and the output t consists of a set of coefficients that describe the phase distortion due to atmospheric turbulence. These output values are then used to make real-time adjustments of the multiple mirror segments to cancel the atmospheric distortion.

Classification and regression problems also can be viewed as special cases of density estimation. The most general and complete description of the data is given by the probability distribution function $p(x, t)$ in the joint input-target space. However, the usual goal is to be able to make good predictions for the target variables when presented with new values of the inputs. In this case, it is convenient to decompose the joint distribution in the form

$$p(x, t) = p(t \mid x)p(x) \tag{23.1}$$

and to consider only the conditional distribution $p(t \mid x)$, in other words the distribution of t *given* the value of x. Thus, classification and regression involve the estimation of *conditional* densities, a problem which has its own idiosyncracies.

The organization of the chapter is as follows. In section 23.2 we present examples of network representations of unconditional and conditional densities. In section 23.3 we discuss the problem of adjusting the parameters of these networks to fit them to data. This problem has a number of practical aspects, including the choice of optimization procedure and the method used to control network complexity. We then discuss a broader perspective on probabilistic network models in section 23.4. The final section presents further information and pointers to the literature.

23.2 Representation

In this section we describe a selection of neural network architectures that have been proposed as representations for unconditional and conditional densities. After a brief discussion of density estimation, we discuss classification and regression, beginning with simple models that illustrate the fundamental ideas and then progressing to more complex architectures. We focus here on representational issues, postponing the problem of learning from data until the following section.

Density Estimation

We begin with a brief discussion of density estimation, utilizing the Gaussian **mixture model** as an illustrative model. We return to more complex density estimation techniques later in the chapter.

Although density estimation can be the main goal of a learning system, as in the diagnosis example mentioned in the Introduction, density estimation models arise more often as components of the solution to a more general classification or regression problem. To return to Eq. (23.1), note that the joint density is composed of $p(t \mid x)$, to be handled by classification or regression models, and $p(x)$, the (unconditional) input density. There are several reasons for wanting to form an explicit model of the input density. First, real-life data sets often have missing components in the input vector. Having a model of the density allows the missing components to be filled in in an intelligent way. This can be useful both for training and for prediction (cf. Bishop [1995]). Second, as we see in Eq. (23.1), a model of $p(x)$ makes possible an estimate of the joint probability $p(x, t)$. This in turn provides us with the necessary information to estimate the inverse conditional density $p(x \mid t)$. The calculation of such inverses is important for applications in control and optimization.

A general and flexible approach to density estimation is to treat the density as being composed of a set of M simpler densities. This approach involves modeling the observed data as a sample from a *mixture density*,

$$p(x \mid w) = \sum_{i=1}^{M} \pi_i \, p(x \mid i, w_i) \tag{23.2}$$

where the π_i are constants known as *mixing proportions*, and the $p(x \mid i, w_i)$ are the *component densities*, generally taken to be from a simple parametric family. A common choice of component density is the multivariate Gaussian, in which case the parameters w_i are the means and covariance matrices of each of the components. By varying the means and covariances to place and orient the Gaussians appropriately, a wide variety of high-dimensional, multimodal data can be modeled. This approach to density estimation is essentially a probabilistic form of clustering.

Gaussian mixtures have a representation as a network diagram, as shown in Fig. 23.1. The utility of such network representations will become clearer as we proceed; for now, it suffices to note that not only mixture models, but also a wide variety of other classical statistical models for density estimation, are representable as simple networks with one or more layers of adaptive weights. These methods include *principal component analysis, canonical correlation analysis, kernel density estimation,* and *factor analysis* [Anderson 1984].

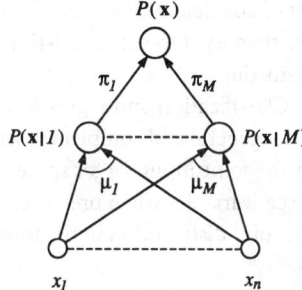

FIGURE 23.1 A network representation of a Gaussian mixture distribution. The input pattern x is represented by numerical values associated with the input nodes in the lower level. Each link has a weight μ_{ij}, which is the jth component of the mean vector for the ith Gaussian. The ith intermediate node contains the covariance matrix Σ_i and calculates the Gaussian conditional probability $p(x \mid i, \mu_i, \Sigma_i)$. These probabilities are weighted by the mixing proportions π_i and the output node calculates the weighted sum $p(x) = \sum_i \pi_i p(x \mid i, \mu_i, \Sigma_i)$.

Linear Regression and Linear Discriminants

Regression models and classification models both focus on the conditional density $p(t \mid x)$. They differ in that in regression the target vector t is a real-valued vector, whereas in classification t takes its values from a discrete set representing the class labels.

The simplest probabilistic model for regression is one in which t is viewed as the sum of an underlying deterministic function $f(x)$ and a Gaussian random variable ϵ,

$$t = f(x) + \epsilon \tag{23.3}$$

If ϵ has zero mean, as is commonly assumed, $f(x)$ then becomes the *conditional mean* $E(t \mid x)$. It is this function that is the focus of most regression modeling. Of course, the conditional mean describes only the first moment of the conditional distribution, and, as we discuss in a later section, a good regression model will also generally report information about the second moment.

In a linear regression model, the conditional mean is a linear function of x: $E(t \mid x) = Wx$, for a fixed matrix W. Linear regression has a straightforward representation as a network diagram in which the jth input unit represents the jth component of the input vector x_j, each output unit i takes the weighted sum of the input values, and the weight w_{ij} is placed on the link between the jth input unit and the ith output unit.

The conditional mean is also an important function in classification problems, but most of the focus in classification is on a different function known as a **discriminant function**. To see how this function arises and to relate it to the conditional mean, we consider a simple two-class problem in which the target is a simple binary scalar that we now denote by t. The conditional mean $E(t \mid x)$ is equal to the probability that t equals one, and this latter probability can be expanded via Bayes rule

$$p(t = 1 \mid x) = \frac{p(x \mid t = 1)p(t = 1)}{p(x)} \tag{23.4}$$

The density $p(t \mid x)$ in this equation is referred to as the *posterior probability* of the class given the input, and the density $p(x \mid t)$ is referred to as the *class-conditional density*. Continuing the derivation, we expand the denominator and (with some foresight) introduce an exponential,

$$p(t = 1 \mid x) = \frac{p(x \mid t = 1)p(t = 1)}{p(x \mid t = 1)p(t = 1) + p(x \mid t = 0)p(t = 0)} \tag{23.5}$$

$$= \frac{1}{1 + \exp\left\{ -\ln\left[\frac{p(x \mid t=1)}{p(x \mid t=0)}\right] - \ln\left[\frac{p(t=1)}{p(t=0)}\right] \right\}}$$

We see that the posterior probability can be written in the form of the *logistic function*:

$$y = \frac{1}{1 + e^{-z}} \tag{23.6}$$

where z is a function of the likelihood ratio $p(x \mid t = 1)/p(x \mid t = 0)$, and the prior ratio $p(t = 1)/p(t = 0)$. This is a useful representation of the posterior probability if z turns out to be simple.

It is easily verified that if the class conditional densities are multivariate Gaussians with identical covariance matrices, then z is a linear function of x: $z = w^T x + w_0$. Moreover, this representation is appropriate for any distribution in a broad class of densities known as the exponential family (which includes the Gaussian, the Poisson, the gamma, the binomial, and many other densities). All of the densities in this family can be put in the following form:

$$g(x; \theta, \phi) = \exp\{(\theta^T x - b(\theta))/a(\phi) + c(x, \phi)\} \tag{23.7}$$

where θ is the *location parameter* and ϕ is the *scale parameter*. Substituting this general form in Eq. (23.5), where θ is allowed to vary between the classes and ϕ is assumed to be constant between classes, we see that z is in all cases a linear function. Thus, the choice of a linear-logistic model is rather robust.

The geometry of the two-class problem is shown in Fig. 23.2, which shows Gaussian class-conditional densities, and suggests the logistic form of the posterior probability.

The function z in our analysis is an example of a discriminant function. In general, a discriminant function is any function that can be used to decide on class membership [Duda and Hart 1973]; our analysis has produced a particular form of discriminant function that is an intermediate step in the calculation of a posterior probability. Note that if we set $z = 0$, from the form of the logistic function we obtain a probability of 0.5, which shows that $z = 0$ is a *decision boundary* between the two classes.

The discriminant function that we found for exponential family densities is linear under the given conditions on ϕ. In more general situations, in which the class-conditional densities are more complex than a single exponential family density, the posterior probability will not be well characterized by the linear-logistic form. Nonetheless, it still is useful to retain the logistic function and focus on *nonlinear* representations for the function z. This is the approach taken within the neural network field.

To summarize, we have identified two functions that are important for regression and classification, respectively: the conditional mean and the discriminant function. These are the two functions that are of concern for simple linear models and, as we now discuss, for more complex nonlinear models as well.

Nonlinear Regression and Nonlinear Classification

The linear regression and linear discriminant functions introduced in the previous section have the merit of simplicity, but are severely restricted in their representational capabilities. A convenient way to see this is to consider the geometrical interpretation of these models. When viewed in the d-dimensional x-space, the linear regression function $w^T x + w_0$ is constant on hyperplanes which are orthogonal to the vector w. For many practical applications, we need to consider much more general classes of function. We therefore seek representations for nonlinear mappings which can approximate any given mapping to

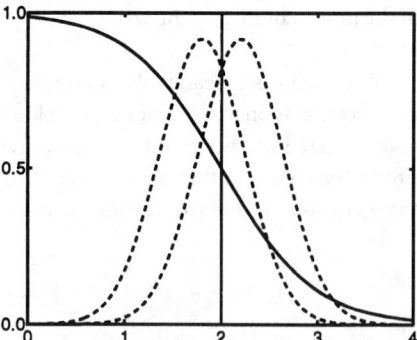

FIGURE 23.2 This shows the Gaussian class-conditional densities $p(x \mid C_1)$ (dashed curves) for a two-class problem in one dimension, together with the corresponding posterior probability $p(C_1 \mid x)$ (solid curve) which takes the form of a logistic sigmoid. The vertical line shows the decision boundary for $y = 0.5$, which coincides with the point at which the two density curves cross.

arbitrary accuracy. One way to achieve this is to transform the original x using a set of M nonlinear functions $\phi_j(x)$ where $j = 1, \ldots, M$, and then to form a linear combination of these functions, so that

$$y_k(x) = \sum_j w_{kj} \phi_j(x) \qquad (23.8)$$

For a sufficiently large value of M, and for a suitable choice of the $\phi_j(x)$, such a model has the desired universal approximation properties. A familiar example, for the case of one-dimensional input spaces, is the simple polynomial, for which the $\phi_j(x)$ are simply successive powers of x and the w are the polynomial coefficients. Models of the form in Eq. (23.8) have the property that they can be expressed as network diagrams in which there is a *single* layer of adaptive weights.

There are a variety of families of functions in one dimension that can approximate any continuous function to arbitrary accuracy. There is, however, an important issue which must be addressed, called the *curse of dimensionality*. If, for example, we consider an Mth-order polynomial then the number of independent coefficients grows as d^M [Bishop 1995]. For a typical medium-scale application with, say, 30 inputs, a fourth-order polynomial (which is still quite restricted in its representational capability) would have over 46,000 adjustable parameters. As we shall see in the section on complexity control, in order to achieve good generalization it is important to have more data points than adaptive parameters in the model, and this is a serious problem for methods that have a power law or exponential growth in the number of parameters.

A solution to the problem lies in the fact that, for most real-world data sets, there are strong (often nonlinear) correlations between the input variables such that the data do not uniformly fill the input space but are effectively confined to a subspace whose dimensionality is called the *intrinsic dimensionality* of the data. We can take advantage of this phenomenon by considering again a model of the form in Eq. (23.8) but in which the basis functions $\phi_j(x)$ are *adaptive* so that they themselves contain weight parameters whose values can be adjusted in the light of the observed dataset. Different models result from different choices for the basis functions, and here we consider the two most common examples. The first of these is called the **multilayer perceptron** (MLP) and is obtained by choosing the basis functions to be given by linear-logistic functions Eq. (23.6). This leads to a multivariate nonlinear function that can be expressed in the form

$$y_k(\boldsymbol{x}) = \sum_{j=1}^{M} w_{kj} g \left(\sum_{i=1}^{d} w_{ji} x_i + w_{j0} \right) + w_{k0} \qquad (23.9)$$

Here w_{j0} and w_{k0} are *bias* parameters, and the basis functions are called *hidden units*. The function $g(\cdot)$ is the logistic sigmoid function of Eq. (23.6). This also can be represented as a network diagram as in Fig. 23.3. Such a model is able to take account of the intrinsic dimensionality of the data because the first-layer weights w_{ji} can adapt and hence orient the surfaces along which the basis function response is constant. It has been demonstrated that models of this form can approximate to arbitrary accuracy any continuous function, defined on a compact domain, provided the number M of hidden units is sufficiently large. The MLP model can be extended by considering several successive layers of weights. Note that the use of nonlinear activation functions is crucial, because if $g(\cdot)$ in Eq. (23.9) was replaced by the identity, the network would reduce to several successive linear transformations, which would itself be linear.

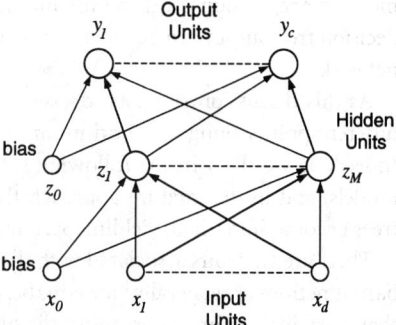

FIGURE 23.3 An example of a feedforward network having two layers of adaptive weights. The bias parameters in the first layer are shown as weights from an extra input having a fixed value of $x_0 = 1$. Similarly, the bias parameters in the second layer are shown as weights from an extra hidden unit, with activation again fixed at $z_0 = 1$.

The second common network model is obtained by choosing the basis functions $\phi_j(\boldsymbol{x})$ in Eq. (23.8) to be functions of the radial variable $\boldsymbol{x} - \boldsymbol{\mu}_j$ where $\boldsymbol{\mu}_j$ is the *center* of the jth basis function, which gives rise to the **radial basis function (RBF) network** model. The most common example uses Gaussians of the form

$$\phi_j(\boldsymbol{x}) = \exp \left\{ -\frac{1}{2}(\boldsymbol{x} - \boldsymbol{\mu}_j)^T \Sigma_j^{-1} (\boldsymbol{x} - \boldsymbol{\mu}_j) \right\} \qquad (23.10)$$

Here both the mean vector $\boldsymbol{\mu}_j$ and the covariance matrix Σ_j are considered to be adaptive parameters. The curse of dimensionality is alleviated because the basis functions can be positioned and oriented in input space such as to overlay the regions of high data density and hence to capture the nonlinear correlations between input variables. Indeed, a common approach to training an RBF network is to use a two-stage procedure [Bishop 1995]. In the first stage, the basis function parameters are determined using the input data alone, which corresponds to a density estimation problem using a mixture model in which the component densities are given by the basis functions $\phi_j(\boldsymbol{x})$. In the second stage, the basis function parameters are frozen and the second-layer weights w_{kj} are found by standard least-squares optimization procedures.

Decision Trees

MLP and RBF networks are often contrasted in terms of the support of the basis functions that compose them. MLP networks are often referred to as "global", given that linear-logistic basis functions are bounded away from zero over a significant fraction of the input space. Accordingly, in an MLP, each input vector generally gives rise to a distributed pattern over the hidden units. RBF networks, on the other hand, are referred to as "local", due to the fact that their Gaussian basis functions typically have support over a local region of the input space. It is important to note, however, that local support does not necessarily mean nonoverlapping support; indeed, there is nothing in the RBF model that prefers basis functions that have nonoverlapping support. A third class of model that does focus on basis functions with nonoverlapping support is the **decision tree** model [Breiman et al. 1984]. A decision tree is a regression or classification model that can be viewed as asking a sequence of questions about the input vector. Each question is implemented as a linear discriminant, and a sequence of questions can be viewed as a recursive partitioning of the input space. All inputs that arrive at a particular leaf of the tree define a polyhedral region in the input space. The collection of such regions can be viewed as a set of basis functions. Associated with each basis function is an output value which (ideally) is close to the average value of the conditional mean (for regression) or discriminant function (for classification; a majority vote is also used). Thus, the decision tree output can be written as a weighted sum of basis functions in the same manner as a layered network.

As this discussion suggests, decision trees and MLP/RBF neural networks are best viewed as being different points along the continuum of models having overlapping or nonoverlapping basis functions. Indeed, as we show in the following section, decision trees can be treated probabilistically as mixture models, and in the mixture approach the sharp discriminant function boundaries of classical decision trees become smoothed, yielding partially overlapping basis functions.

There are tradeoffs associated with the continuum of degree-of-overlap; in particular, nonoverlapping basis functions are generally viewed as being easier to interpret and better able to reject noisy input variables that carry little information about the output. Overlapping basis functions often are viewed as yielding lower variance predictions and as being more robust.

General Mixture Models

The use of mixture models is not restricted to density estimation; rather, the mixture approach can be used quite generally to build complex models out of simple parts. To illustrate, let us consider using mixture models to model a conditional density in the context of a regression or classification problem. A mixture model in this setting is referred to as a "mixtures of experts" model [Jacobs et al. 1991].

Suppose that we have at our disposal an elemental conditional model $p(t \mid x, w)$. Consider a situation in which the conditional mean or discriminant exhibits variation on a local scale that is a good match to our elemental model, but the variation differs in different regions of the input space. We could use a more complex network to try to capture this global variation; alternatively, we might wish to combine local variants of our elemental models in some manner. This can be achieved by defining the following probabilistic mixture:

$$p(t \mid x, w) = \sum_{i=1}^{M} p(i \mid x, v) p(t \mid x, i, w_i) \qquad (23.11)$$

Comparing this mixture to the unconditional mixture defined earlier [Eq. (23.2)], we see that both the mixing proportions and the component densities are now conditional densities dependent on the input vector x. The former dependence is particularly important: we now view the mixing proportion $p(i \mid x, v)$ as providing a probabilistic device for choosing different elemental models ("experts") in different regions of the input space. A learning algorithm that chooses values for the parameters v as well as the values for the parameters w_i can be viewed as attempting to find both a good partition of the input space and a good fit to the local models within that partition.

This approach can be extended recursively by considering mixtures of models where each model may itself be a mixture model [Jordan and Jacobs 1994]. Such a recursion can be viewed as providing a probabilistic interpretation for the decision trees discussed in the previous section. We view the decisions in the decision tree as forming a recursive set of probabilistic selections among a set of models. The total probability of target t given input x is the sum across all paths down the tree,

$$p(t \mid x, w) = \sum_{i=1}^{M} p(i \mid x, u) \sum_{j=1}^{M} p(j \mid x, i, v_i) \cdots p(t \mid x, i, j, \ldots, w_{ij\ldots}) \qquad (23.12)$$

where i and j are the decisions made at the first level and second level of the tree, respectively, and $p(t \mid x, i, j, \ldots, w_{ij\ldots})$ is the elemental model at the leaf of the tree defined by the sequence of decisions. This probabilistic model is a conditional hierarchical mixture. Finding parameter values u, v_i, etc., to fit this model to data can be viewed as finding a nested set of partitions of the input space and fitting a set of local models within the partition.

The mixture model approach can be viewed as a special case of a general methodology known as *learning by committee*. Bishop [1995] provides a discussion of committees; we will also meet them in the section on Bayesian methods later in the chapter.

23.3 Learning from Data

The previous section has provided a selection of models to choose from; we now face the problem of matching these models to data. In principle, the problem is straightforward: given a family of models of interest we attempt to find out how probable each of these models is in the light of the data. We can then select the most probable model [a selection rule known as *maximum a posteriori* (MAP) estimation], or we can select some highly probable subset of models, weighted by their probability (an approach that we discuss in the section on Bayesian methods). In practice, there are a number of problems to solve, beginning with the specification of the family of models of interest. In the simplest case, in which the family can be described as a fixed structure with varying parameters (e.g., the class of feedforward MLPs with a fixed number of hidden units), the learning problem is essentially one of *parameter estimation*. If, on the other hand, the family is not easily viewed as a fixed parametric family (e.g., feedforward MLPs with a variable number of hidden units), then we must solve the *model selection* problem.

In this section we discuss the parameter estimation problem. The goal will be to find MAP estimates of the parameters by maximizing the probability of the parameters given the data $\mathcal{D}$. We compute this probability using Bayes rule,

$$p(w \mid \mathcal{D}) = \frac{p(\mathcal{D} \mid w)\, p(w)}{p(\mathcal{D})} \tag{23.13}$$

where we see that to calculate MAP estimates we must maximize the expression in the numerator (the denominator does not depend on w). Equivalently we can minimize the negative logarithm of the numerator. We thus define the following **cost function** $J(w)$:

$$J(w) = -\ln p(\mathcal{D} \mid w) - \ln p(w) \tag{23.14}$$

which we wish to minimize with respect to the parameters w. The first term in this cost function is a (negative) log **likelihood**. If we assume that the elements in the training set $\mathcal{D}$ are conditionally independent of each other given the parameters, then the likelihood factorizes into a product form. For density estimation we have

$$p(\mathcal{D} \mid w) = \prod_{n=1}^{N} p(x_n \mid w) \tag{23.15}$$

and for classification and regression we have

$$p(\mathcal{D} \mid w) = \prod_{n=1}^{N} p(t_n \mid x_n, w) \tag{23.16}$$

In both cases this yields a log likelihood which is the sum of the log probabilities for each individual data point. For the remainder of this section we will assume this additive form; moreover, we will assume that the log prior probability of the parameters is uniform across the parameters and drop the second term. Thus, we focus on *maximum likelihood* (ML) estimation, where we choose parameter values w_{ML} that maximize $\ln p(\mathcal{D} \mid w)$.

Likelihood-Based Cost Functions

Regression, classification, and density estimation make different probabilistic assumptions about the form of the data and therefore require different cost functions.

Equation (23.3) defines a probabilistic model for regression. The model is a conditional density for the targets t in which the targets are distributed as Gaussian random variables (assuming Gaussian errors ϵ) with mean values $f(x)$. We now write the conditional mean as $f(x, w)$ to make explicit the dependence on the parameters w. Given the training set $\mathcal{D} = \{x_n, t_n\}_{n=1}^{N}$, and given our assumption that the targets t_n are sampled independently (given the inputs x_n and the parameters w), we obtain

$$J(w) = \frac{1}{2} \sum_n \|t_n - f(x_n, w)\|^2 \tag{23.17}$$

where we have assumed an identity covariance matrix and dropped those terms that do not depend on the parameters. This cost function is the standard least-squares cost function, which is traditionally used in neural network training for real-valued targets. Minimization of this cost function is typically achieved via some form of gradient optimization, as we discuss in the following section.

Classification problems differ from regression problems in the use of discrete-valued targets, and the likelihood accordingly takes a different form. For binary classification the Bernoulli probability model $p(t \mid x, w) = y^t (1 - y)^{1-t}$ is natural, where we use y to denote the probability $p(t = 1 \mid x, w)$. This model yields the following log likelihood:

$$J(w) = -\sum_n [t_n \ln y_n + (1 - t_n) \ln(1 - y_n)] \tag{23.18}$$

which is known as the *cross-entropy* function. It can be minimized using the same generic optimization procedures as are used for least squares.

For multiway classification problems in which there are C categories, where $C > 2$, the multinomial distribution is natural. Define t_n such that its elements $t_{n,i}$ are one or zero according to whether the nth data point belongs to the ith category, and define $y_{n,i}$ to be the network's estimate of the posterior probability of category i for data point n; that is, $y_{n,i} \equiv p(t_{n,i} = 1 \mid x_n, w)$. Given these definitions, we obtain the following cost function:

$$J(w) = -\sum_n \sum_i t_{n,i} \ln y_{n,i} \tag{23.19}$$

which again has the form of a cross entropy.

We now turn to density estimation as exemplified by Gaussian mixture modeling. The probabilistic model in this case is that given in Eq. (23.2). Assuming Gaussian component densities with arbitrary covariance matrices, we obtain the following cost function:

$$J(w) = -\sum_n \ln \sum_i \pi_i \frac{1}{|\Sigma_i|^{1/2}} \exp \left\{ -\frac{1}{2}(x_n - \mu_i)^T \Sigma_i^{-1} (x_n - \mu_i) \right\} \tag{23.20}$$

where the parameters w are the collection of mean vectors μ_i, the covariance matrices Σ_i, and the mixing proportions π_i. A similar cost function arises for the generalized mixture models [cf. Eq. (23.12)].

Gradients of the Cost Function

Once we have defined a probabilistic model, obtained a cost function, and found an efficient procedure for calculating the gradient of the cost function, the problem can be handed off to an optimization routine. Before discussing optimization procedures, however, it is useful to examine the form that the gradient takes for the examples that we have discussed in the previous two sections.

The ith output unit in a layered network is endowed with a rule for combining the activations of units in earlier layers, yielding a quantity that we denote by z_i and a function that converts z_i into the output y_i. For regression problems, we assume linear output units such that $y_i = z_i$. For binary classification problems, our earlier discussion showed that a natural output function is the logistic: $y_i = 1/(1 + e^{-z_i})$. For multiway classification, it is possible to generalize the derivation of the logistic function to obtain an analogous representation for the multiway posterior probabilities known as the *softmax function* [cf. Bishop 1995]:

$$y_i = \frac{e^{z_i}}{\sum_k e^{z_k}} \tag{23.21}$$

where y_i represents the posterior probability of category i.

If we now consider the gradient of $J(w)$ with respect to z_i, it turns out that we obtain a single canonical expression of the following form:

$$\frac{\partial J}{\partial w} = \sum_i (t_i - y_i) \frac{\partial z_i}{\partial w} \tag{23.22}$$

As discussed by Rumelhart et al. [1995], this form for the gradient is predicted from the theory of generalized linear models [McCullagh and Nelder 1983], where it is shown that the linear, logistic, and softmax functions are (inverse) *canonical links* for the Gaussian, Bernoulli, and multinomial distributions, respectively. Canonical links can be found for all of the distributions in the exponential family, thus providing a solid statistical foundation for handling a wide variety of data formats at the output layer of a network, including counts, time intervals, and rates.

The gradient of the cost function for mixture models has an interesting interpretation. Taking the partial derivative of $J(w)$ in Eq. (23.20) with respect to μ_i, we find

$$\frac{\partial J}{\partial \mu_i} = \sum_n h_{n,i} \Sigma_i (x_n - \mu_i) \tag{23.23}$$

where $h_{n,i}$ is defined as follows:

$$h_{n,i} = \frac{\pi_i |\Sigma_i|^{-1/2} \exp\left\{-\frac{1}{2}(x_n - \mu_i)^T \Sigma_i^{-1}(x_n - \mu_i)\right\}}{\sum_k \pi_k |\Sigma_k|^{-1/2} \exp\left\{-\frac{1}{2}(x_n - \mu_k)^T \Sigma_k^{-1}(x_n - \mu_k)\right\}} \tag{23.24}$$

When summed over i, the quantity $h_{n,i}$ sums to one and is often viewed as the "responsibility" or "credit" assigned to the ith component for the nth data point. Indeed, interpreting Eq. (23.24) using Bayes rule shows that $h_{n,i}$ is the posterior probability that the nth data point is generated by the ith component Gaussian. A learning algorithm based on this gradient will move the ith mean μ_i toward the data point x_n, with the effective step size proportional to $h_{n,i}$.

The gradient for a mixture model will always take the form of a weighted sum of the gradients associated with the component models, where the weights are the posterior probabilities associated with each of the components. The key computational issue is whether these posterior weights can be computed efficiently. For Gaussian mixture models, the calculation [Eq. (23.24)] is clearly efficient. For decision trees there is a set of posterior weights associated with each of the nodes in the tree, and a recursion is available that computes the posterior probabilities in an upward sweep [Jordan and Jacobs 1994].

Mixture models in the form of a chain are known as **hidden Markov models**, and the calculation of the relevant posterior probabilities is performed via an efficient algorithm known as the Baum–Welch algorithm.

For general layered network structures, a generic algorithm known as backpropagation is available to calculate gradient vectors [Rumelhart et al. 1986]. Backpropagation is essentially the chain rule of calculus realized as a graphical algorithm. As applied to layered networks it provides a simple and efficient method that calculates a gradient in $O(W)$ time per training pattern, where W is the number of weights.

Optimization Algorithms

By introducing the principle of maximum likelihood in section 23.1, we have expressed the problem of learning in neural networks in terms of the minimization of a cost function, $J(w)$, which depends on a vector, w, of adaptive parameters. An important aspect of this problem is that the gradient vector $\nabla_w J$ can be evaluated efficiently (for example, by backpropagation). Gradient-based minimization is a standard problem in unconstrained nonlinear optimization for which many powerful techniques have been developed over the years. Such algorithms generally start by making an initial guess for the parameter vector w and then iteratively updating the vector in a sequence of steps,

$$w^{(\tau+1)} = w^{(\tau)} + \Delta w^{(\tau)} \qquad (23.25)$$

where τ denotes the step number. The initial parameter vector $w^{(0)}$ is often chosen at random, and the final vector represents a minimum of the cost function at which the gradient vanishes. Because of the nonlinear nature of neural network models, the cost function is generally a highly complicated function of the parameters and may possess many such minima. Different algorithms differ in how the update $\Delta w^{(\tau)}$ is computed.

The simplest such algorithm is called *gradient descent* and involves a parameter update which is proportional to the negative of the cost function gradient $\Delta = -\eta \nabla E$ where η is a fixed constant called the learning rate. It should be stressed that gradient descent is a particularly inefficient optimization algorithm. Various modifications have been proposed, such as the inclusion of a *momentum* term, to try to improve its performance. In fact, much more powerful algorithms are readily available, as described in standard textbooks such as Fletcher [1987]. Two of the best known are called *conjugate gradients* and *quasi-Newton* (or *variable metric*) methods. For the particular case of a sum-of-squares cost function, the *Levenberg–Marquardt* algorithm can also be very effective. Software implementations of these algorithms are widely available.

The algorithms discussed so far are called *batch* since they involve using the whole dataset for each evaluation of the cost function or its gradient. There is also a *stochastic* or *on-line* version of gradient descent in which, for each parameter update, the cost function gradient is evaluated using just one of the training vectors at a time (which are then cycled either in order or in a random sequence). Although this approach fails to make use of the power of sophisticated methods such as conjugate gradients, it can prove effective for very large datasets, particularly if there is significant redundancy in the data.

Hessian Matrices, Error Bars, and Pruning

After a set of weights have been found for a neural network using an optimization procedure, it is often useful to examine second-order properties of the fitted network as captured in the Hessian matrix $H = \partial^2 J / \partial w \partial w^T$. Efficient algorithms have been developed to compute the Hessian matrix in time $O(W^2)$ [Bishop 1995]. As in the case of the calculation of the gradient by backpropagation, these algorithms are based on recursive message passing in the network.

One important use of the Hessian matrix lies in the calculation of error bars on the outputs of a network. If we approximate the cost function locally as a quadratic function of the weights (an approximation which is equivalent to making a Gaussian approximation for the log likelihood), then the estimated variance of the ith output y_i can be shown to be

$$\hat{\sigma}_{y_i}^2 = \left(\frac{\partial y_i}{\partial \boldsymbol{w}}\right)^T H^{-1} \left(\frac{\partial y_i}{\partial \boldsymbol{w}}\right) \tag{23.26}$$

where the gradient vector $\partial y_i/\partial \boldsymbol{w}$ can be calculated via backpropagation.

The Hessian matrix also is useful in pruning algorithms. A pruning algorithm deletes weights from a fitted network to yield a simpler network that may outperform a more complex, **overfitted** network (discussed subsequently) and may be easier to interpret. In this setting, the Hessian is used to approximate the increase in the cost function due to the deletion of a weight. A variety of such pruning algorithms is available [cf. Bishop 1995].

Complexity Control

In previous sections we have introduced a variety of models for representing probability distributions, we have shown how the parameters of the models can be optimized by maximizing the likelihood function, and we have outlined a number of powerful algorithms for performing this minimization. Before we can apply this framework in practice there is one more issue we need to address, which is that of model complexity. Consider the case of a mixture model given by Eq. (23.2). The number of input variables will be determined by the particular problem at hand. However, the number M of component densities has yet to be specified. Clearly if M is too small the model will be insufficiently flexible and we will obtain a poor representation of the true density. What is not so obvious is that if M is too large we can

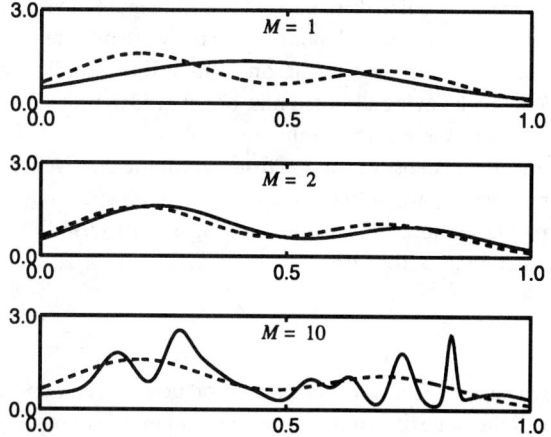

FIGURE 23.4 Effects of model complexity illustrated by modeling a mixture of two Gaussians (shown by the dashed curves) using a mixture of M Gaussians (shown by the solid curves). The results are obtained for 20 cycles of EM.

also obtain poor results. This effect is known as *overfitting* and arises because we have a dataset of finite size. It is illustrated using the simple example of mixture density estimation in Fig. 23.4. Here a set of 100 data points in one dimension has been generated from a distribution consisting of a mixture of two Gaussians (shown by the dashed curves). This dataset has then been fitted by a mixture of M Gaussians by use of the expectation-maximization (EM) algorithm. We see that a model with 1 component ($M = 1$) gives a poor representation of the true distribution from which the data were generated, and in particular is unable to capture the bimodal aspect. For $M = 2$ the model gives a good fit, as we expect since the data were themselves generated from a two-component Gaussian mixture. However, increasing the number of components to $M = 10$ gives a poorer fit, even though this model contains the simpler models as special cases.

The problem is a very fundamental one and is associated with the fact that we are trying to infer an entire distribution function from a finite number of data points, which is necessarily an ill-posed problem. In regression, for example, there are infinitely many functions which will give a perfect fit to the finite number of data points. If the data are noisy, however, the best generalization will be obtained for a function which does not fit the data perfectly but which captures the underlying function from which the data were generated. By increasing the flexibility of the model, we are able to obtain ever better fits to the training data, and this is reflected in a steadily increasing value for the likelihood function at its maximum. Our goal is to model the true underlying density function from which the data were generated since this allows us to make the best predictions for new data. We see that the best approximation to this density occurs for an intermediate value of M.

The same issue arises in connection with nonlinear regression and classification problems. For example, the number M of hidden units in an MLP network controls the model complexity and must be optimized to give the best generalization. In a practical application, we can train a variety of different models having different complexities, compare their generalization performance using an independent validation set, and then select the model with the best generalization. In fact, the process of optimizing the complexity using a validation set can lead to some partial overfitting to the validation data itself, and so the final performance of the selected model should be confirmed using a third independent data set called a *test* set.

Some theoretical insight into the problem of overfitting can be obtained by decomposing the error into the sum of bias and variance terms [Geman et al. 1992]. A model which is too inflexible is unable to represent the true structure in the underlying density function, and this gives rise to a high bias. Conversely, a model which is too flexible becomes tuned to the specific details of the particular data set and gives a high variance. The best generalization is obtained from the optimum tradeoff of bias against variance.

As we have already remarked, the problem of inferring an entire distribution function from a finite data set is fundamentally ill posed since there are infinitely many solutions. The problem becomes well posed only when some additional constraint is imposed. This constraint might be that we model the data using a network having a limited number of hidden units. Within the range of functions which this model can represent there is then a unique function which best fits the data. Implicitly, we are assuming that the underlying density function from which the data were drawn is relatively smooth. Instead of limiting the number of parameters in the model, we can encourage smoothness more directly using the technique of **regularization**. This involves adding penalty term Ω to the original cost function J to give the total cost function $\tilde{J}$ of the form

$$\tilde{J} = J + \nu\Omega \qquad (23.27)$$

where ν is called a regularization coefficient. The network parameters are determined by minimizing $\tilde{J}$, and the value of ν controls the degree of influence of the penalty term Ω. In practice, Ω is typically chosen to encourage smooth functions. The simplest example is called *weight decay* and consists of the sum of the squares of all of the adaptive parameters in the model,

$$\Omega = \sum_i w_i^2 \qquad (23.28)$$

Consider the effect of such a term on the MLP function [Eq. (23.9)]. If the weights take very small values then the network outputs become approximately linear functions of the inputs (since the sigmoidal function is approximately linear for small values of its argument). The value of ν in Eq. (23.27) controls the effective complexity of the model, so that for large ν the model is oversmoothed (corresponding to high bias), whereas for small ν the model can overfit (corresponding to high variance). We can therefore consider a network with a relatively large number of hidden units and control the effective complexity by changing ν. In practice, a suitable value for ν can be found by seeking the value which gives the best performance on a validation set.

The weight decay regularizer [Eq. (23.28)] is simple to implement but suffers from a number of limitations. Regularizers used in practice may be more sophisticated and may contain multiple regularization coefficients [Neal 1994].

Regularization methods can be justified within a general theoretical framework known as *structural risk minimization* [Vapnik 1995]. Structural risk minimization provides a quantitative measure of complexity known as the **VC dimension**. The theory shows that the VC dimension predicts the difference between performance on a training set and performance on a test set; thus, the sum of log likelihood and (some function of) VC dimension provides a measure of generalization performance. This motivates regularization methods [Eq. (23.27)] and provides some insight into possible forms for the regularizer Ω.

Bayesian Viewpoint

In earlier sections we discussed network training in terms of the minimization of a cost function derived from the principle of maximum a posteriori or maximum likelihood estimation. This approach can be seen as a particular approximation to a more fundamental, and more powerful, framework based on Bayesian statistics. In the maximum likelihood approach, the weights w are set to a specific value, w_{ML}, determined by minimization of a cost function. However, we know that there will typically be other minima of the cost function which might give equally good results. Also, weight values close to w_{ML} should give results which are not too different from those of the maximum likelihood weights themselves.

These effects are handled in a natural way in the Bayesian viewpoint, which describes the weights not in terms of a specific set of values but in terms of a probability distribution over all possible values. As discussed earlier [cf. Eq. (23.13)], once we observe the training dataset $\mathcal{D}$ we can compute the corresponding *posterior* distribution using Bayes' theorem, based on a *prior* distribution function $p(w)$ (which will typically be very broad), and a *likelihood* function $p(\mathcal{D} \mid w)$,

$$p(w \mid \mathcal{D}) = \frac{p(\mathcal{D} \mid w) p(w)}{p(\mathcal{D})} \tag{23.29}$$

The likelihood function will typically be very small except for values of w for which the network function is reasonably consistent with the data. Thus, the posterior distribution $p(w \mid \mathcal{D})$ will be much more sharply peaked than the prior distribution $p(w)$ (and will typically have multiple maxima). The quantity we are interested in is the predicted distribution of target values t for a new input vector x once we have observed the data set $\mathcal{D}$. This can be expressed as an integration over the posterior distribution of weights of the form

$$p(t \mid x, \mathcal{D}) = \int p(t \mid x, w) p(w \mid \mathcal{D}) dw \tag{23.30}$$

where $p(t \mid x, w)$ is the conditional probability model discussed in the introduction.

If we suppose that the posterior distribution $p(w \mid \mathcal{D})$ is sharply peaked around a single most-probable value w_{MP}, then we can write Eq. (23.30) in the form:

$$p(t \mid x, \mathcal{D}) \simeq p(t \mid x, w_{MP}) \int p(w \mid \mathcal{D}) dw \tag{23.31}$$

$$= p(t \mid x, w_{MP}) \tag{23.32}$$

and so predictions can be made by fixing the weights to their most probable values. We can find the most probable weights by maximizing the posterior distribution or equivalently by minimizing its negative logarithm. Using Eq. (23.29), we see that w_{MP} is determined by minimizing a regularized cost function of the form in Eq. (23.27) in which the negative log of the prior $-\ln p(w)$ represents the regularizer $\nu\Omega$. For example, if the prior consists of a zero-mean Gaussian with variance ν^{-1}, then we obtain the weight-decay regularizer of Eq. (23.28).

The posterior distribution will become sharply peaked when the size of the dataset is large compared to the number of parameters in the network. For datasets of limited size, however, the posterior distribution has a finite width and this adds to the uncertainty in the predictions for t, which can be expressed in terms of error bars. Bayesian error bars can be evaluated using a local Gaussian approximation to the posterior distribution [MacKay 1992]. The presence of multiple maxima in the posterior distribution also contributes to the uncertainties in predictions. The capability to assess these uncertainties can play a crucial role in practical applications.

The Bayesian approach can also deal with more general problems in complexity control. This can be done by considering the probabilities of a set of alternative models, given the dataset

$$p(\mathcal{H}_i \mid \mathcal{D}) = \frac{p(\mathcal{D} \mid \mathcal{H}_i) p(\mathcal{H}_i)}{p(\mathcal{D})} \tag{23.33}$$

Here different models can also be interpreted as different values of regularization parameters as these too control model complexity. If the models are given the same prior probabilities $p(\mathcal{H}_i)$ then they can be ranked by considering the *evidence* $p(\mathcal{D} \mid \mathcal{H}_i)$, which itself can be evaluated by integration over the model parameters w. We can simply select the model with the greatest probability. However, a full Bayesian treatment requires that we form a linear combination of the predictions of the models in which the weighting coefficients are given by the model probabilities.

In general, the required integrations, such as that in Eq. (23.30), are analytically intractable. One approach is to approximate the posterior distribution by a Gaussian centered on w_{MP} and then to linearize $p(t \mid x, w)$ about w_{MP} so that the integration can be performed analytically [MacKay 1992]. Alternatively, sophisticated Monte Carlo methods can be employed to evaluate the integrals numerically [Neal 1994]. An important aspect of the Bayesian approach is that there is no need to keep data aside in a validation set as is required when using maximum likelihood. In practical applications for which the quantity of available data is limited, it is found that a Bayesian treatment generally outperforms other approaches.

Preprocessing, Invariances and Prior Knowledge

We have already seen that neural networks can approximate essentially arbitrary nonlinear functional mappings between sets of variables. In principle, we could therefore use a single network to transform the raw input variables into the required final outputs. However, in practice for all but the simplest problems the results of such an approach can be improved upon considerably by incorporating various forms of preprocessing, for reasons we shall outline in the following.

One of the simplest and most common forms of preprocessing consists of a simple normalization of the input, and possibly also target, variables. This may take the form of a linear rescaling of each input variable independently to give it zero mean and unit variance over the training set. For some applications, the original input variables may span widely different ranges. Although a linear rescaling of the inputs is equivalent to a different choice of first-layer weights, in practice the optimization algorithm may have considerable difficulty in finding a satisfactory solution when typical input values are substantially different. Similar rescaling can be applied to the output values, in which case the inverse of the transformation needs to be applied to the network outputs when the network is presented with new inputs. Preprocessing is also used to encode data in a suitable form. For example, if we have categorical variables such as red, green, and blue, these may be encoded using a 1-of-3 binary representation.

Another widely used form of preprocessing involves reducing the dimensionality of the input space. Such transformations may result in loss of information in the data, but the overall effect can be a significant improvement in performance as a consequence of the curse of dimensionality discussed in the complexity control section. The finite dataset is better able to specify the required mapping in the lower dimensional space. Dimensionality reduction may be accomplished by simply selecting a subset of the original variables but more typically involves the construction of new variables consisting of linear or nonlinear combinations of the original variables called *features*. A standard technique for dimensionality reduction is principal component analysis [Anderson 1984]. Such methods, however, make use only of the input data and ignore the target values and can sometimes be significantly suboptimal.

Yet another form of preprocessing involves correcting deficiencies in the original data. A common occurrence is that some of the input variables are missing for some of the data points. Correction of this problem in a principled way requires that the probability distribution $p(x)$ of input data be modeled.

One of the most important factors determining the performance of real-world applications of neural networks is the use of *prior knowledge*, which is information additional to that present in the data. As an example, consider the problem of classifying handwritten digits discussed in section 23.1. The most direct approach would be to collect a large training set of digits and to train a feedforward network to map from the input image to a set of 10 output values representing posterior probabilities for the 10 classes. However, we know that the classification of a digit should be independent of its position within the input image. One way of achieving such *translation invariance* is to make use of the technique of *shared weights*. This involves a network architecure having many hidden layers in which each unit takes inputs only from a small patch, called a *receptive field*, of units in the previous layer. By a process of constraining neighboring units to have common weights, it can be arranged that the output of the network is insensitive to translations of the input image. A further benefit of weight sharing is that the number of independent parameters is much smaller than the number of weights, which assists with the problem of model complexity. This approach is the basis for the highly successful U.S. postal code recognition system of LeCun et al. [1989]. An alternative to shared weights is to enlarge the training set artificially by generating virtual examples based on applying translations and other transformations to the original training set [Poggio and Vetter 1992].

23.4 Graphical Models

Neural networks express relationships between variables by utilizing the representational language of graph theory. Variables are associated with nodes in a graph and transformations of variables are based on algorithms that propagate numerical messages along the links of the graph. Moreover, the graphs are often accompanied by probabilistic interpretations of the variables and their interrelationships. As we have seen, such probabilistic interpretations allow a neural network to be understood as a form of a probabilistic model and reduce the problem of learning the weights of a network to a problem in statistics.

Related graphical models have been studied throughout statistics, engineering, and AI in recent years. Hidden Markov models, Kalman filters, and path analysis models are all examples of graphical probabilistic models that can be fitted to data and used to make inferences. The relationship between these models and neural networks is rather strong; indeed, it is often possible to reduce one kind of model to the other. In this section, we examine these relationships in some detail and provide a broader characterization of neural networks as members of a general family of graphical probabilistic models.

Many interesting relationships have been discovered between graphs and probability distributions [Spiegelhalter et al. 1993, Pearl 1988]. These relationships derive from the use of graphs to represent conditional independencies among random variables. In an undirected graph, there is a direct correspondence between conditional independence and graph separation: random variables X_i and X_k are conditionally independent given X_j if nodes X_i and X_k are separated by node X_j (we use the symbol X_i to represent both a random variable and a node in a graph). This statement remains true for sets of nodes. [See Fig. 23.5(a).] Directed graphs have a somewhat different semantics due to the ability of directed graphs to represent induced dependencies. An induced dependency is a situation in which two nodes which are marginally independent become conditionally dependent given the value of a third node. [See Fig. 23.5(b).] Suppose, for example, that X_i and X_k represent independent coin tosses, and X_j represents the sum of X_i and X_k. Then X_i and X_k are marginally independent but are conditionally dependent given X_j. The semantics of independence in directed graphs is captured by a graphical criterion known as *d-separation* [Pearl 1988], which differs from undirected separation only in those cases in which paths have two arrows arriving at the same node [as in Fig. 23.5(b)].

Although the neural network architectures that we have discussed until now all have been based on directed graphs, undirected graphs also play an important role in neural network research. Constraint satisfaction architectures, including the Hopfield network [Hopfield 1982] and the **Boltzmann machine** [Hinton and Sejnowski 1986], are the most prominent examples. A Boltzmann machine is an undirected probabilistic graph that respects the conditional independency semantics previously described [cf. Fig. 23.5(a)]. Each node in a Boltzmann machine is a binary-valued random variable X_i (or, more generally, a discrete-valued random variable). A probability distribution on the 2^N possible configurations of such variables

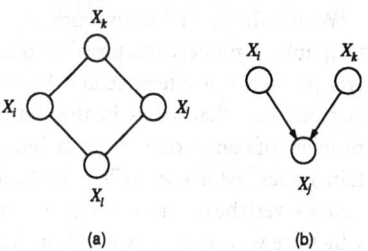

(a)   (b)

FIGURE 23.5   (a) An undirected graph in which X_i is independent of X_j given X_k and X_l, and X_k is independent of X_l given X_i and X_j. (b) A directed graph in which X_i and X_k are marginally independent but are conditionally dependent given X_j.

is defined via an *energy function E*. Let J_{ij} be the weight on the link between X_i and X_j, let $J_{ij} = J_{ji}$, let α index the configurations, and define the energy of configuration α as follows:

$$E_\alpha = -\sum_{i<j} J_{ij} X_i^\alpha X_j^\alpha \tag{23.34}$$

The probability of configuration α is then defined via the Boltzmann distribution:

$$P_\alpha = \frac{e^{-E_\alpha/T}}{\sum_\gamma e^{-E_\gamma/T}} \tag{23.35}$$

where the *temperature T* provides a scale for the energy.

An example of a directed probabilistic graph is the hidden Markov model (HMM). An HMM is defined by a set of *state variables* H_i, where i is generally a time or a space index, a set of output variables O_i, a *probability transition matrix* $A = p(H_i \mid H_{i-1})$, and an *emission matrix* $B = p(O_i \mid H_i)$. The directed graph for an HMM is shown in Fig. 23.6(a). As can be seen from considering the separatory properties of the graph, the conditional independencies of the HMM are defined by the following Markov conditions:

$$H_i \perp \{H_1, O_1, \ldots, H_{i-2}, O_{i-2}, O_{i-1}\} | H_{i-1}, \qquad 2 \le i \le N \tag{23.36}$$

and

$$O_i \perp \{H_1, O_1, \ldots, H_{i-1}, O_{i-1}\} | H_i, \qquad 2 \le i \le N \tag{23.37}$$

where the symbol $\perp$ is used to denote independence.

Figure 23.6(b) shows that it is possible to treat an HMM as a special case of a Boltzmann machine [Luttrell 1989, Saul and Jordan 1995]. The probabilistic structure of the HMM can be captured by defining the weights on the links as the logarithms of the corresponding transition and emission probabilities. The Boltzmann distribution [Eq. (23.35)] then converts the additive energy into the product form of the standard HMM probability distribution. As we will see, this reduction of a directed graph to an undirected graph is a recurring theme in the graphical model formalism.

General mixture models are readily viewed as graphical models [Buntine 1994]. For example, the unconditional mixture model of Eq. (23.2) can be represented as a graphical model with two nodes—a multinomial hidden node, which represents the selected component, a visible node representing x, with a directed link from the hidden node to the visible node (hidden/visible distinction discussed subsequently). Conditional mixture models [Jacobs et al. 1991] simply require another visible node with directed links to the hidden node and the visible nodes. Hierarchical conditional mixture models [Jordan and Jacobs 1994] require a chain of hidden nodes, one hidden node for each level of the tree.

Within the general framework of probabilistic graphical models, it is possible to tackle general problems of inference and learning. The key problem that arises in this setting is the problem of computing the probabilities of certain nodes, which we will refer to as *hidden nodes*, given the observed values of other nodes, which we will refer to as *visible nodes*. For example, in an HMM, the variables O_i are generally treated as visible, and it is desired to calculate a probability distribution on the hidden states H_i. A similar inferential calculation is required in the mixture models and the Boltzmann machine.

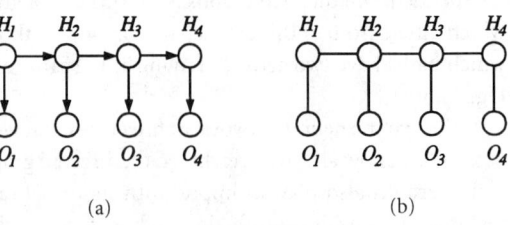

(a)   (b)

FIGURE 23.6   (a) A directed graph representation of an HMM. Each horizontal link is associated with the transition matrix A, and each vertical link is associated with the emission matrix B. (b) An HMM as a Boltzmann machine. The parameters on the horizontal links are logarithms of the entries of the A matrix, and the parameters on the vertical links are logarithms of the entries of the B matrix. The two representations yield the same joint probability distribution.

Generic algorithms have been developed to solve the inferential problem of the calculation of posterior probabilities in graphs. Although a variety of inference algorithms have been developed, they can all be viewed as essentially the same underlying algorithm [Shachter et al. 1994]. Let us consider undirected graphs. A special case of an undirected graph is a *triangulated graph* [Spiegelhalter et al. 1993], in which any cycle having four or more nodes has a chord. For example, the graph in Fig. 23.5(a) is not triangulated but becomes triangulated when a link is added between nodes X_i and X_j. In a triangulated graph, the cliques of the graph can be arranged in the form of a *junction tree*, which is a tree having the property that any

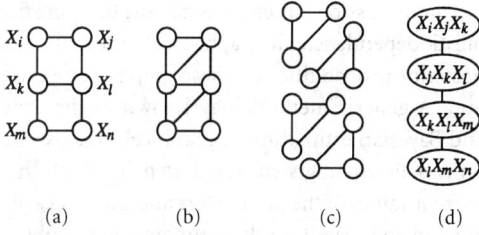

(a)   (b)   (c)   (d)

FIGURE 23.7 The basic structure of the junction tree algorithm for undirected graphs. The graph in (a) is first triangulated (b), then the cliques are identified (c), and arranged into a tree (d). Products of potential functions on the nodes in (d) yield probability distributions on the nodes in (a).

node that appears in two different cliques in the tree also appears in every clique on the path that links the two cliques (the "running intersection property"). This cannot be achieved in nontriangulated graphs. For example, the cliques in Fig. 23.5(a) are $\{X_i, X_k\}, \{X_k, X_j\}, \{X_j, X_l\}$, and it is not possible to arrange these cliques into a tree that obeys the running intersection property. If a chord is added, the resulting cliques are $\{X_i, X_j, X_k\}$ and $\{X_i, X_j, X_l\}$, and these cliques can be arranged as a simple chain that trivially obeys the running intersection property. In general, it turns out that the probability distributions corresponding to triangulated graphs can be characterized as *decomposable*, which implies that they can be factorized into a product of local functions (potentials) associated with the cliques in the triangulated graph.[1] The calculation of posterior probabilities in decomposable distributions is straightforward and can be achieved via a local message-passing algorithm on the junction tree [Spiegelhalter et al. 1993].

Graphs that are not triangulated can be turned into triangulated graphs by the addition of links. If the potentials on the new graph are defined suitably as products of potentials on the original graph, then the independencies in the original graph are preserved. This implies that the algorithms for triangulated graphs can be used for *all* undirected graphs; an untriangulated graph is first triangulated. (See Fig. 23.7.) Moreover, it is possible to convert *directed* graphs to undirected graphs in a manner that preserves the probabilistic structure of the original graph [Spiegelhalter et al. 1993]. This implies that the junction tree algorithm is indeed generic; it can be applied to any graphical model.

The problem of calculating posterior probabilities on graphs is NP-hard; thus, a major issue in the use of the inference algorithms is the identification of cases in which they are efficient. Chain structures such as HMMs yield efficient algorithms, and indeed the classical forward-backward algorithm for HMMs is a special, efficient case of the junction tree algorithm [Smyth et al. 1996]. Decision tree structures such as the hierarchical mixture of experts yield efficient algorithms, and the recursive posterior probability calculation of Jordan and Jacobs [1994] described earlier is also a special case of the junction tree algorithm. All of the simpler mixture model calculations described earlier are therefore also special cases. Another interesting special case is the state estimation algorithm of the Kalman filter [Shachter and Kenley 1989]. Finally, there are a variety of special cases of the Boltzmann machine which are amenable to the exact calculations of the junction tree algorithm [Saul and Jordan 1995].

For graphs that are outside of the tractable categories of trees and chains, the junction tree algorithm often performs surprisingly well, but for highly connected graphs the algorithm can be too slow. In such cases, approximate algorithms such as Gibbs sampling are utilized. A virtue of the graphical framework is that Gibbs sampling has a generic form, which is based on the notion of a *Markov boundary* [Pearl 1988]. A special case of this generic form is the stochastic update rule for general Boltzmann machines.

---

[1]An interesting example is a Boltzmann machine on a triangulated graph. The potentials are products of $\exp(J_{ij})$ factors, where the product is taken over all (i, j) pairs in a particular clique. Given that the product across potentials must be the joint probability, this implies that the partition function [the denominator of Eq. (23.35)] must be unity in this case.

Our discussion has emphasized the unifying framework of graphical models both for expressing probabilistic dependencies in graphs and for describing algorithms that perform the inferential step of calculating posterior probabilities on these graphs. The unification goes further, however, when we consider learning. A generic methodology known as the expectation-maximization algorithm is available for MAP and Bayesian estimation in graphical models [Dempster et al. 1977]. EM is an iterative method, based on two alternating steps: an *E step*, in which the values of hidden variables are estimated, based on current values of the parameters and the values of visible variables, and an *M step*, in which the parameters are updated, based on the estimated values obtained from the E step. Within the framework of the EM algorithm, the junction tree algorithm can readily be viewed as providing a generic E step. Moreover, once the estimated values of the hidden nodes are obtained from the E step, the graph can be viewed as fully observed, and the M step is a standard MAP or ML problem. The standard algorithms for all of the tractable architectures described (mixtures, trees, and chains) are, in fact, instances of this general graphical EM algorithm, and the learning algorithm for general Boltzmann machines is a special case of a generalization of EM known as GEM [Dempster et al. 1977].

What about the case of feedforward neural networks such as the multilayer perceptron? It is, in fact, possible to associate binary hidden values with the hidden units of such a network (cf. our earlier discussion of the logistic function; see also Amari [1995]) and apply the EM algorithm directly. For N hidden units, however, there are 2^N patterns whose probabilities must be calculated in the E step. For large N, this is an intractable computation, and recent research has therefore begun to focus on fast methods for approximating these distributions [Hinton et al. 1995, Saul et al. 1996].

Defining Terms

Boltzmann machine: An undirected network of discrete valued random variables, where an energy function is associated with each of the links, and for which a probability distribution is defined by the Boltzmann distribution.

Classification: A learning problem in which the goal is to assign input vectors to one of a number of (usually mutually exclusive) classes.

Cost function: A function of the adaptive parameters of a model whose minimum is used to define suitable values for those parameters. It may consist of a likelihood function and additional terms.

Decision tree: A network that performs a sequence of classificatory decisions on an input vector and produces an output vector that is conditional on the outcome of the decision sequence.

Density estimation: The problem of modeling a probability distribution from a finite set of examples drawn from that distribution.

Discriminant function: A function of the input vector that can be used to assign inputs to classes in a classification problem.

Hidden Markov model: A graphical probabilistic model characterized by a state vector, an output vector, a state transition matrix, an emission matrix, and an initial state distribution.

Likelihood function: The probability of observing a particular data set under the assumption of a given parametrized model, expressed as a function of the adaptive parameters of the model.

Mixture model: A probability model that consists of a linear combination of simpler component probability models.

Multilayer perceptron: The most common form of neural network model, consisting of successive linear transformations followed by processing with nonlinear activation functions.

Overfitting: The problem in which a model which is too complex captures too much of the noise in the data, leading to poor generalization.

Radial basis function network: A common network model consisting of a linear combination of basis functions, each of which is a function of the difference between the input vector and a center vector.

Regression: A learning problem in which the goal is to map each input vector to a real-valued output vector.

Regularization: A technique for controlling model complexity and improving generalization by the addition of a penalty term to the cost function.

VC dimension: A measure of the complexity of a model. Knowledge of the VC dimension permits an estimate to be made of the difference between performance on the training set and performance on a test set.

References

Amari, S. 1995. The EM algorithm and information geometry in neural network learning. *Neural Comput.* 7(1):13–18.

Anderson, T. W. 1984. *An Introduction to Multivariate Statistical Analysis.* Wiley, New York.

Bengio, Y. 1996. *Neural Networks for Speech and Sequence Recognition.* Thomson Computer Press, London.

Bishop, C. M. 1995. *Neural Networks for Pattern Recognition.* Oxford University Press.

Breiman, L., Friedman, J. H., Olshen, R. A., and Stone, C. J. 1984. *Classification and Regression Trees.* Wadsworth International Group, Belmont, CA.

Buntine, W. 1994. Operations for learning with graphical models. *J. Artif. Intelligence Res.* 2:159–225.

Dempster, A. P., Laird, N. M., and Rubin, D. B. 1977. Maximum-likelihood from incomplete data via the EM algorithm. *J. R. Stat. Soc.* B39:1–38.

Duda, R. O. and Hart, P. E. 1973. *Pattern Classification and Scene Analysis.* Wiley, New York.

Fletcher, R. 1987. *Practical Methods of Optimization*, 2nd ed. Wiley, New York.

Geman, S., Bienenstock, E., and Doursat, R. 1992. Neural networks and the bias/variance dilemma. *Neural Comput.* 4:1–58.

Hertz, J., Krogh, A., and Palmer, R. G. 1991. *Introduction to the Theory of Neural Computation.* Addison–Wesley, Redwood City, CA.

Hinton, G. E., Dayan, P., Frey, B., and Neal, R. 1995. The wake-sleep algorithm for unsupervised neural networks. *Science* 268:1158–1161.

Hinton, G. E. and Sejnowski, T. 1986. Learning and relearning in Boltzmann machines. In *Parallel Distributed Processing: Vol. 1.* D. E. Rumelhart and J. L. McClelland, eds., pp. 282–317. MIT Press, Cambridge, MA.

Hopfield, J. J. 1982. Neural networks and physical systems with emergent collective computational abilities. *Proc. Nat. Acad. Sci.* 79:2554–2558.

Jacobs, R. A., Jordan, M. I., Nowlan, S. J., and Hinton, G. E. 1991. Adaptive mixtures of local experts. *Neural Comput.* 3:79–87.

Jordan, M. I. and Jacobs, R. A. 1994. Hierarchical mixtures of experts and the EM algorithm. *Neural Comput.* 6:181–214.

Le Cun, Y., Boser, B., Denker, J. S., Henderson, D., Howard, R. E., Hubbard, W., and Jackel, L. D. 1989. Backpropagation applied to handwritten zip code recognition. *Neural Comput.* 1(4):541–551.

Luttrell, S. 1989. The Gibbs machine applied to hidden Markov model problems. *Royal Signals and Radar Establishment: SP Res. Note 99*, Malvern, UK.

MacKay, D. J. C. 1992. A practical Bayesian framework for back-propagation networks. *Neural Comput.* 4:448–472.

McCullagh, P. and Nelder, J. A. 1983. *Generalized Linear Models.* Chapman and Hall, London.

Neal, R. M. 1994. *Bayesian Learning for Neural Networks.* Unpublished Ph.D. thesis, Department of Computer Science, University of Toronto, Canada.

Pearl, J. 1988. *Probabilistic Reasoning in Intelligent Systems.* Morgan Kaufmann, San Mateo, CA.

Poggio, T. and Vetter, T. 1992. Recognition and structure from one 2D model view: observations on prototypes, object classes and symmetries. *Artificial Intelligence Lab.*, AI Memo 1347, Massachusetts Institute of Technology, Cambridge, MA.

Rabiner, L. R. 1989. A tutorial on hidden Markov models and selected applications in speech recognition. *Proc. IEEE* 77:257–286.

Rumelhart, D. E., Durbin, R., Golden, R., and Chauvin, Y. 1995. Backpropagation: the basic theory. In *Backpropagation: Theory, Architectures, and Applications*, Y. Chauvin, and D. E. Rumelhart, eds., pp. 1–35. Lawrence Erlbaum, Hillsdale, NJ.

Rumelhart, D. E., Hinton, G. E., and Williams, R. J. 1986. Learning internal representations by error propagation. In *Parallel Distributed Processing: Vol. 1*. D. E. Rumelhart and J. L. McClelland, eds., pp. 318–363. MIT Press, Cambridge, MA.

Sandler, D. G., Barrett, T. K., Palmer, D. A., Fugate, R. Q., and Wild, W. J. 1991. Use of a neural network to control an adaptive optics system for an astronomical telescope. *Nature* 351:300–302.

Saul, L. K., Jaakkola, T., and Jordan, M. I. 1996. Mean field learning theory for sigmoid belief networks. *J. Artif. Intelligence Res.* 4:61–76.

Saul, L. K. and Jordan, M. I. 1995. Boltzmann chains and hidden Markov models. In *Advances in Neural Information Processing Systems 7*, G. Tesauro, D. Touretzky, and T. Leen, eds. MIT Press, Cambridge, MA.

Shachter, R., Andersen, S., and Szolovits, P. 1994. Global conditioning for probabilistic inference in belief networks. In *Uncertainty in Artificial Intelligence: Proc. 10th Conf.*, pp. 514–522. Seattle, WA.

Shachter, R. and Kenley, C. 1989. Gaussian influence diagrams. *Management Sci.* 35(5):527–550.

Smyth, P., Heckerman, D., and Jordan, M. I. in press. Probabilistic independence networks for hidden Markov probability models. *Neural Computation*.

Spiegelhalter, D., Dawid, A., Lauritzen, S., and Cowell, R. 1993. Bayesian analysis in expert systems. *Stat. Sci.* 8(3):219–283.

Tarassenko, L. 1995. Novelty detection for the identification of masses in mammograms. *Proc. 4th IEE Int. Conf. Artif. Neural Networks* Vol. 4, pp. 442–447.

Vapnik, V. N. 1995. *The Nature of Statistical Learning Theory*. Springer–Verlag, New York.

Further Information

In this chapter we have emphasized the links between neural networks and statistical pattern recognition. A more extensive treatment from the same perspective can be found in Bishop [1995]. For a view of recent research in the field, the proceedings of the annual Neural Information Processing Systems (NIPS), MIT Press, conferences are highly recommended.

Neural computing is now a very broad field, and there are many topics which have not been discussed for lack of space. Here we aim to provide a brief overview of some of the more significant omissions, and to give pointers to the literature.

The resurgence of interest in neural networks during the 1980s was due in large part to work on the statistical mechanics of fully connected networks having symmetric connections (i.e., if unit i sends a connection to unit j then there is also a connection from unit j back to unit i with the same weight value). We have briefly discussed such systems; a more extensive introduction to this area can be found in Hertz et al. [1991].

The implementation of neural networks in specialist very large-scale integrated (VLSI) hardware has been the focus of much research, although by far the majority of work in neural computing is undertaken using software implementations running on standard platforms.

An implicit assumption throughout most of this chapter is that the processes which give rise to the data are stationary in time. The techniques discussed here can readily be applied to problems such as time series forecasting, provided this stationarity assumption is valid. If, however, the generator of the data is itself evolving with time, then more sophisticated techniques must be used, and these are the focus of much current research (see Bengio [1996]).

One of the original motivations for neural networks was as models of information processing in biological systems such as the human brain. This remains the subject of considerable research activity, and there is a continuing flow of ideas between the fields of neurobiology and of artificial neural networks. Another historical springboard for neural network concepts was that of adaptive control, and again this remains a subject of great interest.

24

Genetic Algorithms

Stephanie Forrest*
University of New Mexico

24.1 Introduction

A genetic algorithm is a form of evolution that occurs in a computer. Genetic algorithms are useful, both as search methods for solving problems and for modeling evolutionary systems. This chapter describes how genetic algorithms work, gives several examples of genetic algorithm applications, and reviews some mathematical analysis of genetic algorithm behavior.

In genetic algorithms, strings of binary digits are stored in a computer's memory, and over time the properties of these strings evolve in much the same way that populations of individuals evolve under natural selection. Although the computational setting is highly simplified when compared with the natural world, genetic algorithms are capable of evolving surprisingly complex and interesting structures. These structures, called **individuals**, can represent solutions to problems, strategies for playing games, visual images, or computer programs. Thus, genetic algorithms allow engineers to use a computer to evolve problem solutions over time, instead of designing them by hand. Although genetic algorithms are known primarily as a problem-solving method, they can also be used to study and model evolution in various settings, including biological (such as ecologies, immunology, and population genetics), social (such as economies and political systems), and cognitive systems.

24.2 Underlying Principles

The basic idea of a genetic algorithm is quite simple. First, a population of individuals is created in a computer, and then the population is evolved using the principles of variation, selection, and inheritance. Random variations in the population result in some individuals being more fit than others (better suited to their environment). These individuals have more offspring, passing on successful variations to their children, and the cycle is repeated. Over time, the individuals in the population become better adapted to their environment. There are many ways of implementing this simple idea. Here I describe the one invented by Holland [1975, Goldberg 1989].

*Significant portions of this chapter are excerpted with permission from Forrest, S. 1993. Genetic algorithms: principles of adaption applied to computation. *Science* 261 (Aug. 13):872–878. ©1993 American Association for the Advancement of Science.

The idea of using selection and variation to evolve solutions to problems goes back at least to Box [1957], although his work did not use a computer. In the late 1950s and early 1960s there were several independent efforts to incorporate ideas from evolution in computation. Of these, the best known are genetic algorithms [Holland 1962], evolutionary programming [Fogel et al. 1966], and evolutionary strategies [Back and Schwefel 1993]. Rechenberg [Back and Schwefel 1993] emphasized the importance of selection and mutation as mechanisms for solving difficult real-valued optimization problems. Fogel et al. [1966] developed similar ideas for evolving intelligent agents in the form of finite state machines. Holland [1962, 1975] emphasized the adaptive properties of entire populations and the importance of recombination mechanisms such as **crossover**. In recent years, genetic algorithms have taken many forms, and in some cases bear little resemblance to Holland's original formulation. Researchers have experimented with different types of representations, crossover and mutation operators, special-purpose operators, and different approaches to reproduction and selection. However, all of these methods have a family resemblance in that they take some inspiration from biological evolution and from Holland's original genetic algorithm. A new term, *evolutionary computation*, has been introduced to cover these various members of the genetic algorithm family, evolutionary programming, and evolution strategies.

Figure 24.1 gives an overview of a simple genetic algorithm. In its simplest form, each individual in the population is a bit string. Genetic algorithms often use more complex representations, including richer alphabets, diploidy, redundant encodings, and multiple **chromosomes**. However, the binary case is both the simplest and most general. By analogy with genetics, the string of bits is referred to as the **genotype**. Each individual consists only of its genetic material, and it is organized into one (haploid) chromosome. Each bit position (set to 1 or 0) represents one gene. I will use the term bit string to refer both to genotypes and the individuals that they define. A natural question is how genotypes built from simple strings of bits can specify a solution to a specific problem. In other words, how are the binary genes expressed? There are many techniques for mapping bit strings to different problem domains, some of which are described in the following subsections.

The initial population of individuals is usually generated randomly, although it need not be. For example, prior knowledge about the problem solution can be encoded directly into the initial population, as in Hillis [1990]. Each individual is tested empirically in an environment, receiving a numerical evaluation of its merit, assigned by a **fitness function** F. The environment can be almost anything: another computer simulation, interactions with other individuals in the population, actions in the physical world (by a robot for example), or a human's subjective judgment. The fitness function's evaluation typically returns a

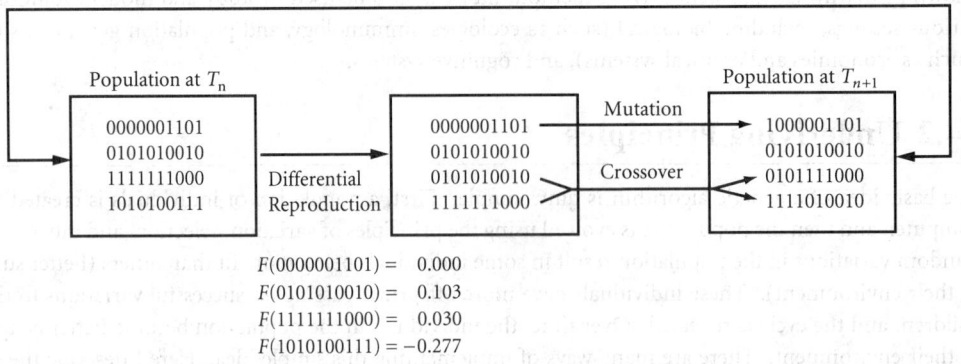

FIGURE 24.1 Genetic algorithm overview: A population of four individuals is shown. Each is assigned a fitness value by the function $F(x, y) = yx^2 - x^4$. (See Fig. 24.3.) On the basis of these fitnesses, the selection phase assigns the first individual (0000001101) one copy, the second (0101010010) two copies, the third (1111111000) one copy, and the fourth (1010100111) zero copies. After selection, the genetic operators are applied probabilistically; the first individual has its first bit mutated from a 0 to a 1, and crossover combines the last two individuals into two new ones. The resulting population is shown in the box labeled $T_{(N+1)}$. (See color version of this figure in the color section of the Handbook.)

single number (usually, higher numbers are assigned to fitter individuals). This constraint is sometimes relaxed so that the fitness function returns a vector of numbers [Fonseca and Fleming 1995], which can be appropriate for problems with multiple objectives. The fitness function determines how each gene (bit) of an individual will be interpreted and thus what specific problem the population will evolve to solve. The fitness function is the primary place where the traditional genetic algorithm is tailored to a specific problem.

Once all individuals in the population have been evaluated, their fitnesses form the basis for selection. Selection is implemented by eliminating low-fitness individuals from the population, and inheritance is implemented by making multiple copies of high-fitness individuals. Genetic operators such as **mutation** (flipping individual bits) and crossover (exchanging substrings of two individuals to obtain new offspring) are then applied probabilistically to the selected individuals to produce a new population (or **generation**) of individuals. The term crossover is used here to refer to the exchange of homologous substrings between individuals, although the biological term crossing over generally implies exchange within an individual. New generations can be produced either synchronously, so that the old generation is completely replaced, or asynchronously, so that generations overlap.

By transforming the previous set of good individuals to a new one, the operators generate a new set of individuals that ideally have a better than average chance of also being good. When this cycle of evaluation, selection, and genetic operations is iterated for many generations, the overall fitness of the population generally improves, as shown in Fig. 24.2, and the individuals in the population represent improved solutions to whatever problem was posed in the fitness function.

There are many details left unspecified by this description. For example, selection can be performed in any of several ways—it could arbitrarily eliminate the least fit 50% of the population and make one copy of all of the remaining individuals, it could replicate individuals in direct proportion to their fitness (fitness-proportionate selection), or it could scale the fitnesses in any of several ways and replicate individuals in direct proportion to their scaled values (a more typical method). Similarly, the crossover operator can pass on both offspring to the new generation, or it can arbitrarily choose one to be passed on; the number of crossover points can be restricted to one per pair, two per pair, or N per pair. These and other variations of the basic algorithm have been discussed extensively in Goldberg [1989], in Davis [1991], and in the Proceedings of the International Conference on Genetic Algorithms. (See Further Information section.)

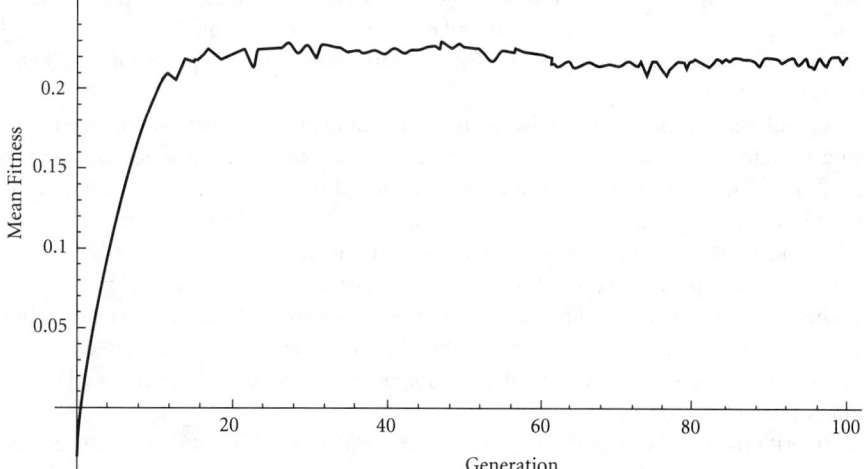

FIGURE 24.2   Mean fitness of a population evolving under the genetic algorithm. The population size is 100 individuals, each of which is 10 bits long (5 bits for x, 5 bits for y, as described in Fig. 24.3), mutation probability is 0.0026/bit, crossover probability is 0.6 per pair of individuals, and the fitness function is $F = yx^2 - x^4$. Population mean is shown every generation for 100 generations.

The genetic algorithm is interesting from a computational standpoint, at least in part, because of the claims that have been made about its effectiveness as a biased sampling algorithm. The classical argument about genetic algorithm performance has three components [Holland 1975, Goldberg 1989]:

- Independent sampling is provided by large populations that are initialized randomly.
- High-fitness individuals are preserved through selection, and this biases the sampling process toward regions of high fitness.
- Crossover combines partial solutions, called building blocks, from different strings onto the same string, thus exploiting the parallelism provided by the population of candidate solutions.

A partial solution is taken to be a hyperplane in the search space of strings and is called a **schema** (see section 24.4). A central claim about genetic algorithms is that schemas capture important regularities in the search space and that a form of *implicit parallelism* exists because one fitness evaluation of an individual comprising l bits implicitly gives information about the 2^l schemas, or hyperplanes, of which it is an instance. The Schema Theorem states that the genetic algorithm operations of reproduction, mutation, and crossover guarantee exponentially increasing samples of the observed best schemas in the next time step. By analogy with the k-armed bandit problem it can be argued that the genetic algorithm uses an optimal sampling strategy [Holland 1975]. See section 24.4 for details.

24.3 Best Practices

The simple computational procedure just described can be applied in many different ways to solve a wide range of problems. In designing a genetic algorithm to solve a specific problem there are two major design decisions: (1) specifying the mapping between binary strings and candidate solutions (this is commonly referred to as the representation problem) and (2) defining a concrete measure of fitness. In some cases the best representation and fitness function are obvious, but in many cases they are not, and in all cases, the particular representation and fitness function that are selected will determine the ultimate success of the genetic algorithm on the chosen problem. Possibly the simplest representation is a *feature list* in which each bit, or gene, represents the presence or absence of a single feature. This representation is useful for learning pattern classes defined by a critical set of features. For example, in spectroscopic applications, an important problem is selecting a small number of spectral frequencies that predict the concentration of some substance (e.g., concentration of glucose in human blood). The feature list approach to this problem assigns 1 bit to represent the presence or absence of each different observable frequency, and high fitness is assigned to those individuals whose feature settings correspond to good predictors for high (or low) glucose levels [Thomas 1993].

Genetic algorithms in various forms have been applied to many scientific and engineering problems, including optimization, automatic programming, machine and robot learning, modeling natural systems, and artificial life. They have been used in a wide variety of optimization tasks, including numerical optimization (see section on function optimization) and combinatorial optimization problems such as circuit design and job shop scheduling (see section on ordering problems). Genetic algorithms have also been used to evolve computer programs for specific tasks (see section on automatic programming) and to design other computational structures, e.g., cellular automata rules and sorting networks. In machine learning, they have been used to design neural networks, to evolve rules for rule-based systems, and to design and control robots. For an overview of genetic algorithms in machine learning, see DeJong [1990a, 1990b] and Schaffer et al. [1992].

Genetic algorithms have been used to model processes of innovation, the development of bidding strategies, the emergence of economic markets, the natural immune system, and ecological phenomena such as biological arms races, host–parasite coevolution, symbiosis, and resource flow. They have been used to study evolutionary aspects of social systems, such as the evolution of cooperation, the evolution of communication, and trail-following behavior in ants. They have been used to study questions in population genetics, such as "under what conditions will a gene for recombination be evolutionarily viable?" Finally, genetic algorithms are an important component in many artificial-life models, including systems that model interactions between species evolution and individual learning. See Further Information section and Mitchell and Forrest [1994] for details about genetic algorithms in modeling and artificial life.

The remainder of this section describes four illustrative examples of how genetic algorithms are used: numerical encodings for function optimization, permutation representations and special operators for sequencing problems, computer programs for automated programming, and endogenous fitness and other extensions for ecological modeling. The first two cover the most common classes of engineering applications. They are well understood and noncontroversial. The third example illustrates one of the most promising recent advances in genetic algorithms, but it was developed more recently and is less mature than the first two. The final example shows how genetic algorithms can be modified to more closely approximate natural evolutionary processes.

Function Optimization

Perhaps the most common application of genetic algorithms, pioneered by DeJong [1975], is multiparameter function optimization. Many problems can be formulated as a search for an optimal value, where the value is a complicated function of some input parameters. In some cases, the parameter settings that lead to the exact greatest (or least) value of the function are of interest. In other cases, the exact optimum is not required, just a near optimum, or even a value that represents a slight improvement over the previously best-known value. In these latter cases, genetic algorithms are often an appropriate method for finding good values.

As a simple example, consider the function $f(x, y) = yx^2 - x^4$. This function is solvable analytically, but if it were not, a genetic algorithm could be used to search for values of x and y that produce high values of $f(x, y)$ in a particular region of $\Re^2$. The most straightforward representation (Fig. 24.3) is to assign regions of the bit string to represent each parameter (variable). Once the order in which the parameters are to appear is determined (in the figure x appears first and y appears second), the next step is to specify the domain for x and y (that is, the set of values for x and y that are candidate solutions). In our example, x and y will be real values in the interval $[0, 1)$. Because x and y are real valued in this example, and

Degray	0 0 0 0 1	1 1 0 1 0	Bit String (Gray Coded)
	0 0 0 0 1	1 0 1 1 1	Base 2
	1	19	Base 10
	0.03	0.59	Normalized

$F(0000111010) = F(0.03, 0.59) = 0.59 \times (0.03)^2 - (0.03)^4 = 0.0005$

FIGURE 24.3 Bit-string encoding of multiple real-valued parameters. An arbitrary string of 10 bits is interpreted in the following steps: (1) segment the string into two regions with the first 5 bits reserved for x and the second 5 bits for y; (2) interpret each 5-bit substring as a Gray code and map back to the corresponding binary code; (3) map each 5-bit substring to its decimal equivalent; (4) scale to the interval $[0, 1)$; (5) substitute the two scaled values for x and y in the fitness function F; (6) return $F(x, y)$ as the fitness of the original string.

we are using a bit representation, the parameters need to be discretized. The precision of the solution is determined by how many bits are used to represent each parameter. In the example, 5 bits are assigned for x and 5 for y, although 10 is a more typical number. There are different ways of mapping between bits and decimal numbers, and so an encoding must also be chosen, and here we use gray coding.

Once a representation has been chosen, the genetic algorithm generates a random population of bit strings, decodes each bit string into the corresponding decimal values for x and y, applies the fitness function ($f(x, y) = yx^2 - x^4$) to the decoded values, selects the most fit individuals [those with the highest $f(x, y)$] for copying and variation, and then repeats the process. The population will tend to converge on a set of bit strings that represents an optimal or near optimal solution. However, there will always be some variation in the population due to mutation (Fig. 24.2).

The standard binary encoding of decimal values has the drawback that in some cases all of the bits must be changed in order to increase a number by one. For example, the bit pattern 011 translates to 3 in decimal, but 4 is represented by 100. This can make it difficult for an individual that is close to an optimum to move even closer by mutation. Also, mutations in high-order bits (the leftmost bits) are more significant than mutations in low-order bits. This can violate the idea that bit strings in successive generations will have a better than average chance of having high fitness, because mutations may often be disruptive. Gray codes address the first of these problems. Gray codes have the property that incrementing or decrementing any number by one is always 1 bit change. In practice, Gray-coded representations are often more successful for multiparameter function optimization applications of genetic algorithms.

Many genetic algorithm practitioners encode real-valued parameters directly without converting to a bit-based representation. In this approach, each parameter can be thought of as a gene on the chromosome. Crossover is defined as before, except that crosses take place only between genes (between real numbers). Mutation is typically redefined so that it chooses a random value that is close to the current value. This representation strategy is often more effective in practice, but it requires some modification of the operators [Back and Schwefel 1993, Davis 1991]. There are a number of other representation tricks that are commonly employed for function optimization, including logarithmic scaling (interpreting bit strings as the logarithm of the true parameter value), dynamic encoding (a technique that allows the number and interpretation of bits allocated to a particular parameter to vary throughout a run), variable-length representations, delta coding (the bit strings express a distance away from some previous partial solution), and a multitude of nonbinary encodings.

This completes our description of a simple method for encoding parameters onto a bit string. Although a function of two variables was used as an example, the strength of the genetic algorithm lies in its ability to manipulate many parameters, and this method has been used for hundreds of applications, including aircraft design, tuning parameters for algorithms that detect and track multiple signals in an image, and locating regions of stability in systems of nonlinear difference equations. See Goldberg [1989], Davis [1991], and the Proceedings of the International Conference on Genetic Algorithms for more detail about these and other examples of successful function-optimization applications.

Ordering Problems

A common problem involves finding an optimal ordering for a sequence of N items. Examples include various NP-complete problems such as finding a tour of cities that minimizes the distance traveled (the traveling salesman problem), packing boxes into a bin to minimize wasted space (the bin packing problem), and graph coloring problems.

For example, in the traveling salesman problem, suppose there are four cities: 1, 2, 3, and 4 and that each city is labeled by a unique bit string.[1] A common fitness function for this problem is the length of the candidate tour. A natural way to represent a tour is as a permutation, so that 3 2 1 4 is one candidate tour and 4 1 2 3 is another. This representation is problematic for the genetic algorithm because mutation and crossover do not necessarily produce legal tours. For example, a crossover between positions two and three in the example produces the individuals 3 2 2 3 and 4 1 1 4, both of which are illegal tours—not all of the cities are visited and some are visited more than once.

Three general methods have been proposed to address this representation problem: (1) adopting a different representation, (2) designing specialized crossover operators that produce only legal tours, and (3) penalizing illegal solutions through the fitness function. Of these, the use of specialized operators has been the most successful method for applications of genetic algorithms to ordering problems such as the traveling salesman problem (for example, see Mühlenbein et al. [1988]), although a number of generic representations have been proposed and used successfully on other sequencing problems. Specialized crossover operators tend to be less general, and I will describe one such method, **edge recombination**, as

---

[1] For simplicity, we will use integers in the following explanation rather than the bit strings to which they correspond.

an example of a special-purpose operator that can be used with the permutation representation already described.

When designing special-purpose operators it is important to consider what information from the parents is being transmitted to the offspring, that is, what information is correlated with high-fitness individuals. In the case of traditional bitwise crossover, the answer is generally short, low-order schemas. (See section 24.4.) But in the case of sequences, it is not immediately obvious what this means. Starkweather et al. [1991] identified three potential kinds of information that might be important for solving an ordering problem and therefore important to preserve through recombination: absolute position in the order, relative ordering (e.g., precedence relations might be important for a scheduling application), and adjacency information (as in the traveling salesman problem). They designed the edge-recombination operator to emphasize adjacency information. The operator is rather complicated, and there are many variants of the originally published operator. A simplified description follows (for details, see Starkweather et al. [1991]). For each pair of individuals to be crossed: (1) construct a table of adjacencies in the parents (see Fig. 24.4) and (2) construct one new permutation (offspring) by combining information from the two parents:

```
3  6  2  1  4  5                    3  6  4  1  2  5
5  2  1  3  6  4    ----------->
```

Original Individuals   New Individual

Adjacency List

Key	Adjacent Keys
1	2, 2, 3, 4
2	1, 1, 3, 6
3	1, 6, 6
4	1, 5, 6
5	2, 4
6	2, 3, 3, 4

FIGURE 24.4 Example of edge-recombination operator. The adjacency list is constructed by examining each element in the parent permutations (labeled Key) and recording its adjacent elements. The new individual is constructed by selecting one parent arbitrarily (the top parent) and assigning its first element (3) to be the first element in the new permutation. The adjacencies of 3 are examined, and 6 is chosen to be the second element because it is a shared adjacency. The adjacencies of 6 are then examined, and of the unused ones, 4 is chosen randomly. Similarly, 1 is assigned to be the fourth element in the new permutation by random choice from {1, 5}. Then 2 is placed as the fifth element because it is a shared adjacency, and then the one remaining element, 5, is placed in the last position.

- Select one parent at random and assign the first element in its permutation to be the first one in the child.
- Select the second element for the child, as follows: If there is an adjacency common to both parents, then choose that element to be the next one in the child's permutation; if there is an unused adjacency available from one parent, choose it; or if (1) and (2) fail, make a random selection.
- Select the remaining elements in order by repeating step 2.

An example of the edge-recombination operator is shown in Fig. 24.4. Although this method has proved effective, it should be noted that it is more expensive to build the adjacency list for each parent and to perform edge recombination operation than it is to use a more standard crossover operator.

A final consideration in the choice of special-purpose operators is the amount of random information that is introduced when the operator is applied. This can be difficult to assess, but it can have a large effect (positive or negative) on the performance of the operator.

Automatic Programming

Genetic algorithms have been used to evolve a special kind of computer program [Koza 1992]. These programs are written in a subset of the programming language Lisp and more recently other languages. Lisp programs can naturally be represented as trees (Fig. 24.5). Populations of random program trees are generated and evaluated as in the standard genetic algorithm. All other details are similar to those described for binary genetic algorithms with the exception of crossover. Instead of exchanging substrings, **genetic programs** exchange subtrees between individual program trees. This modified form of crossover

appears to have many of the same advantages as traditional crossover (such as preserving partial solutions).

Genetic programming has the potential to be extremely powerful, because Lisp is a general-purpose
programming language and genetic programming
eliminates the need to devise an explicit chromosomal representation. In practice, however, genetic
programs are built from subsets of Lisp tailored to
particular problem domains, and at this point considerable skill is required to select just the right set
of primitives for a particular problem. Although the
method has been tested on a wide variety of problems, it has not yet been used extensively in real applications.

The genetic programming method is intriguing
because its solutions are so different from human-
designed programs for the same problem. Humans
try to design elegant and general computer programs,
whereas genetic programs are often needlessly complicated, not revealing the underlying algorithm. For
example, a human-designed program for comput-
ing $\cos 2x$ might be $1 - 2\sin^2 x$, expressed in Lisp as

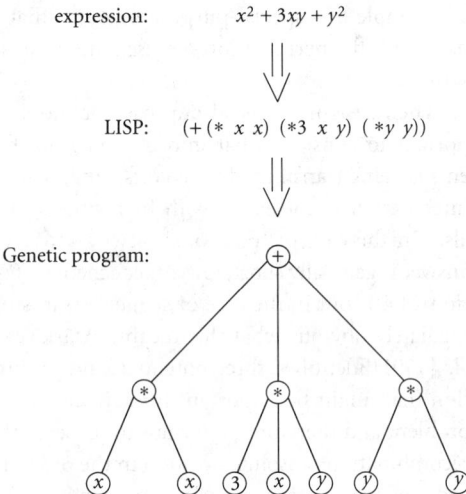

FIGURE 24.5   Tree representation of computer programs: The displayed tree corresponds to the expression $x^2 + 3xy + y^2$. Operators for each expression
are displayed as a root, and the operands for each expression are displayed as children. (*Source:* Forrest, S.
1993a. *Science* 261:872–878. With permission.)

$(-1(*2(*(\sin x)(\sin x))))$, whereas genetic programming discovered the following program (Koza 1992,
p. 241):

$$(\sin(-(-2(*x\,2))(\sin(\sin(\sin(\sin(\sin(\sin(*(\sin(\sin 1))(\sin(\sin 1))))))))))))$$

For anyone who has studied computer programming this is apparently a major drawback because the
evolved programs are inelegant, redundant, inefficient, difficult for a human to read, and do not reveal
the underlying structure of the algorithm. However, genetic programs do resemble the kinds of ad hoc
solutions that evolve in nature through gene duplication, mutation, and modifying structures from one
purpose to another. There is some evidence that the junk components of a genetic program sometimes turn
out to be useful components in other contexts. Thus, if the genetic programming endeavor is successful,
it could revolutionize software design.

Genetic Algorithms for Making Models

The past three examples concentrated on understanding how genetic algorithms can be applied to solve
problems. This subsection discusses how the genetic algorithm can be used to model other systems.
Genetic algorithms have been employed as models of a wide variety of dynamical processes, including
induction in psychology, natural evolution in ecosystems, evolution in immune systems, and imitation
in social systems. Making computer models of evolution is somewhat different from many conventional models because the models are highly abstract. The data produced by these models are unlikely
to make exact numerical predictions. Rather, they can reveal the conditions under which certain qualitative behaviors are likely to arise—diversity of phenotypes in resource-rich (or poor) environments,
cooperation in competitive nonzero-sum games, and so forth. Thus, the models described here are being used to discover qualitative patterns of behavior and, in some cases, critical parameters in which
small changes have drastic effects on the outcomes. Such modeling is common in nonlinear dynamics and in artificial intelligence, but it is much less accepted in other disciplines. Here we describe one
of these examples: ecological modeling. This exploratory research project is still in an early stage of

development. For examples of more mature modeling projects, see Holland et al. [1986] and Axelrod [1986].

The Echo system [Holland 1995] shows how genetic algorithms can be used to model ecosystems. The major differences between Echo and standard genetic algorithms are: (1) there is no explicit fitness function, (2) individuals have local storage (i.e., they consist of more than their genome), (3) the genetic representation is based on a larger alphabet than binary strings, and (4) individuals always have a spatial location. In Echo, fitness evaluation takes place implicitly. That is, individuals in the population (called *agents*) are allowed to make copies of themselves anytime they acquire enough *resources* to replicate their genome. Different resources are modeled by different letters of the alphabet (say, A, B, C, D), and genomes are constructed out of those same letters. These resources can exist independently of the agent's genome, either free in the environment or stored internally by the agent. Agents acquire resources by interacting with other agents through trading relationships and combat. Echo thus relaxes the constraint that an explicit fitness function must return a numerical evaluation of each agent. This **endogenous fitness function** is much closer to the way fitness is assessed in natural settings. In addition to trade and combat, a third form of interaction between agents is mating. Mating provides opportunities for agents to exchange genetic material through crossover, thus creating hybrids. Mating, together with mutation, provides the mechanism for new types of agents to evolve.

Populations in Echo exist on a two-dimensional grid of sites, although other connection topologies are possible. Many agents can cohabit one site, and agents can migrate between sites. Each site is the source of certain renewable resources. On each time step of the simulation, a fixed amount of resources at a site becomes available to the agents located at that site. Different sites can produce different amounts of different resources. For example, one site might produce 10 As and 5 Bs each time step, and its neighbor might produce 5 As, 0 Bs, and 5 Cs. The idea is that an agent will do well (reproduce often) if it is located at a site whose renewable resources match well with its genomic makeup or if it can acquire the relevant resources from other agents at its site.

In preliminary simulations, the Echo system has demonstrated surprisingly complex behaviors, including something resembling a biological arms race (in which two competing species develop progressively more complex offensive and defensive strategies), functional dependencies among different species, trophic cascades, and sensitivity (in terms of the number of different phenotypes) to differing levels of renewable resources. Although the Echo system is still largely untested, it illustrates how the fundamental ideas of genetic algorithms can be incorporated into a system that captures important features of natural ecological systems.

24.4 Mathematical Analysis of Genetic Algorithms

Although there are many problems for which the genetic algorithm can evolve a good solution in reasonable time, there are also problems for which it is inappropriate (such as problems in which it is important to find the exact global optimum). It would be useful to have a mathematical characterization of how the genetic algorithm works that is predictive. Research on this aspect of genetic algorithms has not produced definitive answers. The domains for which one is likely to choose an adaptive method such as the genetic algorithm are precisely those about which we typically have little analytical knowledge—they are complex, noisy, or dynamic (changing over time). These characteristics make it virtually impossible to predict with certainty how well a particular algorithm will perform on a particular problem instance, especially if the algorithm is stochastic, as is the case with the genetic algorithm. In spite of this difficulty, there are fairly extensive theories about how and why genetic algorithms work in idealized settings.

Analysis of genetic algorithms begins with the concept of a search space. The genetic algorithm can be viewed as a procedure for searching the space of all possible binary strings of fixed length l. Under this interpretation, the algorithm is searching for points in the l-dimensional space $\{0, 1\}^l$ that have high fitness. The search space is identical for all problems of the same size (same l), but the locations of good points will generally differ. The surface defined by the fitness of each point, together with the neighborhood relation imposed by the operators, is sometimes referred to as the **fitness landscape**. The longer the bit strings, corresponding to higher values of l, the larger the search space is, growing exponentially with the length of l. For problems with a sufficiently large l, only a small fraction of this size search space can be examined, and thus it is unreasonable to expect an algorithm to locate the global optimum in the space. A more reasonable goal is to search for good regions of the search space corresponding to regularities in the problem domain. Holland [1975] introduced the notion of a *schema* to explain how genetic algorithms search for regions of high fitness. Schemas are theoretical constructs used to explain the behavior of genetic algorithms, and are not processed directly by the algorithm. The following description of schema processing is excerpted from Forrest and Mitchell [1993b].

A schema is a template, defined over the alphabet $\{0, 1, *\}$, which describes a pattern of bit strings in the search space $\{0, 1\}^l$ (the set of bit strings of length l). For each of the l bit positions, the template either specifies the value at that position (1 or 0), or indicates by the symbol $*$ (referred to as don't care) that either value is allowed.

For example, the two strings A and B have several bits in common. We can use schemas to describe the patterns these two strings share:

$$A = 100111$$
$$B = 010011$$
$$**0*11$$
$$****11$$
$$**0***$$
$$**0**1$$

A bit string x that matches a schema s's pattern is said to be an *instance* of s; for example, A and B are both instances of the schemas just shown. In schemas, 1s and 0s are referred to as *defined bits*; the *order* of a schema is the number of defined bits in that schema, and the *defining length* of a schema is the distance between the leftmost and rightmost defined bits in the string. For example, the defining length of $**0**1$ is 3.

Schemas define hyperplanes in the search space $\{0, 1\}^l$. Figure 24.6 shows four hyperplanes, corresponding to the schemas $0****$, $1****$, $*0***$, and $*1***$. Any point in the space is simultaneously an instance of two of these schemas. For example, the point shown in Fig. 24.6 is an instance of both $1****$ and $*0***$ (and also of $10***$).

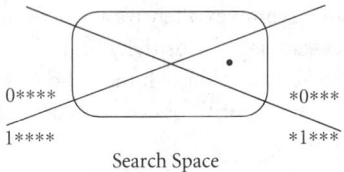

Search Space

The fitness of any bit string in the population gives some information about the average fitness of the 2^l different schemas of which it is an instance, and so an explicit evaluation of a population of M individual strings is also an implicit evaluation of a much larger number of schemas. This is referred to as implicit parallelism. At the explicit level the genetic algorithm searches

FIGURE 24.6 Schemas define hyperplanes in the search space. (*Source:* Forrest, S. and Mitchell, M. 1993b. *Machine Learning* 13:285–319. With permission.)

through populations of bit strings, but the genetic algorithm's search can also be interpreted as an implicit schema sampling process. Feedback from the fitness function, combined with selection and recombination, biases the sampling procedure over time away from those schemas that give negative feedback (low average fitness) and toward those that give positive feedback (high average fitness). Ultimately, the search procedure should identify regularities, or patterns, in the environment that lead to high fitness. Because the space of possible patterns is larger than the space of possible individuals (3^l vs. 2^l), implicit parallelism is potentially advantageous.

An important theoretical result about genetic algorithms is the Schema Theorem [Holland 1975, Goldberg 1989], which states that the observed best schemas will on average be allocated an exponentially increasing number of samples in the next generation. Figure 24.7 illustrates the rapid convergence on fit schemas by the genetic algorithm. This strong convergence property of the genetic algorithm is a two-edged sword. On the one hand, the fact that the genetic algorithm can close in on a fit part of the

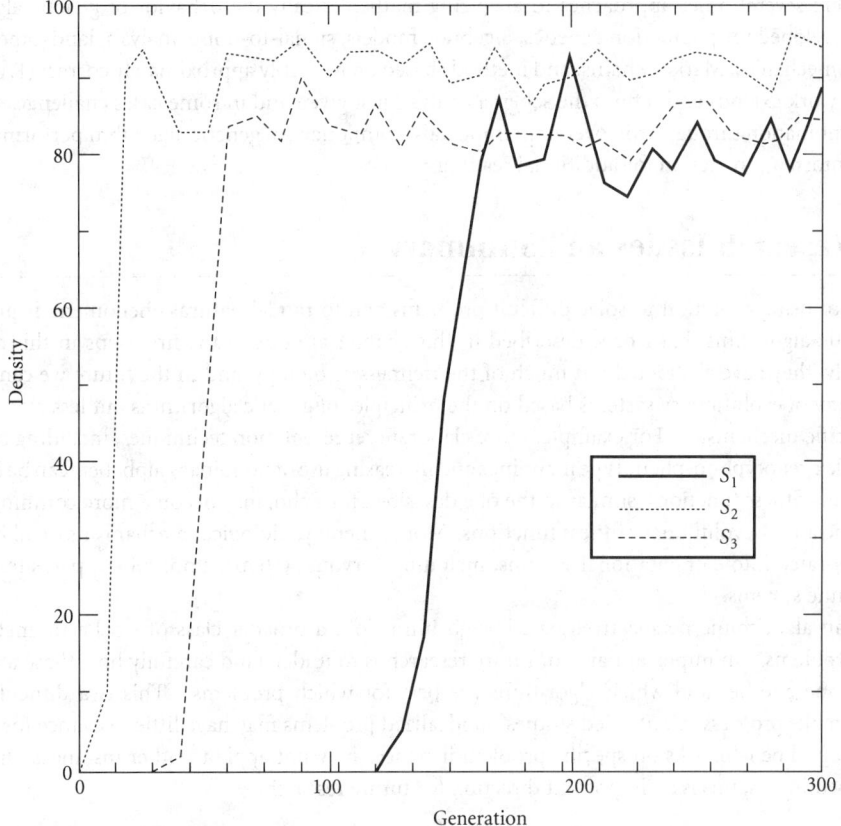

FIGURE 24.7   Schema frequencies over time. The graph plots schema frequencies in the population over time for three schemas:

$s_1 = 1111111111111111***$;

$s_2 = ***************1111111111111111*******************************$;

$s_3 = *******************************1111111111111111111111111111111$.

The function plotted was a royal road function [Forrest and Mitchell 1993a] in which the optimum value is the string of all 1s. (*Source:* Forrest, S. 1993a. *Science* 261:872–878. With permission.)

space very quickly is a powerful property; on the other hand, because the genetic algorithm always operates on finite-size populations, there is inherently some sampling error in the search, and in some cases the genetic algorithm can magnify a small sampling error, causing premature convergence on local optima.

According to the **building blocks hypothesis** [Holland 1975, Goldberg 1989], the genetic algorithm initially detects biases toward higher fitness in some low-order schemas (those with a small number of defined bits), and converges on this part of the search space. Over time, it detects biases in higher-order schemas by combining information from low-order schemas via crossover, and eventually it converges on a small region of the search space that has high fitness. The building blocks hypothesis states that this process is the source of the genetic algorithm's power as a search and optimization method. If this hypothesis about how genetic algorithms work is correct, then crossover is of primary importance, and it distinguishes genetic algorithms from other similar methods, such as simulated annealing and greedy algorithms. A number of authors have questioned the adequacy of the building blocks hypothesis as an explanation for how genetic algorithms work and there are several active research efforts studying schema processing in genetic algorithms. Nevertheless, the explanation of schemas and recombination that I have just described stands as the most common account of why genetic algorithms perform as they do.

There are several other approaches to analyzing mathematically the behavior of genetic algorithms: models developed for population genetics, algebraic models, signal-to-noise analysis, landscape analysis, statistical mechanics, Markov chains, and methods based on probably approximately correct (PAC) learning. This work extends and refines the schema analysis just given and in some cases challenges the claim that recombination through crossover is an important component of genetic algorithm performance. See Further Information section for additional reading.

24.5 Research Issues and Summary

The idea of using evolution to solve difficult problems and to model natural phenomena is promising. The genetic algorithms that I have described in this chapter are one of the first steps in this direction. Necessarily, they have abstracted out much of the richness of biology, and in the future we can expect a wide variety of evolutionary systems based on the principles of genetic algorithms but less closely tied to these specific mechanisms. For example, more elaborate representation techniques, including those that use complex genotype-to-phenotype mappings and increasing use of nonbinary alphabets can be expected. Endogenous fitness functions, similar to the one described for Echo, may become more common, as well as dynamic and coevolutionary fitness functions. More generally, biological mechanisms of all kinds will be incorporated into computational systems, including nervous systems, embryology, parasites, viruses, and immune systems.

From an algorithmic perspective, genetic algorithms join a broader class of stochastic methods for solving problems. An important area of future research is to understand carefully how these algorithms relate to one another and which algorithms are best for which problems. This is a difficult area in which to make progress. Controlled studies on idealized problems may have little relevance for practical problems, and benchmarks on specific problem instances may not apply to other instances. In spite of these impediments, this is an important direction for future research.

Acknowledgments

The author gratefully acknowledges support from the National Science Foundation (Grant IRI-9157644), the Office of Naval Research (Grant N00014-95-1-0364), ATR Human Information Processing Research Laboratories, and the Santa Fe Institute. Ron Hightower prepared Fig. 24.2.

Defining Terms

Building blocks hypothesis: The hypothesis that the genetic algorithm searches by first detecting biases toward higher fitness in some low-order schemas (those with a small number of defined bits) and converging on this part of the search space. Over time, it then detects biases in higher-order schemas by combining information from low-order schemas via crossover and eventually converges on a small region of the search space that has high fitness. The building blocks hypothesis states that this process is the source of the genetic algorithm's power as a search and optimization method [Holland 1975, Goldberg 1989].

Chromosome: A string of symbols (usually in bits) that contains the genetic information about an individual. The chromosome is interpreted by the fitness function to produce an evaluation of the individual's fitness.

Crossover: An operator for producing new individuals from two parent individuals. The operator works by exchanging substrings between the two individuals to obtain new offspring. In some cases, both offspring are passed to the new generation; in others, one is arbitrarily chosen to be passed on; the number of crossover points can be restricted to one per pair, two per pair, or N per pair.

Edge recombination: A special-purpose crossover operator designed to be used with permutation representations for sequencing problems. The edge-recombination operator attempts to preserve adjacencies between neighboring elements in the parent permutations [Starkweather et al. 1991].

Endogenous fitness function: Fitness is not assessed explicitly using a fitness function. Some other criterion for reproduction is adopted. For example, individuals might be required to accumulate enough internal resources to copy themselves before they can reproduce. Individuals who can gather resources efficiently would then reproduce frequently and their traits would become more prevalent in the population.

Fitness function: Each individual is tested empirically in an environment, receiving a numerical evaluation of its merit, assigned by a fitness function F. The environment can be almost anything—another computer simulation, interactions with other individuals in the population, actions in the physical world (by a robot for example), or a human's subjective judgment.

Fitness landscape: The surface defined by the fitness of each point in the search space, together with the neighborhood relation imposed by the operators.

Generation: One iteration, or time step, of the genetic algorithm. New generations can be produced either synchronously, so that the old generation is completely replaced (the time step model), or asynchronously, so that generations overlap. In the asynchronous case, generations are defined in terms of some fixed number of fitness-function evaluations.

Genetic programs: A form of genetic algorithm that uses a tree-based representation. The tree represents a program that can be evaluated, for example, an S-expression.

Genotype: The string of symbols, usually bits, used to represent an individual. Each bit position (set to 1 or 0) represents one gene. The term bit string in this context refers both to genotypes and to the individuals that they define.

Individuals: The structures that are evolved by the genetic algorithm. They can represent solutions to problems, strategies for playing games, visual images, or computer programs. Typically, each individual consists only of its genetic material, which is organized into one (haploid) chromosome.

Mutation: An operator for varying an individual. In mutation, individual bits are flipped probabilistically in individuals selected for reproduction. In representations other than bit strings, mutation is redefined to an appropriate smallest unit of change. For example, in permutation representations, mutation is often defined to be the swap of two neighboring elements in the permutation; in real-valued representations, mutation can be a creep operator that perturbs the real number up or down some small increment.

Schema: A theoretical construct used to explain the behavior of genetic algorithms. Schemas are not processed directly by the algorithm. Schemas are coordinate hyperplanes in the search space of strings.

Selection: Some individuals are more fit than others (better suited to their environment). These individuals have more offspring, that is, they are selected for reproduction. Selection is implemented by eliminating low-fitness individuals from the population, and inheritance is implemented by making multiple copies of high-fitness individuals.

References

Axelrod, R. 1986. An evolutionary approach to norms. *Am. Political Sci. Rev.* 80 (Dec).

Back, T. and Schwefel, H. P. 1993. An overview of evolutionary algorithms. *Evolutionary Comput.* 1:1–23.

Belew, R. K. and Booker, L. B., eds. 1991. *Proc. 4th Int. Conf. Genet. Algorithms.* July. Morgan Kaufmann, San Mateo, CA.

Booker, L. B., Riolo, R. L., and Holland, J. H. 1989. Learning and representation in classifier systems. *Art. Intelligence* 40:235–282.

Box, G. E. P. 1957. Evolutionary operation: a method for increasing industrial productivity. *J. R. Stat. Soc.* 6(2):81–101.

Davis, L., ed. 1991. *The Genetic Algorithms Handbook.* Van Nostrand Reinhold, New York.

DeJong, K. A. 1975. *An analysis of the behavior of a class of genetic adaptive systems.* Ph.D. thesis, University of Michigan, Ann Arbor.

DeJong, K. A. 1990a. Genetic-algorithm-based learning. *Machine Learning* 3:611–638.

DeJong, K. A. 1990b. Introduction to second special issue on genetic algorithms. *Machine Learning.* 5(4):351–353.

Eshelman, L. J., ed. 1995. *Proc. 6th Int. Conf. Genet. Algorithms.* Morgan Kaufmann, San Francisco.

Filho, J. L. R., Treleaven, P. C., and Alippi, C. 1994. Genetic-algorithm programming environments. *Computer* 27(6):28–45.

Fogel, L. J., Owens, A. J., and Walsh, M. J. 1966. *Artificial Intelligence Through Simulated Evolution.* Wiley, New York.

Fonseca, C. M. and Fleming, P. J. 1995. An overview of evolutionary algorithms in multiobjective optimization. *Evolutionary Comput.* 3(1):1–16.

Forrest, S. 1993a. Genetic algorithms: principles of adaptation applied to computation. *Science* 261:872–878.

Forrest, S., ed. 1993b. *Proc. Fifth Int. Conf. Genet. Algorithms.* Morgan Kaufmann, San Mateo, CA.

Forrest, S. and Mitchell, M. 1993a. Towards a stronger building-blocks hypothesis: effects of relative building-block fitness on ga performance. In *Foundations of Genetic Algorithms*, Vol. 2, L. D. Whitley, ed., pp. 109–126. Morgan Kaufmann, San Mateo, CA.

Forrest, S. and Mitchell, M. 1993b. What makes a problem hard for a genetic algorithm? Some anomalous results and their explanation. *Machine Learning* 13(2/3).

Goldberg, D. E. 1989. *Genetic Algorithms in Search, Optimization, and Machine Learning.* Addison Wesley, Reading, MA.

Grefenstette, J. J. 1985. *Proc. Int. Conf. Genet. Algorithms Appl.* NCARAI and Texas Instruments.

Grefenstette, J. J. 1987. *Proc. 2nd Int. Conf. Genet. Algorithms.* Lawrence Erlbaum, Hillsdale, NJ.

Hillis, W. D. 1990. Co-evolving parasites improve simulated evolution as an optimization procedure. *Physica D* 42:228–234.

Holland, J. H. 1962. Outline for a logical theory of adaptive systems. *J. ACM* 3:297–314.

Holland, J. H. 1975. *Adaptation in Natural and Artificial Systems.* University of Michigan Press, Ann Arbor; MI, 1992. 2nd ed. MIT Press, Cambridge, MA.

Holland, J. H. 1992. Genetic algorithms. *Sci. Am.*, pp. 114–116.

Holland, J. H. 1995. *Hidden Order: How Adaptation Builds Complexity.* Addison–Wesley, Reading, MA.

Holland, J. H., Holyoak, K. J., Nisbett, R. E., and Thagard, P. 1986. *Induction: Processes of Inference, Learning, and Discovery.* MIT Press, Cambridge, MA.

Koza, J. R. 1992. *Genetic Programming.* MIT Press, Cambridge, MA.

Männer, R. and Manderick, B., eds. 1992. *Parallel Problem Solving From Nature 2.* North Holland, Amsterdam.

Mitchell, M. 1996. *An Introduction to Genetic Algorithms.* MIT Press, Cambridge, MA.

Mitchell, M. and Forrest, S. 1994. Genetic algorithms and artificial life. *Artif. Life* 1(3):267–289; reprinted 1995. In *Artificial Life: An Overview*, C. G. Langton, ed. MIT Press, Cambridge, MA.

Mühlenbein, H., Gorges-Schleuter, M., and Kramer, O. 1988. *Parallel Comput.* 6:65–88.

Rawlins, G., ed. 1991. *Foundations of Genetic Algorithms.* Morgan Kaufmann, San Mateo, CA.

Schaffer, J. D., ed. 1989. *Proc. 3rd Int. Conf. Genet. Algorithms.* Morgan Kaufmann, San Mateo, CA.

Schaffer, J. D., Whitley, D., and Eshelman, L. J. 1992. Combinations of genetic algorithms and neural networks: a survey of the state of the art. In *Int. Workshop Combinations Genet. Algorithms Neural Networks*, L. D. Whitley and J. D. Schaffer, eds., pp. 1–37. IEEE Computer Society Press, Los Alamitos, CA.

Schwefel, H. P. and Männer, R., eds. 1990. Parallel problem solving from nature. *Lecture Notes in Computer Science.* Springer–Verlag, Berlin.

Srinivas, M. and Patnaik, L. M. 1994. Genetic algorithms: a survey. *Computer* 27(6):17–27.

Starkweather, T., McDaniel, S., Mathias, K., Whitley, D., and Whitley, C. 1991. A comparison of genetic sequencing operators. In *4th Int. Conf. Genet. Algorithms*, R. K. Belew and L. B. Booker, eds., pp. 69–76. Morgan Kaufmann, Los Altos, CA.

Thomas, E. V. 1993. Frequency Selection Using Genetic Algorithms. *Sandia National Lab. Tech. Rep.* SAND93-0010, Albuquerque, NM.

Whitley, L. D., ed. 1993. *Foundations of Genetic Algorithms 2*. Morgan Kaufmann, San Mateo, CA.

Whitley, L. D. and Vose, M., eds. 1995. *Foundations of Genetic Algorithms 3*. Morgan Kaufmann, San Francisco.

Further Information

Review articles on genetic algorithms include Booker et al. [1989], Holland [1992], Forrest [1993a], Mitchell and Forrest [1994], Srinivas and Patnaik [1994] and Filho et al. [1994]. Books that describe the theory and practice of genetic algorithms in greater detail include Holland [1975], Goldberg [1989], Davis [1991], Koza [1992], Holland et al. [1986], and Mitchell [1996]. Holland [1975] was the first book-length description of genetic algorithms, and it contains much of the original insight about the power and breadth of adaptive algorithms. The 1992 reprinting contains interesting updates by Holland. However, Goldberg [1989], Davis [1991], and Mitchell [1996] are more accessible introductions to the basic concepts and implementation issues. Koza [1992] describes genetic programming and Holland et al. [1986] discuss the relevance of genetic algorithms to cognitive modeling.

Current research on genetic algorithms is reported many places, including the Proceedings of the International Conference on Genetic Algorithms [Grefenstette 1985, 1987, Schaffer 1989, Belew and Booker 1991, Forrest 1993b, Eshelman 1995], the proceedings of conferences on Parallel Problem Solving from Nature [Schwefel and Männer 1990, Männer and Manderick 1992], and the workshops on Foundations of Genetic Algorithms [Rawlins 1991, Whitley 1993, Whitley and Vosee 1995]. Finally, the artificial-life literature contains many interesting papers about genetic algorithms.

There are several archival journals that publish articles about genetic algorithms. These include *Evolutionary Computation* (a journal devoted to GAs), *Complex Systems*, *Machine Learning*, *Adaptive Behavior*, and *Artificial Life*.

Information about genetic algorithms activities, public domain packages, etc., is maintained through the WWW at URL http://www.aic.nrl.navy.mil/galist/ or through anonymous ftp at ftp.aic.nrl.navy.mil [192.26.18.68] in /pub/galist.

25

Computer Vision

Daniel Huttenlocher
Cornell University and
   Xerox Palo Alto Research
   Center

25.1 Introduction

The goal of computer vision is to extract information from images. For example, **structure from motion** methods can recover a three-dimensional model of an object from a sequence of views, for use in robot grasping, medical imaging, and graphical modeling; **model-based recognition** methods can determine the best matches of stored models to image data, for use in visual inspection and image database searches; and *visual* **motion analysis** can recover image motion patterns for use in vehicle guidance and processing **digital video.** Computer vision is closely related to the field of image processing. In computer vision the focus is on extracting information from image data, whereas in image processing the focus is on transforming images. For instance, extracting a three-dimensional model from two-dimensional images is more of a computer vision problem than an image processing one, whereas image enhancement is more of an image processing problem than a computer vision one.

Computer vision is an interdisciplinary area, which falls primarily within the field of computer science, but also draws heavily on a number of other areas including image processing, differential and combinatorial geometry, numerical methods, and statistics. Some research in computer vision also has ties with biology and psychophysics; however, computer vision tends to be more concerned with building artificial vision systems than with accurately modeling human or animate systems. One of the main challenges for students of computer vision is developing adequate depth across such a wide range of areas. There are a number of books that provide relatively broad coverage of the field, with more advanced treatments in Faugeras [1993], Haralick and Shapiro [1992], and Horn [1986].

Human visual perception appears to be nearly effortless, in contrast with cognition, which can require substantial conscious effort. However, visual perception tasks are arguably at least as difficult as cognitive ones. For instance, computers can now beat all but the best human chess players and computational mathematics systems are routinely used to solve calculus problems that are too involved for people to do. Yet computational vision systems have only achieved human levels of performance in very restricted domains, such as automated parts inspection (under controlled lighting conditions). One particularly successful area of visual information processing is the development of systems for recognizing printed text. However, such optical character recognition (OCR) systems still make mistakes that a grade school student would not make, even if the child did not know the particular words being recognized. The main problem is that artificial vision systems are brittle, in the sense that small variations in the input may cause enormous changes in the output. Developing vision systems that degrade gracefully is a major challenge of computer vision.

Computer vision systems operate on **digital images**. A digital image is quantized into discrete values, both in space and in intensity. The discrete spatial locations are called **pixels**, and are generally arranged on a square grid, spaced equally apart (although the area covered by each pixel may not actually be square). Each pixel takes on a range of integer values. For a *gray-level* (or *intensity*) image these values are generally between 0 and 255 (8 b). For a *binary* image (or *bitmap*) the values are just 0 and 1. Color images can be represented in several ways; commonly three intensity images are used, one for each of three color channels (e.g., red, green, and blue). Digital images are large: a single gray-level frame from a video camera is about $\frac{1}{3}$ of a megabyte, a 24-b color image of a page scanned at 400 dots/in is about 44 megabytes, and an uncompressed 24-b color video stream is about 30 megabytes/s.

Computer vision methods are often classified into low, middle, and high levels. Although these classes are by no means universal, they still provide a useful way of categorizing computer vision problems. We will consider the following definitions:

- Low-level vision techniques are those that operate directly on images and produce outputs that are other images in the same coordinate system as the input. For example, an **edge detection** algorithm takes an intensity image as input and produces a binary image indicating where edges are present.

- Middle-level vision techniques are those that take images or the results of low-level vision algorithms as input and produce outputs that are something other than pixels in the image coordinate system. For example, a **structure from motion** algorithm takes as input sets of image features and produces as output the three-dimensional coordinates of those features.

- High-level vision techniques are those that take the results of low- or middle-level vision algorithms as input and produce outputs that are abstract data structures. For example, a **model-based recognition** system can take a set of image features as input and return the geometric transformations mapping models in its database to their locations in the image.

There are many applications of computer vision techniques. Traditionally, most computer vision systems have been designed for military and industrial applications. Common military applications include target recognition, visual guidance for autonomous vehicles, and interpretation of reconaissance imagery. Common industrial applications include parts inspection and visual control of automated systems. Over the past few years a number of new applications have emerged in medical imaging and multimedia systems. In medical applications computer vision methods are being used to register preoperative scans with a patient in the operating room. Computer vision techniques are also being used for realistic rendering and virtual reality applications, as well as image database retrieval.

In this chapter we will discuss a few computer vision problems in enough detail to give the reader an idea of some of the issues and to illustrate the kinds of techniques that are used to solve them. The presentation is divided according to low-, middle-, and high-level vision.

25.2 Low-Level Vision

Low-level vision computations operate directly on images and produce outputs that are pixel based and in the image coordinate system. Low-level vision computations include finding intensity edges in an image, representing images at multiple scales based on smoothing the image with different filters, computing visual motion fields, and analyzing the color information in images. In order to illustrate low-level vision methods, we will consider the problems of edge detection and image smoothing in more detail.

Local Edge Detectors

The primary goal of edge detection is to extract information about the geometry of an image for use in higher-level processing. There are many physical events in the world that cause intensity changes, or edges, in an image. Only some of these are geometric: object boundaries produce intensity changes due to a discontinuity in depth or difference in surface color and texture, surface boundaries produce intensity changes due to a difference in surface orientation. Other intensity changes do not directly reflect geometry (though it may be possible to derive some geometric information from them): specular reflections produce sharp intensity changes due to direct reflection of light; shadows and interreflections produce intensity changes due to other objects or parts of the same object.

We will refer to a gray-level image as $I(x, y)$, which denotes intensity as a function of the image coordinate system. Intensity edges correspond to rapid changes in the value of $I(x, y)$; thus it is common to use local differential properties such as the squared **gradient magnitude**,

$$\|\nabla I\|^2 = \left(\frac{\partial I}{\partial x}\right)^2 + \left(\frac{\partial I}{\partial y}\right)^2$$

Simplistically speaking, where the squared gradient magnitude is large, there is an edge. Another local differential operator that is used in edge detection is the Laplacian (see Horn [1986, Ch. 8]),

$$\nabla^2 I = \frac{\partial^2 I}{\partial x^2} + \frac{\partial^2 I}{\partial y^2}$$

This second derivative operator preserves information about which side of an edge is brighter. The zero crossings (sign changes) of $\nabla^2 I$ correspond to intensity edges in the image, and the sign on each side of a zero crossing indicates which side is brighter.

The images used in computer vision systems are digitized both in space and in intensity, producing an array $I[j, k]$ of discrete intensity values. Thus, in order to compute local differential operators, finite difference approximations are used to estimate the derivatives. For a discrete one-dimensional sampled function, represented as a vector of values $F[j]$, the derivative dF/dx can be approximated as $F[j+1] - F[j]$, and the second derivative d^2F/dx^2 can be approximated as $F[j-1] - 2F[j] + F[j+1]$. The squared gradient magnitude, $\|\nabla I\|^2 = (\partial I/\partial x)^2 + (\partial I/\partial y)^2$, can be approximated (at the center of a 2×2 grid of pixels) as

$$\left(\frac{\partial I}{\partial x}\right)^2 + \left(\frac{\partial I}{\partial y}\right)^2 \approx (I[j+1, k+1] - I[j, k])^2 + (I[j, k+1] - I[j+1, k])^2 \qquad (25.1)$$

The Laplacian, $\nabla^2 I$, can be computed in a similar manner using the approximation to the second derivative (see Horn [1986, Ch. 8]).

In practice edges cannot be computed reliably using these kinds of local operators, which consider just a 2×2 or 3×3 window of pixel values. The high degree of variability in images causes such operators to both report edge points where there are none and to miss edge points. For example, Fig. 25.1(b) shows the result of running a local gradient magnitude edge detector on the image shown in Fig. 25.1(a). This detector simply finds local maxima in the gradient magnitude which are larger than some threshold. Note the broken edges and large number of isolated edge points. In contrast, Fig. 25.1(c) shows the edges detected using a nonlocal (or less local) gradient magnitude computation, which is described later in the section on the **Canny edge operator**.

A number of local edge operators have been developed, which can mainly be understood in terms of directional first and second derivatives. A more detailed discussion of some of these operators can be found in Haralick and Shapiro [1992]. For example, the Sobel operator is a directional first derivative, based on the approximation $F[j+1] - F[j-1]$ (as opposed to $F[j+1] - F[j]$ as used earlier). The Sobel operator also uses a simple form of local smoothing. These local edge operators (which use 4×4

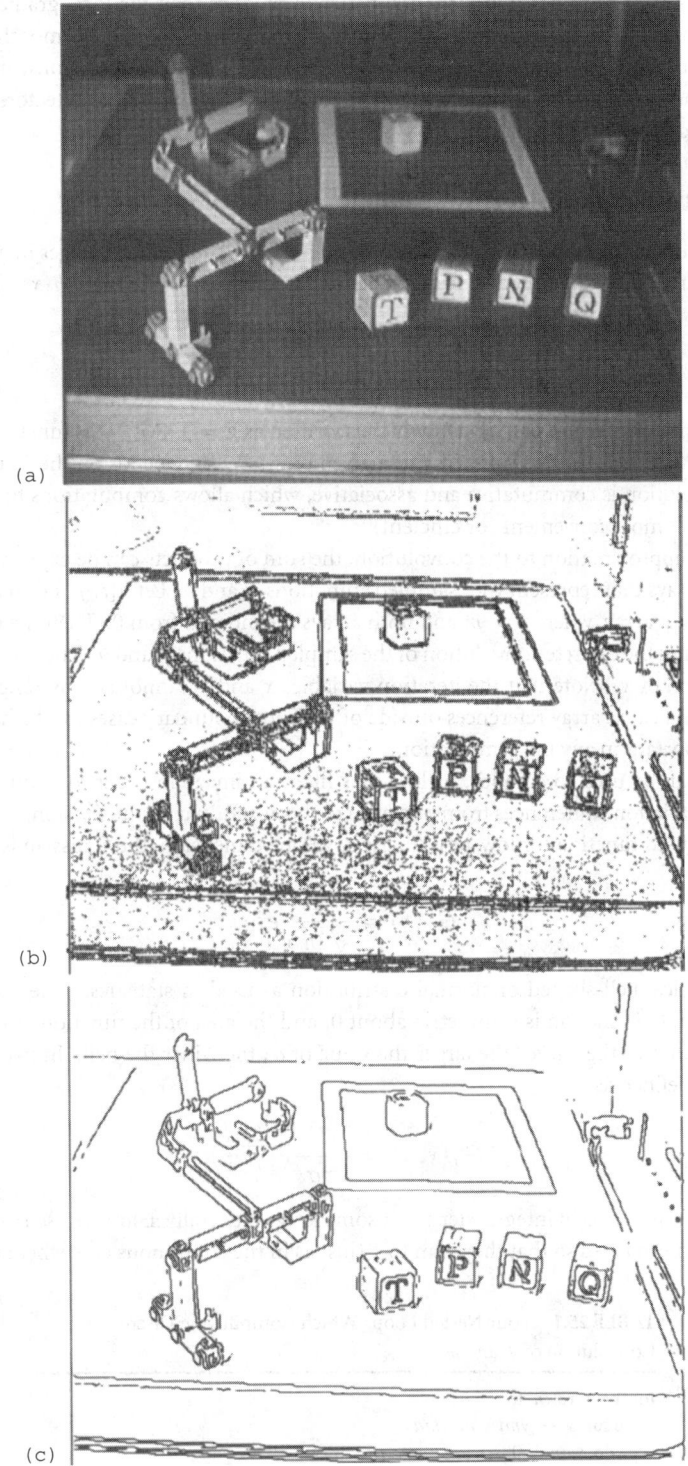

FIGURE 25.1 An example of edge detection: (a) an image, (b) a local gradient magnitude edge operator, (c) the Canny edge operator.

or 5 × 5 windows of pixel values) work slightly better in practice than the local gradient magnitude or Laplacian operators. The main reason is that they do local averaging (or weighted smoothing) of the image as part of the processing. We now turn to a discussion of local smoothing operations, and then put these operations together with the Laplacian and gradient magnitude to obtain edge detectors that work better than local methods such as Sobel.

Image Smoothing and Filtering

The basic operation used to smooth images in computer vision (and to filter images in general) is convolution. Consider the following function $g(x, y)$, defined in terms of $f(x, y)$ and $h(x, y)$,

$$g(x, y) = \int_{-\infty}^{\infty} \int_{-\infty}^{\infty} f(x - \xi, y - \eta) h(\xi, \eta) \, d\xi \, d\eta$$

We say that g is the convolution of f and h, which is written as $g = f \otimes h$. This function can be difficult to understand at first, because the value of g at a given point (x, y) depends on the values of f and h at all points. Convolution is commutative and associative, which allows computations to be rearranged in whatever fashion is most convenient (or efficient).

In the discrete approximation to the convolution, the sum of products can be expressed as four nested loops over two arrays that represent the (sampled) functions f and h. Let $h[i, j]$ be an $m \times m$ array and $f[x, y]$ be an $n \times n$ array, where $n > m$ and both arrays are indexed from 0. Then the code fragment in Table 25.1 computes the discrete convolution of the sampled functions f and h. The notation $\lfloor x \rfloor$ denotes the integer portion of x. Note that the iteration variables x and y cannot simply range between 0 and $n - 1$ as this would cause array references outside of f. These boundary cases can be handled in several ways and are important in any implementation.

Convolution can be used to smooth, or low-pass filter, an image in order to handle the problem of high-frequency variation (differences from one pixel to the next). In computer vision, the **Gaussian** is the most commonly used function for low-pass filtering. In one dimension, the Gaussian is given by

$$G_\sigma(x) = \frac{1}{\sqrt{2\pi}\sigma} e^{\frac{-x^2}{2\sigma^2}}$$

This is the canonical bell-shaped or normal distribution as used in statistics. The maximum value is attained at $G_\sigma(0)$, the function is symmetric about 0, and the area of the function is 1. The parameter σ controls the width of the curve: the larger the value of σ, the wider the bell. In two dimensions, the Gaussian can be defined as

$$G_\sigma(x, y) = \frac{1}{2\pi\sigma^2} e^{-\frac{(x^2+y^2)}{2\sigma^2}}$$

In the discrete case the values at integral steps over some range (generally $\pm 4\sigma$) are used as approximations. These values are normalized so that they sum to 1 (just as in the continuous case where the integral is 1).

TABLE 25.1 Four Nested Loops Which Compute the Discrete Convolution of f and h

```
for x ← xmin to xmax
  do for y ← ymin to ymax
    do sum = 0
      for i ← 0 to m − 1
        do for j ← 0 to m − 1
          do sum ← sum + h[i, j] f[x − ⌊m/2⌋ + i , y − ⌊m/2⌋ + j]
      g[x, y] = sum
```

A direct implementation of discrete convolution as in Table 25.1 requires $O(m^2 n^2)$ operations for an $m \times m$ mask representing the Gaussian and an $n \times n$ image. In the case of Gaussians, the operator is *separable* into the product of two one-dimensional Gaussians, and we can use this fact to speed up the convolution to $O(mn^2)$. The underlying idea is to do a one-dimensional convolution in the x-direction followed by a one-dimensional convolution in the y-dimension. The reason that this works is beyond the scope of this chapter. This is a significant savings, both theoretically and in practice, over the direct implementation. Any smoothing method (or edge detector) that uses separable filtering operators should be implemented in this manner. Gaussian smoothing can be made even faster using an approximation which is nearly independent of σ and thus the size of the mask, m. These methods are $O(n^2 + m)$, and use a form of the central limit theorem to approximate a Gaussian by several repeated convolutions with functions of constant height. Such convolutions can be performed in a constant number of operations for each image pixel (see Wells [1986]). In practice for Gaussian smoothing with σ of more than about 1 the repeated convolution approximation method is the fastest. For smaller values of σ the two one-dimensional convolutions are faster.

The Canny Edge Operator

The Canny [1986] edge detector is based on computing the squared gradient magnitude of the Gaussian smoothed image. Local maxima of the gradient are identified using a process known as *non-maximum suppression* (NMS). The NMS operation enables thin, connected chains of edge pixels to be identified. Conceptually, it is much like following the ridge lines in a mountain range, rather than just finding the (isolated) peaks. This is done by defining local maxima with respect to the gradient direction (the direction of steepest change) rather than in all directions. The NMS operation still leaves many local maxima that are not very large. These are then thresholded based on the gradient magnitude (or strength of the edge) to remove the small maxima. The maxima that pass this threshold are classified as edge pixels. Canny uses a thresholding operation with two thresholds, lo and hi. Any local maximum for which the gradient magnitude $m(x, y)$ is larger than hi is kept as an edge pixel. Moreover, any local maximum for which $m(x, y) > $ lo and some neighbor is an edge pixel is also kept as an edge pixel. Note that this is a recursive definition: any pixel that is above the low threshold and adjacent to an *edge* pixel is itself an *edge* pixel.

The steps of the Canny method are as follows:

1. Gaussian smooth the image, $I_s = G_\sigma \otimes I$.
2. Compute the gradient ∇I_s and the squared magnitude $\|\nabla I_s\|^2$ as in Eq. (25.1.)
3. Perform NMS. Let $(\delta \mathbf{x}, \delta \mathbf{y})$ be the unit vector in the gradient direction, ∇I_s. Compare $\|\nabla I_s(x, y)\|^2$ with $\|\nabla I_s(x + \delta \mathbf{x}, y + \delta \mathbf{y})\|^2$ and $\|\nabla I_s(x - \delta \mathbf{x}, y - \delta \mathbf{y})\|^2$ to see if it is a local maximum.
4. Threshold strong local maxima using $\|\nabla I_s\|^2$ as measure of edge strength, with two thresholds on edge strength, lo and hi, as previously described.

The edges in Fig. 25.1(c) are from the Canny edge detector. In practice, this edge operator or variants of it (see Faugeras [1993, Ch. 4]) are the most useful and widely used.

Multiscale Processing

One problem with image filtering is choosing the *scale* of smoothing (e.g., the value of σ to use for the Gaussian). As the scale increases less of the detailed information in an image is preserved (and spatial localization gets worse). It is often desirable to be able to process an image at multiple scales; for instance, in order to determine the significance of edges by finding the range of scales over which they occur. Witkin [1983] developed the idea of multiscale signals resulting from smoothing with a Gaussian at different scales, which he called the **scale space** representation. For example, given an image $I(x, y)$, the corresponding scale-space function is

$$\mathcal{I}(x, y, \sigma) = I(x, y) \otimes G_\sigma(x, y)$$

where the scale parameter σ ranges from 0 to ∞ [at $\sigma = 0$ the scale-space function is the original function $I(x, y)$].

The use of multiscale representations and the characterization of signals from their edges at multiple scales can be viewed in terms of **wavelet** theory (cf. Mallat and Zhong [1992]). A wavelet function is a function whose integral is zero,

$$\int_{-\infty}^{\infty} h(x) = 0$$

and which has a scaling property

$$h_s(x) = \frac{1}{s} h\left(\frac{x}{s}\right)$$

The wavelet transform of a function f at scale s is then defined as $W_s^h f(x) = f(x) \otimes h_s(x)$, where h is a wavelet function (has integral zero and the scaling property). The derivative of a Gaussian has both the zero integral and scaling properties and thus is a wavelet function. Therefore, the multiscale edge representation of an image can be viewed as a wavelet transform. Wavelets have been used for a number of applications in image processing and analysis, such as image compression.

Multiscale representations using oriented filters are also common in computer vision [Perona 1992]. The gradient magnitude of the Gaussian smoothed image is not sensitive to orientation; it responds equally to edges in all directions. It is possible to design filters that are sensitive primarily to edges at a particular orientation, such as vertical or horizontal edges. The visual systems of many animals perform such orientation-sensitive filtering. For example, in many environments large horizontal edges are of considerable interest (especially moving ones which, could be predators).

Visual Motion and Optical Flow

Given a sequence of images taken over time (e.g., a digitized video stream at 30 frames/s), the goal of visual motion analysis is to recover information about the motion of objects in the image. The true motion of points in the world is not directly observable in an image sequence, only the changes in image intensity or the **optical flow**. For example, consider a rotating uniform sphere; in this case there is no optical flow (no change in brightness patterns in the image) but there is motion of the object. On the other hand, consider a shadow moving across a static scene. In this case there is optical flow, but there is no motion of the objects in the scene.

The optical flow is the vector field that tells us where each point in the image at time t "moved to" at time $t + \delta t$. Let $\mathbf{u}(x, y)$ and $\mathbf{v}(x, y)$ denote the components of this vector field; in other words the vector $(\mathbf{u}(x, y), \mathbf{v}(x, y))$ denotes the instantaneous motion of each point (x, y). One class of methods for computing the optical flow are based on measuring local intensity changes in the image with respect to time (see Horn [1986, Ch. 12]). The *optical flow constraint equation*

$$\frac{\partial I}{\partial x}\mathbf{u} + \frac{\partial I}{\partial y}\mathbf{v} + \frac{\partial I}{\partial t} = 0$$

shows how the components of the optical flow, $\mathbf{u}$ and $\mathbf{v}$, can be computed from local derivatives of the image with respect to space and time (x, y, and t). This equation also illustrates a fundamental problem with measuring motion from local intensity change. The equation can be rewritten in vector form as $(\partial I/\partial x, \partial I/\partial y) \cdot (\mathbf{u}, \mathbf{v}) = -\partial I/\partial t$. In other words, only the component of the optical flow in the direction of the brightness gradient $(\partial I/\partial x, \partial I/\partial y)$ is recoverable. This problem is known as the *aperture problem*; intuitively if a line segment is viewed through a small local window it is possible to tell what the motion of the segment is only in the direction normal to the segment.

There are a number of ways to provide additional constraints so that the flow problem can be solved. One approach is to assume that there is a single body moving rigidly [Horn 1986, Ch. 12]. This is powerful but generally overly restrictive. A second approach is to assume that the motion field varies smoothly in most parts of the image, which allows for nonrigid deformations and multiple objects, but has the drawback of performing poorly at motion boundaries because it blurs together multiple motions. A third approach is to use robust statistical techniques to combine local motion estimates [Black and Anandan 1993]. Another approach is to determine the motion of edge contours, rather than using local intensity differences with respect to time (see Horn [1986, Ch. 9]). As long as the contour is not a line segment, it is possible to extract the complete two-dimensional motion (because there are at least two distinct normals to the edge, which thus span the plane).

In practice, the best techniques for computing the optical flow are **area-based** methods, rather than computing the flow from discrete approximations to partial derivatives. Area-based methods operate by considering a small area, or *window*, around each pixel of an image. For each location (x, y) in the image at one time, I_t, a window of the image is matched against a set of windows in a local neighborhood of the next image I_{t+1}. The best matching window of I_{t+1}, using some match criterion, specifies the motion of the point (x, y). For example, if the best match for the window at (x, y) in I_t is the window at (w, z) in I_{t+1}, then the motion is $\mathbf{u}(x, y) = w - x$ and $\mathbf{v}(x, y) = z - y$. Area-based techniques generally use a simple matching measure for comparing windows, such as the **sum-squared difference (SSD)** of the corresponding pixels in the two windows. There is no ideal choice of window size for computing the optical flow. As the window gets smaller the flow estimate becomes noisy because the windows are not distinctive enough. As the window gets larger the motion boundaries become inaccurate because there are multiple motions in a window.

The pseudocode fragment in Table 25.2 illustrates the area-based computation of $\mathbf{u}$ and $\mathbf{v}$ for two images **img1** and **img2,** where the match window is of width m (size $(2m + 1) \times (2m + 1)$) and the search neighborhood is of width n (size $(2n + 1) \times (2n + 1)$). The window of **img1** centered at each (x, y) location is compared against the windows in the search neighborhood of **img2,** to find the best match for each window using the function **ssd.** The function **ssd** computes the sum-squared difference (or L_2 norm) of two images. The estimates of the optical flow resulting from a simple computation like this must generally be processed further in order to be useful.

As with local differential methods, the motions computed with area-based methods are generally processed by smoothing or aggregating the motion over local regions. Preprocessing of the images can also be used to improve the performance of area-based methods, particularly at motion boundaries where there are different motions in the same window. One such preprocessing method is based on transforming

TABLE 25.2 Area-Based Computation of the Optical Flow $(\mathbf{u}(x, y), \mathbf{v}(x, y))$

```
for x ← xmin to xmax
  do for y ← ymin to ymax
    do min ← ∞
      for i ← −n to n
        do for j ← −n to n
          do diff ← SSD(img1, x, y, img2, x + i, y + j, m)
            if diff < min
              then min ← diff
                umin ← i
                vmin ← j
      u(x, y) = umin
      v(x, y) = vmin

SSD(img1, x1, y1, img2, x2, y2, w)
  sum ← 0
  for i ← −w to w
    do for j ← −w to w
      do sum ← sum + (img1(x1 + i, y1 + i) − img2(x2 + i, y2 + i))²
  return sum
```

the images using local nonparametric measures [Zabih and Woodfill 1994]. Two of these transforms are known as the *rank* and *census* transforms. The idea of both transforms is to replace the image intensities with measures based on order statistics in a local neighborhood. These measures are relatively insensitive to overall changes in intensity and thus are more reliable for use in area-based matching. In the rank transform, each pixel is replaced with the rank of its intensity over a local neighborhood. For example, when the neighborhood is size 15×15, if the point at the center of the neighborhood is brighter than any of the others its rank is 1, if it is darker than any of its neighbors its rank is 225, and if it is the median intensity its rank is 113. In the census transform a bit vector is used to encode information about which of the neighboring pixels are brighter or darker than the center pixel in the window.

25.3 Middle-Level Vision

Recall that we consider middle-level vision techniques to be those that take images or the results of low-level vision algorithms as inputs and produce some output other than pixels in the image coordinate system. One of the goals of middle-level vision is to extract three-dimensional geometric information from images. Extracting three-dimensional geometry from images is often referred to as *shape-from-x*, because there are a number of different sources of information that can be used to recover the three-dimensional structure, or shape, of a scene from two-dimensional images. Shading in an image reveals information about three-dimensional shapes (see Horn [1986, Ch. 11]). For instance, much of the way that the shape of a sphere in a photograph is perceived as being a solid rather than a disk is due to the uniform change in brightness away from the light source. Specular reflections can also provide information about the three-dimensional shapes of objects [Blake and Brelstaff 1988]. Another source of three-dimensional shape information is provided by the change in location of an object from one image to another in a set of two or more images. The main techniques for extracting image shape from multiple images are **stereopsis** and **structure from motion**.

Another goal of middle level vision is to extract structural descriptions of images. Active contour models or **snakes** are often used to fit models to data [Amini et al. 1990, Kass et al. 1988]. Relations between image structures can be identified using *perceptual grouping* methods. Grouping methods are generally concerned with recovering nonaccidental alignments of image primitives, such as colinear line segments or cocircular arcs [Lowe 1985]. We will not discuss grouping methods further in this chapter. First we will consider the shape-recovery methods of stereopsis and structure from motion, and then the fitting of contours using snakes.

Stereopsis

In the basic stereo vision paradigm, there are two cameras observing a scene. The central idea is that objects that are closer to the cameras will move more between the two views than will objects that are farther away. This phenomenon is readily observable by alternately blinking the two eyes. For a given point in the scene, the magnitude of its motion between the two views is referred to as the *disparity*. If the camera system is calibrated in world coordinates then the actual distance to a point can be determined from its disparity. The disparity between two images is often displayed as an intensity image with brighter points corresponding to larger intensities (things that are closer to the cameras). Figure 25.2 shows two intensity images and the resulting disparities from an area-based stereo matcher (discussed further subsequently).

A simple *pinhole* camera model consists of a focal point (or optical center), o, through which all rays of light pass, and an image plane I onto which these rays are projected. The *optical axis* of the camera is the line perpendicular to the image plane I and through the focal point o. The *focal length* f is the distance from the optical center o to the image plane I. For stereo vision there are two cameras at fixed relative positions. We will consider a simple stereo camera geometry in which the optical axes of the two cameras are parallel to one another and are perpendicular to the *baseline* that connects the two camera centers (which are denoted by o_l and o_r). We will also assume that the focal length f of the two cameras is the same. Denote the length of the baseline (distance between the camera centers) by b, and place the

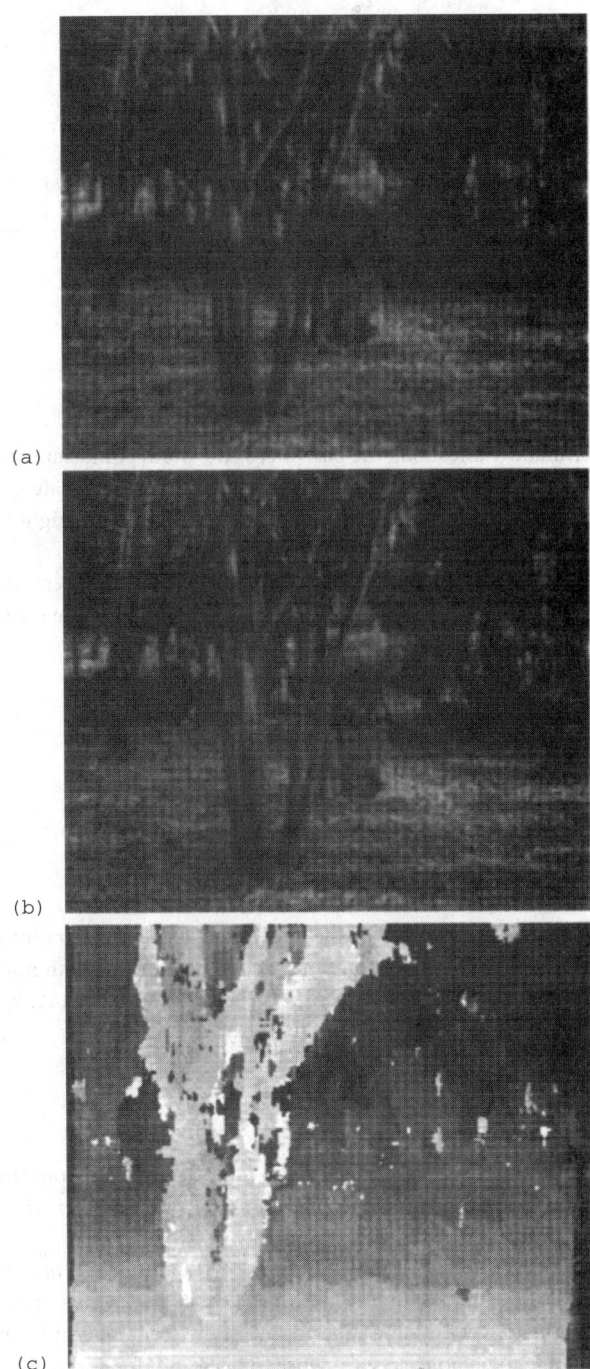

FIGURE 25.2 An example of stereo matching: (a) the left image, (b) the right image, (c) a disparity map with brighter points being larger disparity (closer to the cameras).

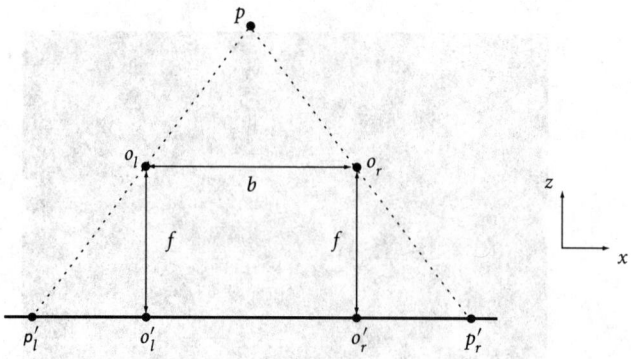

FIGURE 25.3   A simple stereo camera geometry.

origin of the world coordinate frame along the baseline, at the point equidistant between the two camera centers (at distance $b/2$ from each o_l and o_r). Let the origin of the coordinate system for the left image plane L be the projection of its optic axis o_l' (and similarly the origin of the right image plane R is at o_r'). This simple camera model is illustrated in Fig. 25.3.

Consider a point $p = (x, y, z)$ in the world which is projected into L at location $p_l' = (x_l', y_l')$ and into R at location $p_r' = (x_r', y_r')$. From the geometry of the two cameras, it can be seen that

$$\frac{x_l'}{f} = \frac{x + b/2}{z}$$

$$\frac{x_r'}{f} = \frac{x - b/2}{z}$$

$$\frac{y_l'}{f} = \frac{y_r'}{f} = \frac{y}{z}$$

Note that in this simple camera geometry, only the x location of a projected point differs between the left and right images. The y location of a given point in space is the same for both images.

The disparity is defined as the distance between $p_l' = (x_l', y_l')$ and $p_r' = (x_r', y_r')$, which in this case is just $x_l' - x_r'$. From the preceding equations, we see that

$$\frac{x_l' - x_r'}{f} = \frac{b}{z}$$

Thus, if b and f are known, the depth z of the point p can be computed from the disparity. If b and f are unknown then the *relative* depths of points can be computed, but not their absolute distance from the camera.

The basic computation in stereopsis is that of finding corresponding pairs of points, p_l' in L and p_r' in R, which are images of the same point p in the world. Note that a given point p need not have an image in both L and R; there may be some other point in the scene that hides p from view in the left or right image, causing there to be no correspondence. This search for corresponding pairs is simplified by the fact that corresponding pairs of points must lie along certain lines in the two images, known as **epipolar lines**, rather than anywhere in the image. For the simple camera geometry illustrated in Fig. 25.3, the corresponding epipolar lines are those with the same y-coordinates in the two images. Thus, given $p_l' = (x_l', y_l')$, it is necessary only to search for points p_r' in R that have $y_r' = y_l'$. In general, the corresponding epipolar lines are not parallel to one another but the search is still one dimensional. The epipolar lines in general form a pencil of lines in each image plane (a pencil is the set of lines through a given point). The point common to all of the epipolar lines in each image is the intersection of the image with the baseline connecting the two camera centers (for the simple camera geometry this point is at infinity so the epipolar lines are parallel). The correspondence of epipolar lines in the left and right images can be determined through an iterative process that identifies sparse corresponding points in the two images [Deriche et al. 1994].

In practice the recovery of corresponding points in two images L and R is usually accomplished using area-based matching techniques like those described in the section on visual motion. Multiple cameras can be used for more accuracy (which is called multibaseline stereo). Using special hardware, these methods can compute depth maps at video rates. The matching process in stereo is made particularly difficult by the fact that the two images were taken by different cameras, with different gain and bias. However, the matching process is also simplified by the fact that in stereo the search region is one dimensional (along an epipolar line). A useful postprocess in stereo matching is *cross checking* [Hannah 1989], which ensures that if p_r in R is the best match for p_l in L, then the converse is also true. At locations where this is not the case, the match is probably incorrect. There have also been a number of matching techniques developed specifically for stereo, which directly implement constraints such as the fact that the disparity should in general not change quickly, except at object boundaries where there may be depth discontinuities (see Horn [1986, Ch. 6]).

In order to derive surface models from the depth (or disparity) information recovered from stereo, a surface interpolation process is often applied to the data (cf. Grimson [1983] and Terzopoulos [1988]). Such interpolation methods have broad applicability beyond computer vision.

Structure from Motion

Determining the three-dimensional structure, or shape, of an object in the world from a sequence of two-dimensional views is a problem that has been studied extensively (starting with Ullman [1979]). It is generally assumed that some feature points have been extracted and *tracked* through successive frames, so that the correspondence between points in each frame is known. The task is to recover the three-dimensional structure of the feature points in the world and the trajectory of the camera. Much of the early work on structure from motion was concerned with identifying the minimum number of points and frames that are needed to determine the three-dimensional structure of the points. We will not discuss these results here, because in practice it is necessary to have many more than the theoretical minimum number of observations in order to obtain a reliable solution.

To illustrate structure from motion techniques we will consider the case of orthographic projection (where all of the light rays are perpendicular to the image plane rather than going through a single focal point) and all of the points are visible in all of the frames. We will also limit the problem to a static scene and a moving camera, so that the goal is to recover the locations of P points in space and the positions of the camera at each of F time frames. The data provides $2PF$ observations (the x and y coordinates of the image points in each frame). The key issue is to use the redundancy of the observations to cancel out any (unbiased) noise in the locations of the image points and thereby recover the true structure and motion. There are a number of possible techniques for doing this; the approach of Tomasi and Kanade [1992] has been particularly effective.

Let (u_{fp}, v_{fp}) be the image coordinates of the pth point in the fth frame, $1 \leq p \leq P, 1 \leq f \leq F$, where the origin of each frame is the centroid of the P points in that frame. Given the assumption that all of the points are visible in each frame, using the centroid as the origin of each frame omits the need to consider translations of the points. Let $W = [UV]^T$ be a matrix containing the data points, where U and V are the $F \times P$ matrices of the u and v coordinates of each observed point. If there were no error in the observed image data, then the measurement matrix W could be expressed as the matrix product $W = MS$, where $M = [\mathbf{i}_1 \ldots \mathbf{i}_F \mathbf{j}_1 \ldots \mathbf{j}_F]^T$ represents the camera motion [each $(\mathbf{i}_f, \mathbf{j}_f)$ is a pair of orthogonal unit vectors specifying the orientation of the image plane at frame f] and $S = [s_1 \ldots s_P]$ is the shape matrix (the locations of the points in space, with their centroid as the origin). In other words the observed image data, as represented by W, is the product of two parts, the camera motion and the three-dimensional positions of the points.

As M is a $2F \times 3$ matrix and S is a $3 \times P$ matrix, if the measurements were exact $W = MS$ would be of rank at most 3. In reality, the problem is to recover M and S given W despite the fact that there is some error in the observed data. Of course, in the case of even small errors in the locations of the image points, the rank of W will not be 3. However, the singular value decomposition (SVD) can be used to determine the approximate rank of W, which we expect to be 3 (see a numerical methods book such as Press et al. [1988] for a description of the SVD).

The SVD of the measurement matrix is $W = L\Sigma R$ where L is a $2F \times P$ matrix, Σ is a $P \times P$ diagonal matrix of singular values and R is a $P \times P$ matrix. The information of interest corresponds to the three greatest singular values, so the best approximation is

$$\hat{W} = L'\Sigma'R' = \left(L'\Sigma'^{\frac{1}{2}}\right)\left(\Sigma'^{\frac{1}{2}}R'\right) = \hat{M}\hat{S}$$

where L' is the $2F \times 3$ matrix corresponding to the first three columns of L, Σ' is the 3×3 diagonal matrix corresponding to the upper left part of Σ, and R' is the $3 \times P$ matrix corresponding to the first three rows of R. One problem is that this is not a unique factorization as any linear transformation of $\hat{M} = L'\Sigma'^{\frac{1}{2}}$ and $\hat{S} = \Sigma'^{\frac{1}{2}}R'$ yields a valid result. It is possible to solve for the correct motion and shape transformations by noting that the true motion matrix M is composed of unit vectors, and the first F rows are orthogonal to the remaining rows (because each pair of rows $\mathbf{i}$ and $\mathbf{i} + F$ correspond to the two orthogonal unit vectors defining the image plane at frame $\mathbf{i}$). This specifies a unique solution up to a rotational ambiguity, which corresponds to the initial position of the camera with respect to the world. In other words, the overall orientation of the points in the world can be recovered only relative to the initial orientation of the camera.

A number of structures from motion methods have considered the problem of reconstructing the shape of objects up to certain transformations of space (such as affine or projective transformations). A nice book on the subject of solid shape is Koenderink [1990].

Snakes: Active Contour Models

Active contours, or *snakes*, are useful for applications that involve fitting models to image data. An active contour model is a curve that seeks to minimize both internal and external forces which control its shape. The internal forces are generally related to the smoothness of the curve, as reflected by measures such as first and second derivatives of the contour. The external forces are generally based on image measurements, but may also be due to user inputs in interactive curve-fitting applications. The image forces are often based on the gradient magnitude, so that the contour is attracted to edges. In most applications a contour is initially placed near some image structure, and then the constraint forces act on the snake to make it fit the image structure. An iterative update procedure is used to find the position of the snake which (locally) minimizes the forces.

The active contour model was defined in Kass et al. [1988] as a curve $\mathbf{v}(s) = (x(s), y(s))$ with an associated energy functional,

$$\int_0^1 E_{\text{int}}(s) + E_{\text{ext}}(s) \, ds \qquad (25.2)$$

The internal energy is composed of first- and second-order terms that measure the smoothness of the curve, and which make the contour act like a membrane or a thin plate,

$$E_{\text{int}}(s) = (\alpha(s)|\mathbf{v}_s(s)|^2 + \beta(s)|\mathbf{v}_{ss}(s)|^2)/2$$

where the subscripts denote the first and second derivatives of the curve with respect to arclength, and the functions α and β are weights (in practice these functions are often constant). The external energy is related to some image measure such as the gradient magnitude $E_{\text{ext}}(s) = -|\nabla I(x, y)|^2$. The curve that minimizes Eq. (25.2) can be found using variational methods, by deriving a pair of Euler equations in x and y, discretizing them and then solving the discrete equations iteratively until they converge. This technique was developed in Kass et al. [1988]. Related methods have been developed for fitting three-dimensional energy-minimizing models to two- or three-dimensional data.

There are several difficulties with the variational approach to minimizing the energy of an active contour model. First, there is no constraint for the distance between points on the contour, so that many points of the contour can cluster near or on top of one another. More generally it is not possible to specify hard constraints, such as a minimum distance between points on the contour, because the energy functional must be differentiable. It is also difficult to choose the parameters of the minimization, and to determine to what degree the external energy term must be smoothed in order to produce a stable solution.

A different approach to the energy minimization is taken in Amini et al. [1990] where a discrete dynamic programming method is used. One of the main advantages of this method is that it allows the incorporation of hard constraints such as a minimum distance between points on the contour. One of the main disadvantages is that it is fairly computationally intensive. In this approach, the internal energy terms are discretized, $E_{int}(i) = (\alpha_i |\mathbf{v}_i - \mathbf{v}_{i-1}|^2 + \beta_i |\mathbf{v}_{i+1} - 2\mathbf{v}_i + \mathbf{v}_{i-1}|^2)/2$ and $E_{ext}(i) = -|\nabla I(x_i, y_i)|$. The energy over all of the contour points $(\mathbf{v}_0, \ldots, \mathbf{v}_{n-1})$, $\mathbf{v}_i = (x_i, y_i)$ is then

$$\sum_{i=0}^{n-1} E_{int}(i) + E_{ext}(i)$$

The minimization of this sum can be performed using $O(n)$ separate stages, where each stage considers only a local neighborhood around each of three successive contour points, because only the variables indexed by i, $i - 1$, and $i + 1$ must be considered simultaneously. This is used to develop a dynamic programming solution that runs in time $O(m^3 n)$ where m is the size of the local neighborhood around each point. In practice, the dynamic programming methods are easier to implement and more numerically stable than the variational ones.

25.4 High-Level Vision

High-level vision methods are those that make abstract decisions or categorizations based on visual data (generally using the outputs of low- or middle-level vision algorithms). For instance, *object recognition* systems can be used to determine whether or not particular objects are present in a scene, as well as to recover the locations of objects with respect to the camera. *Object tracking* systems can be used to follow a moving object in a video sequence and thus to guide a mobile robot or an autonomous vehicle (cf. Dickmanns and Graefe [1988] and Thorpe et al. [1988]). In this section we will discuss some approaches to object recognition.

Object Recognition

Object recognition systems generally operate by comparing an unknown image to stored object models in order to determine whether any of the models are present in the image. Many systems perform both recognition and localization, identifying an object and recovering its location in the image or in the world. The location of an object is often referred to as its **pose**, and it is generally specified by a transformation mapping the model coordinate system to the image or world coordinate system. One way of categorizing object recognition systems is in terms of the kind of problems they solve. The simplest recognition problems involve identifying two-dimensional objects which are completely unoccluded (i.e., none of the object is hidden from view), appear against a uniform background, and where the lighting conditions are controlled (e.g., there are no shadows or reflections). Many industrial inspection problems fall into this category and can be handled quite accurately with commercially available systems. Recognition problems become more difficult when there are many objects in a scene, when objects may be touching and occluding one another, when the background is highly textured, and when the lighting conditions are unknown. Recognition problems also become more difficult as the number of object models increases. Finally, recognizing three-dimensional objects in a two-dimensional image is more difficult than recognizing two-dimensional objects.

We will primarily consider approaches that can handle images with multiple objects, partly occluded objects, and some amount of background clutter. Most methods that address these kinds of recognition problems are based on extracting geometric information such as intensity edges from an image. This two-dimensional image geometry can then be compared with three-dimensional geometric models. For instance, Kriegman and Ponce [1990] use silhouettes for recognizing objects from intensity edges. A different approach is to compute invariant representations of the image geometry, which remain unchanged under changes in viewpoint [Mundy and Zisserman 1992]. These representations can be used to index into a library of object models. One of the central issues in geometric recognition systems is that of determining which portions of an image correspond to a given object. The recognition problem is often framed as that of recovering a correspondence between local features of an image and an object model. Three major classes of methods can be identified based on how they search for possible matches between model and image features: (1) *correspondence methods* consider the space of possible corresponding features, (2) *transformation space methods* consider the space of possible transformations mapping the model to the image, and (3) *hypothesize and test* methods consider k-tuples of model and data features. A more detailed treatment of geometric search methods can be found in Grimson [1990].

Correspondence Search: Interpretation Tree

Given a set of image features $S = \{s_1, \ldots, s_n\}$ and a set of model features $M = \{m_1, \ldots, m_r\}$, an *interpretation* of the image is a set of pairs $N = \{(m_i, s_j), \ldots\}$ specifying which model features correspond to which image features. If model features may be occluded and there may be extraneous image features, then in principle the set N can be any subset of the set of all pairs of model and image features. The **interpretation tree** approach (see Grimson [1990, Ch. 3]) is a pruned search of this exponential-sized space of possible interpretations (pairings of model and image features). The main idea is to use pairwise relations between features to prune a tree of possible model and image feature pairs, where paths in the tree correspond to interpretations. For concreteness, we consider the case where the features are points in the plane, and the transformation mapping a model to an image is a rigid motion (translation and rotation) plus some allowable error tolerance. As distances are preserved under rigid motion, the distance between pairs of points can be used as a constraint for pruning the search. In order for an interpretation to contain the pairs (m_i, s_j) and (m_p, s_q), the distances $\|m_p - m_i\|$ and $\|s_q - s_j\|$ must be equal (up to the allowable error tolerance).

The search for interpretations is structured as a pruned tree search, in the following manner. Each level of the tree (other than the root) corresponds to a given image feature. Each branch at a given node corresponds to a given model feature or to a special branch called the *null face*. Thus each node of the tree specifies a pair of an image feature (the one at that level of the tree) and a model feature or the null feature (the branch that was taken from the previous level). The search is depth first from the root, and at each node the null face branch is expanded last in the search. A given node is expanded only if it is pairwise consistent with all of the nodes along the path from the current node back to the root of the tree. That is, a given node is paired with each node along the path back to the root, and for each such pair the distance constraint is checked. Only when all of these pairs satisfy the constraint is the node expanded. (The null face branch is always consistent.)

A path from the root to a leaf node of the tree is an interpretation that accounts for zero or more model features (any of the branches may be null branches that do not account for a model feature). A path that accounts for k model features is called a k-interpretation. A threshold k_0 is used to filter out any hypotheses that do not account for enough model features (i.e., the matcher should report only those interpretations for which $k > k_0$). Note that a k-interpretation is guaranteed only to be *pairwise* consistent. Thus an additional step of model verification is performed, which estimates the best transformation for each k-interpretation and checks that this transformation brings each model feature within some error range ϵ of each corresponding image feature.

The HYPER and LFF systems (see Grimson [1990, Ch. 7]) similarly structure recognition as a search for consistent sets of model and image features, using pairwise constraints to prune the search. HYPER starts by matching a privileged model feature against compatible image features. A privileged feature is one that is believed a priori to be more reliable (e.g., for line segment features, longer segments might be considered more reliable). LFF searches for maximal cliques in a graph structure formed from pairwise consistent features.

Transformation Space Search

Transformation space (or pose space) methods are based on searching the space of possible transformations mapping a model to the image or world coordinate system. The idea underlying these methods is to accumulate independent pieces of evidence for a match. Pairs of model and image features, which are part of the same correct match, will specify approximately the same transformation, whereas random pairs of model and image features will tend to result in randomly distributed transformations. Therefore, pairs of features which result in a cluster of similar transformations are assumed to correspond to a match of the model to the data. The validity of this assumption, however, depends on there being a low likelihood that random clusters will be as large as those clusters resulting from correct matches (see Grimson [1990, Ch. 11]). Transformation space search methods compute the transformations that are consistent with each pair of model and image features (or in general each k-tuple of feature pairs). Then the space of possible transformations is searched to find clusters of similar transformations. The exact means of searching the space and identifying clusters depends on the particular transformation space search method.

The generalized **Hough transform** (see Grimson [1990, Ch. 11]) is a transformation search method that operates by voting for buckets in a discrete transformation space. For example, a rigid motion of the plane can be represented using three parameters x, y, and θ corresponding to two translations and a rotation. Each of these parameters is broken into discrete ranges, forming a three-dimensional array of buckets, which tile the space of transformations. Every pair of model and image features then votes for those buckets containing transformations that map the given model feature to the given image feature, up to the allowable sensing error. A bucket that gets many votes corresponds to a possible transformation of the model to the image. There are a number of practical issues with the generalized Hough transform, such as what parameterization of the transformations to use, how to break the space up into a reasonable number of buckets (hierarchical schemes are often used rather than forming an explicit array of buckets), what kind of weighting scheme to use when a given feature pair votes for multiple buckets, and what to do about clusters that occur near bucket boundaries (because then votes may be spread over neighboring buckets instead of all occurring in the same bucket, possibly causing matches to be missed).

Another class of transformation search methods is based on precisely characterizing the regions of transformation space that are specified by each k-tuple of model and image feature pairs. The arrangement of these regions is then searched to find those cells where a large number of regions overlap. For example, consider the case of matching two point sets under translation, where a positional uncertainty of ϵ is allowed for each point. For each pair of model and image points there is a circle of translations of radius ϵ, which places the model point within ϵ of the image point. Translations at which many of these circles overlap are good potential matches of the model and image. Recognition methods that search the arrangement of these transformation space regions (e.g., Cass [1992]) make heavy use of techniques from computational geometry.

k-Tuple Search: Alignment and Linear Combinations

In the absence of sensor uncertainty, k pairs of model and image features exactly determine the transformation mapping a model to an image (where k depends on the kind of feature and the kind of transformation). For example, under translation a single pair of model and image points specifies the transformation. The idea underlying k-tuple search methods is to use the transformations specified by each k-tuple as an hypothesis about the pose of a model in an image (see Grimson [1990, Ch. 7]). If the transformation specified by k pairs of model and image features corresponds to a correct hypothesis, then it will map the other model features onto image features. In this case we say that the transformation *aligns* the model with the image. If the transformation is incorrect then other model features will in general not be mapped onto image features. When there is sensor uncertainty, it is necessary to account for the fact that the transformations computed from k-tuples will not in general bring other model features precisely into correspondence with image features.

Many applications of k-tuple search have considered the case of affine transformations of the plane. An affine transformation of the plane can be represented as $A(x) = Lx + b$, where L is a nonsingular 2×2 matrix, and b is a two-dimensional translation. Three pairs of corresponding points uniquely define such a transformation (which maps any triangle to any other triangle). Thus under an affine transformation each triple of model and image features defines a possible alignment of a model with an image. An affine transformation mapping three model points to three image points also constrains the three-dimensional location of an object. Under an orthographic projection camera model (where all of the light rays are parallel rather than going through an optical center), the three-dimensional position and orientation can be recovered up to a reflective ambiguity [Huttenlocher and Ullman 1990].

The most basic k-tuple search method is called the *alignment* technique, because it simply considers k-tuples of model and image feature pairs, checking the resulting transformations to find those that align a large number of model features with image features. In order to find all possible matches of a model to an image each ordered k-tuple of model and image features must be considered, although in some methods the search may terminate after one or more matches are found. For affine transformations of the plane, each ordered triple of model points and ordered triple of image points defines a basis set which specifies a possible transformation (or two transformations for a three-dimensional object). For each such basis set the transformation mapping the model into the image is computed, and the transformation is evaluated by using it to map the remaining model features into the image. The quality of a transformation is measured by the number of transformed model features for which there are nearby image features. The size of the region to search for each transformed model feature depends on the degree of error in sensing the image points, the spatial configuration of the basis triples, and the location of the given point with respect to the basis points.

An interesting extension of alignment techniques is the linear combinations method [Ullman and Basri 1991], which is based on the idea of forming two-dimensional images of a three-dimensional model as combinations of two-dimensional views of the object. That is, an object is modeled as a set of two-dimensional views, with known correspondence between the points in different views. A new view is recognized by determining whether it is a linear combination of a small number of stored two-dimensional views. This method assumes an orthographic projection camera model and is developed both for point sets and for objects with smooth bounding contours.

Invariants, Indexing, and Geometric Hashing

The central idea underlying the use of **geometric invariants** in model-based recognition is to develop representations of objects that remain unchanged as the viewpoint changes. Such invariants can be found in classical geometry and can also be derived using algebraic techniques (cf. Mundy and Zisserman [1992]). The invariant representations of objects can be used as keys to index into a table of stored objects. In principle it is then possible to look up an object in a large library of stored models in time that is essentially independent of the number of models. One of the key issues that separates such model indexing from hashing in general is the fact that different instances of the same object in various images will not generate exactly the same key or invariant signature, due to sensing uncertainty. Thus in practice exact numerical values cannot be used to hash into a data structure.

A number of methods have been developed for using invariants in recognition, based on geometric, differential, photometric, and even thermal properties of images (see Mundy and Zisserman [1992]). Structural indexing methods use combinations of simple features such as points and segments. Other geometric techniques use invariant properties of curves such as coplanar conics. The main advantage of using curve features is lower combinatorial complexity, and the main disadvantage is the need to extract these features from noisy, cluttered images. Photometric (intensity) information provides a richer description of an image than do geometric features such as points and line segments. The use of photometric invariants in recognition has been investigated in Nayar and Bolle [1993]. In order to illustrate the use of invariants in recognition we consider the **geometric hashing** approach [Lamdan and Wolfson 1988]. As in the previous section we examine the case of two-dimensional affine transformations. The fundamental observation underlying affine-invariant geometric hashing is the fact that three points define a coordinate system or basis with respect to which other points can be encoded in an invariant manner.

The geometric hashing method consists of two basic stages: (1) the construction of a model hash table and (2) the matching of the models to an image. The hash table is used to store a redundant, transformation-invariant representation of each object. Each model is entered into a hash table prior to recognition. For a given model, each ordered triple of model points m_1, m_2, m_3 forms an affine basis with origin $o = m_1$ and axes $u = m_2 - m_1, v = m_3 - m_1$. For each such basis every additional model point m_i is rewritten as (α_i, β_i) such that $m_i - o = \alpha_i u + \beta_i v$. The basis triple (o, u, v) and the point m_i are then stored in the hash table using the affine invariant indices (α_i, β_i). This results in a table with $O(r^4)$ entries for r model points. The table is generally formed using buckets rather than using a hashing scheme, as in practice the sensing uncertainty in real data makes it impossible to use exact values for retrieval. The issue of how to determine appropriate bucket sizes is somewhat complicated.

At recognition time, the hash table is used to determine which models are present in the image. The idea is that when the image points are rewritten in terms of an *image basis* that corresponds to an instance of the model, then the same *model basis* will be retrieved from the table many times. Each ordered triple of image points s_1, s_2, s_3 is used to form a basis, with origin $O = s_1$ and axes $U = s_2 - s_1, V = s_3 - s_1$. For a given image basis, each additional image point s_i is rewritten as (α_i', β_i') such that $s_i - O = \alpha_i' U + \beta_i' V$. The indices (α_i', β_i') are used to retrieve the corresponding entries from the hash table. For each model basis retrieved from the table, a corresponding counter is incremented in a histogram. Once all of the image points have been considered for a given image basis, the histogram contains votes for those model bases that could correspond to the current image basis, (O, U, V). If the peak in the histogram for a given model basis, (o, u, v), is sufficiently high, then this basis is selected as a possible match. When a new image basis is chosen the histogram counts are cleared.

This basic method often does not work well in practice due to the effects of sensing uncertainty on the locations of image features. Several weighted schemes have been developed which work well in practice. These methods enter each basis triple into multiple buckets in the table based on the sensing uncertainty [Rigoutsos and Hummel 1992].

Dense Feature Matching: Hausdorff Distances

The methods considered so far are primarily useful when there are relatively small numbers of model and image features, because they are based on considering subsets of features. A different approach is taken in Huttenlocher et al. [1993], which is based on computing distances between point sets rather than finding correspondences of points in two sets. These methods can be used for large sets of points, such as entire edge maps. The methods are similar to the template matching techniques used in some commercial recognition systems, but they use a new measure of image similarity based on the **Hausdorff** distance, and provide efficient algorithms for searching cluttered images.

Given two point sets $\mathcal{P}$ and $\mathcal{Q}$, with m and n points, respectively, and a fraction, $0 \leq f \leq 1$, the generalized Hausdorff measure is defined in Huttenlocher et al. [1993] and Rucklidge [1995] as

$$h_f(\mathcal{P}, \mathcal{Q}) = f_{p \in \mathcal{P}}^{\text{th}} \min_{q \in \mathcal{Q}} \|p - q\|$$

where $f_{p \in \mathcal{P}}^{\text{th}} g(p)$ denotes the fth quantile of $g(p)$ over the set $\mathcal{P}$. For example, the 1st quantile is the maximum (the largest element), and the $\frac{1}{2}$th quantile is the median. This generalizes the classical Hausdorff distance, which *maximizes* over $p \in \mathcal{P}$. Hausdorff-based measures are asymmetric; for example, $h_f(\mathcal{P}, \mathcal{Q})$ and $h_f(\mathcal{Q}, \mathcal{P})$ can attain very different values as there may be points of $\mathcal{P}$ that are not near any points of $\mathcal{Q}$, or vice versa. This asymmetry is useful in recognition problems, where a hypothesize-and-test paradigm is often employed.

The generalized Hausdorff measure has been used for a number of matching and recognition problems. There are two complementary ways in which the measure has been employed. The first approach is to specify a fixed fraction f, and then determine the distance $d = h_f(\mathcal{P}, \mathcal{Q})$. In other words, find the smallest distance d, such that $k = \lceil fm \rceil$ of the points of $\mathcal{P}$ are within d of points of $\mathcal{Q}$. This has been termed finding the distance for a given fraction. Intuitively, it measures how well the best subset of size $k = \lceil fm \rceil$ of $\mathcal{P}$ matches $\mathcal{Q}$, with smaller distances being better matches. The second approach is to specify a fixed distance d, and then determine the resulting fraction of points that are within that distance. In other words, find the largest f such that $h_f(\mathcal{P}, \mathcal{Q}) \leq d$. Intuitively, this measures what portion of $\mathcal{P}$ is near $\mathcal{Q}$ for some fixed neighborhood size d. This has been termed "finding the fraction for a given distance." It measures how well two sets match, with larger fractions being better matches.

Most applications of the measure are based on the second of the approaches, computing the *Hausdorff fraction*, because in most visual matching problems there is a reasonable prior estimate of the uncertainty in the positional location of image features. For example, a positional error of one pixel is generally introduced by the digitization process. If the feature points are edge features, then there is an uncertainty based on the degree of smoothing of the image. Efficient methods for finding the transformations of one point set such that the Hausdorff fraction is above some threshold (and the distance below some threshold) have been developed for affine transformations of the plane [Huttenlocher et al. 1993, Rucklidge 1995]. When the transformations are restricted to translations the fastest methods use dilation and correlation, whereas for full affine transformations the fastest methods use a hierarchical decomposition of the parameter space. The initial methods for computing Hausdorff distances were combinatorial algorithms using techniques from computational geometry, but current practice does not use these combinatorial techniques.

Appearance-Based Matching: Subspace Methods

Appearance-based recognition methods using **subspace techniques** have proven successful in a number of visual matching and recognition systems (e.g., Murase and Nayar [1995]). Appearance-based methods differ from those that we have already considered in that they operate by directly comparing images or image regions rather than by matching sets of local geometric features. The main advantage of appearance-based methods is that they are useful for tasks in which there is a large database of objects to be searched. The main disadvantage is that, in general, they do not work well with occlusion or with complex scenes and cluttered backgrounds. The most effective applications have been to problems such as the recognition of faces from mug shots, where the faces are generally about the same size and location in each image, and the background is a fixed color [Pentland et al. 1994].

The central idea underlying subspace methods is to represent images in terms of their projection into a relatively low-dimensional space, which captures the important characteristics of the set of objects to be recognized. This low-dimensional space is generally formed using the predominant eigenvectors (or principal components) of a set of known model images, where the k largest eigenvectors serve as the coordinate axes of the space (see Press et al. [1988] for more on eigendecomposition). Each model image is represented in terms of the k coefficients that result from projecting the image into the space. An unknown image is recognized by projecting it into the subspace and finding the closest model(s). The projection of an image can be thought of as a summary, which represents the image using just k numbers.

Subspace methods are attractive for problems in which there is a relatively large database of known objects, because the set of model images can be represented using just k coefficients in each, rather than the thousands of pixels in each image. This both saves storage and speeds the process of finding the closest matching images in the database. Moreover, to the extent that a subspace captures the important characteristics of a given set of images while omitting the unimportant characteristics, it can be used to generalize from a set of models. The main limitation of subspace methods is that when extraneous information from the background of an unknown image is projected into the subspace, it tends to cause incorrect recognition results. This is analogous to the problem that occurs when using the SSD to compare two image windows in motion or stereo, where background pixels included in a matching window can significantly alter the value of the SSD and cause incorrect matches.

Let I denote a two-dimensional image with N pixels, and let $\mathbf{x}$ be its representation as a (column) vector in scan line order. Given a set of training or model images, $\{I_m\}$, $1 \leq m \leq M$, define the matrix $X = [\mathbf{x}_1 - c, \ldots, \mathbf{x}_M - c]$, where $\mathbf{x}_m$ denotes the representation of I_m as a vector, and c is the average of the $\mathbf{x}_m$. The average image is subtracted from each $\mathbf{x}_m$ so that the predominant eigenvectors of XX^T will capture the maximal variation of the original set of images. Generally the $\mathbf{x}_m$ are also normalized prior to forming X, such as making $\|\mathbf{x}_m\| = 1$, to prevent the overall brightness of the images from affecting the results.

The eigenvectors of XX^T are an orthogonal basis in terms of which the $\mathbf{x}_m$ can be rewritten (and other, unknown, images as well). Let λ_i, $1 \leq i \leq N$, denote the ordered (from largest to smallest) eigenvalues of XX^T and let $\mathbf{e}_i$ denote each corresponding eigenvector. Define E to be the matrix of eigenvectors $[\mathbf{e}_1, \ldots, \mathbf{e}_N]$. Then $g_m = E^T(\mathbf{x}_m - c)$ is the rewriting of $\mathbf{x}_m - c$ in terms of the orthogonal basis defined by the eigenvectors of XX^T. It is straightforward to show that $\|\mathbf{x}_m - \mathbf{x}_n\|^2 = \|g_m - g_n\|^2$ [Murase and Nayar 1995], because distances are preserved under an orthonormal change of basis. That is, the SSD can be computed using the squared distance between the eigenspace representations of the two images.

The central idea underlying the use of subspace methods is to *approximate* the $\mathbf{x}_m$, and thus the SSD, using just those eigenvectors corresponding to the few largest eigenvalues. That is, $\mathbf{x}_m \approx \sum_{i=1}^{k} g_{m_i} \mathbf{e}_i + c$, where $k \ll N$. This low-dimensional representation is intended to capture the important characteristics of the set of training images. As this representation uses just the k predominant eigenvectors, it is not necessary to compute all N eigenvalues and eigenvectors of XX^T (which would be quite impractical as N is usually many thousands). Each model I_m is just represented by the k coefficients $(g_{m_1}, \ldots, g_{m_k})$, so comparing it with an unknown image requires only k rather than N comparisons. Generally k is considerably smaller than the number of models M, which is in turn much smaller than the number of pixels N.

Defining Terms

Appearance-based recognition: Recognizing objects based on views, generally using properties such as surface reflectance patterns; often used in contrast with model-based recognition.

Area-based matching: A means of identifying corresponding points in two images, for motion or stereo, by comparing small areas or windows of the two images.

Canny operator: An edge detector based on finding local maxima (peaks and ridges) of the gradient magnitude.

Correlation: The product of one function and shifted versions of another function; the maximum of the correlation can be used to find the best relative shift of two functions.

Digital image: Sampling of an image into discrete units in both space and intensity (or color) for processing with a digital computer; an array of pixels.

Digital video: A sequence of digital images representing a sampled video signal.

Edge detection: Locating significant changes in an intensity (gray-level) image, generally using local differential image properties.

Epipolar geometry: The constraint that the locations of points in one image must lie along a particular line in order to correspond to the same scene point as a given point in another image.

Gaussian smoothing: Convolution of a signal (an image) with a Gaussian function in order to remove high spatial frequency changes from the image (a type of low pass filtering).

Geometric hashing: A model-based recognition technique using a highly redundant representation that is invariant to certain geometric transformations.

Geometric invariants: Properties of a geometric model (e.g., a set of points) that remain unchanged under specified types of geometric transformations (e.g., distances between points under rigid motion).

Gradient magnitude: The magnitude of the gradient vector, or equivalently the square root of the sums of the squares of the local directional derivatives.

Hausdorff matching: A geometric technique for comparing binary images based on computing a variant of the classical Hausdorff distance from point set topology.

Hough transform: A technique used in model-based recognition to accumulate independent pieces of evidence for a match, using local features to vote for possible transformations mapping a model to an image.

Interpretation tree: A model-based recognition technique that uses a pruned exponential tree search to find corresponding sets of model and image features.

Model-based recognition: Approaches to recognizing objects in images that are based on comparing sets of stored model features against features extracted from an unknown image (generally geometric features).

Motion analysis: The recovery of information about how objects are moving in the image or in the world, or about the shape of objects, based on changes in brightness patterns over time.

Optical flow: The local change in image brightness as a function of time.

Pixels: The discrete spatial units of a digital image, generally obtained by conversion of an analog image signal from a camera or scanner; each pixel can generally take on a range of integer values (e.g., 0–255).

Pose recovery: Determining the position and orientation of an object in the world with respect to the camera coordinate system.

Scale space: Representation of a signal (an image) at multiple scales of smoothing.

Snakes: Energy-minimizing contours that combine internal constraints on their shape, such as smoothness of the contour, and external constraints from the image, such as brightness or gradient magnitude.

Stereopsis: The recovery of depth (or relative depth) by finding corresponding points in two or more images of the same scene.

Structure from motion: The recovery of the three-dimensional structure of an object based on a sequence of views as the camera moves with respect to the object.

Subspace methods: Reducing the dimensionality of image matching or recognition problems by representing the images in terms of their projection into a lower-dimensional space.

Sum-squared difference (SSD): A measure used to find the best relative shift of two images (or functions) based on the squared L_2 distance; similar to correlation, but based on minimizing a distance rather than maximizing a product.

Wavelets: Functions with a scaling and zero-sum property that are used to form a multiscale representation of an image (or function).

References

Amini, A. A., Weymouth, T. E., and Jain, R. C. 1990. Using dynamic programming for solving variational problems in vision. *IEEE Trans. Pattern Anal. Machine Intelligence* 12(9):855–867.

Black, M. J. and Anandan, P. 1993. A framework for the robust estimation of optical flow. In *Proc. Int. Conf. Comput. Vision*, pp. 231–236.

Blake, A. and Brelstaff, G. 1988. Geometry from specularities. In *Proc. Int. Conf. Comput. Vision*, pp. 394–403.

Canny, J. 1986. A computational approach to edge detection. *IEEE Trans. Pattern Anal. Machine Intelligence* 8(6):679–697.

Cass, T. A. 1992. Polynomial-time object recognition in the presence of clutter, occlusion, and uncertainty. In *Proc. Eur. Conf. Comput. Vision*, pp. 834–842. 1992.

Deriche, R., Zhang, Z., Luong, Q. T., and Faugeras, O. 1994. Robust recovery of the epipolar geometry for an uncalibrated stereo rig. *Proc. Eur. Conf. Comput. Vision* A:567–576.

Dickmanns, E. and Graefe, V. 1988. Dynamic monocular machine vision. *Machine Vision Appl.* 1:223–240.

Faugeras, O. 1993. *Three Dimensional Computer Vision, A Geometric Viewpoint.* MIT Press, Cambridge, MA.

Grimson, W. E. L. 1983. An implementation of a computational theory of visual surface interpolation. *Comput. Vision, Graphics, Image Proc.* 22(1):39–69.

Grimson, W. E. L. 1990. *Object Recognition by Computer: The Role of Geometric Constraints.* MIT Press, Cambridge, MA.

Hannah, M. J. 1989. A system for digital stereo image matching. *Photogrammetric Eng. Remote Sensing* 55:1765–1770.

Haralick, R. M. and Shapiro, L. G. 1992. *Computer and Robot Vision.* Addison–Wesley, Reading, MA.

Horn, B. K. P. 1986. *Robot Vision.* McGraw–Hill, New York.

Huttenlocher, D. P., Klanderman, G. A., and Rucklidge, W. J. 1993. Comparing images using the Hausdorff distance. *IEEE Trans. Pattern Anal. Machine Intelligence* 15(9):850–863.

Huttenlocher, D. P. and Ullman, S. 1990. Recognizing solid objects by alignment with an image. *Int. J. Comput. Vision* 5(2):195–212.

Kass, M., Witkin, A., and Terzopoulos, D. 1988. Snakes: active contour models. *Int. J. Comput. Vision* 1(3):321–331.

Koenderink, J. J. 1990. *Solid Shape.* MIT Press, Cambridge, MA.

Kriegman, D. J. and Ponce, J. 1990. On recognizing and positioning curved 3-d objects from image contours. *IEEE Trans. Pattern Anal. Machine Intelligence* 12:1127–1137.

Lamdan, Y. and Wolfson, H. J. 1988. Geometric hashing: a general and efficient model-based recognition scheme. In *Proc. Int. Conf. Comput. Vision,* pp. 238–249.

Lowe, D. G. 1985. *Perceptual Organization and Visual Recognition.* Kluwer Academic.

Mallat, S. G. and Zhong, S. 1992. Characterization of signals for multiscale edges. *IEEE Trans. Pattern Anal. Machine Intelligence* 14(7):710–732.

Mundy, J. L. and Zisserman, A. 1992. *Geometric Invariants In Computer Vision.* MIT Press, Cambridge, MA.

Murase, H. and Nayar, S. K. 1995. Visual learning and recognition of 3-d objects from appearance. *Int. J. Comput. Vision.* 14:5–24.

Nayar, S. K. and Bolle, R. M. 1993. Reflectance ratio: a photometric invariant for object recognition. In *Proc. Int. Conf. Comput. Vision,* pp. 280–285.

Pentland, A., Moghaddam, B., and Starner, T. 1994. View-based and modular eigenspaces for face recognition. In *Proc. IEEE Conf. Comput. Vision Pattern Recognition,* pp. 84–91.

Perona, P. 1992. Steerable-scalable kernels for edge detection and junction analysis. In *Proc. Eur. Conf. Comput. Vision,* pp. 3–18.

Press, W. H., Flannery, B. P., Teukolsky, S. A., and Vetterling, W. T. 1988. *Numerical Recipes: The Art of Scientific Computing.* Cambridge University Press, Cambridge, England.

Rigoutsos, I. and Hummel, R. 1992. Massively parallel model matching: geometric hashing on the connection machine. *Computer* (Feb.):33–41.

Rucklidge, W. J. 1995. Locating objects using the Hausorff distance. In *Proc. Int. Conf. Comput. Vision,* pp. 457–464.

Terzopoulos, D. 1988. The computation of visible-surface representations. *IEEE Trans. Pattern Anal. Machine Intelligence* 10(4):417–438.

Thorpe, C. E., Hebert, M., Shafer, S. A., and Kanade, T. 1988. Vision and navigation for the Carnegie-Mellon navlab. *IEEE Trans. Pattern Anal. Machine Intelligence* 10(3):361–372.

Tomasi, C. and Kanade, T. 1992. Shape and motion from image streams under orthography: a factorization method. *Int. J. Comput. Vision* 9(2):137–154.

Ullman, S. 1979. *The Interpretation of Visual Motion.* MIT Press, Cambridge, MA.

Ullman, S. and Basri R. 1991. Recognition by linear combinations of models. *IEEE Trans. Pattern Anal. Machine Intelligence* 13(10):992–1006.

Wells, W. M. 1986. Efficient synthesis of Gaussian filters by cascaded uniform filters. *IEEE Trans. Pattern Anal. Machine Intelligence* 8:234–239.

Witkin, A. P. 1983. Scale-space filtering. In *Proc. Int. J. Conf. Artif. Intelligence,* pp. 1019–1022.

Zabih R. and Woodfill, J. 1994. Non-parametric local tranforms for computing visual correspondence. *Proc. Eur. Conf. Comput. Vision* B:151–158.

Further Information

Much of the material in computer vision is in the form of original research articles, a few of which are cited in the References. In-depth coverage of the field can be found in the books by Faugeras [1993] and Haralick and Shapiro [1992], and good coverage of low-level vision is provided by Horn [1986]. The area of model-based object recognition is covered in Grimson [1990], and the geometric invariants approach to recognition is in Mundy and Zisserman [1992].

26

Understanding Spoken Language

26.1 Introduction

Computers are fast becoming a ubiquitous part of our lives, and our appetite for information is ever increasing. As a result, many researchers have sought to develop convenient human–computer interfaces, so that ordinary people can effortlessly access, process, and manipulate vast amounts of information—any time and anywhere—for education, decision making, purchasing, or entertainment. A speech interface, in a user's own language, is ideal because it is the most natural, flexible, efficient, and economical form of human communication.

After many years of research, spoken input to computers is just beginning to pass the threshold of practicality. The last decade has witnessed dramatic improvement in **speech recognition** (SR) technology, to the extent that high-performance algorithms and systems are becoming available. In some cases, the transition from laboratory demonstration to commercial deployment has already begun. Speech input capabilities are emerging that can provide functions such as voice dialing (e.g., "call home"), call routing (e.g., "I would like to make a collect call"), simple data entry (e.g., entering a credit card number), and preparation of structured documents (e.g., a radiology report).

Defining the Problem

Speech recognition is a very challenging problem in its own right, with a well-defined set of applications. However, many tasks that lend themselves to spoken input, making travel arrangements or selecting a movie, as illustrated in Fig. 26.1, are in fact exercises in interactive problem solving. The solution is often built up incrementally, with both the user and the computer playing active roles in the conversation. Therefore, several language-based input and output technologies must be developed and integrated to

0-8493-2909-4/97/$0.00+$.50

reach this goal. Regarding the former, speech recognition must be combined with **natural language** (NL) processing so that the computer can understand spoken commands (often in the context of previous parts of the dialogue). On the output side, some of the information provided by the computer, and any of the computer's requests for clarification, must be converted to natural sentences, perhaps delivered verbally.

This chapter describes the technologies that are utilized by computers to achieve spoken language understanding. Spoken language understanding by machine has been the focus of much research over the past 10 years around the world. In contrast, spoken **language generation** has not

---

Book me a flight on American to Dallas tomorrow morning.

Send me all his X-rays taken over the past six months.

Wo ist das nächste Italienische Restaurant das die American Express Kreditkarte nimmt?

List all mergers and acquisitions in the entertainment and telecommunication industries in the last three months.

Quel temps fera-t-il à Washington?

Transfer $500 from my savings account to my checking account.

Dove sta la biblioteca vicino a Central Square?

What's that movie with Marilyn Monroe in which two men dressed as women?

FIGURE 26.1 An illustration of the types of queries a user is likely to produce for interactive problem solving.

received nearly as much attention, even though it is a critical component of a fully interactive, conversational system. The remainder of this chapter will focus mainly on the input side. Although such an imbalance in treatment may be viewed as being inappropriate, it is, unfortunately, an accurate reflection of the research landscape.

Spoken language communication is an active process that utilizes many different sources of knowledge, some of them deeply embedded in the linguistic competence of the talker and the listener. For example, the two phrase, "lettuce spray" and "let us pray" differ in subtle ways at the acoustic level. In the first phrase, the /s/ phonemes ending the first word and starting the second merge into a longer acoustic segment than the /s/ in "us." Furthermore, the /p/ in "spray" is *unaspirated*, because it is embedded in a consonant cluster with the preceding /s/ in the syllable onset, thus sounding more like a /b/. Other very similar word sequences are more problematic, because in some cases the acoustic realizations can be essentially identical. For the pair "meter on Main Street," and "meet her on Main Street," *syntactic* constraints would suggest that the first one is not a well-formed sentence. For the contrastive pair, "is the baby crying," and "is the bay bee crying," the acoustic differences would be very subtle, and both phrases are *syntactically* legitimate. However, the second one could be ruled out as implausible on the basis of *semantic* constraints. A popular example among speech researchers is the pair, "recognize speech," and "wreck a nice beach." These two are surprisingly similar acoustically, but it is hard to imagine a **discourse** context that would support both. In practice, acoustics, syntax, semantics, and discourse context should all be utilized to contribute to a goodness score for all competing hypotheses.

Higher level linguistic knowledge can play an important role in helping to constrain the permissible word sequences. Thus, for example, the phoneme sequence /w ɛ r I z I t/ is linguistically much more likely to be "where is it" than "wear is it," simply because the first one makes more sense. On the other hand, if used too aggressively, syntactic and semantic constraints can cause a system to fail even though enough of the content words have been recognized correctly to infer a plausible action to take. Problems arise not only because people often violate syntactic rules in conversational speech, but also because recognition errors can lead to pathological syntactic forms. For instance, if the recognizer's top choice hypothesis is the nonsense phrase, "between four in five o'clock," a parser may fail to recognize the intended meaning. Three alternative and contrastive strategies to cope with such problems have been developed. The first would be to analyze "between four" and "five o'clock" as separate units, and then infer the relationship between them after the fact through plausibility constraints. A second approach would be to permit "in" to substitute for "and" at selected places in the grammar rules, based on the assumption that these are confusable pairs acoustically. The final, and intuitively most appealing method, is to tightly integrate the natural language component into the recognizer search, so that "and" is so clearly preferred over "in" in the preceding situation that the latter is never chosen. Although this final approach seems most logical, it turns out that researchers have not yet solved the problem of computational overload that occurs when a **parser** is used to predict the next word hypotheses. A compromise that is currently popular is to allow the recognizer to propose a list of *N* ordered theories, and have the linguistic analysis examine each theory in turn, choosing the one that appears the most plausible.

System Architecture and Research Issues

Figure 26.2 shows the major components of a typical **conversational system**. The spoken input is first processed through the speech recognition component. The natural language component, working in concert with the recognizer, produces a meaning representation. For information retrieval applications illustrated in this figure, the meaning representation can be used to retrieve the appropriate information in the form of text, tables, and graphics. If the information in the utterance is insufficient or ambiguous, the system may choose to query the user for clarification. Natural language generation and **text-to-speech synthesis** can be used to produce spoken responses that may serve to clarify the tabular information. Throughout the process, discourse information is maintained and fed back to the speech recognition and language understanding components, so that sentences can be properly understood in context.

The development of conversational systems offers a set of significant challenges to speech and natural language researchers, and raises several important research issues. First, the system must begin to deal with conversational, extemporaneously produced speech. Spontaneous speech is often extremely difficult to recognize and understand, since it may contain false starts, hesitations, and words and linguistic constructs unknown to the system.

Second, the system must have an effective strategy for coupling speech recognition with language understanding. Speech recognition systems typically implement linguistic constraints as a statistical grammar that specifies the probability of a word given its predecessors. Although these simple language models have been effective in reducing the search space and improving performance, they do not begin to address the issue of speech understanding. On the other hand, most natural language systems are developed with text input in mind; it is usually assumed that the entire word string is known with certainty. This assumption is clearly false for speech input, where many words are competing for the same time span, and some words may be more reliable than others because of varying signal robustness. Researchers in each discipline need to investigate how to exchange and utilize information so as to maximize overall system performance. In some cases, one may have to make fundamental changes in the way systems are designed.

Similarly, the natural language generation and text-to-speech components on the output side of conversational systems should also be closely coupled in order to produce natural-sounding spoken language. For example, current systems typically expect the language generation component to produce a textual surface form of a sentence (throwing away valuable linguistic and prosodic knowledge) and then require the text-to-speech component to produce linguistic analysis anew. Clearly, these two components would benefit from a shared knowledge base. Furthermore, language generation and dialogue modeling should be intimately coupled, especially for applications over the phone and without displays. For example, if there is too much information in the table to be delivered verbally to the user, a clarification subdialogue may be necessary to help the system narrow down the choices before enumerating a subset.

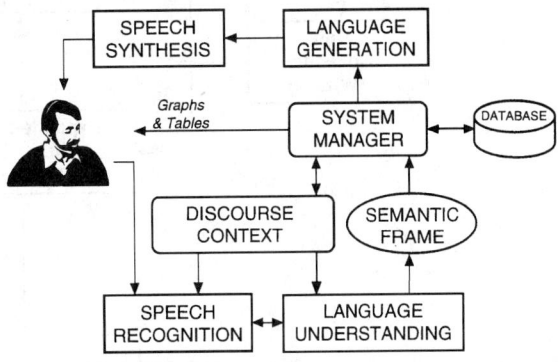

FIGURE 26.2 A generic block diagram for a typical spoken language system.

26.2 Underlying Principles

Procedure for System Development

Figure 26.3 illustrates the typical procedure for system development. For a newly emerging domain or language, an initial system is developed with some limited natural language capabilities, based on the inherent knowledge and intuitions of system developers. Once the system has some primitive capabilities, a **wizard** mode data collection episode can be initiated, in which a human wizard helps the system answer questions posed by naive subjects. The resulting data (both speech and text) are then used for further development and training of both the speech recognizer and the natural language component. As these components begin to mature, it becomes feasible to give the system increasing responsibility in later data collection episodes. Eventually, the system can stand alone without the aid of a wizard, leading to less costly and more efficient data collection possibilities. As the system evolves, its changing behaviors have a profound influence on the subjects' speech, so that at times there is a *moving target* phenomenon. Typically, some of the collected data are set aside for performance evaluation, in order to test how well the system can handle previously unseen material. The remainder of this section provides some background information on speech recognition and language understanding components, as well as the data collection and performance evaluation procedures.

Data Collection

Development of spoken language systems is driven by the availability of representative training data, capturing how potential users of a system would want to talk to it. For this reason, data collection and evaluation have been important areas of research focus. Data collection enables application development and training of the recognizer and language understanding systems; evaluation techniques make it possible to compare different approaches and to measure progress. It is difficult to devise a way to collect realistic data reflecting how a user would use a spoken language system when there is no such system; indeed, the data are needed in order to *build* the system.

Most researchers in the field have now adopted an approach to data collection which uses a system in the loop to facilitate data collection and provide realistic data. At first, some limited natural language

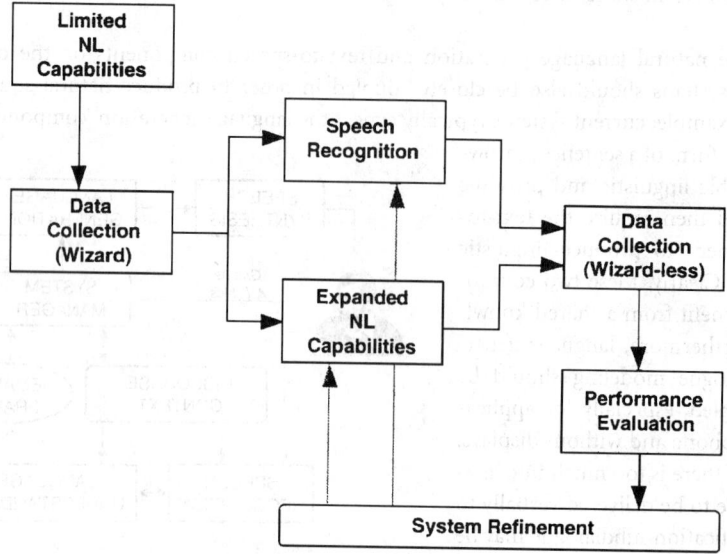

FIGURE 26.3   An illustration of the spoken language system development cycle.

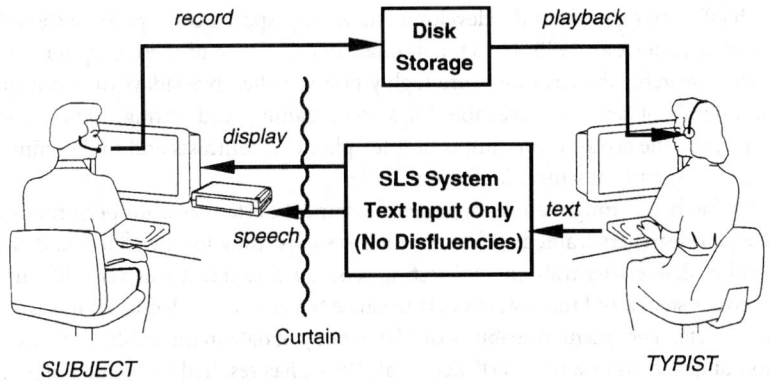

FIGURE 26.4   A person-behind-the-curtain, or wizard, paradigm for data collection.

understanding capabilities are developed for the particular application. In early stages, the data are collected in a simulation mode, where the speech recognition component is replaced by an expert typist. An experimenter in a separate room types in the utterances spoken by the subject, typically after removing false starts and hesitations. The natural language component then translates the typed input into a query to the database, returning a display to the user, perhaps along with a verbal response clarifying what is being shown. In this way, data collection and system development are combined into a single tightly coupled cycle. Since only a transcriber is needed, not an expert wizard, this approach is quite cost effective, allowing data collection to begin quite early in the application development process, and permitting realistic data to be collected. (See Fig. 26.4.) As system development progresses, the simulated portions of the system can be replaced with their real counterparts, ultimately resulting in stand-alone data collection, yielding data that accurately reflect the way system would be used in practice.

Since the subjects brought in for data collection are not true users with clear goals, it is critical to provide a mechanism to help them focus their dialogue with the computer. A popular approach is to devise a set of short scenarios for them to solve. These are necessarily artificial, and the exact wording of the sentences in the scenarios often has a profound influence on the subjects' choices of linguistic constructs. An alternative is to allow the subjects complete freedom to design their own scenarios. This is perhaps somewhat more realistic, but subjects may wander from topic to topic because of a lack of a clearly defined problem.

As the system's dialogue model evolves, data previously collected can become somewhat obsolete, since the users' utterances are markedly influenced by the computer feedback. Hence it is problematic to achieve advances in the dialogue model without suffering from temporary inadequacies in recognition, until the system can bootstrap from new releases of training material.

Speech Recognition

The past decade has witnessed unprecedented progress in speech recognition technology. Word error rates continue to drop by a factor of 2 every two years while barriers to speaker independence, continuous speech, and large vocabularies have all but fallen. There are several factors that have contributed to this rapid progress. First, there is the coming of age of the stochastic modeling techniques known as **hidden Markov modeling (HMM)**. HMM is a doubly stochastic model, in which the generation of the underlying phoneme string and its surface acoustic realizations are *both* represented probabilistically as Markov processes [Rabiner 1986]. HMM is powerful in that, with the availability of training data, the parameters of the model can be trained automatically to give optimal performance. The systems typically operate with the support of an *n*-gram **(statistical)** language model and adopt either a **Viterbi** (time-synchronous) or an **A*** (best fit) search strategy. Although the application of HMM to speech recognition began nearly 20 years ago [Jelinek et al. 1974], it was not until the past few years that it has gained wide acceptance in the research community.

Second, much effort has gone into the development of large speech corpora for system development, training, and testing [Zue et al. 1990, Hirschman et al. 1992]. Some of these corpora are designed for acoustic phonetic research, whereas others are highly task specific. Nowadays, it is not uncommon to have tens of thousands of sentences available for system training and testing. These corpora permit researchers to quantify the acoustic cues important for phonetic contrasts and to determine parameters of the recognizers in a statistically meaningful way.

Third, progress has been brought about by the establishment of standards for performance evaluation. Only a decade ago, researchers trained and tested their systems using locally collected data and had not been very careful in delineating training and testing sets. As a result, it was very difficult to compare performance across systems, and the system's performance typically degraded when it was presented with previously unseen data. The recent availability of a large body of data in the public domain, coupled with the specification of evaluation standards [Pallett et al. 1994], has resulted in uniform documentation of test results, thus contributing to greater reliability in monitoring progress.

Finally, advances in computer technology have also indirectly influenced our progress. The availability of fast computers with inexpensive mass storage capabilities has enabled researchers to run many large-scale experiments in a short amount of time. This means that the elapsed time between an idea and its implementation and evaluation is greatly reduced. In fact, speech recognition systems with reasonable performance can now run in real time using high-end workstations without additional hardware—a feat unimaginable only a few years ago. However, recognition results reported in the literature are usually based on more sophisticated systems that are too computationally intensive to be practical in live interaction. An important research area is to develop more efficient computational methods that can maintain high-recognition accuracy without sacrificing speed.

Historically, speech recognition systems have been developed with the assumption that the speech material is read from prepared text. Spoken language systems offer new challenges to speech recognition technology in that the speech is extemporaneously generated, often containing **disfluencies** (i.e., unfilled and filled pauses such as "umm" and "aah," as well as word fragments) and words outside the system's working vocabulary. Thus far, some attempts have been made to deal with these problems, although this is a research area that deserves greater attention. For example, researchers have improved their system's recognition performance by introducing explicit acoustic models for the filled pauses [Ward 1990, Butzberger et al. 1992]. Similarly, *trash* models have been introduced to detect the presence of unknown words, and procedures have been devised to learn the new words once they have been detected [Asadi et al. 1991].

Most recently, researchers are beginning to seriously address the issue of recognition of telephone quality speech. It is highly likely that the first several spoken language systems to become available to the general public will be accessible via telephone, in many cases replacing presently existing touch-tone menu driven systems. Telephone-quality speech is significantly more difficult to recognize that high-quality recordings, both because the band-width has been limited to under 3.3 kHz and because noise and distortions are introduced in the line. Furthermore, the background environment could include disruptive sounds such as other people talking or babies crying.

Language Understanding

Natural language analysis has traditionally been predominantly syntax driven—a complete syntactic analysis is performed which attempts to account for *all* words in an utterance. However, when working with spoken material, researchers quickly came to realize that such an approach [Bobrow et al. 1990, Seneff 1992b], although providing some linguistic constraints to the speech recognition component and a useful structure for further linguistic analysis, can break down dramatically in the presence of unknown words, novel linguistic constructs, recognition errors, and spontaneous speech events such as false starts. Spoken language tends to be quite informal; people are perfectly capable of speaking, and willing to accept, sentences that are agrammatical.

Due to these problems, many researchers have tended to favor more semantic-driven approaches, at least for spoken language tasks in limited domains. In such approaches, a meaning representation or

semantic frame is derived by *spotting* key words and phrases in the utterance [Ward 1990]. Although this approach loses the constraint provided by syntax, and may not be able to adequately interpret complex linguistic constructs, the need to accommodate spontaneous speech input has outweighed these potential shortcomings. At the present time, almost all viable systems have abandoned their original goal of achieving a complete syntactic analysis of every input sentence, favoring a more robust strategy that can still answer when a full parse is not achieved [Jackson et al. 1991, Seneff 1992a, Stallard and Bobrow 1992]. This can be achieved by identifying parsable phrases and clauses, and providing a separate mechanism for gluing them together to form a complete meaning analysis [Seneff 1992a]. Ideally, the parser includes a probabilistic framework with a smooth transition to parsing fragments when full linguistic analysis is not achievable. Examples of systems that incorporate such *stochastic* modeling techniques can be found in Pieraccini et al. [1992] and Miller et al. [1994].

Speech Recognition/Natural Language Integration

One of the critical research issues in the development of spoken language systems is the mechanism by which the speech recognition component interacts with the natural language component in order to obtain the correct meaning representation. At present, the most popular strategy is the so-called **N-best interface** [Soong and Huang 1990], in which the recognizer can propose its best N complete sentence hypotheses[1] one by one, stopping with the first sentence that is successfully analyzed by the natural language component. In this case, the natural language component acts as a filter on *whole sentence* hypotheses. However, it is still necessary to provide the recognizer with an inexpensive language model that can partially constrain the theories. Usually, a statistical language model such as a bigram is used, in which every word in the lexicon is assigned a probability reflecting its likelihood in following a given word.

In the N-best interface, a natural language component filters hypotheses that span the entire utterance. Frequently, many of the candidate sentences differ minimally in regions where the acoustic information is not very robust. Although confusions such as "an" and "and" are acoustically reasonable, one of them can often be eliminated on linguistic grounds. In fact, many of the top N sentence hypotheses could have been eliminated before reaching the end if syntactic and semantic analyses had taken place early on in the search. One possible control strategy, therefore, is for the speech recognition and natural language components to be tightly coupled, so that only the acoustically promising hypotheses that are linguistically meaningful are advanced. For example, partial theories are arranged on a stack, prioritized by score. The most promising partial theories are extended using the natural language component as a predictor of all possible next-word candidates; any other word hypotheses are not allowed to proceed. Therefore, any theory that completes is guaranteed to parse. Researchers are beginning to find that such a tightly coupled integration strategy can achieve higher performance than an N-best interface, often with a considerably smaller stack size [Goodine et al. 1991, Goodeau 1992, Moore et al. 1995]. The future is likely to see increasing instances of systems making use of linguistic analysis at early stages in the recognition process.

Discourse and Dialogue

Human verbal communication is a two-way process involving multiple, active participants. Mutual understanding is through direct and indirect speech acts, turn taking, clarification, and pragmatic considerations. An effective spoken language interface for information retrieval and interactive transactions must incorporate extensive and complex **dialogue modeling**: initiating appropriate clarification subdialogues based on partial understanding, and taking an active role in directing the conversation toward a valid conclusion. Although there has been some theoretical work on the structure of human–human dialogue [Grosz and Sidner 1990], this has not yet led to effective insights for building human–machine interactive systems.

---

[1] N is a parameter of the system that can be set arbitrarily as a compromise between accuracy and computation.

Systems can maintain an active or a passive role in the dialogue, and each of these extremes has advantages and disadvantages. An extreme case is a system which asks a series of prescribed questions, and requires the user to answer each question in turn before moving on. This is analogous to the interactive voice response systems that are now available via the touch-tone telephone, and users are usually annoyed by their inflexibility. At the opposite extreme is a system that never asks any questions or gives any unsolicited advice. In such cases the user may feel uncertain as to what capabilities exist, and may, as a consequence, wander quite far from the domain of competence of the system, leading to great frustration because nothing is understood. Researchers are still experimenting with setting an important balance between these two extremes in managing the dialogue.

It is absolutely essential that a system be able to interpret a user's queries in context. For instance, if the user says, "I want to go from Boston to Denver," followed with, "show me only United flights," they clearly do not want to see *all* United flights, but rather just the ones that fly from Boston to Denver. The ability to inherit information from preceding sentences is particularly helpful in the face of recognition errors. The user may have asked a complex question involving several restrictions, and the recognizer may have misunderstood a single word, such as a flight number or an arrival time. If a good context model exists, the user can now utter a very short correction phrase, and the system will insert the correction for the misunderstood word correctly, preventing the user from having to reutter the entire sentence, running the risk of further recognition errors.

At this point, it is probably educational to give an example of a real dialogue between a spoken language system and a human. For this purpose, we have selected the Pegasus system, a system developed at Massachusetts Institute of Technology (MIT), which is capable of helping a user make flight plans [Zue et al. 1994]. Pegasus connects, via a modem over the telephone line, to the Eaasy Sabre flight database, offered by American Airlines. As a consequence, users can make explicit flight reservations on real flights using Pegasus. Figures 26.5 and 26.6 contain an example of the log of an actual round-trip booking to illustrate the system's capability. This dialogue shows examples where the system asks directed questions, cases where a great deal of context information is carried over from one query to the next, "please wait" requests where the system is warning the user of possible delays, and instances where the system provides additional information that was not explicitly requested, such as the ticket summary.

Evaluation

Over the past few years, the speech recognition and natural language communities have placed strong emphasis on performance evaluation. With the use of common corpora and metrics, the relative merits of different approaches and systems can be assessed, using both subjective and objective measures.

Spoken language systems can be evaluated along several dimensions. First, the *accuracy* of the system and its various modules can be documented. Thus, for example, a given system's phonetic, word, and sentence accuracy can be measured, as well as linguistic and task completion accuracy. Second, the *coverage* and *habitability* of the system can be evaluated. Coverage can be measured with respect to the lexicon, the language model, and the application back-end. Third, the system's *flexibility* can be established. For example, how easy is it to add new knowledge to the system? How easy is it for a naive user to learn how to communicate with the system? How difficult is it for researchers, and, more especially, for nonexperts, to port the system to a different application or language? Finally, the *efficiency* of the system can be assessed. One such measure may be the task completion time.

Whether one wants to evaluate the accuracy of a spoken language system in part or as a whole, one must first establish what the *reference* should be. For example, determining word accuracy for speech recognizers requires that the reference string of words first be transcribed. Similarly, assessing the appropriateness of a syntactic parse presupposes that we know what the correct parse is. In some cases, establishing the reference is relatively straightforward and can be done almost objectively. In other cases, such as specifying the correct system response, the process can be highly subjective. For example, should the correct answer to the query, "Do you know of any Chinese restaurants?" be simply, "Yes," or a list of the restaurants that the system knows?

USER: I WANT TO GO FROM BOSTON TO SAN FRANCISCO
PEGASUS: What date will you be travelling on?
USER: OCTOBER TWENTIETH
PEGASUS: Here are the flights from Boston to San Francisco on Wednesday October twentieth.

	FLIGHT		FROM	TO	DEPARTURE	ARRIVAL	NSTOPS
1	DL	565	BOS	SFO	7:00 A.M.	11.55 A.M.	1
2	US	31	BOS	SFO	7:00 A.M.	12:03 P.M.	1
3	NW	181	BOS	SFO	7:30 A.M.	10:58 A.M.	0
4	UA	31	BOS	SFO	8:10 A.M.	11:23 A.M.	0
5	NW	353	BOS	SFO	8:15 A.M.	1:18 P.M.	1
6	UA	59	BOS	SFO	9:05 A.M.	12:34 P.M.	0

. . .

You can ask for more flights.
USER: I'D LIKE A NONSTOP UNITED FLIGHT
PEGASUS: Here are the non-stop United flights from Boston to San Francisco on Wednesday October twentieth.

	FLIGHT		FROM	TO	DEPARTURE	ARRIVAL	NSTOPS
1	UA	31	BOS	SFO	8:10 A.M.	11:23 A.M.	0
2	UA	59	BOS	SFO	9:05 A.M.	12:34 P.M.	0
3	UA	93	BOS	SFO	5:10 P.M.	8:23 P.M.	0

. . .

USER: BOOK THE THIRD FLIGHT PLEASE
PEGASUS: Please wait while I determine what fares are available for this flight. The following fares are available. Which fare class would you like?

	ONE WAY	FARE CLASS	MIN-STAY	MAX-STAY	RETURN REQ'D
1	$ 214.	THRIFT DISCOUNT	YES	30	YES
2	$ 234.	COACH DISCOUNT	YES	30	YES
3	$ 294.	THRIFT DISCOUNT	YES	NO	YES
4	$ 304.	COACH DISCOUNT	YES	30	YES

. . .

USER: COACH DISCOUNT
PEGASUS: The following fares are available. Which fare class would you like?

	ONE WAY	FARE CLASS	MIN-STAY	MAX-STAY	RETURN REQ'D
1	$ 234.	COACH DISCOUNT	YES	30	YES
2	$ 304.	COACH DISCOUNT	YES	30	YES
3	$ 336.	COACH DISCOUNT	31	NO	YES

. . .

FIGURE 26.5 An example of an actual verbal booking dialogue using Pegasus. Due to space limitations, irrelevant parts of the system's responses have been omitted.

The outputs of the system modules naturally become more general at the higher levels of the system, since these outputs represent more abstract information. Unfortunately, this makes an automatic comparison with a reference output more difficult, both because the *correct* response may become more ambiguous and because the output representation must become more flexible. The added flexibility that is necessary to express more general concepts also allows a given concept to be expressed in many ways, making the comparison with a reference more difficult.

Objective evaluation of spoken language systems comes with large overhead costs, particularly if it is applied on a common evaluation dataset across a wide community. Researchers must first agree on formal definitions of the *correct* answers, which becomes particularly problematic when discourse context may lead to ambiguities. If a pool of systems are to be evaluated on a common set of dialogues, then the systems used to collect the dialogues must be extremely passive, since users' responses to system queries may not be interpretable in the absence of knowledge about the collection system's half of the conversation. Yet, if systems are not evaluated on common data, it becomes impossible to make objective comparisons of their performance.

A possible alternative is to utilize more subjective evaluations, where an evaluator examines a prior dialogue between a subject and a computer, and decides whether each exchange in the dialogue was

USER: THE CHEAPEST ONE
PEGASUS: I'll book United flight 93 from Boston to San Francisco on Wednesday October twentieth for you.
                   Your reservation has been entered.
                   Now I'll help you with the return flight.
                   What date will you be returning on?
USER: ON SUNDAY
PEGASUS: Here are the United flights from San Francisco to Boston on Sunday October twenty-fourth.

	FLIGHT	FROM	TO	DEPARTURE	ARRIVAL	NSTOPS
1	UA 92	SFO	BOS	7:45 A.M.	4:06 P.M.	0
2	US 20	SFO	BOS	1:30 P.M.	9:54 P.M.	0
3	US 90	SFO	BOS	10:10 P.M.	6:32 A.M.	0

                   . . .
USER: SHOW ME NONSTOP FLIGHTS LEAVING IN THE MORNING
PEGASUS: Here is United flight 92 from San Francisco to Boston on Sunday October twenty-fourth.

	FLIGHT	FROM	TO	DEPARTURE	ARRIVAL	NSTOPS
1	UA 92	SFO	BOS	7:45 A.M.	4:06 P.M.	0

USER: I'LL TAKE THIS FLIGHT
PEGASUS: Please wait while I determine what fares are available for this flight. I'll book United flight 92 from
                   San Francisco to Boston on Sunday October twenty-fourth for you. Your reservation has been
                   entered.
                   Give me a moment to confirm your reservation.
                   Here's your ticket summary.
                   You saved a total of 894 dollars over regular coach fare.
                   Based on fares in effect today, your airline tickets will cost:
                   Each adult: 488.00 USD
                   Fare codes: QE7ONQ, QE7PNQ
                   Adults: 1
                   Total fare: 488.00 USD
                   . . .   Please wait while I retrieve your itinerary. . .

						Price:	488.00
						Total:	488.00

	Date	Flight	Class	From	To	Leave	Arrive
1	OCT20	UA 93	Q	BOS	SFO	510P	823P
2	OCT24	UA 92	Q	SFO	BOS	745A	406P

Can I help you with something else?

FIGURE 26.6   Continuation of the example shown in Fig. 26.5.

effective. A small set of categories, such as correct, incorrect, partially correct, and out of domain, can be used to tabulate statistics on the performance. If the scenario comes with a single unique correct answer, then it is also straightforward to measure how many times users solved their problem successfully, as well as how long it took them to do so. The time is rapidly approaching when real systems will be accessible to the general public via the telephone line, and so the ultimate evaluation will be successful active use of such systems in the real world.

26.3 Best Practices

Spoken language systems are a relatively new technology, having first come into existence in the late 1980s. Prior to that time, computer processing and memory limitations precluded the possibility of real-time speech recognition making it difficult for researchers to conceive of interactive human computer dialogues. All of the systems focus within a narrowly defined area of expertise, and vocabulary sizes are generally limited to under 3000 words. Nowadays, these systems can typically run in real time on standard workstations with no additional hardware.

During the late 1980s, two major government-funded efforts involving multiple sites on two continents provided the momentum to thrust spoken language systems into a highly visible and exciting success story, at least within the computer speech research community. The two programs were the Esprit speech understanding and dialog (SUNDIAL) program in Europe [Peckham 1992] and the Advanced Research Projects Agency (ARPA) spoken language understanding program in the U.S. These two programs were remarkably parallel in that both involved database access for travel planning, with the European one including both flight and train schedules, and the American one being restricted to air travel. The European program was a multilingual effort involving four languages (English, French, German, and Italian), whereas the American effort was, understandably, restricted to English.

The Advanced Research Projects Agency Spoken Language System (SLS) Project

The Air Travel Information Service (ATIS) Common Task

The spoken language systems (SLS) program sponsored by ARPA of the Department of Defense in the U.S. has provided major impetus for spoken language system development. In particular, the program adopted the approach of developing the underlying technologies within a common domain called Air Travel Information Service (ATIS) [Price 1990]. ATIS permits users to verbally query for air travel information, such as flight schedules from one city to another, obtained from a small **relational database** excised from the Official Airline Guide. By requiring that all system developers use the same database, it has been possible to compare the performance of various spoken language systems based on their ability to extract the correct information from the database, using a set of prescribed training and test data, and a set of interpretation guidelines. Indeed, periodic common evaluations have occurred at regular intervals, and steady performance improvements have been observed for all systems. Figure 26.7 shows the error rates for the best ATIS systems, measured in several dimensions over the past four years. Many of the systems currently run in real time on high-end workstations with no additional hardware, although with some performance degradation.

As shown in Fig. 26.7, the speech recognition performance has improved steadily over the past four years. Word error rate (WE) decreased by more than eightfold, while sentence error rate (SE) decreased more than fourfold in this period. In both cases, the reduction in error rate for spontaneous speech has followed the trend set forth for read speech, namely, halving the error every two years. In the most recent formal evaluation of the ARPA-SLS program in the ATIS domain, the best system achieved a word error rate of 2.3% and a sentence error rate of 15.2% [Pallett et al. 1994]. The vocabulary size was more than 2500 words, and the bigram and trigram language models had a **perplexity** of about 20 and 14, respectively.

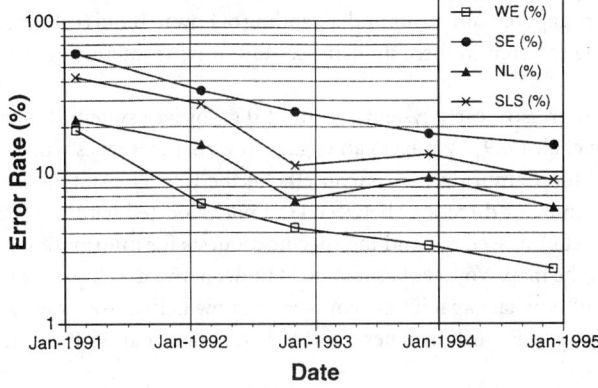

FIGURE 26.7 Best performance achieved by systems in the ATIS domain over the past four years. See text for a detailed description.

Note that all of the performance results quoted in this section are for the so-called *evaluable* queries, i.e., those queries that are within the ATIS domain and for which an appropriate answer is available from the database.

The ARPA-SLS community has carefully defined a common answer specification (CAS) evaluation protocol, whereby a system's performance is determined by comparing its output, expressed as a set of database tuples, with one or more predetermined reference answers [Bates et al. 1991]. The CAS protocol has the advantage that system evaluation can be carried out automatically, once the principles for generating the reference answers have been established and a corpus has

TABLE 26.1 Examples Illustrating Particularly Difficult Sentences Within the ATIS Domain That Systems Are Capable of Handling

GIVE ME A FLIGHT FROM MEMPHIS TO LAS VEGAS AND NEW YORK CITY TO LAS VEGAS ON SUNDAY THAT ARRIVE AT THE SAME TIME
I WOULD LIKE A LIST OF THE ROUND TRIP FLIGHTS BETWEEN INDIANAPOLIS AND ORLANDO ON THE TWENTY SEVENTH OR THE TWENTY EIGHTH OF DECEMBER
I WANT A ROUND TRIP TICKET FROM PHOENIX TO SALT LAKE CITY AND BACK. I WOULD LIKE THE FLIGHT FROM PHOENIX TO SALT LAKE CITY TO BE THE EARLIEST FLIGHT IN THE MORNING AND THE FLIGHT FROM SALT LAKE CITY TO PHOENIX TO BE THE LATEST FLIGHT IN THE AFTERNOON.

been annotated accordingly. Since direct comparison across systems can be performed relatively easily with this procedure, the community has been able to achieve cross fertilization of research ideas, leading to rapid research progress. Figure 26.7 shows that language understanding error rate (NL) has declined by more than threefold in the past four years.[2] This error rate is measured by passing the transcription of the spoken input, after removing partial words, through the natural language component. In the most recent formal evaluation in the ATIS domain, the best natural language system achieved an understanding error rate of only 5.9% on all the evaluable sentences in the test set [Pallett et al. 1994]. Table 26.1 contains several examples of relatively complex sentences that some of the NL systems being evaluated are able to handle.

The performance of the entire spoken language system can be assessed using the same CAS protocol for the natural language component, except with speech rather than text as input. Figure 26.7 shows that this speech understanding error rate (SLS) has fallen from 42.6% to 8.9% over the four-year interval. It is interesting to note that this error rate is considerably less than the sentence recognition error rate, suggesting that a large number of sentences can be understood even though the transcription may contain errors.

Other Advanced Research Projects Agency Projects

The major ARPA sites are Carnegie Mellon University (CMU), MIT, SRI International, and Bolt, Beranek, and Newman (BBN) Systems and Technology. Most of these sites have developed other interesting spoken language systems besides the ATIS task. These systems are at varying stages of completion, and it is fair to say that none of them are as yet sufficiently robust to be deployable. However, each one has revealed interesting new research areas where gaps remain in our understanding of how to build truly practical spoken language systems. Active ongoing research in these domains should provide interesting new developments in the future.

Probably the most impressive early system was the MIT Voyager system, first assembled for English dialogues in 1989 [Zue et al. 1989]. Voyager can engage in verbal dialogues with users about a restricted geographical region within Cambridge, Massachusetts, in the U.S. The system can provide information about distances, travel times, or directions between landmarks located within this area (e.g., restaurants, hotels, banks, libraries, etc.) as well as handling specific requests for information such as address, phone number, or location on the map. Voyager has remained under active development since 1989, particularly in the dimension of multilingual capabilities. Voyager can now help a user solve ground travel planning problems in three languages: English, Japanese, and Italian [Glass et al. 1995]. The process of porting to

---

[2]The error rate for both text (NL) and speech (SLS) input increased somewhat in the 1993 evaluation. This was largely due to the fact that the database was increased from 11 cities to 46 that year, and some of the travel-planning scenarios used to collect the newer data were considerably more difficult.

new languages has motivated researchers to redesign the system so that language-dependent aspects can be contained in external tables and rules.

The CMU office manager system is designed to provide users with voice access to a set of application programs such as a calendar, an on-line Rolodex®, a voice mail system, and a calculator for the office of the future [Rudnicky et al. 1991]. For speech recognition the office manager relies on a version of the Sphinx system [Lee 1989]. Language understanding support is provided either by a finite-state language model or a frame-based representation that extracts the meaning of an utterance by spotting key words and phrases [Ward 1990].

In early 1991, researchers at BBN demonstrated a spoken language interface [Bates et al. 1991] to the dynamic analysis and replanning tool (DART), which is a system for military logistical transportation planning. A spoken language interface to DART has the potential advantage of reducing task completion time by allowing the user to access information more efficiently and naturally than would be possible with the keyboard and mouse buttons. Their spoken language system demonstration centers around the task of database query and information retrieval by combining the BYBLOS speech recognition system [Chow et al. 1987] and the Delphi natural language system [Bates et al. 1990] for speech understanding.

An interesting realistic system, which grew out of the ARPA ATIS effort, is the previously mentioned Pegasus system, which is an extension of an existing ATIS system. The distinguishing feature of Pegasus is that it connects via a modem over the phone line to a real flight reservation system. The system has knowledge of flights to and from some 220 cities worldwide. Pegasus has a fairly extensive dialogue model to help it cope with difficult problems such as date restrictions imposed by discount fares or aborted flight plans due to selections being sold out. A *displayless* version of Pegasus is under development, which will ultimately enable users to make flight reservations by speaking with a computer over the telephone [Seneff et al. 1992].

In 1994, researchers at MIT started the development of Galaxy [Goddeau et al. 1994], a system that enables universal information access using spoken dialogue. Galaxy differs from current spoken language systems in a number of ways. First, it is distributed and decentralized: Galaxy uses a client-server architecture to allow sharing of computationally expensive processes (such as large vocabulary speech recognition), as well as knowledge intensive processes. Second, it is multidomain, intended to provide access to a wide variety of information sources and services while insulating the user from the details of database location and format. It is presently connected to many real, on-line databases, including the National Weather Services, the NYNEX Electronic Yellow Pages, and the World Wide Web. Users can query Galaxy in natural English (e.g., "what is the weather forecast for Miami tomorrow," "how many hotels are there in Boston," and "do you have any information on Switzerland," etc.), and receive verbal and visual responses. Finally, it is extensible; new knowledge domain servers can be added to the system incrementally.

The SUNDIAL Program

Whereas the ARPA ATIS program in the U.S. emphasized competition through periodic common evaluations, the European SUNDIAL program [Peckham 1992] promoted cooperation and plug compatibility by requiring different sites to contribute distinct components to a single multisite system. Another significant difference was that the European program made dialogue modeling an integral and important part of the research program, whereas the American program was focused more strictly on speech understanding, minimizing the effort devoted to usability considerations. The common evaluations carried out in America led to an important breakthrough in forcing researchers to devise robust **parsing** techniques that could makes some sense out of even the most garbled spoken input. At the same time, the emphasis on dialogue in Europe led to some interesting advances in dialogue control mechanisms.

Although the SUNDIAL program formally terminated in 1993, some of the systems it spawned have continued to flourish under other funding resources. Most notable is the Philips Automatic Train Timetable Information System, which is probably the foremost real system in existence today [Eckert et al. 1993]. This system operates in a displayless mode and thus is capable of communicating with the user solely by voice. As a consequence, it is accessible from any household in Germany via the telephone line. The system is presently under field trial, and has been actively promoted through German press releases in order to encourage people to try it. Data are continuously collected from the callers, and can then be used directly to improve system performance. The system runs on a UNIX workstation, and has a vocabulary of 1800 words, 1200 of which are distinct railway station names. The dialogue relies heavily on confirmation requests to permit correction of recognition errors, but the overall success rate for usage is remarkably high.

Other Systems

There are a few other spoken language systems that fall outside of the ARPA ATIS and Esprit SUNDIAL efforts. A notable system is the Berkeley restaurant project (BeRP) [Jurafsky et al. 1994], which acts as a restaurant guide in the Berkeley area. This system is currently distinguished by its neural networks-based recognizer and its probabilistic natural language system. Another novel emergent system is the Waxholm system, being developed by researchers at KTH in Sweden [Blomberg et al. 1993]. Waxholm provides timetables for ferries in the Stockholm archipelago as well as port locations, hotels, camping sites, and restaurants that can be found on the islands. The Waxholm developers are designing a flexible, easily controlled dialogue module based on a scripting language that describes dialogue flow.

26.4 Research Issues and Summary

As we can see, significant progress has been made over the past few years in research and development of systems that can understand spoken language. To meet the challenges of developing a language-based interface to help users solve real problems, however, we must continue to improve the core technologies while expanding the scope of the underlying Human Language Technology (HLT) base. In this section, we outline some of the new research challenges that have heretofore received little attention.

Working in Real Domains

The rapid technological progress that we are witnessing raises several timely questions. When will this technology be available for productive use? What technological barriers still exist that will prevent large-scale HLT deployment? An effective strategy for answering these questions is to develop the underlying technologies within *real* applications, rather than relying on mockups, however realistic they might be, since this will force us to confront some of the critical technical issues that may otherwise elude our attention. Consider, for example, the task of accessing information in the Yellow Pages® of a medium-sized metropolitan area such as Boston, a task that can be viewed as a logical extension of the VOYAGER system developed at MIT. The vocabulary size of such a task could easily exceed 100,000, considering the names of the establishments, street and city names, and listing headings. A task involving such a huge vocabulary presents a set of new technical challenges. Among them are:

- How can adequate acoustic and language models be determined when there is little hope of obtaining a sufficient amount of domain-specific data for training?
- What search strategy would be appropriate for very large vocabulary tasks? How can natural language constraints be utilized to reduce the search space while providing adequate coverage?
- How can the application be adapted and/or customized to the specific needs of a given user?
- How can the system be efficiently ported to a different task in the same domain (e.g., changing the geographical area from Boston to Washington D.C.), or to an entirely different domain (e.g., library information access)?

There are many other research issues that will surface when one is confronted with the need to make human language technology truly useful for solving real problems, some of which will be described in the remainder of this section. Aside from providing the technological impetus, however, working within real domains also has some practical benefits. While years may pass before we can develop unconstrained spoken language systems, we are fast approaching a time when systems with limited capabilities can help users interact with computers with greater ease and efficiency. Working on real applications thus has the potential benefit of shortening the interval between technology demonstration and its ultimate use. Besides, applications that can help people solve problems *will* be used by real users, thus providing us with a rich and continuing source of useful data.

The New Word Problem

Yet another important issue concerns unknown words. The traditional approach to spoken language recognition and understanding research and development is to define the working vocabulary based on domain-specific corpora [Hetherington and Zue 1991]. However, experience has shown that, no matter how large the size of the training corpora, the system will invariably encounter previously unseen words. This is illustrated in Fig. 26.8. For the ATIS task, for example, a 100,000-word training corpus will yield a vocabulary of about 1,000 words. However, the probability of the system encountering an unknown word, is about 0.002. Assuming that an average sentence contains 10 words, this would mean that approximately 1 in 50 sentences will contain an unknown word.

In a *real* domain such as Electronic Yellow Pages, a much larger fraction of the words uttered by users will not be in the system's working vocabulary. This is unavoidable partly because it is not possible to anticipate all of the words that all users are likely to use, and partly because the database is usually changing with time (e.g., new restaurants opening up). In the past, we have not paid much attention to the unknown word problem because the tasks we have chosen assume a closed vocabulary. In the limited cases where the vocabulary has been open, unknown words have accounted for a small fraction of the word tokens in the test corpus. Thus researchers could either construct generic *trash word* models and hope for the best, or ignore the unknown word problem altogether and accept a small penalty on word error rate. In real applications, however, the system must be able to cope with unknown words simply because they will always be present, and ignoring them will not satisfy the user's needs; if a person wants to know how to go from MIT to Lucia's restaurant, they will not settle for a response such as, "I am sorry I don't understand you. Please rephrase the question." The system must be able not only to *detect* new words, taking into account acoustic, phonological, and linguistic evidence, but also to adaptively *acquire* them, both in terms of their orthography and

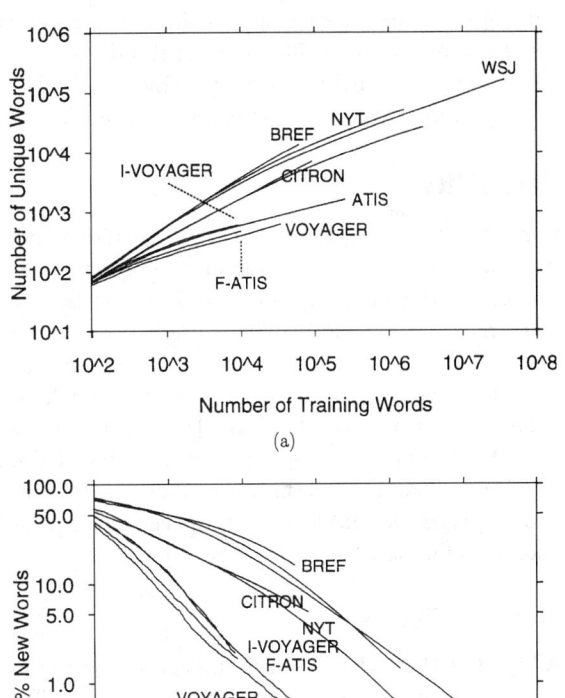

FIGURE 26.8 (a) The number of unique words (i.e., task vocabulary) as a function of the size of the training corpora, for several spoken language tasks and (b) the percentage of unknown words in previously unseen data as a function of the size of the training corpora used to determine the vocabulary empirically. The sources of the data are: F − ATIS = French ATIS, I-VOYAGER = Italian Voyager, BREF = French *La Monde*, NYT = *New York Times*, WSJ = *Wall Street Journal*, and CITRON = Directory Assistance.

linguistic properties. In some cases, fundamental changes in the problem formulation and search strategy may be necessary.

Spoken Language Generation

With few exceptions [Zue et al. 1989, 1994], current research in spoken language systems has focused on the input side, i.e., the understanding of the input queries, rather than the *conveyance* of the information.

Spoken language generation is an extremely important aspect of the human–computer interface problem, especially if the transactions are to be conducted over a telephone. Models and methods must be developed that will generate natural sentences appropriate for spoken output, across many domains and languages [Glass et al. 1994]. In many cases, particular attention must be paid to the interaction between language generation and dialogue management; the system may have to initiate clarification dialogue to reduce the amount of information returned from the backend, in order not to generate unwieldy verbal responses. On the speech side, we must continue to improve speech synthesis capabilities, particularly with regard to the encoding of prosodic and paralinguistic information such as emotion and mood. As is the case on the input side, we must also develop integration strategies for language generation and speech synthesis. Finally, evaluation methodologies for spoken language generation technology must be developed, and comparative evaluation performed.

Portability

Currently, the development of speech recognition and language understanding technologies has been domain specific, requiring a large amount of annotated training data. However, it may be costly, or even impossible, to collect a large amount of training data for certain applications, such as Yellow Pages.

Therefore, we must address the problems of producing a spoken language system in a new domain given at most a small amount of domain-specific training data. To achieve this goal, we must strive to cleanly separate the algorithmic aspects of the system from the application-specific aspects. We must also develop automatic or semiautomatic methods for acquiring the acoustic models, language models, grammars, semantic structures for language understanding, and dialogue models required by a new application. The issue of portability spans across different acoustic environments, databases, knowledge domains, and languages. Real deployment of spoken language technology cannot take place without adequately addressing this issue.

Defining Terms

A* (best first) search: A search strategy for speech recognition in which the theories are prioritized by score, and the best scoring theory is incrementally advanced and returned to the stack. An estimated future score is included to normalize theories. The search is admissible if the estimated future score is an *upper-bound* estimate, in which case it can be guaranteed that the overall best-scoring theory will arrive at the end first.

Conversational system: A computer system that is able to carry on a spoken dialogue with a user in order to solve some problem. Usually there is a database of information that the user is attempting to access, and it may involve explicit goals such as making a reservation.

Dialogue modeling: The part of a conversational system that is concerned with interacting with the user in an effective way. This includes planning what to say next and keeping track of the state of completion of a task such as form filling. Important considerations are the ability to offer help at certain critical points in the dialogue or to recover gracefully from recognition errors. A good dialogue model can help tremendously to improve the usability of the system.

Discourse modeling: The part of a conversational system that is concerned with interpreting user queries in context. Often information that was mentioned earlier must be retained in interpreting a new query. The obvious cases are pronominal reference such as *it* or *this one*, but there are many difficult cases where inheritance is only implicit.

Disfluencies (false starts):   Portions of a spoken sentence that are not fluent language. These can include false starts (a word or phrase that is abruptly ended prior to being fully uttered, and then verbally replaced with an alternative form), filled pauses (such as "umm" and "er"), or agrammatical constructs due to a changed plan midstream. Dysfluencies are particularly problematic for recognition systems.

Hidden Markov modeling (HMM):   A very prevalent recognition framework that begins with an observation sequence derived from an acoustic waveform, and searches through a sequence of states, each of which has a set of hidden observation probabilities and a set of state transition probabilities, to seek an optimal solution. A distinguished *begin* state starts it off, and a distinguished *end* state concludes the search. In recognition, each phoneme is typically associated with an explicit state transition matrix, and each word is encoded as a sequence of specific phonemes. In some cases, phonological pronunciation rules may expand a word's phonetic realization into a set of alternate choices.

Language generation: The process of generating a well-formed expression in English (or some other language) that conveys appropriate information to a user based on diverse sources such as a database, a user query, a partially completed electronic form, and a discourse context (narrow definition for conversational systems).

Natural language understanding: The process of converting an utterance (text string) into a meaning representation (e.g., **semantic frame**).

N-best interface: An interface between a speech recognition system and a natural language system in which the recognizer proposes *N* whole-sentence hypotheses, and the NL system selects the most plausible alternative from among the *N* theories. In an alternative tightly coupled mode, the NL system is allowed to influence partial theories during the initial recognizer search.

***n*-Gram (statistical) language models:**   A powerful mechanism for providing linguistic constraint to a speech recognizer. The models specify the set of follow words with associated probabilities, based on the preceding $n - 1$ words. Statistical language models depend on large corpora of training data within the domain to be effective.

Parser: A program that can analyze an input sentence into a hierarchical structure (a parse tree) according to a set of prescribed rules (a grammar) as an intermediate step toward obtaining a meaning representation (semantic frame).

Perplexity: A measure associated with a statistical language model, characterizing the geometric mean of the number of alternative choices at each branching point. Roughly, it indicates the average number of words the recognizer must consider at each decision point.

Relational database: An electronic database in which a collection of tables contain database entries along with sets of attributes, such that the data can be accessed along complex dimensions using the standard query language (SQL). Such databases make it convenient to look up information based on specifications derived from a semantic frame.

Semantic frame: A meaning representation associated with a user query. For very restricted domains it could be a flat structure of (key: value) pairs. Parsers that retain the syntactic structure can produce semantic frames that preserve the clause structure of the sentence.

Speech recognition: The process of converting an acoustic waveform (digitally recorded spoken utterance) into a sequence of hypothesized words (an orthographic transcription).

Speech synthesis: The process of converting a text string representing a sentence in English (or some other language) into an acoustic waveform that appropriately expresses the phonetics of the text string.

Viterbi search: A search strategy for speech recognition in which all partial theories are advanced lock-stepped in time. Inferior theories are pruned prior to each advance.

Wizard-of-Oz paradigm: A procedure for collecting speech data to be used for training a conversational system in which a human wizard aids the system in answering the subjects' queries. The wizard may simply enter user queries verbatim to the system, eliminating recognition errors, or may play a more active role by extracting appropriate information from the database and formulating canned responses. As the system becomes more fully developed it can play an ever-increasing role in the data collection process, eventually standing alone in a wizardless mode.

References

Asadi, A., Schwartz, R., and Makhoul, J. 1991. Automatic modelling for adding new words to a large vocabulary continuous speech recognition system, pp. 305–308. In *Proc. ICASSP '91*.

Bates, M., Boisen, S., and Makhoul, J. 1990. Developing an evaluation methodology for spoken language systems, pp. 102–108. In *Proc. ARPA Workshop Speech Nat. Lang.*

Bates, M., Ellard, P., and Shaked, V. 1991. Using spoken language to facilitate military transportation planning, pp. 217–220. In *Proc. ARPA Workshop Speech Nat. Lang.* Morgan Kaufmann, San Mateo, CA.

Blomberg, M., Carlson, R., Elenius, K., Granstrom, B., Gustafson, J., Hunnicutt, S., Lindell, R., and Neovius, L. 1993. An experimental dialogue system: Waxholm, pp. 1867–1870. In *Proc. Eurospeech '93*. Berlin, Germany.

Bobrow, R., Ingria, R., and Stallard, R. 1990. Syntactic and semantic knowledge in the DELPHI unification grammar, pp. 230–236. In *Proc. DARPA Speech Nat. Lang. Workshop*.

Butzberger, J., Murveit, H., and Weintraub, M. 1992. Spontaneous speech effects in large vocabulary speech recognition applications, pp. 339–344. In *Proc. ARPA Workshop Speech Nat. Lang.*

Chow, Y. et al. 1987. BYBLOS: the BBN continuous speech recognition system, pp. 89–92. In *Proc. ICASSP*.

Eckert, W., Kuhn, T., Niemann, H., Rieck, S., Scheuer, A., and Schukat-Talamazzini, E. G. 1993. A spoken dialogue system for German intercity train timetable enquiries, pp. 1871–1874. In *Proc. Eurospeech '93*. Berlin, Germany.

Glass, J., Flammia, G., Goodine D., Phillips, M., Polifroni, J., Sakai, S., Seneff, S., and Zue, V. 1995. Multilingual spoken-language understanding in the MIT Voyager system. *Speech Commun.* 17:1–18.

Glass, J., Polifroni, J., and Seneff, S. 1994. Multilingual language generation across multiple domains, pp. 983–976. In *ICSLP '94*.

Goddeau, D. 1992. Using probabilistic shift-reduce parsing in speech recognition systems, pp. 321–324. In *Proc. Int. Conf. Spoken Lang. Process.*

Goddeau, D., Brill, E., Glass, J., Pao, C., Phillips, M., Polifroni, J., Seneff, S., and Zue, V. 1994. Galaxy: a human-language interface to on-line travel information, pp. 707–710. In *Proc. ICSLP '94*.

Goodine, D., Seneff, S., Hirschman, L., and Philips, M. 1991. Full integration of speech and language understanding in the MIT spoken language system, pp. 845–848. In *Proc. Eurospeech*.

Grosz, B. and Sidner, C. 1990. Plans for discourse. In *Intentions in Communication*. MIT Press, Cambridge, MA.

Hetherington, I. L. and Zue, V. 1991. New words: implications for continuous speech recognition. In *Proc. Eur. Conf. Speech Commun. Tech.* Berlin, Germany.

Hirschman, L. et al. 1992. Multi-site data collection for a spoken language corpus, pp. 903–906. In *Proc. Int. Conf. Spoken Lang. Process.*

Jackson, E., Appelt, D., Bear, J., Moore, R., and Podlozny, A. 1991. A template matcher for robust NL interpretation, pp. 190–194. In *Proc. DARPA Speech Nat. Lang. Workshop*.

Jelinek, F., Bahl, L., and Mercer, R. 1974. Design of a linguistic decoder for the recognition of continuous speech, pp. 255–266. In *Proc. IEEE Symp. Speech Recognition*.

Jurafsky, D., Wooters, C., Tajchman, G., Segal, J., Stolcke, A., Fosler, E., and Morgan, N. 1994. The Berkeley restaurant project, pp. 2139–2142. In *Proc. ICSLP '94*. Yokohama, Japan.

Lee, K. F. 1989. *Automatic Speech Recognition: The Development of the SPHINX System*, Kluwer Academic, Boston, MA.

Miller, S., Schwartz, R., Bobrow, R., and Ingria, R. 1994. Statistical language processing using hidden understanding models. *Proc. ARPA Speech Nat. Lang. Workshop*.

Moore, R., Appelt, D., Dowding, J., Gawron, J., and Moran, D. 1995. Combining linguistic and statistical knowledge sources in natural-language processing for ATIS, pp. 261–264. In *Proc. ARPA Spoken Language Systems Workshop*. Austin, TX.

Pallett, D., Fiscus, J., Fisher, W., Garafolo, J., Lund, B., Martin, A., and Pryzbocki, M. 1994. Benchmark tests for the ARPA spoken language program, pp. 5–36. In *Proc. ARPA Spoken Lang. Sys. Tech. Workshop*. Austin, TX.

Peckham, J. 1992. A new generation of spoken dialogue systems: results and lessons from the SUNDIAL project, pp. 33–40. In *Proc. Eurospeech*.

Pieraccini, R., Levin, E., and Lee, C. H. 1992. Stochastic representation of conceptual structure in the ATIS task, pp. 121–124. In *Proc. DARPA Speech Nat. Lang. Workshop*.

Price, P. 1990. Evaluation of spoken language systems: the ATIS domain, pp. 91–95. In *Proc. DARPA Speech Nat. Lang. Workshop*.

Rabiner, L. R. 1989. A tutorial on hidden Markov models and selected applications in speech recognition, pp. 257–285. In *Proc. IEEE* 77(2), February.

Rudnicky, A., Lunati, J.-M., and Franz, A. 1991. Spoken language recognition in an office management domain, pp. 829–832. In *Proc. ICASSP*.

Seneff, S. 1992a. Robust parsing for spoken language systems, pp. 189–192. In *Proc. ICASSP*.

Seneff, S. 1992b. TINA: A natural language system for spoken language applications. *Comput. Linguistics* 18(1):61–86.

Seneff, S., Meng, H., and Zue, V. 1992. Language modelling for recognition and understanding using layered bigrams, pp. 317–320. In *Proc. Int. Conf. Spoken Lang. Process.*

Seneff, S., Zue, V., Polifroni, J., Pao, C., Hetherington, L., Goddeau, D., and Glass, J. 1995. The preliminary development of a displayless Pegasus system, pp. 212–217. In *Proc. ARPA Spoken Lang. Tech. Workshop*. Austin, TX.

Soong, F. and Huang, E. 1990. A tree-trellis based fast search for finding the N-best sentence hypotheses in continuous speech recognition, pp. 199–202. In *Proc. ARPA Workshop Speech Nat. Lang.*

Stallard, D. and Bobrow, R. 1992. Fragment processing in the DELPHI system, pp. 305–310. In *Proc. DARPA Speech Nat. Lang. Workshop*.

Ward, W. 1989. Modelling non-verbal sounds for speech recognition, pp. 47–50. In *Proc. DARPA Workshop Speech Nat. Lang.*

Ward, W. 1990. The CMU air travel information service: understanding spontaneous speech, pp. 127–129. In *Proc. ARPA Workshop Speech Nat. Lang.* Morgan Kaufmann, San Mateo, CA.

Zue, V., Glass, J., Goodine, D., Leung, H., Phillips, M., Polifroni, J., and Seneff, S. 1989. The Voyager speech understanding system: a progress report, pp. 160–167. In *Proc. DARPA Speech Nat. Lang. Workshop*.

Zue, V., Seneff, S., and Glass, J. 1990. Speech database development at MIT: TIMIT and beyond. *Speech Commun.* 9(4):351–356.

Zue, V., Seneff, S., Polifroni, J., Phillips, M., Pao, C., Goddeau, D., Glass, J., and Brill, E. 1994. Pegasus: a spoken language interface for on-line air travel planning. *Speech Commun.* 15:331–340.

Further Information

Fundamentals of Speech Recognition, by Larry Rabiner and Bing-Huang Juang (Prentice–Hall, Englewood Cliffs, NJ, 1993) provides a good description of the basic speech recognition technology.

Natural Language Understanding, by James Allen (2nd ed., Benjamin Cummings, 1995) provides a good description of basic natural language technology.

Proceedings of ICASSP, *Proceedings of Eurospeech*, *Proceedings of ICSLP*, and *Proceedings of DARPA Speech and Natural Language Workshop* all provide excellent coverage of state-of-the-art spoken language systems.

27

Planning and Scheduling

Thomas Dean
Brown University

Subbarao
Kambhampati
Arizona State University

27.1 Introduction

In this chapter, we use the generic term **planning** to encompass both planning and scheduling problems, and the terms *planner* or *planning system* to refer to software for planning or scheduling. Planning is concerned with reasoning about the consequences of acting in order to choose from among a set of possible courses of action. In the simplest case, a planner might enumerate a set of possible courses of action, consider their consequences in turn, and choose one particular course of action that satisfies a given set of requirements.

Algorithmically, a planning problem has as input a set of possible courses of actions, a predictive model for the underlying dynamics, and a performance measure for evaluating courses of action. The output or solution to a planning problem is one or more courses of action that satisfy the specified requirements for performance. Most planning problems are combinatorial in the sense that the number of possible courses of actions or the time required to evaluate a given course of action is exponential in the description of the problem.

Just because there is an exponential number of possible courses of action does not imply that a planner has to enumerate them all in order to find a solution. However, many planning problems can be shown to be NP-hard, and, for these problems, all known exact algorithms take exponential time in the worst case. The computational complexity of planning problems often leads practitioners to consider approximations, computation time vs. solution quality tradeoffs, and heuristic methods.

Planning and Scheduling Problems

We use the travel planning problem as our canonical example of planning (distinct from scheduling). A *travel planning problem* consists of a set of travel options (airline flights, cabs, subways, rental cars, and shuttle services), travel dynamics (information concerning travel times and costs and how time and cost are affected by weather or other factors), and a set of requirements. The requirements for a travel planning problem include an itinerary (be in Providence on Monday and Tuesday, and in Phoenix from Wednesday morning until noon on Friday) and constraints on solutions (leave home no earlier than the Sunday before, arrive back no later than the Saturday after, and spend no more than $1000 in travel-related costs). Planning can be cast either in terms of **satisficing** (find some solution satisfying the constraints) or **optimizing** (find the least cost solution satisfying the constraints).

We use the job-shop scheduling problem as our canonical example of scheduling (distinct from planning). The specification of a *job-shop scheduling problem* includes a set of jobs, where each job is a partially ordered set of tasks of specified duration, and a set of machines, where each machine is capable of carrying out a subset of the set of all tasks. A *feasible solution* to a job-shop scheduling problem is a mapping from tasks to machines over specific intervals of time, so that no machine has assigned to it more than one task at a time and each task is completed before starting any other task that follows it in the specified partial order. Scheduling can also be cast in terms of either satisficing (find a feasible solution) or optimizing (find a solution that minimizes the total time required to complete all jobs).

Distinctions and Disciplines

To distinguish between planning and scheduling, we note that scheduling is primarily concerned with figuring out *when* to carry out actions whereas planning is concerned with *what* actions need to be carried out. In practice, this distinction often blurs and many real-world problems involve figuring out both what and when.

In real job shops, each task need not specify a rigid sequence of steps (the what). For example, drilling a hole in a casting may be accomplished more quickly if a positioning device, called a fixture, is installed on the machine used for drilling. However, the fixture takes time to install and may interfere with subsequent machining operations for other tasks. Installing a fixture for one task may either expedite (the next job needs the same fixture) or retard (the fixture is in the way in the next job) subsequent jobs. In this version of our canonical scheduling problem, planning can take on considerable importance.

We can, however, design a problem so as to emphasize either planning or scheduling. For example, it may be reasonable to let a human decide the what (e.g., the type of machine and specific sequence of machining steps for each task) and a computer program decide the when (e.g., the time and machine for each task). This division of labor has allowed the field of operations research to focus effort on solving problems that stress scheduling and finesse planning, as we previously distinguished the two. Restricting attention to scheduling has the effect of limiting the options available to the planner, thereby limiting the possible interactions among actions and simplifying the combinatorics. In addition, some scheduling problems do not allow for the possibility of events outside the direct control of the planner, so-called *exogenous* events. Planning researchers in artificial intelligence generally allow a wide range of options (specifying both what and when) resulting in a very rich set of interactions among the individual actions in a given course of action and between actions and exogenous events.

The travel planning problem includes as a special case the classic traveling salesperson problem, a problem of considerable interest in operations research. In the traveling salesperson problem, there is a completely connected graph with L vertices corresponding to L distinct cities, an $L \times L$ matrix whose entries encode the distance between each pair of cities, and the objective is to find a minimal-length tour of a specified subset of the cities. The classic traveling salesperson problem involves a very limited set of possible interactions (e.g., you must finish one leg of a tour before beginning the next, and the next leg of a tour must begin at the city in which the previous leg ended). In contrast, variants of the travel planning problem studied in artificial intelligence generally consider a much richer set of possible interactions (e.g., if you start on a multileg air trip it is generally more cost effective to continue with the same airline; travel that extends over a Saturday is less expensive than travel that does not).

Planning of the sort studied in artificial intelligence is similar in some respects to problems studied in a variety of other disciplines. We have already mentioned operations research; planning is also similar to the problem of synthesizing controllers in control theory or the problem of constructing decision procedures in various decision sciences. Planning problems of the sort considered in this chapter differ from those studied in other disciplines mainly in the details of their formulation. Planning problems studied in artificial intelligence typically involve very complex dynamics, requiring expressive languages for their representation, and encoding a wide range of knowledge, often symbolic, but invariably rich and multifaceted.

27.2 Classifying Planning Problems

In this section, we categorize different planning problems according to their inputs: the set of basic courses of action, the underlying dynamics, and the performance measure. We begin by considering models used to predict the consequences of action.

Representing Dynamical Systems

We refer to the environment in which actions are carried out as a **dynamical system**. A description of the environment at an instant of time is called the *state* of the system. We assume that there is a finite, but large set of states S, and a finite set of actions A, that can be executed.

States are described by a vector of *state variables*, where each state variable represents some aspect of the environment that can change over time (e.g., the location or color of an object). The resulting dynamical system can be described as a deterministic, nondeterministic, or stochastic finite-state machine, and time is isomorphic to the integers. In the case of a deterministic finite-state machine, the dynamical system is defined by a **state-transition function** f that takes a state $s_t \in S$ and an action $a_t \in A$ and returns the next state $f(s_t, a_t) = s_{t+1} \in S$.

If there are N state variables each of which can take on two or more possible values, then there are as many as 2^N states and the state-transition function is N dimensional. We generally assume each state variable at t depends on only a small number (at most M) of state variables at $t - 1$. This assumption enables us to *factor* the state-transition function f into N functions, each of dimension at most M, so that $f(s, a) = \langle g_1(s, a), \ldots, g_N(s, a) \rangle$ where $g_i(s, a)$ represents the ith state variable.

In most planning problems, a plan is constructed at one time and executed at a later time. The state-transition function models the evolution of the state of the dynamical system as a consequence of actions carried out by a *plan executor*. We also want to model the information available to the plan executor. The plan executor may be able to observe the state of the dynamical system, partial state information corrupted by noise, or only the current time. We assume that there is a set of possible observations O and the information available to the plan executor at time t is determined by the current state and the *output function* $h: S \to O$, so that $h(s_t) = o_t$. We also assume that the plan executor has a clock and can determine the current time t.

Figure 27.1 depicts the general planning problem. The planner is notated as Γ; it takes as input the current observation o_t and has as output the current **plan** π_t. The planner need not issue a new plan on every state transition and can keep a history of past observations if required. The plan executor is notated as Ψ; it takes as input the current observation o_t and the current plan π_t and has as output the current action a_t.

In the classic formulation of the problem, all planning is done prior to any execution. This formulation is inappropriate in cases where new information becomes available in the midst of execution and replanning is called for. The idea of the planner and plan executor being part of the specification of a planning problem is relatively new in artificial intelligence. The theory that relates to accounting for computations performed during execution is still in its infancy and is only touched upon briefly in this chapter.

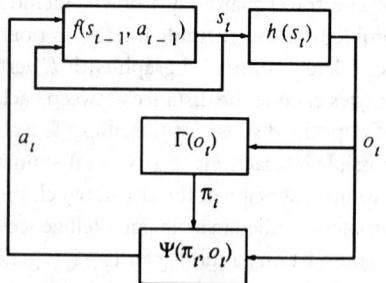

FIGURE 27.1 A block diagram for the general planning problem with state-transition function f, output function h, planner Γ, and plan executor Ψ.

Some physical processes modeled as dynamical systems evolve deterministically; the next state of the system is completely determined by the current state and action. Other processes, said to be *stochastic*, are subject to random changes or are so complex that it is often convenient to model their behavior in statistical terms; the next state of such a system is summarized by a distribution over the set of states.

If the state transitions are governed by a stochastic process, then the state-transition and output functions are random functions and we define the state-transition and output conditional probability distributions as follows:

$$Pr(f(s_t, a_t) \mid s_t, a_t)$$

$$Pr(h(s_t) \mid s_t)$$

In the general case, it requires $O(2^N)$ storage to encode these distributions for boolean state variables. However, in many practical cases, these probability distributions can be factored by taking advantage of independence among state variables.

As mentioned earlier, we assume that the ith state variable at time t depends on a small subset (of size at most M) of the state variables at time $t-1$. Let $\text{Parents}(i, s)$ denote the subset of state variables that the ith state variable depends on in s. We can represent the conditional probability distribution governing state transitions as the following product:

$$Pr(\langle g_1(s_t, a_t), \ldots, g_N(s_t, a_t) \rangle \mid s_t, a_t) = \prod_{i=1}^{N} Pr(g_i(s_t, a_t) \mid \text{Parents}(i, s_t), a_t)$$

This factored representation requires only $O(N2^M)$ storage for Boolean state variables, which is reasonable assuming that M is relatively small.

The preceding descriptions of dynamical systems provide the semantics for a planning system embedded in a dynamic environment. There remains the question of syntax, specifically: how do you represent the dynamical system? In artificial intelligence, the answer varies widely. Researchers have used first-order logic [Allen et al. 1991], dynamic logic [Rosenschein 1981], **state-space operators** [Fikes and Nilsson 1971], and factored probabilistic state-transition functions [Dean and Kanazawa 1989]. In the later sections, we examine some of these representations in more detail.

In some variants of job-shop scheduling the dynamics are relatively simple. We might assume, for example, that if a job is started on a given machine, it will successfully complete in a fixed, predetermined amount of time known to the planner. Everything is under the control of the planner, and evaluating the consequences of a given plan (schedule) is almost trivial from a computational standpoint.

We can easily imagine variants of the travel planning problems in which the dynamics are quite complicated. For example, we might wish to model flight cancellations and delays due to weather and mechanical failure in terms of a stochastic process. The planner cannot control the weather but it can plan to avoid the deleterious effects of the weather (e.g., take a Southern route if a chance of snow threatens to close Northern airports). In this case, there are factors not under control of the planner and evaluating a given travel plan may require significant computational overhead.

Representing Plans of Action

We have already introduced a set of actions $\mathcal{A}$. We assume that these actions are *primitive* in that they can be carried out by the hardware responsible for executing plans. Semantically, a plan π is a mapping from what is known at the time of execution to the set of actions. The set of all plans for a given planning problem is notated Π.

For example, a plan might map the current observation o_t to the action a_t to take in the current state s_t. Such a plan would be independent of time. Alternatively, a plan might ignore observations altogether and map the current time t to the action to take in state s_t. Such a plan is independent of the current state, or at least the observable aspects of the current state.

If the action specified by a plan is dependent on observations of the current state, then we say that the plan is *conditional*. If the action specified by a plan is dependent on the current time, then we say that the plan is *time variant*, otherwise we say it is *stationary*. If the mapping is one-to-one, then we say that the plan is deterministic, otherwise it is nondeterministic and possibly stochastic if the mapping specifies a distribution over possible actions.

Conditional plans are said to run in a **closed loop**, since they enable the executor to react to the consequences of prior actions. Unconditional plans are said to run in an **open loop**, since they take no account of exogenous events or the consequences of prior actions that were not predicted using the dynamical model.

Now that we have the semantics for plans, we can think about how to represent them. If the mapping is a function, we can use any convenient representation for functions, including decision trees, tabular formats, hash tables, or artificial neural networks. In some problems, an unconditional, time-variant, deterministic plan is represented as a simple sequence of actions. Alternatively, we might use a set of possible sequences of actions perhaps specified by a partially ordered set of actions to represent a nondeterministic plan [i.e., the plan allows any total order (sequence) consistent with the given partial order].

Measuring Performance

For a deterministic dynamical system in initial state s_0, a plan π determines a (possibly infinite) sequence of states $h_\pi = \langle s_0, s_1, \ldots \rangle$, called a **history or state-space trajectory**. More generally, a dynamical system together with a plan induces a probability distribution over histories, and h_π is a random variable governed by this distribution. A value function V assigns to each history a real value. In the deterministic case, the performance J of a plan π is the value of the resulting history, $J(\pi) = V(h_\pi)$. In the general case, the performance J of a plan is the expected value according to V over all possible histories, $J(\pi) = E[V(h_\pi)]$, where E denotes taking an expectation.

In artificial intelligence planning (distinct from scheduling), much of the research has focused on goal-based performance measures. A **goal** $\mathcal{G}$ is a subset of the set of states $\mathcal{S}$.

$$V(\langle s_0, s_1, \ldots \rangle) = \begin{cases} 1 & \text{if } \exists i, s_i \in \mathcal{G} \\ 0 & \text{otherwise} \end{cases}$$

Alternatively, we can consider the number of transitions until we reach a goal state as a measure of performance.

$$V(\langle s_0, s_1, \ldots \rangle) = \begin{cases} -\min_i s_i \in \mathcal{G} & \text{if } \exists i, s_i \in \mathcal{G} \\ -\infty & \text{otherwise} \end{cases}$$

In the stochastic case, the corresponding measure of performance is called *expected time to target*, and the objective in planning is to minimize this measure.

Generalizing on the expected-time-to-target performance measure, we can assign to each state a cost using the cost function C. This cost function yields the following value function on histories:

$$V(\langle s_0, s_1, \ldots \rangle) = -\sum_{i=0}^{\infty} C(s_i)$$

In some problems, we may wish to discount future costs using a discounting factor $0 \leq \gamma < 1$,

$$V(\langle s_0, s_1, \ldots \rangle) = -\sum_{i=0}^{\infty} \gamma^i C(s_i)$$

This performance measure is called *discounted cumulative cost*. These value functions are said to be *separable* since the total value of a history is a simple sum or weighted sum (in the discounted case) of the costs of each state in the history.

It should be noted that we can use any of the preceding methods for measuring the performance of a plan to define either a satisficing criterion (e.g., find a plan whose performance is above some fixed threshold) or an optimizing criterion (e.g., find a plan maximizing a given measure of performance).

Categories of Planning Problems

Now we are in a position to describe some basic classes of planning problems. A planning problem can be described in terms of its dynamics, either deterministic or stochastic. We might also consider whether the actions of the planner completely or only partially determine the state of the environment.

A planning problem can be described in terms of the knowledge available to the planner or executor. In the problems considered in this chapter, we assume that the planner has an accurate model of the underlying dynamics, but this need not be the case in general. Even if the planner has an accurate predictive model, the executor may not have the necessary knowledge to make use of that model. In particular, the executor may have only partial knowledge of the system state and that knowledge may be subject to errors in observation (e.g., noisy, error-prone sensors).

We can assume that all computations performed by the planner are carried out prior to any execution, in which case the planning problem is said to be **off-line**. Alternatively, the planner may periodically compute a new plan and hand it off to the executor; this sort of planning problem is said to be **on-line**. Given space limitations, we are concerned primarily with off-line planning problems in this chapter.

Now that we have some familiarity with the various classes of planning problems, we consider some specific techniques for solving them. Our emphasis is on the design, analysis, and application of planning algorithms.

27.3 Algorithms, Complexity and Search

Once we are given a set of possible plans and a *performance function* implementing a given performance measure, we can cast any planning or scheduling problem as a search problem. If we assume that evaluating a plan (applying the performance function) is computationally simple, then the most important issue concerns how we search the space of possible plans. Specifically, given one or more plans currently under consideration, how do we extend the search to consider other, hopefully better, performing plans? We focus on two methods for extending search: *refinement* methods and *repair* methods.

A refinement method takes an existing partially specified plan (schedule) and refines it by adding detail. In job-shop scheduling, for example, we might take a (partial) plan that assigns machines and times to k of the jobs, and extend it so that it accounts for $k + 1$ jobs. Alternatively, we might build a plan in chronological order by assigning the earliest interval with a free machine on each iteration.

A repair method takes a completely specified plan and attempts to transform it into another completely specified plan with better performance. In travel planning, we might take a plan that makes use of one airline's flights and modify it to use the flights of another, possibly less expensive or more reliable airline. Repair methods often work by first analyzing a plan to identify unwanted interactions or bottlenecks and then attempting to eliminate the identified problems.

The rest of this section is organized as follows. In "Complexity Results," we briefly survey what is known about the complexity of planning and scheduling problems, irrespective of what methods are used to solve them. In "Planning with Deterministic Dynamics," we focus on traditional search methods for generating plans of actions given deterministic dynamics. We begin with open-loop planning problems with complete knowledge of the initial state, progressing to closed-loop planning problems with incomplete knowledge of the initial state. In "Scheduling with Deterministic Dynamics," we focus on methods for generating schedules given deterministic dynamics. In both of the last two sections just mentioned we discuss refinement- and repair-based methods. In "Improving Efficiency," we mention related work in machine learning concerned with learning search rules and adapting previously generated solutions to planning and scheduling problems. In "Approximation in Stochastic Domains," we consider a class of planning problems involving stochastic dynamics and address some issues that arise in trying to approximate the value of conditional plans in stochastic domains. Our discussion begins with a quick survey of what is known about the complexity of planning and scheduling problems.

Complexity Results

Garey and Johnson [1979] provide an extensive listing of NP-hard problems, including a great many scheduling problems. They also provide numerous examples of how a hard problem can be rendered easy by relaxing certain assumptions. For example, most variants of job-shop scheduling are NP-hard. Suppose, however, that you can suspend work on one job in order to carry out a rush job, resuming the suspended job on completion of the rush job so that there is no time lost in suspending and resuming. With this assumption, some hard problems become easy. Unfortunately, most real scheduling problems are NP-hard. Graham et al. [1977] provide a somewhat more comprehensive survey of scheduling problems with a similarly dismal conclusion. Lawler et al. [1985] survey results for the traveling salesperson problem, a special case of our travel planning problem. Here again the prospects for optimal, exact algorithms are not good, but there is some hope for approximate algorithms.

With regard to open-loop, deterministic planning, Chapman [1987], Bylander [1991], and Gupta and Nau [1991] have shown that most problems in this general class are hard. Dean and Boddy [1988] show that the problem of evaluating plans represented as sets of partially ordered actions is NP-hard in all but the simplest cases. Bäckström and Klein [1991] provide some examples of easy (polynomial time) planning problems, but these problems are of marginal practical interest.

Regarding closed-loop, deterministic planning, Papadimitriou and Tsitsiklis [1987] discuss polynomial-time algorithms for finding an optimal conditional plan for a variety of performance functions. Unfortunately, the polynomial is in the size of the state space. As mentioned earlier, we assume that the size of the state space is exponential in the number of state variables. Papadimitriou and Tsitsiklis also list algorithms for the case of stochastic dynamics that are polynomial in the size of the state space.

From the perspective of worst-case, asymptotic time and space complexity, most practical planning and scheduling problems are computationally very difficult. The literature on planning and scheduling in artificial intelligence generally takes it on faith that any interesting problem is at least NP-hard. The research emphasis is on finding powerful heuristics and clever search algorithms. In the remainder of this section, we explore some of the highlights of this literature.

Planning with Deterministic Dynamics

In the following section, we consider a special case of planning in which each action deterministically transforms one state into another. Nothing changes without the executor performing some action. We assume that the planner has an accurate model of the dynamics. If we also assume that we are given complete information about the initial state, it will be sufficient to produce unconditional plans that are produced off-line and run in an open loop.

Recall that a state is described in terms of a set of state variables. Each state assigns to each state variable a value. To simplify the notation, we restrict our attention to Boolean variables. In the case of Boolean variables, each state variable is assigned either true or false. Suppose that we have three Boolean state variables: P, Q, and R. We represent the particular state s in which P and Q are true and R is false by the *state-variable assignment*, $s = \{P = \text{true}, Q = \text{true}, R = \text{false}\}$, or, somewhat more compactly, by $s = \{P, Q, \neg R\}$, where $X \in s$ indicates that X is assigned true in s and $\neg X \in s$ indicates that X is assigned false in s.

An action is represented as a *state-space operator* α defined in terms of *preconditions* ($\text{Pre}(\alpha)$) and *postconditions* (also called effects) ($\text{Post}(\alpha)$). Preconditions and postconditions are represented as state-variable assignments that assign values to subsets of the set of all state variables. Here is an example operator α_{eg}:

Operator α_{eg}

$$\text{Preconditions:} \quad P, \neg R$$

$$\text{Postconditions:} \quad \neg P, \neg Q$$

If an operator (action) is applied (executed) in a state in which the preconditions are satisfied, then the variables mentioned in the postconditions are assigned their respective values in the resulting state. If the preconditions are not satisfied, then there is no change in state.

In order to describe the state-transition function, we introduce a notion of consistency and define two operators $\oplus$ and $\ominus$ on state-variable assignments. Let φ and ϑ denote state-variable assignments. We say that φ and ϑ are *inconsistent* if there is a variable X such that φ and ϑ assign X different values; otherwise, we say that φ and ϑ are *consistent*. The operator $\ominus$ behaves like set difference with respect to the variables in assignments. The expression $\varphi \ominus \vartheta$ denotes a new assignment consisting of the assignments to variables in φ that have no assignment in ϑ (e.g., $\{P, Q\} \ominus \{P\} = \{P, Q\} \ominus \{\neg P\} = \{Q\} \ominus \{\} = \{Q\}$). The operator $\oplus$ takes two consistent assignments and returns their union (e.g., $\{Q\} \oplus \{P\} = \{P, Q\}$, but $\{P\} \oplus \{\neg P\}$ is undefined).

The state-transition function is defined as follows:

$$f(s, \alpha) = \begin{cases} s & \text{if } s \text{ and } Pre(\alpha) \text{ are inconsistent} \\ Post(\alpha) \oplus (s \ominus Post(\alpha)) & \text{otherwise} \end{cases}$$

If we apply the operator α_{eg} to a state where the variables P and Q are true, and R is false, we have

$$f(\{P, Q, \neg R\}, \alpha_{eg}) = \{\neg P, \neg Q, \neg R\}$$

We extend the state-transition function to handle sequences of operators in the obvious way,

$$f(s, \langle \alpha_1, \alpha_2, \ldots, \alpha_n \rangle) = f(f(s, \alpha_1), \langle \alpha_2, \ldots, \alpha_n \rangle)$$

$$f(s, \langle \rangle) = s$$

Our performance measure for this problem is goal based. Goals are represented as state-variable assignments that assign values to subsets of the set of all state variables. By assigning values to one or more state variables, we designate a set of states as the goal. We say that a state s *satisfies* a goal ϕ, notated $s \models \phi$, just in case the assignment ϕ is a subset of the assignment s. Given an initial state s_0, a goal ϕ, and a library of operators, the objective of the planning problem is to find a sequence of state-space operators $\langle \alpha_1, \ldots, \alpha_n \rangle$ such that $f(s_0, \langle \alpha_1, \ldots, \alpha_n \rangle) \models \phi$.

Using a state-space operator to transform one state into the next state is called **progression**. We can also use an operator to transform one goal into another, namely, the goal that the planner would have prior to carrying out the action corresponding to the operator. This use of an operator to transform goals is called **regression**. In defining regression, we introduce the notion of an impossible assignment, denoted $\bot$. We assume that if you regress a goal using an operator with postconditions that are inconsistent with the goal, then the resulting regressed goal is impossible to achieve. Here is the definition of regression:

$$b(\phi, \alpha) = \begin{cases} \bot & \text{if } \phi \text{ and } Post(\alpha) \text{ are inconsistent} \\ Pre(\alpha) \oplus (\phi \ominus Post(\alpha)) & \text{otherwise} \end{cases}$$

Conditional Postconditions and Quantification

Within the general operator-based state-transition framework previously described, a variety of syntactic abbreviations can be used to facilitate compact action representation. For example, the postconditions of an action may be *conditional*. A conditional postcondition of the from $P \Rightarrow Q$ means that the action changes the value of the variable Q to true only if the value of P is true in the state where the operator is applied. It is easy to see that an action with such a conditional effect corresponds to two simpler actions, one which has a precondition P and the postcondition Q, and the other which has a precondition $\neg P$ and does not mention Q in its postconditions.

Similarly, when state variables can be typed in terms of objects in the domain to which they are related, it is possible to express preconditions and postconditions of an operator as quantified formulas. As an example, suppose in the travel domain, we have one state variable $loc\,(c)$ which is true if the agent is in city c and false otherwise. The action of flying from city c to city c' has the effect that the agent is now at city c', and the agent is not in any other city. If there are n cities, $c_1, \ldots, c_n$, the latter effect can be expressed either as a set of propositional postconditions $\neg loc\,(c_1), \ldots, \neg loc\,(c_{j-1}), \neg loc\,(c_{j+1}), \ldots, \neg loc\,(c_n)$ where $c' = c_j$, or, more compactly, as the quantified effect $\forall_{z:city\,(z)}\, z \neq c' \Rightarrow \neg loc\,(z)$. Since operators with conditional postconditions and quantified preconditions and postconditions are just shorthand notations for finitely many propositional operators, the transition function, as well as the progression and regression operations, can be modified in straightforward ways to accommodate them. For example, if a goal formula $\{W, S\}$ is regressed through an operator having preconditions $\{P, Q\}$ and postconditions $\{R \Rightarrow \neg W\}$, we get $\{\neg R, S, P, Q\}$. Note that by making $\neg R$ a part of the regressed formula, we ensure that $\neg W$ will not be a postcondition of the operator, thereby averting the inconsistency with the goals.

Representing Partial Plans

Although solutions to the planning problems can be represented by operator sequences, to facilitate efficient methods of plan synthesis, it is useful to have a more flexible representation for partial plans. A *partial plan* consists of a set of *steps*, a set of *ordering constraints* that restrict the order in which steps are to be executed, and a set of *auxiliary constraints* that restrict the value of state variables over particular intervals of time. Each step is associated with a state-space operator. To distinguish between multiple instances of the same operator appearing in a plan, we assign each step a unique integer i and represent the ith step as the pair (i, α_i) where α_i is the operator associated with the ith step.

Figure 27.2 shows a partial plan π_{eg} consisting of seven steps. The plan π_{eg} is represented as follows:

$$\Big\langle\ \{(0, \alpha_0), (1, \alpha_1), (2, \alpha_2), (3, \alpha_3), (4, \alpha_4), (5, \alpha_5), (\infty, \alpha_\infty)\},$$
$$\{(0 \preceq 1), (1 \prec 2), (1 \prec 4), (2 \prec 3), (3 \prec 5), (4 \prec 5), (5 \preceq \infty)\},$$
$$\Big\{\Big(1 \xrightarrow{Q} 2\Big), \Big(3 \xrightarrow{R} \infty\Big)\Big\}\ \Big\rangle$$

An ordering constraint of the form $(i \prec j)$ indicates that step i precedes step j. An ordering constraint of the form $(i \preceq j)$ indicates that step i is contiguous with step j, that is, step i precedes step j and no other steps intervene. The steps are *partially ordered* in that step 2 can occur either before or after step 4. An auxiliary constraint of the form $(i \xrightarrow{P} j)$ is called an *interval preservation constraint* and indicates that

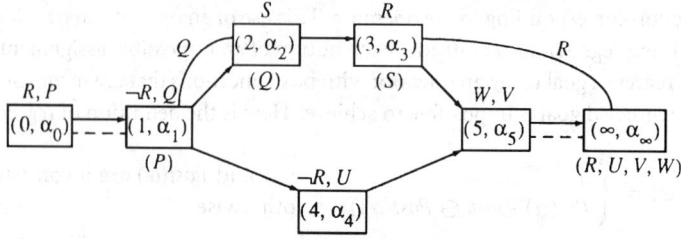

FIGURE 27.2 This figure depicts the partial plan π_{eg}. The postconditions (effects) of the steps are shown above the steps, whereas the preconditions are shown below the steps in parentheses. The ordering constraints between steps are shown by arrows. The interval preservation constraints are shown by arcs, whereas the contiguity constraints are shown by dashed lines.

P is to be preserved in the range between steps i and j (and therefore no operator with postcondition $\neg P$ should occur between steps i and j). In particular, according to the constraint $(3 \xrightarrow{R} \infty)$, step 4 should not occur between steps 3 and ∞.

The set of steps $\{\sigma_1, \sigma_2, \ldots, \sigma_n\}$ with contiguity constraints

$$\{(\sigma_0 \precsim \sigma_1), (\sigma_1 \precsim \sigma_2), \ldots, (\sigma_{n-1} \precsim \sigma_n)\}$$

is called the *header* of the plan π. The last element of the header σ_n is called the *head step*. The state defined by $f(s_0, \langle \alpha_{\sigma_1}, \ldots, \alpha_{\sigma_n} \rangle)$, where α_{σ_i} is the operator associated with σ_i, is called the *head state*. In a similar manner, we can define the *tail, tail step*, and *tail state*.

As an example, the partial plan π_{eg} shown in Fig. 27.2 has the steps 0 and 1 in its header, with step 1 being the head step. The head state (which is the state resulting from applying α_1 to the initial state) is $\{P, Q\}$. Similarly, the tail consists of steps 5 and ∞, with step 5 being the tail step. The tail state (which is the result of regressing the goal conditions through the operator α_5) is $\{R, U\}$.

Refinement Search

A large part of the work on plan synthesis in artificial intelligence falls under the rubric of refinement search. Refinement search can be seen as search in the space of partial plans. The search starts with the empty partial plan, and adds details to that plan until a complete plan results. Semantically, a partial plan can be seen as a shorthand notation for the set of complete plans (action sequences) that are consistent with the constraints. A refinement strategy converts a partial plan π into a set of new plans $\{\pi_1, \ldots, \pi_n\}$ such that all of the potential solutions represented by π are represented by at least one of $\pi_1, \ldots, \pi_n$. Syntactically, this is accomplished by generating each of the children plans (refinements) by adding additional constraints to π.

The following is a general template for refinement search. The search starts with the null plan $\langle \{(0, \alpha_0), (\infty, \alpha_\infty)\}, \{(0 \prec \infty)\}, \{\} \rangle$, where α_0 is a dummy operator with no preconditions and postconditions corresponding to the initial state, and α_∞ is a dummy operator with no postconditions and preconditions corresponding the goal. For example, if we were trying to find a sequence of actions to transform the initial state $\{P, Q, \neg R\}$ into a state satisfying the goal $\{R\}$, then we would have $\text{Pre}(\alpha_0) = \{\}$, $\text{Post}(\alpha_0) = \{P, Q, \neg R\}$, $\text{Pre}(\alpha_\infty) = \{R\}$, and $\text{Post}(\alpha_\infty) = \{\}$.

We define a generic refinement procedure, $\text{Refine}(\pi)$, as follows [Kambhampati et al. 1995]:

1. If an action sequence $\langle \alpha_1, \alpha_2, \ldots, \alpha_n \rangle$ corresponds to a total order consistent with both ordering constraints and the auxiliary constraints of π, and is a solution to the planning problem, then terminate and return $\langle \alpha_1, \alpha_2, \ldots, \alpha_n \rangle$.
2. If the constraints in π are inconsistent, then eliminate π from future consideration.
3. Select a refinement strategy, and apply the strategy to π and add the resulting refinements to the set of plans under consideration.
4. Select a plan π' from those under consideration and call $\text{Refine}(\pi')$.

In step 3, the search selects a refinement strategy to be applied to the partial plan. There are several possible choices here, corresponding intuitively to different ways of splitting the set of potential solutions represented by the plan. In the following sections, we outline four popular refinement strategies employed in the planning literature.

State-Space Refinements

The most straightforward way of refining partial plans involves using progression to convert the initial state into a state satisfying the goal conditions, or using regression to convert a set of goal conditions into a set of conditions that are satisfied in the initial state. From the point of view of partial plans, this corresponds to growing the plan from either the beginning or the end.

Progression (or forward state-space) refinement involves advancing the head state by adding a step σ, such that the preconditions of α_σ are satisfied in the current head state, to the header of the plan. The step

σ may be newly added to the plan or currently present in the plan. In either case, it is made contiguous to the current head step and becomes the new head step.

As an example, one way of refining the plan π_{eg} in Fig. 27.2 using progression refinement would be to apply an instance of the operator α_2 (either the instance that is currently in the plan $(2, \alpha_2)$ or a new instance) to the head state (recall that it is $\{P, Q\}$). This is accomplished by putting a contiguity constraint between $(2, \alpha_2)$ and the current head step $(1, \alpha_1)$ (thereby making the former the new head step).

We can also define a refinement strategy based on regression, which involves regressing the tail state of a plan through an operator. For example, the operator α_3 is applicable (in the backward direction) through this tail state (which is $\{R, U\}$), whereas the operator α_4 is not (since its postconditions are inconsistent with the tail state). Thus, one way of refining π_{eg} using regression refinement would be to apply an instance of the operator α_3 (either the existing instance in step 3 or a new one) to the tail state in the backward direction. This is accomplished by putting a contiguity constraint between $(3, \alpha_3)$ and the current tail step.

From a search control point of view, one of the important questions is deciding which of the many refinements generated by progression and regression refinements are most likely to lead to a solution. It is possible to gain some focus by using *state difference heuristics*, which prefer the refinements where the set difference between the tail state and the head state is the smallest.

Although the state difference heuristic works well enough for regression refinements, it does not provide sufficient focus to progression refinements. The problem is that in a realistic planning problem, there potentially may be many operators that are applicable in the current head state, and only a few of them may be relevant to the goals of the problem. Thus, the strategy of generating all of the refinements and ranking them with respect to the state difference heuristic can be prohibitively expensive. We need a method of automatically zeroing on those operators which are possibly relevant to the goals.

One popular way of generating the list of relevant operators is to use *means-ends analysis*. The general idea is the following. Suppose we have an operator α whose postconditions match a goal of the problem. Clearly, α is a relevant operator. If the preconditions of α are satisfied in the head state of the current partial plan, we can apply it directly. Suppose they are not all satisfied. In such a case, we can consider the preconditions of α as subgoals, look for an operator α' whose postconditions match one of these subgoals, and check if it is applicable to the head state. This type of recursive analysis can be continued to find the set of relevant operators, and focus progression refinement.

Plan-Space Refinements

As we have seen previously in state-space refinements, partial plans are extended by adding new steps and new contiguity constraints. The contiguity constraints are required since without them the head state and tail state are not well defined. State-space refinements have the disadvantage that they completely determine the order and position of every step introduced into the plan. Although it is easy to see whether or not a given step is relevant to a plan, often the precise position at which a step must occur in the final plan is not apparent until all of the steps have been added. In such situations, state-space refinement can lead to premature commitment to the order of steps, causing extensive backtracking.

Plan-space refinement attempts to avoid this premature commitment. The main idea in plan-space refinement is to shift the attention from advancing the world state to establishing goals. A precondition P of a step (i, α_i) in a plan is said to be *established* if there is some step (j, α_j) in the plan that precedes i and causes P to be true, and no step that can possibly intervene between j and i has postconditions that are inconsistent with P. It is easy to see that if every precondition of every step in the plan is established, then that plan will be a solution plan. Plan-space refinement involves picking a precondition P of a step (i, α_i) in the partial plan, and adding enough additional step, ordering, and auxiliary constraints to ensure the establishment of P.

We illustrate the main ideas in precondition establishment through an example. Consider the partial plan at the top in Fig. 27.3. Step 2 in this plan requires a precondition Q. To establish this precondition, we need a step which has Q as its postcondition. None of the existing steps have such a postcondition. Suppose an operator α_3 in the library has a postcondition $R \Rightarrow Q$. We introduce an instance of α_3 as step 3

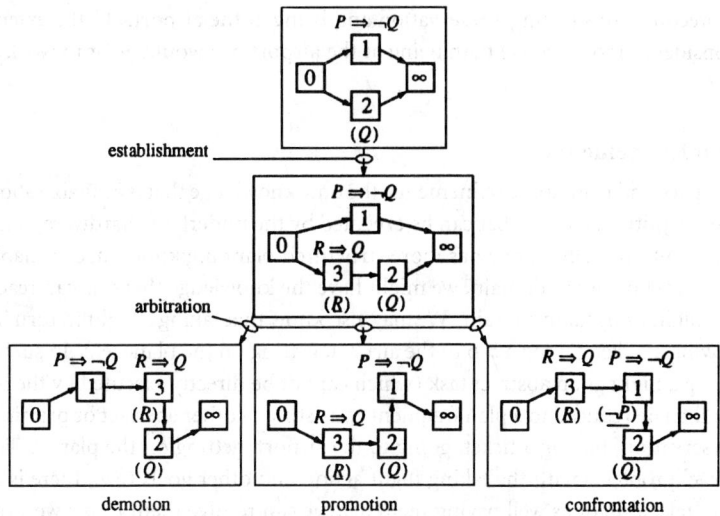

FIGURE 27.3 An example of precondition establishment. This diagram illustrates an attempt to establish Q for step 2. Establishing a postcondition can result in a potential conflict, which requires arbitration to avert the conflict. Underlined preconditions correspond to secondary preconditions.

into the plan. Step 3 is ordered to come before step 2 (and after step 0). Since α_3 makes Q true only when R is true before it, to make sure that Q will be true following step 3, we need to ensure that R is true before it. This can be done by posting R as a precondition of step 3. Since R is not a normal precondition of α_3, and is being posted only to guarantee one of its conditional effects, it is called a *secondary precondition* [Pednault 1988].

Now that we have introduced step 3 and ensured that it produces Q as a postcondition, we need to make sure that Q is not violated by any steps possibly intervening between steps 3 and 2. This phase of plan-space refinement is called *arbitration*. In our example, step 1, which can possibly intervene between steps 3 and 2, has a postcondition $P \Rightarrow \neg Q$ that is potentially inconsistent with Q. To avert this inconsistency, we can either order step 1 to come before step 3 (demotion), or order step 1 to come after step 2 (promotion), or ensure that the offending conditional effect will not occur. This last option, called confrontation, can be carried out by posting $\neg P$ as a (secondary) precondition of step 1. All these partial plans, corresponding to different ways of establishing the precondition Q at step 2 are returned as the refinements of the original plan.

One problem with this precondition-by-precondition establishment approach is that the steps added in establishing a precondition might unwittingly violate a previously established precondition. Although this does not affect the completeness of the refinement search, it can lead to wasted planning effort, and necessitate repeated establishments of the same precondition within the same search branch. Many variants of plan-space refinements avoid this inefficiency by *protecting* their establishments. Whenever a condition P of a step σ is established with the help of the effects of a step σ', an interval preservation constraint $(\sigma' \overset{P}{-} \sigma)$ is added to remember this establishment. If the steps introduced by later refinements violate this preservation constraint, those conflicts are handled much the same way as in the arbitration phase previously discussed. In the example shown in Fig. 27.3, we can protect the establishment of precondition Q by adding the constraint $3 \overset{Q}{-} 2$.

Although the order in which preconditions are selected for establishment does not have any effect on the completeness of a planner using plan-space refinement, it can have a significant impact on the size of the search space explored by the planner (and thereby its efficiency). Thus, any available domain specific information regarding the relative importance of the various types of preconditions can be gainfully exploited. As an example, in the travel domain, the action of taking a flight to go from one place to another may have as its preconditions having a reservation and being at the airport. To the extent that having a reservation is considered more critical than being at the airport, we would want to work on establishing the former first.

Task-Reduction Refinements

In both the state-space and plan-space refinements, the only knowledge that is available about the planning task is in terms of primitive actions (that can be executed by the underlying hardware), and their preconditions and postconditions. Often, one has more structured planning knowledge available in a domain. For example, in a travel planning domain, we might have the knowledge that one can reach a destination by either taking a flight or by taking a train. We may also know that taking a flight in turn involves making a reservation, buying a ticket, taking a cab to the airport, getting on the plane, etc. In such a situation, we can consider taking a flight as an abstract task (which cannot be directly executed by the hardware). This abstract task can then be reduced to a plan fragment consisting of other abstract or primitive tasks (in this case making a reservation, buying a ticket, going to the airport, getting on the plane). This way, if there are some high-level problems with the taking flight action and other goals (e.g., there is not going to be enough money to take a flight as well paying the rent), we can resolve them *before* we work on low-level details such as getting to the airport.

This idea forms the basis for task reduction refinement. Specifically, we assume that in addition to the knowledge about primitive actions, we also have some abstract actions, and a set of schemas (plan fragments) that can replace any given abstract action. Task reduction refinement takes a partial plan π containing abstract and primitive tasks, picks an abstract task σ, and for each reduction schema (plan fragment) that can be used to reduce σ, a refinement of π is generated with σ replaced by the reduction schema (plan fragment). As an example, consider the partial plan on the left in Fig. 27.4. Suppose the operator α_2 is an abstract operator. The central box in Fig. 27.4 shows a reduction schema for step 2, and the partial plan shown on the right of the figure shows the result of refining the original plan with this reduction schema. At this point any interactions between the newly introduced plan fragment and the previously existing plan steps can be resolved using techniques such as promotion, demotion, and confrontation discussed in the context of plan-space refinement. This type of reduction is carried out until all of the tasks are primitive.

In some ways, task reduction refinements can be seen as macrorefinements that package together a series of state-space and plan-space refinements, thereby reducing a considerable amount of search. This, and the fact that in most planning domains, canned reduction schemas are readily available, have made task reduction refinement a very popular refinement choice for many applications.

Hybrid Refinements

Although early refinement planning systems tended to subscribe exclusively to a single refinement strategy, it is possible and often effective to use multiple refinement strategies. As an example, the partial plan π_{eg} shown in Fig. 27.2 can be refined with progression refinement (e.g., by putting a contiguity constraint

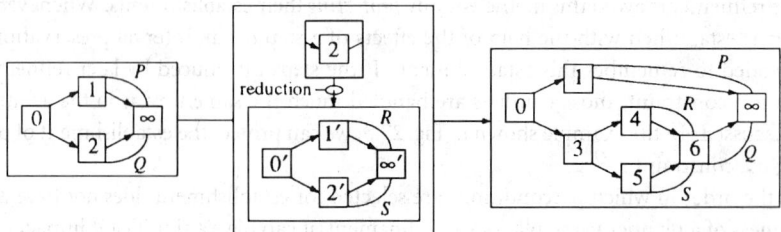

FIGURE 27.4   Step 2 in the partial plan shown on the left is reduced to obtain a new partial plan shown on the right. In the new plan, step 2 is replaced with the (renamed) steps and constraints specified in the reduction shown in the center box.

between step 1 and step 2), with regression refinement (e.g., by putting a contiguity constraint between step 3 and step 5), or plan-space refinement (e.g., by establishing the precondition S of step 3 with the help of the effect step 2). Finally, if the operator α_4 is a nonprimitive operator, we can also use task reduction refinement to replace α_4 with its reduction schema. There is some evidence that planners using multiple refinement strategies intelligently can outperform those using single refinement strategies [Kambhampati 1995]. However, the question as to which refinement strategy should be preferred when is still largely open.

Handling Incomplete Information

Although the refinement methods just described were developed in the context of planning problems where the initial state is completely specified, they can be extended to handle incompletely specified initial states. Incomplete specification of the initial state means that the values of some of the state variables in the initial state are not specified. Such incomplete specification can be handled as long as the state variables are observable (i.e., the correct value of the variable can be obtained at execution time).

Suppose the initial state is incomplete with respect to the value of the state variable ϕ. If ϕ has only a small number of values, K, then we can consider this planning problem to be a collection of K problems, each with the same goal and a complete initial state in which ϕ takes on a specific value. Once the K problems are solved, we can make a K-way conditional plan that gives the correct plan conditional given the observed value of the state variable ϕ. There exist methods for extending refinement strategies so that instead of working on K unconditional plans with significant overlap, a single, multithreaded conditional plan is generated [Peot and Shachter 1991].

Conditional planning can be very expensive in situations in which the unspecified variable ϕ has a large set of possible values or there are several unspecified variables. If there are U unspecified variables each with K possible values, then a conditional plan that covers all possible contingencies has to account for U^K possible initial states. In some cases, we can avoid a combinatorial explosion by performing some amount of on-line planning; first plan to obtain the necessary information, then, after obtaining this information, plan what to do next. Unfortunately, this on-line approach has potential problems.

In travel planning, for example, you could wait until you arrive in Boston's Logan Airport to check on the weather in Chicago in order to plan whether to take a Southern or Northern route to San Francisco. But, if you do wait and it is snowing in Chicago, you may find out that all of the flights taking Southern routes are already sold out. In this case, it would have been better to anticipate the possibility of snow in Chicago and reserve a flight to San Francisco taking a Southern route. Additional complications arise concerning the time when you observe the value of a given variable, the time when you need to know the value of a variable, and whether or not the value of a variable changes between when you observe it and when you need to know a value.

Uncertainty arises not only with respect to initial conditions, but also as a consequence of the actions of the planner (e.g., you get stuck in traffic and miss your flight) or the actions of others (e.g., the airline cancels your flight). In general, uncertainty is handled by introducing *sensing* or *information gathering* actions (operators). These operators have preconditions and postconditions similar to other operators, but some of the postconditions, those corresponding to the consequences of information gathering, are nondeterministic; we will not know the actual value of these postconditions until after we have executed the action [Etzioni et al. 1992].

The approach to conditional planning just sketched theoretically extends to arbitrary sources of uncertainty, but in practice search has to be limited to consider only outcomes that are likely to have a significant impact on performance. Subsequently we briefly consider planning using stochastic models that quantify uncertainty involving outcomes.

Repair Methods in Planning

The refinement methods for plan synthesis described in this section assume access to the complete dynamics of the system. Sometimes, the system dynamics are complex enough that using the full model during plan synthesis can be inefficient. In many such domains, it is often possible to come up with a simplified model of the dynamics that is approximately correct. As an example, in the travel domain, the action of taking a flight from one city to another has potentially many preconditions, including ones such as: having enough money to buy tickets and enough clean clothes to take on the travel. Often, most of these preconditions are trivially satisfied, and we are justified in approximating the set of preconditions to simply ensure that we have a reservation and are at the airport on time. In such problems, a simplified model can be used to drive plan generation using refinement methods, and the resulting plan can then be *tested* with respect to the complete dynamical model of the system. If the testing shows the plan to be correct, we are done. If not, the plan needs to be *repaired* or *debugged*. This repair process involves both adding and deleting constraints from the plan.

If the complete dynamical model is declarative (instead of being a black box), it is possible to extract from the testing phase an *explanation* of why the plan is incorrect (for example, in terms of some of the preconditions that are not satisfied, or are violated by some of the indirect effects of actions). This explanation can then be used to focus the repair activity [Simmons and Davis 1987, Hammond 1989]. Similar repair methods can also be useful in situations where we have probably approximately correct *canned* plans for generic types of goals, and we would like to solve planning problems involving collections of these goals by putting the relevant canned plans together and modifying them.

Scheduling with Deterministic Dynamics

As we mentioned earlier, scheduling is typically concerned with deciding when to carry out a given set of actions so as to satisfy various types of constraints on the order in which the actions need to be performed, and the ways in which different resources are consumed. Artificial intelligence approaches to scheduling typically declaratively represent and reason with the constraints.

Constraint-based schedulers used in a real applications generally employ sophisticated programming languages to represent a range of constraints. For example, many schedulers require temporal constraints that specify precedence, contiguity, duration, and earliest and latest start and completion times for tasks.

In some schedulers, temporal constraints are enforced rigidly, so that they never need to be checked during search. Many scheduling problems also manage a variety of resources. In the job-shop scheduling problem, machines are resources; only one task can be performed on a machine at a time. Other resources encountered in scheduling problems include fuel, storage space, human operators and crew, vehicles, and assorted other equipment. Tasks have constraints that specify their resource requirements and resources have capacity constraints that ensure that a schedule does not over allocate resources.

In addition to constraints on the time of occurrence and resources used by tasks, there are also constraints on the state of the world that are imposed by physics: the dynamics governing the environment. For example, a switch can be in the on or off position but not both at the same time. In some scheduling problems, the dynamical system is represented as a large set of state constraints.

Scheduling and Constraint Satisfaction

Scheduling problems are typically represented in terms of a set of variables and constraints on their values. A schedule is then represented as an assignment of values to all of the variables that satisfies all of the constraints. The resulting formulation of scheduling problems is called a *constraint satisfaction problem* [Tsang 1993].

Formally, a constraint satisfaction problem is specified by a set of n variables $\{x_1, \ldots, x_n\}$, their respective value domains $\Omega_1, \ldots, \Omega_n$, and a set of m constraints $\{C_1, \ldots, C_m\}$. A constraint C_i involves a subset $\{x_{i_1}, \ldots, x_{i_k}\}$ of the set of all variables $\{x_1, \ldots, x_n\}$ and is defined by a subset of the Cartesian product $\Omega_{i_1} \times \cdots \times \Omega_{i_k}$. A constraint C_i is *satisfied* by a particular assignment in which $x_{i_1} \leftarrow v_{i_1}, \ldots, x_{i_k} \leftarrow v_{i_k}$ just in case $\langle v_{i_1}, \ldots, v_{i_k} \rangle$ is in the subset of $\Omega_{i_1} \times \cdots \times \Omega_{i_k}$ that defines C_i. A solution is an assignment of values to all of the variables such that all of the constraints are satisfied. There is a performance function that maps every complete assignment to a numerical value representing the cost of the assignment. An optimal solution is a solution that has the lowest cost.

As an example, consider the following formulation of a simplified version of the job-shop scheduling problem as a constraint satisfaction problem. Suppose we have N jobs, $1, 2, \ldots, N$, each consisting of a single task, and M machines, $1, 2, \ldots, M$. Since there is exactly one task for each job, we just refer to jobs. Assume that each job takes one unit of time and there are T time units, $1, 2, \ldots, T$. Let $z_{ij} = 1$ if the jth machine can handle the ith job and $z_{ij} = 0$ otherwise. The z_{ij} are specified in the description of the problem. Let x_i for $1 \le i \le N$ take on values from $\{j \mid z_{ij} = 1\} \times \{1, 2, \ldots, T\}$, where $x_i = (j, k)$ indicates that the ith job is assigned to the jth machine during the kth time unit. The x_i are assigned values in the process of planning. There are $N(N - 1)$ constraints of the form $x_i \ne x_j$, where $1 \le i, j \le N$, and $i \ne j$. We are searching for an assignment to the x_i that satisfies these constraints.

Refinement-Based Methods

A refinement-based method for solving a constraint satisfaction problem progresses by incrementally assigning values to each of the variables. A partial plan (schedule) π is represented as partial assignment of values to variables $\{x_{\pi_1} \leftarrow v_{\pi_1}, \ldots, x_{\pi_k} \leftarrow v_{\pi_k}\}$, where $\{x_{\pi_1}, \ldots, x_{\pi_k}\}$ is a subset of the set of all variables $\{x_1, \ldots, x_n\}$. The partial assignment π can be seen as a shorthand notation for all of the complete assignments that agree on the assignment of values to the variables in $\{x_{\pi_1}, \ldots, x_{\pi_k}\}$. A partial assignment π is said to be inconsistent if the assignment of values to variables in π already violates one or more constraints. If the partial assignment π is consistent, it can be refined by selecting a variable x_j that is not yet assigned a value in π and extending π to produce a set of refinements each of which assigns x_j one of the possible values from its domain Ω_j. Thus, the set of refinements of π is $\{\pi \cup \{x_j \leftarrow v\} \mid v \in \Omega_j\}$. In the case of satisficing scheduling, search terminates when a complete and consistent assignment is produced. In the case of optimizing scheduling, search is continued with *branch-and-bound* techniques until an optimal solution is found.

From the point of view of efficiency, it is known that the order in which the variables are considered and the order in which the values of the variables are considered during refinement have a significant impact on the efficiency of search. Considering variables with the least number of possible values first is known to provide good performance in many domains. Other ways of improving search efficiency include using *lookahead* techniques to prune inconsistent partial assignments ahead of time, to process the domains of the remaining variables so that any infeasible values are removed, or using dependency directed backtracking techniques to recover from inconsistent partial assignments intelligently. See Tsang [1993] for a description of these techniques and their tradeoffs.

Repair-Based Methods

A repair-based method for solving constraint satisfaction problems is to start with an assignment to all of the variables in which not all of the constraints are satisfied and reassign a subset of the variables so that more of the constraints are satisfied. Reassigning a subset of the variables is referred to as repairing an assignment. Consider the following repair method for solving the simplified job-shop scheduling problem.

We say that two variables x_i and x_j ($i \ne j$) *conflict* if their values violate a constraint; in the simplified job-shop scheduling problem considered here, a constraint is violated if $x_i = x_j$. *Min-conflicts* is a heuristic for repairing an existing assignment that violates some of the constraints to obtain a new assignment that violates fewer constraints. The hope is that by performing a short sequence of repairs as determined by the min-conflicts heuristic we obtain an assignment that satisfies all of the constraints. The min-conflicts heuristic counsels us to select a variable that is in conflict and assign it a new value that minimizes the number of conflicts. See the Johnston and Minton article in [Zweben and Fox 1994] for more on the min-conflicts heuristic.

Min-conflicts is a special case of a more general strategy that proceeds by making *local* repairs. In the job-shop scheduling problem, a local repair corresponds to a change in the assignment of a single variable.

For the traveling salesperson problem, there is a very effective local repair method that works quite well in practice. Suppose that there are five cities A, B, C, D, E, and an existing tour (a path consisting of a sequence of edges beginning and ending in the same city) (A, B), (B, C), (C, D), (D, E), (E, A). Take two edges in the tour, say (A, B) and (C, D), and consider the length of the tour (A, C), (C, B), (B, D), (D, E), (E, A) that results from replacing (A, B) and (C, D) with (A, C) and (B, D). Try all possible pairs of edges [there are $O(L^2)$ such edges where L is the number of cities], and make the replacement (repair) that results in the shortest tour. Continue to make repairs in this manner until no improvement (reduction in the length of the resulting tour) is possible. Lin and Kernighan's algorithm, which is based on this local repair method, generates solutions that are within 10% of the length of the optimal tour on a large class of practical problems [Lin and Kernighan 1973].

Rescheduling and Iterative Repair Methods

Repair methods are typically implemented with iterative search methods; at any point during the scheduling process, there is a complete schedule available for use. This ready-when-you-are property of repair methods is important in applications that require frequent rescheduling, such as job shops in which change orders and new rush jobs are a common occurrence.

Most repair methods employ greedy strategies that attempt to improve the current schedule on every iteration by making local repairs. Such greedy strategies often have a problem familiar to researchers in combinatorial optimization. The problem is that many repair methods, especially those that perform only local repairs, are liable to converge to local extrema of the performance function and thereby miss an optimal solution. In many cases, these local extrema correspond to very poor solutions.

To improve performance and reduce the risk of becoming stuck in local extrema corresponding to badly suboptimal solutions, some schedulers employ stochastic techniques that occasionally choose to make repairs other than those suggested by their heuristics. Simulated annealing [Kirkpatrick et al. 1983] is one example of a stochastic search method used to escape local extrema in scheduling. In simulated annealing, there is a certain probability that the scheduler will choose a repair other than the one suggested by the scheduler's heuristics. These random repairs force the scheduler to consider repairs that at first may not look promising but in the long term lead to better solutions. Over the course of scheduling this probability is gradually reduced to zero. See the article by Zweben et al. in Zweben and Fox [1994] for more on iterative repair methods using simulated annealing.

Another way of reducing the risk of getting stuck in local extrema involves making the underlying search systematic (so that it eventually visits all potential solutions). However, traditional systematic search methods tend to be too rigid to exploit local repair methods such as the min-conflicts heuristic. In general, local repair methods attempt to direct the search by exploiting the local gradients in the search space. This guidance can sometimes be at odds with the commitments that have already been made in the current search branch. Iterative methods do not have this problem since they do not do any bookkeeping about the current state of the search. Recent work on partial-order dynamic backtracking algorithms [Ginsberg and McAllester 1994] provides an elegant way of keeping both systematicity and freedom of movement.

Improving Efficiency

Whereas the previous sections surveyed the methods used to organize the search for plans and discussed their relative advantages, as observed in the section on complexity results most planning problems are computationally hard. The only way we can expect efficient performance is to exploit the structure and idiosyncrasies of the specific applications. One attractive possibility involves dynamically customizing the performance of a general-purpose search algorithm to the structure and distribution of the application problems. A variety of machine learning methods have been developed and used for this purpose. We briefly survey some of these methods.

One of the simplest ways of improving performance over time involves caching plans for frequently occurring problems and subproblems, and reusing them in subsequent planning scenarios. This approach is called *case-based planning (scheduling)* [Hammond 1989, Kambhampati and Hendler 1992] and is motivated by similar considerations to those motivating task-reduction refinements. In storing a previous planning experience, we have two choices: store the final plan or store the plan along with the search decisions that lead to the plan. In the latter case, we exploit the previous experience by *replaying* the previous decisions in the new situation.

Caching typically involves only storing the information about the successful plan and the decisions leading to it. Often, there is valuable information in the search failures encountered in coming up with the successful plan. By analyzing the search failures and using *explanation-based learning* techniques, it is possible to learn *search control rules* that, for example, can be used to advise a planner as to which refinement or repair to pursue under what circumstances. For more about the connections between planning and learning see Minton [1992].

Approximation in Stochastic Domains

In this section, we consider a planning problem involving stochastic dynamics. We are interested in generating conditional plans for the case in which the state is completely observable [the output function is the identity $h(x_t) = x_t$] and the performance measure is expected discounted cumulative cost with discount γ. This constitutes an extreme case of closed-loop planning in which the executor is able to observe the current state at any time without error and without cost.

In this case, a plan is just a mapping from (observable) states to actions $\pi : S \to A$. To simplify the presentation, we notate states with the integers $0, 1, \ldots, |S|$, where $s_0 = 0$ is the initial state. We refer to the performance of a plan π starting in state i as $J(\pi \mid i)$. We can compute the performance of a plan by solving the following set of $|S| + 1$ equations in $|S| + 1$ unknowns,

$$
J(\pi \mid i) = C(i) + \gamma \sum_{j=0}^{|S|} Pr(f(i, \pi(i)) = j \mid i, \pi(i)) J(\pi \mid j)
$$

The objective in planning is to find a plan π from the set of all possible plans Πs such that for all $\pi' \in \Pi$, $J(\pi \mid i) \geq J(\pi' \mid i)$ for $0 \leq i \leq |S|$.

As an aside, we note that the conditional probability distribution governing state transitions, $Pr(f(i, \pi(i)) = j \mid i, \pi(i))$, can be specified in terms of *probabilistic state space operators*, allowing us to apply the techniques of the section on planning with deterministic dynamics. A probabilistic state-space operator α is a set of triples of the form $\langle \phi, \rho, \omega \rangle$ where ϕ is a set of preconditions, ρ is a probability, and ω is a set of postconditions. Semantically, if ϕ is satisfied just prior to α, then with probability ρ the postconditions in ω are satisfied immediately following α. If a proposition is not included in ϕ, then it is assumed not to affect the outcome of α; if a proposition is not included in ω, then it is assumed to be unchanged by α. For example, given the following representation for α:

$$
\alpha = \{\langle \{P\}, 1, \emptyset \rangle, \langle \{\neg P\}, 0.2, \{P\} \rangle, \langle \{\neg P\}, 0.8, \{\neg P\} \rangle\}
$$

if P is true prior to α, nothing is changed following α; but if P is false, then 20% of the time P becomes true and 80% of the time P remains false. For more on planning in stochastic domains using probabilistic state-space operators, see Kushmerick et al. [1994].

There are well-known methods for computing an optimal plan for the problem previously described [Puterman 1994]. Most of these methods proceed using iterative repair-based methods that work by improving an existing plan π using the computed function $J(\pi \mid i)$. On each iteration, we end up with a new plan π' and must calculate $J(\pi' \mid i)$ for all i. If, as we assumed earlier, $|S|$ is exponential in the number of state variables, then we are going to have some trouble solving a system of $|S| + 1$ equations. In the rest of this section, we consider one possible way to avoid incurring an exponential amount of work in evaluating the performance of a given plan.

Suppose that we know the initial state s_0 and a bound $C_{\max}$ ($C_{\max} \geq \max_i C(i)$) on the maximum cost incurred in any state. Let π be any plan, $J_\infty(\pi) = J(\pi \mid 0)$ be the performance of π accounting for an infinite sequence of state transitions, and $J_K(\pi)$ the performance of π accounting for only K state transitions. We can bound the difference between these two measures of performance as follows (see

Fiechter [1994] for a proof):

$$|J_\infty(\pi) - J_K(\pi)| \leq \gamma^K C_{\max}/(1 - \gamma)$$

These result implies that if we are willing to sacrifice a (maximum) error of $\gamma^K C_{\max}/(1 - \gamma)$ in measuring the performance of plans, we need only concern ourselves with histories of length K. So how do we calculate $J_K(\pi)$? The answer is a familiar one in statistics; namely, we estimate $J_K(\pi)$ by sampling the space of K-length histories.

Using a factored representation of the conditional probability distribution governing state transitions, we can compute a random K-length history in time polynomial in K and N (the number of state variables), assuming that M (the maximum dimensionality of a state-variable function) is constant. The algorithm is simply, given s_0, for $t = 0$ to $K - 1$, determine s_{t+1} according to the distribution $Pr(s_{t+1} \mid s_t, \pi(s_t))$. For each history $\langle s_0, \ldots, s_K \rangle$ so determined, we compute the quantity $V(\langle s_0, \ldots, s_K \rangle) = \sum_{j=0}^{K} \gamma^j C(s_j)$ and refer to this as one *sample*.

If we compute enough samples and take their average, we will have an accurate estimate of $J_K(\pi)$. The following algorithm takes two parameters, ϵ and δ, and computes an estimate $\hat{J}_K(\pi)$ of $J_K(\pi)$ such that

$$Pr[J_K(\pi)(1 - \epsilon) \leq \hat{J}_K(\pi) \leq J_K(\pi)(1 + \epsilon)] > 1 - \delta$$

1. $T \leftarrow 0; Y \leftarrow 0$
2. $S \leftarrow 4 \log(2/\delta)(1 + \epsilon)/\epsilon^2$
3. *While* $Y < S$ *do*
   a. $T \leftarrow T + 1$
   b. *Generate a random history* $\langle s_0, \ldots, s_K \rangle$
   c. $Y \leftarrow Y + V(\langle s_0, \ldots, s_K \rangle)$
4. *Return* $J_K(\pi) = S/T$

This algorithm terminates after generating $E[T]$ samples, where

$$E[T] \leq 4 \log(2/\delta)(1 + \epsilon) \left(J_K(\pi)\epsilon^2 \right)^{-1}$$

so that the entire algorithm for approximating $J_\infty(\pi)$ runs in expected time polynomial in $1/\delta$, $1/\epsilon$, $1/(1 - \gamma)$ (see Dagum et al. [1995] for a detailed analysis).

Approximating $J_\infty(\pi)$ is only one possible step in an algorithm for computing an optimal or near-optimal plan. In most iterative repair-based algorithms, the algorithm evaluates the current policy and then tries to improve it on each iteration. In order to have a polynomial time algorithm, we not only have to establish a polynomial bound on the time required for evaluation but also a polynomial bound on the total number of iterations. The point of this exercise is that when faced with combinatorial complexity, we need not give up but we may have to compromise. In practice, making reasonable tradeoffs is critical in solving planning and scheduling problems. The simple analysis demonstrates that we can trade time (the expected number of samples required) against the accuracy (determined by the ϵ factor) and reliability (determined by the δ factor) of our answers.

Practical Planning

There currently are no off-the-shelf software packages available for solving real-world planning problems. Of course, there do exist general-purpose planning systems. The SIPE [Wilkins 1988] and O-Plan [Currie and Tate 1991] systems are examples that have been around for some time and have been applied to a range of problems from spacecraft scheduling to fermentation planning for commercial breweries. In scheduling, several companies have sprung up to apply artificial intelligence scheduling techniques to commercial applications, but their software is proprietary and by no means turn-key. Moreover, these systems are rather large; they are really programming environments meant to support design and not necessarily to provide the basis for stand-alone products.

Why, you might ask, are there not convenient libraries in C, Pascal, and Lisp for solving planning problems much as there are libraries for solving linear programs? The answer to this question is complicated, but we can provide some explanation for why this state of affairs is to be expected. Before you can solve a planning problem you have to understand it and translate it into an appropriate language for expressing operators, goals, and initial conditions. Although it is true, at least in some academic sense, that most planning problems can be expressed in the language of propositional operators that we introduced earlier in this chapter, there are significant practical difficulties to realizing such a problem encoding. This is especially true in problems that require reasoning about geometry, physics, and continuous change.

In most problems, operators have to be encoded in terms of schemas and somehow generated on demand; such schema-based encodings require additional machinery for dealing with variables that draw upon work in automated theorem proving and logic programming. Dealing with quantification and disjunction, although possible in finite domains using propositional schemas, can be quite complex. Finally, in addition to just encoding the problem, it is also necessary to cope with the inevitable combinatorics that arise by encoding expert heuristic knowledge to guide search. Designing heuristic evaluation functions is more an art than a science and, to make matters worse, an art that requires deep knowledge of the particular domain.

The point is that the problem-dependent aspects of building planning systems are monumental in comparison with the problem-independent aspects that we have concentrated upon in this chapter. Building planning systems for real-world problems is further complicated by the fact that most people are uncomfortable turning over control to a completely automated system. As a consequence, the interface between humans and machines is a critical component in planning systems that we have not even touched upon in this brief overview.

To be fair, the existence of systems for solving linear programs does not imply off-the-shelf solutions to any real-world problems either. And, once you enter the realm of mixed integer and linear programs, the existence of systems for solving such programs is only of marginal comfort to those trying to solve real problems given that the combinatorics severely limit the effective use of such systems. The bottom line is that if you have a planning problem in which discrete-time, finite-state changes can be modeled as operators, then you can look for advice in books such as Wilkins's account of applying SIPE to real problems [Wilkins 1988] and look to the literature on heuristic search to implement the basic engine for guiding search given a heuristic evaluation function. But you should be suspicious of anyone offering a completely general-purpose system for solving planning problems. The general planning problem is just too hard to admit to quick off-the-shelf technological solutions.

27.4 Research Issues and Summary

In this chapter, we provide a framework for characterizing planning and scheduling problems that focuses on properties of the underlying dynamical system and the capabilities of the planning system to observe its surroundings. The presentation of specific techniques distinguishes between refinement-based methods that construct plans and schedules piece by piece, and repair-based methods that modify complete plans and schedules. Both refinement- and repair-based methods are generally applied in the context of heuristic search.

Most planning and scheduling problems are computationally complex. As a consequence of this complexity, most practical approaches rely on heuristics that exploit knowledge of the planning domain. Current research focuses on improving the efficiency of algorithms based on existing representations and on developing new representations for the underlying dynamics that account for important features of the domain (e.g., uncertainty) and allow for the encoding of appropriate heuristic knowledge. Given the complexity of most planning and scheduling problems, an important area for future research concerns identifying and quantifying tradeoffs, such as those involving solution quality and algorithmic complexity.

Planning and scheduling in artificial intelligence cover a wide range of techniques and issues. We have not attempted to be comprehensive in this relatively short chapter. Citations in the main text provide attribution for specifically mentioned techniques. These citations are not meant to be exhaustive by any means. General references are provided in the Further Information section at the end of this chapter.

Defining Terms

Closed-loop planner: A planning system that periodically makes observations of the current state of its environment and adjusts its plan in accord with these observations.

Dynamical system: A description of the environment in which plans are to be executed that account for the consequences of actions and the evolution of the state over time.

Goal: A subset of the set of all states such that a plan is judged successful if it results in the system ending up in one of these states.

History or (state-space) trajectory: A (possibly infinite) sequence of states generated by a dynamical system.

Off-line planning algorithm: A planning algorithm that performs all of its computations prior to executing any actions.

On-line planning algorithm: A planning algorithm in which planning computations and the execution of actions are carried out concurrently.

Open-loop planner: A planning system that executes its plans with no feedback from the environment, relying exclusively on its ability to accurately predict the evolution of the underlying dynamical system.

Optimizing: A performance criterion that requires maximizing or minimizing a specified measure of performance.

Plan: A specification for acting that maps from what is known at the time of execution to the set of actions.

Planning: A process that involves reasoning about the consequences of acting in order to choose from among a set of possible courses of action.

Progression: The operation of determining the resulting state of a dynamical system given some initial state and specified action.

Regression: The operation of transforming a given (target) goal into a prior (regressed) goal so that if a specified action is carried out in a state in which the regressed goal is satisfied, then the target goal will be satisfied in the resulting state.

Satisficing: A performance criterion in which some level of satisfactory performance is specified in terms of a goal or fixed performance threshold.

State-space operator: A representation for an individual action that maps each state into the state resulting from executing the action in the (initial) state.

State-transition function: A function that maps each state and action deterministically to a resulting state. In the stochastic case, this function is replaced by a conditional probability distribution.

References

Allen, J. F., Hendler, J., and Tate, A., eds. 1990. *Readings in Planning*. Morgan Kaufmann, San Francisco, CA.

Allen, J. F., Kautz, H. A., Pelavin, R. N., and Tenenberg, J. D. 1991. *Reasoning about Plans*. Morgan Kaufmann, San Francisco, CA.

Bäckström, C. and Klein, I. 1991. Parallel non-binary planning in polynomial time, pp. 268–273. In *Proc. IJCAI 12*, IJCAII.

Bylander, T. 1991. Complexity results for planning, pp. 274–279. In *Proc. IJCAI 12*, IJCAII.

Chapman, D. 1987. Planning for conjunctive goals. *Artif. Intelligence* 32:333–377.

Currie, K. and Tate, A. 1991. O-Plan: the open planning architecture. *Artif. Intelligence* 51(1):49–86.

Dagum, P., Karp, R., Luby, M., and Ross, S. M. 1995. An optimal stopping rule for Monte Carlo estimation. In *Proc. 1995 Symp. Found. Comput. Sci.*

Dean, T., Allen, J., and Aloimonos, Y. 1995. *Artificial Intelligence: Theory and Practice*. Benjamin Cummings, Redwood City, CA.

Dean, T. and Boddy, M. 1988. Reasoning about partially ordered events. *Artif. Intelligence* 36(3):375–399.

Dean, T. and Kanazawa, K. 1989. A model for reasoning about persistence and causation. *Comput. Intelligence* 5(3):142–150.

Dean, T. and Wellman, M. 1991. *Planning and Control*. Morgan Kaufmann, San Francisco, CA.

Etzioni, O., Hanks, S., Weld D., Draper, D., Lesh, N., and Williamson, M. 1992. An approach to planning with incomplete information. In *Proc. 1992 Int. Conf. Principles Knowledge Representation Reasoning*.

Fiechter, C.-N. 1994. Efficient reinforcement learning, pp. 88–97. In *Proc. 7th Annu. ACM Conf. Comput. Learning Theory*.

Fikes, R. and Nilsson, N. J. 1971. Strips: a new approach to the application of theorem proving to problem solving. *Artif. Intelligence* 2:189–208.

Garey, M. R. and Johnson, D. S. 1979. *Computers and Intractibility: A Guide to the Theory of NP-Completeness*. W. H. Freeman, New York.

Georgeff, M. P. 1987. Planning. In *Annual Review of Computer Science*, Vol. 2, J. F. Traub, ed., Annual Review Incorp.

Ginsberg, M. L. and McAllester, D. 1994. GSAT and dynamic backtracking. In *Proc. 1994 Int. Conf. Principles Knowledge Representation and Reasoning*.

Graham, R. L., Lawler, E. L., Lenstra, J. K., Rinnooy, and Kan, A. H. G. 1977. Optimization and approximation in deterministic sequencing and scheduling: a survey. In *Proc. Discrete Optimization*.

Gupta, N. and Nau, D. S. 1991. Complexity results for blocks-world planning, pp. 629–633. In *Proc. AAAI-91*, AAAI.

Hammond, K. J. 1989. *Case-Based Planning*. Academic Press, New York.

Hendler, J., Tate, A., and Drummond, M. 1990. AI planning: systems and techniques. *AI Mag.* 11(2):61–77.

Kambhampati, S. 1995. A comparative analysis of partial-order planning and task-reduction planning. *ACM SIGART Bull* 6(1).

Kambhampati, S. and Hendler, J. 1992. A validation structure based theory of plan modification and reuse. *Artif. Intelligence* 55(2–3):193–258.

Kambhampati, S., Knoblock, C., and Yang, Q. 1995. Refinement search as a unifying framework for evaluating design tradeoffs in partial order planning. *Art. Intelligence* 76(1–2):167–238.

Kirkpatrick, S., Gelatt, C. D., and Vecchi, M. P. 1983. Optimization by simulated annealing. *Science* 220:671–680.

Kushmerick, N., Hanks, S., and Weld, D. 1994. An algorithm for probabilistic planning. In *Proc. AAAI-94*. AAAI.

Lawler, E. L., Lenstra, J. K., Rinnooy Kan, A. H. G., and Shmoys, D. B. 1985. *The Travelling Salesman Problem*. Wiley, New York.

Lin, S. and Kernighan, B. W. 1973. An effective heuristic for the travelling salesman problem. *Operations Res.* 21:498–516.

Minton, S., ed. 1992. *Machine Learning Methods for Planning and Scheduling*. Morgan Kaufmann, San Francisco, CA.

Papadimitriou, C. H. and Tsitsiklis, J. N. 1987. The complexity of Markov chain decision processes. *Math. Operations Res.* 12(3):441–450.

Pednault, E. P. D. 1988. Synthesizing plans that contain actions with context-dependent effects. *Comput. Intelligence* 4(4):356–372.

Penberthy, J. S. and Weld, D. S. 1992. UCPOP: a sound, complete, partial order planner for ADL, pp. 103–114. In *Proc. 1992 Int. Conf. Principles Knowledge Representation and Reasoning*.

Peot, M. and Shachter, R. 1991. Fusion and propagation with multiple observations in belief networks. *Artif. Intelligence* 48(3):299–318.

Puterman, M. L. 1994. *Markov Decision Processes.* Wiley, New York.

Rosenschein, S. 1981. Plan synthesis: a logical perspective, pp. 331–337. In *Proceedings IJCAI 7,* IJCAII.

Simmons, R. and Davis, R. 1987. Generate, test and debug: combining associational rules and causal models, pp. 1071–1078. In *Proceedings IJACI 10,* IJCAII.

Tsang, E. 1993. *Foundations of Constraint Satisfaction.* Academic, San Diego, CA.

Wilkins, D. E. 1988. *Practical Planning: Extending the Classical AI Planning Paradigm.* Morgan Kaufmann, San Francisco, CA.

Zweben, M. and Fox, M. S. 1994. *Intelligent Scheduling.* Morgan Kaufmann, San Francisco, CA.

Further Information

Research on planning and scheduling in artificial intelligence is published in the journals *Artificial Intelligence, Computational Intelligence,* and the *Journal of Artificial Intelligence Research.* Planning and scheduling work is also published in the proceedings of the International Joint Conference on Artificial Intelligence and the National Conference on Artificial Intelligence. Specialty conferences such as the International Conference on Artificial Intelligence Planning Systems and the European Workshop on Planning cover planning and scheduling exclusively.

Georgeff [1987] and Hendler et al. [1990] provide useful summaries of the state of the art. Allen et al. 1990 is a collection of readings that covers many important innovations in automated planning. Dean et al. [1995] and Penberthy and Weld [1992] provide somewhat more detailed accounts of the basic algorithms covered in this chapter. Zweben and Fox [1994] is a collection of readings that summarizes many of the basic techniques in knowledge-based scheduling. Allen et al. [1991] describe an approach to planning based on first-order logic. Dean and Wellman [1991] tie together techniques from planning in artificial intelligence, operations research, control theory, and the decision sciences.

28

Knowledge-Based Systems for Natural Language Processing

Kavi Mahesh
New Mexico State University

Sergei Nirenburg
New Mexico State University

28.1 Introduction

A large variety of information processing applications deal with natural language texts. Many such applications require extracting and processing the meanings of the texts, in addition to processing their surface forms. For example, applications such as intelligent information access, automatic document classification, and **machine translation** benefit greatly by having access to the underlying meaning of a text. In order to extract and manipulate text meanings, a natural language processing (NLP) system must have available to it a significant amount of **knowledge** about the world and the domain of discourse. The knowledge-based approach to NLP concerns itself with methods for acquiring and representing such knowledge and for applying the knowledge to solve well-known problems in NLP such as **ambiguity** resolution. In this chapter, we look at knowledge-based solutions to NLP problems, a range of applications for such solutions, and several exemplary systems built in the knowledge-based paradigm.

A piece of natural language text can be viewed as a set of cues to the meaning conveyed by the text, where the cues are structured according to the rules and conventions of that natural language and the style of its authors. Such cues include the words in a language, their inflections, the order in which they appear in a text, punctuation, and so on. It is well known that such cues do not normally specify a direct mapping to a unique meaning. Instead, they suggest many possible meanings for a given text with various ambiguities and gaps. In order to extract the most appropriate meaning of an input text, the NLP system must put to use several types of knowledge including knowledge of the natural language, of the domain of discourse, and of the world in general.

Every NLP system, even a connectionist or purely statistical one, uses at least some knowledge of the natural language in question. Knowledge of the rules according to which the meaning cues in the text are structured (such as grammatical and **semantic** knowledge of the language) is used by practically every NLP system. The term *knowledge-based NLP system* (KB-NLP) is applied in particular to those systems that, in addition to using linguistic knowledge, also rely on explicitly formulated domain or **world knowledge** to solve typical problems in NLP such as ambiguity resolution and inferencing. In this chapter, we concentrate on KB-NLP systems.

Knowledge-Based Ambiguity Resolution: Some Examples

Not all problems in NLP require world knowledge. For example, see the following sentences:

1. The large can can hold the water.
2. The man deposited his salary in the bank.
3a. The child inspected the new bar.
3b. The officer inspected the new bar.
4. The hat was hidden by three o'clock.

A sentence such as example 1 has several local syntactic ambiguities, all of which can be resolved using grammatical knowledge alone. Sentence 2 has a *lexical* or **word sense** ambiguity where the word "bank" can mean either a financial institution or a river bank among other meanings. This ambiguity can be resolved using statistical knowledge that the financial institution sense is more frequent in the given corpus or domain. It can also be resolved (with a better confidence in the result, perhaps) using world knowledge that salaries are much more likely to be deposited in financial institutions in the modern world than in a river bank. Such knowledge is called a **selectional constraint** or *preference.*

Sentence 3 also has a word sense ambiguity in the word "bar," which can mean (in its noun form) either a rod or a drinking place. The ambiguity can be resolved in favor of the rod meaning (such as in a chocolate bar) in sentence 3a using world knowledge that a child is not allowed to inspect drinking places (or by knowing that the rod sense occurs much more frequently with child than does the drinking-place sense). Sentence 3b, however, is truly ambiguous and is hard to interpret one way or the other in the absence of strong knowledge of the domain of discourse.

Sentence 4 is an example of a semantic ambiguity in the preposition "by." Once again, world knowledge about hiding events informs, the system that the hat must have been hidden before the point in time known as three o'clock rather than by Mr. three o'clock or near the place known as three o'clock.

Knowledge-Based Systems in Natural Language Processing

A majority of KB-NLP systems are domain specific. They use rich knowledge about a particular domain to process texts from that domain. NLP is easier in domain-specific systems for several reasons: some of the ambiguities present in the language in general are eliminated in the chosen domain by overriding preferences for a domain-specific (or technical) interpretation; the range of inferences that can be made to bridge a gap is narrowed by knowledge of the chosen domain; and it is much more tractable to encode detailed knowledge of a narrow domain than it is to represent all of the knowledge in the world. Domain-specific NLP systems can also benefit from expert knowledge about the domain, which may be available in the form of rules, heuristics, or just episodic or case knowledge.

Some applications, however, do not allow us to constrain the domain narrowly. Often, any piece of text must be acceptable, as input to the system or texts from a domain may nevertheless talk about pretty much anything in the world. In such tasks, since it is impossible to provide the system with expert knowledge in every domain, one must build a base of common-sense knowledge about the world and use that knowledge to solve problems such as ambiguity.

Knowledge-based solutions have influenced the field of NLP in significant ways, providing a viable alternative to approaches based only on grammars and other linguistic information. In particular, knowledge-based systems have made it possible to integrate NLP systems with other knowledge-based artificial intelligence (AI) systems such as those for problem solving, engineering design, and reasoning (e.g., Peterson et al. [1994]). Efforts at such integration have moved the focus away from natural language front ends to more tightly coupled systems where NLP and other tasks interact with and aid each other significantly. KB-NLP systems have provided a means for these AI systems to be grounded in real-world input and output in the form of natural languages.

KB-NLP systems have also brought to focus a range of applications that involve NLP of more than one natural language. Knowledge-based systems are particularly desirable in tasks involving multilingual processing such as machine translation, multilingual database query, information access, or multilingual summarization. Since linguistic knowledge tends to differ in significant ways from one language to another, such systems, especially those that deal with more than two languages, benefit greatly by having a common, interlingual **representation** of the meanings of texts. Deriving and manipulating such language-independent meaning representations go hand in hand with the knowledge-based approach to NLP.

Knowledge-based solutions are equally applicable to both language understanding and generation. However, knowledge-based techniques have been explored and applied in practice to language understanding to a far greater extent than to generation. Although we use understanding as the task in this chapter to illustrate the majority of problems and techniques, KB-NLP has been quite influential in the field of generation as well.

28.2 Underlying Principles

NLP can be construed as a search task where the different possible meanings of words (or phrases) and the different ways in which those meanings can be composed with each other define the combinatorics of the space. All knowledge-based systems can be viewed as search systems that use different types of knowledge to constrain the search space and make the search for an optimal or acceptable solution more efficient. Knowledge such as that contained in selectional constraints aids NLP in two ways: it constrains the search by pruning certain branches of the search space that are ruled out as per the knowledge and it also guides the system in the remaining search space along paths that are more likely to yield preferred interpretations. Thus, knowledge helps to both reduce the size of the search problem and to make the search in the reduced space more efficient.

In designing a KB-NLP system one must first determine how to represent the meanings of texts. Once such a *meaning representation* is designed, knowledge about the world, meanings of words in a language, and meanings of texts can all be expressed in well-formed representations built upon the common meaning representation [Mahesh and Nirenburg 1996].

Knowledge in KB-NLP systems is typically partitioned into a **lexical knowledge** base and a world KB. A lexical KB typically contains linguistic knowledge, such as word meanings, the syntactic patterns in which they occur, and special usage and idiosyncratic information, organized around the words in the language. It is, in general, a good practice to keep language-specific knowledge in the lexical KB and domain or world knowledge in a separate world KB, which is sometimes also called an **ontology,** especially in multilingual systems [Carlson and Nirenburg 1990, Mahesh 1996]. In such a design, the world KB can be common across all of the natural languages being processed and can also be shared potentially with other knowledge-based AI systems. In what follows, whenever we refer to a KB or to retrieving knowledge from a KB, we mean the world KB, not a lexicon.

In some systems, such as case-based NLP systems [Cullingford 1978, DeJong 1979, 1982, Lebowitz 1983, Lehnert et al. 1983, Ram 1989, Riesbeck and Martin 1986a, 1986b], in addition to ontological world knowledge, one or more episodic knowledge bases may be employed. An episodic KB, sometimes also known as an *onomasticon* or *case library,* contains remembered instances of ontological **concepts** (such as people, places, organizations, events, etc.) and is typically used for case-based inferences but may also be used for selection tasks.

The application of knowledge in KB-NLP systems can be divided into one of two fundamental operations:

- *Knowledge-based selection.* A very common problem in NLP is ambiguity where the system must choose between two or more possible solutions (meanings of a word, for instance). Knowledge-based selection is a method for choosing one (or a subset) of the possible meanings as per constraints derived from the system's knowledge of how meanings combine with those of other words in (syntactically related parts of) the text. Use of such selectional constraints for word sense **disambiguation** was illustrated in example sentences 2 and 3 given earlier. Knowledge of the relationships between each of the possible meanings of a unit and meanings of other units of the text is used to set up a set of constraints for each possible meaning. These constraints are evaluated using the knowledge base and the results compared with each other. The choice that best meets all of the constraints is selected by the system as the most preferred meaning of the unit.

- *Knowledge-based inference.* Often, there are zero choices to select from or there are gaps in the input where certain parts of the meaning or how certain parts compose with each other is not specified by any cues in the input. Knowledge-based systems apply their knowledge to infer such missing information to fill the gaps in the meaning. Knowledge-based inference is also often necessary to resolve a word sense ambiguity by making inferences about the meanings. For example, in the sentence "The box was in the pen," an understander has to infer that the pen must have been a playpen (not a writing pen) using its knowledge of the default, relative sizes of boxes, writing pens, and playpens (unless it was known in the prior context that it was a playpen).

  Many tasks, such as question answering and information retrieval, also require NLP systems to often make inferences beyond the literal meanings of input texts. The knowledge in the system permits such inferences to be made by means of reasoning and inferencing methods commonly employed in expert systems and other AI systems. Knowledge-based inferences are made essentially by matching partial meanings of input texts with stored knowledge structures and adding the resulting instantiations of the knowledge to the current interpretation of the text. Making such inferences is typically an expensive process due to the cost of retrieving the right knowledge structure from a KB and the typically large search spaces for making inferences from the knowledge structures. KB-NLP systems must often find ways to control inference processes and keep them goal-oriented.

Important issues in knowledge-based selection include:

- It is often the case that the best alternative according to KB selection is inappropriate. When such an error is revealed by later processing, the KB-NLP system must be able to *backtrack* and continue the search. It is also desirable to *recover from an error* by repairing the incorrect meaning rather than continue the search and reinterpret the text from the start. Various methods of error recovery have been implemented in different KB-NLP systems (e.g., Eiselt [1989] and Mahesh [1995]). Backtracking and failure recovery require the system to retain unselected alternatives for possible use later during recovery.

- The algorithm may in fact be made to eliminate those choices whose scores fall below a certain threshold to prune the search space and reduce the size of the search problem.

- KB selection might also use heuristics to guide the search just like in other AI systems. Examples of such heuristics include structural syntactic preferences (such as *minimal attachment* and *right association*, discussed subsequently), those derived from knowledge of the domain, or other search heuristics.

- In addition to doing a best-first search and backtracking, the algorithm can also implement different schemes for dependency analysis to assign blame for an error and make an informed choice during backtracking. Such methods are warranted by the typically large search spaces encountered in trying to analyze the meanings of real-world texts with complex sentences containing multiple, interacting ambiguities.

- NLP involves the application of several different types of knowledge, including syntactic, semantic, and domain or world knowledge. Constraints from each knowledge source are satisfied to various degrees, resulting in different scores for each constraint. All of this evidence must be combined to make an integral decision in resolving an ambiguity. Since there is no optimal ordering for processing different types of knowledge that is always the right one for NLP, flexible architectures such as **blackboard systems** are ideally suited for building KB-NLP systems [Mahesh 1995, Nirenburg and Frederking 1994, Nirenburg et al. 1994, Reddy et al. 1973].

- It is also possible for a selection algorithm to pursue a limited number of parallel interpretations and delay decisions until further information eliminates some of the meanings under consideration. This approach leads to various problems in representing parallel interpretations and controlling their processing.

Methodological Issues

A basic requirement for KB-NLP is that the system must be provided with all of the necessary knowledge. In spite of some advances in automated acquisition of knowledge for the purposes of NLP [Wilks et al. 1995], acquisition must be done manually for the most part. As a result, KB-NLP systems share the knowledge engineering bottleneck with other knowledge-based AI systems. However, for many applications, the types of knowledge necessary for NLP make it easier to acquire the knowledge and to automate acquisition partially than in the case of problem solving or other expert systems.

There has been a lot of attention focused on the problem of designing a standardized knowledge representation language that serves as a common interchange format so that different systems and research groups can share knowledge bases developed by each other. Work on defining such standards is still in preliminary stages. However, several examples of sharable knowledge bases with translators between different formats can be found already in experimental development [Genesereth and Fikes 1992, Gruber 1993, IJCAI 1995].

Knowledge acquisition for KB-NLP involves building a broad coverage world model (or ontology, [Carlson and Nirenburg 1990, Knight and Luk 1994, Mahesh 1996, Mahesh and Nirenburg 1995b]) and a sufficiently large lexicon for each of the desired languages that represents meanings of words using the concepts in the world model. Any such effort to build a nontrivial KB-NLP system requires the collaborative efforts of a number of linguists, native speakers, lexicographers, ontologists, domain experts, and system developers and testers. In order to aid the acquisition of both the world model and the lexicons, it is essential to have a set of tools that support a number of activities including:

- Searching and browsing the hierarchies and the conceptual meanings in the ontology
- Searching and browsing previously acquired lexical entries
- Searching and browsing an on-line dictionary, thesaurus, or corpus
- Graphical editing of the hierarchies and conceptual relationships in the ontology
- Lexical entry creation and editing (with minimal manual typing)
- Various types of consistency checking both within the ontology and with the lexicon and for conformance with overall design and guidelines
- Automatic testing and correction of ontological and lexical entries
- Supporting interactions between ontologists, domain experts, knowledge engineers, and lexicon acquirers, such as through an interface for submitting requests for changes or additions and for logging them
- Database maintenance, version control, any necessary format conversions, and support for distributed maintenance and access

In practice, a system never has all of the knowledge that it might ever need. As such, it is important for KB-NLP systems to be able to reason under uncertainty and make inferences from incomplete knowledge. Since natural languages tend to have a large variety of constructs with many idiosyncracies, it is invaluable for KB-NLP systems to be robust and be able to produce meaningful output even when complete knowledge is not available.

A drawback of the KB-NLP paradigm is that it is often intractable to provide the system with all of the knowledge it may need. As a result, KB-NLP systems often fail to solve a particular instance of an NLP problem as they may not have the complete knowledge necessary for proper solution. There are two trends in current research aimed at solving this problem: multiengine and human assisted systems.

Multiengine systems use a KB-NLP engine as one of several engines that together solve the problem of NLP. For example, a KB-NLP system may work together with a statistical or a purely **syntax**-based system. In such systems, several designs are possible: a dispatcher could be used to examine each input and decide which engine to call; all engines could be run in parallel and an *output selector* used to select one or more of the outputs; or all engines could be run in parallel, in a race, and the first result used [Nirenburg and Frederking 1994, McRoy and Hirst 1990]. In general, a dispatcher or output-selector architecture is easier to implement than one requiring the outputs of several engines to be combined.

Multiengine architectures have been applied especially to machine translation among multiple natural languages [Nirenburg et al. 1992b, 1994, Frederking et al. 1993, Nirenburg and Frederking 1994, Frederking and Nirenburg 1994]. Heterogeneous multiengine architectures where different subsets of techniques and engines can be combined for different languages are particularly suited to multilingual NLP systems. They enable the overall system to have different sets of tools and engines for different languages. Such flexibility is essential for rapid development, since certain tools or engines may be too expensive to build for some languages but are readily available for others.

Human assisted KB-NLP systems produce partial solutions to NLP problems and interact with a human user to incrementally perform the task. Depending on whether the human user is a naive user or an expert linguist, different modes and depths of interaction may be desirable. However, even when the users are expert linguists, these systems are very effective since they provide the experts with fast, sophisticated access to on-line dictionaries, lexicons, corpora, and so on and perform most of the less tricky computation very quickly. These systems also aid the users in visualizing problems, knowledge representations, and intermediate results with ease. With advances in hardware and software technology for building highly usable graphical interfaces, this alternative is becoming highly popular in situations that do not require fully automatic NLP. This also makes it possible to build integrated NLP workstations that support program development, debugging, testing, knowledge acquisition, and end-user interaction all in a single environment with immediate interactions and feedback between the different components (e.g., Nirenburg et al. [1992b] and Nirenburg [1994]).

28.3 Knowledge-Based Natural Language Processing in Practice

Knowledge-based solutions have been applied to solve a variety of problems in NLP. In this section, we illustrate some of the well-known problems and describe methods using knowledge-based selection to solve the problem. We also introduce the reader to several exemplary systems that have addressed the problems. These system descriptions also serve as a broad coverage of some of the most successful KB-NLP systems built so far. Systems using knowledge-based inference can be viewed as applications of NLP and will be described at the end of this section.

Syntactic Ambiguity

A syntactic ambiguity can be of two kinds: category ambiguity and attachment ambiguity. Category (or part of speech) ambiguities often can be resolved using syntactic knowledge of the language. Attachment

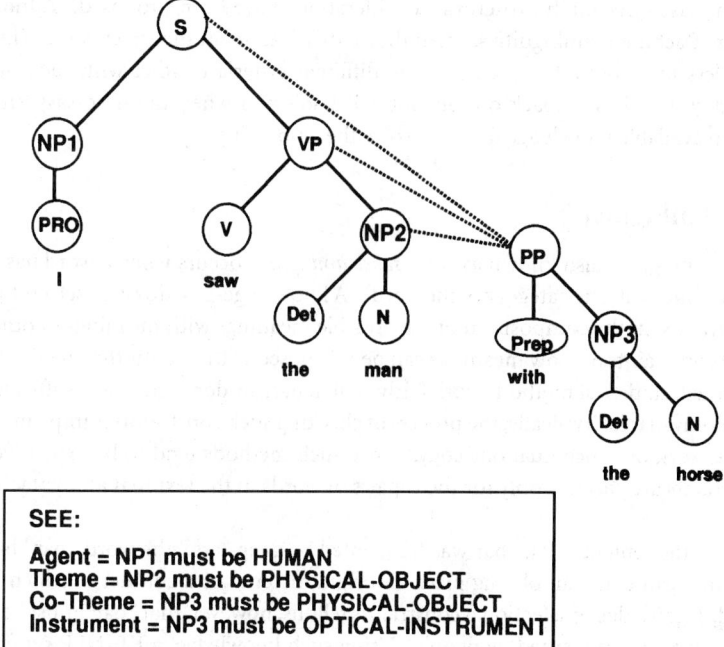

FIGURE 28.1 A PP attachment ambiguity showing an ambiguous parse tree and selectional constraints on attachments.

ambiguities often require semantic and world knowledge to resolve. A commonly occurring type of attachment ambiguity is the prepositional phrase- (PP-) attachment problem where there is ambiguity as to which part of a sentence a prepositional phrase is modifying. Constraints on composing the meanings of the two units being attached (see Fig. 28.1) derived from semantic or world knowledge must often be applied to select one of the possible attachments.

A knowledge-based algorithm for resolving attachment ambiguities sets up constraints derived from semantic and world knowledge for composing the meanings of the child unit (being attached) with meanings of each of the possible parent syntactic units (at which the child could be attached). Such selectional constraints (e.g., Allen [1987] and Wilks [1975]) are typically represented in the form of a permissible range of fillers for slots in frames representing the meanings. The potential filler (i.e., the meaning of the child unit) is compared against this constraint by a fuzzy match function. A popular way to do this fuzzy match is to compute a weighted distance between the two meanings in a semantic or ontological network of concepts. The closer the two are in the network, the higher the score assigned to the particular choice. The algorithm then combines the scores from different constraints for the same choice by applying a mathematical function which may be as simple as addition or multiplication or may be complex variants of root-mean-squared functions and the like. The resulting combined scores are used to select the best choice according to the knowledge of selectional constraints (e.g., Beale et al. [1996]).

For example, in Fig. 28.1, the PP "with the horse" can be attached in any one of the three ways shown in the tree. Because of this ambiguity, it is not clear whether the horse is accompanying the man or is an instrument of seeing. However, as shown in the lower-half of the figure, knowledge of selectional constraints tells us that an instrument of seeing must be an optical instrument. Since a horse is not (typically) an optical instrument, the attachment to the verb phrase (VP) "saw" can be ruled out, resulting in an interpretation where the horse is an accompanier to "the man."

There have been a number of attempts, both in NLP and psycholinguistics, to develop purely syntactic methods for resolving syntactic attachment ambiguities. Several criteria, such as minimal attachment and right association, based on purely structural considerations have been proposed. Although they work in many cases of attachment ambiguities, often they either lead to an incorrect choice (i.e., semantically meaningless or less preferred attachment) or the different criteria conflict with each other. They are nevertheless highly useful as fallback options for a KB solution when the necessary knowledge is not available or when available knowledge fails to resolve the ambiguity.

Word Sense Ambiguity

Lexical semantic ambiguity, also known as *word sense ambiguity,* occurs when a word has more than one meaning (for the same syntactic category of the word). A knowledge-based system selects the meaning that best meets constraints on the composition of the possible meanings with meanings of other words in the text. Selectional constraints on how meanings can be composed with one another are derived from world knowledge, although statistical methods work fairly well in narrow domains where sufficient training data on word senses are available. Typically, the process of checking such constraints is implemented as a search in a semantic network or conceptual ontology. Since such methods tend to be expensive in large-sized networks, constraints are checked only for those pairs of words in the text that are syntactically related to one another.

For example, in the sentence "The bar was frequented by gangsters," the word "bar" has a word sense ambiguity: it can mean either an oblong piece of a rigid material or a counter at which liquors or light meals are served. Knowledge of selectional constraints tells us, however, that "bar" must be a place of some kind in order for it to be frequented by people. Using such knowledge, a KB-NLP system can correctly resolve the word sense ambiguity and determine that "bar" in this case is a liquor-serving place.

Knowledge-Based Systems for Syntactic and Semantic Analysis

KB-NLP systems for sentence processing can be described in terms of the representations and processing mechanisms they use to apply world knowledge to solve problems such as syntactic and semantic ambiguity. In the following descriptions, we pay particular attention to methods for combining linguistic knowledge with world knowledge. It can be seen that many KB-NLP systems combine the two types of knowledge directly in the representations whereas others try to combine them during processing.

Early models of sentence processing built in the KB-NLP paradigm attempted integration of multiple types of knowledge in the lexical entries of words. A good example of a model of this kind is the conceptual analyzer (CA) model [Birnbaum and Selfridge 1981] based on conceptual dependency representations [Schank and Abelson 1977]. Lexical entries in CA contained rules about what to expect before and after the word in various situations and how to combine the meanings of surrounding words with those of the head words. These lexical entries contained both linguistic and world knowledge. Because of the complete reliance on lexical packaging of all types of knowledge, these models suffered from lack of compositionality and problems of scalability. Small and Rieger [1982] built a model called the word expert parser which used a similar technique. Each word was an expert that knew exactly how to combine its meaning with those of other words in the surrounding context.

Many recent models such as Jurafsky's [1992] SAL use integrated representations of all of the different kinds of knowledge in a monolithic knowledge base of integrated constructs called grammatical constructions. Jurafsky's model differs from other integrated models mentioned previously in detailing an explicit decomposition of processing into distinct phases called *access, selection,* and *integration* of alternative interpretations. This is an example of a decomposition of the language processing task that is orthogonal to standard units of analyses such as syntax, semantics, and pragmatics.

A related type of integrated model was proposed by Wilks [1975] based on the use of stored conceptual knowledge in the form of templates. Wilks proposed that these templates store constraints on fillers that act as selectional preferences rather than as restrictions. These templates also included syntactic ordering information and as such were packages that combined syntactic, semantic, and conceptual knowledge in an integrated representation.

A different kind of representation is employed in models where syntactic and semantic knowledge are separable but the mappings between them are encoded a priori in individual lexical entries. In this formalism, there are separate syntactic and semantic processes (typically arranged in a syntax-first sequential architecture) but the mapping from syntactic structures to semantic structures is precomputed and enumerated in the lexicon. Models based on lexical semantics, such as the Diana and Mikrokosmos analyzers for machine translation [Meyer et al. 1990, Nirenburg et al. 1992a, Onyshkevych and Nirenburg 1995] are typically sequential models using such representations.

NL-SOAR [Lehman et al. 1991, Lewis 1993a, 1993b] is a model of sentence understanding based on the SOAR architecture for production systems that are capable of learning by chunking [Laird et al. 1987]. NL-SOAR used architectural constraints from SOAR and a large set of language processing rules to model a range of syntactic phenomena in human sentence processing including structural ambiguity resolution, garden path sentences, and parsing breakdown (such as in center embedded sentences). The model also showed a gradual transition in behavior from deliberative reasoning to immediate recognition. This was made possible by the chunking mechanism in SOAR that produced chunks by combining all of the productions that were involved in selecting a particular interpretation for a sentence. A drawback of this approach was the gradual loss of functional independence between the different knowledge sources. Though syntactic and semantic productions were encoded separately to begin with, the chunking process combined them to produce bigger monolithic units applicable to specific types of sentences. Thus, NL-SOAR starts with separate representations of different types of knowledge but gradually builds its own integrated representations. Another drawback of using the SOAR architecture was its serial nature with the consequence that NL-SOAR could pursue only one interpretation of a sentence at any time. Moreover, since productions could contain any type of knowledge, one could encode productions that are rather specific to a particular type of sentence. In fact, chunking would produce such productions that would be applicable in sentences with particular combinations of ambiguities and syntactic and semantic contexts.

Cardie and Lehnert [1991] have extended a conceptual analyzer (such as the CA described earlier) to handle complex syntactic constructs such as embedded clauses. They show that the conceptual parser can correctly interpret the complex syntactic constructs without a separate syntactic grammar or explicit parse tree representations. This is accomplished by a mechanism called lexically indexed control kernel (LICK), which is essentially a method for dynamically creating a copy of the conceptual parsing mechanism for each embedded clause.

A model called ABSITY that had separate modules for syntax and semantics was developed by Hirst [1988]. ABSITY had separate representations of syntactic and semantic knowledge. A syntactic parser similar to PARSIFAL [Marcus 1980] ran in tandem with a semantic analyzer based on Montague semantics. This model was able to resolve both structural syntactic and lexical semantic ambiguities and produced incremental interpretations. Syntax was provided with semantic feedback so that syntax and semantics could influence each other incrementally. Lexical disambiguation was made possible by the use of individual processes or demons for each word, called *Polaroid Words*, that interacted with each other to select the meaning that is most appropriate to the context. However, semantic analysis was dependent on the parser producing correct syntactic interpretations. This was a consequence of the requirement of strict correspondence between pieces of syntactic and semantic knowledge for syntax–semantics interaction to work in Hirst's model. Rules for semantic composition had to be paired with corresponding syntactic rules for the tandem design to work.

A noticeable characteristic of Hirst's model was the use of separate mechanisms for solving each subproblem in sentence understanding. The model used a parser based on a capacity limit for syntax, a set of continuously interacting processes through marker passing (e.g., Charniak [1983], discussed subsequently) for resolving word sense ambiguities, a *semantic enquiry desk* for semantic feedback to syntax, and strict correspondence between syntactic and semantic knowledge for ensuring consistency of incremental interpretations. This characteristic is inherited completely by a more recent model by McRoy and Hirst [1990], which has more modules than Hirst's original model and appears to be as heterogeneous as its predecessor. This enhanced model is organized in a *race-based* architecture which simulates syntax and semantics running in parallel by associating time costs with each operation. The model is able to resolve a variety of ambiguities by simply selecting whichever alternative minimizes the time cost (hence the name race-based). The model employed a *sausage-machine*-like two-stage parser for syntactic processing [Frazier and Fodor 1978] and ABSITY for semantic analysis. The two were put together through an *attachment processor* and a set of grammar application routines. The attachment processor also consulted three other sets of routines called the grammar consultant routines, knowledge base routines, and argument consultant routines, resulting in a highly complex and heterogeneous model.

COMPERE is a recent model of sentence understanding designed to integrate syntactic and semantic knowledge dynamically during processing [Mahesh 1995, Mahesh and Eiselt 1994]. COMPERE uses an arbitrating process in a blackboard architecture to choose the best interpretation based on both syntactic and semantic preferences. Additional expert modules can be added to the arbitrating process to account for other linguistic phenomena. A key feature of COMPERE is its close, incremental interactions between syntax and semantics so that most ambiguities can be resolved locally at the earliest possible opportunity in order to minimize the combinatorial effects of multiple ambiguities in a sentence.

Mikrokosmos is a knowledge-based machine translation (KBMT) system that has focused particularly on lexical disambiguation [Onyshkevych and Nirenburg 1995, Mahesh and Nirenburg 1995a, Beale et al. 1996]. It is one of the largest ever attempts at building a knowledge-based system expressly for NLP. In Mikrokosmos, the meaning of the input text is extracted and represented as instantiated elements of an independently motivated model of the world, that is, an ontology [Carlson and Nirenburg 1990, Mahesh 1996]. The link between the ontology and the meaning representation is provided by the lexicon, where the meanings of most open class lexical items are defined in terms of their mappings into ontological concepts and their resulting contributions to meaning. The ontology and the lexicon are the two main knowledge sources in the Mikrokosmos system. Information about the nonpropositional components of text meaning such as speech acts, speaker attitudes and intentions, relations among text units, coreferences, etc., is also derived from the lexicon with inputs from other expert *microtheories*.

The Mikrokosmos core semantic analyzer produces a text meaning representation (TMR) starting from the output of a syntactic analysis module. This knowledge-based engine extracts constraints both from lexical knowledge of a language and from an ontological world model and applies efficient search and constraint satisfaction algorithms to derive the TMR that best meets all known constraints on the meaning of the given text [Beale et al. 1996, Mahesh and Nirenburg 1995a]. The TMR is intended to serve as an interlingua input for generating the translation in a target language.

In a real-world text, many words have lexical ambiguities. Both the lexicon and the ontology specify constraints on the relationships between the various concepts under consideration, based on language-specific and world knowledge, respectively. The Mikrokosmos analyzer checks each of these selectional constraints by determining how closely a candidate filler meets the constraints on an interconcept relationship. Closeness is measured by an ontological search algorithm (also known as the *semantic affinity checker*) that computes a weighted distance between pairs of concepts in the network of concepts present in the ontology. Given a pair of concepts (such as "metal rod" and "place," or, "drinking place" and "place," to determine which sense of "bar" is closer to a "place" that can be the destination of a physical motion, for instance), the ontological search process finds the least cost path between the two concepts in the ontology where the cost of each link is determined by the type of that link. For example, taxonomic links have a lower cost than "member of" or "has parts" links, and so on. By combining the scores for each link along a path, the core semantic analyzer selects a combination of word senses that minimizes the total deviation from all known constraints and constructs the TMR resulting from the chosen relationships between the concepts [Beale et al. 1996].

Another class of models has used a symbolic equivalent of spreading activation called *marker passing* in a semantic network to build models of semantic analysis and lexical ambiguity resolution. Among these models, Charniak's [1983] model and Eiselt's [1989] ATLAST model are particularly interesting. Charniak proposed a semantic processor that used marker passing to initially propose semantic alternatives without being influenced by a syntactic processor (that was running in parallel to the marker passer). There was a third module that combined syntactic preferences with the semantic alternatives to select the combined interpretation. ATLAST was a model of lexical semantic and pragmatic ambiguity resolution, as well as error recovery, also using marker passing [Eiselt 1989]. ATLAST divided both syntax and semantics into three different modules called the *capsulizer*, the *proposer*, and the *filter*. The capsulizer was an initial stage that packaged the input sentence into local syntactic units and sent them incrementally to the proposer. The proposer was a marker passer quite akin to the one in Charniak's model. The third module, the filter, combined syntactic and semantic preferences to arrive at the final interpretation. The most important aspect of this model was its ability to recover from its errors without completely reprocessing the sentence by conditionally retaining previously unselected alternatives [Eiselt 1989]. ATLAST was one of the very few models that highlighted the importance of error recovery in shaping the design of a sentence understander. ATLAST's drawback was its limited syntactic capabilities (limited to simple subject-object-verb single-clause sentences) realized in a simple augmented transition-network (ATN) parser (e.g., Allen [1987]) that did not interact in any interesting way with the semantic analyzer.

Models have also been built using connectionist networks with activation and inhibition between nodes being the means of interaction. For example, Waltz and Pollack [1985] built a connectionist model where syntactic and semantic decisions were made in separate networks, but the networks exchanged activation with each other and settled on a combined interpretation at the end. A fundamental problem with these models is the inability of their processing mechanism, spreading activation, to deal with syntactic processing in a large scale. In spite of recent advances in connectionist networks, it is yet to be demonstrated that a connectionist network can process the complex syntax of natural languages and scale up to handle real-world texts.

Knowledge-Based Solutions to Other Natural Language Processing Problems

Problems of reference and coreference in NLP often require world knowledge in addition to different types of linguistic knowledge for effective solution. A KB-NLP system can use instances of various concepts in its meaning representation to keep track of references. A unification method may be used to determine coreference between a pair of instances in the meaning representation when suggested by linguistic cues. This method is particularly feasible in a KB-NLP system whose world knowledge KB has inheritance capabilities. Inherited properties of instances help determine coreference relations whether or not such information is explicitly mentioned in the input text.

Knowledge-based solutions are also applicable to a variety of other problems in NLP such as thematic analysis, topic identification, discourse and context tracking, and temporal reasoning. Knowledge-based methods also play a crucial role in natural language generation for problems such as lexical choice (i.e., selecting a good word or words to realize a given meaning in a language) and text planning (i.e., designing discourse structure, sentence and paragraph boundaries, generating texts with ellipsis and anaphora, and so on, e.g., Viegas and Bouillon [1994] and Viegas and Nirenburg [1995].

Knowledge-Based Inference for Applied Natural Language Processing

Knowledge-based inference mechanisms for NLP offer partial solutions to a number of practical applications involving text processing. Some of the significant applications include database query; information retrieval and question answering; conceptually based document retrieval (such as searching the World Wide Web); information extraction; knowledge acquisition from natural language texts (e.g., Peterson et al. [1994]); automatic summarization; knowledge-based machine translation; interpreting nonliteral expressions such as metonymy, metaphor, and idioms; document classification; intelligent authoring environments; and intelligent agents that communicate in natural languages. Development of KB-NLP systems for many of these applications is still in somewhat early stages.

28.4 Research Issues and Discussion

The development of KB-NLP systems both contributes to and benefits from basic and applied research in several areas including computational linguistics, natural language semantics, knowledge representation, knowledge acquisition, architectures for language processing, and cognitive science. In this section, we briefly examine some of the research issues in knowledge representation and acquisition.

Knowledge Representation for Natural Language Processing

KB-NLP requires several types of knowledge, both linguistic (such as grammatical and lexical knowledge) and world knowledge. A key issue in knowledge representation for NLP is how to represent and combine the different types of knowledge. Should there be a single representation that combines all of the types of knowledge a priori (perhaps in the form of word experts [Small and Rieger 1982], in an episodic memory [Riesbeck and Martin 1986a, 1986b, Schank and Abelson 1977] or in a semantic grammar [Burton 1976]), or should there be separate representations that are compatible with one another? The choice is especially important in a multilingual NLP system where it is highly desirable to keep the linguistic knowledge separate for each language but share the common world knowledge across the entire system.

If the representations of linguistic and world knowledge are kept separate, problems arise due to differences in expressiveness and types of inferences supported by the different representations. Whereas attempts have been made to design uniform representation formats (e.g., predicate-logic-based representations like Knowledge Interchange Format (KIF), Genesereth and Fikes [1992] and Gruber [1993]) for all types of knowledge, in practice it seems best to allow slightly different representations and guarantee compatibility across the different representations through an integrated methodology for acquisition [Mahesh 1996, Mahesh and Nirenburg 1995b, 1996]. Such compatibility can be guaranteed by encoding explicit cross references (between lexical and world knowledge and between grammatical and lexical knowledge). Such a design also allows one to develop a flexible, blackboard architecture [Mahesh 1995, Nirenburg and Frederking 1994, Nirenburg et al. 1994, Reddy et al. 1973] in which different expert microtheories that solve particular problems in NLP using appropriate knowledge come together to achieve the common goals of KB-NLP [Mahesh and Nirenburg 1995a, Onyshkevych and Nirenburg 1995].

The design of a knowledge representation for KB-NLP also affects the accessibility of the represented knowledge for NLP purposes. For example, acquiring lexical entries requires one to find the right concepts in the world knowledge in order to represent meanings of words. Resolving certain word sense ambiguities and interpreting nonliteral expressions requires one to measure distances between concepts in the world knowledge. A representation employed for KB-NLP must be designed expressly to support such operations.

Knowledge Acquisition for Natural Language Processing

Basic research problems in knowledge acquisition for NLP address possible ways of automating the acquisition of knowledge and the development of methodologies for acquiring linguistic and world knowledge together so that they are compatible with one another.

It is clear that different sources, experts, and constraints govern the acquisition of lexical and world knowledge, especially in a multilingual system. It is important to have an integrated, situated methodology for acquisition where the linguistic and world-knowledge acquisition teams interact with each other regularly and negotiate their choices for each piece of representation. Such an incremental methodology accompanied by a set of tools for supporting the interaction and for quality control is inevitable to guarantee compatibility between the knowledge sources. It will also guarantee that all of the acquired knowledge is in fact usable for KB-NLP purposes. Although it is highly desirable to have general-purpose knowledge bases that can be used for both KB-NLP and other tasks, it has not yet been demonstrated convincingly that knowledge that was not acquired for use in NLP, such as Cyc [Lenat and Guha 1990], can in fact be used effectively for practical KB-NLP.

In situated knowledge acquisition for KB-NLP, lexical and world knowledge bases are developed concurrently. Often the best choices for each type of knowledge conflict with one another. Practice shows that negotiations to meet the constraints on both a lexical entry and a concept in the ontology leads to the best choice in each case [Mahesh 1996, Mahesh and Nirenburg 1995b]. It also ensures that every entry in each knowledge base is consistent, compatible with its counterparts, and has a purpose toward the ultimate objectives of KB-NLP. It is also very important to have a well-documented set of guidelines for making choices in both linguistic and world knowledge acquisition [Mahesh 1996, Mahesh and Nirenburg 1995b]. The ideal method of situated development of knowledge sources for multilingual NLP is one where an ontology and at least two lexicons for different languages are developed concurrently. This ensures that world knowledge in the ontology is truly language independent and that representational needs of more than one language are taken into account.

Knowledge acquisition for KB-NLP is very expensive in any nontrivial system. The acquisition of world knowledge is typically underconstrained, even in well-defined domains. The situated methodology described previously offers a practical way of constraining the amount of world knowledge to be acquired. Only those concepts and conceptual relationships that are required to represent and disambiguate word meanings need to be acquired for KB-NLP purposes. This automatically eliminates certain types of knowledge that may well be within the domain of interest but will never be used for KB-NLP purposes (e.g., Mahesh [1996]).

In addition to situated development, knowledge acquisition for KB-NLP can be made more tractable by partial automation and by the use of advanced tools. Various attempts have been made to automatically construct grammars and lexicons by analyzing a large corpus of texts. Attempts are also being made to bootstrap the acquisition process so that the KB-NLP system learns and acquires more knowledge as a result of processing input texts. Although these are promising approaches to reduce the costs of knowledge acquisition, at present some of the most successful KB-NLP systems have been built by careful manual knowledge acquisition supported by sophisticated tools that help achieve parsimony and high quality in the acquired representations.

28.5 Summary

Communication in natural languages is possible only when there is a significant amount of general knowledge about the world that is shared among the different participants. An NLP system can make use of such general knowledge of the world as well as specific knowledge of the particular domain in order to solve a range of hard problems in NLP. In this chapter, we have shown how such knowledge can be applied in practice to resolve syntactic and semantic ambiguities and make necessary inferences. We have described algorithms for these solutions, pointed the reader at some of the best implemented systems, and presented some of the current trends and research issues in this field. We would like to emphasize the key practical issue in KB-NLP, namely, that KB solutions are viable and attractive but are often incomplete or rather expensive in practice. It is a good design strategy to evaluate a KB-NLP solution against its alternatives for a particular problem and make the best choice of single method, or, if appropriate, design a multi-engine or human-assisted NLP system. We hope this chapter has helped the reader make a well-informed decision.

Defining Terms

Ambiguity: A situation where there is more than one possible interpretation for a part of a text.

Attachment: A syntactic relation between two parts of a sentence where one modifies the meaning of the other.

Blackboard system: A system where several independent modules interact and coordinate with each other by accessing and posting results on a public data structure called the blackboard.

Concept: A unit in world knowledge representation that denotes some object, state, event, or property in the world.

Disambiguation: The process of resolving an ambiguity by selecting one (or a subset) of the possible set of interpretations for a part of a text.

Knowledge: A blanket term for any piece of information that can be applied to solve problems.

Lexical knowledge: Knowledge of words in a natural language. Includes knowledge of syntactic constructs in which the word(s) appears in the language, possible meanings of the words, pronunciation, part of speech, possible inflections, and so on.

Machine translation:　Translating a text in one natural language to another natural language by computer.

Ontology:　A model of the world that defines each concept that exists in the world as well as taxonomic and other relationships among the concepts. A knowledge base containing information about the domain of interest.

Representation of knowledge:　An encoding of knowledge that is computationally tractable.

Selectional constraint:　A constraint on the range of concepts with which a concept can have a particular relationship.

Semantics:　The branch of linguistics that deals with the meanings of words and texts.

Syntax:　The branch of linguistics that deals with the rules that explain the kinds of sentence structure permissible in a language.

Word sense:　One of the possible meanings of a word in a language.

World knowledge:　Extra-linguistic knowledge. Knowledge of the world or a particular domain that is not specific to any particular natural language.

References

Allen, J. 1987. *Natural Language Understanding*. Benjamin/Cummings.

Allen, J. 1989. Natural language understanding. Section G: Conclusion. In *The Handbook of Artificial Intelligence*, Vol. IV, A. Barr, P. R. Cohen, and E. A. Feigenbaum, eds., pp. 238–239. Addison–Wesley, Reading, MA.

Barr, A. and Feigenbaum, E. A., eds. 1981. *The Handbook of Artificial Intelligence*. Addison–Wesley, Reading, MA.

Beale, S., Nirenburg, S., and Mahesh, K. 1996. Hunter-gatherer: three search techniques integrated for natural language semantics. In *Proc. 13th Nat. Conf. Artif. Intell.* Portland, OR.

Birnbaum, L. and Selfridge, M. 1981. Conceptual analysis of natural language. In *Inside Computer Understanding*. R. Schank and C. Riesbeck, eds., pp. 318–353. Lawrence Erlbaum Associates, Hillsdale, NJ.

Burton, R. 1976. Semantic Grammar: An Engineering Technique for Constructing Natural Language Understanding Systems. *BBN Rep. No.* 3453, Bolt, Beranek, and Newman, Cambridge, MA.

Cardie, C. and Lehnert, W. 1991. A cognitively plausible approach to understanding complex syntax, pp. 117–124, In *Proc. 9th Nat. Conf. Artif. Intell.* Morgan Kaufmann, San Mateo, CA.

Carlson, L. and Nirenburg, S. 1990. World Modeling for NLP. *Center for Machine Translation, Tech. Rep.* CMU-CMT-90-121, Carnegie Mellon University, Pittsburgh, PA.

Charniak, E. 1983. Passing markers: a theory of contextual influence in language comprehension. *Cognitive Sci.* 7:171–190.

Cullingford, R. 1978. *Script Application: Computer Understanding of Newspaper Stories*. Ph.D. dissertation. Department of Computer Science, Yale University, New Haven, CT. Res. Rep. 116.

DeJong, G. 1979. *Skimming Stories in Real Time: An Experiment in Integrated Understanding*. Ph.D. dissertation. Department of Computer Science, Yale University, New Haven, CT. Res. Rep. 158.

DeJong, G. 1982. An overview of the FRUMP system. In *Strategies for Natural Language Processing*. W. G. Lehnert and M. H. Ringle, eds. Lawrence Erlbaum Associates, Hillsdale, NJ.

Eiselt, K. P. 1989. *Inference Processing and Error Recovery in Sentence Understanding*. Ph.D. dissertation. University of California, Irvine. Tech. Rep. 89-24.

Frazier, L. and Fodor, J. D. 1978. The sausage machine: a new two-stage parsing model. *Cognition* 6:291–325.

Frederking, R., Grannes, D., Cousseau, P., and Nirenburg, S. 1993. An MAT tool and its effectiveness. In *Proc. DARPA Human Language Tech. Workshop*, Princeton, NJ.

Frederking, R. and Nirenburg, S. 1994. Three heads are better than one. In ANLP-94. *Proc. App. Nat. Language Process. Conf.* Stuttgart, Oct.

Genesereth, M. and Fikes, R. 1992. Knowledge interchange format version 3.0 reference manual. Computer Science Department, Stanford University, Stanford, CA.

Grishman, R. 1986. *Computational Linguistics: An Introduction.* Cambridge University Press.

Gruber, T. 1993. Toward Principles for the Design of Ontologies Used for Knowledge Sharing. *Stanford Knowledge Systems Lab. Tech. Rep.* KSL 93-04, Aug. 23.

Hirst, G. 1988. Semantic interpretation and ambiguity. *Artif. Intell.* 34:131–177.

IJCAI. 1995. Ontology workshop *Proc. Workshop Basic Ontological Issues in Knowledge Sharing,* Int. J. Conf. Artif. Intell. Montreal, Canada, Aug.

Jurafsky, D. 1992. An on-line computational model of human sentence interpretation, pp. 302–308. In *AAAI 92, Proc. 10th Nat. Conf. on Artif. Intell.*

Knight, K. and Luk, S. K. 1994. Building a large-scale knowledge base for machine translation. In AAAI-94 *Proc. 12th Nat. Conf. on Artif. Intell.*

Laird, J., Newell, A., and Rosenbloom, P. 1987. Soar: an architecture for general intelligence. *Artif. Intell.* 33:1–64.

Lebowitz, M. 1983. Memory-based parsing. *Artif. Intell.* 21:363–404.

Lehman, J. F., Lewis, R. L., and Newell, A. 1991. Integrating knowledge sources in language comprehension, pp. 461–466. In *Proc. 13th Annu. Conf. Cognitive Sci. Soc.*

Lehnert, W. G., Dyer, M. G., Johnson, P. N., Yang, C. J., and Harley, S. 1983. Boris—an experiment in in-depth understanding of narratives. *Artif. Intell.* 20(1):15–62.

Lenat, D. B. and Guha, R. V. 1990. *Building Large Knowledge-Based Systems.* Addison–Wesley, Reading, MA.

Lewis, R. L. 1993a. An architecturally-based theory of human sentence comprehension, pp. 108–113. In *Proc. 15th Annu. Conf. Cognitive Sci. Soc.* Lawrence Erlbaum Associates, Hillsdale, NJ.

Lewis, R. L. 1993b. *An Architecturally-Based Theory of Human Sentence Comprehension.* Ph.D. dissertation. Computer Science Department, Carnegie Mellon University, Pittsburgh, PA. Tech. Rep. CMU-CS-93-226.

Mahesh, K. 1995. *Syntax-Semantics Interaction in Sentence Understanding.* Ph.D. dissertation. College of Computing, Georgia Institute of Technology, Atlanta. Tech. Rep. GIT-CC-95/10.

Mahesh, K. 1996. Ontology Development: Ideology and Methodology. *Computing Research Lab. Tech. Rep.* MCCS-96-292, New Mexico State University.

Mahesh, K. and Eiselt, K. 1994. Uniform representations for syntax-semantics arbitration. In *Proc. 16th Annu. Conf. Cognitive Sci. Soc.,* pp. 589–594. Lawrence Erlbaum Associates. Hillsdale, NJ.

Mahesh, K. and Nirenburg, S. 1995a. Semantic classification for practical natural language processing. In *Proc. 6th ASIS SIG/CR Classification Res. Workshop: An Interdisciplinary Meeting,* Am. Soc. Inf. Sci. Chicago, IL. Oct.

Mahesh, K. and Nirenburg, S. 1995b. A situated ontology for practical NLP. In IJACI-95, *Proc. Workshop Basic Ontological Issues in Knowledge Sharing.* Int. J. Conf. Artif. Intell. Montreal, Canada, Aug.

Mahesh, K. and Nirenburg, S. 1996. Meaning representation for knowledge sharing in practical machine translation. In *Proc. FLAIRS-96 Special Track Inf. Interchange,* Florida AI Res. Symp. Key West, May 19–22.

Marcus, M. 1980. *A Theory of Syntactic Recognition for Natural Language.* MIT Press, Cambridge, MA.

McRoy, S. W. and Hirst, G. 1990. Race-based parsing and syntactic disambiguation. *Cognitive Sci.* 14:313–353.

Meyer, I., Onyshkevych, B., and Carlson, L. 1990. Lexicographic principles and design for knowledge-based machine translation. *Center for Machine Translation Tech. Rep.* CMU-CMT-90-118, Carnegie Mellon University, Pittsburgh, PA.

Nirenburg, S., ed. 1994. The Pangloss Mark III Machine Translation System. *NMSU CRL, USC ISI and CMU CMT Tech. Rep.*

Nirenburg, S., Carbonell, J., Tomita, M., and Goodman, K. 1992a. *Machine Translation: A Knowledge-Based Approach.* Morgan Kaufmann, San Mateo, CA.

Nirenburg, S. and Frederking, R. 1994. Toward multi-engine machine translation. In *Proc. Hum. Language*

Tech. Conf. Princeton, NJ.

Nirenburg, S., Frederking, R., Farwell, D., and Wilks, Y. 1994. Two types of adaptive MT environments. In COLING-94, *Proc. Int. Conf. Comput. Linguistics.* Kyoto, Aug.

Nirenburg, S., Shell, P., Cohen, A., Cousseau, P., Grannes, D., and McNeilly, C. 1992b. The translator's workstation. In *Proc. 3rd Conf. on Appl. Nat. Language Process.* Trento, Italy, April.

Onyshkevych, B. A. and Nirenburg, S. 1995. A lexicon for knowledge-based MT. *Mach. Trans.* Spec. Issue on building lexicons for machine translation II 10:(1–2):5–57.

Peterson, J., Mahesh, K., and Goel, A. 1994. Situating natural language understanding within experience-based design. *Int. J. Hum.–Comput. Stud.* 41(6):881–913.

Ram, A. 1989. *Question-Driven Understanding: An Integrated Theory of Story Understanding, Memory and Learning.* Ph.D. dissertation. Department of Computer Science, Yale University, New Haven, CT. Res. Rep. 710.

Reddy, D., Erman, L., and Neely, R. 1973. A model and a system for machine recognition of speech. *IEEE Trans. Audio Electroacoustics* AU-21:229–238.

Rich, E. and Knight, K. 1991. *Artificial Intelligence.* 2nd ed. McGraw–Hill, New York.

Riesbeck, C. K. and Martin, C. E. 1986a. Direct memory access parsing. In *Experience, Memory, and Reasoning,* J. L. Kolodner and C. K. Riesbeck, eds., pp. 209–226. Lawrence Erlbaum Associates, Hillsdale, NJ.

Riesbeck, C. K. and Martin, C. E. 1986b. Towards completely integrated parsing and inferencing, pp. 381–387. In *Proc. 8th Annu. Conf. Cognitive Sci. Soc.* Cognitive Science Society.

Russell, S. J. and Norvig, P. 1995. *Artificial Intelligence: A Modern Approach.* Prentice–Hall.

Schank, R. C. and Abelson, R. P. 1977. *Scripts, Plans, Goals, and Understanding.* Lawrence Erlbaum Associates, Hillsdale, NJ.

Schank, R. and Riesbeck, C. K., eds. 1981. *Inside Computer Understanding: Five Programs Plus Miniatures.* Lawrence Erlbaum Associates, Hillsdale, NJ.

Small, S. L. and Rieger, C. 1982. Parsing and comprehending with word experts. In *Strategies for Natural Language Processing,* W. G. Lehnert and M. H. Ringle, eds. Lawrence Erlbaum Associates, Hillsdale, NJ.

Viegas, E. and Bouillon, P. 1994. Semantic lexicons: the cornerstone for lexical choice in natural language generation. In *Proc. 7th Int. Workshop Nat. Language Generation,* Kennebunkport, ME, June.

Viegas, E. and Nirenburg, S. 1995. The semantic recovery of event ellipsis: its computational treatment. In *Proc. Workshop Context Nat. Language Process.* 14th Int. J. Conf. AI. Montreal.

Waltz, D. L. and Pollack, J. B. 1985. Massively parallel parsing: a strongly interactive model of natural language interpretation. *Cognitive Sci.* 9:51–74.

Wilks, Y. 1975. A preferential, pattern-seeking, semantics for natural language inference. *Artif. Intell.* 6(1):53–74.

Wilks, Y., Slator, B., and Guthrie, L. 1995. *Electric Words: Dictionaries, Computers and Meanings.* MIT Press, Cambridge, MA.

Further Information

A good introduction to NLP can be found in *Natural Language Understanding* by James Allen or in *Computational Linguistics, An Introduction* by Ralph Grishman. Knowledge-based approaches to NLP are introduced using practical systems in *Inside Computer Understanding: Five Programs Plus Miniatures* by Roger C. Schank and Christopher K. Riesbeck. A useful account of a knowledge-based machine translation system can be found in *Machine Translation: A Knowledge-Based Approach* by Sergei Nirenburg, Jaime Carbonell, Masaru Tomita, and Kenneth Goodman.

The *Handbook of Artificial Intelligence,* edited by Avron Barr and Edward A. Feigenbaum, is also a useful source, especially for early work in the area. The section on *Natural Language Understanding* by James Allen in Vol. IV is particularly relevant. Recent AI textbooks such as *Artificial Intelligence* by Elaine Rich and Kevin Knight and *Artificial Intelligence: A Modern Approach* by Stuart Russel and Peter Norvig contain good introductory material and pointers to related work.

Important journals in this area include *Computational Linguistics, Cognitive Science, Machine Translation,* and many of the major journals in Artificial Intelligence.

Major associations in the area include the American Association for Artificial Intelligence (AAAI), the Association for Computational Linguistics (ACL), and the Cognitive Science Society. Proceedings of annual conferences of these associations as well as the biannual International Joint Conference on Artificial Intelligence report some of the latest work in this field.

Useful newsgroups on the Internet include comp.ai.nat-lang, comp.ai.nlang-know-rep, and comp.ai. A compilation of frequently asked questions (FAQs) and answers on NLP is posted regularly on these newsgroups.

Many sites on the World Wide Web contain articles, useful information, and free software relevant to this field. Some useful pointers include the ACL NLP/CL Universe at http://www.cs.columbia.edu/~radev/cgi-bin/universe.cgi and the Computing Research Laboratory homepage at http://crl.nmsu.edu/Home.html. An electronic archive of papers on computational linguistics and NLP is located at http://xxx.lanl.gov/cmp-lg/.

29

Logic-Based Deductive Reasoning

James J. Lu
Bucknell University

Erik Rosenthal
University of New Haven

29.1 Introduction

Modern interest in artificial intelligence (AI) is coincident with the development of high-speed digital computers. Shortly after World War II many hoped that truly intelligent machines would soon be a reality. In 1950, Turing, in his now famous article *Computing Machinery and Intelligence*, which appeared in the journal *Mind*, predicted that machines would duplicate human intelligence by the end of the century. In 1956, at a workshop held at Dartmouth College, McCarthy introduced the term *artificial intelligence*, and the race was on. The first attempts at mechanizing reasoning included Newell and Simon's 1956 computer program, the *Logic Theory Machine*, and a computer program developed by Wang that proved theorems in propositional logic.

Early on it was recognized that **automated reasoning** is central to the development of machine intelligence, and central to automated reasoning is automated theorem proving, which may be thought of as mechanical techniques for determining whether a logical formula is satisfiable. The key to automated theorem proving is inference rules that can be implemented as algorithms. The first major breakthrough was Robinson's landmark paper in 1965, in which the resolution principle and the unification algorithm were introduced [Robinson 1965]. That paper marks the beginning of a veritable explosion of research in machine-oriented logics.

The focus of this chapter is on deductive reasoning through mechanical inference techniques. The fundamental principles underlying many different logics will be introduced, and several of the numerous theorem proving techniques that have been developed will be explored. The emphasis on logic-based inference is not unfounded; after all, it is the key to implementing most reasoning systems. It is also the foundation of logic programming.

The underlying logics may generally be classified as **classical**—roughly, the logic described by Aristotle, or as **nonstandard**—logics that were developed somewhat more recently. Most of the alternative logics that have been proposed are extensions of classical logic, and inference methods for them are typically based on classical deduction techniques. Reasoning with uncertainty through **fuzzy logic** and nonmonotonic

0-8493-2909-4/97/$0.00+$.50
© 1997 by CRC Press, Inc.

reasoning through **default logics**, for example, have for the most part been based on variants and extensions of classical proof techniques; see, for example, Reiter's paper in Bobrow [1980] and Lee [1972] and Lifschitz [1995].

This chapter touches on three disciplines: artificial intelligence, automated theorem proving, and symbolic logic. Many researchers have made important contributions to each of them, and it is impossible to describe all logics and inference rules that have considered. We believe that the methodologies described are typical and should give the reader a basis for further exploration of the vast and varied literature on the subject.

29.2 Underlying Principles

We begin with a brief review of **propositional classical logic**. Propositional logic is not adequate for most deductive purposes, and so we also examine first-order logic and generalize to some of the nonstandard logics. An excellent source for a more detailed exposition of the fundamentals of computational logic is the book by Chang and Lee [1973].

Propositional Logic

A *proposition* is a statement that is either true or false (but not both); *carbon is an element* and *Hercules is president* are both examples. A single proposition is often called an *atomic formula* or simply an *atom*. Logical formulas are built from a set $\mathcal{A}$ of atoms, a set of connectives, and a set of logical constants in the following way: Atoms and constants are formulas; if Θ is an n-ary connective and if $\mathcal{F}_1, \mathcal{F}_2, \ldots, \mathcal{F}_n$ are formulas, then so is $\Theta(\mathcal{F}_1, \mathcal{F}_2, \ldots, \mathcal{F}_n)$. The expression

$$(A \vee \neg B) \leftrightarrow (B \rightarrow \textit{true})$$

is an example with one constant *true*, two atoms A and B, one unary connective $\neg$ (*negation*), and three binary connectives $\vee$ (*logical or*), $\leftrightarrow$ (*logical if and only if*), and $\rightarrow$ (*logical implication*). Negated atoms play a special role in most deduction techniques, and the word **literal** is used for an atom or for the negation of an atom.

The semantics (meaning) of a logical formula is characterized by truth values. In classical logic the possible truth values are *true* and *false*; in general, the set of truth values may be any set.

Connectives may be viewed as function symbols appearing in the strings representing formulas, or, less formally, if Δ is the set of truth values, then an n-ary connective is a function from Δ^n to Δ. In classical logic, $\Delta = \{\textit{true, false}\}$; there are four functions from Δ to Δ, and only one is an interesting connective: standard negation. There are 16 binary connectives (i.e., functions from $\Delta \times \Delta$ to Δ). Of particular interest are conjunction $\wedge$ and disjunction $\vee$.

Most **inference rules** assume that formulas have been normalized in some manner. A formula is in **negation normal form** (**NNF**) if conjunction and disjunction are the only binary connectives and if all negations are at the atomic level. A formula is in **conjunctive normal form** (**CNF**) if it is a conjunction of clauses, where a clause is a disjunction of literals; the term **clause form** refers to CNF. Observe that CNF is a special case of NNF.

An **interpretation** is a function from the atom set $\mathcal{A}$ to the set Δ of truth values. In practice, we use the word interpretation for a *partial interpretation* for a formula $\mathcal{F}$: an assignment of a truth value only to the atoms that occur in $\mathcal{F}$. Interpretations may be extended to complex formulas according to the functions represented by the connectives in the logic. A formula in classical logic is **satisfiable** if it evaluates to *true* under some interpretation. For other logics, where the set Δ of truth values is arbitrary, *satisfiability* is determined by a designated subset Δ^*. That is, an interpretation I satisfies a formula $\mathcal{F}$ if I maps $\mathcal{F}$ to a truth value in Δ^*. A formula C is said to be a **logical consequence** of $\mathcal{F}$ if every interpretation I that satisfies $\mathcal{F}$ also satisfies C; in that case, we write $\mathcal{F} \models C$.

Inference and Deduction

To paraphrase Hayes [1977], the meaning and the implementation of a logic meet in the notion of inference. Automated inference techniques may roughly be put into one of two categories: **inference rules** and **rewrite rules**. Inference rules are applied to a formula, producing a conclusion that is conjoined to the original formula. When a rewrite rule is applied to a formula, the result is a new formula in which the original formula may not be present. The distinction is really not that clear: The rewritten formula may be interpreted as a conclusion and conjoined to the original formula. We will consider examples of both. Resolution is an inference rule, and the **tableau method** and **path dissolution** may be thought of as rewrite rules.

Inference rules may be written in the following general form:

$$\frac{premise}{conclusion} \tag{29.1}$$

where *premise* is a set of formulas[1] and *conclusion* is a formula. A *deduction* of a formula C from a given set of formulas S is a sequence

$$C_0, C_1, \ldots, C_n$$

such that $C_0 \in S$, $C_n = C$, and for each i, $1 \le i \le n$, C_i satisfies one of the following conditions:

1. $C_i \in S$.
2. There is an inference rule (*premise/conclusion*) such that *premise* $\subseteq \{C_0, C_1, \ldots, C_{i-1}\}$, and $C_i =$ *conclusion*.

We use the notation $S \vdash C$ to indicate that there is a deduction of C from S.

A simple example of an inference rule is

$$\frac{premise}{chocolate\ is\ good\ stuff}$$

i.e., *chocolate is good stuff* is inferred from any premise. We have several colleagues who can really get behind this particular rule, but it appears to lack something from the automated reasoning point of view. To avoid the possible problems inherent in this rule, there are two standards against which inference rules are judged:

Soundness: Suppose $\mathcal{F} \vdash C$. Then $\mathcal{F} \models C$.
Completeness: Suppose $\mathcal{F} \models C$. Then $\mathcal{F} \vdash C$.

Of the two properties, the first is the more important: The ability to draw valid (and only valid!) conclusions is more critical than the ability to draw all valid conclusions. In practice, most researchers are interested in *refutation* completeness, i.e., the ability to verify that an unsatisfiable formula is in fact unsatisfiable. As we shall see when we consider nonmonotonic reasoning, even soundness may not always be a desirable property.

First-Order Logic

Most serious deduction techniques require **first-order logic**. Typically, one starts with propositional inference rules and then employs some variant of Robinson's **unification algorithm** and the **lifting lemma** [Robinson 1965]. In this section we present the basics of first-order logic.

---

[1] In most settings—certainly in this chapter—a *set of formulas* is essentially the conjunction of the formulas in the set.

Atoms are usually called *predicates* in first-order logic, and predicates are allowed to have arguments. For example, if M is the predicate, "is a man," then $M(x)$ may be interpreted as, "x is a man." Thus, $M(\text{Socrates})$, $M(7)$, and (since function symbols are allowed) $M(f(x))$ are all well formed. In general, predicates may have any (finite) number of arguments, and any *term* may be substituted for any argument. Terms are defined recursively as follows: Variables and constant symbols are terms, and if $t_1, t_2, \ldots, t_n$ are terms and if f is an n-ary function symbol, then $f(t_1, t_2, \ldots, t_n)$ is a term.

First-order formulas are essentially the same as propositional formulas, with the obvious exception that the atoms that appear are predicates. However, first-order formulas may be **quantified**. In the following example, c is a constant, x is a *universally* quantified variable, and y is *existentially* quantified; the unquantified variable z is said to be *quantifier-free* or simply *free*,

$$\forall x \exists y (P(x, y) \lor \neg Q(y, z, c))$$

Interpretations at the first-order level are different because a **domain of discourse** over which the variables may vary must be selected. If $\mathcal{F}$ is a formula with n free variables, if I is an interpretation, and if $\mathcal{D}$ is the corresponding domain of discourse, then I maps $\mathcal{F}$ to a function from $\mathcal{D}^n$ to Δ. A *valuation* is an assignment of variables to elements of $\mathcal{D}$. Under interpretation I and valuation V, a formula $\mathcal{F}$ yields a truth value, and two formulas are said to be *equivalent* if they evaluate to the same truth value under all interpretations and valuations.

Of particular importance to the theoretical development of inference techniques in AI is the class of *Herbrand interpretations*. These are interpretations whose domain of discourse is the **Herbrand universe**, which is built from the variable-free terms in the given formula. It can be defined recursively as follows: Let $\mathcal{F}$ be any formula. Then H_0 is the set of constants that appear in $\mathcal{F}$. If there are no constants, let a be any constant symbol, and let $H_0 = \{a\}$. For $n = 0, 1, 2, \ldots, H_{n+1}$ is the union of H_n and the set of all terms of the form $f(t_1, t_2, \ldots, t_m)$, where $t_i \in H_n$ for $i = 0, 1, 2, \ldots, m$. Then the Herbrand universe is $H = \bigcup_{i=0}^{\infty} H_i$. The importance of Herbrand interpretations is made clear by the following theorem: *A formula $\mathcal{F}$ is unsatisfiable if and only if $\mathcal{F}$ is unsatisfiable under Herbrand interpretations.*

In general, it is possible to transform any first-order formula to an equivalent (i.e., truth preserving) **prenex normal form**: all quantifiers appear in the front of the formula. A formula $\mathcal{F}$ in prenex normal form can be further normalized to a satisfiability preserving **Skolem standard form** $\mathcal{G}$: All existentially quantified variables are replaced by constants or by functions of constants and the universally quantified variables. *Skolemizing* a formula in this manner preserves satisfiability: $\mathcal{F}$ is satisfiable if and only if $\mathcal{G}$ is.[2]

Since the quantifiers appearing in $\mathcal{G}$ are all universal, we can (and typically do) write $\mathcal{G}$ without quantifiers, it being understood that all variables are universally quantified.

A **substitution** is a function that maps variables to terms. Any substitution may be extended in a straightforward way to apply to arbitrary expressions. Given a set of expressions $E_1, \ldots, E_n$, each of which may be a term, an atom, or a clause, a substitution θ is a **unifier** for the set $\{E_1, \ldots, E_n\}$ if $\theta(E_1) = \theta(E_2) = \cdots = \theta(E_n)$. A unifier θ of a set of expressions E is called *the most general unifier* (mgu) if given any unifier γ of E, $\gamma \circ \theta = \gamma$. For example, the two expressions $P(a, y)$, $P(x, f(z))$ are unifiable via the substitution θ_1 which maps y to $f(z)$ and x to a. They are also unifiable via the substitution θ_2 which maps y to $f(a)$, z to a, and x to a. The substitution θ_1 is more general than θ_2. When a substitution is applied to a formula, the resulting formula is called an *instance* of the given formula.

Robinson's unification algorithm [Robinson 1965] provides a means of finding the mgu of any set of unifiable expressions. Robinson proved the lifting lemma in the same paper, and the two together probably represent the most important single advance in the development of automated theorem proving.

---

[2] Perhaps surprisingly, Skolemization does not in general preserve equivalence.

29.3 Best Practices

Classical Logic

Not surprisingly, the most widely adopted logic in AI systems is classical (two-valued) logic. The truth value set Δ is $\{true, false\}$, and the designated truth value set Δ^* is $\{true\}$. Some examples of AI programs based on classical logic include: problem solvers such as Green's program [Green 1969], theorem provers such as OTTER [McCune 1992], Astrachan's METEOR (see Wrightson [1994]), the Boyer and Moore [1979] theorem prover, and the *rewrite rule laboratory* [Kapur and Zhang 1989], later extended by Zhang. There are several deduction-based programming languages such as Prolog; a good source is the book by Sterling and Shapiro [1986]. In this section we describe one inference rule—**resolution**—and two rewrite rules—the tableau method and its generalization, path dissolution.

Resolution

Perhaps the most widely applied inference rule in all of AI is the resolution principle of Robinson [1965]. It assumes that each formula is in CNF (a conjunction of clauses). To define resolution for propositional logic, suppose we have a formula in CNF containing the two clauses in the premise; then the conclusion may be inferred,

$$\frac{(A_1 \vee A_2 \vee \cdots \vee A_m \vee L) \wedge (B_1 \vee B_2 \vee \cdots \vee B_n \vee \neg L)}{(A_1 \vee A_2 \vee \cdots \vee A_m \vee B_1 \vee B_2 \vee \cdots \vee B_n)} \tag{29.2}$$

The conclusion is called the *resolvent*, and the two clauses in the premise are called the *parent clauses*.

It is easy to see why resolution is sound: If an interpretation satisfies the formula, then it must satisfy every clause. Since L and $\neg L$ cannot simultaneously evaluate to *true*, one of the other literals must be *true*. Resolution is also complete; the proof is beyond the scope of this chapter.

The lifting lemma [Robinson 1965] enables the application of resolution to formulas in first-order logic. Roughly speaking, it says that if instances of two clauses can be resolved, then the clauses can be unified and resolved. The effect is that two first-order clauses can be resolved if they contain, respectively, positive and negative unifiable occurrences of the same predicate. To state the first-order resolution inference rule, let L_1 and L_2 be two occurrences of the same predicate, one positive, one negative, and let θ be the mgu of L_1 and L_2. Then

$$\frac{(A_1 \vee A_2 \vee \cdots \vee A_m \vee L) \wedge (B_1 \vee B_2 \vee \cdots \vee B_n \vee \neg L)}{(\theta(A_1) \vee \theta(A_2) \vee \cdots \vee \theta(A_m) \vee \theta(B_1) \vee \theta(B_2) \vee \cdots \vee \theta(B_n))} \tag{29.3}$$

In practice, an implementation based on resolution alone has limited value because unrestricted resolution tends to produce an enormous number of inferences. There are several approaches to controlling the search space. One is the *set of support* strategy, which identifies a subset of the original set of clauses as its set of support, and then insists that at least one parent clause in every resolvent come from the set of support. Another strategy, one that is especially useful for logic programming, is the *linear restriction*, wherein one parent clause must be the most recent resolvent. These strategies may be thought of as control of the deduction process through metalevel restrictions on the search space.

As a simple example, consider the knowledge that "Tweety is a canary," "a canary is a bird," and "a bird flies," encoded as the following set of clauses:

$$Canary(I)$$

$$\neg Canary(x) \vee Bird(x)$$

$$\neg Bird(x) \vee Flies(x)$$

A linear resolution deduction of the fact that Tweety flies can be obtained as in Fig. 29.1. The substitutions θ_1 and θ_2 are both functions that map x to the constant *Tweety*.

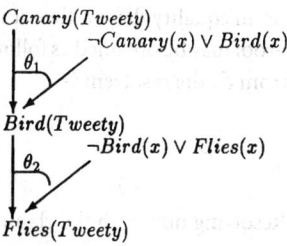

$Canary(Tweety)$
$\quad\quad\neg Canary(x) \vee Bird(x)$
θ_1
$Bird(Tweety)$
$\quad\quad\neg Bird(x) \vee Flies(x)$
θ_2
$Flies(Tweety)$

FIGURE 29.1 A deduction of *Flies(Tweety)*.

Bundy et al. argue that "[logic] provides only a low-level, step by step understanding, whereas a high-level, strategic understanding is also required," and there are restrictions on the search space that attempt to incorporate some kind of understanding [Bundy et al. 1988]. For example, equality and inequality have a special status in many theories. An axiom $a = b$ in a theory typically indicates that a and b may be used interchangeably in any context. Formally, the following equality axioms are implicitly assumed for theories requiring this property:

1. (Reflexivity) $x = x$.
2. (Symmetry) $(x = y) \rightarrow (y = x)$.
3. (Transitivity) $(x = y) \wedge (y = z) \rightarrow (x = z)$.
4. (Substitution 1) $(x_i = y) \wedge P(x_1, \ldots, x_i, \ldots, x_n) \rightarrow P(y_1, \ldots, y, \ldots, y_n)$ for $1 \leq i \leq n$, for each n-ary predicate symbol P.
5. (Substitution 2) $(x_i = y) \rightarrow f(x_1, \ldots, x_i, \ldots, x_n) = f(x_1, \ldots, y, \ldots, x_n)$ for $1 \leq i \leq n$, for each n-ary function symbol f.

The explicit incorporation of these axioms tends to drastically increase the search space, so Robinson and Wos [1969] proposed a specialized inference rule, **paramodulation**, for handling equality. Let $L[t]$ be a literal, let θ be the mgu of r and t, and let $\theta(L[s])$ be the literal obtained from $L[t]$ by replacing one occurrence of $\theta(t)$ in $L[t]$ with $\theta(s)$:

$$\frac{L[t] \vee D_1, (r = s) \vee D_2}{\theta(L[s]) \vee \theta(D_1) \vee \theta(D_2)} \tag{29.4}$$

The conclusion is called the *paramodulant* of the two clauses.

Using the Tweety example, suppose we have the additional knowledge that Tweety is known by the alias "fred" (i.e., *Tweety = fred*). Then the question, "Can fred fly?" may be answered by extending the resolution proof shown in Fig. 29.1 with the paramodulation inference, which substitutes the constant *Tweety* in the conclusion *Flies(Tweety)* with *fred* to obtain *Flies(fred)*.

An inference rule such as paramodulation is semantically based since its definition comes from unique properties of the predicate and function symbols. Paramodulation treats the equality symbol $=$ with a special status that enables it to perform larger inference steps. Other semantically based inference rules can be found in Slagle [1972], Manna and Waldinger [1986], Stickel [1985], and Bledsoe et al. [1985].

Controlling paramodulation in an implementation is difficult. One system designed to handle equality is the RUE[3] system of Digricoli and Harrison [1986]. Its goal directed nature tends to produce better computational behavior than paramodulation. The essential idea, which is illustrated with the following example, is to build the two substitution axioms into resolution. Let S be the set of clauses

$$\{P(f(a)), \neg P(f(b)), (a = b)\}$$

and let $\mathcal{E}$ be the equality axioms; i.e., $\mathcal{E}$ consists of the rules for reflexivity, transitivity, symmetry, and the following two substitution axioms:

$$(x = y) \wedge \neg P(x) \rightarrow P(y)$$

$$(x = y) \rightarrow (f(x) = f(y))$$

---

[3] Resolution with unification and equality.

As an equality theory, the set is unsatisfiable. That is, $S \cup \mathcal{E}$ is not satisfiable; a straightforward resolution proof may be obtained as follows. Apply resolution to the first substitution axiom and the clause $\neg P(f(b))$ from S; the resolvent is

$$x \neq f(b) \vee P(x)$$

Resolving now with the clause $P(f(a))$ yields the resolvent

$$f(a) \neq f(b)$$

Finally, this clause may be resolved with the second substitution axiom to produce the clause $a \neq b$. This resolves with $a = b$ from the set S to complete the proof.

RUE builds into the resolution inference the substitution axioms by observing that the two substitution axioms may be expressed equivalently as

$$P(y) \wedge \neg P(x) \rightarrow x \neq y$$

$$f(x) \neq f(y) \rightarrow x \neq y$$

The inference rule

$$\frac{P(y), \neg P(x)}{x \neq y}$$

is introduced from the first axiom, and from the second, we obtain the inference rule

$$\frac{f(x) \neq f(y)}{x \neq y}$$

RUE further optimizes its computation by allowing the application of both inference rules in a single step. Thus, from the clauses $P(f(a))$ and $\neg P(f(b))$, the RUE resolvent $a \neq b$ may be obtained in a single step. Note that if only the first inference rule is applied, then the RUE resolvent of $P(f(a))$ and $\neg P(f(b))$ would be $f(a) \neq f(b)$.

The Method of Analytic Tableaux and Path Dissolution

Most theorem proving techniques employ clause form; the tableau method and the more general path dissolution do not. Tableau methods were originally developed and studied by a number of logicians, among them Beth [1955], Hintikka [1955], and Smullyan [1995], who built on the work of Gentzen [1969]. It is probably Smullyan who is most responsible for popularizing these methods; his particularly elegant variation on these techniques is known as the *method of analytic tableaux*. More recently, tableau methods have been receiving considerable attention from researchers investigating both automated deduction and logics for artificial intelligence; this includes serious implementors and those whose focus is primarily theoretical. See Beckert and Posegga [1995] for a particularly elegant implementation of the tableau method (eight lines of Prolog code!). *Path dissolution* operates on a complementary pair of literals within a formula by restructuring the formula in such a way that all paths through the link vanish. The tableau method restructures the formula so that the paths through the link are immediately accessible and then marks them closed, in effect deleting them. It does this by selectively expanding the formula toward disjunctive normal form. The sense in which path dissolution generalizes the tableau method is that dissolution need not distinguish between the restructuring and the closure operations.

$$((\overline{C} \wedge A) \vee D \vee E) \wedge (\overline{A} \vee (B \wedge C)) \quad \equiv \quad \begin{array}{c} \overline{C} \\ \wedge \;\; \vee \;\; D \;\; \vee \;\; E \\ A \\ \qquad \wedge \\ \qquad B \\ \overline{A} \;\; \vee \;\; \wedge \\ \qquad C \end{array}$$

FIGURE 29.2

Negation Normal Form

One way to classify deduction systems is by what normalization they require of the formulas upon which they operate. As we have seen, resolution uses conjunctive normal form. Path dissolution and the tableau method cannot be restricted to clause form: Both restructure a formula in such a way that clause form may not be preserved. In essence, both methods work with formulas in negation normal form (NNF). It turns out that NNF formulas can be far more complex than formulas in clause form, and so a careful analysis of NNF is useful before we introduce these proof techniques.

Recall that a formula is in NNF if conjunction and disjunction are the only binary connectives and if all negations are at the atomic level. Propositional formulas in NNF may be described recursively as follows:

1. The constants t and f are NNF formulas.
2. The literals A and $\neg A$ are NNF formulas.
3. If $\mathcal{F}$ and $\mathcal{G}$ are formulas, then so are $\mathcal{F} \wedge \mathcal{G}$ and $\mathcal{F} \vee \mathcal{G}$.

Each formula used in the construction of an NNF formula is called an *explicit subformula*. When a formula contains occurrences of *true* and *false*, the obvious truth-functional reductions apply. For example, if $\mathcal{F}$ is any formula, then $t \wedge \mathcal{F} = \mathcal{F}$. Unless otherwise stated, we will assume that formulas are automatically so reduced.

It is often convenient to write formulas as two-dimensional graphs in a manner that can easily be understood by considering a simple example. In Fig. 29.2, the formula on the left is displayed graphically on the right. For a more detailed exposition, see Murray and Rosenthal [1993].

Naturally, a literal A may occur more than once in a formula. As a result, we use the term *node* for a literal occurrence in a formula. If A and B are nodes in a formula $\mathcal{F}$, and if $\mathcal{F}$ contains the subformula $X \wedge Y$ with A in X and B in Y, then we say that A and B are *c-connected*; *d-connected* nodes are similarly defined. In Fig. 29.2, C is c-connected to each of B, A, $\overline{C}$, D, and E and is d-connected to $\overline{A}$.

Let $\mathcal{F}$ be a formula. A *partial c-path through* $\mathcal{F}$ is a set of nodes such that any two are c-connected, and a *c-path* through $\mathcal{F}$ is a partial c-path that is not properly contained in any partial c-path. The c-paths of the formula of Fig. 29.2 are: $\{\overline{C}, A, \overline{A}\}$, $\{\overline{C}, A, B, C\}$, $\{D, \overline{A}\}$, $\{D, B, C\}$, $\{E, \overline{A}\}$, $\{E, B, C\}$. We similarly define d-path using d-connected nodes in place of c-connected nodes.

The c-paths of a formula are the clauses of one of its disjunctive normal form equivalents. Similarly, the d-paths correspond to the clauses of a CNF equivalent. It is easy to see that a formula is unsatisfiable if and only if every c-path in it is unsatisfiable, and a c-path is unsatisfiable if and only if it contains a *link*: a complementary pair of literals (i.e., an atom and its negative). Most inference mechanisms operate on links; several are path based. The idea of the method of analytic tableaux is to isolate paths containing a link and then to eliminate them. Path dissolution accomplishes this without first isolating the paths in question. Other path-based methods include Andrews's work on matings [Andrews 1976] and Bibel's connection and connection graph methods (see Bibel [1987]).

The Tableau Method

The reader should be forewarned that there is a potentially misleading difference in emphasis between the typical descriptions of the tableau method and the one presented here. Tableau proofs are usually cast as tree structures in which paths may grow through the addition of new *lines* in the tree; the number of paths may increase due to a *splitting* or *branching* operation. The lines and branch points of a tableau proof tree are metalinguistic representations of conjunction and disjunction, respectively. Here we strip them of their special status; a tableau proof tree then becomes merely a single formula. This simplifies the presentation, which is an advantage because of space limitations, and makes the relationship to path dissolution easier to see. Smullyan's book is an excellent source for the traditional description of the tableau method [Smullyan 1995].

Defining the tableau method in terms of formulas requires three rules: *separation*, *dispersion*, and *closure*. It is also convenient to designate certain subformulas as *primary*; they form a tree that corresponds precisely to the proof tree maintained by the tableau method. Initially, the entire formula is the only primary subformula. A separation is performed on any primary subformula whose highest level connective is a conjunction by removing the primary designation from it and bestowing that designation on its conjuncts. There is essentially no cost to this operation, and it may be regarded as automatic whenever a conjunction becomes primary.

A separation may also be performed on a disjunction that is a leaf in the primary tree. (Separating an interior disjunction is not allowed since such an operation would destroy the tree structure.) Separations of such disjunctions should not be regarded as automatic: This operation increases the number of paths in the tree (we call such paths *tree* paths to distinguish them from c-paths); thus, although there is no cost to the operation itself, there is a potential penalty from the extra paths.

The process of dispersing a primary subformula whose highest level connective is a disjunction may now be defined precisely: A copy of the subformula is placed at the end of one path descending from it and separated. For example, suppose that $X = X_1 \vee X_2$ is a primary subformula, and that the leaf Y is a descendent of X. On the left, we show the original tree path from X to Y; on the right is the extension of that path produced by dispersing X,

$$
\begin{array}{ccc}
 & & \vdots \\
 & & X \\
\vdots & & \wedge \\
X & & \vdots \\
\wedge & & \wedge \\
\vdots & & Y \\
\wedge & & \wedge \\
Y & & X_1 \vee X_2
\end{array}
$$

The subformulas X and Y remain primary, and X_1 and X_2 are now designated as primary. Note that if a subformula is eventually dispersed to the ends of every path descending from it, the original copy is no longer required.

The key operations in a tableau deduction are the closures, which close tree paths. Marking a tree path closed is equivalent to deleting it. In terms of the tree of primaries, a tree path may be closed when a separation or a dispersion makes primary a literal that forms a link with one of its ancestors. The literal and all of its descendants are deleted, and any leaves that result are in turn deleted. Since the tree path through the link is removed, the effect is to delete all c-paths through the link.

As an example, the unsatisfiable formula $(((A \wedge B) \vee \overline{C}) \wedge C \wedge (\overline{A} \vee \overline{B}))$ is pictured in Fig. 29.3(a). Boxes are used to designate primary subformulas; initially the entire formula is the only one. Since it is a conjunction, this primary subformula is automatically separated; the result is Fig. 29.3(b).

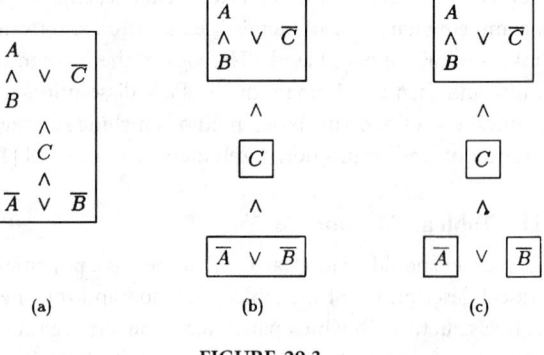

FIGURE 29.3

Note that there is yet only one path in the tree. As a result, a dispersion may be performed by moving (without a duplicate copy) any primary to the leaf position and then separating.

For simplicity, we separate the primary that is already a leaf (Fig. 29.3(c)). The formula, as a formula, is unchanged. But by designating $\overline{A}$ and $\overline{B}$ as primary, the proof tree has split and now contains two tree paths. Only one operation can follow: dispersion of the upper primary. It may be dispersed twice, once for each tree path, allowing the deletion of the original copy at the root; the result is pictured in Fig. 29.4.

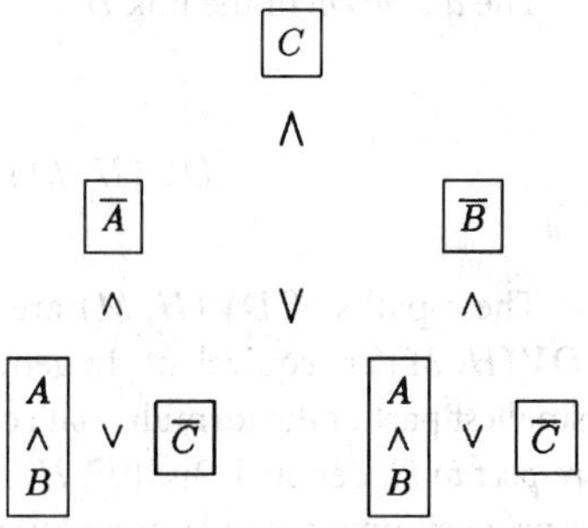

The remainder of the proof is straightforward. Each of the primary subformulas ($A \wedge B$) is separated into two primaries, and each of the four paths in the tree can be closed; two paths contain a $\{C, \overline{C}\}$ link, another contains $\{\overline{A}, A\}$, and the fourth contains $\{\overline{B}, B\}$.

A tree path is the conjunction of its nodes, i.e., of its primary subformulas. We may also view a tree path as a collection of c-paths. For instance, the single tree path in Fig. 29.3(c) that contains $\overline{A}$ is a conjunction of three primaries and contains the c-paths $\{\{A, B, C, \overline{A}\}, \{\overline{C}, C, \overline{A}\}\}$. Note also that dispersion is the source of all literal duplication (the expensive part) with the tableau method. In Fig. 29.4, for example, an extra copy (in this case only one) of $((A \wedge B) \vee \overline{C})$ has been created for all but the last descendent leaf to which it has been dispersed.

FIGURE 29.4

Path Dissolution

Path dissolution operates on a *link*—complementary pair of literals—within a formula by restructuring the formula in such a way that all paths through the link vanish. The tableau method restructures a formula so that the paths through the link are immediately accessible and then marks them closed, in effect deleting them. It does this by selectively expanding the formula toward disjunctive normal form. The sense in which dissolution generalizes the tableau method is that dissolution need not distinguish between the restructuring and the closure operations.

Path dissolution is in general applicable to collections of links; here we restrict attention to single links. Suppose then that we have complementary literals A and $\overline{A}$ residing in conjoined subformulas X and Y, respectively. Consider, for example, the link $\{A, \overline{A}\}$ on the left in Fig. 29.5. Then the formula is $\mathcal{D} = (X \wedge Y)$, where

$$X = \begin{matrix} \overline{C} \\ \wedge \\ A \end{matrix} \vee D \vee E \quad \text{and} \quad Y = \overline{A} \vee \begin{matrix} B \\ \wedge \\ C \end{matrix}$$

The *c-path complement* of a node A with respect to X, written $CC(A, X)$, is defined to be the subformula of X consisting of all literals in X that lie on c-paths that do not contain A; the *c-path extension* of A with respect to X, written $CPE(A, X)$, is the subformula containing all literals in X that lie on paths that *do* contain A.

In Fig. 29.5, $CC(A, X) = (D \vee E); CPE(A, X) = (\overline{C} \wedge A)$.

It is intuitively clear that the paths through ($X \wedge Y$) that do not contain the link are those through $CPE(A, X) \wedge CC(\overline{A}, Y))$ plus those through $(CC(A, X) \wedge CPE(\overline{A}, Y))$ plus those through $(CC(A, X) \wedge CC(\overline{A}, Y))$. The reader is referred to Murray and Rosenthal [1993] for the formal definitions of CC and of CPE and for the appropriate theorems.

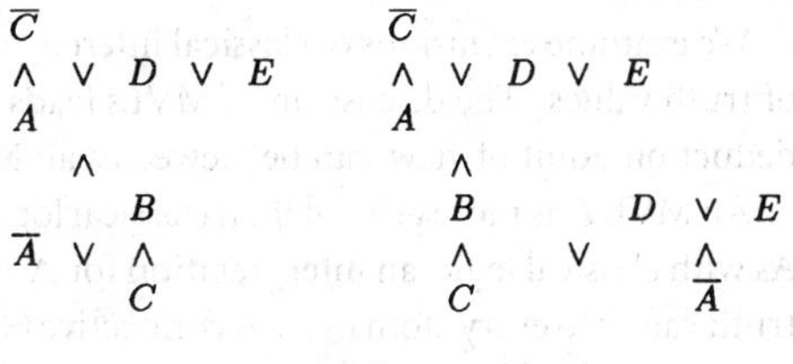

FIGURE 29.5

The *dissolvent* of the link $H = \{A, \overline{A}\}$ in $M = X \wedge Y$ is defined by

$$DV(H, M) = \begin{array}{ccccc} CPE(A, X) & & CC(A, X) & & CC(A, X) \\ \wedge & \vee & \wedge & \vee & \wedge \\ CC(\overline{A}, Y) & & CPE(\overline{A}, Y) & & CC(\overline{A}, Y) \end{array}$$

The c-paths of $DV(H, M)$ are exactly the c-paths of M that *do not* contain the link. Thus M and $DV(H, M)$ are equivalent. In general, M need not be the entire formula; without being precise, M is the smallest part of the formula that contains the link. If $\mathcal{F}$ is the entire formula, then the *dissolvent of $\mathcal{F}$ with respect to H*, denoted Diss $(\mathcal{F}, H)$, is the formula produced by replacing M in $\mathcal{F}$ by $DV(H, M)$. If $\mathcal{F}$ is a propositional formula, then Diss $(\mathcal{F}, H)$ is equivalent. Since the paths of the new formula are all that appeared in $\mathcal{F}$ except those that contained the link, this formula has strictly fewer c-paths than $\mathcal{F}$. As a result, finitely many dissolutions (bounded above by the number of c-paths in the original formula) will yield a linkless equivalent formula. We may therefore say that path dissolution is a strongly complete rule of inference for propositional logic; i.e., if a formula is unsatisfiable, any sequence of dissolution steps will eventually produce the empty clause.

A useful special case of dissolution arises when X consists of A alone; then $CC(A, X)$ is empty, and the dissolvent of the link $\{A, \overline{A}\}$ in the subformula $X \wedge Y$ is $X \wedge CC(\overline{A}, Y)$; i.e., dissolving has the effect of replacing Y by $CC(\overline{A}, Y)$, which is formed by deleting $\overline{A}$ and anything directly conjoined to it. Hence no duplications whatsoever are required. A tableau closure is essentially a dissolution step of this type. Observe that a separation in a tableau proof does not really affect the structure of the formula; it is a bookkeeping device employed to keep track of the primaries in the tree. A dispersion is essentially an application of the distributive laws, which of course can be used by any logical system. As a result, every tableau proof is a dissolution proof, but certainly not vice versa.

Nonclassical Logics

Many departures from classical logic that have been formalized in the AI research program have been aimed at common sense reasoning. Perhaps the two most widely addressed limits of classical logic are its inability to model reasoning with uncertain knowledge and reasoning with incomplete information. A number of nonclassical logics have been proposed; here we consider **multiple-valued logics**, fuzzy logic, and default logic. Alternatives to uncertain reasoning include probablistic reasoning—see Chapter 31, the chapter on Knowledge Representation—and nonmonotonic formalisms for knowledge representation—see Chapter 28 for an in-depth discussion. Inference techniques in nonclassical logics remain important; a common approach is to designate a set of satisfying truth values and then to adapt classical inference techniques.

Multiple-Valued Logics

One of the drawbacks of classical logic is its restricted set Δ of truth values. A logic that generalizes Δ to be an arbitrary set is commonly referred to as a *multiple-valued* (or *many-valued*) logic (MVL). Though the precise boundary of what can be classified as a multiple-valued logic is not clear [Urquhart 1986], logics that are generally agreed to fit that description have been applied to reasoning with uncertainty, to reasoning with inconsistency, to natural language processing, and, to some extent, to nonmonotonic reasoning.

We examine extensions of classical inference techniques to MVLs based on the framework of **signs**: sets of truth values. The discussion of MVLs leads naturally to the examination of fuzzy logic, which, from a deduction point of view, can be viewed as an MVL.

An MVL Λ is more general than classical logic in one respect: The set Δ of truth values may be arbitrary. As with classical logic, an interpretation for Λ is a function from its atom set $\mathcal{A}$ to Δ; i.e., an assignment of truth values to every atom in Λ. A connective Θ of arity n denotes a function $\Theta : \Delta^n \rightarrow \Delta$. Interpretations are extended in the usual way to mappings from the set of formulas to Δ.

Consider, for example, Łukasiewicz's three-valued logic [Łukasiewicz 1970]. The set of truth values is $\Delta = \{0, 1/2, 1\}$, the binary connectives are $\wedge$, $\vee$, and $\rightarrow$, and the only unary connective is $\neg$. The connectives are defined by the truth table, Table 29.1

The designated set of truth values of Łukasiewicz's logic is $\{1\}$. Thus, the consequence relation means that $\mathcal{F} \models C$ if whenever an interpretation I assigns the value 1 to $\mathcal{F}$, then I also assigns 1 to C.

Intuitively, the truth value $1/2$ in Łukasiewicz's logic denotes possible. A variant of this three-valued logic is Kleene's, in which the truth value $1/2$ corresponds to

TABLE 29.1 Truth Table

G	H	$G \wedge H$	$G \vee H$	$G \rightarrow H$	$\neg G$
0	0	0	0	1	1
0	1/2	0	1/2	1	1
0	1	0	1	1	1
1/2	0	0	1/2	1/2	1/2
1/2	1/2	1/2	1/2	1	1/2
1/2	1	1/2	1	1	1/2
1	0	0	1	0	0
1	1/2	1/2	1	1/2	0
1	1	1	1	1	0

undefined [Kleene 1952]. In a more recent three-valued logic proposed by Priest, the value $1/2$ may be thought of as denoting inconsistency.

A good reference for deduction techniques for MVL's is Hähnle's monograph, *Automated Deduction in Multiple-Valued Logics* [Hähnle 1994]. The key to the approach described there is the use of signs—subsets of the set Δ of truth values. If $\mathcal{F}$ is any formula and if S is any sign, the expression $S:\mathcal{F}$ may be interpreted as the assertion, "$\mathcal{F}$ evaluates to a truth value in S." For each interpretation over the MVL Λ, this assertion is either true or false, so $S : \mathcal{F}$ can be treated as a proposition in classical logic. For example, let $\mathcal{F}$ be a formula in Łukasiewicz's logic, and suppose we are interested in determining whether a formula C is a logical consequence of $\mathcal{F}$. One strategy is to determine whether the following disjunction of signed formulas must evaluate to true under all interpretations over Λ:

$$\{0, 1/2\} : \mathcal{F} \vee \{1\} : C$$

The connective $\vee$ is classical. Thus, given any interpretation, for the disjunction to evaluate to 1, either $\mathcal{F}$ evaluates to 0 or to $1/2$ or C evaluates to 1. In particular, if $\mathcal{F}$ evaluates to 1, then so must C, which is to say, C is a logical consequence of $\mathcal{F}$.

For example, let $\mathcal{F}$ be the formula $(p \wedge r) \wedge (p \rightarrow q)$. To show that q is a logical consequence of $\mathcal{F}$, we must determine whether the formula

$$\{0, 1/2\}:\mathcal{F} \vee \{1\} : q[-3pt] \tag{29.5}$$

is a tautology. Using the tableau method, we attempt to find a closed tableau for the negation of Eq. 29.5. It is useful to first drive the signs inward; thus, from the truth table:

$$\{1\} : ((p \wedge r) \wedge (p \rightarrow q)) \wedge \{0, 1/2\} : q = \{1\} : p \wedge \{1\} : r \wedge \{1\} : (\neg p \vee q) \wedge \{0, 1/2\} : q$$

$$= \{1\} : p \wedge \{1\} : r \wedge (\{1\} : \neg p \vee \{1\} : q) \wedge \{0, 1/2\} : q$$

$$= \{1\} : p \wedge \{1\} : r \wedge (\{0, 1/2\} : p \vee \{1\} : q) \wedge \{0, 1/2\} : q$$

The last formula is the initial tableau tree; if the disjunction is dispersed, the tableau becomes

$$\boxed{\{1\} : p}$$
$$\wedge$$
$$\boxed{\{1\} : r}$$
$$\wedge$$
$$\boxed{\{0, 1/2\} : q}$$
$$\wedge$$
$$\boxed{\{0, 1/2\} : p} \quad \vee \quad \boxed{\{1\} : q}$$

To close a tree path, the path must contain a *generalized link*: signed atoms $S_1 : A$ and $S_2 : A$, where S_1 and S_2 are disjoint. There are two tree paths in this tableau; the left path contains the link $\{\{1\} : p, \{0, 1/2\} : p\}$, and the right contains the link $\{\{0, 1/2\} : q, \{1\} : q\}$. Thus, the tableau may be closed and the proof is complete.

Most other classical inference techniques—for example, resolution and path dissolution—may similarly be generalized to MVLs using signed formulas, which may be thought of as formalizing metalevel reasoning about an arbitrary MVL Λ. In Lu et al. [1996], signed formulas provide a framework for adapting most classical inference techniques. The idea is to treat each signed formula $S : \mathcal{F}$ as a proposition in classical logic. Then classical inferences made by restricting attention to the Λ-*consistent interpretations*—those interpretations that assign true to a proposition $S : \mathcal{F}$ if and only if there is a corresponding interpretation over Λ that assigns some truth value in S to $\mathcal{F}$—yield an inference about the MVL Λ.

The language of signed formulas is closely related to the system of annotated logic studied in Kifer and Lozinskii [1992] and in Blair and Subrahmanian [1989]. Those authors had the goal of developing reasoning systems capable of dealing with inconsistent information. An examination of the relationship between logic programming based on signs and annotations can be found in Lu [1996]. Automated reasoning systems that implement reasoning in MVLs based on signed formulas include the 3TAP program of Beckert et al. [1992].

Fuzzy Logic

In recent years, fuzzy logic, which was introduced by Zadeh in 1965, has received considerable attention, largely for engineering applications such as heuristic control theory. There are at least two other views of fuzzy logic—see Dubois and Prade [1995] and Gaines [1977]—both closer to mainstream AI. One is that fuzzy logic can be regarded as an extension of classical logic for handling *uncertain* propositions—propositions whose truth values are derived from the unit interval [0, 1]. But relatively little attention has been paid to inference techniques for fuzzy logic; examples of deduction-based systems that do include Baldwin [1986], Lee [1972], Mukaidono [1982], Weigert et al. [1993].

In fuzzy logic, the set of truth values is $\Delta = [0, 1]$, and so interpretations assign each proposition a value in the interval [0, 1]. As usual, n-ary connectives are functions from $[0, 1]^n$ to [0, 1]. Conjunction $\wedge$ is usually the function min, disjunction $\vee$ is usually the function max, and $\neg$ is usually defined by $\neg \phi = 1 - \phi$. There are several possibilities for the function $\rightarrow$; perhaps the most obvious is $A \rightarrow B \equiv \neg A \vee B$.

The designated set of truth values Δ^* in fuzzy logic is a subinterval of [0, 1] of the form $[\alpha, 1]$ for some $\alpha \geq 0.5$. We call such an interval positive, and correspondingly call an interval of the form $[0, \alpha]$, where $\alpha \leq 0.5$, negative. For example, Weigert et al. [1993] defined a threshold of acceptability, $\tau \geq 0.5$, which in effect specifies Δ^* to be $[\tau, 1]$. On the other hand, Lee and Mukaidono do not explicitly define Δ^*. However, their systems implicitly adopt $\Delta^* = [0.5, 1]$. We begin this section by considering the fuzzy logic developed by Lee [1972] and extended by Mukaidono [1982] and then examine the more recent work of Weigert et al. [1993].

If we restrict attention to fuzzy formulas that use $\wedge$ and $\vee$ interpreted as min and max, respectively, as the only binary connectives and $\neg$ as defined previously as the only unary connective, then, as in the classical case, a formula may be put into an equivalent CNF. The keys are the observations

$$\neg(\mathcal{F} \wedge \mathcal{G}) = \neg\mathcal{F} \vee \neg\mathcal{G} \quad \text{and} \quad \neg(\mathcal{F} \vee \mathcal{G}) = \neg\mathcal{F} \wedge \neg\mathcal{G} \tag{29.6}$$

The resolution inference rule introduced by Lee is the obvious generalization of classical resolution. Let C_1 and C_2 be clauses (i.e., disjunctions of literals), and let L be an atom. Then the resolvent is defined by

$$\frac{L \vee C_1, \neg L \vee C_2}{C_1 \vee C_2} \tag{29.7}$$

Lee proved the following: Let C_1 and C_2 be two clauses, and let $R(C_1, C_2)$ be a resolvent of C_1 and C_2. If I is any interpretation, let $\max\{I(C_1), I(C_2)\} = b$ and $\min\{I(C_1), I(C_2)\} = a > 0.5$. Then $a \leq I(R(C_1, C_2)) \leq b$.

Mukaidono defines an inference to be *significant* if for any interpretation, the truth value of the conclusion is greater than or equal to the truth value of the minimum of the clauses in the premise. Lee's theorem may thus be interpreted to say that an inference using resolution is significant.

Weigert et al. [1993] built on the work of Lee and Mukaidono. They augmented the language by allowing infinitely many negation symbols that they call *fuzzy operators*. A formula is defined as follows: Let A be an atom, let $\mathcal{F}$ and $\mathcal{G}$ be fuzzy formulas, and let $\phi \in [0, 1]$.[4] Then:

1. ϕA is a fuzzy formula (also called a *fuzzy literal*)
2. $\phi(\mathcal{F} \wedge \mathcal{G})$ is a fuzzy formula
3. $\phi(\mathcal{F} \vee \mathcal{G})$ is a fuzzy formula

A simple example of a fuzzy formula is

$$\mathcal{F} = A \wedge 0.3(0.9B \vee 0.2C)$$

Several observations are in order. First, fuzzy operators are represented by real numbers in the unit interval. (That there are uncountably many fuzzy operators should not cause alarm. In practice, considering only rational fuzzy operators is not likely to be a problem. Indeed, with a computer implementation, we are restricted to a finite set of terminating decimals of at most n digits for some not very large n.) In particular, real numbers in the unit interval denote both truth values and fuzzy operators. Second, every formula and subformula is prefixed by a fuzzy operator; any subformula that does not have an explicit fuzzy operator prefix is understood to have 1 as its fuzzy operator.

The semantics of fuzzy operators are given via a kind of fuzzy product.

Definition 29.1. If $\phi, \delta \in [0, 1]$, then $\phi \otimes \delta = (2\phi - 1) \cdot \delta - \phi + 1$.

Observe that $\otimes$ is commutative and associative. Also observe that

$$\phi \otimes \delta = \phi \cdot \delta + (1 - \phi) \cdot (1 - \delta)$$

This last observation provides the intuition behind the fuzzy product $\otimes$: Were ϕ the probability that A_1 is true and were δ the probability that A_2 is true, then $\phi \otimes \delta$ would be the probability that A_1 and A_2 are both true or both false. (This probabilistic analogy is for intuition only; fuzzy logic is not based on probability.)

It turns out that the following generalization of Eq. (29.6) holds: Let $\mathcal{F}$ and $\mathcal{G}$ be fuzzy formulas, and let ϕ be a fuzzy operator. If $\phi > 0.5$, then

$$\phi(\mathcal{F} \wedge \mathcal{G}) = \phi\mathcal{F} \wedge \phi\mathcal{G} \qquad \text{and} \qquad \phi(\mathcal{F} \vee \mathcal{G}) = \phi\mathcal{F} \vee \phi\mathcal{G}$$

If $\phi < 0.5$, then

$$\phi(\mathcal{F} \wedge \mathcal{G}) = \phi\mathcal{F} \vee \phi\mathcal{G} \qquad \text{and} \qquad \phi(\mathcal{F} \vee \mathcal{G}) = \phi\mathcal{F} \wedge \phi\mathcal{G}$$

In particular, every fuzzy formula is equivalent to one in which 1 is the only fuzzy operator applied to nonatomic arguments.

In addition to introducing fuzzy operators, Weigert et al. extended Lee and Mukaidono's work with the *threshold of acceptability*: a real number $\tau \in [0.5, 1]$. Then an interpretation I is said to τ-*satisfy* the formula $\mathcal{F}$ if $I(\mathcal{F}) \geq \tau$. Observe that the threshold of acceptability is essentially a redefinition of Δ^* to $[\tau, 1]$. That is, the threshold of acceptability provides a variable for the definition of the designated set of truth values.

---

[4]Weigert et al. [1993] use λ for the fuzzy operator and Λ for the threshold of acceptability. We use ϕ and τ to avoid confusion with notation used in other parts of the chapter.

The significance of the threshold can be made clear by looking at some simple examples. Let $\tau = 0.7$ and consider each of the following three formulas: $0.8A, 0.2A$, and $0.6A$. Suppose A is 1; that is, $I(A) = 1$ for some interpretation I. Then $I(0.8A) = 0.8 \geq \tau$, so that the first formula is satisfied. The latter two evaluate to 0.2 and to 0.6, and so neither is satisfied. Now suppose A is 0. The first formula evaluates to $0.8 \otimes 0 = 0.2$, and the second evaluates to $0.2 \otimes 0 = 0.8$, so that the second formula is τ-satisfied. In effect, since the fuzzy operator 0.2 is less than $1 - \tau$, $0.2A$ is a negative literal and is τ-satisfied by assigning false to the atom A. The value of the third formula is now $0.6 \otimes 0 = 0.4$. Thus, in either case, the third formula is τ-unsatisfiable. Weigert et al. in fact define a clause to be τ-*empty* if every fuzzy operator of every literal in the clause lies between $1 - \tau$ and τ; it is straightforward to prove that every τ-empty clause is τ-unsatisfiable.

The fuzzy resolution rule relies upon complementary pairs of literals. However, complementarity is a relative notion depending on the threshold τ. Two literals $\phi_1 A$ and $\phi_2 A$ are said to be be τ-*complementary* if $\phi_1 \leq 1 - \tau$ and $\phi_2 \geq \tau$. Resolution for fuzzy logic, which Weigert et al. proved is sound and complete, can now be defined with respect to the threshold τ: Let $\phi_1 A$ and $\phi_2 A$ be τ-complementary, and let C_1 and C_2 be fuzzy clauses. Then

$$\frac{\phi_1 A \vee C_1, \phi_2 A \vee C_2}{C_1 \vee C_2} \tag{29.8}$$

Suppose we now reconsider the example from the Resolution section, which encoded the knowledge that "Tweety is a canary," "A canary is a bird," and "A bird flies." In reality, not all birds fly. A perhaps more realistic representation of this knowledge

   1 *Canary(Tweety)*
   0 *Canary(x)* $\vee$ 1 *Bird(x)*
   0*Bird(x)* $\vee$ 0.8 *Flies(x)*

The first two clauses represent the facts that Tweety is a canary and that all canaries are birds. In other words, the fuzzy operator 1 represents true, and the fuzzy operator 0 represents false. The fuzzy operator 0.8 may be interpreted as *highly likely*; thus, the third clause expresses the notion that most birds fly. Using the threshold 0.7 again and applying the same resolution steps as in Fig. 29.1, we can infer 0.8 *flies(Tweety)*. Observe that this means that the truth value of *flies(Tweety)* must be at least 5/6 since $0.8 \otimes 5/6 = 0.7$.

Nonmonotonic Logics

Common sense reasoning requires the ability to draw conclusions in the presence of incomplete information. Indeed, very few conclusions in our everyday thinking are based on knowledge of every piece of relevant information. Typically, numerous assumptions are required. Even a simple inference such as, *if x is a bird, then x flies*, is based on a host of assumptions regarding x, for instance, that x is not an unusual bird such as an ostrich or a penguin. It follows that logics for common sense reasoning must be capable of modeling reasoning processes that permit incorrect (and reversible) conclusions based on false assumptions. This observation has motivated the development of nonmonotonic logics, whose origins may be traced to foundational works by Clark, McCarthy, McDermott and Doyle, and Reiter; see Gallaire and Minker [1978] and Bobrow [1980].

The name **nonmonotonic logic** highlights the fundamental technical difference from classical logic, which is monotonic in the sense that

$$\mathcal{F}_1 \vdash \gamma \quad \text{and} \quad \mathcal{F}_1 \subseteq \mathcal{F}_2 \quad \text{implies } \mathcal{F}_2 \vdash \gamma \tag{29.9}$$

That is, classical entailment dictates that the set of conclusions from a knowledge base is inviolable: The addition of new knowledge never invalidates previously inferred conclusions. A classically based reasoning agent will therefore never be able to retract a conclusion in light of new, possibly contradictory information. Nonmonotonic logics, on the other hand, need not obey Eq. (29.9).

The investigation of inference techniques for nonmonotonic logics has been limited, but there have been several interesting attempts. Comprehensive studies of nonmonotonic reasoning and default logic include Etherington [1988], Marek and Truszcyński [1993], Besnard [1989], and Moore's autoepistemic logic [Moore 1985]. We will focus on the nonmonotonic formalism of Reiter known as *default logic*; see his paper in Bobrow [1980]. Some of the other proposed systems contain technical differences, but the essence of *nonmonotonicity*—failure to obey Eq. (29.9)—is adhered to by all.

A *default* is an inference (scheme) of the form

$$\frac{\alpha : M\beta_1, \ldots, M\beta_m}{\gamma} \qquad (29.10)$$

where $\alpha, \beta_1, \ldots, \beta_m$, and γ are formulas. The formula α is the *prerequisite* of the default, $\{M\beta_1, \ldots, M\beta_m\}$ is the *jusitification*, and γ is the *consequent*; the M in the justification serve merely to demark the justification. A *default theory* is a pair (D, W), where W is a set of formulas and D is a set of defaults. Intuitively, W may be thought of as the set of knowledge that is known to be true. A default theory (D, W) then enables a reasoner to draw additional conclusions through the defaults in D.

As a motivating example, consider the default rule

$$\frac{Bird(x) : M\,Fly(x)}{Fly(x)} \qquad (29.11)$$

The prerequisite specifies *if an individual x is a bird*, and the justification specifies, *if it is consistent to assume that x flies*, then we may infer by default that x flies. Suppose we have a theory consisting of the clause set $W = \{Canary(Tweety), \neg Canary(x) \vee Bird(x)\}$. Using classical logic, one logical consequence of W is the fact $Bird(Tweety)$. If $S = \{\alpha \mid W \vdash \alpha\}$ represents the reasoner's knowledge, then $S \cup \{Fly(Tweety)\}$ is consistent. That is, $S \cup \{Fly(Tweety)\}$ has a satisfying interpretation in classical logic. The default rule Eq. (29.11) then warrants the inference $Fly(Tweety)$.

It is important to note that formulas and consequences in default logics are classical in the sense that a formula is either true or false. The difference from classical logic lies in the manner in which consequences are derived from a given set of formulas. From W we may conclude $Fly(Tweety)$. However, suppose we add to W the additional knowledge set A consisting of the clauses:

> $Broken\_Wings(Tweety)$
> $Broken\_Wings(x) \rightarrow \neg Fly(x)$

Then $S = \{\alpha \mid W \vdash \alpha\}$ contains, among other things, the fact $\neg Fly(Tweety)$. In this case, $Fly(Tweety)$ is inconsistent with S; that is, $S \cup \{Fly(Tweety)\}$ is not satisfiable. Thus, the condition for the justification part of the default rule (29.11) is not met, and hence the conclusion $Fly(Tweety)$ cannot be drawn. This provides a clear illustration of the nonmonotonic nature of default inference rules such as rule (29.11),

$$W \vdash_D Fly(Tweety), \qquad \text{but} \qquad W \cup A \not\vdash_D Fly(Tweety)$$

where $\vdash_D$ means deduction based on classical inference or the default rules in D.

In the initial version of the example, the conclusion $Fly(Tweety)$ was obtained starting with the set S (which is the set of all classical consequences of W), and then applying the default rule according to the condition that $Bird(Tweety)$ is a consequence of S, and that $\{Fly(Tweety)\} \cup S$ is satisfiable. Now suppose a reasoner holds the following initial set of beliefs

$$S_0 = \{Bird(Tweety), Fly(Tweety), Canary(Tweety), Canary(x) \rightarrow Bird(x)\}$$

This set contains S, and applications of the default rule (29.11) with respect to S_0 yield no additional conclusions. That is, since $Bird(Tweety)$ is a classical consequence of S_0, and $Fly(Tweety) \in S_0$, so that $\{Fly(Tweety)\} \cup S_0$ is consistent, adding the conclusion $Fly(Tweety)$ from the consequent of the default rule produces no changes in the beliefs S_0. A set that has such a property holds a special status in default logic and is called an *extension*. Intuitively, an extension E may be thought of as a set of formulas that *agree* with all the default rules in the logic; i.e., every default whose prerequisites are in E and whose justification is consistent with E must have its consequence in E. Still another way to look at an extension E of W is as a superset of W that is closed under both classical and default inference.

To formally define extension, given a set of formulas E, let $th(E)$ denote the set of all classical consequence of E; that is, $th(E) = \{\alpha \mid E \vdash \alpha\}$. IF (D, W) is a default theory, let $\Gamma(E)$ be the smallest set of formulas that satisfies the following conditions:

1. $W \subseteq \Gamma(E)$.
2. $\Gamma(E) = th(\Gamma(E))$.
3. Suppose $(\alpha : M\beta_1, \ldots, M\beta_m/\gamma) \in D$, $\alpha \in \Gamma(E)$, and $\neg\beta_1, \ldots, \neg\beta_m \notin E$. Then $\gamma \in \Gamma(E)$.

Then E is an extension of (D, W) if $E = \Gamma(E)$.

Observe that the third part of the definition requires $\neg\beta_i \notin E$ for each i. This is, in general, a weaker notion than the requirement that β_i be consistent with E. That is, were E not deductively closed under inference in classical logic, it is possible that one $\neg\beta_i$ is not a member of but is a logical consequence of E; $E \cup \{\beta_i\}$ would then be inconsistent. However, in the case of an extension, which is closed under classical deduction, the two notions coincide.

The simple example just illustrated suggests a natural way of computing extensions, namely, begin with the formulas in W and repeatedly apply each default inference rule until no new inferences are possible. Of course, since default rules may be interdependent, the choice of which default rule to apply first may affect the extension that is obtained. For example, with the following two simple rules, the application of one prevents the application of the other by making the justification of the other inconsistent with the inferred fact:

$$\frac{true : MB}{\neg A} \qquad \frac{true : MA}{\neg B}$$

Marek and Truszcyński introduced the operator R^D to compute extensions of a default logic. Let U be a set of formulas; then

$$R^D(U) = th\left(U \cup \left\{\gamma \,\middle|\, \frac{\alpha}{\gamma} \in D \text{ and } \alpha \in U\right\}\right)$$

Observe that α/γ is justification free. Thus, R^D amounts to the closure of U under classical and justification free default inference.

Repeated applications of the operator R^D is merely function composition and may be written as follows:

$$R^D \uparrow 0(U) = U$$

$$R^D \uparrow (\alpha + 1)(U) = R^D(R^D \uparrow \alpha(U))$$

$$R^D \uparrow \lambda(U) = \bigcup\{R^D \uparrow \alpha(U) \mid \alpha < \lambda\} \qquad \text{for a limit ordinal } \lambda$$

Given a default theory (D, W) and a set of formulas S, the *reduct* of D with respect to S, D_S, is defined to be the set of justification-free inference rules of the form α/γ, where $(\alpha : M\beta_1, \ldots, M\beta_m/\gamma)$ is a default in D, and $\neg\beta_i \notin S$ for each i. The point is, once we know that a justification is satisfiable, then the corresponding justification-free rule is essentially equivalent.

Marek and Truszcyński showed that from the knowledge base W of a default theory, it is possible to use the operator R^D to determine whether a set of formulas is an extension.[5] More precisely, a set of formulas E is an extension of a default theory (D, W) if and only if $E = R^{D_E} \uparrow \omega(W)$.

---

[5] Even for propositional default theory, this checking process is computationally very expensive since it is necessary to choose the set E nondeterministically.

Default logic is intimately connected with nonmonotonic logic programming. Analogs of the many results regarding extensions may be found in nonmonotonic logic programming. The problem of determining whether a formula is contained in some extension of a default theory is called the *extension membership problem*; in general, it is quite difficult because it is not semidecidable (as compared, for example, with first-order classical logic). This makes implementation of nonmonotonic reasoning much harder than the already difficult task of implementing monotonic reasoning. Reiter speculated that a reasonable computational approach to default logic will necessarily allow for incorrect (unsound) inferences [Bobrow 1980]. This issue was also considered in Etherington [1988]. Some recent work on proof procedures for default logic can be found in Barback and Lobo [1995] (resolution based), in the work of Thielscher and Schaub (see Lifschitz [1995]), and in the tableau-based work of Risch and Schwind (see Wrightson [1994]). Work on general nonmonotonic deduction systems include Kraus et al. [1990].

29.4 Research Issues and Summary

Logic-based deductive reasoning has played a central role in the development of artificial intelligence. The scope of AI research has expanded to include, for example, vision and speech [Russell and Norvig 1995], but the importance of logical reasoning remains. At the heart of logic-based deductive reasoning is the ability to perform inference. In this chapter we have discussed but a few of the numerous inference rules that have been widely applied.

There are several directions that researchers commonly pursue in automated reasoning. One is the exploration of new logics. This line of research is more theoretical and intimately tied to the philosophical foundation of the reasoning processes of intelligent agents. Typically, the motivation behind newly proposed logics lies with some aspect of reasoning for which classical two-valued logic may not be adequate. Among the many examples are temporal logic, which attempts to deal with time-oriented reasoning [Allen 1991]; modal logics, which address questions of knowledge and beliefs [Fagin et al. 1992], and alternative MVLs [Ginsberg 1988].

Another area of ongoing research is the development of new inference techniques for existing logics, both classical and nonclassical, for example, Henschen [1979] and McRobbie [1991]. Such techniques might produce better general purpose inference engines or might be especially well suited for some narrowly defined reasoning process.

Inference also plays an important role in *complexity theory*, which is, more or less, the analysis of the running time of algorithms; it is carefully described in other chapters. A fundamental question in complexity theory—indeed, a famous open question in all of computer science—is, "Does the class $\mathcal{NP}$ equal the class $\mathcal{P}$?" It has been shown [Cook 1971] that this question is equivalent to the question, "Is there a fast algorithm for determining whether a formula in classical propositional logic is satisfiable?" (Roughly speaking, *fast* means a running time that is polynomial in the size of the input.)

Implementation of deduction techniques continues to receive a great deal of attention from researchers. Considerable effort has gone into controlling the search space. In recent years, many authors have chosen to replace domain-independent, general purpose control strategies with domain-specific strategies, for example, the work of Bundy, van Harmelen, Hesketh and Smaill, Smith, and Wos. Other implementation issues for theorem provers include the use of discrimination trees by McCune, flatterms by Christian, and parallel representation by Fishman and Minker.

Defining Terms

Artificial intelligence: The field of study that attempts to capture aspects of human intelligence in machines.

Automated deduction, automated reasoning: Deduction techniques that may be mechanized.

Classical logic: The standard logic that employs the usual connectives and the truth values {*true, false*}.

Clause: A disjunction of literals.

Completeness: An inference or rewrite rule is complete if the rule can verify that an unsatisfiable formula is unsatisfable.

Conjunctive normal form (CNF): A conjunction of clauses.

Default logic: A nonmonotonic logic.

Domain of discourse: The set of values to which a first-order variable may be assigned.

First-order logic: Logic in which predicates may have arguments and formulas may be quantified.

Fuzzy logic: An extension of classical logic for handling uncertain propositions.

Herbrand universe: A domain of discourse constructed from the constants and function symbols that appear in a logical formula.

Inference rule: A rule that, when applied to a formula, produces another formula.

Interpretation: A function from the atom set to the set of truth values.

Lifting lemma: A lemma for proving completeness at the first-order level from completeness at the propositional level.

Literal: An atom or the negation of an atom.

Logical consequence: A formula C is a logical consequence of $\mathcal{F}$ if every interpretation that satisfies $\mathcal{F}$ also satisfies C.

Multiple-valued logic: Any logic whose set of truth values is *not* restricted to {*true, false*}.

Negation normal form (NNF): A form for logical formulas in which conjunction and disjunction are the only binary connectives and in which all negations are at the atomic level.

Nonmonotonic logic: A logic in which the addition of new knowledge may invalidate previously inferrable conclusions.

Paramodulation: An specialized inference rule for handling equality.

Path dissolution: An inference mechanism that operates on formulas in NNF.

Prenex normal form: A form for first-order logical formulas in which all quantifiers appear in the front of the formula.

Propositional logic: Logic in which predicates may not have variables as arguments.

Quantifier: A restriction on variables in a first-order formula.

Resolution: An inference rule that operates on sets of clauses.

Rewrite rule: A rule that modifies formulas.

Satisfiable: A formula in classical logic is satisfiable if it evaluates to *true* under some interpretation.

Sign: Any subset of the set of truth values.

Skolem standard form: A form for first-order logical formulas in which all existentially quantified variables are replaced by constants or by functions of constants and the universally quantified variables.

Soundness: An inference (or rewrite rule) is sound if every inferred formula is a logical consequence of the original formula.

Substitution: A function that maps variables to terms.

Tableau method: An inference mechanism that operates on formulas in NNF.

Unification algorithm: An algorithm that finds the most general unifier of a set of terms.

Unifier: A substitution that unifies—makes identical—terms in different predicate occurrences.

References

Allen, J. F. 1991. Time and time again: the many ways to represent time. *J. Intelligent Syst.* 6:341–355.

Andrews, P. B. 1976. Refutations by matings. *IEEE Trans. Comput.* C-25:801–807.

Baldwin, J. 1986. Support logic programming. In *Fuzzy Sets Theory and Applications.* A. Jones, A. Kaufmann, and H. Zimmermann., eds., pp. 133–170. D. Reidel.

Barback, M. and Lobo, J. 1995. A resolution-based procedure for default theories with extensions. In *Nonmonotonic Extensions of Logic Programming,* J. Dix, L. Pereira, and T. Przymusinski., eds. Springer–Verlag, Heidelberg.

Beckert, B., Gerberding, S., Hähnle, R., and Kernig, W. 1992. The tableau-based theorem prover 3TAP for multiple-valued logics. In *Proc. 11th Int. Conf. Automated Deduction*, pp. 758–760. Springer–Verlag, Heidelberg.

Beckert, B. and Posegga, J. 1995. leanTAP: Lean tableau-based deduction. *J. Automated Reasoning* 15(3):339–358.

Besnard, P. 1989. *An Introduction to Default Logic.* Springer–Verlag, Heidelberg.

Beth, E. W. 1955. Semantic Entailment and Formal Derivability. *Mededelingen van de Koninklijke Nederlandse Akad. van Wetenschappen, Afdeling Letterkunde, N.R.* 18(3):309–342.

Bibel, W. 1987. *Automated Theorem Proving.* Vieweg Verlag, Braunschweig.

Blair, H. A. and Subrahmanian, V. S. 1989. Paraconsistent logic programming. *Theor. Comput. Sci.* 68:135–154.

Bledsoe, W. W., Kunen, K., and Shostak, R. 1985. Completeness results for inequality provers. *Artificial Intelligence* 27:255–288.

Bobrow, D. G., ed. 1980. *Artificial Intelligence: Spec. Issue Nonmonotonic Logics.* 13.

Boyer, R. S. and Moore, J. S. 1979. *A Computational Logic.* Academic Press, New York.

Bundy, A., van Harmelen, F., Hesketh, J., and Smaill, A. 1988. Experiments with proof plans for induction. *J. Automated Reasoning* 7(3):303–324.

Chang, C. L. and Lee, R. C. T. 1973. *Symbolic Logic and Mechanical Theorem Proving.* Academic Press, New York.

Cook, S. A. 1971. The complexity of theorem proving procedures, pp. 151–158. In *Proc. 3rd Annu. ACM Symp. Theory Comput.* ACM Press, New York.

Digricoli, V. J. and Harrison, M. C. 1986. Equality based binary resolution. *J. ACM* 33(2):253–289.

Dubois, D. and Prade, H. 1995. What does fuzzy logic bring to AI? *ACM Comput. Surveys* 27(3):328–330.

Etherington, D. W. 1988. *Reasoning with Incomplete Information.* Pitman, London, UK.

Fagin, R., Halpern, J. Y., and Vardi, M. Y. 1992. What can machines know? On the properties of knowledge in distributed systems. *J. ACM* 39(2):328–376.

Fitting, M. 1990. *Automatic Theorem Proving.* Springer–Verlag, Heidelberg.

Gabbay, D. M., Hogger, C. J., and Robinson, J. A., eds. 1993–95. *Handbook of Logic in Artificial Intelligence and Logic Programming.* Vols. 1–4, Oxford University Press, Oxford, UK.

Gaines, B. R. 1977. Foundations of fuzzy reasoning. In *Fuzzy Automata and Decision Processes.* M. M. Gupta, G. N. Saridis, and B. R. Gaines, eds., pp. 19–75. North-Holland.

Gallaire, H. and Minker, J., eds. 1978. *Logic and Data Bases.* Plenum Press.

Genesereth, M. R. and Nilsson, N. J. 1988. *Logical Foundations of Artificial Intelligence.* Morgan Kaufmann, Menlo Park, CA.

Gentzen, G. 1969. Investigations in logical deduction. In *Studies in Logic*, M. E. Szabo, ed., pp. 132–213. Amsterdam.

Ginsberg, M. 1988. Multivalued logics: A uniform approach to inference in artificial intelligence. *Comput. Intelligence* 4(3).

Green, C. 1969. Application of theorem proving to problem solving, pp. 219–239. In *Proc. 1st Int. Conf. Artificial Intelligence.* Morgan Kaufmann, Menlo Park, CA.

Hähnle, R. 1994. *Automated Deduction in Multiple-Valued Logics.* Vol. 10. International series of monographs on computer science. Oxford University Press, Oxford, UK.

Hayes, P. 1977. In defense of logic, pp. 559–565. In *Proc. 5th IJCAI.* Morgan-Kaufman, Palo Alto, CA.

Henschen, L. 1979. Theorem proving by covering expressions. *J. ACM* 26(3):385–400.

Hintikka, K. J. J. 1955. Form and content in quantification theory. *Acta Philosohica Fennica* 8:7–55.

Kapur, D. and Zhang, H. 1989. An overview of RRL: rewrite rule laboratory. In *Proc. 3rd Int. Conf. Rewriting Tech. Its Appl.* LNCS 355:513–529.

Kifer, M. and Lozinskii, E. 1992. A logic for reasoning with inconsistency. *J. Automated Reasoning* 9(2):179–215.

Kleene, S. C. 1952. *Introduction to Metamathematics.* Van Nostrand, Amsterdam.

Kraus, S., Lehmann, D., and Magidor, M. 1990. Nonmonotonic reasoning, preferential models and cumulative logics. *Artificial Intelligence* 44:167–207.

Lee, R. C. T. 1972. Fuzzy logic and the resolution principle. *J. ACM* 19(1):109–119.

Lifschitz, V., ed. 1995. *J. Automated Reasoning: Spec. Issue Common Sense Nonmonotonic Reasoning* 15(1).

Loveland, D. W. 1978. *Automated Theorem Proving: A Logical Basis*. North-Holland, New York.

Lu, J. J. 1996. Logic programming based on signs and annotations. *J. Logic Comput.* (to appear).

Lu, J. J., Murray, N. V., and Rosenthal, E. 1996. A framework for automated reasoning in multiple-valued logics. *J. Automated Reasoning* (to appear).

Łukasiewicz, J. 1970. *Selected Works*, L. Borkowski, ed., North-Holland, Amsterdam.

Manna, Z. and Waldinger, R. 1986. Special relations in automated deduction. *J. ACM* 33(1):1–59.

Marek, V. W. and Truszcyński, M. 1993. *Nonmonotonic Logic: Context-Dependent Reasoning*. Springer–Verlag, Heidelberg.

McCune, W. 1992. Experiments with discrimination-tree indexing and path indexing for term retrieval. *J. Automated Reasoning* 9(2):147–168.

McRobbie, M. A. 1991. Automated reasoning and nonclassical logics: introduction. *J. Automated Reasoning: Spec. Issue Automated Reasoning Nonclassical Logics* 7(4):447–452.

Mendelson, E. 1979. *Introduction to Mathematical Logic*. Van Nostrand Reinhold, Princeton, NJ.

Moore, R. C. 1985. Semantical considerations on nonmonotonic logic. *Artificial Intelligence* 25:27–94.

Mukaidono, M. 1982. Fuzzy inference of resolution style. In *Fuzzy Set and Possibility Theory*, R. Yager, ed., pp. 224–231. Pergamon, New York.

Murray, N. V. and Rosenthal, E. 1993. Dissolution: making paths vanish. *J. ACM* 40(3):502–535.

Robinson, J. A. 1965. A machine-oriented logic based on the resolution principle. *J. ACM* 12:23–41.

Robinson, J. A. 1979. *Logic: Form and Function*. Elsevier North-Holland, New York.

Robinson, G. and Wos, L. 1969. Paramodulation and theorem proving in first-order theories with equality. In *Machine Intelligence*, Vol. IV, B. Melzer and D. Michie, eds., pp. 135–150. Edinburgh University Press, Edinburgh, UK.

Russell, S. and Norvig, P. 1995. *Artificial Intelligence: A Modern Approach*. Prentice-Hall, Englewood Cliffs, NJ.

Slagle, J. 1972. Automatic theorem proving with built-in theories including equality, partial ordering, and sets. *J. ACM* 19(1):120–135.

Smullyan, R. M. 1995. *First-Order Logic*, 2nd ed. Dover, New York.

Sterling, L. and Shapiro, E. 1986. *The Art of Prolog*. ACM Press, Cambridge, MA.

Stickel, M. E. 1985. Automated deduction by theory resolution. *J. Automated Reasoning* 1(4):333–355.

Urquhart, A. 1986. Many-valued logic. In *Handbook of Philosophical Logic*, Vol. III, D. Gabbay and F. Guenthner, eds., pp. 71–116. D. Reidel.

Weigert, T. J., Tsai, J. P., and Liu, X. H. 1993. Fuzzy operator logic and fuzzy resolution. *J. Automated Reasoning* 10(1):59–78.

Wos, L., Overbeek, R., Lusk, E., and Boyle, J. 1992. *Automated Reasoning: Introduction and Applications*. 2nd ed. Prentice–Hall, Englewood Cliffs, NJ.

Wrightson, G., ed. 1994. *J. Automated Reasoning: Spec. Issues Automated Reasoning Analytic Tableaux* 13(2,3).

Zadeh, L. A. 1965. Fuzzy sets. *Inf. Control* 8:338–353.

Further Information

The *Journal of Automated Reasoning* is an excellent reference for current research and advances in logic-based automated deduction techniques for both classical and nonclassical logics. The International Conference on Automated Deduction (CADE) is the major forum for researchers focusing on logic-based deduction techniques; its proceedings are published by Springer–Verlag. Other conferences with an emphasis on computational logic and logic-based reasoning include: the International Logic Programming Conference (ICLP), Logics in Computer Science (LICS), Logic Programming and Nonmonotonic Reasoning

Conference (LPNMR), International Symposium on Multiple-Valued Logics (ISMVL), the International Symposium on Methodologies for Intelligent Systems (ISMIS), and the IEEE International Conference on Fuzzy Systems.

More general conferences on AI include the two major annual meetings: The conference of the AAAI and the International Joint Conference on AI. Each of these conferences regularly publishes logic-based deduction papers. The *Artificial Intelligence* journal is an important source for readings on logics for common sense reasoning and related deduction techniques.

Other journals of relevance include: the *Journal of Logic and Computation*, the *Journal of Computational Intelligence*, the *Journal of Logic Programming*, the *Journal of Symbolic Computation*, *IEEE Transactions on Fuzzy Systems*, *Theoretical Computer Science*, and the *Journal of the Association of Computing Machinery*.

Most of the texts referenced in this chapter provide a more detailed introduction to the field of computational logic. They include: Bibel [1987], Chang and Lee [1973], Fitting [1990], Loveland [1978], Robinson [1979], and Wos et al. [1992]. Good introductory texts for mathematical logic are: Mendelson [1979] and Smullyan [1995].

30

Search

Danny Kopec
Richard Stockton College

T. A. Marsland
University of Alberta

30.1 Introduction

Artificial intelligence (AI) efforts to solve problems with computers—which humans routinely handle by employing innate cognitive abilities, pattern recognition, perception, and experience, invariably must turn to considerations of search. This chapter explores search methods in AI, including both **blind** exhaustive methods and informed **heuristic** and optimal methods, along with some more recent findings. The search methods covered include (for nonoptimal, uninformed approaches) state-space search, **generate and test, means–ends analysis**, problem reduction, **AND/OR trees, depth-first search**, and **breadth-first search**. Under the umbrella of heuristic (informed) methods we discuss hill climbing, **best-first search, bi-directional search**, and the **A* algorithm**. Tree searching algorithms for games have proven to be a rich source of study and provide empirical data about heuristic methods. Included here are the **SSS* algorithm**, variations on the **alpha–beta** minimax algorithm, and the use of **iterative deepening**.

Since many of these methods are computationally intensive, the second half of the chapter focuses on parallel methods. The importance of parallel search is presented through an assortment of relatively recent parallel algorithms including the parallel iterative deepening algorithm (PIDA*), **principal variation splitting (PVSplit)**, and the **young brothers wait concept**. Coincident with the continuing price-performance improvement of small computers is a growing interest in reimplementing some of the heuristic techniques developed for problem solving and planning programs, to see if they can be enhanced or replaced by more algorithmic methods. The application of raw computing power, although an anathema to some, often provides better answers than is possible by reasoning or analogy. Thus, brute force techniques form a good basis against which to compare more sophisticated methods designed to mirror the human deductive process. Parallel methods are important not only for single-agent search, but also through a variety of parallelizations for adversary games. In the latter case there is an emphasis on the problems that pruning poses in unbalancing the work load, and so we cover some of the dynamic tree-splitting methods that have evolved. One source of extra computing power comes through the use of parallel processing on a multicomputer.

0-8493-2909-4/97/$0.00+$.50

30.2 Uninformed Search Methods

Search Strategies

All search methods in computer science share in common three necessities: (1) a world model or database of facts based on a choice of representation providing the current state, as well as other possible states and a goal state; (2) a set of operators that defines possible transformations of states; and (3) a control strategy that determines how transformations amongst states are to take place by applying operators. Reasoning about the current state to identify a state which is closer to a goal is known as *forward reasoning*. Working backwards to a current state from a goal state is called *backward reasoning*. As such it is possible to make distinctions between bottom up and top down approaches to problem solving. Bottom up is often *goal oriented*, that is, reasoning backwards from a goal state to solve intermediary subgoal states. Top down or data-driven reasoning is based on simply being able to get to a state which is defined as closer to a goal state than the current state. Often application of operators to a problem state may not lead directly to a goal state and some **backtracking** may be necessary before a goal state can be found [Barr and Feigenbaum 1981].

State-Space Search

Exhaustive search of a problem space (or search space) is often not feasible or practical due to the size of the problem space. In some instances, however, it is necessary. More often, we are able to define a set of legal transformations of a state space (moves in the world of games) from which those that are more likely to bring us closer to a goal state are selected, whereas others are never explored further. This technique in problem solving is known as *split and prune*. In AI, the technique that emulates this approach is called generate and test. The basic method is shown in Fig. 30.1.

Good generators are complete and will eventually produce all possible solutions, while not proposing redundant ones. They are also informed; that is, they will employ additional information to constrain the solutions they propose.

Means–ends analysis is another state-space technique whose purpose is to reduce the difference (distance) between current state and a goal state. Determining *distance* between any state and a goal state can be facilitated by *difference-procedure tables*, which can effectively prescribe what the next state might be. To perform means-ends analysis follow the procedure in Fig. 30.2.

The technique of *problem reduction* is another important approach in AI. That is, solve a complex or larger problem by identifying smaller manageable problems (or subgoals), which you know can be solved in fewer steps.

For example, Fig. 30.3 shows the sliding block puzzle Donkey, which has been known for over 100 years. Subject to constraints on the movement of pieces in the sliding block puzzle, the task is to slide the blob around the vertical bar with the goal of moving it to the other side. The blob occupies four spaces and needs two adjacent vertical or horizontal spaces in order to be able to move, whereas the vertical bar needs two adjacent empty vertical spaces to move left or right, or one empty space above or below it to move up or down. The horizontal bars' movements are complementary to the vertical bar. Likewise, the circles can move to any empty space around them in a horizontal or vertical line. A relatively uninformed state

```
Repeat
        Generate a candidate solution
        Test the candidate solution
Until a satisfactory solution is found, or
        no more candidate solutions can be generated:
If an acceptable solution is found, announce it;
        Otherwise, announce failure.
```

FIGURE 30.1 Generate and test method.

```
Repeat
        Describe the current state, the goal state,
        and the difference between the two.
        Use the difference between the current state and goal state,
        to select a promising transformation procedure.
        Apply the promising procedure and update the current state.
    Until the GOAL is reached or
        no more procedures are available
    If the GOAL is reached, announce success;
        Otherwise, announce failure.
```

FIGURE 30.2   Means–end analysis.

space search can result in over 800 moves for this problem to be solved, with plenty of backtracking necessary. By problem reduction, resulting in the subgoal of trying to get the blob on the two rows above or below the vertical bar, it is possible to solve this puzzle in just 82 moves!

Another example of a technique for problem reduction is called AND/OR trees. Here the goal is to find a solution path to a given tree by applying the following rules.

A node is solvable if:

1. It is a terminal node (a primitive problem).
2. It is a nonterminal node whose successors are AND nodes that are all solvable.
3. Or it is a nonterminal node whose successors are OR nodes and at least one of them is solvable.

Similarly, a node is unsolvable if:

1. It is a nonterminal node that has no successors (a nonprimitive problem to which no operator applies).
2. It is a nonterminal node whose successors are AND nodes and at least one of them is unsolvable.
3. Or it is a nonterminal node whose successors are OR nodes and all of them are unsolvable.

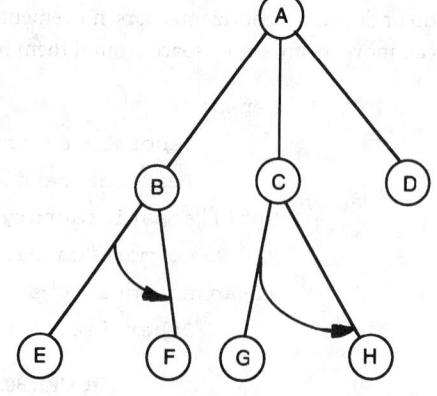

FIGURE 30.3   Problem reduction and the sliding block puzzle.

In Fig. 30.4 nodes B and C serve as exclusive parents to subproblems EF and GH, respectively. One way of viewing the tree is with nodes B, C, and D serving as individual, alternative subproblems representing OR nodes. Node pairs E and F and G and H, respectively, with curved arrowheads connecting them, represent AND nodes. That is, to solve problem B you must solve both subproblems E and F. Likewise, to solve subproblem C, you must solve subproblems G and H. Solution paths would therefore be: {A-B-E-F}, {A-C-G-H}, and {A-D}. In the special case where no AND nodes occur, we have the ordinary graph occurring in a state-space search. However, the presence of AND nodes distinguishes AND/OR trees (or graphs) from ordinary state structures, which call for their own specialized search techniques. Typical problems tackled by AND/OR trees include games or puzzles, and other well-defined state-space goal-oriented problems, such as robot planning, movement through an obstacle course, or setting a robot the task of reorganizing blocks on a flat surface.

FIGURE 30.4   AND/OR tree.

Breadth-First Search

One way to view search problems is to consider all possible combinations of subgoals, by treating the problem as a tree search. Breadth-first search always explores nodes closest to the root node first, thereby visiting all nodes at a given layer first before moving to any longer paths. It pushes uniformly into the search tree. Because of memory requirements, breadth-first search is only practical on shallow trees, or those with an extremely low branching factor. It is, therefore, not used much in practice, except as a basis for such best-first search algorithms as A* and SSS*.

Depth-First Search

Depth-first search (DFS) is one of the most basic and fundamental blind search algorithms. It is used for bushy trees (with high branching factor) where a potential solution does not lie too deeply down the tree. That is, "DFS is a good idea when you are confident that all partial parts either reach dead ends or become complete paths after a reasonable number of steps." In contrast, "DFS is a bad idea if there are long paths, particularly indefinitely long paths, that neither reach dead ends nor become complete paths" [Winston 1992]. To conduct a DFS:

1. Put the start node on the list called open.
2. If open is empty, exit with failure; otherwise continue.
3. Remove the first node from open and put it on a list called closed. Call this node n.
4. If the depth of n equals the depth bound, go to step 2; otherwise continue.
5. Expand node n, generating all immediate successors. Put these at the beginning of open (in predetermined order) and provide pointers back to n.
6. If any of the successors are goal nodes, exit with the solution obtained by tracing back through the pointers; otherwise go to step 2.

DFS always explores the deepest node to the left first. That is, the one which is farthest down from the root of the tree. When a dead end (terminal node) is reached, the algorithm backtracks one level and then tries to go forward again. To prevent consideration of unacceptably long paths, a *depth bound* is often employed to limit the depth of search. At each node immediate successors are generated and a transition made to the leftmost node, where the process continues recursively until a dead end or depth limit is reached. DFS would explore the tree in Fig. 30.5 in the order: I-E-b-F-B-a-G-c-H-C-a-D-A. Here the notation using lowercase letters represents the possible storing of provisional information about the subtree. For example, this could be a lower bound on the value of the tree.

Figure 30.6 enhances depth-first search with a form of *iterative deepening* that can be used in a single agent search like A*. DFS expands an immediate successor of some node **N** in a tree. The next successor (**N.i**) expanded is the one with lowest cost function. Thus, the expected value of node **N.i** is the estimated

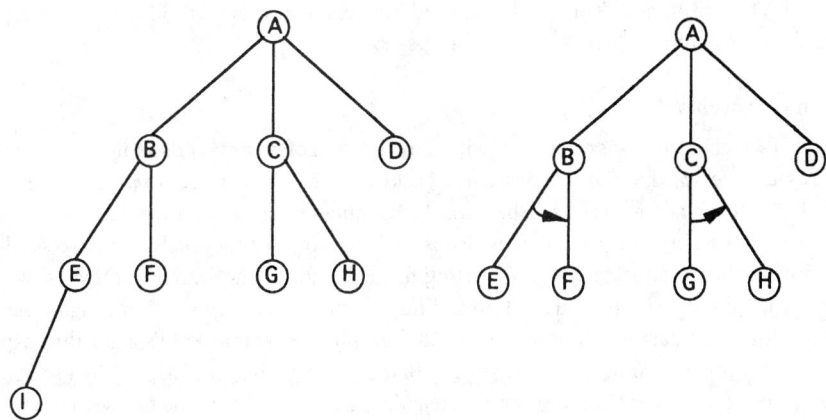

FIGURE 30.5   Tree example for depth-first and breadth-first search.

```
// The A* (DFS) algorithm expands the N.i successors of node N
// in best first order. It uses and sets solved, a global indicator.
// It also uses a heuristic estimate function H(N), and a
// transition cost C(N,N.i) of moving from N to N.i
//
IDA* (N) → cost
     bound ← H(N)
     while not solved
          bound ← DFS (N, bound)
     return bound                            // optimal cost

DFS (N, bound) → value
     if H(N) ≡ 0                             // leaf node
          solved ← true
          return 0
     new_bound ← ∞
     for each successor N.i of N
          merit ← C(N, N.i) + H(N.i)
          if merit ≤ bound
               merit ← C(N, N.i) + DFS (N.i, bound - C(N, N.i))
               if solved
                    return merit
          if merit < new_bound
               new_bound ← merit
     return new_bound
```

FIGURE 30.6   The A* depth first search (DFS) algorithm for use with IDA*.

cost **C(N,N.i)** plus **H(N)**, the known value of node **N**. The basic idea in iterative deepending is that a DFS is started with a depth bound of 1, and this continues with the depth bound increasing by one with each iteration. With each increase in depth the algorithm must reinitiate its depth-first search for the prescribed bound. The idea of iterative deepening, in conjunction with a memory function to retain the best available potential solution paths from iteration to iteration, is credited to Slate and Atkin [1977] who used it in their chess program. Korf [1985] showed how efficient this method is in single-agent search, with his iterative deepening A* (IDA*) algorithm.

Bidirectional Search

To this point all search algorithms discussed (with the exception of means–ends analysis and backtracking) have been based on forward reasoning. Searching backwards from goal nodes to predecessors is relatively easy. Pohl [1971] combined forward and backward reasoning into a technique called *bidirectional search*. The idea is to replace a single search graph, which is likely to grow exponentially, with two smaller graphs: one starting from the initial state and one starting from the goal. The search terminates when the two graphs intersect. This algorithm is guaranteed to find the shortest solution path through a general state-space graph. Empirical data for randomly generated graphs shows that Pohl's algorithm expands only about 1/4 as many nodes as unidirectional search [Barr and Feigenbaum 1981]. Pohl also implemented heuristic versions of this algorithm. However, determining when and how the two searches will intersect is a complex process.

30.3 Heuristic Search Methods

George Polya, via his wonderful book *How To Solve It* [1945], may be regarded as the father of heuristics. Polya's efforts focused on problem solving, thinking, and learning. He developed a short dictionary of heuristic primitives. Polya's approach was both practical and experimental. He sought to develop commonalties in the problem solving process through the formalization of observation and experience.

Present day notions of heuristics are somewhat different from Polya's [Bolc and Cytowski 1992]. Current tendencies seek formal and rigid algorithmic solutions to specific problem domains rather than the development of general approaches which could be appropriately selected and applied to specific problems.

The goal of a heuristic search is to greatly reduce the number of nodes searched in seeking a goal. In other words, problems whose complexity grows combinatorially large may be tackled. Through knowledge, information, rules, insights, analogies and simplification, in addition to a host of other techniques, heuristic search aims to reduce the number of objects that must be examined. Heuristics do not guarantee the achievement of a solution, although good heuristics should facilitate this. Over the years heuristic search has been defined in many different ways:

- It is a practical strategy increasing the effectiveness of complex problem solving [Feigenbaum and Feldman 1963].
- It leads to a solution along the most probable path, omitting the least promising ones.
- It should enable one to avoid the examination of dead ends, and to use already gathered data.

The points at which heuristic information can be applied in a search include:

1. Deciding which node to expand next, instead of doing the expansions in either a strict breadth-first or depth-first order
2. In the course of expanding a node, deciding which successor or successors to generate, instead of blindly generating all possible successors at one time
3. Deciding that certain nodes should be discarded, or pruned, from the search tree

Bolc and Cytowski [1992] add:

... use of heuristics in the solution construction process increases the uncertainty of arriving at a result ... due to the use of informal knowledge (rules, laws, intuition, etc.) whose usefulness have never been fully proven. Because of this, heuristic methods are employed in cases where algorithms give unsatisfactory results or do not guarantee to give any results. They are particularly important in solving very complex problems (where an accurate algorithm fails), especially in speech and image recognition, robotics and game strategy construction.... Heuristic methods allow us to exploit uncertain and imprecise data in a natural way.... The main objective of heuristics is to aid and improve the effectiveness of an algorithm solving a problem. Most important is the elimination from further consideration of some subsets of objects still not examined....

Most modern heuristic search methods are expected to bridge the gap between the completeness of algorithms and their optimal complexity [Romanycia and Pelletier 1985]. Strategies are being modified in order to arrive at a quasioptimal, instead of an optimal, solution with a significant cost reduction [Pearl 1984]. Games, especially two-person, zero-sum games of perfect information such as chess and checkers, have proven to be a very promising domain for studying and testing heuristics.

Hill Climbing

Hill climbing is a depth-first search with a heuristic measure that orders choices as nodes are expanded. The heuristic measure is the estimated remaining distance to the goal. The effectiveness of hill climbing is completely dependent on the accuracy of the heuristic measure. To conduct a hill climbing search of a tree:

```
Form a one-element queue consisting of a zero-length path that
      contains only the root node.
Repeat
      Remove the first path from the queue;
      Create new paths by extending the first path to all the
      neighbors of the terminal node.
         If New Path(s) result in a loop Then
           Reject New Path(s).
      Sort any New Paths by the estimated distances between
      their terminal nodes and the GOAL.
         If any shorter paths exist Then
           Add them to the front of the queue.
Until the first path in the queue terminates at the GOAL node or
      the queue is empty
If the GOAL node is found, announce SUCCESS, otherwise
      announce FAILURE.
```

In this algorithm neighbors refers to children of nodes which have been explored, and terminal nodes are equivalent to leaf nodes. Winston [1992] explains the potential problems affecting hill climbing. They are all related to the issue of local vision vs. global vision of the search space. The *foothills problem* is particularly subject to local maxima where global ones are sought, whereas the *plateau problem* occurs when the heuristic measure does not hint toward any significant gradient of proximity to a goal. The *ridge problem* illustrates just what it is called: you may get the impression that the search is taking you closer to a goal state, when in fact you are traveling along a ridge which prevents you from actually attaining your goal.

Best-First Search

Best-first search (Fig. 30.7) is a general algorithm for heuristically searching any state-space graph, that is, a graph representation for a problem which includes initial states, intermediate states, and goal states. In this sense a directed acyclic graph (DAG), for example, is a special case of a state-space graph. Best-first search is equally applicable to data and goal driven searchers and supports the use of heuristic evaluation functions. It can be used with a variety of heuristics, ranging from a state's *goodness* to sophisticated measures based on the probability of a state leading to a goal which can be illustrated by examples of Bayesian statistical measures.

Similar to the depth-first and breadth-first search algorithms, best-first search uses lists to maintain states: open to keep track of the current fringe of the search and closed to record states already visited. In addition, the algorithm orders states on open according to some heuristic estimate of their proximity to a goal. Thus, each iteration of the loop considers the most promising state on the open list. According to Luger and Stubblefield [1993], just where hill climbing fails with its short-sighted and local vision is where the best-first search improves. The following description of the algorithm closely follows that of Luger and Stubblefield [1993, p. 121]:

> At each iteration, best-first search removes the first element from the open list. If it meets the goal conditions, the algorithm returns the solution path that led to the goal. Each state retains ancestor information to allow the algorithm to return the final solution path.
>
> If the first element on open is not a goal, the algorithm generates it descendants. If a child state is already on open or closed, the algorithm checks to make sure that the state records the

```
Procedure Best_First_Search (Start) → pointer
        OPEN ← {Start}                                        // Initialize
        CLOSED ← { }
        While OPEN ≠ { } Do                                   // States Remain
            remove the leftmost state from OPEN, call it X;
            if X ≡ goal then
                return the path from Start to X
            else
                generate children of X
                for each child of X do
                CASE
                the child is not on open or CLOSED:
                        assign the child a heuristic value
                        add the child to OPEN
                the child is already on OPEN:
                        if the child was reached by a shorter path
                        then give the state on OPEN the shorter path
                the child is already on CLOSED:
                        if the child was reached by a shorter path then
                            remove the state from CLOSED
                            add the child to OPEN
                end_CASE
                put X on closed;
                re-order states on OPEN by heuristic merit (best leftmost)
        return NULL                                           // OPEN is empty
```

FIGURE 30.7 The best-first search algorithm. (*Source:* Based on Luger, J. and Stubblefield, W. 1993. *Artificial Intelligence: Structures and Strategies for Complex Problem Solving*, 2nd ed., p. 121. Benjamin/Cummings, Redwood City, CA. With permission.)

shorter of the two partial solution paths. Duplicate states are not retained. By updating the ancestor history of nodes on open and closed, when they are rediscovered, the algorithm is more likely to find a quicker path to a goal.

Best-first search then heuristically evaluates the states on open, and the list is sorted according to the heuristic values. This brings the *best* state to the front of open. It is noteworthy that these estimates are heuristic in nature and therefore the next state to be examined may be from any level of the state space. Open, when maintained as a sorted list, is often referred to as a *priority queue*.

Our notation in Fig. 30.8 is that the dashed branches represent transitions that were not taken by the search algorithm. The solid lines relate to nodes that were put on the open list, with the thick lines pointing to nodes that have been fully evaluated and moved onto the closed list. Much of the excellent Luger and Stubblefield's [1993] text's fourth chapter is devoted to the best-first search. We provide their description:

[Figure 30.8] shows a hypothetical state space with heuristic evaluations attached to some of its states. The states with attached evaluations are those actually generated in best-first search. The states expanded by the heuristic search algorithm are indicated in BOLD; note that it does not search all of the space. The goal of best-first search is to find the goal state by looking at as few

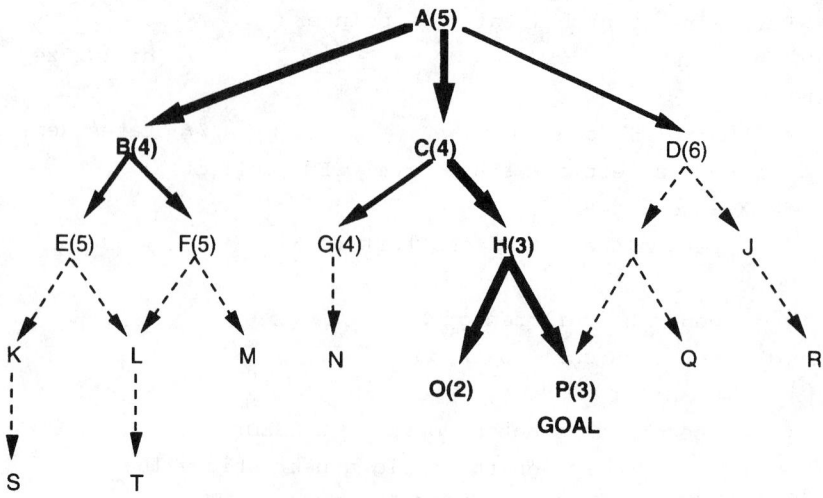

FIGURE 30.8 A hypothetical state space with heuristic evaluations for best-first search, (*Source:* Based on Luger, J. and Stubblefield, W. 1993. *Artificial Intelligence: Structures and Strategies for Complex Problem Solving.* 2nd ed., p. 122. Benjamin/Cummings, Redwood City, CA. With permission.)

states as possible; the more *informed* the heuristic, the fewer states are processed in finding the goal.

A trace of the execution of Procedure Best_First_Search appears below. P is the goal state in this example, so states along the path to P tend to have low heuristic values. The heuristic is fallible: the state O has a lower value than the goal itself and is examined first. Unlike hill climbing, the algorithm recovers from this error and finds the correct goal.

1. Open = {A5}; Closed = { }
2. evaluate A5; Open = {B4, C4, D6}; Closed = {A5}
3. evaluate B4; Open = {C4, E5, F5, D6}; Closed = {B4, A5}
4. evaluate C4; Open = {H3, G4, E5, F5, D6}; Closed = {C4, B4, A5}
5. evaluate H3; Open = {O2, P3, G4, E5, F5, D6}; Closed = {H3, C4, B4, A5}
6. evaluate O2; Open = {P3, G4, E5, F5, D6}; Closed = {O2, H3, C4, B4, A5}
7. evaluate P3; a solution is found!

When the best-first search algorithm is used, the states are sent to the open list in such a way that the most promising one will be expanded next. Because the search heuristic being used for measurement of distance from the goal state may prove erroneous, the alternatives to the preferred state are kept on open. If the algorithm follows an incorrect path, it will retrieve the next best state and shift its focus to another part of the space. In the example of Fig. 30.8, children of state B were found to have poorer heuristic evaluations, than B's sibling C, and so the search shifted there. However, the children of B were kept on open and could be returned to later if other optimal solutions are sought.

The A* Algorithm

The A* algorithm, first described by Hart et al. [1968], attempts to find the minimal cost path joining the start node and the goal in a state-space graph. The algorithm employs an ordered state-space search and an estimated heuristic cost to a goal state, f^* (known as an evaluation function), as does the best-first search of the preceding section, but is unique in how it defines f^*, so that it can guarantee an optimal solution path. The A* algorithm falls into the **branch and bound** class of algorithms, typically employed in operations research to find the shortest path to a solution node in a graph.

The evaluation function $f^*(n)$ estimates the quality of a solution path through node n, based on values returned from two components, $g^*(n)$ and $h^*(n)$. Here $g^*(n)$ is the minimal cost of a path from a start node to n, and $h^*(n)$ is a lower bound on the minimal cost of a solution path from node n to a goal node. As in branch and bound algorithms, for trees g^* will determine the single unique shortest path to node n. For graphs, on the other hand, g^* can err only in the direction of overestimating the minimal cost; if a shorter path is found, its value is readjusted downward. The function h^* is the carrier of *heuristic information*, and the ability to ensure that the value of $h^*(n)$ is less than $h(n)$ [that is, $h^*(n)$ is an underestimate of the actual cost, $h(n)$, of an optimal path from n to a goal node] is essential to the optimality of the A* algorithm. This property, whereby $h^*(n)$ is always less than $h(n)$, is known as the **admissibility condition**. If h^* is zero, then A* reduces to the blind uniform-cost algorithm. If two otherwise similar algorithms, A1 and A2 can be compared to each other with respect to their h^* function, that is, $h1^*$ and $h2^*$, then algorithm A1 is said to be *more informed* than A2 if, whenever a node n (other than a goal node) is evaluated, $h1^*(n) > h2^*(n)$. Considerations for the cost of computing h^* in terms of the overall computational effort involved, and algorithmic utility, determine the *heuristic power* of an algorithm. That is, an algorithm which employs an h^* which is usually accurate, but sometimes inadmissible, may be preferred over an algorithm where h^* is always minimal but hard to effect [Barr and Feigenbaum 1981].

Thus, we can summarize that the A* algorithm is a branch and bound algorithm augmented by the *dynamic programming principle*: the best way through a particular, intermediate node is the best way to that intermediate node from the starting place, followed by the best way from that intermediate node to the goal node. There is no need to consider any other paths to or from the intermediate node [Winston 1992].

30.4 Game-Tree Search

The Alpha–Beta Algorithms

To the human player of two-person games the notion behind the alpha–beta algorithm is understood *intuitively* as:

> If I have determined that a move or a sequence of moves is bad for me (because of a refutation move or variation by my opponent), then I don't need to determine just how bad that move is. Instead I can spend my time exploring other alternatives earlier in the tree.
>
> Conversely, if I have determined that a variation or sequence of moves is bad for my opponent, I don't need to determine how bad it is.

Some of these ideas are illustrated in Fig. 30.9. Here the bold face solid line represents the current solution path. This in turn has replaced a candidate solution, here shown with dotted lines. Everything to the right of the optimal solution path represents alternatives that are simply proved inferior. The path of the current solution is called the principal variation (PV) and nodes on that path are marked as PV nodes. Similarly, the alternatives to PV nodes are *cut* nodes, where only a few successors are examined before a proof of inferiority is found. In time the successor to a cut node will be an *all* node where everything must be examined to prove the cut off at the cut node. The number or bound value by each node represents the return to the root of the cost of the solution path.

In the 40 years since its inception, the alpha–beta minimax algorithm has undergone many revisions and refinements to improve the efficiency of its pruning, so that today it is the primary search engine for two-person games. There have been many landmarks on the way, including Knuth and Moore's [1975] formulation in a negamax framework, Pearl's [1980] introduction of Scout and the special formulation for chess with the principal variation search [Marsland and Campbell 1982] and NegaScout [Reinefeld 1983]. The essence of the method is that the search seeks a path whose value falls between two bounds called alpha and beta, which form a window. With this approach one can also incorporate an artificial narrowing of the alpha–beta window, thus encompassing the notion of *aspiration search*, with a mandatory research

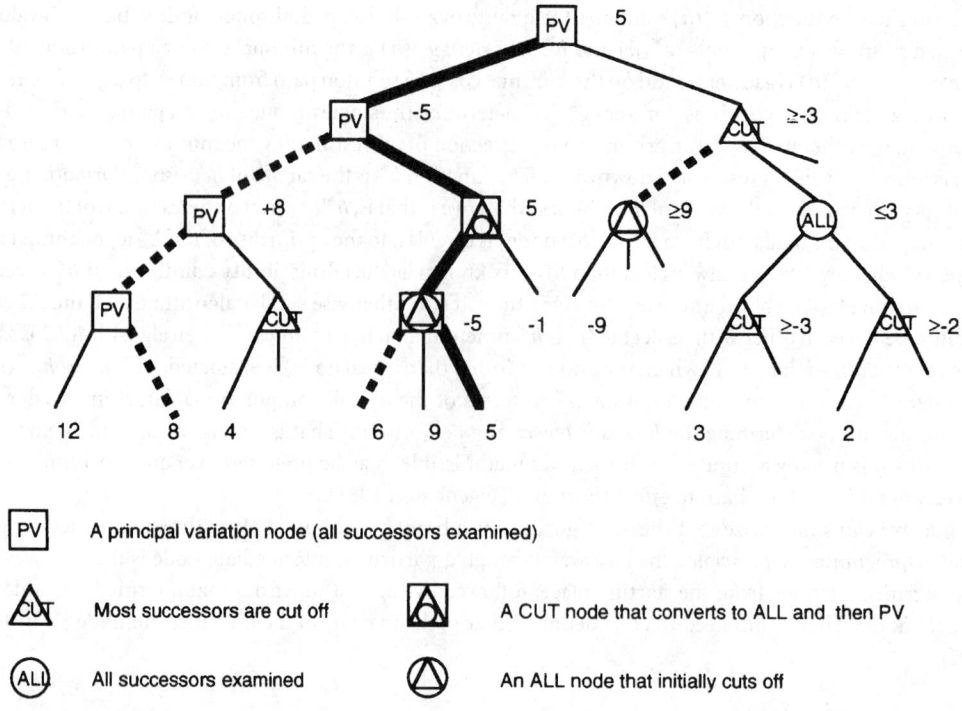

FIGURE 30.9   The PV, cut, and all nodes of a tree, showing its optimal path (bold) and value (5).

on failure to find a value within the corrected bounds. This leads naturally to the incorporation of null window search (NWS) to improve upon Pearl's test procedure.

Here the null window search procedure covers the search at a cut node (Fig. 30.9), where the cutting bound (beta) is negated and increased by 1 in the recursive call. This refinement has some advantage in the parallel search case, but otherwise NWS, Fig. 30.10, is entirely equivalent to the minimal window call in NegaScout. Additional improvements include the use of iterative deepening with *transposition tables* and other move-ordering mechanisms to retain a memory of the search from iteration to iteration. These improvements help ensure that the better subtrees are searched sooner, leading to greater pruning efficiency (more cutoffs) in the later subtrees. Figure 30.10 encapsulates the essence of the algorithm and shows how the first variation from a set of PV nodes, as well as any superior path that emerges later, is given special treatment. Alternates to PV nodes will always be cut nodes, where a few successors will be examined. In a minimal game tree only one successor to a cut node will be examined, and it will be an all node where every thing is examined. In the general case the situation is more complex, as Fig. 30.9 shows.

The SSS* Algorithm

The SSS* algorithm was introduced by Stockman [1979] as a game-searching algorithm that traverses subtrees of the game tree in a best-first fashion similar to the A* algorithm. SSS* was shown to be superior to the original alpha–beta algorithm in the sense that it never looks at more nodes, while occasionally examining fewer [Pearl 1984]. Roizen and Pearl [1983], the source of the following description of SSS*, state:

> ... the aim of SSS* is the discovery of an optimal solution tree.... In accordance with the best-first split-and-prune paradigm, SSS* considers "clusters" of solution trees and splits (or refines) that cluster having the highest upper bound on the merit of its constituents. Every node in the game tree represents a cluster of solution trees defined by the set of all solution trees that

```
ABS (node, alpha, beta, height) → tree_value
      if height ≡ 0
            return evaluate (node)                          // a terminal node
      next ← FirstSuccessor (node)                          // a PV node
      best ← - ABS (next, -beta, -alpha, height -1)
      next ← SelectSibling (next)
      while next ≠ NULL do
            if best ≥ beta then
                  return best                               // a CUT node
            alpha ← max (alpha, best)
            merit ← - NWS (next, -alpha, height -1)
            if merit > best then
                  if (merit ≤ alpha) or (merit ≥ beta) then
                        best ← merit
                  else best ← -ABS (next, -beta, -merit, height -1)
            next ← SelectSibling (next)
      end
      return best                                           //a PV node
end
NWS (node, beta, height) → bound_value
      if height ≡ 0 then
            return Evaluate (note)                          // a terminal node
      next ← FirstSuccessor (node)
      estimate ← - ∞
      while next ≠ NULL do
            merit ← - NWS (next, -beta +1, height -1)
            if merit > estimate then
                  estimate ← merit
            if merit ≥ beta then
                  return estimate                           // a CUT node
            next ← SelectSibling (next)
      end
      return estimate                                       // an ALL node
end
```

FIGURE 30.10   Scout/PVS version of alpha–beta search (ABS) in the Negamax framework.

share that node.... The merit of a partially developed solution tree in a game is determined solely by the properties to the frontier nodes it contains, not by the cost of the paths leading to these nodes. The value of a frontier node is an upper bound on each solution tree in the cluster it represents, ... SSS* establishes upper bounds on the values of partially developed solution trees by seeking the value of terminal nodes, left to right, taking the minimum value of those examined so far. These monotonically nonincreasing bounds are used to order the solution trees so that the tree of highest merit is chosen for development. The development process continues until one solution tree is fully developed, at which point that tree represents the optimal strategy and its value coincides with the minimax value of the root [Roizen and Pearl 1983].

> ... The disadvantage of SSS* lies in the need to keep in storage a record of all contending candidate clusters, which may require large storage space, growing exponentially with search depth [Pearl 1984, p. 245].

Heavy space and time overheads have kept SSS* from being much more than an example of a best-first search, but current research seems destined to now relegate SSS* to a historical footnote. Recently, Plaat et al. [1995] formulated the node-efficient SSS* algorithm into the alpha–beta framework using successive NWS search invocations (supported by perfect transposition tables) to achieve a Memory-enhanced Test (MT) procedure that provides a best-first search. With their MTD(f) function (here MTD(f) refers to a set of drivers using the Memory-enhanced Test (MT) procedure with a parameter f that can range from −infinity to +infinity). Plaat et al. [1995] claim that SSS* can be viewed as a special case of the time-efficient alpha–beta algorithm, instead of the earlier view that alpha–beta is a k-partition variant of SSS*. This is an important contribution that should find wider application for best-first search because of its improved efficiency.

30.5 Parallel Search

The easy availability of low-cost computers has stimulated interest in the use of a multitude of processors for parallel traversals of decision trees. The few theoretical models of parallelism do not accommodate communication and synchronization delays that inevitably impact the performance of working systems. There are several other factors to consider, too, including:

1. How best to employ the additional memory and input/output resources that become available with the extra processors
2. How best to distribute the work across the available processors
3. How to avoid excessive duplication of computation

Some important combinatorial problems have no difficulty with the third point because every eventuality must be considered, but these tend to be less interesting in an artificial intelligence context.

One problem of particular interest is the game-tree search, where it is necessary to compute the value of the tree, while communicating an improved estimate to the other parallel searchers as it becomes available. This can lead to an *acceleration anomaly* when the tree value is found earlier than is possible with a sequential algorithm. Even so, uniprocessor algorithms can have special advantages in that they can be optimized for best pruning efficiency, while a competing parallel system may not have the right information in time to achieve the same degree of pruning, and so do more work (suffer from search overhead). Further, the very fact that pruning occurs makes it impossible to determine in advance how big any piece of work (subtree to be searched) will be, leading to a potentially serious work imbalance and heavy synchronization (waiting for more work) delays.

Although the standard basis for comparing the efficiency of parallel methods is, simply,

$$\text{speedup} = \frac{\text{time taken by a sequential uni-processor algorithm}}{\text{time taken by a } P\text{-processor system}}$$

this basis is often misused, since it depends on the efficiency of the uniprocessor implementation.

The exponential growth of the tree size (solution space) with depth of search makes parallel search algorithms especially susceptible to anomalous speedup behavior. Clearly, acceleration anomalies are among the welcome properties, but more commonly anomalously bad performance is seen, unless the algorithm has been designed with care.

In game playing programs of interest to artificial intelligence, parallelism is not primarily intended to find the answer more quickly, but to get a more reliable result (e.g., based on a deeper search). Here, the emphasis lies on salability instead of speedup. While speedup holds the problem size constant and increases the system size to get a resolute sooner, salability measures the ability to expand the size of both the problem and the system at the same time:

$$\text{scaleup} = \frac{\text{time taken to solve a problem of size } s \text{ by a single processor}}{\text{time taken to solve a } (P \times s) \text{ problem by an } P\text{-processor system}}$$

Thus, scaleup close to unity reflects successful parallelism.

Parallel Single-Agent Search

Single-agent game tree search is important because it is useful for several robot planning activities, such as finding the shortest path through a maze of obstacles. It seems to be more amenable to parallelization than the techniques used in adversary games, because a large proportion of the search space must be fully seen, especially when optimal solutions are sought. This traversal can safely be done in parallel, since there are no cutoffs to be missed. Although move ordering can reduce node expansions, it does not play the same crucial role as in dual-agent game tree search, where significant parts of the search space are often pruned away. For this reason, parallel single-agent search techniques usually achieve better speedups than their counterparts in adversary games.

Most parallel single-agent searches are based on A* or IDA*. As in the sequential case, parallel A* outperforms IDA* on a node count basis, although parallel IDA* needs only linear storage space and runs faster. In addition, cost-effective methods exist (e.g., **parallel window search** described subsequently) that determine nonoptimal solutions with even less computing time.

Parallel A*

Given P processors, the simplest way to parallelize A* is to let each machine work on one of the currently best states on a global openlist (a place holder for nodes that have not yet been examined). This approach minimizes the search overhead, as confirmed in practice by Kumar et al. [1988]. Their relevant experiments were run on a shared memory BBN-Butterfly machine with 100 processors, where a search overhead of less than 5% was observed for the traveling salesperson (TSP) problem.

But elapsed time is more important than the node expansion count, because the global openlist is accessed both before and after each node expansion, so that memory contention becomes a serious bottleneck. It turns out, that a centralized strategy for managing the openlist is only useful in domains where the node expansion time is large compared to the openlist access time. In the TSP problem, near linear time speedups were achieved with up to about 50 processors, when a sophisticated heap data structure was used to significantly reduce the openlist access time [Kumar et al. 1988].

Distributed strategies using local openlists reduce the memory contention problem. But again some communication must be provided to allow processors to share the most promising state descriptors, so that no computing resources are wasted in expanding inferior states. For this purpose a global *blackboard table* can be used to hold state descriptors of the currently best nodes. After selecting a state from its local openlist, each processor compares its f-value (lower bound on the solution cost) to that of the states contained in the blackboard. If the local state is much better (or much worse) than those stored in the blackboard, then node descriptors are sent (or received), so that all active processors are exploring states of almost equal heuristic value. With this scheme, a 69-fold speedup was achieved on an 85-processor BBN Butterfly [Kumar et al. 1988].

Although a blackboard is not accessed as frequently as a global openlist, it still causes memory contention with increasing parallelism. To alleviate this problem, Huang and Davis [1989] proposed a distributed heuristic search algorithm called parallel iterative A* (PIA*), which works solely on local data structures. On a uniprocessor, PIA* expands the same nodes as A*, whereas in the multiprocessor case, it performs a parallel best-first node expansion. The search proceeds by repetitive synchronized iterations, in which processors working on inferior nodes are stopped and reassigned to better ones. To avoid unproductive waiting at the synchronization barriers, the processors are allowed to perform speculative processing. Although Huang and Davis [1989] claim that "this algorithm can achieve almost linear speedup on a large number of processors," it has the same disadvantage as the other parallel A* variants, namely, excessive memory requirements.

Parallel IDA*

IDA* (Fig. 30.6) has proved to be effective, when excessive memory requirements undermine best-first schemes. Not surprisingly, it has also been a popular algorithm to parallelize. Rao et al. [1987] proposed PIDA*, an algorithm with almost linear speedup even when solving the 15-puzzle with its trivial node expansion cost. The 15-puzzle is a popular game made up of 15 tiles that slide within a 4×4 matrix. The object is to slide the tiles through the one empty spot until all tiles are aligned in some goal state. An optimal solution to a hard problem might take 66 moves. PIDA* splits the search space into disjoint parts, so that each processor performs a local cost-bounded depth-first search on its private portion of the state space. When a process has finished its job, it tries to get an unsearched part of the tree from other processors. When no further work can be obtained, all processors detect global termination and compute the minimum of the cost bounds, which is used as a new bound in the next iteration. Note, that more than a P-fold speedup is possible when a processor finds a goal node early in the final iteration. In fact, Rao et al. [1987] report an average speedup of 9.24 with 9 processors on the 15-puzzle! Perhaps more relevant is the all-solution case where no superlinear speedup is possible. Here, an average speedup of $0.93P$ with up to 30 (P) processors on a bus-based multiprocessor architecture (Sequent Balance 21000) was achieved. This suggests that only low multiprocessing overheads (locking, work transfer, termination detection, and synchronization) were experienced.

PIDA* employs a task attraction scheme such as that shown in Fig. 30.11 for distributing the work among the processors. When a processor becomes idle, it asks a neighbor for a piece of the search space. The donor then splits its depth-first search stack and transfers to the requester some nodes (subtrees) for parallel expansion. An optimal splitting strategy would depend on the regularity (uniformity of width and height) of the search tree, though short subtrees should never be given away. When the tree is regular (as in the 15-puzzle) a coarse gained work transfer strategy can be used (e.g., transferring only nodes near the root), otherwise a slice of nodes (e.g., nodes A, B, and C in Fig. 30.11) should be transferred.

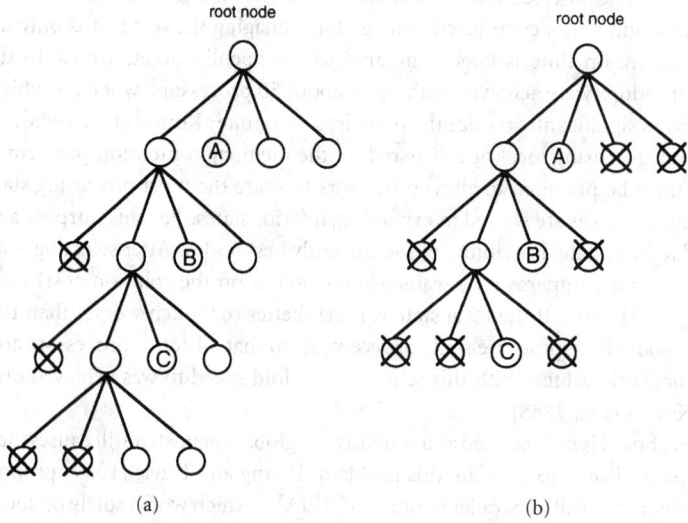

FIGURE 30.11 A work distribution scheme: (a) search tree of sending processor before transferring work nodes: A, B, and C and (b) search tree of receiving processor after accepting work nodes: A, B, and C.

A Comparison with Parallel Window Search

Another parallel IDA* approach borrows from Baudet's [1978] parallel window method for searching adversary games (described subsequently). Powley and Korf [1991] adapted this method to single-agent search, under the title parallel window search (PWS). Their basic idea is to simultaneously start as many iterations as there are processors. This works for a small number of processors, which either expand the tree up to their given thresholds until a solution is found (and the search is stopped), or they completely expand their search space. A global administration scheme then determines the next larger search bound and node expansion starts over again.

Note that the first solution found by PWS need not necessarily be optimal. Suboptimal solutions are often found in searches of poorly ordered trees. There a processor working with a higher cutoff bound finds a goal node in a deeper tree level, while other processors are still expanding shallower tree parts (that may contain cheaper solutions). But according to Powley and Korf [1991], PWS is not primarily meant to compete with IDA*, but it "can be used to find a nearly optimal solution quickly, improve the solution until it is optimal, and then finally guarantee optimality, depending on the amount of time available." Compared to PIDA*, the degree of parallelism is limited, and it remains unclear how to apply PWS in domains where the cost-bound increases are variable.

In summary, PWS and PIDA* complement each other, and so it seems natural to combine them to form a single search scheme that runs PIDA* on groups of processors administered by a global PWS algorithm. The amount of communication needed depends on the work distribution scheme. A fine-grained distribution requires more communication, whereas a coarse-grained work distribution generates fewer messages (but may induce unbalanced work load). Note that the choice of the work distribution scheme also affects the frequency of good acceleration anomalies. Along these lines perhaps the best results have been reported by Reinefeld [1995]. Using asynchronous parallel IDA* (AIDA*) near linear speedup was obtained on a 1024 transputer-based system solving 13 instances of the 19-puzzle. Reinefeld's paper includes a discussion of the communication overheads in both ring and toroid systems, as well as a description of the work distribution scheme.

Adversary Games

In the area of two-person games, early simulation studies with a mandatory work first (MWF) scheme [Akl et al. 1982], and the PVSplit algorithm [Marsland and Campbell 1982], showed that a high degree of parallelism was possible, despite the work imbalance introduced by pruning. Those papers saw that in key applications, (e.g., chess) the game-trees are well ordered because of the wealth of move ordering heuristics that have been developed [Slate and Atkin 1977], and so the bulk of the computation occurs during the search of the first subtree. The MWF approach uses the shape of the critical tree that must be searched. Since that tree is well defined and has regular properties, it is easy to generate. In their simulation Akl et al. [1982] consider the merits of searching the critical game tree in parallel, with the balance of the tree being generated algorithmically and searched quickly by simple tree splitting. Marsland and Campbell [1982], on the other hand, recognized that the first subtree of the critical game tree has the same properties as the whole tree, but its maximum height is one less. This so-called principal variation can be recursively split into parts of about equal size for parallel exploration. PVSplit, an algorithm based on this observation, was tested and analyzed by Marsland and Popowich [1985]. Even so, the static processor allocation schemes such as MWF and PVSplit cannot achieve high levels of parallelism, although PVSplit does very well with up to half a dozen processors. MWF, in particular, ignores the true shape of the average game tree, and so is at its best with shallow searches, where the pruning imbalance from the so-called deep cutoffs has less effect. Other working experience includes the first parallel chess program by Newborn, who later presented performance results [Newborn 1988]. For practical reasons Newborn only split the tree down to some prespecified common depth from the root (typically, 2), where the greatest benefits from parallelism can be achieved. This use of a common depth has been taken up by Hsu [1990] in his proposal for large-scale parallelism. Depth limits are also an important part of changing search modes and in managing transposition tables.

Parallel Aspiration–Window Search

In an early paper on parallel game-tree search, Baudet [1978] suggests partitioning the range of the alpha–beta window rather than the tree. In his algorithm, all processors search the whole tree, but each with a different, nonoverlapping, alpha–beta window. The total range of values is subdivided into P smaller intervals (where P is the number of processors), so that approximately one-third of the range is covered. The advantage of this method is that the processor having the true minimax value inside its narrow window will complete more quickly than a sequential algorithm running with a full window. Even the unsuccessful processors return a result: They determine whether the true minimax value lies below or above their assigned search window, providing important information for rescheduling idle processors until a solution is found.

Its low-communication overhead and lack of synchronization needs are among the positive aspects of Baudet's approach. On the negative side, however, Baudet estimates a maximum speedup of between 5 and 6 even when using infinitely many processors. In practice, parallel window search can only be effectively employed on systems with two or three processors. This is because even in the best case (when the successful processor uses a minimal window) at least the critical game tree must be expanded. The critical tree has about the square root of the leaf nodes of a uniform tree of the same depth, and it represents the smallest tree that must be searched under any circumstances.

Advanced Tree-Splitting Methods

Results from fully recursive versions of PVSplit using the Parabelle chess program [Marsland and Popowich 1985] confirmed the earlier simulations and offered some insight into a major problem: In a P-processor system, $P - 1$ processors are often idle for an inordinate amount of time, thus inducing a high synchronization overhead for large systems. Moreover, the synchronization overhead increases as more processors are added, accounting for most of the total losses, because the search overhead (number of unnecessary node expansions) becomes almost constant for the larger systems. This led to the development of variations that dynamically assign processors to the search of the principal variation. Notable is the work of Schaeffer [1989], which uses a loosely coupled network of workstations, and the Hyatt et al. [1989] independent implementation for a shared-memory computer. These dynamic splitting works have attracted growing attention through a variety of approaches. For example, the results of Feldmann et al. [1990] show a speedup of 11.8 with 16 processors (far exceeding the performance of earlier systems) and Felten and Otto [1988] measured a 101 speedup on a 256 processor hypercube. This latter achievement is noteworthy because it shows an effective way to exploit the 256 times bigger memory that was not available to the uniprocessor. Use of the extra transposition table memory to hold results of search by other processors provides a significant benefit to the hypercube system, thus identifying clearly one advantage of systems with an extensible address space.

These results shows a wide variation not only of methods but also of apparent performance. Part of the improvement is accounted for by the change from a static assignment of processors to the tree search (e.g., from PVSplit) to the dynamic processor reallocation schemes of Hyatt et al. [1989], and also Schaeffer [1989]. These later systems try to dynamically identify all the nodes of Fig. 30.9 and search them in parallel, leaving the cut nodes (where only a few successors might be examined) for serial expansion. In a similar vein Ferguson and Korf (1988) proposed a bound-and-branch method that only assigned processors to the leftmost child of the tree-splitting nodes where no bound (subtree value) exists. Their method is equivalent to the static PVSplit algorithm, and realizes a speedup of 12 with 32 processors for alpha–beta trees generated by Othello programs. This speedup result might be attributed to the smaller average branching factor of about 10 for Othello trees, compared to an average branching factor of about 35 for chess. If that uniprocessor solution is inefficient, for example, by omitting an important node-ordering mechanism such as the use of transposition tables [Reinefeld and Marsland 1994], the speedup figure may look good. For that reason comparisons with a standard test suite from a widely accepted game is often done, and should be encouraged. Most of the working experience with parallel methods for two-person games has centered on the alpha–beta algorithm. Parallel methods for more node-count-efficient sequential methods, such as SSS*, have not been successful until recently, when the potential advantages of using heuristic methods such as hash tables to replace the openlist were exploited [Plaat et al. 1995].

Dynamic Distribution of Work

The key to successful large-scale parallelism lies in the dynamic distribution of work. The young brothers wait concept [Feldmann 1993] is one such scheme in which the parallelism is best described through the help of a definition: The search for a successor $N.j$ of a node N in a game tree must not be started until after the leftmost sibling $N.1.$, the leftmost sibling of $N.j$ is completely evaluated. Thus $N.j$ can be given to another processor if and only if it has not yet been started and the search of $N.1$ is complete. Since this is also the requirement for PVSplit, how then do the two methods differ and what are the tradeoffs? There are two significant differences. The first is at startup and the second is in the potential for parallelism. PVSplit starts much more quickly since all of the processors traverse the first variation (first path from the root to the search horizon of the tree), and then split the work at the nodes on the path as the processors back up the tree to the root. Thus all of the processors are busy from the beginning, but on the other hand, this method suffers from increasingly large synchronization delays as the processors work their way back to the root of the game tree [Marsland and Popowich 1985]. Thus, good performance is possible only with relatively few processors, because the splitting is purely static.

In the work of Feldmann et al. [1990] the startup time for this system is lengthy, because initially only one processor (or a small group of processors) is used to traverse the first path. When that is complete, the right siblings of the nodes on the path can be distributed for parallel search to the waiting processors. If there are a thousand such processors, then perhaps less than 1% would initially be busy. Gradually, the idle processors are brought in to help the busy ones, but this takes time. However, and here comes the big advantage, the system is now much more dynamic in the way it distributes work, so that it is less prone to serious synchronization loss. Further, although many of the nodes in the tree will be cut nodes (which are a poor choice for parallelism because they generate high search overhead), others will be all nodes, where every successor must be examined and they can simply be done in parallel. Usually cut nodes generate a cutoff quite quickly, and so by being cautious about how much work is initially given away once $N.1$ has been evaluated, one can keep excellent control over the search overhead while getting full benefit from the dynamic work distribution that Feldmann's method provides.

Recent Developments

Although there have been several successful implementations involving parallel computing systems, significantly better methods for NP-hard problems, such as game-tree search, remain elusive. Theoretical studies often focus on showing that linear speedup is possible on worst-order game trees. Although not wrong, they make only the trivial point that where exhaustive search is necessary, and where pruning is impossible, then even simple work distribution methods yield excellent results. The true challenge, however, is to consider the case of average game trees, or even better the strongly ordered model (where extensive pruning occurs), which result in asymmetric trees and a significant work distribution problem.

Many people have recognized the intrinsic difficulty of searching game trees under pruning conditions, and one way or another try to recognize dynamically when the critical game tree assumption is being violated, and hence to redeploy the processors. For example, Feldmann et al. [1990] introduced the concept of making *young brothers wait* to reduce search overhead, and the *helpful master* scheme to eliminate the idle time of masters waiting for their slaves' results.

Generalized depth-first searches are fundamental to many AI problems, and Kumar and Rao (1990) have fully examined a method that is well suited to doing the early iterations of single-agent IDA* search. The unexplored part of the trees are marked and are dynamically assigned to any idle processor. In principle, this work distribution method (illustrated in Fig. 30.11) could also be used for deterministic adversary game trees. Finally, we come to the issue of salability and the application of massive parallelism. None of the work discussed so far for game tree search seems to be extensible to arbitrarily many processors. Nevertheless, there have been claims for better methods and some insights into the extra hardware that may be necessary to do the job. Perhaps most confident is Hsu's [1990] recent thesis. His project for the design of the Deep Blue chess program is to manufacture a new VLSI processor in large quantity. Though the original aim was to make a system that was salable to 1000 processors, initially the machine may use 32 nodes of an SP-2 computer each with half a dozen processors that are chess specific. The original design was the major contribution of the thesis, and with it Hsu predicts, based on some simulation studies, a 350-fold speedup. No doubt there will be many inefficiencies to correct before that comes to pass, but in time we will know if massive parallelism solves our game-tree search problems.

Acknowledgment

The authors thank David Kopec for assistance with artwork.

Defining Terms

A* algorithm: A best-first procedure that uses an admissible heuristic estimating function to guide the search process to an optimal solution.

Admissibility condition: The necessity that the heuristic measure never overestimates the cost of the remaining search path, thus ensuring that an optimal solution will be found.

Alpha–beta: The conventional name for the bounds on a depth-first minimax procedure that are used to prune away redundant subtrees in two-person games.

AND/OR tree: A tree which enables the expression of the decomposition of a problem into subproblems enabling alternate solutions to subproblems through the use of AND/OR node labeling schemes.

Backtracking: A component process of many search techniques whereby recovery from unfruitful paths is sought by backing up to a juncture where new paths can be explored.

Best-first search: A heuristic search technique that finds the most promising node to explore next by maintaining and exploring an ordered open node list.

Bidirectional search: A search algorithm that replaces a single search graph, which is likely to grow exponentially, with two smaller graphs: one starting from the initial state and one starting from the goal state.

Blind search: A characterization of all search techniques that are heuristically uninformed. Included amongst these would normally be state-space search, means–ends analysis, generate and test, depth-first search, and breadth-first search.

Branch and bound algorithm: A potentially optimal search technique that keeps track of all partial paths contending for further consideration, always extending the shortest path one level.

Breadth-first search: An uniformed search technique that proceeds level by level visiting all of the nodes at each level (closest to the root node) before proceeding to the next level.

Depth-first search: A search technique that first visits each node as deeply and to the left as possible.

Generate and test: A search technique that proposes possible solutions and then tests them for their feasibility.

Heuristic search: An informed method of searching a state space with the purpose of reducing its size and finding one or more suitable goal states.

Iterative deepening: A successive refinement technique that progressively searches a longer and longer tree until an acceptable solution path is found.

Mandatory work first: A static two-pass process that first traverses the minimal game tree and uses the provisional value found to improve the pruning during the second pass over the remaining tree.

Means–ends analysis: An AI technique that tries to reduce the difference between a current state and a goal state.

Parallel window-aspiration search: A method in which a multitude of processors search the same tree, but each with different (nonoverlapping) alpha–beta bounds.

Principal variation splitting (PVSplit):   A static parallel search method that takes all of the processors down the first variation to some limiting depth, and then splits the subtrees among the processors as they back up to the root of the tree.

SSS*:   A best-first procedure for two-person games.

Young brothers wait concept: A dynamic variation of PVSplit in which idle processors wait until the first path of leftmost subtree has been searched before giving work to an idle processor.

References

Akl, S. G., Barnard, D. T., and Doran, R. J. 1982. Design, analysis and implementation of a parallel tree search machine. *IEEE Trans. Pattern Anal. Mach. Intell.* 4(2):192–203.

Barr, A. and Feigenbaum, E. A. 1981. *The Handbook of Artificial Intelligence V1*. William Kaufmann, Stanford, CA.

Baudet, G. M. 1978. *The Design and Analysis of Algorithms for Asynchronous Multiprocessors*. Ph.D. Thesis, Department of Computing Science, Carnegie Mellon University, Pittsburgh, PA.

Bolc, L. and Cytowski, J. 1992. *Search Methods for Artificial Intelligence*. Academic Press, San Diego, CA.

Feigenbaum, E. and Feldman, J. 1963. *Computers and Thought*. McGraw–Hill, New York.

Feldmann, R. 1993. *Game Tree Search on Massively Parallel Systems*. Ph.D. Thesis, University of Paderborn, Germany.

Feldmann, R., Monien, B., Mysliwietz, P., and Vornberger, O. 1990. Distributed game tree search. In *Parallel Algorithms for Machine Intelligence and Vision*, V. Kumar, P. S. Gopalakrishnan, and L. Kanal, eds., pp. 66–101. Springer–Verlag, New York.

Felten, E. W. and Otto, S. W. 1988. A highly parallel chess program, pp. 1001–1009. In *Proc. Int. Conf. 5th Generation Comput Syst.*

Ferguson, C. and Korf, R. E. 1988. Distributed tree search and its application to alpha-beta pruning, Vol. 1, pp. 128–132. In *Proc. 7th Nat. Conf. Artif. Intell.* Saint Paul, MN. Kaufmann, Los Altos, CA.

Hart, P. E., Nilsson, N. J., and Raphael, B. 1968. A formal basis for the heuristic determination of minimum cost paths. *IEEE Trans. SSC* SSC-4:100–107.

Hsu, F.-H. 1990. Large scale parallelization of alpha–beta search: an algorithmic and architectural study with computer chess. *Carnegie-Mellon University Tech. Rep.* CMU-CS-90-108. Pittsburgh, PA.

Huang, S. and Davis, L. R. 1989. Parallel iterative A* search: an admissible distributed search algorithm. Vol. 1, pp. 23–29. In *Procs. 11th Int. Jt. Conf. Artif. Intell. AI.* Detroit, IL. Kaufmann, Los Altos, CA.

Hyatt, R. M., Suter, B. W., and H. L. Nelson. 1989. A parallel alpha–beta tree searching algorithm. *Parallel Comput.* 10(3):299–308.

Knuth, D. and Moore, R. 1975. An analysis of alpha-beta pruning. *Artif. Intell.* 6(4):293–326.

Korf, R. E. 1985. Depth-first iterative-deepening: an optimal admissible tree search. *Artif. Intell.* 27(1):97–109.

Korf, R. E. 1989. Generalized game trees, Vol. 1, pp. 328–333. In *Proc. 11th Int. Jt. Conf. Artif. Intell.* Detroit, IL. Kaufmann, Los Altos, CA.

Kumar, V. and Rao, V. N. 1990. Scalable parallel formulations of depth-first search. In Parallel Algorithms for Machine Intelligence and Vision. V. Kumar, P. S. Gopalakrishnan, and L. Kanal, eds., pp. 1–41. Springer–Verlag, New York.

Kumar, V., Ramesh, K., and Nageshwara-Rao, V. 1988. Parallel best-first search of state-space graphs: a summary of results, pp. 122–127, In *Procs. 7th Nat. Conf. Artif. Intell. AAAI-88*, Saint Paul, MN. Kaufmann, Los Altos, CA.

Luger, J. and Stubblefield, W. 1993. *Artificial Intelligence: Structures and Strategies for Complex Problem Solving*, 2nd ed. Benjamin/Cummings, Redwood City, CA.

Marsland, T. A. and Campbell, M. 1982. Parallel search of strongly ordered game trees. *ACM Compu. Sur.* 14(4):533–551.

Marsland, T. A. and Popowich, F. 1985. Parallel game-tree search. *IEEE Trans. Pattern Anal. Mach. Intell.* 7(4):442-452.

Newborn, M. M. 1988. Unsynchronized iteratively deepening parallel alpha–beta search. *IEEE Trans. Pattern Anal. and Mach. Intell.* 10(5):687–694.

Nilsson, N. 1971. *Problem-Solving Methods in Artificial Intelligence*. McGraw–Hill, New York.

Pearl, J. 1980. Asymptotic properties of minimax trees and game-searching procedures. *Artif. Intell.* 14(2):113–38.

Pearl, J. 1984. *Heuristics: Intelligent Search Strategies for Computer Problem Solving*. Addison–Wesley, Reading, MA.

Plaat, A., Schaeffer, J., Pijls, W., and de Bruin, A. 1995. Best-first fixed-depth game-tree search in practice, pp. 273–279. In *Proc. IJCAI-95*. Montreal, Canada, Aug. Kaufmann, Los Altos, CA.

Pohl, I. 1971. Bi-dircetional search. In *Machine Intelligence 6*. B. Meltzer and D. Michie, eds., pp. 127–140. American Elsevier, New York.

Polya, G. 1945. *How to Solve It*. Princeton University Press, Princeton, NJ.

Powley, C. and Korf, R. E., 1991. Single-agent parallel window search. *IEEE Trans. Pattern Anal. Mach. Intell.* 13(5):466–477.

Rao, V. N., Kumar, V., and Ramesh, K. 1987. A parallel implementation of iterative-deepening A*, pp. 178–182. In *Proc. 6th Nat. Conf. Artif. Intell.* Seattle, WA.

Reinefeld, A. 1983. An improvement to the scout tree-search algorithm. *Int. Comput. Chess Assoc. J.* 6(4):4–14.

Reinefelid. A. 1995. Salability of massively parallel depth-first search. In *Parallel Processing of Discrete Optimization Problems*. P. M. Pardalos. M. G. C. Resende, and K. G. Ramakrishnan, eds., DIMACS series in discrete mathematics and theoretical computer science, Vol. 22, pp. 305–322. American Mathematical Society, Providence, RI.

Reinefeld, A. and Marsland, T. A. 1994. Enhanced iterative-deepening search. *IEEE Trans. Pattern Anal. Mach. Intell.* 16(7):701–710.

Roizen, I. and Pearl, J. 1983 A minimax algorithm better than alpha–beta?: yes and no. *Artif. Intell.* 21(1-2):199–220.

Romanycia, M. and Pelletier, F. J. 1985. What is heuristic? *Comput. Intelli.* 1:24–36.

Schaeffer, J. 1989. Distributed game-tree search. *J. Parallel Distributed Comput.* 6(2):90–114.

Slate, D. J. and Atkin, L. R. 1977. Chess 4.5—the Northwestern University chess program. In *Chess Skill in Man and Machine*, P. Frey, ed., pp. 82–118. Springer–Verlag, New York.

Stockman, G. 1979. A minimax algorithm better than alpha–beta? *Artif. Intell.* 12(2):179–96.

Winston, P. H. 1992. *Artiffficial Intelligence*, 3rd ed. Addison–Wesley, Reading, MA.

Further Information

The most regularly and consistently cited source of information for this article is the *Journal of Artificial Intelligence*. There are numerous other journals including, for example, *AAAI Magazine, CACM, IEEE Expert, ICCA Journal,* and the *International Journal of Man–Machine Studies,* which frequently publish articles related to this subject area. Also prominent has been the *Machine Intelligence Series* of volumes edited by Donald Michie with various others. An excellent reference source is the three-volume *Handbook of Artificial Intelligence* by Barr and Feigenbaum [1981].

In addition there are numerous national and international and national conferences on AI with published proceedings, headed by the International Joint Conference on AI (IJCAI). Classic books on AI methodology include Feigenbaum and Feldman's [1963] *Computers and Thought* and Nils Nilsson's [1971] *Problem-Solving Methods in Artificial Intelligence*. There are a number of popular and thorough textbooks on AI. Two relevant books on the subject of search in are *Heuristics* [Pearl 1984] and the more recent *Search Methods for Artificial Intelligence* [Bolc and Cytowski 1992].

31

Graphical Models for Probabilistic and Causal Reasoning

Judea Pearl
University of California,
Los Angeles

31.1 Introduction

This chapter surveys the development of graphical models known as Bayesian networks, summarizes their semantical basis, and assesses their properties and applications to reasoning and planning.

Bayesian networks are directed acyclic graphs (DAGs) in which the nodes represent variables of interest (e.g., the temperature of a device, the gender of a patient, a feature of an object, the occurrence of an event) and the links represent causal influences among the variables. The strength of an influence is represented by conditional probabilities that are attached to each cluster of parents–child nodes in the network.

Figure 31.1 illustrates a simple yet typical Bayesian network. It describes the causal relationships among the season of the year (X_1), whether rain falls (X_2) during the season, whether the sprinkler is on (X_3) during that season, whether the pavement would get wet (X_4), and whether the pavement would be slippery (X_5). All variables in this figure are binary, taking a value of either true or false, except the root variable X_1, which can take one of four values: spring, summer, fall, or winter. Here, the absence of a direct link between X_1 and X_5, for example, captures our understanding that the influence of seasonal variations on the slipperiness of the pavement is mediated by other conditions (e.g., the wetness of the pavement).

As this example illustrates, a Bayesian network constitutes a model of the environment rather than, as in many other knowledge representation schemes (e.g., logic, rule-based systems, and neural networks), a model of the reasoning process. It simulates, in fact, the causal mechanisms that operate in the environment and thus allows the investigator to answer a variety of queries, including associational queries, such as "Having observed A, what can we expect of B?"; abductive queries, such as "What is the most plausible explanation for a given set of observations?"; and control queries, such as "What will happen if we intervene and act on the environment?" Answers to the first type of query depend only on probabilistic knowledge of the domain, whereas answers to the second and third types rely on the causal knowledge embedded in the network. Both types of knowledge, associative and causal, can effectively be represented and processed in Bayesian networks.

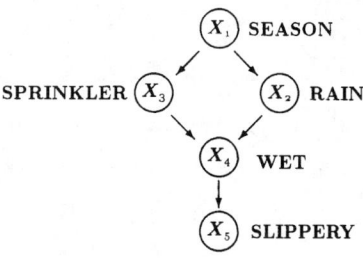

FIGURE 31.1 A Bayesian network representing causal influences among five variables. (*Source:* Arbib, M. A., ed. 1995. Bayesian networks. In *The Handbook of Brain Theory and Neural Networks.* MIT Press, Cambridge, MA.)

The associative facility of Bayesian networks may be used to model cognitive tasks such as object recognition, reading comprehension, and temporal projections. For such tasks, the probabilistic basis of Bayesian networks offers a coherent semantics for coordinating top-down and bottom-up inferences, thus bridging information from high-level concepts and low-level percepts. This capability is important for achieving selective attention, that is, selecting the most informative next observation before actually making the observation. In certain structures, the coordination of these two modes of inference can be accomplished by parallel and distributed processes that communicate through the links in the network.

However, the most distinctive feature of Bayesian networks, stemming largely from their causal organization, is their ability to represent and respond to changing configurations. Any local reconfiguration of the mechanisms in the environment can be translated, with only minor modification, into an isomorphic reconfiguration of the network topology. For example, to represent a disabled sprinkler, we simply delete from the network all links incident to the node sprinkler. To represent a pavement covered by a tent, we simply delete the link between rain and wet. This flexibility is often cited as the ingredient that marks the division between deliberative and reactive agents and that enables the former to manage novel situations instantaneously, without requiring retraining or adaptation.

31.2 Historical Background

Networks employing DAGs have a long and rich tradition, starting with the geneticist Sewall Wright [1921]. He developed a method called *path analysis* [Wright 1934], which later became an established representation of causal models in economics [Wold 1964], sociology [Blalock 1971, Kenny 1979], and psychology [Duncan 1975]. Good [1961] used DAGs to represent causal hierarchies of binary variables with disjunctive causes. *Influence diagrams* represent another application of DAG representation [Howard and Matheson 1981]. Developed for decision analysis, they contain both event nodes and decision nodes. *Recursive models* is the name given to such networks by statisticians seeking meaningful and effective decompositions of contingency tables [Lauritzen 1982, Wermuth and Lauritzen 1983, Kiiveri et al. 1984].

The role of the network in the applications cited was primarily to provide an efficient description for probability functions; once the network was configured, all subsequent computations were pursued by symbolic manipulation of probability expressions. The potential for the network to work as a computational architecture, and hence as a model of cognitive activities, was noted in Pearl [1982], where a distributed scheme was demonstrated for probabilistic updating on tree-structured networks. The motivation behind this particular development was the modeling of distributed processing in reading comprehension [Rumelhart 1976], where both top-down and bottom-up inferences are combined to form a coherent interpretation. This dual mode of reasoning is at the heart of Bayesian updating and in fact motivated Reverend Bayes's original 1763 calculations of posterior probabilities (representing explanations), given prior probabilities (representing causes), and likelihood functions (representing evidence).

Bayesian networks have not attracted much attention in the logic and cognitive modeling circles, but they did in expert systems. The ability to coordinate bidirectional inferences filled a void in expert systems technology of the late 1970s, and it is in this area that Bayesian networks truly flourished. Over the past 10 years, Bayesian networks have become a tool of great versatility and power, and they are now the most common representation scheme for probabilistic knowledge [Shafer and Pearl 1990, Shachter

1990, Oliver and Smith 1990, Neapolitan 1990]. They have been used to aid in the diagnosis of medical patients [Heckerman 1991, Andersen et al. 1989, Heckerman et al. 1990, Peng and Reggia 1990] and malfunctioning systems [Agogino et al. 1988]; to understand stories [Charniak and Goldman 1991]; to filter documents [Turtle and Croft 1991]; to interpret pictures [Levitt et al. 1990]; to perform filtering, smoothing, and prediction [Abramson 1991]; to facilitate planning in uncertain environments [Dean and Wellman 1991]; and to study causation, nonmonotonicity, action, change, and attention. Some of these applications are described in a tutorial article by Charniak [1991]; others can be found in Pearl [1988], Shafer and Pearl [1990], and Goldszmidt and Pearl [1996].

31.3 Bayesian Networks as Carriers of Probabilistic Information

Formal Semantics

Given a DAG G and a joint distribution P over a set $X = \{X_1, \ldots, X_n\}$ of discrete variables, we say that G *represents* P if there is a one-to-one correspondence between the variables in X and the nodes of G, such that P admits the recursive product decomposition

$$P(x_1, \ldots, x_n) = \prod_i P(x_i \mid pa_i) \tag{31.1}$$

where pa_i are the direct predecessors (called *parents*) of X_i in G. For example, the DAG in Fig. 31.1 induces the decomposition

$$P(x_1, x_2, x_3, x_4, x_5) = P(x_1)P(x_2 \mid x_1)P(x_3 \mid x_1)P(x_4 \mid x_2, x_3)P(x_5 \mid x_4) \tag{31.2}$$

The recursive decomposition in Eq. (31.1) implies that, given its parent set pa_i, each variable X_i is conditionally independent of all its other predecessors $\{X_1, X_2, \ldots, X_{i-1}\}\backslash pa_i$. Using Dawid's notation [Dawid 1979], we can state this set of independencies as

$$X_i \perp\!\!\!\perp \{X_1, X_2, \ldots, X_{i-1}\}\backslash pa_i \mid pa_i \quad i = 2, \ldots, n \tag{31.3}$$

Such a set of independencies is called *Markovian*, since it reflects the Markovian condition for state transitions: each state is rendered independent of the past, given its immediately preceding state. For example, the DAG of Fig. 31.1 implies the following Markovian independencies:

$$X_2 \perp\!\!\!\perp \{0\} \mid X_1, \qquad X_3 \perp\!\!\!\perp X_2 \mid X_1, \qquad X_4 \perp\!\!\!\perp X_1 \mid \{X_2, X_3\}, \qquad X_5 \perp\!\!\!\perp \{X_1, X_2, X_3\} \mid X_4 \tag{31.4}$$

In addition to these, the decomposition of Eq. (31.1) implies many more independencies, the sum total of which can be identified from the DAG using the graphical criterion of d-separation [Pearl 1988]:

Definition 31.1 (d-separation). Let a path in a DAG be a sequence of consecutive edges, of any directionality. A path p is said to be d-separated (or blocked) by a set of nodes Z iff:

1. p contains a chain $i \longrightarrow j \longrightarrow k$ or a fork $i \longleftarrow j \longrightarrow k$ such that the middle node j is in Z.
2. Or p contains an inverted fork $i \longrightarrow j \longleftarrow k$ such that neither the middle node j nor any of its descendants (in G) are in Z.

If X, Y, and Z are three disjoint subsets of nodes in a DAG G, then Z is said to d-separate X from Y, denoted $(X \perp\!\!\!\perp Y \mid Z)_G$, iff Z d-separates every path from a node in X to a node in Y.

In Fig. 31.1, for example, $X = \{X_2\}$ and $Y = \{X_3\}$ are d-separated by $Z = \{X_1\}$; the path $X_2 \leftarrow X_1 \rightarrow X_3$ is blocked by $X_1 \in Z$, while the path $X_2 \rightarrow X_4 \leftarrow X_3$ is blocked because X_4 and all its descendants are outside Z. Thus, $(X_2 \parallel X_3 \mid X_1)_G$ holds in Fig. 31.1. However, X and Y are not d-separated by $Z' = \{X_1, X_5\}$, because the path $X_2 \to X_4 \leftarrow X_3$ is rendered active by virtue of X_5, a descendant of X_4, being in Z'. Consequently $(X_2 \parallel X_3 \mid \{X_1, X_5\})_G$ does not hold; in words, learning the value of the consequence X_5 renders its causes $\overline{X}_2$ and X_3 dependent, as if a pathway were opened along the arrows converging at X_4.

The d-separation criterion has been shown to be both necessary and sufficient relative to the set of distributions that are represented by a DAG G [Verma and Pearl 1990, Geiger et al. 1990]. In other words, there is a one-to-one correspondence between the set of independencies implied by the recursive decomposition of Eq. (31.1) and the set of triples (X, Z, Y) that satisfies the d-separation criterion in G. Furthermore, the d-separation criterion can be tested in time linear in the number of edges in G. Thus, a DAG can be viewed as an efficient scheme for representing Markovian independence assumptions and for deducing and displaying all of the logical consequences of such assumptions.

An important property that follows from the d-separation characterization is a criterion for determining when two DAGs are observationally equivalent, that is, every probability distribution that is represented by one of the DAGs is also represented by the other:

Theorem 31.1 [Verma and Pearl 1990].   *Two DAGs are observationally equivalent if and only if they have the same sets of edges and the same sets of v-structures, that is, head-to-head arrows with nonadjacent tails.*

The soundness of the d-separation criterion holds not only for probabilistic independencies but for any abstract notion of conditional independence that obeys the semigraphoid axioms [Verma and Pearl 1990, Geiger et al. 1990]. Additional properties of DAGs and their applications to evidential reasoning in expert systems are discussed in Pearl [1988, 1993a], Pearl et al. [1990], Geiger [1990] Lauritzen and Spiegelhalter [1988], and Spiegelhalter et al. [1993].

Inference Algorithms

The first algorithms proposed for probability updating in Bayesian networks used message-passing architecture and were limited to trees [Pearl 1982] and singly connected networks [Kim and Pearl 1983]. The idea was to assign each variable a simple processor, forced to communicate only with its neighbors, and to permit asynchronous back-and-forth message passing until equilibrium was achieved. Coherent equilibrium can indeed be achieved this way, but only in singly connected networks, where an equilibrium state occurs in time proportional to the diameter of the network.

Many techniques have been developed and refined to extend the tree-propagation method to general, multiply connected networks. Among the most popular are Shachter's [1988] method of node elimination, Lauritzen and Spiegelhalter's [1988] method of clique-tree propagation, and the method of loop-cut conditioning [Pearl 1988, Ch. 4.3].

Clique-tree propagation, the most popular of the three methods, works as follows. Starting with a directed network representation, the network is transformed into an undirected graph that retains all of its original dependencies. This graph, sometimes called a Markov network [Pearl 1988, Ch. 3.1], is then triangulated to form local clusters of nodes (cliques) that are tree structured. Evidence propagates from clique to clique by ensuring that the probability of their intersection set is the same, regardless of which of the two cliques is considered in the computation. Finally, when the propagation process subsides, the posterior probability of an individual variable is computed by projecting (marginalizing) the distribution of the hosting clique onto this variable.

Whereas the task of updating probabilities in general networks is NP-hard [Rosenthal 1977, Cooper 1990], the complexity for each of the three methods cited is exponential in the size of the largest clique found in some triangulation of the network. It is fortunate that these complexities can be estimated prior to actual processing; when the estimates exceed reasonable bounds, an approximation method such as stochastic simulation [Pearl 1987, Henrion 1988] can be used instead. Learning techniques have also been developed for systematic updating of the conditional probabilities $P(x_i \mid pa_i)$ so as to match empirical data [Spiegelhalter and Lauritzen 1990].

System's Properties

By providing graphical means for representing and manipulating probabilistic knowledge, Bayesian networks overcome many of the conceptual and computational difficulties of earlier knowledge-based systems [Pearl 1988]. Their basic properties and capabilities can be summarized as follows:

1. Graphical methods make it easy to maintain consistency and completeness in probabilistic knowledge bases. They also define modular procedures of knowledge acquisition that reduce significantly the number of assessments required [Pearl 1988, Heckerman 1991].

2. Independencies can be dealt with explicitly. They can be articulated by an expert, encoded graphically, read off the network, and reasoned about; yet they forever remain robust to numerical imprecision [Geiger 1990, Geiger et al. 1990, Pearl et al. 1990].

3. Graphical representations uncover opportunities for efficient computation. Distributed updating is feasible in knowledge structures which are rich enough to exhibit intercausal interactions (e.g., explaining away) [Pearl 1982, Kim and Pearl 1983]. And, when extended by clustering or conditioning, tree-propagation algorithms are capable of updating networks of arbitrary topology [Lauritzen and Spiegelhalter 1988, Shachter 1986, Pearl 1988].

4. The combination of predictive and abductive inferences resolves many problems encountered by first-generation expert systems and renders belief networks a viable model for cognitive functions requiring both top-down and bottom-up inferences [Pearl 1988, Shafer and Pearl 1990].

5. The causal information encoded in Bayesian networks facilitates the analysis of action sequences, their consequences, their interaction with observations, their expected utilities, and, hence, the synthesis of plans and strategies under uncertainty [Dean and Wellman 1991, Pearl 1993b, 1994b].

6. The isomorphism between the topology of Bayesian networks and the stable mechanisms, which operate in the environment, facilitates modular reconfiguration of the network in response to changing conditions and permits deliberative reasoning about novel situations.

Recent Developments

Causal Discovery

One of the most exciting prospects in recent years has been the possibility of using the theory of Bayesian networks to discover causal structures in raw statistical data. Several systems have been developed for this purpose [Pearl and Verma 1991, Spirtes et al. 1993], which systematically search and identify causal structures with hidden variables from empirical data. Technically, because these algorithms rely merely on conditional independence relationships, the structures found are valid only if one is willing to accept weaker forms of guarantees than those obtained through controlled randomized experiments: minimality and stability [Pearl and Verma 1991]. Minimality guarantees that any other structure compatible with the data is necessarily less specific, and hence less falsifiable and less trustworthy, than the one(s) inferred. Stability ensures that any alternative structure compatible with the data must be less stable than the one(s) inferred; namely, slight fluctuations in experimental conditions will render that structure no longer compatible with the data. With these forms of guarantees, the theory provides criteria for identifying genuine and spurious causes, with or without temporal information.

Alternative methods of identifying structure in data assign prior probabilities to the parameters of the network and use Bayesian updating to score the degree to which a given network fits the data [Cooper and Herskovits 1990, Heckerman et al. 1994]. These methods have the advantage of operating well under small sample conditions but encounter difficulties coping with hidden variables.

Plain Beliefs

In mundane decision making, beliefs are revised not by adjusting numerical probabilities but by tentatively accepting some sentences as true for all practical purposes. Such sentences, often named *plain beliefs*, exhibit both logical and probabilistic character. As in classical logic, they are propositional and deductively closed; as in probability, they are subject to retraction and to varying degrees of entrenchment [Spohn 1988, Goldszmidt and Pearl 1992].

Bayesian networks can be adopted to model the dynamic of plain beliefs by replacing ordinary probabilities with nonstandard probabilities, that is, probabilities that are infinitesimally close to either zero or one. This amounts to taking an order of magnitude approximation of empirical frequencies and adopting new combination rules tailored to reflect this approximation. The result is an integer-addition calculus, very similar to probability calculus, with summation replacing multiplication and minimization replacing addition. A plain belief is then identified as a proposition whose negation obtains an infinitesimal probability (i.e., an integer greater than zero). The connection between infinitesimal probabilities and nonmonotonic logic is described in Pearl [1994a] and Goldszmidt and Pearl [1996].

This combination of infinitesimal probabilities with the causal information encoded by the structure of Bayesian networks facilitates linguistic communication of belief commitments, explanations, actions, goals, and preferences and serves as the basis for current research on qualitative planning under uncertainty [Darwiche and Pearl 1994, Goldszmidt and Pearl 1992, Pearl 1993b, Darwiche and Goldszmidt 1994]. Some of these aspects will be presented in the next section.

31.4 Bayesian Networks as Carriers of Causal Information

The interpretation of DAGs as carriers of independence assumptions does not necessarily imply causation and will in fact be valid for any set of Markovian independencies along any ordering (not necessarily causal or chronological) of the variables. However, the patterns of independencies portrayed in a DAG are typical of causal organizations and some of these patterns can be given meaningful interpretation only in terms of causation. Consider, for example, two independent events, E_1 and E_2, that have a common effect E_3. This triple represents an intransitive pattern of dependencies: E_1 and E_3 are dependent, E_3 and E_2 are dependent, yet E_1 and E_2 are independent. Such a pattern cannot be represented in undirected graphs because connectivity in undirected graphs is transitive. Likewise, it is not easily represented in neural networks, because E_1 and E_2 should turn dependent once E_3 is known. The DAG representation provides a convenient language for intransitive dependencies via the converging pattern $E_1 \rightarrow E_3 \leftarrow E_2$, which implies the independence of E_1 and E_2 as well as the dependence of E_1 and E_3 and of E_2 and E_3. The distinction between transitive and intransitive dependencies is the basis for the causal discovery systems of Pearl and Verma [1991] and Spirtes et al. [1993]. (See subsection on causal discovery.)

However, the Markovian account still leaves open the question of how such intricate patterns of independencies relate to the more basic notions associated with causation, such as influence, manipulation, and control, which reside outside the province of probability theory. The connection is made in the mechanism-based account of causation.

The basic idea behind this account goes back to structural equations models [Wright 1921, Haavelmo 1943, Simon 1953] and it was adapted in Pearl and Verma [1991] for defining probabilistic causal theories, as follows. Each child–parents family in a DAG G represents a deterministic function

$$X_i = f_i(pa_i, \epsilon_i) \tag{31.5}$$

where pa_i are the parents of variable X_i in G, and $\epsilon_i, 0 < i < n$, are mutually independent, arbitrarily distributed random disturbances. Characterizing each child–parent relationship as a deterministic function, instead of the usual conditional probability $P(x_i \mid pa_i)$, imposes equivalent independence constraints on the resulting distributions and leads to the same recursive decomposition that characterizes DAG models [see Eq. (31.1)]. However, the functional characterization $X_i = f_i(pa_i, \epsilon_i)$ also specifies how the resulting distributions would change in response to external interventions, since each function is presumed to represent a stable mechanism in the domain and therefore remains constant unless specifically altered. Thus, once we know the identity of the mechanisms altered by the intervention and the nature of the alteration, the overall effect of an intervention can be predicted by modifying the appropriate equations in the model of Eq. (31.5) and using the modified model to compute a new probability function of the observables.

The simplest type of external intervention is one in which a single variable, say X_i, is forced to take on some fixed value x_i'. Such *atomic* intervention amounts to replacing the old functional mechanism $X_i = f_i(pa_i, \epsilon_i)$ with a new mechanism $X_i = x_i'$ governed by some external force that sets the value x_i'. If we imagine that each variable X_i could potentially be subject to the influence of such an external force, then we can view each Bayesian network as an efficient code for predicting the effects of atomic interventions and of various combinations of such interventions, without representing these interventions explicitly.

Causal Theories, Actions, Causal Effect, and Identifiability

Definition 31.2. A causal theory is a 4-tuple

$$T = \langle V, U, P(u), \{f_i\} \rangle$$

where:

1. $V = \{X_1, \ldots, X_n\}$ is a set of observed variables.
2. $U = \{U_1, \ldots, U_n\}$ is a set of unobserved variables which represent disturbances, abnormalities, or assumptions.
3. $P(u)$ is a distribution function over $U_1, \ldots, U_n$.
4. $\{f_i\}$ is a set of n deterministic functions, each of the form

$$X_i = f_i(PA_i, u) \quad i = 1, \ldots, n \tag{31.6}$$

where PA_i is a subset of V not containing X_i.

The variables PA_i (connoting parents) are considered the direct of X_i and they define a directed graph G, which may, in general, be cyclic. Unlike the probabilistic definition of parents in Bayesian networks [Eq. (31.1)] PA_i is selected from V by considering functional mechanisms in the domain, not by conditional independence considerations. We will assume that the set of equations in (31.6) has a unique solution for $X_i, \ldots, X_n$, given any value of the disturbances $U_i, \ldots, U_n$. Therefore, the distribution $P(u)$ induces a unique distribution on the observables, which we denote by $P_T(v)$.

We will consider concurrent actions of the form $do(X = x)$, where $X \subseteq V$ is a set of variables and x is a set of values from the domain of X. In other words, $do(X = x)$ represents a combination of actions that forces the variables in X to attain the values x.

Definition 31.3 (effect of actions). The effect of the action $do(X = x)$ on a causal theory T is given by a subtheory T_x of T, where T_x obtains by deleting from T all equations corresponding to variables in X and substituting the equations $X = x$ instead.

The framework provided by Definitions 31.2 and 31.3 permits the coherent formalization of many subtle concepts in causal discourse, such as causal influence, causal effect, causal relevance, average causal effect, identifiability, counterfactuals, exogeneity, and so on. Examples are as follows:

- *X influences Y* in context u if there are two values of X, x, and x', such that the solution for Y under $U = u$ and $do(X = x)$ is different from the solution under $U = u$ and $do(X = x')$.
- *X can potentially influence Y* if there exist both a subtheory T_z of T and a context $U = u$ in which X influences Y.
- Event $X = x$ *is the (singular) cause* of event $Y = y$ if (1) $X = x$ and $Y = y$ are true, and (2) in every context u compatible with $X = x$ and $Y = y$, and for all $x' \neq x$, the solution of Y under $do(X = x')$ is not equal to y.

The definitions are deterministic. Probabilistic causality emerges when we define a probability distribution $P(u)$ for the U variables, which, under the assumption that the equations have a unique solution, induces a unique distribution on the endogenous variables for each combination of atomic interventions.

Definition 31.4 (causal effect).   Given two disjoint subsets of variables, $X \subseteq V$ and $Y \subseteq V$, the causal effect of X on Y, denoted $P_T(y \mid \hat{x})$, is a function from the domain of X to the space of probability distributions on Y, such that

$$P_T(y \mid \hat{x}) = P_{T_x}(y) \tag{31.7}$$

for each realization x of X. In other words, for each $x \in \mathrm{dom}(X)$, the causal effect $P_T(y \mid \hat{x})$ gives the distribution of Y induced by the action $do(X = x)$.

Note that causal effects are defined relative to a given causal theory T, though the subscript T is often suppressed for brevity.

Definition 31.5 (identifiability).   Let $Q(T)$ be any computable quantity of a theory T; Q is identifiable in a class M of theories if for any pair of theories T_1 and T_2 from M, $Q(T_1) = Q(T_2)$ whenever $P_{T_1}(v) = P_{T_2}(v)$.

Identifiability is essential for estimating quantities Q from P alone, without specifying the details of T, so that the general characteristics of the class M suffice. The question of interest in planning applications is the identifiability of the causal effect $Q = P_T(y \mid \hat{x})$ in the class M_G of theories that share the same causal graph G. Relative to such classes we now define the following:

Definition 31.6 (causal-effect identifiability).   The causal effect of X on Y is said to be identifiable in M_G if the quantity $P(y \mid \hat{x})$ can be computed uniquely from the probabilities of the observed variables, that is, if for every pair of theories T_1 and T_2 in M_G such that $P_{T_1}(v) = P_{T_2}(v)$, we have $P_{T_1}(y \mid \hat{x}) = P_{T_2}(y \mid \hat{x})$.

The identifiability of $P(y \mid \hat{x})$ ensures that it is possible to infer the effect of action $do(X = x)$ on Y from two sources of information:

1. Passive observations, as summarized by the probability function $P(v)$
2. The causal graph, G, which specifies, qualitatively, which variables make up the stable mechanisms in the domain or, alternatively, which variables participate in the determination of each variable in the domain.

Simple examples of identifiable causal effects will be discussed in the next subsection.

Acting vs Observing

Consider the example depicted in Fig. 31.1. The corresponding theory consists of five functions, each representing an autonomous mechanism:

$$X_1 = U_1$$

$$X_2 = f_2(X_1, U_2)$$

$$X_3 = f_3(X_1, U_3) \qquad (31.8)$$

$$X_4 = f_4(X_3, X_2, U_4)$$

$$X_5 = f_5(X_4, U_5)$$

To represent the action "turning the sprinkler ON," $do(X_3 = \text{ON})$, we delete the equation $X_3 = f_3(x_1, u_3)$ from the theory of Eq. (31.8) and replace it with $X_3 = \text{ON}$. The resulting subtheory, $T_{X_3=\text{ON}}$, contains all of the information needed for computing the effect of the actions on other variables. It is easy to see from this subtheory that the only variables affected by the action are X_4 and X_5, that is, the descendant, of the manipulated variable X_3.

The probabilistic analysis of causal theories becomes particularly simple when two conditions are satisfied:

1. The theory is recursive, i.e., there exists an ordering of the variables $V = \{X_1, \ldots, X_n\}$ such that each X_i is a function of a subset PA_i of its predecessors

$$X_i = f_i(PA_i, U_i), \quad PA_i \subseteq \{X_1, \ldots, X_{i-1}\} \qquad (31.9)$$

2. The disturbances $U_1, \ldots, U_n$ are mutually independent, that is,

$$P(u) = \prod_i P(u_i) \qquad (31.10)$$

These two conditions, also called Markovian, are the basis of the independencies embodied in Bayesian networks [Eq. (31.1)] and they enable us to compute causal effects directly from the conditional probabilities $P(x_i \mid pa_i)$, without specifying the functional form of the functions f_i, or the distributions $P(u_i)$ of the disturbances. This is seen immediately from the following observations: The distribution induced by any Markovian theory T is given by the product in Eq. (31.1)

$$P_T(x_1, \ldots, x_n) = \prod_i P(x_i \mid pa_i) \qquad (31.11)$$

where pa_i are (values of) the parents of X_i in the diagram representing T. At the same time, the subtheory $T_{x'_j}$, representing the action $do(X_j = x'_j)$ is also Markovian; hence, it also induces a product-like distribution

$$P_{T_{x'_j}}(x_1, \ldots, x_n) = \begin{cases} \prod_{i \neq j} P(x_i \mid pa_i) = \frac{P(x_1, \ldots, x_n)}{P(x_j \mid pa_j)} & \text{if } x_j = x'_j \\ 0 & \text{if } x_j \neq x'_j \end{cases} \qquad (31.12)$$

where the partial product reflects the surgical removal of the

$$X_j = f_j(pa_j, U_j)$$

from the theory of Eq. (31.9) (see Pearl [1993a]).

In the example of Fig. 31.1, the pre-action distribution is given by the product

$$P_T(x_1, x_2, x_3, x_4, x_5) = P(x_1)P(x_2 \mid x_1)P(x_3 \mid x_1)P(x_4 \mid x_2, x_3)P(x_5 \mid x_4) \tag{31.13}$$

whereas the surgery corresponding to the action $do(X_3 = \text{ON})$ amounts to deleting the link $X_1 \rightarrow X_3$ from the graph and fixing the value of X_3 to ON, yielding the postaction distribution

$$P_T(x_1, x_2, x_3, x_4, x_5 \mid do(X_3 = \text{ON})) = P(x_1)P(x_2 \mid x_1)P(x_4 \mid x_2, X_3 = \text{ON})P(x_5 \mid x_4) \tag{31.14}$$

Note the difference between the action $do(X_3 = \text{ON})$ and the observation $X_3 = \text{ON}$. The latter is encoded by ordinary Bayesian conditioning, whereas the former by conditioning a mutilated graph, with the link $X_1 \rightarrow X_3$ removed. Indeed, this mirrors the difference between seeing and doing: after observing that the sprinkler is ON, we wish to infer that the season is dry, that it probably did not rain, and so on; no such inferences should be drawn in evaluating the effects of the deliberate action "turning the sprinkler ON." The amputation of $X_3 = f_3(X_1, U_3)$ from Eq. (31.8) ensures the suppression of any abductive inferences from X_3, the action's recipient.

Note also that Eqs. (31.11–31.14) are independent of T; in other words, the pre-action and postaction distributions depend only on observed conditional probabilities but are independent of the particular functional form of $\{f_i\}$ or the distribution $P(u)$, which generate those probabilities. This is the essence of identifiability as given in Definition 31.6, which stems from the Markovian assumptions (31.9) and (31.10). The next subsection will demonstrate that certain causal effects, though not all, are identifiable even when the Markovian property is destroyed by introducing dependencies among the disturbance terms.

Generalization to multiple actions and conditional actions are reported in [Pearl and Robins 1995]. Multiple actions $do(X = x)$, where X is a compound variable result in a distribution similar to Eq. (31.12), except that all factors corresponding to the variables in X are removed from the product in Eq. (31.11). Stochastic conditional strategies [Pearl 1994b] of the form

$$do(X_j = x_j) \quad \text{with probability} \quad P^*(x_j \mid pa_j^*) \tag{31.15}$$

where pa_j^* is the support of the decision strategy, also result in a product decomposition similar to Eq. (31.11), except that each factor $P(x_j \mid pa_j)$ is *replaced* with $P^*(x_j \mid pa_j^*)$.

The surgical procedure just described is not limited to probabilistic analysis. The causal knowledge represented in Fig. 31.1 can be captured by logical theories as well, for example,

$$
\begin{aligned}
x_2 &\Longleftrightarrow [(X_1 = \text{Winter}) \vee (X_1 = \text{Fall}) \vee ab_2] \wedge \neg ab_2' \\
x_3 &\Longleftrightarrow [(X_1 = \text{Summer}) \vee (X_1 = \text{Spring}) \vee ab_3] \wedge \neg ab_3' \\
x_4 &\Longleftrightarrow (x_2 \vee x_3 \vee ab_4) \wedge \neg ab_4' \\
x_5 &\Longleftrightarrow (x_4 \vee ab_5) \wedge \neg ab_5'
\end{aligned}
\tag{31.16}
$$

where x_i stands for X_i = true, and ab_i and ab_i' stand, respectively, for triggering and inhibiting abnormalities. The double arrows represent the assumption that the events on the right-hand side of each equation are the *only* direct causes for the left-hand side, thus identifying the surgery implied by any action.

It should be emphasized though that the models of a causal theory are not made up merely of truth value assignments which satisfy the equations in the theory. Since each equation represents an autonomous process, the content of each individual equation must be specified in any model of the theory, and this can be encoded using either the graph (as in Fig. 31.1) or the generic description of the theory, as in Eq. (31.8). Alternatively, we can view a model of a causal theory to consist of a mutually consistent set of submodels, with each submodel being a standard model of a single equation in the theory.

Action Calculus

The identifiability of causal effects demonstrated in the last subsection relies critically on the Markovian assumptions (31.9) and (31.10). If a variable that has two descendants in the graph is unobserved, the disturbances in the two equations are no longer independent, the Markovian property (31.9) is violated, and identifiability may be destroyed. This can be seen easily from Eq. (31.12); if any parent of the manipulated variable X_j is unobserved, one cannot estimate the conditional probability $P(x_j \mid pa_j)$, and the effect of the action $do(X_j = x_j)$ may not be predictable from the observed distribution $P(x_1, \ldots, x_n)$. Fortunately, certain causal effects are identifiable even in situations where members of pa_j are unobservable [Pearl 1993a] and, moreover, polynomial tests are now available for deciding when $P(x_i \mid \hat{x}_j)$ is identifiable and for deriving closed-form expressions for $P(x_i \mid \hat{x}_j)$ in terms of observed quantities [Galles and Pearl 1995].

These tests and derivations are based on a symbolic calculus [Pearl 1994b, 1995] to be described in the sequel, in which interventions, side by side with observations, are given explicit notation and are permitted to transform probability expressions. The transformation rules of this calculus reflect the understanding that interventions perform local surgeries as described in Definition 31.3, i.e., they overrule equations that tie the manipulated variables to their preintervention causes.

Let X, Y, and Z be arbitrary disjoint sets of nodes in a DAG G. We denote by $G_{\overline{X}}$ the graph obtained by deleting from G all arrows pointing to nodes in X. Likewise, we denote by $G_{\underline{X}}$ the graph obtained by deleting from G all arrows emerging from nodes in X. To represent the deletion of both incoming and outgoing arrows, we use the notation $G_{\overline{X}\underline{Z}}$. Finally, the expression $P(y \mid \hat{x}, z) \overset{\Delta}{=} P(y, z \mid \hat{x})/P(z \mid \hat{x})$ stands for the probability of $Y = y$ given that $Z = z$ is observed and X is held constant at x.

Theorem 31.2. *Let G be the directed acyclic graph associated with a Markovian causal theory, and let $P(\cdot)$ stand for the probability distribution induced by that theory. For any disjoint subsets of variables X, Y, Z, and W we have the following rules:*

- *Rule 1 (Insertion/deletion of observations):*

$$P(y \mid \hat{x}, z, w) = P(y \mid \hat{x}, w) \quad \text{if } (Y \perp\!\!\!\perp Z \mid X, W)_{G_{\overline{X}}} \qquad (31.17)$$

- *Rule 2 (Action/observation exchange):*

$$P(y \mid \hat{x}, \hat{z}, w) = P(y \mid \hat{x}, z, w) \quad \text{if } (Y \perp\!\!\!\perp Z \mid X, W)_{G_{\overline{X}\underline{Z}}} \qquad (31.18)$$

- *Rule 3 (Insertion/deletion of actions):*

$$P(y \mid \hat{x}, \hat{z}, w) = P(y \mid \hat{x}, w) \quad \text{if } (Y \perp\!\!\!\perp Z \mid X, W)_{G_{\overline{X}, \overline{Z(W)}}} \qquad (31.19)$$

where $Z(W)$ is the set of Z-nodes that are not ancestors of any W-node in $G_{\overline{X}}$.

Each of the inference rules follows from the basic interpretation of the $\hat{x}$ operator as a replacement of the causal mechanism that connects X to its pre-action parents by a new mechanism $X = x$ introduced by the intervening force. The result is a submodel characterized by the subgraph $G_{\overline{X}}$ (named *manipulated graph* in Spirtes et al. [1993]), which supports all three rules.

Corollary 31.1. *A causal effect $q : P(y_1, \ldots, yk \mid \hat{x}_1, \ldots, \hat{x}_m)$ is identifiable in a model characterized by a graph G if there exists a finite sequence of transformations, each conforming to one of the inference rules in Theorem 31.2, which reduces q into a standard (i.e., hat-free) probability expression involving observed quantities.*

Although Theorem 31.2 and Corollary 31.1 require the Markovian property, they also can be applied to non-Markovian, recursive theories because such theories become Markovian if we consider the unobserved variables as part of the analysis, and represent them as nodes in the graph. To illustrate, assume that variable X_1 in Fig. 31.1 is unobserved, rendering the disturbances U_3 and U_2 dependent since these terms now include the common influence of X_1. Theorem 31.2 tells us that the causal effect $P(x_4 \mid \hat{x}_3)$ is identifiable, because

$$P(x_4 \mid \hat{x}_3) = \sum_{x_2} P(x_4 \mid \hat{x}_3, x_2) P(x_2 \mid \hat{x}_3)$$

Rule 3 permits the deletion

$$P(x_2 \mid \hat{x}_3) = P(x_2), \quad \text{because } (X_2 \parallel X_3)_{G_{\overline{X}_3}}$$

whereas rule 2 permits the exchange

$$P(x_4 \mid \hat{x}_3, x_2) = P(x_4 \mid x_3, x_2), \quad \text{because } (X_4 \parallel X_3 \mid X_2)_{G_{\underline{X}_3}}$$

This gives

$$P(x_4 \mid \hat{x}_3) = \sum_{x_2} P(x_4 \mid x_3, x_2) P(x_2)$$

which is a hat-free expression, involving only observed quantities.

In general, it can be shown [Pearl 1995]:

1. The effect of interventions can often be identified (from nonexperimental data) without resorting to parametric models.
2. The conditions under which such nonparametric identification is possible can be determined by simple graphical tests.[1]
3. When the effect of interventions is not identifiable, the causal graph may suggest nontrivial experiments which, if performed, would render the effect identifiable.

The ability to assess the effect of interventions from nonexperimental data has immediate applications in the medical and social sciences, since subjects who undergo certain treatments often are not representative of the population as a whole. Such assessments are also important in artificial intelligence (AI) applications where an agent needs to predict the effect of the next action on the basis of past performance records, and where that action has never been enacted out of free will, but in response to environmental needs or to other agents' requests.

Historical Remarks

An explicit translation of interventions to *wiping out* equations from linear econometric models was first proposed by Strotz and Wold [1960] and later used in Fisher [1970] and Sobel [1990]. Extensions to action representation in nonmonotonic systems were reported in Goldszmidt and Pearl [1992] and Pearl [1993a]. Graphical ramifications of this translation were explicated first in Spirtes et al. [1993] and later in Pearl [1993b]. A related formulation of causal effects, based on event trees and counterfactual analysis,

---

[1]These graphical tests offer, in fact, a complete formal solution to the *covariate-selection* problem in statistics: finding an appropriate set of variables that need be adjusted for in any study which aims to determine the effect of one factor upon another. This problem has been lingering in the statistical literature since Karl Pearson, the founder of modern statistics, discovered (1899) what in modern terms is called the Simpson's paradox; any statistical association between two variables may be reversed or negated by including additional factors in the analysis [Aldrich 1995].

was developed by Robin [(1986, pp. 1422–1425)]. Calculi for actions and counterfactuals based on this interpretation are developed in Pearl [1994b] and Balke and Pearl [1994], respectively.

31.5 Counterfactuals

A counterfactual sentence has the form:

> *If A were true, then C would have been true?*

where A, the counterfactual antecedent, specifies an event that is contrary to one's real-world observations, and C, the counterfactual consequent, specifies a result that is expected to hold in the alternative world where the antecedent is true. A typical example is "If Oswald were not to have shot Kennedy, then Kennedy would still be alive," which presumes the factual knowledge of Oswald's shooting Kennedy, contrary to the antecedent of the sentence.

The majority of the Philosophers who have examined the semantics of counterfactual sentences have resorted to some version of Lewis' *closest world* approach: "*C* if it were *A*" is true, if *C* is true in worlds that are closest to the real world yet consistent with the counterfactual's antecedent A [Lewis 1973]. Ginsberg [1986] followed a similar strategy. Whereas the closest world approach leaves the precise specification of the closeness measure almost unconstrained, causal knowledge imposes very specific preferences as to which worlds should be considered closest to any given world. For example, considering an array of domino tiles standing close to each other, the manifestly closest world consistent with the antecedent "tile i is tipped to the right" would be a world in which just tile i is tipped, and all of the others remain erect. Yet, we all accept the counterfactual sentence "Had tile i been tipped over to the right, tile $i + 1$ would be tipped as well" as plausible and valid. Thus, distances among worlds are not determined merely by surface similarities but require a distinction between disturbed mechanisms and naturally occurring transitions. The local surgery paradigm expounded in the beginning of section 31.4 offers a concrete explication of the closest world approach which respects causal considerations. A world w_1 is closer to w than a world w_2 is, if the set of atomic surgeries needed for transforming w into w_1 is a subset of those needed for transforming w into w_2. In the domino example, finding tile i tipped and $i + 1$ erect requires the breakdown of two mechanism (e.g., by two external actions) compared with one mechanism for the world in which all j-tiles, $j > i$, are tipped. This paradigm conforms to our perception of causal influences and lends itself to economical machine representation.

Formal Underpinning

The structural equation framework offers an ideal setting for counterfactual analysis.

Definition 31.7 (context-based potential response). Given a causal theory T and two disjoint sets of variables, X and Y, the potential response of Y to X in a context u, denoted $Y(x, u)$ or $Y_x(u)$, is the solution for Y under $U = u$ in the subtheory T_x. $Y(x, u)$ can be taken as the formal definition of the counterfactual English phrase: "the value that Y would take in context u, had X been x."[2]

Note that this definition allows for the context $U = u$ and the proposition $X = x$ to be incompatible in T. For example, if T describes a logic circuit with input U it may well be reasonable to assert the counterfactual: "Given $U = u$, Y would be high if X were low," even though the input $U = u$ may

---

[2]The term *unit* instead of *context* is often used in the statistical literature [Rubin 1974], where it normally stands for the identity of a specific individual in a population, namely, the set of attributes u that characterizes that individual. In general, u may include the time of day, the experimental conditions under study, and so on. Practitioners of the counterfactual notation do not explicitly mention the notions of *solution* or *intervention* in the definition of $Y(x, u)$. Instead, the phrase "the value that Y would take in unit u, had X been x," viewed as basic, is posited as the definition of $Y(x, u)$.

preclude X from being low. It is for this reason that one must invoke some motion of intervention (alternatively, a theory change or a *miracle* [Lewis 1973]) in the definition of counterfactuals.

If U is treated as a random variable, then the value of the counterfactual $Y(x, u)$ becomes a random variable as well, denoted as $Y(x)$ of Y_x. Moreover, the distribution of this random variable is easily seen to coincide with the causal effect $P(y \mid \hat{x})$, as defined in Eq. (31.7), i.e.,

$$P((Y(x) = y) = P(y \mid \hat{x})$$

The probability of a counterfactual conditional $x \to y \mid o$ may then be evaluated by the following procedure:

- Use the observations o to update $P(u)$ thus forming a causal theory $T^o = \langle V, U, \{f_i\}, P(u \mid o) \rangle$
- Form the mutilated theory T_x^o (by deleting the equation corresponding to variables in X) and compute the probability $P_{T^o}(y \mid \hat{x})$ which T_x^o induces on Y.

Unlike causal effect queries, counterfactual queries are not identifiable even in Markovian theories, but require that the functional-form of $\{f_i\}$ be specified. In Balke and Pearl [1994] a method is devised for computing sharp bounds on counterfactual probabilities which, under certain circumstances may collapse to point estimates. This method has been applied to the evaluation of causal effects in studies involving noncompliance and to the determination of legal liability.

Applications to Policy Analysis

Counterfactual reasoning is at the heart of every planning activity, especially real-time planning. When a planner discovers that the current state of affairs deviates from the one expected, a plan repair activity needs to be invoked to determine what went wrong and how it could be rectified. This activity amounts to an exercise in counterfactual thinking, as it calls for rolling back the natural course of events and determining, based on the factual observations at hand, whether the culprit lies in previous decisions or in some unexpected, external eventualities. Moreover, in reasoning forward to determine if things would have been different, a new model of the world must be consulted, one that embodies hypothetical changes in decisions or eventualities—hence, a breakdown of the old model or theory.

The logic-based planning tools used in AI, such as STRIPS and its variants or those based on situation calculus, do not readily lend themselves to counterfactual analysis, as they are not geared for coherent integration of abduction with prediction, and they do not readily handle theory changes. Remarkably, the formal system developed in economics and social sciences under the rubric structural equations models does offer such capabilities but, as will be discussed, these capabilities are not well recognized by current practitioners of structural models. The analysis presented in this chapter could serve both to illustrate to AI researchers the basic formal features needed for counterfactual and policy analysis and to call the attention of economists and social scientists to capabilities that are dormant within structural equation models.

Counterfactual thinking dominates reasoning in political science and economics. We say, for example, "If Germany were not punished so severely at the end of World War I, Hitler would not have come to power," or "If Reagan did not lower taxes, our deficit would be lower today." Such thought experiments emphasize an understanding of generic laws in the domain and are aimed toward shaping future policy making, for example, "defeated countries should not be humiliated," or "lowering taxes (contrary to Reaganomics) tends to increase national debt."

Strangely, there is very little formal work on counterfactual reasoning or policy analysis in the behavioral science literature. An examination of a number of econometric journals and textbooks, for example, reveals a glaring imbalance: although an enormous mathematical machinery is brought to bear on problems of estimation and prediction, policy analysis (which is the ultimate goal of economic theories) receives almost no formal treatment. Currently, the most popular methods driving economic policy making are based on so-called *reduced-form* analysis: to find the impact of a policy involving decision variables X on outcome variables Y, one examines past data and estimates the conditional expectation $E(Y \mid X = x)$, where x is the particular instantiation of X under the policy studied.

The assumption underlying this method is that the data were generated under circumstances in which the decision variables X act as exogenous variables, that is, variables whose values are determined outside the system under analysis. However, although new decisions should indeed be considered exogenous for the purpose of evaluation, past decisions are rarely enacted in an exogenous manner. Almost every realistic policy (e.g., taxation) imposes control over some endogenous variables, that is, variables whose values are determined by other variables in the analysis. Let us take taxation policies as an example. Economic data are generated in a world in which the government is reacting to various indicators and various pressures; hence, taxation is endogenous in the data-analysis phase of the study. Taxation becomes exogenous when we wish to predict the impact of a specific decision to raise or lower taxes. The reduced-form method is valid only when past decisions are nonresponsive to other variables in the system, and this, unfortunately, eliminates most of the interesting control variables (e.g., tax rates, interest rates, quotas) from the analysis.

This difficulty is not unique to economic or social policy making; it appears whenever one wishes to evaluate the merit of a plan on the basis of the past performance of other agents. Even when the signals triggering the past actions of those agents are known with certainty, a systematic method must be devised for selectively ignoring the influence of those signals from the evaluation process. In fact, the very essence of *evaluation* is having the freedom to imagine and compare trajectories in various counterfactual worlds, where each world or trajectory is created by a hypothetical implementation of a policy that is free of the very pressures that compelled the implementation of such policies in the past.

Balke and Pearl [1995] demonstrate how linear, nonrecursive structural models with Gaussian noise can be used to compute counterfactual queries of the type: "Given an observation set O, find the probability that Y would have attained a value greater than y, had X been set to x." The task of inferring causes of effects, that is, of finding the probability that $X = x$ is the cause for effect E, amounts to answering the counterfactual query: "Given effect E and observations O, find the probability that E would not have been realized, had X not been x." The technique developed in Balke and Pearl [1995] is based on probability propagation in dual networks, one representing the actual world and the other representing the counterfactual world. The method is not limited to linear functions but applies whenever we are willing to assume the functional form of the structural equations. The noisy OR-gate model [Pearl 1988] is a canonical example where such functional form is normally specified. Likewise, causal theories based on Boolean functions (with exceptions), such as the one described in Eq. (31.16) lend themselves to counterfactual analysis in the framework of Definition 31.7.

Acknowledgments

The research was partially supported by Air Force Grant F49620-94-1-0173, National Science Foundation (NSF) Grant IRI-9420306, and Northrop/Rockwell Micro Grant 94-100.

References

Abramson, B. 1991. ARCO1: an application of belief networks to the oil market. In *Proc. 7th Conf. Uncertainty Artificial Intelligence*. Morgan Kaufmann, San Mateo, CA.

Agogino, A. M., Srinivas, S., and Schneider, K. 1988. Multiple sensor expert system for diagnostic reasoning, monitoring and control of mechanical systems. *Mech. Sys. Sig. Process.* 2:165–185.

Aldrich, J. 1995. Correlations genuine and spurious in Pearson and Yule. Forthcoming *Stat. Sci.*

Andersen, S. K., Olesen, K. G., Jensen, F. V., and Jensen, F. 1989. Hugin—a shell for building Bayesian belief universes for expert systems, pp. 1080–1085. In *11th Int. Jt. Conf. Artificial Intelligence*.

Balke, A. and Pearl, J. 1994. Counterfactual probabilities: computational methods, bounds, and applications. In *Uncertainty in Artificial Intelligence 10*. R. Lopez de Mantaras and D. Poole, eds., pp. 46–54. Morgan Kaufmann, San Mateo, CA.

Balke, A. and Pearl, J. 1995. Counterfactuals and policy analysis in structural models. In *Uncertainty in Artificial Intelligence 11*. P. Besnard and S. Hanks, eds., pp. 11–18. Morgan Kaufmann, San Francisco, CA.

Blalock, H. M. 1971. *Causal Models in the Social Sciences*. Macmillan, London.

Charniak, E. 1991. Bayesian networks without tears. *AI Mag.* 12(4):50–63.

Charniak, E. and Goldman, R. 1991. A probabilistic model of plan recognition. In *Proceedings, AAAI-91*. AAAI Press/MIT Press, Anaheim, CA.

Cooper, G. F. 1990. Computational complexity of probabilistic inference using Bayesian belief networks. *Artificial Intelligence* 42(2):393–405.

Cooper, G. F. and Herskovits, E. 1990. A Bayesian method for constructing Bayesian belief networks from databases, pp. 86–94. In *Proc. Conf. Uncertainty in AI*. Morgan Kaufmann.

Darwiche, A. and Goldszmidt, M. 1994. On the relation between kappa calculus and probabilistic reasoning. In *Uncertainty in Artificial Intelligence*. Vol. 10, R. Lopez de Mantaras and D. Poole, eds., pp. 145–153. Morgan Kaufmann, San Francisco, CA.

Darwiche, A. and Pearl, J. 1994. Symbolic causal networks for planning under uncertainty, pp. 41–47. In *Symp. Notes AAAI Spring Symp. Decision-Theoretic Planning*. Stanford, CA.

Dawid, A. P. 1979. Conditional independence in statistical theory. *J. R. Stat. Soc. Ser. A* 41:1–31.

Dean, T. L. and Wellman, M. P. 1991. *Planning and Control*. Morgan Kaufmann, San Mateo, CA.

Duncan, O. D. 1975. *Introduction to Structural Equation Models*. Academic Press, New York.

Fisher, F. M. 1970. A correspondence principle for simultaneous equations models. *Econometrica* 38:73–92.

Galles, D. and Pearl, J. 1995. Testing identifiability of causal effects. In *Uncertainty in Artificial Intelligence 11*, P. Besnard and S. Hanks, eds., pp. 185–195. Morgan Kaufmann, San Francisco, CA.

Geiger, D. 1990. *Graphoids: A Qualitative Framework for Probabilistic Inference*. Ph.D. Thesis. Department of Computer Science, University of California, Los Angeles.

Geiger, D., Verma, T. S., and Pearl, J. 1990. Identifying independence in Bayesian networks. In *Networks*. Vol. 20, pp. 507–534. Wiley, Sussex, England.

Ginsberg, M. L. 1986. Counterfactuals. *Artificial Intelligence* 30:35–79.

Goldszmidt, M. and Pearl, J. 1992. Rank-based systems: a simple approach to belief revision, belief update, and reasoning about evidence and actions. In *Proc. 3rd Int. Conf. Knowledge Representation Reasoning*. B. Nobel, C. Rich, and M. Swartout, eds., pp. 661–672. Morgan Kaufmann, San Mateo, CA.

Goldszmidt, M. and Pearl, J. 1996. Qualitative probabilities for default reasoning, belief revision, and causal modeling. *Artificial Intelligence* 84(1–2):57–112.

Good, I. J. 1961. A causal calculus, I-II. *Br. J. Philos. Sci.* 11:305–318, 12:43–51.

Haavelmo, T. 1943. The statistical implications of a system of simultaneous equations. *Econometrica* 11:1–12.

Heckerman, D. 1991. Probabilistic similarity networks. *Networks* 20(5):607–636.

Heckerman, D., Geiger, D., and Chickering, D. 1994. Learning Bayesian networks: The combination of knowledge and statistical data, pp. 293–301. In *Proc. 10th Conf. Uncertainty Artificial Intelligence*. Seattle, WA, July. Morgan Kaufmann, San Mateo, CA.

Heckerman, D. E., Horvitz, E. J., and Nathwany, B. N. 1990. Toward normative expert systems: The pathfinder project. *Medical Comput. Sci. Group, Tech. Rep.* KSL-90-08, Section on Medical Informatics, Stanford University, Stanford, CA.

Henrion, M. 1988. Propagation of uncertainty by probabilistic logic sampling in Bayes' networks. In *Uncertainty in Artificial Intelligence 2*. J. F. Lemmer and L. N. Kanal, eds., pp. 149–164. Elsevier Science, North-Holland, Amsterdam.

Howard, R. A. and Matheson, J. E. 1981. Influence diagrams. *Principles and Applications of Decision Analysis*. Strategic Decisions Group. Menlo Park, CA.

Kenny, D. A. 1979. *Correlation and Causality*. Wiley, New York.

Kiiveri H., Speed, T. P., and Carlin, J. B. 1984. Recursive causal models. *J. Australian Math. Soc.* 36:30–52.

Kim, J. H. and Pearl, J. 1983. A computational model for combined causal and diagnostic reasoning in inference systems, pp. 190–193. In *Proceedings IJCAI-83*. Karlsruhe, Germany.

Lauritzen, S. L. 1982. *Lectures on Contingency Tables*, 2nd ed. University of Aalborg Press, Aalborg, Denmark.

Lauritzen, S. L. and Spiegelhalter, D. J. 1988. Local computations with probabilities on graphical structures and their application to expert systems (with discussion). *J. R. Stat. Soc. Ser. B* 50(2):157–224.

Levitt, J. M., Agosta, T. S., and Binford, T. O. 1990. Model-based influence diagrams for machine vision. In M. Hension, R. D. Shachter, L. N. Kanal, and J. F. Lemmer, eds., pp. 371–388. *Uncertainty in Artificial Intelligence* 5. North-Holland, Amsterdam.

Lewis, D. 1973. *Counterfactuals*. Basil Blackwell, Oxford, England.

Neapolitan, R. E. 1990. *Probabilistic Reasoning in Expert Systems: Theory and Algorithms*. Wiley, New York.

Oliver, R. M. and Smith, J. Q., eds. 1990. *Influence Diagrams, Belief Nets, and Decision Analysis*. Wiley, New York.

Pearl, J. 1982. Reverend Bayes on inference engines: a distributed hierarchical approach, pp. 133–136. In *Proc. AAAI Nat. Conf. AI*. Pittsburgh, PA.

Pearl, J. 1987. Bayes decision methods. In *Encyclopedia of AI*, pp. 48–56. Wiley Interscience, New York.

Pearl, J. 1988. *Probabilistic Reasoning in Intelligence Systems*, rev. 2nd printing, Morgan Kaufmann, San Mateo, CA.

Pearl, J. 1993a. From Bayesian networks to causal networks, pp. 25–27. In *Proc. Adaptive Comput. Inf. Process. Semin.* Brunel Conf. Centre, London, Jan. See also *Stat. Sci.* 8(3):266–269.

Pearl, J. 1993b. From conditional oughts to qualitative decision theory. In *Proc. 9th Conf. Uncertainty Artificial Intelligence*. D. Heckerman and A. Mamdani, eds., pp. 12–20. Morgan Kaufmann.

Pearl, J. 1994a. From Adams' conditionals to default expressions, causal conditionals, and counterfactuals. In *Probability and Conditionals*. E. Eells and B. Skyrms, eds., pp. 47–74. Cambridge University Press, Cambridge, MA.

Pearl, J. 1994b. A probabilistic calculus of actions. In *Uncertainty in Artificial Intelligence 10*. R. Lopez de Mantaras and D. Poole, eds., pp. 454–462. Morgan Kaufmann, San Mateo, CA.

Pearl, J. 1995. Causal diagrams for experimental research. *Biometrika* 82(4):669–710.

Pearl, J., Geiger, D., and Verma, T. 1990. The logic and influence diagrams. In *Influence Diagrams, Belief Nets and Decision Analysis*. R. M. Oliver and J. Q. Smith, eds., pp. 67–87. Wiley, New York.

Pearl, J. and Robins, J. M. 1995. Probabilistic evaluation of sequential plans from causal models with hidden variables. In *Uncertainty in Artificial Intelligence 11*. P. Besnard and S. Hanks, eds., pp. 444–453. Morgan Kaufmann, San Francisco, CA.

Pearl, J. and Verma, T. 1991. A theory of inferred causation. In *Principles of Knowledge Representation and Reasoning: Proc. 2nd Int. Conf.* J. A. Allen, R. Fikes, and E. Sandewall, eds., pp. 441–452. Morgan Kaufmann, San Mateo, CA.

Peng, Y. and Reggia, J. A. 1990. *Abductive Inference Models for Diagnostic Problem-Solving*. Springer–Verlag, New York.

Robin, S. M. 1986. A new approach to causal inference in mortality studies with a sustained exposure period—applications to control of the healthy workers survivor effect. *Math. Model* 7:1393–1512.

Rosenthal, A. 1977. A computer scientist looks at reliability computations. In *Reliability and Fault Tree Analysis*. Barlow et al., eds., pp. 133–152. SIAM, Philadelphia, PA.

Rubin, D. B. 1974. Estimating causal effects of treatments in randomized and nonrandomized studies. *J. Educ. Psych.* 66:688–701.

Rumelhart, D. E. 1976. Toward an interactive model of reading. *University of California Tech. Rep. CHIP-56*, University of California, La Jolla.

Shachter, R. D. 1986. Evaluating influence diagrams. *Op. Res.* 34(6):871–882.

Shachter, R. D. 1988. Probabilistic inference and influence diagrams. *Op. Res.* 36:589–604.

Shachter, R. D. 1990. Special issue on influence diagrams. *Networks: Int. J.* 20(5).

Shafer, G. and Pearl, J., eds. 1990. *Readings in Uncertain Reasoning*. Morgan Kaufmann, San Mateo, CA.

Simon, H. A. 1953. Causal ordering and identifiability. In *Studies in Econometric Method.* W. C. Hood and T. C. Koopmans, eds., pp. 49–74. Wiley, New York.

Sobel, M. E. 1990. Effect analysis and causation in linear structural equation models. *Psychometrika* 55(3):495–515.

Spiegelhalter, D. J. and Lauritzen, S. L. 1990. Sequential updating of conditional probabilities on directed graphical structures. *Networks* 20(5):579–605.

Spiegelhalter, D. J., Lauritzen, S. L., Dawid, P. A., and Cowell, R. G. 1993. Bayesian analysis in expert systems. *Stat. Sci.* 8:219–247.

Spirtes, P., Glymour, C., and Schienes, R. 1993. *Causation, Prediction, and Search.* Springer–Verlag, New York.

Spohn, W. 1988. A general non-probabilistic theory of inductive reasoning, pp. 315–322. In *Proc. 4th Workshop Uncertainty Artificial Intelligence.* Minneapolis, MN.

Strotz, R. H. and Wold, H. O. A. 1960. Causal models in the social sciences. *Econometrica* 28:417–427.

Turtle, H. R. and Croft, W. B. 1991. Evaluation of an inference network-based retrieval model. *ACM Trans. Inf. Sys.* 9(3).

Verma, T. and Pearl, J. 1990. Equivalence and synthesis of causal models. In *Uncertainty in Artificial Intelligence 6*, pp. 220–227. Elsevier Science, Cambridge, MA.

Wermuth, N. and Lauritzen, S. L. 1983. Graphical and recursive models for contingency tables. *Biometrika* 70:537–552.

Wold, H. 1964. *Econometric Model Building.* North-Holland, Amsterdam.

Wright, S. 1921. Correlation and causation. *J. Agric. Res.* 20:557–585.

Wright, S. 1934. The method of path coefficients. *Ann. Math. Stat.* 5:161–215.

32

Qualitative Reasoning

Kenneth D. Forbus
Northwestern University

32.1 Introduction

Qualitative reasoning is the area of artificial intelligence (AI) which creates representations for continuous aspects of the world, such as space, time, and quantity, which support reasoning with very little information. Typically, it has focused on scientific and engineering domains, hence its other name, qualitative physics. It is motivated by two observations. First, people draw useful and subtle conclusions about the physical world without differential equations. In our daily lives we figure out what is happening around us and how we can affect it, working with far less data, and less precise data, than would be required to use traditional, purely quantitative methods. Creating software for robots that operate in unconstrained environments and modeling human cognition require understanding how this can be done. Second, scientists and engineers appear to use qualitative reasoning when initially understanding a problem, when setting up more formal methods to solve particular problems, and when interpreting the results of **quantitative simulations**, calculations, or measurements. Thus, advances in qualitative physics should lead to the creation of more flexible software that can help engineers and scientists.

Qualitative physics began with de Kleer's investigation on how qualitative and quantitative knowledge interacted in solving a subset of simple textbook mechanics problems [de Kleer 1977]. After roughly a decade of initial explorations, the potential for important industrial applications led to a surge of interest in the mid-1980s, and the area has been growing steadily, with rapid progress. Qualitative representations have made their way into commercial supervisory control software for curing composite materials, and the first product known to have been designed using qualitative physics techniques appeared on the market in 1994 [Shimomura et al. 1995]. Given the strong potential for industrial applications that is only starting to be realized, and its potential importance in understanding human cognition, work in qualitative modeling is likely to remain an important area in artificial intelligence.

This chapter first surveys the state of the art in qualitative representations and in qualitative reasoning techniques. The application of these techniques to various problems is discussed subsequently.

32.2 Qualitative Representations

As with many other representation issues, there is no single, universal right or best qualitative representation. Instead, there exists a spectrum of choices, each with its own advantages and disadvantages for particular tasks. What all of them have in common is that they provide notations for describing and reasoning about continuous properties of the physical world. Two key issues in qualitative representation are resolution and **compositionality**. We discuss each in turn.

Resolution concerns the level of information detail in a representation. Resolution is an issue because one goal of qualitative reasoning is to understand how little information suffices to draw useful conclusions. Low-resolution information is available more often than precise information ("the car heading toward us is slowing down" vs. "the derivative of the car's speed along the line connecting us is −28 km/h"), but conclusions drawn with low-resolution information are often ambiguous. The role of ambiguity is important: the prediction of alternative futures (i.e., "the car will hit us" vs. "the car won't hit us") suggests that we may need to gather more information, analyze the matter more deeply, or take action, depending on what alternatives our qualitative reasoning uncovers. High-resolution information is often needed to draw particular conclusions [i.e., a finite element analysis of heat flow within a notebook computer design to ensure that the central processing unit (CPU) will not cook the battery], but qualitative reasoning with low-resolution representations reveals what the interesting questions are. Qualitative representations comprise one form of tacit knowledge that people, ranging from the person on the street to scientists and engineers, use to make sense of the world.

Compositionality concerns the ability to combine representations for different aspects of a phenomenon or system to create a representation of the phenomenon or system as a whole. Compositionality is an issue because one goal of qualitative physics is to formalize the modeling process itself. Many of today's AI systems are based on handcrafted knowledge bases that express information about a specific artifact or system needed to carry out a particular narrow range of tasks involving it. By contrast, a substantial component of the knowledge of scientists and engineers consists of principles and laws that are broadly applicable, both with respect to the number of systems they explain and the kinds of tasks they are relevant for. Qualitative physics is developing the ideas and organizing techniques for knowledge bases with similar expressive and inferential power, called **domain theories**.

The remainder of this section surveys the fundamental representations used in qualitative reasoning for quantity, mathematical relationships, **modeling assumptions**, causality, space, and time.

Representing Quantity

Qualitative reasoning has explored tradeoffs in representations for continuous parameters ranging in resolution from sign algebras to the hyperreals. Most of the research effort has gone into understanding the properties of low-resolution representations, since the properties of high-resolution representations tend to already be well understood due to work in mathematics.

The lowest resolution representation for continuous parameters is the status abstraction, which represents a quantity by whether or not it is *normal* [Abbott et al. 1987]. It is a useful representation for certain diagnosis and monitoring tasks because it is the weakest representation that can express the difference between something working and not working. The next step in resolution is the sign algebra, which represents continuous parameters as either −, +, or 0, according to whether the sign of the underlying continuous parameter is negative, positive, or zero. The sign algebra is surprisingly powerful: Since a parameter's derivatives are themselves parameters whose values can be represented as signs, some of the main results of the differential calculus (e.g., the mean value theorem) can be applied to reasoning about sign values [de Kleer and Brown 1984]. This allows sign algebras to be used for qualitative reasoning about dynamics, including expressing properties such as oscillation and stability. The sign algebra is the weakest representation that supports such reasoning.

Representing continuous values via sets of ordinal relations (also known as the **quantity space** representation) is the next step up in resolution [Forbus 1984]. For example, the temperature of a fluid might be represented in terms of its relationship between the freezing point and boiling point of the material that comprises it. Like the sign algebra, quantity spaces are expressive enough to support qualitative reasoning about dynamics. (The sign algebra can be modeled by a quantity space with only a single comparison point, zero.) Unlike the sign algebra, which draws values from a fixed finite algebraic structure, quantity spaces provide variable resolution because new points of comparison can be added to refine values. The temperature of water in a kettle on a stove, for instance, will likely be defined in terms of its relationship with the temperature of the stove as well as its freezing and boiling points. There are two kinds of comparison points used in defining quantity spaces. **Limit points** are derived from general properties of a domain as applicable to a specific situation. Continuing with the kettle example, the particular ordinal relationships used were chosen because they determine whether or not the **physical processes** of freezing, boiling, and heat flow occur in that situation. The precise numerical value of limit points can change over time, e.g., the boiling point of a fluid is a function of its pressure. **Landmark** values are constant points of comparison introduced during reasoning to provide additional resolution [Kuipers 1986]. To ascertain whether an oscillating system is overdamped, underdamped, or critically damped, for instance, requires comparing successive peak values. Noting the peak value of a particular cycle as a landmark value, and comparing it to the landmarks generated for successive cycles in the behavior, provides a way of making this inference.

Intervals are a well-known variable-resolution representation for numerical values and have been heavily used in qualitative reasoning. A quantity space can be thought of as partial information about a set of intervals. If we have complete information about the ordinal relationships between limit points and landmark values, these comparison points define a set of intervals that partition a parameter's value. This natural mapping between quantity spaces and intervals has been exploited by a variety of systems that use intervals whose endpoints are known numerical values to refine predictions produced by purely qualitative reasoning. Fuzzy intervals also have been used in similar ways, e.g., in reasoning about control systems.

Order of magnitude representations stratify values according to some notion of scale. They can be important in resolving ambiguities and in simplifying models because they enable reasoning about what phenomena and effects may safely be ignored in a given situation. For instance, heat losses from turbines are generally ignored in the early stages of power plant design, because the energy lost is very small relative to the energy being produced. Several stratification techniques have been used in the literature, including hyperreal numbers, numerical thresholds, and logarithmic scales. Three issues faced by all these formalisms are (1) the conditions under which many small effects can combine to produce a significant effect, (2) the soundness of the reasoning supported by the formalism, and (3) the efficiency of using them. Understanding the properties of these formalisms and their tradeoffs is still an area of active research.

Although many qualitative representations of number use the reals as their basis, two other bases have been used with interesting results. One is the hyperreals, otherwise known as the infinitesimal calculus. Aside from order of magnitude representations, hyperreals have been used in modeling **comparative analysis**, dynamics, and time. The other basis for qualitative representations of number is finite algebras. One motivation for using finite algebras is that observations are often naturally categorized into a finite set of labels, i.e., very small, small, normal, large, very large. Research on such algebras is aimed at solving problems such as how to increase the compositionality of such representations, e.g., how to propagate information across different resolution scales.

Representing Mathematical Relationships

Like number, a variety of qualitative representations of mathematical relationships have been developed, often by adopting and adapting systems developed in mathematics. Abstractions of the analytic functions are commonly used to provide the lower resolution and compositionality desired. For example, **conflu-ences** are differential equations over the sign algebra [de Kleer and Brown 1984]. An equation such as $V = IR$ can be expressed as the confluence

$$[V] = [I] + [R]$$

where $[Q]$ denotes taking the sign of Q. Differential equations also can be expressed in this manner, for instance,

$$[F] = \partial_V$$

which is a qualitative version of $F = MA$ (assuming M is always positive). Thus, any system of algebraic and differential equations with respect to time can be described as a set of confluences.

Many of the algebraic operations taken for granted in manipulating analytic functions over the reals are not valid in weak algebras [Struss 1988]. Since qualitative relationships most often are used to prop-agate information, this is not a serious limitation. In situations where algebraic solutions themselves are desirable, mixed representations that combine algebraic operations over the reals and move to qualitative abstractions when appropriate are a promising approach [Williams 1991].

Another low-resolution representation of equations uses monotonic functions over particular ranges, i.e.,

$$M + (\text{force, acceleration})$$

states that force depends only on the acceleration, and the function relating them is increasing monotonic [Kuipers 1986]. Compositionality is achieved by using qualitative proportionalities [Forbus 1984] to express partial information about functional dependency, e.g.,

$$\text{force} \propto_{Q+} \text{acceleration}$$

states that force depends on acceleration and is increasing monotonic in its dependence on acceleration, but may depend on other factors as well. Additional constraints on the function which determines force can be added by additional **qualitative proportionalities**, e.g.,

$$\text{force} \propto_{Q-} \text{mass}$$

states that force also depends on mass, and is decreasing monotonic in this dependence. Qualitative proportionalities must be combined via closed-world assumptions to ascertain all of the effects on a quantity. Similar primitives can be defined for expressing relationships involving derivatives, to define a complete language of compositional qualitative mathematics for ordinary differential equations. As with confluences, few algebraic operations are valid for combining monotonic functions, mainly composition of functions of identical sign, i.e.,

$$M + (f, g) \wedge M + (g, h) \Rightarrow M + (f, h))$$

In addition to resolution and compositionality, another issue arising in qualitative representations of mathematical relationships is causality. There are three common views on how mathematical relationships interact with the causal relationships people use in common sense reasoning. One view is that there is no relationship between them. The second view is that mathematical relationships should be expressed with primitives that also make causal implications. For example, qualitative proportionalities include a causal interpretation, i.e., a change in acceleration causes a change in force, but not the other way around. The third view is that acausal mathematical relationships give rise to causal relationships via the particular process of using them. For example, confluences have no built-in causal direction, but are used in causal reasoning by identifying the flow of information through them while reasoning with a presumed flow of causality in the physical system they model. One method for imposing causality on a set of acausal constraint equations is by computing a causal ordering [Iwasaki and Simon 1986] that imposes directionality on a set of equations, starting from variables considered to be exogenous within the system.

Each view of causality has its merits. For tasks where causality is truly irrelevant, ignoring causality might be the best approach. To create software that can span the range of human common sense reasoning, something like a combination of the second and third views appears necessary because the appropriate notion of causality varies. In reasoning about chemical phenomena, for instance, changes in concentration are always caused by changes in the amounts of the constituent parts and never the other way around. In electronics, on the other hand, it is often convenient to consider voltage changes as being caused by changes in current in one part of a circuit and to consider current changes as being caused by changes in voltage in another part of the same circuit [Forbus and Gentner 1986]. Understanding why different domain idealizations lead to different notions of causality is one of the interesting research issues in qualitative physics.

Ontology

Ontology concerns how to carve up the world, i.e., what kinds of things there are and what sorts of relationships can hold between them. Ontology is central to qualitative physics because one of its main goals is formalizing the art of building models of physical systems. A key choice in any act of modeling is figuring out how to construe the situation or system to be modeled in terms of the available models for classes of entities and phenomena. No single ontology will suffice for the span of reasoning about physical systems that people do. What is being developed instead is a catalog of ontologies, describing their properties and interrelationships and specifying conditions under which each is appropriate. Whereas some ontologies are currently well understood, at this writing the catalog contains many gaps.

An example of ontologies will make this point clearer. Consider the representation of liquids. Broadly speaking, the major distinction in reasoning about fluids is whether one individuates fluid according to a particular collection of particles or by location [Hayes 1985]. The former are called Eulerian, or piece of stuff, ontologies. The latter are called Lagrangian, or contained stuff, ontologies. It is the contained stuff view of liquids we are using when we treat a river as a stable entity, even though the particular set of molecules that comprises it is changing constantly. It is the piece of stuff view of liquids we are using when we think about the changes in a fluid as it flows through a steady-state system, such as a working refrigerator. Ontologies multiply as we try to capture more of human reasoning. For instance, the piece of stuff ontology can be further divided into three cases, each with its own rules of inference: (1) molecular collections, which describe the progress of an arbitrary piece of fluid that is small enough to never split apart but large enough to have extensive properties; (2) slices which, like molecular collections, never subdivide but unlike them are large enough to interact directly with their surroundings; and (3) pieces of stuff large enough to be split into several pieces (e.g., an oil slick). Similarly, the contained stuff ontology can be further specialized according to whether or not individuation occurs simply by container (abstract contained stuffs) or by a particular set of containing surfaces (bounded stuffs). Abstract contained stuffs provide a low-resolution ontology appropriate for reasoning about system-level properties in complex systems (e.g., the changes over time in a lubricating oil subsystem in a propulsion plant), whereas bounded stuffs contain the geometric information needed to reason about the interactions of fluids and shape in systems such as pumps and internal combustion engines.

Cutting across the ontologies for particular physical domains are systems of organization for classes of ontologies. The most commonly used ontologies are the device ontology [de Kleer and Brown 1984] and the process ontology [Forbus 1984]. The device ontology is inspired by network theory and system dynamics. Like those formalisms, it construes physical systems as networks of devices whose interactions occur solely through a fixed set of ports. Unlike those formalisms, it provides the ability to write and reason automatically with device models whose governing equations can change over time.

The process ontology is inspired by studies of human mental models and observations of practice in thermodynamics and chemical engineering. It construes physical systems as consisting of entities whose changes are caused by physical processes. Process ontologies thus postulate a separate ontological category for causal mechanisms, unlike device ontologies, where causality arises solely from the interaction of the parts. Another difference between the two classes of ontologies is that in the device ontology the system of devices and connections is fixed over time, whereas in the process ontology entities and processes can come into existence and vanish over time. Each is appropriate in different contexts: For most purposes, an electronic circuit is best modeled as a network of devices, whereas a chemical plant is best modeled as a collection of interacting processes.

State, Time, and Behaviors

A qualitative state is a set of propositions that characterize a qualitatively distinct behavior of a system. A qualitative state describing a falling ball, for instance, would include information about what physical processes are occurring (e.g., motion downwards, acceleration due to gravity) and how the parameters of the ball are changing (e.g., its position is getting lower and its downward velocity is getting larger). A qualitative state can abstractly represent an infinite number of quantitative states: Although the position and velocity of the ball are different at each distinct moment during its fall, until the ball collides with the ground the qualitative state of its motion is unchanged.

Qualitative representations can be used to partition behavior into natural units. For instance, the time over which the state of the ball falling holds is naturally thought of as an interval, ending when the ball collides with the ground. The collision itself can be described as yet another qualitative state, and the fact that falling leads to a collision with the ground can be represented via a transition between the two states. If the ball has a nonzero horizontal velocity and there is some obstacle in its direction of travel, another possible behavior is that the ball will collide with that object instead of the ground. In general, a qualitative state can have transitions to several next states, reflecting ambiguity in the qualitative representations. Returning to our ball example, and assuming that no collisions with obstacles occur, notice that the qualitative state of the ball falling occurs again once the ball has reached its maximum height after the collision. If continuous values are represented by quantity spaces and the sources of comparisons are limit points, then a finite set of qualitative states is sufficient to describe every possible behavior of a system. A collection of such qualitative states and transitions is called an **envisionment** [de Kleer 1977]. Many interesting dynamical conclusions can be drawn from an envisionment. For instance, oscillations correspond to cycles of states. Unfortunately, the fixed resolution provided by limit points is not sufficient for other dynamical conclusions, such as ascertaining whether or not the ball's oscillation is damped. If comparisons can include landmark values, such conclusions can sometimes be drawn, e.g., by comparing the maximum height on one bounce to the maximum height obtained on the next bounce. The cost of introducing landmark values is that the envisionment no longer need be finite; every cycle in a corresponding fixed-resolution envisionment could give rise to an infinite number of qualitative states in an envisionment with landmarks.

A sequence of qualitative states occurring over a particular span of time is called a behavior. Behaviors can be described using purely qualitative knowledge, purely quantitative knowledge, or a mixture of both. If every continuous parameter is quantitative, the numerical aspects of behaviors coincide with the notion of trajectory in a state-space model. If qualitative representations of parameters are used, a single behavior can represent a family of trajectories through state space.

An idea closely related to behaviors is histories [Hayes 1985]. Histories can be viewed as local behaviors, that is, how a single individual or property varies through time. A behavior is equivalent to a global history, that is, the union of all of the histories for the participating individuals. The distinction is important for two reasons. First, histories are the dual of situations in the situation calculus; histories are bounded spatially and extended temporally, whereas situations are bounded temporally and global spatially. Using histories avoids the frame problem, instead trading it for the more tractable problems of generating histories locally and determining how they interact when they intersect in space and time. The second reason is that history-based simulation algorithms can be more efficient than state-based simulation algorithms, since no commitments need to be made concerning irrelevant information.

In a correct envisionment, every possible behavior of the physical system corresponds to some path through the envisionment. Since envisionments reflect only local constraints, the converse is not true; that is, an arbitrary path through an envisionment may not represent a physically possible behavior. All such paths must be tested against global constraints, such as energy conservation, to ensure their physical validity. Since the typical uses of an envisionment are to test whether an observed behavior is plausible or to propose possible behaviors, this limitation is not serious. A more serious limitation is that envisionments are often exponential in the size of the system being modeled. This means that, in practice, envisionments often are not generated explicitly, and instead possible behaviors are searched in ways similar to those used in other areas of AI.

Many tasks require integrating qualitative states with other models of time, such as numerical models. Including precise information (e.g., algebraic expressions or floating-point numbers) about the endpoints of intervals in a history does not change their essential character.

Space and Shape

Qualitative representations of space and shape involve quantization, just as qualitative representations of continuous one-dimensional parameters do. However, problem-independent purely qualitative spatial representations suffice for fewer tasks than in the one-dimensional case, because of the increased ambiguity in higher dimensions [Forbus et al. 1991]. Consider, for example, deciding whether a protrusion can fit snugly inside a hole. If we have detailed information about their shapes we can derive an answer. If we consider a particular set of protrusions and a particular set of holes, we can construct a qualitative representation of these particular protrusions and holes that would allow us to derive whether or not a specific pair would fit, based on their relative sizes. But if we first compute a qualitative representation for each protrusion and hole in isolation, in general the rules of inference that can be derived for this problem will be very weak. Work in qualitative spatial representations thus tends to take two approaches. The first approach is to explore what aspects do lend themselves to qualitative representations. The second approach is to use a quantitative representation as a starting point and compute problem-specific qualitative representations to reason with. We summarize each in turn.

There are several purely qualitative representations of space and shape that have proven useful. Topological relationships between regions in two-dimensional space have been formalized, with transitivity inferences similar to those used in temporal reasoning identified for various vocabularies of relations [Cohn and Randall 1992]. The beginnings of a rich qualitative mechanics have been developed. This includes qualitative representations for vectors using the sign of the vector's quadrant to reason about possible directions of motion [Nielsen 1988] and using relative inclination of angles to reason about linkages [Kim 1992].

The use of quantitative representations to ground qualitative spatial reasoning can be viewed as a model of the ways humans use diagrams and models in spatial reasoning. "For this reason such work is also known as **diagrammatic reasoning**." One form of diagram representation is the occupancy array that encodes the location of an object by cells in a (two- or three-dimensional) grid (cf. Funt [1980]). These representations simplify the calculation of spatial relationships between objects (e.g., whether or not one object is above another), albeit at the cost of making the object's shape implicit. Another form of diagram representation uses symbolic structures with quantitative, e.g., numerical, algebraic, or interval (cf. Forbus et al. [1991]). These representations simplify calculations involving shape and spatial relationships, without the scaling and resolution problems that sometimes arise in array representations. However, they require a set of primitive shape elements that spans all of the possible shapes of interest, and identifying such sets for particular tasks can be difficult. For instance, many intuitively natural sets of shape primitives are not closed with respect to their complement, which can make characterizing free space difficult.

Diagram representations are used for qualitative spatial reasoning in two ways. The first is as a decision procedure for spatial questions. This mimics one of the roles diagrams play in human perception. Often these operations are combined with domain-specific reasoning procedures to produce an analog style of inference, where for instance the effects of perturbations on a structure are mapped into the diagram, the effect on the shapes in the diagram noted, and the results mapped back into a physical interpretation. The second way uses the diagram to construct a problem-specific qualitative vocabulary, imposing new spatial entities representing physical properties, such as the maximum height a ball can reach or regions of free space that can contain a motion. This is the **metric diagram/place vocabulary** model of qualitative spatial reasoning.

The best developed area in qualitative spatial representation is the representation of kinematic mechanisms. The possible motions of objects are represented by qualitative regions in configuration space representing the legitimate positions of parts of mechanisms [Faltings 1990]. Whereas in principle a single high-dimensional configuration space could be used to represent a mechanism's possible motions (each dimension corresponding to a degree of freedom of a part of the mechanism), in practice a collection of configuration spaces, one two-dimensional space for each pair of parts that can interact is used. These techniques suffice to analyze a wide variety of kinematic mechanisms [Joscowicz and Sacks 1993].

Another important class of spatial representations concerns qualitative representations of spatially distributed phenomena, such as flow structures and regions in phase space. These models use techniques from computer vision to recognize or impose qualitative structure on a continuous field of information. This qualitative structure, combined with domain-specific models of how such structures tie to the underlying physics, enables them to interpret physical phenomena in much the same way that a scientist examining the data would (cf. Nishida [1994], Yip [1991], and Zhao [1994]).

Compositional Modeling, Domain Theories, and Modeling Assumptions

There is almost never a single correct model for a complex physical system. Most systems can be modeled in a variety of ways, and different tasks can require different types of models. The creation of a system model for a specific purpose is still something of an art. Qualitative physics has developed formalisms that combine logic and mathematics with qualitative representations to help automate the process of creating and refining models. The compositional modeling methodology [Falkenhainer and Forbus 1991], which has become standard in qualitative physics, works like this: Models are created from domain theories, which describe the kinds of entities and phenomena that can occur in a physical domain. A domain theory consists of a set of **model fragments**, each describing a particular aspect of the domain. Creating a model is accomplished by instantiating an appropriate subset of model fragments, given some initial specification of the system (e.g., the propositional equivalent of a blueprint) and information about the task to be performed. Reasoning about appropriateness involves the use of modeling assumptions. Modeling assumptions are the control knowledge used to reason about the validity or appropriateness of using model fragments. Modeling assumptions are used to express the relevance of model fragments. Logical constraints between modeling assumptions comprise an important component of a domain theory.

An example of a modeling assumption is assuming that a turbine is isentropic. Here is a model fragment that illustrates how this assumption is used:

```
(defEquation Isentropic-Turbine
    ((turbine ?g ?in ?out)(isentropic ?g))
    (:= (spec-s ?in) (spec-s ?out)))
```

in other words, when a turbine is isentropic, the specific entropy of its inlet and outlet are equal. Other knowledge in the domain theory puts constraints on the predicate isentropic,

```
(for-all (?self (turbine ?self))
    (iff (= (nu-isentropic ?self) 1.0)
    (isentropic ?self)))
```

that is, a turbine is isentropic exactly when its isentropic thermal efficiency is 1. Even though no real turbine is isentropic, assuming that turbines are isentropic simplifies early analyses when creating a new design. In later design phases, when tighter performance bounds are required, this assumption is retracted and the impact of particular values for the turbine's isentropic thermal efficiency are explored. The consequences of choosing particular modeling assumptions can be quite complex; the fragments shown here are less than one-fourth of the knowledge expressing the consequences of assuming that a turbine is isentropic in a typical knowledge base.

Modeling assumptions can be classified in a variety of ways. An ontological assumption describes which onotology should be used in an analysis. For instance, reasoning about the pressure at the bottom of a swimming pool is most simply performed using a contained stuff representation, whereas describing the location of an oil spill is most easily performed using a piece of stuff representation. A perspective assumption describes which subset of phenomena operating in a system will be the subject. For example, in analyzing a steam plant one might focus on a fluid perspective, a thermal perspective, or both at once. A grain assumption describes how much detail is included in an analysis. Ignoring the implementation details of subsystems, for instance, is useful in the conceptual design of an artifact, but the same implementation details may be critical for troubleshooting that artifact. The relationships between these classes of assumptions can be complicated and domain dependent; for instance, it makes no sense to include a model of a heating coil (a choice of granularity) if the analysis does not include thermal properties (a choice of perspective).

Relationships between modeling assumptions provide global structure to domain theories. Assumptions about the nature of this global structure can significantly impact the efficiency of model formulation, as discussed subsequently. In principle, any logical constraint could be imposed between modeling assumptions. In practice, two kinds of constraints are the most common. The first are implications, such as one modeling assumption requiring or forbidding another. For example,

```
(for-all (?s (system ?s))
    (implies (consider (black-box ?s))
            (for-all (?p (part-of ?p ?s)) (not (consider ?p))))))
```

says that if one is considering a subsystem as a black box, then all of its parts should be ignored. Similarly,

```
(for-all (?l (physical-object ?l))
    (implies (consider (pressure ?l))
            (consider (fluid-properties ?l))))
```

states that if an analysis requires considering something's pressure, then its fluid properties are relevant.

The second kind of constraint between modeling assumptions is assumption classes. An assumption class expresses a choice required to create a coherent model under particular conditions. For example,

```
(defAssumptionClass (turbine ?self)
    (isentropic ?self)
    (not (isentropic ?self)))
```

states that when something is modeled as a turbine, any coherent model including it must make a choice about whether or not it is modeled as isentropic. The choice may be constrained by the data so far (e.g., different entrance and exit specific entropies), or it may be an assumption that must be made in order to complete the model. The set of choices need not be binary. For each valid assumption class there must be exactly one of the choices it presents included in the model.

32.3 Qualitative Reasoning Techniques

A wide variety of qualitative reasoning techniques have been developed which use the qualitative representations just outlined.

Model Formulation

Methods for automatically creating models for a specific task are one of the hallmark contributions of qualitative physics. These methods formalize knowledge and skills typically left implicit by most of traditional mathematics and engineering. To be sure, many models used in qualitative reasoning are still entirely handcrafted, using system-specific laws and implicit task-specific simplifications. However, the state of the art in model formulation algorithms is advancing rapidly enough that this practice should soon become quite rare.

The simplest model formulation algorithm is to instantiate every possible model fragment from a domain theory, given a propositional representation of the particular scenario to be reasoned about. This algorithm is adequate when the domain theory is very focused and thus does not contain much irrelevant information. It is inadequate for broad domain theories and fails completely for domain theories that include alternative and mutually incompatible perspectives (e.g., viewing a contained liquid as a finite object vs. an infinite source of liquid). It also fails to take task constraints into account. For example, it is possible in principle to analyze the cooling of a cup of coffee using quantum mechanics. Even if it were possible in practice to do so, for most tasks simpler models suffice. Just how simple a model can be and remain adequate depends on the task. If I want to know if the cup of coffee will still be drinkable after an hour, a qualitative model suffices to infer that its final temperature will be that of its surroundings. If I want to know its temperature within 5% after 12 min have passed, a macroscopic quantitative model is a better choice. In other words, the goal of model formulation is to create the simplest adequate model of a system for a given task.

More sophisticated model formulation algorithms search the space of modeling assumptions, since they control which aspects of the domain theory will be instantiated. The model formulation algorithm of Falkenhainer and Forbus [1991] instantiated all potentially relevant model fragments and used an assumption-based truth maintenance system to find all legal combinations of modeling assumptions that sufficed to form a model that could answer a given query. The simplicity criterion used was to minimize the number of modeling assumptions. This algorithm is very simple and general but has two major drawbacks: (1) full instantiation can be very expensive, especially if only a small subset of the model fragments are eventually used and (2) the number of consistent combinations of model fragments tends to be exponential for most problems. The rest of this section describes algorithms that overcome these problems.

Efficiency in model formulation can be gained by imposing additional structure on domain theories. Under at least one set of constraints, model formulation can be carried out in polynomial time [Nayak 1994]. The constraints are (1) the domain theory can be divided into independent assumption classes and (2) within each assumption class, the models can be organized by a (perhaps partial) simplicity ordering of a specific nature, forming a lattice of *causal approximations*. Nayak's algorithm computes a simplest model, in the sense of simplest within each local assumption class, but does not necessarily produce the globally simplest model.

Conditions that ensure the creation of *coherent* models, that is, models which include sufficient information to produce an answer of the desired form, provide powerful constraints on model formulation. For example, in generating "what if" explanations of how a change in one parameter might affect particular other properties of the system, a model must include a complete causal chain connecting the changed parameter to the other parameters of interest. This insight can be used to treat model formulation as a best-first search for a set of model fragments providing the simplest complete causal chain [Rickel and Porter 1994]. A novel feature of this algorithm is that it also selects models at an appropriate time scale. It does this by choosing the slowest time-scale phenomenon that provides a complete causal model, since this provides accurate answers that minimize extraneous detail.

As with other AI problems, knowledge can reduce search. One kind of knowledge that experienced modelers accumulate concerns the range of applicability of various modeling assumptions and strategies for how to reformulate when a given model proves inappropriate. Model formulation often is an iterative process. For instance, an initial qualitative model often is generated to identify the relevant phenomena, followed by the creation of a narrowly focused quantitative model to answer the questions at hand. Similarly, domain-specific error criterion can determine that a particular model's results are internally inconsistent, causing the reasoner to restart the search for a good model. Formalizing the decision making needed in iterative model formulation is an area of active research.

Causal Reasoning

Causal reasoning explains an aspect of a situation in terms of others in such a way that the aspect being explained can be changed if so desired. For instance, a flat tire is caused by the air inside flowing out, either through the stem or through a leak. To refill the tire, we must both ensure that the stem provides a seal and that there are no leaks. Causal reasoning is thus at the heart of diagnostic reasoning as well as explanation generation.

The techniques used for causal reasoning depend on the particular notion of causality used, but they all share a common structure. First, causality involving factors within a state are identified. Second, how the properties of a state contribute to a transition (or transitions) to another state are identified, to extend the causal account over time. Since causal reasoning often involves qualitative simulation, we turn to simulation next.

Simulation

The new representations of quantity and mathematical relationships of qualitative physics expand the space of simulation techniques considerably. We start by considering varieties of purely qualitative simulation, and then describe several simulation techniques that integrate qualitative and quantitative information.

Understanding *limit analysis*, the process of finding state transitions, is key to understanding qualitative simulation. Recall that a qualitative state consists of a set of propositions, some of them describing the values of continuous properties in the system. (For simplicity, in this discussion we will assume that these values are described as ordinal relations, although the same method works for sign representations and representations richer than ordinals.) Two observations are critical: (1) the phenomena which cause changes in a situation often depend on ordinal relationships between parameters of the situation and (2) knowing just the sign of the derivatives of the parameters involved in these ordinal relationships suffices to predict how they might change over time. The effects of these changes, when calculated consistently, describe the possible transitions to other states.

An example will make this clearer. Consider again a pot of water sitting on a stove. Once the stove is turned on, heat begins to flow to the water in the pot because the stove's temperature is higher than that of the water. The causal relationship between the temperature inequality and the flow of heat means that to predict changes in the situation, we should figure out their derivatives and any other relevant ordinal relationships that might change as a result. In this qualitative state, the derivative of the water's temperature is positive, and the derivative of the stove's temperature is constant. Thus, one possible state change is that the water will reach thermal equilibrium with the stove and the flow of heat will stop. That is not the only possibility, of course. We know that boiling can occur if the temperature of the water begins to rise above its boiling temperature. That, too, is a possible transition that would end the state. Which of these transitions occurs depends on the relationship between the temperature of the stove and the boiling temperature of water.

This example illustrates several important features of limit analysis. First, surprisingly weak information (i.e., ordinal relations) suffice to draw important conclusions about broad patterns of physical behavior. Second, limit analysis with purely qualitative information is fundamentally ambiguous: It can identify what transitions might occur but cannot by itself determine in all cases which transition will occur. Third, like other qualitative ambiguities, higher resolution information can be brought in to resolve the ambiguities as needed. Returning to our example, any information sufficient to determine the ordinal relationship between the stove temperature and boiling suffices to resolve this ambiguity. If we are designing an electric kettle, for instance, we would use this ambiguity as a signal that we must ensure that the heating element's temperature is well above the boiling point, and if we are designing a drink warmer, its heating element should operate well below the boiling point.

Qualitative simulation algorithms vary along four dimensions: (1) their initial states, (2) what conditions they use to filter states or transitions, (3) whether or not they generate new landmarks, and (4) how much of the space of possible behaviors they explore. *Envisioning* is the process of generating an envisionment, e.g., generating all possible behaviors. Two kinds of envisioning algorithms have been used in practice: *attainable* envisioners produce all states reachable from a set of initial states, and total envisioners produce a complete envisionment. *Behavior generation* algorithms start with a single initial state, generate landmark values, and use a variety of task-dependent constraints as filters and termination criteria (e.g., resource bounds, energy constraints).

Higher resolution information can be integrated with qualitative simulation in several ways. One method for resolving ambiguities in behavior generation is to provide numerical envelopes to bound mathematical relationships. These envelopes can be dynamically refined to provide tigher situation-specific bounds. Such systems are called **semiquantitative simulators**.

A different approach to integration is to use qualitative reasoning to automatically construct a numerical simulator that has integrated explanation facilities. These *self-explanatory simulators* [Forbus and Falkenhainer 1990] use traditional numerical simulation techniques to generate behaviors, which are also tracked qualitatively. The concurrently evolving qualitative description of the behavior is used both in generating explanations and in ensuring that appropriate mathematical models are used when applicability thresholds are crossed. Self-explanatory simulators can be compiled in polynomial time for efficient execution, even on small computers, or created in an interpreted environment.

Comparative Analysis

Comparative analysis answers a specific kind of "what if" questions, namely, the changes that result from changing the value of a parameter in a situation. Given higher resolution information, traditional analytic or numerical sensitivity analysis methods can be used to answer these questions, but (1) such reasoning is commonly carried out by people who have neither the data nor the expertise to carry out such analyses and (2) purely quantitative techniques tend not to provide good explanations. Sometimes purely qualitative information suffices to carry out such reasoning, using techniques such as *exaggeration* [Weld 1990]. Consider for instance the effect of increasing the mass of a block in a spring-block oscillator. If the mass were infinite the block would not move at all, corresponding to an infinite period. Thus, we can conclude that increasing the mass of the block will increase the period of the oscillator.

One paradox concerning comparative analysis with purely qualitative representations as an explanation for human common sense reasoning is that the set of unambiguous, sound inferences appears to be smaller than the set of common sense conclusions. It may be that human reasoning in this area is often unsound, or relies on experiential, higher resolution information. Recently it has been suggested that comparative analysis may be an important form of inference with diagrams, since the concrete nature of diagrams avoids such ambiguities [Pisan 1995].

Teleological Reasoning

Teleological reasoning connects the structure and behavior of a system to its goals. (By its goals, we are projecting the intent of its designer or the observer, since purposes often are ascribed to components of evolved systems.) To describe how something works entails ascribing a function to each of its parts and to explain how these functions together achieve the goals. Teleological reasoning is accomplished by a combination of abduction and recognition. Abduction is necessary because most components and behaviors can play several functional roles [de Kleer 1984]. A turbine, for instance, can be used to generate work in a power generation system and to expand a gas in a liquefication system. Recognition is important because it explains patterns of function in a system in terms of known, commonly used abstractions. A complex power-generation system with multiple stages of turbines and reheating and regeneration, for instance, still can be viewed as a Rankine cycle after the appropriate aggregation of physical processes involved in its operation [Everett 1995].

Data Interpretation

There are two ways that the representations of qualitative physics have been used in data interpretation problems. The first is to explain a temporal sequence of measurements in terms of a sequence of qualitative states; the second is to create a qualitative model of phase space by interpreting the results of successive numerical simulation experiments. The underlying commonality in these problems is the use of qualitative descriptions of physical constraints to formulate compatibility constraints that prune the set of possible interpretations. We describe each in turn.

In measurement interpretation tasks, numerical and symbolic data are partitioned into intervals, each of which can be explained by a qualitative state or sequence of qualitative states. By using precomputed envisionments or performing limit analysis on line, possible transitions between states used as interpretations can be found for filtering purposes. Specifically, if a state S1 is a possible interpretation for interval I1, then at least one transition from S1 must lead to a state which is an interpretation for the next interval. This compatibility constraint, applied in both directions, can provide substantial pruning. Additional constraints that can be applied include likelihoods of particular states occurring, likelihood of particular transitions occurring, and estimates of durations for particular states. Algorithms have been developed which can use all these constraints to maintain a *best* interpretation of a set of incoming measurements that operate in polynomial time [de Coste 1991].

In phase space interpretation tasks, an analytic model or numerical simulation is used to gather information about the possible behaviors of a system given a set of initial parameters. The geometric patterns these behaviors form in phase space are described using vision techniques to create a qualitative characterization of the behavior. Initially simulations are performed on a coarse grid to create an initial description of phase space. This initial description is then used to guide additional numerical simulation experiments, using rules that express physical properties visually [Yip 1991].

Planning

The ability of qualitative physics to provide predictions with low-resolution information and to determine what manipulations might achieve a desired effect makes it a useful component in planning systems involving the physical world. A tempting approach is to carry out qualitative reasoning entirely in a planner, by *compiling* the domain theory and physics into operators and inference rules. Unfortunately, such straightforward translations tend to have poor combinatorics. A different approach is to treat actions as another kind of state transition in qualitative simulation. This can be effective if qualitative reasoning is interleaved with execution monitoring [Drabble 1993] or used with a mixture of backward and forward reasoning with partial states.

Spatial Reasoning

What distinguishes qualitative spatial reasoning from other forms of spatial reasoning is the extraction and explicit representation of qualitative descriptions of shape and space. Otherwise, the processing techniques are mainly borrowed from research in vision and robotics. Recently this flow has begun to reverse, with vision researchers adopting qualitative representations because they are more robust to compute from the data and are more appropriate for many tasks [Kuipers and Byun 1991].

32.4 Applications of Qualitative Physics

Given that qualitative physics only began as a research enterprise in the late 1980s, it should not be surprising that there are to date few fielded applications. Applications in supervisory process control have been successful enough to be embedded in several commercial systems. Qualitative reasoning techniques were also used in the design of the Mita Corporation's DC-6090 photocopier [Shimomura et al. 1995], which came to market in 1994. Here we briefly summarize some representative application-oriented projects in varying stages of fruition.

Monitoring and Diagnosis

Monitoring and diagnosis, although often treated as distinct problems, are in many applications deeply intertwined. Since these tasks also have deep theoretical commonalities, they are described together here. Monitoring a system requires summarizing its behavior at a level of description that is useful for taking action. Qualitative representations correspond to the natural descriptions applied by system operators and designers and thus can help provide new opportunities for automation. An important benefit of using qualitative representations is that the concepts the software uses can be made very similar to those of people who interact with the software, thus improving the human–computer communication bandwidth. Diagnosis tasks impose similar requirements. It is rarely beneficial to spend the resources required to construct a very detailed quantitative model of the way a particular part has failed when the goal is to isolate a problem. Qualitative models often provide sufficient resolution for fault isolation. Qualitative models also provide the framework for organizing fault detection (i.e., noticing that a problem has occurred) and for working around a problem, even when these tasks require quantitative information.

Operative diagnosis tasks are those where the system being monitored must continue being operated in spite of faults. One example of operative diagnosis is diagnosing engine trouble in civilian commercial aircraft. FaultFinder [Abbott et al. 1987], under development at NASA Langley Research Center, is intended to detect engine trouble and provide easily understood advice to pilots, whose information processing load is already substantial. Faultfinder prototypes compare engine data with a numerical simulation to detect the onset of a problem. A causal model, using low-resolution qualitative information (essentially, working vs. not working) is used to construct failure hypotheses, to be communicated to the pilot in a combination of natural language and graphics.

Many process control tasks involve monitoring. It has been demonstrated that qualitative representations can be used to provide more robust control than statistical process control in curing composite parts [LeClair et al. 1989]. This technique is called *qualitative process automation* (QPA). In the early stage of curing a composite part, the temperature of the furnace needs to be kept relatively low because the part is outgassing. Keeping the furnace low during the entire curing process is inefficient, however, because lower temperatures means longer cure times. Therefore, it is more productive to keep temperature low until outgassing stops and then increase it to finish the cure process more quickly. Statistical process control methods use a combination of analytic models and empirical tests to figure out an optimal pattern of high/low cooking times. QPA incorporates a qualitative description of behavior into the controller, allowing it to detect the change in qualitative regime and control the furnace accordingly. The use of qualitative distinctions in supervisory control provided both faster curing times and higher yield rates than traditional techniques. QPA-inspired supervisory control techniques are now in regular use in curing certain kinds of composite components and have been incorporated into commercial control software.

Many alarm conditions are specified as thresholds, indicating when a system is approaching a dangerous mode of operation or when a component is no longer behaving normally. Alarms are insufficient for fault detection, since they do not reflect the lack of normal behaviors. Experienced operators gain a feel for a system and can sometimes spot potential problems long before they have become serious enough to trigger an alarm. Some of this expertise can be replicated by using a combination of causal models and statistical reasoning over historical data concerning the system in question [Doyle 1995].

In some applications a small set of fault models can be pre-enumerated. A set of models, which includes the nominal model of the system plus models representing common faults, can then be used to track the behavior of a system with a qualitative or semiquantitative simulator. Any fault model whose simulation is inconsistent with the observed behavior can thus be ruled out.

Relying on a pre-existing library of fault models can limit the applicability of automatic monitoring and diagnosis algorithms. One approach to overcoming this limitation is to create algorithms that require only models of normal behavior. Most consistency-based diagnosis algorithms take this approach. The problem with this approach is that the ways a system can fail are still governed by natural laws, which impose more constraint than logical consistency. This extra constraint can be exploited by using a domain theory to generate explanations that could account for the problem, via abduction. These explanations are useful because they make additional predictions which can be tested and which also can be important for reasoning about safety in operative diagnosis (e.g., if a solvent tank's level is dropping because it is leaking, then where is the solvent going?).

Design

Engineering design activities are divided into conceptual design, the initial phase when the overall goals, constraints, and functioning of the artifact are established, and detailed design, when the results of conceptual design are used to synthesize a constructable artifact or system. Most computer-based design tools, such as computer-aided design (CAD) systems and analysis programs, facilitate detailed design. Yet many of the most costly mistakes occur during the conceptual design phase. The ability to reason with partial information makes qualitative reasoning one of the few technologies that provides substantial leverage during the conceptual design phase. Qualitative reasoning can also help automate aspects of detailed design.

The best example is the Mita Corporation's DC-6090 photocopier [Shimomura et al. 1995]. It is an example of a *self-maintenance machine*, in which redundant functionality is identified at design time so that the system can dynamically reconfigure itself to temporarily overcome certain faults. An envisionment including fault models, created at design time, was used as the basis for constructing the copier's control software. In operation, the copier keeps track of which qualitative state it is in, so that it can produce the best quality copy it can.

In some fields experts formulate general design rules and methods expressed in natural language. Qualitative representations can enable these rules, and methods can be further formalized so that they can be used in software. In chemical engineering, for instance, design methods for distillation plants (including Seader and Westerberg and Nath and Motard) have been formalized using qualitative representations, and designs comparable to those in the chemical engineering research literature have been generated automatically.

Automatic analysis and synthesis of kinematic mechanisms have received considerable attention. Complex fixed-axis mechanisms, such as mechanical clocks, can be simulated qualitatively, and a simplified dynamics can be added to produce convincing animations. Initial forays into conceptual design of mechanisms have been made, and qualitative kinematics simulation has been demonstrated to be competitive with conventional approaches in some linkage optimization problems.

Qualitative reasoning also is being used to reason about the effects of failures and operating procedures. Such information can be used in failure modes and effects analysis (FMEA). For example, potential hazards in a chemical plant design can be identified by perturbing a qualitative model of the design with various faults and using qualitative simulation to ascertain the possible indirect consequences of each fault. FMEA software for electrical system design is now being fielded for use in automotive design [Price et al. 1995].

Intelligent Tutoring Systems and Learning Environments

One of the original motivations for the development of qualitative physics was its potential applications in intelligent tutoring systems (ITSs) and intelligent learning environments (ILEs). Qualitative representations provide a formal language for a student's *mental models* [Gentner and Stevens 1983], and thus they facilitate communication between software and student. For example, a sequence of qualitative models can be designed that helps students learn complex domains such as electronics more easily. Student protocols can be analyzed in qualitative terms to diagnose misconceptions.

Qualitative representations are being used in software for teaching plant operators and engineers. They provide a level of explanation for how things work that facilitates teaching control. For example, systems for teaching the operation of power generation plants, including nuclear plants, are under construction in various countries. Teaching software often uses hierarchies of models to help students understand a typical industrial process and design controllers for it. Qualitative representations also can help provide teaching software with the physical intuitions required to help find students' problems. For instance, qualitative representations are used to detect physically impossible designs in an ILE for engineering thermodynamics.

Qualitative representations can be particularly helpful in teaching domains where quantitative knowledge is either nonexistent, inaccurate, or incomplete. For example, efforts underway to create ITSs for ecology in Brazjl, to support conservation efforts, are using qualitative representations to explain how environmental conditions affect plant growth.

Cognitive Modeling

Since qualitative physics was inspired by observations of how people reason about the physical world, one natural application of qualitative physics is cognitive simulation, i.e., the construction of programs whose primary concern is accurately modeling some aspect of human reasoning, as measured by comparison with psychological results. Suprisingly little has been done in this area. Most of the research has been concerned with modeling scientific discovery, e.g., how analogy can be used to create new physical theories and modeling scientific discovery [Falkenhainer 1990]. Given the increasing importance of human–computer interaction, a better understanding of how people reason qualitatively could have substantial economic as well as scientific benefits.

32.5 Research Issues and Summary

Qualitative reasoning is rapidly moving from an area of pure research to a mature subfield with a mixture of basic and applied activities, including fielded applications. The substantial increases in available computing power, combined with the now urgent need to make software that is more articulate, suggests that the importance of qualitative reasoning will continue to grow.

Although there is a substantial research base to draw upon, there are many open problems and areas that require additional research. There are still many unanswered questions about purely qualitative representations (e.g., what is the minimum information that is required to guarantee that all predicted behaviors generated from an envisionment are physically possible?), but the richest vein of research concerns the integration of qualitative knowledge with other kinds of knowledge: numerical, analytic, teleological, etc. The work on modeling to date, although a solid foundation, is still very primitive; better model formulation algorithms, well-tested conventions for structuring domain theories, and robust methods for integrating the results of multiple models are needed. Substantial domain theories for a broad range of scientific and engineering knowledge need to be created. And finally, there are many domains where traditional mathematics has intruded, but where the amount and/or precision of the data available has not enabled it to be very successful. These areas are ripe for qualitative modeling. Examples where such efforts are underway include medicine, organizational theory, economics, and ecology.

Defining Terms

Comparative analysis: A particular form of what if question, i.e., how a physical system changes in response to the perturbation of one of its parameters.

Compositional modeling: A methodology for organizing domain theories so that models for specific systems and tasks can be automatically formulated and reasoned about.

Confluence: An equation involving sign values.

Diagrammatic reasoning: Spatial reasoning, with particular emphasis on how people use diagrams.

Domain theory: A collection of general knowledge about some area of human knowledge, including the kinds of entities involved and the types of relationships that can hold between them, and the mechanisms that cause changes (e.g., physical processes, component laws, etc.). Domain theories range from purely qualitative to purely quantitative to mixtures of both.

Envisionment: A description of all possible qualitative states and transitions between them for a system. *Attainable envisionments* describe all states reachable from a particular initial state; *total envisionments* describe all possible states.

Landmark: A comparison point indicating a specific value achieved during a behavior, e.g., the successive heights reached by a partially elastic bouncing ball.

Limit point: A comparison point indicating a fundamental physical boundary, such as the boiling point of a fluid. Limit points need not be constant over time, e.g., boiling points depend on pressure.

Metric diagram: A quantitative representation of shape and space used for spatial reasoning, the computer analog to or model of the combination of diagram/visual apparatus used in human spatial reasoning.

Model fragment: A piece of general domain knowledge that is combined with others to create models of specific systems for particular tasks.

Modeling assumption: A proposition expressing control knowledge about modeling, such as when a model fragment is relevant.

Physical process: A mechanism that can cause changes in the physical world, such as heat flow, motion, and boiling.

Place vocabulary: A qualitative description of space or shape that is grounded in a quantitative representation.

Qualitative proportionality: A qualitative relationship expressing partial information about a functional dependency between two parameters.

Qualitative simulation: The generation of predicted behaviors for a system based on qualitative information. Qualitative simulations typically include branching behaviors due to the low resolution of the information involved.

Quantity space: A set of ordinal relationships that describes the value of a continuous parameter.

Semiquantitative simulation: A qualitative simulation that uses quantitative information, such as numerical values or analytic bounds, to constrain its results.

References

Abbott, K., Schutte, P., Palmer, M., and Ricks, W. 1987. Faultfinder: a diagnostic expert system with graceful degradation for onboard aircraft application. In *14th Int. Symp. Aircraft Integrated Monitoring Syst.*

Cohn, A. and Randall, Z.1992. An interval logic for space based on 'connection,' pp. 394–398. *Proc. 10th European Conf. Artif. Intell.*

de Coste, D. 1991. Dynamic across-time measurement interpretation. *Artif. Intell.* 51:273–341.

de Kleer, J. 1977. Multiple representations of knowledge in a mechanics problem solver, pp. 299–304. *Proc. IJCAI-77.*

de Kleer, J. 1984. How circuits work. *Artif. Intell.* 24:205–280.

de Kleer, J. and Brown, J. 1984. A qualitative physics based on confluences. *Artif. Intell.* 24:7–83.

Doyle, R. 1995. Determining the loci of anomalies using minimal causal models, pp. 1821–1827. *Proc. IJCAI-95.*

Drabble, B. 1993. Excalibur: a program for planning and reasoning with processes. *Artif. Intell.* 62(1):1–40.

Everett, J. 1995. A theory of mapping from structure to function applied to engineering thermodynamics, pp. 1837–1843. *Proc. IJCAI-95.*

Falkenhainer, B. 1990. A unified approach to explanation and theory formation. In *Computational Models of Scientific Discovery and Theory Formation.* Shrager and Langley, eds. Morgan Kaufmann, San Mateo, CA. Also In Sharlik and Dietterich (eds.), *Readings in Machine Learning.* Morgan Kaufmann, San Mateo, CA.

Falkenhainer, B. and Forbus, K. 1991. Compositional modeling: finding the right model for the job. *Artif. Intell.* 51:95–143.

Faltings, B. 1990. Qualitative kinematics in Mechanisms. *Artif. Intell.*, 44(1):89–119.

Faltings, B. and Struss, P., eds. 1992. *Recent Advances in Qualitative Physics*. MIT Press, Cambridge, MA.

Forbus, K. 1984. Qualitative process theory. *Artif. Intell.* 24:85–168.

Forbus, K. and Falkenhainer, B. 1990. Self explanatory Simulations: an integration of qualitative and quantitative knowledge, pp. 380–387. *Proc. AAAI-90*.

Forbus, K. and Gentner, D. 1986. Learning physical domains: towards a theoretical framework. In *Machine Learning: An Artificial Intelligence Approach*. R. Michalski, J. Carbonell, and T. Mitchell, eds. Vol. 2, pp. 311–348. Morgan-Kaufmann, San Mateo, CA.

Forbus, K., Nielsen, P., and Faltings, B. 1991. Qualitative spatial reasoning: the CLOCK project. *Artif. Intell.* 51:417–471.

Funt, B. 1980. Problem-solving with diagrammatic representations. In *Diagrammatic Reasoning*. J. Glasgow, B. Karan, and N. Narayanan, eds., pp. 33–68. AAAI Press/MIT Press, Cambridge, MA.

Gentner, D. and Stevens, A. eds. 1983. *Mental Models*. Erldaum, Hillsdale, NJ.

Glasgow, J., Karan, B., and Narayanan, N., eds. 1995. *Diagrammatic Reasoning*. AAAI Press/MIT Press, Cambridge, MA.

Hayes, P. 1985. Naive physics 1: ontology for liquids. In *Formal Theories of the Commonsense World* R. Hobbs and R. Moore, eds. Ablex, Norwood, NJ.

Hollan, J., Hutchins, E., and Weitzman, L. 1984. STEAMER: an interactive inspectable simulation-based training system. *AI Mag.* 5(2):15–27.

Iwasaki, Y. and Simon, H. 1986. Theories of causal observing: reply to de Kleer and Brown. *Artif. Intell.* 29(1):63–68.

Iwasaki, Y., Tessler, S., and Law, K. 1995. Qualitative structural analysis through mixed diagrammatic and symbolic reasoning. In *Diagrammatic Reasoning*. J. Glasgow, B. Karan, and N. Narayanan, eds., pp. 711–729. AAAI Press/MIT Press, Cambridge, MA.

Joscowicz, L. and Sacks, E. 1993. Automated modeling and kinematic simulation of mechanisms. *Computer Aided Design* 25(2).

Kim, H. 1992. Qualitative kinematics of linkages. In *Recent Advances in Qualitative Physics*. B. Faltings and P. Struss, eds. MIT Press, Cambridge, MA.

Kuipers, B. 1986. Qualitative simulation. *Artif. Intell.* 29:289–338.

Kuipers, B. 1994. *Qualitative Reasoning: Modeling and Simulation with Incomplete Knowledge*. MIT Press, Cambridge, MA.

Kuipers, B. and Byun, Y. 1991. A robot exploration and mapping strategy based on semantic hierarchy of spatial reasoning. *J. Robotics Autonomous Syst.* 8:47–63.

Le Clair, S., Abrams, F., and Matejka, R. 1989. Qualitative process automation: self directed manufacture of composite materials. *Artif. Intell. Eng. Design Manuf.* 3(2):125–136.

Milne, R. and Trave-Massuyes, L. 1993. Application oriented qualitative reasoning, pp. 145–156. *7th Int. Workshop Qualitative Reasoning About Phys. Syst.*

Nayak, P. 1994. Causal approximations. *Artif. Intell.* 70:277–334.

Nielsen, P. 1988. A qualitative approach to mechanical constraint. *Proc. AAAI-88*.

Nishida, T. 1994. Qualitative reasoning for automated explanation for chaos, pp. 1211–1216. *Proc. AAAI-94*.

Pisan, Y. 1995. A visual routines based model of graph understanding, pp. 692–697. In *Proc. 17th Annu. Conf. Cognitive Sci. Soc.* Lawrence Erlbaum Associates, Hillsdale, NJ.

Price, C., Pugh, D., Wilson, M., and Snooke, N. 1995. The FLAME system: automating electrical failure modes and effects analysis (FMEA), pp. 90–95. *Proc. Annu. Reliability Maintainability Symp.* IEEE.

Rickel, J. and Porter, B. 1994. Automated modeling for answering prediction questions: selecting the time scale and system boundary, pp. 1191–1198. *Proc. AAAI-94*.

Shimomura, Y., Tanigawa, S., Umeda, Y., and Tomiyama, T. 1995. Development of self-maintenance photocopiers, pp. 171–180. *Proc. IAAI-95*.

Struss, P. 1988. Mathematical aspects of qualitative reasoning. *Int. J. Artif. Intell. Eng.* 3(3):156–169.

Weld, D. 1990. *Theories of Comparative Analysis*. MIT Press, Cambridge, MA.

Weld, D. and de Kleer, J., eds. 1990. *Readings in Qualitative Reasoning about Physical Systems.* Morgan Kaufmann, Los Altos, CA.

Williams, B. 1991. A theory of interactions: unifying qualitative and quantitative algebraic reasoning. *Artif. Intell.* 51(1–3):39–94.

Yip, K. 1991. *KAM: A System for Intelligently Guiding Numerical Experimentation by Computer.* Artificial intelligence series. MIT Press, Cambridge, MA.

Zhao, F. 1994. Extracting and representing qualitative behaviors of complex systems in phase space. *Artif. Intell.* 69:51–92.

Further Information

A good introduction to qualitative physics is Weld and de Kleer [1990], which provides access to many of the classic papers in the field. Faltings and Struss [1992] provide a sample of more recent papers. An excellent textbook on the QSIM approach to qualitative physics is Kuipers [1994]. For an introduction to diagrammatic reasoning, see Glasgow et al. [1995]. Milne and Trave-Massuyes [1993] provide an extensive survey of application-oriented qualitative reasoning work. There are a variety of qualitative reasoning resources on the World Wide Web, including extensive bibliographies, papers, and software.

Papers on qualitative reasoning routinely appear in *Artificial Intelligence, Journal of Artificial Intelligence Research (JAIR), AI in Engineering Design and Manufacturing* (AIEDAM), and *IEEE Expert.* Many papers first appear in the proceedings of the American Association for Artificial Intelligence (AAAI), the International Joint Conferences on Artificial Intelligence (IJCAI), and the European Conference on Artificial Intelligence (ECAI). Every year there is an International Qualitative Reasoning Workshop, whose proceedings document the latest developments in the area. Proceedings for a particular workshop are available from its organizers.

33

Robotics

F. L. Lewis
University of Texas at Arlington

M. Fitzgerald
University of Texas at Arlington

K. Liu
University of Texas at Arlington

33.1 Introduction

The word *robot* was introduced by the Czech playright Karel Čapek in his 1920 play *Rossum's Universal Robots*. The word *robota* in Czech means simply work. In spite of such practical beginnings, science fiction writers and early Hollywood movies have given us a romantic notion of robots and expectations that they will revolutionize several walks of life including industry. However, many of the more far-fetched expectations from robots have failed to materialize. For instance, in underwater assembly and

0-8493-2909-4/97/$0.00+$.50
© 1997 by CRC Press, Inc.

oil mining, teleoperated robots are very difficult to manipulate due to sea currents and low visibility, and have largely been replaced or augmented by automated smart quick-fit couplings that simplify the assembly task. However, through good design practices and painstaking attention to detail, engineers have succeeded in applying robotic systems to a wide variety of industrial and manufacturing situations where the environment is *structured* or predictable. Thus, the first successful commercial implementation of process robotics was in the U.S. automobile industry; the word *automation* was coined in the 1940s at Ford Motor Company, a contraction of automatic motivation.

As machines, robots have precise motion capabilities, repeatability, and endurance. On a practical level, robots are distinguished from other electromechanical motion equipment by their dexterous manipulation capability in that robots can work, position, and move tools and other objects with far greater dexterity than other machines found in the factory. The capabilities of robots are extended by using them as a basis for *robotic workcells*. Process robotic workcells are integrated functional systems with grippers, **end effectors**, sensors, and process equipment organized to perform a controlled sequence of jobs to execute a process. Robots must coordinate with other devices in the workcell such as machine tools, conveyors, part feeders, cameras, and so on. Sequencing jobs to correctly perform automated tasks in such circumstances is not a trivial matter, and robotic workcells require sophisticated planning, sequencing, and control systems.

Today, through developments in computers and artificial intelligence (AI) techniques (and often motivated by the space program), we are on the verge of another breakthrough in robotics that will afford some levels of autonomy in *unstructured environments*. For applications requiring increased autonomy it is particularly important to focus on the design of the data structures and command-and-control information flow in the robotic system. Therefore, this chapter focuses on the design of robotic workcell *systems*. A distinguishing feature of robotics is its multidisciplinary nature: to successfully design robotic systems one must have a grasp of electrical, mechanical, industrial, and computer engineering, as well as economics and business practices. The purpose of this chapter is to provide a background in these areas so that design of robotic systems may be approached from a position of rigor, insight, and confidence.

The chapter begins by discussing layouts and architectures for robotic workcell design. Then, components of the workcell are discussed from the bottom up, beginning with robots, sensors, and conveyors/part feeders, and progressing upwards in abstraction through task coordination, job sequencing, and resource dispatching, to task planning, assignment, and decomposition. Concepts of user interface and exception handling/fault recovery are included.

33.2 Robot Workcells

In factory automation and elsewhere it was once common to use layouts such as the one in Fig. 33.1, which shows an assembly line with distinct workstations, each performing a dedicated function. Robots have been used at the workstation level to perform operations such as assembly, drilling, surface finishing, welding, palletizing, and so on. In the assembly line, parts are routed sequentially to the workstations by conveyors. Such systems are very expensive to install, require a cadre of engineering experts to design and program, and are extremely difficult to modify or reprogram as needs change. In today's high-mix low-volume (HMLV) manufacturing scenario, these characteristics tolled the death knell for such rigid antiquated designs.

In the assembly line, the robot is *restricted* by placing it into a rigid sequential system. Robots are versatile machines with many capabilities, and their potential can be significantly increased by using them as a basis for robotic workcells such as the one in Fig. 33.2 [Decelle 1988, Jamshidi et al. 1992, Pugh 1983]. In the robotic workcell, robots are used for part handling, assembly, and other process operations. The workcell is designed to make full use of the workspace of the robots, and components such as milling machines, drilling machines, vibratory part feeders, and so on are placed within the robots' workspaces to allow servicing by the robots. Contrary to the assembly line, the physical layout does not impose a priori a fixed sequencing of the operations or jobs. Thus, as product requirements change, all that is required is to reprogram the workcell in software. The workcell is ideally suited to emerging HMLV conditions in manufacturing and elsewhere.

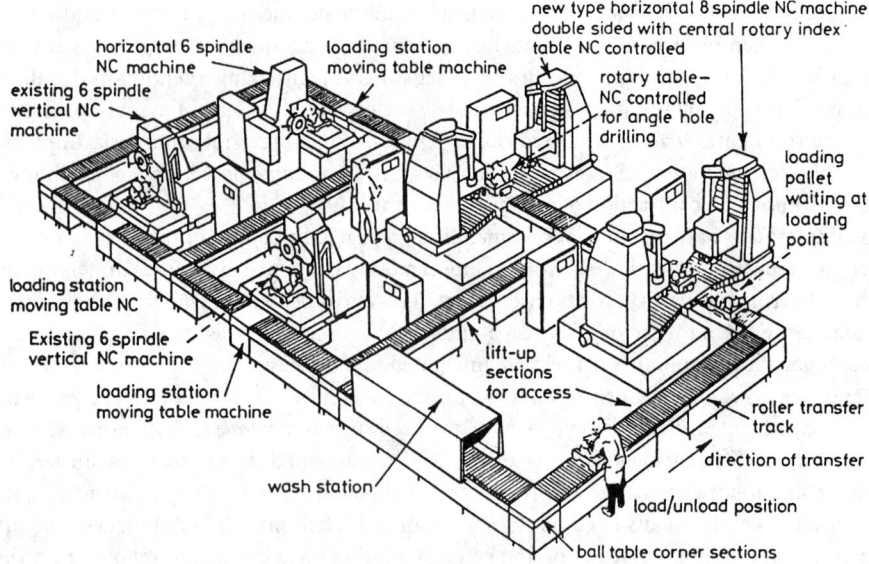

FIGURE 33.1 Antiquated sequential assembly line with dedicated workstations. (*Source:* Courtesy of Edkins, M. 1983. Linking industrial robots and machine tools. In *Robotic Technology*. A. Pugh, ed. Peregrinus, London. With permission.)

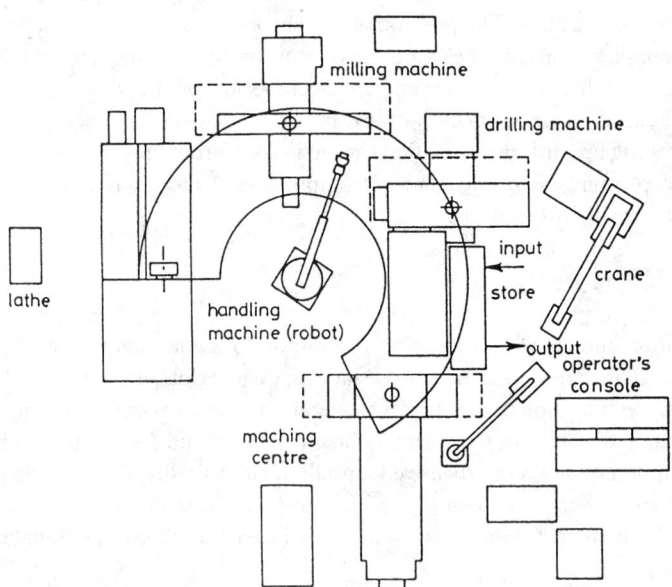

FIGURE 33.2 Robot workcell. (*Source:* Courtesy of Edkins, M. 1983. Linking industrial robots and machine tools. In *Robotic Technology*. A. Pugh, ed. Peregrinus, London. With permission.)

The rising popularity of robotic workcells has taken emphasis away from hardware design and placed new emphasis on innovative *software techniques and architectures* that include planning, coordination, and control (PC&C) functions. Research into individual robotic devices is becoming less useful; what is needed are rigorous design and analysis techniques for integrated multirobotic systems.

33.3 Workcell Command and Information Organization

In this section we define some terms, discuss the design of intelligent control systems, and specify a planning, coordination, and control structure for robotic workcells. The remainder of the chapter is organized around that structure. The various architectures used for modeling AI systems are relevant to this discussion, although here we specialize the discussion to intelligent control architecture.

Intelligent Control Architectures

Many structures have been proposed under the general aegis of the so-called intelligent control (IC) architectures [Antsaklis and Passino 1992]. Despite frequent heated philosophical discussions, it is now becoming clear that most of the architectures have much in common, with apparent major differences due to the fact that different architectures focus on different aspects of intelligent control or different levels of abstraction. A general IC architecture based on work by Saridis is given in Fig. 33.3, which illustrates the principle of decreasing precision with increasing abstraction [Saridis 1996]. In this figure, the organization level performs as a manager that schedules and assigns tasks, performs task decomposition and planning, does path planning, and determines for each task the required job sequencing and assignment of resources. The coordination level performs the prescribed job sequencing, coordinating the workcell agents or resources; in the case of shared resources it must execute dispatching and conflict resolution.

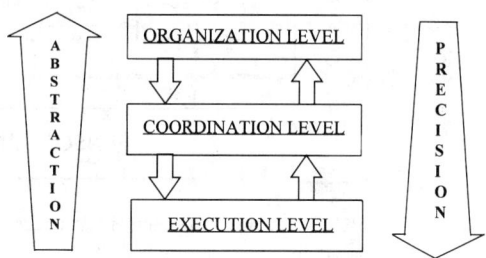

FIGURE 33.3 Three-level intelligent control architecture from work by Saridis.

The *agents* or *resources* of the workcell include robot manipulators, grippers and tools, conveyors and part feeders, sensors (e.g., cameras), mobile robots, and so on. The execution level contains a closed-loop controller for each agent that is responsible for the real-time performance of that resource, including trajectory generation, motion and force feedback servo-level control, and so on. Some permanent built-in motion sequencing may be included (e.g., stop robot motion prior to opening the gripper).

At each level of this hierarchical IC architecture, there may be several systems or nodes. That is, the architecture is not strictly hierarchical. For instance, at the execution level there is a real-time controller for each workcell agent. Several of these may be coordinated by the coordination level to sequence the jobs needed for a given task. At each level, each node is required to sense conditions, make decisions, and give commands or status signals. This is captured in the sense/world-model/execute (SWE) paradigm of Albus [1992], shown in the NASREM configuration in Fig. 33.4; each node has the SWE structure.

Behaviors and Hybrid Systems Design

In any properly designed IC system, the supervisory levels should not destroy the capabilities of the systems supervised. Thus, design should proceed in the manner specified by Brooks [1986], where *behaviors* are built in at lower levels, then selected, activated, or modified by upper-level supervisors. From the point of view of still higher level nodes, the composite performance appears in terms of new more complex or emergent behaviors. Such *subsumption* design proceeds in the manner of adding layers to an onion, as depicted loosely in Fig. 33.5.

Near or slightly below the interfaces between the coordination level and the execution level one must face the transition between two fundamentally distinct worlds. Real-time servo-level controller design and control may be accomplished in terms of *state-space systems*, which are time-varying dynamical systems (either continuous time or discrete time) having continuous-valued states such as temperatures, pressures, motions, velocities, forces, and so on. On the other hand, the coordinator is not concerned about such

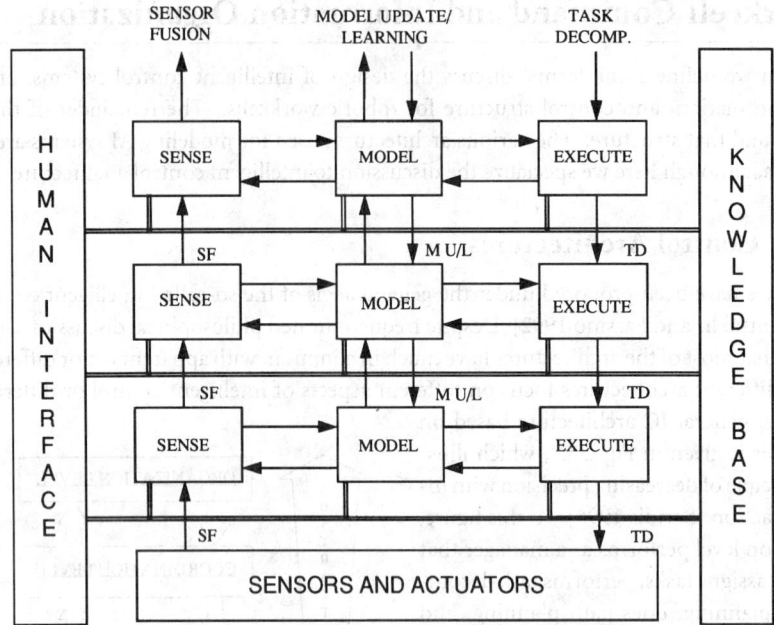

FIGURE 33.4 Three-element structure at all levels of the IC architecture: the NASREM paradigm.

details, but speaks in terms of *discrete events* such as "perform this job" or "check this condition." The theory of *hybrid systems* is concerned with the interface between continuous-state systems and discrete event systems.

These concepts are conveniently illustrated by figures such as Fig. 33.6, where a closed-loop real-time feedback controller for the plant having dynamics $\dot{x} = f(x, u)$ is shown at the execution level. The function of the coordinator is to select the details of this real-time feedback control structure; that is, the outputs $z(t)$, control inputs $u(t)$, prescribed reference trajectories $r(t)$, and controllers K to be switched in at the low level. Selecting the outputs amounts to selecting which sensors to read; selecting the control inputs amounts to selecting to which actuators the command signals computed by the controller should be sent. The controller K is selected from a library of stored predesigned controllers.

A specific combination of (z, u, r, K) defines a *behavior* of the closed-loop system. For instance, in a mobile robot, for path-following behavior one may select: as outputs, the vehicle speed and heading; as controls, the speed and steering inputs; as the controller, an adaptive **proportional-integral-derivative (PID) controller**; and as reference input, the prescribed path. For wall-following behavior, for instance, one

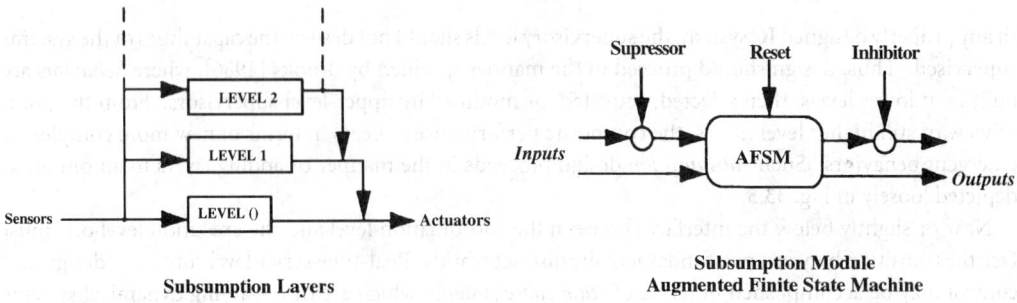

FIGURE 33.5 Behavior-based design after the subsumption technique of Brooks.

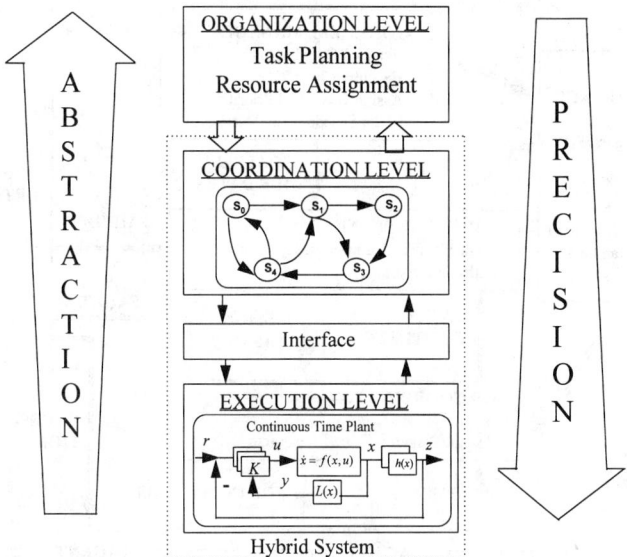

FIGURE 33.6   Hybrid systems approach to defining and sequencing the plant behaviors.

simply selects as output the sonar distance from the wall, as input the steering command, and as reference input the prescribed distance to be maintained. These distinct closed-loop behaviors are sequenced by the coordinator to perform the prescribed job sequence.

Workcell Planning, Coordination, and Control Structure

A convenient planning, coordination, and control structure for robotic workcell design and operation is given in Fig. 33.7, which is modified from the next generation controller (NGC) paradigm. This is an operational PC&C architecture fully consistent with the previous IC structures. In this figure, the term *virtual agent* denotes the agent plus its low-level servocontroller and any required built-in sequencing coordinators. For instance, a *virtual robot* includes the manipulator, its commercial controller with servo-level joint controllers and trajectory generator, and in some applications the gripper controller plus an agent internal coordinator to sequence manipulator and gripper activities. A *virtual camera* might include the camera(s) and framegrabber board, plus software algorithms to perform basic vision processing such as edge detection, segmentation, and so on; thus, the virtual camera could include a *data abstraction*, which is a set of data plus manipulations on that data.

The remainder of the chapter is structured after this PC&C architecture, beginning at the execution level to discuss robot manipulator kinematics, dynamics and control; end effectors and tooling; sensors; and other workcell components such as conveyors and part feeders. Next considered is the coordination level including sequencing control and dispatching of resources. Finally, the organization level is treated including task planning, path planning, workcell management, task assignment, and scheduling.

Three areas are particularly problematic. At each level there may be *human operator interfaces*; this complex topic is discussed in a separate section. An equally complex topic is *error detection and recovery*, also allotted a separate section, which occurs at several levels in the hierarchy. Finally, the strict NGC architecture has a component known as the *information* or *knowledge base*; however, in view of the fact that all nodes in the architecture have the SWE structure shown in Fig. 33.4, it is clear that the knowledge base is distributed throughout the system in the world models of the nodes. Thus, a separate discussion on this component is not included.

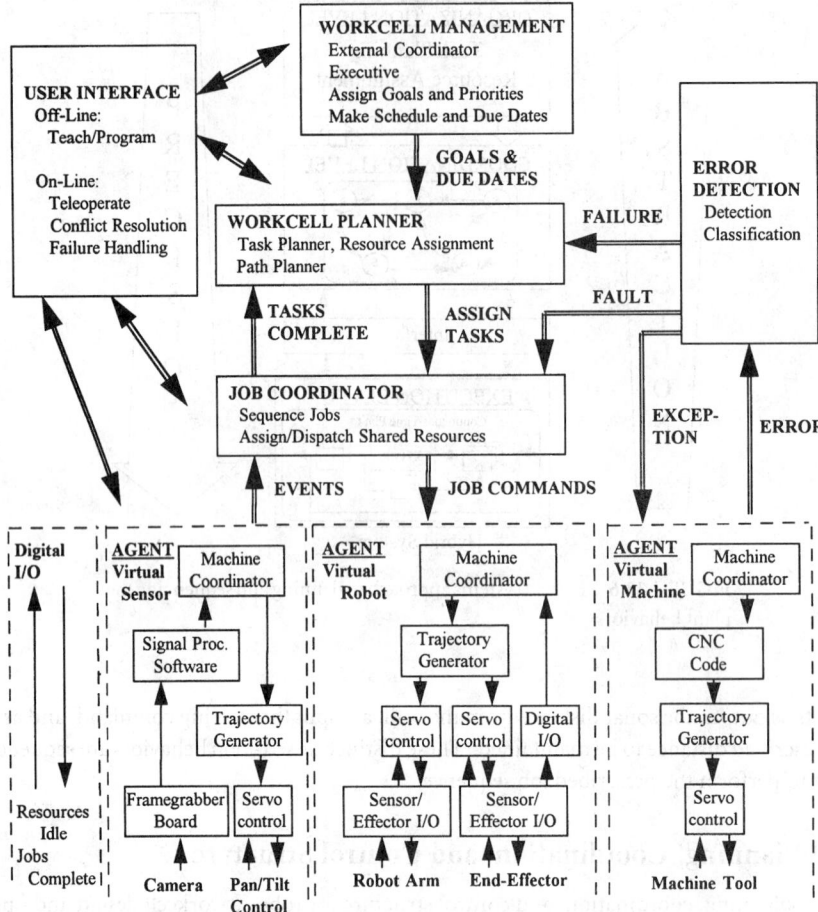

FIGURE 33.7   Robotic workcell planning, coordination, and control operational architecture.

33.4 Commercial Robot Configurations and Types

Robots are highly reliable, dependable, and technologically advanced factory equipment. The majority of the world's robots are supplied by established companies using reliable off-the-shelf component technologies. All commercial industrial robots have two physically separate basic elements, the manipulator arm and the controller. The basic architecture of all commercial robots is fundamentally the same, and consists of digital servocontrolled electrical motor drives on serial-link kinematic machines, usually with no more than six axes (degrees of freedom). All are supplied with a proprietary controller. Virtually all robot applications require significant design and implementation effort by engineers and technicians. What makes each robot unique is how the components are put together to achieve performance that yields a competitive product. The most important considerations in the application of an industrial robot center on two issues: manipulation and integration.

Manipulator Performance

The combined effects of kinematic structure, axis drive mechanism design, and real-time motion control determine the major manipulation performance characteristics: reach and dexterity, payload, quickness, and precision. Caution must be used when making decisions and comparisons based on manufacturers' published performance specifications because the methods for measuring and reporting them are not standardized across the industry. Usually motion testing, simulations, or other analysis techniques are used to verify performance for each application.

Reach is characterized by measuring the extent of the *workspace* described by the robot motion and *dexterity* by the angular displacement of the individual joints. Some robots will have unusable spaces such as dead zones, singular poses, and wrist-wrap poses inside of the boundaries of their reach.

Payload weight is specified by the manufacturers of all industrial robots. Some manufacturers also specify inertial loading for rotational wrist axes. It is common for the payload to be given for extreme velocity and reach conditions. Weight and inertia of all tooling, workpieces, cables and hoses must be included as part of the payload.

Quickness is critical in determining throughput but difficult to determine from published robot specifications. Most manufacturers will specify a maximum speed of either individual joints or for a specific kinematic tool point. However, *average speed* in a working cycle is the quickness characteristic of interest.

Precision is usually characterized by measuring **repeatability**. Virtually all robot manufacturers specify static position repeatability. **Accuracy** is rarely specified, but it is likely to be at least four times larger than repeatability. Dynamic precision, or the repeatability and accuracy in tracking position, velocity, and acceleration over a continuous path, is not usually specified.

Common Kinematic Configurations

All common commercial industrial robots are serial-link manipulators, usually with no more than six kinematically coupled axes of motion. By convention, the axes of motion are numbered in sequence as they are encountered from the base on out to the wrist. The first three axes account for the spatial positioning motion of the robot; their configuration determines the shape of the space through which the robot can be positioned. Any subsequent axes in the kinematic chain generally provide rotational motions to orient the end of the robot arm and are referred to as *wrist axes*. There are two primary types of motion that a **robot axis** can produce in its driven link—either **revolute** or **prismatic**. It is often useful to classify robots according to the orientation and type of their first three axes. There are four very common commercial robot configurations: articulated, type I **selectively compliant assembly robot arm (SCARA)**, type II SCARA, and Cartesian. Two other configurations, cylindrical and spherical, are now much less common.

Articulated Arms.   The variety of commercial articulated arms, most of which have six axes, is very large (Fig. 33.8). All of these robot's axes are revolute. The second and third axes are parallel and work together

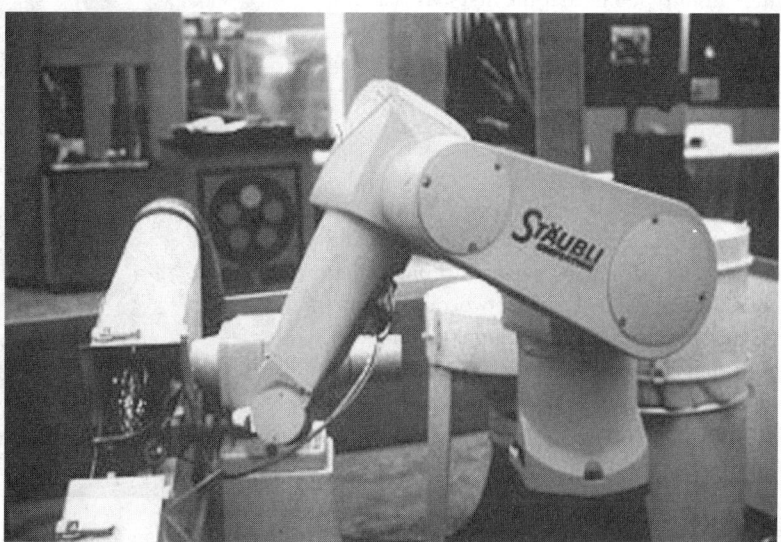

FIGURE 33.8   Articulated arm; six-axis arm grinding from a casting. (*Source:* Courtesy of Staubli Unimation, Inc.).

to produce motion in a vertical plane. The first axis in the base is vertical and revolves the arm to sweep out a large work volume. Many different types of drive mechanisms have been devised to allow wrist and forearm drive motors and gearboxes to be mounted close to the first and second axis of rotation, thus minimizing the extended mass of the arm. The workspace efficiency of well-designed articulated arms, which is the degree of quick dexterous reach with respect to arm size, is unsurpassed by other arm configurations when five or more degrees of freedom are needed. A major limiting factor in articulated arm performance is that the second axis has to work to lift both the subsequent arm structure and the payload. Historically, articulated arms have not been capable of achieving accuracy as well as other arm configurations, as all axes have joint angle position errors which are multiplied by link radius and accumulated for the entire arm.

Type I SCARA. The type I SCARA (selectively compliant assembly robot arm) uses two parallel revolute joints to produce motion in the horizontal plane (Fig. 33.9). The arm structure is weight-bearing but the first and second axes do no lifting. The third axis of the type I SCARA provides work volume by adding a vertical or z axis. A fourth revolute axis will add rotation about the z axis to control orientation in the horizontal plane. This type of robot is rarely found with more than four axes. The type I SCARA is used extensively in the assembly of electronic components and devices, and it is used broadly for the assembly of small- and medium-sized mechanical assemblies.

Type II SCARA. The type II SCARA, also a four-axis configuration, differs from type I in that the first axis is a long vertical prismatic z stroke, which lifts the two parallel revolute axes and their links (Fig. 33.10). For quickly moving heavier loads (over approximately 75 lb) over longer distance (more than about 3 ft), the type II SCARA configuration is more efficient than the type I.

Cartesian Coordinate Robots. Cartesian coordinate robots use orthogonal prismatic axes, usually referred to as x, y, and z, to translate their end effector or payload through their rectangular workspace (Fig. 33.11). One, two, or three revolute wrist axes may be added for orientation. Commercial robot companies supply several types of Cartesian coordinate robots with workspace sizes ranging from a few cubic inches to tens of thousands of cubic feet, and payloads ranging to several hundred pounds. Gantry robots, which have an elevated bridge structure, are the most common Cartesian style and are well suited to material handling applications where large areas and/or large loads must be serviced. They are particularly useful in applications such as arc welding,

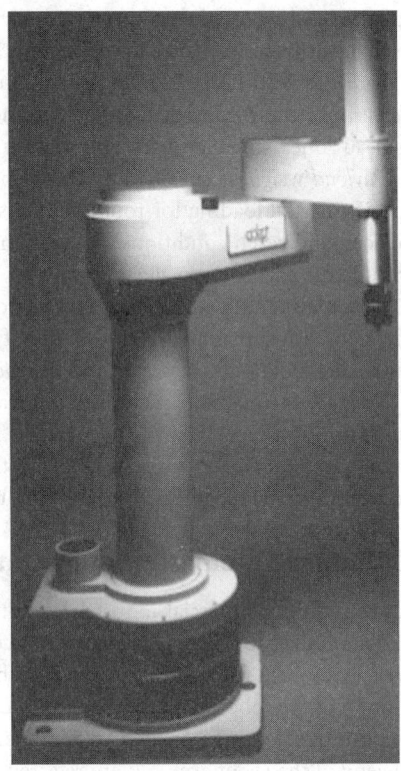

FIGURE 33.9 Type I SCARA arm. High-precision, high-speed midsized SCARA I. (*Source:* Courtesy of Adept Technologies, Inc. With permission.)

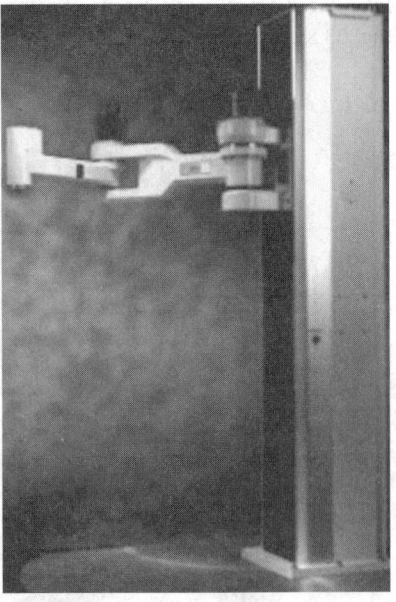

FIGURE 33.10 Type II SCARA. (*Source:* Courtesy of Adept Technologies, Inc. With permission.)

FIGURE 33.11   Cartesian robot. Three-axis robot constructed from modular single-axis motion modules. (*Source:* Courtesy of Adept Technologies, Inc. With permission.)

waterjet cutting, and inspection of large complex precision parts. Modular Cartesian robots are also commonly available from several commercial sources. Each module is a self-contained completely functional single-axis actuator; the modules may be custom assembled for special-purpose applications.

Spherical and Cylindrical Coordinate Robots.   The first two axes of the spherical coordinate robot are revolute and orthogonal to one another, and the third axis provides prismatic radial extension. The result is a natural spherical coordinate system with a spherical work volume. The first axis of cylindrical coordinate robots is a revolute base rotation. The second and third are prismatic, resulting in a natural cylindrical motion. Commercial models of spherical and cylindrical robots (Fig. 33.12) were originally very common and popular in machine tending and material handling applications. Hundreds are still in use but now there are only a few commercially available models. The decline in use of these two configurations is attributed to problems arising from use of the prismatic link for radial extension/retraction motion; a solid boom requires clearance to fully retract.

Drive Types of Commercial Robots

The vast majority of commercial industrial robots use electric servomotor drives with speed reducing transmissions. Both ac and dc motors are popular. Some servohydraulic articulated arm robots are available now for painting applications. It is rare to find robots with servopneumatic drive axes. All types of mechanical transmissions are used, but the tendency is toward low- and zero-backlash type drives. Some robots use direct drive methods to eliminate the amplification of inertia and mechanical backlash associated with other drives. Joint angle position sensors, required for real-time servo-level control, are generally considered an important part of the drive train. Less often, velocity feedback sensors are provided.

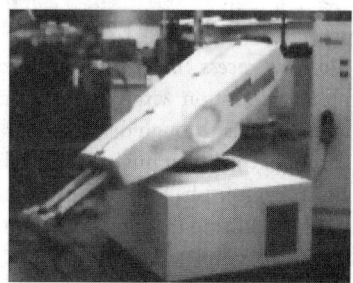

(a)

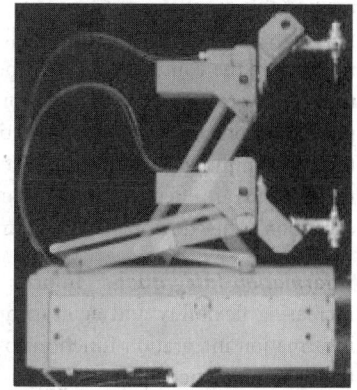

(b)

FIGURE 33.12   Spherical and cylindrical robots. (a) Hydraulic powered spherical robot. (*Source:* Courtesy of Kohol Systems, Inc. With permission.) (b) Cylindrical arm using scissor mechanism for radial prismatic motion. (*Source:* Courtesy of Yamaha Robotics. With permission.)

Commercial Robot Controllers

Commercial robot controllers are specialized multiprocessor computing systems that provide four basic processes allowing integration of the robot into an automation system: motion trajectory generation and following, motion/process integration and sequencing, human user integration, and information integration.

Motion Trajectory Generation and Following.   There are two important controller related aspects of industrial robot motion generation. One is the extent of manipulation that can be programmed, the other is the ability to execute controlled programmed motion. A unique aspect of each robot system is its real-time servo-level motion control. The details of real-time control are typically not revealed to the user due to safety and proprietary information secrecy reasons. Each robot controller, through its operating system programs, converts digital data from higher level coordinators into coordinated arm motion through precise computation and high-speed distribution and communication of the individual axis motion commands, which are executed by individual joint servocontrollers. Most commercial robot controllers operate at a sample period of 16 ms. The real-time motion controller invariably uses classical independent-joint proportional-integral-derivative control or simple modifications of PID. This makes commercially available controllers suitable for point-to-point motion, but most are not suitable for following continuous position/velocity profiles or exerting prescribed forces without considerable programming effort, if at all.

Motion/Process Integration and Sequencing.   Motion/process integration involves coordinating manipulator motion with process sensors or other process controller devices. The most primitive process integration is through discrete digital input/output (I/O). For example, a machine controller external to the robot controller might send a 1-b signal indicating that it is ready to be loaded by the robot. The robot controller must have the ability to read the signal and to perform logical operations (if then, wait until, do until, etc.) using the signal. Coordination with sensors (e.g., vision) is also often provided.

Human Integration.   The controller's human interfaces are critical to the expeditious setup and programming of robot systems. Most robot controllers have two types of human interface available: computer style CRT/keyboard terminals for writing and editing program code off line, and *teach pendants*, which are portable manual input terminals used to command motion in a telerobotic fashion via touch keys or joy sticks. Teach pendants are usually the most efficient means available for positioning the robot, and a memory in the controller makes it possible to play back the taught positions to execute motion trajectories. With practice, human operators can quickly teach a series of points which are chained together in playback mode. Most robot applications currently depend on the integration of human expertise during the programming phase for the successful planning and coordination of robot motion. These interface mechanisms are effective in unobstructed workspaces where no changes occur between programming and execution. They do not allow human interface during execution or adaptation to changing environments.

Information Integration.   Information integration is becoming more important as the trend toward increasing flexibility and agility impacts robotics. Many commercial robot controllers now support information integration functions by employing integrated personal computer (PC) interfaces through the communications ports (e.g., RS-232), or in some through direct connections to the robot controller data bus.

33.5 Robot Kinematics, Dynamics, and Servo-level Control

In this section we shall study the kinematics, dynamics, and servocontrol of robot manipulators; for more details see Lewis et al. [1993]. The objective is to turn the manipulator, by proper design of the control system and trajectory generator, into an *agent with desirable behaviors*, which behaviors can then be selected by the job coordinator to perform specific jobs to achieve some assigned task. This agent, composed of the robot plus servo-level control system and trajectory genarator, is the *virtual robot* in Fig. 33.7; this philosophy goes along with the subsumption approach of Brooks (Fig. 33.5).

Kinematics and Jacobians

Kinematics of Rigid Serial-Link Manipulators

The kinematics of the robot manipulator are concerned only with relative positioning and not with motion effects.

Link A Matrices.   Fixed-base serial-link rigid robot manipulators can be considered as a sequence of joints held together by links. Each joint i has a **joint variable** $\mathbf{q}_i$, which is an angle for revolute joints (units of degrees) and a length for prismatic or extensible joints (units of length). The *joint vector* of an n-link robot is defined as $\mathbf{q} = [\mathbf{q}_1 \, \mathbf{q}_2 \, \cdots \, \mathbf{q}_n]T \in \Re^n$; the joints are traditionally numbered from the base to the end effector, with link 0 being the fixed base. A robot with n joints has n degrees of freedom, so that for complete freedom of positioning and orientation in our 3D space $\Re^3$ one needs a six-link arm.

For analysis purposes, it is considered that to each link is affixed a coordinate frame. The *base frame* is attached to the manipulator base, link 0. The location of the coordinate frame on the link is often selected according to the *Denavit–Hartenberg* (DH) convention [Lewis et al. 1993]. The relation between the links is given by the *A matrix for link i*, which has the form

$$A_i(\mathbf{q}_i) = \begin{bmatrix} R_i & \mathbf{p}_i \\ 0 & 1 \end{bmatrix} \tag{33.1}$$

where $R_i(\mathbf{q}_i)$ is a 3×3 rotation matrix ($R_i^{-1} = R_i^T$) and $\mathbf{p}_i(\mathbf{q}_i) = [x_i \, y_i \, z_i]^T \in \Re^3$ is a translation vector. R_i specifies the rotation of the coordinate frame on link i with respect to the coordinate frame on link $i - 1$; $\mathbf{p}_i$ specifies the translation of the coordinate frame on link i with respect to the coordinate frame on link $i - 1$. The 4×4 *homogeneous transformation A_i* thus specifies completely the orientation and translation of link i with respect to link $i - 1$.

The A matrix $A_i(\mathbf{q}_i)$ is a function of the joint variable, so that as $\mathbf{q}_i$ changes with robot motion, A_i changes correspondingly. A_i is also dependent on the parameters link twist and link length, which are fixed for each link. The A matrices are often given for a specific robot in the manufacturers handbook.

Robot T Matrix.   The position of the end effector is given in terms of the base coordinate frame by the *arm T matrix* defined as the concatenation of A matrices

$$T(\mathbf{q}) = A_1(\mathbf{q}_1)A_2(\mathbf{q}_2) \cdots A_n(\mathbf{q}_n) \equiv \begin{bmatrix} R & \mathbf{p} \\ 0 & 1 \end{bmatrix} \tag{33.2}$$

This 4×4 homogeneous transformation matrix is a function of the joint variable vector $\mathbf{q}$. The 3×3 cumulative rotation matrix is given by $R(\mathbf{q}) = R_1(\mathbf{q}_1)R_2(\mathbf{q}_2) \cdots R_n(\mathbf{q}_n)$.

Joint Space Versus Cartesian Space.   An n-link manipulator has n degrees of freedom, and the position of the end effector is completely fixed once the joints variables $\mathbf{q}_i$ are prescribed. This position may be described either in joint coordinates or in Cartesian coordinates. The joint coordinates position of the end effector is simply given by the value of the n-vector $\mathbf{q}$. The Cartesian position of the end effector is given in terms of the base frame by specifying the orientation and translation of a coordinate frame affixed to the end effector in terms of the base frame; this is exactly the meaning of $T(\mathbf{q})$. That is, $T(\mathbf{q})$ gives the Cartesian position of the end effector.

The Cartesian position of the end effector may be completely specified in our 3D space by a six vector; three coordinates are needed for translation and three for orientation. The representation of Cartesian translation by the arm $T(\mathbf{q})$ matrix is suitable, as it is simply given by $\mathbf{p}(\mathbf{q}) = [x \, y \, z]^T$. Unfortunately, the representation of Cartesian orientation by the arm T matrix is inefficient in that $R(\mathbf{q})$ has nine elements. More efficient representations are given in terms of quaternions or the *tool configuration vector*.

Kinematics and Inverse Kinematics Problems. The robot *kinematics problem* is to determine the Cartesian position of the end effector once the joint variables are given. This is accomplished simply by computing $T(\mathbf{q})$ for a given value of $\mathbf{q}$.

The **inverse kinematics** problem is to determine the required joint angles q_i to position the end effector at a prescribed Cartesian position. This corresponds to solving Eq. (33.2) for $\mathbf{q} \in \mathfrak{R}^n$ given a desired orientation R and translation $\mathbf{p}$ of the end effector. This is not an easy problem, and may have more than one solution (e.g., think of picking up a coffee cup, one may reach with elbow up, elbow down, etc.). There are various efficient techniques for accomplishing this. One should avoid the functions arcsin, arccos, and use where possible the numerically well-conditioned arctan function.

Robot Jacobians

Transformation of Velocity and Acceleration. When the manipulator moves, the joint variable becomes a function of time t. Suppose there is prescribed a generally nonlinear transformation from the joint variable $\mathbf{q}(t) \in \mathfrak{R}^n$ to another variable $y(t) \in \mathfrak{R}^p$ given by

$$y(t) = h(\mathbf{q}(t)) \tag{33.3}$$

An example is provided by the equation $y = T(\mathbf{q})$, where $y(t)$ is the Cartesian position. Taking partial derivatives one obtains

$$\dot{y} = \frac{\partial h}{\partial \mathbf{q}}\dot{\mathbf{q}} \equiv J(\mathbf{q})\dot{\mathbf{q}} \tag{33.4}$$

where $J(\mathbf{q})$ is the *Jacobian* associated with $h(\mathbf{q})$. This equation tells how the joint velocities are transformed to the velocity $\dot{y}$.

If $y = T(\mathbf{q})$ the Cartesian end effector position, then the associated Jacobian $J(\mathbf{q})$ is known as the **manipulator Jacobian**. There are several techniques for efficiently computing this particular Jacobian; there are some complications arising from the fact that the representation of orientation in the homogeneous transformation $T(\mathbf{q})$ is a 3×3 rotation matrix and not a three vector. If the arm has n links, then the Jacobian is a $6 \times n$ matrix; if n is less than 6 (e.g., SCARA arm), then $J(\mathbf{q})$ is not square and there is not full positioning freedom of the end effector in 3D space. The **singularities** of $J(\mathbf{q})$ (where it loses rank), define the limits of the robot workspace; singularities may occur within the workspace for some arms.

Another example of interest is when $y(t)$ is the position in a *camera coordinate frame*. Then $J(\mathbf{q})$ reveals the relationships between manipulator joint velocities (e.g., joint incremental motions) and incremental motions in the camera image. This affords a technique, for instance, for moving the arm to cause desired relative motion of a camera and a workpiece. Note that, according to the velocity transformation (33.4), one has that incremental motions are transformed according to $\Delta y = J(\mathbf{q})\Delta q$.

Differentiating Eq. (33.4) one obtains the *acceleration transformation*

$$\ddot{y} = J\ddot{\mathbf{q}} + \dot{J}\dot{\mathbf{q}} \tag{33.5}$$

Force Transformation. Using the notion of virtual work, it can be shown that forces in terms of $\mathbf{q}$ may be transformed to forces in terms of y using

$$\tau = J^T(\mathbf{q})\mathbf{F} \tag{33.6}$$

where $\tau(t)$ is the force in joint space (given as an n-vector of torques for a revolute robot), and $\mathbf{F}$ is the force vector in y space. If y is the Cartesian position, then $\mathbf{F}$ is a vector of three forces $[\mathbf{f}_x\ \mathbf{f}_y\ \mathbf{f}_z]^T$ and three torques $[\tau_x\ \tau_y\ \tau_z]^T$. When $J(\mathbf{q})$ loses rank, the arm cannot exert forces in all directions that may be specified.

Robot Dynamics and Properties

The robot dynamics considers motion effects due to the control inputs and inertias, Coriolis forces, gravity, disturbances, and other effects. It reveals the relation between the control inputs and the joint variable motion $\mathbf{q}(t)$, which is required for the purpose of servocontrol system design.

Robot Dynamics. The dynamics of a rigid robot arm with joint variable $\mathbf{q}(t) \in \Re^n$ are given by

$$M(\mathbf{q})\ddot{\mathbf{q}} + V_m(\mathbf{q}, \dot{\mathbf{q}})\dot{\mathbf{q}} + \mathbf{F}(\mathbf{q}, \dot{\mathbf{q}}) + \mathbf{G}(\mathbf{q}) + \boldsymbol{\tau}_d = \boldsymbol{\tau} \tag{33.7}$$

where M is an inertia matrix, V_m is a matrix of Coriolis and centripetal terms, $\mathbf{F}$ is a friction vector, $\mathbf{G}$ is a gravity vector, and $\boldsymbol{\tau}_d$ is a vector of disturbances. The n-vector $\boldsymbol{\tau}(t)$ is the control input. The dynamics for a specific robot arm are not usually given in the manufacturers specifications, but may be computed from the kinematics A matrices using principles of Lagrangian mechanics.

The dynamics of any actuators can be included in the robot dynamics. For instance, the electric or hydraulic motors that move the joints can be included, along with any gearing. Then, as long as the gearing and drive shafts are noncompliant, the form of the equation with arm-plus-actuator dynamics has the same form as Eq. (33.7). If the actuators are not included, the control $\boldsymbol{\tau}$ is a torque input vector for the joints. If joint dynamics are included, then $\boldsymbol{\tau}$ might be, for example, a vector of voltage inputs to the joint actuator motors.

The dynamics may be expressed in Cartesian coordinates. The *Cartesian dynamics* have the same form as Eq. (33.7), but appearances there of $\mathbf{q}(t)$ are replaced by the Cartesian position $y(t)$. The matrices are modified, with the manipulator Jacobian $J(\mathbf{q})$ becoming involved. In the Cartesian dynamics, the control input is a six vector of forces, three linear forces and three torques.

Robot Dynamics Properties. Being a Lagrangian system, the robot dynamics satisfy many physical properties that can be used to simplify the design of servo-level controllers. For instance, the inertia matrix $M(\mathbf{q})$ is symmetric positive definite, and bounded above and below by some known bounds. The gravity terms are bounded above by known bounds. The Coriolis/centripetal matrix V_m is linear in $\dot{\mathbf{q}}$, and is bounded above by known bounds. An important property is the **skew-symmetric property** of **rigid-link robot arms**, which says that the matrix $(\dot{M} - 2V_m)$ is always skew symmetric.

This is a statement of the fact that the fictitious forces do no work, and is related in an intimate fashion to the *passivity* properties of Lagrangian systems, which can be used to simplify control system design. Ignoring passivity can lead to unacceptable servocontrol system design and serious degradations in performance, especially in teleoperation systems with transmission delays.

State-Space Formulations and Computer Simulation. Many commercially available controls design software packages, including MATLAB, allow the simulation of state-space systems of the form $\dot{\mathbf{x}} = f(\mathbf{x}, u)$ using, for instance, Runge–Kutta integration. The robot dynamics can be written in state-space form in several different ways. One state-space formulation is the position/velocity form

$$\begin{aligned}
\dot{\mathbf{x}}_1 &= \mathbf{x}_2 \\
\dot{\mathbf{x}}_2 &= -M^{-1}(\mathbf{x}_1)[V_m(\mathbf{x}_1, \mathbf{x}_2)\mathbf{x}_2 + \mathbf{F}(\mathbf{x}_1, \mathbf{x}_2) + \mathbf{G}(\mathbf{x}_1) + \boldsymbol{\tau}_d] + M^{-1}(\mathbf{x}_1)\boldsymbol{\tau}
\end{aligned} \tag{33.8}$$

where the control input is $u = M^{-1}(\mathbf{x}_1)\boldsymbol{\tau}$, and the state is $\mathbf{x} = [\mathbf{x}_1^T \ \mathbf{x}_2^T]^T$, with $\mathbf{x}_1 = \mathbf{q}$, and $\mathbf{x}_2 = \dot{\mathbf{q}}$ both n-vectors. In computation, one should not invert $M(\mathbf{q})$; one should either obtain an analytic expression for M^{-1} or use least-squares techniques to determine $\dot{\mathbf{x}}_2$.

Robot Servo-level Motion Control

The objective in robot servo-level motion control is to cause the manipulator end effector to follow a prescribed trajectory. This can be accomplished as follows for any system having the dynamics Eq. (33.7), including robots, robots with actuators included, and robots with motion described in Cartesian coordinates.

Generally, design is accomplished for robots including actuators, but with motion described in joint space. In this case, first, solve the inverse kinematics problem to convert the desired end effector motion $y_d(t)$ (usually specified in Cartesian coordinates) into a desired joint-space trajectory $\mathbf{q}_d(\mathbf{t}) \in \Re^n$ (discussed subsequently). Then, to achieve tracking motion so that the actual joint variables $\mathbf{q}(t)$ follow the prescribed trajectory $\mathbf{q}_d(t)$, define the *tracking error $e(t)$* and *filtered tracking error $r(t)$* as

$$e(t) = \mathbf{q}_d(t) - \mathbf{q}(t) \tag{33.9}$$

$$r(t) = \dot{e} + \Lambda e(t) \tag{33.10}$$

with Λ a positive definite design parameter matrix; it is common to select Λ diagonal with positive elements.

Computed Torque Control.   One may differentiate Eq. (33.10) to write the robot dynamics Eq. (33.7) in terms of the filtered tracking error as

$$M\dot{r} = -V_m r + f(\mathbf{x}) + \boldsymbol{\tau}_d - \boldsymbol{\tau} \tag{33.11}$$

where the *nonlinear robot function* is given by

$$f(\mathbf{x}) = M(\mathbf{q})(\ddot{\mathbf{q}}_d + \Lambda \dot{e}) + V_m(\mathbf{q}, \dot{\mathbf{q}})(\dot{\mathbf{q}}_d + \Lambda e) + F(\mathbf{q}, \dot{\mathbf{q}}) + G(\mathbf{q}) \tag{33.12}$$

Vector $\mathbf{x}$ contains all of the time signals needed to compute $f(\cdot)$, and may be defined for instance as $\mathbf{x} \equiv [e^T \ \dot{e}^T \ \mathbf{q}_d^T \ \dot{\mathbf{q}}_d^T \ \ddot{\mathbf{q}}_d^T]^T$. It is important to note that $f(\mathbf{x})$ contains all the potentially unknown robot arm parameters including payload masses, friction coefficients, and Coriolis/centripetal terms that may simply be too complicated to compute.

A general sort of servo-level tracking controller is now obtained by selecting the control input as

$$\boldsymbol{\tau} = \hat{f} + K_v r - v(t) \tag{33.13}$$

with $\hat{f}$ an *estimate* of the nonlinear terms $f(\mathbf{x})$, $K_v r = K_v \dot{e} + K_v \Lambda e$ an *outer proportional-plus-derivative (PD) tracking loop*, and $v(t)$ an auxiliary signal to provide robustness in the face of disturbances and modeling errors. The *multiloop control structure* implied by this scheme is shown in Fig. 33.13. The nonlinear inner loop that computes $\hat{f}(x)$ provides *feedforward compensation* terms that improve the

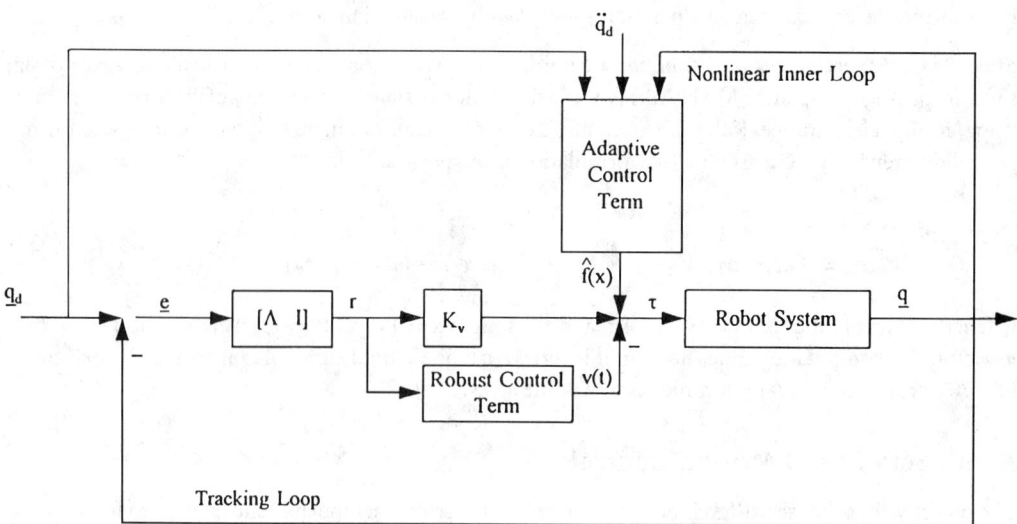

FIGURE 33.13   Robot servo-level tracking controller.

tracking capabilities of the PD outer loop, including an *acceleration feedforward* term $M(\mathbf{q})\ddot{\mathbf{q}}_d$, *friction compensation* $\mathbf{F}(\mathbf{q}, \dot{\mathbf{q}})$, and a *gravity compensation* term $\mathbf{G}(\mathbf{q})$.

This controller is a variant of **computed-torque control**, since the torque required for trajectory following is computed in terms of the tracking error and the additional nonlinear robot terms in $f(\mathbf{x})$. An integrator may be added in the outer tracking loop to ensure zero steady-state error, obtaining a PID outer loop.

Commercial Robot Controllers.   Commercial robot controllers do not implement the entire computed torque law. Most available controllers simply use a PD or PID control loop around each joint, dispensing entirely with the inner nonlinear compensation loop $\hat{f}(\mathbf{x})$. It is not clear exactly what is going on in most commercially available controllers, as they are proprietary and the user has no way to modify the joint tracking loops. However, in some controllers (e.g., Adept Hyperdrive), there appears to be some inner-loop compensation, where some of the terms in $f(\mathbf{x})$ are included in $\tau(t)$. For instance, acceleration feedforward may be included. To implement nonlinear feedback terms that are not already built-in on commercial controllers, it is usually necessary to perform hardware modifications of the controller.

Adaptive and Robust Control.   There are by now many advanced control techniques for robot manipulators that either estimate the nonlinear robot function or compensate otherwise for uncertainties in $f(\mathbf{x})$. In **adaptive control** the estimate $\hat{f}$ of the nonlinear terms is updated online in real-time using additional internal controller dynamics, and in **robust control** the robustifying signal $v(t)$ is selected to overbound the system modeling uncertainties. In **learning control**, the nonlinearity correction term is improved over each repetition of the trajectory using the tracking error over the previous repetition (this is useful in repetitive motion applications including spray painting). Neural networks (NN) or fuzzy logic (FL) systems can be used in the inner control loop to manufacture the nonlinear estimate $\hat{f}(\mathbf{x})$ [Lewis et al. 1995]. Since both NN and FL systems have a universal approximation property, the restrictive **linear in the parameters assumption** required in standard adaptive control techniques is not needed, and no *regression matrix* need be computed. FL systems may also be used in the outer PID tracking loop to provide additional robustness.

Though these advanced techniques significantly improve the tracking performance of robot manipulators, they cannot be implemented on existing commercial robot controllers without hardware modifications.

Robot Force/Torque Servocontrol

In many industrial applications it is desired for the robot to exert a prescribed force normal to a given surface while following a prescribed motion trajectory tangential to the surface. This is the case in surface finishing, etc. A hybrid position/force controller can be designed by extension of the principles just presented.

The robot dynamics with environmental contact can be described by

$$M(\mathbf{q})\ddot{\mathbf{q}} + V_m(\mathbf{q}, \dot{\mathbf{q}})\dot{\mathbf{q}} + \mathbf{F}(\mathbf{q}, \dot{\mathbf{q}}) + \mathbf{G}(\mathbf{q}) + \tau_d = \tau + J^T(\mathbf{q})\lambda \qquad (33.14)$$

where $J(\mathbf{q})$ is a constraint Jacobian matrix associated with the contact surface geometry and λ (the so-called Lagrange multiplier) is a vector of contact forces exerted normal to the surface, described in coordinates relative to the surface.

The hybrid position/**force control** problem is to follow a prescribed motion trajectory $\mathbf{q}_{1_d}(t)$ tangential to the surface while exerting a prescribed contact force $\lambda_d(t)$ normal to the surface. Define the filtered motion error $r_m = \dot{e}_m + \Lambda e_m$, where $e_m = \mathbf{q}_{1_d} - \mathbf{q}_1$ represents the motion error in the plane of the surface and Λ is a positive diagonal design matrix. Define the force error as $\bar{\lambda} = \lambda_d - \lambda$, where $\lambda(t)$ is the normal force measured in a coordinate frame attached to the surface. Then a hybrid position/force controller has the structure

$$\tau = \hat{f} + K_v L(\mathbf{q}_1)r_m + J^T[\lambda_d + K_f\bar{\lambda}] - v \qquad (33.15)$$

where $\hat{f}$ is an estimate of the nonlinear robot function (33.12) and $L(\cdot)$ is an extended Jacobian determined from the surface geometry using the implicit function theorem.

This controller has the basic structure of Fig. 33.13, but with an additional *inner force control loop*. In the hybrid position/force controller, the nonlinear function estimate inner loop $\hat{f}$ and the robustifying term $v(t)$ can be selected using adaptive, robust, learning, neural, or fuzzy techniques. A simplified controller that may work in some applications is obtained by setting $\hat{f} = 0$, $v(t) = 0$, and increasing the PD motion gains $K_v \Lambda$ and K_v and the force gain K_f.

It is generally not possible to implement force control on existing commercial robot controllers without hardware modification and extensive low-level programming.

Motion Trajectory Generation

In the section on servo-level motion control it was shown how to design real-time servo-level control loops for the **robot joint** actuators to cause the manipulator to follow a prescribed joint-space trajectory $q_d(t)$ and, if required by the job, to exert forces normal to a surface specified by a prescribed force trajectory $\lambda_d(t)$. Unfortunately, the higher level **path planner** and job coordinator in Fig. 33.7 do not specify the position and force trajectories in the detail required by the servo-level controllers. Most commercial robot controllers operate at a sampling period of 16 ms, so that they require specific desired motion trajectories $q_d(t)$ sampled every 16 ms. On the other hand, the path planner wishes to be concerned at the level of abstraction only with general path descriptions sufficient to avoid obstacles or accomplish desired high-level jobs (e.g., move to prescribed final position, then insert pin in hole).

Path Transformation and Trajectory Interpolation

Joint Space Versus Cartesian Space Prescribed Trajectories. The job coordinator in Fig. 33.7 passes required path-following commands to the virtual robot in the form of discrete events to be accomplished, which could be in the form of commands to "move to a specified final position passing through prescribed *via points.*" These prescribed path via points are given in Cartesian coordinates y, are usually not regularly spaced in time, and may or may not have required times of transit associated with them. Via points are given in the form $(y_i, \dot{y}_i, t_i)$, with y_i the required Cartesian position at point i and $\dot{y}_i$ the required velocity. The time of transit t_i may or may not be specified. The irregularly spaced Cartesian-space via points must be *interpolated* to produce joint-space trajectory points regularly spaced at every sampling instant, often every 16 ms. It should be clearly understood that the path and the joint trajectory are both prescribed for each coordinate: the path for three Cartesian position coordinates and three Cartesian orientation coordinates, and the trajectory for each of the n manipulator joints. If n is not equal to 6, there could be problems in that the manipulator might not be able to exactly reach the prescribed via points. Thus, in its planning process the path planner must take into account the limitations of the individual robots.

Two procedures may be used to convert prescribed Cartesian path via points into desired joint-space trajectory points specified every 16 ms. One may either: (1) use the arm inverse kinematics to compute the via points in joint-space coordinates and then perform trajectory interpolation in joint space, or (2) perform interpolation on the via points to obtain a Cartesian trajectory specified every 16 ms, and then perform the inverse kinematics transformation to yield the joint-space trajectory $q_d(t)$ for the servo-level controller. The main disadvantage of the latter procedure is that the full inverse kinematics transformation must be performed every 16 ms. The main disadvantage of the former procedure is that interpolation in joint space often has strange effects, such as unexpected motions or curvilinear swings when viewed from the point of Cartesian space; one should recall that the path planner selects via points in Cartesian space, e.g., to avoid obstacles, often assuming linear Cartesian motion between the via points. The latter problem may be mitigated by spacing the Cartesian path via points more closely together. Thus, procedure 1 is usually selected in robotic workcell applications.

Trajectory Interpolation. A trajectory specified in terms of via points, either in joint space or Cartesian space, may be interpolated to obtain connecting points every 16 ms by many techniques, including interpolation by cubic polynomials, second- or third-order splines, minimum-time techniques, etc. The interpolation must be performed separately for each coordinate of the trajectory (e.g., n interpolations if done in joint space). Cubic interpolation is not recommended as it can result in unexpected swings or overshoots in the computed trajectory.

The most popular technique for trajectory interpolation may be linear functions with parabolic blends (LFPB). Let us assume that the path via points are specified in joint space, so that the inverse kinematics transformation from the Cartesian path via points obtained from the path planner has already been performed. Then, the path is specified in terms of the via points $(\mathbf{q}(t_i), \dot{\mathbf{q}}(t_i), t_i)$; note that the time of transit t_i of point i is specified; the transit times need not be uniformly spaced. Within each path segment connecting two via points, one uses constant acceleration or deceleration to obtain the required transit velocity, then zero acceleration during the transit, then constant acceleration or deceleration to obtain the prescribed final position and velocity at the next via point. Sample LFPB trajectories are given in Fig. 33.14. Note that LFPB results in quadratic motion, followed by linear motion, followed by quadratic motion. The maximum acceleration/deceleration is selected taking into account the *joint actuator torque limits*.

There are standard formulas available to compute the LFPB trajectory passing through two prescribed via points, for instance the following. In Fig. 33.14 two design parameters are selected: the *blend time* t_b and the *maximum velocity* v_M. Then the joint-space trajectory passing through via points i and $(i+1)$, shown in Fig. 33.14(c), is given by

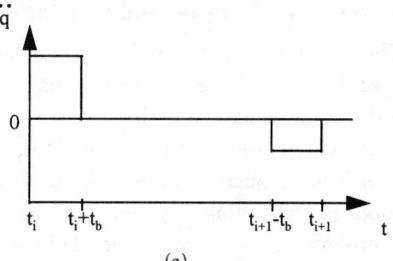

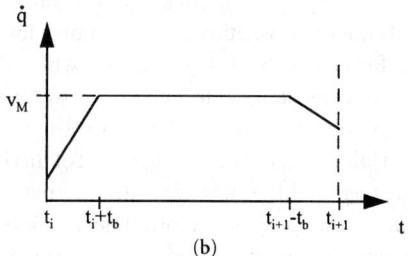

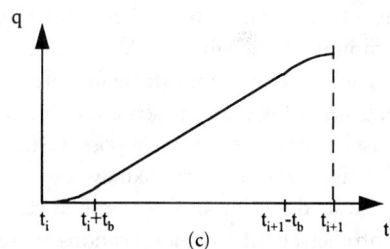

FIGURE 33.14 LFPB trajectory: (a) acceleration profile, (b) velocity profile, and (c) position profile.

$$\mathbf{q}_d(t) = \begin{cases} a + (t - t_i)b + (t - t_i)^2 c, & t_i \le t < t_i + t_b \\ d + v_M t, & t_i + t_b \le t < t_{i+1} - t_b \\ e + (t - t_{i+1})f + (t - t_{i+1})^2 g, & t_{i+1} - t_b \le t < t_{i+1} \end{cases} \quad (33.16)$$

It is not difficult to determine that the coefficients required to pass through the ith and $(i + 1)$st via points are given by

$$a = \mathbf{q}(t_i), \qquad b = \dot{\mathbf{q}}(t_i), \qquad c = \frac{v_M - \dot{\mathbf{q}}(t_i)}{2t_b}$$

$$d = \frac{\mathbf{q}(t_i) + \mathbf{q}(t_{i+1}) - v_M t_{i+1}}{2}$$

$$e = \mathbf{q}(t_{i+1}), \qquad f = \dot{\mathbf{q}}(t_{i+1})$$

$$g = \frac{v_M t_{i+1} + \mathbf{q}(t_i) - \mathbf{q}(t_{i+1}) + 2t_b[\dot{\mathbf{q}}(t_{i+1}) - v_M]}{2t_b^2}$$

$$(33.17)$$

One must realize that this interpolation must be performed for each of the n joints of the robot. Then, the resulting trajectory n-vector is passed as a prescribed trajectory to the servo-level controller, which functions as in Robot Servo-level Motion Control subsection to cause trajectory-following arm motion.

Types of Trajectories and Limitations of Commercial Robot Controllers

The two basic types of trajectories of interest are motion trajectories and force trajectories. Motion specifications can be either in terms of motion from one prescribed point to another, or in terms of following a prescribed position/velocity/acceleration motion profile (e.g., spray painting).

In robotic assembly tasks point-to-point motion is usually used, without prescribing any required transit time. Such motion can be programmed with commercially available controllers using standard robot programming languages (section 33.12). Alternatively, via points can usually be taught using a telerobotic teach pendant operated by the user (section 33.11); the robot memorizes the via points, and effectively plays them back in operational mode. A speed parameter may be set prior to the motion that tells the robot whether to move more slowly or more quickly. Trajectory interpolation is automatically performed by the robot controller, which then executes PD or PID control at the joint servocontrol level to cause the desired motion. This is by far the most common form of robot motion control.

In point-to-point motion control the commercial robot controller performs trajectory interpolation and joint-level PD servocontrol. All of this is transparent to the user. Generally, it is very difficult to modify any stage of this process since the internal controller workings are proprietary, and the controller hardware does not support more exotic trajectory interpolation or servo-level control schemes. Though some robots by now do support following of prescribed position/velocity/acceleration profiles, it is generally extremely difficult to program them to do so, and especially to modify the paths once programmed. Various tricks must be used, such as specifying the Cartesian via points $(y_i, \dot{y}_i, t_i)$ in very fine time increments, and computing t_i such that the desired acceleration is produced.

The situation is even worse for force control, where additional sensors must be added to sense forces (e.g., wrist force-torque sensor, see section 33.7), kinematic computations based on the given surface must be performed to decompose the tangential motion control directions from the normal force control directions, and then very tedious low-level programming must be performed. Changes in the surface or the desired motion or force profiles require time-consuming reprogramming. In most available robot controllers, hardware modifications are required.

33.6 End Effectors and End-of-Arm Tooling

End effectors and end-of-arm tooling are the devices through which the robot manipulator interacts with the world around it, grasping and manipulating parts, inspecting surfaces, and so on [Wright and Cutkosky 1985]. End effectors should not be considered as accessories, but as a major component in any workcell; proper selection and/or design of end effectors can make the difference between success and failure in many process applications, particularly when one includes reliability, efficiency, and economic factors. End effectors consist of the *fingers*, the *gripper*, and the *wrist*. They can be either standard commercially available mechanisms or specially designed tools, or can be complex systems in themselves (e.g., welding tools or dextrous hands). Sensors can be incorporated in the fingers, the gripper mechanism, or the wrist mechanism. All end effectors, end-of-arm tooling, and supply hoses and cables (electrical, pneumatic, etc.) must be taken into account when considering the manipulator payload weight limits of the manufacturer.

Part Fixtures and Robot Tooling

In most applications the end effector design problem should not be decoupled from the part fixturing design problem. One should consider the wrist, gripper, fingers, and part fixturing as a *single system*. Integrated design can often yield innovative solutions to otherwise intractable problems; nonintegrated design can often lead to unforseen problems and unexpected failure modes. Coordinated design of fixtures and end effectors can often avoid the use of high-level expensive sensors (e.g., vision) and/or complex feedback control systems that required overall coordinated control of the robot arm motion, the gripper action, and the part pose. An ideal example of a device that allows simplified control strategies is the **remote-center-of-compliance (RCC)** wrist in Fig. 33.17(b), if correctly used.

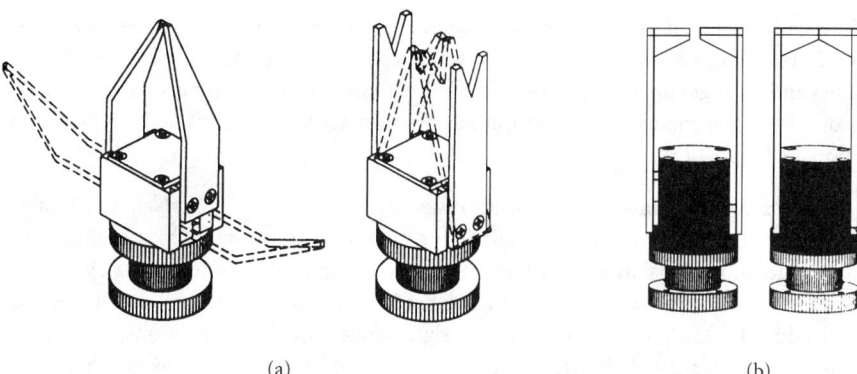

FIGURE 33.15 Angular and parallel motion robot grippers: (a) angular motion gripper and (b) parallel motion gripper, open and closed. (*Source:* Courtesy of Robo-Tech Systems, Gastonia, NC. With permission.)

Grippers and Fingers

Commercial catalogs usually allow one to purchase end effector components separately, including fingers, grippers, and wrists. Grippers can be actuated either pneumatically or using servomotors. Pneumatic actuation is usually either open or closed, corresponding to a binary command to turn the air pressure either off or on. Grippers often lock into place when the fingers are closed to offer failsafe action if air pressure fails. Servomotors often require analog commands and are used when finer gripper control is required. Available gripping forces span a wide range up to several hundred pounds force.

Gripper Mechanisms. *Angular motion grippers*, see Fig. 33.15(a), are inexpensive devices allowing grasping of parts either externally or internally (e.g., fingers insert into a tube and gripper presses them outward). The fingers can often open or close by 90°. These devices are useful for simple pick-and-place operations. In electronic assembly or tasks where precise part location is needed, it is often necessary to use *parallel grippers*, see Fig. 33.15(b), where the finger actuation affords exactly parallel closing motion. Parallel grippers generally have a far smaller range of fingertip motion that angular grippers (e.g., less than 1 in). In some cases, such as electronic assembly of parts positioned by wires, one requires *center seeking grippers*, see Fig. 33.16(a), where the fingers are closed until one finger contacts the part, then that finger stops and the other finger closes until the part is grasped.

There are available many grippers with advanced special-purpose mechanisms, including Robo-Tech's Versagrip III shown in Fig. 33.16(b), a 3-fingered gripper whose fingers can be rotated about a longitudinal

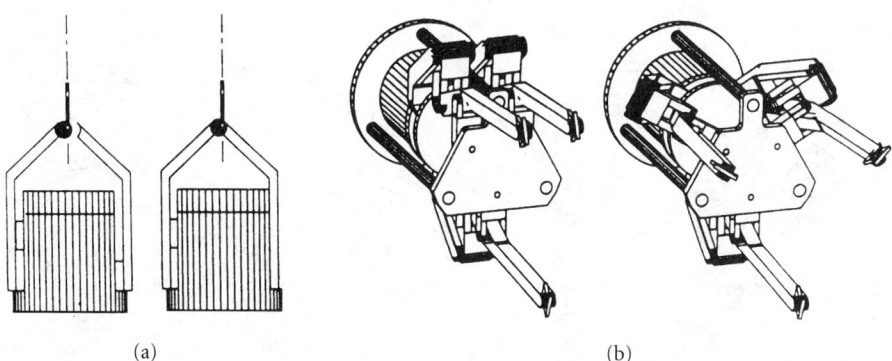

FIGURE 33.16 Robot grippers: (a) center seeking gripper showing part contact by first finger and final closure by second finger and (b) Versagrip III adjustable three-finger gripper. (*Source:* Courtesy of Robo-Tech Systems, Gastonia, NC. With permission.)

axis to offer a wide variety of 3-fingered grasps depending on the application and part geometry. Finger rotation is affected using a fine motion servomotor that can be adjusted as the robot arm moves.

The gripper and/or finger tips can have a wide variety of sensors including binary part presence detectors, binary closure detectors, analog finger position sensors, contact force sensors, temperature sensors, and so on (section 33.7).

The Grasping Problem and Fingers. The study of the *multifinger grasping problem* is a highly technical area using mathematical and mechanical engineering analysis techniques such as rolling/slipping concepts, friction studies, force balance and center of gravity studies, etc. [Pertin-Trocaz 1989] These ideas may be used to determine the required gripper mechanisms, number of fingers, and finger shapes for a specific application. Fingers are usually specially designed for particular applications, and may be custom ordered from end-effector supply houses. Improper design and selection of fingers can doom to failure an application of an expensive robotic system. By contrast, innovative finger and contact tip designs can solve difficult manipulation and grasping problems and greatly increase automation reliability, efficiency, and economic return.

Fingers should not be thought of as being restricted to anthropomorphic forms. They can have vacuum contact tips for grasping smooth fragile surfaces (e.g., auto windshields), electromagnetic tips for handling small ferrous parts, compliant bladders or wraparound air bladders for odd-shaped or slippery parts, Bernoulli effect suction for thin fragile silicon wafers, or membranes covering a powder to distribute contact forces for irregular soft fragile parts [Wright and Cutkosky 1985].

Multipurpose grippers are advantageous in that a single end effector can perform multiple tasks. Some multipurpose devices are commercially available; they are generally expensive. The ideal multipurpose end effector is the *anthropomorphic dextrous hand*. Several dextrous robot hands are now available and afford potential applications in processes requiring active manipulation of parts or handling of many sorts of tooling. Currently, they are generally restricted to research laboratories since the problems associated with their expense, control, and coordination are not yet completely and reliably solved.

Robot Wrist Mechanisms

Wrist mechanisms couple the gripper to the robot arm, and can perform many functions. Commercial *adapter plates* allow wrists to be mounted to any commercially available robot arm. As an alternative to expensive multipurpose grippers, *quick change wrists* allow end effectors to be changed quickly during an application, and include quick disconnect couplings for mechanical, electrical, pneumatic and other connections. Using a quick change wrist, required tools can be selected from a magazine of available tools/end effectors located at the workcell. If fewer tools are needed, an alternative is provided by inexpensive *pivot gripper wrists*, such as the 2-gripper-pivot device shown in Fig. 33.17(a), which allows one of two grippers

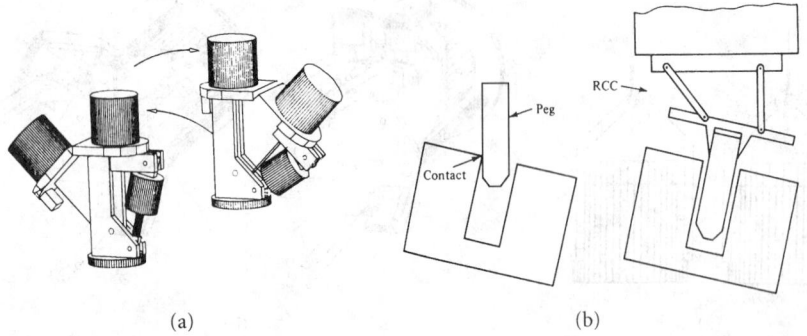

(a)   (b)

FIGURE 33.17   Robot wrists. (a) Pivot gripper wrist. (*Source:* Courtesy of Robo-Tech Systems, Gastonia, NC. With permission.) (b) Remote-center-of-compliance (RCC) wrist. (*Source:* Courtesy of Lord Corporation, Erie, PA. With permission.)

to be rotated into play. With this device, one gripper can unload a machine while the second gripper subsequently loads a new blank into the machine. Other *rotary gripper wrists* allow one of several (up to six or more) grippers to be rotated into play. With these wrists, the grippers are mounted in parallel and rotate much like the chamber of an old-fashioned western Colt 45 revolver; they are suitable if the grippers will not physically interfere with each other in such a parallel configuration.

Safety wrists automatically deflect, sending a fault signal to the machine or job coordinator, if the end-of-arm tooling collides with a rigid obstacle. They may be reset automatically when the obstacle is removed.

Part positioning errors frequently occur due to robot end effector positioning errors, part variations, machine location errors, or manipulator repeatability errors. It is unreasonable and expensive to require the robot joint controller to compensate exactly for such errors. *Compliant wrists* offset positioning errors to a large extent by allowing small passive part motions in response to forces or torques exerted on the part. An example is pin insertion, where small positioning errors can result in pin breakage or other failures, and compensation by gross robot arm motions requires sophisticed (e.g., expensive) force-torque sensors and advanced (e.g., expensive) closed-loop feedback force control techniques. The compliant wrist allows the pin to effectively adjust its own position in response to sidewall forces so that it slides into the hole. A particularly effective device is the remote-center-of-compliance (RCC) wrist, Fig. 33.17(b), where the rotation point of the wrist can be adjusted to correspond, e.g., to the part contact point [Groover et al. 1986]. Compliant wrists allow successful assembly where vision or other expensive sensors would otherwise be needed.

The wrist can contain a wide variety of sensors, with possibly the most important class being the *wrist force-torque sensors* (section 33.7), which are quite expensive. A general rule-of-thumb is that, for economic and control complexity reasons, robotic force/torque sensing and control should be performed at the lowest possible level; e.g., fingertip sensors can often provide sufficient force information for most applications, with an RCC wrist compensating for position inaccuracies between the fingers and the parts.

Robot/Tooling Process Integration and Coordination

Many processes require the design of sophisticated end-of-arm tooling. Examples include spray painting guns, welding tools, multipurpose end effectors, and so on. Indeed, in some processes the complexity of the tooling can rival or exceed the complexity of the robot arm that positions it. Successful coordination and sequencing of the robot manipulator, the end effector, and the end-of-arm tooling calls for a variety of considerations at several levels of abstraction in Fig. 33.7.

There are two philosophically distinct points of view that may be used in considering the robot manipulator plus its end-of-arm tooling. In the first, the robot plus tooling is viewed as a *single virtual agent* to be assigned by an upper-level organizer/manager and commanded by a midlevel job coordinator. In this situation, all machine-level robot/tool coordination may be performed by the internal virtual robot machine coordinator shown in Fig. 33.7. This point of view is natural when the robot must perform sophisticated trajectory motion during the task and the tool is unintelligent, such as in pick-and-place operations, surface finishing, and grinding. In such situations, the end effector is often controlled by simple binary on/off or open/close commands through digital input/output signals from the machine coordinator. Many commercially available robot controllers allow such communications and support coordination through their programming languages (section 33.12).

In the second viewpoint, one considers the manipulator as a dumb platform that positions the tooling or maintains its relative motion to the workpiece while the tooling performs a job. This point of view may be taken in the case of processes requiring sophisticated tooling such as welding. In this situation, the robot manipulator and the tooling may be considered as *two separate agents* which are coordinated by the higher level job coordinator shown in Fig. 33.7.

A variety of processes fall between these two extremes, such as assembly tasks which require some coordinated intelligence by both the manipulator and the tool (insert pin in hole). In such applications both machine-level and task-level coordination may be required. The decomposition of coordination commands into a portion suitable for machine-level coordination and a portion for task-level coordination is not easy. A rule-of-thumb is that any coordination that is invariant from process to process should be apportioned to the lower level (e.g., do not open gripper while robot is in motion). This is closely connected to the appropriate definition of robot/tooling behaviors in the fashion of Brooks et al. [1986].

33.7 Sensors

Sensors and actuators [Tzou and Fukuda 1992] function as *transducers*, devices through which the workcell planning, coordination, and control system interfaces with the hardware components that make up the workcell. Sensors are a vital element as they convert states of physical devices into signals appropriate for input to the workcell PC&C control system; inappropriate sensors can introduce errors that make proper operation impossible no matter how sophisticated or expensive the PC&C system, whereas innovative selection of sensors can make the control and coordination problem much easier.

The Philosophy of Robotic Workcell Sensors

Sensors are of many different types and have many distinct uses. Having in mind an analogy with biological systems, *proprioceptors* are sensors internal to a device that yield information about the internal state of that device (e.g., robot arm joint-angle sensors). *Exteroceptors* yield information about other hardware external to a device. Sensors yield outputs that are either analog or digital; digital sensors often provide information about the status of a machine or resource (gripper open or closed, machine loaded, job complete). Sensors produce inputs that are required at all levels of the PC&C hierarchy, including uses for:

- Servo-level feedback control (usually analog proprioceptors)
- Process monitoring and coordination (often digital exteroceptors or part inspection sensors such as vision)
- Failure and safety monitoring (often digital, e.g., contact sensor, pneumatic pressure-loss sensor)
- Quality control inspection (often vision or scanning laser).

Sensor output data must often be processed to convert it into a form meaningful for PC&C purposes. The sensor plus required signal processing is shown as a *virtual sensor* in Fig. 33.7; it functions as a *data abstraction*, that is, a set of data plus operations on that data (e.g., camera, plus framegrabber, plus signal processing algorithms such as image enhancement, edge detection, segmentation, etc.). Some sensors, including the proprioceptors needed for servo-level feedback control, are integral parts of their host devices, and so processing of sensor data and use of the data occurs within that device; then, the sensor data is incorporated at the servocontrol level or machine coordination level. Other sensors, often vision systems, rival the robot manipulator in sophistication and are coordinated by the job coordinator, which treats them as valuable shared resources whose use is assigned to jobs that need them by some priority assignment (e.g., dispatching) scheme. An interesting coordination problem is posed by so-called *active sensing*, where, e.g., a robot may hold a scanning camera, and the camera effectively takes charge of the coordination problem, directing the robot where to move it to effect the maximum reduction in entropy (increase in information) with subsequent images.

Types of Sensors

This section summarizes sensors from an operational point of view. More information on functional and physical principles can be found in Fraden [1993], Fu et al. [1987], Groover et al. [1986], and Snyder [1985].

Tactile Sensors.   Tactile sensors [Nichols and Lee 1989] rely on physical contact with external objects. Digital sensors such as limit switches, microswitches, and vacuum devices give binary information on whether contact occurs or not. Sensors are available to detect the onset of slippage. Analog sensors such as spring-loaded rods give more information. Tactile sensors based on rubberlike carbon- or silicon-based *elastomers* with embedded electrical or mechanical components can provide very detailed information about part geometry, location, and more. Elastomers can contain resistive or capacitive elements whose electrical properties change as the elastomer compresses. Designs based on LSI technology can produce *tactile grid pads* with, e.g., 64×64 *forcel* points on a single pad. Such sensors produce *tactile images* that have properties akin to digital images from a camera and require similar data processing. Additional tactile sensors fall under the classification of force sensors discussed subsequently.

Proximity and Distance Sensors. The noncontact proximity sensors include devices based on the Hall effect or inductive devices based on the electromagnetic effect that can detect ferrous materials within about 5 mm. Such sensors are often digital, yielding binary information about whether or not an object is near. Capacitance-based sensors detect any nearby solid or liquid with ranges of about 5 mm. Optical and ultrasound sensors have longer ranges.

Distance sensors include time-of-flight range finder devices such as sonar and lasers. The commercially available Polaroid sonar offers accuracy of about 1 in up to 5 ft, with angular sector accuracy of about $15°$. For $360°$ coverage in navigation applications for mobile robots, both scanning sonars and ring-mounted multiple sonars are available. Sonar is typically noisy with spurious readings, and requires low-pass filtering and other data processing aimed at reducing the false alarm rate. The more expensive laser range finders are extremely accurate in distance and have very high angular resolution.

Position, Velocity, and Acceleration Sensors. Linear position-measuring devices include linear potentiometers and the sonar and laser range finders just discussed. Linear velocity sensors may be laser- or sonar-based Doppler-effect devices.

Joint-angle position and velocity proprioceptors are an important part of the robot arm servocontrol drive axis. Angular position sensors include potentiometers, which use dc voltage, and *resolvers*, which use ac voltage and have accuracies of ± 15 min. Optical encoders can provide extreme accuracy using digital techniques. *Incremental optical encoders* use three optical sensors and a single ring of alternating opaque/clear areas, Fig. 33.18(a), to provide angular position relative to a reference point and angular velocity information; commercial devices may have 1200 slots per turn. More expensive *absolute optical*

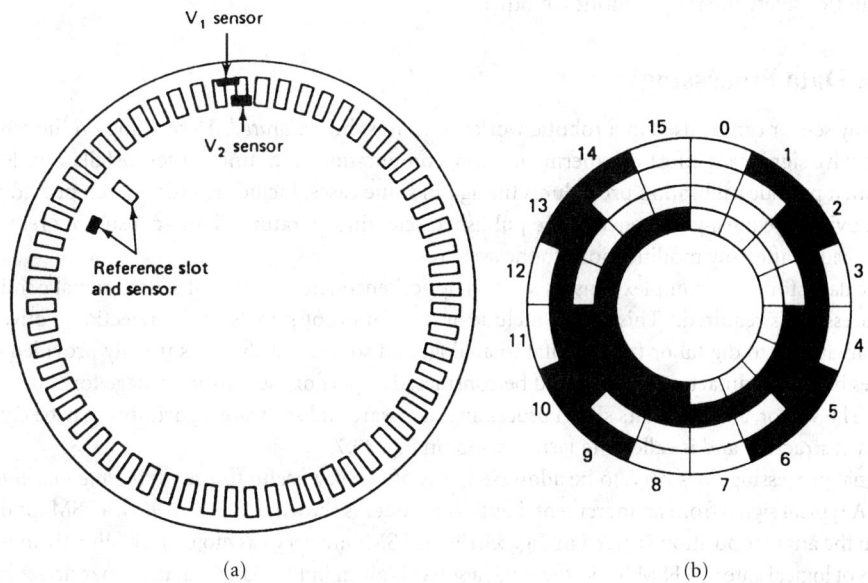

(a)   (b)

FIGURE 33.18 Optical encoders: (a) incremental optical encoder and (b) absolute optical encoder with $n = 4$ using Grey code (Snyder, W. E. 1985. *Industrial Robots: Computer Interfacing and Control.* Prentice–Hall, Englewood Cliffs, NJ. With permission.)

encoders, Fig. 33.18(b), have n concentric rings of alternating opaque/clear areas and require n optical sensors. They offer increased accuracy and minimize errors associated with data reading and transmission, particularly if they employ the *Grey code*, where only one bit changes between two consecutive sectors. Accuracy is $360°/2^n$, with commercial devices having $n = 12$ or so.

Gyros have good accuracy if repeatability problems associated with drift are compensated for. Directional gyros have accuracies of about $\pm1.5°$; vertical gyros have accuracies of $0.15°$ and are available to measure multiaxis motion (e.g., pitch and roll). Rate gyros measure velocities directly with thresholds of $0.05°/s$ or so.

Various sorts of accelerometers are available based on strain gauges (next paragraph), gyros, or crystal properties. Commercial devices are available to measure accelerations along three axes.

Force and Torque Sensors.   Various torque sensors are available, though they are often not required; for instance, the internal torques at the joints of a robot arm can be computed from the motor armature currents. Torque sensors on a drilling tool, for instance, can indicate when tools are dull. Linear force can be measured using load cells or strain gauges. A strain gauge is an elastic sensor whose resistance is a function of applied strain or deformation. The piezoelectric effect, the generation of a voltage when a force is applied, may also be used for force sensing. Other force sensing techniques are based on vacuum diodes, quartz crystals (whose resonant frequency changes with applied force), etc.

Robot arm force-torque wrist sensors are extremely useful in dextrous manipulation tasks. Commercially available devices can measure both force and torque along three perpendicular axes, providing full information about the Cartesian force vector **F**. Transformations such as Eq. (33.6) allow computation of forces and torques in other coordinates. Six-axis force-torque sensors are quite expensive.

Photoelectric Sensors.   A wide variety of photoelectric sensors are available, some based on fiber optic principles. These have speeds of response in the neighborhood of 50 μs with ranges up to about 45 mm, and are useful for detecting parts and labeling, scanning optical bar codes, confirming part passage in sorting tasks, etc.

Other Sensors.   Various sensors are available for measuring pressure, temperature, fluid flow, etc. These are useful in closed-loop servocontrol applications for some processes such as welding, and in job coordination and/or safety interrupt routines in others.

Sensor Data Processing

Before any sensor can be used in a robotic workcell, it must be *calibrated*. Depending on the sensor, this could involve significant effort in experimentation, computation, and tuning after installation. Manufacturers often provide calibration procedures though in some cases, including vision, such procedures may not be obvious, requiring reference to the published scientific literature. Time-consuming recalibration may be needed after any modifications to the system.

Particularly for more complex sensors such as optical encoders, significant sensor signal conditioning and processing is required. This might include amplification of signals, noise rejection, conversion of data from analog to digital or from digital to analog, and so on. Hardware is usually provided for such purposes by the manufacturer and should be considered as part of the sensor package for robot workcell design. The sensor, along with its signal processing hardware and software algorithms may be considered as a data abstraction and is called the *virtual sensor* in Fig. 33.7.

If signal processing does need to be addressed, it is often very useful to use *finite state machine (FSM) design*. A typical signal from an incremental optical encoder is shown in Fig. 33.19(a); a FSM for decoding this into the angular position is given in Fig. 33.19(b). FSMs are very easy to convert directly to hardware in terms of logical gates. A FSM for sequencing a sonar is given in Fig. 33.20(a); the sonar driver hardware derived from this FSM is shown in Fig. 33.20(b).

A particular problem is obtaining angular velocity from angular position measurements. All too often the position measurements are simply differenced using a small sample period to compute velocity. This is

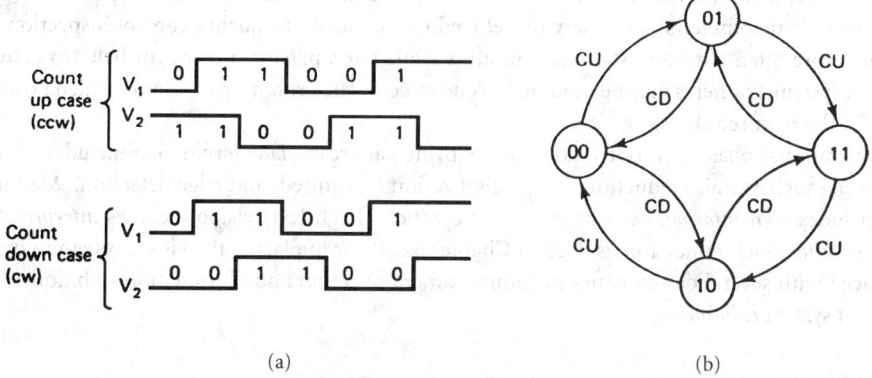

(a)   (b)

FIGURE 33.19   Signal processing using FSM for optical encoders: (a) phase relations in incremental optical encoder output and (b) finite state machine to decode encoder output into angular position. [Snyder 1985].

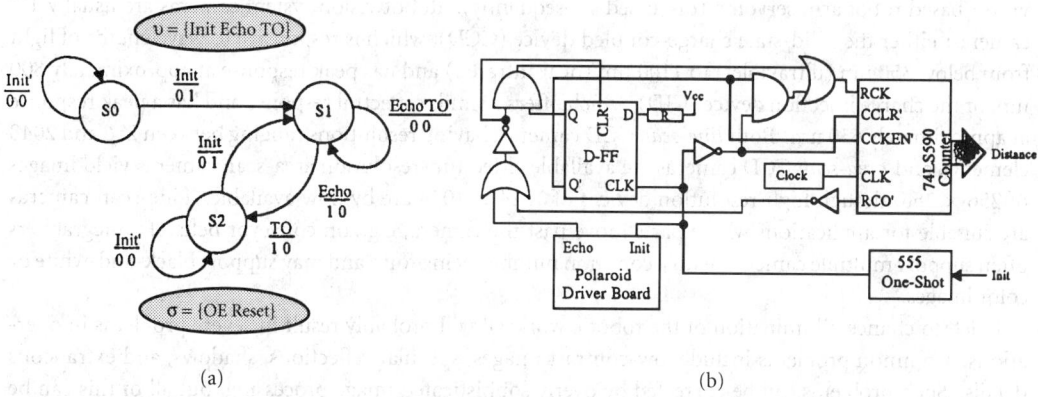

(a)   (b)

FIGURE 33.20   Hardware design from FSM: (a) FSM for sonar transducer control on a mobile robot and (b) sonar driver control system from FSM.

guaranteed to lead to problems if there is any noise in the signal. It is almost always necessary to employ a low-pass-filtered derivative where velocity samples v_k are computed from position measurement samples $\mathbf{p}_k$ using, e.g.,

$$v_k = \alpha v_{k-1} + (1 - \alpha)(\mathbf{p}_k - \mathbf{p}_{k-1})/T \qquad (33.18)$$

where T is the sample period and α is a small filtering coefficient. A similar approach is needed to compute acceleration.

Vision for Robotics

Computer vision is covered in Chapter 25 of this Handbook; the purpose of this section is to discuss some aspects of vision that are unique to robotics [Fu et al. 1987, Lee and Li 1991, Lee and Blenis 1994]. Industrial robotic workcells often require vision systems that are reliable, accurate, low cost, and rugged yet perform sophisticated image processing and decision making functions. Balancing these conflicting demands is not always easy. There are several commercially available vision systems, the most sophisticated of which may be the Adept vision system, which supports multiple cameras. However, it is sometimes necessary to design one's own system. Vision may be used for three purposes in robotic workcells: inspection and quality control, robotic manipulation, and servo-level feedback control. In quality control inspection systems the cameras are often affixed to stationary mounts while parts pass on a conveyor belt. In *active vision* inspection systems, cameras may be mounted as end effectors of a robot manipulator, which positions the camera for the required shots.

The operational phase of robot vision has six principal areas. *Low-level vision* includes *sensing* and *preprocessing* such as noise reduction, image digitization if required, and edge detection. Medium-level vision includes *segmentation, description*, and *recognition*. High-level vision includes *interpretation* and *decision making*. Such topics are disussed in Chapter 32. Prior to placing the vision system in operation, one is faced with several design issues including camera selection and illumination techniques, and the problem of system *calibration*.

Cameras and Illumination

Typical commercially available vision systems conform to the RS-170 standard of the 1950s, so that frames are acquired through a framegrabber board at a rate of 30 frames/s. Images are scanned; in a popular U.S. standard, each complete scan or *frame* consists of 525 lines of which 480 contain image information. This sample rate and image resolutions of this order are adequate for most applications with the exception of vision-based robot arm servoing (discussed subsequently). Robot vision system cameras are usually TV cameras: either the solid-state charge-coupled device (CCD), which is responsive to wavelengths of light from below 350 nm (ultraviolet) to 1100 nm (near infrared) and has peak response at approximately 800 nm, or the charge injection device (CID), which offers a similar spectral response and has a peak response at approximately 650 nm. Both *line-scan* CCD cameras, having resolutions ranging between 256 and 2048 elements, and *area-scan* CCD cameras are available. Medium-resolution area-scan cameras yield images of 256×256, though high-resolution devices of 1024×1024 are by now available. Line-scan cameras are suitable for applications where parts move past the camera, e.g., on conveyor belts. Framegrabbers often support multiple cameras, with a common number being four, and may support black-and-white or color images.

If left to chance, illumination of the robotic workcell will probably result in severe problems in operations. Common problems include low-contrast images, specular reflections, shadows, and extraneous details. Such problems can be corrected by overly sophisticated image processing, but all of this can be avoided by some proper attention to details at the workcell design stage. Illumination techniques include spectral filtering, selection of suitable spectral characteristics of the illumination source, diffuse-lighting techniques, backlighting (which produces easily processed silhouettes), structured-lighting (which provides additional depth information and simplifies object detection and interpretation), and directional lighting.

Coordinate Frames and Camera Perspective Transformation

A typical robot vision system is depicted in Fig. 33.21, which shows a gimball-mounted camera. There are illustrated the base frame (or world frame) (X, Y, Z), the gimball platform, the camera frame (x, y, z), and the *image plane* having coordinates of (ξ, υ).

Image Coordinates of a Point in Base Coordinates. The primary tools for analysis of robot vision systems are the notion of *coordinate transforms* and the camera *perspective transformation*. Four-by-four *homogeneous transformations* (discussed earlier) are used, as they provide information on translations, rotations, scaling, and perspective.

Four homogeneous transformations may be identified in the vision system, as illustrated in Fig. 33.22. The gimball transformation G represents the base frame in coordinates affixed to the gimball platform. If the camera is mounted on a robot end effector, G is equal to T^{-1}, with T the robot arm T matrix detailed in earlier in section 33.5; for a stationary-mounted camera G is a constant matrix capturing the camera platform mounting offset $r_0 = [X_0 \ Y_0 \ Z_0]^T$. The pan/tilt transformation R represents the gimball

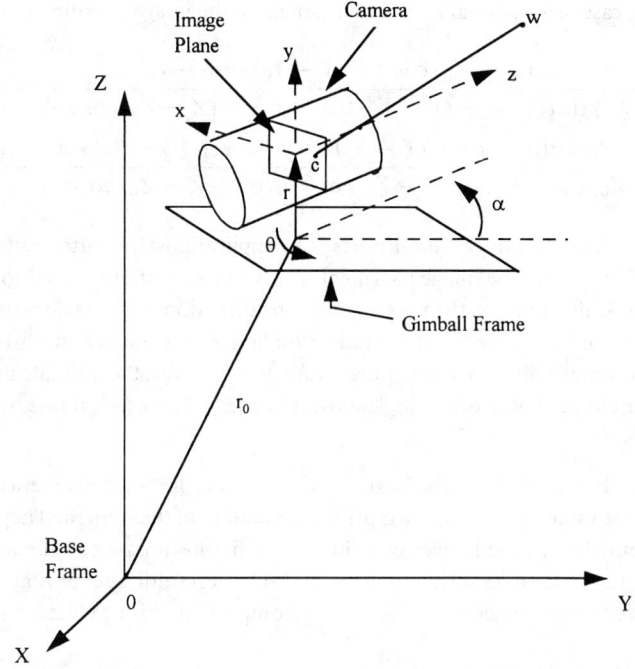

FIGURE 33.21   Typical robot workcell vision system.

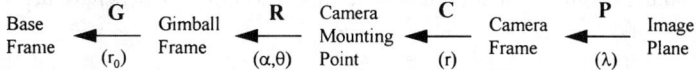

FIGURE 33.22   Homogeneous transformations associated with the robot vision system.

platform with respect to the mounting point of the camera. This rotation transformation is given by

$$R = \begin{bmatrix} \cos\theta & \sin\theta & 0 & 0 \\ -\sin\theta\cos\alpha & \cos\theta\cos\alpha & \sin\alpha & 0 \\ \sin\theta\sin\alpha & -\cos\theta\sin\alpha & \cos\alpha & 0 \\ 0 & 0 & 0 & 1 \end{bmatrix} \tag{33.19}$$

with θ the pan angle and α the tilt angle. C captures the offset $r = [r_x\ r_y\ r_z]^T$ of the camera frame with respect to the gimbal frame. Finally, the perspective transformation

$$\begin{bmatrix} 1 & 0 & 0 & 0 \\ 0 & 1 & 0 & 0 \\ 0 & 0 & 1 & 0 \\ 0 & 0 & -\frac{1}{\lambda} & 1 \end{bmatrix} \tag{33.20}$$

projects a point represented in camera coordinates (x, y, z) onto a position (ξ, υ) in the image, where λ is the camera focal length.

In terms of these constructions, the image position of a point w represented in base coordinates as (X, Y, Z) is given by the *camera transform equation*

$$c = PCRGw \tag{33.21}$$

which evaluates in the case of a stationary-mounted camera to the image coordinates

$$
\xi = \lambda \frac{(X - X_0)\cos\theta + (Y - Y_0)\sin\theta - r_x}{-(X - X_0)\sin\theta\sin\alpha + (Y - Y_0)\cos\theta\sin\alpha - (Z - Z_0)\cos\alpha + r_z + \lambda}
$$
$$
\upsilon = \lambda \frac{-(X - X_0)\sin\theta\cos\alpha + (Y - Y_0)\cos\theta\cos\alpha + (Z - Z_0)\sin\alpha - r_y}{-(X - X_0)\sin\theta\sin\alpha + (Y - Y_0)\cos\theta\sin\alpha - (Z - Z_0)\cos\alpha + r_z + \lambda}
$$

(33.22)

Base Coordinates of a Point in Image Coordinates.   In applications, one often requires the inverse of this transformation; that is, from the image coordinates (ξ, υ) of a point one wishes to determine its base coordinates (X, Y, Z). Unfortunately, the perspective transformation P is a *projection* which loses depth information z, so that the inverse perspective transformation P^{-1} is not unique. To compute unique coordinates in the base frame one therefore requires either two cameras, the ultimate usage of which leads to *stereo imaging*, or multiple shots from a single moving camera. Many techniques have been developed for accomplishing this.

Camera Calibration.   Equation (33.22) has several parameters, including the camera offsets r_0 and r and the focal length λ. These values must be known prior to operation of the camera. They may be measured, or they may be computed by taking images of points w_i with known base coordinates (X_i, Y_i, Z_i). To accomplish this, one must take at least six points w_i and solve a resulting set of nonlinear simultaneous equations. Many procedures have been developed for accomplishing this by efficient algorithms.

High-Level Robot Vision Processing

Besides scene interpretation, other high-level vision processing issues must often be confronted, including decision making based on vision data, relation of recognized objects to stored CAD data of parts, recognition of faults or failures from vision data, and so on. Many technical papers have been written on all of these topics.

Vision-Based Robot Manipulator Servoing

In standard robotic workcells, vision is not often used for servo-level robot arm feedback control. This is primarily due to the facts that less expensive lower level sensors usually suffice, and reliable techniques for vision-based servoing are only now beginning to emerge. In vision-based servoing the standard frame rate of 30 ft/s is often unsuitable; higher frame rates are often needed. This means that commercially available vision systems cannot be used. Special purpose cameras and hardware have been developed by several researchers to address this problem, including the vision system in Lee and Blenis [1994].

Once the hardware problems have been solved, one has yet to face the design problem for real-time servocontrollers with vision components in the feedback loop. This problem may be attacked by considering the nonlinear dynamical system (33.7) with measured outputs given by combining the camera transformation (33.21) and the arm kinematics transformation (33.2) [Ghosh et al. 1994].

33.8 Workcell Planning

Specifications for workcell performance vary at distinct levels in the workcell planning, coordination, and control architecture in Fig. 33.7. At the machine servocontrol level, motion specifications are given in terms of continuous trajectories in joint space sampled every 16 ms. At the job coordination level, motion specifications are in terms of Cartesian path via points, generally nonuniformly spaced, computed to achieve a prescribed task. Programming at these lower levels involves tedious specifications of points, motions, forces, and times of transit. The difficulties involved with such low-level programming have led to requirements for *task-level programming*, particularly in modern robot workcells which must be flexible and reconfigurable as products vary in response to the changing desires of customers. The function of the workcell planner is to allow task-level programming from the workcell manager by performing *task planning and decomposition* and *path planning*, thereby automatically providing the more detailed specifications required by the job coordinator and servo-level controllers.

Workcell Behaviors and Agents

The virtual machines and virtual sensors in the (PC&C) architecture of Fig. 33.7 are constructed using the considerations discussed in previous sections. These involve commercial robot selection, robot kinematics and servo-level control, end effectors and tooling, and sensor selection and calibration. The result of design at these levels is a set of workcell *agents*—robots, machines, or sensors—each with a set of behaviors or *primitive actions* that each workcell agent is capable of. For instance, proper design could allow a robot agent to be capable of behaviors including accurate motion trajectory following, tool changing, force-controlled grinding on a given surface, etc. A camera system might be capable of identifying all Phillips screw heads in a scene, then determining their coordinates and orientation in the base frame of a robot manipulator.

Given the workcell agents with their behaviors, the higher level components in Fig. 33.7 must be able to assign tasks and then decompose them into a suitable sequencing of behaviors. In this and the next section are discussed the higher level PC&C components of workcell planning and job coordination.

Task Decomposition and Planning

A *task* is a specific goal that must be accomplished, often by an assigned *due date*. These goals may include completed robotic processes (e.g., weld seam), finished products, and so on. Tasks are accomplished by a sequence of *jobs* that, when completed, result in the achievement of the final goal. *Jobs* are specific primitive activities that are accomplished by a well-defined set of resources or agents (e.g., drill hole, load machine). Once a resource has been assigned to a job it can be interpreted as a behavior. To attain the task goal, jobs must usually be performed in some *partial ordering*, e.g., tasks *a* and *b* are immediate prerequisties for task *c*. The jobs are not usually completely ordered, but have some possibility for concurrency.

At the workcell management level, tasks are assigned, along with their due dates, without specifying details of resource assignment or selection of specific agents. At the job coordination level, the required specifications are in terms of sequences of jobs, with resources assigned, selected to achieve the assigned goal tasks. The function of the workcell planner is to convert between these two performance specification paradigms. In the task planning component, two important transformations are made. First, assigned *tasks are decomposed into the required job sequences*. Second, workcell agents and *resources are assigned* to accomplish the individual jobs. The result is a task plan that is passed for execution to the job coordinator.

Task Plan. A *task plan* is a sequence of jobs, along with detailed resource assignments, that will lead to the desired goal task. Jobs with assigned resources can be interpreted as behaviors. Plans should not be overspecified—the required job sequencing is usually only a partial ordering, often with significant concurrency remaining among the jobs. Thus, some decisions based on real-time workcell status should be left to the job coordinator; among these are the final detailed sequencing of the jobs and any *dispatching* and *routing* decisions where shared resources are involved.

Computer Science Planning Tools

There are well-understood techniques in computer science that can be brought to bear on the robot task planning problem [Fu et al. 1987]. Planning and scheduling is covered in Chapter 27, decision trees in Chapter 22, search techniques in Chapter 30, and decision making under uncertainty in Chapter 31; all of these are relevant to this discussion. However, the structure of the robotic workcell planning problem makes it possible to use some refined and quite rigorous techniques in this chapter, which are introduced in the next subsections.

Task planning can be accomplished using techniques from problem solving and learning, especially learning by analogy. By using *plan schema* and other *replanning* techniques, it is possible to modify existing plans when goals or resources change by small amounts. *Predicate logic* is useful for representing knowledge in the task planning scenario and many problem solving software packages are based on production systems.

Several task planning techniques use *graph theoretic* notions that can be attacked using search algorithms such as A^*. *State-space search techniques* allow one to try out various approaches to solving a problem until a suitable solution is found: the set of states reachable from a given initial state forms a graph. A plan is often represented as a finite state machine, with the states possibly representing jobs or primitive actions. *Problem reduction* techniques can be used to decompose a task into smaller subtasks; in this context it is often convenient to use AND/OR graphs. *Means–ends analysis* allows both forward and backward search techniques to be used, solving the main parts of a problem first and then going back to solve smaller subproblems.

For workcell assembly and production tasks, product data in CAD form is usually available. Assembly task planning involves specifying a sequence of assembly, and possibly process, steps that will yield the final product in finished form. *Disassembly planning techniques* work backwards from the final product, performing part disassembly transformations until one arrives at the initial raw materials. Care must be taken to account for part obstructions, etc. The relationships between parts should be specified in terms of symbolic spatial relationships between *object features* (e.g., place $block_1 - face_2$ against $wedge_2 - face_3$ and $block_1 - face_1$ against $wedge_2 - face_1$ or place pin in slot). *Constructive solid geometric* techniques lead to graphs that describe objects in terms of features related by set operations such as intersection, union, etc.

Industrial Engineering Planning Tools

In industrial engineering there are well-understood design tools used for product assembly planning, process planning, and resource assignment; they should be used in workcell task planning. The *bill of materials* (BOM) for a product is a computer printout that breaks down the various subassemblies and component parts needed for the product. It can be viewed as a matrix B whose elements $B(i, j)$ are set to 1 if subassembly j is needed to produce subassembly i. This matrix is known as *Steward's Sequencing Matrix;* by studying it one can decompose the assembly process into hierarchically interconnected subsystems of subassemblies [Warfield 1973], thereby allowing parallel processing and simplification of the assembly process. The *assembly tree* [Wolter et al. 1992] is a graphical representation of the BOM.

The *resource requirements matrix* is a matrix R whose elements $R(i, j)$ are set equal to 1 if resource j is required for job i. The resources may include machines, robots, fixtures, tools, transport devices, and so on. This matrix has been used by several workers for analysis and design of manufacturing systems; it is very straightforward to write down given a set of jobs and available resources. The *subassembly tree* is an assembly tree with resource information added.

Task Matrix Approach to Workcell Planning

Plans can often be specified as finite state machines. However, in the robot workcell case, FSM are neither general enough to allow versatile incorporation of workcell status and sensor information, nor specific enough to provide all of the information needed by the job coordinator.

A very general robot workcell task plan can be completely specified by four *task plan matrices* [Lewis and Huang 1995]. The *job sequencing matrix* and *job start matrix* are independent of resources and carry the job sequencing information required for task achievement. Resources are subsequently added by constructing the *resource requirements matrix* and the *resource release matrix*. The function of the task planner is to construct these four matrices and pass them to the job coordinator, who uses them for job coordination, sequencing, and resource dispatching. The task plan matrices are straightforward to construct and are easy to modify in the event of goal changes, resource changes, or failures; that is, they accommodate task planning as well as *task replanning*.

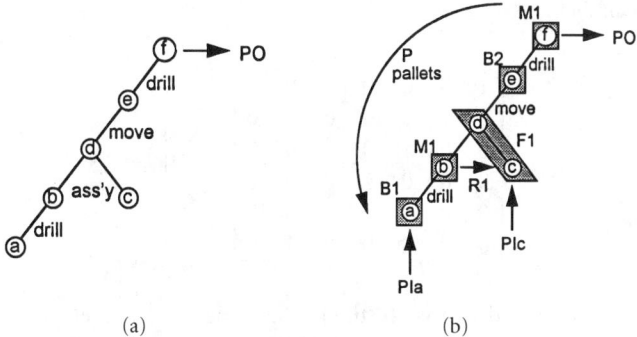

FIGURE 33.23 Product information for task planning: (a) assembly tree with job sequencing information and (b) subassembly tree with resource information added to the jobs.

The task planning techniques advocated here are illustrated through an assembly design example, which shows how to select the four task plan matrices. Though the example is simple, the technique extends directly to more complicated systems using the notions of *block matrix (e.g., subsystem) design*. First, job sequencing is considered, then the resources are added.

Workcell Task Decomposition and Job Sequencing

In Fig. 33.23(a) is given an assembly tree which shows the required sequence of actions (jobs) to produce a product. This sequence may be obtained from stored product CAD data through disassembly techniques, etc. The assembly tree contains information analogous to the BOM; it does not include any resource information. Part a enters the workcell and is drilled to produce part b, then assembled with part c to produce part d, which is again drilled (part e) to result in part f, which is the cell output (*PO* denotes 'product out'). The assembly tree imposes only a *partial ordering* on the sequence of jobs. It is important not to overspecify the task decomposition by imposing additional temporal orderings that are not required for job sequencing.

Job Sequencing Matrix. Referring to Fig. 33.23(a), define the *job vector* as $\mathbf{v} = [\mathbf{a}\ \mathbf{b}\ \mathbf{c}\ \mathbf{d}\ \mathbf{e}\ \mathbf{f}]^T$. The Steward's sequencing matrix F_v for the assembly tree in Fig. 33.23(a) is then given by

$$
F_v = \begin{array}{c} \\ \mathbf{a} \\ \mathbf{b} \\ \mathbf{c} \\ \mathbf{d} \\ \mathbf{e} \\ \mathbf{f} \\ \mathbf{PO} \end{array}
\begin{array}{c} \begin{array}{cccccc} \mathbf{a} & \mathbf{b} & \mathbf{c} & \mathbf{d} & \mathbf{e} & \mathbf{f} \end{array} \\
\begin{bmatrix}
0 & 0 & 0 & 0 & 0 & 0 \\
1 & 0 & 0 & 0 & 0 & 0 \\
0 & 0 & 0 & 0 & 0 & 0 \\
0 & 1 & 1 & 0 & 0 & 0 \\
0 & 0 & 0 & 1 & 0 & 0 \\
0 & 0 & 0 & 0 & 1 & 0 \\
0 & 0 & 0 & 0 & 0 & 1
\end{bmatrix} \end{array}. \tag{33.23}
$$

In this matrix, an entry of 1 in position (i, j) indicates that job j must be completed prior to starting job i. F_v is independent of available resources; in fact, regardless of the resources available, F_v will not change.

Sequencing State Vector and Job Start Equation. Define a *sequencing state vector* x, whose components are associated with the vector $[\mathbf{a}\ \mathbf{b}\ \mathbf{c}\ \mathbf{d}\ \mathbf{e}\ \mathbf{f}\ \mathbf{PO}]^T$, that checks the conditions of the rules needed for job sequencing. The components of $\mathbf{x}$ may be viewed as situated between the nodes in the assembly tree.

Then, the *job start equation* is

$$
\mathbf{v}_s =
\begin{bmatrix}
1 & 0 & 0 & 0 & 0 & 0 & 0 \\
0 & 1 & 0 & 0 & 0 & 0 & 0 \\
0 & 0 & 1 & 0 & 0 & 0 & 0 \\
0 & 0 & 0 & 1 & 0 & 0 & 0 \\
0 & 0 & 0 & 0 & 1 & 0 & 0 \\
0 & 0 & 0 & 0 & 0 & 1 & 0
\end{bmatrix}
\begin{bmatrix}
\mathbf{x}_1 \\
\mathbf{x}_2 \\
\mathbf{x}_3 \\
\mathbf{x}_4 \\
\mathbf{x}_5 \\
\mathbf{x}_6 \\
\mathbf{x}_7
\end{bmatrix}
\equiv S_v \mathbf{x}
\tag{33.24}
$$

where $\mathbf{v}_s$ is the job start command vector. In the job start matrix S_v, an entry of 1 in position (i, j) indicates that job i can be started when component j of the sequencing state vector is active.

In this example, the matrix S_v has 1s in locations (i, i) so that S_v appears to be redundant. This structure follows from the fact that the assembly tree is an *upper semilattice*, wherein each node has a unique node above it; such a structure occurs in the manufacturing *re-entrant flowline with assembly*. In the more general *job shop* with variable part routings the semilattice structure of the assembly tree does not hold. Then, S_v can have multiple entries in a single column, corresponding to different routing options; nodes corresponding to such columns have more than one node above them.

Adding the Resources

To build a job dispatching coordination controller for shop-floor installation to perform this particular assembly task, the resources available must now be added. The issue of required and available resources is easily confronted as a *separate engineering design issue* from job sequence planning. In Fig. 33.23(b) is given a *subassembly tree* for the assembly task, which includes resource requirements information. This information would in practice be obtained based on the resources and behaviors available in the workcell and could be assigned by a user during the planning stage using interactive software. The figure shows that part input *PIc* and part output (*PO*) do not require resources, pallets (*P*) are needed for part a and its derivative subassemblies, buffers (*B*1, *B*2) hold parts a and e, respectively, prior to drilling, and both drilling operations need the same machine (*M*1). The assembly operation is achieved by fixturing part c in fixture *F*1 while robot *R*1 inserts part b.

Note that drilling machine *M*1 represents a *shared resource*, which performs two jobs, so that dispatching decision making is needed when the two drilling jobs are simultaneously requested, in order to avoid possible problems with *deadlock*. This issue is properly faced by the job coordinator in real-time, as shown in section 33.9, not by the task planner. Shared resources impose *additional temporal restrictions* on the jobs that are not present in the job sequencing matrix; these are concurrency restrictions of the form: both drilling operations may not be performed simultaneously.

Resource Requirements (RR) Matrix. Referring to Fig. 33.23(b), define the *resource vector* as $\mathbf{r} = [R1A\ F1A\ B1A\ B2A\ PA\ M1A]^T$, where A denotes available. In the RR matrix F_r, a 1 in entry (i, j) indicates that resource j is needed to activate sequencing vector component $\mathbf{x}_i$ (e.g., in this example, to accomplish job i). By inspection, therefore, one may write down the RR matrix

$$
F_r =
\left[
\begin{array}{ccccc|c}
0 & 0 & 1 & 0 & 1 & 0 \\
0 & 0 & 0 & 0 & 0 & 1 \\
0 & 1 & 0 & 0 & 0 & 0 \\
1 & 0 & 0 & 0 & 0 & 0 \\
0 & 0 & 0 & 1 & 0 & 0 \\
0 & 0 & 0 & 0 & 0 & 1 \\
0 & 0 & 0 & 0 & 0 & 0
\end{array}
\right]
\tag{33.25}
$$

Row 3, for instance, means that resource *F1A*, the fixture, is needed as a precondition for firing $\mathbf{x}_3$; which matrix S_v associates with job c. Note that column 6 has two entries of 1, indicating that *M*1 is a shared resource that is needed for two jobs b and f. As resources change or machines fail, the RR matrix is easily modified.

Resource Release Matrix. The last issue to be resolved in this design is that of *resource release*. Thus, using manufacturing engineering experience and Fig. 33.23(b), select the *resource release matrix* S_r in the resource release equation

$$\mathbf{r}_s = \begin{bmatrix} R1A_s \\ F1A_s \\ B1A_s \\ B2A_s \\ PA_s \\ M1A_s \end{bmatrix} = \begin{bmatrix} 0 & 0 & 0 & 0 & 1 & 0 & 0 \\ 0 & 0 & 0 & 1 & 0 & 0 & 0 \\ 0 & 1 & 0 & 0 & 0 & 0 & 0 \\ 0 & 0 & 0 & 0 & 0 & 1 & 0 \\ 0 & 0 & 0 & 0 & 0 & 0 & 1 \\ 0 & 0 & 0 & 1 & 0 & 0 & 1 \end{bmatrix} \begin{bmatrix} \mathbf{x}_1 \\ \mathbf{x}_2 \\ \mathbf{x}_3 \\ \mathbf{x}_4 \\ \mathbf{x}_5 \\ \mathbf{x}_6 \\ \mathbf{x}_7 \end{bmatrix} \equiv S_r \mathbf{x} \qquad (33.26)$$

where subscript s denotes a command to the workcell to start resource release. In the resource release matrix S_r, a 1 entry in position (i, j) indicates that resource i is to be released when entry j of $\mathbf{x}$ has become high (e.g., in this example, on completion of job j). It is important to note that rows containing multiple ones in S_r correspond to columns having multiple ones in F_r. For instance, the last row of S_r shows that $M1A$ is a shared resource, since it is released after *either* $\mathbf{x}_4$ is high or $\mathbf{x}_7$ is high; that is, after either job b or job f is complete.

Petri Net from Task Plan Matrices

It will be shown in section 33.9 that the four task plan matrices contain all of the information needed to implement a matrix-based job coordination controller on, for instance, a programmable logic workcell controller. However, there has been much discussion of uses of *Petri nets* in task planning. It is now shown that the four task plan matrices correspond to a Petri net (PN). The job coordinator would not normally be implemented as a Petri net; however, it is straightforward to derive the PN description of a manufacturing system from the matrix controller equations, as shown by the next result.

Theorem 35.1 (Petri net from task plan matrices).   *Given the four task plan matrices F_v, S_v, F_r, S_r, define the* activity completion matrix F *and the* activity start matrix S *as*

$$F = [\, F_v \quad F_r \,], \qquad S = \begin{bmatrix} S_v \\ S_r \end{bmatrix} \qquad (33.27)$$

Define X as the set of elements of sequencing state vector $\mathbf{x}$, and A (activities) as the set of elements of the job and resource vectors $\mathbf{v}$ and $\mathbf{r}$. Then (A, X, F, S^T) is a Petri net.

The theorem identifies F as the input incidence matrix and S^T as the output incidence matrix of a PN, so that the PN incidence matrix is given by

$$W = S^T - F = \begin{bmatrix} S_v^T - F_v & S_r^T - F_r \end{bmatrix} \qquad (33.28)$$

Based on the theorem, the PN in Fig. 33.24 is easily drawn for this example. In the figure, initial markings have been added; this is accomplished by determining the *number of resources available* in the workcell. The inputs u_{D1}, u_{D2} are required for dispatching the shared resource $M1$, as discussed in section 33.9. This theorem provides a formal technique for constructing a PN for a workcell task plan and allows all of the PN analysis tools to be used for analysis of the workcell plan. It formalizes some work in the literature (e.g., top-down and bottom-up design [Zhou et al. 1992]).

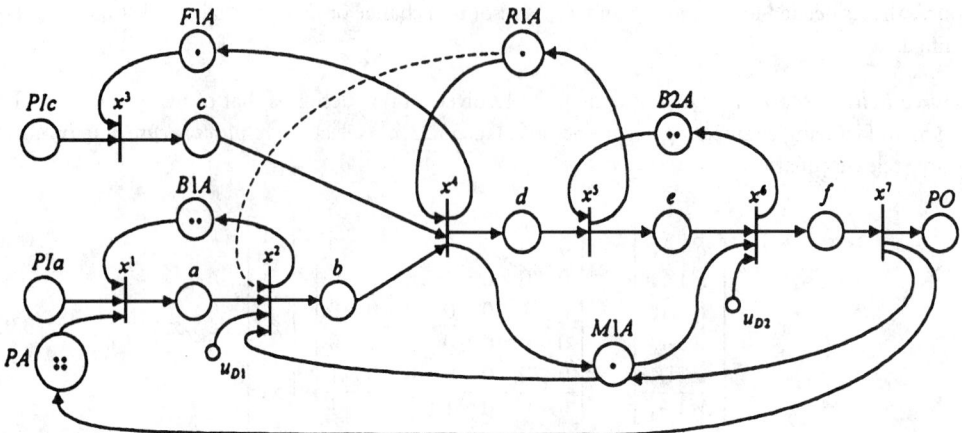

FIGURE 33.24 Petri net representation of workcell with shared resource.

Behaviors.   All of the PN transitions occur along the *job paths*. The places in the PN along the job paths correspond to jobs with assigned resources and can be interpreted as behaviors. The places off the task paths correspond to resource availability.

Path Planning

The path planning problem [Latombe 1991] may be decomposed into motion path planning, grasp planning, and error detection and recovery; only the first is considered here. *Motion path planning* is the process of finding a continuous path from an initial position to a prescribed final position or goal without collision. The output of the path planner for robotic workcells is a set of path *via points* which are fed to the machine trajectory generator (discussed previously). *Off-line path planning* can be accomplished if all obstacles are stationary at known positions or moving with known trajectories. Otherwise, *on-line or dynamic* path planning is required in real time; this often requires techniques of *collision or obstacle avoidance*. In such situations, paths preplanned off line can often be modified to incorporate collision avoidance. This subsection deals with off-line path planning except for the portion on the potential field approach, which is dynamic planning. See Zhou [1996] for more information.

Initial and final positions may be given in any coordinates, including the robot's joint space. Generally, higher level workcell components think in terms of Cartesian coordinates referred to some world frame. The Cartesian position of a robot end effector is given in terms of three position coordinates and three angular orientation coodinates; therefore, the general 3D path planning problem occurs in $\Re^6$. If robot joint-space initial and final positions are given one may work in *configuration space*, in which points are specified by the joint variable vector q having coordinates q_i, the individual joint values. For a six-degrees-of-freedom arm, configuration space is also isomorphic to $\Re^6$. Path planning may also be carried out for initial and final values of *force/torque*. In 3D, linear force has three components and torque has three components, again placing the problem in $\Re^6$. Hybrid position/force planning is also possible. In this subsection path planning techniques are illustrated in $\Re^2$, where it is convenient to think in terms of planning paths for mobile robots in a plane.

If the number of degrees of freedom of a robot is less than six, there could be problems in that the manipulator may not be able to reach the prescribed final position and the via points generated in the planning process. Thus, the path planner must be aware of the limitations of the individual robots in its planning process; in fact, it is usually necessary to select a specific robot agent for a task prior to planning the path in order to take such limitations into account.

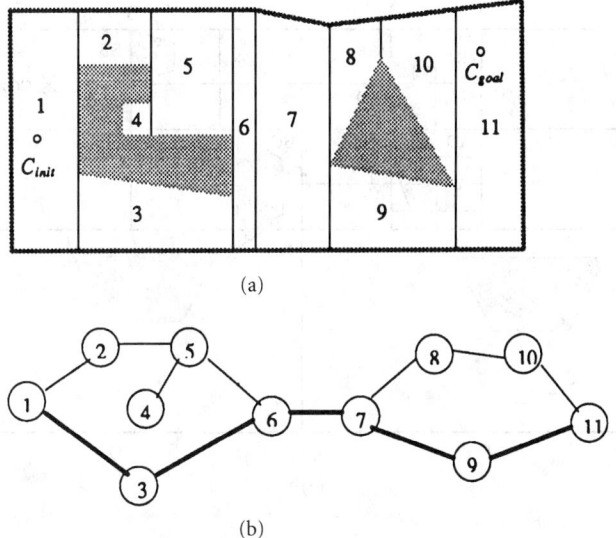

(a)

(b)

FIGURE 33.25 Cell decomposition approach to path planning: (a) free space decomposed into cells using the vertical-line-sweep method and (b) connectivity graph for the decomposed space. (*Source:* Courtesy of Zhou, C. 1996. Planning and intelligent control. In *CRC Handbook of Mechanical Engineering*. F. Kreith, ed. CRC Press, Boca Raton, FL.)

Cell Decomposition Approach

In the *cell decomposition approach to path planning*, objects are enclosed in polygons. The object polygons are expanded by an amount equal to the radius of the robot to ensure collision avoidance; then, the robot is treated simply as a moving point. The free space is decomposed into simply connected free-space regions within which any two points may be connected by a straight line. When the Euclidean metric is used to measure distance, convex regions satisfy the latter requirement. A sample cell decomposition is shown in Fig. 33.25. The decomposition is not unique; the one shown is generated by sweeping a vertical line across the space. Based on the decomposed space, a *connectivity graph* may be constructed, as shown in the figure. To the graph may be added weights or costs at the arcs or the nodes, corresponding to distances traveled, etc. Then, graph search techniques may be used to generate the shortest, or otherwise least costly, path.

Road Map Based on Visibility Graph

In the road map approach the obstacles are modeled as polygons expanded by the radius of the robot, which is treated simply as a moving point. A **visibility graph** is a nondirected graph whose nodes are the vertices of the polygons and whose links are straight line segments connecting the nodes without intersecting any obstacles. A *reduced visibility graph* does not contain links that are dominated by other links in terms of distance. Figure 33.26 shows a reduced visibility graph for the free space. Weights may be assigned to the arcs or nodes and graph search techniques may be used to generate a suitable path. The weights can reflect shortest distance, path smoothness, etc.

Road Map Based on Voronoi Diagram. A **Voronoi diagram** is a diagram where the path segment lines have equal distance from adjacent obstacles. In a polygonal space, the Voronoi diagram consists of straight lines and parabolas: when both adjacent object segments are vertices or straight lines, the equidistant line is straight, when one object is characterized by a vertex and the other by a straight line, the

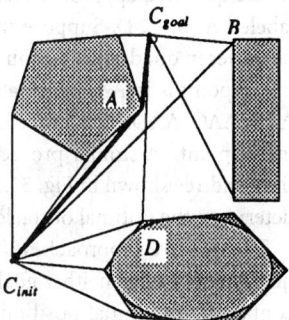

FIGURE 33.26 Road map based on visibility graph. (*Source:* Courtesy of Zhou, C. 1996. Planning and intelligent control. In *CRC Handbook of Mechanical Engineering*. F. Kreith, ed. CRC Press, Boca Raton, FL.)

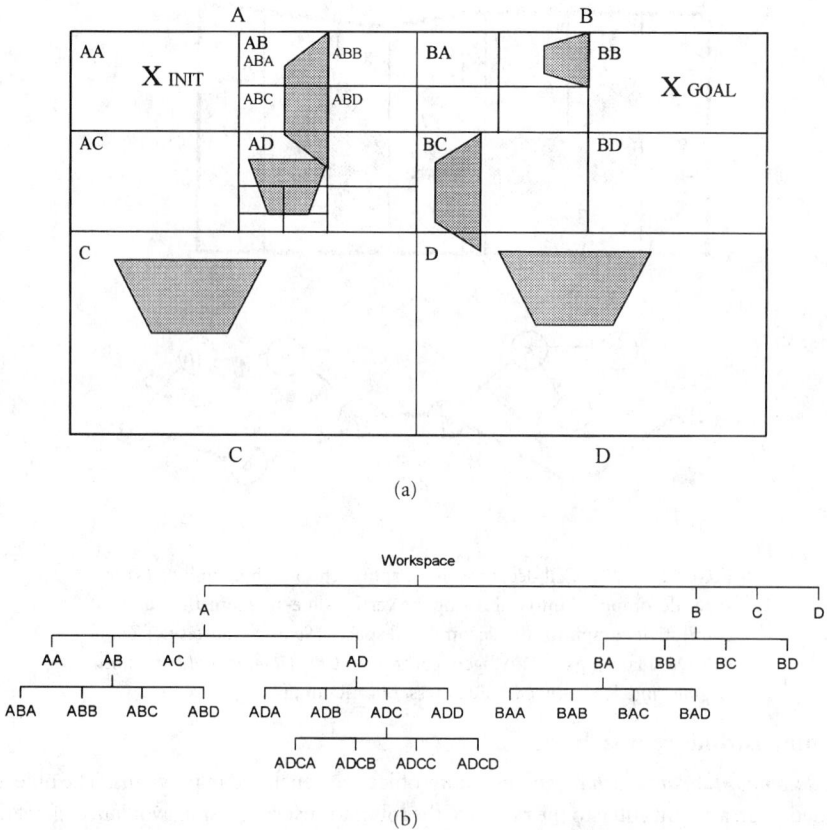

FIGURE 33.27 Quadtree approach to path planning: (a) quadtree decomposition of the work area and (b) quadtree constructed from space decomposition.

equidistant line is parabolic. In the Voronoi approach, generated paths are generally longer than in the visibility graph approach, but the closest point of approach (CPA) to obstacles is maximized.

Quadtree Approach

In the quadtree approach, Fig. 33.27(a) rectangular workspace is partitioned into four equal quadrants labeled A, B, C, D. Suppose the initial point is in quadrant A with the goal in quadrant B. If there are obstacles in quadrant A, it must be further partitioned into four quadrants AA, AB, AC, AD. Suppose the initial point is in AA, which also contains obstacles; then AA is further partitioned into quadrants AAA, AAB, AAC, AAD. This procedure terminates when there are no obstacles in the quadrant containing the initial point. A similar procedure is effected for the goal position. Based on this space decomposition, the quadtree shown in Fig. 33.27(b) may be drawn. Now, tree search methods such as A^* may be used to determine the optimal obstacle-free path.

The quadtree approach has the advantage of partitioning the space only as finely as necessary. If any quadrant contains neither goal, initial point, nor obstacles, it is not further partitioned. If any quadrant containing the initial position or goal contains no obstacles, it is not further partitioned. In 3D, this approach is called *octree*.

Maneuvering Board Solution for Collision Avoidance of Moving Obstacles

The techniques just discussed generate a set of via points between the initial and final positions. If there are moving obstacles within the free-space regions, one may often modify the paths between the via points online in real-time to avoid collision. If obstacles are moving with constant known velocities in the free space, a technique used by the U.S. Navy based on the *maneuvering board* can be used for on-line obstacle avoidance. Within a convex free-space region, generated for instance by the **cell decomposition** approach, one makes a *relative polar plot* with the moving robot at the center and other moving objects plotted as straight lines depending on their relative courses and speeds. A steady bearing and decreasing range (SBDR) indicates impending collision. Standard graphical techniques using a parallel ruler allow one to alter the robot's course and/or speed to achieve a prescribed CPA; these can be converted to explicit formulas for required course/speed changes. An advantage of this technique for mobile robots is that the coordinates of obstacles in the relative polar plot can be directly measured using onboard sonar and/or laser range finders. This technique can can be modified into a *navigational technique* when some of the stationary obstacles have fixed absolute positions, such obstacles are known as *reference landmarks*.

Potential Field Approach

The potential field approach [Arkin 1989] is especially popular in mobile robotics as it seems to emulate the reflex action of a living organism. A fictitious attractive potential field is considered to be centered at the goal position [Fig. 33.28(a)]. Repulsive fields are selected to surround the obstacles [Fig. 33.28(b)]. The sum of the potential fields [Fig. 33.28(c)] produces the robot motion as follows. Using $\mathbf{F}(\mathbf{x}) = ma$, with m the vehicle mass and $\mathbf{F}(\mathbf{x})$ equal to the sum of the forces from the various potential fields computed at the current vehicle position $\mathbf{x}$, the required vehicle acceleration $a(\mathbf{x})$ is computed. The resulting motion avoids obstacles and converges to the goal position. This approach does not produce a global path planned a priori. Instead, it is a real-time on-line motion control technique that can deal with moving obstacles,

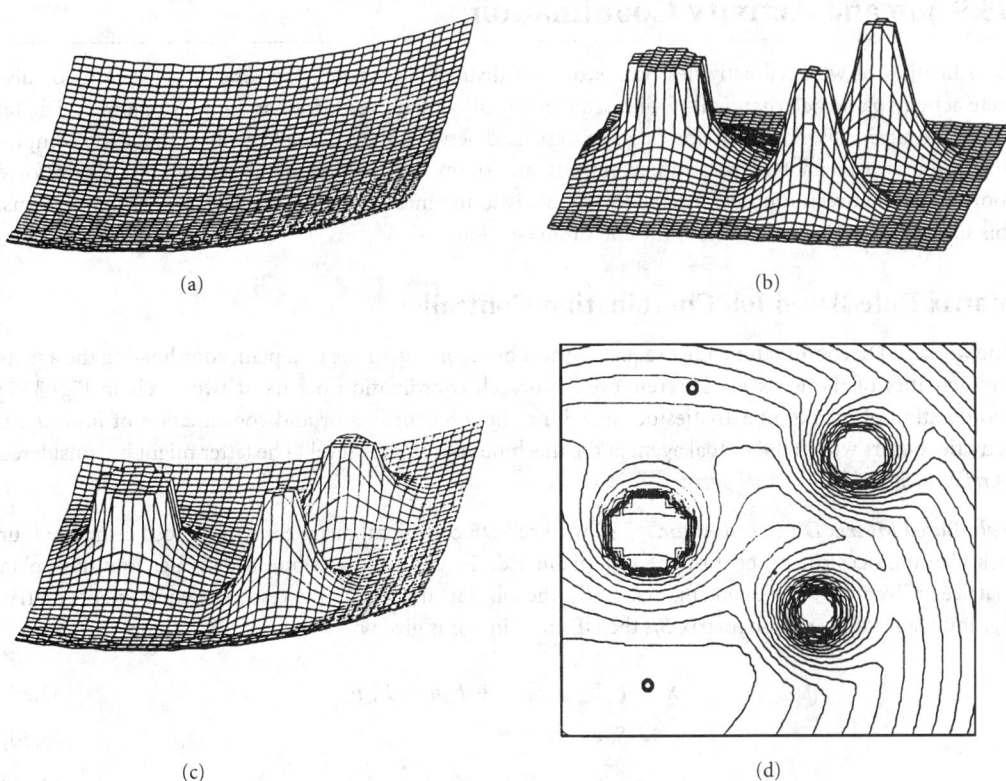

(a)   (b)   (c)   (d)

FIGURE 33.28 Potential field approach to navigation: (a) attractive field for goal at lower left corner, (b) repulsive fields for obstacles, (c) sum of potential fields, and (d) contour plot showing motion trajectory. (*Source:* Courtesy of Zhou, C. 1996. Planning and intelligent control. In *CRC Handbook of Mechanical Engineering*. F. Kreith, ed. CRC Press, Boca Raton, FL. With permission.)

particularly if combined with maneuvering board techniques. Various methods have been proposed for selecting the potential fields; they should be limited to finite influence distances, or else the computation of the total force $\mathbf{F(x)}$ requires knowledge of all obstacle relative positions.

The potential field approach is particularly convenient as the force $\mathbf{F}$ may be computed knowing only the *relative positions* of the goal and obstacles from the vehicle; this information is directly provided by onboard sonar and laser readings. The complete potential field does not need to be computed, only the force vector of each field acting on the vehicle. A problem with the potential field approach is that the vehicle may become trapped in *local minima* (e.g., an obstacle is directly between the vehicle and the goal); this can be corrected using various techniques, including adding a dither force to get the vehicle out of these false minima. The potential field approach can be combined with Lyapunov analysis techniques to integrate the path planning and trajectory following servocontrol functions of a mobile robot [Jagannathan et al. 1994].

In fact, Lyapunov functions and potential fields may simply be added in an overall controls design technique.

Emergent Behaviors.   The responses to individual potential fields can be interpreted as *behaviors* such as seek goal, avoid obstacle, etc. Potential fields can be selected to achieve specialized behaviors such as *docking* (i.e., attaining a goal position with a prescribed angle of approach) and remaining in the center of a corridor (simply define repulsive fields from each wall). The sum of all of the potential fields yields an *emergent behavior* that has not been preprogrammed (e.g., seek goal while avoiding obstacle and remaining in the center of the hallway). This makes the robot exhibit behaviors that could be called intelligent or self-determined.

33.9 Job and Activity Coordination

Coordination of workcell activities occurs on two distinct planes. On the *discrete event* (DE) or discrete activity plane, *job coordination* and sequencing, along with resource handling, is required. Digital input/output signals, or sequencing *interlocks*, are used between the workcell agents to signal job completion, resource availability, errors and exceptions, and so on. On a lower plane, servo-level motion/force coordination between multiple interacting robots is sometimes needed in special-purpose applications; this specialized topic is relegated to the end of this section.

Matrix Rule-Based Job Coordination Controller

The workcell DE coordinators must sequence the jobs according to the task plan, coordinating the agents and activities of the workcell. Discrete event workcell coordination occurs at two levels in Fig. 33.7: coordination of *interagent* activities occurs within the job coordinator and coordination of *intra-agent* activities occurs within the virtual agent at the machine coordinator level. The latter might be considered as *reflex actions* of the virtual agent.

Rule-Based Matrix DE Coordinator.   The workcell DE coordinators are easily produced using the four task plan matrices constructed by the task planner designed in the section on Petri net from task plan matrices. Given the job sequencing matrix F_v, the job start matrix S_v, the resource requirements matrix F_r, and the resource release matrix S_r, the DE coordinator is given by

$$\bar{\mathbf{x}} = F_v \bar{v}_c + F_r \bar{r}_c + F_u \bar{u} + F_D \bar{u}_D \tag{33.29}$$

$$\mathbf{v}_s = S_v x \tag{33.30}$$

$$\mathbf{r}_s = S_r x \tag{33.31}$$

where Eq. (33.29) is the *controller state equation*, Eq. (33.30) is the *job start equation*, and Eq. (33.31) is the *resource release equation*. This is a set of *logical equations* where all matrix operations are carried out in the matrix or/and algebra; addition of elements is replaced by OR, and multiplication of elements is replaced by AND. Overbars denote logical negation, so that Eq. (33.29) is a rule base composed of AND statements (e.g., if job b is completed and job c is completed and resource $R1$ is available, then set state component $\mathbf{x}_4$ high), and Eqs. (33.30) and (33.31) are rule bases composed of OR statements (e.g., if state component $\mathbf{x}_4$ is high or state component $\mathbf{x}_7$ is high, then release resource $M1$).

For complex tasks with many jobs, the matrices in the DE controller can be large. However, they are sparse. Moreover, for special manufacturing structures such as the re-entrant flow line, the matrices in Eq. (33.29) are *lower block triangular*, and this special structure gets around problems associated with the NP-hard of general manufacturing job shops. Finally, as rule bases, the DE controller equations may be fired using standard efficient techniques for forward chaining, backward chaining (Rete algorithm), and so on.

The structure of the DE job coordination controller is given in Fig. 33.29. This shows that the job coordinator is simply a *closed-loop feedback control system* operating at the DE level. At each time increment, workcell status signals are measured including the job complete status vector $\mathbf{v}_c$ (where entries of 1 denote jobs complete), the resource availability vector $\mathbf{r}_c$ (where entries of 1 denote resources available), and the part input vector $\mathbf{u}$ (where entries of 1 denote parts coming into the workcell). These signals are determined using workcell digital interlocks and digital input/output between the agents. Based on this workcell status information, the DE controller computes which jobs to start next and which resources to release. These commands are passed to the workcell in the job start vector $\mathbf{v}_s$ (where entries of 1 denote jobs to be started) and the resource release vector $\mathbf{r}_s$ (where entries of 1 denote resources to be released).

Deadlocks and Resource Dispatching Commands. Outer feedback loops are required to compute the *dispatching input* u_D. A separate dispatching input is required whenever a column of F_r contains multiple ones, indicating that the corresponding resource is a *shared resource* required for more than one job. In a manufacturing workcell, if care is not taken to assign shared resources correctly, *system deadlock* may occur. In deadlock, operation of a subsystem ceases as a *circular blocking* of resources has developed [Wysk et al. 1991]. In a circular blocking, each resource is waiting for each other resource, but none will ever again become available. In the example of Fig. 33.24, a circular blocking occurs if machine $M1$ is waiting at b to be unloaded by $R1$, but $R1$ already has a part at d and the buffer $B2$ is full at e. On the other hand, buffer $B2$ cannot be unloaded at e since $M1$ already has a part at b.

There are well-established algorithms in industrial engineering for job dispatching [Panwalker and Iskander 1977], including first-in–first-out, earliest due date, last buffer first serve, etc. *Kanban systems* are pull systems where no job can be started unless a kanban card is received from a downstream job (possibly indicating buffer space or resource availability, or part requirements). A generalized kanban system that guarantees deadlock avoidance can be detailed in terms of the DE controller matrices, for it can be shown that all the *circular waits* of the workcell for a particular task are given in terms of the graph defined by $S_r F_r$, with S_r the resource release matrix and F_r the resource requirements matrix. Only circular waits can develop into circular blockings. Based on this, in Lewis and Huang [1995] a procedure known as maximum work-in-process (MAXWIP) is given that guarantees dispatching with no deadlock.

Process Integration, Digital I/O, and Job Coordination Controller Implementation

Information integration is the process by which the activities and status of the various workcell agents interact. Subcategories of information integration include *sensor integration* and *sensor/actuator integration*. Motion/process integration involves coordinating manipulator motion with process sensor or process controller devices. The most primitive process integration is through discrete digital I/O, or sequencing interlocks. For example a machine controller external to the robot controller might send a 1-b signal indicating that it is ready to be loaded by the robot. The DE matrix-based job coordination controller provides an ideal technique for information integration. The workcell status signals required in Fig. 33.29, the job completion vector and resource availability vector, are given by digital output signals provided by sensors in the worccell (e.g., gripper open, part inserted in machine). The workcell command signals in

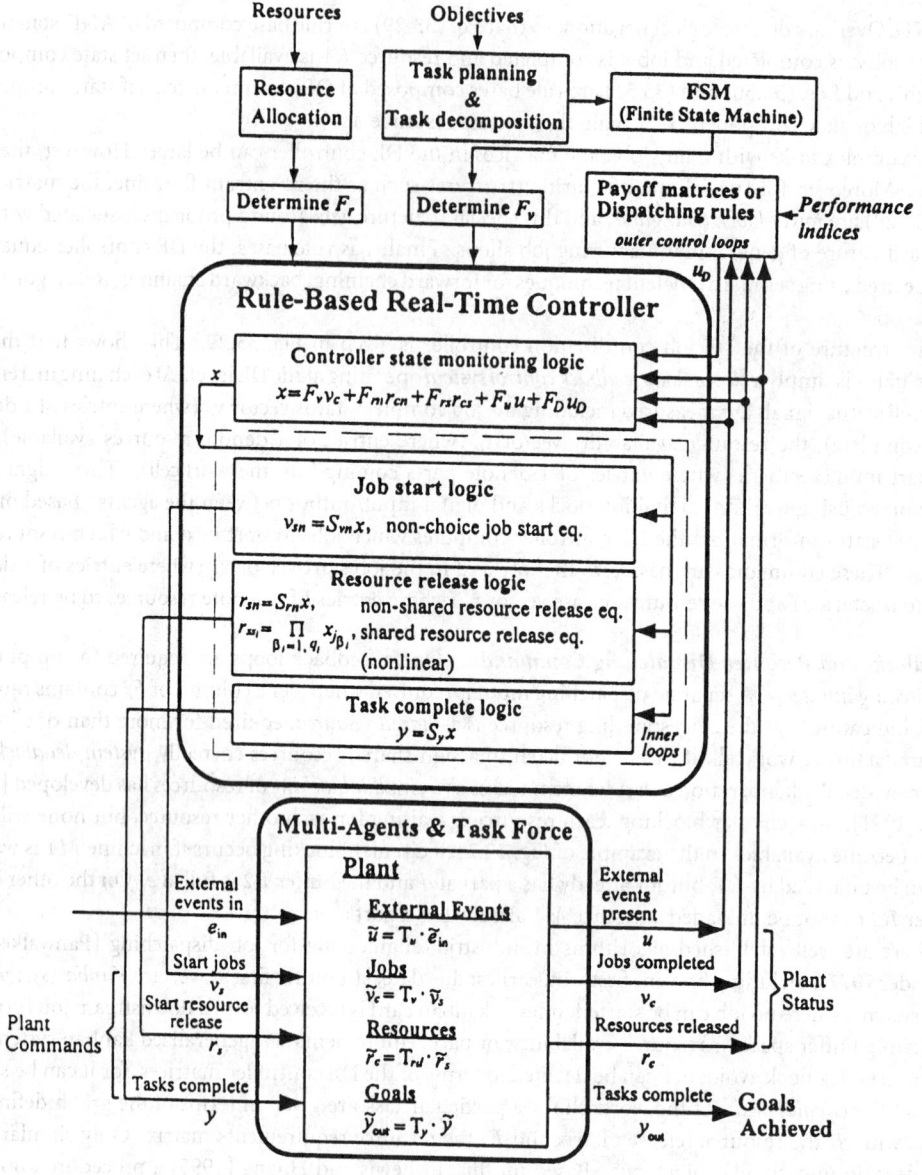

FIGURE 33.29   Matrix-based DE feedback controller for job sequencing and resource allocation.

Fig. 33.29, the job start vector and resource release vector, are given by digital *discrete input* signals to the workcell agents (e.g., move robot to next via point, pick up part, load machine).

   The DE job coordination controller is nothing but a rule base, and so may be implemented either on commercial progammable logic controllers (PLC) or on commercial robot controllers. In many cases, the DE controller matrices are block triangular, so that portions of the controller can be implemented hierarchically in separate subsystems (e.g., coordination between a single robot and a camera for some jobs). If this occurs, portions of the controller may be implemented in the machine coordinator in Fig. 33.7, which often resides within the robot arm controller. Many robot controllers now support information integration functions by employing integrated PC interfaces through the communications ports, or in some through direct connections to the robot controller data bus. Higher level interactive portions of the DE controller should be implemented at the job coordination level, which is often realized on a dedicated

PLC. Vision-guided high-precision pick and place and assembly are major applications in the electronics and semiconductor industries. Experience has shown that the best integrated vision/robot performance has come from running both the robot and the vision system internal to the same computing platform, since data communication is much more efficient due to data bus access, and computing operations are coordinated by one operating system.

Coordination of Multiple Robots

Coordination of multiple robots can be accomplished either at the discrete event level or the servocontrol level. In both cases it is necessary to avoid collisions and other interactions that impede task completion. If the arms are not interacting, it is convenient to coordinate them at the DE level, where collision avoidance may be confronted by assigning any intersecting workspace where two robots could collide as *a shared resource*, accessible by only a single robot at any given time. Then, techniques such as those in the section on the task matrix approach to workcell planning may be used (where $M1$ was a shared resource). Such approaches are commonly used in coordination control of automated guided vehicles (AGV) [Gruver et al. 1984].

Two-Arm Motion/Force Coordination. In specialized robotic applications requiring two-arm interaction, such as coordinated lifting or process applications, it may be necessary to coordinate the motion and force exertion of the arms on the joint servocontrol level [Hayati et al. 1989]. In such cases, the two robot arms may be considered in Fig. 33.7 as a *single virtual agent* having specific behaviors as defined by the feedback servocontroller.

There are two basic approaches to two-arm servocontrol. In one approach, one arm is considered as the *master*, whose commanded trajectory is in terms of motion. The other, *slave*, arm is commanded to maintain prescribed forces and torques across a payload mass, which effectively constrains its relative motion with respect to the master arm. By this technique, the motion control and force/torque control problems are relegated to different arms, so that the control objectives are easily accomplished by servo-level feedback controller design. Another approach to two-arm coordination is to treat both arms as equals, coordinating to maintain prescribed linear and angular motions of the center-of-gravity (c.g.) of a payload mass, as well as prescribed internal forces and torques across the payload. This approach involves complex analyses to decompose the payload c.g. motion and internal forces into the required motion of each arm; kinematic transformations and Jacobians are needed.

33.10 Error Detection and Recovery

The material in this section is modfied from Zhou [1996]. In the execution of a task, errors can occur. The errors can be classified into several categories: hardware error, software error, and operational error. The hardware errors include errors in mechanical and electrical mechanisms of the robot, such as failure in the drive system or sensing system. Software errors can be bugs in the application program or control software. Timing with cooperative devices can also be called software error. Operational errors are errors in the robot environment that are external to the robot system such as jamming of parts or collision with obstacles.

During this discussion one should keep in mind the PC&C structure in Fig. 33.7. Error handling can be classified into two activities, *error detection* and *error recovery*. Error detection is composed of error sensing, interpretation, and classification. Error recovery is composed of decision making and corrective job assignment. While corrective jobs are being performed, the assigned task may be interrupted, or may continue to run at a reduced capability (e.g., one of two drilling machines may be down).

Error Detection

The sensors used in error detection can include all those discussed in section 33.7 including tactile sensors for sensing contact errors, proximity sensors for sensing location or possible collision, force/torque sensors for sensing collision and jamming, and vision for sensing location, orientation and error existence. Once an error is sensed, it must be interpreted and classified. This may be accomplished by servo-level state observers, logical rule-based means, or using advanced techniques such as neural networks.

Error Recovery

The occurrence of an error usually causes interruption of the normal task execution. Error recovery can be done at three levels, where errors can be called *exceptions, faults, and failures*. At the lowest level the exception will be corrected automatically, generally in the real-time servocontrol loops, and the task execution continued. An example is jamming in the pin insertion problem, where a force/torque wrist sensor can indicate jamming as well as provide the information needed to resolve the problem. At the second level, the error is a fault that has been foreseen by the task planner and included in the task plan passed to the job coordinator.

The vector $\mathbf{x}$ in Eq. (33.29) contains *fault states*, and logic is built into the task plan matrices to allow corrective job assignment. Upon detection of an error, jobs can be assigned to correct the fault, with the task subsequently continued from the point where the error occurred. At this level, the error detection/recovery logic can reside either in the machine coordinator or in the job coordinator.

At the highest level of recovery, the error was not foreseen by the task planner and there is no error state in the task plan. This results in a failure, where the task is interrupted. Signals are sent to the planner, who must correct the failure, sometimes with external resources, and replan the task, passing another plan to the coordinator. In the worst case, manual operator intervention is needed. It can be seen that the flow of error signals proceeds upwards and of commands proceeds downwards, exactly as in the NASREM architecture in Fig. 33.4.

At the lowest servocontrol level, additional sensory information is generally required for error recovery, as in the requirement for a wrist force/torque sensor in pin insertion. At the mid-level, additional logic is needed for error recovery. At the highest level, task replanning capabilities are needed.

33.11 Human Operator Interfaces

Human operator integration is critical to the expeditious setup, programming, maintenance, and sometimes operation of the robotic workcell. Especially important for effective human integration are the available human I/O devices, including the information available to the operator in graphical form and the modes of real-time control operation available for human interaction. Teaching, programming, and operational efforts are dramatically influenced by the type of user interface I/O devices available.

Levels of User Interface

Discounting workcell design and layout, operator interfaces occur at several levels in Fig. 33.7 and may be classified into off-line and on-line activities. Off-line interfaces occur in task definition and setup, often consisting of teaching activities. In workcell management, user inputs include assignment of tasks, due dates, and so on. At the workcell planning level, user functions might be required in task planning, both in task decomposition/job sequencing and in resource assignment. Off-line CAD programs are often useful at this level. In path planning, the user might be required to teach a robot specific path via points for job accomplishment. Finally, if failures occur, a human might be required to clear the failure, reset the workcell, and restart the job sequence.

On-line user interfaces may occur at the discrete event level and the servocontrol level. In the former case, a human might perform some of the jobs requested by the job coordinator, or may be required to perform corrective jobs in handling foreseen faults. At the servocontrol level, a human might perform teleoperator functions, or may be placed in the inner feedback control loop with a machine or robotic device.

Mechanisms for User Interface

Interactive 3D CAD. Computer integrated manufacturing operations require off-line programming and simulation in order to layout production facilities, model and evaluate design concepts, optimize motion of devices, avoid interference and collisions, minimize process cycle times, maximize productivity, and ensure maximum return on investment. Graphical interfaces, available on some industrial robots, are very effective for conveying information to the operator quickly and efficiently. A graphical interface is most important for design and simulation functions in applications which require frequent reprogramming and setup changes. Several very useful off-line programming software systems are available from third party suppliers (CimStation [SILMA 1992], ROBCAD, IGRIP). These systems use CAD and/or dynamics computer models of commercially available robots to simulate job execution, path motion, and process activities, providing rapid programming and virtual prototyping functions. Interactive off-line CAD is useful for assigning tasks at the management level and for task decomposition, job sequencing, and resource assignment at the task planning level.

Off-Line Robot Teaching and Workcell Programming. Commercial robot or machine tool controllers may have several operator interface mechanisms. These are generally useful at the level of off-line definition or teaching of jobs, which can then be sequenced by the job coordinator or machine coordinator to accomplish assigned tasks. At the lowest level one may program the robot in its operating language, specifying path via points, gripper open/close commands, and so on. Machine tools may require programming in CNC code. These are very tedious functions, which can be avoided by object-oriented and open architecture approaches in well-designed workcells, where such functions should be performed automatically, leaving the user free to deal with other higher level supervisory issues. In such approaches, macros or subroutines are written in machine code which *encapsulate the machine behaviors* (e.g., set speed, open gripper, go to prescribed point). Then, higher level software passes specific parameters to these routines to execute behaviors with specific location and motion details as directed by the job coordinator.

Many robots have a teach pendant, which is a low-level teleoperation device with push buttons for moving individual axes and other buttons to press commanding that certain positions should be memorized. On job execution, a playback mode is switched in, wherein the robot passes through the taught positions to sweep out a desired path. This approach is often useful for teaching multiple complex poses and Cartesian paths.

The job coordinator may be implemented on a programmable logic controller (PLC). PLC programming can be a tedious and time-consuming affair, and in well-designed flexible reconfigurable workcells an object-oriented approach is used to avoid reprogramming of PLCs. This might involve a programming scheme that takes the task plan matrices in section 33.9 as inputs and automatically implements the coordinator using rule-based techniques (e.g., forward chaining, Rete algorithm).

Teleoperation and Man-in-the-Loop Control. Operator interaction at the servocontrol level can basically consist of two modes. In man-in-the-loop control, a human provides or modifies the feedback signals that control a device, actually operating a machine tool or robotic device. In teleoperation, an inner feedback loop is closed around the robot, and a human provides motion trajectory and force commands to the robot in a master/slave relationship. In such applications, there may be problems if extended communications distances are involved, since delays in the communications channel can destabilize a teleoperation system having force feedback unless careful attention is paid to designing the feedback loops to maintain *passivity*. See Lewis [1996] for more details.

33.12 Robot Workcell Programming

The robotic workcell requires programming at several levels [Leu 1985]. At the lower levels one generally uses commercial programing languages peculiar to device manufacturers of robots and CNC machine tools. At the machine coordination level, robot controllers are also often used with discrete I/O signals and decision making commands. At the job coordination level prorammable logic controllers (PLCs) are often used in medium complexity workcells, so that a knowledge of PLC programming techniques is required. In modern manufacturing and process workcells, coordination may be accomplished using general purpose computers with programs written, for instance, in C.

Robot Programming Languages

Subsequent material in this section is modified from Bailey [1996]. Each robot manufacturer has its own proprietary programming language. The variety of motion and position command types in a programming language is usually a good indication of the robot's motion generation capability. Program commands which produce complex motion should be available to support the manipulation needs of the application. If palletizing is the application, then simple methods of creating position commands for arrays of positions are essential. If continuous path motion is needed, an associated set of continuous motion commands should be available. The range of motion generation capabilities of commercial industrial robots is wide. Suitability for a particular application can be determined by writing test code.

The earliest industrial robots were simple sequential machines controlled by a combination of servomotors, adjustable mechanical stops, limit switches, and PLCs. These machines were generally programmed by a record and play-back method with the operator using a teach pendant to move the robot through the desired path. MHI, the first robot programming language, was developed at Massachusetts Institute of Technology (MIT) during the early 1960s. MINI, developed at MIT during the mid-1970s was an expandable language based on LISP. It allowed programming in Cartesian coordinates with independent control of multiple joints. VAL and VAL II [Shimano et al. 1984], developed by Unimation, Inc., were interpreted languages designed to support the PUMA series of industrial robots. **A manufacturing language (AML)** was a completely new programming language developed by IBM to support the R/S 1 assembly robot. It was a subroutine-oriented, interpreted language which ran on the Series/1 minicomputer. Later versions were compiled to run on IBM compatible personal computers to support the 7535 series of SCARA robots. Several additional languages [Gruver et al. 1984, Lozano-Perez 1983] were introduced during the late 1980s to support a wide range of new robot applications which were developed during this period.

V+, A Representative Robot Language

V+, developed by Adept Technologies, Inc., is a representative modern robot programming language with several hundred program instructions and reserved keywords. V+ will be used to demonstrate important features of robot programming. Robot program commands fall into several categories, as detailed in the following subsections.

Robot Control.   Program instructions required to control robot motion specify location, trajectory, speed, acceleration, and obstacle avoidance. Examples of V+ robot control commands are as follows:

> MOVE: Move the robot to a new location.
> DELAY: Stop the motion for a specified period of time.
> SPEED: Set the speed for subsequent motions.
> ACCEL: Set the acceleration and deceleration for subsequent motions.
> OPEN: Open the hand.
> CLOSE: Close the hand.

System Control.   In addition to controlling robot motion, the system must support program editing and debugging, program and data manipulation, program and data storage, program control, system definitions and control, system status, and control/monitoring of external sensors. Examples of V+ control instructions are as follows:

EDIT:	Initiate line-oriented editing.
STORE:	Store information from memory onto a disk file.
COPY:	Copy an existing disk file into a new program.
EXECUTE:	Initiate execution of a program.
ABORT:	Stop program execution.
DO:	Execute a single program instruction.
WATCH:	Set and clear breakpoints for diagnostic execution.
TEACH:	Define a series of robot location variables.
CALIBRATE:	Initiates the robot positioning system.
STATUS:	Display the status of the system.
ENABLE:	Turn on one or more system switches.
DISABLE:	Turn off one or more system switches.

Structures and Logic. Program instructions are needed to organize and control execution of the robot program and interaction with the user. Examples include familiar commands such as FOR, WHILE, IF as well as commands like the following:

WRITE:	Output a message to the manual control pendant.
PENDANT:	Receive input from the manual control pendant.
PARAMETER:	Set the value of a system parameter.

Special Functions. Various special functions are required to facilitate robot programming. These include mathematical expressions such as COS, ABS, and SQRT, as well as instructions for data conversion and manipulation, and kinematic transformations such as the following:

BCD:	Convert from real to binary coded decimal.
FRAME:	Compute the reference frame based on given locations.
TRANS:	Compose a transformation from individual components.
INVERSE:	Return the inverse of the specified transformation.

Program Execution. Organization of a program into a sequence of executable instructions requires scheduling of tasks, control of subroutines, and error trapping/recovery. Examples include the following:

PCEXECUTE:	Initiate the execution of a process control program.
PCABORT:	Stop execution of a process control program.
PCPROCEED:	Resume execution of a process control program.
PCRETRY:	After an error, resume execution at the last step tried.
PCEND:	Stop execution of the program at the end of the current execution cycle.

Example Program. This program demonstrates a simple pick and place operation. The values of position variables *pick* and *place* are specified by a higher level executive that then initiates this subroutine:

1	.PROGRAM move.parts()	
2	;	Pick up parts at location "pick" and put them down at "place"
3	parts = 100	; Number of parts to be processed
4	height1 = 25	; Approach/depart height at "pick"
5	height2 = 50	; Approach/depart height at "place"
6	PARAMETER.HAND.TIME = 16	; Setup for slow hand
7	OPEN	; Make sure hand is open
8	MOVE start	; Move to safe starting location
9	For i = 1 TO parts	; Process the parts
10	APPRO pick, height1	; Go toward the pick-up
11	MOVES pick	; Move to the part
12	CLOSEI	; Close the hand

13	DEPARTS height1	; Back away
14	APPRO place, height2	; Go toward the put-down
15	MOVES place	; Move to the destination
16	OPENI	; Release the part
17	DEPARTS height2	; Back away
18	END	; Loop for the next part
19	TYPE "ALL done.", /I3, parts, "parts processed"	
20	STOP	; End of the program
21	.END	

Off-Line Programming and Simulation. Commercially available software packages (discussed in section 33.11) provide support for off-line design and simulation of 3D workcell layouts including robots, end effectors, fixtures, conveyors, part positioners, and automatic guided vehicles. Dynamic simulation allows off-line creation, animation, and verification of robot motion programs. However, these techniques are limited to verification of overall system layout and preliminary robot program development. With support for data exchange standards [e.g., **International Graphics Exchange Specification (IGES)**, **Virtual Data Acquisition and File Specification (VDAFS)**, **Specification for Exchange of Text (SET)**], these software tools can pass location and trajectory data to a robot control program, which in turn can provide the additional functions required for full operation (operator guidance, logic, error recovery, sensor monitoring/control, system management, etc.).

33.13 Mobile Robots and Automated Guided Vehicles

A topic which has always intrigued computer scientists is that of **mobile robots** [Zheng 1993]. These machines move in generally unstructured environments and so require enhanced decision making and sensors; they seem to exhibit various anthropomorphic aspects since vision is often the sensor, decision making mimics brain functions, and mobility is similar to humans, particularly if there is an onboard robot arm attached. Here are discussed mobile robot research and factory automated guided vehicle (AGV) systems, two widely disparate topics.

Mobile Robots

Unfortunately, in order to focus on higher functions such as decision making and high-level vision processing, many researchers treat the mobile robot as a dynamical system obeying Newton's laws $F = ma$ (e.g., in the potential field approach to motion control, discussed earlier). This simplified dynamical representation does not correspond to the reality of moving machinery which has nonholonomic constraints, unknown masses, frictions, Coriolis forces, drive train **compliance**, wheel slippage, and backlash effects. In this subsection we provide a framework that brings together three camps: computer science results based on the $F = ma$ assumption, nonholonomic control results that deal with a kinematic steering system, and full servo-level feedback control that takes into account all of the vehicle dynamics and uncertainties.

Mobile Robot Dynamics

The full dynamical model of a rigid mobile robot (e.g., no flexible modes) is given by

$$M(\mathbf{q})\ddot{\mathbf{q}} + V_m(\mathbf{q}, \dot{\mathbf{q}})\dot{\mathbf{q}} + F(\mathbf{q}, \dot{\mathbf{q}}) + \mathbf{G}(\mathbf{q}) + \boldsymbol{\tau}_d = B(\mathbf{q})\boldsymbol{\tau} - A^T(\mathbf{q})\boldsymbol{\lambda} \qquad (33.32)$$

which should be compared to Eqs. (33.7) and (33.14). In this equation, M is an inertia matrix, V_m is a matrix of Coriolis and centripetal terms, $\mathbf{F}$ is a friction vector, $\mathbf{G}$ is a gravity vector, and $\boldsymbol{\tau}_d$ is a vector of disturbances. The n-vector $\boldsymbol{\tau}(t)$ is the control input. The dynamics of the driving and steering motors should be included in the robot dynamics, along with any gearing. Then, τ might be, for example, a vector of voltage inputs to the drive actuator motors.

The vehicle variable $\mathbf{q}(t)$ is composed of Cartesian position (x, y) in the plane plus orientation θ. If a robot arm is attached, it can also contain the vector of robot arm joint variables. A typical mobile robot with no onboard arm has $\mathbf{q} = [x \; y \; \theta]^T$, where there are three variables to control, but only two inputs, namely, the voltages into the left and right driving wheels (or, equivalently, vehicle speed and heading angle).

The major problems in control of mobile robots are the fact that there are more degrees of freedom than control inputs, and the existence of nonholonomic constraints.

Nonholonomic Constraints and the Steering System

In Eq. (33.12) the vector of constraint forces is λ and matrix $A(\mathbf{q})$ is associated with the constraints. These may include nonslippage of wheels and other *holonomic* effects, as well as the *nonholonomic constraints*, which pose one of the major problems in mobile robot control. Nonholonomic constraints are those which are nonintegrable, and include effects such as the impossibility of sideways motion (think of an automobile). In research laboratories, it is common to deal with omnidirectional robots that have no nonholonomic constraints, but can rotate and translate with full degrees of freedom; such devices do not correspond to the reality of existing shop floor or cross-terrain vehicles which have nonzero turn radius.

A general case is where all kinematic equality constraints are independent of time and can be expressed as

$$A(\mathbf{q})\dot{\mathbf{q}} = 0 \qquad (33.33)$$

Let $S(\mathbf{q})$ be a full-rank basis for the nullspace of $A(\mathbf{q})$ so that $AS = 0$. Then one sees that the linear and angular velocities are given by

$$\dot{\mathbf{q}} = S(\mathbf{q})\mathbf{v(t)} \qquad (33.34)$$

where $\mathbf{v(t)}$ is an auxiliary vector. In fact, $\mathbf{v(t)}$ often has physical meaning, consisting of two components: the commanded vehicle speed and the heading angle. Matrix $S(\mathbf{q})$ is easily determined independently of the dynamics (33.32) from the wheel configuration of the mobile robot. Thus, Eq. (33.34) is a *kinematic* equation that expresses some simplified relations between motion $\mathbf{q}(t)$ and a fictitious *ideal speed and heading* vector $\mathbf{v}$. It does not include dynamical effects, and is known in the nonholonomic literature as the *steering system*. In the case of omnidirectional vehicles $S(\mathbf{q})$ is 3×3 and Eq. (33.34) corresponds to the Newton's law model $F = ma$ used in, e.g., potential field approaches.

There is a large literature on selecting the command $\mathbf{v(t)}$ to produce desired motion $\mathbf{q}(t)$ in nonholonomic systems; the problem is that $\mathbf{v}$ has two components and $\mathbf{q}$ has three. Illustrative references include the chapters by Yamamato and Yun and by Canudas de Wit et al. in Zheng [1993], as well as Samson and Ait-Abderrahim [1991]. There are basically three problems considered in this work: following a prescribed path, tracking a prescribed trajectory (e.g., a path with prescribed transit times), and stabilization at a prescribed final docking position (x, y) and orientation θ. Single vehicle systems as well as multibody systems (truck with multiple trailers) are treated. The results obtained are truly remarkable and are in the vein of a path including the forward/backward motions necessary to park a vehicle at a given docking position and orientation. All of the speed reversals and steering commands are automatically obtained by solving certain coupled nonlinear equations. This is truly the meaning of intelligence and autonomy.

Conversion of Steering System Commands to Actual Vehicle Motor Commands

The steering system command vector obtained from the nonholonomic literature may be called $\mathbf{v}_c(\mathbf{t})$, the ideal desired value of the speed/heading vector $\mathbf{v(t)}$. Under the so-called perfect velocity assumption the actual vehicle velocity $\mathbf{v(t)}$ follows the command vector $\mathbf{v}_c(\mathbf{t})$, and can be directly given as control input to the vehicle. Unfortunately, in real life this assumption does not hold. One is therefore faced with the problem of obtaining drive wheel and steering commands for an actual vehicle from the steering system command $\mathbf{v}_c(\mathbf{t})$.

To accomplish this, premultiply Eq. (33.32) by $S^T(\mathbf{q})$ and use Eq. (33.34) to obtain

$$\overline{M}(\mathbf{q})\dot{\mathbf{v}} + \overline{V}_m(\mathbf{q}, \dot{\mathbf{q}})\mathbf{v} + \overline{\mathbf{F}}(\mathbf{v}) + \overline{\tau}_d = \overline{B}(\mathbf{q})\tau \tag{33.35}$$

where gravity plays no role and so has been ignored, the constraint term drops out due to the fact that $AS = 0$, and the overbar terms are easily computed in terms of original quantities. The true model of the vehicle is thus given by combining both Eqs. (33.34) and (33.35). However, in the latter equation it turns out that $(\overline{B})(\mathbf{q})$ is square and invertible, so that standard computed torque techniques (see section on robot servo-level motion control) can be used to compute the required vehicle control τ from the steering system command $\mathbf{v}_c(\mathbf{t})$. In practice, correction terms are needed due to the fact that $\mathbf{v} \neq \mathbf{v}_c$; they are computed using a technique known as *integrator backstepping* [Fierro and Lewis 1995].

The overall controller for the mobile robot is similar in structure to the multiloop controller in Fig. 33.13, with an inner nonlinear **feedback linearization** loop (e.g., computed torque) and an outer tracking loop that computes the steering system command. The robustifying term is computed using backstepping. Adaptive control and neural net control inner loops can be used instead of computed torque to reject uncertainties and provide additional dynamics learning capabilities. Using this multiloop control scheme, the idealized control inputs provided, e.g., by potential field approaches, can also be converted to actual control inputs for any given vehicle. A major criticism of potential field approaches has been that they do not take into account the vehicle nonholonomic constraints.

Automated Guided Vehicle Systems

Though research in mobile robots is intriguing, with remarkable results exhibiting intelligence at the potential field planning level, the nonholonomic control level, and elsewhere, few of these results make their way into the factory or other unstructured environments. There, reliability and repeatability are the main issue of concern.

Navigation and Job Coordination.   If the environment is unstructured one may either provide sophisticated planning, decision making, and control schemes or one may force structure onto the environment. Thus, in most AGV systems the vehicles are guided by wires buried in the floor or stripes painted on the floor. Antennas buried periodically in the floor provide check points for the vehicle as well as transmitted updates to its commanded job sequence.

A single computer may perform scheduling and routing of multiple vehicles. Design of this coordinating controller is often contorted and complex in actual installed systems, which may be the product of several engineers working in an ad hoc fashion over several years of evolution of the system. To simplify and unify design, the discrete event techniques in the task matrix approach section may be used for planning. Track intersections should be treated as shared resources only accessible by a single vehicle at a time, so that on-line dispatching decisions are needed. The sequencing controller is then implemented using the approach in section 33.9.

Sensors, Machine Coordination, and Servo-level Control.   Autonomous vehicles often require extensive sensor suites. There is usually a desire to avoid vision systems and use more reliable sensors including contact switches, proximity detectors, laser rangefinders, sonar, etc. Optical bar codes are sometimes placed on the walls; these are scanned by the robot so it can update its absolute position. Integrating this multitude of sensors and performing coordinated activities based on their readings may be accomplished using simple decision logic on low-level microprocessor boards. Servo-level control consists of simple PD loops that cause the vehicle to follow commanded speeds and turn commands. Distance sensors may provide information needed to maintain minimum safe intervehicular spacing.

Defining Terms

Accuracy: The degree to which the actual and commanded position (of, e.g., a robot manipulator) correspond.

Adaptive control: A large class of control algorithms where the controller has its own internal dynamics and so is capable of learning the unknown dynamics of the robot arm, thus improving performance over time.

A manufacturing language (AML): A robot programming language.

Automatic programming of tools (APT): A robot programming language.

Cell decomposition: An approach to path planning where the obstacles are modeled as polygons and the free space is decomposed into cells such that a straight line path can be generated between any two points in a cell.

Compliance: The inverse of stiffness, useful in end effectors and tooling whenever a robot must interact with rigid constraints in the environment.

Computed torque control: An important and large class of robot arm controller algorithms that relies on subtracting out some or most of the dynamical nonlinearities using feedforward compensation terms including, e.g., gravity, friction, Coriolis, and desired acceleration feedforward.

End effector: Portion of robot (typically at end of chain of links) designed to contact the external world.

Feedback linearization: A modern approach to robot arm control that formalizes computed torque control mathematically, allowing formal proofs of stability and design of advanced algorithms using Lyapunov and other techniques.

Force control: A class of algorithms allowing control over the force applied by a robot arm, often in a direction normal to a prescribed surface while the position trajectory is controlled in the plane of the surface.

Forward kinematics: Identification of Cartesian task coordinates given robot joint configuration.

International Graphics Exchange Specification (IGES): A data exchange standard.

Inverse kinematics: Identification of possible robot joint configurations given desired Cartesian task coordinates.

Joint variables: Scalars specifying position of each joint, one for each degree of freedom. The joint variable for a revolute joint is an angle in degrees; the joint variable for a prismatic joint is an extension in units of length.

Learning control: A class of control algorithms for repetitive motion applications (e.g., spray painting) where information on the errors during one run is used to improve performance during the next run.

Linearity in the parameters: A property of the robot arm dynamics, important in adaptive controller design, where the nonlinearities are linear in the unknown parameters such as unknown masses and friction coefficients.

Manipulator Jacobian: A configuration-dependent matrix relating joint velocities to Cartesian coordinate velocities.

Mechanical part feeders: Mechanical devices for feeding parts to a robot with a specified frequency and orientation. They are classified as vibratory bowl feeders, vibratory belt feeders, and programmable belt feeders.

Mobile robot: A special type of manipulator which is not bolted to the floor but can move. Based on different driving mechanisms, mobile robots can be further classified as wheeled mobile robots, legged mobile robots, treaded mobile robots, underwater mobile robots, and aerial vehicles.

Path planning: The process of finding a continuous path from an initial robot configuration to a goal configuration without collision.

Prismatic joint: Sliding or telescoping robot joint that produces relative translation of the connected links.

Proportional-integral-derivative (PID) control: A classical servocontrol feedback algorithm where the actual system output is subtracted from the desired output to obtain a tracking error. Then, a weighted linear combination of the tracking error, its derivative, and its integral are used as the control input to the system.

Remote-center-of-compliance (RCC):   A compliant wrist or end effector designed so that task-related forces and moments produce deflections with a one-to-one correspondence (i.e., without side effects). This property simplifies programming of assembly and related tasks.

Repeatability: The degree to which the actual positions resulting from two repeated commands to the same position (of, e.g., a robot manipulator) correspond.

Revolute joint: Rotary robot joint producing relative rotation of the connected links.

Robot axis: A direction of travel or rotation usually associated with a degree of freedom of motion.

Robot joint: A mechanism that connects the structural links of a robot manipulator together while allowing relative motion.

Robot link: The rigid structural elements of a robot manipulator that are joined to form an arm.

Robust control: A large class of control algorithms where the controller is generally nondynamic, but contains information on the maximum possible modeling uncertainties so that the tracking errors are kept small, often at the expense of large control effort. The tracking performance does not improve over time so the errors never go to zero.

SCARA: Selectively compliant assembly robot arm.

Singularity: Configuration for which the manipulator Jacobian has less than full rank.

Skew symmetry: A property of the dynamics of rigid-link robot arms, important in controller design, stating that $\dot{M} - \frac{1}{2}V_m$ is skew symmetric, with M the inertia matrix and V_m the Coriolis/centripetal matrix. This is equivalent to stating that the internal forces do no work.

Specification for Exchange of Text (SET):   A data exchange standard.

Task coordinates: Variables in a frame most suited to describing the task to be performed by manipulator. They are generally taken as Cartesian coordinates relative to a base frame.

Virtual Data Acquisition and File Specification (VDAFS):   A data exchange standard.

Visibility graph: A road map approach to path planning where the obstacles are modeled as polygons. The visibility graph has nodes given by the vertices of the polygons, the initial point, and the goal point. The links are straight line segments connecting the nodes without intersecting any obstacles.

Voronoi diagram: A road map approach to path planning where the obstacles are modeled as polygons. The Voronoi diagram consists of line as having an equal distance from adjacent obstacles; it is composed of straight lines and parabolas.

References

Albus, J. S. 1992. A reference model architecture for intelligent systems design. *An Introduction to Intelligent and Autonomous Control*. P. J. Antsaklis and K.M. Passino, eds., pp. 27–56. Kluwer, Boston, MA.

Antsaklis, P. J. and Passino, K. M. 1992. *An Introduction to Intelligent and Autonomous Control*. Kluwer, Boston, MA.

Arkin, R. C. 1989. Motor schema-based mobile robot navigation. *Int. J. Robotic Res.* 8(4):92–112.

Bailey, R. 1996. Robot programming languages. In *CRC Handbook of Mechanical Engineering*. F. Kreith, ed. CRC Press, Boca Raton, FL.

Brooks, R. A. 1986. A robust layered control system for a mobile robot. *IEEE. J. Robotics Automation.* RA-2(1):14–23.

Decelle, L. S. 1988. Design of a robotic workstation for component insertions. *AT&T Tech. J.* 67(2):15–22.

Edkins, M. 1983. Linking industrial robots and machine tools. *Robotic Technology*. A. Pugh, ed. IEE control engineering ser. 23, Pergrinus, London.

Fierro, R. and Lewis, F. L. 1995. Control of a nonholonomic mobile robot: backstepping kinematics into dynamics, pp. 3805–3810. *Proc. IEEE Conf. Decision Control*. New Orleans, LA., Dec.

Fraden, J. 1993. *AIP Handbook Of Modern Sensors, Physics, Design, and Applications*. American Institute of Physics.

Fu, K. S., Gonzalez, R. C., and Lee, C. S. G. 1987. *Robotics*. McGraw–Hill, New York.

Ghosh, B., Jankovic, M., and Wu, Y. 1994. Perspective problems in systems theory and its application in machine vision. *J. Math. Sys. Estim. Control.*

Groover, M. P., Weiss, M., Nagel, R. N., and Odrey, N. G. 1986. *Industrial Robotics*. McGraw–Hill, New York.

Gruver, W. A., Soroka, B. I., and Craig, J. J. 1984. Industrial robot programming languages: a comparative evaluation. *IEEE Trans. Syst., Man, Cybernetics* SMC-14(4).

Jagannathan, S., Lewis, F., and Liu, K. 1994. Motion control and obstacle avoidance of a mobile robot with an onboard manipulator. *J. Intell. Manuf.* 5:287–302.

Jamshidi, M., Lumia, R., Mullins, J., and Shahinpoor, M. 1992. *Robotics and Manufacturing: Recent Trends in Research, Education, and Applications*, Vol. 4. ASME Press, New York.

Hayati, S., Tso, K., and Lee, T. 1989. Dual arm coordination and control. *Robotics* 5(4):333–344.

Latombe, J. C. 1991. *Robot Motion Planning*, Kluwer Academic, Boston, MA.

Lee, K.-M. and Li, D. 1991. Retroreflective vision sensing for generic part presentation. *J. Robotic Syst.* 8(1):55–73.

Lee, K.-M. and Blenis, R. 1994. Design concept and prototype development of a flexible integrated vision system. *J. Robotic Syst.* 11(5):387–398.

Leu, M. C. 1985. Robotics software systems. *Rob. Comput. Integr. Manuf.* 2(1):1–12.

Lewis, F. 1996. In *CRC Handbook of Mechanical Engineering*, ed. F. Kreith. CRC Press, Boca Raton, FL.

Lewis, F. L., Abdallah, C. T., and Dawson, D. M. 1993. *Control of Robot Manipulators*. Macmillan, New York.

Lewis, F. L. and Huang, H.-H. 1995. Manufacturing dispatching controller design and deadlock avoidance using a matrix equation formulation, pp. 63–77. *Proc. Workshop Modeling, Simulation, Control Tech. Manuf.* SPIE Vol. 2596. R. Lumia, organizer. Philadelphia, PA. Oct.

Lewis, F. L., Liu, K., and Yeşildirek, A. 1995. Neural net robot controller with guaranteed tracking performance. *IEEE Trans. Neural Networks* 6(3):703–715.

Lozano-Perez, T. 1983. Robot programming. *Proc. IEEE* 71(7):821–841.

Nichols, H. R. and Lee, M. H. 1989. A survey of robot tactile sensing technology. *Int. J. Robotics Res.* 8(3):3–30.

Panwalker, S. S. and Iskander, W. 1977. A survey of scheduling rules. *Operations Res.* 26(1):45–61.

Pertin-Trocaz, J. 1989. Grasping: a state of the art. In *The Robotics Review 1*. O. Khatib, J. Craig and T. Lozano-Perez, eds., pp. 71–98. MIT Press, Cambridge, MA.

Pugh, A., ed. 1983. *Robotic Technology*. IEE control engineering ser. 23, Pergrinus, London.

Samson, C. and Ait-Abderrahim, K. 1991. Feedback control of a nonholonomic wheeled cart in Cartesian space, pp. 1136–1141. *Proc. IEEE Int. Conf. Robotics Automation*. April.

Saridis, G. N. 1996. Architectures for intelligent control. In *Intelligent Control Systems*. M. M. Gupta and R. Sinha, eds. IEEE Press.

Shimano, B. E., Geschke, C. C., and Spalding, C.H., III. 1984. Val-II: a new robot control system for automatic manufacturing, pp. 278–292. *Proc. Int. Conf. Robotics*, March 13–15.

SILMA. 1992. SILMA CimStation Robotics Technical Overview, SILMA Inc., Cupertino, CA.

Snyder, W. E. 1985. *Industrial Robots: Computer Interfacing and Control*. Prentice–Hall, Englewood Cliffs, NJ.

Spong, M. W. and Vidyasagar, M. 1989. *Robot Dynamics and Control*. Wiley, New York.

Tzou, H. S. and Fukuda, T. 1992. *Precision Sensors, Actuators, and Systems*. Kluwer Academic, Boston, MA.

Warfield, J. N. 1973. Binary matrices in system modeling. *IEEE Trans. Syst. Man, Cybernetics*. SMC-3(5):441–449.

Wolter, J., Chakrabarty, S., and Tsao, J. 1992. Methods of knowledge representation for assembly planning, pp. 463–468. *Proc. NSF Design and Manuf. Sys. Conf.* Jan.

Wright, P. K. and Cutkosky, M. R. 1985. Design of grippers. In *The Handbook of Industrial Robotics*, S. Nof, ed. Chap. 21. Wiley, New York.

Wysk, R. A., Yang, N. S., and Joshi, S. 1991. Detection of deadlocks in flexible manufacturing cells. *IEEE Trans. Robotics Automation* 7(6):853–859.

Zheng., Y. F., ed. 1993. *Recent Trends in Mobile Robots.* World Scientific, Singapore.

Zhou, C. 1996. Planning and intelligent control. In *CRC Handbook of Mechanical Engineering.* F. Kreith, ed. CRC Press, Boca Raton, FL.

Zhou, M.-C., DiCesare, F., and Desrochers, A. D. 1992. A hybrid methodology for synthesis of Petri net models for manufacturing systems. *IEEE Trans. Robotics Automation* 8(3):350–361.

Further Information

For further information one is referred to the chapter on "Robotics" by F. L. Lewis in the *CRC Handbook of Mechanical Engineering*, edited by F. Kreith, CRC Press, 1996. Also useful are robotics books by Craig (1989), Lewis, Abdallah, and Dawson (1993), and Spong and Vidyasagar (1989).

IV

Computational Science

Joe Thompson, Section Adviser
Mississippi State University

T HE EMERGING AREA OF COMPUTATIONAL SCIENCE unites computational simulation, experimental investigations, and theoretical pursuits as three fundamental modes of scientific discovery. It uses scientific visualization, made possible by computational simulation, as the window into the analysis of physical phenomena and processes, providing a virtual microscope/telescope for inquiry and investigation at an unprecedented level of detail.

789

34

Geometry-Grid Generation

Bharat K. Soni
Mississippi State University

Nigel P. Weatherill
Mississippi State University

34.1 Introduction

With the advent and rapid development of supercomputers and high-performance workstations, computational field simulation (CFS) is rapidly emerging as an essential analysis tool for science and engineering problems. In particular, CFS has been extensively utilized in analyzing fluid mechanics, heat and mass transfer, biomedics, geophysics, electromagnetics, semiconductor devices, atmospheric and ocean science, hydrodynamics, solid mechanics, civil engineering related transport phenomena, and other physical field problems in many science and engineering firms and laboratories.

The basic equations governing the general physical field can be represented as a set of nonlinear partial differential equations pursuant to a particular set of boundary conditions. For computational simulation, the field is decomposed into a collection of elemental areas (2D)/volumes (3D). The governing equations associated with the field under consideration are then approximated by a set of algebraic equations on these elemental volumes and are numerically solved to get discrete values which approximate the solution of the pertinent governing equations over the field. This discretization of the field (domain, region) into finite-elemental areas/volumes is referred to as grid generation, and the collection of discretized elemental areas/volumes is called the grid.

The numerical solution process associated with general CFS applications first involves discretization of the integral or differential form of the governing set of partial differential equations (PDEs) formulated in continuum. The discretization of these equations is usually influenced by the grid strategy under consideration and the solution strategy to be employed. In general, the solution strategies are classified as: finite difference, finite volume, and finite element. In the case of finite differences, the derivatives in the PDEs are represented by algebraic difference expressions obtained by performing Taylor series expansions of the associated solution variables at several neighbors of the point of evaluation [Thompson and Mastin 1983]. The differential forms of the governing equations are utilized in this case. However, the integral forms of the governing equations are used in the cases of finite-element and finite-volume strategies. Here the solution process involves representation of the solution variables over the cell in terms of selected functions, and then these functions are integrated over the volume (in case of finite-element) or the associated fluxes through cell sides (edges) are balanced (in case of finite volume).

The finite-element approach itself comes in two basic forms—the *variational*, where the PDEs are replaced by a more fundamental integral variational principle (from which they arise through the calculus of variations), or the *weighted residual* (Galerkin) approach in which the PDEs are multiplied by certain functions and then integrated over the cell.

In the finite-volume approach the fluxes through the cell sides (which separate discontinuous solution values) are best calculated with a procedure which represents the dissolution of such a discontinuity during the time step (Riemann solver).

The finite-difference approach, using the discrete points, is associated by many with rectangular Cartesian grids since such a regular lattice structure provides easy identification of neighboring points to be used in the representation of derivatives, whereas the finite-element approach has always been, by the nature of its construction on discrete cells, considered well-suited for irregular regions since a network of cells can be made to fill any arbitrarily shaped region and each cell is an entity unto itself, the representation being on a cell, not across cells. In view of the discretization strategy employed, the grids can be classified as structured, unstructured, or hybrid.

Structured Grids

Let $\mathbf{r} = (x, y, z)$ and $\Omega = (\xi, \eta, \zeta)$ denote the coordinates in the physical and computational space. The structured grid is presented by a network of lines of constant ξ, η, and ζ such that a one-to-one mapping can be established between physical and computational space. The computational space is made up of uniformly distributed points within a square in two dimensions or a cube in three dimensions as demonstrated in Fig. 34.1. The structured grid involving identity transformation between physical and computational space (that is, $x = \xi$, $y = \eta$, $z = \zeta$) is called a Cartesian grid. However, the body-fitted grid generated by utilizing discrete/analytic arbitrary transformations between physical and computational space is classified as a curvilinear grid. A grid around a cylinder demonstrating the Cartesian grid and curvilinear two-dimensional grid demonstrating O, C, and H type strategies and their respective correspondence with the computational domain are displayed in Figs. 34.2(a)–(d).

The curvilinear grid points conform to the solid surfaces/boundaries and hence provide the most economical and accurate way for specifying boundary conditions. For example, in the O-type grid the boundary of the cylinder is specified at $\eta = 1$ boundary, and $\xi = 1$ and $\xi = \xi_{max}$ boundaries represent the same physical boundary (commonly referred to as a cut line). In the C-type grid the cylinder boundary is mapped into only a part of the ξ boundary, as shown in the Fig. 34.2(c). Here, the boundary segment in the front of the airfoil in the computational domain and at the tail of the airfoil represent the cut line

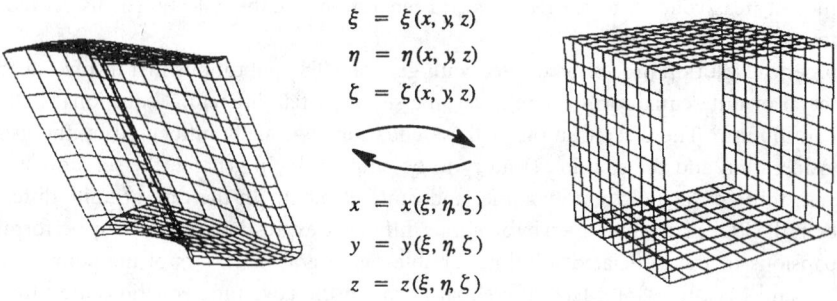

$$\xi = \xi(x, y, z)$$
$$\eta = \eta(x, y, z)$$
$$\zeta = \zeta(x, y, z)$$

$$x = x(\xi, \eta, \zeta)$$
$$y = y(\xi, \eta, \zeta)$$
$$z = z(\xi, \eta, \zeta)$$

FIGURE 34.1 Physical to computational space mapping.

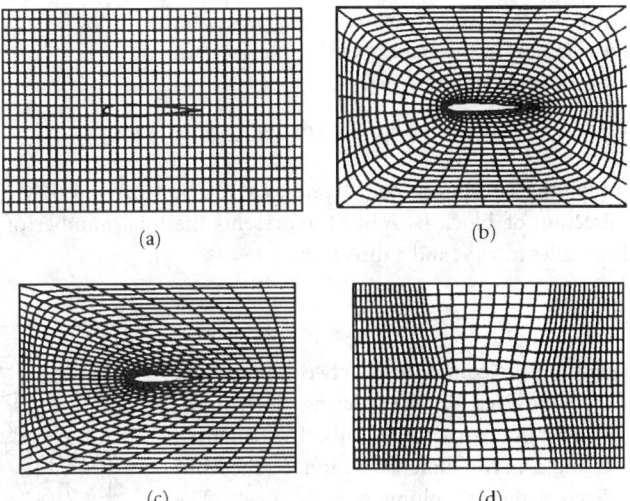

FIGURE 34.2 (a) Cartesian grid, (b) O-type grid, (c) C-type grid, and (d) H-type grid.

boundary. However, in the H-type grid an airfoil is mapped into the middle of the computational domain as shown in Fig. 34.2(d).

For complicated geometrical configurations, the physical region is divided into subregions, within each of which a structured grid is generated. These subgrids are commonly referred to as composite grids, and the generation process is referred to as domain decomposition [Thompson 1987]. The resulting subgrids may be patched together at common interfaces, may be overlapped, or may be overlaid. Self-explanatory pictorial views of domain decomposition strategies allowing patched blocks, overlapping blocks, and overlaid blocks are presented in Figs. 34.3(a) and (b). Overlaid grids are also called chimera grids after the composite monster of Greek mythology.

The use of composite grids allowing patched and overlapped blocks is very common in computational fluid dynamics where geometrical configurations representative of the physical space are extremely complex and/or distinct sets of underlying partial differential equations are to be simulated in different regions. An application of a chimera (overlaid) grid simplifies an overall grid generation and solution process when temporally deforming/moving geometrical components are involved in the simulation. The transfer of solution information at the block interface is very critical for successful simulation. In case of chimera and nonmatching blocks, the interface treatment involves interpolation, and these interpolated physical

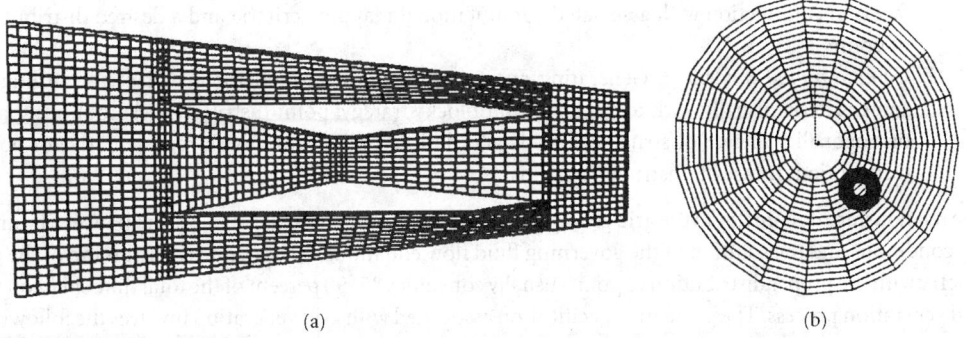

FIGURE 34.3 (a) Multiblock grid showing matching, nonmatching, and overlapping strategies and (b) overlaid chimera grid.

variables may not satisfy the conservation requirement associated with the overall simulation process, resulting in spurious oscillations in the vicinity of the interface.

In general, the composite grids are presented as

$$\mathbf{r}_{ijkl}(i = 1, \ldots, I\text{MAX}(l), \quad j = 1, \ldots, J\text{MAX}(l), \quad k = 1, \ldots, K\text{MAX}(l), \quad l = 1, \ldots, N\text{BLKS})$$

where i, j, and k identify three curvilinear coordinates, $I\text{MAX}(l)$, $J\text{MAX}(l)$, $K\text{MAX}(l)$ denote the grid dimensions in each direction of block l. NBLKS represents the total number of blocks and vector $\mathbf{r}$ contains the physical variables in x, y, and z direction.

Unstructured Grids

In structured grid generation the connections between points are automatically defined from the given (i, j, k) ordering. Such ordering (structure) does not exist in unstructured grids. Unstructured grids are composed of triangles in two dimensions and tetrahedrons in three dimensions. Each elemental volume is called a cell. The grid information is presented by a set of coordinates (nodes) and the connectivity between the nodes. The connectivity table specifies connections between nodes and cells. A pictorial view of a simple unstructured grid is presented in Fig. 34.4.

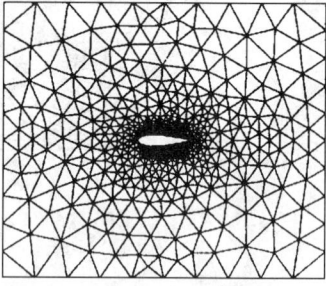

FIGURE 34.4   Unstructured grid.

Hybrid Grid

A grid formed by a combination of structured–unstructured grids and/or allowing polygonal cells with different numbers of sides is called a hybrid or generalized grid. An usual practice is to generate structured grids near solid components up to desired distance and fill in remaining regions with unstructured (triangular/tetrahedron) grids. A typical hybrid grid is displayed in Fig. 34.5.

Generation Process

Regardless of which grid strategy is being considered, creation of a computational grid requires the following:

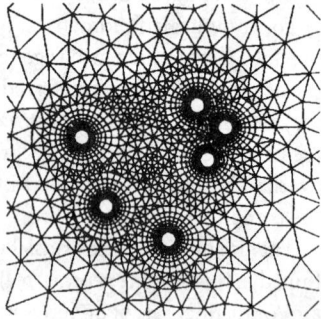

FIGURE 34.5   Hybrid grid.

1. *Computational mapping:* Establishing an appropriate mapping from physical to computational space allowing proper multiblock strategies in case of structured and hybrid grids or establishing an ordering of nodes in case of unstructured grids and hybrid grids.
2. *Geometry generation:* Defining an accurate numerical description of all solid components (surfaces) in conjunction with associated computational mapping criteria and a desired distribution of points.
3. *Computational modeling:* Generating an *appropriate* grid around these surfaces according to some criteria, usually with a specified multiblock strategy, point distribution, smoothness, and orthogonality in case of structured grids and desired background mesh representative of point distribution in case of unstructured grids.

The relationship of geometry to the grid generation process is analogous to the relationship between boundary conditions and the solution of the governing fluid flow equations. An accurate construction of the geometry with the proper distribution of points usually consumes 85–90 percent of the total time spent on the grid generation process. The geometry specification associated with grid generation involves the following:

1. Determine the desired distribution of grid points. This depends upon the expected flow characteristics.

2. Evaluate boundary segments and surface patches to be defined in order to resolve an accurate mathematical description of the geometry in question.
3. Select the geometry tools to be utilized to define these boundary segments/surface patches.
4. Follow an appropriate logical path to blend the aforementioned tasks obtaining the desired discretized mathematical description of the geometry with properly distributed points.

The accuracy of the numerical algorithm depends not only on the formal order of approximations but also on the distribution of grid points in the computational domain. The grid employed can have a profound influence on the quality and convergence rate of the solution. Grid adaption offers the use of excessively fine, computationally expensive, grids by directing the grid points to congregate so that a functional relationship on these points can represent the physical solution with sufficient accuracy.

Underlying principles and methodologies for grid generation and adaption follow.

34.2 Underlying Principles

Terminology and Grid Characteristics

The differencing and solution techniques involving Cartesian (regular) grids are well developed and well understood. The use of structured curvilinear grids (nonorthogonal in most cases) in the numerical solution of PDEs is not, in principle, any more difficult than using Cartesian grids. This is accomplished by transforming the pertinent PDEs from the physical space to computational space. The following notations will be utilized in the development of structured grids and these transformations; detailed mathematical analysis can be found in [Thompson et al. 1985]:

$$
\begin{aligned}
\mathbf{r} = (x, y, z) &\equiv (x_1, x_2, x_3) &&= \text{physical space} \\
\Omega = (\xi, \eta, \zeta) &\equiv (\xi^1, \xi^2, \xi^3) &&= \text{computational space} \\
\mathbf{a}_i = \mathbf{r}_{\xi^i}, \ i &= 1, 2, 3 &&= \text{covariant base vectors} \\
\mathbf{a}^i = \nabla \xi^i, \ i &= 1, 2, 3 &&= \text{contravariant base vectors} \\
g_{ij} = \mathbf{a}_i \cdot \mathbf{a}_j \ (i = 1, 2, 3), \ (j = 1, 2, 3) &&= \text{covariant metric tensor components} \\
g^{ij} = \mathbf{a}^i \cdot \mathbf{a}^j \ (i = 1, 2, 3), \ (j = 1, 2, 3) &&= \text{contravariant metric tensor components} \\
\sqrt{g} = \mathbf{a}_1 \cdot (\mathbf{a}_2 \times \mathbf{a}_3) &&= \text{Jacobian of transformation}
\end{aligned}
$$

Also it can be shown that for any tensor $\mathbf{u}$

$$
\nabla \cdot \mathbf{u} = \frac{1}{\sqrt{g}} \sum_{i=1}^{3} [(\mathbf{a}_j \cdot \mathbf{a}_k) \cdot \mathbf{u}]_{\xi^i}, \quad (i, j, k) \text{ cyclic} \tag{34.1}
$$

and

$$
\nabla \cdot \mathbf{u} = \frac{1}{\sqrt{g}} \sum_{i=1}^{3} (\mathbf{a}_j \cdot \mathbf{a}_k) \cdot \mathbf{u}_{\xi^i}, \quad (i, j, k) \text{ cyclic} \tag{34.2}
$$

Although the Eqs. (34.1) and (34.2) are equivalent expressions, the numerical representations of these two forms may not be equivalent. The form given by Eq. (34.1) is called conservative form and that of Eq. (34.2) is called the nonconservative form. Equation (34.1) or (34.2) is utilized for transforming derivatives from physical to computational space. With moving grids the time derivatives also must be transformed using the following equation:

$$
\left(\frac{\partial \mathbf{u}}{\partial t} \right)_{\mathbf{r}} = \left(\frac{\partial \mathbf{u}}{\partial t} \right)_{\Omega} - \dot{\mathbf{r}} \cdot \nabla \mathbf{u} \tag{34.3}
$$

where $\dot{\mathbf{r}}$ indicates the associated grid speed.

The discretization process associated with the finite-volume scheme employed in unstructured or hybrid grids is also very straightforward. Here, the integral equations are utilized. The edge-based data structure is used for connectivity information and can be easily utilized to compute areas of cells and associated fluxes. The area, for example, of a region bounded by a two-dimensional boundary $\partial\Omega$ is

$$A = \int_{\partial\Omega} x \, dy \tag{34.4}$$

which can be approximated to

$$A = \sum_{\text{edges}} x \, \Delta y \tag{34.5}$$

where x and Δy are interpreted as edge quantities. Also, the governing equations are discretized in similar fashion. For example, consider an integral form of the Navier-Stokes equation without body force as follows:

$$\frac{\partial}{\partial t} \int_v \mathbf{Q} \, dv + \oint_{\partial\Omega} \mathbf{F}(Q) \cdot \mathbf{n} \, ds = \oint_{\partial\Omega} \mathbf{F}^v(Q) \cdot \mathbf{n} \, ds \tag{34.6}$$

where $\mathbf{n}$ is the outward normal to the control volume with components n_x and n_y in the x and y directions. The domain of interest is discretized into a set of nonoverlapping polygons (unstructured or hybrid grid), and the cell-averaged variables are stored at the cell center. For each cell the semidiscretized form of the governing Eqs. (34.6) can be written as

$$V_i \frac{\partial Q}{\partial t} = -\sum_{j=1}^{k} F_{ij} \cdot n_j \, ds + \sum_{j=1}^{k} F_{ij}^v \cdot n_j \, ds \tag{34.7}$$

where j varies over the cell faces, and F_{ij} and F_{ij}^v are the convective and viscous part of the numerical flux at jth edge of cell i. Detailed analysis and description of this discretization process can be found in Weatherill [1990], and Koomullil et al. [1996b].

On structured grids the definition of an order of a difference representation is integrally tied to point distribution functions, commonly referred to as stretching functions. The order is determined by the error behavior as the spacing varies with the points fixed in a certain distribution, either by increasing the number of points or by changing a parameter in the distribution. Actually, a global order is not meaningful in case of nonuniform structured grids. The order is relevant only locally in regions where the spacing does in fact decrease as the point distribution changes. Looking at the truncation error analysis involving nonuniform structured curvilinear grids, the following grid requirements can be outlined: (1) The structured grid system must be either a right-handed system [$\sqrt{g} > 0$ for all (i, j, k)] or a left-handed system [$\sqrt{g} < 0$ for all (i, j, k)]. (2) Application of the distribution function with bounds on higher order derivatives does not change the formal order of approximation. The following functions involving exponential and hyperbolic tangent or hyperbolic sine stretching functions are widely utilized as distribution functions for distributing N points:

$$x(\xi) = \frac{e^{\alpha(s-1)} - 1}{e^{\alpha} - 1} \quad \text{or} \quad x(\xi) = 1 - \frac{\tanh(\alpha(1-s))}{\tanh(\alpha)} \tag{34.8}$$

where

$$s = (\xi - 1)/(N - 1) \quad 1 \le \xi \le N.$$

(3) Numerical derivative evaluation of the distribution function is preferred, i.e., instead of analytical

definition of x_ξ, where $x_\xi = [(x_{i+1} - x_{i-1})/2]$. (4) The truncation error is inversely proportional to the sine of the angle between grid lines. This in turn indicates that the mildly skewed grid does not increase truncation error significantly. In fact, as a rule of thumb, nonorthogonality resulting in an angle between grid lines not less than $45°$ does not increase truncation errors significantly.

Geometry Preparation

The geometry preparation is the most critical and labor intensive part of the overall grid generation process. Most of the geometrical configurations of interest to practical engineering problems are designed in the computer-aided design/computer-aided manufacturing (CAD/CAM) system. There are numerous geometry output formats which require the designer to spend a great deal of time manipulating geometrical entities in order to achieve a useful sculptured geometrical description with appropriate distribution for grid points. Also, there is a danger of loosing fidelity of the underlying geometry in this process. The desired point distribution on the boundary segment/surface patch is achieved by computing a concentration array of unit length. The concentration array is computed using specified spacing and by selecting an exponential or hyperbolic tangent stretching function. The geometry preparation involves the discrete-sculptured definitions of all outer boundaries/surfaces associated with the domain of interest. In case of unstructured or hybrid grids, the geometry preparation involves the definition of discrete points on the boundaries associated with the domain. The following definitions will be utilized in this development.

Definition 34.1. Given a set of points on a curve with physical Cartesian coordinates (x_i, y_i, z_i), $i = i1, i1 + 1, \ldots, i2$. A number sequence $r = (r_1, r_2, \ldots, r_n)$, with $0 \leq r_j \leq 1$, $r_1 = 0, r_n = 1, r_i \leq r_j$ for all $i < j$ represents the distribution of points, such that there exists a one-to-one correspondence between the element r_j of r and the triplet (x_j, y_j, z_j). This number sequence r is called a curve distribution. For example, the normalized chord length with $r_{i1} = 0$ and

$$
r_i = \frac{\sum_{u=i1+1}^{i} \sqrt{(x_u - x_{u-1})^2 + (y_u - y_{u-1})^2 + (z_u - z_{u-1})^2}}{\sum_{u=i1+1}^{i2} \sqrt{(x_u - x_{u-1})^2 + (y_u - y_{u-1})^2 + (z_u - z_{u-1})^2}}
\tag{34.9}
$$

where $i = i1 + 1, \ldots, i2$ satisfies the definition of a curve distribution.

Definition 34.2. Given a set of points on a surface with physical Cartesian coordinates (x_{ij}, y_{ij}, z_{ij}), $i = i1, i1 + 1, \ldots, i2$; $j = j1, j1 + 1, \ldots, j2$. A mesh (s_{ij}, t_{ij}), $i = i1, i1 + 1, \ldots, i2$; $j = j1, j1 + 1, \ldots, j2$, is called a surface distribution mesh if $(s_{i1j}, \ldots, s_{i2j})$ for $j = j1, \ldots, j2$ represents the curve distribution for the curve $((x_{ij}, y_{ij}, z_{ij}), i = i1, i1+1, \ldots, i2)$ for $j = j1, j1, \ldots, j2)$ and $(t_{ij1}, \ldots, t_{ij2})$ for $i = i1, \ldots, i2$ represents the curve distribution for the curve $((x_{ij}, y_{ij}, z_{ij}), j = j1, j1+1, \ldots, j2)$, for $i = i1, i1 + 1, \ldots, i2)$.

Also, there exists a one-to-one correspondence between the physical domain and the surface distribution mesh and between the surface distribution mesh and the computational domain. These relations are demonstrated in Fig. 34.6.

Definition 34.3. Let $(X_{ijk}, Y_{ijk}, Z_{ijk})$, $i = 1, 2, \ldots, N$; $j = 1, 2, \ldots, M$; $k = 1, 2, \ldots, L$ be a single block three-dimensional structured grid. A mesh $(s_{ijk}, t_{ijk}, q_{ijk})$, $i = 1, 2, \ldots, N$; $j = 1, 2, \ldots, M$; $k = 1, 2, \ldots, L$ is called a volume distribution mesh if $(\overline{s}_{ij\overline{k}}, \overline{t}_{ij\overline{k}})$ $i = 1, 2, \ldots, N$; $j = 1, 2, \ldots, M$ represents the surface distribution mesh for the surface $(\overline{X}_{ij\overline{k}}, \overline{Y}_{ij\overline{k}}, \overline{Z}_{ij\overline{k}})$, $\overline{k} = 1, 2, \ldots, L$, $(t_{\overline{i}jk}, q_{\overline{i}jk})$ represents the surface distribution mesh for the surface $(X_{\overline{i}jk}, Y_{\overline{i}jk}, Z_{\overline{i}jk})$, for all $\overline{i}$ and $(s_{i\overline{j}k}, q_{i\overline{j}k})$ represents the surface distribution mesh for the surface $(\overline{X}_{i\overline{j}k}, \overline{Y}_{i\overline{j}k}, \overline{Z}_{i\overline{j}k})$, for all $\overline{j}$.

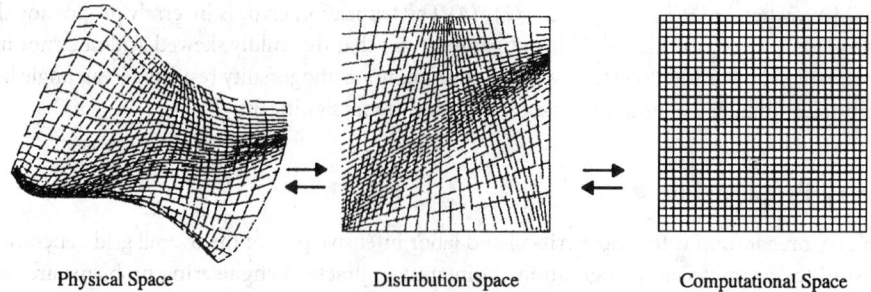

Physical Space   Distribution Space   Computational Space

FIGURE 34.6 Relationship between physical space, distribution mesh, and computational space.

Structured Grid Generation

Algebraic Generation Methods

An algebraic 3D generation system based on transfinite interpolation (using either Lagrange or Hermite interpolation) is widely utilized for grid generation [Gordan and Thiel 1982, Soni 1992b]. The interpolation, in general complete transfinite interpolation from all boundaries, can be restricted to any combination of directions or lesser degrees of interpolation, and the form (Lagrange, Hermite, or incomplete Hermite) can be different in different directions or in different blocks. The blending functions can be linear or, more appropriately, based on the distribution surface/volume mesh. Hermite interpolation, based on cubic blending functions, allows orthogonality at the boundary. Incomplete Hermite uses quadratic functions and, hence, can give orthogonality at any one of two opposing boundaries, whereas Lagrange, with its linear functions, does not give orthogonality.

The transfinite interpolation is accomplished by the appropriate combination of 1D projectors F for the type of interpolation specified. (Each projector is simply the 1D interpolation in the direction indicated.) For interpolation from all sides of the section, if all three directions are indicated and the section is a volume, this interpolation is from all six sides, and the combination of projectors is the Boolean sum of the three projectors,

$$F_1 \oplus F_2 \oplus F_3 \equiv F_1 + F_2 + F_3 - F_1 F_2 - F_2 F_3 - F_3 F_1 + F_1 F_2 F_3 \tag{34.10}$$

With interpolation in only the two directions j and k, or if the section is a surface on which ξ^i is constant, the combination is the Boolean sum of F_j and F_k

$$F_j \oplus F_k \equiv F_j + F_k - F_j F_k, \quad (i, j, k) \text{ cyclic} \tag{34.11}$$

With interpolation in only a single direction i, or if the section is a line on which ξ^i varies, the interpolation is between the two sides on which ξ^i is constant using only the single projector F_i.

Blocks can be divided into subblocks for the purpose of generation of the algebraic grid. Point distributions on the sides of the subblocks can either be specified or automatically generated by transfinite interpolation from the edge of the side. This allows additional control over the grid in general configurations and is particularly useful in cases where point distributions need to be specified in the interior of a block or to prevent grid overlap in highly curved regions. This also allows points in the interior of the field to be excluded if desired, e.g., to represent holes in the field.

Elliptic Generation Method

An elliptic grid generation system [Thompson 1987b] commonly used is

$$\sum_{i=1}^{3} \sum_{l=1}^{3} g^{il} r_{\xi^i \xi^l} + \sum_{l=1}^{3} g^{ll} P_l r_{\xi^l} = 0 \tag{34.12}$$

where the g^{il}, the elements of the contravariant metric tensor, can be evaluated as

$$g^{il} = \frac{1}{g}(g_{jm}g_{kn} - g_{jn}g_{km}), \quad (i = 1, 2, 3), \quad (l = 1, 2, 3) \tag{34.13}$$

$$(i, j, k) \text{ cyclic}, \quad (l, m, n) \text{ cyclic}$$

where g is the square of the Jacobian.

The P_n are the *control functions*, which serve to control the spacing and orientation of the grid lines in the field. The first and second coordinate derivatives are normally calculated using second-order central differences. One-sided differences dependent on the sign of the control function P_n (backward for $P_n < 0$ and forward for $P_n > 0$) are useful to enhance convergence with very strong control functions. The control functions are evaluated either directly from the initial algebraic grid and then smoothed or by interpolation from the boundary point distributions.

The three components of the elliptic grid generation system provide a set of three equations that can be solved simultaneously at each point for the three control functions $P_n(n = 1, 2, 3)$, with the derivatives here represented by central differences.

The elliptic generation system is solved by point successive over relaxation (SOR) iteration using a field of locally optimum acceleration parameters. These optimum parameters make the solution robust and capable of convergence with strong control functions.

Control functions can also be evaluated on the boundaries using the specified boundary point distribution in the generation system, with certain necessary assumptions (orthogonality at the boundary) to eliminate some terms, then they can be interpolated from the boundaries into the field. More general regions can, however, be treated by interpolating elements of the control functions separately. Thus, control functions on a line on which ξ^n varies can be expressed as

$$P_n = A_n + \frac{S_n}{\varrho_n} \tag{34.14}$$

where A_n is the logarithmic derivative of the arc length, S_n is the arc length spacing, and ϱ_n is the radius of curvature of the surface on which ξ^n is constant.

A second-order elliptic generation system allows either the point locations on the boundary or the coordinate line slope at the boundary to be specified, but not both. It is possible, however, to iteratively adjust the control functions in the generation system until not only a specified line slope, but also the spacing of the first coordinate surface off the boundary is achieved, with the point locations on the boundary specified. The extent of the orthogonality into the field can also be controlled. This orthogonality feature is also applicable on specified grid surfaces within the field, allowing grid surfaces in the field to be kept fixed while retaining continuity of slope of the grid line crossing the surface. This is quite useful in controlling the skewness of grid lines in some troublesome areas. Alternatively, boundary orthogonality can be achieved through Neumann boundary conditions, which allow the boundary points to move over a surface spline. The boundary locations are located by Newton iteration on the spline to be at the foot of normals to the adjacent field points.

Hyperbolic Generation Method

It is also possible to base a grid generation system on hyperbolic PDEs rather than elliptic equations. In this case, the grid is generated by numerically solving a hyperbolic system [Steger and Chaussee 1980, Chan and Steger 1992], marching in the direction of one curvilinear coordinate between two boundary curves in two dimensions or between two boundary surfaces in three dimensions. The hyperbolic system, however, allows only one boundary to be specified and is therefore of interest only for use in calculation on physically unbounded regions where the precise location of a computational outer boundary is not important. The hyperbolic grid generation system has the advantage of being generally faster than elliptic generation systems, but, as just noted, is applicable only to certain configurations. Hyperbolic generation

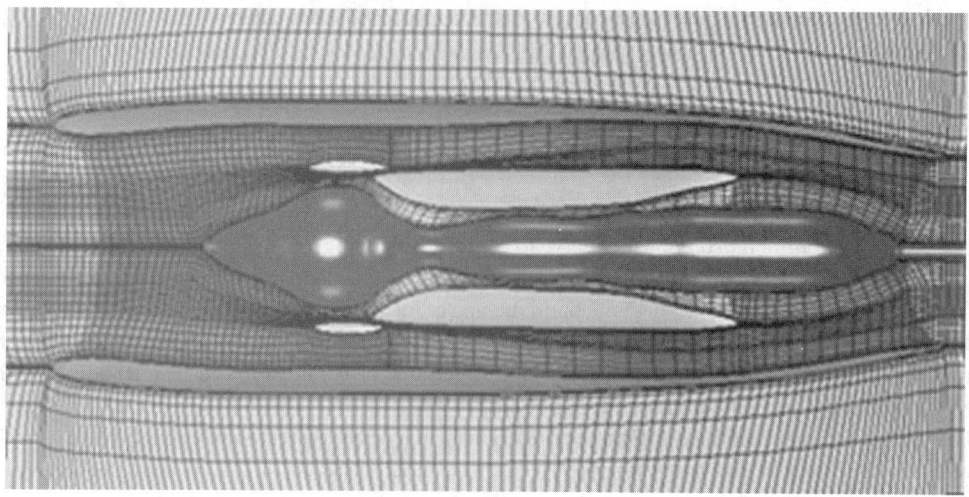

FIGURE 34.7 Multiblock grid.

systems can be used to generate orthogonal grids. In two dimensions, the condition of orthogonality is simply $g_{12} = 0$.

If either the cell area $\sqrt{g}$ or the cell diagonal length (squared) $g_{11} + g_{22}$ is a specified function of the curvilinear coordinates, i.e.,

$$\sqrt{g} = F(\xi, \eta) \qquad \text{or} \qquad g_{11} + g_{22} = F(\xi, \eta) \tag{34.15}$$

then the system consisting of $g_{12} = 0$ and either of the preceding two equations is hyperbolic. Since the system is hyperbolic, a noniterative marching solution can be constructed proceeding in one coordinate direction away from a specified boundary.

Multiblock Systems

Although in principle it is possible to establish a correspondence between any physical region and a single empty logically rectangular block for general 3D configurations, the resulting grid is likely to be much too skewed and irregular to be usable when the boundary geometry is complicated. A better approach with complicated physical boundaries is to segment the physical region into contiguous subregions, each bounded by six curved sides (four in 2D) and each of which transforms to a logically rectangular block in the computational region. Each subregion has its own curvilinear coordinate system, irrespective of that in the adjacent subregions (see Fig. 34.7). This then allows both the grid generation and numerical solutions on the grid to be constructed to operate in a logically rectangular computational region, regardless of the shape or complexity of the full physical region. The full region is treated by performing the solution operations in all of the logically rectangular computational blocks. With the composite framework, PDE solution procedures written to operate on logically rectangular regions can be incorporated into a code for general configurations in a straightforward manner, since the code only needs to treat a rectangular block. The entire physical field then can be treated in a loop over all the blocks. Transformation relations for PDEs are covered in detail in Thompson et al. [1985]. Discretization error related to the grid is covered in Thompson and Mastin [1983]. The evaluation and control of grid quality is an ongoing area of active research [Gatlin et al. 1991].

Grid lines at the interfaces may meet with complete continuity, with or without slope continuity, or may not meet at all. Complete continuity of grid lines across the interface requires that the interface [Thompson 1987a] be treated as a branch cut on which the generation system is solved just as it is in the interior of blocks. The interface locations are then not fixed, but are determined by the grid generation system. This is most easily handled in coding by providing an extra layer of points surrounding each block. Here, the grid points on an interface of one block are coincident in physical space with those of another interface of the same or another block, and also the grid points on the surrounding layer outside the first interface are coincident with those just inside the other interface, and vice versa. This coincidence can be maintained during the course of an iterative solution of an elliptic generation system by setting the values on the surrounding layers equal to those at the corresponding interior points after each iteration. All of the blocks are thus iterated to convergence, so that the entire composite grid is generated at once. The same procedure is followed by PDE solution codes on the block-structured grid.

Chimera Grids

The *chimera* (overlaid) grids [Belk 1995, Meakin 1991, Benek et al. 1985] are composed of completely independent component grids which may even overlap other component boundary elements, creating holes in the component grids. This requires flagging procedures to locate grid points that lie out of the field of computation, but such holes can be handled even in tridiagonal solvers by placing 1s at the corresponding positions on the matrix diagonal and all 0s off the diagonal. These overlaid grids also require interpolation to transfer data between grids, and that subject is the principal focus of effort with regard to the use of this type of composite grid.

Adaptive Grid Generation

With structured grids, the adaptive strategy based on redistribution is by far the most simple to implement, requiring only the regeneration of the grid and interpolation of flow properties at the new grid points at each adaptive stage without modification of the flow solver unless time accuracy is desired. Time accuracy can be achieved, as far as the grid is concerned, by simply transforming the time derivatives, thus adding convectivelike terms that do not alter the basic conservation form of the PDEs.

Adaptive redistribution of points traces its roots to the principle of equidistribution of error [Brackbill 1993, Soni et al. 1993] by which a point distribution is set so as to make the product of the spacing and a weight function constant over the points

$$w \Delta x = \text{const}$$

With the point distribution defined by a function ξ_i, where ξ varies by a unit increment between points, the equidistribution principle can be expressed as

$$w x_\xi = \text{const} \tag{34.16}$$

This one-dimensional equation can be applied in each direction in an alternating fashion. A direct extension to multiple dimensions using algebraic, variational, and elliptic systems has been developed.

The weight function is usually formulated by utilizing scaled gradients and curvatures of the solution variables considered for adaption.

The control of the characteristics and distribution of a grid system can be achieved by varying the values of the control functions P_l in Eq. (34.12). The application of the one-dimensional form of Eq. (34.12) with Eq. (34.16) results in the definition of control functions in three dimensions,

$$P_i = \frac{W_{\xi^i}}{W} \quad (i = 1, 2, 3) \tag{34.17}$$

These control functions were generalized by Eiseman [1985] as

$$P_i = \sum_{j=1}^{3} \frac{g^{ij}}{g^{ii}} \frac{(W_i)_{\xi^i}}{W_i} \quad (i = 1, 2, 3) \tag{34.18}$$

In order to conserve the geometrical characteristics of the existing grid the definition of control functions is extended as

$$P_i = (P_{\text{initial geometry}})_i + c_i(P_{wt}) \quad (i = 1, 2, 3) \tag{34.19}$$

where $(P_{\text{initial geometry}})$ is the control function based on initial grid geometry, P_{wt} is the control function based on gradient of flow parameter, and c_i is the constant weight factors.

These control functions are evaluated based on the current grid at the adaptation step. This can be formulated as

$$P_i^{(n)} = P_i^{(n-1)} + c_i(P_{wt})^{(n-1)} \quad (i = 1, 2, 3) \tag{34.20}$$

where

$$P_i^{(1)} = (P_{\text{initial geometry}})_i^{(0)} + c_i(P_{wt})^{(0)} \quad (i = 1, 2, 3) \tag{34.21}$$

Unstructured Grid Generation

The Delaunay Triangulation

Dirichlet, in 1850, first proposed a method whereby a domain could be systematically decomposed into a set of packed convex polyhedra. For a given set of points in space, $\{P_k\}, k = 1, \ldots, K$, the regions $\{V_k\}, k = 1, \ldots, K$, are the territories which can be assigned to each point P_k, such that V_k represents the space closer to P_k than to any other point in the set. Clearly, these regions satisfy

$$V_k = \{P_i : |p - P_i| < |p - P_j|\} \; \forall \; j \neq i \tag{34.22}$$

This geometrical construction of tiles is known as the Dirichlet tessellation or Voronoi diagram. This tessellation of a closed domain results in a set of nonoverlapping convex polyhedra, called Voronoi regions, covering the entire domain. If all point pairs which have some segment of a Voronoi boundary in common are joined, the result is a triangulation of the convex hull of the set of points $\{P_k\}$. This triangulation is known as the *Delaunay triangulation* [Baker 1990, George and Hermeline 1992]. The definition is valid for n-dimensional space.

From the preceding discussion, it is apparent that in two dimensions a line segment of the Voronoi diagram is equidistant from the two points it separates. Hence, the vertices of the Voronoi diagram must be equidistant from each of the three nodes which form the Delaunay triangles. Clearly, it is possible to construct a circle, centered at a Voronoi vertex, which passes through the three points, which form a triangle. Furthermore, it is evident that, given the definition of Voronoi line segments and regions, no circle can contain any point. This latter condition is referred to as the in-circle criterion.

Advancing Front Procedure

The advancing front procedure is based on the method originally proposed in Peraire et al. [1987] for two dimensions and then extended to three dimensions in [Peraire et al. 1988, 1990]. The advocated approach is regarded as a generalization of the advancing front technique [George 1971, Lo 1985] with the distinctive feature that elements, i.e., triangles or tetrahedra, and points are generated simultaneously. This enables the generation of elements of variable size and stretching and differs from the approach followed in tetrahedral generators which are based on Delaunay concepts [Baker 1990, Cavendish et al. 1985], which generally connect grid points which have already been distributed in space.

The generation problem consists of subdividing an arbitrarily complex domain into a consistent assembly of elements. The consistency of the generated mesh is guaranteed if the generated elements cover the entire domain and the intersection between elements occurs only on common points, sides, or triangular faces in the three-dimensional case. The final mesh is constructed in a bottom-up manner. The process starts by discretising each boundary curve. Nodes are placed on the boundary curve components and then contiguous nodes are joined with straight line segments. In later stages of the generation process, these segments will become sides of some triangles. The length of these segments must therefore, be consistent with the desired local distribution of mesh size. This operation is repeated for each boundary curve in turn.

The next stage consists of generating triangular planar faces. For each two-dimensional region or surface to be discretized, all of the edges produced when discretizing its boundary curves are assembled into the so-called initial front. The relative orientation of the curve components with respect to the surface must be taken into account in order to give the correct orientation to the sides in the initial front. The front is a dynamic data structure which changes continuously during the generation process. At any given time, the front contains the set of all of the sides which are currently available to form a triangular face. A side is selected from the front and a triangular element is generated. This may involve creating a new node or simply connecting to an existing one. After the triangle has been generated, the front is updated and the generation proceeds until the front is empty. The size and shape of the generated triangles must be consistent with the local desired size and shape of the final mesh. In the three-dimensional case, these triangles will become faces of the tetrahedra to be generated later.

The geometrical characteristics of a general mesh are locally defined in terms of certain mesh parameters. If $N = (2 \text{ or } 3)$ is the number of dimensions, then the parameters used are a set of N mutually orthogonal directions $\boldsymbol{\alpha}_i$, $i = 1, \ldots, N$, and N associated element sizes δ_i, $i = 1, \ldots, N$. Thus, at a certain point, if all N element sizes are equal, the mesh in the vicinity of that point will consist of approximately equilateral elements. To aid the mesh generation procedure, a transformation T which is a function of $\boldsymbol{\alpha}_i$ and δ_i is defined. This transformation is represented by a symmetric $N \times N$ matrix and maps the physical space onto a space in which elements, in the neighborhood of the point being considered, will be approximately equilateral with unit average size. This new space will be referred to as the normalized space. For a general mesh this transformation will be a function of position. The transformation T is the result of superimposing N scaling operations with factors $1/\delta_i$ in each $\boldsymbol{\alpha}_i$ direction. Thus,

$$T(\boldsymbol{\alpha}_i, \delta_i) = \sum_{i=1}^{N} \frac{1}{\delta_i} \boldsymbol{\alpha}_i \otimes \boldsymbol{\alpha}_i \qquad (34.23)$$

where $\otimes$ denotes the tensor product of two vectors.

Grid Adaption Methods

For the solution adaptive grid generation procedure, an error indicator is required that detects and locates appropriate features in the flowfield. In order to provide flexibility in isolating varying features, multiple error indicators are used. Each can isolate a particular type of feature. The error indicators are usually set to the negative and positive components of the gradient in the direction of the velocity vector as given by

$$\begin{aligned} e_1 &= \min[\mathbf{V} \cdot \nabla(\mathbf{u}), 0] \\ e_2 &= \max[\mathbf{V} \cdot \nabla(\mathbf{u}), 0] \end{aligned} \qquad (34.24)$$

and the magnitude of the gradient in all directions normal to the velocity vector as given by

$$e_3 = |\nabla \mathbf{u} - \mathbf{V}(\mathbf{V} \cdot \nabla \mathbf{u})/\mathbf{V} \cdot \mathbf{V}| \qquad (34.25)$$

Where $\mathbf{V}$ is the velocity vector and $\mathbf{u}$ is any suitable flow property. Typically, density is used as the basis for the error indicator. The first two error indicators represent expansions and compressions in the flow direction and the third represents gradients normal to the flow direction. The indicators can be scaled by the relative element size. Length scaling can improve detection of weak features on a coarse grid with the present procedure. Each error indicator is treated independently, allowing particular features in the

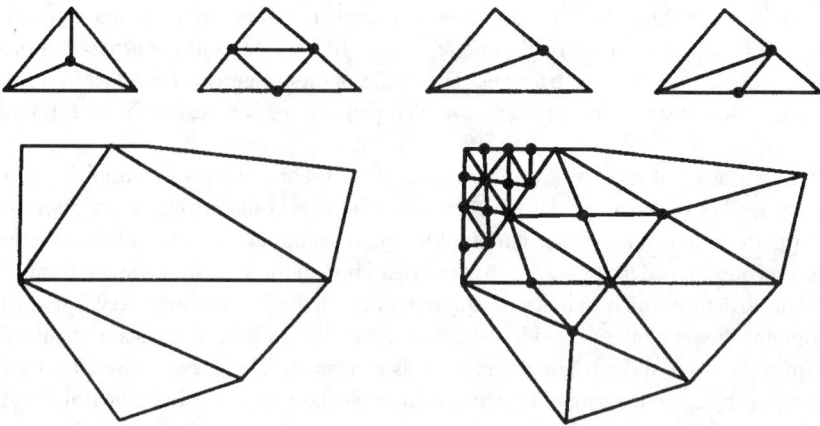

FIGURE 34.8 Types of h-refinement in two dimensions.

flowfield to be isolated. For each error indicator, an error is determined from

$$e_{\lim} = e_m + c_{\lim} \cdot e_s \tag{34.26}$$

where $e_{\lim}$ is the error limit, e_m is the mean of the error indicator, e_s is the standard deviation of the error indicator, and $c_{\lim}$ is a constant. Typically a value near 1 is used for the constant. The error indicators are used to control the local reduction in relative element size during grid generation.

One of the advantages of an unstructured grid is that it provides a natural environment for grid adaptation using h-refinement or mesh enrichment. Points can be added to the mesh with the consequence that new elements are formed and only local modifications to the connectivity matrix need to be made. In addition, no modification or special treatments are required within the solution algorithm provided that on enriching the mesh the distribution of elements and points remains smooth. Once the regions for enrichment are determined and individual elements are identified there are a number of strategies for adding points. The most suitable methods attempt to ensure smoothness of the enriched mesh, and in this respect local refinement strategies can prove to be useful. Some examples of point enrichment are given in Fig. 34.8.

In addition to h-refinement, node movement has been found to be necessary for an efficient implementation of grid adaptation. Node movement can be applied in the form

$$\mathbf{r}_0^{n+1} = \mathbf{r}_0^n + \omega_i \frac{\sum_{i=1}^{M} C_{i0}(\mathbf{r}_i^n - \mathbf{r}_0^{ni})}{\sum_{i=1}^{M} c_{i0}} \tag{34.27}$$

where $\mathbf{r} = (x, y)$, $\mathbf{r}_0^{n+1}$ is the position of node 0 at relaxation level $n + 1$, C_{i0} is the adaptive weight function between nodes i and 0, and ω is the relaxation parameter. An adaptive weight function C_{i0} is used which takes the form

$$C_{i0} = k_1 + k_2 \left| \frac{\phi_i - \phi_0}{\phi_i + \phi_0} \right| \tag{34.28}$$

where ϕ is the driving variable e.g., pressure, density, Mach number, etc.; k_1 and k_2 are constants, k_1 acts to damp out noise; and k_2 amplifies the gradients along edges. In practice, this is implemented in a form which guarantees positive area cells after movement, even in regions close to a wall, which for viscous grids can have very small volumes.

34.3 Best Practices

In the last few years, numerical grid generation has evolved as an essential tool in obtaining the numerical solution of the partial differential equations of fluid mechanics. A multitude of techniques and computer codes have been developed to support multiblock structured and unstructured grid generation associated with complex configurations. Structured grid generation methodologies can be grouped in two main categories: direct methods, where algebraic interpolation techniques are utilized, and indirect methods, where a set of partial differential equations is solved. Both of these techniques are utilized either separately or in combination, to efficiently generate grids in the aforementioned codes.

In algebraic methods the most widely used technique is transfinite interpolation. Historically, application of algebraic methods in grid generation has progressed as follows. In the 1970s (and early 1980s) the algebraic methods based on Lagrange and Hermite (mostly cubic) interpolation methods in tensor product form or Coon's patching, commonly referred to as transfinite interpolation technique form, and parametric spline (mostly cubic natural splines) were utilized to construct an initial grid for the iterative grid evaluation associated with a set of partial differential equations (mostly elliptic equations). In the 1980s (and 1990–1991), the development of high-powered graphics workstations along with the application of Bezier B-spline curve/surface definition in a control point form revolutionized the grid generation process with graphically interactive generation strategies and grid quality (smoothness-orthogonality with precise distribution improvements) with fast and efficient parametric curve/surface description based on basis splines. The parametric based nonuniform rational B-spline (NURBS) is a widely utilized representation for geometrical entities in computer-aided geometric design (CAGD) and CAD/CAM systems [Yu 1995]. The convex hull, local support, shape preserving forms, and variation diminishing properties of NURBS are extremely attractive in engineering design applications and in geometry-grid generation. In fact, the NURBS representation is becoming the de facto standard for the geometry description in most of the modern grid generation systems. Recently, the research concentration in algebraic grid generation is placed on utilizing CAGD techniques for efficient and accurate geometric modeling (boundary/surface grid generation). The development of NASA initial graphics exchange specification (IGES) standard [Blake and Chou 1992]; NURBS data structure in the National Grid Project [Thompson 1993]; DT-NURBS library and its implementation in various grid systems; the grid systems NGP, ICEM, GRIDGEN, IGG, GENIE++; and computer aided grid interface (CAGI) system is just a partial list of the outcome of this research concentration.

The best practice in grid generation is to transform all geometrical entities associated with the complex configuration under consideration into parametric control points based on NURBS representation allowing standard data structure. The grid generation algorithms are then tailored to exhibit NURBS representation in the generation process. An overall generation process is usually based on utilizing the best features of direct and indirect methods in case of structured grids and Delaunay triangulation and advancing front methods in case of unstructured grids.

Structured Grid Generation

The following observations and evaluations on the structured grid generation methodologies are important to consider before the development of overall grid generation process: algebraic systems are fast and economical; precise spacing control (a well-distributed grid) is always achieved with algebraic systems; grid generation by elliptic systems is always smooth; algebraic systems may cause grids to overlap; however, elliptic systems resist grid line overlapping; the control functions can be formulated to achieve boundary orthogonality and spacing control (near solid boundary surfaces) by elliptic generation systems; the control functions can be formulated to accomplish field orthogonality in a given computational direction (ξ, ς, or η) and spacing control by elliptic generation systems by iteratively updating various terms in the generation system. This is very time consuming especially in three-dimensional problems. Algebraic systems require a high degree of understanding and visual user interaction. However, elliptic systems can be readily adaptable for generalization. This is extremely useful in grid adaptations. The hyperbolic systems preserve the orthogonality at the solid boundary and the point distribution in the field. However, their applicability is restricted to external flows where the accurate geometrical shape of the outer boundaries/surfaces is not important as long as their location is a certain distance away from the body. Also in three-dimensional applications of hyperbolic systems the grid quality is directly influenced by the characteristics of the surfaces associated with the computational domain.

Transfinite Interpolation Method

In general, the algebraic methods are based on utilizing tensor product form of interpolation (in case of surface generation) and transfinite interpolation (in case of 2D or full 3D volume grid generation [Gordon and Thiol 1989]). Define a one-dimensional interpolation projector as follows:

$$T[r(s)] = \sum_{k=0}^{q} \sum_{j=0}^{p} \phi_{j,k}(s) r_j^{(k)} \tag{34.29}$$

where the parameter s is such that $0 \leq s \leq 1$, $\phi_{j,k}$ are the blending functions, and $r_j(k)$ is the kth derivative of the variable r at parametric location $s_j (0 = s_0 \leq s_1 \leq s_2 \cdots \leq s_p = 1)$. The following example clarifies the Eq. (34.29).

Example 34.1. Let

$$q = 0, \qquad p = 1$$
$$\phi_{0,0}(s) = (1 - s), \phi_{1,0}(s) = s$$

then

$$T[r(s)] = \phi_{0,0}(s) r_0^0 + \phi_{1,0}(s) r_1^0 = (1 - s) r_0^0 + s r_1^0$$

defines a straight line between two points r_0^0 and r_1^0.

Example 34.2. Let

$$q = 1, \qquad p = 1$$
$$\phi_{0,0}(s) = 2s^3 - 3s^2 + 1$$
$$\phi_{1,0}(s) = -2s^3 + 3s^2$$
$$\phi_{0,1}(s) = s^3 - 2s^2 + 1 \tag{34.30}$$
$$\phi_{1,1}(s) = s^3 - s^2$$
$$T[r(s)] = \phi_{0,0}(s) r_0^0 + \phi_{1,0}(s) r_1^0 + \phi_{0,1}(s) r_0^1 + \phi_{1,1}(s) r_1^1$$

defines a Hermite cubic polynomial between two points r_0^0 and r_1^0 with specified slopes r_0^1 and r_1^1 at the respective end points. The linear interpolation and Hermite cubic interpolation are widely applied in grid generation.

For 2D grid generation the transfinite interpolation (TFI) method is defined as follows:

$$T[r(s_{ij}, t_{ij})] = T_I[r(s_{ij}, t_{ij})] \oplus T_J[r(s_{ij}, t_{ij})]$$

where T_I is a one-dimensional interpolation projector applied in the $i(\xi)$ direction keeping t_{ij} fixed and T_J is the one-dimensional interpolation projector applied in the $j(\eta)$ direction keeping s_{ij} fixed. $T[r] = (T_I + T_J - T_I T_J)[r]$, and (s_{ij}, t_{ij}) is the distribution mesh for the 2D grid configuration under consideration. The $T_I T_J$ represents the tensor product of interpolation in both the I and J directions.

The transfinite interpolation method for 3D grid generation can be defined as

$$T[r(s_{ijk}, t_{ijk}, q_{ijk})] = T_I[r(s_{ijk}, t_{ijk}, q_{ijk})] \oplus T_J[r(s_{ijk}, t_{ijk}, q_{ijk})] \oplus T_K[r(s_{ijk}, t_{ijk}, q_{ijk})] \quad (34.31)$$

where T_I is a one-dimensional interpolation projector applied in the $t(\xi)$ direction keeping t_{ijk} and q_{ijk} fixed, T_J and T_K are similarly defined. Here $(s_{ijk}, t_{ijk}, q_{ijk})$ represents volume distribution mesh associated with 3D grid generation.

For example, if (s_{ij}, t_{ij}) represents an $N \times M$ size distribution mesh, $((X_{i1}, Y_{i1}), (X_{iM}, Y_{iM}))$, $i = 1, 2, \ldots, N$; and $((X_{1j}, Y_{1j}), (X_{Nj}, Y_{Nj}))$, $j = 1, 2 \ldots, M$ boundaries are known, T_I is selected as a linear interpolation projector and T_J is selected as a Hermite interpolation projector and then

$$T_I[r(s_{ij}, t_{ij})] = (1 - s_{ij})r_{1j} + s_{ij}r_{Nj} \quad (34.32)$$

and

$$T_J[r(s_{ij}, t_{ij})] = \Phi_{00}(t_{ij})r_{i1} + \Phi_{1,0}(t_{ij})r_{iM} + \Phi_{0,1}(t_{ij})\frac{\partial r}{\partial \eta}\bigg|_{r_{i1}} + \Phi_{1,1}(t_{ij})\frac{\partial r}{\partial \eta}\bigg|_{r_{iM}} \quad (34.33)$$

and the respective 2D grid can be evaluated as

$$r_{ij} = T_I + T_J - T_I T_J$$

where

$$T_I T_J[r(s_{ij}, t_{ij})] = (1 - s_{ij}) \left[\Phi_{0,0}(t_{ij})\gamma_{11} + \Phi_{1,0}(t_{ij})\gamma_{1M} + \Phi_{0,1}(t_{ij})\frac{\partial r}{\partial \eta}\bigg|_{r_{11}} + \Phi_{1,1}(t_{ij})\frac{\partial r}{\partial \eta}\bigg|_{1M} \right]$$

$$+ \left[s_{ij} \; \Phi_{0,0}(t_{ij})r_{N1} + \Phi_{1,0}(t_{ij})r_{NM} + \Phi_{0,0}(t_{ij})\frac{\partial r}{\partial \eta}\bigg|_{r_{NM}} + \Phi_{1,1}(t_{ij})\frac{\partial r}{\partial \eta}\bigg|_{NM} \right] \quad (34.34)$$

An important factor in applying the Hermite interpolation projectors in TFI formulation is the evaluation of slopes and twist vectors (cross derivatives). The slope vector $\bar{\mathbf{r}}_\xi$ can be evaluated by solving

$$\mathbf{r}_\xi \cdot \mathbf{r}_\eta = (\sqrt{g_{11}g_{22}}) \cos(\theta_1)$$
$$\|\mathbf{r}_\xi \times \mathbf{r}_\zeta\| = A \quad (34.35)$$

where θ_1 is the desired angle between grid lines ξ and η, and A is the desired area of the cell. The metric terms g_{11} and g_{22} can be evaluated using the desired change in arc length or from the appropriate algebraic grid (precise spacing control property of the algebraic grid can be exploited here). The system (34.35) can be uniquely solved to evaluate $\mathbf{r}_\xi$ and $\mathbf{r}_\eta$, and $\mathbf{r}_\zeta$ can be evaluated similarly. The twist vectors $\mathbf{r}_{\xi\eta}$ and $\mathbf{r}_{\xi\xi}$ can be evaluated by solving

$$(\mathbf{r}_\xi \cdot \mathbf{r}_\eta)_\eta = [(\sqrt{g_{11}g_{22}}) \cos(\theta_1)]_\eta$$
$$\|\mathbf{r}_\xi \times \mathbf{r}_\eta\|_\eta = (A)_\eta \quad (34.36)$$

and

$$(\mathbf{r}_\xi \cdot \mathbf{r}_\eta)_\xi = [(\sqrt{g_{11}g_{22}}) \cos(\theta_1)]_\xi$$
$$\|\mathbf{r}_\xi \times \mathbf{r}_\xi\|_\eta = (A)_\xi \quad (34.37)$$

respectively. The other cross derivatives can be evaluated similarly. Observe that if orthogonality is desired, that is, $\theta_1 = 90°$, then the right-hand side will have zeros except for A, A_ξ, and A_η where A_ξ and A_η represent the change in desired volumes in all areas in the ξ and η directions. This concept can be easily extended to three-dimensional configurations.

Elliptic Grid Generation

A multitude of general purpose elliptic generation systems have appeared [Thompson 1987b]. Most of these algorithms are based on an iterative adjustment of control functions to achieve boundary orthogonality. The following analysis is provided to illustrate this development.

Consider

$$(g_{ij})_{\xi^k} \equiv (\text{derivative of } g_{ij} \text{ with respect } \xi^k)$$

$$\equiv \mathbf{r}_{\xi^i \xi^k} \cdot \mathbf{r}_{\xi^j} + \mathbf{r}_{\xi^i} \cdot \mathbf{r}_{\xi^i \xi^k} \tag{34.38}$$

$$i = 1, 2, 3; \qquad j = 1, 2, 3; \qquad \text{and} \qquad k = 1, 2, 3$$

Using Eq. (34.12), the following statement can be obtained:

$$\mathbf{r}_{\xi^i \xi^j} \cdot \mathbf{r}_{\xi^k} = \frac{(g_{ik})_{\xi^j} - (g_{ij})_{\xi^k} + (g_{jk})_{\xi^i}}{2}$$

$$i = 1, 2, 3; \qquad j = 1, 2, 3; \qquad k = 1, 2, 3$$

The three-dimensional elliptic grid generation system presented in Eq. (34.12) can be rewritten by taking the dot product with r_{ξ^q}, $q = 1, 2, 3$ as

$$\sum_{i=1}^{3} \sum_{j=1}^{3} g^{ij} \mathbf{r}_{\xi^i \xi^j} \cdot \mathbf{r}_{\xi^q} + \sum_{k=1}^{3} \Phi_k g^{kk} \mathbf{r}_{\xi^k} \cdot \mathbf{r}_{\xi^q} = 0 \tag{34.39}$$

This can be written in terms of metric terms and their derivatives as

$$\sum_{i=1}^{3} \sum_{j=1}^{3} g^{ij} \frac{((g_{iq})_{\xi^j} - (g_{ij})_{\xi^q} + (g_{jq})_{\xi^i})}{2} + \sum_{k=1}^{3} \Phi_k g^{kk} g_{kq} = 0 \quad q = 1, 2, 3 \tag{34.40}$$

Now

$$g_{ii} = \bar{\mathbf{r}}_{\xi^i} \cdot \bar{\mathbf{r}}_{\xi^i} = \|\bar{\mathbf{r}}_{\xi^i}\|^2$$

represents an increment of arc length on a coordinate line along which ξ^i varies and

$$g_{ij} = \bar{\mathbf{r}}_{\xi^i} \cdot \bar{\mathbf{r}}_{\xi^j} = |\bar{\mathbf{r}}_{\xi^i}| \cdot |\bar{\mathbf{r}}_{\xi^j}| \cdot \cos\theta \quad i \neq j$$

represents a measure of orthogonality between grid lines along which ξ^i and ξ^j varies. These quantities can be evaluated if the desired increment in the arc length and desired angles between grid lines are known. Looking at the precise control of spacing property of the algebraic grid [Soni 1992a, b], the quantities g_{ij} can be evaluated from the well–defined algebraic grid, and using

$$g_{ij} = (\sqrt{g_{ii}})(\sqrt{g_{jj}}) \cos\theta \tag{34.41}$$

where θ is the desired angle between ξ^i, ξ^j grid lines, the quantities g_{ij}, $i \neq j$, can be evaluated. Once all g_{ij} are known, then Eq. (34.40) can be solved for the forcing functions Φ_k, $k = 1, 2, 3$.

If orthogonality is enforced, i.e., $\theta = 90°$ or $g_{ij} = 0$ for $i \neq j$, then Φ_k, $k = 1, 2, 3$, can be formulated as

$$\hat{\Phi}_k = \frac{1}{2} \frac{d}{d\xi^k} \left(\ln \frac{g_{kk}}{g_{ii} g_{jj}} \right), \qquad (i, j, k) \text{ cyclic} \quad k = 1, 2, 3 \tag{34.42}$$

and

$$\Phi_k = -\frac{(g_{ii})(g_{jj})\hat{\Phi}_k}{g}, \qquad (i, j, k) \text{ cyclic} \quad k = 1, 2, 3 \tag{34.43}$$

A usual practice is to utilize Eq. (34.43) in the following form:

$$\Phi_k = -\frac{\mathbf{r}_{\xi^k} \cdot \mathbf{r}_{\xi^k \xi^k}}{|\mathbf{r}_{\xi^k}|^2} \frac{\mathbf{r}_{\xi^k} \cdot \mathbf{r}_{\xi^i \xi^i}}{|\mathbf{r}_{\xi^i}|^2} \frac{\mathbf{r}_{\xi^k} \cdot \mathbf{r}_{\xi^j \xi^j}}{|\mathbf{r}_{\xi^j}|^2} \tag{34.44}$$

In fact, the firm term of the definition of Φ_k provides the distribution control and the remaining two terms contribute toward the curvature control.

To understand the iterative evaluation of the control function, consider a two-dimensional elliptic system:

$$g_{22}(\mathbf{r}_{\xi\xi} - \phi\mathbf{r}_\xi) - 2g_{12}\mathbf{r}_{\xi\eta} + g_{11}(\mathbf{r}_{\eta\eta} - \psi\mathbf{r}_\eta) = 0 \tag{34.45}$$

The control functions ϕ and ψ can be formulated as

$$\Phi = -\frac{\mathbf{r}_\xi \cdot \mathbf{r}_{\xi\xi}}{|\mathbf{r}_\xi|^2} - \frac{\mathbf{r}_\xi \cdot \mathbf{r}_{\eta\eta}}{|\mathbf{r}_\eta|^2} \tag{34.46}$$

and

$$\Psi = -\frac{\mathbf{r}_\eta \cdot \mathbf{r}_{\eta\eta}}{|\mathbf{r}_\eta|^2} - \frac{\mathbf{r}_\eta \cdot \mathbf{r}_{\xi\xi}}{|\mathbf{r}_\xi|^2} \tag{34.47}$$

During the evaluation of ϕ: (1) the quantities $\mathbf{r}_\xi$ and $\mathbf{r}_{\xi\xi}$ can be evaluated by utilizing appropriate finite-difference approximations, (2) the $\mathbf{r}_\eta$ are evaluated by solving

$$\mathbf{r}_\xi \cdot \mathbf{r}_\eta = 0 \qquad \text{and} \qquad \|\mathbf{r}_\xi \times \mathbf{r}_\eta\| = (\Delta A) \tag{34.48}$$

where (ΔA) is the desired cell area, (3) the $\mathbf{r}_{\eta\eta}$ quantities are calculated using the finite-difference approximation on the current grid. These quantities are updated at every iteration. Another approach is to utilize well-distributed algebraic grid characteristics to solve following equations in order to evaluate $\mathbf{r}_{\eta\eta}$:

$$(\mathbf{r}_\xi \cdot \mathbf{r}_\eta)_\xi = 0 \qquad \text{and} \qquad \|\mathbf{r}_\xi \times \mathbf{r}_\eta\|_\xi = (\Delta A)_\xi \tag{34.49}$$

and

$$(\mathbf{r}_\xi \cdot \mathbf{r}_\eta)_\eta = 0 \qquad \text{and} \qquad \|\mathbf{r}_\xi \times \mathbf{r}_\eta\|_\eta = (\Delta A)_\eta \tag{34.50}$$

where $(\Delta A)_\xi$ and $(\Delta A)_\eta$ represent the change of cell area in the ξ and η directions, respectively (they can be computed using finite-difference approximation from a well-distributed algebraic grid or by utilizing desired cell areas on the boundaries). The control functions are usually evaluated on the boundaries and then interpolated in the interior. The distribution mesh can be utilized as a parametric space for doing this interpolation.

The Delaunay Algorithm

The algorithm widely utilized to generate the Delaunay triangulation is based upon the work of Bowyer. The algorithm is based on the in-circle criterion, and is a sequential process with each point introduced into an existing Delaunay satisfying structure, which is broken and then reconnected to form a new Delaunay triangulation [Baker 1990, George and Hermeline 1992]. The algorithm is applicable in two- and three-dimensions and in step-by-step format is as follows:

ALGORITHM I.

1. *Define a set of points which forms a convex hull within which all points will lie.*
2. *Introduce a new point anywhere within the convex hull.*
3. *Determine all vertices of the Voronoi diagram to be deleted. A point which lies within a circle (sphere), centered at a vertex of the Voronoi diagram and which passes through its three (four) forming points, results in the deletion of that vertex. This follows from the in-circle criterion of the Voronoi construction.*
4. *Find the forming points of all the deleted Voronoi vertices. These are the contiguous points to the new point.*
5. *Determine the neighboring Voronoi vertices to the deleted vertices which have not themselves been deleted.*
6. *Determine the forming points of the new Voronoi vertices. The forming points of new vertices must include the new point together with two (three) points which are contiguous to the new point and form an edge (face) of a neighboring Voronoi diagram data structure, overwriting the entries of the deleted vertices.*
7. *Repeat steps (2–6) for the next point.*

In the preceding algorithm, the interpretation for three dimensions is included in parentheses. This algorithm has been used for the construction of the triangulation in two and three dimensions. It does not differ in content from that used in earlier work, but its implementation has made use of highly efficient search procedures and, hence, the computational time is considerably less than that used in earlier work.

For grid generation purposes the boundary of the domain is defined by points and associated connectivities. It will be assumed that the grid points on the boundary reflect appropriate variations in geometrical slope and curvature. Ideally any method which automatically creates points should ensure that the boundary point distribution is extended into the domain in a spatially smooth manner. An algorithm which achieves this in both two and three dimensions is the following:

ALGORITHM II.

1. *Compute the point distribution function for each boundary point $r_0 = (x, y, z)$ (i.e., for point 0)*

$$dp_0 = \frac{1}{M} \sum_{i=1}^{M} |r_1 - r_0|$$

   where $|\;|$ is the Euclidean distance and it is assumed that point 0 is surrounded by M points, $i = 1, M$.
2. *Generate the Delaunay triangulation of the boundary points.*
3. *Initialize the number of interior field points created, $N = 0$.*
4. *For all tetrahedra within the domain:*
   a. *Define a prospective point Q to be at the centroid of the tetrahedron.*
   b. *Derive the point distribution dp for the point Q, by interpolating the point distribution function from the nodes of the tetrahedron, dp_m, $m = 1, \ldots, 4$.*
   c. *Compute the distances d_m, $m = 1, \ldots, 4$, from the prospective point, Q, to each of the four points of the tetrahedron.*
   If $\{d_m < a dp_m\}$ for any $m = 1, \ldots, 4$ then
   *   reject the point :- Return to the beginning of step 4.*
   If $\{d_m > a dp_m\}$ for any $m = 1, \ldots, 4$ then
   *   compute the distance s_j, $(j = 1, \ldots, N)$, from the prospective point Q, to other points to*
   *   be inserted, P_j, $j = 1, N$.*
   *   If $\{s_j < \beta dp_m\}$ then*
   *   reject the point :- Return to the beginning of step 4.*
   *   If $\{s_j > \beta dp_m\}$ then*

accept the point Q for insertion by the Delaunay triangulation algorithm and include Q in the list P_j, $j = 1, N$.

d. *Assign the interpolated value of the point distribution function dp to the new node P_N.*

e. *Next tetrahedra.*

5. *If $N = 0$ go to step 7.*

6. *Perform the Delaunay triangulation of the derived points, P_j, $j = 1, N$. Go to step 3.*

7. *Smooth the mesh.*

In the preceding algorithm, the term tetrahedron and triangle are interchangeable. The coefficient α controls the grid point density by changing the allowable shape of formed tetrahedra, whereas β has an influence on the regularity of the triangulation by not allowing points within a specified distance of each other to be inserted in the same sweep of the tetrahedra within the field. The effects of the parameters α and β are demonstrated in the following examples, which for convenience are presented for domains in two dimensions.

The interpolation of the boundary point distribution function is linear throughout the field. This can be modified to provide a weighting toward the boundaries so as to ensure greater point density in such regions. The implementation of such a procedure involves a scaling of the point distribution of the nodes, which form an element on the boundary. It should be noted that this point creation algorithm can be implemented very efficiently within the Delaunay triangulation procedure. In particular, if a point is accepted for insertion, then in the Delaunay algorithm a tetrahedron is known which contains this point, since by the very nature of the procedure the tetrahedron from which the point was created is known. However, after the insertion of one point the tetrahedron numbering can be changed, and if the tetrahedra formed from the inserted points overlap, then the tetrahedron numbers which have been flagged for each new point can be then incorrect. However, the exclusion zone, controlled by the parameter β, ensures that the points created from one sweep through the tetrahedra are sufficiently spatially separated that on the insertion of each point the resulting tetrahedra do not overlap and, hence, the original tetrahedron numbers associated with each new point are valid. Hence, in this way β improves the regularity of the tetrahedra and also ensures that no search is required to find a circle which includes the point.

The procedure outlined creates points consistent with the point distribution on the boundaries. Simple modifications provide greater flexibility.

Point Creation by the Use of Sources. In somewhat of an analogous way to point sources used as control functions with elliptic partial differential equations, it is possible to define line and point sources to provide grid control for unstructured meshes. Local point spacing, at position r, can be defined as

$$dp(r) = A_j e^{B_j |R_j - r|}$$

where A_j and B_j are the user specified amplification and decay parameters of the sources j, $j = 1, M$, and R_j is the position of each point source. Grid point creation is then performed as outlined in Algorithm II but in step 4b the appropriate point distribution function at the centroid is determined by Eq. (34.2). Various forms of implementation of this can be devised. One simple modification is to define the point spacing as

$$dp(r) = \min \left(dp_{\text{boundary}}, A_j e^{B_j |R_j - r|} \right)$$

In this case dp_{boundary} is the point distribution from the boundary spacing. This then provides the desired point clustering in regions influenced by the sources but farther away the boundary point distribution has a dominant effect.

Point Creation Controlled by a Background Mesh. Another way to control the point spacing in the domain is to use a background mesh. A mesh is overlaid over the domain and at each node a point spacing is specified. To encompass this approach within the framework of Algorithm III, the point distribution function *dp* for a prospective point is obtained from the interpolated spacing from the background mesh. Within Algorithm II step 4b is replaced by the interpolation of dp from the background mesh. The effect is similar to that achieved by sources.

Boundary Integrity. This problem is widely recognized as a problem inherent to the generation of boundary conforming grids using the Delaunay triangulation. Several approaches to its solution have been proposed. The procedure followed here is an extension of earlier work and is closely related to the work of George. Both these approaches involve the addition of points to *block* the penetration of tetrahedra through the boundary surface. In the former approach, points were added on the boundary, which were then connected to the surrounding points using the Delaunay triangulation procedure. In two dimensions and in some cases in three dimensions, this approach proves to be adequate. Hence, the approach used in two dimensions uses this technique. However, in three dimensions, for some severe cases, it proves to be difficult to completely ensure the reconstruction of surface triangles. Hence, a different approach has been devised for three dimensions, which also involves the addition of points, but these are connected directly to the tetrahedral construction rather than using the Delaunay triangulation. A finite number of direct connections can be formulated for all types of tetrahedral penetration and hence in the proposed procedure the recovery of the surface can be guaranteed.

The necessary and sufficient conditions for a face with nodes (P, Q, R) to be present in the triangulation $\{\tau_l\}, l = 1, T$ are as follows:

1. The nodes P, Q, and R exist in the tetrahedral construction $\{\tau_l\}, l = 1, T$.
2. The cyclic combinations of P, Q, and R that is, PQ, QR, RP occur in one of the tetrahedra in the construction $\{\tau_l\}, l = 1, T$.
3. The combination (P, Q, R) exists in one of the tetrahedra in $\{\tau_l\}, l = 1, T$.

Hence, to recover an arbitrary set of triangular faces, these three conditions must be met for each face. The first condition appears to be self-evident.

If a boundary face is not in τ, this is because edges and faces of the tetrahedra $\{\tau\}$ intersect the required face. Since a face is formed from edges, and it is assumed that the points P, Q and R are present, it is necessary to firstly recover edges PQ, QR, RP and then the face (P, Q, R). This is achieved, for a given face PQR, by firstly finding the tetrahedra which are intersected by the edges PQ, QR, and RP. These tetrahedra are then modified and new tetrahedra created so that the required edges are present. Once edges are present, a similar procedure follows to recover the face. If the edges PQ, QR, and RP exist but the face (P, Q, R) does not exist, then all tetrahedra which possess at least one edge which intersects the face (P, Q, R) are determined. These tetrahedra are then modified accordingly to recover the missing faces.

Edge Swapping. In circumstances where it is required to have a complete surface, although there is not a fixed constraint that a given set of faces is recovered, it is possible to ensure boundary faces are coincident with faces within the tetrahedral construction by swapping edges within the boundary surface triangulation. If in the tetrahedral construction faces ABC and BCD exist, but in the surface triangulation faces ACD and ABD exist, then the two can be made to agree if in the surface triangulation edge AD is replaced by BC. There are conditions under which such a transformation is not allowed and these must be checked.

Edge swapping can be incorporated as an option. However, if it can be used, it can greatly reduce the amount of work to be carried out in the edge and face recovery routines.

Boundary Edge Recovery. The procedure to recover a missing edge of a boundary face involves two steps. Firstly, it is necessary to identify the faces, edges, and points of tetrahedra which the edge intersects. Second, local transformations involving tetrahedra are performed to recover the edge. The intersections of edges with tetrahedra can be readily computed. Once this has been performed, a set of transformations are used to recover edges.

Boundary Face Recovery. After the recovery of all boundary face edges, the boundary faces can then be recovered. Clearly, although the edges of all boundary faces are present, this does not imply that boundary faces are present, since other tetrahedra can penetrate the interior of a face but not the boundary face edges. If a boundary face is not present, it is necessary to determine all tetrahedra which intersect the face. One, two, three, or four edges and associated segments of a tetrahedra can intersect a face. Hence, for each missing face, all tetrahedra which have an edge or edges which intersect the face are determined and each of the tetrahedra are then classified accordingly.

Removal of Added Points. Most of the transformations used to recover the edges and faces in both 2D and 3D grids involve the creation of one or more points. These added surface points are used purely as part of the boundary recovery procedure and are removed after the boundary is complete.

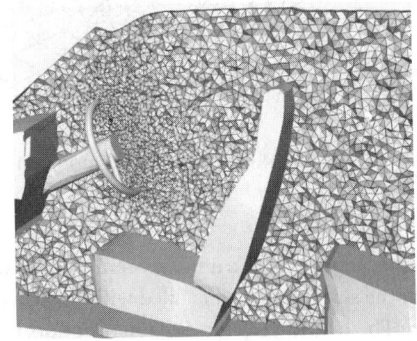

The mechanics of node removal involve taking each added point in turn and finding all elements connected to it. These elements are deleted leaving an empty polyhedron, which is then triangulated in a direct manner by finding point connections which lead to the optimum-shaped tetrahedral construction. This is a rapid process since this operation is performed locally for a relatively small number of points.

A pictorial view of an unstructured grid for automotive application is presented in Figs. 34.9 and 34.10. The grid is generated by advancing front local reconnection algorithm [Marcum 1995].

FIGURE 34.9 Volume grid of Ford Explorer.

Hybrid Grid Generation

A hybrid grid consists of structured grid in part of the physical domain and unstructured grid in rest of the field. In general, a hybrid grid can be defined as an agglomeration of cells having polygons with a different number of sides. This necessitates a hybrid grid generation process as a combination of structured and unstructured grid methodologies. The truncation error of a triangular cell is inversely proportional to the sine of the minimum angle of the triangle. Therefore, to reduce the truncation error the number of triangles in the boundary layer has to increase and, consequently, it will increase the total number of cells in the field. This will lead to more memory and central processing unit (CPU) time for the flow solver. The structured grid in the boundary layer will help to overcome these difficulties.

The basic steps involved in hybrid grid generation are the following. The first step is to decompose the complex geometries into simple geometric entities. A structured grid is generated around these geometric entities using an advancing layer-type method based on the surface normals together with the application of a local elliptic solver [Koomullil et al. 1996a]. The second step involves the trimming of the overlapping structured grid from different solid bodies in the domain, by comparing the aspect ratio. Cells having

FIGURE 34.10 Surface grid of Ford Explorer interior.

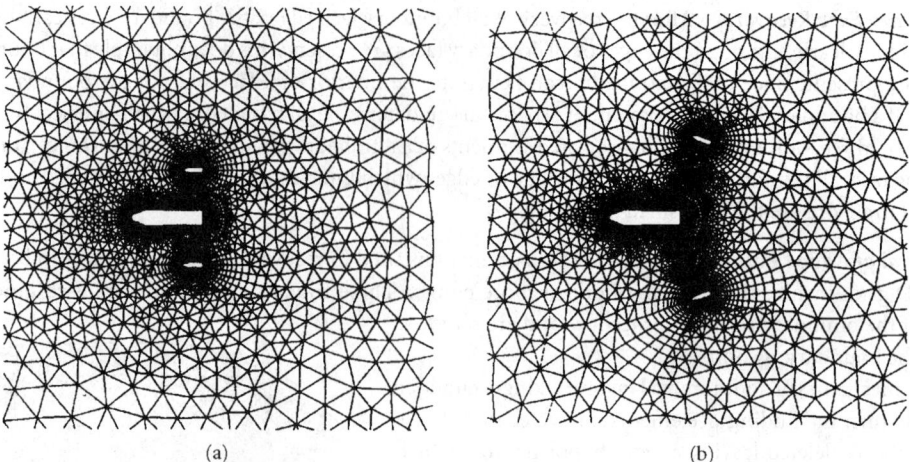

FIGURE 34.11 Hybrid grid around a launch vehicle with boosters: (a) grid at an initial state and (b) grid after booster separation.

aspect ratio less than unity are removed. In the third step the void in the physical domain after the trimming of the structured grid is filled with unstructured grid.

The hybrid grid for dynamic motion-type geometries can be generated quickly and efficiently using the approach present in [Koomullil et al. 1996a]. An example of a hybrid grid for the moving geometry problem is illustrated in Fig. 34.11. The strap-on separation from the main launch vehicle booster is shown in this figure. The relative position of the strap-ons and booster rockets at different times and the hybrid grid around this configuration is shown in Figs. 34.11(a) and (b).

Grid Adaption: Construction of Weight Functions

Application of the equidistribution law results in grid spacing inversely proportional to the weight function, and hence, the weight function determines the grid point distribution. Ideally, the weight would be the local truncation error ensuring a uniform distribution of error. Determination of this function is one of the most challenging areas of adaptive grid generation. The overall solution is only as accurate as the least accurate region. Thus, excessive resolution in certain regions does not increase the accuracy of the overall solution.

Evaluation of higher order derivatives from discrete data is progressively less accurate and subject to noise. However, lower order derivatives must be nonzero in regions of wide variations of higher order derivatives, and are proportional to the rate of variation. Therefore, it is possible to employ lower order derivatives as a proxy for the truncation error.

Analysis of the weight functions explored to date indicates that density or velocity derivatives are not independently sufficient to represent the different types and strengths of flow features. Density, or pressure for that manner, varies insufficiently in the boundary layer to be used to construct weight functions for representation of these features. Whereas velocity derivatives for viscous flows by themselves are dominated by the boundary layer, additional variables must be included to represent other flow features. The weight function [Thornburg and Soni 1994] consists of relative derivatives of density and the three conservative velocities,

$$W_{ijk}^k = 1.0 + \frac{|\hat{q}_{ijk}|}{|\hat{q}_{ijk} + e|_{\max}} + \frac{|(\hat{q}_{\xi^k})_{ijk}/\hat{q}_{ijk}|}{|(\hat{q}_{\xi^k})_{ijk}/\hat{q}_{ijk} + e|_{\max}} + \frac{|(\hat{q}_{\xi^k\xi^k})_{ijk}/\hat{q}_{ijk}|}{|(\hat{q}_{\xi^k\xi^k})_{ijk}/\hat{q}_{ijk} + e|_{\max}}$$

where $\hat{q} = (\varrho, \varrho u, \varrho v, \varrho w)$. The relative derivatives are necessary to detect features of varying intensity, so that weaker, but important structures such as vortices are accurately reflected in the weight function. One-sided differences are used at boundaries, and no-slip boundaries require special treatment since the velocity is zero. This case is handled in the same manner as zero velocity regions in the field. A small value, epsilon in the preceding equation, is added to all normalizing quantities. Also it appears that the Boolean sum construction method of Thornburg et al. [1996] would balance the weight functions more evenly, as several features are reflected in multiple variables, whereas some are reflected in only one.

34.4 Grid Systems

A multitude of general purpose grid generation codes to address complex three-dimensional structured–unstructured grid generation needs are newly available in the public domain or as proprietary commercial code. A brief description of the widely utilized candidate general-purpose codes follows.

The National Grid Project (NGP) system of Mississippi State University is an interactive geometry and grid generation system for block-structured and tetrahedral grids. The system reads CAD data via IGES and converts all surface patches to NURBS. A carpet, composed of interfacing NURBS patches, is then laid over the CAD patches to correct for gaps and overlaps. The system also has internal CAD capability for the construction or repair of surfaces. Surface grids are generated on the NURBS carpet, and can be projected onto the original CAD patches. Both the surface grids and the subsequent volume grid can be generated as block structured via elliptic, hyperbolic, or TFI methods, or as unstructured via Delaunay or advancing front procedure. A pictorial view demonstrating the graphical interface is displayed in Fig. 34.12.

The ICEM–CFD [Akdag and Wulf 1992] system is a commercial code which offers block-structured grids, tetrahedral grids, and unstructured hexahedral grids. The system interfaces with numerous CAD systems and has been connected to a number of flow solvers.

The GRIDGEN [Steinbrenner and Chawner 1993] system is a graphically interactive block-structured commercial code. The user constructs curves, which are in turn used to build the topological surface and volume components. The user then selects curves as the boundaries of surface grids, and finally surfaces as the boundaries of volume grids (blocks). With this system, grid generation is a user-in-the-loop task.

EAGLEView [Soni et al. 1992] is a graphical system that allows interactive construction of geometry and block-structured grids with journaling capability.

GridPro/az3000 [Eiseman 1995] is a commercial block-structured code topology input language (TIL) to define both the surface and the block-structured grid. The language includes components (objects) that can be invoked, and therefore admits the formation of element libraries.

CFD-GEOM [Hufford et al. 1995] is an interactive geometric modeling and grid generation system for block structured grids, tetrahedral (advancing front) grids, and hybrid grids. All elements are linked so

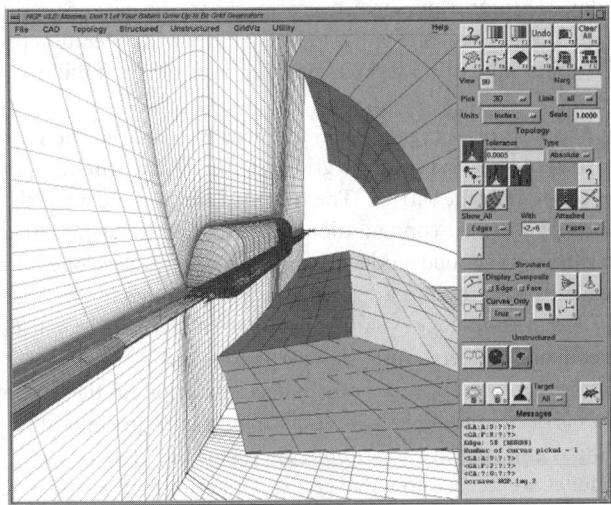

FIGURE 34.12 NGP code interface demonstrating multiblock grid.

that updates are propagated throughout the database. The geometry is NURBS based, reads IGES files, and has some internal CAD capabilities. The system also has macrolibrary capability.

The GEMS [Dener et al. 1994] block-structured grid generation system of SAMTEK-ITC in Turkey is based on object-oriented programming and C++ that uses case-based reasoning and reinforcement learning to capture CFD expertise. The system selects the case that is best suited for a particular geometry from among known ones.

The 3DGRAPE/AL [Sorenson and Alta 1995] system of NASA Ames is a block-structured grid generator that now includes the specification of arbitrary intersection angles at boundary surfaces, as well as the orthogonality pioneered by Steger–Sorenson.

The GENIE++ [Soni et al. 1992] block-structured grid generation system of Mississippi State was also introduced in the late 1980s and has been continually enhanced over the years. This system uses TFI with elliptic smoothing and includes various splining methods.

The RAPID [Smith et al. 1995] system of NASA Langley is specialized to a class of airplane configurations.

VGRID [Parikh and Pirzadeh 1992] of NASA Langley is a tetrahedral grid generator which uses advancing front with a Cartesian background grid to control resolution.

TGrid of Fluent is a tetrahedral grid generator based on the Delaunay approach.

The first general-purpose-domain connectivity codes for chimera grids were the PEGSUS (from the Air Force Arnold Engineering Development Center) and CMPGRD (from IBM) codes in the late 1980s [Meakin 1995], which continue to be enhanced. Advances in CMPGRD are detailed in Henshaw et al. [1992]. Later codes are DCF3D of NASA Ames and Overset Methods (MEAKIN 1991) and BEGGAR of the Air Force Wright Laboratory at Eglin [Belk 1995, Maple and Belk 1994]. A detailed description of these codes and a comprehensive review of existing codes and technology can be found in Thompson [1996]

34.5 Research Issues and Summary

The first step of the CFS process is the construction of a discrete approximation of the general region of interest. This representation could be multiblock structured, unstructured, or hybrid. Only for the simplest of applications can a grid be generated quickly or easily. In fact, geometry-grid generation is by far the most time consuming aspect of the entire CFS process. Rapid turnaround, reliable accuracy, and affordability are the three key requirements to be addressed for CFS to play its rightful role in supporting multidisciplinary design environment. To this end, industry is targeting grid generation in 1 h for complex configuration, i.e., reliable 1-h grid generation turnaround for one-time geometries when run by designers. The system must include CAD-to-grid links, which resolve tolerance issues and produce grids with a quality good enough for the CFS solver. The designer has to feel that the grid generation process is under control and is predictable. The following critical barriers must be overcome to fulfill the aforestated industrial requirements.

The time-consuming aspects of grid generation are usually related to the geometry definition, i.e., the input of the geometry information into the grid system. Today, trimming the surfaces associated with their intersection is a significant barrier. The trimmed surfaces are a widely utilized entity in the construction of complex geometrical configurations in CAD/CAM systems. Algorithms which utilize surface triangulation techniques and solid modeling schemes need to be developed. Reliable methodologies for representing triangulated surfaces into a single surface represented by the NURBS are needed. The CAD/CAM technology and methodology evolution based on solid modeling will also reduce this present barrier of addressing geometries undergoing design perturbations.

Automatic (noninteractive) algorithms for domain decomposition for the development of multiblock structured grids pose a barrier for addressing multidisciplinary design applications involving geometry optimization. The solution grid adaptive algorithms, at present, are limited to simple three-dimensional configurations. Techniques are needed to enhance the applicability of adaptive schemes pertaining to complex configurations. Parallel and distributed processing of grid generation algorithms is also essential for these multidisciplinary applications. Algorithms need to be developed to improve the quality of unstructured surface grids since they highly influence the quality of unstructured volume grids. Hybrid-generalized grid techniques are promising, especially for multidisciplinary CFS applications that include dynamic motion. This technology does not exist for full three-dimensional configurations.

References

Akdag, V. and Wulf, A. 1992. Integrated geometry and grid generation system for complex configurations, p. 161. In *Proc. Software Syst. Surface Modeling Grid Generation Workshop*. R. E. Smith, ed. NASA Conf. Pub. 3143, NASA Langley Research Center, Hampton, VA.

Baker, T. J. 1990. Unstructured mesh generation by a generalized Delaunay algorithm, pp. 20.1–20.10. In *Appl. Mesh Generation to Complex 3-D Configurations*. AGARD Conf. Proc. No. 464.

Belk, D. M. 1995. The role of overset grids in the development of the general purpose CFD code, p. 193. In *Proc. Surface Modeling, Grid Generation Related Issues Comput. Fluid Dyn. Workshop*, NASA Conf. Publ. 3291, NASA Lewis Research Center, Cleveland, OH, May.

Benek, J. A., Buning, P. G., and Steger, J. L. 1985. A 3D chimera grid embedding technique. AIAA Paper 85-1523.

Blake, M. W. and Chou, J. J. 1992. The NASA-IGES geometry data exchange standard. *Proc. Workshop Sponsored by NASA*. Washington, DC, Langley Research Center, Hampton, VA, April.

Brackbill, J. U. 1993. An adaptive grid with directional control. *J. Comput. Physics*. 108:38.

Cavendish, J. C., Field, D. A., and Frey, W. H. 1985. An approach to automatic three dimensional finite element mesh generation. *Int. J. Num. Methods Eng.* 21:329–348.

Dener, C., Koc, E., and Sirin, I. 1994. Extentions to GEMS for automatic grid generation and intelligent topology definition. *Numerical Grid Generation in Computational Field Simulation and Related Fields*. N. P. Weatherill, P. R. Eiseman, J. Hauser, and J. F. Thompson, eds., p. 453. Proc. 4th Int. Grid Con. Pineridge Press, Ltd. Swansea, Wales, UK.

Dirichlet, G. L. 1850. Uber die Reduction der positiven quadratischen formen mit drei understimmten, ganzen Zahlen. *J. Reine Angew. Math.* 40(3):209–227.

Eiseman, P. R. 1985. Grid generation for fluid mechanics computations. *Annu. Rev., Fluid Mech.* 17:487–522.

Eiseman, P. R. 1995. Multiblock grid generation with automatic zoning, p. 143. In *Proc. Surface Modeling, Grid Generation Related Issues Comput. Fluid Dyn. Workshop*, NASA Conf. Pub. 3291, NASA Lewis Research Center, Cleveland, OH, May.

Gatlin, B., Thompson, J. F., Yoon, Y.-H., Luong, P. V., Ganapathiraju, D., and Wolverton, M. K. 1991. Extensions to the EAGLE grid code for quality control and efficiency. *29th AIAA Aerospace Sci. Meeting*, AIAA Paper 91-0148, Reno, NV, Jan.

George, A. J. 1971. *Computer Implementation of the Finite Element Method*. Ph.D. Thesis, Stanford University, STAN-CS-71-208.

George, P. L. and Hermeline, F. 1992. Delaunay's mesh of a conven polyhedron in dimension D; application for arbitrary polyhedra. *Int. J. Num. Methods Eng.* 33:975–995.

Gordon, W. J. and Thiel, L. C. 1982. Transfinite mappings and their application to grid generation. In *Numerical Grid Generation*. J. F. Thompson, ed. North Holland, Amsterdam.

Henshaw, W. D., Chessire, G., and Henderson, M. E. 1992. On constructing three-dimensional overlapping grids with CMPGRD, p. 415. In *Proc. Software Systems Surface Modeling Grid Generation Workshop*, R. E. Smith, ed. NASA Conf. Pub. 3143, NASA Langley Research Center, Hampton, VA.

Hufford, G. S., Harrand, V. J., Patel, B. C., and Mitchell, C. R. 1995. Evaluation of grid generation technologies from an applied perspective, p. 401. In *Proc. Surface Modeling, Grid Generation Related Issues Comput. Fluid Dyn. Workshop*, NASA Conf. Pub. 3291, NASA Lewis Research Center, Cleveland, OH, May.

Koomullil, R. P., Soni, B. K., and Huang, C. 1996a. Flow simulations on generalized grids, pp. 527–536. In *5th Int. Conf. Num. Grid Generation Comput. Fluid Dyn. Related Fields*. B. K. Soni, J. F. Thompson, J. Hauser and P. Eiseman, eds. Mississippi State University, April 1–5.

Koomullil, R. P., Soni, B. K., and Huang, C. 1996b. Navier–Stokes Simulation on Hybrid Grids. *34th Aerospace Sci. Meeting,* AIAA Paper 96-767, Reno, NV, Jan. 15–18.

Lo, S. H. 1985. A new mesh generation scheme for arbitrary planar domains. *Int. J. Num. Methods in Eng.* 21:1403–1426.

Lohner, R. and Parikh, P. 1988. Three-dimensional grid generation by the advancing-front method. *Int. J. Num. Methods Fluids* 8:1135–1149.

Maple, R. C. and Belk, D. M. 1994. Automated setup of blocked, patched and embedded grids in the beggar flow solver. In *Numerical Grid Generation in Computational Field Simulations and Related Fields,* N. P. Weatherill, P. R. Eiseman, J. Hauser, and J. F. Thompson, eds., p. 151. Proc. 4th Int. Grid Conf. Pineridge Press Limited: Swansea, Wales, UK.

Marcum, D. L. 1995. Generation of unstructured grids for viscous flow applications. *33rd Aerospace Sci. Meeting and Exhibit,* AIAA Paper 95-0212, Reno, NV, Jan. 9–12.

Meakin, R. L. 1991. A new method for establishing intergrid communication among systems of overset grids. In *10th AIAA Comput. Fluid Dyn. Conf.,* AIAA Paper 91-1586, Honolulu, HI, June.

Meakin, R. L. 1995. Grid related issues for static and dynamic geometry problems using systems of overset structured grids, p. 181. In *Proc. Surface Modeling, Grid Generation Related Issues Comput. Fluid Dyn. Workshop.* NASA Conf. Pub. 3291, NASA Lewis Research Center, Cleveland, OH, May.

Parikh, P. and Pirzadeh, S. 1992. Recent advanced in unstructured grid generation, p. 435. In *Proc. Software Syst. Surface Modeling Grid Generation Workshop.* R. E. Smith, ed. NASA Conf. Pub. 3143, NASA Langley Research Center, Hampton, VA.

Peraire, J., Morgan, K., and Peiro, J. 1990. Unstructured finite element mesh generation and adaptive procedures for CFD, pp. 18.1–18.12. In *Appl. Mesh Generation Complex 3-D Configurations,* AGARD Conf. Proc. No. 464.

Peraire, J., Peiro, J., Formaggia, L., Morgan, K., and Zeinkiewica, O. C. 1988. Finite element euler computations in three dimensions. *Int. J. Num. Methods Eng.* 26.

Peraire, J., Vahdati, M., Morgan, K., and Zienkiewicz, O. C. 1987. Adaptive remeshing for compressible flow computations. *J. Complex Physics* 72:449–466.

Soni, B. K. 1992a. Grid generation: algebraic and partial differential equations techniques revisited. In *Proc. Comput. Fluid Dyn. '92,* Hirsch, Periaux, and Kordulla, eds. Vol. 2, pp. 929–936, Sept.

Soni, B. K. 1992b. Grid generation for internal flow configurations. *Comput. Math. Appl.* 24(5/6):191–201.

Soni, B. K., Thompson, J. F., Stokes, M. L., and Shih, M.-H. 1992. GENIE^{++}, EAGLEView and TIGER: general and special purpose graphically interactive grid systems. *30th AIAA Aerospace Sci. Meeting,* AIAA Paper 92-0071, Reno, NV, Jan.

Soni, B. K. Weatherill, N. P., and Thompson, J. F. 1993. Grid adaptive strategies in CFD. In *Advances in Hydro–Science & Engineering.* S. S. Y. Wang, ed., pp. 1.A:201–208. University of Mississippi Press.

Sorenson, R. L. and Alta, S. J. 1995. 3D GRAPE/AL: the Ames-Langley technology upgrade, p. 447. In *Proc. Surface Modeling, Grid Generation Related Issues Comput. Fluid Dyn. Workshop.* NASA Conf. Pub. 3291, NASA Lewis Research Center, Cleveland, Ohio, May.

Steger, J. L. and Chaussee, D. S. 1980. Generation of body-fitted coordinates using hyperbolic pratial differential equations. *SIAM J. Sci. and Stat. Comput.* 1:431–443.

Steinbrenner, J. P. and Chawner, J. R. 1993. Incorporation of a hierarchical grid component structure into GRIDGEN. *31st AIAA Aerospace Sci. Meeting.* AIAA Paper 93-0429. Reno, NV. Jan.

Thompson, J. F. 1985. A survey of dynamically adaptive grids in numerical solution of partial differential equations. *Appl. Numerical Math.* 1:3–27.

Thompson, J. F. 1987a. A composite grid generation code for general 3-D regions. *25th AIAA Aerospace Sci. Meeting.* AIAA Paper 87-0275. Reno, NV, Jan. 1987.

Thompson, J. F. 1987b. A general three-dimensional elliptic grid generation system on a composite block structure. *Comput. Methods Appl. Mech. Eng.* 64:377–411.

Thompson, J. F. 1993. The national grid project. *Comput. Syst. Eng.* 3(1–4):393–399.

Thompson, J. F. 1996. A reflection on grid generation in the 90's: trends, needs, and influences. In *Int. Num. Grid Generation Comput. Field Simulations.* B. K. Soni, J. F. Thompson, J. Hauser, and P. R. Eiseman, eds., 1029. Proc. 5th Int. Grid Generation Conf. ERC Press.

Thompson, J. F. and Mastin, C. W. 1983. Order of difference expressions curvilinear coordinate systems. *J. Fluids Eng.* 50:215.

Thompson, J. F., Warsi, Z. U. A., and Mastin, C. W. 1985. *Numerical Grid Generation: Foundations and Applications.* North Holland, Amsterdam.

Thornburg, H., Soni, B., and Boyalakuntla, K. 1996. A structured based solution–adaptive technique for complex separated flows. *J. Appl. Math. Comput.* (to appear).

Voronoi, G. 1908. Nouvelles applications des parametres continus a la theorie des formes quadratiques, Rescherches sur les parallelloedres primitifs. *J. Reine Angew. Math.* 134.

Weatherill, N. P. 1990. Mixed structured-unstructured meshes for aerodynamic flow simulation. *Aeronautical J.* 94(934):111–123.

Yu, T.-U. 1995. *CAGD Techniques in Grid Generation.* Ph.D. dissertation, Computational Engineering Program, Mississippi State University.

Further Information

For complete in-depth literature on grid generation, the reader is referred to Thompson et al. [1985] and five proceedings associated with the 1985, 1988, 1992, 1994, and 1996 International Conferences on Numerical Grid Generation in Computational Fluid Dynamics and Related Areas. (The first four proceedings were published by Pineridge Press and the fifth conference proceedings was published by the Engineering Research Center at Mississippi State University.) The literature on surface grid generation and practical applications can also be found in the NASA conference *Proceedings on Surface Modeling, Grid Generation, and Related Issues in Computational Fluid Dynamics Solutions* of 1992 and 1995.

In view of the importance of and worldwide interest in grid generation, the organizing committee of the 5th International Conference has proposed an establishment of the International Society of Grid Generation (ISGG). The ISGG will be a focal point for assimilating the progress and advances realized in grid generation by publishing a quarterly electronic journal of grid generation and a newsletter and by maintaining a grid generation Internet index to the grid generation literature, researchers, test cases, and information on public domain and commercial geometry-grid systems. The ISGG can also be the focal point for organizing future grid generation related workshops and conferences. The organization committee feels that the time is right for the emergence of a formal society and journal for grid generation. Additional information on the ISGG can be obtained from Bharat Soni, NSF Engineering Research Center, Mississippi State University, by e-mailing: bsoni@erc.msstate.edu.

35

Scientific Visualization

William R. Sherman
*National Center for
Supercomputing
Applications*

Alan B. Craig
*National Center for
Supercomputing
Applications*

M. Pauline Baker
*National Center for
Supercomputing
Applications*

Colleen Bushell
*National Center for
Supercomputing
Applications*

35.1 Introduction

The field of scientific visualization is broad and requires technical knowledge and an understanding of many communication issues. This chapter provides information about its evolution, its uses in computational science, and the creative process involved. Also included are descriptions of various software tools currently available and examples of work which illustrate various visualization techniques. Relevant concerns, such as visual perception, representation, audience communication, and information design, are discussed throughout the chapter and are referenced for further investigation. An overview of current research efforts provides insight into the future directions of this field.

35.2 Historic Overview

Visualization didn't begin after the advent of computers. There has always been a need for people to visualize information. At the dawn of human history, humans began spreading pigment on surfaces to convey events that took place and later to indicate quantities of goods. From that time on, the medium of choice for representing such information has continued to evolve (Fig. 35.1).

In general, visualization efforts required that the creators of the image represent their data by hand. Often this was a painstaking process that involved an artistic ability to mentally envision a pictorial representation of a phenomenon and the manual skills required to transpose the mental image into a suitable medium. The researcher had to be a capable artist and craftsman as well as a scientist. Usually, the visualizer would render the representation onto paper. However, other media for visualization were used as well.

As the scientific method developed, certain forms of visualization became accepted practices. As a scientist observed a phenomenon, it could be recorded onto an XY plot, representing the relationship between

0-8493-2909-4/97/$0.00+$.50
© 1997 by CRC Press, Inc.

two quantities. A line was often drawn through the data to show the probable continuous pattern. Error bars were added to represent uncertainty in the data (Fig. 35.2). We can now render detailed, data-based visual images by machine to show both quantitative relationships and qualitative overviews. How this process is accomplished, and its value to the scientific method, demand investigation.

The Motivation for Computer-Generated Visualization

The advent of the digital computer brought about the ability to collect, create, and store more information than previously. It also brought about a new method of science: *computational* [Kaufmann and Smarr 1993].

Computational science is the process of simulating a relevant subset of the laws of nature using a supercomputer. The laws of nature are described by a set of equations, which yield numeric solutions regarding the science being studied. Because these computational simulations of nature produce vasts amounts of numeric information, a scientist may not be able to see, much less interpret, all the results. Fortu-

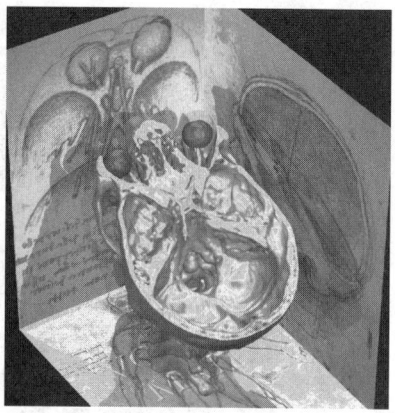

FIGURE 35.1 A combination of scientific representations from the fifteenth century and today demonstrates that the craft of visualization has been practiced for many years. (Courtesy of U. Tiede, T. Schiemann, and K. H. Hohne, IMDM, University of Hamburg, Germany [Tiede et al. 1996].)

nately, as the computational power of computers has increased, allowing these complex simulations to be calculated, so has the graphical power of computers increased. Thus, we also have access to a medium capable of creating and presenting all this information in a way useful to the researcher.

There are tradeoffs in how numeric information can be visualized. Interactive visualization gives the researcher the ability to control specific portions of the dataset to examine and to control the type and parameters of the visual output. Interactivity, though, may limit the percentage of the data that can be examined at a given moment and limit the types of representation available. Alternatively, the data display may be created as a batch process, allowing complex representations which are not possible in real time. The researcher may take advantage of both methods by beginning with interactive exploration and, when an interesting region of the data is located, producing a detailed animation.

Another consideration when creating visualization is whether to render a view of the entire dataset (a qualitative overview) or to precisely represent a subset of the data that the scientist can analyze (a quantitative study). Both are important in computational science. The qualitative overview can give the scientist a sense of the entire simulation, which can help in comparisons with observed nature. Because it gives an overall understanding of the dataset, it provides a sense of context when looking at the details.

The details are in the quantitative representation provided by a precise mathematical description. Qualitative information is helpful to this process, but high-resolution quantitative displays are essential. Representations such as contour plots and two-dimensional (2-D) vector diagrams are precise and aid in data analysis. Quantitative representations, such as these, provide the ability to pore over a particular subset of the data, even to the point of measuring phenomenon from the display. Because there is a limit to the amount of useful

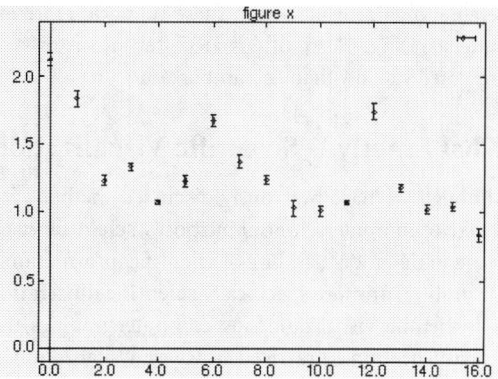

FIGURE 35.2 An *XY* plot with error bars.

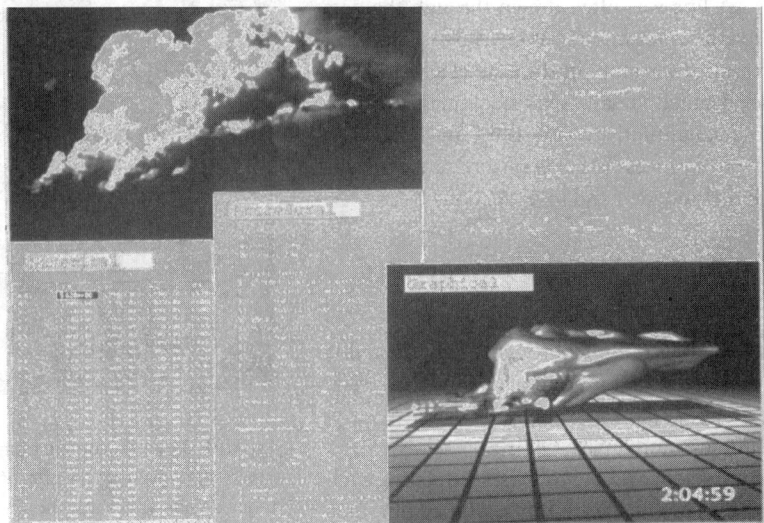

FIGURE 35.3   Computational science consists of observations, equations, algorithms, numerical solutions, and graphical representations. (Courtesy of Matthew Arrott; NCSA.)

information that can be rendered on a screen, focusing on a subset allows the data to be more completely displayed.

The Process of Computational Science in Relation to Visualization

To understand the role of scientific visualization in the computational science process, we must first review the scientific process itself. Figure 35.3 depicts the steps involved in the computational science process [Arrott and Latta 1992]. Computational science begins with observations of some natural phenomenon, in this particular case the formation of a severe thunderstorm. The scientist then expresses the observations in mathematics—the language of science. These equations can be manipulated by the researcher, though the problems are generally of a complexity sufficient to require solution on a supercomputer. In today's computing environment, the mathematical representation of the phenomenon is not a suitable means of input to typical computing systems. The computer requires the phenomenon be simulated in discrete steps of space and time, whereas mathematics allows a continuous representation. The mathematical representation is therefore translated into a programming language, implementing appropriate algorithms for a discrete numerical solution.

The resultant solution is typically a numeric value or set of values—a **dataset**. While the scientist may be able to gain insight from these numeric representations, a more intuitive visual form often aids in understanding. Also, others are often much more likely to understand the scientist's work when it is presented visually than as numbers only.

What Exactly Is Scientific Visualization?

Although we have previously discussed methods by which scientific data were represented before the advent of computers, for our purposes, **scientific visualization** is the use of data-driven computer graphics to aid in the understanding of scientific information. We will refer to an artist's rendering of a concept as illustration, or more specifically **scientific illustration**.

Is scientific visualization just computer graphics, then? Computer graphics is the medium in which modern visualization is practiced; however, visualization is more than simply computer graphics. Visualization uses graphics as the tool to represent data and concepts from computer simulations and collected data. Visualization is actually the process of selecting and combining representations and packaging this information in understandable presentations.

Other Modes of Presenting Information

There are ways to present information other than visually, of course. The primary of these is aurally, but one also might imagine the use of haptic (force, texture, temperature, etc.) display [Brooks et al. 1990], or even display to our other senses. Currently, sound (sonification) is increasingly recognized as an important method of information display for special types of data [Scaletti and Craig 1991].

Perhaps the field of representing information would be more appropriately termed **perceptualization**. The goal is, after all, to increase the information observer's perception of what is taking place in the data.

Application Areas

The use of scientific visualization to represent data is as broad as science itself. It spans the range of scales from the atomic and subatomic worlds to the vastness of the universe. It encompasses the study of complicated molecules and the building of complicated machinery. It looks at dynamic systems of living creatures and at the dynamics of whole ecosystems. Each of the areas touched by scientific visualization has representations that are particular to itself. Yet, there is much overlap in the techniques used by visualization developers due to commonality in the underlying mathematical expressions of natural systems.

Often a variety of what would seem unrelated sciences share similar or identical computational techniques. For example, computational fluid dynamics is used to study atmospheric effects, ocean currents, cosmology, mixing, injection molding, blood flow, and aeronautics. Finite-element analysis is used for solid and structural mechanics, fracture mechanics, crash-worthiness, heat transfer, electromagnetic fields, soil mechanics, metal forming, etc. Beyond these, there are many other fields that benefit from similar computational algorithms, including molecular modeling, population dynamics, diffusion, wave theory, and n-body problems.

Evolution of Scientific Visualization

The representation of the numeric output of simulations has developed from the simple printing of characters on paper, to vector display and plotter graphics, to three-dimensional (3-D) static images, to animated 2-D and then 3-D renderings of a simulation over time. The level of interaction has increased from the creation of visualization animations in batch mode, to real-time viewpoint control of fixed geometries, to interactive rendering allowing modification of the simulation in real time, and now to interacting with the simulation and representations in an immersive virtual environment.

As the underlying tools have improved, so have the idioms for representing information. New idioms are developed, and old ones are used in new ways. Many advances in representation are able to occur because of the advance of computing power and of computer input and output enhancements. Faster computing means more graphical computations can be done to create images, and higher-resolution displays allow for more detail to be presented.

These advances bring higher expectations. When 3-D pictures, animated 2-D pictures, and then finely rendered 3-D animations were first used by researchers to present their work, the visualization might have been considered the highlight of the presentation rather than the underlying science. The broader the audience, the more likely for this to be the case, giving rise to a situation where the scientists' credibility to the audience may be more correlated with the beauty of their images than with the underlying scientific theory. If scientists do not use the latest methods of computer graphics and animation to present their work, then it may not receive the attention it deserves. We are not arguing that this is how it *should* work, but it is important for scientists to be aware of the impact that visualization has on the communication of their research.

35.3 Underlying Principles

In this section, we will look at the various reasons for using scientific visualization and the effect they have on how a visualization is produced. We will also examine the basic concepts of visualization production and some of the considerations a producer should think about. Why are these important? Because scientific visualization is a means of communication. Sometimes the communication is between the raw numeric data and the researcher, and sometimes it is between the researcher and a group of people. Either way, for effective communication, it is important that both the producer of the visuals (or the tools used to create the visuals) and the audience have a grasp on what happens to the information as it passes from numbers to pictures.

The Goal of Scientific Visualization

Recall that the reason scientific visualization is employed as a tool is to more readily gain insight into a natural process. There are other similar goals that may be accomplished with scientific visualization. For instance, the goal might be to demonstrate a scientific concept to others, in which case the medium, representation, and the degree of detail chosen vary with the audience. A presentation shown to other scientists in the field will differ widely from one shown to the public or to government bodies.

The amount and the level of explanation required in a visualization is based on the experience of the intended audience. Also, design choices should be made which determine the amount of interactivity possible for wider audiences. For example, presentations designed for a mass audience are typically designed as noninteractive video animations. Alternatively, by utilizing computer delivered media, such as CD-ROMs, or multimedia presentations over the Internet, a limited dataset can be presented with a limited selection of visualization options, allowing for some experimentation by the audience.

When the audience is only the individual scientist, the primary goal is to uncover patterns in the data. Still, the goals of the study can vary. The goal might be to compare the patterns in the simulated data with patterns observed in nature. The closer these patterns match, the more confidence the scientist has in the theory expressed in the computational algorithm. Or the goal might be to discover new patterns that give clues to a better mathematical expression of the process. This is more frequently the case when the process is less well understood and the data are collected rather than simulated. For example, in analysis of the stock market, researchers might look for patterns that give rise to the ability to determine profitable opportunities.

The Basic Steps of the Scientific Visualization Process

Though it is possible to jump right into using visualization as a tool, there are several important steps that occur in the process of creating an effective visualization. At one level, scientific visualization can be thought of analytically as simply a transfer function between numbers and images, bearing in mind that this transfer may be irreversible and cause distortion of meaning.

At another level, visualization involves a barrage of procedures, each of which influences the final outcome and its ability to convey meaningful information. That is, the process of visualization includes consideration for data filtering, representation, potential inaccuracy, and human perception.

Data Filtering

Seldom is it possible to make pictures straight from the data source (a computational simulation or data acquisition system). Typically, work needs to be done on the data before they can be appropriately visualized. This can include cleaning up the data and performing operations on them that yield more useful data. Examples of cleaning up data are removing noise, replacing missing values, clamping values to be within the range of interest, etc. New numeric forms of the data often are derived in order to produce a dataset which will lead to greater insight. For example, the vertical vorticity of a fluid flow can be calculated from the horizontal wind velocity.

The medium of presentation may also be the cause of data filtering. From a practical point of view, it is often necessary to adjust the data so that they can be conveniently produced within the constraints of a particular medium. For example, a standard NTSC video device displays at a rate of 30 frames per second. Thus, the dataset must be adjusted to produce 30 images for each second of the animation. This is done by interpolating the data over time. To fill spatial gaps in the visual imagery, interpolation is also frequently done over space.

No matter what the medium, whenever any form of interpolation or other data-filtering operation is used, there are problems that can arise which may cause the imagery to be misleading to viewers who don't know or understand what has happened to the data.

Representation Issues

As noted at the beginning of this chapter, computational science involves choosing appropriate representations of the phenomenon being studied (e.g., numeric, mathematical, etc.). The representations of the data appropriate for computer manipulation are not necessarily the most conducive to human understanding. Representing numeric data so that they can be more readily perceived by the audience involves mapping those numbers to a geometric form, sonic waves, etc.

Visual representation of information requires a certain literacy on the part of the developer and the viewer [Keates]. Minimal information will be communicated if the viewer is not able to understand the visual language the developer has chosen. The language elements (symbols) of visualization come primarily from the adoption and evolution of symbols used in other visual domains, particularly scientific illustration. When new symbols are created, care must be taken to ensure that adequate explanation is given.

Beyond representing the numerical output of the simulation, it is desirable to indicate information about the simulation itself. This includes items such as a representation of the grid of the computational domain, the coordinate system, scale information, and the resolution of the computation.

When choosing which representational idiom to use, it is important to consider several issues before coming to a decision. What type of information is being investigated? Is the primary goal to convey quantitative or qualitative information? How detailed should the representations be? Making these determinations depends to a great extent on the characteristics of the audience for which the visualization is intended. The resolution of the display affects the ability to present quantitative information and thus is a factor in which idioms can be chosen.

The goal of the visualization limits the medium of delivery. The medium, in turn, puts constraints on the possible choice of representation and interaction. So, for example, if motion is important to show some aspect of the data, then a medium that can support time-varying imagery needs to be used (e.g., film, videotape, interactive computer graphics). If the delivery medium is constrained to be a single image, then one must find a means to represent motion statically.

The ability to communicate with the audience relies on a well-designed presentation of information. A common problem is to give equal visual importance to all the elements in a scene, making it difficult to comprehend. We can learn techniques for making effective imagery from experts in information presentation such as graphic designers and instructional designers.

Accuracy

It is good practice for scientists to question the accuracy and validity of any information they are presented. All too often, compelling visualizations are used without the audience really questioning what they are seeing. Today, visualizations sometimes accompany peer-reviewed publications without being subjected to the same critical examination as the paper.

Where does inaccuracy come from, and what forms does it take? Whenever data change representation, the possibility for the introduction of error exists. Illusions are a danger in any medium. This is especially true when representing three-dimensional imagery on a two-dimensional display. For example, parallax can lead to misreading sizes of objects. The bias of the visualization producer often can affect the accuracy of the presentation. This does not necessarily imply that they might deliberately misrepresent the information, but many of the choices made during the production can add up to a presentation that gives an inaccurate view of the results. Some of these might include the choice of which representation to focus on, and which to leave in the background, or the selection of viewpoint or color and lighting that can make objects look ominous or insignificant.

High production values often lead to a sense of quality. The quality of the imagery does not necessarily reflect the quality of the underlying science or representations. The computer graphics techniques used should not get more emphasis than the science (i.e., should not have "glitz" merely for its own sake). Adding glitz can make a visualization appealing but can also occlude the important elements of the presentation. High production values and accuracy are two separate factors and should not be confused. The overuse of glitz in visualization is satirically treated in the animation *The Dangers of Glitziness and Other Visualization Faux Pas* (Fig. 35.4) [Lytle 1993].

Labels can be used in any visualization, both as a tool for showing features of the visual representations and as a means to help clarify potentially confusing or unclear items. By adding labels, the viewer can be shown the size of the domain, the range of the values, the coarseness of the simulation, etc. Labels make the visualization more clear, understandable, and, therefore, a more useful means of communication.

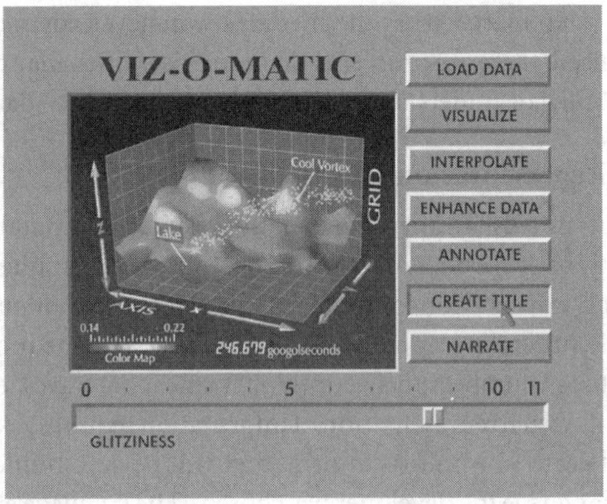

FIGURE 35.4   The Viz-o-matic animation pokes fun at the tendency to overuse graphic elements at the expense of accurate portrayal of the data. (Courtesy of Wayne Lytle; CTC.)

Human Perception

One filter that will always have an effect on how data are viewed is the human perceptual system [Weintraub and Walker 1969]. Our perception of the world does not exactly match with physical reality. In fact, there are many elements of the real world that we cannot directly perceive at all. For example, human vision covers only a small range of the electromagnetic spectrum; human hearing perceives only air pressure changes within a specific range of frequencies; human olfactory nerves have limited precision and no real ability to determine directionality of smells. Each sense has its own perceptual anomalies.

It is likely that some animals are able to perceive phenomena that humans cannot consciously perceive. For example, many species of birds can sense the earth's magnetic field and use it as a navigational aid (magnetoperception). Fortunately, we are able to use instruments that can sense elements of the physical world that we are unable to sense. These instruments then translate the sensory input into a display that humans are able to interpret. Visualization often involves the mapping of information from imperceptible forms to something we can interpret and analyze.

The field of *human factors* studied the relationship between people and machines and has a deep body of research to draw upon. It will be beneficial for anyone creating visualizations to familiarize themselves with the work of this field [Wickens et al. 1994].

All human perception, however, suffers from the additional problem that each of our brains interprets the incoming signals differently. Our experiences through life have trained our perceptual systems uniquely. Many biases are fairly consistent throughout a culture because the individuals within that culture will have had many similar experiences. Symbols in a culture lead to an understanding of what lies ahead or within; a skull and crossbones has a specific meaning in western culture, as does a railroad crossing sign. Colors are also culturally biased. Red can mean danger or stop; yellow, caution; and green, go. But for some, green also implies money or envy. In science, colors often are used to represent a scale of information, but this might not be understood by the rest of the culture. Which colors should represent high and low values, and which should signify the interesting data?

It is important to take human perceptual biases into account when designing an information display, but one also must recognize that these biases are not necessarily universal. Thus, a scale or legend should always be used to illustrate how information is being mapped.

35.4 The Practice of Scientific Visualization

In this section, we examine the process and visual components of visualization, we discuss several types of tools available for producing visualization, and we look at examples of how visualization has been applied to particular sciences and evolved over the course of several years.

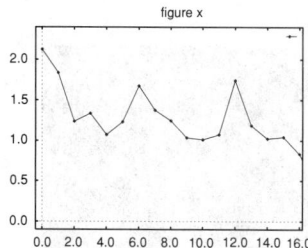 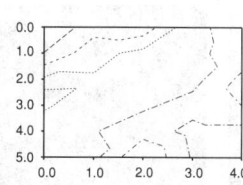 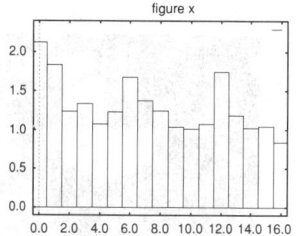

FIGURE 35.5 Simple 2-D graph representations convey both quantitative and qualitative information.

Representation Techniques

One of the most important elements in creating scientific visualizations is the choice of visual representation, or *visual idioms*. This section surveys a variety of commonly used visual idioms, each of which is appropriate in different situations. The visual representation is created by combining the elements of form, color, and motion that together show features of the data. This representation can range from very realistic renderings of real physical objects to abstract glyphs used to combine many pieces of information into a single idiom.

The traditional forms of data display should not be dismissed. Quantitative data can readily be retrieved from such representations as the XY plot, the contour map, and the bar chart (Fig. 35.5).

Realism Continuum

Accurate representation of information does not necessitate that the display be realistic. Information can be represented in a very abstract form, and its contents can be read accurately by someone that knows what the form symbolizes. Sometimes it is useful, though not always possible, to show a realistic representation of what the data symbolize, as in many architectural visualizations and prototype designs (Fig. 35.6).

In contrast, it is often more useful to create an abstract representation. This is especially true when there is not a direct physical counterpart to the concept being displayed. This is a convenient way of representing many variables simultaneously. The drawback with using very abstract representational schemes is that it takes time for the viewer to learn and become fluent with the symbols. In Fig. 35.7, symbols (or glyphs) represent different aspects of current weather conditions.

Color

Color adds to the appeal of a visualization. More importantly, color is used to convey information about data. Some fields, including atmospheric sciences, chemistry, and seismic analysis, have developed common conventions about color use within their application areas. Colors must be chosen carefully, with a view to the goal of the visual analysis. For example, visual displays often are used to identify the quantitative value at a point. Color can be used for this task by assigning colors to specific data values. The number of colors should be limited to about seven, and the colors should be easily distinguished from each other. Alternatively, the visualization task might be to determine the overall structure of the data slice, so that a color map that is continuously varying can be more effective.

Figure 35.8 shows the use of a variety of color palettes to color the same 2-D slice of data. The top two color maps use a wide range of hues—these are variations of a "rainbow" palette. These color

FIGURE 35.6 A realistic graphic representation of a front-end loader rendered from CAD data. (Courtesy of Caterpillar; Mark Bajuk; NCSA.)

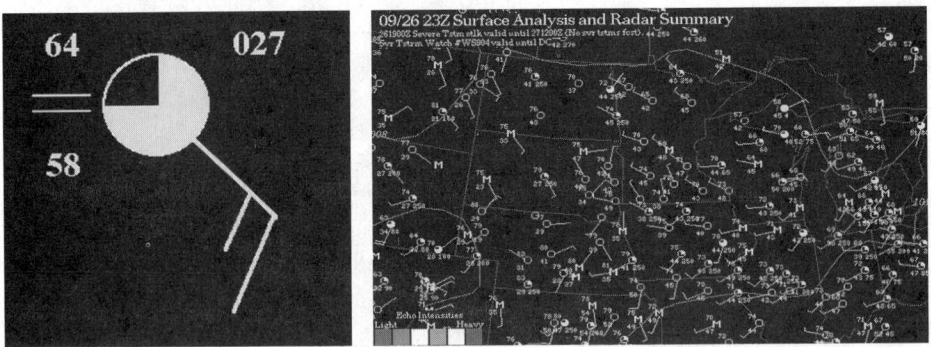

FIGURE 35.7 Weather maps often include an abstract symbol which depicts wind speed and direction, cloud coverage, etc.

maps introduce discontinuities into the image that may not be present in the data. For example, the edge from yellow to green or yellow to red suggests a distinct change in the data, which may or may not be there. The distinct bands may be useful if we're trying to isolate all the data values within a particular range. However, if we're looking for overall structure, a smoothly changing palette with a restricted hue, such as the "fire" palette shown in the lower right, is more appropriate.

Form

Two-Dimensional Plane. The simplest form to present on a sheet of paper or a computer screen is a 2-D plane. Depending on the dimensionality of the data, there can be tradeoffs

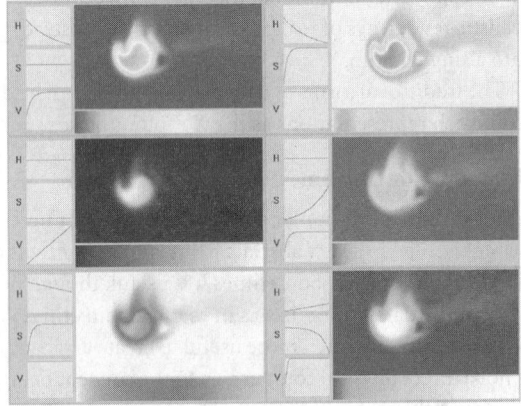

FIGURE 35.8 (See color version of this figure in the color section of the Handbook.)(Courtesy of NCSA, Yale University.)

between representing data purely in two dimensions and giving a 3-D view of the data. In a 2-D representation, we do not have to deal with perspective projection or other potentially confusing 3-D cues, thus making it much easier to make quantitative determinations from the rendering. Of course, this works best for data which inherently lie on a 2-D grid. Higher-dimensional datasets can be examined a slice at a time by cutting 2-D planes through it.

One can represent data on a 2-D raster image (also known as a false-color map). An example of this is the visualization of a 2-D fluid flow problem involving a jet of high-density material into a lower-density material.

The range of values is mapped to a range of colors. The choice of colors requires considerable thought in order to produce a meaningful image. Choosing colors arbitrarily may lead to false impressions of what the data indicate. It is not always necessary to employ a continuous range of color values. Sometimes a few colors chosen to indicate discrete states can convey information more effectively. In Fig. 35.9, discrete shades of yellow and green represent the age of the forest since the last fire in the region, red indicates a current fire, and black represents areas of nonvegetation.

There are several other idioms for representing information on a 2-D plane, many of which extend into 3-D forms. Contour lines and vector plots are two examples. Contour lines can be drawn on a 2-D plane of data indicating where specific values of the data are located. Typically, multiple contours are drawn in a single visualization to show the range of values. Vector values on a 2-D plane of data often are represented

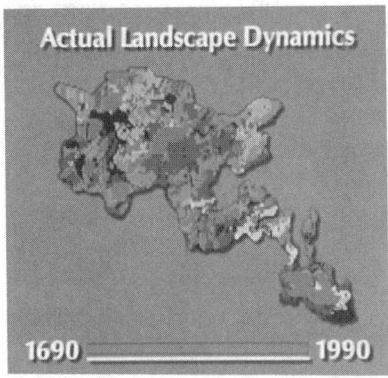

FIGURE 35.9 (See color version of this figure in the color section of the Handbook.) Discrete color mapping is used to depict the age of the forest in Yellowstone National Park [Kovacic et al. 1990]. (Courtesy of D. Kovacic, A. Craig, and R. Patterson; UIUC, NCSA.)

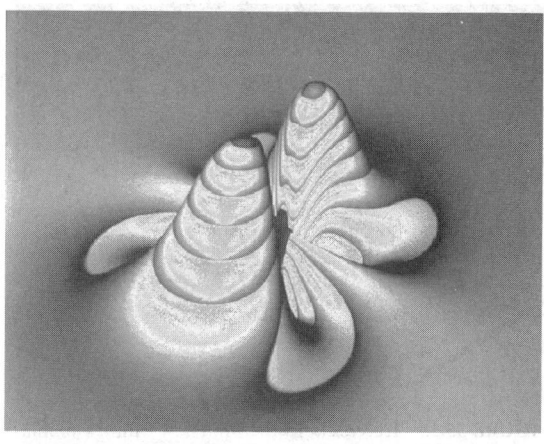

FIGURE 35.10 Visualization of gravitational effects of colliding black holes. (Courtesy of Mark Bajuk, Edward Seidel; NCSA [Anninos et al. 1993].)

as arrows that point in the direction of the vector, with a length proportional to the magnitude. To view the entire domain, arrows are placed at the location of each data element.

Height. If a goal is to see correlations between two variables in the dataset, one can add the element of height to the aforementioned color mapping idiom. One variable is mapped to the color, and the other is used to determine the height of the surface (e.g., Fig. 35.10). One must be careful in viewing such representations so as to avoid inaccuracies due to poor viewpoint selection. For example, in the extreme case, viewing a height–color rendering from directly above causes the height information to be lost.

Volume Rendering. The technique of color-mapping a single variable of a 2-D dataset can be extended to a 3-D dataset using a computer graphics technique called volume rendering [Drebin et al. 1988]. In looking at 3-D data, a viewpoint must be selected, and then an image is rendered by traversing through the elements of the dataset and assigning color and transparency values. A common usage is to volume-render medical data, such as the dog heart in Fig. 35.11. In this example, the less dense material is assigned a high transparency value, so we can see through to the higher density material of the muscle and bone [Moran and Potter 1992].

Vectors: Arrows, Tracers, Streamlines, etc. There are several common idioms used for displaying vector data. A standard representation of a 2-D vector field is to simply draw an arrow from each cell in the data in the direction of the vector, with the length proportional to the magnitude (Fig. 35.12).

Just as techniques such as colored smoke are used to visualize fluid flows in the laboratory, computer graphics equivalents

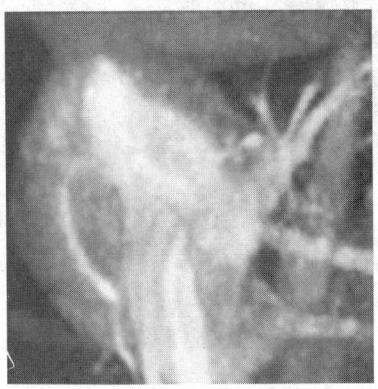

FIGURE 35.11 Volume rendering of a dog's heart. (See color version of this figure in the color section of the Handbook.) (Courtesy of E. Hoffman, P. Moran, C. Potter; NCSA.)

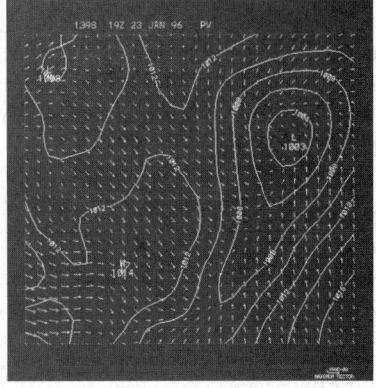

FIGURE 35.12 Wind vector field and pressure contours.

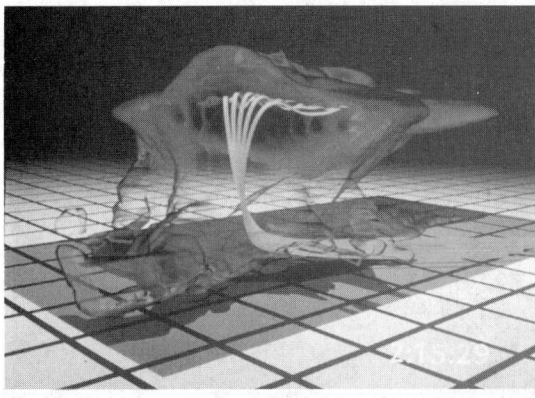

FIGURE 35.13 Tracers clarify the wind flow within a simulation of a severe thunderstorm. (See color version of this figure in the color section of the Handbook.) (Courtesy of Robert Wilhelmson et al. NCSA.)

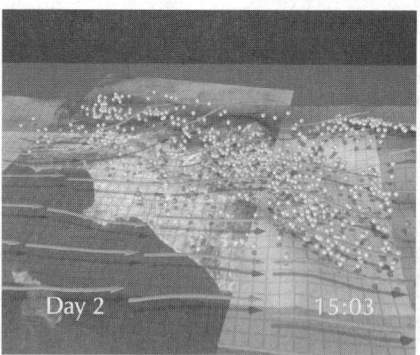

FIGURE 35.14 Streaklines and particles depict smog in Los Angeles. Each green particle represents 10 tons of reactive organic gases; each yellow particle, carbon monoxide; and each red particle, nitrogen oxide. (See color version of this figure in the color section of the Handbook.) (Courtesy of W. Sherman, M. McNeill et al. NCSA.)

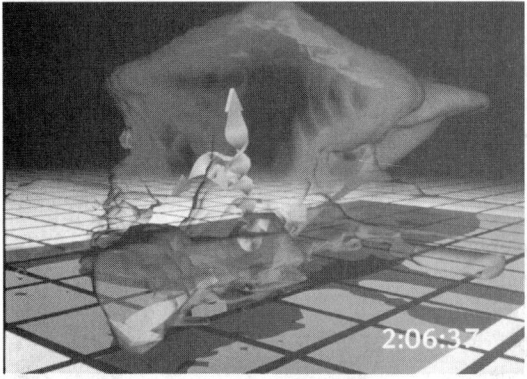

FIGURE 35.15 A ribbon's rate of twist indicates the streamwise vorticity. (See color version of this figure in the color section of the Handbook.) (Courtesy of Robert Wilhelmson et al. NCSA.)

FIGURE 35.16 Isosurfaces depict regions of electron-density change. (Courtesy of Jeffrey Thingvold; NCSA.)

of smoke may be used to display flowfields in computational simulations [Merzkirch 1987]. There are several idioms that give an effective qualitative view of the data. The simplest utilizes particles to indicate the flow. By adding lines showing the history of the paths taken by the particles, we derive a different technique often called tracers (Fig. 35.13). These are also referred to as streamlines, especially when the particle itself is not represented. A similar idiom, the streakline, shows a path through the vector field for a static moment of time (Fig. 35.14). In other words, it shows a portion of the path a particle would take if it were traveling infinitely fast.

As with many geometric representations, additional information (age, velocity, temperature, etc.) can be mapped onto a tracer, streamline, or streakline. This information can be represented with color, transparency, texture, or twist. Figure 35.15 shows a good example of representing the streamwise vorticity of a fluid flow simulation. In this representation, the twist along a streamline is proportional to the amount of vorticity at the particular location of the domain.

Isosurfaces. Extending the isoline (or contour line) to three dimensions gives us the isosurface [Lorensen and Cline 1987]. An isosurface is a surface of constant value in three dimensions. All points inside the surface have values below the threshold level, and all outside have values above the threshold level, or vice versa. For example, the isosurfaces shown in the cloud surface shown in Fig. 35.16 represent changes in electron density.

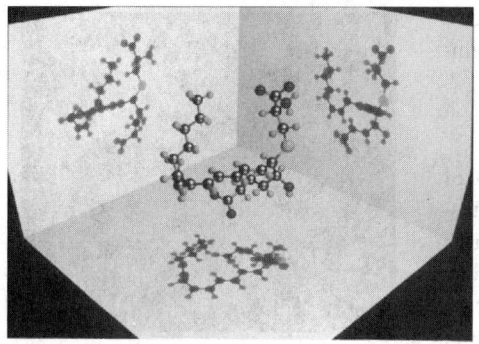

FIGURE 35.17 A computer-generated ball-and-stick model of a leukotriene molecule. Colored shadows aid in perceiving the 3-D shape. (See color version of this figure in the color section of the Handbook.) (Courtesy of D. Herron, Eli Lilly & Co.; J. Thingvold, W. Sherman, NCSA.)

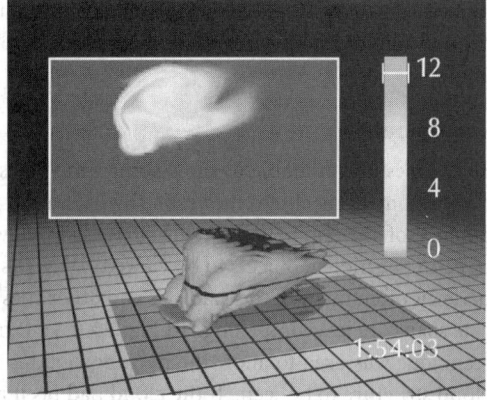

FIGURE 35.18 A cutting plane displays additional detail of a particular slice of a 3-D dataset. (See color version of this figure in the color section of the Handbook.) (Courtesy of Robert Wilhelmson et al. NCSA.)

Forms from Traditional Science. It often is desirable to choose representational idioms with which the audience is already familiar. For example, chemists are familiar with representing molecular structure with ball-and-stick models. These can be replicated in computer graphics. Using computer-generated imagery, however, removes certain physical constraints, and additional information may be given. For example, Fig. 35.17 shows a picture of a leukotriene molecule in which colored, orthographic shadow views are projected onto a three-walled stage to give the viewer a better clue to the 3-D shape.

Cutting Plane. Sometimes it is useful when looking at a 3-D dataset to cut the 3-D cube with what is called a cutting plane, revealing data along a single 2-D slice of that cube. Figure 35.18 shows a cutting plane, together with the 3-D isosurface. The black line in the isosurface indicates where the plane intersects it. The scale alongside the cutting place shows the mapping between the level of water content and the colors.

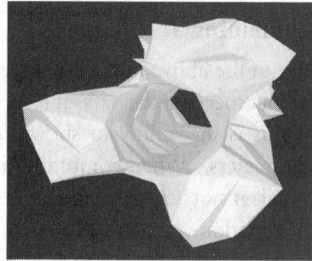

FIGURE 35.19 Alpha-shape representation of gramicidin A molecule. (Courtesy of H. Edelsbrunner, P. Fu, UIUC, NCSA.)

Alpha Shapes. A technique called alpha shapes allows one to represent the concept of shape applied to a collection of points in space [Edelsbrunner and Mucke 1994]. Alpha is the name of the parameter that, when varied, affects the complexity of the resultant shape. Large values of alpha produce a simple representation of the shape—a convex hull. As the value of alpha is continuously decreased, more complex shapes are created, revealing cavities, tunnels, and voids inherent in the dataset. For example, Fig. 35.19 shows a geometric representation of a molecule of gramicidin A, the first clinically used antibiotic (predating penicillin).

Motion

Motion can be a very important cue to the viewer for understanding the data being portrayed. It provides a qualitative overview of dynamic data by showing how the system evolves over time. Of course, the medium in which the imagery is presented affects the ability to indicate motion. In a book, for example, motion must be indicated by techniques such as motion blur, tracers, or series of still images from a time sequence (small multiples) [Tufte 1990] (Fig. 35.31).

Motion also can be used to aid the viewer in discerning the three-dimensionality of the object. Figure 35.20 shows an example of a 3-D set of molecules. The image in this book looks flat. However, when the molecules are rotating on a screen, a strong sense of the 3-D structure is conveyed. This representation also demonstrates how motion through the simulation can be represented in a single static geometry. In this visualization, the collection of small spheres represents the positions the atoms take over the course of the entire simulation.

Transparency

Sometimes it is desirable to show something that's inside or behind something else in the scene. By increasing the transparency of the geometry, the visualizer can allow internal structure or hidden objects to be viewed. When using transparency, often many of the shadings and other cues that indicate shape are less compelling. In Fig. 35.15, a transparency technique was used to allow the twisting ribbons inside the cloud structure to be seen. If the cloud had been made uniformly transparent, it would be difficult to see its shape. Here, the transparency is dependent on how nearly perpendicular the surface is to the viewpoint. This makes it easy to see the twisting ribbons in the center, while maintaining a well-defined edge to the surface.

Combining Techniques

Any of the above idioms may be combined to illustrate different aspects of the dataset (e.g., see Fig. 35.21). The benefit of this is the ability to show correlations between various parameters. When combining idioms, however, care must be taken not to overload the display or occlude important information.

The Visualization Process

This subsection provides an overview of the process of creating various kinds of visualizations. We begin by looking at basic still images. We then add features such as choreography, real-time interaction, and, finally, simulation control from within the visualization process.

Still Imagery

The basic method for transforming numeric data into visual output is depicted in Fig. 35.22. In this diagram, information flow follows the arrows, with dashed lines indicating optional parameters. At the top, data are provided from some numeric source such as a mathematically based computational model, a conceptually based computational model, or a collection of observed values. Data filtering involves a wide range of operations. Data must be translated into a form that is required by the visualization software. Additionally, the filtering process is used to sift out the most relevant aspects and discard unnecessary values.

After the data are filtered, they are mapped to some geometric form. At that point, decisions must be made about the materials and the characteristics of what these

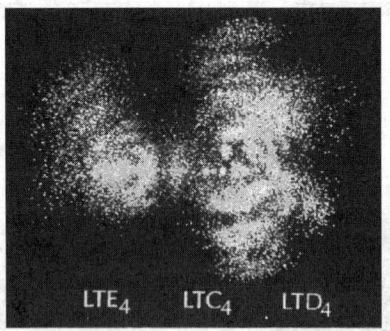

FIGURE 35.20 This image shows all of simulation time in a single geometric form and uses motion in the animation to aid the viewer in discerning the 3-D structure. (Courtesy of D. Herron, Eli Lilly & Co.; J. Thingvold, W. Sherman, NCSA.)

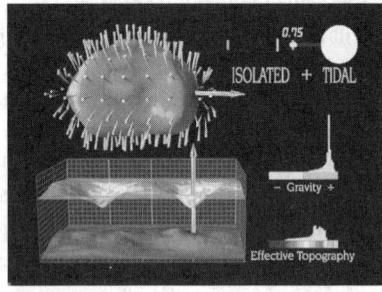

FIGURE 35.21 Multiple techniques combine to show gravity pull on the surface of Mars's moon Phobos. (Courtesy of Wayne Lytle, CTC.)

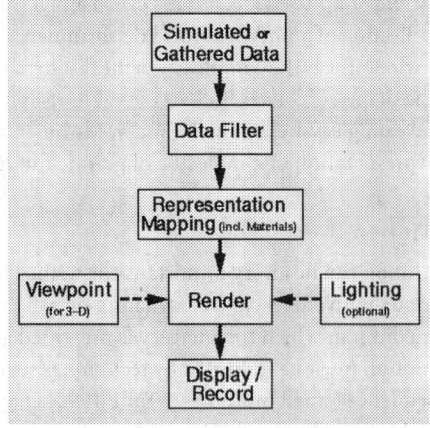

FIGURE 35.22 The basic visualization process. (Courtesy of W. Sherman.)

objects will look like—are they shiny, are they smooth, are they rough, are they dull, are they red, are they green, are they semitransparent? These parameters can also be driven by the computational model within the constraints imposed by the visualization software. Lighting information—i.e., how many lights, how diffuse they are, their relative location, etc.—is then combined with the viewpoint and information about the geometry by a computer program called the renderer. The renderer takes this information and computes the 2-D image which the eyes see. For a stereoscopic view, two images are rendered, from the point of view of each eye. The resultant images are then either recorded on film or video tape or displayed on a computer screen or other output device.

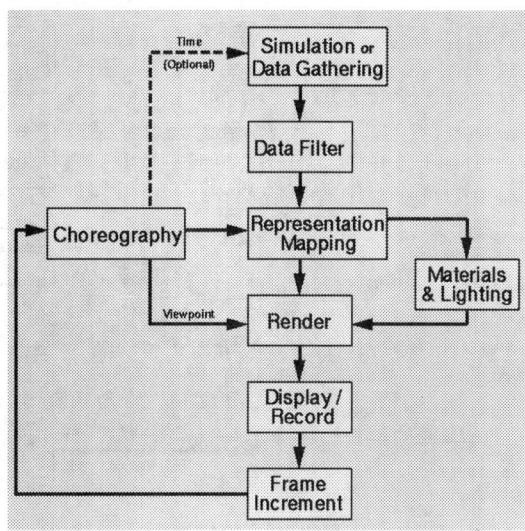

FIGURE 35.23   Process of creating a scientific visualization animation. (Courtesy of W. Sherman.)

Animation

The basic methodology for creating a computer animation is an extension of the process used to create the still image. The major extension is the concept of choreography. Now, the same steps are followed as in Fig. 35.22, but the viewpoint information can be choreographed, along with the objects, their representations, and the passage of simulation time itself, resulting in the diagram in Fig. 35.23. Animation is the result of rendering the scene repeatedly while varying the choreographic information and then displaying the images in rapid succession.

Choreography is that part of the process which controls the viewpoint of the scene and the movement of objects within the scene. Choreography can also be used to control how representations of the data change during an animation, such as the transparency of a material of a particular idiom.

Another important element of the dataset that can be controlled by the choreographer is *time*. Usually, time is considered to be constantly moving, at least as long as the animation is playing. However, there are two notions of time in a visualization animation. One is the passage of time of the viewer while watching the animation. The other, for visualizations of time-varying datasets, is the notion of time within the data. There is no requirement that data time move at a steady pace through the animation. It often can be insightful to hold time constant while other parameters of the representation are choreographed, revealing additional information. Moving time at different rates may reveal features that otherwise would be hidden.

Interaction

Animated sequences are made of several still images displayed in rapid succession. Often these images take longer to create than the fraction of a second that they will be displayed. This results in a situation where the choreography for all the action is planned ahead of time and is then unchangeable once the animation is complete. This is not generally the optimal method for a researcher who wants to probe a dataset interactively.

To have a more insightful experience with the data, scientists want to be able to spend more time looking at specific pieces. They want to be able to look at them from different angles, change representation parameters, and watch them at different rates of speed. To allow this kind of interaction, the images that are animated on the screen must be created as fast as they are displayed (i.e., in *real time*). Thus, the rendering stage in our visualization figure becomes a real-time (RT) process, and the display–record stage is now display only, with the option of recording (Fig. 35.24).

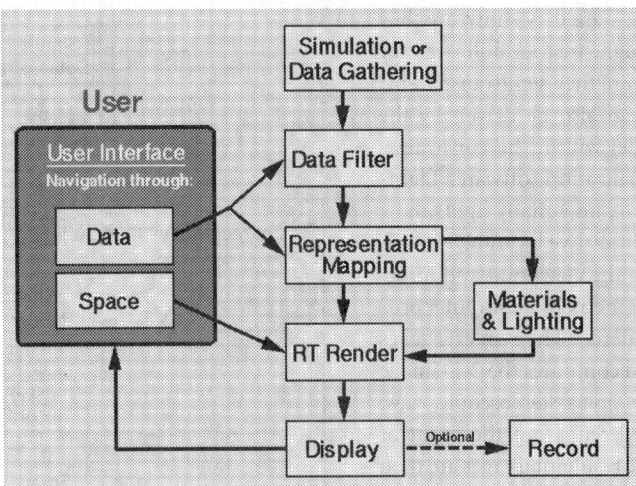

FIGURE 35.24   Interactive visualization allows the user to control the data filter, the representation, and the viewpoint. (Courtesy of W. Sherman.)

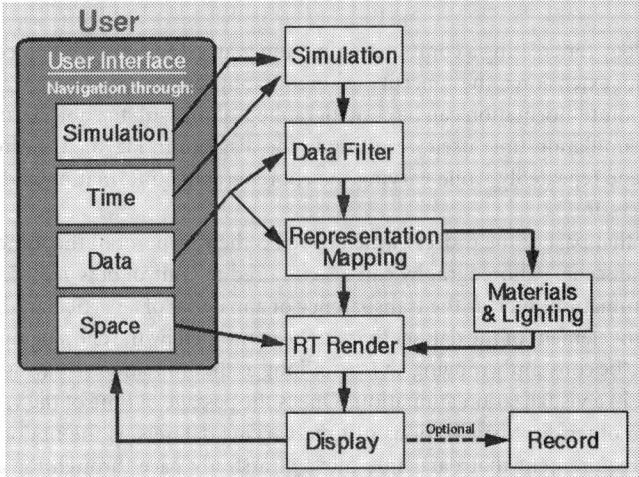

FIGURE 35.25   Interactive steering allows the user to manipulate the simulation in addition to the visualization. (Courtesy of W. Sherman.)

The most significant change in the interactive visualization process is that the choreography now happens in real time too; instead of an animator creating the choreography, the scientists themselves choreograph the scene through the package's user interface (UI). At this level of interaction, users are able to guide themselves through space and time and control the representation of the data (Fig. 35.24).

Interactive Steering

Interactive steering is the ability for the user to alter the course of the simulation in real time [Haber 1989]. Figure 35.25 further extends the visualization flow chart to include user control of the simulation—including direct control over passage of time. It is vital that both the simulation and graphics system provide real-time performance and allow for user input. The user can then interactively steer the simulation by modifying variables and data in the simulation, as shown in the figure. The user can still steer the processing of the data into graphical representations and steer the viewpoint of the scene.

Visualization Tools

A number of tools are available for creating visualizations of information. The choices have improved with the evolution of computer technology. Visualization tools provide varying elements of the visualization processes shown in the figures. They can be categorized as follows: plotting libraries, turnkey packages, dataflow packages, and animation packages. There is also the option of creating the visualization by writing the software in-house.

Each package (and type of package) has certain advantages and limitations that may make one more suitable for a particular application or situation. We will examine some examples of each type of visualization tool.

Plotting Libraries

Historically, computer-graphics-based visualization originated with the use of alphanumeric printouts to mimic a bar graph, a shaded "color" map, or values from which contours can be hand-plotted. Computer visualization began to take shape with the flexibility brought about by new output devices, such as pen plotters and vector displays. These new devices made the creation of computer-generated contour and vector plots more feasible.

Software libraries were developed that enabled researchers to generate charts, graphs, and plots without the need for reinventing the graphics themselves. These packages have evolved as the computers and computer I/O have improved and are still available today. Since these packages are basically programming interfaces, they do not typically involve any form of interaction other than through programming and (perhaps) a command-line interface.

The limits on the interactivity of these tools suggest that they are of limited functionality. Although this is partially true, there are also some benefits. Because scientists often want to work with quantitative representations of their data, plotting libraries can provide the easiest means for producing the appropriate output. Print output has much higher resolution than screen output, and these libraries can produce high-resolution plots specifically for output to such printers. Screen-based interactive packages often limit their resolution to what is available on the screen, thus limiting their ability to produce quantitative visualization.

Turnkey Visualization Packages

A turnkey visualization package is a program designed specifically for doing visualization and contains controls (widgets) for most options a user would want to exercise when visualizing the data. This is accomplished through the use of pulldown menus or popup windows with control panels.

A number of commercial and no-cost visualization software packages are available. Most packages are available for Unix workstations, and many are also available for PC compatibles and Apple Macintoshes. Commercial packages include Fluent's CFD software package [Fluent], NASA/Sterling Software's FAST [FAST], and Fortner Research's visualization suite [Fortner]. Freely available packages include VIS-5D from the University of Wisconsin [Vis-5D], Sci-An from Florida State University [Sci-An], and Gnu-Plot from Dartmouth College [Gnuplot].

There are definite advantages to using turnkey visualization software. The primary advantage is that one can often read one's data directly into the packages and immediately begin manipulating the controls to look at various aspects of them. This assumes that the data are initially in a format readable by the package and that the controls are intuitive enough for someone to experiment with and get good results.

These packages are designed as interactive tools; their ability to manipulate the view and the visualization parameters is key to putting the visualization process into the hands of individual researchers. Also, because the target user is a scientist, the user interface is usually designed for ease of learning and use.

Off-the-shelf packages also have some limitations, however. Foremost is that their flexibility is often limited. For instance, the user won't be able to add visualization representations that are not provided by the software. Also, if too many representations are included, then the user interface may suffer from an overwhelming number of control selections, making it hard for the user to find the control needed to perform a specific operation.

The desire for interactivity also imposes limits on these packages. To be interactive, data must be easily retrievable, and the representational forms should not be rapidly changing. This means that displaying features of time-dependent data may be too slow to be very interactive or may not be allowed by the package at all. To have interactive real-time rendering, many constraints are put on the amount of complexity the images can have. Generally, very beautiful imagery requires more complexity than can be achieved in real time. One solution for this is to allow the user to experiment with representations and choreography of the data; when a satisfactory situation is selected, information can be sent to a noninteractive rendering system to produce nicer images.

Dataflow Packages

Like the turnkey visualization packages above, dataflow visualization packages are designed as tools to be used directly by the scientist. However, dataflow packages are much more modular, with each stage of the visualization process represented as an independent unit. These units are then connected in the appropriate manner to allow data to be passed (or *flow*) from one unit to the next. This style of interface is inherently more flexible and also provides a map of the visualization process the system is using. These packages also are designed to be more extensible, allowing the user to add features and functionality that are not provided off the shelf.

Dataflow software currently requires the availability of a Unix workstation. Most packages now run across different brands of workstations. Despite the narrowing gap between PC and workstation, not many run on PCs. Three factors that contribute to this situation are the requirement for doing a considerable amount of processing of the data, the use of large amounts of memory, and the need for large-screen displays with reasonable graphics performance.

The dataflow concept consists in breaking down tasks into small programs, each of which does one thing. Each task is represented as a *module* (a software building block, or "black box") that performs the specified operation. Each module has a defined set of required inputs and outputs for passing information between modules. Figure 35.26 shows a simple connection network of some modules. In this example, data are retrieved with the Read HDF module and flow to the Isosurface module, which creates a geometric representation of the data, passing the new information to the Render module, which renders a pixel map, which in turn is passed to the Display module, which puts the image on the screen.

Though the dataflow concept had been described for doing other tasks (such as image manipulation), AVS (Application Visualization System) was probably the first package to apply the dataflow concept to the task of doing visualization [Upson et al. 1989, AVS]. AVS is available on most Unix workstations. IRIS Explorer [IRIS Explorer], originally developed for Silicon Graphics workstations, is also now available on other workstations. IBM's Data Explorer [IBM Data Explorer] is another example of commercially available software. Khoros [Khoros] is one freely available package of this nature. SCIrun, a powerful new implementation of the dataflow paradigm, is also freely available to some institutions [Johnson and Parker 1995]. The user interfaces for most of these programs are surprisingly similar, and familiarity with one package makes learning another easier.

Advantages of systems designed around the dataflow paradigm are numerous. As with turnkey packages, real-time rendering coupled with user control of the viewpoint and visualization parameters allows researchers to experience their data first hand. Flexibility is enhanced by the user's ability to connect modules together, creating a network of modules specific to the task at hand.

Flexibility is raised another degree by having the option to create new modules that perform operations not provided by the base collection of modules provided with the system. New modules may be

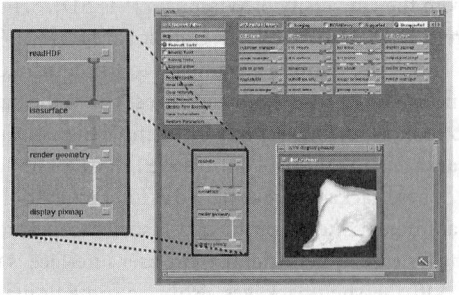

FIGURE 35.26 Dataflow packages such as AVS connect modules to customize a visualization. (See color version of this figure in the color section of the Handbook.) (Courtesy of W. Sherman, NCSA.)

written in common programming languages, such as C or Fortran. By utilizing routines provided with the system, the new modules will exhibit the same user interface and data handling as the standard modules. Since many people typically share the same desire for specific extensions, sharing of modules is common practice, thus making these packages even more valuable.

There are some disadvantages to this form of visualization package. For many, the primary disadvantage is the fact that most dataflow packages are available only on Unix workstations. Another is the difficulty with which this type of package handles data that change over time as part of a real-time visualization. Typically, data flowing down the network is for a particular time step of the computational simulation, and the modules that provide the representation of particle flow through a vector field will not be able to get the

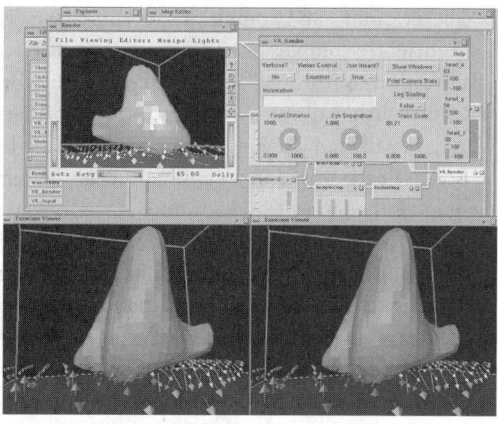

FIGURE 35.27 Customization capabilities allow the package to be extended to new display paradigms, such as in this stereo image linked to a virtual-reality viewing device. (See color version of this figure in the color section of the Handbook.) (Courtesy of W. Sherman, NCSA.)

data fast enough to give smooth-flowing animation of those particles in a changing dataset (though they generally can handle particles flowing through a time-static dataset).

And last, the dataflow interface *is* a type of programming; thus, the end user at one level must learn how to program the system. While easier than writing code in C or Fortran, this process is more complicated than in the turnkey approach. The degree of difficulty increases as the networks become complex and the application has greater requirements, such as looping time-varying data that are flowing through the system.

The modularity of the dataflow paradigm also allows for the addition of a new style of interfacing with the user. New input devices such as Spaceballs™ and 3-D trackers can allow the system to control the viewpoint of the environment. New output devices can provide a head-tracked, stereo view, allowing *virtual reality* to become a potential method of interacting with a visualization [Sherman 1993] (Fig. 35.27).

Animation

Today, presentation-quality visualization animations typically use standard computer animation packages [Fangmeier 1988]. Until recently, computer animation packages were the only off-the-shelf software available that provided 3-D rendering of any sort. Thus, these were often used to provide the rendering stage of the visualization process.

Computer animation packages are designed for all types of computer animation, not just for use in scientific visualization. However, it can be said that whatever they are used for, they are used to show (or visualize) the world the animator has conceptualized. For scientific visualization, the objects in the scene are typically derived from data. An animation package typically has separate components that allow the modeling of objects; the choreography of the objects, lights, and viewpoint; and the ability to design visual material qualities for each of the objects. These are all tied together into an image or series of images by the rendering component.

The primary advantage of using animation packages is that they are extremely flexible. The user is able to create objects of any type or size to represent the subject of the visualization and any supporting objects that might be required to give context to the representations. In a visualization environment, most of the objects are created by writing specialized programs that convert data from their native format to a format suitable for the animation package. This flexibility does have limits, however, such as when there is a need to represent the data using a method not available in the package. For example, a volume rendering technique may be required, and the renderer for the package may support only polygonal techniques.

Another advantage of animation packages is the ability to create very high-quality imagery. It is important to note, though, that the image quality is also dependent on the work done by the visualization creator.

The flexibility and complexity of animation packages also can be a disadvantage. These tools can be sufficiently complex that to use them as a visualization tool requires an outside expert to bring the visualization to fruition. Such an expert would be able to write the custom programs that convert data from their original state to an appropriate format. The fact that the consultant doing the visualization work is usually not an expert in the field of study being represented also can be a disadvantage.

Because each image in an animation can be very detailed, rendered using computationally expensive techniques that give rise to high-quality scenes, the time required to create the image is unlikely to be short enough to allow for what is referred to as real-time rendering. In fact, it can take several minutes to several hours to produce an image that will be viewed for one-thirtieth of a second (at video rates). The time required is a disadvantage even in situations where months are available to produce the animation, because of the delay between the time a small change is made and the time the scientist can view the resulting work.

Another disadvantage of using an animation package is the increased potential loss of data integrity that can occur in the process of translating the data from their original form to a form suitable for the animation package. The likelihood of data integrity problems is higher in this environment for two reasons: more custom work is done to translate the data, and the animation package may have options to make the output look good at the expense of giving an accurate view of the object being rendered. In an off-the-shelf visualization package, the techniques used to bring data to the screen are often well-known ones that a user can research and learn how they may affect the resulting image. Custom code used with animation packages may not be as heavily scrutinized if it is used only once for a particular application. Also, animation packages have features such as automatic smoothing that take the rough spots out of the data. This changed representation probably will not give an accurate view of the original information.

WYOS: Write Your Own Software

Before visualization and animation packages were available, tools to look at data were custom programmed for the task at hand. This is still done for certain visualizations—especially ones that handle large amounts of time-varying data or require new computer graphics techniques that have not been implemented in off-the-shelf software (e.g., certain volume visualization techniques). Now, however, users generally choose to use off-the-shelf software, perhaps with a few modifications.

Some examples of visualizations requiring custom renderers include *L.A.—The Movie* [Hussey et al. 1986], which used special techniques for overlaying satellite imagery onto a topological map of the Los Angeles region, and *The Deluge* [Sims 1989], which brings a still image to life by using a renderer written specifically for that task.

The only advantage to writing your own application software is in situations where that is the only way you can obtain the visualization you need.

Tools Summary

In summary, many tools have been developed which aid in visualizing scientific data. These tools are improving, making visualization an easier task to perform. However, even with a very easy user interface, the skill of the visualization developer is the overriding issue. No tool, no matter how easy to use, can replace the skill and insight of a visualization expert.

Examples of Scientific Visualizations

We have explained the components of scientific visualization and examined how these are used to create output that the researcher can use. In order to have a better understanding of how visualization is typically used in practice, and to look at how representational idioms have evolved, we will describe a selection of visualizations used as a means for scientific study.

The Study of a Severe Thunderstorm

The primary example of scientific visualization we will examine in this chapter is the computational simulation of severe thunderstorms. The simulation of meteorological processes has a long history in the computational sciences and consequently can be used to demonstrate the evolution of many visualization techniques. In particular, we will look at the research performed by Robert Wilhelmson's research group at NCSA and the Department of Atmospheric Sciences at UIUC and the visualizations done by his group and the NCSA Visualization Group.

Evolution of Atmospheric Simulations and Visualization. Wilhelmson's group has been studying the process of thunderstorm development computationally for 25 years. Over this period, several developments have been made in scientific visualization, and these are reflected in the many visualizations made of his work.

In the 1950s, the simulations run on digital computers were so simple that the researchers could literally examine every number produced.

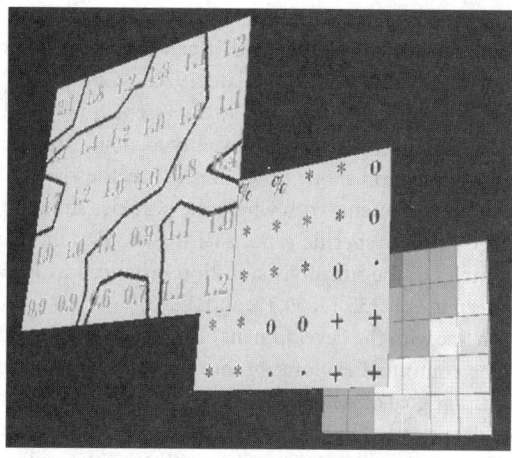

FIGURE 35.28 Early visualization attempts included hand-drawn contours and shaded images constructed from a judicious choice of ASCII characters.

In the 1960s, digital computers had become fast enough to be used to execute 2-D simulations of atmospheric conditions over time. At first, the data were examined simply by printing the numbers in a 2-D array on paper. The researcher could then actually draw contours through the data by hand based on the printed numbers. An alternative early technique was to print a 2-D array of letters, with each letter representing a range of data values (this was often referred to as "gray-shading," with each character effectively representing a different level of gray) (Fig. 35.28).

About this time, a package for producing standard visualizations of 2-D data was released by the National Center for Atmospheric Research (NCAR). This package, known as *NCAR-graphics* (UCAR) could produce representations such as the image of contour levels within the developing storm front shown in Fig. 35.29. In general, all the visual output of the simulations were of a static slice of time.

The 1970s saw the true advent of 3-D simulation models for all areas of atmospheric science. With the expansion into three dimensions, it became more difficult to interpret and explain the results. By this time, NCAR-graphics was the primary visualization package used by atmospheric sciences. With the addition of the ability to produce 3-D isocontours (i.e., isosurfaces) (Fig. 35.30), Wilhelmson was able to view storms from a real-world perspective and from a radar perspective, among others.

Other techniques that began to be used in the late 1970s include the use of color to visually separate different representations, giving the ability to view them simultaneously. Color also was used to draw different layers of isosurfaces within a 3-D structure. In addition to color, Wilhelmson began to look at the change of the system over time; initially, this was limited to an alternation between two images on a raster-frame storage device.

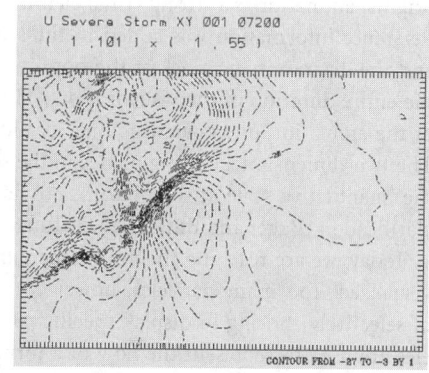

FIGURE 35.29

FIGURE 35.30 An early isosurface representing the late stage of the storm.

In the early 1980s, the use of flow tracers (particles) to visualize flow through a field began in earnest—though some had been done in the late 1970s. Computing particle location was a new way to represent flow through a fluid substance, providing the ability to view a large region of the simulation at one time. In practice, the visualization creator experiments with where the particles should be released to show specific aspects of the simulation. Several examples of the storm visualization produced in 1989 can be seen in Figs. 35.13, 35.15, and 35.18.

In line with the development of visualization technology is the continued exploration of effective information design. The goal of visualization is to communicate aspects of the data clearly. To achieve this clarity, one must be aware of basic design principles and have the ability to effectively critique the design quality of visualization. Issues of visual

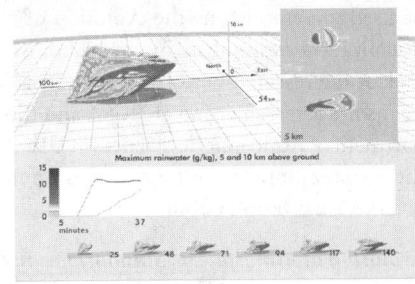

FIGURE 35.31   The thunderstorm visualization was redesigned with less contrast between the grid and the ground and small multiples added to represent changes in time. (Courtesy of Yale University, NCSA.)

hierarchy, composition, choreography, color theory, and typography cannot be disregarded. Attempts to demonstrate scale, provide a sense of context, combine qualitative and quantitative information simultaneously, direct the viewer's eye, create multiple representations, and point out intentional manipulations in the data are tasks that the creator must work through in order to produce quality information design [Tufte 1990].

The storm visualization shown in Fig. 35.18 was redesigned to illustrate several issues of information design. In the new design (Fig. 35.31), the contrast between the elements of the scene was adjusted so that visual emphasis is placed on the important details of the visualization. The grid floor in the original animation is a strong visual element and distracts from the dynamics of the cloud development. Reference information was added to the scene, such as the dimensions of the computational domain and the diagrammatic clock at the bottom. This clock includes a stripe of small images that illustrate the entire duration of the visualization. This gives contextual information by indicating which portion of the animation one is viewing and in addition shows the numerical time step for specific reference. The two-dimensional cutting planes in the scene in conjunction with the three-dimensional cloud, and the quantitative scale at the bottom, provide multiple representations of the data for comparison and clarification [Baker and Bushell 1995, Tufte 1996].

Today, we are reaching the point where the flow of large numbers of particles can be computed and visualized. Too many particles, however, can make it hard to see what is going on, so the technique of selectively chosing to display specific particles is being explored. In the picture in Fig. 35.32, particles are used to represent the flow of a tornado. To highlight just the tornado, only particles with high vertical vorticity, within a certain region, are shown. The amount of data and objects required to represent it also brings about a logistical problem of where all this information is stored. Large disk drives are required to create visualizations of massive numbers of particles, and files must be carefully managed to stay within the constraints of the storage.

Virtual reality (VR) also is being experimented with as a means of gaining insights from simulated scientific experiments [Baker and Wickens 1995]. VR experiences have been designed for the VROOM [Brown 1994] and SuperVROOM [Korab and Brown 1995] venues at SIGGRAPH '94 and SuperComputing '95, respectively. In these experiences, the domain of the data is represented by a bounding box, with various tools available to represent the data, including

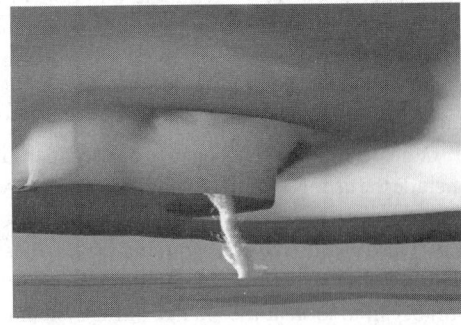

FIGURE 35.32   Thousands of particles show the shape of a tornado. (Courtesy of Matthew Arrott.)

isosurfaces, and particle flow. In a CAVE™ [Cruz-Neira et al. 1992] VR system, the user can easily navigate the data space and has the ability to control the representations. In the particle flow representation, the user can control the release of particles with a hand-held device. As the user releases the particles, a flow simulation program is begun that updates the particle positions. This program (and any of the representation modules) can execute on a separate larger supercomputer for better interactivity.

It is hard to judge how much additional insight can be gained by immersive interactive visualizations (i.e., virtual reality). This stems from the deep knowledge atmospheric scientists already have of their simulations. Major concerns of the scientist in using virtual reality include: Would I use a device that requires me to be standing or walking for an extended period? Analyzing the data requires knowing where one is with respect to the cloud, which is easier when viewing the cloud from the outside—in which case, what is the advantage over the desktop? To be truly immersed, the user should feel the wind, hear the thunder, see the lightning, none of which is accurately representable with current VR technology.

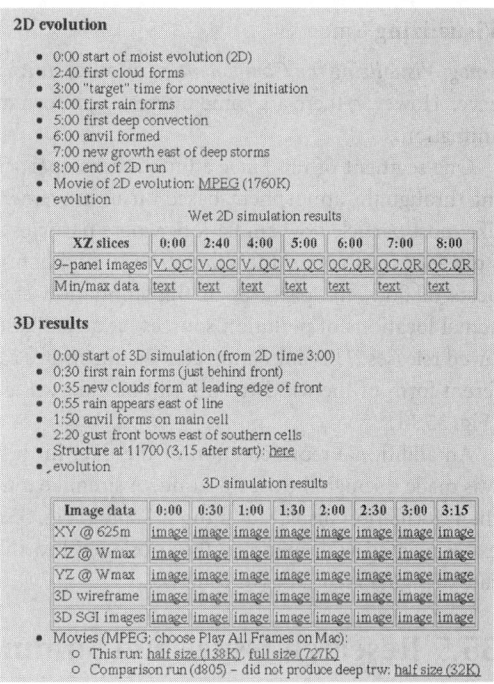

FIGURE 35.33 Portion of a WWW page used to document a research project. (Courtesy of Brian Jewett; NCSA.)

Scientists have become proficient at deciphering the 3-D content from flat 2-D displays from years of experience. Because of this, they see less need for the improved 3-D cues that VR provides. However, they are continuing to explore the use of virtual reality as a visualization tool. Since it is a new medium, there are many technological and user interface improvements ahead; VR still may prove to provide unique insight not obtainable through other means.

The World Wide Web (Fig. 35.33) is viewed as a very useful tool for disseminating research information. It can be used by the public to access weather forecast information or by other atmospheric researchers to directly examine the simulation data themselves. It can be the complete tool for a researcher—the notebook that contains all the diary entries over the course of a study, including explanations, data, images, animations, and interactive visualization tools. It can be made available not only to the researcher but also to colleagues, advisors, and (if desired) the scientific community [Jewett and Wilhelmson 1995].

What Does the Scientist Use Today? With all that is available to the scientist today, what tools are actually used? To be specific, everything. Line graphs, 2-D contours, and 2-D vector plots still provide the basis for closely investigating the results of a simulation. Three-dimensional stills and animations are excellent tools for illustrating concepts. Interactive visualization tools help to find out what's in the dataset and to locate interesting regions that demand close inspection.

Animation, while a good tool for presenting the overall contents of a research study, poses problems when it comes to publishing the study. Scientific publishing still exists primarily as a print medium, and thus it is difficult to include animation. Indeed, even submitting an animation for inclusion in the review process is difficult. The World Wide Web is beginning to provide a solution to these problems. The entire research submission, including animations and interactive tools, can be included as one comprehensive document for the review. Once reviewed, such a submission can be published on CD-ROM, while still remaining available online where it can be periodically updated.

Visualizing Smog

Smog: Visualizing the Components is another animation visualization in the domain of atmospheric science. However, there are some interesting information presentation features that are highlighted in this animation.

One segment of the *Smog* animation has particles that represent different forms of pollutants flowing through the atmosphere, based on air movement data. Unlike most particle flow animations where particles are released from locations within the simulation to highlight specific aspects of flow, the particles in this animation emanate from actual locations of pollution sources, at a rate based on measured releases. This animation also demonstrates a slightly different form of the wind-tracing ribbons known as streaklines (Fig. 35.14).

An additional representation of the data in this animation was made through the use of sound. A sirenlike tone reinforces the information displayed in the *XY* plot (Fig. 35.34). It also experimented with mapping the ozone level to the repetition rate of digital recordings of coughs.

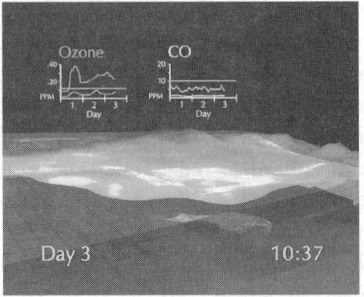

FIGURE 35.34 Los Angeles smog visualization. (Courtesy of NCSA.)

35.5 Research Issues and Summary

Though great strides have been taken in developing scientific visualization as a tool in computational science, a multitude of issues remain for further research and development. As computer power increases, the need for effective visualization tools increases due to the larger volume of information being produced. Fortunately, the increase in computer power also allows us to develop more compelling visualization tools with more flexibility, better performance, and higher fidelity.

When studying the field of visualization, it is important to recognize that visualization is being done effectively in a variety of fields outside of the realm of computational science. Medical imaging, factory automation, and financial researchers, to name a few, are taking advantage of the power of visual representation of information.

Additionally, there is a growing wealth of knowledge about the visual representation of information, and how we perceive and interpret visual information, being published by researchers in fields such as cognitive psychology, perceptual psychology, human factors, etc. As visualization researchers and tool developers incorporate new findings, we may take better advantage of visualization as a powerful analysis and communication tool. The art and design community also is demonstrating new ways of presenting and representing information. The use of sound in representing scientific data is becoming more important. Research is being done to find what types of information are best represented in audio.

In order to move beyond the current level of visualization tools, we must continue to develop the basic technologies and algorithms involved in the image-generation aspects of the process—for example, improved rendering algorithms, improved user interface, improved representations, etc.

In the ideal world, the computational scientist would not be required to think about human visual perception, optical illusions, how cultural biases affect the interpretation of imagery, or how color is understood. If research in the appropriate areas can be integrated into the tools which scientists use, we can raise the level of effectiveness of the visual analysis and presentation of science and, indeed, the science itself. Research is currently under way [Baker 1994] to develop tools which do automatic idiom selection. Based on input from the scientists regarding the type of data, the insight they are trying to achieve, etc., the tool will generate representations that have a high likelihood of being useful.

Visualization is rapidly moving beyond the researcher's workstation and videotape presentations. New media and communications systems are allowing the scientist to collaborate remotely over networks and share data, imagery, etc. The Internet and the World Wide Web are allowing the scientist to publish findings, including visual representations, to a worldwide audience. These new media bring a variety of new concerns and research issues. Some examples include the size and resolution of imagery that it is possible to share. Current network speeds render it impractical to share multiple large, high-resolution images. Animations also are just beginning to be practical over the network. As data compression techniques and sharing of executable code over the network improve, the scientist will truly be able to explore this new medium.

Current work in VR techniques is enabling scientists to see and interact with their research in many new ways. Stereoscopic imagery, wide fields of view, and tracking of the viewer's head enable the scientist to become immersed within the simulation and become a part of the system. Current systems suffer from a lack of resolution, tracker lag, etc. As these improve, virtual reality promises to be a powerful tool for computational scientists. Many scientists who have currently abandoned virtual reality systems with their clumsy, low-resolution head-mounted displays are finding that the less intrusive, higher-resolution projection-based displays such as the CAVE^{TM} are transforming virtual reality from a novelty to a useful tool for analysis and display.

In summary, the field of scientific visualization is sufficiently mature to allow the scientist to harness the power of the eye–brain connection for data analysis and presentation. At the same time, it is breaking new ground with respect to online collaborative computing, worldwide publication, and fully immersive interaction through virtual reality.

Acknowledgments

We would like to thank Robert Wilhelmson for sharing his wealth of experience and historical perspective on visualization in computational science; also, from the Atmospheric Science Group, Crystal Shaw and Brian Jewett for additional support. Thanks also to the reviewers and editors for helpful comments and suggestions to improve this chapter.

Many of the ideas in this chapter were the result of our interactions with a variety of scientists. We have had the opportunity to work on a wide variety of challenging visualization problems with numerous world-class researchers.

We gained a much greater insight into the field of scientific visualization as a direct result of interaction with the participants of the Representation Project, including the guest speakers from a wide variety of disciplines and institutions. In particular, the members of the former Visualization Group at NCSA including Boss Dan Brady, Matthew Arrott, Mark Bajuk, Ingrid Kallick, Mike Krogh, Mike McNeill, Gautam Mehrotra, Jeff Thingvold, and Deanna Spivey. This group was intellectually stimulating, an endless source of talent and ideas, and most of all, our friends. We feel honored to have been part of such a magical team.

Defining Terms

Alpha shapes: A technique that allows one to represent the concept of shape applied to a collection of points in space.

Choreography: In computer animation, the timing and sequencing of activity and representation.

Computer graphics: The medium in which modern visualization is practiced.

Dataset: A value or set of values that is the input or output of a computational simulation.

Glyph: Generally, a symbol used to represent information. For example, in visualizing vector fields, an arrow often is used to show the direction and magnitude of the vector value at each location.

Interactive steering: The practice of dynamically modifying the parameters of a running simulation, guided by a real-time visualization of the simulation's progress.

Isosurface: The shape defined within a volume of scalar values on which all the values are equal to some constant.

Parallax: The difference in the apparent position of an object caused by a change in the point of observation.

Perceptualization: A term perhaps more suitable than "visualization," in recognition of the efficacy of using auditory and tactile techniques for representing and communicating about scientific data.

Scientific illustration: The traditional use of graphics created by an artist to show scientific concepts.

Scientific visualization: The use of computer-generated graphics, often animated, interactive, and three-dimensional, to represent scientific data and concepts.

Simulation: A computer model of natural phenomena.

Streakline: A line showing the path taken by all particles that pass through a given location in a vector field.

Streamline: A line drawn in a vector field such that, at any instant, the tangent to the line at any point on the line is the direction of the flow. Often restricted to fields with steady flow, in which case the streamline shows the path of a tracer particle.

Tracer: An animated symbol, usually a sphere, showing the path that would be taken by a particle in a vector field.

Vector field: An n-dimensional collection of vector values arranged in space, such as the wind velocity over a two-dimensional surface.

Virtual reality: A medium composed of highly interactive computer simulations that utilizes data about the participant's position and replaces or augments one or more of their senses—giving the feeling of being immersed, or being present, in the simulation [Sherman and Craig 1996].

Visual idiom: A technique for representing scientific data that has a commonly accepted interpretation.

References

Anninos, P., Bajuk, M., Bernstein, Seidel, E., Smarr, L., and Hobill, D. The evolution of distorted black holes. Physics Computing '93.

Arrott, M. and Latta, S. 1992. Perspectives on visualization, pp. 61–65. *IEEE Spectrum*, Sept.

AVS. AVS Home Page, http://www.avs.com/.

Bajuk, M. 1992. Camera evidence: visibility analysis through a multi-camera viewpoint,

Baker, M. P. 1994. KnowVis: an experiment in automating visualization, p. 456. In *Proc. Decision-Support 2001*, Toronto, Ontario, Sept.

Baker, M. P. and Bushell, C. 1995. After the storm: considerations for information visualization. *IEEE Trans. Comput. Graphics Appl.* 15(3):12–15 (May).

Baker, M. P. and Wickens, C. D. 1995. Human factors in virtual environments for the visual analysis of scientific data. *NCSA Tech. Rep.* 032, Aug.

Brooks, F. P., Jr., Ming O.-Y., Batter, J. J., and Kilpatrick, P. J. 1990. Project GROPE—haptic displays for scientific visualization. *Comput. Graphics (Proc. SIGGRAPH)* 24(4):177–185 (Aug.).

Brown, M., ed. 1994. *Comput. Graphics*, SIGGRAPH '94 Visual Proceedings, Aug., Orlando, FL.

Cruz-Neira, C., Sandin, D., DeFanti, T, Kenyon, R., and Hart, J. 1992. The CAVE audio visual experience automatic virtual environment. *Comm. ACM* 35(6):64–72. (June). URL: http://www.ncsa.uiuc.edu/EVL/docs/html/CAVE.html.

Drebin, R. A., Carpenter, L., and Hanrahan, P. 1988. Volume rendering. *Comput. Graphics (Proc. SIGGRAPH)* 22(4):64–75 (Aug.).

Edelsbrunner, H. and Mucke, E. P. 1994. Three-dimensional alpha shapes. *ACM Trans. Graphics* 13:43–72.

Fangmeier, S. M. 1988. The scientific visualization process, pp. 26–38. SIGGRAPH '88 Course Notes, Course 20: Computer Graphics in Science.

Fast. Flow Analysis Software Toolkit (FAST) home page. http://www.nas.nasa.gov/NAS/FAST/fast.html. See also Sterling Software WWW Home Page. http://www.sterling.com/.

Fluent. Fluent Incorporated Home Menu Page. Fluent CFD software. http://www.fluent.com/.

Fortner. Fortner Research LLC. http://www.langsys.com/langsys/.

Gnuplot. Gnuplot. http://www.cs.dartmouth.edu/gnuplot_info.html.

Haber, R. B. 1989. Scientific visualization and the rivers project at the National Center for Supercomputing Applications. *Computer* 22(8):84–89 (Aug.).

Herron, D. K., Bollinger, N. G., Chaney, M. O., Varshavsky, A. D., Yost, J. B., Sherman, W. R., and Thingvold, J. A. 1995. Visualization and comparison of molecular dynamics simulations or leukotriene C(4), leukotriene D(4), and leukotriene E(4). *J. Mol. Graphics* 13:337–341, Elsevier Science, New York.

Hussey, K., Mortensen, B., and Hall, J. 1986. Jet Propulsion Lab Animation: *L.A.—The Movie*. Visualization in scientific computing. *ACM SIGGRAPH Video Review*, No. 28.

IBM Data Explorer. IBM Visualization Data Explorer (DX). http://www-i.almaden.ibm.com/dx/.

IRIS Explorer. IRIS Explorer Center. http://www.nag.co.uk:70/1h/Welcome_IEC.

Jewett, B. F. and Wilhelmson, R. B. 1995. Use of HTML and web tools in atmospheric sciences research. URL: http://redrock.ncsa.uiuc.edu/AOS/publications/IIPS96/web-atmossci.html.

Johnson, C. R. and Parker, S. G. 1995. Applications in computational medicine using SCIRun: a computational steering programming environment, H. W. Meuer, ed., pp. 2–19. In *Supercomputing '95*. URL: http://www.cs.utah.edu/~sci/

Kaufmann, W. J., III, and Smarr, L. L. 1993. *Supercomputing and Science*. Scientific American Library, New York.

Keates, J. F. *Understanding Maps*, p. 39.

Khoros. The Khoral Research Home Page. http://www.khoros.unm.edu/home.html.

Korab, H. and Brown, M., eds. 1995. Virtual environments and distributed computing at SC '95. ACM/IEEE Supercomputing '95, Dec., San Diego, CA.

Kovacic, D., Craig, A., Patterson, R., Romme, W., and Despain, D. 1990. *Fire Dynamics in the Yellowstone Landscape, 1690–1990: An Animation*. Model driven visual simulation, Proc. Resource Technology 90, Second International Symposium on Advanced Technology in Natural Resources Management.

Lorensen, W. E., and Cline, H. 1987. Marching cubes: a high resolution 3D surface construction algorithm, *Comput. Graphics (Proc. SIGGRAPH)* 21(4):163–169 (July).

Lytle, W. 1993. The dangers of glitziness and other visualization faux pas. Animation, Cornell Theory Center, *SIGGRAPH 93 Electronic Theater*. No. 91.

Marshall, R., Kempf, J., and Dyer, S. 1990. Visualization methods and simulation steering for a 3D turbulence model of Lake Erie. Symposium on Interactive 3D Graphics. *ACM SIGGRAPH* 24(2):89–97 (Mar.).

Merzkirch, W. 1987. *Flow Visualization*, 2nd ed. Academic Press, Orlando, FL.

Moran, P. J. and Potter, C. S. 1992. Tiller: a tool for analyzing 4-d data. In *Visual Data Interpretation*, Proc. SPIE, 1668:124–128.

NCAR. NCAR Graphics Home Page. http://ngwww.ucar.edu/.

Scaletti, C. and Craig, A. B. 1991. Using sound to extract meaning from complex data. In *Extracting Meaning from Complex Data: Processing, Display, Interaction II*, Proc. SPIE, 1459:207–219.

SciAn. SciAn—Scientific Visualization and Animation Package. http://www.scri.fsu.edu/~mimi/scian.html.

Sherman, W., McNeill, M., Arrott, M., Bajuk, M., and Corson, A. 1990. Animation: Smog: Visualizing the Components. SIGGRAPH '90 Film & Video Theater. *ACM SIGGRAPH Video Rev.*, No. 62.

Sherman, W. R. 1990. Integrating virtual environments into the dataflow paradigm. Fourth Eurographics Workshop on Visualization in Scientific Computing, Abingdon, U.K. Apr.

Sherman, W. R. and Craig, A. B. 1997. *Working with Virtual Reality*. Morgan Kaufmann, San Francisco.

Sims, C. 1989. Animation: Leonardo's Deluge SIGGRAPH '89 Computer Graphics Theater. *ACM SIGGRAPH Video Rev.*, No. 52.

Tiede, U., Schiemann, T. and Hohne, K. H. 1996. Visualizing the visible human. In *IEEE Computer Graphics & Applications* 16(1).

Tufte, E. R. 1996. *Envisioning Information*. Graphics Press, Cheshire, CT.

Tufte, E. R. 1996. *Visual Explanations*. Graphics Press, Cheshire, CT.

Upson, C., Faulhaber, T., Kamins, D., Laidlaw, D., Vroom, J., Gurwitz, R., and van Dam, A. 1989. The application visualization system: a computational environment for scientific visualization. *IEEE Trans. Comput. Graphics Appl.* 9(4):30–42 (July).

Vis5D. Vis5D Home Page. http://www.ssec.wisc.edu/~billh/vis5d.html.

Weintraub, D. J. and Walker, E. H. 1969. *Perception*. Brooks/Cole, Belmont, CA.

Weintraub, D. J. and Walker, E. H. 1969. *Perception.* Brooks/Cole, Belmont, CA.

Wickens, C., Merwin, and Lin. 1994. The human factors implications of graphics enhancements for the visualization of scientific data. In *1994 Human Factors*, Vol. 36. Also in *Proceedings of the 38th Annual Meeting of the Human Factors Society (1994)*, pp. 44–61.

Wickens, C. and Seidler, K. 1995. Information access and utilization. In *Emerging Needs and Opportunities for Human Factors Research*, R. Nickerson, ed., National Academy Press.

Wilhelmson, R., 2nd. 1993. PATHFINDER: Probing ATmospHeric Flows in an INteractive and Distributed EnviRonment. In *Proc. 9th Conf. on Interactive Information and Processing Systems for Meteorology, Oceanography, and Hydrology*, Anaheim, CA, Jan.

Further Information

Brown, J. R., Earnshaw, R., Jern, M., and Vince, J. 1995. *Visualization: Using Computer Graphics to Explore Data and Present Information.* Wiley, New York.

Chambers, J., Cleveland, W., Kleiner, B., and Tukey, P. 1983. *Graphical Methods for Data Analysis.* Wadsworth International Group, Belmont, CA.

Dent, B. D. 1990. *Cartography Schematic Map Design.* William C. Brown.

Dondis, D. A. 1973. *A Primer of Visual Literacy.* MIT Press, Cambridge, MA.

Friedhoff, R. M. and Benzon, W. 1989. *Visualization.* W. H. Freeman, New York.

Gallager, R. S., ed. 1995. *Computer Visualization.* CRC Press.

Hearn, D. and Baker, M. P. 1994. *Computer Graphics*, 2nd ed. Addison–Wesley.

Huff, D. 1954. *How to Lie with Statistics.* Norton, New York.

Kaufmann, W. J., III, and Smarr, L. L. 1993. *Supercomputing and Science.* Scientific American Library, New York.

Keates, J. F. *Understanding Maps.*

Keller, P. R. and Keller, M. M. 1993. *Visual Cues.* IEEE Press.

Lauer, D. A. 1990. *Design Basics.* Harcourt Brace Jovanovich.

MacEachren, A. M. 1995. *How Maps Work.* Guilford Press, New York.

McCormick, B. H., DeFanti, T. A., and Brown, M. D. 1987. Visualization in scientific computing. *Comput. Graphics* 21(6): (Nov.).

Tufte, E. R. 1983. *The Visual Display of Quantitative Information.* Graphics Press, Cheshire, CT.

Wickens, C. 1992. *Engineering Psychology and Human Performance*, 2nd ed. HarperCollins, New York.

36

Computational Structural Mechanics

Ahmed K. Noor
University of Virginia

36.1 Introduction

Structural mechanics deals with (1) the idealization of actual structures and their environments and (2) prediction of response, failure, life, and performance of structures. In the last three decades the discipline of computational structural mechanics (CSM) has emerged as an insightful blend between structural mechanics, on the one hand, and other disciplines such as computer science, numerical analysis, and approximation theory, on the other hand. This rapidly evolving discipline is having a major impact on the development of structures technology, as well as on its application to various engineering systems. Development of the modern finite-element method during the 1950s marks the beginning of CSM.

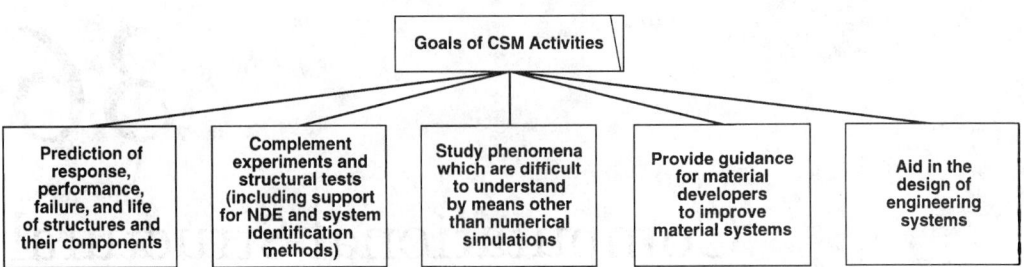

FIGURE 36.1 Five major goals of CSM activities (NDE refers to nondestructive evaluation techniques).

Finite-element technology is the backbone of many structural analysis software systems which are widely used by government, academia, and industry to solve complex structures problems.

The five major goals of CSM activities are shown in Fig. 36.1. In support of these goals, current activities of CSM cover the study of phenomena, through numerical simulations, at a wide range of length scales ranging from the microscopic level to the structural level. Today, no important design can be completed without CSM, nor can any new theory be validated without it.

A number of survey papers and monographs have been written on various aspects of CSM (see, for example, Noor and Atluri [1987], Noor and Venneri [1990, 1995], and Ju [1995]). Also, a number of workshops and symposia have been devoted to CSM and proceedings have been published (for example, Grandhi et al. [1989], Noor et al. [1992], Onaté et al. [1992], Ladevéze and Zienkiewicz [1992], Stein [1993], and Storaasli and Housner [1994]). The present chapter attempts to present, in a concise manner, the broad spectrum of problems covered by CSM along with the basic principles, formulations, and solution techniques for these problems. A brief history is given of the development of software systems used for the modeling and analysis of structures. The research areas in CSM, which have high potential for meeting future technological needs, are identified.

36.2 Classification of Structural Mechanics Problems

Structural Characteristics and Source Variables

The functions which govern the response, failure, life, and performance of structures can be grouped into four categories, namely:

> *Kinematic variables:* e.g., displacements, velocities, strains, and strain rates
> *Kinetic variables:* e.g., stresses and internal forces (or stress resultants)
> *Material characteristics:* e.g., material stiffness and compliance coefficients, and
> *Source variables:* which include environmental effects and external forces (e.g., mechanical, aerodynamic, thermal, optical, and electromagnetic forces)

The relations between the external forces and response quantities are shown in Table 36.1.

TABLE 36.1 Relations Between External Forces and Response Quantities

Quantities	Relation	Type of Relation
External forces		
	Balance equations	Physical (conservation) laws
Stresses		
	Constitutive relations	Semi-empirical based on experiments
Strains		
	Strain–displacement relations	Geometric–based on logic
Displacements		

Different Classes of Structural Mechanics Problems

A number of classifications can be made for structural mechanics problems depending on: (1) presence or absence of uncertainties about the structural characteristics and source variables, (2) nature of the functions which govern the behavior of the structure (e.g., time dependence or independence), (3) the functional form of the relations between the source variables and response quantities (e.g., linear or nonlinear), and (4) the geometric characteristics of the structure and its components. A general classification, incorporating the aforementioned factors, is shown in Fig. 36.2. Additional classifications can be made based on:

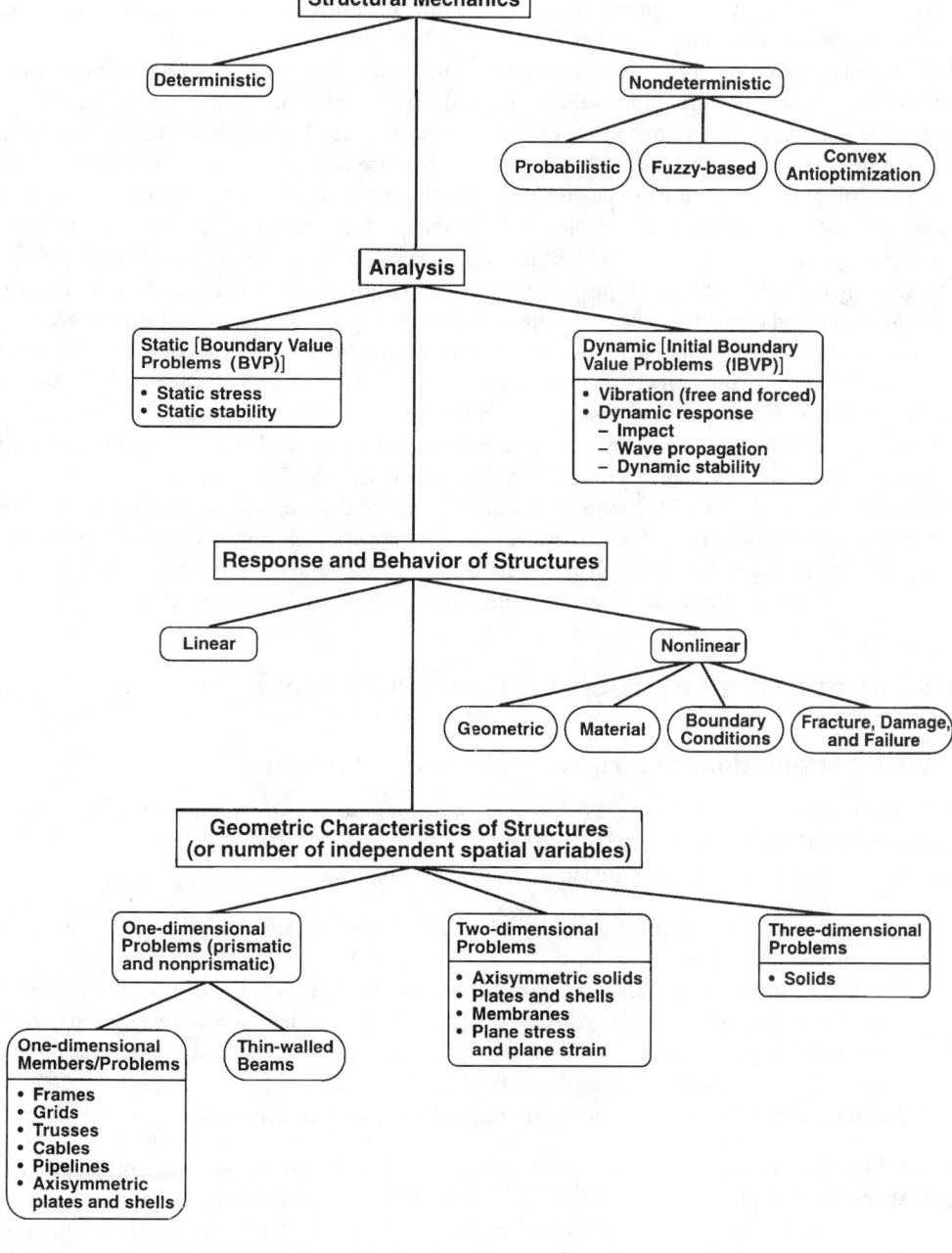

FIGURE 36.2 Classification of structural mechanics problems.

(1) material response (e.g., homogeneous or nonhomogeneous, isotropic or anisotropic), (2) nature of source variables (e.g., conservative or nonconservative), and (3) coupling or noncoupling between source variables and response quantities (e.g., whether the changes in the source variables with structural deformations are pronounced or not).

Deterministic and Nondeterministic Methods

Deterministic methods of structural mechanics have become quite elaborate and include sophisticated mathematical models, highly refined computational methods, and optimization techniques. However, there is a growing realization among engineers during the past 25 years that unavoidable uncertainties in geometry, material properties, boundary conditions, loading, and operational environment must be taken into account to produce meaningful designs.

Three possible approaches for handling uncertainty can be identified, depending on the type of uncertainty and the amount of information available about the structural characteristics and the operational environment. The three approaches are: *probabilistic analysis, fuzzy-logic approach,* and *set theoretical, convex* (or antioptimization) *approach*. A discussion of the three approaches and their combinations is given in Elishakoff [1995]. In the probabilistic analysis, the structural characteristics and/or the source variables are assumed to be random variables (or functions), and the joint probability density functions of these variables are selected. The main objective of the analysis is the determination of the reliability of the system. Reliability is defined as the probability that the structure will adequately perform its intended mission for a specified interval of time, when operating under specified environmental conditions.

If the uncertainty is due to vague definition of structural and/or operational characteristics, imprecision of data and subjectivity of opinion or judgment, then fuzzy logic-based treatment is appropriate. The distinction between randomness and fuzziness is the fact that whereas randomness describes the uncertainty in the occurrence of an event (e.g., damage or failure of the structure), fuzziness describes the ambiguity of the event (e.g., imprecisely defined criteria for failure or damage—see Ross [1995]).

When the information about the structural and/or operational characteristics is fragmentary (e.g., only a bound on a maximum possible response function is known), then convex modeling can be used. Convex modeling produces the maximum or least favorable response, and the minimum or most favorable response, of the structure under the constraints within the set-theoretical description.

36.3 Formulation of Structural Mechanics Problems

Different Formulations of Structural Mechanics Problems

Several classifications can be made of the different formulations used for structural mechanics problems. Two of the major classifications are discussed next.

First, with respect to the approach used in deriving governing equations:

1. Differential equation formulation. This is the classical approach which is referred to as the *strong statement* of the problem.
2. Variational formulation. This is based on application of the principle of virtual displacements, and is referred to as the *weak statement* of the problem. A variety of variational principles for static and dynamic problems can be derived from the principle of virtual displacements (see, for example, Washizu [1982] and Geradin [1980]).
3. Integral, integro-differential or boundary integral equation formulation.

Some of the discretization techniques can be applied directly to the strong statement of the problem. The most notable example is the finite-difference method. Other techniques, such as finite elements, are typically used with the weak statement. Boundary element methods are generally used with the third formulation.

Second, with respect to fundamental unknowns in the governing equations or the governing functional:

1. Single-field formulation in which the fundamental unknowns belong to one field. The most commonly used formulation is the displacement formulation.
2. Multifield formulation in which the fundamental unknowns belong to more than one field, i.e., stresses and displacements. A detailed discussion of multifield formulations is given in Noor [1980].

Description of the Motion of a Structure

For nonlinear structural problems, three basic choices need to be made, namely, the approach used for describing the motion, the kinematic description (i.e., deformation measure), and the kinetic description (i.e., stress measure). Typically, the choice of the deformation measure determines the stress measure since the two have to be work conjugates. The two most commonly used approaches for describing the motion are as follows.

First is a *total Lagrangian description,* in which all the kinematic and kinetic variables are referred to the initial (undeformed) configuration. The Green–Lagrange strain tensor is used as the measure of deformation, and the associated second Piola–Kirchhoff stress tensor is used as the stress measure. For large-rotation problems, it is useful to separate the rigid-body movement from stretching by using *local corotational frames* and the polar decomposition method. This eliminates the problems associated with approximating finite rotations using trigonometric functions (or series expansions thereof).

Second is an *updated Lagrangian description,* in which the kinematic and kinetic variables are referred to the current configuration. For small-strain problems, the Almansi strain is used as the measure of deformation and the associated Cauchy stress is used as the stress measure. For large-strain problems, it is convenient to use the velocity strain (or rate of deformation) and the Jaumann stress rate as the deformation and stress measures.

A detailed discussion of the aforementioned formulations and their combinations, along with the appropriate deformation and stress measures is given in Bathe [1996] and Cescotto et al. [1979].

36.4 Steps Involved in the Application of Computational Structural Mechanics to Practical Engineering Problems

Major Steps in the Application of Computational Structural Mechanics

The application of CSM to contemporary structural problems involves a sequence of five steps, namely (Fig. 36.3):

- Observation of response phenomena of interest.
- Development of computational models for the numerical simulation of these phenomena. This in turn includes: (1) selecting mathematical models which represent the environmental effects and external forces and describe the phenomena, analyzing the models to ensure that the problem is properly formulated, and testing the range of validity of the models; and (2) development of a discrete model, computational strategy, and numerical algorithms to approximate the mathematical model. Successful computational models for structures are based on a thorough familiarity with the response phenomena being simulated and a good understanding of the mathematical models available to describe them.
- Development and assembly of software and/or hardware to implement the computational model.
- Postprocessing and interpretation of the predictions of the computational model.
- Utilization of the computational model in the analysis and design of engineering structures.

Within this general framework, CSM is being used today in a broad range of practical applications. To

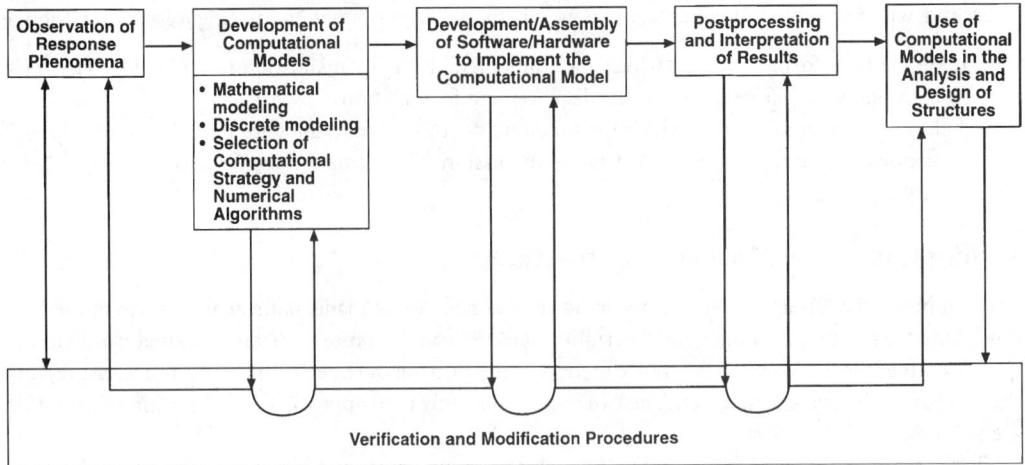

FIGURE 36.3   Application of CSM to practical structural problems.

date, large structural calculations are performed which account for complicated geometry, complex loading history, and material behavior. The applications span several industries, including aerospace, automotive, naval, nuclear, and microelectronics.

Selection of the Mathematical Models

The mathematical models are idealized representations of the real structure and its environment. Proper selection of the mathematical models is strongly influenced by the goals of the computation. The models will represent reality only if they take into consideration all factors which affect the conclusions drawn from them. The mathematical models are described by their governing equations (in one of the forms described in the previous section).

It is useful to view a particular mathematical model as one in a sequence (or hierarchy) of models of progressively increasing complexity. The effect of the model selection on the accuracy of the response predictions for simple structures, boundary conditions, and joints is discussed in Szabo and Babuska [1991]. A number of comparisons can be made between the predictions of the model being formulated and more elaborate models, i.e., models which account for a greater number of potentially relevant phenomena.

Discretization Techniques

Because of the complexity of the governing equations of the mathematical models used in representing real structures, exact solutions can only be obtained in very few cases, and one has to resort to approximate or numerical discretization techniques. A variety of approximate and numerical discretization techniques have been applied to structural mechanics problems. Two possible classifications of these techniques are shown in Fig. 36.4 and are based on: (1) the formulation used, namely, differential equation, variational, or integral-equation formulation; and (2) modifications made in the form of the governing equations (replacement of the governing equations by an equivalent form).

The finite-element method is the most commonly used discretization technique to date. Extensive literature exists on various aspects of the finite-element method. A list of monographs, books, and conference proceedings on the method is given in Noor [1991]. The boundary element method is a computational tool for the boundary integral equation formulation. The method works with values of the dependent variables on the boundary only, and therefore, is well suited to problems involving a large volume to surface ratio (Kane et al. [1993], Banerjee [1994]). In a number of applications, hybrid combinations of analytical and numerical discretization techniques were shown to be more effective than individual techniques, and

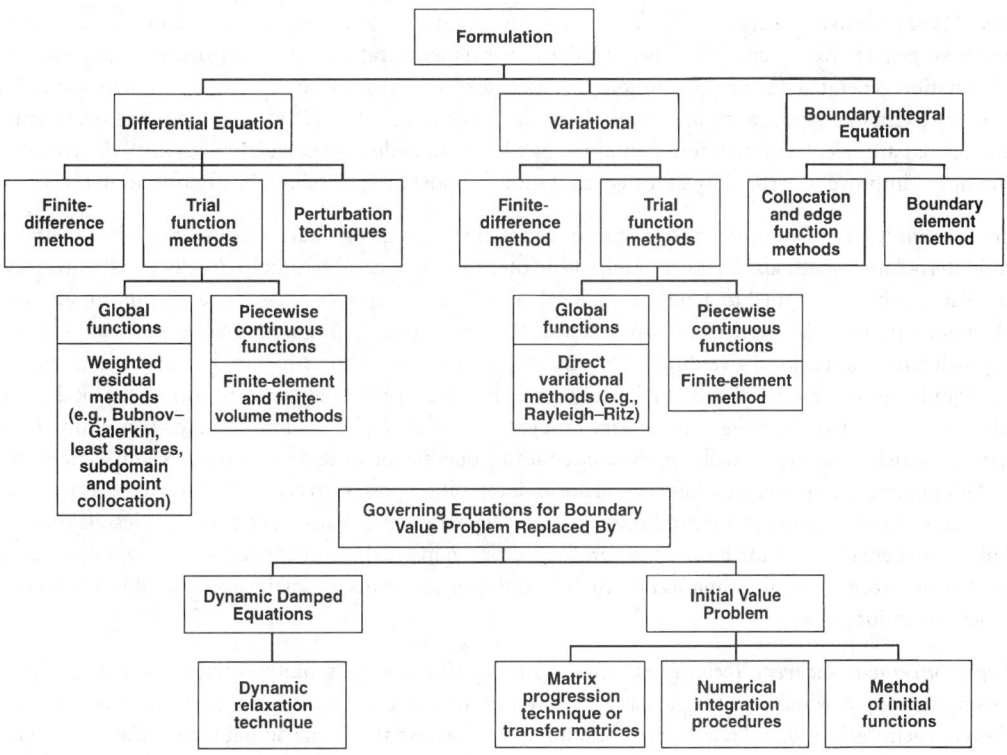

FIGURE 36.4 Approximate and numerical discretization techniques for structural mechanics problem.

resulted in dramatic savings in the computational effort (see, for example, Noor and Andersen [1992] and Noor [1994]).

Model and Mesh Generation

The reliability of the predictions of numerical discretization techniques (e.g., finite differences, finite elements, finite volumes, and boundary elements), and the computational effort involved in obtaining them, is very much influenced by the selection of the mesh and the procedure for generating it. Considerable effort has been devoted to the development of robust mesh generation procedures capable of producing controlled meshes over domains of arbitrary complex geometry. Finite-element mesh generation activities have been approached from both geometric modeling and adaptive finite-element viewpoints. Overviews and classifications of finite-element mesh generation methods are given in Shephard and Wentorf [1994], Shephard [1988], Ho-Le [1988], George [1991], Sabin [1991], and Mackerle [1993].

Among the recent activities on model generation are: (1) application of knowledge-based analysis-assistance tools, which allow a simple description of the analysis objectives and generate the corresponding discrete models appropriate for these objectives (Turkiyyah and Fenves [1996]); and (2) the development of paving and plastering techniques for automated generation of quadrilateral and hexahedral finite-element grids. These techniques generate well-formed elements (with reasonably small distortion metric), and are based on iteratively layering or paving rows of elements to the interior of a region's boundary (see Blacker and Stephenson [1991]).

Quality Assessment and Control of Numerical Solutions

Assessment of the reliability of computational models has been the focus of intense efforts in recent years. These efforts can be grouped into three general categories (see, for example, Noor and Babuska [1987],

Noor [1992], Demkowicz et al. [1992], Krizek and Neittaanmaki [1987]): a posteriori error estimation, superconvergent recovery techniques, and adaptive improvement strategies. A posteriori error estimates use information generated during the solution process to assess the discretization errors. In superconvergent recovery techniques, more accurate values of certain response quantities (e.g., derivatives of fundamental unknowns) are calculated than those obtained by direct finite-element calculations. Adaptive strategies attempt to improve the reliability of the computational model by controlling the discretization error.

Error Estimation. Two broad classes of error estimation schemes are currently used: residual methods and interpolation methods. Residual methods involve the use of local residuals, usually as data in a local auxiliary problem designed to generate the local error to an acceptable accuracy. A significant amount of computation may be required in implementing these methods. In interpolation methods the available approximate solution for a given mesh (or time step) is used to estimate higher derivatives locally (e.g., local gradients or second derivatives). The higher derivatives are used in turn to determine the local error. Although these error estimates can be very crude, they are portable: a subroutine for computing local estimates can be added to virtually any existing code that operates on unstructured meshes with some effort.

Although significant progress has been made in developing a posteriori error estimates for linear elliptic problems, the error estimates for nonlinear and time-dependent problems are considerably less developed. This is particularly true for bifurcation problems, problems with multiple scales, and problems with resonance. Work on error estimation for highly nonlinear problems has mainly been a subject of ad hoc experimentation.

Superconvergent Recovery Techniques. Superconvergent recovery techniques refer to simple postprocessing techniques which provide increased accuracy of the sought quantities at some isolated points (e.g., Gauss-Legendre, Jacobi, or Lobatto); in a subdomain; or even in the whole domain (Krizek and Neittaanmaki [1987]). In the latter two cases, the techniques are referred to as local- and global-superconvergent recovery techniques, respectively. Recent work included development of local-superconvergent patch derivative techniques for both interior and boundary (or material interface) points (Babuska and Miller [1984a, 1984b], Zienkiewicz and Zhu [1992], Zienkiewicz et al. [1993], Tabbara et al. [1994]). It was shown in Zienkiewicz and Zhu [1992] and Zienkiewicz et al. [1993] that the superconvergent recovery technique can be used to obtain a posteriori error estimates for the finite-element solution.

Adaptive Strategies. Different strategies have been used for adaptive improvement of the numerical solutions, including: (1) mesh refinement (or derefinement) schemes, h methods; (2) moving mesh (node redistribution) schemes, r methods; (3) subspace enrichment schemes (selection of the local order of approximation), p methods; (4) mesh superposition schemes (overlapping local finite-element meshes on the global one), s methods (see, for example, Fish and Markolefas [1993, 1994]); and (5) hybrid (or combined) schemes. Examples of these schemes are: (1) simultaneous selection of the meshes and local order of approximation, h–p methods; recent theoretical results have shown that the fastest possible convergence rates can be attained by optimally decreasing the mesh size h and increasing the degree of the polynomial degree p in a special way; and (2) simultaneous selection of the meshes and node redistribution, h–r method. These methods can be effective in shock problems since an r-method might align the mesh along discontinuities prior to a mesh refinement.

36.5 Overview of Static, Stability, and Dynamic Analysis

Flow charts for the basic components of the solution methods for static, stability, and dynamic problems are given in Figs. 36.5 and 36.6. In this section a brief summary is given of the fundamental equations and solution techniques used in static, stability, and dynamic analysis. A single-field displacement formulation is used, and the spatial discretization of the structure is assumed to have been performed. The external load vector and associated displacement vector will henceforth be denoted by $\{F^{ext}\}$ and $\{X\}$, respectively. For linear problems $\{X\}$ is a linear function of the components of $\{F^{ext}\}$, and for nonlinear problems $\{X\}$ is a nonlinear function of these components.

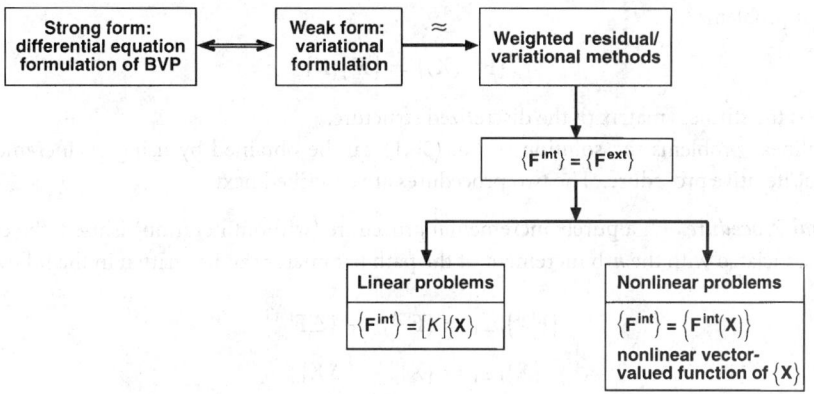

FIGURE 36.5 Basic components of solution methods for static problems.

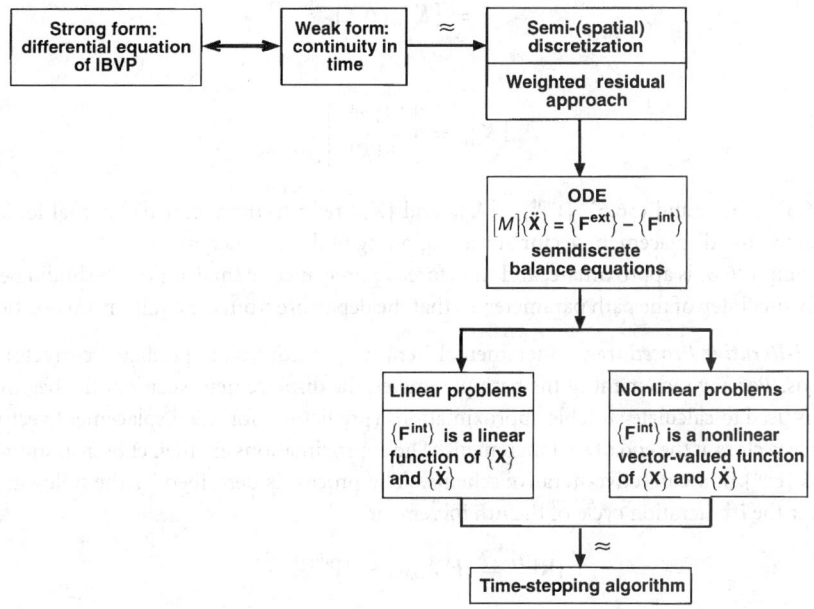

FIGURE 36.6 Basic components of solution methods for dynamic problems.

Static Analysis

The equilibrium equations for the discretized structure can be written in the following form:

$$\{F^{int}(X)\} = \{F^{ext}\} \tag{36.1}$$

where $\{F^{int}(X)\}$ is the vector of internal forces, which is a vector-valued function of the displacements $\{X\}$. For conservative loading $\{F^{ext}\}$ is independent of $\{X\}$, and for nonconservative loading it is a function of $\{X\}$.

The equilibrium path is usually expressed in terms of one or more variable path parameters (typically taken as load, displacement, or arc-length parameters in the solution space). For simplicity, in subsequent discussion the loading is assumed to be conservative and proportional to a single path parameter p. The displacement vector is therefore a function of p, i.e., $\{X\} = \{X(p)\}$.

For linear problems

$$\{\mathbf{F}^{int}(\mathbf{X})\} = [K]\{\mathbf{X}\} \tag{36.2}$$

where $[K]$ is the stiffness matrix of the discretized structure.

For nonlinear problems the solution of Eq. (36.1) can be obtained by using an incremental or an incremental/iterative procedure. The two procedures are described next.

Incremental Procedure.   If a purely incremental procedure (without iteration) is used, the equilibrium equations associated with the nth increment of the path parameter can be written in the following form:

$$\{\mathbf{F}^{int}\}_{n+1} = \{\mathbf{F}^{int}\}_n + \{\Delta\mathbf{F}^{int}\}_n \tag{36.3}$$

$$\{\mathbf{X}\}_{n+1} = \{\mathbf{X}\}_n + \{\Delta\mathbf{X}\}_n \tag{36.4}$$

where

$$\{\Delta\mathbf{F}^{int}\}_n = \{\mathbf{F}^{ext}\}_{n+1} - \{\mathbf{F}^{int}\}_n \tag{36.5}$$

$$\cong [K]_n\{\Delta\mathbf{X}\}_n \tag{36.6}$$

and

$$[K]_n = \left[\frac{\partial\{\mathbf{F}^{int}\}}{\partial\{\mathbf{X}\}}\right]_n \tag{36.7}$$

In Eqs. (36.3), (36.4), and (36.6), $\{\mathbf{F}^{int}\}_n$, $[K]_n$ and $\{\mathbf{X}\}_n$ refer to the vector of internal forces, tangent stiffness matrix, and displacement vector at the beginning of the nth increment.

Note that Eq. (36.6) is approximate, and therefore, a purely incremental approach should be used with a sufficiently small step of the path parameter, so that the departure from the equilibrium position is small.

Incremental-Iterative Procedures.   Incremental-iterative procedures are predictor-corrector continuation methods. For any increment of the path parameter, the displacement vector at the beginning of the increment is used to calculate suitable approximations (predictors) for the displacement vectors and the internal force vectors at the end of that increment. The approximations are then chosen as initial estimates for $\{\mathbf{X}\}$ and $\{\mathbf{F}^{int}\}$ in a corrective-iterative scheme. The process is described by the following recursive equations for the ith iteration cycle of the nth increment.

$$\{\mathbf{R}\}^{(i)} = \{\mathbf{F}^{ext}\}_{n+1} - \{\mathbf{F}^{int}\}_{n+1}^{(i-1)} \tag{36.8}$$

$$= [K]\{\Delta\mathbf{X}\}_n^{(i)} \tag{36.9}$$

and

$$\{\mathbf{X}\}_{n+1}^{(i)} = \{\mathbf{X}\}_{n+1}^{(i-1)} + \{\Delta\mathbf{X}\}_n^{(i)} \tag{36.10}$$

where $\{\mathbf{R}\}$ is the residual force vector.

The iterational process is continued until convergence. As a test for convergence, a number of error norms can be used. Two of the error norms are described next. First is the modified Euclidean (spectral) norm,

$$|e| = \frac{1}{n}|\Delta\mathbf{X}|/|\mathbf{X}| \leq \text{tolerance} \tag{36.11}$$

where n is the total number of displacement parameters in the model. Second is the energy norm,

$$|e| = \frac{\left[\{\Delta\mathbf{X}\}^{(i)}\right]^T\left[\{\mathbf{F}^{ext}\}_{n+1} - \{\mathbf{F}^{int}\}_{n+1}^{(i-1)}\right]}{\left[\{\Delta\mathbf{X}\}^{(1)}\right]^T\left[\{\mathbf{F}^{ext}\}_{n+1} - \{\mathbf{F}^{int}\}_n\right]} \leq \text{tolerance} \tag{36.12}$$

In Eqs. (36.8–36.10) superscripts $(i - 1)$ and (i) refer to the values of the vectors at the beginning and end of the ith iteration cycle and $[K]$ is an approximation to the tangent stiffness matrix. A number of different iterative processes are distinguished by the choice of $[K]$.

Newton–Raphson Technique. The matrix $[K]$ is selected to be the tangent stiffness matrix based on the solution at the end of iteration cycle $(i - 1)$, i.e.,

$$[K] = \left[\frac{\partial \{\mathbf{F}^{\text{int}}\}}{\partial \{\mathbf{X}\}}\right]^{(i-1)}_{n+1} = [K]^{(i-1)}_{n+1} \tag{36.13}$$

Modified Newton Method. The matrix $[K]$ is selected to be the tangent stiffness matrix associated with increment m of the path parameter, where $m \leq n$, i.e.,

$$[K] = [K]_m \tag{36.14}$$

Broyden–Fletcher–Goldfarb–Shanno (BFGS) Method. A secant approximation to the stiffness matrix is used in successive iterations, through updating the inverse of the stiffness matrix using vector products. This is equivalent to updating the stiffness matrix, based on iteration history, as follows:

$$[K] = [K]^{(i-1)}_{n+1} \tag{36.15}$$

Dynamic Analysis

The balance equations of the discretized structure can be written in the form of a system of ordinary differential equations in time as follows:

$$[M]\{\ddot{\mathbf{X}}\} = \{\mathbf{F}^{\text{ext}}\} - \{\mathbf{F}^{\text{int}}\} \tag{36.16}$$

where $[M]$ is the global mass matrix and $\{\ddot{\mathbf{X}}\}$ is the acceleration vector. The explicit characterization of $[M]$, $\{\ddot{\mathbf{X}}\}$, and $\{\mathbf{F}^{\text{int}}\}$ depends on the particular structure under consideration. When damping or viscous effects exist, the vector $\{\mathbf{F}^{\text{int}}\}$ is a function of both the displacement vector and the velocity vector, i.e.,

$$\{\mathbf{F}^{\text{int}}\} = \{\mathbf{F}^{\text{int}}(\mathbf{X}, \dot{\mathbf{X}})\} \tag{36.17}$$

For linear problems, $\{\mathbf{F}^{\text{int}}\}$ is a linear function of both $\{\mathbf{X}\}$ and $\{\dot{\mathbf{X}}\}$, i.e.,

$$\{\mathbf{F}^{\text{int}}\} = [K]\{\mathbf{X}\} + [C]\{\dot{\mathbf{X}}\} \tag{36.18}$$

where $[C]$ is the global damping matrix.

For nonlinear problems $\{\mathbf{F}^{\text{int}}\}$ is a nonlinear function of $\{\mathbf{X}\}$ and $\{\dot{\mathbf{X}}\}$. At any time instant, the tangent stiffness and tangent damping matrices are given by

$$[K] = \left[\frac{\partial \{\mathbf{F}^{\text{int}}\}}{\partial \{\mathbf{X}\}}\right] \tag{36.19}$$

$$[C] = \left[\frac{\partial \{\mathbf{F}^{\text{int}}\}}{\partial \{\dot{\mathbf{X}}\}}\right] \tag{36.20}$$

The approaches used for obtaining the response-time history of the structure [solution of Eq. (36.16)] can be divided into two general categories, namely, direct integration techniques and modal superposition methods. The application of the two approaches to nonlinear dynamic analysis is described next.

Direct Integration Techniques

Direct temporal integration techniques are time-stepping (or step-by-step) strategies in which the response vectors at an initial time are used to generate the corresponding vectors at subsequent times. The techniques are based on: (1) satisfying the balance equations only at discrete time intervals and (2) assuming the functional dependence of the response vectors within each time interval. A variety of approximations for the response vectors within each time interval have been applied to structural dynamics problems. The approximations for the velocity and displacement vectors in the nth time step can be expressed as follows:

$$\{\dot{\mathbf{X}}\}_{n+1} = \frac{\alpha_1}{\Delta t}\{\ddot{\mathbf{X}}\}_{n+1} + L(\{\dot{\mathbf{X}}\}_n, \{\ddot{\mathbf{X}}\}_n, \ldots) \tag{36.21}$$

$$\{\mathbf{X}\}_{n+1} = \frac{\alpha_2}{(\Delta t)^2}\{\ddot{\mathbf{X}}\}_{n+1} + M(\{\mathbf{X}\}_n, \{\dot{\mathbf{X}}\}_n, \{\ddot{\mathbf{X}}\}_n, \ldots) \tag{36.22}$$

where Δt is the time-step size, α_1 and α_2 are coefficients used in the approximation, L and M are functions of the response vectors, and subscripts n and $n + 1$ refer to the values of the vectors at the beginning and end of the nth time step.

Temporal integration techniques can be classified into two general categories, namely, explicit and implicit techniques. In explicit methods the response at the end of a time step is evaluated using the balance equations at the beginning of the time step [α_1 and α_2 in Eqs. (36.21) and (36.22) are both zero]. By contrast, implicit methods use the balance equations at the end of the time step, with either α_1 and/or α_2 in Eqs. (36.21) and (36.22) as nonzero.

Explicit techniques generally require fewer computations per time step than implicit techniques and can easily handle complex nonlinearities. However, the time-step size must often be very small to ensure numerical stability. By contrast, for linear problems, the time-step size in implicit techniques is only restricted by accuracy requirements. However, based on available information, unconditional stability of implicit methods in linear problems does not necessarily extend to nonlinear problems. Explicit and implicit techniques can also be classified into *single-step* and *multistep methods* according to whether the response at any instant is related to the response at one or more previous times. Detailed discussions and assessments of different explicit and implicit techniques are given in Belytschko and Hughes [1983] and Belytschko et al. [1987]. Mixed explicit-implicit techniques are described in Belytschko and Hughes [1983].

Herein, the application of two of the widely used temporal integration techniques, central difference explicit scheme and the Newmark implicit scheme is described. Both are single-step methods.

Central Difference Explicit Scheme. The central difference scheme is based on the following approximations for the velocity and the acceleration vectors:

$$\{\dot{\mathbf{X}}\}_{n+\frac{1}{2}} = \frac{1}{\Delta t}(\{\mathbf{X}\}_{n+1} - \{\mathbf{X}\}_n) \tag{36.23}$$

$$\{\ddot{\mathbf{X}}\}_n = \frac{1}{(\Delta t)^2}[\{\mathbf{X}\}_{n-1} - 2\{\mathbf{X}\}_n + \{\mathbf{X}\}_{n+1}] \tag{36.24}$$

where Δt is the time-step size, and subscripts n, $n + \frac{1}{2}$, and $n + 1$ refer to the values of the vectors at the beginning, middle, and end of the nth time step.

Based on Eqs. (36.25) and (36.26), the update formulas for the velocity and displacement vectors are

$$\{\dot{\mathbf{X}}\}_{n+\frac{1}{2}} = \{\dot{\mathbf{X}}\}_{n-\frac{1}{2}} + \Delta t[M]^{-1}(\{\mathbf{F}^{\text{ext}}\}_n - \{\mathbf{F}^{\text{int}}\}_n) \tag{36.25}$$

$$\{\mathbf{X}\}_{n+1} = \{\mathbf{X}\}_n + \Delta t\{\dot{\mathbf{X}}\}_{n+\frac{1}{2}} \tag{36.26}$$

If a lumped (diagonal) mass matrix is used, then Eq. (36.25) uncouples.

Initially (at $t = 0$), Eqs. (36.25) and (36.26) are replaced by

$$\{\dot{\mathbf{X}}\}_{\frac{\Delta t}{2}} = \{\dot{\mathbf{X}}\}_0 + \frac{\Delta t}{2}[M]^{-1}\left(\{\mathbf{F}^{\text{ext}}\}_0 - \{\mathbf{F}^{\text{int}}\}_0\right) \tag{36.27}$$

and

$$\{\mathbf{X}\}_{\Delta t} = \{\mathbf{X}\}_0 + \frac{\Delta t}{2}\{\dot{\mathbf{X}}\}_{\frac{\Delta t}{2}} \tag{36.28}$$

where subscript 0 in Eqs. (36.27) and (36.28) refers to the value of the vector at $t = 0$.

Note that the central difference scheme is only conditionally stable. Therefore, the time step Δt must be smaller than a critical value Δt_{cr}, which for linear problems is calculated from the smallest period of vibration T_{min}. Specifically,

$$\Delta t \leq \Delta t_{\text{cr}} = \frac{T_{\text{min}}}{\pi} = \frac{2}{\omega_{\text{max}}}$$

For nonlinear problems experience has shown that a 10–20% reduction in the time step is usually sufficient to maintain stability.

Newmark's Method. Newmark's method is based on the following approximations for the displacement and velocity vectors:

$$\{\mathbf{X}\}_{n+1} = \{\mathbf{X}\}_n + \Delta t\{\dot{\mathbf{X}}\}_n + (\Delta t)^2\left[\left(\frac{1}{2} - \beta\right)\{\ddot{\mathbf{X}}\}_n + \beta\{\ddot{\mathbf{X}}\}_{n+1}\right] \tag{36.29}$$

$$\{\dot{\mathbf{X}}\}_{n+1} = \{\dot{\mathbf{X}}\}_n + \Delta t[(1 - \gamma)\{\ddot{\mathbf{X}}\}_n + \gamma\{\ddot{\mathbf{X}}\}_{n+1}] \tag{36.30}$$

where β and γ are free parameters of Newmark's method. The particular choice $\beta = 1/6, \gamma = 1/2$ corresponds to a linear approximation of the acceleration over the nth time step. The choice $\beta = 1/4, \gamma = 1/2$ corresponds to a constant average acceleration (trapezoidal rule), which is unconditionally stable for linear systems. The central difference scheme corresponds to $\beta = 0, \gamma = 1/2$.

If Eqs. (36.29) and (36.30) are combined with the balance equations, Eq. (36.16), a set of simultaneous algebraic equations result at each time step. For nonlinear problems, the equations are nonlinear and are expressed by the following process for the nth time step:

1. Predict velocities and displacements using the explicit approximations

$$\{\tilde{\dot{\mathbf{X}}}\}_{n+1} = \{\dot{\mathbf{X}}\}_n + \Delta t(1 - \gamma)\{\ddot{\mathbf{X}}\}_n \tag{36.31}$$

$$\{\tilde{\mathbf{X}}\}_{n+1} = \{\mathbf{X}\}_n + \Delta t\{\dot{\mathbf{X}}\}_n + (\Delta t)^2\left(\frac{1}{2} - \beta\right)\{\ddot{\mathbf{X}}\}_n \tag{36.32}$$

where a tilde ($\sim$) refers to the predicted value of the vector.

2. Obtain corrections to the displacements, velocities, and accelerations by the following iterative process:

$$[\overset{*}{K}]\{\Delta\mathbf{X}\}_{n+1}^{(i)} = \{\mathbf{R}\}_{n+1}^{(i)} \tag{36.33}$$

$$\{\mathbf{X}\}_{n+1}^{(i+1)} = \{\mathbf{X}\}_{n+1}^{(i)} + \{\Delta\mathbf{X}\}_{n+1}^{(i)} \tag{36.34}$$

$$\{\dot{\mathbf{X}}\}_{n+1}^{(i+1)} = \{\dot{\mathbf{X}}\}_{n+1}^{(i)} + \gamma\Delta t\{\ddot{\mathbf{X}}\}_{n+1}^{(i+1)} \tag{36.35}$$

$$\{\ddot{\mathbf{X}}\}_{n+1}^{(i+1)} = \frac{1}{\beta(\Delta t)^2}\left[\{\mathbf{X}\}_{n+1}^{(i+1)} - \{\tilde{\mathbf{X}}\}_{n+1}\right] \tag{36.36}$$

For $i = 0$, the following values are used for the displacement, velocity, and acceleration vectors

$$\{\mathbf{X}\}_{n+1}^{(0)} = \{\tilde{\mathbf{X}}\}_{n+1}$$

$$\{\dot{\mathbf{X}}\}_{n+1}^{(0)} = \{\dot{\tilde{\mathbf{X}}}\}_{n+1}$$

$$\{\ddot{\mathbf{X}}\}_{n+1}^{(0)} = 0$$

where superscript (0) refers to the value of the vector associated with $i = 0$, $[\overset{*}{K}]$ and $\{\mathbf{R}\}$ in Eq. (36.33) are the effective stiffness matrix and the residual vector given by

$$[\overset{*}{K}] = [K] + \frac{\gamma}{\beta \Delta t}[C] + \frac{1}{\beta(\Delta t)^2}[M] \tag{36.37}$$

$$\{\mathbf{R}\}_{n+1}^{(i)} = \{\mathbf{F}^{\text{ext}}\}_{n+1} - [M]\{\ddot{\mathbf{X}}\}_{n+1}^{(i)} - \{\mathbf{F}^{\text{int}}\}_{n+1}^{(i)} \tag{36.38}$$

Superscripts i and $i + 1$ refer to the values of the vectors at the beginning and end of the ith iteration cycle. As for nonlinear static problems, the choice of $[\overset{*}{K}]$ depends on the iterative procedure used.

Modal Superposition Method

In this method the structural response at any time is expressed as a linear combination of a number of preselected modes (or basis vectors). This is expressed by the following transformation:

$$\{\mathbf{X}\} = [\Gamma]\{\psi\} \tag{36.39}$$

where $[\Gamma]$ is the transformation matrix whose columns are the modes (or basis vectors) and $\{\psi\}$ is a vector of unknown coefficients (amplitudes of the preselected modes).

The number of modes (or basis vectors) is usually selected to be much smaller than the number of degrees of freedom of the discretized structure (components of the vector $\{\mathbf{X}\}$). A Bubnov–Galerkin technique is used to approximate the balance equations of the discretized structure, Eq. (36.16), by a much smaller system of equations in $\{\psi\}$. The resulting equations have the following form:

$$[\Gamma]^T[M][\Gamma]\{\ddot{\psi}\} = [\Gamma]^T\{\mathbf{F}^{\text{ext}}\} - [\Gamma]^T\{\mathbf{F}^{\text{int}}(\psi)\} \tag{36.40}$$

For linear problems, if the basis vectors are selected to be the free vibration modes of the structure, Eq. (36.40) uncouples in the components of $\{\psi\}$. For nonlinear problems the free vibration modes and frequencies of the structure change with time.

The mode superposition technique is effective when the response can be adequately represented by few modes and the time integration has to be carried out over many time steps (e.g., earthquake loading), and the cost of calculating the required modes is reasonable.

Energy Balance in Transient Dynamic Analysis

Energy balance (conservation) in transient analysis is a reflection of the accuracy of the time integrator. Furthermore, conservation properties are intimately related to stability. The construction of stable integrators is often approached by enforcing conservation laws.

In linear problems, instability can be easily detected by the spurious growth in velocities. On the other hand, in large structural problems with material nonlinearities the energy generated by an instability may be rapidly dissipated as the material becomes inelastic and the erroneousness of the results does not become obvious to the user. This kind of instability has been termed *arrested instability* in Belytschko and Hughes [1983] and the energy balance check described subsequently was suggested to detect it.

Let $\{\Delta\mathbf{X}\}_n$ be the increment of the displacement vector from time t_n to time t_{n+1}, and let the internal energy, and the increment of external work be approximated by the trapezoidal rule as follows:

$$U_j = \sum_{I=1}^{j-1} \frac{1}{2}\{\Delta\mathbf{X}\}_I^T \left(\{\mathbf{F}^{int}\}_I + \{\mathbf{F}^{int}\}_{I+1}\right) \tag{36.41}$$

$$\Delta W_j = \frac{1}{2}\{\Delta\mathbf{X}\}_j^T \left(\{\mathbf{F}^{ext}\}_j + \{\mathbf{F}^{ext}\}_{j+1}\right) \tag{36.42}$$

The kinetic energy T_j at time t_j is given by

$$T_j = \frac{1}{2}\{\dot{\mathbf{X}}\}_j^T [M]\{\dot{\mathbf{X}}\}_j \tag{36.43}$$

If the mass matrix is positive definite, then the energy and the displacements of the structure at time t_{n+1} are bounded if the following inequality is satisfied:

$$T_{n+1} + U_{n+1} \leq (1+\varepsilon)(T_n + U_n) + \Delta W_n \tag{36.44}$$

where ε is a small number. Equation (36.44) provides an energy balance check provided $U_n \geq 0$.

Detailed discussion of the stability, convergence, and decay of energy is given in Hughes [1976]. Examples of the construction of integrators through energy conservation are given in Hughes et al. [1978] and Haug et al. [1977].

Stability Analysis

A number of instability phenomena can occur in structures depending on the structural characteristics and operational environment. These instabilities can be either time dependent or time invariant. Time-dependent instability analysis can be performed by using Lyapunov's method (see Kratzig and Eller [1992] and Kounadis and Kratzig [1995]). Time-invariant instabilities can be in the form of bifurcation points (simple or multiple) or limit points in the solution path. The *critical loads* associated with these points are referred to as bifurcation buckling and limit loads, respectively.

The algorithmic tools for *time-invariant instability analysis* encompass three distinct aspects: (1) nonlinear (or linear) analysis of the prebuckling state, using the nonlinear equations—Eq. (36.1) [or the linearized version, Eq. (36.2)]; (2) determination of the critical points (e.g., bifurcation and/or limit points) on the equilibrium path; and (3) tracing the postcritical response. Critical points can be detected by the singularity of the tangent stiffness matrix, Eq. (36.19). Bifurcation loads associated with linear prebuckling state can be determined by solving a linear eigenvalue problem as described in the succeeding subsection. Special numerical algorithms are available for overcoming the difficulties associated with commonly used iterative techniques near critical points (see, for example, Riks [1984] and Crisfield [1983]).

Eigenvalue Problems

Undamped free vibration and linear (bifurcation) buckling problems of structures can be represented by the general linear matrix eigenvalue equation

$$[K]\{\mathbf{X}\} = \lambda[B]\{\mathbf{X}\} \tag{36.45}$$

where $[K]$ is the stiffness matrix of the discretized structure, $\{\mathbf{X}\}$ is the displacement vector, $[B]$ is the mass matrix for free vibration problems or the negative of the geometric stiffness matrix for buckling problems, and λ is the eigenvalue square of vibration frequency or buckling load parameter. Typically, most of the elements of $[K]$ and $[B]$ are zero and only a few pairs $(\lambda, \{\mathbf{X}\})$ are wanted.

Although the governing equations for both the free vibration and linear buckling problems are similar, the properties of the two matrices $[K]$ and $[B]$ are different. For vibration problems, the stiffness matrix can be positive definite, positive semidefinite, or indefinite. For an unsupported structure the stiffness matrix is indefinite. If the structure is stable (except for rigid motions), the stiffness matrix will be positive semidefinite. Also, some lumping procedures can produce mass matrices, which are positive semidefinite because they have zero mass elements on the diagonal. The mass matrix can be positive definite or semidefinite (if some of the diagonal mass elements are zero, due to the lumping procedure used).

For buckling problems, the stiffness matrix is positive definite, provided the deformation state is stable (which may be assessed by a buckling analysis). The geometric stiffness matrix may be indefinite.

The preferred eigenvalue extraction techniques in use to date are sampling techniques. They create a linear operator (or matrix) and apply it to a sequence of carefully constructed vectors. From these transformed vectors the dominant eigenvectors and their eigenvalues can be approximated. Examples of these techniques are subspace iteration (or simultaneous iteration) and Lanczos techniques. The details of these methods are described in Bathe [1996], Parlett [1987], and Hughes [1987].

The eigensolver algorithms typically generate a set of eigenpairs (eigenvalues and associated eigenvectors) for the lowest eigenstates of the system, i.e., for the smallest eigenvalues (in absolute order). Often, it is necessary to compute eigenpairs for cases other than the set of smallest eigenvalues. This may be performed by introducing a shift α, which defines the shifted eigenvalue

$$\lambda = \bar{\lambda} + \alpha$$

The shifted eigenvalue problem is

$$[[K] - \alpha[B]]\{\mathbf{X}\} = \bar{\lambda}[B]\{\mathbf{X}\} \tag{36.46}$$

Equation (36.46), with a nonzero α, can be used to compute the eigenvalues for an unsupported structure (one for which $[K]$ is singular).

Sensitivity Analysis

Sensitivity analysis refers to methods of calculating the rates of change of: (1) response quantities (e.g., displacements, stresses, vibration frequencies, and buckling loads) with respect to changes in the structure characteristics (e.g., geometric and material parameters of the structure) and (2) the optimum design variable values with respect to changes in the structure parameters (e.g., applied loads and allowable stresses). The two types of calculations are usually designated by response and optimum design sensitivity analysis, and the rates of change are referred to as sensitivity coefficients.

A number of techniques have been developed for evaluating the sensitivity coefficients of response quantities using either the governing discrete equations or continuum equations of the structure. The techniques used with the discrete equations can be grouped into three categories: *analytical, direct differentiation methods; finite difference methods;* and *semianalytical or quasianalytical methods* (see Kleiber and Hisada [1993] and Hinton and Sienz [1994]).

Methods for computing sensitivity coefficients for linear structural response have been developed for over 20 years (see, for example, Haftka and Adelman [1989], Haber et al. [1993], and Choi [1993]). However, only recently have attempts been made to extend the domain of sensitivity analysis to (1) nonlinear structural response and to path-dependent problems, for which the sensitivity coefficients depend also on the deformation history (e.g., viscoplastic response and frictional contact—see, for example, Kleiber et al. [1994], Kulkarni and Noor [1995], and Karaoglan and Noor [1995]); and (2) structural systems exhibiting probabilistic uncertainties.

Because of the importance of its role in structural optimization and in assessing the effect of uncertainties in the input parameters on the structural response, some commercial software systems have incorporated response sensitivity analysis into their systems. Also, an automatic differentiation facility has been developed for evaluating the derivatives of functions defined by computer programs exactly to within machine precision. The facility, automatic differentiation of Fortran (ADIFOR), is described in

Chinchalkar [1994] and Carle et al. [1994]. The use of ADIFOR to evaluate the sensitivity coefficients from incremental/iterative forms of three-dimensional fluid flow problems is discussed in Sherman et al. [1994], and the additional facilities needed for ADIFOR to become competitive with hand-differentiated codes are listed in Carle et al. [1994].

Strategies and Numerical Algorithms for New Computing Systems

In recent years, intense efforts have been devoted to the development of efficient computational strategies and numerical algorithms which exploit the capabilities of new computing systems (in particular, the vector and parallel processing of the powerful high-performance computers—see, for example, Onaté et al. [1992]). Efficient direct and iterative numerical algorithms have been developed for solution of large sparse linear systems of equations (see Wang and Bruch [1993], Law and Mackay [1993], Qin and Mackay [1993], Vaughan [1991], and Papadrakakis [1993]).

Most parallel strategies are related to the *divide-and-conquer* paradigm based on breaking a large problem into a number of smaller subproblems which may be solved separately on individual processors. The degree of independence of the subproblems is a measure of the effectiveness of the algorithm since it determines the amount and frequency of communication and synchronization. The numerical algorithms developed for structural analysis can be classified into three major categories: namely, *elementwise algorithms, nodewise algorithms, and domainwise algorithms.*

The elementwise parallel algorithms include element-by-element equation solvers and parallel frontal equation solvers. The nodewise parallel equation solvers include node-by-node iterative solvers as well as column-oriented direct solvers. The domainwise algorithms include nested dissection-based (substructuring) techniques and domain-decomposition methods. The first two categories of numerical algorithms allow only small granularity of the parallel tasks and require frequent communication among the processors. By contrast, the third category allows a larger granularity, which can result in improved performance for the algorithm.

Nested dissection ordering schemes have been found to be effective in reducing both the storage requirements and the total computational effort required of direct factorization [Noor et al. 1978]. The performance of nested dissection-based linear solvers depends on balancing the computational load across processors in a way that minimizes interprocessor communication. Several nested dissection ordering schemes have been developed which differ in the strategies used in partitioning the structure and selecting the separators. Among the proposed partitioning strategies are: recursive bisection strategies (e.g., spectral graph bisection, recursive coordinate bisection, and recursive graph bisection [Pothen et al. 1990, Hendrickson and Leland 1992]); combinatorial and design-optimization based strategies (e.g., simulated annealing algorithm, genetic algorithm and neural-network-based techniques [Khan and Topping 1993]); and heuristic strategies (e.g., methods based on geometric projections and mappings; and algorithms based on embedding the problem in Euclidean space [Bui and Jones 1993]). For highly irregular and/or three-dimensional structures the effectiveness of nested dissection-based schemes may be reduced. However, this is also true for most other parallel numerical algorithms. Scalable parallel computational strategies for nonlinear, postbuckling, and dynamic contact/impact problems are presented in Watson and Noor [1996a, 1996b].

36.6 Brief History of the Development of Computational Structural Mechanics Software

Development of CSM software spans a period of less than 40 years, and may be divided into four stages, each stage lasting approximately 8–10 years. In the first stage (during the 1950s and 1960s), the aircraft industry pioneered development of in-house finite-element programs. These programs were generally based on the force method of analysis and were used to automate analysis of highly redundant structural components. Subsequent efforts in industry and academia led to development of simple two- and three-dimensional finite elements based on the displacement formulation. The variational process for formulating the elemental matrices was also introduced in this period.

In the second stage, general-purpose finite-element programs, such as NASTRAN, ASKA, ANSYS, STARDYNE, MARC, SAP, SESAM, and SAMCEF, were released for public use in the U.S. and Europe. These programs brought a significant technology base that led to development of numerous commercial finite-element software systems. This development included mixed and hybrid finite-element models with the fundamental unknowns consisting of stress and displacement parameters, efficient numerical algorithms for the solution of algebraic equations and extraction of eigenvalues, and substructuring and modal synthesis techniques for handling large problems. The finite-element method's success in linear static problems has encouraged bolder applications to nonlinear and transient response problems.

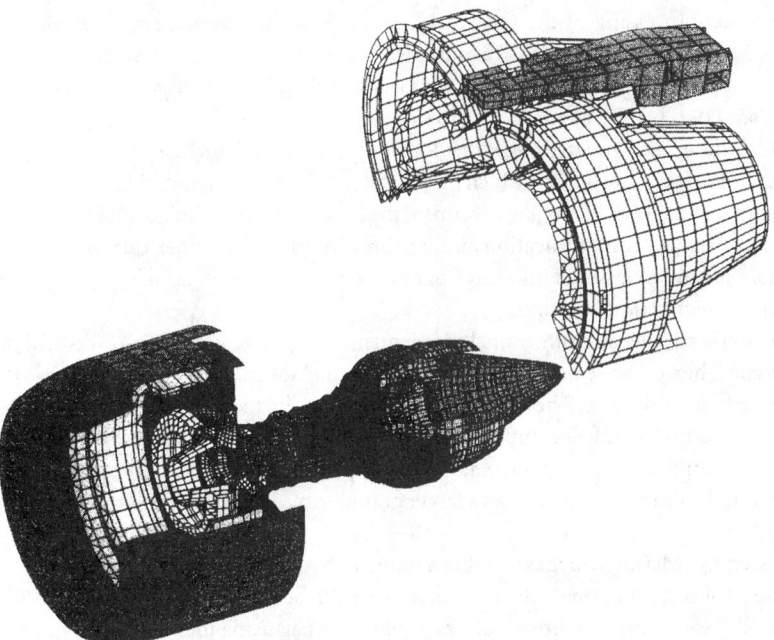

FIGURE 36.7   MSC/NASTRAN finite-element model of a G.E. Engine: 180,000 degrees of freedom. (Courtesy of GE Aircraft Engines.)

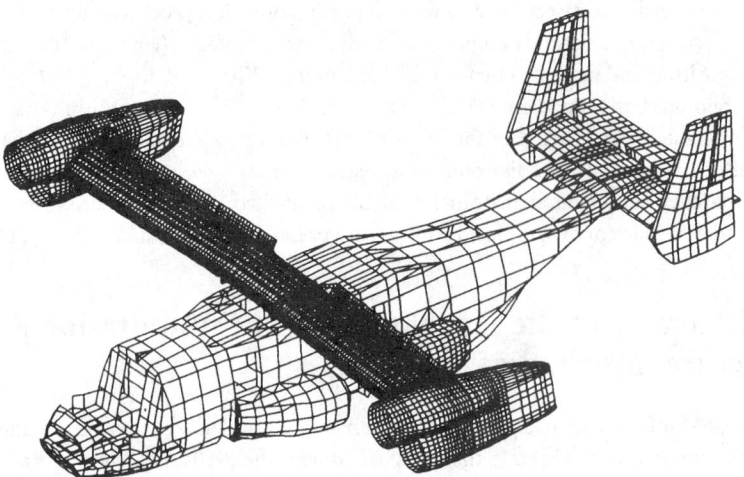

FIGURE 36.8   MSC/NASTRAN finite-element dynamics model of the V-22 Osprey Tiltrotor: 134,982 degrees of freedom (22,497 grid points), 44,006 elements. (Courtesy of Bell-Boeing.)

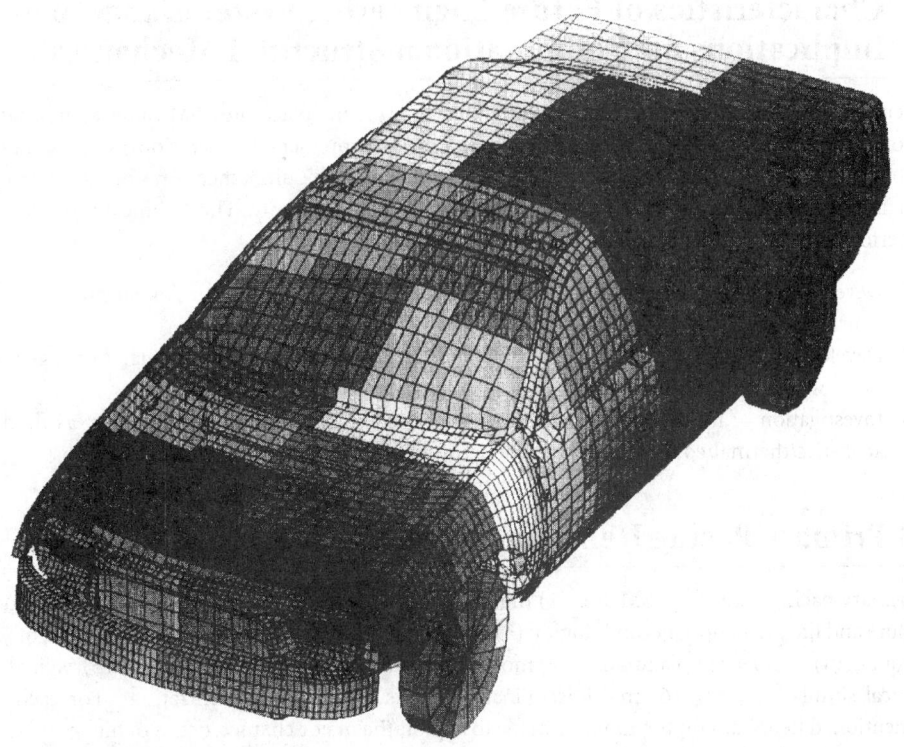

FIGURE 36.9 DYNA 3D finite element car model used in crash simulation: 27,000 shell elements, 162,000 degrees of freedom. (Courtesy of Lawrence Livermore.)

The third stage involved refining the commercial software codes and expanding their technology base. Design optimization techniques were also developed in this stage, as were pre- and postprocessing software and computer-aided design systems. The technology development included singular elements for fracture mechanics applications, boundary-element techniques, coupling of finite elements with other discretization techniques such as boundary elements, and quality assurance methods for both software and finite-element models.

The fourth stage included in the adaptation of CSM software to new computing systems (vector, multiprocessor, and massively parallel machines), development of efficient computational strategies and numerical algorithms for these machines, widespread availability of CSM software on personal computers and workstations, and the addition of substantial nonstructural modeling and analysis capabilities to CSM software systems such as MSC/NASTRAN and ANSYS. The latter capabilities were added because future flight vehicles and high-performance engineering systems (e.g., health monitoring aircraft and microsized spacecraft) will require significant interactions between CSM and other disciplines such as aerodynamics, controls, acoustics, electromagnetics, and optics.

Technology development in the fourth stage included introduction of advanced material models, development of stochastic models and probabilistic methods for structural analysis and design, and development of facilities for quality assessment and control of numerical simulations.

The four stages of CSM software development parallel the four stages of the computing environment's evolution: noncentralized mainframes, centralized mainframes, mainframe computing with timesharing, and distributed computing and networking. A summary of the major characteristics of currently used finite-element systems is given in Mackerle and Fredriksson [1988], and a guide to information sources on finite-element methods is given in Mackerle [1991]. Commercial finite-element programs for structural analysis have a rich variety of elements, and are widely used for performing structural calculations on large components and/or entire structures (see, for example, Figs. 36.7–36.9).

36.7 Characteristics of Future Engineering Systems and Their Implications on Computational Structural Mechanics

The demands that future high-performance engineering systems place on CSM differ somewhat from those of current systems. The radically different and more unpredictable operational environments for many of the systems (e.g., future flight vehicles) are one reason for this difference. Another is the stringency of design requirements for high performance, light weight, and economy. The technical needs for future high-performance engineering systems include:

1. Development of new high-performance material systems, such as smart/intelligent material systems
2. Development of novel structural concepts, such as structural tailoring and smart/intelligent structures, with active and/or passive adaptive control of dynamic deformations
3. Investigation of more complex phenomena and interdisciplinary couplings such as fluid flow/ acoustics/thermal/control/electromagnetic/optics, and structural couplings.

36.8 Primary Pacing Items and Research Issues

The primary pacing items for CSM are: (1) high-fidelity modeling of the structure and its components; (2) failure and life prediction methodologies; (3) hierarchical, integrated methods and adaptive modeling techniques; (4) nondeterministic analysis, modeling methods, and risk assessment; (5) validation of numerical simulations; and (6) multidisciplinary analysis and design optimization. For each of the aforementioned items, attempts should be made to exploit the major characteristics of high-performance computing technologies, as well as the future computing environment. The six primary pacing items are described subsequently. Note that some of the tasks within the pacing items are of a generic nature. Others are specific to certain components of future engineering systems (e.g., propulsion systems or airframes).

High-Fidelity Modeling of the Structure

The reliability of the predictions of response, failure, and life of structures is critically dependent on: (1) the accurate characterization and modeling of material behavior and (2) high-fidelity modeling of the critical details of the structure and its components (e.g., joints, damping, and for large deformations, frictional contact between the different parts of the structure). The simple material models used to date are inadequate for many of the future applications, especially those involving severe environment (e.g., high temperatures). Needed work on material modeling can be grouped in two general areas: (1) modeling the response and damage of advanced material systems in the actual operating environment of future engineering systems and (2) numerical simulation of manufacturing (fabrication) processes.

Advanced material systems include new polymer composites, metal-matrix composites, ceramic composites, carbon/carbon, and advanced metallics. The length scale selected in the model must be adequate for capturing the response phenomena of interest (e.g., micromechanics, mesomechanics, and macromechanics). For materials used in high-temperature applications, work is needed on the modeling of damage accumulation and propagation to fracture; modeling of thermoviscoplastic response, thermal-mechanical cycling, and ratcheting; and prediction of long-term material behavior from short-term data, which are particularly important.

Failure and Life Prediction Methodologies

Practical numerical techniques are needed for predicting the life, as well as the failure initiation and propagation in structural components made of new, high-performance materials in terms of measurable and controllable parameters. Examples of these materials are high-temperature materials; piezoelectric composites; and electronic, optical, and smart materials. For some of the materials, accurate constitutive descriptions, failure criteria, damage theories, and fatigue data are needed, along with more realistic characterization of interface phenomena (such as contact and friction). The constitutive descriptions may require investigations at the microstructure level or even the atomic level, as well as carefully designed and conducted experiments. Failure and life prediction of structures made of these materials is difficult and numerical models often constructed under restricting assumptions may not capture the dominant and underlying physical failure mechanisms. Moreover, material failure and structural response (such as instability) often couple in the failure mechanism.

Hierarchical, Integrated Multiple Methods and Adaptive Modeling Techniques

The effective use of numerical simulations for predicting the response, life, performance, and failure of future engineering systems requires strategies for treating phenomena occurring at disparate spatial and time scales, using reasonable computer resources. The strategies are based on using multiple mathematical models in different regions of the structure to take advantage of efficiencies gained by matching the model to the expected response in each region. To achieve the full potential of hierarchical modeling, there should be minimal reliance on a priori assumptions about the response. This is accomplished by adding adaptivity to the strategy. The key tasks of the research in this area are: (1) simple design-oriented models for use in the early stages of the design process; (2) rational selection of a set of nested mathematical models for different regions, and discretization techniques for use in conjunction with the mathematical models, which in turn requires the availability of a capability for holistic modeling from micro to structural response with varying degrees of accuracy; (3) simulation of local phenomena through global/local methodologies; (4) automated (or semiautomated) coupling of different mathematical/discrete models; (5) error estimation and adaptive modeling strategies; (6) efficient methods for coupling different components (e.g., engine airframe and rotor/engine frame); and (7) sensitivity analysis to assess the sensitivity of the response to each of the parameters neglected in the current mathematical model.

Nondeterministic Analysis, Modeling, and Risk Assessment

The new methodology developed for treating different types of uncertainties in geometry, material properties, boundary conditions, loading, and operational environment in the structural analysis formulation of structural components needs to be extended to the design and risk assessment of engineering systems. The ability to quantify inherent uncertainties in the response of the structure is obviously of great advantage. However, the principal benefit of using any nondeterministic method consists of the insights into engineering, safety, and economics that are gained in the process of arriving at those quantitative results and carrying out reliability analyses. As future engineering structures become more complicated, modeling of failure mechanisms will account for uncertainties from the beginning of the design process, and potential design improvements will be evaluated to assess their effects on reducing overall risk. The results combined with economic considerations will be used in systematic cost-benefit analyses (perhaps also done on a nondeterministic basis) to determine the structural design with the most acceptable balance of cost and risk.

Validation of Numerical Simulations

In addition to selecting a benchmark set of structures for assessing new computational strategies and numerical algorithms, a high degree of interaction and communication is needed between computational modelers and experimentalists. This is done on four different levels, namely, (1) laboratory tests on small specimens to obtain material data, (2) component tests to validate computational models, (3) full-scale tests to validate the modeling of details, and for flight vehicles (4) flight tests to validate the entire modeling process.

Multidisciplinary Analysis and Design Optimization

The realization of new complex engineering systems requires integration between the structures discipline and other traditionally separate disciplines such as aerodynamics, propulsion, and control. This is mandated by significant interdisciplinary interactions and couplings which need to be accounted for in predicting response, as well as in the optimal design of these structures. Examples are the couplings between the aerodynamic flow field, structural heat transfer, and structural response of high-speed aircraft and propulsion systems and the couplings between the control system and structural response in control-configured aircraft and spacecraft. This activity also includes design optimization with multiobjective functions (e.g., performance, durability, integrity, reliability, and cost), and multiscale structural tailoring (micro, local, and global levels). For propulsion systems, it also includes design with damping for high-cycle fatigue, low-cycle fatigue, and acoustic fatigue.

Typically, in the design process questions arise regarding influence of design variable changes on system behavior. Answers to these questions, quantified by the derivatives of behavior with respect to the design variables or by parametric studies, guide design improvements toward a better overall system. In large applications, this improvement process is executed by numerical optimization, combined with symbolic/artificial intelligence (AI) techniques, and human judgment aided by data visualization. Efficiency of the computations that provide data for such a process is decisive for the depth, breadth, and rate of progress achievable, and hence, ultimately, is critical for the final product quality.

Related Tasks

For CSM to impact the design process, the following tasks need to be addressed by the research community: (1) development of automated or semiautomated model (and mesh) generation facilities; (2) pre- and postdata processing and use of advanced visualization technology, including multimedia and virtual reality facilities; and (3) adaptation of object-oriented technology and AI tools (knowledge-based expert systems and neural networks) to CSM.

CSM's wide acceptance can affect the design and operation of future engineering systems and structures in three ways. It can provide a better understanding of the phenomena associated with response, failure, and life, thereby identifying desirable structural design attributes. CSM can verify and certify designs and also allow low-cost modifications to be made during the design process. Finally, it can improve the design team's productivity and allow major improvements and innovations during the design phase, enabling fully integrated design in an integrated product and process development (IPPD) environment. Such an environment allows computer simulation of the entire life cycle of the engineering system, including material selection and processing, multidisciplinary design, automated manufacturing and fabrication, quality assurance, certification, operation, health monitoring and control, retirement, and disposal.

Acknowledgment

The present work is partially supported by NASA Cooperative Agreement NCCW-0011 and by Air Force Office of Scientific Research Grant AFOSR-F49620-93-1-0184.

References

Babuska, I. and Miller, A. 1984a. The postprocessing approach in the finite element method—part 1: calculation of displacements, stresses and other higher derivatives of the displacements. *Int. J. Num. Methods. Eng.* 20:1085–1109.

Babuska, I. and Miller, A. 1984b. The postprocessing approach in the finite element method—part 2: the calculation of stress intensity factors. *Int. J. Num. Methods. Eng.* 20:1110–1129.

Banerjee, P. K. 1994. *The Boundary Element Methods in Engineering*, 2nd ed. McGraw–Hill, New York.

Bathe, K. J. 1996. *Finite Element Procedures*. Prentice–Hall, Englewood Cliffs, NJ.

Belytschko, T., Engelmann, B. E., and Liu, W. K. 1987. A review of recent developments in time integration. In *State-of-the-Art Surveys on Computational Mechanics*, A. K. Noor and J. T. Oden, eds., pp. 185–199. American Society of Mechanical Engineers, New York.

Belytschko, T. and Hughes, T. J. R. 1983. *Computational Methods for Transient Analysis.* Elsevier Science, Amsterdam.

Blacker, T. D. and Stephenson, M. B. 1991. Pacing: a new approach to automated quadrilateral mesh generation. *Int. J. Num. Methods. Eng.* 32:811–847.

Bui, T. N. and Jones, C. 1993. A heuristic for reducing fill-in in sparse matrix factorization. *Proc. 6th SIAM Conf. Parallel Process. Sci. Comput.*

Carle, A., Green, L. L., Bischof, C. H., and Newman, P. A. 1994. Application of automatic differentiation in CFD. *Proc. 25th AIAA Fluid Dynamics Conf.* Colorado Springs, CO, June 20–23, AIAA Paper 94-2197.

Cescotto, S., Frey, F., and Fonder, G. 1979. Total and updated Lagrangian descriptions in nonlinear structural analysis: a unified approach. In *Energy Methods in Finite Element Analysis*, R. Glowinski, E. Y. Rodin, and O. C. Zienkiewicz, eds., pp. 283–296. Wiley, New York.

Chinchalkar, S. 1994. The application of automatic differentiation to problems in engineering analysis. *Comp. Methods. Appl. Mech. Eng.* 118:197–207.

Choi, K. K. 1993. Design sensitivity of nonlinear structures—II. In *Structural Optimization: Status and Promise*, M. P. Kamat, ed., pp. 407–446. American Institute of Aeronautics and Astronautics, Washington, DC.

Crisfield, M. A. 1983. An arc-length method including line searches and accelerations. *Int. J. Num. Methods Eng.* 19:1269–1289.

Demkowicz, L., Oden, J. T., and Babuska, I. 1992. *Reliability in Computational Mechanics. Comput. Methods Appl. Mech. Eng.* Spec. issue 101(1–3).

Elishakoff, I. 1995. Essay on uncertainties in elastic and viscoelastic structures: from A. M. Freudenthal's criticisms to modern convex modeling. *Comput. Struct.* 56(6):871–895.

Fish, J. and Markolefas, S. 1993. Adaptive s-method for linear elastostatics. *Comput. Methods. Appl. Mech. Eng.* 104:363–396.

Fish, J. and Markolefas, S. 1994. Adaptive global-local refinement strategy based on the interior error estimates of the h-method. *Int. J. Num. Methods Eng.* 37:827–838.

George, P. L. 1991. *Automatic Mesh Generation. Application to Finite Element Methods.* Wiley, New York.

Geradin, M. 1980. Variational methods of structural dynamics and their finite element implementation. In *Advanced Structural Dynamics*, J. Donea, ed., pp. 1–41. Applied Science, Ltd., London.

Grandhi, R. V., Stroud, W. J., and Venkayya, V. B., eds. 1989. *Computational Structural Mechanics and Multidisciplinary Optimization*, AD Vol. 16, American Society of Mechanical Engineers, New York.

Haber, R. B., Tortorelli, D. A., and Vidal, C. A. 1993. Design sensitivity analysis of nonlinear structures—I: large deformation hyperelasticity and history dependent material response. In *Structural Optimization: Status and Promise*, M. P. Kamat, ed., pp. 369–406. American Institute of Aeronautics and Astronautics, Washington, DC.

Haftka, R. T. and Adelman, H. M. 1989. Recent developments in structural sensitivity analysis. *Struct. Optimization* 1(3):137–151.

Haug, E., Nguyen, Q. S., and de Rouvray, A. L. 1977. An improved energy conserving implicit time integration algorithm for nonlinear dynamic structural analysis. In *Trans. 4th Conf. Struct. Mech. Reactor Tech.*, T. A. Jaeger and B. A. Boley, eds. IASMiRT, Paper No. M-5/3.

Hendrickson, B. and Leland, R. 1992. An Improved Spectral Graph Partitioning Algorithm for Mapping Parallel Computations. *Sandia National Lab.* TR SAND92-1460.

Hinton, E. and Sienz, J. 1994. Aspects of adaptive finite-element analysis and structural optimization. In *Advances in Structural Optimization*, B. H. V. Topping and M. Papadrakakis, eds., pp. 1–25. Civil-Comp, Ltd., Edinburgh, Scotland.

Ho-Le, K. 1988. Finite element mesh generation methods: review and classification. *Comput. Aided Design* 1(20):27–38.

Hughes, T. J. R. 1976. Stability, convergence and growth and decay of energy of the average acceleration method in nonlinear structural dynamics. *Comput. Struct.* 6:313–324.

Hughes, T. J. R. 1987. *The Finite Element Method.* Prentice–Hall, Englewood Cliffs, NJ.

Hughes, T. J. R., Caughey, T. K., and Liu, W. K. 1978. Finite element methods for nonlinear elastodynamics which conserve energy. *J. Appl. Mech.* 45:366–370.

Ju, J. W., ed. 1995. *Numerical Methods in Structural Mechanics,* ASME Press, New York.

Kane, J. H., Maier, G., Tosaka, N., and Atluri, S. N., eds. 1993. *Advances in Boundary Element Techniques.* Springer–Verlag, New York.

Karaoglan, L. and Noor, A. K. 1995. Dynamic sensitivity analysis of frictional contact/impact response of axisymmetric composite structures. *Comp. Methods Appl. Mech. Eng.* 128(1–2):169–190.

Khan, A. I. and Topping, B. H. V. 1993. Subdomain generation for parallel finite element analysis. *Comput. Syst. Eng.* 4(4–6):473–488.

Kleiber, M., Hien, T. D., and Postek, E. 1994. Incremental finite-element sensitivity analysis for nonlinear mechanics applications. *Int. J. Num. Methods. Eng.* 37(19):3291–3308.

Kleiber, M. and Hisada, T., eds. 1993. *Design-Sensitivity Analysis.* Atlanta Technology, Atlanta, GA.

Kounadis, A. N. and Kratzig, W. B. 1995. *Nonlinear Stability of Structures: Theory and Computational Techniques.* Springer–Verlag, Vienna.

Kratzig, W. B. and Eller, C. 1992. Numerical algorithms for unstable dynamic shell responses. *Comput. Struct.* 44(1–2):263–271.

Krizek, M. and Neittaanmaki, P. 1987. On superconvergence techniques. *Acta Applicandae Mathematicae* 9:175–198.

Kulkarni, M. and Noor, A. K. 1995. Sensitivity analysis of the nonlinear dynamic viscoplastic response of two-dimensional structures with respect to material parameters. *Int. J. Num. Methods. Eng.* 38(2):183–198.

Ladevéze, P. and Zienkiewicz, O. C., eds. 1992. *New Advances in Computational Structural Mechanics* (*Proc. Eur. Conf. New Advances Comput. Struc. Mechanics*). Giens, France, April 2–5, 1991. Elsevier, Amsterdam.

Law, K. H. and Mackay, D. R. 1993. A parallel row-oriented sparse solution method for finite element structural analysis. *Int. J. Num. Methods. Eng.* 36:2895–2919.

Mackerle, J. 1991. *Finite Element Methods, A Guide to Information Sources.* Elsevier, Amsterdam.

Mackerle, J. 1993. Mesh generation and refinement for FEM and BEM—A bibliography. *Finite Elements Anal. Design* 15:177–188.

Mackerle, J. and Fredriksson, B. 1988. *Handbook of Finite Element Software.* Studentlitteratur, Lund, Sweden.

Noor, A. K. 1980. Mixed methods of analysis. In *Structural Mechanics Software Series,* N. Perrone and W. Pilkey, eds., Vol. III, pp. 263–305. University Press of Virginia, Charlottesville.

Noor, A. K. 1991. Bibliography of books and monographs on finite element technology. *Appl. Mech. Rev.,* 44(6):307–317.

Noor, A. K., ed. 1992. *Adaptive, Multilevel and Hierarchical Computational Strategies.* Proc. Symp. ASME Winter Annu. Meeting, Nov. 8–13, Anaheim, CA. AMD Vol. 157, American Society of Mechanical Engineers, New York.

Noor, A. K. 1994. Recent advances and applications of reduction methods. *Appl. Mech. Rev.* 47(5): 125–146.

Noor, A. K. and Andersen, C. M. 1992. Hybrid analytical techniques for the nonlinear analysis of curved beams. *Comput. Struct.* 43(5):823–830.

Noor, A. K. and Atluri, S. N. 1987. Advances and trends in computational structural mechanics. *AIAA J.* 25(7):977–995.

Noor, A. K. and Babuska, I. 1987. Quality assessment and control of finite element solutions. *Finite Elements Anal. Design* 3:1–26.

Noor, A. K., Housner, J. M., Starnes, J. H., and Hopkins, D. A. compilers. 1992. *Computational Structures Technology for Airframes and Propulsion Systems.* NASA CP-3142.

Noor, A. K., Kamel, H. A., and Fulton, R. E. 1978. Substructuring techniques—status and projections. *Comput. Struct.* 8:621–632.

Noor, A. K. and Venneri, S. L. 1990. Advances and trends in computational structural technology. *Comput. Syst. Eng.* 1(1):23–36.

Noor, A. K. and Venneri, S. L., eds. 1995. Computational structures technology. *Flight-Vehicle Materials, Structures and Dynamics*, Vol. 6. American Society of Mechanical Engineers, New York.

Onaté, E., Periaux, J., and Samuelsson, A., eds. 1992. *The Finite Element Method in the Nineteen Ninety's: A Book Dedicated to O. C. Zienkiewicz*. Springer–Verlag, New York.

Papadrakakis, M., ed. 1993. *Solving Large-Scale Problems in Mechanics: The Development and Application of Computational Solution Methods*. Wiley, Chichester, UK.

Parlett, B. N. 1987. The state-of-the-art in extracting eigenvalues and eigenvectors in structural mechanics. In *State-of-the-Art Surveys on Computational Mechanics*. A. K. Noor and J. T. Oden, eds., pp. 201–218. American Society of Mechanical Engineers, New York.

Pothen, A., Simon, H. D., and Liou, K. 1990. Partitioning sparse matrices with eigenvectors of graphs. *SIAM J. Matrix Anal. App.* 11(3):430–452.

Qin, J. and Mackay, D. R. 1993. A new parallel-vector finite element analysis software on distributed memory computers. *Proc. 34th AIAA/ASME/ASCE/AHS/ASC Struct. Structural Dynamics Mater. Conf.* La Jolla, CA, April 15–22.

Riks, E. 1984. Bifurcation and stability. A numerical approach. In *Proc. Int. Conf. Innovative Methods Nonlinear Probl.* W. K. Liu, T. Belytschko, and K. C. Park, eds., pp. 313–344. Pineridge Press, Swansea, UK.

Ross, T. 1995. *Fuzzy Logic with Engineering Applications*. McGraw–Hill, New York.

Sabin, M. 1991. Criteria for comparison of automatic mesh generation methods. *Advances Eng. Software & Workstations* 13(5–6):220–225.

Shephard, M. S. 1988. Approaches to the automatic generation and control of finite element meshes. *Appl. Mech. Rev.* 41:169–185.

Shephard, M. S. and Wentorf, R. 1994. Toward the implementation of automated analysis idealization control. *Appl. Num. Math.* 14(1–3).

Sherman, L. L., Taylor, A. C., III, Green, L. L., Newman, P. A., Hou, G. J.-W., and Korivi, V. M. 1994. First- and second-order aerodynamic sensitivity derivatives via automatic differentiation with incremental iterative methods. *5th AIAA/USAF/NASA/ISSMO Symp. Multidisciplinary Anal. Optimization*, AIAA Paper 94-4262. Panama City Beach, FL. Sept. 7–9.

Stein, E., ed. 1993. *Progress in Computational Analysis of Inelastic Structures*. Springer–Verlag, Vienna.

Storaasli, O. O. and Housner, J. M., eds. 1994. *Large-Scale Structural Analysis for High-Performance Computers and Workstations, Comput. Syst. Eng.* Spec. issue 5(4–6).

Szabo, B. and Babuska, I. 1991. *Finite Element Analysis*. Wiley, New York.

Tabbara, M., Blacker, T., and Belytschko, T. 1994. Finite element derivative recovery by moving least square interpolants. *Comput. Meth. Appl. Mech. Eng.* 117(1–2):211–223.

Turkiyyah, G. M. and Fenves, S. J. 1996. Knowledge-based assistance for finite element modeling. *IEEE Expert Intell. Syst. Their Appl.* (June)11(3):23–32.

Vaughan, C. T. 1991. Structural analysis on massively parallel computers. *Comput. Systems Eng.* 2(2/3):261–267.

Wang, K. P. and Bruch, J. C. 1993. A highly efficient iterative parallel computational method for finite element systems. *Eng. Comput.* 10:195–204.

Washizu, K. 1982. *Variational Methods in Elasticity and Plasticity*, 3rd ed. Pergamon Press, Oxford, UK.

Watson, B. C. and Noor, A. K. 1996a. Large-scale contact/impact simulation and sensitivity analysis on distributed-memory computers. *Comp. Methods Appl. Mech. Eng.* to appear.

Watson, B. C. and Noor, A. K. 1996b. Sensitivity analysis for large-deflection and postbuckling responses on distributed-memory computers. *Comp. Methods. Appl. Mech. Eng.* 129:393–409.

Zienkiewicz, O. C. and Zhu, J. Z. 1992. The superconvergent patch recovery and *a posteriori* error estimates—part 1: the recovery technique. *Int. J. Num. Methods. Eng.* 33:1331–1364.

Zienkiewicz, O. C., Zhu, J. Z., and Wu, J. 1993. Superconvergent patch recovery techniques—some further tests. *Commun. Num. Methods Eng.* 9(3):251–258.

Further Information

Information about CSM software is available on the Internet including structural analysis and design programs, and commercial finite-element programs. A number of publications on finite-element practice is available from the National Agency for Finite Element Methods and Standards (NAFEMS), Department of Trade and Industry, National Engineering Laboratory, East Kilbride, Glasgow G75 OQU, United Kingdom, including *A Finite Element Primer* (1986) and *Guidelines to Finite Element Practice* (1984). A finite element bibliography, including books and conference proceedings published since 1967, is under preparation for the World Wide Web. The WWW address is: http://ohio.ikp.liu.se/fe/index.html.

37

Computational Fluid Dynamics

David A. Caughey
Cornell University

37.1 Introduction

The use of computer-based methods for the prediction of fluid flows has seen tremendous growth in the past several decades. Fluid dynamics has been one of the earliest, and most active fields for the application of numerical techniques. This is due to the essential nonlinearity of most fluid flow problems of practical interest—which makes analytical, or closed-form, solutions virtually impossible to obtain—combined with the geometrical complexity of these problems. In fact, the history of computational fluid dynamics can be traced back virtually to the birth of the digital computer itself, with the pioneering work of John von Neumann and others in this area. Von Neumann was interested in using the computer not only to solve engineering problems, but to understand the fundamental nature of fluid flows themselves. This is possible because the complexity of fluid flows arises, in many instances, not from complicated or poorly understood formulations, but from the nonlinearity of partial differential equations that have been known for more than a century. A famous paragraph written by von Neumann in 1946 serves to illustrate this point. He wrote [Goldstine and von Neumann 1963]:

> Indeed, to a great extent, experimentation in fluid mechanics is carried out under conditions where the underlying physical principles are not in doubt, where the quantities to be observed are completely determined by known equations. The purpose of the experiment is not to verify a proposed theory but to replace a computation from an unquestioned theory by direct measurements. Thus, wind tunnels are, for example, used at present, at least in part, as computing devices of the so-called analogy type . . . to integrate the nonlinear partial differential equations of fluid dynamics.

The present article provides some of the basic background in fluid dynamics required to understand the issues involved in numerical solution of fluid flow problems, then outlines the approaches that have been successful in attacking problems of practical interest.

37.2 Underlying Principles

In this section, we will provide the background in fluid dynamics required to understand the principles involved in the numerical solution of the governing equations. The formulation of the equations in

0-8493-2909-4/97/$0.00+$.50

generalized, curvilinear coordinates and the geometrical issues involved in the construction of suitable grids also will be discussed.

Fluid-Dynamical Background

As can be inferred from the quotation of von Neumann presented in the introduction, fluid dynamics is fortunate to have a generally accepted mathematical framework for describing most problems of practical interest. Such diverse problems as the high-speed flow past an aircraft wing, the motions of the atmosphere responsible for our weather, and the unsteady air currents produced by the flapping wings of a housefly all can be described as solutions to a set of partial differential equations known as the Navier–Stokes equations. These equations express the physical laws corresponding to conservation of mass, Newton's second law of motion relating the acceleration of fluid elements to the imposed forces, and conservation of energy, under the assumption that the stresses in the fluid are linearly related to the local rates of strain of the fluid elements. This latter assumption is generally regarded as an excellent approximation for everyday fluids such as water and air—the two most common fluids of engineering and scientific interest.

We will describe the equations for problems in two space dimensions, for the sake of economy of notation; here, and elsewhere, the extension to problems in three dimensions will be straightforward unless otherwise noted. The Navier–Stokes equations can be written in the form

$$\frac{\partial \mathbf{w}}{\partial t} + \frac{\partial \mathbf{f}}{\partial x} + \frac{\partial \mathbf{g}}{\partial y} = \frac{\partial \mathbf{R}}{\partial x} + \frac{\partial \mathbf{S}}{\partial y} \tag{37.1}$$

where $\mathbf{w}$ is the vector of conserved quantities

$$\mathbf{w} = \{\rho, \rho u, \rho v, e\}^T \tag{37.2}$$

where ρ is the fluid density, u and v are the fluid velocity components in the x and y directions, respectively, and e is the total energy per unit volume. The inviscid flux vectors $\mathbf{f}$ and $\mathbf{g}$ are given by

$$\mathbf{f} = \{\rho u, \rho u^2 + p, \rho u v, (e + p)u\}^T \tag{37.3}$$

$$\mathbf{g} = \{\rho v, \rho u v, \rho v^2 + p, (e + p)v\}^T \tag{37.4}$$

where p is the fluid pressure, and the flux vectors describing the effects of viscosity are

$$\mathbf{R} = \{0, \tau_{xx}, \tau_{xy}, u\tau_{xx} + v\tau_{xy} - q_x\}^T \tag{37.5}$$

$$\mathbf{S} = \{0, \tau_{xy}, \tau_{yy}, u\tau_{xy} + v\tau_{yy} - q_y\}^T \tag{37.6}$$

The viscous stresses appearing in these terms are related to the derivatives of the components of the velocity vector by

$$\tau_{xx} = 2\mu \frac{\partial u}{\partial x} - \frac{2}{3}\mu \left\{ \frac{\partial u}{\partial x} + \frac{\partial v}{\partial y} \right\} \tag{37.7}$$

$$\tau_{xy} = \mu \left\{ \frac{\partial u}{\partial y} + \frac{\partial v}{\partial x} \right\} \tag{37.8}$$

$$\tau_{yy} = 2\mu \frac{\partial v}{\partial y} - \frac{2}{3}\mu \left\{ \frac{\partial u}{\partial x} + \frac{\partial v}{\partial y} \right\} \tag{37.9}$$

and

$$q_x = -k\frac{\partial T}{\partial x} \tag{37.10}$$

$$q_y = -k\frac{\partial T}{\partial y} \tag{37.11}$$

represent the x and y components of the heat flux vector, according to Fourier's law. In these equations μ and k represent the coefficients of viscosity and thermal conductivity, respectively.

If the Navier–Stokes equations are nondimensionalized by normalizing lengths with respect to a representative length L and velocities by a representative velocity V_∞, and normalizing the fluid properties (such as density and coefficient of viscosity) by their values in the freestream, an important nondimensional parameter, the *Reynolds number*

$$\mathbf{Re} = \frac{\rho_\infty V_\infty L}{\mu_\infty} \tag{37.12}$$

appears as a parameter in the equations. In particular, the viscous stress terms on the right-hand side of Eq. (37.1) are multiplied by the reciprocal of the Reynolds number. Physically, the Reynolds number can be interpreted as an inverse measure of the relative importance of the contributions of the viscous stresses to the dynamics of the flow; i.e., when the Reynolds number is large, the viscous stresses are small almost everywhere in the flowfield.

The computational resources required to solve the complete Navier–Stokes equations are enormous, particularly when the Reynolds number is large and regions of **turbulent flow** must be resolved. In 1970 Howard Emmons of Harvard University estimated the computer time required to solve a simple turbulent pipe-flow problem, including direct computation of all turbulent eddies containing significant energy [Emmons 1970]. For a computational domain consisting of approximately 12 diameters of a pipe of circular cross section, the computation of the solution at a Reynolds number based on the pipe diameter of $\mathbf{Re}_d = 10^7$ would require approximately 10^{17} seconds on a 1970s mainframe computer. Of course, much faster computers are now available, but even at a computational speed of 100 gigaflops—i.e., 10^{11} floating-point operations per second—such a calculation would require more than 3000 *years* to complete.

Because the resources required to solve the complete Navier–Stokes equations are so large, it is common to make approximations to bring the required computational resources to a more modest level for problems of practical interest. Expanding slightly on a classification introduced by Chapman [1979], the sequence of fluid-mechanical equations can be organized into the hierarchy shown in Table 37.1.

Table 37.1 summarizes the physical assumptions and time periods of most intense development for each of the stages in the fluid-mechanical hierarchy. Stage IV represents an approximation to the Navier–Stokes equations in which only the largest, presumably least isotropic scales of turbulence are resolved; the subgrid scales are modeled. Stage III represents an approximation in which the solution is decomposed into time-averaged or ensemble-averaged and fluctuating components for each variable. For example, the velocity components and pressure can be decomposed into

$$u = U + u' \tag{37.13}$$

$$v = V + v' \tag{37.14}$$

$$p = P + p' \tag{37.15}$$

where U, V, and P are the average values of u, v, and p, respectively, taken over a time interval that is long compared to the turbulence time scales, but short compared to the time scales of any nonsteadiness of the

TABLE 37.1   The Hierarchy of Fluid-Mechanical Approximations, Following Chapman [1979]

Stage	Model	Equations	Time Frame
I	Linear potential	Laplace's equation	1960s
IIa	Nonlinear potential	Full potential equation	1970s
IIb	Nonlinear inviscid	Euler equations	1980s
III	Modeled turbulence	Reynolds-averaged Navier–Stokes equations	1990s
IV	Large-eddy simulation (LES)	Navier–Stokes equations with subgrid turbulence model	1980s–1990s
V	Direct numerical simulation (DNS)	Fully resolved Navier–Stokes equations	1980s–1990s

averaged flowfield. If we let $\langle u \rangle$ denote such a time average of the u-component of the velocity, then, e.g.,

$$\langle u \rangle = U \tag{37.16}$$

When a decomposition of the form of Eqs. (37.13–37.15) is substituted into the Navier–Stokes equations and the equations are averaged as described above, the resulting equations describe the evolution of the mean-flow quantities (such as U, V, P, etc.). These equations, called the **Reynolds-averaged Navier–Stokes equations**, are nearly identical to the original Navier–Stokes equations written for the mean-flow variables because terms that are linear in $\langle u' \rangle$, $\langle v' \rangle$, $\langle p' \rangle$, etc., are identically zero. The nonlinearity of the equations, however, introduces terms that depend upon the fluctuating components. In particular, terms proportional to $\langle \rho u' v' \rangle$, $\langle \rho u'^2 \rangle$, and $\langle \rho v'^2 \rangle$ appear in the averaged equations. Dimensionally, these terms are equivalent to stresses; in fact, these quantities are called the Reynolds stresses. Physically, the Reynolds stresses are the turbulent counterparts of the molecular viscous stresses, and appear as a result of the transport of momentum by the turbulent fluctuations. The appearance of these terms in the equations describing the mean flow means that the mean flow cannot be determined without some knowledge of these fluctuating components.

In order to solve the Reynolds-averaged Navier–Stokes equations, the Reynolds stresses must be related to the mean flow at some level of approximation using a phenomenological model. The simplest procedure is to try to relate the Reynolds stresses to the local mean-flow properties. Since the turbulence that is responsible for these stresses is a function not only of the local mean-flow state, but also of the flow history as well, such an approximation cannot have broad generality, but such local, or algebraic, turbulence models, based on the analogy of the mixing length from the kinetic theory of gases, are useful for many flows of engineering interest, especially for flows in which boundary-layer separation does not occur.

A more general procedure is to develop partial differential equations for the Reynolds stresses themselves, or for quantities that can be used to determine the scales of the turbulence, such as the turbulent kinetic energy k and the dissipation rate ε. The latter, so-called k–ε model is widely used in engineering analyses of turbulent flows. These differential equations can be derived by taking higher-order moments of the Navier–Stokes equations. For example, if the x-momentum equation is multiplied by v and then averaged, the result will be an equation for the Reynolds-stress component $\langle \rho u' v' \rangle$. Again because of the nonlinearity of the equations, however, yet higher moments of the fluctuating components (e.g., terms proportional to $\langle \rho u'^2 v' \rangle$, $\langle \rho u' v'^2 \rangle$, etc.) will appear in the equations for the Reynolds stresses. This is an example of the problem of closure of the equations for the Reynolds stresses; i.e., to solve the equations for the Reynolds stresses, these third-order quantities must be known. Equations for the third-order quantities can be derived by taking yet higher-order moments of the Navier–Stokes equations, but these equations will involve fourth-order moments. Thus, at some point the taking of moments must be terminated and models must be developed for the unknown higher-order quantities. It is hoped that more nearly universal models may be developed for these higher-order quantities, but the superiority of second-order models, in which the equations for the Reynolds-stress components are solved, subject to modeling assumptions for the required third-order quantities, has yet to be established for most practical problems.

For many design purposes, especially in aerodynamics, it is sufficient to represent the flow as that of an inviscid, or ideal, fluid. This is appropriate for flows at high Reynolds numbers that contain only negligibly small regions of separated flow. The equations describing inviscid flows can be obtained as a simplification of the Navier–Stokes equations in which the viscous terms are neglected altogether. This results in the **Euler equations** of inviscid, compressible flow,

$$\frac{\partial \mathbf{w}}{\partial t} + \frac{\partial \mathbf{f}}{\partial x} + \frac{\partial \mathbf{g}}{\partial y} = 0 \tag{37.17}$$

This approximation corresponds to stage IIb in the hierarchy described in Table 37.1.

The Euler equations constitute a hyperbolic system of partial differential equations, and their solutions contain features that are absent from the Navier–Stokes equations. In particular, while the viscous diffusion terms appearing in the Navier–Stokes equations guarantee that solutions will remain smooth, the absence of these dissipative terms from the Euler equations allows them to have solutions that are discontinuous across surfaces in the flow. Solutions to the Euler equations must be interpreted within the context of generalized (or weak) solutions, and this theory provides the mathematical framework for developing the properties of any discontinuities that may appear. In particular, jumps in dependent variables (such as density, pressure, and velocity) across such surfaces must be consistent with the original conservation laws upon which the differential equations are based. For the Euler equations, these jump conditions are called the Rankine–Hugoniot conditions.

Solutions of the Euler equations for flows containing shock waves can be computed using either **shock fitting** or **shock capturing** methods. In the former, the shock surfaces must be located and the Rankine–Hugoniot jump conditions enforced explicitly. In shock capturing methods, **artificial viscosity** terms are added in the numerical approximation to provide enough dissipation to allow the shocks to be captured automatically by the scheme, with no special treatment in the vicinity of the shock waves. The numerical viscosity terms usually act to smear out the discontinuity over several grid cells. Numerical viscosity also is used when solving the Navier–Stokes equations for flows containing shock waves, because usually it is impractical to resolve the shock structure defined by the physical dissipative mechanisms.

In many cases, the flow can further be approximated as steady and irrotational. In these cases, it is possible to define a velocity potential Φ such that the velocity vector $\mathbf{V}$ is given by

$$\mathbf{V} = \nabla \Phi \tag{37.18}$$

and the fluid density is given by the isentropic relation

$$\rho = \left(1 + \frac{\gamma - 1}{2} \mathbf{M}_\infty^2 [1 - (u^2 + v^2)] \right)^{\frac{1}{\gamma - 1}} \tag{37.19}$$

where

$$u = \frac{\partial \Phi}{\partial x} \tag{37.20}$$

$$v = \frac{\partial \Phi}{\partial y} \tag{37.21}$$

and $\mathbf{M}_\infty$ is a reference Mach number corresponding to the freestream state in which $\rho = 1$ and $u^2 + v^2 = 1$. The steady form of the Euler equations then reduces to the single equation

$$\frac{\partial \rho u}{\partial x} + \frac{\partial \rho v}{\partial y} = 0 \tag{37.22}$$

Equation (37.19) can be used to eliminate the density from Eq. (37.22), which then can be expanded to the form

$$(a^2 - u^2)\frac{\partial^2 \Phi}{\partial x^2} - 2uv\frac{\partial^2 \Phi}{\partial x \, \partial y} + (a^2 - v^2)\frac{\partial^2 \Phi}{\partial y^2} = 0 \qquad (37.23)$$

where a is the local speed of sound, which is a function only of the fluid speed $V = \sqrt{u^2 + v^2}$. Thus, Eq. (37.23) is a single equation that can be solved for the velocity potential Φ.

Equation (37.23) is a second-order quasilinear partial differential equation whose type depends on the sign of the discriminant $1 - \mathbf{M}^2$, where $\mathbf{M}$ is the *local* Mach number. The equation is elliptic or hyperbolic according as the Mach number is less than or greater than unity. Thus, the nonlinear potential equation contains a mathematical description of the physics necessary to predict the important features of transonic flows. It is capable of changing type, and the conservation form of Eq. (37.22) allows surfaces of discontinuity, or shock waves, to be computed. Solutions at this level of approximation, corresponding to stage IIa, are considerably less expensive to compute than solutions of the full Euler equations, since only one dependent variable need be computed and stored. The jump relation corresponding to weak solutions of the potential equation is different than the Rankine–Hugoniot relations, but is a good approximation to the latter when the shocks are not too strong.

Finally, if the flow can be approximated by small perturbations to some uniform reference state, Eq. (37.23) can further be simplified to

$$\left(1 - \mathbf{M}_\infty^2\right)\frac{\partial^2 \phi}{\partial x^2} + \frac{\partial^2 \phi}{\partial y^2} = 0 \qquad (37.24)$$

where ϕ is the *perturbation* velocity potential defined according to

$$\Phi = x + \phi \qquad (37.25)$$

if the uniform velocity in the reference state is assumed to be parallel to the x-axis and be normalized to have unit magnitude. Further, if the flow can be approximated as incompressible, i.e., in the limit as $\mathbf{M}_\infty \to 0$, Eq. (37.23) reduces to

$$\frac{\partial^2 \Phi}{\partial x^2} + \frac{\partial^2 \Phi}{\partial y^2} = 0 \qquad (37.26)$$

Since Eqs. (37.24, 37.26) are linear, superposition of elementary solutions can be used to construct solutions of arbitrary complexity. Numerical methods to determine the singularity strengths for aerodynamic problems are called **panel methods**, and constitute stage I in the hierarchy of approximations.

It is important to realize that, even though the time periods for development of some of the different models overlap, the applications of the various models may be for problems of significantly differing complexity. For example, DNS calculations were performed as early as the 1970s, but only for the simplest flows—homogeneous turbulence—at very low Reynolds numbers. Flows at higher Reynolds numbers and nonhomogeneous flows are being performed only now, whereas calculations for three-dimensional flows with modeled turbulence were performed as early as the mid 1980s.

Treatment of Geometry

For all numerical solutions to partial differential equations, including those of fluid mechanics, it is necessary to discretize the boundaries of the flow domain. For the panel method of stage I, this is all that is required, since the problem is linear and superposition can be used to construct solutions satisfying the boundary conditions. For nonlinear problems, it is necessary to discretize the entire flow domain. This can be done using either structured or unstructured grid systems. *Structured grids* are those in which the grid points can be ordered in a regular Cartesian structure; i.e., the points can be given indices (i, j) such that the nearest neighbors of the (i, j) point are identified by the indices $(i \pm 1, j \pm 1)$. The grid cells for these meshes are thus quadrilateral in two dimensions and hexahedral in three dimensions. *Unstructured grids* have no regular ordering of points or cells, and a connectivity table must be maintained to identify which points and edges belong to which cells. Unstructured grids most often consist of triangles (in two dimensions), tetrahedra (in three dimensions), or combinations of these and quadrilateral and hexahedral cells, respectively. In addition, grids having purely quadrilateral cells may also be unstructured, even though they have a locally Cartesian structure—e.g., when multilevel grids are used for adaptive refinement.

Implementations on structured grids are generally more efficient than those on unstructured grids, since indirect addressing is required for the latter, and efficient implicit methods often can be constructed that take advantage of the regular ordering of points (or cells) in structured grids. A great deal of effort has been expended to generate both structured and unstructured grid systems, much of which is closely related to the field of computational geometry.

Structured Grids

A variety of techniques are used to generate structured grids for use in fluid-mechanical calculations. These include relatively fast algebraic methods, including those based on transfinite interpolation and conformal mapping, as well as more costly methods based on the solution of either elliptic or hyperbolic systems of partial differential equations for the grid coordinates. These techniques are discussed in a review article by Eiseman [1985].

For complex geometries, it often is not possible to generate a single grid that conforms to all the boundaries. Even if it is possible to generate such a grid, it may have undesirable properties, such as excessive skewness or a poor distribution of cells, which could lead to poor stability or accuracy in the numerical algorithm. Thus, structured grids for complex geometries are generally constructed as combinations of simpler grid blocks for various parts of the domain. These grids may be allowed to overlap, in which case they are referred to as *Chimera* grids, or be required to share common surfaces of intersection, in which case they are identified as *multiblock* grids. In the latter case, grid lines may have varying degrees of continuity across the interblock boundaries, and these variations have implications for the construction and behavior of the numerical algorithm in the vicinity of those boundaries. Grid generation techniques based on the solution of systems of partial differential equations are described by Thompson et al. [1985].

Numerical methods to solve the equations of fluid mechanics on structured grid systems are implemented most efficiently by taking advantage of the ease with which the equations can be transformed to a generalized coordinate system. The expression of the system of conservation laws in the new, body-fitted coordinate system reduces the problem to one effectively of Cartesian geometry. The transformation will be described here for the Euler equations.

Consider the transformation of Eq. (37.17) to a new coordinate system (ξ, η). The local properties of the transformation at any point are contained in the Jacobian matrix of the transformation, which can be defined as

$$J = \left\{ \begin{matrix} x_\xi & x_\eta \\ y_\xi & y_\eta \end{matrix} \right\} \tag{37.27}$$

for which the inverse is given by

$$J^{-1} = \left\{ \begin{matrix} \xi_x & \xi_y \\ \eta_x & \eta_y \end{matrix} \right\} = \frac{1}{h} \left\{ \begin{matrix} y_\eta & -x_\eta \\ -y_\xi & x_\xi \end{matrix} \right\} \tag{37.28}$$

where $h = x_\xi y_\eta - y_\xi x_\eta$ is the determinant of the Jacobian matrix. Subscripts in Eqs. (37.27) and (37.28) denote partial differentiation.

It is natural to express the fluxes in conservation laws in terms of their contravariant components. Thus, if we define

$$\{\mathbf{F}, \mathbf{G}\}^T = J^{-1}\{\mathbf{f}, \mathbf{g}\}^T \tag{37.29}$$

then the transformed Euler equations can be written in the compact form

$$\frac{\partial h\mathbf{w}}{\partial t} + \frac{\partial h\mathbf{F}}{\partial \xi} + \frac{\partial h\mathbf{G}}{\partial \eta} = 0 \tag{37.30}$$

if the transformation is independent of time (i.e., if the grid is not moving). If the grid is moving or deforming with time, the equations can be written in a similar form, but additional terms must be included that allow for the fluxes induced by the motion of the mesh.

The Navier–Stokes equations can be transformed in a similar manner, although the transformation of the viscous contributions is somewhat more complicated and will not be included here. Since the nonlinear potential equation is simply the continuity equation (the first of the equations that comprise the Euler equations), the transformed potential equation can be written as

$$\frac{\partial}{\partial \xi}(\rho h U) + \frac{\partial}{\partial \eta}(\rho h V) = 0 \tag{37.31}$$

where

$$\begin{Bmatrix} U \\ V \end{Bmatrix} = J^{-1} \begin{Bmatrix} u \\ v \end{Bmatrix} \tag{37.32}$$

are the contravariant components of the velocity vector.

Unstructured Grids

Unstructured grids generally have greater geometric flexibility than structured grids, because of the relative ease of generating triangular or tetrahedral tesselations of two- and three-dimensional domains. Advancing-front methods and Delaunay triangulations are the most frequently used techniques for generating triangular/tetrahedral grids. Unstructured grids also are easier to adapt locally so as to better resolve localized features of the solution.

37.3 Best Practices

Panel Methods

The earliest numerical methods used widely for making fluid-dynamical computations were developed to solve linear potential problems, described as stage I calculations in the previous section. Mathematically, panel methods are based upon the fact that Eq. (37.26) can be recast as an integral equation giving the solution at any point (x, y) in terms of the freestream speed U (here assumed unity) and angle of attack α and the line integrals of singularities distributed along the curve C representing the body surface:

$$\Phi(x, y) = x \cos\alpha + y \sin\alpha + \int_C \sigma \ln\left(\frac{r}{2\pi}\right) ds + \int_C \delta(\partial/\partial n) \ln\left(\frac{r}{2\pi}\right) ds \tag{37.33}$$

In this equation, σ and δ represent the source and doublet strengths distributed along the body contour, respectively, r is the distance from the point (x, y) to the boundary point, and n is the direction of the outward normal to the body surface. When the point (x, y) is chosen to lie on the body contour C, Eq. (37.33) can be interpreted as giving the solution Φ at any point on the body in terms of the singularities distributed along the surface. This effectively reduces the dimension of the problem by one (i.e., the two-dimensional problem considered here becomes essentially one-dimensional, and the analogous procedure applied to a three-dimensional problem results in a two-dimensional equation requiring the evaluation only of integrals over the body surface).

Equation (37.33) is approximated numerically by discretizing the boundary C into a collection of panels (line segments in this two-dimensional example) on which the singularity distribution is assumed to be of some known functional form, but of an as yet unknown magnitude. For example, for a simple nonlifting body, the doublet strength δ might be assumed to be zero, while the source strength σ is assumed to be constant on each segment. The second integral in Eq. (37.33) is then zero, while the first integral can be written as a sum over all the elements of integrals that can be evaluated analytically as

$$\Phi(x, y) = x \cos \alpha + y \sin \alpha + \sum_{i=1}^{N} \sigma_i \int_{C_i} \ln\left(\frac{r}{2\pi}\right) ds \qquad (37.34)$$

where C_i is the portion of the body surface corresponding to the ith segment and N is the number of segments, or panels, used.

The source strengths σ_i must be determined by enforcing the boundary condition

$$\frac{\partial \Phi}{\partial n} = 0 \qquad (37.35)$$

which specifies that the component of velocity normal to the surface be zero (i.e., that there be no flux of fluid across the surface). This is implemented by requiring that Eq. (37.35) be satisfied at a selected number of control points. For the example of constant-strength source panels, if one control point is chosen on each panel, the requirement that Eq. (37.35) be satisfied at each of the control points will result in N equations of the form

$$\sum_{i=1}^{N} A_{i,j} \sigma_i = \mathbf{U} \cdot \hat{\mathbf{n}}, \quad j = 1, 2, \ldots, N \qquad (37.36)$$

where $A_{i,j}$ are the elements of the influence-coefficient matrix that give the normal velocity at control point j due to sources of unit strength on panel i, and $\mathbf{U}$ is a unit vector in the direction of the freestream. Equations (37.36) constitute a system of N linear equations that can be solved for the unknown source strengths σ_i. Once the source strengths have been determined, the velocity potential, or the velocity itself, can be computed directly at any point in the flowfield using Eq. (37.34). A review of the development and application of panel methods is provided by Hess [1990].

A major advantage of panel methods, relative to the more advanced methods required to solve the nonlinear problems of stages II–V, is that it is necessary to describe (and to discretize into panels) only the surface of the body. While linearity is a great advantage in this regard, it is not clear that the method is computationally more efficient than the more advanced nonlinear field methods. This results from the fact that the influence-coefficient matrix in the system of equations that must be solved to give the source strengths for each panel is not sparse; i.e., the velocities at each control point are affected by the sources on all the panels. In contrast, the solution at each mesh point in a finite-difference formulation (or each mesh cell in a finite-volume formulation) is related to the values of the solution at only a few neighbors, resulting in a very sparse matrix of influence coefficients that can be solved very efficiently using iterative methods. Thus, the primary advantage of the panel method is the geometrical one associated with the reduction in dimension of the problem. For nonlinear problems the use of finite-difference, finite-element, or finite-volume methods requires discretization of the entire flowfield, and the associated mesh generation task has been a major pacing item limiting the application of such methods.

Nonlinear Methods

For nonlinear equations, superposition of elementary solutions is no longer a valid technique for constructing solutions, and it becomes necessary to discretize the solution throughout the entire domain, not just the boundary surface. In addition, since the equations are nonlinear, some sort of iteration must be used to compute successively better approximations to the solution. This iterative process may approximate the physics of an unsteady flow process or may be chosen to provide more rapid convergence to the solution for steady-state problems. Solutions for both the nonlinear potential and Euler equations are generally determined using a finite-difference, finite-volume, or finite-element method. In any of these techniques a grid, or network of points and cells, is distributed throughout the flowfield. In a finite-difference method the derivatives appearing in the original differential equation are approximated by discrete differences in the values of the solution at neighboring points (and times, if the solution is unsteady), and substitution of these into the differential equation yields a system of algebraic equations relating the values of the solution at neighboring grid points. In a finite-volume method, the unknowns are taken to be representative of the values of the solution in the control volumes formed by the intersecting grid surfaces, and the equations are constructed by balancing fluxes across the bounding surfaces of each control volume with the rate of change of the solution within the control volume. In a finite-element method, the solution is represented using simple interpolating functions within each mesh cell, or element, and equations for the nodal values are obtained by integrating a variational or residual formulation of the equations over the elements. The algebraic equations relating the values of the solution in neighboring cells can be very similar (or even identical) in appearance for all three methods, and the choice of method often is primarily a matter of taste. Stable and efficient finite-difference and finite-volume methods were developed earlier than finite-element methods for compressible flow problems, but finite-element methods capable of treating very complex compressible flows now are available. A review of recent progress in the development of finite-element methods for compressible flows and remaining issues is given by Glowinski and Pironneau [1992].

Since the nonlinear potential, Euler, and Navier–Stokes equations all are nonlinear, the algebraic equations resulting from these discretization procedures also are nonlinear, and a scheme of successive approximation usually is required to solve them. As mentioned earlier, however, these equations tend to be highly local in nature, and efficient iterative methods have been developed to solve them in many cases.

Nonlinear Potential-Equation Methods

The primary advantage of solving the potential equation rather than the Euler (or Navier–Stokes) equations derives from the fact that the flowfield can be described completely in terms of a single scalar function, the velocity potential Φ. The formulation of numerical schemes to solve Eq. (37.22) is complicated by the fact that, as noted earlier, the equation changes type according to whether the local Mach number is subsonic or supersonic. Differencing schemes for the potential equation must, therefore, be type-dependent—i.e., they must change their form depending on whether the local Mach number is less than or greater than unity. These methods usually are based upon central, or symmetric, differencing formulas that are appropriate for the elliptic case (corresponding to subsonic flows); they are then modified by adding an upwind bias to maintain stability in hyperbolic regions (where the flow is supersonic). This directional bias can be introduced into the difference equations either by adding an artificial viscosity proportional to the third derivative of the velocity potential Φ in the streamwise direction, or by replacing the density at each point in supersonic zones with its value at a point slightly upstream of the actual point. Mathematically, these two approaches can be seen to be equivalent, since the upwinding of the density evaluation also effectively introduces a correction proportional to the third derivative of the potential in the flow direction.

It is important to introduce such artificial viscosity (or compressibility) terms in such a way that their effect vanishes in the limit as the mesh is refined. In this way, the numerical approximation will approach the differential equation in the limit of zero mesh spacing, and the method is said to be consistent with the original differential equation. In addition, for flows with shock waves, it is important that the numerical approximation be **conservative**; this guarantees that the properties of discontinuous solutions will be consistent with the jump relations of the original conservation laws. The shock jump relations corresponding to Eq. (37.22), however, are different from the Rankine–Hugoniot conditions describing shocks within the framework of the Euler equations. Since entropy is everywhere conserved in the potential theory, and since there is a finite entropy jump across a Rankine–Hugoniot shock, it is clear that the jump relations must be different. For weak shocks, however, the differences are small and the economies afforded by the potential formulation make computations based on this approximation attractive for many transonic problems.

Perturbation techniques can be used to demonstrate that the effect of entropy jump across the shocks is more important than the rotationality introduced by these weak shocks. Thus, it makes sense to develop techniques that allow for the entropy jump, but are still within the potential formulation. Such techniques have been developed (see, e.g., Hafez [1985]), but have been relatively little used, as developments in techniques to solve the Euler equations have overtaken these approaches.

The nonlinear difference equations resulting from discrete approximations to the potential equation generally are solved using iterative, or relaxation, techniques. The equations are linearized by computing approximations to all but the highest (second) derivatives from the preceding solution in an iterative sequence, and a correction is computed at each mesh point in such a way that the equations are identically satisfied. It is useful in developing these iterative techniques to think of the iterative process as a discrete approximation to a continuous time-dependent process [Garabedian 1956]. Thus, the iterative process approximates an equation of the form

$$\beta_0 \frac{\partial \Phi}{\partial t} + \beta_1 \frac{\partial^2 \Phi}{\partial \xi \, \partial t} + \beta_2 \frac{\partial^2 \Phi}{\partial \eta \, \partial t} = \frac{a^2}{\rho} \left\{ \frac{\partial}{\partial \xi} (\rho h U) + \frac{\partial}{\partial \eta} (\rho h V) \right\} \qquad (37.37)$$

The parameters β_0, β_1, and β_2, which are related to the mix of old and updated values of the solution used in the difference equations, can then be chosen to ensure that the time-dependent process converges to a steady state in both subsonic and supersonic regions. The formulation of transonic potential flow problems and their solution is described by Caughey [1982].

Even when the values of the parameters are chosen to provide rapid convergence, many hundreds of iterations may be necessary to achieve convergence to acceptable levels, especially when the mesh is very fine. This slow convergence is a characteristic of virtually all iterative schemes, and is a result of the fact that the representation of the difference equations must be highly local if the scheme is to be computationally efficient. As a result of this locality, the reduction of the low-wave-number component of the error to acceptable levels often requires many iterations. A powerful technique for circumventing this difficulty with the iterative solution of numerical approximations to partial differential equations, called the multigrid technique, has been applied with great success to problems in fluid mechanics.

The multigrid method relies for its success on the fact that after a few cycles of any good iterative technique, the error remaining in the solution is relatively smooth, and can be represented accurately on a coarser grid. Application of the same iterative technique on the coarser grid soon makes the error on this grid smooth as well, and the grid-coarsening process can be repeated until the grid contains only a few cells in each coordinate direction. The corrections that have been computed on all coarser levels are then interpolated back to the finest grid, and the process is repeated. The accuracy of the converged solution on the fine grid is determined by the accuracy of the approximation on that grid, since the coarser levels are used only to effect a more rapid convergence of the iterative process.

A particularly efficient multigrid technique for steady transonic potential-flow problems has been developed by Jameson [1979]. It uses a generalized alternating-direction implicit smoothing algorithm in conjunction with a full-approximation-scheme multigrid algorithm. In theory, for a wide class of problems, the work (per mesh point) required to solve the equations using a multigrid approach is independent of the number of mesh cells. In many practical calculations, 10 or fewer multigrid cycles may be required even on very fine grids.

Euler-Equation Methods

As noted earlier, the Euler equations constitute a hyperbolic system of partial differential equations, and numerical methods for solving them rely heavily upon the rather well-developed mathematical theory of such systems. As for the nonlinear potential equation, discontinuous solutions corresponding to flows with shock waves play an important role. Shock-capturing methods are much more widely used than shock-fitting methods, and for these methods it is important to use schemes that are conservative.

As mentioned earler, it is necessary to add artificial, or numerical, dissipation to stabilize the Euler equations. This can be done by adding dissipative terms explicitly to an otherwise nondissipative central difference scheme or by introducing upwind approximations in the flux evaluations. Both approaches are highly developed. The most widely used central difference methods are those modeled after the approach of Jameson et al. [1981]. This approach introduces an adaptive blend of second and fourth differences of the solution in each coordinate direction; a local switching function, usually based on a second difference of the pressure, is used to reduce the order of accuracy of the approximation to first order locally in the vicinity of shock waves and to turn off the fourth differences there. The second-order terms are small in smooth regions where the fourth-difference terms are sufficient to stabilize the solution and ensure convergence to the steady state. More recently, Jameson [1992] has developed improved symmetric limited positive (SLIP) and upstream limited positive (USLIP) versions of these blended schemes (see also Tatsumi et al. [1995]).

Much effort has been directed toward developing numerical approximations for the Euler equations that capture discontinuous solutions as sharply as possible without overshoots in the vicinity of the discontinuity. For purposes of exposition of these methods, we consider the one-dimensional form of the Euler equations, which can be written

$$\frac{\partial \mathbf{w}}{\partial t} + \frac{\partial \mathbf{f}}{\partial x} = 0 \tag{37.38}$$

where $\mathbf{w} = \{\rho,\ \rho u,\ e\}^T$ and $\mathbf{f} = \{\rho u,\ \rho u^2 + p,\ (e + p)u\}^T$. Not only is the exposition clearer for the one-dimensional form of the equations, but the implementation of these schemes for multidimensional problems also generally is done by dimensional splitting in which one-dimensional operators are used to treat variations in each of the mesh directions.

For smooth solutions, Eq. (37.38) are equivalent to the quasilinear form

$$\frac{\partial \mathbf{w}}{\partial t} + \mathbf{A}\frac{\partial \mathbf{w}}{\partial x} = 0 \tag{37.39}$$

where $\mathbf{A} = \{\partial \mathbf{f}/\partial \mathbf{w}\}$ is the Jacobian of the flux vector with respect to the solution vector. The eigenvalues of $\mathbf{A}$ are given by $\lambda = u,\ u + a,\ u - a$, where $a = \sqrt{\gamma p/\rho}$ is the speed of sound. Thus, for subsonic flows, one of the eigenvalues will have a different sign than the other two. For example, if $0 < u < a$, then $u - a < 0 < u < u + a$. The fact that various eigenvalues have different signs in subsonic flows means that simple one-sided difference methods cannot be stable. One way around this difficulty is to split the flux vector into two parts, the Jacobian of each of which has eigenvalues of only one sign. Such an approach has been developed by Steger and Warming [1981]. They used a relatively simple splitting that has discontinuous derivatives whenever an eigenvalue changes sign; an improved splitting has been developed by van Leer [1982] that has smooth derivatives at the transition points.

Each of the characteristic speeds can be identified with the propagation of a wave. If a mesh surface is considered to represent a discontinuity between two constant states, these waves constitute the solution to a Riemann (or shock-tube) problem. A scheme developed by Godunov [1959] assumes the solution to be piecewise constant over each mesh cell, and uses the fact that the solution to the Riemann problem can be given in terms of the solution of algebraic (but nonlinear) equations to advance the solution in time. Because of the assumption of piecewise constancy of the solution, Godunov's scheme is only first-order accurate. Van Leer [1979] has shown how it is possible to extend these ideas to a second-order monotonicity-preserving scheme using the so-called monotone upwind scheme for systems of conservation laws (MUSCL) formulation. The efficiency of schemes requiring the solution of Riemann problems at each cell interface for each time step can be improved by the use of approximate solutions to the Riemann problem [Roe 1986].

More recent ideas to control oscillation of the solution in the vicinity of shock waves include the concept of total-variation-diminishing (TVD) schemes, first introduced by Harten (see, e.g., Harten [1983, 1984]), and essentially nonoscillatory (ENO) schemes, introduced by Osher and his coworkers (see, e.g., Harten et al. [1987] and Shu and Osher [1988]).

Hyperbolic systems describe the evolution in time of physical systems undergoing unsteady processes governed by the propagation of waves. This feature frequently is used in fluid mechanics, even when the flow to be studied is steady. In this case, the unsteady equations are solved for long enough times that the steady state is reached asymptotically—often to within roundoff error. To maintain the hyperbolic character of the equations, and to keep the numerical method consistent with the physics it is trying to predict, it is necessary to determine the solution at a number of intermediate time levels between the initial state and the final steady state. Such a sequential process is said to be a time marching of the solution.

The simplest practical methods for solving hyperbolic systems are **explicit** in time. The size of the time step that can be used to solve hyperbolic systems using an explicit method is limited by a constraint known as the Courant–Friedrichs–Lewy or **CFL condition**. Broadly interpreted, the CFL condition states that the time step must be smaller than the time required for a signal to propagate across a single mesh cell. Thus, if the mesh is very fine, the allowable time step also must be very small, with the result that many time steps must be taken to reach an asymptotic steady state.

Multistage, or Runge–Kutta, methods have become extremely popular for use as explicit time-stepping schemes. After discretization of the spatial operators, using finite-difference, finite-volume, or finite-element approximations, the Euler equations can be written in the form

$$\frac{d\mathbf{w}_i}{dt} + Q(\mathbf{w}_i) + D(\mathbf{w}_i) = 0 \tag{37.40}$$

where $\mathbf{w}_i$ represents the solution at the ith mesh point, or in the ith mesh cell, and Q and D are discrete operators representing the contributions of the Euler fluxes and numerical dissipation, respectively. An m-stage time integration scheme for these equations can be written in the form

$$\mathbf{w}_i^{(k)} = \mathbf{w}_i^{(0)} - \alpha_k \, \Delta t \left[Q\big(\mathbf{w}_i^{(k-1)}\big) + D\big(\mathbf{w}_i^{(k-1)}\big)\right], \quad k = 1, 2, \ldots, m \tag{37.41}$$

with $\mathbf{w}_i^{(0)} = \mathbf{w}_i^n$, $\mathbf{w}_i^{(m)} = \mathbf{w}_i^{n+1}$, and $\alpha_m = 1$. The dissipative and dispersive properties of the scheme can be tailored by the sequence of α_i chosen; note that, for nonlinear problems, this method may be only first-order accurate in time, but this is not necessarily a disadvantage if one is interested only in the steady-state solution. The principal advantage of this formulation, relative to versions that may have better time accuracy, is that only two levels of storage are required regardless of the number of stages used. This approach was first introduced for problems in fluid mechanics by Graves and Johnson [1978], and has been further developed by Jameson et al. [1981]. In particular, Jameson and his group have shown how to tailor the stage coefficients so that the method is an effective smoothing algorithm for use with multigrid (see, e.g., Jameson and Baker [1983]).

Implicit techniques also are highly developed, especially when structured grids are used. Approximate factorization of the implicit operator usually is required to reduce the computational burden of solving a system of linear equations for each time step. Methods based on alternating-direction implicit (ADI) techniques date back to the pioneering work of Briley and McDonald [1974] and Beam and Warming [1976]. An efficient diagonalized ADI method has been developed within the context of the multigrid method by Caughey [1987], and a lower-upper symmetric Gauss–Seidel method has been developed by Yoon and Kwak [1994]. The multigrid implementations of these methods are based on the work of Jameson [1983].

Navier–Stokes Equation Methods

As described earlier, the relative importance of viscous effects is characterized by the value of the Reynolds number. If the Reynolds number is not too large, the flow remains smooth, and adjacent layers (or laminae) of fluid slide smoothly past one another. When this is the case, the solution of the Navier–Stokes equations is not too much more difficult than solution of the Euler equations. Greater resolution is required to resolve the large gradients in the boundary layers near solid boundaries, especially as the Reynolds number becomes large, so more mesh cells are required to achieve acceptable accuracy. In most of the flowfield, however, the flow behaves as if it were nearly inviscid, so methods developed for the Euler equations are appropriate and effective. The equations must, of course, be modified to include the additional terms resulting from the viscous stresses, and care must be taken to ensure that any artificial dissipative effects are small relative to the physical viscous dissipation in regions where the latter is important. The solution of the Navier–Stokes equations for laminar flows, then, is somewhat more costly in terms of computer resources, but not significantly more difficult from an algorithmic point of view, than solution of the Euler equations. Unfortunately, most flows of engineering interest occur at large enough Reynolds numbers that the flow in the boundary layers near solid boundaries becomes turbulent.

Turbulence Models

Solution of the Reynolds-averaged Navier–Stokes equations requires modeling of the Reynolds stress terms. The simplest models, based on the original mixing-length hypothesis of Prandtl, relate the Reynolds stresses to the local properties of the mean flow. The Baldwin–Lomax model [Baldwin and Lomax 1978] is the most widely used model of this type and gives good correlation with experimental measurements for wall-bounded shear layers so long as there are no significant regions of separated flow.

The most complete commonly used turbulence models include two additional partial differential equations that determine characteristic length and time scales for the turbulence. The most widely used of these techniques is called the k–ε model, since it is based on equations for the turbulence kinetic energy (usually given the symbol k) and the turbulence dissipation rate (usually given the symbol ε). This method has grown out of work by Launder and Spaulding [1972]. Another variant, based on a formulation by Kolmogorov, calculates a turbulence frequency ω instead of the dissipation rate (and hence is called a k–ω model). More complete models that compute all elements of the Reynolds stress tensor are in an active state of development. The common base for most of these models is the work of Launder et al. [1975], with more recent contributions by Lumley [1978], Speziale [1987], and Reynolds [1987]. These models, and their implementation within the context of CFD, are described in the book by Wilcox [1993].

More limited models, based on a single equation for a turbulence scale, also have been developed. These include the models of Baldwin and Barth [1991] and of Spalart and Allmaras [1992]. These models are applicable principally to boundary-layer flows of aerodynamic interest.

Large-Eddy Simulations

The difficulty of developing generally applicable phenomenological turbulence models on the one hand, and the enormous computational resources required to resolve all scales in turbulent flows at large Reynolds number on the other, have led to the development of large-eddy simulation (LES) techniques. In this approach, the largest length and time scales of the turbulent motions are resolved, but the smaller (subgrid) scales are modeled. This is an attractive approach because it is the largest scales that contain the preponderance of turbulent kinetic energy and that are responsible for most of the mixing. At the same time, the smaller scales are believed to be more nearly isotropic and independent of the larger scales, and thus are less likely to behave in problem-specific ways—i.e., it should be easier to develop universal models for these smaller scales.

A filtering technique is applied to the Navier–Stokes equations which results in equations having a form similar to the original equations, but with additional terms representing a subgrid-scale tensor that describes the effect of the modeled terms on the larger scales. The solution of these equations is not well posed if there is no initial knowledge of the subgrid scales; the correct statistics are predicted for the flow, but it is impossible to reproduce a specific realization of the flow without this detailed initial condition. Fortunately, for most engineering problems, it is only the statistics that are of interest.

LES techniques date back to the pioneering work of Smagorinsky [1963], who developed an eddy-viscosity subgrid model for use in meteorological problems. The model turned out to be too dissipative for large-scale meteorological problems in which large-scale, predominantly two-dimensional motions are affected by three-dimensional turbulence. Smagorinsky's model finds wide application in engineering problems, however. Details of the LES approach are discussed by Rogallo and Moin [1984], and more recent developments and applications are described by Lesieur and Métais [1996].

Direct Numerical Simulations

Direct numerical simulations generally use spectral or pseudospectral approximations for the spatial discretization of the Navier–Stokes equations (see, e.g., Gottlieb and Orszag [1977] or Hussaini and Zang [1987]). The difference between spectral and pseudospectral methods is in the way that products are computed; the advantage of spectral methods, which are more expensive computationally, is that aliasing errors are removed exactly [Orszag 1972].

A description of the issues involved and some results are given by Rogallo and Moin [1984]. Direct numerical simulations are particularly valuable for the insight that they provide into the fundamental nature of turbulent flows. The computational resources required for DNS calculations grow so rapidly with increasing Reynolds number that they are unlikely to be directly useful for engineering predictions, but they will remain an invaluable tool providing insight needed to construct better turbulence models for use in LES and with the Reynolds-averaged equations.

37.4 Research Issues and Summary

Computational fluid dynamics continues to be a field of intense research activity. The development of accurate algorithms based on unstructured grids for problems involving complex geometries, and the increasing application of CFD techniques to unsteady problems, including aeroelasticity and acoustics, are examples of areas of great current interest. Algorithmically, there continues to be fruitful work on the incorporation of adaptive grids that automatically increase resolution in regions where it is required to maintain accuracy, and on the development of inherently multidimensional high-resolution schemes. Finally, the continued expansion of computational capability allows the application of DNS and LES methods to problems of higher Reynolds number and increasingly realistic flow geometries.

Defining Terms

Artificial viscosity: Terms added to a numerical approximation that provide artificial—i.e., nonphysical—dissipative mechanisms to stabilize a solution or to allow shock waves to be captured automatically by the numerical scheme.

CFL condition: The Courant–Friedrichs–Lewy (CFL) condition is a stability criterion that limits the time step of an explicit time-marching scheme for hyperbolic systems of differential equations. In the simplest one-dimensional case, if Δx is the physical mesh spacing and $\rho(\mathbf{A})$ is the spectral radius of the Jacobian matrix $\mathbf{A}$ corresponding to the fastest wave speed for the problem, then the time step for explicit schemes must satisfy the constraint $\Delta t \leq K \, \Delta x / \rho(\mathbf{A})$, where K is a constant. For the simplest explicit schemes, $K = 1$, which implies that the time step must be less than that required for the fastest wave to cross the cell.

Conservative: A numerical approximation is said to be conservative if it is based on the conservation (or divergence) form of the differential equations, and the net flux across a cell interface is the same when computed from either direction; in this way the properties of discontinuous solutions are guaranteed to be consistent with the jump relations for the original integral form of the conservation laws.

Direct numerical simulation (DNS):   A solution of the complete Navier–Stokes equations in which all length and time scales, down to those describing the smallest eddies containing significant turbulent kinetic energy, are fully resolved.

Euler equations: The equations describing the inviscid flow of a compressible fluid. These equations constitute a hyperbolic system of differential equations; the Euler equations are nondissipative, and weak solutions containing discontinuities (which can be viewed as approximations to shock waves) must be allowed for many practical problems.

Explicit method: A method in which the solution at each point for the new time level is given explicitly in terms of values of the solution at the previous time level; to be contrasted with an *implicit method*, in which the solution at each point at the new time level also depends on the solution at one or more neighboring points at the new time level, so that an algebraic system of equations must be solved at each time step.

Finite-difference method: A numerical method in which the solution is computed at a finite number of points in the domain; the solution is determined from equations that relate the solution at each point to its values at selected neighboring points.

Finite-element method: A numerical method for solving partial differential equations in which the solution is approximated by simple functions within each of a number of small elements into which the domain has been divided.

Finite-volume method: A numerical method for solving partial differential equations, especially those arising from systems of conservation laws, in which the rate of change of quantities within each mesh volume is related to fluxes across the boundaries of the volume.

Implicit method: See **explicit method**.

Large-eddy simulation: A numerical solution of the Navier–Stokes equations in which the largest, energy-carrying eddies are completely resolved, but the effects of the smaller eddies, which are more nearly isotropic, are accounted for by a subgrid model.

Mach number: The ratio $\mathbf{M} = V/a$ of the fluid velocity V to the speed of sound a. This nondimensional parameter characterizes the importance of compressibility in the dynamics of the fluid motion.

Mesh generation: The generation of mesh systems suitable for the accurate representation of solutions to partial differential equations.

Panel method: A numerical method to solve Laplace's equation for the velocity potential of a fluid flow. The boundary of the flow domain is discretized into a set of nonoverlapping facets, or panels, on each of which the strength of some elementary solution is assumed constant. Equations for the normal velocity at control points on each panel can be solved for the unknown singularity strengths to give the solution. In some disciplines this approach is called the *boundary integral element method* (BIEM).

Reynolds-averaged Navier–Stokes equations: Equations for the mean quantities in a turbulent flow obtained by decomposing the fields into mean and fluctuating components and averaging the Navier–Stokes equations. Solution of these equations for the mean properties of the flow requires knowledge of various correlations (the Reynolds stresses), of the fluctuating components.

Shock capturing: A numerical method in which shock waves are treated by smearing them out with artificial dissipative terms in a manner such that no special treatment is required in the vicinity of the shocks; to be contrasted with *shock fitting* methods in which shock waves are treated as discontinuities with the jump conditions explicitly enforced across them.

Shock fitting: See **shock capturing**.

Shock wave: Region in a compressible flow across which the flow properties change almost discontinuously; unless the density of the fluid is extremely small, the shock region is so thin relative to other significant dimensions in most practical problems that it is a good approximation to represent the shock as a surface of discontinuity.

Turbulent flow: Flow in which unsteady fluctuations play a major role in determining the effective mean stresses in the field; regions in which turbulent fluctuations are important inevitably appear in fluid flow at large Reynolds numbers.

Upwind method: A numerical method for CFD in which upwinding of the difference stencil is used to introduce dissipation into the approximation, thus stabilizing the scheme. This is a popular mechanism for the Euler equations, which have no natural dissipation, but is also used for Navier–Stokes algorithms, especially those designed to be used at high Reynolds number.

Visualization: The use of computer graphics to display features of solutions to CFD problems.

References

Abid, R., Vatsa, V. N., Johnson, D. A., and Wedan, B. W. 1990. Prediction of separated transonic wing flows with non-equilibrium algebraic turbulence models. *AIAA J.* 28:1426–1431.

Baldwin, B. S. and Barth, T. J. 1991. A one-equation turbulence transport model for high Reynolds number wall-bounded flows. *AIAA Paper* 91-0610, 29th Aerospace Sciences Meeting, Reno, NV.

Baldwin, B. S. and Lomax, H. 1978. Thin layer approximation and algebraic model for separated turbulent flows. *AIAA Paper* 78-257, 16th Aerospace Sciences Meeting, Huntsville, AL.

Beam, R. M. and Warming, R. F. 1976. An implicit finite-difference algorithm for hyperbolic systems in conservation law form. *J. Comput. Phys.* 22:87–110.

Briley, W. R. and McDonald, H. 1974. Solution of the three-dimensional compressible Navier–Stokes equations by an implicit technique. In *Lecture Notes in Physics*, Vol. 35, pp. 105–110. Springer–Verlag, New York.

Caughey, D. A. 1982. The computation of transonic potential flows. *Ann. Rev. Fluid Mech.* 14:261–283.

Caughey, D. A. 1987. Diagonal implicit multigrid solution of the Euler equations. *AIAA J.* 26:841–851.

Chapman, D. R. 1979. Computational aerodynamics: review and outlook. *AIAA J.* 17:1293–1313.

Eiseman, P. R. 1985. Grid generation for fluid mechanics computations. *Ann. Rev. Fluid Mech.* 17:487–522.

Emmons, H. W. 1970. Critique of numerical modeling of fluid-mechanics phenomena. *Ann. Rev. Fluid Mech.* 2:15–36.

Garabedian, P. R. 1956. Estimation of the relaxation factor for small mesh size. *Math. Tables Aids Comput.* 10:183–185.

Glowinski, R. and Pironneau, O. 1992. Finite element methods for Navier–Stokes equations. *Ann. Rev. Fluid Mech.* 24:167–204.

Godunov, S. K. 1959. A finite-difference method for the numerical computation of discontinuous solutions of the equations of fluid dynamics. *Mat. Sb.* 47:357–393.

Goldstine, H. H. and von Neumann, J. 1963. On the principles of large scale computing machines. In *John von Neumann, Collected Works*, A. H. Taub, ed. Vol. 5, p. 4. Pergamon Press, New York.

Gottleib, D. and Orszag, S. A. 1977. *Numerical Analysis of Spectral Methods: Theory and Application*, CBMS-NSF Reg. Conf. Ser. Appl. Math. 26. SIAM, Philadelphia.

Graves, R. A. and Johnson, N. E. 1978. Navier–Stokes solutions using Stetter's method. *AIAA J.* 16:1013–1015.

Hafez, M. M. 1985. Numerical algorithms for transonic, inviscid flow calculations. In *Advances in Computational Transonics*, W. G. Habashi, ed., pp. 23–58. Pineridge Press, Swansea.

Harten, A. 1983. High resolution schemes for hyperbolic conservation laws. *J. Comput. Phys.* 49:357–393.

Harten, A. 1984. On a class of total-variation stable finite-difference schemes. *SIAM J. Numer. Anal.* 21:1–23.

Harten, A., Engquist, B., Osher, S., and Chakravarthy, S. 1987. Uniformly high order accurate, essentially non-oscillatory schemes III. *J. Comput. Phys.* 71:231–323.

Hess, J. L. 1990. Panel methods in computational fluid dynamics. *Ann. Rev. Fluid Mech.* 22:255–274.

Hussaini, M. Y. and Zang, T. A. 1987. Spectral methods in fluid dynamics. *Ann. Rev. Fluid Mech.* 19:339–367.

Jameson, A. 1979. A multi-grid scheme for transonic potential calculations on arbitrary grids, pp. 122–146. In *Proc. AIAA 4th Comput. Fluid Dynamics Conf.* Williamsburg, VA.

Jameson, A. 1983. Solution of the Euler Equations by a Multigrid Method. *MAE Rep.* 1613, Department of Mechanical and Aerospace Engineering, Princeton University.

Jameson, A. 1992. *Computational Algorithms for Aerodynamic Analysis and Design. Tech. Rep. INRIA 25th Anniversary Conference on Computer Science and Control,* Paris. Princeton University Rep. MAE 1966, Princeton, NJ, Dec.

Jameson, A. and Baker, T. J. 1983. Solution of the Euler equations for complex configurations, pp. 293–302. In *Proc. AIAA Comput. Fluid Dynamics Conf.*

Jameson, A., Schmidt, W., and Turkel, E. 1981. Numerical solution of the Euler equations by finite volume methods using Runge–Kutta time stepping schemes. *AIAA Paper* 81-1259, AIAA Fluid and Plasma Dynamics Conf., Palo Alto, CA.

Launder, B. E., Reese, G. J., and Rodi, W. 1975. Progress in the development of a Reynolds-stress turbulence closure. *J. Fluid Mech.* 68(3):537–566.

Launder, B. E. and Spaulding, D. B. 1972. *Mathematical Models of Turbulence.* Academic Press, London.

Lesieur, M. and Métais, O. 1996. New trends in large-eddy simulations of turbulence. *Ann. Rev. Fluid Mech.* 28:45–82.

Lumley, J. L. 1978. Computational modeling of turbulent flows. *Adv. Appl. Mech.* 18:123–176.

Orszag, S. A. 1972. Comparison of pseudo-spectral and spectral approximation. *Stud. Appl. Math.* 51:253–259.

Reynolds, W. C. 1987. Fundamentals of turbulence for turbulence modeling and simulation, pp. 1–66. In *Lecture Notes for von Karman Institute,* AGARD Lecture Series No. 86, NATO, New York.

Roe, P. L. 1986. Characteristic-based schemes for the Euler equations. *Ann. Rev. Fluid Mech.* 18:337–365.

Rogallo, R. S. and Moin, P. 1984. Numerical simulation of turbulent flows. *Ann. Rev. Fluid. Mech.* 16:99–137.

Shu, C. and Osher, S. 1988. Efficient implementation of essentially non-oscillatory shock-capturing schemes. *J. Comp. Phys.* 77:439–471.

Smagorinsky, J. 1963. General circulation experiments with the primitive equations. *Mon. Weather Rev.* 91:99–164.

Spalart, P. R. and Allmaras, S. R. 1992. A one-equation turbulence model for aerodynamic flows. *AIAA Paper* 92-0439, 30th Aerospace Sciences Meeting, Reno, NV.

Speziale, C. G. 1987. On nonlinear k–ℓ and k–ε models of turbulence. *J. Fluid Mech.* 178:459–475.

Steger, J. L. and Warming, R. F. 1981. Flux vector splitting of the inviscid gasdynamic equations with application to finite-difference methods. *J. Comput. Phys.* 40:263–293.

Tatsumi, S., Martinelli, L., and Jameson, A. 1995. Flux-limited schemes for the compressible Navier–Stokes equations. *AIAA J.* 33:252–261.

Thompson, J. F., Warsi, Z. U. A., and Mastin, C. W. 1985. *Numerical Grid Generation.* North Holland, New York.

van Leer, B. 1974. Towards the ultimate conservative difference scheme, II. Monotonicity and conservation combined in a second-order accurate scheme. *J. Comput. Phys.* 14:361–376.

van Leer, B. 1979. Towards the ultimate conservative difference scheme, V. A second-order sequel to Godunov's scheme. *J. Comput. Phys.* 32:101–136.

van Leer, B. 1982. Flux-vector splitting for the Euler equations. In *Lecture Notes in Phys.* 170:507–512.

Wilcox, D. C. 1993. *Turbulence Modeling for CFD.* DCW Industries, La Cañada, CA.

Yoon, S. and Kwak, D. 1994. Multigrid convergence of an LU scheme, pp. 319–338. In *Frontiers of Computational Fluid Dynamics—1994,* D. A. Caughey and M. M. Hafez, eds. Wiley-Interscience, Chichester, U.K.

Further Information

Several organizations sponsor regular conferences devoted completely, or in large part, to computational fluid dynamics. The American Institute of Aeronautics and Astronautics (AIAA) sponsors the AIAA Computational Fluid Dynamics Conferences in odd-numbered years, usually in July; the proceedings of this conference are published by AIAA. In addition, there typically are many sessions on CFD and its applications at the AIAA Aerospace Sciences Meeting, held every January, and the AIAA Fluid and Plasma Dynamics Conference, which is held every summer, in conjunction with the AIAA CFD Conference in those years when the latter is held. The Fluids Engineering Conference of the American Society of Mechanical Engineers, held every summer, also contains sessions devoted to CFD. In even-numbered years, the International Conference on Numerical Methods in Fluid Dynamics is held, alternating between Europe and America; the proceedings of this conference are published in the *Lecture Notes in Physics* series by Springer-Verlag. The International Symposium on Computational Fluid Dynamics, sponsored by the CFD Society of Japan, is held in odd-numbered years, alternating between the U.S. and Asia; in September 1997 this meeting will be held in Beijing, China.

The *Journal of Computational Physics* contains many articles on CFD, especially covering algorithmic issues. The *AIAA Journal* also has many articles on CFD, including aerospace applications. The *International Journal for Numerical Methods in Fluids* contains articles emphasizing the finite-element method applied to problems in fluid mechanics. The journals *Computers and Fluids* and *Theoretical and Computational Fluid Dynamics* are devoted exclusively to CFD, the latter journal emphasizing the use of CFD to elucidate basic fluid-mechanical phenomena. The first issue of the *CFD Review*, which attempts to review important developments in CFD, was published in 1995. The *Annual Review of Fluid Mechanics* also contains a number of review articles on topics in CFD.

The following textbooks provide excellent coverage of many aspects of CFD:

Anderson, D. A., Tannehill, J. C., and Pletcher, R. H. 1984. *Computational Fluid Mechanics and Heat Transfer*. Hemisphere, Washington.

Hirsch, C. 1988, 1990. *Numerical Computation of Internal and External Flows, Vol. I: Fundamentals of Numerical Discretization, Vol. II: Computational Methods for Inviscid and Viscous Flows*. Wiley, Chichester.

Wilcox, D. C. 1993. *Turbulence Modeling for CFD*. DCW Industries, La Cañada, CA.

An up-to-date summary on algorithms and applications for high-Reynolds-number aerodynamics is found in:

Caughey, D. A. and Hafez, M. M., eds. *Frontiers of Computational Fluid Dynamics—1994*. J. Wiley Chichester, U.K.

38

Computational Reacting Flow

Pasquale Cinnella
Mississippi State University

Carey F. Cox
Mississippi State University

38.1 Introduction

Numerical simulations of fluid flows dominated by chemical reactions and other nonequilibrium phenomena are a recent addition to the field of computational science. In essence, a double challenge awaits the scientist trying to understand reacting flows: fluid-dynamic phenomena and chemistry are intimately coupled and have to be resolved at the same time in order for useful predictions of flow behavior to be made. A cursory examination of the problems facing practitioners of Computational Fluid Dynamics (CFD), described in Chapter 37, shows that even computations based on an ideal-gas model can require significant human and hardware resources. Taking chemistry and nonequilibrium phenomena into consideration can easily result in a manifold increase of those resource requirements.

The path from computations based on the ideal-gas model to reacting flow models (and to **thermochemical nonequilibrium** models, in some cases) might seem deceivingly simple. In summary, detailed species mass balances replace a global mass balance, and more detailed energy balances are needed. However, both the flow physics and the computational requirements are enormously affected by the inclusion of nonequilibrium phenomena. For example, combustion or dissociation reactions alter significantly the properties of the fluid, especially its energy balance. Moreover, the consideration of a large number of chemical species in place of one global mass balance results in computational requirements that typically scale with the *cube* of the number of species involved. The physical complications result in some uncertainty about how to extend to reacting flows the algorithms derived for the simulation of an ideal gas. The increase in computational requirements forces the scientist into tradeoffs between physical complexity and geometric resolution. These cautionary notes notwithstanding, the simulation of reacting flows has made very impressive strides in recent years.

0-8493-2909-4/97/$0.00+$.50
© 1997 by CRC Press, Inc.

The most common applications of reacting-flow simulations are in the areas of combustion and hypersonics. In the former, laminar or turbulent mixing, ignition delays, the possibility of detonation, and the conditions that make the flame diffusion-dominated or kinetics-dominated are examples of problems that require accurate physical modeling and reliable algorithms in order to be properly studied. In the latter, high temperatures behind strong shock waves and strong gradients in the boundary layer result in nonequilibrium phenomena ranging from dissociation to ionization to surface ablation. Other areas where simulations of nonequilibrium flows play an important role include lasers, studying the structure of shock waves, and biochemical problems such as the growth of biofilms in piping systems or even human arteries.

In the following, the basic principles that govern nonequilibrium flows are reviewed, and the most widely accepted practices for their computation are discussed. An example of the differences that can be expected between ideal-gas behavior and reactive flows is given as a case study, and the most important current research issues are mentioned. Finally, the reader is referred to the additional sources listed for more details on specific subjects.

38.2 Underlying Principles

A gas mixture whose temperature is high enough to allow for the onset and/or sustenance of chemical reactions will typically deviate from the ideal-gas behavior. Internal energy modes which behave nonlinearly with temperature will be activated, and, in extreme cases encountered in hypersonic and laser applications, local thermal nonequilibrium will be established.

Internal Energy

In recent years, a great deal of effort has been spent on physically accurate modeling of *real-gas effects*, which is the somewhat misleading terminology used to indicate departures from the low-temperature ideal-gas laws. A useful review of the current literature is presented by Gnoffo et al. [1989].

A general representation for the species internal energy per unit mass, e_s, suitable for the modeling of problems where thermal nonequilibrium plays a key role, is the following:

$$e_s = \tilde{e}_s(T) + e_{n_s}\left(T_{n_s}\right) \tag{38.1}$$

In the above, $\tilde{e}_s(T)$ is the contribution due to translation and those internal modes which can be assumed to be in equilibrium at the translational temperature T. Moreover, $e_{n_s}(T_{n_s})$ is the contribution due to internal modes that are in thermal nonequilibrium, meaning that they are not in equilibrium at the translational temperature T but may be assumed to satisfy a **Boltzmann distribution** at a different temperature T_{n_s}.

Equation (38.1) is valid for a very broad range of physical conditions and is susceptible to simplification and reduction to simpler models when the flow regime is less extreme. This formulation is very general, and it allows the utilization of many practical thermodynamic models for the determination of the specific functional form of internal energy, specific heats, and related properties. Examples of specific models currently used can be found in the references at the end of the chapter.

In the following, a gas mixture composed of N species will be considered, and two major representations of the thermodynamic state of the mixture will be studied. The first one is a *nonequilibrium model*, wherein the first M species contain a nonequilibrium contribution e_{n_s}. A simplification of the previous thermodynamic model may be achieved for flows where there is enough time for the internal modes to reach equilibrium at the translational temperature T. This *equilibrium model* will feature only one contribution to the internal energy in Eq. (38.1), namely $\tilde{e}_s$. The usual ideal-gas model is a special case of the equilibrium model, when vibrational and electronic modes are neglected, and translational and fully excited rotational contributions are included in $\tilde{e}_s$.

The mixture value for the internal energy per unit mass, e, may be written using a standard mass-fraction-averaged summation, as follows:

$$e = \sum_{s=1}^{N} Y_s e_s = \sum_{s=1}^{N} Y_s \tilde{e}_s + \sum_{s=1}^{M} Y_s e_{n_s} \tag{38.2}$$

where the species mass fraction Y_s has been introduced, defined as the ratio of the species density ρ_s to the mixture density $\rho = \sum_{s}^{N} \rho_s$.

A very important subcase of the equilibrium model is *local chemical equilibrium*, wherein it is assumed that the kinetic rates in the flowfield are fast enough to bring the chemical reactions that occur in the system virtually instantaneously to their equilibrium values, dictated by the **laws of mass action**. This assumption has played a significant role in the modeling of real-gas effects, due to the simplification that it brings in the analysis.

Equations of State

Most flows of practical interest will involve low to moderate pressures and densities. In these instances, *Dalton's law* will yield the mixture pressure as the sum of species partial pressures:

$$p = \sum_{s=1}^{N} \rho_s R_s T = \rho R T \tag{38.3}$$

where R_s is the species gas constant, and the mixture gas constant R is defined as follows:

$$R = \sum_{s=1}^{N} Y_s R_s \tag{38.4}$$

Equation (38.3) is also known as the *Thermal Equation of State*.

Unlike the ideal-gas case, the state relationship of the pressure to the specific internal energy cannot, in general, be written directly, but it occurs implicitly through the translational temperature. A *caloric equation of state* such as Eq. (38.2) will relate internal energy, or portions thereof, to temperature. Iterative procedures are necessary for the determination of the temperature, due to the inherently nonlinear character of this equation [Cinnella and Grossman 1991]. Once T is found, the pressure is evaluated from Eq. (38.3).

Chemistry Models

When chemically reacting flows are investigated, a classification of chemical phenomena according to the time available for the completion of reactions is usually employed for order-of-magnitude estimates of the flowfield properties. If reaction times are very large compared with the time scale at which the fluid dynamics is evolving, then reactions have virtually no time to occur, and a *frozen* flow can be assumed. Incidentally, perfect-gas results are a particular class of frozen-chemistry simulations. The other limiting case occurs when the reaction times are very small compared to the fluid-dynamic scales, which results in the reactions having enough time to reach their local chemical equilibrium values, given by the laws of mass action.

The real situation, however, is the general case of *finite-rate chemistry*, wherein in at least portions of the flowfield and/or at some point in the time evolution of the flow, the reaction times are comparable to the fluid-dynamic scales. In this instance, it becomes necessary to simulate the actual kinetic behavior of the chemical system.

Finite-Rate Chemistry

A general simulation of chemical effects can be achieved for a system containing N species where J reactions take place:

$$\nu'_{1,j} X_1 + \nu'_{2,j} X_2 + \cdots + \nu'_{N,j} X_N \rightleftharpoons \nu''_{1,j} X_1 + \nu''_{2,j} X_2 + \cdots + \nu''_{N,j} X_N, \qquad j = 1, \ldots, J \qquad (38.5)$$

where the $\nu'_{s,j}$ and the $\nu''_{s,j}$ are stoichiometric coefficients of species X_s in the jth reaction. Then the rate of production of the sth species, w_s, may be written as a summation over the reactions:

$$w_s \equiv \frac{d\rho_s}{dt} = \mathcal{M}_s \sum_{j=1}^{J} (\nu''_{s,j} - \nu'_{s,j}) \left[k_{f,j} \prod_{l=1}^{N} \left(\frac{\rho_l}{\mathcal{M}_l} \right)^{\nu'_{l,j}} - k_{b,j} \prod_{l=1}^{N} \left(\frac{\rho_l}{\mathcal{M}_l} \right)^{\nu''_{l,j}} \right], \qquad s = 1, \ldots, N$$

$$(38.6)$$

where $\mathcal{M}_s$ is the species molecular mass. For reaction j, the forward and backward reaction rates, $k_{f,j}$ and $k_{b,j}$, are assumed to be known functions of temperature, and they are related by thermodynamics:

$$k_{f,j}(T) = k_{b,j}(T) K_{e,j}(T) \qquad (38.7)$$

where $K_{e,j}$ is the equilibrium constant, which is a known function of the thermodynamic state.

The finite-rate chemistry model described remains valid in conjunction with all of the thermodynamic models discussed. However, the presence of thermal nonequilibrium in the flowfield is likely to exercise some effect on the reaction rates, especially when diatomic or polyatomic molecules are involved, whose vibrational modes may be excited. Several attempts to model the interaction between chemical and thermal nonequilibrium have been recorded. More details are given in the references at the end of the chapter.

Local Chemical Equilibrium

There are many problems that render a finite-rate computation a nontrivial task. The most important one is the uncertainty in the **reaction paths** and in the reaction rates. Although the composition of a given mixture in chemical equilibrium as a function of thermodynamic variables can be determined quite accurately by theoretical and experimental means, there is less knowledge of the actual mechanism with which a given reaction occurs. Moreover, the numerical simulation of flows in chemical nonequilibrium requires the inclusion of N species continuity equations in the algorithms instead of only one global mass conservation equation, as will be discussed in the following. Also, the finite-rate equations become extremely stiff when the flow is close to chemical equilibrium [Oran and Boris 1987]. All of the above reasons render a numerical simulation based on chemical equilibrium very appealing from a computational standpoint when compared with finite-rate calculations. In addition, for a large number of flows of practical interest, results of considerable accuracy can be obtained using this simplifying assumption.

A close examination of the governing equations, including the combined first and second laws of thermodynamics, indicates that local-chemical-equilibrium calculations should be performed at constant density and internal energy [Liu and Vinokur 1989]. The equations to be solved for the determination of the N species densities and the temperature will be: N_e mass conservation equations written for an equivalent number of *elemental species*, $N - N_e$ equations corresponding to a possible formulation of the laws of mass action, and one energy equation, of the type given in Eq. (38.2), specialized for thermal equilibrium. Several methods for the solution of these nonlinear equations have been utilized, and a fairly recent review may be found in [Liu and Vinokur 1989].

Practical Chemistry Models

It was previously mentioned that a detailed knowledge of the kinetic behavior of a chemically reacting gas mixture is necessary for finite-rate chemistry investigations. In recent years, much attention has been paid in the scientific community to the kinetics of air and of hydrogen–air mixtures, spurred by the drive toward hypersonic flight and the necessary propulsive tools to achieve it. A compilation of the most recent data for the kinetics of air may be found in [Park 1991], where attention is given to thermal nonequilibrium and its influence on the reaction rates. Detailed studies of hydrogen–air mixtures have been published recently [Oldenborg et al. 1990]. Moreover, computerized databases are now available, involving several thousand reactions and hundreds of chemical species [Mallard et al. 1993].

When local-chemical-equilibrium calculations are performed, the actual reaction paths are no longer necessary. Chemical equilibrium is dictated by thermodynamics; the entropy has to be a maximum for a system at constant density and internal energy [Liu and Vinokur 1989].

Governing Equations

The governing equations for flows in thermochemical nonequilibrium represent the conservation and/or time variation of physical quantities such as mass of a species in the mixture, momentum, nonequilibrium energy contributions, and energy. They are established either for a control volume of arbitrary size (finite or infinitesimal) or for a generic point in the flowfield. The former approach results in an integrodifferential form, the latter in a nonlinear differential system. In the following, the integral form of the equations will be introduced for consistency with the finite-volume-based numerical techniques to be analyzed later. The differential form of the equations can be readily obtained from the integral form and has been extensively discussed in the literature (e.g., see Lee [1985]).

The governing equations written in integral form are valid both in regions of smooth flows and across discontinuities, where they can be shown to reduce to the Rankine-Hugoniot **jump conditions** in the limit of infinitesimal volumes. The control volume employed for the derivation is allowed to move and deform with time, although many important applications take advantage of the simpler case of fixed control volumes. For an arbitrary volume $\mathcal{V}$, closed by a boundary $\mathcal{S}$, the governing equations in integral form read

$$\frac{\partial}{\partial t} \iiint_{\mathcal{V}} Q \, d\mathcal{V} + \oiint_{\mathcal{S}} (S - S_v) \cdot n \, d\mathcal{S} = \iiint_{\mathcal{V}} W \, d\mathcal{V} \tag{38.8}$$

where Q is the vector of conserved variables, W is the vector of source terms, and S and S_v are the inviscid and viscous flux vectors, respectively. The unit vector n is normal to the infinitesimal area $d\mathcal{S}$ and points outward.

Thermochemical Nonequilibrium Equations

In conjunction with the nonequilibrium model, and utilizing a Cartesian frame of reference (x, y, z) whose unit vectors are i, j, and k, respectively, the vectors Q and W in the previous equation read

$$Q = \begin{pmatrix} \rho_1 \\ \rho_2 \\ \vdots \\ \rho_N \\ \rho u \\ \rho v \\ \rho w \\ \rho_1 e_{n_1} \\ \vdots \\ \rho_M e_{n_M} \\ \rho e_0 \end{pmatrix} \qquad W = \begin{pmatrix} w_1 \\ w_2 \\ \vdots \\ w_N \\ \sum_s \rho_s g_{sx} \\ \sum_s \rho_s g_{sy} \\ \sum_s \rho_s g_{sz} \\ Q_1 \\ \vdots \\ Q_M \\ \left(\sum_s \rho_s g_s \right) \cdot u \end{pmatrix} \tag{38.9}$$

and the flux vectors S, S_v read

$$
S = \begin{pmatrix} \rho_1(u - u_V) \\ \rho_2(u - u_V) \\ \vdots \\ \rho_N(u - u_V) \\ \rho u(u - u_V) + p i \\ \rho v(u - u_V) + p j \\ \rho w(u - u_V) + p k \\ \rho_1 e_{n_1}(u - u_V) \\ \vdots \\ \rho_M e_{n_M}(u - u_V) \\ \rho e_0(u - u_V) + p u \end{pmatrix}, \qquad
S_v = \begin{pmatrix} -\rho_1 v_1 \\ -\rho_2 v_2 \\ \vdots \\ -\rho_N v_N \\ \tau_{xx} i + \tau_{xy} j + \tau_{xz} k \\ \tau_{yx} i + \tau_{yy} j + \tau_{yz} k \\ \tau_{zx} i + \tau_{zy} j + \tau_{zz} k \\ -\rho_1 e_{n_1} v_1 - q_{n1} \\ \vdots \\ -\rho_M e_{n_M} v_M - q_{nM} \\ \Theta \end{pmatrix}
\tag{38.10}
$$

where

$$
\Theta = (\tau_{xx} u + \tau_{yx} v + \tau_{zx} w) i + (\tau_{xy} u + \tau_{yy} v + \tau_{zy} w) j
$$

$$
+ (\tau_{xz} u + \tau_{yz} v + \tau_{zz} w) k - q - q^R - \sum_{s=1}^{M} q_{ns} - \sum_{s=1}^{N} \rho_s h_s v_s
\tag{38.11}
$$

In these formulas, the *mass-averaged velocity* u for the mixture has been utilized, whose Cartesian components are (u, v, w), respectively. This velocity can be defined in terms of species velocities u_s as follows:

$$
\rho u = \sum_{s=1}^{N} \rho_s u_s
\tag{38.12}
$$

The difference between species and mass-averaged velocities is a species *diffusion velocity*

$$
v_s = u_s - u, \qquad s = 1, \ldots, N
\tag{38.13}
$$

The velocity u_V that appears in the inviscid flux vector is the control-surface velocity, which equals zero when the control volume is fixed with time. The first N equations are species continuity equations, relating the time rate of change of species densities ρ_s to convective and diffusive transport and to the creation/destruction of the species due to chemical reactions. The species rates of production w_s have been defined by Eq. (38.6), when finite-rate chemistry models were discussed. Following species continuity are the three components of the momentum equation. The source term $\sum_s \rho_s g_s$ represents body forces, e.g., gravity and electric fields, and the viscous fluxes involve the components of the *shear stress tensor*. After the momentum, M nonequilibrium energy equations are written, describing the creation and evolution in time and space of the nonequilibrium components of the internal energy. The viscous fluxes associated with these equations describe the heat transfer due to the transport of nonequilibrium energy by diffusion, $\rho_s e_{n_s} v_s$, as well as the conductive heat transfer q_{ns}. The source terms for the nonequilibrium energy equations, Q_s, will be described in more detail below. Finally, the global energy equation describes the time evolution of the total internal energy per unit volume $\rho e_0 = \rho e + \rho(u^2 + v^2 + w^2)/2$. The viscous flux in this case represents viscous dissipation; convective heat transfer due to both equilibrium and nonequilibrium portions of the internal energy, q and $\sum_s^M q_{ns}$, respectively; radiative heat transfer, q^R; and diffusive heat transfer, $\sum_s^N \rho_s h_s v_s$, where h_s is the species enthalpy, $h_s = e_s + R_s T$.

In addition to the equations presented, thermal and caloric equations of state will relate pressure and temperatures to equilibrium and nonequilibrium internal energy components, as already discussed, and the viscous flux vector entries will be expressed in terms of the entries in the vector of conserved variables.

Simplifications in the physical model employed for the analysis of nonequilibrium flows will bring about corresponding changes in the governing equations. For instance, the equilibrium model can be handled simply by dropping all of the nonequilibrium energy equations, along with the nonequilibrium contributions to the global energy equation. The whole process is equivalent to setting M equal to zero. Inviscid flows are modeled by dropping the viscous flux vector in conjunction with any of the thermodynamic models discussed.

Local-Chemical-Equilibrium Equations

The previous form of the governing equations is written for finite-rate chemistry problems. When the important special subcase of the equilibrium model that deals with local chemical equilibrium is considered, the vectors that appear in Eq. (38.8) are further simplified. Global continuity replaces the N species continuity equations, the nonequilibrium energy contributions disappear, and the final form of Q and W reads

$$Q = \begin{pmatrix} \rho \\ \rho u \\ \rho v \\ \rho w \\ \rho e_0 \end{pmatrix} \qquad W = \begin{pmatrix} 0 \\ \sum_s \rho_s g_{sx} \\ \sum_s \rho_s g_{sy} \\ \sum_s \rho_s g_{sz} \\ \left(\sum_s \rho_s g_s\right) \cdot u \end{pmatrix} \tag{38.14}$$

where only body forces are present in the vector of source terms W. Inviscid and viscous fluxes read

$$S = \begin{pmatrix} \rho(u - u_V) \\ \rho u(u - u_V) + pi \\ \rho v(u - u_V) + pj \\ \rho w(u - u_V) + pk \\ \rho e_0(u - u_V) + pu \end{pmatrix} \qquad S_v = \begin{pmatrix} 0 \\ \tau_{xx}i + \tau_{xy}j + \tau_{xz}k \\ \tau_{yx}i + \tau_{yy}j + \tau_{yz}k \\ \tau_{zx}i + \tau_{zy}j + \tau_{zz}k \\ \Theta \end{pmatrix} \tag{38.15}$$

where

$$\Theta = (\tau_{xx}u + \tau_{yx}v + \tau_{zx}w)i + (\tau_{xy}u + \tau_{yy}v + \tau_{zy}w)j$$
$$+ (\tau_{xz}u + \tau_{yz}v + \tau_{zz}w)k - q - q^R \tag{38.16}$$

and diffusion has been neglected.

The closure to the mathematical problem of the governing equations for flows in local chemical equilibrium again will be provided by thermal and caloric equations of state. With temperature and species composition known from the equilibrium solver, the previous equations are tantamount to a general equation of state relating pressure to density and internal energy

$$p = p(\rho, e) \tag{38.17}$$

Thermal-Nonequilibrium Models

Several nonequilibrium models have been utilized for the study of hypersonic flow and lasers. The most common ones are reviewed by Gnoffo et al. [1989]. Additional references are listed at the end of the chapter.

By far, the most common source of nonequilibrium behavior is vibrational energy. Two major physical phenomena contribute to the source terms Q_s in Eq. (38.9), when it is assumed that vibrational nonequilibrium is present. The first one is the change in vibrational energy caused by inelastic collisions, that is, creation or destruction of vibrating particles due to chemical reactions; the second one is due to elastic collisions and is tantamount to energy exchanges between the different internal energy modes. The most important contribution to the latter is the relaxation of nonequilibrium vibrational energy toward the equilibrium levels at the local translational temperature. Additional contributions include energy exchanges among the different vibrating species [Lee 1985].

Viscous Fluxes and Transport Properties

The most common assumptions that are made when dealing with **transport phenomena** concern the treatment of the viscous stress tensor. Generally only *Newtonian* fluids are considered, where there is a linear relationship between stress and rate of deformation. Moreover, *bulk viscosity* effects are neglected. Under these assumptions, the stress-tensor components are expressed as

$$\tau_{ij} = \mu \left(\frac{\partial u_i}{\partial x_j} + \frac{\partial u_j}{\partial x_i} \right) - \frac{2}{3} \mu \left(\frac{\partial u_1}{\partial x_1} + \frac{\partial u_2}{\partial x_2} + \frac{\partial u_3}{\partial x_3} \right) \delta_{ij}, \qquad i, j = 1, 2, 3 \qquad (38.18)$$

where μ is the viscosity coefficient and δ_{ij} is the Kronecker delta. The indices i and j refer to the x, y, and z space directions and velocity components.

For mixtures of gases in thermal equilibrium, the heat flux vector q is modeled as the thermal conductivity λ times the temperature gradient, a result known as *Fourier's law*. The extension of this approach to flows in thermal nonequilibrium is usually done by considering similar contributions for the nonequilibrium temperatures. The resulting expressions read

$$q = -\lambda \nabla T, \qquad q_{n_s} = -\lambda_{n_s} \nabla T_{n_s}, \qquad s = 1, \ldots, M \qquad (38.19)$$

Mass diffusion is a physical phenomenon that arises mostly because of the presence of gradients of mass or mole concentrations in the mixture. Assuming that the mixture behaves like a binary mixture yields the so-called *Fick's law of diffusion*,

$$\rho_s v_s = -\rho D_s \nabla \left(\frac{\rho_s}{\rho} \right), \qquad s = 1, \ldots, N \qquad (38.20)$$

where D_s is a diffusion coefficient.

Theoretical formulas for the viscosity coefficients of individual species and mixtures have been given as a result of the asymptotic analysis that leads to the **Navier–Stokes equations**, but they are very cumbersome to use. Curve-fit functional expressions have been proposed, such as the very popular one due to Blottner et al. [1971]. Another very common approach is the extension of Sutherland's law for a perfect gas to a generic component. Values of the viscosity coefficient for the mixture usually are recovered by means of *Wilke's rule* [Wilke 1950], which is an extension of a Sutherland-type equation to multicomponent systems, obtained on the basis of kinetic theory and several simplifying assumptions.

Values of the thermal conductivity to be used in Eq. (38.19) also are obtained from an asymptotic analysis for small deviations from equilibrium. Curve-fit expressions also have been used for thermal conductivity, and a Sutherland-type formula has been proposed, similarly to what has been done for the viscosity coefficient [Gnoffo et al. 1989]. Results for the mixture thermal conductivity can be obtained again by means of Wilke's rule.

The simplest modeling of the diffusion coefficients is obtained by assuming a constant **Schmidt number**, where only one global diffusion coefficient is used. Multicomponent effects can be partially taken into account by more complex choices of diffusion coefficients, relating those with binary diffusion coefficients [Park 1991].

The numerical treatment of transport properties for mixtures of gases out of thermal equilibrium usually consists of fairly simple assumptions, due to the lack of a solid theoretical framework. Diffusion coefficients and viscosity are left unaltered, and the thermal conductivity is modified in a straightforward manner: only the portions of the specific heats that are in thermal equilibrium are included in the equations for λ; the remaining portions contribute to the determination of λ_{n_s}.

The inclusion of turbulence in the viscous flux vectors is obtained in its simplest form by the addition of turbulent viscosity, thermal conductivity, and diffusion coefficients to their laminar counterparts. The turbulent contributions are then modeled by means of algebraic treatments—foremost among which is the model originated by Baldwin and Lomax [1978]—or by introducing and solving additional partial differential equations, for example for turbulent kinetic energy and dissipation. In general, the models for turbulence in the presence of thermochemical nonequilibrium have been straightforward extensions of the ones derived for an ideal gas. Recent departures from this tendency are methods based on statistical techniques such as the *probability distribution function* [Pope 1985].

38.3 Best Practices

In order for a computer simulation to be possible, the governing equations presented in the previous section must be *discretized*, that is, reduced to a set of algebraic equations to be solved. A very convenient discretization approach is the finite-volume technique, whereby the integral form of the governing equations is solved for the unknown volume averages of conserved variables in some small, but finite, control volume. The advantage of this method is its use of the integral form of the equations, which allows a correct treatment of discontinuities and consistently treats general grid topologies. Other discretizations are also possible: for example, spectral techniques and the more usual finite-difference approach [Fletcher 1988].

The Finite-Volume Technique

The finite-volume technique takes advantage of the integral form of the governing equations to achieve a discretized algebraic approximation that can be handled by a digital computer. The physical domain of interest is partitioned into a large set of subdomains, volume-averaged values for the conservative variables Q are defined for each subdomain (computational cell), and the governing equations described in Eq. (38.8) are written for each cell in terms of the volume averages. The surface integral that involves inviscid and viscous fluxes is partitioned into separate contributions from each of the faces that compose the boundary of the computational cell, and each contribution is written in terms of surface-averaged values of the fluxes. Then, the area-averaged values are related to the unknown volume-averaged values, typically by extrapolation. The resulting set of algebraic equations will be the discretized form of the original problem that is actually solved by means of a digital computer.

Extrapolation Strategies

Relating the area-averaged values of inviscid and viscous fluxes to the volume-averaged values of the conserved variables is the key operation of any finite-volume strategy. A generalized extrapolation procedure to recover values at a face starting from quantities at the center of a volume can be formulated for a generic vector q, and this results in two typically different left- and right-extrapolated vectors, q^- and q^+, respectively (which are both needed in general). In order to minimize instabilities and nonmonotonic behavior of the solution, high-order extrapolation formulas will involve limiters. A review of commonly used limiters and their properties is given by Sweby [1984], and general extrapolation formulas are presented by Cinnella and Grossman [1991].

Two different techniques have emerged in recent years for the treatment of the inviscid flux vector. The first one identifies q with the flux vector itself. Consequently, a vector of inviscid fluxes is created from volume-averaged values of the variables in a cell, and, subsequently, face values are created by extrapolation. The second method, called MUSCL extrapolation (monotone upstream-centered schemes for conservation laws), identifies q with the vector of conserved variables, Q. The extrapolation produces face values for the conserved variables, and then the fluxes are constructed from these values. It is useful to point out that for many applications, the extrapolation of *primitive* or even *characteristic* variables is preferred to the extrapolation of conserved variables.

Flux-Split Algorithms: Popular Schemes

In recent years, *flux-split techniques* have obtained wide acceptance as an accurate means of discretizing the inviscid fluxes. Originally developed for a perfect gas, they have been extended to flows in chemical equilibrium and to mixtures out of chemical and thermal equilibrium. They are found to be very accurate and robust when used for transonic, supersonic, and hypersonic flows, and they are fully compatible with conservative finite-volume, shock-capturing approaches.

Probably the two most popular schemes in the flux-vector splitting category are the ones due to Steger and Warming [1981] and to Van Leer [1982]. The basic premise is to split the inviscid flux vector in one space dimension into two parts, containing the information that propagates downstream and upstream, respectively. The two parts are constructed using the extrapolation strategies already outlined consistently with the direction of propagation. Another very popular scheme, less robust for hypersonic flows, but more accurate, is the flux-difference splitting technique due to Roe [1981]. It consists of an approximate Riemann solver, where an arbitrary discontinuity is supposed to exist at the cell surface between the left and the right state, and an approximate solution for this situation is written in terms of waves propagating upstream and downstream. In this instance, the inviscid flux vector is not split but rather is reconstructed from the upstream and downstream contributions that constitute the left and the right states in the **Riemann problem**. A complete summary of the nonequilibrium version of these techniques can be found in [Cinnella and Grossman 1991], and derivations of similar algorithms are presented in the references at the end of the chapter.

Extensions to more space directions usually are made by superimposing pseudo-one-dimensional problems, where the extrapolation formulas to get left and right states keep track of the relevant direction only. In recent times, much effort has been directed toward the design of truly multidimensional algorithms, which would be at least approximately independent of the grid orientation, as will be discussed shortly.

Additional Algorithms

The algorithms presented do not exhaust the category of flux-split techniques. Osher-type formulations exist for perfect gases, and their extension to reacting flows has been attempted. More recently, new kinds of flux splitting have been emerging, wherein the treatment of the pressure gradient plays a central role, and they seem to show promise of improvement over the older schemes. Again, the original development is for perfect gases, although the extension to flows in thermochemical nonequilibrium is fairly straightforward. More details are given in the references at the end of the chapter.

All of the previously mentioned upwind algorithms originate from the **Euler equations** for gas dynamics. The Euler equations, however, can be obtained as moments of the Boltzmann equation, established in the kinetic theory of gases, provided that the velocity distribution function is Maxwellian. This connection between the Boltzmann equation and the Euler equations has led to a new class of upwind *kinetic* schemes which are based on the principle that an upwind scheme at the Boltzmann level leads to an upwind scheme at the Euler level. Eppard and Grossman [1993] have extended the kinetic schemes to flows with chemical and thermal nonequilibrium by means of nonequilibrium kinetic theory.

Although the Roe scheme has been very successful, it does have well-documented failings, such as expansion shocks and the carbuncle phenomenon [Quirk 1992]. In an attempt to address these failings, the positively conservative scheme of Harten, Lax, van Leer, and Einfeldt (HLLE) has been recently introduced [Einfeldt et al. 1991]. The HLLE scheme, although very robust, is also highly dissipative at contact discontinuities and shear layers. Modifications to this algorithm to increase its resolution are in progress (e.g., see Obayashi and Wada [1994]).

Truly Multidimensional Algorithms

To obtain accurate numerical simulations for complex flow problems, solvers must be capable of calculating strong and weak shock waves, shear and contact discontinuities, flame fronts, mixing layers, and boundary layers. Although there has been considerable progress in these areas over the last decade, some improvements are still necessary. The current generation of Euler and Navier–Stokes solvers are limited by a noticeable degradation in the accuracy of resolution of these fluid-dynamic phenomena when they are oriented obliquely to the computational grid. This problem arises in central-difference codes because of the inability to properly tune the numerical dissipation and in upwind codes because the current technology is essentially one-dimensional in nature and must be implemented in a dimensionally split approach for calculations in two and three spatial dimensions. Consequently, excessive numbers of grid points and associated large amounts of computational time and memory are required to accurately resolve the critical flow features. This problem is magnified for calculations involving chemical and thermal nonequilibrium, since additional rate equations must be solved in a coupled fashion and the species densities and nonequilibrium energies must be stored at each grid point.

One approach toward the development of upwind solvers with multidimensional behavior involves rotated or *multidirectional* Riemann solvers (e.g., see Van Leer [1993]). These schemes require the choice of a dominant upwinding direction and a local Riemann solution with left and right states that are functions of the upwinding angle. Solvers of this type have succeeded in calculating shocks oblique to the grid with nearly the same resolution as shocks aligned with the grid. However, for three-dimensional flows with complex fluid phenomena, the task of choosing only one pertinent upwinding direction is a formidable one. Although these schemes represent an improvement over directionally split schemes, they are not the final answer and still leave the CFD community in need of truly multidimensional ideas.

Low-Speed Reacting Flows

The upwind methods examined thus far are well suited for the calculation of fast flows (subsonic or higher), but some convergence and/or efficiency problems arise when the Mach number is decreased to low values. Several applications in combustion and flame propagation involve low-speed flows where compressibility effects are important. The investigation of these flowfields is complicated by the fact that chemical reactions and the ensuing multiple space and time scales are an essential part of the physics.

Two major factors contribute to the inefficiency of standard methods for low Mach numbers: the ill-coupling of pressure and velocity in the continuity and momentum equations; and the **stiffness** of the inviscid fluxes, as measured by the ratio between the maximum and the minimum eigenvalue of the flux Jacobians (which is inversely proportional to the Mach number). *Preconditioning* techniques attempt to improve the performance of high-speed algorithms applied to low-speed compressible and reactive problems by rescaling the eigenvalues (characteristic speeds) of the problem and rendering their ratio nonsingular as the Mach number goes to zero. In the low-speed, compressible arena, proposed solutions that alter the steady-state equations generally have been discarded in favor of techniques that alter the time evolution of the problem. Unsteady problems have been considered by introducing pseudotime derivatives, which allow the calculations to converge to a time-accurate unsteady residual. Techniques that simplify the governing equations by means of small-Mach-number expansions have been proposed, but they have the distinct disadvantage of not being accurate for slightly larger Mach numbers. A review of preconditioning techniques for the compressible equations is provided by Turkel [1992].

A different solution to the low-speed difficulties of flux-split algorithms is represented by the family of SIMPLE and SIMPLER algorithms. These techniques have been applied to reactive flow problems, but their overall accuracy has been repeatedly questioned [Oran and Boris 1987].

Viscous Fluxes

The viscous vectors usually are discretized using standard second-order accurate central differences. In most applications of engineering interest, the thin-layer version of the Navier–Stokes equations is actually discretized, whereby the viscous terms in the space directions that are not normal to a solid surface are neglected. On the other hand, when the full Navier–Stokes equations are analyzed, some mixed second-order derivative terms appear in the formulation, and they usually are treated in a diagonal-dominance-preserving fashion [Chakravarthy et al. 1985].

Source Terms

When using the finite-volume technique, the inclusion of source terms in the discretized equations is usually a trivial matter. Unless radiative transfer is included, source terms are local in nature and do not require the evaluation of space derivatives or gradients. The vector of source terms W, introduced in Eqs. (38.8) and (38.9), is algebraically summed to the unsteady and flux contributions.

Significant complications usually arise when radiative heat transfer is included in the source terms for the energy equations, because those contributions are *nonlocal*; they depend on the thermochemical properties of the surrounding medium. Drastic simplifications of the radiative source terms are common in these cases [Vincenti and Kruger 1986].

Time Integration

After fluxes and source terms have been discretized and related to the vector of unknown volume-averaged variables, the remaining problem to be solved is how to advance the numerical solution in time, given a known initial state. The time integration schemes that commonly are used fall into two categories, *explicit* and *implicit*. Explicit schemes evaluate fluxes and source terms at the old time step prior to moving to a new time, whereas implicit techniques utilize linearized estimates of fluxes and source terms at the new time step when advancing the solution in time.

Two classes of problems usually arise when dealing with time integration. The first class contains the *unsteady* problems, where an accurate time integration is essential for the overall validity of the simulation. Moreover, some unsteady problems could involve moving boundaries and/or moving and deforming control volumes. The second class involves *steady* problems, where time accuracy is of no concern. A pseudotransient problem usually is solved in this case, where the solution is advanced in a pseudotime until convergence to a steady state is reached. Most algorithms will be able to handle both steady and unsteady problems, provided that the discretized form of the *geometric conservation law* [Thomas and Lombard 1979] is enforced in the calculation of the time derivative of the control volumes for cases when the latter are deforming.

Explicit Schemes

A popular and computationally inexpensive scheme that is second-order accurate in time is the m-stage, Jameson-style, explicit Runge–Kutta method. That and other explicit algorithms used in computational fluid dynamics applications are discussed in [Fletcher 1988]. Unfortunately, all explicit schemes have the major drawback of becoming extremely inefficient for stiff problems. In these conditions, the time step necessary for stability can become orders of magnitude smaller than the value which would be necessary for an efficient resolution of the overall gas-dynamic transient. For this reason, time-implicit schemes are advocated, as they show an increase in efficiency that can be dramatic.

Hybrid schemes also have been investigated, whereby the source terms are treated implicitly and the fluxes remain explicit. These techniques show some promise, but they are still the object of current research.

Implicit Schemes

Implicit time integration schemes are very popular, especially for steady problems, due to their unconditional stability, the consequent increase in time step per iteration, and the overall efficiency that they make possible. Moreover, for reacting flow problems it has been shown that treating the source terms in an implicit fashion is tantamount to rescaling the governing equations so that all of the time scales in a flow problem become of the same order of magnitude [Bussing and Murman 1985]. Consequently, implicit algorithms are a viable solution to stiffness problems.

For cases where time accuracy is important, as in unsteady problems, second-order-accurate schemes in time have been developed, such as the implicit two-step Runge–Kutta scheme developed by Iannelli and Baker [1988], or three-time-level schemes, such as the popular one proposed by Beam and Warming [1978]. Examples abound of problems where even first-order algorithms such as the Euler implicit are deemed satisfactory. Incidentally, for steady problems, the Euler implicit technique is almost universally accepted as the most efficient approach.

After any one of the implicit techniques is applied to the discretized equations, an extremely large linear problem is obtained, whose unknowns are the volume-averaged values at each control volume and whose left-hand side comprises the Jacobian matrices of fluxes and source terms. For three-dimensional problems, the storage requirements for an exact solution of this linear problem are too restrictive. Approximate solutions have been devised, namely, solutions in a plane in conjunction with relaxation in the third space direction, or some sort of approximate factorization that reduces the original problem to a series of smaller problems.

All of the techniques previously discussed apply to flows in local chemical equilibrium as well as flows in thermochemical nonequilibrium. The difference is in the size of the problem, due to the increased number of equations that are solved per control volume in the latter case and the fact that in the former case some thermodynamic properties have to be obtained from the composition solver.

Numerical Results: A Case Study

Extensive numerical results have been obtained in recent years for applications involving chemically reacting flows, and the reader is referred to the end of the chapter for numerous references discussing simulations in one, two, and three space dimensions, both steady and unsteady.

In the following, the differences between ideal-gas behavior and reactive flows are demonstrated through a case study involving two different flow simulations. The results presented were obtained as part of a Research Experiences for Undergraduates (REU) program at Mississippi State University in the summer of 1994.2 The computer code utilized for the computations is a three-dimensional, multiblock, inviscid solver, which runs with both perfect-gas and local-chemical-equilibrium models [Cox and Cinnella 1994]. Both perfect and reactive gases utilized here are models for air, and the latter includes both molecular and atomic nitrogen and oxygen as well as nitric oxide. A modified HLLE scheme was utilized [Obayashi and Wada 1994] in conjunction with high-order accurate flux extrapolation.

The first case studied represents an incident shock wave moving against an obstacle, namely a pyramid. The base of the pyramid is 1.0 m, the height is 0.5 m, and the ambient temperature is 298 K. The speed of the shock wave is Mach 5.5, and its interaction with the obstacle is recorded at regular time intervals. The time is kept the same for both physical models (air and reactive), so that a one-on-one comparison can be made of the flow evolution. A representative set of results is shown in Figs. 38.1 (a)–(g). The commercially available FAST (Fluid Analysis Software Toolkit) software [Clucas et al. 1993] is used for visualizing and animating all of the computations. In the figures, the temperature for both perfect-gas and reactive flows is shown at regular intervals. The temperatures associated with the perfect-gas model are much higher than their reactive counterparts; the reactive flow is dominated by chemical dissociation of the diatomic molecules, which significantly reduces the temperature because dissociation reactions are **endothermic**. The overall flow is fairly complicated; a reflected, curved shock wave forms in front of the obstacle, and when the flow reaches the tip of the pyramid and spills over, a strong expansion wave appears.

The second test case represents a greatly simplified version of the inlet of a supersonic engine. The flow of air as it enters the engine is studied with both perfect-gas and reactive models. The figures show Mach-number contours; the starting Mach number is 5. The inlet temperature was artificially raised to a value of 1750 K to enhance deviations from perfect-gas behavior. The inlet is composed of two ramps; the upper one makes a 20° angle and the lower one makes a 10° angle with the horizontal direction. The purpose of these ramps is to slow down the air entering the engine and increase its pressure and temperature are to prepare it for combustion with the fuel. The mechanisms for slowing down the flow are through oblique shock waves, which are formed at the ramps, as seen in Figs. 38.2 (a)–(g). These shock waves end up intersecting each other and forming a complicated diamond pattern, as seen in the figures,

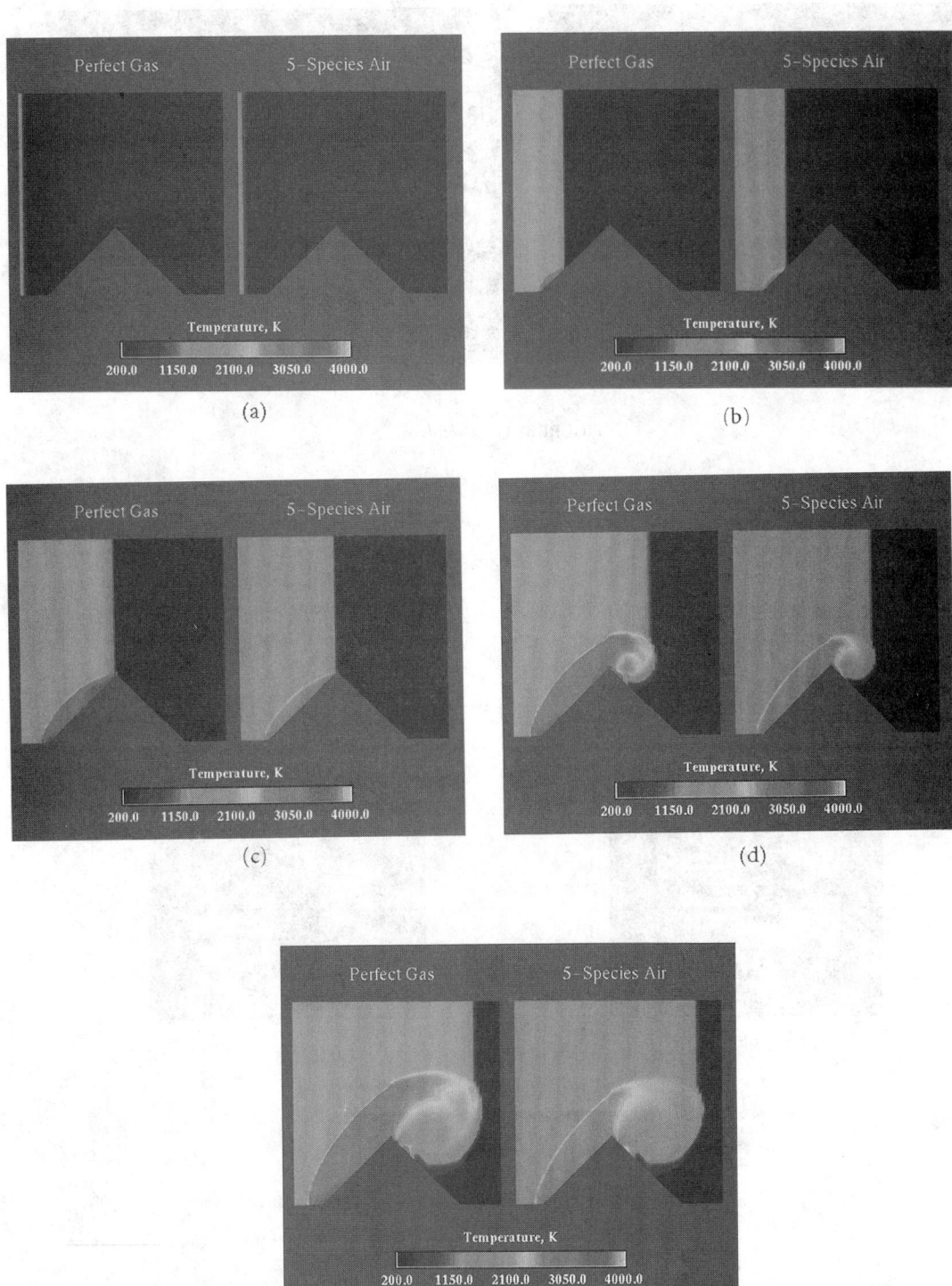

FIGURE 38.1 A shock wave impinges on a pyramid. Temperature contours for a perfect-gas model and a five-species reactive-air model. (See color version of this figure in the color section of the Handbook.)

(Continued)

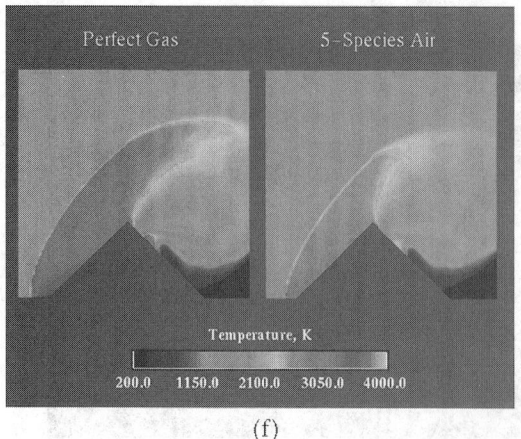

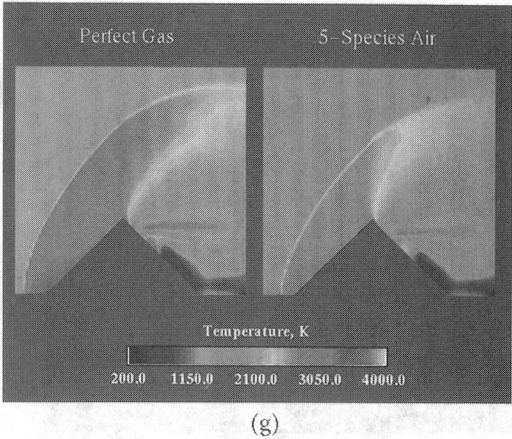

(f)

(g)

FIGURE 38.1 (*Continued*)

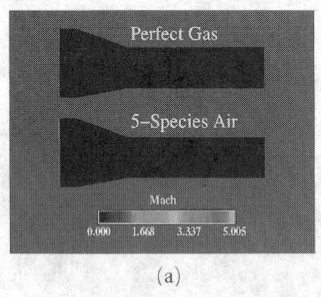

(a)

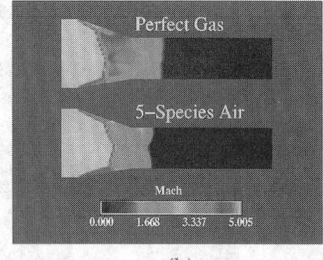

(b)

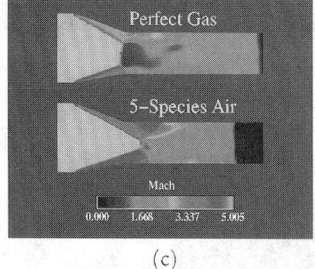

(c)

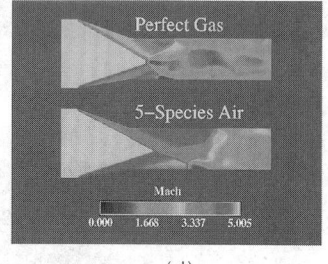

(d)

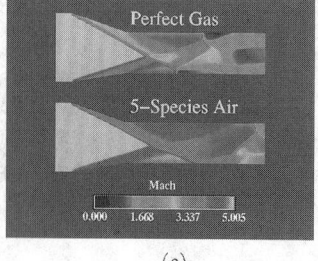

(e)

FIGURE 38.2 Mach 5 flow in a supersonic inlet. Mach-number contours for a perfect-gas model and a five-species reactive-air model. (See color version of this figure in the color section of the Handbook.)

(*Continued*)

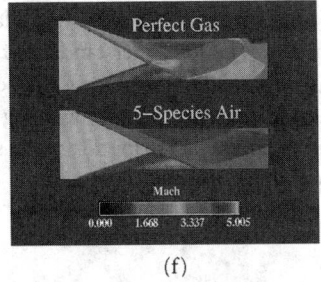

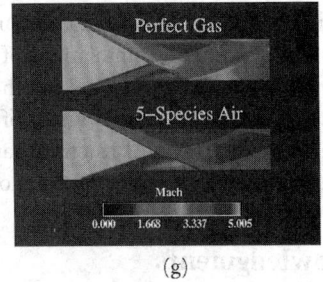

(f)   (g)

FIGURE 38.2 (*Continued*)

which is typical in these cases. Moreover, at the corners where the inlet becomes straight again, expansion waves form, which also interact with the shock waves. Throughout the calculation, the five-species air model develops more rapidly than the perfect-gas model; in every frame its features are better defined, and it reaches steady state faster. The speed difference is due to two main reasons: the angles that the shocks make with the ramps are slightly smaller for the reactive case (a well-known result), and the overall temperatures are much lower again.

38.4 Summary and Research Issues

The present chapter has attempted to review the state of the art in the area of numerical methods for physically challenging flows such as those in thermochemical nonequilibrium. The exposition has been essentially limited to finite-volume techniques and upwind algorithms, although an effort has been made to mention other approaches.

The thermodynamic foundation for the study of mixtures of reacting gases was discussed first. These flows can be either in *thermal equilibrium,* wherein all internal energy modes are in equilibrium at the translational temperature, or in *thermal nonequilibrium,* wherein some modes satisfy Boltzmann distributions at different temperatures. Moreover, three categories of flows were identified based on chemical activity: *frozen flow,* when reaction time scales are much slower than fluid-dynamic time scales; *local chemical equilibrium,* when reaction time scales are much faster than fluid-dynamic time scales; and the general case of *finite-rate flow,* when reaction time scales are of the same order of magnitude as fluid-dynamic time scales. *Governing equations* for the general case of flows in thermochemical nonequilibrium were presented, and the most common simplifying assumptions were introduced. The treatment of *transport properties* for mixtures of reacting gases was detailed.

Numerical methods used to obtain practical solutions to the governing equations were presented. The discussion centered on *finite-volume* techniques, whereby the physical domain is discretized by means of small control volumes, and *upwind algorithms,* whereby information is extrapolated from the prominent flow direction. It was pointed out that inviscid fluxes usually are discretized using upwind algorithms via *flux-split techniques,* and viscous fluxes usually are discretized using central differencing. Among flux-split techniques, the most popular are the flux-vector split formulations of Steger and Warming and of Van Leer and the flux-difference split formulation of Roe. More recent upwind algorithm developments were also discussed, including the HLLE scheme and truly multidimensional techniques. Since these upwind methods were originally developed for high-speed compressible flows, *preconditioning* techniques are necessary to improve their performance with low-speed problems. Flows in thermochemical nonequilibrium generally entail multiple time scales, due to the various reaction times; consequently, particular attention must be paid to *time discretization.* A few of the most commonly used algorithms were mentioned. Finally, two sample simulations were presented, depicting the differences between perfect-gas behavior and reacting flows. The presence of chemistry is shown to have a significant effect on the flow physics.

Future work in this field is very promising, due to the availability of larger and faster computers and to advances in physical modeling and numerical algorithms. Parallel computers are likely to play a major role in making numerical simulations of reacting flows affordable. A better understanding of turbulent combustion and nonequilibrium aerothermodynamics is necessary, and some progress is being made in these fields. Lastly, the *open* field of biochemical and biomedical applications involving one- and multiphase fluids, chemical reactions, and interface exchange phenomena is ripe for more systematic advances in the state-of-the-art simulation capabilities; see [Board 1993].

Acknowledgments

Derrick P. Tucker and Cindy G. Callahan ran the two test cases, respectively. The National Science Foundation partially funds the Research Center, and Eglin Air Force Base funded the original development of the code utilized for the computations. Computer time on a Cray-YMP was provided by the Mississippi Center for Supercomputing Research.

Defining Terms

Boltzmann distribution: In statistical thermodynamics, a function that describes the distribution of energy for gas particles (i.e., what fraction of gas particles has a certain amount of internal energy). It is valid for "normal" conditions, away from the gas triple point or extremely low temperatures.

Endothermic reactions: Chemical reactions that result in products whose energetic content is higher than that of the reactants. They require energy (usually heat) to occur. Their opposite is **exothermic reactions** (e.g., combustion processes), which result in the release of heat.

Euler equations: The governing equations for inviscid fluid flows.

Jump conditions: Relationships (usually algebraic) that are satisfied across discontinuities by the flow variables.

Laws of mass action: A set of algebraic equations that involve the concentrations of chemical species in chemical equilibrium at a given thermodynamic state. Also known as Le Châtelier's principle.

Navier–Stokes equations: The governing equations for viscous fluid flows.

Reaction path: In chemistry, the actual path followed by chemical reactants to yield reaction products, including intermediate steps and/or compounds.

Riemann problem: A mathematical problem, typically in one space dimension, whose solution is the flow that evolves in time given two uniform but different initial fluid conditions, separated by an interface.

Schmidt number: One of many nondimensional ratios used in fluid dynamics to estimate the relative importance of two competing phenomena in a fluid flow. Specifically, the ratio of diffusion forces to viscous forces.

Stiffness: For a system of hyperbolic partial differential equations, stiffness is encountered when the characteristic times that describe the evolution of the solution are of different orders of magnitude.

Thermochemical nonequilibrium: Simultaneous *chemical nonequilibrium*, which occurs when chemical reactions are present in the gas system, and *thermal nonequilibrium*, which occurs when the internal energy of the gas ceases to be described by a single Boltzmann distribution at the translational temperature.

Transport phenomena: The common name given to three distinct physical phenomena (viscosity, thermal conductivity, and mass diffusion), which arise as a consequence of the random motion of gas particles that accompanies the macroscopic motion of the gas as a whole.

References

Baldwin, B. S. and Lomax, H. 1978. Thin layer approximation and algebraic model for separated turbulent flows. *AIAA Paper* 78-257.

Beam, R. M. and Warming, R. F. 1978. An implicit factored scheme for the compressible Navier–Stokes equations. *AIAA J.* 16(4):393–402.

Blottner, F. G., Johnson, M., and Ellis, M. 1971. Chemically reacting viscous flow program for multi-component gas mixtures. *Sandia Lab. Rep.* SC-RR-70-754.

Board, J. A., Jr. 1993. Grand challenges in biomedical computing. In *High-Speed Performance in Biomedical Research*, T. Pilkington et al., eds., pp. 479–502. CRC Press, Boca Raton, FL.

Bussing, T. R. A. and Murman, E. M. 1985. A finite-volume method for the calculation of compressible chemically reacting flows. *AIAA Paper* 85-0331.

Chakravarthy, S. R., Szema, K.-Y., Goldberg, U. C., Gorski, J. J., and Osher, S. 1985. Application of a new class of high accuracy TVD schemes to the Navier–Stokes equations. *AIAA Paper* 85-0165.

Cinnella, P. and Grossman, B. 1991. Flux-split algorithms for hypersonic flows. In *Computational Methods in Hypersonic Aerodynamics*, T. K. S. Murthy, ed., pp. 153–202. Computational Mechanics Publications, Southampton, U.K.

Clucas, J., Walatka, P., and McCabe, R. K. 1993. *FAST Programmer's Manual.* NASA Ames Research Center: NAS Division, RND Branch.

Cox, C. F. and Cinnella, P. 1994. General solution procedure for flows in local chemical equilibrium. *AIAA J.* 32(3):519–527.

Einfeldt, B., Munz, C. D., Roe, P. L., and Sjogreen, B. 1991. On Gudonov-type methods near low densities. *J. Comput. Phys.* 92(2):273–295.

Eppard, W. M. and Grossman, B. 1993. An upwind kinetic flux-vector splitting method for flows in chemical and thermal nonequilibrium. *AIAA Paper* 93-0894.

Fletcher, C. A. J. 1988. *Computational Techniques for Fluid Dynamics.* Springer–Verlag, New York.

Gnoffo, P. A., Gupta, R. N., and Shinn, J. L. 1989. Conservation equations and physical models for hypersonic air flows in thermal and chemical nonequilibrium. *NASA Tech. Paper* 2867.

Iannelli, G. S. and Baker, A. J. 1988. A stiffly-stable implicit Runge–Kutta algorithm for CFD applications. *AIAA Paper* 88-0416.

Lee, J.-H. 1985. Basic governing equations for the flight regimes of aeroassisted orbital transfer vehicles. *Prog. Astronaut. Aeronaut.* 96:3–53.

Liu, Y. and Vinokur, M. 1989. Equilibrium gas flow computations. I. Accurate and efficient calculations of equilibrium gas properties. *AIAA Paper* 89-1736.

Mallard, W. G., Westley, F., Herron, J. T., Hampson, R. F., and Frizzell, D. H. 1993. *NIST Chemical Kinetics Database: Version 5.0.* National Institute of Standards and Technology, Gaithersburg, MD.

Obayashi, S. and Wada, Y. 1994. Practical formulation of a positively conservative scheme. *AIAA J.* 32(5):1093–1095.

Oldenborg, R., Chinitz, W., Friedman, M., Jaffe, R., Jachimowski, C., Rabinowitz, M., and Schott, G. 1990. Hypersonic combustion kinetics. *NASA Tech. Memo.* 1107.

Oran, E. S. and Boris, J. P. 1987. *Numerical Simulation of Reactive Flow.* Elsevier, Amsterdam.

Park, C. 1991. *Nonequilibrium Hypersonic Aerothermodynamics.* Wiley, New York.

Pope, S. B. 1985. PDF methods for turbulent reactive flows. *Prog. Energy Comb. Sci.* 11:119–192.

Quirk, J. J. 1992. A contribution to the great Riemann solver debate. *ICASE Rep.* 92-64. NASA Langley Research Center, Langley, VA.

Roe, P. L. 1981. Approximate Riemann solvers, parameter vectors, and difference schemes. *J. Comput. Phys.* 43:357–372.

Steger, J. L. and Warming, R. F. 1981. Flux vector splitting of the inviscid gas dynamic equations with applications to finite-difference methods. *J. Comput. Phys.* 40:263–293.

Sweby, P. K. 1984. High resolution schemes using flux limiters for hyperbolic conservation laws. *SIAM J. Numer. Anal.* 21(5):995–1011.

Thomas, P. D. and Lombard, C. K. 1979. Geometric conservation law and its application to flow computations on moving grids. *AIAA J.* 17(10):1030–1037.

Turkel, E. 1992. Review of preconditioning methods for fluid dynamics. *ICASE Rep.* 92-47. NASA Langley Research Center, Langley VA.

Van Leer, B. 1982. Flux-vector splitting for the Euler equations. In *Lecture Notes in Physics*, Vol. 170, pp. 507–512. Springer–Verlag, New York.

Van Leer, B. 1993. Progress in multidimensional upwinding. In *Lecture Notes in Physics*, Vol. 414, pp. 1–26, Springer–Verlag, New York.

Vincenti, W. G. and Kruger, C. H., Jr. 1986. *Introduction to Physical Gas Dynamics*. Robert E. Krieger, Malabar, FL.

Wilke, C. R. 1950. A viscosity equation for gas mixtures. *J. Chem. Phys.* 18(4):517–519.

Further Information

General Reference

Anderson, J. D., Jr. 1989. *Hypersonic and High Temperature Gas Dynamics*. McGraw–Hill, New York.

Bertin, J. J., Glowinski, R., and Periaux, J., eds. 1989. *Hypersonics*. Birkhäuser, Boston.

Clarke, J. F. and McChesney, M. 1964. *The Dynamics of Real Gases*. Butterworth, London. Second ed. under the title *The Dynamics of Relaxing Gases*, 1975.

Dervieux, A. and Larrouturou, B., eds. 1990. *Numerical Combustion*. Lecture notes in physics. Springer–Verlag, New York.

Thermodynamic Models, Including Nonequilibrium

Candler, G. V. and MacCormack, R. W. 1988. The computation of hypersonic ionized flows in chemical and thermal nonequilibrium. *AIAA Paper* 88-0511.

Marrone, P. V. and Treanor, C. E. 1963. Chemical relaxation with preferential dissociation from excited vibrational levels. *Phys. Fluids* 6(9):1215–1221.

Viscous Fluxes and Turbulence

Bird, R. B., Stewart, W. E., and Lightfoot, E. N. 1960. *Transport Phenomena*. Wiley, New York.

Clavin, P. 1994. Premixed combustion and gasdynamics. *Annu. Rev. Fluid Mech.* 26:321–352.

Hirschfelder, J. O., Curtiss, C. F., and Bird, R. B. 1954. *Molecular Theory of Gases and Liquids*. Wiley, New York.

Upwind Algorithms

Abgrall, R. and Montagné, J. L. 1989. Generalization of Osher's Riemann solver to mixture of perfect gas and real gas. *Rech. Aerosp.* 1989(4):1–13.

Glaister, P. 1988. An approximate linearised Riemann solver for the three-dimensional Euler equations for real gases using operator splitting. *J. Comput. Phys.* 77:361–383.

Grossman, B. and Cinnella, P. 1990. Flux-split algorithms for flows with nonequilibrium chemistry and vibrational relaxation. *J. Comput. Phys.* 88(1):131–168.

Liu, Y. and Vinokur, M. 1989. Upwind algorithms for general thermochemical nonequilibrium flows. *AIAA Paper* 89-0201.

Multidimensional Algorithms

Dadone, A. and Grossman, B. 1992. A rotated upwind scheme for the Euler equations. *AIAA J.* 30(10):2219–2226.

Davis, S. F. 1984. A rotationally-biased upwind difference scheme for the Euler equations. *J. Comput. Phys.* 56:65–92.

Rumsey, C. L., Van Leer, B., and Roe, P. L. 1991. Effect of a multidimensional flux function on the monotonicity of Euler and Navier–Stokes computations. *AIAA Paper* 91-1530-CP.

Struijs, R., Deconinck, H., De Palma, P., Roe, P. L., and Powell, K. G. 1991. Progress on multidimensional upwind Euler solvers for unstructured grids. *AIAA Paper* 91-1550-CP.

Low-Speed Reactive Flows

Bhattacharjee, S., Altenkirch, R. A., Srikantaiah, N., and Vedhanayagam, M. 1990. A theoretical description of flame spreading over solid combustibles in a quiescent environment at zero gravity. *Combust. Sci. and Tech.* 69:1–15.

Withington, J. P., Shuen, J.-S., and Yang, V. 1991. A time-accurate, implicit method for chemically reacting flows at all Mach numbers. *AIAA Paper* 91-0581.

Time Integration and Stiffness

Chang, S.-H. 1989. On the application of subcell resolution to conservation laws with stiff source terms. *NASA Tech. Memo.* 102384.

LeVeque, R. J. and Yee, H. C. 1988. A study of numerical methods for hyperbolic conservation laws with stiff source terms. *NASA Tech. Memo.* 100075.

Taylor, L. K. and Whitfield, D. L. 1991. Unsteady three-dimensional incompressible Euler and Navier–Stokes solver for stationary and dynamic grids. *AIAA Paper* 91-1650.

Radiative Heat Transfer

Anderson, J. D., Jr. 1969. An engineering survey of radiating shock layers. *AIAA J.* 7(9):1665–1675.

Cinnella, P. and Elbert, G. J. 1992. Two-dimensional radiative heat transfer calculations for flows in thermo-chemical nonequilibrium. *AIAA Paper* 92-0121.

Hartung, L. C. and Hassan, H. A. 1992. Radiation transport around axisymmetric blunt body vehicles using a modified differential approach. *AIAA Paper* 92-0119.

Structure of Shock Waves

Lumpkin, F. E., III. 1991. Accuracy of the Burnett equations for hypersonic real gas flows. *AIAA Paper* 91-0771.

Zong, X., MacCormack, R. W., and Chapman, D. R. 1991. Stabilization of the Burnett equations and application to high-altitude hypersonic flows. *AIAA Paper* 91–0070.

Biochemical Applications

Characklis, W. G. and Marshall, K. C., eds. 1990. *Biofilms.* Wiley, New York.

Szego, S., Cinnella, P., and Cunningham, A. B. 1993. Numerical simulation of biofilm processes in closed conduits. *J. Comput. Phys.* 108(2):246–263.

39

Computational Electromagnetics

J. S. Shang
Wright Laboratory

Nomenclature

$\mathbf{B}$ = Magnetic flux density
C = Coefficient matrix of flux-vector formulation
$\mathbf{D}$ = Electric displacement
$\mathbf{E}$ = Electric field strength
F = Flux vector component
$\mathbf{H}$ = Magnetic flux intensity
i, j, k = Index of discretization
$\mathbf{J}$ = Electric current density
n = Index of temporal level of solution
S = Similar matrix of diagonalization
t = Time
U = Dependent variables
V = Elementary-cell volume
x, y, z = Cartesian coordinates
Δ = Forward difference operator
λ = Eigenvalue
ξ, η, ζ = Transformed coordinates
∇ = Gradient, backward difference operator

39.1 Introduction

Computational electromagnetics (CEM) is a natural extension of the analytical approach in solving the Maxwell equations. In spite of the fundamental difference between representing the solution in a

912

0-8493-2909-4/97/$0.00+$.50

continuum and in a discretized space, both approaches satisfy all pertaining theorems rigorously. The analytic approach to electromagnetics is elegant, and the results can describe the specific behavior as well as the general patterns of a physical phenomenon in a given regime. However, exact solutions to the Maxwell equations are usually unavailable. Some of the closed-form results that exist have restrictive underlying assumptions that limit their range of validity. Solutions of CEM generate only a point value for a specific simulation, but complexity of the physics or of the field configuration is no longer a limiting factor. The numerical accuracy of CEM is an issue to be addressed. Nevertheless, with the advent of high-performance computing systems, CEM is becoming a mainstay for engineering applications.

CEM in the present context is focused on simulation methods for solving the Maxwell equations in the time domain. First of all, time dependence is the most general form of the Maxwell equations, and the dynamic electromagnetic field is not confined to a time-harmonic phenomenon. Therefore, CEM in the time domain has the widest range of engineering applications. In addition, several new numerical algorithms for solving the first-order hyperbolic partial differential equations, as well as coordinate transformation techniques, were introduced recently to the CEM community. These finite-difference and finite-volume numerical algorithms were devised specifically to mimic the physics involving directional wave propagation. Meanwhile, very complex shapes associated with the field can be easily accommodated by incorporating a coordinate transformation technique. These methodologies have the potential to radically change future research in electromagnetics.

In order to use CEM effectively, it will be beneficial to understand the fundamentals of numerical simulation and its limitations. The inaccuracy incurred by a numerical simulation is attributable to the mathematical model for the physics, the numerical algorithm, and the computational accuracy. In general, differential equations in computational electromagnetics consist of two categories: the first-order divergence–curl equations and the second-order curl–curl equations [Elliott 1966, Harrington 1961, 1968]. In specific applications, further simplifications into the frequency-domain or the Helmholtz equations and the potential formulation have been accomplished. Poor numerical approximations to physical phenomena can result, however, from solving overly simplified governing equations. Under these circumstances, no meaningful quantification of errors for the numerical procedure can be achieved. Physically incorrect values and inappropriate implementation of initial and/or boundary conditions are another major source of error. The placement of the far-field boundary and the type of initial or boundary conditions have also played an important role. These concerns are easily appreciated in the light of the fact that the governing equations are identical, but the different initial/boundary conditions generate different solutions.

Numerical accuracy is also controlled by the algorithm and computing system adopted. The error induced by the discretization consists of the roundoff and the truncation error. The roundoff error is contributed by the computing system and is problem-size-dependent. Since this error is random, it is the most difficult to evaluate. One anticipates that this type of error will be a concern for solution procedures involving large-scale matrix manipulation such as the method of moments and the implicit numerical algorithm for finite-difference or finite-volume methods. The truncation error for time-dependent calculations appears as dissipation and dispersion, which can be assessed and alleviated by mesh-system refinements.

Finally, numerical error can be the consequence of a specific formulation. The error becomes pronounced when a special phenomenon is investigated or when a discontinuous and distinctive stratification of the field is encountered, such as a wave propagating through the interface between media of different characteristic impedances, for which the solution is piecewise continuous. Only in a strongly conservative formulation can the discontinuous phenomenon be adequately resolved. Another example is encountered in radar cross-section simulation, where the scattered-field formulation has been shown to be superior to the total-field formulation.

The Maxwell equations in the time domain consist of a first-order divergence–curl system and are difficult to solve by conventional numerical methods. Nevertheless, the pioneering efforts by Yee and others have attained impressive achievements [Yee 1966, Taflove 1992]. Recently, numerical techniques in CEM have been further enriched by development in the field of computational fluid dynamics (CFD). In CFD, the Euler equations, which are a subset of the Navier–Stokes equations, have the same partial-differential-system classification as that of the time-dependent Maxwell equations. Both are hyperbolic systems and constitute initial-value problems [Sommerfeld 1949]. For hyperbolic partial differential equations, the solutions need not be analytic functions. More importantly, the initial values together with any possible discontinuities are continued along a time–space trajectory, which is commonly referred to as the characteristic. A series of numerical schemes have been devised in the CFD community to duplicate the directional information-propagation feature. These numerical procedures are collectively designated as the characteristic-based method, which in its most elementary form is identical to the Riemann problem [Roe 1986]. The characteristic-based method when applied to solve the Maxwell equations in the time domain has exhibited many attractive attributes. A synergism of the new numerical procedures and scalable parallel-computing capability will open up a new frontier in electromagnetics research. For this reason, a major portion of the present chapter will be focused on introducing the characteristic-based finite-volume and finite-difference methods [Shang 1993, Shang and Gaitonde 1993, Shang and Fithen 1994].

39.2 Governing Equations

The time-dependent Maxwell equations for the electromagnetic field can be written as [Elliott 1966, Harrington 1961]:

$$\frac{\partial \mathbf{B}}{\partial t} + \nabla \times \mathbf{E} = \mathbf{0} \tag{39.1}$$

$$\frac{\partial \mathbf{D}}{\partial t} - \nabla \times \mathbf{H} = -\mathbf{J} \tag{39.2}$$

$$\nabla \cdot \mathbf{B} = 0 \tag{39.3}$$

$$\nabla \cdot \mathbf{D} = \rho$$

The only conservation law for electric charge and current densities is

$$\frac{\partial \rho}{\partial t} + \nabla \cdot \mathbf{J} = 0 \tag{39.4}$$

where ρ and $\mathbf{J}$ are the charge and current density, respectively, and represent the source of the field. The constitutive relations between the magnetic flux density and intensity and between the electric displacement and field strength are $\mathbf{B} = \mu \mathbf{H}$ and $\mathbf{D} = \epsilon \mathbf{E}$. Equation (39.4) is regarded as a fundamental law of electromagnetics, derived from the generalized Ampere's circuit law and Gauss's law. Since Eqs. (39.1) and (39.2) contain the information on the propagation of the electromagnetic field, they constitute the basic equations of CEM.

The above partial differential equations also can be expressed as a system of integral equations. The following expression is obtained by using Stoke's law and the divergence theorem to reduce the surface and volume integrals to a circuital line and surface integrals, respectively [Elliott 1966]. These integral relationships hold only if the first derivatives of the electric displacement $\mathbf{D}$ and the magnetic flux density $\mathbf{B}$ are continuous throughout the control volume:

$$\oint \mathbf{E} \cdot d\mathbf{L} = -\iint \frac{\partial \mathbf{B}}{\partial t} \cdot d\mathbf{S} \tag{39.5}$$

$$\oint \mathbf{H} \cdot d\mathbf{L} = \iint \left(\mathbf{J} + \frac{\partial \mathbf{D}}{\partial t} \right) \cdot d\mathbf{S} \tag{39.6}$$

$$\iint \mathbf{D} \cdot d\mathbf{S} = \iiint \rho \, d\mathbf{V} \tag{39.7}$$

$$\iint \mathbf{B} \cdot d\mathbf{S} = 0 \tag{39.8}$$

The integral form of the Maxwell equations is rarely used in CEM. They are, however, invaluable as a validation tool for checking the global behavior of field computations.

The second-order curl–curl form of the Maxwell equations is derived by applying the curl operator to get

$$\nabla \times \nabla \times \mathbf{E} + \frac{1}{c^2} \frac{\partial^2 \mathbf{E}}{\partial t^2} = -\frac{\partial(\mu \mathbf{J})}{\partial t} \tag{39.9}$$

$$\nabla \times \nabla \times \mathbf{B} + \frac{1}{c^2} \frac{\partial^2 \mathbf{B}}{\partial t^2} = \nabla \times (\mu \mathbf{J}) \tag{39.10}$$

The outstanding feature of the curl–curl formulation of the Maxwell equations is that the electric and magnetic fields are decoupled. The second-order equations can be further simplified for harmonic fields. If the time-dependent behavior can be represented by a harmonic function $e^{i\omega t}$, the separation-of-variables technique will transform the Maxwell equations into the frequency domain [Elliott 1966, Harrington 1961]. The resultant partial differential equations in spatial variables become elliptic:

$$\nabla \times \nabla \times \mathbf{E} - k^2 \mathbf{E} = -i\omega(\mu \mathbf{J})$$
$$\nabla \times \nabla \times \mathbf{B} - k^2 \mathbf{B} = \nabla \times (\mu \mathbf{J}) \tag{39.11}$$

where $k = \omega/c$ is called the propagation constant or the wave number, and ω is the angular frequency of a component of a Fourier series or a Fourier integral [Elliott 1966, Harrington 1961]. The above equations are frequently the basis for finite-element approaches [Rahman et al. 1991].

In order to complete the description of the differential system, initial and/or boundary values are required. For Maxwell equations, only the source of the field and a few physical boundary conditions at the media interfaces are pertinent [Elliott 1966, Harrington 1961]:

$$\mathbf{n} \times (\mathbf{E}_1 - \mathbf{E}_2) = 0$$
$$\mathbf{n} \times (\mathbf{H}_1 - \mathbf{H}_2) = \mathbf{J}_s$$
$$\mathbf{n} \cdot (\mathbf{D}_1 - \mathbf{D}_2) = \rho_s \tag{39.12}$$
$$\mathbf{n} \cdot (\mathbf{B}_1 - \mathbf{B}_2) = 0$$

where the subscripts 1 and 2 refer to media on two sides of the interface, and $\mathbf{J}_s$ and ρ_s are the surface current and charge densities, respectively.

Since all computing systems have finite memory, all CEM computations in the time domain must be conducted on a truncated computational domain. This intrinsic constraint requires a numerical far-field condition at the truncated boundary to mimic the behavior of an unbounded field. This numerical boundary unavoidably induces a reflected wave to contaminate the simulated field. In the past, absorbing boundary conditions at the far-field boundary have been developed from the radiation condition [Sommerfeld 1949, Enquist and Majda 1977, Higdon 1986, Mur 1981]. In general, a progressive order-of-accuracy procedure can be used to implement the numerical boundary conditions with increasing accuracy [Enquist and Majda 1977, Higdon 1986]. On the other hand, the characteristic-based methods which satisfy the physical domain of influence requirement can specify the numerical boundary condition readily. For this formulation, the reflected wave can be suppressed by eliminating the undesirable incoming numerical data. Although the accuracy of the numerical far-field boundary condition depends on the coordinate system, in principle this formulation under ideal circumstances can effectively suppress artificial wave reflections.

39.3 Characteristic-Based Formulation

The fundamental idea of the characteristic-based method for solving the hyperbolic system of equations is derived from the eigenvalue–eigenvector analyses of the governing equations. For Maxwell equations in the time domain, every eigenvalue is real but not all of them are distinct [Shang 1993, Shang and Gaitonde 1993, Shang and Fithen 1994]. In a time–space plane, the eigenvalue actually defines the slope of the characteristic or the phase velocity of the wave motion. All dependent variables within the time–space domain bounded by two intersecting characteristics are completely determined by the values along these characteristics and by their compatibility relationship. The direction of information propagation is also clearly described by these two characteristics [Sommerfeld 1949]. In numerical simulation, the well-posedness requirement on initial or boundary conditions and the stability of a numerical approximation are also ultimately linked to the eigenvalues of the governing equation [Anderson et al. 1984, Richtmyer and Morton 1967]. Therefore, characteristic-based methods have demonstrated superior numerical stability and accuracy to other schemes [Roe 1986, Shang 1993]. However, characteristic-based algorithms also have an inherent limitation in that the governing equation can be diagonalized only in one spatial dimension at a time. The multidimensional equations are required to split into multiple one-dimensional formulations. This limitation is not unusual for numerical algorithms, such as the approximate factored and the fractional-step schemes [Shang 1993, Anderson et al. 1984]. A consequence of this restriction is that solutions of the characteristic-based procedure may exhibit some degree of sensitivity to the orientation of the coordinate selected. This numerical behavior is consistent with the concept of optimal coordinates.

In the characteristic formulation, data on the wave motion are first split according to the direction of phase velocity and then transmitted in each orientation. In each one-dimensional time–space domain, the direction of the phase velocity degenerates into either a positive or a negative orientation. They are commonly referred to as the right-running and the left-running wave components [Sommerfeld 1949, Roe 1986]. The sign of the eigenvalue is thus an indicator of the direction of signal transmission. The corresponding eigenvectors are the essential elements for diagonalizing the coefficient matrices and for formulating the approximate Riemann problem [Roe 1986]. In essence, knowledge of eigenvalues and eigenvectors of the Maxwell equations in the time domain becomes the first prerequisite of the present formulation.

The system of governing equations cast in the flux-vector form in the Cartesian frame becomes [Shang 1993, Shang and Gaitonde 1993, Shang and Fithen 1994]

$$\frac{\partial U}{\partial t} + \frac{\partial F_x}{\partial x} + \frac{\partial F_y}{\partial y} + \frac{\partial H_z}{\partial z} = -J \tag{39.13}$$

where U is the vector of dependent variables. The flux vectors are formed by the inner product of the coefficient matrix and the dependent variable: $F_x = C_x U$, $F_y = C_y U$, and $F_z = C_z U$, with

$$U = \{B_x, B_y, B_z, D_x, D_y, D_z\}^T \tag{39.14}$$

and

$$F_x = \{0, -D_z/\epsilon, D_y/\epsilon, 0, B_z/\mu, -B_y/\mu\}^T$$

$$F_y = \{D_z/\epsilon, 0, -D_x/\epsilon, -B_z/\mu, 0, B_x/\mu\}^T \tag{39.15}$$

$$F_z = \{-D_y/\epsilon, D_x/\epsilon, 0, B_y/\mu, -B_x/\mu, 0\}^T$$

The coefficient matrices, or the Jacobians of the flux vectors C_x, C_y, and C_z are [Shang 1993]:

$$C_x = \begin{bmatrix} 0 & 0 & 0 & 0 & 0 & 0 \\ 0 & 0 & 0 & 0 & 0 & -\frac{1}{\epsilon} \\ 0 & 0 & 0 & 0 & \frac{1}{\epsilon} & 0 \\ 0 & 0 & 0 & 0 & 0 & 0 \\ 0 & 0 & \frac{1}{\mu} & 0 & 0 & 0 \\ 0 & -\frac{1}{\mu} & 0 & 0 & 0 & 0 \end{bmatrix} \tag{39.16a}$$

$$C_y = \begin{bmatrix} 0 & 0 & 0 & 0 & 0 & \frac{1}{\epsilon} \\ 0 & 0 & 0 & 0 & 0 & 0 \\ 0 & 0 & 0 & -\frac{1}{\epsilon} & 0 & 0 \\ 0 & 0 & -\frac{1}{\mu} & 0 & 0 & 0 \\ 0 & 0 & 0 & 0 & 0 & 0 \\ \frac{1}{\mu} & 0 & 0 & 0 & 0 & 0 \end{bmatrix}$$ (39.16b)

$$C_z = \begin{bmatrix} 0 & 0 & 0 & 0 & -\frac{1}{\epsilon} & 0 \\ 0 & 0 & 0 & \frac{1}{\epsilon} & 0 & 0 \\ 0 & 0 & 0 & 0 & 0 & 0 \\ 0 & \frac{1}{\mu} & 0 & 0 & 0 & 0 \\ -\frac{1}{\mu} & 0 & 0 & 0 & 0 & 0 \\ 0 & 0 & 0 & 0 & 0 & 0 \end{bmatrix}$$ (39.16c)

where ϵ and μ are the permittivity and permeability, which relate the electric displacement to the electric field intensity and the magnetic flux density to the magnetic field intensity, respectively.

The eigenvalues of the coefficient matrices C_x, C_y, and C_z in the Cartesian frame are identical and contain multiplicities [Shang 1993, Shang and Gaitonde 1993]. Care must be exercised to ensure that all associated eigenvectors

$$\lambda = \left\{ +\frac{1}{\sqrt{\epsilon\mu}}, -\frac{1}{\sqrt{\epsilon\mu}}, 0, +\frac{1}{\sqrt{\epsilon\mu}}, -\frac{1}{\sqrt{\epsilon\mu}}, 0 \right\}$$ (39.17)

are linearly independent. The linearly independent eigenvectors associated with each eigenvalue are found by reducing the matrix equation, $(C - \bar{\bar{I}}\lambda)U = 0$, to the Jordan normal form [Shang 1993, Shang and Fithen 1994].

Since the coefficient matrices C_x, C_y, and C_z can be diagonalized, there exist nonsingular similar matrices S_x, S_y, and S_z such that

$$\Lambda_x = S_x^{-1} C_x S_x$$

$$\Lambda_y = S_y^{-1} C_y S_y$$ (39.18)

$$\Lambda_z = S_z^{-1} C_z S_z$$

where the Λs are the diagonalized coefficient matrices. The columns of the similar matrices S_x, S_y, and S_z are simply the linearly independent eigenvectors of the coefficient matrices C_x, C_y, and C_z, respectively.

The fundamental relationship between the characteristic-based formulation and the Riemann problem can be best demonstrated in the Cartesian frame of reference. For the Maxwell equations in this frame of reference and for an isotropic medium, all the similar matrices are invariant with respect to temporal and spatial independent variables. In each time–space plane (x–t, y–t, and z–t), the one-dimensional governing equation can be given just as in the x–t plane:

$$\frac{\partial U}{\partial t} + C_x \frac{\partial U}{\partial x} = 0$$ (39.19)

Substitute the diagonalized coefficient matrix to get

$$\frac{\partial U}{\partial t} + S_x \Lambda_x S_x^{-1} \frac{\partial U}{\partial x} = 0$$ (39.20)

Since the similar matrix in the present consideration is invariant with respect to time and space, it can be brought into the differential operator. Multiplying the above equation by the left-hand inverse of S_x, S_x^{-1}, we have

$$\frac{\partial \left(S_x^{-1} U \right)}{\partial t} + \Lambda_x \frac{\partial \left(S_x^{-1} U \right)}{\partial x} = 0 \qquad (39.21)$$

One immediately recognizes the group of variables $S_x^{-1} U$ as the characteristic, and the system of equations is decoupled [Shang 1993]. In scalar-variable form and with appropriate initial values, this is the Riemann problem [Sommerfeld 1949, Courant and Hilbert 1965]. This differential system is specialized to study the breakup of a single discontinuity. The piecewise continuous solutions separated by the singular point are also invariant along the characteristics. Equally important, stable numerical operators can now be easily devised to solve the split equations according to the sign of the eigenvalue. In practice it has been found if the multidimensional problem can be split into a sequence of one-dimensional equations, this numerical technique is applicable to those one-dimensional equations [Roe 1986, Shang 1993].

The gist of the characteristic-based formulation is also clearly revealed by the decomposition of the flux vector into positive and negative components corresponding to the sign of the eigenvalue:

$$\Lambda = \Lambda^+ + \Lambda^-, \qquad F = F^+ + F^- \qquad (39.22)$$

$$F^+ = S\Lambda^+ S^{-1}, \qquad F^- = S\Lambda^- S^{-1} \qquad (39.23)$$

where the superscripts $+$ and $-$ denote the split vectors associated with positive and negative eigenvalues, respectively.

The characteristic-based algorithms have a deep-rooted theoretical basis for describing the wave dynamics. They also have however an inherent limitation in that the diagonalized formulation is achievable only in one dimension at a time. All multidimensional equations are required to be split into multiple one-dimensional formulations. The approach yields accurate results so long as discontinuous waves remain aligned with the computational grid. This limitation is also the state-of-the-art constraint in solving partial differential equations [Roe 1986, Shang 1993, Anderson et al. 1984].

39.4 Maxwell Equations in a Curvilinear Frame

In order to develop a versatile numerical tool for computational electromagnetics in a wide range of applications, the Maxwell equations can be cast in a general curvilinear frame of reference [Shang and Gaitonde 1993, Shang and Fithen 1994]. For efficient simulation of complex electromagnetic field configurations, the adoption of a general curvilinear mesh system becomes necessary. The system of equations in general curvilinear coordinates is derived by a coordinate transformation [Anderson et al. 1984, Thompson 1982]. The mesh system in the transformed space can be obtained by numerous grid generation procedures [Thompson 1982]. Computational advantages in the transformed space are also realizable. For a body-oriented coordinate system, the interface between two different media is easily defined by one of the coordinate surfaces. Along this coordinate parametric plane, all discretized nodes on the interface are precisely prescribed without the need for an interpolating procedure. The outward normal to the interface, which is essential for boundary-value implementation, can be computed easily by $n = \nabla S / \|\nabla S\|$. In the transformed space, computations are performed on a uniform mesh space, but the corresponding physical spacing can be highly clustered to enhance the numerical resolution.

As an illustration of the numerical advantage of solving the Maxwell equations on nonorthogonal curvilinear, body-oriented coordinates, a simulation of the scattered electromagnetic field from a re-entry vehicle has been performed [Shang and Gaitonde 1994]. The aerospace vehicle, X24C-10D, has a complex geometrical shape (Fig. 39.1). In addition to a blunt leading-edge spherical nose and a relatively flat delta-shaped underbody, the aft portion of the vehicle consists of five control surfaces—a central fin, two middle fins, and two strakes. A body-oriented, single-block mesh system enveloping the configuration is adopted. The numerical grid system is generated by using a hyperbolic grid generator for the near-field mesh adjacent to the solid surface, and a transfinite technique for the far field. The two mesh systems are merged by the Poisson averaging technique [Thompson 1982, Shang and Gaitonde 1994]. In this manner, the composite grid sys-

FIGURE 39.1 Radar-wave fringes on X24C-10D, grid 181 × 59 × 162, TE excitation, $L/\lambda = 9.2$.

tem is orthogonal in the near field but less restrictive in the far field. All solid surfaces of the X24C-10D are mapped onto a parametric surface in the transformed space, defined by $\eta = 0$. The entire computational domain is supported by a 181 × 59 × 162 grid system, where the first coordinate index denotes the number of cross sections in the numerical domain. The second index describes the number of cells between the body surface and the far-field boundary, while the third index gives the cells used to circumscribe each cross-sectional plane. The electromagnetic excitation is introduced by a harmonic incident wave traveling along the x-coordinate. The fringe pattern of the scattered electromagnetic waves on the X24C-10D is presented in Fig. 39.1 for a characteristic-length-to-wavelength ratio $L/\lambda = 9.2$. A salient feature of the scattered field is brought out by the surface curvature: the smaller the radius of surface curvature, the broader the diffraction pattern. The numerical result exhibits highly concentrated contours at the chine (the line of intersection between upper and lower vehicle surfaces) of the forebody and the leading edges of strakes and fins.

For the most general coordinate transformation of the Maxwell equations in the time domain, a one-to-one relationship between two sets of temporal and spatial independent variables is required. However, for most practical applications, the spatial coordinate transformation is sufficient:

$$\xi = \xi(x, y, z)$$
$$\eta = \eta(x, y, z) \tag{39.24}$$
$$\zeta = \zeta(x, y, z)$$

The governing equation in the strong conservation form is obtained by dividing the chain-rule-differentiated equations by the Jacobian of coordinate transformation and by invoking metric identities [Shang and Gaitonde 1993, Anderson et al. 1984]. The time-dependent Maxwell equations on a general curvilinear frame of reference and in the strong conservative form are

$$\frac{\partial U}{\partial t} + \frac{\partial F_\xi}{\partial \xi} + \frac{\partial F_\eta}{\partial \eta} + \frac{\partial F_\zeta}{\partial \zeta} = -J \tag{39.25}$$

where the dependent variables are now defined as

$$U = U(B_x V, B_y V, B_z V, D_x V, D_y V, D_z V) \tag{39.26}$$

Here V is the Jacobian of the coordinate transformation and is also the inverse local cell volume. If the

Jacobian has nonzero values in the computational domain, the correspondence between the physical and the transformed space is uniquely defined [Anderson et al. 1984, Thompson 1982]. Since systematic procedures have been developed to ensure this property of coordinate transformations, detailed information on this point is not repeated here [Anderson et al. 1984, Thompson 1982]. We have

$$V = \begin{bmatrix} \xi_x & \eta_x & \zeta_x \\ \xi_y & \eta_y & \zeta_y \\ \xi_z & \eta_z & \zeta_z \end{bmatrix} \tag{39.27}$$

and ξ_x, η_x, ζ_x, etc. are the metrics of coordinate transformation and can be computed easily from the definition given by Eq. (39.24). The flux-vector components in the transformed space have the following form:

$$F_\xi = \begin{bmatrix} 0 & 0 & 0 & 0 & -\frac{\xi_z}{\epsilon V} & \frac{\xi_y}{\epsilon V} \\ 0 & 0 & 0 & \frac{\xi_z}{\epsilon V} & 0 & -\frac{\xi_x}{\epsilon V} \\ 0 & 0 & 0 & -\frac{\xi_y}{\epsilon V} & \frac{\xi_x}{\epsilon V} & 0 \\ 0 & \frac{\xi_z}{V\mu} & -\frac{\xi_y}{V\mu} & 0 & 0 & 0 \\ -\frac{\xi_z}{V\mu} & 0 & \frac{\xi_x}{V\mu} & 0 & 0 & 0 \\ \frac{\xi_y}{V\mu} & -\frac{\xi_x}{V\mu} & 0 & 0 & 0 & 0 \end{bmatrix} \begin{Bmatrix} B_x \\ B_y \\ B_z \\ D_x \\ D_y \\ D_z \end{Bmatrix} \tag{39.28}$$

$$F_\eta = \begin{bmatrix} 0 & 0 & 0 & 0 & -\frac{\eta_z}{\epsilon V} & \frac{\eta_y}{\epsilon V} \\ 0 & 0 & 0 & \frac{\eta_z}{\epsilon V} & 0 & -\frac{\eta_x}{\epsilon V} \\ 0 & 0 & 0 & -\frac{\eta_y}{\epsilon V} & \frac{\eta_x}{\epsilon V} & 0 \\ 0 & \frac{\eta_z}{V\mu} & -\frac{\eta_y}{V\mu} & 0 & 0 & 0 \\ -\frac{\eta_z}{V\mu} & 0 & \frac{\eta_x}{V\mu} & 0 & 0 & 0 \\ \frac{\eta_y}{V\mu} & -\frac{\eta_x}{V\mu} & 0 & 0 & 0 & 0 \end{bmatrix} \begin{Bmatrix} B_x \\ B_y \\ B_z \\ D_x \\ D_y \\ D_z \end{Bmatrix} \tag{39.29}$$

$$F_\zeta = \begin{bmatrix} 0 & 0 & 0 & 0 & -\frac{\zeta_z}{\epsilon V} & \frac{\zeta_y}{\epsilon V} \\ 0 & 0 & 0 & \frac{\zeta_z}{\epsilon V} & 0 & -\frac{\zeta_x}{\epsilon V} \\ 0 & 0 & 0 & -\frac{\zeta_y}{\epsilon V} & \frac{\zeta_x}{\epsilon V} & 0 \\ 0 & \frac{\zeta_z}{V\mu} & -\frac{\zeta_y}{V\mu} & 0 & 0 & 0 \\ -\frac{\zeta_z}{V\mu} & 0 & \frac{\zeta_x}{V\mu} & 0 & 0 & 0 \\ \frac{\zeta_y}{V\mu} & -\frac{\zeta_x}{V\mu} & 0 & 0 & 0 & 0 \end{bmatrix} \begin{Bmatrix} B_x \\ B_y \\ B_z \\ D_x \\ D_y \\ D_z \end{Bmatrix} \tag{39.30}$$

After the coordinate transformation, all coefficient matrices now contain metrics which are position-dependent, and the system of equations in the most general frame of reference possesses variable coefficients. This added complexity of the characteristic formulation of the Maxwell equations no longer permits the system of one-dimensional equations to be decoupled into six scalar equations and reduced to the true Riemann problem [Shang 1993, Shang and Gaitonde 1993, Shang and Fithen 1994] like that on the Cartesian form.

39.5 Eigenvalues and Eigenvectors

As previously mentioned, eigenvalue and the eigenvector analyses are the prerequisites for characteristic-based algorithms. The analytic process to obtain the eigenvalues and the corresponding eigenvectors of the Maxwell equations in general curvilinear coordinates is identical to that in the Cartesian frame. In each of the temporal–spatial planes $t-\xi$, $t-\eta$, and $t-\zeta$, the eigenvalues are easily found by solving the sixth-degree characteristic equation associated with the coefficient matrices [Sommerfeld 1949, Courant and Hilbert

1965]

$$\lambda_\xi = \left\{ -\frac{\alpha}{V\sqrt{\epsilon\mu}}, -\frac{\alpha}{V\sqrt{\epsilon\mu}}, \frac{\alpha}{V\sqrt{\epsilon\mu}}, \frac{\alpha}{V\sqrt{\epsilon\mu}}, 0, 0 \right\} \tag{39.31}$$

$$\lambda_\eta = \left\{ -\frac{\beta}{V\sqrt{\epsilon\mu}}, -\frac{\beta}{V\sqrt{\epsilon\mu}}, \frac{\beta}{V\sqrt{\epsilon\mu}}, \frac{\beta}{V\sqrt{\epsilon\mu}}, 0, 0 \right\} \tag{39.32}$$

$$\lambda_\zeta = \left\{ -\frac{\gamma}{V\sqrt{\epsilon\mu}}, -\frac{\gamma}{V\sqrt{\epsilon\mu}}, \frac{\gamma}{V\sqrt{\epsilon\mu}}, \frac{\gamma}{V\sqrt{\epsilon\mu}}, 0, 0 \right\} \tag{39.33}$$

where $\alpha = \sqrt{\xi_z^2 + \xi_y^2 + \xi_x^2}$, $\beta = \sqrt{\eta_z^2 + \eta_y^2 + \eta_x^2}$, and $\gamma = \sqrt{\zeta_z^2 + \zeta_y^2 + \zeta_x^2}$

One recognizes that the eigenvalues in each one-dimensional time–space plane contain multiplicities, and hence the eigenvectors do not necessarily have unique elements [Shang 1993, Courant and Hilbert 1965]. Nevertheless, linearly independent eigenvectors associated with each eigenvalue still have been found by reducing the coefficient matrix to the Jordan normal form [Shang 1993, Shang and Fithen 1994]. For reasons of wide applicability and internal consistency, the eigenvectors are selected in such a fashion that the similar matrices of diagonalization reduce to the same form as in the Cartesian frame. Furthermore, in order to accommodate a wide range of electromagnetic field configurations such as antennas, waveguides, and scatterers, the eigenvalues are no longer identical in the three time–space planes. This complexity of formulation is essential to facilitate boundary-condition implementation on the interfaces of media with different characteristic impedances.

From the eigenvector analysis, the similarity transformation matrices for diagonalization in each time-space plane are formed by using eigenvectors as the column arrays as shown in the following equations. For an example, the first column of the similar matrix of diagonalization,

$$\left[-\frac{\sqrt{\mu}\xi_y}{\sqrt{\epsilon}\alpha}, \frac{\sqrt{\mu}(\xi_x^2 + \xi_z^2)}{\sqrt{\epsilon}\xi_x\alpha}, \frac{\sqrt{\mu}\xi_y\xi_z}{\sqrt{\epsilon}\xi_x\alpha}, \frac{\xi_y}{\xi_x}, 0, 1 \right]$$

in the t–ξ plane is the eigenvector corresponding to the eigenvalue $\lambda_\xi = -\alpha/V\sqrt{\epsilon\mu}$. We have

$$S_\xi = \begin{bmatrix} -\dfrac{\sqrt{\mu}\xi_y}{\sqrt{\epsilon}\alpha} & \dfrac{\sqrt{\mu}\xi_z}{\sqrt{\epsilon}\alpha} & \dfrac{\sqrt{\mu}\xi_y}{\sqrt{\epsilon}\alpha} & -\dfrac{\sqrt{\mu}\xi_z}{\sqrt{\epsilon}\alpha} & 1 & 0 \\[2mm] \dfrac{\sqrt{\mu}(\xi_x^2+\xi_z^2)}{\sqrt{\epsilon}\xi_x\alpha} & \dfrac{\sqrt{\mu}\xi_y\xi_z}{\sqrt{\epsilon}\xi_x\alpha} & -\dfrac{\sqrt{\mu}(\xi_x^2+\xi_z^2)}{\sqrt{\epsilon}\xi_x\alpha} & -\dfrac{\sqrt{\mu}\xi_y\xi_z}{\sqrt{\epsilon}\xi_x\alpha} & \dfrac{\xi_y}{\xi_x} & 0 \\[2mm] -\dfrac{\sqrt{\mu}\xi_y\xi_z}{\sqrt{\epsilon}\xi_x\alpha} & -\dfrac{\sqrt{\mu}(\xi_x^2+\xi_y^2)}{\sqrt{\epsilon}\xi_x\alpha} & \dfrac{\sqrt{\mu}\xi_y\xi_z}{\sqrt{\epsilon}\xi_x\alpha} & \dfrac{\sqrt{\mu}(\xi_x^2+\xi_y^2)}{\sqrt{\epsilon}\xi_x\alpha} & \dfrac{\xi_z}{\xi_x} & 0 \\[2mm] -\dfrac{\xi_z}{\xi_x} & -\dfrac{\xi_y}{\xi_x} & -\dfrac{\xi_z}{\xi_x} & -\dfrac{\xi_y}{\xi_x} & 0 & 1 \\[2mm] 0 & 1 & 0 & 1 & 0 & \dfrac{\xi_y}{\xi_x} \\[2mm] 1 & 0 & 1 & 0 & 0 & \dfrac{\xi_z}{\xi_x} \end{bmatrix} \tag{39.34}$$

$$S_\eta = \begin{bmatrix} -\dfrac{(\eta_y^2+\eta_z^2)\sqrt{\mu}}{\sqrt{\epsilon}\eta_y\beta} & -\dfrac{\eta_x\eta_z\sqrt{\mu}}{\sqrt{\epsilon}\eta_y\beta} & \dfrac{(\eta_y^2+\eta_z^2)\sqrt{\mu}}{\sqrt{\epsilon}\eta_y\beta} & \dfrac{\eta_x\eta_z\sqrt{\mu}}{\sqrt{\epsilon}\eta_y\beta} & \dfrac{\eta_x}{\eta_y} & 0 \\[2mm] \dfrac{\eta_x\sqrt{\mu}}{\sqrt{\epsilon}\beta} & -\dfrac{\eta_z\sqrt{\mu}}{\sqrt{\epsilon}\beta} & -\dfrac{\eta_x\sqrt{\mu}}{\sqrt{\epsilon}\beta} & \dfrac{\eta_z\sqrt{\mu}}{\sqrt{\epsilon}\beta} & 1 & 0 \\[2mm] \dfrac{\eta_x\eta_z\sqrt{\mu}}{\sqrt{\epsilon}\eta_y\beta} & \dfrac{(\eta_x^2+\eta_y^2)\sqrt{\mu}}{\sqrt{\epsilon}\eta_y\beta} & -\dfrac{\eta_x\eta_z\sqrt{\mu}}{\sqrt{\epsilon}\eta_y\beta} & -\dfrac{(\eta_x^2+\eta_y^2)\sqrt{\mu}}{\sqrt{\epsilon}\eta_y\beta} & \dfrac{\eta_z}{\eta_y} & 0 \\[2mm] 0 & 1 & 0 & 1 & 0 & \dfrac{\eta_x}{\eta_y} \\[2mm] -\dfrac{\eta_z}{\eta_y} & -\dfrac{\eta_x}{\eta_y} & -\dfrac{\eta_z}{\eta_y} & -\dfrac{\eta_x}{\eta_y} & 0 & 1 \\[2mm] 1 & 0 & 1 & 0 & 0 & \dfrac{\eta_z}{\eta_y} \end{bmatrix} \tag{39.35}$$

$$S_\zeta = \begin{bmatrix} \frac{\sqrt{\mu}(\zeta_y^2+\zeta_z^2)}{\sqrt{\epsilon}\zeta_z\gamma} & \frac{\sqrt{\mu}\zeta_x\zeta_y}{\sqrt{\epsilon}\zeta_z\gamma} & -\frac{\sqrt{\mu}(\zeta_y^2+\zeta_z^2)}{\sqrt{\epsilon}\zeta_z\gamma} & -\frac{\sqrt{\mu}\zeta_x\zeta_y}{\sqrt{\epsilon}\zeta_z\gamma} & \frac{\zeta_x}{\zeta_z} & 0 \\[2mm] -\frac{\sqrt{\mu}\zeta_x\zeta_y}{\sqrt{\epsilon}\zeta_z\gamma} & -\frac{\sqrt{\mu}(\zeta_x^2+\zeta_z^2)}{\sqrt{\epsilon}\zeta_z\gamma} & \frac{\sqrt{\mu}\zeta_x\zeta_y}{\sqrt{\epsilon}\zeta_z\gamma} & \frac{\sqrt{\mu}(\zeta_x^2+\zeta_z^2)}{\sqrt{\epsilon}\zeta_z\gamma} & \frac{\zeta_y}{\zeta_z} & 0 \\[2mm] -\frac{\sqrt{\mu}\zeta_x}{\sqrt{\epsilon}\gamma} & \frac{\sqrt{\mu}\zeta_y}{\sqrt{\epsilon}\gamma} & \frac{\sqrt{\mu}\zeta_x}{\sqrt{\epsilon}\gamma} & -\frac{\sqrt{\mu}\zeta_y}{\sqrt{\epsilon}\gamma} & 1 & 0 \\[2mm] 0 & 1 & 0 & 1 & 0 & \frac{\zeta_x}{\zeta_z} \\[2mm] 1 & 0 & 1 & 0 & 0 & \frac{\zeta_y}{\zeta_z} \\[2mm] -\frac{\zeta_y}{\zeta_z} & -\frac{\zeta_x}{\zeta_z} & -\frac{\zeta_y}{\zeta_z} & -\frac{\zeta_x}{\zeta_z} & 0 & 1 \end{bmatrix} \tag{39.36}$$

Since the similar matrices of diagonalization, S_ξ, S_η, and S_ζ, are nonsingular, the left-hand inverse matrices S_ξ^{-1}, S_η^{-1}, and S_ζ^{-1} are easily found. Although these left-hand inverse matrices are essential to the diagonalization process, they provide little insight for the following flux-vector splitting procedure. The rather involved results are omitted here, but they can be found in [Shang and Fithen 1994].

39.6 Flux-Vector Splitting

An efficient flux-vector splitting algorithm for solving the Euler equations was developed by Steger and Warming [1987]. The basic concept is equally applicable to any hyperbolic differential system for which the solution need not be analytic [Sommerfeld 1949, Courant and Hilbert 1965]. In most CEM applications, discontinuous behavior of the solution is associated only with the wave across an interface between different media, a piecewise continuous solution. Even if a jump condition exists, the magnitude of the finite jump across the interface is much less drastic than the shock waves encountered in supersonic flows. Nevertheless, the salient feature of the piecewise continuous solution domains of the hyperbolic partial differential equation stands out: The coefficient matrices of the time-dependent, three-dimensional Maxwell equations cast in the general curvilinear frame of reference contain metrics of coordinate transformation. Therefore, the equation system no longer has constant coefficients even in an isotropic and homogeneous medium. Under this circumstance, eigenvalues can change sign at any given field location due to the metric variations of the coordinate transformation. Numerical oscillations have appeared in results calculated using the flux-vector splitting technique when eigenvalues change sign. A refined flux-difference splitting algorithm has been developed to resolve fields with jump conditions [Van Leer 1982, Anderson et al. 1985]. The newer flux-difference splitting algorithm is particularly effective at locations where the eigenvalues vanish. Perhaps more crucial for electromagnetics, the polarization of the medium, making the basic equations become nonlinear, occurs only in the extremely high-frequency range [Elliott 1966, Harrington 1961]. In general the governing equations are linear; at most, the coefficients of the differential system are dependent on physical location and phase velocity. For this reason, the difference between the flux-vector splitting [Steger and Warming 1987] and flux-difference splitting [Van Leer 1982, Anderson et al. 1985] schemes, when applied to the time-dependent Maxwell equations, is not of great importance.

The basic idea of the flux-vector splitting of Steger and Warming is to process data according to the direction of information propagation. Since diagonalization is achievable only in each time–space plane, the direction of wave propagation degenerates into either the positive or the negative orientation. This designation is consistent with the notion of the right-running and the left-running wave components. The flux vectors are computed from the point value, including the metrics at the node of interest. This formulation for solving hyperbolic partial differential equations not only ensures the well-posedness of the differential system but also enhances the stability of the numerical procedure [Roe 1986, Shang 1993, Anderson et al. 1984, Richtmyer and Morton 1967]. Specifically, the flux vectors F_ξ, F_η, and F_ζ will be split according to the sign of their corresponding eigenvalues. The split fluxes are differenced by an upwind algorithm to allow for the zone of dependence of an initial-value problem [Roe 1986, Shang 1993, Shang and Gaitonde 1993, Shang and Fithen 1994].

From the previous analysis, it is clear that the eigenvalues contain multiplicities, and hence the split flux of the three-dimensional Maxwell equations is not unique [Shang and Gaitonde 1993, Shang and Fithen

1994]. All flux vectors in each time–space plane are split according to the signs of the local eigenvalues:

$$F_\xi = F_\xi^+ + F_\xi^-$$

$$F_\eta = F_\eta^+ + F_\eta^-$$

$$F_\zeta = F_\zeta^+ + F_\zeta^-$$

(39.37)

The flux-vector components associated with the positive and negative eigenvalues are obtainable by a straightforward matrix multiplication:

$$F_\xi^+ = S_\xi \lambda_\xi^+ S_\xi^{-1} U$$

$$F_\xi^- = S_\xi \lambda_\xi^- S_\xi^{-1} U$$

$$F_\eta^+ = S_\eta \lambda_\eta^+ S_\eta^{-1} U$$

$$F_\eta^- = S_\eta \lambda_\eta^- S_\eta^{-1} U$$

$$F_\zeta^+ = S_\zeta \lambda_\zeta^+ S_\zeta^{-1} U$$

$$F_\zeta^- = S_\zeta \lambda_\zeta^- S_\zeta^{-1} U$$

(39.38)

It is also important to recognize that even if the split flux vectors in each time–space plane are non-unique, the sum of the split components must be unambiguously identical to the flux vector of the governing equation (39.25). This fact is easily verifiable by performing the addition of the split matrices to reach the identities in Eqs. (39.28), (39.29), and (39.30). In addition, if one sets the diagonal elements of metrics, ξ_x, η_y, and ζ_z equal to unity and the off-diagonal elements equal to zero, the coefficient matrices will recover the Cartesian form:

$$F_\xi^+ =
\begin{bmatrix}
\frac{\xi_y^2+\xi_z^2}{2\sqrt{\epsilon\mu}V\alpha} & \frac{-\xi_x\xi_y}{2\sqrt{\epsilon\mu}V\alpha} & \frac{-\xi_x\xi_z}{2\sqrt{\epsilon\mu}V\alpha} & 0 & \frac{-\xi_z}{2\epsilon V} & \frac{\xi_y}{2\epsilon V} \\
\frac{-\xi_x\xi_y}{2\sqrt{\epsilon\mu}V\alpha} & \frac{\xi_x^2+\xi_z^2}{2\sqrt{\epsilon\mu}V\alpha} & \frac{-\xi_y\xi_z}{2\sqrt{\epsilon\mu}V\alpha} & \frac{\xi_z}{2\epsilon V} & 0 & \frac{-\xi_x}{2\epsilon V} \\
\frac{-\xi_x\xi_z}{2\sqrt{\epsilon\mu}V\alpha} & \frac{-\xi_y\xi_z}{2\sqrt{\epsilon\mu}V\alpha} & \frac{\xi_x^2+\xi_y^2}{2\sqrt{\epsilon\mu}V\alpha} & \frac{-\xi_y}{2\epsilon V} & \frac{\xi_x}{2\epsilon V} & 0 \\
0 & \frac{\xi_z}{2V\mu} & \frac{-\xi_y}{2V\mu} & \frac{\xi_y^2+\xi_z^2}{2\sqrt{\epsilon\mu}V\alpha} & \frac{-\xi_x\xi_y}{2\sqrt{\epsilon\mu}V\alpha} & \frac{-\xi_x\xi_z}{2\sqrt{\epsilon\mu}V\alpha} \\
\frac{-\xi_z}{2V\mu} & 0 & \frac{\xi_x}{2V\mu} & \frac{-\xi_x\xi_y}{2\sqrt{\epsilon\mu}V\alpha} & \frac{\xi_x^2+\xi_z^2}{2\sqrt{\epsilon\mu}V\alpha} & \frac{-\xi_y\xi_z}{2\sqrt{\epsilon\mu}V\alpha} \\
\frac{\xi_y}{2V\mu} & \frac{-\xi_x}{2V\mu} & 0 & \frac{-\xi_x\xi_z}{2\sqrt{\epsilon\mu}V\alpha} & \frac{-\xi_y\xi_z}{2\sqrt{\epsilon\mu}V\alpha} & \frac{\xi_x^2+\xi_y^2}{2\sqrt{\epsilon\mu}V\alpha}
\end{bmatrix}
\begin{Bmatrix}
B_x \\ B_y \\ B_z \\ D_x \\ D_y \\ D_z
\end{Bmatrix}$$

(39.39)

$$F_\xi^- =
\begin{bmatrix}
\frac{-(\xi_y^2+\xi_z^2)}{2\sqrt{\epsilon\mu}V\alpha} & \frac{\xi_x\xi_y}{2\sqrt{\epsilon\mu}V\alpha} & \frac{\xi_x\xi_z}{2\sqrt{\epsilon\mu}V\alpha} & 0 & \frac{-\xi_z}{2\epsilon V} & \frac{\xi_y}{2\epsilon V} \\
\frac{\xi_x\xi_y}{2\sqrt{\epsilon\mu}V\alpha} & \frac{-(\xi_x^2+\xi_z^2)}{2\sqrt{\epsilon\mu}V\alpha} & \frac{\xi_y\xi_z}{2\sqrt{\epsilon\mu}V\alpha} & \frac{\xi_z}{2\epsilon V} & 0 & \frac{-\xi_x}{2\epsilon V} \\
\frac{\xi_x\xi_z}{2\sqrt{\epsilon\mu}V\alpha} & \frac{\xi_y\xi_z}{2\sqrt{\epsilon\mu}V\alpha} & \frac{-(\xi_x^2+\xi_y^2)}{2\sqrt{\epsilon\mu}V\alpha} & \frac{-\xi_y}{2\epsilon V} & \frac{\xi_x}{2\epsilon V} & 0 \\
0 & \frac{\xi_z}{2V\mu} & \frac{-\xi_y}{2V\mu} & \frac{-(\xi_y^2+\xi_z^2)}{2\sqrt{\epsilon\mu}V\alpha} & \frac{\xi_x\xi_y}{2\sqrt{\epsilon\mu}V\alpha} & \frac{\xi_x\xi_z}{2\sqrt{\epsilon\mu}V\alpha} \\
\frac{-\xi_z}{2V\mu} & 0 & \frac{\xi_x}{2V\mu} & \frac{\xi_x\xi_y}{2\sqrt{\epsilon\mu}V\alpha} & \frac{-(\xi_x^2+\xi_z^2)}{2\sqrt{\epsilon\mu}V\alpha} & \frac{\xi_y\xi_z}{2\sqrt{\epsilon\mu}V\alpha} \\
\frac{\xi_y}{2V\mu} & \frac{-\xi_x}{2V\mu} & 0 & \frac{\xi_x\xi_z}{2\sqrt{\epsilon\mu}V\alpha} & \frac{\xi_y\xi_z}{2\sqrt{\epsilon\mu}V\alpha} & \frac{-(\xi_x^2+\xi_y^2)}{2\sqrt{\epsilon\mu}V\alpha}
\end{bmatrix}
\begin{Bmatrix}
B_x \\ B_y \\ B_z \\ D_x \\ D_y \\ D_z
\end{Bmatrix}$$

(39.40)

$$
F_\eta^+ = \begin{bmatrix}
\frac{\eta_y^2+\eta_z^2}{2\sqrt{\epsilon\mu}\beta V} & \frac{-\eta_x\eta_y}{2\sqrt{\epsilon\mu}\beta V} & \frac{-\eta_x\eta_z}{2\sqrt{\epsilon\mu}\beta V} & 0 & \frac{-\eta_z}{2\epsilon V} & \frac{\eta_y}{2\epsilon V} \\[2mm]
\frac{-\eta_x\eta_y}{2\sqrt{\epsilon\mu}\beta V} & \frac{\eta_x^2+\eta_z^2}{2\sqrt{\epsilon\mu}\beta V} & \frac{-\eta_y\eta_z}{2\sqrt{\epsilon\mu}\beta V} & \frac{\eta_z}{2\epsilon V} & 0 & \frac{-\eta_x}{2\epsilon V} \\[2mm]
\frac{-\eta_x\eta_z}{2\sqrt{\epsilon\mu}\beta V} & \frac{-\eta_y\eta_z}{2\sqrt{\epsilon\mu}\beta V} & \frac{\eta_x^2+\eta_y^2}{2\sqrt{\epsilon\mu}\beta V} & \frac{-\eta_y}{2\epsilon V} & \frac{\eta_x}{2\epsilon V} & 0 \\[2mm]
0 & \frac{\eta_z}{2V\mu} & \frac{-\eta_y}{2V\mu} & \frac{\eta_y^2+\eta_z^2}{2\sqrt{\epsilon\mu}\beta V} & \frac{-\eta_x\eta_y}{2\sqrt{\epsilon\mu}\beta V} & \frac{-\eta_x\eta_z}{2\sqrt{\epsilon\mu}\beta V} \\[2mm]
\frac{-\eta_z}{2V\mu} & 0 & \frac{\eta_x}{2V\mu} & \frac{-\eta_x\eta_y}{2\sqrt{\epsilon\mu}\beta V} & \frac{\eta_x^2+\eta_z^2}{2\sqrt{\epsilon\mu}\beta V} & \frac{-\eta_y\eta_z}{2\sqrt{\epsilon\mu}\beta V} \\[2mm]
\frac{\eta_y}{2V\mu} & \frac{-\eta_x}{2V\mu} & 0 & \frac{-\eta_x\eta_z}{2\sqrt{\epsilon\mu}\beta V} & \frac{-\eta_y\eta_z}{2\sqrt{\epsilon\mu}\beta V} & \frac{\eta_x^2+\eta_y^2}{2\sqrt{\epsilon\mu}\beta V}
\end{bmatrix}
\begin{Bmatrix} B_x \\ B_y \\ B_z \\ D_x \\ D_y \\ D_z \end{Bmatrix}
\tag{39.41}
$$

$$
F_\eta^- = \begin{bmatrix}
\frac{-(\eta_y^2+\eta_z^2)}{2\sqrt{\epsilon\mu}\beta V} & \frac{\eta_x\eta_y}{2\sqrt{\epsilon\mu}\beta V} & \frac{\eta_x\eta_z}{2\sqrt{\epsilon\mu}\beta V} & 0 & \frac{-\eta_z}{2\epsilon V} & \frac{\eta_y}{2\epsilon V} \\[2mm]
\frac{\eta_x\eta_y}{2\sqrt{\epsilon\mu}\beta V} & \frac{-(\eta_x^2+\eta_z^2)}{2\sqrt{\epsilon\mu}\beta V} & \frac{\eta_y\eta_z}{2\sqrt{\epsilon\mu}\beta V} & \frac{\eta_z}{2\epsilon V} & 0 & \frac{-\eta_x}{2\epsilon V} \\[2mm]
\frac{\eta_x\eta_z}{2\sqrt{\epsilon\mu}\beta V} & \frac{\eta_y\eta_z}{2\sqrt{\epsilon\mu}\beta V} & \frac{-(\eta_x^2+\eta_y^2)}{2\sqrt{\epsilon\mu}\beta V} & \frac{-\eta_y}{2\epsilon V} & \frac{\eta_x}{2\epsilon V} & 0 \\[2mm]
0 & \frac{\eta_z}{2V\mu} & \frac{-\eta_y}{2V\mu} & \frac{-(\eta_y^2+\eta_z^2)}{2\sqrt{\epsilon\mu}\beta V} & \frac{\eta_x\eta_y}{2\sqrt{\epsilon\mu}\beta V} & \frac{\eta_x\eta_z}{2\sqrt{\epsilon\mu}\beta V} \\[2mm]
\frac{-\eta_z}{2V\mu} & 0 & \frac{\eta_x}{2V\mu} & \frac{\eta_x\eta_y}{2\sqrt{\epsilon\mu}\beta V} & \frac{-(\eta_x^2+\eta_z^2)}{2\sqrt{\epsilon\mu}\beta V} & \frac{\eta_y\eta_z}{2\sqrt{\epsilon\mu}\beta V} \\[2mm]
\frac{\eta_y}{2V\mu} & \frac{-\eta_x}{2V\mu} & 0 & \frac{\eta_x\eta_z}{2\sqrt{\epsilon\mu}\beta V} & \frac{\eta_y\eta_z}{2\sqrt{\epsilon\mu}\beta V} & \frac{-(\eta_x^2+\eta_y^2)}{2\sqrt{\epsilon\mu}\beta V}
\end{bmatrix}
\begin{Bmatrix} B_x \\ B_y \\ B_z \\ D_x \\ D_y \\ D_z \end{Bmatrix}
\tag{39.42}
$$

$$
F_\zeta^+ = \begin{bmatrix}
\frac{\zeta_y^2+\zeta_z^2}{2\sqrt{\epsilon\mu}V\gamma} & \frac{-\zeta_x\zeta_y}{2\sqrt{\epsilon\mu}V\gamma} & \frac{-\zeta_x\zeta_z}{2\sqrt{\epsilon\mu}V\gamma} & 0 & \frac{-\zeta_z}{2\epsilon V} & \frac{\zeta_y}{2\epsilon V} \\[2mm]
\frac{-\zeta_x\zeta_y}{2\sqrt{\epsilon\mu}V\gamma} & \frac{\zeta_x^2+\zeta_z^2}{2\sqrt{\epsilon\mu}V\gamma} & \frac{-\zeta_y\zeta_z}{2\sqrt{\epsilon\mu}V\gamma} & \frac{\zeta_z}{2\epsilon V} & 0 & \frac{-\zeta_x}{2\epsilon V} \\[2mm]
\frac{-\zeta_x\zeta_z}{2\sqrt{\epsilon\mu}V\gamma} & \frac{-\zeta_y\zeta_z}{2\sqrt{\epsilon\mu}V\gamma} & \frac{\zeta_x^2+\zeta_y^2}{2\sqrt{\epsilon\mu}V\gamma} & \frac{-\zeta_y}{2\epsilon V} & \frac{\zeta_x}{2\epsilon V} & 0 \\[2mm]
0 & \frac{\zeta_z}{2V\mu} & \frac{-\zeta_y}{2V\mu} & \frac{\zeta_y^2+\zeta_z^2}{2\sqrt{\epsilon\mu}V\gamma} & \frac{-\zeta_x\zeta_y}{2\sqrt{\epsilon\mu}V\gamma} & \frac{-\zeta_x\zeta_z}{2\sqrt{\epsilon\mu}V\gamma} \\[2mm]
\frac{-\zeta_z}{2V\mu} & 0 & \frac{\zeta_x}{2V\mu} & \frac{-\zeta_x\zeta_y}{2\sqrt{\epsilon\mu}V\gamma} & \frac{\zeta_x^2+\zeta_z^2}{2\sqrt{\epsilon\mu}V\gamma} & \frac{-\zeta_y\zeta_z}{2\sqrt{\epsilon\mu}V\gamma} \\[2mm]
\frac{\zeta_y}{2V\mu} & \frac{-\zeta_x}{2V\mu} & 0 & \frac{-\zeta_x\zeta_z}{2\sqrt{\epsilon\mu}V\gamma} & \frac{-\zeta_y\zeta_z}{2\sqrt{\epsilon\mu}V\gamma} & \frac{\zeta_x^2+\zeta_y^2}{2\sqrt{\epsilon\mu}V\gamma}
\end{bmatrix}
\begin{Bmatrix} B_x \\ B_y \\ B_z \\ D_x \\ D_y \\ D_z \end{Bmatrix}
\tag{39.43}
$$

$$
F_\zeta^- = \begin{bmatrix}
\frac{-(\zeta_y^2+\zeta_z^2)}{2\sqrt{\epsilon\mu}V\gamma} & \frac{\zeta_x\zeta_y}{2\sqrt{\epsilon\mu}V\gamma} & \frac{\zeta_x\zeta_z}{2\sqrt{\epsilon\mu}V\gamma} & 0 & \frac{-\zeta_z}{2\epsilon V} & \frac{\zeta_y}{2\epsilon V} \\[2mm]
\frac{\zeta_x\zeta_y}{2\sqrt{\epsilon\mu}V\gamma} & \frac{-(\zeta_x^2+\zeta_z^2)}{2\sqrt{\epsilon\mu}V\gamma} & \frac{\zeta_y\zeta_z}{2\sqrt{\epsilon\mu}V\gamma} & \frac{\zeta_z}{2\epsilon V} & 0 & \frac{-\zeta_x}{2\epsilon V} \\[2mm]
\frac{\zeta_x\zeta_z}{2\sqrt{\epsilon\mu}V\gamma} & \frac{\zeta_y\zeta_z}{2\sqrt{\epsilon\mu}V\gamma} & \frac{-(\zeta_x^2+\zeta_y^2)}{2\sqrt{\epsilon\mu}V\gamma} & \frac{-\zeta_y}{2\epsilon V} & \frac{\zeta_x}{2\epsilon V} & 0 \\[2mm]
0 & \frac{\zeta_z}{2V\mu} & \frac{-\zeta_y}{2V\mu} & \frac{-(\zeta_y^2+\zeta_z^2)}{2\sqrt{\epsilon\mu}V\gamma} & \frac{\zeta_x\zeta_y}{2\sqrt{\epsilon\mu}V\gamma} & \frac{\zeta_x\zeta_z}{2\sqrt{\epsilon\mu}V\gamma} \\[2mm]
\frac{-\zeta_z}{2V\mu} & 0 & \frac{\zeta_x}{2V\mu} & \frac{\zeta_x\zeta_y}{2\sqrt{\epsilon\mu}V\gamma} & \frac{-(\zeta_x^2+\zeta_z^2)}{2\sqrt{\epsilon\mu}V\gamma} & \frac{\zeta_y\zeta_z}{2\sqrt{\epsilon\mu}V\gamma} \\[2mm]
\frac{\zeta_y}{2V\mu} & \frac{-\zeta_x}{2V\mu} & 0 & \frac{\zeta_x\zeta_z}{2\sqrt{\epsilon\mu}V\gamma} & \frac{\zeta_y\zeta_z}{2\sqrt{\epsilon\mu}V\gamma} & \frac{-(\zeta_x^2+\zeta_y^2)}{2\sqrt{\epsilon\mu}V\gamma}
\end{bmatrix}
\begin{Bmatrix} B_x \\ B_y \\ B_z \\ D_x \\ D_y \\ D_z \end{Bmatrix}
\tag{39.44}
$$

39.7 Finite-Difference Approximation

Once the detailed split fluxes are known, formulation of the finite-difference approximation is straightforward. From the sign of an eigenvalue, the stencil of a spatially second- or higher-order-accurate windward differencing can be easily constructed to form multiple one-dimensional difference operators [Shang 1993, Anderson et al. 1984, Richtmyer and Morton 1967]. In this regard, the forward difference and the backward difference approximations are used for the negative and the positive eigenvalues, respectively. The split flux vectors are evaluated at each discretized point of the field according to the signs of the eigenvalues. For the present purpose, a second-order accurate procedure is given:

$$
\begin{aligned}
\text{If} \quad \lambda < 0, \quad & \Delta U_i = [-3U_i + 4U_{i+1} - U_{i+2}]/2 \\
\text{If} \quad \lambda > 0, \quad & \nabla U_i = [\ 3U_i - 4U_{i-1} + U_{i-2}]/2
\end{aligned}
\tag{39.45}
$$

The necessary metrics of the coordinate transformation are calculated by central-differencing, except at the edges of computational domain, where one-sided differences are used. Although the fractional-step or the time-splitting algorithm [Shang 1993, Anderson et al. 1984, Richtmyer and Morton 1967] has demonstrated greater efficiency in data storage and a higher data-processing rate than predictor–corrector time integration procedures [Shang 1993, Shang and Gaitonde 1993, Shang and Fithen 1994], it is limited to second-order accuracy in time. With respect to the fractional-step method, the temporal second-order result is obtained by a sequence of symmetrically cyclic operators [Shang 1993, Richtmyer and Morton 1967]:

$$
U^{n+2} = L_\xi L_\eta L_\zeta L_\zeta L_\eta L_\xi U^n
\tag{39.46}
$$

where L_ξ, L_η, and L_ζ are the difference operators for one-dimensional equations in the ξ, η, and ζ coordinates, respectively.

In general, second-order and higher temporal resolution is achievable through multiple-time-step schemes [Anderson et al. 1984, Richtmyer and Morton 1967]. However, one-step schemes are more attractive because they have less memory requirements and don't need special startup procedures [Shang and Gaitonde 1993, Shang and Fithen 1994]. For future higher-order accurate solution development potential, the Runge–Kutta family of single-step, multistage procedures is recommended. This choice is also consistent with the accompanying characteristic-based finite-volume method [Shang and Gaitonde 1993]. In the present effort, the two-stage, formally second-order accurate scheme is used:

$$
\begin{aligned}
U_0 &= U_n \\
U_1 &= U_0 - \Delta U\,(U_0) \\
U_2 &= U_0 - 0.5\,(\Delta U\,(U_1) + \Delta U\,(U_0)) \\
U_{n+1} &= U_2
\end{aligned}
\tag{39.47}
$$

where ΔU comprises the incremental values of dependent variables during each temporal sweep. The resultant characteristic-based finite-difference scheme for solving the three-dimensional Maxwell equations in the time domain is second-order accurate in both time and space.

The most significant feature of the flux-vector splitting scheme lies in its ability to easily suppress reflected waves from the truncated computational domain. In wave motion, the compatibility condition at any point in space is described by the split flux vector [Shang 1993, Shang and Gaitonde 1993, Shang and Fithen 1994]. In the present formulation, an approximated no-reflection condition can be achieved by setting the incoming flux component equal to zero:

$$
\text{either} \quad \lim_{r \to \infty} F^+(\xi, \eta, \zeta) = 0 \quad \text{or} \quad \lim_{r \to \infty} F^-(\xi, \eta, \zeta) = 0
\tag{39.48}
$$

The one-dimensional compatibility condition is exact when the wave motion is aligned with one of the coordinates [Shang 1993]. This unique attribute of the characteristic-based numerical procedure in removing a fundamental dilemma in computational electromagnetics will be demonstrated in detail later.

39.8 Finite-Volume Approximation

The finite-volume approximation solves the governing equation by discretizing the physical space into contiguous cells and balancing the flux-vectors on the cell surfaces. Thus in discretized form, the integration procedure reduces to evaluation of the sum of all fluxes aligned with surface-area vectors

$$\frac{\Delta U}{\Delta t} + \frac{\Delta F}{\Delta \xi} + \frac{\Delta G}{\Delta \eta} + \frac{\Delta H}{\Delta \zeta} - J = 0 \tag{39.49}$$

In the present approach, the continuous differential operators have been replaced by discrete operators. In essence, the numerical procedure needs only to evaluate the sum of all flux vectors aligned with surface-area vectors [Shang and Gaitonde 1993, Shang and Fithen 1994, Van Leer 1982, Anderson et al. 1985]. Only one of the vectors is required to coincide with the outward normal to the cell surface, and the rest of the orthogonal triad can be made to lie on the same surface. The metrics, or more appropriately the direction cosines, on the cell surface are uniquely determined by the nodes and edges of the elementary volume. This feature is distinct from the finite-difference approximation. The shape of the cell under consideration and the stretching ratio of neighbor cells can lead to a significant deterioration of the accuracy of finite-volume schemes [Leonard 1988].

The most outstanding aspect of finite-volume schemes is the elegance of its flux-splitting process. The flux-difference splitting for Eq. (39.25) is greatly facilitated by a locally orthogonal system in the transformed space [Van Leer 1982, Anderson et al. 1985]. In this new frame of reference, eigenvalues and eigenvectors as well as metrics of the coordinate transformation between two orthogonal systems are well known [Shang 1993, Shang and Gaitonde 1993]. The inverse transformation is simply the transpose of the forward mapping. In particular, the flux vectors in the transformed space have the same functional form as that in the Cartesian frame. The difference between the flux vectors in the transformed and the Cartesian coordinates is a known quantity and is given by the product of the surface outward normal and the cell volume, $V(\nabla S / \|\nabla S\|)$ [Shang and Gaitonde 1993]. Therefore, the flux vectors can be split in the transformed space according to the signs of the eigenvalues but without detailed knowledge of the associated eigenvectors in the transformed space. This feature of the finite-volume approach provides a tremendous advantage over the finite-difference approximation in solving complex problems in physics.

The present formulation adopts Van Leer's kappa scheme in which solution vectors are reconstructed on the cell surface from the piecewise data of neighboring cells [Van Leer 1982, Anderson et al. 1985]. The spatial accuracy of this scheme spans a range from first-order to third-order upwind biased approximations,

$$U^+_{i+\frac{1}{2}} = U_i + \frac{\phi}{4} \left[(1 - \kappa) \nabla + (1 + \kappa) \Delta \right] U_i$$

$$U^-_{1+\frac{1}{2}} = U_i - \frac{\phi}{4} \left[(1 + \kappa) \nabla + (1 - \kappa) \Delta \right] U_{i+1} \tag{39.50}$$

where $\Delta U_i = U_i - U_{i-1}$ and $\nabla U_i = U_{i+1} - U_i$ are the forward and backward differencing discretizations. The parameters ϕ and κ control the accuracy of the numerical results. For $\phi = 1$, $\kappa = -1$ a two-point windward scheme is obtained. This method has an odd-order leading truncation-error term; the dispersive error is expected to dominate. If $\kappa = \frac{1}{3}$, a third-order upwind-biased scheme will emerge. In fact both upwind procedures have discernible leading phase error. This behavior is a consequence of using the two-stage time integration algorithm, and the dispersive error can be alleviated by increasing the temporal resolution. For $\phi = 1$, $\kappa = 0$ the formulation recovers the Fromm scheme [Van Leer 1982, Anderson et al. 1985]. If $\kappa = 1$, the formulation yields the spatially central scheme. Since the fourth-order dissipative term is suppressed, the central scheme is susceptible to parasitic odd–even point decoupling [Anderson et al. 1984, 1985].

The time integration is carried out by the same two-stage Runge–Kutta method as in the present finite-difference procedure [Shang 1993, Shang and Gaitonde 1993]. The finite-volume procedure is therefore second-order accurate in time and up to third-order accurate in space [Shang and Gaitonde 1993, Shang and Fithen 1994]. For the present purpose, only the second-order upwinding and the third-order upwind biased options are exercised. The second-order windward schemes in the form of the flux-vector splitting finite-difference and the flux-difference splitting finite-volume scheme are formally equivalent [Shang and Gaitonde 1993, Shang and Fithen 1994, Van Leer 1982, Anderson et al. 1985, Leonard 1988].

39.9 Summary and Research Issues

The technical merits of the characteristic-based methods for solving the time-dependent, three-dimensional Maxwell equations can best be illustrated by the following two illustrations. In Fig. 39.2, the exact electrical field of a traveling wave is compared with numerical results. The numerical results were generated at the maximum allowable time-step size defined by the Courant–Friedrichs–Lewy (CFL) number of 2 ($\lambda \, \Delta x / \Delta t = 2$) [Anderson et al. 1984, Richtmyer and Morton 1967]. The numerical solutions presented are at instants when a right-running wave reaches the midpoint of the computational domain and exits the numerical boundary respectively. For this one-dimensional simulation, the characteristic-based scheme using the single-step upwind explicit algorithm exhibits the shift property, which indicates a perfect translation of the initial value in space [Anderson et al. 1984]. As the impulse wave moves through the initially quiescent environment, the numerical result duplicates the exact solution at each and every discretized point, including the discontinuous incoming wavefront. Although this highly desirable property of a numerical solution is achievable only under very restrictive conditions and is not preserved for multidimensional problems [Anderson et al. 1984, Richtmyer and Morton 1967], the ability to simulate the nonanalytic solution behavior in the limit is clearly illustrated.

In Fig. 39.3, another outstanding feature of the characteristic-based method is highlighted by simulating the oscillating electric dipole. For the radiating electric dipole, the depicted temporal calculations are sampled at the instant when the initial pulse has traveled a distance of 2.24 wavelengths from the dipole.

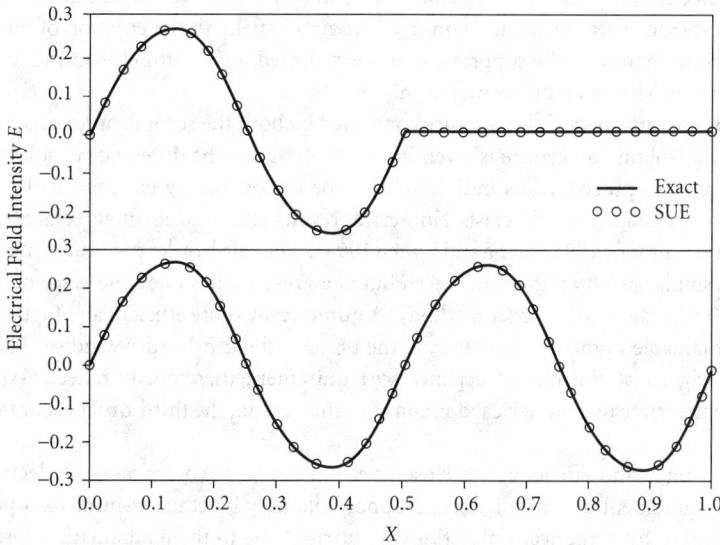

FIGURE 39.2 Perfect-shift property of a one-dimensional wave computation, CFL = 2.

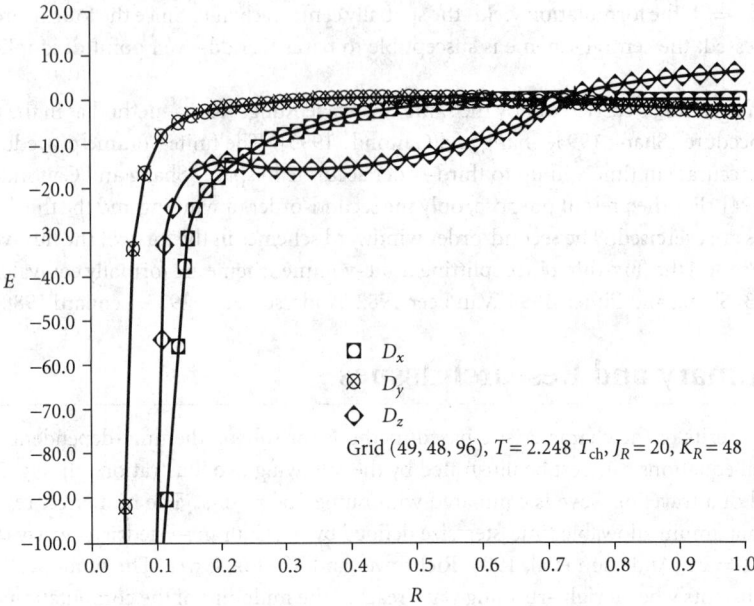

FIGURE 39.3   Instantaneous distributions of oscillating-dipole electric field.

The numerical results are generated on a $48 \times 48 \times 96$ mesh system with the second-order scheme. Under that condition each wavelength is resolved by 15 mesh points and the difference between numerical results by the finite-volume and the finite-difference method is negligible. However, on an irregular mesh like the present spherical polar system, the finite-volume procedure has shown a greater numerical error than the finite-difference procedure on highly stretched grid systems [Shang and Fithen 1994, Anderson et al. 1985]. Under the present computational conditions, both numerical procedures yield uniformly excellent comparison with the theoretical result. Significant error appeared only in the immediate region of the coordinate origin. At that location, the solution has a singularity that behaves as the inverse cube of the radial distance. The most interesting numerical behavior, however, is revealed in the truncated far field. The no-reflection condition at the numerical boundary is observed to be satisfied within the order of the truncation error. For a spherically symmetric radiating field, the orientation of the wave is aligned with the radial coordinate, and the suppression of the reflected wave within the numerical domain is the best achievable by the characteristic formulation.

The corresponding magnetic field intensity computed by both the second-order accurate finite-difference and the finite-volume procedure is given in Fig. 39.4. Again the difference in solution between the two distinct numerical procedures is indiscernible. For the oscillating electric dipole, only the x and y components of the magnetic field exist. Numerical results attain excellent agreement with theoretical values [Shang and Gaitonde 1993, Shang and Fithen 1994]. The third-order accurate finite-volume scheme also produces a similar result on the same mesh but at a greater allowable time-step size (a CFL value of 0.87 is used vs. 0.5 for the second-order method). A numerically more efficient and higher-order accurate simulation is obtainable in theory. However, at the present, the third-order windward biased algorithm cannot reinforce rigorously the zone-of-dependence requirement; therefore the reflected-wave suppression is incomplete in the truncated numerical domain. For this reason, the third-order accurate results are not included here.

Numerical accuracy and efficiency are closely related issues in computational electromagnetics. A high-accuracy requirement of a simulation is supportable only by efficient numerical procedures. The inaccuracies incurred by numerical simulations are attributable to the mathematical formulation of the problem, to the algorithm, to the numerical procedure, and to computational inaccuracy. A basic approach to relieve the accuracy limitation must be derived from using high-order schemes or spectral methods.

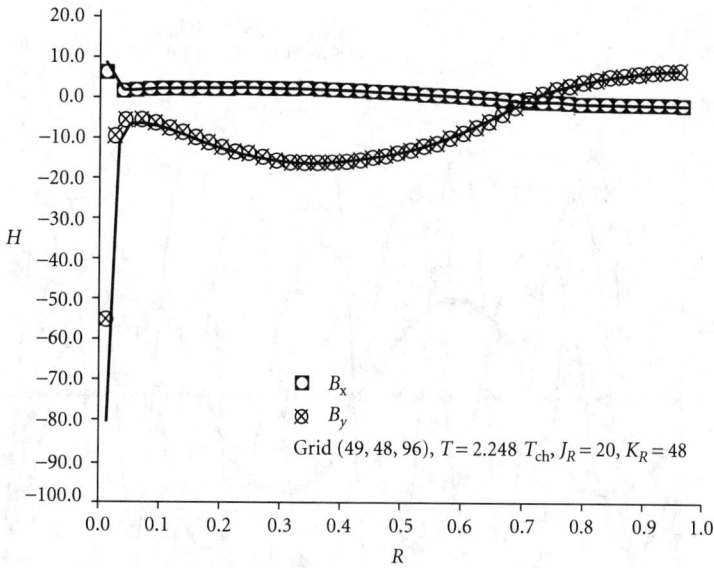

FIGURE 39.4 Instantaneous distributions of oscillating-dipole magnetic field.

The numerical efficiency of CEM can be enhanced substantially by using scalable multicomputers [Shang et al. 1993]. The effective use of a distributed-memory, message-passing homogeneous multicomputer still requires a judicious tradeoff between a balanced work load and interprocessor communication. A characteristic-based finite-volume computer program has been successfully mapped onto distributed-memory systems by a rudimentary domain decomposition strategy [Shang et al. 1993]. For example, a square waveguide, at five different frequencies up to the cutoff, was simulated.

Figure 39.5 displays the x-component of the magnetic field intensity within the waveguide. The simulated transverse electric mode, $TE_{1,1}$, $E_x = 0$, which has a half period of π along the x and y coordinates, is generated on a $24 \times 24 \times 128$ mesh system. Since the entire field is described by simple harmonic functions, the remaining field components are similar and only half the solution domain along the z-coordinate is presented to minimize repetition. In short, the agreement between the closed-form and numerical solutions is excellent at each frequency. In addition, the numerical simulations duplicate the physical phenomenon at the cutoff frequency, below which there is no phase shift along the waveguide and the wave motion ceases [Elliott 1966, Harrington 1961]. For simple harmonic wave motion in an isotropic medium, the numerical accuracy can be quantified. At a grid-point density of 12 nodes per wavelength, the L2 norm [Richtmyer and Morton 1967] has a nearly uniform magnitude of order 10^{-4}. The numerical results are fully validated by comparison with theory. However, further efforts are still required to substantially improve the parallel and scalable numerical efficiency. In fact, this is the most promising area in CEM research.

The pioneering efforts in CEM usually employed the total-field formulation on staggered-mesh systems [Yee 1966, Taflove 1992]. That particular combination of numerical algorithm and procedure has been proven to be very effective. In earlier RCS calculations using the present numerical procedure, the total-field formulation was also utilized [Shang and Gaitonde 1994, Shang et al. 1993]. However, for three-dimensional scatterer simulation, the numerical accuracy requirement for RCS evaluations becomes extremely stringent. In the total-field formulation, the dynamic range of the field variables has a substantial difference from the exposed and the shadow region, and the incident wave must also traverse the entire computation domain. Both requirements impose severe demands on the numerical accuracy of simulation. In addition, the total field often contains only the residual of partial cancelations of the incident and the diffracted waves. The far-field electromagnetic energy distribution becomes a secular problem—a small difference between two variables of large magnitude. An alternative approach via the

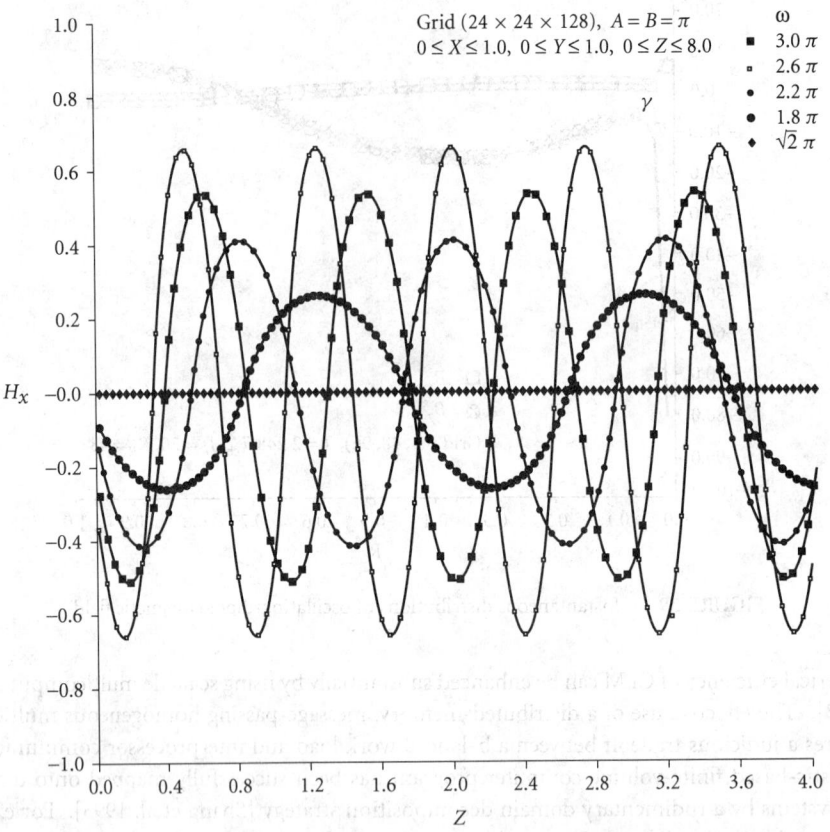

FIGURE 39.5 Cutoff frequency of a square waveguide, TE$_{1,1}$.

scattered-field formulation for RCS calculations appears to be very attractive. Particularly in this formulation, the numerical dissipation of the incident wave that must propagate from the far-field boundary to the scatterer is completely eliminated from the computations. In short, the incident field can be directly specified on the scatterer's surface. The numerical advantage over the total-field formulation is substantial.

The total-field formulation can be cast in the scattered-field form by replacing the total field with scattered field variables [Elliott 1966, Harrington 1961]:

$$U_s = U_t - U_i \qquad (39.51)$$

Since the incident field U_i must satisfy the Maxwell equations identically, the equations of the scattered field remain unaltered from the total-field formulation. Thus, the scattered-field formulation can be considered as a dependent-variable transform of the total-field equations. In the present approach, both formulations are solved by a characteristic-based finite-volume scheme.

The comparison of horizontal polarized RCS of a perfect electrically conducting (PEC) sphere, $\sigma(\theta, 0.0)$, from the total-field and scattered-field formulations at $ka = 5.3$ (where k = wave number and a = diameter of the sphere) is presented in Fig. 39.6. The validating datum is the exact solution for the scattering of a plane electromagnetic wave by a PEC sphere, which is commonly referred to as the Mie series [Elliott 1966]. Both numerical results are generated under identical computational conditions. The location of the truncated far-field boundary is prescribed at 2.5 wavelengths from the center of the PEC sphere. Numerical results of the total-field formulation reveal far greater error than for the scattered-field formulation. The additional source of error is incurred when the incident wave must propagate from the far-field boundary to the scatterer. In the scattered-field formulation, the incident field data are

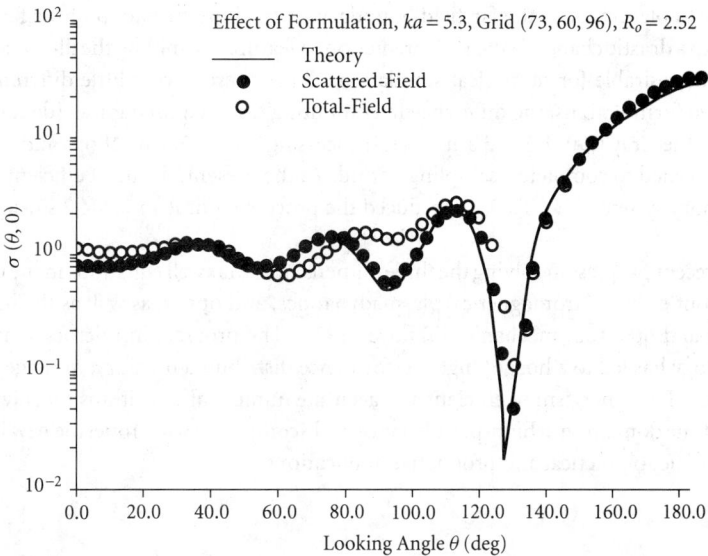

FIGURE 39.6 Comparison of total-field and scattered-field RCS calculations of $\sigma(\theta, 90°)$, $ka = 5.3$.

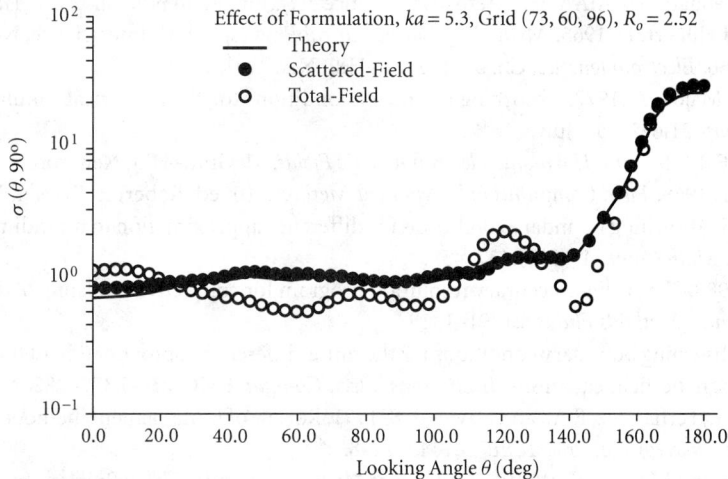

FIGURE 39.7 Comparison of total-field and scattered-field RCS calculations of $\sigma(\theta, 90°)$, $ka = 5.3$.

described precisely by the boundary condition on the scatterer surface. Since the far-field electromagnetic energy distribution is derived from the near-field parameters [Taflove 1992, Sommerfeld 1949, Shang and Gaitonde 1993], the advantage of describing the data incident on a scatterer without error is tremendous. Numerical errors of the total-field calculations are evident in the exaggerated peaks and troughs over the entire viewing-angle displacement.

In Fig. 39.7, the vertically polarized RCS $\sigma(\theta, 90.0°)$ of the $ka = 5.3$ case substantiates the previous observation. In fact, the numerical error of the total-field calculation is excessive in comparison with the result of the scattered-field formulation. Since the results are not obtained for the optimal far-field placement for RCS computation, the results of the scattered-field formulation overpredict the theoretical values by 2.7 percent. The deviation of the total-field result from the theory, however, exceeds 25.6 percent and becomes unacceptable. In addition, computations by the total-field formulation exhibit a strong sensitivity to placement of the far-field boundary. A small perturbation of the far-field boundary placement leads to a drastic change in the RCS prediction: a feature resembling the ill-posedness condition, which is highly undesirable for numerical simulation. Since there is very little difference in computer coding for the two formulations, the difference in computing time required for an identical simulation is insignificant. On the Cray C90, 1,505.3 s at a data-processing rate of 528.8 Mflops and an average vector length of 62.9 is needed to complete a sampling period. At the present, the most efficient calculation on a distributed memory system, Cray T3D, has reduced the processing time to 1,204.2 s using 76 computing nodes.

In summary, recent progress in solving the three-dimensional Maxwell equations in the time domain has opened a new frontier in electromagnetics, plasmadynamics, and optics, as well as the interface between electrodynamics and quantum mechanics [Taflove 1992]. The progress in microchip and interconnect network technology has led to a host of high-performance distributed memory, message-passing parallel computer systems. The synergism of efficient and accurate numerical algorithms for solving the Maxwell equations in the time domain with high-performance multicomputers will propel the new interdisciplinary simulation technique to practical and productive applications.

References

Anderson, D. A., Tannehill, J. C., and Pletcher, R. H. 1984. *Computational Fluid Mechanics and Heat Transfer*. Hemisphere, New York.

Anderson, W. K., Thomas, J. L., and Van Leer, B. 1985. A comparison of finite-volume flux splittings for the Euler equations., AIAA 23rd Aerospace Science Meeting, Reno NV, Jan. *AIAA Paper* 85-0122.

Courant, R. and Hilbert, D. 1965. *Methods of Mathematical Physics*, Vol. II. Interscience, New York.

Elliott, R. A. 1966. *Electromagnetics*, Ch. 5. McGraw–Hill, New York.

Enquist, B. and Majda, A. 1977. Absorbing boundary conditions for the numerical simulation of waves. *Math Comp.* 31:629-651, July.

Harrington, R. F. 1961. *Time-Harmonic Electromagnetic Fields*. McGraw–Hill, New York.

Harrington, R. F. 1968. *Field Computation by Moment Methods*, 4th ed. Robert E. Krieger, Malabar, FL.

Higdon, R. 1986. Absorbing boundary conditions for difference approximation to multidimensional wave equation. *Math Comp.* 47(175):437–459.

Leonard, B. P. 1988. Simple high-accuracy resolution program for convective modeling of discontinuities. *Int. J. Numer. Methods Fluids* 8:1291–1318.

Mur, G. 1981. Absorbing boundary conditions for the finite-difference approximation of the time-domain electromagnetic-field equations. *IEEE Trans. Elect. Compat.* EMC-23(4):377–382, Nov.

Rahman, B. M. A., Fernandez, F. A., and Davies, J. B. 1991. Review of finite element methods for microwave and optical waveguide. *Proc. IEEE* 79:1442, 1448.

Richtmyer, R. D. and Morton, K. W. 1967. *Difference Methods for Initial-Value Problem*. Interscience, New York.

Roe, P. L. 1986. Characteristic-based schemes for the Euler equations. *Ann. Rev. Fluid Mech.* 18:337–365.

Shang, J. S. 1993. A fractional-step method for solving 3-D, time-domain Maxwell equations. AIAA 31st Aerospace Science Meeting, Reno NV, Jan. *AIAA Paper* 93-0461; *J. Comput. Phys.* Vol. 118(1):109–119, Apr. 1995.

Shang, J. S. and Fithen, R. M. 1994. A comparative study of numerical algorithms for computational electromagnetics. AIAA 25th Plasmadynamics and Laser Conference, Colorado Springs, June 20–23, *AIAA Paper* 94-2410.

Shang, J. S. and Gaitonde, D. 1993. Characteristic-based, time-dependent Maxwell equation solvers on a general curvilinear frame. AIAA 24th Fluid Dynamics, Plasmadynamics, and Laser Conference, Orlando FL, July. *AIAA Paper* 93-3178; *AIAA J.* 33(3):491–498, Mar. 1995.

Shang, J. S. and Gaitonde, D. 1994. Scattered electromagnetic field of a reentry vehicle. AIAA 32nd Aerospace Science Meeting, Reno NV, Jan. *AIAA Paper* 94-0231; *J. Spacecraft and Rockets* 32(2):294–301, Mar.–Apr. 1995.

Shang, J. S., Hill, K. C., and Calahan, D. 1993. Performance of a characteristic-based, 3-D, time-domain Maxwell equations solver on a massively parallel computer. AIAA 24th Plasmadynamics & Lasers Conference, Orlando FL, July 6–9 *AIAA Paper* 3179; *Appl. Comput. Elect. Soc.* 10(1):52–62, Mar. 1995.

Sommerfeld, A. 1949. *Partial Differential Equations in Physics*, Ch. 2. Academic Press, New York.

Steger, J. L. and Warming, R. F. 1987. Flux vector splitting of the inviscid gas dynamics equations with application to finite difference methods. *J. Comput. Phys.* 20(2):263–293, Feb.

Taflove, A. 1992. Re-inventing electromagnetics: supercomputing solution of Maxwell's equations via direct time integration on space grids. *Comput. Systems Eng.* 3(1–4):153–168.

Thompson, J. F. 1982. *Numerical Grid Generation*. Elsevier Science, New York.

Van Leer, B. 1982. Flux-vector splitting for the Euler equations. *TR* 82-30, ICASE, Sept., pp. 507–512, *Lecture Notes in Physics*, Vol. 170.

Yee, K. S. 1966. Numerical solution of initial boundary value problems involving Maxwell's equations. In *Isotropic Media, IEEE Trans. Ant. Prop.* 14(3):302–307.

40

Computational Ocean Modeling

Lakshmi Kantha
Naval Research Laboratory and University of Colorado

Steve Piacsek
Naval Research Laboratory

40.1 Introduction

Oceanography is a relatively young field, barely a century old; major discoveries—such as the reason for the western intensification of currents such as the Gulf Stream and Kuroshio, and the existence of a deep sound channel in which acoustic energy can travel for thousands of kilometers with little attenuation—were not made till the 1940s. Even today, our knowledge of the circulation in the global oceans is rather sketchy. Computational ocean modeling is even younger; the very first comprehensive numerical global ocean model was formulated by Kirk Bryan [1969] in the late sixties. However, the advent of supercomputers has led to a phenomenal growth in the field, especially in the last decade. In a brief article such as this, it is impossible to provide a detailed account of the many different versions of the ocean models that exist at present. Instead we will attempt to provide a bird's-eye view of the field and a detailed account of a selected few. The objective is to provide a road map that enables an interested reader to consult appropriate sources for details of a particular approach.

Oceans act as thermal flywheels and moderate our long-term weather. They are also huge reservoirs of CO_2 and have long memory and therefore play a crucial role in determining the climatic conditions on our planet on a variety of time scales. A better understanding of the oceans is also important for other reasons, including defense and commerce needs of nations. They are a source of protein, and might be able to supply part of our energy and mineral needs in the coming century. However, the oceans are data-poor in general. It is only in the last decade or so that satellite-borne sensors such as infrared radiometers, microwave imagers, and altimeters have begun to fill in the data gaps, especially in the poorly

0-8493-2909-4/97/$0.00+$.50
© 1997 by CRC Press, Inc.

explored southern-hemisphere oceans. Since collection of in situ data in the oceans is quite expensive, and since satellite-borne sensors provide information mostly on the near-surface layers of the ocean, it is often thought that ocean models are central to understanding the way the oceans function. The hope is that comprehensive ocean models in combination with sparse in situ and relatively abundant remotely sensed data will provide the best means of studying and monitoring the oceans. Herein lies the importance of ocean modeling. For prediction purposes, of course, numerical ocean models are quite indispensable.

40.2 Underlying Principles

The choice of a particular ocean model or modeling approach depends very much on the intended application and on the computational and pre- and postprocessing capabilities available. A judicious compromise is essential for success. With this in mind, numerical ocean models can be classified in many different ways.

Global or Regional

The former necessarily requires supercomputing capabilities, whereas it may be possible to run the latter on powerful modern workstations. Even then, the resolution demanded (grid sizes in the horizontal and vertical) is critical. A doubling of the resolution in a three-dimensional model often requires an order of magnitude increase in computing (and analysis) resources. It is therefore quite easy to overwhelm even the most modern supercomputer (or workstation), whether it be a coarse-grained multiple CPU vector processor such as a Cray C-90 or a modern massively parallel machine such as a CM5 or Cray T3D, irrespective of whether the model is global or regional. Regional models have to contend with the problem of how to inform the model about the state of the rest of the ocean, in other words, of prescribing suitable conditions along the open boundaries. Often the best solution is to nest the fine-resolution regional model in a coarse-resolution model of the basin.

Deep Basin or Shallow Coastal

The prevailing physical processes and the underlying driving mechanisms are essentially different for the two cases. Circulation in shallow coastal regions is highly variable, driven primarily by synoptic wind and other rapidly changing surface forcing (and, near river outflows, by buoyancy differences between the fresh river water and saline ambient shelf water). Wind mixing at the surface and benthic processes are important, and a numerical model that has reliable mixing physics and that resolves the benthic boundary layer is therefore better suited for coastal applications. A model such as the one developed at Princeton University, which employs a bottom-following, sigma-coordinate vertical grid and incorporates an advanced turbulence closure [Blumberg and Mellor 1987; Kantha and Piacsek 1993], may be essential for such applications.

Deep basins, on the other hand, are comparatively sluggish, and the horizontal density gradients, especially below the wind-mixed upper layers, are a dominant factor in the circulation. The upper mixed layer can often be modeled less rigorously, especially for applications that do not require consideration of air–sea interaction processes. The popular z-level Geophysical Fluid Dynamics Laboratory Modular Ocean Model (MOM), with or without an upper mixed layer, is a good candidate for modeling the ocean basins (and deep marginal and semienclosed seas) on a variety of time scales ranging from synoptic to climate. The very first global ocean model [Bryan 1969], on which many modern global models are based [Philander et al., 1987; Semtner and Chervin 1992], was a z-level model without an upper mixed layer.

In a z-level model, a number of horizontal levels are defined in the water column, and the equations are written for the oceanic variables at each level and each point on the model grid in the horizontal and solved. This is an Eulerian approach. Another equally viable approach is a semi-Lagrangian approach [Hurlburt and Thompson 1980] that divides the ocean vertically into a number of layers and models the variation in properties such as the thickness and density of each layer at each grid point on the horizontal grid. More recently developed isopycnal models [Oberhuber 1993, Bleck and Smith 1990] belong to this category. Since mixing in the deep oceans is primarily along **isopycnals** (density surfaces), isopycnal models perform a better job of depicting interior mixing and are ideal for long-term simulations of circulation in the global oceans.

Rigid Lid or Free Surface

Oceanic response to surface forcing can often be divided into two parts: fast barotropic response mediated by external Kelvin and gravity waves on the sea surface, and relatively slow baroclinic adjustment via internal gravity and other waves. On long time scales, it is the internal adjustment that is important to model and it is possible to suppress the external gravity waves by imposing a *rigid lid* on the free surface. This permits larger time-stepping of the model, and models used for climatic-type simulations are usually the rigid-lid kind. The very first global ocean model [Bryan 1969] was a rigid-lid model. At each time step an elliptic Poisson equation for the stream function has to be solved. This is difficult to carry out efficiently on vector and parallel processors and for complicated basin shapes (including islands). Also, under synoptic forcing, the convergence of the iterative solver slows down. For these reasons, free-surface models are becoming more popular for nonclimatic simulations. A mode-splitting technique must then be employed to circumvent the severe limitation on time-stepping that would otherwise be imposed. For shallow-water applications, such as storm-surge and tide modeling, free-surface dynamics must be retained. To diminish the drawbacks of a rigid-lid model, Dietrich et al. [1987] and Dukowicz and Smith [1994] have developed versions in which one works with the pressure on the rigid lid, rather than the barotropic stream function, leaving the domain multiply connected and with better matrix inversion characteristics.

Comprehensive or Purely Dynamical

Since the density gradients are overwhelmingly important and the density below the upper layers in the global oceans changes very slowly, it is often possible to ignore completely the changes in density with time. The model then becomes purely dynamical and can be used to explore the consequences of changing wind forcing at the surface. Purely dynamical layered models, originally developed at Florida State University [Metzger et al. 1992; Wallcraft 1991], belong to this category and are essentially isopycnal models without the thermodynamic component [Hurlburt and Thompson 1980]. Their principal advantage is that it is often possible to select a limited number of layers in the vertical (as few as two) and still include salient dynamical processes. This enables very high horizontal resolutions necessary for resolving mesoscale eddies in the oceans to be afforded. The highest-resolution global model at present is the Naval Research Laboratory $\frac{1}{8}°$ eddy-resolving global model [Metzger et al. 1992] that needs a 16-processor Cray C-90 for multiyear simulations.

Even with modern computing power, it is necessary to sacrifice either vertical or horizontal resolution for many (especially global) simulations. The layered (and isopycnal) models sacrifice vertical resolution, whereas the z-level models, employing large numbers of levels in the vertical, are necessarily comparatively coarse-grained in the horizontal. The highest-resolution dynamical–thermodynamic z-level model at present is the $\frac{1}{6}°$ Semtner model at Los Alamos (Dukowicz and Smith 1994; see Semtner and Chervin 1992 for a description of the basic model) and it stretches the capability of a 256-processor CM5 to the limit.

With Applications to Short-Term Simulations or Long-Term Climate Studies

On climatic time scales, it is extremely important to model correctly the thermohaline circulation driven by the formation of dense deep-water masses during strong wintertime cooling in subpolar seas, especially in the Atlantic. These water masses flow along the ocean bottom, and several centuries are needed for a water particle that sank (say) in the north Atlantic to surface again in the Indian or the Pacific Ocean. Because of this long memory of the deep oceans, it is necessary to make multicentury simulations, and, irrespective of whether isopycnal or *z*-level models are employed, the horizontal and vertical resolutions that can be afforded are necessarily coarse. Accurate ocean simulations on climatic time scales belong to the category of grand challenge problems requiring a teraflop (10^{12} floating-point operations per second) computing capability that has been the holy grail of the computer industry.

Quasigeostrophic (QG) or Primitive-Equation-Based

In the seventies and early eighties, the limited computing power available led some to explore simplifications to the governing equations to be solved. QG models assume that there is a near-balance between the Coriolis acceleration and pressure gradient in the dynamical equations in the rotating coordinate frame of reference in which most ocean models are formulated. The resulting simplification enables higher vertical and horizontal resolutions to be achieved. QG models have strong limitations with respect to the accuracy with which physical processes are depicted and are becoming obsolete in the modern high-computing-power environment. We will not discuss QG models in this article, but instead refer the reader to Holland [1986]. Neither shall we discuss intermediate models, which are in between QG and primitive equation (PE) models in complexity.

Barotropic or Baroclinic

In the former, the density gradients are neglected and therefore the currents become independent of the depth in the water column. Many phenomena, such as tidal sea-surface elevation fluctuations and storm surges, can be simulated quite adequately by a **barotropic** model, which is a two-dimensional (in the horizontal) model based on the vertically integrated equations of motion. The advantage is that a barotropic model requires an order of magnitude less computing resources than a comparable baroclinic model. However, when it is important to model the vertical structure of currents, or the density field, a fully three-dimensional **baroclinic** model is necessary.

Purely Physical or Physical–Chemical–Biological

Often there is a need to model the fate of chemical and biological constituents in the ocean, and to do so it is essential to include not only the dynamical equations governing the circulation and other physical variables, but also the conservation equations for chemical and biological variables. Modeling the fate of inorganic CO_2 in the oceans (a problem germane to global warming) and modeling the primary production in the upper layers of the ocean are two such examples. The former requires solving for at least two more variables, the total CO_2 and alkalinity, whereas the simplest biological model must solve for at least three additional quantities, the nutrient, phytoplankton, and zooplankton concentrations (the so-called NPZ model). The governing equations are transport equations with appropriate source and sink terms, whose parametrization is not always quite straightforward. This not only implies additional complexity but also requires considerably more computing (and data) resources.

Process-Studies-Oriented or Application-Oriented

Models that are used to study some salient processes (for example western boundary currents and gyre circulation) can often be considerably simplified and hence made less computationally (and observational data-) intensive, since it is often possible to isolate and retain only the relevant physical process and ignore the rest in the model. Also, such models may not require extensive observational data for model initialization or forcing. They are most often run free in a predictive (or prognostic) mode. On the other hand, application-oriented models such as those used for ocean prediction purposes require extensive observational data for realistic initialization, forcing, and data assimilation. Assimilation into the model by one means or other of observational data is indispensable for **nowcast**, forecast, and **hindcast** applications, and data-assimilative models often employ approaches very much similar to those employed by numerical weather prediction (NWP) models in the atmosphere.

With and Without Coupling to Sea Ice

Global ocean models do not at present include sea ice. Ice-ocean-coupled basin models of the Arctic exist, however; for coupling to a z-level ocean model, see Hibler and Bryan [1987], and to an isopycnal model, see Oberhuber [1993]. Sea ice insulates the ocean from the cold atmosphere and mediates the exchange of heat and momentum between the two, and therefore such models involve solving dynamical and thermodynamic equations for the sea ice cover and its coupling to the underlying ocean.

Coupled to the Atmosphere or Uncoupled

Finally, for accurate simulation of long-time-scale processes, it is essential to couple ocean models with atmospheric models. Such coupled models are being increasingly used for such things as forecasting **El Niño** events. Most often, either the atmosphere or the ocean is highly simplified in such models, although modern supercomputers are enabling comprehensive atmospheric general circulation models to be coupled to global ocean models, at least for simulations of interannual variability. Truly comprehensive coupled models with applications to long-term climate studies require teraflop computing capability that might be routinely available in the coming century.

All numerical ocean models solve one form or other of the same governing equations for oceanic motions, written in a rotating coordinate frame of reference. These equations are essentially Navier–Stokes equations (or more appropriately **Reynolds-averaged** equations for mean term quantities, since the flow is invariably turbulent), but with the buoyancy and Coriolis force terms (fictitious accelerations due to the noninertial nature of the rotating coordinate frame) prominent in the dynamical balance. In addition, an equation of state relating the density of seawater to its temperature and salinity, and conservation equations for temperature and salinity, are also solved. In those models which model turbulence explicitly, equations for turbulence quantities such as the turbulence velocity scale (or equivalently turbulence kinetic energy) and turbulence macroscale are also solved. If chemical or biological components are included, conservation equations for relevant species with appropriate source and sink terms are solved as well.

Global and basin-scale models are formulated in spherical coordinates, but regional models are usually cast in rectangular coordinates instead. For simplicity we present the governing equations in rectangular Cartesian coordinates (spherical coordinate version can be found for example in Semtner [1986]). The x_1-axis is usually taken to be in the zonal direction (positive to the east), the x_2-axis is in the meridional direction (positive to the north), and the z-axis is in the vertical direction, positive upwards, with the origin located at the sea surface. Using tensorial notation and treating the horizontal coordinates separately from the vertical (indices take values 1 and 2 only), the governing equations consist of the continuity equation

$$\frac{\partial U_k}{\partial x_k} + \frac{\partial W}{\partial z} = 0 \tag{40.1}$$

where U_k denotes the horizontal components of velocity and W the vertical, and the momentum equations

$$\frac{\partial U_j}{\partial t} + \frac{\partial}{\partial x_k}(U_k U_j) + \frac{\partial}{\partial z}(W U_j) + f\varepsilon_{j3k}U_k = -\frac{\partial P}{\partial x_j} + \frac{\partial \Phi}{\partial x_j} + \frac{\partial}{\partial z}\left(K_M \frac{\partial U_j}{\partial z}\right) + F_j \tag{40.2}$$

$$\frac{\partial P}{\partial z} = -\frac{\rho}{\rho_0}g \tag{40.3}$$

where f is the Coriolis parameter ($2\,\Omega\cos\phi$), Ω the earth's angular velocity, ϕ the latitude, P the kinematic pressure, K_M the vertical mixing coefficient, ρ the in situ density, ρ_0 the reference density, g the gravitational acceleration, and Φ the gravitational potential.

The transport equations for **potential temperature** Θ and salinity S are

$$\frac{\partial \Theta}{\partial t} + \frac{\partial}{\partial x_k}(U_k \Theta) + \frac{\partial}{\partial z}(W\Theta) = \frac{\partial}{\partial z}\left(K_H \frac{\partial \Theta}{\partial z} \right) + S_\Theta + F_\Theta \tag{40.4}$$

$$\frac{\partial S}{\partial t} + \frac{\partial}{\partial x_k}(U_k S) + \frac{\partial}{\partial z}(WS) = \frac{\partial}{\partial z}\left(K_H \frac{\partial S}{\partial z} \right) + F_s \tag{40.5}$$

where K_H is the vertical mixing coefficient for scalar quantities, and S_Θ denotes a volumetric heat source such as due to penetrative solar heating. The equation of state is given by

$$\rho = \rho(\Theta, S) \tag{40.6}$$

Various simplifications have been made in deriving the above equations. The ocean is considered incompressible, a very good approximation for most applications. It is also considered to be in hydrostatic balance in the vertical, and hence the only terms remaining in the vertical component of the momentum equations are the buoyancy and pressure gradient terms. The hydrostatic approximation involves neglecting the vertical acceleration and regarding the fluid as essentially motionless in the vertical. In addition, the Boussinesq approximation has been used. This involves considering the ocean to be of constant density except when buoyancy forces are computed, thus assuming $\rho = \rho_0$ in all except the terms involving gravitation. Also, terms containing the horizontal component of rotation are neglected. The resulting equations are sufficiently accurate for most ocean circulation modeling. The equation of state employed is the so-called UNESCO equation [Pond and Pickard 1979] and is in the form of polynomial expansions in temperature and salinity. The pressure P is given by

$$P = g\eta + \frac{g}{\rho_0} \int_z^0 \rho(x_j, z', t)\, dz' \tag{40.7}$$

where η is the sea surface height. The terms F_j, F_Θ, F_S are horizontal mixing terms corresponding to unresolved subgrid-scale processes and are most often parametrized simply as Laplacian diffusion terms:

$$F_{j,\Theta,S} = A_{M,H} \frac{\partial^2}{\partial x_k \partial x_k}(U_j, \Theta, S) \tag{40.8}$$

where $A_{M,H}$ are horizontal mixing coefficients. A more rigorous form for these terms can be found in Blumberg and Mellor [1987]. The values for these coefficients are most often chosen as constants in a rather ad hoc manner based on purely numerical considerations. While the vertical mixing coefficients K_M and K_H can be rigorously modeled by turbulence closure theories [for example, Kantha and Clayson 1994, Mellor and Yamada 1982], there does not exist a similar approach for these terms. One approach, widely used in atmospheric modeling, is that due to Smagorinsky [1963], which is similar to the classical mixing-length theory of turbulence. Here the mixing coefficient is assumed to be proportional to the mean strain rate, so that

$$A_M = C(\Delta x_1)(\Delta x_2)\left[\left(\frac{\partial U_i}{\partial x_j} + \frac{\partial U_j}{\partial x_i} \right)\left(\frac{\partial U_i}{\partial x_j} + \frac{\partial U_j}{\partial x_i} \right) \right]^{1/2}, \quad A_H \sim A_M \tag{40.9}$$

where C is the Smagorinsky coefficient, with a value of around 0.04, and Δx_1 and Δx_2 are the grid sizes. This approach assumes that the subgrid scales fall within the Kolmogoroff inertial subrange, an assumption not always satisfied. A practical consequence of using this model is that strong horizontal shear is accompanied by strong horizontal mixing, which tends to smear out thermal fronts. A more general approach is to assume that the mixing coefficients are a sum of a constant background value and the Smagorinsky value given by Eq. (40.9) and to choose the values assigned to each appropriately [for example, Kantha 1995]. Another approach is to assign a constant cell Reynolds number $R_N = |U_j||\Delta x_j|/A_M$ and determine the value of the mixing coefficient thus [Choi and Kantha 1995]. Some modelers have used biharmonic form

$$\frac{\partial^2}{\partial x_k \partial x_k} \frac{\partial^2}{\partial x_k \partial x_k}$$

to model these terms [O'Brien 1985]. In this form, the terms serve principally to control the so-called $2\,\Delta x$ noise in the numerical solutions. Suffice it to say that modeling horizontal diffusion terms is still rather ad hoc.

The oceans are driven by momentum, heat, and salt fluxes at the air–sea interface. The boundary conditions at the sea surface $(z = \eta)$ are therefore

$$K_M \left(\frac{\partial U_j}{\partial z} \right) = \tau_{0j}$$

$$K_H \left(\frac{\partial}{\partial z}(\Theta, S) \right) = (q_H, q_S) \tag{40.10}$$

$$W = \frac{\partial \eta}{\partial t} + U_j \frac{\partial \eta}{\partial x_j}$$

where τ_{0j} is the kinematic shear stress acting at the free surface due to the action of winds and waves (taken mostly as equal to the kinematic wind stress) and $q_{H,S}$ are the kinematic heat and salt fluxes. The value of q_H is determined by the net heat balance at the air–sea interface due to short-wave and long-wave solar heating, back radiation by the ocean surface, and the turbulent sensible and latent heat exchanges, and that of q_S is determined by the difference between evaporation and precipitation. Accurate parametrization of these air–sea fluxes has been the subject of intense research for several decades (for example, the 1992 multinational Tropical Ocean Global Atmosphere/Coupled Ocean Atmosphere Response Experiment).

The conditions at the ocean bottom $(z = -H)$ are of no mass transfer through the bottom

$$W = -U_j \frac{\partial H}{\partial x_j}$$

and

$$K_M \left(\frac{\partial U_j}{\partial z} \right) = \tau_{bj}$$

$$K_H \left(\frac{\partial}{\partial z}(\Theta, S) \right) = (0, 0) \tag{40.11}$$

The last of the above conditions implies no heat or salt transfer through the ocean bottom. The bottom stresses are usually parametrized using a quadratic drag law with $c_d \sim 0.0025$:

$$\tau_{bj} = c_d |U_{bj}| U_{bj} \tag{40.12}$$

or by assuming that the lowest model grid point falls within the logarithmic-law region and using the well-known logarithmic relationship between the mean velocity and the friction velocity to derive the drag coefficient [Blumberg and Mellor 1987, Kantha and Piacsek 1993, for example].

When additional quantities such as turbulence velocity and length scales are modeled explicitly, corresponding conservation equations need to be solved along with appropriate boundary conditions at the ocean surface and bottom [Blumberg and Mellor 1987, Kantha and Clayson 1994, Kantha and Piacsek 1993]. The same holds for modeling of chemical and biological quantities.

If the lateral boundary is closed, then it is straightforward to apply the lateral boundary condition; the component of velocity perpendicular to the boundary is zero, and there is no lateral mass, heat, or salt flux through the boundary. If it is open, on the other hand, as is usual in regional models of coastal and marginal seas, then it is necessary to prescribe open boundary conditions. This is a difficult problem, since complete information on various flow properties must be specified, and this depends to a large extent on how well the flow at the boundary is known. The best strategy is to nest the model in a coarser-resolution model of the basin. In many cases, this is not feasible, and hence it is not possible to inform the model about what the rest of the ocean is doing. The best under these conditions is some form of Sommerfeld radiation boundary condition on dynamical quantities, which assures that disturbances approaching the boundary from the inside are radiated out and not bottled up [Blumberg and Kantha 1985, Kantha et al. 1990, Roed and Cooper 1986]. This is usually of the form

$$\frac{\partial \zeta}{\partial t} + C\frac{\partial \zeta}{\partial x_n} = 0 \qquad (40.13)$$

where ζ is a variable such as the sea surface height, n denotes direction normal to the boundary, and C is the phase speed of the approaching disturbance. Proper prescription of C is important to the success of the radiative boundary condition and has been the subject of much research [Blumberg and Kantha 1985, Orlonski 1976].

If there is inflow at the lateral boundary, temperature and salinity of the incoming flow must be prescribed. If there is outflow, on the other hand, these quantities are simply advected out:

$$\frac{\partial (\Theta, S)}{\partial t} + U_n\frac{\partial (\Theta, S)}{\partial x_n} = 0 \qquad (40.14)$$

40.3 Best Practices

In many applications, ocean simulation is an initial-value problem and therefore the initial state of the modeled ocean must be prescribed as accurately as possible. This is difficult to do in practice, since observational data are insufficient to specify the state of the ocean at any given point in time. The best alternative is to prescribe some sort of a climatological average as the initial state and spin up the ocean from that state under prescribed surface forcing. Here appeal is made to databases such as [Levitus 1982] (updated in 1995) and the U.S. Navy GDEM that contain distributions of climatological average temperature and salinity in the global oceans derived from historical archives of in situ observations.

Prescription of surface forcing is itself a major problem. Both momentum and buoyancy fluxes need to be specified at the sea surface, and both are determined by processes in the adjacent atmospheric boundary layer. For determining the long-term average state of the oceans, it is once again possible to appeal to climatological databases such as that due to Hellerman and Rosenstein [1983] and the Comprehensive Ocean-Atmosphere Data Set (COADS) database (see [Woodruff et al. 1987]; the database has been updated to 1990s recently), which provide gridded monthly average values of wind stress over the global oceans derived from historical marine surface observations. However, for many applications, including ocean prediction, it is necessary to provide surface forcing on a daily or even a multihourly basis. This is impossible to do from observations alone (although satellite-sensed air–sea fluxes may help in the future), and one has little choice but to appeal to six-hourly analyses and predictions of NWP centers, even though the accuracy of the surface forcing so derived depends very much on the skill of the particular NWP model and methodology.

Finally, the governing equations are discretized and the resulting finite-difference equations (or a set of algebraic equations) are solved to determine the evolution with time of the various oceanic variables such as temperature, salinity, and velocity at each model grid point. There are two principal means of discretization in the horizontal direction: finite differences [for example, Blumberg and Mellor 1987, Choi and Kantha 1995, Semtner and Chervin 1992] and finite elements [Le Provost et al. 1994], the latter being quite popular with civil engineers. The principal advantage of the latter is the flexibility the finite-element grid provides in assigning localized high resolution where needed. The reader is referred to Kantha [1995] for an example of a finite-difference approach to modeling the barotropic equations for global tides and to Le Provost et al. [1994] for a finite-element approach to the same problem. Advanced discretization techniques such as adaptive grids and nonorthogonal coordinate systems that are quite routinely used in conventional computational fluid dynamics are still in the developmental stage in ocean modeling; their efficacy is still largely unproven.

The finite-difference grid can be staggered or nonstaggered, with the former being more accurate. There are several possibilities [Mesinger and Arakawa 1976], including the so-called Arakawa C-grid, where the velocity component U_1 is displaced half a grid to the west and U_2 half a grid to the south of the grid center where all scalar quantities such as η, Θ, and S reside. The C-grid has better wave propagation characteristics if the grid size is smaller than the Rossby radius of deformation. The Arakawa B-grid, where both velocities are displaced half a grid point to the west, is better if it is larger [Semtner 1986]. With increased computing power and hence finer resolution, C-grid is becoming more popular.

Explicit or implicit methods can be used for time-stepping the equations; the latter are more efficient but require more complex simultaneous solution at all model grid points. The former are more easily adapted to massively parallel processors and are being increasingly used despite the limitation imposed by numerical stability considerations. The maximum time step that can be taken in an explicit scheme (for a staggered grid) is given by the Courant–Friedrichs–Lewy (CFL) condition, which is of the form

$$\Delta t \leq 0.5(\Delta x_e / C_e) \qquad (40.15)$$

where

$$\Delta x_e = (1/(\Delta x_1)^2 + 1/(\Delta x_2)^2)^{-1/2}$$

is the effective grid size, which is smaller than the grid size in the individual directions, and C_e is the effective gravity-wave speed, which is the sum of the gravity-wave speed and the advection velocity. In the barotropic problem, for example, $C_e = \max[|U_j| + \sqrt{gH}]$.

Explicit inclusion of the free-surface dynamics in a model requires that a mode-splitting technique [Blumberg and Mellor 1987, Kantha and Piacsek 1993, Madala and Piacsek 1977] be employed to overcome the severe limitations on the solution due to stability considerations imposed by fast-moving external gravity waves on the free surface. This technique consists essentially of splitting the solution into barotropic and baroclinic modes, with the barotropic part solved at the time step dictated by external gravity waves and the baroclinic part at a much larger time step, 20 to 50 times larger. This approach takes into account the fact that internal baroclinic adjustments are much slower.

It is the discretization of the vertical coordinate that is the most distinguishing feature of various ocean models. Several choices are possible, including that of no discretization (for a barotropic model). We will describe these next.

Barotropic Models

If the density gradients are neglected in the governing equations, or alternatively the ocean is considered to be of uniform density, the current distribution in the vertical becomes independent of depth (away from regions of frictional influence such as the surface and the bottom). Under these conditions, it is possible to ignore the transport equations for Θ and S and integrate the governing equations for continuity and momentum over the water column to arrive at a vertically integrated set of equations that govern the sea-surface elevation η and the vertically averaged velocity components $\overline{U}_j$:

$$\frac{\partial \eta}{\partial t} + \frac{\partial}{\partial x_k}(\overline{U}_k D) = 0$$

$$\frac{\partial}{\partial t}(\overline{U}_j D) + \frac{\partial}{\partial x_k}(\overline{U}_j \overline{U}_k D) + f\varepsilon_{j3k}(\overline{U}_k D) = -gD\frac{\partial \eta}{\partial x_j} - D\frac{\partial P_a}{\partial x_j} \tag{40.16}$$

$$+ g\, D\frac{\partial \xi}{\partial x_j} + (\tau_{0j} - \tau_{bj}) + D\overline{F}_j$$

where $D = H + \eta$ is the total depth of the water column.

The bottom friction is now determined using the column average velocity $\overline{U}_j$. Note the presence of tidal potential terms involving ξ on the right-hand side of the momentum equations that contain astronomical forcing terms due to the gravitational forces of the moon and the sun. Note also the terms due to atmospheric pressure and wind stress forcing. The astronomical forcing can be prescribed a priori from a knowledge of the ephemerides of the sun and the moon [see, for example, Kantha 1995, Schwiderski 1980]. The atmospheric forcing terms are also known and can be prescribed as a function of time during the model run. This set of equations can be used to solve for the sea surface height (SSH) and depth-averaged currents due to phenomena such as tides and storm surges.

Figure 40.1 shows an example of the application of **barotropic** equations to the problem of deducing the tidal SSH in the global oceans. The reader is referred to Kantha [1995] for details, but, briefly, the equations are cast in spherical coordinates, and the tidal potential terms are expressed as a sum of a series containing various tidal components such as the semidiurnal M_2, with a period of 12.42 h, and the diurnal K_1, with a period of 23.93 h (the atmospheric forcing terms are zeroed out for this application). The resulting equations are solved on a $\frac{1}{5}^{\circ}$ latitude–longitude C-grid covering the global oceans (excluding the Arctic) for each tidal component. The bottom depths over the model grid are derived from a digital database (ETOP05 from NOAA) containing world topography at $\frac{1}{12}^{\circ}$ resolution. However, for the results to be accurate enough for certain applications such as altimetry, inevitable errors that result from inaccurate knowledge of bottom depths and friction coefficients have to be overcome by data assimilation. Tidal SSHs can be derived in the deeper parts of the oceans quite accurately from measurements of SSH fluctuations by a satellite-borne microwave altimeter. The tidal SSH data derived from the currently operational NASA/CNES TOPEX/Poseidon precision **altimeter** [Desai and Wahr 1995] have been assimilated into the model as well as those from coastal tide gauges around the world's coastlines. A simple data assimilation scheme has been used where, at each time step, the model-predicted SSH is replaced by a weighted sum of the model SSH and the observed SSH, with weights determined a priori. The result is tidal SSH that is accurate to within a few centimeters over the global oceans, including shallow coastal and semienclosed seas. This information is useful for many applications, such as an accurate determination of the subtidal SSH variability from altimetric data, gravimetry, and determination of tidal dissipation. Figure 40.1 shows the M_2 **coamplitude** and **cophase** (with respect to Greenwich) distributions of the tidal SSH and the tidal-current ellipses over the global oceans. Figure 40.2 shows the accuracy attained by this data-assimilative tidal model in the form of scatterplots of comparison of modeled and observed tides from an independent set of accurate tide and bottom-pressure gauges over the global oceans, whose locations are also shown.

Barotropic models such as these can also be used to study the response of the SSH to atmospheric pressure forcing [Kantha et al. 1994, Ponte 1994]. It is often assumed that the ocean responds instantaneously to pressure forcing as an inverse barometer with roughly one centimeter of increase (decrease) for every millibar of drop (rise) in atmospheric pressure. This is not always true, and the departures from the inverse-barometer response are quite important to satellite ocean altimetry [Kantha et al. 1994].

Finally, a very important application of barotropic models is for prediction of storm surge effects along a coastline due to approaching hurricanes. The strong hurricane-force winds (augmented by the pressure drop in the eye of the hurricane) pile up water against the coast that often leads to an increase in sea level of several meters and consequent inundation of structures along the coastline. Hurricane Camille in 1969

CU Global Tide Model
version 1.3
topex assimilation

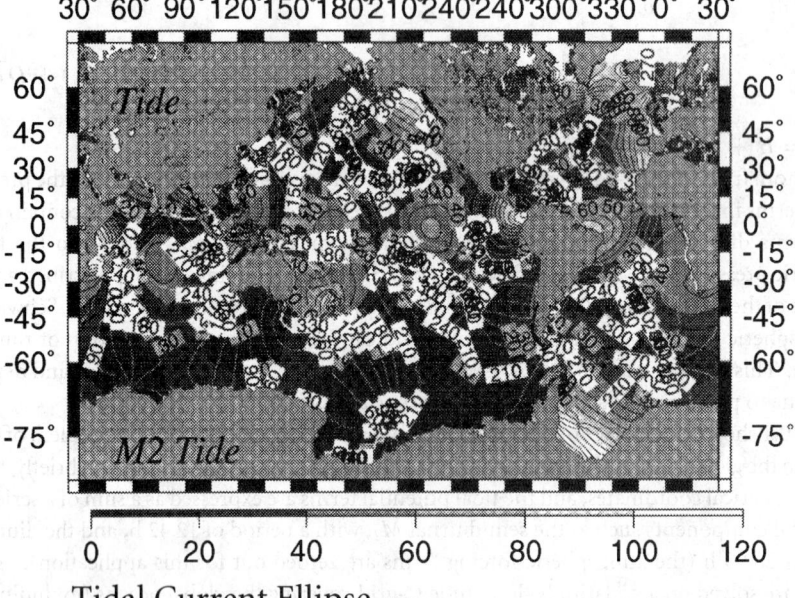

Tidal Current Ellipse
Log10(.1 cm/s) = 0
Thin = Counter Clockwise
Thick = Clockwise

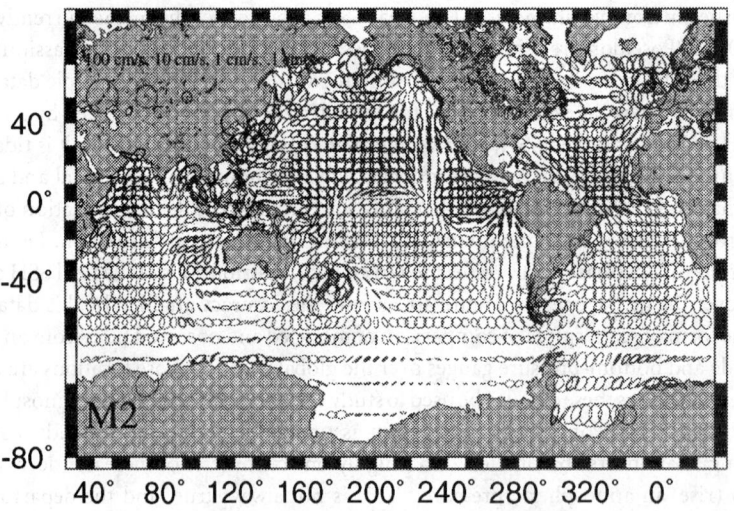

FIGURE 40.1 A map of the distribution of coamplitude and cophase (top), and tidal-current ellipses plotted every 25th point in each direction (bottom) for the M_2 tidal component in the global oceans. Note the logarithmic scale for ellipses.

GU Global Tide Model Error Distribution

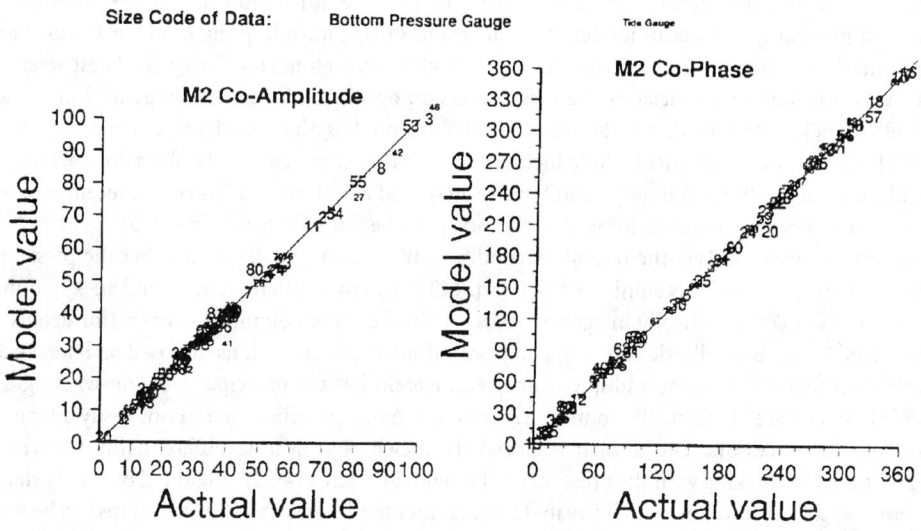

Size Code of Data: Bottom Pressure Gauge   Tide Gauge

Pelagic Tide Gauges
Bottom Pressure Gauge
Coastal Tide Gauge

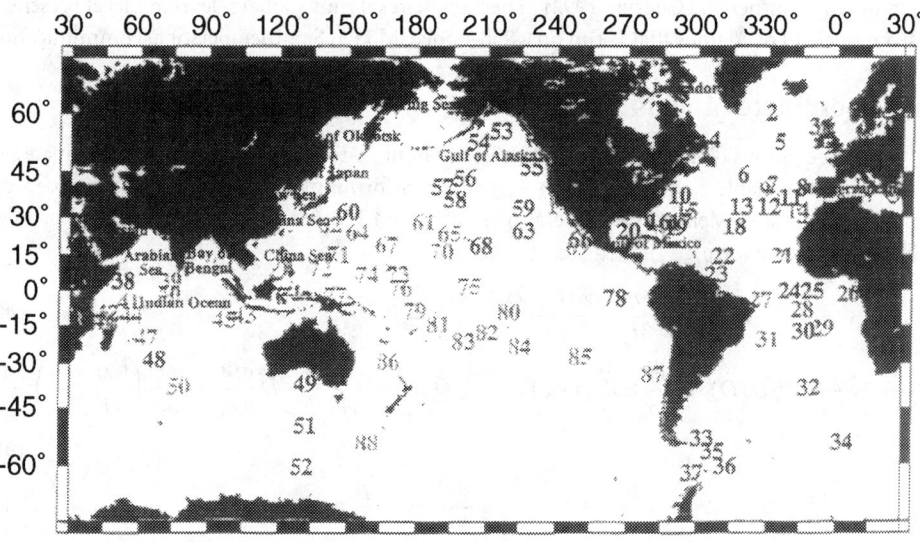

FIGURE 40.2   Scatterplots (top) of modeled M_2 coamplitudes and cophases vs. those observed at pelagic tidal stations, the locations of which are shown at the bottom (darker numbers: bottom pressure gauges; lighter ones: coastal tide gauges).

caused a storm surge of nearly 8 m along the Mississippi coast, leading to widespread destruction and devastation. Provided the local bathymetry is known accurately and the characteristics of the hurricane (such as the wind stress distribution and forward velocity) can be deduced reasonably well from NWP forecasts, it is possible to predict the resulting storm surge quite accurately using a barotropic model driven by the wind stress and atmospheric pressure terms on the right-hand side.

z-Level Models

The Bryan–Cox–Semtner z-level model [Bryan 1969, Cox 1985, Semtner and Chervin 1992] is the oldest and the most popular global ocean model. Several versions exist, including the Modular Ocean Model (MOM) from the Geophysical Fluid Dynamics Laboratory in Princeton, New Jersey, the latest version of which (MOM2) includes free-surface dynamics. A version optimized for massively parallel processors, called POPS (Parallel Ocean Prediction System), is available from Los Alamos Scientific Laboratory [Smith et al. 1992]. Recent improvements include inclusion of a free surface in a variety of versions around the world [Killworth et al. 1991, Dukowicz and Smith 1994], and adoption of a C-grid. A recent review of the current state of ocean modeling using z-level models can be found in Semtner [1995].

Imposition of a rigid lid (via the boundary condition $W = 0$ at $z = 0$) means that the pressure at $z = 0$ enters as an unknown. It is eliminated from Eq. (40.2) by cross-differentiation, and an equation for the stream function ψ for vertically integrated transport in the water column is derived (for details, see [Semtner 1986]). This is an elliptic equation and is solved subject to conditions imposed on lateral ocean boundaries, which are in general multiply connected. Herein lies the principal problem with rigid-lid models. While they are efficient, the solution technique is more complicated and not easily adapted to vector and parallel processors. The problem is alleviated somewhat by not cross-differentiating to derive the stream function, but working with the pressure on the rigid lid (section 40.2, "Rigid Lid or Free surface").

Numerous applications of the z-level Bryan–Cox–Semtner model and its various versions can be found in the literature (for example in the *Journal of Physical Oceanography* and the *Journal of Geophysical Research, Oceans*). It has been used extensively to study the seasonal, interannual, and climatic variations in the global oceans [for example, Philander et al. 1986, 1987]. It is also a central part of the ocean analysis system for the tropical oceans [Leetma and Ji 1989], where a best estimate of the state of these oceans is determined by assimilation of observational data into a tropical-ocean version of the model. The most recent application can be found in [Semtner and Chervin 1992]. The highest-resolution global z-level model at present is the $\frac{1}{6}°$ POPS model at Los Alamos that is run on a 256-node CM5 (A. Semtner, personal communication).

Sigma-Coordinate Models

The governing equations (40.1) to (40.5) can be cast in a bottom-topography-following coordinate system by defining a new variable $\sigma = (z - \eta)/(H + \eta)$ and transforming the equations to the new coordinate system ([Blumberg and Mellor 1987]; see also [Kantha and Piacsek 1993] for the general orthogonal curvilinear coordinate form):

$$\frac{\partial \eta}{\partial t} + \frac{\partial (U_k D)}{\partial x_k} + \frac{\partial \omega}{\partial \sigma} = 0 \tag{40.17}$$

$$\frac{\partial (U_j D)}{\partial t} + \frac{\partial}{\partial x_k}(U_k U_j D) + \frac{\partial}{\partial \sigma}(\omega U_j) + f \varepsilon_{j3k} U_k D = -D \frac{\partial P}{\partial x_j} + D \frac{\partial \Phi}{\partial x_j} + \frac{\partial}{\partial \sigma}\left(\frac{K_M}{D}\frac{\partial U_j}{\partial \sigma}\right) + D F_j \tag{40.18}$$

$$\frac{\partial P}{\partial \sigma} = -\frac{\rho}{\rho_0} g D \tag{40.19}$$

$$\frac{\partial (\Theta D)}{\partial t} + \frac{\partial}{\partial x_k}(U_k \Theta D) + \frac{\partial}{\partial \sigma}(\omega \Theta) = \frac{\partial}{\partial \sigma}\left(\frac{K_H}{D}\frac{\partial \Theta}{\partial \sigma}\right) + D S_\Theta + D F_\Theta \tag{40.20}$$

$$\frac{\partial (SD)}{\partial t} + \frac{\partial}{\partial x_k}(U_k SD) + \frac{\partial}{\partial \sigma}(\omega S) = \frac{\partial}{\partial \sigma}\left(\frac{K_H}{D}\frac{\partial S}{\partial \sigma}\right) + D F_S \tag{40.21}$$

where $D = H + \eta$, the total depth of the water column, and ω is the pseudo vertical velocity in the new coordinate system, zero at the ocean surface ($\sigma = 0$) and the bottom ($\sigma = -1$).

These equations, along with corresponding conservation relations for turbulence quantities, form the basis of the popular **sigma-coordinate** Princeton model developed by George Mellor's group at Princeton

University [Blumberg and Mellor 1987; see also Mellor 1991, Kantha and Piacsek 1993]. In this coordinate system, the number of levels is the same everywhere in the ocean, irrespective of the depth of the water column. It is therefore possible to resolve the bottom boundary layer where needed. This set of equations is best suited to modeling the shallow coastal oceans, although there is no inherent barrier to its application to deep basins. The principal problem is in applying it over sharply changing topography such as the continental slope separating the shelf from the deep basin. Here, unless the topographic gradients are suitably reduced by a nonlinear smoother, the errors in the calculation of pressure gradients induced by horizontal gradients of density can lead to spurious along-slope currents [Haney 1991]. While the problem due to strong topographic changes manifests itself in one form or another in all ocean models, the problem is particularly serious in sigma-coordinate models.

Many applications of this model can be found in the literature (for example in the *Journal of Physical Oceanography* and the *Journal of Geophysical Research, Oceans*). An application of a modified version developed at the University of Colorado, incorporating an improved mixed-layer formulation [Kantha and Clayson 1994] and involving assimilation of altimetric data, is given in section 40.4. This version has also been converted to CM5 and applied to the Straits of Sicily, and its Cray T3D version is being applied to the North Pacific Ocean.

Layered Models

In layered models, the ocean is divided into several (N) layers in the vertical, and Eq. (40.1) to (40.3) are integrated over each layer ($n = 1, \ldots, N$) to obtain expressions for the thickness of and velocity in each layer. For example, Wallcraft [1991] obtains

$$\frac{\partial h^n}{\partial t} + \frac{\partial}{\partial x_k}\left(h^n U_k^n\right) = w^n - w^{n-1}$$

$$\frac{\partial\left(h^n U_j^n\right)}{\partial t} + \left[\frac{\partial}{\partial x_k}\left(h^n U_k^n\right) + U_k^n \frac{\partial}{\partial x_k}\right] U_j^n + f\varepsilon_{j3k} h^n U_k^n$$

$$= -h^n \sum_{k=1}^{N} G_k^n \frac{\partial}{\partial x_k}\left(h^n - h_0^n\right) + \left(\tau_j^{n-1} - \tau_j^n\right) + A_M \frac{\partial^2}{\partial x_k \partial x_k}\left(h^n U_j^n\right)$$

$$+ \max\left(0, -w^{n-1}\right)U_j^{n-1} + \max(0, w^k)U_j^{n+1} - [\max(0, -w^n)$$

$$+ \max(0, w^{n-1})]U_j^n + \max(0, -c_{de}w^{n-1})\left(U_j^{n-1} - U_j^n\right)$$

$$+ \max(0, -c_{de}w^n)\left(U_j^{n+1} - U_j^n\right) \tag{40.22}$$

where h^n is the thickness and U_j^n the velocity of the nth layer, w^k is the vertical velocity at the kth interface, and h_0 is the layer thickness at rest. The Nth layer contains the model basin topography, and its thickness is the total depth of the water column minus the sum of the thicknesses of the remaining layers. Finally,

$$G_k^n = \begin{cases} g, & k \geq n \\ g\left[1 - \dfrac{\rho^n - \rho^k}{\rho_0}\right], & k < n \end{cases}$$

$$\tau_j^n = \begin{cases} \tau_w, & n = 0 \\ c_{dn}\left|U_j^n - U_j^{n+1}\right|\left(U_j^n - U_j^{n+1}\right), & n = 1, \ldots, N-1 \\ c_{db}\left|U_j^N\right|U_j^N, & n = N \end{cases} \tag{40.23}$$

The factor c_d is the drag coefficient, c_{de} is the drag due to entrainment of fluid from one layer to the adjacent one, τ_w is the wind stress, and ρ^n is the density of the nth layer. Note that the layer densities do not change with time, only their thickness does at each model grid point. The conditional statements have to do with entrainment and detrainment at each interface between two adjacent layers, the details of which can be found in Wallcraft [1991].

The thinning of a layer to vanishing thickness is a major problem in layered models that leads to numerical difficulties. The traditional solution has been to make each layer thick enough, but this distorts the representation of the oceanic vertical structure. An alternative solution is to entrain fluid into the thinning layer from below to thicken it. Such entrainment has to be balanced by global detrainment in the layer so as to keep the density of each layer constant in space and time. For details of this and the model numerics, see Wallcraft [1991].

It is essential to select the number and rest thicknesses of layers carefully in layered models. Since topographic variations are contained in the bottommost layer only, these models are generally incapable of simulating circulation in coastal and shallow seas. They are, however, excellent at capturing the important lowest-order dynamics of the basin circulation and are therefore widely used for process-oriented studies. They are also being increasingly used for a variety of applications. One example is the six-layer, $\frac{1}{8}^\circ$ global model at the Naval Research Laboratory at Stennis Space Center, Mississippi, the SSH from which is shown in two parts, the Atlantic and Indian Oceans in Fig. 40.3(a) and the Pacific Ocean in Fig. 40.3(b). Realistic depiction of mesoscale activity, especially in regions of strong ocean currents—such as the Gulf Stream in the Atlantic, Kuroshio in the Pacific, the Brazil/Malvinas Current off Brazil, the Agulhas Current off Africa, and the Circumpolar Current around the continent of Antarctica—are noteworthy. The SSH variability from a layered model like this, driven by synoptic winds from a NWP center such as Fleet Numerical Meteorology and Oceanography Center, compares well with the variability indicated by altimeters such as the U.S. Navy's GEOSAT.

A simple subset of the layered model is the so-called reduced-gravity model (also called $1\frac{1}{2}$-layer model), where the water column is assumed to consist of two layers: an active top layer of thickness H and a quiescent bottom layer of infinite thickness, with a density interface between the two of intensity $\Delta\rho$. It is remarkable that this very simple model often captures the essential dynamics of the circulation; for example, a reduced-gravity model of the Gulf of Mexico demonstrated conclusively that it is the instability of the Loop Current that is responsible for the shedding of the Loop Current eddies [Hurlburt and Thompson 1980]. The governing equations are identical to the barotropic equations (40.15), except that the gravity parameter g is replaced by $g' = g\,(\Delta\rho/\rho_0)$, the reduced gravity (whose value is two orders of magnitude smaller than g; hence the name reduced-gravity model), with H now denoting the rest thickness of the upper layer and η denoting the deflection of the interface.

Isopycnal Models

Isopycnal models are similar to the layered models discussed above but are fully dynamical and thermodynamic. Despite the numerical problems associated with surfacing and vanishing of layers, they are well suited to simulate basin dynamics. Considerable progress has been made over the last decade in isopycnal modeling, and with the inclusion of adequate upper-mixed-layer physics they are also becoming quite practical. Examples of applications can be found in Oberhuber [1993], Bleck and Boudra [1986], Bleck and Smith [1990]. Since they principally deal with isopycnals (surfaces of equal density) and do not consider temperature and salinity separately, but instead treat density as the prognostic variable, they are not well suited to handling situations where temperature and salinity must be computed separately. A linear equation of state and identical diffusion characteristics for temperature and salinity are implicit in these models. This is valid over a majority of the global oceans, if one excludes regions such as those near river outfows and sea-ice formation.

Data Assimilation

Inevitable errors in initial conditions and imperfect parametrization of physical processes make a model ocean diverge rapidly from the real ocean. This is simply due to the extreme sensitivity of this system to

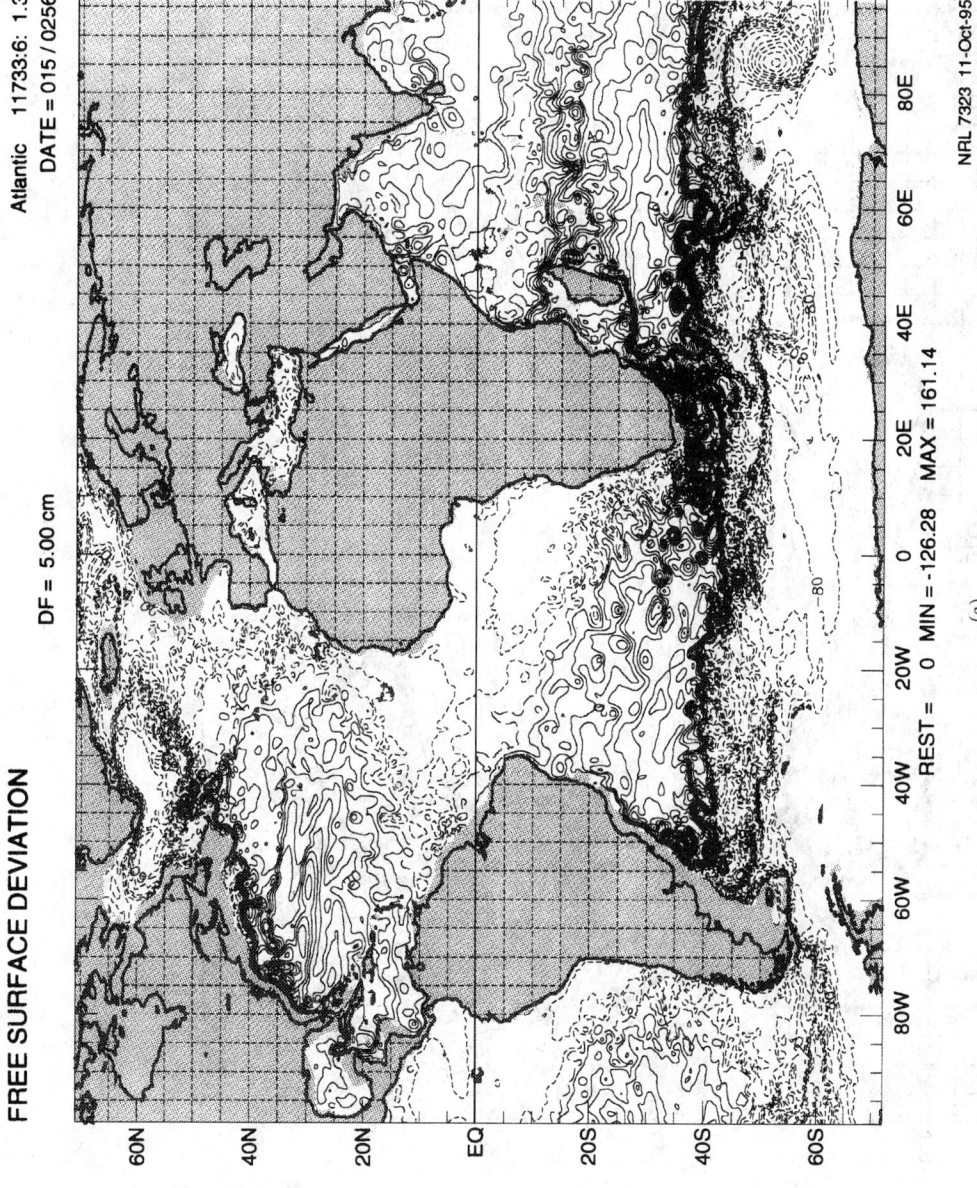

FIGURE 40.3   Sea surface from the six-layer Naval Research Laboratory $\frac{1}{8}^{\circ}$ global model for the Atlantic and the Indian Oceans (a) and for the Pacific Ocean (b). Note the eddy-resolving capability of this model displayed in the realistic mesoscale activity in regions of strong currents such as the Aghulas around Africa.

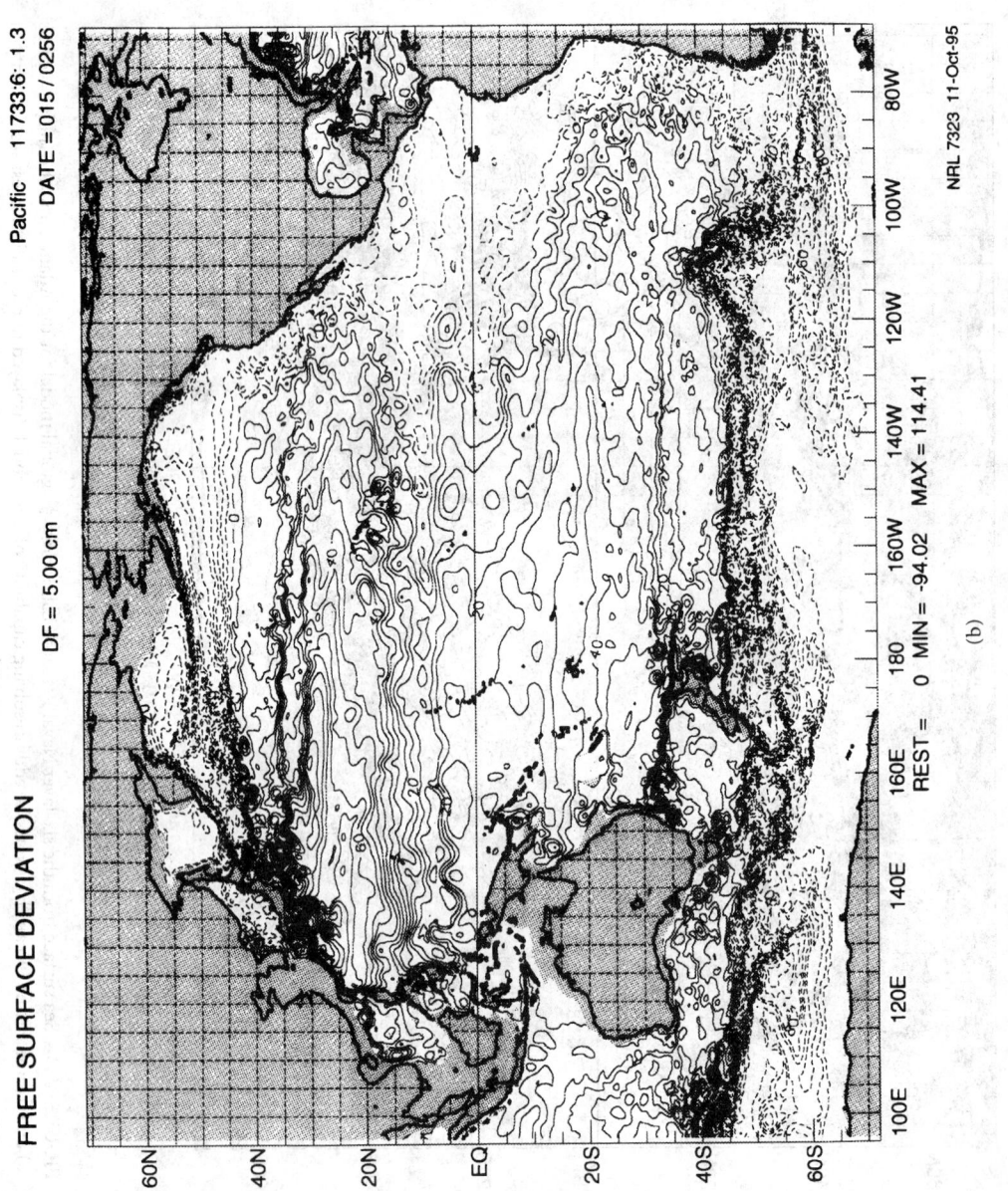

FIGURE 40.3 *continued*

even minute changes in initial conditions, typical of chaotic nonlinear systems. It is therefore essential to employ observational data in ocean models to retain the modeled ocean state close to the real state. The situation is no different from that in modeling the state of the atmosphere for NWP purposes, except that the time scales for loss of predictability is weeks for the oceans compared to days for the atmosphere. The process of employing observed data from the real ocean (atmosphere) to keep the modeled ocean (atmosphere) realistic is called data assimilation [for example, Anderson and Moore 1986] and consists of combining the modeled fields with observed data at various points in the domain to produce the best possible estimate of the real state of the ocean over the entire model domain. Exactly how this is best done has been the subject of considerable research in the atmospheric community [Bengtsson et al. 1981] over the past few decades, and more recently in the oceanic community as well [Haidvogel and Robinson 1989].

NWP centers use predominantly the so-called analysis–forecast cycle of assimilation. Here, the current state of the modeled atmosphere as predicted by the previous forecast is combined with observations of the atmosphere by radiosondes and surface stations all over the world, by an analysis–initialization process, to produce initial fields of various model variables suitable for describing the initial state for the next model forecast. The forecast skill depends very much on the accuracy of the initial state so derived, since errors in the latter tend to get magnified with time during the forecast.

Another possible assimilation method is the so-called continuous assimilation [Bengtsson 1975], where a numerical model is kept running and current by assimilation of observed data as and when they become available. A forecast can then be initiated by a similar model running forward *free* (without any data assimilation). The principal advantage of this method over the analysis–forecast cycle is that the model derives benefit from all past data as opposed to a single set of observations at a particular time. Also, the shock of data insertion due to inevitable mismatch between the model and observed states is less severe. Since such a mismatch can lead to severe noise superimposed on the true state of the forecast atmosphere (ocean), often making the forecast worthless, considerable effort has been expended in devising means to minimize such a mismatch, resulting in a procedure called *initialization* in NWP terminology. Continuous assimilation tends to reduce this shock and is therefore often preferable. The reader is referred to Bengtsson et al. [1981] and Haidvogel and Robinson [1989] for a discussion of assimilation philosophies.

The method of combining data into a model can vary from the simplest one (called data insertion), in which the model-predicted values are just replaced by observed values, to Kalman filters [Gelb 1988] (which blend the model and observed values optimally, taking into account the model error and observational error statistics), adjoint techniques [Thacker and Long 1987], and variational methods [Derber and Rosati 1989]. It is also possible to use nudging techniques in which appropriate Newtonian damping terms that damp the variable to the observed value with a predetermined time scale are introduced into the governing equations. The most commonly employed method is optimal interpolation [see Choi and Kantha 1995, for example], since methods such as Kalman filters and adjoint techniques are computationally expensive and at present still impractical for applications in NWP and ocean prediction.

It is beyond the scope of this article to go into details of data assimilation methods. Instead, the reader is referred to the above two references (and more recent work in the literature, especially on NWP), with a reminder that most assimilation methods replace the model-predicted values by a weighted combination of model-predicted value and observed values during the assimilation step, with the weight either determined a priori by statistical methods such as optimal interpolation or updated at each assimilation step by a method such as Kalman filtering. For examples of oceanic data assimilation, the reader is referred to Derber and Rosati [1989], Glenn and Robinson [1995], and Choi and Kantha [1995].

Computational Issues

Ocean models make a large demand on computer resources, CPU time, core memory, and disk storage, because ocean eddies are much smaller than weather systems, and the resolution needed is therefore much finer. Fine resolution also forces one to take smaller time steps in explicit models on account of CFL constraints. Even in "implicit" ocean models, the advection terms are treated explicitly, thus imposing a time-step limitation.

For explicit free-surface models, the time step is limited by the step of the fast-moving surface gravity waves, and one has to take a large number of small time steps to integrate over a simulation or forecast period (mode splitting helps alleviate this problem). For implicit ocean models, which filter out these gravity modes, the CPU-time requirement is governed by the rate of convergence of the iterative method used to solve the resulting Poisson equation.

Explicit model codes are usually readily vectorizable and parallelizable and generally need few additional arrays to store the auxiliary variables that may be needed to speed up the computations. In contrast, the vectorization/parallelization of the Helmholtz solvers associated with implicit codes is usually a nontrivial problem and, for some schemes that have been used up to now on serial machines, not at all feasible. The extra work resulting from the iterative or matrix inversion solution can often increase the total CPU time so that it is comparable to, or even exceeds, that for explicit codes, especially on vector/parallel computers. In addition, there are almost always extra arrays needed during this stage of the computations. The two- or four-color versions of the successive overrelaxation (SOR) and the conjugate-gradient method are two techniques that are well suited to vectorization and parallelization in implicit codes.

In the early days of supercomputers, the core memory available was usually so small that all the arrays needed for computations in ocean models, especially global ones, would not fit within the core, and elaborate methods were employed to make efficient use of high-speed disks to transfer arrays into and out of core as needed. GFDL models (MOM2 for example) still retain such an architecture. With high-speed memory becoming much cheaper, modern supercomputers have core memories measured in gigawords (Gw), and many ocean models can now reside in memory, although the need for out-of-core models has not totally vanished, especially for very high resolutions and global coverage. In-core models such as the Los Alamos POPS are, however, better suited to efficient massive parallelization than the out-of-core ones such as MOM, because of the considerable disk I/O involved.

Disk/tape storage requirements for storing ocean model results are also often in tens to hundreds of gigabytes and depend on the length of the simulation and on how often and how many variables are required to be stored for later analyses. Disk storage and postprocessing requirements often constrain the temporal resolution and the details of the analyses carried out on the results of an ocean model.

Data-assimilative ocean models require even more resources than the free-running ones, with the additional memory and CPU-time requirements depending very much on the method of assimilation. It is not unusual for assimilation to more than double the CPU time requirements, even for simple OI-type schemes. Methods such as Kalman filters and adjoint methods are even more demanding. Generally, data assimilation on massively parallel computers requires considerable investment of time and effort for efficient implementation. We will give some typical CPU-time and memory requirements for large ocean models and for diverse computers, to cover a spectrum of configurations and to familiarize the reader with resource requirements of computational ocean modeling. The $\frac{1}{8}^\circ$ six-layer NRL global model $(2051 \times 1145 \times 6$ grid) requires 1.8 Gw of memory and 2 CPU hours per month of simulation on a 256-node CM5-E. The $\frac{1}{16}^\circ$ Pacific model $(1977 \times 1313 \times 6$ grid) requires 385 Megawords (Mw) and 17 single-processor CPU hours per month on a Cray C-90. The $\frac{1}{5}^\circ$ global explicit barotropic tidal model discussed in section 40.3 under "Barotropic Models" $(1801 \times 729$ grid) employs a time step of 13 s, assimilates 4000 data points every time step, and requires 65 Mw of memory and 22 single-processor CPU hours for a 10-day simulation on a 16-processor Cray C-90, assimilating 4000 data points every time step. A 15-level northern hemisphere Arctic ice–ocean model $(360 \times 360 \times 15$ grid) requires 40 Mw of memory and 25 CPU hours per month on a Cray C-90. A 30-level sigma-coordinate model of the eastern Pacific $(163 \times 229 \times 30$ grid) requires 42 Mw of memory and 4 CPU hours per month on Cray C-90. The small Gulf of Mexico sigma-coordinate nowcast–forecast model $(85 \times 86 \times 22$ grid) discussed in the next section requires 30 Mw of memory and 6 CPU hours for a month-long simulation in the nowcast mode, and 4 CPU hours in the forecast mode, the additional time requirements for the simple OI-based data assimilation being in this case about 50%. An idea of the storage requirements can be obtained from the fact that even this small model required 2 Gbytes to store the model output at 5-day intervals for a 10-year-long simulation without any data assimilation, and postprocessing of this output required numerous hours on a powerful Sun Sparc workstation.

40.4 Nowcast/Forecast in the Gulf of Mexico (a Case Study)

An important application of ocean models is in prediction of the current (nowcast) and future (forecast) state of the ocean. Given the fact that more than 50% of the burgeoning human population lives within 100 miles of a coastline and hence uses/abuses the coastal oceans, such predictions, especially in the coastal and marginal seas, might be particularly useful for societal needs such as sea-level predictions, mapping of currents, and pollution tracking. We will provide one such example from a marginal semienclosed sea in the north Atlantic, the Gulf of Mexico. The offshore oil fields of this "mini-ocean basin" account for roughly half the U.S. domestic oil production, and the Louisiana–Texas (LATEX) continental shelf is dotted with thousands of oil platforms. Exploration and production are expanding steadily into deeper waters, waters as deep as 1000 m.

The major oceanic phenomenon in the Gulf is the so-called Loop Current variability. Every second, about 28 million cubic meters of subtropical waters enter the Gulf through the Yucatan Straits between Mexico and Cuba and leave it through the Florida Straits between Florida and Cuba to eventually become the Gulf Stream. The extent of penetration of this so-called Loop Current into the Gulf is highly variable. Occasionally the Loop Current becomes unstable and sheds off a huge anticyclonic (clockwise) eddy, anywhere from 100 to 350 km in diameter, that pinches off the Loop Current and moves into the western Gulf. This Loop Current eddy (LCE) is the principal mechanism for renewal of waters in the western Gulf [Hurlburt and Thompson 1980]. The path of LCEs is also highly variable, and occasionally a LCE traverses the Gulf in close proximity to the LATEX continental shelf. Because of the strong currents (often as much as 4 knots, 2 m s^{-1} in magnitude) associated with LCEs, this is the second major source of concern (the first being hurricanes in late summer and fall) to production and exploration activities in the Gulf. A capability to accurately forecast the movement of a LCE is valuable to the oil industry.

A forecast of the path an LCE takes is possible with the use of a numerical model of the Gulf. However, accurate information on the initial location of the LCE once it is shed and the corresponding Gulf-wide oceanic state is crucial to the forecast skill. An accurate nowcast is therefore essential, and this requires a data-assimilative numerical model. Since in situ data, even in the Gulf, are sparse and often nonexistent, remotely sensed data need to be relied upon for this purpose. Since the sea surface temperature from IR sensors is not always useful in locating a LCE (especially in summer) and since altimetry can almost always detect such an eddy if it happens to straddle its track, altimetric SSH anomalies can be assimilated into the ocean model to provide a reliable nowcast of not only the eddy location but also the initial state of the Gulf. Forecasts can then be made from this nowcast and the path of the LCE predicted.

The methodology employed by Choi and Kantha [1995] for producing a nowcast (in a hindcast mode, that is, prediction of past events for which data are available for verification) is called the *continuous* or *four-dimensional* assimilation method and has its origins in NWP. The model is run from a time in the past to the present, assimilating altimetric data track by track [see Choi and Kantha 1995 for details]. Altimetric SSH anomalies derived from NASA/CNES TOPEX and ESA ERS1 altimeters are converted to anomalies in the temperature in the water column and assimilated into the model using simple optimal interpolation. In the particular example shown here, the model was run starting at the beginning of January 1993 (day 1 corresponds to Jan. 1) to produce a nowcast for day 240, at which a LCE pinches off and separates from the Loop Current. The model is then run forward free (without assimilation) to produce a forecast over the next 40 days, assuming nonchanging winds over the period. The forecast skill is assessed by comparison with the nowcast, which was also carried out over the rest of 1993.

Figure 40.4 shows a comparison of the forecast and nowcast SSH fields over the Gulf at days 260 (top) and 280 (bottom). The forecasts are shown on the left and the corresponding nowcasts on the right. There is a close correspondence between forecast LCE position and the nowcast position, suggesting that the LCE path is being predicted reasonably well. The error, however, between forecast and nowcast LCE positions is larger at day 280 than at day 260. This particular experiment suggests that the forecast has some skill to about 30 days or so, beyond which the predicted path (forecast) deviates increasingly from the actual path (nowcast for the corresponding day). Since altimetric data are available within several days of their collection by the sensor, this experiment suggests that with some skill forecasts can be made two

Gulf of Mexico
Model Run Day 260
(J. K. Choi & L. H. Kantha)

Forecast (Without Assimilation)   Nowcast (With Assimilation)

(unit: m)   (unit: m)

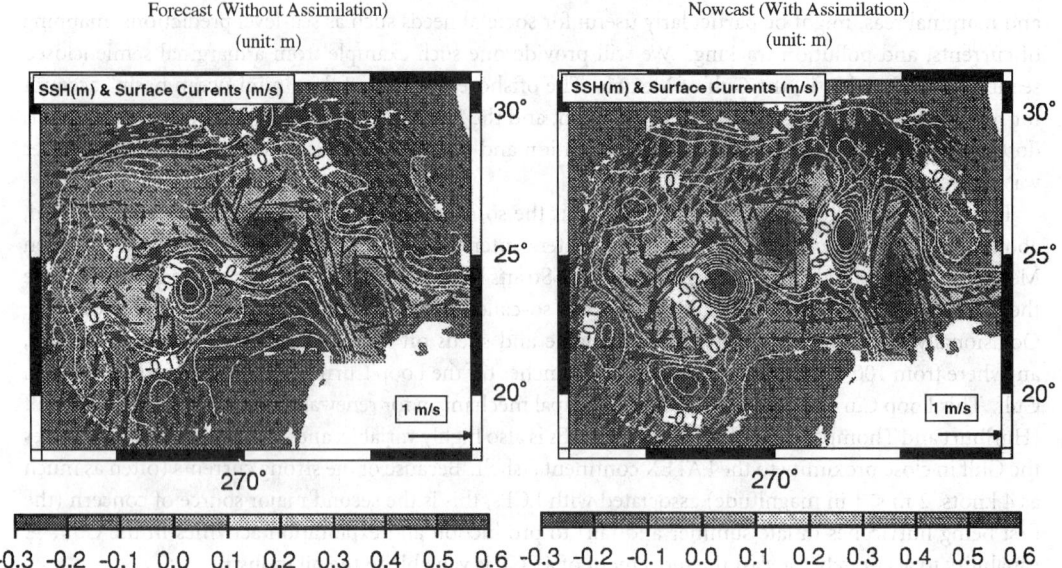

Model Run Day 280
(J. K. Choi & L. H. Kantha)

Forecast (Without Assimilation)   Nowcast (With Assimilation)

(unit: m)   (unit: m)

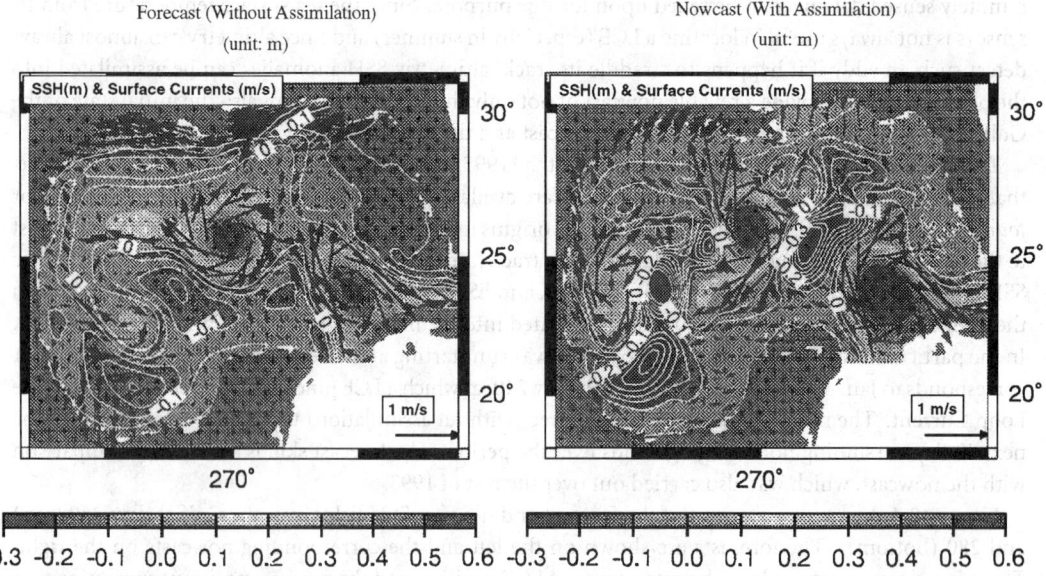

FIGURE 40.4   The sea surface elevation and currents from the forecast (left) and the nowcast (right) at days 260 (top) and 280 (bottom) from a three-dimensional circulation model of the Gulf of Mexico assimilating altimetric data from TOPEX and ERS1. The forecast was started at day 240. Compare the forecast with the corresponding nowcast to assess the model skill.

to three weeks in advance. If this is proven correct, this nowcast–forecast capability might be useful to drilling/exploration activities in the Gulf. It is in applications such as this that an ocean model, acting in concert with routine ocean monitoring via satellite-borne sensors, can prove useful.

40.5 Research Issues and Summary

We have provided a thumbnail sketch of ocean modeling as it is practiced today. As we said earlier, the field has undergone a phenomenal growth in recent years, and it is impossible to do justice to the subject in a short review like this. The reader is encouraged to pursue a particular model or approach of interest via the references cited.

The major issue in ocean modeling is the dearth of data for model initialization, forcing, assimilation, and of course verification or skill assessment. In situ data are rather sparse and, given the cost of ship time, likely to remain so. Therefore, increasing reliance will be placed on remote sensing to fill in gaps. However, this approach itself has limitations, and it is not clear what might fill the gap. Smart autonomous vehicles, a product of the Cold War, roaming the world oceans, and buoys sprinkled into the global oceans, telemetering data via communication satellites, may one day provide more in situ data than we currently acquire. Combined with multiteraflop computing capabilities of the coming century, an ocean observing and monitoring system consisting of satellites, moored arrays, buoys, and autonomous vehicles might one day finally enable us to set up realistic ocean prediction systems to satisy the needs of the coming generations. Ocean modeling will play a central role in all this.

Acknowledgments

Lakshmi Kantha acknowledges with pleasure the support provided by The Minerals Management Service of the Department of the Interior through an interagency agreement with the U.S. Navy through contract N00014-92C-6011, administered by Walter Johnson of MMS and Donald Johnson of the Naval Research Laboratory. Lakshmi Kantha was also supported by the NOMP program of the Office of Naval Research under contract N00014-95-1-0343, administered by Tom Curtin, and by the Coastal Sciences Section of the Office of Naval Research under contract N00014-92-J-1766, administered by Thomas Kinder.

Defining Terms

Altimeter: A microwave device measuring the time delay between an emitted microwave signal and its return by reflection from the sea surface. When the position of the instrument in space is independently determined, it enables sea surface topography to be measured to an accuracy of a few centimeters along the satellite track.

Baroclinic: Conditions in which the vertical shear is generated because of the horizontal gradients of density.

Barotropic: Conditions in which there are no variations in currents in the vertical direction.

Coamplitude and cophase: Lines of maximum tidal amplitude and lines of the time of occurrence of maximum tide, referred to either local or universal time.

Coriolis force: A fictitious force needed to allow for the noninertial nature of a rotating coordinate system.

Data assimilation: The process of blending observational data into numerical models.

El Niño: A frequent phenomenon in the tropical Pacific, occurring at 3–7-year intervals, when the eastern Pacific gets anomalously warm and sets off changes in the tropical atmosphere that affect weather all over the globe.

Gravimetry: The science of precise measurement of the earth's gravity.

Hindcast: A forecast excercise conducted for a period in the past to take account of the availability of accurate observational data for forcing, assimilation, and verification.

Inverse barometer effect: The effect where the changes in the atmospheric pressure are compensated exactly by the ocean by inverse changes in its height so that no oceanic motions are induced.

Isopycnal: A surface on which the density is constant.

Kelvin waves: Waves that run along the ocean margins (with the coast to the right in the northern hemisphere) at the speed of the shallow-water gravity wave. These waves are important for oceanic adjustment to changing surface forcing.

Nowcast: An estimate of the present state, often by an optimal blend of model and data.

Potential temperature: The temperature attained by a parcel of water brought adiabatically to a reference depth.

Reynolds averaging: The process of obtaining equations for mean quantities in a turbulent flow by considering each quantity to consist of a mean and a fluctuating component and taking averages over time or realizations.

Sigma coordinates: A coordinate system where the vertical coordinate is normalized by local depth; it is bottom-fitting or topographically conformal.

Synoptic forcing: Multihourly forcing from atmospheric models run at NWP centers, obtained in the past from a synopsis of weather charts.

Western intensification (boundary current):   Strong currents found at the western boundaries of the ocean basins or eastern sides of continents because the effect of Earth's rotation variation with latitude (the so-called β-effect).

References

Andersen, O. B., Woodworth, P. L., and Flather, R. A. 1995. Intercomparison of recent ocean tidal models. *J. Geophys. Res.* 100:25261–25282.

Anderson, D. L. T. and Moore, A. M. 1986. Data assimilation. In *Advanced Physical Oceanographic Numerical Modelling*, J. J. O'Brien, ed., pp. 437–464. Reidel, Dordrecht.

Bengtsson, L., Ghil, M., and Kallen, E., ed. 1981. *Dynamic Meteorology: Data Assimilation Methods*, p. 330. Springer–Verlag, New York.

Bleck, R. and Smith, L. T. 1990. A wind-driven isopycnic coordinate model of the north and equatorial Atlantic Ocean. Part I: Model development and supporting experiments. *J. Geophys. Res.* 95:3273–3285.

Blumberg, A. F. and Kantha, L. H. 1985. Open boundary conditions for circulation models. *J. Hydraulic Eng.* 111:237–255.

Blumberg, A. F. and Mellor, G. L. 1987. A description of a three-dimensional coastal ocean circulation model. In *Three-dimensional Coastal Ocean Models*, N. Heaps, ed., pp. 1–16. American Geophysical Union, Washington, DC.

Bryan, K. 1969. A numerical model for the study of the circulation of the world oceans. *J. Comput. Phys.* 4:347–359.

Choi, J.-K. and Kantha, L. H. 1995. A nowcast/forecast experiment using TOPEX/Poseidon and ERS-1 altimetric data assimilation into a three-dimensional circulation model of the Gulf of Mexico. Abstract, XXI IAPSO Meeting, Hawaii, Aug. 5–12.

Cox, M. D. 1985. An eddy-resolving numerical model of the ventilated thermocline. *J. Phys. Oceanogr.* 15:1312–1324.

Derber, J. and Rosati, A. 1989. A global oceanic data assimilation system. *J. Phys. Oceanogr.* 19:1333–1347.

Desai, S. D. and Wahr, J. M. 1995. Empirical ocean tide models estimated from TOPEX/POSEIDON altimetry. *J. Geophys. Res.* 100:25205–25228.

Dietrich, D. E., Marietta, M. G., and Roach, P. J. 1987. An ocean modeling system with turbulent boundary layers and topography. *Int. J. Numer. Methods Fluids* 7:833–855.

Dukowicz, J. K. and Smith, R. D. 1994. Implicit free-surface model for the Bryan–Cox–Semtner ocean model. *J. Geophys. Res.* 99:7991.

Dukowicz, J. K., Smith, R. D., and Malone, R. C. 1993. A reformulation and implementation of the Bryan–Cox–Semtner ocean model on the Connection Machine. *J. Atmos. Ocean. Technol.* 10:195.

Gelb, A., ed. 1988. *Applied Optimal Estimation*, p. 374. MIT Press, Cambridge, MA.

Gill, A. E. 1982. *Atmosphere–Ocean Dynamics*. p. 666. Academic Press, New York.

Glenn, S. M. and Robinson, A. R. 1995. Verification of an operational Gulf Stream forecasting model. In *Quantitative Skill Assessment for Coastal Ocean Models*, D. R. Lynch and A. M. Davies, eds., pp. 469–499. American Geophysical Union, Washington, DC.

Haidvogel, D. B. and Robinson, A. R. 1989. In *Data Assimilation*, Special issue, *Dyn. Atmos. Oceans*. 13:171–515.

Haney, R. L. 1991. On the pressure gradient force over steep topography in sigma-coordinate ocean models. *J. Phys. Oceanogr.* 21:610–619.

Heaps, N., ed. 1987. *Three-Dimensional Coastal Ocean Models*. p. 208. American Geophysical Union, Washington, DC.

Hellerman, S. and Rosenstein, M. 1983. Normal monthly wind stress over the world ocean with error estimates. *J. Phys. Oceanogr.* 13:1093–1104.

Hibler, W. D., III and Bryan, K. 1987. Diagnostic ice–ocean model. *J. Phys. Oceanogr.* 17:987–1015.

Holland, W. R. 1986. Quasi-geostrophic modeling of eddy-resolved ocean circulation. In *Advanced Physical Oceanographic Numerical Modeling*, J. J. O'Brien, ed., pp. 203–231. Reidel, Dordrecht.

Hurlburt, H. E. and Thompson, J. D. 1980. A numerical study of Loop Current intrusions and eddy-shedding. *J. Phys. Oceanogr.* 10:1611.

Kantha, L. H. 1995. Barotropic tides in the global oceans from a nonlinear tidal model assimilating altimetric tides. 1. Model description and results. *J. Geophys. Res.* 100:25283–25308.

Kantha, L. H. and Clayson, C. A. 1994. An improved mixed layer model for geophysical applications. *J. Geophys. Res.* 99:25235–25266.

Kantha, L. H. and Piacsek, S. A. 1993. Ocean models. In *Computational Science Education Project*, Oak Ridge Nat. Lab., Dept. Energy Rep. CSEP.

Kantha, L. H., Blumberg, A. F., and Mellor, G. L. 1990. Computing phase speeds at an open boundary. *J. Hydraulic Eng.* 116:592–597.

Kantha, L., Whitmer, K., and Born, G. 1994. The inverted barometer effect in altimetry: a study in the North Pacific. *TOPEX/Poseidon Res. News* 2:18–23.

Killworth, P. D., Stainforth, D., Webb, D. J., and Paterson, S. M. 1991. The development of a free surface Bryan–Cox–Semtner ocean model. *J. Phys. Oceanogr.* 21:1333–1348.

Kowalik, Z. and Murty, T. S. 1993. *Numerical Modeling of Ocean Dynamics*, p. 481. World Scientific, Singapore.

Le Provost, C., Genco, M. L., Lyard, F., Vincent, P., and Canceil, P. 1994. Spectroscopy of the world tides from a finite element hydrodynamical model. *J. Geophys. Res.* 99:24777–24797.

Levitus, S. 1982. Climatological atlas of the world ocean. *NOAA Professional Paper* 13, Geophys. Fluid Dyn. Lab., Princeton, NJ. 173 pp.

Madala, R. V. and Piacsek, S. A. 1977. A model for baroclinic oceans. *J. Comput. Phys.* 22:167.

Mellor, G. L. 1991. User's guide for a three-dimensional, primitive equation, numerical ocean model. *Princeton University Rep.* 91. 35 pp.

Mellor, G. L. and Yamada, T. 1982. Development of a turbulence closure model for geophysical fluid problems. *Rev. Geophys.* 20:851–875.

Mesinger, F. and Arakawa, A. 1976. *Numerical Methods Used in Atmospheric Models*, Vol. 1. Global Atmospheric Research Program Publication 17. 64 pp.

Metzger, E. J., Hurlburt, H. E., Kindle, J. C., Serkes, Z., and Pringle, J. M. 1992. Hindcasting of wind-driven anomalies using a reduced-gravity global ocean model. *Mar. Technol. Soc. J.* 26:23–32.

Oberhuber, J. M. 1993. Simulation of the Atlantic circulation with a coupled sea ice–mixed layer–isopycnal general circulation model. Part I: Model description. *J. Phys. Oceanogr.* 23:808–829.

O'Brien, J. J. 1985. *Advanced Physical Oceanographic Numerical Modeling*. Reidel, New York.

Orlonski, I. 1976. A simple boundary condition for unbounded hyperbolic flows. *J. Comput. Phys.* 21:251–269.

Pond, S. and Pickard, G. L. 1979. *Introductory Dynamical Oceanography*, 2nd ed. p. 329. Pergamon Press, New York.

Ponte, R. M. 1994. Understanding the relation between wind driven sea level variability and atmospheric pressure. *J. Geophys. Res.* 99:8033–8040.

Roed, L. P. and Cooper, C. K. 1986. Open boundary conditions in numerical ocean models. In *Advanced Physical Oceanographic Numerical Modeling*, J. J. O'Brien, ed., pp. 411–436. Reidel, Dordrecht.

Schwiderski, E. W. 1980. On charting global ocean tides. *Rev. Geophys.* 18:243–268.

Semtner, A. J. 1986. Finite-difference formulation of a world ocean model. In *Advanced Physical Oceanographic Numerical Modeling*, J. J. O'Brien, ed., pp. 187–202. Reidel, Dordrecht.

Semtner, A. J. 1995. Modeling ocean circulation. *Science* 269:1379–1385.

Semtner, A. J., Jr. and Chervin, R. M. 1992. Ocean general circulation from a global eddy-resolving model. *J. Geophys. Res.* 97:5493–5550.

Smagorinskiy, J. 1963. General circulation experiments with primitive equations: I. The basic experiment. *Mon. Weather Rev.* 91:99–164.

Smith, R. D., Dukowicz, J. K., and Malone, R. C. 1992. Parallel ocean general circulation modeling. *Phys. D* 60:38.

Thacker, W. C. and Long, R. B. 1987. Fitting dynamics to data. *J. Geophys. Res.* 93:1227–1240.

Wallcraft, A. J. 1991. The Navy layered ocean model users guide. *NOARL Rep.* 35, 21 pp.

Warren, B. A. and Wunsch, C., eds. 1981. *Evolution of Physical Oceanography.* p. 623. MIT Press, Cambridge, MA.

Further Information

This review article has been necessarily sketchy. The reader is therefore encouraged to consult the various references cited for more details. The monograph on numerical ocean modeling edited by James O'Brien [1985] is still the best starting point, especially since the models described there have remained essentially unchanged, undergoing only small evolutionary changes such as adaptation to massively parallel computers and inclusion of better mixing algorithms. A good starting point in coastal ocean modeling is the American Geophysical Union (AGU) volume edited by N. Heaps [1987]. Kowalik and Murty [1993] is an excellent "cookbook" for details of numerics such as finite-differencing and the split-mode technique. Reference can be made to Haidvogel and Robinson [1989] for a good description of data assimilation methods. Textbooks by Pond and Pickard [1979] and Gill [1982] are good starting points for exploring the dynamics of the oceans. The Henry Stommel 60th Birthday volume on physical oceanography [Warren and Wunsch 1981] is a good followup.

There is no specific journal for ocean modeling; instead, modeling advances are published in journals such as the *Journal of Physical Oceanography* of the American Meteorological Society (AMS) and *Journal of Geophysical Research (Oceans)* of the American Geophysical Union (AGU). The *Journal of Hydraulic Engineering* of the American Society of Civil Engineers publishes modeling papers mostly related to coastal and estuarine studies. The *Journal of Continental Shelf Research* specializes in coastal research, including coastal modeling. Purely computational advances often appear in journals such as the *Journal of Computational Physics.* Semiyearly meetings of the AGU, meetings of the AMS and biennial Ocean Sciences, and quadrennial meetings of the International Union of Geodesy and Geophysics (IUGG) are examples of venues where latest advances in ocean modeling are presented and critiqued.

The GFDL z-level deep-water-basin model (ftp.gfdl.gov or 140.208.1.2; directory pub/GFDL_MOM2), Princeton sigma-coordinate shallow-water coastal model (ftp.gfdl.gov or 140.208.1.2; directory pub/glm), and University of Miami isopycnal model (ftp_mount.ee.umn.edu; directory pub/ocean) are all available from the respective anonymous ftp sites. Readers are encouraged to offload the model codes and experiment with them. A good starting point for hands-on ocean modeling is the Ocean Models chapter of the Computational Science Education Project [Kantha and Piacsek 1993], available on the World Wide Web at http://compsci.cas.vanderbilt.edu/csep.html. It contains model code, graphics, animation, and exercises on simple ocean models that serve as a good introduction to the field. Net surfers might be interested in browsing through http://www.cast.msstate.edu/Tides2D and http://www.cast.msstate.edu/Altimetry for recent examples related to modeling and altimetry. These also provide a glimpse into the world of electronic publishing of the coming century.

41

Computational Biology

David T. Kingsbury
Johns Hopkins University
School of Medicine

41.1 Introduction

In the past five years computational biology has emerged as an analog to the development of molecular biology as a discipline in its own right. In the late 1950s and early 1960s a group of scientists began to apply the tools of multiple disciplines—genetics, microbiology, physics, biochemistry, and biophysics—to analyze biological problems in a new way. The power of this approach was so great that it emerged as the discipline of molecular biology. Now the application of mathematical and computational tools to all areas of biology is producing equally exciting results and is providing insights into biological problems too complex for traditional analysis.

Biology, regardless of the subspecialty, is overwhelmed with a large amount of very complex data. However, what sets biology apart from other data-rich fields is the *complexity* of its data. In contrast to other data-rich fields, biology generally remains a scientific "cottage industry," the data being generated in a highly distributed mode, with no standard format or syntax.

Thus, all areas of the biological sciences have urgent needs for the development of organized and accessible storage of biological data. While this is referred to as biological database development, traditional database technology such as transaction-oriented relational database systems is often inadequate to serve many areas of the biological sciences, due to the complexity of biological data and the absence of a standardized data-structuring strategy. It is clear that collaboration between computer scientists and biologists will be necessary to design information platforms which accommodate the needs for variation in the representation of biological data, the distributed nature of the data acquisition system, the variable demands placed on different data sets, and the absence of adequate algorithms for data comparison, which are characteristic of biological science.

The recent dramatic advances in commercially available hardware and approaches to building hardware clusters have had a profound effect on computational biology. Traditional general-purpose hardware was inadequate for many of the most computationally intense problems in the biological sciences. These

0-8493-2909-4/97/$0.00+$.50
© 1997 by CRC Press, Inc.

computational problems were best handled by special-purpose equipment designed by teams of biologists and chip and circuit designers. This condition has essentially disappeared as high-performance general-purpose instruments have become more widely available. However, not only hardware limitations have affected the productivity of the computational biologist. There is a continuing need for new algorithm development to cover many tasks, especially real-time data analysis and comparisons between objects and images. Imaging technology is central to almost all of biology, and data representation through image construction remains an elusive but astoundingly powerful tool.

During the last decade there were dramatic advances in instrumentation and related methodologies for both light and electron microscopy. The advances lie not simply in higher resolution, but rather in a broader size range of structures that can be analyzed, more powerful methods for putting together the pieces of three-dimensional puzzles of cell form, and the addition of dynamic details of biological form and function, ranging from the subcellular to the physiological level. These new approaches are computationally demanding. Existing computational resources, which were typically set up for entirely different processing needs, are proving inadequate for dealing with the massive data flow. An effort to develop new computational approaches is underway in a few laboratories around the world. However, it is important that new software be developed within the context of the experimental research driving the needs; that is, there must be close collaboration between those developing the software and the groups carrying out research on static and dynamic structures.

X-ray crystallography and *NMR* are the major experimental methods for deducing macromolecular structures at atomic resolution. Both methods produce extremely large amounts of data and are entirely dependent upon the availability of powerful computers and sophisticated processing algorithms for the interpretation of raw data. In addition, there are fundamental scientific problems in both areas that require major computational advances. Substantial opportunities exist for combining structural information from several experimental techniques. This may provide the basis for a structural solution where only partial data are available from any single technique. With improved computational tools, combining physical data from a variety of sources may become commonplace. These developments will allow solutions to be obtained for structural problems which would otherwise be intractable. Analysis of errors in structures based upon experimental data from several sources also represents a significant computational challenge.

Advances in X-ray and NMR data analysis will lead directly to rapid developments in the field of protein-folding which will be synergistic with developments in other areas of biology itself, and especially computational biology. Common problems of data representation, search strategy, pattern recognition, and data visualization appear in many fields. There is a particularly exciting synergistic relationship between the protein-folding field and the fields of structure determination by X-ray crystallography and 2-D NMR. Each field will benefit from rapid advances in the others. Improved folding algorithms provide a new way to attack the phase problem in crystallography, and new, more carefully refined protein structures provide rich new insights into protein folding.

Computational neurobiology gives us the hope of interpreting the mass of anatomical and physiological information about the nervous system that is now available in functional terms. Better interpretation of these data will permit neurobiology to make contact with other fields such as psychology and artificial intelligence. This work will make specific, testable predictions in the areas of sensory perception (visual, olfactory, and auditory), memory, learning, and motor control. Above all, it will lead to the integration of all these aspects to provide an eventual understanding of the total functioning of the nervous system. Such integration can be expected to provide new insights that will lead to improvements in the treatment of diseases of the nervous system at all levels, from neuropharmacology to psychotherapy. In addition, studies of this kind may be expected to contribute to major advances in artificial intelligence and practical robotics.

The area of *genome analysis* continues to be a major focal point in computational biology, and much progress has been made over the past few years. Robust approaches to both **linkage mapping** and **physical mapping** have been developed, and significant effort has been invested in new algorithmic approaches to computing these problems. However, considerable effort is still required to make genetic linkage maps effective tools for genetic research. To be useful in common situations, more markers must be identified and mapped to produce higher-resolution maps. In many cases marker analysis requires the ability to analyze small families and consider quantitative traits. To be fully useful in a meaningful quantitative sense, this analysis will require powerful computer simulation and modeling. Major algorithmic advances have been made in the area of sequence assembly and clone assembly in physical mapping. As biologists continue to pursue the rapid sequencing of many genomes, strategies for the assembly of random fragments remain another major problem. Common to all of the problem areas mentioned is the need for good visualization of data. Visualization is necessary because the map and sequence analysis phase for a molecular biologist is equivalent to exploratory analysis for a statistician. It is at this point that the experimentalist gains the feeling for, and understanding of, a physical or linkage map or sequence, which may then guide many months of experimental work. The complexity inherent in biological systems is so great that very sophisticated methods of analysis are required. These are the tools which must be readily accessible to molecular and cellular biologists untrained in computer technology.

Ecology and evolutionary biology encompass a broad range of levels of biological organization, from the organism through the population to communities and whole ecosystems. This complexity demands computational solutions. The need for enhanced computational ability is most evident when one attempts to couple large numbers of individual units into highly interactive and largely parallel networks, whether at the tissue, community, or ecosystem level of organization. The proliferation of information from remote sensing introduces the need for geographical information systems that provide a framework for classifying information, spatial statistics for analyzing patterns, and dynamic simulation models that allow the integration of information across multiple spatial, temporal, and organizational scales.

What follows is an examination of several specific areas of computational biology, with a particular emphasis on those areas related to molecular biology, and a short development of the experimental paradigm and highlights of the current computational challenges regarding the requirements for further development of that area. There are common themes which appear in several of the sections, and these themes deserve special attention, since they appear to be limiting the development of the entire field, regardless of the specific area of research. (This review will not attempt to cover the important areas of computational neuroscience and ecology and evolutionary biology in any further detail. Both fields are rich in computational challenges and theory and, like some of the areas covered here, deserve a chapter of their own.)

41.2 Databases

Biology is inherently information-rich because of the complexity and variety of living systems. Understanding these systems requires information about their organization, structure, and function at a multitude of levels from the macroscopic to the molecular. Moreover, each species (and in many species each organism) represents the potential for a unique solution to the problems of life processes and the organism's interaction with the environment. The full understanding of biology requires extending organismic complexity to include the relationships of species and organisms in their ecological niches, as well as the evolution of the biosphere over time.

Until recently, much of this information was accessible to scientific inquiry only at high levels of abstraction, so that the inherent richness of information was not reflected in the volume of data available. This situation has changed rapidly over the last decade in a number of biological fields, among them molecular biology, neurobiology, ecology, and taxonomy. This change has been made especially dramatic by the widespread use of the World Wide Web (WWW). Emerging scientific paradigms will require even more data organized into large and dynamic databases to support ongoing biological research. Indeed, some aspects of modern biology (e.g., the Human Genome Project, or protein structure–function studies) are now utterly dependent upon database and computer technology. In many cases current data collections are not well organized for ease of retrieval but remain central to work in a given field. For example, several important macromolecular databases are maintained as flat files, poorly delimited and not accessible to ad hoc queries because of the absence of well-structured fields. Because of the importance of these data, they are the subject of active database development by numerous investigators. The reliance of modern molecular biology on databases is exemplified by its heavy dependence on the collection of DNA-sequence databases (GenBank, EMBL, DDBJ, and GSDB). At the present time it is impossible to publish new information about a gene sequence without (1) providing evidence of prior examination of the gene-sequence database and reporting the relationship with existing sequences, if any exist, and (2) providing evidence of submission of the new sequence to a public database.

As a result of this need, a substantial number of people now devote their careers to data management and computational analysis in biological disciplines. However, despite significant and vigorous efforts, the present generation of biological databases will fail in the next decade without significant continued development. They were not (and could not have been) designed to deal with the volume, complexity, and diversity of the data which will need to be accessible for future biological research.

The explosion of data is derived, in large part, from the desire of increasing numbers of investigators to have shared data repositories. The pressure on databases is severe in a quantitative sense, but is equally daunting in terms of the diversity and the interrelationships which must be represented among the data. These problems are further complicated by the way data are generated in biological research, which is geographically dispersed and with little standardization. Taken together, these problems pose unique transdisciplinary challenges for database design. Indeed, it is important to note that the technology is not yet in hand which can support the design of adequate databases for much of the biology of the next decade. In fact, the application of the term "database" itself tends to trivialize the problem and is misleading. Database technology and theory as it currently exists is an inadequate paradigm for what is needed now and in the future to represent and organize biological information. For example, biological data include mixtures of measurements, images, and interpretation, including extensive collections of metadata. Ideally each of these data types would be available to ad hoc queries. At present only a few extended relational systems have addressed these needs, and without complete success.

Computational science has, however, begun to deliver technologies which have tremendous promise for future biological databases. Examples include the object-oriented design paradigm [Kent 1981], object-based development environments (e.g., OPM [Chen and Markowitz 1994]), extended relational systems [Stonebreaker and Rowe 1984], and semantic search models [King 1988]. Biologists are becoming increasingly sophisticated in their use of computers and in their abilities to state their research requirements in terms of informational and computational strategies. Biological science is now posing questions which not only require computational solutions, but also provide problems of fundamental interest to computer science researchers, creating the possibility of effective interdisciplinary work. Fortunately, there is a growing community of transdisciplinary workers whose expertise is centered at the interface of biology and computing, and who can provide much of the insight into how the two fields can interact productively.

One of the principal challenges in scientific computing in the next decade will be the development of database systems which can handle the inherent complexity of biological information. The existence and availability of such databases will transform the way the science is done, and make possible completely new paradigms of biological research. Meeting this challenge will require the construction of databases in fields where none are available, significant research and development in database and knowledge-base technology, and the provision of a robust and widely available computational infrastructure for biological science through new algorithmic approaches to data analysis and tools to embed database access into the analysis routines. The beginnings of such a system have been developed [Costain and Marinescu 1995].

Access and Communication

The emergence of the World Wide Web has had an enormous impact on biological databases. Because the subdomains of biology are fragmented, it has been necessary to develop many distinct and customized databases. The fact that there is little semantic consistency between these databases has raised a significant barrier to linking them through standard query mechanisms. In the past two years a number of investigators have built hypertext-based linkages between a number of different databases, bringing together a richer data resource. One representative of the many examples is ProtWeb, a collection of molecular- and structural-biology databases developed by Dan Jacobsen at Johns Hopkins School of

Medicine (http://www.gdb.org/hopkins.html). Complex searches are accommodated through sophisticated forms interfaces and links built on the fly through a series of C and Perl routines.

As powerful as the WWW-based systems are, they still lack the potential for supporting complex ad hoc queries that would be achieved through a true federation of biological databases. The need for such a federation has been recognized most acutely in the genomics community, and the outlines for such a federation have been developed [Robbins 1994]. It was suggested that for minimum technical linkage all of the participating databases present similar **APIs** to the Net. All of the databases in the federation should also be relational systems that support SQL queries. Ideally these databases should (1) be self-documenting, (2) be stable, and (3) conform to an agreed-upon federation-wide semantics. The problem with this strategy is that in many cases it places the goals of the federation in conflict with the rapidly changing nature of biological research and the needs of the specialty user community. In balance the lack of interoperability places the future of large programs like the Human Genome Project at great risk—coordination between genome-relevant community databases is essential.

Representation/Data Modeling

To enhance the continued development of shared data resources we must increase database expressiveness, study representation of biological knowledge, and automate modeling of database schemas. Biological knowledge is extremely rich and diverse; it includes raw experimental data (images, numbers, symbols), interpretations of experimental data (descriptions of biological objects with complex properties and internal structures), descriptions of experimental methods (complex procedures), and theoretical knowledge (documents, equations, descriptions of processes such as gene expression). Existing database technology provides a small number of simple representational primitives (such as relations) allowing databases to capture only a small piece of this rich biological semantics. Research is needed to provide richer representational capabilities, and to study how to represent the wide range of biological knowledge. Further, because biological databases will model a large number of complex entities, biological database schemas will be correspondingly complex. Researchers must investigate automated methods of managing database schemas to increase the efficiency of the database design process.

41.3 Imaging, Microscopy and Tomography

Image reconstruction with light and electron microscopy and other imaging techniques is evolving as a powerful tool for the characterization of biological structures in three dimensions over a wide range of scales. Biological structures amenable to one or more techniques of imaging include single macromolecules and macromolecular assemblies, subcellular organelles, whole cells, and tissues. Characterization of the spatial organization of biological structures is critical for determining the functionality of these structures. Many fundamental problems in biology are open to study by microscopy and other imaging techniques. The recent advances in computer-controlled scanning microscopies, high-resolution atomic-force microscopy, and cryoelectron microscopy will add substantially to the level of detail which will be attained by these methods.

The continuing advances and widespread applications of **transmission electron microscopy** at conventional, intermediate, and high voltages (with and without energy filtration), as well as confocal light microscopy, are leading to the production of vast amounts of data that need to be processed to extract the structural information and biologically important details. Each of these techniques can produce three-dimensional images of biologically important structures. Successes in these studies do not imply that the technical problems have been solved; currently, there are substantial computational challenges to meet before achieving the goal of making efficient use of these techniques. These challenges include developing improved image processing and reconstruction algorithms.

In transmission electron microscopy there have evolved three distinct methodologies for producing three-dimensional information: (1) electron crystallography of two-dimensional lattices and the processing of images of symmetric structures; (2) the analysis of multiple images of isolated asymmetric macromolecular assemblies; and (3) electron tomography, or three-dimensional reconstruction from a tilt series of images obtained from a single structure. Confocal light microscopy (CLM), a tool used in cell biology, produces three-dimensional images by scanning light focused to a single point over a three-dimensional grid in the specimen, and then imaging the light onto a point detector. Compared to conventional microscopies, CLM has a somewhat higher resolution and a much smaller depth of field, minimizing mixing of image data from different depths in the specimen.

The successes of these methods have been substantial. For instance, electron microscope tomography has been used to solve the structure of a single transcription unit, a dendrite, and of chromosome fibers [Belmont et al. 1993]. Three-dimensional reconstruction from images of single molecules has been used to study mammalian ribosomal subunits, calcium release channels, and the flagellar motor (basal bodies) [Schuster and Kahn 1994]. Three-dimensional reconstruction of symmetric structures has been used to characterize a large number of structures, including many viruses. Examples of the achievements of electron crystallography include the three-dimensional structure of gap junctions and the high-resolution structure of purple membrane. Confocal microscopy has been used to image chromosomes in three dimensions, and its use with immunofluorescence localization is a powerful method for determining the spatial distribution of molecules within a cell and (more recently) resolving the electrophoretic separation of DNA fragments in high-speed, high-throughput DNA sequencers [Huang et al. 1992]. Specimen preparation methods have also improved. The use of frozen specimens in transmission electron microscopy (cryoelectron microscopy) has provided a method for determining the structure of assemblies, such as enveloped viruses, that are readily distorted by standard preparative techniques.

Use of these imaging techniques entails three fundamental problems. The first is the vast quantity of data being produced which require complex processing. This is a problem common to all imaging techniques within and outside biology. For example, cryoelectron-microscopic analysis of icosahedral viruses involves sample preparation followed by microscopy followed further by a series of computational steps, many of which are very computation-intensive [Cheng et al. 1995]. Because of the limited computing power available, only a limited number of independent images can be analyzed. However, three-dimensional images are intrinsically complex, and the amount of information required to characterize an image in detail is very high. For instance, construction of the three-dimensional image of a molecule from electron-microscope images of single particles routinely requires thousands of particle images. This means that the actual time to produce a three-dimensional image is weeks to months, due in large part to the user-mediated steps that remain in the analysis. For efforts like this to prove maximally fruitful, it is crucial that each step be automated as much as possible.

The second fundamental problem common to all imaging techniques is the existence of a point spread function due to the instrumental broadening that is intrinsic to each form of imaging. For instance, transmission electron microscopy loses a cone of data in the Fourier transform of the image, and the restoration of this cone represents a difficult, open problem. In **scanning confocal microscopy**, the point spread function is greatly extended in the direction parallel to the optical axis, and narrowing it could improve the resulting three-dimensional images.

Third, the results of any imaging method must be quantitated and displayed. The problems of image enhancement and visualization are completely general, although each technique may benefit more from specific display modes than others. Quantitative comparison of images also provides substantial challenges, especially in the presence of noise. Comparison of two images of flexible objects (e.g., cells or chromosomes) represents a substantially greater challenge.

One of the recurring problems common to all imaging techniques is the existence of artifacts due to incomplete data collection. These artifacts may seriously interfere with the interpretation of the recon-struction, and may even lead to incorrect conclusions. This problem is most serious in electron microscopy, where a full range of viewing angles is not usually accessible, and data corresponding to a cone or wedge in Fourier space cannot be collected. In confocal microscopy, the resolution in the z-direction (parallel to the optical axis) is much lower than in the x and y directions, as reflected by a nonisotropic instrumental point spread function.

Although several restoration algorithms have been in existence for some time, only the recent increases in computational speed have made their practical implementation possible. Two algorithms with potential application to signal restoration have attracted special attention due to their success in other fields—projection onto convex sets (POCS) [Bellon and Lanzavecchia 1995] and maximum entropy (ME) [Schmeider et al. 1995]. Both methods are extremely computation-intensive, making their application to realistic-sized image volumes ($64 \times 64 \times 64$ or $128 \times 128 \times 128$) exceedingly cumbersome on most computers. A full development of these algorithms into something useful for detailed images of biological systems will require many computation cycles, to allow many different parameter values to be tested (ME) or a variety of different constraints to be used (POCS). Thus, a serious attempt to make three-dimensional image restoration viable for biological images will always require the highest available computational speed, along with advances in the theory and design of algorithms.

41.4 Determination of Structures from X-Ray Crystallography and NMR

The three-dimensional structures of proteins and nucleic acids are essential elements in the pathway relating gene sequence to function. The problem of predicting three-dimensional structure from a sequence remains unsolved at present. X-ray crystallography and NMR are the major experimental methods for deducing these structures at atomic resolution. While the rate at which these structures are being determined is increasing, it lags significantly behind the rate at which new sequences are being accumulated. Each newly determined structure increases our knowledge in two ways. First, it adds to the database of known structures which can be used in knowledge-based methods in subsequent structure determinations of other macromolecules. Equally important, each new structure provides insights into the fundamental biological processes that support all life.

NMR and X-ray crystallography both produce extremely large amounts of raw data and are entirely dependent upon the availability of powerful computers and sophisticated processing algorithms for the interpretation of those data. In addition, there are fundamental scientific problems in both areas that require major computational advances. Both NMR and crystallography make use of constraint refinement to optimize the fit of experimental data to working models, and both fields use large scale-simulations to correlate the molecular models to known biological properties. Like investigators in three-dimensional microscopy, crystallographers and NMR spectroscopists are using maximum-entropy reconstruction as a major tool in computational analysis.

Determination of Macromolecular Structures from NMR Data

NMR methods for determining three-dimensional structures of macromolecules in solution have become increasingly important over the past several years. Major limitations on the speed and ease of analyzing the NMR data are the difficulty of assigning individual resonances in the spectra to particular protons in the molecule and then of calculating the full three-dimensional structure using distance geometry, molecular dynamics, or algorithms. For example, when determining the structure of a 100–150-residue protein, it may take as much as two weeks of NMR spectrometer time to collect the raw data, two or three days to calculate the spectra, and months or years to fully interpret the results. The complexity of this process depends on both the size of the protein and the extent of peak overlap within the spectra. Since this is the critical bottleneck in obtaining the structural information, it limits the size of molecules which can be considered. One critical element is the solubility of biological macromolecules. The balance between solubility and the sensitivity of modern NMR equipment places the current lower limit on concentration at around 0.5 mM. Many proteins, especially those of high molecular mass, aggregate at such high concentrations, leading to broad spectral lines of no value in structural studies. The solution to this problem lies with the development of enhanced computational approaches, such as maximum-entropy reconstruction of specially collected data sets. This approach requires approximately 100 times the computational work of the traditional discrete-Fourier-transform processing. The refinement of this approach will require the continued collaboration of structural biologists and computer scientists [Schmieder et al. 1995]. Future advances in algorithms, software development, and availability of more powerful computers will make a major impact on the time required to interpret NMR data.

A critical step in interpreting the data is to use the relationships between protons signified by two-dimensional (or three-dimensional) crosspeaks to assign the resonances to particular protons in the molecule. Assignment is frequently the rate-limiting step in structure determination. There are a number of different strategies for assignment, and approaches to automate this process are under active investigation. It is clear that several approaches will be necessary to deal with the problems associated with ambiguous data. Both the **sequential** and **mainchain-directed** assignment procedures make use of patterns of *J*-correlated and distance-correlated relationships. In both cases the procedure is still largely manual, although there have been recent attempts to automate parts of the analysis. The development of computer-assisted or fully automatic pattern recognition techniques would make a major impact on the time required to make the assignments, as well as the size of molecules which can be studied. This is particularly true as the complexity of the original NMR data increases.

Once protons have been assigned, a three-dimensional structure can be estimated using the peak areas from the **NOESY** spectrum to determine distance constraints. Extensive computing is required to calculate structures using distance geometry, molecular dynamics, or **Kalman filtering** techniques. Several refinements of the structure are needed, and a family of structures is usually generated. Recently, back-calculation of the NOESY spectrum has been used to try to refine the structure. The value of this procedure and the effect of using different techniques to calculate NMR structures still need to be established. However, both the development and application of this technique require major computing power.

X-Ray Structure Determination of Macromolecules

X-ray crystallography provides a fine example of how the availability of supercomputers, through the National Science Foundation-funded centers, has dramatically sped up the rate at which X-ray structures can be determined. There are four major phases in crystal structure determination: crystal growth; solution of the phase problem; interpretation of the resulting electron density in terms of an atomic model; and, finally, refinement of the atomic model. Up until very recently, the refinement of protein structures using X-ray data required substantial manual intervention and took one or more years to get to a satisfactory stage. With the incorporation of **simulated annealing** methods into the refinement procedure, the time required for this phase has dropped to one or two months using minisupercomputers. It should be noted that while the established protocols for simulated annealing can now be run fairly comfortably on workstations, the original development and testing of the method required the availability of supercomputers such as the Cray.

The fundamental problem in protein crystallography is the **phase problem,** which, with the improvement in refinement techniques, has become the major bottleneck in the structure determination process. The current method of multiple isomorphous replacement (MIR) relies on measuring data using crystals of the protein soaked in different metal compounds. The soaking procedure does not always yield suitable crystals, and in some cases the lack of metal derivatives delays the determination of the structure by several years. Attempts to solve the phase problem without resorting to MIR or other experimental techniques fall into two classes:

1. *Ab initio phasing.* This does not rely on any additional experimental information other than that provided by the X-ray data set on the native protein crystal. Methods that have been successful for small molecules break down in the range of 100 atoms or so, well below the range of 1000 atoms or more in protein molecules. The availability of more computing power is gradually leading to the extension of some of these methods to larger polypeptides, and intensive computational effort may yield success for the smaller proteins (approximately 50 amino acids). The ab initio solution

of virus structures may be aided by taking advantage of their high degree of noncrystallographic symmetry. If an adequate starting model is available, then these techniques allow the extension of phases directly over a wide resolution range [Rossman et al. 1985]. Novel approaches, such as those based on maximum entropy, may also provide computational tools that will have a substantial impact on obtaining phases for diffraction data.

2. *Use of prior knowledge.* In cases where the three-dimensional structure of a closely related protein is known, the phase problem can be solved by using the known structure as a starting point for refinement *(molecular replacement)*. Computation-intensive methods such as simulated annealing have proved to be particularly valuable here, since significant distortions may have to be introduced into the starting structure and the optimization procedure needs to escape from the local minimum of the starting structure. The inclusion of prior knowledge into phase determination is likely to become extremely important in the near future, because it is becoming clear that proteins are built up from smaller subunits or segments that are commonly shared between large numbers of proteins. As the database of known structures increases, it should be possible to use segments of structures in search procedures to solve the phase problem. There is a need here to develop and apply novel search procedures and pattern recognition algorithms, many of which will also be applicable in the next stage of the structure determination.

Noisy electron density maps are difficult to interpret manually because of false or missing connections in the electron density. It is not uncommon for a structure determination project to arrive at this stage rapidly, once crystals are obtained, and then to spend a long time getting a better electron density map (the phase problem). Computing is essential in two ways. First, phase bias from the model has to be minimized by optimizing relative contributions from calculated and observed diffraction data. Second, rapid interactive refinement of difficult-to-interpret regions must be performed to help to solve the problem. This requires high-speed supercomputing and a fast network link between the supercomputer and a high-performance graphics workstation.

Even in the case of a well-defined electron density map, fitting an atomic model to these maps is a time-consuming process. The problem lies in automatic recognition of characteristic patterns of macro-molecular structure, such as alpha helices, which cannot be achieved by existing algorithms. Because of the complexity of the three-dimensional pattern recognition, further advances in methodology and computing speed are required.

41.5 Protein Folding

Protein folding, recognized for many years as one of biology's core problems, still occupies a central role in biology. Folding converts the linear, one-dimensional information of the amino acid sequence into the biologically active three-dimensional structure. Folding may be thought of as a final unsolved aspect of the genetic code, and it is clear that progress on the folding problem will have tremendous theoretical and practical implications for biology. As described above, there have been dramatic advances in crystallography and two-dimensional NMR, and the virtual explosion of protein sequence data reemphasizes the importance of working on the folding problem. Even limited progress in this area could have a tremendous payoff. The *protein folding problem* can be considered either as:

- the problem of understanding the actual kinetic pathway by which a protein folds, or
- the problem of predicting the final folded conformation.

Obviously, a detailed structural understanding of folding intermediates would lead to a prediction of the final folded structure. However, experimental studies of folding intermediates have been very difficult because the intermediates are present in vanishingly small amounts. While there are very few experimental data that can serve to guide the theoretical work, the rich database of known structures provides an excellent guide regarding the final folded conformation.

At first glance, the protein folding problem may seem to have a tantalizing simplicity (a given string of 100 amino acids contains all the data needed to determine the final folded structure), but the problem is extraordinarily complex. If each residue in a polypeptide chain can adopt 10 distinct conformations, then the protein could adopt 10^{100} distinct structures, which leads to an extraordinarily difficult search problem. Many different strategies have been used in approaching the folding problem. Some methods have relied on detailed physical models of the polypeptide chain and have tried to carefully simulate the interactions (hydrogen bonds, van der Waals contacts, electrostatic interactions, etc.) that stabilize the chain. Other methods rely on the structural database that has accumulated over the past decades. In some sense, it appears that "structure is more conserved than sequence," so that the structural database is a useful guide when modeling new proteins.

Current approaches to the folding problem can generally be placed into one of two methods: direct determination of the folded confirmations, or a template-based method. Direct methods seek to determine the lowest acceptable energy point in a suitably defined conformational space. Template-based methods compare the sequence of the unknown with a collection of solved structures and select a limited number of highly scored possibilities [Luthy et al. 1992, Sali et al. 1994].

Recently an additional approach [Srinivasan and Rose 1995] was described, based on a direct prediction method which was built on the observation that globular proteins are organized as a structural hierarchy [Rose 1979]. The basis of this approach is that neighboring chain sites interact to form primitive folding modules, which interact in an iterative fashion, resulting in larger molecular structures. In this scheme all folding events are local at a critical step in the self-assembly process.

One core problem with direct methods is the difficulty of searching through the astronomical number of possible structures. A naive calculation may suggest that there are 10^{100} conformations, yet it is clear that only "a few" of these are of low enough energy to be plausible structures or plausible folding intermediates. The multiple-minima problem arises repeatedly in studies of folding. Both physical approaches (based on a detailed molecular potential surface) and pattern recognition schemes (based on analogy) encounter the same problems with multiple minima.[1] Stochastic search algorithms have proved especially useful in handling problems with multiple minima, and the method of **simulated annealing** is frequently used. This method corresponds to a simulation of the molecular dynamics under the influence of random thermal forces. Other search algorithms involve buildup or stochastic buildup based on genetic algorithms. In order to estimate the difficulty of the multiple-minima problem in protein folding, it is possible to draw upon some parts of statistical physics for help. Simple lattice models have been used to estimate excluded-volume effects after polymer collapse, and these models indicate that the number of allowed conformations may be far smaller than suggested by the initial naive estimates. This result is very encouraging, since it suggests that the search problem can focus on a much smaller region of conformational space.

Detailed atomic models have been used to study protein folding and stability. The models are based upon well-understood principles of physical chemistry and have been used in conjunction with molecular dynamics and Monte Carlo methods to study the underlying forces that determine the stability of folded proteins. The application of free-energy perturbation theory has been a particularly exciting development. These approaches are beginning to provide a much better understanding of the key forces that are involved in protein folding and stability, such as the true role of electrostatic interactions and the origin of the hydrophobic effect. Continued close interactions between experimental biologists and computational/theoretical researchers have also been extremely important for this field. Theory can help design new experiments, and better data can allow the refinement of basic physical models.

Although not directly linked to the protein folding problem, an extremely important area of computational biology involves the molecular modeling of protein function. A molecular understanding of enzyme catalysis can now be approached through a combination of molecular dynamics and quantum

---

[1]The same search problem occurs in other parts of computational biology, including sequence comparison problems in genomics, neural networks, and immune-network modeling.

chemistry. There have been many applications of this approach, and important insights about the mechanism of triose phosphate isomerase have recently come from such simulations. Another exciting area is the recent development of computational approaches to modeling electron transfer. This a fundamental and inherently quantum-mechanical process involved in energy transduction and photosynthesis. Signal transduction is another extremely important and active area of research. This involves problems of docking and protein–protein recognition. Allosteric transitions are a frequent consequence of such interactions, and new methods for studying protein motions on long time scales are being developed [Gilson et al. 1994].

Much recent enthusiasm has been driven by the ability to clone and express a protein of choice, followed by deliberate mutation of individual residues or larger segments. Perhaps the simplest application of folding that can be envisioned would be to predict the structural perturbation caused by a single-residue mutational change in a protein of known structure. Most often, such mutations do not result in major rearrangements of the chainfold, as shown by the work of Kossiakoff [Eigenbrot et al. 1993] and others. While the structural effect of single-site changes has been successfully predicted in some cases [Desjarlais and Berg 1992], there are many conspicuous failures, and clearly more work is needed. At the level of larger segments, attempts to predict antibody hypervariable loops have also met with partial success [Tramontano and Lesk 1995; Pan et al. 1995]. Recent experiments suggest that deletion of entire loops may be tolerated while partial deletions of the same segments are not. Such results suggest the existence of quasi-independent modules, which would simplify calculations.

41.6 Genomics

Genomics is the study of DNA and its products at the genome level. This includes both experimental and theoretical aspects of the problem. For the computer scientist, the interest lies in the discrete domains and output of the system; for the biologist, the goal is to reveal the function of a sequence, either of the DNA or, more commonly, of the protein gene product. The biological effort includes genetic mapping, physical mapping (restriction maps, ordered clones, X-ray diffraction maps, cytogenetic maps, and sequence assembly), and sequence analysis. There is a natural division of the work into two major computational areas: support for the construction and representation of various maps, and analysis of the data produced. This overlaps at the boundary with computation involved in the analysis of protein folding and interacts with work to assign function to gene products. The systematic effort to map genomes in the presence of variability and error also mandates the use of new computational techniques to assist in the design of efficient experiments. Beyond this, there are four major areas in which the use of computers is indispensable: database searching for DNA sequence analysis, maximum-likelihood calculations for genetic linkage mapping, heuristics for the NP-hard problem of DNA mapping and sequencing, and general bookkeeping (e.g., laboratory information management systems).

Genetic Mapping

Genetic mapping deals with the inheritance of certain genetic **markers** within the pedigree of families. These markers can be genes, sequences associated with genetic disease, or arbitrary probes determined to be of significance [e.g., RFLP probes, anonymous DNA segments, sequence tagged sites (STSs), or expressed site tags (ESTs)]. The sequence of such markers and their probabilistic distance (measured in centimorgans) along the genome can often be determined with fair accuracy by the use of maximum-likelihood methods. Inheritance of traits across the pedigrees of multiple families is determined by hybridizing each family member's genome against the predetermined probes. Eventually, the genetic map most likely to produce the observed data is constructed. Only a few years ago, the knowledge of the mathematics involved and the computational complexity of algorithms based on that mathematics limited analysis to no more than five or six markers. As knowledge of approximations to the formulas and likelihood estimation has improved, software capable of producing maps for 60 markers or more [Magness et al. 1993, Matise et al. 1994] has been developed. Recent advances in this area include the identification of a large number of STS and EST probes and software capable of tractably producing maps based on hybridization pedigree data [Cinkosky and Fickett 1993]. Progress in this area has used mathematical methods such as combinatorics, graph theory, and statistics, and computer science methods such as search theory. Although significant progress has been made over the last few years, considerable effort is still required to make genetic linkage maps effective tools for genetic research. Most maps that have been produced to date are of low resolution, being based on the STS content of mega-yeast artificial chromosome (YAC) ordered libraries. However, some high-resolution maps involving fingerprinting of cosmid libraries are appearing [Stallings et al. 1990]. To be useful in common situations, more markers must be identified and mapped to produce higher-resolution maps. Furthermore, tools to address more complex situations (such as manic-depressive disease, which is likely to involve multiple genes) are badly needed.

Physical Mapping, Including Sequence Assembly

The basic problem in **physical mapping** is to build representations of the sequence information in a genome, with the ultimate goal of producing the sequence of base pairs that constitutes the entire genome [Lander and Waterman 1988]. Most current work can be characterized as a "shotgun assembly" process: fragments of a genetic sequence are randomly sampled, and then experiments are performed on the fragments in order to determine which pairs of fragments come from overlapping regions [Iris 1994]. This overlap information is then used to piece together, or assemble, the fragments into an ordered layout of the fragments that covers the original genetic sequence (see Fleischmann et al. [1995] for a good example). This problem must be solved at a hierarchy of size levels driven by current biotechnology. An assemblage of YACs averaging approximately 500–1500 kb is sampled to represent a chromosome; an assemblage of cosmids (35 kb) represents each YAC, and an assemblage of inserts (500 bp) represents each cosmid. At the bottommost level, the actual sequence of each fragment is determined by electrophoresis, and the resulting assembly consequently gives the sequence of the cosmid, in turn giving the sequence of the YAC and ultimately the genome.

This problem, called **sequence assembly**, has been extensively examined [Indury and Waterman 1995], and is somewhat distinct from higher-level problems where it is not possible to obtain a sequence directly [Dyer et al. 1995]. Instead, experiments are performed to infer when two fragments are very likely to overlap the sampled original. For example, oligonucleotide hybridizations determine if a fragment contains a particular sequence; restriction digests give the lengths of the segments when the fragment is cleaved at all occurrences of a given short sequence. The assemblages for these higher-level problems are called ordered clone libraries. Assembly problems pose an interesting suite of combinatorial optimization problems. Overlap relationships are typically modeled as an overlap graph and the assembly problems can be viewed as constrained versions of classical graphs problems. Alternatively, they can be viewed as finding the maximum-likelihood interval graph explaining the overlaps.

The problem is further compounded by the fact that all the data are inaccurate (e.g., digest lengths are off by up to 10%, electrophoretically determined sequences contain 1–2% incorrect base assignments, etc.). Moreover, the data are partial, not all regions of the original are represented in the sample, and the orientation of fragments is frequently unknown. Other issues involve the amount and type of information gathered to infer overlaps. More information implies more confidence in the veracity of an overlap, but some false positives will occur by chance. Consequently, it is likely that when building ordered clone libraries, a variety of different experiments yielding heterogeneous types of information will be performed and must be used simultaneously to detect overlaps. A key analysis problem is to accurately assess the statistical significance of overlaps under some stochastic model. Another issue involves how much to sample. Statistical and biological considerations show that for large problems with moderate overlap information, one will never achieve coverage without an impractical amount of sampling. This is the gambler's dilemma: at some point one must stop rolling the dice and move to another strategy to complete an assembly project.

While preliminary solutions to genome analysis have been of great use, many challenges remain.

1. The scale of the problems is such that they are computationally intractable. Better algorithms and the exploitation of parallelism are required.
2. Each assembly problem has a somewhat different combinatorial structure due to variations in the experimental methods used to infer overlaps. There is clearly a central generalized assembly problem, but each variation requires its own optimization in order to best lever the combinatorial structure. However, many of these projects are one-time efforts. The challenge is to build software that is both general and efficient.
3. The resulting assemblages are large, and software is needed that permits one to visualize and navigate a solution. Further engineering is required to allow investigators to manipulate solutions according to their expert knowledge, and to maintain versions of the data and a record of the work. A most critical need is the ability to automatically read autoradiograms (image processing), or a better technique to replace autoradiography.
4. Finally, the central assembly problem involves NP-hard combinatorial problems for which heuristics work well on typical data. But as the scale and number of the problems to be solved increase, we will need ever more trustworthy solutions [Goldberg et al. 1995].

Sequence Analysis

Database searching has become an essential part of modern molecular biology. Database searching is our most effective means of identifying the biochemical function of a newly sequenced protein or gene. It is hard to imagine the number of hours of trial and error that are eliminated by the advent of this technique.

Matching a Defined Pattern

The need for speed in database searches has led to using heuristic methods. Current research topics in this area include the improvement of the sensitivity of searching, which can be severely reduced by the overall nucleotide composition of the genomes involved. Database searches are generally conducted with a global alignment algorithm that finds the "best" alignment over the entire length of both sequences. Recent advances in our understanding of domain structure of proteins and the intron–exon organization of genes has made it very desirable to develop a fast, sensitive local alignment search algorithm to identify regions within a pair of sequences that show the highest similarity. Currently, the most promising approach to this is an implementation of one of the rigorous dynamic programming local alignment algorithms on very highly parallel hardware [Lim et al. 1993].

Several other searching techniques are widely used by experimental molecular biologists as aids for guiding and interpreting experiments. These include things as simple as finding the highly hydrophilic and hydrophobic regions in a protein sequence or regions capable of forming helices on the surface of a protein. Finding specific patterns of codon usage in a newly sequenced gene can yield insights into its expression, and finding the intron–exon junction is necessary before the correct protein sequence can be derived from a genomic DNA sequence. There are several important research topics in searching for signals. First is the need for procedures for easily and clearly specifying very arbitrary, complex patterns in a sequence. Faster algorithms for finding these patterns are needed as well. In many cases, the patterns, or biological signals, which are being sought are too complex to be readily identified by visual inspection of sequences. This is especially true if some variation is permissible in the signal pattern. Thus, another important research topic is better algorithms for selecting the most likely patterns to be associated with a signal in a sequence. For example, given several genes known to contain exons, from a single organism, how do we discover the pattern which signals the beginning and end of each exon? These algorithms need to function effectively when separate instances of pattern are not identical. Farber et al. [1992] and Uberbacher and his associates [Uberbacher and Mural 1991] have applied sophisticated *neural network* programming approaches to this problem, but even these are limited by the basic problem of pattern recognition in genomic DNA.

Alignment

Sequence alignment is an important type of searching. It is, basically, the search for the most similar juxtaposition of sequences or regions within sequences. Good alignments are necessary if our inferences about the homology of genes are to be accurate. Even more crucially dependent on good alignments are methods for reconstructing phylogenies based on sequence data [Steel et al. 1994]. Finally, some problems in identifying signal patterns are appreciably simplified given well-aligned sequences. Fortunately, the state of the art in pairwise sequence alignment is well advanced. There are rigorous algorithms for both global and local alignments of pairs of sequences. These algorithms allow both flexible and sophisticated treatment of insertion/deletion gaps. One possible topic for research in this area is the context-sensitive treatment of these gaps. This would include cases where an insertion/deletion would change the reading frame of a coding region in a gene sequence or change the relative positions of amino acids known to be essential to protein function.

Multiple sequence alignment techniques are not as far advanced as the pairwise techniques. The rigorous pairwise algorithm can be, and has been, extended to multiple sequence problems. However, this approach requires computer time and memory proportional to the product of the lengths of the sequences. Recent advances have reduced this requirement by a large constant factor. However, even with this improvement, this approach soon exhausts even the fastest present-day computers. If a phylogeny of the sequences is available independent of the sequences themselves, it may be used to convert a sequence alignment problem into a series of pairwise problems. However, since a frequent reason for doing a multiple sequence alignment is to generate a phylogeny, this is not a general solution. Thus, a variety of heuristic algorithms are used for most multiple sequence alignment problems.

Several kinds of research are needed here. First, faster and more rigorous algorithms are required. Where algorithms are not completely rigorous, we need to characterize their performance so that we know how close to an optimal solution we can come. We also need to identify what sequence features may cause an algorithm to perform badly.

Defining Terms

API: Application programming interface. An API provides a wide and varied set of functions, messages, and structures that give applications access to the features and capabilities of the underlying operating system. The API consists of a set of standard interfaces to such features as window management, graphics device interface, system services, multimedia services, and remote procedure calls.

Clone assembly: The essence of experimental physical mapping is the ordering and placement of fragmented regions of chromosomes in an overlapping contiguous stretch of DNA that covers the entire chromosomal interval. The DNA fragments are derived from the original DNA and each fragment propagated independently (cloned) to obtain working quantities. Fragment overlaps are determined by hybridization experiments with unique DNA sequences that will permit the unique identification of a chromosomal region. This is a computationally difficult problem.

Kalman filtering: A digital-image averaging procedure for the enhancement of the signal-to-noise ratio in noisy images. The Kalman filter is a recursive version of the true averaging procedure, and takes the form $y_1 = (i - 1)y_i - 1/i$, in which the filter parameters are not constant but vary with the frame number i so that the latest averaged image y_n always equals $(x_1 + x_2 + \cdots + x_n)/n$. This gives a straightforward average over n frames, with a maximum signal-to-noise ratio, without having to prespecify n.

Linkage and linkage mapping: The proximity of two or more markers (e.g., genes, DNA sequences) on a chromosome; the closer the markers are, the lower the probability that they will be separated during repair or replication, and therefore, the greater the probability they will be inherited together. A map of the relative positions of genetic markers on a chromosome, determined on the basis of how often they are inherited together, is referred to as a linkage map.

Marker: An identifiable physical location on a chromosome (e.g., a specific identifiable sequence such as a restriction-enzyme cutting site, or a gene) whose inheritance can be monitored. Markers can be expressed regions of DNA (genes) or a segment of DNA with no known coding function but whose inheritance can be followed with molecular techniques.

Mainchain-directed alignment: A technique for predicting protein structure from NMR data, based on aligning only mainchain residues and ignoring sidechain structures at the primary alignment stage. This technique ignores the complexity of the sidechain packing but suffers from distortions based on the effects of sidechain groups.

NOESY: Nuclear Overhauser effect (NOE) enhancement spectroscopy. The NOE is a consequence of relaxation of dipole-coupled spins induced in magnetic resonance. The enhancement factor in NMR is the deviation from thermal equilibrium induced in one of a pair of dipole-coupled spins following the selective radiation of one of the pairs.

Phase problem: In X-ray diffraction structure determination a collimated beam of monochromatic X-rays is directed through an object. The X-rays are scattered in all directions by the electrons of every atom in the object. The magnitude of the scattering is proportional to the size of the electron complement of atoms in the target. Because of the variable composition of the target, the scattering of X-rays appears continuous. However, when regular structures are radiated, and known heavy-metal ions are included, it is possible through a series of least-squares calculations to reassemble the phased waves of the diffracted X-rays and compute a clean electron density map.

Physical map: A map of the locations of identifiable regions on DNA (e.g., specific identifiable sequences or genes), *regardless of inheritance*. Distance is measured in base pairs. The physical map with the highest possible resolution is the entire nucleotide sequence of the chromosome.

Scanning confocal microscopy: A technique which—unlike conventional light microscopy, which focuses an optical image of a specimen on the image plane of a collection system—involves the physical scanning of the specimen with a diffraction-limited point of light. In some cases this may be a one- or two-dimensional array of points. In such microscopy the result of the interaction of the scanned light beam(s) with successive regions of the specimen is measured and recorded. With such instruments digital signal processing may significantly enhance the signal-to-noise ratio, yielding three-dimentional images of extraordinary quality.

Sequence assembly: One of the most common methods of sequencing DNA is "shotgun" sequencing in which a DNA strand is read as a series of random substrings of length 350 to 500. The reconstruction of the sequence of the whole molecule from these random strings is referred to as sequence assembly. There is great complexity in this process, since the reactions to produce the sequencing substrate may be obtained from either strand. Therefore, when comparing two fragments one must take into account that they could be derived from the same strand or from different strands; in the latter case it is necessary to take the reverse complement of one of them before making the comparison.

Sequential alignment: A technique for evaluating the degree of secondary structure alignment by sequential evaluation of the root-mean-square deviation between backbone atoms. This involves identification of secondary structure, application of a clustering function to locate collections of such structures, and examining the more extended alignments outside of the initial regions.

Simulated annealing: This technique is a derivative of the Metropolis algorithm and applies statistical mechanics to optimization to many-body systems. In brief, the Metropolis algorithm is used to provide a simulation of a number of atoms in equilibrium at a given temperature. In each step of the algorithm an atom is given a small random displacement, and the resulting change ΔE in the energy of the system is computed. If the $\Delta E \leq 0$, the displacement is accepted and the value is the basis for the next round. Through a series of probabilistic and cost functions the algorithm optimizes at a given temperature. The simulated annealing procedure applies this function to a system which has been "melted," and the temperature is lowered until the system is frozen. It is essential that at each stage the system proceed long enough to reach a steady state.

Stochastic search algorithm: A form of optimization searching based on random sampling (Monte Carlo) methods where points v from a set V are chosen at random with probability $1/|V|$. The minimum values of $f(v)$ are recorded as the random sampling proceeds, and the sampling does not terminate arbitrarily as might occur in a deterministic search. Simulated annealing is an example of a stochastic algorithm.

Transmission electron microscopy (TEM):   A technique in which a suitably prepared sample is placed in a beam of electrons being controlled in an electric field. A moving electron has a wavelength which is inversely proportional to its momentum (mass times velocity). Therefore, the higher the accelerating voltage of the TEM, the smaller the wavelength and the higher the resolution. The modern TEM consists of an electron source, an imaging lens, and an image-recording system, all housed in a column under high vacuum. Electrons are emitted from a heated tungston filament held at a large negative potential and accelerated through voltages greater than 80 kV. The column is equipped with condenser electromagnetic lenses for focusing.

References

Bellon, P. L. and Lanzavecchia, S. 1995. A direct Fourier method (DFM) for X-ray tomographic reconstructions and the accurate simulations of sinograms. *Int. J. Bio-Med. Comput.* 38(1):55–69.

Belmont, A., Zhai, S., and Thilenius, Y. 1993. Lamin B distribution and association with peripheral chromatin revealed by optical sectioning and electron microscopy tomography. *J. Cell Biol.* 123(6, Pt 2):1671–1685.

Chen, I. A. and Markowitz, V. M. 1994. The object–protocol model. *Lawrence Berkeley Lab. Tech. Rep.* LBL-32738.

Cheng, R. H., Kuhn, R. J., Olson, N. H., Rossmann, M. G., Choi, H. K., Smith, T. J., and Baker, T. S. 1995. Nucleocapsid and glycoprotein organization in an enveloped virus. *Cell* 80(4):621–630.

Cinkosky, M. J. and Fickett, J. W. 1993. *SIGMA User Manual.* Los Alamos National Laboratory, Los Alamos, NM.

Constain, C. and Marinescu, D. 1995. SOCRATES: an environment for high performance computing. In *Proceedings of the 4th IEEE Workshop on Enabling Technologies: Infrastructure for Collaborative Enterprises.* IEEE Press, New York.

Desjarlais, J. R. and Berg, J. M. 1992. Redesigning the DNA-binding specificity of a zinc finger protein: a database guided approach. *Proteins: Struct. Funct. Genet.* 12:101–104.

Dyer, M., Frieze, A., and Suen, S. 1995. Ordering clone libraries in computational biology. *J. Comput. Biol.* 2:207–218.

Eigenbrot, C., Randal, M., Presta, L., Carter, P., and Kossiakoff, A. A. 1993. X-ray structures of the antigen-binding domains from three variants of humanized anti-p185HER2 antibody 4D5 and comparison with molecular modeling. *J. Mol. Biol.* 229:969–995.

Farber, R., Lapedes, A., and Sirotkin, K. 1992. Determination of eukaryotic protein coding regions using neural networks and information theory. *J. Mol. Biol.* 226:471–479.

Fleichmann, R. D. et al. 1995. Whole-genome random sequencing and assembly of *Haemophilus influenzae* Rd. *Science* 269:496–512.

Gilson, M. K., Straatsma, T. P., McCammon, J. A., Ripcoll, D. R., Faerman, C. H., Axelsen, P. H., Silman, I., and Sussman, J. L. 1994. Open "back door" in a molecular dynamics simulation of acetylcholinesterase. *Science* 263:1276–1278.

Goldberg, P. W., Golumbic, M. C., Kaplan, H., and Shamir, R. 1995. Four strikes against physical mapping of DNA. *J. Comput. Biol.* 2:139–152.

Huang, X. C., Quesada, M. A., and Mathies, R. A. 1992. DNA sequencing using capillary array electrophoresis. *Anal. Chem.* 64(18):2149–2154.

Indury, R. M. and Waterman, M. S. 1995. A new algorithm for DNA sequence assembly. *J. Comput. Biol.* 2:291–306.

Iris, F. J. M. 1994. Optimized methods for large-scale shotgun sequencing in Alu-rich genomic regions, pp. 199–210. In M. D. Adams, C. Fields, and J. C. Venter, eds. *Automated DNA Sequencing and Analysis,* Academic Press, London.

Kent, W. 1981. Limitations of record-based information models. *ACM Trans. Database Systems* 6(4).

King, R. 1988. My cat is object oriented. In *Object-Oriented Concepts, Databases, and Applications,* W. Kim, and F. H. Lochovsky, eds. Addison–Wesley, Reading, MA.

Lander, E. S. and Waterman, M. S. 1988. Genomic mapping by fingerprinting random clones: a mathematical analysis. *Genomics* 2:231–239.

Lim, H. A., Fickett, J. W., Cantor, C. R., and Robbins, R. J., eds. 1993. In *Proc. 2nd Int. Conf. on Bioinformatics, Supercomputing and Complex Genome Analysis,* St. Petersburg, FL, July 1992. World Scientific, Singapore.

Luthy, R., Bowie, J. U., and Eisenberg, D. 1992. Assessment of protein models with three-dimensional profiles. *Nature (London)* 356:83–85.

Magness, C., Xu, Y., and Green, P. 1993. *SEGMAP—a program for computing and displaying YAC-based STS-content maps.* Washington University School of Medicine, St. Louis.

Matise, T. C., Perlin, M., and Chakravarti, A. 1994. Automated construction of genetic linkage maps using an expert system (MultiMap): a human genome linkage map. *Nature Genet.* 6:384–390.

Pan, Y., Yuhasz, S. C., and Amzel, L. M. 1995. Anti-idiotypic antibodies: biological function and structural studies. *FASEB J.* 9:43–49.

Robbins, R. J., ed. 1994. Report on the Invitational DOE Workshop on Genome Informatics, 26–27 April 1993, Baltimore, Maryland; Genome Informatics I: Community Databases. *J. Comput. Biol.* 1:173–190.

Rose, G. D. 1979. Hierarchic organization of domains in globular proteins. *J. Mol. Biol.* 134:447–470.

Rossman, M. G., Arnold, E., Erickson, J. W., Frankenberger, E. A., Griffith, J. P., Hecht, H. J., Johnson, J. E., Kamer, G., Luo, M., Mosser, A. G., Rueckert, R. R., Sherry, B., and Vriend, G. 1985. Structure of a human common cold virus and functional relationship to other picornaviruses. *Nature (London),* 317:145–153.

Sali, A., Shakhnovich, E. I., and Karplus, M. 1994. How does a protein fold? *Nature (London)* 369:248–251.

Schmieder, P., Hoch, J. C., Stern, A. S., and Wagner, G. 1995. Maximum entropy reconstruction of non-linearly sampled data. Keystone Symposium on Frontiers of NMR in Molecular Biology-IV, Keystone, Colorado, Apr. 3–9, 1995. *J. Cell. Biochem. Suppl.* 21b, p. 76.

Schuster, S. C. and Khan, S. 1994. The bacterial flagellar motor. *Annu. Rev. Biophys. Biomol. Struct.* 23:509–539.

Srinivasan, R. and Rose, G. D. 1995. LINUS: a hierarchic procedure to predict the fold of a protein. *Proteins: Struct. Funct. Genet.* 22:81–99.

Stallings, R. L., Torney, D. C., Hildebrand, C. E., Longmire, J. L., Deaven, L. L., Jett, J. H., Dogett, N. A., and Moyzis, R. K. 1990. Physical mapping of human chromosomes by repetitive sequence fingerprinting. *Proc. Natl. Acad. Sci. USA* 87:6218–6222.

Steel, M. A., Szekely, L. A., and Hendy, M. D. 1994. Reconstructing trees when sequence sites evolve at variable rates. *J. Comput. Biol.* 1:153–163.

Stonebraker, M. and Rowe, L. A. 1984. Portals: a new application program interface. In *Proc. Int. Conf. on Very Large Databases (VLDB).* Singapore.

Tramontano, A. and Lesk, A. M. 1995. Common features of the conformations of antigen-binding loops in immunoglobulins and application to modeling loop conformations. *Proteins* 13:231–245.

Uberbacher, E. and Mural, R. 1991. Locating protein coding regions in human DNA sequences by a multiple sensor–neural network approach. *Proc. Natl. Acad. Sci. USA* 88:11261–11265.

Further Information

The Journal of Computational Biology is a regular source of computational molecular biology. Volume 2, Number 2 (Summer 1995) is an unusually rich source of material related to genome analysis, sequence alignment, and fragment assembly. The *Journal of Structural Biology* and *Proteins: Structure, Function and Genetics* are also sources of current work. For a comprehensive treatment of the mathematics of molecular biology the reader is directed to *Introduction to Computational Biology: Maps, Sequences and Genomes,* by Michael Waterman, published by Chapman and Hall, 1995.

V

Database and Information Retrieval

Raghu Ramakrishnan, Section Adviser
University of Wisconsin—Madison

T HE SUBJECT AREA OF DATABASE AND INFORMATION retrieval addresses the general problem of storing large amounts of data in such a way that it is reliable, up to date, and efficiently
retrieved. This problem is prominent in a wide range of applications in industry, government, and
academic research. Availability of such data on the Internet and in forms other than text (e.g., audio and
video) makes this problem increasingly complex.

977

42

Data Models

Avi Silberschatz*
Bell Laboratories, Lucent Technologies, Inc.

Henry F. Korth
Bell Laboratories, Lucent Technologies, Inc.

S. Sudarshan
Indian Institute of Technology—Bombay

42.1 Introduction

Underlying the structure of a database is the concept of a *data model*. A data model is a collection of conceptual tools for describing the real-world entities to be modeled in the database and the relationships among these entities. Data models differ in the primitives available for describing data and in the amount of semantic detail that can be expressed.

The various data models that have been proposed fall into three different groups: object-based logical models, record-based logical models, and physical data models. Physical data models are used to describe data at the lowest level. Only a few physical data models are in use; two widely known ones are the unifying model [Batory and Gotlieb 1982] and the frame memory model [March et al. 1981]. Physical data models capture aspects of database system implementation that are not covered in this article. We shall focus instead on the object-based and record-based logical models.

42.2 Object-Based Models

Object-based logical models are used in describing data at the conceptual and view levels. The object-based models use the concepts of **entities** or **objects** and **relationships** among them rather than the implementation-based concepts of the record-based models. They provide flexible structuring capabilities and allow data constraints to be specified explicitly. Several object-based models are in use; some of the more widely known ones are:

- The entity–relationship model.
- The object-oriented model.
- The semantic data model.
- The functional data model.

*Portions of this chapter were reproduced from Korth, H. et al. 1997. *Database System Concepts*, 3rd ed. McGraw-Hill, New York. With permission.

In this article, we examine the *entity–relationship model* and the *object-oriented model* as representatives of the class of the object-based logical models. The entity–relationship model has gained acceptance in database design and is widely used in practice. The object-oriented model includes many of the concepts of the entity–relationship (E–R) model, but represents executable code as well as data. It is rapidly gaining acceptance in practice. Below are brief descriptions of both models.

The Entity–Relationship Model

The E–R data model derives from the perception of the world, or more specifically, of a particular enterprise in the world, as consisting of a set of basic objects called *entities* and *relationships* among these objects. It facilitates database design by allowing the specification of an *enterprise schema*, which represents the overall logical structure of a database. The E–R data model is one of several semantic data models; that is, it attempts to represent the meaning of the data.

Basics

There are three basic notions that the E–R data model employs: entity sets, relationship sets, and attributes. An *entity* is a "thing" or "object" in the real world that is distinguishable from all other objects. For example, each person in the universe is an entity. Each entity is described by a collection of features, called **attributes**. For example, the entity John Smith may have as attributes a name, Social Security number, and date of birth. The values for some attributes may uniquely identify an entity. For example, the Social Security number 890-12-3456 identifies an entity, since it uniquely identifies one particular person. (We assume that the enterprise we are dealing with is in the U.S.A., and that all the person entities with which we deal have unique Social Security numbers.) An entity may be concrete, such as a person or a book, or it may be abstract, such as a bank account, or a holiday, or a concept.

An *entity set* is a set of entities of the same type that share the same properties (attributes). The set of all persons working at a bank, for example, can be defined as the entity set *employee*, and the entity John Smith may be a member of the *employee* entity set. Similarly, the entity set *account* might represent the set of all accounts in a particular bank. A database thus includes a collection of entity sets each of which contains any number of entities of the same type.

Attributes are descriptive properties possessed by all members of an entity set. The designation of attributes expresses that the database stores similar information concerning each entity in an entity set; however, each entity will have its own value for each attribute. Possible attributes of the *employee* entity set are *employee-name*, *social-security-number*, and *employee-address*. Possible attributes of the *account* entity set are *account-number* and *account-balance*. For each attribute there is a set of permitted values, called the *domain* (or *value set*) of that attribute. The domain of the attribute *employee-name* might be the set of all text strings of a certain length. Similarly, the domain of attribute *account-number* might be the set of all positive integers.

Entities in an entity set are distinguished according to their attribute values. A set of attributes that suffices to distinguish all entities in an entity set is chosen, and called a *primary key* of the entity set. For the *employee* entity set, *social-security-number* serves as a primary key, since in the enterprise modeled by the database, no two people may have the same social security number, and thus *social-security-number* distinguishes employees from each other.

A *relationship* is an association among several entities. Thus, an *employee* entity and a *department* entity might be related by a *works-for* relationship; for example, John Smith works for the bank's credit department. Just as all *employee* entities are grouped into an *employee* entity set, all *works-for* relationship instances are grouped into a *works-for* relationship set. A relationship set may also have descriptive attributes. For example, consider a relationship set *depositor* between the *customer* and *account* entity sets. We could associate an attribute *last-access* to specify the date of the most recent access to the account. The relationship sets *works-for* and *depositor* are examples of a binary relationship set, that is, one which involves two entity sets. Most of the relationship sets in a database system are binary.

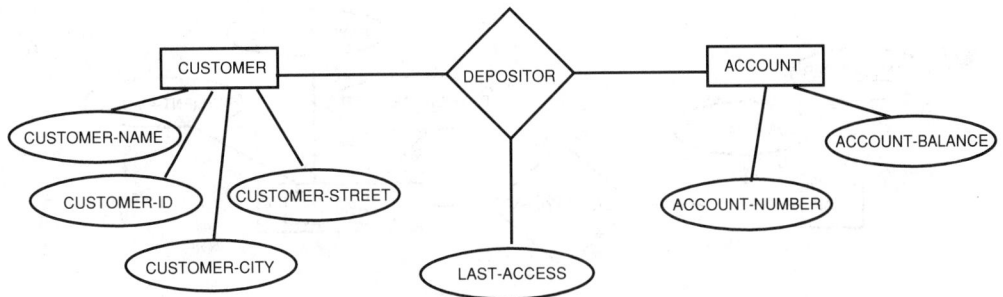

FIGURE 42.1   E–R diagram.

The overall logical structure of a database can be expressed graphically by an E–R *diagram*. Such a diagram consists of the following major components:

- *Rectangles*, which represent entity sets.
- *Ellipses*, which represent attributes.
- *Diamonds*, which represent relationship sets.
- *Lines*, which link entity sets to relationship sets, and link attributes to both entity sets and relationship sets.

An entity–relationship diagram for a portion of our simple banking example is shown in Fig. 42.1.

Although the basic E–R concepts can model most database features, some aspects of a database may be more aptly expressed by certain extensions to the basic E–R model. Commonly used extended E–R features include specialization, generalization, higher- and lower-level entity sets, attribute inheritance, and aggregation. A full explanation of these features is beyond the scope of this chapter; we refer readers to the references listed at the end for additional information.

Representing Data Constraints

In addition to entities and relationships, the E–R model represents certain constraints to which the contents of a database must conform. One important constraint is *mapping cardinalities*, which express the number of entities to which another entity can be associated via a relationship set. Therefore, relationships can be classified as many-to-many, many-to-one, or one-to-one. A many-to-many *works-for* relationship between employee and department exists if a department may have one or more employees and an employee may work for one or more departments. A many-to-one *works-for* relationship between *employee* and *department* exists if a department may have one or more employees but an employee must work for only department. A one-to-one *works-for* relationship exists if a department were required to have exactly one employee, and an employee required to work for exactly one department. In an E–R diagram, an arrow is used to indicate the type of relationship, as shown in Fig. 42.2.

Another important class of constraints is *existence dependencies*. If the existence of entity *x* depends on the existence of entity *y*, then *x* is said to be *existence-dependent* on *y*. To illustrate, consider Fig. 42.3, which depicts the entity set *loan*, and the entity set *payment*, which keeps information about all the payments that were made in connection with a particular loan. The *payment* entity set is described by the attributes

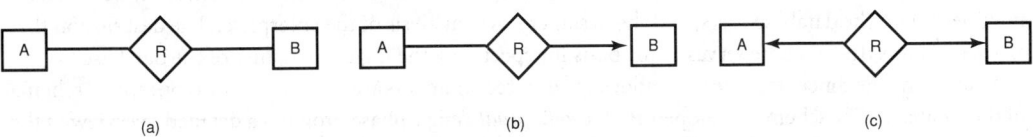

FIGURE 42.2   Relationships between *A* and *B*: (a) many-to-many, (b) many-to-one, (c) one-to-one.

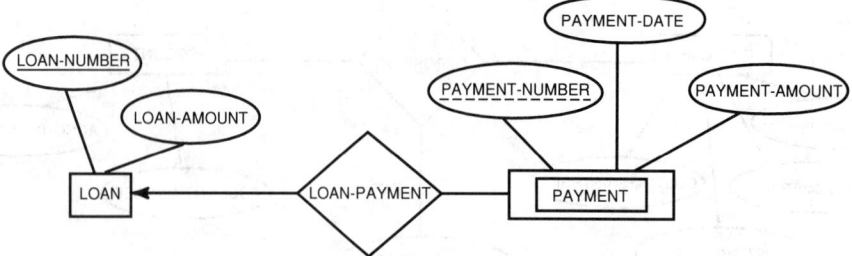

FIGURE 42.3   E–R diagram with existence dependency.

payment-number, payment-date, and *payment-amount.* We form a relationship set *loan-payment* between these two entity sets, which is one-to-many from *loan* to *payment.* Every *payment* entity must be associated with a *loan* entity. If a *loan* entity is deleted, then all of its associated *payment* entities must be deleted also. In contrast, *payment* entities can be deleted from the database without affecting any *loan.* The entity set *loan,* therefore, is dominant and *payment* is subordinate in the *loan-payment* relationship set.

It is possible that an entity set does not have sufficient attributes to form a primary key. Such an entity set is termed a *weak entity set.* An entity set which has a primary key is termed a *strong entity set.* To illustrate, consider again the entity set *payment,* which has the three attributes: *payment-number, payment-date,* and *payment-amount.* Although each *payment* entity is distinct, payments for different loans may share the same payment number. Thus, this entity set does not have a primary key and is therefore a weak entity set. In order for a weak entity set to be meaningful, it must be part of a one-to-many relationship set. The concepts of strong and weak entity sets are related to existence dependencies; a member of a strong entity set is by definition a dominant entity, while a member of a weak entity set is a subordinate entity.

Although a weak entity set does not have a primary key, we nevertheless need a means of distinguishing among all those entities in the entity set that depend on one particular strong entity. The *discriminator* of a weak entity set is a set of attributes that allows this distinction to be made. For example, the discriminator of the weak entity set *payment* is the attribute *payment-number,* since for each loan a payment number uniquely identifies one single payment for that loan. The discriminator of a weak entity set may also be referred to as the *partial key* of the entity set.

The primary key of a weak entity set is formed by the primary key of the strong entity set on which it is existence-dependent, plus its discriminator. In the case of the entity set *payment,* its primary key is {*loan-number, payment-number*}, where *loan-number* identifies the dominant entity of a *payment,* and *payment-number* distinguishes *payment* entities within the same loan.

In an E–R diagram, weak entity sets are depicted by double rectangles; primary keys are underlined with a solid line, while discriminators are underlined with a dashed line.

Use of the E–R Model in Database Design

A high-level data model, such as the E–R model, serves the database designer by providing a conceptual framework in which to specify, in a systematic fashion, the data requirements of the database users and how the database will be structured to fulfill these requirements. The initial phase of database design, then, is to fully characterize the data needs of the prospective database users. The outcome of this phase will be a *specification of user requirements.* The initial specification of user requirements may be based on interviews with the database users, and the designer's own analysis of the enterprise. The description that arises from this design phase serves as the basis for specifying the logical structure of the database.

By applying the concepts of the E–R model, the user requirements are translated into a conceptual schema of the database. The schema developed at this *conceptual design* phase provides a detailed overview of the enterprise. Stated in terms of the E–R model, the conceptual schema specifies all entity sets, relationship sets, attributes, and mapping constraints. The schema can be reviewed to confirm that all data requirements are indeed satisfied and are not in conflict with each other. The design can also be examined to remove any redundant features. The focus at this point is on describing the data and their relationships rather than on physical storage details.

A fully developed conceptual schema will also indicate the functional requirements of the enterprise. In a *specification of functional requirements*, users describe the kinds of operations (or transactions) that will be performed on the data. Example operations include modifying or updating data, searching for and retrieving specific data, and deleting data. A review of the schema for meeting functional requirements can be made at the conceptual design stage.

The process of moving from a conceptual schema to the actual implementation of the database involves two final design phases. Although these final phases extend beyond the role of data models, we present a brief description of the final mapping from model to physical implementation. In the *logical design* phase, the high-level conceptual schema is mapped onto the implementation data model of the database management system (DBMS) that will be used. The resulting DBMS-specific database schema is then used in the subsequent *physical design* phase, in which the physical features of the database are specified. These features include the form of file organization and the internal storage structures.

To illustrate the conceptual design process, we present an extremely simplified banking enterprise example, and apply the E–R data model to translate user requirements into a conceptual design schema that is depicted as an E–R diagram.

A bank is organized into departments; each department has a unique name and a manager. A department may have several employees, and an employee may work for one or more departments. Bank employees are identified by their Social Security numbers; the bank also records employees' names and addresses. An employee may serve as the personal banker for one or more customers; only one such personal banker is assigned to a customer. Accounts are identified by a unique account number, and each account's balance is recorded. The bank identifies customers with a customer-id number. It stores the name and the city and street address of each customer. The *depositor* relationship identifies which customers own which accounts. Loans are assigned a loan number. The bank stores the amount of each loan, and information about loan payments.

The E–R diagram of Fig. 42.4 depicts the entity sets, relationship sets, attributes, and data constraints that describe the banking enterprise.

Because the E–R model is extremely useful in mapping the meanings and interactions of real-world enterprises onto a conceptual schema, a number of database design tools draw on E–R concepts. Further, the simplicity and pictorial clarity of the E–R diagramming technique may well account in large part for the widespread use of the E–R model.

Although the basic E–R concepts can model most database features, some aspects of a database may be more aptly expressed by certain extensions to the basic E–R model. Commonly used extended E–R features include specialization, generalization, higher- and lower-level entity sets, attribute inheritance, and aggregation. An explanation of these features is beyond the scope of this article. Further discussion of the E–R model appears in the classic paper by Chen [1976], which introduced the E–R model.

Deriving a Relational Database Design from the E–R Model

A database which conforms to an E–R diagram can be represented by a collection of tables. For each entity set and each relationship set in the database, there is a unique table which is assigned the name of the corresponding entity set or relationship set. Each table has a number of columns, which again have unique names. The conversion of database representation from an E–R diagram to a table format is the basis for deriving a relational database design. We cover the relational database model in section 42.3. Although important differences exist between a relation and a table, informally, a relation can be considered to be a table. The column headers of a table representing an entity set correspond to the attributes of the entity, and the primary key of the entity becomes the primary key of the relation. The column headers of a table

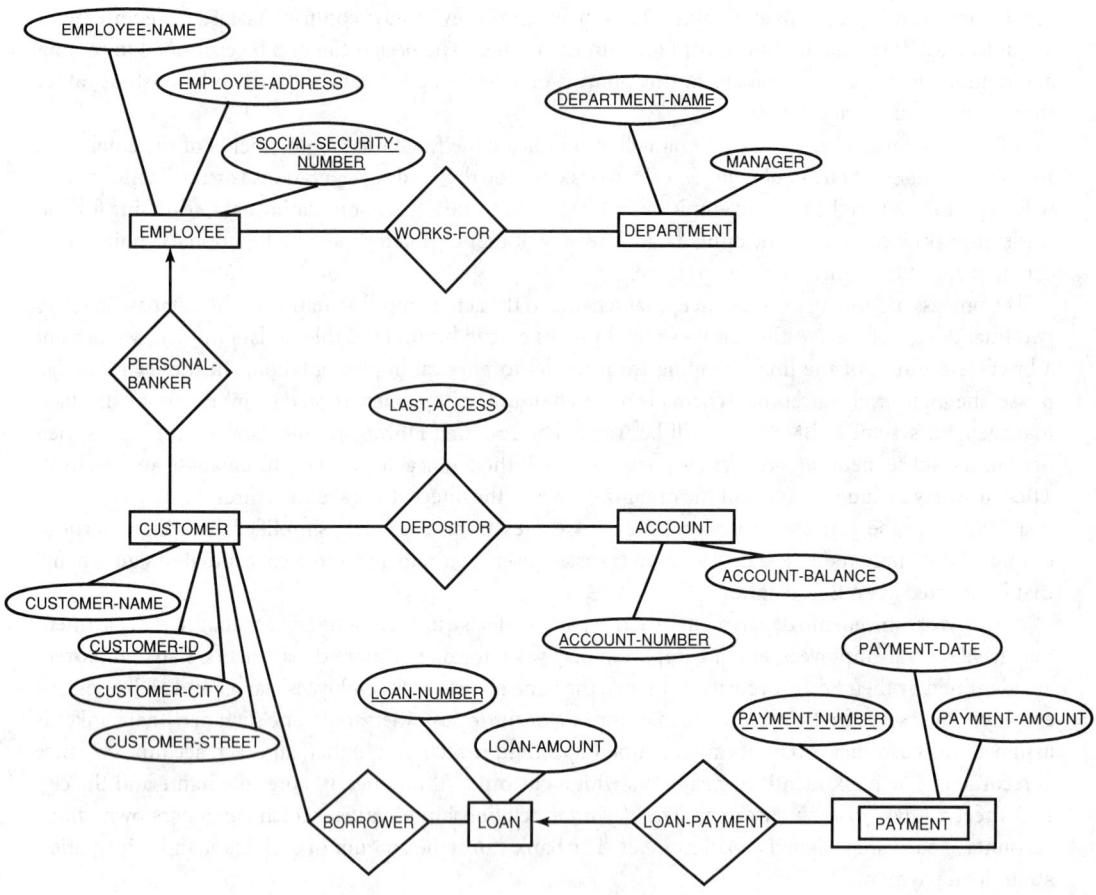

FIGURE 42.4   E–R diagram for banking enterprise.

representing a relationship set correspond to the primary-key attributes of the participating entity sets and the attributes of the relationship set. The combined primary keys of the participating entity sets form the primary key of the table. For such a table, the primary keys of the participating entity sets are called **foreign keys** of the table. The rows of the tables correspond to individual members of the entity or relationship set.

TABLE 42.1   The *Customer* Table

Customer-Name	Customer-Id	Customer-Street	Customer-City
Oliver	654-32-1098	Main	Harrison
Harris	890-12-3456	North	Rye
Marsh	456-78-9012	Main	Harrison
Pepper	369-12-1518	North	Rye
Ratliff	246-80-1214	Park	Pittsfield
Brill	121-21-2121	Putnam	Stamford
Evers	135-79-1357	Nassau	Princeton

Tables 42.1, 42.2, 42.3, and 42.4 show the sets of tables that correspond, respectively, to the *customer* entity set, the *borrower* relationship set, and the *loan* and *payment* entity sets of Fig. 42.1. Note that the *loan-payment* relationship set is not converted to a table; rather, the *payment* table contains the information needed to express its association with the *loan* table.

Object-Oriented Model

The object-oriented data model is an adaptation of the object-oriented programming paradigm to database systems. The object-oriented approach to programming was first introduced by the language Simula 67, which was designed for programming simulations. More recently, the languages C++ and Smalltalk have become the most widely known object-oriented programming languages. Database applications in such areas as computer-aided design, software engineering, and document processing do not fit the set of assumptions made for older, data-processing-style applications. The object-oriented data model has been proposed to deal with some of these new types of applications. The model is based on the concept of encapsulating data, and code that operates on those data, in an object.

Basics

Like the E–R model, the object-oriented model is based on a collection of objects. Entities, in the sense of the E–R model, are represented as *objects* with attribute values represented by *instance variables* within the object. The value stored in an instance variable may itself be an object. Thus, a containment relationship, the *is-part-of* relationship, is established among objects. Additionally, objects can contain objects to an arbitrarily deep level of nesting. At the bottom of this hierarchy are objects such as integers, character strings, and other data types that are built into the object-oriented system and serve as the foundation of the object-oriented model. The set of built-in object types varies from system to system.

In addition to representing data, objects have the ability to initiate operations. An object may send a *message* to another object, causing that object to execute a *method* in response. Methods are procedures, written in a general-purpose programming language, which manipulate the object's local instance variables and send messages to other objects. Messages provide the only means by which an object can be accessed. Therefore, the internal representation of an object's data need not influence the implementation of any other object. Different objects may respond differently to the same message. For example, an office-memo object may respond to a *print* message by formatting its text and then printing it on a laser printer, while a C++-source-code object may respond to a *print* message by simply sending its text to a line printer. This encapsulation of code and data has proven useful in developing higher modular systems. It corresponds to the programming-language concept of abstract data types.

The only way in which one object can access the data of another object is by invoking a method of that other object. This is called *sending a message* to the object. Thus, the call interface of the methods of an object defines its externally visible part. The internal part of the object—the instance variables and method code—are not visible externally. The result is two levels of data abstraction.

To illustrate the concept, consider an object representing a bank account. Such an object contains instance variables *account-number* and *account-balance*, representing the account number and account

TABLE 42.2 The *Borrower* Table

Customer-Id	Loan-Number
654-32-1098	L-259
654-32-1098	L-630
890-12-3456	L-401
456-78-9012	L-700
369-12-1518	L-199
246-80-1214	L-467
246-80-1214	L-115
121-21-2121	L-183
135-79-1357	L-118
135-79-1357	L-225
135-79-1357	L-210

TABLE 42.3 The *Loan* Table

Loan-Number	Loan-Amount
L-259	1000
L-630	2000
L-401	1500
L-700	1500
L-199	500
L-467	900
L-115	1200
L-183	1300
L-118	2000
L-225	2500
L-210	2200

TABLE 42.4 The *Payment* Table

Loan-Number	Payment-Number	Payment-Date	Payment-Amount
L-259	5	11 May 1994	50
L-630	11	17 May 1994	75
L-401	22	23 May 1994	300
L-700	69	28 May 1994	500
L-199	103	3 June 1994	900
L-259	6	7 June 1994	50
L-115	53	7 June 1994	125
L-199	104	13 June 1994	200
L-259	7	17 June 1994	100

balance. It contains a method *pay-interest*, which adds interest to the balance. Assume that the bank had been paying 6 percent interest on all accounts but now is changing its policy to pay 5 percent if the balance is less than $1000 or 6 percent if the balance is $1000 or greater. Under most data models, this would involve changing code in one or more application programs. Under the object-oriented model, the only change is made within the *pay-interest* method. The external interface to the object remains unchanged.

Classes

Objects that contain the same types of values and the same methods are grouped together into **classes**. A class may be viewed as a type definition for objects. This combination of data and code into a type definition is similar to the programming-language concept of abstract data types. Thus, all *employee* objects may be grouped into an *employee* class. Classes themselves can be grouped into a hierarchy of classes; for example, the *employee* class and the *customer* classes may be grouped into a *person* class. The hierarchy of classes allows sharing of common methods. It also allows several distinct views of objects: an employee, for an example, may be viewed in the role of either person or employee, whichever is more appropriate.

Object-Oriented Database Programming Languages

There are two approaches to creating an object-oriented database language: the concepts of object orientation can be added to existing database languages, or existing object-oriented languages can be extended to deal with databases by adding concepts such as persistence and collections. Extended relational languages take the former approach; persistent programming languages take the latter. Persistent extensions to C++ have made significant progress in recent years. A major challenge in the ongoing development of object-oriented database languages is the integration of persistence seamlessly and orthogonally with existing language constructs.

42.3 Record-Based Models

The record-based data models use the concept of a record as the fundamental building block. They provide structures for grouping similar records, and for defining connections between records. Record-based logical models are used in describing data at the conceptual and view levels. In contrast to object-based data models, they are used both to specify the overall logical structure of the database and to provide a higher-level description of the implementation.

Record-based models are so named because the database is structured in fixed-format records of several types. Each record type defines a fixed number of fields, or attributes, and each field is usually of a fixed length. The use of fixed-length records simplifies the physical-level implementation of the database. This is in contrast to many of the object-based models in which objects may contain other objects to an arbitrary depth of nesting. The richer structure of these databases often leads to variable-length records at the physical level. The most important record-based models are the relational model, the network model, and the hierarchical model.

The Relational Model

The relational model was developed in the late 1960s and early 1970s by E. F. Codd. The 1970s saw the development of several experimental database systems based on the relational model and the emergence of a formal theory to support the design of relational databases. The commercial application of relational databases began in the late 1970s but was limited by the poor performance of early relational systems. During the 1980s numerous commercial relational systems with good performance became available. Simultaneously, simple database systems based loosely on the relational approach were introduced for single-user personal computers. In the latter part of the 1980s, efforts were made to integrate collections of personal computer databases with large mainframe databases.

The relational model has since established itself as the primary data model for commercial data-processing applications. The first database systems were based on either the network model or the hierarchical model (see the following two sections). Those two older models are tied more closely to the underlying implementation of the database than is the relational model. The relational model is now being applied outside data processing in systems for computer-aided design and other environments.

Basics

The power of the relational data model lies in its rigorous mathematical foundations and a simple user-level paradigm for representing data. Mathematically speaking, a **relation** is a subset of the Cartesian product of an ordered list of domains. For example, let E be the set of all employee identification numbers, D the set of all department names, and S the set of all salaries. An employment relation is a set of 3-tuples (e, d, s) where $e \in E$, $d \in D$, and $s \in S$. A tuple (e, d, s) represents the fact that employee e works in department d and earns salary s.

At the user level, a relation is represented as a *table*. The table has one column for each domain and one row for each tuple. Each column has a name, which serves as a column header, and is called an **attribute** of the relation. The list of attributes for a relation is called the **relation schema**. A depiction of the *customer* relation, including attribute values for individual tuples, is shown in Table 42.1.

Relational-Database Design

The process of designing a conceptual-level schema for a relational database involves the selection of a set of relation schemas. There are often many possible choices that the database designer might make. A proper balance must be struck among three criteria for a good design:

1. Minimization of redundant data.
2. Ability to represent all relevant relationships among data items.
3. Ability to test efficiently data dependencies which require certain attributes to be unique identifiers.

To illustrate these criteria for a good design, consider a database of employees, departments, and managers. Let us assume that a department has only one manager, but a manager may manage one or more departments. If we use a single schema (employee, department, manager), then we must repeat the manager of a department once for each employee. Thus we have redundant data.

We can avoid redundancy by using two schemas (employee, manager) and (manager, department). However, consider a manager, Martin, who manages both the sales and the service departments. If Clark works for Martin, we cannot represent that fact that Clark works in the service department but not the sales department. Thus we cannot represent all relevant relationships among data. If instead, we chose the two schemas (employee, department) and (manager, department) we would avoid this difficulty, and at the same time avoid redundancy.

There are several types of data dependencies. The most important of these are *functional dependencies*. A functional dependency is a constraint that the value of a tuple on one attribute or set of attributes determines its value on another. For example, the constraint that a department has only one manager could be stated as "department functionally determines manager." Because functional dependencies represent facts about the enterprise being modeled, it is important that the system check newly inserted data to ensure no functional dependency is violated (as in the case of a second manager being inserted for some department). Such checks ensure that the update does not make the information in the database inconsistent. The cost of this check depends on the design of the database.

There is a formal theory of relational-database design that allows us to construct automatically designs which have minimal redundancy consistent with meeting the requirements of representing all relevant relationships and allowing efficient testing of functional dependencies. Details can be found in Ullman [1988] and Abiteboul et al. [1995].

Relational Languages

As we have seen, the relational data model is based on a collection of tables. The user of the database system may query these tables, insert new tuples, delete tuples, and update (modify) tuples. There are several languages for expressing these operations. The tuple relational calculus and the domain relational calculus are nonprocedural languages that represent the basic power required in a relational query language. Both of these languages are based on statements written in mathematical logic. The relational algebra is a procedural language that is equivalent in power to both forms of the relational calculus. The algebra defines the basic operations used within relational query languages.

The relational algebra and the relational calculi are terse, formal languages that are inappropriate for casual users of a database system. Commercial database systems have therefore used languages with more "syntactic sugar." The three most influential commercial languages are SQL, QBE, and Quel. Of these three, SQL has clearly established itself as *the* standard relational database language. There are numerous versions of SQL. In 1986, the American National Standards Institute (ANSI) and the International Standards Organization (ISO) published an SQL standard, called SQL-86 [ANSI 1986]. An extended standard for SQL, SQL-89, was published in 1989, and database systems today typically support at least the features of SQL-89 [ANSI 1989]. The current version of the ANSI–ISO SQL standard is the SQL-92 standard [ANSI 1992]. Further versions of the SQL standard are currently under development.

The Network Model

The network data model is an abstraction of the design concepts used in the implementation of database systems. As a result, the model is tied more closely to physical-level design than is the relational model.

Records and Links

In the relational model, the data and the relationships among data are represented by a collection of tables. The network model differs from the relational model in that data are represented by collections of *records* and relationships among data are represented by *links*.

A record is in many respects similar to an entity in the entity–relationship model. Each record is a collection of fields (attributes), each of which contains only one data value. All records for entities of a given entity set will have the same fields. Thus, all of the records have the same type and are represented as a *record type*. A relationship in the E–R model may be represented in the network model by a link. A link is an association between precisely two records. Thus, a link can be viewed as a restricted (binary) form of relationship in the sense of the E–R model. There are several restrictions usually imposed on links, depending on whether the relationship is many-to-many, many-to-one, or one-to-one.

In the CODASYL DBTG standard for network databases [CODASYL 1971], all links are treated as many-to-one relationships, since multiple records can point to a single record. The structure consisting of two record types that are linked together is referred to in the DBTG model as a *DBTG set*. Each DBTG set has one record type designated as the *owner* of the set and the other record type designated as the *member* of the set. A DBTG set can have any number of *set occurrences*.

Representing Many-to-Many Relationships

In order to model many-to-many relationships, a record type is defined to represent the relationship and two links are used. For example, consider the E–R diagram of Fig. 42.5. Since *works-for* is a many-to-many relationship, we define a record type *works-for*, as well as record types *department* and *employee*. The two links required define a many-to-one relationship between *works-for* and *department* and a many-to-one relationship between *works-for* and *employee*.

The reason these restrictions are included in the DBTG standard involves implementation considerations. The direct implementation of many-to-many relationships requires the use of variable-length records. In our example, each *department* record would have a pointer to the *employee* record for each employee who works for the department, and each *employee* record would have a pointer to the *department*

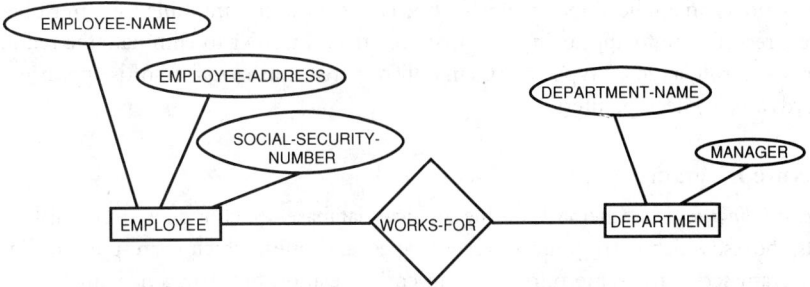

FIGURE 42.5 Employee E–R diagram.

record of each department in which the employee works. A many-to-one relationship can be implemented by a circular list using exactly one pointer per record. The result is that all records of a record type have the same length, allowing a more efficient file structure.

Data-Structure Diagrams

A *data-structure diagram* is a schema representing the design of a network database. Such a diagram consists of two basic components:

- *Boxes*, which correspond to record types.
- *Lines*, which correspond to links.

A data-structure diagram serves the same purpose as an entity–relationship diagram; namely, it specifies the overall logical structure of the database. For every E–R diagram, there is a corresponding data-structure diagram. Figure 42.6 depicts an E–R diagram and its corresponding data-structure diagram.

The Hierarchical Model

The hierarchical model is similar to the network model in the sense that data and relationships among data are represented by records and links, respectively. The primary distinction between the two models is that links in the hierarchical model must form a tree structure, while the network model allows the links to form arbitrary graphs.

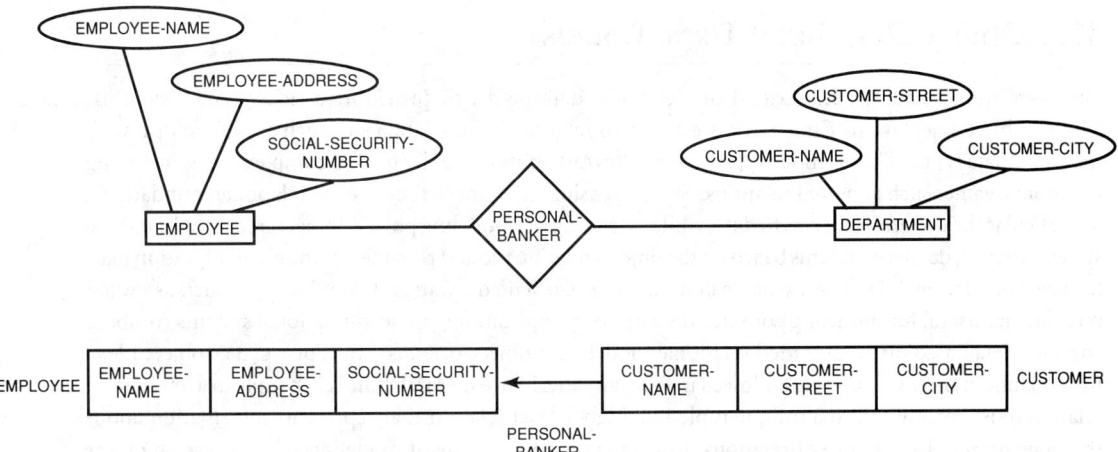

FIGURE 42.6 Data-structure diagram.

In order to model an application for which a tree is not sufficient, multiple trees must be used. This may require a record type to appear in more than one tree. In order to eliminate the redundancy that would otherwise result, a record type appears in only one tree, and *virtual* records appear in other trees. Virtual records are, in effect, pointers.

Tree-Structure Diagrams

A *tree-structure diagram* is a schema for a hierarchical database. Such a diagram consists of two basic components: boxes, which correspond to record types, and lines, which correspond to links. A tree-structure diagram serves the same purpose as an entity–relationship diagram; namely, it specifies the overall logical structure of the database. For every entity–relationship diagram, there is a corresponding tree-structure diagram.

A tree-structure diagram is similar to a data-structure diagram in the network model. The main difference is that in the former record types are organized in the form of an arbitrary graph, while in the latter record types are organized in the form of a *rooted tree*. Figure 42.7 presents a tree-structure diagram that corresponds to the data-structure diagram of Fig. 42.6. The top node is the root, and many-to-one links cannot point downwards in the tree.

The database schema is thus represented as a collection of tree-structure diagrams. For each

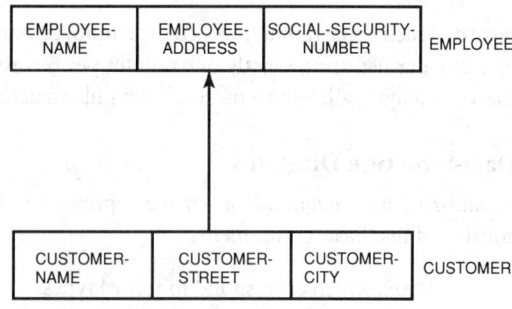

FIGURE 42.7   Tree-structure diagram.

such diagram, there exists a *single* instance of a database tree. The root of this tree is a dummy node. The children of that node are actual instances of the appropriate record type. Each such instance may, in turn, have several instances of various record types, as specified in the corresponding tree-structure diagram.

Summary of Record-Based Models

The relational model differs from the network and hierarchical models in that it does not use pointers or links. Instead, the relational model relates records by the values they contain. This freedom from the use of pointers allows a formal mathematical foundation to be defined, and frees programmers from worrying about implementation details.

42.4 Object–Relational Data Models

Object–relational data models extend the relational data model by providing a richer type system including object orientation. Constructs are added to relational query languages such as SQL to deal with the added data types. The extended type systems allow attributes of tuples to have complex types, including nonatomic values such as nested relations. Such extensions attempt to preserve the relational foundations, in particular the declarative access to data, while extending the modeling power. Object–relational database systems (that is, database systems based on the object–relation model) provide a convenient migration path for users of relational databases who wish to use object-oriented features. Complex types such as nested relations are useful for modeling complex data in many applications. Object–relational systems combine complex data based on an extended relational model with object-oriented concepts such as object identity and inheritance. Relations are allowed to form an inheritance hierarchy, and each tuple in a lower-level relation must correspond to a unique tuple in a higher-level relation that represents information about the same object. Inheritance of relations provides a convenient way of modeling roles, where an object can acquire and relinquish roles over a period of time.

Several object-oriented extensions to SQL have been proposed in the recent past. POSTGRES [Stonebraker et al. 1987] was an early implementation of an object–relational system. Illustra is the commercial object–relational system which is the successor of POSTGRES. The Iris database system from Hewlett-Packard [Fishman et al. 1990] provides object-oriented extensions on top of a relational database system. Iris supports a language called Object SQL (OSQL) which is an object-oriented extension of SQL. The O_2 database system supports an object-oriented query language, described in Bancilhon et al. [1992]. Kim [1995] contains a collection of papers on modern database systems, including descriptions of OSQL and ZQL[C++], which integrates declarative querying with C++. Standardization efforts for object-oriented extensions to SQL are being carried out as part of the SQL-3 standard that is currently being developed.

Defining Terms

Attribute: (1) A descriptive feature of an entity or relationship in the entity–relationship model. (2) The name of a column header in a table, or, in relational-model terminology, the name of a domain used to define a relation.

Class: A set of objects in the object-oriented model that contain the same types of values and the same methods; also, a type definition for objects.

Entity: A distinguishable item in the real-world enterprise being modeled by a database schema.

Foreign key: A set of attributes in a relation schema that serves as a primary key for another relation schema.

Functional dependency: A rule stating that given values for some set of attributes, the value for some other set of attributes is uniquely determined. X functionally determines Y if whenever two tuples in a relation have the same value on X, they must also have the same value on Y.

Instance variables: Attributes within objects.

Key: (1) A set of attributes in the entity–relationship model that serves as a unique identifier for entities. (2) A set of attributes in a relation schema that functionally determines the entire schema. (3) Candidate key—a minimal key. (4) Primary key—a candidate key chosen as the primary means of accessing the entity set or relation.

Message: The means by which an object invokes a method in another object.

Method: Procedures within an object that operate on the instance variables of the object and/or send messages to other objects.

Object: Data and behavior (methods) representing an entity.

Persistence: The ability of information to survive (persist) despite failures of all kinds, including crashes of programs, operating systems, networks, and hardware.

Relation: (1) A subset of a Cartesian product of domains. (2) Informally, a table.

Relation schema: A type definition for relations consisting of attribute names and a specification of the corresponding domains.

Relationship: An association among several entities.

Subclass: A class that lies below some other class (a superclass) in a class inheritance hierarchy; a class that contains a subset of the objects in a superclass.

References

Abiteboul, S., Hull, R., and Viannu, V. 1995. *Foundations of Databases*. Addison–Wesley, Reading, MA.

Agrawal, R. and Gehani, N. 1989. Ode (Object Database and Environment): the language and the data model. pp. 36–45. In *Proc. ACM SIGMOD Int. Conf. on the Management of Data*.

ANSI 1986. American national standard for information systems: database language SQL. *FDT*, ANSI X3,135-1986. American National Standards Institute, New York.

ANSI 1989. Database language SQL with integrity enhancement. ANSI X3,135-1989. American National Standards Institute, New York. Also available as ISO/IEC Document 9075:1989.

ANSI 1992. American national standard for information systems: database language SQL. ANSI X3,135-1992. American National Standards Institute, New York.

Bancilhon, F., Delobel, C., and Kannelakis, P., eds. 1992. *Building an Object-Oriented Database System: The Story of O₂*. Morgan Kaufmann, San Mateo, CA.

Banerjee, J., Chou, H. T., Garza, J. F., Kim, W., Woelk, D., Ballou, N., and Kim, H. J. 1987. Data model issues for object-oriented applications. *ACM Trans. Office Inf. Systems.* 5(1):3–26, Jan.

Batory, D. S. and Gotlieb, C. C. 1982. A unifying model of physical databases. *ACM Trans. Database Systems* 7(4):509–539, Dec.

Cardenas, A. F. 1985. *Data Base Management Systems*, 2nd ed. Allyn and Bacon, Boston.

Carey, M. J., DeWitt, D., Graefe, G., Haight, D., Richardson, J., Schuh, D., Shekita, E., and Vandenberg, S. 1990. The EXODUS extensible DBMS project: an overview. In [Zdonik and Maier 1990], pp. 474–499.

Chamberlin, D. D. et al. 1981. A history and evaluation of System R. *Comm. ACM* 24(10):632–646, Oct.

Chen, P. P. 1976. The entity–relationship model: toward a unified view of data. *ACM Trans. Database Systems* 1(1):9–36, Jan.

CODASYL 1971. CODASYL Data Base Task Group April 71 report, ACM, New York.

CODASYL 1978. CODASYL *Data Description Language Journal of Development*. Material Data Management Branch, Dept. of Supply and Services, Ottawa, Ontario.

Codd, E. F. 1970. A relational model for large shared data banks. *Comm. ACM* 13(6):377–387, June.

Cox, B. J. 1986. *Object-Oriented Programming: An Evolutionary Approach*. Addison–Wesley, Reading, MA.

Date, C. J. 1995. *An Introduction to Database Systems*, 6th ed. Addison–Wesley, Reading, MA.

Date, C. J. and Darwen, H. 1993. *A Guide to the SQL Standard*, 3rd ed. Addison–Wesley, Reading, MA.

Dittrich, K., ed. 1988. *Advances in Object-Oriented Database Systems*. Lecture Notes in Computer Science 334. Springer–Verlag.

Dogac, A., Ozsu, M., Biliris, A., and Selis, T. 1994. *Advances in Object-Oriented Database Systems*. Computer and Systems Sciences, F 130. Springer–Verlag.

ElMasri, R. and Navathe, S. B. 1994. *Fundamentals of Database Systems*, 2nd ed. Benjamin Cummings, Redwood City, CA.

Fishman, D., Beech, D., Cate, H., Chow, E., Connors, T., Davis, J., Derrett, N., Hoch, C., Kent, W., Lyngbaek, P., Mahbod, B., Neimat, M., Ryan, T., and Shan, M. 1990. IRIS: an object-oriented database management system. In [Zdonik and Maier 1990], pp. 216–226.

Gardarin, G. and Valduriez, P. 1989. *Relational Databases and Knowledge Bases*. Addison–Wesley, Reading, MA.

Goldberg, A. and Robson, D. 1983. *Smalltalk-80: The Language and Its Implementation*. Addison–Wesley, Reading, MA.

Hudson, S. E. and King, R. 1987. Object-oriented database support for software environments, pp. 491–593. In *Proc. ACM SIGMOD Int. Conf. on the Management of Data*.

Kapp, D. and Leben, J. 1978. *IMS Programming Techniques*. Van Nostrand Reinhold, New York.

Kim, W. 1990. *Introduction to Object-Oriented Databases*. MIT Press, Cambridge, MA.

Kim, W., ed. 1995. *Modern Database Systems: The Object Model, Interoperability and Beyond*. ACM Press/Addison–Wesley, New York.

Kim, W., Ballou, N., Banerjee, J., Chou, H. T., Garga, J. F., and Woelk, D. 1988. Integrating an object-oriented programming system with a database system. In *Proc. Int. Conf. on Object-Oriented Programming Systems, Languages, and Applications*.

Kim, W. and Lochovsky, F., eds. 1989. *Object-Oriented Concepts, Databases, and Applications*. Addison–Wesley, Reading, MA.

Kroenke, D. and Dolan, K. 1988. *Database Processing*, 3rd ed. Science Research Associates, Chicago.

Lamb, C., Landis, G., Orenstein, J., and Weinreb, D. 1991. The ObjectStore database system. *Comm. ACM* 34(10):51–63, Oct.

Maier, D. 1983. *The Theory of Relational Databases*. Computer Science Press, Rockville, MD.

Maier, D., Stein, J., Otis, A., and Purdy, A. 1986. Development of an object-oriented DBMS., pp. 472–482. In *Proc. Int. Conf. on Object-Oriented Programming Systems, Languages, and Applications.*

March, S. T., Severance, D. G., and Wilens, M. 1981. Frame memory: a storage architecture to support rapid design and implementation of efficient databases. *ACM Trans. on Database Systems* 6(3):441–463, Sept.

McGee, W. C. 1977. The information management system IMS/VS part I: general structure and operation. *IBM Systems J.* 16(2):84–168, June.

Melton, J. and Simon, A. R. 1993. *Understanding The New SQL: A Complete Guide.* Morgan Kaufmann, San Francisco.

Olle, T. W. 1978. *The* CODASYL *Approach to Data Base Management.* Wiley, New York.

Peterson, G. E. 1987. *Object-Oriented Computing.* IEEE Computer Society Press.

Silberschatz, A., Korth, H. F., and Sudarshan, S. 1996. *Database System Concepts,* 3rd ed. McGraw–Hill, New York.

Smith, J. M. and Smith, D. C. P. 1977. Database abstractions: aggregation and generalization, *ACM Trans. Database Systems* 2(2):105–133, Mar.

Stonebraker, M., ed. 1986. *The Ingres Papers.* Addison–Wesley, Reading, MA.

Stonebraker, M., Anton, J., and Hanson, E. 1987. Extending a database system with procedures, *ACM Trans. Database Systems* 12(3):350–376, Sept.

Stroustrup, B. 1992. *The C++ Programming Language,* 2nd ed. Addison–Wesley, Reading, MA.

Taylor, R. W. and Frank, R. L. 1976. CODASYL data base management systems. *ACM Comput. Surveys* 8(1):67–103, Mar.

Tsichritzis, D. C. and Lochovsky, F. H. 1976. Hierarchical data-base management: a survey. *ACM Comput. Surveys* 8(1):67–103.

Tsichritzis, D. C. and Lochovsky, F. H. 1982. *Data Models.* Prentice–Hall, Englewood Cliffs, NJ.

Ullman, J. D. 1988. *Principles of Database and Knowledge-Base Systems,* Vol. I. Computer Science Press, Rockville, MD.

Zdonik, S. and Maier, D. 1990. *Readings in Object-Oriented Database Systems.* Morgan Kaufmann, San Mateo, CA.

Zloof, M. M. 1977. Query-by-Example: A data base language. *IBM Systems J.* 16(4):324–343.

Further Information

The Entity–Relationship Model

The entity–relationship data model was introduced by Chen [1976]. Various data manipulation languages for the E–R model have been proposed, though none is in widespread commercial use. The concepts of generalization, specialization, and aggregation were introduced by Smith and Smith [1977]. Basic textbook discussions are offered by Tsichritzis and Lochovsky [1982], ElMasri and Navathe [1994], and Silberschatz et al. [1996].

The Object-Oriented Model

Object-oriented programming is discussed in Goldberg and Robson [1983], Cox [1986], and Peterson [1987]. Stroustrup [1992] describes the C++ programming language. There are numerous implemented object-oriented database systems, including (in alphabetical order) Cactis, developed at the University of Colorado [Hudson and King 1987], E/Exodus, developed at the University of Wisconsin [Carey et al. 1990], Gemstone [Maier et al. 1986], Iris, developed at Hewlett-Packard [Fishman et al. 1990], O_2 [Bancilhon et al. 1992], ObjectStore [Lamb et al. 1991], Ode, developed at AT&T Bell Labs [Agrawal and Gehani 1989], Ontos, Open-OODB, Orion [Banerjee et al. 1987, Kim et al. 1988], UniSQL, Versant, and others.

Overviews of object-oriented database research include Dittrich [1988], Kim and Lochovsky [1989], Kim [1990], and Zdonik and Maier [1990]. Recent advances in object-oriented database systems are discussed in Dogac et al. [1994].

The Relational Model

The relational model was proposed by E. F. Codd of the IBM San Jose Research Laboratory in the late 1960s [Codd 1970]. Following Codd's original paper, several research projects were formed with the goal of constructing practical relational database systems, including System R at the IBM San Jose Research Laboratory [Chamberlin et al. 1981], Ingres at the University of California at Berkeley [Stonebraker 1986], and Query-by-Example at the IBM T. J. Watson Research Center [Zloof 1977].

The official standards for SQL-89 and SQL-92 are available as ANSI [1989] and ANSI [1992] respectively. Textbook descriptions of the SQL-92 language include Date and Darwen [1993] and Melton and Simon [1993].

General discussion of the relational data model appears in most database texts, including Abiteboul et al. [1995], Date [1995], Ullman [1988], ElMasri and Navathe [1994], and Silberschatz et al. [1996]. Texts devoted exclusively to the relational data model include Gardarin and Valduriez [1989] and Maier [1983].

The Network and Hierarchical Models

In the late 1960s, several commercial database systems emerged that relied on the network model. These systems led to the first database standard specification, called the CODASYL DBTG 1971 report [CODASYL 1971]. Since then, a number of changes have been proposed, the last official one in 1978 [CODASYL 1978]. A survey paper on the DBTG model is presented by Taylor and Frank [1976].

Two influential database systems relying on the hierarchical model are IBM's Information Management System (IMS) [McGee 1977] and MRI's System 2000. A survey paper on the hierarchical data model is presented by Tsichritzis and Lochovsky [1976]. Textbook discussions on the network and hierarchical models are offered by Tsichritzis and Lochovsky [1982], Cardenas [1985], Olle [1978], Kapp and Leben [1978], Kroenke and Dolan [1988], Ullman [1988], Date [1995], and Silberschatz et al. [1996].

Tuning Database Design for High Performance

Dennis Shasha*
*New York University and
AT&T Bell Laboratories*

43.1 Introduction

In fields ranging from arbitrage to tactical missile defense, speed of access to data can determine success or failure. Database tuning is the activity of making a database system run faster. Like optimization activities in other areas of computer science and engineering, database tuning must work within the constraints of its underlying technology. Just as compiler optimizers, for example, cannot directly change the underlying hardware, database tuners cannot change the underlying database management system.

The tuner can, however, modify the design of tables, select new indexes, rearrange transactions, tamper with the operating system, or buy hardware. The goals are to eliminate bottlenecks, decrease the number of accesses to disks, and guarantee response time, at least in a statistical sense.

Understanding how to do this well requires deep knowledge of the interaction among the different components of a database management system. Further, these interactions change with technology. For example, parallel servers require an understanding of commit overheads and the costs of redistributing a table to a set of sites. Tuning, then, is for well-informed generalists. This chapter introduces a principled foundation for tuning, focusing on principles that are likely to hold true for years to come.

43.2 Underlying Principles

To understand the principles of tuning, you must understand the range of database applications and what affects performance.

*Supported in part by the National Science Foundation, Grant CCR-93-20577. E-mail: shasha@cs.nyu.edu.

What Databases Do

At a high level of abstraction, databases are used for two purposes: on-line transaction processing and decision support. **On-line transaction processing** typically involves access to a small number of records, generally to modify them. A typical such transaction records a sale or updates a bank account. These transactions use indexes to access their few records without scanning through an entire table.

Decision support queries, by contrast, read many records and typically form some kind of summary result. Typical decision support queries are "find the total sales of widgets in the last quarter in the northeast" or "calculate the available inventory per unit item." These queries typically scan large portions of tables, occasionally build indexes on the fly, or perform sorts.

Performance Spoilers

Having divided the database applications into two broad areas, we can now discuss what slows them down.

1. *Imprecise data searches.* These occur typically when a selection retrieves a small number of data from a large table, yet must search the entire table to find those data. Establishing an index is often helpful in this case, though other actions, including reorganizing the table, may also have an effect.

2. *Random vs. sequential disk accesses.* As we will see, sequential disk bandwidth runs between 10 and 30 times faster than random-access disk bandwidth. (The variation depends on technology and on tunable parameters such as the degree of prefetching and size of pages.) Index accesses tend to be random whereas scans are sequential. Thus, removing an index can often improve performance, because either the index is never used for reading (and therefore constitutes only a burden for updates) or the index is used for reading and behaves poorly.

3. *Many short data interactions, either over a network or to the database.* This may occur, for example, if a language like C++ accesses a database repeatedly from within a "for" loop rather than once before the loop begins.

4. *Delays due to lock conflicts.* These occur either when update transactions execute too long or when several transactions want to access the same datum, but are delayed because of locks.

As mentioned in the introduction, avoiding such performance problems requires changes at all levels of a database system. We will discuss tactics used at several of these levels and their interactions—hardware, concurrency control subsystem, indexes, and conceptual level. There are other levels such as recovery and query rewriting that we mostly defer to the reference [Shasha 1992].

43.3 Best Practices

Understanding how to tune each level of a database system (see Fig. 43.1 for the levels in a relational system) requires understanding the factors leading to good performance at that level. Each of the following subsections discusses these factors before discussing tuning tactics.

Tuning Hardware

Each processing unit consists of a processor, one or more disks, and some memory. Assuming a 100-MIPS (million instructions per second) processor, disks will be the bottleneck for on-line transaction-processing applications until the processor has at least 10–20 disks. Each transaction spends far more time waiting for head movement on disk than in the processor.

Decision-support queries, by contrast, often entail massive scans of a table, so even a 100-MIPS processor will constitute the bottleneck once a few disks are delivering data at their optimal rates of 3–10 Mbyte/s. Thus, decision-support sites need fewer disks per processor than transaction-processing sites for the

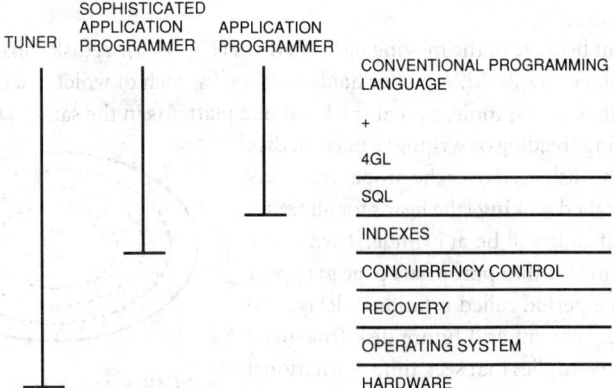

FIGURE 43.1 Architecture of relational database systems: responsibilities of people with different skills.

purposes of matching aggregate disk bandwidth to processor speed.[1]

Solid-state random access memory (RAM) obviates the need to go to disk. Database systems reserve a portion of RAM as a *buffer*, whose logical role is illustrated in Fig. 43.2. In all applications, the buffer usually holds frequently accessed pages (*hot* pages, in database parlance) including the first few levels of indexes. Increasing the amount of RAM buffer tends to be particularly helpful in on-line transaction applications where disks are the bottleneck.

The read **hit ratio** in a database is the portion of database reads that are satisfied by the buffer. Hit ratios of 90% or higher are common in on-line transaction applications but less common in decision support applications. Even in transaction processing applications, hit ratios tend to level off as one increases the buffer size if there are one or more tables that are accessed unpredictably and are much larger than available RAM (e.g., sales records).

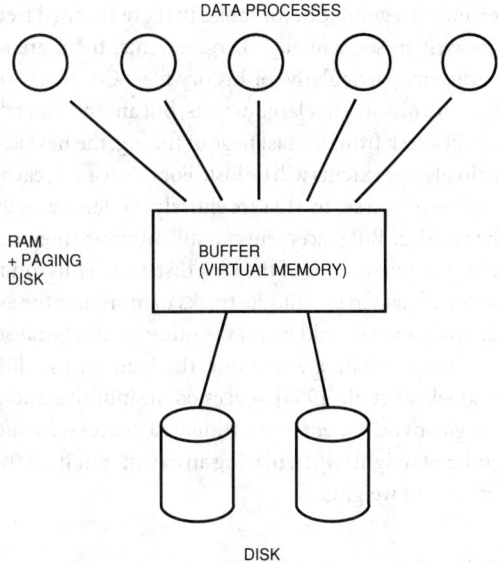

FIGURE 43.2 Buffer architecture. The buffer is in virtual memory, though its greater part should be in random-access memory.

Tuning the Operating System

The operating system, in combination with the lower levels of the database system, determines such features as the layout of files on disk as well as the assignment and safe use of transaction priorities.

---

[1] This point requires a bit more explanation. There are two reasons one might need more disks: (1) for disk bandwidth (the number of bytes coming from the disk per second); or (2) for space. Disk bandwidth is usually the issue in on-line transaction processing. Airline reservations systems, for example, often run their disks at less than 50% utilization. Decision support applications tend to run into the space issue more frequently, because scanning allows disks to deliver their optimal bandwidth.

File Layout

File layout is important because of the moving parts on disks (Fig. 43.3). A disk consists of a set of platters each of which resembles a CD-ROM. A platter holds a set tracks, each of which is a concentric circle. The platters are held together on a spindle, so that track i of one platter is in the same *cylinder* as track i of all other platters. Accessing (reading or writing) a page on disk requires (1) moving the disk head over the proper track, say track t, an operation called **seeking** (the heads for all tracks move together, so all heads will be at cylinder t when the seek is done); (2) waiting for the appropriate page to appear under the head, a time period called rotational delay; and (3) accessing the data. Present and future disk (magnetic and optical) technology implies that seek time > rotational delay > access time.

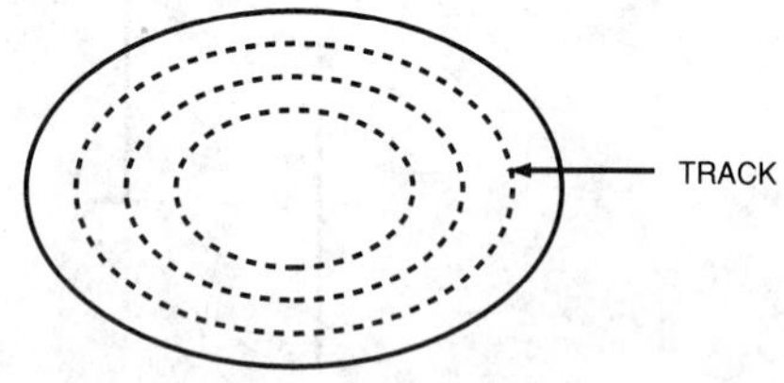

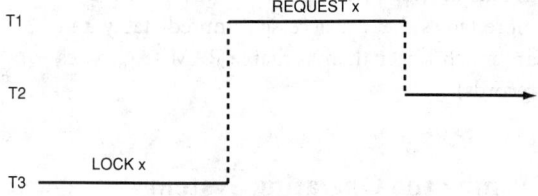

FIGURE 43.3 Anatomy of a disk: surface of a platter on a disk. The concentric dashed circles are called tracks.

As noted above, if one could eliminate the overhead caused by seeks and rotational delay, the aggregate bandwidth could increase by a factor of 10 or more. Making this possible requires laying out the data to be read sequentially along tracks.[2]

Recognizing the advantage of sequential reads on properly laid-out data, most database systems require administrators to lay out tables in relatively large *extents* (consecutive portions of disk). Having a few large extents is a good idea for tables that are scanned frequently or (like database recovery logs or history files) are written sequentially. Large extents, then, are a necessary condition for good performance, but not sufficient, particularly for history files. Consider, for example, the scenario in which a database log is laid out on a disk in a few large extents, but another hot table is also on that disk. The accesses to the hot table may entail a seek from the last page of the log; the next access to the log will entail another seek. So, much of the gain of large extents will be lost. For this reason, each log or history file should be the only hot file on its disk.

When accesses to a file are entirely random (as is the case in on-line transaction processing), seeks cannot be avoided. But placement can still minimize their cost, since seek time is roughly proportional to a constant plus the square root of the seek distance. Thus, on magnetic disks, it is good to place the most frequently accessed data in the middle tracks to minimize the average seek time. On optical disks, the most frequently accessed data should be on the outer tracks, because those tracks hold more data than inner tracks.

Once the data are laid out, the load on the different disks may be unbalanced. The Bubba system [Copeland et al. 1988] suggested an intuitive and practical technique for balancing disk load. Assign a *weight* to each extent proportional to its access frequency. Now, place the extents on the disks in descending order of weight; when placing an extent, put it on the least-loaded disk, where load is measured as the sum of current weights.

Priorities

Since some transactions are more important to a given application than others, many users would like to assign priorities to transactions according to the rule: if a transaction t is more important than t', then t should have higher priority. Unfortunately, this can lead to a counterintuitive phenomenon known as *priority inversion*.

FIGURE 43.4 Priority inversion. T1 waits for a lock that only T3 can release, but T2 runs instead.

Priority inversion occurs because of an interaction between priorities and locking. The situation in Fig. 43.4 illustrates the following sequence of events: a low-priority transaction acquires an exclusive lock

---

[2]Informed readers will realize that the physical layout of sequential data on tracks is not always contiguous—whether it is or not depends on the relative speed ratios of the controller and the disk. The net effect is that there is a layout that eliminates rotational and seek time delay for table scans.

on a datum x; a high-priority transaction preempts and runs until it is blocked on x, causing the low-priority transaction to take over. Many people think this is priority inversion, but it isn't: the only way for the high-priority transaction to continue executing is for the low-priority one to continue executing and eventually release its lock.[3] The priority inversion occurs next: the middle-priority transaction enters the system, preempting the low priority one, and executes for a long time. We call this inversion because the middle transaction effectively slows down the high-priority transaction by preventing the low-priority transaction from completing its execution and thereby releasing its lock on x.

The solution that some vendors choose is known as *priority inheritance*: the database system boosts the priority of the low priority transaction to that of the highest priority transaction that is waiting for one of its locks. In the case of Fig. 43.4, the low-priority transaction would have its priority raised to that of the highest-priority transaction. So the middle-priority transaction does not execute until the lock on x is released.

If your database system does not support priority inheritance or some other mechanism to overcome priority inversion, you should probably run all your database transactions at the same priority.

Tuning Concurrency Control

As the chapter on Concurrency Control and Recovery in this handbook explains, database systems attempt to give users the illusion that each transaction executes in isolation from all others. The ANSI SQL standard, for example, makes this explicit with its concept of degrees of isolation [Gray and Reuter 1993]. Full isolation or **serializability** is the guarantee that each transaction that completes will appear to execute one at a time *except that its performance may be affected by other transactions*. This ensures, for example, that in an accounting database in which every update (e.g. sale, purchase, etc.) is recorded as a double-entry transaction, any transaction that sums assets, liabilities, and owners' equity will find that assets equal the sum of the other two. There are less stringent notions of isolation that are appropriate when users don't require such a high degree of consistency.

The concurrency-control algorithm in predominant use is two-phase locking, sometimes with optimizations for data structures. Two-phase locking has **read** (or *shared*) and **write** (or *exclusive*) locks. Two transactions may both hold a shared lock on a datum. If one transaction holds an exclusive lock on a datum, however, then no other transaction may hold any lock on that datum; in this case, the two transactions are said to **conflict**. The notion of datum (the basic unit of locking) is deliberately left unspecified in the field of concurrency control, because the same algorithmic principles apply regardless of the size of the datum, whether a page, a record, or a table. The performance may differ, however. For example, Oracle, IBM DB2, and Informix hold locks at the record level, whereas Sybase through version 11 holds locks on pages. This implies that two transactions accessing a Sybase page may conflict even if they don't access the same data records. Normally, this occurs so rarely that it isn't a factor, but it is a factor for insertions to heap data structures and to indexes on sequential keys, as we will see in the subsection on indexes.

Rearranging Transactions

Tuning concurrency control entails trying to reduce the number and duration of conflicts. This often entails understanding application semantics. Consider, for example, the following code for a purchase application of item i for price p for a company in bankruptcy (for which the cash cannot go below 0):

PURCHASE TRANSACTION (p, i)

```
1   BEGIN TRANSACTION
2       if cash < p then roll back transaction
3       inventory(i) := inventory(i) + p
4       cash := cash − p
5   END TRANSACTION
```

---

[3] Another option would be for the high-priority transaction to force the low-priority one to roll back and release its locks, but this solution is never used in a commercial database management system, because of the wasted work it requires.

From a concurrency-control-theoretical point of view, this code does the right thing. For example, if the cash remaining is 100, and purchase P1 is for item i with price 50, and purchase P2 is for item j with price 75, then one of these will roll back.

From the point of view of performance, however, this transaction design is very poor, because every transaction must acquire an exclusive lock on cash from the beginning to avoid deadlock. (Otherwise, many transactions will obtain shared locks on cash and none will be able to obtain an exclusive lock on cash.) That will make cash a bottleneck and have the effect of serializing the purchases. Since inventory is apt to be large, accessing inventory(i) will take at least one disk access, taking about 20 ms. Since the transactions will serialize on cash, only one transaction will access inventory at a time. This will limit the number of purchase transactions to about 50 per second. Even a company in bankruptcy may find this rate to be unacceptable.

A surprisingly simple rearrangement helps matters greatly:

REDESIGNED PURCHASE TRANSACTION (p, i)

```
1   BEGIN TRANSACTION
2       inventory(i) := inventory(i) + p
3       if cash < p then roll back transaction
4       else cash := cash − p
5   END TRANSACTION
```

Cash is still a hot spot, but now each transaction will avoid holding cash while accessing inventory. Since cash is so hot, it will be in the RAM buffer. The lock on cash can be released as soon as the commit occurs (in some systems, such as Informix, this lock is released even before the commit is written to the log).

Other techniques are available that "chop" transactions into independent pieces to shorten lock times further, but they are quite technical. We refer interested readers to [Shasha et al. 1995].

Living Dangerously

Many applications live with less than full isolation due to the high cost of holding locks during user interactions. Consider the following full-isolation transaction from an airline reservation application:

AIRLINE RESERVATION TRANSACTION (p, i)

```
1   BEGIN TRANSACTION
2       Retrieve list of seats available.
3       Reservation agent talks with customer regarding availability.
4       Secure seat.
5   END TRANSACTION
```

The performance of a system built from such transactions would be intolerably slow, because each customer would hold a lock on all available seats for a flight while chatting with the reservations agent. This solution does, however, guarantee two conditions: (1) no two customers will be given the same seat, and (2) any seat that the reservation agent identifies as available in view of the retrieval of seats will still be available when the customer asks to secure it.

Because of the poor performance, however, the following is done instead:

LOOSELY CONSISTENT AIRLINE RESERVATION TRANSACTION (p, i)

```
1   Retrieve list of seats available.
2   Reservation agent talks with customer regarding availability.
3   BEGIN TRANSACTION
4       Secure seat.
5   END TRANSACTION
```

This design relegates lock conflicts to the secure step, thus guaranteeing that no two customers will be given the same seat. It does allow the possibility, however, that a customer will be told that a seat is available, will ask to secure it, and will then find out that it is gone. This has actually happened to a particularly garrulous colleague of mine.

Indexes

Access methods, also known as **indexes**, are discussed in another chapter. Here we review the basics, then discuss tuning considerations. An **index** is a data structure plus a method of arranging the data tuples in the table (or other kind of collection object) being indexed. Let's discuss the data structure first.

Data Structures

Two data structures are used in practice: B-trees and Hash structures. Since B-trees are used much more often (one vendor's tuning book puts it this way: "When in doubt, use a B-tree"), we will discuss them first. Our discussion will concentrate on tuning aspects. You will find B-tree algorithms in the chapter on Access Methods.

A **B-tree** (strictly speaking a B+ tree) is a balanced tree whose nodes contain a sequence of key–pointer pairs [Comer 1979]. The keys are sorted by value. The pointers at the leaves point to the tuples in the indexed table (Fig. 43.5).

B-trees are self-reorganizing through operations known as splits and merges (though occasional reorganizations for the purpose of reducing the number of seeks do take place). Further, they support many different query types well: equality queries (find the employee record of the person having a specific social security number), min–max queries (find the highest-paid employee in the company), and range queries (find all salaries between $70,000 and $80,000).

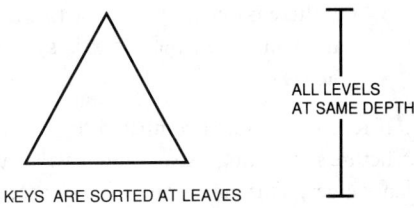

FIGURE 43.5 B-tree essentials. A B-tree is a multiary tree such that every leaf is at the same distance from the root and the keys (attributes being indexed) are sorted at the leaves.

Because an access to disk secondary memory costs 10 to 30 ms if it requires a seek (as index accesses will), the performance of a B-tree depends critically on the number of nodes in the average path from root to leaf. (The root will tend to be in RAM, but the other levels may or not be, and the farther down the tree one goes, the less likely they are to be in RAM.) The number of nodes in the path is known as the number of levels. One technique that database management systems use to minimize the number of levels is to make each interior node have as many children as possible (1000 or more for many B-tree implementations). The maximum number of children a node can have is called its *fanout*. Because a B-tree node consists of key-pointer pairs, the bigger the key is, the lower the fanout.

For example, a B-tree with a million records and a fanout of 1000 requires three levels (including the level where the records are kept). A B-tree with a million records and a fanout of 10 requires seven levels. If we increase the number of records to a billion, the numbers of levels increase to four and ten respectively. This is why accessing data through indexes on large keys is slower than accessing data through small keys on most systems (the exceptions are those few systems that have good compression).

Hash structures, by contrast, are a method of storing key–value pairs based on a pseudorandomizing function called a *hash function*. The hash function can be thought of as the root of the structure. Given a key, the hash function returns a location that contains either a page address (usually on disk) or a directory location that holds a set of page addresses. That page either contains the key and associated record or is the first page of a linked list of pages, known as an *overflow chain* leading to the record(s) containing the key. (You can keep overflow chaining to a minimum by using only half the available space in a hash setting.)

In the absence of overflow chains, hash structures can answer equality queries (e.g., find the employee with Social Security number 156-87-9864) in one disk access, making them the best data structures for

that purpose. The hash function will return arbitrarily different locations on key values that
are close but unequal, e.g., Smith and Smythe.
As a result, records containing such close keys
will likely be on different pages. This explains
why hash structures are completely unhelpful
for range and min–max queries and why they
give good concurrent performance for sequential keys.

Clustering and Sparse Indexes

The data-structure portion of an index has pointers at its leaves to either data pages or data records
as shown in Fig. 43.6.

- If there is at most one pointer from the
  data structure to each data page, then
  the index is said to be **sparse**.
- If there is one pointer to each record in
  the table, then the index is said to be
  dense.

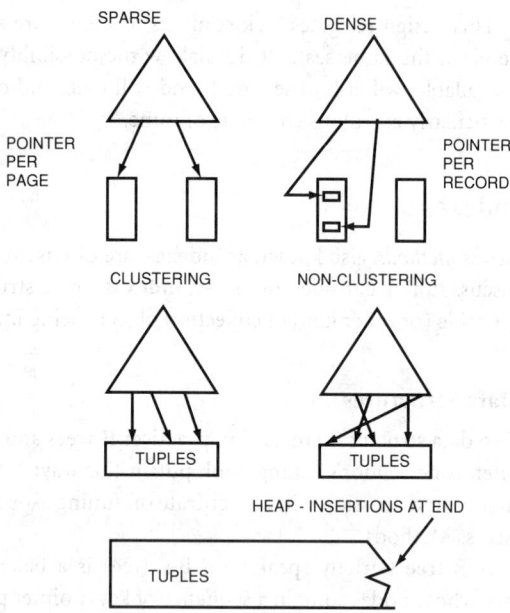

FIGURE 43.6   Index possibilities.

If records are small compared to pages, then there will be many records per data page and the data
structure supporting a sparse index will usually have one fewer level than the data structure supporting a
dense index. This means one less disk access if the table is large. By contrast, if records are almost as large
as pages, then a sparse index will rarely have better disk access properties than dense indexes.

The main virtue of dense indexes is that they can support certain read queries within the data structure
itself. For example, if there is a dense index on the keywords of a document retrieval system, one can count
the records containing some term, e.g., "derivatives scandals," without accessing the records themselves.
(Count information is useful for that application, because queriers frequently reformulate a query when
they discover that it would retrieve too many documents.) A secondary virtue is that a query that makes
use of several dense indexes can identify all relevant tuples before accessing the data records; instead, one
can just form intersections and unions of pointers to data records.

A **clustering index** on an attribute (or set of attributes) X is an index that puts records close to one
another if their X-values are *near* one another. What "near" means depends on the data structure. On
B-trees, two X-values are near if they are close in their sort order. For example, 50 and 51 are near, as are
Smith and Sneed. In hash structures, two X-values are near only if they are identical.

Sparse indexes must be clustering, but clustering indexes need not be sparse. In fact, clustering indexes
are sparse in some systems (e.g., SYBASE, ORACLE hash structures) and dense in others (e.g., ORACLE B-
trees, DB2). Because a clustering index implies a certain table organization and the table can be organized
in only one way at a time, there can be at most one clustering index per table.

A *nonclustering index* (sometimes called a *secondary* index) is an index on an attribute (or set of attributes)
Y that puts no constraint on the table organization. The table can be clustered according to some other
attribute X or can be organized as a heap, as we discuss below. A nonclustering index is always dense—there
is one leaf pointer per record. There can be many nonclustering indexes per table.

A **heap** is the simplest table organization of all. Records are ordered according to their time of entry
(Fig. 43.6). That is, new insertions are added to the last page of the data structure. For this reason,
inserting a record requires a single page access. That is the good news. The bad news is that concurrent
insert transactions will tend to lock one another out unless your system offers record locking, because all
inserts access the same page. This is a frequent cause of concurrency control bottlenecks in systems that
have page level locking. (Perhaps because it has page-level locking, Sybase Version 11 [Sybase Web Site]

has introduced a feature that splits the head of a heap into several parts with successive insertions applying to different parts according to a round-robin protocol.)

Nonclustering indexes are useful if each query retrieves significantly fewer records than there are pages in the file. We use the word "significant" for the following reason: a table scan can often save time by reading many pages at a time, provided the table is stored on contiguous tracks. For example, INGRES 5.0 normally reads 8 pages at a time on a scan. Therefore, even if the scan and the index both read all the pages of the table, the scan may complete 2 to 10 times faster than if it read one page at a time.

Consider the following two examples, adapted from a Wall Street application, concerning a query that retrieves many records.

1. Suppose a table T has 2 million records and an average of 74,000 are retrieved by a query on A. Suppose there are 22 records per page. Is a nonclustering index on A a help or a hindrance? *Evaluation.* The total number of pages is about 90,000. So nearly every page will contain a record, and using the index will entail random accesses. By contrast, scanning the table will entail few seeks. So using the index will give worse performance than scanning the table.
2. Consider the same situation as above, except that there are only two records per page. *Evaluation.* In this case, there are one million pages, and a query on the nonclustering index will touch only every thirteenth page on the average, so the index will help.

In summary, nonclustering indexes work best if the average query using the index will access far fewer records than there are data pages. Large records and high selectivity both contribute to the usefulness of nonclustering indexes.

Locks and Data Structures

Many database instructors discuss data structures in the absence of concurrency considerations and then discuss concurrency control in the absence of data structure considerations. The two are closely related, however, as the following scenario illustrates.

A *sequential key* is an attribute or set of attributes that are monotonic with time. A typical sequential key is a timestamp or a sequential counter. If one designs a B-tree based on a sequential key, then all insert operations will go down the same path (using rightmost pointers all the way). Since these inserts must exclusively lock the leaf node, they will serialize on that node if the inserts come from different transactions. Systems that have record level locking (e.g., Oracle 7.x) will hold such node locks only for a short time (as "latches"), whereas systems with page-level locking (e.g., Sybase System 11) will hold such locks until the end of the transaction. In either case, but especially in the page-level case, this serialization can severely hurt performance. Thus, B-trees are a bad idea for sequential keys if there is heavy insert traffic and the inserts come from different transactions. If they all come from the same transaction, say a batch transaction, then they will not conflict. By contrast, a hash structure works well for sequential keys, since consecutive keys will usually go to different leaf pages, causing few lock conflicts.

Here we have an interaction among index types, insert patterns, and locking—a tuner must take all this into consideration.

Face-Lifts and Related Tips

Though indexes don't develop wrinkles that you can see with the naked eye, the scan performance of clustering indexes suffers from undulations after a time. Figure 43.6 shows the reason: when first built, a clustering index on some attribute A gains its power for range and prefix queries on A from the fact that the records of the underlying table are ordered according to their A-values. Thus, to answer the range query "find records whose A-values lie between 500 and 700," the system needs only to scan a few consecutive cylinders on disk. The net result is that the records are retrieved at optimal disk rates.

Unfortunately, insertions and updates can cause pages to overflow. Some systems attempt to put overflow records from page p on p's cylinder but this is not always possible. Consider a range query on a clustering index having many overflows. Records can no longer be read at optimal rates necessarily. Instead, scans must perform seeks to jump from parent page to overflow page and back again.[4] The net effect is that the clustering index performs worse after a while than when it was first built. If that happens with a clustering index, it is good to rebuild it. Since rebuilding an index can be expensive, smart system administrators often create clustering indexes with low page utilization, so that many insertions can be accommodated before resorting to overflow pages.

Controlling the Use of Indexes

Query optimizers may decline to use an index for a variety of reasons.

- Users of the Oracle 6 rule-based query optimizer—were able to suppress the use of the index on attribute A by rewriting the qualification $A = 6$ to $A + 0 = 6$. The optimizer would not use an index on A if A was combined with some other term in an arithmetic expression on its side of an equation.
- Data-sensitive query optimizers will not use an index on a small table (e.g. smaller than what can fit on a track), because it is usually quicker to read the entire table. The sizes of data tables are stored in a system catalog. A user with a large table may be surprised to learn (by looking at the plan of a query) that the optimizer has chosen to scan a large table rather than use an apparently useful index. The reason often turns out to be that the system catalog is out of date. An Update Statistics command (and, sometimes, a recompilation of affected queries) will ensure that the proper indexes are used.

Final Remarks Concerning Indexes

The main point to remember is that the use of indexes is a two-edged sword: I have seen an index reduce the time to execute a query from hours to a few seconds in one application, yet increase batch load time by a factor of 80 in another application. Add them with care.

Tuning-Table Design

Table design is the activity of deciding which attributes should appear in which tables in a relational system. The Conceptual Database Design chapter discusses this issue, emphasizing the desirability of arriving at a **normalized** schema. Performance considerations sometimes suggest choosing a nonnormalized schema, however. More commonly, performance considerations may suggest choosing one normalized schema over another or they may even suggest the use of redundant tables.

To Normalize or Not to Normalize

Consider the normalized schema consisting of two tables: *Sale*(sale_id, customer_id, product, quantity) and *Customer*(customer_id, customer_location).

If we frequently want sales per customer location or sales per product per customer location or even sales per salesperson per customer location, then this table design requires a join on customer_id for each of these queries. A denormalized alternative is to add customer_location to *Sale*, yielding *Sale*(sale_id, customer_id, product, quantity, customer_location) and *Customer*(customer_id, customer_location). In this alternative, we still would need the *Customer* table to avoid anomalies such as the inability to store the location of a customer who has not yet bought anything.

Comparing these two schemas, we see that the denormalized schema requires more space and more work on insertion of a sale. (Typically, the data-entry operator would type in the customer_id, product, and quan-

---

[4]Clever query processors may do better by recording the fact that they must visit an overflow page, but defer the visit until they arrive on the cylinder having that page.

tity; the system would generate a sale_id and do a join on customer_id to get customer_location.) On the other hand, the denormalized schema is much better for finding the products sold at a particular location.

The tradeoff of space plus insertion cost vs. improved speeds for certain queries is the characteristic one in deciding when to use a denormalized schema. Good practice suggests starting with a normalized schema and then denormalizing sparingly.

Redundant Tables

The previous example illustrates a special situation that we can sometimes exploit by implementing wholly redundant tables. Such tables store the aggregates we want. For example:

Sale(sale_id, customer_id, product, quantity) *Customer*(customer_id, customer_location) *Customer_Agg* (customer_id, totalquantity) *Loc_Agg* (customer_location, totalquantity). This reduces the query time but imposes an update (and small space) overhead. The tradeoff is worthwhile in situations where many aggregate queries are issued, since even the denormalized schema requires a sort or hash to compute the aggregate.

Aggregate-intensive situations often result from decision-support or *data-mining* applications. Data mining is the activity of looking for profitable patterns in data. For example, any customer who buys disposable diapers will often also buy lipstick (this is a real pattern discovered by Wal-Mart). A typical installation may use a transaction-oriented database system such as for on-line updates and replicate the data to a back-end query engine for data mining, often with many redundant tables.

Tuning Normalized Schemas

Even restricting our attention to normalized schemas without redundant tables, we find tuning opportunities because many normalized schemas are possible. Consider a bank whose *Account* relation has the normalized schema (account_id is the key):

- *Account*(account_id, balance, name, street, postal_code)

Consider the possibility of replacing this by the following pair of normalized tables:

- *AccountBal*(account_id, balance)
- *AccountLoc*(account_id, name, street, postal_code)

The second schema results from **vertical partitioning** of the first (all nonkey attributes are partitioned). The second schema has the following benefits for simple account update transactions that access only the id and the balance:

- A sparse clustering index on account_id of *AccountBal* may be a level shorter than it would be for the *Account* relation, especially if the name, street, and postal_code fields are long relative to account_id and balance. The reason is that the leaves of the data structure in a sparse index point to data pages. If *AccountBal* has far fewer pages than the original table, then there will be far fewer leaves in the data structure.
- More account_id–balance pairs will fit in memory, thus increasing the hit ratio. Again, the gain is large if *AccountBal* tuples are much smaller than *Account* tuples.

On the other hand, consider the further decomposition:

- *AccountBal*(account_id, balance)
- *AccountStreet*(account_id, name, street)
- *AccountPost*(account_id, postal_code)

Though still normalized, this schema probably wouldn't work well for this application, since queries (e.g. monthly statements, account update) require both street and postal_code or neither. Vertical partitioning, then, is a technique to be used for users who have intimate knowledge of the application.

43.4 New Challenges: Object-Oriented and Parallel Database Systems

Except for our discussion of table design, the tuning considerations raised so far pertain directly to object-oriented database systems. All of the above applies to parallel relational database servers. These relatively new kinds of database systems simply introduce new considerations.

Object-Oriented Database Systems

Object orientation provides many potential benefits to the designer of a database system—encapsulation, reusability, extensible type systems, and so on. Object-oriented database systems (OODBs) combine the interface of an object-oriented language (usually C++ or Smalltalk) with such database amenities as persistence (the ability to retain data after a program completes), concurrency control, and recovery. OODBs are used mainly in applications whose data cannot be modeled easily as records with fixed numbers of fields, e.g., circuit design, hypermedia, graphical layout databases, and the like.

From the performance point of view, OODBs have two principal defining characteristics:

1. They implement pointers directly, in contrast to relational systems, which simulate pointers by using foreign keys. The potential advantage of following a pointer in a few machine cycles rather than traversing an index with a foreign key gives rise to claims of 1000-fold speedups over relational systems. Whether such performance improvements can be realized for an entire application or not depends on how often pointers must be dereferenced. In applications characterized by nodes and edges (like circuit design), the speedup will tend to be significant, especially if the working set of an application fits in RAM.
2. They provide ordered data types such as arrays, in contrast to relational systems where relations impose no order on tuples. This can be a significant advantage in systems that must hold data in time order, for example, or require the ability to address sequences.

The main tuning consideration pertaining to object-oriented systems has to do with pointers. An object may have data members such as simple character arrays or numerics, but may also have pointers to other complex objects such as images, known as *subobjects*. Two questions arise with respect to these subobjects:

- When should subobjects be clustered with their parent objects, and when should they be separated? Separating the subobjects is a good idea when the subobjects are large and seldom accessed together with the parent object—as suggested in the discussion of vertical partitioning for relations.
- Should subobjects be prefetched when parent objects are fetched? The answer to this question depends on whether the accesses between parents and subobjects are correlated or not.

Object-oriented systems offer other facilities, such as support for very long transactions and distributed buffer. New-generation relational systems (conformant with SQL3) have also begun to support such facilities. We defer the discussion of tuning such facilities, as well as sample applications of OODBs, to [Shasha 1992].

Parallel Database Systems

In large enterprise database systems, the future belongs to parallel database servers.[5] The architecture of new mainframes (often called "enterprise servers" by the vendors) consists of collections of commodity microprocessors (the same ones that run personal computers) connected by a proprietary network.

---

[5]In fact, even classical mainframes like the IBM 3090 were six-way multiprocessors. The difference between such mainframes and new ones is that classical mainframes were proprietary from the ground up and today's degree of multiprocessing—modern database machines have up to 1000 processors—was not feasible then.

The reason is simple: this architecture constitutes the cheapest way to get high speed while riding the commodity technology curve. Further, even though multiprocessors suffer the processor overhead of communication, this overhead becomes nearly insignificant now that microprocessors have become so fast (e.g., up to 500 million instructions per second for a DEC Alpha) and so cheap.

This said, parallelism still does not imply location independence. The performance of local cache accesses can be a factor of 1000 times faster than remote Random Access Memory access. This speed difference bubbles up to the programmer level: database designers must lay out data to maximize cache locality. That is the major new tuning challenge.

The database tuner can influence locality in three ways:

1. For each table T, the tuner can choose the subset S of processing nodes on which to *partition T*. That is, one divides the records of T into disjoint sets and puts each set into a different node. (In practice, systems divide tables into many more disjoint sets than there are nodes in order to obtain better load balancing.) Fortunately for the tuner, folklore and some preliminary research suggests that partitioning medium-to-large tables over all nodes is usually a good idea.

2. Given S, the next question is *how* to partition the data among those nodes. There are three main methods:

   - The *round-robin* technique holds that tuples should be inserted to the nodes in S in circular order, so all nodes hold n tuples before any node holds $n + 1$ tuples for all non-negative n. This maintains the numbers of tuples in each node in balance. (Deletions can disturb this balance, however.)

   - The *range-partitioning* technique on some attribute (or attributes) A maintains the invariant that all tuples at any site s_i have A-values falling in a disjoint interval from the tuples of any site s_j where $i \neq j$. This strategy works very well for min–max as well as equality queries on A and localizes the work of range and prefix queries on A to a few nodes, at the cost of an unbalanced load. Insertions can easily cause some nodes to have many more tuples than others.

   - The *hash-partitioning* technique on some attribute (or attributes) A works as follows. Number the nodes in S from 0 to $|S| - 1$. Construct a hash function h from the domain of A to the range 0 to $|S| - 1$. Map tuple t to $h(t.A)$. This strategy works very well for equality selections based on A (the selection can be routed to one node based on the hash function, as is also possible for range-partitioning). Other queries, even min–max ones, require all processors to work in parallel. Suitable tricks such as establishing far more buckets than there are nodes and putting many buckets on each node can ensure that the load on the different nodes is more or less balanced. For this reason, hash partitioning is the most popular partitioning strategy.

3. Suppose T is partitioned based on A and there are many equality selections on a secondary key B. It seems wasteful to search all nodes for a matching record, even if each node has a local index. To avoid this, the tuner can construct an index on B partitioned over several or all nodes. Such an index allows an equality query on B to probe an index leaf page to discover the node holding the tuple—reducing the search to two nodes. If even this proves to be too expensive, one may choose to set up a redundant copy of T partitioned on B. The notion of setting up a redundant table for the sake of its partitioning key is an enticing possibility, but, like all redundancy strategies, should be used sparingly, since it increases space requirements as well as update times.

43.5 Case Studies—Typical Tuning Actions

Often, tuning consists in applying the techniques cited above, such as the selection and placement of indexes or the splitting up of transactions to reduce locking conflicts. At other times, tuning consists in recognizing fundamental inefficiencies and attacking them.

1. Simple problems are often the worst. I've seen a situation where the database was very slow because the computer supporting the database was also the mail router. Offloading nondatabase applications is the simplest method of speeding up database applications.

2. Another simple problem having a simple solution concerns the recovery log. Putting the log on a disk by itself is a very good way to increase performance by a third or more for high-update transactions. For reliability purposes, it is good to mirror the log on two disks.

3. Batch-load performance problems often have to do with indexes. Recall that every insertion in a batch load must update all indexes. Since it is the case for at least one modern database management system that each index adds a factor of 1.5 to the cost of inserting a row, inserting a set of rows into a table with five indexes will take more than 7 times longer than inserting those rows into a table with no indexes—even if none of the indexes checks for duplicates.

4. Inefficient queries are a major source of performance problems. A good place to look for such inefficiencies is in trigger code. Since procedural languages for triggers resemble standard programming languages, bad habits sometimes emerge. Consider for example a trigger that loops over all records inserted by an update statement. If the loop has an expensive multitable join operation, it is important to pull that join out of the loop if possible. I've seen a 10-fold speedup for a critical update operation following such a change.

5. Some administrators think that the more users you put on a database system, the better. Unfortunately, this can cause electronic traffic jams that can be as annoying as automotive ones. Reduce the maximum-number-of-database-processes parameter if your database monitoring tools suggest that there are too many lock waits, or if operating-system statistics suggest that the amount of paging is too high. A transaction processing monitor such as CICS, Tuxedo, or Encina may help reduce the paging load by multiplexing many users onto fewer threads, particularly if the users are often inactive.

6. There are many ways to partition load to avoid performance bottlenecks in a large enterprise. One approach is to distribute the data across many sites connected by wide-area networks. This can result, however, in performance and administrative overheads unless networks are extremely reliable. Another approach is to distribute queries over time. For example, banks typically send out $\frac{1}{20}$ of their monthly statements every working day rather than send out all of them at the end of the month.

43.6 Summary and Research Results

Database tuning is based on a few principles and a body of knowledge. Some of that knowledge depends on the specifics of systems (e.g., which index types each system offers), but most of it is independent of version number, vendor, and even data model (e.g., hierarchical, relational, or object-oriented). This chapter has attempted to provide a taste of the principles that govern effective database tuning.

Various research and commercial efforts have attempted to automate the database tuning process. Given information about table sizes and access patterns, Oracle's RDB Expert [RDB Web Site] can give advice about storage requirements, data compression, buffer cache utilization, load balancing, physical placement on disk, and index selection. Tuners would do well to exploit these tools as much as possible. Human expertise then comes into play only when deep application knowledge is necessary (e.g., in the determination of locking isolation levels or in rewriting queries or table designs) or when these tools don't work as advertised (the problems are all NP-complete).

The COMFORT project [Weikum et al. 1994] at ETH in Zurich has studied lower level automatic tuning strategies that could be incorporated directly into a database management system. Using data from Swiss banks as well as synthesized data, the project has proposed paging strategies and load control strategies that demonstrably improve performance. Alex Thomasian and Kyung Ryu of IBM have independently verified the load-control work with a theoretical model [Thomasian and Ryu 1991].

Technological advances such as parallelism and object orientation add other challenges to the already challenging task of diagnosing performance problems and finding solutions. Doing this well may not require a good bedside manner, but good tuning can put a sick database system back on its feet.

Acknowledgment

The reviewer of this chapter improved the presentation greatly.

Defining Terms

B-tree: The most used data structure in database systems. A B-tree is a balanced tree structure that permits fast access for a wide variety of queries. In virtually all database systems, the actual structure is a B+-tree in which all key–pointer pairs are at the leaves.

Clustering index: A data structure plus an implied table organization. For example, if there is a clustering index based on a B-tree on last name, then all records with the last names that are alphabetically close will be packed onto as few pages as possible.

Conflict (between locks): An incompatibility relationship between two lock types. Read locks are compatible (nonconflicting) with read locks, meaning different transactions may have read locks on the same data items. A write lock, however, conflicts with all kinds of locks.

Decision support: Queries that help planners decide what to do next, e.g., which products to push, which factories require overtime, and so on.

Denormalization: The activity of changing a schema to make certain relations denormalized for the purpose of improving performance (usually by reducing the number of joins). Should not be used for relations that change often or in cases where disk space is scarce.

Dense index: An index in which the underlying data structure has a pointer to each record among the data pages. Clustering indexes can be dense in some systems (e.g., ORACLE). Nonclustering indexes are always dense.

Hash structure: A tree structure whose root is a function, called the hash function. Given a key, the hash function returns a page that contains pointers to records holding that key or is the root of an overflow chain. Should be used when selective equality queries and updates are the dominant access patterns.

Heap: In the absence of a clustering index, the tuples of a table will be laid out in their order of insertion. Such a layout is called a heap. (Some systems, such as RDB, reuse the space in the interior of heaps, but most do not.)

Hit ratio: The number of logical accesses satisfied by the database buffer divided by the total number of logical accesses.

Index: A data organization to speed the execution of queries on tables or object-oriented collections. It consists of a data structure, e.g., a B-tree or hash structure, and a table organization.

Locking: The activity of obtaining and releasing read locks and write locks (see corresponding entries) for the purposes of concurrent synchronization (concurrency control) among transactions.

Nonclustering index: A dense index that puts no constraints on the table organization, also known as a secondary index. For contrast, see **clustering index**.

Normalized: A relation R is normalized if every functional dependency "X functionally determines A," where A and the attributes in X are contained in R (but A does not belong to X), has the property that X is the key or a superset of the key of R. X functionally determines A if any two tuples with the same X values have the same A value. X is a key if no two records have the same values on all attributes of X.

On-line transaction processing: The class of applications where the transactions are short, typically ten disk I/Os or fewer per transaction, the queries are simple, typically point and multipoint queries, and the frequency of updates is high.

Read lock: If a transaction T holds a read lock on a data item x, then no other transaction can obtain a write lock on x.

Seek: Moving the read/write head of a disk to the proper track.

Serializability: The assurance that each transaction in a database system will appear to execute in isolation of all others. Equivalently, the assurance that a concurrent execution of committed transactions will appear to execute in serial order as far as their input/output behaviors are concerned.

Sparse index: An index in which the underlying data structure contains exactly one pointer to each data page. Only clustering indexes can be sparse.

Track: A narrow ring on a single platter of a disk. If the disk head over a platter does not move, then a track will pass under that head in one rotation. The implication is that reading or writing a track does not take much more time than reading or writing a portion of a track.

Transaction: A program fragment delimited by Commit statements having database accesses that are supposed to appear as if they execute alone on the database. A typical transaction may process a purchase by increasing inventory and decreasing cash.

Two-phase locking: An algorithm for concurrency control whereby a transaction acquires a write lock on x before writing x and holds that lock until after its last write of x; acquires a read or write lock on x before reading x and holds that lock until after its last read of x; and never releases a lock on any item x before obtaining a lock on any (perhaps different) item y. Two-phase locking can encounter deadlock. The database system resolves this by rolling back one of the transactions involved in the deadlock.

Vertical partitioning: A method of dividing each record (or object) of a table (or collection of objects) so that some attributes, including a key, of the record (or object) are in one location and others are in another location, possibly another disk. For example, the account id and the current balance may be in one location and the account id and the address information of each tuple may be in another location.

Write lock: If a transaction T holds a write lock on a datum x, then no other transaction can obtain any lock on x.

Transaction: Unit of work within a database application that should appear to execute atomically (i.e., either all its updates should be reflected in the database or none should; it should appear to execute in isolation).

References

Comer, D. 1979. The ubiquitous B-tree. *ACM Comput. Surveys* 11(2):121–137.

Copeland, G., Alexander, W., Bougherty, E., and Keller, T. 1988. Data placement in Bubba, pp. 99–108. In *Proc. ACM SIGMOD Conf.*, May.

Gray, J. and Reuter, A. 1993. *Transaction Processing: Concepts and Techniques*. Morgan Kaufmann, San Mateo, CA.

RDB Web Site, ongoing. http://www.oracle.com/info/products/rdbfamily.html.

Shasha, D. 1992. *Database Tuning: A Principled Approach*. Prentice–Hall, Englewood Cliffs, NJ.

Shasha, D., Llirbat, F., Simon, E., and Valduriez, P. 1995. Transaction chopping: algorithms and performance studies. *ACM Trans. Database Systems*, Oct., pp. 325–363.

Sybase Web Site, ongoing. http://www.sybase.com/index/toc.html.

Thomasian, A. and Ryu, K. 1991. Performance analysis of two-phase locking. *IEEE Trans. Software Eng.*, 17(5):68–76 (May).

Weikum, G., Hasse, C., Moenkeberg, A., and Zabback, P. 1994. The COMFORT automatic tuning project. *Inf. Systems* 19(5):381–432.

Further Information

Whereas the remarks of this chapter apply to most database systems, each vendor will give you valuable specific information in the form of tuning guides or administrator's manuals. The guides vary in quality, but they are particularly useful for telling you how to monitor such aspects of your system as the relationship between buffer space and hit ratio, the number of deadlocks, the disk load, and so on.

The popular journal *DBMS*, published by Miller Freeman, Inc., of San Mateo, California, features useful articles about database tuning for specific releases of specific systems from time to time.

A very nice discussion of object orientation outside the database context can be found in Bjarne Stroustrup's recent book *The Design and Evolution of C++* published by Addison–Wesley.

A performance-oriented general textbook on databases is Pat O'Neil's book *Database* published by Morgan Kaufmann.

Jim Gray has produced some beautiful viewgraphs of the technology trends and applications leading to parallel database architectures.

Steve Rozen's Ph.D. thesis at New York University, *Automating Physical Database Design: An Extensible Approach,* presented some useful heuristics underlying an automatic tuning tool.

44

Access Methods

Betty Salzberg
Northeastern University

44.1 Introduction

Although main memories are becoming larger, there are still many large databases which cannot fit entirely in main memory. In addition, because main memory is larger and processing is faster, new applications are storing and displaying image data as well as text, sound, and video. This means that the data stored can be measured in terabytes. Few main memories hold a terabyte of data. So data still have to be transferred from a magnetic disk to main memory. Such a transferral is called a **disk access**.

Disk access speeds have improved. But they have not and cannot improve as rapidly as central processing unit (CPU) speed. Disk access requires mechanical movement. To move a **disk page** from a magnetic disk to main memory, first one must move the **arm** of the disk drive to the correct **cylinder**. A cylinder is the collection of **tracks** at a fixed distance from the center of the disk drive. The disk arm moves toward or away from the center to place the read/write **head** over the correct track on one of the disks. As the disks rotate, the correct part of the track moves under the head. Only then can the page be transferred to the main memory of the computer. A disk drive is pictured in Fig. 44.1.

The fastest disks today have an average access time of 10 ms. This is at a time when CPU operations are measured in nanoseconds. Therefore, the access of one disk page is at least one million times slower than adding two integers in the CPU. In addition, to request a disk page, the CPU has to perform several thousand instructions, and often the operating system must make a process switch. Thus, although the development of efficient access methods is not a new topic, it is becoming increasingly important.

In addition, new application areas are requiring more complex disk access methods. Some of the data being stored in large databases are multidimensional, requiring that the records

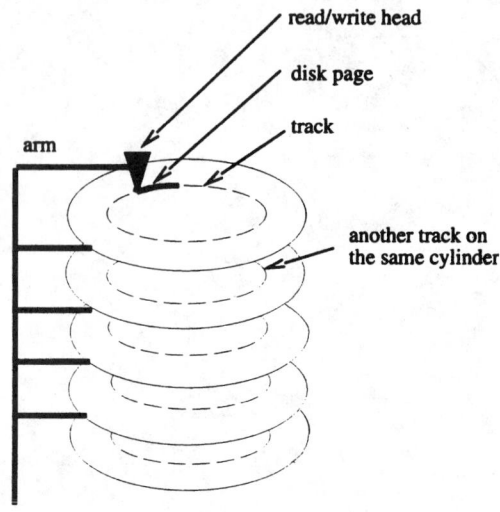

FIGURE 44.1 A disk drive.

0-8493-2909-4/97/$0.00+$.50
© 1997 by CRC Press, Inc.

stored in one disk page refer to points in two- or three-dimensional space, which are close to each other in that space. Data mining, or discovery of patterns over time, requires access methods which are sensitive to the time dimension. The use of video requires indexing that will allow retrieval by pictorial subject matter. The increasingly large amount of textual data being gathered electronically requires new thinking about information retrieval. Other chapters in this book will look at video and text databases, but here we will treat spatial and temporal data as well as the usual linear business data.

44.2 Underlying Principles

Because disk access is so expensive in elapsed time and in CPU operations, and because the minimum number of bytes which can be transferred at one time is now 4096 bytes, or 4 kilobytes, the main goal of access methods is to group or cluster on one disk page data which will be requested by an application in a short time frame. Then after a page is placed in main memory, in order to read a requested item, there will be a good chance that when a subsequent item is requested, a separate disk access will not be necessary.

Since application logic cannot be predicted in general, the clustering will follow some kind of logical relationship or some kind of specific knowledge about the applications which are run. For example, if it is known that address labels are printed out in order of the zip code, one can attempt to store records with the same zip code in the same page.

In addition to disk page clustering, there are several other important principles of access methods. One is that disk space should be used wisely. Suppose an application accesses customer records individually and randomly by social security number. A very simple solution which would require only one disk access per customer would be to reserve one disk page for each possible social security number. With 1000 customers, 1×10^9 disk pages would be reserved. Since each disk page is 4 kilobytes, this is about 4 terabytes of disk space. Although this seems like a ridiculous example, papers have been written which propose access methods and do not take into account the amount of disk space used. Total disk space used should always be considered in proposing a disk access method.

Another principle regards the number of pages needed to be accessed to find one particular record. In general, one aims to minimize the number of pages accessed which contain *none* of the data wanted. This is one of the reasons why binary search trees are not used for large databases. Binary search trees can be very unbalanced, and the mapping from parts of the tree to disk pages is not defined. One could, for example, read in the page which had the root of binary search tree in it and then follow the tree in memory until it led to another disk page, and so forth. If one of the records which is accessed relatively often thus required reading 20 disk pages every time it is requested, this is a poor access method for this application. In general, if pagination of the access method cannot be specified, it will perform poorly.

Last, insertion and deletion of records should modify as few disk pages as possible. Most good access methods modify only one page when a new record is inserted most of the time. That is, the record is inserted into a page which has empty space. This page is the only one modified. Occasionally, one or two other pages have to be modified. For a good access method, it is *never* necessary to modify a large number of pages when one record is inserted or deleted.

We summarize the principles we have discussed in a list. We will often refer to these in the rest of the chapter. The **properties of good access methods** are as follows:

1. *Clustering:* Data should be clustered in disk pages according to anticipated queries.
2. *Space usage:* Total disk space usage should be minimized.
3. *Few answer-free pages in search:* Search should not touch many pages having no relevant data.
4. *Local insertion and deletion:* Insertion and deletion should modify only one page most of the time and occasionally two or three. Insertion and deletion should never modify a large number of pages.

Most access methods have **data pages** and also have pages which contain no data, which are called **index pages**. The index pages contain **references** to other index pages and/or references to data pages or

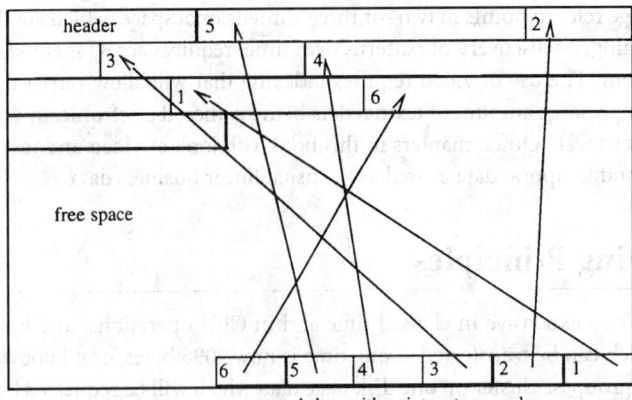

FIGURE 44.2   A typical disk page layout. New records are placed in the
free space after old records (growing down) and new slot numbers and
pointers are placed before the old ones (growing up). Variable length
records are accommodated.

references to records within data pages. A reference to a data page or an index page is a disk page address.
A reference to a record may be only the disk page address of the page where the record is kept (and some
other means must be used to locate the record in the page, usually the value of one of the attributes of the
record), it may be the combination of a disk page address and a record slot number within that disk page
(see Fig. 44.2), or it may be some attribute value which can be used in another index to find the given
record. (We freely exchange the words *index* and *access method*.)

If the index pages reference data pages but never individual records within those data pages, the access
method is called a **sparse index** or sometimes a **primary index**. In this case, the location for insertion of
a new record is determined by the access method and usually (but not always) depends on the value of a
primary key for the record. (It could instead depend on some other attribute which is not primary, for
example, the zip code in an address.) Thus, *primary indices* are not always associated with *primary keys*.

If there are references in the index to individual records, so that a data page with 20 data records in it
has at least 20 separate references in the index, the index is called a **dense index** or a **secondary index**.
Insertion of new records can then occur anywhere. Often secondary indices are associated with data pages
where the data are placed in order of insertion.

We shall use the terms primary index and secondary index. A data collection often has one primary
index and several secondary indices. The primary index determines the placement of the records and
the secondary indices allow lookup by attributes other than the ones used for placement. Most primary
indices can be converted to secondary indices by replacing the data records with references to the same
data records and actually storing the data records elsewhere. Thus, the question: "Is this access method a
primary index or a secondary index?" does not always make sense.

Indices are sometimes called **clustering indices**. In commercial products, this usually means that the
data are loaded into data pages initially in the order specified by a secondary index. When new data are
inserted in the database, such an index loses the clustering property. Such an index clusters data *statically*
and not *dynamically*. We shall not again use the term clustering index in this chapter as we believe it has
been abused and frequently misunderstood in both industry and academia.

Next in this chapter, we shall present one exceptionally good access method, the B$^+$-tree [Bayer and
McCreight 1972], and use it as an example against which other access methods are compared. The B$^+$-
tree has all of the good properties previously listed. We shall then present the hash table (and some
proposed variants) and briefly review some of the proposed access methods for spatial and temporal
data.

44.3 Best Practices

The B⁺-Tree

The B⁺-tree [Bayer and McCreight 1972] is the most widely used access method in databases today. A picture of a B⁺-tree is shown in Fig. 44.3. Each node of the tree is a disk page and, hence, contains 4096 bytes. The leaves of the tree when it is used as a primary index contain the data records or, in the case of a secondary B⁺-tree, references to the data records which lie elsewhere. The leaves of the tree are all at the same level of the tree.

The index entries contain values and pointers. Search begins at the root. The search key is compared with the values in the index entries. In Fig. 44.3, the pointer associated with the largest index entry value smaller than or equal to the search key is followed. To search for coconut, for example, first the pointer at the root associated with caramel is followed. Then at the next level, the pointer associated with chocolate is followed. Search for a single record visits only one node at each level of the tree.

In the remainder of this section, we shall assume the B⁺-tree is being used as a primary index. This is certainly not always the case. Many commercial database management systems have no primary B⁺-tree indices. Even those which do offer the B⁺-tree as a primary index must offer the B⁺-tree as a secondary index as well, as it will be necessary in most cases to have more than one index on each relation, and only one of them can be primary.

The main reasons that the B⁺-tree is the most widely used index is (1) that it clusters data in pages in the order of one of the attributes (or a concatenation of attributes) of the records and (2) it maintains that clustering dynamically, never having to be reorganized to retain disk page clustering. Thus, the B⁺-tree satisfies property 1 for good access methods. It clusters data for the anticipated queries.

Storage utilization for B⁺-trees in both index and data pages is 69% on average [Yao 1978]. Total space usage is approximately 1/0.69 times the space needed to store the data records packed in pages. All of the ·data are stored in the leaves of the tree. The space needed above the level of the leaves is less than 1/100 of the space needed for the leaves. Thus, the B⁺-tree satisfies property 2 for good access methods. It does not use too much total disk space.

In addition, B⁺-trees have no pathological behavior. That is, no records require many accesses to be found. Searches for a record given its key follow one path from the root of the tree to a leaf visiting only

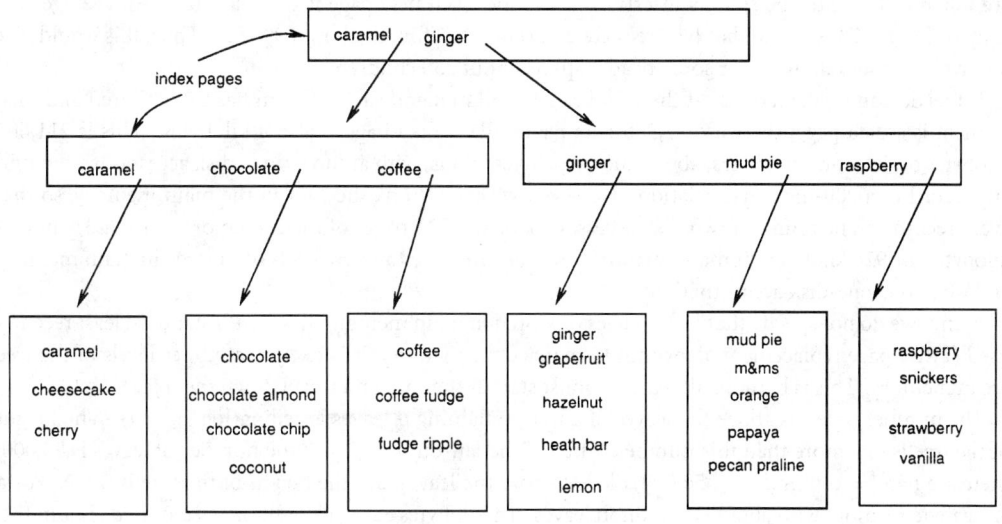

FIGURE 44.3   A B⁺-tree.

one page at each level of the tree. If the record is in the database, it is in the leaf page visited. All of the leaves are at the same level of the tree.

All records can thus be found with at most as many disk accesses as the height of the tree and the height of the tree is always small. The reason some records can be found with less disk accesses is that the upper levels of the tree are often still in main memory when subsequent requests for data are made. Thus, B^+-trees satisfy property 3 for good access methods. Few answer-free pages (in this case index pages) are accessed in any search.

Insertion and deletion algorithms enable the B^+-tree to retain its clustering and to maintain the property that all leaves are at the same level of the tree. Thus, search for one record remains efficient and search for ranges remains efficient. Insertion in a leaf which has enough empty space requires writing only one disk page. (Although this seems obvious, we will see that the R-tree discussed subsequently does not have this property.) Deletion from a page which does not require it to become sparse also writes only one disk page. The probability that more than one existing index or data page must be updated is low. In the worst case, one page is updated and a new page created at each level of the tree. This is an extremely rare event. Thus, the B^+-tree satisfies property 4 for good access methods. The B^+-tree has local insertion and deletion algorithms.

Fan-Out Calculations

The height of the tree is small because the **fan-out** is large. The fan-out is the ratio of the size of the data page collection to the size of the index page collection. The secret of the B^+-tree is that each index page has hundreds of children.

Some data structure textbooks suggest a B-tree where data records are stored in all pages of the index. But when data records are stored high up in the tree, the fan-out is too small. A data record may have 100 or 200 bytes. This would limit the number of children each high-level page in the index could have. This variation has never been used for database management systems for this reason. It is also the reason we use the notation B^+-tree, which has historically stood for the variation of B-trees, which has all of the data in the leaves, the only variation of the B-tree in use.

A disk page address is usually 4 bytes. A key value may be about 8 bytes if it is an alphanumeric key. This means each index term is 12 bytes. But there are 4096 bytes in an index page. Theoretically, each page could have $4096/12 = 341$ children. (Actually there is some header information.) But when pages are full and they must be split as insertions are made, each new page is only half-full. An average case analysis [Yao 1978] showed that B^+-trees are on average $\ell_n 2$ full, or about 69% full. Thus, this would give our average node, allowing for some header space, about 230 children.

Let us do some calculations. If the root has 230 children and each of them has 230 children and each of them is a data page, there are 52,900 data pages. Each data page is also 4 kilobytes. This is 211,600 kilobytes or, in round numbers, about 200 megabytes. Thus, with at most three disk accesses we can find any record in a 200-megabyte relation. However, we can also fix the root in the main memory, so that every record can be found in two disk accesses. If we use 231 pages of main memory space (231 times 4 kilobytes, or 924 kilobytes of main memory), we can store the top two levels of the tree in main memory and we have a one-disk access method.

Even if we do not specify that index pages be stored in main memory, if we use a standard least recently used (LRU) page replacement algorithm to manage the memory, it is likely that upper levels of the tree are in memory. This is because all searches must start at the root and travel down the tree.

The number of levels above the leaves of a tree containing n leaves is ceiling($\log_{\text{fan-out}} n$). The height of the tree is one more than this number. Thus, if the fan-out is 230 and the number of leaves is 50,000, there are two (= ceiling($\log_{230} 50,000$)) levels above the leaves and the height of the tree is three. For a 4-gigabyte relation, with about one million leaves of 4 kilobytes each, a height of 4 is required. But in this case, the level right below the root may be small. Suppose the level below the root has only 20 pages. If each of these is a root of a subtree of height three with 50,000 leaves, we have one million leaves in total. But storing in memory these 20 pages at the level below the root plus the root takes up only 84 kilobytes

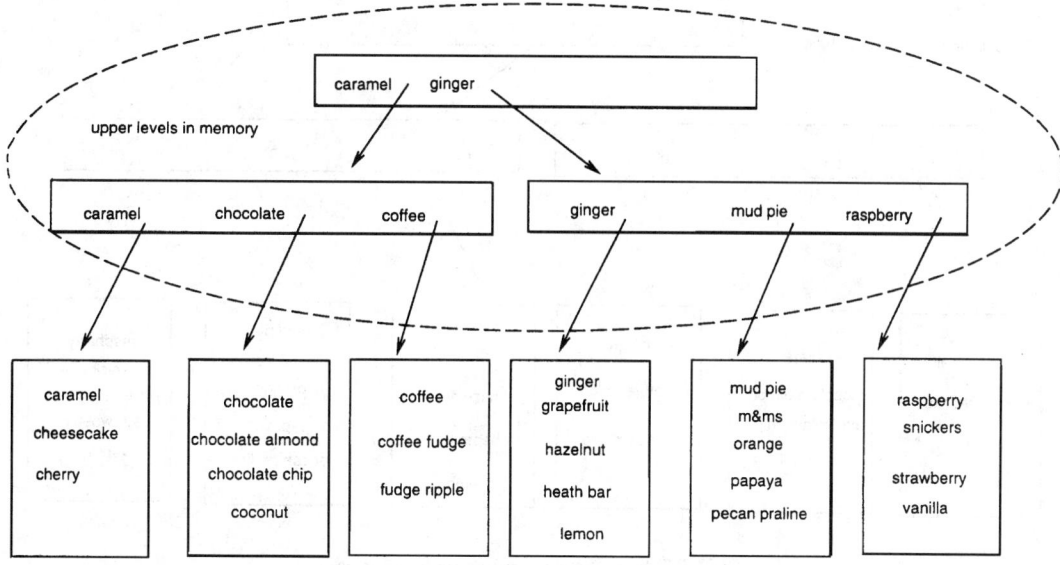

FIGURE 44.4 The number of disk accesses to find a record in a B$^+$-tree depends on how much of the tree can be kept in main memory. Adding one more level, as shown here, does not necessarily add one more disk access.

(21 times 4 kilobytes). It is likely that these two levels will stay in main memory. Then, even with a height of 4, the number of disk accesses for search for one record is only two.

Thus, tree height is not the same as number of disk accesses required to find a record. The number of disk accesses required to find a record depends on what part of the top levels of the tree are already stored in main memory. This is illustrated in Fig. 44.4, where we have shown the top two levels of the sample tree in main memory, making this B$^+$-tree a one-disk-access method.

Key Compression and Binary Search Within Pages

Since the fan-out determines the height of the tree, which in turn limits the number of disk accesses for record search, it is worthwhile to enhance fan-out. This would be particularly desirable if the size of the attributes or concatenated attributes used as the key in the B$^+$-tree is large. A smaller key means a larger fan-out.

One technique often used for enhancing B$^+$-tree performance is key compression. The prefix B$^+$-tree [Bayer and Unterauer 1977] is used in many implementations. Here, what is stored in the index pages is not the full key but only a **separator**. A separator holds only enough characters to differentiate one page from the next. Binary search still can be used within the index pages. Separators are illustrated in Fig. 44.5. (It is not true that separators must be changed when records are deleted from the database. It is probably not worthwhile to make a disk access to modify the separator when a record is deleted. The old separator still allows correct search, even though it is not as short as it might be.)

Although schemes to omit some of the first characters of the key when they are shared by some of the entries have been considered, unless they are shared by *all* of the index entries in an index page, such schemes have not been used. This is because compressing prefixes of keys makes binary search impossible, and search within an index page becomes linear. Since index page search is a frequent operation, prefix compression is not worthwhile. Also, it is predicted that disk page sizes will get larger, making linear search even worse. Figure 44.5 illustrates storing a common prefix (the first few characters used in *every key in a page*) only once.

Either the index terms can be organized within index pages using a binary search tree, or they can be kept in order using an array. The usual practice for index pages is to use an array, which saves space and allows more fan-out. Index pages are updated relatively infrequently. When a new index term is inserted,

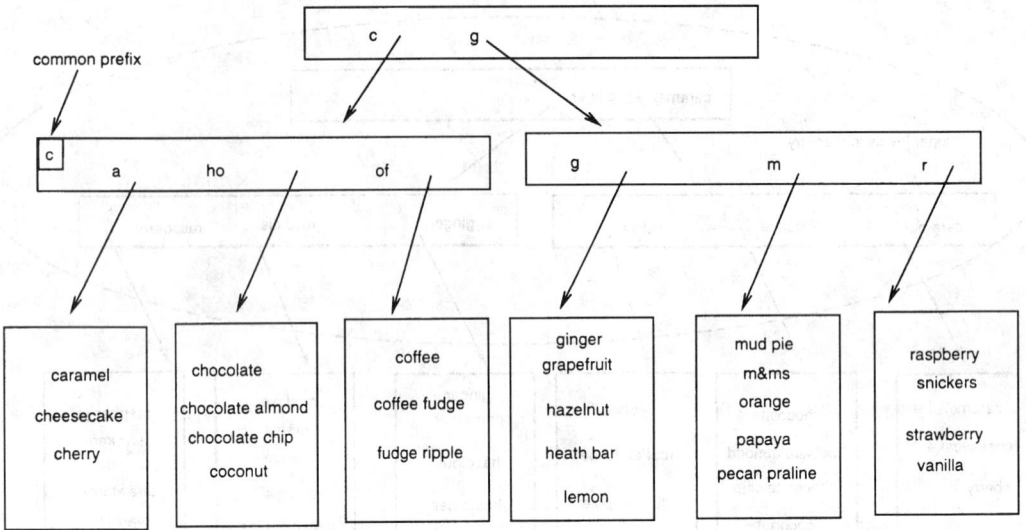

FIGURE 44.5   Separators in a prefix B⁺-tree.

the array elements after the insertion must of course be moved to a new position in the page. Binary search is performed on the array in each index page visit during a record search operation.

In data pages, the keys can be stored separately from the rest of the record, with a pointer to the rest of the record. Then the keys also can be stored in order in an array, and binary search can be used.

Insertion and Deletion

Usually, insertion of a record simply searches for the correct leaf page in the tree and updates that leaf page. Occasionally [with a probability of $2/\max(\text{recs})$ where $\max(\text{recs})$ is the maximum number of records in a leaf page] the leaf is full, and it must be split. A leaf split in a B⁺-tree is illustrated in Fig. 44.6. In Fig. 44.6, the record with key lime is entered in the tree of Fig. 44.5. In this example, we assume that only five records may reside in a leaf. Thus, we split the leaf where lime would be inserted.

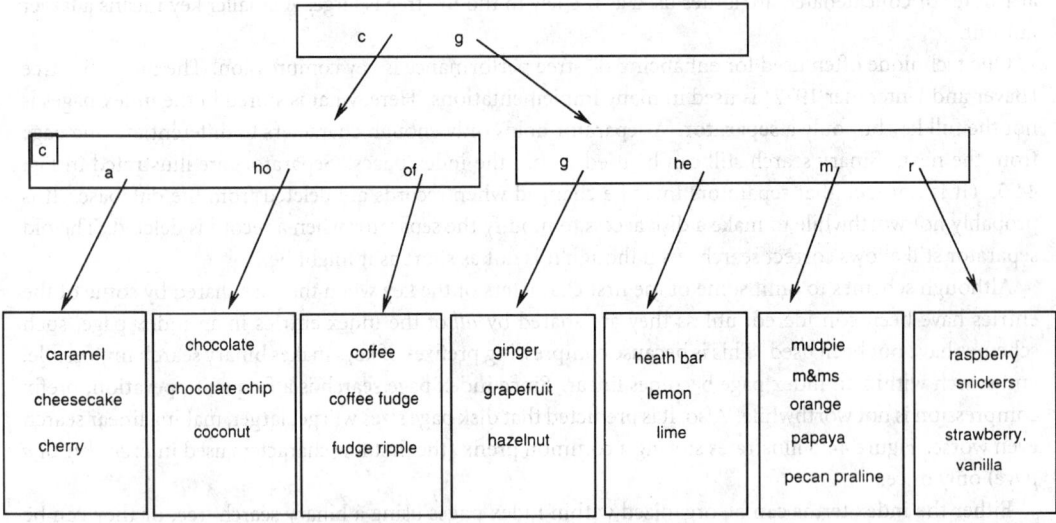

FIGURE 44.6   A leaf split in a B⁺-tree.

Half of the records stay in the old leaf node and half of the records are copied to a newly allocated leaf node. An entry is placed in the parent to separate the old and new nodes. With a much lower probability, the parent may need to be split as well, and this can percolate up the tree on a path to the root. The probability that an insertion causes a split of both the leaf and its parent is $(2/\max(\text{recs})) \times (2/\max(\text{index entries}))$.

For example, if records are 200 bytes and there are 4 kilobyte pages, max(rec) might be 20 allowing for header space in the data page. Then, after a split, 10 more records must be inserted in the page before a split is needed again (assuming no deletes). Thus, the probability of a split is at worst 1/10. If max(index entries) is equal to 300, the probability of an insertion causing an index page split is $(1/10) \times (1/150)$ or 1/1500.

For record deletion, similarly, one need only remove the record from the data page in most cases. Sometimes pages are considered sparse and are merged with their siblings. In fact, many commercial products do not consider any pages sparse unless they are completely empty [Mohan and Levine 1992]. An analysis of this issue, which concludes that node consolidation in B^+-trees is not useful unless the nodes are completely empty, can be found in Johnson and Shasha [1989].

Most textbooks, however, consider anything less than 50% to be sparse in order to maintain the property that all nodes are at least 50% full. This is not a good idea in practice, even if one does not go so far as to wait until nodes are empty before consolidating them, because it may cause **thrashing**.

If the threshold for node sparseness is 50%, so that nodes are not allowed to ever fall below 50% full, and if a record is inserted into an overflowing page, then the page is *split* to get a 50% utilization, then a record is deleted from the page, the B^+-tree may needlessly thrash between splitting and consolidating the same node. This issue is discussed in detail in Maier and Salveter [1981].

A B^+-tree node consolidation is pictured in Fig. 44.7. (The records with keys m&ms, orange, and heath bar have been removed from Fig. 44.6.) We do not change the separator he, although a shorter separator could have been chosen. In some cases, search will be incorrect if a new separator is chosen. Also, using the old separator makes the node consolidation algorithm *simpler*, which is an important consideration for access methods.

The case pictured deallocates a page. It also is possible to move records from a full sibling to a sparse one. In both cases the parent must be updated to reflect the change. In rare cases, sparse node consolidation could percolate to higher levels, because deletion of an index term could cause a parent to become sparse.

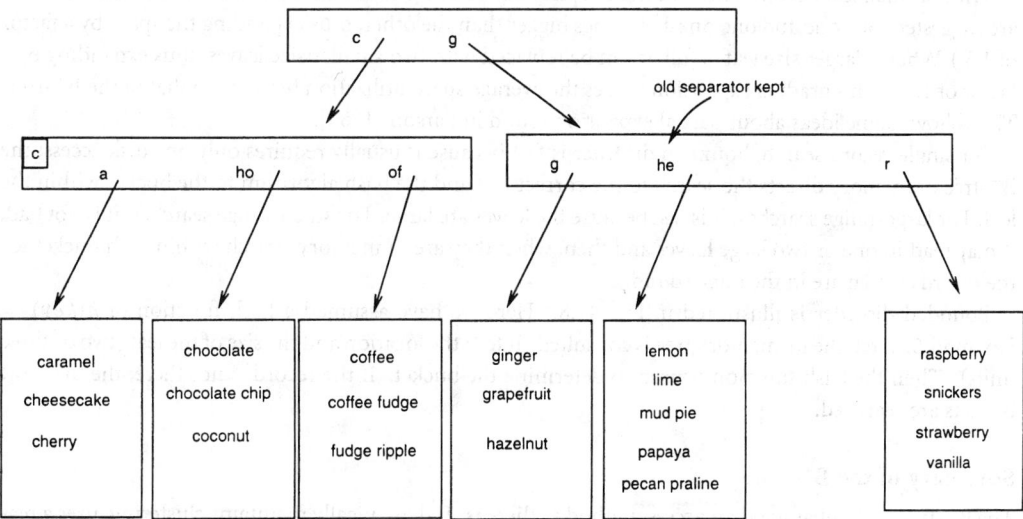

FIGURE 44.7 A B^+-tree node consolidation.

Range Searches and Reorganizing

B^+-trees are very good for small range searches. Each leaf of the tree contains all of the records whose keys are within a given range. If the range is small, it is worthwhile to use the B^+-tree to access only a few disk pages to get all of the records wanted. However, if the range is large, it might be better to read in all of the leaf pages at once and not use the B^+-tree to find out which ones satisfy the query.

The reason for this is that after there have been many splits of the leaves in a B^+-tree, the leaf pages are not stored on the disk in order. Within each page, the records are in key order. But two adjacent pages in terms of key order could be in very different places on the disk. The disk arm might need to move a long distance to access the pages which were required. On the other hand, sequential reading on the disk, where the arm sits over one cylinder and reads all of the tracks on that cylinder and then moves to the next cylinder and reads all of the tracks on it and so forth, is much faster than reading the same amount of data one page at a time, moving the disk arm for each page.

Calculations in Salzberg [1988] show that in circumstances where a fairly large-size range is required (say 10% of the data), it is better to read in all of the data in no particular order than to make separate disk accesses for pages which satisfy the query. This is even worse when the B^+-tree is used as a secondary index. The range must be quite small to make use of the index worthwhile. Here each record in a range could be stored in a separate disk page.

Because primary B^+-trees get out of order in this sense, it has sometimes been considered worthwhile to reorganize them. This can be done while keeping the B^+ trees on-line, as shown in Smith [1990], a description of an algorithm written and programmed by F. Putzolu. One leaf is written at a time. Sometimes the data in two leaves are interchanged. This is all done as a background process, and it is done transactionally (i.e., with lock protection so that searches will be correct even when reorganization is in progress).

Bounded Disorder

Efforts have been made to improve the B^+-tree. Most of these have not been implemented in commercial systems because it is not considered worth the trouble. Mostly, the B^+-tree is good enough. However, we will outline a particularly nice attempt in this area [Lomet 1988].

The basic idea of bounded disorder is to keep a small B^+-tree in memory and then vary the size of the leaves, using consecutive disk pages for one leaf. To find a particular record within a leaf, a hash algorithm is used to find the correct **bucket** within the leaf holding that record. There is one hash overflow bucket stored with each leaf. In the variation we illustrate, buckets are two or three consecutive disk pages.

When a small leaf becomes full, it can be replaced with a larger leaf. (Only two different sizes for leaves are suggested, one one and one one-half times bigger than the other, thus expanding the space by a factor of 1.5.) When a larger size leaf is full, it can be replaced with two smaller size leaves, thus expanding by a factor of 1.33. This gradual expansion makes the average space utilization better than that of the B^+-tree. This follows some ideas about partial expansion found in Larson [1980].

For single record search, bounded disorder is fast because it usually requires only one disk access; the B^+-tree in memory directs the search to the correct leaf and the hash algorithm to the bucket within the leaf. For large-range searches, it is fast because the leaves are large. For small-range searches, it is not bad; it may read in one or two large leaves and then, when they are in memory, search within each bucket for the records which are in the queried range.

Bounded disorder is illustrated in Fig. 44.8. Here we have assumed a hash function of $h(key) = key$ mod 5. First, the in-memory tree is consulted. It tells the location and the size of the leaf (two or three units). Then the hash function is used to determine the bucket. If the record is not there, the overflow buckets are searched.

Summary of the B^+-Tree

The B^+-tree is about as good an access method as there is. It dynamically maintains clustering, uses a reasonable total amount of disk space, never has pathological search cases, and has local insertion and deletion algorithms, usually modifying only one page. All other access methods we shall describe do not do as well.

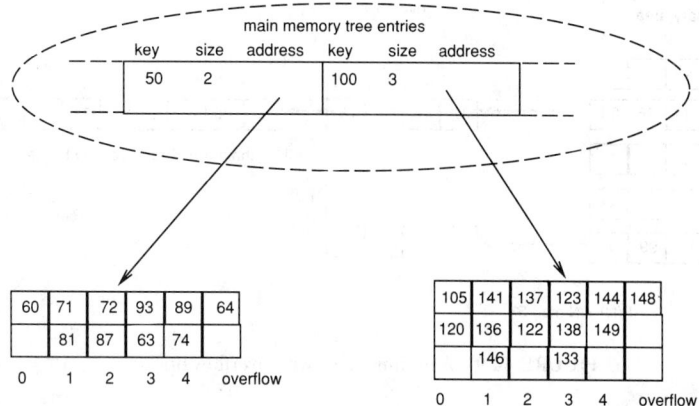

FIGURE 44.8 Bounded disorder. A small main memory B^+-tree directs the search to a large leaf. A hash function (here $h(key) = key \bmod 5$) yields the bucket within the leaf.

Hashing

The second most used access method is **hashing**. In hashing, a function applied to a database key determines the location of the record. For example, if the set of keys were known to be the integers between 1 and 1000 and the data records all had the same size, contiguous disk space for 1000 records could be allocated, and the identity function would yield the offset from the beginning of the allocated space. However, usually database keys are not consecutive integers.

Hashing algorithms start with a function which maps a database key to a number. In order to make the result uniformly distributed, certain functions have been found to be especially effective. The best coverage of hashing functions can still be found in Knuth [1968].

Basically, the database key is transformed to a number N by some method such as adding together the ASCII (American Standard Code for Information Interchange) codes of the letters of the key. Then N is multiplied by a large prime number P_1 and then added to another number P_2, and the result is taken modulo P_3, where P_3 corresponds to the number of consecutive pages on the disk allocated for the **primary area** of the hash table. Thus, $f(N) = (N \times P_1 + P_2) \bmod P_3$. Knuth [1968] discusses how the parameters are chosen.

For a primary hashing index, the data records themselves are stored in the primary area. For a secondary hashing index, the primary area is filled with addresses of data records which are stored elsewhere. In the remainder of our discussion, we shall assume all indices are primary indices.

A page (or sometimes a set of consecutive pages) corresponding to one value of the hash function is called a bucket. When all records fit into their correct bucket, hashing is a one-disk-access method. The hash value gives the bucket address as an offset from the beginning of the primary area. That bucket is accessed and the record is there. The same algorithm is used to insert a new record into a bucket as long as there is room for it.

When there is no longer room in a bucket, an overflow bucket is allocated from the **overflow area**. Let us assume each overflow bucket is a disk page. When an overflow bucket is allocated, its address is placed in the primary bucket to which it corresponds (or in the last overflow bucket allocated to this hash number), forming a chain of overflow buckets for each hash value). (Most overflow chains should be empty if the average search time is to be reasonable.)

Then the search algorithm is as follows: Hash the database key to get the address of the correct bucket in the primary area. Search in the bucket. If the record is not there, search in the overflow bucket(s). A basic hashing index with overflow buckets is illustrated in Fig. 44.9.

We shall define the **hash table fill factor** as the total space needed for the data divided by the space used by the primary area. As long as the hash table fill factor is a certain amount below one, there is not much

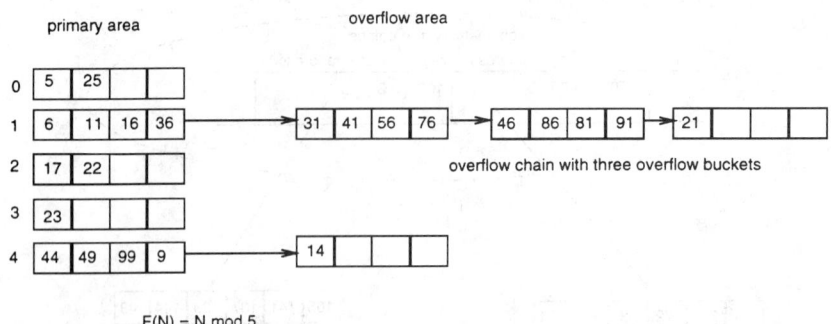

FIGURE 44.9   A hashing table with overflow buckets.

overflow and search time is fast. For example, if buckets large enough for 50 records are used and the hash table fill factor is near 70% (which would be comparable to B+-tree leaf space utilization), search is very close to one disk access on average [Knuth 1968].

However, one basic drawback of hashing is that the fill factor grows as the database grows. Hashing does not adjust well to growth of the database. The database can even become several times as large as the allocated primary area. In this case, the number of disk accesses needed on average can become quite high. Even without the database becoming large, hashing functions which produced very long chains of overflow buckets are possible and have occurred. Thus, hashing does not always satisfy property 3 for good access methods (efficient search).

If hashing must be reorganized and a new table constructed because the old table is too small (which is done in most systems that use hashing), property 4 is violated: insertion is not local when massive reorganization is required.

Further, by its nature, hashing does not cluster data which are related, unless all of the records which are related have the same key. Hashing functions which effectively create a uniform distribution are never order preserving. If such an order-preserving hashing function existed, it could do sorting of n elements in $O(n)$ time, by hashing each one into its proper place in an array [Lomet 1991]. Thus, hashing does not satisfy property 1: clustering. In fact, it is virtually useless to use a hashing structure for a range query. Each possible value in the range would have to be hashed.

Since hashing does not support range queries and since it does not adjust well to growth, it is not used as often as B+-trees. When memories were small, a B+-tree access from a B+-tree of height four meant at least three and sometimes four disk accesses, because not even the root could always be kept in main memory. Now that memories are larger, single-record search in B+-trees is more likely to be one or two disk accesses even when the height of the tree is four. More importantly for most relations, a B+-tree of height three is likely to be a one-disk access method. This makes B+-trees competitive with hashing for single-record search.

Before leaving this topic, we will look at two proposed variations on hashing to support database growth. Although many papers have been written on this topic, most are refinements of these two. Most commercial systems which provide hashing have only hashing with overflow buckets.

Linear Hashing

Linear hashing [Litwin 1980] and extendible hashing [Fagin et al. 1979] both add one new bucket to the primary area at a time. Linear hashing (in the variation we explain here) adds a new bucket when the hash table fill factor becomes too large. Extendible hashing adds one whenever an overflow bucket would otherwise occur. We explain linear hashing first.

In linear hashing, data are placed in a bucket according to the last k bits or the last $k + 1$ bits of the hash function value of the key. A pointer keeps track of the boundary between the values whose last k bits are used and the values whose last $k + 1$ bits are used. The current fill factor for the hash table is stored as the two values (bytes of data, bytes of primary space).

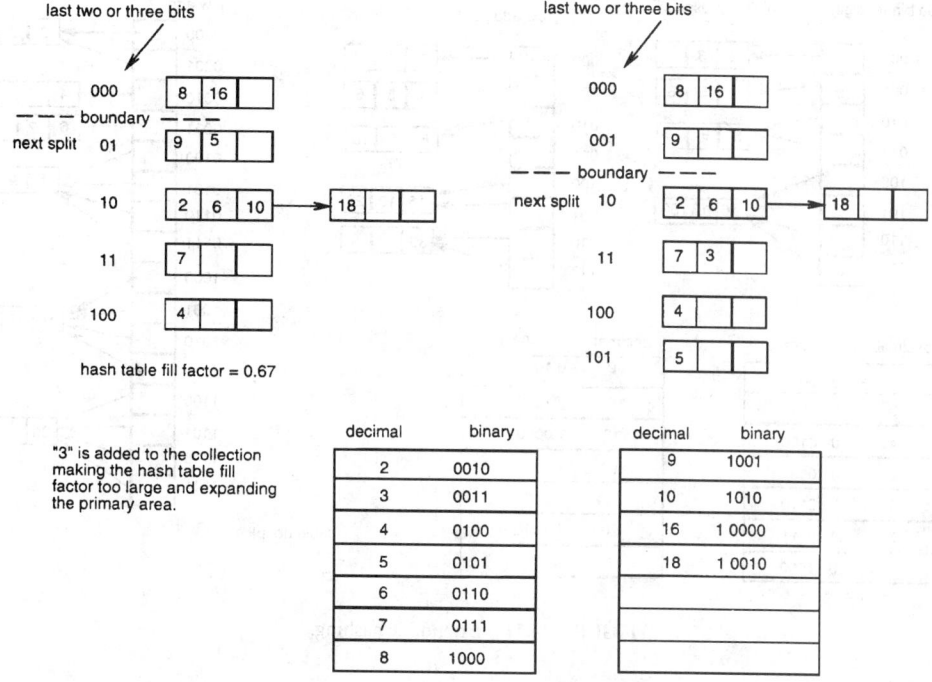

FIGURE 44.10 Linear hashing.

When the insertion of a new record causes the fill factor to go over a limit, the data in the k-bucket on the boundary between $k + 1$ and k bits is split into two buckets, each of which are placed according to the last $k + 1$ bits. There is now one more bucket in the primary area and the boundary has moved down by one bucket. When all buckets use $k + 1$ bits, an insertion causing the fill factor to go over the limit starts another expansion, so that some buckets begin to use $k + 2$ bits (k is incremented).

There is no relationship between the bucket obtaining the insertion causing the fill factor to go over the limit and the bucket which is split. Linear hashing also has overflow bucket chains.

We give an example of linear hashing in Fig. 44.10. We assume that k is 2. We assume that each bucket has room for three records and the limit for the fill factor is 0.667. We show how the insertion of one record causes the fill factor limit to be exceeded and how a new bucket is added to the primary area.

The main advantage of linear hashing is that insertion never causes massive reorganization. Search can still be long if long overflow chains exist. Range searches and clustering still are not enabled. There are two main criticisms of linear hashing: (1) some of the buckets are responsible for twice as many records on average as others, causing overflow chains to be likely even when the fill factor is reasonable and (2) in order for the addressing system to work, massive amounts of consecutive disk pages must be allocated.

Actually, the second objection is not more of a problem here than for other access methods. File systems have to allocate space for growing data collections. Usually this is done by allocating an **extent** to a relation when it is created and specifying how large new extents should be when the relation grows. Extents are large amounts of consecutive disk space. The information about the extents should be a very small table which is kept in main memory while the relation is in use. (Some file systems are not able to do this and are unsuitable for large relations.)

Many papers have been written about expanding linear hashing tables by a factor of less than two. The basic idea here is that, for example (as was used in the bounded disorder method previously mentioned), at the first expansion, what took two units of space expands to three units of space. At the second expansion, what was in three buckets expands to four buckets. At this point, the file is twice as big as it was before the first expansion. In this way, no buckets are responsible for twice as much data as any other buckets; the factors are 1.5 or 1.33. This idea originated in Larson [1980].

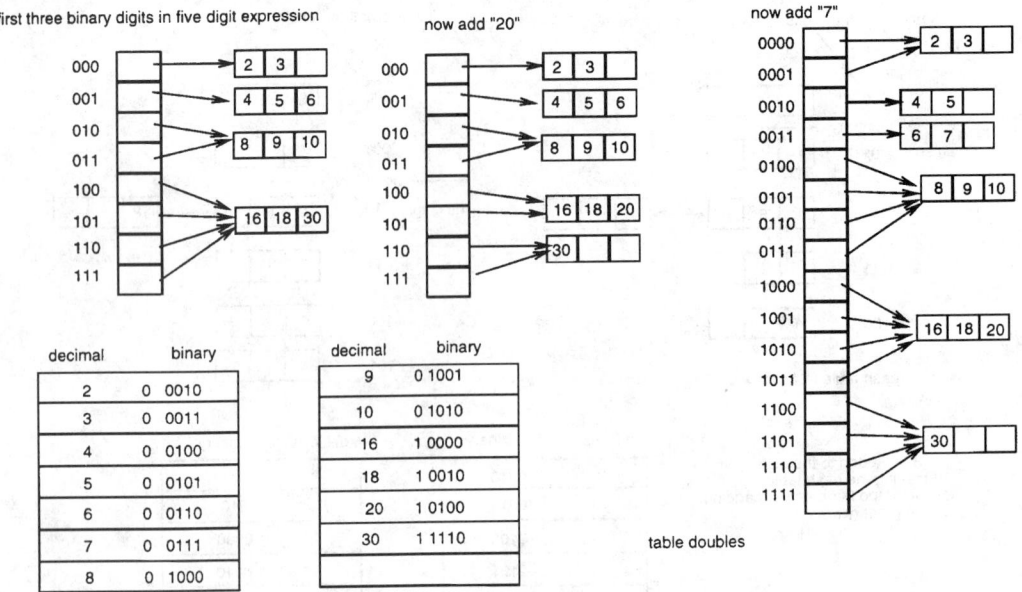

FIGURE 44.11   Extendible hashing.

Extendible Hashing

Another variant on hashing which also allows the primary area to grow one bucket at a time is called extendible hashing [Fagin et al. 1979]. Extendible hashing does not allow overflow buckets. Instead, when an insertion would cause a bucket to overflow, its contents are split between a new bucket and an old bucket and a table keeping track of where the data are is updated.

The table is based on the first k bits of a hash number. A bucket B can belong to 2^j table entries where $j < k$. In this case, all of the numbers whose first $k - j$ bits match those in the table will be in B. For example, if k is 3, there are eight entries in the table. They could refer to eight different buckets. Or two of the entries with the same two first bits could refer to the same bucket. Or four of the entries with the same first bit could refer to the same bucket. We illustrate extendible hashing in Fig. 44.11.

The insertion of a new record which would cause an overflow either causes the table to double or else it causes some of the entries to be changed. For example, a bucket which was referred to by four entries might have its contents split into two buckets, each of which was referred to by two entries. Both cases are illustrated in Fig. 44.11.

The advantage of extendible hashing is that it never has more than two disk accesses for any record. Often, the table will fit in memory, so that it becomes a one-disk-access method. There are no overflow chains to follow.

The main problems with this variation on hashing are total space utilization and need for massive reorganization (of the table). Suppose the buckets can hold 50 records and there are 51 records with the identical first 13 bits in the hash number. Then there are at least $2^{14} = 16,384$ entries in the table. It does not matter how many other records there are in the database.

Like the other variations on hashing, extendible hashing does not support range queries. All hashing starts with a *hashing function*, which will randomize the keys before applying the rest of the algorithm.

Spatial Methods

New application areas in geography, meteorology, astronomy, and geometry require spatial access and make nearest-neighbor queries. For spatial access methods, data should be clustered in disk pages by nearness in the application area space, for example, in latitude and longitude. Then the question: "find

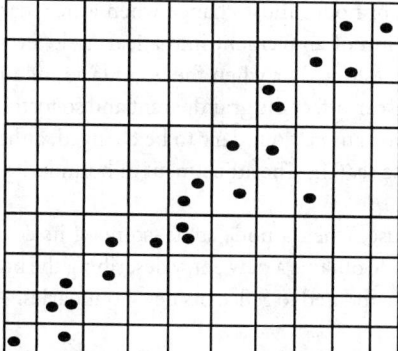

Correlated Data: A disk page is reserved for each cell in the grid. Many disk pages are empty.

The Grid File: A grid index where each cell in the grid has the address of a disk page containing some data. Many of the cells contain the same address. The index can become larger than the data collection.

FIGURE 44.12 A grid and a grid index.

all of the desert areas in photographs whose center is within 100 miles of the equator" would have its answer in a smaller number of disk pages than if the data were organized alphabetically, for example, by name.

However, it is a difficult problem to organize data spatially and still maintain the four properties of good access methods. For example, one way to organize space is to make a grid and assign each cell of the grid to one disk page. However, if the data are correlated as in Fig. 44.12, most of the disk pages will be empty. In this case, $O(n^k)$ disk space is needed for n records in k-dimensional space. Using a grid as an index has similar problems; only the constant in the asymptotic expression is changed. A grid index is also pictured in Fig. 44.12.

A proposal was made for a grid index or *grid file* in Nievergelt et al. [1984]. Because it uses $O(n^k)$ space, in the worst case it can use too much total disk space for the index; thus, it violates property 2. Range searches can touch very many pages of the index just to find one data page; thus, it violates property 3. Insertion or deletion can cause massive reorganization; thus, it violates property 4. One main problem with the index is that the index is not paginated (no specification of which disk pages correspond to which parts of the grid is made). Thus, a search over a part of it, which may even be small, can touch many disk pages of the index.

Tree indices have had more success. In particular, the R-tree and Z-ordering have been used commercially, especially in geographic information systems.

R-Tree and R*-Tree

The R-tree [Guttman 1984] organizes the data in disk pages (nodes of the R-tree) corresponding to a brick or a rectangle in space. It was originally suggested for use with spatial objects. Each object is represented by the coordinates of its smallest enclosing rectangle (or brick) with sides parallel to the coordinate axes. Thus, any two-dimensional object is represented by four numbers: its lowest x-coordinate, its highest x-coordinate, its lowest y-coordinate, and its highest y-coordinate. Then when a number of such objects are collected, their smallest enclosing brick becomes the boundary of the disk page (a leaf node) containing the records (or in the case of a secondary index, pointers to the records).

At each level, the boundaries of the spaces corresponding to nodes of the R-tree can overlap. Thus, search for objects involves backtracking.

When a new item is inserted, at each level of the tree the node where the insertion would cause the least increase in the corresponding brick is chosen. Often, the new item can be inserted in a data page without increasing the area at all.

But sometimes the area, and hence the boundaries, of nodes must change when a new data item is inserted. This is an example of a case where the insertion of an element into a leaf node, even though there is room and no splits occur, causes updates of ancestors. For when the boundaries of a leaf node change, its entry in its parent must be updated. This also can affect the grandparent and so forth if further enclosing boundaries are changed. A deletion could also cause a boundary to be changed, although this could be ignored at the parent level without causing false search. The adjustment of boundaries can thus violate property 4, local insertion and deletion.

Node splits are similar to those of the B^+-tree because, when a node splits, some of its contents are moved to a newly allocated disk page. A parent index node obtains a new entry describing the boundaries of the new child and the entry referring to the old child is updated to reflect its new boundaries. An R-tree split is illustrated in Fig. 44.13.

Since an R-tree node can be split in many different ways, an algorithm is presented which first chooses two *seed elements* in the node to be split. One seed element is to remain in the old node, and the other seed is to be placed in the new node. After that, each remaining element is tested to see in which of the two nodes its insertion would cause the greatest increase in area (or volume). These elements are placed in the node which causes the least increase in area, until one of the nodes has reached a threshold in the number of elements. Then all remaining elements are placed in the other node.

Sparse nodes are consolidated by deallocating the sparse node, removing its reference from its parent, adjusting upward boundaries, and reinserting the orphaned elements from the sparse node (each of which requires a search, an insertion, and a possible boundary adjustment). Thus, property 4 is violated by sparse node consolidation.

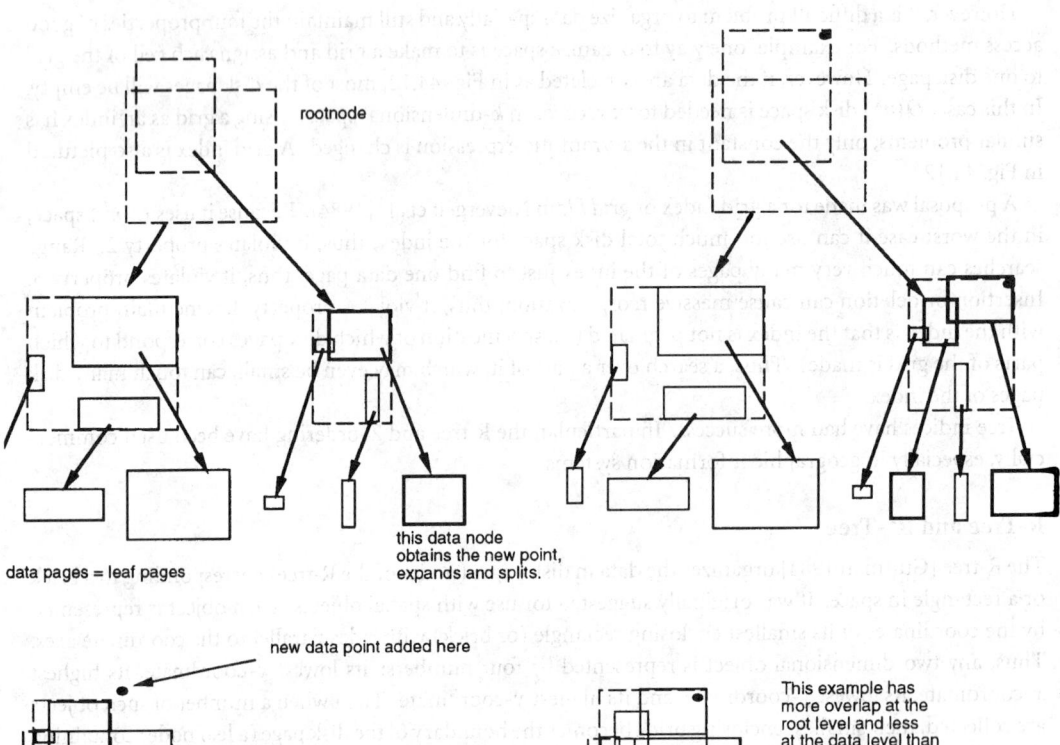

rootnode

this data node
obtains the new point,
expands and splits.

data pages = leaf pages

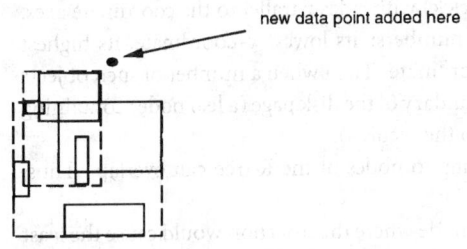

new data point added here

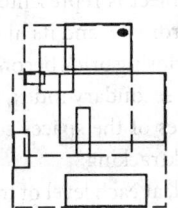

This example has
more overlap at the
root level and less
at the data level than
before the new point
was added.

FIGURE 44.13   An R-tree split.

Another drawback of the R-tree is its sensitivity to dimension. This is not as severe as the problems of grid files; the space usage is $O(nk)$ since each of the $2k$ coordinates of each child must be stored in its parent. This affects the fan-out.

Keeping boundaries does have the good property of stopping searches in areas which have no points. For example, with the information in the root of the R-tree, one knows the outer boundaries of the entire data collection. A search for a point not in this space can be stopped without accessing any lower level pages. Since the root is likely to be in main memory, this means negative searches (searches that retrieve no data) can in some cases be very efficient. The tradeoff between keeping boundaries and searching in space where there are no points is a general one, not confined to the R-tree. Every search method must make a decision whether to keep boundaries of existing data in index terms, making the index larger but making negative searches more efficient, or having a smaller index but risking accessing several pages when making searches in areas where there are no data.

The main problem with the R-tree, however, is that the boundaries of space covered by nodes at a given level of the tree overlap. This means that a search for a data item could do backtracking and thus visit extra index pages and data pages which have no relevant data. This violates property 3. Attempts to decrease the amount of overlap have generated additional suggestions.

For example, the R*-tree is a collection of such suggestions [Beckmann et al. 1990]. In the R*-tree, insertion of a new element follows the path where the least new overlap of areas would occur at each level of the tree rather than the least area increase. Splits first sort the elements by lower value, and then separately by upper value, and measure the perimeter of the resulting possible splits to choose a best *axis* for the split, and then along that axis the position of least overlap is chosen.

The R*-tree also suggests forced reinsertions of elements when an insertion is made which would otherwise cause a split. This causes property 4 to be violated more seriously than in the original R-tree because one insertion of a data item may cause a number of pages to be retrieved and written. However, the claim is that there is less overlap as a result and that therefore searches require less backtracking.

hB$^\Pi$-Trees

One spatial access method which has local insertion and deletion algorithms is the hB$^\Pi$-tree [Evangelidis et al. 1996]. This is a combination of the hB-tree [Lomet and Salzberg 1990a] and a concurrency method for a generalized tree called a Π-tree [Lomet and Salzberg 1992]. The method is described for point data. (Like any other point-data method, it can be used for spatial data by representing spatial elements by the coordinates of their smallest enclosing box.)

The basic idea of Π-tree concurrency is (1) to do a node split as an atomic action, locking only the node which is split, and keeping a pointer to the new sibling in the old sibling; (2) to do posting of information to the parent as a separate atomic action, locking only the parent and briefly the child to make sure the action is still necessary; and (3) to do node consolidation as a separate atomic action locking only the parent and the two children being consolidated. Only possibly the node split is part of a database transaction. The other actions are done asynchronously. Here we will concentrate on the access method aspects and not the concurrency, which is treated in another chapter of this book.

A k-d-tree [Bentley 1979] is used in index nodes to describe the spaces of its children and simultaneously to describe the spaces of the siblings it is pointing to. In data nodes, a k-d-tree also describes the space now covered by a sibling the data node points to. An illustration of a data node in an hB$^\Pi$-tree is given in Fig. 44.14(a).

Search is as in the B$^+$-tree. There is exactly one path from the root to the leaves for any point in the space. There is no backtracking. Insertion of an element never writes on more than one page if there is room for the element in the page, unlike the R-tree.

The splitting discipline of the hB$^\Pi$-tree is similar to the B$^+$-tree or the R-tree. When a page becomes full, some of the contents is moved to a new sibling and an index term will be posted (in this case asynchronously) in the parent. Unlike the R-tree, the spaces of the two siblings never overlap. At each level of the tree, the spaces corresponding to the pages partition the whole space.

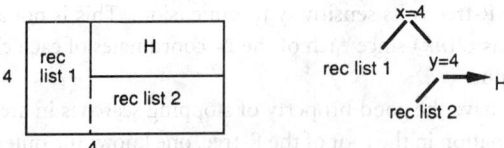

a) data node organization in hB–Pi tree. A local kd–tree indicates
the spaces where there are record lists and the addresses of
siblings that have been extracted. Here an upper corner of the
space has been split off and is in another node with disk address "H."

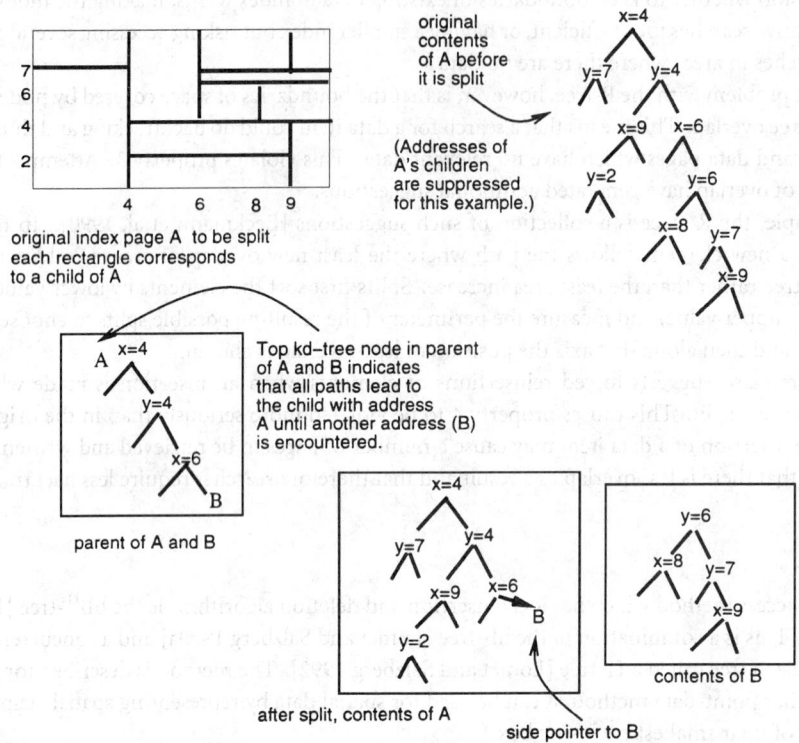

b) hB–Pi tree index split with full path posted

FIGURE 44.14   An hB$^\Pi$-tree split.

Splitting data nodes is the same in the hB$^\Pi$-tree as in the hB-tree. Usually one axis can be chosen and between one-third and two-thirds of the data are on one side of the hyperplane described by some one value on that axis. In Lomet and Salzberg [1990a] it is proved that, in any case, a corner can be chosen that contains between one-third and two-thirds of the data. The information posted to the parent consists of at most k k-d-tree nodes and usually only one. The hB$^\Pi$-tree is not as sensitive to increases in dimension as the R-tree (and transitively, much less sensitive to increases in dimension than the grid file).

Several variations on splitting index nodes and posting are suggested in Evangelidis et al. [1997]. We shall briefly describe *full paths* and *split anywhere*. Split anywhere is the split policy of the hB-tree [Lomet and Salzberg 1990a]. Index nodes contain k-d-trees. Index nodes contain no data. To split an index node, one follows the path from the root to a subtree having more than two-thirds of the total contents of the node. The split is made there. The subtree is moved to a newly allocated disk page. All of the k-d-tree nodes on the path from the root of the original k-d-tree to the extracted subtree are copied to the parent(s). (This is the full path.) A split of an hB$^\Pi$-tree index node with full path posting is illustrated in

Fig. 44.14(b). In this figure, we begin with a full index node A containing a k-d-tree. A subtree is extracted and placed in a newly allocated sibling node B.

The choice of not keeping boundaries of existing data elements, but instead partitioning the space, means that searches for ranges outside the areas where data exist can touch several pages containing no data. This seems to be a general tradeoff: if boundaries of existing data are kept, as in the R-tree, the total space usage of the access structure grows at least linearly in the number of dimensions of the space and the insertion and deletion algorithms become nonlocal. But some searches (especially those that retrieve no data) will be more efficient.

Z-Ordering

A tried and true method for spatial access is bit interleaving or Z-ordering. Here the bits of each coordinate of a data point are interleaved to form one number. Then the record corresponding to that point is inserted into a B$^+$-tree according to that number. Z-ordering is illustrated in Fig. 44.15. One reference for Z-ordering is Orenstein and Merrett [1984].

The Z-ordering forms a path in space. Points are entered into the B$^+$-tree in the order of that path. Leaves of the B$^+$-tree correspond to connected segments of the path. Thus, clustering is good, although some points which are far apart in space may be clustered together when the path jumps to another area,

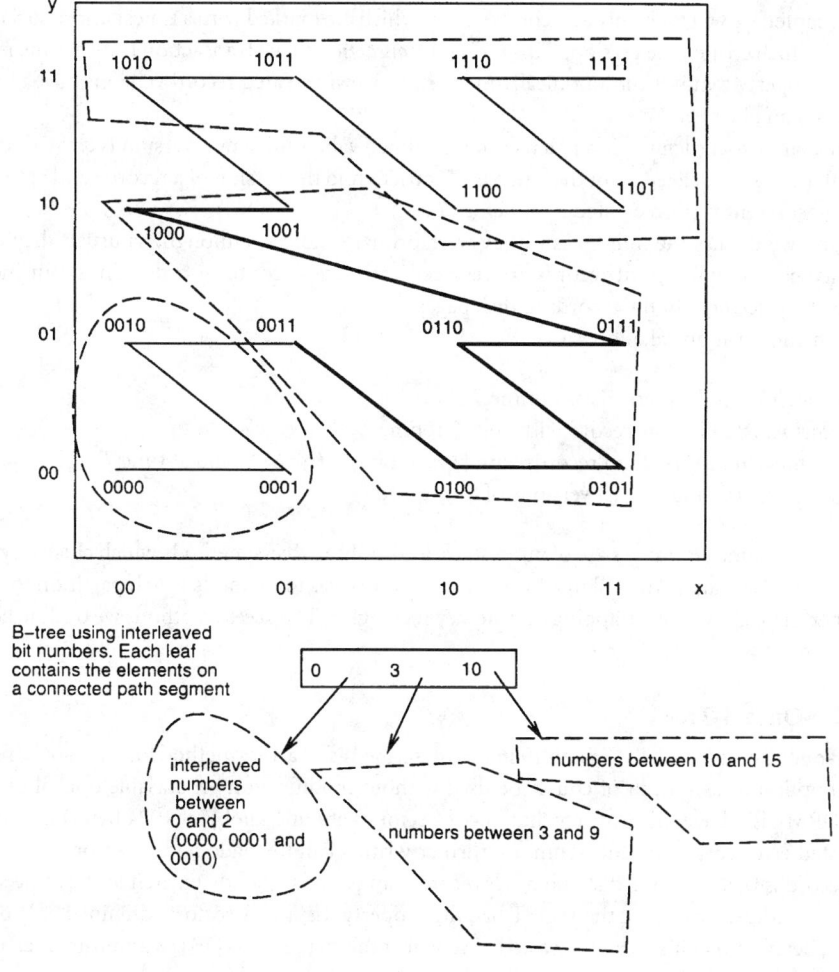

FIGURE 44.15   Z-ordering.

and some close-by points in space are far apart on the path. The disk space usage is also good since each point is stored once and a standard B$^+$-tree is used. Insertion and deletion and exact-match search are also efficient. In addition, since a well-known method (the B$^+$-tree) is used, existing software in file systems and databases can be adapted to this method.

There are two problems. One is that the bits chosen for bit interleaving can have patterns which inhibit good clustering. For example, the first 13 bits of the first attribute could be identical in 95% of the records. Then if two attributes are used, clustering is good only for the second attribute.

The other problem is the range query. Ranges correspond to many disjoint segments of the path and many different B$^+$-tree leaves. How can these segments be determined? In Orenstein and Merrett [1984], a recursive algorithm finds all segments completely contained in the search area and then obtains all B-tree leaves intersecting those segments. (This may require visiting a number of index pages and data pages which may in fact have no points in the search area, but this is true of all spatial access methods as pages whose space intersects the border of the query space may or may not contain answers to the query.)

Temporal Methods

To do *data mining*, for example, to discover trends in buying patterns, or for many legal and financial applications, all old versions of records are maintained. This is in contrast to the usual policy in databases of replacing records with their new versions, or *updating in place*. In order to maintain a database of current and old versions of records, special access structures are necessary.

In this chapter, we will look only at record versions which are marked with a timestamp associated to the transaction which created the version. This is called *transaction time*. Transaction time has the interesting and useful property that it is monotonically increasing. Newly created record versions have more recent timestamps than older versions.

We also assume that each version of a record is assumed valid until a new version is created or until the record with that key is deleted from the database. Thus, to find the version of a record valid at time T, one finds the most recent version created at or before T.

As before, we discuss the indices as if they were primary indices, although, as usual, they can all be regarded as secondary indices if records are replaced with references to records. Thus, our indices will determine the placement of the records in disk pages.

We assume four canonical queries:

1. *Time slice:* Find all records as of time T.
2. *Exact match:* Find the record with key K at time T.
3. *Key range/time slice:* Find records with keys in range (K_1, K_2) valid at time T.
4. *Past versions:* Find all past versions of this record.

Because we assume key ranges are of interest, we look only at access methods which cluster by time and by key range in disk pages. We will in addition restrict ourselves to methods which partition the time-key space rather than allowing overlapping of time-key rectangles. The access methods we outline here are all variations on the write-once B-tree.

The Write-Once B-Tree

The write-once B-tree (WOBT) [Easton 1986] was described as an access method for write-once read-many (WORM) optical disks. It can of course be used without modification on erasable optical or magnetic disks. WORM disks have the property that once a sector of about 1 kilobyte (1024 bytes) is written, with its associated error correcting checksum, no further writing can be made on the sector. In particular, if say one record is written to the disk, no other records can be subsequently written in the same sector.

(The original specification of the WOBT had the property that many sectors contained only one record each, and therefore much space was wasted. We will speak here of a WOBT without this restriction, to illustrate the basic idea.)

Basically, data pages or nodes in a WOBT span a key-time rectangle. Such a key-time rectangle on the data level is pictured in Fig. 44.16. The set of all of the pages on a given level of the WOBT partition the key-time space. The WOBT has good clustering for the canonical queries. It satisfies property 1.

Search for a record with key K and time T follows exactly one path down the tree to the one data page where K is in the key range and T is in the time range. Only one page on each level is visited. The WOBT satisfies property 3 of good access methods.

The WOBT splits as a B$^+$-tree. Assume a new record version is to be inserted in a data page including its key and the current time. All

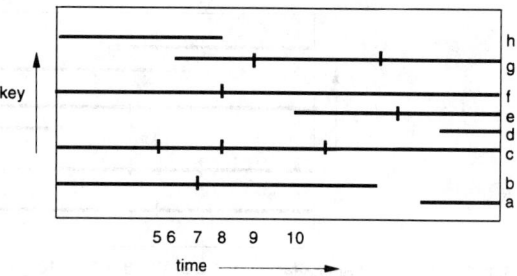

FIGURE 44.16 A time-key rectangle covered by a data page. Line segments represent distinct record versions. At time instant 5, a new version with key c is created. At time instant 6, a record with key g is inserted. (*Source:* Salzberg, B. 1994. On indexing spatial and temporal data. *Information Systems* 19(6):447–465. Elsevier Science Ltd. With permission.)

of the current record versions in that page are then copied to one or two newly allocated pages. Only one page is used if the number of current records in the page falls below a threshold. Two pages are used if the number of record versions in the page which are still valid is large. If two pages are used, they are distinguished by key range: all of the current records with key value over or equal to some value K_0 are placed in one page and those with key value less than K_0 are placed in the other page. WOBT data node splits are illustrated in Fig. 44.17(a) (by time) and Fig. 44.18(a) (by time, then by key).

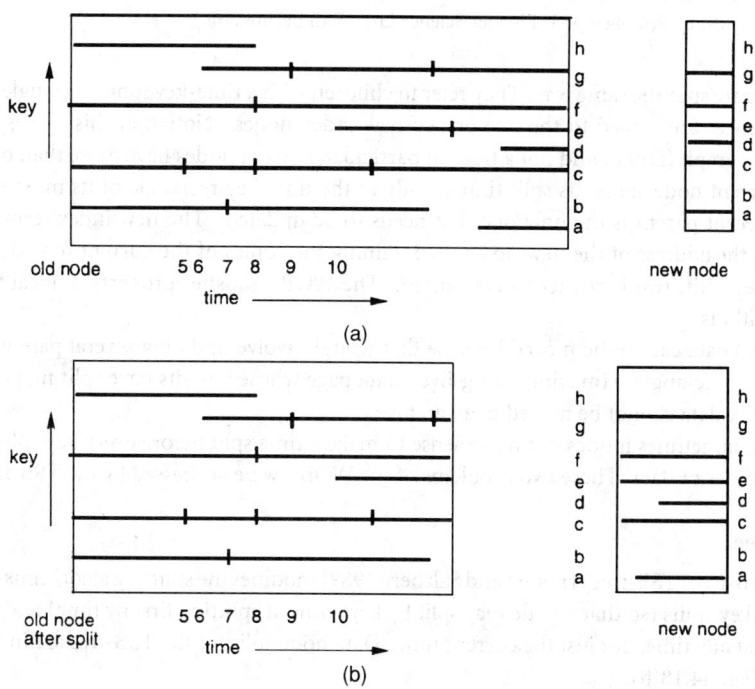

FIGURE 44.17 WOBT and TSB-tree time splits. (a) The WOBT splits at current time, copying current records into a new node. (b) The TSB tree can choose other times to split. (*Source:* Salzberg, B. 1994. On indexing spatial and temporal data. *Information Systems* 19(6):447–465. Elsevier Science Ltd. With permission.)

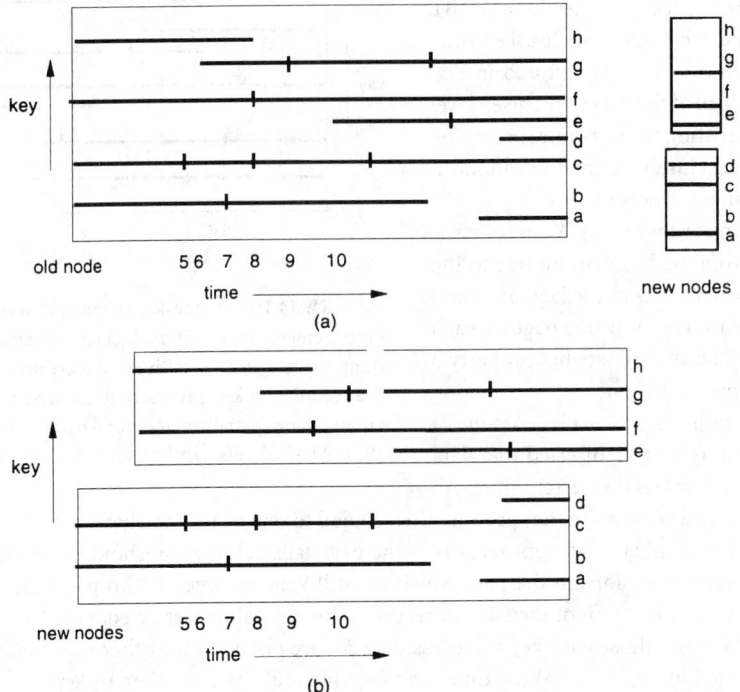

FIGURE 44.18 WOBT and TSB-tree key splits. (a) The WOBT splits data nodes
first by time and then sometimes also by key. (b) The TSB-tree can split by key alone.
(*Source:* Salzberg, B. 1994. On indexing spatial and temporal data. *Information
Systems* 19(6):447–465. Elsevier Science Ltd. With permission.)

Index nodes are split the same way. They refer to children with a time-key range rectangle. The current
children references are copied to the one or two new index nodes. Note that this makes the WOBT a
directed acyclic graph (DAG), and not a tree. In particular, current nodes have more than one parent.

When a current node splits, its split time is only in the time-key rectangle of its most recent parent.
So the most recent parent is the only one that needs to be updated. The new index term indicates the
split time and the address of the new node(s) containing the copies of the current records. In case of a
(time-and-) key split, two index terms are posted. The WOBT satisfies property 4, local insertion and
deletion algorithms.

However, old data cannot be moved because that would involve updating several parents referring to
the old time-key rectangle. (Imagine a long-lived data page whose parents have split many times.) This
means that older data cannot be moved to an archive.

In addition, sometimes it does not make sense to make a time split before every key split. This makes
unnecessary copies of data. These two problems of the WOBT were addressed by the TSB-tree.

The TSB-Tree

The time-split B-tree (TSB-tree) [Lomet and Salzberg 1989] modifies the splitting algorithms of the WOBT.
It allows pure key splits (so that a node may split by key without splitting first by time) and it allows data
nodes to split at any time, not just the current time. Data node splits of the TSB-tree are indicated in Fig.
44.17(b) and Fig. 44.18(b).

In the TSB-tree, index nodes may split only by a time before or at the earliest begin time of any current
child. This way, current children have only one parent and old data can be migrated to an archive. Key
splits for index nodes must be by a key boundary of a current child. This is also needed to assure that
current children have only one parent. TSB-tree index node splits are illustrated in Fig. 44.19.

The total space utilization of the TSB-tree is significantly smaller than that of the WOBT. This is important since all past versions of records are kept and since copies of records are made. An analysis of the space usage is in [Lomet and Salzberg 1990b]. Both the TSB-tree and the WOBT use $O(N)$ space, where N is the number of distinct versions. The constant is smaller for the TSB-tree. However, this sometimes comes at a price. Because pure key splits are allowed, it is possible to split a data node so that in some of the earlier time instants not many record versions are valid in one or both of the new nodes. This is illustrated in Fig. 44.18b.

Both TSB-tree and the WOBT have no node-consolidation algorithm. This problem is solved by two other extensions of the WOBT.

Other Extensions of the Write Once B-Tree

The persistent B-tree [Lanka and Mays 1991] does time splits by current time only, as does the WOBT. It specifies a node consolidation algorithm by splitting a sparse current node and one of its siblings at current time and copying the combined current record versions to a new node. If there are too many of them, two new nodes can be used instead. It makes the mistake we discussed earlier with the B$^+$-tree of having a 50% threshold for node consolidation, thereby allowing thrashing.

It also has two features which are not useful in general: (1) extra nodes on the path from the root to the leaves and (2) an unbalanced overall structure caused by having a *superroot* called *root**. Having a superroot usually increases the average height of the tree. (The WOBT, but not the TSB-tree, also has a superroot.)

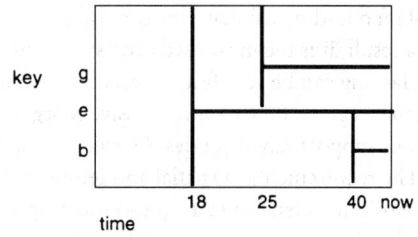

a) original index page: rectangles represent key–time space of children

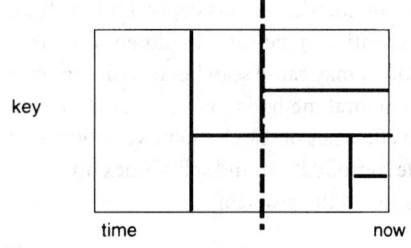

b) split by begin time of oldest current child

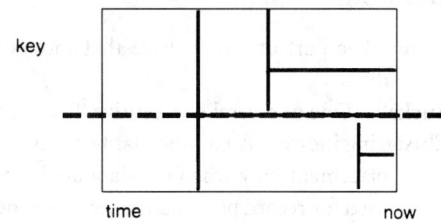

c) split by a current key boundary

FIGURE 44.19 TSB-tree index-node splits. (*Source:* Salzberg, B. 1994. On indexing spatial and temporal data. *Information Systems* 19(6):447–465. Elsevier Science Ltd. With permission.)

The multiversion B-tree [Becker et al. 1993] eliminates the extra nodes in the search path and uses a smaller than 50% threshold for node consolidation but does not eliminate the superroot.

Both the persistent B-tree and the multiversion B-tree always split by current time. This implies that a lower limit on the number of record versions valid at a given time in a node's time interval can be guaranteed, but old record versions cannot be migrated to an archive.

A good compromise of all of the previous methods would be to allow pure-key splitting as in the TSB-tree, but only when the resulting minimum number of valid records in the time intervals of each new node is above a certain level. Index nodes should be split by earliest begin time of current children as in the TSB-tree. (This includes the case when they are split for node consolidation.) Node consolidation should be supported as in the persistent B-tree but with a lower threshold to avoid thrashing as in the multiversion B-tree.

44.4 Research Issues and Summary

Good access methods should cluster data according to anticipated queries, use only a reasonable amount of total disk space, be efficient in search, and have local insertion and deletion algorithms. We have seen that for one-dimensional data, usually used in business applications, the B^+-tree has all of these properties. As a result, it is the most used access method today.

Hashing can be very fast, especially for a large and nearly static collection of data. However, as hashed databases grow, they require massive reorganization to regain their good performance. In addition, they do not support range queries. Hashing is the second most used access method today.

The requirements of spatial and temporal indexing, on the other hand, lead to subtle problems. Grid-style solutions tend to take up too much space, especially in large dimensions. R-tree-like solutions with overlapping can have poor search performance due to backtracking. R-trees are also somewhat sensitive to larger dimensions as all boundary coordinates of each child are stored in their parent. Access methods based on interleaved bits depend on the bit patterns of the data. Methods such as the hB^Π-tree, which is not sensitive to increases in dimension but where index terms do not keep boundaries of existing data in children, may cause searches to visit too many data nodes without data in the query area.

Temporal methods (using transaction time) trade off total space usage (numbers of copies of records) with efficiency of search. Some variation of the WOBT, which allows pure key splits some of the time, does node consolidation, and splits index nodes by earliest begin time of current children, is a good compromise solution to the problem.

Defining Terms

Arm: The part of a disk drive that moves back and forth toward and away from the center of the disks.

Bucket: One or several consecutive disk pages corresponding to one value of a hashing function.

Clustering index: A commercial term used often to denote a secondary index which is used for data placement only when the data are loaded in the database. After initial loading, the index can be used for record placement only when there is still space left in the correct page. Records never move from the page where they are originally placed. Clustering indices tend not to be clustering after a number of insertions in the database. This term is avoided in this chapter for this reason.

Cylinder: The set of tracks on a collection of disks on a disk drive which are the same distance from the center of the disks. (One track on each side of each disk.) Reading information which is stored on the same cylinder of a disk drive is fast because the disk arm does not have to move.

Data page: A disk page in an access method which contains data records.

Dense index: A secondary index.

Disk access: The act of transferring information from a magnetic disk to the main memory of a computer or the reverse. This involves the mechanical movement of the disk arm so that it is placed at the correct cylinder and the rotation of the disk so that the correct disk page falls under a read/write head on the disk arm.

Disk page: The smallest unit of transfer from (or to) a disk to (or from) main memory. In most systems today, this is 4 kilobytes, or 4096 bytes. It is expected that the size of a disk page will grow in the future so that many systems may begin to have 8-kilobyte disk pages or even 32-kilobyte disk pages. The reason for the size increase is that main memory space and CPU speed are increasing faster than disk access speed.

Extent: The large amount of consecutive disk space assigned to a relation when it is created and subsequent such chunks of consecutive disk space assigned to the relation as it grows. Some file systems cannot assign extents and thus are unsuitable for many access methods.

Fan-out: The ratio of the size of the data page collection to the size of the index page collection. In B^+-tree N-like access methods, this is approximately the average number of children of an index node, and sometimes fan-out is used in this sense.

Hash table fill factor: The total space needed for the data divided by the space used by the primary area.

Hashing: In hashing, a function maps a database key to the location (address) of the record having that key. (A secondary hashing method maps the database key to the location containing the address of the record.)

Head: The head on a disk arm is where the bits are moved off and onto the disk (read and write). Much effort has been made to allow disk head placement to be more precise so that the density of bits on the disk (number of bits per track and number of tracks per disk of a fixed diameter) can become larger.

Index page: A disk page in an access method or indexing method which does not contain any data records.

Overflow area: That part of a hashing access method where records are placed when there is no room for them in the correct bucket in the primary area.

Primary area: That part of the disk which holds the buckets of a hashing method accessible with one disk access, using the hashing function.

Primary index: A primary index determines the physical placement of records. It does not contain references to individual records. A primary index can be converted to a secondary index by replacing all of the data records with references to data records. Many database systems have no primary indices.

Reference: A reference to an index page or to a data page is a disk address. A reference to a data record can be (1) just the address of the disk page where the record is, with the understanding that some other criteria will be used to locate the record, (2) a disk page address and a slot number within that disk page, or (3) some collection of attribute values from the record which can be used in another index to find the record.

Secondary index: An index that contains a reference to every data record. Secondary indices do not determine data record placement. Sometimes secondary indices refer to data which are placed in the database in insertion order. Sometimes secondary indices refer to data which are placed in the database according to a separate primary index based on other attributes of the record. Sometimes secondary indices refer to data which are loaded into the database originally in the same order as specified by the secondary index but not thereafter. Many database management systems have only secondary indices.

Separator: A prefix of a (possibly former) database key which is long enough to differentiate one page on a lower level of a B$^+$-tree from the next. These are used instead of database keys in the index pages of a B$^+$-tree.

Sparse index: A primary index.

Thrashing: When repeated deletions and insertions of records cause the same data page to be repeatedly split and then consolidated, the access method is said to be thrashing. Commercial database systems prevent thrashing by setting the threshold for B$^+$-tree node consolidation at 0% full; only empty nodes are considered sparse and are consolidated with their siblings.

Track: A circle on one side of one disk with its center in the middle of the disk. Tracks tend to hold on the order of 100,000 bytes. The set of tracks at the same distance from the center, but on different disks or on different sides of the disk, form a cylinder on a given disk drive.

References

Bayer, R. and McCreight, E. 1972. Organization and maintenance of large ordered indices. *Acta Informatica* 1(3):173–189.

Bayer, R. and Unterauer, K. 1997. Prefix B-trees. *ACM Trans. Database Syst.* 2(1):11–26.

Becker, B., Gschwind, S., Ohler, T., Seeger, B., and Widmayer, P. 1993. On optimal multiversion access structures. In *Proc. Symp. Large Spatial Databases.* Lecture notes in computer science 692, pp. 123–141. Springer–Verlag, Berlin.

Beckmann, N., Kriegel, H.-P., Schneider, R., and Seeger, B. 1990. The R*-tree: an efficient and robust access method for points and rectangles, pp. 322–331. In *Proc. ACM SIGMOD.*

Bentley, J. L. 1979. Multidimensional binary search trees in database applications. *IEEE Trans. Software Eng.* 5(4):333–340.

Comer, D. 1979. The ubiquitous B-tree. *Comput. Surv.* 11(4):121–137.

Easton, M. C. 1986. Key-sequence data sets on indelible storage. *IBM J. Res. Dev.* 30(3):230–241.

Evangelidis, G., Lomet, D., and Salzberg, B. 1997. The hB$^\Pi$-tree: a multiattribute index supporting concurrency, recovery and node consolidation. *J. Very Large Databases* 6(1).

Fagin, R., Nievergelt, J., Pippenger, N., and Strong, H. R. 1979. Extendible hashing—a fast access method for dynamic files. *Trans. Database Syst.* 4(3):315–344.

Guttman, A. 1984. R-trees: a dynamic index structure for spatial searching, pp. 47–57. In *Proc. ACM SIGMOD*.

Johnson, T. and Shasha, D. 1989. B-trees with inserts and deletes: why free-at-empty is better than Merge-at-half. *J. Comput. Syst. Sci.* 47(1):45–76.

Knuth, D. E. 1968. *The Art of Computer Programming*. Addison–Wesley, Reading, MA.

Lanka, S. and Mays, E. 1991. Fully persistent B$^+$-trees, pp. 426–435. In *Proc. ACM SIGMOD*.

Larson, P. 1980. Linear hashing with partial expansions, pp. 224–232. In *Proc. Very Large Database*.

Litwin, W. 1980. Linear hashing: a new tool for file and table addressing, pp. 212–223. In *Proc. Very Large Database*.

Lomet, D. 1988. A simple bounded disorder file organization with good performance. *Trans. Database Syst.* 13(4):525–551.

Lomet, D. 1991. Grow and post index trees: role, techniques and future potential. In *2nd Symp*. Large Spatial Databases (SSD91), *Advances in Spatial Databases*. Lecture notes in computer science 525, pp. 183–206. Springer–Verlag. Berlin.

Lomet, D. and Salzberg, B. 1989. Access methods for multiversion data, pp. 315–324. In *Proc. ACM SIGMOD*.

Lomet, D. and Salzberg, B. 1990a. The hB-tree: A multiattribute indexing method with good guaranteed performance. *Trans. Database Syst.* 15(4):625–658.

Lomet, D. and Salzberg, B. 1990b. The performance of a multiversion access method, pp. 353–363. In *Proc. ACM SIGMOD*.

Lomet, D. and Salzberg, B. 1992. Access method concurrency with recovery, pp. 351–360. In *Proc. ACM SIGMOD*.

Maier, D. and Salveter, S. C. 1981. Hysterical B-trees. *Inf. Process. Lett.* 12:199–202.

Mohan, C. and Levine, F. 1992. ARIES/IM: an efficient and high concurrency index management method using write-ahead logging, pp. 371–380. In *Proc. ACM SIGMOD*.

Nievergelt J., Hinterberger, H., and Sevcik, K. C. 1984. The grid file: an adaptable, symmetric, multikey file structure. *Trans. Database Syst.* 9(1):38–71.

Orenstein, J. A. and Merrett, T. 1984. A class of data structures for associative searching, pp. 181–190. In *Proc. ACM SIGMOD/SIGACT Principles Database Syst. (PODS)*.

Salzberg, B. 1988. *File Structures: An Analytic Approach*. Prentice–Hall, Englewood Cliffs, NJ.

Smith, G. 1990. Online reorganization of key-sequenced tables and files. (Description of software designed and implemented by F. Putzolu.) *Tandem Syst. Rev.* 6(2):52–59.

Yao, A. C. 1978. On random 2-3 trees. *Acta Informatica* 9:159–170.

Further Information

The best information about new access methods for new applications can be obtained by attending the annual conference of the Association for Computing Machinery Special Interest Group on Management of Data (ACM SIGMOD). This is held every year in an American city. The meeting is announced in the newsgroup dbworld. To subscribe to dbworld, send a message by email to listproc@cs.wisc.edu with the words "subscribe dbworld" and your full name. The reason to attend conferences in person is that you will hear information which is not published and you will get an impression of which articles are the best ones to read in detail.

The second best source for information about new access methods is the *Proceedings of the Annual Conference of the ACM SIGMOD*. This can be found in most university libraries. If you join ACM and join SIGMOD, it will be sent to you as part of your membership.

Proceedings of the International Conference on Very Large Data Bases is another reasonably good source for new results in access methods. The VLDB meeting is also announced on dbworld and is held yearly in different cities in the world. Recent meetings have been in Barcelona, Vancouver, Dublin, Santiago, and Zurich, for example.

The quarterly journal *ACM Transactions on Database Systems* is of high quality and publishes some access methods articles.

The textbook, *Transaction Processing: Techniques and Concepts* by Jim Gray and Andreas Reuter has chapters on file structures and access methods in a modern setting. This book was published in 1993 by Morgan Kaufman.

The author of this chapter, Betty Salzberg, has written a textbook *File Structures: An Analytic Approach.* Many topics touched upon in this chapter are elaborated with exercises and examples. This book was published in 1988 by Prentice–Hall.

45

Query Optimization

Yannis E. Ioannidis*
*University of
Wisconsin–Madison*

45.1 Introduction

Imagine yourself standing in front of an exquisite buffet filled with numerous delicacies. Your goal is to try them all out, but you need to decide in what order. What order of tastes will maximize the overall pleasure of your palate?

Although much less pleasurable and subjective, that is the type of problem that query optimizers are called to solve. Given a query, there are many plans that a database management system (DBMS) can follow to process it and produce its answer. All plans are equivalent in terms of their final output but vary in their cost, i.e., the amount of time that they need to run. What is the plan that needs the least amount of time?

Such *query optimization* is absolutely necessary in a DBMS. The cost difference between two alternatives can be enormous. For example, consider the following database schema, which will be used throughout this chapter:

    emp(name,age,sal,dno)
    dept(dno,dname,floor,budget,mgr,ano)
    acnt(ano,type,balance,bno)
    bank(bno,bname,address)

*Partially supported by the National Science Foundation under Grants IRI-9113736 and IRI-9157368 (PYI Award) and by grants from DEC, IBM, HP, AT&T, Informix, and Oracle.

TABLE 45.1

Parameter Description	Parameter Value
Number of emp pages	20,000
Number of emp tuples	100,000
Number of emp tuples with sal>100 K	10
Number of dept pages	10
Number of dept tuples	100
Indices of emp	Clustered B+-tree on emp.sal (3 levels deep)
Indices of dept	Clustered hashing on dept.dno (average bucket length of 1.2 pages)
Number of buffer pages	3
Cost of one disk page access	20 ms

Further, consider the following very simple SQL query:

> **select** name, floor
> **from** emp, dept
> **where** emp.dno=dept.dno **and** sal>100 K.

Assume the characteristics in Table 45.1 for the database contents, structure, and run-time environment:
Consider the following three different plans:

P1: Through the B+-tree find all tuples of emp that satisfy the selection on emp.sal. For each one, use the hashing index to find the corresponding dept tuples. (Nested loops, using the index on both relations.)

P2: For each dept page, scan the entire emp relation. If an emp tuple agrees on the dno attribute with a tuple on the dept page and satisfies the selection on emp.sal, then the emp–dept tuple pair appears in the result. (Page-level nested loops, using no index.)

P3: For each dept tuple, scan the entire emp relation and store all emp–dept tuple pairs. Then, scan this set of pairs and, for each one, check if it has the same values in the two dno attributes and satisfies the selection on emp.sal. (Tuple-level formation of the cross product, with subsequent scan to test the join and the selection.)

Calculating the expected I/O costs of these three plans shows the tremendous difference in efficiency that equivalent plans may have. P1 needs 0.32 s, P2 needs a bit more than an hour, and P3 needs more than a whole day. Without query optimization, a system may choose plan P2 or P3 to execute this query, with devastating results. Query optimizers, however, examine "all" alternatives, so they should have no trouble choosing P1 to process the query.

The path that a query traverses through a DBMS until its answer is generated is shown in Fig. 45.1. The system modules through which it moves have the following functionality:

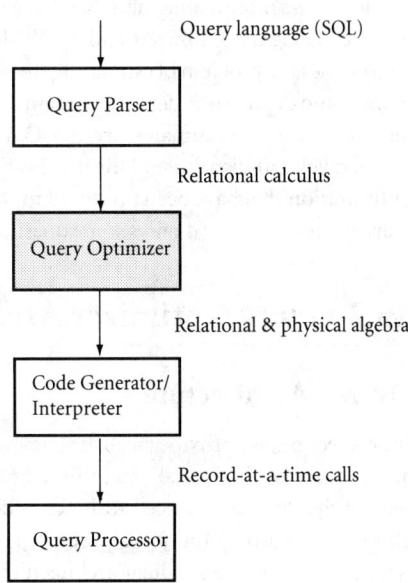

FIGURE 45.1 Query flow through a DBMS.

- The *Query Parser* checks the validity of the query and then translates it into an internal form, usually a relational calculus expression or something equivalent.
- The *Query Optimizer* examines all algebraic expressions that are equivalent to the given query and chooses the one that is estimated to be the cheapest.
- The *Code Generator* or the *Interpreter* transforms the access plan generated by the optimizer into calls to the query processor.
- The *Query Processor* actually executes the query.

Queries are posed to a DBMS by interactive users or by programs written in general-purpose programming languages (e.g., C/C++, Fortran, PL/I) that have queries embedded in them. An interactive (ad hoc) query goes through the entire path shown in Fig. 45.1. On the other hand, an embedded query goes through the first three steps only once, when the program in which it is embedded is compiled (*compile time*). The code produced by the Code Generator is stored in the database and is simply invoked and executed by the Query Processor whenever control reaches that query during the program execution (*run time*). Thus, independent of the number of times an embedded query needs to be executed, optimization is not repeated until database updates make the access plan invalid (e.g., index deletion) or highly suboptimal (e.g., extensive changes in database contents). There is no real difference between optimizing interactive or embedded queries, so we make no distinction between the two in this chapter.

The area of query optimization is very large within the database field. It has been studied in a great variety of contexts and from many different angles, giving rise to several diverse solutions in each case. The purpose of this chapter is to primarily discuss the core problems in query optimization and their solutions and only touch upon the wealth of results that exist beyond that. More specifically, we concentrate on optimizing a single *flat SQL query* with "and" as the only Boolean connective in its qualification (also known as *conjunctive query, select–project–join query*, or *nonrecursive Horn clause*) in a centralized relational DBMS, assuming that full knowledge of the run-time environment exists at compile time. Likewise, we make no attempt to provide a complete survey of the literature, in most cases providing only a few example references. More extensive surveys can be found elsewhere [Jarke and Koch 1984, Mannino et al. 1988].

The rest of the chapter is organized as follows. Section 45.2 presents a modular architecture for a query optimizer and describes the role of each module in it. Section 45.3 analyzes the choices that exist in the shapes of relational query access plans, and the restrictions usually imposed by current optimizers to make the whole process more manageable. Section 45.4 focuses on the dynamic programming search strategy used by commercial query optimizers and briefly describes alternative strategies that have been proposed. Section 45.5 defines the problem of estimating the sizes of query results and/or the frequency distributions of values in them and describes in detail histograms, which represent the statistical information typically used by systems to derive such estimates. Section 45.6 discusses query optimization in noncentralized environments, i.e., parallel and distributed DBMSs. Section 45.7 briefly touches upon several advanced types of query optimization that have been proposed to solve some hard problems in the area. Finally, section 45.8 summarizes the chapter and raises some questions related to query optimization that still have no good answer.

45.2 Query Optimizer Architecture

Overall Architecture

In this section, we provide an abstraction of the query optimization process in a DBMS. Given a database and a query on it, several execution plans exist that can be employed to answer the query. In principle, all the alternatives need to be considered so that the one with the best estimated performance is chosen. An abstraction of the process of generating and testing these alternatives is shown in Fig. 45.2, which is essentially a modular architecture of a query optimizer. Although one could build an optimizer based on this architecture, in real systems the modules shown do not always have boundaries so clear-cut as in Fig. 45.2. Based on Fig. 45.2, the entire query optimization process can be seen as having two

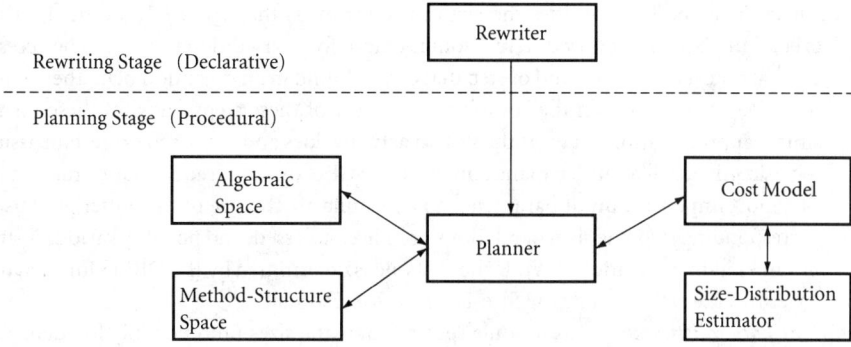

FIGURE 45.2 Query optimizer architecture.

stages: *rewriting* and *planning*. There is only one module in the first stage, the *Rewriter*, whereas all other modules are in the second stage. The functionality of each of the modules in Fig. 45.2 is analyzed below.

Module Functionality

Rewriter: This module applies transformations to a given query and produces equivalent queries that are hopefully more efficient, e.g., replacement of views with their definition, flattening out of nested queries, etc. The transformations performed by the Rewriter depend only on the declarative, i.e., static, characteristics of queries and do not take into account the actual query costs for the specific DBMS and database concerned. If the rewriting is known or assumed to always be beneficial, the original query is discarded; otherwise, it is sent to the next stage as well. By the nature of the rewriting transformations, this stage operates at the *declarative* level.

Planner: This is the main module of the ordering stage. It examines all possible execution plans for each query produced in the previous stage and selects the overall cheapest one to be used to generate the answer of the original query. It employs a *search strategy*, which examines the space of execution plans in a particular fashion. This space is determined by two other modules of the optimizer, the *Algebraic Space* and the *Method–Structure Space*. For the most part, these two modules and the search strategy determine the cost, i.e., running time, of the optimizer itself, which should be as low as possible. The execution plans examined by the Planner are compared based on estimates of their cost so that the cheapest may be chosen. These costs are derived by the last two modules of the optimizer, the *Cost Model* and the *Size-Distribution Estimator*.

Algebraic Space: This module determines the action execution orders that are to be considered by the Planner for each query sent to it. All such series of actions produce the same query answer but usually differ in performance. They are usually represented in relational algebra as formulas or in tree form. Because of the algorithmic nature of the objects generated by this module and sent to the Planner, the overall planning stage is characterized as operating at the *procedural* level.

Method–Structure Space: This module determines the implementation choices that exist for the execution of each ordered series of actions specified by the Algebraic Space. This choice is related to the available join methods for each join (e.g., nested loops, merge scan, and hash join), if supporting data structures are built on the fly, if/when duplicates are eliminated, and other implementation characteristics of this sort, which are predetermined by the DBMS implementation. This choice is also related to the available indices for accessing each relation, which is determined by the physical schema of each database stored in its catalogs. Given an algebraic formula or tree from the Algebraic Space, this module produces all corresponding complete execution plans, which specify the implementation of each algebraic operator and the use of any indices.

Cost Model: This module specifies the arithmetic formulas that are used to estimate the cost of execution plans. For every different join method, for every different index type access, and in general for every distinct kind of step that can be found in an execution plan, there is a formula that gives its cost. Given the complexity of many of these steps, most of these formulas are simple approximations of what the system actually does and are based on certain assumptions regarding issues like buffer management, disk–CPU overlap, sequential vs random I/O, etc. The most important input parameters to a formula are the size of the buffer pool used by the corresponding step, the sizes of relations or indices accessed, and possibly various distributions of values in these relations. While the first one is determined by the DBMS for each query, the other two are estimated by the Size-Distribution Estimator.

Size-Distribution Estimator: This module specifies how the sizes (and possibly frequency distributions of attribute values) of database relations and indices as well as (sub)query results are estimated. As mentioned above, these estimates are needed by the Cost Model. The specific estimation approach adopted in this module also determines the form of statistics that need to be maintained in the catalogs of each database, if any.

Description Focus

Of the six modules of Fig. 45.2, three are not discussed in any detail in this chapter: the Rewriter, the Method–Structure Space, and the Cost Model. The Rewriter is a module that exists in some commercial DBMSs (e.g., DB2-Client/Server and Illustra), although not in all of them. Most of the transformations normally performed by this module are considered an advanced form of query optimization and not part of the core (planning) process. The Method–Structure Space specifies alternatives regarding join methods, indices, etc., which are based on decisions made outside the development of the query optimizer and do not really affect much of the rest of it. For the Cost Model, for each alternative join method, index access, etc., offered by the Method–Structure Space, either there is a standard straightforward formula that people have devised by simple accounting of the corresponding actions (e.g., the formula for the tuple-level nested-loop join) or there are numerous variations of formulas that people have proposed and used to approximate these actions (e.g., formulas for finding the tuples in a relation having a random value in an attribute). In either case, the derivation of these formulas is not considered an intrinsic part of the query optimization field. For these reasons, we do not discuss these three modules any further until section 45.7, where some Rewriter transformations are described. The following three sections provide a detailed description of the Algebraic Space, the Planner, and the Size-Distribution Estimator modules, respectively.

45.3 Algebraic Space

As mentioned above, a flat SQL query corresponds to a select–project–join query in relational algebra. Typically, such an algebraic query is represented by a *query tree* whose leaves are database relations and nonleaf nodes are algebraic operators like selections (denoted by σ), projections (denoted by π), and joins[1] (denoted by $\bowtie$). An intermediate node indicates the application of the corresponding operator on the relations generated by its children, the result of which is then sent further up. Thus, the edges of a tree represent data flow from bottom to top, i.e., from the leaves, which correspond to data in the database, to the root, which is the final operator producing the query answer. Figure 45.3 gives three examples of query trees for the query

> **select** name, floor
> **from** emp, dept
> **where** emp.dno=dept.dno **and** sal>100 K

---

[1] For simplicity, we think of the cross-product operator as a special case of a join with no join qualification.

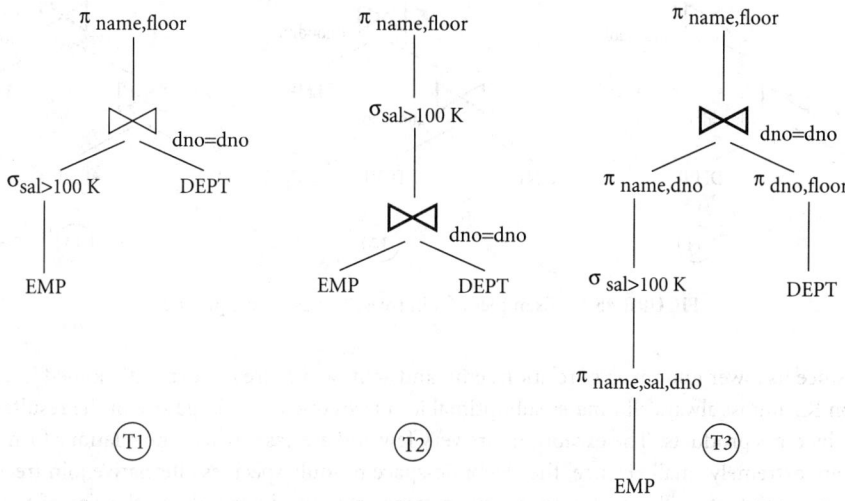

FIGURE 45.3 Examples of general query trees.

For a complicated query, the number of all query trees may be enormous. To reduce the size of the space that the search strategy has to explore, DBMSs usually restrict the space in several ways. The first typical restriction deals with selections and projections:

> *R1*: Selections and projections are processed on the fly and almost never generate intermediate relations. Selections are processed as relations are accessed for the first time. Projections are processed as the results of other operators are generated.

For example, plan P1 of section 45.1 satisfies restriction R1: the index scan of emp finds emp tuples that satisfy the selection on emp.sal on the fly and attempts to join only those; furthermore, the projection on the result attributes occurs as the join tuples are generated. For queries with no join, R1 is moot. For queries with joins, however, it implies that all operations are dealt with as part of join execution. Restriction R1 eliminates only suboptimal query trees, since separate processing of selections and projections incurs additional costs. Hence, the Algebraic Space module specifies alternative query trees with join operators only, selections and projections being implicit.

Given a set of relations to be combined in a query, the set of all alternative join trees is determined by two algebraic properties of join: commutativity ($R_1 \bowtie R_2 \equiv R_2 \bowtie R_1$) and associativity [$(R_1 \bowtie R_2) \bowtie R_3 \equiv R_1 \bowtie (R_2 \bowtie R_3)$]. The first determines which relation will be inner and which outer in the join execution. The second determines the order in which joins will be executed. Even with the R1 restriction, the alternative join trees that are generated by commutativity and associativity are very large, $\Omega(N!)$ for N relations. Thus, DBMSs usually further restrict the space that must be explored. In particular, the second typical restriction deals with cross products.

> *R2*: Cross products are never formed, unless the query itself asks for them. Relations are combined always through joins in the query.

For example, consider the following query:

> **select** name, floor, balance
> **from** emp, dept, acnt
> **where** emp.dno=dept.dno **and** dept.ano=acnt.ano

Figure 45.4 shows the three possible join trees (modulo join commutativity) that can be used to combine the emp, dept, and acnt relations to answer the query. Of the three trees in the figure, tree T3 has a cross

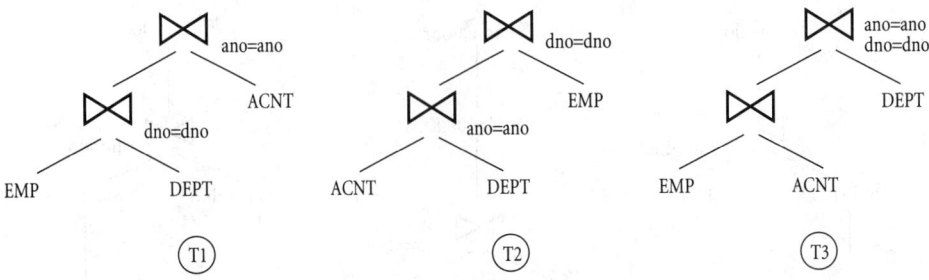

FIGURE 45.4   Examples of join trees; T3 has a cross product.

product, since its lower join involves relations emp and acnt, which are not explicitly joined in the query. Restriction R2 almost always eliminates suboptimal join trees due to the large size of the results typically generated by cross products. The exceptions are very few and are cases where the relations forming cross products are extremely small. Hence, the algebraic-space module specifies alternative join trees that involve no cross product.   The exclusion of unnecessary cross products reduces the size of the space to be explored, but that still remains very large. Although some systems restrict the space no further (e.g., Ingres and DB2-Client/Server), others require an even smaller space (e.g., DB2/MVS). In particular, the third typical restriction deals with the shape of join trees:

> *R3*: The inner operand of each join is a database relation, never an intermediate result.

For example, consider the following query:

> **select** name, floor, balance, address
> **from** emp, dept, acnt, bank
> **where** emp.dno=dept.dno **and** dept.ano=acnt.ano **and** acnt.bno=bank.bno

Figure 45.5 shows three possible cross-product-free join trees that can be used to combine the emp, dept, acnt, and bank relations to answer the query. Tree T1 satisfies restriction R3, whereas trees T2 and T3 do not, since they have at least one join with an intermediate result as the inner relation. Because of their shape (Fig. 45.5), join trees that satisfy restriction R3, e.g., tree T1, are called *left-deep*. Trees that have their

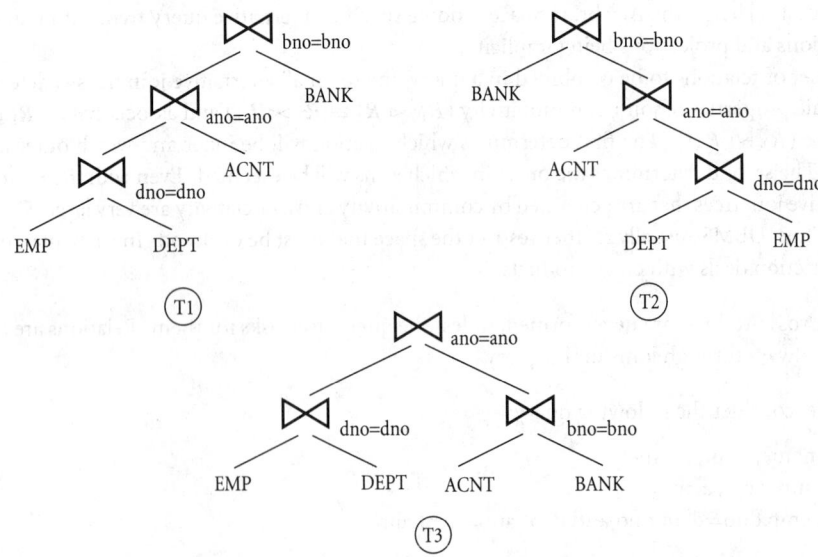

FIGURE 45.5   Examples of left-deep (T1), right-deep (T2), and bushy (T3) join trees.

outer relation always being a database relation, e.g., tree T2, are called *right-deep*. Trees with at least one join between two intermediate results, e.g., tree T3, are called *bushy*. Restriction R3 is of a more heuristic nature than R1 and R2 and may well eliminate the optimal plan in some cases. It has been claimed that most often the optimal left-deep tree is not much more expensive than the optimal tree overall. The typical arguments used are two:

- Having original database relations as inners increases the use of any preexisting indices.
- Having intermediate relations as outers allows sequences of nested-loops joins to be executed in a pipelined fashion.[2]

Both index usage and pipelining reduce the cost of join trees. Moreover, restriction R3 significantly reduces the number of alternative join trees to $O(2^N)$ for many queries with N relations. Hence, the Algebraic Space module of the typical query optimizer specifies only join trees that are left-deep.

In summary, typical query optimizers make restrictions R1, R2, and R3 to reduce the size of the space they explore. Hence, unless otherwise noted, our descriptions follow these restrictions as well.

45.4 Planner

The role of the Planner is to explore the set of alternative execution plans, as specified by the Algebraic Space and the Method–Structure Space, and find the cheapest one, as determined by the Cost Model and the Size-Distribution Estimator. The following three subsections deal with different types of search strategies that the Planner may employ for its exploration. The first one focuses on the most important strategy, dynamic programming, which is the one used by essentially all commercial systems. The second one discusses a promising approach based on randomized algorithms, and the third one talks about other search strategies that have been proposed.

Dynamic Programming Algorithms

Dynamic programming was first proposed as a query optimization search strategy in the context of System R [Astrahan et al. 1976] by Selinger et al. [1979]. Commercial systems have since used it in various forms and with various extensions. We present this algorithm pretty much in its original form [Selinger et al. 1979], only ignoring details that do not arise in flat SQL queries, which are our focus.

The algorithm is essentially a dynamically pruning, exhaustive search algorithm. It constructs all alternative join trees (that satisfy restrictions R1–R3) by iterating on the number of relations joined so far, always pruning trees that are known to be suboptimal. Before we present the algorithm in detail, we need to discuss the issue of *interesting order*. One of the join methods that is usually specified by the Method–Structure-Space module is *merge scan*. Merge scan first sorts the two input relations on the corresponding join attributes and then merges them with a synchronized scan. If any of the input relations, however, is already sorted on its join attribute (e.g., because of earlier use of a B+-tree index or sorting as part of an earlier merge-scan join), the sorting step can be skipped for the relation. Hence, given two partial plans during query optimization, one cannot compare them based on their cost only and prune the more expensive one; one has to also take into account the sorted order (if any) in which their result comes out. One of the plans may be more expensive but may generate its result sorted on an attribute that will save a sort in a subsequent merge-scan execution of a join. To take into account these possibilities, given a query, one defines its *interesting orders* to be orders of intermediate results on any relation attributes that participate in joins. (For more general SQL queries, attributes in order-by and group-by clauses give rise to interesting orders as well.) For example, in the query of section 45.3, orders on the attributes emp.dno, dept.dno, dept.ano, acnt.ano, acnt.bno, and bank.bno are interesting. During optimization of this query, if any intermediate result comes out sorted on any of these attributes, then the partial plan that gave this result must be treated specially.

---

[2]A similar argument can be made in favor of right-deep trees regarding sequences of hash joins.

Using the above, we give below a detailed English description of the dynamic programming algorithm optimizing a query of N relations:

Step 1: For each relation in the query, all possible ways to access it, i.e., via all existing indices and including the simple sequential scan, are obtained. (Accessing an index takes into account any query selection on the index key attribute.) These partial (single-relation) plans are partitioned into equivalence classes based on any interesting order in which they produce their result. An additional equivalence class is formed by the partial plans whose results are in no interesting order. Estimates of the costs of all plans are obtained from the Cost Model module, and the cheapest plan in each equivalence class is retained for further consideration. However, the cheapest plan of the no-order equivalence class is not retained if it is not cheaper than all other plans.

Step 2: For each pair of relations joined in the query, all possible ways to evaluate their join using all relation access plans retained after step 1 are obtained. Partitioning and pruning of these partial (two-relation) plans proceeds as above.

⋮

Step i: For each set of $i - 1$ relations joined in the query, the cheapest plans to join them for each interesting order are known from the previous step. In this step, for each such set, all possible ways to join one more relation with it without creating a cross product are evaluated. For each set of i relations, all generated (partial) plans are partitioned and pruned as before.

⋮

Step N: All possible plans to answer the query (the unique set of N relations joined in the query) are generated from the plans retained in the previous step. The cheapest plan is the final output of the optimizer to be used to process the query.

For a given query, the above algorithm is guaranteed to find the optimal plan among those satisfying restrictions R1–R3. It often avoids enumerating all plans in the space by being able to dynamically prune suboptimal parts of the space as partial plans are generated. In fact, although in general still exponential, there are query forms for which it generates only $O(N^3)$ plans [Ono and Lohman 1990].

An example that shows dynamic programming in its full detail takes too much space. We illustrate its basic mechanism by showing how it would proceed on the simple query below:

> **select** name, mgr
> **from** emp, dept
> **where** emp.dno=dept.dno **and** sal>30 K **and** floor=2

Assume that there is a B+-tree index on emp.sal, a B+-tree index on emp.dno, and a hashing index on dept.floor. Also assume that the DBMS supports two join methods: nested loops and merge scan. (Both types of information should be specified in the Method–Structure-Space module.) Note that, based on the definition, potential interesting orders are those on emp.dno and dept.dno, since these are the only join attributes in the query. The algorithm proceeds as follows:

Step 1: All possible ways to access emp and dept are found. The only interesting order arises from accessing emp via the B+-tree on emp.dno, which generates the emp tuples sorted and ready for the join with dept. The entire set of alternatives, appropriately partitioned, is shown in Table 45.2. Each partial plan is associated with some hypothetical cost; in reality, these costs are obtained from the Cost-Model module. Within each equivalence class, only the cheapest plan is retained for the next step, as indicated by the boxes surrounding the corresponding costs in the table.

Step 2: Since the query has two relations, this is the last step of the algorithm. All possible ways to join emp and dept are found, using both supported join methods and all partial plans for individual relation access retained from step 1. For the nested-loops method, which relation is inner and which is outer is also specified. Since this is the last step of the algorithm, there is no issue of interesting orders. The entire set of alternatives is shown in Table 45.3 in a way

TABLE 45.2

Relation	Interesting Order	Plan Description	Cost
emp	emp.dno	Access through B+-tree on emp.dno.	$\boxed{700}$
	–	Access through B+-tree on emp.sal. Sequential scan.	$\boxed{200}$ 600
dept	–	Access through hashing on dept.floor. Sequential scan.	$\boxed{50}$ 200

similar to step 1. Based on hypothetical costs for each of the plans, the optimizer produces as output the plan indicated by the box surrounding the corresponding cost in the table.

As the above example illustrates, the choices offered by the Method–Structure Space in addition to those of the Algebraic Space result in an extraordinary number of alternatives that the optimizer must search through. The memory requirements and running time of dynamic programming grow exponentially with query size (i.e., number of joins) in the worst case, since all viable partial plans generated in each step must be stored to be used in the next one. In fact, many modern systems place a limit on the size of queries that can be submitted (usually around fifteen joins), because for larger queries the optimizer crashes due to its very high memory requirements. Nevertheless, most queries seen in practice involve less than ten joins, and the algorithm has proved to be very effective in such contexts. It is considered the standard in query optimization search strategies.

Randomized Algorithms

To address the inability of dynamic programming to cope with really large queries, which appear in several novel application fields, several other algorithms have been proposed recently. Of these, randomized algorithms, i.e., algorithms that "flip coins" to make decisions, appear very promising.

TABLE 45.3

Join Method	Outer/Inner	Plan Description	Cost
Nested loops	emp/dept	• For each emp tuple obtained through the B+-tree on emp.sal, scan dept through the hashing index on dept.floor to find tuples matching on dno.	1800
		• For each emp tuple obtained through the B+-tree on emp.dno and satisfying the selection on emp.sal, scan dept through the hashing index on dept.floor to find tuples matching on dno.	3000
	dept/emp	• For each dept tuple obtained through the hashing index on dept.floor, scan emp through the B+-tree on emp.sal to find tuples matching on dno.	2500
		• For each dept tuple obtained through the hashing index on dept.floor, probe emp through the B+-tree on emp.dno using the value in dept.dno to find tuples satisfying the selection on emp.sal.	$\boxed{1500}$
Merge scan	–	• Sort the emp tuples resulting from accessing the B+-tree on emp.sal into L_1. • Sort the dept tuples resulting from accessing the hashing index on dept.floor into L_2. • Merge L_1 and L_2.	2300
		• Sort the dept tuples resulting from accessing the hashing index on dept.floor into L_2. • Merge L_2 and the emp tuples resulting from accessing the B+-tree on emp.dno and satisfying the selection on emp.sal.	2000

The most important class of these optimization algorithms is based on plan transformations instead of the plan construction of dynamic programming, and includes algorithms like *Simulated Annealing, Iterative Improvement*, and *Two-Phase Optimization*. These are generic algorithms that can be applied to a variety of optimization problems and are briefly described below as adapted to query optimization. They operate by searching a graph whose nodes are all the alternative execution plans that can be used to answer a query. Each node has a cost associated with it, and the goal of the algorithm is to find a node with the globally minimum cost. Randomized algorithms perform *random walks* in the graph via a series of *moves*. The nodes that can be reached in one move from a node S are called the *neighbors* of S. A move is called *uphill* move (*downhill*) if the cost of the source node is lower (higher) than the cost of the destination node. A node is a *global minimum* if it has the lowest cost among all nodes. It is a *local minimum* if, in all paths starting at that node, any downhill move comes after at least one uphill move.

Algorithm Description

Iterative Improvement (II) [Nahar et al. 1986, Swami 1989, Swami and Gupta 1988] performs a large number of *local optimizations*. Each one starts at a random node and repeatedly accepts random downhill moves until it reaches a local minimum. II returns the local minimum with the lowest cost found.

Simulated Annealing (SA) performs a continuous random walk, accepting downhill moves always and uphill moves with some probability, trying to avoid being caught in a high-cost local minimum [Ioannidis and Kang 1990, Ioannidis and Wong 1987, Kirkpatrick et al. 1983,]. This probability decreases as time progresses and eventually becomes zero, at which point execution stops. Like II, SA returns the node with the lowest cost visited.

The *Two-Phase Optimization* (2PO) algorithm is a combination of II and SA [Ioannidis and Kang 1990]. In phase 1, II is run for a small period of time, i.e., a few local optimizations are performed. The output of that phase, which is the best local minimum found, is the initial node of the next phase. In phase 2, SA is run starting from a low probability for uphill moves. Intuitively, the algorithm chooses a local minimum and then searches the area around it, still being able to move in and out of local minima, but practically unable to climb up very high hills.

Results

Given a finite amount of time, these randomized algorithms have performance that depends on the characteristics of the cost function over the graph and the connectivity of the latter as determined by the neighbors of each node. They have been studied extensively for query optimization, being mutually compared and also compared against dynamic programming [Ioannidis and Kang 1990, Ioannidis and Wong 1987, Kang 1991, Swami 1989, Swami and Gupta 1988]. The specific results of these comparisons vary depending on the choices made regarding issues of the algorithms' implementation and setup, but also choices made in other modules of the query optimizer, i.e., the Algebraic Space, the Method–Structure Space, and the Cost Model. In general, however, the conclusions are as follows. First, up to about ten joins, dynamic programming is preferred over the randomized algorithms because it is faster and it guarantees finding the optimal plan. For larger queries, the situation is reversed, and despite the probabilistic nature of the randomized algorithms, their efficiency makes them the algorithms of choice. Second, among randomized algorithms, II usually finds a reasonable plan very quickly, while given enough time, SA is able to find a better plan than II. 2PO gets the best of both worlds and is able to find plans that are as good as those of SA, if not better, in much shorter time.

Other Search Strategies

To complete the picture on search strategies we briefly describe several other algorithms that people have proposed in the past, deterministic, heuristic, or randomized. Ibaraki and Kameda were the ones that proved that query optimization is an NP-complete problem even if considering only the nested-loops join method [Ibaraki and Kameda 1984]. Given that result, there have been several efforts to obtain algorithms that solve important subcases of the query optimization problem and run in polynomial time. Ibaraki and Kameda themselves presented an algorithm (referred to as IK here) that takes advantage of the special form of the cost formula for nested loops and optimizes a tree query of N joins in $O(N^2 \log N)$ time. They also presented an algorithm that is applicable to even cyclic queries and finds a good (but not always optimal) plan in $O(N^3)$ time.

The KBZ algorithm uses essentially the same techniques, but it is more general and more sophisticated and runs in $O(N^2)$ time for tree queries [Krishnamurthy et al. 1986]. As with IK, the applicability of KBZ depends on the cost formulas for joins to be of a specific form. Nested loops and hash join satisfy this requirement but, in general, merge scan does not.

The AB algorithm mixes deterministic and randomized techniques and runs in $O(N^4)$ time [Swami and Iyer 1993]. It uses KBZ as a subroutine, which needs $O(N^2)$ time, and essentially executes it $O(N^2)$ times on randomly selected spanning trees of the query graph. Through an interesting separation of the cost of merge scan into a part that affects optimization and a part that does not, AB is applicable to all join methods despite the dependence on KBZ.

In addition to SA, II, and 2PO, *Genetic Algorithms* [Goldberg 1989] form another class of generic randomized optimization algorithms that have been applied to query optimization. These algorithms simulate a biological phenomenon: a random set of solutions to the problem, each with its own cost, represents an initial population; pairs of solutions from that population are matched (*cross over*) to generate offspring that obtain characteristics from both parents, and the new children may also be randomly changed in small ways (*mutation*); between the parents and the children, those with the least cost (*most fit*) survive in the next generation. The algorithm ends when the entire population consists of copies of the same solution, which is considered to be optimal. Genetic algorithms have been implemented for query optimization with promising results [Bennett et al. 1991].

Another interesting randomized approach to query optimization is pure, uniformly random generation of access plans [Galindo-Legaria et al. 1994]. Truly uniform generation is a hard problem but has been solved for tree queries. With an efficient implementation of this step, experiments with the algorithm have shown good potential, since there is no dependence on plan transformations or random walks.

In the artificial intelligence community, the A* heuristic algorithm is extensively used for complex search problems. A* has been proposed for query optimization as well and can be seen as a direct extension to the traditional dynamic programming algorithm [Yoo and Lafortune 1989]. Instead of proceeding in steps and using all plans with n relations to generate all plans with $n + 1$ relations together, A* proceeds by expanding one of the generated plans at hand at a time, based on its expected proximity to the optimal plan. Thus, A* generates a full plan much earlier than dynamic programming and is able to prune more aggressively in a branch-and-bound mode. A* has been proposed for query optimization and has been shown quite successful for not very large queries.

Finally, in the context of extensible DBMSs, several unique search strategies have been proposed, which are all rule-based. Rules are defined on how plans can be constructed or modified, and the Planner follows the rules to explore the specified plan space. The most representative of these efforts are those of Starburst [Haas et al. 1990, Lohman 1988] and Volcano/Exodus [Graefe and DeWitt 1987, Graefe and McKenna 1993]. The Starburst optimizer employs constructive rules, whereas the Volcano/Exodus optimizers employ transformation rules.

45.5 Size-Distribution Estimator

The final module of the query optimizer that we examine in detail is the Size-Distribution Estimator. Given a query, it estimates the sizes of the results of (sub)queries and the frequency distributions of values in attributes of these results.

Before we present specific techniques that have been proposed for estimation, we use an example to clarify the notion of frequency distribution. Consider the simple relation *OLYMPIAN* on the left in Table 45.4, with the frequency distribution of the values in its Department attribute on the right.

TABLE 45.4

Name	Salary	Department	Department	Frequency
Zeus	100 K	General Management	General Management	2
Poseidon	80 K	Defense	Defense	2
Pluto	80 K	Justice	Education	1
Aris	50 K	Defense	Domestic Affairs	2
Ermis	60 K	Commerce	Agriculture	1
Apollo	60 K	Energy	Commerce	1
Hefestus	50 K	Energy	Justice	1
Hera	90 K	General Management	Energy	3
Athena	70 K	Education		
Aphrodite	60 K	Domestic Affairs		
Demeter	60 K	Agriculture		
Hestia	50 K	Domestic Affairs		
Artemis	60 K	Energy		

One can generalize the above and discuss distributions of frequencies of combinations of arbitrary numbers of attributes. In fact, to calculate/estimate the size of any query that involves multiple attributes from a single relation, multiattribute joint frequency distributions or their approximations are required. Practical DBMSs, however, deal with frequency distributions of individual attributes only, because considering all possible combinations of attributes is very expensive. This essentially corresponds to what is known as the *attribute value independence assumption*, and, although rarely true, it is adopted by all current DBMSs.

Several techniques have been proposed in the literature to estimate query result sizes and frequency distributions, most of them contained in the extensive survey by Mannino et al. [1988] and elsewhere [Christodoulakis 1989]. Most commercial DBMSs (e.g., DB2, Informix, Ingres, Sybase, Microsoft SQL server) base their estimation on *histograms*, so our description mostly focuses on those. We then briefly summarize other techniques that have been proposed.

Histograms

In a *histogram* on attribute a of relation R, the domain of a is partitioned into *buckets*, and a uniform distribution is assumed within each bucket. That is, for any bucket b in the histogram, if a value $v_i \in b$, then the frequency f_i of v_i is approximated by $\sum_{v_j \in b} f_j / |b|$. A histogram with a single bucket generates the same approximate frequency for all attribute values. Such a histogram is called *trivial* and corresponds to making the *uniform distribution assumption* over the entire attribute domain. Note that, in principle, any arbitrary subset of an attribute's domain may form a bucket and not necessarily consecutive ranges of its natural order.

Continuing with the example of the *OLYMPIAN* relation, we present in Table 45.5 two different histograms on the Department attribute, both with two buckets. For each histogram, we first show which frequencies are grouped in the same bucket by enclosing them in the same shape (box or circle), and then we show the resulting approximate frequency, i.e., the average of all frequencies enclosed by identical shapes.

There are various classes of histograms that systems use or researchers have proposed for estimation. Most of the earlier prototypes, and still some of the commercial DBMSs, use trivial histograms, i.e., make the uniform distribution assumption [Selinger et al. 1979]. That assumption, however, rarely holds in real data, and estimates based on it usually have large errors [Christodoulakis 1984, Ioannidis and Christodoulakis 1991]. Excluding trivial ones, the histograms that are typically used belong to the class of *equiwidth* histograms [Kooi 1980]. In those, the number of consecutive attribute values or the size of the range of attribute values associated with each bucket is the same, independent of the frequency of each attribute value in the data. Since these histograms store a lot more information than trivial histograms (they typically have 10–20 buckets), their estimations are much better. Histogram H1 above is equiwidth, since the first bucket contains four values starting from A–D and the second bucket contains also four values starting from E–Z.

TABLE 45.5

Department	Histogram H1		Histogram H2	
	Frequency in Bucket	Approximate Frequency	Frequency in Bucket	Approximate Frequency
Agriculture	[1]	1.5	[1]	1.33
Commerce	[1]	1.5	[1]	1.33
Defense	[2]	1.5	[2]	1.33
Domestic Affairs	[2]	1.5	(2)	2.5
Education	(1)	1.75	[1]	1.33
Energy	(3)	1.75	(3)	2.5
General Management	(2)	1.75	[2]	1.33
Justice	(1)	1.75	[1]	1.33

Although we are not aware of any system that currently uses histograms in any other class than those mentioned above, several more advanced classes have been proposed and are worth discussing. *Equidepth* (or *equiheight*) histograms are essentially duals of equiwidth histograms [Kooi 1980, Piatetsky-Shapiro and Connell 1984]. In those, the sum of the frequencies of the attribute values associated with each bucket is the same, independent of the number of those attribute values. Equiwidth histograms have a much higher worst-case and average error for a variety of selection queries than equidepth histograms. Muralikrishna and DeWitt [1988] extended the above work for multidimensional histograms that are appropriate for multiattribute selection queries.

In *serial* histograms [Ioannidis and Christodoulakis 1993], the frequencies of the attribute values associated with each bucket are either all greater or all less than the frequencies of the attribute values associated with any other bucket. That is, the buckets of a serial histogram group frequencies that are close to each other with no interleaving. Histogram H1 in Table 45.5 is not serial, as frequencies 1 and 3 appear in one bucket and frequency 2 appears in the other, while histogram H2 is. Under various optimality criteria, serial histograms have been shown to be optimal for reducing the worst-case and the average error in equality selection and join queries [Ioannidis 1993, Ioannidis and Christodoulakis 1993, Ioannidis and Poosala 1995].

Identifying the optimal histogram among all serial ones takes exponential time in the number of buckets. Moreover, since there is usually no order correlation between attribute values and their frequencies, storage of serial histograms essentially requires a regular index that will lead to the approximate frequency of every individual attribute value. Because of all these complexities, the class of *end-biased* histograms has been introduced. In those, some number of the highest frequencies and some number of the lowest frequencies in an attribute are explicitly and accurately maintained in separate individual buckets, and the remaining (middle) frequencies are all approximated together in a single bucket. End-biased histograms are serial, since their buckets group frequencies with no interleaving. Identifying the optimal end-biased histogram, however, takes only slightly over linear time in the number of buckets. Moreover, end-biased histograms require little storage, since usually most of the attribute values belong in a single bucket and do not have to be stored explicitly. Finally, in several experiments it has been shown that most often the errors in the estimates based on end-biased histograms are not too far off from the corresponding (optimal) errors based on serial histograms. Thus, as a compromise between optimality and practicality, it has been suggested that the optimal end-biased histograms should be used in real systems.

Other Techniques

In addition to histograms, several other techniques have been proposed for query result size estimation [Christodoulakis 1989, Mannino et al. 1988]. Those that, like histograms, store information in the database typically approximate a frequency distribution by a parametrized mathematical distribution or a polynomial. Although requiring very little overhead, these approaches are typically inaccurate because most often real data do not follow any mathematical function. On the other hand, those based on *sampling* primarily operate at run time [Haas and Swami 1992, 1995, Lipton et al. 1990, Olken and Rotem 1986] and compute their estimates by collecting and possibly processing random samples of the data. Although producing highly accurate estimates, sampling is quite expensive, and therefore its practicality in query optimization is questionable, especially since optimizers need query result size estimations frequently.

45.6 Noncentralized Environments

The preceding discussion focuses on query optimization for sequential processing. This section touches upon issues and techniques related to optimizing queries in noncentralized environments. The focus is on the Method–Structure-Space and Planner modules of the optimizer, as the remaining ones are not significantly different from the centralized case.

Parallel Databases

Among all parallel architectures, the shared-nothing and the shared-memory paradigms have emerged as the most viable ones for database query processing. Thus, query optimization research has concentrated on these two. The processing choices that either of these paradigms offers represent a huge increase over the alternatives offered by the Method–Structure-Space module in a sequential environment. In addition to the sources of alternatives that we discussed earlier, the Method–Structure-Space module offers two more: the number of processors that should be given to each database operation (*intraoperator parallelism*) and placing operators into groups that should be executed simultaneously by the available processors (*interoperator parallelism*, which can be further subdivided into *pipelining* and *independent parallelism*). The *scheduling* alternatives that arise from these two questions add at least another superexponential factor to the total number of alternatives and make searching an even more formidable task. Thus, most systems and research prototypes adopt various heuristics to avoid dealing with a very large search space. In the two-stage approach [Hong and Stonebraker 1991], given a query, one first identifies the optimal sequential plan for it using conventional techniques like those discussed in section 45.4, and then one identifies the optimal parallelization/scheduling of that plan. Various techniques have been proposed in the literature for the second stage, but none of them claims to provide a complete and optimal answer to the scheduling question, which remains an open research problem. In the segmented execution model, one considers only schedules that process memory-resident right-deep segments of (possibly bushy) query plans one-at-a-time (i.e., no independent interoperator parallelism). Shekita et al. [1993] combined this model with a novel heuristic search strategy with good results for shared-memory. Finally, one may be restricted to deal with right-deep trees only [Schneider and DeWitt 1990].

In contrast to all the search-space reduction heuristics, Lanzelotte et al. [1993] dealt with both deep and bushy trees, considering schedules with independent parallelism, where all the pipelines in an execution are divided into phases, pipelines in the same phase are executed in parallel, and each phase starts only after the previous phase ended. The search strategy that they used was a randomized algorithm, similar to 2PO, and proved very effective in identifying efficient parallel plans for a shared-nothing architecture.

Distributed Databases

The difference between distributed and parallel DBMSs is that the former are formed by a collection of independent, semiautonomous processing sites that are connected via a network that could be spread over a large geographic area, whereas the latter are individual systems controlling multiple processors that are in the same location, usually in the same machine room. Many prototypes of distributed DBMSs have been implemented [Bernstein et al. 1981, Mackert and Lohman 1986], and several commercial systems are offering distributed versions of their products as well (e.g., DB2, Informix, Sybase, Oracle).

Other than the necessary extensions of the Cost-Model module, the main differences between centralized and distributed query optimization are in the Method–Structure-Space module, which offers additional processing strategies and opportunities for transmitting data for processing at multiple sites. In early distributed systems, where the network cost was dominating every other cost, a key idea was using semijoins for processing in order to only transmit tuples that would certainly contribute to join results [Bernstein et al. 1981, Mackett and Lohman 1986]. An extension of that idea is using Bloom filters, which are bit vectors that approximate join columns and are transferred across sites to determine which tuples *might* participate in a join so that only these may be transmitted [Mackett and Lohman 1986].

45.7 Advanced Types of Optimization

In this section, we attempt to provide a brief glimpse of advanced types of optimization that researchers have proposed over the past few years. The descriptions are based on examples only; further details may be found in the references provided. Furthermore, there are several issues that are not discussed at all due to lack of space, although much interesting work has been done on them, e.g., nested query optimization, rule-based query optimization, query optimizer generators, object-oriented query optimization, optimization with materialized views, heterogeneous query optimization, recursive query optimization, aggregate query optimization, optimization with expensive selection predicates, and query-optimizer validation.

Semantic Query Optimization

Semantic query optimization is a form of optimization mostly related to the Rewriter module. The basic idea lies in using integrity constraints defined in the database to rewrite a given query into *semantically equivalent* ones [King 1981]. These can then be optimized by the Planner as regular queries, and the most efficient plan among all can be used to answer the original query. As a simple example, using a hypothetical SQL-like syntax, consider the following integrity constraint:

> **assert** sal-constraint **on** emp:
> sal>100 K **where** job="Sr. Programmer".

Also consider the following query:

> **select** name, floor
> **from** emp, dept
> **where** emp.dno=dept.dno **and** job="Sr. Programmer".

Using the above integrity constraint, the query can be rewritten into a semantically equivalent one to include a selection on sal:

> **select** name, floor
> **from** emp, dept
> **where** emp.dno=dept.dno **and** job="Sr. Programmer" **and** sal>100 K.

Having the extra selection could help tremendously in finding a fast plan to answer the query if the only index in the database is a B+-tree on emp.sal. On the other hand, it would certainly be a waste if no such index exists. For such reasons, all proposals for semantic query optimization present various heuristics or rules on which rewritings have the potential of being beneficial and should be applied and which should not.

Global Query Optimization

So far, we have focused our attention to optimizing individual queries. Quite often, however, multiple queries become available for optimization at the same time, e.g., queries with unions, queries from multiple concurrent users, queries embedded in a single program, or queries in a deductive system. Instead of optimizing each query separately, one may be able to obtain a global plan that, although possibly suboptimal for each individual query, is optimal for the execution of all of them as a group. Several techniques have been proposed for global query optimization [Sellis 1988]. As a simple example of the problem of global optimization consider the following two queries:

> **select** name, floor
> **from** emp, dept
> **where** emp.dno=dept.dno **and** job="Sr. Programmer,"
> **select** name
> **from** emp, dept
> **where** emp.dno=dept.dno **and** budget>1 M.

Depending on the sizes of the emp and dept relations and the selectivities of the selections, it may well be that computing the entire join once and then applying separately the two selections to obtain the results of the two queries is more efficient than doing the join twice, each time taking into account the corresponding selection. Developing Planner modules that would examine all the available global plans and identify the optimal one is the goal of global/multiple query optimizers.

Parametric/Dynamic Query Optimization

As mentioned earlier, embedded queries are typically optimized once at compile time and are executed multiple times at run time. Because of this temporal separation between optimization and execution, the values of various parameters that are used during optimization may be very different during execution. This may make the chosen plan invalid (e.g., if indices used in the plan are no longer available) or simply not optimal (e.g., if the number of available buffer pages or operator selectivities has changed, or if new indices have become available). To address this issue, several techniques [Cole and Graefe 1994, Graefe and Ward 1989, Ioannidis et al. 1992] have been proposed that use various search strategies (e.g., randomized algorithms [Ioannidis et al. 1992] or the strategy of Volcano [Cole and Graefe 1994]) to optimize queries as much as possible at compile time, taking into account all possible values that interesting parameters may have at run time. These techniques use the actual parameter values at run time and simply pick the plan that was found optimal for them with little or no overhead. Of a drastically different flavor is the technique of Rdb/VMS [Antoshenkov 1993], where by dynamically monitoring how the probability distribution of plan costs changes, plan switching may actually occur during query execution.

45.8 Summary

To a large extent, the success of a DBMS lies in the quality, functionality, and sophistication of its query optimizer, since that determines much of the system's performance. In this chapter, we have given a bird's-eye view of query optimization. We have presented an abstraction of the architecture of a query optimizer and focused on the techniques currently used by most commercial systems for its various modules. In addition, we have provided a glimpse of advanced issues in query optimization, whose solutions have not yet found their way into practical systems, but could certainly do so in the future.

Although query optimization has existed as a field for more than twenty years, it is very surprising how fresh it remains in terms of being a source of research problems. In every single module of the architecture of Fig. 45.2, there are many questions for which we do not have complete answers, even for the most simple, single-query, sequential, relational optimizations. When is it worthwhile to consider bushy trees instead of just left-deep trees? How can one model buffering effectively in the system's cost formulas? What is the most effective means of estimating the cost of operators that involve random access to relations (e.g., nonclustered index selection)? Which search strategy can be used for complex queries with confidence, providing consistent plans for similar queries? Should optimization and execution be interleaved in complex queries so that estimate errors do not grow very large? Of course, we do not even attempt to mention the questions that arise in various advanced types of optimization.

We believe that the next twenty years will be as active as the previous twenty and will bring many advances to query optimization technology, changing many of the approaches currently used in practice. Despite its age, query optimization remains an exciting field.

Acknowledgments

I would like to thank Minos Garofalakis, Joe Hellerstein, Navin Kabra, and Vishy Poosala for their many helpful comments.

References

Antoshenkov, G. 1993. Dynamic query optimization in Rdb/VMS, pp. 538–547. In *Proc. IEEE Int. Conf. on Data Engineering*, Vienna, Austria, Mar.

Astrahan, M. M. et al. 1976. System R: a relational approach to data management. *ACM Trans. Database Sys.* 1(2):97–137, June.

Bennett, K., Ferris, M. C., and Ioannidis, Y. 1991. A genetic algorithm for database query optimization, pp. 400–407. In *Proc. 4th Int. Conf. on Genetic Algorithms*, San Diego, CA, July.

Bernstein, P. A., Goodman, N., Wong, E., Reeve, C. L., and Rothnie, J. B. 1981. Query processing in a system for distributed databases (SDD-1). *ACM Trans. Database Syst.* 6(4):602–625, Dec.

Christodoulakis, S. 1984. Implications of certain assumptions in database performance evaluation. *ACM Trans. Database Syst.* 9(2):163–186, June.

Christodoulakis, S. 1989. On the estimation and use of selectivities in database performance evaluation. *Research Report CS-89-24.* Dept. of Computer Science, University of Waterloo, June.

Cole, R. and Graefe G. 1994. Optimization of dynamic query evaluation plans, pp. 150–160. In *Proc. ACM-SIGMOD Conf. on the Management of Data*, Minneapolis, MN, June.

Galindo-Legaria, C., Pellenkoft, A., and Kersten, M. 1994. Fast, randomized join-order selection—why use transformations?, pp. 85–95. In *Proc. 20th Int. VLDB Conf.*, Santiago, Chile, Sept. Also available as *CWI Tech. Report CS-R9416.*

Goldberg, D. E. 1989. *Genetic Algorithms in Search, Optimization, and Machine Learning.* Addison–Wesley, Reading, MA.

Graefe, G. and DeWitt, D. 1987. The exodus optimizer generator, pp. 160–172. In *Proc. ACM-SIGMOD Conf. on the Management of Data*, San Francisco, CA, May.

Graefe, G. and McKenna, B. 1993. The Volcano optimizer generator: extensibility and efficient search. In *Proc. IEEE Data Engineering Conf.*, Vienna, Austria, Mar.

Graefe, G. and Ward, K. 1989. Dynamic query evaluation plans, pp. 358–366. In *Proc. ACM-SIGMOD Conference on the Management of Data*, Portland, OR, May.

Haas, L. et al. 1990. Starburst mid-flight: as the dust clears. *IEEE Trans. Knowledge and Data Eng.* 2(1):143–160, Mar.

Haas, P. and Swami, A. 1992. Sequential sampling procedures for query size estimation, pp. 341–350. In *Proc. 1992 ACM-SIGMOD Conf. on the Management of Data*, San Diego, CA, June.

Haas, P. and Swami, A. 1995. Sampling-based selectivity estimation for joins using augmented frequent value statistics. In *Proc. 1995 IEEE Conf. on Data Engineering*, Taipei, Taiwan, Mar.

Hong, W. and Stonebraker, M. 1991. Optimization of parallel query execution plans in xprs, pp. 218–225. In *Proc. 1st Int. PDIS Conf.*, Miami, FL, Dec.

Ibaraki, T. and Kameda, T. 1984. On the optimal nesting order for computing n-relational joins. *ACM Trans. Database Syst.* 9(3):482–502, Sept.

Ioannidis, Y. 1993. Universality of serial histograms, pp. 256–267. In *Proc. 19th Int. VLDB Conf.*, Dublin, Ireland, Aug.

Ioannidis, Y. and Christodoulakis, S. 1991. On the propagation of errors in the size of join results, pp. 268–277. In *Proc. 1991 ACM-SIGMOD Conf. on the Management of Data*, Denver, CO, May.

Ioannidis, Y. and Christodoulakis, S. 1993. Optimal histograms for limiting worst-case error propagation in the size of join results. *ACM Trans. Database Syst.* 18(4):709–748, Dec.

Ioannidis, Y. and Kang, Y. 1990. Randomized algorithms for optimizing large join queries, pp. 312–321. In *Proc. ACM-SIGMOD Conf. on the Management of Data*, Atlantic City, NJ, May.

Ioannidis, Y. and Poosala, V. 1995. Balancing histogram optimality and practicality for query result size estimation, pp. 233–244. In *Proc. 1995 ACM-SIGMOD Conf. on the Management of Data*, San Jose, CA, May.

Ioannidis, Y. and Wong, E. 1987. Query optimization by simulated annealing, pp. 9–22. In *Proc. ACM-SIGMOD Conf. on the Management of Data*, San Francisco, CA, May.

Ioannidis, Y., Ng, R., Shim, K., and Sellis, T. K. 1992. Parameteric query optimization, pp. 103–114. In *Proc. 18th Int. VLDB Conf.*, Vancouver, BC, Aug.

Jarke, M. and Koch, J. 1984. Query optimization in database systems. *ACM Comput. Surveys* 16(2):111–152, June.

Kang, Y. 1991. *Randomized Algorithms for Query Optimization.* Ph.D. thesis, University of Wisconsin, Madison, May.

King, J. J. 1981. Quist: a system for semantic query optimization in relational databases, pp. 510–517. In *Proc. 7th Int. VLDB Conf.*, Cannes, France, Aug.

Kirkpatrick, S., Gelatt, C. D., Jr., and Vecchi, M. P. 1983. Optimization by simulated annealing. *Science* 220(4598):671–680, May.

Kooi, R. P. 1980. *The Optimization of Queries in Relational Database.* Ph.D. thesis, Case Western Reserve University, Sept.

Krishnamurthy, R., Boral, H., and Zaniolo, C. 1986. Optimization of nonrecursive queries, pp. 128–137. In *Proc. 12th Int. VLDB Conf.*, Kyoto, Japan, Aug.

Lanzelotte, R., Valduriez, P., and Zait, M. 1993. On the effectiveness of optimization search strategies for parallel execution spaces, pp. 493–504. In *Proc. 19th Int. VLDB Conf.*, Dublin, Ireland, Aug.

Lipton, R. J., Naughton, J. F., and Schneider, D. A. 1990. Practical selectivity estimation through adaptive sampling, pp. 1–11. In *Proc. 1990 ACM-SIGMOD Conf. on the Management of Data*, Atlantic City, NJ, May.

Lohman, G. 1988. Grammar-like functional rules for representing query optimization alternatives, pp. 18–27. In *Proc. ACM-SIGMOD Conf. on the Management of Data.*, Chicago, IL, June.

Mackert, L. F. and Lohman, G. M. 1986. R^* validation and performance evaluation for distributed queries, pp. 149–159. In *Proc. 12th Int. VLDB Conf.*, Kyoto, Japan, Aug.

Mannino, M. V., Chu, P., and Sager, T. 1988. Statistical profile estimation in database systems. *ACM Comput. Surveys* 20(3):192–221, Sept.

Muralikrishna, M. and DeWitt, D. J. 1988. Equi-depth histograms for estimating selectivity factors for multi-dimensional queries, pp. 28–36. In *Proc. 1988 ACM-SIGMOD Conf. on the Management of Data*, Chicago, IL, June.

Nahar, S., Sahni, S., and Shragowitz, E. 1986. Simulated annealing and combinatorial optimization, pp. 293–299. In *Proc. 23rd Design Automation Conf.*

Olken, F. and Rotem, D. 1986. Simple random sampling from relational databases, pp. 160–169. In *Proc. 12th Int. VLDB Conf.*, Kyoto, Japan, Aug.

Ono, K. and Lohman, G. 1990. Measuring the complexity of join enumeration in query optimization, pp. 314–325. In *Proceedings of the 16th Int. VLDB Conf.*, Brisbane, Australia, Aug.

Piatetsky-Shapiro, G. and Connell, C. 1984. Accurate estimation of the number of tuples satisfying a condition, pp. 256–276. In *Proc. 1984 ACM-SIGMOD Conf. on the Management of Data*, Boston, MA, June.

Schneider, D. and DeWitt, D. 1990. Tradeoffs in processing complex join queries via hashing in multiprocessor database machines, pp. 469–480. In *Proc. of the 16th Int. VLDB Conf.*, Brisbane, Australia, Aug.

Selinger, P. G., Astrahan, M. M., Chamberlin, D. D., Lorie, R. A., and Price, T. G. 1979. Access path selection in a relational database management system, pp. 23–34. In *Proc. ACM-SIGMOD Conf. on the Management of Data*, Boston, MA, June.

Sellis, T. 1988. Multiple query optimization. *ACM Trans. Database Syst.* 13(1):23–52, Mar.

Shekita E., Young, H., and Tan, K.-L. 1993. Multi-join optimization for symmetric multiprocessors, pp. 479–492. In *Proc. 19th Int. VLDB Conf.*, Dublin, Ireland, Aug.

Swami, A. 1989. Optimization of large join queries: combining heuristics and combinatorial techniques, pp. 367–376. In *Proc. ACM-SIGMOD Conf. on the Management of Data*, Portland, OR, June.

Swami, A. and Gupta, A. 1988. Optimization of large join queries, pp. 8–17. In *Proc. ACM-SIGMOD Conf. on the Management of Data*, Chicago, IL, June.

Swami, A. and Iyer, B. 1993. A polynomial time algorithm for optimizing join queries. In *Proc. IEEE Int. Conf. on Data Engineering*, Vienna, Austria, Mar.

Yoo, H. and Lafortune, S. 1989. An intelligent search method for query optimization by semijoins. *IEEE Trans. Knowledge and Data Eng.* 1(2):226–237, June.

46

Concurrency Control and Recovery

46.1 Introduction

Many service-oriented businesses and organizations, such as banks, airlines, catalog retailers, hospitals, etc., have grown to depend on fast, reliable, and correct access to their "mission-critical" data on a constant basis. In many cases, particularly for global enterprises, 7×24 access is required; that is, the data must be available seven days a week, twenty-four hours a day. *Database management systems* (DBMSs) are often employed to meet these stringent performance, availability, and reliability demands. As a result, two of the core functions of a DBMS are (1) to protect the data stored in the database and (2) to provide correct and highly available access to those data in the presence of concurrent access by large and diverse user populations, despite various software and hardware failures. The responsibility for these functions resides in the **concurrency control** and **recovery** components of the DBMS software. *Concurrency control* ensures that individual users see consistent states of the database even though operations on behalf of many users may be interleaved by the database system. *Recovery* ensures that the database is fault-tolerant; that is, that the database state is not corrupted as the result of a software, system, or media failure. The existence of this functionality in the DBMS allows applications to be written without explicit concern for concurrency and fault tolerance. This freedom provides a tremendous increase in programmer productivity and allows new applications to be added more easily and safely to an existing system.

For database systems, correctness in the presence of concurrent access and/or failures is tied to the notion of a **transaction**. A transaction is a unit of work, possibly consisting of multiple data accesses and updates, that must **commit** or **abort** as a single atomic unit. When a transaction *commits*, all updates it performed on the database are made permanent and visible to other transactions. In contrast, when a transaction *aborts*, all of its updates are removed from the database and the database is restored (if necessary) to the state it would have been in if the aborting transaction had never been executed. Informally, transaction

---

*Portions of this chapter are reprinted with permission from Franklin, M., Zwilling, M., Tan, C., Carey, M., and DeWitt, D., Crash recovery in client-server EXODUS. In *Proc. ACM Int. Conf. on Management of Data (SIGMOD'92)*, San Diego, June 1992. © 1992 by the Association for Computing Machinery, Inc. (ACM).

executes are said to respect the **ACID properties** [Gray and Reuter 1993]:

Atomicity: This is the "all-or-nothing" aspect of transactions discussed above—either all operations of a transaction complete successfully, or none of them do. Therefore, after a transaction has completed (i.e., committed or aborted), the database will not reflect a partial result of that transaction.

Consistency: Transactions preserve the consistency of the data—a transaction performed on a database that is internally consistent will leave the database in an internally consistent state. Consistency is typically expressed as a set of declarative *integrity constraints.* For example, a constraint may be that the salary of an employee cannot be higher than that of his or her manager.

Isolation: A transaction's behavior is not impacted by the presence of other transactions that may be accessing the same database concurrently. That is, a transaction sees only a state of the database that could occur if that transaction were the only one running against the database and produces only results that it could produce if it was running alone.

Durability: The effects of *committed* transactions survive failures. Once a transaction commits, its updates are guaranteed to be reflected in the database even if the contents of volatile (e.g., main memory) or nonvolatile (e.g., disk) storage are lost or corrupted.

Of these four transaction properties, the concurrency control and recovery components of a DBMS are primarily concerned with preserving *atomicity, isolation,* and *durability.* The preservation of the *consistency* property typically requires additional mechanisms such as compile-time analysis or run-time triggers in order to check adherence to integrity constraints.[1] For this reason, this chapter focuses primarily on the A, I, and D of the ACID transaction properties.

Transactions are used to structure complex processing tasks which consist of multiple data accesses and updates. A traditional example of a transaction is a money transfer from one bank account (say account A) to another (say B). This transaction consists of a withdrawal from A and a deposit into B and requires four accesses to account information stored in the database: a read and write of A and a read and write of B. The data accesses of this transaction are as follows:

```
TRANSFER( )
    01 A_bal := Read(A)
    02 A_bal := A_bal − $50
    03 Write(A, A_bal)
    04 B_bal := Read(B)
    05 B_bal := B_bal + $50
    06 Write(B, B_bal)
```

The value of A in the database is read and decremented by $50, then the value of B in the database is read and incremented by $50. Thus, TRANSFER preserves the invariant that the sum of the balances of A and B prior to its execution must equal the sum of the balances after its execution, regardless of whether the transaction commits or aborts.

Consider the importance of the atomicity property. At several points during the TRANSFER transaction, the database is in a temporarily inconsistent state. For example, between the time that account A is updated (statement 3) and the time that account B is updated (statement 6) the database reflects the decrement of A but not the increment of B, so it appears as if $50 has disappeared from the database. If the transaction reaches such a point and then is unable to complete (e.g., due to a failure or an unresolvable conflict, etc.), then the system must ensure that the effects of the partial results of the transaction (i.e., the update to A) are removed from the database—otherwise the database state will be incorrect. The durability property, in contrast, only comes into play in the event that the transaction successfully commits. Once the user is

---

[1] In the case of triggers, the recovery mechanism is typically invoked to abort an offending transaction.

notified that the transfer has taken place, he or she will assume that account B contains the transferred funds and may attempt to use those funds from that point on. Therefore, the DBMS must ensure that the results of the transaction (i.e., the transfer of the $50) remain reflected in the database state even if the system crashes.

Atomicity, consistency, and durability address correctness for **serial execution** of transactions, where only a single transaction at a time is allowed to be in progress. In practice, however, database management systems typically support **concurrent execution**, in which the operations of multiple transactions can be executed in an interleaved fashion. The motivation for concurrent execution in a DBMS is similar to that for multiprogramming in operating systems, namely, to improve the utilization of system hardware resources and to provide multiple users a degree of fairness in access to those resources. The *isolation* property of transactions comes into play when concurrent execution is allowed.

Consider a second transaction that computes the sum of the balances of accounts A and B:

REPORTSUM( )
      01 A_bal := **Read**(A)
      02 B_bal := **Read**(B)
      03 Print(A_bal + B_bal)

Assume that initially, the balance of account A is $300 and the balance of account B is $200. If a REPORTSUM transaction is executed on this state of the database, it will print a result of $500. In a database system restricted to *serial execution* of transactions, REPORTSUM will also produce the same result if it is executed after a TRANSFER transaction. The atomicity property of transactions ensures that if the TRANSFER aborts, all of its effects are removed from the database (so REPORTSUM would see A = $300 and B = $200), and the durability property ensures that if it commits, then all of its effects remain in the database state (so REPORTSUM would see A = $250 and B = $250).

Under concurrent execution, however, a problem could arise if the isolation property is not enforced. As shown in Fig. 46.1, if REPORTSUM were to execute after TRANSFER has updated account A but before it has updated account B, then REPORTSUM could see an inconsistent state of the database. In this case, the execution of REPORTSUM sees a state of the database in which $50 has been withdrawn from account A but has not yet been deposited in account B, resulting in a total of $450—it seems that $50 has disappeared from the database. This result is not one that could be obtained in any serial execution of TRANSFER and REPORTSUM transactions. It occurs because in this example, REPORTSUM accessed the database when it was in a temporarily inconsistent state. This problem is sometimes referred to as the *inconsistent retrieval* problem. To preserve the isolation property of transactions the DBMS must prevent the occurrence of this and other potential anomalies that could arise due to concurrent execution. The formal notion of correctness for concurrent execution in database systems is known as **serializability** and is described in section 46.2.

Although the transaction processing literature often traces the history of transactions back to antiquity (such as Sumerian tax records) or to early contract law [Gray 1981, Gray and Reuter 1993, Korth 1995], the roots of the transaction concept in information systems are typically traced back to the early 1970s and the work of Bjork [1973] and Davies [1973]. Early systems such as IBM's IMS addressed related issues, and a systematic treatment and understanding of ACID transactions was developed several years later by

TRANSFER	REPORTSUM
01 A_bal := **Read**(A)	
02 A_bal := A_bal - $50	
03 **Write**(A,A_bal)	
	01 A_bal := **Read**(A) /* value is $250 */
	02 B_bal := **Read**(B) /* value is $200 */
	03 Print(A_bal + B_bal) /* result = $450 */
04 B_bal := **Read**(B)	
05 B_bal := B_bal + $50	
06 **Write**(B,B_bal)	

FIGURE 46.1   An incorrect interleaving of TRANSFER and REPORTSUM.

members of the IBM System R group [Gray et al. 1975, Eswaran et al. 1976] and others [e.g., Rosenkrantz et al. 1977, Lomet 1977]). Since that time, many techniques for implementing ACID transactions have been proposed and a fairly well accepted set of techniques has emerged. The remainder of this chapter contains an overview of the basic theory that has been developed as well as a survey of the more widely known implementation techniques for concurrency control and recovery. A brief discussion of work on extending the simple transaction model is presented at the end of the chapter.

It should be noted that issues related to those addressed by concurrency control and recovery in database systems arise in other areas of computing systems as well, such as file systems and memory systems. There are, however, two salient aspects of the ACID model that distinguish transactions from other approaches. First is the incorporation of both isolation (concurrency control) and fault-tolerance (recovery) issues. Second is the concern with treating arbitrary groups of *write* and/or *read* operations on multiple data items as atomic, isolated units of work. While these aspects of the ACID model provide powerful guarantees for the protection of data, they also can induce significant systems implementation complexity and performance overhead. For this reason, the notion of ACID transactions and their associated implementation techniques have remained largely within the DBMS domain, where the provision of highly available and reliable access to "mission critical" data is a primary concern.

46.2 Underlying Principles

Concurrency Control

Serializability

As stated in the previous section, the responsibility for maintaining the isolation property of ACID transactions resides in the concurrency-control portion of the DBMS software. The most widely accepted notion of correctness for concurrent execution of transactions is *serializability*. Serializability is the property that an (possibly interleaved) execution of a group of transactions has the same effect on the database, and produces the same output, as some serial (i.e., noninterleaved) execution of those transactions. It is important to note that serializability does not specify any *particular* serial order, but rather, only that the execution is equivalent to *some* serial order. This distinction makes serializability a slightly less intuitive notion of correctness than transaction initiation time or commit order, but it provides the DBMS with significant additional flexibility in the scheduling of operations. This flexibility can translate into increased responsiveness for end users.

A rich theory of database concurrency control has been developed over the years [see Papadimitriou 1986, Bernstein et al. 1987, Gray and Reuter 1993], and serializability lies at the heart of much of this theory. In this chapter we focus on the simplest models of concurrency control, where the operations that can be performed by transactions are restricted to *read(x)*, *write(x)*, *commit*, and *abort*. The operation *read(x)* retrieves the value of a data item from the database, *write(x)* modifies the value of a data item in the database, and *commit* and *abort* indicate successful or unsuccessful transaction completion respectively (with the concomitant guarantees provided by the ACID properties). We also focus on a specific variant of serializability called *conflict serializability*. Conflict serializability is the most widely accepted notion of correctness for concurrent transactions because there are efficient, easily implementable techniques for detecting and/or enforcing it. Another well-known variant is called *view serializability*. View serializability is less restrictive (i.e., it allows more legal schedules) than conflict serializability, but it and other variants are primarily of theoretical interest because they are impractical to implement. The reader is referred to [Papadimitriou 1986] for a detailed treatment of alternative serializability models.

Transaction Schedules

Conflict serializability is based on the notion of a **schedule** of transaction operations. A schedule for a set of transaction executions is a partial ordering of the operations performed by those transactions, which shows how the operations are interleaved. The ordering defined by a schedule can be partial in the sense that it is only required to specify two types of dependencies:

- All operations of a given transaction for which an order is specified by that transaction must appear in that order in the schedule. For example, the definition of REPORTSUM above specifies that account A is read before account B.
- The ordering of all **conflicting operations** from different transactions must be specified. Two operations are said to conflict if they both operate on the same data item and at least one of them is a *write( )*.

The concept of a schedule provides a mechanism to express and reason about the (possibly) concurrent execution of transactions. A *serial* schedule is one in which all the operations of each transaction appear consecutively. For example, the serial execution of TRANSFER followed by REPORTSUM is represented by the following schedule:

$$r_0[A] \rightarrow w_0[A] \rightarrow r_0[B] \rightarrow w_0[B] \rightarrow c_0 \rightarrow r_1[A] \rightarrow r_1[B] \rightarrow c_1 \qquad (46.1)$$

In this notation, each operation is represented by its initial letter, the subscript of the operation indicates the *transaction number* of the transaction on whose behalf the operation was performed, and a capital letter in brackets indicates a specific data item from the database (for read and write operations). A transaction number (tn) is a unique identifier that is assigned by the DBMS to an execution of a transaction. In the example above, the execution of TRANSFER was assigned tn 0 and the execution of REPORTSUM was assigned tn 1. A right arrow ($\rightarrow$) between two operations indicates that the left-hand operation is ordered before the right-hand one. The ordering relationship is transitive; the orderings implied by transitivity are not explicitly drawn.

For example, the interleaved execution of TRANSFER and REPORTSUM shown in Fig. 46.1 would produce the following schedule:

$$r_0[A] \rightarrow w_0[A] \rightarrow r_1[A] \rightarrow r_1[B] \rightarrow c_1 \rightarrow r_0[B] \rightarrow w_0[B] \rightarrow c_0 \qquad (46.2)$$

The formal definition of serializability is based on the concept of equivalent schedules. Two schedules are said to be *equivalent* ($\equiv$) if:

- they contain the same transactions and operations, and
- they order all conflicting operations of nonaborting transactions in the same way.

Given this notion of equivalent schedules, *a schedule is said to be serializable if and only if it is equivalent to some serial schedule*. For example, the following concurrent schedule is serializable because it is equivalent to schedule (46.1):

$$r_0[A] \rightarrow w_0[A] \rightarrow r_1[A] \rightarrow r_0[B] \rightarrow w_0[B] \rightarrow c_0 \rightarrow r_1[B] \rightarrow c_1 \qquad (46.3)$$

In contrast, the interleaved execution of schedule (46.2) is *not* serializable. To see why, notice that in any serial execution of TRANSFER and REPORTSUM either *both* writes of TRANSFER will precede *both* reads of REPORTSUM or vice versa. However, in schedule (46.2) $w_0[A] \rightarrow r_1[A]$ but $r_1[B] \rightarrow w_0[b]$. Schedule (46.2), therefore, is not equivalent to any possible serial schedule of the two transactions so it is not serializable. This result agrees with our intuitive notion of correctness, because recall that schedule (46.2) resulted in the apparent loss of $50.

Testing for Serializability

A schedule can easily be tested for serializability through the use of a *precedence graph*. A precedence graph is a directed graph that contains a vertex for each *committed* transaction execution in a schedule (noncommitted executions can be ignored). The graph contains an edge from transaction execution T_i to transaction execution T_j ($i \neq j$) if there is an operation in T_i that is constrained to precede an operation of T_j in the schedule. A schedule is serializable if and only if its precedence graph is *acyclic*. Figure 46.2(a) shows the precedence graph for schedule (46.2). That graph has an edge $T_0 \rightarrow T_1$ because the

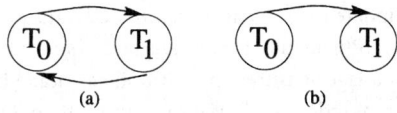

FIGURE 46.2 Precedence graphs for (a) non-serializable and (b) serializable schedules.

schedule contains $w_0[A] \rightarrow r_1[A]$ and an edge $T_1 \rightarrow T_0$ because the schedule contains $r_1[B] \rightarrow w_0[b]$. The cycle in the graph shows that the schedule is nonserializable. In contrast, Fig. 46.2(b) shows the precedence graph for schedule (46.1). In this case, all ordering constraints are from T_0 to T_1, so the precedence graph is acyclic, indicating that the schedule is serializable.

There are a number of practical ways to implement conflict serializability. These and other implementation issues are addressed in section 46.3. Before discussing implementation issues, however, we first survey the basic principles underlying database recovery.

Recovery

Coping with Failures

Recall that the responsibility for the atomicity and durability properties of ACID transactions lies in the recovery component of the DBMS. For recovery purposes it is necessary to distinguish between two types of storage: (1) **volatile storage**, such as main memory, whose state is lost in the event of a system crash or power outage, and (2) **nonvolatile storage**, such as magnetic disks or tapes, whose contents persist across such events. The recovery subsystem is relied upon to ensure correct operation in the presence of three different types of failures (listed in order of likelihood):

- *Transaction failure:* When a transaction that is in progress reaches a state from which it cannot successfully commit, all updates that it made must be removed from the database in order to preserve the atomicity property. This is known as *transaction rollback*.

- *System failure:* If the system fails in a way that causes the loss of volatile memory contents, recovery must ensure that: (1) the updates of all transactions that had committed prior to the crash are reflected in the database and (2) all updates of other transactions (aborted or in progress at the time of the crash) are removed from the database.

- *Media failure:* In the event that data are lost or corrupted on the nonvolatile storage (e.g., due to a disk-head crash), then the on-line version of the data is lost. In this case, the database must be restored from an archival version of the database and brought up to date using operation logs.

In this chapter we focus on the issues of rollback and crash recovery, the most frequent uses of the DBMS recovery subsystem. Recovery from media crashes requires substantial additional mechanisms and complexity beyond what is covered here. Media recovery is addressed in the recovery-related references listed at the end of this chapter.

Buffer Management Issues

The process of removing the effects of an incomplete or aborted transaction for preserving atomicity is known as *UNDO*. The process of reinstating the effects of a committed transaction for durability is known as *REDO*. The amount of work that a recovery subsystem must perform for either of these functions depends on how the DBMS buffer manager handles data that are updated by in-progress and/or committing transactions [Haerder and Reuter 1983, Bernstein et al. 1987]. Recall that the buffer manager is the DBMS component that is responsible for coordinating the transfer of data between main memory (i.e., volatile storage) and disk (i.e., nonvolatile storage). The unit of storage that can be written atomically to nonvolatile storage is called a *page*. Updates are made to copies of pages in the (volatile) buffer pool, and those copies are written out to nonvolatile storage at a later time. If the buffer manager allows an update made by an *uncommitted* transaction to overwrite the most recent committed value of a data item on nonvolatile storage, it is said to support a *STEAL* policy (the opposite is called *NO-STEAL*). If the buffer manager ensures that all updates made by a transaction are reflected on nonvolatile storage before the transaction is allowed to commit, then it is said to support a *FORCE* policy (the opposite is *NO-FORCE*).

Support for the STEAL policy implies that in the event that a transaction needs to be rolled back (due to transaction failure or system crash), UNDOing the transaction will involve restoring the values of any nonvolatile copies of data that were overwritten by that transaction back to their previous committed state. In contrast, a NO-STEAL policy guarantees that the data values on nonvolatile storage are valid, so they do not need to be restored. A NO-FORCE policy raises the possibility that some committed data values may be lost during a system crash because there is no guarantee that they have been placed on nonvolatile storage. This means that substantial REDO work may be required to preserve the durability of committed updates. In contrast, a FORCE policy ensures that the committed updates *are* placed on nonvolatile storage, so that in the event of a system crash, the updates will still be reflected in the copy of the database on nonvolatile storage.

From the above discussion, it should be apparent that a buffer manager that supports the combination of NO-STEAL and FORCE would place the fewest demands on UNDO and REDO recovery. However, these policies may negatively impact the performance of the DBMS during normal operation (i.e., when there are no crashes or rollbacks) because they restrict the flexibility of the buffer manager. NO-STEAL obligates the buffer manager to retain updated data in memory until a transaction commits or to write those data to a temporary location on nonvolatile storage (e.g., a swap area). The problem with a FORCE policy is that it can impose significant disk write overhead during the critical path of a committing transaction. For these reasons, many buffer managers support the STEAL and NO-FORCE (**STEAL/NO-FORCE**) policies.

Logging

In order to deal with the UNDO and REDO requirements imposed by the STEAL and NO-FORCE policies respectively, database systems typically rely on the use of a **log**. A log is a sequential file that stores information about transactions and the state of the system at certain instances. Each entry in the log is called a **log record**. One or more log records are written for each update performed by a transaction. When a log record is created, it is assigned a **log sequence number** (LSN) which serves to uniquely identify that record in the log. LSNs are typically assigned in a monotonically increasing fashion so that they provide an indication of relative position in the log. When an update is made to a data item in the buffer, a log record is created for that update. Many systems write the LSN of this new log record into the page containing the updated data item. Recording LSNs in this fashion allows the recovery system to relate the state of a data page to logged updates in order to tell if a given log record is reflected in a given state of a page.

Log records are also written for transaction management activities such as the commit or abort of a transaction. In addition, log records are sometimes written to describe the state of the system at certain periods of time. For example, such log records are written as part of the **checkpointing** process. Checkpoints are taken periodically during normal operation to help bound the amount of recovery work that would be required in the event of a crash. Part of the checkpointing process involves the writing of one or more *checkpoint records*. These records can include information about the contents of the buffer pool and the transactions that are currently active, etc. The particular contents of these records depend on the method of checkpointing that is used. Many different checkpointing methods have been developed, some of which involve quiescing the system to a consistent state, while others are less intrusive. A particularly nonintrusive type of checkpointing is used by the ARIES recovery method [Mohan et al. 1992] that is described in section 46.3.

For transaction update operations there are two basic types of logging: *physical* and *logical* [Gray and Reuter 1993]. Physical log records typically indicate the location (e.g., position on a particular page) of modified data in the database. If support for UNDO is provided (i.e., a STEAL policy is used), then the value of the item prior to the update is recorded in the log record. This is known as the *before image* of the item. Similarly the *after image* (i.e., the new value of the item after the update), is logged if REDO support is provided. Thus, physical log records in a DBMS with STEAL/NO-FORCE buffer management contain both the old and new data values of items. Recovery using physical log records has the property that recovery actions (i.e., UNDOs or REDOs) are *idempotent*, meaning that they have the same effect no matter how many times they are applied. This property is important if recovery is invoked multiple times, as will occur if a system fails repeatedly (e.g., due to a power problem or a faulty device).

Logical logging (sometimes referred to as *operational logging*) records only high-level information about operations that are performed, rather than recording the actual changes to items (or storage locations) in the database. For example, the insertion of a new tuple into a relation might require many physical changes to the database such as space allocation, index updates, and reorganization, etc. Physical logging would require log records to be written for all of these changes. In contrast, logical logging would simply log the fact that the insertion had taken place, along with the value of the inserted tuple. The REDO process for a logical logging system must determine the set of actions that are required to fully reinstate the insert. Likewise, the UNDO logic must determine the set of actions that make up the inverse of the logged operation.

Logical logging has the advantage that it minimizes the amount of data that must be written to the log. Furthermore, it is inherently appealing because it allows many of the implementation details of complex operations to be hidden in the UNDO/REDO logic. In practice however, recovery based on logical logging is difficult to implement because the actions that make up the logged operation are not performed atomically. That is, when a system is restarted after a crash, the database may not be in an *action consistent* state with respect to a complex operation—it is possible that only a subset of the updates made by the action had been placed on nonvolatile storage prior to the crash. As a result, it is difficult for the recovery system to determine which portions of a logical update are reflected in the database state upon recovery from a system crash. In contrast, physical logging does not suffer from this problem, but it can require substantially higher logging activity.

In practice, systems often implement a compromise between physical and logical approaches that has been referred to as *physiological logging* [Gray and Reuter 1993]. In this approach log records are constrained to refer to a single page, but may reflect logical operations on that page. For example, a physiological log record for an insert on a page would specify the value of the new tuple that is added to the page, but would not specify any free-space manipulation or reorganization of data on the page resulting from the insertion; the REDO and UNDO logic for insertion would be required to infer the necessary operations. If a tuple insert required updates to multiple pages (e.g., data pages plus multiple index pages), then a separate physiological log record would be written for each page updated. Physiological logging avoids the action consistency problem of logical logging, while reducing, to some extent, the amount of logging that would be incurred by physical logging. The ARIES recovery method is one example of a recovery method that uses physiological logging.

Write-ahead Logging (WAL)

A final recovery principle to be addressed in this section is the **write-ahead logging** (WAL) protocol. Recall that the contents of volatile storage are lost in the event of a system crash. As a result, any log records that are not reflected on nonvolatile storage will also be lost during a crash. WAL is a protocol that ensures that in the event of a system crash, the recovery log contains sufficient information to perform the necessary UNDO and REDO work when a STEAL/NO-FORCE buffer management policy is used. The WAL protocol ensures that:

1. All log records pertaining to an updated page are written to nonvolatile storage before the page itself is allowed to be overwritten in nonvolatile storage.
2. A transaction is not considered to be committed until all of its log records (including its commit record) have been written to stable storage.

The first point ensures that UNDO information required due to the STEAL policy will be present in the log in the event of a crash. Similarly, the second point ensures that any REDO information required due to the NO-FORCE policy will be present in the nonvolatile log. The WAL protocol is typically enforced with special support provided by the DBMS buffer manager.

46.3 Best Practices

Concurrency Control

Two-phase Locking

The most prevalent implementation technique for concurrency control is locking. Typically, two types of locks are supported, *shared* (S) locks and *exclusive* (X) locks. The compatibility of these locks is defined by the *compatibility matrix* shown in Table 46.1. The compatibility matrix shows that two different transactions are allowed to hold S locks simultaneously on the same data item, but that X locks cannot be held on an item simultaneously with any other locks (by other transactions) on that item. S locks are used for protecting *read* access to data (i.e., multiple concurrent readers are allowed), and X locks are used for protecting *write* access to data. As long as a transaction is holding a lock, no other transaction is allowed to obtain a conflicting lock. If a transaction requests a lock that cannot be granted (due to a lock conflict), that transaction is *blocked* (i.e., prohibited from proceeding) until all the conflicting locks held by other transactions are released.

TABLE 46.1 Compatibility Matrix for S and X Locks

	S	X
S	y	n
X	n	n

S and X locks as defined in Table 46.1 directly model the semantics of conflicts used in the definition of conflict serializability. Therefore, locking can be used to enforce serializability. Rather than testing for serializability after a schedule has been produced (as was done in the previous section), the blocking of transactions due to lock conflicts can be used to prevent nonserializable schedules from *ever* being produced.

A transaction is said to be **well formed** with respect to *reads* if it always holds an S or an X lock on an item while reading it, and well formed with respect to *writes* if it always holds an X lock on an item while writing it. Unfortunately, restricting all transactions to be well formed is not sufficient to guarantee serializability. For example, a nonserializable execution such as that of schedule (46.2) is still possible using well formed transactions. Serializability can be enforced, however, through the use of **two-phase locking** (2PL). Two-phase locking requires that all transactions be well formed and that they respect the following rule: *Once a transaction has released a lock, it is not allowed to obtain any additional locks.* This rule results in transactions that have two phases:

1. a *growing phase* in which the transaction is acquiring locks, and
2. a *shrinking phase* in which locks are released.

The two-phase rule dictates that the transaction shifts from the growing phase to the shrinking phase at the instant it first releases a lock.

To see how 2PL enforces serializability, consider again schedule (46.2). Recall that the problem arises in this schedule because $w_0[A] \rightarrow r_1[A]$ but $r_1[B] \rightarrow w_0[B]$. This schedule could not be produced under 2PL, because transaction 1 (REPORTSUM) would be blocked when it attempted to read the value of A because transaction 0 would be holding an X lock on it. Transaction 0 would not be allowed to release this X lock before obtaining its X lock on B, and thus it would either abort or perform its update of B before transaction 1 is allowed to progress. In contrast, note that schedule (46.1) (the serial schedule) would be allowed in 2PL. 2PL would also allow the following (serializable) interleaved schedule:

$$r_1[A] \rightarrow r_0[A] \rightarrow r_1[B] \rightarrow c_1 \rightarrow w_0[A] \rightarrow r_0[B] \rightarrow w_0[B] \rightarrow c_0 \qquad (46.4)$$

It is important to note, however, that two-phase locking is sufficient but not necessary for implementing serializability. In other words, there are schedules that are serializable but would not be allowed by two-phase locking. Schedule (46.3) is an example of such a schedule.

In order to implement 2PL, the DBMS contains a component called a *lock manager*. The lock manager is responsible for granting or blocking lock requests, for managing queues of blocked transactions, and for unblocking transactions when locks are released. In addition, the lock manager is also responsible for dealing with **deadlock** situations. A deadlock arises when a set of transactions is blocked, each waiting for another member of the set to release a lock. In a deadlock situation, none of the transactions involved can make progress. Database systems deal with deadlocks using one of two general techniques: avoidance or detection. Deadlock avoidance can be achieved by imposing an order in which locks can be obtained on data, by requiring transactions to predeclare their locking needs, or by aborting transactions rather than blocking them in certain situations.

Deadlock detection, on the other hand, can be implemented using *timeouts* or explicit checking. Timeouts are the simplest technique; if a transaction is blocked beyond a certain amount of time, it is assumed that a deadlock has occurred. The choice of a timeout interval can be problematic, however. If it is too short, then the system may infer the presence of a deadlock that does not truly exist. If it is too long, then deadlocks may go undetected for too long a time. Alternatively the system can explicitly check for deadlocks using a structure called a *waits-for graph*. A waits-for graph is a directed graph with a vertex for each active transaction. The lock manager constructs the graph by placing an edge from a transaction T_i to a transaction T_j ($i \neq j$) if T_i is blocked waiting for a lock held by T_j. If the waits-for graph contains a cycle, all of the transactions involved in the cycle are waiting for each other, and thus they are deadlocked. When a deadlock is detected, one or more of the transactions involved is rolled back. When a transaction is rolled back its locks are automatically released, so the deadlock will be broken.

Isolation Levels

As should be apparent from the previous discussion, transaction isolation comes at a cost in potential concurrency. Transaction blocking can add significantly to transaction response time.[2] As stated previously, serializability is typically implemented using two-phase locking, which requires locks to be held at least until all necessary locks have been obtained. Prolonging the holding time of locks increases the likelihood of blocking due to data contention.

In some applications, however, serializability is not strictly necessary. For example, a data analysis program that computes aggregates over large numbers of tuples may be able to tolerate some inconsistent access to the database in exchange for improved performance. The concept of *degrees of isolation* or *isolation levels* has been developed to allow transactions to trade concurrency for consistency in a controlled manner [Gray et al. 1975, Gray and Reuter 1993, Berenson et al. 1995]. In their 1975 paper, Gray et al. defined four degrees of consistency using characterizations based on locking, dependencies, and anomalies (i.e., results that could not arise in a serial schedule). The degrees were named degree 0–3, with degree 0 being the least consistent, and degree 3 intended to be equivalent to serializable execution.

The original presentation has served as the basis for understanding relaxed consistency in many current systems, but it has become apparent over time that the different characterizations in that paper were not specified to an equal degree of detail. As pointed out in a recent paper by Berenson et al. [1995], the SQL-92 standard suffers from a similar lack of specificity. Berenson et al. have attempted to clarify the issue, but it is too early to determine if they have been successful. In this section we focus on the locking-based definitions of the isolation levels, as they are generally acknowledged to have "stood the test of time" [Berenson et al. 1995]. However, the definition of the degrees of consistency requires an extension to the previous description of locking in order to address the *phantom problem*.

An example of the phantom problem is the following: assume a transaction T_i reads a set of tuples that satisfy a query predicate. A second transaction T_j inserts a new tuple that satisfies the predicate. If T_i then executes the query again, it will see the new item, so that its second answer differs from the first. This behavior could never occur in a serial schedule, as a "phantom" tuple appears in the midst of a transaction; thus, this execution is anomalous. The phantom problem is an artifact of the transaction model, consisting of reads and writes to *individual* data that we have used so far. In practice, transactions include *queries* that dynamically define sets based on predicates. When a query is executed, all of the tuples that satisfy the predicate at that time can be locked as they are accessed. Such individual locks, however, do not protect against the later addition of further tuples that satisfy the predicate.

---

[2]Note that other, non-blocking approaches discussed later in this section also suffer from similar problems.

One obvious solution to the phantom problem is to lock predicates instead of (or in addition to) individual items [Eswaran et al. 1976]. This solution is impractical to implement, however, due to the complexity of detecting the overlap of a set of arbitrary predicates. Predicate locking can be approximated using techniques based on locking clusters of data or ranges of index values. Such techniques, however, are beyond the scope of this chapter. In this discussion we will assume that predicates can be locked without specifying the technical details of how this can be accomplished (see [Gray and Reuter 1993, Mohan 1992] for detailed treatments of this topic).

The locking-oriented definitions of the isolation levels are based on whether or not read and/or write operations are well formed (i.e., protected by the appropriate lock), and if so, whether those locks are *long duration* or *short duration*. Long-duration locks are held until the end of a transaction (EOT) (i.e., when it commits or aborts); short-duration locks can be released earlier. Long-duration write locks on data items have important benefits for recovery, namely, they allow recovery to be performed using *before images*. If long-duration write locks are not used, then the following scenario could arise:

$$w_0[A] \rightarrow w_1[A] \rightarrow a_0 \tag{46.5}$$

In this case restoring A with T_0's before image of it will be incorrect because it would overwrite T_1's update. Simply ignoring the abort of T_0 is also incorrect. In that case, if T_1 were to subsequently abort, installing its before image would reinstate the value written by T_0. For this reason and for simplicity, locking systems typically hold long-duration locks on data items. This is sometimes referred to as *strict* locking [Bernstein et al. 1987].

Given these notions of locks, the degrees of isolation presented in the SQL-92 standard can be obtained using different lock protocols. In the following, all levels are assumed to be well formed with respect to writes and to hold long duration *write* (i.e., exclusive) locks on updated data items. Four levels are defined (from weakest to strongest)[3]:

> READ UNCOMMITTED: This level, which provides the weakest consistency guarantees, allows transactions to read data that have been written by other transactions that have not committed. In a locking implementation this level is achieved by being ill formed with respect to reads (i.e., not obtaining read locks). The risks of operating at this level include (in addition to the risks incurred at the more restrictive levels) the possibility of seeing updates that will eventually be rolled back and the possibility of seeing some of the updates made by another transaction but missing others made by that transaction.

> READ COMMITTED: This level ensures that transactions only see updates that have been made by transactions that have committed. This level is achieved by being well formed with respect to reads on individual data items, but holding the read locks only as short-duration locks. Transactions operating at this level run the risk of seeing *nonrepeatable* reads (in addition to the risks of the more restrictive levels). That is, a transaction T_0 could read a data item twice and see two different values. This anomaly could occur if a second transaction were to update the item and commit in between the two reads by T_0.

> REPEATABLE READ: This level ensures that reads to individual data items are repeatable, but does not protect against the phantom problem described previously. This level is achieved by being well formed with respect to reads on individual data items, and holding those locks for long duration.

> SERIALIZABLE: This level protects against all of the problems of the less restrictive levels, including the phantom problem. It is achieved by being well formed with respect to reads on *predicates* as well as on individual data items and holding all locks for long duration.

---

[3]It should be noted that two-phase locks can be substituted for the long-duration locks in these definitions without impacting the consistency provided. Long-duration locks are typically used, however, to avoid the recovery-related problems described previously.

A key aspect of this definition of degrees of isolation is that as long as all transactions execute at the READ UNCOMMITTED level or higher, they are able to obtain at least the degree of isolation they desire without interference from any transactions running at lower degrees. Thus, these degrees of isolation provide a powerful tool that allows application writers or users to trade off consistency for improved concurrency. As stated earlier, the definition of these isolation levels for concurrency-control methods that are not based on locking has been problematic. This issue is addressed in depth in [Berenson et al. 1995].

It should be noted that the discussion of locking so far has ignored an important class of data that is typically present in databases, namely, *indexes*. Because indexes are auxiliary information, they can be accessed in a non-two-phase manner without sacrificing serializability. Furthermore, the hierarchical structure of many indexes (e.g., B-trees) makes them potential concurrency bottlenecks due to high contention at the upper levels of the structure. For this reason, significant effort has gone into developing methods for providing highly concurrent access to indexes. Pointers to some of this work can be found in the Further Information section at the end of this chapter.

Hierarchical Locking

The examples in the preceeding discussions of concurrency control primarily dealt with operations on a single granularity of data items (e.g., tuples). In practice, however, the notions of conflicts and locks can be applied at many different granularities. For example, it is possible to perform locking at the granularity of a page, a relation, or even an entire database. In choosing the proper granularity at which to perform locking there is a fundamental tradeoff between potential concurrency and locking overhead. Locking at a fine granularity, such as an individual tuple, allows for maximum concurrency, as only transactions that are truly accessing the same tuple have the potential to conflict. The downside of such fine-grained locking, however, is that a transaction that accesses a large number of tuples will have to acquire a large number of locks. Each lock request requires a call to the lock manager. This overhead can be reduced by locking at a coarser granularity, but coarse granularity raises the potential for *false conflicts*. For example, two transactions that update different tuples residing on the same page would conflict under page-level locking but not under tuple-level locking.

The notion of hierarchical or multigranular locking was introduced to allow concurrent transactions to obtain locks at different granularities in order to optimize the above tradeoff [Gray et al. 1975]. In hierarchical locking, a lock on a granule at a particular level of the granularity hierarchy implicitly locks all items included in that granule. For example, an S-lock on a relation implicitly locks all pages and tuples in that relation. Thus, a transaction with such a lock can read any tuple in the relation without requesting additional locks. Hierarchical locking introduces additional lock modes beyond S and X. These additional modes allow transactions to declare their *intention* to perform an operation on objects at lower levels of the granularity hierarchy. The new modes are IS, IX, and SIX for *intention shared, intention exclusive,* and *shared with intention exclusive.* An IS (or IX) lock on a granule provides no privileges on that granule, but indicates that the holder intends to obtain S (or X) locks on one or more finer granules. An SIX lock combines an S lock on the entire granule with an IX lock. SIX locks support the common access pattern of scanning the items in a granule (e.g., tuples in a relation) and choosing to update a fraction of them based on their values.

Similarly to S and X locks, these lock modes can be described using a compatibility matrix. The compatibility matrix for these modes is shown in Table 46.2. In order for transactions locking at different granularities to coexist, all transactions must follow the same hierarchical locking protocol starting from the root of the granularity hierarchy. This protocol is shown in Table 46.3. For example, to read a single record,

TABLE 46.2 Compatibility Matrix for Regular and Intention Locks

	IS	IX	S	SIX	X
IS	y	y	y	y	n
IX	y	y	n	n	n
S	y	n	y	n	n
SIX	y	n	n	n	n
X	n	n	n	n	n

TABLE 46.3 Hierarchical Locking Rules

To Get	Must Have on All Ancestors
IS or S	IS or IX
IX, SIX, or X	IX or SIX

a transaction would obtain IS locks on the database, relation, and page, followed by an S lock on the specific tuple. If a transaction wanted to read all or most tuples on a page, then it could obtain IS locks on the database and relation, followed by an S lock on the entire page. By following this uniform protocol, potential conflicts between transactions that ultimately obtain S and/or X locks at different granularities can be detected.

A useful extension to hierarchical locking is known as *lock escalation.* Lock escalation allows the DBMS to automatically adjust the granularity at which transactions obtain locks, based on their behavior. If the system detects that a transaction is obtaining locks on a large percentage of the granules that make up a larger granule, it can attempt to grant the transaction a lock on the larger granule so that no additional locks will be required for subsequent accesses to other objects in that granule. Automatic escalation is useful because the access pattern that a transaction will produce is often not known until run time.

Other Concurrency Control Methods

As stated previously, two-phase locking is the most generally accepted technique for ensuring serializability. Locking is considered to be a *pessimistic* technique because it is based on the assumption that transactions are likely to interfere with each other and takes measures (e.g., blocking) to ensure that such interference does not occur. An important alternative to locking is **optimistic concurrency control**. Optimistic methods [e.g., Kung and Robinson 1981] allow transactions to perform their operations without obtaining any locks. To ensure that concurrent executions do not violate serializability, transactions must perform a *validation phase* before they are allowed to commit. Many optimistic protocols have been proposed. In the algorithm of [Kung and Robinson 1981], the validation process ensures that the reads and writes performed by a validating transaction did not conflict with any other transactions with which it ran concurrently. If during validation it is determined a conflict had occurred, the validating transaction is aborted and restarted.

Unlike locking, which depends on *blocking* transactions to ensure isolation, optimistic policies depend on transaction *restart.* As a result, although they don't perform any blocking, the performance of optimistic policies can be hurt by data contention (as are pessimistic schemes)—a high degree of data contention will result in a large number of unsuccessful transaction executions. The performance tradeoffs between optimistic and pessimistic have been addressed in numerous studies [see Agrawal et al. 1987]. In general, locking is likely to be superior in resource-limited environments because blocking does not consume cpu or disk resources. In contrast, optimistic techniques may have performance advantages in situations where resources are abundant, because they allow more executions to proceed concurrently. If resources are abundant, then the resource consumption of restarted transactions will not significantly hurt performance. In practice, however resources are typically limited, and thus concurrency control in most commercial database systems is based on locking.

Another class of concurrency control techniques is known as **multiversion concurrency control** [e.g., Reed 1983]. As updating transactions modify data items, these techniques retain the previous versions of the items on line. Read-only transactions (i.e., transactions that perform no updates) can then be provided with access to these older versions, allowing them to see a consistent (although possibly somewhat out-of-date) snapshot of the database. Optimistic, multiversion, and other concurrency control techniques (e.g., timestamping) are addressed in further detail in [Bernstein et al. 1987].

Recovery

The recovery subsystem is generally considered to be one of the more difficult parts of a DBMS to design for two reasons: First, recovery is required to function in failure situations and must correctly cope with a huge number of possible system and database states. Second, the recovery system depends on the behavior of many other components of the DBMS, such as concurrency control, buffer management, disk management, and query processing. As a result, few recovery methods have been described in the literature in detail. One exception is the ARIES recovery system developed at IBM [Mohan et al. 1992]. Many details about the ARIES method have been published, and the method has been included in a number of DBMSs. Furthermore, the ARIES method involves only a small number of basic concepts. For these reasons, we focus on the ARIES method in the remainder of this section. The ARIES method is related to many other recovery methods such as those described in [Bernstein et al. 1987, Gray and Reuter 1993]. A comparison with other techniques appears in [Mohan et al. 1992].

Overview of ARIES

ARIES is a fairly recent refinement of the **write-ahead-logging** (WAL) protocol. Recall that the WAL protocol enables the use of a STEAL/NO FORCE buffer management policy, which means that pages on stable storage can be overwritten at any time and that data pages do not need to be forced to disk in order to commit a transaction. As with other WAL implementations, each page in the database contains a **log sequence number** (LSN) which uniquely identifies the log record for the latest update that was applied to the page. This LSN (referred to as the *pageLSN*) is used during recovery to determine whether or not an update for a page must be redone. LSN information is also used to determine the point in the log from which the REDO pass must commence during restart from a system crash. LSNs are often implemented using the physical address of the log record in the log to enable the efficient location of a log record given its LSN.

Much of the power and relative simplicity of the ARIES algorithm is due to its REDO paradigm of *repeating history*, in which it redoes updates for *all* transactions—including those that will eventually be undone. Repeating history enables ARIES to employ a variant of the *physiological logging* technique described earlier: it uses *page-oriented REDO* and a form of *logical UNDO*. Page-oriented REDO means that REDO operations involve only a single page and that the affected page is specified in the log record. This is part of physiological logging. In the context of ARIES, logical UNDO means that the operations performed to undo an update do not need to be the exact inverses of the operations of the original update.

In ARIES, logical UNDO is used to support fine-grained (i.e., tuple-level) locking and high-concurrency index management. For an example of the latter issue, consider a case in which a transaction T1 updates an index entry on a given page P1. Before T1 completes, a second transaction T2 could split P1, causing the index entry to be moved to a new page (P2). If T1 must be undone, a physical, page-oriented approach would fail because it would erroneously attempt to perform the UNDO operation on P1. Logical UNDO solves this problem by using the index structure to find the index entry, and then applying the UNDO operation to it in its new location. In contrast to UNDO, page-oriented REDO can be used because the repeating history paradigm ensures that REDO operations will always find the index entry on the page referenced in the log record—any operations that had affected the location of the index operation at the time the log record was created will be replayed before that log record is redone.

ARIES uses a three-pass algorithm for restart recovery. The first pass is the *analysis* pass, which processes the log forward from the most recent checkpoint. This pass determines information about dirty pages and active transactions that is used in the subsequent passes. The second pass is the *REDO* pass, in which history is repeated by processing the log forward from the earliest log record that could require REDO, thus ensuring that all logged operations have been applied. The third pass is the *UNDO* pass. This pass proceeds backwards from the end of the log, removing from the database the effects of all transactions that had not committed at the time of the crash. These passes are shown in Fig. 46.3. (Note that the relative ordering of the starting point for the REDO pass, the endpoint for the UNDO pass, and the checkpoint can be different than that shown in the figure.) The three passes are described in more detail below.

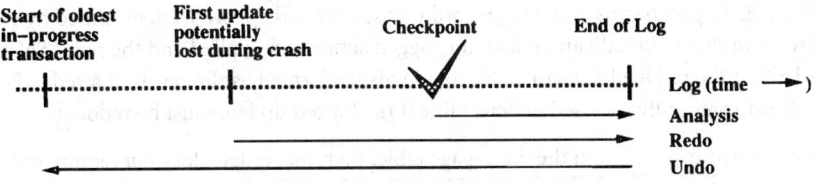

FIGURE 46.3 The three passes of ARIES restart.

ARIES maintains two important data structures during normal operation. The first is the *transaction table*, which contains status information for each transaction that is currently running. This information includes a field called the *lastLSN*, which is the LSN of the most recent log record written by the transaction. The second data structure, called the *dirty-page table*, contains an entry for each "dirty" page. A page is considered to be dirty if it contains updates that are not reflected on stable storage. Each entry in the dirty-page table includes a field called the *recoveryLSN*, which is the LSN of the log record that caused the associated page to become dirty. Therefore, the *recoveryLSN* is the LSN of the earliest log record that might need to be redone for the page during restart. Log records belonging to the same transaction are linked backwards in time using a field in each log record called the *prevLSN* field. When a new log record is written for a transaction, the value of the *lastLSN* field in the transaction-table entry is placed in the *prevLSN* field of the new record and the new record's LSN is entered as the *lastLSN* in the transaction-table entry.

During normal operation, checkpoints are taken periodically. ARIES uses a form of fuzzy checkpoints which are extremely inexpensive. When a checkpoint is taken, a checkpoint record is constructed which includes the contents of the transaction table and the dirty-page table. Checkpoints are efficient, since no operations need be quiesced and no database pages are flushed to perform a checkpoint. However, the effectiveness of checkpoints in reducing the amount of the log that must be maintained is limited in part by the earliest *recoveryLSN* of the dirty pages at checkpoint time. Therefore, it is helpful to have a background process that periodically writes dirty pages to non-volatile storage.

Analysis

The job of the analysis pass of restart recovery is threefold: (1) it determines the point in the log at which to start the REDO pass, (2) it determines which pages could have been dirty at the time of the crash in order to avoid unnecessary I/O during the REDO pass, and (3) it determines which transactions had not committed at the time of the crash and will therefore need to be undone.

The analysis pass begins at the most recent checkpoint and scans forward to the end of the log. It reconstructs the transaction table and dirty-page table to determine the state of the system as of the time of the crash. It begins with the copies of those structures that were logged in the checkpoint record. Then, the contents of the tables are modified according to the log records that are encountered during the forward scan. When a log record for a transaction that does not appear in the transaction table is encountered, that transaction is added to the table. When a log record for the commit or the abort of a transaction is encountered, the corresponding transaction is removed from the transaction table. When a log record for an update to a page that is not in the dirty-page table is encountered, that page is added to the dirty-page table, and the LSN of the record which caused the page to be entered into the table is recorded as the *recoveryLSN* for that page. At the end of the analysis pass, the dirty-page table is a conservative (since some pages may have been flushed to nonvolatile storage) list of all database pages that could have been dirty at the time of the crash, and the transaction table contains entries for those transactions that will actually require undo processing during the UNDO phase. The earliest *recoveryLSN* of all the entries in the dirty-page table, called the *firstLSN*, is used as the spot in the log from which to begin the REDO phase.

REDO

As stated earlier, ARIES employs a redo paradigm called *repeating history*. That is, it redoes updates for *all* transactions, committed or otherwise. The effect of repeating history is that at the end of the REDO pass, the database is in the same state with respect to the logged updates that it was in at the time that the crash occurred. The REDO pass begins at the log record whose LSN is the *firstLSN* determined by analysis and scans forward from there. To redo an update, the logged action is reapplied and the *pageLSN* on the page is set to the LSN of the redone log record. No logging is performed as the result of a redo. For each log record the following algorithm is used to determine if the logged update must be redone:

- If the affected page is not in the dirty-page table, then the update does *not* require redo.
- If the affected page is in the dirty-page table, but the *recoveryLSN* in the page's table entry is *greater than* the LSN of the record being checked, then the update does *not* require redo.

- Otherwise, the LSN stored on the page (the *pageLSN*) must be checked. This may require that the page be read in from disk. If the *pageLSN* is *greater than or equal to* the LSN of the record being checked, then the update does *not* require redo. Otherwise, the update *must* be redone.

UNDO

The UNDO pass scans backwards from the end of the log. During the UNDO pass, all transactions that had not committed by the time of the crash must be undone. In ARIES, undo is an *unconditional* operation. That is, the *pageLSN* of an affected page is not checked, because it is always the case that the undo must be performed. This is due to the fact that the *repeating of history* in the REDO pass ensures that all logged updates have been applied to the page.

When an update is undone, the undo operation is applied to the page and is logged using a special type of log record called a *compensation log record* (CLR). In addition to the undo information, a CLR contains a field called the *UndoNxtLSN*. The *UndoNxtLSN* is the LSN of the next log record that must be undone for the transaction. It is set to the value of the *prevLSN* field of the log record being undone. The logging of CLRs in this fashion enables ARIES to avoid ever having to undo the effects of an undo (e.g., as the result of a system crash during an abort), thereby limiting the amount of work that must be undone and bounding the amount of logging done in the event of multiple crashes. When a CLR is encountered during the backward scan, no operation is performed on the page, and the backward scan continues at the log record referenced by the *UndoNxtLSN* field of the CLR, thereby jumping over the undone update and all other updates for the transaction that have already been undone (the case of multiple transactions will be discussed shortly). An example execution is shown in Fig. 46.4.

In Fig. 46.4, a transaction logged three updates (LSNs 10, 20, and 30) before the system crashed for the first time. During REDO, the database was brought up to date with respect to the log (i.e., 10, 20, and/or 30 were redone if they weren't on nonvolatile storage), but since the transaction was in progress at the time of the crash, they must be undone. During the UNDO pass, update 30 was undone, resulting in the writing of a CLR with LSN 40, which contains an *UndoNxtLSN* value that points to 20. Then, 20 was undone, resulting in the writing of a CLR (LSN 50) with an *UndoNxtLSN* value that points to 10. However, the system then crashed for a second time before 10 was undone. Once again, history is repeated during REDO, which brings the database back to the state it was in after the application of LSN 50 (the CLR for 20). When UNDO begins during this second restart, it will first examine the log record 50. Since the record is a CLR, no modification will be performed on the page, and UNDO will skip to the record whose LSN is stored in the *UndoNxtLSN* field of the CLR (i.e., LSN 10). Therefore, it will continue by undoing the update whose log record has LSN 10. This is where the UNDO pass was interrupted at the time of the second crash. Note that no extra logging was performed as a result of the second crash.

In order to undo multiple transactions, restart UNDO keeps a list containing the next LSN to be undone for each transaction being undone. When a log record is processed during UNDO, the *prevLSN* (or *UndoNxtLSN*, in the case of a CLR) is entered as the next LSN to be undone for that transaction. Then the UNDO pass moves on to the log record whose LSN is the most recent of the next LSNs to be redone. UNDO continues backward in the log until all of the transactions in the list have been undone up to and including their first log record. UNDO for *transaction rollback* works similarly to the UNDO pass of the restart algorithm as described above. The only difference is that during transaction rollback, only a single transaction (or part of a transaction) must be undone. Therefore, rather than keeping a list of LSNs to

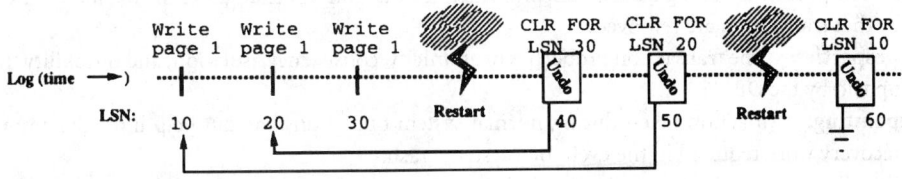

FIGURE 46.4 The use of CLRs for UNDO.

be undone for multiple transactions, rollback can simply follow the backward chain of log records for the transaction to be rolled back.

46.4 Research Issues and Summary

The model of ACID transactions that has been described in this chapter has proven to be quite durable in its own right, and serves as the underpinning for the current generation of database and transaction processing systems. This chapter has focused on the issues of concurrency control and recovery in a centralized environment. It is important to note, however, that the basic model is used in many types of distributed and parallel DBMS environments and the mechanisms described here have been successfully adapted for use in these more complex systems. Additional techniques, however, are needed in such environments. One important technique is *two-phase commit*, which is a protocol for ensuring that all participants in a distributed transaction agree on the decision to commit or abort that transaction.

While the basic transaction model has been a clear success, its limitations have also been apparent for quite some time [e.g., Gray 1981]. Much of the ongoing research related to concurrency control and recovery is aimed at addressing some of these limitations. This research includes the development of new implementation techniques, as well as the investigation of new and extended transaction models.

The ACID transaction model suffers from a lack of flexibility and the inability to model many types of interactions that arise in complex systems and organizations. For example, in collaborative work environments, strict isolation is not possible or even desirable [Korth 1995]. Workflow management systems are another example where the ACID model, which works best for relatively simple and short transactions, is not directly appropriate. For these types of applications, a richer, multilevel notion of transactions is required.

In addition to the problems raised by complex application environments, there are also many computing environments for which the ACID model is not fully appropriate. These include environments such as mobile wireless networks, where large periods of disconnection are expected, and loosely coupled wide-area networks (the Net is an extreme example) in which the availability of systems is relatively low. The techniques that have been developed for supporting ACID transactions must be adjusted to cope with such highly variable situations. New techniques must also be developed to provide concurrency control and recovery in nontraditional environments such as heterogeneous systems, dissemination-oriented environments, and others.

A final limitation of ACID transactions in their simplest form is that they are a general mechanism, and hence, do not exploit the semantics of data and/or applications. Such knowledge could be used to significantly improve system performance. Therefore, the development of concurrency control and recovery techniques that can exploit application-specific properties is another area of active research.

As should be obvious from the preceding discussion, there is still a significant amount of work that remains to be done in the areas of concurrency control and recovery for database systems. The basic concepts, however, such as serializability theory, two-phase locking, write-ahead logging, etc. will continue to be a fundamental technology, both in their own right and as building blocks for the development of more sophisticated and flexible information systems.

Defining Terms

Abort: The process of rolling back an uncommitted transaction. All changes to the database state made by that transaction are removed.

ACID properties: The transaction properties of atomicity, consistency, isolation, and durability that are upheld by the DBMS.

Checkpointing: An action taken during normal system operation that can help limit the amount of recovery work required in the event of a system crash.

Commit: The process of successfully completing a transaction. Upon commit, all changes to the database state made by a transaction are made permanent and visible to other transactions.

Concurrency control: The mechanism that ensures that individual users see consistent states of the database even though operations on behalf of many users may be interleaved by the database system.

Concurrent execution: The (possibly) interleaved execution of multiple transactions simultaneously.

Conflicting operations: Two operations are said to conflict if they both operate on the same data item and at least one of them is a *write( )*.

Deadlock: A situation in which a set of transactions is blocked, each waiting for another member of the set to release a lock. In such a case none of the transactions involved can make progress.

Log: A sequential file that stores information about transactions and the state of the system at certain instances.

Log record: An entry in the log. One or more log records are written for each update performed by a transaction.

Log sequence number (LSN): A number assigned to a log record, which serves to uniquely identify that record in the log. LSNs are typically assigned in a monotonically increasing fashion so that they provide an indication of relative position.

Multiversion concurrency control: A concurrency control technique that provides read-only transactions with conflict-free access to previous versions of data items.

Nonvolatile storage: Storage, such as magnetic disks or tapes, whose contents persist across power failures and system crashes.

Optimistic concurrency control: A concurrency control technique that allows transactions to proceed without obtaining locks and ensures correctness by validating transactions upon their completion.

Recovery: The mechanism that ensures that the database is fault-tolerant; that is, that the database state is not corrupted as the result of a software, system, or media failure.

Schedule: A schedule for a set of transaction executions is a partial ordering of the operations performed by those transactions, which shows how the operations are interleaved.

Serial execution: The execution of a single transaction at a time.

Serializability: The property that a (possibly interleaved) execution of a group transactions has the same effect on the database, and produces the same output, as some serial (i.e., non-interleaved) execution of those transactions.

STEAL/NO-FORCE: A buffer management policy that allows committed data values to be overwritten on nonvolatile storage and does not require committed values to be written to nonvolatile storage. This policy provides flexibility for the buffer manager at the cost of increased demands on the recovery subsystem.

Transaction: A unit of work, possibly consisting of multiple data accesses and updates, that must commit or abort as a single atomic unit. Transactions have the ACID properties of *atomicity, consistency, isolation*, and *durability*.

Two-phase locking (2PL): A locking protocol that is a sufficient but not a necessary condition for serializability. Two-phase locking requires that all transactions be well formed and that once a transaction has released a lock, it is not allowed to obtain any additional locks.

Volatile storage: Storage, such as main memory, whose state is lost in the event of a system crash or power outage.

Well formed: A transaction is said to be well formed with respect to reads if it always holds a shared or an exclusive lock on an item while reading it, and well formed with respect to writes if it always holds an exclusive lock on an item while writing it.

Write-ahead logging: A protocol that ensures all log records required to correctly perform recovery in the event of a crash are placed on nonvolatile storage.

References

Agrawal, R., Carey, M., and Livny, M. 1987. Concurrency control performance modeling: alternatives and implications. *ACM Trans. Database Systems* 12(4), Dec.

Berenson, H., Bernstein, P., Gray, J., Melton, J., Oneil, B., and Oneil, P. 1995. A critique of ANSI SQL Isolation Levels. In *Proc. ACM SIGMOD Int. Conf. on the Management of Data*, San Jose, CA, June.

Bernstein, P., Hadzilacos, V., and Goodman, N. 1987. *Concurrency Control and Recovery in Database Systems*. Addison–Wesley, Reading, MA.

Bjork, L. 1973. Recovery scenario for a DB/DC system. In *Proc. ACM Annual Conf.* Atlanta.

Davies, C. 1973. Recovery semantics for a DB/DC system. In *Proc. ACM Annual Conf.* Atlanta.

Eswaran, L., Gray, J., Lorie, R., and Traiger, I. 1976. The notion of consistency and predicate locks in a database system. *Commun. ACM* 19(11), Nov.

Gray, J., Lorie, R., Putzolu, G., and Traiger, I. 1975. Granularity of locks and degrees of consistency in a shared database. In *IFIP Working Conf. on Modelling of Database Management Systems*.

Gray, J. 1981. The transaction concept: virtues and limitations. In *Proc. Seventh International Conf. on Very Large Databases*, Cannes.

Gray, J. and Reuter, A. 1993. *Transaction Processing: Concepts and Techniques*. Morgan Kaufmann, San Mateo, CA.

Haerder, T. and Reuter, A. 1983. Principles of transaction-oriented database recovery. *ACM Comput. Surveys* 15(4).

Korth, H. 1995. The double life of the transaction abstraction: fundamental principle and evolving system concept. In *Proc. Twenty-First International Conf. on Very Large Databases*, Zurich.

Kung, H. and Robinson, J. 1981. On optimistic methods for concurrency control. *ACM Trans. Database Systems* 6(2).

Lomet, D. 1977. Process structuring, synchronization and recovery using atomic actions. *SIGPLAN Notices* 12(3), Mar.

Mohan, C. 1990. ARIES/KVL: a key-value locking method for concurrency control of multiaction transactions operating on B-tree indexes. In *Proc. 16th Int. Conf. on Very Large Data Bases*, Brisbane, Aug.

Mohan, C., Haderle, D., Lindsay, B., Pirahesh, H., and Schwarz, P. 1992. ARIES: a transaction method supporting fine-granularity locking and partial rollbacks using write-ahead logging. *ACM Trans. Database Systems* 17(1), Mar.

Papadimitriou, C. 1986. *The Theory of Database Concurrency Control*. Computer Science Press, Rockville, MD.

Reed, D. 1983. Implementing atomic actions on decentralized data. *ACM Trans. Comput. Systems* 1(1), Feb.

Rosenkrantz, D., Sterns, R., and Lewis, P. 1977. System level concurrency control for distributed database systems. *ACM Trans. Database Systems* 3(2).

Further Information

For many years, what knowledge that existed in the public domain about concurrency control and recovery was passed on primarily though the use of multiple-generation copies of a set of lecture notes written by Jim Gray in the late seventies ("Notes on Database Operating Systems" in *Operating Systems: An Advanced Course* published by Springer–Verlag, Berlin, 1978). Fortunately, this state of affairs has been supplanted by the publication of *Transaction Processing: Concepts and Techniques* by Jim Gray and Andreas Reuter (Morgan Kaufmann, San Mateo, CA, 1993). This latter book contains a detailed treatment of all of the topics covered in this chapter, plus many others that are crucial for implementing transaction processing systems.

An excellent treatment of concurrency control and recovery theory and algorithms can be found in *Concurrency Control and Recovery in Database Systems* by Phil Bernstein, Vassos Hadzilacos, and Nathan Goodman (Addison–Wesley, Reading, MA, 1987). Another source of valuable information on concurrency control and recovery implementation is the series of papers on the ARIES method by C. Mohan and others at IBM, some of which are referenced in this chapter. The book *The Theory of Database Concurrency Control* by Christos Papadimitriou (Computer Science Press, Rockville, MD, 1986) covers a number of serializability models.

The performance aspects of concurrency control and recovery techniques have been only briefly addressed in this chapter. More information can be found in the recent books *Performance of Concurrency Control Mechanisms in Centralized Database Systems* edited by Vijay Kumar (Prentice–Hall, Englewood Cliffs, NJ, 1996) and *Recovery in Database Management Systems*, edited by Vijay Kumar and Meichun Hsu (Prentice–Hall, Englewood Cliffs, NJ, in press). Also, the performance aspects of transactions are addressed in *The Benchmark Handbook: For Database and Transaction Processing Systems* (2nd ed.), edited by Jim Gray (Morgan Kaufmann, San Mateo, CA, 1993).

Finally, extensions to the ACID transaction model are discussed in *Database Transaction Models*, edited by Ahmed Elmagarmid (Morgan Kaufmann, San Mateo, CA, 1993). Papers containing the most recent work on related topics appear regularly in the ACM SIGMOD Conference and the International Conference on Very Large Databases (VLDB), among others.

47

Database Performance Measurement

Patrick O'Neil
University of
Massachusetts—Boston

47.1 Introduction

Our concern in this chapter will be with database benchmarks—performance tests by which different database hardware/software platforms are compared in terms of price and performance. The reader would benefit from having read other chapters of this Handbook as a prelude to what follows. Specifically, it would be useful to have already read Chapter 46 on Concurrency Control and Recovery, Chapter 43 on Tuning Database Design for High Performance, and Chapter 45 on Query Optimization.

As the computer field matures there is a desire to treat computerized work as a commodity resource—to rate the performance of various hardware/software platforms in the same way that electricity is measured in kilowatt hours or automobile engines in horsepower and mileage. To create such ratings, carefully defined tests are required that will measure performance in different application domains on different platforms. These performance measurement tests are known as *benchmarks*. To give an early example, the *Whetstone* benchmark [Curnow and Wichman 1976] was created to rate a variety of machines in how efficiently they performed a program of scientific calculations, including a substantial amount of floating-point arithmetic. To compare different machines fairly, the program was written in a high-level language, originally ALGOL-60 and later Fortran. The Whetstone benchmark measured a subset of capabilities of the CPU, together with how well the compiler translated high-level computational statements into efficient machine instructions.

As we will see, benchmarks exist for a large number of different database and transactional application domains. In what follows, we will refer to database processing and transaction processing under the single term *database processing*, although these terms are often differentiated by computer vendors (see [Gray 1993]). Most database benchmarks assume a large collection of persistent data, which must reside on disk or tape (with current technology), and which cannot be made memory-resident all at once. Because

0-8493-2909-4/97/$0.00+$.50
© 1997 by CRC Press, Inc.

of this, benchmark performance tuning often deals with rather sophisticated interrelationships among CPU, memory, and I/O. Even more complex considerations of concurrency and logging for recovery come into play with benchmarks of update transactions. The degree of sophistication and differentiation in database benchmarks is motivated by database customer demand. This demand is driven by the enormous dollar investment in database systems and tools, which currently amounts to about seven billion dollars a year. Because of the volume of investment, customers making acquisition decisions are eager to adopt any objective measures that might make these decisions easier.

Ideally, a good database benchmark will allow a prospective customer to compare ratings of different platforms in terms of their relative price/performance ratios before deciding on an acquisition. Other beneficiaries of such benchmarks include hardware designers creating new computer architectures and software engineers writing database systems. Benchmarks enable these groups to focus on improving specific performance measures that the industry has discovered to be important. Application designers also benefit by being able to plan large applications with a better understanding of what kind of performance they can expect to achieve: the throughput, the response time, and the appropriate cost per user. For all of these aspects the ultimate focus is on improving customer value; in what follows we concentrate our benchmark discussion on the customer's desire to make a platform acquisition that meets expected database needs at the lowest price.

47.2 Underlying Principles

There are a large number of database applications, leading to a wide variety of benchmarks to evaluate different kinds of work. Every potential database customer has a specific application to implement, but many of these generalize well to standard industry usage. We differentiate between a **custom benchmark** and a **generic benchmark**. A custom benchmark is a test created by a particular customer based on a specific application and measured on a number of platforms to aid in an acquisition decision. A generic benchmark is created to represent a commonly perceived paradigm of use in an application domain. The advantage of an intelligently designed custom benchmark is that it will measure precisely the capabilities that the customer wants; however, the cost of performing such a design and porting the benchmark to a number of different platforms can be daunting. Generic benchmarks, on the other hand, are created to be generally useful for a large group of potential customers with applications that are relatively standard across an application domain. Because they focus on a particular application domain, generic benchmarks are also known as *domain-specific* benchmarks. In addition, there is a commonly used approach of creating an inexpensive custom benchmark by starting with the framework of a generic benchmark, then modifying or adding tests to better reflect a specific need of the customer. The result is often referred to as a **hybrid benchmark**.

Benchmarks and Application Domains

The oldest database benchmark still used today is the *Wisconsin benchmark* [Bitton et al. 1983]; it systematically measures the elapsed time for a relational database system to perform a set of queries requested one after another by a single user. This mode of use places it in the **single-user benchmark** category. Relational database system products were first being introduced at the time the Wisconsin benchmark was published, and it had the rather general aim of validating performance for the new query capabilities offered by such systems. A surprisingly large number of product weaknesses became visible as a result, leading to furious vendor activity and ultimately to improved performance.

When the Wisconsin benchmark was published, however, it seemed to have little applicability to the work performed by large high-performance transaction systems, and this spurred commercial practitioners to develop a second benchmark, known as the **DebitCredit benchmark** [Anon. 1985]. DebitCredit dealt with concurrent order entries by clerks or tellers on thousands of terminals in large applications known as on-line transaction processing (**OLTP**) systems. It was true then and it remains true today that this type of application, and its smaller cousins with only dozens of order-entry clerks, is the application area where the most money is spent by customers on hardware and software. Since benchmarks are created to assist customers making purchase decisions, the descendants of DebitCredit have become the most sophisticated benchmarks currently available.

More recently, other database benchmarks have been created to evaluate database performance supporting complex queries by decision makers on a decision support system (**DSS**), object-oriented database system (**OODBS**) capabilities on workstations to support applications such as electronic computer-aided design (ECAD), text retrieval by library researchers on a document retrieval service (DRS), and so on. We will discuss these domain-specific benchmarks in the current chapter.

Leaving aside for a moment the wide variation in application domains, it must seem that even in a specific domain such as OLTP it is difficult to compare the performance of the wide variety of database hardware/software platforms, ranging from IBM mainframe computers running DB2 to personal computers running dBase. To compare different platforms in an unbiased manner, two requirements are common for sophisticated database benchmarks. First, such benchmarks have a list of domain-specific qualifications that must be met by any platform to be tested. For example, current OLTP benchmarks require that the platform measured must provide the standard **ACID properties** of transactions, including *isolation* of update effects under concurrent **workload** and *durability* in the event of system crash. Second, most of the benchmarks in common use have taken a cue from the original DebitCredit benchmark by requiring two specific measures to be reported for all platforms: (1) a measure of peak performance, for example in transactions per second (**tps**), and (2) a measure of price/performance to achieve this measure, for example in dollars per transaction per second (**$/tps**).

Together, these two benchmark requirements have had a surprisingly normalizing effect. For a small and inexpensive platform, the benchmark qualifications may turn out to be impossible: for example, if ACID transactions are not supported or top throughput rates required by a customer cannot be achieved. On the other hand, some customers have learned from normalized benchmark results such as this that they have been paying more for a traditional database platform than they should—that a less expensive system is perfectly suitable for their needs. The celebrated phenomenon of *computer downsizing* owes an important debt to benchmarks, which spurred some companies to switch their OLTP and other database work to less expensive platforms. As we will indicate a bit later, however, benchmarks are not currently able to measure all system costs, and a certain amount of skepticism is appropriate in estimating potential savings from downsizing based solely on benchmark price/performance results.

Scalar Measures of Performance vs Vector Measures of Performance

There is a good deal of controversy about the idea of using a single performance measure—a *scalar* measure such as tps—to report peak performance and price/performance in a benchmark of a complex system. The original Wisconsin benchmark did not do this, but instead reported a *vector* of (elapsed-time) performance results: the benchmarks had 32 queries, each with some number of measurements to report under varying conditions. The *SPEC* benchmarks [Gray 1993, Chapter 9] make a particular point of measuring multiple capabilities of a system, providing measures for integer and floating-point CPU performance with "20 colors" (i.e., 20 test results).

> SPEC strongly discourages a monochromatic view of performance characterization .... Decision makers should not depend on a single benchmark or a single performance number in selecting systems.

But the SPEC benchmarks measure CPU performance and are not actually database-oriented. Many database customers seem to be demanding simple scalar performance measures, which would be comparable to the commonly requested but flawed MIPS rating (millions of instructions per second) that was popular for many years in reporting performance of IBM and DEC mainframe machines. (See section 2.2 of [Hennessy and Patterson 1990] for a discussion.) Indeed, even SPEC results are commonly reported in terms of *SPEC marks*, the geometric mean of the 20 SPEC results. DeWitt, one of the original authors of the Wisconsin benchmark makes the point that benchmarks to measure multiuser performance, intended by him and coauthors as successors of the Wisconsin benchmark, were probably hurt in their acceptance by the fact that they did not reduce each system measure to a single number. This made it more difficult to compare two systems, a capability in demand, "regardless of the superficiality of the comparison" [Gray 1993, Chapter 4].

The value of vector measures of performance in a benchmark is that they allow users to distinguish different performance effects in terms of their functional origins. We refer to this benchmark feature as **separability** of performance effects. Separability can have important value when a planned site application does not exercise all of the functionality of the generic benchmark being used in the acquisition decision, so that extrapolation of the generic benchmark results becomes necessary. But no matter how superficial it seems, it is not inappropriate for potential database customers to want simple scalar measures of comparison between systems. After all, an acquisition decision is usually extremely simplistic: I have this much money to spend and I am going to acquire one of these systems; I need to decide which one is best for my needs. There has to be a clear winner in this decision, and a vector of measures, which might show one system the winner in one component and another in a different component, is not very useful as an aid to the decision without a good deal of guidance. On the other hand, the problem with most benchmarks which offer simple ratings is that they might not match well with the actual workload to be performed at a customer site, and will therefore not be good substitutes for custom benchmarks created to faithfully reflect that workload.

The *Set Query* benchmark [O'Neil 1993], a descendent of the Wisconsin benchmark, was created to measure performance of complex queries in decision support system (DSS) applications, and also attempted to offer guidance on how to use the vector of measures (CPU, I/O, and elapsed time) that resulted. To set the ground rules, an argument was made that the queries of this benchmark achieved **functional coverage**, meaning that any query in a site-specific workload performing this type of application would find a representative query in the benchmark. Since a workload is defined as the list of queries and their expected rates of submission on the target system, it was proposed that a reasonable custom rating of a database query system could be estimated by taking a convex combination of measures for queries measured in the benchmark, weighted by the frequency of corresponding queries in the workload, and estimating from this the CPU and I/O demand for the site. As a result, any published Set Query numbers could be used to estimate a custom benchmark rating merely by application of some arithmetic.

Unfortunately, the Set Query benchmark, like the earlier Wisconsin benchmark, took the simple approach of assuming a single-user environment. Because of the complex interaction between I/O and CPU in most database applications, a relatively complex queuing model must be used to estimate response time and throughput based on resource use of individual database actions making up a multiuser mix. (We define the *database actions* of a benchmark to be the individual atomic requests making up the benchmark: ad hoc queries in the Wisconsin benchmark or embedded SQL programs that bind several SQL statements together in a single transaction in DebitCredit.) It is enticing to assume on most database systems that one can tune an application so that DSS queries share resources effectively without any surprising bottlenecks; this is because queries can share data access and do not typically contend with one another for *locks*. But a single-user benchmark is not convincing when extrapolated to a multiuser workload, and database practitioners have good reason to be skeptical about balanced resource use until it has been demonstrated in the multiuser case.

Benchmarks that provide vector ratings of a platform have generally been of single-user type (although the recently released *TPC-D* benchmark is a multiuser benchmark that requires a vector of query timing intervals in its full-disclosure report). Multiuser benchmarks that run different types of database actions in a workload have the weights of the different database actions preset. The original DebitCredit benchmark and its direct descendants, the *TPC-A* and *TPC-B* benchmarks (covered below and in Chapter 2 of [Gray 1993]), actually contain only a single type of database transaction profile. Since the workload represented has all its weight on a single type of action, the problem of vector measurement does not arise. This rather unusual workload is justified by the aim of the benchmark, to measure common bottlenecks of transactional systems under stress, rather than attempting to measure many different types of work. However, TPC-A and TPC-B have become somewhat outmoded and have been replaced by benchmarks that combine numerous different database actions with different weights in the workload they measure. Since the weights are all preset, there is some concern that the benchmark workload is unrepresentative for some sites, even though the workload contains all the right types of work. The weights are not susceptible to recalculated weighting because in the multiuser run to test the concurrent-queuing effect, everything is "all mixed together in a bucket." In an ideal world, a database customer should be able to vary workload weights to generate resource measures that reflect a multiuser situation. This is our ultimate aim when we speak of separability of performance effects.

Criteria for Domain-Specific Benchmarks

In [Gray 1993, section 1.3], Gray provides the following list of criteria that a standard domain-specific database benchmark should meet.

1. *Relevance:* It should measure the peak performance and price/performance of systems when performing typical operations within that problem domain.
2. *Portability:* It should be easy to implement on many different systems and architectures.
3. *Scalability:* It should be applicable to small and large computer systems. It should be possible to scale the benchmark up to larger systems and parallel systems as computer performance and architecture improve.
4. *Simplicity:* It should be understandable; otherwise it will lack credibility.

We defer consideration of the first criterion. Criteria 2 and 3, portability and scalability, emphasize two different dimensions in which a domain-specific benchmark should be able to rate different platforms. Insistence on wide applicability allows potential customers to evaluate price/performance on old and new systems, responding quickly to innovative database capabilities for a particular application domain. The point made earlier, that a particular application domain may impose requirements (such as transactional ACID properties) on any usable database system, is not in conflict with criteria supporting wide applicability. The desire to port to different architectures and scale to large or small systems is not to be taken to an unrealistic level; any domain-specific benchmark is assumed to require features needed by realistic applications in that domain, and architectures that do not support these features are excluded from measurement.

Criterion 4, simplicity, says that customers should be able to follow the assumptions and implementation of a benchmark sufficiently far to believe that the ratings generated will be meaningful for their needs. This is not a completely obvious statement. There are many examples in commercial settings of administrators hiring experts to perform work that is not understood by the managers paying the bills. The assumption here is that costs of platform acquisition fall in a different category. Just as bank managers make it their business to thoroughly understand accounting rules and money flow, MIS managers should not let the details of price/performance escape them in making a platform acquisition. As we will see, the current benchmark state of the art is pushing the envelope of comprehension. We can expect to see how well the simplicity criterion holds up in the next few years.

To return to criterion 1, relevance, there are two points being made, and they both deserve discussion. The first is that there should be a single measure of performance (to measure peak performance). The second is that there is a way to cost the system to report price/performance at the peak rating. We will consider these points separately.

As we have already mentioned, the concept of a single measure of performance (a scalar rather than a vector measure) gives rise to difficulties in many application domains. A typical application often has a number of different kinds of work (database actions) being performed, and two database platforms might show different winners for different actions. The problem is a common one in taxonomy: a generic benchmark is an ideal, meant to span a generally understood application domain, but some application domains are too variegated in their database actions to be well represented by a preset weighting in a benchmark. At the same time, we must reduce our benchmark results to a single measure of performance in order to support an acquisition decision. While we do not challenge this assumption of the relevance criterion, we believe there is a need to advance the state of the art by which ratings are generated so as to be able to create customized weightings in a multiuser environment.

The second point made by the relevance criterion, the need to properly cost a database system, is one which receives a good deal of attention from Gray. It is important to note that database benchmarks do not attempt to reflect all costs of an application implementation at a site; they may not even reflect the most important costs. The standard price for a platform is calculated as the five-year cost of hardware (purchase assumed to depreciate to zero in five years) and software (purchase or license) together with all hardware and software support required. The performance, as we have seen, is calculated as peak throughput per second during run time.

But in reality, total costs for a database system stem from a number of other factors, including hardware and software reliability (the system may be fast and cheap, but break down a lot), development risks (in difficult programming environments it might be impossible ever to get complex applications working), cost of programming (based on typical remuneration for programmers with the needed expertise), ease of performing needed utilities (one database might take 12 hours to load, and another five days), operational support costs (an expensive computer operations staff might be needed), and so on. Many of these costs are difficult to calculate in a general way, but they can still be crucial, as the following analogy shows. Imagine having to make the tradeoff decision between writing a complex application system in assembly language and writing it in a higher-level language such as C or C++. Individual assembly-language programs generally have better run-time efficiency than programs written in C++ if they are written by very knowledgeable programmers. But there are problems trying to write a large system in assembly language. There is a good deal more code needed, and it will take longer to write. Programmers who can write it are rarer, and the resulting code is not portable to other architectures, which increases the system's long-term risk. Clearly, a benchmark run by a vendor with a reason to put the assembly language in the best light can demonstrate the inherent run-time advantage. But the decision to implement the application system in assembly language will lead to greater programmer costs and increase the complexity and risks of the project.

Hardware costs have been decreasing for many years, so that now the major costs of most application projects is in salaries for programmers and operational support staff. Clearly we would like to include personnel costs in a benchmark, but there seem to be insuperable difficulties in doing so. Given that a database system vendor usually performs the benchmark, how can one set the rules to guarantee that a typical programming staff with no prior experience with the code will be used in measuring the cost of implementation? This is clearly an area for future research in benchmarking.

47.3 Best Practices

The Wisconsin and DebitCredit benchmarks revolutionized the database industry. The Wisconsin benchmark was easy to measure, and competitive ratings of different hardware/software platforms were measured by individual customers, spurring database-system vendors to release ratings to demonstrate improved performance in new releases. The DebitCredit benchmark was much more difficult to measure properly, but demand for objective measures in transactional systems was so great that most vendors felt they had to publish DebitCredit ratings for their products, even when these ratings were rather embarrassing. As DebitCredit results were released, it became clear that vendors often changed the rules of the DebitCredit benchmark to minimize bottlenecks and show their products off in the best light. For a history of Debit-Credit measurements in the period following its publication in 1985, the interested reader is referred to [Serlin 1993].

In what follows we trace a bit of the history of benchmarking, then describe the database application domains for which benchmarks currently exist along with the benchmarks available in these domains.

Transaction Processing Performance Council (TPC) Benchmarks

In reaction to the great general interest in DebitCredit, Omri Serlin called for the formation of a vendor consortium known as the Transaction Processing Performance Council (TPC), with the intention of standardizing the rules under which benchmarks were run. The TPC soon released a pair of standards based on the DebitCredit benchmark, the TPC-A and TPC-B benchmarks, which were used for several years. More recently, the TPC has replaced TPC-A and TPC-B with the more complex transactional benchmark, *TPC-C* (covered below [Raab 1993]). This benchmark, together with another recently released TPC-D benchmark for decision support systems [TPC-D 1995], represents the current TPC state of the art.

Unfortunately, current TPC benchmarks are somewhat difficult to grasp and expensive to run. They are arguably pushing the envelope of the simplicity criterion mentioned earlier. The relatively simple TPC-B benchmark, though decommissioned in the sense that the TPC will no longer validate results, is still being used by many vendors as a quick check for performance bottlenecks. Performance consultants such as Tom Sawyer [see Sawyer 1993] are the ones that actually validate results, and they will continue to do this for TPC-B as long as there is a demand.

The TPC benchmarks are the ones that draw the greatest public interest. This is not to say that the other benchmarks we cover are not used: many are utilized heavily by vendors for their internal measurements, to detect performance "bugs" in new product releases, and as a means of focusing on performance areas that need improvement. But results of non-TPC benchmarks are rarely published, presumably because companies selling database hardware and software want to control public debate on performance. The TPC benchmarks are the standards that have been chosen to avoid the "benchmark wars" that accompanied the introduction of the Wisconsin and DebitCredit benchmarks. Marketing departments of companies whose products perform well on the TPC benchmarks reinforce the popularity of the TPC benchmarks with full-page ads in the trade press. The TPC, with all database vendors represented, offers a number of important advantages to its members. First, vendors want a chance to tune their product for emerging benchmarks, and so a slow process of adoption is to be preferred. Another consideration is that some benchmark characteristics have been perceived as making it unfairly difficult for some vendors to show good results. As an example, the DebitCredit benchmark originally required 95% of all transactions to be completed in one second; this seemed unfair to some of the smaller vendors who had difficulty meeting this requirement, and they argued successfully that TPC-A should specify 90% in two seconds, since this level of response was perfectly acceptable to most users. Finally, it is important that the TPC benchmark rules of execution and full disclosure be extremely carefully defined, as this limits the possibility that one vendor will change the rules to its own advantage. The TPC is a clearinghouse for these kinds of considerations.

More recently it has been found that new benchmarks being created and performed outside the TPC framework are often not publishable under common license terms for some database systems [Carey et al. 1993]. Because of this, there is often a dearth of published benchmark results in various application domains, and the ones available might very well not fit the purchaser's needs. Many customers attempt to extrapolate TPC benchmark ratings (see [Raab 1993] for a short description of appropriate use). Unfortunately, such attempts often extrapolate benchmark results beyond what is appropriate, with problematic results. It is advisable to seek expert help if there is any doubt that a proposed custom application fits a generic benchmark description.

Application Domains and Benchmarks

This section provides an overview of the different application domains, and within each domain an explanation of the domain-specific benchmarks that are available. We usually attempt to make clear the strengths and weaknesses of the benchmarks presented. While it is the easiest thing in the world to criticize an existing benchmark, it is not always easy to do a better job. Often there are criticisms of existing benchmarks on both sides of an issue (too complex and too simplistic), and there may be no way to please everyone. Of course this can also mean that a generic benchmark is trying to cover too large an application domain.

OLTP Domain: DebitCredit, TP1, TPC-A, TPC-B, and TPC-C Benchmarks

OLTP (on-line transaction processing) applications support large numbers of terminal users (order entry clerks, bank tellers, airport reservation agents) concurrently entering fundamental business data for an enterprise (purchase orders, account deposits and withdrawals, airplane ticketing information and seat reservations). Applications of this kind exercise the most sophisticated capabilities of a transactional system, and bottlenecks have historically arisen for the following reasons: lock contention on popular records that many users want to update simultaneously (called **hotspots**), inappropriate lock granularity (e.g., writing records to a sequential file with **page locks**), poorly performing disk buffers (used to amortize I/Os of popular pages for multiple reads and updates), a very large number of communication threads to communicate with on-line users, and the need to write journals to disk for durability at transaction commit time.

The DebitCredit benchmark in [Anon. 1985] took as its model a stylized representation of a large bank and a multiuser workload consisting of a single, rather lightweight database transaction which stressed the bottlenecks just listed. DebitCredit logic was triggered by reading a terminal message, and then performed triple-entry bookkeeping on three different tables (one of them small enough to exhibit lock contention), wrote a history record to a sequential file, sent a message back to the terminal, and committed. The nominal throughput assumed was 100 tps, but this was scalable to larger volume with linearly enlarged tables. The dollar cost calculated for the DebitCredit configuration ignored terminal and communication cost, but a large number of communication threads was required, since each terminal user was assumed to have a **think time** (during which the user submitted no request) of 100 s. The response-time requirement had 95% of transactions returning the terminal message within 1 s. A common alternative to the DebitCredit benchmark published by vendors was known as the **TP1 benchmark** (see section 2.5 of [Serlin 1993]), which removed the multithread requirement and drove the logic with a small number of batch threads. This was much easier to run and gave better price/performance numbers (and thus many vendors preferred it).

The TPC released the TPC-A benchmark in November 1989, and the TPC-B benchmark in August 1990. The TPC-A benchmark standardized DebitCredit with some variations from the original, and TPC-B standardized TP1. These benchmarks had the important effect of making all vendors follow the same rules in reporting ratings. See Table 2.3 of [Serlin 1993] for a detailed comparison of features between DebitCredit and TPC-A. The response-time criterion was relaxed in TPC-A from DebitCredit, as mentioned earlier, and the smallest table was made smaller, which made the lock contention test more difficult. LAN configurations were introduced, and terminals and local communications equipment were included in the cost calculation. This change was particularly significant. In fact, terminal and communications costs swamped all other costs with a 100-s think time for DebitCredit, so the think time for TPC-A was reduced to 10 s, where fewer terminals were needed for the throughput volume assumed. Even so, terminal and communication costs made up nearly 50% of the total cost for most TPC-A configurations, assuming the cheapest possible terminals. This is a significant point for most customers who are considering acquisition, especially since the assumed 10-s think time is unrealistically small. A certain amount of blame for this relatively high communication cost may be attached to the lightweight simplicity of the TPC-A transaction profile, since most terminal users of commercial applications trigger much greater resource use with each request. The lightweight TPC-A transaction weakness was one of the motivations for creating the TPC-C benchmark.

As time passed, most database-system vendors responded to the TPC challenge by fixing the common OLTP bottlenecks, the TPC-A and TPC-B benchmarks became outmoded, and they were replaced by the TPC-C benchmark (see Chapter 3 of [Gray 1993]). As of 1995, the TPC-A and TPC-B ratings are no longer being reported by the TPC, and all TPC-C benchmarks need to be audited by a TPC-approved auditor to be officially accepted. TPC-C models a relatively realistic warehouse order application, with eight different tables, including tables for warehouses, items carried, stock on hand, customers, orders, and line items within an order. There are five different transaction profiles occurring in the workload, including the high-frequency action of placing a new order, a read-only transaction to query the status of a customer's most recent order, and a mini batch transaction to "deliver" stock items for a set of newly entered orders. The TPC-C application model is much more realistic than the older TPC-A benchmark, but there is some reason to question that the specific workload used is broadly representative of the generic OLTP application class. There are probably many real OLTP applications that have an entirely different mix of transaction types. Of course the same is true for any particular mix of actions one might select.

The DebitCredit and TPC-A benchmarks are no longer used. However, the TPC-B benchmark is relatively easy and inexpensive to run, and it is still being used by vendors interested in a quick test for transaction bottlenecks. The TPC-C benchmark is quite complex and difficult to run. Running the TPC-C benchmark for the first time at an installation is estimated to cost about $30,000 in resources and consulting, and only vendors who stand to benefit from publicizing the results can be expected to run it. Furthermore, one of the aims of TPC-C was not achieved: to demonstrate that terminals and communication are not as large a fraction of OLTP acquisition costs as TPC-A seemed to indicate. While average TPC-C transactions are about ten times the "weight" of TPC-A transactions, the terminals assumed are more modern and expensive, so the percentage of total cost for terminals and communication remains high. In 1995, the TPC dropped costing of terminals from the required full-disclosure report for TPC-C (Revision 3.0). The TPC is also engaged in an effort to release a server-only version of TPC-C (a bit like what TPC-B was compared to TPC-A, while trying to avoid some of the credibility pitfalls that TPC-B encountered). In spite of any shortcomings that may exist in TPC-C, nothing else competes with it as an OLTP benchmarking tool.

DSS Domain: Wisconsin, AS³AP, Set Query, TPC-D Benchmarks

Decision support system (DSS) applications support relatively small numbers of users performing read-only queries to analyze fundamental enterprise data of the kind created by OLTP applications. Update activity may be undesirable in a DSS, and an extract of operational data is often placed in a ***data warehouse*** where subsequent OLTP updates will not take place concurrently (a minibatch refresh updating the warehouse data might be performed under controlled circumstances). As a result, update transactions will not contend with long-running queries, which can therefore run with maximum efficiency. The tradeoff is that data will not remain up-to-the-second with the latest transaction updates, but this is often acceptable. For business enterprises, DSS applications usually support marketing analysis, e.g., factor analysis of sales data by customer categories, product packaging, sales channel, geographic or temporal placement, and various other regression factors. The queries involved commonly examine a great many data and data indexes, and thus use significantly more resources than OLTP transactions, with consequently longer response times and a reduction in the number of users that can be supported.

A wide diversity of basic functions are involved in performing DSS queries, with much greater variation in performance between platforms than what is measured with OLTP. Such functions include multipage disk reads with 10-to-1 efficiency advantage over single page I/O, the ability to resolve conjunctions of several low selectivity predicates (color = 'red' and license_ state = 'NY' and make = 'Chevrolet' and year = 1994) by combining index extracts, and the availability of sophisticated algorithms to perform joins. These basic capabilities are used in query plans, and represent the "bag of tricks" available to the query optimizer.

Parallelism, the ability to spread the work of a single query over a number of CPUs, is another important capability in DSS applications. The wide variation in query efficiency between platforms is particularly significant because DSS applications are more common than they used to be, perhaps amounting to 20% of customer dollar investment in new applications.

The Wisconsin benchmark [Bitton et al. 1983] has a number of well-publicized deficiencies when used for modern acquisition decisions. One of the deficiencies relates to the small size of the tables and the fact that if the tables are *scaled up* (increased in size), the meaning of some of the attributes (columns) of the tables is changed. But this is not a serious problem and is addressed in [DeWitt 1993], which shows that the scaled Wisconsin benchmark can measure how close parallel database systems come to providing certain desirable performance properties. These properties include linear **speedup** (where an *n*-fold increase in hardware will allow a query to complete in $1/n$ of the previous elapsed time) and linear **scaleup** (where a constant response time can be maintained as the size of workload increases with database size, by adding a proportional number of processors and disks). The scaleup definition was originally motivated by the need to perform overnight batch jobs in a constant time window. As database sizes increased, this requirement forced customers to purchase additional hardware to stay within the time window. Database parallelism is driven by a customer desire for better response time (since the response time for complex queries on million-row tables is commonly measured in minutes) and price/performance (since the function of CPU cost vs. power curves upward strongly, and one can save money using multiple small CPUs, appropriately interconnected, instead of one large one). While OLTP platforms achieve parallelism by partitioning work for distinct users on separate CPUs, individual DSS queries often require so many computer resources that partitioning queries on separate CPUs is insufficient, and the database platforms must apply parallelism within a single query task, a property known as *intraquery parallelism*. The criterion of *scalability*, given in Gray's list of desirable benchmark criteria, is assumed to imply this kind of parallelism for large queries [see Gray 1993, section 2].

The AS^{3}AP benchmark [Turbyfill et al. 1993] was created to improve on the Wisconsin benchmark by building scalability considerations into the fundamental design, rather than as an afterthought. It introduces the concept of *equivalent database size*, the maximum scaled size of an AS^{3}AP database for which the system under test can complete the set of AS^{3}AP operations in (just) under 12 h. This size can be a difficult measure to determine, and an accurate size value may require multiple passes, which adds undesirable complexity. However, the size measure reflects the phenomenon of scaleup quite precisely, maintaining a constant response time with larger tables as the platform increases its capacity. The AS^{3}AP benchmark includes a number of multiuser tests, but unfortunately, in every test, all but one of the user threads perform single row updates or single row retrievals, so that complex queries accessing large numbers of data are never performing concurrently. Multiuser tests of concurrent complex queries are difficult to perform, but these are the most rewarding from the standpoint of building a realistic model. The AS^{3}AP benchmark includes the most complete set of join tests allowing separability of performance effects of any DSS benchmark, with joins of up to four tables. Careful analysis of join test results have in the past exposed deficiencies in some database-system join strategies.

The Set Query benchmark [O'Neil 1993] is a single-user benchmark with several suites of queries that were characterized as complex when the benchmark was released in 1991. It has since been suggested that these queries are rather basic, not at all complex in comparison with those of TPC-D. Proponents claim that these benchmark queries provide functional coverage of common DSS usage, and that appropriate weighting of the queries as generically measured on any given platform will permit a customer to estimate resource use for a custom application. As mentioned earlier, the Set Query benchmark measures the queries only in a single-user setting. The Set Query benchmark is unique in that it separates resource use by different queries, measuring I/O of different types as well as their CPU use and elapsed time. A careful analysis of these results allows an analyst to determine how well the database query system implements some basic query capabilities—the "bag of tricks" used by the query optimizer—including multipage disk reads, combining index extracts for multiple predicates, and sophisticated algorithms to perform joins. However, the Set Query benchmark does not challenge the optimizer in generating optimal access plans for complex queries involving multitable joins.

The TPC-D benchmark [TPC-D 1995] was released by the TPC in April 1995, after a long period of development. The TPC-D benchmark can be thought of as executing multiuser queries to analyze the sort of operational business data entered by TPC-C (i.e., relatively realistic warehouse order data). Many of the seventeen DSS queries of TPC-D potentially reference all the data collected over a long span in several large tables. For example, the query called Q6 considers all the line items shipped in a given year and lists the amount by which total revenue would have increased if certain discounts had been eliminated. While most of the TPC-D activity is in queries, there are two update functions that insert new sales information and delete old sales information from the database. These update functions can be executed either concurrently with the queries or separately in a batch run when queries are not being performed. The purpose of the update functions is to reflect a realistic cost for creating a large number of secondary indices. The TPC-D multiuser test runs multiple threads of queries (with update functions, either concurrent or separately), and is known as the *update test*. The time interval to perform each query and update statement of each stream must be reported in the benchmark results, together with the throughput metric which combines all activity in a single rating. Because individual query time intervals are reported, it is claimed that the queries can be weighted to give custom metric ratings, as with the Set Query benchmark. The units reported for the TPC-D ratings are in queries per hour (QPH) and dollars per query per hour ($/QPH).

As explained at the beginning of this section, there are no recent published results for the Wisconsin, AS^{3}AP, or Set Query benchmark, although these benchmarks have been used internally by many vendors to track product performance. The TPC-D benchmark is the standard for published ratings for the foreseeable future. The TPC-D benchmark's greatest strength is that it measures multiuser performance of complex queries running in separate query streams, the only DSS benchmark to do so. It also attempts to extrapolate custom ratings using a vector of query-interval measurements in a multiuser test, and this feature deserves careful study as further experience is gained with vendor results. A number of potential weaknesses have also been pointed out by some benchmark practitioners. The most important is that TPC-D seems to inappropriately emphasize some areas of performance over others. For example, more than half of TPC-D queries contain a *group by* clause, which may have the effect of de-emphasizing other important features that would be stressed in common practice. As another example, all values used in TPC-D queries are constants—no host variable substitutions are used. This tends to overemphasize ad hoc queries at the expense of embedded SQL, where more sophistication in precompilation is required. The decision to include updates as part of the load seems to militate against the read-only environment found in many data warehouses, although warehouses without concurrent update are currently increasing in importance. On the other hand, François Raab, who designed the TPC-D benchmark, argues that the future will see more frequent updates to replicate OLTP changes to data warehouses. As for complexity, some think that the TPC-D benchmark definition is too complex, and others that it is not complex enough to cover the diversity of practical cases that appear in real DSS applications.

OODBS, Engineering Workstation: The OO1 and OO7 Benchmarks

Object-oriented database systems (OODBS) depart from the older relational model by incorporating a number of concepts from object-oriented programming languages: rows of tables are replaced by objects in classes, allowing hierarchical class inheritance; the objects support composite structures, permitting complex nesting of sets of objects inside other objects. (See Chapter 52 on Object Database Systems for details.) For certain types of engineering workstation applications such as computer-aided software engineering (CASE) and computer-aided design (CAD), some of the properties of an OODBS are an important aid to performance. In typical applications, small sets of objects are sometimes accessed and held buffer-resident in a local workstation over an extended period, where editing tools perform interactive reads and updates that require extremely high performance. At other times, sequences of objects are traversed in a data-specific order.

The OO1 benchmark [Cattell 1993] is a relatively simple single-user workstation benchmark that emphasizes main-memory-resident access to data and claims to be independent of the underlying data model: object-oriented, relational, network, hierarchical, or custom application-specific database systems are all permitted. The benchmark database must be located on a machine different from the user's, and is defined as having two logical records, *parts* and *connections*, with a 90% bias toward locality of connection between parts. The benchmark contains ratings on three types of operations: *lookups*, where randomly chosen parts are accessed, *traversals*, where parts connected to a given randomly chosen part are all accessed, out to a connection distance of seven, and *inserts*, where a number of parts and their connections to existing parts are entered in the database. Each simple operation measured must be called from a workstation program to simulate an application where engineering calculations mix with database operations. All measures are performed 10 times, starting the first time with no data specifically cached (cold start) and ending with optimal cache residence (warm start). Different random parts are used for lookup in successive measures, however, so caching is not complete after a single set of measures. The database is scalable, and there are three different sizes used to vary the effectiveness of caching. The sizes are denoted *small* (where the entire database fits in memory cache), *large*, and *huge*. The benchmark requires the underlying database to support the ACID properties, but does not insist on protected access to the data.

The OO7 benchmark [Carey et al. 1993] was created to provide a comprehensive measure of some of the capabilities of OODBS systems that were not covered by OO1. It has the same sort of model as OO1, but greater data complexity and functional coverage in its tests. In particular, OO7 measures the following types of operations: sparse and dense traversals involving cached and noncached data; updates of indexed and unindexed object fields, repeated updates, sparse updates, and creation and deletion of objects; and exercise of query capabilities with several different kinds of queries. The benchmark reference reports a vector of measurements, and seems to provide excellent separability of performance effects for a number of commercial products.

The strength of the OO1 benchmark lies in its simplicity and its realism (due to early interviews with users during its design), particularly for ECAD applications. Weaknesses include complaints about allowing unprotected access to the data (introduction to Chapter 6 of [Stonebraker 1994]). The OO7 benchmark attempts to generalize the measurement of OODBS capabilities and cover a number of capabilities not treated by OO1. For example, many OODBS commercial offerings have turned out to be deficient in query capabilities, since they lack some of the structure expected for traditional SQL. The OO7 benchmark attempts to measure this area of performance, although it permits hand-coded queries which would demand exorbitant programmer time. At the last minute, the creators of the OO7 benchmark were unable to present some results in their report because a vendor they were measuring threatened legal action for what it said was an unsatisfactory benchmark process. It is noteworthy that license requirements for many database system products today (not just OODBS products) make it a condition of license that licensees not report any performance results without the permission of the vendor. Clearly this has a chilling effect on independent researchers who want to devise new benchmarks and publish measurement ratings.

Full-Text Retrieval (FTR): Full-Text Document Retrieval Benchmark

Full-text retrieval (FTR) systems often deal with large collections of documents, such as newspaper articles, legal briefs, or research articles accumulated from multiple sources over a number of years. The data model usually involves an understanding of how large documents are partitioned into smaller document parts: for example, newspaper articles might be divided into paragraphs and sentences. Retrieval usually involves Boolean search on all words in a document (rather than just a small number of keywords, the reason we speak of *full*-text retrieval), and there are predicates to specify proximity distance between given words. Thus we might ask to retrieve all research articles with the words WINDOWS and (DATABASE within five words of SERVER). FTR systems often generate synonyms for search terms [thus DATABASE might have the synonym (DATA followed by BASE)], and retrieved document lists are often *ranked* in terms of *relevance.* Clearly, most of these capabilities do not fit in the relational model. All the words of a document can be thought of as being in the same multivalued table column, and there is no relational equivalent to proximity or relevance ranking. For an excellent reference work on FTR concepts, see [Salton 1988] and Chapter 50 in this book.

The FTR benchmark [DeFazio 1993] was created to represent the workload generated by a generic document retrieval service (DRS). It is modeled after the TPC benchmarks, with a multiuser workload and a performance rating in search transactions per minute (tpm-S), where the database size must scale up with increasing rating. The database is scalable by specifying the number of document partitions, each partition containing 1 Gbyte of textual documents between 1000 and 50,000 bytes in length, with an average size of 5000 bytes, together with whatever indexing data are used by the system under test. The tpm-S performance rating is measured in terms of search transactions, which typically use indexing information to determine an answer set of documents fitting a given search expression; there are also retrieval transactions, which retrieve individual documents determined by earlier search transactions, and the generated workload must contain 10 retrieval transactions for each search transaction. The search expression to specify a search transaction is randomly generated following strict rules: the number of words specified in the search expression is randomly selected from 1 to 50; each successive term in the search expression uses either an equal match or a proximity operator, at random, and words included in these operators are chosen from three different lists of words determined by their frequency of appearance in documents; the terms are joined with randomly chosen Boolean connectors.

The FTR benchmark was based on a good deal of experience with real document retrieval applications. Synonym generation and relevance ranking, capabilities that are supported by several commercial products, are not included in the benchmark. The average number of words appearing in a search expression (25) seems high, but it may indeed be realistic.

47.4 Research Issues and Summary

A tremendous amount of money is spent every year on database and transaction processing systems. Database benchmarks have been created to help prospective database customers compare different platforms in terms of needed capability and cost/performance. The task is not a simple one, and a number of rather sophisticated concepts have been developed in the short period since the Wisconsin benchmark was first published in 1983. The criteria for domain-specific benchmarks, relevance, portability, scalability, and simplicity are commonly accepted in the field, but a number of issues remain unresolved, such as scalar versus vector measures.

The Transaction Processing Performance Council has expanded its role since its founding, and it now provides benchmark for DSS as well as OLTP systems: TPC-D and TPC-C. While these two benchmarks cover the most important application domains, they do not address many other domains covered in the previous section. The fact that results of non-TPC benchmarks usually do not appear in print (following the initial publication of the benchmark) means that customers are often unable to find relevant measures for new application areas. As a result, benchmark results are frequently used in acquisition decisions for which they are inappropriate. This is a problem area that database benchmarking practitioners should strive to clean up.

Some of the other issues that seem to deserve future research effort include the following.

It is important to find a way to incorporate more of the actual costs of system ownership and use in benchmark measures, including the cost of programming an application, the risk factors in a development project traceable to the platform, etc. Current benchmarks are defined with sufficient safeguards so that vendors, who have strong motivation to see their system perform well, can undergo a fairly straightforward benchmark audit to guarantee there are no distortions in the results. This aim seems much more problematic when personnel costs and risk factors enter, and no one has yet suggested a solution, or even a very good metric for these additional factors.

We would like to find a way to perform multiuser benchmarks which still allow us to achieve separability of performance effects, so that the individual actions in the benchmark workload mix can be differentiated in their effects. This would be approachable as a technical problem if we could develop an analytic framework whereby multiuser performance for different mixes of workload actions could be predicted from individual single-user measures of those actions. Such an analytic framework seems possible for many types of database work. The attempt at vector measurement and extrapolation in the TPC-D benchmark seems to be a good basis for future work.

New types of generic benchmarks will certainly be needed in the future. A great deal of concern is expressed that new types of data are being introduced that do not correspond to classical numeric or text values held in relational columns. One such data type is *binary large objects* (BLOBS) which are commonly used to hold bit-mapped graphical data. These new types of data are often passive, in the sense that they are located using normal data retrieval on classical field types and then displayed or used on some peripheral device. But probably the most important changes in application domains will occur as new types of retrieval are performed on these data types, so that they cease to be passively located. Some foreshadowing of this occurs in the SEQUOIA 2000 benchmark [Stonebraker et al. 1994], where geographic search on nontraditional data is supported by nontraditional access methods. It is presumed that other database applications in the future will benefit from such new capabilities.

Defining Terms

$/tps: A benchmark rating: dollar cost for a system divided by its tps rating.

ACID properties: The characteristic properties of transactional systems: atomicity, consistency, isolation, and durability. Tests of these properties are often specified in benchmarks, e.g., in the TPC-C specification [Raab 1993, paragraph 3.1].

Custom benchmark: A benchmark created for the purpose of evaluating different database platforms for a specific workload to be implemented at a site.

Data warehouse: Usually refers to a system using an extract of operational data for purposes of efficient query-only processing. The data are not up-to-the-minute, but this is considered of secondary importance for the DSS function being performed.

DebitCredit benchmark: Published in [Anon. 1985], the precursor of the TPC-A benchmark, which also included sort and scan tests. See also **TP1 benchmark**.

DSS: Decision support system. See coverage of this application domain in the section 47.3.

Functional coverage: Applied to query benchmarks, meaning that any query in a site-specific workload in a given application domain should find a representative query performing the same type of function in the benchmark.

Generic benchmark: A benchmark created to represent a paradigm of database application that is commonly in use by a number of different sites.

Hotspot: A bottleneck in a system, usually resulting from the popularity of particular data in the database as a result of concurrency contention.

Hybrid benchmark: A customization of a generic benchmark for a specific purpose.

Locks: A standard method by which ACID transactions avoid isolation anomalies. A transaction attempting to read or update a datum must first be assigned a read or update lock, to hold until commit. Two transactions cannot simultaneously hold locks on the same datum unless they are both read locks. Lock contention for popular data often leads to a bottleneck.

OLTP: On-line transaction processing. See description in section 47.3 under "OLTP Domain".

OODBS: Object-oriented database system.

Page locks: See **locks**. Some database systems assign locks on entire pages containing a datum being read or updated by a transaction. This can cause unnecessary contention, since other data on the same page will then be unavailable for conflicting access by other transactions.

Scaleup: A database-system property for a given type of workload where a constant response time can be maintained as the size of workload increases with database size by adding a proportional number of processors and disks.

Separability (of performance effects): The ability to differentiate the resource use of different actions in a benchmark workload.

Single-user benchmark: A benchmark where only a single user makes requests.

Synthetic data: Data generated probabilistically to have certain good statistical properties. An advantage cited in section 4.2 of [Gray 1993] in comparison with *empirical data* is the ability with synthetic data to scale (increase the size of the database) easily.

Think time: The time taken by an emulated terminal in a benchmark simulation between receiving a reply from one request and sending the next request.

TP1 benchmark: Batch version of DebitCredit [Serlin 1993]. Sometimes referred to loosely (for example in the Chapter 6 introduction to [Stonebraker 1994] as the third performance test in [Anon. 1985] that is not Sort or Scan.

tps: Benchmark rating unit for the number of transactions per second a system can provide.

Transaction processing: Subsumed under the term *database processing* in this article.

Workload: For a database application at a specific site, the set of database operations performed together with their frequency of execution.

References

Anon. 1985. A measure of transaction processing power, *Datamation*, Feb. 1985. Also in [Stonebraker 1994], Ch. 6.

Bitton, D., DeWitt, D. J., and Turbyfill, C. 1983. Benchmarking database systems, a systematic approach. In *Proc. Intl. Conf. on VLDB*, vol. 9, pp. 8–19.

Carey, M. J., DeWitt, D. J., and Naughton, J. F. 1993. The OO7 benchmark. In *Proc. ACM SIGMOD Conf. 1993*, 12–21. A more complete writeup is available under the same title as a CS Tech. Report from the Univ. of Wisconsin—Madison.

Cattell, R. G. G. 1993. An Engineering Database Benchmark. In [Gray 1993].

Curnow, H. and Wichman, B. A. 1976, A synthetic benchmark (referred to as the Whetstone benchmark). *Comput. J.* 19(1):43–49.

DeFazio, S. 1993. Full-text document retrieval benchmark. In [Gray 1993].

DeWitt, D. J. 1993. The Wisconsin benchmark: past, present, and future. In [Gray 1993].

Gray, J., ed. 1993. *The Benchmark Handbook for Database and Transaction Processing Systems*, 2nd ed. Morgan Kaufmann. Contains numerous benchmark descriptions as cited here, together with TPC-A, TPC-B, and TPC-C results in Appendix B.

Hennessy, J. L. and Patterson, D. A. 1990. *Computer Architecture: A Quantitative Approach*. Morgan Kaufmann.

O'Neil, P. 1993. The Set Query benchmark. In [Gray 1993].

Raab, F. 1993. Overview of the TPC benchmark C^{TM}: A complex OLTP benchmark. In [Gray 1993].

Salton, G. 1988 (corrected ed. 1989). *Automatic Text Processing*. Addison–Wesley.

Sawyer, T. 1993. Doing your own benchmark. In [Gray 1993].

Serlin, O. 1993. The history of DebitCredit and the TPC. In [Gray 1993].

Serlin, O. Various dates. *Fault-Tolerant Systems Newsletter (FTSN)*. ITOM International, Los Altos, CA.

Shasha, D. 1992. *Database Tuning: A Principled Approach*. Prentice–Hall.

Stonebraker, M., ed. 1994. *Readings in Database Systems*, 2nd ed. Morgan Kaufmann.

Stonebraker, M., Frew, J., Gardels, K., and Meredith, J. 1994. The sequoia 2000 storage benchmark. In *Readings in Database Systems*, 2nd ed. M. Stonebraker, ed. Morgan Kaufmann. Originally in *Proc. ACM SIGMOD Conf. 1993*, pp. 2–11.

Turbyfill, C., Orji, C., and Bitton, D. 1993. AS^3AP: An ANSI SQL Standard scaleable and portable benchmark for relational database systems. In [Gray 1993]. Ch. 5.

TPC-D 1995. DSS benchmark released by TPC, April 1995 all TPC benchmark descriptions and results are available on the TPC Home Page, http://www.tpc.org.

Further Information

The most important reference work in the database benchmarking field is [Gray 1993]. Other discussions of database benchmarks occur in Ch. 6 of [Stonebraker 1994] and Ch. 6 of [Shasha 1992]. For a discussion of benchmark auditing, see the article by Tom Sawyer [1994]. Those interested in the early history of database benchmarking should read the article by Omri Serlin [1993] or refer to individual issues of the *Fault Tolerant Systems Newsletter* (published by ITOM International, Los Altos, CA, owned by Omri Serlin) for the period.

A number of other benchmarks exist that are not covered in the current chapter, and more are published every year. Some of these involve new application domains. For example, the performance of *geographic search* in earth science (ES) applications is measured in the SEQUOIA 2000 storage benchmark [Stonebraker et al. 1994]. The interested reader who wishes to keep up with the field should follow the yearly proceedings of the ACM SIGMOD and VLDB conferences, where many of these benchmarks originally appeared.

Program implementations of some of the benchmarks covered here are available in machine-readable form.

A distribution containing sample TPC-A and TPC-B programs, the Wisconsin benchmark, the Set Query benchmark, and OO1 is available from Morgan Kaufmann Publishers (at a rather high cost, several hundred dollars). The software provided includes data generators, database creation and loading programs, SQL implementations of the benchmark operations, and programs to execute and time the benchmarks. Phone (415) 578-9911, or fax (415) 578-0672

An OO7 implementation is available by anonymous ftp from the OO7 directory of ftp.cs.wisc.edu.

48

Distributed and Parallel Database Systems

M. Tamer Özsu
University of Alberta

Patrick Valduriez
INRIA, Rocquencourt

48.1 Introduction

The maturation of database management system (DBMS) technology has coincided with significant developments in distributed computing and parallel processing technologies. The end result is the emergence of **distributed database management systems** and **parallel database management systems**. These systems have started to become the dominant data management tools for highly data-intensive applications.

The integration of workstations in a distributed environment enables a more efficient function distribution in which application programs run on workstations, called *application servers*, while database functions are handled by dedicated computers, called *database servers*. This has led to the present trend in distributed system architecture, where sites are organized as specialized servers rather than as general-purpose computers.

A parallel computer, or multiprocessor, is itself a distributed system made of a number of nodes (processors and memories) connected by a fast network within a cabinet. Distributed database technology can be naturally revised and extended to implement *parallel database systems*, i.e., database systems on parallel computers [DeWitt and Gray 1992, Valduriez 1993]. Parallel database systems exploit the parallelism in data management [Boral 1988] in order to deliver high-performance and high-availability database servers at a much lower price than equivalent mainframe computers [DeWitt and Gray 1992, Valduriez 1993].

In this paper, we present an overview of the distributed and parallel DBMS technologies, highlight the unique characteristics of each, and indicate the similarities between them. This discussion should help establish their unique and complementary roles in data management.

48.2 Underlying Principles

A distributed database (DDB) is a collection of multiple, logically interrelated databases distributed over a computer network. A distributed database management system (distributed DBMS) is then defined as the software system that permits the management of the distributed database and makes the distribution transparent to the users [Özsu and Valduriez 1991a]. These definitions point to two identifying architectural principles. The first is that the system consists of a (possibly empty) set of *query sites* and a nonempty set of *data sites*. The data sites have data storage capability, while the query sites do not. The latter only run the user interface routines in order to facilitate the data access at data sites. The second is that each site (query or data) is assumed to consist *logically* of a single independent computer. Therefore, each site has its own primary and secondary storage, runs its own operating system (which may be the same or different at different sites), and has the capability to execute applications on its own. The sites are interconnected by a computer network rather than a multiprocessor configuration. The important point here is the emphasis on loose interconnection between processors which have their own operating systems and operate independently.

The database is physically distributed across the data sites by *fragmenting* and *replicating* the data [Ceri et al. 1987]. Given a relational-database schema, fragmentation subdivides each relation into horizontal or vertical partitions. *Horizontal fragmentation* of a relation is accomplished by a selection operation which places each tuple of the relation in a different partition according to a fragmentation predicate (e.g., an Employee relation may be fragmented according to the location of the employees). *Vertical fragmentation* divides a relation into a number of fragments by projecting over its attributes (e.g., the Employee relation may be fragmented so that the Emp_number, Emp_name, and Address information is in one fragment, and Emp_number, Salary, and Manager information is in another fragment). Fragmentation is desirable because it enables the placement of data in close proximity to their place of use, thus potentially reducing transmission cost, and it reduces the size of relations that are involved in user queries.

Based on the user access patterns, each of the fragments may also be replicated. This is preferable when the same data are accessed from applications that run at a number of sites. In this case, it may be more cost-effective to duplicate the data at a number of sites rather than continuously moving them from site to site.

When the above architectural assumptions of a distributed DBMS are relaxed, one gets a parallel database system. The differences between a parallel DBMS and a distributed DBMS are somewhat unclear. In particular, shared-nothing parallel DBMS architectures, which we discuss below, are quite similar to the loosely interconnected distributed systems. Parallel DBMSs exploit recent multiprocessor computer architectures in order to build high-performance and high-availability database servers at a much lower price than equivalent mainframe computers.

A parallel DBMS can be defined as a DBMS implemented on a multiprocessor computer. This includes many alternatives, ranging from the straightforward porting of an existing DBMS, which may require only rewriting the operating system interface routines, to a sophisticated combination of parallel processing and database-system functions into a new hardware–software architecture. As always, we have the traditional tradeoff between *portability* (to several platforms) and *efficiency*. The sophisticated approach is better able to fully exploit the opportunities offered by a multiprocessor at the expense of portability.

The solution, therefore, is to use large-scale parallelism to magnify the raw power of individual components by integrating these in a complete system along with the appropriate parallel database software. Using standard hardware components is essential in order to exploit the continuing technological improvements with minimal delay. Then, the database software can exploit the three forms of parallelism inherent in data-intensive application workloads. **Interquery parallelism** enables the parallel execution of multiple queries generated by concurrent transactions. **Intraquery parallelism** makes the parallel execution of multiple independent operations (e.g., select operations) possible within the same query. Both interquery and intraquery parallelism can be obtained by using *data partitioning*, which is similar to horizontal fragmentation. Finally, with **intraoperation parallelism**, the same operation can be executed as many suboperations using *function partitioning* in addition to data partitioning. The set-oriented mode of database languages (e.g., SQL) provides many opportunities for intraoperation parallelism.

There are a number of identifying characteristics of the distributed and parallel DBMS technology.

1. The distributed/parallel database is a database, not some "collection" of files that can be individually stored at each node of a computer network. This is the distinction between a DDB and a collection of files managed by a distributed file system. To form a DDB, distributed data should be

logically related, where the relationship is defined according to some structural formalism (e.g., the relational model), and access to data should be at a high level via a common interface.

2. The system has the full functionality of a DBMS. It is neither, as indicated above, a distributed file system, nor is it a transaction-processing system. Transaction processing is only one of the functions provided by such a system, which also provides query processing, structured organization of data, and other functions that transaction-processing systems do not necessarily deal with.

3. The distribution (including fragmentation and replication) of data across multiple sites/processors is not visible to the users. This is called **transparency**. The distributed/parallel database technology extends the concept of *data independence*, which is a central notion of database management, to environments where data are distributed and replicated over a number of machines connected by a network. This is provided by several forms of transparency: *network* (and, therefore, *distribution*) *transparency*, *replication transparency*, and *fragmentation transparency*. Transparent access means that users are provided with a single logical image of the database even though it may be physically distributed, enabling them to access the distributed database as if it were a centralized one. In its ideal form, full transparency would imply a query-language interface to the distributed/parallel DBMS which is no different from that of a centralized DBMS. Transparency concerns are more pronounced in the case of distributed DBMSs. There are fundamental reasons for this. First of all, the multiprocessor system on which a parallel DBMS is implemented is controlled by a single operating system. Therefore, the operating system can be structured to implement some aspects of DBMS functionality, thereby providing some degree of transparency. Secondly, software development on parallel systems is supported by parallel programming languages which can provide further transparency.

In a distributed DBMS, data and the applications that access them can be localized at the same site, eliminating (or reducing) the need for remote data access that is typical of teleprocessing-based timesharing systems. Furthermore, since each site handles fewer applications and a smaller portion of the database, contention for resources and for data access can be reduced. Finally, the inherent parallelism of distributed systems provides the possibility of interquery parallelism and intraquery parallelism.

If the user access to the distributed database consists only of querying (i.e., read-only access), then provision of interquery and intraquery parallelism would imply that as much of the database as possible should be replicated. However, since most database accesses are not read-only, the mixing of read and update operations requires support for distributed transactions (as discussed in a later section).

Higher performance is probably the most important objective of parallel DBMSs. In these systems, higher performance can be obtained through several complementary solutions: database-oriented operating system support, parallelism, optimization, and load balancing. Having the operating system constrained and "aware" of the specific database requirements (e.g., buffer management) simplifies the implementation of low-level database functions and therefore decreases their cost. For instance, the cost of a message can be significantly reduced to a few hundred instructions by specializing the communication protocol. Parallelism can increase throughput (using interquery parallelism) and decrease transaction response times (using intraquery and intraoperation parallelism).

Distributed and parallel DBMSs are intended to improve reliability, since they have replicated components and thus eliminate single points of failure. The failure of a single site or processor, or the failure of a communication link which makes one or more sites unreachable, is not sufficient to bring down the entire system. This means that although some of the data may be unreachable, with proper system design users may be permitted to access other parts of the distributed database. The "proper system design" comes in the form of support for distributed transactions. Providing transaction support requires the implementation of distributed concurrency control and distributed reliability (commit and recovery) protocols, which are reviewed in a later section.

In a distributed or parallel environment, it should be easier to accommodate increasing database sizes or increasing performance demands. Major system overhauls are seldom necessary; expansion can usually be handled by adding more processing and storage power to the system.

Ideally, a parallel DBMS (and to a lesser degree a distributed DBMS) should demonstrate two advantages: **linear scaleup** and **linear speedup**. Linear scaleup refers to sustained performance for a linear increase in both database size and processing and storage power. Linear speedup refers to a linear increase in performance for a constant database size, and a linear increase in processing and storage power. Furthermore, extending the system should require minimal reorganization of the existing database.

The price/performance characteristics of microprocessors and workstations make it more economical to put together a system of smaller computers with the equivalent power of a single big machine. Many commercial distributed DBMSs operate on minicomputers and workstations in order to take advantage of their favorable price/performance characteristics. The current reliance on workstation technology has come about because most of the commercial distributed DBMSs operate within local-area networks for which the workstation technology is most suitable. The emergence of distributed DBMSs that run on wide-area networks may increase the importance of mainframes. On the other hand, future distributed DBMSs may support hierarchical organizations where sites consist of clusters of computers communicating over a local-area network with a high-speed backbone wide-area network connecting the clusters.

48.3 Distributed and Parallel Database Technology

Distributed and parallel DBMSs provide the same functionality as centralized DBMSs except in an environment where data are distributed across the sites on a computer network or across the nodes of a multiprocessor system. As discussed above, the users are unaware of data distribution. Thus, these systems provide the users with a *logically integrated* view of the *physically distributed* database. Maintaining this view places significant challenges on system functions. We provide an overview of these new challenges in this section. We assume familiarity with basic database management techniques.

Architectural Issues

There are many possible distribution alternatives. The currently popular **client–server architecture** [Orfali et al. 1994], where a number of client machines access a single database server, is the most straightforward one. In these systems, which can be called *multiple-client–single-server*, the database management problems are considerably simplified, since the database is stored on a single server. The pertinent issues relate to the management of client buffers and the caching of data and (possibly) locks. The data management is done *centrally* at the single server.

A more distributed and more flexible architecture is the *multiple-client–multiple-server* architecture, where the database is distributed across multiple servers which have to communicate with each other in responding to user queries and in executing transactions. Each client machine has a "home" server to which it directs user requests. The communication of the servers among themselves is transparent to the users. Most current database management systems implement one or the other type of the client–server architectures.

A truly distributed DBMS does not distinguish between client and server machines. Ideally, each site can perform the functionality of a client and a server. Such architectures, called *peer-to-peer*, require sophisticated protocols to manage the data distributed across multiple sites. The complexity of required software has delayed the offering of peer-to-peer distributed DBMS products.

Parallel-system architectures range between two extremes, the **shared-nothing** and the **shared-memory** architectures. A useful intermediate point is the **shared-disk** architecture.

In the shared-nothing approach, each processor has exclusive access to its main memory and disk unit(s). Thus, each node can be viewed as a local site (with its own database and software) in a distributed database system. The difference between shared-nothing parallel DBMSs and distributed DBMSs is basically one of implementation platform; therefore most solutions designed for distributed databases may be reused in parallel DBMSs. In addition, shared-nothing architecture has three main virtues: cost, extensibility, and availability. On the other hand, it suffers from higher complexity and (potential) load-balancing problems.

Examples of shared-nothing parallel database systems include the Teradata's DBC and Tandem's Non-StopSQL products as well as a number of prototypes such as BUBBA [Boral et al. 1990], EDS [EDS, 1990], GAMMA [DeWitt et al. 1990], GRACE [Fushimi et al. 1986], PRISMA [Apers et al. 1992] and ARBRE [Lorie et al. 1989].

In the shared-memory approach, any processor has access to any memory module or disk unit through a fast interconnect (e.g., a high-speed bus or a crossbar switch). Several new mainframe designs such as the IBM 3090 or Bull's DPS8, and symmetric multiprocessors such as Sequent and Encore, follow this approach. Shared memory has two strong advantages: simplicity and load balancing. These are offset by three problems: cost, limited extensibility, and low availability.

Examples of shared-memory parallel database systems include XPRS [Stonebraker et al. 1988], DBS3 [Bergsten et al. 1991], and Volcano [Graefe 1990], as well as portings of major RDBMSs on shared-memory multiprocessors. In a sense, the implementation of DB2 on an IBM 3090 with six processors was the first example. All the shared-memory commercial products (e.g., INGRES and ORACLE) today exploit interquery parallelism only (i.e., no intraquery parallelism).

In the shared-disk approach, any processor has access to any disk unit through the interconnect, but exclusive (nonshared) access to its main memory. Each processor can then access database pages on the shared disk and copy them into its own cache. To avoid conflicting accesses to the same pages, global locking and protocols for the maintenance of cache coherency are needed. Shared disk has a number of advantages: cost, extensibility, load balancing, availability, and easy migration from uniprocessor systems. On the other hand, it suffers from higher complexity and potential performance problems.

Examples of shared-disk parallel DBMS include IBM's IMS/VS Data Sharing product and DEC's VAX DBMS and Rdb products. The implementation of ORACLE on DEC's VAXcluster and NCUBE computers also uses the shared-disk approach, since it requires minimal extensions of the RDBMS kernel. Note that all these systems exploit interquery parallelism only.

Query Processing and Optimization

Query processing is the process by which a declarative query is translated into low-level data manipulation operations. SQL is the standard query language that is supported in current DBMSs. **Query optimization** refers to the process by which the "best" execution strategy for a given query is found from among a set of alternatives.

In centralized DBMSs, the process typically involves two steps: *query decomposition* and *query optimization*. Query decomposition takes an SQL query and translates it into one expressed in relational algebra. In the process, the query is analyzed semantically so that incorrect queries are detected and rejected as easily as possible, and correct queries are simplified. Simplification involves the elimination of redundant predicates which may be introduced as a result of query modification to deal with views, security enforcement, and semantic integrity control. The simplified query is then restructured as an algebraic query.

For a given SQL query, there is more than one possible algebraic query. Some of these algebraic queries are "better" than others. The quality of an algebraic query is defined in terms of expected performance. The traditional procedure is to obtain an initial algebraic query by translating the predicates and the target statement into relational operations as they appear in the query. This initial algebraic query is then transformed, using algebraic transformation rules, into other algebraic queries until the "best" one is found. The "best" algebraic query is determined according to a cost function which calculates the cost of executing the query according to that algebraic specification. This is the process of query optimization.

In distributed DBMSs, two more steps are involved between query decomposition and query optimization: *data localization* and *global query optimization*.

The input to data localization is the initial algebraic query generated by the query decomposition step. The initial algebraic query is specified on global relations irrespective of their fragmentation or distribution. The main role of data localization is to localize the query's data using data distribution information. In this step, the fragments which are involved in the query are determined and the query is transformed into one that operates on fragments rather than global relations. As indicated earlier, fragmentation is defined through fragmentation rules which can be expressed as relational operations (horizontal fragmentation by selection, vertical fragmentation by projection). A distributed relation can be reconstructed by applying the inverse of the fragmentation rules. This is called a *localization program*. The localization program for a horizontally (vertically) fragmented query is the union (join) of the fragments. Thus, during the data localization step each global relation is first replaced by its localization program, and then the resulting fragment query is simplified and restructured to produce another "good" query. Simplification and restructuring may be done according to the same rules used in the decomposition step. As in the decomposition step, the final fragment query is generally far from optimal; the process has only eliminated "bad" algebraic queries.

The input to the third step is a fragment query, that is, an algebraic query on fragments. The goal of query optimization is to find an execution strategy for the query which is close to optimal. Remember that finding the optimal solution is computationally intractable. An execution strategy for a distributed query can be described with *relational algebra operations* and *communication primitives* (send/receive operations) for transferring data between sites. The previous layers have already optimized the query—for example, by eliminating redundant expressions. However, this optimization is independent of fragment characteristics such as cardinalities. In addition, communication operations are not yet specified. By permuting the ordering of operations within one fragment query, many equivalent query execution plans may be found. Query optimization consists of finding the "best" one among candidate plans examined by the optimizer.[1] The query optimizer is usually seen as three components: a search space, a cost model, and a search strategy. The *search space* is the set of alternative execution plans to represent the input query. These plans are equivalent in the sense that they yield the same result, but they differ in the execution order of operations and the way these operations are implemented. The *cost model* predicts the cost of a given execution plan. To be accurate, the cost model must have accurate knowledge about the parallel execution environment. The *search strategy* explores the search space and selects the best plan. It defines which plans are examined and in which order.

In a distributed environment, the cost function, often defined in terms of time units, refers to computing resources such as disk space, disk I/Os, buffer space, CPU cost, communication cost, and so on. Generally, it is a weighted combination of I/O, CPU, and communication costs. Nevertheless, a typical simplification made by distributed DBMSs is to consider communication cost as the most significant factor. This is valid for wide-area networks, where the limited bandwidth makes communication much more costly than it is in local processing. To select the ordering of operations it is necessary to predict execution costs of alternative candidate orderings. Determining execution costs before query execution (i.e., static optimization) is based on fragment statistics and the formulas for estimating the cardinalities of results of relational operations. Thus the optimization decisions depend on the available statistics on fragments. An important aspect of query optimization is *join ordering*, since permutations of the joins within the query may lead to improvements of several orders of magnitude. One basic technique for optimizing a sequence of distributed join operations is through use of the semijoin operator. The main value of the semijoin in a distributed system is to reduce the size of the join operands and thus the communication cost. However, more recent techniques, which consider local processing costs as well as communication costs, do not use semijoins because they might increase local processing costs. The output of the query optimization layer is an optimized algebraic query with communication operations included on fragments.

Parallel query optimization exhibits similarities with distributed query processing. It takes advantage of both intraoperation parallelism, which was discussed earlier, and interoperation parallelism.

Intraoperation parallelism is achieved by executing an operation on several nodes of a multiprocessor machine. This requires that the operands have been previously partitioned, i.e., horizontally fragmented, across the nodes. The way in which a base relation is partitioned is a matter of physical design. Typically, partitioning is performed by applying a hash function on an attribute of the relation, which will often be the join attribute. The set of nodes where a relation is stored is called its *home*. The *home of an operation* is the

---

[1]The difference between an optimal plan and the best plan is that the optimizer does not, because of computational intractability, examine all of the possible plans.

set of nodes where it is executed, and it must be the home of its operands in order for the operation to access its operands. For binary operations such as join, this may imply repartitioning one of the operands. The optimizer may even sometimes find that repartitioning both the operands is useful. Parallel optimization to exploit intraoperation parallelism can make use of some of the techniques devised for distributed databases.

Interoperation parallelism occurs when two or more operations are executed in parallel, either as a dataflow or independently. We designate as *dataflow* the form of parallelism induced by *pipelining*. *Independent* parallelism occurs when operations are executed at the same time or in arbitrary order. Independent parallelism is possible only when the operations do not involve the same data.

Concurrency Control

Whenever multiple users access (read and write) a shared database, these accesses need to be synchronized to ensure database consistency. The synchronization is achieved by means of **concurrency control algorithms** which enforce a correctness criterion such as **serializability**. User accesses are encapsulated as **transactions** [Gray 1981], whose operations at the lowest level are a set of read and write operations to the database. Concurrency control algorithms enforce the **isolation** property of transaction execution, which states that the effects of one transaction on the database are isolated from other transactions until the first completes its execution.

The most popular concurrency control algorithms are **locking**-based. In such schemes, a lock, in either shared or exclusive mode, is placed on some unit of storage (usually a page) whenever a transaction attempts to access it. These locks are placed according to lock compatibility rules such that *read–write*, *write–read*, and *write–write* conflicts are avoided. It is a well-known theorem that if lock actions on behalf of concurrent transactions obey a simple rule, then it is possible to ensure the serializability of these transactions: "No lock on behalf of a transaction should be set once a lock previously held by the transaction is released." This is known as **two-phase locking** [Gray 1979], since transactions go through a growing phase when they obtain locks and a shrinking phase when they release locks. In general, releasing of locks prior to the end of a transaction is problematic. Thus, most of the locking-based concurrency control algorithms are *strict* in that they hold on to their locks until the end of the transaction.

In distributed DBMSs, the challenge is to extend both the serializability argument and the concurrency control algorithms to the distributed execution environment. In these systems, the operations of a given transaction may execute at multiple sites where they access data. In such a case, the serializability argument is more difficult to specify and enforce. The complication is due to the fact that the serialization order of the same set of transactions may be different at different sites. Therefore, the execution of a set of distributed transactions is serializable if and only if

1. the execution of the set of transactions at each site is serializable, and
2. the serialization orders of these transactions at all these sites are identical.

Distributed concurrency control algorithms enforce this notion of *global serializability*. In locking-based algorithms there are three alternative ways of enforcing global serializability: centralized locking, primary-copy locking, and distributed locking algorithm.

In *centralized locking*, there is a single lock table for the entire distributed database. This lock table is placed, at one of the sites, under the control of a single lock manager. The lock manager is responsible for setting and releasing locks on behalf of transactions. Since all locks are managed at one site, this is similar to centralized concurrency control and it is straightforward to enforce the global serializability rule. These algorithms are simple to implement, but suffer from two problems. The central site may become a bottleneck, both because of the amount of work it is expected to perform and because of the traffic that is generated around it; and the system may be less reliable, since the failure or inaccessibility of the central site would cause system unavailability.

Primary-copy locking is a concurrency control algorithm that is useful in replicated databases where there may be multiple copies of a datum stored at different sites. One of the copies is designated as a primary copy, and it is this copy that has to be locked in order to access that item. The set of primary copies for each datum is known to all the sites in the distributed system, and the lock requests on behalf of transactions are directed to the appropriate primary copy. If the distributed database is not replicated, copy locking degenerates into a distributed locking algorithm. Primary-copy locking was proposed for the prototype distributed version of INGRES.

In *distributed* (or *decentralized*) *locking*, the lock management duty is shared by all the sites in the system. The execution of a transaction involves the participation and coordination of lock managers at more than one site. Locks are obtained at each site where the transaction accesses a datum. Distributed locking algorithms do not have the overhead of centralized locking ones. However, both the communication overhead to obtain all the locks and the complexity of the algorithm are greater. Distributed locking algorithms are used in System R* and in NonStop SQL.

One side effect of all locking-based concurrency control algorithms is that they cause **deadlocks**. The detection and management of deadlocks in a distributed system is difficult. Nevertheless, the relative simplicity and better performance of locking algorithms make them more popular than alternatives such as *timestamp-based algorithms* or *optimistic concurrency control*. Timestamp-based algorithms execute the conflicting operations of transactions according to their timestamps, which are assigned when the transactions are accepted. Optimistic concurrency control algorithms work from the premise that conflicts among transactions are rare and proceed with executing the transactions up to their termination, at which point a validation is performed. If the validation indicates that serializability would be compromised by the successful completion of that particular transaction, then it is aborted and restarted.

Reliability Protocols

We indicated earlier that distributed DBMSs are potentially more reliable because there are multiples of each system component, which eliminates single points of failure. This requires careful system design and the implementation of a number of protocols to deal with system failures.

In a distributed DBMS, four types of failures are possible: *transaction failures*, *site (system) failures*, *media (disk) failures*, and *communication-line failures*. Transactions can fail for a number of reasons. Failure can be due to an error in the transaction caused by input data, as well as the detection of a present or potential deadlock. The usual approach to take in cases of transaction failure is to abort the transaction, resetting the database to its state prior to the start of the database.

Site (or system) failures are due to a hardware failure (e.g., processor, main memory, power supply) or a software failure (bugs in system or application code). The effect of system failures is the loss of main memory contents. Therefore, any updates to the parts of the database that are in the main memory buffers (also called **volatile database**) are lost as a result of system failures. However, the database that is stored in secondary storage (also called **stable database**) is safe and correct. To achieve this, DBMSs typically employ **logging protocols**, such as *write-ahead logging*, which record changes to the database in system logs and move these log records and the volatile database pages to stable storage at appropriate times. From the perspective of distributed transaction execution, site failures are important, since the failed sites cannot participate in the execution of any transaction.

Media failures refer to the failure of secondary storage devices that store the stable database. Typically, these failures are addressed by duplexing storage devices and maintaining archival copies of the database. Media failures are frequently treated as problems local to one site and therefore are not specifically addressed in the reliability mechanisms of distributed DBMSs.

The three types of failures described above are common to both centralized and distributed DBMSs. Communication failures, on the other hand, are unique to distributed systems. There are a number of types of communication failures. The most common ones are errors in the messages, improperly ordered messages, lost (or undelivered) messages, and line failures. Generally, the first two of these are considered to be the responsibility of the computer network protocols and are not addressed by the distributed DBMS. The last two, on the other hand, have an impact on the distributed DBMS protocols and therefore need to be considered in the design of these protocols. If one site is expecting a message from another site and this message never arrives, this may be because (1) the message is lost, (2) the line(s) connecting the two sites are broken, or (3) the site which is supposed to send the message has failed. Thus, it is not always possible to distinguish between site failures and communication failures. The waiting site simply times out and has to assume that the other site is incommunicado. Distributed DBMS protocols have to deal with this uncertainty. One drastic result of line failures may be *network partitioning* in which the sites form groups where communication within each group is possible but communication across groups is not. This is difficult to deal with in that it may not be possible to make the database available for access while at the same time guaranteeing its consistency.

Two properties of transactions are maintained by reliability protocols: **atomicity** and **durability**. Atomicity requires that either all the operations of a transaction are executed or none of them are (all or nothing). Thus, the set of operations contained in a transaction is treated as one atomic unit. Atomicity is maintained even in the face of failures. Durability requires that the effects of successfully completed (i.e., committed) transactions endure subsequent failures.

The enforcement of atomicity and durability requires the implementation of *atomic commitment protocols* and *distributed recovery protocols*. The most popular atomic commitment protocol is **two-phase commit**. The recoverability protocols are built on top of the local recovery protocols, which are dependent upon the supported mode of interaction (of the DBMS) with the operating system.

Two-phase commit (2PC) is a very simple and elegant protocol that ensures the atomic commitment of distributed transactions. It extends the effects of local atomic commit actions to distributed transactions by insisting that all sites involved in the execution of a distributed transaction agree to commit the transaction before its effects are made permanent (i.e., all sites terminate the transaction in the same manner). If all the sites agree to commit a transaction, then all the actions of the distributed transaction take effect; if one of the sites declines to commit the operations at that site, then all of the other sites are required to abort the transaction. Thus, the fundamental 2PC rule states:

1. If even one site declines to commit (which means it votes to abort) the transaction, the distributed transaction has to be aborted at each site where it executes.
2. If all the sites vote to commit the transaction, the distributed transaction is committed at each site where it executes.

The simple execution of the 2PC protocol is as follows. There is a *coordinator* process at the site where the distributed transaction originates, and *participant* processes at all the other sites where the transaction executes. Initially, the coordinator sends a "prepare" message to all the participants, each of which independently determines whether or not it can commit the transaction at that site. Those that can commit send back a "vote-commit" message, while those that are not able to commit send back a "vote-abort" message. Once a participant registers its vote, it cannot change it. The coordinator collects these messages and determines the fate of the transaction according to the 2PC rule. If the decision is to commit, the coordinator sends a "global-commit" message to all the participants; if the decision is to abort, it sends a "global-abort" message to those participants who had earlier voted to commit the transaction. No message needs to be sent to those participants that had originally voted to abort, since they can assume, according to the 2PC rule, that the transaction is eventually going to be globally aborted. This is known as the *unilateral abort* option of the participants.

There are two rounds of message exchanges between the coordinator and the participants, hence the name 2PC protocol. There are a number of variations of 2PC, such as the linear 2PC and distributed 2PC, that have not found much favor among distributed-DBMS vendors. Two important variants of 2PC are the *presumed abort 2PC* and *presumed commit 2PC* [Mohan and Lindsay 1983]. These are important because they reduce the message and I/O overhead of the protocols. Presumed abort protocol is included in the X/Open XA standard and has been adopted as part of the ISO standard for open distributed processing.

One important characteristic of 2PC protocol is its *blocking* nature. Failures can occur during the commit process. As discussed above, the only way to detect these failures is by means of a timeout of the process waiting for a message. When this happens, the process (coordinator or participant) that times out follows a **termination protocol** to determine what to do with the transaction that was in the middle of the commit process. A nonblocking commit protocol is one whose termination protocol can determine what to do with a transaction in case of failures under any circumstance. In the case of 2PC, if a site failure occurs at the coordinator site and one participant site while the coordinator is collecting votes from the participants, the remaining participants cannot determine the fate of the transaction among themselves, and they have to remain blocked until the coordinator or the failed participant recovers. During this period, the locks that are held by the transaction cannot be released, which reduces the availability of the database.

Assume that a participant times out after it sends its commit vote to the coordinator, but before it receives the final decision. In this case, the participant is said to be in the READY state. The termination protocol for the participant is as follows. First, note that the participant cannot unilaterally reach a termination decision. Since it is in the READY state, it must have voted to commit the transaction. Therefore, it cannot now change its vote and unilaterally abort it. On the other hand, it cannot unilaterally decide to commit the transaction, since it is possible that another participant may have voted to abort it. In this case the participant will remain blocked until it can learn from someone (either the coordinator or some other participant) the ultimate fate of the transaction. If we consider a centralized communication structure where the participants cannot communicate with one another, the participant that has timed out has to wait for the coordinator to report the final decision regarding the transaction. Since the coordinator has failed, the participant will remain blocked. In this case, no reasonable termination protocol can be designed.

If the participants can communicate with each other, a more distributed termination protocol may be developed. The participant that times out can simply ask all the other participants to help it reach a decision. If during termination all the participants realize that only the coordinator site has failed, they can elect a new coordinator, which can restart the commit process. However, in the case where both a participant site and the coordinator site have failed, it is possible for the failed participant to have received the coordinator's decision and terminated the transaction accordingly. This decision is unknown to the other participants; thus if they elect a new coordinator and proceed, there is the danger that they may decide to terminate the transaction differently from the participant at the failed site. The above case demonstrates the blocking nature of 2PC. There have been attempts to devise nonblocking commit protocols (e.g., three-phase commit), but the high overhead of these protocols has precluded their adoption.

The inverse of termination is recovery. When the failed site recovers from the failure, what actions does it have to take to recover the database at that site to a consistent state? This is the domain of distributed recovery protocols. Consider the recovery side of the case discussed above, in which the coordinator site recovers and the recovery protocol now has to determine what to do with the distributed transaction(s) whose execution it was coordinating. The following cases are possible:

1. The coordinator failed before it initiated the commit procedure. Therefore, it will start the commit process upon recovery.
2. The coordinator failed while in the READY state. In this case the coordinator has sent the "prepare" command. Upon recovery, the coordinator will restart the commit process for the transaction from the beginning by sending the "prepare" message one more time. If the participants had already terminated the transaction, they can inform the coordinator. If they were blocked, they can now resend their earlier votes and resume the commit process.
3. The coordinator failed after it informed the participants of its global decision and terminated the transaction. Thus, upon recovery, it does not need to do anything.

Replication Protocols

In replicated distributed databases,[2] each logical data item has a number of physical instances. For example, the salary of an employee (*logical datum*) may be stored at three sites (*physical copies*). The issue in this

---

[2]Replication is not a significant concern in parallel DBMSs, because the data are normally not replicated across multiple processors. Replication may occur as a result of data shipping during query optimization, but this is not managed by the replica control protocols.

type of a database system is to maintain some notion of consistency among the copies. The most discussed consistency criterion is **one-copy equivalence**, which asserts that the values of all copies of a logical datum should be identical when the transaction that updates it terminates.

If replication transparency is maintained, transactions will issue read and write operations on a logical datum x. The replica control protocol is responsible for mapping operations on x to operations on physical copies of x ($x_1, \ldots, x_n$). A typical replica control protocol that enforces one-copy serializability is known as the **read-once–write-all** (ROWA) protocol. ROWA maps each read on x [Read(x)] to a read on one of the physical copies x_i [Read(x_i)]. The copy which is read is insignificant from the perspective of the replica control protocol and may be determined by performance considerations. On the other hand, each write on the logical datum x is mapped to a set of writes on *all* copies of x.

The ROWA protocol is simple and straightforward, but it requires that all copies of all logical data that are updated by a transaction be accessible for the transaction to terminate. Failure of one site may block a transaction, reducing database availability.

A number of alternative algorithms have been proposed which reduce the requirement that all copies of a logical datum be updated before the transaction can terminate. They relax ROWA by mapping each write to only a subset of the physical copies.

This idea of possibly updating only a subset of the copies, but nevertheless successfully terminating the transaction, has formed the basis of quorum-based voting for replica control protocols. The majority consensus algorithm can be viewed from a slightly different perspective: It assigns equal votes to each copy, and a transaction that updates that logical datum can successfully complete as long as it has a majority of the votes. Based on this idea, an early **quorum-based voting algorithm** [Gifford 1979] assigns a (possibly unequal) vote to each copy of a replicated datum. Each operation then has to obtain a *read quorum* (V_r) or a *write quorum* (V_w) to read or write a datum, respectively. If a given datum has a total of V votes, the quorums have to obey the following rules:

1. $V_r + V_w > V$ (a datum is not read and written by two transactions concurrently, avoiding the read–write conflict);
2. $V_w > V/2$ (two write operations from two transactions cannot occur concurrently on the same datum, avoiding write–write conflict).

The difficulty with this approach is that transactions are required to obtain a quorum even to read data. This significantly and unnecessarily slows down read access to the database. An alternative quorum-based voting protocol that overcomes this serious performance drawback [Abbadi et al. 1985] has also been proposed. However, this protocol makes unrealistic assumptions about the underlying communication system. It requires that failures that change the network's topology be detected by all sites instantaneously, and that each site have a view of the network consisting of all the sites with which it can communicate. In general, communication networks cannot guarantee to meet these requirements. The single-copy-equivalence replica control protocols are generally considered to be restrictive in the availability they provide. Voting-based protocols, on the other hand, are considered too complicated, with high overheads. Therefore, these techniques are not used in current distributed-DBMS products. More flexible replication schemes have been investigated where the type of consistency between copies is under user control. A number of *replication servers* have been developed or are being developed with this principle. Unfortunately, there is no clear theory that can be used to reason about the consistency of a replicated database when the more relaxed replication policies are used. Work in this area is still in its early stages.

48.4 Research Issues

Distributed and parallel DBMS technologies have matured to the point where fairly sophisticated and reliable commercial systems are now available. As expected, there are a number of issues that have yet to be satisfactorily resolved. In this section we provide an overview of some of the more important research issues.

Data Placement

In a parallel database system, proper data placement is essential for load balancing. Ideally, interference between concurrent parallel operations can be avoided by having each operation work on an independent dataset. These independent datasets can be obtained by *declustering* (horizontal partitioning) the relations according to a function (hash function or range index) applied to some placement attribute(s), and allocating each partition to a different disk. As with horizontal fragmentation in distributed databases, declustering is useful for obtaining interquery parallelism, by having independent queries working on different partitions, and intraquery parallelism, by having a query's operations working on different partitions. Declustering can be single-attribute or multiattribute. In the latter case [Ghandeharizadeh et al. 1992], an exact match query requiring the equality of all attributes can be processed by a single node without communication. The choice between hashing and range index for partitioning is a design issue: hashing incurs less storage overhead but provides direct support for exact-match queries only, while range index can also support range queries. Initially proposed for shared-nothing systems, declustering has been shown to be useful for shared-memory designs as well, by reducing memory access conflicts [Bergsten et al. 1991].

Full declustering, whereby each relation is partitioned across all the nodes, causes problems for small relations or systems with large numbers of nodes. A better solution is *variable declustering*, where each relation is stored on a certain number of nodes as a function of the relation size and access frequency [Copeland et al. 1988]. This can be combined with multirelation clustering to avoid the communication overhead of binary operations.

When the criteria used for data placement change to the extent that load balancing degrades significantly, dynamic reorganization is required. It is important to perform such dynamic reorganization on line (without stopping the incoming of transactions) and efficiently (through parallelism). By contrast, existing database systems perform static reorganization for database tuning [Shasha 1992]. Static reorganization takes place periodically when the system is idle, to alter data placement according to changes in either database size or access patterns. In contrast, dynamic reorganization does not need to stop activities and adapts gracefully to changes. Reorganization should also remain transparent to compiled programs that run on the parallel system. In particular, programs should not be recompiled because of reorganization. Therefore, the compiled programs should remain independent of data location. This implies that the optimizer does not know the actual disk nodes where a relation is stored or where an operation will actually take place. The set of nodes where a relation is stored when a certain operation is to be executed is called its *home*. Similarly, the set of nodes where the operation will be executed is called the home of the operation. However, the optimizer needs abstract knowledge of the home (e.g., relation R is hashed on A over 20 nodes), and the run-time system makes the association between the home and the actual nodes.

A serious problem in data placement is how to deal with skewed data distributions which may lead to nonuniform partitioning and negatively affect load balancing. Hybrid architectures with nodes of different memory and processing power can be exploited usefully here. Another solution is to treat nonuniform partitions appropriately, e.g., by further declustering large partitions. Separation between logical and physical nodes is also useful, since a logical node may correspond to several physical nodes.

A final complicating factor in data placement is data replication for high availability. A naive approach is to maintain two copies of the same data, a primary and a backup copy, on two separate nodes. However, in case of a node failure, the load of the node having the copy may double, thereby hurting load balancing. To avoid this problem, several high-availability data replication strategies have been proposed and recently compared [Hsiao and DeWitt 1991]. An interesting solution is Teradata's interleaved declustering, which declusters the backup copy on a number of nodes. In failure mode, the load of the primary copy is balanced among the backup-copy nodes. However, reconstructing the primary copy from its separate backup copies may be costly. In normal mode, maintaining copy consistency may also be costly. A better solution is Gamma's chained declustering, which stores the primary and the backup copy on two adjacent nodes. In failure mode, the load of the failed node and the backup nodes are balanced among all remaining nodes by using both primary- and backup-copy nodes. In addition, maintaining copy consistency is cheaper. An open issue remains how to perform data placement taking into account data replication. As with the fragment allocation in distributed databases, this should be considered an optimization problem.

Network Scaling Problems

The database community does not have a full understanding of the performance implications of all the design alternatives that accompany the development of distributed DBMSs. Specifically, questions have been raised about the scalability of some protocols and algorithms as the systems become geographically distributed [Stonebraker 1989] or as the number of system components increases [Garcia-Molina and Lindsay 1990]. Of specific concern is the suitability of the distributed transaction processing mechanisms (i.e., the two-phase locking, and, particularly, two-phase commit protocols) in wide-area-network-based distributed database systems. As mentioned before, there is a significant overhead associated with these protocols, and implementing them over a slow wide-area network may be difficult [Stonebraker 1989].

Scaling issues are only one part of a more general problem, namely that we do not have a good handle on the role of the network architectures and protocols in the performance of distributed DBMSs. Almost all the performance studies of which we are aware assume a very simple network cost model—sometimes as unrealistic as using a fixed communication delay that is independent of all network characteristics such as load, message size, network size, and so on. In general, the performance of the proposed algorithm and protocols in different local-area network architectures is not well understood, nor is their comparative behavior in moving from local-area networks to wide-area networks. The proper way to deal with scalability issues is to develop general and sufficiently powerful performance models, measurement tools and methodologies. Such work on centralized DBMSs has been going on for some time, but has not yet been sufficiently extended to distributed DBMSs.

Even though there are plenty of performance studies of distributed DBMSs, these usually employ simplistic models, artificial workloads, or conflicting assumptions or consider only a few special algorithms. This does not mean that we do not have some understanding of the tradeoffs. In fact, certain tradeoffs have long been recognized, and even the earlier systems have considered them in their design. However, these tradeoffs can mostly be spelled out only in qualitative terms; their quantification requires more research on performance models.

Distributed and Parallel Query Processing

As discussed earlier, global query optimization generates an optimal execution plan for the input fragment query by making decisions regarding operation ordering, data movement between sites, and the choice of both distributed and local algorithms for database operations. There are a number of problems related to this step. They have to do with the restrictions imposed on the cost model, the focus on a subset of the query language, the tradeoff between optimization cost and execution cost, and the optimization–reoptimization interval.

The cost model is central to global query optimization, since it provides the necessary abstraction of the distributed DBMS execution system in terms of access methods, as well as an abstraction of the database in terms of physical schema information and related statistics. The cost model is used to predict the execution cost of alternative execution plans for a query. A number of important restrictions are often associated with the cost model, limiting the effectiveness of optimization in improving throughput. Work in extensible query optimization [Freytag 1987] can be useful in parametrizing the cost model, which can then be refined after much experimentation. Even though query languages are becoming increasingly powerful (e.g., new versions of SQL), global query optimization typically focuses on a subset of the query language, namely select–project–join (SPJ) queries with conjunctive predicates. This is an important class of queries for which good optimization opportunities exist. As a result, a good deal of theory has been developed for join and semijoin ordering. However, there are other important queries that warrant optimization, such as queries with disjunctions, unions, fixpoint, aggregations, or sorting. A promising solution is to separate the language understanding from the optimization itself, which can be dedicated to several optimization "experts."

There is a necessary tradeoff between optimization cost and quality of the generated execution plans. Higher optimization costs are probably acceptable to produce "better" plans for repetitive queries, since this would reduce the query execution cost and amortize the optimization cost over many executions. However, high optimization cost is unacceptable for ad hoc queries which are executed only once. The optimization cost is mainly incurred by searching the solution space for alternative execution plans. In a distributed system, the solution space can be quite large because of the wide range of distributed execution strategies. Therefore, it is critical to study the application of efficient search strategies that avoid the exhaustive search approach. Global query optimization is typically performed prior to the execution of the query; hence it is called static. A major problem with this approach is that the cost model used for optimization may become inaccurate because of changes in the fragment sizes or the database reorganization which is important for load balancing. The problem, therefore, is to determine the optimal intervals of recompilation/reoptimization of the queries taking into account the tradeoff between optimization and execution cost.

The crucial issue in search strategy is the join-ordering problem, which is NP-complete in the number of relations [Ibaraki and Kameda 1984]. A typical approach to solving the problem is to use dynamic programming [Selinger et al. 1979], which is a *deterministic* strategy. This strategy is almost exhaustive and assures that the best of all plans is found. It incurs an acceptable optimization cost (in time and space) when the number of relations in the query is small. However, this approach becomes too expensive when the number of relations is greater than five or six. For this reason, there has been recent interest in *randomized* strategies, which reduce the optimization complexity but do not guarantee the best of all plans. Randomized strategies investigate the search space in a way which can be fully controlled so that optimization ends after a given optimization time budget has been reached. Another way to cut off optimization complexity is to adopt a heuristic approach. Unlike deterministic strategies, randomized strategies allow the optimizer to trade optimization time for execution time [Ioannidis and Wong 1987, Swami and Gupta 1988, Ioannidis and Kang 1990].

Distributed Transaction Processing

There are still topics worthy of investigation in the area of distributed transaction processing. We have already discussed the scaling problems of transaction management algorithms. Additionally, replica control protocols, more sophisticated transaction models, and nonserializable correctness criteria require further attention. The field of data replication needs further experimentation; research is required on replication methods for computation and communication; and more work is necessary to enable the systematic exploitation of application-specific properties. Experimentation is required to evaluate the claims that are made by algorithm and system designers, and we lack a consistent framework for comparing competing techniques.

One of the difficulties in quantitatively evaluating replication techniques lies in the absence of commonly accepted failure incidence models. For example, Markov models that are sometimes used to analyze the availability achieved by replication protocols assume the statistical independence of individual failure events and the rarity of network partitions relative to site failures. Currently, we do not know that either of these assumptions is tenable, nor are we aware how sensitive Markov models are to these assumptions. The validation of the Markov models by simulation requires empirical measurement, since simulations often embody the same assumptions that underlie the Markov analysis. There is a need, therefore, for empirical studies to monitor failure patterns in real-life production systems, with the purpose of constructing a simple model of typical failure loads.

To achieve the twin goals of data replication, namely availability and performance, it is necessary to provide integrated systems in which the replication of data goes hand in hand with the replication of computation and communication (including I/O). Only data replication has been studied intensively; relatively little has been done in the replication of computation and communication.

In addition to replication, and related to it, work is required on more elaborate transaction models, especially those that exploit the semantics of the application [Elmagarmid 1992, Weihl 1989]. Greater availability and improved performance, as well as concurrency, can be achieved with such models. As database technology enters new application domains such as engineering design, software development, and office information systems, the nature of and requirements for transactions change. Thus, work is needed on more complicated transaction models as well as on correctness criteria different from serializability.

Complex transaction models are important in distributed systems for a number of reasons. The most important is that the new application domains that distributed DBMSs will support in the future (e.g., engineering design, office information systems, cooperative work, etc.) require transaction models that incorporate more abstract operations that execute on complex data. Furthermore, these applications have a different sharing paradigm than the typical database access to which we are accustomed. For example, computer-assisted cooperative work environments require participants to cooperate in accessing shared resources rather than competing for them, as is usual in typical database applications. These changing requirements necessitate the development of new transaction models and accompanying correctness criteria.

Object-oriented DBMSs are now being investigated as potential candidates for meeting the requirements of such "advanced" applications. These systems encapsulate operations (methods) with data. Therefore, they require a clear definition of their update semantics, and transaction models that can exploit the semantics of the encapsulated operations [Özsu 1994].

48.5 Summary

Distributed and parallel DBMSs have become a reality in the last few years. They provide the functionality of centralized DBMSs, but in an environment where data are distributed over the sites of a computer network or the nodes of a multiprocessor system. Distributed databases have enabled the natural growth and expansion of databases by the simple addition of new machines. The price/performance characteristics of these systems are favorable in part due to the advances in computer network technology. Parallel DBMSs are perhaps the only realistic approach to meet the performance requirements of a variety of important applications which place significant throughput demands on the DBMS. In order to meet these requirements, distributed and parallel DBMSs need to be designed with special consideration for the protocols and strategies. In this article, we provide an overview of these protocols and strategies.

There are a number of related issues that we did not cover. Two important topics that we omitted are multidatabase systems and distributed object-oriented databases. Most information systems evolve independently of other systems, with their own DBMS implementations. Later requirements to "integrate" these autonomous and possibly heterogeneous systems pose significant difficulties. Systems which provide access to independently designed and implemented, and possibly heterogeneous, databases are called *multidatabase systems* [Sheth and Larson 1990].

The penetration of database management technology into areas (e.g., engineering databases, multimedia systems, geographic information systems, image databases) which relational database systems were not designed to serve has given rise to a search for new system models and architectures. A primary candidate for meeting the requirements of these systems is the object-oriented DBMS [Dogac et al. 1994]. The distribution of object-oriented DBMSs gives rise to a number of issues generally categorized as distributed object management [Özsu et al. 1994]. We have ignored both multidatabase system and distributed object management issues in this paper.

Defining Terms

Atomicity: The property of transaction processing whereby either all the operations of a transaction are executed or none of them are (all or nothing).

Client–server architecture: A distributed/parallel DBMS architecture where a set of client machines with limited functionality access a set of servers which manage data.

Concurrency control algorithm: An algorithm that synchronizes the operations of concurrent transactions that execute on a shared database.

Data independence: The immunity of application programs and queries to changes in the physical organization (physical data independence) or logical organization (logical data independence) of the database and vice versa.

Deadlock: An occurrence where each transaction in a set of transactions circularly waits on locks that are held by other transactions in the set.

Distributed database management system: A database management system that manages a database that is distributed across the nodes of a computer network and makes this distribution transparent to the users.

Durability: The property of transaction processing whereby the effects of successfully completed (i.e., committed) transactions endure subsequent failures.

Interquery parallelism: The parallel execution of multiple queries generated by concurrent transactions.

Intraoperation parallelism: The execution of one relational operation as many suboperations.

Intraquery parallelism: The parallel execution of multiple, independent operations possible within the same query.

Isolation: The property of transaction execution which states that the effects of one transaction on the database are isolated from other transactions until the first completes its execution.

Linear scaleup: Sustained performance for a linear increase in both database size and processing and storage power.

Linear speedup: Linear increase in performance for a constant database size and linear increase in processing and storage power.

Locking: A method of concurrency control where locks are placed on database units (e.g., pages) on behalf of transactions that attempt to access them.

Logging protocol: The protocol which records, in a separate location, the changes that a transaction makes to the database before the change is actually made.

One-copy equivalence: A replica control policy which asserts that the values of all copies of a logical datum should be identical when the transaction that updates that item terminates.

Parallel database management system: A database management system that is implemented on a tightly coupled multiprocessor.

Query optimization: The process by which the "best" execution strategy for a given query is found from among a set of alternatives.

Query processing: The process by which a declarative query is translated into low-level data manipulation operations.

Quorum-based voting algorithm: A replica control protocol where transactions collect votes to read and write copies of data. They are permitted to read or write data if they can collect a quorum of votes.

Read-once–write-all protocol: The replica control protocol which maps each logical read operation to a read on one of the physical copies and maps a logical write operation to a write on all of the physical copies.

Serializability: The concurrency control correctness criterion which requires that the concurrent execution of a set of transactions should be equivalent to the effect of some serial execution of those transactions.

Shared-disk architecture: A parallel DBMS architecture where any processor has access to any disk unit through the interconnect but exclusive (nonshared) access to its main memory.

Shared-memory architecture: A parallel DBMS architecture where any processor has access to any memory module or disk unit through a fast interconnect (e.g., a high-speed bus or a crossbar switch).

Shared-nothing architecture: A parallel DBMS architecture where each processor has exclusive access to its main memory and disk unit(s).

Stable database: The portion of the database that is stored in secondary storage.

Termination protocol: A protocol by which individual sites can decide how to terminate a particular transaction when they cannot communicate with other sites where the transaction executes.

Transaction: A unit of consistent and atomic execution against the database.

Transparency: Extension of data independence to distributed systems by hiding the distribution, fragmentation, and replication of data from the users.

Two-phase commit: An atomic commitment protocol which ensures that a transaction is terminated the same way at every site where it executes. The name comes from the fact that two rounds of messages are exchanged during this process.

Two-phase locking: A locking algorithm where transactions are not allowed to request new locks once they release a previously held lock.

Volatile database: The portion of the database that is stored in main memory buffers.

References

Abbadi, A. E., Skeen, D., and Cristian, F. 1985. An efficient, fault-tolerant protocol for replicated data management, pp. 215–229. In *Proc. 4th ACM SIGACT–SIGMOD Symp. on Principles of Database Systems*, Portland, OR, Mar.

Apers, P., van den Berg, C., Flokstra, J., Grefen, P., Kersten, M., and Wilschut, A. 1992. Prisma/DB: a parallel main-memory relational DBMS. *IEEE Trans. Data and Knowledge Eng.* 4(6):541–554.

Bell, D. and Grimson, J. 1992. *Distributed Database Systems*. Addison–Wesley, Reading, MA.

Bergsten, B., Couprie, M., and Valduriez, P. 1991. Prototyping DBS3, a shared-memory parallel database system, pp. 226–234. In *Proc. Int. Conf. on Parallel and Distributed Information Systems*, Miami, Dec.

Bernstein, P. A., Hadzilacos, V., and Goodman, N. 1987. *Concurrency Control and Recovery in Database Systems*. Addison–Wesley, Reading, MA.

Boral, H. 1988. Parallelism and data management, pp. 362–373. In *Proc. 3rd Int. Conf. on Data and Knowledge Bases*, Jerusalem, June.

Boral, H., Alexander, W., Clay, L., Copeland, G., Danforth, S., Franklin, M., Hart, B., Smith, M., and Valduriez, P. 1990. Prototyping Bubba, a highly parallel database system. *IEEE Trans. Knowledge and Data Eng.* 2(1):4–24, Mar.

Ceri, S. and Pelagatti, G. 1984. *Distributed Databases: Principles and Systems*. McGraw–Hill, New York.

Ceri, S., Pernici, B., and Wiederhold, G. 1987. Distributed database design methodologies. *Proc. IEEE* 75(5):533–546, May.

Copeland, G., Alexander, W., Bougherty, E., and Keller, T. 1988. Data placement in Bubba, pp. 99–108. In *Proc. ACM SIGMOD Int. Conf. on Management of Data*, Chicago, May.

DeWitt, D. J., Ghandeharizadeh, S., Schneider, D. A., Bricker, A., Hsiao, H.-I., and Rasmussen, R. 1990. The GAMMA database machine project. *IEEE Trans. Knowledge and Data Eng.* 2(1):44–62, Mar.

DeWitt, D. and Gray, J. 1992. Parallel database systems: the future of high-performance database systems. *Commun. ACM* 35(6):85–98 June.

Dogac, A., Özsu, M. T., Biliris, A., and Sellis, T., eds. 1994. *Advances in Object-Oriented Database Systems*. Springer–Verlag, Berlin.

EDS: European Declarative System Database Group. 1990. EDS—Collaborating for a high-performance parallel relational database. In *Proc. ESPRIT Conf.*, Brussels, Nov.

Elmagarmid, A. K., ed. 1992. *Transaction Models for Advanced Database Applications*. Morgan Kaufmann, San Mateo, CA.

Freytag, J.-C. 1987. A rule-based view of query optimization, pp. 173–180. In *Proc. ACM SIGMOD Int. Conf. on Management of Data*, San Francisco.

Freytag, J.-C., Maier, D., and Vossen, G. 1993. *Query Processing for Advanced Database Systems*. Morgan Kaufmann, San Mateo, CA.

Fushimi, S., Kitsuregawa, M., and Tanaka, H. 1986. An overview of the system software of a parallel relational database machine GRACE, pp. 209–219. In *Proc. 12th Int. Conf. on Very Large Data Bases*, Kyoto, Aug.

Garcia-Molina, H. and Lindsay, B. 1990. Research directions for distributed databases. *IEEE Q. Bull. Database Eng.* 13(4):12–17, Dec.

Ghandeharizadeh, S., DeWitt, D., and Quresh, W. 1992. A performance analysis of alternative multi-attributed declustering strategies, pp. 29–38. In *Proc. ACM SIGMOD Int. Conf. on Management of Data*, San Diego, CA, June.

Gifford, D. K. 1979. Weighted voting for replicated data, pp. 150–159. In *Proc. 7th ACM Symp. on Operating System Principles*, Pacific Grove, CA, Dec.

Graefe, G. 1990. Encapsulation of parallelism in the volcano query processing systems, pp. 102–111. In *Proc. ACM SIGMOD Int. Conf.*, Atlantic City, NJ, May.

Gray, J. N. 1979. Notes on data base operating systems, pp. 393–481. In *Operating Systems: An Advanced Course*, R. Bayer, R. M. Graham, and G. Seegmüller, eds. Springer–Verlag, New York.

Gray, J. 1981. The transaction concept: virtues and limitations, pp. 144–154. In *Proc. 7th Int. Conf. on Very Large Data Bases*, Cannes, France, Sept.

Gray, J. and Reuter, A. 1993. *Transaction Processing: Concepts and Techniques*. Morgan Kaufmann, San Mateo, CA.

Hsiao, H.-I. and De Witt, D. 1991. A performance study of three high-availability data replication strategies, pp. 18–28. In *Proc. Int. Conf. on Parallel and Distributed Information Systems*, Miami, Dec.

Ibaraki, T. and Kameda, T. 1984. On the optimal nesting order for computing N-relation joins. *ACM Trans. Database Syst.* 9(3):482–502, Sept.

Ioannidis, Y. and Kang, Y. C. 1990. Randomized algorithms for optimizing large join queries, pp. 312–321. In *Proc. ACM SIGMOD Int. Conf. on Management of Data*.

Ioannidis, Y. and Wong, E. 1987. Query optimization by simulated annealing, pp. 9–22. In *Proc. ACM SIGMOD Int. Conf. on Management of Data*.

Lorie, R., Daudenarde, J.-J., Hallmark, G., Stamos, J., and Young, H. 1989. Adding intra-parallelism to an existing DBMS: early experience. *IEEE Bull. Database Eng.* 12(1):2–8, Mar.

Mohan, C. and Lindsay, B. 1983. Efficient commit protocols for the tree of processes model of distributed transactions, pp. 76–88. In *Proc. 2nd ACM SIGACT–SIGMOD Symp. on Principles of Distributed Computing*.

Orfali, R., Harkey, D., and Edwards, J. 1994. *Essential Client–Server Survival Guide*. Wiley, New York.

Özsu, M. T. 1994. Transaction models and transaction management in object-oriented database management systems, pp. 147–183. In *Advances in Object-Oriented Database Systems*, A. Dogac, M. T. Özsu, A. Biliris and T. Sellis, eds. Springer–Verlag, Berlin.

Özsu, M. T. and Valduriez, P. 1991a. *Principles of Distributed Database Systems*. Prentice–Hall, Englewood Cliffs, NJ.

Özsu, M. T. and Valduriez, P. 1991b. Distributed database systems: where are we now? *IEEE Comput.* 24(8):68–78 Aug.

Özsu, M. T., Dayal, U., and Valduriez, P., eds. 1994. *Distributed Object Management*. Morgan Kaufmann, San Mateo. CA.

Selinger, P. G., Astrahan, M. M., Chamberlin, D. D., Lorie, R. A., and Price, T. G. 1979. Access path selection in a relational database management system, pp. 23–34. In *Proc. ACM SIGMOD Int. Conf. on Management of Data*, Boston, May.

Shasha, D. 1992. *Database Tuning: A Principled Approach*. Prentice–Hall, Englewood Cliffs, NJ.

Sheth, A. and Larson, J. 1990. Federated databases: architectures and integration. *ACM Comput. Surv.* 22(3):183–236, Sept.

Stonebraker, M. 1989. Future trends in database systems. *IEEE Trans. Knowledge and Data Eng.* 1(1):33–44, Mar.

Stonebraker, M., Katz, R., Patterson, D., and Ousterhout, J. 1988. The design of XPRS, pp. 318–330. In *Proc. 14th Int. Conf. on Very Large Data Bases*, Los Angeles, Sept.

Swami, A. and Gupta, A. 1988. Optimization of large join queries, pp. 8–17. In *Proc. ACM SIGMOD Int. Conf. on Management of Data*.

Valduriez, P. 1993. Parallel database systems: open problems and new issues. *Distributed and Parallel Databases* 1(2):137–165, Apr.

Weihl, W. 1989. Local atomicity properties: modular concurrency control for abstract data types. *ACM Trans. Prog. Lang. Syst.* 11(2):249–281, Apr.

Further Information

There are two current textbooks on distributed and parallel databases. One is our book [Özsu and Valduriez 1991a], and the other is [Bell and Grimson 1992]. The first serious book on this topic was [Ceri and Pelagatti 1984], which is now quite dated. Our paper [Özsu and Valduriez 1991b], which is a companion to our book, discusses many open problems in distributed databases. Two excellent papers on parallel database systems are [DeWitt and Gray 1992, Valduriez 1993].

There are a number of more specific texts. On query processing, [Freytag et al. 1993] provides an overview of many of the more recent research results. [Elmagarmid 1992] has descriptions of a number of advanced transaction models. [Gray and Reuter 1993] is an excellent overview of building transaction managers. Another classical textbook on transaction processing is [Bernstein et al. 1987]. These books cover both concurrency control and reliability.

49

Database Security and Privacy

Sushil Jajodia
George Mason University

49.1 Introduction

With rapid advancements in computer and network technology, it is possible for an organization to collect, store, and retrieve vast amounts of data of all kinds quickly and efficiently. This, however, represents a threat to the organizations as well as individuals. Consider the following incidents of security and privacy problems:

- On Nov. 2, 1988, Internet came under attack from a program containing a *worm*. The program affected an estimated 2000–3000 machines, bringing them to a virtual standstill.

- In 1986, a group of West German hackers broke into several military computers, searching for classified information, which was then passed to the KGB.

- According to a U.S. General Accounting Office study, authorized users (or *insiders*) were found to represent the greatest threat to the security of the Federal Bureau of Investigation's National Crime Information Center. Examples of misuse included insiders disclosing sensitive information to outsiders in exchange for money or using it for personal purposes (such as determining if a friend or a relative has a criminal record).

- Another U.S. General Accounting Office study uncovered improper accesses of taxpayer information by authorized users of the Internal Revenue Service (IRS). The report identified instances where IRS employees manipulated taxpayer records to generate unauthorized refunds and browsed tax returns that were unrelated to their work, including those of friends, relatives, neighbors, or celebrities.

The essential point of these examples is that databases of today no longer contain only data used for day-to-day data processing; they have become information systems that store everything, whether it is vital or not to an organization. Information is of strategic and operational importance to any organization; if the concerns related to security are not properly resolved, security violations may lead to losses of information that may translate into financial losses or losses whose values are obviously high by other measures (e.g., national security).

These large information systems also represent a threat to personal privacy since they contain a great amount of detail about individuals. Admittedly, the information collection function is essential for an organization to conduct its business; however, indiscriminate collection and retention of data can represent an extraordinary intrusion on the privacy of individuals.

To resolve these concerns, security or privacy issues must be carefully thought out and integrated into a system very early in its developmental life cycle. Timely attention to system security generally leads to effective measures at lower cost. A complete solution to security and privacy problems requires the following three steps:

- *Policy:* The first step consists of developing a security and privacy policy. The policy precisely defines the requirements that are to be implemented within the hardware and software of the computing system, as well as those that are external to the system such as physical, personnel, and procedural controls. The policy lays down broad goals without specifying how to achieve them. In other words, it expresses what needs to be done rather than how it is going to be accomplished.
- *Mechanism:* The security and privacy policy is made more concrete in the next step, which proposes the mechanism necessary to implement the requirements of the policy. It is important that the mechanism perform the intended functions.
- *Assurance:* The last step deals with the assurance issue. It provides guidelines for ensuring that the mechanism meets the policy requirements with a high degree of assurance. Assurance is directly related to the effort that would be required to subvert the mechanism. Low-assurance mechanisms may be easy to implement, but they are also relatively easy to subvert. On the other hand, high-assurance mechanisms can be notoriously difficult to implement.

Since most commercial database management systems (DBMSs) and database research have security rather than privacy as their main focus, we devote most of this chapter to the issues related to security. We conclude with a brief discussion of the issues related to privacy in database systems.

49.2 General Security Principles

There are three high-level objectives of security in any system:

- *Secrecy* aims to prevent unauthorized disclosure of information. The terms *confidentiality* or *nondisclosure* are synonyms for secrecy.
- *Integrity* aims to prevent unauthorized modification of information or processes.
- *Availability* aims to prevent improper denial of access to information. The term *denial of service* is often used as a synonym for denial of access.

These three objectives apply to practically every information system. For example, payroll system secrecy is concerned with preventing an employee from finding out the boss's salary; integrity is concerned with preventing an employee from changing his or her salary in the database; availability is concerned with ensuring that the paychecks are printed and distributed on time as required by law. Similarly, military command and control system secrecy is concerned with preventing the enemy from determining the target coordinates of a missile; integrity is concerned with preventing the enemy from altering the target coordinates; availability is concerned with ensuring that the missile does get launched when the order is given.

49.3 Access Controls

The purpose of access controls is to ensure that a user is permitted to perform certain operations on the database only if that user is authorized to perform them. Commercial DBMSs generally provide access controls that are often referred to as **discretionary access controls** (as opposed to the **mandatory access controls** which will be described later in the chapter).

Access controls are based on the premise that the user has been correctly identified to the system by some **authentication** procedure. Authentication typically requires the user to supply his or her claimed identity (e.g., user name, operator number, etc.) along with a password or some other authentication token. Authentication may be performed by the operating system, the DBMS, a special authentication server, or some combination thereof. Authentication is not discussed further in this chapter; we assume that a suitable mechanism is in place to ensure proper access controls.

Discretionary Access Controls

Most commercial DBMSs provide security by controlling modes of access by users to data. These controls are called discretionary since any user who has discretionary access to certain data can pass the data along to other users. Discretionary policies are used in commercial systems because of their flexibility; this makes them suitable for a variety of environments with different protection requirements.

There are many different administrative policies that can be applied to issue authorizations in systems that enforce discretionary protection. Some examples are *centralized* administration, where only a few privileged users may grant and revoke authorizations; *ownership-based* administration, where the creator of an object is allowed to grant and revoke accesses to the object; and *decentralized* administration, where other users, at the discretion of the owner of an object, may also be allowed to grant and revoke authorizations on the object.

Granularity and Modes of Access Control

Access controls can be imposed in a system at various degrees of granularity. In relational databases, some possibilities are the entire database, a single relation, or some rows or columns within a relation. Access controls are also differentiated by the operation to which they apply. For instance, among the basic SQL (Structured Query Language) operations, access control modes are distinguished as SELECT access, UPDATE access, INSERT access, and DELETE access. Beyond these access control modes, which apply to individual relations or parts thereof, there are also privileges which confer special authority on selected users. A common example is the DBA privilege for database administrators.

Data-Dependent Access Control

Database access controls can also be established based on the contents of the data. For example, some users may be limited to seeing salaries which are less than $30,000. Similarly, managers may be restricted to seeing the salaries only for employees in their own departments. *Views* and *query modification* are two basic techniques for implementing data-dependent access controls in relational databases.

Granting and Revoking Access

The granting and revocation operations allow users with authorized access to certain information to selectively and dynamically grant or restrict any of those access privileges to other users. In SQL, granting of access privileges is accomplished by means of the GRANT statement, which has the following general form:

```
GRANT    privileges
[ON      relation]
TO       users
[WITH    GRANT OPTION]
```

Possible privileges users can exercise on relations are *select* (select tuples from a relation), *insert* (add tuples to a relation), *delete* (delete tuples from a relation), and *update* (modify existing tuples in a relation). These access modes apply to a relation as a whole, with the exception of the update privilege, which can be further refined to refer to specific columns inside a relation. When a privilege is given with the grant option, the recipient can in turn grant the same privilege, with or without grant option, to other users.

The GRANT command applies to base relations within the database as well as views. Note that it is not possible to grant a user the grant option on a privilege without allowing the grant option itself to be further granted.

Revocation in SQL is accomplished by means of the REVOKE statement, which has the following general format:

REVOKE   privileges
[ON   relation]
FROM   users

The meaning of REVOKE depends upon who executes it, as explained next.

A grant operation can be modeled as a tuple of the form $\langle s, p, t, ts, g, go \rangle$ stating that user s has been granted privilege p on relation t by user g at time ts. If $go =$ yes, s has the grant option and, therefore, s is authorized to grant other users privilege p on relation t, with or without grant option. For example, tuple $\langle$Bob, *select*, T, 10, Ann, *yes*$\rangle$ indicates that Bob can select tuples from relation T, and grant other users authorizations to select tuples from relation T, and that this privilege was granted to Bob by Ann at time 10. Tuple $\langle C$, *select*, T, 20, B, *no*$\rangle$ indicates that user C can select tuples from relation T and that this authorization was granted to C by user B at time 20; this authorization, however, does not entitle user C to grant other users the select privilege on T.

The semantics of the revocation of a privilege from a user (revokee) by another user (revoker) is to consider as valid the authorizations that would have resulted had the revoker never granted the revokee the privilege. As a consequence, every time a privilege is revoked from a user, a recursive revocation may take place to delete all of the authorizations which would have not existed had the revokee never received the authorization being revoked.

To illustrate this concept, consider the sequence of grant operations for privilege p on relation t illustrated in Fig. 49.1(a), where every node represents a user, and an arc between node u_1 and node u_2 indicates that u_1 granted the privilege on the relation to u_2. The label of the arc indicates the time the privilege was granted. For the sake of simplicity, we make the assumption that all authorizations are granted with the grant option. Suppose now that Bob revokes the privilege on the relation from David at some time later

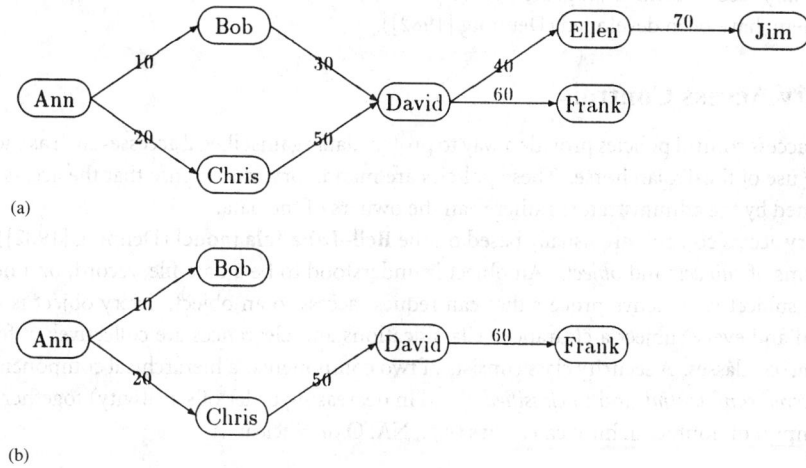

(a)

(b)

FIGURE 49.1   Bob revokes the privilege from David.

than 70. According to the semantics of recursive revocation, the resulting authorization state has to be as if David had never received the authorization from Bob, and the time of the original granting is the arbiter of this recursion. That is, if David had never received the authorization from Bob, he could not have granted the privilege to Ellen (his request would have been rejected by the system at time 40). Analogously, Ellen could not have granted the authorization to Jim. Therefore, the authorizations granted by David to Ellen and by Ellen to Jim must also be deleted. Note that the authorization granted by David to Frank does not have to be deleted since David could have granted it even if he had never received the authorization from Bob (because of the authorization from Chris at time 50). The set of authorizations holding in the system after the revocation is shown in Fig. 49.1(b).

Limitation of Discretionary Access Controls

Whereas discretionary access control mechanisms are adequate for preventing unauthorized disclosure of information to honest users, malicious users who are determined to seek unauthorized access to the data must be restricted by other devices. The main drawback of discretionary access controls is that although it allows an access only if it is authorized, it does not impose restrictions on further dissemination of information by a user once the user obtains it. This weakness makes discretionary controls vulnerable to **Trojan horse** attacks. A Trojan horse is a computer program with an apparent or actual useful function, but which contains additional *hidden* functions that surreptitiously exploit the access gained by legitimate authorizations of the invoking process. To understand how a Trojan horse can leak information to unauthorized users despite discretionary access control, consider the following example.

Suppose a user Burt (the bad guy) wants to access a file called **my_data** owned by Vic (the victim). To achieve this, Burt creates another file **stolen_data** and gives Vic the write authorization to **stolen_data** (Vic is not informed about this). Moreover, Burt modifies the code of an application generally used by Vic to include a Trojan horse containing two hidden operations, the first operation reads **my_data** and the second operation copies **my_data** into **stolen_data**. When Vic executes the application the next time, the application executes on behalf of Vic and, as a result, the personal information in **my_data** is copied to **stolen_data**, which can then be read by Burt.

This simple example illustrates how easily the restrictions stated by the discretionary authorizations can be bypassed and, therefore, the lack of assurance that results from the authorizations imposed by discretionary policies. For this reason discretionary policies are considered unsafe and not satisfactory for environments with stringent protection requirements.

To overcome this weakness further restrictions, beside the simple presence of the authorizations for the required operations, should be imposed on the accesses. To this end, the idea of *mandatory* (or *nondiscretionary*) access controls, together with a protection mechanism called the **reference monitor** for enforcing them, have been developed (Denning [1982]).

Mandatory Access Controls

Mandatory access control policies provide a way to protect data against illegal accesses such as those gained through the use of the Trojan horse. These policies are mandatory in the sense that the accesses allowed are determined by the administrators rather than the owners of the data.

Mandatory access controls are usually based on the **Bell–LaPadula model** (Denning [1982]), which is stated in terms of *subjects* and *objects*. An object is understood to be a data file, record, or a field within a record. A subject is an active process that can request access to an object. Every object is assigned a classification and every subject a clearance. Classifications and clearances are collectively referred to as *security* or *access* classes. A security class consists of two components: a hierarchical component (usually, *top secret, secret, confidential,* and *unclassified,* listed in decreasing order of sensitivity) together with a set (possibly empty) of nonhierarchical categories (e.g., NATO or Nuclear).[1]

---

[1]Although this discussion is couched within a military context, it can easily be adapted to meet nonmilitary security requirements.

Security classes are partially ordered as follows: Given two security classes L_1 and L_2, $L_1 \geq L_2$ if and only if the hierarchical component of L_1 is greater than or equal to that of L_2 and the categories in L_1 contain those in L_2. Since the set inclusion is not a total order, neither is $\geq$.

The Bell–LaPadula model imposes the following restrictions on all data accesses:

The simple security property: A subject is allowed a read access to an object only if the former's clearance is identical to or higher (in the partial order) than the latter's classification.

The $\star$-property: A subject is allowed a write access to an object only if the former's clearance is identical to or lower than the latter's classification.

These two restrictions are intended to ensure that there is no direct flow of information from high objects to low subjects.[2] The Bell–LaPadula restrictions are mandatory in the sense that the reference monitor checks security classes of all reads and writes and enforces both restrictions automatically. The $\star$-property is specifically designed to prevent a Trojan horse operating on behalf of a user from copying information contained in a high object to another object having a lower or incomparable classification.

Covert Channels

It turns out that a system may not be secure even if it always enforces the two Bell–LaPadula restrictions correctly. A secure system must guard against not only the direct revelation of data but also violations that do not result in the direct revelation of data yet produce illegal information flows. **Covert channels** fall into the violations of the latter type. They provide indirect means by which information by subjects within high-security classes can be passed to subjects within lower security classes.

To illustrate, suppose a distributed database uses two-phase commit protocol to commit a transaction. Further, suppose that a certain transaction requires a *ready-to-commit* response from both a secret and an unclassified process to commit the transaction; otherwise, the transaction is aborted. From a purely database perspective, there does not appear to be a problem, but from a security viewpoint, this is sufficient to compromise security. Since the secret process can send one bit of information by agreeing either to commit or not to commit a transaction, both secret and unclassified processes may cooperate to compromise security as follows: The unclassified process generates a number of transactions; it always agrees to commit a transaction, but the secret process by selectively causing transaction aborts can establish a covert channel to the unclassified process.

Polyinstantiation

The application of mandatory policies in relational databases requires that all data stored in relations be classified. This can be done by associating security classes with a relation as a whole, with individual tuples (rows) in a relation, with individual attributes (columns) in a relation, or with individual elements (attribute values) in a relation. In this chapter we assume that each tuple of a relation is assigned a classification.

The assignment of security classes to tuples introduces the notion of a *multilevel* relation. An example of a multilevel relation is shown in Table 49.1. Since the security class of the first tuple is secret, any user logged in at a lower security class will not be shown this tuple.

Multilevel relations suffer from a peculiar integrity problem known as **polyinstantiation** (Abrams et al. [1995]). Suppose an unclassified user (i.e., a user who is logged in at an unclassified security class) wants to enter a tuple in a multilevel relation in which each tuple is labeled either secret or unclassified. If the same key is already occurring in a secret tuple, we cannot prevent the unclassified user from inserting the unclassified tuple without leakage of one bit of information by inference. In other words the

TABLE 49.1 A Multilevel Relation

STARSHIP	DESTINATION	SECURITY_CLASS
Voyager	Rigel	Secret
Enterprise	Mars	Unclassified

---

[2] The terms *high* and *low* are used to refer to two security classes such that the former is strictly higher than the latter in the partial order.

classification of the tuple has to be treated as part of the relation key. Thus unclassified tuples and secret tuples will always have different keys, since the keys will have different security classes.

To illustrate this further, consider the multi-level relation of Table 49.2, which has the key

TABLE 49.2   A Polyinstantiated Multilevel Relation

STARSHIP	DESTINATION	SECURITY_CLASS
Voyager	Rigel	Secret
Voyager	Mars	Unclassified

STARSHIP, SECURITY_CLASS. Suppose a secret user inserts the first tuple in this relation. Later, an unclassified user inserts the second tuple of Table 49.2 This later insertion cannot be rejected without leaking the fact to the unclassified user that a secret tuple for the Voyager already exists. The insertion is therefore allowed, resulting in the relation of Table 49.2. Unclassified users see only one tuple for the Voyager, viz., the unclassified tuple. Secret users see two tuples. There are two different ways these two tuples might be interpreted as follows:

- There are two distinct starships named Voyager going to two distinct destinations. Unclassified users know of the existence of only one of them, viz., the one going to Mars. Secret users know about both of them.
- There is a single starship named Voyager. Its real destination is Rigel, which is known to secret users. There is an unclassified cover story alleging that the destination is Mars.

Presumably, secret users know which interpretation is intended.

The main drawback of mandatory policies is their rigidity, which makes them unsuitable for many application environments. In particular, in most environments there is a need for a decentralized form of access control to designate specific users who are allowed (or who are forbidden) access to an object. Thus, there is a need for access control mechanisms that are able to provide the flexibility of discretionary access control and, at the same time, the high assurance of mandatory access control. The development of a high-assurance discretionary access control mechanism poses several difficult challenges. Because of this difficulty, the limited research effort that has been devoted to this problem has yielded no satisfactory solutions.

49.4 Assurance

In order that a DBMS meets the U.S. Department of Defense (DoD) requirements, it must also be possible to *demonstrate* that the system is secure. To this end, designers of secure DBMSs follow the concept of a **trusted computing base**[3](**TCB**) (also known as a *security kernel*), which is responsible for all security-relevant actions of the system. TCB mediates all database accesses and cannot be bypassed; it is small enough and simple enough so that it can be formally verified to work correctly; it is isolated from the rest of the system so that it is tamperproof.

DoD established a metric against which various computer systems can be evaluated for security. It developed a number of *levels,* A1, B3, B2, B1, C2, C1, and D, and for each level, it listed a set of requirements that a system must have to achieve that level of security. Briefly, systems at levels C1 and C2 provide discretionary protection of data, systems at level B1 provide mandatory access controls, and systems at levels B2 or above provide increasing assurance, in particular against covert channels. The level A1, which is most rigid, requires verified protection of data. The D level consists of all systems which are not secure enough to qualify for any of levels A, B, or C.

Although these criteria were designed primarily to meet DoD requirements, they also provide a metric for the non-DoD world. Most commercial systems which implement security would fall into the C1 or D levels. The C2 level requires that decisions to grant or deny access can be made at the granularity of individual users. In principle, it is reasonably straightforward to modify existing systems to meet C2 or even B1 requirements. This has been successfully demonstrated by several operating system and DBMS

---

[3]The reference monitor resides inside the trusted computing base.

vendors. It is not clear how existing C2 or B1 systems can be upgraded to B2 because B2 imposes modularity requirements on the system architectures. At B3 or A1 it is generally agreed that the system would need to be designed and built from scratch.

For obvious reasons the DoD requirements tend to focus on secrecy of information. Information integrity, on the other hand, is concerned with unauthorized or improper modification of information, such as caused by the propagation of viruses which attach themselves to executables. The commercial world also must deal with the problem of authorized users who misuse their privileges to defraud the organization. Many researchers believe that we need some notion of mandatory access controls, possibly different from the one based on the Bell–LaPadula model, in order to build high-integrity systems. Consensus on the nature of this mandatory access controls has been illusive.

49.5 General Privacy Principles

In this section, we describe the basic principles for achieving information privacy. These principles are made more concrete when specific mechanisms are proposed to support them:

- *Proper acquisition and retention* are concerned with what information is collected and after it is collected how long it is retained by an organization.
- *Integrity* is concerned with maintaining information on individuals that is correct, complete, and timely. The source of the information should be clearly stated, especially when the information is based on indirect sources.
- *Aggregation and derivation of data* are concerned with ensuring that any aggregations or derivations performed by an organization on its information are necessary to carry out its responsibilities. Aggregation is the combining of information from various sources. Derivation goes one step further; it uses different pieces of data to deduce or create new or previously unavailable information from the aggregates. Aggregation and derivation are important and desirable effects of collecting data and storing them in databases; they become a problem, however, when legitimate data are aggregated or used to derive information that is either not authorized by law or not necessary to the organizations. Aggregates and derived data pose serious problems since new information can be derived from available information in several different ways. Nonetheless, it is critical that data be analyzed for possible aggregation or derivation problems. With a good understanding of the ways problems may arise, it should be possible to take steps to eliminate them.
- *Information sharing* is concerned with authorized or proper disclosure of information to outside organizations or individuals. Information should be disclosed only when specifically authorized and used solely for the limited purpose specified. This information should be generally prohibited from being redisclosed by requiring that it be either returned or properly destroyed when no longer needed.
- *Proper access* is concerned with limiting access to information and resources to authorized individuals who have a demonstrable need for it in order to perform official duties. Thus, information should not be disclosed to those that either are not authorized or do not have a need to know (even if they are authorized).

Privacy protection is a personal and fundamental right of all individuals. Individuals have a right to expect that organizations will keep personal information confidential. One way to ensure this is to require that organizations collect, maintain, use, and disseminate identifiable personal information and data only as necessary to carry out their functions. In the U.S., Federal privacy policy is guided by two key legislations:

Freedom of Information Act of 1966: It establishes an openness in the Federal government by improving the public access to the information. Under this act, individuals may make written requests for copies of records of a department or an agency that pertain to them.

The Privacy Act of 1974: It provides safeguards against the invasion of personal privacy by the Federal government. It permits individuals to know what records pertaining to them are collected, maintained, used, and disseminated.

49.6 Relationship Between Security and Privacy Principles

Although there appears to be a large overlap in principle between security and privacy, there are significant differences between their objectives.

Consider the area of secrecy. Although both security and privacy seek to prevent unauthorized observation of data, security principles do not concern themselves with whether it is proper to gather a particular piece of information in the first place and, after it is collected, how long it should be retained. Privacy principles seek to protect individuals by limiting what is collected and, after it is collected, by controlling how it is used and disseminated. As an example, the IRS is required to collect only the information that is both necessary and relevant for tax administration and other legally mandated or authorized purposes. IRS must dispose of personally identifiable information at the end of the retention periods required by law or regulation.

Security and privacy have different goals when new, more general information is deduced or created using available information. The objective of security controls is to determine the sensitivity of the derived data; any authorized user can access this new information. Privacy concerns, on the other hand, dictate that the system should not allow aggregation or derivation if the new information is either not authorized by law or not necessary to carry out the organization's responsibilities.

There is one misuse—denial of service—that is of concern to security but not privacy. In denial of service misuse, an adversary seeks to prevent someone from using features of the computer system by tying up the computer resources.

49.7 Research Issues

Current research efforts in the database security area are moving in three main directions. We refer the reader to Bertino et al. [1995] for a more detailed discussion and relevant citations.

Discretionary Access Controls

The first research direction concerns discretionary access control in relational DBMSs. Recent efforts are attempting to extend the capabilities of current authorization models so that a wide variety of application authorization policies can be directly supported. Related to these extensions is the problem of developing appropriate tools and mechanisms to support those models. Examples of these extension are models that permit negative authorizations, role-based and task-based authorization models, and temporal authorization models.

One extension introduces a new type of revoke operation. In the current authorization models, whenever an authorization is revoked from a user, a recursive revocation takes place. A problem with this approach is that it can be very disruptive. Indeed, in many organizations the authorizations users possess are related to their particular tasks or functions within the organization. If a user changes his or her task or function (for example, if the user is promoted), it is desirable to remove only the authorizations of this user, without triggering a recursive revocation of all of the authorizations granted by this user.

To support this concept, a new type of revoke operation, called *noncascading* revoke, has been introduced. Whenever a user, say Bob, revokes a privilege from another user, say David, a noncascading revoke operation would not revoke authorizations granted by David; instead, they are respecified as if they had been granted by Bob, the user issuing revocation. The semantics of the revocation without cascade is to produce the authorization state that would have resulted if the revoker (Bob) had granted the authorizations that had been granted by revokee (David).

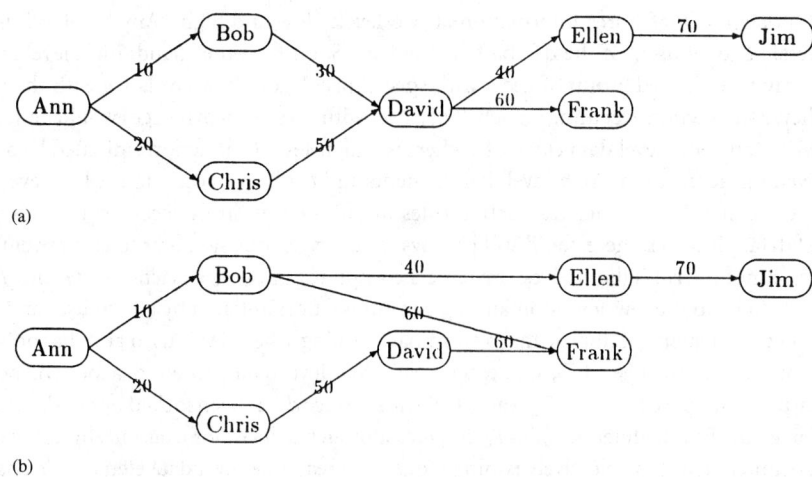

FIGURE 49.2 Bob revokes the privilege from David without cascade.

To illustrate how noncascading revocation works, consider the sequence of authorizations shown in Fig. 49.2(a). Suppose now that Bob invokes the noncascading revoke operation to the privilege granted to David. Figure 49.2(b) illustrates the authorization state after revocation. The authorizations given by David to Ellen and Frank are respecified with Bob as the grantor and Jim retains the authorization given him by Ellen.

Another extension of current authorization models concerns negative authorizations. Most DBMSs use a *closed world* policy. Under this policy, the lack of an authorization is interpreted as a negative authorization. Therefore, whenever a user tries to access a table, if a positive authorization (i.e., an authorization permitting access) is not found in the system catalogs, the user is denied the access.

This approach has a major problem in that the lack of a given authorization for a user does not guarantee that he or she will not acquire the authorization any time in the future. That is, anyone possessing the right to administer an object can grant any user the authorization to access that object. The use of explicit negative authorizations can overcome this drawback. An explicit negative authorization expresses a *denial* for a user to access a table under a specified mode. Conflicts between positive and negative authorizations are resolved by applying the *denials-take-precedence* policy under which negative authorizations override positive authorizations. That is, whenever a user has both a positive and a negative authorization for a given privilege on the same table, the user is prevented from using that privilege on the table. The user is denied access even if a positive authorization is granted *after* a negative authorization has been granted. There are more flexible models in which negative authorizations do not always take precedence over positive authorizations (Bertino et al. [1995]).

Negative authorizations can also be used for temporarily blocking possible positive authorizations of a user and for specifying exceptions. For example, it is possible to grant an authorization to all members of a group except one specific member by granting the group the positive authorization for the privilege on the table and the given member the corresponding negative authorization.

Mandatory Access Controls

The second research direction deals with extending the relational model to incorporate mandatory access controls. Several results have been reported for relational DBMSs, some of which have been applied to commercial products.

When dealing with multilevel secure DBMSs, there is a need to revise not only the data models but also the transaction processing algorithms. In this section, we show that the two most popular concurrency control algorithms, *two-phase locking* and *timestamp ordering*, do not satisfy the secrecy requirements.

Consider a database that stores information at two levels: low and high. Any low-level information is made accessible to all users of the database by the DBMS; on the other hand, high-level information is available only to a selected group of users with special privileges. In accordance with the mandatory security policy, a transaction executing on behalf of a user with no special privileges would be able to access (read and write) only low-level data elements, whereas a high-level transaction (initiated by a high user) would be given full access to the high-level data elements and read-only access to the low-level elements.

It is easy to see that the previous transaction rules would prevent direct access by unauthorized users to high-level data. However, there could still be ways for an ingenious saboteur to circumvent the intent of these rules, if not the rules themselves. Imagine a conspiracy of two transactions: T_L and T_H. T_L is a transaction confined to the low-level domain; T_H is a transaction initiated by a high user and, therefore, able to read all data elements. Suppose that a two-phase locking scheduler is used and that only these two transactions are currently active. If T_H requests to read a low-level data element d, a lock will be placed on d for that purpose. Suppose that next T_L wants to write d. Since d has been locked by another transaction, T_L will be forced by the scheduler to wait. T_L can measure such a delay, for example, by going into a busy loop with a counter. Thus, by selectively issuing requests to read low-level data elements, transaction T_H could modulate delays experienced by transaction T_L, effectively sending signals to T_L. Since T_H has full access to high-level data, by transmitting such signals, it could pass on to T_L the information that the latter is not authorized to see. The information channel thus created is known as a **signaling channel**.

Note that we can avoid a signaling channel by aborting the high transactions whenever a low-transaction wants to acquire a conflicting lock on a low data item. However, the drawback with this approach is that a malicious low transaction can starve a high transaction by causing it to abort repeatedly.

The standard timestamp-ordering technique also possesses the same secrecy-related flaw. Let T_L, T_H, and d be as before. Suppose that timestamps are used instead of locks to synchronize concurrent transactions. Let $ts(T_L)$ and $ts(T_H)$ be the (unique) timestamps of transactions T_L and T_H. Let $rts(d)$ be the read timestamp of data element d. (By definition, $rts(d) = \max(rts(d), ts(T))$, where T is the last transaction that read d.) Suppose that $ts(T_L) < ts(T_H)$ and T_H reads d. If, after that, T_L attempts to write d, then T_L will be aborted. Since a high-transaction can selectively cause a (cooperating) low transaction to abort, a signaling channel can be established.

Since there does not appear to be a completely satisfactory solution for single-version multilevel databases, researchers have been looking in alternative directions for solutions. One alternative is to maintain multiple versions of data instead of a single version. Using this alternative, transaction T_H will be given older versions of low-level data, thus eliminating both the signaling channels and starvations. The other alternative is to use correctness criteria that are weaker than serializability, yet they preserve database consistency in some meaningful way.

Authorization for Advanced Database Management Systems

A third direction concerns the development of adequate authorization models for advanced DBMSs, like object-oriented DBMSs or active DBMSs. These DBMSs are characterized by data models that are richer than the relational model. Advanced data models often include notions such as inheritance hierarchies, composite objects, versions, and methods. Therefore, authorization models developed for relational DBMSs must be properly extended to deal with the additional modeling concepts.

Authorization models developed in the framework of relational DBMSs need substantial extensions to be suitable for object-oriented DBMSs (OODBMSs). The main requirements driving such extensions can be summarized as follows. First, the authorization model must account for all semantic relationships which may exist among data (i.e., inheritance, versioning, or composite relationship). For example, in order to execute some operation on a given object (e.g., an instance), the user may need to have the authorization to access other objects (e.g., the class to which the instance belongs). Second, administration of authorizations becomes more complex. In particular, the ownership concept does not have a clear interpretation in the context of object-oriented databases. For example, a user can create an instance from a class owned by some other user. As a result, it is not obvious who should be considered the owner of the instance and administer authorizations to access the instance. Finally, different levels of authorization granularity must be supported. Indeed, in object-oriented database systems, objects are the units of access. Therefore, the authorization mechanism must allow users to associate authorizations with single objects. On the other hand, such fine granularity may decrease performance when accessing sets of objects, as in the case of queries. Therefore, the authorization mechanisms must allow users to associate authorizations with classes, or even class hierarchies, if needed. Different granularities of authorization objects are not required in relational DBMSs, where the tuples are always accessed in a set-oriented basis, and thus authorizations can be associated with entire relations or views.

Some of those problems have been been addressed by recent research. However, work in the area of authorization models for object-oriented databases is still at a preliminary stage. Of the OODBMSs, only Orion and Iris provide authorization models comparable to the models provided by current relational DBMSs.

With respect to mandatory controls, the Bell–LaPadula model is based on the subject-object paradigm. Application of this paradigm to object-oriented systems is not straightforward. Although this paradigm has proven to be quite effective for modeling security in operating systems as well as relational databases, it appears somewhat forced when applied to object-oriented systems. The problem is that the notion of an object in the object-oriented data model does not correspond to the Bell–LaPadula notion of an object. The former combines the properties of a passive information repository, represented by attributes and their values, with the properties of an active entity, represented by methods and their invocations. Thus, the object of the object-oriented data model can be thought of as the object and the subject of the Bell–LaPadula paradigm fused into one. Moreover, as with relational databases, the problem arises of assigning security classifications to information stored inside objects. This problem is made more complex by the semantic relationships among objects which must be taken into consideration in the classification. For example, the access level of an instance cannot be lower than the access level of the class containing the instance; otherwise, it would not be possible for a user to access the instance.

Some work has been performed on applying the Bell–LaPadula principles to object-oriented systems. A common characteristic to the various models is the requirement that objects must be single level (i.e., all attributes of an object must have the same security level). A model based on single-level objects has the important advantage of making the security monitor small enough that it can be easily verified. However, entities in the real world are often multilevel: some entities may have attributes with different levels of security. Since much modeling flexibility would be lost if multilevel entities could not be represented in the database, most of the research work on applying mandatory policies to object-oriented databases has dealt with the problem of representing these entities with single-level objects.

Defining Terms

Authentication: The process of verifying the identity of users.

Bell–LaPadula model: A widely used formal model of mandatory access control. It requires that the simple security property and the $\star$-property be applied to all subjects and objects.

Covert channel: Any component or feature of a system that is misused to encode or represent information for unauthorized transmission, without violating access control policy of the system.

Discretionary access controls: Means of restricting access. Discretionary refers to the fact that the users at their discretion can specify to the system who can access their files.

Mandatory access controls: Means of restricting access. Mandatory refers to the fact that the security restrictions are applied to all users. Mandatory access control is usually based on the Bell–LaPadula security model.

Polyinstantiation: A multilevel relation containing two or more tuples with the same primary key values but differing in security classes.

Reference monitor: Mechanism responsible for deciding if an access request of a subject for an object should be granted or not. In the context of multilevel security, it contains security classes of all subjects and objects and enforces two Bell–LaPadula restrictions faithfully.

Signaling channel: A means of information flow inherent in the basic model, algorithm, or protocol and, therefore, implementation invariant.

Trojan horse: A malicious computer program that performs some apparently useful function but contains additional hidden functions that surreptitiously leak information by exploiting the legitimate authorizations of the invoking process.

Trusted computing base: Totality of all protection mechanisms in a computer system, including all hardware, firmware, and software that is responsible for enforcing the security policy.

References

Abrams, M. D., Jajodia, S., and Podell, H. J., eds. 1995. *Information Security: An Integrated Collection of Essays.* IEEE Computer Society Press.

Adam, N. R. and Wortmann, J. C. 1989. Security-control methods for statistical databases: a comparative study. *ACM Comput. Sur.* 21(4):515–556.

Amoroso, E. 1994. *Fundamentals of Computer Security Technology.* Prentice–Hall, Englewood Cliffs, NJ.

Bertino, E., Jajodia, S., and Samarati, P. 1995. Database security: research and practice. *Inf. Syst.* 20(7):537–556.

Castano, S., Fugini, M., Martella, G., and Samarati, P. 1994. *Database Security.* Addison–Wesley, Reading, MA.

Cheswick, W. R. and Bellovin, S. M. 1994. *Firewalls and Internet Security.* Addison–Wesley, Reading, MA.

Denning, D. E. 1982. *Cryptography and Data Security.* Addison–Wesley, Reading, MA.

Kaufman, C., Perlman, R., and Speciner, M. 1995. *Network Security: Private Communication in a Public World.* Prentice–Hall, Englewood Cliffs, NJ.

Further Information

In this chapter, we have mainly focused on the security issues related to DBMSs. It is important to note, however, that the security measures discussed here constitute only a small aspect of overall security. As an increasing number of organizations become dependent on access to their data over the Internet, network security is also critical.

The most popular security measure these days is a *firewall* (Cheswick and Bellovin [1994]). A firewall sits between an organization's internal network and the Internet. It monitors all traffic from outside to inside and blocks any traffic that is unauthorized. Although firewalls can go a long way to protect organizations against the threat of intrusion from the Internet, they should be viewed only as the first line of defense. Firewalls are not immune to penetrations; once an outsider is successful in penetrating a system, firewalls typically do not provide any protection for internal resources. Moreover, firewalls do not protect against security violations from *insiders*, who are an organization's authorized users. Most security experts believe that insiders are responsible for a vast majority of computer crimes.

For general reference on computer security, refer to Abrams et al. [1995], Amoroso [1994], and Denning [1982]. Text by Castano et al. [1994] is specific to database security. Kaufman et al. [1995] deals with security for computer networks. Security in statistical databases is covered in Denning [1982] and in the survey by Adam and Wortman [1989].

50

Text Databases and Information Retrieval

Ellen Riloff
University of Utah

Lee Hollaar
University of Utah

50.1 Introduction

For thousands of years of human civilization, "text" has referred to words written down on paper or tablet. But the modern computer age has made it possible for words to be saved not only on a perceptible medium, such as paper, but also in the memory of our computers. One of the main advantages of having text available on line is that we can organize it and search it using computers to find information that we need.

The field of information retrieval (IR) is the study of techniques for organizing and retrieving information from text databases.[1] Given this broad definition, information retrieval encompasses a wide variety of problems. The traditional query-based retrieval task is analogous to someone walking into a library and asking the librarian to recommend books about a certain topic. The person formulates a **query** by specifying the topic of interest. The librarian then tells the person which books discuss this topic—that is, which books satisfy the query. An information retrieval system essentially plays the role of the librarian. A person types a query into the retrieval system, and the computer tells the person which books or documents are relevant to their query. Many IR systems produce a ranked list of potentially relevant documents, where the first item on the list is thought to be most relevant and the last is thought to be least relevant.

Another branch of information retrieval is concerned with static information needs: essentially, a single query that will be of interest to one or more people for some period of time. Text categorization, text routing, and text filtering systems are all concerned with long-term information needs. **Text categorization** systems label texts automatically based on a set of predefined categories. For example, computer science abstracts might be categorized by subject areas such as operating systems, compilers, algorithm design, automata theory, programming languages, artificial intelligence, or information retrieval. Presumably, many people interested in computer science would find these categorizations useful. **Text routing** systems accept a set of user profiles, or categories of interest, and automatically route texts that satisfy a profile to the corresponding user. **Text filtering** systems allow only certain texts to pass through a filter. The filter specifies

---

[1] "Information retrieval" is sometimes used to describe tasks that involve retrieving nontextual information as well, such as pictures. In this chapter, however, we will use the term information retrieval to be synonymous with text retrieval.

which topics are, or are not, of interest to a user, and only the appropriate texts are passed on to the user. Text routing and text filtering systems are usually applied to incoming streams of data, such as newswire feeds.

In this chapter, we present an overview of the field of information retrieval and discuss techniques that have been developed for building IR systems. In the first section, we discuss some of the phenomena that occur in natural language and make information retrieval a challenging task. We hope that this section will motivate the reader to better understand the issues involved in building an effective IR system, and to appreciate the strengths and weaknesses of different approaches. In the following section, we present some of the methods that are commonly used to represent, retrieve, and classify textual information. Finally, we conclude with a discussion of the state of the art in information retrieval and challenges for the future.

50.2 Underlying Principles

Information retrieval is a difficult task. Query-based IR systems must be able to accept a query about any topic and find texts that contain the information specified in the query. The query can be about any topic in the world and may be completely general or tediously specific. Many text databases (sometimes called **text corpora**[2]) are very large, and IR systems must operate in real time, so they must be able to search large volumes of text quickly and efficiently. Furthermore, the search is conducted on unconstrained text that has all of the ambiguities and idiosyncrasies inherent in natural language. In this section, we briefly discuss some of the issues that IR systems must face.

Synonymy

Synonymy occurs when different words or phrases mean essentially the same thing. For example, suppose that a user is looking for newspaper articles that discuss corporate investments. One simple strategy would be to retrieve all texts that contain the word "invest." However, this strategy will fail to retrieve many texts that mention a corporate investment because there are a lot of different ways to refer to investments. A company might "invest in," "finance," "back," "fund," "support," "capitalize," "provide capital for," "pay for," "buy into," "put up money for," or "speculate on" a company. The object might be referred to as an "investment," a "venture," a "risk," or a "speculation."

Natural language is filled with many words and phrases that have similar meanings, and it is often impossible for users to list all of the synonymous terms that might be relevant to their query. To address this problem, some information retrieval systems automatically expand a query using a thesaurus or on-line dictionary to include additional terms that are potentially relevant.

Polysemy

Polysemy occurs when a single word has more than one possible meaning. For example, the word "shot" can refer to:

a shooting	Example: HE SHOT THE SHERIFF.
an attempt	Example: I TOOK A SHOT AT PLAYING THE LOTTERY.
an injection	Example: THE DOCTOR GAVE ME A SHOT.
pellets	Example: HE FILLED THE GUN WITH BIRD SHOT.
a quantity of liquor	Example: HE DRANK A SHOT OF TEQUILA.
a photograph	Example: I TOOK A NICE SHOT OF THE EIFFEL TOWER.

Although the word "shot" may seem to be exceptionally ambiguous, many common words have multiple meanings, so this phenomenon is not unusual. For example, the word "take" has 102 different meanings in the Random House College Dictionary, Revised Edition, 1980. Although many of the semantic differences

---

[2]A single text database is called a corpus.

between dictionary entries are subtle, the sheer number of different entries indicates that "take" can have a wide variety of meanings in different contexts. Indeed, it is an interesting exercise to look up any common word in the dictionary; one is almost always surprised to find more meanings than expected. Word–sense disambiguation techniques have been investigated to address the issue of polysemy, although few IR systems currently use them.

Phrases

Expressions consisting of multiple words often have a meaning that is substantially different from the meanings of the words individually. For example, the phrase "passed away" means that someone died but the words "passed" and "away" have very different meanings on their own. Similarly, an "operating system" is more than just a system that is currently operating. Composite expressions are prevalent in natural language, and a variety of phrase-based indexing techniques have been developed to deal with them [see Croft et al. 1991, Fagan 1989, Dillon and Gray 1983].

Object Recognition

Certain types of information require special procedures to identify them. For example, dates come in a variety of forms, such as January 2, 1995; January 2; Jan. 2; Jan. 2, 1995; 1/2/95; 2.1.95. As another example, proper names are often difficult to recognize. Looking for capitalized phrases is helpful, but this solution is not sufficient because the first word of every sentence is also capitalized. Many proper names also contain both capitalized and uncapitalized words, such as "Bank of America" or "Smith, Jones, and Carlson Associates." Furthermore, many texts are not available in mixed case. Some systems use specialized procedures to recognize certain types of items that are common, such as dates, company names, and person names. These procedures typically rely on clues such as titles (e.g., Mr.) and suffixes (e.g., Corp.), as well as more complex pattern recognition.

Semantics and Role Relationships

Some information can be identified only through semantics. For example, suppose a user is interested in finding information about murders of government officials. This seemingly simple query contains two types of semantic information. First, the system must know which people are government officials, such as presidents, senators, mayors, and cabinet ministers. Second, the system must be able to identify role relationships that depend on local context. For example, a government official is typically the *victim* of a murder, not the perpetrator. It is impossible to identify whether someone is a victim or a perpetrator only by looking at the person's name (e.g., John Smith). The phrase "John Smith was killed" specifies that John was the victim, but the phrase "John Smith killed" specifies that John was the perpetrator. As another example, consider a query for texts about corporate mergers. It is impossible to determine whether a company is participating in a merger solely by looking at its name. The local context contains phrases that tell the reader whether the company is merging with another company, whether it is doing something completely unrelated, or whether it is mentioned only in reference to something, such as a product.

Recognizing role relationships requires natural language processing to analyze the text conceptually and identify the appropriate relationships. Role relationships have been found to make a substantial difference in performance for some text classification tasks [Riloff and Lehnert 1994].

Computable Values

Determining whether information is relevant sometimes depends on a specific calculation. For example, suppose a business executive is interested in newspaper articles about corporate takeovers that occurred after January 1990. The IR system must identify the date of the takeover and calculate whether it occurred before or after January 1990. Similarly, the executive might be interested only in corporate takeovers that involved more than 5 million dollars. The IR system must then identify the appropriate revenue amount (e.g., $1,800,000) and calculate whether that is more or less than 5 million dollars. These types of computations require special functions that must be provided to the IR system.

Heterogeneous Texts

Texts come in all shapes and sizes. There are two dimensions that are particularly important for IR systems: length and cohesiveness. Some texts are short (memos) and some texts are very long (books). Short documents usually discuss only a single topic. For example, abstracts are a good example of short texts that are succinct and cohesive. Longer documents, however, usually discuss many different topics in separate sections. Even medium-length documents, such as newspaper and magazine articles, typically discuss several topics in different paragraphs. For example, a newspaper article about an earthquake might contain separate paragraphs about the damage done by the earthquake, the geology of the area, a scientific analysis of earthquakes, background information about other natural disasters that recently affected the region, and a report on seismic activity in other parts of the world. Many IR systems use normalization techniques to allow for the varying lengths of documents. Until recently, however, most of the standard IR collections have consisted of short, cohesive documents (often technical abstracts). Newer collections, such as the TREC database, are much larger[3] and contain considerably longer and more heterogeneous documents. As a result, the problem of subtopic identification [Callan 1994, Hearst and Plaunt 1993, Salton et al. 1993] is receiving growing attention in the field.

IR systems are often characterized in terms of the text representation and retrieval strategy that they use. In the next section, we discuss various methods that have been developed to generate effective text representations and describe some well-known text retrieval and classification algorithms.

50.3 Best Practices

Text Representation

The purpose of an information retrieval system is to search text databases for relevant documents. However, the search must be done in real time and text databases are often very large, so it is usually too expensive to search the full texts in response to each query. The text database is therefore usually preprocessed into a more easily searchable form. This preprocessed form is called the **text representation** because the representation of the documents is what the retrieval system actually searches.

In this section, we overview some techniques that are commonly used to generate effective text representations. Traditional approaches to information retrieval [Salton 1971, Turtle and Croft 1991, Van Rijsbergen 1979] are word-based techniques that treat each text as a collection of words and retrieve documents by searching for relevant words or phrases. Inverted files, stopword lists, and stemming algorithms are frequently used in word-based systems. Phrase-based indexing techniques are sometimes used as well to represent more complex indexing terms. Experiments with richer text representations have been conducted using artificial intelligence techniques, such as natural language processing and case-based reasoning.

Inverted Files

An **inverted file** is a data structure for efficiently **indexing** texts by their words. One can view an inverted file as a list of words where each word is followed by the identifier of every text that contains the word. The number of occurrences of each word in a text is usually stored in the inverted file as well. For example, an inverted file might look something like Fig. 50.1.

In Fig. 50.1, only one text (doc-0583) contains the word "aardvark," and the word "aardvark" appears eight times in the text. Two texts (doc-0071 and doc-8914) contain the word "abracadabra," and five texts contain the word "abrupt." The word "abracadabra" appears three times in text doc-0071 and ten times

---

[3]The TREC database contains about 3.5 gigabytes of text.

aardvark	(doc-0583 8)
abracadabra	(doc-0071 3) (doc-8914 10)
abrupt	(doc-0103 2) (doc-0498 8) (doc-3938 5) (doc-8734 3) (doc-9877 10)
...	

FIGURE 50.1 Sample inverted file.

in text doc-8914. The inverted file allows an IR system to quickly determine what documents contain a given set of words and how often each word appears in a document. Additional information may also be stored in the inverted file, such as the location of each word in the text. Phrase recognition is sometimes approximated by computing the proximity of two words (whether they are adjacent or have less than *N* words between them). This requires that the location of each word be stored in the file, sometimes causing the inverted file to be several times larger than the document database itself. Inverted files can be implemented using data structures such as a sorted array, hash table, b-tree, or combinations of these structures [Frakes and Baeza-Yates 1992].

Stopword Lists

Some words are extremely common and occur in a large majority of documents. For example, words such as "a," "the," "and," "of," and "by" appear in almost every text. The entries for these words in the inverted file take considerable space. Furthermore, word-based information retrieval techniques usually do not benefit from the presence of these words, because they search for each word independently of the others and the most common words typically do not have much semantic content on their own. Therefore the most frequently occurring words, called **stopwords**, are usually thrown away and are not stored in the inverted file. Most information retrieval systems maintain a stoplist of words that should be ignored when constructing the inverted file.

The number of stopwords varies, but the stoplist usually contains a hundred words or more. If the corpus is limited to a single subject area, then common words in that subject are often placed on the stoplist. For example, the word "computer" might be a stopword for a computer science database, since it will appear in virtually every text. But stopwords can make it difficult to locate phrases, where particular stopwords have significance. Most retrieval systems that use stopwords are unable to locate the famous phrase "To be, or not to be." When its longer form ("To be, or not to be, that is the question") is used, all documents containing the word "question" will be found. Furthermore, some types of words that are almost always on stoplists, such as prepositions and auxiliary verbs, have been shown to be very important for some text classification tasks [Riloff 1995].

Superimposed Codewords

An alternative text representation to inverted files is representing each document using *codewords* that represent the different words in that document. Codewords have most commonly been used in retrieval system implementations on parallel machines [Stanfill and Kahle 1986] or special-purpose hardware [Lee and Lochovsky 1990]. While the representation of a document can be constructed by concatenating binary codes that represent the words in that document, most implementations use *superimposed codewords*. Superimposed codewords result in a fixed-length bit vector representing the document, regardless of the actual document length, a benefit for many parallel or hardware-based systems.

Superimposed codewords are formed by ORing the codewords for all words (except stopwords) in the document. Each codeword is formed by hashing the word with a number of different algorithms, and setting a bit in the codeword corresponding to each hash result. To determine if a query term occurs in the document, the query term is hashed using the same algorithms. If there is an instance where a bit corresponding to a hash result is not set, the word does not occur in the document. In parallel implementations, each node of the parallel computer holds the codeword for a document, and all nodes check the bit corresponding to a hash algorithm simultaneously. Only those nodes that have the bit set remain active, and those nodes still active after all hashings of the query term contain that term.

There are two major difficulties with superimposed codewords. First, queries that contain *don't-cares* cannot be handled. (A don't-care is an indication that any letter or substring is acceptable at an indicated point of a query term, such as allowing any suffix on a specified term.) Only completely specified terms can be hashed to find the bits of the corresponding word.

Second, and more troublesome, is the problem with *false drops* that results because the superimposed codeword for a document is the OR of the codewords for each distinct word in the document. This results in bit patterns being set that represent words not present in the document. For example, if the codeword for "shark" is (5,9,27) and the codeword for "worm" is (8,15,31), a document containing both words will have a superimposed codeword of (5,8,9,15,27,31) plus bits corresponding to the other words in the document. If the query term "lawyer" has a codeword of (8,27,31), documents containing both "shark" and "worm" will be retrieved when documents about "lawyer" are requested. False drops can be minimized by having very long codewords in comparison with the number of hashes performed, but can never be eliminated.

Stemming Algorithms

Word-based IR techniques retrieve documents by searching for texts that contain relevant words. However, looking for an *exact* match is not sufficient, because many words have several morphological variants. Inflectional variants are different syntactic forms for a single part of speech, such as verb tenses (kick, kicks, kicked, kicking), and singular or plural noun forms (book, books). Derivational variants may form a different part of speech or produce a different meaning. For example, some nouns can be used as an adjective by replacing an "-e" ending with a "-y," such as "noise" → "noisy" and "rose" → "rosy." Suffixes such as "-tion" and "-ship" can also produce new word forms, for example, "motive" → "motivation" and "friend" → "friendship."

Stemming algorithms are often used to reduce words to a root form called a stem. Some stemming algorithms use morphological analysis to create true root forms, but others use simpler techniques that merely strip off characters at the end of a word. The stems, instead of the original words, are then used to construct the inverted file. The two main advantages of stemming algorithms are space efficiency and retrieval generality. The size of the inverted file can be reduced dramatically because many different words are indexed under the same stem and require only a single entry in the inverted file. For example, the words "kicks," "kicked," and "kicking" would all be indexed under the stem "kick." Furthermore, generality is enhanced because the query terms no longer have to match a text exactly. The stemming algorithm is also applied to the query so only the stems must match. For example, if the query contains the word "education," then the stemmed form "educat" is matched against the inverted file and all documents that contain words with the same stem ("educate," "educated," "educating," "education") are retrieved.

However, stemming algorithms can also produce false hits during retrieval. Stemming algorithms are far from perfect, and even morphologically based algorithms often produce the same stem for words that have very different meanings. For example, "gorgeous" is different than "gorge," "army" is different than "arm," and "hardly" is different than "hard." Stemming has produced mixed results when integrated into information retrieval systems, so the jury is still out on whether the advantages of stemming algorithms outweigh the disadvantages [Frakes and Baeza-Yates 1992, Harman 1991, Rilof 1995].

Phrase-Based Indexing

Word-based retrieval systems are sometimes augmented with phrase-based indexing techniques so that multiword phrases can be represented explicitly in a query. Phrases often have a meaning different from their composite words. For example, if a user wants to retrieve documents about "operating systems," then the entire phrase should be used to represent the correct concept. If the words "operating" and "systems" are used independently, many irrelevant texts will likely be retrieved.

However, recognizing phrases requires part-of-speech tagging and phrase bracketing, which are not supported by most IR systems. Therefore techniques have been developed to approximate phrase recognition. One method for phrase-based indexing is to use proximity measures [see Croft et al. 1991] to specify the acceptable distance between words. Two or more words are considered to be a phrase if they are separated by no more than *N* words in the text. For example, if $N = 0$ then the words must be adjacent. If $N = 1$ then they can have only one word between them. The main advantage of this approach is that the inverted file can be easily adapted for phrase recognition by simply including the location of the words in each text. However, the inverted file will be substantially larger. Another nonsyntactic approach to phrase-based indexing uses frequency and structural information (the words must be in the same sentence or paragraph) as well as proximity measures [Fagan 1989].

There have also been several approaches to phrase-based indexing that use syntactic analysis. For example, Dillon and Gray [1983] used syntactic patterns to select appropriate phrases. A suffix dictionary and disambiguation rules were used to assign syntactic categories to words. Multiword groups that matched predefined syntactic patterns were then chosen as candidate phrases. The phrases were purged of stopwords, stemmed, and sorted. The sorting procedure produced some significant errors—"library science" is very different from a "science library"[4]—but the system performed comparably to a stem-based system and achieved slightly higher precision results at most recall levels.

Natural Language Processing

Experiments with richer text representations have also been conducted using natural language processing (NLP) techniques. NLP representations range from syntactic structures to complex knowledge structures. For example, the TTP parser generates syntactic parse trees using the Linguistic String Grammar, and has been used to generate syntactic phrases as query terms [Strzalkowski 1993]. The syntactic structures are head–modifier pairs, for example, a head noun plus its left or right adjunct, a main verb plus its object, or a main verb plus its subject. In general, syntactic approaches to IR have performed at best only slightly better than traditional word-based methods. One reason may be that most systems that do syntactic analysis still operate within the same paradigm as traditional IR systems. Although the indexing terms are slightly more complex (noun phrases or head–modifier pairs), these systems also retrieve documents based on the statistical properties of the individual indexing terms, using similar retrieval models (such as the vector-space model).

Natural language processing has also been used to generate semantic text representations. For example, the FERRET system [Mauldin 1991] is a conceptual information retrieval system that parses each text into a conceptual dependency representation [Schank 1975] and uses case frame matching to compare documents with queries. This approach has the advantage that the system can support language independence and uses semantic representations, but it also requires a large dictionary of case frames (over 12,000 were used), which were hand-coded. The DR-LINK system [Liddy and Myaeng 1994] is another conceptual information retrieval system that integrates many different levels of natural language processing. This system includes separate modules to process proper nouns, complex nominals, conceptual relations, and discourse structure. DR-LINK produces a conceptual graph for each text, which is then matched against a conceptual graph generated for the query.

A compromise between word-based approaches and in-depth natural language processing is a technique called *information extraction* [Lehnert and Sundheim 1991], which has been applied to the problem of text classification. Given a training corpus, local linguistic expressions called *relevancy signatures* [Riloff and Lehnert 1994] are generated automatically to represent domain-specific expressions (for example, "*X* was kidnapped" is a common expression in the terrorism domain). Relevancy signatures can then be used to classify new texts with respect to the target domain. Semantic features can also be incorporated into augmented relevancy signatures to support more fine-grained semantic distinctions during classification.

Retrieval Strategies

Boolean Keyword Systems

Most commercial information retrieval systems are based on Boolean operations (AND, OR, and NOT) on user-specified search keywords. The keywords can be particular words, phrases, or patterns indicating

---

[4]Proximity-based phrase indexing algorithms do not usually take word order into account either.

specific characters to be matched and locations where any character or substring is acceptable (don't-cares). The Boolean information retrieval system finds all documents that precisely match the query specified by the user.

Using an AND expression may result in relevant documents not being found if they do not contain all of the terms in an expression. Sometimes, no document in the database contains all the terms in the expression, so nothing is found. Using an OR expression may result in far too many documents being found than the user wishes to review.

In a strictly Boolean system, all documents in the database will either match the query or not. There is no "goodness" ranking of the documents. Many Boolean systems extend the matching by attempting to rank the documents according to some metric, such as giving a higher rank to a document that contains many instances of a search term. Search terms can also be given a weighting factor, indicating their relative importance to other search terms. The computed score of each document that matches the Boolean query is then used to order the documents for display to the user, with the documents receiving the higher scores (presumably of more interest to the user) displayed first.

The Vector-Space Model

The vector-space model for information retrieval [Salton et al. 1975, Salton 1989] is an attempt to find documents that are the closest to a specified query. Each document and query is viewed as a vector in an N-dimensional space, where N is the number of different terms of interest in the database. In the basic vector-space model, the length in a dimension is either 1 if the term corresponding to that dimension is present, or 0 if it isn't. Alternatively, the length can be a function of the relative importance of a term, with a longer length used, for example, with terms that occur frequently in a document or better specify the desired documents in a query. To simplify processing, it is generally assumed that there is no correlation between terms, so that the specification or occurrence of a term affects only its dimension.

To process a query, a vector is formed based on the terms specified in the query and their importance. That vector is then compared with the vector for each document. A number of measures of vector similarity have been proposed [Salton 1989], such as the cosine between the vectors. The documents are then ranked according to their vector similarity to the query. An advantage of the vector-space model over Boolean retrieval is that documents do not have to contain all query terms to be retrieved. Disadvantages include vectors of tens of thousands of dimensions (and increasing as documents with new terms are added to the database), the difficulty in accurately determining the correlation between terms, and the need to compare the query vector with the vectors for each document for every new query.

Probabilistic Models

Another type of retrieval strategy is based on probability estimates of whether a document satisfies a user's query. The INQUERY system [Turtle and Croft 1991] is a well-known IR system based on this model. INQUERY uses Bayesian inference networks to represent both the documents in the text database and the query. A document network is constructed with one layer of nodes to represent the documents, another layer of nodes for the text representation (e.g., words in the documents), and a third layer of nodes for more complex concepts (e.g., phrases). A separate query network is constructed dynamically from the query to represent the information need. The query network contains a layer of query nodes and concept nodes, which are then connected to the concept nodes in the document network. Prior probabilities are assigned, based on statistics from the text corpus, and conditional probabilities are propagated to the interior nodes. When the query network is hooked up to the document network, the belief associated with the query is calculated for every document in the corpus and the documents are ranked by these belief ratings.

Bayesian networks can be computationally expensive, but simplifications can allow them to be evaluated efficiently [Turtle and Croft 1991], and they have performed well in competitive performance evaluations such as TREC. One of the strengths of this approach is that different types of text representations can be easily incorporated into a single model.

AI Approaches

Although word-based approaches to information retrieval have largely dominated the field, there has also been research on building intelligent information retrieval systems using techniques from artificial intelligence (AI). AI encompasses a wide variety of methods, many of which have been applied to information retrieval, including natural language processing, rule-based systems, case-based reasoning, and machine learning. We have already discussed research concerning applications of natural language processing to IR, so we will briefly overview other areas of AI in this section. (Additional discussion of AI approaches to information retrieval can be found in section III of this handbook.)

Many AI systems rely on a knowledge base to provide background information needed to reason about a topic. Knowledge-based methods can achieve good performance and often provide intuitive explanations for users. However, building an appropriate knowledge base is time-consuming, so these systems are typically applied to problems that have long-term information needs for a particular domain. For example, CONSTRUE [Hayes and Weinstein 1991] is a knowledge-based text categorization system that achieved high recall and precision for a categorization problem involving 674 categories. CONSTRUE relied on a manually constructed rule base, which took several person-years to build, although the developers claim that knowledge-engineering tools that they developed will substantially reduce the time required to develop new rule bases. RUBRIC [Tong et al. 1983] is a knowledge-based system that was developed for query-based IR tasks, but it only operated in a single domain (terrorism). Users could specify a query by selecting concepts from a hierarchy of topics and subtopics that were implemented as production rules.

Case-based reasoning (CBR) is an area of AI that solves a new problem by considering previous problem-solving episodes (cases) and applying these cases to the current situation. For example, a student trying to schedule classes for his first year in college might begin with the schedule of a freshman from the previous year and then adapt it, rather than deriving a schedule from scratch. In some ways, case-based reasoning techniques are complementary to traditional IR methods. Most CBR systems perform an in-depth analysis to retrieve the best cases from a relatively small knowledge base (called the case base). Most IR techniques, on the other hand, perform a shallow analysis (word statistics) to retrieve the best texts from a relatively large text database. The goal of both types of systems, however, is to retrieve relevant items from a database.

Some work has been done on combining these methods to exploit their complementary strengths. For example, Daniels and Rissland combined a case-based reasoning system with the INQUERY retrieval engine using relevance feedback mechanisms [Daniels and Rissland 1995]. Their system was applied to two legal domains, where each case represents the facts associated with a legal case. Given a case description from a user, the CBR system analyzes the case and retrieves the best N cases from its case base. The full texts associated with these cases (the case opinions) are then given to INQUERY's relevance feedback mechanism as examples of relevant texts, and INQUERY retrieves other texts from a large text database that are potentially relevant to the original case.

As an example of a less knowledge-based CBR method, Masand et al. [1992] experimented with a memory-based reasoning approach to text classification using a database of 49,652 newswire articles that had been classified into eight categories. To classify a new text, the "nearest neighbors" were selected from the database using a traditional information retrieval system, and each category was assigned a weight by summing the similarity measures of the neighbors.

Machine learning techniques have also been applied to problems in information retrieval. The goal of machine learning algorithms is to enable a system to improve its performance on a task, either by acquiring new knowledge or by improving its skills (for example, solving a problem more efficiently). Learning algorithms have been incorporated into IR systems for both the standard query-based retrieval task and text classification tasks. For example, the FIGLEAF system [Lehnert et al. 1995] used an inductive decision-tree algorithm to learn how to classify texts in a medical domain.

There have also been hybrid approaches to IR that combine several different AI techniques. For example, Riloff developed a text classification system that combined natural language processing, case-based reasoning, and statistics to classify new texts automatically by considering the classifications of previous texts that were semantically similar [Riloff and Lehnert 1994]. Although this system used knowledge-based methods, the primary knowledge base used by the NLP and CBR modules (the dictionary of domain-specific case frames) could be generated automatically with only minimal human effort, using a dictionary construction system called AutoSlog [Riloff 1993].

Evaluation

Information retrieval systems are usually evaluated using two measures: **recall** and **precision**. Intuitively, recall measures the percentage of relevant documents in the database that the system successfully identifies as being relevant. Recall is calculated as

$$\text{Recall} = \frac{\text{\# relevant documents retrieved}}{\text{\# relevant documents in database}} \tag{50.1}$$

For example, 100% recall indicates that the IR system successfully found every relevant document in the database, while 30% recall indicates that the IR system only found 3 of every 10 relevant documents in the database. Precision, on the other hand, measures how reliable a system is when it does retrieve a document. This metric provides a sense of how often the system produces false hits. Precision is calculated as

$$\text{Precision} = \frac{\text{\# relevant documents retrieved}}{\text{total \# documents retrieved}} \tag{50.2}$$

For example, 100% precision indicates that every document retrieved by the system actually was relevant, while 30% precision indicates that only 3 out of every 10 documents retrieved by the system were truly relevant. Note that a system can achieve very high recall and very low precision at the same time; for example, a system that simply retrieves every text in the database will achieve 100% recall but presumably with extremely low precision. Conversely, a system can achieve very high precision with very low recall; for example, a system might retrieve only 5 out of 100 relevant documents (5% recall), but if it *only* retrieved these 5 documents, then it was always correct when it retrieved something (100% precision).

Another common metric used in the IR community is the E-measure, which is attributed to Van Rijsbergen [1979]. The E-metric combines recall and precision into a single measure and accepts a β-value to adjust the relative weighting of recall and precision. The formula for the E-metric is:

$$E(\beta) = 1 - \frac{(\beta^2 + 1.0)PR}{\beta^2 P + R} \tag{50.3}$$

where R is recall and P is precision. For example, $\beta = 1.0$ gives recall and precision equal weighting, $\beta = 0.5$ makes precision twice as important as recall, and $\beta = 2.0$ makes recall twice as important as precision.[5]

IR systems typically exhibit a recall–precision tradeoff. In order to increase precision, the criterion for judging relevance has to be more strict so the system is more reliable and produces fewer false hits. Consequently, however, the system usually retrieves fewer texts and recall suffers. Conversely, in order to increase recall, the criterion for judging relevance has to be more lenient, so more texts are retrieved. But the side effect is usually a greater percentage of false hits.

There are some applications for which high recall is critically important (for example, legal domains), and there are some applications for which high precision is critically important (time-dependent tasks). Therefore, IR systems are usually evaluated by measuring precision at multiple recall levels. A common approach is to chart the precision scores for 11 recall levels, ranging from 0 to 100 in increments of 10. Table 50.1 shows a sample recall–precision table.

TABLE 50.1 Example Recall–Precision Table

Recall (%)	Precision (%)
0	70.2
10	63.4
20	55.3
30	46.9
40	33.8
50	26.7
60	19.3
70	14.4
80	11.2
90	7.6
100	2.4
Average	31.9

[5]The F-measure is also used sometimes, which is simply $1 - E$.

Table 50.1 indicates that the system achieved 70.2% precision at 0% recall[6], and 2.4% precision at 100% recall. The average precision over all recall levels was 31.9%. In practice, most retrieval systems produce a ranked list of documents for which the texts at the top of the list are thought to be more relevant than those at the bottom. For each recall level, the precision score is calculated by stepping down the ranked list until N% of the relevant texts have been included. For example, 10% of the relevant texts might appear in the first 200 entries. These first 200 entries are then used as the basis for calculating the precision of the system at the 10% recall level.

Although recall and precision are the predominant measures used to evaluate IR systems, it is often difficult to quantify relevance, and there are some important aspects of performance that these measures do not capture. In real applications, the true recall is often impossible to calculate because someone would have to read every text in the database to know how many of them are actually relevant to a query. Modern text databases often contain thousands or even millions of documents, and it is simply too expensive for people to read them all.

For example, the TREC evaluations are based on a large database containing several gigabytes of text, so calculating the true recall figures for each topic is impossible. Instead, only the documents retrieved by each system were manually judged to be relevant or irrelevant. Recall was estimated by collecting the union of all of the relevant documents retrieved by the systems. The assumption behind this approach is that most relevant documents in the database were likely to be found by at least one system. However, there is no evidence indicating that this assumption was valid; it is possible that there were a significant number of relevant documents in the database that were not retrieved by any of the systems. An experiment performed by other researchers reported that, on average, no more than 20% of the relevant texts in a database were retrieved by their IR system even though the users believed that the system had retrieved at least 75% of the relevant texts [Blair and Maron 1985].

Another complicating factor is that relevance is in the eye of the beholder. Given any query, users will have different opinions about whether a document is relevant to that query. Some documents will clearly be relevant to the query and almost all users will agree that they should have been retrieved. But many documents fall into gray areas and these documents will seem relevant to some people but not to others. For example, consider a query for news articles describing recent corporate takeovers. Most people will agree that an article about a corporate takeover that happened the previous day is relevant to the query. But how about an article about a corporate takeover that happened last week? Last month? Last year? How about articles about corporate mergers? How about an article that discusses a recent stock-market plunge and briefly mentions a corporate takeover as a possible cause? How about a personal-interest piece about a well-known personality which casually refers to a recent corporate takeover in which he was involved? The issue of whether a text is relevant to any particular query is often a matter of opinion and ultimately depends on the need of the user. Little work has been done on query analysis or on developing tools to help users construct effective queries, but this is an area that could provide much needed guidance to both IR practitioners and users.

Finally, an important issue that has received little attention in the literature is that not all false hits are equally bad from a user's perspective. Informal experiments have shown that users are more satisfied with a system that produces false hits for reasons that seem to make sense to them, than they are with a system that produces false hits for seemingly stupid reasons. Users become frustrated when a system retrieves texts that are clearly irrelevant, but are more understanding when a system retrieves texts that are "in the right ballpark," even if they are ultimately judged to be irrelevant as well.

As text databases grow and information retrieval systems are applied to a broader range of texts and queries, evaluating IR systems is likely to become even more complicated but also more important. Although quantitative measures are useful for providing feedback on the performance of a system, qualitative measures and user considerations are also crucially important to get a sense of how a system is likely to perform in a real-world scenario.

---

[6]This does not mean that 0 texts were retrieved but that the actual recall percentage rounded down to 0.

Relevance Feedback

Relevance feedback [Salton 1989, 1971] is a technique used by some IR systems to improve performance on a query by asking the user for feedback about the retrieved texts. First, a user gives the system a query and the system retrieves texts that it believes to be relevant. Next, the user reads some of the retrieved documents and labels them as either relevant or irrelevant to the query. Some relevance feedback strategies ask the user to label only relevant texts, and others ask the user to label both relevant and irrelevant texts. The labeled texts are then given to a relevance feedback algorithm which uses this information to modify the original query. Most relevance feedback algorithms add new terms to the original query (this process is called **query expansion**), modify weights associated with the original query terms, or do a combination of both.

Originally, relevance feedback techniques were designed as an extension to the vector-space model to allow users to modify their query vectors in view of the results of a previous query. The user judges documents as relevant or irrelevant, and a new query vector is formed by moving the past query vector toward the relevant documents' vectors, and possibly away from irrelevant documents' vectors. Sometimes, the relevant documents' vectors are not located in a single cluster, but in two or more clusters instead. In this case, it may be desirable to split the query vector into separate vectors, each of which is moved toward a different cluster.

Relevance feedback has been shown to improve the performance of retrieval systems [Salton 1989] and is not limited to retrieval systems using the vector-space model. For example, relevance feedback has been used in systems based on probabilistic models. It can also be implemented on a Boolean keyword system by allowing users to select relevant documents, or paragraphs within documents, and then searching for documents that also contain the terms in the selected portions of text. However, it is hard to determine the effect of query modification, and the modification can substantially alter the operation and performance of the retrieval system. Furthermore, it may be difficult for users to understand why a particular document has received a high score in response to a relevance feedback query, since the modified query may contain many terms that the user did not include.

Passage-Based Retrieval

The explosion of on-line corpora has had a major impact on both the quantity and quality of available texts. Until recently, the corpora used by IR systems typically contained small, cohesive texts such as technical abstracts or short articles. Short texts are usually succinct, discuss only a single subject, and contain a high proportion of content words that are directly relevant to the subject. Longer texts, however, often discuss multiple topics in different paragraphs or sections and contain a wider variety of words.

IR systems that use word-based statistics to identify relevant documents have a more difficult time handling longer documents because the density of relevant words is smaller. For example, consider two texts that are each 20 pages long and contain 50 relevant terms. On first glance, both texts might appear to be equally relevant. However, suppose the 50 relevant terms in the first text are scattered throughout its 20 pages. This implies that the main topic of the text is probably the relevant subject. Now suppose that the 50 relevant terms in the second text are all contained on the fifth paragraph on the fourth page and nowhere else. This implies that the second text mentions the relevant subject only in one paragraph, so it is not the main topic of the text. For example, a magazine article about a Nobel Prize winner in chemistry might contain a paragraph about the small town in Poland where she was born even though the main topic of the article is chemistry, not Poland.

To alleviate this problem, researchers have begun to develop "passage-based" retrieval algorithms that determine whether a document is relevant by dividing the document into subsections, or "passages," and searching for relevant terms in each passage. Intuitively, if many different passages of a text contain relevant words, then the targeted subject is probably the topic of the text. But if only a single passage of the text contains relevant words, then the targeted subject is probably only a subtopic of the text. One of the advantages of passage-based retrieval is that the potentially relevant portions of the text are easily identified and can be displayed to the user so he or she does not have to browse through the whole document.

The main issue involved in passage-based retrieval is how to divide a text into meaningful sections. In general, most methods fall into one of three categories: document-level structuring, window structuring, and semantic structuring. Document-level structuring divides a text into passages based on document cues, such as sentences, paragraphs, or marked section headers. Window structuring divides a text into passages using a fixed-length window of words; for example, every sequence of 50 words is represented as a separate passage. Callan [1994] found that document-level structuring performed poorly because of a wide variation in the size and content shift associated with paragraph breaks, and got consistently better results by combining window structuring with full-text retrieval. To minimize the chances of splitting a small relevant block into two separate passages, he used overlapping passages so that each window overlapped with half of the words of the next window.

Passage-based retrieval using semantic structuring attempts to divide texts into sections based on subtopic identification. The TextTiling algorithm [Hearst and Plaunt 1993] automatically segments a text into "coherent" units by identifying approximate topic shifts based on lexical connectivity. A text is first divided into blocks of approximately 3–5 sentences, and then adjacent blocks are assigned similarity values by comparing the words that appear in each text and the relative frequencies of these words in the blocks and in the full text. The similarity values are then graphed and smoothed, and the valleys in the graph are identified as locations of potential topic shifts. To evaluate its potential usefulness for passage-based retrieval, the performance of the SMART information retrieval system [Salton 1971] was compared using full texts and using segments generated by TextTiling. These experiments showed that the segmented texts produced better retrieval results than the full texts. Several different methods for combining the scores of multiple segments were also evaluated: choosing only the documents associated with the top-scoring segments, choosing a fixed number of texts mostly from the highly ranked segments, or choosing documents according to the combined scores for multiple segments. Combining the scores of multiple highly ranked segments proved to be the most effective strategy.

Text Classification

All information retrieval systems are designed to classify texts. In the traditional query-based IR task, an IR system must classify each text in the database as being either relevant or irrelevant to a query. Boolean keyword systems simply retrieve all texts that are classified as relevant, while ranking systems order the texts according to their degree of relevance. However, query-based IR systems are designed for *short-term information needs*; the system is expected to receive a wide variety of queries over its lifetime, and the types of queries that it will receive cannot be defined in advance. Consequently, query-based IR systems must be as general as possible so that they can handle queries about any topic whatsoever.

A different breed of IR systems is concerned with *long-term information needs*, where one or more specific topics are expected to be of interest to a user, or group of users, for some period of time. Systems that are designed for long-term information needs are referred to broadly as **text classification** systems. Text classification includes several different tasks, such as text categorization, text routing, and text filtering. Because text classification systems focus on only one or a few specific topics, more specialized techniques can be used to build them.

The goal of text categorization systems is to classify texts automatically with respect to a predefined set of categories. For example, a company might want to build a text categorization system for different types of business transactions, such as mergers, acquisitions, joint ventures, and corporate takeovers. Text classification was the focus of some of the earliest work in information retrieval. For example, Maron [1961] developed a text categorization system for 32 categories based on the conditional probability of each category given certain words. Borko and Bernick [1963] followed up Maron's work by applying factor analysis to a correlation matrix composed of the same words. This approach involved deriving the categories themselves as well as classifying texts.

More recent work on text classification has focused on AI approaches, many of which were described earlier [see Hayes and Weinstein 1991, Masand et al. 1992, Riloff and Lehnert 1994]. The AIR/X system [Fuhr et al. 1991] tackled a text classification task of a different variety, which was essentially a form of automatic indexing. AIR/X automatically indexed texts using terms called "descriptors" that came from a controlled (predetermined) vocabulary. The system used weighting and probabilistic classification trees to assign one or more descriptors to a text. Another, slightly different text classification system was NLDB [Rau and Jacobs 1991], which automatically segments text databases for IR applications. NLDB uses lexical analysis, pattern matching, and company-name recognition to assign keywords and topic indicators to texts. The keywords are segmented into different categories so a user can search the database more effectively.

Text routing and text filtering are also active areas of research. Text routing systems are given a set of "profiles" from users and then automatically route texts that match one of the profiles to the appropriate user. Text filtering systems are given a list of topics that are or are not of interest to someone and allow only the texts that satisfy the filter to pass through to the user. For example, a potential application of text filtering is an e-mail or newsgroup filter that only shows you messages about certain topics. Both types of systems are usually applied to incoming streams of data, rather than static text databases.

Conclusions

Information retrieval encompasses a wide variety of problems and methods that we have tried to outline in this section. Traditional word-based methods have dominated the field since the 1960s, but the IR community is facing new challenges posed by larger and more heterogeneous text corpora, which have led to an explosion of new approaches and methodologies. Richer text representations, passage-based retrieval, and AI approaches to intelligent information retrieval are relatively new directions for IR that reflect the rapid changes that are taking place in the field.

50.4 Research Issues and Summary

In this overview of the field of information retrieval, we have tried to convey a sense of history about its evolution over the past few decades. However, research in information retrieval has experienced a surge that reflects a growing awareness of the critical role that text-processing technology will play in our future. In the past few years, the amount of text available on line has grown by staggering proportions, and we have every reason to believe that this trend will continue, perhaps at an even faster pace. But as our information repositories grow, our ability to use the information effectively decreases at the same time. The scenario is rather like an office worker trying to clear his desk, but every time he throws one paper in the trash can his boss comes along and puts ten more papers on his desk. It is simply a losing battle to try to manage so much information by hand; the only way that we will be able to benefit from the explosion of on-line information is through automated methods and computer technology.

Managing vast information repositories is a multifaceted challenge. We must be able to retrieve information in response to specific queries, as described in this chapter, but there are also countless possibilities for other forms of information retrieval and management. Text categorization, text filtering, text routing, and text extraction are some of the related applications that have already surfaced, but many additional tasks surely lurk on the horizon.

Much of the recent activity in information retrieval has been instigated by government-sponsored performance evaluations. The first Text Retrieval Conference (TREC-1), held in 1993, was initiated by the U.S. government to assess the state of the art in information retrieval and to bring together researchers working on different approaches to IR. TREC meetings have been held annually since then, with about 30 laboratories from academic and industrial research sites participating each year. For each conference, each participating site develops its own IR system, and all of the systems are evaluated on the same data sets. This design allows a clean comparison to be made of the strengths and weaknesses of different approaches, since everyone is working with exactly the same texts and topics. In addition, one of the main contributions of the TREC conferences has been to bring greater visibility to text-routing applications because the TREC participants are evaluated on both query-based IR tasks as well as text routing tasks. This emphasis on text routing has, in turn, brought greater visibility to related tasks such as text categorization and text filtering.

The future also holds great promise for integrating information retrieval techniques with natural language processing systems. In general, the strengths of these technologies are largely complementary. IR systems typically do shallow text analysis (word statistics), which allows them to process large amounts of text quickly and efficiently. However, the accuracy of these systems often suffers because of a lack of semantic analysis, especially for complex information requests. Natural language processing systems, on the other hand, usually perform deeper text analysis (generating syntactic and semantic structures), which allows them to produce richer meaning representations of text. However, NLP techniques are more computationally expensive and therefore have more trouble scaling up to large text collections.

The Tipster program, also sponsored by the U.S. government, was started expressly for the purpose of bringing IR and NLP technologies together. Like the TREC evaluations, the Tipster project involves researchers from different sites who build text-processing systems that are compared on uniform test data. The Tipster *detection* sites build information retrieval systems, and the Tipster *extraction* sites build information extraction (NLP) systems. Ultimately, the goal of this project is to integrate these technologies into a single text-processing architecture that can achieve intelligent information retrieval.

Some of the grand challenges for this field include intelligent information retrieval, automated construction of large-scale hypertext systems, and multimedia information retrieval (retrieving images and pictures as well as text). Towards this end, the National Science Foundation (NSF), the Department of Defense Advanced Research Projects Agency (ARPA), and the National Aeronautics and Space Administration (NASA) recently initiated a joint project on *digital libraries* to advance the state of the art in information retrieval, processing, and management. Six four-year projects were funded under this initiative at major research universities across the country, with collaborative efforts from other organizations such as libraries, government agencies, and schools.

So what does the future hold? Most people would probably agree that it is impossible to predict which information retrieval applications will ultimately achieve the greatest success or what new problems lie ahead. The one thing that is certain is that information retrieval technology will become an increasingly integral part of our lives. The recent prominence of the Internet in popular culture promises to bring the power of digital technology into homes throughout the country. But the extent to which information retrieval technology weaves its way into our everyday life will largely depend on the new ideas and approaches put forth by the next generation of IR researchers. Perhaps one day in the not too distant future we will be able to navigate vast libraries of digital information with ease and be able to find even the most obscure pieces of information in the comfort of our homes.

Defining Terms

Indexing: The process of transforming a text database into another representation that can be searched quickly. For example, most word-based systems use automated indexing techniques to create an inverted file so each text is indexed by its words.

Inverted file: A data structure for quickly indexing texts by the words they contain.

Precision: The number of relevant texts retrieved divided by the total number of texts retrieved.

Query: The specification given to an information retrieval system that describes the topic in which a user is interested.

Query expansion: The process of adding new terms to a query automatically, for example, in response to relevance feedback.

Recall: The number of relevant texts retrieved divided by the total number of relevant texts in a corpus.

Relevance feedback: The process of asking a user for feedback concerning which retrieved texts are relevant and which ones are not, and then feeding that information back into the retrieval system to improve its performance.

Stemming: The process of reducing words to a shorter form called a *stem*, which may involve morphological analysis or simply stripping off characters. For example, the word "education" may be stemmed as "educat" or "educate."

Stopword: A word that occurs frequently in a large majority of documents. For example, "a," "the," "and," "by," and "of" are common stopwords.

Text categorization: See **text classification**.

Text classification: A task that involves labeling texts as either relevant or irrelevant to a particular category. For example, binary classification tasks require each text to be labeled as either relevant or irrelevant to a single category, while multiclass tasks (sometimes called *text categorization* problems) require each text to be assigned one or more different category labels.

Text corpora: Text databases. A single text database is called a *text corpus*.

Text filtering: A task that involves filtering a set of texts so that only relevant texts which satisfy the filter are passed on to a user.

Text representation: The form that is used to represent a document. For example, a document might be represented as a set of words or as a set of linguistic structures.

Text routing: A task that involves categorizing texts with respect to a set of user profiles and sending relevant texts to the appropriate user.

References

Blair, D. C. and Maron, M. E. 1985. An evaluation of retrieval effectiveness for a full-text document-retrieval system. *Commun. ACM* 28(3):289–299.

Borko, H. and Bernick, M. 1963. Automatic document classification. *J. ACM* 10(2):151–162.

Callan, J. P. 1994. Passage-level evidence in document retrieval, pp. 302–310. In *Proc. 17th Annual Int. ACM SIGIR Conf. on Research and Development in Information Retrieval.*

Croft, W. B., Turtle, H. R., and Lewis, D. D. 1991. The use of phrases and structured queries in information retrieval, pp. 32–45. In *Proc. 14th Annual Int. ACM SIGIR Conf. on Research and Development in Information Retrival.*

Daniels, J. J. and Rissland, E. L. 1995. A case-based approach to intelligent information retrieval, pp. 238–245. In *Proc. 18th Annual Int. ACM SIGIR Conf. on Research and Development in Information Retrieval.*

Dillon, M. and Gray, A. 1983. FASIT: a fully automatic syntactically based indexing system. *J. Am. Soc. Inf. Sci.* 34(2):99–108.

Fagan, J. 1989. The effectiveness of a nonsyntactic approach to automatic phrase indexing for document retrieval. *J. Am. Soc. Inf. Sci.* 40(2):115–132.

Frakes, W. B. and Baeza-Yates, R., eds. 1992. *Information Retrieval: Data Structures and Algorithms.* Prentice–Hall, Englewood Cliffs, NJ.

Fuhr, N., Hartmann, S., Lustig, G., Schwantner, M., and Tzeras, K. 1991. AIR/X–a rule-based multistage indexing system for large subject fields, pp. 606–623. In *Proc. RIAO 91.*

Harman, D. 1991. How effective is suffixing? *J. Am. Soc. Inf. Sci.* 42(1):7–15.

Hayes, P. J. and Weinsteins, S. P. 1991. Construe-TIS: a system for content-based indexing of a database of news stories, pp. 49–64. In *Proc. Second Annual Conf. on Innovative Applications of Artificial Intelligence.* AAAI Press.

Hearst, M. A. and Plaunt, C. 1993. Subtopic structuring for full-length document access, pp. 59–68. In *Proc. 16th Annual Int. ACM SIGIR Conf. on Research and Development in Information Retrieval.*

Lee, D. L. and Lochovsky, F. H. 1990. HYTREM—a hybrid text-retrieval machine for large databases. *IEEE Trans. Comput.* 39(1):111–123.

Lehnert, W. G. and Sundheim. B. 1991. A performance evaluation of text analysis technologies. *AI Mag.* 12(3):81–94.

Lehnert, W., Soderland, S., Aronow, D., Feng, F., and Shmueli, A. 1995. Inductive text classification for medical applications. *J. Exp. Theor. Artif. Intell.* 7(1):271–302.

Liddy, E. D. and Myaeng. S. H. 1994. DR-LINK: a system update for TREC-2, pp. 85–99. In *Proc. Second Text Retrieval Conference (TREC2)*, D. K. Harman, ed. NIST Special Publication 500-215.

Maron, M. 1961. Automatic indexing: an experimental inquiry. *J. ACM* 8:404–417.

Masand, B., Linoff, G., and Waltz, D. 1992. Classifying news stories using memory based reasoning, pp. 59–65. In *Proc. 15th Annual Int. ACM SIGIR Conf. on Research and Development in Information Retrieval.*

Mauldin, M. 1991. Retrieval performance in FERRET: a conceptual information retrieval system, pp. 347–355. In *Proc. 14th Annual Int. ACM SIGIR Conf. on Research and Development in Information Retrieval.*

Rau, L. F. and Jacobs, P. S. 1991. Creating segmented databases from free text for text retrieval, pp. 337–346. In *Proc. 14th Annual Int. ACM SIGIR Conf. on Research and Development in Information Retrieval.*

Riloff, E. 1993, Automatically constructing a dictionary for information extraction tasks, pp. 811–816. In *Proc. Eleventh National Conf. on Artificial Intelligence.* AAAI Press/MIT Press.

Riloff, E. 1995. Little words can make a big difference for text classification, pp. 130–136. In *Proc. 18th Annual Int. ACM SIGIR Conf. on Research and Development in Information Retrieval.*

Riloff, E. and Lehnert, W. 1994. Information extraction as a basis for high-precision text classification. *ACM Trans. Inf. Systems* 12(3):296–333.

Salton, G., ed. 1971. *The SMART Retrieval System: Experiments in Automatic Document Processing.* Prentice–Hall, Englewood Cliffs, NJ.

Salton, G. 1989. *Automatic Text Processing: The Transformation, Analysis, and Retrieval of Information by Computer.* Addison–Wesley, Reading, MA.

Salton, G., Wong, A., and Yang, C. S. 1975. A vector space model for automatic indexing. *Commun. ACM* 18(11):613–620.

Salton, G. Allan, J., and Buckley, C. 1993. Approaches to passage retrieval in full text information systems. In *Proc. 16th Annual International ACM SIGIR Conf. on Research and Development in Information Retrieval.*

Schank, R. C. 1975. *Conceptual Information Processing,* Ch. 3, pp. 22–82. North Holland.

Stanfill, C. and Kahle, B. 1986. Parallel free-text search on the connection machine system. *Commun. ACM* 29(12):1229–1239.

Strzalkowski, T. 1993. Robust text processing in automated information retrieval, pp. 9–19. In *Proc. ACL Workshop on Very Large Corpora: Academic and Industrial Perspectives.*

Tong, R. M., Shapiro, D. G., McCune, B. P., and Dean. J. S. 1983. A rule-based approach to information retrieval: some results and comments, pp. 411–415. In *Proc. Third National Conference on Artificial Intelligence.* William Kaufmann.

Turtle. H. and Croft, W. B. 1991. Efficient probabilistic inference for text retrieval, pp. 644–661. In *Proc. of RIAO 91.*

Van Rijsbergen, C. J. 1979. *Information Retrieval,* 2nd ed. Butterworths, London.

Further Information

A good introduction to the basic structures and techniques used to build information retrieval systems is *Information Retrieval: Data Structures and Algorithms,* William B. Frakes and Ricardo Baeza-Yates, editors.

The TREC evaluations have been instrumental in providing large, realistic corpora for developing IR systems and in providing a forum for comparative analyses of a variety of different approaches to IR. The first Text Retrieval Conference (TREC1) was held in 1993, and the second Text Retrieval Conference (TREC2) was held in 1994. The proceedings of the TREC conferences are available from the National Institute of Standards and Technology as:

Harman, D., ed. 1993. *Proc. First Text Retrieval Conference (TREC1).* National Institute of Standards and Technology Special Publication 200-207.

Harman, D., ed. 1994. *Proc. Second Text Retrieval Conference (TREC2).* National Institute of Standards and Technology Special Publication 500-215.

The Tipster program has been run in conjunction with the TREC conferences, but involves a smaller number of IR researchers as well as researchers in information extraction. The ultimate goal of Tipster is to demonstrate how text detection systems (IR systems) can be integrated with text extraction systems to produce more effective text processing technology. The proceedings for the first phase of the Tipster program are available as:

1993. *Proc. TIPSTER Text Program (Phase I).* Morgan Kaufmann, San Francisco.

The primary venue for the presentation of new research contributions to IR is the Annual International ACM SIGIR Conference on Research and Development in Information Retrieval (commonly referred to as "SIGIR"). Proceedings of this conference can be obtained through the ACM Special Interest Group in Information Retrieval, which also publishes a quarterly newsletter called *SIGIR Forum.*

Archival journals that often publish research papers in IR are the *ACM Transactions on Information Systems* and *Information Processing and Management.*

51

Rules in Database Systems

Stefano Ceri
Politecnico di Milano

Raghu Ramakrishnan
*University of
Wisconsin—Madison*

51.1 Introduction

Extending database systems with deductive and active rules is one of the major steps in the recent evolution of database technology. By means of rules, database designers express both declarative and procedural knowledge about their applications; rules introduce a significant increase of the expressive power of database languages, resulting in the enhancement of processing capabilities within database servers.

Deductive database systems are database management systems whose query language and (usually) storage structure are designed around a logical model of data. As relations are naturally thought of as the "value" of a logical predicate, and relational languages such as SQL are syntactic sugarings of a limited form of logical expression, it is easy to see deductive database systems as an advanced form of relational systems.

Deductive systems are not the only class of systems with a claim to being an extension of relational systems. The deductive systems do, however, share with the relational systems the important property of being *declarative*, that is, of allowing the user to query or update by saying *what* he or she wants, rather than *how* to perform the operation. Declarativeness is now being recognized as an important driver of the success of relational systems.

Deductive rules were influenced by work in logic programming and offer a rich query language, which extends SQL in many important directions (including support for aggregation, negation, and recursion). Deductive rules express declarative knowledge: they represent queries in a style that reflects the meaning of the query itself and does not depend on the query evaluation strategy. Thus, they have pure, declarative semantics. In addition, rules may be interpreted as integrity constraints.

The increased power of deductive languages, in comparison with conventional database query languages such as SQL, is important in a variety of application domains, including decision support, financial analysis, scientific modeling, various applications of transitive closure (e.g., bill of materials, path problems), and

0-8493-2909-4/97/$0.00+$.50
© 1997 by CRC Press, Inc.

language analysis and parsing. (See Ramakrishnan [1994] for a collection of articles on applications of deductive systems.) Deductive database systems are best suited for applications in which a large number of data must be accessed and complex queries must be supported.

Active rules were influenced by work in artificial intelligence (and specifically by production rules and expert systems). They describe reactive computations which occur automatically in response to data manipulation events (Widom and Ceri [1996]). Active rules express knowledge in terms of event–condition–action computations: when given events occur on the database, if the rule's conditions hold, then the rule's actions are executed. Thus, the semantics of active rules is procedural, and in particular it depends on various features of rules, which influence their execution model (in particular, their scheduling).

The potential uses of reactive behavior are very significant. Active rules are a vehicle for supporting data derivations (Ceri and Widom [1994]), integrity maintenance (Ceri and Widom [1990]), workflow management (Dayal et al. [1990]), replication management, and more. For instance, when active rules maintain data integrity, a user-defined transaction may cause the loss of integrity; system-defined active rules take the responsibility of reacting to the integrity violations, either by repairing them or by rolling back the transaction. More in general, active rules may impose the so-called "business rules" to user-defined applications, thereby incorporating some domain-specific knowledge, e.g., about bond marketing, retail trading, production scheduling, and so on (Loucopoulos [1994]).

Deductive and active rules are supported by many commercial products and research prototypes. Deductive rules are supported by several research prototypes (e.g., Coral, Aditi, Glue-Nail, LDL, Lola, XSB; see Ramakrishnan and Ullman [1995]). The forthcoming system Validity, developed by Bull (Friesen et al. [1994]), is perhaps the first commercial version of a deductive and object-oriented database (DOOD) system.

Active rules are supported by most relational database systems, including Allbase, DB2, Illustra, Informix, Ingres, Oracle, RDB, Sybase, and others. Active rules in these systems (called triggers) support basic functionalities and are inspired by the still evolving SQL3 standard, which is not published yet. Thus, triggers of different systems are syntactically and semantically different. Triggers are also supported in several object-oriented database prototypes (e.g., Chimera, Naos, Ode, Reach, Samos, Sentinel, and many others; see Widom and Ceri [1996]).

This article is organized as follows. The second section is devoted to deductive rules, which are considered essentially from two viewpoints: the expressive power of deductive databases, which is progressively extended to encompass generalized recursion, negation, and aggregation; and the techniques for the efficient evaluation of deductive rules. The third section is devoted to active rules; it presents the main features of active-rule languages, provides an operational semantics for an abstract active-rule execution engine, gives some examples of active rules, and introduces the main problems in active-rule design. The fourth section aims at giving a joint description of deductive and active rules and finally introduces knowledge independence, a new design abstraction which is common to both deductive and active rules.

51.2 Deductive Databases

The origins of deductive databases can be traced back to work in automated theorem proving and, later, logic programming. For an interesting survey of the early development of the field, see Minker et al. [1984].

Deductive database systems divide their information into two categories:

1. *Data*, or facts, that are normally represented by a predicate with constant arguments (by a *ground atom*). For example, the fact *parent(joe, sue)*, means that Sue is a parent of Joe. Here, *parent* is the name of a predicate, and this predicate is represented *extensionally*, that is, by storing in the database a relation of all the true tuples for this predicate. Thus, (*joe, sue*) would be one of the tuples in the stored relation.

2. *Rules*, or program, which are normally written in Prolog-style notation as

$$p :- q_1, \ldots, q_n.$$

This rule is read declaratively as "q_1 and q_2 and ... and q_n implies p." Here p (the *head*) and the q_i's (the *subgoals* of the *body*) are *atomic formulas* (also referred to as *literals*), consisting of a predicate applied to *terms*, which are either constants, variables, or function symbols applied to terms. Programs in which terms are either constants or variables often are referred to as *datalog* programs. The data are often referred to as the *EDB* (*extensional database*) and the rules as the *IDB* (*intensional database*). Following Prolog convention, we use names beginning with lowercase letters for predicates, function symbols, and constants, whereas variables have names beginning with uppercase letters. In later sections we also consider programs that contain features like negation and aggregation (e.g., sum) operations applied to subgoals.

Example 51.1. Consider the following program:

$$sg(X, Y) :- flat(X, Y).$$

$$sg(X, Y) :- up(X, U), sg(U, V), down(V, Y).$$

Here, *sg* is a predicate ("same-generation"), and the head of each of the two rules is the atomic formula $p(X, Y)$. X and Y are variables. The other predicates found in the rules are *flat*, *up*, and *down*. These are presumably stored extensionally, while the relation for *sg* is intensional, that is, defined only by the rules. Intensional predicates play a role similar to views in conventional database systems, although we expect that in deductive applications there will be large numbers of intensional predicates and rules defining them, far more than the number of views defined in typical database applications.

The first rule can be interpreted as saying that individuals X and Y are at the same generation if they are related by the predicate flat, that is, if there is a tuple (X, Y) in the relation for flat. The second rule says that X and Y are also at the same generation if there are individuals U and V such that:

1. X and U are related by the *up* predicate.
2. U and V are at the same generation.
3. V and Y are related by the *down* predicate.

These rules thus define the notion of being at the same generation recursively. Since common implementations of SQL do not support general recursions such as this example without going to a host-language program, we see one of the important extensions of deductive systems: the ability to support declarative, recursive queries.

The optimization of recursive queries has been an active research area and often has focused on some important classes of recursion. We say that a predicate p *depends upon* a predicate q—not necessarily distinct from p—if some rule with p in the head has a subgoal whose predicate either is q or (recursively) depends on q. If p depends upon q and q depends upon p, then p and q are said to be *mutually recursive*. A program is said to be *linear recursive* if each rule contains at most one subgoal whose predicate is mutually recursive with the head predicate.[1]

Optimization Techniques for Deductive DBs

In a sense, deductive systems are an attempt to adapt Prolog, which has a "small-data" view of the world, to a "large-data" world. (Equally, one could think of deductive systems as an attempt to extend relational database systems; indeed, this is the more common view.) Prolog's depth-first evaluation strategy can lead to infinite loops in the computation, but this problem is adequately addressed by memoing extensions to Prolog evaluation. However, the tuple-at-a-time Prolog evaluation is inefficient when large datasets are involved and access to disk must be minimized.

---

[1] Sometimes, a more restrictive definition is used, requiring that no two distinct predicates can be mutually recursive or even that there be at most one recursive rule in the program. We shall not worry about such distinctions.

The key to accessing disk data efficiently is to utilize the set-oriented nature of typical database operations and to tailor both the clustering of data on disk and the management of buffers in order to minimize the number of pages fetched from disk. The goal of deductive databases is to deal with a superset of relational algebra that includes support for recursion in a way that permits efficient handling of disk data. Evaluation strategies should retain Prolog's goal-directed flavor but be more *set-at-a-time*. There are two aspects to set orientation:

- The run-time computation should utilize traditional relational operations such as selects, projects, joins, and unions; thus, conventional database processing techniques can be utilized.

- The overall computation should be organized so as to make as many operations as possible (logically) concurrent, thereby creating more flexibility in terms of reordering operations. In particular, it is desirable to generate and process *sets of goals* rather than proceed one (sub)goal at a time.

These considerations have led to the development of optimization techniques in the deductive database area that differ significantly from the traditional approaches to optimizing logic programs. Whereas for nonrecursive rules the optimization problem is similar to that of conventional relational optimization, the presence of recursive rules opens up a variety of new options and problems. There is an extensive literature on the subject, and we shall attempt here to give only the most basic ideas and motivation.

An important problem is that, frequently, a query asks not for the entire relation corresponding to an intensional predicate but for a small subset. An example would be a query like *sg(john, Z)*, that is, "who is at the same generation as John?" asked of the predicate defined in Example 51.1. It is important that we answer this query by examining only the part of the database that involves individuals somehow connected to John.

A top-down, or backward-chaining, search would start from the query as a goal and use the rules from head to body to create more goals, and none of these goals would be irrelevant to the query, although some might cause us to explore paths that happen to "dead end," because data that would lead to a solution to the query happen not to be in the database. Prolog evaluation is the best-known example of top-down evaluation. However, the Prolog algorithm, like all purely top-down approaches, suffers from some problems. It is prone to recursive loops, it may perform repeated computation of some subgoals, and it is often hard to tell that all solutions to the query goal have been found.

On the other hand, a bottom-up (or forward-chaining) search, working from the bodies of the rules to the heads, would cause us to infer *sg* facts that would never even be considered in the top-down search. Yet, bottom-up evaluation is desirable because it avoids the problems of looping and repeated computation that are inherent in the top-down approach. Also, bottom-up approaches allow us to use set-at-a-time operations like relational joins, which may be made efficient for disk-resident data, while the pure top-down methods use tuple-at-a-time operations.

Magic Sets

Magic sets is a technique that allows us to rewrite the rules for each *query form* (i.e., which arguments of the predicate are bound to constants, and which are variable) so that the advantages of top-down and bottom-up methods are combined. That is, we get the focus inherent in top-down evaluation combined with the freedom from looping, easy termination testing, and efficient evaluation of bottom-up evaluation. Magic-sets is a rule-rewriting technique. We shall not give the method, of which many variations are known and used in practice. The following example should suggest the idea.

Example 51.2. Given the rules of Example 51.1, together with the query *sg(john, Z)*, a typical magic-sets transformation of the rules would be

$$sg(X, Y) :- magic\_sg(X), flat(X, Y).$$
$$sg(X, Y) :- magic\_sg(X), up(X, U), sg(U, V), down(V, Y).$$
$$magic\_sg(U) :- magic\_sg(X), up(X, U).$$
$$magic\_sg(john).$$

Intuitively, the *magic_sg* facts correspond to queries or subgoals. The definition of the *magic_sg* predicate mimics how goals are generated in a top-down evaluation. The set of *magic_sg* facts is used as a *filter* in the rules defining *sg* to avoid generating facts that are not answers to some subgoal. Thus, a purely bottom-up, forward-chaining evaluation of the rewritten program achieves a restriction of search similar to that achieved by top-down evaluation of the original program.

The original paper on magic sets was (Bancilhon et al. [1986]), and its extension to general programs was in Beeri and Ramakrishnan [1987]. Independently, the article (Rohmer et al. [1986]) described the *Alexander method*, which is essentially the "generalized supplementary magic sets method" of Beeri and Ramakrishnan [1987] for the case of left-to-right evaluation within rules. There are a number of other approaches to optimizing rules that have similar effects without rewriting rules, and several variations and extensions of the idea have been proposed. In addition, complementary techniques and refinements have been proposed.

An important point to note is that the magic-sets technique, while originally developed to deal with recursive queries, is clearly applicable to nonrecursive queries as well. Indeed, it has been adapted to deal with SQL queries (which contain features such as grouping, aggregation, arithmetic conditions, and multiset relations that are not present in pure logic queries) and found to be superior to techniques used in commercial database systems for *nonrecursive* "nested" SQL queries (Mumick et al. [1990]). It is especially useful for optimizing complex decision-support queries and is finding its way into commercial relational database systems.

The survey (Ramakrishnan and Ullman [1995]) offers additional pointers to work on optimization of deductive database queries.

Iterative Fixpoint Evaluation

Most rule-rewriting techniques like magic sets expect implementation of the rewritten rules by a bottom-up technique, where, starting with the facts in the database, we repeatedly evaluate the bodies of the rules with whatever facts are known (including facts for the intensional predicates) and infer what facts we can from the heads of the rules. This approach is called *naive fixpoint evaluation*.

We can improve the efficiency of this algorithm by a simple trick. If in some round of the repeated evaluation of the bodies we discover a new fact f, then we must have used, for at least one of the subgoals in the utilized rule, a fact that was discovered on the previous round. For if not, then f itself would have been discovered in a previous round. We may thus reorganize the substitution of facts for the subgoals so that at least one of the subgoals is replaced by a fact that was discovered in the previous round.

Example 51.3. Consider the same-generation rules of Example 51.1. The first rule has a body, $flat(X, Y)$, that never changes, so after the first round, it can never yield any new *sg* facts. The second rule's body can have only new facts for the $sg(U, V)$ subgoal; the $up(X, U)$ and $down(V, Y)$ subgoals are extensional and do not change during the iteration. Thus, we can, on each round, use only the new *sg* facts from the previous round, along with the full *up* and *down* relations. Because in general, only a small fraction of the *sg* facts will be new on any one round, we significantly reduce the amount of work required.

A number of researchers have independently proposed this evaluation technique and developed variants. The formulation presented in Balbin and Ramamohanarao [1987] is probably the most widely used. It is now known widely as *seminaive evaluation*.

The fixpoint evaluation of a logic program also can be refined by taking certain algebraic properties of the program into consideration. Such refinements, and techniques for detecting when they are applicable, have been investigated by several researchers.

Deductive Database Queries with Negation and Aggregation

We turn back to the expressive power of datalog and introduce two powerful extensions to the language: negation and aggregation.

Negation

A deductive database query language can be enhanced by permitting negated subgoals in the bodies of rules. However, we then lose an important property of our rules. When rules have the form introduced in section 51.2, there is a unique *minimal model* of the rules and data. A *model* of a program is a set of facts such that for any rule, replacing body literals by facts in the model results in a head fact that is also in the model. Thus, in the context of a model, a rule can be understood as saying, essentially, "if the body is true, the head is true." A *minimal model* is a model such that no subset is a model. The existence of a unique minimal model, or least model, is clearly a fundamental and desirable property. Indeed, this least model is the one computed by naive or seminaive evaluation, as discussed in the sub-subsection on Iterative Fixpoint Evaluation above. Intuitively, we expect the programmer had in mind the least model.

In 1976, van Emden and Kowalski [1976] showed that the least fixpoint of a Horn-clause logic program coincided with its least Herbrand model. This provided a firm foundation for the semantics of logic programs, especially for deductive databases, since fixpoint computation is the operational semantics associated with deductive databases (at least, of those implemented using bottom-up evaluation).

However, in the presence of negated literals, a program may not have a least model.

Example 51.4. The program

$$p(a) \leftarrow \neg p(b)$$

has two minimal models: $\{p(a)\}$ and $\{p(b)\}$.

The meaning of a program with negation is usually given by some "intended" model.[2] The challenge is to develop algorithms for choosing an intended model that

1. makes sense to the user of the rules, and
2. allows us to answer queries about the model efficiently. In particular, it is desirable that it work well with the magic-sets transformation, in the sense that we can modify the rules by some suitable generalization of magic sets, and the resulting rules will allow (only) the relevant portion of the selected model to be computed efficiently. (Alternatively, other efficient evaluation techniques must be developed.)

We note that relying upon such an intended model in general results in a treatment of negation that differs from classical logic. In Example 51.4, we just saw that choosing one of the two minimal models over the other cannot be justified in terms of classical logic, since the rule is logically equivalent to $p(a) \vee p(b)$. One important class of negation that has been extensively studied is stratified negation (see articles in Minker [1988]). A program is *stratified* if there is no recursion through negation. Programs in this class have a very intuitive semantics and also can be efficiently evaluated. The following example describes a stratified program.

Example 51.5. Consider the following program $P2$:

$$r1 : anc(X, Y) \quad \leftarrow par(X, Y).$$

$$r2 : anc(X, Y) \quad \leftarrow par(X, Z), anc(Z, Y).$$

$$r3 : nocyc(X, Y) \leftarrow anc(X, Y), \neg anc(Y, X).$$

Intuitively, this program is stratified because the definition of the predicate *nocyc* depends (negatively on) the definition of *anc*, but the definition of *anc* does not depend on the definition of *nocyc* at all.

---

[2] Clark's *completed program* and Reiter's *closed world assumption* approaches do not fall into this category.

A bottom-up evaluation of $P2$ would first compute a fixpoint of rules $r1$ and $r2$ (the rules defining *anc*). Rule $r3$ is applied only when all the *anc* facts are known.

Several approaches to negation have been proposed, extending the class of programs for which a reasonable semantics is defined and extending optimization techniques to cover these classes of programs.

Set Grouping and Aggregation

The SQL query language allows aggregate operators like *sum* and *count* to be used in conjunction with *partitioning* a relation into *groups* of tuples. A similar construct has been explored in deductive databases; the presence of recursion and the ability to store *nested* relations makes this construct more powerful (and expensive) than SQL's variant. The following example illustrates the use of a *grouping* or *aggregation* construct $\langle \ \rangle$:

$$set\_of\_grades(Class, \langle Grade \rangle) \leftarrow student(Name, Class, Grade).$$

We first (conceptually) create a set of tuples for *set_of_grades* using the rule

$$set\_of\_grades(Class, Grade) \leftarrow student(Name, Class, Grade).$$

Now for each value of *Class* (in general, each value of those arguments of the head that are not enclosed in the $\langle \ \rangle$), we create a set containing all the corresponding values for *Grade*. For each value of *Class*, let this set be called S_{Class}; we then create a fact $set\_of\_grades(Class, S_{Class})$.

Aggregate operations such as *count*, *sum*, *min*, and *max* can be combined with $\langle \ \rangle$:

$$max\_grade\_given(Class, max\langle Grade \rangle) \leftarrow student(Name, Class, Grade).$$

As before, for each value of *Class* we create a set. But now we apply the aggregate operation *max* to the set and create a head fact using this value rather than the set itself.[3] A number of important practical problems, such as bill of materials (generating various summaries of the contents of a complex part in a part–subpart hierarchy) and shortest paths, involve a combination of aggregation and recursion.

We observe that before any head fact can be derived, all body facts that can contribute to the multiset created in the head fact must be available. This introduces a situation that is very similar to negation, and several approaches (to defining semantics and to optimization) used for negation carry over to grouping.

Prototypes of Deductive Database Systems

There have been a number of implementations of deductive databases. These include the Aditi system (Univ. of Melbourne), COL (INRIA), ConceptBase (Univ. of Aachen), CORAL (Univ. of Wisconsin), EKS-V1 (ECRC), LogicBase (Simon Fraser Univ.), DECLARE (MAD Intelligent Systems), Hy+ (Univ. of Toronto), X4 (Univ. of Karlsruhe), LDL and LDL++ (MCC), LOGRES (Polytechnic of Milan), LOLA (Technical Univ. of Munich), Glue-Nail (Stanford Univ.), Starburst (IBM Almaden), and XSB (SUNY—StonyBrook). A discussion of these systems and pointers for further reading can be found in Ramakrishnan and Ullman [1995].

51.3 Active Rules

Active rules, also called production rules or triggers, originally were introduced in the context of expert systems and in particular languages such as OPS5 (Brownston et al. [1985]); they are now being tightly

---

[3]The $\langle \ \rangle$ construct is a generalization of SQL's *group-by* construct. It is defined to generate a nested set of values in LDL, where it was originally proposed. Defining it to generate a nested *multiset* of values, as in CORAL, brings it closer to the SQL *group-by* construct.

integrated into database management (Chakravarthy and Lomet [1992], Sellis [1989], Widom [1993], Widom and Ceri [1996]). They follow the *event–condition–action* paradigm; a seamless integration of active rules within databases occurs by mapping events to data manipulation operations, by expressing conditions as database queries, and by including database manipulations within the activities that can be performed by actions. Thus, active rules are a vehicle for providing reactive behaviors to databases.

Active-Rule Components

Rules in active database systems are defined in the *data definition language* (DDL) of DBMSs and typically concern all applications. In their most general form, active database rules consist of three parts:

- *Event* causes the rule to be *triggered*.
- *Condition* is checked when the rule is triggered.
- *Action* is executed when the rule is triggered and its condition is true.

The event, condition, and action for a rule usually are specified as part of a *create* command provided for rules. A corresponding *drop* command is necessary to remove rules. A *modify* command may be provided as well, so that components of a rule can be changed without deleting the rule and creating a new one. A pair of commands provided by some systems is *deactivate* and *activate*; these commands allow certain rules to be "turned off" temporarily.

Typically, the *event* part of an active rule indicates a data manipulation operation that occurs within the database; thus, in a relational database system, events are either one of the three SQL data modification operations—*insert, delete,* or *update*—on a particular table. In an object-oriented database system, a data modification event might be specified as the creation, deletion, or modification of an object or as the invocation of a particular method of an object.

The *condition* specifies an additional condition to be checked once the rule is triggered and before the action is executed. Normally, the condition specifies that a certain predicate holds either on the database or on a query using the database system's query language. In the first case, the condition's evaluation produces a Boolean value; in the second case, it produces a relational answer and is interpreted as a true condition when the query produces a nonempty answer.

The *action* is executed when the rule is triggered and its condition is true. Actions normally include data modification operations (e.g., SQL *insert, delete,* or *update* operations), data retrieval operations (e.g., SQL *select* operations), or other database commands, including operations for data definition, operations for transaction control (e.g., *rollback, commit*), operations for granting and revoking privileges, and so on. When the rule language is integrated with a database programming language, actions may also call application procedures.

Rule execution is *instance-oriented* if a rule is executed once for each database instance triggering the rule or satisfying the rule's condition; rule execution is *set-oriented* if a rule is executed once for all database instances triggering the rule or satisfying the rule's condition. For example, consider a relational active database rule R_1 that is triggered by deletions from a particular table. If rule execution is instance-oriented, then R_1's condition is evaluated and its action is executed once for each deleted tuple. If rule execution is set-oriented, then R_1's condition is evaluated and its action is executed once for the entire set of deleted tuples.

The execution semantics for active database rules sometimes requires that one rule be selected from a set of eligible rules. For this reason, an active database rule language may include a mechanism for declaring rule *priorities*. Priorities may be specified by ordering the set of rules, by declaring relative priorities between pairs of rules, or by assigning numerical priority values to each rule.

Examples of Active Rules

In the following, we present two representative examples of active rules. The first example presents a classical rule for enforcing referential integrity, written in the Starburst rule language (Widom and

Finkelstein [1990]). This active rule is set-oriented; it triggers whenever some departments (*dept*) are deleted by an application and deletes the tuples from the *emp* relation whose department number matches with the numbers of deleted departments:

```
create rule cascade on dept
when deleted
then delete from emp
     where dept-no in
         (select dept-no
          from deleted)
```

The second example, a bit more complex, is written in Oracle 7. It implements a business rule for the automatic management of reorders in an inventory. The trigger is tuple-oriented; it activates at each update of the *parts_on_hand* attribute of the *inventory* table. The condition consists in checking that the quantity remaining in the inventory is below a reorder point. If the condition is satisfied, then an action is executed which is encoded as a PL/SQL program (PL/SQL is a procedural extension of SQL supported by Oracle). Although the precise description of the program is beyond our interests, the program indeed checks that there are no pending orders for the peculiar part and then issues an order for that part.

```
CREATE TRIGGER reorder
AFTER UPDATE OF parts_on_hand ON inventory
WHEN (NEW.parts_on_hand < NEW.reorder_point)
FOR EACH ROW
   DECLARE NUMBER X
   BEGIN
      SELECT COUNT(*) INTO X   /* X=1 if already ordered */
      FROM pending_orders      /* X=0 otherwise */
      WHERE part_no = NEW.part_no;
      IF X=0 THEN
         INSERT INTO pending_orders
            VALUES (NEW.part_no, NEW.reorder_quantity, SYSDATE)
      END IF;
   END;
```

The reader will appreciate that the two examples are rather different in style; indeed, the former is an example of a trigger as supported by a research prototype which aims at a rather declarative style and clean semantics, whereas the latter is a trigger as currently supported by a very popular product.

Semantics of Active Rules

There is not a unique semantics for active rule systems; each different product, research prototype, or even theoretical contribution is characterized by a different semantics from all others. Thus, we give an abstract semantics, which grossly describes the behavior of an active-rule system and as such can be considered as representative of many proposals. A thorough comparative analysis of the semantics of active rule systems can be found in Fraternali and Tanca [1996].

Definition 51.1. An *active database* is a pair $\langle\langle E, R\rangle\rangle$ where E is the extensional database and R is the active-rule set, i.e., the set of all the active rules defined for E. Rules of R may be partially ordered; if r_i and r_j are two rules of R and $r_i < r_j$, then r_i has *higher priority* than r_j.

The *evolution of the active database* is obtained by alternating *user-specified transitions* and *rule processing*. The entire sequence of data manipulation operations performed by user-specified transitions and by active-rule processing is committed or aborted as a *transaction*; we assume the usual ACID properties of transactions. In the context of a single transaction, each rule is *triggered* by the execution of any operation in its event set.

Definition 51.2. The *rule-processing algorithm* consists of iterating the following steps:

1. If there is no triggered rule, then exit.
2. Select one of the triggered rules, which is detriggered.
3. Evaluate the condition of the selected rule.
4. If the condition is true, then execute the action of the selected rule.

The rule selected at step 2 is nondeterministically chosen among the triggered rules with highest priority.

Definition 51.3. An *active state* is a pair $\langle D, T \rangle$ where D is a database state and T is the set of triggered rules.

Rule processing can be represented as a sequence of active states. A rule, r, links active states $\langle D_1, T_1 \rangle$ and $\langle D_2, T_2 \rangle$ when r, belonging to the set of triggered rules T_1, is selected at step 2 of the rule-processing algorithm, its condition is true on database state D_1, and its action executes on database state D_1 yielding database state D_2 and a new set of triggered rules, T_2. If the rule-processing algorithm terminates by exiting at step 1, a quiescent state is reached, in which no rule is triggered:

Definition 51.4. A *quiescent state* is an active state $\langle D, \emptyset \rangle$ in which the set of triggered rules is empty.

Rule processing is initiated by system-specific mechanisms; rules may be distinguished as *immediate* or *deferred*. Processing of immediate rules normally is initiated either immediately before or immediately after the operation triggering them, whereas processing of deferred rules is initiated when the user issues the *commit work* statement. In addition, *detached rules* run asynchronously, e.g., in the context of a different transaction. These assumptions apply to many active database prototypes and commercial systems (Widom and Ceri [1996]) (in particular, they apply to the set-oriented trigger languages of several relational products, such as Oracle, Informix, and Sybase (Widom and Ceri [1996])). They also apply to tuple-oriented iterative expert-system languages, such as OPS5 (Brownston et al. [1985]).

Advanced Features

Several features enrich the semantic options offered when writing a collection of active rules.

- The rule language usually includes a mechanism for referencing values associated with transitions, called *transition values*. For example, *inserted*, *deleted*, and *updated* might be reserved words that, when used in a rule condition or action, denote the data that were inserted, deleted, or modified in the triggering transition. In the case of a rule triggered by data modification, keyword *new* might be used to reference the new value of the modified data, and keyword *old* might be used to reference the old value.

- Some systems allow a rich collection of events. In these systems, elementary events may include *data retrieval events* (e.g., a *select* operation on a particular table), *temporal events* specifying that a rule should be triggered at an absolute time or at periodic intervals (e.g., **on friday at 5PM**), and *application-defined* events, specified by an application of them (e.g., *high-pressure*) and then used within the event part of active rules. Then, an application may notify the database system of the occurrence of an event, thereby triggering all associated rules. Triggering events also may be *composite*, i.e., they may be combinations of single events or other composite events. Useful operators for combining events are logical operators, sequence, and temporal composition.

Very complex composite events are possible if an event specification language is based on regular expressions or a context-free grammar.

- Rules are *detached* (also called *decoupled*) when triggering occurs within one transaction and then condition evaluation and/or action execution take place in another transaction. Decoupled rules can be useful when a long series of rules are triggered in order to decompose the resulting large transaction into a set of smaller ones. Decoupled rules can further be subdivided into *dependent decoupled*, where the separate transaction isn't spawned unless the original transaction commits, and *independent decoupled*, where the separate transaction is spawned regardless of whether the original transaction commits. In addition, a *causality* between the transactions might be specified, such as requiring that the spawned transaction be later than the original transaction in the serialization ordering.

- Two distinct *event consumption modes* are possible for each rule; this feature is relevant when a given rule is considered multiple times in the context of the same transaction.
  - Events can be *consumed* after the consideration of a rule. In this case, each instance of an event is considered by a rule only at its next execution and then disregarded.
  - Events can be *preserved* after the consideration of a rule. In this case, all instances of the event since the transaction started are considered at each execution of that rule.

- Events can be composed, yielding their *net-effect computation*; such computation is relevant when the same object is the target of multiple operations. The computation of net effect is performed as follows:
  - A sequence of *creation* and *deletion* primitives on the same object, possibly with an arbitrary number of intermediate *modification* primitives on that object, has a null net effect.
  - A sequence of *creation* and several *modification* primitives on the same object has the net effect of a single *creation* operation.
  - A sequence of several *modifications* and *deletion* primitives on the same object has the net effect of a single *deletion* operation on that object.

Designing Active-Rule Applications

Applications of active-rule systems can be classified as internal, extended, and external.

- We regard as *internal application* of active rules their use for implementing system-supported facilities instead of using other built-in mechanisms of database systems. Such facilities include management of integrity constraints, computation of views and derived data, or authorization support.

- We regard as *extended applications* the use of active rules for supporting software systems which are devoted to specific tasks, thus becoming a *hidden implementation mechanism* for these systems. Examples of external applications include workflow management, e-mail managers (capable of intelligent message storage), document retrieval and classification systems (capable of automatically shipping documents to requestors based on their interest), version management (e.g., in the context of engineering databases), and replication management (e.g., in the context of federated databases).

- Finally, we consider all other applications of active rules as *external*; these respond to generic requirements. A number of external applications of active rules are reported, e.g., in the context of software engineering and testing, container packing and loading, order entry and processing, stock-market control, production control, air-traffic control, and so on.

Properties of Rule Behavior

Formal properties may help to define rule-processing behavior. In particular, it is useful to know a priori, through a compile-time analysis of rules, whether the rule execution is guaranteed the properties of:

- *Termination*, i.e., ensuring that rule processing initiated by any application will eventually terminate.
- *Confluence*, i.e., ensuring, in addition to termination, that rule processing initiated by any application will eventually produce the same final database state.
- *Observable determinism*, i.e., ensuring, in addition to confluence, that all visible actions performed by rules activated by any application will be the same.

Determining the most general conditions which guarantee the above properties is addressed in Aiken et al. [1992] and van der Voort and Siebes [1993].

Several applications of active databases can be specified declaratively; these specifications lead to the derivation of rules with a known structure and a clear objective, for which some degree of rule analysis can be performed automatically. This approach is proposed for repairing integrity constraints after violations (in Ceri and Widom [1990]) or for incremental maintenance of views (in Ceri and Widom [1994]). Declarative specifications are given by the designer, possibly through an interactive process. Then, specifications are translated into active rules, which are installed in the active database. For some problems, the derivation of rules from specifications is *fully automatic*. For other problems, in which several active behaviors are possible, the designer must provide directives to a rule analyzer; rule derivation is *semiautomatic*.

Rule analysis techniques may be used to validate the rules produced by the generator. For instance, conservative analysis may identify a collection of rules which are potentially nonterminating (e.g., rules that may trigger each other indefinitely); in this context, the designer may be asked to validate the rule set and ensure that its execution on actual data would always terminate; the designer may also make some changes to the rules in order to achieve this property.

One advantage of using an automatic generation mechanism for producing rules of given, known structure is that *rule optimization* can be successfully applied to them. A variety of standard techniques are possible, such as replacing several rules with one equivalent rule, thereby factoring activities common to several rules, or replacing the references to entire tables with references to smaller data structures describing changes produced by transactions.

Products and Prototypes

Active rules are supported by most relational database systems, including Allbase, DB2, Illustra, Informix, Ingres, Oracle, RDB, Sybase, and others. Active rules in these systems (called triggers) support basic functionalities and are inspired by the still-evolving SQL3 standard, which is not published yet [ISO–ANSI 1994]. Thus, triggers of different systems are syntactically and semantically different.

Several research prototypes are relational in nature, including Starburst (Widom and Finkelstein [1990]), Postgres (Stonebraker et al. [1990]), A-RDL (Simon et al. [1992]), and Ariel (Hanson [1992]). Triggers also are supported in several object-oriented database prototypes (e.g., Hipac [Chakravarthy et al. 1989], Chimera (Ceri et al. [1994]), Naos (Collet et al. [1994]), Ode (Agrawal and Gehani [1989]), Reach (Buchmann et al. [1992]), Samos (Gatziv et al. [1991]), Sentinel (Anwar et al. [1993]), and many others.

Researchers in the field consider relational triggers as the first step in the evolution of active rules in databases and propose several ways in which their functionalities should be extended. Important dimensions are the introduction of transactional coupling modes, of event consumption modes, and of complex events and event calculus and the ability to access events and past database states from within rules. All these features make research prototypes much more powerful than relational systems with triggers.

51.4 Common Features of Deductive and Active Rules

Although deductive and active rules appear to be quite different programming paradigms (and indeed they have been so considered by most of the researchers in the field), recent research efforts have shown that deductive and active rules have several features in common and that it is possible to accommodate both formalisms within the same framework. Widom [1994] notes that all rule languages consist of at least some form of *antecedent* and some form of *consequent*; in addition, some rule systems include an *activator*. A generic rule, albeit simplified, is represented as the sequence below, where square brackets denote optionality:

$$[\text{Activator} \rightarrow] \text{ Antecedent} \rightarrow \text{Consequent}$$

In essence, the proposed model characterizes deductive and active databases as follows:

- The antecedent of deductive rules corresponds to a query on the database that produces certain tuples: the consequent states that the tuples produced by the antecedent should contribute to tuples in the relation specified by the consequent; deductive rules have no activator.
- Classical active rules have an activator; the antecedent is a query that produces certain tuples and that is evaluated on a database state, possibly the one causing the activation; and the consequent executes commands that may cause a change of state by using the tuples produced by the antecedent.

In all cases, rule processing consists of repeatedly executing rules whose antecedent is true until there is no change, i.e., until a fixpoint is reached. Each rule execution ensures that the consequent becomes true. This feature can be recognized both in the *naive evaluation* of section 51.2 under Optimization Techniques for Deductive DBs: Iterative Fixpoint Evaluation and in the rule processing algorithm of section 51.3 under Semantics of Active Rules.

The model is completed by showing that certain systems (such as RDL and Ariel) have no activator, although they behave similarly to active rules; on the other hand, rules in the Postgres system have more power than the classical active rule sketched above, because their consequent can alter the activator events (through the *instead of* clause). Thus, the model accommodates datalog, RDL, Ariel, Starburst, and Postgres along an active–deductive continuum.

More recently, Zaniolo [1995] has illustrated a model of "declarative" active rules where the semantics of active rules is given by means of *stable models*, a classical approach to the definition of rule semantics which originally was developed in the context of deductive databases. The work in Widom [1994] and Zaniolo [1995] is representative of a reconciliation trend between deductive and active rules, which also is occurring in the arena of developing applications, as illustrated in the next section.

51.5 Knowledge Independence and Rule Design

The importance of deductive and active rules is that they may express a lot of the semantics that is normally encoded in every application. This feature can be regarded as an additional level of independence for applications, called *knowledge independence* (Friesen et al. [1994]). Traditionally, databases provide physical independence (from the actual storage implementation) and logical independence (from the structure of the relational schema). The additional level of independence is achieved by factoring "knowledge" out of the applications and expressing it in the form of rules.

The power of knowledge independence is that rules are automatically shared by all applications. Rules are specified and designed by *database administrators* and are logically part of the database schema. One of the major consequences of knowledge independence is that integrity constraints and their monitoring are automatically imposed on all applications. As with all other levels of independence, extracting knowledge from applications and factoring knowledge into rules makes knowledge evolution much more feasible and controllable: it is sufficient to change the rules without changing the content of the application code.

Although rule technology increasingly is being incorporated in database systems, these features often remain unused. Researchers and developers agree that one of the main difficulties in the development of rule applications is the lack of design methods and of suitable design tools; consequently, application designers lack appropriate experience and culture. Some recent research has focused on design issues and techniques, e.g., on detecting that certain constraints are not satisfiable (Bry et al. [1988]), or deriving integrity-restoring active rules from constraint specifications (Ceri and Widom [1990]), or introducing the notion of modularization within active rules (Baralis et al. [1996]). A complete, computer-assisted methodology for designing database applications with objects and rules, covering their analysis, design, prototyping, implementation, and maintenance, is being developed and tested within the IDEA Project, sponsored by the EC [Ceri and Fraternali, to appear].

51.6 Conclusion

We have reviewed several results in the fields of deductive databases and active rules. In the area of deductive databases, efficient evaluation methods have been developed for the "large data" environment that are sound and complete with respect to an intuitive declarative semantics for large classes of programs with powerful features like negation and aggregation. Systems based upon these methods are being developed and offer good support for rule-based applications. There is also ongoing work that seeks to combine the powerful query-language capability of deductive databases with features from object-oriented systems, and this likely will lead to a new generation of more powerful systems that bring database languages and programming languages closer to each other.

In the area of active rules, there are many ongoing efforts. Two standard bodies (ANSI and ISO) are proposing trigger standardization, and triggers are being added to almost all relational products. Object-oriented databases also are starting to be extended with active-rule capabilities. Finally, more powerful research prototypes are being developed. We consider the relational products that already support triggers to be the first generation of active databases, with a considerable amount of processing power but also severe limitations. We expect that designers of commercial products will extend them to incorporate more and more features and also will remove some of their current limitations, so as to raise the level of abstraction at which they can be programmed. We also expect that design methods and tools will play a major role in facilitating the effective use of active database technology.

Acknowledgments

Stefano Ceri is supported by the Esprit Project P6333, "Idea," and by grant VDS 1/94 from ENEL. The work of R. Ramakrishnan was supported by a Packard Foundation Fellowship in Science and Engineering; a Presidential Young Investigator Award, with matching grants from DEC, IBM, Tandem, and Xerox; and a grant from NASA.

References

Agrawal, R. and Gehani, N. 1989. Ode (Object Database and Environment): the language and the data model, pp. 36–45. In *Proc. ACM SIGMOD Int. Conf. on Management of Data*, Portland, OR, May.

Aiken, A., Widom, J., and Hellerstein, J. M. 1992. Behavior of database production rules: termination, confluence, and observable determinism, pp. 59–68. In *Proc. ACM SIGMOD Int. Conf. on Management of Data*, M. Stonebraker, ed., San Diego, CA, May.

Anwar, E., Maugis, L., and Chakravarthy, S. 1993. A new perspective on rule support for object-oriented databases, pp. 99–108. In *Proc. ACM SIGMOD Int. Conf. on Management of Data*, Washington, May.

Balbin, I. and Ramamohanarao, K. 1987. A generalization of the differential approach to recursive query evaluation. *J. Logic Programming* 4(3): Sept.

Bancilhon, F., Maier, D., Sagiv, Y., and Ullman, J. D. 1986. Magic sets and other strange ways to implement logic programs, pp. 1–15. In *Proc. ACM Symp. on Principles of Database Systems*, Cambridge, MA, Mar.

Baralis, E., Ceri, S., and Paraboschi, S. 1996. Modularization techniques for active rule design. *ACM Trans. Database Systems*.

Beeri, C. and Ramakrishnan, R. 1987. On the power of magic, pp. 269–283. In *Proc. ACM Symp. on Principles of Database Systems*, San Diego, CA, Mar.

Brownston, L., Farrell, R., Kant, E., and Martin, N. 1985. *Programming Expert Systems in OPS5: An Introduction to Rule-Based Programming.* Addison–Wesley, Reading, MA.

Bry, F., Decker, H., and Manthey, R. 1988. A uniform approach to constraint satisfaction and constraint satisfiability in deductive databases, pp. 487–505. In *Proc. First Int. Conf. on Extending Database Technology.* Lecture Notes on Comput. Sci. 303.

Buchmann, A., Branding, H., Kudrass, T., and Zimmerman, J. 1992. REACH: a REal-time, ACtive, and Heterogeneous mediator system. *IEEE Data Eng. Bull., Special Issue on Active Databases* 15(4):44–47, Dec.

Ceri, S. and Fraternali, P. to appear. *Designing Database Applications with Objects and Rules: the IDEA Methodology.* Addison–Wesley, Reading, MA.

Ceri, S. and Widom, J. 1990. Deriving production rules for constraint maintenance, pp. 566–577. In *Proc. Sixteenth Int. Conf. on Very Large Data Bases,* D. McLeod, R. Sacks-Davis, and H. Schek, eds., Brisbane, Australia, Aug.

Ceri, S. and Widom, J. 1994. Deriving incremental production rules for deductive data. *Inf. Systems* 19(6):467–490, Nov.

Ceri, S., Fraternali, P., Paraboschi, S., and Tanca, L. 1994. Active rule management in chimera. In *Active Database Systems,* J. Widom and S. Ceri, eds., Morgan-Kaufmann, San Mateo, CA.

Chakravarthy, S. and Lomet, D., eds., 1992. *Special Issue on Active Databases, IEEE Data Eng. Bull.* 15(4), Dec.

Chakravarthy, S., Blaustein, B., Buchmann, A., Carey, M., Dayal, U., Goldhirsch, D., Hsu, M., Jauhari, R., Ladin, R., Livny, M., McCarthy, D., McKee, R., and Rosenthal, A. 1989. HiPAC: a research project in active, time-constrained database management. *Tech. Rep.* XAIT-89-02, Xerox Advanced Information Technology, Cambridge, MA, July.

Collet, C., Coupaye, T., and Svensen, T. 1994. NAOS—efficient and modular reactive capabilities in an object-oriented system, pp. 132–143. In *Proc. Twentieth Int. Conf. on Very Large Data Bases,* Santiago, Chile, Sept.

Dayal, U., Hsu, M., and Ladin, R. 1990. Organizing long-running activities with triggers and transactions, pp. 204–214. In *Proc. ACM SIGMOD Int. Conf. on Management of Data,* H. Garcia-Molina and H. V. Jagadish, eds., Atlantic City, NJ, May.

Diaz, O., Jaime, A., and Paton, N. 1993. DEAR: A DEbugger for Active Rules in an object-oriented context, pp. 180–193. In *Proc. First Workshop on Rules in Database Systems,* N. W. Paton and M. H. Williams, eds., WICS, Edinburgh, Scotland, Aug. Springer–Verlag, Berlin.

Diaz, O., Patom, N., and Gray, P. 1991. Rule management in object-oriented databases: a uniform approach, pp. 317–326. In *Proc. Seventeenth Int. Conf. on Very Large Data Bases,* Barcelona, Spain, Sept.

Fraternali, P. and Tanca, L. 1996. A structured approach for the definition of the semantics of active databases. *ACM Trans. Database Systems,* Mar.

Friesen, O., Villars, G. G., Lefebvre, A., and Vieille, L. 1994. Applications of deductive object-oriented databases (DOOD) using Datalog Extended Language (DEL). In *Applications of Logic Databases,* R. Ramakrishnan, ed., Kluwer, Boston, MA.

Gatziu, S., Geppert, A., and Dittrich, K. R. 1991. Integrating active concepts into an object-oriented database system. In *Proc. Third Int. Workshop on Database Programming Languages,* Nafplion, Greece, Aug.

Hanson, E. 1992. Rule condition testing and action execution in Ariel, pp. 49–58. In *Proc. ACM SIGMOD Int Conf. on Management of Data,* M. Stonebraker, ed., San Diego, CA, May.

ISO–ANSI 1994. Working draft, database language SQL. *SQL3 Document* X3H2-94-080 *and* SOU-003.

Loucopoulos, P. 1994. Requirements engineering: conceptual modeling and CASE perspectives. *Tech. Rep., COMETT/FORMITT Course on Conceptual Modeling, Databases and CASE,* Lausanne, Switzerland, Oct.

Minker, J., ed., 1988. *Foundations of Deductive Databases and Logic Programming.* Morgan-Kaufmann, Los Altos, CA.

Minker, J., Gallaire, H., and Nicolas, J.-M. 1984. Logic and databases: a deductive approach. *ACM Comput. Surveys* 16(2):153–185.

Mumick, I. S., Finkelstein, S., Pirahesh, H., and Ramakrishnan, R. 1990. Magic is relevant. In *Proc. ACM SIGMOD Int. Conf. on Management of Data,* Atlantic City, NJ, May.

Ramakrishnan, R., ed. 1994. *Applications of Logic Databases.* Kluwer Academic, Boston, MA.

Ramakrishnan, R. and Ullman, J. D. 1995. A survey of deductive database systems. *J. Logic Programming* 23(2):125–149.

Rohmer, J., Lescoeur, R., and Kerisit, J. M. 1986. The Alexander method—a technique for the processing of recursive axioms in deductive database queries. *New Generation Comput.* 4:522–528.

Sellis, T., ed. 1989. *Special Issue on Rule Management and Processing in Expert Database Systems, SIGMOD Record* 18(3), Sept.

Simon, E., Kiernan, J., and de Maindreville, C. 1992. Implementing high level active rules on top of a relational DBMS, pp. 315–326. In *Proc. Eighteenth Int. Conf. on Very Large Data Bases,* Vancouver, Aug.

Stonebraker, M., Jhingran, A., Goh, J., and Potamianos, S. 1990. On rules, procedures, caching and views in data base systems, pp. 281–290. In *Proc. ACM SIGMOD Int. Conf. on Management of Data,* Atlantic City, NJ, May.

van der Voort, L. and Siebes, A. 1993. Termination and confluence of rule execution. In *Proc. Second Int. Conf. on Information and·Knowledge Management,* Washington, Nov.

van Emden, M. H. and Kowalski, R. A. 1993. The semantics of predicate logic as a programming language. *J. ACM* 23(4):733–742, Oct.

Widom, J. 1993. Research issues in active database systems: report from the closing panel at RIDE-ADS '94, pp. 180–193. In *Proc. First Workshop on Rules in Database Systems,* N. W. Paton and M. H. Williams, eds., WICS, Edinburgh, Aug. Springer–Verlag, Berlin.

Widom, J. 1994. Deductive and active databases: two paradigms or ends of a spectrum? *SIGMOD Record* 23(3):41–43, Sept.

Widom, J. and Ceri, S. 1996. *Active Database Systems: Triggers and Rules for Advanced Data Processing.* Morgan-Kaufmann, San Mateo, CA.

Widom, J. and Finkelstein, S. J. 1990. Set-oriented production rules in relational database systems, pp. 259–270. In *Proc. ACM SIGMOD Int. Conf. on Management of Data,* Atlantic City, NJ, May.

Zaniolo, C. 1995. Active database rules with transaction-conscious stable-model semantics. In *Proc. Fourth Int. Conf. on Deductive and Object-Oriented Databases,* T. W. Ling, A. Mendelzon, and L. Vieille, eds., Singapore, Dec. LNCS 1013. Springer–Verlag, Berlin.

52

Object Database Systems

François Bancilhon
O2 Technology

52.1 History

Object databases (just like relational databases) started first as a research field and quickly evolved into a field dominated by industrial development. The research effort started at the beginning of the 1980s. The seminal paper [Copeland and Maier 1983] introduced the concept of an object-based data model and seamless integration with a programming language, Smalltalk. From there, a number of prototypes were built. Examples are Orion [Kim et al. 1990], O2 [Bancilhon et al. 1993], ODE [Agrawal and Gehani 1989], Iris [Fishman et al. 1989], and Exodus [Carey and DeWitt 1987].

In 1987 and 1988, the first products came out: GemStone by Servio Logic and Vbase by Ontologic. GemStone is still alive and marketed by Gemstone; Vbase was withdrawn from the market and replaced by Ontos. In 1989, a paper was published which represented the consensus of the research community on the definition of an object database system. The manifesto [Atkinson et al. 1989] defined an object database system by a set of 14 rules, 6 for the database aspect and 8 for the object-oriented aspect.

A second generation of products came out at the beginning of the 1990s: ObjectStore by Object Design, Versant/DB by Versant, Objectivity/DB by Objectivity, O2 by O2 Technology, Orion by Itasca, Matisse by Intelletic, and Ontos by Ontologic. In 1992, the Object Data Management Group (ODMG), a standards group, was created at the initiative of some object database vendors with the goal of defining and promoting a portability standard for object databases. Several drafts of the standard were published starting in 1993, the latest one being ODMG 1.2 in 1995 [Cattel 1995]. Compliant products began to appear in 1995.

Finally a new breed of system, the so-called "object–relational systems," came out a couple of years later: Illustra by Illustra and UniSQL/X by UniSQL.

52.2 Delimiting the Scope: Object Databases, Object–Relational Databases, Object-to-Relational Interoperability

An *object database* system is a database system that supports an object model. An *object–relational database* system is a database system that supports an object model while remaining compatible with the relational model. Both types of systems make sense: they are different technically and address different problems.

The object–relational approach addresses the relational-database market and attempts to transition relational-database users from the relational world to the object world by adding some object features to a relational system while remaining strictly compliant with the relational model. Thus, object–relational-system designers started from the relational model (as defined by the SQL92 standard) and added some object features to it.

The object approach addresses the object market by providing database solutions to the object users. Thus object database system designers started from existing object models (C++, Smalltalk, and OMG) and provid database functionality in combination with these models. To summarize, the object–relational systems target SQL programmers, and object database systems target C++ and Smalltalk program mers.

Object-to-relational interoperability technology addresses yet another issue: that of allowing object developers to interact with relational databases and of allowing relational-database users to interact with object applications. This implies the need to establish bridges between products. Examples of this technology are SQL92 and ODBC interfaces on top of object database systems, C++ and Smalltalk interfaces on top of relational database systems, object view mechanisms on a relational schema and flat relational views of object schemas, data and schema migration tools that transform relational to object schemas and vice versa, and tools that transform relational databases to object databases and vice versa. All object database systems need such technology, and these tools are or will be at some point provided by such systems.

This survey does not address at all the issue of object–relational systems, nor object-to-relational technology. It concentrates exclusively on object database systems.

52.3 ODMG

ODMG was created at the initiative of a set of object database vendors with the goal of defining and promoting a portability standard for object databases. ODMG is chaired by Rick Cattell. It has a set of eight voting members (O_2 Technology, Versant, Poet, GemStone, Objectivity, Object Design, UniSQL, IBEX). Voting members have to ship an object database product that matches the ODMG definition (a database engine with a seamless language binding). Other members are reviewing members (ADB, AMS, Andersen Consulting, CERN, EDS, Fujitsu, Hewlett Packard, Lockheed Martin, Micram, Microsoft, Mitre, Ontos, Persistence, Sybase, Texas Instruments, and Unidata) and academic members (David DeWitt, David Maier, Stan Zdonick, Mike Carey, Eliot Moss, and Marv Solomon).

A portability standard ensures application developers that they can port their application from one compliant system to another at minimal cost. It allows for comparisons and avoids users being locked into a specific vendor. The initial standard was produced by five vendors in about 18 months. Version 1 of the standard was published in October 1993. A new version of the standard (1.1) was issued in 1994, and recently an enhanced version (1.2) came out in December 1995 [ODMG 1995].

The standard covers five different points:

- *Object model.* It describes the specific object model supported by the ODMG system. This model is an extension of the OMG model.
- *Object definition language* (ODL). It is used to specify the schema of an object database. It has a specific syntax, which is an extension of the IDL language from OMG, but also a C++ binding, i.e., you can use C++ to define your database schema.

- *Object query language* (OQL). It can be used as a standalone language for interactive queries, or embedded in a programming language (C++ or Smalltalk). It is an SQL-like language.
- *C++ and Smalltalk bindings:* they specify how one can write C++ and Smalltalk applications on top of an ODMG database.

A lot has been said on the pros and cons of the ODMG standard. The initial versions were far from perfect. But as new versions come out, the overall specification is strengthening and improving rapidly.

Here are some of the initial and remaining problems. There were a number of contradictions between the data model and the language-binding chapters, but most of them have disappeared in Version 1.2. A more formal definition of the semantics of the query language would be useful, and this is being developed. The initial definition of the Smalltalk binding was very weak, and that has been partially corrected in Version 1.2. And finally, the standard needs some certification process to be complete, and this is being worked on. As a result, ODMG is sufficient to serve as a specification for an object-database-system implementation. The technical choices which were made are sound and reasonable. The overall philosophy is good, and implemented versions exist.

52.4 Object Database: Definition

An object database system is a database system, and as such it supports the same type of functionality as a traditional database:

- transactions and concurrency,
- recovery,
- client/server, distribution, and replication,
- version management,
- schema evolution.

We will not elaborate on these aspects, since they are well-known database concepts which do not depend on the object nature of the system.

An object database system is furthermore a database system supporting an object data model. Though incomplete, this definition is a reasonable first approximation. It needs to be extended in two ways: (1) object databases need more than just a data model: they need a data model, a behavior model, a naming model, and persistence model; (2) object databases are integrated with programming languages in a completely new way.

52.5 Object Database Model: Data and Behavior

All object databases support an object model. These object models support the following features [Atkinson et al. 1989]:

Complex objects. Objects are built from atomic objects (integer, real, Boolean, string, bit, etc.) by recursively applying object constructors: tuple, set, bags, lists, and arrays.

Relationships. The tuple constructor allows for the definition of relationships between objects. These relationships can be one-way or two-way; they can be 1–1, 1–n, or n–p.

Object identity. Each object has its own identity, which is independent of its value. Through object identity, objects can share subcomponents, and objects can even be cyclic. The ODMG model supports both objects (which have an identity) and literals (which do not).

Encapsulation. An object consists of an identity, a value (a literal), and a set of operations (the methods). The only way an outside operator can update or read an object is through the methods; thus the internal structure of the object is not visible from the outside.

Classes. Objects with the same internal structure and the same set of methods can be grouped in classes. Thus a class has an interface description and an implementation description.

Extents. The extent of a class is a collection (a set, hopefully) containing all the objects of the class in question. There has been quite a debate around the notion of extents and the support they should be given in an object database system. One radical solution consists of requiring that every class have an extent which is both automatically maintained by the system and accessible to the programmers through some name (see section 52.7). The arguments for this solution are that users find it practical to access extents and that it resembles the relational approach (the extents mimic the relation). The other extreme is to assume that the programmer is in charge of maintaining the extents of classes if he/she wants to. The argument for this solution is that in many cases it does not make sense to store the extent (think of the **Address** or the **Date** class in section. 52.9) and in other cases it makes sense to have multiple extents (think of several users sharing a class definition but not necessarily an extent). Between these two ends of the spectrum, there are of course a number of intermediate solutions, such as allowing the user to declare extents or multiple extents maintained by the system.

Inheritance. Classes can be specialized by adding new attributes, new relationships, and new methods. Inheritance can be single or multiple. The set of classes forms a directed acyclic graph.

Message passing. The binding of the name of the method is actually done at run time (late binding) so as to ensure that the proper implementation of the method takes into account the actual class to which the object belongs.

52.6 Object Database Model: Persistence

In a traditional database, programming is done using a programming language and embedding queries in the program to access or update the database. The programming language has its own type system (the equivalent of a data model), and the database system has its own data model (equivalent to a type system). A protocol explains how to map data from one of these models to the other one. This can be done for instance by doing the communication at the atomic type level. The difference between these two models is the "impedance mismatch" which forces the developer to convert data in and out of the database.

Furthermore, there is a "wall" between the database system and the application program. The database contains only persistent data, i.e., every modification done to the database by a committed transaction remains effective until another transaction eventually modifies it again. The application program only manipulates transient data, i.e., all the data it operates on disappear after the end of the execution of the program. Interaction between the application and the database is always done through read and write statements by moving data from one space to the other. Thus, the program can read data from the database, insert new data in the database, or update existing data in the database. The only thing one can do to a database object is bring it to the work area; the only way one can modify it is by rewriting it from the program work area (with, of course, the exception of the update statement of SQL).

Object databases have a very different approach to this issue. The data model of the database and the type system of the programming language are the same (in other words, every object of every type can be stored as such in the database). This solves the impedance mismatch: no data conversion is needed, and the developer does not have to be aware of two different data models.

Furthermore, the dichotomy between the database and the programming language is replaced by a more general notion of persistence. In this approach, the application program manipulates both persistent and transient data. Persistent data are updated in transactions, and persistent data updates are final once committed. Transient data have the life span of the application. And finally, the same operations can be done on a transient and a persistent object.

The semantics of these features is defined by a model of persistence. This model specifies how objects are created in each space and whether and how they can change their persistence status. Persistence models can be static (the persistence status of the object does not change during its lifetime) or dynamic (the persistence status can be modified).

52.7 Object Database Model: Naming

The naming model specifies the entry points in the database. There are two styles of naming in object databases. One is inherited from the relational world, and the other from the programming-language world. In a relational database, the naming model consists of giving a name to every relation in the database and providing this name as the sole and only access to the database. Some object databases have copied this approach and provided names for the extent of every class in the schema as the sole access to the database. In classical programming languages, one can associate names to variables and assign them values. Some object databases have chosen this naming model: the schema contains a number of names, and to these names, objects of any type can be attached.

52.8 Language Bindings

Instead of inventing "yet another programming language," most object database system designers found it more realistic and attractive to use existing object-oriented programming languages. This yields an approach which is very different from the classical one. Language bindings provide the ability (1) to write the methods of a class in a given programming language and (2) to develop an application using an object database. The language bindings specify how (1) to declare persistence, (2) to manage persistent objects, (3) to embed queries, (4) to represent and manipulate all the features of the data model, and (5) to manage databases and transactions.

We now turn to a presentation of the ODMG standard.

52.9 The ODMG Standard: Object Database Model and C++ Binding

The ODMG object model, following the OMG object model, supports the notion of class, of objects with attributes and methods, and of inheritance and specialization. It also offers the classical types to deal with date, time, and character strings. To illustrate this, let us first define the elementary objects of a schema without establishing connections between them:

```
class Person{
  String name;
  Date birthdate;

// Methods:
  Person();              // Constructor: a new Person is born
  int age();             // Returns an atomic type
};

class Employee: Person{  // A subclass of Person
  float salary;
};

class Student: Person{   // A subclass of Person
  String grade;
};
```

```
class Address{
    int number;
    String street;
};

class Building{
    Address address;          // A complex value (Address) embedded
                              // in this object
};

class Apartment{
    int number;
};
```

The ODMG extensions to the OMG data model include relationships and collections. An object refers to another object through a **Ref**. A **Ref** behaves as a C++ pointer, but with more semantics. Firstly, a **Ref** is a persistent pointer. Futhermore, referential integrity can be expressed in the schema and maintained by the system. This is done by declaring the relationship as symmetric. For instance, we can say that a **Person** lives in an **Apartment**, and that this **Apartment** is used by this **Person**, in the following way:

```
class Person{
    Ref<Apartment>  lives_in   inverse is_used_by;
};

class Apartment{
    Ref<Person>     is_used_by inverse lives_in;
};
```

The keyword **inverse** is the only ODMG extension to the standard C++ class definitions. It is, of course, optional. It ensures the referential integrity constraint: if a **Person** moves to another **Apartment**, the attribute **is_used_by** is automatically reset to **NULL** until a new **Person** takes this **Apartment** again. Moreover, if an **Apartment** object is deleted, the corresponding **lives_in** attribute is automatically reset to **NULL**, thereby avoiding dangling references.

ODMG introduces a set of predefined generic classes for the management of very large collections: **Set<T>**, **Bag<T>** (a multiset, i.e., a set allowing duplicates), **Varray<T>** (variable-size array), **List<T>** (variable-size and insertable array). A *collection* is a container of elements of the same class. As usual, polymorphism is obtained through the class hierarchy. For instance a **Set<Ref<Person>>** may contain **Person**s as well as **Employee**s, if the class **Employee** is a subclass of the class **Person**:

```
Set<  Ref<Person> > Persons;       // The Person class extent.
Set<  Ref<Apartment> > Apartments; // The Apartment class extent.
Set<  Ref<Apartment> > Vacancy;    // The set of vacant apartments.
List< Ref<Apartment> > Directory;  // The list of apartments.
                                   // ordered by their number
                                   // of rooms.
```

The notion of 1–1 relationships defined previously has to be extended to 1–*n* and *n*–*m* relationships, with the same guarantee of referential integrity. For example, a **Person** has two **parents** and possibly several **children**; in a **building**, there are many **apartments**.

```
class Person{
   Set  < Ref<Person> > parents  inverse children; // 2 parents.
   List < Ref<Person> > children inverse parents;  // Ordered by
                                                    // birthdate
};

class Building{
   List< <Ref<Apartment> > apartments inverse building;
   // Ordered by apartment number
};

class Apartment{
   int number;
   Ref<Building> building inverse apartments;
};
```

Let us now define our example schema completely:

```
class Person{
   String name;
   Date    birthdate;
   Set  < Ref<Person> > parents  inverse children;
   List < Ref<Person> > children inverse parents;

   Ref<Apartment> lives_in      inverse is_used_by;

// Methods:
   Person();                         // Constructor: a new
                                     // Person is born
   int age();                        // Returns an atomic
                                     // type
   void marriage( Ref<Person> spouse); // This Person gets a
                                     // spouse
   void birth( Ref<Person> child);   // This Person gets a
                                     // child
   Set< Ref<Person> > ancestors;     // Set of ancestors of
                                     // this Person
   virtual Set<String> activities(); // A redefinable method
};

class Employee: Person{             // A subclass of Person
   float salary;
// Method
   virtual Set<String> activities();  // The method is
                                     // redefined
};
```

```
class Student: Person{                      // A subclass of
                                            // Person
   String grade;
// Method
   virtual Set<String> activities();        // The method is
                                            // redefined

};

class Address{
   int number;
   String street;
};

class Building{
   Address address;        // A complex value (Address) embedded
                           // in this object
   List< <Ref<Apartment> > apartments inverse building;
// Method
   Ref<Apartment> less_expensive();
};

class Apartment{
   int number;
   Ref<Building> building;
   Ref<Person> is_used_by inverse lives_in;
};

Set< Ref<Person> >     Persons;    // All persons and employees
Set< Ref<Apartment> > Apartments;// The Apartment class extent
Set< Ref<Apartment> > Vacancy;    // The set of vacant apartments
List< Ref<Apartment> > Directory; // The list of apartments
                                  // ordered by number of rooms
```

To implement the above schema, we write the body of each method. These bodies can be easily written in C++. In fact, because **Ref<T>** is equivalent to a pointer (**T***), manipulating persistent objects through **Ref**s is done in exactly the same way as through normal pointers.

To run applications on a database instantiating such a schema, ODMG provides classes to deal with *databases* (with open and closed methods) and *transactions* (with start, commit, and abort methods). When an application creates an object, it can create a transient object which will disappear at the end of the program, or a persistent object which will survive when the program ends and can be shared by many other programs possibly running at the same time. Here is an example of a program to create a new persistent apartment and let **john** move into it:

```
Transaction move;
move.begin();
   Ref< Apartment > home = new(database) Apartment;
   Ref< Person >    john = database->lookup_object(''john'');
                              // Retrieve a named object
```

```
    Apartments.insert_element(home); // Put this new apartment
                                     // in the class extent
    john->lives_in = home;  // Persistent objects are
                            // handled as standard C++ objects
  move.commit();
```

52.10 The ODMG Standard: Query Language

Object databases, like traditional databases, need query languages. The SQL query language cannot be used directly, because it lacks support for (1) complex objects (ability to query nested structures and arbitrary collections) and (2) method invocation. ODMG adopted the OQL language. OQL has an SQL flavor but supports complex objects and methods. Recent improvements (OQL 1.1 and 1.2) have brought it closer to the SQL92 standard. We use the database described in the previous section, and we give an overview of the most relevant features through examples.

As explained above, one can enter a database through a named object. We use the notation **.** (or indifferently **->**), which enables us to follow pointers or attribute names. For instance, if we have a person **p** and we want to know the name of the street where this person lives, the OQL query is

```
    p.lives_in.building.address.street
```

This example has treated 1–1 relationships; let us now look at *n–p* relationships. Assume we want the names of the children of the person **p**. We cannot write **p.children.name**, because **children** is a list of references, so the interpretation of the result of this query would be undefined. Intuitively, the result should be a collection of names, but we need an unambiguous notation to traverse such a multiple relationship, and we use the select-from-where clause to handle collections, just as in SQL:

```
    select c.name from p.children c
```

The result of this query is a value of type **Bag<String>**. If we want to get a set, we simply drop duplicates, as in SQL, by using the **distinct** keyword:

```
    select distinct c.name from p.children c
```

Now we have a way to navigate from an object towards any object, following any relationship and entering any complex subvalues of the object. For instance, we want the set of addresses of the children of each person of the database. We know the collection named **Persons** contains all the persons of the database. We have now to traverse two collections: **Persons** and **Person.children**. As in SQL, the select-from operator allows us to query more than one collection. These collections then appear in the **from** part. In OQL, a collection in the **from** part can be derived from a previous one by following a path which starts from it, and the answer is

```
    select c.lives_in.building.address
    from Persons p,
         p.children c
```

This query inspects all children of all persons. Its result is a value whose type is **Bag<Address>**.

Of course, the **where** clause can be used to define any predicate, which then serves to select only the data matching the predicate. For instance, suppose we want to restrict the previous query to the people living on Main Street and having at least two children. Moreover we are only interested in the addresses of the children who do not live in the same apartment as their parents. The query is

```
select c.lives_in.building.address
from  Persons p,
        p.children c
where   p.lives_in.building.address.street = ''Main Street''
and
        count(p.children) >= 2 and
        c.lives_in != p.lives_in
```

In the **from** clause, collections which are not directly related can also be declared. As in SQL, this allows us to compute *joins* between these collections. For instance, to get the people living in a street and having the same name as their street, we do the following: the **building** extent is not defined in the schema, so we have to compute it from the **Apartments** extent. To compute this intermediate result, we need a select-from operator again. This shows that in a query where a collection is expected, it can be computed recursively by a select-from-where operator, without any restriction. So the join is done as follows:

```
select p
from Persons p,
     (select distinct a.building from a in Apartments) b
where p.name = b.address.street
```

This query highlights the need for an optimizer. In this case, the inner **select** subquery must be computed only once and not for each person.

A major difference between OQL and SQL is that object query languages must manipulate complex values. OQL can therefore create any complex value as a final result, or inside the query as an intermediate calculation. To build a complex value, OQL uses the constructors **struct**, **set**, **bag**, **list**, and **array**. For example, to obtain the addresses of the children of each person, along with the address of this person, we use the following query:

```
select struct(me: p.name,
              my_address:p.lives_in.building.address,
              my_children:
                          (select struct(name: c.name,
                            address: c.lives_in.building.address)
                          from p.children c))
from Persons p
```

This gives for each person the name, the address, and the name and address of each of his or her children, and the type of the resulting value is

```
struct result_type{
   String me;
   Address my_address;
   Bag<struct{String name; Address address}> my_children;
}
```

OQL can also create complex objects. For this purpose, it uses the name of a class as a constructor. Attributes of the object of this class can be initialized explicitly by any valid expression. For instance, to create a new building with two apartments, if there is a type name in the schema, called **List_apart**, defined by **typedef List<<Ref<Apartment> > List_apart**, then the query is

```
Building(address: struct(number: 10, street: ``Main street''),
         apartments: List_apart(Apartment(number: 1),
                                        Apartment(number: 2)))
```

OQL allows us to call a method with or without parameters anywhere the result type of the method matches the expected type in the query. The notation for calling a method is exactly the same as for accessing an attribute or traversing a relationship in the case where the method has no parameter. If it has parameters, these are given between parentheses.

This flexible syntax frees the user from knowing whether the property is stored (an attribute) or computed (a method). For instance, to get the age of the oldest child of the person Paul, we write the following query:

```
select max(select c.age from c in p.children)
from Persons p,
where p.name = ``Paul''
```

Of course, a method can return a complex object or a collection, and then its call can be embedded in a complex path expression. Suppose, for instance, that inside a building **b**, we want to know who inhabits the least expensive apartment. The following path expression gives the answer:

```
b.less_expensive.is_used_by.name
```

Although **less_expensive** is a method, we "traverse" it as if it were a relationship. The following will give the activities of each person:

```
select p.activities from Persons p
```

Here **activities** is a method which has three incarnations. Depending on the kind of person the current **p** is, the right incarnation is called.

OQL is a purely functional language: all operators can be composed freely as long as the type system is respected. That is why the language is so simple and its manual so short. This philosophy is different from that of SQL, which is an ad-hoc language whose composition rules are not orthogonal. Adopting complete orthogonality allows us not to restrict the power of expression and makes the language easier to learn without losing the SQL style for the simplest queries.

Among the operators offered by OQL, but not yet introduced, we can mention the set operators (union, intersect, except), the universal (for all) and existential (exists) quantifiers, the sort and group-by operators, and the aggregate operators (count, sum, min, max, and avg).

To illustrate this free composition of operators, let us write a rather complex query. We want to know the name of the street where employees live who have the smallest salary on average, compared to employees living on other streets. We proceed step by step and then do it all at once. We can use the **define** OQL instruction to evaluate temporary results.

Build the extent of the class **Employee** (not supported directly by the schema):

```
define Employees as
   select (Employee) p from Persons p
   where ``has a job'' in p.activities
```

Group the employees by street, and compute the average salary in each street:

```
define salary_map as
    group   e in Employees
    by      (street: e.lives_in.building.address.street)
    with    (average_salary: avg(select x.salary
                                      from x in partition))
```

The group-by operator splits the employees into partitions according to the criterion (the name of the street where this person lives). The **with** clause computes, in each partition, the average of the salaries of the employees belonging to this partition. The result of the query is of type **Set<struct{String street; float average_salary;}>**.

Sort this set by salary:

```
define sorted_salary_map as
    sort s in salary_map by s.average_salary
```

The result is now of type **List<struct{String street; float average_salary;}.** Now get the smallest salary (the first in the list) and take the corresponding street name. This is the final result:

```
sorted_salary_map[0].street
```

In a single query we could have written

```
(sort s in(
    group   e in (select (Employee) p from p in Persons
                    where ''has a job'' in p.activities)
    by      (street: e.lives_in.building.address.street)
    with    (average_salary:
                    avg(select x.salary from x in partition)))
    by s.average_salary)[0].street
```

52.11 Performance

Object databases are currently targeting specific application areas, mainly those manipulating complex data. A number of benchmarks [Anderson et al. 1990] have been proposed to represent this type of work load; the best known are the OO1 benchmark [Cattell and Skeen 1992] and the OO7 benchmarks [Carey et al. 1993]. OO1 was mainly used to compare object databases with relational databases. It shows that on this type of workload, object database systems outperform relational systems. OO7 was mainly used to compare object database systems with one another. More discussion of benchmarks and performance appears in Chapter 47.

Performance improvement over relational systems for this type of workload can be easily explained:

- The database engine is specifically designed to support the object model. It offers native support for complex objects and avoids the join operation of relational systems when building these objects.
- Object databases were designed with a client-centered architecture in mind. Thus, they transfer the workload onto the client as much as possible, and support client caches for objects.

52.12 Conclusion

It is now 10 years since the beginning of the object database field, and it is clear the area has reached a first level of maturity. Through 5 years of design, prototyping, and implementation and 7 years of usage and experimentation in the marketplace, the technology is now mature and in full use. A multiplicity of available products confirms this status, as does the convergence of most products towards the ODMG standard.

References

Agrawal, R. and Gehani, N. H. 1989. ODE (Object Database and Environment): the language and the data model. In *Proc. ACM SIGMOD Conf.*

Anderson, T., Berre, A., Allison, M., Porter, H., and Schneider, B. 1990. The hypermodel benchmark. In *Proc. EDBT Conf.*, Venice, March.

Atkinson, M., Bancilhon, F., DeWitt, D., Dittrich, K., Maier, D., and Zdonik, S. 1989. The object-oriented database system manifesto. In *Proc. First DOOD Conf.* Kyoto, Dec.

Bancilhon, F., Delobel, C., and Kanellakis, P. 1993. *Building an Object-Oriented Database System, the Story of O_2*. Morgan Kaufman.

Carey, M. and DeWitt, D. 1987. An overview of the Exodus project. *IEEE Trans. Database Eng.* 10(2), June.

Carey, M. J., DeWitt D. J., and Naughton, J. F. 1993. The 007 benchmark. In *Proc. ACM SIGMOD Conf.*, Washington, DC.

Cattell, R. G. G. 1994. *Object Data Management*. Addison–Wesley.

Cattell, R. G. G., ed. 1995. *The Object Database Standard: ODMG*. Morgan Kaufman.

Cattell, R. and Skeen, J. 1992. Object operation benchmark. *ACM Trans. Database Systems* 17(1), March.

Copeland, G. and Maier, D. 1983. Making Smalltalk a database system. In *Proc. ACM SIGMOD Conf.*, June.

Fishman, D. et al. 1989. IRIS: an object-oriented database management system. *ACM Trans. Office Inf. Syst.* 5(1), Jan.

Gemstone 1995. *GemStone Reference Manual*. Gemstone, Beaverton, OR.

Kim, W., Garza, J. F., Ballou, N., and Woelk, D. 1990. Architecture of the ORION next generation database system. *IEEE Trans. Data and Knowledge Eng.* 2(1), March.

O_2 1995. A technical overview of the O_2 system. *O_2 Technology Tech. Rep.* O_2 Technology, Versailles.

Objectivity 1994. *Objectivity/DB User Manual*. Objectivity, Inc., Mountain View, CA.

ObjectStore 1994. *ObjectStore User Manual*. Object Design, Inc., Burlington, MA.

Ontos 1993. *ONTOS Reference Manual*. Ontos, Inc., Burlington, MA.

POET 1995. *POET Programmer's and Reference Guide*. Poet Software, Hamburg, Germany.

UniSQL 1995. *UniSQL/X Database Management System Product Description*. UniSQL Inc., Austin, TX.

53

The SQL Language: A Case Study

Jim Melton
Sybase, Inc.

53.1 Introduction

SQL is a *data sublanguage* used with a *host language* for access to relational databases. It is not a complete programming language, but depends on input/output and control facilities of the host language; SQL has an English-like syntax, and its statements read a bit like English. SQL is the subject of both *de jure* and *de facto* standards—ANSI (American National Standards Institute) has published three generations of the SQL standard, as has ISO (International Organization for Standardization). X/Open, a consortium, has also published an SQL specification.

In 1974, Dr. E. F. Codd published a seminal paper [Codd 1974] providing a mathematical foundation for *logical* representation and manipulation of data, independent of physical representation, relationships, and other implementation considerations. Shortly afterwards, Don Chamberlin and Raymond Boyce published the first paper on what became SQL [Chamberlin and Boyce 1974], arising from research prototypes on data languages (SQUARE and SEQUEL) and on IBM's research relational database System R. The relational model uses *relation, attribute*, and *tuple* for SQL's **table, column**, and **row**; SQL does not correspond perfectly to the relational model, most significantly in SQL's lack of complete prohibition of duplicate rows in a table (though SQL permits users to restrict their tables to containing only unique rows).

In 1978, ANSI approved work to standardize a data definition language for network databases; a new Technical Committee, X3H2, was formed for this project. In 1986, a complete network database language standard was published as Database Language NDL. X3H2 members recognized the relational model's importance, too—after working on a derivation of SEQUEL called RDL ("Relational Database Language"), a proposal to use IBM's SQL specification was accepted. X3H2, in cooperation with the corresponding ISO group, spent another year refining the SQL specification, which was published in 1986.

SQL-86 omitted support for **referential integrity**, but a revised standard was published three years later with a minimal referential-integrity facility. In 1992, a significant new version of the language was published. While SQL-86 and SQL-89 did not have adequate features for real applications, SQL-92 contained language features and conformance requirements to allow significant applications. The fourth generation of SQL (the project is called "SQL3") is currently being prepared for publication, perhaps as early as 1998; it adds significant new facilities to SQL, including support for the object paradigm, and divides the specification into several parts that can be processed more or less independently.

53.2 Underlying Concepts

SQL's most fundamental concept is the **table**. Tables have one or more columns, each of which has a name and a data type. Data in a table are stored in rows that have columns corresponding to those of the table. Each column of a table has a single data type for all rows in that table. (A column in a row is sometimes called a "cell," though the SQL standard does not use or define that term.) Figure 53.1 illustrates these concepts.

SQL provides a number of data types, broken into the categories of *numeric, string,* and *datetime.* Table 53.1 shows each category, the further breakdown of those categories, and the specific data types.

All data in an SQL database belong to one of those data types. However, database designers occasionally need to capture the fact that some particular datum is not stored—it might be inapplicable in a given row, or it might be unknown when the row is created. SQL provides a concept of *null* to permit representation of this situation. The concept of null doesn't have a data type itself, but the cell in which a null is stored always has one of the SQL data types.

In addition to representing data, SQL databases are *self-describing*; that is, they contain tables with *metadata* that describe the tables in the database (and describing the tables containing the metadata). While the SQL standard doesn't define the word "database," it does define the words **catalog** and **schema**. A catalog is a named collection of schemas, including the special schema that contains the metadata for all objects in the catalog. A schema is a named collection of tables (and their columns), character sets, and other SQL-defined objects. Catalog names *qualify* schema names, allowing multiple schemas with the same name to exist in different catalogs; similarly, schema names qualify the names of tables and other objects.

Part of the power of SQL lies in the helpers that it provides database and application designers. SQL databases can contain **constraints**, including semantic integrity constraints that instruct the database

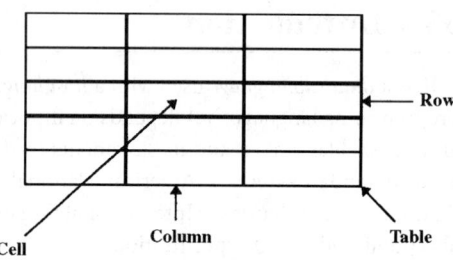

FIGURE 53.1   Illustration of table concepts.

TABLE 53.1   SQL Data Types

Numeric	
Exact numeric	Represents values exactly
INTEGER and **SMALLINT**	System-defined precisions
DECIMAL and **NUMERIC**	User-defined precisions
Approximate numeric	"Floating point" numbers
REAL and **DOUBLE PRECISION**	System-defined precisions
FLOAT	User-defined precision
String	
Character string	In specific character sets
CHARACTER,... VARYING	User-specified character set
NATIONAL CHARACTER	System-defined character set
Bit string	Zeros and ones
BIT,... VARYING	Fixed- or varying-length
Datetime and interval	
DATE, **TIME**, and **TIMESTAMP**	Specific dates and times
INTERVAL (with precision)	Difference between datetimes

system how to enforce business rules associated with the data stored in the database, as well as referential integrity constraints that tell the database system how to keep its data internally consistent when changes are made by applications. If an application attempts to violate a semantic integrity constraint (for example, a rule that says "all salaries must be greater than 0"), then it is notified of the error and the statement attempting that violation is not executed. Attempted violations of some referential constraints (e.g., a rule prohibiting elimination of departments having employees) are handled similarly. However, referential constraints can be more sophisticated—a database designer might permit resignation of a project's manager, but require resolution of the status of the project. One design results in the project's automatic deletion, a second design assigns the project to someone responsible for orphaned projects, and the third design leaves the project in an unassigned state pending explicit action at a later time. Each of these designs results in automatic resolution without execution of any SQL statements by the application.

A similar feature, called *triggers*, allows a database designer to force specific database actions whenever certain tables are accessed in specified ways. For example, a trigger could be defined to add a row to a log table whenever changes are made to the salary column of an employee table, or to adjust the budgets for departments whenever new projects are assigned to them.

When rows are created in a table, an application programmer can choose to provide a value for every column in each created row. Alternatively, some rows might have an obvious default value; for example, employees might be hired as members of the Staff department often enough that the application assumes that assignment if no specific department is provided. SQL allows the database designer to specify a default value for each column in a table; if no default value is specified, then a default of null is implied.

It sometimes happens that database designers find themselves using a particular combination of data type, constraint, and default value frequently (perhaps in various tables). SQL allows the definition of a *domain* to give a name to that combination; the domain name can then be used in place of the data type (and constraint and default value) when defining columns in tables. For example, the name **MONEY** might be applied to a domain providing a data type of **DECIMAL(8,2)**—decimal with 8 total digits of precision, two of them after the decimal point—with a constraint saying that the value must never be negative, and a default value of null. Columns such as **SALARY** and **BUDGET** could then be defined to be **MONEY**, providing a convenient shorthand as well as ensuring consistency of specification.

SQL programmers have several alternatives for using the language. The most widely used alternative is to *embed* SQL statements into programs written in ordinary third-generation programming languages (3GLs). This technique, called **embedded SQL**, requires the application programmer to write the application in a 3GL (the SQL standard supports Ada, C, COBOL, Fortran, MUMPS, Pascal, and PL/I; SQL implementations often support other languages). Each embedded SQL statement starts with "**EXEC SQL**". In a typical SQL implementation, this embedded SQL program is processed by a *preprocessor* that extracts the SQL statements and (conceptually, at least) replaces them with a "call statement" to invoke the (conceptual or literal) **procedure** that the system creates to contain the SQL statement. The SQL statement (contained in that procedure) is then compiled and optimized by the SQL system to prepare it for later execution; the remaining application program is compiled in the normal way. When the program executes, the optimized SQL statements are executed as specified by the 3GL code.

In some SQL implementations and the SQL standard, it is possible to write actual SQL procedures (each containing a single SQL statement), collecting related procedures together into a **module**. Called *module language*, this technique permits applications to be written in a more modular fashion—database-related operations are coded in "pure SQL" and processed by an SQL compiler, while other application operations are coded in the appropriate 3GL and processed by that language's compiler. The SQL procedures are invoked through actual "call statements" by the application program. The two techniques are completely isomorphic with one another. In implementations that support both techniques, the choice of which to use is often a matter of taste or of organization policy.

In many applications, such as traditional mainframe applications, the SQL statements to be executed are well known when the application is written. Embedded (or module language) SQL is appropriate for such applications. In other situations, such as *ad hoc* query generators, graphical database browsers, or client–server systems with widely varying users, the SQL statements that will be executed are often not known until execution time, when the user formulates a question. A technique called *dynamic SQL* allows SQL statements to be formulated at run time, prepared for execution by the database system, and executed on demand. Dynamic SQL is typically slower than *static* SQL because of its inability to precompile and optimize statements. Of course, the benefits of flexibility often make this a worthwhile cost.

53.3 Retrieving Information from a Database

Arguably, the most basic operation that is performed on an SQL database is the retrieval of information stored in tables in that database. Information may be retrieved from the database directly into the application, or it may be retrieved for use within an SQL statement for a variety of purposes.

The **SELECT** expression is the foundation for retrieval of SQL information. With a little variation in syntax, the **SELECT** expression can be used as an SQL statement to retrieve the information into the application or to define a view, but is used most often in the form of a **subquery** within an SQL statement. The format of a **SELECT** expression is

```
SELECT select-list
    FROM table, table...
    WHERE predicates
    GROUP BY grouping-columns
    HAVING predicates
```

The **WHERE**, **GROUP BY**, and **HAVING** clauses are all optional. The result of the **SELECT** expression is always a *virtual table*; SQL exhibits *closure* in that operations on tables produce new tables.

A SELECT expression is evaluated according to the following rules (effectively, that is; products must provide this effect, but may—and usually do—provide significant optimizations):

1. First, all rows in the table or tables specified in the **FROM** clause are retrieved; if more than one table is specified, then the Cartesian product of all tables is retrieved, producing new, extended rows. (Two tables with N and M columns and n and m rows have a Cartesian product with $N + M$ columns and $n \cdot m$ rows; each row in one table is "matched" with every row from the other table.)

2. The predicates in the **WHERE** clause (if present) are applied to the rows produced by the preceding step. All rows that do not satisfy the predicates are eliminated from the working set of rows.

3. If a **GROUP BY** clause is present, then rows are grouped together according to equal values in the column or columns (**grouping-columns**) identified in that clause.

4. If a **HAVING** clause is present, then its predicates are applied to the groups; all groups that do not satisfy the predicates are eliminated. (A **HAVING** clause without a **GROUP BY** clause effectively makes the result of the **WHERE** clause a single group.)

5. Finally, the **select-list** is used to determine the columns produced as the result of the **SELECT** expression. If groups have been formed by the presence of a **GROUP BY** clause or a **HAVING** clause, then the **select-list** can include only columns used as grouping columns, "statistical operations" (sum, average, maximum, and minimum) on other columns, and count operations on the resulting table. These statistical and count operations also can be used without groups having been formed.

To illustrate the effects of a **SELECT** expression, consider the two tables in Fig. 53. 2: Evaluation of the **SELECT** expression

```
SELECT EMP_NAME
    FROM EMPLOYEES
    WHERE EMP_ID = 100
```

EMPLOYEES

EMP_ID INTEGER	EMP_NAME VARCHAR(20)	DEPT_ID CHAR(3)
100	Joe	SLS
200	Mary	EXC
300	Bob	ENG
400	Sally	SLS
500	José	SLS
600	Krishna	ENG
700	Barbara	ENG
800	Jack	ENG
900	Bill	SLS
1000	Andie	EXC

DEPARTMENTS

DEPT_ID CHAR(3)	DEPT_NAME VARCHAR(20)
ENG	Engineering
SLS	Sales
EXC	Executive

FIGURE 53.2 Sample SQL tables.

will produce a (virtual) table with one column, named **EMP_NAME**, and one row; the value of the one column in that row will be **Joe**. In some situations in SQL, this expression can be treated as a scalar value; in such places, it is called a *scalar subquery*.

A different query,

```
SELECT EMP_NAME, DEPT_NAME
  FROM EMPLOYEES, DEPARTMENTS
 WHERE EMP_ID = 100 AND
         EMPLOYEES.DEPT_ID = DEPARTMENTS.DEPT_ID
```

will similarly produce a table, this one with two columns, named **EMP_NAME** and **DEPT_NAME**, and one row; the values of the two columns will be **Joe** and **Sales**. This expression can be treated as a single row and is sometimes called a *row subquery*. In order to "match" employees with information for the department for which they work, the **WHERE** clause has an extra predicate specifying the match rules.

A third query,

```
SELECT EMP_NAME, DEPT_NAME
  FROM EMPLOYEES AS E, DEPARTMENTS AS D
 WHERE DEPT_ID = 'SLS' AND
         E.DEPT_ID = D.DEPT_ID
```

will similarly produce a table, this one with two columns, still named **EMP_NAME** and **DEPT_NAME**, and four rows. This expression can be treated as the table it produces and is sometimes called a *table subquery*. Both the second and third examples *joined* the **EMPLOYEES** table with the **DEPARTMENTS** table to produce the requested information. Joining tables together means matching rows of one table with rows of another table, usually according to some meaningful criteria; in this case, the criterion is equality of **DEPT_ID** columns in the two tables. The third query uses a technique called *correlation names* to identify the tables more concisely; while merely a convenience in this example, correlation names are sometimes necessary to allow distinguishing among multiple instances of the same table when joined with itself.

It is worth noting that SQL supports several sorts of joins, including so-called inner joins (where the result includes only rows that have a match between the two tables being joined) and *outer joins* (in which the result may include rows from one or both tables that have no match in the other table—columns corresponding to those from the table with no match are filled in with nulls).

To form a **SELECT** *statement*, the target of the retrieved information must be given:

```
SELECT select-list
   INTO target-list
          FROM table, table...
          WHERE predicates
          GROUP BY grouping-columns
          HAVING predicates
```

The **target-list** is a list of host language variables (or, in module language, a list of parameters of the containing procedure); there must be as many targets as there are columns in the **select-list**, and the data types of each target must match the data type of its corresponding **select-list** column.

A problem arises if the **SELECT** statement produces a virtual table of more than one row. The host language variables into which the columns of the result (consider any 3GL) can only accept a single value. If the **SELECT** statement were to produce multiple rows, then only one of them could be retrieved into the list of targets. As this illustrates, SQL may be characterized as a "set-at-a-time" language, while 3GLs are typically "datum-at-a-time" languages. (Using an analogy taken from electrical engineering, this is often called an *impedance mismatch* between SQL and the host languages.) The **SELECT** statement, then, is really a "single-row **SELECT** statement." If it produces more than one row, an error is signaled.

Of course, it's not always sufficient to select just columns from the virtual table that the other clauses produce. SQL supports a variety of *expressions* that can also be used. Though it's probably not meaningful in practice, the following **SELECT** expression is valid:

```
SELECT EMP_ID/100
   FROM EMPLOYEES
     WHERE EMP_NAME = 'Bill'
```

This scalar subquery will return the value 9. Analogously, the SELECT expression:

```
SELECT EMP_NAME || `` || DEPT_NAME || '' || DEPT_ID
   FROM EMPLOYEES, DEPARTMENTS
   WHERE DEPT_ID = 'SLS'
```

will return four rows, but each row will have a single column containing the employee's name, the corresponding department name, and the department identifier, separated from one another by a single space. Numeric operators that SQL provides are **+** (addition), **-** (subtraction), ***** (multiplication), and **/** (division); the string operator is **||** (concatenation); datetime and interval operators are analogous to numeric operators with the appropriate restrictions based on the data type. SQL also provides several functions for each data type. Functions that return a numeric type include **LENGTH** (of a string), **EXTRACT** (from a datetime datum), and **POSITION** (of a substring in a string); the functions that return a string include **SUBSTRING** (to extract a substring from a string) and **TRIM** (to eliminate leading and/or trailing blanks or other characters); the functions that return a datetime value are **CURRENT_DATE**, **CURRENT_TIME**, and **CURRENT_TIMESTAMP** (which return what their names suggest).

SQL is fairly strict about how items of different data types can be used together. For example, although SQL permits expressions that mix exact numeric and approximate numeric data items, it would not permit an expression that attempted to multiply a bit string by an integer, even though a bit string can be considered to represent a numeric value. To allow such operations where the application programmer requires them, SQL provides a **CAST** function that allows explicit data-type conversion. A bit string could be "converted" to an integer with an expression like:

```
CAST (BIT_COLUMN AS INTEGER)
```

Most "meaningful" data conversions are supported by **CAST** with the obvious restrictions (for example, a character string can be **CAST** to a numeric type, but an error will be signaled if nonnumeric characters are encountered).

As just seen, the difference between SQL's set-at-a-time semantics and the datum-at-a-time character of the host languages makes it infeasible to write a simple **SELECT** statement to retrieve more than one row. Yet many applications need exactly this capability. SQL resolves this impedance mismatch by providing a device called a *cursor*, which allows the application to identify sets of rows, but to process them one at a time. The set of rows is identified by a *cursor declaration* that specifies the **SELECT** expression:

```
DECLARE cursor-name CURSOR FOR
   SELECT select-list
   FROM tables
   WHERE predicates
```

The cursor declaration is just that—a declaration. It is not executed at all. However, when the application program *opens* the cursor, the **SELECT** expression is evaluated. Once the cursor has been opened, rows can be *fetched* through the cursor until the last row has been retrieved (signaled by a special status returned to the application program). Finally, the cursor is *closed*:

```
OPEN cursor-name
label:
FETCH FROM cursor-name
   INTO target-list
3GL statements to process the data
if not end-of-data, then loop to label
CLOSE cursor-name
```

The **FETCH** statement provides additional capabilities, invoked by inserting syntax immediately after the keyword **FETCH**. **NEXT** provides the default behavior of fetching the next row following the row that was just processed, while **PRIOR** retrieves the row *preceding* the row just processed. **FIRST** positions the cursor at the first row in the virtual table identified by the **SELECT** expression, and **LAST** positions it at the very last row. **ABSOLUTE**, followed by a numeric expression, positions the cursor on the row identified by the value of that expression, and **RELATIVE**, also followed by a numeric expression, moves the cursor forward or backward by a number of rows corresponding to the value of that expression.

Consider the following scenario: An application is required to retrieve and print the name of each employee and the name of the department for which they work. Because the number of employees might be rather large and vary over time, it would not be feasible to attempt to retrieve the information into some fixed-size host language data structure, like an array. Instead, a cursor provides the most reasonable approach to this problem:

```
/* Sample C program to demonstrate cursor usage */
#include...
char ename[21], dname[21];
EXEC SQL DECLARE emp_cursor CURSOR FOR
```

```
    SELECT emp_name, dept_name
    FROM employees, departments
    ORDER BY dept_name;
/* Open the cursor in preparation for fetch/print loop */
EXEC SQL OPEN emp_cursor ;
/* Error handling is omitted from this example */
while (1 == 1) {
EXEC SQL FETCH FROM emp_cursor
        INTO :ename, :dname ;
   printf (...);
   /* if end of data, then get out of the loop */
   if (SQLSTATE == "02000") break;
}
EXEC SQL CLOSE emp_cursor;
```

The **DECLARE CURSOR** in this example shows another feature of cursors: the ability to order the results of the cursor. Set-oriented statements in SQL have no ordering capability, simply because it is not normally needed; because the entire set of data is processed in precisely the same way by the statement, the sequence in which rows are processed is unimportant. However, when interacting with the application, as with a cursor, that order may very well be significant.

53.4 Inserting Information into a Database

However useful it might be to retrieve information from a database, that information must first be somehow placed in the database. In SQL, the **INSERT** statement is used to insert information into tables. While it is theoretically possibly to insert information into a view, by translating the **INSERT** into a series of **INSERT**s on the physical tables underlying the view, SQL provides rather little support for that feature. Views can be arbitrarily complex, and while some limited views are theoretically updatable, it's not so easy to deal with other views. As a result, SQL can insert into (and update and delete from) only a very limited subset of views.

The **INSERT** statement has three alternative formats. The first of these is used when the information to be inserted is provided in the form of literal values in the SQL statement itself or from host variables in the application program:

```
INSERT INTO table-name
  ( column-name, column-name... )
  literal-or-variable, literal-or-variable...
```

In this form of **INSERT**, the parenthesized list of **column-names** is optional; if it's not specified, then the system assumes that you meant to write such a list containing every column of the table, in the order in which they are defined in the table. The number of column names and the number of literals or variables must be the same, and the data types of each column must match the data type of the literal or host variable. A single row is inserted into the identified table. For example:

```
INSERT INTO departments
  ( dept_id, dept-name )
  'MNT', 'Maintenance'
```

The second form is used when it is necessary to retrieve information from one table and insert it into a second table:

```
INSERT INTO table-name
   ( column-name, column-name... )
   select-expression
```

In this second form of **INSERT**, the select expression is evaluated and zero or more rows in some table are identified. Those rows are then *copied* to the table identified by the **table-name**. Of course, the list of column names is optional, and the system will provide a default list as in the first form. In this form, however, the cardinality of the **select-expression**'s **select-list** must match the number of column names, and the data types of corresponding columns must match. Of course, the **select-expression** might select all columns from the source table, or only certain columns; in any case, the target table's rows are taken from those selected from the source table.

The third form of **INSERT** is used to create a row in a table using the default values for every column in the row:

```
INSERT INTO table-name
   (column-name, column-name... )
   DEFAULT VALUES
```

In this last form, a single row is inserted into the identified table, with every column in that row taking its default value (which, of course, is null if no explicit default value has been defined).

53.5 Updating Information in a Database

In addition to retrieving information from a database and inserting new information into it, real applications require that data already in a database be modified. In fact, after retrieval, updating information is probably the most common operation performed on SQL databases.

SQL provides two categories of UPDATE statements—one for set-oriented update operations, and a second for cursor-oriented updates. The first category, sometimes called a *searched update* because of its self-contained operation of locating and updating rows in tables, exemplifies SQL's set-oriented nature. By using the searched update statement, an application is able to change many rows of a table with one statement—without the programmer having to write a loop of any sort. The SQL implementation takes all responsibility for determining the most appropriate ways to access the rows, taking into consideration issues like their location on disk, physical access paths, and so forth. SQL shows its nature as a *declarative* language, rather than a procedural one, in statements like this one.

The format of the searched **UPDATE** is

```
UPDATE table-name
         SET column-name = update-value, column-name = update-value...
     WHERE predicates
```

The **WHERE** clause in the searched **UPDATE** is optional; if it is omitted, then *all* rows of the table are affected. The **update-values** can be expressions with a data type suitable for the corresponding column (including scalar subqueries), but they can also be the keyword **NULL** or the keyword **DE-FAULT**. In the cases of **NULL** or **DEFAULT**, the corresponding column in each identified row is set to null or to its default value (which, of course, might itself be null). When an expression is used for an

update-value, the expression can use values in the row being updated. For example,

```
UPDATE employees
    SET salary = salary * 1.05
  WHERE dept_id = 'ENG'
```

will give a 5% raise in salary to every member of the engineering department (assuming that, unlike our example table, the **EMPLOYEES** table actually has a **SALARY** column).

Another example illustrates the use of literals and subqueries as update values:

```
UPDATE employees
    SET emp_name = 'Fred',
            dept_id = ( SELECT dept_id
                        FROM departments
                        WHERE dept_name = 'Engineering' )
    WHERE emp_id = :eid
```

In this example, the application provides the employee identification number through a host variable, changes that employee's name to Fred, and sets its department identifier from a value retrieved from another table using a **SELECT** expression.

The second form of **UPDATE** is called the *positioned update*; the word "positioned" is used to imply that the statements refer to the row on which a cursor is currently positioned. The format of the positioned **UPDATE** statement is

```
UPDATE table-name
    SET column-name = update-value, column-name = update-value...
  WHERE CURRENT OF cursor-name
```

The **WHERE** clause in this form identifies the cursor whose position identifies the row to be updated. (The **table-name** is required even though the cursor clearly identifies the table to be updated: the table on which the cursor's **SELECT** expression is defined; this is nothing more than an eccentricity in SQL's definition.) The same options for update value apply to the positioned **UPDATE** that applied for the searched **UPDATE**.

Using the earlier example of printing information about all employees, the addition of a positioned **UPDATE** statement would permit giving raises to certain employees:

```
/* Sample C program to demonstrate cursor usage */
#include...
char ename[21], dname[21];
int eid;
EXEC SQL DECLARE emp_cursor CURSOR FOR
    SELECT emp_id, emp_name, dept_name
    FROM employees, departments
    ORDER BY dept_name ;
/* Open the cursor in preparation for fetch/print loop */
EXEC SQL OPEN emp_cursor ;
/* Error handling is omitted from this example */
while (1 == 1) {
```

```
    EXEC SQL FETCH FROM emp_cursor
            INTO :eid, :ename, :dname ;
    printf (...);
    if (eid == 200) {
            EXEC SQL UPDATE employees
                SET salary = salary + 500.00
                WHERE CURRENT OF CURSOR ;
    }
    /* if end of data, then get out of the loop */
    if ( SQLSTATE == ''02000'' ) break;
}
EXEC SQL CLOSE emp_cursor ;
```

This example gives Mary a $500 raise (well, it's dollars if that happens to be the currency represented in the **SALARY** column) while printing every employee's ID, name, and department information.

53.6 Removing Information from a Database

Of course, not all data that are put into a database remain there forever; some data become obsolete and must be removed, while other insertion operations are wrong and the incorrect rows must be deleted. SQL provides two forms of the **DELETE** statement, analogous to the two forms of the **UPDATE** statement, to allow applications to remove data from tables.

The first form, called the *searched delete*, allows applications to remove (possibly many) rows from a table according to criteria specified in the statement. The format of the searched **DELETE** statement is

```
DELETE FROM table-name
    WHERE predicates
```

The **WHERE** clause here, as in the searched **UPDATE** statement, is optional; if absent, then all rows in the specified table are deleted:

```
DELETE FROM employees
```

This statement has the unfortunate effect of firing all of the employees. **DELETE** statements without **WHERE** clauses are obviously something to be used with great caution.

If the **WHERE** clause is present, then all rows matching the criteria of the predicates are deleted from the specified table. An example illustrates how all salespeople can be removed from the database:

```
DELETE FROM employees
    WHERE dept_id = 'SLS'
```

The second form of the **DELETE** statement is called the *positioned delete* and deletes from the identified table only the row on which the specified cursor is positioned. Its format is

```
DELETE FROM table-name
    WHERE CURRENT OF cursor-name
```

To illustrate its use, consider the earlier example of printing information about all employees. However, rather than giving Mary a raise, she is to be removed from the database:

```
/* Sample C program to demonstrate cursor usage */
#include...
char ename[21], dname[21];
int eid;
EXEC SQL DECLARE emp_cursor CURSOR FOR
   SELECT emp_id, emp_name, dept_name
   FROM employees, departments
   ORDER BY dept_name ;
/* Open the cursor in preparation for fetch/print loop */
EXEC SQL OPEN emp_cursor ;
/* Error handling is omitted from this example */
while (1 == 1) {
  EXEC SQL FETCH FROM emp_cursor
          INTO :eid, :ename, :dname ;
  printf (...);
  if (eid == 200) {
  EXEC SQL DELETE FROM employees
      WHERE CURRENT OF CURSOR ;
  }
  /* if end of data, then get out of the loop */
  if ( SQLSTATE == ''02000'' ) break;
}
EXEC SQL CLOSE emp_cursor ;
```

53.7 Metadata Management

Manipulation and management of data in an SQL database depend, of course, on the existence of the database. SQL does not specify how a database itself is created; there are simply too many different reasonable (and commercially successful) implementation techniques to encourage standardization on any one or set of them. As mentioned earlier, the SQL standard does not even define the term "database." In the context of the SQL standard, the catalog may be the closest analog to a database. Catalogs contain schemas, including the schema (called the *information schema*) that describes all other schemas (and their contained objects) in the catalog, just as one would expect a single database to have a single information schema in addition to its other schemas. The SQL standard does not provide statements for creating and destroying catalogs, either. It explicitly leaves that to "implementation-defined" means.

However, the standard does provide a **CREATE SCHEMA** statement that allows users to define new schemas, as well as a **DROP SCHEMA** statement to allow the destruction of schemas and their contents. Creation of a schema is normally accompanied by the creation of one or more objects within the schema, such as tables. Schemas may belong to the creator (that is, to the authorization identifier that caused the invocation of **CREATE SCHEMA**) or to a specified authorization identifier. (Authorization identifiers are the way that SQL identifies users of the database; every SQL statement is executed under the privileges of exactly one authorization identifier.)

The format of a **CREATE SCHEMA** is

```
CREATE SCHEMA schema-name
  AUTHORIZATION auth-id
  schema-element...
```

If **schema-name** is not present, then the schema is given a name equivalent to **auth-id**, and if the **AUTHORIZATION** clause is not present, then the schema's owner is the authorization identifier executing the **CREATE SCHEMA**; however, they must not both be missing. The **schema-elements** can be **CREATE TABLE** statements, **CREATE VIEW** statements, or several other alternatives.

A **CREATE TABLE** statement has this syntax:

```
CREATE TABLE table-name (
  table-element, table-element... )
```

Each **table-element** can be a column definition or a table constraint definition. (Incidentally, SQL offers-several sorts of *temporary tables*, tables whose contents are not persistently stored, but which are available only during a single database session; temporary tables are not covered in this summary of the language.)

Column definitions look like this:

```
column-name data-type
  default-clause
  column-constraint
  collation
```

The **column-name** must be unique within the table, and the **data-type** can be either one of SQL's data types or the name of a domain. The other clauses are optional; the **default-clause** provides an explicit default for the column, and a collation instructs the database system how to sort character string columns.

A **column-constraint** is a constraint that gives the database system information about valid values for the column. For example, specification of **NOT NULL** prohibits nulls from appearing the column, while **UNIQUE** means that no two rows in the table are allowed to have the same value in that column.

A **table-constraint** definition also provides information about valid values, but is permitted to cover more than a single column. For example, **NOT NULL (EMP_ID, EMP_NAME)** means that the combination of the **EMP_ID** and **EMP_NAME** columns are prohibited from being null, but either one of them is permitted to be null; similarly, **UNIQUE (EMP_NAME, DEPT_ID)** means that no department is allowed to have two employees with the same name working for it—the combination of **EMP_NAME** and **DEPT_ID** must be unique (note that this does not prohibit an employee from working for more than one department, but it does require that those two departments must have different names). One interesting feature of table constraints is that they are required to be satisfied *for every row in the table*; the unexpected implication of this is that empty tables can never cause a table constraint to fail. Therefore, the table definition

```
CREATE TABLE EMPLOYEES (
  EMP_ID          INTEGER,
  EMP_NAME        CHARACTER VARYING (20),
  DEPT_ID         CHARACTER(3),
  CHECK (SELECT COUNT(*) FROM EMPLOYEES > 5) )
```

will fail if the **EMPLOYEES** table contains one, two, three, or four employees, but will *not* fail if the **EMPLOYEES** table contains no employees at all.

SQL permits constraints to be specified at the schema level, too; these higher-level constraints are called *assertions*, and they are allowed to govern relationships between data in multiple tables. The syntax for a **CREATE ASSERTION** is

```
CREATE ASSERTION assertion-name
   CHECK ( predicates )
```

As with any constraint, an assertion requires that the predicates be satisfied ("true") at the end of every SQL statement; if the predicates are not satisfied, the SQL statement fails.

Tables are dropped if they are not needed. To drop a table from a schema implies eliminating all the columns as well as all the data stored in the table:

```
DROP TABLE table-name
   drop-behavior
```

If **drop-behavior** is **CASCADE**, then all columns in the table are dropped, followed by the destruction of the table. If **drop-behavior** is **RESTRICT**, the presence of existing columns in the table will cause the **DROP** to fail; this provides a sort of safety mechanism to require that applications really mean to drop what they want to drop.

A special sort of constraint, briefly mentioned earlier, is the referential integrity constraint. There are actually two sorts of referential integrity constraints. One provides a *key* for a table, ensuring that the value of a column or set of columns in the table is always unique and never null; this feature is called *primary key*. The syntax can be provided as a column constraint as long as only one column makes up the primary key; a table constraint can be used to make a primary key out of one or more columns:

```
CREATE TABLE employees (
    emp_id          INTEGER PRIMARY KEY,
    emp_name        CHARACTER VARYING(20),
    dept_id         CHARACTER(3) )
```

is equivalent to

```
CREATE TABLE employees (
    emp_id          INTEGER,
    emp_name        CHARACTER VARYING(20),
    dept_id         CHARACTER(3),
    PRIMARY KEY (emp_id) )
```

The second sort of referential integrity constraint specifies the relationship that one table, the *referencing table*, must have to another table, the *referenced table*. The mechanism used is called a *foreign key*:

```
CREATE TABLE employees (
    emp_id          INTEGER PRIMARY KEY,
    emp_name        CHARACTER VARYING(20),
    dept_id         CHARACTER(3) REFERENCES departments )
```

or, using a table constraint,

```
CREATE TABLE employees (
   emp_id          INTEGER PRIMARY KEY,
   emp_name        CHARACTER VARYING(20),
   dept_id         CHARACTER(3),
   FOREIGN KEY (dept_id) REFERENCES departments )
```

If the column in the referenced table (**DEPARTMENTS**, in this case) to which a match is required is not the primary key of that table, then the column name can be included in parentheses after the table name:

```
CREATE TABLE employees (
   emp_id          INTEGER PRIMARY KEY,
   emp_name        CHARACTER VARYING(20),
   dept_id         CHARACTER(3) REFERENCES departments (dept_id) )
```

The semantics of a foreign key (or the **REFERENCES** column constraint) is that every row in the referencing table (the table with the FOREIGN KEY or **REFERENCES** constraint) must correspond to a row in the referenced table that has the same value in the referenced column or columns as the referencing row has in the corresponding columns.

The sort of foreign key constraints shown so far are those that *restrict* the values in the referencing and referenced tables. If an **INSERT** statement attempts to add a row to the referencing table without there being a corresponding row in the referenced table, an error is signaled; conversely, if a **DELETE** statement attempts to delete a row from the referenced table while there remain corresponding rows in the referencing table, an error is signaled.

A more powerful sort of foreign key constraint allows the database designer to specify the actions that the database system must perform in order to make things right when a statement like the just-mentioned **DELETE** or **INSERT**—or an **UPDATE** with a similar effect—is executed. Choices of behavior are **CASCADE** (meaning that the same action is taken on the referencing rows), **SET NULL** (meaning that referencing columns are set to null), and **SET DEFAULT** (meaning that referencing columns are set to their default values):

```
CREATE TABLE employees (
   emp_id          INTEGER PRIMARY KEY,
   emp_name        CHARACTER VARYING(20),
   dept_id         CHARACTER(3)
     REFERENCES departments (dept_id)
     ON DELETE SET NULL )
```

The **ON DELETE SET NULL** clause instructs the database that, if an employee's department is eliminated, then the **DEPT_ID** field for that employee is to be set to null. Similarly,

```
CREATE TABLE employees (
   emp_id           INTEGER PRIMARY KEY,
   emp_name         CHARACTER VARYING(20),
   dept_id          CHARACTER(3)
     REFERENCES departments (dept_id)
     ON UPDATE CASCADE )
```

means that any change to the **DEPT_ID** code in the **DEPARTMENTS** table will cause the records of all employees assigned to that department to be updated to the same new value.

The **CREATE VIEW** statement establishes new views that can derive information from persistent tables and/or from other views. It provides a way to easily capture many types of common queries so that application program **SELECT** statements are simpler. For example, if an application system frequently required the ability to identify employees and their department names, the **SELECT** expression

```
SELECT emp_name, dept_name
  FROM employees AS e, departments AS d
  WHERE e.dept_id = d.dept_id
```

could be captured in a view

```
CREATE VIEW emp_dept_names AS
  SELECT emp_name, dept_name
    FROM employees AS e, departments AS d
    WHERE e.dept_id = d.dept_id
```

The application queries then could simply be

```
SELECT emp_name, dept_name
  FROM emp_dept_names
  WHERE emp_name = :host_variable
```

53.8 Other Aspects of SQL

Two other significant areas deserve some discussion. The first of these is the security structure of SQL. As mentioned earlier, every schema belongs to a single authorization identifier (which will be simply called "a user"). In addition, *every* object in that schema, such as tables and views, belongs to the user who owns the schema. The owner of an object, perhaps obviously, has all possible privileges on that object. That owner may choose to *grant* some or all of those privileges to other users; furthermore, those grants can provide or deny the grantees the ability to make further grants of the privileges to still other users.

Therefore, if an authorization identifier **MANAGEMENT** is the owner of the **EMPLOYEES** table, the following SQL statement might be useful:

```
GRANT SELECT ON employees TO supervisor
```

Assuming that **SUPERVISOR** identifies another user, then that user will be able to execute **SELECT** expressions (and statements) that reference the **EMPLOYEES** table; however, that user will *not* be able to hire, fire, or modify employees unless **INSERT**, **DELETE**, or **UPDATE** privileges are also granted.

Once granted, privileges can also be revoked:

```
REVOKE SELECT ON employees FROM supervisor
```

This statement will make **SUPERVISOR** unable to execute a **SELECT** that references the **EMPLOY-EES** table. Only the grantor of a privilege is able to revoke it.

The other area that must be covered is the notion of errors: signaling them and handling them. SQL defines two *status variables* or *status parameters*, called **SQLCODE** and **SQLSTATE**. **SQLSTATE** is the recommended variable to use because it is more thoroughly standardized. **SQLSTATE** is a five-character variable. The first two characters, called the *class code*, help identify the principal cause of any errors that arise during execution of an SQL statement; for example, **07** is a class code meaning "data exception." The last three characters, called the *subclass code*, further identify the cause of an error; for example, **07012** means "data exception—division by zero." The characters in a class code or subclass code must be taken from the digits (0–9) and the 26 basic upper-case letters of the Latin alphabet (A–Z). If a class code value begins with 0–4 or A–H, then it is defined in the SQL standard or a closely related standard; if it begins with 5–9 or I–Z, then it is defined by the SQL implementation that returned the error. For class codes in the range defined by the standard, the same rules apply to the subclass code value's first letter; however, for class codes in the implementation-defined range, all subclass code values other than **000** are also implementation-defined (**000** means "no subclass code specified" in all cases).

Embedded SQL programs have one additional way to handle errors—a **WHENEVER** statement. The format of **WHENEVER** is

```
WHENEVER condition action
```

The **condition** is either **SQLERROR** or **NOT FOUND**. **SQLERROR** means to take the **action** whenever an **SQLSTATE** value with a class code other than **00** (successful completion), **01** (warnings), or **02** (no data) results from an SQL statement's execution. The **action** can be either **CONTINUE** (meaning "go on to the next statement") or **GOTO** (also spelled **GO TO**, it means to change the program's execution flow by branching to the specified label or statement number). The condition **NOT FOUND** means that the specified action is to be taken if the resulting **SQLSTATE** value is **02** (no data). It is worth mentioning that many implementations also provide **SQLWARNING** as an acceptable condition in this statement.

An unusual aspect of the **WHENEVER** statement is that it is *declarative* and not truly executable. In an application program, the behavior specified by a **WHENEVER** statement applies to every SQL statement *lexically* following the **WHENEVER**, until an overriding **WHENEVER** statement is encountered. Consider the cursor example earlier, with some **WHENEVER** statements inserted:

```
/* Sample C program to demonstrate cursor usage */
#include...
char ename[21], dname[21];
int eid;
EXEC SQL DECLARE emp_cursor CURSOR FOR
   SELECT emp_id, emp_name, dept_name
   FROM employees, departments
   ORDER BY dept_name ;
EXEC SQL WHENEVER SQLERROR GOTO error_place ;
/* Open the cursor in preparation for fetch/print loop */
EXEC SQL OPEN emp_cursor ;
while (1 == 1) {
  EXEC SQL FETCH FROM emp_cursor
           INTO :eid, :ename, :dname ;
  printf (...);
  EXEC SQL WHENEVER SQLERROR CONTINUE ;
  if (eid == 200) {
    EXEC SQL DELETE employees
            WHERE CURRENT OF CURSOR ;
```

```
    }
    /* if end of data, then get out of the loop */
    if ( SQLSTATE == ''02000'' ) break;
}
EXEC SQL CLOSE emp_cursor ;
error_place:
```

The first **WHENEVER** statement instructs the system to go to **error_place** whenever any errors occur on the statements that *lexically* follow—until the second **WHENEVER** statement overrides it by saying to simply **CONTINUE** . That means that the **FETCH** statement in the loop will always cause a branch to **error_place** if an error occurs, while the **UPDATE** statement in the same loop will not branch to **error_place**, but will ignore the error and continue execution.

Defining Terms

The SQL standard defines a number of important terms. Among the terms most central to SQL are these:

Catalog: A catalog is a named collection of schemas. The name of a catalog is used to qualify the names of the schemas in that catalog.

Column: A column is a multiset of values that may vary over time. All values of the same column are of the same data type and are values in the same table. A value of a column is the smallest unit of data that can be selected from a table or updated in a table.

Constraint: A constraint is a rule that defines valid states of data in an SQL database by constraining the values of data, or the relationships between values of data, in tables.

Embedded SQL: Embedded SQL is a technique of placing SQL statements in programs written in one of several more conventional programming languages, thus producing the effect of a conventional program "calling" procedures written in SQL alone.

Module: A module is a unit of SQL language, containing declarations of certain SQL entities as well as one or more procedures, each containing a single SQL statement.

Procedure: A procedure is a subroutine written in SQL, contained in a module, that is invoked by means of a *call* from a program written in a conventional programming language.

Referential integrity: Referential integrity places requirements on data so that values stored in one or more columns of rows of one table, called the referencing table, are identical to values stored in corresponding columns of rows of another table, called the referenced table.

Row: A row is an instance of a row type. A row type is a sequence of {name, datatype} pairs. A row is the smallest unit that can be inserted into or deleted from a table.

Schema: A schema is a named conceptual entity that contains tables, constraints, and other SQL objects. In SQL, data are considered to be associated with, but distinct from, schemas. The name of a schema is used to qualify the names of the objects in that schema.

SQL statement: An SQL statement is the smallest executable unit of SQL. SQL statements are used to create, alter, and remove metadata objects, as well as to insert, update, and delete data in tables and to retrieve data from tables.

Subquery: A subquery is an expression that identifies data from tables. Subqueries are used as expressions in the context of SQL statements.

Table: A table is a collection of rows. Every row in a table has the same row type. The *degree* of a table is the number of columns that it has. The *cardinality* of a table is the number of rows in that table.

View: A view is a *virtual table*—a table that is not physically stored in the database, but whose contents are derived only when needed by an SQL statement.

References

ANSI 1986. *ANSI X3. 135-1986, American National Standard for Information Systems—Database Language SQL*. American National Standards Institute.

ANSI 1989a. *ANSI X3. 135-1989, American National Standard for Information Systems—Database Language SQL with Integrity Enhancement*. American National Standards Institute.

ANSI 1989b. *ANSI X3. 168-1989, American National Standard for Information Systems—Database Language Embedded SQL*. American National Standards Institute.

ANSI 1992. *ANSI X3. 135-1992, American National Standard for Information Systems—Database Language SQL*. American National Standards Institute.

Cannan, S. and Otten, G. 1993. *SQL—The Standard Handbook*. McGraw–Hill.

Chamberlin, D. D. and Boyce, R. F. 1974. SEQUEL: a structured English query language, pp. 249–264. In *Proc. ACM SIGFIDET Workshop* (May).

Codd, E. F. 1974. A relational model of data for large shared data banks. *Commun. ACM*. 13(6):377–387.

Date, C. J. and Darwen, H. 1993. *A Guide to the SQL Standard*, 3rd ed. Addison–Wesley.

ISO 1987. *ISO 9075:1987, Database Languages—SQL*. International Organization for Standardization.

ISO 1989. *ISO/IEC 9075:1989, Information technology—Database languages—SQL*. International Organization for Standardization.

ISO 1992. *ISO/IEC 9075:1992, Information technology—Database languages—SQL*. International Organization for Standardization.

ISO 1995a. *ISO/IEC JTC1/SC21 N9462 through N9467, Working Draft 9075, Information technology—Database languages—SQL, Parts 1 through 6*. International Organization for Standardization, 1995.

ISO 1995b. *ISO/IEC 9075-3:1995, Information technology—Database languages—SQL, Part 3, Call-Level Interface (SQL/CLI)*. International Organization for Standardization.

ISO 1996. *ISO/IEC DIS 9075-4, Information technology—Database languages—SQL, Part 4, Persistent Stored Modules (SQL/PSM)*. International Organization for Standardization.

Melton, J. and Simon, A. R. 1993. *Understanding the New SQL: A Complete Guide*. Morgan Kauffman.

X/Open 1992. *CAE Specification—Structured Query Language (SQL)*. X/Open Company Ltd.

Further Information

The most popular *de facto* standard for SQL can be found in X/Open [1992].

Three popular books address the SQL-92 standard, covering programming in SQL, providing critiques of various language aspects, and explaining the language in detail: Melton and Simon [1993], Date and Darwen [1993], and Cannan Otten [1993].

VI

Graphics

Donald H. House, Section Adviser
Texas A&M University

COMPUTER GRAPHICS is the study and realization of complex processes for representing physical and conceptual objects visually on a computer screen. These processes include the internal modeling of objects, rendering, hidden surface elimination, color, shading, projection, and representing motion. Fundamental to all graphics applications are the processes of modeling and rendering. The reconstruction of scanned images is another major area.

54

Overview of Three-Dimensional Computer Graphics

Donald H. House
Texas A&M University

54.1 Introduction

The name *three-dimensional computer graphics* has been used freely in the computer graphics community for many years now [Foley et al. 1990, Glassner 1995, Hill 1990, Rogers 1985, Watt and Watt 1992]. It is something of a misnomer, because the graphics themselves are not in any sense three-dimensional (3D). Rather, the way that the graphics are generated is dependent upon the construction of a virtual 3D model in the computer, which is then imaged via a virtual camera, usually implying a simulation of a real physical illumination process. The term *three-dimensional* merely emphasizes the fact that a simulation of a 3D world underlies the image-making process and also that the images produced often display the kinds of foreshortening distortions apparent in photographs or perspective drawings of real 3D scenes. This chapter is devoted to outlining the various aspects of the process of generating 3D computer graphic images. It is meant to give the reader an overview, or "big picture," that can be filled in by reading Chapters 55 through 62 of the Handbook, which provide more detailed information on specific aspects of the process.

54.2 Organization of a Three-Dimensional Computer Graphics System

A three-dimensional graphics system can be thought of as having three major components, each of which performs a distinct and clearly defined key role in the process of image generation. These three components are responsible for *scene specification, rendering,* and *image storage and display.* Figure 54.1 gives a schematic view of the process used in 3D graphics, showing the role that each of these components plays. Each of these major components can itself be broken down into groups of important subcomponents.

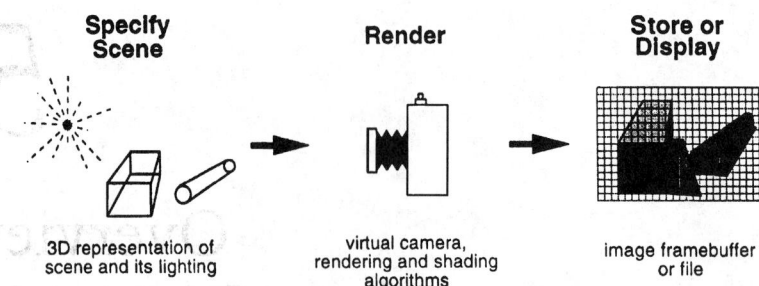

FIGURE 54.1   The 3D graphics process.

Scene Specification

The *scene specification* section of a 3D graphics system is responsible for providing an internal representation of the virtual scene that is eventually to be imaged. This requires both an interface to allow user specification and modification of the scene definition and a set of internal data structures that store and organize the scene so that it can be accessed by the user interface and the rendering system. This can be a highly interactive program, providing access to a variety of modeling tools via an interactive user interface; it may be script-driven, providing a scene description language that the user communicates in; or it can be as simple as a program that reads basic geometric information from a tightly formatted scene description file. In any case, the scene specification system will need to support some concept of a geometric coordinate system and provide some way of describing the geometry of the scene to be imaged. Scene description systems also will provide a way in which the user can specify what (virtual) materials objects are made of and how the scene is lit.

Coordinate Systems

Key to the geometric structure of a 3D graphics system is a compact means for storing and utilizing descriptions of **local coordinate systems**. The local coordinate systems are used in the definition of the various components of a model describing the geometry and other characteristics of the scene, much as the local coordinates used on a plan are used in describing the design of a real object. For example, the coordinates on the plan for a complete airplane will necessarily be much different from the coordinates used on the plan for the airplane's wheel assembly.

Consistent with the usual representation of 3D coordinates in mathematics and engineering, most current books, articles, and implementations of 3D graphics systems use right-handed coordinate systems [Foley et al. 1990]. This gives a natural organization with respect to the display screen, with the x-coordinate measuring horizontal distance across the screen, the y-coordinate measuring vertical distance up the screen, and the z-coordinate providing the third spatial dimension as distance in front of the screen. However, in the early development of computer graphics, coordinate systems were often left-handed [Foley and van Dam 1982]. In screen space, the difference is that the positive z or depth coordinate is measured into the screen. Of course, for modeling, a local coordinate system can be positioned and oriented anywhere in space and is not usually aligned with the screen. Figure 54.2 shows the ordering of right-handed and left-handed coordinate systems.

A local coordinate system is usually defined in terms of a small set of intuitive

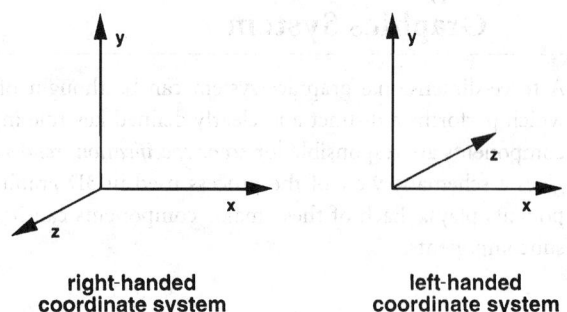

FIGURE 54.2   Right- and left-handed coordinate systems.

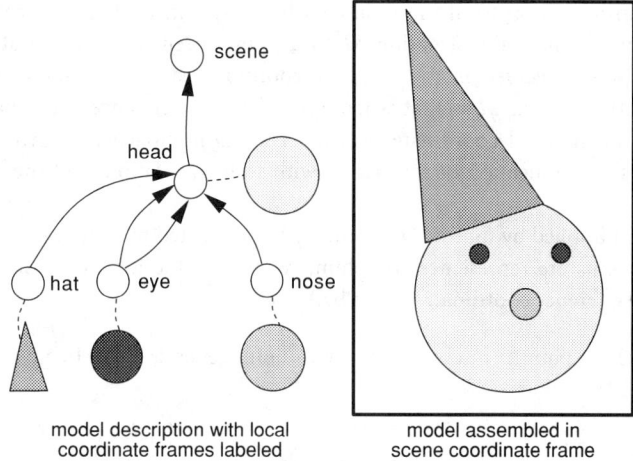

model description with local
coordinate frames labeled

model assembled in
scene coordinate frame

FIGURE 54.3 A clown described by a hierarchy of local coordinate frames.

geometric operations—the **affine transformations**. These are

1. translation—a change in the position of the origin of the local system,
2. scaling—a change in the scale of measurement in the local system,
3. rotation—a change in the orientation of the local system, and
4. shear—transformation from an orthogonal coordinate system to a nonorthogonal system or vice versa via shearing deformations.

All of these elements of the local-coordinate-system definition are specified with respect to the origin, scale, orientation, and shear of some external reference system, which might itself be specified relative to some other external system. In this way, local coordinates can be nested within each other, providing the possibility for models to be described in a **hierarchical** fashion. For example, the simple clown model of Fig. 54.3 is described in terms of a hierarchy of coordinate frames, which allows for the design and modeling of the head, eye, nose, and hat in their own separate local coordinate frames but then places the two eyes, the nose, and the hat on the head with respect to the head's frame. Finally, the assembled head is placed and oriented in the scene with respect to a local reference frame for the scene. The reference frame at the top of such a hierarchy is usually referred to as the **global coordinate system**.

The basic geometric unit is the 3D point, which is typically represented in a 3D graphics system as a 3-vector and stored as a an array of three elements, representing the x, y, and z components of the point. Orientation vectors, like normals to surfaces and directions in space, are also represented by 3-vectors. Thus, the point (x, y, z) is given by the vector

$$\begin{bmatrix} x \\ y \\ z \end{bmatrix}$$

(or in some systems by its transpose $[x \ y \ z]$).

A local-coordinate-system is usually specified by a four-dimensional (4D) *homogeneous transformation matrix* of the form

$$M = \begin{bmatrix} a_{11} & a_{12} & a_{13} & a_{14} \\ a_{21} & a_{22} & a_{23} & a_{24} \\ a_{31} & a_{32} & a_{33} & a_{34} \\ 0 & 0 & 0 & 1 \end{bmatrix}$$

which specifies a transformation from the local coordinate system to its reference coordinate system. In other words, applying the transformation M to a point specified in the local coordinate system will yield the same point specified in the reference coordinate system. Another way of thinking of the same transformation matrix M is that when applied to the reference coordinate system it aligns it with and scales it to the local coordinate system. The 4D homogeneous form of the transformation matrix M allows the unification of translation with scaling, rotation, and shear in a single matrix representation.

The transformation implied by matrix M is actually implemented by a three-step process. Assuming that 3D geometric points are represented as column vectors in the local coordinate system, they are transformed into the reference coordinate system by:

1. extending the 3D point **p** into a 4-vector **v** in homogeneous space by giving it a fourth, or w, coordinate of 1:

$$\mathbf{p} = \begin{bmatrix} x \\ y \\ z \end{bmatrix} \implies \mathbf{v} = \begin{bmatrix} x \\ y \\ z \\ 1 \end{bmatrix}$$

2. premultiplying this extended vector by the matrix M yielding a transformed 4-vector $\mathbf{v}'$.

$$M\mathbf{v} = \mathbf{v}' = \begin{bmatrix} x' \\ y' \\ z' \\ 1 \end{bmatrix}$$

3. converting the resulting 4-vector $\mathbf{v}'$ into the transformed 3D point $\mathbf{p}'$ by discarding its w-coordinate:

$$\mathbf{v}' = \begin{bmatrix} x' \\ y' \\ z' \\ 1 \end{bmatrix} \implies \mathbf{p}' = \begin{bmatrix} x' \\ y' \\ z' \end{bmatrix}$$

Inspection of matrix M will show that it is defined to always send the original w-coordinate to itself, thus making the third step legitimate. (In earlier computer graphics systems, it was usual for points to be represented by row vectors instead of column vectors and for step 2 to be done by *postmultiplying* the homogeneous row vector **v** by the transpose of matrix M.)

The basic transformations of translation, rotation, scaling, and shear are given by the following matrices, which assume that points are represented as column vectors in a right-handed coordinate system and that transformations will be done by premultiplication of the vector (extended to homogeneous coordinates) by the matrix. Use of left-handed coordinates instead of right-handed coordinates will affect the rotations only as indicated below. If row vectors and postmultiplication are being used to represent points and their transforms, the matrices must be transposed.

- *Translation* by Δx in the x-direction, Δy in the y-direction, and Δz in the z-direction:

$$T(\Delta x, \Delta y, \Delta z) = \begin{bmatrix} 1 & 0 & 0 & \Delta x \\ 0 & 1 & 0 & \Delta y \\ 0 & 0 & 1 & \Delta z \\ 0 & 0 & 0 & 1 \end{bmatrix}, \quad T(\Delta x, \Delta y, \Delta z) \begin{bmatrix} x \\ y \\ z \end{bmatrix} = \begin{bmatrix} x + \Delta x \\ y + \Delta y \\ z + \Delta z \end{bmatrix}$$

- *Scaling* of s_x in the x-direction, s_y in the y-direction, and s_z in the z-direction:

$$S(s_x, s_y, s_z) = \begin{bmatrix} s_x & 0 & 0 & 0 \\ 0 & s_y & 0 & 0 \\ 0 & 0 & s_z & 0 \\ 0 & 0 & 0 & 1 \end{bmatrix}, \quad S(s_x, s_y, s_z) \begin{bmatrix} x \\ y \\ z \end{bmatrix} = \begin{bmatrix} s_x x \\ s_y y \\ s_z z \end{bmatrix}$$

- *Rotation* through angle θ around the x, y, or z axis, with right-handed rotation around the axis taken as a positive rotation (i.e., aligning the thumb of the right hand with the axis, the fingers grasp the axis in the direction of positive rotation; note that if left-handed coordinates are being used, the signs of the *sine* terms in R_x and R_y should be reversed, but R_z is unaffected):

$$R_x(\theta) = \begin{bmatrix} 1 & 0 & 0 & 0 \\ 0 & \cos\theta & -\sin\theta & 0 \\ 0 & \sin\theta & \cos\theta & 0 \\ 0 & 0 & 0 & 1 \end{bmatrix}, \quad R_x(\theta) \begin{bmatrix} x \\ y \\ z \end{bmatrix} = \begin{bmatrix} x \\ y\cos\theta - z\sin\theta \\ y\sin\theta + z\cos\theta \end{bmatrix}$$

$$R_y(\theta) = \begin{bmatrix} \cos\theta & 0 & \sin\theta & 0 \\ 0 & 1 & 0 & 0 \\ -\sin\theta & 0 & \cos\theta & 0 \\ 0 & 0 & 0 & 1 \end{bmatrix}, \quad R_y(\theta) \begin{bmatrix} x \\ y \\ z \end{bmatrix} = \begin{bmatrix} x\cos\theta + z\sin\theta \\ y \\ -x\sin\theta + z\cos\theta \end{bmatrix}$$

$$R_z(\theta) = \begin{bmatrix} \cos\theta & -\sin\theta & 0 & 0 \\ \sin\theta & \cos\theta & 0 & 0 \\ 0 & 0 & 1 & 0 \\ 0 & 0 & 0 & 1 \end{bmatrix}, \quad R_z(\theta) \begin{bmatrix} x \\ y \\ z \end{bmatrix} = \begin{bmatrix} x\cos\theta - y\sin\theta \\ x\sin\theta + y\cos\theta \\ z \end{bmatrix}$$

- *Shear* parallel to the (x, y) plane as a function of z, or parallel to the (y, z) plane as a function of x, or parallel to the (z, x) plane as a function of y:

$$H_{xy}(h_{xz}, h_{yz}) = \begin{bmatrix} 1 & 0 & h_{xz} & 0 \\ 0 & 1 & h_{yz} & 0 \\ 0 & 0 & 1 & 0 \\ 0 & 0 & 0 & 1 \end{bmatrix}, \quad H_{xy}(h_{xz}, h_{yz}) \begin{bmatrix} x \\ y \\ z \end{bmatrix} = \begin{bmatrix} x + h_{xz}z \\ y + h_{yz}z \\ z \end{bmatrix}$$

$$H_{yz}(h_{yx}, h_{zx}) = \begin{bmatrix} 1 & 0 & 0 & 0 \\ h_{yx} & 1 & 0 & 0 \\ h_{zx} & 0 & 1 & 0 \\ 0 & 0 & 0 & 1 \end{bmatrix}, \quad H_{yz}(h_{yx}, h_{zx}) \begin{bmatrix} x \\ y \\ z \end{bmatrix} = \begin{bmatrix} x \\ y + h_{yx}x \\ z + h_{zx}x \end{bmatrix}$$

$$H_{zx}(h_{xy}, h_{zy}) = \begin{bmatrix} 1 & h_{xy} & 0 & 0 \\ 0 & 1 & 0 & 0 \\ 0 & h_{zy} & 1 & 0 \\ 0 & 0 & 0 & 1 \end{bmatrix}, \quad H_{zx}(h_{xy}, h_{zy}) \begin{bmatrix} x \\ y \\ z \end{bmatrix} = \begin{bmatrix} x + h_{xy}y \\ y \\ z + h_{zy}y \end{bmatrix}$$

More complex coordinate transformations, involving combinations of the basic transformations, can be obtained by specifying them as a sequence of operations, each described by a 4D transformation matrix. The product of these matrices will yield a compound transformation matrix that has the same effect as each transformation applied separately. For example, a rotation of 30° around the x axis followed by a translation to the point $(10, 20, -10)$ would be given by

$$M = T(10, 20, -10)\,R_x(30°)$$

Geometric Modeling

Virtually all 3D graphics systems provide the ability to work with simple geometric primitives that can be specified as lists of 3D points. These primitives include points, lines, and polygons. Points can be arranged together to indicate a *sampled* surface, lines to form a *wireframe* representation, and polygons to form *polyhedral* surfaces. More sophisticated modelers will provide **parametric surfaces**, which are defined via an underlying piecewise polynomial formulation [Rogers and Adams 1990, Bartels et al. 1987]. Polynomial coefficients are adjusted to give the surface a specific shape, and these coefficients are often given intuitive form by encoding them via simple geometric devices, such as control polyhedra.

A typical surface formulation is a *biparametric surface*, which describes a surface in three spatial dimensions (x, y, z) via a set of three functions of two parameters u and v:

$$x = X(u, v), \qquad y = Y(u, v), \qquad z = Z(u, v)$$

This concept is illustrated in Fig. 54.4. The rectangular grid on the left of the figure, defined in the (u, v) parametric coordinate system, is mapped, via the functions X, Y, and Z, into a 3D surface like that shown on the right. A set of points on a parametric surface can be obtained algorithmically by looping over a collection of sample points on the (u, v) plane. Simple geometric primitives and parametric surfaces are described more fully in Chapter 55—Geometric Primitives.

Implicit surfaces are a common alternative to parametric surfaces. Here, surfaces are defined as the set of points satisfying a mathematical expression of the form

$$F(x, y, z) = 0$$

Thus, these surfaces are defined implicitly. Any point (x, y, z) in 3D space can be tested to determine whether or not it is above $[F(x, y, z) > 0]$, below $[F(x, y, z) < 0]$, or on $[F(x, y, z) = 0]$ the surface. However, it is not generally easy to algorithmically generate a set of points guaranteed to be on the surface, without resorting to iterative search or relaxation techniques. Implicit formulations are especially useful for defining solids, where the implicit equation can be evaluated to determine whether a point is inside, outside, or on the surface of the solid. For example, the well-known equation for a sphere of radius r centered at the point (x_0, y_0, z_0) can be written implicitly as

$$(x - x_0)^2 + (y - y_0)^2 + (z - z_0)^2 - r^2 = 0$$

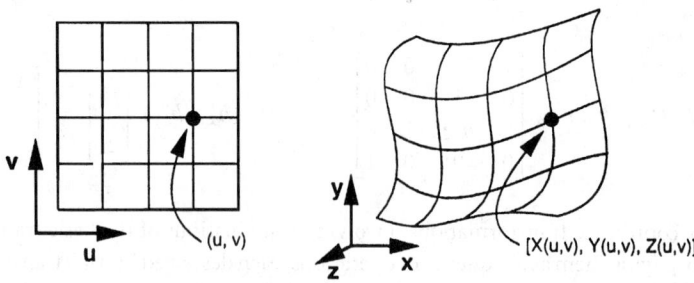

FIGURE 54.4   A biparametric surface.

Implicit techniques can be generalized to describe surfaces or solids defined algorithmically in the form of a programmed function. This technique is useful in describing a variety of natural-looking forms [Ebert 1994]. Functional techniques are described in Chapter 57—Advanced Geometric Modeling.

Some geometric data come naturally in the form of a set of scalar values distributed in a 3D field. For example, data from medical scanners, such as MR or CT devices, consist of a set of material density values distributed on a regular lattice within a volume of space. This type of data has its own specialized set of modeling and visualization techniques that are described thoroughly in Chapter 60—Volume Visualization.

Materials

In the context of a 3D graphics system, a **material** is an attribute of a geometric object that provides a description of how the surface of the object will appear when viewed from a particular direction under a particular illumination. In physical terms, what we need to define here is how a surface reflects (or transmits) light as a function of incident angle, reflection (or refraction angle), and wavelength. A function providing these relationships is known as the material's **bidirectional reflectance distribution function** or BRDF.

In practical terms for computer graphics applications, it is usually enough to approximate the BRDF for a material with a collection of parameters and maps. A usual material specification system will provide parameters for the specification of a material's color, diffuse reflectance factor, specular reflectance factor, specularity, transmissivity, and refraction index. These factors and their use in lighting calculations are described in detail in Chapter 57—Mainstream Rendering Techniques.

A material specification will also often include the capability to provide both **texture maps** and **bump maps**. A texture map provides a pattern of color that is to be applied to the surface of an object during the rendering process. These can be anything from a digital image that will be projected onto the surface, to a regular geometric pattern like a checkerboard or polka-dot design. A bump map is a pattern of perturbations to the normal vector to a surface that simulates the effect that bumps would have on the appearance of the

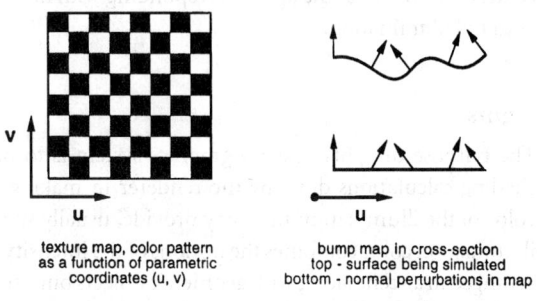

texture map, color pattern
as a function of parametric
coordinates (u, v)

bump map in cross-section
top - surface being simulated
bottom - normal perturbations in map

FIGURE 54.5 Texture and bump maps.

surface. Figure 54.5 illustrates texture and bump maps. Some systems also provide for **displacement maps**, which are used to locally perturb both the surface normal and the surface itself. It is usual to relate texture and other map coordinates directly to the parametric coordinates of a parametric surface. For nonparametric surfaces, specific texture coordinates must be provided by the user or by some algorithmic technique. For example, many modelers allow the user to provide texture coordinates along with 3D geometric coordinates for each vertex in a polygonal surface.

Within the field of computer graphics, color is becoming as complex a topic as it is in physics, psychology, and art. However, from the point of view of usual practice, color is most often represented in 3D graphics systems in one of two related color systems, the *RGB* and the *HSV* systems.

The RGB or "red–green–blue" system is the usual system used for storage or display of color. This is because this color system relates directly to the three-electron-gun RGB organization of color CRT displays. An RGB color is stored as a triple of three numbers giving the relative amount of each of the three color primaries—red, green, and blue. Because the RGB system organizes color into three *primaries*, and allows us to scale each primary independently, we can think of all of the colors that are represented by the system as being organized in the shape of a cube, as shown in Fig. 54.6. We call this the RGB color cube or the RGB color space (when we add coordinate axes to measure R, G, and B levels). Note that the corners of the RGB color cube represent pure black and pure white; the three primary colors red, green, and blue; and the three secondary colors yellow, cyan, and magenta. The diagonal from the black corner to the white corner represents all of the gray levels. Other locations within the cube correspond with all of the other colors that can be displayed.

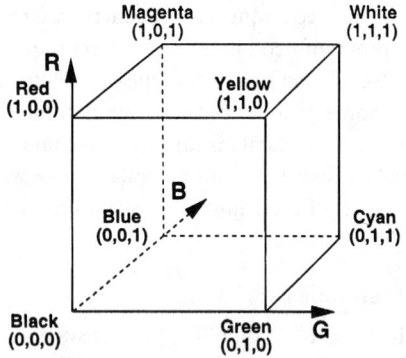

FIGURE 54.6   RGB color cube.

The HSV or "hue–saturation–value" color system represents colors using three measures that relate directly to how artists often think about color. It provides separate measures of *hue* (corresponding to dominant color name), *saturation* (purity of color), and *value* (brightness on gray scale). Its structure is derived directly from the RGB system, and in fact there is a simple translation from RGB to HSV and back. If the RGB color cube of Fig. 54.6 is viewed along its white–black diagonal, it presents a hexagonal silhouette. "Peeling" off layers of the face of the RGB cube visible along the white–black diagonal and projecting these cross sections onto a flat surface result in a series of smaller and smaller hexagonal cross sections. If these cross sections are then stacked up, they form a hexagonal cone. The HSV system is the cone-shaped space derived in this way, shown in Fig. 54.7. Figure 54.8 shows how the coordinates of the HSV color space are organized. The hue or h coordinate is an angular measurement (usually in degrees) around the face of the cone, with red at 0°, green at 120°, and blue at 240°. The saturation or s-coordinate is measured from the center of the face of the cone out to its perimeter, with 0 at the center corresponding with gray and 1 at the perimeter corresponding with a fully saturated color of the chosen hue. The value or v-coordinate is measured from the apex of the cone to its face along the central axis, with 0 at the apex corresponding with no illumination (black) and 1 at the face corresponding with full illumination.

Lights

The purpose of lights in a 3D graphics system is to provide the illumination source for the simulated shading calculations done by the renderer in making an image. Thus, all light sources must define a color of the illumination that they provide, usually specified in RGB or HSV coordinates. Note that this illumination color combines the notions of the intensity of the light and its chromatic attributes. Lights are arranged in a scene along with geometric objects but usually carry no geometric properties other than their position and, for directional lights, a direction of orientation. Some rendering algorithms work with light sources with other geometric properties such as shape and area, but most 3D graphics systems work with only two types of lights—*infinite lights* and *point lights*. Figure 54.9 illustrates infinite and point light sources and some of their variants.

An infinite light is one that is so far away from the geometry being illuminated that light rays can be assumed to be parallel to each other, like the rays of light from the sun illuminating the earth. For this type of light, no position needs to be specified. All that matters geometrically is the direction of the light rays.

A point light source is assumed to have no area, with light emitted in all directions from the geometric position of the light. Simple variants of point light sources include the addition of conic or other shading devices with the light specification, so that the light shines only in a particular direction. A further variation is to have the intensity of light rays fall off gradually as a function

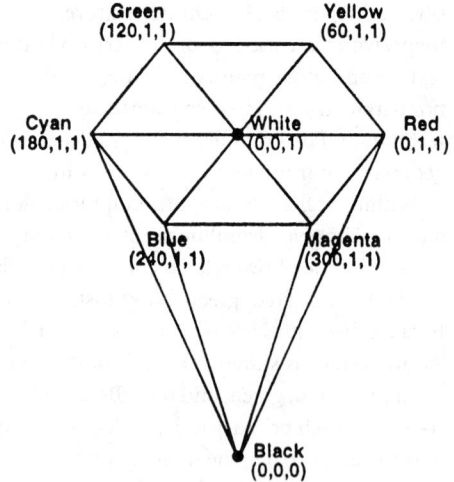

FIGURE 54.7   HSV color cone.

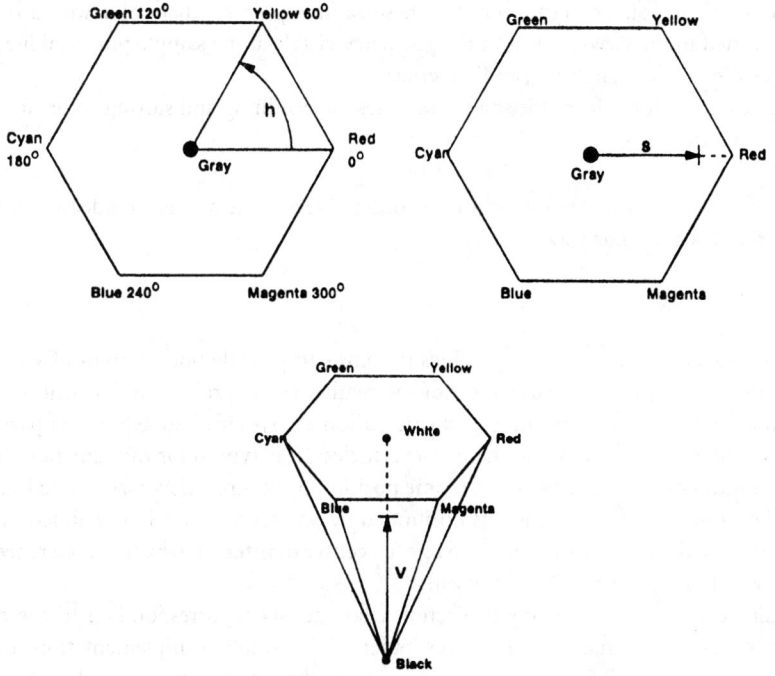

FIGURE 54.8 HSV parameterization of hue, saturation, and value.

of angular distance from the central directional axis of the light. With these variations, a point light can provide a reasonable approximation to an unshaded incandescent bulb, a shaded desk or studio lamp, a flashlight, or a spotlight.

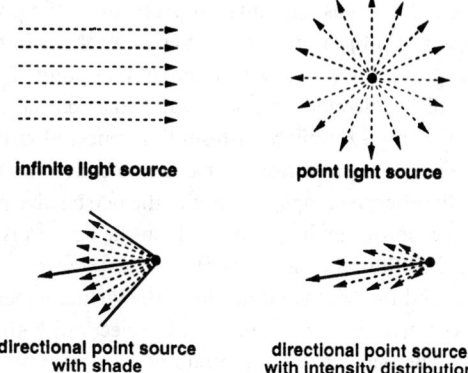

FIGURE 54.9 Infinite and point light sources and variants.

Rendering

Rendering is simply the process of transforming a 3D scene description into a two-dimensional (2D) image. It is generally done by a simulation of the physical process that occurs in a camera when a picture is recorded on film. Making this process algorithmic, so that it can be simulated efficiently and accurately in a computer, is the essence of the rendering problem. Chapter 57—Mainstream Rendering Techniques—describes practical approaches to rendering in some detail, so a brief synopsis will suffice here.

Briefly, the main steps in the rendering process are:

1. point of view—orienting the 3D scene as if it were being viewed from a particular point in space,
2. projection—associating points in the 3D scene with their images on a 2D virtual image plane or screen by projecting the 3D scene description onto the plane,
3. visible-surface determination—deciding which surfaces projected onto the image plane would actually be visible from the present viewpoint,
4. sampling—fixing a set of sample points across the virtual image plane, usually corresponding in some way with the pixels that will be used to store the calculated image, and associating these sample points with visible points on the scene's 3D geometry,

5. shading calculation—determining, for these sample points, what color would be reflected or transmitted to the viewpoint from the geometry visible at the sample point, taking into account the scene's geometry, lighting, and materials,

6. image construction—from the shaded samples, determining and storing colors for each pixel in the output image.

These steps are not necessarily completed in this order. Nevertheless, every renderer will have to solve each of these problems in some way.

Camera

The role of the virtual camera in a 3D graphics system is to provide both a point of view from which to render an image and the basic parameters of the mathematical projection that will be used to form the virtual image. The camera's position and orientation are specified sometimes as part of the scene description system and sometimes elsewhere. Nevertheless, it is typical for the camera to be positioned in the global coordinate system, usually with some positioning controls that correspond to the operation of a real studio camera. Often, a camera is positioned by one set of controls and aimed by another set. Aiming is usually made easier by the option to select a **center of interest**, which is a reference point in the scene toward which the camera will orient itself.

Theoretically, cameras can have any projection characteristics, corresponding to the entire variety of lens types, either real or imagined. However, practical 3D graphics implementations usually implement only the standard parallel or perspective projections that are common in architectural and design drafting.

A perspective projection is one in which all light rays coming from the scene converge at a common point, known as the **center of projection**. If a projection plane is interposed between the scene and the center of projection, the point at which a ray from the scene through the center of projection intersects the projection plane is the *image* of that point. Figure 54.10 shows the geometry of this projection. The parallel projections can be considered as special cases of the perspective projections, where the center of projection is infinitely far from the scene and virtual screen. Thus, rays between points on the scene and the center of projection are parallel to each other when intersecting the screen.

In order to completely specify the perspective projection, the position of the camera (i.e., the center of projection), the direction in which the camera is aimed (i.e., its central ray of projection), the camera's up direction, the distance of the virtual screen from the center of projection along the central projection ray, and the width and height of the virtual screen must be known. This assumes that the virtual screen is centered on the central ray of projection, with its surface normal aligned with the central ray (i.e., it is perpendicular to the central ray). It is also possible to build fancier cameras, where the screen can be moved off center and oriented skewed to the central ray.

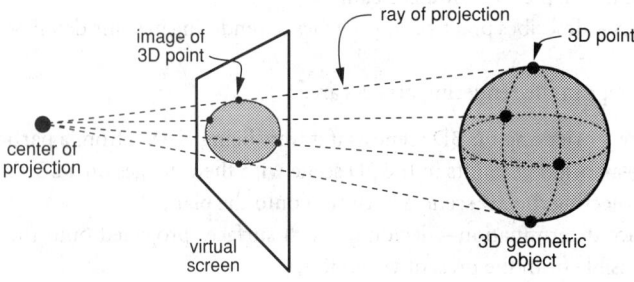

FIGURE 54.10 Geometry of a perspective projection.

Renderer

The renderer in a 3D graphics system is essentially the engine that drives the picture-making process. We can think of the renderer as viewing the scene through the lens of the virtual camera and constructing an image of what it sees, by first sampling points on the scene geometry and calling on the shader to calculate colors for each sample, and then combining these sampled colors into the pixels of the image. There are so many approaches to rendering, and the subject is so complex, that we will direct the interested reader to Chapter 57—Mainstream Rendering Techniques—for more information.

Shader

The shader is the algorithm that uses the information collected by the renderer about a point sample on the scene geometry, its material attributes, and the available lighting to calculate a color for the sample. Generally this is done by a more or less approximate physical simulation of how light is reflected toward the camera from the position on the surface at which the sample is being taken. Again, the reader is referred to Chapter 57—Mainstream Rendering Techniques—for more detailed information on shading and how it is done in a typical graphics system.

Image Construction

The final step in the rendering process is the construction of a digital image from the set of shaded samples across the virtual screen. This is done in any of a variety of ways, all of which are forms of low-pass filtering and resampling, providing a smooth blending and interpolation of the color samples into image pixel values [Wolberg 1990]. In practical terms, the digital image pixel grid is superimposed over the virtual screen, so that its pixels become associated with locations on the screen. Then the color of each pixel in the grid is calculated by taking a weighted average of the shaded samples in the vicinity of the pixel.

Storage and Display

For a 3D graphics system to be useful, there must be a way to turn the results of calculations into tangible images that can be both viewed and stored for archiving and transmission. Thus, a 3D graphics system is organized around a model of a digital image data structure, one or more image file formats, and a notion of the kind of display device that will be used to view images.

Image

The *pixmap* is the basic data structure for in-memory storage of digital images. A pixmap is simply a 2D array of pixel values, with each pixel's value stored in units of one or more bits. Typical pixel sizes are 1, 8, 24, and 32 bits.

A pixmap that allocates only one bit per pixel is known as a *bitmap* and can be used to store only monochrome images (i.e., each pixel is either full on or full off). Pixmaps with 8 bits per pixel can be used to store up to 256 levels of gray for a shaded gray-tone image, or, if the image is colored, the 8 bits are usually used to index a *lookup table*, which is simply an array containing up to 256 RGB colors that are used in the image. Pixmaps of this type are limited to pictures with a palette of no more than 256 distinct colors, although these colors can be drawn from a much larger set of possible colors. The size of this set is determined by the number of bits per entry in the lookup table. This scheme is often supported by hardware as described below in the discussion of *framebuffers*.

Pixmaps with 24 bits per pixel normally allocate 8 bits, or one byte, to each of the three RGB color primaries, giving a color resolution of 256 levels per primary, or 16,777,216 distinct colors. This color resolution is well beyond the ability of the human eye to distinguish color differences, so that even color gradations in these images appear to be as smooth as they would be in a continuous-tone color image.

On a high-end graphics computer, it is not unusual to allocate more than 24 bits per pixel in a pixmap. The extra space can be used to store colors at higher than 8-bit resolution, which is often handy to avoid roundoff errors in image-processing operations. A common configuration is a 32-bit pixel, where only 8 bits are used for each color primary, and the additional 8 bits are used to store an *alpha* value. The alpha value is used in image-compositing operations as a measure of pixel opacity. For purposes of compositing images together, pixels with high alpha values are treated as if they were opaque and pixels with low alpha values are treated as if they were transparent. Other uses for extra bits in a pixmap are to store aspects of the geometric information of the original model, such as surface normal, object id, material type, 3D position, etc. This information can be used in postprocessing of the image to do things like modify shading, or to add embellishments to the image that give a notion of the underlying form and structure of the geometry of the imaged objects.

Display Devices

The display device most frequently used in conjunction with a 3D graphics system is the *CRT* or *cathode-ray tube*. A CRT works on exactly the same principle as a simple vacuum tube. A schematic diagram of the organization of a monochrome CRT is shown in Fig. 54.11. Electrons traveling from the negatively charged cathode toward the positively charged plate are focused into a beam by focusing coils. The plate end of the CRT is a glass screen coated with phosphor. The grid control voltage adjusts the intensity of the beam and thus determines the brightness of the glowing phosphor dot where the beam hits the screen. Steering or deflection coils push the beam left/right and up/down so that it can be precisely directed to any desired spot on the screen.

A color CRT works like a monochrome CRT, but the tube has three separately controllable electron beams or *guns*. The screen has dots of red-, green-, and blue-colored phosphors, and each of the three beams is calibrated to illuminate only one of the phosphor colors. Thus, even though beams of electrons have no color, we can think of the CRT as having red, green, and blue electron guns. Colors are made using the RGB system, as optical mixing of the colors of the tiny adjacent dots takes place in the eye. Typically, the colored phosphors are arranged in triangular patterns known as *triads*, and an opaque *shadow mask* is positioned between the electron guns and the phosphors to ensure that each gun excites only the phosphors of the appropriate color.

A CRT can be used to display a picture in two different ways. The electron beam can be directed to "draw" a line-drawing on the screen—much like a high-speed electronic Etch-a-Sketch. The picture is drawn over and over on the screen at very high speed, giving the illusion of a permanent image. This type of device is known as a *vector display* and was quite popular for use in computer graphics and computer-aided design up until the early 1980s. By far the most popular type of CRT-based display device today is the *raster display*. These work by scanning the electron beam across the screen in a regular pattern of *scanlines* to "paint" out a picture, as shown in Fig. 54.12. The resulting pattern of scanlines is known as a **raster**.

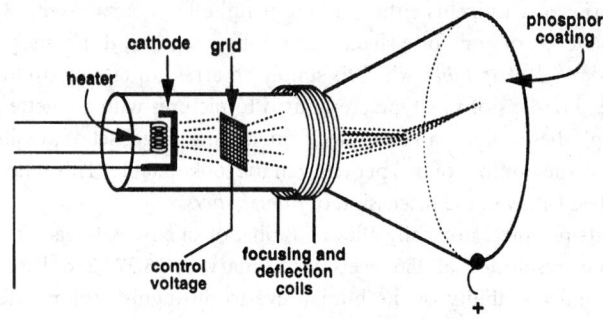

FIGURE 54.11 Schematic diagram of a CRT.

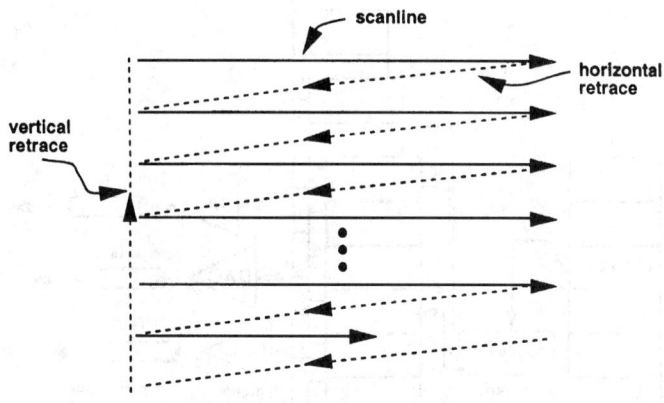

FIGURE 54.12 Raster scan pattern.

As a scanline is traced across the screen by the beam, the beam is modulated proportional to the intended brightness of the corresponding point on the picture. After a scanline is drawn, the beam is turned off and brought back to the starting point of the next scanline. As opposed to a vector display, which essentially makes a line drawing on the screen, a raster display can be used to paint out a shaded image.

The NTSC broadcast TV standard that is used throughout most of America uses 585 scanlines with 486 of these in the visible raster. The extra scanlines are used to transmit additional information, like control signals and closed-caption titling. The NTSC standard specifies a *framerate* of 30 frames per second, with each *frame* (single image) broadcast as two *interlaced fields*. The first of each pair of fields contains every even-numbered scanline, and the second contains every odd-numbered scanline. In this way, the screen is *refreshed* 60 times every second, giving the illusion of a solid flicker-free image. Actually, most of the screen is blank (or dark) most of the time. High-quality color CRTs for computer graphics greatly exceed the resolution and framerate of the NTSC standard, offering noninterlaced framerates of 60 or more frames per second with 1000 or more scanlines per frame.

Framebuffers

A *framebuffer* is the hardware interface between the pixmap data structure of a digital image and a CRT display. It is simply an array of computer memory, large enough to hold the color information for one or more frames (i.e., screenful) and display hardware to convert the current frame into control signals to drive a CRT. The color framebuffer schematized in Fig. 54.13 holds an 8-bit per pixel color image in a pixmap. The circuitry that controls the electron gun on the CRT loops through each row of the image array, fetching each pixel value in turn and using it to index an array of 256×3 high-speed hardware registers arranged as a lookup table. The values fetched from each of the three indexed registers are converted to voltages by digital-to-analog converters (DACs) and used to control the grid voltages of the CRT's three RGB electron guns. The timing has to be such that the memory fetches and conversion to grid voltages are synchronized exactly with the trace of the beam across the corresponding screen scanline, so that the correct position on the screen is associated with the appropriate pixel from the framebuffer.

A full-color-resolution framebuffer is shown in schematic form in Fig. 54.14. This type of device will have at least 24 bits per pixel (8 bits per color primary), driving three color guns, either directly or (as shown in the figure) through a separate lookup table per color primary. In this case, the lookup table is not used to increase the color resolution but instead can be used to correct nonlinearities or to obtain certain effects like overlay planes or pseudocoloring. Higher-end framebuffers may have more than 24 bits allocated per pixel, for hardware handling of such tasks as image compositing, depth buffering for hidden-surface resolution, double buffering for real-time animated display, and overlays.

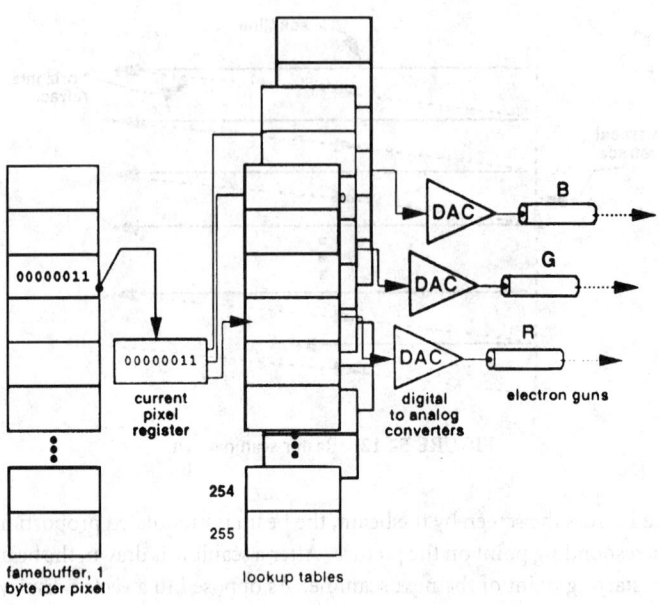

FIGURE 54.13   8-bit color framebuffer with lookup table.

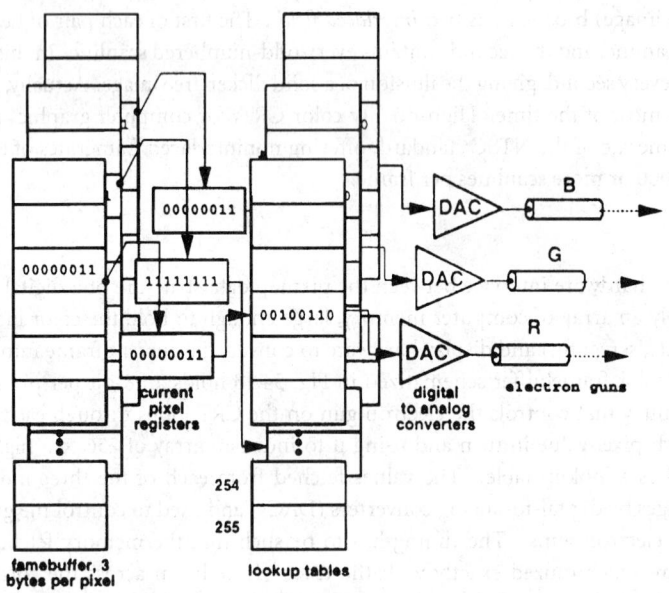

FIGURE 54.14   24-bit color framebuffer with three lookup tables.

Image Files

Due to the potential for using huge amounts of space, image file storage is a very important issue in computer graphics. A TV-resolution image has about $\frac{1}{3}$ million pixels—so a full-color RGB TV image will contain $3 \times \frac{1}{3} = 1$ million bytes of color information. Now, at 1800 frames (or images) per minute in a computer animation, we can expect to use up most of a 2-gigabyte disk for each minute of animation. Fortunately, we can do somewhat better than this by various file compression techniques, but disk storage space remains a crucial issue. Related to the space issue is the speed-of-access issue—that is, the bigger an image file, the longer it takes to read, write, and display.

For purposes of this overview, we will look closely at only a very simple, but very widely used image file format that has no intrinsic notion of compression. The *PPM*, or *portable pixmap*, format was devised to be an intermediate format for use in developing file-format conversion systems [Murray and vanRyper 1994]. Anyone familiar with computer graphics knows that the number of popular image file formats is immense. These include GIF, Targa, RLA, SGI, PICT, RLE, RLB, and many more. Converting images from one format to another is one of the common tasks in computer graphics, because different software packages and hardware units require different file formats. If there were N different file formats, and we wanted to be able to convert any one of these formats into any of the other formats, we would have to have $N \times (N - 1)$ conversion programs. The PPM idea is that we have one format that serves as a common source or target for all format conversions. We then write only N programs to convert all other formats into PPM and N more programs to convert PPM files into all other formats. In this way, we need write only $2 \times N$ programs to build a complete image conversion library.

The PPM format is not intended to be an archival format, so it does not need to be too storage-efficient. Although it is one of the simplest formats, PPM will nevertheless serve to illustrate some common features of image files. Most file formats are variants of the following organization. The file will typically contain some indication of the file type (which has come to be known as the format's *magic number*), a block of header or control information, and the image description data. Most (but not all) formats have a magic number, which identifies the file type. Often the magic number is not a number at all but rather a string of characters. The header block contains various descriptive information necessary to interpret the data in the image data block or can contain such archival information as creation date, image name, etc. The image data block is some form of encoding of the pixmap that describes the image. Some formats are much more complex than this, but this is the basic layout of a high percentage of formats.

In the PPM format, the magic number is either P1, P2, P3, P4, P5, or P6. P1 and P4 indicate that the image data are in a bitmap. These files are called PBM (portable bitmap) files. P2 and P5 are used to indicate gray-scale images or PGM (portable graymap) files. P3 and P6 are used to indicate full-color PPM (portable pixmap) files. The lower numbers—P1, P2, P3—indicate that the image data are stored as ASCII characters; i.e., all numbers are stored as character strings. This has the advantage that you can read the file in a text editor. The higher numbers—P4, P5, P6—indicate that image data are stored in a more compact binary encoding. We will look here only at P6-type files.

The header for a PPM file consists of the information shown in Fig. 54.15, stored as ASCII characters in consecutive bytes in the file. In the header, all whitespace is ignored, so the program that writes the file can freely intersperse spaces and line breaks. The im-

```
P6              -- magic number
# comment       -- any number of comment lines starting with #
200 300         -- image width & height
255             -- max color value
```

FIGURE 54.15 PPM P6 file header block layout.

age width and height determine the length of a scanline and the number of scanlines. The maximum color value cannot exceed 255, but it may be less if less than 8 bits of color information per primary are available.

The PPM P6 data block begins with the first pixel of the top scanline of the image (upper left-hand corner), and pixel data are stored in scanline order from left to right in 3-byte chunks giving the R, G, B values for each pixel, encoded as binary numbers. There is no separator between scanlines, and none is needed, as the image width given in the header block exactly determines the number of pixels per scanline. Figure 54.16(a) shows a red cube on a mid-gray background, and Fig. 54.16(b) gives the first several lines of a hexadecimal dump of the contents of the corresponding PPM file. Each line of this dump has the hexadecimal byte count on the left, followed by 16 bytes of hexadecimal information from the file, and ends with the same 16 bytes of information displayed in ASCII (nonprinting characters are displayed as !). Except for the first line of the dump, which contains the file header information, the ASCII information is meaningless, because the image data in the file are binary encoded. A line in the dump containing only a * indicates a sequence of lines all containing exactly the same information as the line above.

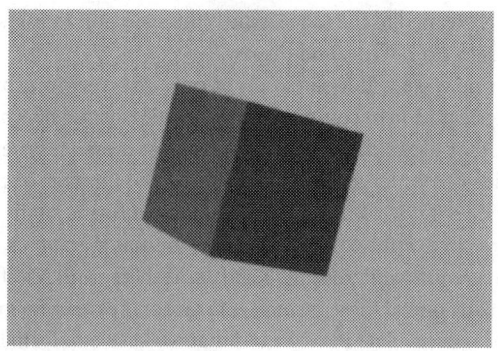

a) red cube on a midgrey (0.5 0.5 0.5) background

```
000000  5036 0a33 3030 2032 3030 0a32 3535 0a7f   P6!300 200!255!!
000010   7f7f 7f7f 7f7f 7f7f 7f7f 7f7f 7f7f 7f7f   !!!!!!!!!!!!!!!!!
*
009870  886d 6d92 5959 8a69 6984 7676 817d 7d80   !mm!YY!ii!vv!}}!
009880   7f7f 7f7f 7f7f 7f7f 7f7f 7f7f 7f7f 7f7f   !!!!!!!!!!!!!!!!!
*
009bf0   7f80 7e7e 9d41 41b5 0909 b211 11a9 2424   !!~~!AA!!!!!!!$$
009c00  9f3b 3b94 5454 8b68 6886 7272 827b 7b80   !;;!TT!hh!rr!{{!
009c10   7f7f 7f7f 7f7f 7f7f 7f7f 7f7f 7f7f 7f7f   !!!!!!!!!!!!!!!!!
*
009f70   7f7f 7f7f 7f82 7979 a72b 2bb9 0000 ba00   !!!!!!yy!++!!!!!
009f80  00ba 0000 b901 01b7 0606 b20f 0faf 1d1d   !!!!!!!!!!!!!!!!
009f90  a532 3297 4d4d 8d62 6286 7272 827a 7a80   !22!MM!bb!rr!zz!
009fa0  7e7e 7f7f 7f7f 7f7f 7f7f 7f7f 7f7f 7f7f   ~~!!!!!!!!!!!!!!
009fb0   7f7f 7f7f 7f7f 7f7f 7f7f 7f7f 7f7f 7f7f   !!!!!!!!!!!!!!!!!
*
00a2f0   7f7f 7f7f 7f7f 7f7f 7f8a 6969 b014 14ba   !!!!!!!!!!!ii!!!!
00a300  0000 ba00 00ba 0000 ba00 00ba 0000 ba00   !!!!!!!!!!!!!!!!!
00a310  00ba 0000 ba00 00b9 0505 b60d 0daf 1d1d   !!!!!!!!!!!!!!!!
00a320  a62f 2f9d 4141 915b 5b88 6d6d 8279 7980   !///!AA![[!mm!yy!
00a330  7e7e 7f7f 7f7f 7f7f 7f7f 7f7f 7f7f 7f7f   ~~!!!!!!!!!!!!!!
00a340   7f7f 7f7f 7f7f 7f7f 7f7f 7f7f 7f7f 7f7f   !!!!!!!!!!!!!!!!!
*
```

b) dump of start of PPM P6 red-cube image file

FIGURE 54.16   Red cube image and corresponding PPM P6 image file data: (a) red cube on a mid-gray (0.5 0.5 0.5) background, (b) dump of start of PPM P6 red-cube image file.

54.3 Research Issues and Summary

In this chapter, we have taken a quick, broad-brush look at 3D graphics systems, from scene specification to image display and storage. The attempt has been to lay a basic foundation and to provide certain essential details necessary for further study. A practical example of the implementation and use of a 3D graphics system appears in Chapter 62—Renderman®: An Interface for Image Synthesis.

As research in graphics tends to be specialized, readers are directed to the Research Issues and Summary sections of Chapters 56 through 61 of this Handbook for information on important and interesting open research areas. However, we note here some broad areas of research that are both timely and important to the development of the field. In the area of rendering, there are two areas that seem to be generating much current interest: extending solutions to the global illumination problem to handle a wider variety of material types and developing nonphotorealistic techniques. The entire fields of virtual reality and volumetric modeling are just getting off the ground and promise to be strong research areas for many years. Within the subfield of computer animation, three areas are of very strong current interest.

These are physically based modeling and simulation [Barzel 1992], artificial-intelligence and artificial-life techniques for character animation and choreography, and higher-order interactive techniques that exploit the capabilities of new 3D position and motion tracking devices. Finally, in the area of modeling, there is much room for improvement in interactive modeling tools and techniques, again possibly exploiting 3D position and motion trackers. And there is a continuing search for compact, powerful ways to represent natural forms.

Defining Terms

Affine transformation: A coordinate-system transformation where each transformed coordinate is a linear sum of the three original coordinates plus a constant.

Bidirectional reflectance distribution function: A function defined for a reflecting material that, given arrival and departure directions, gives the fraction of light energy of a particular wavelength arriving at a surface from the arrival direction that is reflected from the surface in the departure direction.

Bump map: A pattern of surface normal displacements, simulating the undulations of a bumpy surface, that is to be mapped onto a geometric surface during rendering.

Center of interest: A point in space toward which the virtual camera is always aimed.

Center of projection: The point in space at which all rays of projection for a perspective projection converge. This often is considered to be the position of the virtual camera.

Displacement map: A pattern of surface position displacements to create the undulations of a bumpy surface that is to be mapped onto a geometric surface during rendering.

Global coordinate system: A coordinate system with respect to which all other coordinate systems in the definition of a 3D scene are defined.

Hierarchical model: A geometric model defined within a set of nested reference coordinate frames, whose references form a directed tree or acyclic graph from a set of leaf frames, where basic geometric objects are defined, to a root frame in which the entire scene is defined.

Homogeneous coordinate system: In a 3D graphics system, this is a 4D coordinate system into which points are transformed before being multiplied by a homogeneous transformation matrix. The system is called homogeneous because all 3D points lie on the same hyperplane perpendicular to the fourth coordinate axis. Typically, this coordinate is 1 for all points.

Implicit surface: A surface defined implicitly as the set of points satisfying an equation of the form $F(x, y, z) = 0$.

Infinite light: A light source taken to be infinitely far from the model being illuminated, so that all of its rays reaching the model can be considered to be parallel.

Local coordinate system: A coordinate system defined with respect to some reference coordinate system, usually used to define a part or subassembly within a scene definition.

Material: The collective set of properties of the surface of an object that determines how it will reflect or transmit light.

Parametric surface: A surface defined explicitly by a set of functions of the form $X(\mathbf{u})$, $Y(\mathbf{u})$, $Z(\mathbf{u})$ that returns (x, y, z) coordinates of a point on the surface as a function of the set of parameters $\mathbf{u}$. Most commonly, $\mathbf{u} = (u, v)$, which yields a biparametric surface parametrized by the parametric coordinates u and v.

Pixel: A square or rectangular uniformly colored area on a raster display that forms the basic unit or picture element of a digital image.

Point light: A light source that radiates light uniformly in all directions from a single geometric point in space.

Projection: A transformation typically from a higher-dimensional space to a lower-dimensional space. In computer graphics the most commonly used projection is the camera projection, which projects 3D scene geometry onto the plane of a 2D virtual screen, as one of the key steps in the rendering process.

Raster: An array of scanlines, painted across a CRT screen, which taken together form a rectangular 2D image. Often the term raster is used to refer to the 2D array of pixel values stored digitally in a framebuffer.

Surface normal: A vector perpendicular to the tangent plane to a surface at a point. If the surface is planar, the three coefficients of the surface normal vector are the three scaling coefficients of the plane equation.

Texture map: A pattern of color to be mapped onto a geometric surface during rendering.

Transformation matrix: For a 3D system, this is a 4×4 matrix that, when multiplied by a point in homogeneous coordinates, gives the coordinates of the point in a transformed homogeneous coordinate system. The 4D homogeneous form of the matrix allows the unification of 3D translation, rotation, scaling, and shear into one operator.

References

Bartels, R., Beatty, J., and Barsky, B. 1987. *An Introduction to Splines for Use in Computer Graphics and Geometric Modeling.* Morgan Kaufmann, Los Altos, CA.

Barzel, R. 1992. *Physically-Based Modeling for Computer Graphics.* Academic Press, San Diego.

Ebert, D. S., ed. 1994. *Texturing and Modeling: A Procedural Approach.* AP Professional, Boston.

Foley, J. and van Dam, A. 1982. *Fundamentals of Interactive Computer Graphics.* Addison–Wesley, Reading, MA.

Foley, J., van Dam, A., Feiner, S., and Hughes, J. 1990. *Computer Graphics Principles and Practice.* Addison–Wesley, Reading, MA.

Glassner, A., ed. 1990. *Graphics Gems.* Academic Press, San Diego.

Glassner, A. 1995. *Principles of Digital Image Synthesis.* Morgan Kaufmann, San Francisco.

Gonzalez, R. and Woods, R. 1992. *Digital Image Processing.* Addison–Wesley, Reading, MA.

Hill, F. 1990. *Computer Graphics.* Macmillan, New York.

Murray, J. and vanRyper, W. 1994. *Encyclopedia of Graphics File Formats.* O'Reilly, Sebastopol, CA.

Press, W., Teukolsky, S., Vetterling, W., and Flannery, B. 1992. *Numerical Recipes in C, The Art of Scientific Computing,* 2nd ed. Cambridge University Press, Cambridge.

Rogers, D. 1985. *Procedural Elements for Computer Graphics.* McGraw–Hill, New York.

Rogers, D. and Adams, J. 1990. *Mathematical Elements for Computer Graphics.* McGraw–Hill, New York.

Russ, J. 1992. *The Image Processing Handbook.* CRC Press, Boca Raton, FL.

Watt, A. and Watt, M. 1992. *Advanced Animation and Rendering Techniques.* Addison–Wesley, Reading, MA.

Wolberg, G. 1990. *Digital Image Warping.* IEEE Computer Society Press, Los Alamitos, CA.

Further Information

The reader seeking further information on three-dimensional graphics systems should refer first to Chapters 55 through 62 of this Handbook, which provide much of the detail that this overview intentionally skips. The Further Information sections of these chapters provide pointers to the best source books on each of the specialized topics covered.

Beyond this volume, the primary source book for a broad coverage of the field is *Computer Graphics Principles and Practice* by Foley, van Dam, Feiner, and Hughes, published by Addison–Wesley. For a host of practical information and implementation tips, the five-volume series *Graphics Gems*, published by Academic Press, is an invaluable source. Information on image file formats can be found in the *Encyclopedia of Graphics File Formats* by Murray and vanRyper, published by O'Reilly. Fine practical guides to image-processing techniques are *The Image Processing Handbook* by Russ, published by CRC Press, and *Digital Image Processing* by Gonzalez and Woods, published by Addison–Wesley. The mathematics of computer graphics is given a very lucid treatment in *Mathematical Elements of Computer Graphics* by Rogers and Adams, published by McGraw–Hill, and there is no better reference to practical approaches to the implementation of numerical algorithms than *Numerical Recipes* (in various computer-language editions) by Press, Teukolsky, Vetterling, and Flannery, published by Cambridge University Press. Finally, the recent two-volume set *Principles of Digital Image Synthesis* by Glassner, published by Morgan Kaufmann, provides an excellent comprehensive coverage of the theoretical groundings of the field.

Persons interested in keeping up with the latest research in the field should turn to the ACM SIGGRAPH Conference Proceedings, published each year as a special issue of the ACM journal *Computer Graphics*. Other important conferences with published proceedings are *Eurographics* sponsored by the European Association for Computer Graphics, *Graphics Interface* sponsored by the Canadian Human–Computer Communications Society, and *Computer Graphics International* sponsored by the Computer Graphics Society. The IEEE journal *Computer Graphics and Applications* provides an applications-oriented view of recent developments, as well as publishing news and articles of general interest. ACM's *Transactions on Graphics* carries significant research papers, often with a focus on geometric modeling. Other important journals include *The Visual Computer*, IEEE's *Transactions on Visualization and Computer Graphics*, and the *Journal of Visualization and Computer Animation*.

55

Geometric Primitives

Alyn P. Rockwood
Arizona State University

55.1 Introduction

Geometric primitives are rudimentary for creating the sophisticated objects seen in computer graphics. They provide uniformity and standardization in addition to enabling hardware support.

Initially, definition of geometric primitives was driven by the capabilities of the hardware. Only simple primitives were available, e.g., points, line segments, triangles. In addition to the hardware constraints, other driving forces in the development of a geometric primitive have been either its general applicability to a broad range of needs or its satisfying ad hoc, but useful applications. The triangular facet is an example of a primitive that is simple to generate, easy to support in hardware, and widely used to model many graphics objects. An example of a specific primitive can be drawn from flight simulation in the case of *light strings*, which are instances of variable-intensity, directional points of light used to model airport and city lights at night. It is not a common primitive, but it is supported by a critical and profitable application.

As hardware and CPU increased in capability, the sophistication of the primitives grew as well. The primitives became somewhat less dependent on hardware; software primitives became more common, although for raw speed hardware primitives still dominate.

One direction for the sophistication of graphics primitives has been in the geometric order of the primitive. Initially, primitives were discrete or first-order approximations of objects, that is, collections of points, lines, and planar polygons. This has been extended to higher-order primitives represented by polynomial or rational curves and surfaces in any dimension.

The other direction for the sophistication of primitives has been in attributes that are assigned to the geometry. Color, transparency, associated bitmaps and textures, surface normals, and labels are examples of attributes attached to a primitive and used in its display.

0-8493-2909-4/97/$0.00+$.50
© 1997 by CRC Press, Inc.

This summary of graphics primitives is in rough chronological order of development which basically corresponds to increasing complexity. It concentrates on common primitives. It is beyond the scope of this review to include anything but occasional allusions to the plentiful special-purpose developments.

55.2 Screen Specification

To locate the graphics primitive in the viewing window, a local coordinate system is defined. By convention the origin is at the bottom left corner of the window. The positive x-axis extends horizontally from it, while the positive y-axis extends vertically. For 3D objects, the z-axis is imagined to extend into the screen away from the viewer. In the 3D case, it is necessary to transform the object to the screen via a set of viewing transformations (see Chapter 54).

Unlike pen and paper, we cannot draw a straight line between two points. The screen is a discrete grid; individual pixels must be illuminated in some pattern to indicate the desired line segment or other graphics primitive. A screen has from 80 to 120 pixels per inch, with high-resolution screens exhibiting 300 per inch. Most screens are about 1024 pixels wide by 780 pixels high. The problem of rendering a graphics primitive on a raster screen is called *scan conversion* and is discussed in Chapter 57. It is mentioned because it is closely related to the geometry of the object drawn and related drawing attributes. It is the scan conversion method that is embedded in hardware to accelerate the display of the graphics in a system. The expense and efficacy of graphics hardware depend on careful selection of the primitives for the facility desired.

55.3 Simple Primitives

Text

There are two standard ways to represent textual characters for graphics purposes. The first method is to save a representation of the letter as a bitmap, called a *font cache*. This method allows fast reproduction of the character on screen. Usually the font cache has more resolution than needed, and the character is downsampled to the display pixels. Even on high-resolution devices such as a quality laser printer, the discrete nature of the bitmap can be apparent, creating jagged edges. When transformations are applied to bitmaps such as rotations or shearing, aliasing problems can also be apparent. See the examples in Fig. 55.1.

To improve the quality of transformed characters and to compress the amount of data needed to transfer text, a second method of representing characters was developed using polygons or curves. When the text is displayed, the curves or polygons are scan-converted; thus the quality is constant regardless of the transformation (Fig. 55.2). The transformation is applied to the curve or polygon basis before scan conversion. Postscript™ is a well-known product for text transferal. In a "Postscript" printer, for instance, the definition of the fonts resides in the printer where it is scan-converted. Transfer across the network requires only a few parameters to describe font size, type, style, etc. Those printers which do not have resident Postscript databases and interpreters must transfer bitmaps with resulting loss of quality and time.

FIGURE 55.1 Bitmap characters. Note jagged edges and sampling artifacts. Compare Fig. 55.2.

Postscript is based on parametrically defined curves called Bézier curves (see section 55.8 below).

Fonts are designed in the bitmap case by simply scanning script from existing print, while special font design programs exist to design fonts with curves.

FIGURE 55.2 Curved representation of characters scan-converted.

1214 Graphics*Graphics*

Lines and Polylines

The first and still one of the most prevalent of the graphics primitives is the *line* (actually a line segment), typically specified by "move to $(x1, y1)$" and then "draw to $(x2, y2)$" commands, where $(x1, y1)$ and $(x2, y2)$ are the end points of the line segments. The endpoints can also be given as 3D points and then projected to the screen through viewing transforms (see Chapter 54).

The line can be easily extended to a *polyline* graphics primitive, that is to a piecewise polygonal path. It is usually stored as an array of 2D or 3D points. The first element of the array tells how many points are in the polyline. The following array elements are then the arguments of the line drawing commands. The first entry invokes the "move to" command; successive array entries use the "draw to" command as an operator.

Attributes for lines and polylines are used to modify their appearance without changing position. Popular attributes include *thickness*, *color*, and *style* (i.e., continuous, dotted, dashed, and variable dash lengths). In more advanced systems each vertex of a polyline is associated to a color and the line segment is drawn by linearly blending the one color to the other along the segment. This is called *color blending*.

Attributes are set either by extending the array to include attribute fields (polylines), or they are set *modally* (lines and polylines); that is, there is an independent command that sets the mode of the line drawing until changed explicitly by another command. Default values control the attributes that are not explicitly set. "Continuous" is the default for style, for example.

Impressive pictures can be produced from these simple line-based primitives. Figure 55.3 shows a turbine engine done with polylines.

Because of the discrete, pixel-based nature of the raster display screen, lines are prone to alias (see Chapter 58); that is, they may break apart or merge with one another to form distracting moiré patterns. They are also susceptible to jagged edges, a signature of older graphics displays. Serious line-drawing systems found it therefore important to add hardware that would "antialias" their lines while drawing them. See Chapter 58 for more details. Figure 55.3 used hardware antialiasing.

A *marker* is another primitive; it is either a point or a small square, triangle, or circle, often placed at the vertices of a polyline to indicate specific details such as distinguishing between lines of a chart. Markers are themselves usually made by predetermined lines or polylines. This unifies the display technology for the hardware, making them faster to draw. Even the point is often a very short line. A single command

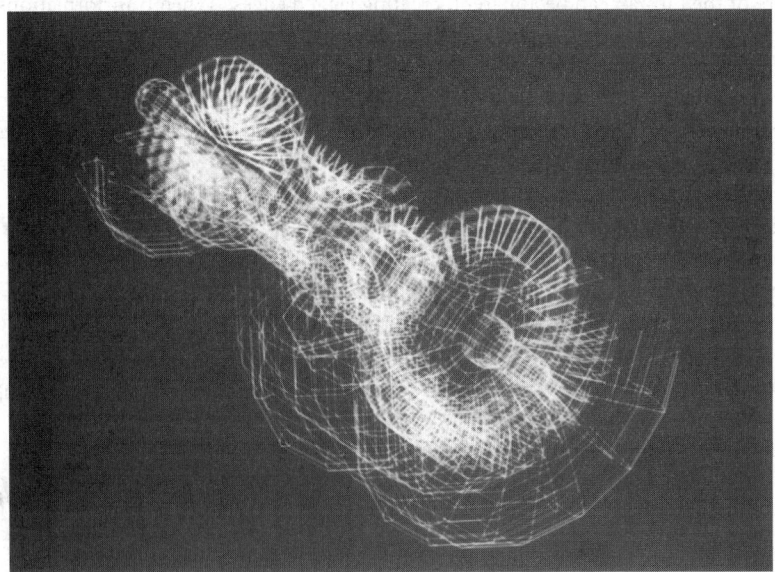

FIGURE 55.3 A polyline display of a turbine engine (created on an Evans and Sutherland Multi-Picture System in 1986).

is then used to center the marker. Style attributes don't usually apply to markers, but color and thickness attributes can be used to advantage.

Elliptical Arcs

Elliptical arcs in 2D may be specified in many equivalent ways. One way is to give the center position, major and minor axes, and start and end angles. Both angles are given counterclockwise from the major axis. Elliptical arcs may have all the attributes given to lines.

Such an arc may be a closed figure if the angles are properly chosen. In this case, it makes sense to have the ability to *fill* the ellipse with a given pattern or color using a scan conversion algorithm. Even in the case of partial arcs, the object is closed by a line between the end points of the arc so that it may be filled, if wished.

In 3D the plane of the elliptical arc must also be specified by the unit normal of the plane. While 2D arcs are common, 3D arcs are limited to high-end systems that can justify the cost of the hardware. Viewing transforms must compute the arc that is the image on the screen and then scan-convert that arc (usually elliptical, since perspective takes conics to conics).

It should be mentioned that arcs can also be represented by a polyline with enough segments; thus a software primitive for the arc which induces the properly segmented and positioned polyline may be a cost-effective *macro* for elliptical arcs. This macro should consider the effects of zoom and perspective to avoid revealing the underlying polygonality of the arc. It is surprising how small a number of line segments, properly chosen, can give the impression of a smooth arc.

One of the most commonly used elliptical arcs is, of course, the circular arc and its closure, the circle.

55.4 Wireframes

Given hardware or software macro arc primitives in a system, complex curves can be generated by piecing the arcs and lines end to end. Several computer aided design (CAD) systems exist that can generate a rich set of line-based models in both 2D and 3D. They are called **wireframe** models. Figure 55.3 is a wireframe model of a turbine engine.

Wireframe models are popular in engineering applications, for instance, because of their visual precision. Drafting is also a natural application. They have other advantages in 3D because of the ability to see through them to parts of the object that are behind.

Too many lines can be confusing, however; to improve wireframe models a hidden-line routine may be employed (see Chapter 57). This requires derivation of a surface implied by the line. Yet line models can be ambiguous. Figure 55.4 shows a classical example of an object for which the implied surface can be legitimately interpreted in many ways.

Finally, wireframe objects do not support realism. Most objects have a surface that is colored and reflects light according to physical laws of irradiance. This leads us to the next type of primitive.

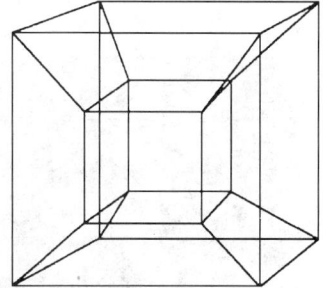

FIGURE 55.4 An ambiguous wireframe model. Where do the surfaces belong?

55.5 Polygons

Closing a polyline by matching start and end points creates a **polygon**. It is probably the most commonly employed primitive in graphics, because it is easy to define an associated surface. Not all polygons have surfaces (they may be a primitive used in wireframe modeling), but most systems that admit polygons will support both *filled* and *empty* polygons. Filling a polygon is an important example of scan conversion. There are many ways to do this (see Chapter 57). The different ways to fill become attributes of the polygon. The filled polygon is the basis for the numerous hidden-surface algorithms (see Chapter 57). Because of their usefulness for defining surfaces, polygons almost always appear as 3D primitives which subsume the 2D case.

The most sophisticated polygon primitive allows nonconvex polygons that contain other polygons, called *holes* or *islands* depending on whether they are filled or not. The scan conversion routine selectively fills the appropriate portions of the polygon depending on whether they are holes or islands. This complex polygon is probably made as a macro out of simple polygon primitives. Triangulation routines exist, for example, that reduce such polygons to simple triangular facets.

55.6 The Triangular Facet

If the polygon is the most popular primitive, then the simple triangle is certainly the most popular polygon. Figure 55.5 shows an object simple object composed of triangular facets.

As mentioned, the triangular facet often supports more complex polygons. It is fast to draw and supports many diverse and powerful methods (see Chapter 57). Another major advantage is that they are always flat; they do not leave the plane due to numerical problems, data errors, or nonplane-preserving transformations. At this writing, graphics workstations exist that can render 10 million triangular facets per second with hidden surfaces removed and smooth shaded display between neighboring triangles implemented. It is certain that this number will continue to increase. With such capacity several dozens of the objects in Fig. 55.5 could be smoothly animated.

In order to increase the speed of polygon rendering and decrease the size of the database, triangular and quadrilateral facets can be stored and processed as meshes; i.e., collections of facets that share edges and vertices. The triangular mesh, for instance, is given by defining three vertices for the "lead" triangle, and then giving a new vertex for each successive triangle (Fig. 55.6). The succeeding triangles use the last two vertices in the list with the newly given one to form the next triangle. Such a mesh contains about one-third the data and uses the shared edges to increase processing speed. Most graphics objects have large contiguous areas that can take advantage of meshing. The quadrilateral mesh requires two additional vertices be used, with the last two given, and therefore has less savings.

Because of its benefits and ubiquity, many other primitives are defined in software as configurations of the triangular facet. One example is the faceted sphere. With enough facets the sphere will look smooth. It is very commonly encountered in engineering applications and molecular modeling.

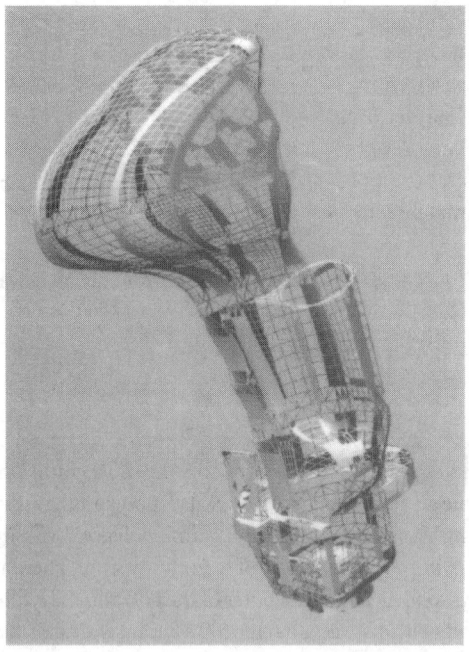 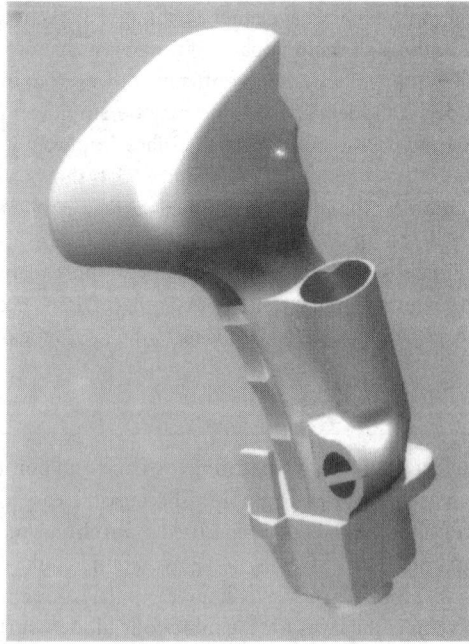

FIGURE 55.5   A model composed of simple triangular facets (polygon model).

Most of the attributes associated with polygons, and triangular facets in particular, deal with the rendering of the facet (see Chapter 57), e.g., color blending between vertices, pointers into a bitmap for texturing, normal vectors (from an underlying surface), reflectance parameters, transparency, and color.

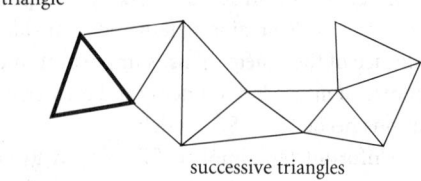

Start triangle

successive triangles

FIGURE 55.6   A mesh of triangular facets.

55.7 Implicit Modeling

Implicit Primitives

The need to model mechanical parts drove the definition of implicitly defined primitives. These modeling primitives naturally became graphics primitives to serve the design industry. They have since been used in other fields such as animation.

An implicit function f maps a point $\mathbf{x}$ in $\mathbf{R}^n$ to a real number, i.e., $f(\mathbf{x}) = r$. Usually $n = 3$. By absorbing r into the function we can view the implicit function as mapping to zero, i.e., $g(\mathbf{x}) = f(\mathbf{x}) - r = 0$. The importance of implicit functions in modeling is that the function divides space into three parts: $f(\mathbf{x}) < 0$, $f(\mathbf{x}) > 0$, and $f(\mathbf{x}) = 0$. The last case defines the surface of an **implicit object**. The other two cases define respectively the *inside* and *outside* of the implicit object. Hence implicit objects are useful in *solid modeling* where it is necessary to distinguish the inside and outside of an object. Modeling objects by polygons does not, for instance, distinguish between the inside and outside of the object. It is, in fact, quite possible to describe an object in which inside and outside are ambiguous. This facility to determine inside and outside is further enhanced by using Boolean operations on the implicit objects to define more complex objects (see the subsection on CSG objects below).

Another advantage of implicit objects is that the surface normal of an implicitly defined surface is given simply as the gradient of the function: $N(\mathbf{x}) = \nabla f(\mathbf{x})$. Furthermore, many common engineering surfaces are easily given as implicit surfaces; thus the *plane* (not the polygon) is defined by $\mathbf{n} \cdot \mathbf{x} + \mathbf{d} = 0$, where $\mathbf{n}$ is the surface normal and $\mathbf{d}$ is the perpendicular displacement of the plane to the origin. The *ellipsoid* is defined by $(x/a)^2 + (y/b)^2 + (z/c)^2 - r^2 = 0$, where $\mathbf{x} = (x, y, z)$. General *quadrics*, which include ellipsoids, paraboloids, and hyperboloids, are defined by $\mathbf{x}^T M \mathbf{x} + \mathbf{b} \cdot \mathbf{x} + \mathbf{d} = 0$, where M is a 3-by-3 matrix and $\mathbf{b}$ and $\mathbf{d}$ vectors in $\mathbf{R}^3$. The quadrics include such important forms as the cylinder, cone, sphere, paraboloids, and hyperboloids of one and two sheets. Other implicit forms used are the *torus*, *blends* (transition surfaces between other surfaces), and *superellipsoids* defined by $(x/a)^k + (y/b)^k + (z/c)^k - R = 0$ for any integer k.

CSG Objects

An important extension to implicit modeling arises from applying set operations such as union, intersection, and difference to the sets defined by implicit objects. The intersection of six half spaces defines a cube, for example. This method of modeling is called *constructive solid geometry* (CSG) [Foley 1990]. All set operations can be reduced to some combination of just union, intersection, and complementation. Because these create an algebra on the sets that is isomorphic to Boolean algebra, corresponding to multiply, add, and negate, respectively, the operations are often referred to as *Booleans*.

A convenient form for visualizing and storing a CSG object is to use a tree in which the nodes are implicit objects and the branches indicate the operation. Traversal of the tree indicates the order of the binary operations and to which sets they pertain. Figure 55.7 shows a CSG tree for a simple model.

Figure 55.8 demonstrates an object made exclusively from Boolean parts of plane quadrics, a part torus, and

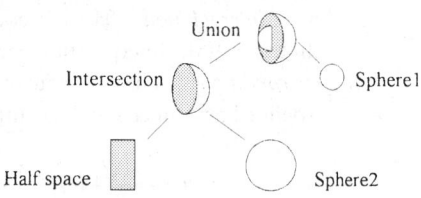

Union

Intersection

Sphere1

Half space

Sphere2

FIGURE 55.7   A CSG model and its tree.

blended transition surfaces. For any point in space it is straightforward to determine whether it is inside, outside, or on the
surface of the object. This is important in determining volume,
center of mass, and moments and for performing Boolean operations needed by CSG models.

Unfortunately, implicit forms tend to be difficult to render,
except for ray tracing. Algorithms for polygonizing implicits and
CSG models tend to be quite complex [Bloomenthal 1988]. The
implicit object gives information about a point relative to the
surface in space, but no information is given as to where on the
surface a point is located; there is no local coordinate system.
This makes it difficult to tessellate into rendering elements such
as triangular facets. In the case of ray tracing, however, the parametric form of a ray $\mathbf{x}(t) = (x(t), y(t), z(t))$ composed with
implicit function $f(\mathbf{x}(t)) = 0$ leads to a root-finding solution

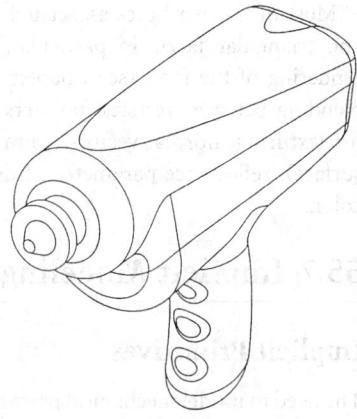

FIGURE 55.8   A solid model of a drill.

of the intersection points on the surface which are critical points
in the ray-tracing algorithm. Determining whether points are part of a CSG model is simply exclusion
testing on the CSG tree.

55.8 Parametric Curves

An important class of geometric primitives are formed by **parametrically defined curves and surfaces**.
These constitute a flexible set of modeling primitives that are locally parametrized; thus in space the curve
is given by $\mathbf{x}(t) = (x(t), y(t), z(t))$ and the surface by $\mathbf{s}(u, v) = (x(u, v), y(u, v), z(u, v))$. In this section
and the next one, we will give examples of only the most popular types of parametric curves and surfaces.
There are many variations on the parametric forms (see Farin [1992]).

Bézier Curves

The general form of a Bézier curve of degree n is

$$\mathbf{f}(t) = \sum_{i=0}^{n} \mathbf{b}_i B_i^n(t) \tag{55.1}$$

where $\mathbf{b}_i$ are vector coefficients, the *control points*, and

$$B_i(t) = \binom{n}{i} t^i (1 - t^{n-i})$$

where $\binom{n}{i}$ is the binomial coefficient. The $B_i^n(t)$ are called *Bernstein functions*. They form a basis for the
set of polynomials of degree n. Bézier curves have a number of characteristics which are derived from the
Bernstein functions and which define their behavior.

>*End-point interpolation:*   The Bézier curve interpolates the first and last control points, $\mathbf{b}_0$ and $\mathbf{b}_n$.
>In terms of the interpolation parameter t, $\mathbf{f}(0) = \mathbf{b}_0$ and $\mathbf{f}(1) = \mathbf{b}_n$.
>
>*Tangent conditions:*   The Bézier curve is cotangent to the first and last segments of the *control polygon*
>(defined by connecting the control points) at the first and last control points; specifically
>
>$$\mathbf{f}'(0) = (\mathbf{b}_1 - \mathbf{b}_0)n \quad \text{and} \quad \mathbf{f}'(1) = (\mathbf{b}_n - \mathbf{b}_{n-1})n$$
>
>*Convex hull:*   The Bézier curve is contained in the convex hull of its control points for $0 \leq t \leq 1$.

Affine invariance: The Bézier curve is affinely invariant with respect to its control points. This means that any linear transformation or translation of the control points defines a new curve which is just the transformation or translation of the original curve.

Variation diminishing: The Bézier curve does not undulate any more than its control polygon; it may undulate less.

Linear precision: The Bézier curve has linear precision: If all the control points form a straight line, the curve also forms a line.

Figure 55.9 shows a Bézier curve with its control polygon. Notice how it follows the general shape. This together with the other properties makes it desirable for shape design.

Evaluation of the Bézier curve function at a given value t produces a point $\mathbf{f}(t)$. As t varies from 0 to 1, the point $\mathbf{f}(t)$ traces out the curve segment. One way to evaluate Eq. (55.1) is by direct substitution. This is probably the worst way. There are several better methods available for evaluating the Bézier curve. One method is the *de Casteljau algorithm*. This method not only provides a general, relatively fast and robust algorithm, but it gives insight into the behavior of Bézier curves and leads to several important operations on the curves.

To formalize de Casteljau's algorithm we need to use a recursive scheme. The control points are the input. Thereafter each point is superscripted by its level of recursion. Finally, for any point,

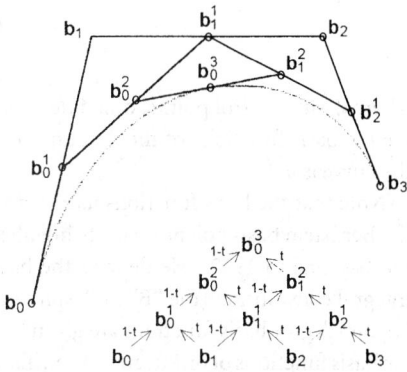

FIGURE 55.9   Bézier curve, control polygon, and de Casteljau algorithm.

$$\mathbf{b}_i^j(t) = (1 - t)\mathbf{b}_i^{j-1} + t\mathbf{b}_{i+1}^{j-1} \quad \text{for } i = 0, \ldots, n, \quad j = 0, \ldots, n - i$$

Note that $\mathbf{b}_0^n(t) = \mathbf{f}(t)$; it is a point on the curve.

One of the most important devices for evaluating curves is the systolic array. It is a triangular arrangement of vectors in which each row reflects the levels of recursion of the de Casteljau algorithm. The first row consists of the Bézier control points. Each successive row corresponds to the points produced by iterating with de Casteljau's algorithm.

The point $\mathbf{b}_0^3$ is the point on the curve for some value of the parameter t. Any point in the systolic array may be computed by linearly interpolating the two points in the preceding row with the parameter t; thus for example:

$$\mathbf{b}_1^2 = \mathbf{b}_1^1(1 - t) + \mathbf{b}_2^1 t$$

One of the most important operations on a curve is that of subdividing it. The de Casteljau algorithm not only evaluates a point on the curve, it also subdivides a curve into two parts as a bonus. The control points of the two new curves are given as the legs of the systolic array. In Fig. 55.10 is a cubic Bézier curve after three iterations of the de Casteljau algorithm, with the parameter $t = 0.5$. By using the left and right legs of the systolic array as control points, we obtain two separate Bézier curves which together replicate the original. We have subdivided the curve at $t = 0.5$.

Subdivision permits existing designs to be refined and modified; for example, by incorporating additional curves

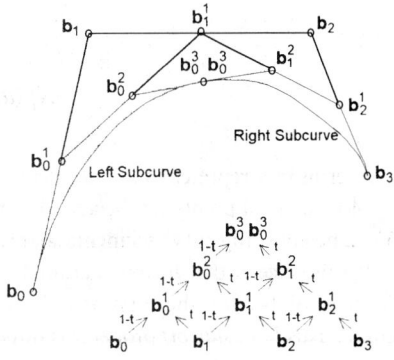

FIGURE 55.10   Subdividing the curve.

into an object. One method of intersecting a Bézier curve with a line is to recursively subdivide the curve, testing for intersections of the curve's control polygons with the line. This process is continued until a sufficiently fine intersection is attained.

B-Spline Curves

A single *B-spline* curve segment is defined much like a Bézier curve. It looks like

$$\mathbf{d}(t) = \sum_{i=0}^{n} \mathbf{d}_i N_i(t)$$

where $\mathbf{d}_i$ are control points, called *de Boor points*. The $N_i(t)$ are the basis functions of the B-spline curve. The degree of the curve is n.

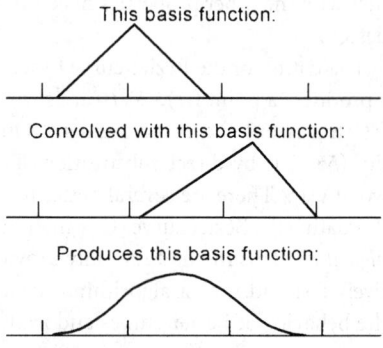

Note that the basis functions used here are different from the Bernstein basis polynomials. Schoenberg first introduced the B-spline in 1949. He defined the basis functions using integral convolution (the "B" in B-spline stands for "basis"). Higher-degree basis functions are given by convolving multiple basis functions of one degree lower. Linear basis functions are just "tents" as shown in Fig. 55.11. When convolved together they make piecewise parabolic "bell" curves.

The tent basis function (which has a degree of one) is nonzero over two intervals, the parabola is nonzero over three intervals, and so forth. This gives the region of influence for

FIGURE 55.11 Defining basis functions by convolution.

different degree B-spline control points. Notice also that each convolution results in higher-order continuity between segments of the basis function. When the control points (de Boor points) are weighted by these basis functions, the B-spline curve results.

The major advantage of the B-spline form is in piecing curves together to form a *spline*. If two B-spline curve segments share $n - 1$ control points, they will fit together at the junction point with C^{n-1} continuity. The picture in Fig. 55.12 shows the case for a cubic B-spline with two segments:

- The first curve has control points $\mathbf{d}_0$, $\mathbf{d}_1$, $\mathbf{d}_2$, and $\mathbf{d}_3$.
- The second curve has control points $\mathbf{d}_1$, $\mathbf{d}_2$, $\mathbf{d}_3$, and $\mathbf{d}_4$.

Instead of integrating to evaluate the basis functions, a recursive formula has been derived:

$$N_i^n(u) = \frac{u - u_{i-1}}{u_{i+n-1} - u_{i-1}} N_i^{n-1}(u) + \frac{u_{i+n} - u}{u_{i+n} - u_i} N_{i+1}^{n-1}(u)$$

where

$$N_i^0(u) = \begin{cases} 1 & \text{if } u_{i-1} \leq u \leq u_i \\ 0 & \text{otherwise} \end{cases}$$

The terms in u represent a *knot sequence*, the spans over which the de Boor points influence the B-spline.

More control points can be added to make a longer and more elaborate spline curve. As seen in Fig. 55.12 neighboring curve segments share n control points.

It can be seen that for any parameter value u only four basis functions are nonzero; thus only four control points affect the curve at u. If a control point is moved it influences only a limited portion of the curve. This *local support* property is important for modeling.

If the first n and last n control points are made to correspond, then the curve's end points will match; it will form a closed curve. This is called a *periodic B-spline*.

Any point on the curve is a convex combination of the control points, i.e., it must be in the convex hull of the control points associated with the nonzero basis functions.

Like the Bézier curve, the B-spline curve also satisfies a variation-diminishing property, and is affinely invariant. Linear precision follows, as in the Bézier-curve case, from the convex-hull property.

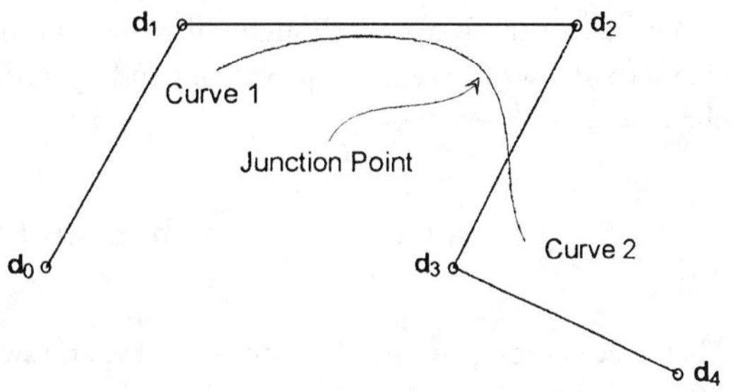

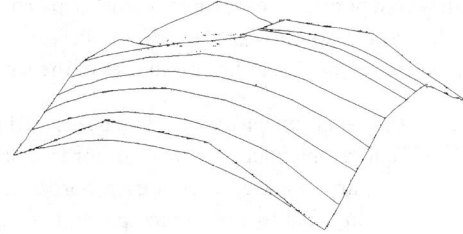

FIGURE 55.12 B-spline curve with two segments.

The recursive form for evaluating B-splines via basis functions is seldom used in practice. The best way to evaluate a B-spline curve is to use the de Boor algorithm.

Formally, the de Boor algorithm is written as

$$\mathbf{d}_i^k(u) = \frac{u_{i+n-k} - u}{u_{i+n-k} - u_{i-1}} \mathbf{d}_{i-1}^{k-1}(u) + \frac{u - u_{i-1}}{u_{i+n-k} - u_{i-1}} \mathbf{d}_i^{k-1}(u)$$

We see that the de Boor points form a systolic array; each point is defined in terms of preceding points. Thus we may write an iterative procedure to evaluate a point on a B-spline curve in much the same way as de Casteljau's algorithm above evaluates a point on a Bézier curve. Only the weighting factors differ. The last point produced in the method is the point on the curve.

55.9 Parametric Surfaces

Bézier Surfaces

Imagine moving the control points of the Bézier curve in three dimensions. As they move in space, new curves are generated. If they are moved smoothly, then the curves formed create a surface, which may be thought of as a bundle of curves. If each of the control points is moved along a Bézier curve of its own, then a Bézier surface patch is created; if a B-spline curve is extruded, then a B-spline surface results (Fig. 55.13).

In the Bézier case this can be written by changing the control points in the Bézier formula into Bézier curves; thus a surface is defined by

FIGURE 55.13 Sweeping a curve to make the surface.

$$\mathbf{s}(u, v) = \sum_{i=0}^{n} \mathbf{b}_i(u) B_i(v) \tag{55.2}$$

Notice that we have one parameter for the control curves and one for the "swept" curve. It is convenient to write the control curves as Bézier curves of the same degree. If we let the ith control curve have control points $\mathbf{b}_{ij}$, then the surface given in Eq. (55.2) above can be written as

$$\mathbf{s}(u, v) = \sum_{i=0}^{n} \left(\sum_{j=0}^{m} \mathbf{b}_{ij} B_j(u) \right) B_i(v) \tag{55.3}$$

where m is the degree of the control curves.

A surface can always be thought of as nesting one set of curves inside another. From this simple characteristic we derive many properties and operations for surfaces. Simple algebra changes Eq. (55.3) above into

$$s(u, v) = \sum_{i=0}^{n} \left(\sum_{j=0}^{m} \mathbf{b}_{ij} B_j(u) \right) B_i(v) = \sum_{i=0}^{n} \sum_{j=0}^{m} \mathbf{b}_{ij} B_i(v) B_j(u) \qquad (55.4)$$

That is, even though we started with one curve and swept it along the other, there is no preferred direction. The surface patch could have been written as:

$$s(u, v) = \sum_{i=0}^{n} \mathbf{b}_j(v) B_j(u), \qquad \text{where} \quad \mathbf{b}_j(v) = \sum_{i=0}^{n} \mathbf{b}_{ij} B_i(v)$$

The curve is simply swept in the other direction.

The set of control points forms a rectangular control mesh. A 3-by-3 (bicubic) control mesh is shown in Fig. 55.14 with the surface. There are 16 control points in the bicubic control mesh. In general there will be $(n + 1) \times (m + 1)$ control points. By convention we associate the i-index with the u-parameter, and the j-index with the v-parameter. Hence if we take:

$$\mathbf{b}_{i0}, \quad i = 1, \dots, n$$

we get the Bézier curve

$$\mathbf{b}(u, 0) = \sum_{i=0}^{n} \mathbf{b}_{i0} B_i^n(u)$$

Each marginal set of control points defines a Bézier curve (the four border curves), and each of these curves is a boundary of the Bézier surface patch. Such a surface is shown in white in Fig. 55.14.

Many of the properties of the Bézier surface are derived directly from those of the Bézier curve, especially those curves that form the boundaries of the patch:

End-point interpolation: The Bézier surface patch passes through all four corner control points.

Tangent conditions: The four border curves of the Bézier surface patch are cotangent to the first and last segments of each border control polygon, at the first and last control points. The normal to the surface patch at each vertex may be found from the cross product of the tangents.

Convex hull: The Bézier surface patch is contained in the convex hull of its control mesh for $0 < u < 1$ and $0 < v < 1$.

Affine invariance: The Bézier surface patch is affinely invariant with respect to its control mesh. This means that any linear transformation or translation of the control mesh defines a new patch which is just the transformation or translation of the original patch.

Evaluation of the Bézier Surface Patch

As with the properties described above, the evaluation of a Bézier surface patch can also be derived from the Bézier curve. If we want to evaluate a point on the patch at parameter value (u, v), we apply the de Casteljau algorithm in a nested fashion to Eq. (55.3). That is, we first evaluate the control curves in the u direction, which reduce to control points in the v direction. These points are again evaluated with de Casteljau's algorithm.

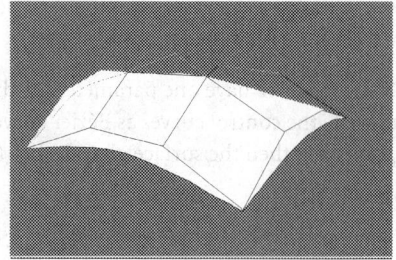

FIGURE 55.14 Bicubic surface with control mesh.

Subdivision of the Bézier Surface Patch

Again, as with the Bézier curve, we can apply the de Casteljau algorithm in nested fashion to subdivide a Bézier surface patch. When a surface patch is subdivided, it yields
four subpatches that share a corner at the (u, v) subdivision
point. Recall that when a curve was subdivided, the new
curve's control points appeared as the legs of a systolic array. In the surface case we subdivide each row of the control
mesh, producing points of the systolic array for each. Each
point on each leg of every row's systolic array now becomes
a control point for a columnar set. A biquadratic case is
shown in Fig. 55.15. Here we see that three points in each

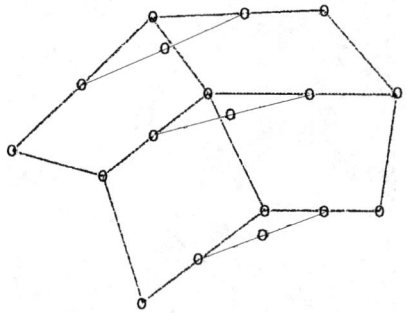

FIGURE 55.15 First-level subdivision.

row produce five after subdivision. Now we consider the points in columns, subdividing the columns with
de Casteljau's algorithm. The points in the legs of their systolic arrays become the control points of the
new subpatches. In our example rows with three control points produce five "leg" points, i.e., five columns
of three points. Each column then produces five control points, so a 3-by-3 grid generates a 5-by-5 grid
after subdivision. The control meshes of the four new patches are produced as shown in Fig. 55.16. The
central row and column of control points are shared by each 3-by-3 subpatch as shown in Fig. 55.17. Note
that the order of the scheme does not matter. Columns might have been taken first, and then rows.

Subdivision is a basic operation on surfaces.
Many "divide and conquer" algorithms are based on it. To clip a surface to a viewing window
we can use the convex-hull property and subdivision, for example. The convex-hull test can
determine if a patch is entirely contained in the
window. If not, it is subdivided, and the sub-
patches are then tested. Recursion is applied until the pixel level.

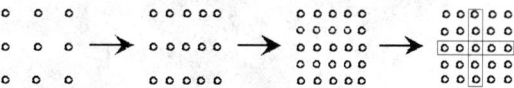

FIGURE 55.16 Control points: progression for subdivision.

B-Spline Surfaces

As with the Bézier surface, the B-spline surface is defined as a nested bundle of curves, thus yielding

$$\mathbf{s}(u, v) = \sum_{i=0}^{L+n-1} \sum_{j=0}^{M+m-1} \mathbf{d}_{ij} N_i(u) N_j(v)$$

where

> n, m are the degrees of the B-splines,
> L, M are the number of segments, so there are $L \times M$ patches.

All operations used for B-spline curves carry over to the
surface via the nesting scheme implied by its definition. We
recall that B-spline curves are especially convenient for obtaining continuity between polynomial segments. This convenience is even stronger in the case of B-spline surfaces;
the patches meet with higher-order continuity depending
on the degree of the respective basis functions. B-splines
define quilts of patches with "local support" design flexibility. Finally, since B-spline curves are more compact to store
than Bézier curves, the advantage is "squared" in the case of
surfaces.

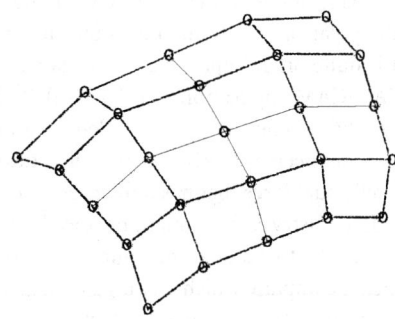

FIGURE 55.17 The subdivided patch.

FIGURE 55.18   Turbine engine modeled by B-spline surfaces.

These advantages are tempered by the fact that operations are typically more efficient on Bézier curves. Conventional wisdom says that it is best to design and represent surfaces as B-splines, and then convert to Bézier form for operations. Figure 55.18 shows an object made of many surface patches.

55.10 Standards

Several movements have occurred to standardize primitives. Some of the standards grew out of the mechanical CAD/CAM industry, because of the obvious connection between it and graphics. IGES is a popular one. Although it cannot be said that IGES has become commonplace in the graphics world, it is quite often important to translate between graphics models and these standards when dealing with CAD/CAM applications. GKS [ANSI 1985], PHIGS, and PHIGS+ [ANSI 1988] are standards developed by a broad consortium of academics and industrial users. In spite of major efforts and backing, they have not been as directly successful as the standards evolved in the industry. They have, however, been an intellectual force that has influenced many of the industrial efforts.

In industry each company has developed and pushed for its particular set of graphics standards. Perhaps the most successful at this time is GL (for Graphics Library), which was developed by Silicon Graphics, Inc., a company which based its computer workstation product on high-powered graphics. It has been licensed by IBM and many other companies. It has evolved into Open GL, which is supported by many manufacturers and threatens to become the standard.

55.11 Research Issues and Summary

There have been efforts to cast higher-order primitives like parametric surfaces into graphics hardware, but the best approach seems to be to convert these into polygons and then render [Rockwood et al. 1990]. There may be hardware support for this process, but the polygon processing remains at the heart of graphics primitives. This trend is likely to continue into the future if for no other reason than its own inertia. Special needs will continue to drive the development of specialized primitives.

One new trend that may affect development is that of volume rendering (see Chapter 60). Although it is currently quite expensive to render, hardware improvements and cheaper memory costs should make it increasingly more viable. As a technique it subsumes many of the current methods, usually with better quality, as well as enabling the visualization of volume-based objects. Volume-based primitives, i.e., tetrahedra, cuboids, and curvilinear volume cubes, will receive more attention and research.

Defining Terms

Implicit objects: Defined by implicit functions, they define solid objects of which outside and inside can be distinguished. Common engineering forms such as the plane, cylinder, sphere, torus, etc. are defined simply by implicit functions.

Parametrically defined curves and surfaces: Higher-order surface primitives used widely in industrial design and graphics. Parametric surfaces such as B-spline and Bézier surfaces have flexible shape attributes and convenient mathematical representations.

Polygon: A closed object consisting of vertices, lines, and usually an interior. When pieced together it gives a piecewise (planar) approximation of objects with a surface. Triangular facets are the most common form and form the basis of most graphics primitives.

Wireframe: Simplest and earliest form of graphics model, consisting of line segments and possibly elliptical arcs that suggest the shape of an object. It is fast to display and has advantages in precision and "see through" features.

References

Adobe Systems Inc. 1985. *Postscript Language Reference Manual.* Addison–Wesley, Reading, MA.

ANSI (American National Standards Institute) 1985. *American National Standards for Information Processing Systems—Computer Graphics—Graphical Kernel System (GKS) Functional Description.* ANSI X3.124-1985. ANSI, New York.

ANSI (American National Standards Institute) 1988. *American National Standards for Information Processing Systems—Programmer's Hierarchical Interactive Graphics Systems (PHIGS) Functional Description.* ANSI X3.144-1988. ANSI, New York.

Bézier, P. Mathematical and practical possibilities of UNISURF. In *Computer Aided Geometrical Design*, R. E. Barnhill and R. Riesenfeld, eds. Academic Press, New York.

Bloomenthal, J. 1988. Polygonisation of implicit surfaces. *Comput. Aided Geom. Des.* 5:341–345.

Boehm, W., Farin, G., and Kahman, J. 1984. A survey of curve and surface methods in CAGD. *Comput. Aided Geom. Des.* 1(1):1–60, July.

Farin, G. 1992. *Curves and Surfaces for Computer Aided Geometric Design.* Academic Press, New York.

Faux, I. D. and Pratt, M. J. 1979. *Computational Geometry for Design and Manufacture.* Wiley, New York.

Foley, J. D. 1990. *Computer Graphics: Principles and Practice.* Addison–Wesley, Reading, MA.

Rockwood, A., Dawis, T., and Heaton, K. *Real Time Rendering of Trimmed Surfaces*, Computer Graphics.

56

Advanced Geometric Modeling

David S. Ebert
*University of Maryland
Baltimore County*

56.1 Introduction

In the past thirty years, geometric modeling techniques in computer graphics have evolved significantly as the field matured and attempted to portray the complexities of nature. Earlier geometric models, such as polygonal models, patches, points, and lines, are insufficient to represent the complexities of natural objects and intricate man-made objects in a manageable and controllable fashion. Higher-level modeling techniques have been developed to provide an *abstraction* of the model, encode classes of objects, and allow high-level control and specification of the models. Most of these advanced modeling techniques can be considered *procedural* modeling techniques: code segments or algorithms are used to abstract and encode the details of the model, instead of explicitly storing vast numbers of low-level primitives. The use of algorithms unburdens the modeler/animator from low-level control, provides great flexibility and allows amplification of their efforts through parametric control: a few parameters to the model yield large amounts of geometric detail (Smith [1984] referred to this as *database amplification*). This amplification allows a savings in storage of data and user specification time. The modeler has the flexibility to capture the *essence* of the object or phenomena being modeled without being constrained by the laws of physics and nature. He can include as much physical accuracy, and also as much artistic expression, as he wishes into the model.

This survey examines several types of procedural advanced geometric modeling techniques, including **fractals**, **grammar-based models**, **volumetric procedural models**, **implicit surfaces**, and **particle systems**. Most of these techniques are used to model natural objects and phenomena because the inherent complexity of nature renders traditional modeling techniques impractical. These techniques can also be classified into **surface-based modeling** techniques and **volumetric modeling** techniques. Fractals, grammar-based models, and implicit surfaces[1] are surface-based modeling techniques. Volumetric procedural models and particle systems are volumetric modeling techniques.

---

[1]Implicit surfaces are rendered as surfaces, although the actual model is volumetric.

0-8493-2909-4/97/$0.00+$.50
© 1997 by CRC Press, Inc.

56.2 Fractals

Fractals and chaos theory have rapidly grown in popularity since the early 1960s [Peitgen et al. 1992]. Mathematicians in the late nineteenth century and early twentieth century, including Cantor, Sierpiński, and von Koch, "discovered" fractal mathematics, but considered these formulas to be "mathematical monsters" that defied normal mathematical principles. Benoit Mandelbrot, who coined the term *fractal,* was the first person to realize that these mathematical formulas were a geometry for describing nature.

Fractals [Peitgen et al. 1992] have a precise mathematical definition, but in computer graphics their definition has been extended to generally refer to models with a large degree of *self-similarity:* subpieces of the object appear to be scaled down, possibly translated and rotated versions of the original object.[2] Along these lines, Musgrave [Ebert et al. 1994] defines a fractal as "a geometrically complex object, the complexity of which arises through the repetition of form over some range of scale." Many natural objects exhibit this characteristic, including mountains, coastlines, trees, plants (e.g., cauliflower), water, and clouds. In describing fractals, the amount of "roughness," "detail," or amount of space filled by the fractal can be mathematically characterized by its *fractal dimension* (self-similarity dimension), D. The fractal dimension is related to the common integer dimensionality of geometric objects: a line is 1D, a plane is 2D, a sphere is 3D. Fractal objects have noninteger dimensionality. An easy way to explain fractal dimension is to define it in terms of the recursive subdivision technique usually used to create simple fractals. If the original object is subdivided into a pieces using a reduction factor of s, the dimension D is related by the power law [Peitgen et al. 1992]

$$a = \frac{1}{s^D}$$

which yields

$$D = \frac{\log a}{\log(1/s)}$$

Normal geometric objects produce fractal (self-similarity) dimensions that are integers. Fractals produce noninteger, fractional, fractal dimensions. The following examples will illustrate this. If a cube is subdivided into 27 equal pieces, each one is scaled down by a factor of one-third, yielding

$$D = \frac{\log 27}{\log \left(1/\frac{1}{3}\right)} = \frac{\log 27}{\log 3} = 3$$

Conversely, a fractal object such as the *von Koch snowflake*[3] has a noninteger fractal dimension. The von Koch snowflake can be constructed by taking each side of an equilateral triangle, recursively dividing it into three equal pieces, and replacing the middle piece with two equal length pieces rotated to form two sides of an equilateral triangle as illustrated in Fig. 56.1. Analyzing the self-similarity dimension of this object yields a noninteger value:

$$D = \frac{\log 4}{\log \left(1/\frac{1}{3}\right)} = \frac{\log 4}{\log 3} \approx 1.2618595$$

The von Koch curve has a fractal dimension between one and two, indicating that it is more space-filling than a line, but not as much as a two-dimensional object. This property is characteristic of fractal curves. By definition, a fractal has a noninteger self-similarity dimension.

Fractals can generally be classified as deterministic and nondeterministic (also called random fractals), depending on whether they contain randomness. In computer graphics, deterministic fractals are closely related to the grammar-based L-systems described in the following section. Random fractals have been used extensively in computer graphics to model self-similar, complex, natural objects, including terrain,

---

[2] Mathematically speaking, the self-similarity must be infinite for the set to be a true fractal.
[3] Named for the mathematician, Helga von Koch, who "discovered" it in 1904.

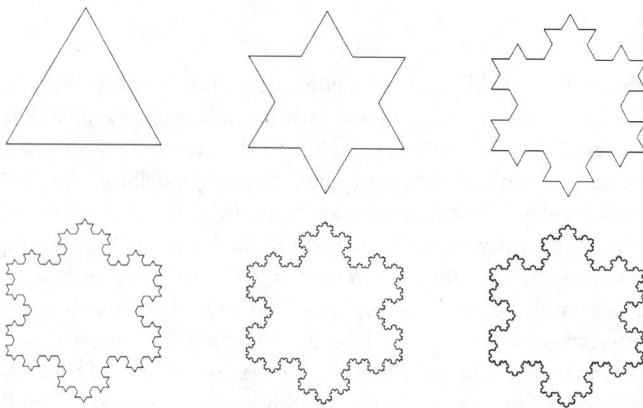

FIGURE 56.1   The von Koch snowflake after 0, 1, 2, 3, 4, and 5 iterations.

mountains, clouds, water, and even entire planets [Ebert et al. 1994]. Indeed, fractals are the most common technique used in graphics for modeling mountains. Most fractal terrain generation algorithms work through recursive subdivision and pseudorandom perturbation. An original surface is defined and divided equally into subparts. New vertices are added and pseudorandomly displaced from the original surface, with a displacement magnitude that decreases at each iteration. Therefore, the first iteration gives the large peaks on the surface, and later subdivisions add smaller-scale detail. A common algorithm for mountains by Fournier, Fussell, and Carpenter uses this technique by starting with a triangle, adding new vertices at the midpoint of each edge with a random height displacement, generating new triangles from these points, and repeating this subdivision with decreased height displacements at each iteration. Typically, only parameters for controlling the random number generator, the level of subdivision, and the "roughness" of the surface are needed to define extremely complex mountains and terrain. Others have used simulations of fractional Brownian motion (fBm) and statistical simulations of noise to produce the height displacements for the mountains. Musgrave [Ebert et al. 1994] uses a nonrecursive algorithm to displace the vertices of a regular grid with values iteratively calculated based on fBm and noise functions (at each iteration, the frequency is increased and the magnitude decreased) to create realistic terrain models. Recent work on fractal terrain generation has included eroding the fractal terrain to simulate natural erosion processes and the use of *multifractal* models. Multifractal terrain models allow the fractal dimension of the terrain to vary across the surface. This variation allows rougher areas and smoother areas, providing more realistic natural terrain. An example of the realistic landscapes that can be produced by these techniques can be seen in Fig. 56.2.

Recent work in fractals has included the simulation of diffusion-limited aggregation (DLA) models, the previously mentioned use of multifractals, and the use of fractal models to add complex details into models. DLA is a process based on random walks (fBm motion) of particles. Several initial sticky particles are placed in space. A large number of additional particles are moved on random walks; if they touch one of the sticky particles, they stick and may become sticky also. This process continues until all the additional particles attach to or move far enough away from the original particles. DLA models are being used to model a wide range of random processes from the formation of dendrite clusters to the formation of galaxies.

There are two common applications where geometric details are added with fractals. One is the addition of realistic, detailed, fractal terrain to coarse digital elevation data to provide realistic, higher-resolution terrain models. Another is the use of fractals to add small levels of geometric detail to standard geometric models. This allows less geometry in the model, with the procedural fractal functions being applied at rendering time to add an appropriate level of detail [Hart 1995].

There are many open areas of research in fractal modeling. Better erosion models that take into account different rock hardnesses, better rain distribution models, deposition of material in addition to erosion,

FIGURE 56.2 A fractal terrain model by Ken Musgrave. (See color version of this figure in the color section of the Handbook.) © 1992 *Slickrock*. F. Kenton Musgrave. Used by permission.

wind erosion, and the use of nonheight fields to allow rock overhangs will improve the realism of fractal terrains. Multifractals, diffusion-limited aggregation, and fractal detail addition have just begun to be explored and show great potential for geometric modeling.

56.3 Grammar-Based Models

Grammar-based models also allow natural complexity to be specified with a few parameters. Smith [1984] introduced grammar-based models to graphics, calling them *graftals*. The most commonly used grammar-based model, an L-system (named for Aristid Lindenmayer), was originally developed as a mathematical theory of plant development [Prusinkiewicz and Lindenmayer 1990]. An L-system is a formal language, a parallel graph grammar, where all the rules are applied in parallel to provide a final "word" describing the object. This parallel application of the production rules distinguishes L-systems from Chomsky grammars. Like Chomsky grammars, there are context-free L-systems (0L), and context-sensitive L-systems (1L and 2L).

Grammar-based models have been used by many authors, including Fowler, Lindenmayer, Prusinkiewicz, and Smith, to produce remarkably realistic models and images of trees, plants, and seashells. These models describe natural structures algorithmically and are closely related to deterministic fractals in their self-similarity, but fail to meet the precise mathematical definition of a fractal. Many deterministic fractals can be defined with L-systems, but not all L-systems meet the definition of a fractal.

As with most formal languages, an L-system can be described by an alphabet for the grammar, the grammar production rules, and an initial axiom. In plant modeling, alphabet symbols represent botanical structures (usually letters) and branching commands (usually "[ ]" denotes the beginning and end of a branch). We can add denotation for angular movement by introducing a "+" for clockwise rotations and "−" for counterclockwise rotations. The following simple L-system can produce a basic tree:

Alphabet: $a, [,], +, -$

Production Rule: $a \rightarrow a[+a]a[-a-a]a$

Initial Axiom: a

In the L-system, each terminal symbol represents a part of the object (e.g., a branch element, internode, apex) or a directional command (e.g., turn left 30 deg) to be interpreted by a three-dimensional drawing

FIGURE 56.3 Trees produced after 1, 2, and 3 derivations with the PGF software by Prusinkiewicz.

mechanism (turtle graphics). A "word" for a tree would contain subwords describing each branch, its length, size, and branching angle, when it develops, and its connection in the tree. For example, if we interpret each "$[+a]$" as defining a right branch at 30 deg, each "$[-a]$" as defining a left branch at 30 deg, and each "a" as a tree segment (internode or apex), the above grammar can be interpreted graphically as the trees in Fig. 56.3 after 1, 2, and 3 derivations and symbolically as the following:

Derivation	Word
0	a
1	$a[+a]a[-a-a]a$
2	$a[+a]a[-a-a]a[+a[+a]a[-a-a]a]a[+a]a[-a-a]a[-a[+a]a[-a-a]a$
	$-a[+a]a[-a-a]a]a[+a]a[-a-a]a$

To allow more complex plant and plant growth models, L-systems have been extended to include context sensitivity, word age information, and stochastic rule evaluation. Context sensitivity allows the relationships between parts of plants to be incorporated into the model. 1L-systems have one-sided contexts: either a right or a left context, yielding production rules of the form

$$a_l < a \to F$$

and

$$a > a_r \to F$$

The production rule is applied if and only if either its right context, aa_r, is satisfied or its left context, $a_l a$, is satisfied (either if a is preceded by a_l or if a is followed by a_r). 2L systems have both a left context and a right context, yielding production rules of the form

$$a_l < a > a_r \to F$$

Stochastic L-systems assign a probability to the application of a rule. This allows randomness into the creation of the plant, so that each plant is slightly different. Deterministic L-systems produce identical plants each time they are evaluated and would, therefore, create an unrealistic field of identical flowers. Stochastic rule evaluation is added into the grammar by associating a probability, p, with each production rule as follows:

$$a \xrightarrow{p} F$$

Parametric L-systems allow word age information and conditional expressions based on parameter values to be included into the model, permitting developmental growth models for plants. A parametric

L-system has rules of the form

$$F(t) : t > 2 \rightarrow F(t-1)H(t+7)$$

where H is another parametric production rule. Parametric L-systems can be combined with stochastic L-systems and context-sensitive L-systems to achieve a powerful, flexible grammar.

Recent work in L-systems allows better developmental models, more advanced biologically based growth models, incorporation of more growth parameters, and environmental effects. Extensions of L-systems allow the death of the buds, dropping of leaves, and cutting of branches with the inclusion of erasing and cutting operators. Botanically based flowering structures (inflorescences), branching structures (sympodial and monopodial), and arrangement of lateral plant organs (phyllotaxis) [Fowler et al. 1992] have been incorporated into L-systems to more accurately model plants. **Tropism** effects (gravity, wind, growth toward light), pruning, amount of light, and availability of nutrients have also been incorporated into these grammars. These natural effects and growth processes not only affect the structure (topology) of the tree, but also affect the branching angles, petal and seed location and shape, and thickness of each branch segment. When generating the geometric plant model from the L-system grammar, these effects need to be included in determining the geometry and size of each structure in the plant. Figure 56.4 shows a realistic image of a horse chestnut tree generated by a modified L-system that takes into account the competition of branches for light.

FIGURE 56.4 Horse chestnut tree created with a modified L-system that takes into account branch competition for light. (See color version of this figure in the color section of this Handbook.) © 1995 by R. Měch and P. Prusinkiewicz. University of Calgary, Canada. Used by permission.

Ongoing L-systems research includes environmentally-sensitive L-systems [Prusinkiewicz et al. 1995] and the use of L-systems for modeling other growth processes and artificial life. Additionally, better developmental models can be simulated and more accurate modeling of natural growth factors can be included.

56.4 Volumetric Procedural Models

Another procedural modeling technique, volumetric procedural modeling (also called hypertextures, volume density functions, and fuzzy blobbies), uses algorithms to define and animate three-dimensional volumetric objects and natural phenomena [Ebert et al. 1994]. These techniques have been used to model natural phenomena such as fire (Stam and Inakage), gases such as smoke, clouds, and fog (Ebert, Perlin, Sakas, Stam), and water (Ebert, Perlin). The volumetric procedures take as input a point location in space, time (if animating), and parameters that describe the object being modeled, and return the density and color of the object for that location in space. Complex volumetric phenomena can therefore be described with a few parameters.

An extremely simple volumetric procedural model for a spherical volume object is given below. This function, **spherical_vpm**, defines a sphere of radius **outer_radius**, with a solid center of radius **inner_radius**. The region between **inner_radius** and **outer_radius** is a semi-solid area that increases in density as the inner radius is approached. The following function, suggested by Perlin to create a *soft sphere* [Ebert et al. 1994], returns a density value when a point in space and the sphere definition are given as input:

```
typedef struct rgb_td { float r,g,b} rgb_td;
typedef struct xyz_td { float x,y,z} xyz_td;
typedef struct vol_td
    {
    float  density;
    rgb_td color;
    } vol_td;

vol_td spherical_vpm(xyz_td pnt,xyz_td center,float inner_radius,
                  float outer_radius, float density_factor)
{
    float outer_radius_2, inner_radius_2, distance_2;
    vol_td vol;

    distance_2 = (pnt.x - center.x)*(pnt.x - center.x)
              + (pnt.y - center.y)* (pnt.y - center.y)
              + (pnt.z - center.z)*(pnt.z - center.z);

    /* compute outer and inner radius squared values */
    outer_radius_2 = outer_radius * outer_radius;
    inner_radius_2 = inner_radius * inner_radius;

  if (distance_2 < inner_radius_2)
    { /* inside inner radius */
     vol.density =1.0*density_factor;
     vol.color.r = 1.0; vol.color.g = 0.0; vol.color.b = 0.0;
    }
```

```
        else if (distance_2 > outer_radius_2)
          { /* outside of the sphere */
          vol.density = 0.0;
          vol.color.r = 0.0; vol.color.g = 0.0; vol.color.b = 0.0;
          }
        else
          { /* in the soft area of the sphere */
          vol.density = density_factor*(distance-inner_radius_2)/
               (outer_radius_2-inner_radius_2);
          vol.color.r=vol.color.g = 1.0*vol.density;vol.color.b=0;
          }
        return(vol);
    }
```

Many authors have used these techniques to describe a wide range of natural objects. Perlin has success-fully created realistic rock arches, woven fabric, smoke, and fur [Ebert et al. 1994], basing his procedures on a statistical simulation of turbulence and random noise to give natural-looking complexity to the objects. Perlin's turbulence function has been used as a building block for volumetric procedural objects, solid textures, and many other applications in computer graphics. This turbulence function defines a three-dimensional turbulence space by summing octaves of random noise, increasing the frequency and decreas-ing the amplitude at each step. The C function below is one implementation of Perlin's turbulence function:

```
    float turbulence(xyz_td pnt, float pixel_size)
    {
      float t, scale;
      t=0;
      for(scale=1.0; scale >pixel_size; scale/=2.0)
        {
          pnt.x = pnt.x/scale; pnt.y = pnt.y/scale;
          pnt.z = pnt.z/scale;
          t+=calc_noise(pnt)* scale ;
        }
      return(t);
    }
```

This function takes as input a three-dimensional point location in space, **pnt**, and an indication of the number of octaves of noise to sum, **pixel_size**,[4] and returns the turbulence value for that location in space. This function has a fractal characteristic in that it is self-similar and sums the octaves of random noise, doubling the frequency while halving the amplitude at each step. The heart of the **turbulence** function is the **calc_noise** function used to simulate uncorrelated random noise. Many authors have used various implementations of the noise function (see [Ebert et al. 1994] for several possible implementations). One implementation is the **calc_noise** function given below, which uses linear interpolation of a $64 \times 64 \times 64$ grid of random numbers[5]:

---

[4]This variable name is used in reference to the projected area of the pixel in the three-dimensional turbulence space for antialiasing.

[5]The actual implementation uses a 65^3 table with the 64th entry equal to the 0th entry for quicker interpolation.

```
#define SIZE        64
#define SIZE_1      65
double drand48();
float calc_noise();
float noise[SIZE+1][SIZE+1][SIZE+1];

/*
 *********************************************************************
 *                          Calc_noise
 *********************************************************************
 * This is basically how the trilinear interpolation works:
 * interpolate down left front edge of the cube first, then the
 * right front edge of the cube(p_l, p_r). Next, interpolate down
 * the left and right back edges (p_l2, p_r2). Interpolate across
 * the front face between p_l and p_r (p_face1) and across the
 * back face between p_l2 and p_r2 (p_face2). Finally, interpolate
 * along line between p_face1 and p_face2.
 *********************************************************************
 */
float
calc_noise(xyz_td pnt)
{
   float t1;
   float p_l,p_l2, /* value lerped down left side of face1 &
                    * face 2 */
         p_r,p_r2, /* value lerped down right side of face1 &
                    * face 2 */
         p_face1, /* value lerped across face 1 (x-y plane ceil
                   * of z) */
         p_face2, /* value lerped across face 2 (x-y plane floor
                   * of z) */
         p_final; /* value lerped through cube (in z) */

   extern float noise[SIZE_1][SIZE_1][SIZE_1];
   float         tnoise;
   register int x, y, z,px,py,pz;
   int           i,j,k, ii,jj,kk;
   static int    firstime =1;

   /* During first execution, create the random number table of
    * values between 0 and 1, using the Unix random number
    * generator drand48(). Other random number generators may be
    * substituted. These noise values can also be stored to a
    * file to save time.
    */
```

```
    if (firsttime)
      { for (i=0; i<SIZE; i++)
        for (j=0; j<SIZE; j++)
          for (k=0; k<SIZE; k++)
            {
              noise[i][j][k] = (float)drand48();
              /* A crude way to make element[64]=element[0] for
               * easier linear interpolation */
              if(i==0) noise[SIZE][j][k] = noise[i][j][k];
              if(j==0) noise[i][SIZE][k] = noise[i][j][k];
              if(k==0) noise[i][j][SIZE] = noise[i][j][k];
            }
        firsttime=0;
      }

  px = (int)pnt.x;
  py = (int)pnt.y;
  pz = (int)pnt.z;
  x = px &(SIZE-1); /* make sure the values are in the table */
  y = py &(SIZE-1); /* Effectively, replicates the table
                     * throughout space */
  z = pz &(SIZE-1);

  t1 = pnt.y - py;
  p_l = noise[x][y][z+1] +t1*(noise[x][y+1][z+1]
        - noise[x][y][z+1]);
  p_r = noise[x+1][y][z+1] +t1*(noise[x+1][y+1][z+1]
        - noise[x+1][y][z+1]);
  p_l2 = noise[x][y][z] +t1*(noise[x][y+1][z] - noise[x][y][z]);
  p_r2 = noise[x+1][y][z] +t1*(noise[x+1][y+1][z]
        - noise[x+1][y][z]);
  t1 = pnt.x - px;
  p_face1 = p_l + t1 * (p_r - p_l);
  p_face2 = p_l2 + t1 * (p_r2 -p_l2);
  t1 = pnt.z - pz;
  p_final = p_face2 + t1*(p_face1 -p_face2);
  return(p_final);
  }
```

[Ebert et al. 1994] has used similar functions to model and animate steam, fog, smoke, clouds, and solid marble. The turbulence function and random noise functions allow a simple simulation of turbulent flow processes. To simulate steam rising from a teacup, a volume of gas is placed over the teacup. The basic gas is defined by the following function:

```
float basic_gas(xyz_td pnt, float density, float density_scalar,
                float exponent)
{
  float turb, density;

  turb =turbulence(pnt);
  density = pow(turb*density_scalar, exponent);
  return(density);
}
```

This function creates a three-dimensional gas space controlled by the values **density_scalar** and **exponent**. **density_scalar** controls the denseness of the gas, while **exponent** controls the sharpness and sparseness of the gas (from continuously varying to sharp individual plumes). This function is then shaped to create steam over the center of the teacup by spherically attenuating the density toward the edge of the cup and linearly attenuating the density as the distance from the top of the cup increases, simulating the gas dissipation as it rises. The following procedure will produce an image of steam rising from a teacup, as in in Fig. 56.5.

```
float steam(xyz_td pnt, xyz_td pnt_world, float exponent,
            float density_scalar,xyz_td tea_center,float radius)
{
  float   turb, dist, dist_sq, density_max, fall_off, offset;
  xyz_td  diff;

  turb = turbulence(pnt);
```

FIGURE 56.5 Steam rising from a teacup. (See color version of this figure in the color section of the Handbook.) © 1991 by David S. Ebert.

```
        density = pow(turb*density_scalar, exponent);

        /* determine distance from center of the teacup squared. */
        XYZ_SUB(diff,tea_center, pnt_world);
        dist_sq = DOT_XYZ(diff,diff);
        /* calculate relative distance from center with some
         * randomness */
        density_max = dist_sq/(radius*radius);
        density_max += .2*noise(pnt);

        /* Use a cosine function to spherically attenuate the
         * density */
        if(density_max >= .25) /* ramp off if > 25% from center */
           { /* get table index 0:RAMP_SIZE-1 */
               density_max = MAX((density_max -.25)*4/3, 1.0);
               fall_off = (cos(density_max*M_PI)+1.0)/2.0;
               density *=fall_off;

           }
        /* Use exponential attenuation to decrease density as steam
         * rises */
        dist = pnt_world.y - tea_center.y;
        if(dist > 0.0)
           { dist = (dist +noise(pnt)*.1)/radius.y;
             if(dist > .05)
                { offset = (dist -.05)*1.111111;
                  offset = 1 - (exp(offset)-1.0)/1.718282;
                  density = density*offset;
                }

           }
     return(density);

   }
```

These procedural techniques can be easily animated by adding time as a parameter to the algorithm [Ebert et al. 1994]. They allow the use of simple simulations of natural complexity (noise, turbulence) to speed computation, but also allow the incorporation of physically based parameters where appropriate and feasible. This flexibility is one of the many advantages of procedural techniques.

Volumetric procedural models require volume rendering techniques to create images of these objects. Most authors use a modification of volume ray tracing. Several authors have incorporated physically based models for shadowing and illumination into rendering algorithms for these models [Ebert and Parent 1990, Stam and Fiume 1995].

Volumetric procedural modeling is still in its infancy and has many research problems to address. Efficient rendering of these models is still an important issue, as is the development of a larger toolbox of useful primitive functions. The incorporation of more physically based models will increase the accuracy and realism of the water, gas, and fire simulations. Finally, the development of an interactive volumetric procedural modeling system will speed the development of volumetric procedural modeling techniques.

56.5 Implicit Surfaces

While previously discussed techniques have been used primarily for modeling the complexities of nature, implicit surfaces (also called blobby molecules [Blinn 1982], metaballs [Nishimura et al. 1985], and soft objects [Wyvill et al. 1986]) are used in modeling organic shapes, complex man-made shapes, and "soft" objects that are difficult to animate and describe using more traditional techniques. Implicit surfaces are surfaces of constant value, **isosurfaces**, created from blending primitives (functions or skeletal elements) represented by implicit equations of the form $F(x, y, z) = 0$, and were first introduced into computer graphics by Blinn [1982] to produce images of electron density clouds. A simple example of an implicit surface is the sphere defined by the equation

$$F(x, y, z) : x^2 + y^2 + z^2 - r^2 = 0$$

Implicit surfaces are a more concise representation than parametric surfaces and provide greater flexibility in modeling and animating soft objects.

For modeling complex shapes, several basic implicit surface primitives are smoothly blended to produce the final shape. For the blending function, Blinn used an exponential decay of the field values, whereas Wyvill [Wyvill et al. 1986, 1993] uses the cubic function

$$F_{\text{cub}}(r) = -\frac{4}{9}\frac{r^6}{R^6} + \frac{17}{9}\frac{r^4}{R^4} - \frac{22}{9}\frac{r^2}{R^2} + 1$$

This cubic blending function, whose values range from 1 when $r = 0$ to 0 at $r = R$, has several advantages for complex shape modeling. First, its value drops off quickly to zero (at the distance R), reducing the number of primitives that must be considered in creating the final surface. Second, it has zero derivatives at $r = 0$ and $r = R$ and is symmetrical about the contour value 0.5, providing for smooth blends between primitives. Finally, it can provide volume-preserving primitive blending. Figure 56.6(a) shows a graph of this blending function, and Fig. 56.6(c) shows the blending of two spheres using this function.

For implicit surface primitives, Wyvill uses procedures that return a functional (field) value for the field defined by the primitive. Field primitives, such as lines, points, polygons, circles, splines, spheres, and ellipsoids, are combined to form a basic skeleton for the object being modeled. The surfaces resulting from these skeletal elements can be seen in Fig. 56.6(b). The object is then defined as an offset (isosurface) from this series of blended skeletal elements. Skeletons are an intuitive representation and are easily displayed and animated.

Modeling and animation of implicit surfaces is achieved by controlling the skeletal elements and blending functions, providing complex models and animations from a few parameters (another example of data

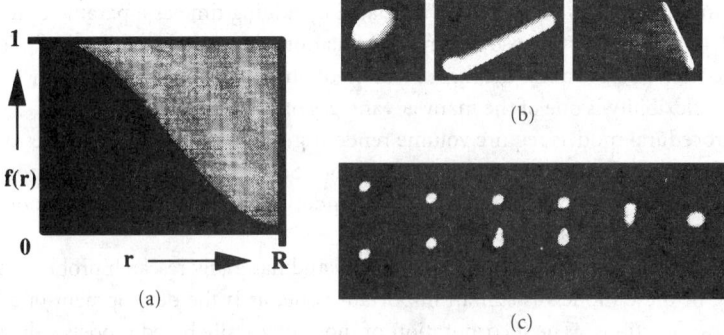

FIGURE 56.6   (a) Blending function, (b) surfaces produced by skeletal elements point, line and polygon, and (c) blended spheres. © 1995 Brian Wyvill. Used by permission.

FIGURE 56.7 Ten years in implicit surface modeling. The locomotive labeled 1985 shows a more traditional soft object created by implicit surface techniques. The locomotive labeled 1995 shows the results achievable by incorporating constructive solid geometry techniques with implicit surface models. (See color version of this figure in the color insert section of the Handbook.) © 1995 by Brian Wyvill. Used by permission.

amplification). Deformation, path following, warping, squash and stretch, gravity, and metamorphosis effects can all be easily achieved with implicit surfaces. Very high-level animation control is achieved by animating the basic skeleton, with the surface defining the character following naturally. The animator does not have to be concerned with specifying the volume-preserving deformations of the character as it moves.

There are two common approaches to rendering implicit surfaces. One approach is to directly ray-trace the implicit surfaces, requiring the modification of a standard ray tracer. The second approach is to polygonalize the implicit surfaces [Ning and Bloomenthal 1993, Wyvill et al. 1993] and then use traditional polygonal rendering algorithms on the result. Uniform-voxel space polygonization can create large numbers of unnecessary polygons to accurately represent surface details. More complicated tessellation and shrinkwrapping algorithms have been developed which create appropriately sized polygons [Wyvill et al. 1993].

Recent work in implicit surfaces [Wyvill and Gascuel 1995] has extended their use to character modeling and animation, human figure modeling, and representation of rigid objects through the addition of constructive solid geometry (CSG) operators. Implicit surface modeling techniques have advanced significantly in the past 10 years, as can be seen by comparing the locomotives in Fig. 56.7. The development of better blending algorithms, which solve the problems of unwanted primitive blending and surface bulging, is an active area of research [Bloomenthal 1995]. Advanced animation techniques for implicit surfaces, including higher-level animation control, surface collision detection, and shape metamorphosis animation, are also active research areas. Finally, the development of interactive design systems for implicit surfaces will greatly expand the use of this modeling technique.

56.6 Particle Systems

Particle systems are different from the previous four techniques in that their abstraction is in control of the animation and specification of the object. A particle-system object is represented by a large collection (cloud) of very simple geometric particles that change stochastically over time. Therefore, particle systems do use a large database of geometric primitives to represent natural objects ("fuzzy objects"), but the animation, location, birth, and death of the particles representing the object are controlled algorithmically. As with the other procedural modeling techniques, particle systems have the advantage of database amplification, allowing the modeler/animator to specify and control this extremely large cloud of geometric particles with only a few parameters.

Particle systems were first used in computer graphics by Reeves [1983] to model a wall of fire for the movie *Star Trek II: The Wrath of Khan* (see Fig. 56.8). Since particle systems are a volumetric modeling technique, they are most commonly used to represent volumetric natural phenomena such as fire, water, clouds, snow, and rain [Reeves 1983]. An extension of particle systems, *structured particle systems,* has also been used to model grass and trees [Reeves and Balu 1985].

A particle system is defined by both a collection of geometric particles and the algorithms that govern their creation, movement, and death. Each geometric particle has several attributes, including its initial position, velocity, size, color, transparency, shape, and lifetime.

To create an animation of a particle system object the following steps are performed at each time step [Reeves 1983]:

1. New particles are generated and assigned their attributes.
2. Particles that have existed in the system past their lifetime are removed.
3. Each remaining particle is moved and transformed by the particle-system algorithms as prescribed by their individual attributes.
4. These particles are rendered, using special-purpose rendering algorithms, to produce an image of the particle system.

The creation, death, and movement of particles are controlled by stochastic procedures, allowing complex, realistic motion to be created with a few parameters. The creation procedure for particles is controlled by parameters defining either the mean number of particles created at each time step and its

FIGURE 56.8   An image from *Star Trek II: The Wrath of Khan* showing a wall of fire created with a particle system. (See the color version of this figure in the color section of the Handbook.)
© 1987 Pixar. Used by permission.

variance, or the mean number of particles created per unit of screen area at each time step and its variance.[6] The actual number of particles created is stochastically determined to be within *mean + variance* and *mean − variance*. The initial color, velocity, size, and transparency are also stochastically determined by mean and variance values. The initial shape of the particle system is defined by an origin, a region about this origin in which new generated particles are placed, angles defining the orientation of the particle system, and the initial direction of movement for the particles.

The movement of particles is also controlled by stochastic procedures (stochastically determined velocity vectors). These procedures move the particles by adding their velocity vector to their position vector. Random variations can be added to the velocity vector at each frame, and acceleration procedures can be incorporated to simulate effects such as gravity, vorticity, and conservation of momentum and energy. The simulation of physically based forces allows realistic motion and complex dynamics to be displayed by the particle system, while being controlled by only a few parameters. Besides the movement of particles, their color and transparency can also change dynamically to give more complex effects. The death of particles is controlled very simply by removing particles from the system whose lifetimes have expired or that have strayed more than a given distance from the origin of the particle system.

An example of the effects achievable by such a particle system can be seen in Fig. 56.8, an image from the Genesis Demo sequence from *Star Trek II: The Wrath of Khan*. In this image, a two-level particle system was used to create the wall of fire. The first-level particle system generated concentric, expanding rings of particle systems on the planet's surface. The second-level particle system generated particles at each of these locations, simulating explosions. During the Genesis Demo sequence, the number of particles in the system ranged from several thousand initially to over 750,000 near the end.

Reeves extended the use of particle systems to model fields of grass and forests of trees, calling this new technique structured particle systems [Reeves and Blau 1985]. In structured particle systems, the particles are no longer an independent collection of particles, but rather form a connected, cohesive three-dimensional object and have many complex relationships among themselves. Each particle represents an element of a tree (e.g., branch, leaf) or part of a blade of grass. These particle systems are therefore similar to L-systems and graftals, specifically probabilistic, context-sensitive L-systems. Each particle is similar to a letter in an L-system alphabet, and the procedures governing the generation, movement, and death of particles are similar to the production rules. However, they differ from L-systems in several ways. First, the goal of structured particle systems is to model the visual appearance of whole collections of trees and grass, and not to correctly model the detailed geometry of each plant. Second, they are not concerned with biological correctness or modeling growth of plants. Structured particle systems construct trees by recursively generating subbranches, with stochastic variations of parameters such as branching angle, thickness, and placement within a value range for each type of tree. Additional stochastic procedures are used for placement of the trees on the terrain, random warping of branches, and bending of branches to simulate tropism. A forest of such trees can therefore be specified with a few parameters for distribution of tree species and several parameters defining the mean values and variances for tree height, width, first branch height, length, angle, and thickness of each species.

Both regular particle systems and structured particle systems pose special rendering problems because of the large number of primitives. Regular particle systems have been rendered simply as point light sources (or linear light sources for antialiased moving particles) for fire effects, accumulating the contribution of each particle into the frame buffer and compositing the particle system image with the surface rendered image (as in Fig. 56.8). No occlusion or interparticle illumination is considered. Structured particle systems are much more difficult to render, and specialized probabilistic rendering algorithms have been developed to render them [Reeves and Blau 1985]. Illumination, shadowing, and hidden-surface calculations need to be performed for the particles. Since stochastically varying objects are being modeled, approximately correct rendering will provide sufficient realism. Probabilistic and approximate techniques are used to determine the shadowing and illumination of each tree element. The particle's distance into the tree from the light

---

[6]These values can be varied over time as well.

source determines its amount of **diffuse shading** and probability of having **specular highlights**. Self-shadowing is simulated by exponentially decreasing the **ambient illumination** as the particle's distance within the tree increases. External shadowing is also probabilistically calculated to simulate the shadowing of one tree by another tree. For hidden-surface calculations, an initial depth sort of all trees and a **painter's algorithm** is used. Within each tree, again, a painter's algorithm is used, along with a back-to-front bucket sort of all the particles. This will not correctly solve the hidden-surface problem in all cases, but will give realistic, approximately correct images.

Efficient rendering of particle systems is still an open research problem. Although particle systems allow complex scenes to be specified with only a few parameters, they sometimes require rather slow, specialized rendering algorithms. Simulation of fluids [Miller and Pearce 1989], cloth [Breen et al. 1994], and surface modeling with oriented particle systems [Szeliski and Tonnesen 1992] are recent, promising extensions of particle systems. Sims [1990] demonstrated the suitability of highly parallel computing architectures to particle-system simulation. Particle systems, with their ease of specification and good dynamical control, have great potential when combined with other modeling techniques such as implicit surfaces [Witkin and Heckbert 1994] and volumetric procedural modeling.

56.7 Research Issues and Summary

Advanced modeling techniques will continue to play an important role in computer graphics. As computers become more powerful, the complexity that can be rendered will increase; however, the capability of humans to specify more geometric complexity (millions of primitives) will not. Therefore, procedural techniques, with their capability to amplify the user's input, are the only viable alternative. These techniques will evolve in their capability to specify and control incredibly realistic and detailed models with a small number of user-specified parameters. More work will be done in allowing high-level control and specification of models in user-understandable terms, while more complex algorithms and improved physically based simulations will be incorporated into these procedures. Finally, the automatic generation of procedural models through artificial evolution techniques, similar to those of Sims [1994], will greatly enhance the capabilities and uses of these advanced modeling techniques.

Defining Terms

Ambient illumination: An approximation of the global illumination on the object, usually modeled as a constant amount of illumination per object.

Diffuse shading: The illumination of an object where light is reflected equally in all directions, with the intensity varying based on surface orientation with respect to the light source. This is also called Lambertian reflection, since it is based on Lambert's law of diffuse reflection.

Fractal: Generally refers to a complex geometric object with a large degree of self-similarity and a noninteger fractal dimension that is not equal to the object's topological dimension.

Grammar-based modeling: A class of modeling techniques based on formal languages and formal grammars where an alphabet, a series of production rules, and initial axioms are used to generate the model.

Implicit surfaces: Isovalued surfaces created from blending primitives which are modeled with implicit equations.

Isosurface: A surface defined by all the points where the field value is the same.

L-system: A parallel graph grammar in which all the production rules are applied simultaneously.

Painter's algorithm: A hidden-surface algorithm that sorts primitives in a back-to-front order, then "paints" them into the frame buffer in this order, overwriting previously "painted" primitives.

Particle system: A modeling technique that uses a large collection (thousands) of particles to model complex natural phenomena, such as snow, rain, water, and fire.

Phyllotaxis: The regular arrangement of plant organs, including petals, seeds, leaves, and scales.

Specular highlights: The bright spots or highlights on objects caused by angular-dependent illumination. Specular illumination is dependent on the surface orientation, the observer location, and the light source location.

Surface-based modeling: Refers to techniques for modeling the three-dimensional surfaces of objects.

Tropism: An external directional influence on the branching patterns of trees.

Volumetric modeling: Refers to techniques that model objects as three-dimensional volumes of material, instead of being defined by surfaces.

Volumetric procedural models: Use algorithms to define the three-dimensional volumetric representation of an object.

References

Blinn, J. F. 1982. A generalization of algebraic surface drawing. *ACM Trans. Graphics* 1(3):235–256.

Bloomenthal, J. 1995. *Skeletal Design of Natural Forms.* Ph.D. thesis, Department of Computer Science, University of Calgary.

Breen, D. E., House, D. H., and Wozny, M. J. 1994. Predicting the drape of woven cloth using interacting particles, pp. 365–372. In *Proc. SIGGRAPH '94 (Orlando, Florida, July 24–29, 1994)*, A. Glassner, ed., Computer Graphics Proceedings, Annual Conf. Ser. ACM SIGGRAPH. ACM Press.

Ebert, D. and Parent, R. 1990. Rendering and animation of gaseous phenomena by combining fast volume and scanline a-buffer techniques. *Comput. Graphics (Proc. SIGGRAPH)* 24:357–366.

Ebert, D., Musgrave, F. K., Peachey, D., Perlin, K., and Worley, S. 1994. *Texturing and Modeling: A Procedural Approach.* AP Professional, Boston.

Foley, J. D., van Dam, A., Feiner, S. K., and Hughes, J. F. 1990. *Computer Graphics: Principles and Practices,* 2nd ed. Addison–Wesley, Reading, MA.

Fowler, D. R., Prusinkiewicz, P., and Battjes, J. 1992. A collision-based model of spiral phyllotaxis. *Comput. Graphics (Proc. SIGGRAPH)* 26:361–368.

Hart, J. 1995. Procedural models of geometric detail. In *SIGGRAPH '95: Course 33 Notes.* ACM SIGGRAPH.

Mandelbrot, B. B. 1983. *The Fractal Geometry of Nature.* W. H. Freeman, New York.

Miller, G. and Pearce, A. 1989. Globular dynamics: a connected particle system for animating viscous fluids. *Comput. and Graphics* 13(3):305–309.

Ning, P. and Bloomenthal, J. 1993. An evaluation of implicit surface tilers. *IEEE Comput. Graphics Appl.* 13(6):33–41.

Nishimura, H., Hirai, A., Kawai, T., Kawata, T., Kawa, I. S., and Omura, K. 1985. Object modelling by distribution function and a method of image generation (in Japanese). In *Journals of Papers Given at the Electronics Communication Conference '85*, J68-D(4).

Peitgen, H.-O., Jürgens, H., and Saupe, D. 1992. *Chaos and Fractals: New Frontiers of Science.* Springer–Verlag, New York.

Prusinkiewicz, P. and Lindenmayer, A. 1990. *The Algorithmic Beauty of Plants.* Springer-Verlag.

Prusinkiewicz, P., Hammel, M., and Mech, R. 1995. The artificial life of plants. In *SIGGRAPH '95: Course Notes.* ACM SIGGRAPH.

Reeves, W. T. 1983. Particle systems—a technique for modeling a class of fuzzy objects. *ACM Trans. Graphics* 2:91–108.

Reeves, W. T. and Blau, R. 1985. Approximate and probabilistic algorithms for shading and rendering structured particle systems. *Comput. Graphics (Proc. SIGGRAPH)* 19:313–322.

Sims, K. 1990. Particle animation and rendering using data parallel computation. *Comput. Graphics (Proc. SIGGRAPH)* 24:405–413.

Sims, K. 1994. Evolving virtual creatures, pp. 15–22. *Proc. of SIGGRAPH '94 (Orlando, Florida, July 24–29, 1994)*, A. Glassner, ed. Computer Graphics Proc., Annual Conf. Series. ACM SIGGRAPH, ACM Press.

Smith, A. R. 1984. Plants, fractals and formal languages. *Computer Graphics (Proc. SIGGRAPH)* 18: 1–10.

Stam, J. and Fiume, E. 1995. Depicting fire and other gaseous phenomena using diffusion processes, pp. 129–136. In *Proc. SIGGRAPH '95 (Los Angeles, California, August 6–11, 1995)*, R. Cook, ed. Computer Graphics Proc., Annual Conf. Series. ACM SIGGRAPH, ACM Press.

Szeliski, R. and Tonnesen, D. 1992. Surface modeling with oriented particle systems. *Comput. Graphics (Proc. SIGGRAPH)* 26:185–194.

Watt, A. and Watt, M. 1992. *Advanced Animation and Rendering Techniques: Theory and Practice.* Addison–Wesley, Reading, MA.

Witkin, A. P. and Heckbert, P. S. 1994. Using particles to sample and control implicit surfaces, pp. 269–278. In *Proc. SIGGRAPH '94 (Orlando, Florida, July 24–29, 1994)*, A. Glassner, ed., Computer Graphics Proc. Annual Conf. Series. ACM SIGGRAPH, ACM Press.

Wyvill, B. and Gascuel, M.-P., 1995. *Implicit Surfaces '95, The First International Workshop on Implicit Surfaces.* INRIA, Eurographics.

Wyvill, G., McPheeters, C., and Wyvill, B. 1986. Data structure for soft objects. *The Visual Computer* 2(4):227–234.

Wyvill, B., Bloomenthal, J., Wyvill, G., Blinn, J., Hart, J., Bajaj, C., and Bier, T. 1993. Modeling and animating implicit surfaces. In *SIGGRAPH '93: Course 25 Notes.*

Further Information

There are many sources of further information on advanced modeling techniques. Two of the best resources are the proceedings and course notes of the annual ACM SIGGRAPH conference. The SIGGRAPH conference proceedings usually feature a section on the latest, and often best, results in modeling techniques. The course notes are a very good source for detailed, instructional information on a topic. Several courses at SIGGRAPH '92, '93, '94, and '95 contained notes on procedural modeling, fractals, particle systems, implicit surfaces, L-systems, artificial evolution, and artificial life.

Standard graphics texts, such as *Computer Graphics: Principles and Practice* by Foley, van Dam, Feiner, and Hughes [Foley et al. 1990] and *Advanced Animation and Rendering Techniques* by Watt and Watt [1992], contain introductory explanations to these topics. The reference list contains references to excellent books and, in most cases, the most comprehensive sources of information on the subject. Additionally, the book *The Fractal Geometry of Nature*, by Benoit Mandelbrot [1983], is a classic reference for fractals. For implicit surfaces, a book by Bloomenthal, Wyvill et al. will be published by Morgan Kaufmann in 1996.

Another good source of reference material is specialized conference and workshop proceedings on modeling techniques. For example, the proceedings of the Eurographics '95 Workshop on Implicit Surfaces contains state-of-the-art implicit surfaces techniques.

57

Mainstream Rendering
Techniques

Alan Watt
University of Sheffield

57.1 Introduction

Rendering is the name given to the process in three-dimensional graphics whereby a geometric description of an object is converted into a two-dimensional image-plane representation that "looks real."

Three methods of rendering are now firmly established. The first and most common method is to use a simulation of light–object interaction in conjunction with **polygon mesh** objects, and we have called this approach *rendering polygon mesh objects*. Although the light–object simulation is independent of the object representation, the combination of empirical light–object interaction and polygon mesh representation has emerged as the most popular rendering technique in computer graphics. Because of its ubiquity and importance we shall devote most of this chapter to this approach.

This approach to rendering suffers from a significant disadvantage. The reality of light–object interaction is simulated as a crude approximation—albeit an effective and cheap simulation. In particular, objects are considered to exist in isolation with respect to a light source or sources, and no account is taken of light interaction between objects themselves. This means in practice that although we simulate the reflection of light incident on an object from a light source, we resolutely ignore the effects that the reflected light has on the scene when it travels onwards from its first reflection to encounter, perhaps, other objects, and so on. Thus common phenomena that depend on light reflecting from one object onto another, like shadows and objects reflecting in each other, cannot be produced by such a model.

Such models for this reason are called **local reflection models** to distinguish them from **global reflection models**, which attempt to follow the adventures of light emanating from a source as it hits objects, is reflected, hits other objects, and so on. The reason local reflection models "work"—in the sense that they produce visually acceptable, or even impressive, results—is that in reality the reflected light in a scene that emanates from first-hit incident light predominates. However, the subtle interactions that one normally encounters in an environment that are due to object–object interaction are important, and this motivation led to the development of the two **global reflection** models: **ray tracing** and **radiosity**.

Ray tracing simulates global interaction by explicitly tracking infinitely thin beams, or rays, of light as they travel through the scene from object to object. Radiosity, on the other hand, considers light reflecting in all directions from the surface of an object and calculates how light radiates from one surface to another as a function of the geometric relationship between surfaces—their proximity, relative orientation, etc. Ray tracing operates on points in the scene, radiosity on finite areas called patches.

Ray tracing and radiosity formed popular research topics in the eighties. Both methods are very much more expensive than polygon mesh rendering, and a common research motivation was efficiency—particularly in the case of ray tracing.

For reasons that will become clear later, ray tracing and radiosity can each simulate only one aspect of global interaction. Ray tracing deals with specular interaction and is fine for scenes consisting of shiny, mutually reflective objects. On the other hand, radiosity deals with diffuse or dull surfaces and is used mostly to simulate interiors of rooms. In effect the two methods are mutually exclusive—ray tracing cannot simulate diffuse interaction, and radiosity cannot cope with specular interaction. This fact led to another major research effort, which was to incorporate specular interaction in the radiosity method. In the nineties, although research continues into the radiosity method, work on ray tracing appears to have died out. At the moment most of the commercial implementations of these methods are of the early algorithms and we will give a basic treatment of these in this chapter.

57.2 Rendering Polygon Mesh Objects

Introduction

The overall process of rendering polygon mesh objects can be broken down into a sequence of geometric transformations and pixel processes that have been established for at least two decades as a de facto standard. Although not the only way to produce a shaded image of a three-dimensional object, the particular processes we shall describe represent a combination of popularity and ease of implementation. There is no established name for this group of processes which has emerged for rendering objects represented by a polygon mesh—by far the most popular form of representation. The generic term *rendering pipeline* applies to any set of processes that is used to render objects in three-dimensional graphics.

Ignoring any transformations that are involved in positioning many objects to make up a scene—modeling transformations—we can list and summarize these processes as follows:

> *Viewing transformation:* A process that is invoked to generate a representation of the object or scene as seen from the viewpoint of an observer positioned somewhere in the scene and looking towards some aspect of it. This involves a simple transformation that transforms the object from its database representation to one that is represented in a coordinate system related to the viewer's position and viewing direction. It establishes the size of the object, according to its distance from the viewer, and what parts of it are seen, according to the viewing direction.
>
> *Clipping:* The need for clipping is easily exemplified by considering a viewpoint that is embedded amongst objects in the scene. Objects and parts of objects, for example, behind the viewer need to be eliminated from consideration. Clipping is nontrivial because, in general, it involves removing parts of polygons and creating new ones. It means "cutting chunks" off the objects.
>
> *Projective transformation:* This transformation generates a two-dimensional image on the image or viewing plane from the three-dimensional view-space representation of the object.
>
> *Shading algorithm:* The orientation of the polygonal facets that represent the object are compared with the position of a light source (or sources) and a reflected light intensity calculated for each point on the surface of the object. In practice "each point on the surface" means those pixels onto which the polygonal facet projects. Thus it is convenient to calculate the set of pixels onto

which a polygon projects and drives this process from pixel space—a process that is usually called *rasterization*. Shading algorithms use a local **reflection model** and an *interpolative method* to distribute the appropriate light intensity amongst pixels inside a polygon. It is the computational efficiency and visual efficacy of the shading algorithm that has supported the popularity of the polygon mesh representation. (The polygon mesh representation has many drawbacks—its major advantage is simplicity.)

Hidden-surface removal: Those surfaces that cannot be seen from the viewpoint need to be removed from consideration. In the seventies much research was carried out on the best way to remove hidden surfaces, but the Z-buffer algorithm, with its easy implementation, is the de facto algorithm, with others being used only in specialized contexts. However, it does suffer from inefficiency and produces aliasing artifacts in the final image.

The above processes are not carried out in a sequence but are merged together in a way that depends on the overall rendering strategy. The use of the Z-buffer algorithm, as we shall see, conveniently allows polygons to be fetched from the database in any order. This means that the units which the whole rendering process operates on are single polygons that are passed through the processes one at a time. The entire process can be seen as a black box with a polygon input as a set of vertices in three-dimensional world space. The output is a shaded polygon in two-dimensional screen space as a set of pixels onto which the polygon has been projected.

Although, as we have implied, the processes themselves have become a kind of standard, rendering systems vary widely in detail, particularly in differences amongst subprocesses such as rasterization and the kind of viewing system that is used.

The marriage of interpolative shading with the polygon mesh representation of objects has served, and continues to serve, the graphics community well. It does suffer from a significant disadvantage, which is that antialiasing measures are not easily incorporated in it (except by the inefficient device of calculating a virtual image at a resolution much higher than the final screen resolution). Antialiasing measures are described elsewhere in the text.

The first two processes, viewing transformation and clipping, are geometric processes that operate on the vertex list of a polygon, producing a new vertex list. At this stage polygons are still represented by a list of vertices where each vertex is a coordinate in a three-dimensional space with an implicit link between vertices in the list. The projective transformation is also a geometric process, but it is embedded in the pixel-level processes. The shading algorithm and hidden-surface removal algorithm are pixel-level processes operating in screen space (which, as we shall see, is considered for some purposes to possess a third dimension). For these processes the polygon becomes a set of pixels in two-dimensional space. However, some aspects of the shading algorithm require us to return to three-dimensional space. In particular, calculating light intensity is a three-dimensional calculation. This functioning of the shading algorithm in both screen space and a three-dimensional object space is the source of certain visual artifacts. These arise because the projective transformation is nonlinear. Such subtleties will not be considered here—but see [Watt and Watt 1992] for more information on this point.

Viewing and Clipping

When viewing transformations are considered in computer graphics, an analogy is often made with a camera and the term "virtual camera" is employed. There are certainly direct analogies to be made between a camera, that records a two-dimensional projection of a real scene on a film, and a computer graphics system. However, it needs to be borne in mind that these concern "external" attributes such as the position of the camera and the direction that it is pointing in. There are implementations in a computer graphics system (notably the near and far clip planes) that are not available in a camera and facilities in a camera that are not usually imitated in a computer graphics system (notably depth of field and lens distortion effects). The analogy is a general one, and its utility disappears when details are considered.

The facilities that are incorporated in a viewing system can vary widely, and in this section we will look at a system which will suffice for most general-purpose rendering systems. The ways in which the attributes of the viewing system are represented, which have ramifications for the design of a user interface for a viewing system, also vary widely.

We can discuss our requirements by considering just three attributes (Fig. 57.1). First we establish a viewpoint and a viewing direction. A viewpoint, which we can also use as a center of projection, is a single point in world space, and a viewing direction is a vector in this space. Second we position a view plane somewhere in this space, and it is convenient to constrain the view-plane normal to be the viewing direction and its position to be such that a line from the viewpoint, in the viewing direction, passes through the center of the view plane. The distance of the viewpoint from the object being viewed and the distance of the view plane from the viewpoint determine the size of the projection of the object on the view plane. Also these two distances determine the degree of "perspective effect" in the object projection, since we are going to be using a perspective projection. This arrangement defines a new three-dimensional space, known as view space. Normally we would take the origin of this space to be the viewpoint, and the coordinate axes are oriented by the viewing direction.

Normally we assume that the view plane is of finite

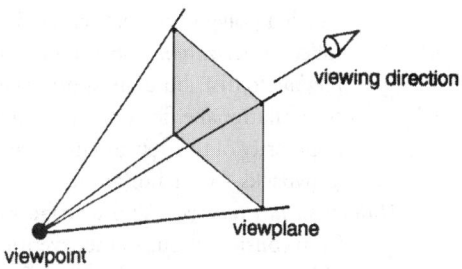

FIGURE 57.1 The three basic attributes required in a viewing system.

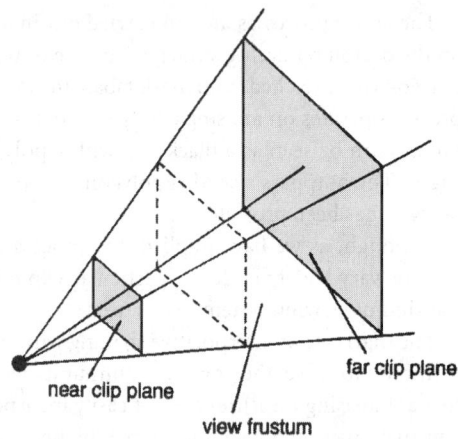

FIGURE 57.2 View frustum formed by near and far planes.

extent and rectangular—because its contents are eventually going to be mapped onto the display device. Additionally we can add *near* and *far clip planes* (Fig. 57.2) to further constrain those elements of the scene that are projected onto the view plane—a caprice of computer graphics not available in a camera.

Such a setup, as can be seen from the figure, defines a so-called view volume, and consideration of this gives the motivation for clipping. Clipping means that the part of the scene which lies outside the view frustum should be discarded from the rendering process. We perform this operation in three-dimensional view space, clipping polygons to the view volume. This is a nontrivial operation, but it is vital in scenes of any complexity where only a small proportion of the scene is finally going to appear on the screen. In simple single-object applications, where the viewpoint is not going to be inside the bounds of the scene and we do not implement a near and a far clip plane, we could project all the scene onto the view plane and perform the clipping operation in two-dimensional space.

Now we are in a position to define **viewing** and **clipping** as those operations that transform the scene from world space into view space, at the same time discarding that part of the scene or object that lies outside the view frustum.

We will deal separately with the transformation into the view space and clipping. First we consider the viewing transformation. A useful practical facility that we should consider is the addition of another vector to specify the rotation of the view plane about its axis (the view-direction vector). Returning to our camera analogy, this is equivalent to allowing the user to rotate the camera about the direction that it is pointing in. A user of such a system has to specify:

1. a viewpoint or camera position **C**, which forms the origin of view space. This point is also the center of projection (see below).

2. a viewing direction vector **N** (the positive z-axis in view space)—a vector normal to the view plane.
3. an "up" vector **V** that orients the camera about the view direction.
4. an (optional) vector **U** to denote the direction of increasing x in the eye coordinate system. This establishes a right- or left-handed coordinate system (**UVN**). This system is represented in Fig. 57.3.

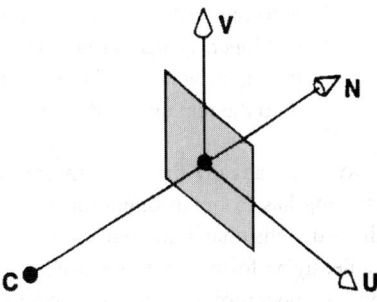

FIGURE 57.3 UVN coordinate system embedded in the view plane.

The transformation required to take an object from world space into view space, T_{view}, can be split into a translation T and a change of basis B:

$$T_{\text{view}} = T B,$$

where

$$T = \begin{bmatrix} 1 & 0 & 0 & -C_x \\ 0 & 1 & 0 & -C_y \\ 0 & 0 & 1 & -C_z \\ 0 & 0 & 0 & 1 \end{bmatrix}$$

It can be shown [Fiume 1989] that B is given by

$$B = \begin{bmatrix} U_x & U_y & U_z & 0 \\ V_x & V_y & V_z & 0 \\ N_x & N_y & N_z & 0 \\ 0 & 0 & 0 & 1 \end{bmatrix}$$

The only problem now is specifying a user interface for the system and mapping whatever parameters are used by the interface into **U**, **V**, and **N**. A user needs to specify **C**, **N**, and **V**. **C** is easy enough. **N**, the viewing direction or view-plane normal, can be entered, say, using two angles in a spherical coordinate system. **V** is more problematic. For example, a user may require "up" to have the same sense as "up" in the world coordinate system. However, this cannot be achieved by setting

$$\mathbf{V} = (0, 0, 1)$$

because **V** must be perpendicular to **N**. A useful strategy is to allow the user to specify, through a suitable interface, an approximate value for **V**, having the program alter this to a correct value.

Clipping and Culling

Clipping and culling means discarding, at an early stage in the rendering process, polygons or parts of polygons that will not appear on the screen. Polygons that need to be discarded fall into three categories:

1. Complete objects that lie outside the view volume should be removed in their entirety without invoking any tests at the level of individual polygons. This can be done by comparing a bounding volume, such as a sphere, with the view-volume extents and removing or not the entire set of polygons that represent an object.
2. Complete polygons that face away from a viewer need not invoke the expense of a clipping procedure. These are called back-facing polygons and in any (convex) object account on average for 50 percent of the object polygons. These can be eliminated by a simple geometric test, which is termed **culling** or *back-face removal*.

3. Polygons that straddle a view-volume plane have to be clipped against the view frustum and the resulting fragment, which comprises a new polygon, passed onto the remainder of the process.

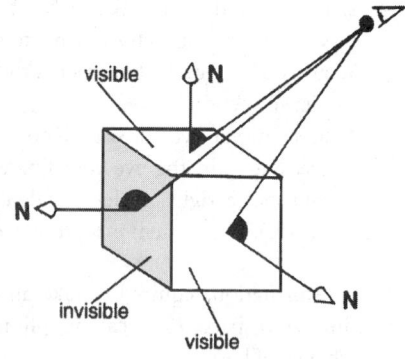

We have already discussed the reasons why, in general, clipping has to be an operation that is carried out in the three-dimensional domain of view space, and culling is conveniently performed in this space also. Culling is a pure geometric operation that discards polygons on the basis of the direction of their surface normal compared to the viewing direction. Clipping is an algorithmic operation because some process has to be invoked that produces a new polygon from the polygon that is clipped by one of the view volume planes.

FIGURE 57.4   Back-face removal or culling.

We deal with the simple operation of culling first. If we are considering a single convex object, then culling performs complete hidden-surface removal. If we are dealing with objects that are partially concave, or if there is more than one object in the scene, then a general hidden-surface removal algorithm is required. This is because in these cases the event of one polygon partially obscuring another arises, a situation impossible with a single convex object.

Determining whether a single polygon is visible from a viewpoint involves a simple geometric test (Fig. 57.4). We compare the angle between the (true) normal of each polygon and a line-of-sight vector. If this is greater than 90 deg, then the polygon cannot be seen. This condition can be written as

$$\text{visibility} := \mathbf{N}_p \cdot \mathbf{L}_{os} > 0$$

where $\mathbf{N}_p$ is the polygon normal and $\mathbf{L}_{os}$ is a vector representing a line from the polygon to the viewpoint. A polygon normal can be calculated from any three (noncollinear) vertices by taking a cross product of vectors parallel to the two edges defined by the vertices.

The most popular clipping algorithm, like most of the algorithms used in rendering, goes back over 20 years and is the Sutherland–Hodgman re-entrant polygon clipper [Sutherland and Hodgman, 1974]. We will describe, for simplicity, its operation in two-dimensional space, but it is easily extended to three dimensions. A polygon is tested against a clip boundary by testing each polygon edge against a single infinite clip boundary. This structure is shown in Fig. 57.5.

We consider the innermost loop of the algorithm, where a single edge is being tested against a single clip boundary. In this step the process outputs zero, one, or two vertices to add to the list of vertices defining the clipped polygon. Figure 57.6 shows the four possible cases. An edge is defined by vertices **S** and **F**. In the first case the edge is inside the clip boundary and the existing vertex **F** is added to the output list. In the second case the edge crosses the clip boundary and a new vertex **I** is calculated and output. The third case shows an edge that is completely outside the clip boundary. This produces no output. (The intersection for the edge that caused the excursion outside is calculated in the previous iteration, and the intersection for the edge that causes the incursion inside is calculated in the next iteration.) The final case again produces a new vertex, which is added to the output list.

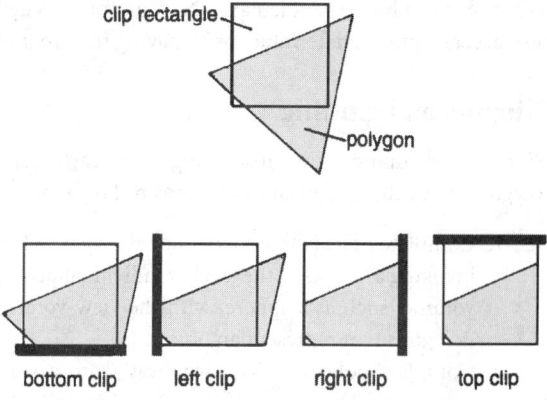

FIGURE 57.5   Sutherland–Hodgman clipper clips each polygon against each edge of each rectangle.

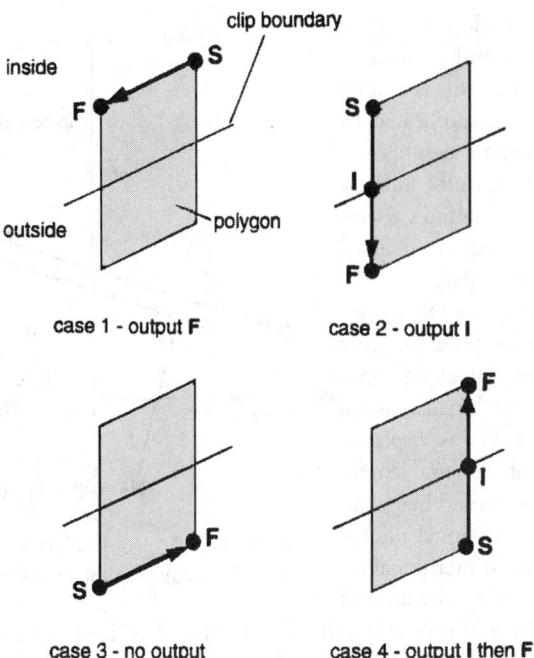

FIGURE 57.6 Sutherland–Hodgman clipper—within the polygon loop each edge of a polygon is tested against each clip boundary.

To calculate whether a point or vertex is inside, outside, or on the clip boundary we use a dot-product test. Figure 57.7 shows clip boundary **C** with an outward normal $\mathbf{N}_c$ and a line with end points **S** and **F**. We represent the line parametrically as

$$\mathbf{P}(t) = \mathbf{S} + (\mathbf{F} - \mathbf{S})t \tag{57.1}$$

where

$$0 \le t \le 1$$

We define an arbitrary point on the clip boundary as **X** and consider a vector from **X** to any point on the line. The dot product of this vector and the normal allows us to distinguish whether a point on the line is outside, inside, or on the clip boundary. In the case shown in Fig. 57.7,

$$\mathbf{N}_c \cdot (\mathbf{S} - \mathbf{X}) > 0 \quad \Rightarrow \quad \mathbf{S} \text{ is outside the clip region}$$

$$\mathbf{N}_c \cdot (\mathbf{F} - \mathbf{X}) < 0 \quad \Rightarrow \quad \mathbf{F} \text{ is inside the clip region}$$

and

$$\mathbf{N}_c \cdot (\mathbf{P}(t) - \mathbf{X}) = 0$$

defines the point of intersection of the line and the clip boundary. Solving (Eq. 57.1) for t enables the intersecting vertex to be calculated and added to the output list.

In practice the algorithm is written recursively. As soon as a vertex is output, the procedure calls itself with that vertex, and no intermediate storage is required for the partially clipped polygon. This structure makes the algorithm eminently suitable for hardware implementation.

Projective Transformation and Three-Dimensional Screen Space

A projective transformation takes the object representation in view space and produces a projection on the view plane. This is a fairly simple procedure somewhat complicated by the fact that we have to retain a depth value for each point for eventual use in the hidden-surface removal algorithm. Sometimes, therefore, the space of this transformation is referred to as three-dimensional screen space.

A perspective projection is the more popular or common choice in computer graphics because it incorporates foreshortening. In a perspective projection relative dimensions are not preserved, and a distant line is displayed smaller than a nearer line of the same length. This familiar effect enables human beings to perceive depth in a two-dimensional photograph or a stylization of three-dimensional reality. A perspective projection is characterized

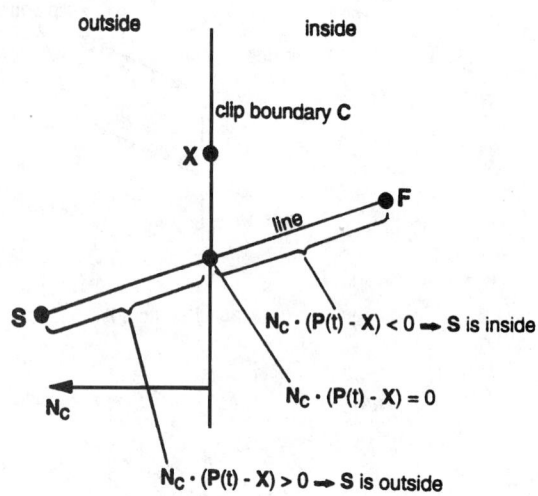

FIGURE 57.7 Dot-product test determines whether a line is inside or outside a clip boundary.

by a point known as the center of projection, the same point as the viewpoint in our discussion, and the projection of three-dimensional points onto the view plane is the intersection of the lines from each point to the center of projection. This familiar idea is shown in Fig. 57.8.

Figure 57.9 shows how a perspective projection is derived. Point $P(x_e, y_e, z_e)$ is a three-dimensional point in the view coordinate system. This point is to be projected onto a view plane normal to the z_e-axis and positioned at distance D from the origin of this system. Point P' is the projection of this point in the view plane and has two-dimensional coordinates (x_s, y_s) in a view-plane coordinate system with the

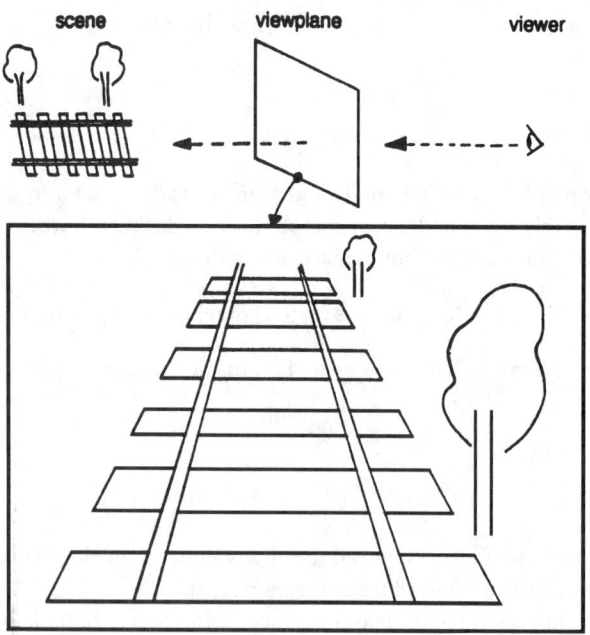

FIGURE 57.8 The "perspective effect."

origin at the intersection of the z_e-axis and the view plane. In this system we consider the view plane to be the view surface or screen.

Similar triangles give

$$x_s = D\frac{x_e}{z_e}$$

$$y_s = D\frac{y_e}{z_e}$$

Screen space is defined to act within a closed volume—the viewing frustum that delineates the volume of space which is to be rendered. For the purposes of this text we will consider a simplified view volume that constrains some of the dimensions that would normally be found in a more general viewing system. A simple view volume can be specified as follows: Suppose we have a square window— that area of the view plane that is mapped onto the view surface or screen—of size $2h$ arranged symmetrically about the viewing direction. The four planes defined by

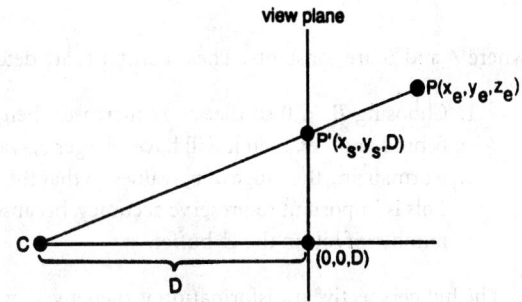

FIGURE 57.9 Perspective projection.

$$x_e = \pm h\frac{z_e}{D}$$

$$y_e = \pm h\frac{z_e}{D}$$

—together with the two additional planes, called the near and far clipping planes, respectively (perpendicular to the viewing direction), defined by

$$z_e = D$$

$$z_e = F$$

—make up the definition of the viewing frustum as shown in Fig. 57.10. Additionally we invoke the constraint that the view plane and the near clip plane are to be coincident. This simple system is based on a treatment given in an early classic textbook on computer graphics [Newman and Sproull 1973].

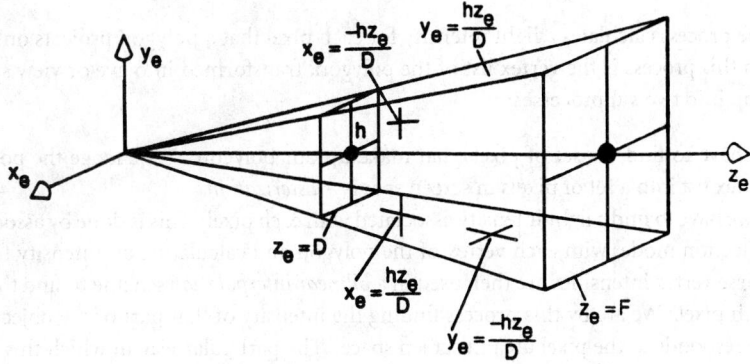

FIGURE 57.10 The six planes that define the view frustum.

This deals with transforming the x and y coordinates of points in view space, and we shall now discuss the transformation of the third component of screen space, namely z_e. In order to perform hidden-surface calculations (in the Z-buffer algorithm), depth information has to be generated on arbitrary points, in practice pixels, within the polygon by interpolation. This is possible in screen space only if, in moving from view space to screen space, lines transform into lines and planes transform into planes. It can be shown [Newman and Sproull 1973] that these conditions are satisfied provided the transformation of z takes the form

$$z_s = A + \frac{B}{z_e}$$

where A and B are constants. These constants are determined from the following constraints:

1. Choosing $B < 0$ so that as z_e increases then so does z_s. This preserves depth. If one point is behind another, then it will have a larger z_e-value; if $B < 0$, it will also have a larger z_s-value.
2. Normalizing the range of z_s-values so that the range $z_e \in [D, F]$ maps into the range $z_e \in [0, 1]$. This is important to preserve accuracy, because a pixel depth is going to be represented by a fixed number of bits in the Z-buffer.

The full perspective transformation is then given by

$$x_s = D \frac{x_e}{h z_e}$$

$$y_s = D \frac{y_e}{h z_e}$$

$$z_s = F \frac{1 - D/z_e}{F - D}$$

where the additional constant, h, appearing in the transformation for x_s and y_s ensures that these values fall in the range $[-1, 1]$ over the square screen.

It is instructive to consider the relationship between z_e and z_s a little more closely; although as we have seen they both provide a measure of the depth of a point, interpolating along a line in eye space is not the same as interpolating along this line in screen space. As z_e approaches the far clipping plane, z_s approaches 1 more rapidly. Objects in screen space thus get pushed and distorted towards the back of the viewing frustum. This difference can lead to errors when interpolating quantities, other than position, in screen space.

Shading Algorithm

This part of the process calculates a light intensity for each pixel that a polygon projects onto. The input information to this process is the vertex list of the polygon, transformed into eye or view space, and the process splits up into two subprocesses:

First we have to find the set of pixels that make up our polygon. We change the polygon from a vertex list into a set of pixels in screen space—*rasterization*.

Second we have to find a light intensity associated with each pixel. This is done by associating a local reflection model with each vertex of the polygon and calculating an intensity for the vertex. These vertex intensities are then used in a *bilinear interpolation* scheme to find the intensity of each pixel. We are by this process finding the intensity of that part of the object surface that corresponds to the pixel area in screen space. The particular way in which this is done leads to the (efficiency–quality) hierarchy of **flat shading**, **Gouraud shading**, and **Phong shading**.

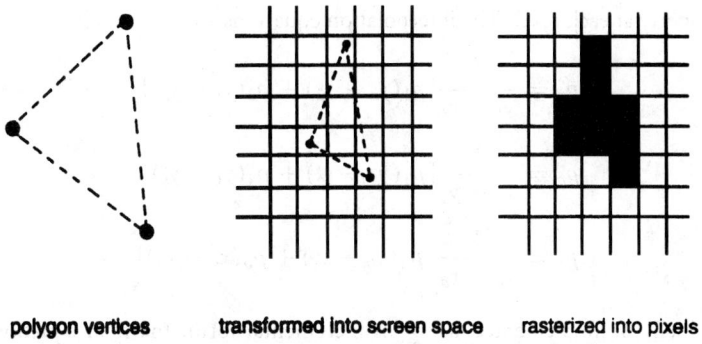

polygon vertices   transformed into screen space   rasterized into pixels

FIGURE 57.11   Different representations of a polygon in the rendering process.

Rasterization

Rasterization, or finding the set of polygons onto which the polygon projects, has to be done carefully because adjacent polygons must fit together accurately after they have been mapped into pixel sets. If it is not done accurately, holes can result in the image—probably the most common defect seen in rendering software. As shown in Fig. 57.11, we see that the precise geometry of the polygon will map into a set of fully and partially covered pixels. We have to decide which of the partially covered pixels are to be considered part of the polygon. Deciding on the basis of the area of coverage is extremely expensive and would wipe out the efficiency advantage of the bilinear interpolation scheme that is used to find a pixel intensity. It is better to map the vertices in some way to the nearest pixel coordinate and set up a consistent rule for deciding the fate of partially covered pixels. This is the crux of the matter. If the rules are not consistent and carefully formulated, then the rounding process will produce holes or unfilled pixels between polygons that share the same scan line. Note that the process will cause a shape change in the polygon, which we ignore because the polygon is already an approximation, and to some extent an arbitrary representation of the "real" surface of the object.

Sometimes called *scan-line conversion*, rasterization proceeds by moving a horizontal line through the polygon in steps of a pixel height. For a current scan line, interpolation (see next section) between the appropriate pairs of polygon vertices will yield x_{start} and x_{end}, the start and end points of the portions of the scan line crossing the polygons (using real arithmetic). The following scheme is a simple set of rules that converts these values into a run of pixels:

1. Round x_{start} up.
2. Round x_{end} down.
3. If the fractional part of x_{end} is 0, then subtract 1 from it.

Applying the rasterization process to a complete polygon implies embedding this operation in a structure that keeps track of the edges that are to be used in the current scan-line interpolation, and this is normally implemented using a linked-list approach.

Bilinear Interpolation

As we have already mentioned, light intensity values are assigned to the set of pixels that we have now calculated, not by individual calculation but by interpolating from values calculated only at the polygon vertices. At the same time we interpolate depth values for each pixel to be used in the hidden-surface determination. So in this sub-subsection we consider the interpolation of a pixel property from vertex values independent of the nature of the property. Referring to Fig. 57.12, the interpolation proceeds by moving a scan line down through the pixel set and obtaining start and end values for a scan line by interpolating between the appropriate pair of vertex properties. Interpolation along a scan line then yields a value for the property at each pixel. The interpolation equations are

$$p_a = \frac{1}{y_1 - y_2}[p_1(y_s - y_2) + p_2(y_1 - y_s)]$$

$$p_b = \frac{1}{y_1 - y_4}[p_1(y_s - y_4) + p_4(y_1 - y_s)] \qquad (57.2)$$

$$p_s = \frac{1}{x_b - x_a}[p_a(x_b - x_s) + p_b(x_s - x_a)]$$

These would normally be implemented using an incremental form, the final equation, for example, becoming

$$p_s := p_s + \Delta p$$

with the constant value Δp calculated once per scan line.

Local Reflection Models

Given that we find pixel intensities by an interpolation process, the next thing to discuss is how to find the reflected light intensity at the vertices of a polygon, those values from which the pixel intensity values are derived. This is done by using a simple local reflection model—the one most commonly used is the Phong reflection model [Phong 1975] (not to be confused with Phong shading, which is a vector interpolation scheme). A local reflection model calculates a value for the reflected light intensity at a point on the surface of an object—in this case that point is a polygon vertex—due to incident light from a source, which for reasons we will shortly examine is usually a point light source. The model is a linear combination of three components—diffuse, specular, and ambient. We assume that the behavior of reflected light at a point on a surface can be simulated by assuming that the

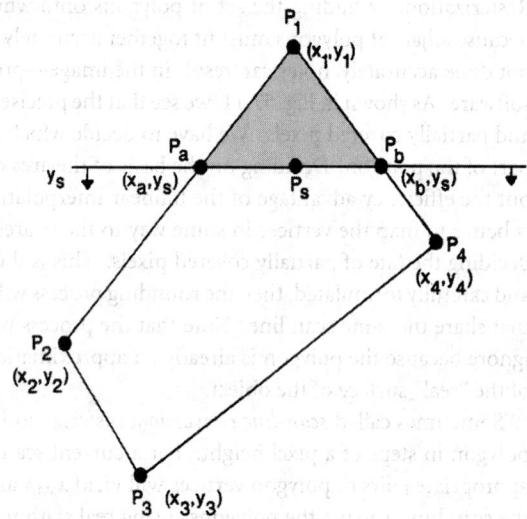

FIGURE 57.12 Notation used in property interpolation within a polygon.

surface is some combination of a perfect diffuse surface together with an (imperfect) specular or mirrorlike surface. The light scattered from a perfect diffuse surface is the same in all directions, and the reflected light intensity from such a surface is given by Lambert's cosine law, which is, in computer graphics notation,

$$I_d = I_i k_d \mathbf{L} \cdot \mathbf{N}$$

where $\mathbf{L}$ is the light direction vector and both $\mathbf{L}$ and $\mathbf{N}$ are unit vectors as shown in Fig. 57.13(a), k_d is a diffuse reflection coefficient, and I_i is the intensity of a (point) light source. The specular contribution is a function of the angle between the viewing direction $\mathbf{V}$ and the mirror direction $\mathbf{R}$:

$$I_s = I_i k_s (\mathbf{R} \cdot \mathbf{V})^n$$

where n is an index that simulates surface roughness and k_s is a specular reflection coefficient. For a perfect mirror n would be infinity and reflected light would be constrained to the mirror direction. For small integer values of n a reflection lobe is generated, where the thickness of the lobe is a function of the surface roughness [Fig. 57.13(b)]. The effect of the specular reflection term in the model is to produce a

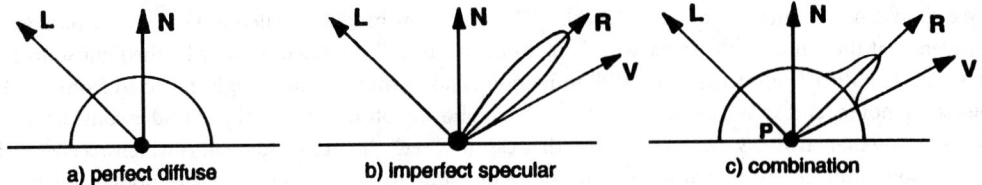

FIGURE 57.13 Components of the Phong local reflection model.

so-called highlight on the rendered object. This is basically a reflection of the light source spread over an area of the surface to an extent that depends on the value of n. The color of the specularly reflected light is different from that of the diffuse reflected light—hence the term "highlight." In simple models of specular reflection the specular component is assumed to be the color of the light source. If, say, a green surface is illuminated with white light, then the diffuse reflection component is green but the highlight is white.

Adding the specular and diffuse components together gives a very approximate imitation to the behavior of reflected light from a point on the surface of an object. Consider Fig. 57.13(c). This is a cross section of the overall reflectivity response as a function of the orientation of the view vector $\mathbf{V}$. The cross section is in a plane that contains the vector $\mathbf{L}$ and the point $\mathbf{P}$ and it thus slices through the specular bump. The magnitude of the reflected intensity, the sum of the diffuse and specular terms, is the distance from $\mathbf{P}$ along the direction $\mathbf{V}$ to where $\mathbf{V}$ intersects the profile.

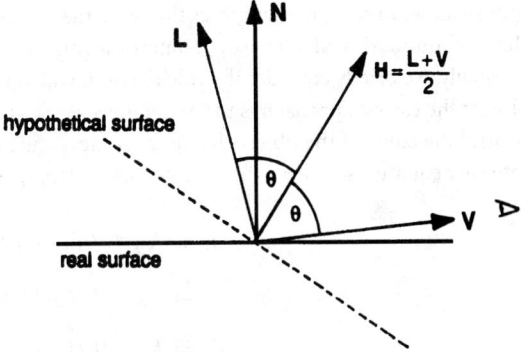

FIGURE 57.14 $\mathbf{H}$ is the unit normal to a hypothetical surface oriented in a direction that bisects the angle between $\mathbf{L}$ and $\mathbf{V}$.

An ambient component is usually added to the diffuse and specular term. Such a component illuminates surfaces that, because we generally use a point light source, would otherwise be rendered black. These are surfaces that are visible from the viewpoint but not from the light source.

Adding the diffuse, specular, and ambient components together, we have

$$I = K_a + I_i(k_d \mathbf{L} \cdot \mathbf{N} + k_s (\mathbf{R} \cdot \mathbf{V})^n) \tag{57.3}$$

where K_a is the constant ambient term. The expense of Eq. (57.3) can be considerably reduced by making some geometric assumptions and approximations. Firstly, if the light source and the viewpoint are considered to be at infinity, then $\mathbf{L}$ and $\mathbf{V}$ are constant over the domain of the scene. The vector $\mathbf{R}$ is expensive to calculate, and although Phong gives an efficient method for calculating $\mathbf{R}$, it is better to use a vector $\mathbf{H}$. This appears to have been first introduced by Blinn [1977]. The specular term then becomes a function of $\mathbf{N} \cdot \mathbf{H}$ rather than $\mathbf{R} \cdot \mathbf{V}$. $\mathbf{H}$ is the unit normal to a hypothetical surface that is oriented in a direction halfway between the light direction vector $\mathbf{L}$ and the viewing vector $\mathbf{V}$ (Fig. 57.14).

$$\mathbf{H} = \frac{\mathbf{L} + \mathbf{V}}{2}$$

Together with a shading algorithm, this simple model is responsible for the look of most shaded computer graphics images, and it has been in use constantly since 1975. Its main disadvantages are that objects look as if they were made from some kind of plastic material, which is either shiny or dull. Also, in reality the magnitude of the specular component is not independent, as Eq. (57.3) implies, of the direction of the incoming light. Consider, for example, the glare from a (nonshiny) road surface when you are driving along a road in the direction of the setting sun: this does not occur when the sun is overhead.

We should now return to the term "local." The reflection model is called a local model because it considers that the point on the surface under consideration is illuminated directly by the light source in the scene. No other (indirect) source of illumination is taken into account. Light reflected from nearby objects is ignored, and so we see no reflections of neighboring objects in the object under consideration. It also means that shadows, which are areas that cannot "see" the light source and which receive their illumination indirectly from another object, cannot be modeled. In a scene using this model, objects are illuminated as if they were floating in a dark space illuminated only by the light source.

When shadows are added into rendering systems that use local reflection models, these are purely geometric. That is, the area that the shadow occupies on the surface of an object, due to the intervention of another object between it and the light source, is calculated. The reflected light intensity within this area is then arbitrarily reduced. There is no way of calculating, using local reflection models, how much indirect light illuminates the shadowed area. There are visual consequences of this which should be considered when including shadows in an "add-on" manner. These may detract from the real appearance of the final rendered image rather than add to it. First, because shadows are important to us in reality, we easily spot shadows in computer graphics that have the wrong intensity. This is compounded by the hard-edged shadow boundaries calculated by geometric algorithms. In reality shadows normally have soft subtle edges.

Finally, we briefly consider the role of color, which up to now we have said nothing about. For colored objects the easiest approach is to model the specular highlights as white (for a white light source) and to control the color of the objects by appropriate setting of the diffuse reflection coefficients. We use three intensity equations to drive the monitor's red, green, and blue inputs:

$$I_r = K_a + I_i(k_{dr}\mathbf{L} \cdot \mathbf{N} + k_s(\mathbf{N} \cdot \mathbf{H})^n)$$

$$I_g = K_a + I_i(k_{dg}\mathbf{L} \cdot \mathbf{N} + k_s(\mathbf{N} \cdot \mathbf{H})^n)$$

$$I_b = K_a + I_i(k_{db}\mathbf{L} \cdot \mathbf{N} + k_s(\mathbf{N} \cdot \mathbf{H})^n)$$

where the specular coefficient k_s is common to all three equations, but the diffuse component varies according to the object's surface color.

It should be mentioned at this stage that this three-sample approach to color is a crude approximation and accurate treatment of color requires far more than three samples. This means that to accurately model the behavior of reflected light we would have to evaluate many more than three equations. We would have to sample the spectral energy distribution of the light source as a function of wavelength and the reflectivity of the object as a function of wavelength and apply Eq. (57.3) at each wavelength sample. The solution then obtained would have to be converted back into three intensities to drive the monitor. The colors that we would get from such an approach would certainly be different from the three-sample implementation. Except in very specialized applications this problem is completely ignored.

We now discuss shading options. These options differ in where the reflection model is applied and how calculated intensities are distributed amongst pixels. There are three options: flat shading, Gouraud shading, and Phong shading, in order of increasing expense and increasing image quality.

Flat Shading

Flat shading is the option where we invoke no interpolation within the polygon and shade each pixel within the polygon set with the same intensity. The reflection model is used once only. The (true) normal for the polygon (in eye or view space) is inserted into Eq. (57.3) and the calculated intensity is applied to each polygon. The efficiency advantages are obvious—the entire interpolation procedure is avoided, and shading reduces to rasterization plus a once-only intensity calculation per polygon. The (visual) disadvantage is that the polygon edges remain glaringly visible and we render not the surface that the polygon mesh represents but the polygon mesh itself. As far as image quality is concerned this is more disadvantageous than the fact that there is no variation in light intensity amongst the polygon pixels. Flat shading is used as a fast preview facility.

Gouraud Shading

Both Gouraud and Phong shadings exhibit two strong advantages—in fact they are their *raison d'être*. They use the interpolation scheme already described and so are efficient, and they diminish or eliminate the visibility of the polygon edges. In a Gouraud- or Phong-shaded object these are now visible only along silhouette edges. This elegant device meant their enduring success and the idea originated by Gouraud [1971] and cleverly elaborated by Phong was one of the major breakthroughs in three-dimensional computer graphics.

In Gouraud shading, intensities are calculated at each vertex and inserted into the interpolation scheme. The trick is in the normals used at a polygon vertex. Using the true polygon normal would not work (be-

cause all the vertex normals would be parallel and the reflection model would evaluate the same intensity at each). What we have to do is to calculate a normal at each vertex that somehow relates back to the original surface and Gouraud vertex normals are calculated by considering the average of the true polygon normals of those polygons that contribute to the vertex (Fig. 57.15). This calculation is normally regarded as part of the setting up of the object, and these vectors are stored as part of the object database (although there is a problem when polygons are clipped—new vertex normals then have to be calculated as part of the rendering process). Because polygons now share vertex normals, the interpolation process ensures that there is no change in intensity across the edge between two polygons, and in this way the polygonal structure of the object representation is rendered invisible. (However, an optical illusion, known as Mach banding, persists along the edges with Gouraud shading.)

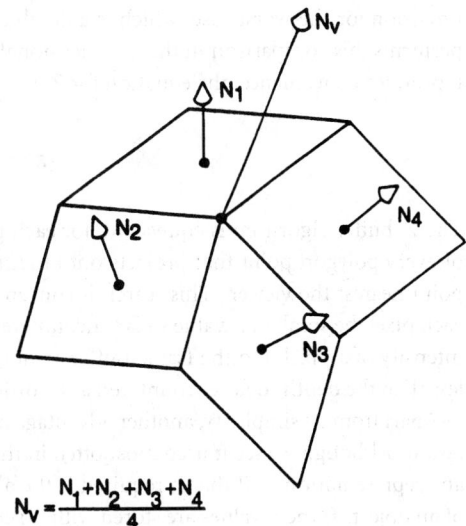

$$N_v = \frac{N_1 + N_2 + N_3 + N_4}{4}$$

FIGURE 57.15 The concept of a vertex normal.

Gouraud shading is used extensively and gives excellent results for the diffuse component. However, calculating reflected light intensity only at the vertices leads to problems with the specular component. The easiest case to consider is that of a highlight which, if it was visible, would be within the polygon boundaries—meaning it does not extend to the vertices. In this case the Gouraud scheme would simply miss the highlight completely.

Phong Shading

Phong shading [Phong 1975] was developed to overcome the problems of Gouraud shading and specular highlights, and in this scheme the property to be interpolated is the vertex normals themselves, with each vector component now inserted into three versions of Eq. (57.3). It is a strange hybrid with an interpolation procedure running in pixel or screen space controlling vector interpolation in three-dimensional view space (or world space). But it works very well. We are estimating the normal to the surface at a point that corresponds to the pixel under consideration in screen space, or at least estimating it to within the limitations and approximations that have been imposed by the polygonal representation and the interpolation scheme. We can then apply the reflection model at each pixel, and a unique reflected light intensity is now calculated for each pixel. We may end up with a result that is different from what would be obtained if we had access to the true surface normal at the point on the real surface that corresponded to the pixel, but it does not matter, because the quality of Phong shading is so good that we cannot perceive any erroneous effects on the monitor.

Phong shading is much slower than Gouraud shading because the interpolation scheme is three times as lengthy, and also the reflection model [Eq. (57.3)] is now applied at each pixel. A good rule of thumb is that Phong shading has five times the cost of Gouraud shading.

Hidden-Surface Removal

As already mentioned, we shall describe the Z-buffer as the de facto hidden-surface removal algorithm. That it has attained this status is due to its ease of implementation—it is virtually a single "if" statement—and its ease of incorporation into a polygon-based renderer. Screen space algorithms (the Z-buffer falls into this category) operate by associating a depth value with each pixel. In our polygon renderer the depth values are available only at a vertex, and the depth values for a pixel are obtained by using the same interpolation scheme as for intensity in Gouraud shading.

Hidden-surface removal eventually comes down to a point-by-point depth comparison. Certain algorithms operate on area units, scan line segments, or even complete polygons, but they must contain a provision for the worst case, which is a depth comparison between two pixels. The Z-buffer algorithm performs this comparison in three-dimensional screen space. We have already defined this space and we repeat, for convenience, the equation for Z_s:

$$z_s = F \frac{1 - D/z_e}{F - D}$$

The Z-buffer algorithm is equivalent, for each pixel (x_s, y_s), to a search through the associated z-values of every polygon point that projects onto that pixel to find that point with the minimum z-value—the point nearest the viewer. This search is conveniently implemented by using a Z-buffer, which holds for each pixel the smallest z-value so far encountered. During the processing of a polygon we either write the intensity of a pixel into the frame buffer or not, depending on whether the depth of the current pixel is less than the depth so far encountered as recorded in the Z-buffer.

Apart from its simplicity, another advantage of the Z-buffer is that it is independent of object representation. Although we see it used most often in the context of polygon mesh rendering, it can be used with any representation—all that is required is the ability to calculate a z-value for each point on the surface of an object. If the z-values are stored with pixel values, separately rendered objects can be merged into a multiple object scene using Z-buffer information on each object.

The main disadvantage of the Z-buffer is the amount of memory it requires. The size of the Z-buffer depends on the accuracy to which the depth value of each point (x, y) is to be stored, which is a function of scene complexity. Usually 20–32 bits is deemed sufficient for most applications. Recall in a previous section that we discussed the compression of z_s-values. This means that a pair of distinct points with different z_e-values can map into identical z_s-values. Note that for frame buffers with less than 24 bits per pixel, say, the Z-buffer will in fact be *larger* than the frame buffer. In the past Z-buffers have tended to be part of the main memory of the host processor, but now graphics terminals are available with dedicated Z-buffers, and this represents the best solution.

57.3 Rendering Using Ray Tracing

Ray tracing is a simple and elegant algorithm whose appearance in computer graphics is usually attributed to Whitted [1980]. It combines in a single algorithm

- hidden-surface removal,
- reflection due to direct illumination—the same factor we have calculated in the previous method using a local model,
- reflection due to indirect illumination—that is, reflection due to light striking the object which itself has been reflected from another object,
- transmission of light through transparent or partially transparent objects,
- shading due to object–object interaction—global illumination,
- the computation of (hard-edged) shadows.

It does this by tracing rays—infinitesimally thin beams of light—in the reverse direction of light propagation; that is, it traces light rays from the eye into the scene and from object to object. In this way it "discovers" the way in which light interacts between objects and is able to produce visualizations such as objects reflecting in other objects and the distortion of an object viewed through another (transparent or glass) object due to refraction. Rays are traced from the eye or viewpoint because we are interested only in those rays that pass through the view plane. If we traced rays from the light source, then (theoretically) we would have to trace an infinity of rays.

Ray-tracing algorithms exhibit a strong visual signature because a basic ray tracer can simulate only one aspect of the global interaction of light in an environment—specular reflection and specular transmission. Thus ray-traced scenes always look ray-traced, because they tend to consist of objects that exhibit mirrorlike reflection, in which you can see the (perfect) reflections of other objects. Simulating nonperfect specular reflection is computationally impossible with the normal ray-tracing approach, because this means that at a hit point a single incoming ray will produce a multiplicity of reflected rays instead of just one. The same argument applies to transparent objects. A single incident ray can produce only a single transmitted or refracted ray, and such behavior would happen only in a perfect material that did not scatter light passing through it. With transparent objects the refractive effect can be simulated, but the material looks like "perfect glass." Thus perfect surfaces and perfect glass, behavior that does not occur in practice, betray the underlying rendering algorithm.

A famous development, called distributed ray tracing [Cook et al. 1984], addressed exactly this problem, using a Monte Carlo approach to simulate the specular reflection and specular transmission spread without invoking a combinatorial explosion. The algorithm produces shiny objects that look real (that is, their surfaces look rough or imperfect), blurred transmission through glass, and blurred shadows. The modest cost of this method involved initiating 16 rays per pixel instead of one. This is still a considerable increase in an already expensive algorithm, and most ray tracers still utilize the perfect specular interaction model.

The algorithm is conceptually easy to understand and is also easy to implement using a recursive procedure. A pictorial representation is given in Fig. 57.16. The algorithm operates in three-dimensional world space, and for each pixel in screen space we calculate an initial ray from the viewpoint through the center of the pixel. The ray is injected into the scene and will either hit an object or not. (In the case of a closed environment, some object will always be encountered by an initial ray, even if it is just the background such as a wall.) When it hits an object its spawns two more rays, a reflected ray and a transmitted ray, which refracts into the object if the object is partially transparent. These rays travel onwards and themselves spawn other rays at their next hits. The process is sometimes represented as a binary tree, with a light–surface hit at each node in the tree. This process can be implemented as a recursive procedure which for each ray invokes an intersection test that spawns a transmitted and a reflected ray by calling itself twice with parameters representing the reflected and the transmitted or refracted directions of the new rays. At the heart of the recursive control procedure is an intersection test. This procedure is supplied with a ray, compares the ray geometry with all objects in the scene, and returns the nearest surface that the ray intersects. If the ray is an initial ray, then this effectively implements hidden-surface removal. Intersection tests account for most of the computational overheads in ray tracing, and much research effort has gone into ways of reducing this cost.

Grafted onto this basic recursive process, which follows specular interaction through the scene, is the computation of direct reflection and shadow computation. This means that at each node or surface hit we calculate these two contributions. Direct reflection is calculated by applying, for each light source, a Phong reflection model (or some other local model) at the node under consideration. The direct reflection contribution is diminished if the point is in shadow with respect to a light source. Thus at any hit point or node there are three contributions to the light intensity passed back up through the recursion: a reflected-ray contribution, a transmitted-ray contribution, and a local contribution unaltered or modified by the shadow computation.

Shadow computation is easily implemented by injecting the light direction vector, used in the local contribution calculation, into the intersection test to see if it is interrupted by any intervening objects. This ray is called a shadow feeler. If **L** is so interrupted, then the current surface point lies in shadow.

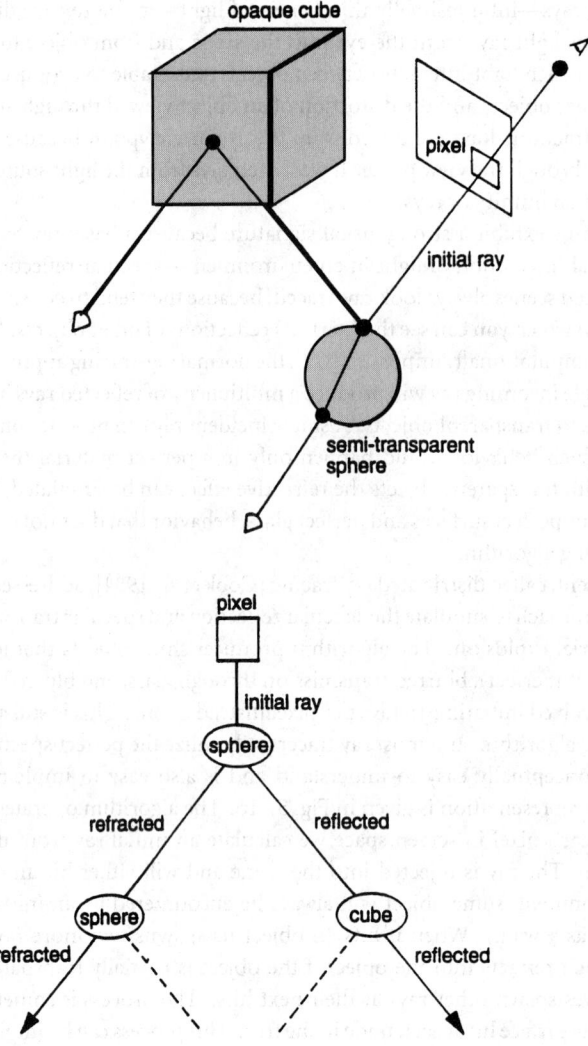

FIGURE 57.16   A representation of a ray-tracing algorithm.

If a wholly opaque object lies in the path of the shadow feeler, then the local contribution is reduced to the ambient value. An attenuation in the local contribution is calculated if the intersecting object is partially transparent. Note that it is no longer appropriate to consider **L** as a constant vector (light source at infinity) and the so-called shadow feelers are rays whose direction is calculated at each hit point. Because light sources are normally point sources, this procedure produces, like add-on shadow algorithms, hard-edged shadows. (Strictly speaking, a shadow feeler intersecting partially transparent objects should be refracted. It is not possible to do this, however, in the simple scheme described. The shadow feeler is initially calculated as the straight line between the surface intersection and the light source. This is an easy calculation, and it would be difficult to trace a ray from this point to the light source and include refractive effects.)

Finally note that, as the number of light sources increases from one, the computational overheads for shadow testing rapidly predominate. This is because the "main" rays are traced only to an average depth of between one and two. However, each ray–surface intersection spawns n shadow feelers (where n is the number of light sources), and the object intersection cost for a shadow feeler is exactly the same as for a main ray.

Intersection Testing

We have mentioned that intersection testing forms the heart of a ray tracer and accounts for most of the cost of the algorithm. In the eighties much research was devoted to this aspect of ray tracing, and there is a large body of literature on the subject which dwarfs the work devoted to improvements in the ray-traced image such as, for example, distributed ray tracing.

Intersection testing means finding if a ray intersects an object and then, if so, the point of intersection. Expressing the problem in this way hints at the usual approach, which is to try and cut down the overall cost by using some scheme, like a bounding volume, which prevents the intersection test searching through all the polygons in an object if the ray cannot hit the object.

The cost of ray tracing and the different approaches that are possible depend much on the way in which objects are represented. For example, if a voxel representation is used and the entire space is labeled with object occupancy, then discretizing the ray into voxels and stepping along it from the start point will reveal the first object that the ray hits. Contrast this with a brute-force intersection test, which must test a ray against every object in the scene to find the hit nearest to the ray start point.

57.4 Rendering Using the Radiosity Method

The radiosity method arrived in computer graphics in the mid-eighties a few years after ray tracing. Most of the early development work was carried out at Cornell University under the guidance of D. Greenberg—a major figure in the development of the technique. The emergence of the hemicube algorithm and later the progressive refinement algorithm established the method and enabled it to leave research laboratories and become a practical rendering tool. Nowadays many commercial systems are available, and most are implementations of these early algorithms.

The radiosity method provides a solution to diffuse interaction, which, as we have discussed, cannot be easily incorporated in ray tracing, but at the expense of dividing the scene into large patches (over which the radiosity is constant). This approach cannot cope with (sharp) specular reflections, and essentially we have two global methods: ray tracing, which simulates global specular reflection, and transmission and radiosity, which simulates global diffuse interaction.

In terms of the global phenomena that they simulate the methods are mutually exclusive, and predictably a major research bias has involved the unification of the two methods into a single global solution. Research is still actively pursued into many aspects of the method—particularly form-factor determination and scene decomposition into elements or patches.

Basic Theory

The radiosity method works by dividing the environment into largish elements called patches. For every pair of patches in the scene, a parameter F_{ij}, called a form factor, which depends on the geometric relationship between patches i and j, is evaluated. This factor is used to determine the strength of diffuse light interaction between pairs of patches, and a large system of equations is set up which on solution yields the radiosity for each patch in the scene.

The radiosity method is an object-space algorithm, solving for the intensity at discrete points or surface patches within an environment and not for pixels in an image-plane projection. The solution is thus independent of viewer position. This complete solution is then injected into a renderer that computes a particular view by removing hidden surfaces and forming a projection. This phase of the method does not require much computation (intensities are already calculated), and different views are easily obtained from the general solution. The method is based on the assumption that all surfaces are perfect diffusers or ideal Lambertian surfaces.

Radiosity, B, is defined as the energy per unit area leaving a surface patch per unit time and is the sum of the emitted and the reflected energy:

$$B_i \, dA_i = E_i A_i + R_i \int_i B_j F_{ji} \, dA_j$$

Expressing this equation in words, we have for a single patch i

$$\text{radiosity} \times \text{area} = \text{emitted energy} + \text{reflected energy}$$

E_i is the energy emitted from a patch, and emitting patches are, of course, light sources. The reflected energy is given by multiplying the incident energy by R_i the reflectivity of the patch. The incident energy is that energy that arrives at patch i from all other patches in the environment; that is, we integrate over the environment, for all $j (j \neq i)$, the term $B_j F_{ji} dA_j$. This is the energy leaving each patch j that arrives at patch i.

For a discrete environment the integral is replaced by a summation and constant radiosity is assumed over small discrete patches, and it can be shown that

$$B_i = E_i + R_i \sum_{j=1}^{n} B_j F_{ij}$$

Such an equation exists for each surface path in the enclosure and the complete environment produces a set of n simultaneous equations of the form

$$\begin{bmatrix} 1 - R_1 F_{11} & -R_1 F_{12} & \cdots & -R_1 F_{1n} \\ -R_2 F_{21} & 1 - R_2 F_{22} & \cdots & -R_2 F_{2n} \\ \vdots & \vdots & & \vdots \\ -R_n F_{n1} & -R_n F_{n2} & \cdots & 1 - R_n F_{nn} \end{bmatrix} \begin{bmatrix} B_1 \\ B_2 \\ \vdots \\ B_n \end{bmatrix} = \begin{bmatrix} E_1 \\ E_2 \\ \vdots \\ E_n \end{bmatrix}$$

The E_i's are nonzero only at those surfaces that provide illumination, and these terms represent the input illumination to the system. The R_i's are known, and the F_{ij}'s are a function of the geometry of the environment. The reflectivities are wavelength-dependent terms, and the above equation should be regarded as a monochromatic solution, a complete solution being obtained by solving for however many color bands are being considered. We can note at this stage that $F_{ii} = 0$ for a plane or convex surface—none of the radiation leaving the surface will strike itself. Also, from the definition of the form factor the sum of any row of form factors is unity.

Since the form factors are a function only of the geometry of the system, they are computed once only. Solving this set of equations produces a single value for each patch, and this information is then input to a modified Gouraud renderer to give an interpolated solution across all patches.

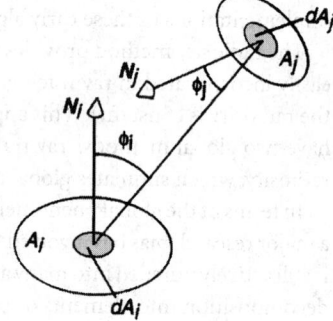

FIGURE 57.17 Parameters used in the definition of a form factor.

Form-Factor Determination

A significant early development was a practical method to evaluate form factors. The algorithm is both an approximation and an efficient method of achieving a numerical estimation of the result. The form factor between patches i and j is defined as

$$F_{ij} = \frac{\text{radiative energy leaving surface } A_i \text{ that strikes } A_j \text{ directly}}{\text{radiative energy leaving } A_i \text{ in all directions in the hemispherical space surrounding } A_i}$$

This is given by

$$F_{ij} = \frac{1}{A_i} \int_{A_i} \int_{A_j} \frac{\cos \phi_i \cos \phi_j}{\pi r^2} dA_j dA_i$$

where the geometric conventions are illustrated in Fig. 57.17. Now it can be shown that this patch-to-patch form factor can be approximated by the differential-area-to-finite-area form factor

$$F_{dA_i A_j} = \int_{A_j} \frac{\cos \phi_i \cos \phi_j}{\pi r^2} \, dA_j$$

where we are now considering the form factor between the elemental area da_i and the finite area A_j, and it is this approximation that is calculated by the hemicube algorithm. The factors that enable the approximation to a single integral and its veracity are quite subtle and are outside the scope of this treatment.

A good intuition of the workings of the algorithm can be gained from a pictorial visualization (Fig. 57.18). Figure 57.18(a) is a representation of the property known as the Nusselt analog. In the example patches A, B, and C all have the same form factor with respect to patch i. Patch B is the projection of A onto a hemicube, centered on patch i, and C is the projection of A onto a hemisphere. This property is the foundation of the hemicube algorithm [Cohen and Greenberg 1985], which places a hemicube on each patch i [Fig. 57.18(b)]. The hemicube is subdivided into elements, and associated with each element is a precalculated *delta form factor*. The hemicube is placed on patch i, and then patch j is projected onto it. In practice this involves a clipping operation, since a patch can in general project onto three faces of the hemicube. To evaluate F_{ij} involves simply summing the values of the delta form factors onto which patch j projects [Fig. 57.18(c)].

Another aspect of form-factor determination solved by the hemicube algorithm is the intervening-patch problem. This means that normally we cannot evaluate the form-factor relationship between a pair of patches independently of one or more patches that happen to be situated between them. The hemicube algorithm solves this by making the hemicube a kind of Z-buffer in addition to its role as five projection planes. This is accomplished as follows. For the patch i under consideration, every other patch in the scene is projected onto the hemicube. For each projection the distance from patch i to the patch being projected is compared with the smallest distance associated with previously projected patches, which is stored in hemicube elements. If a projection from a nearer patch occurs, then the identity of that patch and its distance from patch i are stored in

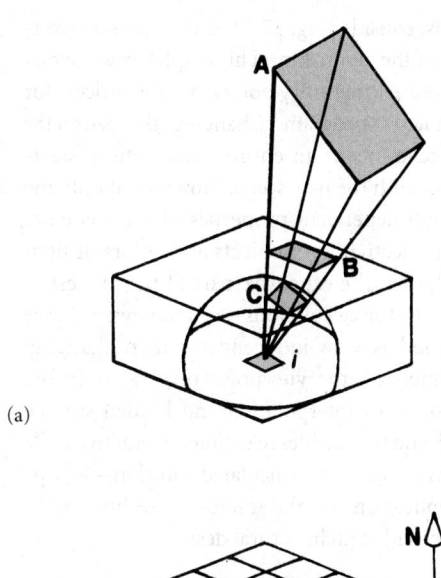

(a)

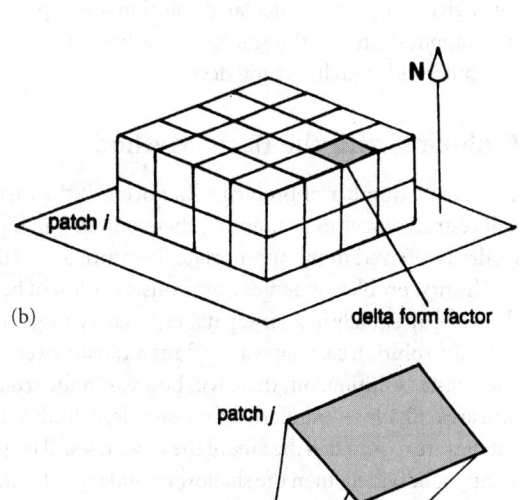

(b)

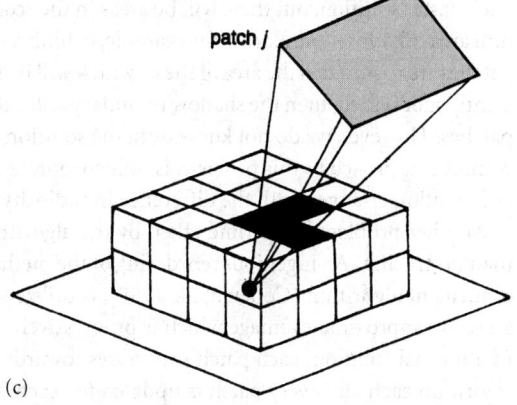

(c)

FIGURE 57.18 Visualization of the properties used in the hemicube algorithm for form-factor evaluation; (a) Nusselt analogue: patches A, B, and C have the same form factor with respect to patch i; (b) delta form factors are precalculated for the hemicube; (c) the hemicube is positioned over patch i. Each patch j is projected onto the hemicube. F_{ij} is calculated by summing the delta form factors of the hemicube elements onto which j projects.

the hemicube elements onto which it projects. When all patches are projected, the form factor F_{ij} is calculated by summing the delta form factors that have registered patch j as a nearest projection.

Finally, consider Fig. 57.19, which gives an overall view of the algorithm. This emphasizes the fact that there are three entry points into the process for an interactive program. Changing the geometry of the scene means an entire recalculation, starting afresh with the new scene. However, if only the wavelength-dependent properties of the scene are altered (reflectivities of objects and colors of light sources), then the expensive part of the process— the form-factor calculations—is unchanged. Since the method is view-independent, then changing the position of the viewpoint involves only the final process of interpolation and hidden-surface removal, and this enables real-time, interactive walk-throughs using the precalculated solution—a popular application of the radiosity technique in computer-aided architectural design.

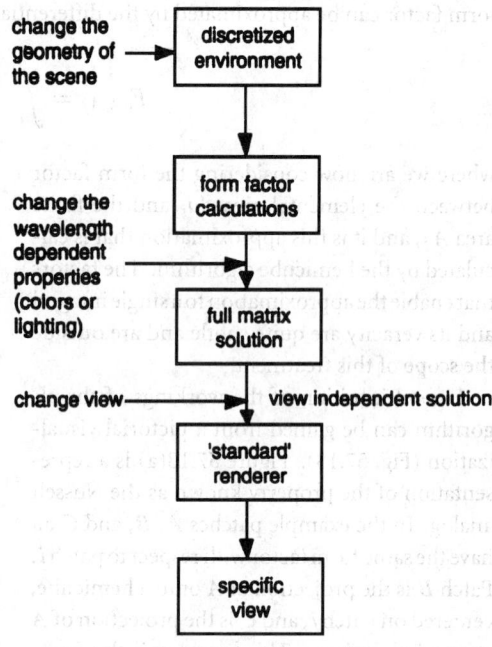

FIGURE 57.19 Processes and interactive entry points in a radiosity algorithm.

Problems with the Basic Method

There are a number of problems that occur if the method is implemented without elaboration. Here we will restrict ourselves to pointing these out and giving a pointer to the solutions that have emerged. The reader is referred to the appropriate literature for further details.

The first problem emerges from consideration of how to divide the environment into patches. Dividing the scene up equally into large patches, as far as the geometry of the objects allows, will not suffice. The basic radiosity solution calculates a constant radiosity over the area of a patch. Larger patches means less patches and a faster solution, but there will be areas in the scene that will exhibit a fast change in reflected light per unit area. Shadow boundaries, for example, exhibit a shape that depends on the obscuring object, and small patches are required in the area of these, which will in general be curves. If the shadowed surface is not sufficiently subdivided, then the shadow boundary will exhibit steps in the shape of the (square or rectangular) patches. However, we do not know until the solution is computed where such areas are in the scene, and a common approach to this problem is to incorporate a variable subdivision into the solution, subdividing as a solution emerges until the difference in radiosity between adjacent patches falls below a threshold.

Another problem is the time taken by the algorithm—approximately one order of magnitude longer than ray tracing. An ingenious reordering of the method for solving the equation set, called the progressive refinement algorithm [Cohen et al. 1988], is utilized to ameliorate this problem and provide a user with an early (approximate) image which is progressively refined, that is, it becomes more and more accurate. In a normal solution, each patch progresses towards a final value in turn. In the progressive refinement algorithm each and every patch is updated for each iteration, and all patches proceed towards their final value simultaneously. The scene is at first dark and then gets lighter and lighter as the patches increase their radiosity. Early, approximate solutions are quickly computed by distributing an arbitrary ambient component amongst all patches. The early solutions are "fully" lit but incorrect. Progressive refinement means displaying more and more accurate radiosities on each patch and reducing the approximate ambient component at the same time. A user can thus see from the start some solution, which then proceeds to get more and more accurate. The process can then be interrupted at some stage if the image is not as required or has some visually obvious error.

57.5 The Future of Rendering

Like many developments in computer science, the evolution of rendering techniques has been influenced by many factors other than image quality. One of the strongest influences has been the development of graphics workstations with hardware implementing various polygon-based rendering techniques. This has led to the perception of a graphics pipeline as a black box like any other piece of hardware such as a floating-point unit.

The significant problem that rendering now faces is coping with high-quality moving imagery. In this context "high quality" means both high complexity in terms of scene elements and photorealistic rendering of such scenes. Although time is not currently too much of a problem in computer animation and film special effects—where the production is off line—we are now faced with the real-time demands of virtual reality (VR). The current image-quality–cost tradeoff means that most medium-priced VR systems produce extremely low-quality images. It is normal to dismiss this problem by saying that falling hardware costs will eventually result in equipment that will cope with high-quality VR, but it is unlikely that this will happen in the next five to ten years. And it must be remembered that further developments in rendering will place even greater demands on rendering hardware. The tendency is always for improvements to cost more; the quality spectrum from flat shading to radiosity is exactly reflected in the rendering time.

The main problem for VR applications is the black-box renderer, where any slight change in the viewpoint (due, say, to a user moving his head) causes the entire original scene description to be injected into the process and a new frame to be calculated completely independently of any previous computations. So how can the standard approach to rendering be altered to cope with these new demands? A number of schemes are under investigation—all of which flow from relaxing the accuracy, in some way, of the final image.

An approach long used in flight simulators is to switch the level of detail at which objects are rendered as a function of the projected area of the object on the screen or the viewing distance. There is no point in rendering an object, consisting of a large number of polygons, at its finest level of detail, if the object projects onto only a few pixels. Traditional solutions in flight simulator systems have involved constructing the object database at different levels of detail, switching to a more detailed representation as the object projection becomes larger and vice versa. In its simplest manifestation, however, detail switches on and off—a clearly undesirable effect. Invoking a scheme for smoothly changing from one level of detail to another starts to increase the cost again.

Another idea is *priority rendering* [Regan and Pose 1994], where different objects in the scene are rendered at different rates. Nearby objects are rendered at full animation rate, whereas objects further away from the user are rendered at a lower rate. Thus the "geometric correctness" of the scene projected into an HMD varies with the user's virtual position with respect to all the virtual objects in the scene.

Finally there is the precalculation theme. Can we store enough precalculated solutions of, say, an interior to be able to compose a frame for any viewpoint in the room? For example, in a virtual museum or virtual art-gallery application we could store a set of environment maps precalculated from a set of viewpoints arranged in a grid contained in a plane positioned at head height. A stationary user positioned at any one of the viewpoints can gaze in any direction, and an appropriate view can be constructed from the environment map. Movement through the environment would then consist of "hopping" from one viewpoint to another. This idea is explored in recent work [Chen 1995] which uses, as an environment map, 360-deg cylindrical panoramic images.

Defining Terms

Ambient: The constant term used in local reflection models to simulate global illumination. It illuminates parts of a surface which cannot be seen from the light source but which are visible from the viewpoint.

Antialiasing: The term given to measures designed to eliminate defects in computer graphics images, the most common being the effect produced when a curved edge in the image is displayed as jagged due to the finite extent of the pixels. The term is somewhat inappropriate, as "aliasing" is a specific signal-processing term and many computer graphics defects are not aliases in that sense.

Clipping: It is common to define a view frustum, a volume emanating from the viewpoint, against which objects are clipped. For example, objects behind the viewer need to be eliminated from consideration.

Culling: A process to remove whole polygons that cannot be seen from the viewpoint and do not therefore need to be considered by the hidden-surface removal algorithm.

Diffuse: Local reflection models separately evaluate a diffuse, a specular, and an ambient component. The diffuse component is the light that is reflected from a point equally in all directions, simulating a matte or plastic-like surface.

Flat shading: A shading option where polygons are allocated the same shade—there is no variation within a polygon. This makes the underlying polygonal structure visible.

Global reflection models: Models which attempt to model indirect illumination. At a surface point they consider both light coming directly from light sources and light incoming that has been reflected from other objects.

Gouraud shading: An interpolative shading method for polygons where the parameter that is interpolated is the reflected light intensity at the vertices.

Hemicube: An algorithmic device that enables an efficient calculation of the form-factor values in the radiosity method. A hemicube is an efficient approximation to a hemisphere.

Hidden-surface removal: The general algorithm which removes hidden surfaces and deals with such cases where one object partially obscures another. A hidden surface algorithm will in general eliminate those fragments of a polygon which are not visible because they are behind another object.

Local reflection models: Models that simulate the reflection of light incident directly on an object from a light source. Unlike global models, they take no account of (indirect) light reflected from another object.

Phong shading: An interpolative shading method where the parameter that is interpolated is the vertex normal at the vertices. This produces an interpolated normal for every pixel onto which the polygon projects, and a reflection model is evaluated at each pixel in the image plane.

Polygon mesh: The most common form of object representation is to build a set of planar facets or polygons that (approximately, in general) represent the surface of an object.

Progressive refinement: An elaboration of the original radiosity method that enables a visualization of the solution to emerge as the equations are being solved. The solution, originally approximated with an ambient term, is gradually made more and more accurate.

Projective transformation: Usually the final geometric transformation, it produces the perspective foreshortening desired in most applications.

Radiosity: A global reflection model which divides the scene into large elements called patches, calculates a parameter that reflects the geometric relationship between all pairs of patches, and sets up a system of equations whose solution yields a constant radiosity for each patch. These radiosity values are input to a Gouraud-type interpolation to produce a rendered image. The geometric extent of the patches means that the basic method can deal only with diffuse interaction.

Ray tracing: A global reflection model that casts infinitesimally thin rays into the scene from each pixel and follows or traces these as they are reflected and transmitted by objects that they encounter. Such a ray tracer can find out only about specular interaction.

Shading algorithm: A general term that describes that part of the process that makes a geometric description look like a solid three-dimensional object.

Specular: The specular component is the light reflected from a point in the mirror direction—as if the surface were a perfect mirror. In practice this component is "empirically spread" to simulate a practical glossy surface.

Viewing transformation: Generates a representation of the object or scene as seen from the viewpoint of an observer positioned somewhere in the scene and looking towards some aspect of it.

Virtual camera: A common analogy used for the series of geometric transformations that form a two-dimensional image on the view surface. These transformations have the same effect as a (perfect) pinhole camera. The analogy is particulary useful in animation where the camera is to be "choreographed."

References

Blinn, J. 1977. Models of light reflection for computer synthesized pictures, pp. 192–198. *Comput. Graphics (Proc. SIGGRAPH)*.

Chen, S. E. 1995. Quicktime VR—An image based approach to virtual environment navigation, pp. 29–38. *Comput. Graphics (Proc. SIGGRAPH)*.

Cohen, M. F. and Greenberg, D. P. 1985. A radiosity solution for complex environments, pp. 31–40. *Comput. Graphics (Proc. SIGGRAPH)*.

Cohen, M. F., Chen, S. E., Wallace, J. R., and Greenberg, D. P. 1988. A progressive refinement approach to fast radiosity image generation, pp. 75–84. *Comput. Graphics (Proc. SIGGRAPH)*.

Cook, R. L., Porter, T., and Carpenter, L. 1984. Distributed ray tracing, pp. 137–145. *Comput. Graphics (Proc. SIGGRAPH)*.

Fiume, E. L. 1989. *The Mathematical Structure of Computer Graphics*. Academic Press, San Diego, CA.

Gouraud, H. 1971. Illumination for computer generated pictures. *Commun. ACM* 18(60):628–678

Newman, W. and Sproull, R. 1973. *Principles of Interactive Computer Graphics*. McGraw–Hill, New York.

Phong, B. T. 1975. Illumination for computer generated pictures. *Commun. ACM* 18(6):311–317.

Regan, M. and Pose, R. 1994. Priority rendering with a virtual reality address recalculation pipeline. *Comput. Graphics (Proc. SIGGRAPH)*.

Sutherland, I. E. and Hodgman, G. W. 1974. Re-entrant polygon clipping. *Commun. ACM* 17(1):32–42.

Watt, A. and Watt, M. 1992. *Advanced Animation and Rendering Techniques*. ACM Press, New York.

Whitted, T. 1980. An improved illumination model for shaded display. *Commun. ACM* 26(6):342–349.

Further Information

The above references are mostly the original sources of the algorithms that are commonly incorporated in rendering engines. A would-be implementer, however, would be best directed to a general textbook, such as [Watt and Watt 1992] in the above list, or the encyclopedic *Computer Graphics—Principles and Practice* by Foley et al.

Undoubtedly the best source of rendering techniques and their development is the annual ACM SIGGRAPH conference (proceedings published by ACM Press). Browsing through past proceedings gives a feel for the fascinating development and history of image synthesis.

58

Sampling, Reconstruction, and Antialiasing

George Wolberg

City College of New York

58.1 Introduction

This chapter reviews the principal ideas of sampling theory, reconstruction, and antialiasing. Sampling theory is central to the study of sampled-data systems, e.g., digital image transformations. It lays a firm mathematical foundation for the analysis of sampled signals, offering invaluable insight into the problems and solutions of sampling. It does so by providing an elegant mathematical formulation describing the relationship between a continuous signal and its samples. We use it to resolve the problems of image reconstruction and aliasing. Reconstruction is an interpolation procedure applied to the sampled data. It permits us to evaluate the discrete signal at any desired position, not just the integer lattice upon which the sampled signal is given. This is useful when implementing geometric transformations, or warps, on the image. Aliasing refers to the presence of unreproducibly high frequencies in the image and the resulting artifacts that arise upon undersampling.

Together with defining theoretical limits on the continuous reconstruction of discrete input, sampling theory yields the guidelines for numerically measuring the quality of various proposed filtering

0-8493-2909-4/97/$0.00+$.50
© 1997 by CRC Press, Inc.

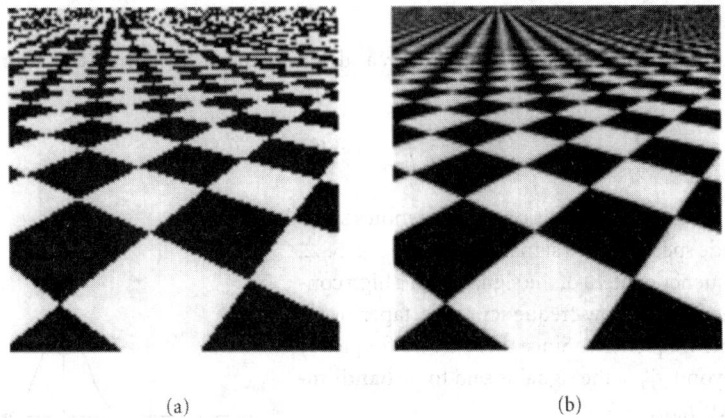

FIGURE 58.1 Oblique checkerboard: (a) unfiltered, (b) filtered.

techniques. This proves most useful in formally describing reconstruction, aliasing, and the filtering necessary to combat the artifacts that may appear at the output.

In order to better motivate the importance of sampling theory and filtering, we demonstrate its role with the following examples. A checkerboard texture is shown projected onto an oblique planar surface in Fig. 58.1. The image exhibits two forms of artifacts: jagged edges and moiré patterns. Jagged edges are prominent toward the bottom of the image, where the input checkerboard undergoes magnification. It reflects poor reconstruction of the underlying signal. The moiré patterns, on the other hand, are noticeable at the top, where minification (compression) forces many input pixels to occupy fewer output pixels. This artifact is due to aliasing, a symptom of undersampling.

Figure 58.1(a) was generated by projecting the center of each output pixel into the checkerboard and sampling (reading) the value of the nearest input pixel. This point sampling method performs poorly, as is evident by the objectionable results of Fig. 58.1(a). This conclusion is reached by sampling theory as well. Its role here is to precisely quantify this phenomenon and to prescribe a solution. Figure 58.1(b) shows the same mapping with improved results. This time the necessary steps were taken to preclude artifacts. In particular, a superior reconstruction algorithm was used for interpolation to suppress the jagged edges, and antialiasing filtering was carried out to combat the symptoms of undersampling that gave rise to the moiré patterns.

58.2 Sampling Theory

Both reconstruction and antialiasing share the twofold problem addressed by sampling theory:

1. Given a continuous input signal $g(x)$ and its sampled counterpart $g_s(x)$, are the samples of $g_s(x)$ sufficient to exactly describe $g(x)$?
2. If so, how can $g(x)$ be reconstructed from $g_s(x)$?

The solution lies in the frequency domain, whereby spectral analysis is used to examine the spectrum of the sampled data.

The conclusions derived from examining the reconstruction problem will prove to be directly useful for resampling and indicative of the filtering necessary for antialiasing. Sampling theory thereby provides an elegant mathematical framework in which to assess the quality of reconstruction, establish theoretical limits, and predict when it is not possible.

Sampling

Consider a 1D signal $g(x)$ and its spectrum $G(f)$, as determined by the Fourier transform:

$$G(f) = \int_{-\infty}^{\infty} g(x)e^{-i2\pi f x}\, dx \tag{58.1}$$

Note that x represents spatial position and f denotes spatial frequency.

The magnitude spectrum of a signal is shown in Fig. 58.2. It shows the frequency content of the signal, with a high concentration of energy in the low-frequency range, tapering off toward the higher frequencies. Since there are no frequency components beyond f_{max}, the signal is said to be **bandlimited** to frequency f_{max}.

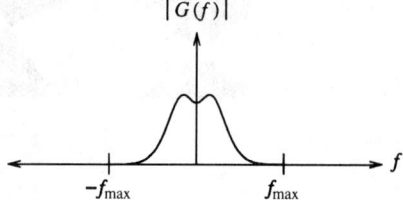

FIGURE 58.2 Spectrum $G(f)$.

The continuous output $g(x)$ is then digitized by an ideal impulse sampler, the comb function, to get the sampled signal $g_s(x)$. The ideal 1D sampler is given as

$$s(x) = \sum_{n=-\infty}^{\infty} \delta(x - nT_s) \tag{58.2}$$

where δ is the familiar impulse function and T_s is the sampling period. The running index n is used with δ to define the impulse train of the comb function. We now have

$$g_s(x) = g(x)s(x) \tag{58.3}$$

Taking the Fourier transform of $g_s(x)$ yields

$$G_s(f) = G(f) * S(f) \tag{58.4}$$

$$= G(f) * \left[\sum_{n=-\infty}^{n=\infty} f_s \delta(f - nf_s) \right] \tag{58.5}$$

$$= f_s \sum_{n=-\infty}^{n=\infty} G(f - nf_s) \tag{58.6}$$

where f_s is the sampling frequency and $*$ denotes convolution. The above equations make use of the following well-known properties of Fourier transforms:

1. Multiplication in the spatial domain corresponds to convolution in the frequency domain. Therefore, Eq. (58.3) gives rise to a convolution in Eq. (58.4).
2. The Fourier transform of an impulse train is itself an impulse train, giving us Eq. (58.5).
3. The spectrum of a signal sampled with frequency f_s ($T_s = 1/f_s$) yields the original spectrum replicated in the frequency domain with period f_s [Eq. (58.6)].

This last property has important consequences. It yields a spectrum $G_s(f)$, which, in response to a sampling period $T_s = 1/f_s$, is *periodic in frequency* with period f_s. This is depicted in Fig. 58.3. Notice, then, that a small sampling period is equivalent to a high sampling frequency, yielding spectra replicated far apart from each other. In the limiting case when the sampling period approaches zero ($T_s \to 0$, $f_s \to \infty$), only a single spectrum appears—a result consistent with the continuous case.

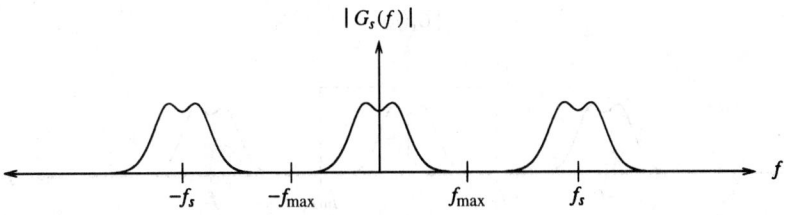

FIGURE 58.3   Spectrum $G_s(f)$.

58.3 Reconstruction

The above result reveals that the sampling operation has left the original input spectrum *intact*, merely replicating it periodically in the frequency domain with a spacing of f_s. This allows us to rewrite $G_s(f)$ as a sum of two terms, the low-frequency (baseband) and high-frequency components. The *baseband* spectrum is exactly $G(f)$, and the high-frequency components, $G_{high}(f)$, consist of the remaining replicated versions of $G(f)$ that constitute harmonic versions of the sampled image:

$$G_s(f) = G(f) + G_{high}(f) \qquad (58.7)$$

Exact signal reconstruction from sampled data requires us to discard the replicated spectra $G_{high}(f)$, leaving only $G(f)$, the spectrum of the signal we seek to recover. This is a crucial observation in the study of sampled-data systems.

Reconstruction Conditions

The only provision for exact reconstruction is that $G(f)$ be undistorted due to overlap with $G_{high}(f)$. Two conditions must hold for this to be true:

1. The signal must be bandlimited. This avoids spectra with infinite extent that are impossible to replicate without overlap.
2. The sampling frequency f_s must be greater than twice the maximum frequency f_{max} present in the signal. This minimum sampling frequency, known as the **Nyquist rate**, is the minimum distance between the spectra copies, each with bandwidth f_{max}.

The first condition merely ensures that a sufficiently large sampling frequency exists that can be used to separate replicated spectra from each other. Since all imaging systems impose a bandlimiting filter in the form of a point spread function, this condition is always satisfied for images captured through an optical system. Note that this does not apply to synthetic images, e.g., computer-generated imagery.

The second condition proves to be the most revealing statement about reconstruction. It answers the problem regarding the sufficiency of the data samples to exactly reconstruct the continuous input signal. It states that exact reconstruction is possible only when $f_s > f_{Nyquist}$, where $f_{Nyquist} = 2 f_{max}$. Collectively, these two conclusions about reconstruction form the central message of sampling theory, as pioneered by Claude Shannon in his landmark papers on the subject [Shannon 1948, 1949].

Ideal Low-Pass Filter

We now turn to the second central problem: Given that it is theoretically possible to perform reconstruction, how may it be done? The answer lies with our earlier observation that sampling merely replicates the spectrum of the input signal, generating $G_{high}(f)$ in addition to $G(f)$. Therefore, the act of reconstruction requires us to completely suppress $G_{high}(f)$. This is done by multiplying $G_s(f)$ with $H(f)$,

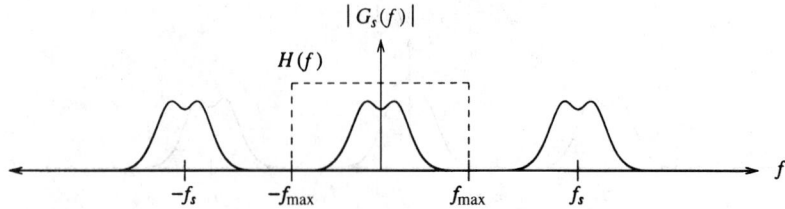

FIGURE 58.4   Ideal low-pass filter $H(f)$.

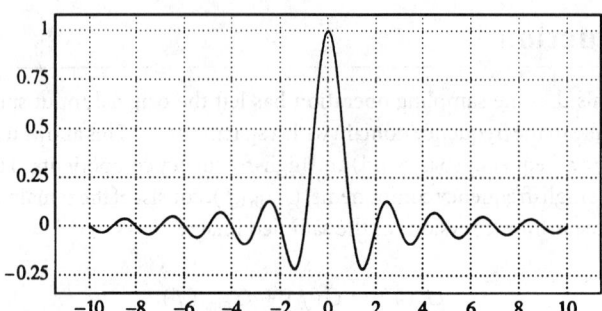

FIGURE 58.5   The sinc function.

given as

$$H(f) = \begin{cases} 1, & |f| < f_{max} \\ 0, & |f| \geq f_{max} \end{cases} \tag{58.8}$$

$H(f)$ is known as an *ideal low-pass filter* and is depicted in Fig. 58.4, where it is shown suppressing all frequency components above f_{max}. This serves to discard the replicated spectra $G_{high}(f)$. It is ideal in the sense that the f_{max} cutoff frequency is strictly enforced as the transition point between the transmission and complete suppression of frequency components.

Sinc Function

In the spatial domain, the ideal low-pass filter is derived by computing the inverse Fourier transform of $H(f)$. This yields the sinc function shown in Fig. 58.5. It is defined as

$$\text{sinc}\,(x) = \frac{\sin(\pi x)}{\pi x} \tag{58.9}$$

The reader should note the reciprocal relationship between the height and width of the ideal low-pass filter in the spatial and frequency domains. Let A denote the amplitude of the sinc function, and let its zero crossings be positioned at integer multiples of $1/2W$. The spectrum of this sinc function is a rectangular pulse of height $A/2W$ and width $2W$, with frequencies ranging from $-W$ to W. In our example above, $A = 1$ and $W = f_{max} = 0.5$ cycles/pixel. This value for W is derived from the fact that digital images must not have more than one half cycle per pixel in order to conform to the Nyquist rate.

The sinc function is one instance of a large class of functions known as cardinal splines, which are interpolating functions defined to pass through zero at all but one data sample, where they have a value of one. This allows them to compute a continuous function that passes through the uniformly spaced data samples.

Since multiplication in the frequency domain is identical to convolution in the spatial domain, sinc (x) represents the convolution kernel used to evaluate any point x on the continuous input curve g given only

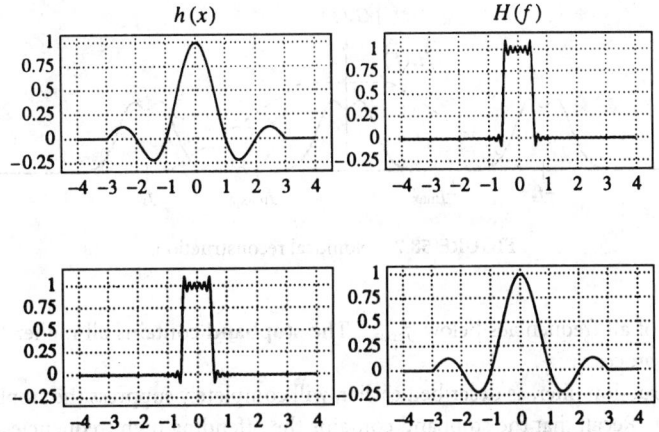

FIGURE 58.6 Truncation in one domain causes ringing in the other domain.

the sampled data g_s:

$$g(x) = \text{sinc}(x) * g_s(x) = \int_{-\infty}^{\infty} \text{sinc}(\lambda)\, g_s(x - \lambda)\, d\lambda \qquad (58.10)$$

Equation (58.10) highlights an important impediment to the practical use of the ideal low-pass filter. The filter requires an infinite number of neighboring samples (i.e., an infinite filter support) in order to precisely compute the output points. This is, of course, impossible owing to the finite number of data samples available. However, truncating the sinc function allows for approximate solutions to be computed at the expense of undesirable "ringing," i.e., ripple effects. These artifacts, known as the **Gibbs phenomenon**, are the overshoots and undershoots caused by reconstructing a signal with truncated frequency terms. The two rows in Fig. 58.6 show that truncation in one domain leads to ringing in the other domain. This indicates that a truncated sinc function is actually a poor reconstruction filter because its spectrum has infinite extent and thereby fails to bandlimit the input.

In response to these difficulties, a number of approximating algorithms have been derived, offering a tradeoff between precision and computational expense. These methods permit local solutions that require the convolution kernel to extend only over a small neighborhood. The drawback, however, is that the frequency response of the filter has some undesirable properties. In particular, frequencies below f_{max} are tampered, and high frequencies beyond f_{max} are not fully suppressed. Thus, nonideal reconstruction does not permit us to exactly recover the continuous underlying signal without artifacts.

Nonideal Reconstruction

The process of nonideal reconstruction is depicted in Fig. 58.7, which indicates that the input signal satisfies the two conditions necessary for exact reconstruction. First, the signal is bandlimited, since the replicated copies in the spectrum are each finite in extent. Second, the sampling frequency exceeds the Nyquist rate, since the copies do not overlap. However, this is where our ideal scenario ends. Instead of using an ideal low-pass filter to retain only the baseband spectrum components, a nonideal reconstruction filter is shown in the figure.

The filter response $H_r(f)$ deviates from the ideal response $H(f)$ shown in Fig. 58.4. In particular, $H_r(f)$ does not discard all frequencies beyond f_{max}. Furthermore, that same filter is shown to attenuate some frequencies that should have remained intact. This brings us to the problem of assessing the quality of a filter.

The accuracy of a reconstruction filter can be evaluated by analyzing its frequency-domain characteristics. Of particular importance is the filter response in the passband and stopband. In this problem, the

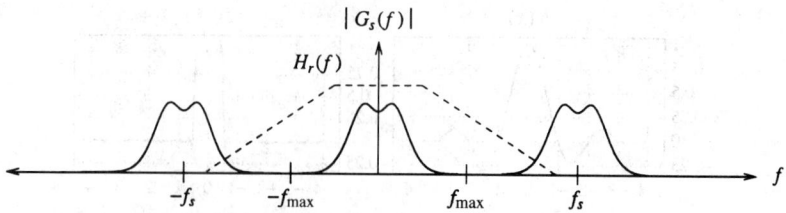

FIGURE 58.7   Nonideal reconstruction.

passband consists of all frequencies below f_{max}. The **stopband** contains all higher frequencies arising from the sampling process.

An ideal reconstruction filter, as described earlier, will completely suppress the stopband while leaving the passband intact. Recall that the stopband contains the offending high frequencies that, if allowed to remain, would prevent us from performing exact reconstruction. As a result, the sinc filter was devised to meet these goals and serve as the ideal reconstruction filter. Its kernel in the frequency domain applies unity gain to transmit the passband and zero gain to suppress the stopband.

The breakdown of the frequency domain into passband and stopband isolates two problems that can arise due to nonideal reconstruction filters. The first problem deals with the effects of imperfect filtering on the passband. Failure to impose unity gain on *all* frequencies in the passband will result in some combination of image smoothing and image sharpening. Smoothing, or blurring, will result when the frequency gains near the cutoff frequency start falling off. Image sharpening results when the high-frequency gains are allowed to exceed unity. This follows from the direct correspondence of visual detail to spatial frequency. Furthermore, amplifying the high passband frequencies yields a sharper transition between the passband and stopband, a property shared by the sinc function.

The second problem addresses nonideal filtering on the stopband. If the stopband is allowed to persist, high frequencies will exist that will contribute to aliasing (described later). Failure to fully suppress the stopband is a condition known as **frequency leakage**. This allows the offending frequencies to fold over into the passband range. These distortions tend to be more serious, since they are visually perceived more readily.

In the spatial domain, nonideal reconstruction is achieved by centering a finite-width kernel at the position in the data at which the underlying function is to be evaluated, i.e., reconstructed. This is an interpolation problem which, for equally spaced data, can be expressed as

$$f(x) = \sum_{k=0}^{K-1} f(x_k)h(x - x_k) \tag{58.11}$$

where h is the reconstruction kernel that weighs K data samples at x_k. Equation (58.11) formulates interpolation as a convolution operation. In practice, h is nearly always a symmetric kernel, i.e., $h(-x) = h(x)$. We shall assume this to be true in the discussion that follows.

The computation of one interpolated point is illustrated in Fig. 58.8. The kernel is centered at x, the location of the point to be interpolated. The value of that point is equal to the sum of the values of the discrete input scaled by the corresponding values of the reconstruction kernel. This follows directly from the definition of convolution.

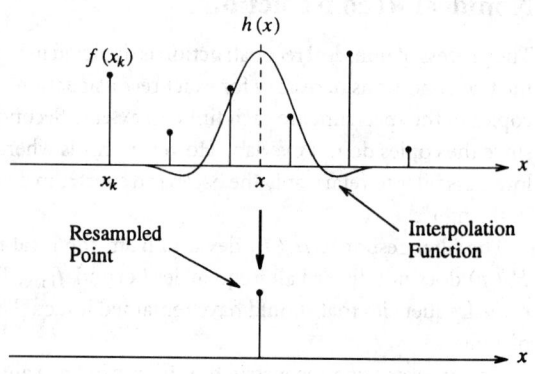

FIGURE 58.8   Interpolation of a single point.

58.4 Reconstruction Kernels

The numerical accuracy and computational cost of reconstruction are directly tied to the convolution kernel used for low-pass filtering. As a result, filter kernels are the target of design and analysis in the creation and evaluation of reconstruction algorithms. They are subject to conditions influencing the tradeoff between accuracy and efficiency. This section reviews several common nonideal reconstruction filter kernels in the order of their complexity: box filter, triangle filter, cubic convolution, and windowed sinc functions.

Box Filter

The box filter kernel is defined as

$$h(x) = \begin{cases} 1, & -0.5 < x \le 0.5 \\ 0, & \text{otherwise} \end{cases} \tag{58.12}$$

Various other names are used to denote this simple kernel, including the sample-and-hold function and Fourier window. The kernel and its Fourier transform are shown in Fig. 58.9.

Convolution in the spatial domain with the rectangle function h is equivalent in the frequency domain to multiplication with a sinc function. Due to the prominent side lobes and infinite extent, a sinc function makes a poor low-pass filter. Consequently, this filter kernel has a poor frequency-domain response relative to that of the ideal low-pass filter. The ideal filter, drawn as a dashed rectangle, is characterized by unity gain in the passband and zero gain in the stopband. This permits all low frequencies (below the cutoff frequency) to pass and all higher frequencies to be suppressed.

Triangle Filter

The triangle filter kernel is defined as

$$h(x) = \begin{cases} 1 - |x|, & 0 \le |x| < 1 \\ 0, & 1 \le |x| \end{cases} \tag{58.13}$$

The kernel h is also referred to as a tent filter, roof function, Chateau function, or Bartlett window.

This kernel corresponds to a reasonably good low-pass filter in the frequency domain. As shown in Fig. 58.10, its response is superior to that of the box filter. In particular, the side lobes are far less prominent, indicating improved performance in the stopband. Nevertheless, a significant amount of spurious high-frequency components continues to leak into the passband, contributing to some aliasing. In addition, the passband is moderately attenuated, resulting in image smoothing.

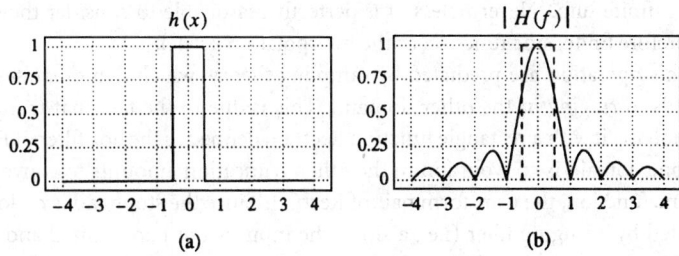

FIGURE 58.9   Box filter: (a) kernel, (b) Fourier transform.

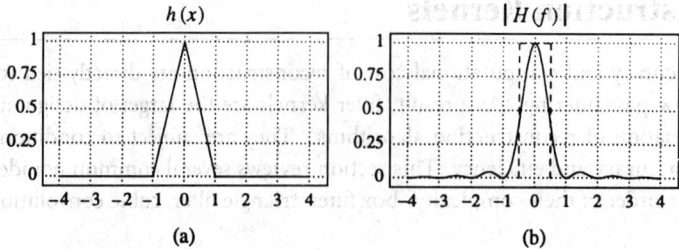

FIGURE 58.10   Triangle filter: (a) kernel, (b) Fourier transform.

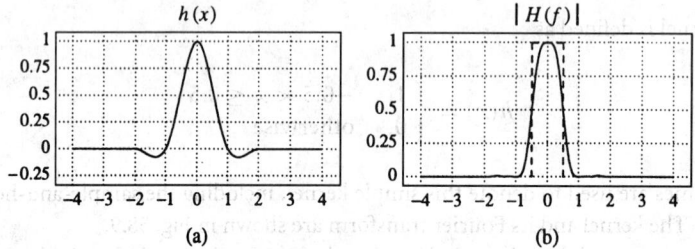

FIGURE 58.11   Cubic convolution: (a) kernel ($a = -0.5$), (b) Fourier transform.

Cubic Convolution

The cubic convolution kernel is a third-degree approximation to the sinc function. It is symmetric, space-invariant, and composed of piecewise cubic polynomials:

$$h(x) = \begin{cases} (a+2)|x|^3 - (a+3)|x|^2 + 1, & 0 \leq |x| < 1 \\ a|x|^3 - 5a|x|^2 + 8a|x| - 4a, & 1 \leq |x| < 2 \\ 0, & 2 \leq |x| \end{cases} \qquad (58.14)$$

where $-3 < a < 0$ is used to make h resemble the sinc function.

Of all the choices for a, the value -1 is preferable if visually enhanced results are desired. That is, the image is sharpened, making visual detail perceived more readily. However, the results are not mathematically precise, where precision is measured by the order of the Taylor series. To maximize this order, the value $a = -0.5$ is preferable. A cubic convolution kernel with $a = -0.5$ and its spectrum are shown in Fig. 58.11.

Windowed Sinc Function

Sampling theory establishes that the sinc function is the ideal interpolation kernel. Although this interpolation filter is exact, it is not practical, since it is an infinite impulse response (IIR) filter defined by a slowly converging infinite sum. Nevertheless, it is perfectly reasonable to consider the effects of using a truncated, and therefore finite, sinc function as the interpolation kernel.

The results of this operation are predicted by sampling theory, which demonstrates that truncation in one domain leads to ringing in the other domain. This is due to the fact that truncating a signal is equivalent to multiplying it with a rectangle function Rect(x), defined as the box filter of Eq. (58.12). Since multiplication in one domain is convolution in the other, truncation amounts to convolving the signal's spectrum with a sinc function, the transform pair of Rect(x). Since the stopband is no longer eliminated, but rather attenuated by a ringing filter (i.e., a sinc), the input is not bandlimited and aliasing artifacts are introduced. The most typical problems occur at step edges, where the Gibbs phenomena becomes noticeable in the form of undershoots, overshoots, and ringing in the vicinity of edges.

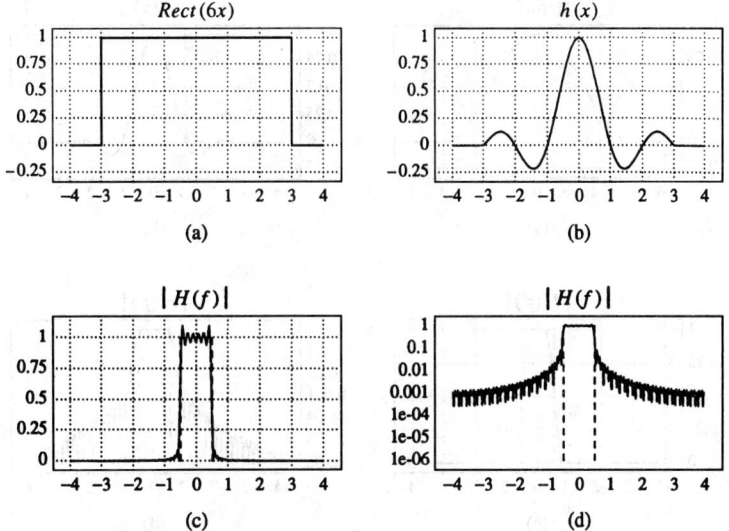

FIGURE 58.12 (a) Rectangular window; (b) windowed sinc; (c) spectrum; (d) log plot.

The Rect function above served as a window, or kernel, that weighs the input signal. In Fig. 58.12(a), we see the *Rect* window extended over three pixels on each side of its center, i.e., Rect(6x) is plotted. The corresponding windowed sinc function $h(x)$ is shown in Fig. 58.12(b). This is simply the product of the sinc function with the window function, i.e., sinc (x) Rect(6x). Its spectrum, shown in Fig. 58.12(c), is nearly an ideal low-pass filter. Although it has a fairly sharp transition from the passband to the stopband, it is plagued by ringing. In order to more clearly see the values in the spectrum, we use a logarithmic scale for the vertical axis of the spectrum in Fig. 58.12(d). The next few figures will use this same four-part format.

Ringing can be mitigated by using a different windowing function exhibiting smoother falloff than the rectangle. The resulting windowed sinc function can yield better results. However, since slow falloff requires larger windows, the computation remains costly.

Aside from the rectangular window mentioned above, the most frequently used window functions are Hann, Hamming, Blackman, and Kaiser. These filters identify a quantity known as the ripple ratio, defined as the ratio of the maximum side-lobe amplitude to the main-lobe amplitude. Good filters will have small ripple ratios to achieve effective attenuation in the stopband. A tradeoff exists, however, between ripple ratio and main-lobe width. Therefore, as the ripple ratio is decreased, the main-lobe width is increased. This is consistent with the reciprocal relationship between the spatial and frequency domains, i.e., narrow bandwidths correspond to wide spatial functions.

In general, though, each of these smooth window functions is defined over a small finite extent. This is tantamount to multiplying the smooth window with a rectangle function. While this is better than the Rect function alone, there will inevitably be some form of aliasing. Nevertheless, the window functions described below offer a good compromise between ringing and blurring.

Hann and Hamming Windows

The Hann and Hamming windows are defined as

$$\text{Hann/Hamming}(x) = \begin{cases} \alpha + (1 - \alpha) \cos \frac{2\pi x}{N-1}, & |x| < \frac{N-1}{2} \\ 0, & \text{otherwise} \end{cases} \tag{58.15}$$

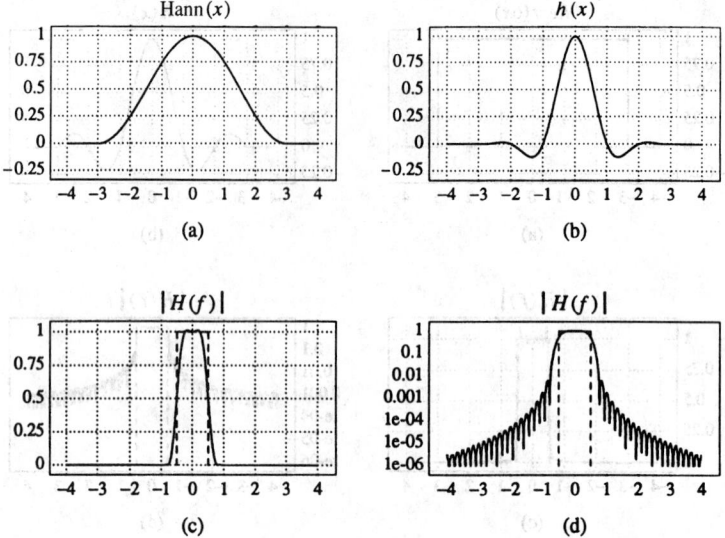

FIGURE 58.13 (a) Hann window; (b) windowed sinc; (c) spectrum; (d) log plot.

where N is the number of samples in the windowing function. The two windowing functions differ in their choice of α. In the Hann window $\alpha = 0.5$, and in the Hamming window $\alpha = 0.54$. Since they both amount to a scaled and shifted cosine function, they are also known as the raised cosine window. The Hann window is illustrated in Fig. 58.13. Notice that the passband is only slightly attenuated, but the stopband continues to retain high-frequency components in the stopband, albeit less than that of Rect(x).

Blackman Window

The Blackman window is similar to the Hann and Hamming windows. It is defined as

$$\text{Blackman}(x) = \begin{cases} 0.42 + 0.5 \cos \frac{2\pi x}{N-1} + 0.08 \cos \frac{4\pi x}{N-1}, & |x| < \frac{N-1}{2} \\ 0, & \text{otherwise} \end{cases} \tag{58.16}$$

The purpose of the additional cosine term is to further reduce the ripple ratio. This window function is shown in Fig. 58.14.

Kaiser Window

The Kaiser window is defined as

$$\text{Kaiser}(x) = \begin{cases} \frac{I_0(\beta)}{I_0(\alpha)}, & |x| < \frac{N-1}{2} \\ 0, & \text{otherwise} \end{cases} \tag{58.17}$$

where I_0 is the zeroth-order Bessel function of the first kind, α is a free parameter, and

$$\beta = \alpha \left[1 - \left(\frac{2x}{N-1} \right)^2 \right]^{1/2} \tag{58.18}$$

The Kaiser window leaves the filter designer much flexibility in controlling the ripple ratio by adjusting the parameter α. As α is increased, the level of sophistication of the window function grows as well. Therefore, the rectangular window corresponds to a Kaiser window with $\alpha = 0$, while more sophisticated windows such as the Hamming window correspond to $\alpha = 5$.

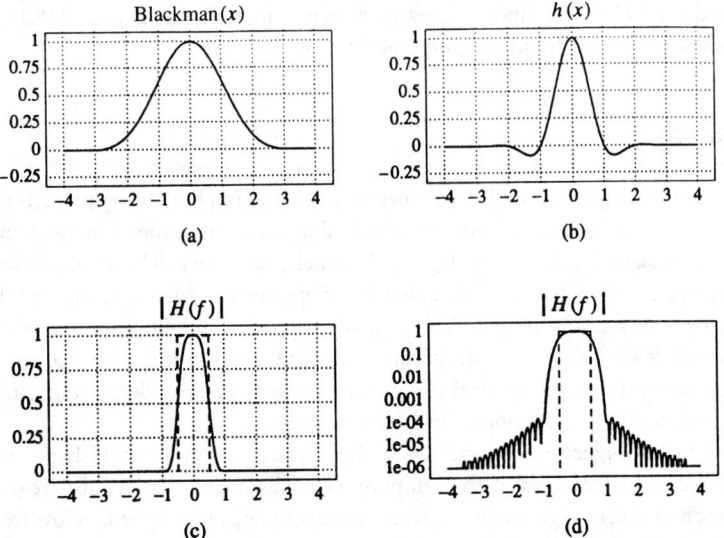

FIGURE 58.14   (a) Blackman window; (b) windowed sinc; (c) spectrum; (d) log plot.

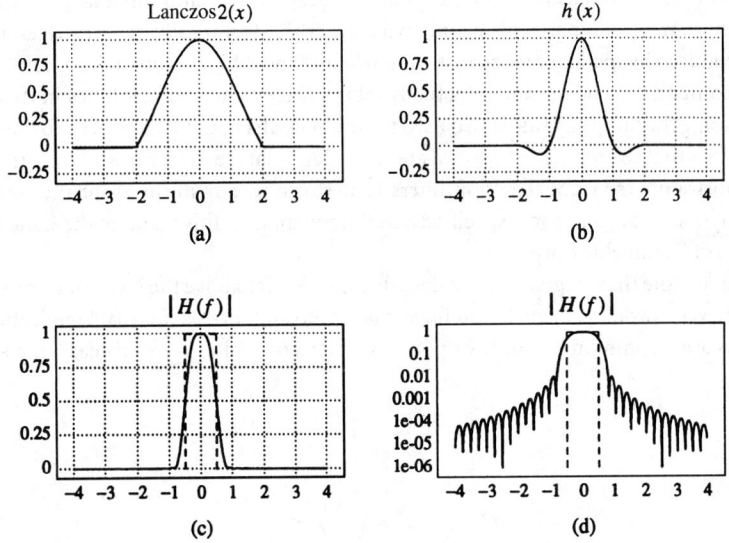

FIGURE 58.15   (a) Lanczos2 window; (b) windowed sinc; (c) spectrum; (d) log plot.

Lanczos Window

The Lanczos window is based on the sinc function rather than cosines as used in the previous methods. The two-lobed Lanczos window function is defined as

$$\text{Lanczos2}(x) = \begin{cases} \frac{\sin(\pi x/2)}{\pi x/2}, & 0 \le |x| < 2 \\ 0, & 2 \le |x| \end{cases} \tag{58.19}$$

The Lanczos2 window function, shown in Fig. 58.15, is the central lobe of a sinc function. It is wide enough to extend over two lobes of the ideal low-pass filter, i.e., a second sinc function. This formulation can be generalized to an N-lobed window function by replacing the value 2 in Eq. (58.19) with the value N. Larger N results in superior frequency response.

58.5 Aliasing

If the two reconstruction conditions outlined earlier are not met, sampling theory predicts that exact reconstruction is *not* possible. This phenomenon, known as **aliasing**, occurs when signals are not bandlimited or when they are undersampled, i.e., $f_s \leq f_{\text{Nyquist}}$. In either case there will be unavoidable overlapping of spectral components, as in Fig. 58.16. Notice that the irreproducible high frequencies fold over into the low frequency range. As a result, frequencies originally beyond f_{max} will, upon reconstruction, appear in the form of much *lower* frequencies. In comparison with the spurious high frequencies retained by nonideal reconstruction filters, the spectral components passed due to undersampling are more serious, since they actually corrupt the components in the original signal.

Aliasing refers to the higher frequencies becoming aliased and indistinguishable from the lower-frequency components in the signal if the sampling rate falls below the Nyquist frequency. In other words, undersampling causes high-frequency components to appear as spurious low frequencies. This is depicted in Fig. 58.17, where a high-frequency signal appears as a low-frequency signal after sampling it too sparsely. In digital images, the Nyquist rate is determined by the highest frequency that can be displayed: one cycle every two pixels. Therefore, any attempt to display higher frequencies will produce similar artifacts.

There is sometimes a misconception in the computer graphics literature that jagged (staircased) edges are always a symptom of aliasing. This is only partially true. Technically, jagged edges arise from high frequencies introduced by inadequate reconstruction. Since these high frequencies are not corrupting the low-frequency components, no aliasing is actually taking place. The confusion lies in that the suggested remedy of increasing the sampling rate is also used to eliminate aliasing. Of course, the benefit of increasing the sampling rate is that the replicated spectra are now spaced farther apart from each other. This relaxes the accuracy constraints for reconstruction filters to perform ideally in the stopband, where they must suppress all components beyond some specified cutoff frequency. In this manner, the same nonideal filters will produce less objectionable output.

It is important to note that a signal may be densely sampled (far above the Nyquist rate) and continue to appear jagged if a zero-order reconstruction filter is used. Box filters used for pixel replication in real-time hardware zooms are a common example of poor reconstruction filters. In this case, the signal is clearly

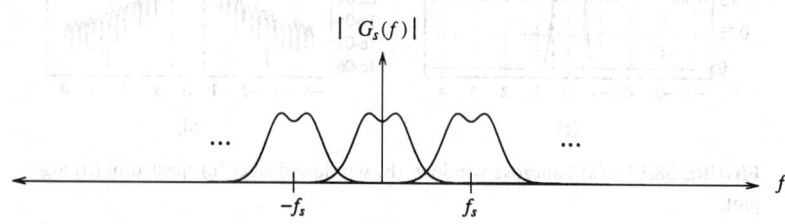

FIGURE 58.16   Overlapping spectral components give rise to aliasing.

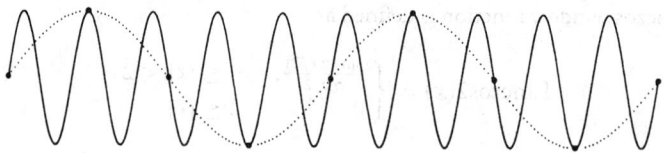

FIGURE 58.17   Aliasing artifacts due to undersampling.

not aliased but rather poorly reconstructed. The distinction between reconstruction and aliasing artifacts becomes clear when we notice that the appearance of jagged edges is improved by blurring. For example, it is not uncommon to step back from an image exhibiting excessive blockiness in order to see it more clearly. This is a defocusing operation that attenuates the high frequencies admitted through nonideal reconstruction. On the other hand, once a signal is truly undersampled, there is no postprocessing possible to improve its condition. After all, applying an ideal low-pass (reconstruction) filter to a spectrum whose components are already overlapping will only blur the result, not rectify it.

58.6 Antialiasing

The filtering necessary to combat aliasing is known as **antialiasing**. In order to determine corrective action, we must directly address the two conditions necessary for exact signal reconstruction. The first solution calls for low-pass filtering *before* sampling. This method, known as **prefiltering**, bandlimits the signal to levels below f_{max}, thereby eliminating the offending high frequencies. Notice that the frequency at which the signal is to be sampled imposes limits on the allowable bandwidth. This is often necessary when the output sampling grid must be fixed to the resolution of an output device, e.g., screen resolution. Therefore, aliasing is often a problem that is confronted when a signal is forced to conform to an inadequate resolution due to physical constraints. As a result, it is necessary to bandlimit, or narrow, the input spectrum to conform to the allotted bandwidth as determined by the sampling frequency.

The second solution is to point-sample at a higher frequency. In doing so, the replicated spectra are spaced farther apart, thereby separating the overlapping spectra tails. This approach theoretically implies sampling at a resolution determined by the highest frequencies present in the signal. Since a surface viewed obliquely can give rise to arbitrarily high frequencies, this method may require extremely high resolution. Whereas the first solution adjusts the bandwidth to accommodate the fixed sampling rate f_s, the second solution adjusts f_s to accommodate the original bandwidth. Antialiasing by sampling at the highest frequency is clearly superior in terms of image quality. This is, of course, operating under different assumptions regarding the possibility of varying f_s. In practice, antialiasing is performed through a combination of these two approaches. That is, the sampling frequency is increased so as to reduce the amount of bandlimiting to a minimum.

Point Sampling

The naive approach for generating an output image is to perform **point sampling**, where each output pixel is a single sample of the input image taken independently of its neighbors (Fig. 58.18). It is clear that information is lost between the samples and that aliasing artifacts may surface if the sampling density is not sufficiently high to characterize the input. This problem is rooted in the fact that intermediate intervals between samples, which should have some influence on the output, are skipped entirely.

The Star image is a convenient example that overwhelms most resampling filters due to the infinitely high frequencies found toward the center. Nevertheless, the extent of the artifacts is related to the quality of the filter and the actual spatial transformation. Figure 58.19 shows two examples of the moiré effects that can appear when a signal is undersampled using point sampling. In Fig. 58.19(a), one out of every two pixels in the Star image was discarded to reduce its dimension. In Fig. 58.19(b), the artifacts of undersampling are more pronounced, as only one out of every four pixels is retained. In order to see the small images more clearly, they are magnified using cubic spline reconstruction. Clearly, these examples show that point sampling behaves poorly in high-frequency regions.

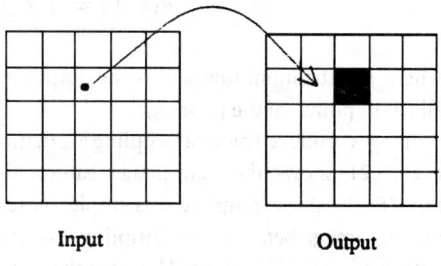

FIGURE 58.18 Point sampling.

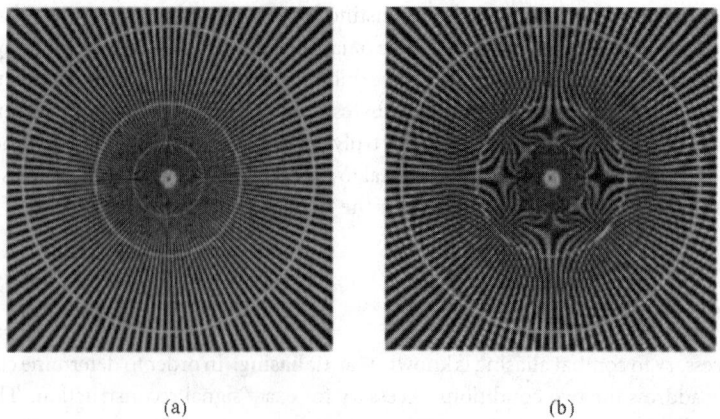

FIGURE 58.19   Aliasing due to point sampling: (a) 1/2 and (b) 1/4 scale.

Aliasing can be reduced by point sampling at a higher resolution. This raises the Nyquist limit, accounting for signals with higher bandwidths. Generally, though, the display resolution places a limit on the highest frequency that can be displayed, and thus limits the Nyquist rate to one cycle every two pixels. Any attempt to display higher frequencies will produce aliasing artifacts such as moiré patterns and jagged edges. Consequently, antialiasing algorithms have been derived to bandlimit the input *before* resampling onto the output grid.

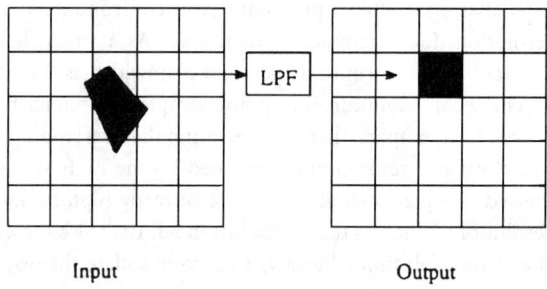

FIGURE 58.20   Area sampling.

Area Sampling

The basic flaw in point sampling is that a discrete pixel actually represents an area, not a point. In this manner, each output pixel should be considered a window looking onto the input image. Rather than sampling a point, we must instead apply a low-pass filter (LPF) upon the projected area in order to properly reflect the information content being mapped onto the output pixel. This approach, depicted in Fig. 58.20, is called **area sampling**, and the projected area is known as the **preimage**. The low-pass filter comprises the **prefiltering** stage. It serves to defeat aliasing by bandlimiting the input image prior to resampling it onto the output grid. In the general case, prefiltering is defined by the convolution integral

$$g(x, y) = \iint f(u, v)h(x - u, y - v)\, du\, dv \tag{58.20}$$

where f is the input image, g is the output image, h is the filter kernel, and the integration is applied to all $[u, v]$ points in the preimage.

Images produced by area sampling are demonstrably superior to those produced by point sampling. Figure 58.21 shows the Star image subjected to the same downsampling transformation as that in Fig. 58.19. Area sampling was implemented by applying a box filter (i.e., unweighted averaging) to the Star image before point sampling. Notice that antialiasing through area sampling has traded moiré patterns for some blurring. Although there is no substitute for high-resolution imagery, filtering can make lower resolution less objectionable by attenuating aliasing artifacts.

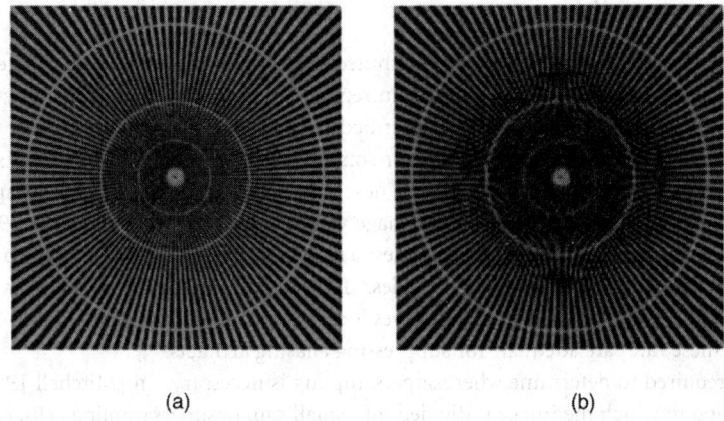

FIGURE 58.21 Aliasing due to area sampling: (a) 1/2 and (b) 1/4 scale.

Supersampling

The process of using more than one regularly spaced sample per pixel is known as **supersampling**. Each output pixel value is evaluated by computing a weighted average of the samples taken from their respective preimages. For example, if the supersampling grid is three times denser than the output grid (i.e., there are nine grid points per pixel area), each output pixel will be an average of the nine samples taken from its projection in the input image. If, say, three samples hit a green object and the remaining six samples hit a blue object, the composite color in the output pixel will be one-third green and two-thirds blue, assuming a box filter is used.

Supersampling reduces aliasing by bandlimiting the input signal. The purpose of the high-resolution supersampling grid is to refine the estimate of the preimages seen by the output pixels. The samples then enter the prefiltering stage, consisting of a low-pass filter. This permits the input to be resampled onto the (relatively) low-resolution output grid without any offending high frequencies introducing aliasing artifacts. In Fig. 58.22 we see an output pixel subdivided into nine subpixel samples which each undergo inverse mapping, sampling the input at nine positions. Those nine values then pass through a low-pass filter to be averaged into a single output value.

Supersampling was used to achieve antialiasing in Fig. 58.1 for pixels near the horizon. There are two problems, however, associated with straightforward supersampling. The first problem is that the newly designated high frequency of the prefiltered image continues to be fixed. Therefore, there will always be sufficiently higher frequencies that will alias. The second problem is cost. In our example, supersampling will take nine times longer than point sampling. Although there is a clear need for the additional computation, the dense placement of samples can be optimized. Adaptive supersampling is introduced to address these drawbacks.

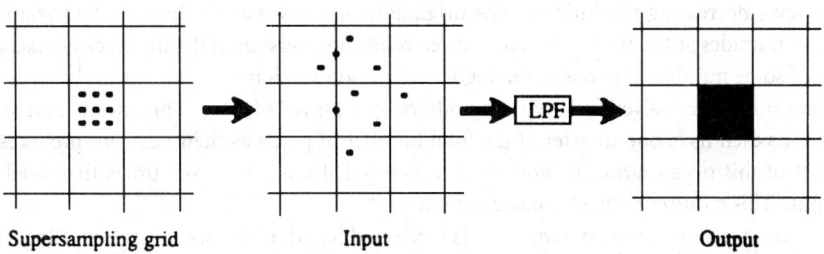

Supersampling grid   Input   Output

FIGURE 58.22 Supersampling.

Adaptive Supersampling

In **adaptive supersampling**, the samples are distributed more densely in areas of high intensity variance. In this manner, supersamples are collected only in regions that warrant their use. Early work in adaptive supersampling for computer graphics is described in [Whitted 1980]. The strategy is to subdivide areas between previous samples when an edge, or some other high-frequency pattern, is present. Two approaches to adaptive supersampling have been described in the literature. The first approach allows sampling density to vary as a function of local image variance [Lee et al. 1985, Kajiya 1986]. A second approach introduces two levels of sampling densities: a regular pattern for most areas and a higher-density pattern for regions demonstrating high frequencies. The regular pattern simply consists of one sample per output pixel. The high density pattern involves local supersampling at a rate of 4 to 16 samples per pixel. Typically, these rates are adequate for suppressing aliasing artifacts.

A strategy is required to determine where supersampling is necessary. In [Mitchell 1987], the author describes a method in which the image is divided into small square supersampling cells, each containing eight or nine of the low-density samples. The entire cell is supersampled if its samples exhibit excessive variation. In [Lee et al. 1985], the variance of the samples is used to indicate high frequency. It is well known, however, that variance is a poor measure of visual perception of local variation. Another alternative is to use contrast, which more closely models the nonlinear response of the human eye to rapid fluctuations in light intensities [Caelli 1981]. Contrast is given as

$$C = \frac{I_{\max} - I_{\min}}{I_{\max} + I_{\min}} \tag{58.21}$$

Adaptive sampling reduces the number of samples required for a given image quality. The problem with this technique, however, is that the variance measurement is itself based on point samples, and so this method can fail as well. This is particularly true for subpixel objects that do not cross pixel boundaries. Nevertheless, adaptive sampling presents a far more reliable and cost-effective alternative to supersampling.

58.7 Prefiltering

Area sampling can be accelerated if constraints on the filter shape are imposed. Pyramids and preintegrated tables are introduced to approximate the convolution integral with a constant number of accesses. This compares favorably against direct convolution, which requires a large number of samples that grow proportionately to preimage area. As we shall see, though, the filter area will be limited to squares or rectangles, and the kernel will consist of a box filter. Subsequent advances have extended their use to more general cases with only marginal increases in cost.

Pyramids

Pyramids are multiresolution data structures commonly used in image processing and computer vision. They are generated by successively bandlimiting and subsampling the original image to form a hierarchy of images at ever decreasing resolutions. The original image serves as the base of the pyramid, and its coarsest version resides at the apex. Thus, in a lower-resolution version of the input, each pixel represents the average of some number of pixels in the higher-resolution version.

The resolution of successive levels typically differs by a power of two. This means that successively coarser versions each have one-quarter of the total number of pixels as their adjacent predecessors. The memory cost of this organization is modest: $1 + 1/4 + 1/16 + \cdots = 4/3$ times that needed for the original input. This requires only 33% more memory.

To filter a preimage, one of the pyramid levels is selected based on the size of its bounding square box. That level is then point sampled and assigned to the respective output pixel. The primary benefit of this approach is that the cost of the filter is constant, requiring the same number of pixel accesses independent of the filter size. This performance gain is the result of the filtering that took place while creating the pyramid. Furthermore, if preimage areas are adequately approximated by squares, the direct convolution methods amount to point-sampling a pyramid. This approach was first applied to texture mapping in [Catmull 1974] and described in [Dungan et al. 1978].

There are several problems with the use of pyramids. First, the appropriate pyramid level must be selected. A coarse level may yield excessive blur, while the adjacent finer level may be responsible for aliasing due to insufficient bandlimiting. Second, preimages are constrained to be squares. This proves to be a crude approximation for elongated preimages. For example, when a surface is viewed obliquely the texture may be compressed along one dimension. Using the largest bounding square will include the contributions of many extraneous samples and result in excessive blur. These two issues were addressed in [Williams 1983] and [Crow 1984], respectively, along with extensions proposed by other researchers.

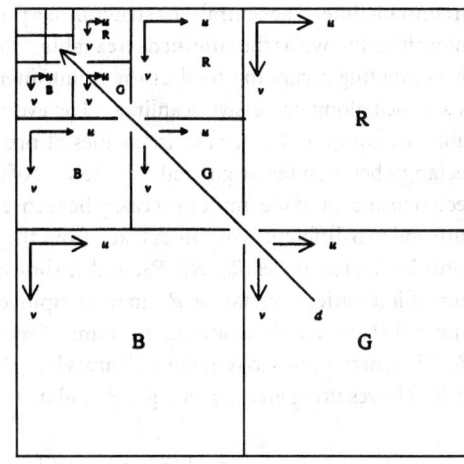

FIGURE 58.23   Mip-map memory organization.

Williams proposed a pyramid organization called *mip map* to store color images at multiple resolutions in a convenient memory organization [Williams 1983]. The acronym "mip" stands for "multum in parvo," a Latin phrase meaning "much in little." The scheme supports trilinear interpolation, where both intra- and interlevel interpolation can be computed using three normalized coordinates: u, v, and d. Both u and v are spatial coordinates used to access points within a pyramid level. The d coordinate is used to index, and interpolate between, different levels of the pyramid. This is depicted in Fig. 58.23.

The quadrants touching the east and south borders contain the original red, green, and blue (RGB) components of the color image. The remaining upper left quadrant contains all the lower-resolution copies of the original. The memory organization depicted in Fig. 58.23 clearly supports the earlier claim that the memory cost is 4/3 times that required for the original input. Each level is shown indexed by the $[u, v, d]$ coordinate system, where d is shown slicing through the pyramid levels. Since corresponding points in different pyramid levels have indices which are related by some power of two, simple binary shifts can be used to access these points across the multiresolution copies. This is a particularly attractive feature for hardware implementation.

The primary difference between mip maps and ordinary pyramids is the trilinear interpolation scheme possible with the $[u, v, d]$ coordinate system. Since they allow a continuum of points to be accessed, mip maps are referred to as pyramidal parametric data structures. In Williams's implementation, a box filter was used to create the mip maps, and a triangle filter was used to perform intra- and interlevel interpolation. The value of d must be chosen to balance the tradeoff between aliasing and blurring. Heckbert suggests

$$d^2 = \max\left(\left(\frac{\partial u}{\partial x}\right)^2 + \left(\frac{\partial v}{\partial x}\right)^2, \left(\frac{\partial u}{\partial y}\right)^2 + \left(\frac{\partial v}{\partial y}\right)^2 \right) \tag{58.22}$$

where d is proportional to the span of the preimage area, and the partial derivatives can be computed from the surface projection [Heckbert 1983].

Summed-Area Tables

An alternative to pyramidal filtering was proposed by Crow in [Crow 1984]. It extends the filtering possible in pyramids by allowing rectangular areas, oriented parallel to the coordinate axes, to be filtered in constant time. The central data structure is a preintegrated buffer of intensities, known as the **summed-area table**. This table is generated by computing a running total of the input intensities as the image is scanned along successive scanlines. For every position P in the table, we compute the sum of intensities of pixels contained in the rectangle between the origin and P. The sum of all intensities in any rectangular area of the input may easily be recovered by computing a sum and two differences of values taken from the table. For example, consider the rectangles R_0, R_1, R_2, and R shown in Fig. 58.24. The sum of intensities in rectangle R can be computed by considering the sum at $[x1, y1]$, and discarding the sums of rectangles R_0, R_1, and R_2. This corresponds to removing all areas lying below and to the left of R. The resulting area is rectangle R, and its sum S is given as

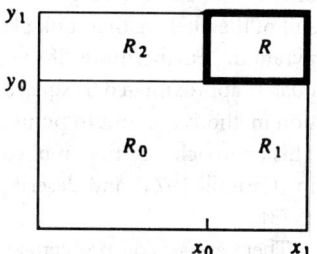

FIGURE 58.24 Summed-area table calculation.

$$S = T[x1, y1] - T[x1, y0] - T[x0, y1] + T[x0, y0] \qquad (58.23)$$

where $T[x, y]$ is the value in the summed-area table indexed by coordinate pair $[x, y]$.

Since $T[x1, y0]$ and $T[x0, y1]$ both contain R_0, the sum of R_0 was subtracted twice in Eq. (58.23). As a result, $T[x0, y0]$ was added back to restore the sum. Once S is determined, it is divided by the area of the rectangle. This gives the average intensity over the rectangle, a process equivalent to filtering with a Fourier window (box filtering).

There are two problems with the use of summed-area tables. First, the filter area is restricted to rectangles. This is addressed in [Glassner 1986], where an adaptive, iterative technique is proposed for obtaining arbitrary filter areas by removing extraneous regions from the rectangular bounding box. Second, the summed-area table is restricted to box filtering. This, of course, is attributed to the use of unweighted averages that keeps the algorithm simple. In [Perlin 1985] and [Heckbert 1986], the summed-area table is generalized to support more sophisticated filtering by repeated integration.

It is shown that by repeatedly integrating the summed-area table n times, it is possible to convolve an orthogonally oriented rectangular region with an nth-order box filter (B-spline). The output value is computed by using $(n + 1)^2$ weighted samples from the preintegrated table. Since this result is independent of the size of the rectangular region, this method offers a great reduction in computation over that of direct convolution. Perlin called this a selective image filter because it allows each sample to be blurred by different amounts.

Repeated integration has rather high memory costs relative to pyramids. This is due to the number of bits necessary to retain accuracy in the large summations. Nevertheless, it allows us to filter rectangular or elliptical regions, rather than just squares as in pyramid techniques. Since pyramid and summed-area tables both require a setup time, they are best suited for input that is intended to be used repeatedly, i.e., stationary background scenes or texture maps. In this manner, the initialization overhead can be amortized over each use.

58.8 Example: Image Scaling

In this section, we demonstrate the role of reconstruction and antialiasing in image scaling. The resampling process will be explained in one dimension rather than two, since resampling is carried out on each axis independently. For example, the horizontal scanlines are first processed, yielding an intermediate image, which then undergoes a second pass of interpolation in the vertical direction. The result is independent of the order: processing the vertical lines before the horizontal lines gives the same results.

A skeleton of a C program that resizes an image in two passes is given below. The input image is assumed to have **INheight** rows and **INwidth** columns. The first pass visits each row and resamples them to have width **OUTwidth.** The second pass visits each column of the newly formed intermediate image and resamples them to have height **OUTheight:**

```
INwidth    =  input image width (pixels/row);
INheight   =  input image height (rows/image);
OUTwidth   =  output image width (pixels/row);
OUTheight  =  output image height (rows/image);
filter     =  convolution kernel to use to filter image;
offset     =  inter-pixel offset (stride);

allocate an intermediate image of size OUTwidth by INheight;

offset = 1;
for(y=0; y<INheight; y++) {        /* process rows */
     src = pointer to row y of input image;
     dst = pointer to row y of intermediate image;
     resample1D(src, dst, INwidth, OUTwidth, filter, offset);
}

offset = OUTwidth;
for(x=0; x<w; x++) {              /* process columns */
     src = pointer to column x of intermediate image;
     dst = pointer to column x of output image;
     resample1D(src, dst, INheight, OUTheight, filter, offset);
}
```

Function **resample1D** is the workhorse of the resizing operation. The inner workings of this function will be described later. In addition to the input and output pointers and dimensions, **resample1D** must be passed **filter,** an integer code specifying which convolution kernel to apply. In order to operate on both rows and columns, the parameter **offset** is given to denote the distance between successive pixels in the scanline. Horizontal scanlines (rows) have **offset = 1** and vertical scanlines (columns) have **offset = OUTwidth.**

There are two operations which **resample1D** must be able to handle: magnification and minification. As mentioned earlier, these two operations are closely related. They both require us to project each output sample into the input, center a kernel, and convolve. The only difference between magnification and minification is the shape of the kernel. The magnification kernel is fixed at $h(x)$, whereas the minification kernel is $ah(ax)$, for $a < 1$. The width of the kernel for minification is due to the need for a low-pass filter to perform antialiasing. That filter now has a narrower response than that of the interpolation function. Consequently, we exploit the following well-known Fourier transform pair:

$$h(ax) \longleftrightarrow \frac{1}{a}H\left(\frac{f}{a}\right) \tag{58.24}$$

This equation expresses the reciprocal relationship between the spatial and frequency domains. Notice that multiplying the spatial axis by a factor of a results in dividing the frequency axis and the spectrum values by that same factor. Since we want the spectrum values to be left intact, we use $ah(ax)$ as the convolution kernel for blurring, where $a > 1$. This implies that the shape of the kernel changes as a function of scale factor when we are downsampling the input. This was not the case for magnification.

A straightforward method to perform 1D resampling is given below. It details the inner workings of the **resample1D** function outlined earlier. In addition, a few interpolation functions are provided.

More such functions can easily be added by the user:

```
#define PI              3.1415926535897931160E0
#define SGN(A)          ((A) > 0 ? 1 : ((A) < 0 ? -1 : 0 ))
#define FLOOR(A)        ((int) (A))
#define CEILING(A)      ((A)==FLOOR(A) ? FLOOR(A) : SGN(A)+FLOOR(A))
#define CLAMP(A,L,H)    ((A)<=(L) ? (L) : (A)<=(H) ? (A) : (H))

resample1D(IN, OUT, INlen, OUTlen, filtertype, offset)
unsigned char *IN, *OUT;
int INlen, OUTlen, filtertype, offset;
{
    int i;
    int left, right;   /* kernel extent in input */
    int pixel;         /* input pixel value */
    double u, x;       /* input (u) , output (x) */
    double scale;      /* resampling scale factor */
    double fwidth;     /* filter width (support) */
    double fscale;     /* filter amplitude scale */
    double weight;     /* kernel weight */
    double acc;        /* convolution accumulator */

    scale = (double) OUTlen / INlen;

    switch(filtertype) {
    case 0: filter = boxFilter; /* box filter (nearest
                                        neighbor) */
            fwidth = .5;
            break;
    case 1: filter = triFilter; /* triangle filter (linear
                                        intrp) */
            fwidth = 1;
            break;
    case 2: filter = cubicConv; /* cubic convolution
                                        filter */
            fwidth = 2;
            break;
    case 3: filter = lanczos3;  /* Lanczos3 windowed sinc
                                        function */
            fwidth = 3;
            break;
    case 4: filter = hann4;      /* Hann windowed sinc
                                        function */
            fwidth = 4;          /* 8-point kernel */
            break;
```

```
        if(scale < 1.0) {       /* minification: h(x) -> h(x*scale)*
                                    scale */
            fwidth = fwidth / scale;   /* broaden filter */
            fscale = scale;            /* lower amplitude */
        } else      fscale = 1.0;

    /* project each output pixel to input, center kernel, and
       convolve */
        for(x=0; x<OUTlen; x++) {
            /* map output x to input u: inverse mapping */
            u = x / scale;

            /* left and right extent of kernel centered at u */
            if(u - fwidth < 0) {
                    left = FLOOR (u - fwidth);
            else left = CEILING(u - fwidth);
            right = FLOOR(u + fwidth);

            /* reset acc for collecting convolution products */
            acc = 0;

            /* weigh input pixels around u with kernel */
            for(i=left; i <= right; i++) {
                    pixel = IN[ CLAMP(i, 0, INlen-1)*offset];
                    weight = (*filter)((u - i) * fscale);
                    acc += (pixel * weight);
            }

            /* assign weighted accumulator to OUT */
            OUT[x*offset] = acc * fscale;
        }
}

/* ~~~~~~~~~~~~~~~~~~~~~~~~~~~~~~~~~~~~~~~~~~~~~~~~~~~~~~~~~~~~~~~~
 * boxFilter:
 *
 * Box (nearest neighbor) filter.
 */
double
boxFilter(t)
double t;
{
    if((t > -.5) && (t <= .5)) return(1.0);
    return(0.0);
}

/* ~~~~~~~~~~~~~~~~~~~~~~~~~~~~~~~~~~~~~~~~~~~~~~~~~~~~~~~~~~~~~~~~
```

```
 * triFilter:
 *
 * Triangle filter (used for linear interpolation).
 */
double
triFilter(t)
double t;
{
      if(t < 0) t = -t;
      if(t < 1.0) return(1.0 - t);
      return(0.0);
}

/* ~~~~~~~~~~~~~~~~~~~~~~~~~~~~~~~~~~~~~~~~~~~~~~~~~~~~~~~~~~~~~~
 * cubicConv:
 *
 * Cubic convolution filter.
 */
double
cubicConv(t)
double t;
{
      double A, t2, t3;

      if(t < 0) t = -t;
      t2 = t * t;
      t3 = t2 * t;

      A = -1.0;          /* user-specified free parameter */
      if(t < 1.0) return((A+2)*t3 - (A+3)*t2 + 1);
      if(t < 2.0) return(A*(t3 - 5*t2 + 8*t - 4));
      return(0.0);
}

/* ~~~~~~~~~~~~~~~~~~~~~~~~~~~~~~~~~~~~~~~~~~~~~~~~~~~~~~~~~~~~~~
 * sinc:
 *
 * Sinc function.
 */
double
sinc(t)
double t;
{
      t *= PI;
      if(t != 0) return(sin(t) / t);
      return(1.0);
```

```
}

/*  ~~~~~~~~~~~~~~~~~~~~~~~~~~~~~~~~~~~~~~~~~~~~~~~~~~~~~~~~~~~~~~
 *  lanczos3:
 *
 *  Lanczos3 filter.
 */
double
lanczos3(t)
double t;
{
        if(t < 0) t = -t;
        if(t < 3.0) return(sinc(t) * sinc(t/3.0));
        return(0.0);
}

/*  ~~~~~~~~~~~~~~~~~~~~~~~~~~~~~~~~~~~~~~~~~~~~~~~~~~~~~~~~~~~~~~
 *  hann:
 *
 *  Hann windowed sinc function. Assume N (width) = 4.
 */
double
hann4(t)
double t;
{
        int N = 4; /* fixed filter width */

        if(t < 0) t = -t;
        if(t < N) return(sinc(t) * (.5 + .5*cos(PI*t/N)));
        return(0.0);
}
```

There are several points worth mentioning about this code. First, the filter width **fwidth** of each of the supported kernels is initialized for use in interpolation (for magnification). We then check to see if the scale factor **scale** is less than one to rescale **fwidth** accordingly. Furthermore, we set **fscale,** the filter amplitude scale factor, to 1 for interpolation, or **scale** for minification. We then visit each of **OUTlen** output pixels, and project them back into the input, where we center the filter kernel. The kernel overlaps a range of input pixels from **left** to **right.** All pixels in this range are multiplied by a corresponding kernel value. The products are added in an accumulator **acc** and assigned to the output buffer.

Note that the **CLAMP** macro is necessary to prevent us from attempting to access a pixel beyond the extent of the input buffer. By clamping to either end, we are effectively replicating the border pixel for use with a filter kernel that extends beyond the image.

In order to accommodate the processing of rows and columns, the variable **offset** is introduced to specify the interpixel distance. When processing rows, **offset = 1.** When processing columns, **offset** is set to the width of a row.

This code can accommodate a polynomial transformation by making a simple change to the evaluation of **u.** Rather than computing **u = x/scale**, we may let u be expressed by a polynomial. The method of forward differences is recommended to simplify the computation of polynomials [Wolberg 1990].

The code given above suffers from three limitations, all dealing with efficiency:

1. A division operation is used to compute the inverse projection. Since we are dealing with a linear mapping function, the new position at which to center the kernel may be computed incrementally. That is, there is a constant offset between each projected output sample. Accordingly, **left** and **right** should be computed incrementally as well.

2. The set of kernel weights used in processing the first scanline applies equally to all the remaining scanlines as well. There should be no need to recompute them each time. This matter is addressed in the code supplied by [Schumacher 1992].

3. The kernel weights are evaluated by calling the appropriate filter function with the normalized distance from the center. This involves a lot of run-time overhead, particularly for the more sophisticated kernels that require the evaluation of a sinc function, division, and several multiplies.

Additional sophisticated algorithms to deal with these issues are given in [Wolberg 1990].

58.9 Research Issues and Summary

The computer graphics literature is replete with new and innovative work addressing the demands of sampling, reconstruction, and antialiasing. Nonuniform sampling has become important in computer graphics because it facilitates variable sampling density and it allows us to trade structured aliasing for noise. Recent work in adaptive sampling and nonuniform reconstruction is discussed in [Glassner 1995]. Excellent surveys in nonuniform reconstruction, which is also known as scattered-data interpolation, can be found in [Franke and Nielson 1991] and [Hoschek and Lasser 1993]. These problems are also of direct consequence to image compression. The ability to determine a unique minimal set of samples to completely represent a signal within some specified error tolerance remains an active area of research. The solution must be closely coupled with a nonuniform reconstruction method. Although traditional reconstruction methods are well understood within the framework described in this chapter, the analysis of nonuniform sampling and reconstruction remains challenging.

We now summarize the basic principles of sampling theory, reconstruction, and antialiasing that have been presented in this chapter. We have shown that a continuous signal may be reconstructed from its samples if the signal is bandlimited and the sampling frequency exceeds the Nyquist rate. These are the two necessary conditions for image reconstruction to be possible. Since sampling can be shown to replicate a signal's spectrum across the frequency domain, ideal low-pass filtering was introduced as a means of retaining the original spectrum while discarding its copies. Unfortunately, the ideal low-pass filter in the spatial domain is an infinitely wide sinc function. Since this is difficult to work with, nonideal reconstruction filters are introduced to approximate the reconstructed output. These filters are nonideal in the sense that they do not completely attenuate the spectra copies. Furthermore, they contribute to some blurring of the original spectrum. In general, poor reconstruction leads to artifacts such as jagged edges.

Aliasing refers to the phenomenon that occurs when a signal is undersampled. This happens if the reconstruction conditions mentioned above are violated. In order to resolve this problem, one of two actions may be taken. Either the signal can be bandlimited to a range that complies with the sampling frequency, or the sampling frequency can be increased. In practice, some combination of both options is taken, leaving some relatively unobjectionable aliasing in the output.

Examples of the concepts discussed thus are concisely depicted in Figs. 58.25 through 58.27. They attempt to illustrate the effects of sampling and low-pass filtering on the quality of the reconstructed signal and its spectrum. The first row of Fig. 58.25 shows a signal and its spectrum, bandlimited to 0.5 cycle/pixel. For pedagogical purposes, we treat this signal as if it were continuous. In actuality, though, it is a 256-sample horizontal cross section taken from a digital image. Since each pixel has four samples contributing to it, there is a maximum of two cycles per pixel. The horizontal axes of the spectrum account for this fact.

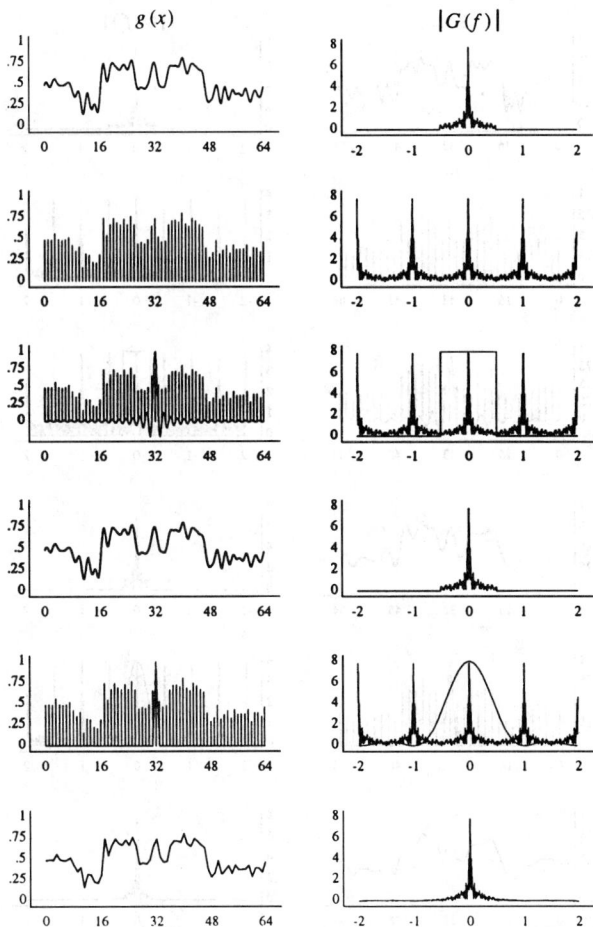

FIGURE 58.25 Sampling and reconstruction (with an adequate sampling rate).

The second row shows the effect of sampling the signal. Since $f_s = 1$ sample/pixel, there are four copies of the baseband spectrum in the range shown. Each copy is scaled by $f_s = 1$, leaving the magnitudes intact. In the third row, the 64 samples are shown convolved with a sinc function in the spatial domain. This corresponds to a rectangular pulse in the frequency domain. Since the sinc function is used here for image reconstruction, it must have an amplitude of unity value in order to interpolate the data. This forces the height of the rectangular pulse in the frequency domain to vary in response to f_s.

A few comments on the reciprocal relationship between the spatial and frequency domains are in order here, particularly as they apply to the ideal low-pass filter. We again refer to the variables A and W as the sinc amplitude and bandwidth. As a sinc function is made broader, the value $1/2W$ is made to change, since W is decreasing to accommodate zero crossings at larger intervals. Accordingly, broader sinc functions cause more blurring, and their spectra reflect this by reducing the cutoff frequency to some smaller W. Conversely, narrower sinc functions cause less blurring, and W takes on some larger value. In either case, the amplitude of the sinc function or its spectrum will change. That is, we can fix the amplitude of the sinc function so that only the rectangular pulse of the spectrum changes height $A/2W$ as W varies. Alternatively, we can fix $A/2W$ to remain constant as W changes, forcing us to vary A. The choice depends on the application.

When the sinc function is used to interpolate data, it is necessary to fix A to 1. Therefore, as the sampling density changes, the positions of the zero crossings shift, causing W to vary. This makes the amplitude of

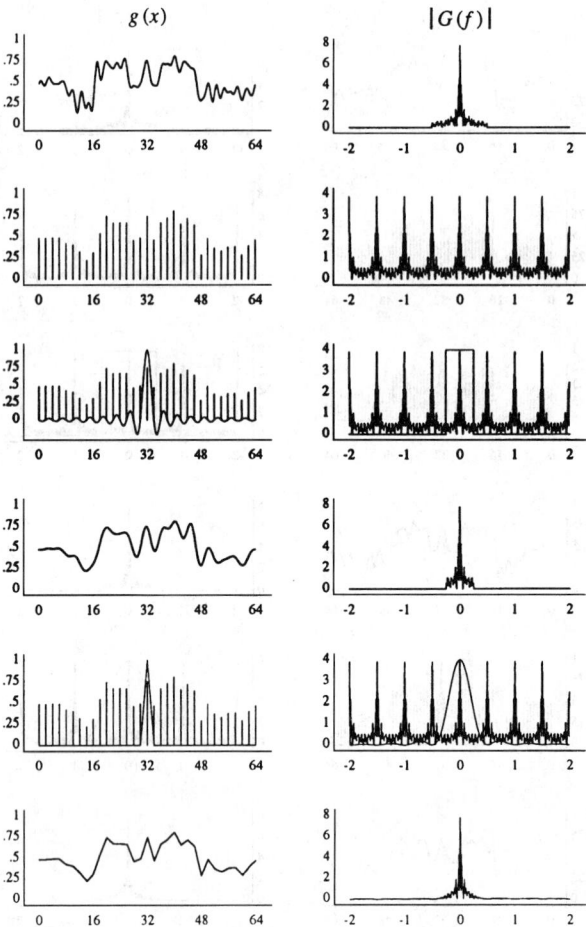

FIGURE 58.26 Sampling and reconstruction (with an inadequate sampling rate).

the spectrum's rectangular pulse change. On the other hand, if the sinc function is applied to bandlimit, not interpolate, the input signal, then it is important to fix $A/2W$ to 1 so that the passband frequencies remain intact. Since W is once again varying, A must change proportionately to keep $A/2W$ constant. Therefore, this application of the ideal low-pass filter requires the amplitude of the sinc function to be responsive to W.

In the examples presented below, our objective is to interpolate (reconstruct) the input, and so $A = 1$ regardless of the sampling density. Consequently, the height of the spectrum of the reconstruction filter changes. To make the Fourier transforms of the filters easier to see, we have not drawn the frequency response of the reconstruction filters to scale. Therefore, the rectangular pulse function in the third row of Fig. 58.25 actually has height $A/2W = 1$. The fourth row of the figure shows the result after applying the ideal low-pass filter. As sampling theory predicts, the output is identical to the original signal. The last two rows of the figure illustrate the consequences of nonideal reconstruction filtering. Instead of using a sinc function, a triangle function corresponding to linear interpolation was applied. In the frequency domain this corresponds to the square of the sinc function. Not surprisingly, the spectrum of the reconstructed signal suffers in both the passband and the stopband.

The identical sequence of filtering operations is performed in Fig. 58.26. In this figure, though, the sampling rate has been lowered to $f_s = 0.5$, meaning that only one sample is collected for every two output pixels. Consequently, the replicated spectra are multiplied by 0.5, leaving the magnitudes at 4.

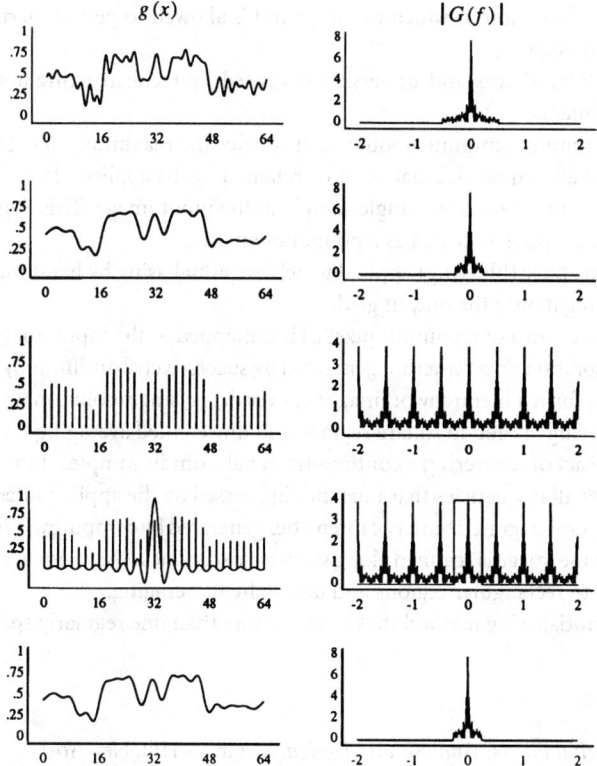

FIGURE 58.27 Antialiasing filtering, sampling, and reconstruction stages.

Unfortunately, this sampling rate causes the replicated spectra to overlap. This, in turn, gives rise to aliasing, as depicted in the fourth row of the figure. Applying the triangle function to perform linear interpolation also yields poor results.

In order to combat these artifacts, the input signal must be bandlimited to accommodate the low sampling rate. This is shown in the second row of Fig. 58.27, where we see that all frequencies beyond $W = 0.25$ are truncated. This causes the input signal to be blurred. In this manner we have traded aliasing for blurring, a far less objectionable artifact. Sampling this function no longer causes the replicated copies to overlap. Convolving with an ideal low-pass filter now properly isolates the bandlimited spectrum.

Defining Terms

Adaptive supersampling: Supersampling with samples distributed more densely in areas of high intensity variance.

Aliasing: Artifacts due to undersampling a signal. This condition prevents the signal from being reconstructed from its samples.

Antialiasing: The filtering necessary to combat aliasing. This generally requires bandlimiting the input before sampling to remove the offending high frequencies that will fold over in the frequency spectrum.

Area sampling: An antialiasing method that treats a pixel as an area, not a point. After projecting the pixel to the input, all samples in the preimage are averaged to compute a representative sample.

Bandlimiting: The act of truncating all frequency components beyond some specified frequency. Useful for antialiasing, where offending high frequencies must be removed to prevent aliasing.

Frequency leakage: A condition in which the stopband is allowed to persist, permitting it to fold over into the passband range.

Gibbs phenomenon: Overshoots and undershoots caused by reconstructing a signal with truncated frequency components.

Nyquist rate: The minimum sampling frequency. It is twice the maximum signal frequency.

Passband: Consists of all frequencies that must be retained by the applied filter.

Point sampling: Each output pixel is a single sample of the input image. This approach generally leads to aliasing because a pixel is treated as a point, not an area.

Prefilter: The low-pass filter (blurring) applied to achieve antialiasing by bandlimiting the input image prior to resampling it onto the output grid.

Preimage: The projected area of an output pixel as it is mapped to the input image.

Pyramid: A multiresolution data structure generated by successively bandlimiting and subsampling the original image to form a hierarchy of images at ever decreasing resolutions. Useful for accelerating antialiasing. Filtering limited to square regions and unweighted averaging.

Reconstruction: The act of recovering a continuous signal from its samples. Interpolation.

Stopband: Consists of all frequencies that must be suppressed by the applied filter.

Summed-area table: Preintegrated buffer of intensities generated by computing a running total of the input intensities as the image is scanned along successive scanlines. Useful for accelerating antialiasing. Filtering limited to rectangular regions and unweighted averaging.

Supersampling: An antialiasing method that collects more than one regularly spaced sample per pixel.

References

Antoniou, A. 1979. *Digital Filters: Analysis and Design*. McGraw–Hill, New York.

Caelli, T. 1981. *Visual Perception: Theory and Practice*. Pergamon Press, Oxford.

Castleman, K. 1996. *Digital Image Processing*. Prentice–Hall, Englewood Cliffs, NJ.

Catmull, E. 1974. *A Subdivision Algorithm for Computer Display of Curved Surfaces*. Ph.D. thesis, Department of Computer Science, University of Utah. Tech. Rep. UTEC-CSc-74-133.

Crow, F. C. 1984. Summed-area tables for texture mapping. *Comput. Graphics (Proc. SIGGRAPH)* 18(3):207–212.

Dungan, W., Jr., Stenger, A., and Sutty, G. 1978. Texture tile considerations for raster graphics. *Comput. Graphics. (Proc. SIGGRAPH)* 12(3):130–134.

Fant, K. M. 1986. A nonaliasing, real-time spatial transform technique. *IEEE Comput. Graphics Appl.* 6(1):71–80.

Franke, R. and Nielson, G. M. 1991. Scattered data interpolation and applications: a tutorial and survey. In *Geometric Modelling: Methods and Their Application*. H. Hagen and D. Roller, eds., pp. 131–160. Springer–Verlag, Berlin.

Glassner, A. 1986. Adaptive precision in texture mapping. *Comput. Graphics (Proc. SIGGRAPH)* 20(4):297–306.

Glassner, A. 1995. *Principles of Digital Image Synthesis*. Morgan Kaufmann, San Francisco.

Gonzalez, R. C. and Woods, R. 1992. *Digital Image Processing*. Addison–Wesley, Reading, MA.

Heckbert, P. 1983. Texture mapping polygons in perspective. Tech. Memo 13, NYIT Computer Graphics Lab.

Heckbert, P. 1986. Filtering by repeated integration. *Comput. Graphics (Proc. SIGGRAPH)* 20(4):315–321.

Hoschek, J. and Lasser, D. 1993. *Computer Aided Geometric Design*. A K Peters, Wellesley, MA.

Jain, A. K. 1989. *Fundamentals of Digital Image Processing*. Prentice–Hall, Englewood Cliffs, NJ.

Kajiya, J. 1986. The Rendering Equation. *Comput. Graphics Proc., SIGGRAPH* 20(4):143–150.

Lee, M., Redner, R. A., and Uselton, S. P. 1985. Statistically optimized sampling for distributed ray tracing. *Comput. Graphics (Proc. SIGGRAPH)* 19(3):61–67.

Mitchell, D. P. 1987. Generating antialiased images at low sampling densities. *Comput. Graphics (Proc. SIGGRAPH)* 21(4):65–72.

Mitchell, D. P. and Netravali, A. N. 1988. Reconstruction filters in computer graphics. *Comput. Graphics* (*Proc. SIGGRAPH*) 22(4):221–228.

Perlin, K. 1985. Course notes. SIGGRAPH'85 State of the Art in Image Synthesis Seminar Notes.

Pratt, W. K. 1991. *Digital Image Processing*, 2nd ed. J. Wiley, New York.

Russ, J. C. 1992. *The Image Processing Handbook*. CRC Press, Boca Raton, FL.

Schumacher, D. 1992. General filtered image rescaling. In *Graphics Gems III*. David Kirk, ed., Academic Press, New York.

Shannon, C.E. 1948. A mathematical theory of communication. *Bell System Tech. J.* 27:379–423 (July 1948), 27:623–656 (Oct. 1948).

Shannon, C. E. 1949. Communication in the presence of noise. *Proc. Inst. Radio Eng.* 37(1):10–21.

Whitted, T. 1980. An improved illumination model for shaded display. *Commun. ACM* 23(6):343–349.

Williams, L. 1983. Pyramidal parametrics. *Comput. Graphics* (*Proc. SIGGRAPH*) 17(3):1–11.

Wolberg, G. 1990. *Digital Image Warping*. IEEE Comput. Soc. Press, Los Alamitos, CA.

Further Information

The material contained in this chapter was drawn from [Wolberg 1990]. Additional image processing texts that offer a comprehensive treatment of sampling, reconstruction, and antialiasing include [Castleman 1996], [Glassner 1995], [Gonzalez and Woods 1992], [Jain 1989], [Pratt 1991], and [Russ 1992].

Advances in the field are reported in several journals, including *IEEE Transactions on Image Processing*, *IEEE Transactions on Signal Processing*, *IEEE Transactions on Acoustics, Speech, and Signal Processing*, and *Graphical Models and Image Processing*. Related work in computer graphics is also reported in *Computer Graphics* (ACM SIGGRAPH Proceedings), *IEEE Computer Graphics and Applications*, and *IEEE Transactions on Visualization and Computer Graphics*.

59

Computer Animation

59.1 Introduction

The main goal of computer animation is to synthesize a desired motion effect, which is a mixing of natural phenomena, perception, and imagination. The animator designs an object's dynamic behavior with his mental representation of causality. He/she imagines how it moves, gets out of shape, or reacts when it is pushed, pressed, pulled, or twisted. So the animation system has to provide the user with motion control tools able to translate his/her wishes from his/her own language. Computer animation methods may also help to understand physical laws by adding motion control to data in order to show their evolution over time. Visualization has become an important way of validating new models created by scientists. When the model evolves over time, computer simulation is generally used to obtain the evolution, and computer animation is a natural way of visualizing the results obtained from the simulation.

To produce a computer animation sequence, the animator has two principal techniques available. The first is to use a model that creates the desired effect. A good example is the growth of a green plant. The second is used when no model is available. In this case, the animator produces "by hand" the real-world motion to be simulated. Until recently most computer-generated films were produced using the second approach: traditional computer animation techniques like keyframe animation, spline interpolation, etc. Then, animation languages, scripted systems, and director-oriented systems were developed. In the next generation of animation systems, motion control tends to be performed automatically using AI and robotics techniques. In particular, motion is planned at a task level and computed using physical laws. More recently, researchers have developed models of behavioral animation and simulation of autonomous creatures.

State Variables and Evolution Laws

Computer animation may be defined as a technique in which the illusion of movement is created by displaying on a screen, or recording on a recording device, a series of individual states of a dynamic scene.

1300

Formally, any computer animation sequence may be defined as a set of objects characterized by state variables evolving over time. For example, a human figure is normally characterized using its joint angles as state variables. To improve computer animation, attention needs to be devoted to the design of evolution laws [Magnenat Thalmann and Thalmann 1985]. Animators must be able to apply any evolution law to the state variables which drive animation.

Classification of Methods

Zeltzer [1985] classifies animation systems as being either guiding, animator-level, or task-level systems. In *guiding systems*, the behaviors of animated objects are explicitly described. Historically, typical guiding systems were BBOP, TWIXT, and MUTAN. In *animator-level systems*, the behaviors of animated objects are algorithmically specified. Typical systems are GRAMPS, ASAS, and MIRA. More details on these systems may be found in [Magnenat Thalmann and Thalmann 1990]. In *task-level systems*, the behaviors of animated objects are specified in terms of events and relationships. There is no general-purpose task-level system available now, but it should be mentioned that JACK [Badler et al. 1993] and HUMANOID [Boulic et al. 1995] are steps towards task-level animation.

Magnenat Thalmann and Thalmann [1991] propose a new classification of computer animation scenes involving synthetic actors both according to the method of controlling motion and according to the kinds of interactions the actors have. A *motion control method* specifies how an actor is animated and may be characterized according to the type of information to which it is privileged in animating the synthetic actor. For example, in a keyframe system for an articulated body, the privileged information to be manipulated is joint angles. In a forward dynamics-based system, the privileged information is a set of forces and torques; of course, in solving the dynamic equations, joint angles are also obtained in this system, but we consider these as derived information. In fact, any motion control method will eventually have to deal with geometric information (typically joint angles), but only geometric motion control methods are explicitly privileged to this information at the level of animation control.

The nature of privileged information for the motion control of actors falls into three categories: geometric, physical, and behavioral, giving rise to three corresponding categories of motion control method.

- The first approach corresponds to methods heavily relied upon by the animator: **performance animation**, *shape transformation*, **parametric keyframe animation**. *Animated objects are locally controlled.* Methods are normally driven by geometric data. Typically the animator provides a lot of geometric data corresponding to a local definition of the motion.

- The second way guarantees a realistic motion by using physical laws, especially *dynamic simulation*. The problem with this type of animation is controlling the motion produced by simulating the physical laws which govern motion in the real world. The animator should provide physical data corresponding to the complete definition of a motion. The motion is obtained by the dynamic equations of motion relating the forces, torques, and constraints and the mass distribution of objects. As trajectories and velocities are obtained by solving the equations, we may consider *actor motions as globally controlled*. Functional methods based on biomechanics are also part of this class.

- The third type of animation is called **behavioral animation** and takes into account the relationship between each object and the other objects. Moreover the control of animation may be performed at a task level, but we may also consider the *animated objects as autonomous creatures*. In fact, we will consider as a behavioral motion control method any method which drives the behavior of objects by providing high-level directives indicating a specific behavior without any other stimulus.

59.2 Underlying Principles and Best Practices

Geometric and Kinematics Methods

In this group of methods, the privileged information is of a geometric or kinematics nature. Typically, motion is defined in terms of coordinates, angles, and other shape characteristics, or it may be specified using velocities and accelerations, but no force is involved. Among the techniques based on geometry and kinematics, we will discuss performance animation, keyframing, morphing, inverse kinematics, and procedural animation. Although these methods have been mainly concerned with determining the displacement of objects, they may also be applied in calculating deformations of objects.

Motion Capture and Performance Animation

Performance animation or motion capture consist of measurement and recording of direct actions of a real person or animal for immediate or delayed analysis and playback. The technique is especially used today in production environments for 3D character animation. It involves mapping of measurements onto the motion of the digital character. This mapping can be direct (e.g., human arm motion controlling a character's arm motion) or indirect (e.g., mouse movement controlling a character's eye and head direction). Maiocchi [1995] gives more details about performance animation.

We may distinguish three kinds of systems: mechanical, optical, and magnetic.

Mechanical systems or digital puppetry allows animation of 3D characters through the use of any number of real-time input devices: mouse, joysticks, DataGloves, keyboard, dial boxes. The information provided by manipulation of such devices is used to control the variation of parameters over time for every animating feature of the character.

Optical motion capture systems are based on small reflective sensors called markers attached to an actor's body and on several cameras focused on performance space. By tracking positions of markers, one can get locations for corresponding key points in the animated model, e.g., we attach markers at joints of a person and record the position of markers from several different directions. We then reconstruct the 3D position of each key point at each time. The main advantage of this method is freedom of movement; it does not require any cabling. There is, however, one main problem: occlusion, i.e., lack of data resulting from hidden markers—for example when the performer lies on his back. Another problem comes with the lack of an automatic way of distinguishing reflectors when they get very close to each other during motion. These problems may be minimized by adding more cameras, but at a higher cost, of course. Most optical systems operate with four or six cameras. Good examples of optical systems are the ELITE and the VICON systems.

Magnetic motion capture systems require the real actor to wear a set of sensors, which are capable of measuring their spatial relationship to a centrally located magnetic transmitter. The position and orientation of each sensor are then used to drive an animated character. One problem is the need for synchronizing receivers. The data stream from the receivers to a host computer consists of 3D positions and orientations for each receiver. For human body motion, eleven sensors are generally needed: one on the head, one on each upper arm, one on each hand, one in the center of the chest, one on the lower back, one on each ankle, and one on each foot. To calculate the rest of the necessary information, the most common way is the use of inverse kinematics. The two most popular magnetic systems are Polhemus Fastrack and Ascension Flock of Birds.

Motion capture methods offer advantages and disadvantages. Let us consider the case of human walking. A walking motion may be recorded and then applied to a computer-generated 3D character. It will provide a very good motion, because it comes directly from reality. However, motion capture does

TABLE 59.1   Applications of VR Devices in Computer Animation

VR Device	Input Data	Application
DataGlove	Positions, orientations, trajectories, gestures, commands	Hand animation
DataSuit	Body positions, gestures	Body animation
6D mouse	Positions, orientations	Shape creation, keyframe
SpaceBall	Positions, orientations, forces	Camera motion
MIDI keyboard	Multidimensional data	Facial animation
Stereo display	3D perception	Camera motion, positioning
Head-mounted display (EyePhone)	Camera positions and trajectories	Camera motion
Force transducers	Forces, torques	Physics-based animation
Real-time video input	Shapes	Facial animation
Real-time audio input	Sounds, speech	Facial animation (speech)

not bring any really new concept to animation methodology. For any new motion, it is necessary to record the reality again. Moreover, motion capture may not be appropriate, especially in real-time simulation activities, where the situation and actions of people cannot be predicted ahead of time, and in dangerous situations, where one cannot involve a human actor.

VR-Based Animation

When motion capture is used on line, it is possible to create applications based on a full 3D interaction metaphor in which the specifications of deformations or motion are given in real time. This new concept drastically changes the way of designing animation sequences. Thalmann [1993] calls all techniques based on this new way of specifying animation *VR-based animation techniques.* He also calls *VR devices* all interactive devices allowing communication with virtual worlds. They include classic devices like head-mounted display systems and DataGloves as well as all 3D mice or SpaceBalls. He also considers as VR devices MIDI keyboards, force-feedback devices, and multimedia capabilities like real-time video input devices and even audio input devices. During the animation creation process, the animator should enter a lot of data into the computer. Table 59.1 shows VR devices and the corresponding data and applications.

Keyframe

This is an old technique consisting of the automatic generation of intermediate frames, called *inbetweens*, based on a set of keyframes supplied by the animator. Originally, the inbetweens were obtained by interpolating the keyframe images themselves. As linear interpolation produces undesirable effects such as lack of smoothness in motion, discontinuities in the speed of motion, and distortions in rotations, spline interpolation methods are used. Splines can be described mathematically as piecewise approximations of cubic polynomial functions. Two kinds of splines are very popular: interpolating splines with C1 continuity at knots, and approximating splines with C2 continuity at knots. For animation, the most interesting splines are the interpolating splines: cardinal splines, Catmull–Rom splines, and Kochanek–Bartels [1984] splines (see section 59.3).

A way of producing better images is to interpolate parameters of the model of the object itself. This technique is called **parametric keyframe animation**, and it is commonly used in most commercial animation systems. In a parametric model, the animator creates keyframes by specifying the appropriate set of parameter values. Parameters are then interpolated and images are finally constructed individually from the interpolated parameters. Spline interpolation is generally used for the interpolation.

Morphing

Morphing is a technique which has attracted much attention recently because of its astonishing effects. It is derived from shape transformation and deals with the metamorphosis of an object into another object over time. While three-dimensional object modeling and deformation is a solution to the morphing problem, the complexity of objects often makes this approach impractical.

The difficulty of the three-dimensional approach can be effectively avoided with a two-dimensional technique called image morphing. Image morphing manipulates two-dimensional images instead of three-dimensional objects and generates a sequence of inbetween images from two images. Image morphing techniques have been widely used for creating special effects in television commercials, music videos, and movies.

The problem of image morphing is basically how an inbetween image is effectively generated from two given images. A simple way for deriving an inbetween image is to interpolate the colors of each pixel between two images. However, this method tends to wash away the features on the images and does not give a realistic metamorphosis. Hence, any successful image morphing technique must interpolate the features between two images to obtain a natural inbetween image.

Feature interpolation is performed by combining warps with the color interpolation. A warp is a two-dimensional geometric transformation and generates a distorted image when it is applied to an image. When two images are given, the features on the images and their correspondences are specified by an animator with a set of points or line segments. Then, warps are computed to distort the images so that the features have intermediate positions and shapes. The color interpolation between the distorted images finally gives an inbetween image. More detailed processes for obtaining an inbetween image are described by Wolberg [1990].

In generating an inbetween image, the most difficult part is to compute warps for distorting the given images. Hence, the research in image morphing has concentrated on deriving warps from the specified feature correspondence. Image morphing techniques can be classified into two categories, mesh-based and feature-based methods, in terms of their ways for specifying features. In mesh-based methods, the features on an image are specified by a nonuniform mesh. Feature-based methods specify the features with a set of points or line segments. Lee and Shin [1995] have given a good survey of digital warping and morphing techniques.

Inverse Kinematics

The *direct kinematics* problem consists of finding the positions of end points (e.g., hand, foot) with respect to a fixed reference coordinate system as a function of time without regard to the forces or the moments that cause the motion. Efficient and numerically well-behaved methods exist for the transformation of position and velocity from joint space (joint angles) to Cartesian coordinates (end of the limb). Parametric keyframe animation is a primitive application of direct kinematics.

The use of *inverse kinematics* permits direct specification of end-point positions. Joint angles are automatically determined. This is the key problem, because the independent variables in an articulated system are joint angles. Unfortunately, the transformation of position from Cartesian to joint coordinates generally does not have a closed-form solution. However, there are a number of special arrangements of the joint axes for which closed-form solutions have been suggested in the context of animation [Girard and Maciejewski 1985].

Procedural Animation

Procedural animation corresponds to the creation of a motion by a procedure specifically describing the motion. Procedural animation should be used when the motion can be described by an algorithm or a formula. For example, consider the case of a clock based on the pendulum law:

$$\alpha = A \sin(\omega t + \phi)$$

A typical animation sequence may be produced using a program such as:

```
create CLOCK (...);
for FRAME:=1 to NB_FRAMES
    TIME:=TIME+1/24;
    ANGLE:= A*SIN (OMEGA*TIME+PHI);
    MODIFY (CLOCK, ANGLE);
    draw CLOCK;
    record CLOCK
    erase CLOCK
```

Procedural animation may be specified using a programming language or interactive system like the extensible director-oriented systems MIRANIM [Magnenat Thalmann and Thalmann 1990].

Character Deformations

Modeling and deformation of 3D characters and especially human bodies during the animation process are important but difficult problems. Researchers have devoted significant efforts to the representation and deformation of the human body shape. Broadly, we can classify their models into two categories: the surface model and the multilayered model.

The surface model [Magnenat Thalmann and Thalmann 1987] is conceptually simple, containing a skeleton and outer skin layer. The envelope is composed of planar or curved patches. One problem is that this model requires the tedious input of the significant points or vertices that define the surface. Another main problem is that it is hard to control the realistic evolution of the surface across joints. Surface singularities or anomalies can easily be produced. Simple observation of human skin in motion reveals that the deformation of the outer skin envelope results from many other factors besides the skeleton configuration.

The multilayered model [Chadwick et al. 1989] contains a skeleton layer, intermediate layers which simulate the physical behavior of muscle, bone, fat tissue, etc.; and a skin layer. Since the overall appearance of a human body is very much influenced by its internal muscle structures, the layered model is the most promising for realistic human animation. The key advantage of the layered methodology is that once the layered character is constructed, only the underlying skeleton need be scripted for an animation; consistent yet expressive shape deformations are generated automatically. Jianhua and Thalmann [1995] describe a highly effective multilayered approach for constructing and animating realistic human bodies. **Metaballs** are employed to simulate the gross behavior of bone, muscle, and fat tissue. They are attached to the proximal joints of the skeleton, arranged in an anatomically based approximation. The skin surfaces are automatically constructed using cross-sectional sampling. Their method, simple and intuitive, combines the advantages of implicit, parametric, and polygonal surface representation, producing very realistic and robust body deformations. By applying smooth blending twice (metaball potential-field blending and B-spline basis blending), the data size of the model is significantly reduced.

Physics-Based Methods

Dynamic Simulation

Kinematics-based systems are generally intuitive but lack dynamic integrity. The animation does not seem to respond to basic physical facts like gravity or inertia. Only modeling of objects that move under the influence of *forces* and *torques* can be realistic. Methods based on parameter adjustment are the most popular approach to dynamics-based animation and correspond to *nonconstraint methods*. There is an alternative: the *constraint-based methods*: the animator states in terms of constraints the properties the model is supposed to have, without needing to adjust parameters to give it those properties. In dynamic-based simulation, there are also two problems to be considered: the *forward dynamics* problem and the *inverse-dynamics* problem. The forward dynamics problem consists of finding the trajectories of some point (e.g., an end effector in an articulated figure) with regard to the forces and torques that cause the

FIGURE 59.1   A motion calculated using dynamic simulation.

motion. The inverse-dynamics problem is much more useful and may be stated as follows: determine the forces and torques required to produce a prescribed motion in a system. Nonconstraint methods have been mainly used for the animation of articulated figures [Armstrong et al. 1987]. There are a number of equivalent formulations which use various motion equations: Newton–Euler formulation (see section 59.3), Lagrange formulation, Gibbs–Appell formulation, D'Alembert formulation. These formulations are popular in robotics, and more details about the equations and their use in computer animation may be found in [Thalmann 1990]. Figure 59.1 shows an example of animation based on dynamics.

Concerning constraint-based methods, Witkin and Kass [1988] propose a new method, called **spacetime constraints**, for creating character animation. In this new approach, the character motion is created automatically by specifying *what* the character has to be, *how* the motion should be performed, what the character's *physical structure* is, and what physical *resources* are available to the character to accomplish the motion. The problem to solve is a problem of constrained optimization. Cohen [1992] takes this concept further and uses a subdivision of spacetime into discrete pieces, or *spacetime windows*, over which subproblems can be formulated and solved. The sensitivity of highly nonlinear constrained optimization to starting values and solution algorithms can thus be controlled to a great extent by the user.

Physics-Based Deformations

Physics-Based Models.   Realistic simulation of deformations may only be performed using physics-based animation. The most well-known model is the Terzopoulos elastic model [Terzopoulos et al. 1987]. In this model, the fundamental equation of motion corresponds to an equilibrium between internal forces (inertia, resistance to stretching, dissipative force, resistance to bending) and external forces (e.g., collision forces, gravity, seaming and attaching forces, wind force). Gourret et al. [1989] proposed a finite element method to model the deformations of human flesh due to flexion of members and/or contact with objects. The method is able to deal with penetrating impacts and true contacts. Simulation of impact with penetration can be used to model the grasping of ductile objects, and requires decomposition of objects into small geometrically simple objects. All the advantages of physical modeling of objects can also be transferred to human flesh. For example, the hand grasp of an object is expected to lead to realistic flesh deformation as well as an exchange of information between the object and the hand which will not only be geometrical.

Collision Detection and Response.   In computer animation, collision detection and response are obviously important. Some works have addressed this subject. Hahn [1988] treated bodies in resting contact as a series of frequently occurring collisions. Baraff [1989] presented an analytical method for finding forces between contacting polyhedral bodies, based on linear programming techniques. He also proposed a formulation of the contact forces between curved surfaces that are completely unconstrained in their tangential movement. Bandi and Thalmann [1995] introduced an adaptive spatial subdivision of the object space based on octree structure and presented a technique for efficiently updating this structure periodically during the simulation. Volino and Magnenat Thalmann [1994] described a new algorithm for detecting self-collisions on highly discretized moving polygonal surfaces. It is based on geometrical shape regularity properties that permit avoiding many useless collision tests.

FIGURE 59.2 Cloth animation.

A Case Study: Cloth Animation. Weil [1986] pioneered cloth animation using an approximated model based on relaxation of the surface. Kunii and Gotoda [1990] used a hybrid model incorporating physical and geometrical techniques to model garment wrinkles. Aono [1990] simulated wrinkle propagation on a handkerchief using an elastic model. Terzopoulos et al. [1987] applied their general elastic model to a wide range of objects including cloth. Lafleur et al. [1991] and Carignan et al. [1992] have described complex interaction of clothes with synthetic actors in motion, which marked the beginning of a new era in cloth animation. However, there were still a number of restrictions on the simulation conditions of the geometrical structure and the mechanical situations, imposed by the simulation model or the collision detection. More recently, Volino et al. [1995] proposed a mechanical model to deal with irregular triangular meshes, handle high deformations despite rough discretization, and cope with complex interacting collisions. Figure 59.2 shows an example of cloth animation.

Behavioral Methods

Task-Level Animation

Similarly to a task-level robotic system, actions in a task-level animation system are specified only by their effects on objects. Task-level commands are transformed into low-level instructions such as a script for algorithmic animation or key values in a parametric keyframe approach. Typical examples of tasks for synthetic actors are:

- walk from an arbitrary point A to another point B;
- pick up an object at location A and move it to location B;
- speak a sentence or make a high-level expression.

Similarly to robotics systems, we may divide task planning for synthetic actors into three phases: world modeling, task specification, and code generation.

The Grasping Task

In the computer animation field, interest in human grasping appeared with the introduction of the synthetic actors. Magnenat Thalmann et al. [1988a] describe one of the first attempts to facilitate the task of animating actors' interaction with their environment. However, the animator has to position the hand and decide the contact points of the hand with the object. Rijpkema and Girard [1991] present a full description of a grasping system that allows either an automatic or an animator-chosen grasp. The main idea is to approximate the objects with simple primitives. The mechanisms to grasp the primitives are known in advance and constitute what they call the knowledge database. Recently, Mas and Thalmann [1994] have presented a hand control and automatic grasping system using an inverse-kinematics-based method. In particular, their system can decide to use a pinch when the object is too small to be grasped by more than two fingers or to use a two-handed grasp when the object is too large.

Figure 59.3 shows an example of object-grasping scene.

FIGURE 59.3   Object-grasping scene.

The Walking Task

Zeltzer [1982] describes a walk controller invoking 12 local motor programs to control the actions of the legs and the swinging of the arms. Bruderlin and Calvert [1989] propose a hybrid approach to human locomotion which combines goal-oriented and dynamic motion control. Knowledge about a locomotion cycle is incorporated into a hierarchical control process. Decomposition of the locomotion determines forces and torques that drive the dynamic model of the legs by numerical approximation techniques. McKenna and Zeltzer [1990] describe an efficient forward dynamic simulation algorithm for articulated figures which has a computational complexity linear in the number of joints. Boulic et al. [1990] propose a model built from experimental data based on a wide range of normalized velocities. At the first level, global spatial and temporal characteristics (normalized length and step duration) are generated. At the second level, a set of parametrized trajectories produce both the position of the body in the space and the internal body configuration. The model is designed to keep the intrinsic dynamic characteristics of the experimental model. Such an approach also allows personalization of the walking action in an interactive real-time context in most cases. Figure 59.4. shows an example walking sequence.

Artificial and Virtual Life

This kind of research is strongly related to the research efforts in behavioral animation introduced by Reynolds [1987] in his distributed behavioral model simulating flocks of birds, herds of land animals, and fish schools. For birds, the simulated flock is an elaboration of a particle system with the simulated birds being the particles. A flock is assumed to be the result of the interaction between the behaviors of individual birds. Working independently, the birds try both to stick together and to avoid collisions with one another and with other objects in their environment. In a module of behavioral animation, the positions, velocities, and orientations of the actors are known from the system at any time. The animator

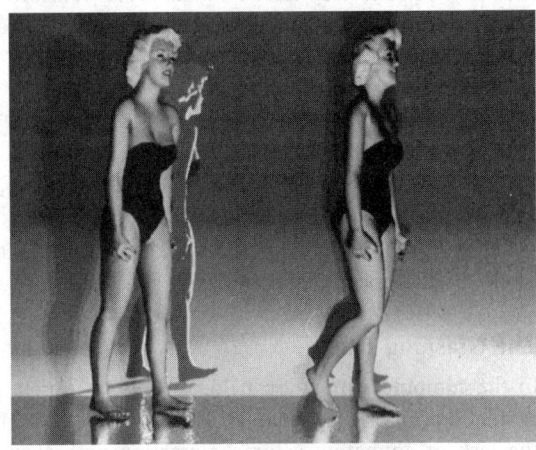

FIGURE 59.4   Walking sequence.

may control several global parameters, e.g., weight of the obstacle avoidance component, weight of the convergence to the goal, weight of the centering of the group, maximum velocity, maximum acceleration, minimum distance between actors. The animator provides data about the leader trajectory and the behavior of other birds relative to the leader. The computer-generated film *Stanley and Stella* was produced using this distributed behavioral model.

Wilhelms [1990] proposes a system based on a network of sensors and effectors. Ridsdale [1990] proposes a method that guides lower-level motor skills from a connectionist model of skill memory, implemented as collections of trained neural networks. We should also mention the huge literature about autonomous agents which represents a background theory for behavioral animation. More recently, genetic algorithms were also proposed to automatically generate morphologies for artificial creatures and the neural systems for controlling their muscle forces.

L-System-Based Behavioral Animation

Another approach for behavioral animation is based on timed and parametrized **L-systems** [Noser et al. 1992] with conditional and pseudostochastic productions. With this production-based approach a user may create any realistic or abstract shape, play with fascinating tree structures, and generate any concept of growth and life development in the resulting animation. To extend the possibilities for more realism in the pictures, external forces have been added, which interact with the L-structures and allow a certain physical modeling. External forces can also have an important impact in the evolution of objects. Tree structures can be elastically deformed and animated by time- and place-dependent vector force fields. The elasticity of each articulation can be set individually by productions. So, the bending of branches can be made dependent on the branches' thickness, making animation more realistic. The force fields, too, can be set and modified with productions. Force can directly affect L-structures. It is possible to simulate the displacement of objects in any vector force field dependent on time and position. An object movement is determined by a class of differential equations, which can be set and modified by productions. The mass of the turtle, who represents the object, can be set as well, by using a special symbol of the grammar. This vector-force-field approach is particularly convenient for simulating the motion of objects in fluids (air, water).

Virtual Sensors

Perception through Virtual Sensors

In a typical behavioral animation scene, the actor perceives the objects and the other actors in the environment, which provides information on their nature and position. This information is used by the behavioral model to decide the action to take, which results in a motion procedure. In order to implement perception, virtual humans should be equipped with visual, tactile, and auditory sensors. These **virtual sensors** should be used as a basis for implementing everyday human behavior such as visually directed locomotion, handling objects, and responding to sounds and utterances. For synthetic audition [Noser and Thalmann 1995], one needs to model a sound environment where the synthetic actor can directly access positional and semantic sound source information of audible sound events. Simulating the haptic system corresponds roughly to a collision detection process. But the most important perceptual subsystem is the vision system as described in the next sub-subsection.

Virtual Vision

The concept of synthetic vision was first introduced by Renault et al. [1990] as a main information channel between the environment and the virtual actor. Reynolds [1993] more recently described an evolved, vision-based behavioral model of coordinated group motion. Tu and Terzopoulos [1994] proposed artificial fishes with perception and vision. In the Renault method, each pixel of the vision input has the semantic information giving the object projected on the pixel, and numerical information giving the distance to the object. So it is easy to know, for example, that there is a table just in front at 3 meters.

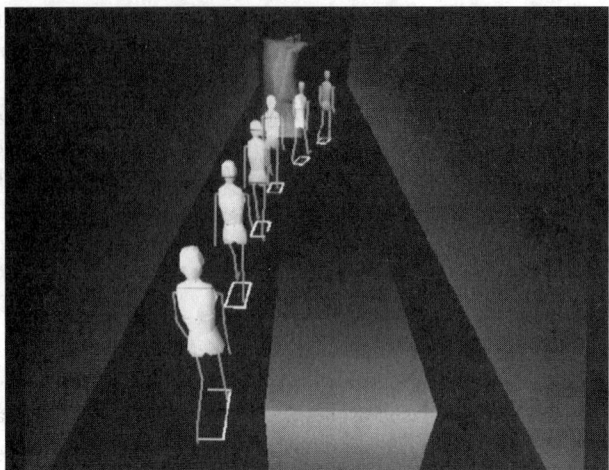

FIGURE 59.5   Obstacle avoidance using virtual vision.

The synthetic actor perceives his environment from a small window in which the environment is rendered from his point of view. As he can access Z-buffer values of the pixels, the color of the pixels, and his own position, he can locate visible objects in his 3D environment. Figure 59.5 shows an example of obstacle avoidance using virtual vision.

More recently, Noser et al. [1995] proposed the use of an octree as the internal representation of the environment seen by an actor because it offers several interesting features. The octree has to represent the visual memory of an actor in a 3D environment with static and dynamic objects. Objects in this environment can grow, shrink, move, or disappear. To illustrate the capabilities of the synthetic vision system, the authors have developed several examples: the actor going out of a maze, walking on sparse foot locations, and playing tennis.

Facial Animation

Realistic animation of a human face is extremely difficult to render by a computer because there are numerous specific muscles and important interactions between the muscles and the bone structure. This complexity leads to what is commonly called facial expression. The properties of facial expressions have been studied for 25 years by psychologist P. Ekman, who proposed a parametrization of muscles with their relationships to emotions: the FACS system [Ekman and Friesen 1978].

The most well-known specialist in facial animation is F. I. Parke of the New York Institute of Technology. He has identified two main approaches to applying computer graphics techniques to facial animation:

1. Using an image keyframe system: this means that a certain number of facial images are specified and the inbetweens are calculated by computer.
2. Using parametrized facial models [Parke 1982]: In this case the animator can create any facial image by specifying the appropriate set of parameter values.

Platt and Badler [1981] have designed a model that is based on underlying facial structure. Points are simulated in the skin, the muscles, and the bones by a set of three-dimensional networks. The skin is the outside level, represented by a set of 3D points that define a surface which can be modified. The bones represent an initial level that cannot be moved. Between the two levels, muscles are groups of points with elastic arcs. Magnenat-Thalmann et al. [1988b] introduce a way of controlling the human face and synchronizing speech based on the concept of abstract muscle action (AMA) procedures. An AMA

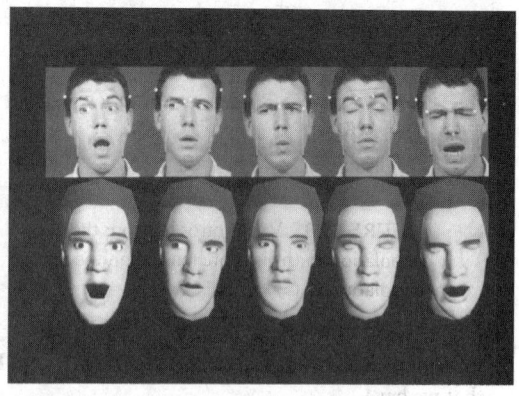

FIGURE 59.6   Facial expression.

FIGURE 59.7   Example of facial animation based on real performance.

procedure is a specialized procedure which simulates specific action of a face muscle. These AMA procedures are specific to the simulated action and are not independent, i.e., they are order-dependent. These procedures work on certain regions of the human face which must be defined when the face is constructed. Each AMA procedure is responsible for a facial parameter corresponding approximately to a muscle, for example, vertical jaw, closed upper lip, closed lower lip, raised lip, etc. Kalra et al. [1992] propose the simulation of muscle actions based on **rational free form deformations** (RFFDs), an extension of the classical free form deformation (FFD) method. Physically, RFFD corresponds to weighted deformations applied to an imaginary parallelepiped of clear, flexible plastic in which are embedded the object(s) to be deformed. The objects are also considered to be flexible so that they are deformed along with the plastic that surrounds them. Figure 59.6 shows an example of facial expression.

Waters [1987] represents the action of muscles using primary motivators on a nonspecific deformable topology of the face. The muscle actions themselves are tested against FACS which employs action units directly to one muscle or a small group of muscles. Any differences found between real action units and those performed on the computer model are easily corrected by changing the parameters for the muscles until reasonable results are obtained. Two types of muscles are created: linear/parallel muscles that pull, and sphincter muscles that squeeze.

Recently several authors have proposed new facial animation techniques which are based on the information derived from human performance. Williams [1990] used a texture-map-based technique with points on the surface of the real face. Mase and Pentland [1990] applied optical flow and principal-direction analysis for lip reading. Waters and Terzopoulos [1991] modeled and animated faces using scanned data obtained from a radial laser scanner. Saji et al. [1992] introduced a new method called *lighting switch photometry* to extract 3D shapes from the moving human face. Pandzic et al. [1994] also describe the underlying approach for recognizing and analyzing the facial movements of a real performance. The output in the form of parameters describing the facial expressions can then be used to drive one or more applications running on the same or on a remote computer. This enables the user to control the graphics system by means of facial expressions. Figure 59.7 shows an example.

59.3 Algorithms

Kochanek–Bartels Spline Interpolation

This method consists of interpolating splines with three parameters for local control—tension, continuity, and bias. Consider a list of points P_i and the parameter t along the spline to be determined. A point V is obtained from each value of t from only the two nearest given points along the curve (one behind P_i, one

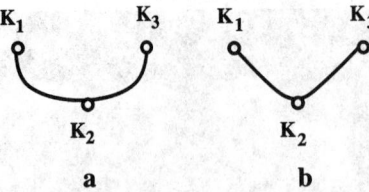

FIGURE 59.8 Variation of tension: the interpolation in (b) is more tense than the interpolation in (a).

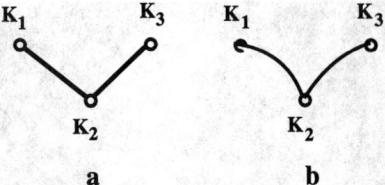

FIGURE 59.9 Variation of continuity: the interpolation in (b) is more discontinuous than the interpolation in (a).

in front of $\mathbf{P}_{i+1}$). But the tangent vectors $\mathbf{D}_i$ and $\mathbf{D}_{i+1}$ at these two points are also necessary. This means that we have

$$\mathbf{V} = THC^T \tag{59.1}$$

where T is the matrix $[t^3 \; t^2 \; t \; 1]$, H is the Hermite matrix, and C is the matrix $[\mathbf{P}_i, \mathbf{P}_{i+1}, \mathbf{D}_i, \mathbf{D}_{i+1}]$. The Hermite matrix is given by

$$H = \begin{bmatrix} 2 & -2 & 1 & 1 \\ -3 & 3 & -2 & -1 \\ 0 & 0 & 1 & 0 \\ 1 & 0 & 0 & 0 \end{bmatrix} \tag{59.2}$$

This equation shows that the tangent vector is the average of the source chord $\mathbf{P}_i - \mathbf{P}_{i-1}$ and the destination chord $\mathbf{P}_{i+1} - \mathbf{P}_i$. Similarly, the source derivative (tangent vector) $\mathbf{DS}_i$ and the destination derivative (tangent vector) $\mathbf{DD}_i$ may be considered at any point $\mathbf{P}_i$.

Using these derivatives, Kochanek and Bartels propose the use of three parameters to control the splines—tension, continuity, and bias.

The tension parameter t controls how sharply the curve bends at a point $\mathbf{P}_i$. As shown in Fig. 59.8, in certain cases a wider, more exaggerated curve may be desired, while in other cases the desired path may be much tighter.

The continuity c of the spline at a point $\mathbf{P}_i$ is controlled by the parameter c. Continuity in the direction and speed of motion is not always desirable. Animating a bouncing ball, for example, requires the introduction of a discontinuity in the motion of the point of impact, as shown in Fig. 59.9.

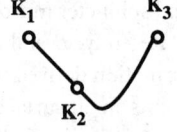

FIGURE 59.10 A biased interpolation at K_2.

The direction of the path as it passes through a point $\mathbf{P}_i$ is controlled by the bias parameter b. This feature allows the animator to have a trajectory anticipate or overshoot a key position by a certain amount, as shown in Fig. 59.10.

Equations combining the three parameters may be obtained:

$$\mathbf{DS}_i = 0.5[(1-t)(1+c)(1-b)(\mathbf{P}_{i+1} - \mathbf{P}_i) + (1-t)(1-c)(1+b)(\mathbf{P}_i - \mathbf{P}_{i-1})] \tag{59.3}$$

$$\mathbf{DD}_i = 0.5[(1-t)(1-c)(1-b)(\mathbf{P}_{i+1} - \mathbf{P}_i) + (1-t)(1+c)(1+b)(\mathbf{P}_i - \mathbf{P}_{i-1})] \tag{59.4}$$

A spline is then generated using Eq. (59.1) with $\mathbf{DD}_i$ and $\mathbf{DS}_{i+1}$ instead of $\mathbf{D}_i$ and $\mathbf{D}_{i+1}$.

Dynamic Simulation

Direct Dynamic

The most popular algorithm in dynamic simulation is the Armstrong–Green algorithm, based on Newton–Euler formulations. To get a simplified form in order to avoid the inversion of matrices larger than three by three, two hypotheses have been made. The first assumes a linear relationship between the linear acceleration of the link and the amount of angular acceleration it undergoes. The second assumes a linear relationship between the linear acceleration and the reactive force on its parent link.

The algorithm can be divided into two opposite processes, inbound and outbound. The inbound calculates some matrices and vectors from the geometric and physical structure of the system, and propagates force and torque along each link from the leaves to the root of the hierarchy. The outbound calculates kinematics quantities, the linear and angular acceleration of each link, then goes on to get the linear and angular velocity by numerical integration for updating the whole structure. It is a recursive algorithm along the whole time of dynamic simulation. The kinematic results of the previous time step are used as initial values for the next-step calculation.

Inverse Dynamics

The inverse dynamics problem is to find at each joint the force and torque that generate the desired motion of the structure. Among many formulations, we select a kind of recursive formulation based on Newton–Euler equations for their computational efficiency. The best advantage is that the computation time is linearly proportional to the number of joints and independent of the limb configuration. It is based on two basic equations,

$$F = ma \tag{59.5}$$

$$N = J\alpha + \omega \times J\omega \tag{59.6}$$

The first is Newton's second law that describes the linear motion of the rigid body. F is the total vector force on the body with mass m. The second is Euler's equation that governs the rotation of the rigid body. N is the total vector torque on the body with the inertia matrix J about its center of mass. The linear acceleration a, angular acceleration α, and angular velocity ω define the kinematics of the rigid body.

To construct the formulations for an articulated body model, we should write a series of equations for each link, using constrained forces and torque to guarantee their connection. The details of this formulation for inverse dynamics are listed below. We can see it has two opposite processes. The forward recursion goes from the inertial coordinate frame to the end-effector coordinate frame to propagate kinematics information forward. The backward recursion propagates the forces and moments exerted on each link in the opposite direction.

Recursive Newton–Euler Algorithm

All vectors with subscript i are expressed in the body coordinate frame of the ith link. The following items except m_i are vectors:

$\dot{q}_{ij}, \ddot{q}_{ij} =$ the first and second derivatives of generalized variable for the jth DOF of the ith joint,
$\omega^i =$ angular velocity of the ith link,
$\alpha^i =$ angular acceleration of the ith link,
$a^i =$ linear acceleration of the ith link at the origin of its local coordinate frame,
$a_c^i =$ linear acceleration of the ith link at the center of mass,
$g =$ gravitational acceleration vector in the inertial coordinate frame,
$g_i =$ gravitation of the ith link,
$f_i =$ force exerted on the ith link by the $(i-1)$st link,
$n_i =$ moment exerted on the ith link by the $(i-1)$st link,
$\tau_i =$ joint generalized actuator torque at the ith joint,
$p_i =$ vector from the origin of the ith link to the origin of the $(i+1)$st link,
$s_i =$ vector from the origin of the ith link to its center of mass,
$z_{ij} =$ rotation vector of the jth DOF of the ith joint,
$m_i =$ mass of link i,
$F_i =$ total external force exerted on the ith link,
$N_i =$ total external torque exerted on the ith link.

The following items are 3-by-3 matrices:

J_i = inertia matrix of the ith link.

A_i^j = pure rotational matrix from the jth link coordination frame to the ith.

Finally, the integer nd_i is the number of DOF at joint i.

The algorithm works as follows:

Initialization:

$$\omega^0 = \alpha^0 = v^0 = 0$$
f_{n+1} = force required at the end effector
n_{n+1} = moment required at the end effector

Forward recursion:

For $i = 1, \ldots, n$ do:

$$\omega^i = A_i^{i-1}\omega^{i-1} + \sum_{j=1}^{nd_i} z_{ij}\dot{q}_{ij}$$

$$\alpha^i = A_i^{i-1}\alpha^{i-1} + \sum_{j=1}^{nd_i} z_i\ddot{q}_{ij} + A_i^{i-1}\omega^{i-1} \times \sum_{j=1}^{nd_i} z_{ij}\dot{q}_{ij}$$

$$a^i = A_i^{i-1}(a^{i-1} + \alpha^{i-1} \times p_{i-1} + \omega^{i-1} \times (\omega^{i-1} \times p_{i-1}))$$

$$a_c^i = a^i + \alpha^i \times s_i + \omega^i \times (\omega^i \times s_i)$$

$$g_i = A_i^0(m_i g)/^*g = (0.0, 0.0, -9.8)^*/$$

$$F_i = m_i a_c^i$$

$$N_i = j_i\alpha^i + \omega^i \times J_i\omega^i$$

Backward recursion:

For $i = n, \ldots, 1$ do:

$$f_i = F_i + A_i^{i+1}f_{i+1} - g_i$$

$$n_i = N_i - A_i^{i+1}n_{i+1} + s_i \times (F_i - g_i) + p_i \times A_i^{i+1}f_{i+1}$$

$$\tau_i = n_i \cdot z_i$$

end;

Principle of Behavioral Animation

A simulation is produced in a synchronous way by a behavioral loop such as

```
t_global ← 0.0
        code to initialize the animation environment
while (t_global < t_final) {
        code to update the scene
        for each actor
                code to realize the perception of the environment
                code to select actions based on sensorial input, actual state and specific behavior
        for each actor
        code executing the above selected actions
        t_global ← t_global + t_interval
}
```

The global time t_global serves as the synchronization parameter for the different actions and events. Each iteration represents a small time step. The action to be performed by each actor is selected by its behavioral model for each time step. The action selection takes place in three phases. First, the actor perceives the objects and the other actors in the environment, which provides information on their nature and position.

This information is used by the behavioral model to decide the action to take, which results in a motion procedure with its parameters: e.g., grasp an object or walk with a new speed and a new direction. Finally, the actor performs the motion.

59.4 Research Issues and Summary

Computer animation should not be considered just as a tool to enhance spatial perception by moving the virtual camera or rotating objects. More sophisticated animation techniques than keyframe animation need to be widely used. Computer animation tends to be more and more based on physics and dynamic simulation methods. In the future, the application of computer animation to the scientific world will also become very common in many areas: fluid dynamics, molecular dynamics, thermodynamics, plasma physics, astrophysics, etc. Real-time complex animation systems will be developed taking advantage of VR devices and simulation methods. Integration of simulation methods with VR-based animation will lead to systems allowing the user to interact with complex time-dependent phenomena, providing interactive visualization and interactive animation. Moreover, real-time synthetic actors will be part of *virtual worlds*, and people will communicate with them through broadband multimedia networks. This development will be made possible only by developing new approaches to real-time motion.

Defining Terms

Behavioral animation: Behavior of objects is driven by providing high-level directives indicating a specific behavior without any other stimulus.

L-systems: Production-based approach to generate any concept of growth and life development in the resulting animation.

Metaballs: Shapes based on implicit surfaces; they are employed to simulate the gross behavior of bone, muscle, and fat tissue.

Morphing or shape transformation: Metamorphosis of an object into another object over time. Image morphing manipulates two-dimensional images instead of three-dimensional objects and generates a sequence of inbetween images from two images.

Parametric keyframe animation: The animator creates keyframes by specifying the appropriate set of parameter values. Parameters are then interpolated, and finally images are individually constructed from the interpolated parameters.

Performance animation: Measurement and recording of direct actions of a real person or animal for immediate or delayed analysis and playback.

Procedural animation: The creation of a motion by a procedure describing the motion specifically. Procedural animation may be specified using a programming language or an interactive system.

Rational free form deformations: Extension of the classical free form deformations. Physically, they correspond to weighted deformations applied to an imaginary parallelepiped of clear, flexible plastic in which are embedded the object(s) to be deformed.

Spacetime constraints: Method for creating automatic character motion by specifying what the character has to be, how the motion should be performed, what the character's *physical structure* is, and what physical *resources* are available to the character to accomplish the motion.

Virtual sensors: Sensors used as a basis for implementing everyday human behavior such as visually directed locomotion, handling objects, and responding to sounds and utterances. Virtual humans should be equipped with visual, tactile, and auditory sensors.

References

Aono, M. 1990. A wrinkle propagation model for cloth, pp. 96–115. In *Proc. Computer Graphics Int. '90*. Springer–Verlag, Tokyo.

Armstrong, W. W., Green, M., and Lake, R. 1987. Near real-time control of human figure models. *IEEE Comput. Graphics Appl.* 7(6):28–38.

Badler, N. I., Phillips, C. B., and Webber, B. L. 1993. *Simulating Humans: Computer Graphics Animation and Control.* Oxford University Press, New York.

Bandi, S. and Thalmann, D. 1995. An adaptive spatial subdivision of the object space for fast collision of animated rigid bodies, pp. 259–270. In *Proc. Eurographics '95*, Maastricht, Aug.

Baraff, D. 1989. Analytical methods for dynamic simulation of non-penetrating rigid bodies, *Comput. Graphics (Proc. SIGGRAPH)* 23(3):223–232.

Boulic, R., Thalmann, D., and Magnenat-Thalmann, N. 1990. A global human walking model with real time kinematic personification. *Visual Comput.* 6(6).

Boulic, R., Capin, T., Huang, Z., Moccozet, L., Molet, T., Kalra, P., Lintermann, B., Magnenat-Thalmann, N., Pandzic, I., Saar, K., Schmitt, A., Shen, J., and Thalmann, D. 1995. The HUMANOID environment for interactive animation of multiple deformable human characters, pp. 337–348. In *Proc. Eurographics '95*, Maastricht, Aug.

Bruderlin, A. and Calvert, T. W. 1989. Goal directed, dynamic animation of human walking. *Comput. Graphics (Proc. SIGGRAPH)* 23(3):233–242

Carignan, M., Yang, Y., Magnenat Thalmann, N., and Thalmann, D. 1992. Dressing animated synthetic actors with complex deformable clothes, *Comput. Graphics (Proc. SIGGRAPH)* 26(2):99–104.

Chadwick, J. E., Haumann, D. R., and Parent, R. E. 1989. Layered construction for deformable animated character. *Comput. Graphics* 23(3):243–252.

Cohen, M. F. 1992. Interactive spacetime control for animation. *Comput. Graphics (Proc. SIGGRAPH)* 92:293–302.

Ekman, P. and Friesen, W. 1978. *Facial Action Coding System.* Consulting Psychologists Press, Palo Alto, CA.

Girard, M. and Maciejewski, A. A. 1985. Computational modeling for computer generation of legged figures. *Comput. Graphics (Proc. SIGGRAPH)* 19(3):263–270.

Gourret, J. P., Magnenat-Thalmann, N., and Thalmann, D. 1989. Simulation of object and human skin deformations in a grasping task. *Comput. Graphics (Proc. SIGGRAPH)* 23(3):21–30.

Hahn, J. K. 1988. Realistic animation of rigid bodies. *Comput. Graphics (Proc. SIGGRAPH)* 22(4):299–308.

Jianhua, S. and Thalmann, D. 1995. Interactive shape design using metaballs and splines. In *Proc. Implicit Surfaces 95*, Eurographics, Grenoble.

Kalra, P., Mangili, A., Magnenat-Thalmann, N., and Thalmann, D. 1992. Simulation of facial muscle actions based on rational free form deformations. In *Proc. Eurographics '92*, Cambridge, U.K.

Kochanek, D. H. and Bartels, R. H. 1984. Interpolating splines with local tension, continuity, and bias control. *Comput. Graphics (Proc. SIGGRAPH)* 18:33–41.

Kunii, T. L. and Gotoda, H. 1990. Modeling and animation of garment wrinkle formation processes, pp. 131–147. In *Proc. Computer Animation '90*, Springer–Verlag, Tokyo.

Lafleur, B., Magnenat Thalmann, N., and Thalmann, D. 1991. Cloth animation with self-collision detection, pp. 179–187. In *Proc. IFIP Conf. on Modeling in Computer Graphics.* Springer–Verlag, Tokyo.

Lee, S. and Shin, S. Y. 1995. Warp generation and transition control in image morphing. In *Interactive Computer Animation*, N. Magnenat Thalmann, and D. Thalmann, eds. Prentice–Hall.

Magnenat Thalmann, N. and Thalmann, D. 1990. *Computer Animation: Theory and Practice*, 2nd ed. Springer–Verlag, Tokyo.

Magnenat Thalmann, N. and Thalmann, D. 1991. Complex models for animating synthetic actors. *IEEE Comput. Graphics Appl.* 11, Sept.

Magnenat Thalmann, N. and Thalmann, D. 1985. 3D Computer animation: more an evolution problem than a motion problem. *IEEE Comput. Graphics and Appl.* 5(10):47–57.

Magnenat Thalmann, N., Laperrière, R., and Thalmann, D. 1988a. Joint-dependent local deformations for hand animation and object grasping, pp. 26–33. In *Proc. Graphics Interface '88*, Edmonton.

Magnenat-Thalmann, N., Primeau, E., and Thalmann, D. 1988b. Abstract muscle action procedures for human face animation. *Visual Comput.* 3(5):290–297.

Magnenat Thalmann, N. and Thalmann, D. 1987. The direction of synthetic actors in the film Rendez-vous a Montreal. *IEEE Comput. Graphics Appl.* 7(12):9–19.

Maiocchi, R. 1995. 3D Character animation using motion capture. In *Interactive Computer Animation*, N. Magnenat Thalmann and D. Thalmann, eds., Prentice–Hall.

Mas, R. and Thalmann, D. 1994. A hand control and automatic grasping system for synthetic actors, pp. 167–178. In *Proc. Eurographic '94.*

Mase, K. and Pentland, A. 1990. Automatic lipreading by computer. *Trans. Inst. Elec. Inform. and Commum. Eng.* J73-D-II (6):796–803.

McKenna, M. and Zeltzer, D. 1990. Dynamic simulation of autonomous legged locomotion. *Computer Graphics (Proc. SIGGRAPH)* 24(4):29–38.

Noser, H. and Thalmann, D. 1995. Synthetic vision and audition for digital actors, pp. 325–336. In *Proc. Eurographics '95*, Maastricht, Aug.

Noser, H., Turner, R., and Thalmann, D. 1992. Interaction between L-systems and vector force-field, pp. 747–761. In *Proc. Comput. Graphics International '92*, Tokyo.

Noser, H., Renault, O., Thalmann, D., and Magnenat Thalmann, N. 1995. Navigation for digital actors based on synthetic vision, memory and learning. *Comput. and Graphics* 19(1):7–19.

Pandzic, I. S., Kalra, P., and Magnenat-Thalmann, N. 1994. Real time facial interaction. *Displays* 15(3):157–163.

Parke, F. I. 1982. Parameterized models for facial animation. *IEEE Comput. Graphics Appl.* 2(9):61–68.

Platt, S. and Badler, N. 1981. Animating facial expressions, *Comput. Graphics (Proc. SIGGRAPH)* 15(3):245–252.

Renault, O., Magnenat-Thalmann, N., and Thalmann, D. 1990. A vision-based approach to behavioral animation. *J. Visualization and Comput. Animation* 1(1):18–21.

Reynolds, C. 1987. Flocks, herds, and schools: a distributed behavioral model. *Comput. Graphics (Proc. SIGGRAPH)* 21(4):25–34.

Reynolds, C. W. 1993. An evolved, vision-based behavioral model of coordinated group motion, pp. 384–392. In *From Animals to Animats, Proc. 2nd Int. Conf. on Simulation of Adaptive Behavior*, J. A. Meyer et al., eds. MIT Press.

Ridsdale, G. 1990. Connectionist modeling of skill dynamics. *J. Visualization and Comput. Animation* 1(2):66–72.

Rijpkema, H. and Girard, M. 1991. Computer animation of knowledge-based grasping. *Comput. Graphics (Proc. SIGGRAPH)* 25(4):339–348.

Saji, H., Hioki, H., Shinagawa, Y., Yoshida, K., and Kunii, T. L. 1992. Extraction of 3D shapes from the moving human face using lighting switch photometry, pp. 69–86. In *Creating and Animating the Virtual World*. N. Magnenat Thalmann and D. Thalmann, eds., Springer–Verlag, Tokyo.

Terzopoulos, D., Platt, J. C., Barr, A. H., and Fleischer, K. 1987. Elastically deformable models. *Comput. Graphics (Proc. SIGGRAPH)* 21(4):205–214.

Thalmann, D. 1990. Robotics methods for task-level and behavioral animation, pp. 129–147. In *Scientific Visualization and Graphics Simulation*. D. Thalmann, ed., Wiley, Chichester, U.K.

Thalmann, D. 1993. Using virtual reality techniques in the animation process. In *Virtual Reality Systems*, R. Earnshaw, M. Gigante, and M. H. Jones, eds., Academic Press.

Tu, X. and Terzopoulos, D. 1994. Artificial fishes: physics, locomotion, perception, behavior. *Comput. Graphics (Proc. SIGGRAPH)* 42–48.

Volino, P. and Magnenat Thalmann, N. 1994. Efficient self-collision detection on smoothly discretised surface animations using geometrical shape regularity. *Comput. Graphics Forum (Proc. Eurographics)* 13(3):155–166.

Volino, P., Courchesnes, M., and Magnenat Thalmann, N. 1995. Versatile and efficient techniques for simulating cloth and other deformable objects. *Comput. Graphics (Proc. SIGGRAPH).*

Waters, K. 1987. A muscle model for animating three-dimensional facial expression. *Comput. Graphics (Proc. SIGGRAPH)* 21(4)17–24.

Waters, K. and Terzopoulos, D. 1991. Modeling and animating faces using scanned data. *J. Visualization and Comput. Animation* 2(4):123–128.

Weil, J. 1986. The synthesis of cloth objects. *Comput. Graphics (Proc. SIGGRAPH)* 20(4):49–54.

Wilhelms, J. 1990. A "notion" for interactive behavioral animation control. *IEEE Comput. Graphics Appl.* 10(3):14–22.

Williams, L. 1990. Performance driven facial animation. *Comput. Graphics (Proc SIGGRAPH)* 24(4):235–242.

Witkin, A. and Kass, M. 1988. Spacetime constraints. *Computer Graphics (Proc. SIGGRAPH)* 22(4):159–168.

Wolberg, G. 1990. *Digital Image Warping*. IEEE Computer Soc. Press.

Zeltzer, D. 1982. Motor control techniques for figure animation. *IEEE Comput. Graphics Appl.* 2(9):53–59.

Zeltzer, D. 1985. Towards an integrated view of 3D computer animation. *Visual Comput.* 1(4):249–259.

Further Information

Several textbooks on computer animation have been published:

Magnenat Thalmann, N. and Thalmann, D., eds. 1990. *Computer Animation: Theory and Practice*, 2nd ed., Springer–Verlag, Tokyo.

Vince, J. 1992. *3-D Computer Animation*. Addison–Wesley.

Mealing, S. 1992. *The Art and Science of Computer Animation*. Intellect, Oxford, U.K.

Magnenat Thalmann, N. and Thalmann, D., eds. 1996. *Interactive Computer Animation*. Prentice–Hall.

There is one journal dedicated to computer animation: *The Journal of Visualization and Computer Animation*, published by John Wiley and Sons, Chichester, U.K., since 1990.

Although computer animation is always represented in major computer graphics conferences like SIG-GRAPH, Computer Graphics International (CGI), Pacific Graphics, and Eurographics, there are also two annual conferences dedicated to computer animation:

1. Computer Animation, organized each year in Geneva by the Computer Graphics Society. Proceedings are published by IEEE Computer Society Press.
2. Eurographics Workshop on Animation and Simulation, organized each year by Eurographics.

Volume Visualization

Arie E. Kaufman
State University of New York at Stony Brook

60.1 Introduction

Volume data are 3D entities that may have information inside them, might not consist of surfaces and edges, or might be too voluminous to be represented geometrically. *Volume visualization* is a method of extracting meaningful information from volumetric data using interactive graphics and imaging, and it is concerned with volume data representation, modeling, manipulation, and rendering [Kaufman 1991]. Volume data are obtained by sampling, simulation, or modeling techniques. An example of a sampled volume data is a sequence of 2D slices obtained from magnetic resonance imaging (MRI) or computed tomography (CT) that is 3D reconstructed into a volume model and visualized for diagnostic purposes or for planning of treatment or surgery. The same technology is often used with industrial CT for nondestructive inspection of composite materials or mechanical parts. Similarly, confocal microscopes produce data which is visualized to study the morphology of biological structures. In many computational fields, such as in computational fluid dynamics, the results of simulation typically running on a supercomputer are often visualized as volume data for analysis and verification. Recently, many traditional geometric computer graphics applications, such as computer aided design (CAD) and simulation, have been exploiting the advantages of volume techniques called *volume graphics* for modeling, manipulation, and visualization.

Over the years many techniques have been developed to visualize 3D data. Since methods for displaying geometric primitives were already well established, most of the early methods involve approximating a surface contained within the data using geometric primitives. When volumetric data are visualized using a surface rendering technique, a dimension of information is essentially lost. In response to this, volume rendering techniques were developed that attempt to capture the entire 3D data in a single 2D image. Volume rendering conveys more information than surface rendering images, but at the cost of increased algorithm complexity, and consequently increased rendering times. To improve interactivity in volume rendering, many optimization methods as well as several special-purpose volume rendering machines have been developed.

We begin with an introduction to volumetric data. Section 60.3 covers briefly surface rendering techniques for volume data. Section 60.4 discusses in detail volume rendering techniques, including image-order, object-order, and domain techniques. Optimization methods for volume rendering are discussed in section 60.5, and special-purpose volume rendering hardware is described in section 60.6. Section 60.7 introduces global illumination of volumetric data, including volumetric ray tracing and volumetric radiosity. Irregular grid rendering is briefly discussed in section 60.8. Volume graphics is introduced in section 60.9, including several volume modeling techniques, such as voxelization, texture mapping, amorphous phenomena, block operations, constructive solid modeling, and volume sculpting.

60.2 Volumetric Data

Volumetric data is typically a set S of samples (x, y, z, v), representing the value v of some property of the data, at a 3D location (x, y, z). If the value is simply a 0 or a 1, with a value of 0 indicating background and a value of 1 indicating the object, then the data is referred to as binary data. The data may instead be multivalued, with the value representing some measurable property of the data, including, for example, color, density, heat, or pressure. The value v may even be a vector, representing, for example, velocity at each location.

In general, the samples may be taken at purely random locations in space, but in most cases the set S is isotropic containing samples taken at regularly spaced intervals along three orthogonal axes. When the spacing between samples along each axis is a constant, but there may be three different spacing constants for the three axes, the set S is anisotropic. Since the set of samples is defined on a regular grid, a 3D array (called also *volume buffer, cubic frame buffer, 3D raster*) is typically used to store the values, with the element location indicating position of the sample on the grid. For this reason, the set S will be referred to as the array of values $S(x, y, z)$, which is defined only at grid locations. Alternatively, either rectilinear, curvilinear (structured), or unstructured grids, are employed (e.g., Speray and Kennon [1990]). In a *rectilinear* grid the cells are axis-aligned, but grid spacings along the axes are arbitrary. When such a grid has been nonlinearly transformed while preserving the grid topology, the grid becomes *curvilinear*. Usually, the rectilinear grid defining the logical organization is called *computational space*, and the curvilinear grid is called *physical space*. Otherwise the grid is called *unstructured* or *irregular*. An unstructured or irregular volume data is a collection of cells whose connectivity has to be specified explicitly. These cells can be of an arbitrary shape such as tetrahedra, hexahedra, or prisms.

The array S only defines the value of some measured property of the data at discrete locations in space. A function $f(x, y, z)$ may be defined over R^3 in order to describe the value at any continuous location. The function $f(x, y, z) = S(x, y, z)$ if (x, y, z) is a grid location, otherwise $f(x, y, z)$ approximates the sample value at a location (x, y, z) by applying some interpolation function to S. There are many possible interpolation functions. The simplest interpolation function is known as *zero-order* interpolation, which is actually just a nearest neighbor function. The value at any location in R^3 is simply the value of the closest sample to that location. With this interpolation method there is a region of constant value around each sample in S. Since the samples in S are regularly spaced, each region is of uniform size and shape. The region of constant value that surrounds each sample is known as a *voxel* with each voxel being a rectangular cuboid having six faces, twelve edges, and eight corners [Kaufman and Sobierajski 1994].

Higher order interpolation functions can also be used to define $f(x, y, z)$ between sample points. One common interpolation function is a piecewise function known as *first-order interpolation,* or *trilinear interpolation.* With this interpolation function, the value is assumed to vary linearly along directions parallel to one of the major axes. Let the point P lie at location (x_p, y_p, z_p) within the regular hexahedron, known as a *cell*, defined by samples A–H. For simplicity, let the distance between samples in all three directions be 1, with sample A at $(0, 0, 0)$ with a value of v_A, and sample H at $(1, 1, 1)$ with a value of v_H. The value v_P, according to trilinear interpolation, is then

$$
\begin{aligned}
v_P = \; & v_A(1 - x_p)(1 - y_p)(1 - z_p) + v_E(1 - x_p)(1 - y_p)z_p \\
& + v_B x_p(1 - y_p)(1 - z_p) + v_F x_p(1 - y_p)z_p \\
& + v_C(1 - x_p)y_p(1 - z_p) + v_G(1 - x_p)y_p z_p \\
& + v_D x_p y_p(1 - z_p) + v_H x_p y_p z_p
\end{aligned} \tag{60.1}
$$

In general, A will be at some location (x_A, y_A, z_A), and H will be at (x_H, y_H, z_H). In this case, x_p in Eq. (60.1) would be replaced by $(x_P - x_A)/(x_H - x_A)$ with similar substitutions made for y_p and z_p.

60.3 Surface Rendering Techniques

Several surface rendering techniques have been developed which approximate a surface contained within volumetric data using geometric primitives, which can be rendered using conventional graphics accelerator hardware. A surface can be defined by applying a binary segmentation function $B(v)$ to the volumetric data. $B(v)$ evaluates to 1 if the value v is considered part of the object, and evaluates to 0 if the value v is part of the background. The surface is then the region where $B(v)$ changes from 0 to 1. If a zero-order interpolation function is being used, then the surface is simply the set of faces which are shared by voxels with differing values of $B(v)$. If a higher order interpolation function is being used, then the surface passes between sample points according to the interpolation function.

For zero-order interpolation functions, the natural choice for a geometric primitive is the 3D rectangular cuboid, since the surface is a set of faces, and each face is a rectangle. An early algorithm for displaying human organs from computed tomograms [Herman and Liu 1979] uses the square as the geometric primitive. To simplify the projection calculation and decrease rendering times, the assumption is made that the sample spacing in all three directions is the same. A software Z-buffer algorithm is then used to project the shaded squares onto the image plane to create the final image.

With continuous interpolation functions, a surface, known as an *isovalued surface* or an *isosurface*, may be defined by a single value. Several methods for extracting and rendering isosurfaces have been developed, a few are briefly described here. The marching cubes algorithm [Lorensen and Cline 1987] was developed to approximate an isovalued surface with a triangle mesh. The algorithm breaks down the ways in which a surface can pass through a cell into 256 cases, reduces by symmetry to only 15 topologies. For each of these 15 cases, a generic set of triangles representing the surface is stored in a lookup table. Each cell through which a surface passes maps to one of the 15 cases, with the actual triangle vertex locations being determined using linear interpolation on the cell vertices. A normal value is estimated for each triangle vertex, and standard graphics hardware can be utilized to project the triangles, resulting in a smooth shaded image of the isovalued surface.

When rendering a sufficiently large data set with the marching cubes algorithm, millions of triangles may be generated; many of them map to a single pixel when projected onto the image plane. This fact led to the development of surface rendering algorithms that use 3D points with normals (called *smart points*) as the geometric primitive. One such algorithm is dividing cubes [Cline et al. 1988], which subdivides each cell through which a surface passes into subcells. The number of divisions is selected such that the subcells project onto a single pixel on the image plane. Another algorithm which uses 3D smart points as the geometric primitive is the trimmed voxel lists method [Sobierajski et al. 1993]. Instead of subdividing, this method uses only one 3D point per visible surface cell, projecting that point on up to three pixels of the image plane to ensure coverage in the image.

60.4 Volume Rendering Techniques

Whereas representing a surface contained within a volumetric data set using geometric primitives can be useful in many applications, there are several main drawbacks to this approach. First, geometric primitives can only approximate surfaces contained within the original data. Adequate approximations may require an excessive amount of geometric primitives. Therefore, a tradeoff must be made between accuracy and space requirements. Second, since only a surface representation is used, much of the information contained within the data is lost during the rendering process. For example, in CT scanned data useful information is contained not only on the surfaces, but within the data as well. Also, amorphous phenomena, such as clouds, fog, and fire cannot be adequately represented using surfaces, and therefore must have a volumetric representation, and must be displayed using volume rendering techniques.

In the next subsections various volume rendering techniques are explored. *Volume rendering* is the process of creating a 2D image directly from 3D volumetric data. Although several of the methods described in these subsections render surfaces contained within volumetric data, these methods operate on the actual data samples, without the intermediate geometric primitive representations used by the algorithms in section 60.3.

Volume rendering can be achieved using an *object-order,* an *image-order,* or a *domain-based* technique. Object-order volume rendering techniques use a *forward mapping* scheme (also called feed forward or projection) where the volume data is mapped onto the image plane. In image-order algorithms, a *backward mapping* scheme (also called feed backward) is used where rays are cast from each pixel in the image plane through the volume data to determine the final pixel value. In a domain-based technique the spatial volume data is first transformed into an alternative domain, such as compression, frequency, and wavelet, and then a projection is generated directly from that domain.

Object-Order Techniques

Object-order techniques involve mapping the data samples onto the image plane. One way to accomplish a projection of a surface contained within the volume is to loop through the data samples, projecting each sample which is part of the object onto the image plane. For this algorithm, the data samples are binary voxels, with a value of 0 indicating background and a value of 1 indicating the object. Also, the data samples are on a grid with uniform spacing in all three directions.

If an image is produced by projecting all voxels with a value of 1 to the image plane in an arbitrary order, we are not guaranteed a correct image. If two voxels project to the same pixel on the image plane, the one that was projected later will prevail, even if it is farther from the image plane than the earlier projected voxel. This problem can be solved by traversing the data samples in a *back-to-front* order. For this algorithm, the strict definition of back-to-front can be relaxed to require that if two voxels project to the same pixel on the image plane, the first processed voxel must be farther away from the image plane than the second one. This can be accomplished by traversing the data plane-by-plane, and row-by-row inside each plane. For arbitrary orientations of the data in relation to the image plane, some axes may be traversed in an increasing order, whereas others may be considered in a decreasing order. The traversal can be accomplished with three nested loops, indexing on x, y, and z. Although the relative orientations of the data and the image plane specify whether each axis should be traversed in an increasing or decreasing manner, the ordering of the axes in the traversal is arbitrary.

An alternative to back-to-front projection is a *front-to-back* method in which the voxels are traversed in the order of increasing distance from the image plane. Although a back-to-front method is easier to implement, a front-to-back method has the advantage that once a voxel is projected onto a pixel, other voxels which project to the same pixel are ignored, since they would be hidden by the first voxel. Another advantage of front-to-back projection methods is that if the axis which is most parallel to the viewing direction is chosen to be the outermost loop of the data traversal, meaningful partial image results can be displayed to the user. This allows the user to better interact with the data and terminate the image generation if, for example, an incorrect view direction was selected. Partial image results can be displayed to the user during a back-to-front method also, but the value of a pixel may change many times during image generation. With a front-to-back method, once a pixel value is set, its value remains unchanged.

Clipping planes orthogonal to the three major axes, and clipping planes parallel to the view plane are easy to implement using either a back-to-front or a front-to-back algorithm. For orthogonal clipping planes, the traversal of the data is limited to a smaller rectangular region within the full data set. To implement clipping planes parallel to the image plane, data samples whose distance to the image plane is less than the distance between the cut plane and the image plane are ignored. This ability to explore the whole data set is a major difference between volume rendering techniques and the surface rendering techniques described in section 60.3. In surface rendering techniques, the geometric primitive representation of the object needs to be changed in order to implement cut planes, which could be a time-consuming process. In a back-to-front method, cut planes can be achieved by simply modifying the bounds of the data traversal, and utilizing a condition when placing depth values in the image plane pixels.

For each voxel, its distance to the image plane could be stored in the pixel to which it maps along with the voxel value. At the end of a data traversal a 2D array of depth values, called a Z-buffer, is created, where the value at each pixel in the Z-buffer is the distance to the closest nonempty voxel. A 2D discrete shading technique can then be applied to the image, resulting in a shaded image suitable for display. The 2D discrete shading techniques described here take as input a 2D array of depth values and a 2D array of projected voxel values, and produce as output a 2D image of intensity values. The simplest 2D discrete shading method is known as depth shading, or *depth-only shading* [Vannier et al. 1983], where only the Z-buffer is used and the intensity value stored in each pixel of the output image is inversely proportional to the depth of the corresponding input pixel. This produces images where features far from the image plane appear dark, while close features are bright. Since surface orientation is not considered in this shading method, most details such as surface discontinuities and object boundaries are lost.

A more accurately shaded image can be obtained by passing the 2D depth image to a gradient-shader [Gordon and Reynolds 1985], which can take into account the object surface orientation and the distance from the light at each pixel to produce a shaded image. This method evaluates the gradient at each (x, y) pixel location in the input image by

$$\nabla z = \left(\frac{\delta z}{\delta x}, \frac{\delta z}{\delta y}, -1 \right) \tag{60.2}$$

where $z = D(x, y)$ is the depth stored in the Z-buffer associated with pixel (x, y). The estimated gradient vector at each pixel is then used as a normal vector for shading purposes.

The value $\delta z / \delta x$ can be approximated using a backward difference $D(x, y) - D(x - 1, y)$, a forward difference $D(x + 1, y) - D(x, y)$, or a central difference $(D(x + 1, y) - D(x - 1, y))/2$. Similar equations are used for approximating $\delta z / \delta y$. In general, the central difference is a better approximation of the derivative, but along object edges where, for example, pixels (x, y) and $(x + 1, y)$ belong to two different objects, a backward difference would provide a better approximation. A context sensitive normal estimation method [Yagel et al. 1992b] was developed to provide more accurate normal estimations by detecting image discontinuities. In this method, two pixels are considered to be in the same context if their depth values, and the first derivative of the depth at these locations do not greatly differ. The gradient vector at some pixel p is then estimated by considering only those pixels which lie within a user-defined neighborhood, and belong to the same context as p. This ensures that sharp object edges, and slope changes are not smoothed out in the final image.

The previous rendering methods consider only binary data samples where a value of 1 indicates the object and a value of 0 indicates the background. Many forms of data acquisition (e.g., CT) produce data samples with 8, 12, or even more bits of data per sample. If these data samples represent the values at some sample points, and the values vary according to some convolution applied to the data samples which can reconstruct the original 3D signal, then a scalar field which approximates the original 3D signal has been defined.

One way to reconstruct the original signal is, as described previously, to define a function $f(x, y, z)$ which determines the value at any location in space. This technique is typically employed by backward-mapping (image-order) algorithms. In forward-mapping algorithms, the original signal is reconstructed by spreading the value at a data sample into space. Westover describes a splatting algorithm [Westover 1990] for approximating smooth object-ordered volume rendering, in which the value of the data samples represents a density. Each data sample $s = (x_s, y_s, z_s, \rho(s))$, $s \in S$, has a function C defining its contribution to every point (x, y, z) in the space

$$C_s(x, y, z) = h_v(x - x_s, y - y_s, z - z_s)\rho(s) \tag{60.3}$$

where h_v is the volume reconstruction kernel and $\rho(s)$ is the density of sample s which is located at (x_s, y_s, z_s). The contribution of a sample s to an image plane pixel (x, y) can then be computed by integration:

$$C_s(x, y) = \rho(s) \int_{-\infty}^{\infty} h_v(x - x_s, y - y_s, u) \, du \tag{60.4}$$

where the u coordinate axis is parallel to the view ray. Since this integral is independent of the sample density, and depends only on its (x, y) projected location, a footprint function F can be defined as follows:

$$F(x, y) = \int_{-\infty}^{\infty} h_v(x, y, u) \, du \tag{60.5}$$

where (x, y) is the displacement of an image sample from the center of the sample's image plane projection. The weight w at each pixel can then be expressed as:

$$w(x, y)_s = F(x - x_s, y - y_s) \tag{60.6}$$

where (x, y) is the pixel location, and (x_s, y_s) is the image plane location of the sample s.

A footprint table can be generated by evaluating the integral in Eq. (60.5) on a grid with a resolution much higher than the image plane resolution. All table values lying outside of the footprint table extent have zero weight and therefore need not be considered when generating an image. A footprint table for a data sample s can be centered on the projected image plane location of s, and be sampled in order to determine the weight of the contribution of s to each pixel on the image plane. Multiplying this weight by $\rho(s)$ then gives the contribution of s to each pixel.

Computing a footprint table can be difficult due to the integration required. Discrete integration methods can be used to approximate the continuous integral, but generating a footprint table is still a costly operation. Luckily, for orthographic projections, the footprint of each sample is the same except for an image plane offset. Therefore, only one footprint table needs to be calculated per veiw. Since this still would require too much computation time, only one generic footprint table is built for the kernel. For each view, a view-transformed footprint table is created from the generic footprint table. The generic footprint table can be precomputed; therefore, it does not matter how long the computation takes.

Generating a view transformed footprint table from the generic footprint table can be accomplished in three steps. First, the image plane extent of the projection of the reconstruction kernel is determined. Next a mapping is computed between this extent and the extent that surrounds the generic footprint table. Finally, the value for each entry in the view transformed footprint table is determined by mapping the location of the entry to the generic footprint table, and sampling. The extent of the reconstruction kernel is either a sphere, or is bounded by a sphere, so that the extent of the generic footprint table is always a circle. If the grid spacing of the data samples is uniform along all three axes, then the reconstruction kernel is a sphere and the image plane extent of the reconstruction kernel will be a circle. The mapping from this extent to the extent of the generic footprint table is simply a scaling operation. If the grid spacing differs along the three axes, then the reconstruction kernel is an ellipsoid and the image plane extent of the reconstruction kernel will be an ellipse. In this case, a mapping from this ellipse to the circular extent of the generic footprint table must be computed.

There are three modifiable parameters in this algorithm which can greatly affect image quality. First, the size of the footprint table can be varied. Small footprint tables produce blocky images, whereas large footprint tables may smooth out details and require more space. Second, different sampling methods can be used when generating the view transformed footprint table from the generic footprint table. Using a nearest neighbor approach is fast, but may produce aliasing artifacts. On the other hand, using bilinear interpolation produces smoother images at the expense of longer rendering times. The third parameter which can be modified is the reconstruction kernel itself. The choice of, for example, a cone function, Gaussian function, sync function, or bilinear function affects the final image.

Drebin et al. [1988] developed a technique for rendering volumes that contain mixtures of materials, such as CT data containing bone, muscle, and flesh. In this method, various assumptions about the volume data are made. First, it is assumed that the scalar field was sampled above the Nyquist frequency, or a low-pass filter was used to remove high frequencies before sampling. The volume contains either several scalar fields, or one scalar field representing the composition of several materials. If the latter is the case, it is assumed that material can be differentiated either by the scalar value at each point, or by additional information about the composition of each volume element.

The first step in this rendering algorithm is to create new scalar fields from the input data, known as material percentage volumes. Each material percentage volume is a scalar field representing only one material. Color and opacity are then associated with each material, with composite color and opacity obtained by linearly combining the color opacity for each material percentage volume. A matte volume, that is, a scalar field on the volume with values ranging between 0 and 1, is used to slice the volume or perform other spatial set operations. Actual rendering of the final composite scalar field is obtained by transforming the volume so that one axis is perpendicular to the image plane. The data is then projected plane by plane in a back-to-front manner and composited to form the final image.

Image-Order Techniques

Image-order volume rendering techniques are fundamentally different from object-order rendering techniques. Instead of determining how a data sample affects the pixels on the image plane, in an image-order technique we determine for each pixel on the image plane, the data samples that contribute to it.

One of the first image-order volume rendering techniques, which may be called *binary ray casting* [H. K. Tuy and L. T. Tuy 1984], was developed to generate images of surfaces contained within binary volumetric data without the need to explicitly perform boundary detection and hidden-surface removal. For each pixel on the image plane, a ray is cast from that pixel to determine if it intersects the surface contained within the data. For parallel projections, all rays are parallel to the view direction, where as for perspective projections, rays are cast from the eye point according to the view direction and the field of view. If an intersection does occur, shading is performed at the intersection point, and the resulting color is placed in the pixel. In order to determine the first intersection along the ray a stepping technique is used where the value is determined at regular intervals along the ray until the object is intersected. Data samples with a value of 0 are considered to be the background whereas those with a value of 1 are considered to be part of the object. A zero-order interpolation technique is used, so that the value at a location along the ray is 0 if that location is not in any voxel of the data; otherwise it is the value of the closest data sample. For a step size d, the ith point sample p_i would be taken at a distance $i \times d$ along the ray. For a given ray, either all point samples along the ray have a value of 0 (the ray missed the object entirely), or there is some sample p_i taken at a distance $i \times d$ along the ray, such that all samples p_j, $j < i$, have a value of 0, and sample p_i has a value of 1. Point sample p_i is then considered to be the first intersection along the ray. In this algorithm, the step size d must be chosen carefully. If d is too large, small features in the data may not be detected. On the other hand, if d is small, the intersection point is more accurately estimated at the cost of higher computation time.

There are several optimizations which can be made to this algorithm. First, the number of steps which must be made along each ray can be reduced by traversing only the part of the ray contained within the bounding box of the data. A second optimization involves the representation of the data in memory. This algorithm was originally developed on a machine with only 32K of random access memory (RAM), so data compression was a critical issue. Instead of simply storing the data as a binary array of 0s and 1s, a

scan-line representation can be used. For each scan line
in the data, a list of end points can be stored which represent the segments belonging to the object. This representation is compact, yet does not add too much time to
the intersection calculation.

The previous algorithm deals with the display of surfaces within binary data. A more general algorithm can
be used to generate surface and composite projections
of multivalued data. Instead of traversing a continuous
ray and determining the closest data sample for each step
with a zero-order interpolation function, a discrete representation of the ray could be traversed. This discrete

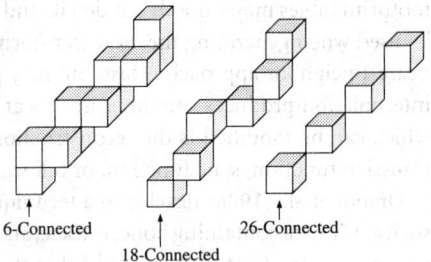

6-Connected   18-Connected   26-Connected

FIGURE 60.1 Examples of 6-, 18-, and 26-
connected paths.

ray is generated using a 3D Bresenham-like algorithm or a 3D line scan-conversion (voxelization) algorithm
[Kaufman and Shimony 1986, Kaufman 1991] (see the subsection on voxelization in section 60.9). As in
the previous algorithms, for each pixel in the image plane, the data samples that contribute to it need to be
determined. This could be done by casting a ray from each pixel in the direction of the viewing ray. This
ray would be discretized (voxelized), and the contribution from each voxel along the path is considered
when producing the final pixel value. This technique is referred to as *discrete ray casting* [Yagel et al. 1992a].

In order to generate a 3D discrete ray using a voxelization algorithm, the 3D discrete topology of 3D
paths has to be understood. There are three types of connected paths: 6 connected, 18 connected, and
26 connected, which are based on the three adjacency relationships between consecutive voxels along
the path. An example of these three types of connected paths is given in Fig. 60.1. Assuming a voxel is
represented as a box centered at the grid point, two voxels are said to be 6 connected if they share a face,
they are 18 connected if they share a face or an edge, and they are 26 connected if they share a face, an
edge, or a vertex. A 6-connected path is a sequence of voxels, $v_1, v_2, \ldots, v_N$, where for each pair of voxels
$v_i, v_{i+1} (1 \leq i < N)$, v_i and v_{i+1} are 6 connected. Similar definitions exist for 18- and 26-connected paths.

In discrete ray casting, a ray is discretized into a 6-, 18-, or 26-connected path, and only the voxels along
this path are considered when determining the final pixel value. If a surface projection is required, the path
is traversed until the first voxel, which is part of the object, is encountered. This voxel is then shaded and
the resulting color value is stored in the pixel. The 6-connected paths contain almost twice as many voxels
as 26-connected paths, and so an image created using 26-connected paths would require less computation,
but a 26-connected path may miss an intersection that would be detected using a 6-connected path.

To produce a shaded image, the distance to the closest intersection is stored at each pixel in the image,
and then this image is passed to a 2D discrete shader, such as those described previously. However, better
results can be obtained by performing a 3D discrete shading operation at the intersection point. One 3D
discrete shading method, known as *normal-based contextual shading* [Chen et al. 1985], can be employed
to estimate the normal when zero-order interpolation is used. The normal for a face of a voxel that is on
the surface of the object is determined by examining the orientation of that face, and the orientation of
the four faces on the surface that are edge connected to that face. Since a face of a voxel can have only six
possible orientations, the error in the approximated normal can be significant. More accurate results can
be obtained using a technique known as *gray-level shading* [Cline et al. 1988, Hoehne and Bernstein 1986,
Lorensen and Cline 1987]. If the intersection occurs at location (x, y, z) in the data, then the gray-level
gradient at that location can be approximated with a central difference,

$$G_x = \frac{f(x+1, y, z) - f(x-1, y, z)}{2D_x}$$

$$G_y = \frac{f(x, y+1, z) - f(x, y-1, z)}{2D_y} \qquad (60.7)$$

$$G_z = \frac{f(x, y, z+1) - f(x, y, z-1)}{2D_z}$$

where (G_x, G_y, G_z) is the gradient vector, and D_x, D_y, and D_z are the distances between neighboring samples in the x, y, and z directions, respectively. The gradient vector is used as a normal vector for shading calculation, and the intensity value obtained from shading is stored in the image. A normal estimation can be performed at point sample p_i, and this information, along with the light direction, and the distance $i \times d$ can be used to shade p_i.

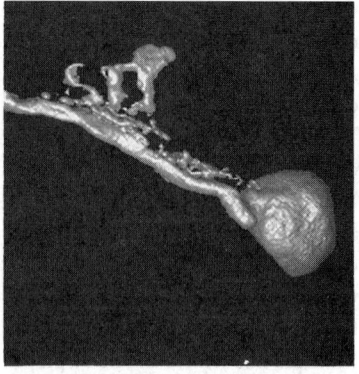

(a)

Actually, stopping at the first opaque voxel and shading there is only one of many operations which can be performed on the voxels along a discrete path or continuous ray. Instead, the whole ray could be traversed, storing in the image plane pixel the maximum value encountered along the ray. Figure 60.2(a) is a first opaque, or surface, projection of a bullfrog sympathetic ganglion cell, which was reconstructed from confocal microscope data, whereas Fig. 60.2(b) is a maximum projection of the same cell. Figure 60.2 was generated using the polygon assisted ray casting (PARC) algorithm, which is described in section 60.5. As opposed to a surface projection, a maximum projection is capable of revealing some internal parts of the data. Another option is to store the sum (simulating X-ray) or average of all values along the ray. More complex techniques, which are described subsequently, may involve defining an opacity and color for each scalar value, and then accumulating intensity along the ray according to some compositing function, revealing 3D structure information and 3D internal features. (See Fig. 60.2(c).)

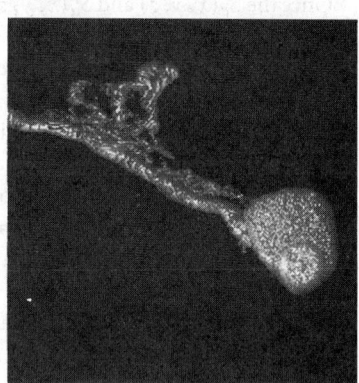

(b)

The previous two rendering techniques, binary ray casting and discrete ray casting, use zero-order interpolation in order to define the scalar value at any location in R^3. One advantage to using zero-order interpolation is simplicity and speed, since many of the calculations required can be done using integer arithmetic. One disadvantage, though, is the aliasing effects in the image. Higher order interpolation functions can be used to create a more accurate image, but generally at the cost of algorithm complexity and computation time. The next three algorithms described in this section all use higher order interpolation functions.

When creating a composite projection of a data set, there are two important parameters, the color at a sample point, and the opacity at that location. An image-order volume rendering algorithm developed by Levoy [1988] states that given an array of data samples S, two new arrays S_c and S_α, which define the color and opacity at each grid location, can be generated using preprocessing techniques. The interpolation functions $f(x, y, z)$, $f_c(x, y, z)$, and $f_\alpha(x, y, z)$, which specify the sample value, color, and opacity at any location in R^3, are then defined. Often, f_c and f_α are referred to as transfer functions.

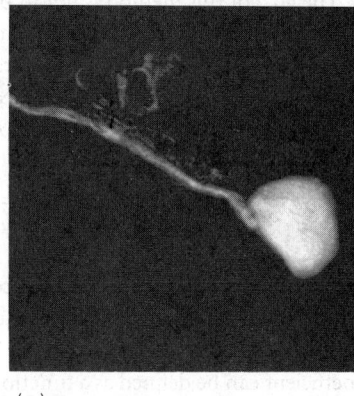

(c)

FIGURE 60.2 (a) A surface projection of a nerve cell, (b) a maximum projection of a nerve cell, and (c) a composited projection of a nerve cell.

Generating the array S_c of color values involves performing a shading operation, such as gray-level shading, at each data sample in the original array S. For this purpose, the Phong illumination model, for example, could be used. The normal at each data sample is the unit gradient vector at that location. The gradient vector at any location can be computed by partially differentiating the interpolation function with respect to x, y, and z to get each component of the gradient. If the interpolation function is not first derivative continuous, aliasing artifacts will occur in the image due to the discontinuous normal vector. A smoother set of gradient vectors can be obtained using a central differencing method similar to the one described earlier in this section.

Calculating the array S_α is essentially a surface classification operation. There are different ways to classify surfaces within a scalar field, and each way requires a new mapping from $S(x, y, z)$ to $S_\alpha(x, y, z)$. When an isosurface at some constant value v with an opacity α_v ought to be viewed, $S_\alpha(x, y, z)$ is simply assigned to α_v if $S(x, y, z)$ is v, otherwise $S_\alpha(x, y, z) = 0$. This would produce aliasing artifacts, which can be reduced by setting $S_\alpha(x, y, z)$ close to α_v if $S(x, y, z)$ is close to v. The best results are obtained when the thickness of the transition region is constant throughout the volume. This can be approximated by having the opacity fall off at a rate inversely proportional to the magnitude of the local gradient vector. Multiple iscsurfaces can be displayed in a single image by separately applying the classification mappings, then combining the opacities.

Once the $S_c(x, y, z)$ and $S_\alpha(x, y, z)$ arrays have been determined, rays are cast from the pixels, through these two arrays, sampling at evenly spaced locations. To determine the value at a location, the trilinear interpolation functions f_c and f_α are used. Once these point samples along the ray have been computed, a fully opaque background is added in, and then the values in a back-to-front order are composited to produce a single color that is placed in the pixel.

Two rendering techniques for displaying volumetric data, known as the V-buffer method, were developed by Upson and Keeler [1988]. One of the methods for visualizing the scalar field is an image-order ray casting technique. In this method, rays are cast from each pixel on the image plane into the volume. For each cell in the volume along the path of this ray, the scalar value is determined at the point where the ray first intersects the cell. The ray is then stepped along until it traverses the entire cell, with calculations for scalar values, shading, opacity, texture mapping, and depth cuing performed at each stepping point. This process is repeated for each cell along the ray, accumulating color and opacity, until the ray exits the volume, or the accumulated opacity reaches unity. At this point, the accumulated color and opacity for that pixel are stored, and the next ray is cast.

The goal of this method is not to produce a realistic image, but instead to provide a representation of the volumetric data which can be interpreted by a scientist or an engineer. For this purpose, the user is given the ability to modify certain parameters in the shading equations, which will lead to an informative, rather than physically accurate, shaded image. A simplified shading equation is used where the perceived intensity as a function of wavelength $I(\lambda)$ is defined as

$$I(\lambda) = K_a(\lambda)I_a + K_d(\lambda) \sum_j [(\mathbf{N} \cdot \mathbf{L}_j)I_j] \tag{60.8}$$

In this equation, K_a is the ambient coefficient, I_a is the ambient intensity, K_d is the diffuse coefficient, $\mathbf{N}$ is the normal approximated by the local gradient, $\mathbf{L}_j$ is the vector to the jth light source, and I_j is the intensity of the jth light source. In order to highlight certain features in the final image, the diffuse coefficient can be defined as a function of not only wavelength, but also scalar value and solid texture,

$$K_d(\lambda, S, M) = K(\lambda)T_d(\lambda, S(x, y, z)M(\lambda, x, y, z)) \tag{60.9}$$

where K is the actual diffuse coefficient, T_d is the color transfer function, S is the sample array, and M is the solid texture map. The color transfer function is defined for red, green, and blue, and maps scalar value to intensity. In this method the following intensity integral is approximated when accumulating along the ray:

$$I(\lambda) = \int_w \left[\tau(d)O(s)\left[K_a(\lambda)I_a + K_d(\lambda, S, M) \sum[(\mathbf{N} \cdot \mathbf{L}_j)I_j] \right] + (1 - \tau(d)bc(\lambda)) \right] d\mathbf{u} \tag{60.10}$$

where $\tau(d)$ represents atmospheric attenuation as a function of distance d, $O(s)$ is the opacity transfer function, *bc* is the background color, and **u** is the vector in the direction of the view ray. The opacity transfer function is similar to the color transfer function in that it defines opacity as a function of scalar value. Different color and opacity transfer functions can be defined to highlight different features in the volume.

The second method for visualizing the scalar field is a cell-by-cell processing technique [Upson and Keeler 1988], where within each cell an image-order ray casting technique is used, thus making this a hybrid technique. In this method, each cell in the volume is processed in a front-to-back order. Processing begins on the plane closest to the viewpoint, and progresses in a plane-by-plane manner. Within each plane, processing begins with the cell closest to the viewpoint, then continues in order of increasing distance from the viewpoint. Each cell is processed by first determining for each scan line in the image plane, which pixels are affected by the cell. Then, for each pixel an integration volume is determined. Within the bounds of the integration volume, an intensity calculation similar to Eq. (60.10) is performed according to

$$I(\lambda) = \int_x \int_y \int_z \left[\tau(d)O(s) \Big[K_a(\lambda)I_a + K_d(\lambda, S, M) \sum [(\mathbf{N} \cdot \mathbf{L}_j)I_j] \Big] \right.$$

$$\left. + (1 - \tau(d)bc(\lambda)) \right] \mathrm{d}x \, \mathrm{d}y \, \mathrm{d}z \qquad (60.11)$$

This process continues in a front-to-back order, until all cells have been processed, with intensity accumulated into pixel values. Once a pixel opacity reaches unity, a flag is set and this pixel is not processed further. Because of the front-to-back nature of this algorithm, incremental display of the image is possible.

In order to simulate light coming from translucent objects, volumetric data with data samples representing density values can be considered as a field of density emitters [Sabella 1988]. A density emitter is a tiny particle that both emits and scatters light. The amount of density emitters in any small region within the volume is proportional to the scalar value in that region. These density emitters are used to correctly model the occlusion of deeper parts of the volume by closer parts, but both shadowing and color variation due to differences in scattering at different wavelengths are ignored. These effects are ignored because it is believed that they would complicate the image, detracting from the perception of density variation. Similar to the V-buffer method, rays are cast from the eye point, through each pixel on the image plane, and into the volume. The intensity I of light for a given pixel is calculated according to

$$I = \int_{t_1}^{t_2} e^{-\tau \int_{t_1}^t \rho^\gamma(\lambda) \, \mathrm{d}\lambda} \rho^\gamma(t) \, \mathrm{d}t \qquad (60.12)$$

In this equation, the ray is traversed from t_1 to t_2, accumulating at each location t the density $\rho^\gamma(t)$ at that location attenuated by the probability

$$e^{-\tau \int_{t_1}^t \rho^\gamma(\lambda) \, \mathrm{d}\lambda}$$

that this light will be scattered before reaching the eye. The parameter τ is modifiable, and controls the attenuation, with higher values of τ specifying a medium, which darkens more rapidly. The parameter γ is also modifiable, and controls the spread of density values. Low γ values produce a diffuse cloud appearance, whereas higher γ values highlight dense portions of the data. For each ray, three values in addition to I maybe computed: the maximum value encountered along the ray, the distance at which that maximum occurred, and the center of gravity of density emitters along the ray. By mapping these values to different color parameters (such as hue, saturation, and lightness), interesting effects can be achieved.

Krueger [1991] showed that the various existing volume rendering models can be described as special cases of an underlying transport theory model of the transfer of particles in inhomogeneous media. The basic idea is that a beam of *virtual* particles is sent through the volume, with the user selecting the particle properties and the laws of interaction between the particles and the data. The image plane then contains the *scattered* virtual particles, and information about the data is obtained from the scattering pattern. If, for example, the virtual particles are chosen to have the properties of photons, and the laws of interaction are governed by optical laws, then this model essentially becomes a generalized ray tracer. Other virtual particles and interaction laws can be used, for example, to identify periodicities and similar hidden symmetries of the data. Using Krueger's transport theory model, the intensity of light I at a pixel can be described as a path integral along the view ray,

$$I = \int_{p_{\text{near}}}^{p_{\text{far}}} Q(p) e^{-\int_{p_{\text{near}}}^{p} \sigma_a(p') + \sigma_{sc}(p') \, dp'} \, dp \qquad (60.13)$$

The emission at each point p along the ray is scaled by the optical depth to the eye to produce the final intensity value for a pixel. The optical depth is a function of the total extinction coefficient, which is composed of the absorption coefficient σ_a, and the scattering coefficient σ_{sc}. The generalized source $Q(p)$ is defined as

$$Q(p) = q(p) + \sigma_{sc(p)} \int \rho_{sc}(\omega' \to \omega) I(S, \omega') \, d\omega' \qquad (60.14)$$

This generalized source consists of the emission at a given point $q(p)$, and the incoming intensity along all directions scaled by the scattering phase ρ_{sc}. Typically, a low-albedo approximation is used to simplify the calculations, reducing the integral in Eq. (60.14) to a sum over all light sources.

Domain Volume Rendering

In domain rendering the spatial 3D data is first transformed into another domain, such as compression, frequency, and wavelet domain, and then a projection is generated directly from that domain or with the help of information from that domain. The frequency-domain rendering applies the Fourier slice projection theorem, which states that a projection of the 3D data volume from a certain view direction can be obtained by extracting a 2D slice perpendicular to that view direction out of the 3D Fourier spectrum and then inverse Fourier transforming it. This approach obtains the 3D volume projection directly from the 3D spectrum of the data, and therefore reduces the computational complexity for volume rendering from $O(N^3)$ to $O(N^2 \log N)$ [Dunne et al. 1990, Malzbender 1993]. A major problem of frequency-domain volume rendering is the fact that the resulting projection is a line integral along the view direction which does not exhibit any occlusion and attenuation effects. Totsuka and Levoy [1993] proposed a linear approximation to the exponential attenuation [Sabella 1988] and an alternative shading model to fit the computation within the frequency-domain rendering framework.

The compression-domain rendering performs volume rendering from compressed scalar data without decompressing the entire data set, and therefore reduces the storage, computation, and transmission overhead of otherwise large volume data. For example, Ning and Hesselink [1993] first applied vector quantization in the spatial domain to compress the volume and, then directly rendered the quantized blocks using regular spatial-domain volume rendering algorithms. Fowler and Yagel [1994] combined differential pulse-code modulation and Huffman coding, and developed a lossless volume compression algorithm, but their algorithm is not coupled with rendering. Yeo and Liu [1995] applied discrete cosine transform based compression technique on overlapping blocks of the data. Chiueh et al. [1994] applied 3D Hartley transform to extend the Joint Photographic Experts Group (JPEG) still image compression algorithm for the compression of subcubes of the volume, and performed frequency-domain rendering on the subcubes before compositing the resulting subimages in the spatial domain. Each of the 3D Fourier coefficients in each subcube is then quantized, linearly sequenced through a 3D zig zag order, and then entropy encoded. In this way, they alleviated the problem of lack of attenuation and occlusion in frequency-domain rendering while achieving high-compression ratios, fast rendering speed compared to spatial volume rendering, and improved image quality over conventional frequency-domain rendering techniques. Figure 60.3 shows a CT scan of a lobster that was rendered out of the compressed frequency domain.

Rooted in time-frequency analysis, wavelet theory [Daubechies 1992] has gained popularity in the recent years. A wavelet is a fast decaying function with zero averaging. The nice features of wavelets are that they have local property in both the spatial and the frequency domains, and can be used to fully represent the volumes with a small number of wavelet coefficients. Muraki [1993] first applied wavelet transform to volumetric data sets, Gross et al. [1995] found an approximate solution for the volume rendering equation using orthonormal wavelet functions, and Westermann [1994] combined volume rendering with wavelet-based compression. However, all of these algorithms have not focused on the acceleration of volume rendering using wavelets. The greater potential of wavelet domain, based on the elegant multiresolution hierarchy provided by the wavelet transform, is still far from fully utilized for volume rendering. A possible research and development challenge is to exploit the local frequency variance provided by wavelet transform and accelerate the volume rendering in homogeneous areas.

FIGURE 60.3 Compression domain volume rendering of a CT scan of a lobster.

60.5 Volume Rendering Optimizations

Volume rendering can produce informative images that can be useful in data analysis, but a major drawback of the techniques previously described is the time required to generate a high-quality image. In this section, several volume rendering optimizations are described that decrease rendering times, and therefore increase interactivity and productivity. Other optimizations have been discussed briefly earlier in the chapter, along with the original algorithms. Another way to speed up volume rendering is to employ special-purpose hardware accelerators for volume rendering, which are described in section 60.6.

Object-order volume rendering typically loops through the data, calculating the contribution of each volume sample to pixels on the image plane. This is a costly operation for moderate- to large-sized data sets (e.g., 128 megabytes for a 512^3 sample data set, with 1 byte per sample), leading to rendering times that are noninteractive. Viewing the intermediate results in the image plane may be useful, but these partial image results are not always representatives of the final image. For the purpose of interaction, it is useful to be able to generate a lower quality image in a shorter amount of time. For data sets with binary sample values, bits could be packed into bytes such that each byte represents a $2 \times 2 \times 2$ portion of the data [H. K. Tuy and L. T. Tuy 1984]. The data would be processed bit by bit to generate the full resolution image, but a lower resolution image could be generated by processing the data byte by byte. If more than 4 b of the byte are set, the byte is considered to represent an element of the object, otherwise it represents the background. This will produce an image with one-half the linear resolution in approximately one-eighth the time.

A more general method for decreasing data resolution is to build a pyramid data structure, which for an original data set of N^3 data samples, consists of a sequence of log N volumes. The first volume is the original data set, whereas the second volume is created by averaging each $2 \times 2 \times 2$ group of samples of the original data set to create a volume of one-eighth the resolution. The third volume is created from the second volume in a similar fashion, with this process continuing until all log N volumes have been created. An efficient implementation of the splatting algorithm, called hierarchical splatting [Laur and Hanrahan 1994], uses such a pyramid data structure. According to the desired image quality, this algorithm scans the appropriate level of the pyramid in a back-to-front order. Each element is splatted onto the image plane using the appropriate-sized splat. The splats themselves are approximated by polygons which can efficiently be rendered by graphics hardware.

Image-order volume rendering involves casting rays from the image plane into the data, and sampling along the ray in order to determine pixel values. The idea of pyramid can also be used here. Actually, Wang and Kaufman [1994b] have proposed the use of multiresolution hierarchy at arbitrary resolutions.

In discrete ray casting, the ray would be discretized, and the contribution from each voxel along the path is considered when producing the final pixel value. It would be quite computationally expensive to discretize every ray cast from the image plane. Fortunately, this is unnecessary for parallel projections. Since all the rays are parallel, one ray can be discretized into a 26-connected line and used as a template for all other rays. This technique, developed by Yagel and Kaufman [1992], is called *template-based volume rendering*. If this template were used to cast a ray from each pixel in the image plane, some voxels in the data may contribute to the image twice while others may not be considered at all. To solve this problem, the rays are cast instead from a *base plane*, that is, the plane of the volume buffer most parallel to the image plane. This ensures that each data sample can contribute at most once to the final image, and all data samples could potentially contribute. Once all of the rays have been cast from the base plane, a simple final step of resampling is needed, which uses bilinear interpolation to determine the pixel values on the image plane from the ray values that have been calculated on the base plane. This template-based ray casting can be extended to allow higher order interpolation [Yagel 1991]. The template for higher order interpolation consists of connected cells, as opposed to the connected voxel template used for zero-order interpolation. Another extension to template-based ray casting allows for screen space supersampling to improve image quality [Yagel 1991]. This is accomplished by allowing rays to originate at subpixel locations.

Lacroute and Levoy [1994] extended the previous ideas with enhancements to an algorithm called shear-warp factorization. It is based on an algorithm that factors the viewing transformation into a 3D shear parallel to the data slices, a projection to form an intermediate but distorted image, and a 2D warp to form an undistorted final image. The algorithm is extended in three ways. First, a fast object-order rendering algorithm based on the factorization algorithms with preprocessing and some loss of image quality has been developed. Shear-warp factorization has the property that rows of voxels in the volume are aligned with rows of pixels in the intermediate image. Consequently, a scan line-based algorithm has been constructed that traverses the volume and the intermediate image in synchrony, taking advantage of the spatial coherence present in both. Spatial data structures based on run-length encoding for both the volume and the intermediate image are used. An implementation running on an SGI Indigo workstation renders a low-resolution 256^3 voxel data set in 1 s. The second extension is shear-warp factorization for perspective viewing transformations. Third, a data structure for encoding spatial coherence in unclassified volumes (i.e., scalar fields with no precomputed opacity) has been introduced. When combined with the shear-warp rendering algorithm this data structure supports classification and rendering a 256^3 voxel volume in 3 s. The method extends to support mixed volumes and geometry and is parallelizable [Lacroute 1995].

One obvious optimization for both discrete and continuous ray casting which has already been discussed is to limit the sampling to the segment of the ray which intersects the data, since samples outside of the data evaluate to 0 and do not contribute to the pixel value. If the data itself contains many zero-valued data samples, or a segmentation function is applied to the data that evaluates to 0 for many samples, the efficiency of ray casting can be greatly enhanced by further limiting the segment of the ray in which samples are taken. One algorithm of this sort is known as polygon assisted ray casting (PARC) [Avila et al. 1992]. This algorithm approximates objects contained within a volume using a crude polyhedral representation. The polyhedral representation is created so that it completely contains the objects. Using conventional graphics hardware, the polygons are projected twice to create two Z-buffers. The first Z-buffer is the standard closest distance Z-buffer, whereas the second is a farthest distance Z-buffer. Since the object is completely contained within the representation, the two Z-buffer values for a given image plane pixel can be used as the starting and ending points of a ray segment on which samples are taken.

The PARC algorithm is part of the *VolVis* volume visualization system [Avila et al. 1992, 1994], which provides a multialgorithm progressive refinement approach for interactivity. By using available graphics hardware, the user is given the ability to interactively manipulate a polyhedral representation of the data.

When the user is satisfied with the placement of the data, light sources, and view, the Z-buffer information is passed to the PARC algorithm, which produces a ray cast image. In a final step, this image is further refined by continuing to follow the PARC rays, which intersected the data according to a volumetric ray tracing algorithm [Sobierajski and Kaufman 1994b] in order to generate shadows, reflections, and transparency (see section 60.7). The ray tracing algorithm uses various optimization techniques, including uniform space subdivision and bounding boxes, to increase the efficiency of the secondary rays. Surface rendering, as well as transparency with color and opacity transfer functions, is incorporated within a global illumination model.

60.6 Special-Purpose Volume Rendering Hardware

The high-computational cost of direct volume rendering makes it impossible for sequential implementations and general-purpose computers to deliver the targeted level of performance. This situation is aggravated by the continuing trend towards higher and higher resolution datasets. For example, to render a dataset of 1024^3 16-b voxels at 30 frames/s requires 2 gigabytes of storage, a memory transfer rate of 60 gigabytes and approximately 300 billion instructions per second, assuming 10 instructions per voxel per projection. To address this challenge, researchers have tried to achieve interactive display rates on supercomputers and massively parallel architectures (e.g., Schroder and Stoll [1992], Silva and Kaufman [1994], and Vezina et al. [1992]). However, most algorithms require very little repeated computation on each voxel and data movement actually accounts for a significant portion of the overall performance overhead. Today's commercial supercomputer memory systems do not have and will not have in the near future adequate latency and memory bandwidth for efficiently transferring the required large amounts of data. Furthermore, supercomputers seldom contain frame buffers and, due to their high cost, are frequently shared by many users.

The same way as the special requirements of traditional computer graphics lead to high-performance graphics engines, volume visualization naturally lends itself to special-purpose volume renderers that separate real-time image generation from general-purpose processing. This allows for standalone visualization environments that help scientists to interactively view their data on a single user workstation, either augmented by a volume rendering accelerator or connected to a dedicated visualization server. Furthermore, a volume rendering engine integrated in a graphics workstation is a natural extension of raster based systems into 3D volume visualization.

Several researchers have proposed special-purpose volume rendering architectures [Kaufman 1991, ch. 6]. More recent research focuses on accelerators for ray casting of regular datasets. Ray casting offers room for algorithmic improvements while still allowing for high-image quality. Recent architectures [Hesser et al. 1995] include VOGUE, VIRIM, and Cube.

VOGUE [Knittel and Strasser 1994], a modular add-on accelerator, is estimated to achieve 2.5 frames/s for 256^3 datasets. For each pixel a ray is defined by the host computer and sent to the accelerator. The VOGUE module autonomously processes the complete ray, consisting of evenly spaced resampling locations, and returns the final pixel color of that ray to the host. Seveal VOGUE modules can be combined to yield higher performance implementations. For example, to achieve 20 projections/s of 512^3 datasets requires 64 boards and a 5.2 gigabytes/s ring-connected cubic network.

VIRIM [Guenther et al. 1994] is a flexible and programmable ray casting engine. The hardware consists of two separate units, the first being responsible for 3D resampling of the volume using lookup tables to implement different interpolation schemes. The second unit performs the ray casting through the resampled dataset according to user programmable lighting and viewing parameters. The underlying ray casting model allows for arbitrary parallel and perspective projections and shadows. An existing hardware implementation for the visualization of $256 \times 256 \times 128$ datasets at 10 frames/s requires 16 processing boards.

The Cube project aims at the realization of high-performance volume rendering systems for large datasets and pioneered several hardware architectures. Cube-1, a first generation hardware prototype, was based on a specially interleaved memory organization [Kaufman and Bakalash 1988], which has also been used in all subsequent generations of the Cube architecture. This interleaving of the n^3 voxel enables conflict-free access to any ray parallel to a main axis of n voxels. A fully operational printed circuit board (PCB) implementation of Cube-1 is capable of generating orthographic projections of 16^3 datasets from a finite number of predetermined directions in real time. Cube-2 was a single-chip very large-scale integration (VLSI) implementation of this prototype [Bakalash et al. 1992].

To achieve higher performance and to further reduce the critical memory access bottleneck, Cube-3 introduced several new concepts [Pfister et al. 1994, Pfister, Wessels, and Kaufman 1995]. A high-speed global communication network aligns and distributes voxels from the memory to several parallel processing units and a circular cross-linked binary tree of voxel combination units composites all samples into the final pixel color. Estimated performance for arbitrary parallel and perspective projections is 30 frames/s for 512^3 datasets. Cube-4 [Pfister, Kaufman, and Wessels 1995] has only simple and local interconnections, thereby allowing for easy scalability of performance. Instead of processing individual rays, Cube-4 manipulates a group of rays at a time. As a result, the rendering pipeline is directly connected to the memory. Accumulating compositors replace the binary compositing tree. A pixel bus collects and aligns the pixel output from the compositors. Cube-4 is easily scalable to very high resolution of 1024^3 16-b voxels and true real-time performance implementations of 30 frames/s [Pfister and Kaufman 1996].

The choice of whether one adopts a general-purpose or a special-purpose solution to volume rendering depends upon the circumstances. If maximum flexibility is required, general purpose appears to be the best way to proceed. However, an important feature of graphics accelerators is that they are integrated into a much larger environment where software can shape the form of input and output data, thereby providing the additional flexibility that is needed. A good example is the relationship between the needs of conventional computer graphics and special-purpose graphics hardware. Nobody would dispute the necessity for polygon graphics acceleration despite its obvious limitations. The exact same argument can be made for special-purpose volume rendering architectures.

60.7 Volumetric Global Illumination

Speed and accuracy of the final image are both important, yet often conflicting aspects of the rendering process. For this reason, a comprehensive volume rendering system, such as VolVis, includes a range of rendering algorithms from the fast, rough approximation of the final image, to the comparatively slow, accurate rendering within a global illumination model. Also, every rendering algorithm should support several levels of accuracy, giving the user an even greater amount of control over the speed and accuracy of the final image.

Standard volume rendering techniques typically employ only a local illumination model for shading, and therefore produce images without global effects. Including a global illumination model within a visualization system has several advantages. First, global effects are often desirable in scientific applications. For example, by placing mirrors in the scene, a single image can show several views of an object in a natural, intuitive manner leading to a better understanding of the 3D nature of the scene. Also, complex geometric surfaces are often easier to render when represented volumetrically than when represented by high-order functions or geometric primitives, and global effects using ray tracing or radiosity are desirable for such applications called volume graphics applications (see section 60.9). Volumetric ray tracing is described in the next section and volumetric radiosity is discussed in the succeeding section.

Volumetric Ray Tracing

A 3D raster ray tracing (RRT) method, developed by Yagel et al. [1992], produces realistic images of volumetric data using a global illumination model. The RRT algorithm is a discrete ray tracing algorithm similar to the discrete ray casting algorithm described previously. Discrete primary rays are cast from the image plane, through the data to determine pixel values. Secondary rays are recursively spawned when a ray encounters a voxel belonging to an object in the data. To save time, the view-independent parts of the illumination equation can be precomputed and added to the voxel color, thereby avoiding the calculation of this quanity during the ray tracing. Also, 2 b per light source per voxel can be precomputed, indicating whether the light is definitely visible, possibly visible, or definitely invisible from that voxel. Shadow rays need only be cast during the ray tracing if the bits indicate that the light is possibly visible through a translucent object. Actually, all view-independent attributes (including normal, texture, and antialiasing) can be precomputed and stored in each voxel.

There are several advantages to using RRT instead of conventional ray tracing. One such advantage is that sampled or computed data, possibly intermixed with voxelized geometric data, can be ray traced directly without having to approximate the sampled data using geometric primitives. Another advan-

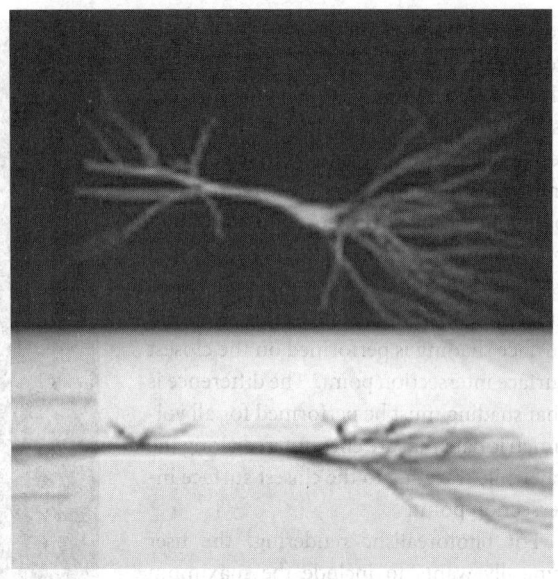

FIGURE 60.4 A maximum-value projection of a cell casting a shadow on the floor using the VolVis volumetric ray tracer.

tage is that there is only one primitive to deal with: the voxel, which greatly simplifies ray–object intersection calculations. Unlike conventional ray tracing that computes expensive continuous ray–object intersections, RRT traverses discrete rays through discrete data and therefore it is basically insensitive to scene complexity and object complexity. RRT is also very effective for ray tracing voxelized geometric models, such as constructive solid geometry (CSG) models. This is an example for the emerging field of volume graphics [Kaufman et al. 1993] in which geometric scenes are modeled using voxelized objects and efficiently rendered using a volume rendering algorithm such as RRT. Volume graphics is discussed in section 60.9.

A volumetric ray tracer [Sobierajski and Kaufman 1994b] is intended to produce much more accurate, informative images. In classical ray tracing, the rendering algorithm is designed to generate images that are accurate according to the laws of optics. A volumetric ray tracer should handle volumetric data as well as classical geometric objects, and strict adherence to the laws of optics is not always desirable. For example, a user may wish to generate an image with no shadows or to view the maximum value along the segment of a ray passing through a volume, instead of the optically correct composited value. Figure 60.4 illustrates the importance of including global effects in a maximum-value projection of a hippocampal pyramidal neuron data set which was obtained using a laser-scanning confocal microscope. Since maximum-value projections do not give depth information, a floor is placed below the cell, and a light source above the cell. This results in a shadow of the cell on the floor, adding back depth information lost by the maximum-value projection.

In order to incorporate both volumetric and geometric objects into one scene, the standard ray tracing intensity equation must be expanded. Since the illumination equation in classical ray tracing is evaluated only at surface locations, volumetric data cannot be incorporated into the scene. This problem can be solved by extending the standard illumination equation to include volumetric effects. The intensity of light, $I_\lambda(x, \omega)$, for a given wavelength λ arriving at a position x from the direction ω can be computed by:

$$I_\lambda(x, \omega) = I_{v\lambda}(x, x') + \tau_\lambda(x, x') I_{s\lambda}(x', \omega) \qquad (60.15)$$

where x' is the first surface intersection point encountered along the ray ω originating at x. $I_{s\lambda}(x', \omega)$ is the intensity of light at this surface location, and can be computed with a standard ray tracing illumination equation [Whitted 1980]. $I_{v\lambda}(x, x')$ is the volumetric contribution to the intensity along the ray from x to x', and $\tau_\lambda(x, x')$ is the attenuation of $I_{s\lambda}(x', \omega)$ by any intervening volumes. These values are determined using volume rendering techniques, based on a transport theory model of light propagation [Kruger 1991]. The basic idea is similar to classical ray tracing, in that rays are cast from the eye into the scene, and surface shading is performed on the closest surface intersection point. The difference is that shading must be performed for all volumetric data that are encountered along the ray while traveling to the closest surface intersection point.

For photorealistic rendering, the user typically wants to include the maximum amount of shading effects that can be calculated within a given time limit. For visualization, however, the user may find it necessary to view volumetric data with no shading effects, such as when using a maximum-value projection. For example,

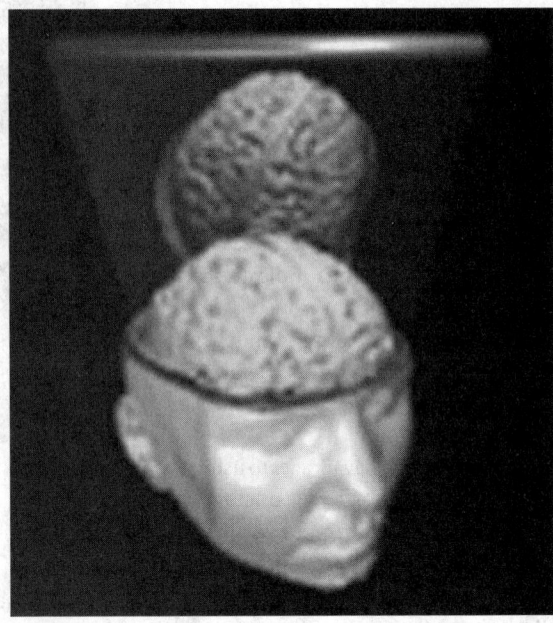

FIGURE 60.5 A ray traced image of a human head. Shadowing effects were not included in this image in order to produce a clear reflection in the mirror.

in Fig. 60.4 no shading effects were included for the maximum-value projection of the cell, while all parts of the illumination equation were considered when shading the geometric polygon. In another example, the user may place a mirror behind a volumetric object in a scene in order to capture two views in one image, but may not want the volumetric object to cast a shadow on the mirror. This can be accomplished easily by turning off the shadowing calculations for the mirror, as shown in Fig. 60.5. The head data was obtained using magnetic resonance imaging, with the brain segmented from the same dataset. The mirror is a voxelized polygon, which was created using the nonbinary voxelization technique described subsequently.

Volumetric Radiosity

The ray tracing algorithm described in the previous section can be used to capture specular interactions between objects in a scene. In reality, most scenes are dominated by diffuse interactions, which are not accounted for in the standard ray tracing illumination model, but accounted for by a radiosity algorithm for volumetric data [Sobierajski and Kaufman 1994a]. In volumetric radiosity, the basic *patch* element of classical radiosity is replaced by a voxel. As opposed to previous methods that use participating media to augment geometric scenes [Hottel and Sarofim 1967, Rushmeier and Torrance 1987], this method moves the radiosity equations into volumetric space, and renders scenes consisting solely of volumetric data. Each voxel can emit, absorb, scatter, reflect, and transmit light. Both *isotropic* and *diffuse* emission of light are allowed, where isotropic implies directional independence, and diffuse implies Lambertian reflection (i.e., dependent on normal or gradient). Light is scattered isotropically, and is reflected diffusely by a voxel. Light entering a voxel that is not absorbed, scattered, or reflected by the voxel is transmitted unchanged.

In order to cope with the high number of voxel interactions required, a hierarchical technique similar to Hanrahan et al. [1991] can be used. The basic hierarchical concept is that the radiosity contribution from some voxel v_i to another voxel v_j is similar to the radiosity contribution from v_i to v_k if the distance between v_j and v_k is small, and the distance between v_i and v_j is large. For each volume, a hierarchical radiosity structure is built by combining each subvolume of eight voxels at one level to form one voxel at the next higher level. An iterative algorithm [Cohen et al. 1988] is then used to shoot voxel radiosities, where several factors govern the highest level in the hierarchy at which two voxels can interact. These factors include the distance between the two voxels, the radiosity of the shooting voxel, and the reflectance and scattering coefficients of the voxel receiving the radiosity. This hierarchical technique can reduce the number of interactions required to converge on a solution by more than four orders of magnitude.

After the view-independent radiosities have been calculated, a view-dependent image is generated using a ray casting technique, where the final pixel value is determined by compositing radiosity values along the ray. Figure 60.6 shows a scene containing a volumetric sphere, polygon, and light source. The light source is isotropically emitting light, and both the sphere and the polygon

FIGURE 60.6 A volumetric radiosity projection of a voxelized sphere and polygon.

are diffusely reflecting light. The light source is above the sphere, and therefore the top-half of the sphere is directly illuminated. The bottom-half of the sphere is indirectly illuminated by light diffusely reflected from the red polygon.

60.8 Irregular Grid Rendering

All of the algorithms previously discussed handle only regular gridded data. Irregular gridded data comes in a large variety [Speray and Kennon 1990], including curvilinear data or unstructured (scattered) data, where no explicit connectivity is defined between cells (one can even be given a scattered collection of points that can be turned into an irregular grid by interpolation [Max 1990]). For rendering purposes, manifold (locally homeomorphic to R^3) grids composed of convex cells are usually necessary. In general, the most convenient grids for rendering purposes are tetrahedral grids and hexahedral grids. One disadvantage of hexahedral grids is that the four points on the side of a cell may not necessarily lie on a plane forcing the rendering algorithm to approximate the cells by convex ones during rendering. Tetrahedral grids have several advantages, including easier interpolation, simple representation (especially for connectivity information because the degree of the connectivity graph is bounded, allowing for compact data structure representation), and the fact that any other grid can be interpolated to a tetrahedral one (with the possible introduction of Steiner points). Among their disadvantages is the fact that the size of the datasets tend to grow as cells are decomposed into tetrahedra. In the case of curvilinear grids, an accurate (and naive) decomposition will make the cell complex contain five times as many cells.

As compared to regular grids, operations for irregular grids are more complicated and the effective visualization methods are more sophisticated in all fronts. Shading, interpolation, point location, etc., are all harder (and some even not well defined) for irregular grids. One notable exception is isosurface generation [Lorensen and Cline 1987], that even in the case of irregular grids is fairly simple to compute given suitable interpolation functions. Slicing operations are also simple [Speray and Kennon 1990].

Volume rendering irregular grids is a hard operation, and there are several different approaches to this problem. The simplest and most inefficient is to resample the irregular grid to a regular grid. In order to achieve the necessary accuracy, a high enough sampling rate has to be used that in most cases will make the resulting regular grid volume too large for storage and rendering purposes, not mentioning the time to perform the resampling.

Extending the simple volumetric point sampling ray tracing to irregular grids is a challenge. For ray tracing, it is necessary to depth-sort samples along rays emanating from each screen pixel. In the case of irregular grids, it is not trivial to perform this sorting operation. Garrity [1990] proposed a scheme where the cells are convex and connectivity information is available. He proposed to preprocess the grid, finding the external cells to help locate the boundary elements during ray tracing, and using the connectivity information for cell skipping. The actual resampling and shading that is simple in the regular grid is not trivial here and has to be carefully considered, usually taking into account the specific application at hand (actually the development of accurate illumination models for volume rendering has irregular grid rendering as one of its main uses [Max 1995]). Simple ray casting is too inefficient, as there is a large amount of interpixel and interscan line coherency in ray casting. Giertsen [1992] proposed a sweep-plane approach to ray casting that uses different forms of *caching* to speed up ray casting irregular grids.

Another approach for rendering irregular grids is the use of feed forward methods, where the cells are projected onto the screen one by one accumulating their contributions incrementally to the final image (e.g., Max et al. [1990], Wilhems and van Gelder [1991], and Williams [1992]). One major advantage of these methods is the ability to use the graphics hardware on graphics workstations to compute the volumetric lighting models (usually simplified) in order to speed up rendering. Another advantage is that the user can see the rendering as it progresses. One problem with this method is generating the ordering for the cell projections. In general, such ordering does not even exist and cells have to be partitioned into multiple cells for projection. The partitioning is (in general) view dependent, but some types of irregular grids (like Delaunay triangulations in space) are acyclic and do not need any partitioning.

60.9 Volume Graphics

The 3D raster representation seems to be more natural for empirical imagery than for geometric objects, due to its ability to represent interiors and digital samples. Nonetheless, the advantages of this representation are also attracting traditional surface-based applications that deal with the modeling and rendering of synthetic scenes made out of geometric models. The geometric model is *voxelized (3D scan converted)* into a set of voxels that best approximate the model. Each of these voxels is then stored in the volume buffer together with the voxel precomputed view-independent attributes. The voxelized model can be either binary (see Kaufman [1987a, 1987b], Kaufman and Shimong [1986], and Kaufman [1991, pp. 280–301]) or volume sampled [Wang and Kaufman 1993], which generates alias-free density voxelization of the model. Some surface-based application examples are the rendering of fractals [Hart et al. 1989], hypertextures [Perlin and Hoffert 1989], fur [Kajiya and Kay 1989], gases [Ebert and Parent 1990], and other complex models, including CAD models [Wang and Kaufman 1994b, 1995] and terrain models for flight simulators [Cohen and Shaked 1993, Kaufman et al. 1994, Wright and Hsieh 1992]. Furthermore, in many applications involving sampled data, such as medical imaging, the data need to be visualized along with synthetic objects that may not be available in digital form, such as scalpels, prosthetic devices, injection needles, radiation beams, and isodose surfaces. These geometric objects can be voxelized and intermixed with the sampled organ in the volume buffer [Kaufman et al. 1990].

Volume graphics [Kaufman et al. 1993], which is an emerging subfield of computer graphics, is concerned with the synthesis, modeling, manipulation, and rendering of volumetric geometric objects, stored in a volume buffer of voxels. Unlike volume visualization which focuses primarily on sampled and computed datasets, volume graphics is concerned primarily with modeled geometric scenes and commonly with those that are represented in a regular volume buffer. As an approach, volume graphics has the potential to greatly advance the field of 3D graphics by offering a comprehensive alternative to traditional surface graphics.

In the next subsections we describe the volumetric approach to several common volume graphics modeling techniques. We describe the generation of object primitives (voxelization), 3D antialiasing, texture and photo mapping, solid-texturing, modeling of amorphous phenomena, modeling by block operations, constructive solid modeling, and volume sculpting. Then, volume graphics is contrasted with surface graphics, and the corresponding advantages are discussed.

Voxelization

An indispensable stage in volume graphics is the synthesis of voxel-represented objects from their geometric representation. This stage, which is called *voxelization*, is concerned with converting geometric objects from their continuous geometric representation into a set of voxels that "best" approximates the continuous object. As this process mimics the scan-conversion process that pixelizes (rasterizes) 2D geometric objects, it is also referred to as *3D scan conversion*. In 2D rasterization the pixels are directly drawn onto the screen to be visualized and filtering is applied to reduce the aliasing artifacts. However, the voxelization process does not render the voxels but merely generates a database of the discrete digitization of the continuous object.

Intuitively, one would assume that a proper voxelization simply selects all voxels which are met (if only partially) by the object body. Although this approach could be satisfactory in some cases, the objects it generates are commonly too coarse and include more voxels than are necessary. For example, when a 2D curve is rasterized into a connected sequence of pixels, the discrete curve does not cover the entire continuous curve, but it is connected and concisely and successfully separates both sides of the curve [Cohen-Or and Kaufman 1995].

One practical meaning of separation is apparent when a voxelized scene is rendered by casting discrete rays from the image plane to the scene. The penetration of the background voxels (which simulate the discrete ray traversal) through the voxelized surface causes the appearance of a hole in the final image of the rendered surface. Another type of error might occur when a 3D flooding algorithm is employed either to fill an object or to measure its volume, surface area, or other properties. In this case the nonseparability of the surface causes a leakage of the flood through the discrete surface.

Unfortunately, the extension of the 2D definition of separation to the third dimension and to voxel surfaces is not straightforward since voxelized surfaces cannot be defined as an ordered sequence of voxels and a voxel on the surface does not have a specific number of adjacent surface voxels. Furthermore, there are important topological issues, such as the separation of both sides of a surface, which cannot be well defined by employing 2D terminology. The theory that deals with these topological issues is called *3D discrete topology*. We sketch some basic notions and informal definitions used in this field next.

Fundamentals of Three-Dimensional Discrete Topology

The 3D discrete space is a set of integral grid points in 3D Euclidean space defined by their Cartesian coordinates (x, y, z). A voxel is the unit cubic volume centered at the integral grid point. The voxel value is mapped onto $\{0, 1\}$: the voxels assigned 1 are called the black voxels representing opaque objects, and those assigned 0 are the white voxels representing the transparent background. In the 3D antialiasing subsection we describe nonbinary approaches where the voxel value is mapped onto the interval $[0,1]$ representing either partial coverage, variable densities, or graded opacities. Because of its larger dynamic range of values, this approach supports 3D antialiasing and thus supports higher quality rendering.

Two voxels are 26 *adjacent* if they share either a vertex, an edge, or a face (see Fig. 60.1). Every voxel has 26 such adjacent voxels: 8 share a vertex (corner) with a center voxel, 12 share an edge, and 6 share a face. Accordingly, face-sharing voxels are defined as 6 *adjacent,* and edge-sharing and face-sharing voxels are defined as 18 *adjacent.* The prefix N is used to define the adjacency relation, where $N = 6, 18$, or 26. A sequence of voxels having the same value (e.g., black) is called an N-*path* if all consecutive pairs are N-adjacent. A set of voxels W is N-*connected* if there is an N-path between every pair of voxels in W. An N-*connected component* is a maximal N-connected set.

Given a 2D discrete 8 connected black curve, there are sequences of 8 connected white pixels (8 component) that pass from one side of the black component to its other side without intersecting it. This phenomenon is a discrete disagreement with the continuous case where there is no way of penetrating a closed curve without intersecting it. To avoid such a scenario, it has been the convention to define *opposite* types of connectivity for the white and black sets. Opposite types in 2D space are 4 and 8, whereas in 3D space 6 is opposite to 26 or to 18.

Assume that a voxel space, denoted by Σ, includes one subset of black voxels S. If $\Sigma - S$ is not N-connected, that is, $\Sigma - S$ consists of at least two white N-connected components, then S is said to be

N-separating in Σ. Loosely speaking, in 2D, an 8 connected black path that divides the white pixels into two groups is 4 separating, and a 4 connected black path that divides the white pixels into two groups is 8 separating. There are no analogous results in 3D space.

Let W be an N-separating surface. A voxel $p \in W$ is said to be an *N-simple voxel* if $W - p$ is still N-separating. An N-separating surface is called *N-minimal* if it does not contain any N-simple voxel. A *cover* of a continuous surface is a set of voxels such that every point of the continuous surface lies in a voxel of the cover. A cover is said to be a *minimal cover* if none of its subsets is also a cover. The cover property is essential in applications that employ space subdivision for fast ray tracing [Glassner 1984]. The subspaces (voxels) which contain objects have to be identified along the traced ray. Note that a cover is not necessarily separating, whereas on the other hand, as previously mentioned, it may include simple voxels. In fact, even a minimal cover is not necessarily N-minimal for any N [Cohen-Or and Kaufman 1995].

Binary Voxelization

An early technique for the digitization of solids was spatial enumeration, which employs point or cell classification methods in either an exhaustive fashion or by recursive subdivision [Lee and Requicha 1982]. However, subdivision techniques for model decomposition into rectangular subspaces are computationally expensive and thus inappropriate for medium- or high-resolution grids. Instead, objects should be directly voxelized, preferably generating an N-separating, N-minimal, and covering set, where N is application dependent. The voxelization algorithms should follow the same paradigm as the 2D scan-conversion algorithms; they should be incremental, accurate, use simple arithmetic (preferably integer only), and have a complexity that is not more than linear with the number of voxels generated.

The literature of 3D scan conversion is relatively small. Danielsson [1970] and Mokrzycki [1988] developed independently similar 3D curve algorithms where the curve is defined by the intersection of two implicit surfaces. Voxelization algorithms have been developed for 3D lines, 3D circles, and a variety of surfaces and solids, including polygons, polyhedra, and quadric objects [Kaufman and Shimony 1986]. Efficient algorithms have been developed for voxelizing polygons using an integer-based decision mechanism embedded within a scan-line filling algorithm [Kaufman 1987a], for parametric curves, surfaces, and volumes using an integer-based forward differencing technique [Kaufman 1987b], and for quadric objects such as cylinders, spheres, and cones using weaving algorithms by which a discrete circle/line sweeps along a discrete circle/line [Kaufman 1991, pp. 280–301]. Figure 60.7 consists of a variety of objects (polygons, boxes, cylinders) voxelized using these methods. These pioneering attempts should now be followed by enhanced voxelization algorithms that, in addition to being efficient and accurate, will also adhere to the topological requirements of separation, coverage, and minimality.

Three-Dimensional Antialiasing

The previous subsection discussed binary voxelization, which generates topologically and geometrically consistent models, but exhibits object space aliasing. These algorithms have used a straightforward method of sampling in space, called *point sampling*. In point sampling, the continuous object is evaluated at the voxel center, and the value of 0 or 1 is assigned to the voxel. Because of this binary classification of the voxels, the resolution of the 3D raster ultimately determines the precision of the discrete model. Imprecise modeling results in jagged surfaces, known as *object-space aliasing* (see Fig. 60.7). In this section, a 3D object-space antialiasing technique is presented. It performs antialiasing once, on a 3D view-independent representation, as part of the modeling stage. Unlike antialiasing of 2D scan-converted graphics, where the main focus is on generating aesthetically pleasing displays, the emphasis in antialiased 3D voxelization is on producing alias-free 3D models that are stored in the view-independent volume buffer for various volume graphics manipulations, including but not limited to the generation of aesthetically pleasing displays.

To reduce object-space aliasing, a *volume sampling* technique has been developed [Wang and Kaufman 1993], which estimates the density contribution of the geometric objects to the voxels. The density of a

FIGURE 60.7 A volumetric model of terrain enhanced with photo mapping of satellite images. The buildings are synthetic voxel models raised on top of the terrain. The voxelized terrain has been mapped with aerial photos during the voxelization stage.

voxel is attenuated by a filter weight function which is proportional to the distance between the center of the voxel and the geometric primitive. To improve performance, precomputed lookup tables of densities for a predefined set of geometric primitives can be used to select the density value of each voxel. For each voxel visited by the binary voxelization algorithm, the distance to the predefined primitive is used as an index into a lookup table of densities.

Since the voxelized geometric objects are represented as volume rasters of density values, they can essentially be treated as sampled or simulated volume datasets, such as 3D medical imaging datasets, and one of many volume rendering techniques for image generation can be employed. One primary advantage of this approach is that volume rendering or volumetric global illumination carries the smoothness of the volume-sampled objects from object space over into its 2D projection in image space [Wang and Kaufman 1994a]. Hence, the silhouette of the objects, reflections, and shadows are smooth. Furthermore, by not performing any geometric ray-object intersections or geometric surface normal calculations, the bulk of the rendering time is saved. In addition, CSG operations between two volume-sampled geometric models are accomplished at the voxel level after voxelization, thereby reducing the original problem of evaluating a CSG tree of such operations down to a fuzzy Boolean operation between pairs of nonbinary voxels [Wang and Kaufman 1994b]. (See section on block operations and constructive solid modeling.) Volume-sampled models are also suitable for intermixing with sampled or simulated datasets, since they can be treated uniformly as one common data representation. Furthermore, volume-sampled models lend themselves to alias-free multiresolution hierarchy construction [Wang and Kaufman 1994b].

Texture Mapping

One type of object complexity involves objects that are enhanced with texture mapping, photo mapping, environment mapping, or solid texturing. Texture mapping is commonly implemented during the last stage of the rendering pipeline, and its complexity is proportional to the object complexity. In volume graphics, however, texture mapping is performed during the voxelization stage, and the texture color is stored in each voxel in the volume buffer.

In photo mapping, six orthogonal photographs of the real object are projected back onto the voxelized object. Once this mapping is applied, it is stored with the voxels themselves during the voxelization stage, and therefore does not degrade the rendering performance. Texture and photo mapping are also viewpoint-independent attributes implying that once the texture is stored as part of the voxel value, texture mapping need not be repeated. This important feature is exploited, for example, by voxel-based flight simulators (see Fig. 60.7) and in CAD systems (see Fig. 60.8).

A central feature of volumetric representation is that unlike surface representation it is capable of representing inner structures of objects, which can be revealed and explored with appropriate manipulation and rendering techniques. This capability is essential for the exploration of sampled or computed objects. Synthetic objects are also likely to be solid rather than hollow. One method for modeling various solid types is solid texturing, in which a function or a 3D map models the color of the objects in 3D (see Fig. 60.8). During the voxelization phase each voxel be-

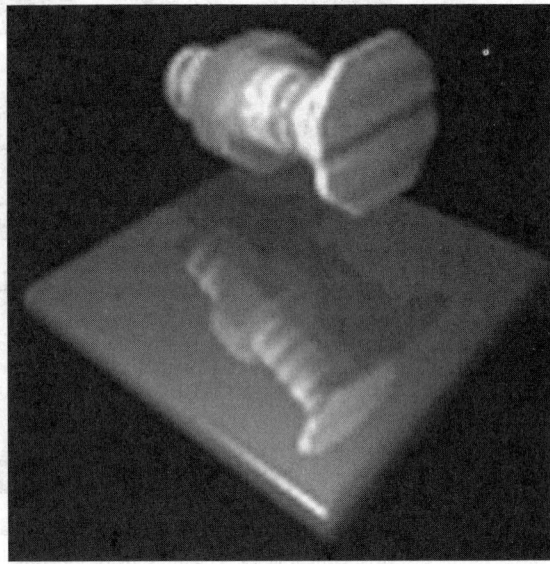

FIGURE 60.8   Volume-sampled bolt and nut generated by a sequence of CSG operations on hexagonal, cylindrical, and helix primitives, reflected on a volume-sampled mirror.

longing to the objects is assigned a value by the texturing function or the 3D map. This value is then stored as part of the voxel information. Again, since this value is view independent, it does not have to be recomputed for every change in the rendering parameters.

Amorphous Phenomena

Although translucent objects can be represented by surface methods, these methods cannot efficiently support the modeling and rendering of amorphous phenomena (e.g., clouds, fire, smoke) that are volumetric in nature and lack any tangible surfaces. A common modeling and rendering approach is based on a function that for any input point in 3D calculates some object features such as density, reflectivity, or color. These functions can then be rendered by ray casting, which casts a ray from each pixel into the function domain. Along the passage of the ray, at constant intervals the function is evaluated to yield a sample. All samples along each ray are combined to form the pixel color. Some examples for the use of this or similar techniques are the rendering of fractals [Hart et al. 1989], hypertextures [Perlin and Hoffert 1989], fur [Kajiya and Kay 1989], and gases [Ebert and Parent 1990].

The process of function evaluation at each sample point in 3D has to be repeated for each image generated. In contrast, the volumetric approach allows the precomputation of these functions at each grid point of the volume buffer. The resulting volumetric dataset can then be rendered from multiple viewpoints without recomputing the modeling function. As in other volume graphics techniques, accuracy is traded for speed, due to the resolution limit. Instead of accurately computing the function at each sample point, some type of interpolation from the precomputed grid values is employed.

Block Operations and Constructive Solid Modeling

The presortedness of the volume buffer naturally lends itself to grouping operations that can be exploited in various ways. For example, by generating multiresolution volume hierarchy, time critical and space critical volume graphics, applications can be better supported. The basic idea is similar to that of level-of-detail surface rendering which has proliferated recently (e.g., Eck et al. [1995], Rossignac and Borrel [1993], and Schroeder et al. [1992]), in which the perceptual importance of a given object in the scene determines its appropriate level-of-detail representation. One simple approach is the 3D *mip-map* approach [Levoy and Whitaker 1990, Sakas and Hartig 1992], where every level of the hierarchy is formed by averaging 8 voxels from the previous level. A better approach is based on sampling theory, in which an object is modeled with a sequence of alias-free volume buffers at different resolutions using the volume-sampled voxelization approach [He et al. 1995]. To accomplish this, high frequencies that exceed the Nyquist frequency of the corresponding volume buffer are filtered out by applying an ideal low-pass filter (*sinc*) with infinite support. In practice, the ideal filter is approximated by filters with finite support. Low-sampling resolution of the volume buffer corresponds to a lower Nyquist frequency, and therefore requires a low-pass filter with wider support for good approximation. As one moves up the hierarchy, low-pass filters with wider and wider support are applied. Compared to the level-of-detail hierarchy in surface graphics, the multiresolution volume buffers are easy to generate and to spatially correspond neighboring levels, and are free of object-space aliasing. Furthermore, arbitrary resolutions can be generated, and errors caused by a nonideal filters do not propagate and accumulate from level to level. Depending on the required speed and accuracy, a variety of low-pass filters (zero order, cubic, Gaussian) can be applied.

An intrinsic characteristic of the volume buffer is that adjacent objects in the scene are also represented by neighboring memory cells. Therefore, rasters lend themselves to various meaningful grouping-based operations, such as *bitblt* in 2D, or *voxblt* in 3D [Kaufman 1992]. These include transfer of volume buffer rectangular blocks (cuboids) while supporting voxel-by-voxel operations between source and destination blocks. Block operations add a variety of modeling capabilities which aid in the task of image synthesis and form the basis for the efficient implementation of a 3D *room manager*, which is the extension of window management to the third dimension.

Since the volume buffer lends itself to Boolean operations that can be performed on a voxel-by-voxel basis during the voxelization stage, it is advantageous to use CSG as the modeling paradigm. Subtraction, union, and intersection operations between two voxelized objects are accomplished at the voxel level, thereby reducing the original problem of evaluating a CSG tree during rendering time down to a 1D Boolean operation between pairs of voxels during a preprocessing stage.

For two point-sampled binary objects the Boolean operations of CSG or voxblt are trivially defined. However, the Boolean operations applied to volume-sampled models are analogous to those of fuzzy set theory (cf. Dubois and Prade [1980]). The volume-sampled model is a density function $d(x)$ over R^3, where d is 1 inside the object, 0 outside the object, and $0 < d < 1$ within the *soft* region of the filtered surface. Some of the common operations, intersection, complement, difference, and union, between two objects A and B are defined as follows:

$$d_{A \cap B}(x) \equiv \min(d_A(x), d_B(x)) \tag{60.16}$$

$$d_{\bar{A}}(x) \equiv 1 - d_A(x) \tag{60.17}$$

$$d_{A-B}(x) \equiv \min(d_A(x), 1 - d_B(x)) \tag{60.18}$$

$$d_{A \cup B}(x) \equiv \max(d_A(x), d_B(x)) \tag{60.19}$$

The law of set theory no longer true is the excluded-middle law (i.e. $A \cap \bar{A} \neq \phi$ and $A \cup \bar{A} \neq Universe$). The use of the min and max functions causes discontinuity at the region where the soft regions of the two objects meet, since the density value at each location in the region is determined solely by one of the two overlapping objects.

Complex geometric models can be generated by performing the CSG operations in Eqs. (60.16–60.19) between volume-sampled primitives. Volume-sampled models can also function as matte volumes [Drebin et al. 1988] for various matting operations, such as performing cut aways and merging multiple volumes into a single volume using the union operation. However, in order to preserve continuity on the cut-away boundaries between the material and the empty space, one should use an alternative set of Boolean operators based on algebraic sum and algebraic product [Dubois and Prade 1980, Goodman and

Sequin 1986]:

$$d_{A\cap B}(x) \equiv d_A(x)d_B(x) \tag{60.20}$$

$$d_{\bar{A}}(x) \equiv 1 - d_A(x) \tag{60.21}$$

$$d_{A-B}(x) \equiv d_A(x) - d_A(x)d_B(x) \tag{60.22}$$

$$d_{A\cup B}(x) \equiv d_A(x) + d_B(x) - d_A(x)d_B(x) \tag{60.23}$$

Unlike the min and max operators, algebraic sum and product operators result in $A \cup A \neq A$, which is undesirable. A consequence, for example, is that during modeling using sweeping, the resulting model is sensitive to the sampling rate of the swept path [Wang and Kaufman 1994b].

Once a CSG model has been constructed in voxel representation, it is rendered in the same way any other volume buffer is. This makes, for example, volumetric ray tracing of constructive solid models straightforward [Sobierajski and Kaufman 1994b]. (See Fig. 60.8.)

Volume Sculpting

Surface-based sculpting has been studied extensively (e.g., Coquillart [1990]), whereas volume sculpting has been recently introduced for clay or waxlike sculptures [Galyean and Hughes 1991] and for comprehensive detailed sculpting [Wang and Kaufman 1995]. The latter approach is a free-form interactive modeling technique based on the metaphor of sculpting and painting a voxel-based solid material, such as a block of marble or wood. There are two motivations for this approach. First, modeling topologically complex and highly detailed objects is still difficult in most CAD systems. Second, sculpting has been shown to be useful in volumetric applications. For example, scientists and physicians often need to explore the inner structures of their simulated or sampled datasets by gradually removing material.

Real-time human interaction could be achieved in this approach, since the actions of sculpting (e.g., carving, sawing) and painting are localized in the volume buffer, a localized rendering can be employed to reproject only those pixels that are affected. Carving is the process of taking a pre-existing volume-sampled tool to chip or chisel the object bit by bit. Since both the object and the tool are represented as independent volume buffers, the process of sculpting involves positioning the tool with respect to the object and performing a Boolean subtraction between the two volumes. Sawing is the process of removing a whole chunk of material at once, much like a carpenter sawing off a portion of a wood piece. Unlike carving, sawing requires generating the volume-sampled tool on-the-fly, using a user interface. To prevent object space aliasing and to achieve interactive speed, 3D splatting is employed.

Surface Graphics Versus Volume Graphics

Contemporary 3D graphics has been employing an object-based approach at the expense of maintaining and manipulating a display list of geometric objects and regenerating the frame-buffer after every change in the scene or viewing parameters. This approach, termed *surface graphics*, is supported by powerful geometry engines, which have flourished in the past decade, making surface graphics the state of the art in 3D graphics.

Surface graphics strikingly resembles vector graphics that prevailed in the 1960s and 1970s, and employed vector drawing devices. Like vector graphics, surface graphics represents the scene as a set of geometric primitives kept in a display list. In surface graphics, these primitives are transformed, mapped to screen coordinates, and converted by scan-conversion algorithms into a discrete set of pixels. Any change to the scene, viewing parameters, or shading parameters requires the image generation system to repeat this process. Like vector graphics that did not support painting the interior of 2D objects, surface graphics generates merely the surfaces of 3D objects and does not support the rendering of their interior.

Instead of a list of geometric objects maintained by surface graphics, volume graphics employs a 3D volume buffer as a medium for the representation and manipulation of 3D scenes. A 3D scene is discretized earlier in the image generation sequence, and the resulting 3D discrete form is used as a database of the scene for manipulation and rendering purposes, which in effect decouples discretization from rendering. Furthermore, all objects are converted into one uniform metaobject, the voxel. Each voxel is atomic and represents the information about at most one object that resides in that voxel.

Volume graphics offers similar benefits to surface graphics, with several advantages that are due to the decoupling, uniformity, and atomicity features. The rendering phase is viewpoint independent and insensitive to scene complexity and object complexity. It supports Boolean and block operations and constructive solid modeling. When 3D

TABLE 60.1 Comparison Between Vector Graphics and Raster Graphics and Between Surface Graphics and Volume Graphics

2D	Vector Graphics	Raster Graphics
Scene/object complexity	−	+
Block operations	−	+
Sampled data	−	+
Interior	−	+
Memory and processing	+	−
Aliasing	+	−
Transformations	+	−
Objects	+	−
3D	Surface Graphics	Volume Graphics

sampled or simulated data are used, such as that generated by medical scanners (e.g., CT, MRI) or scientific simulations (e.g., computational fluid dynamics (CFD)), volume graphics is suitable for their representation, too. It is capable of representing amorphous phenomena and both the interior and exterior of 3D objects. These features of volume graphics as compared with surface graphics are discussed in detail in the next subsection. Several weaknesses of volume graphics are related to the discrete nature of the representation, for instance, transformations and shading are performed in discrete space. In addition, this approach requires substantial amounts of storage space and specialized processing. These weaknesses are discussed in detail subsequently.

Table 60.1 contrasts vector graphics with raster graphics. A primary appeal of raster graphics is that it decouples image generation from screen refresh, thus making the refresh task insensitive to the scene and object complexities. In addition, the raster representation lends itself to block operations, such as bitblt and quadtree. Raster graphics is also suitable for displaying 2D sampled digital images, and thus provides the ideal environment for mixing digital images with synthetic graphic. Unlike vector graphics, raster graphics provides the capability to present shaded and textured surfaces, as well as line drawings. These advantages, coupled with advances in hardware and the development of antialiasing methods, have led raster graphics to supersede vector graphics as the primary technology for computer graphics. The main weaknesses of raster graphics are the large memory and processing power it requires for the frame buffer and the discrete nature of the image. These difficulties delayed the full acceptance of raster graphics until the late 1970s when the technology was able to provide cheaper and faster memory and hardware to support the demands of the raster approach. In addition, the discrete nature of rasters makes them less suitable for geometric operations such as transformations and accurate measurements, and once discretized the notion of objects is lost.

The same appeal that drove the evolution of the computer graphics world from vector graphics to raster graphics, once the memory and processing power became available, is driving a variety of applications from a surface-based approach to a volume-based approach. Naturally, this trend first appeared in applications involving sampled or computed 3D data, such as 3D medical imaging and scientific visualization, in which the datasets are in volumetric form. These diverse empirical applications of volume visualization still provide a major driving force for advances in volume graphics.

The comparison in Table 60.1 between vector graphics and raster graphics strikingly resembles a comparison between surface graphics and volume graphics. Actually, Table 60.1 itself is used also to contrast surface graphics and volume graphics. The next subsection discusses the features of volume graphics whereas the succeeding one discusses the weaknesses of volume graphics relative to surface graphics.

Volume Graphics Features

One of the most appealing attributes of volume graphics is its insensitivity to the complexity of the scene, since all objects have been preconverted into a finite-size volume buffer. Although the performance of the preprocessing voxelization phase is influenced by the scene complexity [Kaufman and Shimony 1986, Kaufman 1987b, Kaufman 1991, pp. 280–301], rendering performance depends mainly on the constant resolution of the volume buffer and not on the number of objects in the scene. Insensitivity to the scene complexity makes the volumetric approach especially attractive for scenes consisting of a large number of objects.

In volume graphics, rendering is decoupled from voxelization and all objects are first converted into one metaobject, the voxel, which makes the rendering process insensitive to the complexity of the objects. Thus, volume graphics is particularly attractive for objects that are hard to render using conventional graphics systems. Examples of such objects include curved surfaces of high order and fractals which require the expensive computation of an iterative function for each volume unit [Norton 1982]. Constructive solid models are also hard to render by conventional methods, but are straightforward to render in volumetric representation (discussed subsequently).

Antialiasing and texture mapping are commonly implemented during the last stage of the conventional rendering pipeline, and their complexity is proportional to object complexity. Solid texturing, which employs a 3D texture image, has also a high complexity proportional to object complexity. In volume graphics, however, antialiasing, texture mapping, and solid texturing are performed only once, during the voxelization stage, where the color is calculated and stored in each voxel. The texture can also be stored as a separate volumetric entity which is rendered together with the volumetric object, as in the VolVis software system for volume visualization [Avila et al. 1992].

The textured objects in Figs. 60.7 and 60.8 have been assigned texture during the voxelization stage by mapping each voxel back to the corresponding value on a texture map or solid. Once this mapping is applied, it is stored with the voxels themselves during the voxelization stage, which does not degrade the rendering performance. In addition, texture mapping and photo mapping are also viewpoint-independent attributes, implying that once the texture is stored as part of the voxel value, texture mapping need not be repeated.

In anticipation of repeated access to the volume buffer (such as in animation), all viewpoint-independent attributes can be precomputed during the voxelization stage, stored with the voxel, and be readily accessible for speeding up the rendering. The voxelization algorithm can generate for each object voxel its color, its texture color, its normal vector (for visible voxels), antialiasing information [Wang and Kaufman 1993], and information concerning the visibility of the light sources from that voxel. Actually, the viewpoint-independent parts of the illumination equation, can also be precomputed and stored as part of the voxel value.

Once a volume buffer with precomputed view-independent attributes is available, a rendering algorithm such as a discrete ray tracing or a volumetric ray tracing algorithm can be engaged. Either ray tracing approach is especially attractive for complex surface scenes and constructive solid models, as well as 3D sampled or computed datasets (discussed subsequently). Figures 60.4, 60.5, and 60.8 show examples of objects that were ray traced in discrete voxel space. In spite of the complexity of these scenes, volumetric ray tracing time was approximately the same as for much simpler scenes and significantly faster than traditional space-subdivision ray tracing methods. Moreover, in spite of the discrete nature of the volume buffer representation, images indistinguishable from the ones produced by conventional surface-based ray tracing can be generated by employing accurate ray tracing, auxiliary object information, or screen supersampling techniques.

Sampled datasets, such as in 3D medical imaging (see Figs. 60.5 and 60.9), volume microscopy (see Figs. 60.2 and 60.4), and geology, and simulated datasets, such as in computational fluid dynamics, chemistry, and materials simulation are often reconstructed from the acquired sampled or simulated points into a regular grid of voxels and stored in a volume buffer. Such datasets provide for the majority of applications using the volumetric approach. Unlike surface graphics, volume graphics naturally and directly supports the representation, manipulation, and rendering of such datasets, as well as provides the volume buffer medium for intermixing sampled or simulated datasets with geometric objects [Kaufman et al. 1990], as can be seen in Figs. 60.4–60.6 and 60.9. For compatibility between the sampled/computed data and the voxelized geometric object, the object can be volume sampled [Wang and Kaufman 1993] with the same, but not necessarily the same, density frequency as the acquired or simulated datasets. In volume sampling the continuous object is filtered during the voxelization stage generating alias-free 3D density primitives.

Volume graphics also naturally supports the rendering of translucent volumetric data-sets. (See Fig. 60.9.)

A central feature of volumetric represen-tation is that, unlike surface representation, it is capable of representing inner structures of objects, which can be revealed and ex-plored with the appropriate volumetric ma-nipulation and rendering techniques. Nat-ural objects as well as synthetic objects are likely to be solid rather than hollow. The inner structure is easily explored using vol-ume graphics and cannot be supported by surface graphics (see Figs. 60.2(b) and (c), 60.3–60.6 and 60.9.) Moreover, while translu-cent objects can be represented by surface methods, these methods cannot efficiently support the translucent rendering of volu-metric objects, or the modeling and render-ing of amorphous phenomena (e.g., clouds, fire, smoke) that are volumetric in nature

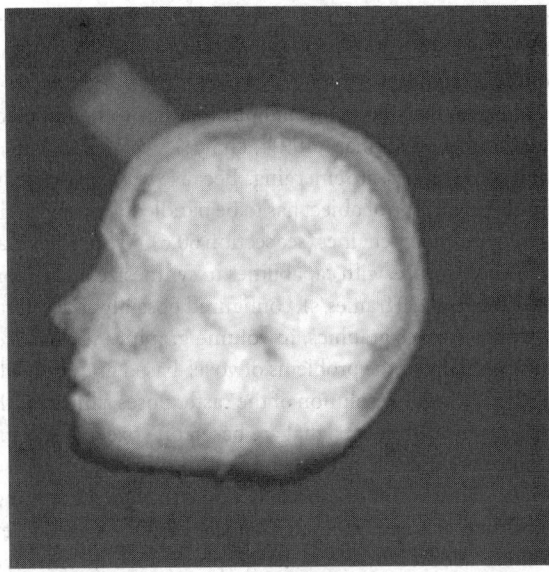

FIGURE 60.9 Intermixing of a volume-sampled cylinder with an MRI head using a union operation.

and do not contain any tangible surfaces [Ebert and Parent 1990, Kajiya and Kay 1989, Perlin and Hoffert 1989].

An intrinsic characteristic of rasters is that adjacent objects in the scene are also represented by neighbor-ing voxels. Therefore, rasters lend themselves to various meaningful block-based operations which can be performed during the voxelization stage. For example, the 3D counterpart of the bitblt operations, termed voxblt (voxel block-transfer), can support transfer of cuboidal voxel blocks with a variety of voxel-by-voxel operations between source and destination blocks [Kaufman 1992]. This property is vary useful for voxblt and CSG. Once a CSG model has been constructed in voxel representation, it is rendered like any other volume buffer. This makes rendering of constructive solid models straightforward.

The spatial presortedness of the volume buffer voxels lends itself to other types of grouping or aggre-gation of neighboring voxels. For example, the terrain image shown in Fig. 60.7 was generated by the voxel-based Hughes Aircraft Co. flight simulator [Wright and Hsieh 1992]. It simulates a flight over voxel-represented terrain enhanced with satellite or aerial photo mapping with additional synthetic raised objects, such as buildings, trees, vehicles, aircraft, clouds, and the like. Since the information below the terrain surface is invisible, terrain voxels can be actually represented as tall cuboids extending from sea level to the terrain height. The raised and moving objects, however, have to be represented in a more conventional voxel-based form.

Similarly, voxels can be aggregated into supervoxels in a pyramidlike hierarchy. For example, in a voxel-based flight simulator, the best resolution can be used for takeoff and landing. As the aircraft ascends, fewer and fewer details need to be processed and visualized, and a lower resolution suffices. Furthermore, even in the same view, parts of the terrain close to the observer are rendered at high reso-lution which decreases toward the horizon. A hierarchical volume buffer can be prepared in advance or on-the-fly by subsampling or averaging the appropriate size neighborhoods of voxels (see also He et al. [1995]).

Weaknesses of Volume Graphics

A typical volume buffer occupies a large amount of memory. For example, for a medium resolution of 512^3, 2 bytes per voxel, the volume buffer consists of 256 megabytes. However, since computer memories are significantly decreasing in price and increasing in their compactness and speed, such large memories are becoming commonplace. This argument echoes a similar discussion when raster graphics emerged as a technology in the mid-1970s. With the rapid progress in memory price and compactness, it is safe to predict that, as in the case of raster graphics, the memory will soon cease to be a stumbling block for volume graphics.

The extremely large throughput that has to be handled requires a special architecture and processing attention (see Kaufman [1991], ch. 6]). *Volume engines,* analogous to the currently available geometry (polygon) engines, are emerging. Because of the presortedness of the volume buffer and the fact that only a simple single type of object has to be handled, volume engines are conceptually simpler to implement than current geometry engines (see section 60.6). We predict that, consequently, volume engines will materialize in the near future, with capabilities to synthesize, load, store, manipulate, and render volumetric scenes in real time (e.g., 30 frames/s), configured possibly as accelerators or cosystems to existing geometry engines.

Unlike surface graphics, in volume graphics the 3D scene is represented in discrete form. This is the cause of many of the problems of voxel-based graphics, which are similar to those of 2D rasters [Eastman 1990]. The finite resolution of the raster poses a limit on the accuracy of some operations, such as volume and area measurements, that are based on voxel counting.

Since the discrete data is sampled during rendering, a low-resolution volume yields high-aliasing artifacts. This becomes especially apparent when zooming in on the 3D raster. When naive rendering algorithms are used, holes may appear between voxels. Nevertheless, this can be alleviated in ways similar to those adopted by 2D raster graphics, such as employing either reconstruction techniques, a higher resolution volume buffer, or volume sampling.

Manipulation and transformation of the discrete volume are difficult to achieve without degrading the image quality or losing some information. Rotation of rasters by angles other than 90° is especially problematic since a sequence of consecutive rotations will distort the image. Again, these can be alleviated in ways similar to the 2D raster techniques.

Once an object has been voxelized, the voxels comprising the discrete object do not retain any geometric information regarding the geometric definition of the object. Thus, it is advantageous, when exact measurements are required (e.g., distance, area), to employ conventional modeling where the geometric definition of the object is available. A voxel-based object is only a discrete approximation of the original continuous object where the volume buffer resolution determines the precision of such measurements. On the other hand, several measurement types are more easily computed in voxel space (e.g., mass property, adjacency detection, and volume computation).

The lack of geometric information in the voxel may inflict other difficulties, such as surface normal computation. In voxel-based models, a discrete shading method is commonly employed to estimate the normal from a context of voxels. A variety of image-based and object-based methods for normal estimation from volumetric data has been devised (see Yagel et al. [1992b] and Kaufman [1991, ch. 4]) and some have been previously discussed. Most methods are based on fitting some type of a surface primitive to a small neighborhood of voxels.

A partial integration between surface and volume graphics is conceivable as part of an object-based approach in which an auxiliary object table, consisting of the geometric definition and global attributes of each object, is maintained in addition to the volume buffer. Each voxel consists of an index to the object table. This allows exact calculation of normal, exact measurements, and intersection verification for discrete ray tracing [Yagel et al. 1992a]. The auxiliary geometric information might be useful also for revoxelizing the scene in case of a change in the scene itself.

60.10 Conclusions

Many of the important concepts and computational methods of volume visualization have been presented. Surface rendering algorithms for volume data were briefly described in which an intermediate representation of the data is used to generate an image of a surface contained within the data. Object order, image order, and domain volume rendering techniques were presented for generating images of surfaces within the data, as well as volume rendered images that attempt to capture all three dimensions of information in the 2D image. Several optimization techniques and volume engines that aim at decreasing the rendering time for volume visualization as well as realistic global illumination rendering were also described.

Although volumetric representations and visualization techniques seem more natural for sampled or computed data sets, their advantages are also attracting traditional geometric-based applications. This trend implies an expanding role for volume visualization, and it has thus the potential to revolutionize the field of computer graphics, by providing an alternative to surface graphics, called volume graphics. We have introduced recent trends in volume visualization that brought about the emergence of volume graphics. As summarized in Table 60.1, volume graphics has advantages over surface graphics by being viewpoint independent, insensitive to scene and object complexities, and it lends itself to the realization of block operations, CSG modeling, and hierarchical representation. It is suitable for the representation of sampled or simulated datasets and their intermixing with geometric objects, and it supports the visualization of internal structures. The problems associated with the volume buffer representation, such as memory size, processing time, aliasing, and lack of geometric representation, echo problems encountered when raster graphics emerged as an alternative technology to vector graphics and can be alleviated in similar ways.

The progress so far in volume graphics, in computer hardware, and memory systems, coupled with the desire to reveal the inner structures of volumetric objects, suggests that volume visualization and volume graphics may develop into major trends in computer graphics. Just as raster graphics in the 1970s superseded vector graphics for visualizing surface, volume graphics has the potential to supersede surface graphics for handling and visualizing volumes, as well as for modeling and rendering synthetic scenes composed of surfaces.

Acknowledgments

Portions of this chapter appeared in *Computer Visualization* (R. S. Gallagher, ed.), published by CRC Press, Boca Raton, FL, 1995, ch. 6, pp. 171–202 [Kaufman and Sobierajski 1995]. Special thanks are due to Lisa Sobierajski, Rick Avila, Roni Yagel, Dany Cohen, Sid Wang, Taosong He, Hanspeter Pfister, Clandio Silva and Lichan Hong who contributed to this paper, coauthored with me related papers [Avila et al. 1994, Kaufman et al. 1993, Kaufman and Sobierajski 1995], and helped with the VolVis software. VolVis can be obtained by sending email to: volvis@cs.sunysb.edu. This work has been supported by the National Science Foundation under grants CCR-9205047 and MIP-9527694, and a grant from the Department of Energy under PICS grant. The MRI head data in Figs. 60.5 and 60.9 is courtesy of Siemens Medical Systems, Inc., Iselin, NJ. Figure 60.7 is courtesy of Hughes Aircraft Company, Long Beach, CA. This image has been voxelized using voxelization algorithms, a voxel-based modeler, and a photo-mapper developed at Stony Brook Visualization Lab. The confocal microscope data in Figs. 60.2 and 60.4 are courtesy of Howard Hughes Medical Institute at Stony Brook, NY. The CT lobster data in Fig. 60.3 is courtesy of AVS, Waltham, MA.

References

Avila, R., Sobierajski, L., and Kaufman, A. 1992. Towards a comprehensive volume visualization system, pp. 13–20. In *Visualization '92 Proc.* Oct.

Avila, R., He, T., Hong, L., Kaufman, A., Pfister, H., Silva, C., Sobierajski, L., and Wang, S. 1994. VolVis: a diversified volume visualization system, pp. 31–38. In *Visualization '94 Proc.* Washington, DC, Oct.

Bakalash, R., Kaufman, A., Pacheco, R., and Pfister, H. 1992. An extended volume visualization system for arbitrary parallel projection. *Proc. Eurographics Workshop Graphics Hardware.* Cambridge, UK, Sept.

Chen, L. S., Herman, G. T., Reynolds, R. A., and Udupa, J. K. 1985. Surface shading in the cuberille environment. *IEEE Comput. Graphics Appl.* 5(12):33–43.

Chiueh, T., He, T., Kaufman, A., and Pfister, H. 1994. Compression Domain Volume Rendering. Computer Science, Tech. Rep. 94.01.04, SUNY Stony Brook, Jan.

Cline, H. E., Lorensen, W. E., Ludke, S., Crawford, C. R., and Teeter, B. C. 1988. Two algorithms for the three-dimensional reconstruction of tomograms. *Med. Phys.* 15(3):320–327.

Cohen, M. F., Chen, S. E., Wallace, J. R., and Greenberg, D. P. 1988. A progressive refinement approach to fast radiosity image generation. *Comput. Graphics (Proc. SIGGRAPH)* 22(4):75–84.

Cohen, D. and Shaked, A. 1993. Photo-realistic imaging of digital terrain. *Comput. Graphics Forum* 12(3):363–374.

Cohen-Or, D. and Kaufman, A. 1995. Fundamentals of surface voxelization. *CVGIP: Graphics Models Image Process.* 56(6):453–461.

Coquillart, S. 1990. Extended free-form deformation: a sculpturing tool for 3D geometric modeling. *Comput. Graphics* 24(4):187–196.

Danielsson, P. E. 1970. Incremental curve generation. *IEEE Trans. Comput.* C-19:783–793.

Daubechies, I. 1992. *Ten Lectures on Wavelets.* CBMS-NSF Reg. Conf. Ser. Appl. Math. SIAM.

Drebin, R. A., Carpenter, L. and Hanrahan, P. 1988. Volume rendering. *Comput. Graphics (Proc. SIGGRAPH)* 22(4):65–74.

Dubois, D. and Prade, H. 1980. *Fuzzy Sets and Systems: Theory and Applications.* Academic Press.

Dunne, S., Napel, S., and Rutt, B. 1990. Fast reprojection of volume data, pp. 11–18. In *Proc. 1st Conf. Visualization Biomed. Comput.* Atlanta, GA.

Eastman, C. M. 1990. Vector versus raster: a functional comparison of drawing technologies. *IEEE Comput. Graphics Appl.* 10(5):68–80.

Ebert, D. S. and Parent, R. E. 1990. Rendering and animation of gaseous phenomena by combining fast volume and scanline A-buffer techniques. *Comput. Graphics* 24(4):357–366.

Eck, M., DeRose, T., Duchamp, T., Hoppe, H., Lounsbery, M., and Stuetzle, W. 1995. Multiresolution analysis of arbitrary meshes, pp. 173–182. In *SIGGRAPH '95 Conf. Proc.* Aug.

Fowler, J. and Yagel, R. 1994. Lossless compression of volume data, pp. 43–50. In *Proc. Symp. Volume Visualization,* Washington, DC, Oct.

Galyean, T. A. and Hughes, J. F. 1991. Sculpting: an interactive volumetric modeling technique. *Comput. Graphics* 25(4):267–274.

Garrity, M. P. 1990. Raytracing irregular volume data. *Comput. Graphics (Proc. SIGGRAPH)* 24(5):35–40

Giertsen, C. 1992. Volume visualization of sparse irregular meshes. *IEEE Comput. Graphics Appl.* 12(2):40–48.

Glassner, A. S., 1984. Space subdivision for fast ray tracing. *IEEE Comput. Graphics Appl.* 4(10):15–22.

Goodman, J. R. and Sequin, C. H. 1981. Hypertree: a multiprocessor interconnection topology. *IEEE Trans. Comput.* C-30(12):923–933.

Gordon, D. and Reynolds, R. A. 1985. Image space shading of 3-dimensional objects. *Comput. Vision, Graphics Image Process.* 29:361–376.

Gross, M. H., Koch, R., Lippert, L., and Dreger, A. 1995. A new method to approximate the volume rendering equation using wavelet bases and piecewise polynomials. *Computers & Graphics* 19(1):47–62.

Guenther, T., Poliwoda, C., Reinhard, C., Hesser, J., Maenner, R., Meinzer, H., and Baur, H. 1994. VIRIM: a massively parallel processor for real-time volume visualization in medicine, pp. 103–108. In *Proc. 9th Eurographics Hardware Workshop.* Oslo, Norway, Sept.

Hanrahan, P., Salzman, D. and Aupperle, L. 1991. A rapid hierarchical radiosity algorithm. *Comput. Graphics (Proc. SIGGRAPH)* 25(4):197–206.

Hart, J. C., Sandin, D. J., and Kaufman, L. H. 1989. Ray tracing deterministic 3-D fractals. *Comput. Graphics* 23(3):289–206.

He, T., Hong, L., Kaufman, A., Varshney, A., and Wang, S. 1995. Voxel-based object simplification, pp. 296–303. In *IEEE Visualization '95 Proc.* Los Alamitos, CA, Oct.

Herman, G. T. and Liu, H. K. 1979. Three-dimensional display of human organs from computed tomograms. *Comput. Graphics Image Process.* 9:1–21.

Hesser, J., Maenner R., Knittel, G., Strasser, W., Pfister, H., and Kaufman, A. 1995. Three architectures for volume rendering. *Comput. Graphics Forum* 14(3):111–122.

Hoehne, K. H. and Bernstein, R. 1986. Shading 3D-images from CT using gray-level gradients. *IEEE Trans. Med. Imaging* MI-5(1):45–47.

Hottel, H. C. and Sarofim, A. D. 1967. *Radiative Transfer*. McGraw–Hill, New York.

Kajiya, J. T. and Kay, T. L. 1989. Rendering fur with three dimensional textures. *Comput. Graphics* 23(3):271–280.

Kaufman, A. 1987a. An algorithm for 3D scan-conversion of polygons, pp. 197–208. In *Proc. EURO-GRAPHICS '87*. Amsterdam, Netherlands, Aug.

Kaufman, A. 1987b. Efficient algorithms for 3D scan-conversion of parametric curves, surfaces, and volumes. *Comput. Graphics* 21(4):171–179.

Kaufman, A. 1991. *Volume Visualization*. IEEE Computer Society Press Tutorial, Los Alamitos, CA.

Kaufman, A. 1992. The *voxblt* engine: a voxel frame buffer processor. In *Advances in Graphics Hardware III*. A. A. M. Kuijk, ed., pp. 85–102. Springer–Verlag, Berlin.

Kaufman, A. and Bakalash, R. 1988. Memory and processing architecture for 3-D voxel-based imagery. *IEEE Comput. Graphics Appl.* 8(6):10–23. Also 1989. *Nikkei Comput. Graphics* 3(30):148–160, in Japanese.

Kaufman, A., Cohen, D., and Yagel, R. 1993. Volume graphics. *IEEE Comput.* 26(7):51–64. Also *Nikkei Comput. Graphics* 1(88):148–155 and 2(89):130–137, in Japanese (1994).

Kaufman, A. and Shimony, E. 1986. 3D scan-conversion algorithms for voxel-based graphics, pp. 45–76. In *Proc. ACM Workshop Interactive 3D Graphics*. Chapel Hill, NC, Oct.

Kaufman, A. and Sobierajski, L. 1995. Continuum volume display. In *Computer Visualization*, R. S. Gallagher, ed., pp. 171–202. CRC Press, Boca Raton, FL.

Kaufman, A., Yagel, R., and Cohen, D. 1990. Intermixing surface and volume rendering. In *3D Imaging in Medicine: Algorithms, Systems, Applications*. K. H. Hoehne, H. Fuchs, and S. M. Pizer, eds., pp. 217–227, June.

Knittel, G. and Strasser, W. 1994. A compact volume rendering accelerator, pp. 67–74. In *Volume Visualization Symp. Proc.* Washington DC, Oct.

Kruger, W. 1991. The application of transport theory to visualization of 3-D scalar data fields. *Comput. Phys.* (July & Aug.):397–406.

Lacroute, P. 1995. Real-time volume rendering on shared memory multiprocessors using the shear-warp factorization, pp. 15–21. In *Parallel Rendering Symp.* Oct.

Lacroute, P. and Levoy, M. 1994. Fast volume rendering using a shear-warp factorization of the viewing transformation. *Comput. Graphics* 28(3):451–458.

Laur, D. and Hanrahan, P. 1994. Hierarchical splatting: a progressive refinement algorithm for volume rendering. *Comput. Graphics* 25(4):285–288.

Lee, Y. T. and Requicha, A. A. G. 1982. Algorithms for computing the volume and other integral properties of solids: I-known methods and open issues; II-a family of algorithms based on representation conversion and cellular approximation. *Commun. ACM* 25(9):635–650.

Levoy, M. 1988. Display of surfaces from volume data. *Comput. Graphics Appl.* 8(5):29–37.

Levoy, M. and Whitaker, R. 1990. Gaze-directed volume rendering. *Comput. Graphics (Proc. Symp. Interactive 3D Graphics)* 24(2):217–223.

Lorensen, W. E. and Cline, H. E. 1987. Marching cubes: a high resolution 3D surface construction algorithm. *Comput. Graphics* 21(4):163–170.

Malzbender, T. 1993. Fourier volume rendering. *ACM Trans. Graphics* 12(3):233–250.

Max, N. 1995. Optical models for direct volume rendering. *IEEE Trans. Visualization Comput. Graphics* 1(2):99–108.

Max, N., Hanrahan, P., and Crawfis, R. 1990. Area and volume coherence for efficient visualization of 3D scalar functions. *Comput. Graphics* 24(5):27–34.

Mokrzycki, W. 1988. Algorithms of discretization of algebraic spatial curves on homogeneous cubical grids. *Comput. Graphics* 12(3/4):477–487.

Muraki, S. 1993. Volume data and wavelet transform. *IEEE Comput. Graphics Appl.* 13(4):50–56.

Ning, P. and Hesselink, L. 1993. Fast volume rendering of compressed data, pp. 11–18. In *Visualization '93 Proc.* Oct.

Norton, V. A. 1982. Generation and rendering of geometric fractals in 3-D. *Comput. Graphics* 16(3):61–67.

Perlin, K. and Hoffert, E. M. 1989. Hypertexture. *Comput. Graphics* 23(3):253–262.

Pfister, H., Kaufman, A., and Chiueh, T. 1994. Cube-3: a real-time architecture for high-resolution volume visualization, pp. 75–82. In *Volume Visualization Symp. Proc.* Washington, DC, Oct.

Pfiser, H., Kaufman, A., and Wessels, F. 1995. Towards a scalable architecture for real-time volume rendering. *10th Eurographics Workshop Graphics Hardware Proc.* Maastricht, The Netherlands, Aug.

Pfister, H., Wessels, F., and Kaufman, A. 1995. Sheared interpolation and gradient estimation for real-time volume rendering. *Computers & Graphics* 19(5):667–677.

Pfister, H. and Kaufman A. 1996. Cube-4: A scalable architecture for real-time volume rendering. In *Volume Visualization Symp. Proc.* San Francisco, CA, Oct.

Rossignac, J. and Borrel, P. 1993. Multi-resolution 3D approximations for rendering complex scenes. In *Modeling in Computer Graphics*. B. Falcidieno and T. L. Kunni, eds., pp. 455–465. Springer–Verlag.

Rushmeier, H. E. and Torrance, K. E. 1987. The zonal method for calculalting light intensities in the presence of a participating medium. *Comput. Graphics* 21(4):293–302.

Sabella, P. 1988. A rendering algorithm for visualizing 3D scalar fields. *Comput. Graphics (Proc. SIGGRAPH)* 22(4):160–165.

Sakas, G. and Hartig, J. 1992. Interactive visualization of large scalar voxel fields, pp. 29–36. In *Proc. Visualization '92*. Boston, MA, Oct.

Schroder, P. and Stoll, G. 1992. Data parallel volume rendering as line drawing, pp. 25–32. In *Workshop Volume Visualization*. Boston, MA, Oct.

Schroeder, W. J., Zarge, J. A., and Lorensen, W. E. 1992. Decimation of triangle meshes. *Comput. Graphics* 26(2):65–70.

Silva, C. and Kaufman, A. 1994. Parallel performance measures for volume ray casting, pp. 196–203. In *Visualization '94 Proc.* Washington, DC, Oct.

Sobierajski, L., Cohen, D., Kaufman, A., Yagel, R., and Acker, D. 1993. A fast display method for volumemtric data. *Visual Comput.* 10(2):116–124.

Sobierajski, L. and Kaufman, A. 1994a. Volumetric radiosity. Technical Report 94.01.05, Computer Science, SUNY Stony Brook, Jan.

Sobierajski, L. and Kaufman, A. 1994b. Volumetric ray tracing, pp. 11–18. In *Volume Visualization Symp. Proc.* Washington, DC, Oct.

Speray, D. and Kennon, S. 1990. Volume probes: interactive data exploration on arbitrary grids. *Comput. Graphics* 24(5):5–12.

Totsuka, T. and Levoy, M. 1993. Frequency domain volume rendering. *Comput. Graphics (Proc. SIGGRAPH)*:271–278.

Tuy, H. K. and Tuy, L. T. 1984. Direct 2-D display of 3-D objects. *IEEE Comput. Graphics Appl.* 4(10):29–33.

Upson, C. and Keeler, M. 1988. V-BUFFER: visible volume rendering. *Comput. Graphics (Proc. SIGGRAPH)* 22(4):59–64

Vannier, M. W., Marsh, J. L., and Warren, J. O. 1983. Three-dimensional computer graphics for craniofacial surgical planning and evaluation. *Comput. Graphics* 17(3):263–273.

Vezina, G., Fletcher, P. A., and Robertson, P. K. 1992. Volume Rendering on the MasPar MP-1, pp. 3–8. In *Workshop Volume Visualization*. Boston, MA, Oct.

Wang, S. and Kaufman, A. 1993. Volume sampled voxelization of geometric primitives, pp. 78–84. In *Visualization '93 Proc.* San Jose, CA, Oct.

Wang, S. and Kaufman, A. 1994a. 3D Antialiasing. Computer Science Tech. Rep. 94.01.03. SUNY Stony Brook, Jan.

Wang, S. and Kaufman, A. 1994b. Volume-sampled 3D modeling. *IEEE Comput. Graphics Appl.* 14(5):26–32.

Wang, S. and Kaufman, A. 1995. Volume sculpting, pp. 151–156. In *ACM Symp. Interactive 3D Graphics*. Monterey, CA, April.

Westermann, R. 1994. A multiresolution framework for volume rendering, pp. 51–58. In *Symp. Volume Visualization*. Washington, DC, Oct.

Westover, L. 1990. Footprint evaluation for volume rendering. *Comput. Graphics (Proc. SIGGRAPH)* 24(4):144–153.

Whitted, T. 1980. An improved illumination model for shaded display. *Commun. ACM* 23(6):343–349.

Wilhems, J. and vanGelder, A. 1991. A coherent projection approach for direct volume rendering. *Comput. Graphics (SIGGRAPH '91 Proc.)* 25:275–284.

Williams, P. L. 1992. Interactive splatting of nonrectilinear volumes, pp. 37–44. In *Proc. Visualization '92*. Oct.

Wright, J. and Hsieh, J. 1992. A voxel-based forward projection algorithm for rendering surface and volumetric data, pp. 340–348. In *Proc. Visualization '92*. Boston, MA, Oct.

Yagel, R. 1991. *Efficient Methods for Computer Graphics*. Ph.D. dissertation. SUNY Stony Brook, Dec.

Yagel, R., Cohen, D., and Kaufman, A. 1992a. Discrete ray tracing. *IEEE Comput. Graphics Appl.* 12(5):19–28.

Yagel, R., Cohen, D., and Kaufman, A. 1992b. Normal estimation in 3D discrete space. *Visual Comput.* (June) 8(5–6):278–291.

Yagel, R. and Kaufman, A. 1992. Template-based volume viewing. *Comput. Graphics Forum* 11(3):153–167.

Yeo, B. and Liu, B. 1995. Volume rendering of DCT-based compressed 3D scalar data. *IEEE Trans. Visualization Comput. Graphics* 1(1):29–43.

61

Virtual Reality

Steve Bryson
MRJ, Inc./NASA Ames Research Center

61.1 Introduction

Virtual reality, also known as virtual environments or virtual worlds, is a new paradigm in computer–human interaction, in which three-dimensional computer-generated worlds, called virtual environments, are created which have the effect of containing objects that have their own location in three-dimensional space. The user's perception of this computer-generated world is as similar to the perception of the real world as the technology will allow, providing appropriate depth and three-dimensional structure cues. User perception in virtual reality can be via a variety of senses, including sight, sound, touch, and force. Virtual environments are often, but not necessarily, immersive, providing the effect of surrounding the user with virtual objects. Objects in the virtual environment are often autonomous and/or interactive. The user interacts with the virtual environment using several interaction techniques, with a stress on direct manipulation in three-dimensional space via interface metaphors from the real world, such as grab and point, where appropriate. In order to create the effect of interactive three-dimensional objects, the virtual environment must be processed and presented at a near-real-time rate of 10 frames/s or greater. The three-dimensional perception and interaction in the virtual environment, its real-world-like interface, and its inherently near-real-time response property make virtual reality a natural interface for three-dimensional applications, including training for real-world tasks.

More precisely, we define **virtual reality** as the use of computer systems and interfaces to create the effect of an interactive three-dimensional environment, called the virtual environment, which contains objects which have spatial presence. By spatial presence, we mean that objects in the environment effectively have the property of spatial location relative to and independent of the user in three-dimensional space. We call the effect of creating a three-dimensional environment which contains objects with a sense of spatial presence the **virtual reality effect**. The essence of virtual reality can be summed up in the idea of three-dimensional "things" in the virtual environment rather than (possibly animated) "pictures of things."

We are defining virtual reality as an interface: there is no statement of content in this definition. In particular there is nothing in the definition of virtual reality which implies an attempt to mimic or otherwise create the illusion of the real world in the computer-generated environment. While some applications such as real-world task training tasks may require mimicking the real world, other applications such as entertainment or scientific visualization use environments which do not attempt to duplicate the real world.

There has been some confusion about the meaning of the phrase "virtual reality," which some people take to be an oxymoron. "Virtual" means "having the effect of being something without actually being that thing," while the definition of "reality" appropriate for our purposes is "having the property of concrete existence." Thus the phrase "virtual reality" translates as "having the effect of concrete existence without actually having concrete existence." This definition is, to some extent, actually achieved in virtual reality systems and distinguishes virtual reality from conventional computer graphics.

Virtual reality is a young, interdisciplinary, growing research field. It is not possible to survey all interesting activities in virtual reality in this short article, nor is it possible to detail particular technologies without this article rapidly going out of date. I will therefore only survey the issues that arise in the design of a virtual reality system, with an emphasis on application development. Many of the results and principles described in this article are the result of experience rather than careful study.

61.2 Underlying Principles

The virtual reality effect is attained through the use of a combination of computer and human–computer interface technologies. While the virtual reality effect can be created using any one or combination of sensory modalities including sight, sound, touch, and force, we shall illustrate the virtual reality effect using the visual modality as an example. Later we shall comment on how the same effect is created using other sensory modalities.

The effect of the spatial presence of an object is provided for the visual sense by rendering the object using conventional three-dimensional computer graphics from the current point of view of the user. **Head tracking** technology is used to measure the position and orientation of the user's head. The computer uses this head information to compute the position and orientation of the center of each of the user's eyes. These eye data are used to set the point of view and perspective projection for the rendering of the three-dimensional virtual scene, once for each eye to create a stereoscopic effect. The graphics is displayed to the user, optionally in stereo, in a way which allows the user to move around to get varying views of the virtual environment.

Consider the case of a single motionless virtual object in the virtual environment: while the image presented to the user changes as the user moves about, that change approximately matches the expectations of the user's cognitive system. The result is that the user's cognitive system interprets the changing images as a motionless object being viewed from a moving point of view. We shall call this effect **spatial constancy**. It is the spatial constancy of virtual objects which results in the object's sense of spatial presence. The virtual scene must be rendered from the user's current point of view with a rate sufficient to provide spatial information about objects in the environment. For purposes of spatial constancy, a rate of 10 frames/s is sufficient, though a higher frame rate is required for interaction with environments which contain rapidly moving objects. Note that at 10 frames/s the environment still appears jerky and the frames are clearly discontinuous. Experience has shown, however, that 10 frames/s is sufficient for the effect of spatial constancy. For lower frame rates the effect of spatial constancy fails and the user perceives the scene as a succession of disconnected images. In general, the human cognitive system is very forgiving about endowing objects with the property of spatial constancy: slight delays in rendering, slightly inaccurate tracking, noticeably jerky motion, and low-quality graphics do not interfere with the effect of spatial constancy.

Interaction in the virtual environment is performed through the measurement of the user's body motions and the interpretation of that motion as intentions by the computer system. Interaction in virtual environments is very different from the transactional, discrete interaction in conventional human–

computer interfaces: interaction in virtual environments is inherently continuous, typically involving spatial manual tasks such as picking, placing, and tracking. This continuous interaction requires accurate tracking and very fast response from the computer system in order to provide the user with appropriate feedback as to the state of the interaction.

The virtual reality effect is critically dependent on the virtual reality system providing a view of the virtual environment which corresponds as closely as possible with the user's head position and orientation as the user moves about. This requirement implies that there must be a minimal graphics frame rate and that the image presented to the user must correspond as closely as possible to the user's current head position and orientation, which implies a short delay between when that position and orientation are sampled and when the resulting rendered scene appears to the user. Thus there are two performance issues critical to the success of a virtual reality system: frame rate (analogous to bandwidth or throughput) and delay (analogous to latency).

Two considerations determine the required frame rate:

- Experience has shown that for the effect of spatial constancy to operate, the virtual scene must be rendered from the user's point of view with a frame rate of at least 10 frames/s. Failure to meet the 10 frame/s requirement will result in the failure of the virtual reality effect.

- If the virtual environment contains moving objects, the Shannon–Nyquist limit requires that the user "sample," in other words see, that virtual object with a frequency at least twice that of the highest frequency of motion of the object. In actual practice the display rate should be at least four times that of the highest frequency of motion. This puts an application-dependent lower limit on the acceptable frame rate. Further, low frame rates have a noticeable impact on the ability to perform spatial manipulation tasks (Bryson 1993, Burdea and Coiffet 1994). Failure to meet this frame-rate requirement will result in an impaired ability of the user to correctly perceive motions and interact with objects in the environment.

Two considerations determine the acceptable delay:

- Delays in head tracking can result in motion or simulator sickness, as the images seen by the user do not correspond with head motion as sensed by the user's vestibular system. How strongly a given delay induces motion sickness is determined by many factors, including head motion frequency and field of view of the virtual reality display. Larger fields of view and high-frequency head motions induce greater motion sickness for a given delay. Experience has indicated that delays of 0.1 s or less are acceptable for head motions limited by reasonable frequencies and wide fields of view.

- Delays in response to hand tracking impair the ability to perform manual tasks such as pick and place and tracking. The highest allowable delay is determined by the accuracy with which tasks must be performed and the frequency of motion of objects with which the user must interact. For example, the accuracy of tracking tasks, where the user's hand must track a nonperiodic target, has been shown to depend linearly on both the delay and the target object's frequency of motion.

These considerations are summarized in the *virtual reality performance requirements:*

- *The graphical frame rate (animation rate) must be greater than two to three times the highest frequency of motion of objects in the environment and in all cases must be greater than 10 frames/s.*

- *The end-to-end delay in response to user input must be small for interaction with objects which have high-frequency motion and in all cases must be less than 0.1 s.*

The displays used in virtual reality come in several forms. The most famous form is that of the **head-mounted display**, which places a display in front of the eyes, usually a separate display for each eye. Head-mounted displays are usually worn on the head and move with the user, so the user always sees the virtual environment (though such displays are sometimes externally supported), providing a strong sense of presence. Issues that arise with head-mounted displays include comfort, image quality, and field of view and will be discussed in more detail below. *Stationary displays* are fixed displays for the virtual environment. Stationary displays may be conventional workstation screens or large projection screens, and may be vertical or horizontal or surround the user for a sense of immersion. Stereo display is typically provided in stationary displays via a time-multiplexed stereo signal, where a polarized image for each eye is displayed alternately, with polarized glasses worn by the user determining which image gets seen by which eye.

The virtual reality effect is attained through the use of head tracking and displays to provide various cues about the user's position and orientation in a three-dimensional environment. The human factors of visual perception, particularly depth perception and personal motion cues, are critical in the successful design of the virtual environment. Human depth cues include the following:

- *Head-motion parallax:* The relative motion of objects at various depths as the user's head moves about in the environment. Head-motion parallax is performed via head tracking.
- *Plane of focus (accommodation):* The location of the focus plane in the environment. As this chapter is written, plane of focus is not supported as a depth cue by any available virtual reality system.
- *Stereopsis:* The differences in relative positions of objects in the virtual environment as seen by the user's two eyes. Stereopsis is supported by providing a different image to each eye, either using a separate display for each eye or single display with images for each typically eye appearing sequentially in time.
- *Occlusion:* If one object blocks the view of another object, the first object is closer. Occlusion is supported by conventional three-dimensional computer graphics systems via hidden surface rendering algorithms.
- *Perspective:* The relative location of an object in the user's field of view as determined by classical perspective transformations. Perspective includes such cues as apparent size and the fact that objects that appear lower in the field of view are perceived to be closer. Perspective is supported by conventional three-dimensional computer graphics systems. Wide-angle displays significantly enhance this cue.
- *Textures:* The appearance of a known texture at different depths gives strong depth cues. Textures are supported by higher-end conventional three-dimensional computer graphics systems.
- *Atmospheric effects:* Blurring and fog effects due to distance. Atmospheric effects are supported by higher-end conventional three-dimensional computer graphics systems.

Self-location and *self-motion* cues are dominated by perception of the object's motion in the user's peripheral field. Thus wide-angle displays significantly enhance the sense of location and the accurate detection of self-motion in the virtual environment.

Hand tracking supports interaction in the virtual environment. Hand tracking takes two forms: position and orientation tracking and gesture (command) tracking. Position and orientation tracking is performed with much the same technology as that used in head tracking. Gesture tracking is performed via a variety of technologies: buttons are often used for a small number of gestures, while measurement of the user's finger joint angles can be used to infer the hand gesture that the user is performing. This hand gesture can then be interpreted as a command to the system. An example is the interpretation of a closed fist as a "grab and move" command. The user's finger joint angles are measured via an instrumented glove-type device.

An understanding of the human factors of manual interaction is critical in the successful design of a virtual environment. One basic result is *Fitts' law*, which states that the shortest time to reach an object is proportional to the log of the ratio of the object's distance to its size. Another result is in the study of manual tracking of a randomly moving target. If error is measured as the mean square distance from the target to the location of the user's cursor, that tracking error is:

- linearly dependent on the frequency of motion of the target, with target motion frequencies greater that 5 Hz being essentially untrackable,

- linearly dependent on end-to-end delays between the user's motions and the resulting motions of the user's cursor,
- linearly dependent (according to preliminary results) on the inverse of the frame rate of the display, even when the display is completely up to date when first shown.

Thus for manual tracking errors depend on both delay and frame rate. As any realistic virtual reality system will have both delays and a finite frame rate, applications which require manual tracking should have frame rates which are as high as possible and delays which are as small as possible.

The virtual environment may appear at many scales: Application requirements entirely determine the scale at which objects in the environment should appear. Some applications, such as real-world simulations or training applications, will have a naturally fixed scale. Other applications, such as a molecular structure application, will naturally have the scale set so that very small objects will appear very large. Other applications will have no natural environment scale at all, leaving the scale setting up to the user.

The above observation about scale generalizes to all aspects of the virtual environment: while virtual reality uses metaphors from the real world in the design of environments, there is no need beyond application requirements for behavior in the virtual environment to match behavior in the real world. Virtual reality applications can be tailored to perform tasks which would be either more difficult or impossible in the real world. Thus effective performance of application tasks should be the guiding principle in virtual environment application design. This focus on the application task as the guiding design principle has led to the abandonment of a conventional interface layer such as the menus and sliders in conventional graphical user interfaces. As this chapter is written it remains to be seen if another layer of conventionality will appear in virtual reality beyond the basic "pick up and move" metaphor of the direct manipulation interface. The desire for a conventional interface is at odds with the opportunity for the creation of application-specific objects which also act as interface objects.

A significant variation on virtual reality is *augmented reality*. In augmented reality the virtual environment is superimposed on the real world. Augmented reality uses either see-through displays, which place the virtual environment in a semitransparent window on the real world, or mixing of video images of the real world with computer-generated images of the virtual world. The dominant application of augmented reality is *information overlays*, which display information about the real world in the virtual environment. These information overlays may match the three-dimensioal position and orientation of real-world objects. This matching of virtual and real objects requires very high accuracy in the tracking of the user's position and orientation as well as models of real-world behaviors. In this chapter we shall treat augmented reality as a subset of virtual reality.

61.3 Best Practices

Display of the Virtual Environment

One of the guiding principles of virtual reality is to provide the same types of information about the virtual environment as are available for the real world. Thus there are several sensual modalities used to present the virtual environment, including visual, auditory, and haptic (touch and force) displays.

Visual Display

Visual display is one of the most important components of a virtual reality system. There are a number of quality considerations which arise in the selection of a visual display for virtual reality. In addition to the usual display quality considerations from conventional graphics such as color, contrast, brightness, and refresh rate, the following issues arise in virtual reality displays:

> *Resolution:* Defined as the angle subtended by a pixel as viewed by the user. Note that this sense of resolution is very different from the conventional sense used in computer graphics. Resolution

will be determined by the pixel spacing on the screen and field of view as determined by the optics. A rough estimate of the angular resolution of a pixel can be obtained by dividing the horizontal field of view by the number of pixels in the horizontal direction.

Pixel spacing: Many screen technologies, such as shadow-masked color cathode-ray tubes and liquid-crystal displays, have significant space between the pixels. This space is magnified by wide-field optics and can significantly deteriorate the image.

Field of view: A strong determinant of the immersiveness of a display. The field of view is determined by the physical screen size and the optics.

Optical quality: Because head-coupled displays are very close to the eyes, they require optics to provide a focused image. These optics also determine the field of view and may induce strong optical distortion.

There are a variety of display technologies available which are appropriate to virtual reality. They fall roughly into two classes: **head-coupled**, which move with the user's head as the user moves about, and **stationary**, which do not. Head-coupled displays place a screen in front of each of the user's eyes, and are often **head-mounted**, rigidly attached to the user's head. Head-coupled displays provide a strong sense of immersion, because the user sees the virtual scene no matter which way the user's head is turned. Head-mounted displays require focusing optics, which can often provide a wide field of view. Some head-mounted display designs use folded optics to provide a more balanced design. Augmented reality displays often use a half-silvered mirror for display, allowing the virtual scene to be overlaid on the real world.

A significant issue for displays is ergonomics, particularly user comfort. Head-mounted displays can be uncomfortable, leading to rejection by the user community. Stationary displays provide both enhanced comfort and a sharable experience, but at the cost of immersiveness. Which display is chosen will generally depend on the application requirements.

Audio Displays

Sound output is an important modality of display in virtual reality. There are roughly two levels of sophistication of sound output: non-spatially-localized (possibly stereo) and three-dimensional spatially localized sound display.

Non-spatially-localized sound display involves using conventional sound rendering techniques, such as sampling and waveform synthesis, to provide nonspatial sound cues such as those found in conventional graphics applications. Nonspatial sounds are typically used to provide feedback as to the occurrence of events such as object collision in the virtual environment and (potentially) data display by varying the characteristics of a continuous sound.

Three-dimensional spatial sound uses various techniques to render sound whose source has a perceived location in three-dimensional space. This sound source location may or may not correspond to a visual or haptic object at the same location in three-dimensional space. Spatially localized sound provides the cues used by the human auditory perceptual system, which are interpreted as the sound source being located in three-dimensional space. These cues include variations in volume from ear to ear, phase differences due to the time difference between when the sound arrives at the two ears, and the distortion of the sound due to the structures on the exterior of the human ear (pinnae).

The simplest method of providing three-dimensional spatially located sound is to surround the user with an array of speakers. A particular sound source is given a perceived spatial location by appropriately balancing the volume of that sound from each speaker, taking into consideration the user's head position as measured by the head-tracking technology. The use of multiple external speakers provides appropriate volume to the user's ears, though the other cues are not well provided.

A more sophisticated method provides spatially localized sound via headphones and appropriate signal processing of the sound source (Begault 1993). Careful measurements are made of the differences in volume, phase, and sound distortion by placing a small microphone in a human ear while moving a sound source about in space. These measured parameters are stored, associated with a locaiton in space, and used to construct a convolution function as a function of three-dimensional position relative to the user. When a new sound is generated and associated with a three-dimensional location in the virtual environment, the convolution function modifies that sound for each ear in the same way that the sound would be modified as it reached the ear from a real-world sound source.

Haptic Displays

Haptics refers to the senses of force and touch. Haptic displays use various types of hardware to provide force or touch feedback when the user encounters an object in the virtual environment. At the time this chapter is written, haptic displays are highly experimental, with few commercial products available. Haptic displays are covered more thoroughly in Burdea and Coiffet (1994). The comments in this section are highly provisional and reflect an active research topic.

The primary purpose of haptic displays is to give the user the effect of "touching" objects in the virtual environment. The experience of touching an object in the real world is extemely complex, involving the texture of, temperature of, vibration (if any) of, and forces exerted by the object. Reflecting this complextiy, haptic displays fall into two classes:

- *Surface displays:* including texture, vibration, and temperature displays,
- *Deep displays:* including force displays such as pneumatic robotics, compressed air actuators, and "memory metal" devices.

Surface displays have been very difficult to build. The most common texture displays have typically involved small pin-type actuators, which are raised or lowered to give the effect of a smooth or rough surface, much like a graphical bitmap. Small vibrators and heat elements mounted in the fingertips of gloves have been used as a substitute for texture display with marginal success.

Deep displays involve exerting some kind of force on the user, the technology for which is well developed in the field of robotics. Thus deep displays are a good deal more mature than surface displays. As a general rule, the larger the force and the volume over which that force is to be exerted, the more difficult and cumbersome the haptic technology will be. As of this writing, a commercial product is available which delivers good force feedback to the fingertip over a volume about 0.5 m on a side.

There are two dominant technologies for force displays:

- Pneumatic actuators, used for whole-arm or whole-body forces. Pneumatic actuators tend to be very powerful but very cumbersome.
- Stepper-motor-based actuators, with the motors connected either directly to the actuator's lever arm or to a collection of strings or wires. Motor-based actuators tend to be smaller and more usable but have a smaller range of forces and a smaller working volume.

Position Tracking

There are two classes of tracking technologies: position and/or orientation trackers and angle trackers. Position and/or orientation trackers typically provide the position and/or orientation of an object at a single point in three-dimensional space. Angle trackers provide the angle between two different objects. Position and orientation sensors are typically used to track the user's head and hand, while angle sensors are usually used to track the angle of bend of a body joint such as the user's fingers.

Position Tracker Data

The data returned by position and orientation trackers fall into two classes: position and orientation. Position data are usually a three-dimensional vector relative to some known reference location. Orientation data are somewhat more complex, owing to the fact that three numbers are not sufficient to describe all orientations in three-dimensional space. Three numbers are sufficient to describe all but two orientations (which two depend on the type and coordinate system of the orientation data), while four numbers are required to describe all orientations. Another option is to use the matrix description of an orientation, which requires nine numbers. There are three common methods of describing orientations:

Euler angles use an ordered triple (roll, pitch, yaw) of numbers: *roll* = rotation around the front-facing axis; *pitch* = rotation around the (new) horizontal axis; and *yaw* = rotation around the (new) vertical axis. Note that these rotations are concatenated. Because they describe rotations using three numbers, Euler angles do not describe all orientations: there are two orientations, characterized by pitch = $\pm 90°$, which are not correctly described by Euler angles. When the tracker is placed in one of these orientations, the tracker is said to be in *gimbal lock.*

Rotation matrices are 3×3 matrices, which describe the rotation of an object. Rotation matrices are standard in conventional computer graphics and are described in (Foley et al. 1990).

Quaternions are an ordered quadruple (w, x, y, z) of numbers which describe all orientations. Physically, quaternions represent a rotation of an angle given by arcsin $(w/2)$ around the axis specified by (x, y, z). In actual use, quaternions are usually translated to rotation matrices using the following formulas:

$$\begin{pmatrix} 1 - 2y^2 - 2z^2 & 2xy + 2wz & 2xz - 2wy \\ 2xy - 2wz & 1 - 2x^2 - 2z^2 & 2yz + 2wx \\ 2xz + 2wy & 2yz - 2wx & 1 - 2x^2 - 2y^2 \end{pmatrix}$$

Quaternions avoid the gimbal-lock problem and require smaller amounts of data storage than rotation matrices.

Position Tracker Technologies

There are a variety of tracker technologies which return position and/or orientation in three-dimensional space. Each of these technologies has its strengths and weaknesses:

- *Electromagnetic trackers:* Electromagnetic trackers use a source containing three orthogonal coils to sequentially produce three oriented radio-frequency electromagnetic fields, each component of which is read by three orthogonal coils in a sensor. These measurements result in nine numbers providing the strength of each of the three fields in three directions, which are used to reconstruct the position and orientation of the sensor relative to the source. Electromagnetic trackers do not require a clear line of sight from the source to the sensor, making them useful for both head and hand tracking, and are readily available commercially. Electromagnetic trackers have limited range (typically 1 to 3 m as of 1996) and are susceptible to electromagnetic distortion and noise in the physical environment, particularly from display monitors.

- *Acoustic trackers:* Acoustic trackers use an ultrasonic sound pulse, which is picked up by an array of receivers. The time of flight of the sound pulse from the source to each receiver is used to reconstruct the position and orientation of the receiver array relative to the source. Acoustic trackers are inexpensive and commercially available. They require a clear line of sight from the source to each receiver, limiting their operational envelope and making them potentially inappropriate for hand tracking. They are, however, appropriate for head tracking when using a stationary desktop display, where the user is physically always looking at a desktop display monitor so that line of sight is assured. Acoustic trackers have a limited range (1 to 2 m) and are very susceptible to acoustic noise and echoes in the physical environment.

- *Mechanical linkage trackers:* Mechanical linkage trackers use a jointed physical structure for position and orientation tracking. These structures have appropriately situated joints, each of which has an angle sensor. By measuring the angle of each joint and knowing the length of each segment, the position and orientation of the end relative to the base of the jointed structure can be determined. Mechanical linkage trackers have the advantage of very high accuracy, allowing a minimum of filtering, resulting in low delays. They have the disadvantage of a usually cumbersome physical structure which can get in the way of task performance.

- *Video tracking:* Video tracking uses multiple video cameras to track objects in the physical space, usually targets placed on the user's head and hands. The location of these targets is identified on the video images, and these image locations are used to reconstruct the three-dimensional position of the targets relative to the cameras. Orientation can be inferred by using multiple targets. Video tracking has the advantage of providing a very fast, accurate signal. It has the disadvantage of requiring a clear line of sight from camera to target, and so is more suited for head tracking than hand tracking in general circumstances. Other disadvantages include the complexity of the camera and video processing setup.

- *Inertial tracking:* Inertial tracking uses gyroscopes and accelerometers to measure the accelerations of the tracker, which are integrated to provide the position, orientation, and velocity. The primary advantage of inertial tracking is that it does not rely on the proximity of source and sensor and so can in principle track over very large volumes. The disadvantage is that errors in the acceleration measurements accumulate over time. As of 1996 practical accelerometers result in errors which become unacceptable within 30 to 60 s, so inertial tracking will not be further discussed here.

- *The global positioning system:* The global positioning system (GPS) is a system of satellites used for navigation. While the simplest civilian use of GPS results in position inaccuracies of a few meters, the use of *differential GPS*, which uses a GPS receiver in a known location to correct for errors in other, nearby moving GPS receivers, can provide accuracies of a few millimeters. This technology is experimental as of 1996 but shows great potential for many virtual reality applications.

Position Tracker Errors

There are two types of errors associated with trackers:

Static error: The difference between the actual position of the tracker and the position returned by the tracking system. Static error occurs in both position and orientation and is typically a function of the position of the tracker. A position error datum is typically a three-dimensional vector. Limits on the magnitude of this vector are usually provided by the tracker manufacturer. Orientation error data are usually expressed in terms of Euler angles, with limits on the error components provided by the tracker manufacturer. The source of static error will depend on the tracker technology.

Dynamic error: The difference between the history of the actual tracker position over time and the position data returned by the tracker system. The dominant sources of dynamic error are delays due to the time required to process the hardware tracker data and due to filtering to eliminate noise in the tracker signal. A second source of dynamic error is the suppression of high frequencies of motion due to the filtering. The hand and head, however, rarely have significant frequencies of motion above 5 Hz, and most filters do not suppress frequencies in this range.

Error Correction

Tracker errors can be corrected when a model of what the signal should be exists. The specific methods depend on whether the correction is to static or dynamic error.

- Static-error correction can be performed when the error is approximately static over time by directly measuring the error by taking tracker data at a known location and building a lookup table of error values indexed by measured position. The error at a given measured position can then be interpolated from the error table and added to the measured position.

- Dynamic-error correction requires a model of the motion of the tracker sensor. Such models are available for head tracking and are based on models of head motion supplemented by a noise factor. Such models are usually implemented via Kalman filters.

Using Head Tracker Data

Head tracker data are usually converted into a 4×4 homogeneous matrix containing both position and orientation information, inverted and multiplied onto the transformation stack of the geometry engine rendering the virtual environment. If V is a vertex to be rendered, and M_{head} is the 4×4 position and orientation matrix describing the user's head, then the transform of V is given by $M_1 M_2 M_3 \cdots M_n M_{\text{head}}^{-1} V$, where $M_1, M_2, M_3, \ldots, M_n$ are the various local transformations of V.

Using Hand Tracker Data for Direct Manipulation

Hand tracker data are used for either selecting or picking up and manipulating an object. There are several methods of selecting an object using hand tracker data:

- *Pointwise collision:* The user's hand position is represented by a simple point, the location of the hand tracker. Alternatively, when finger joint angle information is available, the point may be the tip of one of the user's fingers. An object is selected when that point is within a specified distance from the object.
- *Geometrical collision:* A geometrical model of the user's hand is constructed and used to detect polygon-by-polygon collisions with objects in the virtual environment. This type of selection is more compute-intensive but can more accurately mimic real-world object grasping.
- *Raycasting:* A ray is drawn from the user's hand position in a direction determined by the orientation of the user's hand and finger joint angles (if they are measured). If this ray intersects an object in the environment, then that object is considered selected.

Once an object is selected, it may be "picked up" and moved about by the user's hand. Following (Robinett and Holloway 1992), this picking-up operation is performed by constraining the object's local transformation to be held constant relative to the user's hand transformation as that hand transformation changes. Let the hand transformation relative to the world coordinate system be M_{hand}, and let M_{object} be the object transformation relative to the world coordinate system. Then the orientatin of the object relative to the hand is $M_{\text{hand}}^{-1} M_{\text{object}}$. If M^{old} is the matrix from the last measured transformation of the hand or object, then, given a new hand transformation, the new object transformation is given by

$$M_{\text{object}} = M_{\text{hand}} M_{\text{hand}}^{\text{old}}{}^{-1} M_{\text{object}}^{\text{old}}$$

Gesture Recognition

Many virtual reality systems measure the finger joint angles of the user's hand. These finger joint angle data can be used to infer user commands based on hand gestures. There are two types of hand gestures: static and dynamic. In both cases gesture recognition is a problem of pattern recognition: recent or current measured hand data are matched against values which define a gesture.

A static gesture is defined as a particular configuration of the finger joint angles. The configuration is identified by demanding that each finger joint angle fall within a specified range. For example, the "fist" gesture may be defined as all finger joint angles greater than, say, $60°$. When detected, the application may interpret the fist gesture as a "grab and move" command. The use of many static gestures implies that the matching ranges are small, which makes recognition more sensitive to errors in the joint angle data.

Dynamic gestures are defined by the recent history of the finger angles and the hand position and orientation. The pattern recognition problem for dynamic gestures is complex and subtle and is best addressed by neural network techniques.

The measurement of finger joint angles takes place via a variety of technologies, including optical fibers which have been degraded so that the transmitted light is attenuated when the fiber bends, strain sensors which provide the strain on the sensor as the sensor is bent, and direct angle measurement via optical encoders or variable resistors. All of these technologies require calibration, which will depend on the details of the technology and the way in which the sensors are mounted on the hand. The sensor mountings vary from individual to individual and vary over time as the user's hand moves, introducing errors of as much as $10°$ into the calibration.

Significant issues arise in the definition and interpretation of gestures. While gestures are common in the real world, they are mostly nonconventional with different individuals making the "same" gestures differently. One guiding design principle is to reserve hand gestures for manual tasks such as pick and place, implying two conventional gestures: "fist" for picking up and moving objects, and "point with index finger" for selecting objects. Such use of small numbers of gestures allows large pattern-matching ranges and relative insensitivity to joint angle measurement errors.

61.4 Software Architectures

Because virtual environments are computer-generated, the software architecture used to maintain and operate the environment is of very high importance. The virtual reality performance requirements of frame rates greater than 10 frames/s and delays less that 0.1 s demand very high performance in terms of all tasks that must take place for the presentation of the virtual environment. We shall classify these tasks as shown in Fig. 61.1

First the user state is read, typically measuring head and hand position and orientation and command gesture. Then the environment state is computed. In some applications such as simple environment walkthroughs with static environment contents, very little computation may take place. Other applications may require a great deal of computation, which may include data access

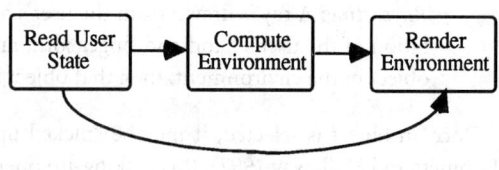

FIGURE 61.1

from mass storage or network communications. One example is a complex walkthrough with moving environment objects, which requires scene culling and other graphics optimization calculations. Another example requiring large amounts of computations is a visualization application, in which the large amounts of computation may result from the movement of an object in the environment, and may include access to large amounts of data from mass storage or a network.

After the environment is computed, it is rendered from the user's point of view as measured by the head tracking technology. In applications which contain a large amount of computation, the computation times may not satisfy the virtual reality frame-rate requirement of greater than 10 frames/s. In addition, the user's head may move, requiring a rerendering of the virtual environment from the new point of view even if the environment contents have not changed. User data obtained at the start of the computation phase will be out of date by the time the rendering phase starts if the rendering waits for the computation. This delay may violate the virtual reality delay requirement of less than 0.1 s. The use of out-of-date head tracking data for rendering will result in a swimming image, which typically induces motion sickness.

For these reasons, it is highly desirable to decouple the computation and rendering phases of the virtual environment, allowing the rendering process to run as fast as it can using the most recent head tracker data. This is accomplished by using one process for the rendering and another for the computation. The most efficient architecture is to have these processes on a single hardware platform implemented as lightweight processes communicating via shared memory, an architecture supported by many workstation vendors. Another option is to have the processes on separate platforms communicating over a network. The rendering and computation processes may themselves be multiple processes. The user data may be read in the rendering process, communicating the user data to the compute process, or the user data may be read in an additional process. An example using shared memory and reading the user state in one process is shown in Fig. 61.2

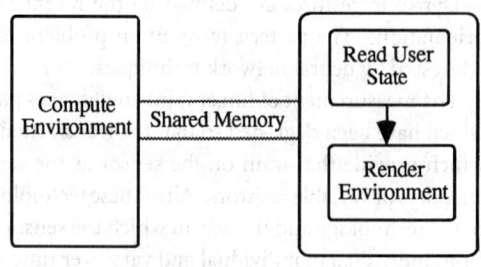

FIGURE 61.2

Polling vs. Events

There are two types of user input for the control of the virtual environment: discrete commands, and continuous motions such as those that arise when directly manipulating an object. This distinction is reflected in the choice of software architecture used to access the data from the interaction hardware. Discrete commands have the property that they must be executed in the order in which they were given by the user. Continuous commands for direct manipulation have the property that they should minimize delay, so that, for example, the position of the user's hand is reflected as accurately as possible (up to specified coordinate transformations) by the position of the environment object showing the position of the user's hand.

We define the *event* interaction architecture as a queue of commands which is read by the application program. Commands may be put on this cue in response to hardware interrupts, polling by a deeper level of the software, or some other method. For our purposes the principal property of an event queue is that it preserves the history of the command sequence, assuring that commands appear on the queue in the order in which they were given by the user. For a wide class of discrete commands, event queues are critical for the correct execution of user commands. Event queues, however, introduce delays and have no built-in sense of simultaneity.

Direct manipulation requires that the scene presented to the user reflect, as closely as possible, the state of the user at the present time, with no sense of history. Polling offers an effective method of taking a "snapshot" of the user's current state. Sampling the user data just before they are used minimizes delays. Simultaneity in the virtual environment can be defined as a single poll of all interaction devices.

Time-Critical Structures

Meeting the virtual reality frame-rate requirement can be a challenging aspect of virtual reality application design. This challenge is made more difficult by the often unpredictable nature of the virtual environment contents and activities, particularly in application environments which allow many user options. In many cases, applications may naturally take more computation or rendering time than the virtual reality frame-rate requirements would allow. In such cases algorithms must be built into the application which determines how much time is available and perform the application task as well as possible in that time. The result is typically a "graceful degradation" of the quality of the application for the purpose of maintaining virtual reality frame rates and responsiveness. While at first sight such degradation may seem to be highly undesirable, many users have indicated a clear preference for increased responsiveness at the cost of quality, so long as the loss of quality is appropriately understood. Designing algorithms which meet a specified time budget and a software architecture which supports these algorithms is known as the problem of *time-critical structures*. Time-critical structures falls naturally into two classes: *time-critical rendering* and *time-critical computing*.

Time-Critical Rendering

The time requirements of a rendering process depend on the method of rendering used, but for most graphics architectures this time has two components: the vertex projection rate, or the time required to compute the screen coordinates of a three-dimensional vertex, and the pixel fill rate, or the time required to fill in a solid object in screen space such as a polygon. Given a particular graphics architecture, the vertex projection time is determined primarily by the number of vertices that must be rendered, which is related to the scene complexity and the way in which the vertices are rendered (an object built out of disconnected triangles will have to render more vertices than the same object built out of triangle strips). Conventional methods of reducing the number of vertices to be projected include optimization techniques such as scene culling, so that only vertices that are potentially visible to the user are sent to the graphics system, and the use of triangle and rectangle strips. These techniques do not degrade the quality of the virtual environment and should be used regardless of the time budget, so they are not considered time-critical. The pixel fill rate is determined in part by the shading algorithm used, including the lighting conditions.

Phong shading, for example, results in a slower rasterization than Gouraud shading, which is slower than flat shading.

Time-critical rendering techniques may use, for example, changes in scene complexity to limit the number of vertices and changes in lighting or in shading algorithm to affect the pixel fill rate. All of these choices can result in variations in scene quality, so they must be designed and implemented with great care. The appropriate choice of these parameters can be made through the use of cost and benefit functions for the rendering of an object (Funkhouser and Sequin 1993). The cost is roughly measured by the time required to render the object, while the benefit is related to such parameters as the object's size on the screen, whether the object appears in the center of the user's field of view is measured by head tracking, the state of motion of the object, and direct indications by the user. The benefit function measures the desirability of a certain object complexity, measured in terms of polygons, in a particular circumstance. The characterization of the benefit function is an area of ongoing research. Given cost and benefit functions, time-critical rendering becomes the problem of assigning a time budget to each object so that the ratio of total benefit to total cost for all time-critical objects in the environment is maximized. The most common way of varying the cost of an object is by reducing the number of vertices in the object, usually performed by selecting among several precomputed representations of the object at varying levels of detail.

Time-Critical Computing

Time-critical computing is the problem of designing computational algorithms which meet a specified time budget by degrading the quality of the output of the computation. As the computations that take place in an application are very application-dependent, time-critical algorithms will be very application-dependent. We will therefore confine ourselves to general comments, illustrated by a few examples.

When the results of the computation do not change very much from frame to frame, the results of a given frame's computation can be used to estimate the cost and benefit of an object in the next frame. Great care must be used, however, when determining what to do if the computation is too costly. For example, objects are of interest if they appear in the user's field of view and are not of interest if they are not. If the computation determines the motion of the object, that computation must be accurately performed in all frames to determine if the object appears in a later frame.

The cost of a computation can be controlled in a variety of ways depending on the computation algorithm: solutions of differential equations can be chosen for higher speed at the cost of accuracy; the computation of extended objects can be truncated after an allotted time; and linear approximations to more complex behavior can be implemented.

Navigation

At its most basic, navigation is the problem of controlling the user's point of view in the virtual environment. Head tracking is used for navigation over small distances and for controlling the orientation of the user's viewpoint. The effective envelope of head tracking is determined by the technology used to track the user's head. Many virtual environments are, however, considerably larger than the useful envelope of the head tracker. In these cases a method must be devised to allow the user to "move about" over large distances.

Movement over distances greater than those supported by head tracking is typically implemented by the addition of an additional transformation which determines the relationship of the head tracking coordinate system to the virtual environment's graphical world coordinate system. Changing this new transformation has the effect of moving the user's point of view as if the user were in a vehicle, so we call this new transformation a *vehicle transformation*. As navigation via head tracking is a well-understood problem, we define the problem of navigation as that of controlling the vehicle transformation. Several methods of controlling the vehicle transformation have appeared in various virtual reality systems:

- *Point and fly*: A three-dimensional orientation tracking device, usually a hand tracker but sometimes the head tracker, is used to indicate a direction in space. When the appropriate command is given, usually through a hand gesture or a button press, the vehicle transformation is translated in the indicated direction, resulting in the effect of flying through the scene. Variations on this method include using a continuous parameter such as the distance from the hand to the body or the angle of bend of a finger when using a dataglove-type device to control the speed of motion. Joystick-type devices can also be used to control both the speed and direction of motion. The primary advantage of the point-and-fly navigation method is its intuitive nature. Its primary disadvantages are poor control and the time required to travel large distances.

- *Teleportation:* When the desired location is known, the vehicle transformation can simply be set to that location. This location can be accessed from a list, indicated by a direct command, or indicated using a miniature representation of the virtual environment. This method has the advantage of speed and the disadvantage of a practical limit on the number of available target locations.

- *Direct manipulation of the environment space:* Rather than creating the effect of the user moving through the environment, limited navigation can be accomplished by providing the effect of the user moving the environment itself. One implementation is to use a hand tracking device to "grab open space," which results in the vehicle transformation being changed by the inverse of hand manipulations. This move may allow the user to control both translation and orientation of the environment. Repeated moves allow the user to manipulate the environment over moderate distances. The advantage of this method is accurate control over the destination of the navigation move and easy control over the orientation of the environment. The disadvantages include a limited operating range.

- *Variable scale:* This method solves the navigation problem by using a scale factor in the vehicle tranformation to shrink the virtual environment so that everything in the environment is within the effective envelope of the head and hand trackers. The hand tracker can then be used to indicate a new desired location, which determines the origin of subsequent scale operations. When the scale is increased again, it is around the new origin at the desired destination, so the user has the experience of ending up at the desired location. Variable scaling is best suited for applications which do not have a preferred scale. This method has the advantages of rapid operation over potentially very large distance scales, and fine control at a user-selectable scale. The disadvantages include the complexity of the navigation operation (scale, select new location, scale).

- *Manipulation of miniatures:* This method uses a miniature model of the environment which contains a representation of the user. Direct manipulation of the user's representation in the model controls the vehicle transformation, resulting in the effect of the user moving to the representation's location in the full-scale environment. It has been found (Pausch et al. 1995) that the most effective and least disorienting way of setting the vehicle transformation using the miniature is to interpolate the vehicle transformation between its original scale, orientation, and translation and that of the representation in the model. This results in the user's experience of the modeling expanding to become the full-scale environment seen from the desired new location and orientation (or equivalently the experience of zooming into the model). This method is appropriate for applications which have a preferred scale, such as an architectural walkthrough. Manipulation of miniatures has the advantage of an intuitive and simple control interface. The disadvantages include the requirement of the model of the environment and the limitations to effective navigation over varying scales using the limited scale of the model.

Virtual Objects

Careful design of the objects in the virtual environment is critical to implementation of a successful application. Such objects fall roughly into two classes:

- *Application objects:* Objects which are directly related to application tasks. These objects will have properties and behaviors directly relevant to the application task. Interaction with these objects will depend on the individual object and its task.

- *Interface objects (widgets):* Objects which exist solely for user control of the environment. Interface objects can provide a layer of a conventional interface between applications. Extensions of conventional graphical user interfaces such as menus and sliders inserted into the virtual environment are examples of interface objects. Using raycasting to select menu items (Jacoby and Ellis 1993) has been shown to be a fruitful implementation of menus in the virtual environment.

Some interface objects will be intimately connected to the application, so the distinction between application and interface objects can become very weak.

The extensive use of direct manipulation in the virtual environment raises the issue of arm fatigue: constantly holding the hand in front of the face when performing extended tasks can be tiring. For non-see-through displays, where the user's physical hand is not visible, one solution to the fatigue problem is to offset the hand cursor from the actual hand position so that when the hand is held low by the user's side the cursor appears in front of the user's face in the virtual environment. Experience has shown that people adapt very quickly to such offsets and these offsets do not impair task performance.

Many objects in the virtual environment will be autonomous, with behavior reflecting such things as simulated physics, information displays, or (in the case of simulated humans) complex volition. The maintenance of autonomous objects and their interactions with the user and other objects can place significant computational burdens on the virtual environment. Such autonomous behaviors will be highly application-dependent and will take advantage of the behavioral modeling literature from conventional computer graphics (Foley et al. 1990).

61.5 Environment Design Concepts

The ability to tailor a virtual environment for an application combined with the lack of a conventional interface layer makes the design of the virtual environment a challenging task. The use of metaphor in the design process has proven to be a useful guide. For our purposes a *metaphor* is a mapping from an application task to a task which is well understood by the user. The design of the virtual environment should be driven by the application, with metaphors from the application domain as the guiding design principle.

Metaphors in the virtual environment can appear at several levels:

- *Overall environment metaphor:* What is the driving metaphor of the application? How does that metaphor determine the overall appearance of the environment? The overall environment metaphor may include user navigation metaphors.

- *Object interaction metaphors:* What is the metaphor for interaction with objects in the virtual environment? There may be several classes of objects, with each class having its own metaphor. For example, an environment may have application objects which are picked up and moved and fall when they are released. The same environment may also have interaction objects which can also be picked up and moved but do not fall when released.

- *Individual object metaphors:* Individual objects in the environment may have their own metaphors which determine their appearance and behavior. A data display object may have a numerical appearance, while a training simulation object may faithfully mimic a real-world object.

When considering all of these levels of metaphor, the following questions should be asked:

- *Is there a metaphor intrinsic to the application?* An example of an intrinsic overall environment metaphor is "walking about the interior of a building" for an architectural walkthrough application. An example of an intrinsic object metaphor is a bat in a virtual baseball application.

- *Is there a metaphor from the language of the target user community?* The user community for the application may have conventions which can be the basis for design metaphors. These metaphors need not be understood by people outside the target user community. A common example of such conventions is the use of menus, sliders, and buttons for user interfaces.

61.6 Distributed Virtual Reality

The implementation of a virtual reality application distributed across several hardware platforms is an important capability, allowing physically separated users to share an environment in a collaborative setting and allowing remote access to large, high-capability systems. The virtual reality performance requirements of greater than 10 frames/s and delays of less than 0.1 s put extreme demands on the network systems involved in such distribution.

Several issues arise in the design of a distributed virtual environment:

- *Network capability:* Primarily the network's bandwidth and latency characteristics. Local area networks (LANs) have typically higher capability than wide area networks (WANs), as LANs usually have a higher bandwidth and lower latency. WANs will have latencies which will generally increase as more nodes are traversed.

- *Minimizing network traffic:* The amount of traffic involved in the operation of the virtual environment should be minimized. This is usually accomplished by transmitting only changes in the environment. Further minimization can be attained when environment changes can be modeled by each system. For example, when the position and velocity of, and forces acting on, an object are known, that object's motion can be predicted so long as no other force acts on that object. In this case only changes in the forces acting on the object need to be transmitted, avoiding constant messages indicating changes in position of the object.

- *Scaling with number of participants:* Some applications are designed for only a small number of participants, while others are designed for an unlimited number of participants. Architectures such as client–server and peer-to-peer unicast that require sequential maintenance of each participant will not have good scaling behavior. Multicast network protocols alleviate this problem somewhat. The NPSNET system (Macedonia et al. 1995) takes advantage of locality information, building multicast groups out of participants which are near each other in the virtual environment. In this case participants who are not near each other are assumed to be far apart.

There are several models of distribution that can be used for virtual reality:

- *Client–server:* In a client–server architecture the environment is maintained on a single computer system (which may itself be distributed over several components), with the state of the environment sent to client workstations which display the environment to the users. As the environment is maintained by a single system, issues of consistency are easily dealt with. User interactions are transmitted to the server, where they are interpreted and change the state of the environment.

- *Peer-to-peer:* In the peer-to-peer architecture, all systems maintain a model of the environment and changes in the environment are sent to each of the other systems. The messages may be sent to the other systems individually by address or via a multicast network message.

A natural issue that arises in any distributed application is that of standards. The lack of standard user interfaces and behaviors in virtual environment design impedes the implementation of standards for distributed virtual environments. Standards have been adapted for particular applications, most notably the Distributed Interactive Simulation standard developed for military simulators. Standards for three-dimensional graphics have also been applied to virtual reality. The Virtual Reality Modeling Language (VRML)(Pesce 1995) is one notable example of a graphics standard, which at the time of this writing is being extended to include simple behaviors.

61.7 Application Evaluation and Design

Virtual reality is based on a paradigm of highly responsive, inherently three-dimensional interaction and display. While this virtual reality paradigm can be highly advantageous when appropriately used, meeting the virtual reality performance requirements can be very difficult to achieve in a real application setting. Further, the new interaction concepts of virtual reality combined with the unusual and often flawed hardware interfaces can impede the acceptance of a virtual reality application by the target user community. Thus careful consideration must be given to the following two questions: "Is virtual reality valuable to the application?" and "Is virtual reality viable for the application?" We shall briefly consider approaches to these questions. As with all application design processes, we strongly recommend that input from the target user community be used when addressing these issues.

Is virtual reality valuable for the application? Virtual realtiy interfaces use inherently three-dimensional interaction and display technologies and concepts. Therefore a virtual reality interface is appropriate to an application if that application has a three- (or more-) dimensional spatial aspect. Classes of examples include: simulations of real-world activities, objects, experiences, or phenomena; abstract simulations such as scientific visualization; abstract experiences for artistic or entertainment purposes; or information displays which can be usefully mapped into a three-dimensional environment such as networking visualization. Examples of classes of applications which would probably not directly benefit from a virtual reality interface include: text-based applications; inherently two-dimensional applications such as image processing (though some image-processing techniques benefit from a three-dimensional representation); and applications which have no inherent spatial display content.

The value of a virtual reality interface for a particular application can be estimated in more detail by considering display and interaction separately:

- *Display: How would the three-dimensional display be used? What would be the role of head tracking?* These questions can be addressed by sketching out the expected appearance of the application environment as experienced by the user, including considerations of how the user would move about in the environment.
- *Interaction: How would the three-dimensional interaction capabilities be used? What is the role of direct manipulation?* These questions can be addressed by identifying representative application tasks that require three-dimensional interaction and sketching out how these tasks are performed.

If the answers to these questions are positive, then a virtual reality interface is probably useful to the application.

Is virtual reality viable for the application? Once the value of a virtual reality interface for an application has been established, it must be established that it is possible to implement that application in a way consistent with both the virtual reality performance requirements and with the current state of the art in both the computational and graphics capabilities and the virtual reality interface hardware. These issues fall into several categories:

- *Can the graphics for the application environment be rendered with a frame rate of 10 frames/s or higher?* Are there more polygons in the scene than can be rendered with available hardware in 0.1 s? Does the rendering require computationally intensive techniques such as volume rendering, raycasting, or sophisticated lighting techniques such as radiosity? While some high-end hardware platforms can perform some of these tasks with the performance required, will such platforms be available to the target user community? Failure to meet the graphics frame-rate requirement may result in a system which does not support head tracking and so does not take advantage of the virtual reality interface.
- *Can direct-manipulation-based application tasks be performed in less than 0.1 s?* Direct-manipulation-based tasks are those that occur in response to a user's continuous motion. Examples include the movement of an object in the virtual environment, which may trigger complex collision detection computations, or a data probe in a scientific visualization environment, where motion triggers the computation of new data displays. These computations may require access to data stored on disk or information over a network. Simple estimates of the computation time required can be made by counting the number of objects required for a collision detection computation or counting floating-point operations in a data display application. If it is found that the application

task probably cannot be performed in 0.1 s or less, possibilities of limiting the task, changing how data are represented, or time-critical techniques which quickly provide approximations can be examined.

- *Do the currently available virtual reality interface hardware devices provide the required accuracy or qualtiy?* At any given time virtual reality interface devices provide only a particular level of quality. Do the available/affordable display devices provide sufficient image quality, resolution, field of view, and/or level of comfort required by this application? Do the trackers appropriate for the application provide the required accuracy over an appropriate range in the environment in which they will actually be used? Is the interface hardware convenient to use? Does the interface provide sufficient benefit to justify any inconvenience? Accuracy issues are particularly critical for augmented reality applications.

61.8 Case Studies

Architectural Walkthrough

An architectural walkthrough is an application which simulates walking through a building. This application is usually used to evaluate the building before it is actually built. The overall environment metaphor is that of being in a building, and there may or may not be individual interactive objects in the building. The navigation metaphor is largely determined by the available interface hardware, and is typically the point-and-fly paradigm. Collision with walls is not typically implemented. Time-critical graphics has been used in these walkthroughs to maintain a constant frame rate.

The Virtual Wind Tunnel

The virtual wind tunnel (Bryson et al. 1995) is an application of virtual reality to the visualization of simulated airflow. The overall environment metaphor is an aircraft body with airflow around it. Various tools are available to visualize the airflow, including streamlines, isosurfaces, and cutting planes. These tools are fully interactive, with movement by the user causing the recomputation of the visualization geometry. The recomputation and display of the visualization geometry allows real-time exploration of complex airflows, providing rapid understanding of the simulation. The virtual wind tunnel uses two asynchronous process groups, one for the computation of the visualization geometry and the other for rendering. Visualization geometry computation uses parallel processing when available. These two processes are typically in the same lightweight process group and communicate via shared memory, but they may be on separate systems communicating over a network in a client–server mode. Several users may be connected to the same server, providing the virtual wind tunnel with a shared-use mode.

Time-critical computation in the virtual wind tunnel is implemented to maintain the responsiveness of the visualization tools when they are moved and to maintain animation rates for time-varying flows. The time-critical computation at a given frame is based on the assignment of a time budget to each visualization geometry computation. The ratio of the total computation time for all visualization geometry computations from the previous frame to the allotted time then multiplies these time budgets, providing a new time budget for each visualization geometry computation. It is then up to the individual computation to decide how to meet the new time budget, either by reducing the size of the visualization computed or by reducing the accuracy of the visualization by switching to a simpler, faster algorithm.

61.9 Research Issues

Essentially all aspects of virtual reality involve research issues. Critical research issues fall into roughly the following categorizations:

- *Human factors:* What is the impact on task perfomance of tracker inaccuracies, frame rates, end-to-end delays, the lack of full sensory feedback, and other technological distortions of experience in virtual reality? What is the appropriate approach to three-dimensional interfaces for task performance? What is the classification of tasks which are appropriate for virtual reality? What enhancements to the virtual environment can be inserted to aid task performance? How do we resolve the tension between application-specific and standard conventional interfaces?
- *Software:* Time-critical rendering and computation concepts, operating systems which support high-resolution scheduling and time-critical operations, and software structures for the rapid design of environments including complex behaviors need to be more fully developed.
- *Hardware:* Technologies need to be developed which deliver immersive visual displays with acceptable ergonomics, with the ideal being the form factor of sunglasses. While three-dimensional computer workstations, networks, and mass storage systems are a mature technology, they are usually developed to maximize throughput with little attention to latency. Developing hardware systems with both high throughput and low latency is a critical need for virtual reality.
- *Design:* Design methodologies for virtual environment applications are currently lacking. A classification of applications and useful approaches for virtual environments is required.

61.10 Summary

Virtual reality is an interface paradigm which relies on the effect of presenting the user with a computer-generated three-dimensional world which contains interactive objects that have three-dimensional locations independent of the user. Virtual reality is an inherently three-dimensional approach to human–computer interfaces which stresses responsive interaction with the virtual environment, which mimics real-world interaction. The three-dimensional aspects and high-performance requirements of virtual reality provide a platform for simulating real-world tasks, interactive entertainment, and exploration of complex data. These requirements also place significant stresses on the performance characteristics of the systems supporting the virtual environment.

Interface hardware for virtual reality involves several sensory modalities including visual, audio, touch, and force. The visual and audio display technologies are relatively mature, with the touch and force technologies requiring further development as of 1996. Virtual reality systems rely on tracking the user in three dimensions using technology which introduces errors and limitations. Working with these errors and limitations is one of the primary challenges of virtual reality application development.

Software systems for virtual reality are typically oriented towards specific applications or application domains. The high-performance requirements of virtual reality place unusual demands on the computation, rendering, and data management software that underlies a virtual reality application. Effective use of multiple processes and optimized, carefully written codes are critical for the success of a virtual reality application. Time-critical structures which gracefully degrade quality in order to maintain performance are important components in virtual reality software systems.

Virtual environment design raises further opportunities and challenges. The appropriate use of three-dimensional display and interface for a specific application must be approached on a task-by-task basis. The use of interface metaphors from the real world and the application domain naturally leads to the design of application-specific interfaces. The implementation and design of standard interfaces in this context is a challenging and open issue.

In spite of its difficulties, virutal reality allows the development of environments tailored to the best way for a user to perform an application task.

Defining Terms

Head-coupled display: A display which moves with the user's head, typically allowing the user to see the screen regardless of the position and orientation of the user's head.

Head-mounted display: A display which is rigidly mounted on the user's head. A subset of head-coupled displays.

Head tracking: The measurement of the position and orientation of the user's head, usually used for rendering the three-dimensional scene from the current point of view.

Spatial constancy: The property of having a spatial location when viewed from a moving point of view.

Spatial presence: The property of having a spatial location relative to and independent of a viewer in three-dimensional space.

Virtual reality: Use of computer systems and interfaces to create the effect of an interactive environment, called the virtual environment, which contains objects which have spatial presence. Also known as virtual environments or virtual worlds.

References

Begault, D. R. 1993. *3-D Sound for Virtual Reality and Multimedia.* Academic Press Professional, Cambridge MA.

Bryson, S. 1993. Impact of lag and frame rate on various tracking tasks. In *Proc. SPIE Conf. on Stereoscopic Displays and Applications*, San Jose, CA.

Bryson, S., Johan, S., Globus, A., Meyer, T., and McEwen, C. 1995. Initial user reaction to the virtual windtunnel. AIAA 95-0114. In *33rd AIAA Aerospace Sciences Meeting and Exhibit,* Reno, NV.

Burdea, G. and Coiffet, P. 1994. *Virtual Reality Technology.* Wiley, New York.

Foley, J., van Dam, A., Feiner, S. K., and Hughes, J. 1990. *Computer Graphics: Principles and Practice.* Addison–Wesley, Reading, MA.

Funkhouser, T. A. and Sequin, C. H. 1993. Adaptive display algorithm for interactive frame rates during visualization of complex virtual environments. In *ACM SIGGRAPH '93 Conf. Proc.* Anaheim, CA, Aug.

Jacoby, R. H. and Ellis, S. R. 1993. Using virtual menus in a virtual environment. In *Proc. Symp. Electronic Imaging Science & Technology*, Vol. 1668. International Society for Optical Engineering/Society for Imaging Science & Technology.

Macedonia, M. R., Zyda, M. J., Pratt, D. R., and Barham, P. T. 1995. Exploiting reality with multicast groups: a network architecture for large scale virtual environments. In *Proc. 1995 IEEE Virtual Reality Annual Int. Symp.* IEEE Computer Society Press, Research Triangle Park, NC, Mar.

Pausch, R., Burnette, T., Broackway, D., and Weiblen, M. E. 1995. Navigation and locomotion in virtual worlds via flight into hand-held miniatures. In *ACM SIGGRAPH '95 Conf. Proc.* Los Angeles, CA, July.

Pesce, M. 1995. *VRML: Browsing and Building Cyberspace.* New Riders Publishing, Indianapolis. IN.

Robinett, W. and Holloway, R. 1992. Implementation of flying, scaling and grabbing in virtual worlds. In *Proc. 1992 Symp. on Interactive 3D Graphics,* Boston, MA.

Further Information

The following books provide surveys of the virtual reality field:

Durlach, N. and Mavor, A. S., eds. 1995. *Virtual Reality: Scientific and Technological Challenges.* National Academy Press, Washington, DC.

Burdea, G. and Coiffet, P. 1994. *Virtual Reality Technology.* Wiley, New York.

Badler, N. I., Phillips, C. B., and Webber, B. L. 1993. *Simulating Humans: Computer Graphics Animation and Control.* Oxford University Press, New York.

Kalawsky, R. S. 1993. *The Science of Virtual Reality and Virtual Environments.* Addison–Wesley. Wokingham, England.

The following books provide a summary of human factors issues:

Boff, K. R., Kaufman, L., and Thomas, J. P. 1986. *Handbook of Perception and Human Performance,* Vols. 1, 2. Wiley, New York.

Ellis, S. R, Kaiser, M, and Grunwald, A. J., eds. 1993. *Pictorial Communications in Real and Virtual Environments*, 2nd ed. Taylor and Francis, Bristol, PA.

The following proceedings contain many important research papers of interest:

Proc. IEEE 1993 Symp. on Research Frontiers in Virtual Reality. IEEE Computer Society Press, San Jose CA, Oct. 1993.

Proc. 1993 IEEE Virtual Reality Annual Int. Symp. IEEE Press, Seattle, WA, Sept. 1993.

Proc. 1995 IEEE Virtual Reality Annual Int. Symp. IEEE Computer Society Press, Research Triangle Park, NC, Mar. 1995.

Proc. 1996 IEEE Virtual Reality Annual Int. Symp. IEEE Computer Society Press, Santa Clara, CA, Mar. 1996.

Singh, G., Feiner, S. K., and Thalmann, D., eds. 1994. *Virtual Reality Software and Technology 1994 Proc.* World Scientific, Singapore.

62

RenderMan®: An Interface for Image Synthesis

Anthony A. Apodaca*
Pixar

62.1 What is RenderMan?

The RenderMan interface is a standard interface between modeling programs and rendering programs capable of producing photorealistic quality images. In this context, a modeling program is any program which is capable of generating a detailed description of a 3D environment, including the positions, shapes, and material characteristics of objects in a scene; the positions and qualities of light sources; and the positions and parameters of virtual cameras. A photorealistic quality image is one in which the scene is captured as though viewed through a physical camera, with few or no artifacts of the computational process, such as visible geometric approximations, arbitrary limits on material texture or scene complexity, aliasing in spatial or temporal domains, or pixel color quantization. The standard interface is the "language" with which the modeling program describes the scene to the rendering program, in sufficient detail that the rendering program is able to generate the highest-quality image possible.

*RenderMan is a registered trademark of Pixar. *The RenderMan Interface Specification* and its bindings are copyright Pixar, and excerpts are used with permission.

RenderMan is, therefore, a *scene description interface*. It is designed and intended to be used to transfer descriptions of the visual characteristics of scenes (still or animated) from a large variety of modeling programs to a large variety of rendering programs, including both scanline and Z-buffer-based renderers and ray tracers and other global-illumination renderers. The interface does not specify *how* an image is to be rendered, but instead specifies *what* image is desired, and has features which are designed to accommodate the needs of both high-quality batch-oriented renderers and low-overhead real-time renderers.

Although it holds a powerful description of the parameters of a scene, RenderMan is not a 3D programming environment or a modeling database. It lacks many features that would be required of such systems, such as 2D drawing primitives, user interface capabilities, and nonvisual information about the objects in scenes, such as object interrelationships and user database records.

The RenderMan interface has two popular language bindings. The primary interface is to C. RenderMan renderers are available as C subroutine libraries, where each entry point has the library prefix **Ri**. The second interface is a bytestream archive (or **metafile**) know as RIB, where each directive is spelled identically except that it omits the **Ri** prefix. There is an almost one-to-one correspondence between C subroutines and RIB directives, the differences generally being due to C language syntax and linking issues. In the text of this chapter, all descriptive text will use the C library subroutine name, and all examples will be presented as RIB archive excerpts (for compactness of representation). For each RenderMan interface routine, the details of the C and RIB versions of that routine are described in the *RenderMan Interface Specification* [Pixar 1989].

RenderMan was developed in 1989 by Pixar, in association with a large number of other companies interested in graphics standards. While it is not an official graphics standard (like PHIGS, for example), it remains the only proposed standard which contains provisions for high-quality image synthesis, and it is a model for the functionality embedded in many proprietary interfaces to commercially available rendering programs.

62.2 Basic RenderMan Program Structure

A RenderMan program[1] uses RenderMan interface calls to describe four features of the virtual scene: the objects in the scene; the lighting of the scene; the camera viewing the scene; and the image capturing the view of the scene. Like other modern computer graphics **APIs**, RenderMan has a hierarchical **graphics state**. That is, it maintains a database of all of the attributes which can apply to geometric objects. At the time a geometric object is defined, it inherits its attributes from the database at that moment. So, to define an object, one first indicates where the object will be located in the scene, what color it will be, and any other material characteristics it will have and finally one specifies the object itself. The state is hierarchical, which means that there is a graphics state stack which can be pushed to save the current graphics state and popped to restore an old graphics state.

The RenderMan graphics state is divided into two parts: *options* and *attributes*. Options are those parts of the graphics state which are global to the entire image being computed, such as the resolution of the image or the position of the camera. Attributes are those parts of the graphics state which can change from one object to the next, such as the color of the object. A RenderMan program, similarly, is divided into two phases: first, setting up the global options, and second, describing the scene with all its attributes and objects. The boundary between these two phases is **RiWorldBegin**, which freezes the options and starts the scene description. **RiWorldEnd** signals the renderer that the scene description is complete and the image should be written out.

The following RIB archive demonstrates a very simple RenderMan program. The order of calls in a RenderMan program obeys one very simple rule: all the data necessary to render an object are specified before that object. This rule includes all camera information, lighting, modeling transformations, and

---

[1] In this chapter, a *RenderMan program* will refer to any sequence of calls to a RenderMan interface, whether compiled into a program written in a standard programming language or as a data stream encoded in a RIB archive.

object attributes.

```
##RenderMan RIB
# Image options...
Format 512 486 1.0
Quantize "rgba" 255 0 255 0.5
Display "foo.tif" "file" "rgba"
# Virtual camera options...
Clipping 1.0 1250.0
Projection "perspective" "fov" [25]
Translate 0 0 10.0
Rotate 45 1 0 0
# Scene description...
WorldBegin
        LightSource "distantlight" 1 "intensity" [1]
        Color [.2 .4 .8]
        Surface "plastic" "Ks" [.4]
        Sphere 1 -1 1 360
WorldEnd
```

Specifying Image Parameters

All of the parameters which define the manner of rendering to be used, determine output pixel quality levels and other such constraints, and provide the file format specifications of the image to be generated are known collectively as *image options*. Image options generally occur first in the RenderMan program before the virtual camera description.

RenderMan's standard image options include the name and file format of the image (set by **RiDisplay**), the image resolution (set with **RiFormat**), and the output pixel bit depth (set with **RiQuantize**). Individual renderers also typically have some set of renderer-specific rendering mode controls, and these are set with **RiOption** in a manner described in the documentation of the particular renderer.

Setting Up the Virtual Camera

The RenderMan virtual camera model has several basic parameters, such as position and viewing direction, projection and field-of-view controls, which are collectively known as *camera options*. In orienting the camera, RenderMan is similar to other modern graphics APIs: the camera is not an "object" in the scene, but rather the scene is laid out in front of a fixed camera. The camera is defined to be at the origin of *camera* coordinates, looking out over the positive *z*-axis, with *x* to the right and *y* up (a left-handed coordinate system). This is often confusing to the novice. However, since the camera's full relationship to the scene is fixed in space before any objects are presented to be imaged, this follows the basic rule that all information necessary to render an object precedes it in the RenderMan program.

The camera description goes at the head of the program, immediately after the image description, and immediately before **RiWorldBegin**. The camera description proceeds in two phases. First, the camera projection is specified with **RiProjection**. Typically, either an *orthographic* or a *perspective* projection is used, although some renderers may provide other alternatives. In the case of perspective projections, the camera description includes a perspective field-of-view parameter, which specifies the full-angle viewing frustum angle. It is also a good idea to set the near and far clipping planes to reasonable values, because most renderers can minimize floating-point quantization errors in internal calculations if they know the depth extent of the scene prior to the start of rendering. **RiClipping** should be called just before **RiProjection**. After **RiProjection** is called, the graphics state is considered to be in *camera* coordinates.

Second, the position and orientation of the camera are specified by providing the *camera-to-world* transformation matrix. For modelers which contain camera descriptions as objects in a global *world* coordinate system, the following routine **AimCamera** can be used to generate the appropriate camera view transformation:

```
#include <values.h>
#include <math.h>
#include <ri.h>


/*
 * Create camera transformation for a camera located at "from"
 * in world-space, looking at a point "to" in world-space.
 * Assumes the Y axis is up, and that the camera should not be
 * rolled. If world-space is right-handed, "rh" should be set
 * to nonzero.
 */
void AimCamera (RtPoint from, RtPoint to, RtInt rh)
{
    RtPoint axis; RtFloat mag, spin;
    double D, E;

    /* Create a normalized viewing axis vector */
    axis[0] = to[0] - from[0];
    axis[1] = to[1] - from[1];
    axis[2] = to[2] - from[2];
    mag = sqrt(axis[0]*axis[0]+axis[1]*axis[1]+axis[2]*axis[2]);
    axis [0] /= mag; axis[1] /= mag; axis[2] /= mag;

    /* If world space is right-handed, reflect left-handed camera
       space. */
    if (rh) RiScale(-1., 1., 1.);

    if (axis[0]==0. && axis[1]==0.) {
        /* Looking straight in or out */
        if (axis[2] < 0.) RiRotate(180., 0., 1., 0.);
    } else {
        /* Preroll camera to ensure Y stays "up" after main
           rotation */
        E = (axis[0]*axis[0] + axis[1]*axis[1]*axis[2]);
        D = (-axis[0] *axis[1] * (1. - axis[2]));
        if (E != 0. || D != 0.) {
            spin = (180./M_PI)*acos(E /sqrt(E*E + D*D));
```

```
            RiRotate(spin, 0., 0., 1.);
    }

    /* Rotate (Z . Axis) degrees around vector (Z ^ Axis) */
    mag = (180./M_PI)*acos(axis[2]);
    RiRotate(-mag, -axis[1], axis[0], 0.);
}
RiTranslate(-from[0], -from[1], -from[2]);
}
```

Scene Description

Once the image and camera options are specified, the actual scene description can begin. **RiWorld-Begin** freezes the options, marks the current coordinate system as *world* coordinates, and starts the scene description. RenderMan scenes are made up of three parts: the geometric primitives themselves; attributes which describe the appearance of those primitives; and the light sources which illuminate those primitives. These are organized in a hierarchy of objects and subobjects whose descriptions are managed through the hierarchical graphics state. In fact, there are two levels of graphics state hierarchies in RenderMan: the transformation stack and the attribute stack. The transformation stack is pushed using **RiTransformBegin**. Subsequent transformations modify the **current transformation matrix** until the stack is popped with **RiTransformEnd**, at which point the saved transformation matrix is restored. **RiAttributeBegin** pushes the attribute stack, which saves all of the graphical attributes of objects (colors, shaders, geometric approximation parameters, etc.), as well as simultaneously pushing the transformation stack. These values are all restored with **RiAttributeEnd**. Since the transform stack is a subset of the attribute stack, these two stacks can be nested in any order necessary. By saving and restoring the current attributes and transformation matrix in this way, complex subobjects can modify the current graphics state arbitrarily, without worry of affecting the state of objects following them in the RenderMan program.

Understanding Transformation Matrices

At any point in the RenderMan program, the RenderMan state machine contains a current transformation matrix. This matrix transforms objects from the "current" coordinate system to one of the standard coordinate systems. Any geometric object which is specified in the RenderMan program is positioned relative to the then-current coordinate system. There are a large number of RenderMan routines which modify the current transformation matrix and thereby move subsequent objects around.

There are two equally valid styles of "reading" a sequence of transformations, and it is useful to pick the one which seems most natural to you in order to analyze RenderMan programs. Both ways lead to the same resulting camera view. However, it is very easy to get confused if you don't realize the difference between the two styles.

In the first style, you imagine your eyes to be the camera (or a viewer in some other fixed coordinate system), and a hand in front of your face is the "current coordinate system." You read sequences of transformations from the top down, and each one moves your hand around just as the transformation says. In this style, rotations spin your hand, and directions are always based on the current orientation of your hand, so if you rotate and then translate, you move your hand in a different direction than if you only translated. When you then place an object, it goes into your hand. The camera sees the object just as your eyes see your hand.

In the second style, you imagine the object to be at the origin of a large gridded space. You read sequences of transformations from the bottom up, and each one moves the object relative to the grid. In this style, directions are always relative to the stationary grid, so rotating the object revolves it about the origin but does not change directions of travel. When you reach the top of the sequence, the object is in its final place. The camera is then placed at the origin and looks at the object from there.

RenderMan transformation matrices are 4-by-4 postmultiplication matrices stored in row-major order. That is, to transform a 3D point by a matrix, a row vector on the left is multiplied by a matrix on the right. The translation component of the matrix is "on the bottom," in elements 12, 13, and 14 of the 16-element C array.

62.3 Geometric Primitives

RenderMan supports a powerful set of geometric primitives. RenderMan's primitives are higher-level than those in many graphics APIs, which generally support only polygons of various varieties. RenderMan's curved surface primitives not only are more concise but also produce higher-quality images than meshes of polygons, and, in many situations, renderers can render these primitives more efficiently as well.

Polygons

Due to their overwhelming popularity and generality, several classes of polygons are supported by Render-Man. RenderMan distinguishes between simple convex polygons (**RiPolygon**) and general polygons (**RiGeneralPolygon**). A general polygon is any polygon which has concavities or which has multiple loops (such as holes or islands). Thus, while **RiPolygon** need only specify the polygon's vertex data, **RiGeneralPolygon** additionally specifies how many vertices are in each loop. This distinction between types is entirely for the benefit of the renderer, as rendering algorithms which know *a priori* that a polygon is convex are often faster than those which deal with more general cases. Any modeler which is unsure of the convexity of its polygons should simply use the general polygon interface, as it is likely to be faster overall than individually testing convexity in order to use the convex interface when appropriate.

In addition to individual polygon primitives, RenderMan supports polyhedra with shared vertices through the concept of points–polygons. A points–polygons primitive is defined by providing a single list of all of the shared vertices and also providing an index array which identifies which vertices (in which order) make up each individual polygon. As before, **RiPointsPolygons** specifies convex polygons with shared vertices, and **RiPointsGeneralPolygons** specifies general polygons by augmenting the data with an array specifying the number of loops in each polygon.

Parameter Lists

In RenderMan, 3D vertices are specified through a data structure known as a *parameter list*. Parameter lists are pairs of values in a variable-length argument list, where the first datum of the pair is the name of the parameter, and the second datum is a pointer to an array which contains the data for that parameter. Note that each "value" in the data array is often a sequence of floating-point numbers. For example, in polygons the vertex lists themselves are identified by the parameter **"P"**. In a polygon which had five vertices, this would be followed by an array of 15 floating-point numbers, which are the five 3D points specified in the current coordinate system.

The RenderMan interface predefines six variables that may be associated with the vertices of any polygon:

"P",	the 3D position of the vertex;
"N",	the smooth-shading surface normal for that vertex;
"Cs",	the color of the vertex;
"Os",	the opacity of the vertex;
"s" and **"t"**,	the texture coordinates of the vertex.

All RenderMan renderers understand these variables and are able to generate textured, Phong-shaded, and Gouraud-shaded images of the polygons by interpolating these values across the surface of each polygon when it is shaded. A RenderMan program can put any combination of these variables on any polygon in its model independently (obviously, **"P"** must be supplied for all polygons).

Quadrics

The RenderMan interface contains seven types of quadric primitives: **RiSphere**, **RiCone**, **Ri-Disk**, **RiCylinder**, **RiHyperboloid** (of one sheet), and **RiTorus**. In each case, the interface provides parameters for specifying a bounded region of the (potentially infinite) mathematical quadric surface. All the quadric primitives are specified about the origin or z-axis of the local coordinate system and are rotationally symmetric about the z-axis. For this reason, models which contain quadrics will typically have a transformation matrix before each to position it appropriately.

Quadric surfaces are *parametric surfaces*. That is, each point on the surface can be located by an equation in two variables. Limiting the range of these variables is the primary means of specifying the bounded quadric-surface regions. Because they are parametric surfaces, it is possible to interpolate other variables along their surface using bilinear interpolation of their parameters **"u"** and **"v"**. Hence, many of the parameter-list variables described above may be attached to quadric surfaces as well. **"Cs"**, **"Os"**, **"s"**, and **"t"** are all legal on quadrics. In each case, a quadric takes four data values in the array of any parameter, one for each of the four corners of the rectangular parametric range. Notice that, in many cases, the parametric corners will be coincident in 3D space (such as at the pole of a sphere); however, they will always retain their rectangularity in the 2D parametric space, and the interpolation remains consistent despite the singularity.

Patch Primitives

RenderMan provides a generous set of curved (higher-order) parametric surface primitives. The simplest of these is the bilinear **RiPatch**, and the most general is the nonuniform rational B-spline surface, **RiNuPatch**. In each case, the surface is a parametric surface of two variables, **"u"** and **"v"**, and the modeler provides the control vertices which define the surface. Control points are specified in the parameter list, just as polygon vertices are, but each primitive requires a different number of such control points, depending on its type. All the other parameter-list values that are legal on polygons are also legal on patch primitives.

RiPatch can be used in two ways: to create bilinear surfaces, or to create bicubic surfaces. Bilinear surfaces have four control points, and the surface created has straight lines between the four points and a curved surface spanning the space in between. Bilinear patches require four data values in each parameter list array.

Bicubic surfaces, on the other hand, have sixteen control points. The interpretation of the control points on a bicubic surface is dependent on the **basis matrix**. RenderMan supports an arbitrary basis matrix[2] in each parametric direction, through the **RiBasis** call, but the four most popular basis matrices (Bezier, B-spline, Catmull–Rom, and Hermite) are predefined. In any case, bicubic patches also require four data values in each parameter-list array (except the control points, of course, which require sixteen).

Patch control points can be specified with **"P"**, in which case they are 3D points and the patch is *polynomial*, or alternatively with a variable parameter-list **"Pw"**, in which case they are 4D *homogeneous* points and the patch is *rational*. Rational patches can take on a larger variety of shapes than polynomial patches can and so are often preferred when patches are used to approximate other "modeler-specific" primitives that RenderMan does not already know about.

Patch Meshes

For surfaces which are made up of a rectangular grid of patches (either bilinear or bicubic), RenderMan supports **RiPatchMesh**. The advantage of using a patch mesh over a set of independent patches is similar to using polyhedra: most of the control points are shared, so the data structures have less redundant

---

[2] Basis-matrix splines and spline surfaces are explained in greater detail in Chapter 55 of this volume.

data and are smaller, and the renderer does less redundant work to transform and manipulate them. Patch meshes can be declared *periodic* in one or both directions. Periodic meshes wrap, or close, in that direction (like a cylinder or torus) and thus share vertices across that edge as well.

Nurbs

RiNuPatch is the most general patch primitive, because it can be used to specify both uniform and nonuniform, and both polynomal rational, B-spline surfaces. Nonuniform, rational B-spline patches are sometimes called *nurbs* [Farin 1992]. RenderMan nu-patches can be of arbitrary order in either parametric direction, and also require an array known as the *knot* vector in each parametric direction to complete their description.

Nu-patches are unique among RenderMan primitives in that they can be "cut up" by the use of **RiTrimCurve**s. Trim curves define holes in the patch where the renderer simply pretends the patch does not exist. Trim curves are also nurbs, in this case 2D nurb curves, and they are drawn in the parameter space of the nu-patch.

Trim curves are an attribute of nu-patch primitives and are stored in the graphics state. Once a trim curve is defined, it affects all of the nu-patches that follow, until it is replaced by a new trim curve (or a null trim curve). It is also saved and restored on the attribute stack when the stack is pushed and popped. It is not uncommon, therefore, for RenderMan programs to push the stack, define a trim curve, define the single nu-patch that uses it, and then pop the stack.

Constructive Solid Geometry

RenderMan supports solid modeling via constructive solid geometry (CSG), providing Boolean operators between hierarchical subobjects in a scene. **RiSolidBegin** begins a node in a CSG tree and specifies the Boolean operator at that node. The Boolean operators include **"union"**, **"intersection"**, and **"difference"**. If the Boolean operator is **"primitive"**, then the node is a leaf and must contain a single or group of geometric primitives which (when taken together) define a closed surface. **RiSolidEnd** ends the node and pops the CSG hierarchy stack.

62.4 Shading Attributes

RenderMan's graphics state provides a rich set of attributes which allow the RenderMan program to control the surface appearance of objects in the scene. High-quality images require scene descriptions which have very high-quality surface appearance descriptions; therefore RenderMan programs should put as much effort into defining the appearances of objects as they do into defining the shapes and placement of them.

Color and Material Properties

The most obvious shading attribute is the surface color, set with **RiColor**. RenderMan also provides a colored opacity, set with **RiOpacity**, which allows translucent colored filters to be defined. However, the most powerful shading attribute is the set of three **shaders** that are applied to the object. A shader is a small program which describes how the material that an object is made from interacts with light and the environment to control how it looks. RenderMan has a set of built-in shaders which cover a wide range of surface appearances that are typical of computer graphics images. In addition, RenderMan has a **Shading Language**, which allows users to write their own shaders in order to model the characteristics of arbitrary materials.

The main shader is the **RiSurface** shader. Surface shaders determine the color of the surface of the object, based on its position, the light-source positions, and any other shading attributes of the object. RenderMan has built-in shaders for **"matte"**, **"plastic"**, and **"metal"** surfaces, which are implementations of the typical standard computer graphics lighting model described in most textbooks. These shaders take parameters such as **"Kd"** and **"roughness"**, which are familiar in that context.

The other object shaders are the **RiDisplacement** shader and the **RiAtmosphere** shader. Displacement shaders manipulate the shape of the object to add divots, dents, or other surface imperfections. In some renderers, displacements actually move the surface points, while in others the appearance of the displacements is merely approximated through the use of bump mapping. Atmosphere shaders manipulate the apparent color of the object by modeling the effect of the substance that lies between the object and the camera (for example, **"fog"**).

All shaders have parameters, and these are specified in the RenderMan program with parameter lists on the shader definition calls. Like geometric primitive parameter lists, this is an arbitrarily long list of name–value pairs; however, in this case, there is only one data value in each value array. Each shader has a default value for every parameter, so any parameter not set explicitly in the parameter list of the call receives the default values. RenderMan predefines a set of commonly used shader parameters (such as **"Kd"**), and this list can be extended arbitrarily to include any parameters of the user-written shaders. In order to add a parameter name to the list of recognized parameters, **RiDeclare** is used to indicate the Shading Language declaration for any parameter referenced in the RenderMan program.

Texture Coordinates

Texture maps are very commonly used to enhance the visual appearance of objects in computer graphics scenes. Specifying exactly where on the surface a texture map should appear and how it should be aligned is a key issue when using texture maps. RenderMan programs have two facilities for describing texture placement. First, each parametric primitive has a default 2D parameter-space mapping. These u, v parametric coordinates run from 0.0 to 1.0 along the whole length of the parametric rectangle of the primitive. **RiTextureCoordinates** defines an affine transformation between u, v and the texture mapping coordinates s, t. This allows textures to be scaled, translated, rotated, and sheared in the 2D parametric space of the primitive. Note that polygons are not parametric primitives, and therefore **RiTextureCoordinates** has no effect on them.

Second, as described above, any primitive can have **"s"** and **"t"** parameters bound to their vertices. These will override the default s, t mapping above. This control is essentially the same as **RiTexture-Coordinates** on four-vertex primitives such as quadrics and isolated patches but is more powerful for polygons and patch meshes, because each vertex has control of its own texture coordinates. Programs which **project** texture coordinates onto objects through orthographic, spherical, or other programmatic mappings will often use this style.

In addition, as is described later in this chapter, shaders written in the RenderMan Shading Language can take complete control over texture placement. This is the most powerful and general solution.

Shading Rate

RenderMan renderers have a pair of shading attributes known as **RiShadingRate** and **RiShadingInterpolation**. These attributes control how frequently the full lighting and shading calculations happen on each object. For example, classic Gouraud shading calculates a shading value on object vertices and interpolates these colors for any pixel between the vertices. Classic Phong shading calculates a shading value at every pixel or subpixel sample point and never needs to interpolate colors. RenderMan renderers can compute the shading at any rate between these extremes. The shading rate defines how many pixels any shading value should cover. Pixels within this region will get either a copy of the nearest shading value or an interpolation of nearby shading values depending on the shading interpolation switch. It is typical for RenderMan programs to request that shading be computed once per pixel (**RiShadingRate(1.0)**) and copy that color to all subpixel sample points within that area (**RiShadingInterpolation("constant")**). Because lighting and shading are often computationally very expensive in high-quality images, shading rate is a very powerful switch to control overall renderer run times.

62.5 Lights

Light sources are defined in the RenderMan program using the **RiLightSource** command. This command takes as arguments the type of light source and a parameter list, which controls parameters specific to that type of light. Each light source created by the RenderMan program also has a *handle*, which is an identifier that permits the RenderMan program to refer again to each particular light source.

There are four standard types of light source provided by all RenderMan renderers. An **"ambient-light"** is a directionless light which is uniformly intense at all points in the scene. A **"distant-light"** is a directional light which emanates from infinity, and whose direction and intensity are both constant at all points in the scene. A **"pointlight"** is a light which, like a small light bulb, is located within the boundaries of the scene, and whose direction and intensity vary at different points in the scene. A **"spotlight"**, as the name suggests, is a pointlight which emanates only within a specified cone. Light sources can also be programmed in the Shading Language.

Pointlights and spotlights have a variable intensity based on distance from the light. These lights follow the inverse-square law of intensity, meaning that the intensity of the light arriving at some point a distance r from the light will have only $1/r^2$ the brightness of the light arriving at distance 1.0. This behavior is physically realistic for pointlight sources but is often a source of confusion when modelers set up the lights in their scene. Pointlight sources which are 10.0 units away from the center of the scene usually have an intensity of 100.0 or greater in order to compensate for this effect. It is not unusual to see RenderMan programs set up lights with intensities of 1,000,000 or more, depending on the modeling units and the distance of the lights from the important objects in the scene.

Global Lights and Private Lights

Any light which is supposed to affect the whole scene must be placed before any of the objects in the scene (in order to satisfy the rule that all data affecting an object must be defined before the object). Typically, these lights are placed immediately after the **RiWorldBegin**. If lights are placed here, they will be "on" for all objects in the scene. Like geometric primitives, lights inherit their position from the current transformation at the time they are described, so these global lights will be in *world* coordinates. The standard lights have **"from"** and **"to"** parameters that let you place the lights at the appropriate places, relative to the world.

Any light which is supposed to affect only a particular object might be thought of as an attribute of that object. It is not unusual to place such a light in the object hierarchy immediately before the object it illuminates. When the attribute block of the current object is popped, the light will be popped with it, and so that light will not affect any of the remaining geometry in the RenderMan program.

Often, the scene contains lights which are naturally part of an object in the scene but which are supposed to illuminate the whole scene globally (e.g., a car headlight). The easiest approach for handling these global lights is to duplicate the entire internal hierarchy of the containing object at the beginning of the scene, but provide only the transformations and light sources. Once the duplicate hierarchy is complete, these lights will be in the scene but turned off. They can be turned back on using **RiIlluminate**.

Light Switches

The list of lights which are "on" and "off" is an attribute in the graphics state. When a light is first created, it is immediately turned on. When the graphics state stack is popped, we are returning to a state where the light was not yet on, and so it is automatically turned off. However, RenderMan supplies a light-switch mechanism with which any light can be turned on and off arbitrarily. **RiIlluminate** takes as arguments the handle of the light (which was assigned when the light was created) and a switch value. This way, lights can be turned off to keep them from affecting an object which they shouldn't, or they can be turned on again after the graphics state where they were created is popped.

62.6 Simulating Global Illumination

The RenderMan interface is intended to support the use of scanline, Z-buffer, and Reyes architecture [Cook et al. 1987] renderers which cannot directly compute global illumination. However, global illumination effects, such as mirror reflections, refractions, shadows, and diffuse interreflection, are all essential to accurate image synthesis for certain lighting environments or with certain surface materials. The standard technique for simulating these effects has, since the late 1970s, been special-purpose texture maps. In particular, texture maps which simulate mirror reflections and shadows are common. Sometimes these texture maps are simply photographs of interesting lighting environments. However, the more interesting case is when the map is generated by one or more rendering prepasses immediately prior to the rendering of the final image. There are standard techniques which the RenderMan program can use to generate these maps via multipass rendering.

Reflection maps are images of the scene that are seen reflected in a flat mirrorlike surface, such as a hanging mirror, a window, or a shiny floor. A quick examination of basic optics reveals that the mirror reflection image is in fact a simple image of an inverted, or *reflected*, version of the original scene. In mathematical terms, this means that the world transformation has a negative scale in one axis. During the prepass, the scene is reflected around the plane of the mirror object. This involves a slightly tricky transformation which brings the plane of the mirror into the x,y plane, negates z, and then undoes the first transformation. After doing this, the camera is looking at the mirror scene as though it was Alice peering "Through the Looking Glass." Figure 62.1 demonstrates the principle. The mirror scene is then rendered from the original camera, with one caveat. The mirror itself, and any objects normally behind the mirror, must be removed from the mirror scene before rendering. They would obviously not appear in the reflection. The image is converted into a standard texture map using **RiMakeTexture**. The resulting reflection map contains the desired mirror reflection pixels in exactly those pixels where the mirror occurs in the final image, plus extraneous stuff in those pixels where the mirror doesn't appear. During final image generation, the mirror object copies pixels directly from the reflection map using the normalized device coordinates (**"NDC"**) of the output pixel as the texture-map lookup coordinates, and the correct result is obtained.

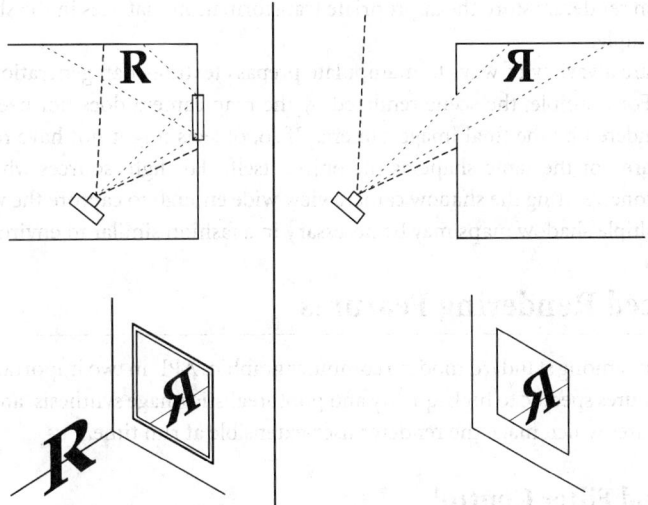

FIGURE 62.1 Creating a mirrored scene for generating a reflection map. On the top left, a camera views a scene which includes a mirror. Below is the image it produces. On the top right, the camera instead views the *mirror scene*. Notice that the image it produces (below) contains the required reflection.

Environment maps are images of the scene that are seen reflected in nonplanar objects, such as spheres and teapots. Nonplanar objects distort their reflections, so simple flat image mappings like reflection maps do not work. Instead, an environment map contains a view of the entire environment as seen from the reflective object. This requires rendering six images of the scene, through the faces of a cube centered on the reflective object. Therefore, the prepass is to move the camera to the center of the reflective object and render six square 90-degree images of the whole scene (minus the reflective object itself), one for each face of the cube. These images are stitched together to form an environment map with **RiMakeCubeFaceEnvironment**. During the final image generation, the renderer computes a reflection vector for each point on the reflective object, based on the viewing vector and the surface normal. This reflection vector is intersected with the cube to determine which pixel of which face image would be seen along that vector, and the environment map color at that pixel is the reflection color.

It is important to realize that this method is only approximate, as the cube contains only the "correct" data for a single point (the center of the object). The approximation assumes that parallax and occlusion differences will be minor, and as long as the point being shaded is relatively close to the center (in comparison to the distance from the camera to the point), this is generally true. However, the closer to planar the reflective object is, the worse the approximation becomes, and environment maps fail completely on planar objects. Reflection maps should be used for those cases.

Shadow maps are images of the scene as seen from the light source. However, rather than store object color in each pixel, shadow maps store the distance to the closest object at each pixel. During the prepass, the camera is moved to the position of the light source and oriented so that it looks down the primary pointing axis of the light. The virtual camera options are set so that the shadow camera can see everything that is potentially illuminated by that light source (for example, a perspective projection frustum angle matching the cone angle of a spotlight). The full scene is then rendered through the shadow camera, and the resulting depth image is converted into the shadow map for that light source using **RiMakeShadow**. During final image generation, the light source accesses the shadow map for each 3D point that it illuminates. The coordinates of each point to be lit are transformed into the coordinate system of the shadow camera and then projected onto the shadow map, and the depth is compared to the value in the map. Any point which is farther away than the map value is occluded by the (closer) object in the map and hence is in shadow. RenderMan renderers store the appropriate transformation matrices in the shadow map, so that this calculation is simple.

Naturally, there are a variety of ways to manipulate prepass texture-map generation to get interesting rendering effects. For example, the scene rendered by the map camera does not need to be exactly the same scene as is rendered by the final image camera. If so, objects might not have reflections or might cast shadows that are not the same shape as the object itself. For light sources which have extremely wide illumination cones, setting the shadow camera view wide enough to capture the whole scene may be problematic, so multiple shadow maps may be necessary in a fashion similar to environment maps.

62.7 Advanced Rendering Features

RenderMan is unique among standard modern computer graphics APIs in two important ways: it supports a wide variety of features specific to high-quality and photorealistic image synthesis, and it supports a very powerful set of features which make the renderer user-extensible at run time.

Antialiasing and Filter Control

An image which is not rendered using some form of antialiasing will generally not be considered high-quality. There are two stages to antialiasing an image. First, you must sample the colors in the scene well enough to ensure that you haven't missed any small objects or any sharp changes in color. RenderMan's sampling rate is controlled by **RiPixelSamples**. This specifies the number of scene samples acquired per pixel. Generally, 2×2 sampling is a minimum for basic antialiasing, and higher sampling rates will often be required. Some RenderMan renderers may have additional controls which modify or adapt various parameters of the sampling pattern in order to acquire the best possible samples.

Second, you must blend the pixel samples together in a way which emphasizes the correct color for the pixel without letting the neighboring pixels' color be so different that an alias occurs.[3] The pixel reconstruction postfilter is specified by **RiPixelFilter**. RenderMan renderers supply a set of common filters (box, Gaussian, sinc, etc.), or the RenderMan program may supply its own. Box filtering generally makes images which appear too blurry, whereas sinc filtering (the theoretically optimal filter) often has unpleasant ringing artifacts. The default filter (Gaussian width 2) is generally a nice compromise.

Depth of Field

Depth of field occurs in a real camera because a lens is not able to focus at all distances simultaneously. It can focus only at one depth at a time, and all other depths are slightly out of focus. The amount of blurriness depends on both the actual distance and on two important parameters of the lens, the f-stop and the focal length. In RenderMan renderers, this effect can be simulated with the **RiDepthOfField** camera option. **RiDepthOfField** takes three parameters: the f-stop and focal length of the virtual camera, and the distance to the objects in focus. Renderers which can simulate depth of field effects will consider the "lens" to be at the origin of *camera* coordinates, with focal length and focal distance both measured in *camera*-space distance units. If depth of field is not used, the camera is called a "pinhole" camera and all objects are in focus, independent of depth.

Motion Blur

In a real camera, objects which move quickly leave a streak across the film, known as a **motion blur**. Because the shutter of a real camera is open for a measurable amount of time, moving objects appear on the film in all the places that they passed by while the shutter was open. This effect can be simulated in RenderMan images in two steps. First, the shutter speed is specified. This is done with the camera option **RiShutter**, which specifies the times at which the camera shutter opens and then closes.

Second, the objects in the scene which are to move must have their motion described in a *motion block*. A motion block contains a sequence of the same RenderMan command, which specifies a single object attribute which changes over time. For example, a sphere which is moving horizontally might have a translational component of its position which is changing over time, as in the following example:

```
MotionBegin [0.0 0.5]
Translate 1.0 0.0 0.0
Translate 2.0 0.0 0.0
MotionEnd
Rotate 45 0 1 0
Sphere 1 -1 1 360
```

This program excerpt specifies that at time 0.0, the sphere is translated 1.0 units in x, but by time 0.5, the sphere has moved to 2.0 units in x. If a spinning sphere were also desired, a second motion block containing the moving rotation would be used. Compound motions are always specified with a series of motion blocks that each contain a single moving command.

The RenderMan specification states that any object attribute can be blurred, including the parameters and vertices of the objects themselves. In practice, however, correct treatment of motion blur is both cumbersome to implement and computationally expensive. Therefore, each RenderMan renderer will have various restrictions on what types of commands may be blurred. In general, most RenderMan

---

[3] Sampling and filtering are explained in greater detail in Chapter 58 of this volume.

renderers will be able to blur transformations from a value at shutter open to a value at shutter close. More than this is renderer-specific.

Named Coordinate Systems

When writing programs in the RenderMan Shading Language, it is often most convenient for certain calculations to take place in some particular coordinate system. For example, when calculating coordinates in an environment map, it is easier in a coordinate system centered and oriented coincident with the camera that created the map than it is in some other random coordinate system. RenderMan has several built-in coordinate systems, such as *camera* and *world*, but often these are not the optimal ones for such special purposes. To handle those cases, a RenderMan program can mark a coordinate system at any time. **RiCoordinateSystem** gives the current coordinate system a name, with which it can be referenced by subsequent shaders. There is no requirement for there to be any geometry or shaders at that node of the transformation hierarchy.

User-Defined Vertex Variables

A RenderMan program can extend the graphics state describing the shading attributes of objects in two ways. First, the shaders attached to the objects can have arbitrary parameters in their parameter lists. Thus, while some graphics APIs would have a specific graphics state for, for example, the coefficient of specular reflection *Ks*, RenderMan does not. Instead, the surface shader will have **Ks** as a parameter if it is appropriate to that particular type of surface.

Parameters attached to shaders are *uniform*, that is, the same everywhere on the surface of the primitive. If the RenderMan program needs to create a graphics state which varies over the surface of a primitive, in the same way that **s** and **t** do, it can create a user-defined *varying* vertex variable with **RiDeclare**. For example, a finite-element modeler might need to place "stress" on the vertices of each of its polygons, while an engineering analysis modeler might need to place both "density" and "temperature" in order for the correct shading to occur. By declaring a vertex variable and specifying that it is of type **varying**, it is then legal to attach this variable on the vertices of polygons and control points of parametric primitives just as **s** or **Cs** is attached. In this way, RenderMan programs can calculate and pass arbitrary data into the shaders, and it will be interpolated across the face of the geometric primitives along with the predefined vertex variables.

62.8 RenderMan Shading Language

One of the great advantages of using RenderMan is the fact that you can describe the appearance characteristics of objects with as much detail and subtlety as you typically describe the shapes and positions of those objects. The RenderMan Shading Language is a special-purpose programming language for describing appearance characteristics [Hanrahan and Lawson 1990]. Shading Language programs, called **shaders**, can be used to model materials and effects in a physically realistic way, or in an "unrealistic" artistic style. Shading Language shaders can be used in RenderMan programs in place of the predefined standard shaders described earlier in RenderMan calls such as **RiSurface**, **RiDisplacement**, **RiLightSource**, and **RiAtmosphere**.

Shading Language Syntax

The Shading Language is a C-like programming language. It has a basic block structure syntax, variable declarations, and arithmetic operators similar to those of C, as is demonstrated by the following example shader. The only significant syntactic difference is that shader formal parameters have default values associated with them. This allows shaders to be called from RenderMan programs using the name–value parameter list syntax:

```
surface example (float Ka = 0.5;
    point axis = point "object" (1,0,0);)
{
    color fore;
    uniform point up;
    float i;

    fore = color "hsl" (.5, 0, .4);
    up = axis ^ point "camera" (0, 0, 1);
    if (Ka > .2) {
        for(i=0; i<3; i+=1) {
            fore *= Ka;
        }
    }
    Ci = Cs * ambient() + fore * diffuse(normalize(up));
}
```

The Shading Language's variable types and built-in functions, however, reflect its special nature. There is a single scalar type, **float**, and two vector types, **point** and **color**, each of which were chosen for their utility in describing material characteristics. The vector types have standard linear-algebraic operators defined for them, such as vector addition, scalar multiplication, and dot and cross products. In addition, constant values of the vector types have coordinate-system names associated with them, a feature which makes geometric calculations significantly easier to describe. There are no pointers, arrays, or structured data types in the Shading Language. Variables in the Shading Language can be declared to be **uniform**, in which case the value is the same everywhere on the surface, or **varying**, in which case every point on the surface may have a different value. These storage declarations are for run-time efficiency only and are optional. All parameters default to **uniform**, and all local variables default to **varying**.

The language provides a very rich set of built-in library functions for arithmetic, color, and geometric calculations. For example, standard C math library functions such as **sin**, **pow**, **sqrt** and utility functions such as **mod**, **min**, **floor** are provided. Color manipulations such as **mix**, color **spline**, and color transformations from one color system into another are provided. Geometric functions such as **length**, **normalize**, **faceforward** (which ensures that surface normals point toward the camera), and point and vector transformations from one coordinate system into another are provided.

In addition, a large set of standard computer graphics illumination functions such as **specular**, **diffuse**, and **fresnel** and commonly used area-operator and antialiasing functions such as **deriv** and **smoothstep** are part of the standard library.

Finally, the Shading Language provides functions to access four varieties of standard image texture maps: **texture**, **environment**, **bump**, and **shadow**. Applications of these texture maps are automatically filtered and antialiased by these functions, so even the simplest shaders get very high-quality results.

Simple Surface Shaders

The function of a surface shader is to calculate the color of arbitrary points on the surface of an object. In order to do this, shaders must have access to a relatively extensive global state which fully describes the location of the points being shaded. RenderMan's shading global state contains **P**, the point being shaded, **u** and **v**, the parametric coordinates of the primitive at that point, and **Cs** and **Os**, the color and opacity of the object, as well as the viewing vector **I**, the normal vector **N**, light vectors **L**, and other information about the overall graphics state and the local geometry of the primitive at the point

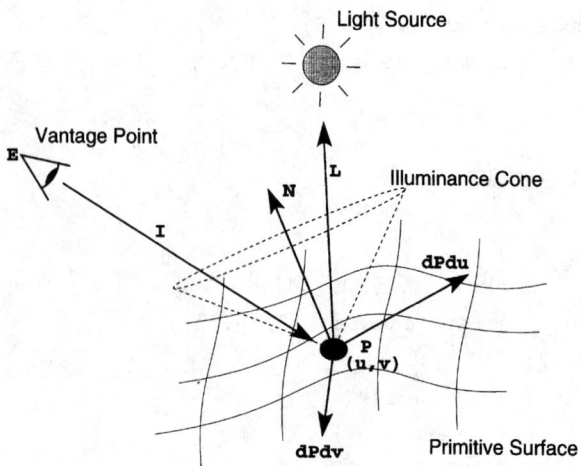

FIGURE 62.2 Geometric global state available to surface shaders. (*Source:* Pixar [1989, p. 110]. Pixar, Richmond, CA. Used with permission.)

being shaded. Figure 62.2 illustrates the relationship of some of the geometric information which is in the global state. Any parameter which affects the surface color, including surface attributes, incident light colors and intensities, angular dependencies, etc. must be taken into account by the surface shader. The resulting computed surface color and surface opacity are placed in **Ci** and **Oi**, respectively.

The richness of the provided function library and global state makes it easy to write simple shaders. For example, the standard computer graphics purely diffuse surface of Eq. (62.1) (known as **"matte"** in RenderMan) and the computer graphics plastic of Eq. (62.1a) (**"plastic"**) are each programmed with only a few lines of Shading Language code, as seen in the following shaders:

$$C = C_{\text{diff}}\left(K_a I_a + K_d \sum_i C_{L_i} (\mathbf{N} \cdot \mathbf{L}_i) \right) \tag{62.1}$$

```
surface matte (float Ka=1; float Kd=1;)
{
        point Nf;
        Nf = faceforward (N,I);
        Oi = Os;
        Ci = Os * Cs * (Ka*ambient() + Kd*diffuse (Nf));
}
```

$$C = C_{\text{diff}}\left(K_a I_a + K_d \sum_i C_{L_i} (\mathbf{N} \cdot \mathbf{L}_i) \right) + C_{\text{spec}} K_s \sum_i C_{L_i} (\mathbf{R}_i \cdot \mathbf{V})^{1/\text{roughness}} \tag{62.1a}$$

```
surface plastic (float Ka=1; float Kd=0.5; float Ks=0.5;
        float roughness=0.1; color specularcolor=color(1,1,1);)
{
        point Nf;
        Nf = faceforward (N,I);
        Oi = Os;
```

```
        Ci = Os * ( Cs * (Ka*ambient() + Kd*diffuse(Nf)) +
             specularcolor * Ks * specular(Nf, -I, roughness));
    }
```

The most basic and nearly universal methodology to writing shaders, as with any program or subroutine, is *divide* and *conquer*. For shading writing, this means dividing the shading calculation up into four phases. The first phase is pattern generation. For most shaders, this is the most interesting part of the calculation. Phase two is layering. Most interesting patterns cannot be described by a single function. Often there are layers of patterns, subtle patterns atop more gross patterns, or tiny patterns inside larger patterns. It is usually helpful to handle each layer separately and merge them together into the final pattern.

The third phase is illumination. For many shaders, copying the illumination from plastic or metal is correct. For other shaders, the illumination model itself is patterned or may be totally nonstandard because the particular surface reflects light in a way quite unlike plastic. The fourth phase is compositing. Some objects have physical layers with totally different characteristics, such as an asphalt road with shallow puddles of oil and water on it, and shaders might calculate the patterns and illumination of each layer independently and then composite them together.

Pattern Generation

The majority of the effort in writing most shaders is devoted to pattern (or texture) generation. Very few surfaces in the natural world exhibit smooth, featureless behavior in more than a couple of parameters. Most surfaces have subtle patterns, defects, asymmetries, or other interesting features in almost every parameter of their material description. Writing shaders for the highest-quality images requires careful attention to such details.

The simplest way to apply patterns to surfaces is through texture mapping. Generally speaking, a texture map is any image which is applied to a surface which contains a surface variation of some parameter of the material. Historically, texture maps have been used to apply variations of surface color, transparency, surface bumpiness, diffuse and specular reflection coefficients, and other parameters of the standard computer graphics lighting equation. Such texture maps are easily applied in Shading Language through the **texture** call. There are no restrictions on the number of textures that can be used on any individual surface or in the scene as a whole.

Image texture maps are always accessed by computing 2D texture coordinates for all points on the 3D surface of the object and using these to index into the pixels of the texture image. There are a variety of ways that these 2D coordinates can be computed (and with an entire language at your disposal, computing intricate 3D-to-2D mappings is at times a very interesting task), but most fall into two categories: parametric mappings and projections. Parametric mappings take the existing 2D u, v coordinate system of the parametric object and manipulate it in some fashion to create a 2D s, t texture coordinate system for the object. Projections create 2D texture coordinates by manipulating the 3D position of the object, such as projecting the position onto an enclosing sphere or cylinder. Decals are often applied with orthographic projection of a plane.

Texture maps have also been used to simulate other effects, such as reflections, refractions, and shadows. RenderMan has separate texture functions for these special-purpose uses (**environment** and **shadow**), since the natural arguments for these uses are not 2D coordinates but rather direction vectors or positions.

Image texture maps have certain limitations which make them inappropriate for use in all situations. First, photographs, when used as textures, include in their color the effects of the lighting in the environment in which they were shot (such as uneven or colored lights, specular highlights, or self-shadowing of a bumpy surface). This is unfortunate because the shader desires to compute lighting itself based on the light sources in the virtual scene. Second, images have a finite resolution and size. This limits the ability of the virtual camera to zoom in or out without revealing the underlying resolution of the texture maps themselves. For this reason, it is often more appropriate for patterns to be generated procedurally.

Procedural Pattern Generation

Procedural texture patterns fall into three categories: regular periodic patterns, stochastic patterns, and perturbed regular patterns.

Regular periodic patterns have no random components. They are strict (although often complex) mappings of 2D (parametric or projected) coordinates or 3D coordinates into a repetitive pattern. Examples of these types of patterns are infinite checkerboards, barber-pole spirals, and rubber-stamp repetition of simple shapes. Because of the purely mathematical mappings required, simple equations using periodic functions such as **sin** often suffice.

An extremely valuable building block for periodic functions is the modulo function **mod**.

The equation

$$y = \frac{\textbf{mod}\ (x, a)}{a} \tag{62.2}$$

generates a sawtooth wave, where y runs from 0 to 1 as x runs from 0 to a, and again as x runs from a to $2a$, etc. If we have a function **func**, which generates an interesting pattern based on inputs from 0 to 1, the equation

$$y = \textbf{func}\left(\frac{\textbf{mod}\ (x, a)}{a}\right) \tag{62.3}$$

will generate a repetition of that interesting pattern every a units.

Similarly, the equation

$$y = \textbf{step}(x, a) - \textbf{step}(x, b) \tag{62.4}$$

generates a unit-high pulse between a and b, and this can be used to isolate a single occurrence of a periodic signal.

Stochastic, or pseudorandom, patterns are completely irregular. They are pseudorandom, and not completely random, for two reasons. First, we usually desire patterns where the pattern at one point of the surface is at least partially related to the pattern at an adjacent part of the surface (if for no other reason than for antialiasing). Second, we usually desire patterns which are repeatable during animation, which means that the pattern at a point on the surface is very similar to the pattern at that same point at some later time. The basic building block for most stochastic patterns is **noise**, a function first described by [Perlin 1985].

RenderMan's **noise** function takes as input a 1D, 2D, or 3D value, and returns a pseudorandom value between 0 and 1, with the following properties:

- it is continuous and smooth;
- it is approximately band-limited;
- the dominant frequencies are all below 1 Hz;
- it is equal to 0.5 at all integer inputs;
- it crosses 0.5 at most one time between any two integers.

Knowing these features about **noise** makes it possible to use **noise** in a stochastic version of spectral synthesis. By scaling the inputs to **noise**, you scale the frequency band of the resulting output. Adding offsets to the inputs results in phase shifts. Scaling the results naturally scales amplitude. Popular constructs such as $1/f$ "fractal" noise, fractional Brownian motion, and turbulence are all constructed by adding together several iterations of **noise** with different frequencies and amplitudes. The following shader code illustrates using **noise** to generate a simple marble pattern:

```
    f = 1; marble = 0;
    for (i = 0; i < 6; i += 1) {
        marble += noise(P * f) * 1/f;
        f *= 2;
    }
```

Another useful type of pattern is a combination of regular and stochastic patterns, known as perturbed regular patterns. This combines a basic recognizable regular structure with just enough random irregularity to keep it interesting. One such approach is known as *bombing*. Bombing uses random values to perturb the centers, sizes, orientations, or some other parameter of an otherwise regular pattern. For example, consider a Shading Language procedure that divides parameter space into a regular grid of cells, where each cell contains a circle at its center. An enhanced shader determines, for every point it shades, which cell that point is in, and it adds a small pseudorandom center offset and size variation to the circle based on the *cell number*. The regular pattern of circles becomes a random polka-dot pattern.

Another useful perturbed regular pattern is perturbed access to texture maps or regular procedural patterns. Consider the following shader code:

```
    ss = s + 0.02 * (noise(P + (1.2,0.3,8.1)) - 0.5);
    tt = t + 0.10 * (noise(P + (3.6,2.1,9.3)) - 0.5);
    Ci = texture(texturename, ss, tt);
```

By adding small pseudorandom perturbations to the texture coordinates, the texture will be subtly warped when applied to the surface. This distortion can be used to hide the edge artifacts that result when a texture is repeated many times across a surface, or to make a photograph of a natural material appear less recognizable when used multiple times in the same scene.

Light-Source Shaders

The function of a light-source shader is to calculate the color and intensity of light leaving a source and arriving at an arbitrary point in space. The global graphics state available to light-source shaders is subtly different from that available to the surface shader because it is optimized for emitting light rather than receiving it. In addition, light sources use one of two special-purpose loop constructs provided in the Shading Language, **solar** and **illuminate**, which determine whether the light is a distant or a local light.

Distant lights—lights which are so far away from the scene that the light they emit arrives in parallel beams—are described by the **solar** loop. The **solar** loop can be thought of as *for all points to illuminate...*, and inside the loop, the special light vector **L** indicates which direction, and **Ps** indicates which point in the scene, to illuminate. Any effects that modulate the amount of light arriving at that point, including directional dependencies, gels, angular attenuation, etc., are computed by the light-source shader and the resulting light intensity is placed in **Cl**.

Local lights—lights whose position in the scene affects their illumination—are described using the **illuminate** loop. As before, the **illuminate** loop can be thought of as *for all points to illuminate...*, and the shader uses **L** and **Ps** to compute **Cl**. Local lights (such as the predefined **spotlight** light source, or the slightly simpler **conelight** shader below) are often dependent on distance or angle from an aiming vector:

```
    light conelight (float intensity = 1;
    float coneangle = radians (30);
        point from = point "shader" (0,0,0);
    point to = point "shader"
        (0,0,1);)
```

```
    {
        float atten, cosangle, llen;
        uniform point Axis = (to-from)/length(to-from);

        illuminate( from, Axis, coneangle ) {
            llen = length(L);
            cosangle = (L/llen) . Axis;
            atten = (1/llen.llen) * step(cos(coneangle), cosangle);
            Cl = atten * intensity;
        }
    }
```

Texture maps or procedural patterns can also be used to modulate the light intensity of light sources. For example, a slide projector is a light source which computes a perspective projection of the point **Ps** onto a source texture map, and sets **Cl** to the color found there. A mirror ball might use bombing to determine which spots on a spherical projection will be bright and which will be dark. In addition, lights often cast shadows using **shadow** maps, computed in an earlier rendering pass. The **shadow** call returns occlusion information for a point in space, so a light source can identify which points are hidden from it and hence should not receive light.

62.9 Summary

The RenderMan interface is a comprehensive scene description interface for 3D image synthesis. Render-Man's strength lies in its large feature set devoted specifically to high-quality image synthesis. Features such as a large variety of curved-surface primitives, advanced virtual camera simulation with depth of field and motion blur, an imaging pipeline with integral antialiasing and filtering, and the Shading Language (which supports procedural texture generation, special-purpose texture mapping, user-defined vertex variables, and arbitrary light reflectance functions) give it descriptive power far beyond other current graphics APIs.

This focus on image quality has caused most implementations of RenderMan renderers to concentrate on high-end, batch-oriented, single-image generation. It should be noted, however, that RenderMan shares many underlying mental models with those other standard computer graphics APIs, which are generally oriented toward interactive and real-time hardware graphics implementations. This makes RenderMan a descendant of a long line of speed-oriented APIs, and it makes RenderMan viable as a model for a future real-time rendering API where speed–quality tradeoffs are more important than pure speed.

In some sense, RenderMan codifies a wide variety of computer graphics tricks and techniques which have been developed over some 20 years of research into rendering. At the same time, it provides a standard interface to a programming environment which encourages new tricks and techniques to be developed. RenderMan was the first public graphics interface to be so focused on state-of-the-art techniques and to be user-extensible to such a large extent. Its paradigms have now found their way into a generation of vendor-proprietary computer graphics systems. The RenderMan interface specification truly revolutionized the design of computer graphics APIs and forever changed the expectations that creators and viewers of computer graphics images have of renderers.

Defining Terms

API: Applications programming interface. The complete interface that an application program has to a utility subroutine library, including all of the subroutine calls themselves, any associated type definitions, global variables, and any related programs or file-based resources which are necessary to use the library.

Cubic basis matrix: A 4 × 4 matrix which represents the style of cubic spline. Specifically, it converts (by vector–matrix multiplication) four coordinates representing a cubic spline from the user-defined cubic-basis-function representation into the monomial (or power) basis.

Current transformation matrix: The 4 × 4 matrix which converts (by vector–matrix multiplication) 3D points from the currently active local coordinate system into the *world* coordinate system. Transformation subroutines modify this matrix, thereby moving the currently active local coordinate system and all the subsequent objects that are put into it.

Graphics state: A complete database of the mode, graphical options and attributes, transformations, and any other information which a renderer maintains between subroutine calls.

Metafile: A data file which contains enough data to exactly recreate an image created through the subroutine library version of a graphics API. This is often a transaction log of the subroutine calls themselves.

Motion blur: The blur or streak effect left on the film when an object moves while the camera shutter is open.

Shader: A program module which computes some facet of the color calculation for a pixel in the image, such as the intensity of a light source, or the color of light reflected toward the camera by a surface.

Shading Language: A special-purpose programming language specifically designed to make it easy for users to write shaders and for rendering programs to execute the shaders that are written.

Texture coordinate projection: The process of mapping a 3D point into 2D so that every point in the 3D space can access some point on a texture map. Common projections include mapping points onto the 2D latitude–longitude coordinates of an enclosing sphere or cylinder or orthographic or perspective projection of points onto the x,y coordinates of an arbitrarily oriented rectangle in space.

References

Cook, R. L., Carpenter, L., and Catmull, E. 1987. The Reyes image rendering architecture. In *Comput. Graphics (Proc. SIGGRAPH)* 21(4):95–102.

Ebert, D. S., ed. 1994. *Textures and Modeling: A Procedural Approach*. Academic Press, Boston.

Farin, G. 1992. *NURB Curves and Surfaces*. A. K. Peters, London.

Hanrahan, P. and Lawson, J. 1990. A language for shading and lighting calculations. In *Comput. Graphics (Proc. SIGGRAPH)* 24(4):289–298.

Neider, J., Davis, T., and Woo, M. 1993. *OpenGL Programming Guide*. Addison–Wesley, Menlo Park, CA.

Perlin, K. 1985. An image synthesizer. In *Comput. Graphics (Proc. SIGGRAPH)* 19(3):287–296.

Pixar. 1989. *The RenderMan Interface Specification: Version 3.1*. Pixar, Richmond, CA.

Upstill, S. 1990. *The RenderMan Companion: A Programmer's Guide to Realistic Computer Graphics*. Addison–Wesley, Menlo Park, CA.

Watt, A. and Watt, M. 1992. *Advanced Animation and Rendering Techniques: Theory and Practice*. ACM Press, New York.

Further Information

The RenderMan Interface Specification, published by Pixar, is the official specification of RenderMan. It contains the exact syntax and detailed semantics of each C library call, each RIB directive, and the Shading Language. Be warned, however: it is quite terse and technical.

A more approachable reference is *The RenderMan Companion: A Programmer's Guide to Realistic Computer Graphics,* by Steve Upstill. Full of excellent examples and images, this book is probably more useful to the casual RenderMan user (although it uses only the C binding, and has no RIB examples).

Advanced Animation and Rendering Techniques: Theory and Practice, by Alan Watt and Mark Watt, describes geometric modeling and high-quality rendering techniques in a thorough yet very readable way.

Textures and Modeling: A Procedural Approach, by Ebert, Musgrave, Peachey, Perlin, and Worley, is the first textbook to explore procedural texture-map generation, and it covers a lot of very interesting territory.

Pixar provides technical information about the RenderMan Interface, and information about products based on RenderMan, on its Web site at http://www.pixar.com. In addition, the newsgroup devoted to RenderMan issues, comp.graphics.rendering.renderman, is very active, and the quality of the discussion is quite high. The *RenderMan FAQ* published monthly in the newsgroup is an excellent source of timely information on RenderMan products and issues.

OpenGL, another computer graphics interface specification aimed specifically at real-time graphics, is described in the *OpenGL Programming Guide,* by Jackie Neider, Tom Davis, and Mason Woo.

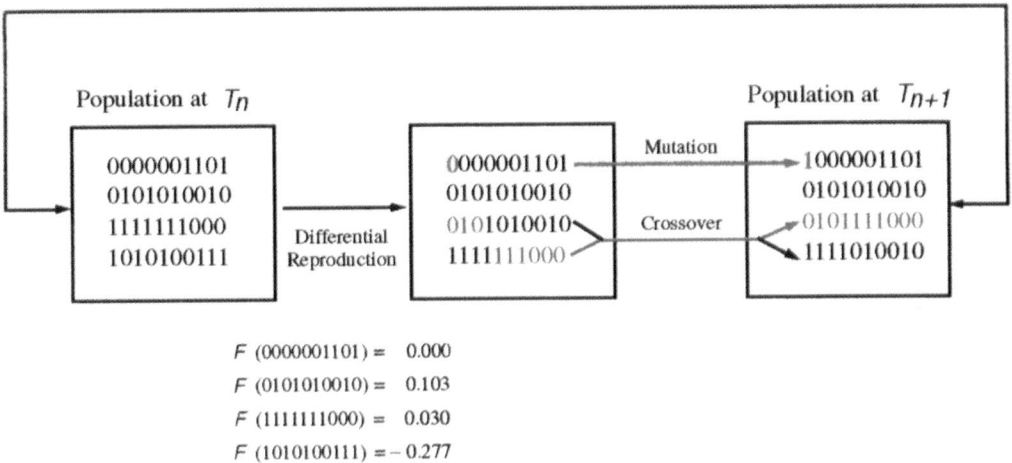

$$F \ (0000001101) = \quad 0.000$$
$$F \ (0101010010) = \quad 0.103$$
$$F \ (1111111000) = \quad 0.030$$
$$F \ (1010100111) = -0.277$$

PLATE 24.1   Genetic algorithm overview: A population of four individuals is shown. Each is assigned a fitness value by the function $F(x,y) = yx^2 - x^4$. On the basis of these fitnesses, the selection phase assigns the first individual (0000001101) one copy, the second (0101010010) two, the third (1111111000) one copy, and the fourth (1010100111) zero copies. After selection, the genetic operators are applied probabilistically; the first individual has its first bit mutated from a 0 to a 1, and crossover combines the last two individuals into two new ones. The resulting population is shown in the box labeled $T_{(N+1)}$.

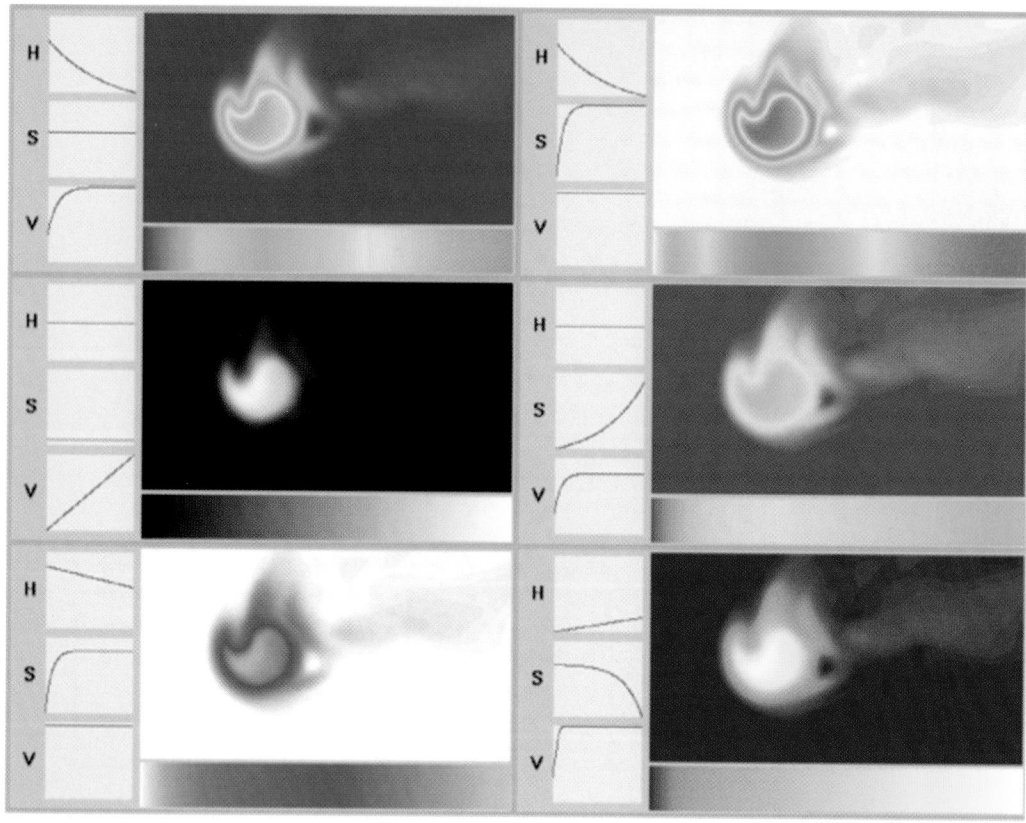

PLATE 35.1   A variety of color palettes used to color the same two-dimensional slice of data. The top two color maps use a wide range of hues and introduce discontinuities into the image that may not be present in the data. If we are looking for overall structure, a smoothly changing palette with a restricted hue — such as the "fire" palette shown in the lower right — is more appropriate. (Courtesy of NCSA, Yale University).

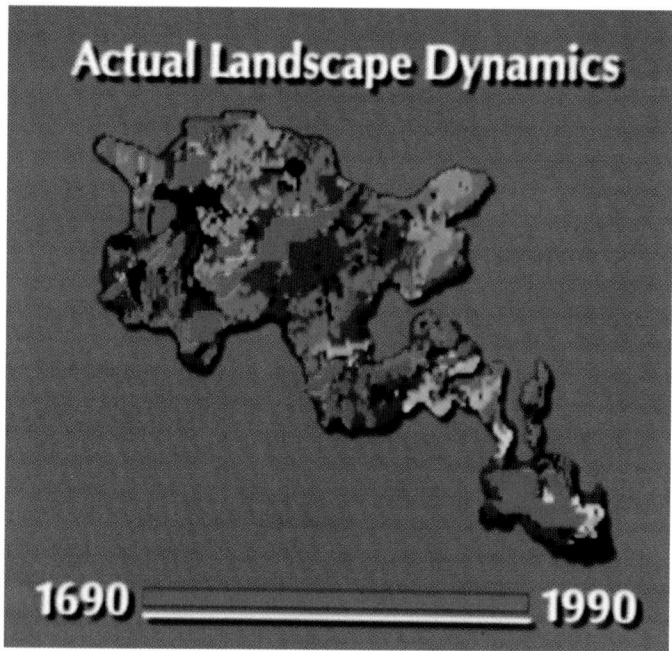

PLATE 35.2   Discrete color mapping is used to depict the age of the forest in Yellowstone National Park [Kovacic et al. 1990]. (Courtesy of D. Kovacic, A. Craig, and R. Patterson; UIUC, NCSA.)

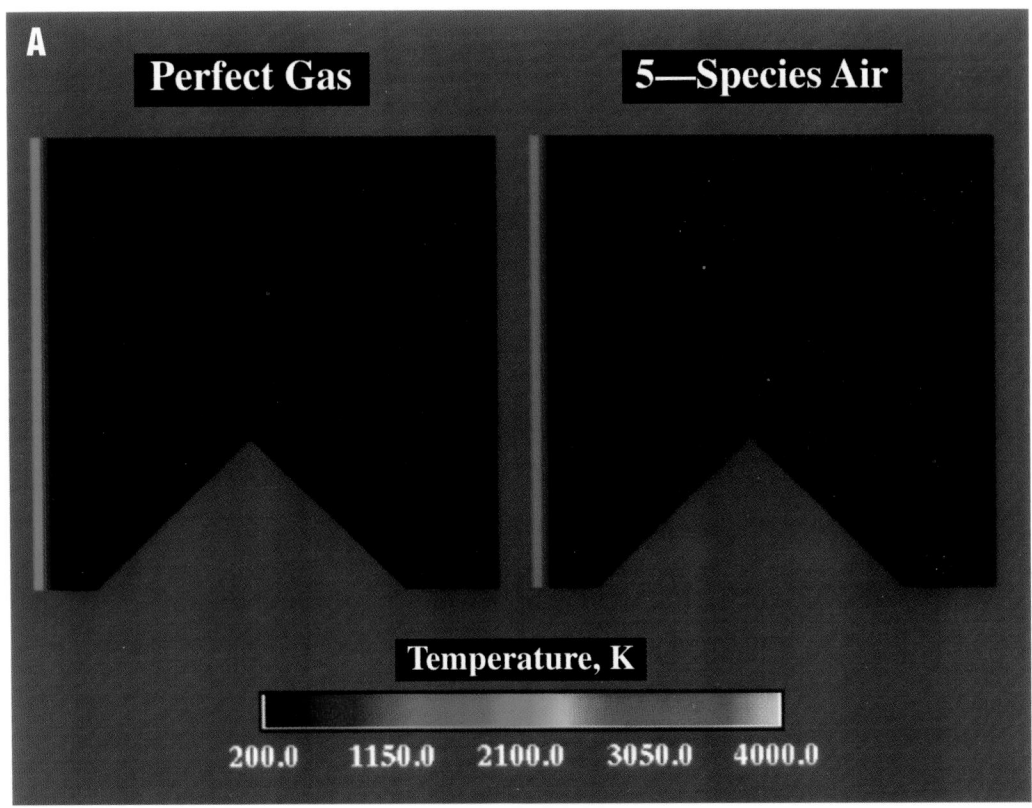

PLATE 38.1A

PLATE 38.1 (A through G) A shock wave impinges on a pyramid. Temperature contours for a perfect gas model and a 5-species reactive air model.

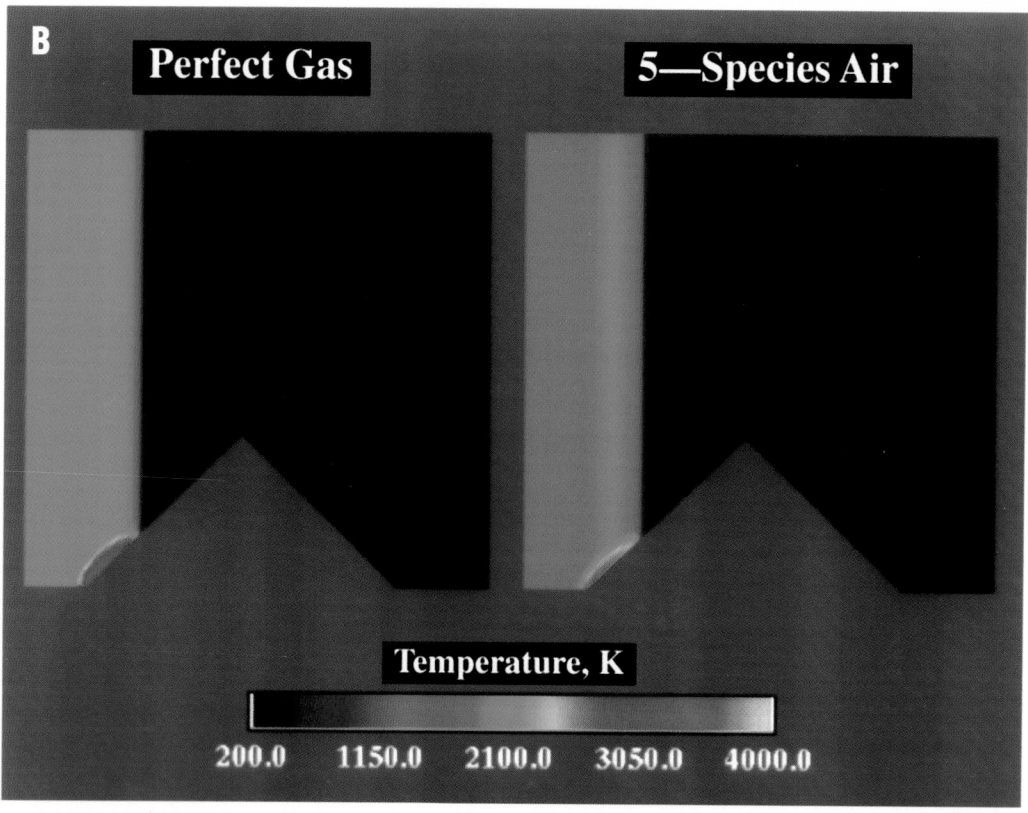

PLATE 38.1B

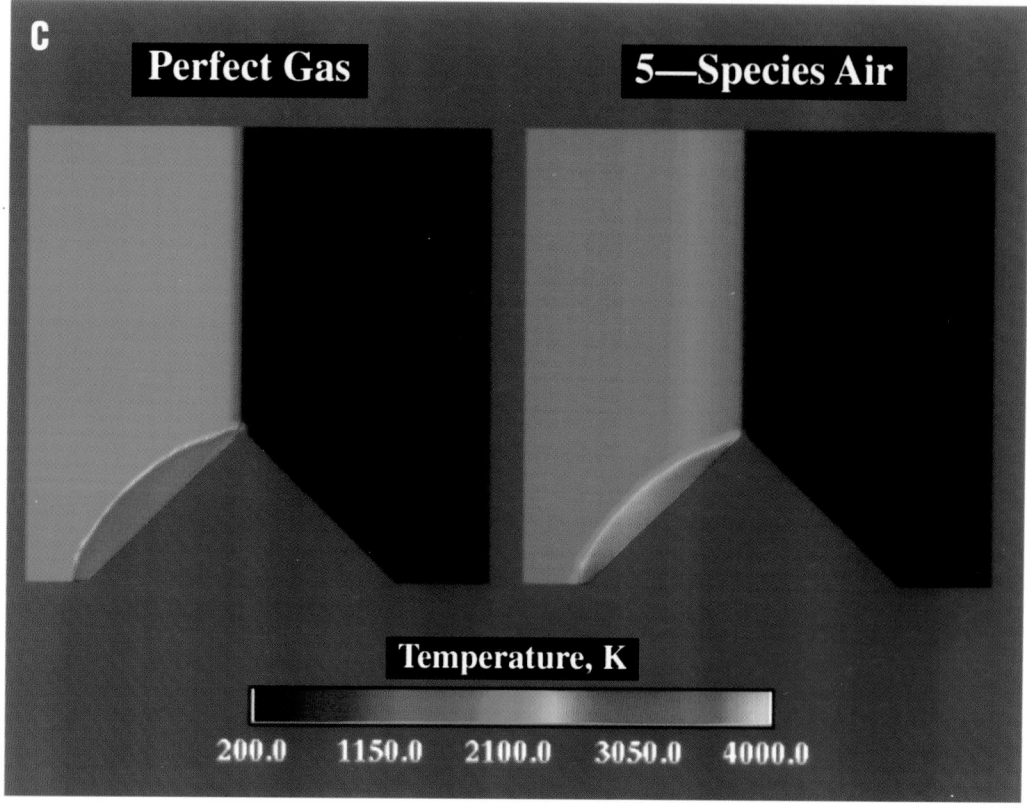

PLATE 38.1C

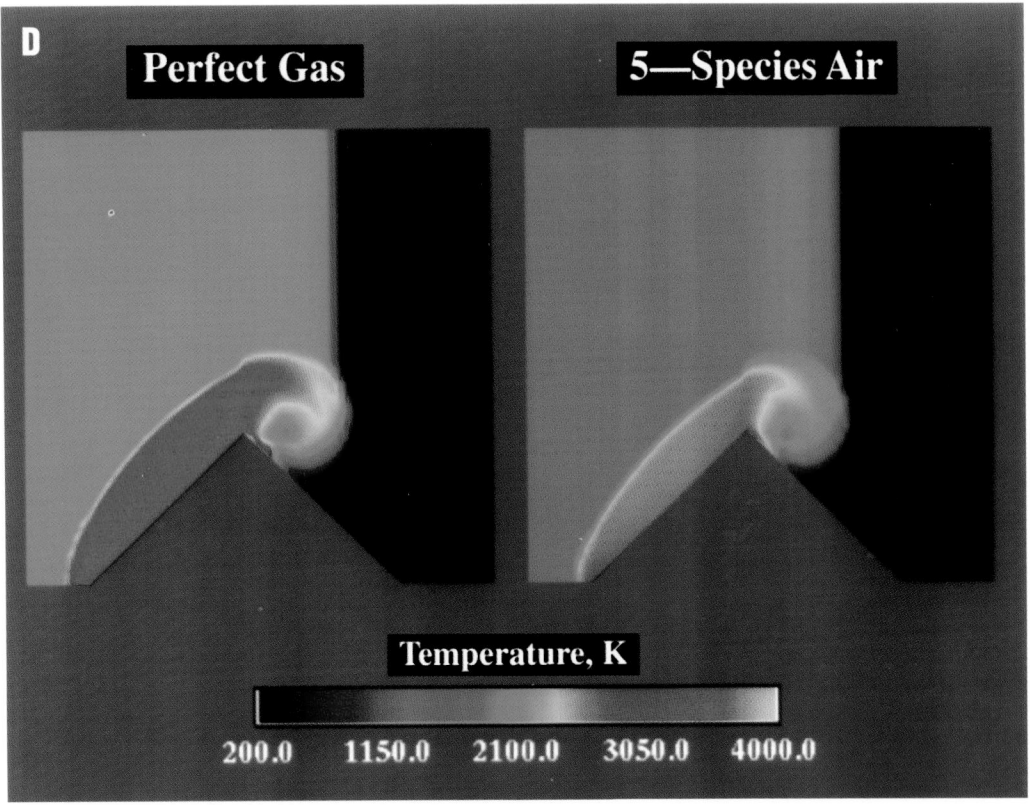

PLATE 38.1D

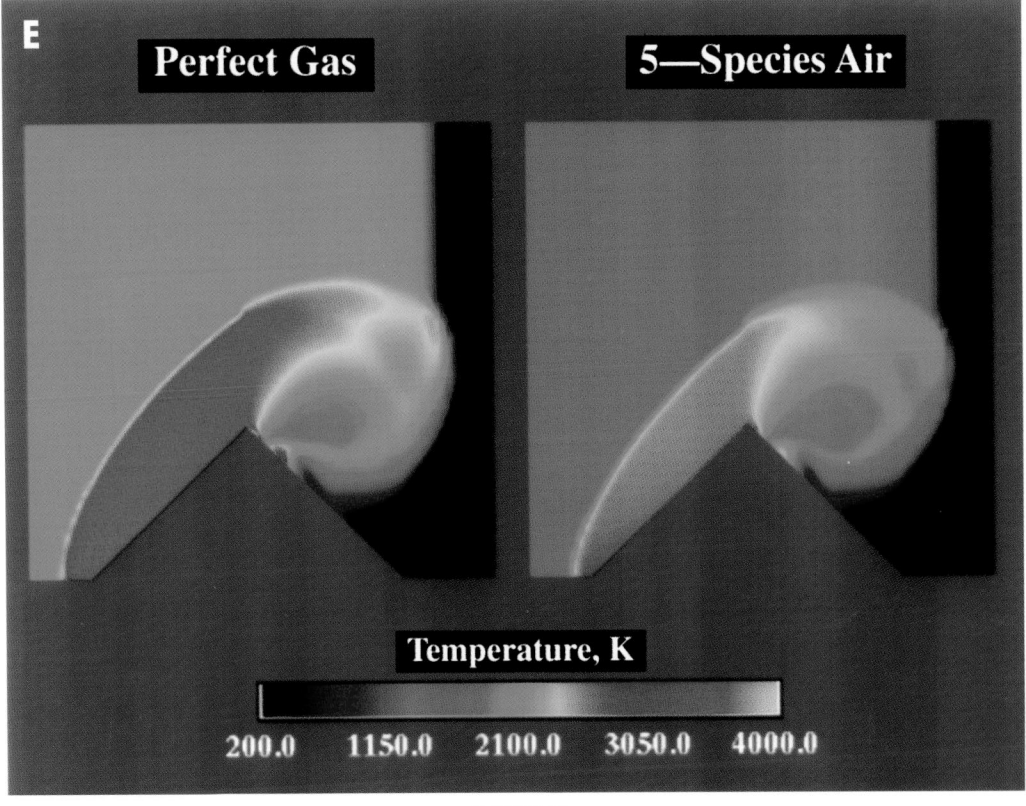

PLATE 38.1E

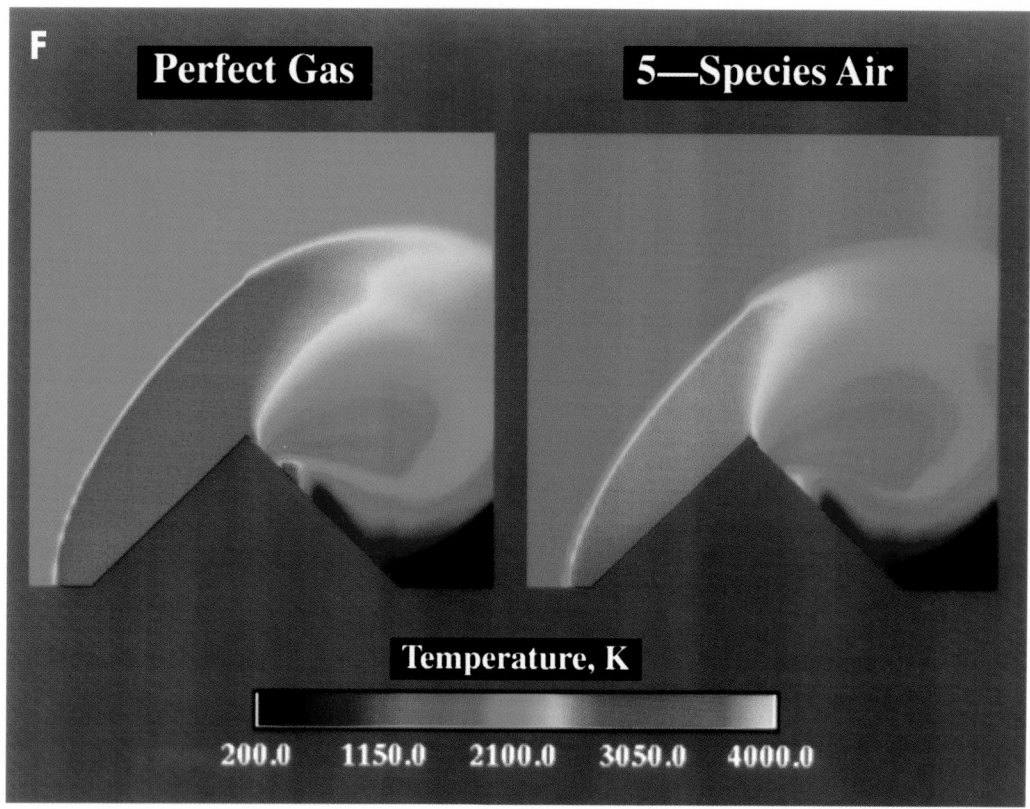

PLATE 38.1F

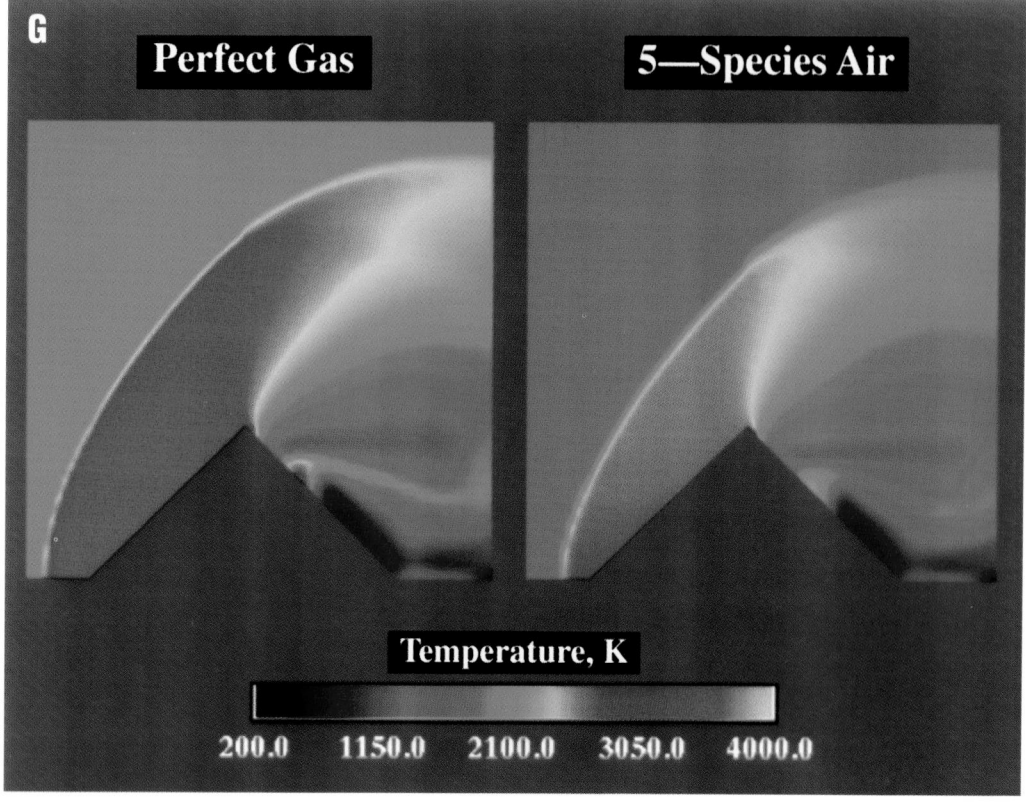

PLATE 38.1G

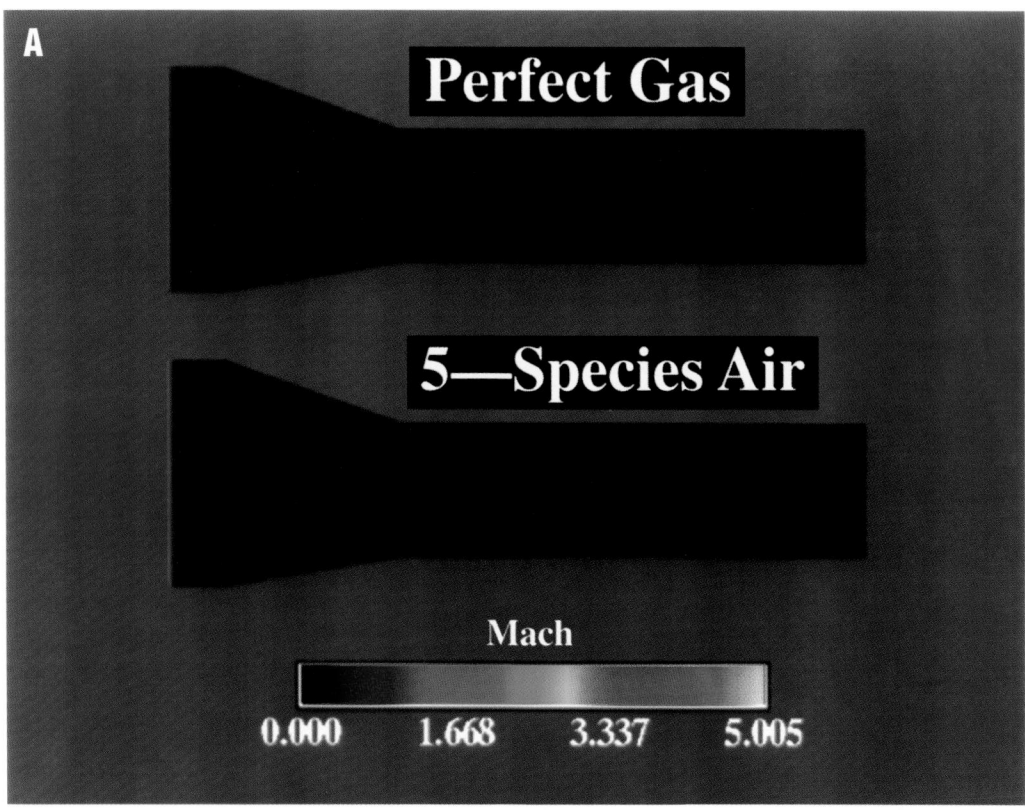

PLATE 38.2A

PLATE 38.2 (A through G) Mach 5 flow in a supersonic inlet. Mach number contours for a perfect gas model and a 5-species reactive air model.

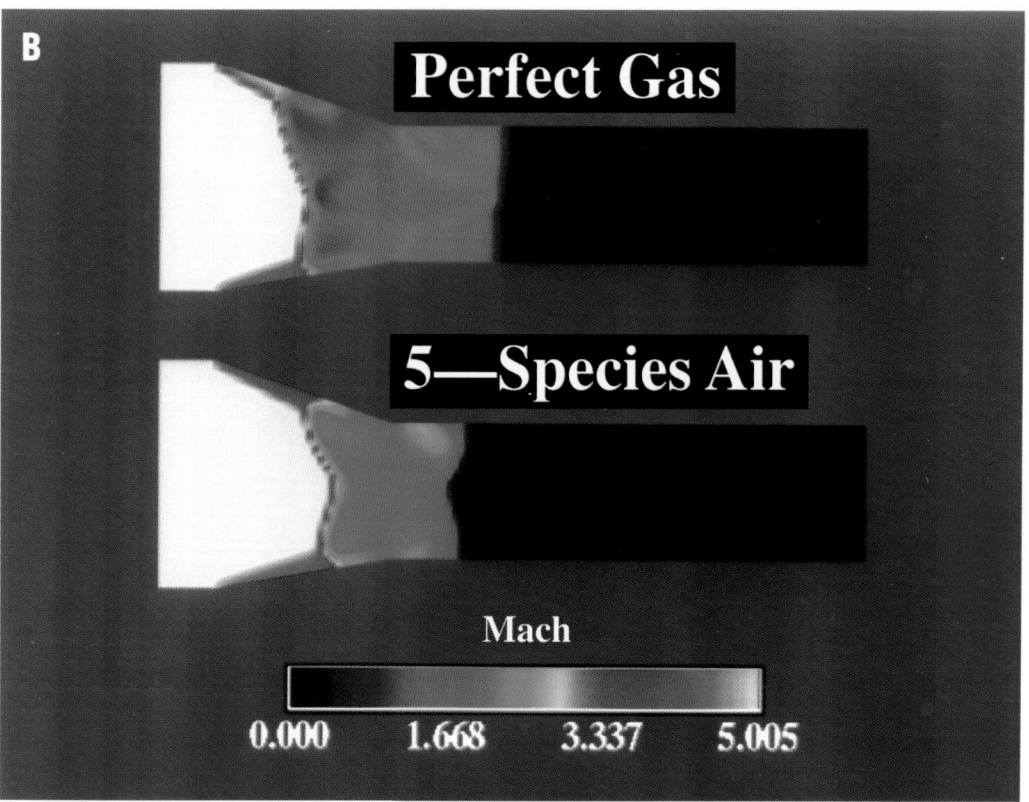

PLATE 38.2B

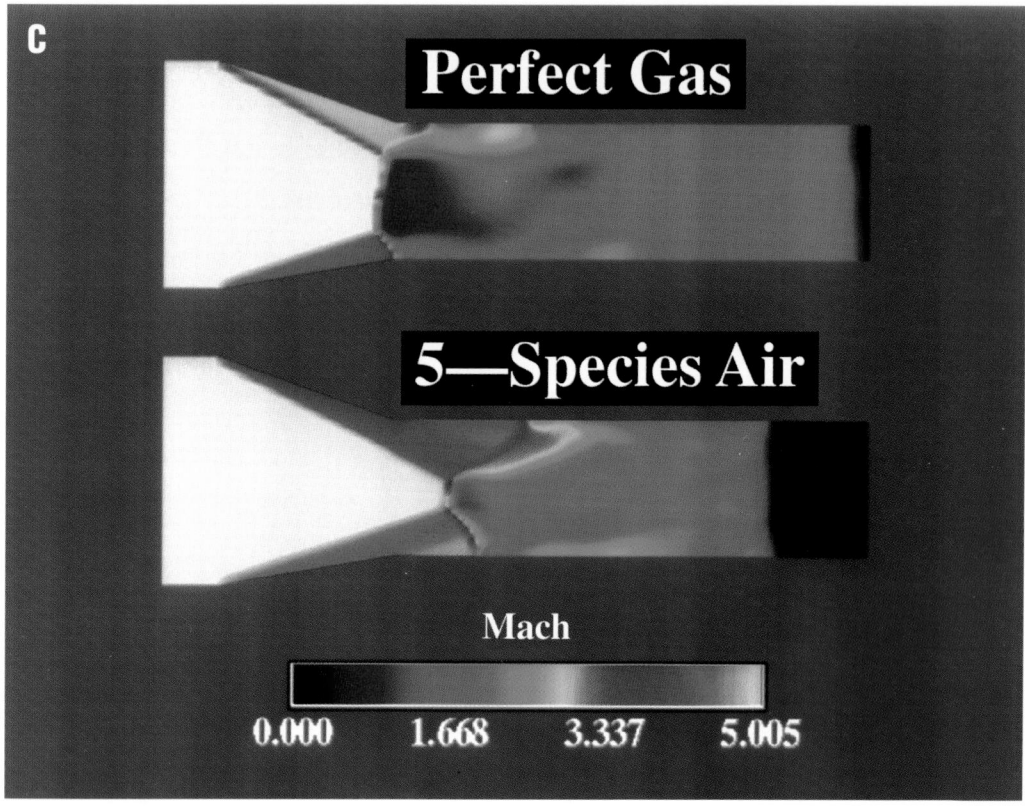

PLATE 38.2C

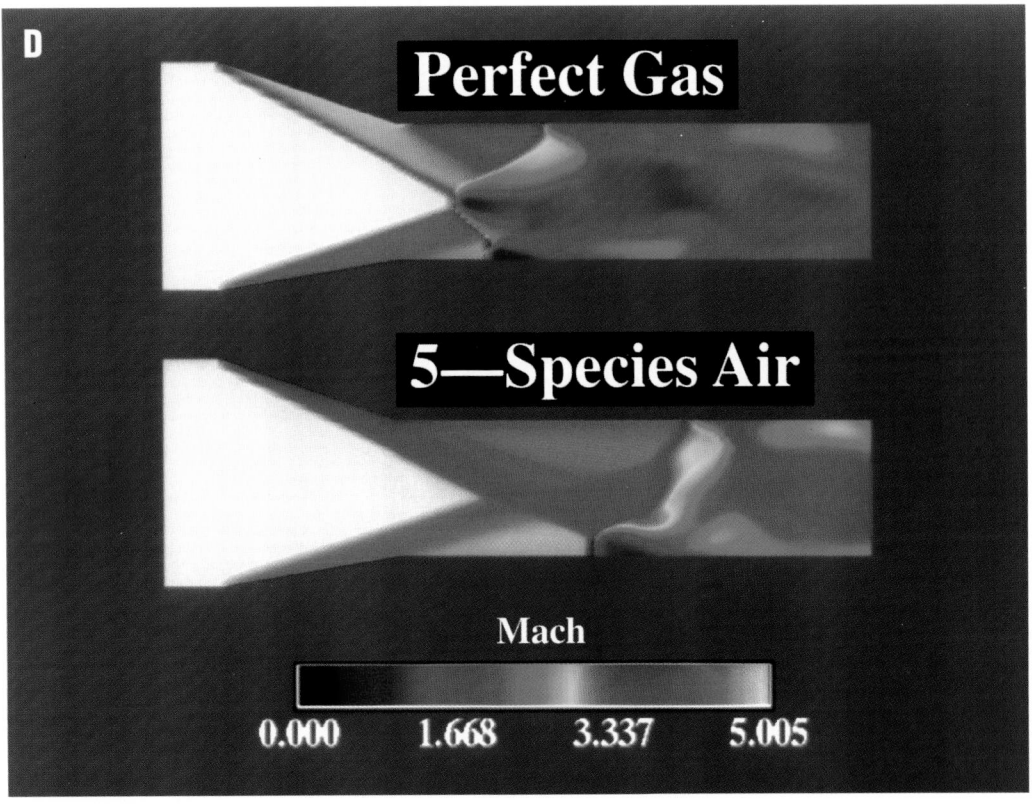

PLATE 38.2D

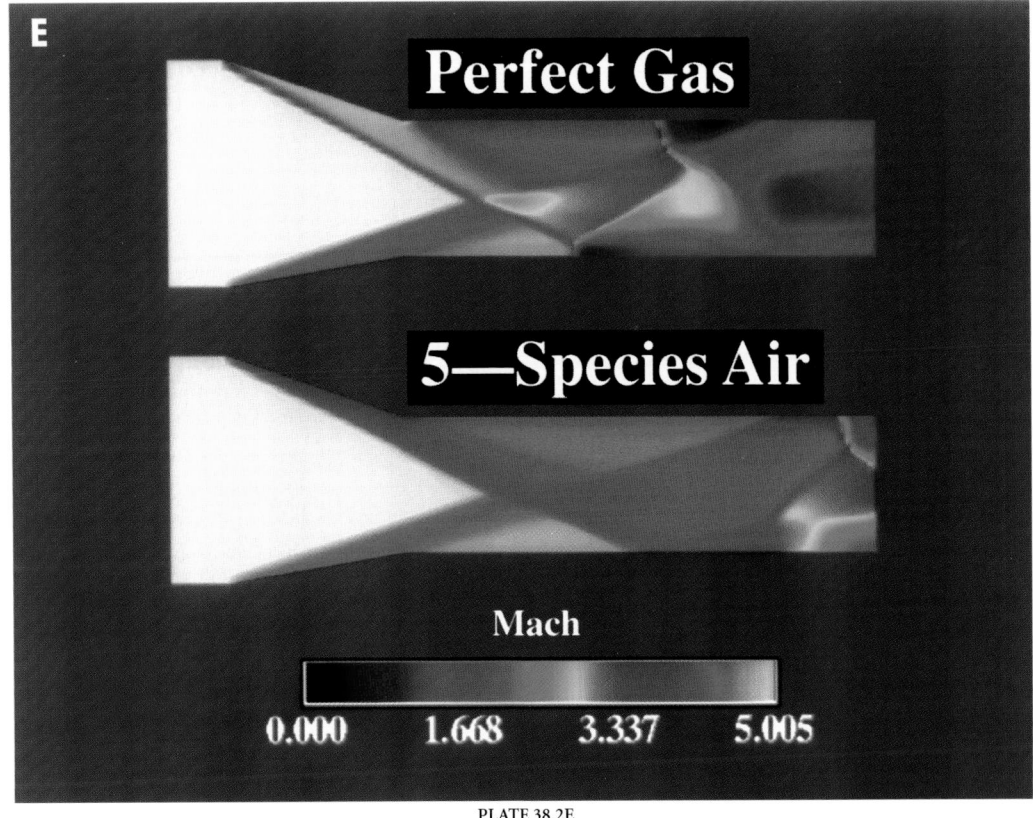

PLATE 38.2E

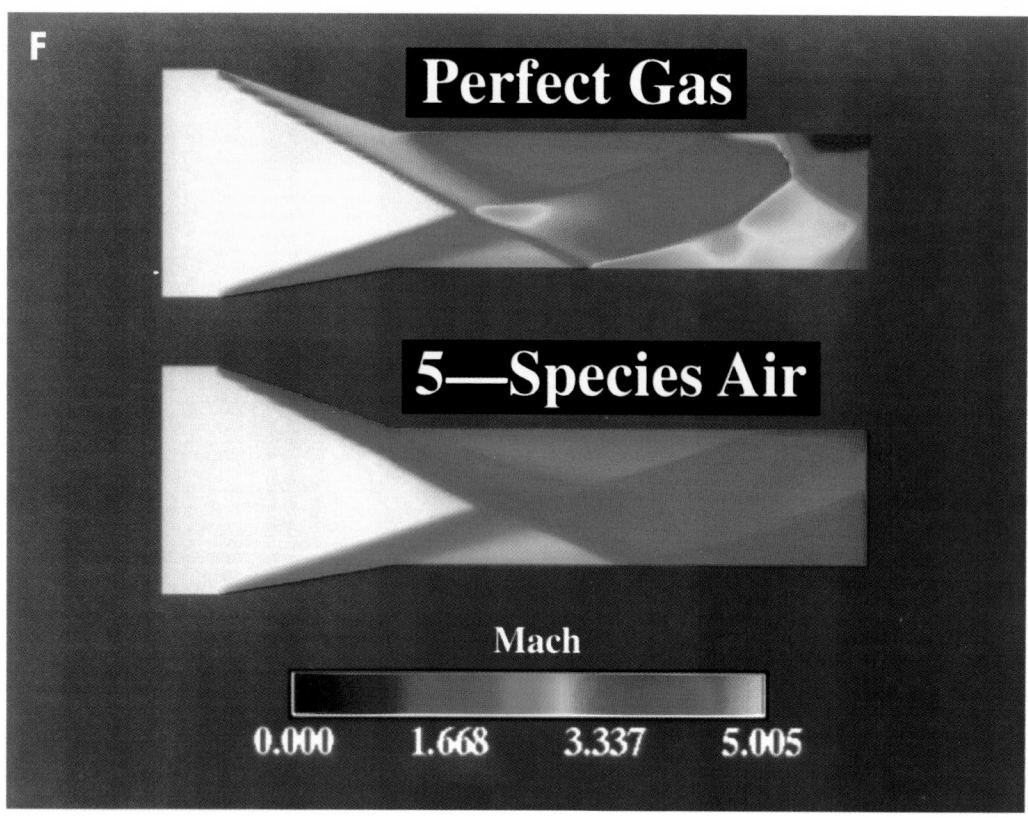

PLATE 38.2F

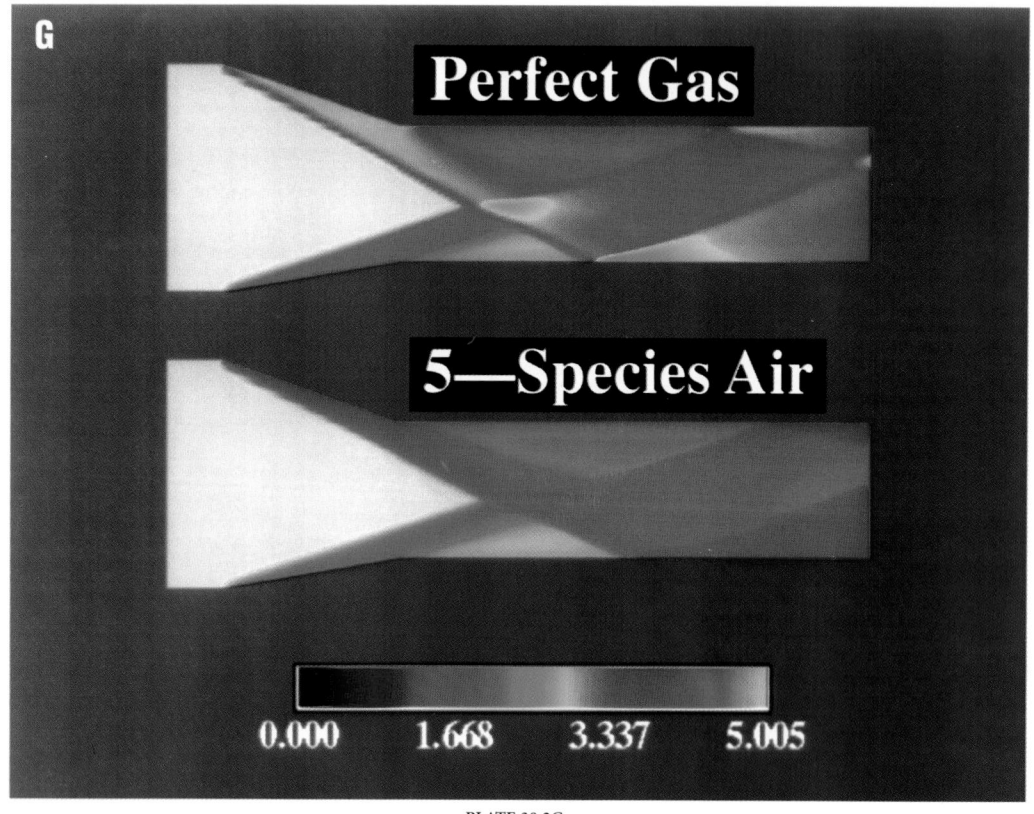

PLATE 38.2G

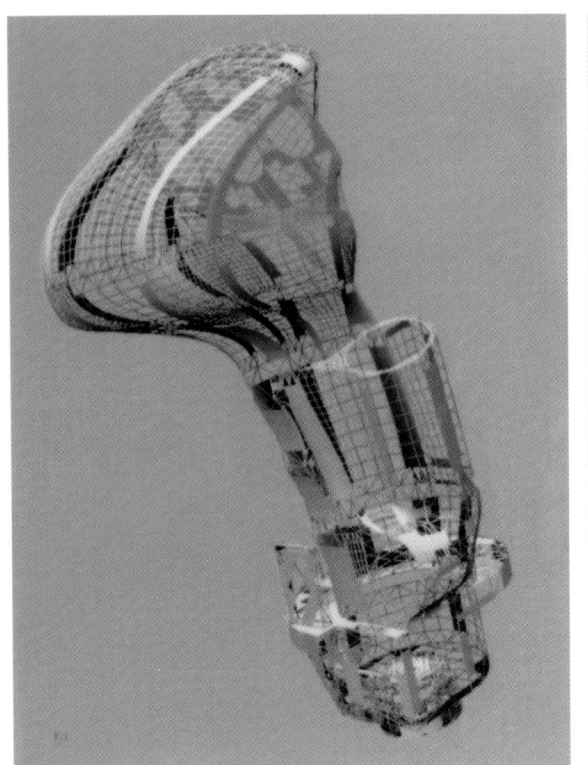 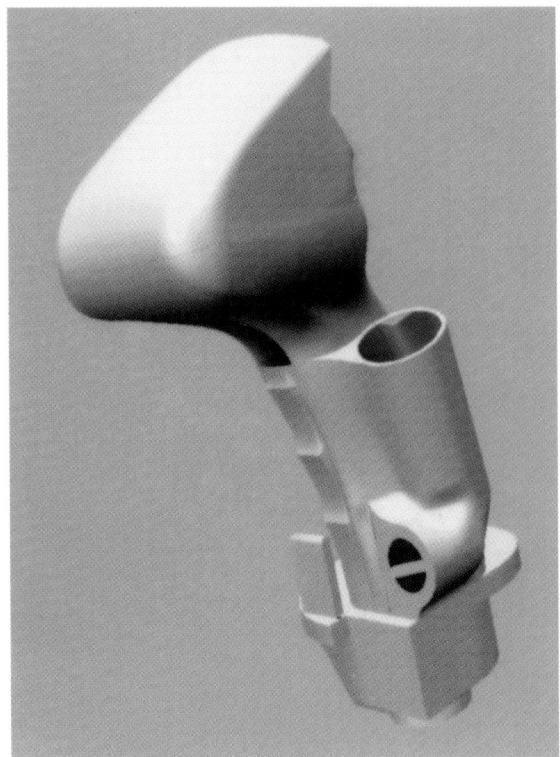

PLATE 55.1 A model composed of simple triangular facets (polygon model).

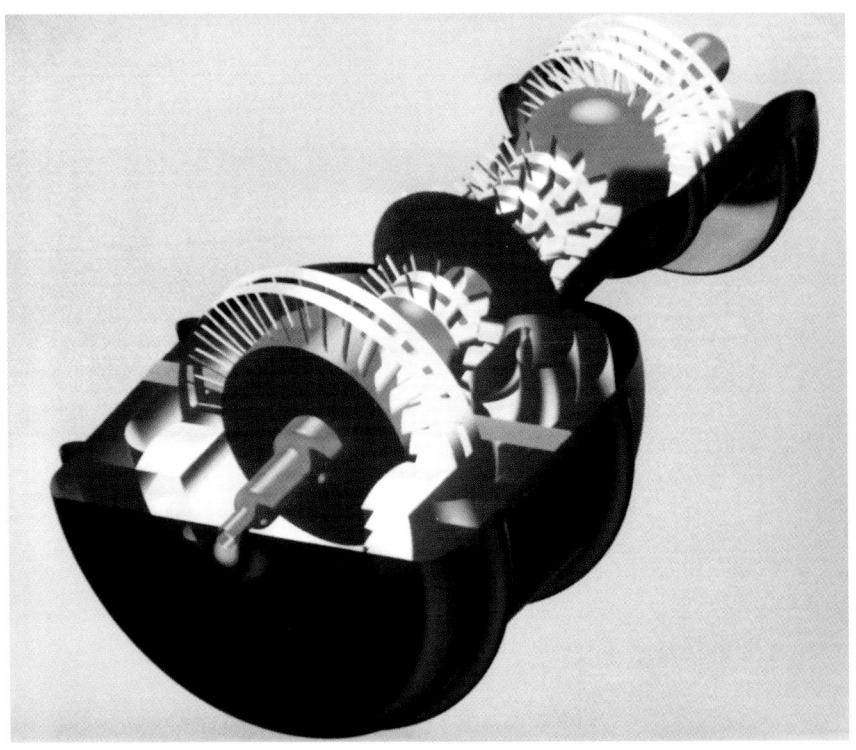

PLATE 55.2 Turbine engine modeled by B-spline surfaces.

PLATE 56.1 A fractal terrain model by Ken Musgrave. (© 1992 *Slickrock*. F. Kenton Musgrave. Used by permission).

PLATE 56.2 Horse chestnut tree created with a modified L-system that takes into account branch competition for light. (© 1995 by R. Měch and P. Prusinkiewicz, University of Calgary, Canada. Used by permission).

PLATE 56.3 Steam rising from a teacup. (© 1991 by David S. Ebert).

PLATE 56.4 Ten years in implicit surface modeling. The locomotive labeled 1985 shows a more traditional soft object created by implicit surface techniques. The locomotive labeled 1995 shows the results achievable by incorporating constructive solid geometry techniques with implicit surface models. (© 1995 by Brian Wyvill. Used by permission).

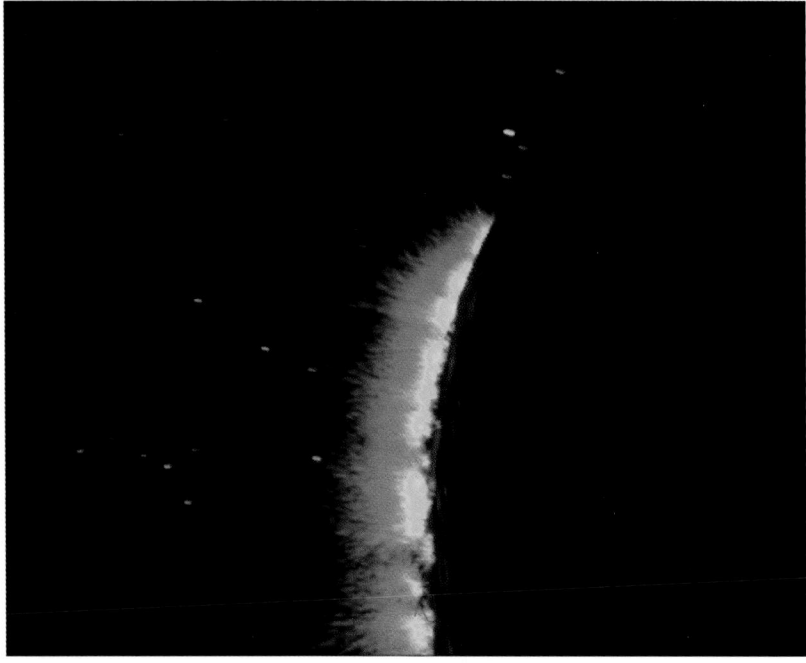

PLATE 56.5 An image from *Star Trek II: The Wrath of Khan* showing a wall of fire created with a particle system. (© 1987 Pixar. Used by permission).

```
 ●  annotddt: cproto.c                                    ? 凹

Annotate ▪    Search ▪    Select ▪   Commands ▪   File ▪   Edit ▪    Move ▪
──────────────────────────────────────────────────────────────────────────
 ✄           if (proto_macro && define_macro) {
                 printf("#if __STDC__ || defined(__cplusplus)\n");
                 printf("#define %s(s) s\n", macro_name);
 (STOP)           printf("#else\n");
                 printf("#define %s(s) ()\n", macro_name);
                 printf("#endif\n\n");
             }

 ∂⌣→ ┃       init_parser();
             if (optind == argc) {
                 if (func_style != FUNC_NONE) {
                     proto_style = PROTO_NONE;
                     variables_out = FALSE;
                     file_comments = FALSE;
                 }
                 process_file(stdin, "stdin");
                 pop_file();
             } else {
──────────────────────────────────────────────────────────────────────────
ReadOnly      Insert Indent                    .../cproto/cproto.c    451/502
```

PLATE 112.1 FIELD program editor. The area on the left is an annotation window showing annotations tying the editor to the other tools. The arrow, for example, indicates the current debugger line of focus while the stop sign indicates a breakpoint.

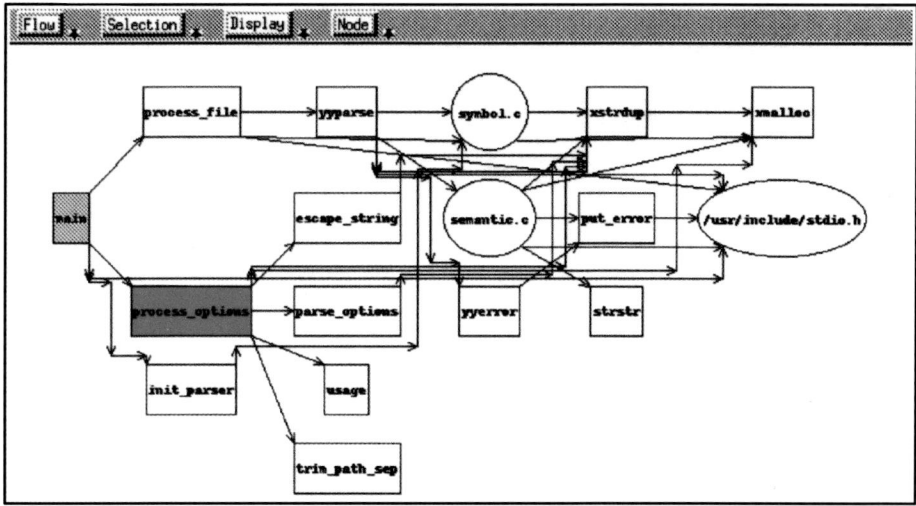

PLATE 112.2 Call graph display from FIELD based on information in the underlying database. The high-lighting reflects the node currently executing the node that is selected. Rectangles indicate functions, ovals represent files.

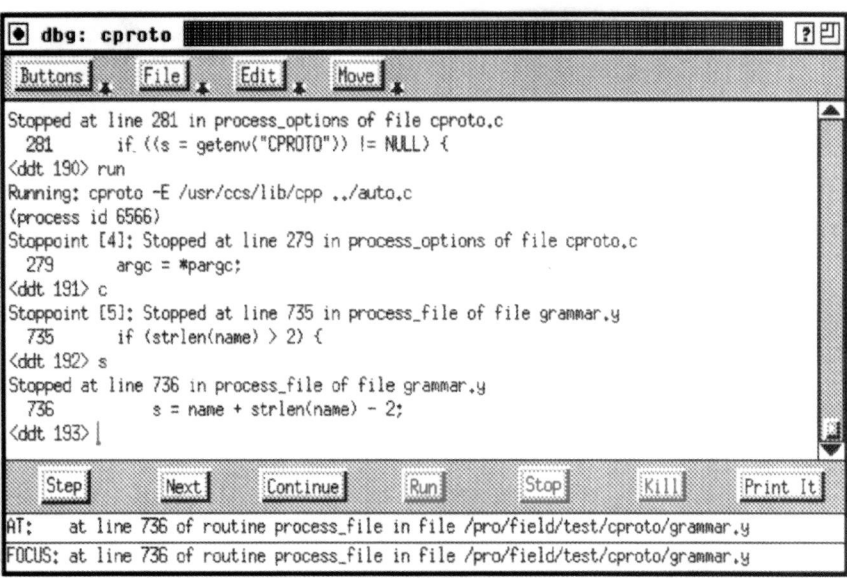

PLATE 112.3 FIELD debugging tool shows visual interface with user-definable buttons as well as a textual transcript of debugging actions.

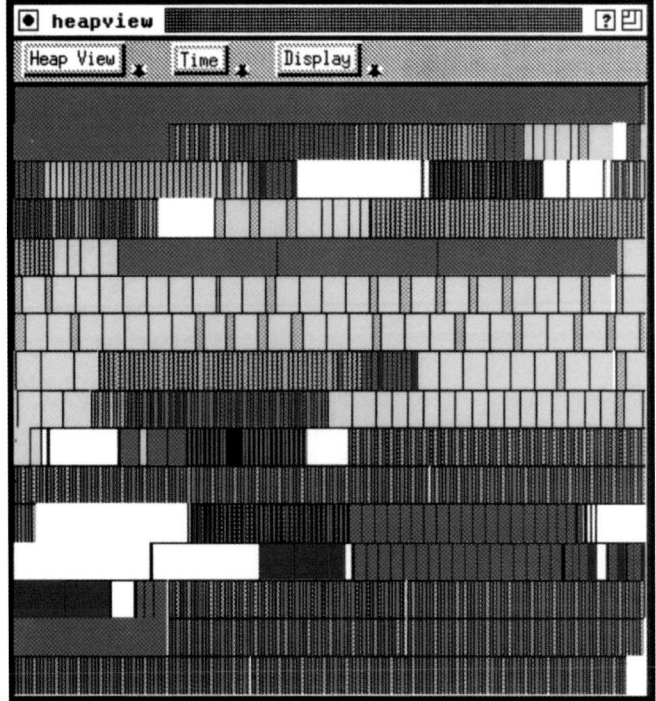

PLATE 112.4 The FIELD heap visualization tool. This diagram shows an abstract view of heap memory in which different colors indicate different allocation sources. The region at the bottom is a memory leak.

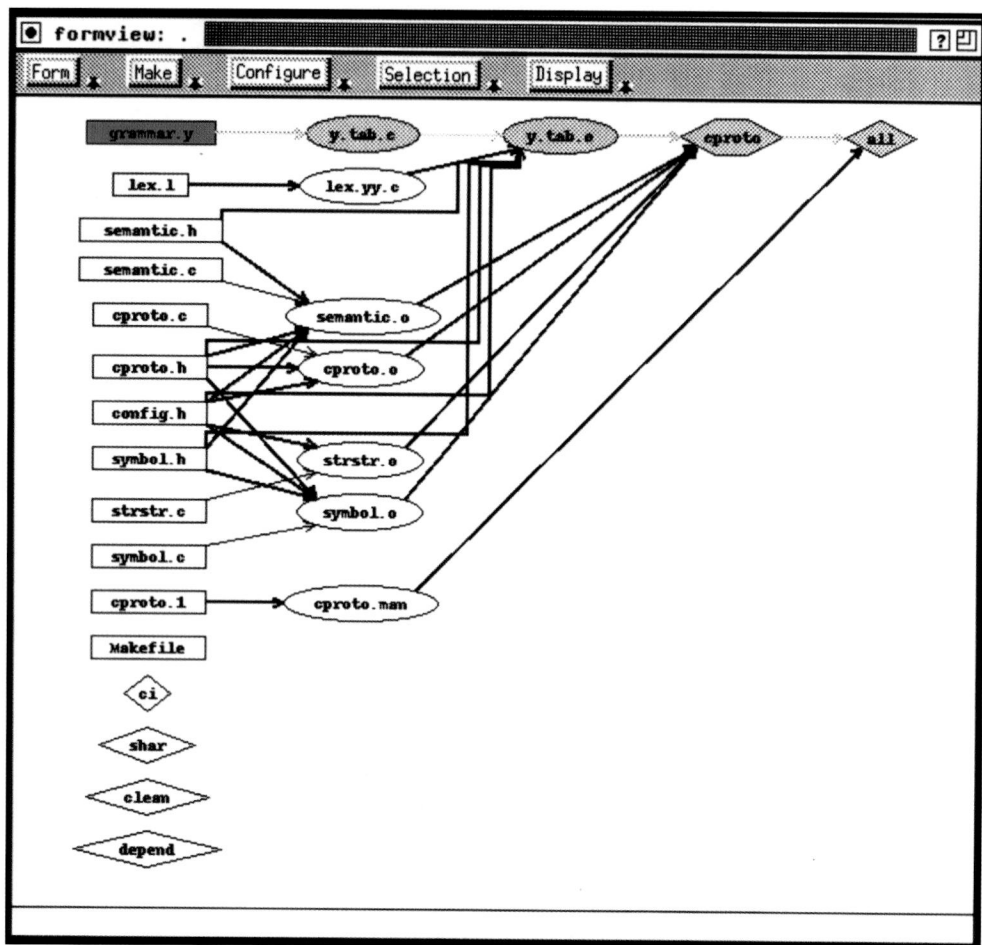

PLATE 112.5 Graphical dependency visualization in the FIELD interface to the UNIX tool make. The color highlighting shows which items have been modified by the programmer and which need to be updated. Elliptical nodes represent intermediate objects, hexagons systems, and diamonds virtual make targets. The thick arrows represent user-defined dependencies while the thin ones were induced by make.

Human–Computer
Interaction

John M. Carroll, Section Adviser
Virginia Polytechnic Institute and State University

THIS AREA, the study of how humans and computers interact, has the goal of improving the quality of the interaction and the effectiveness of those who use technology. This includes the conception, design, implementation, risk analysis, and effects of user interfaces and tools on those who use them in their work. The organizational environment in which people work with technology adds an important dimension to the design of effective user interfaces.

63

Task Analysis and the Design of Functionality

David Kieras
University of Michigan

63.1 Introduction

Task analysis is the process of understanding the user's task thoroughly enough to help design a computer system that will effectively support users in doing the task. By **task** is meant the user's job or work activities, what the user is attempting to accomplish. By *analysis* is meant a relatively systematic approach to understanding the user's task that goes beyond unaided intuitions or speculations, and attempts to document and describe exactly what the task involves. The *design of functionality* is a stage of the design of computer systems in which the user-accessible functions of the computer system are chosen and specified. The basic thesis of this chapter is that the successful design of **functionality** requires a task analysis early enough in the system design to enable the developers to create a system that effectively supports the user's task. Thus, the proper goal of the design of functionality is to choose functions that are *useful* in the user's task and, which together with a good user interface, result in a system that is **usable**, being easy to learn and easy to use.

The user's task is not just to interact with the computer, but to get a job done. Thus, understanding the user's task involves understanding the user's **task domain** and the user's larger job goals. Many systems are designed for ordinary people, who presumably lack specialized knowledge, and so the designers might believe that they understand the user's task adequately without any further consideration. This belief is often not correct; the tasks of ordinary people are often poorly understood by developers. But, in contrast, many economically significant systems are intended for expert users, and understanding their tasks is absolutely critical. For example, a system to assist a petroleum geologist must be based on an understanding of the knowledge and goals of the petroleum geologist. To be useful, such a system will require functions that produce information useful for the geologist; to be usable, the system will have to provide these functions in a way that the frequent and most important activities of the geologist are well supported. Thus, for success, the developer must design not just the **user interface**, but also the functionality behind the interface.

The purpose of this chapter is to provide some background and beginning "how to" information on how to conduct a task analysis and how to approach the design of functionality. The next portion of this chapter discusses why task analysis and the design of functionality are critical stages in software development and how typical development processes interfere with these stages. Then some background on methods for human–machine system design will be presented that have developed in the field of **human factors** over the last few decades. The final section provides some how-to-do-it information in the form of a summary of existing methods and a newly developing method that is especially suitable for computer system design. Other aspects and a general overview of the user interface design process are provided in Chapter 73; the social and organizational aspects of human-computer systems and their development are discussed in Chapter 64. The conventional techniques for ensuring the usability of computer user interfaces are discussed in Chapter 65, whereas presentations of the engineering techniques for interface design and evaluation are presented in sources other than this handbook, especially in Kieras [1988, in press] and John and Kieras [in press a, in press b].

Principles: The Critical Role of Task Analysis and Design of Functionality

In many software development organizations, some group, such as a marketing department or government procurement agents, prepares a list of requirements for the system to be developed. Such requirements specify the system functions, at least in part. The designers and developers then further specify the functions, possibly adding or deleting functions from the list, and then begin to design an implementation for them. Typically, only at this point, or later, is the user interface design begun. Ideally, the design of the user interface will use appropriate techniques to arrive at a usable design, but these techniques normally are based on whatever conception of the user and the user's tasks have already been determined, and the interface is designed in terms of the functions that have already been specified. Thus, even when a usability process is followed, it might well arrive at only a local optimum defined by an inadequate characterization of the user's needs and the corresponding functions.

That is, the focus of usability methods tends to be on relatively low-level questions (such as menu structure) and on how to conduct usability tests to identify usability problems; the problem is posed as developing a usable interface to functions that have already been chosen. Typically, the system requirements have been prepared by a group which then "throws it over the wall" to a development group who arrives at the overall design and functional implementation, and hopefully then throws it over the wall to a usability group to "put a good interface on it."

The problem is that if the initial requirements and system functions are poorly chosen, the rest of the development will probably fail to produce a usable product. It is a truism in human–computer interaction that if customers need the functionality, they will buy and use even a clumsy, hard-to-use product, and if the functionality is poorly chosen, no amount of effort spent on user interface design will result in a usable system, and it might not even be useful at all. This is by no means a rare occurrence; it is easy to find cases of poorly chosen functionality that undermine whatever usability properties the system otherwise possesses. Some examples follow.

The Interface Is Often Not the Problem

An important article with this title by Goransson et al. [1987] presents several brief case studies that involve failures of functionality design masquerading as usability problems. The most clear cut is a database system in a business organization that was considered to be too difficult to use. The interface was improved to make the system reasonably easy to use, but it then became clear that nobody in the organization needed the data provided by the system! Apparently the original system development did not include an analysis of the needs of the organization or the system users. The best way to improve the usability of the system would have been to simply remove it.

Half a Loaf Is Worse than None

The second major version of an otherwise easy-to-use basic word processing application included a multiple-column feature; however, it was not possible to mix the number of columns on a page. Note that documents having a uniform layout of 2 or 3 columns throughout do not exist in the real world; rather, real multicolumn documents always mix the number of columns on at least one page. For example, a common pattern is a title page with a single column for the title that spans the page, followed by the body of the document in two-column format. The application could produce such a document only if two separate documents were prepared, printed, and then physically cut-and-pasted together! In other words, the multiple-column feature of this second version was essentially useless for preparing real documents. A proper task analysis would have determined the kinds and structures of multiple-column documents that users would be likely to prepare. Using this information during the product development would have led to either more useful functionality (like the basic page-layout features in the third major release of the product), or a decision not to waste resources on a premature implementation of incomplete functionality.

Why Doesn't It Do That?

A first-generation hand-held digital diary device provided calendar and date book functions equivalent to paper calendar books, but included no clock, no alarms, and no awareness of the current date, although such functions would have been minor additions to the hardware. In addition, there was no facility for scheduling repeating meetings, making such scheduling remarkably tedious. The only short cut was to use a rather clumsy copy-paste function, but it did not work for the meeting time field in the meeting information. A task analysis of typical user's needs would have identified all of these as highly desirable functions. Including them would have made the first generation of these devices much more viable.

Progress Is Not Necessarily Monotonic

The second version of a personal digital assistant also had a problem with recurring meetings. In the first version, a single interface dialogue was used for specifying recurring meetings, and it was possible to select multiple days per week for the repeating meeting. Thus, the user could easily specify a weekly repeating meeting schedule of the sort common in academics, e.g., a class that meets at the same time every Monday, Wednesday, and Friday for a semester. However, in the second version, which attempted many interface improvements, this facility moved down a couple of menu levels and became both invisible in the interface (unless a certain option was selected) and undocumented in the user manual. If any task analysis was done in connection with either the original or second interface design, it did not take into account the type of repeating meeting patterns needed in academic settings, a major segment of the user population that includes many valuable early adopter customers.

The Role of Task Analysis in Development

Problems with misdefined functionality arise because, first, there is a tendency to assume that the requirements specifications for a piece of software can and should contain all that is necessary to design and implement the software and, second, the common processes for preparing these specifications often fail to include a real task analysis. Usually the requirements are simply a list of desirable features or functions, chosen haphazardly, and without critical examination on how they will fit together to support the user. Blithely assuming that because some user need is mentioned in the requirements does not mean that the final system will include the right functions to make the system either useful or usable for meeting that need. The result is often a serious waste: pointless implementation effort and unused functionality. Thus, understanding the user's task is the most important step in system and interface design. The results of task analyses can be used in several different phases in the development process:

Development of Requirements. This stage of development is emphasized in this chapter. A task analysis should be conducted before developing the system requirements to guide the choice and design of the system functionality; the ultimate usability of the product is actually determined at this stage. The goal of task analysis at this point is to find out what the user needs to be able to accomplish, so that the functionality of the system can be designed so that the user can accomplish the required tasks easily. Although some later revision is likely to be required, these critical choices can be made before the system implementation or user interface is designed.

User Interface Design and Evaluation.   Task analysis results are needed during interface design to effectively design and evaluate the user interface. The usability process itself, and user testing, requires information about the user's tasks. Task analysis results can be used to choose benchmark tasks for user testing that will represent important uses of the system. Usage scenarios valuable during interface design can be chosen that are properly representative of user activities. A task analysis will help identify the portions of the interface that are most important for the user's tasks. Once an interface is designed and is undergoing evaluation, the original task analysis can be supplemented with an additional analysis of how the task would be done with the proposed interface. This can suggest usability improvements either by modifying the interface or improving the fit of the functionality to the more specific form of the user's task entailed by a proposed interface. In fact, some task analysis methods are very similar to user testing; the difference is that in user testing, one seeks to identify problems that the users have with an interface while performing selected tasks; in task analysis, one tries to understand how users will perform their task given a specific interface. Thus, a task analysis might identify usability problems, but task analysis does not necessarily require user testing.

Follow-Up After Installation.   Task analysis can be conducted on fielded or in-place systems to compare systems or identify potential problems or improvements. When a fully implemented system is in place, it is possible to conduct a fully detailed task analysis; these results could be used to compare the demands of different systems, identify problems that should be corrected in a new system, or to determine properties of the task that should be preserved in a new system.

63.2 Research and Application Background

The Contribution of Human Factors to Task Analysis

Task analysis developed in the discipline of human factors, which has a long history of concern with the design process, and how human-centered issues should be incorporated into system design. Much of human factors has been concerned with human participation in extremely large and complex systems, usually military systems, but in fact task analysis methods have been applied to a broad range of systems, ranging from hand-held radios, to radar consoles, to whole aircraft, chemical process plant control rooms, and very complex multiperson systems such as a warship combat information center [see Beevis et al. 1992]. The breadth of this experience is no accident. Historically, human factors is the only discipline with an extended record of involvement and concern with human-system design, long predating the newer field of human–computer interaction (HCI). Unfortunately, comprehensive collections of task analysis techniques have only recently appeared (e.g., Kirwan and Ainsworth [1992] and Beevis et al. [1992]). One purpose of this chapter is to summarize some of these techniques and concepts. But there are some key differences between the interfaces of the traditional kinds of systems treated in the human factors experience and the interfaces to computer-based systems.

Task Properties of Both Types of Systems.   At a general level, the task analysis of both traditional and computer-based interfaces involves collecting and representing the answers to the following primary overall questions:

1. What does the system do as a whole?
2. Where does the human operator fit into the system? What role will the operator play?
3. What specific tasks must the operator perform in order to play that role?

A large-scale general difference between the two kinds of systems shows up in response to these first three questions. Namely, the function or mission of the system as a whole often goes well beyond the immediate goals of the human operators, especially for high-complexity systems. For example, the mission of a system such as AEGIS is to defend a naval task force against all forms of airborne attack. Although each human operator, of course, has an interest in the success of this mission, the individual operator has a much more constrained set of goals, such as monitoring and identifying radar targets. HCI has tended to emphasize the design problem only at the single-operator level, as if only the third question was important. This myopic focus is another manifestation of the problem previously discussed, that the choice and design of functionality is often not considered to be part of the process of designing for usability.

Task Properties of Traditional Systems. At the level of the detailed design of the interface, much of human factors expertise involves the detailed design of traditional interfaces constructed using conventional precomputer technology, often termed *knobs and dials* interfaces. A key design convention of traditional interfaces is that a single display device (e.g., a pressure gauge) displays a single system parameter or sensor value, and a single control device (e.g., a pump power switch) controls a single parameter or component of the system. Such interfaces require many individual physical devices, and so tend to be distributed in space, resulting in control rooms in which every surface is covered with displays and controls. Thus, many of the interface design issues involve the visual and spatial properties of the devices and their layout. For example, some contributors to problems with nuclear power plant control rooms are gauges that cannot be seen by the operator, and misleadingly symmetric switch layouts that have mirror-imaged on and off positions [Woods et al. 1987]. On the other hand, a useful design principle is to place related controls and displays together, and organize them so that the spatial arrangement corresponds to the order in which they are used (e.g., left to right). The entire array of controls and displays is supposed to be always (or readily) visible and accessible simply by looking or moving, and thus constitutes a large external memory for the operator. But the contents of this external memory are usually at the level of individual system parameters and components. This constraint is a product of the available interface and control technology used in traditional systems.

Thus, in addition to the first three task analysis questions, a task analysis of a traditional system seeks answers to the following fourth question:

4. What system parameters and components must be accessible to allow the operator to perform the tasks?

Once these four preliminary questions are answered, an interface can be designed or evaluated. If an interface design does not already exist, then human factors guidelines and principles can be used to guide the creation of a design. Once an interface design is available, a further, more detailed task analysis can then determine how the operator would access the controls and displays involved in order to carry out the tasks. Because of the spatiality of such interfaces, visual and spatial issues receive considerable emphasis. The sequential or temporal properties of the procedures followed by the operators are thus closely related to the spatial layout of the interface.

Task Properties of Computer-Based Systems. In contrast to the myriad controls and displays of traditional interfaces, computer-based systems often have few *in-place objects* (see Beevis et al. [1992]), meaning that the interface for even an extremely complex system can consist of only a single screen, keyboard, and mouse. The operator must thus control the system by sequentially organized activities with this small number of mechanical or visual objects that tend to stay in the same place and compete for the limited keyboard and display space. Time has been traded for space; temporal layout of activity has been traded for spatial layout. The operator often must remember more, because the limited display space means that only a little information is visible at any one time, and complicated procedures are often needed to bring up other information. The user's procedures have relatively little spatial content, and instead become relatively arbitrary sequences of actions on the small set of objects, such as typing different command strings on a keyboard. These procedures can become quite elaborate sequential, repeating, and hierarchical patterns.

On the other hand, the power of the computer behind the interface permits the displayed information to be potentially more useful and more relevant to the user's task than the single-sensor/single-display traditional approach. For example, the software can combine several sensor readings and show a derived value that is what the user really wants to know. Likewise, the software can control several system components in response to a user command, radically simplifying the operator's procedures. Thus, the available computer technology means that the interface displays and controls can be chosen much more flexibly than the traditional knob and dial technology. In conjunction with the greater complexity of such systems, it is now both possible and critically important to choose the system display and control functionality on the basis of what will work well for the user, rather than simply sorting through the system components and parameters to determine the relevant ones.

Thus a task analysis for a computer-based system must articulate what services, or functions, the computer should provide the operator, rather than what fixed components and parameters the operator must have access to, leading to the following additional question for computer-based systems:

5. What display and control functions should the computer provide to support the operator in performing the tasks?

In other words, the critical step in computer-based system design is the *choice of functionality*. Once the functions are chosen, the constraints of computer interfaces mean that the procedural requirements of the interface are especially prominent; if the functions are well chosen, the procedures that the operator must follow will be simple and consistent. Thus, the focus on spatial layout in traditional systems is replaced by a concern with the choice of functionality and interface procedures in computer-based systems. The importance of this combination of task analysis, choice of functionality, and the predominance of the procedural aspects of the interface are the basis for the recommendations later in this chapter.

Contributions of Human–Computer Interaction to Task Analysis

The field of human–computer interaction is a new and highly interdisciplinary field and still lacks a consensus on scientific, practical, and philosophical foundations. Consequently, a variety of ideas have been discussed concerning how developers should approach the problem of understanding what a new computer system needs to do for its users. Although many original researchers and practitioners in HCI had their roots in human factors, during the last decade HCI lost touch with the pragmatic, experience-based approaches from human factors, and has recently been influenced by several other disciplines, which may have strengthened the scientific base of HCI, but with as yet unknown practical benefits.

There are roughly two groups of these disciplines. The first is **cognitive psychology**, which is a branch of scientific psychology that is concerned with human cognitive abilities such as comprehension, problem solving, and learning. The second is a mixture of ideas from the social sciences, such as social-organizational psychology, ethnography, and anthropology. Whereas the contribution from these fields has been important in developing the scientific basis of HCI, these fields either have little experience with humans in a work context, as is the case with cognitive psychology, or no experience with practical system design problems, as with the social sciences. On the other hand, the human factors discipline is almost completely oriented toward solving practical design problems in an ad hoc manner, and is almost completely atheoretic in content. Thus, the disciplines with the broad and theoretical science base lack experience in solving design problems, and the discipline with this practical knowledge lacks a comprehensive scientific foundation.

The current state of task analysis in HCI is thus rather confused; there has been an unfortunate tendency to *reinvent* task analysis under a variety of guises, as each theoretical approach presents its own insights about how to understand a work situation and design a system to support it. The resulting hodgepodge of newly minted ideas is bewildering enough to HCI specialists, but it would be impenetrably obscure to the software developer who merely wants to develop a better system and not become conversant with a variety of academic positions that are often ill formulated and relate poorly to each other. For this reason, this chapter focuses on the tried and true pragmatic methodologies from human factors and a closely related methodology based on the most clearly articulated of the newer theoretical approaches, the **GOMS model**.

New Social-Science Approaches to Task Analysis. At the time of this writing, a few ideas new to HCI that are relevant to task analysis are in the limelight, and so should be briefly mentioned; see Baecker et al. [1995] for a useful sampling and overview. A general methodological approach is *ethnography,* which is the set of methods used by anthropologists to immerse oneself in a culture and document its structure. An approach based on anthropology called *situated cognition* emphasizes understanding human activity in its larger and social context. The proponents of this approach have had some successes, apparently due to their insistence on observing and documenting what people are actually doing in a specific situation. In the context of common industrial practice in system design, this insistence might seem novel and noteworthy, but of course, the reader will recognize such activity as meeting this chapter's definition of task analysis. As presented in the literature, situated cognition makes many other claims about the nature of science and human activity that seem to be unjustified, beside the point for practical application, and perhaps even self-contradictory (see Nardi [1995], for a readable critique).

Another theoretical approach is *activity theory,* which originated in the former Soviet Union as a comprehensive psychological theory, with some interesting differences from conventional western or American psychology. A serious failing in conventional psychology is a gap between the theory of humans as individual intellects and actors and the theory of humans as members of a social group or organization. This leaves HCI as an applied science without an articulated scientific basis for moving between designing a system that works well for an individual user and designing a system that meets the needs of an group. Activity theory, a comprehensive psychological approach from a different tradition, might fill this gap, but based on the recent survey collected in Nardi [1995], this potential is not yet realized. The social aspects of activity theory were not well developed during the period of Soviet repression, and practical applications of the theory have not yet been demonstrated.

Contributions from Cognitive Psychology. The contribution of cognitive psychology to HCI is both more limited and more successful within its limited scope. Cognitive psychology treats an individual human as an information processor who is considered to acquire information from the environment, transform it, store it, retrieve it, and act on it. This *information-processing* approach, also called the *computer metaphor,* has an obvious application to how humans would interact with computer systems. In a cognitive approach to HCI, the interaction between human and computer is viewed as two interacting information-processing systems with different capabilities, and in which one, the human, has goals to accomplish, and the other, the computer, is an artificial system which should be designed to facilitate the human's efforts. The relevance of cognitive psychology research is that it directly addresses two important aspects of usability: how difficult it is for the human to learn how to interact successfully with the computer, and how long it takes the human to conduct the interaction. The underlying topics in cognitive psychology, human learning, problem solving, and human skilled behavior, have been intensively researched for decades.

The application of cognitive psychology research results to human–computer interaction was first systematically presented by Card, Moran, and Newell (1983) at two levels of analysis. The lower level analysis is the Model Human Processor, a summary of about a century's worth of research on basic human perceptual, cognitive, and motor abilities in the form of an engineering model that could be applied to produce quantitative analysis and prediction of task execution times. The higher level analysis was the GOMS model, which is a description of the *procedural knowledge* involved in doing a task. GOMS stands for the following: The user has goals (G) that can be accomplished with the system. Operators (O) are the basic actions such as keystrokes performed in the task. Methods (M) are the procedures, consisting of sequences of operators, that will accomplish the goals. Selection rules (S) determine which method is appropriate for accomplishing a goal in a specific situation. In the Card et al. formulation, the new user of a computer system will use various problem-solving and learning strategies to figure out how to accomplish tasks using the computer system, and then with additional practice, these results of problem solving will become procedures that the user can routinely invoke to accomplish tasks in a smooth, skilled manner. Thus the properties of the procedures will govern both the ease of learning and ease of use of the computer system.

In the research program stemming from the original proposal, approaches to representing GOMS models based on cognitive psychology theory have been developed and validated empirically, along with the corresponding techniques and computer-based tools for representing, analyzing, and predicting human performance in human–computer interaction situations (see Olson and Olson [1990] and John and Kieras [in press a, in press b] for reviews).

The significance of the GOMS model for task analysis is that it provides a way to describe the task procedures in a way that has a theoretically rigorous and empirically validated scientific relationship to human cognition and performance. Space limitations preclude any further presentation of how GOMS can be used to express and evaluate a detailed interface design (see John and Kieras [in press a, in press b] for overviews). Later in this chapter a technique based on GOMS will be used to couple task analysis with the design of functionality.

63.3 Best Practices: How to Do a Task Analysis

The basic idea of conducting a task analysis is to understand the user's activity in the context of the whole system, either an existing or a future system. Whereas understanding human activity is the subject of scientific study in psychology and the social sciences, the conditions under which systems must be designed usually preclude the kind of extended and intensive research necessary to document and account for human behavior in a scientific mode. Thus, a task analysis for system design must be rather more informal, and primarily heuristic in flavor compared to scientific research. The task analysts must do their best to understand the user's task situation well enough to influence the system design given the limited time and resources available.

The Role of Formalized Methods for Task Analysis

Despite the fundamentally informal character of task analysis, many formal and quasiformal systems for task analysis have been proposed and have been widely recommended. Several will be summarized. It is critical to understand that these systems do not in themselves analyze the task or produce an understanding of the task. Rather, they are ways to represent the results of task analysis. They have the important benefit of helping the analyst observe and think carefully about the user's actual task activity, both specifying what kinds of task information are likely to be useful to analyze, and providing a heuristic test for whether the task has actually been understood. That is, a good test for understanding something is whether one can represent it or document it, and constructing such a representation can be a good approach to trying to understand it. A formal representation of a task helps by representing the results of the task analysis in a form that can help document the analysis, so that it can be inspected, criticized, and revised. Finally, some of the more formal representations can be used as the basis for computer simulations or mathematical analyses to obtain quantitative predictions of task performance, but it must be understood that such results are no more correct than the original, and informally obtained, task analysis underlying the representation.

An Informal Task Analysis Is Better than None

Most of the task analysis methods to be surveyed require significant time and effort; spending these resources would usually be justified, given the near-certain failure of a system that fails to meet the actual needs of users. However, the current reality of software development is that developers often will not have adequate time and support to conduct a full-fledged task analysis. Under these conditions, what can be recommended? Based on sources such as Gould [1988] and Grudin [1991], perhaps the most serious problem to remedy is that the developers often have no contact with actual users. Thus, if nothing more systematic is possible, the developers should spend some time in informal observation of real users actually doing real work. The developers should observe unobtrusively, but ask for explanation or clarification as needed, perhaps trying to learn the job themselves, but they do not make any recommendations, and do not discuss the system design. The goal of this activity is simply to try to gain some experience-based intuitions about the nature of the user's job, and what real users do and why. See Gould [1988] for additional discussion. Such informal, intuition-building contact with users will provide tremendous benefits at relatively little cost. The more elaborate methods presented here provide more detail and more systematic documentation, and will permit more careful and exact design and evaluation than casual observation, but some informal observation of users is infinitely better than no attempt at task analysis at all.

Collecting Task Data

Task analysis requires information about the user's situation and activities, but simply collecting data about the user's task is not necessarily a task analysis. In a task analysis, the goal is to understand the properties of the user task that can be used to specify the design of a system; this requires synthesis and interpretation beyond the data. The data collection methods summarized here are those that have been found to produce useful information about tasks (see Kirwan and Ainsworth [1992] and Gould [1988]); the task analytic methods summarized in the next section are approaches that help analysts perform the synthesis and interpretation.

Observation of User Behavior. In this fundamental family of methods, the analyst observes actual user behavior, usually with minimal intrusion or interference, and describes what has been observed in a thorough, systematic, and documented way. This type of task data collection is most similar to user testing, except that, as previously discussed, the goal of task analysis is to understand the user's task, not just to identify problems that the user might have with a specific system design.

The setting for the user's activity can be the actual situation (e.g., in the field) or a laboratory simulation of the actual situation. Either all of the user's behavior can be recorded, or it could be sampled periodically to cover more time while reducing the data collection effort. The user's activities can be categorized, counted, and analyzed in various ways. For example, the frequency of different activities could be tabulated, or the total time spent in different activities could be determined. Both such measures contribute valuable information on which task activities are most frequent or time consuming, and thus important to address in the system design. Finer grain recording and analysis can provide information on the exact timing and sequence of task activities, which can be important in the detailed design of the interface. Videotaping users is a simple recording approach that supports both very general and very detailed analysis at low cost; consumer-grade equipment is often adequate.

A more intrusive method of observation is to have users *think aloud* about a task while performing it, or to have two users discuss and explain to each other how to do the task while performing it. The verbalization can disrupt normal task performance, but such *verbal protocols* are believed to be a rich source of information about the user's mental processes such as inferences and decision making. The pitfall for the inexperienced is that the protocols can be extremely labor-intensive to analyze, especially if the goal is to reconstruct the user's cognitive processes. The most fruitful path is to transcribe the protocols, isolate segments of content, and attempt to classify them into an informative set of categories.

A final technique in this class is *walkthroughs* and *talkthroughs*, in which the users or designers carry out a task and describe it as they do so. The results are similar to a think-aloud protocol, but with more emphasis on the procedural steps involved. An important feature is that the interface or system need not exist; the users or designers can describe how the task would or should be carried out.

Critical Incidents and Major Episodes. Instead of attempting to observe or understand the full variety of activity in the task, the analyst chooses incidents or episodes which are especially informative about the task and the system, and attempts to understand what happens in these; this is basically a case-study approach. Often the critical incidents are accidents, failures, or errors, and the analysis is based on retrospective reports from the people involved and any records produced during the incident. However, the critical incident might be a major episode of otherwise routine activity that serves especially well to reveal the problems in a system. For example, observation of a highly skilled operator performing a very specialized task revealed that most of the time was spent doing ordinary file maintenance; understanding why led to major improvements in the system (Brooks, personal communication).

Questionnaires.   Questionnaires are a fixed set of questions that can be used to collect some types of user and task information on a large scale quite economically. The main problem is that the accuracy of the data is unknown compared to observation, and can be susceptible to memory errors and social influences. Despite the apparent simplicity of a questionnaire, designing and implementing a successful one is not easy, and can require an effort comparable to interviews or workplace observation. The would-be task analyst should consult sources on questionnaire design before proceeding.

Structured Interviews.   Interviews involve talking to users or domain experts about the task. Typically some unstructured interviews might be done first, in which the analyst simply seeks any and all kinds of comments about the task. Structured interviews can then be planned; a series of predetermined questions for the interview is prepared to ensure a more systematic, complete, and consistent collection of information from the main interviewees.

Interface Surveys.   An interface survey collects information about an existing, in-place, or designed interface. Several examples are as follows. Control and display surveys determine what system parameters are shown to the user, and what components can be controlled. Labeling and coding surveys can determine whether there are confusing labels or inconsistent color codes present in the interface. Operator modifications surveys assess changes made to the interface by the users, such as added notes or markings, that can indicate problems in the interface. A concrete example: the author's new Power Macintosh has a yellow Post-It note on the front giving the key combination command–control–power that replaces the previous dedicated reset button. Finally, sightline surveys determine what parts of the interface can be seen from the operator's position; such surveys have found critical problems in nuclear power plant control rooms. Sightlines would not seem important for computer interfaces, but a common interface design problem is that the information required during a task is not on the screen at the time it is required; an analog to a sightline survey would identify these problems.

Representing Systems and Tasks

Once the task data is collected, the problem for the analyst is to determine how to represent the task data, which requires a decision about what aspects of the task are important and how much detail to represent. The key function of a representation is to make the task structure visible or apparent in some way that supports the analyst's understanding of the task. By examining a task representation, an analyst hopes to identify problems in the task flow, such as critical bottlenecks, inconsistencies in procedures, excessive workloads, and activities that could be better supported by the system. Traditionally, a graphical representation, such as a flowchart or diagram, has been preferred, but as the complexity of the system and the operating procedures increase, diagrammatic representations lose their advantage.

Task Decomposition.   One general form of task analysis is often termed *task decomposition*. This is not a well-defined method at all, but merely reflects a philosophy that tasks usually have a complex structure, and a major problem for the analyst will be to decompose the whole task situation into subparts for further analysis, some of which will be critical to the system design, and others possibly less important. For example, one powerful approach is to consider how a task might be decomposed into a hierarchy of subtasks and the procedures for executing them, leading to a popular form of analysis called (somewhat too broadly) *hierarchical task analysis* (HTA). However, another approach would be to decompose the task situation into considerations of how the controls are labeled, how they are arranged, and how the displays are coded. This is also a task decomposition, and might also have a hierarchical structure, but the emphasis is on describing aspects of the displays in the task situation. Obviously, depending on the specific system and its interface, some aspects of the user's task situation may be far more important to analyze than others. Developing an initial task decomposition can help identify what is involved overall in the user's task, and thus allow the analyst to choose what aspects of the task merit intensive analysis.

Level of Detail.   The question of how much detail to represent in a task analysis is difficult to answer. At the level of whole tasks, Kirwan and Ainsworth [1992] suggest a probability × cost rule: if the probability of inadequate performance multiplied by the cost of inadequate performance is low, then the task is probably not worthwhile to analyze. But even if a task has been chosen as important, the level of detail at which to describe the particular task still must be chosen. Some terminology must be clarified at this point: task decompositions can be viewed as a standard inverted tree structure, with a single item, the overall task, at the top and the individual actions (such as keystrokes or manipulating valves) or interface objects (switches, gauges) at the bottom. A *high-level* analysis deals only with the *low-detail* top parts of the tree, whereas a *low-level* analysis includes all of the tree from the top to the *high-detail* bottom.

The cost of task analysis rises quickly as more detail is represented and examined, but on the other hand, many critical design issues appear only at a detailed level. For example, at a high enough level of abstraction, the Unix operating system interface is essentially just like the Macintosh operating system interface; both interfaces provide the functionality for invoking application programs and copying, moving, and deleting files and directories. The notorious usability problems of Unix relative to other systems only appear at a level of detail that the cryptic, inconsistent, and clumsy command structure and generally poor feedback come to the surface. "The devil is in the details." Thus, a task analysis capable of identifying usability problems in an interface design typically involves working at a low, fully detailed, level that involves individual commands and mouse selections. The opposite consideration holds for the design of functionality, as will be discussed further subsequently. When choosing functionality, the question is how the user will carry out tasks using a set of system functions, and it is important to avoid being distracted by the details of the interface.

Task Analysis at the Whole-System Level

When large systems are being designed, an important component of task analysis is to consider how the system, consisting of all of the machines and all of the humans, is supposed to work as a whole in order to accomplish the overall system goal. This kind of very high-level analysis can be done even with very large systems, such as military systems involving multiple machines and humans. The purpose of the analysis is to determine what role in the whole system the individual human operators will play. Various methods for whole system analysis have been in routine use for some time. Briefly, these are as follows (see Beevis et al. [1992] and Kirwan and Ainsworth [1992]):

Mission and Scenario Analysis. Mission and scenario analysis is an approach to starting the system design from a description of what the system has to do (the mission), especially using specific concrete examples or scenarios. See Brooks (this volume) for related discussion.

Function-Flow Diagrams. Function-flow diagrams are constructed to show the sequential or information-flow relationships of the functions performed in the system. Beevis et al. [1992] provide a set of large-scale examples such as naval vessels.

Petri Nets. Petri nets also represent the causal and sequential relationships between the functions performed in a system, but in a rigorous formalism. Various methodologies for modeling and simulating the system performance can also include timing information.

Function Allocation. Function allocation is a set of fairly informal techniques for deciding which system functions should be performed by machines and which by people. Usually mentioned in this context is the *Fitts' list* that describes what kinds of activities can be best performed by humans vs machines. However, according to surveys described in Beevis et al. [1992], this classic technique is rarely used in real design problems since it is simply not specific enough to drive design decisions. Rather, functions are typically allocated in an ad hoc manner, often simply maintaining whatever allocation was used in the predecessor system, or following the rule that whatever can be automated should be, even though it is known that automation often produces safety or vigilance problems for human operators.

In the military systems analyzed heavily in human factors, the overall system goal is normally rather larger scale and well above the level of concerns of the human operators. For example, in designing a new naval fighter aircraft, the system goals might be stated in terms such as "enable air superiority in naval operations under any conditions of weather and all possible combat theaters through the year 2000." At this level, the users of the system as a whole are military strategists, and commanders, not the pilot, and the system as a whole will involve not just the pilot, but other people in the cockpit (such as a radar intercept operator) and also maintenance and ground operations personnel. Thus the humans involved in the system will have a variety of goals and tasks, depending on their role in the whole system.

At first glance, this level of analysis would appear to have little to do with computer systems; we often think of computer users as isolated individuals carrying out their tasks by interacting with their individual computers. However, when considering the needs of an organization, the mission level of analysis is clearly important; the system is supposed to accomplish something as a whole, and the individual humans all play roles defined by their relationship with each other and with the machines in the system. HCI has begun to consider higher levels of analysis, as in the emerging area of computer-supported collaboration, but perhaps the main reason why the mission level of analysis is not common parlance in HCI is that HCI has a cultural bias that organizations revolve around the humans, with the computers playing only a supporting role. Such a bias would explain the movement mentioned earlier toward incorporating more social-science methodology into system design. In contrast, in military systems, the human operators are often viewed as parts in the overall system, whose ultimate user is the commanding officer, leading to a concern with how the humans and machines fit together. Regardless of the perspective taken to the whole system, at some point in the analysis, the activities of the individual humans that actually interact directly with the equipment begin to appear, and it is then both possible and essential to identify the goals that they, as individual operators, will need to accomplish. At this point, task analysis methodology begins to overlap with the concerns of computer user interface design.

Representing the User's Task

Once the whole system and the roles of the individual users and operators has been characterized, the main focus of task analytic work is to identify more specific properties of the situation and activities of the human operator or user. These can be summarized as what the user needs to know, what the user must do, what the user sees and interacts with, or what the user might do wrong.

Representing What the User Needs to Know

The goal of this type of task analysis is to represent what knowledge the human needs to have in order to effectively operate the system. Clearly, the human needs to know how to operate the equipment; such procedural knowledge is treated under its own heading subsequently. But the operator might need additional procedural knowledge that is not directly related to the equipment, as well as additional nonprocedural conceptual background knowledge. For example, a successful fighter aircraft pilot must know more than just the procedures for operating the aircraft and the onboard equipment; the pilot must have additional procedural skills such as combat tactics, navigation, and communication protocols; also, an understanding of the aircraft mechanisms and overall military situation is valuable in dealing with unanticipated and novel situations.

Information on what the user needs to know is clearly useful for specifying the content of operator training and operator qualifications. It can also be useful in choosing system functionality in that large benefits can be obtained by implementing system functions that make it unnecessary for users to know concepts or skills that are difficult to learn; such simplifications typically are accompanied by simplifications in the operating procedures as well. Aircraft computer systems that automate navigation and fuel conservation tasks are an obvious example.

In some cases where user knowledge is mostly procedural in content, it can be represented in a straightforward way, such as decision-action tables that describe what interpretation should be made of a specific situation, as in an equipment trouble-shooting guide. However, the required knowledge can be extremely hard to identify if it does not have a direct and overt relationship to "what to do" operating procedures. An example is the petroleum geologist, who after staring at a display of complex data for some time, comes up with a decision about where to drill, and probably cannot provide a rigorous explanation for how the decision was made. Understanding how and why users make such decisions is difficult because there is very little behavior prior to producing the result; it is all "in the head," a *purely cognitive task*. Analyzing a purely cognitive task is essentially a cognitive psychology research project, and so the available techniques for identifying the required knowledge for a task are all research techniques, which as previously noted, are often impractical to use in system design settings.

An intensive effort to identify the required user knowledge may be justified in some cases. For example, expert systems in applied artificial intelligence are designed using the results of rather informal and labor-intensive knowledge acquisition methods (see Boose [1992]) that are similar to cognitive psychology research techniques and task analysis techniques. Gott [1988] surveys cases in which an intensive effort to identify the knowledge required for tasks can produce large improvements in training programs for highly demanding cognitive tasks such as electronics troubleshooting. Finally, Diaper [1989] contains some methods for cognitive analysis.

However, as reported by Essens et al. [1994], although cognitive task analysis techniques have demonstrated successes in developing training programs, there have been few demonstrations of successful design of computer systems for decision aiding. Landauer [1995] notes that decision-support systems have yet to produce a convincing track record of improving human productivity; this result would be expected given the general lack of task analysis in system design and the great difficulty of task analysis in the case of heavily cognitive tasks.

Because of the difficulty and breadth of cognitive task analysis methods, they will not be presented any further in this chapter; the reader is referred to Essens et al. [1994], who present a general framework for designing systems intended to aid human decision making. The framework attempts to synthesize cognitive task analysis techniques with conventional task analysis and system design practices. Given that support for cognitively demanding tasks is touted as being one of the main contributions of computers, it is critical that more progress be made in the future on how to conduct task analysis and system design for purely cognitive tasks.

Representing What the User Has to Do

A major form of task analysis is describing the actions or activities carried out by the human operator while tasks are being executed. Such analyses have many uses; the description of how a task is currently conducted, or would be conducted with a proposed design, can be used for prescribing training, assisting in the identification of design problems in the interface, or as a basis for quantitative or simulation modeling to obtain predictions of system performance. Depending on the level of detail chosen for the analysis, the description might be very high level, or might be fully detailed, describing the individual valve operations or keystrokes needed to carry out a task. The following are the major methods for representing procedures.

Operational Sequence Diagrams. Operational sequence diagrams and related techniques show the sequence of the operations (actions) carried out by the user (or the machine) to perform a task, represented graphically as a flowchart using standardized symbols for the types of operations. Such diagrams are often partitioned, showing the user's actions on one side and machine's on the other, to show the pattern of operation between the user and the machine.

Timeline Analysis. Timeline analyses simply display activities, or some characteristic of them, as a function of time during task execution. For example, a workload profile for an airliner cockpit would show a large variety and intensity of activities during landing and takeoff, but not during cruising. After constructing a timeline display, the analyst looks for workload peaks (such as the operator having to remember too many things), or conflicts, such as the operator having to use two widely separated controls at the same time.

Hierarchical Task Analysis. Hierarchical task analysis involves describing a task as a hierarchy of tasks and subtasks, emphasizing the procedures that operators will carry out, using several specific forms of description. The term hierarchical is somewhat misleading, since many forms of task analysis produce hierarchical descriptions; a better term might be *procedure hierarchy task analysis*. The results of an HTA are typically represented either graphically, as a sort of annotated tree diagram of the task structure, or in a more compact tabular form. This is one of most widely used forms of task analysis.

HTA descriptions involve goals, tasks, operations, and plans. A goal is a desired state of affairs (e.g., a chemical process proceeding at a certain rate). A task is a combination of a goal and a context (e.g., get a chemical process going at a certain rate given the initial conditions in the reactor). Operations are activities for attaining a goal (e.g., procedures for introducing reagents into the reactor, increasing the temperature, and so forth). Plans specify which operations should be applied under what conditions (e.g., which procedure to follow if the reactor is already hot). Plans usually appear as annotations to the tree-structure diagram that explain which portions of the tree will be executed under what conditions. Each operation in turn might be decomposed into subtasks, leading to a hierarchical structure.

GOMS Models. GOMS models, introduced earlier, are closely related to hierarchical task analysis; and in fact Kirwan and Ainsworth [1992] include GOMS as a form of HTA. GOMS models describe a task in terms of a hierarchy of goals and subgoals, methods which are sequences of operators (actions) that when executed will accomplish the goals, and selection rules that choose which method should be applied to accomplish a particular goal in a specific situation. However, both in theory and in practice, GOMS models are different from HTA. The concept of GOMS models grew out of research on human problem solving and cognitive skill, whereas HTA appears to have originated out of the pragmatic common-sense observation that tasks often involve subtasks, and eventually involve carrying out sequences of actions. Because of its more principled origins, GOMS models are more disciplined than HTA descriptions. The contrast is perhaps most clear in the difficulty HTA descriptions have in expressing the flow of control: the procedural structure of goals and subgoals must be deduced from the plans, which appear only as annotations to the sequence of operations. In contrast, GOMS models represent plans and operations in a uniform format using only methods and selection rules. An HTA plan would be represented as simply a higher order method that carries out lower level methods or actions in the appropriate sequence, along with a selection rule for when the higher order method should be applied.

Representing What the User Sees and Interacts With

The set of objects that the user interacts with during task execution are clearly closely related to the procedures that the user must follow, in that a full procedural description of the user's task will (or should) refer to all objects in the task situation that the user must observe or manipulate. However, it can be useful to attempt to identify and describe the relevant objects and event independently of the procedures in which they are used. Such a task analysis can identify some potentially serious problems or design issues quite rapidly. For example, studies of nuclear power plant control rooms [Woods et al. 1987] found that important displays were located in positions such that they could not be read by the operator. A task decomposition can be applied to break the overall task situation down into smaller portions, and the interface survey technique previously mentioned can then determine the various objects in the task situation. Collecting additional information, e.g., from interviews or walkthroughs, can then lead to an assessment of whether and under what conditions the individual controls or displays are required for task execution. There are then a variety of guidelines in human factors for determining whether the controls and displays are adequately accessible.

A related form of task analysis is concerned with the layout in space of the displays, controls, or other people that the operator must interact with. *Link analysis* is a straightforward methodology for tabulating the pairwise accessibility relationships between people and objects in the task environment, and weighting them by the frequency and difficulty of access (e.g., greater distance). Alternative arrangements can be easily explored in order to minimize the difficulty of the most frequent access paths. For example, a combat information center on a warship was found to be laid out in such a way that the movement and communication patterns involved frequent crossing of paths, sightlines, and so forth. A simple rearrangement of workstation positions greatly reduced the amount of interference. An analog for computer interfaces would be analyzing the transitions between different portions of the interface such as dialogues or screen objects. A design could be improved by making the most frequent transitions short and direct.

Representing What the User Might Do Wrong

Human factors practitioners and researchers have developed a variety of techniques for analyzing situations in which errors have happened, or might happen. The goal is to determine whether human errors will have serious consequences, and to try to identify where they might occur and how likely they are to occur. The design of the system or the interface can then be modified to try to reduce the likelihood of human errors, or mitigate the consequences of them. Some key techniques can be summarized.

Event Trees.   In an event tree, the possible paths, or sequences of behaviors, through the task are shown as a tree diagram; each behavior outcome is represented either as success/failure or a multiway branch, e.g., for the type of diagnosis made by an operation in response to a system alarm display. An event tree can be used to determine the consequences of human errors, such as misunderstanding an alarm. Each path can be given a predicted probability of occurrence based on estimates of the reliability of human operators at performing each step in the sequence (these estimates are controversial; see Reason [1990] for discussion).

Failure Modes and Effects Analysis.   The analysis of human failure modes and their effects is modeled after a common hardware reliability assessment process. The analyst considers each step in a procedure and attempts to list all of the possible failures an operator might commit, such as to omit the action or perform it too early, too late, too forcefully, and so forth. The consequences of each such failure mode can then be worked out, and again a probability of failure could be predicted.

Fault Trees.   In a fault tree analysis, the analyst starts with a possible system failure, and then documents the logical combination of human and machine failures that could lead to it. The probability of the fault occurring can then be estimated, and possible ways to reduce the probability can be determined.

It is mysterious why none of these techniques have been applied in computer user interface design to any visible extent. At most, user interface design guides contain a few general suggestions for how interfaces could be designed to reduce the chance of human error. Incorporating these more refined techniques into interface design would be a valuable contribution.

A Methodology for Using a Task Analysis in Functionality and Interface Design

The task analysis techniques previously described work well enough to have been developed and applied in actual system design contexts. However, they have mainly developed in the analysis of traditional interface technologies rather than computer user interfaces. As previously discussed, although the general concepts of task analysis hold for computer interfaces, there are some key differences and a clear need to address computer user interfaces more directly. In summary, the problems with traditional analysis methods are as follows:

> *Representing a mass of procedural detail.* Computer interface procedures tend to be complicated, repetitious, and hierarchical. The primarily graphical and tabular representations traditionally used for procedures become unwieldy when the amount of detail is large.
>
> *Representing procedures that differ in level of analysis and type.* For example, in hierarchical task analysis, a plan is represented differently from a procedure, even though a plan is simply a kind of higher order procedure.
>
> *Moving from a task analysis to a functional design to an interface design.* To a great extent, human factors practice uses different representations for different stages of the design process (see Kirwan and Ainsworth [1992] and Beevis et al. [1992]). It would be desirable to have a single representation that spans these stages even if it only covers part of the task analysis and design issues.

The subsection to follow describes how GOMS models could be used to represent a high-level task analysis that can be used to help choose the desirable functionality for a system. Because GOMS models have a programming-language like form, they can represent large quantities of procedural detail in a uniform notation that works from a very high level down to the lowest level of the interface design.

High-Level GOMS Analysis

Using high-level GOMS models is an alternative to the conventional requirements development and interface design process discussed in the introduction to this chapter. The approach is to drive the choice of functionality from the high-level procedures for doing the tasks, choosing functions that will produce simple procedures for the user. By considering the task at a high level, these decisions can be made independently of, and prior to, the interface design, thereby improving the chances that the chosen functionality will enable a highly useful and usable product once a good interface is developed. Key interface design decisions, such as whether a color display is needed, can be made explicit and given a well-founded basis, such as how color-coding could be used to make the task easier.

The methodology involves choosing the system functionality based on high-level GOMS analysis of how the task would be done using a proposed set of functions. The analyst can then begin to elaborate the design by making some interface design decisions and writing the corresponding lower level methods. If the design of the functionality is sound, it should be possible to expand the high-level model into a more detailed GOMS model that also has simple and efficient methods. If desired, the GOMS model can be fully elaborated down to the keystroke level of detail that can produce usability predictions (see Kieras [1988, in press] and John and Kieras [in press a, in press b]).

GOMS models involve goals and operators at all levels of analysis, with the lowest level being the so-called keystroke level, individual keystrokes or mouse movements. The lowest level goals will have methods consisting of keystroke-level operators, and might be basic procedures such as moving an object on the screen, or selecting a piece of text. However, in a high-level GOMS model, the goals may refer only to parts of the user's task that are independent of the specific interface, and may not specify operations in the interface. For example, a possible high-level goal would be **Add a Footnote**, but not **Select INSERT FOOTNOTE from EDIT menu**. Likewise, the operators must be well above the keystroke level of detail, and not be specific interface actions. The lowest level of detail a operator may have is to invoke a system function, or perform a mental decision or action such as choosing which files to delete or thinking of a file name. For example, an allowable operator would be **Invoke the database update function**, but not **Click on the UPDATE button**.

The methods in a high-level GOMS model describe the order in which mental actions or decisions, submethods, and invocations of system functions are executed. The methods should document what information the user needs to acquire in order to make any required decisions and to invoke the system functions, and also should represent where the user might detect errors and how they might be corrected with additional system functions. All too often, support for error detection and correction by the user is either missing or is a clumsy add-on to a system design; by including it in the high-level model for the task, the designer may be able to identify ways in which errors can be prevented, detected, and corrected, early and easily.

An Example of High-Level GOMS Analysis

The domain for this example is electronic circuit design and computer-aided design systems for electronic design (ECAD). A task analysis of this domain would reveal many very complex activities on the part of electronic circuit designers. Of special interest are several tasks for which computer support is feasible: After a circuit is designed, its correct functioning must be verified, and then its manufacturing cost estimated, power and cooling requirements determined, the layout of the printed circuit boards designed, automated assembly operations specified, and so forth.

This example involves computer support for the task of verifying the circuit design by using computer simulation to replace the traditional slow and costly *breadboard* prototyping. For many years now, computer-based tools for this process have been available and undergoing development, based on techniques for simulating the behavior of the circuit using an abstract mathematical representation of the circuit. For purposes of this example, attention will be limited to the somewhat simpler domain of digital circuit design. Here the components are black-box modules (i.e., integrated circuits) with known behaviors, and the circuit consists simply of these components with their terminals interconnected with wires.

The high-level functionality design of such a system will be illustrated with an ideal task-driven design example, describing how such systems *should* have been designed. Then the design of a typical actual system will be presented in terms of the high-level analysis and compared with the task-driven design.

Task-Driven Design Example. Given the basic functionality concept of using a simulation of a circuit, the top-level method to accomplish this goal is the first method shown in Fig. 63.1, which accomplishes the goal: **Verify circuit with ECAD system**. This first method needs some explanation of the notation, which is the *Natural* GOMS language (NGOMSL) described in Kieras [1988, in press] for representing GOMS models in a readable format. The first line introduces a method for accomplishing the top-level user goal. It will be executed whenever the goal is asserted, and terminates when the **return with goal accomplished** operator in step 4 is executed. Step 1 represents the user's blackbox mental activity of thinking up the original idea for the circuit design; this **think-of** operator is just a place holder; no attempt is made to represent the extraordinarily complex cognitive processes involved. Step 2 asserts a subgoal to specify the circuit for the ECAD system; the method for this subgoal appears next in Fig. 63.1. Step 3 of the top-level method is a high-level operator for invoking the functionality of the circuit simulator and getting the results; no commitment is made at this point as to how this will be done in the interface. Step 4 documents that at this point the user will decide whether the job is complete or not based on the output of the simulator. Step 5 is another placeholder for a complex cognitive process of deciding what modification to make to the circuit. Step 6 invokes the functionality to modify the circuit, which would probably be much like that involved in step 2, but which will not be further elaborated in this example. Finally, the loop in step 7 shows that the top-level task is iterative.

The next step in the example is to consider the method for entering a circuit into the ECAD system. In this domain, schematic circuit drawings are the conventional representations for a circuit, and so the basic functionality that needs to be provided is a tool for drawing circuit diagrams. This is reflected in the method for the goal **Enter circuit into ECAD system**.

This method starts with invoking the tool, and then has a simple iteration consisting of thinking of something to draw, accomplishing the goal of drawing it, and repeating until done. The method gets more interesting when the goal of drawing an object is considered, because in this domain there are some fundamentally different kinds of objects, and the information requirements for drawing them are different. Only two kinds of objects will be considered here.

A selection rule is needed in a GOMS model to choose what method to apply depending on the kind of object, and then a separate method is needed for each kind of object; the selection rule set in Fig. 63.1 thus accomplishes a general goal by asserting a more specific goal, which then triggers the corresponding method. The method for drawing a component requires a decision about the type of component (e.g., what specific multiplexer chip should be used) and where in the diagram the component should be placed to produce a visually clear diagram. Drawing a connecting wire requires deciding which two points in the circuit the wire should connect, and also how the wire should be routed to produce a clear appearance.

At this point, the analysis has documented some possibly difficult and time-consuming activities that the user will have to do; candidates for additional system functions to simplify these activities could be considered. For example, the step of thinking of an appropriate component might be quite difficult, due to the huge number of integrated circuits available, and likewise thinking of the starting and ending points for the wire involves knowing which input or output function goes with each of the many pins on the chips. Some kind of on-line documentation or database to provide this information in a form that meshes well with the task might be valuable. Likewise, a welcome function might be some automation to choose a good routing for wires.

Although the analysis of the desirable functions has just begun, it is worthwhile to consider what errors the user might make and how the system functionality will support identifying and correcting them. In

```
Method for goal: Verify circuit with ECAD system
    Step 1.   Think-of circuit idea.
    Step 2.   Accomplish Goal: Enter circuit into ECAD system.
    Step 3.   Run simulation of circuit with ECAD system.
    Step 4.   Decide: If circuit performs correct function,
              then return with goal accomplished.
    Step 5.   Think-of modification to circuit.
    Step 6.   Make modification with ECAD system.
    Step 7.   Go to 3.
Method for goal: Enter circuit into ECAD system
    Step 1.   Invoke drawing tool.
    Step 2.   Think-of object to draw next.
    Step 3.   If no more objects, then Return with goal accomplished.
    Step 4.   Accomplish Goal: draw the next object.
    Step 5.   Go to 2.
Selection rule set for goal: Drawing an object
    If object is a component, then accomplish Goal: draw a component.
    If object is a wire, then accomplish Goal: draw a wire.
    ...
    Return with goal accomplished
Method for goal: Draw a component
    Step 1.   Think-of component type.
    Step 2.   Think-of component placement.
    Step 3.   Invoke component-drawing function with type and placement.
    Step 4.   Return with goal accomplished.
Method for goal: Draw a wire
    Step 1.   Think-of starting and ending points for wire.
    Step 2.   Think-of route for wire.
    Step 3.   Invoke wire drawing function with starting point,
              ending point, and route.
    Step 4.   Return with goal accomplished.
```

FIGURE 63.1 Preliminary high-level methods for ECAD system.

this domain, the errors the user might make can be divided into *semantic* errors, in which the specified circuit does not do what it is supposed to do, and *syntactic* errors, in which the specified circuit is invalid, regardless of the function. The semantic errors can only be detected by the user evaluating the behavior of the circuit, and discovering that the idea for the circuit was incorrect or incorrectly specified. Notice that the iteration in the top-level method provides for detecting and correcting semantic errors. Syntactic errors arise in the digital circuit domain because there are certain connection patterns that are incorrect just in terms of what is permissible and meaningful with digital logic circuits. Two common cases are disallowed connections (e.g., shorting an output terminal to ground) or missing connections (e.g., leaving an input terminal unconnected).

It is important to correct syntactic errors prior to running the simulator because their presence will cause the simulator to fail or produce invalid results. Functions to detect syntactic errors can be implemented

```
Method for goal: Enter circuit into ECAD system
   Step 1.   Invoke drawing tool.
   Step 2.   Think-of object to draw next.
   Step 3.   Decide: If no more objects, then go to 6.
   Step 4.   Accomplish Goal: draw the next object.
   Step 5.   Go to 2.
   Step 6.   Accomplish Goal: Proofread drawing.
   Step 7.   Return with goal accomplished.

Method for goal: Proofread drawing
   Step 1.   Find missing connection in drawing.
   Step 2.   Decide: If no missing connection, return with goal accomplished.
   Step 3.   Accomplish Goal: Draw wire for connection.
   Step 4.   Go to 1.

Method for goal: Draw a wire
   Step 1.   Think-of starting and ending points for wire.
   Step 2.   Think-of route for wire.
   Step 3.   Invoke wire drawing function with starting point,
             ending point, and route.
   Step 4.   Decide: If wire is not disallowed, return with goal accomplished.
   Step 5.   Correct the wire.
   Step 6.   Return with goal accomplished.
```

FIGURE 63.2   Revised methods incorporating error detection and correction steps.

fairly easily; the design question is exactly how this functionality should be defined so that it most helps the user. Note that disallowed connections appear at the level of individual wires, whereas missing connections only show up at the level of the entire drawing. Thus, the functions to assist in detecting and correcting disallowed connections should be operating while the user is drawing wires, and those for missing connections while the user is working on the whole drawing. Figure 63.2 shows the corresponding revisions and additions to some of the methods in Fig. 63.1.

The method for entering the circuit now incorporates a method invoked in step 6 to proofread the drawing, which is done by finding the missing connections in the drawing and adding a wire for each one. The method for drawing a wire now has step 4 that checks for the wire being disallowed immediately after it is drawn; if there is a problem, the wire will be corrected before proceeding.

At this point in the design of the functionality, the analysis has documented what information the user needs to get about syntactic errors in the circuit diagram, and where this information can be used to detect and correct the error immediately. Some thought can now be given to what functionality might be useful to support the user in finding syntactically missing and disallowed connections. One obvious candidate that comes to mind is to use color coding; for example, perhaps unconnected input terminals could start out as red in color on the display, and then turn green when validly connected. Likewise, perhaps as soon as a wire is connected at both ends, it could turn red if it is a disallowed connection, and green if it is legal. This use of color should work very well, since properly designed color coding is known to be an extremely effective way to aid visual search. In addition, the color coding calls the user's attention to the problems but without forcibly interrupting the user. This design rationale for using color is an interesting contrast to actual ECAD displays, which generally make profligate use of color in ways that lack any obvious value in performing the task. Figure 63.3 presents a revision of the methods to incorporate this preliminary interface design decision.

```
Method for goal: Proofread drawing
   Step 1.   Find a red terminal in drawing.
   Step 2.   Decide: If no red terminals, return with goal accomplished.
   Step 3.   Accomplish Goal: Draw wire at red terminal.
   Step 4.   Go to 1.

Method for goal: Draw a wire
   Step 1.   Think-of starting and ending points for wire.
   Step 2.   Think-of route for wire.
   Step 3.   Invoke wire drawing function with starting point,
             ending point, and route.
   Step 4.   Decide: If wire is now green, return with goal accomplished.
   Step 5.   Decide: If wire is red, think-of problem with wire.
   Step 6.   Go to 1.
```

FIGURE 63.3   Methods incorporating color codes for syntactic drawing errors.

At this point, the functionality design also has clear implications for how the system implementation needs to be designed, in that the system must be able to perform the required syntax-checking computations on the diagram quickly enough to update the display while the drawing is in progress. Thus, performing the task analysis for the design of the functionality has not only helped guide the design to a fundamentally more usable approach, but also produces some critical implementation specifications very early in the design.

An Actual Design Example.   The preceding example of how the design of functionality can be aided by working out a high-level GOMS model of the task seems straightforward and unremarkable. A good design is usually intuitively right, and, once presented, seems obvious. However, at least the first few generations of ECAD tools did not implement such an intuitively obvious design at all, probably because nothing was done that resembled the kind of task and functionality analysis just presented. Rather, a first version of the system was probably designed and implemented whose methods were the obvious ones shown in Fig. 63.1: the user will draw a schematic diagram in the obvious way, and then run the simulator on it. However, once the system was in use, it became obvious that errors could be made in the schematic diagram that would cause the simulation to fail or produce misleading results. The solution was to simply provide a set of functions to check the diagram for errors, but to do so in an opportunistic, ad hoc fashion, involving minimum implementation effort, that failed to take into account the impact on the user's task. Figure 63.4 shows the resulting method, which was actually implemented in some popular ECAD systems.

The top level is the same as in the previous example, except for step 3, which checks and corrects the circuit after the entire drawing is completed. The method for checking and correcting the circuit first involves invoking a checking function, which was designed to produce a series of error messages that the user would process one at a time. For ease in implementation, the checking function does not work in terms of the drawing, but in terms of the abstract circuit representation, the *netlist*, and so reports the site of the syntactically illegal circuit feature in terms of the name of the node in the netlist. However, the only way the user has to examine and modify the circuit is in terms of the schematic diagram. Thus, the method for processing each error message first involves locating the corresponding point in the circuit diagram, and then making a modification to the diagram. To locate the site of the problem on the circuit diagram, the user invokes an identification function and provides the netlist node name; the function then highlights the corresponding part of the circuit diagram, which the user can then locate on the screen. In other words, to check the diagram for errors, the user must wait until the entire diagram is completely drawn, and then invoke a function whose output must be manually transferred into another function that finally identifies the location of the error!

```
Method for goal: Verify circuit with ECAD tool
   Step 1.   Think-of circuit idea.
   Step 2.   Accomplish Goal: Enter circuit into ECAD tool.
   Step 3.   Accomplish Goal: Check and correct circuit.
   Step 4.   Run simulation of circuit with ECAD tool.
   Step 5.   Decide: If circuit performs correct function,
             then return with goal accomplished.
   Step 6.   Think-of modification to circuit.
   Step 7.   Make modification in ECAD tool.
   Step 8.   Go to 3.

Method for goal: Check and correct circuit
   Step 1.   Invoke checking function.
   Step 2.   Look at next error message.
   Step 3.   If no more error messages, Return with goal accomplished.
   Step 4.   Accomplish Goal: Process error message.
   Step 5.   Go to 2.

Method for goal: Process error message
   Step 1.   Accomplish Goal: Locate erroneous point in circuit.
   Step 2.   Think-of modification to erroneous point.
   Step 3.   Make modification to circuit.
   Step 4.   Return with goal accomplished.

Method for goal: Locate erroneous point in circuit
   Step 1.   Read type of error, netlist node name from error message.
   Step 2.   Invoke identification function.
   Step 3.   Enter netlist node name into identification function.
   Step 4.   Locate highlighted portion of circuit.
   Step 5.   Return with goal accomplished.
```

FIGURE 63.4 Methods for an actual ECAD system.

Obviously, the design of the functionality in this version of the system will inevitably result in a far less usable system than in the task-driven design; instead of getting immediate feedback at the time and place of an error, the user must finish drawing the circuit and then engage in a convoluted procedure to identify the errors in the drawing. Although the interface has not yet been specified, the inferior usability of the actual design relative to the task-driven design is clearly indicated by the additional number of methods and method steps, and the time-consuming nature of many of the additional steps. In contrast to the task-driven design, this actual design seems preposterous, and could be dismissed as a silly example except for the fact that at least one major vendor of ECAD software used exactly this design.

In summary, the task-driven design was based on an analysis of how the user would do the task and what functionality would help the user do it easily. The result was that users could detect and correct errors in the diagram while drawing the diagram, and so could always work directly with the natural display of the circuit structure. In addition, good use was made of color display capabilities, which often go to waste. The actual design probably arose because user errors were not considered until very late in the development process, and the response was minimal add-ons of functionality, leaving the initial functionality decisions intact. The high-level GOMS model makes the difference between the two designs clear by showing the overall structure of the interaction. Even at a very high level of abstraction, a poor design of functionality can result in task methods that are obviously inefficient and clumsy. Thus, high-level GOMS models can capture critical insights from a task analysis to help guide the initial design of a system and its functionality.

63.4 Concluding Summary

The claim that a task analysis is a critical step in system design is well illustrated by the introductory examples, in which entire systems were seriously weakened by failure to consider what users actually need to do and what functionality is needed to support them, and also by the final example, which shows how rather than the usual ad hoc design of functionality, a task analysis can directly support a choice of functions that results in a useful and usable system. Although there are serious practical problems in performing task analysis, the experience of human factors shows that these problems can be overcome, even for rather large and complex systems. The numerous methods developed by human factors for collecting and representing task data are ready to be used and adapted to the problems of computer interface design. The additional contributions of cognitive psychology have resulted in procedural task analyses that can help evaluate designs rapidly and efficiently. Methods for routinely analyzing social groups and arriving at designs for systems that will effectively support them remain at this time in the realm of research. System developers thus have a powerful set of concepts and tools already available, and can anticipate even more comprehensive task analysis methods in the future.

Acknowledgment

The concept of high-level GOMS analysis was developed in conjunction with Ruven Brooks of the Schlumberger Austin Product Center, who also provided helpful comments on this chapter.

Defining Terms

Cognitive psychology: A branch of psychology that is concerned with rigorous empirical and theoretical study of human cognition, the intellectual processes having to do with knowledge acquisition, representation, and application.

Functionality: The set of user-accessible functions performed by a computer system; the kinds of services or computations performed that the user can invoke, control, or observe the results of.

GOMS model: A theoretical description of human procedural knowledge in terms of a set of goals; operators (basic actions); methods, which are sequences of operators that accomplish goals; and selection rules that select methods appropriate for goals. The goals and methods typically have a hierarchical structure. GOMS models can be thought of as programs that the user learns and then executes in the course of accomplishing task goals.

Human factors: Originating when psychologists were asked to tackle serious equipment design problems during World War II, this discipline is concerned with designing systems and devices so that they can be effectively used by humans. Much of human factors is concerned with psychological factors, but important other areas are biomechanics, anthropometrics, work physiology, and safety.

Task: This term is not very well defined, and is used differently in different contexts, even within human factors and human–computer interaction. Here it refers to purposeful activities performed by users, either a general class of activities, or a specific case or type of activity.

Task domain: The set of knowledge, skills, and goals possessed by users that is specific to a kind of job or task.

Usability: The extent to which a system can be used effectively to accomplish tasks. A multidimensional attribute of a system, covering ease of learning, speed of use, resistance to user errors, intelligibility of displays, and so forth.

User interface: The portion of a computer system that the user directly interacts with, consisting not just of physical input and output devices, but also the contents of the displays, the observable behavior of the system, and the rules and procedures for controlling the system.

References

Baecker, R. M., Grudin, J., Buxton, W. A. S., and Greenberg, S., eds. 1995. *Readings in Human–Computer Interaction. Toward the Year 2000.* Morgan Kaufmann, San Francisco.

Beevis, D., Bost, R., Doering, B., Nordo, E., Oberman, F., Papin, J.-P., Schuffel, I. H., and Streets, D. 1992. *Analysis Techniques for Man–Machine System Design.* Defense Research Group Rep. AC/243(P8)TR/7. NATO HQ, Brussels, Belgium.

Boose, J. H. 1992. Knowledge acquisition. In *Encyclopedia of Artificial Intelligence*, 2nd ed., pp. 719–742. Wiley, New York.

Card, S. K., Moran, T. P., and Newell, A. 1983. *The Psychology of Human–Computer Interaction.* Lawrence Erlbaum Associates, Hillsdale, NJ.

Diaper, D., ed. 1989. *Task Analysis for Human–Computer Interaction.* Ellis Horwood, Chichester, U.K.

Essens, P. J. M. D., Fallesen, J. J., McCann, C. A., Cannon-Bowers, J., and Dorfel, G. 1994. COADE: A Framework for Cognitive Analysis, Design, and Evaluation. *Tech. Rep.* TNO Human Factors Research Institute, Soesterberg, Netherlands.

Goransson, B., Lind, M., Pettersson, E., Sandblad, B., and Schwalbe, P. 1987. The interface is often not the problem. In *Proceedings of CHI+GI.* ACM, New York.

Gott, S. P. 1988. Apprenticeship instruction for real-world tasks: the coordination of procedures, mental models, and strategies. In *Review of Research in Education.* E. Z. Rothkopf, ed. AERA, Washington, DC.

Gould, J. D. 1988. How to design usable systems. In *Handbook of Human–Computer Interaction.* M. Helander, ed., pp. 757–789. North-Holland, Amsterdam.

Grudin, J. 1991. Systematic sources of suboptimal interface design in large product development organizations. *Hum.–Comput. Interaction* 6:147–196.

John, B. E. and Kieras, D. E. in press a. Using GOMS for user interface design and evaluation: which technique? *ACM Trans. Comput.–Hum. Interaction.*

John, B. E. and Kieras, D. E. in press b. The GOMS family of user interface analysis techniques: comparison and contrast. *ACM Trans. Comput.–Hum. Interaction.*

Kieras, D. E. 1988. Towards a practical GOMS model methodology for user interface design. In *Handbook of Human–Computer Interact.* M. Helander, ed., pp. 135–158. North-Holland Elsevier, Amsterdam.

Kieras, D. E. in press. A guide to GOMS model usability evaluation using NGOMSL. In M. Helander, T. Landauer, and P. Prabhu, eds. *The Handbook of Human–Computer Interaction*, 2nd ed. North-Holland, Amsterdam.

Kirwan, B. and Ainsworth, L. K. 1992. *A Guide to Task Analysis.* Taylor and Francis, London.

Landauer, T. 1995. *The Trouble with Computers: Usefulness, Usability, and Productivity.* MIT Press, Cambridge, MA.

Nardi, B., ed. 1995. *Context and Consciousness: Activity Theory and Human–Computer Interaction.* MIT Press, Cambridge, MA.

Olson, J. R. and Olson, G. M. 1990. The growth of cognitive modeling in human–computer interaction since GOMS. *Hum.–Comput. Interaction* 5:221–265.

Reason, J. 1990. *Human Error.* Cambridge University Press, Cambridge.

Woods, D. D., O'Brien, J. F., and Hanes, L. F. 1987. Human factors challenges in process control: the case of nuclear power plants. In *Handbook of Human Factors.* G. Salvendy, ed. Wiley, New York.

Further Information

The reference list contains useful sources for following up this chapter. Landauer's book provides excellent economic arguments on how many systems fail to be useful and usable. The most useful sources on task analysis are the books by Kirwan and Ainsworth and by Diaper, and the report by Beevis et al. The Beevis et al. report is probably the single best source on task analysis methods, but may prove difficult to acquire. A readable introduction to GOMS modeling is B. John's article, "Why GOMS?" in *Interactions*, Vol. 2, No. 4, 1995. The John and Kieras and the Kieras references provide detailed overviews and methods.

64

The Organizational Contexts of Development and Use

Jonathan Grudin
*University of California,
Irvine*

M. Lynne Markus
*The Claremont Graduate
School*

64.1 Introduction

Human–computer interaction has focused on individual users and their relationships to systems. Much of the progress in the field has come about by looking for commonalities across the increasing number and diversity of computer-supported tasks. The personal computer (PC) of the 1980s was the perfect laboratory for this effort, and the initial difficulty in networking PCs together only helped shield PC users from group and organizational influences.

The large, expensive systems that preceded the PC had few resources to devote to usability and less reason to worry about it. The users of these systems were people who used the output, typically paper reports. They did not interact directly with computers: that task was left to programmers and computer operators, who acquired the necessary technical competence. With spreadsheets and word processors on PCs, however, to a much greater extent user and operator were synonymous. These users did not see themselves as computer professionals. They had less desire to master technical aspects of systems, and the emerging shrinkwrap software market allowed them to seek out more usable software.

Today, PCs and workstations are networked, intranetworked, and internetworked; once again, all computer use is being carried out in organizational contexts. The implications of this move—from the three key elements of human–computer interaction: user, system, and use, to larger contexts—are described in

1424

0-8493-2909-4/97/$0.00+$.50

other chapters. In this chapter, we examine what has always been a fundamental unit: organization. Organizations affect human–computer interaction in two important ways. First, systems and applications are developed in organizations, and the context of development influences the development process. Second, interactive systems and applications are used in organizations, and successful use is often affected by a range of organizational factors, which has implications for those developing, introducing, and using systems.

64.2 The Need for Organizational Analysis in Human–Computer Interface Design

A case study by Markus and Keil [1994] illustrates the benefit of a careful organizational analysis by demonstrating that a well-executed interface design project can produce a highly usable system that is not useful, and not used. The setting is "CompuSys," a major computer company and employer of leading interface designers. The project was initiated to redesign the interface to a system developed for internal use by the sales organization, an expert system of the sort that achieved prominence in the mid-1980s through the success of Digital's XCON [Barker and O'Connor 1989]. Sales representatives frequently made errors working out details of complex customer system configurations, such as omitting minor but necessary components. CompuSys swallowed the cost of repairing these errors. An expert system for product configuration was built; it was accurate, but was only used with a fraction of orders.

Costly errors continued. Why wasn't the system used? The sales force complained about its usability. A project employing many advanced interface design techniques, including iterative design and user feedback, led to a major redesign of the clearly awkward interface. Users from five pilot sites were trained on the new system. Millions of dollars later, a new system with a much improved interface was introduced. But the new system was not used much more than the one it replaced.

Why? The system design was based on the following model of a typical sales process: (1) a customer identifies system requirements, (2) the sales representative works out a system configuration to meet the requirements, (3) the price is calculated. This seemed logical to the designers, but it was wrong. More often, customers had a fuzzy sense of their problem and a concrete budget for the system. They indicated how much they had to spend and the sales representative would try to identify an adequate system that could be acquired for that amount. The expert system did not support reasoning back from price to configuration—it reasoned only in one direction, from configuration to price. The new interface made it much easier to work from configuration to the price, but it missed the point. The system was usable but not useful. At the end of the project, the developers did not understand why the usable system was neglected. An organizational analysis—actually an interorganizational analysis—uncovered the *counterintuitive* work process.

To avoid such costly mistakes requires greater analysis or awareness of the organizational context and actual work processes than these designers had, even after direct interviews with the sales force. The awareness might be obtained in different ways: through a more sophisticated survey, ethnographic observation, contextual inquiry, or more extensive participation by users. Intuitions could be enhanced through familiarity with the literature on organizational theory and practice. The next section of this chapter summarizes some of the insights from this literature, drawing from Grudin and Markus [in press].

64.3 Organizations and Their Components

Organizations are often defined as collections of people with a common purpose or task. How does this differ from groups? Is it only a matter of size and structure—are organizations (usually) groups of groups?

Organizations arguably have a longer lifespan than groups. Loss or change in one or two members often changes a group completely, and it would seem odd to consider a group the same following entire replacement of its personnel, but organizations (such as Ford Motor Company or the University of California) can continue despite extensive internal reorganization or change. A group may be more than its members, but an organization is much more so.

Organizations also differ from groups in that they often have distinct public and legal identities and engage in a range of activities as a result, whereas groups are more likely to have a single focus. Although rock groups may be anomalous in some ways, we could argue that the Rolling Stones as a group play music; the Rolling Stones as an organization plans concert tours, handles arrangements, invests money, and so forth. People invest in organizations, the annual reports of many organizations are scrutinized, organizations often engage in more competitive activities than we associate with groups. And for these reasons, organizations have possibly been studied more intensely than groups (except for rock groups!).

If we are trying to understand a human–computer dialogue, we would first move one level up and learn what the purpose and context of that interaction was. Only then would we consider the components: an individual with a display, keyboard, and mouse. Similarly, to understand an organization, we first examine the whole of which it is a part—the role that the organization plays in networks of organizations. Then we examine its component parts: groups and individuals with structured relationships and dynamic interactions.

Organizations in Context

Organizations operate in a larger societal context or environment than can usefully be considered a network of organizations. Consider the Boeing Airplane Company. We think of Boeing as making airplanes, but in fact Boeing manufactures very little of the aircraft apart from the wings. Boeing designs the plane, it assembles the plane, and the components are made by scores of organizations around the world. Much of Boeing's work consists of managing this network of organizations, which includes vendors in most countries to which Boeing sells planes. This of course helps ingratiate Boeing to their governments.

Governments are another kind of organization with which the company interacts, as are airlines, financial institutions, unions, passengers, competitors, and so forth. Each organization in this network of interacting organizations performs a different role. Each organization has its own goals or interests that are sometimes mutually compatible and sometimes in conflict. Some organizations are more powerful than the others, controlling more resources, and thus are more likely to win when there are conflicts of interests. If the Chinese government decides to make a foreign policy point by canceling plans for a large order, there is little Boeing can do.

Similarly, the software industry is a network of organizations with different roles involved in the production, sale, support, and use of various products and services. The 1980s saw a proliferation of organizations mediating between developers and users. These include consulting companies, standards organizations, value-added resellers, third-party developers, subcontractors, advertising agencies, professional organizations, magazine companies, and others.

There are opportunities for conflict as well as cooperation in these relationships. Some organizations both cooperate and compete, as in well-known joint ventures between Apple and IBM, for example. Organizations define success differently: for some it is the bottom line, the annual return to stockholders; for other organizations, such as universities, hospitals, governments, social-service agencies, and voluntary organizations, success may mean knowledge creation and transfer, health, social welfare, or member satisfaction. The interaction of organizations with different goals and interests leads to complex, rich dynamics.

Organizational Components

Apart perhaps from quite small organizations, an organization consists of people deployed in various work groups, which may be organized into departments, divisions, or other structural units. A computer manufacturer can be organized by product category (PCs, workstations, minicomputers) or by function (hardware, systems software, applications software, support); they can have sales organized by region or by product line, and so on.

There is no general agreement on optimal organizational structure. Although in some industries, virtually all organizations in the same size class have a similar organizational structure, others are marked by a great deal of local variation. Large organizations often shift from one structure to another, based perhaps on shifts in the external environment, or as a result of the personnel at hand, or perhaps even as a way of releasing adrenaline and enhancing vitality. Choices may not be fixed, but they can determine what the organization can do easily and well. For instance, a computer manufacturer organized by function may have a harder time delivering new PC models to market than a similar company with a product organization, but may have an easier time establishing consistency and interoperability across all product lines.

A specific example of particular interest to many in the human–computer interaction field is the deployment of usability specialists or human factors engineers in a very large research and development division. Should they be grouped together in a central usability laboratory, where they have the benefits of pooled resources and management that understands their roles? One drawback is possible isolation in a usability ghetto. Or should each specialist become a member of a product team, gaining the benefit of close interaction throughout the product cycle and a team that comes to understand and trust them? One drawback is that a usability specialist, managed as a member of a software team by someone with little understanding of usability, may become a software engineer. Both approaches and others are tried.

An ambitious analysis of organizational components has been developed by Mintzberg [1984, 1989]. Mintzberg identifies five basic organizational components or groups: the strategic apex (executives), the middle line (management), the operating core (production workers, whether surgeons in a hospital or assembly line workers), the technostructure (those who define the work processes in an organization), and the support staff (technical support, custodial, kitchen, and so on). Mintzberg argues that these functions will to some degree compete for control, and that five organizational forms can be found, each marking the dominance of one of these groups, along with some hybrids. Mintzberg extends his analysis to look at a wealth of accompanying and interacting factors, such as the approach to measuring output, the organizational environment, and so on. One of us once worked for a 500-person organization that was created with only a loose mission, and it was fascinating to see the correspondence of behavior to Mintzberg's model as each of the five groups jockeyed for control in this uniquely undefined setting.

Thus even when an organization as a whole has a clear set of goals and interests, individuals, groups, and subunits within the organization may not share them fully. Newspaper reporters and newspaper editors may differ, hardware and software engineers may engage in rivalry that is friendly or not so friendly—and both groups may have less than full respect for their colleagues in marketing or upper management. In many user organizations, the information systems staff differs in age and outlook from other groups, which can lead to breakdowns in communication or cooperation.

Employees may be subject to different performance measures and rewards. For example, a software unit may be evaluated on development schedules and budgets, whereas a hardware group may be evaluated largely on manufacturing cost. If so, it may be hard for the units to cooperate to achieve the total organization's goals. In the case of CompuSys, errors in computer system configuration originated with the sales department, which was rewarded for sales volume, but the manufacturing unit bore the costs associated with fixing incorrect orders. Given these incentives and rewards, the sales department devoted little time and energy to ensuring that orders were correct at the outset.

Good management, good communication of organizational purpose, an appropriate set of measures, rewards, and punishments might keep everyone singing the same tune. However, the complexity of large organizations makes it difficult to design efficient and effective management structures. Generally speaking, in organizations of any size and structural complexity, subunits exhibit a fair degree of goal displacement, pursuing goals and values that make sense to the subunit, but that do not necessarily advance the overall interests of the organization. And there is not agreement as to the appropriate level of conflict within an organization; one ideal would be to eliminate internal conflict and competition, but others endorse a management style that fosters a level of internal jockeying for resources.

Groups within organizations can develop distinct cultures (assumptions, beliefs, and language systems) that render clear communication with members of other groups and units more difficult. The same terms are often used in very different ways in different parts of an organization, creating problems for efforts to create a common database. Mark and Mambrey [1996] describe an example in which typists wanted documents categorized by author and completion date, whereas authors wanted categorization by topic. An example noted by Grudin and Markus [in press]: It is a sale to sales when a customer verbally commits to an order, but it is not a sale to legal until a contract is signed; it is not a sale to manufacturing until a purchase order is entered into the manufacturing control system, and the accounting department only acknowledges a sale when an invoice has been prepared. Language differences contribute to widely differing points of view on key organizational decisions. Conflict as well as cooperation occurs in organizational decision making, producing behavioral dynamics inside organizations that are as rich as those observable when organizations interact.

Ways in Which Organizations Differ

It is useful to think of at least three levels of reality operating simultaneously in organizations. The first can be called rational, technical, or economic reality. It takes a stated goal as given and looks for an efficient and effective means of achieving it. The second level of reality can be called socioemotional task reality. People have social needs and organizations are one place in which they attempt to meet them. In addition, people habituate to particular ways of thinking and behaving due in part to their membership in organizational groups and subunits. These ways of thinking and acting may differ dramatically from what an outside observer would see as the rational optimum. The third level of reality is structural/political, focusing on goals and interests created by resources and positions in units and task chains (often called business processes) or interorganizational networks. Keeping all three realities in mind helps in seeing and understanding the complex dynamics of organizational behavior.

This has been a highly simplified overview of organizational behavior. Next, we show how these organizational issues can have enormous implications for the adoption, deployment, use, and consequences of information technology in organizations. However, first we will review some major categories of differences between organizations and their component work groups and subunits that can significantly shape the use of information technology:

- Headcount
- Economic resources, particularly *slack* (uncommitted resources)
- Geography (scope of operations) and space (e.g., in buildings)
- Age of organization, demographic profile, experience including experience working together
- Stated or implicit goals (e.g., least cost producer vs. product innovator)
- Structure/basis of organization (product, geographic, function, technology, time)
- Culture (beliefs, assumptions, language systems, characteristic behavior patterns)
- Management style (measures, rewards, promotion patterns, etc.)
- Information technology infrastructure (prior investments, commitments, governance)

With so many variable factors, it is not surprising that organizations and subunits react quite differently to a given technology.

64.4 Organizational Modeling, Formal and Informal

When a system or application is developed for use within a single organization, the software is likely to better mesh with the behavior of the organization if pertinent aspects of the organization can be formally modeled; that is, represented in a form the computer can manipulate. Formal modeling of organizational processes and data have been a major concern in the development of large systems.

Before addressing this history, though, it is useful to contrast the development of a system for a single organization with other development situations. In particular, when developing a shrinkwrap software product, intended for use in a range of organizations, the initial developers have less motivation for organizational modeling. Rather, those tailoring a system for a given customer might consider organizational modeling. This is significant because commercial software has been the focus of most work in human–computer interaction (HCI). Most in HCI have no familiarity with organizational modeling; many have little experience with databases.

In the area of groupware and computer-supported cooperative work, formal modeling is appearing. Some is at the level of group behavior rather than organizational behavior, but workflow management systems (e.g., Abbott and Sarin [1994] and Marshak [1994]) involve a broad enough context to be considered organizational. The workflow tool developers do not model organizations; it is assumed the customer (or a consultant) will carry out the modeling.

Workflow management systems are often considered in the context of business process re-engineering (BPR). Both are predicated on the idea of creating detailed models of organizational processes. BPR looks to rationalize such processes; workflow management systems look to incorporate them in software and support them. Workflow management systems appear to be useful for high-volume, relatively routine business activities; whether current systems have the flexibility to support other activity is unclear. Bowers et al. [1995] is a nice study of a workflow system.

When we consider *user* organizations and the design and development of interactive software, formal modeling and its limitations arise. In considering the effects of *development* organizations on the design and development of interactive software, formal modeling does not arise, because our examination there deals with the need for people—designers and developers—to contend with the organizational environment, not a program. Fostering *awareness* of organizational influences is the objective, not modeling it formally. Awareness is a kind of informal modeling, of course, a point that arises in contexts of system use as well.

64.5 Organizational Contexts of Development

Developers are well aware that organizations constrain them through time pressures, approval processes, formal specifications, and other practices. What is often less evident is that these constraints differ markedly, and systematically, across organizations. To quote Mahoney [1988], "we speak of the computer industry as if it were a monolith rather than a network of interdependent industries with separate interests and concerns." An organization's structure and practices can have effects on the human–computer interfaces that it produces, and these may be major effects or subtle effects. Differences across segments of the industry affect what techniques can or should be applied, and what tools will or will not be useful. The history of segmentation within the field left traces in systems development practices, and the history differs in North America and Europe. (The following account draws on Grudin [1991a], Grudin and Poltrock [1995], and Grudin and Markus [in press], as well as the sources cited in the text.)

To illustrate the effects of organizational context, consider a key relationship in the design of interactive systems: the relationship between the developers and the users. Adapted from Grudin [1991a], Fig. 64.1 shows, for three development contexts, the times at which development teams are identified and the actual users are identified for a new application. From the top, an organization putting out a project for competitive bidding must produce a preliminary design to give possible contractors a specification to bid on. The users are identified well before the development team is known, and to prevent any favoritism toward a particular contractor, interaction between user and development organization may be curtailed or prohibited. For in-house development of a system by the information systems group within a user organization, both parties are often identifiable from the outset. The developers of a novel commercial product work under still different conditions: After marketing or management has done some analysis and high-level specification, the development team is formed. But the users are not truly known until the product is marketed. As we will see, these and other differences in conditions can greatly affect the process of developing interactive systems.

The Emergence of Distinct Development Contexts in the U.S.

An examination of the historical emergence of development contexts illuminates their differences and explains aspects of systems development practices that can adversely affect human–computer interfaces.

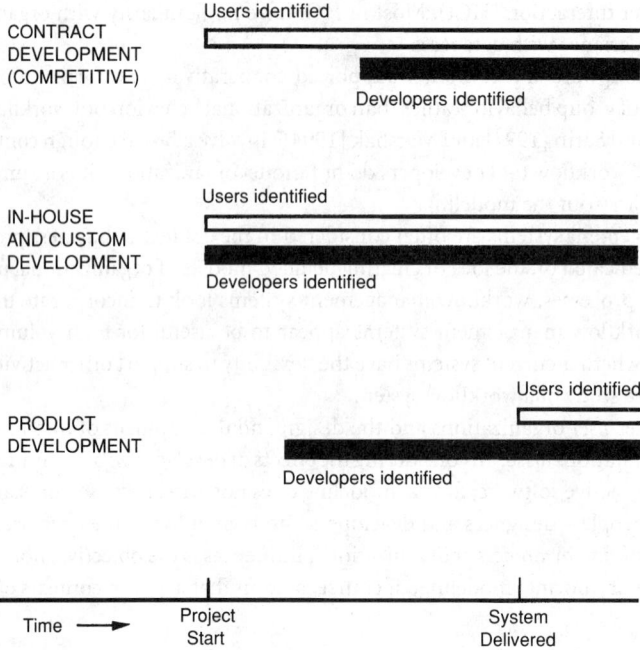

FIGURE 64.1 Identification of users and developers in different development contexts.

Most major early systems were contracted by the U.S. government. The government defined the requirements, and geographically distant organizations bid and were awarded contracts. Different contracts might be awarded for design, development, and maintenance. Legal restrictions governed their communication. This created tremendous organizational barriers to direct contact among developers and users. However, this was of less import because little software was interactive. Most computers were batch processors: a program and data were loaded and the computer computed, and perhaps produced a printed output.

Stage models of systems development evolved naturally in this environment. In the U.S., the waterfall model of Royce [1970], appearing at the dawn of the discipline of software engineering, was enormously influential. This design-it-first, specification-based approach, with little provision for feedback or iteration, became the dominant development methodology.

The design of usable human–computer dialogues requires abandoning the design-it-first approach. Boehm [1988] introduced spiral models explicitly to incorporate prototyping and iterative design for interactive software development. Nevertheless, the separation of developers and users remains a major organizational obstacle to interface design in contract development contexts. Winkler and Buie [1995] provide a good summary of the challenges and possible approaches, such as developing standards that can be included in requests for proposals (RFPs). The primary focus, however, is on raising awareness in user and developer organizations. Perhaps more important than trying to specify the interface in detail is to specify an interface development *process*. Winkler and Buie identify the primary interface development problems in government contracting to be organizational, not technical, in nature.

As business use of computers expanded, in-house development of software in large user organizations became the principal focus of system development. (In Europe, in-house development has always predominated.) Most processing remained batch rather than interactive, with end users primarily engaged in data entry and system operator tasks, but the waterfall development method approach was challenged from the outset in this organizational context, more effectively in Europe, where the stage models did not have the same prominence [Friedman 1989, Hirschheim et al. 1995].

Barriers to interaction among users and developers are less compelling for internal development. Organizational and cultural barriers still hinder user involvement in development—young, highly paid information systems staff may clash culturally with other workers. But developers and users have the same employer and often work in greater proximity.

Methodologies developed to increase user involvement in in-house projects include the sociotechnical approach [Mumford 1983, 1993] and Scandinavian participatory design approaches [Bjerknes et al. 1987, Greenbaum and Kyng 1991]. These stress the need for developers to obtain a detailed understanding of actual work processes and worker preferences. Participatory design also emphasizes educating prospective users about technical possibilities. Early work focused on workplace democracy; when interactive systems arrived, the focus shifted to developing better software tools for workers.

With the arrival of PCs, commercial, off-the-shelf product development focused attention on highly interactive systems and applications. Product development organizations present a different set of constraints on developers. As with contract development, users and developers are in different organizations. Unlike contract development, the development organization is responsible for defining the requirements. Competitive markets mean more emphasis on human–computer interfaces and greater time pressures.

Inadequacies of the waterfall development methodology were clear in this context. Rapid prototyping and iterative design, with heavy user involvement, were promoted from the outset. These principles were widely accepted, but not widely practiced, due to organizational constraints detailed in Grudin [1991b] and Poltrock and Grudin [1994]. When motivated to seek user involvement, which is not always the case, developers can have difficulty identifying appropriate users, obtaining access to them, and motivating users who after all work in other organizations and may never use the product that emerges. If these barriers are overcome, there are often obstacles to using the information that is received. Development schedules rarely provide time for iteration, and changes in the interface can be visible and unsettling to management, as well as to those working on product documentation and training, which are tied closely to the interface.

To overcome these obstacles, product developers have sought to adapt and extend participatory design and related techniques. A notable example is contextual inquiry and modeling [Holtzblatt and Beyer 1993, Holtzblatt and Jones 1993, Beyer and Holtzblatt 1995].

A fourth organizational context for development, custom development, blends characteristics of each of the three previously discussed. The development organization is distinct from the user organization and works under contract. The contract is not necessarily competitively bid and the development organization may be selected based on geographic proximity. Rather than prohibited, long-term, close involvement can serve to mutual advantage, infusing some aspects of internal development into the process. The development organization often focuses on a market niche, where similar custom jobs will amortize their investment. The more successful they are at finding customers with the same needs, the closer their situation comes to resemble product development; in fact, their software may evolve into a commercial product.

Custom development seems likely to thrive as demands for software become increasingly diverse and specialized. Grudin [1996] argues that custom development could be particularly promising for HCI tools that focus on documenting upstream design process for subsequent use in redesign or maintenance, because, for example, it is a context in which design, redesign, and maintenance are likely to be done by the same people.

Figure 64.2 summarizes responses to stage or waterfall models arising in contract development (software engineering), in-house development (information systems), and product or package development (human–computer interaction). For a richer view of the history, see Hirschheim et al. [1995].

This only suggests differences among development contexts and the effects of organizations. There exist organizations to mediate between a developer and a user organization. This activity takes place in economic, cultural, political, and societal conditions to which organizations contribute. Other organizational factors influence how technology is developed and used: structures and processes, size, intended market, geographical placement, age, application novelty, function, culture, environment. In a small startup company, all employees may see one another regularly; in a large organization, the software, documentation, and training developers may work in different states or countries. In the U.S., court decisions regarding *look and feel* copyrights or patents can affect the process and product of development; in Europe, codetermination laws can affect the development process by requiring user involvement.

This list is not exhaustive, but it serves as an indicator that organizational factors within a development environment have a strong bearing on systems development, and thoughtful developers will consider those

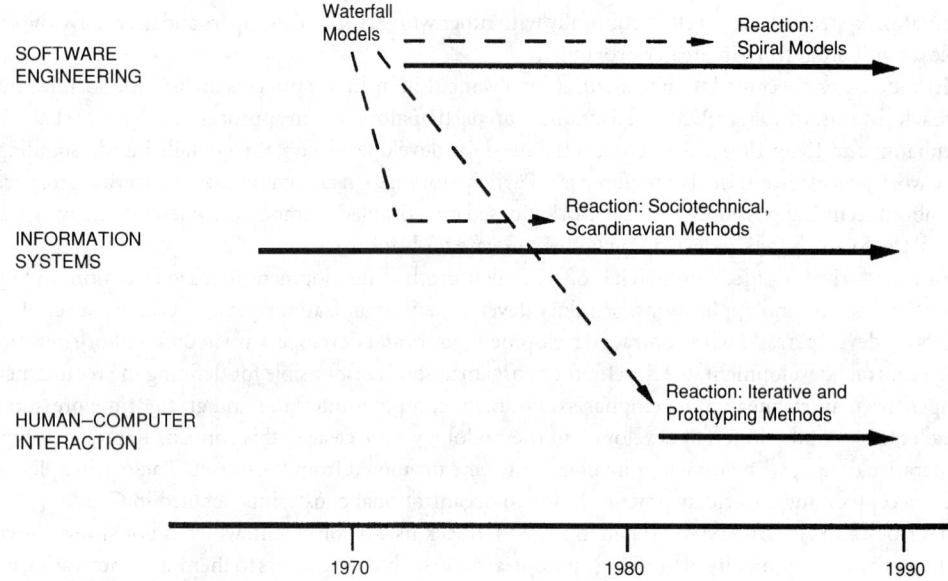

FIGURE 64.2   Approaches to systems development in different development contexts.

factors carefully. Knowledge of organizational influences can alert developers to obstacles and biases that can result in painful mistakes. It can also guide them to techniques that are suited to their context. Keil and Carmel [1995] have closely analyzed successful projects in two contexts, corresponding to what we have called in-house and product development. They report, for example, on 11 different approaches to establishing links among customers and developers and find that the methods used and their levels of effectiveness vary across contexts.

64.6 Organizational Contexts of Use

We now examine efforts to model, understand, or use intermediaries to represent user organizations. The emergence of different contexts of system use and development is relevant here as well. A very good treatment of this topic is found in Hirschheim et al. [1995]. This work to some degree builds on Friedman [1989] which is equally absorbing.

Both of these works focus virtually exclusively on internal or in-house systems development, with perhaps some attention to custom development. This is understandable: in-house development dominated systems development for decades, and this has been even more true in Europe. In the U.S., competitive contract development was more influential in the early days and commercial off-the-shelf software has attracted considerable attention recently.

It is important to keep in mind that close attention to human–computer interaction bloomed in commercial software development, not in in-house systems development. In the latter context, it makes intuitive sense to try to model aspects of the organization, but the focus has not been on improving human–computer interaction.

Hirschheim et al. note that those applications were growing more complex over time, but do not note the dramatic shifts in the user populations, the specific new demands that highly interactive systems introduce, or strong ties between the features of certain systems development methodologies and the development contexts in which they originate.

For example, they note that at one time user requirements elicitation was considered a largely noncontroversial task of asking users what they needed in their jobs. This may always have been a questionable assumption, but it was less so for batch processing than for interactive systems. Elsewhere, they note that software engineering focused on development steps "under the assumption that systems requirements are given" without noting that in government contracting that was often the situation, however problematic it might be. And so forth.

Nevertheless, Hirschheim et al. [1995] is an excellent introduction to organizational and data modeling, as long as one keeps in mind that the history and context of these modeling and development approaches placed less weight on human–computer interaction concerns than the reader might. The distinctions that are introduced include the following:

- Information systems (IS) can be limited to the technical aspects or defined to include humans. IS design can be viewed as designing technical systems that have social consequences or as primarily addressing organizational and social complexity.

- IS development involves changing object systems that are defined differently by different individuals; the choice of a systems development methodology is a major factor among the many that affect a person's definition of the object system.

- Rather than say that a problem exists, it may be better to say that one is constructed among stakeholders with different perceptions.

- Information systems development can be marked by uncertainty regarding the means employed, the effects that will result, or whether the right problem was attacked. Systems development methodologies can be categorized as process oriented or data oriented. Object systems can be perceived and modeled as static, dynamic, or hybrid.

- User participation in system design can be seen as expedient in collecting needed information or overcoming resistance to change, as a prerequisite for creating shared meanings necessary for design, or as a moral right.

- Most methodologies focus on one application; only a few embrace organization-wide planning.

The authors use these and other distinctions to provide alternative views of a range of methodologies. These include structured analysis and design methodologies, which focus on modeling organizational processes and data flows, and data-oriented analysis and design approaches, which represent organizations as a structured collection of facts and associations among them. Object-oriented analysis and design define software units that encapsulate methods and data. In parallel with the evolution of IS development, the authors consider the evolution of data modeling approaches emerging from the techniques that led to the development of database technologies.

The authors identify seven generations of information system design methodologies that replaced unstructured, ad hoc approaches to design. The sequencing is not strictly chronological or convincing, but it is a useful exercise. They are: (1) formal life cycle approaches (from mid-1960s); (2) structured approaches (mid-1970s), emphasizing user requirements; (3) prototyping and evolutionary approaches (late 1970s), tied here to greater competition but not especially to greater prominence of the human–computer interface; (4) sociotechnical and participative approaches (late 1970s, early 1980s) in which users take at least some control and redesign work situations; (5) sense making and problem formulation approaches (early 1980s), involving users and developers from very early in a project; (6) trade union led approaches (mid-1970s through mid-1980s), to support democratic planning and later to design tools to augment workers' skills; and (7) emancipatory approaches (future), a rather vaguely defined approach devised by the authors.

What emerges from this examination of modeling, interpreted through a sense of the contexts in which these approaches have been introduced, is a sense of the trajectory of change and the pressures driving it. When information systems consist largely of entering, organizing, storing, and transforming data, and when the data contain records such as employee names with their ages, salaries, managers, departments, and so forth, then data-centered IS design might flourish in ways that are less likely when the system focuses on real-time computer-mediated communication. A system designed to compute projectile trajectories might give rise to wide agreement on the definition of terms and relevant concepts, whereas a workflow system that must model a group of people and objects that have not previously been considered together will inevitably confront conflicting views of work, conflicting terminology, and other issues. Of particular relevance here, software applications in which over 50% of the program code is devoted to the human–computer interface inevitably introduce a range of considerations and challenges that did not exist for methodologies that often consciously ignored or relegated to a single process substep the establishment of the interface.

Although one does not leave the survey of current practice with a prescription for organizational modeling, one does leave it with a sense of the richness and importance of the problem, and an awareness of the steps taken in the direction of solutions. The growing focus on sense making, on recognizing the multitude of perspectives, and the constant shift in the sense of the target suggests that formal modeling of groups and organizations will encounter limits. And yet as people strive to build systems to support organizations through contact with most or all members, the effort to flexibly model the organizations will inevitably continue.

64.7 What Are the Organizational Issues in Interactive System Use?

This discussion is organized along broad system life cycle phases used by Grudin and Markus [in press]: initiation (idea origination, project funding), acquisition (system acquired or built), implementation and use (internal deployment, use or rejection), and impacts and consequences (system effects experienced; steps taken to augment, mitigate, or otherwise manage them).

Initiation Phase

The source of a technology investment idea can have important influences on downstream events. The source can be external or internal; if internal, it can be from high-ranking line management[1] (possibly responding to advertising or business fads) or lower, perhaps from the technical staff. Approval may require consent of one line manager or an internal technology approval board. Justification can serve a positive purpose of building shared understanding, but the understanding may be inaccurate if distortion is needed to meet preset criteria. See Dean [1987] for a discussion of the technology justification process and Cohen et al. [1972] for a classic paper on decision making in organizations. Suboptimal outcome possibilities include underutilized systems as well as systems rejected because they benefited few users (e.g., only the decision maker [Grudin 1994]). Kling [1978] describes a welfare case tracking system that was adopted despite its lack of operational benefits because the system made the agency look good to potential funders. This was considered a success.

During the initiation phase, the project schedule, funding level, and allocation of funds to different project activities are usually set. Often these decisions have lasting and not entirely positive downstream influences. For instance, an organizational decision to limit access to Lotus Notes to one particular work unit can create difficulties later for activities that cross the boundaries between work units. Or the decision makers may allocate enough resources to acquire or develop a technology but significantly underfund training and support for users, creating predictably negative consequences when the technology is up and running (see Walton [1989]).

In short, the initiation phase of the life cycle can involve political negotiations between people and groups who propose new technologies and people and groups who control essential resources (both technical and financial). These negotiations can result in the selection of inappropriate solutions to organizational problems, or they can result in perceptions and decisions that subsequently have negative effects on system features, use, and impacts.

---

[1] *Line* refers to managers in the core businesses or functions of the organization, as opposed to those in staff functions. In most organizations, IS is viewed as a staff function. This is frequently true even in software development firms, where the IS people who provide support for the internal operations of the business are organizationally separate from line software developers who work on the products the company sells.

Acquisition

A project leader is usually appointed to oversee acquisition of a software product. Several difficulties can arise in this phase.

As noted earlier, the project team may define the capabilities required by the interactive system in purely technical terms [Stinchcombe 1990]. The project team may focus on software features, hardware, and networking requirements, but neglect to provide for the training and support of users and necessary changes in organizational aspects such as their job descriptions, their performance evaluations, and their rewards. Likely result: limited user acceptance and poor quality use [Walton 1989].

The project team is often subjected to influence attempts by interested parties: some may desire to build the technology in-house, even with perfectly acceptable packages available. Some may wish to customize a package to unique organizational needs rather than to change the way the organization works. Project team members may have different personal preferences for a technology vendor, a software package, or a platform, or demand that the solution incorporate certain features. To meet schedule and budgets, certain aspects of the project included in the original proposal may be postponed or dropped altogether. In the acquisition phase the innovation becomes more real and more resistant to change. Errors made in initiation can be fixed, or a well-reasoned decision steered astray [Markus and Soh 1993, Soh and Markus 1995].

Finally, another outcome of the acquisition phase is a new set of social relations among various groups inside and outside the organization who will work with or support the technology during the *use* phase: linkages among technology vendors, third-party developers, in-house developers, in-house computer operations and support personnel, or an outsourcing firm, and users and their managers. Thorny support issues may remain unresolved due to the incentives built into outsourcer's service level agreements as negotiated during the acquisition phase. User skills may stagnate or decline, because a one-shot training program did not address the need for advanced training or for initial training for new hires. In short, what results from the acquisition phase may or may not be adequate for the organization's future needs.

Implementation (Introduction) and Use

Activities occurring during the implementation and use phase can quite substantially alter (for better or worse) the capabilities previously acquired. Technology that is made or bought and thrown "over the wall" to users and their managers usually results in a failure of the technology to yield appropriate organizational benefits. Sometimes line managers and users see some potential in the technologies tossed at them and adopt (and support) these technologies as their own.

There is no guarantee that technology will be used in ways consistent with either the initiator's or the project team's vision. The average user of even a modestly complex information technology like the digital telephone system uses only a small fraction of the technology's features [Manross and Rice 1986]. Once they acquire some level of proficiency, users often stop learning new features unless a new release, a conversion, or a major change in work requirements demands new learning [Tyre and Orlikowski 1994]. Often users will enact time consuming and inefficient workarounds [Gasser 1986], rather than invest the time, energy, and pain required to learn a more efficient procedure.

Sometimes users take simple information technologies and *overuse* them, getting more out of them than designers ever intended. Use unanticipated when the technology was first acquired, rather than the initially planned uses, often results in what are subsequently called organizational *transformations:* radical improvements in business processes, first-in-the-world new products and services, and so forth.

The technical term for the process we have been describing, in which users take a technology and redefine it, using it differently than developers, initiators, and implementors intended, is *reinvention* [Rogers 1995] or *emergence* [Markus and Robey 1988]. Reinvention is significant because it means that the use and hence the impacts of a technology can never fully be determined during the acquisition phase. Even when the project team involves users in design and fashions a careful technology implementation plan, both users and developers may fail to see the organizational implications of a technology until the technology itself is real, installed, and running in an organization.[2] What happens when users get their hands on a technology can never be fully predicted or controlled. Sometimes what emerges during use is much less than vendors, initiators, and implementors had hoped; sometimes it is much more.

Impacts and Performance

Some experts estimate that most benefits obtained from an innovation come from subsequent modifications and enhancements, rather than from the initial change itself [Stinchcombe 1990]. Thus, for example, Frito-Lay did not reap full advantages from its hand-held computer project until it developed analytic tools to improve product promotion decision making and changed the organizational level at which promotion decisions were made [Harvard 1993].

On the other hand, a major reason for the failure of information technology (IT) investments to pay off in terms of improved organizational performance is a tendency for organizations to make nonvalue added improvements in their IT environments [Baily 1986, Baily and Gordon 1988].[3] The people at VeriFone note, "If you're just using e-mail, there's no reason to have a Pentium. You don't need a Ferrari to drive to the supermarket" [Harvard 1994].

Positive organizational impacts from information technology are said to fall into four categories: new products and services enabled by IT, improved business processes, better organizational decision making attributable to databases and analytic tools, and increased organizational flexibility attributable to communication, collaboration, and coordination technologies [Sambamurthy and Zmud 1992]. However, two issues regarding the impacts of technology must be borne in mind.

First, positive organizational impacts due to technology investments do not always result in improved organizational performance, measured in terms important to various organizational stakeholders [Soh and Markus 1995]. Lack of performance improvement despite positive impacts can occur if the innovation is quickly duplicated by competitors or if the improvements only bring company performance up to existing customer expectations [Arthur 1990, Clemons 1991].

Second, positive organizational impacts are almost invariably accompanied by negative impacts on some dimensions of organizational life [Pool 1983, Rogers 1995]. For instance, the improved organizational efficiency and flexibility attributed to electronic communications technologies such as e-mail may be accompanied by depersonalization, stress, overload, and accountability politics [Sproull and Kiesler 1991, Markus 1994b]. And no matter how much people in an organization value the improvements in organizational functioning, they may still mourn the passing of traditional ways that gave meaning and quality to their working lives.

64.8 Conclusions

Human–computer interaction could postpone reckoning with organizational issues when PCs were computational islands. Today we are networked and on the Internet: The day of reckoning has arrived. Designers, developers, acquirers, users, and researchers must all be cognizant of group and organizational issues to a degree previously unnecessary. The HCI and IS fields are quickly merging. Organizational issues will affect many of us working on human–computer interaction, through the organizational contexts of system introduction and adoption, and the organizational contexts of system development. With this knowledge, frustration often gives away to challenge, and challenge evolves into adventure.

---

[2]Social scientists repeatedly warn that user participation in design does not ensure success, since participation can lead to incrementalism or recreation of the status quo [Walton 1989, Leonard-Barton 1988, 1990, Markus and Keil 1994].
[3]This issue and its relationship to usability is explored in detail by Landauer [1995].

References

Abbott, K. R. and Sarin, S. K. 1994. Experiences with workflow management: issues for the next generation, pp. 113–120. *Proc. CSCW '94.*

Arthur, W. B. 1990. Positive feedback in the economy. *Sci. Am.* (Feb.):92–99.

Baily, M. N. 1986. What has happened to productivity growth? *Science* 234:443–451.

Baily, M. N. and Gordon, R. J. 1988. The productivity slowdown, measurement issues and the explosion of computer power. In *Brookings Papers on Economic Activity.* W. C. Brainard and G. L. Perry, eds. The Brookings Institute, Washington, DC.

Barker, V. and O'Connor, D. 1989. Expert systems for configuration at Digital: XCON and beyond. *Commun. ACM* 32(3):298–318.

Beyer, H. and Holtzblatt, K. 1995. Apprenticing with the customer. *Commun. ACM* 38(5):45–52.

Bjerknes, G., Ehn, P., and Kyng, M., eds. 1987. *Computers and Democracy—a Scandinavian Challenge.* Gower, Aldershot, UK.

Boehm, B. 1988. A spiral model of software development and enhancement. *IEEE Comput.* 21(5):61–72.

Bowers, J., Button, G., and Sharrock, W. 1995. Workflow from within and without: technology and cooperative work on the print industry shopfloor, pp. 51–66. *Proc. ECSCW '95.*

Bridges, W. 1991. *Managing Transitions,* Addison–Wesley. Reading, MA.

Clemons, E. K. 1991. Evaluation of strategic investments in information technology. *Commun. ACM* 34(1):22–36.

Cohen, M. D., March J. G., and Olsen, J. P. 1972. A garbage can model of organizational choice. *Adm. Sci. Q.* 17:1–25.

Dean, J. W., Jr. 1987. Building for the future: the justification process for new technology. In *New Technology as Organizational Innovation,* J. M. Pennings and A. Buitendam, eds., pp. 35–58. Ballinger, Cambridge, MA.

Friedman, A. L. 1989. *Computer Systems Development: History, Organization and Implementation.* Wiley, Chichester, UK.

Gasser, L. 1986. The integration of computing and routine work. *ACM Trans. Office Inf. Syst.* 4(3):205–225.

Greenbaum, J. and Kyng, M., eds. 1991. *Design at Work: Cooperative Design of Computer Systems.* Lawrence Erlbaum Associates, Hillsdale, NJ.

Grudin, J. 1991a. Interactive systems: bridging the gaps between developers and users. *IEEE Comput.* 24(4):59–69; republished in *Readings in Human–Computer Interaction: Toward the Year 2000.* R. M. Baecker, J. Grudin, W. A. S. Buxton, and S. Greenberg, eds. Morgan Kaufmann, San Mateo, CA, 1995.

Grudin, J. 1991b. Systematic sources of suboptimal interface design in large product development organizations. *Hum.–Comput. Interaction* 6(2):147–196.

Grudin, J. 1994. Groupware and social dynamics: eight challenges for developers. *Commun. ACM* 37(1):92–105.

Grudin, J. 1996. Evaluating opportunities for design capture. In *Design Rationale: Concepts, Techniques, and Use.* T. Moran and J. Carroll, eds., pp. 453–470. Lawrence Erlbaum, Hillsdale, NJ.

Grudin, J. and Markus, M. L. in press. Organizational issues in development and implementation of interactive systems. In *Handbook of Human–Computer Interaction.* M. Helander and T. Landauer, eds. 2nd ed. Springer–Verlag.

Grudin, J. and Poltrock, S. 1995. Software engineering and the CHI & CSCW communities. In *Software Engineering and Human–Computer Interaction.* R. N. Taylor and J. Coutaz, eds. Lecture notes in computer science 896, pp. 93–112. Springer–Verlag, Berlin.

Harvard. 1993. Frito-Lay, Inc.: A Strategic Transition Case (D) 9-193-004. *Harvard Business School.* Cambridge, MA.

Harvard. 1994. VeriFone: The Transaction Automation Company, Case 9-195-088. *Harvard Business School.* Cambridge, MA.

Hirschheim, R., Klein, H. K., and Lyytinen, K. 1995. *Information Systems Development and Data Modeling: Conceptual and Philosophical Foundations.* Cambridge University Press, Cambridge, UK.

Holtzblatt, K. and Beyer, H. 1993. Making customer-centered design work for teams. *Commun. ACM* 36(10):92–103.

Holtzblatt, K. and Jones, S. 1993. Contextual inquiry: a participatory technique for system design. In *Participatory Design: Principles and Practices*. D. Schuler and A. Namioka, eds. Lawrence Erlbaum Associates, Hillsdale, NJ.

Keil, M. and Carmel, E. 1995. Customer-developer links in software development. *Commun. ACM* 38(5):33–44.

Kling, R. 1978. Automated welfare client-tracking and service integration: the political economy of computing. *Commun. ACM* 21(6):484–493.

Kling, R. and Scacchi, W. 1982. The Web of Computing: Computer Technology as Social Organization. In *Advances in Computers*. M. C. Yovits, ed., pp. 1–89. Academic Press, Orlando, FL.

Landauer, T. K. 1995. *The Trouble with Computers: Usefulness, Usability, and Productivity*. MIT Press, Cambridge, MA.

Leonard-Barton, D. 1988. Implementation as mutual adaptation of technology and organization. *Res. Policy*. 17:251–267.

Leonard-Barton, D. 1990. Implementing new production technologies: exercises in corporate learning. In *Managing Complexity in High Technology Organizations*. Mary Ann Von Glinow and Susan Albers Mohrman, eds., pp. 160–215. Oxford University Press, New York.

Mahoney, M. S. 1988. The history of computing in the history of technology. *Ann. Hist. Comput.* 10:113–125.

Manross, G. G. and Rice, R. E. 1986. Don't hang up: organizational diffusion of the intelligent telephone. *Inf. Manage.* 10(3):161–175.

Mark, G. and Mambrey, P. 1996. Models and metaphors in groupware: toward a group-centered design. Unpublished manuscript.

Markus, M. L. 1994a. Electronic mail as the medium of managerial choice. *Organ. Sci.* 5(4):502–527.

Markus, M. L. 1994b. Finding a happy medium: explaining the negative effects of electronic mail on social life at work. *ACM Trans. Inf. Sys.* 12(2):119–149.

Markus, M. L. and Keil, M. 1994. If we build it they will come: designing information systems that users want to use, *Sloan Manage. Rev.* (Summer):11–25.

Markus, M. L. and Robey, D. 1988. Information technology and organizational change: causal structure in theory and research. *Manage Sci.* 34(5):583–598.

Markus, M. L. and Soh, C. 1993. Banking on information technology: converting IT spending into firm performance. In *Perspectives on the Strategic and Economic Value of Information Technology Investment*. R. D. Banker, R. J. Kauffman, and M. A. Mahmood, eds., pp. 364–392. Idea Group, Middletown, PA.

Marshak, D. S. 1994. Workflow white paper: an overview of workflow software, pp. 15–42. *Proc. Workflow '94*.

Mintzberg, H. 1984. A typology of organizational structure. In *Organizations: A Quantum View*. D. Miller and P. H. Friesen, eds., pp. 68–86. Prentice–Hall, Englewood Cliffs, NJ.

Mintzberg, H. 1989. *Mintzberg on Management*. Free Press, New York.

Mumford, E. 1983. *Designing Human Systems*. Manchester Business School.

Mumford, E. 1993. The participation of users in systems design: an account of the origin, evolution, and use of the ETHICS method. In *Participatory Design: Principles and Practice*. D. Schuler and A. Namioka, eds., pp. 257–270. Lawrence Erlbaum Associates, Hillsdale, NJ.

Poltrock, S. E. and Grudin, J. 1994. Organizational obstacles to interface design and development: two participant observer studies. *ACM Trans. Comput.–Hum. Interaction* 1(1):52–80.

Pool, I. d. S. 1983. *Forecasting the Telephone: A Retrospective Technology Assessment of the Telephone*. Ablex, Norwood, NJ.

Rogers, E. M. 1995. *Diffusion of Innovations*, 4th ed. Free Press, New York.

Royce, W. W. 1970. Managing the development of large software systems: concepts and techniques, pp. 1–9. *Proc. IEEE Wescon*.

Sambamurthy, V. and Zmud, R. W. 1992. Managing IT for Success: The Empowering Business Partnership, Financial Executives Research Foundation, Morristown, NY.

Soe, L. L. 1994. *Substitution and Complementarity in the Diffusion of Multiple Electronic Communication Media: An Evolutionary Approach.* Unpublished Ph.D. dissertation. University of California, Los Angeles.

Soh, C. and Markus, M. L. 1995. How IT creates business value: a process theory synthesis. *Proc. Int. Conf. Inf. Sys.*, pp. 29–41. Amsterdam, The Netherlands.

Sproull, L. and Kiesler, S. 1991. *Connections: New Ways of Working in the Networked Organization.* MIT Press, Cambridge, MA.

Stinchcombe, A. L. 1990. *Information and Organizations.* University of California Press, Berkeley, CA.

Tyre, M. J. and Orlikowski, W. J. 1994. Windows of opportunity: temporal patterns of technological adaptation in organizations. *Organ. Sci.* 5(1).

Walton, R. E. 1989. *Up and Running: Integrating Information Technology and the Organization.* Harvard Business School Press, Boston, MA.

Winkler, I. and Buie, E. 1995. HCI challenges in government contracting. *SIGCHI Bull.* 27(4):35–37.

65

Usability Engineering

65.1 Introduction

Usability engineering [Nielsen 1994b] is not a one-shot event where the user interface is fixed up before the release of a product. Rather, usability engineering is a set of activities that ideally take place throughout the lifecycle of the product, with significant activities happening at the early stages before the user interface has even been designed. The need to have multiple usability engineering stages supplement each other was recognized early in the field, though not always followed by development projects [Gould and Lewis 1985].

Usability cannot be seen in isolation from the broader corporate product development context where one-shot projects are fairly rare. Indeed, usability applies to the development of entire product families and extended projects where products are released in several versions over time. In fact, this broader context only strengthens the arguments for allocating substantial usability engineering resources as early as possible, since design decisions made for any given product have ripple effects due to the need for subsequent products and versions to be backward compatible. Consequently, some usability engineering specialists [Grudin et al. 1987] believe that "human factors involvement with a particular product may ultimately have its greatest impact on future product releases." Of course, having to plan for future versions is also a primary reason to follow up the release of a product with field studies of its actual use.

Table 65.1 shows a summary of the lifecycle stages discussed in this chapter. It is important to note that a usability engineering effort can still be successful even if it does not include every possible refinement at all of the stages.

The lifecycle model emphasizes that one should not rush straight into design. The least expensive way for usability activities to influence a product is to do as much as possible before design is started, since it

0-8493-2909-4/97/$0.00+$.50
© 1997 by CRC Press, Inc.

will then not be necessary to change the design to comply with the usability recommendations. Also, usability work done before the system is designed may make it possible to avoid developing unnecessary features. Several of the predesign usability activities might be considered part of a market research or product planning process as well, and may sometimes be performed by marketing groups. However, traditional market research does not usually employ all of the methods needed to properly inform usability design, and the results are often poorly communicated to developers. But there should be no need for duplicate efforts if management successfully integrates usability and marketing activities [Wichansky et al. 1988]. One outcome of such integration could be the consideration of product usability attributes as features to be used by marketing to differentiate the product. Also, marketing efforts based on usability studies can sell the product on the basis of its benefits as perceived by users (*what* it can do that they want) rather than its features as perceived by developers (*how* does it do it).

TABLE 65.1 Stages of the Usability Engineering Lifecycle

1. Know the user
   a. Individual user characteristics
   b. The user's current and desired tasks
   c. Functional analysis
   d. International use
2. Competitive analysis
3. Setting usability goals
4. Parallel design
5. Participatory design
6. Coordinated design of the total interface
7. Heuristic evaluation
8. Prototyping
9. User testing
10. Iterative design
11. Collect feedback from field use.

65.2 Know the User

The first step in the usability process is to study the intended users and use of the product. At a minimum, developers should visit a customer site so that they have a feel for how the product will be used. Individual user characteristics and variability in tasks are the two factors with the largest impact on usability, so they need to be studied carefully. When considering users, one should keep in mind that they often include installers, maintainers, system administrators, and other support staff in addition to the people who sit at the keyboard. The concept of *user* should be defined to include everybody whose work is affected by the production in some way, including the users of the system's end product or output even if they never see a single screen.

Even though "know the user" is the most basic of all usability guidelines, it is often difficult for developers to get access to users. Grudin [1990, 1991a, 1991b] analyzes the obstacles to such access, including:

- The need for the development company to protect its developers from being known to customers, since customers may bypass established technical support organizations and call developers directly, sidetracking them from their main job
- The reluctance of sales representatives to let anybody else from the company talk to their customers, fearing that the developers or usability people may offend the customer or create dissatisfaction with the current generation of products
- User organizations only making users available for a short time, either because they are highly paid executives or because they are unionized and dislike being studied

All of these issues are real and need to be addressed when trying to get to know the user. No universal solutions are available, except to recommend an explicit effort to get direct access to representative users and not be satisfied with indirect access and hearsay. It is amazing how much time is wasted on certain development projects by arguing over what users *might* be like or what they *may* want to do. Instead of discussing such issues in a vacuum, it is much better (and actually less time consuming) to get hard facts from the users themselves.

Individual User Characteristics

It is necessary to know the class of people who will be using the system. In some situations this is easy since it is possible to identify these users as concrete individuals. This is the case when the product is going to be used in a specific department in a particular company. For other products, users may be more widely scattered such that it is possible to visit only a few, representative customers. Alternatively, the products might be aimed toward the entire population or a very large subset.

By knowing the users' work experience, educational levels, ages, previous computer experience, and so on, it is possible to anticipate their learning difficulties to some extent and to better set appropriate limits for the complexity of the user interface. Certainly one also needs to know the reading and language skills of the users. For example, very young children have no reading ability, so an entirely nontextual interface is required. Also, one needs to know the amount of time users will have available for learning and whether they will have the opportunity for attending training courses: The interface must be made much simpler if users are expected to use it within minimum training.

The users' work environment and social context also need to be known. As a simple example, the use of audible alarms, beeps, or more elaborate sound effects may not be appropriate for users in open office environments. In a field interview I once did, a secretary complained strongly that she wanted the ability to shut off the beep because she did not want others to think that she was stupid because her computer beeped at her all the time.

A great deal of the information needed to characterize individual users may come from market analysis or as a side benefit of the observational studies one may conduct as part of the task analysis. One may also collect such information directly through questionnaires or interviews. In any case, it is best not to rely totally on written information since new insights are almost always achieved by observing and talking to actual users in their own working environment.

Task Analysis

A task analysis [Diaper 1989, Fath and Bias 1992, Johnson 1992] is extremely important as early input to the system design. The users' overall goals should be studied as well as how they currently approach the task, what their information needs are, and how they deal with exceptional circumstances or emergencies. For example, systematic observation of users talking to their clients may reveal input and output needs for a transactions-processing system. Sometimes, interviewing or observing the users' clients or others who interact with them can also provide additional task analysis insights [Garber and Grunes 1992].

The users' model of the task should also be identified, since it can be used as a source for metaphors for the user interface. Also, seek out and observe especially effective users and user strategies and *workarounds* as hints of what a new system could support. Such lead users are often a major source of innovations [von Hippel 1988]. Finally, one should try to identify the weaknesses of the current situation: points where users fail to achieve goals, spend excessive time, or are made uncomfortable. These weaknesses present opportunities for improvements in the product being developed.

A typical outcome of a task analysis is a list of all of the things users want to accomplish with the system (the goals), all of the information they will need to achieve these goals (the preconditions), the steps that need to be performed and the interdependencies between these steps, all of the various outcomes and reports that need to be produced, the criteria used to determine the quality and acceptability of these results, and finally the communication needs of the users as they exchange information with others while performing the task or preparing to do so.

When interviewing users for the purpose of collecting task information, it is always a good idea to ask them to show concrete examples of their work products rather than keeping the discussion on an abstract level. Also, it is preferable to supplement such interviews with observations of some users working on real problems, since users will often rationalize their actions or forget about important details or exceptions when they are interviewed.

Often, a task analysis can be decomposed in a hierarchical fashion [Greif 1991], starting with the larger tasks and goals of the organization and breaking each of them down into smaller subtasks, that can again be further subdivided. Typically, each time a user says, "then I do this," an interviewer could ask two questions: "*Why* do you do it?" (to relate the activity to larger goals) and "*How* do you do it?" (to decompose the activity into subtasks that can be further studied). Other good questions to ask include, "why do you not do this in such and such a manner?" (mentioning some alternative approach you can think of), "Do errors ever occur when doing this?," and "How do you discover and correct these errors?" [Nielsen et al. 1986].

Finally, users should be asked to describe exceptions from their normal work flow. Even though users cannot be expected to remember *all* of the exceptions that have ever occurred, and even though it will be impossible to predict all of the future exceptions, there is considerable value to having a list indicating the *range* of exceptions that must be accommodated. Users should also be asked for remarkable instances of notable successes and failures, problems, what they liked best and least, what changes they would like, what ideas they have for improvements, and what currently annoys them. Even though not all such suggestions may be followed in the final design, they are a rich source of inspiration.

Functional Analysis

A new computer system should not be designed simply to propagate suboptimal ways of doing things that may have been instituted because of limitations in previous technologies. Therefore, one should not analyze just the way users currently do the task, but also the underlying functional reason for the task: What is it that really needs to be done, and what are merely surface procedures which can, and perhaps should, be changed.

As a simple example, initial observations of people reading printed manuals could show them frequently turning pages to move through the document. A naive implementation of on-line documentation might take this observation to indicate the need for really good and fast paging or scrolling mechanisms. A functional analysis would show that manual users really turn pages this much because they want to find specific information, but they have a hard time locating the correct page. Based on this analysis, one could design an on-line documentation interface that first allowed users to specify their search needs, then used an outline of the document to show locations with high search scores, and finally allowed users to jump directly to these locations, highlighting their search terms to make it easier to judge the relevance of the information [Egan et al. 1989]. Of course, there is a limit to how drastically one can change the way users currently approach their task, so the functional analysis should be coordinated with a task analysis.

International Use

A final point related to knowing the users is to plan for any international use of the product from the very beginning of the usability engineering lifecycle [del Galdo and Nielsen 1996]. Some products are only intended for use in a single country, but many development projects need to consider foreign users. Traditionally, internationalization and localization was done after shipping the domestic version of the product, but true international usability requires that international users are considered throughout the lifecycle. Consider, for example, the design of the addressing feature in an e-mail program. If the program is going to be used in a country with an extremely strong sense of hierarchy in the workplace, then customers may require that the addressing feature sort the message recipients by rank, so fitting the program to that culture cannot be done simply by translating the menu items.

65.3 Competitive Analysis

Much of what you need to learn about user interface design for a new product can be gleaned from studies of your competitors' products [Nielsen 1995]. The best prototype of your next product is your own old product since you will presumably want to repeat everything that was good about it and avoid everything that was bad about it. The second-best prototypes are the competing products. Your competitors have invested significant resources in designing and implementing what they believe to be good user interfaces. You should take advantage of those investments.

Please note that I am not suggesting that you violate copyright by cloning your competitors' interface designs. What I *do* suggest is that you can learn a lot by analyzing other products that are designed to solve the same (or related) problems as your own future product. You can see what works and what does not work in these other designs, and you can learn how users approach the tasks by seeing how they work with the competing products. Competitive usability analysis should be performed very early in the usability engineering lifecycle. I would recommend performing competitive usability analysis after the first stages of customer visits, requirements gathering, and defining the product vision, but before you move on to actually designing and prototyping your own user interface.

For competitive usability analysis, I normally recommend acquiring the three or four leading competing products. Often, these products can be bought at a nominal price on the open market, especially in the case of PC software. Even if you are developing for high-end workstations or mainframes, much can still be learned from the design of lower end software even if there is some difference in the supported feature set. To design and develop a prototype yourself will normally take at least a week of engineering time, even for quite low-fidelity prototypes, so a home-grown prototype will cost a minimum of $4000 if the loaded cost of an engineer (whether usability engineer or development engineer) is $100 per hour. Normally one can buy at least 10 commercial software packages for the same money, and these packages will be very high-fidelity prototypes since they are fully functional (even if the features are not exactly the same as the ones you want in your product).

Some classes of products are substantially more expensive than PC software, but it is still often possible to buy evaluation copies, single-user licenses, or other cheap versions of high-end systems. Considering that a competitive usability analysis normally consists of having three or four users use the system for 1 or 2 h each, there is no need to buy the most elaborate version of the competing systems.

If even the cheapest versions of the competing systems are too expensive, you can rely on paper prototyping. Briefly, this method consists of showing users paper printouts of some of the screens from a user interface and asking them to describe what they would do with each screen. A selection of screendumps can usually be acquired from your competitors' sales brochures and so a few hours at a good trade show should suffice to collect more than enough material for an informative usability test.

Often you will have too many competitors to perform a usability study of them all. I usually find that one learns the most from studying four or five competing user interfaces. Three criteria should be used to select the systems that will be subjected to a competitive usability analysis:

- What products have an especially good reputation for good user interface design?
- What products show examples of interesting features or design ideas that you want for your own product?
- Who is the market leader?

Furthermore, you can consider pragmatic issues such as the price of a product and the difficulty of installing and running it on the equipment in your competitive analysis laboratory. If you do not have a competitive analysis laboratory, I highly recommend getting one: buy a computer from each of the major platform families, making sure that it is a high-end model with plenty of memory, a large hard disk, and a compact disc–read-only memory (CD-ROM) drive. You do not want to save on equipment purchases for the competitive analysis laboratory because you will not have the expertise to make each model run optimally: just buy nice big models in vendor-supported configurations. Also buy a good screendump utility for each machine because you will want to include shots of the competing user interfaces in your internal reports and presentations.

The first steps of a competitive usability analysis simply consist of familiarizing yourself with the products and checking how they have designed the features you are contemplating for your product. You can also make lists of user interface elements (commands, features, and attributes) to make sure that you do not overlook something important when designing your own interface.

The next step in competitive usability analysis is a brief usability test where a small number of users are exposed to the various products and asked to perform a few sample tasks. As always with user testing it is important to recruit test users who are representative of the intended user population (that is, the actual end users and not their managers or information systems (IS) support staff—unless, of course, the product is *intended* to have IS personnel as its main users). The tasks should also be chosen to represent the intended usage of the product. For competitive usability analysis, you should select users and tasks that are representative for your future product and not users and tasks that are representative for the other products. After all, the goal is not to evaluate whether the other companies have done a good job designing for *their* customer base but to see what you can learn from their efforts when applied to *your* customer base.

65.4 Goal Setting

Usability is not a one-dimensional attribute of a system. Usability comprises several components, including learnability, efficiency of use, user error rates, and subjective satisfaction, that can sometimes conflict. Normally, not all usability aspects can be given equal weight in a given design project, and so you will have to make your priorities clear on the basis of your analysis of the users and their tasks. For example, learnability would be especially important if new employees were constantly being brought in on a temporary basis, and the ability of infrequent users to return to the system would be especially important for a reconfiguration utility that was used once every three or four months.

The different usability parameters can be operationalized and expressed in measurable ways. Before starting the design of a new interface, it is important to discuss the usability metrics of interest to the project and to specify the goals of the user interface in terms of measured usability [Chapanis and Budurka 1990]. One may not always have the resources available to collect statistically reliable measures of the usability metrics specified as goals, but it is still better to have some idea of the level of usability to strive for.

For each usability attribute of interest, several different levels of performance can be specified as part of a goal-setting process [Whiteside et al. 1988]. One would at least specify the minimum level which would be acceptable for release of the product, but a more detailed goal specification can also include the planned level one is aiming for as well as the current level of performance. Additionally, it can help to list the current value of the usability attribute as measured for ex-

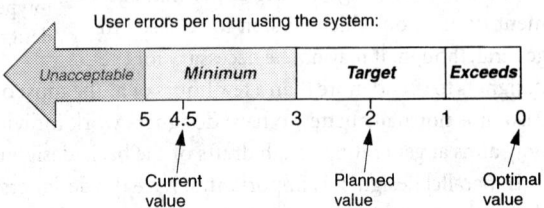

FIGURE 65.1 An example of a usability goal line in a notation similar to that used by Rideout [1991].

isting or competing interfaces, and one can also list the theoretically best possible value, even though this value will typically not be attained. Figure 65.1 shows one possible notation, called a *usability goal line*, for representing the range of specification levels for one usability goal.

In the example in Fig. 65.1, the number of user errors per hour is counted. When using the current system, users make an average of 4.5 errors/h and the planned number of user errors is 2.0/h. Furthermore, the theoretical optimum is obviously to have no errors at all. If the new interface is measured at anything between 1.0 and 3.0 user errors/h, it will be considered on target with respect to this usability goal. A performance in the interval of 3–5 would be a danger signal that the usability goal was not met, even though the new interface could still be released on a temporary basis since a minimal level of usability had been achieved. It would then be necessary to develop a plan to reduce user errors in future releases. Finally, more than 5.0 user errors/h would make this particular product sufficiently unusable to make a release unacceptable.

Usability goals are reasonably easy to set for new versions of existing systems or for systems that have a clearly defined competitor on the market. The minimum acceptable usability would normally be equal to the current usability level, and the target usability could be derived as an improvement that was sufficiently large to induce users to change systems. For completely new systems without any competition, usability goals are much harder to set. One approach is to define a set of sample tasks and ask several usability specialists how long it ought to take users to perform them. One can also get an idea of the minimum acceptable level by asking the users, but unfortunately users are notoriously fickle in this respect; countless projects have failed because developers believed users' claims about what they wanted, only to find that the resulting product was not satisfactory in real use.

Parallel Design

It is often a good idea to start the design with a parallel design process, in which several different designers work out preliminary designs [Nielsen et al. 1993, 1994, Nielsen and Faber 1996]. The goal of parallel design is to explore different design alternatives before one settles on a single approach that can then be developed in further detail and subjected to more detailed usability activities. Figure 65.2 is a conceptual illustration of the relation between parallel and iterative design.

Typically, one can have three or four designers involved in parallel design. For critical products, some large computer companies have been known to devote entire teams to developing multiple alternative designs almost to the final product stage, before upper management decided on which version to release. In general, though, it may not be necessary for the

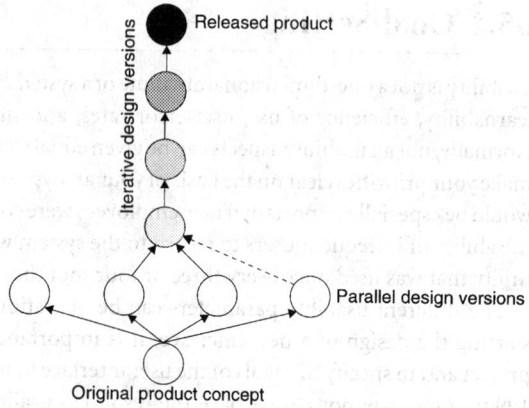

FIGURE 65.2 Conceptual illustration of the relation between parallel and iterative design. Normally, the first prototype would be based on ideas from several of the parallel design sketches.

designers to spend more than a few hours or at the most one or two days on developing their initial designs. Also, it is normally better to have designers work individually rather than in teams, since parallel design only aims at generating rough drafts of the basic design ideas.

In parallel design, it is important to have the designers (or the design teams) work independently, since the goal is to generate as much diversity as possible. Therefore, the designers should not discuss their designs with each other until after they have produced their draft interface designs.

When the designers have completed the draft designs, one will often find that they have approached the problem in at least two drastically different ways that would give rise to fundamentally different user interface models. Even those designers who are basing their designs on the same basic approach almost always have different details in their designs. Usually, it is possible to generate new combined designs after having compared the set of initial designs, taking advantage of the best ideas from each design. If several fundamentally different designs are available, it is preferable to pursue each of the main lines of design a little further in order to arrive at a small number of prototypes that can be subjected to usability evaluation before the final approach is chosen.

A variant of parallel design is called *diversified parallel design* and is based on asking the different designers to concentrate on different aspects of the design problem. For example, one designer could design an interface that was optimized for novice users, at the same time as another designer designed an interface optimized for expert users and a third designer explored the possibilities of producing an entirely nonverbal interface. By explicitly directing the design approach of each designer, diversified parallel design drives each of these approaches to their limit, leading to design ideas that might never have emerged in a unified design. Of course, some of these diversified design ideas may have to be modified to work in a single, integrated design.

It is especially important to employ parallel design for novel systems where little guidance is available for what interface approaches work the best. For more traditional systems, where competitive products are available, the competitive analysis previously discussed can serve as initial parallel designs, but it might still be advantageous to have a few designers create additional parallel designs to explore further possibilities.

The parallel design method might at first seem to run counter to the principle of cost-effective usability engineering, since most of the design ideas will have to be thrown away without even being implemented. In reality, though, parallel design is a very cheap way of exploring the design space, exactly *because* most of the ideas will not need to be implemented, the way they might be if some of them were not tried until later as part of the iterative design. The main financial benefit of parallel design is its parallel nature, which allows several design approaches to be explored at the same time, thus compressing the development schedule for the product and bringing it to market more rapidly. Studies have shown that about a third of the profits are lost when products ship as little as half a year late [House and Price 1991], and so anything that can speed up the development process should be worth the small additional cost of designing in parallel rather than in sequence.

Participatory Design

Participatory design is discussed further in Chapter 66 and elsewhere and will not be covered here.

65.5 Coordinating the Total Interface

Consistency is one of the most important usability characteristics. Consistency should apply across the different media which form the total user interface, including not just the application screens but also the documentation, the on-line help system, and any on-line or videotaped tutorials [Perlman 1989] as well as traditional training classes. For example, in one case studied by Poltrock [1996], training materials described an obsolete way of using an interface because the training department had not been informed about the introduction of a redesigned, and presumably better, interface.

Consistency is not just measured at a single point in time but should apply over successive releases of a product so that new releases are consistent with their predecessors. Also, since very few companies produce only a single product, efforts should be made to promote consistency across entire product families. Corporate user interface standards are one common way of promoting that goal. In spite of the general desirability of consistency, it is obviously not the only desirable usability characteristic, and consistency may sometimes conflict with other interface desiderata [Grudin 1989]. It is necessary to maintain some flexibility so that bad design is not forced upon users for the sake of consistency alone.

To achieve consistency of the total interface it is necessary to have some centralized authority for each development project to coordinate the various aspects of the interface. Typically, this coordination can be done by a single person, but on very large projects or to achieve corporatewide consistency, a committee structure may be more appropriate. Also, interface standards are an important approach to achieving consistency. In addition to such general standards, a project can develop its own ad hoc standard with elements such as a dictionary of the appropriate terminology to be used in all screen designs as well as in the other parts of the total interface.

In addition to formal coordination activities, it is helpful to have a shared culture in the development groups with common understanding of what the user interface should be like. Many aspects of user interface design (especially the dynamics) are hard to specify in written documents but can be fairly easily understood from looking at existing products following a given interface style. Actually, prototyping also helps achieve consistency, since the prototype is an early statement of the kind of interface toward which the project is aiming. Having an explicit instance of parts of the design makes the details of the design more salient for developers and encourages them to follow similar principles in subsequent design activities [Bellantone and Lanzetta 1991].

Furthermore, consistency can be increased through technological means such as code sharing or a constraining development environment. When several products use the same code for parts of their user interface, then those parts of the interface automatically will be consistent. Even if identical code cannot be used, it is possible to constrain developers by providing development tools and libraries that encourage user interface consistency by making it easier to implement interfaces that follow given guidelines [Tognazzini 1989, Wiecha et al. 1989].

65.6 Heuristic Evaluation

Guidelines list well-known principles for user interface design which should be followed in the development project. In any given project, several different levels of guidelines should be used: *general guidelines* applicable to all user interfaces, *category-specific guidelines* for the kind of system being developed (e.g., guidelines for window-based administrative data processing or for voice interfaces accessed through telephone keypads), and *product-specific guidelines* for the individual product. All of these guidelines can be used as background for heuristic evaluation.

Heuristic evaluation [Nielsen and Molich 1990, Nielsen 1994a] is the most popular of the usability inspection methods [Nielsen and Mack 1994]. Heuristic evaluation is done by looking at an interface and trying to come up with an opinion about what is good and bad about the interface. Heuristic evaluation should be done as a systematic inspection of a user interface design for usability. The goal of heuristic evaluation is to find the usability problems in a user interface design so that they can be attended to as part of an iterative design process. Heuristic evaluation involves having a small set of evaluators examine the interface and judge its compliance with recognized usability principles (the heuristics). A recommended set of heuristics is listed in Table 65.2.

In principle, individual evaluators can perform a heuristic evaluation of a user interface on their own, but the experience from several projects indicates that any single evaluator will miss most of the usability problems in an interface. Averaged over six projects, single evaluators found only 35% of the usability problems in the interfaces. However, since different evaluators tend to find different problems, it is possible to achieve substantially better performance by aggregating the evaluations from several evaluators. There is a nice payoff from using more than one evaluator, and it is recommended to use about five evaluators,

TABLE 65.2   List of 10 Heuristics for Good User Interface Design

Visibility of system status: The system should always keep users informed about what is going on, through appropriate feedback within reasonable time.

Match between system and the real world: The system should speak the users' language, with words, phrases, and concepts familiar to the user, rather than system-oriented terms. Follow real-world conventions, making information appear in a natural and logical order.

User control and freedom: Users often choose system functions by mistake and will need a clearly marked emergency exit to leave the unwanted state without having to go through an extended dialogue. Support undo and redo.

Consistency and standards: Users should not have to wonder whether different words, situations, or actions mean the same thing. Follow platform conventions.

Error prevention: Even better than good error messages is a careful design which prevents a problem form occurring in the first place.

Recognition rather than recall: Make objects, actions, and options visible. The user should not have to remember information from one part of the dialogue to another. Instructions for use of the system should be visible or easily retrievable whenever appropriate.

Flexibility and efficiency of use: Accelerators—unseen by the novice user—may often speed up the interaction for the expert user such that the system can cater to both inexperienced and experienced users. Allow users to tailor frequent actions.

Aesthetic and minimalist design: Dialogues should not contain information which is irrelevant or rarely needed. Every extra unit of information in a dialogue competes with the relevant units of information and diminishes their relative visibility.

Help users recognize, diagnose, and recover from errors: Error messages should be expressed in plain language (no codes), precisely indicate the problem, and constructively suggest a solution.

Help and documentation: Even though it is better if the system can be used without documentation, it may be necessary to provide help and documentation. Any such information should be easy to search, focused on the users' tasks, list concrete steps to be carried out, and not be too large.

and certainly at least three. The exact number of evaluators to use would depend on a cost-benefit analysis, and more evaluators should obviously be used in cases where usability is critical or when large payoffs can be expected due to extensive or mission-critical use of a system.

Heuristic evaluation is performed by having each individual evaluator inspect the interface alone. Only after all evaluations have been completed are the evaluators allowed to communicate and have their findings aggregated. This procedure is important in order to ensure independent and unbiased evaluations from each evaluator. The results of the evaluation can be recorded either as written reports from each evaluator or by having an observer present during the evaluation sessions and having the evaluators vocalize their comments as they go through the interface. Written reports have the advantage of presenting a formal record of the evaluation, but require an additional effort from the evaluators and also need to be read and aggregated by an evaluation manager. Using on observer adds to be overhead of each evaluation session but reduces the workload on the evaluators and provides the opportunity for having the result of the evaluation available fairly soon after the last evaluation session since the observer only needs to understand and organize his or her own notes and not a set of reports written by others. Furthermore, the observer can assist the evaluators in operating the interface in case of problems with, e.g., an unstable prototype, and help if the evaluators have limited domain expertise and need to have certain aspects of the interface explained.

Typically, a heuristic evaluation session for an individual evaluator lasts one or two hours. Longer evaluation sessions might be necessary for larger or very complicated interfaces with a substantial number of dialogue elements, but it is likely that it would be better to split up the evaluation in several smaller sessions, each concentrating on a part of the interface.

During the evaluation session, the evaluator goes through the interface several times and inspects the various dialogue elements and compares them with a list of recognized usability principles. These heuristics are general rules that seem to describe common properties of usable interfaces and it is possible to apply lists of heuristics that are more specialized for the particular application domain than the general list given in Table 65.2. In addition to the checklist of heuristics to be considered for all dialogue elements, the evaluator obviously is also allowed to consider any additional usability principles or results that come to mind that may be relevant for any specific dialogue element.

In principle, the evaluators decide on their own how they want to proceed with evaluating the interface. A general recommendation would be that they go through the interface at least twice, however. The first pass would be intended to get a feel for the flow of the interaction and the general scope of the system. The second pass then allows the evaluator to focus on specific interface elements while knowing how they fit the larger whole.

Since the evaluators are not *using* the system as such (to perform a real task), it is possible to perform heuristic evaluation of user interfaces that exist on paper only and have not yet been implemented [Nielsen 1990]. This makes heuristic evaluation suitable for use early in the usability engineering lifecycle.

65.7 Prototyping

One should not start full-scale implementation efforts based on early user interface designs. Instead, early usability evaluation can be based on prototypes of the final systems that can be developed much faster and much more cheaply, and which can thus be changed many times until a better understanding of the user interface design has been achieved.

In traditional models of software engineering most of the development time is devoted to the refinement of various intermediate work products, and executable programs are produced at the last possible moment. A problem with this *waterfall* approach is that there will then be no user interface to test with real users until this last possible moment, since the intermediate work products do not explicitly separate out the user interface in a prototype with which users can interact. Experience also shows that it is not possible to involve the users in the design process by showing them abstract specifications documents, since they will not understand them nearly as well as concrete prototypes.

The entire idea behind prototyping is to save on the time and cost to develop something that can be tested with real users. These savings can only be achieved by somehow reducing the prototype compared with the full system: either cutting down on the number of features in the prototype or reducing the level of functionality of the features such that they *seem* to work but do not actually *do* anything.

Reducing the number of features is called *vertical prototyping* since the result is a narrow system that does include in-depth functionality, but only for a few selected features. A vertical prototype can thus only test a limited part of the full system, but it will be tested in depth under realistic circumstances with real user tasks. For example, for a test of a website, in-depth functionality would mean that a user would actually access a set of documents with real content from the information providers.

Reducing the level of functionality is called *horizontal prototyping* since the result is a surface layer that includes the entire user interface to a full-featured system but with no underlying functionality. A horizontal prototype is a simulation [Life et al. 1990] of the interface where no real work can be performed. In the Web example, this would mean that users should be able to execute all navigation and search commands but without retrieving any real documents as a result of these commands. Horizontal prototyping makes it possible to test the entire user interface, even though the test is of course somewhat less realistic, since users cannot perform any real tasks on a system with no functionality. The main advantages of horizontal prototypes are that they can often be implemented fast with the use of various prototyping and screen design tools and that they can be used to assess how well the entire interface hangs together and feels as a whole.

Finally, one can reduce both the number of features and the level of functionality to arrive at a scenario that is only able to simulate the user interface as long as the test user follows a previously planned path. Scenarios are extremely easy and cheap to build, while at the same time not being particularly realistic. Scenarios are discussed further in other chapters.

In addition to reducing the proportion of the system that is implemented, prototypes can be produced faster by the following.

- Placing less emphasis on the efficiency of the implementation. For example, it will not matter how much disk space the prototype uses since it will only be used for a short time. Similarly, test users may be able to cope with slow response times that would never be acceptable in the final product. Note, however, that response times are an important aspect of usability and that test users may get very frustrated and make errors if the prototype is *too* slow. Of course, efficiency measures of the users' performance will be invalid if the prototype slows them down too much, so inefficient prototypes are better suited for early evaluation of interface concepts than for measurement studies.

- Accepting less reliable or poorer quality code. Even though bugs and crashes do distract users during testing, they can often be compensated for by the experimenter.

- Using simplified algorithms that cannot handle all of the special cases (such as leap years) that normally require a disproportionately large programming effort to get right.

- Using a human expert operating behind the scenes to take over certain computer operations that would be too difficult to program. This approach is often referred to as the *Wizard of Oz technique* after the "pay no attention to that man behind the curtain" scene in this story. Basically, the user interacts normally with the computer, but the users' input is not relayed directly to the program. Instead, the input is transmitted to the wizard who, using another computer, transforms the users' input into an appropriate format. A famous early Wizard of Oz study was the listening typewriter [Gould et al. 1983] simulation of a speech recognition interface where the users' spoken input was typed into a word processor by a human typist located in another room. When setting up a Wizard of Oz simulation, experience with previously implemented systems is helpful in order to place realistic bounds on the wizard's abilities [Maulsby et al. 1993].

- Using a different computer system than the eventual target platform. Often, one will have a computer available that is faster or otherwise more advanced than the final system and which can therefore support more flexible prototyping tools and require less programming tricks to achieve the necessary response times.

- Using low-fidelity media [Virzi 1989] that are not as elaborate as the final interface but still represent the essential nature of the interaction. For example, a prototype hypermedia system could use scanned still images instead of live video for illustrations.

- Using fake data and other content. For example, a prototype of a hypermedia system that was intended to include heavy use of video could use existing video material, even though it did not exactly match the topic of the text, in order to get a feel for the interaction techniques needed to deal with live images. A similar technique is used in the advertising industry, where so-called ripomatics are used as rudimentary television commercials with existing shots from earlier commercials to demonstrate concepts to clients before they commit to pay for the shooting of new footage.

- Using paper mockups instead of a running computer system. Such mockups are usually based on printouts of screen designs, dialogue boxes, pop-up menus, etc., that have been drawn up in some standard graphics or desktop publishing package. They are made into functioning prototypes by having a human play computer and find the next screen or dialogue element from a big pile of paper whenever the user indicates some action. This human needs to be an expert in the way the program is intended to work since it is otherwise difficult to keep track of the state of the simulated computer system and find the appropriate piece of paper to respond to the users' stated input.

- Paper mockups have the further advantage that they can be shown to larger groups on overhead projectors [Rowley and Rhoades 1992] and used in conditions where computers may not be available, such as customer conference rooms. Portable computers with screen projection attachments confer some of the same advantages to computerized prototypes, but also increase the risk of something going wrong.

- Relying on a completely imaginary prototype where the experimenter describes a possible interface to the user orally, posing a series of "what if (the interface did this or that) ..." questions as the user steps though an example task. This verbal prototyping technique has been called *forward scenario simulation* [Cordingley 1989] and is more akin to interviews or brainstorming than a true prototyping technique.

A prototype is a form of design specification, and the final implementation of a user interface is often performed with the prototype as a major way of communicating the design to developers. Unfortunately, the prototype can be *over*specified in some aspects that are not really intended to be part of the design. Whenever something is made concrete, there is a need to instantiate a multitude of representational details that might not have been explicitly designed by anybody. For example, a screen design will have to use certain colors and fonts, even though the designer's focus may have been on the wording and positioning of the dialogue elements. Basically, one needs to be aware that not every aspect of the prototype should be replicated in the final system, and the designers should inform developers about which aspects of the prototype are intentional and which are arbitrary.

65.8 User Testing

The most basic advice with respect to interface evaluation is simply to *do it*, and especially to conduct some user testing. The benefits of employing some reasonable usability engineering methods to evaluate a user interface rather than releasing it without evaluation are much larger than the incremental benefits of using exactly the right methods for a given project.

User testing with real users is the most fundamental usability method and is in some sense irreplaceable, since it provides direct information about how people use computers and what their exact problems are with the concrete interface being tested. Even so, other usability engineering methods [Nielsen 1994b] can serve as good supplements to gather additional information or to gain usability insights at a lower cost. In particular, user testing can often be combined with heuristic evaluation (discussed previously) or other usability inspection methods for greater efficiency: the heuristic evaluation will find many usability problems that should be cleaned up before presenting the design to real users.

The three main rules of user testing are:

- Get real users
- Have them do real tasks
- Shut up while they are trying

The Test Users

Your test users should accurately represent the system's intended users. You cannot simply test with the engineer in the neighboring office, despite his or her suspicious resemblance to a real person. Actually, much can be learned from testing with other engineers and it is also possible to gain substantial insight into potential usability problems by having other experts inspect a design, but a user test should always involve real users who have no special software skills.

Sometimes, during development, the specific individuals who will use the completed system can be identified. This is typically the case when a company is developing a system internally for use by a given department. This makes representative users easy to find, although it may be difficult to have them spend time on user testing instead of their primary job. Internal test users are often recruited through the users' management, who agree to provide a certain number of people. Unfortunately, managers often tend to select their most able staff members for such tests, either to make their department look good or because these staff members have the most interest in new technology. Thus, you should explicitly ask managers to choose a broad test sample based on characteristics such as experience and seniority.

In other cases, the system is designed for a certain user type, such as lawyers, secretaries in a dental clinic, or warehouse managers in small manufacturing companies. These groups can be more or less homogeneous, although it still may be desirable to involve test users from several different locations. Sometimes, existing customers will help with the test because doing so gives them an early look at new software and improves the quality of the resulting product, which they will be using.

But sometimes, no existing customers will be available, making it difficult to gain access to representative users. Test users can then be recruited from temporary employment agencies, or students in the application domain may be attending a local university or trade school. You may also recruit users who are currently unemployed by placing a classified advertisement under job openings. Of course, it will be necessary to pay all users thus recruited.

Test Tasks

Your users should perform specific tasks and not just try out the system. The experience of doing real work is different from that of simply dabbling with the software. For example, a user who plays around and discovers a menu with lots of obscure choices may try one or two commands and will then sound pleasantly surprised if something nice happens. The same user will be completely baffled when viewing that menu in the context of having to perform a specific task that does not map to the menu structure presented.

Test tasks should closely represent the uses to which the installed system will be put. Also, the tasks should provide reasonable coverage of the most important parts of the interface. You can design the test tasks based on a task analysis or on a product-identity statement that lists the product's intended uses. Information that helps you learn how users actually use systems—such as logging the frequency of use for specific commands in existing, similar systems, or direct field observation—can also help construct more representative test task sets for user testing.

The tasks must be small enough to be completed within the time limits of your user test, but they should not be so small they become trivial. For example, a good test task for a spreadsheet might be to enter sales figures for six regions through each of four quarters, using the sample numbers given in the task description. A second test task could be to obtain totals and percentages from the data entered, and a third might be to construct a bar chart showing trends across the six regions.

You should give all test users written task descriptions. This ensures they all receive identical information and also lets them refer to the description during the experiment. After the user receives the task descriptions and has a chance to read them, the tester should allow questions. Normally, task descriptions are distributed in printed form, but they can also be shown on line in a help window. The latter approach works best in computer-paced tests that require users to perform many tasks.

Role of Observers

During testing, the tester should not interface with users, but should let them discover the solutions to problems on their own. Not only does this lead to more valid and interesting test results, it also prevents users from feeling that they are so stupid the tester must solve the problems for them. On the other hand, the tester should not let users struggle endlessly with a task if they are clearly frustrated. In such cases, the tester can gently provide a hint or two to keep the test moving.

Mainly, though, observers should follow one simple rule during a user test: shut up and let the user do the talking. It is common for observers to offer help too quickly. It is human to want to assist a person who is struggling with a system (especially if you designed it), but doing so ruins the study.

Test Stages

A usability test typically has four stages:

1. Preparation
2. Introduction
3. The test itself
4. Debriefing

In preparation for the test, the tester should make sure the test room is prepared, the computer system is in the start specified in the test plan, and that all test materials, instructions, and questionnaires are available. For example, all files needed for the test tasks should be restored to their original content, and any files created during earlier tests should be moved to another computer or at least to another directory. To minimize users' discomfort and confusion, this preparation should be completed before their arrival. Also, any screen savers should be switched off, as should any other system components—such as e-mail notifiers—that might otherwise interrupt the test.

During the introduction, the tester welcomes the test users, gives a brief explanation of the test's purpose, explains the computer setup if it is unfamiliar to the users, and introduces the test procedure. After the introduction, the tester distributes any written instructions for the test, including the first test task, then asks before the start of the test if there are any questions regarding the test procedure, the instructions, or tasks.

Normally, to obtain the most feedback, the tester asks users to think aloud continuously during the test itself. By verbalizing their thoughts, test users help us understand how they view the computer system, which makes it easy to identify their major misconceptions. You get a very direct understanding of what parts of the dialogue cause the most problems, because the thinking-aloud method shows how users interpret each interface item. Thinking aloud should not be used if the test aims at gathering performance data, however, because users may be slowed by having to verbalize.

Thinking aloud feels unnatural to most people, and some test users find it difficult to make a steady stream of comments as they use a system. The tester may need to continuously prompt the user to think aloud by asking questions like, "What are you thinking now?" or, when a user spends more than a second or two on a particular window or dialogue box, "What do you think that message means?"

If the user asks a question like, "Can I do such-and-such?" the tester should not answer, but should instead keep the user talking with a counter-question like, "What do you think will happen if you do so?" If the user acts surprised after a system action but does not otherwise say anything, the tester may prompt the user with a question like, "Is that what you expected would happen?" Of course, following the general principle of not interfering in the users' use of the system, the tester should not use prompts like, "What do you think the message on the bottom of the screen means?" if the user has not appeared to notice that message yet.

After the test, the tester debriefs users and asks them to fill out any user-satisfaction questionnaires. To eliminate tester comments influencing the results, the questionnaires should be distributed before any further discussion of the system. During debriefing, ask users for comments about the system and suggested improvements. Such suggestions may not always lead to specific design changes; you will often find that different users make completely contradictory suggestions, but overall, this type of user suggestion can serve as a rich source of additional ideas to consider in the redesign.

Ethical Issues

Although usability test subjects normally escape actual bodily harm—even from irate developers resenting the users' mistreatment of their beloved software—test participation can still be quite distressing. Users feel a tremendous pressure to perform, even when told the study's purpose is to test the system and not the user. Also, users inevitably make errors and are slow to learn the system, especially when testing early designs that may be burdened with severe usability problems. Users can easily feel inadequate or stupid as they experience these difficulties. Knowing they are being observed, and possibly recorded, makes the feeling of performing inadequately even more unpleasant. On rare occasions, users have been known to cry during usability testing.

The tester is responsible for making the users feel as comfortable as possible during and after the test. Specifically, the tester must never laugh at the users or in any way indicate they are slow at discovering how to operate the system. During the test introduction, the tester should stress that the system is being tested, not the user. To reinforce this, test users should never be referred to as subjects, guinea pigs, or similar terms. More appropriate terms include participant and test user.

Severity Ratings

From whatever evaluation methods are used, a major result will be a list of the usability problems in the interface as well as hints for features to support successful user strategies. It is normally not feasible to solve all of the problems, and so one will need to prioritize them. Priorities are best based on experimental data about the impact of the problems on user performance (e.g., how many people will experience the problem and how much time each of them will waste because of it), but sometimes it is necessary to rely on intuitions only.

Severity ratings are usually gathered by sending a group of usability specialists a list of the usability problems discovered in the interface and asking them to rate the severity of each problem. Sometimes, the severity raters are given access to use the system while making their estimates, and sometimes they are asked to judge the problems based only on written description. Note that the latter approach is possible because the severity raters are supposed to be usability specialists. They should therefore be able to visualize the interface based on the written description (and possibly some screendumps) in a way that regular users would normally not be able to do. Typically, evaluators need only spend about 30 min to provide their severity ratings, though more time may of course be needed if the list of usability problems is extremely long. It is important to note that each usability specialist should provide the individual severity ratings independently of the other evaluators.

Two common approaches to severity ratings are either to have a single scale or to use a combination of several orthogonal scales. A single rating scale for the severity of usability problems might be:

- 0 = This is not a usability problem at all.
- 1 = Cosmetic problem only; need not be fixed unless extra time is available on project.
- 2 = Minor usability problem; fixing this should be given low priority.

TABLE 65.3 Table to Estimate the Severity of Usability Problems Based on the Frequency with Which the Problem is Encountered by Users and the Impact of the Problems on Those Users Who Do Encounter it

Impact of problem on the users who experience it	Proportion of users experiencing the problem	
	Few	*Many*
Small	Low severity	Medium severity
Large	Medium severity	High severity

- 3 = Major usability problem; important to fix, so should be given high priority.
- 4 = Usability catastrophe; imperative to fix this before product can be released.

Alternatively, severity can be judged as a combination of the two most important dimensions of a usability problem: how many users can be expected to have the problem and what is the extent to which those users who do have the problem are hurt by it. A simple example of such a rating scheme is given in Table 65.3. Of course, both dimensions in the table can be estimated at a finer resolution, using more categories than the two shown here for each dimension. Both the proportion of users experiencing a problem and the impact of the problem can be measured directly in user testing. A fairly large number of test users would be needed to measure reliably the frequency and impact of rare usability problems, but from a practical perspective, these problems are less important than more commonly occurring usability problems, so it is normally acceptable to have lower measurement quality for rare problems.

If no user test data is available, the frequency and impact of each problem can be estimated heuristically by usability specialists, but such estimates are probably best when made on the basis of at least a small number of user observations.

One can add a further severity dimension by judging whether a given usability problem will be a problem only the first time it is encountered or whether it will persistently bother users. For example, consider a set of pulldown menus where all of the menus are indicated by single words in the menubar except for a single menu that is indicated by a small icon (as, for example, the Apple menu on the Macintosh). Novice users of such systems can often be observed not even trying to pull down this last menu, simply because they do not realize that the icon is a menu heading. As soon as somebody shows the users that there is a menu under the icon (or if they read the manual), they immediately learn to overcome this small inconsistency and have no problems finding the last menu in future use of the system. This problem is thus not a persistent usability problem and would normally be considered less severe than a problem that also reduced the usability of the system for experienced users.

Usability Laboratories

Many user tests and other usability engineering activities take place in specially equipped usability laboratories [Nielsen 1994c]. Figure 65.3 shows one of Sun Microsystem's usability laboratories: the participant room and the control room. I should stress from the beginning that special laboratories are a convenience but not an absolute necessity for usability testing. It is possible to convert a regular office temporarily into a usability laboratory, and it is possible to perform usability testing with no more equipment than a notepad.

In Sept. 1993, I surveyed 13 usability laboratories from a variety of companies [Nielsen 1994c]. The median floor space of the laboratories was 63 m^2 (678 ft^2), and the median size of the test rooms was 13 m^2 (144 ft^2). The smallest laboratory was 35 m^2 (377 ft^2) with only 9 m^2 (97 ft^2) for the test user. The largest laboratory was 237 m^2 and had 7 rooms, allowing a variety of tests to take place simultaneously [Lund 1994]. The largest single test room was 40 m^2 (430 ft^2) and was found in a telephone company with a need to test groupware interfaces with many users. Even though the survey was conducted in the end of 1993, it seems that more recently built usability laboratories have about the same characteristics as the ones in the survey.

FIGURE 65.3 View of the control room in a usability laboratory. The participant room is visible through the one-way mirror.

Having a permanent usability laboratory decreases the overhead of usability testing (once it is set up, that is!) and may thus encourage increased usability testing in an organization. Having a special room and special equipment dedicated to usability testing means that there will be fewer scheduling problems associated with each test and also makes it possible to run tests without disturbing other groups.

Usability laboratories typically have soundproof, one-way mirrors separating the observation room from the test room to allow the experimenters, other usability specialists, and the developers to discuss user actions without disturbing the user. Users are not so stupid that they do not know that there are observers behind a wall with a large mirror in a test room, so one might as well briefly show the users the observation room before the start of the test. Knowing who and what are behind the mirror is much less stressful for the users than having to imagine it. People usually come to ignore unseen observers during the test, even though they know they are there.

Having an executive observation area in the back of the main observation area allows a third group of observers (e.g., the development team) to discuss the test without disturbing the primary experimenters and the usability specialists.

Typically, a usability laboratory is equipped with several video cameras under remote control from the observation room: the average number of cameras in each test room was 2.2 in my survey, with 2 cameras being the typical number and a few labs using 1 or 3. These cameras can be used to show an overview of the test situation and to focus in on the users' face, the keyboard, the manual and the documentation, and the screen. A producer in the observation room then typically mixes the signal from these cameras to a single video stream that is recorded, and possibly time stamped for later synchronization with an observation log entered into a computer during the experiment. Such synchronization makes it possible to later find the video segment corresponding to a certain interesting user event without having to review the entire videotape.

In many ways, the most important equipment in a usability laboratory is the "do not enter" sign on the door since it makes it possible to conduct the usability test without interruptions. As long as one has a room with a do-not-disturb sign, one can conduct usability tests without any further equipment (you won't even need a computer if you are doing paper prototyping!). The second most important piece of equipment may be high-quality microphones: since there is normally a good deal of background noise from the computer, it will be impossible to hear what the user is saying unless professional microphones are used and unless the user is actually wearing the microphone.

65.9 Iterative Design

Based on the usability problems and opportunities disclosed by the empirical testing, one can produce a new version of the interface. Some testing methods such as thinking aloud provide sufficient insight into the nature of the problems to suggest specific changes to the interface in many cases. Log files of user interaction sequences often help by showing where the user paused or otherwise wasted time, and what errors were encountered most frequently. It often also helps if one is able to understand the underlying cause of the usability problem by relating it to established usability principles such as those listed in Table 65.2. In other cases alternative potential solutions need to be designed based solely on knowledge of usability guidelines, and it may be necessary to test several possible solutions before making a decision. Familiarity with the design options, insight gained from watching users, creativity, and luck are all needed at this point.

Some of the changes made to solve certain usability problems may fail to solve the problems. A revised design may even introduce new usability problems [Bailey 1993]. This is yet another reason for combining iterative design and evaluation. In fact, it is quite common for a redesign to focus on improving one of the usability parameters (for example, reducing the users' error rate), only to find that some of the changes have adversely impacted other usability parameters (for example, transaction speed).

In some cases, solving a problem may make the interface worse for those users who do not experience the problem. Then a tradeoff analysis is necessary as to whether to keep or change the interface, based on a frequency analysis of how many users will have the problem compared to how many will suffer because of the proposed solution. The time and expense needed to fix a particular problem is obviously also a factor in determining priorities. Often, usability problems can be fixed by changing the wording of a menu item or an error message. Other design fixes may involve fundamental changes to the software (which is why they should be discovered as early as possible) and will only be implemented if they are judged to impact usability significantly.

Furthermore, it is likely that additional usability problems appear in repeated tests after the most blatant problems have been corrected. There is no need to test initial designs comprehensively since they will be changed anyway. The user interface should be changed and retested as soon as a usability problem has been detected and understood, so that those remaining problems that have been masked by the initial glaring problems can be found.

I surveyed four projects that had used iterative design and had tested at least three user interface versions [Nielsen 1993]. The median improvement in usability per iteration was 38%, though with extremely high variability. In fact, in 5 of the 12 iterations studied, there was at least one usability metric that had gotten *worse* rather than better. This result certainly indicates the need to keep iterating past such negative results and to plan for at least three versions, since version two may not be any good. Also, the study showed that considerable additional improvements could be achieved after the first iteration, again indicating the benefits of planning for multiple iterations.

During the iterative design process it may not be feasible to test each successive version with actual users. The iterations can be considered a good way to evaluate design ideas simply by trying them out in a concrete design. The design can then be subjected to heuristic analysis and shown to usability experts and consultants or discussed with expert users (or teachers in the case of learning systems). One should not waste users by performing elaborate tests of every single design idea, since test subjects are normally hard to come by and should therefore be conserved for the testing of major iterations. Also, users get worn out as appropriate test subjects as they get more experience with the system and stop being representative of novice users seeing the design for the first time. Users who have been involved in participatory design are especially inappropriate as test subjects, since they will be biased.

65.10 Follow-Up Studies of Installed Systems

The main objective of usability work after the release of a product is to gather usability data for the next version and for new, future products: In the same way that existing and competing products were the best prototypes for the product in the initial competitive analysis phase, a newly released product can be viewed as a prototype of future products. Studies of the use of the product in the field assess how real users use the interface for naturally occurring tasks in their real-world working environment and can therefore provide much insight that would not be easily available from laboratory studies.

Sometimes, field feedback can be gathered as part of standard marketing studies on an ongoing basis. As an example, an Australian telephone company collected customer satisfaction data on a routine basis and found that overall satisfaction with the billing service had gone up from 67% to 84% after the introduction of a redesigned bill printout format developed according to usability engineering principles [Sless 1991]. If the trend in customer satisfaction had been the opposite, there would have been reason to doubt the true usability of the new bill outside the laboratory, but the customer satisfaction survey confirmed the laboratory results.

Alternatively, one may have to conduct specific studies to gather follow-up information about the use of released products. Basically, the same methods can be used for this kind of field study as for other field studies and task analysis, especially including interviews, questionnaires, and observational studies. Furthermore, since follow-up studies are addressing the usability of an existing system, logging data from instrumented versions of the software becomes especially valuable for its ability to indicate how the software is being used across a variety of tasks.

In addition to field studies where the development organization actively seeks out the users, information can also be gained from the more passive technique of analyzing user complaints, modification requests, and calls to help lines. Even when a user complaint at first sight might seem to indicate a programming error (for example, data lost), it can sometimes have its real roots in a usability problem, causing users to operate the system in dangerous or erroneous ways. Defect-tracking procedures are already in place in many software organizations and may only need small changes to be useful for usability engineering purposes [Rideout 1991]. Furthermore, information about common learnability problems can be gathered from instructors who teach courses in the use of the system.

Finally, economic data on the impact of the system on the quality and cost of the users' work product and work life are very important and can be gathered through surveys, supervisors' opinions, and statistics for absenteeism, etc. These data should be compared with similar data collected before the introduction of the system.

References

Bailey, G. 1993. Iterative methodology and designer training in human–computer interface design, pp. 198–205. In *Proc. ACM INTERCHI'93 Conf.* Amsterdam, The Netherlands, April 24–29.

Bellantone, C. E. and Lanzetta, T. M. 1991. Works as advertised: observations and benefits of prototyping, pp. 324–327. In *Proc. Hum. Factors Soc. 35th Annu. Meet.*

Benel, D. C. R., Ottens, D., Jr., and Horst, R. 1991. Use of an eyetracking system in the usability laboratory, pp. 461–465. In *Proc. Hum. Factors Soc. 35th Annu. Meet.*

Chapanis, A. and Budurka, W. J. 1990. Specifying human–computer interface requirements. *Behav. Inf. Tech.* 9(6):479–492.

Cordingley, E. 1989. Knowledge elicitation techniques for knowledge based systems. In *Knowledge Elicitation: Principles, Techniques, and Applications.* D. Diaper, ed., pp. 89–172. Ellis Horwood, Chichester, UK.

del Galdo, E. and Nielsen, J., eds. 1996. *International User Interfaces*, Wiley, New York.

Diaper, D. ed. 1989. *Task Analysis for Human–Computer Interaction*. Ellis Horwood, Chichester, UK.

Egan, D. E., Remde, J. R., Gomez, L. M., Landauer, T. K., Eberhardt, J., and Lochbaum, C. C. 1989. Formative design-evaluation of SuperBook. *ACM Trans. Inf. Syst.* 7(1):30–57.

Fath, J. L. and Bias, R. G. 1992. Taking the task out of task analysis, pp. 379–383. In *Proc. Hum. Factors Soc. 36th Annu. Meet.*

Garber, S. R. and Grunes, M. B. 1992. The art of search: a study of art directors, pp. 157–163. In *Proc. ACM CHI'92 Conf.* Monterey, CA, May 3–7.

Gould, J. D. and Lewis, C. H. 1985. Designing for usability: key principles and what designers think. *Commun. ACM* 28(3):300–311.

Gould, J. D., Conti, J., and Hovanyecz, T. 1983. Composing letters with a simulated listening typewriter. *Commun. ACM* 26(4):295–308.

Greif, S. 1991. Organisational issues and task analysis. In *Human Factors for Informatics Usability*. B. Shackel and S. Richardson, eds., pp. 247–266. Cambridge University Press, Cambridge, UK.

Grudin, J. 1989. The case against user interface consistency. *Commun. ACM* 32(10):1164–1173.

Grudin, J. 1990. Obstacles to user involvement in interface design in large product development organizations, pp. 219–224. In *Proc. IFIP INTERACT'90 3rd Int. Conf. Hum.–Comput. Interaction.* Cambridge, UK, Aug. 27–31.

Grudin, J. 1991a. Interactive systems: bridging the gaps between developers and systems. *IEEE Comput.* 24(4):59–69.

Grudin, J. 1991b. Systematic sources of suboptimal interface design in large product development organizations. *Hum.–Comput. Interaction* 6(2):147–196.

Grudin, J., Ehrlich, S. F., and Shriner, R. 1987. Positioning human factors in the user interface development chain, pp. 125–131. In *Proc. ACM CHI+GI'87 Conf.* Toronto, Canada, April 5–9.

House, C. H. and Price, R. L. 1991. The return map: tracking product teams. *Harvard Bus. Rev.* (Jan.–Feb.):92–100.

Johnson, P. 1992. *Human Computer Interaction: Psychology, Task Analysis and Software Engineering*, McGraw–Hill, London, UK.

Life, M. A., Narborough-Hall, C. S., and Hamilton, W. I., eds. 1990. *Simulation and the User Interface*. Taylor & Francis, London, UK.

Lund, A. M. 1994. Ameritech's usability laboratory: from prototpye of final design. *Behav. Inf. Tech.* 13(1&2):67–80.

Maulsby, D., Greenberg, S., and Mander, R. 1993. Prototyping an intelligent agent through Wizard of Oz, pp. 277–284. In *Proc. ACM INTERCHI'93 Conf.* Amsterdam, The Netherlands. April 24–29.

Nielsen, J. 1990. Paper versus computer implementations as mockup scenarios for heuristic evaluation, pp. 315–320, In *Proc. IFIP INTERACT'90 3rd Int. Conf. Hum.–Comput. Interaction.* Cambridge, UK, Aug. 27–31.

Nielsen, J. 1993. Iterative user interface design. *IEEE Comput.* 26(11):32–41.

Nielsen, J. 1994a. Heuristic evaluation. In *Usability Inspection Methods*. J. Nielsen and R. L. Mack, eds., pp. 25–62. Wiley, New York.

Nielsen, J. 1994b. *Usability Engineering*, paperback ed. AP Professional, Boston, MA.

Nielsen, J. 1994c. Usability laboratories. *Behav. Inf. Tech.* 13(1&2):3–8.

Nielsen, J. 1995. A home-page overhaul using other Web sites. *IEEE Software* 12(3):75–78.

Nielsen, J., Desurvire, H., Kerr, R., Rosenberg, D., Salomon, G., Molich, R., and Stewart, T. 1993. Comparative design review: an exercise in parallel design, pp. 414–417. In *Proc. ACM INTERCHI'93 Conf.* Amsterdam, The Netherlands, April 24–29.

Nielsen, J. and Faber, J. M. 1996. Improving system usability through parallel design. *IEEE Comput.* 29(3):29–35.

Nielsen, J., Fernandes, T., Wagner, A., Wolf, R., and Ehrlich, K. 1994. Diversified parallel design: contrasting design approaches. In *ACM CHI'94 Conf. Companion.* Boston, MA, April 24–28.

Nielsen, J. and Mack, R. L. 1994. *Usability Inspection Methods*. Wiley, New York.

Nielsen, J., Mack, R. L., Bergendorff, K. H., and Grischkowsky, N. L. 1986. Integrated software in the professional work environment: evidence from questionnaires and interviews, pp. 162–167. In *Proc. ACM CHI'86 Conf.* Boston, MA, April 13–17.

Nielsen, J. and Molich, R. 1990. Heuristic evaluation of user interfaces, pp. 249–256. In *Proc. ACM CHI'90 Conf.* Seattle, WA, April 1–5.

Perlman, G. 1989. Coordinating consistency of user interfaces, code, online help, and documentation with multilingual/multitarget software specification. In *Coordinating User Interfaces for Consistency*, J. Nielsen, ed., pp. 35–55. Academic Press, Boston, MA.

Poltrock, S. E. 1996. Participant-observer studies of user interface design and development. In *Human–computer Interface Design: Success Cases, Emerging Methods, and Real-World Context*, M. Rudisill, T. McKay, C. Lewis, and P. Polson, eds. Morgan Kaufmann, San Francisco, CA.

Rideout, T. 1991. Changing your methods from the inside. *IEEE Software* 8(3):99–100, 111.

Rowley, D. E. and Rhoades, D. G. 1992. The cognitive jogthrough: a fast-paced user interface evaluation procedure, pp. 389–395. In *Proc. ACM CHI'92 Conf.* Monterey, CA. May 3–7.

Sless, D. 1991. Designing a new bill for Telecom Australia. *Inf. Design J.* 6(3):255–257.

Tognazzini, B. 1989. Achieving consistency for the Macintosh. In *Coordinating User Interfaces for Consistency*, J. Nielsen, ed., pp. 57–73. Academic Press, Boston, MA.

Virzi, R. A. 1989. What can you learn from a low-fidelity prototype? pp. 224–228. In *Proc. Hum. Factors Soc. 33rd Annu. Meet.* Denver, CO, Oct. 16–20.

von Hippel, E. 1988. *The Sources of Innovation.* Oxford University Press, New York.

Whiteside, J., Bennett, J., and Holtzblatt, K. 1988. Usability engineering: our experience and evolution. In *Handbook of Human–computer Interaction.* M. Helander, ed., pp. 791–817. North-Holland, Amsterdam.

Wichansky, A. M., Abernethy, C. N., Antonelli, D. C., Kotsonis, M. E., and Mitchell, P. P. 1988. Selling ease of use: human factors partnerships with marketing, pp. 598–602. In *Proc. Hum. Factors Soc. 32nd Annu. Meet.*

Wiecha, C., Bennett, W., Boies, S., and Gould, J. 1989. Tools for generating consistent user interfaces. In *Coordinating User Interfaces for Consistency*, J. Nielsen, ed., pp. 107–130. Academic Press, Boston, MA.

Further Information

Usability engineering is the main topic of the annual meetings of the Usability Professionals' Association. For further information contact its office: Usability Professionals' Association, 4020 McEwen, Suite 105, Dallas, TX 75244, USA. Tel. +1-214-233-9107, ext. 206, Fax +1-214-490-4219, email upadallas@aol.com.

66

User Interface Design Activities

Ruven Brooks
*Schlumberger Austin
Product Center*

66.1 The Nature of User-Interface Design

Design is a Discovery Process

For purpose of this discussion, design will be considered to be a human problem solving process whose goal is to specify an artifact that meets a set of constraints as well as possible. Design situations can range from those in which the constraints are fairly well known in advance to those in which many of the most important constraints must be discovered or elaborated during the design process. Design situations at the well-constrained end of the spectrum are sometimes referred to as being *routine* design or as just *specification* without much problem solving flavor. An example of this type of design in the user interface area might be the layout of data entry screens for an existing database in which the layout had to follow a strict set of design standards. Many user interface design activities seem to fall nearer the other end of the spectrum, in which some or many of the important constraints are not known in advance but need to be discovered as the design activity proceeds. User interfaces are a place in software systems where differences in application functionality and intended user community are almost certainly to be reflected in differences in the code that has to be written. This makes it less likely that developers will, over time, have an opportunity to solve the same design problem repeatedly and build up a store of knowledge about what the problems are likely to be. Hence, most user interface design efforts are likely to involve a great deal of constraint discovery.

What may make this constraint discovery difficult is that it may depend on specific design decisions so that some constraints will be discovered only if certain features are included in the design. The partial design may not meet these new constraints very well, and it will then be necessary to try alternatives. In the extreme case, constraints can only be discovered after an attempt is made to construct the artifact; for example, only after writing some of the code may it be discovered that an operating system has a built-in, undocumented limitation when certain features are used in combination. To the extent that design problems contain unknown constraints, trial-and-error is a fundamental characteristic of designing; indeed, Petroski [1985] talks about "failure driven design."

Design is Fundamentally Opportunistic

Experimental studies of experienced designers show that they work in an **opportunistic** fashion, working on those parts of a design on which they can make progress most easily and leaving parts of the design that require more effort until last. The portions which are deferred may include those aspects of the design that require further information gathering, those that require detailed reasoning, and those that are tedious but straightforward to do. Their overall behavior is thus as much a function of the design they are working on as it is driven by any kind of general solution strategy. This behavior is shown both by novice and expert designers [Guindon 1990, Visser 1990].

A priori, one might assume that better designs would result if designers proceeded in a more disciplined fashion, such as maintaining a uniform level of abstraction across the whole design and proceeding *top down*. There are two arguments as to why designers do not and cannot proceed in this fashion. The first is based on the observations just made about design discovery. Since discovery of important constraints may happen only when the design reaches the most concrete and detailed level, the advantage of deferring design commitments of the top-down approach may be more than offset by the damaging effects of delaying constraint discovery if the design problem is not well defined.

A second reason why designers might follow an opportunistic approach instead of a systematic deepening one has to do with managing the knowledge required to do design. Knowledge to be used for design must be readily cognitively available: the designer must have the knowledge in immediate memory or must be able to read it with little effort. If the designer acquires some knowledge, say, by reading a textbook or requirements document but then decides to defer work on the part of the design to which the knowledge applies, the knowledge will have to be reacquired when that part of the design is finally reached. Although the effort to reacquire the knowledge may be substantially less than that needed to acquire it in the first place, it still can be quite substantial. (An argument might be made that having to relearn or review knowledge increases the odds that knowledge will be permanently retained. In educational contexts, this is a highly desirable effect, but, for designing software, the effects can be negative. Remembering the details of each and every design project increases the odds of confusing them. The ability to forget can be very useful). For both these reasons, designers are likely to persist in their opportunistic behavior, despite efforts to encourage them to follow other approaches.

66.2 Selecting Design Activities

Although the occurrence of a discovery process and the following of an opportunistic approach to performing design work are both inherent properties of design in many domains, the actual course of events that takes place during a design process will be very much determined by the particular design problem and by the activities of the designers during this process. Waiting until a preliminary implementation is available before showing the design to the users may have a very different impact on the design process than deciding to show the users paper prototypes. Furthermore, the impact will vary considerably depending on the particular interface being designed. Developing a new user interface for data entry for a legacy application for internal users may profit from very different design activities than developing the next release of an existing application sold to *power users* which may be different, again, from what is needed for developing a *walk up and use* community information service. Design activities which are very beneficial for one situation can have minimum impact or negative impact in another; showing hostile users a paper prototype may encourage their belief that the developers have no real intent of implementing a system.

66.3 Identifying the Primary Design Risks

Given that all design activities have a cost, all design projects need to make decisions about which design activities are most likely to have a positive, cost-effective impact on the design process. The discovery nature of the design process makes it difficult to make these decisions by calculating the cost/benefit ratios for different activities. An alternative approach is to characterize a design situation as to the risks of the situation, those aspects of the situation that are most likely to cause the design process to fail to produce satisfactory designs [Boehm 1986]. For example, suppose that a new system is being built to replace a legacy application with a community of expert users who have had years to learn and tailor the existing interface. If the new system is being built using a widely used, commercial toolkit, it is unlikely that the project will fail because learning to program with the toolkit is too difficult. On the other hand, there is a clear danger that the new interface, despite its superior implementation technology, will be less efficient for the expert users. For example, in an effort to make the menu structure more logical, the developers might put rarely used items in the same menus with frequently used ones, forcing the expert users to go through many more levels of menus than they did previously.

Given limited resources for the design process, it makes sense to focus them on the areas of greatest risk. In this example, spending staff time on task analysis and on usability work is more likely to avoid a cause of project failure than equivalent resources spent on a new, in-house programming interface for the commercial package.

Different design activities address different areas of risk; for example, participatory design reduces the risk that the users will fail to buy in to the design but it does nothing if the main design problem is staying within the performance constraints of the available hardware. Not every design activity is worthwhile for every design problem; instead, design activities need to be selected to suit the risks of the situation and the available resources. The following section discusses some of the more widely used design activities and suggests the design situations in which they are most likely to be beneficial.

66.4 Types of Design Activities

Task Analysis

A task analysis is a description of all of the activities performed by a system user or users in accomplishing their work goals. This description may use a variety of different notations and may also include quantitative information about the frequency and duration of tasks. Chapter 63 discusses different approaches to task analysis and gives an example of how to perform one. Often, a task analysis is part of an overall requirements analysis. A basic task analysis can just be a surface narrative of all of the things a user does. "To open a new account, the user first enters the client name and address information. A new account number is assigned by going to the files and finding the largest existing number and...." A description at this level does not distinguish which aspects of the current system can be or need be changed in designing the new system, so many task analysis methodologies include some effort at abstraction away from this basic description. For example, if there are a number of places in the system in which users need to assign unique numbers, the task description may be reorganized to show that number assignment is a distinct activity. In turn, this may lead to the designers of the new system providing a common user interface function for number assignment. An important point to make here is that the amount of time or effort devoted to task analysis depends on the design risks. If the goal of the project is to port an existing application, basic functionality unchanged, to a new platform with a different look and feel, then a surface task analysis that just catalogs all of the user activities is probably the optimum use of resources. On the other hand, if the project is redesigning the entire system or if it is trying to use the user interface to unify several separate applications, then it is worth trying to spend additional time on abstracting underlying goals or more general tasks from what can be observed currently.

Scenarios

A scenario is a "narrative description of what people do and experience as they try to make use of computer systems and applications" [Carroll 1995]. Scenarios are also sometimes known as use cases [Jacobsen 1995]. Scenarios may vary as to content and level of detail depending how they are being used in the design process. A first use of scenarios is early in the design process as part of the task analysis; the intent of using them then is to provide a view of the structure of activities that is more integrated than the description of the

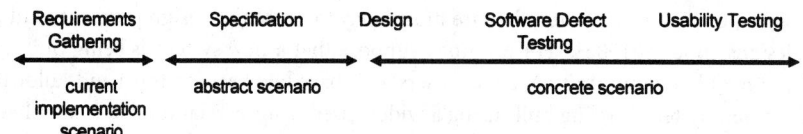

FIGURE 66.1   Usage of different types of scenarios.

individual tasks or actions that compose them. An example of this use of scenarios might be: "the releasing clerk verifies that the amount shown on line 27 has been calculated correctly." The actual calculation procedure would, presumably, be spelled out elsewhere in the task analysis. Note that the language of the scenario is the language of the current implementation. Later in the design process, after the task analysis has been refined, the scenarios may be modified to reflect the higher levels of abstraction of the task analysis. In the example, the language might change to "The calculations of the amount claimed are verified." Scenarios at this level are frequently used as the basis for starting a new design, since they capture the underlying requirements of the activity while not specifying the implementation. They are not as useful, however, for interacting with the customers or users of the product, since they may find it difficult to relate the abstract expressions to their situation. (See Fig. 66.1.)

A final point of use of scenarios is after the new design has been completed, as a means of expressing the intended use of the new design. Scenarios used at this stage are, again expressed in concrete language, that of the new design. Scenarios of this type are frequently combined with prototypes as a means of communicating the design to **stakeholders**. Such scenarios can serve double duty as test scripts once the product is completed.

As with other techniques for improving design, scenarios are not without their disadvantages and drawbacks. First, construction of scenarios which are rich enough in detail to capture all of the relevant aspects of use of an artifact can be a very time-consuming activity. If a system of a completely new type is being constructed, the effort spent writing scenarios that envision all of the ways the system might be used could be better spent on constructing a prototype to find out the ways in which the system will, in fact, be used. To the extent that scenarios capture only some of the uses of an artifact, a second problem is *scenario dropout*, requirements that get lost from the design because they are not part of any of the scenarios that are used to guide the design.

Finally, scenarios shown to the customers and stakeholders in a design may lead to unrealistic expectations about the delivered product. For design purposes, it is very reasonable to either have the scenarios contain much more functionality than will actually be implemented or to gloss over aspects of the design that are not novel. Even careful explanations to the customers about what will and will not be included in the actual product may fail to prevent disappointment when the delivered product does not provide the expected functionality.

Prototypes

A prototype is an artifact that contains some features which are identical to those in a proposed design. This definition is intended to include both software and paper prototypes; a drawing of a screen layout is a prototype in this sense because it shares the features of screen item position and labeling with the proposed software. Prototypes can range from **low-fidelity** prototypes in which the prototype is implemented in a different medium than the product such as the screen layout just mentioned to **high-fidelity** prototypes that are implemented on the same hardware and software platforms as the eventual product and which lack only capacity and performance characteristics of the actual products.

Prototypes are useful in avoiding several different types of **design risks**. They can be very useful in the early requirements definition phase of a project to elicit desired features and to obtain agreement on what features the product should contain. This is particularly likely to be the case if the product performs a function that is brand new to the users or if it has an entirely new type of user interface from their existing software. If the project is using a new or unfamiliar software package or library, a prototype can be used to familiarize developers with the capabilities of the software before committing to a design using these capabilities. A high-fidelity prototype can be used for early usability testing before the underlying structure of the system is completed. Finally, prototypes can be used to maintain customer confidence during a long development cycle in which not much appears to be taking place.

Despite these obvious advantages, there are substantial risks in the use of prototypes. Creating a high-fidelity prototype requires a substantial amount of effort: in the extreme, as much as developing the actual product. If the project runs short of resources, there is a temptation to use the prototype as the basis for further development. This may not be a bad route to follow if the prototype has been carefully designed with this possibility in mind, with a carefully designed interface to the part of the system that will get replaced in the transition to product, but if the prototype has been built in the most expedient fashion possible, then the resulting product may be difficult to maintain or enhance.

Even in situations where there are adequate resources, developing-high fidelity prototypes may have other drawbacks. One is that developers may start to rely on user reaction to the prototype as the major source of user input and may neglect to do other design activities which they are less comfortable with. This may cause entire sets of useful capabilities to be left out of the product because the users assumed that the prototype had all of the essential features of the product and focused their reaction on minor changes.

A related danger of high-fidelity prototypes is that they will produce unreasonable user expectations. Seeing what appears to be a fully functional implementation, users may not appreciate the additional work that is needed to produce a fully functional product and have doubts about the motives of the engineering team or they may propose unrealistic schedule accelerations. Alternately, if the prototype is missing many of the features of the eventual product, the product customers may loose confidence in the ability of the design team to understand the requirements or, even, to produce the product.

Low-fidelity prototypes such as paper drawings or scripted screen snapshots avoid the danger of sliding into development but they are not without their problems in regard to customer expectations. Being shown a paper prototype several years into a multiyear project may raise doubts about whether the project is on schedule. Additionally, low-fidelity prototypes may not be suitable for assessing those aspects of a proposed design that are closely tied to system performance; a paper prototype would be extremely difficult to use to determine whether a 3-D interaction technique will work.

Rather than just assuming that a high-fidelity prototype is the best thing to do if resources are available, a more effective approach to deciding whether and at what level to prototype is to be very explicit about what role the prototype is to play in the project and to tailor the prototype appropriately. If it is decided that the main purpose of the prototype is to gain user support by showing them how much easier the new interface will be to use, then building a complete high-fidelity prototype is not cost effective; instead, only a few of the particularly difficult operations with the existing interface should be selected for prototyping. If the goal of the prototyping effort is to find out what operations the users will find most time consuming with the new interface, then it might be better not to prototype at all, but to spend effort on developing a user model that can be used to predict performance times [John 1995]. If the main purpose of the prototype is only to give the developers experience with a new software package, then to avoid some of the dangers previously cited, the wisest thing to do might be to prototype an entirely different application than the one the project is constructing! Prototype construction should be a planned activity with clear goals and purposes.

Use of Style Guides and Standards in Design

One design risk that occurs in projects with multiple developers is that the user interfaces for different parts of the project will have a different appearance and perform similar operations in different ways. One way to avoid this is to develop project-specific style guides. Typically, the project-specific guide will be based on other, more general style guides, such as those for a window system or for an industry. For example, applications in the petroleum industry might have style guides based on the *POSC E & P User Interface Style Guide* [POSC 1994] which is, in turn, based on the *OSF/Motif Style Guide* [OSF 1993]. The reason for this layering of style guides is that the platform or industry style guides are often very general and offer a great

many acceptable alternatives; more specific standards are needed to guarantee similarity across a project. (See Fig. 66.2.)

The use of a style guide to ensure similarity is often seen as a
relatively low-cost solution and is sometimes assigned to a single individual. This impression is likely to be false. Creating a
style guide implies creating or designing a style. Although this
can be done by abstractly considering the range of applications
to be covered, a more successful approach is to select a small

Product Family Style Guide
Industry Style Guide (POSC)
Platform Style Guide - (OSF/Motif)

FIGURE 66.2   Style guide sets.

set of applications to serve as reference points for style development. The style guide is created by working
out the interface design for these applications and abstracting from the design decisions made during
this process. For this reason, writing the style guide requires an implementation effort, and appropriate
resources need to be assigned. Since those aspects of an interface that are specified by the style guide will
often appear in every component of the system, the usability of those aspects is critical. Usability studies
or user modeling ought to be done with the reference applications to ensure that usability objectives are
met for the proposed style. These studies are particularly cost effective since they can resolve usability
issues across an entire product family.

A problem with any style guide is ensuring that developers follow it. In making style decisions, developers
are more likely to rely on examples than on statements of style [Tetzlaff and Schwartz 1991], particularly
if the source code is available. The reference applications can often serve this function. In addition to
sample applications, it is useful to have style lists that can be used to audit a design for style compliance.

Object-Oriented Analysis and Design

Object-oriented analysis, design, and programming are related, but not identical concepts. From the
viewpoint of user interface design, the first two are of greatest interest. Object-oriented programming
is described in Chapter 96. In object-oriented software design, the design is represented as a collection
of classes, which are specified by the data which are internal to them and the operations which can be
performed on them. The classes are arranged in a hierarchy with each class inheriting properties from
classes above it in the hierarchy. There are a variety of different notations for expressing an object-
oriented design, as well as suggested procedures and software tools for constructing one. In contrast,
nonobject-oriented designs represent the design as a collection of procedures and functions that are
organized by control flow relationships. The association between data structures and sets of procedures
is implicit.

Apart from the advantages of object-oriented approaches for system design, object-oriented design is of
particular interest for user interface design for several reasons. First, object-oriented analyses are claimed to
lead more easily to object-oriented designs. To whatever extent object-oriented analysis does a better job of
capturing user requirements, then these gains are multiplied with a compatible design approach. Second,
in the popular object style of user interface, the user invokes computations by performing operations on
graphical objects such as moving them on the screen or clicking on them with a pointer button. (The object
style is often held to be one aspect of *direct manipulation* interfaces, although it is possible to construct
object interfaces that most observers would agree are not direct manipulation interfaces.) Such interfaces
are easily described with an object-oriented design notation. Finally, most of the available object-oriented
languages also contain extensive graphical user interface facilities (Smalltalk, Visual C++, Visual Basic).
To the extent that object-oriented designs can be most easily implemented in object-oriented languages,
this is an attractive pairing.

Although these claims may, in fact, be true a large portion of the time, the situations in which they fail
need to be considered when deciding whether to use object-oriented techniques. First, the presumption of
a close relationship between the object-oriented task analysis and the system design may not always hold.
If the intent of the design activity is to re-engineer work practices, then functionality and information will
be distributed very differently in the design than in the analysis. The objects in the design can no longer
be directly derived from the objects in the analysis. One approach to coping with this problem is to do

two analyses, one based on the current implementation and one based on projecting how the new system will affect work activities (van der Veer, van Vliet, and Lenting, 1995). The design is then derived from the second analysis.

Second, the relationship between design objects and screen objects can be a very complex one. In the implementation of a word processor, one possible design is to have each paragraph represented by an object which, in turn, contains an instance of a paragraph style object. Several paragraph objects might point to the same style instance. From a software architecture standpoint, this approach makes it easy to share styles across paragraphs and to rearrange paragraphs while retaining their styles.

From a user interface standpoint, though, this may not be a very desirable model since users would have to be aware of which paragraphs share a style before making changes. A better model for the users might be one in which styles do not appear to be shared.

Rather than change the internal model to match the user model, the best design solution might be one in which the shared model was used internally but behind-the-scene instance creation could give the user the appearance of the nonshared model. The design details of such a system would have to be carefully worked out to avoid situations in which the user model breaks down and the internal model governs visible behavior. In situations such as this, it is still possible to capture the relationships in an object-oriented design by explicitly including both types of objects and using the methods of the objects to describe the relationship between them; the user-sets-paragraph-properties method would describe when new paragraph-style objects should be created. What the example does illustrate is that the relationship between screen objects and internal design objects will often be quite complex. Any benefits of object-oriented design for these kinds of design are more likely to come from general improvements in system structuring than from any implicit correspondence between the two kinds of objects.

Finally, the graphical object capabilities of object-oriented programming languages may not be much of an advantage in implementing an object-oriented design. A major problem in all object-oriented design, not just interface design, is that naive implementations of object-oriented designs may have performance problems. Such problems may be somewhat more likely for user interface designs, since interactive operation is frequently an essential element of the designs. Frequently, performance issues can be addressed either by modifying the implementation or by reworking the object representation, but they can be an important reason why transformation of an object-oriented design to an object-oriented implementation can be less straightforward than the language features would suggest.

The overall picture of the use of object-oriented analysis and design in user interface construction is that it may contain complexities and subtleties that are not as prominent in other areas of software construction. This is not to argue that these methods will not work for user interface design or that other approach will be superior, but rather, that more skill and effort may be needed to apply the methods than in other areas.

Participatory Design

Participatory design is a term used to refer to methodologies that seek to achieve a high degree of user involvement and participation in the design process. One of the major origins of **participatory design (PD)** techniques was in Scandinavia in joint projects between trade unions and universities whose goal was to achieve a higher degree of workplace democracy with more worker control over their own activities, not to improve user interfaces per se [Carmel et al. 1993]. From this background, some PD techniques are as much concerned with work redesign as they are with the design of the computing system, but the term is now used as well to cover methodologies whose focus is just on the computing system.

A primary motivation for these methodologies grows out of the discovery nature of the design process described earlier, particularly as regards user needs. Only rarely will it be the case that these needs are completely understood and can be completely recorded before design of the new system begins. What is usually the case is that substantial discovery of user requirements occurs during the design process itself. Unless the developers are themselves the only users of the system, the design process must of necessity have constant access to the future users of the system under development.

A key issue that many of these methodologies seek to address is that the users frequently differ from the developers in knowledge of computing technology, in work methods and practices, in social and organizational standing, and in knowledge of their own requirements. Furthermore, the direct users of a system are not the only ones whose input is required for a successful design; additional, conflicting requirements may come from customers of the users (if the users are providing a direct service), from managers of the users, and from those who will install and maintain the system. (The term **stakeholders** is sometimes used to include all of these different interests.) The goal of many PD methods is to ensure that information from all of these sources is considered in the design process. The range of actual activities to attain this goal is quite wide since one of the tenets of PD is that developers should be creative in their efforts to involve users. A good overview of these techniques is given in Muller and Kuhn [1993]. The following examples are intended to give an idea of this range.

A primary format for contact between users and developers is joint meetings. Joint application design (JAD), which was largely developed in industrial contexts in North America, includes a set of practices for running meetings that are intended to ensure that the meetings do actually serve their intended purpose and do not, for example, become sidetracked into discussions of current management or of details of new software or hardware. Among these practices are the use of a facilitator who is neither a user nor a developer to moderate the meeting, the use of agendas or plans to guide the meeting, and documentation of what takes place during the meeting, including disagreements, instead of just recording the meeting outcomes.

Knowing how a product will be used is frequently critical in deciding what features the product will have. Asking users how they will use a product often does not yield the desired information. If the product should have features or capabilities which are unfamiliar to them, they may have difficulty in envisioning how the new capabilities could be applied. Just asking users what they do may be equally unsuccessful; users may describe only those features of their tasks that they feel will be important to the designers and they are likely to omit those aspects of their jobs that are necessary for the work to proceed but which are not official parts of the work description.

Two techniques to avert these problems are visiting the users in their normal work context and having the developers perform users' jobs. Visits to users can range from formal tours to longer stays in which the developers try to blend into the surroundings, so that the normal behavior of users can be observed. Having the developer do the user's work is most often accomplished by having a developer sit beside a user and be coached through the user's normal activities. It may also be possible to have the developer participate in at least the beginnings of a training program and perform the job as a trainee or novice. For highly technical tasks, neither of these approaches may be feasible—software developers are not going to be able to do even the simplest neurosurgery—but for many tasks, the developers will be able to do enough of the task to give them far better information than they might have received just by interviewing or watching users.

User interface designers are constantly exhorted to know their users, and participatory design methodologies may offer a much better chance of success than design efforts that are driven only by a marketing document and **back-end usability testing**. Nevertheless, participatory design techniques can be among the most risky design activities. A first source of problems is that, as previously mentioned, users may differ in work goals, work styles, organizational status, and organizational perspective from developers. What may be perfectly acceptable, even, desirable behavior on the part of developers within their own community can become a source of conflict outside that community. For example, when developers make requests of each other for new system features or capabilities, they automatically assume that it may not be possible to fulfill all of the requests and that it will be necessary to assign priorities. When working with users, in order to give the users the maximum say as to what gets built, the developers may also ask the users to prioritize their requests. Depending on the history of the user group, this open request for input may instead be interpreted as the developers finding excuses for not building what the users want and relations may deteriorate. Even though the developers were well intentioned, they might have been better off not involving the users.

A second set of problems grows out of the need for balancing the desires of all of the stakeholders in the design. Consider the common case of a new system that management intends will reduce staff requirements. If the current users are aware of this intent, getting them to contribute to design meetings, no matter how well structured they are for user input, is likely to be a very difficult task.

The danger in both of these examples goes beyond the failure of the participatory design technique itself. In the first example, the deterioration of relations may make it difficult to find users who are willing to try the new system. Management may then conclude that this is because the developers have done a poor job of development. Successful use of participatory design techniques may depend heavily on the interpersonal and organizational skills of the developers, an area in which software developers do not have a particularly good reputation. Unless these skills are present or can be acquired, a heavy focus on participatory design activities is not likely to be successful.

Iterative Design and Design Rationale

Because of the way discovery takes place during the design process, the first version of a design is never good enough. All user interfaces need to be improved by evaluating their deficiencies and redesigning to overcome them. A major set of decisions in selecting design activities concerns how and when the design/evaluate/redesign cycle takes place.

A first set of decisions concerns the goals and criteria that will be used to evaluate a design. For many products, the goal for the user interface is the vague, implicit one that the product not get the reputation for having a bad user interface; it should be "easy to learn–easy to use". A better practice, less likely to lead to major opinion conflicts in the project, is to develop explicit operational criteria for quality of the user interface. For example, if a general goal for the product is that it be easy to learn, then an equivalent operational goal is that a group of users with specified background and training materials should be able to learn to perform a criterion set of tasks within a given time limit. Similarly, for the easy to use goal, an operational criterion might be based on the time it takes for specified user experience levels to execute a set of task scenarios.

A second, major set of decisions concerns how to determine whether the design meets the criteria and when in the product development cycle the evaluation takes place. Evaluation methods fall in two classes, empirical and model based. Chapter 66.3 discusses empirical methods. John [1995] describes how a model-based technique, GOMS modeling, can be used in design evaluation.

Evaluation and subsequent redesign can take place in a number of ways in the product development cycle. One possibility is that no evaluation or redesign take place before the product reaches the users for the first time. Once the system is in place, user reactions and requests form the basis for the next iteration on the design. Obviously, this can be a very risky approach; if the initial version is bad enough, there may not be an opportunity for another iteration. It has other, more subtle, disadvantages as well. User comments or complaints provide only partial information on how a design can be improved. Not all users are motivated to provide feedback and those that do may not be representative of the user population as a whole. Furthermore, user comments often take the form of suggestions for features that are minor modifications of features already in the product. Unless developers carefully ferret out the problems the users are trying to solve that underlie the suggestions, product evolution can get trapped in a succession of minor improvements. Another problem with just relying on user comments for feedback is that they do not provide quantitative information on product quality. A ploy used by advocates of usability testing is to ask developers or managers to estimate the amount of time it will take users to do particular tasks. When compared with measured data, the estimates are almost invariably too low. Since users are also often not very good at estimating the amount of time it takes them to perform a task, in the absence of other usability data, the risk is again that the product will be perceived as satisfactory when it still has substantial deficiencies from a usability standpoint.

One improvement on this practice is to conduct usability evaluation on the product once it is completed, either before or after it is released to customers. Evaluation after the product is substantially complete but before it is released can serve two purposes. First, it can ensure that the product meets minimum usability standards to be released at all; this is sometimes referred to as *summative* evaluation. Although waiting until this point to discover that a product does not measure up may seem almost as risky as doing no evaluation, it is still a substantially better practice for two reasons. First, in many situations, it may be still possible to stretch the schedule to address the show stopper problems. This can at least ensure that the product survives until the next release. Second, in actual practice, most products go through multiple versions or releases.

Evaluation results on one release of the product become part of the requirements for the next release and influence the design. Evaluation used this way is sometimes referred to as **formative evaluation**. These evaluations have the advantage that they can often be done economically on actual users performing actual tasks in real-world contexts. Depending on product schedules, the released version of the product can also serve as part of the prototype for a version under development, for example, by using the current software for parts of the system that will be unchanged and paper prototypes for the proposed revisions.

The lowest risk approach is to traverse the design cycle one or more times before the product is ever released to users. In comparison with approaches that can make use of an existing version of the product, this approach may incur a considerable overhead cost in construction of prototypes, development of test scenarios, conducting of evaluations, and possibly, in lengthening the entire project schedule. Landauer [1995] argues that these costs are small in comparison with the savings to end users of an improved user interface, but for many actual projects the issue is not whether the iteration is cost effective for improving the user interface but whether bounded resources are best spent on improving the interface, or, say, reducing the time until a badly needed new system is installed.

A stronger argument that sometimes can be made for the resources to do early iteration is to base it on the risks that stem from having substantial user interface problems. For example, porting an existing product to a platform with a different user interface system on which competing, similar products are already running runs the risk that the interface will retain too much similarity to the interface on the platform on which it was originally developed. This will cause additional learning effort for users on the new platform. Planning in advance for several evaluation/redesign cycles before the product is actually released is probably critical to the product's initial survival. On the other hand, for a product that offers substantial new functionality not related to the user interface, it may be difficult to justify the effort for large amounts of evaluation and redesign.

An important issue in all types of usability evaluation is how to use the results in the design process. Typically, the results will be of two forms: quantitative results, such as times and error rates, and qualitative results, such as problem or issues lists. The main role of quantitative results is to determine how close the product comes to meeting its usability objectives and determining and justifying further expenditure on improving the user interface. The qualitative results are used to identify which particular features or areas need work. Both types of results are useful; a long list of usability deficiencies is much harder to dismiss if it is accompanied by numbers showing that the product is far from meeting its usability objectives.

Some care is required in using qualitative evaluation results. If developers are just provided with an unprioritized list of problems, most developers will try to address as large a number of items on the list as possible and will fix those problems that are easiest to fix first. These may or may not be the problems that have the greatest impact on users. A better approach is to rate separately the effort required to repair a problem and the impact the problem is having on the users. In making the impact ratings, the overall usability objectives of the product are taken into account; a usability problem in learning a system that is intended for constant, routine usage should be given a lower rating than an equally severe problem that affects use after the system has been learned. The combined score is then used to schedule the order in which problems are addressed. Although this approach may result in fewer problems being repaired in the next release, it ensures that those repairs will be those that have the most impact on users.

Design Rationale

Design of software can be viewed as a succession of decisions. Each of these decisions can be characterized by the information available and by the criteria used. **Design rationale** is a set of methods for capturing the criteria and information behind each design decision. These methods have two goals. The first is to make this decision basis available over time. During the course of redesign, decisions made in the previous design period are often revisited. Understanding the basis for the original decision can be very useful. What seems on the surface to be a still reasonable decision can, on closer examination, turn out to be based on criteria or information that is now invalid, or what seems to be a very bad decision can turn out to be still valid after all. The second goal of design rationale methods is to make the decision process explicit during the current design activity. Proponents of design rationale argue that, in many decision situations, neither the design makers nor all of the participants realize all of the assumptions that may lie behind a design decision. By making the assumptions visible, designers and other participants may become aware of those which are doubtful or untrue, and improve the design accordingly.

The two main approaches to design rationale are IBIS [Conklin and Yakemovic 1996] and QOC [Maclean et al. 1996]. One of the differences between them is that IBIS seeks to capture the rationale as an automatic byproduct of the design process by stipulating the use of a particular notation for the design process. QOC takes the contrasting position that just the recording of all individual design decisions is not of much value and that a coherent design rationale is a product of the design process which must be constructed in addition to the design itself. The IBIS rationale contains the reasons behind each decision in the design process, including those that were later reversed or abandoned; the QOC rationale captures only the reason behind those features or characteristics that made it into the final product.

Both approaches have potential problems and associated costs. Since the IBIS process relies entirely on the information captured during designing, there is the danger that no rationale is recorded for important design decisions that were made implicitly. In the case of QOC, there is the possibility that the design rationale constructors do not adequately capture the thinking of the designers. In terms of cost, for IBIS, there is the cost of training and enforcing the use of the design capture tool for the entire design team. For QOC, it is the cost of assigning staff to the task of constructing the design rationale.

A major, unanswered question about all design rationale techniques is how valuable the captured design rationales will be in recovering historical design decisions. Given limited cognitive resources, developers are unlikely to read the design rationale from beginning to end like a novel. Instead, they are more likely to access it on an as needed basis to clear up those aspects of the current design that can not be understood with other available knowledge such as requirements documents, scenarios, etc. This search task can be quite time consuming for a large project that has involved many people over an extended period of time. Since developers will be able to guess the rationale behind the vast majority of design decisions, the effort to track down the rationale behind a decision will only be worthwhile in the case where the decision is a major one and for which the rationale cannot be guessed. In practice, this may well be a rare case.

Even if the benefits of design rationale for historical design recovery turn out to be minimal, there is still a case for using design rationale techniques for making the design process explicit. This is particularly likely to be beneficial in projects in which the designers are less likely to understand each other's decision making process, for example, when they come from different disciplines or different corporate or cultural backgrounds.

66.5 Designing a Design Methodology

This chapter started with two assertions about the nature of user interface design. First, it is perhaps more likely than other types of software design to contain constraints that are unknown at the time the design process begins. Second, like almost all other types of software design, designers are likely to carry out design work in an opportunistic fashion. These two factors, combined with the variability of user interface design problems, make it unlikely that one single design methodology will work for all types of user interface design. Instead, design practices need to be selected with regard to the aspects of the design problem that are most likely to lead to problems in creating a successful design. To illustrate this approach further, consider the following user interface design problems:

- A walk-up-and-use ticket vending machine for a public transit system
- A revolutionary new interface for working with relational databases
- A graphical user interface for expert financial decision makers who are more highly paid than software developers and who are already experienced users of an existing system

For the ticket vending application, an extremely high premium is placed on requiring minimum or no learning to use the system; it must accommodate a constant stream of new users of widely varying backgrounds and ability levels who have never seen the system before. Not quite as obvious is that the system must also be efficient to use; if it takes too long for daily commuters to use the system and there are conventional ticket sales available, they will use these in place of the machine system. Given the relatively shallow functionality of the system and the wide variability of the users, activities such as performing a task analysis or constructing usage scenarios are not likely to yield design information in proportion to resource cost. Instead, highest priority for use of resources for this design problem should probably be given to constructing a series of prototypes of the system that can be set in public places and tried by as wide a range of potential users as possible.

For the database interface problem, the biggest design risk is that the revolutionary new interface will address only some of the problems in using a relational database and will, in fact, be more difficult to use than existing interfaces for tasks the existing interfaces already do well. An activity that can help to guard against this danger would be the collection of a range of scenarios covering both tasks that are easy to do with existing interfaces and those that are more difficult. With these scenarios, it may be possible to use user modeling techniques such as GOMS to analyze proposed designs and to reduce the amount of usability testing needed.

Developing the interface for the financial users is probably the most challenging of the three problems. Although a high level of user involvement in the design process would seem to be very desirable in this case, the relative compensation rates of the users and the developers makes it more cost effective to maximize the use of developer time. One technique that meets this requirement would be to have the developers spend time as apprentice users. Although the cost of the user time for this is still substantial, it is more likely to be effective than, say, having the users participate in a series of design meetings. Development of a scenario collection is also likely to be worthwhile as a method for capturing the experiences of the apprentice users. In contrast, a large effort spent on developing a style guide is not likely to have equivalent returns; since the system will be used on a regular, intensive basis, the users are more likely to be troubled by style compliance that interferes with efficient task operation than they are by style irregularities that present first time learning difficulties. As these examples illustrate, user interface design is an area in which it is difficult to adhere to a single set of techniques or activities across all situations. Instead, successful designs within cost constraints are more likely to result from thoughtful selection of the activities that are included in the design process.

Defining Terms

Back-end usability testing: Usability testing of a product after design and implementation are substantially complete. This can be done either with the intent of making a decision on whether to release the product or as a starting point for requirements gathering for the next release of the product.

Design rationale: Statements of the reasons behind particular design decisions. Approaches to creating and using design rationales differ in the methods used to construct the statements, the form of the statements, and in the way in which the statements are subsequently used.

Design risk: An event occurring during a design activity that produces problems in successfully completing the design. Sometimes these risks can be anticipated and plans can be made for coping with them.

Formative evaluation: Evaluation of a design that takes place during the design process, before final implementation is completed. In the case of user interface design, this may take the form of building a user performance model for the design or of conducting a usability test on a prototype or early implementation of the design.

High-fidelity prototype: A prototype constructed in a medium and environment very similar to that of the eventual product. Often, this takes the form of running code.

Low-fidelity prototype: A prototype constructed in a medium and environment sharing only a few characteristics with the eventual product. An example of a low-fidelity prototype is the use of drawings of screen layouts to walk a user through the product design.

Opportunistic design: In empirical observations of designers at work, the sequence of design steps that a designer actually takes appears to be strongly determined by specific events that have occurred earlier in the design activity rather than by any overall strategy. Often, this results in the designer working on some aspect of the problem as soon as it presents itself. Opportunistic design is a term used to refer to this behavior.

Participatory design (PD): Any of a broad range of techniques designed to increase stakeholder participation in the design process.

Stakeholders: Everyone who has an interest in a software development project. Typically, this includes the end users of the product, the purchasers of the product (who may not be the users), the managers of the users, and the developers and their managers.

References

Boehm, B. 1986. A spiral model of software development and enhancement. *Software Eng. Notes* 11(4).

Booch, G. 1994. *Object-Oriented Analysis and Design with Applications*, 2nd ed. Benjamin Cummings, Redwood City, CA.

Carmel, E., Whitaker, D., and George, J. F. 1993. PD and joint application design: a transatlantic comparison. *Commun. ACM* 36(4):40–48.

Carroll, J. 1995. The scenario perspective on system development. In *Scenario-Based Design: Envisioning Work and Technology in System Development*. J. Carroll, ed. Wiley, New York.

Conklin, E. J. and Yakemovic, K. C. B. 1996. A process-oriented approach to design rationale. In *Design Rationale: Concepts, Techniques, and Use*. J. Carroll and T. Moran, eds. Lawrence Erlbaum Associates, Hillsdale, NJ.

Guindon, R. 1990. Knowledge exploited by experts during software system design. *Int. J. Man-Machine Stud.* 33(3):279–304.

Jacobson, I. 1995. The use-case construct in object-oriented software engineering. In *Scenario-Based Design: Envisioning Work and Technology in System Development*. J. Carroll, ed. Wiley, New York.

John, B. 1995. Why GOMS? *Interactions* 3(4):80–89.

Landauer, T. K. 1995. *The Trouble With Computers*. MIT Press, Cambridge, MA.

MacLean, A., Young, R. M., Bellotti, V. M. E., and Moran, T. P. 1996. Questions, options, and criteria: elements of design space analysis. In *Design Rationale: Concepts, Techniques, and Use*. J. Carroll and T. Moran, eds. Lawrence Erlbaum Associates, Hillsdale, NJ.

Moran, T. P. and Carroll, J. M., eds. 1996. *Design Rationale: Concepts, Techniques, and Use*. Lawrence Erlbaum Associates, Hillsdale, NJ.

Muller, M. and Kuhn, S., eds. 1993. Participatory design. *Commun. ACM Spec. Issue* 36(4).

Open Software Foundation. 1993. *OSF/Motif Style Guide*. Prentice–Hall, Englewood Cliffs, NJ.

Petroski, H. 1985. *To Engineer Is Human*. St. Martin's Press, New York.

POSC 1994. *POSC E & P User Interface Style Guide*, Petrotechnical Open Software Corp. P T R Prentice–Hall, Englewood Cliffs, NJ.

Tetzlaff, L. and Schwartz, D. R. 1991. The use of guidelines in interface design, pp. 329–333. *Proc. ACM CHI'91 Conf. Human Factors Comput. Sys.*

van der Veer, G., van Vliet, J. C., and Lenting, B. F. 1995. Designing complex systems: a structured activity. In *DIS '95 Symp. Designing Interactive Systems: Processes, Practices, Methods & Techniques*. Assn. Computing Machinery, New York.

Visser, W. 1990. More or less following a plan during design: opportunistic deviations in specification. *Int. J. Man-Machine Stud.* 33(3):247–278.

67

International User-Interface Standardization

Wolfgang Dzida
GMD—German National Research Center for Information Technology

67.1 Introduction

Two types of standards are distinguished in the **user interface** technology: standard user interfaces and standards for user interfaces [Stewart 1990]. The first type establishes a de-facto standard for user-interface implementation, either provided by a leading software producer (for instance, OPEN LOOK [SUN 1990] and SAA/CUA [IBM 1992]) or defined by consensus within the software industry (see OSF/Motif [OSF 1994, Berlage 1995]). The second type comprises a series of standards [ISO 9241 1996, Pts. 10–17] which are devoted to aspects of human–computer dialogues such as menus or **direct manipulation** but do not include any guidance for implementation, nor do they involve toolboxes or programming interfaces.

As usual in the standardization of interface components for system-to-system interaction, the existing technologies of different companies get integrated by consensus—for example, so-called computer-aided design (CAD) frameworks to link CAD tools or protocols to ensure that different applications can interoperate at runtime. This kind of standardization is aimed at reducing both development costs and time to market, an approach mainly technology-driven. One may therefore call this type of standard a *technical* standard. The second type of standard, the standards for user interfaces, may appropriately be called *ergonomic* standards. Hence, the design object is far more than purely technical, since the interface component is not located between technical components but between user and system. Standards for user interfaces hold for a variety of user-interface products in all kinds of systems and applications.

The major difference between these two types of standards is that ergonomic standards do not merely rest on the technical state of the art, but rather their development is predominantly determined by the state of evidence obtained from ergonomic research [Dzida 1989]. The impetus for developing ergonomic standards is to achieve an acceptable level of ergonomic **quality** according to the current state of knowledge in the community of ergonomists. State-of-the-art knowledge, however, cannot be provided by consensus among industrial partners. The OSF/Motif standard is thus a surprising exception to the rule that interfaces between humans and technical systems are chiefly determined by the ergonomic tradition of thought, empirical evidence, and agreement between ergonomic experts. The OSF standard therefore does not automatically provide an ergonomic interface; for a discussion of further objections to standard interfaces see Potosnak [1988].

Standards may also be distinguished by their scope of validity, which ranges from locally defined proprietary guidelines to international standards or directives. Within a software company the user-interface designer may be required to stick firmly to the OSF/Motif style guide and the associated toolkit, thereby assuring that for different applications similar interfaces are implemented. Consistency in design can be the main benefit for the user; for instance, the Common User Access (CUA) concept within IBM attempts to achieve software compatibility for user interfaces of all IBM computer systems, from micros to mainfraimes [Pangalos 1992].

The international standards pursue more ambitious objectives than proprietary guidelines. They provide a minimum level of quality (i.e., usability) that should be respected within an international market. Accordingly, the European Council Directive [ECD 1990] requires by law that interface design adheres to such a minimum level of ergonomic quality, so as to establish a harmonized level of work conditions across all member states of the European Union [CEC 1990].

A lot of people are concerned about whether too much standardization in the user-interface area might stifle the designer's creativity and competition between companies in the same way that copyright might stifle them [Samuelson 1995]. Some of this criticism may apply to early approaches to ergonomic standardization and may still be valid for standard interfaces. However, the international ergonomic standards developed by the International Standards Organization [ISO TC 159/SC 4/WG 5 1996] cannot be viewed that way. This group includes nearly all industrial countries involved in the development of information technology, with each of the countries being invited to contribute recommendations and to vote for or against the drafts elaborated by experts in the field.

Sometimes standards have also been suspected of freezing a certain state of technical development. The innovative progress in user interface technology, however, should not be impeded by the ergonomic type of standards, since standardized product attributes are neutral toward any specific implementation. However, standard user interfaces may have a conservative effect in the long run, unless new standard interfaces can outrival the established ones. The freezing of a state of technology may in fact be favored, if a standard interface is widely accepted as a reference for high quality.

This chapter focuses on the international software-ergonomic standards and their relevance in the quality assurance of software products. A methodology of testing products for compliance with international standards is of central concern, since a standard can contribute to an improvement of quality only if the methods of quality definition and testing are well accepted. This raises another issue of internationalization: agreement on a recognized methodical approach enabling designers as well as customers to check whether a user interface is at least in conformance with the minimum level of ergonomic quality prescribed by the standards. Although ergonomic standards suggest a procedure for requirements specification as well as for conformance testing, the "level of specification of the procedure is a matter of negotiation between the involved parties" [ISO 9241-14 1996, par. 4.3]. Relying only on such negotiations, however, will make it impossible to compare the quality of different products.

The approach presented in this chapter is an abstraction of a variety of conformance testing procedures. The approach is called criterion-oriented conformance testing, drawing on the recommendations of Weiss [1972]. The method has strong linkages to a research tradition known as the **evaluation** of existing or innovative social programs; see, for instance, Rossi and Freeman [1989]. The criterion-oriented approach can also be applied in the evaluation of software products. Based on the long tradition in evaluation research it should be possible to adopt this approach in usability testing and to establish an international consensus on methods of conformance testing during the **design-use cycle**.

67.2 Underlying Principles

Evaluation of **usability** has achieved a high level of development during the last decade. This chapter describes some methodological foundations which generally apply to the ergonomic evaluation of software and which may be of particular concern when testing products for conformity with standards. Key methodological questions are what portion of the user interface is under study, what are the leading design objectives (principles of quality), how is usability embedded into a general quality model of software, and how are usability requirements structured in the international standards, so as to easily find them and to operationalize them for design decisions or **compliance tests**. Finally, the liability of ergonomic standards in an international market is addressed, especially from the perspective of the European Union.

User-Interface Reference Models

To structure the rather complex user interface, which couples the user to the application program, one may specify areas of an interactive system particularly apt for software-ergonomic design and evaluation. A number of human–computer interaction reference models have been published; for a survey see Spring et al. [1991]. Some of the reference models are devoted to a functional specification, some describe system architectures, and others provide conceptual models [Norman 1988] for users. Conceptual models include a specific interface model, layer models [Fähnrich and Ziegler 1985], and linguistic models of interaction [Moran 1981]. An interface may be described by a set of rules (or attributes) that determine the interchange of data (information) between user and computer. The interface model was developed by the International Federation for Information Processing [IFIP] Working Group 6.5; for the original model see IFIP [1981], and for its formalized description see Dzida [1987, 1988].

However, designers of interactive systems confine their concept of an interface to only one component (i.e., the input/output interface), which is characterized by facilities of data input as well as attributes of data presentation (such as grouping and coding of data, or echoing keystrokes and mouse clicks). The designers' conceptual models (e.g., the MVC model [Goldberg 1990] or the PAC model [Coutaz 1987]) do not necessarily fit well with the users' conceptual model of an interface. Incompatibility was uncovered particularly for object-oriented models of interactive systems [Wegener 1995], which aimed at taking better account of the reusability of interface components. Abstract data types representing the input/output interface (e.g., push buttons, menus, icons) can be well separated as far as they are neutral toward the application. But data types for the **dialogue** have been treated either as abstractions of input/output or application. Separate data types for the dialogue thus appear to be irrelevant for interface designers, although this concept remains significant from the user's perspective. As a consequence, the designer may develop an interactive system having a concept of interface in mind which may be more restricted than the user's model.

The IFIP Reference Model

The IFIP user-interface reference model (Fig. 67.1) had some impact on the structure of international software-ergonomic standards, insofar as ergonomic principles of design could be grouped according to the conceptual components of the user interface.

The interaction of the user with the system is enabled by an interface, which can be structured into three interface components, each representing a separate aspect of interaction: input/output; the conduct of dialogue; and access to tools, services, or data. This structure can be amended by a fourth component for involving a specific type of interaction; message exchange in an organization or within a computer network (see the organizational interface). An advantage of the model is its focus on four relatively independent concepts appropriate to structure the interface. The focus facilitates communication between users and designers on usability issues.

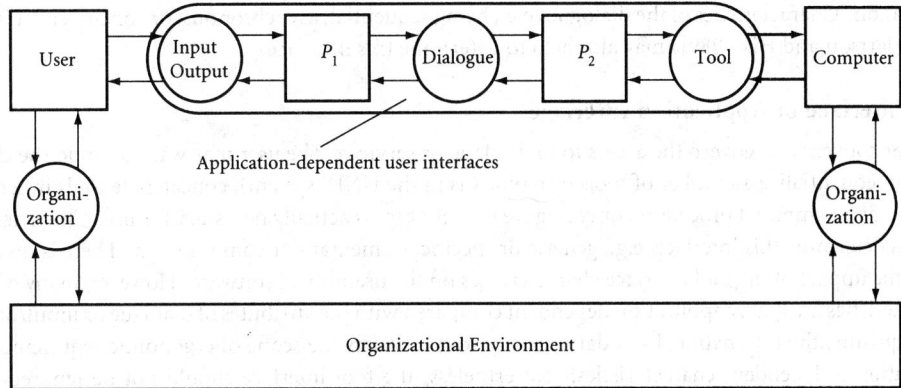

FIGURE 67.1 User-interface reference model of IFIP WG 6.5. (*Source:* Dzida, W. 1987. On tools and interfaces. In *Psychological Issues of Human Computer Interaction in the Work Place.* M. Frese, E. Ulich, and W. Dzida, eds., pp. 339–355. North-Holland, Amsterdam. With permission.)

In Fig. 67.1 circles indicate interfaces, rectangles represent activities (processes), and arrows point to the direction of dataflow.[1] From the user's point of view the four interface components are of central concern, but the user may also be interested in the processes (e.g., P_1 and P_2) representing software components of the user interface provided by a user-interface management system (UIMS) or application program. For instance, P_1 may realize a user input and then react by echoing the input; P_2 may prompt the user for further data input or signal an input error. P_1 and P_2 do not necessarily induce a data processing by the application program, but they do change the display state. A separation of display-related interaction and task-related interaction can be introduced this way. This is a necessary conceptual separation, as the user is interested in the effects of an input (i.e., whether it solely changes the attributes of the screen or causes a transition from the current data state of a task at hand into a new data state). A feature of the model that relies on the Petri net notation is the assimilation of three components into one interface, indicating that the user is virtually faced with one interface that presents three of the interface aspects simultaneously before or after running the application program. (See Fig. 67.1.)

Before design principles are discussed for the user-interface components their functions and attributes should be described.

Input/Output Interface

This part is often referred to as the *surface* of the software. Rules for user input and system output govern the interface—for instance, movement and positioning of the cursor by mouse or arrow keys, placement of a pop-up menu or an icon, size of a window, highlighting of a menu option, and color codings. The design of this part of the user interface is predominantly captured by today's tool boxes or UIMSs. The user interface is sometimes confined to this interface component, i.e., to the tangible surface characteristics. This is, of course, too narrow a concept.

Dialogue Interface

Interaction with the system is dialoguelike, as the user receives information and can control the system by means of an interface language. This language includes command names, letters of short-cut keys, symbols, direct manipulation, voice input, and gestures. Furthermore, the data exchange necessary for conducting a task characterizes the dialogue, which includes activities such as prompting, interrupting, switching to another window, resuming an editing process, and changing the font size. Also, data exchange is necessary after task accomplishment, such as error messages, warnings, system messages, and help

---

[1] The syntax of the model uses a type of Petri nets called channel agency net [Reisig 1985].

infomation. Characteristics of the dialogue (e.g., being sequential, asynchronous, or concurrent [Hartson 1989; Hartson and Hix 1989]) may also help to determine this interface.

Tool Interface or Application Interface

Rules or conventions govern the access to tools, data, or services. The user may want to undo the change of data, sequentialize a number of tools by a pipe (as in the UNIX system), concatenate tools in terms of a macro or a command procedure, or configure the set of tools actually necessary for use. Characteristics of tools determine this interface, e.g., generic or specific, elementary or compounded. There is no doubt about the impact of the tool interface characteristics on the usability of software. However, many of these characteristics are highly application dependent compared with the attributes of dialogue or input/output. Consequently, the international standardization group restricts the scope of ergonomic requirements to application-independent characteristics. Nevertheless, the tool interface should not be ignored when evaluating the usability of software beyond the ergonomic standard requirements.

Notably, the IFIP user-interface reference model has influenced the Seeheim model of interactive systems [Pfaff 1985], which is an approach to develop a system architecture that is still effective in usability engineering for the development of UIMSs [Olsen 1992].

Principles of Design

The interface components of the reference model can help apply the principles of ergonomic design. They represent factors (objectives) of usability, the achievement of which is verified when a number of conceptually coherent user requirements has been satisfied.

The following principles of information presentation pertain to the input/output component of the interface [ISO 9241-12 1996]:

- Clarity
- Discriminability
- Conciseness
- Consistency
- Recognizability (detectability)
- Legibility
- Comprehensibility

The following priciples of ergonomic dialogue design are published in ISO 9241-10 [1996]; for the empirical basis of these principles see Dzida et al. [1978]:

- Suitability for the task
- Self-descriptiveness
- Controllability
- Conformity with user expectations
- Error tolerance
- Suitability for individualization
- Suitability for learning

Although the international ergonomic standards rarely involve ergonomic requirements for the design of the tool interface, a list of such principles is presented as a suggestion; see also ISO 9126 [1994] and McCall et al. [1977]. In software development it is indispensible to adhere to these principles, so that the user can use the software effectively (see effectiveness as a factor in the quality model, Fig. 67.2; see also the definition of usability):

- Functionality
- Reliability

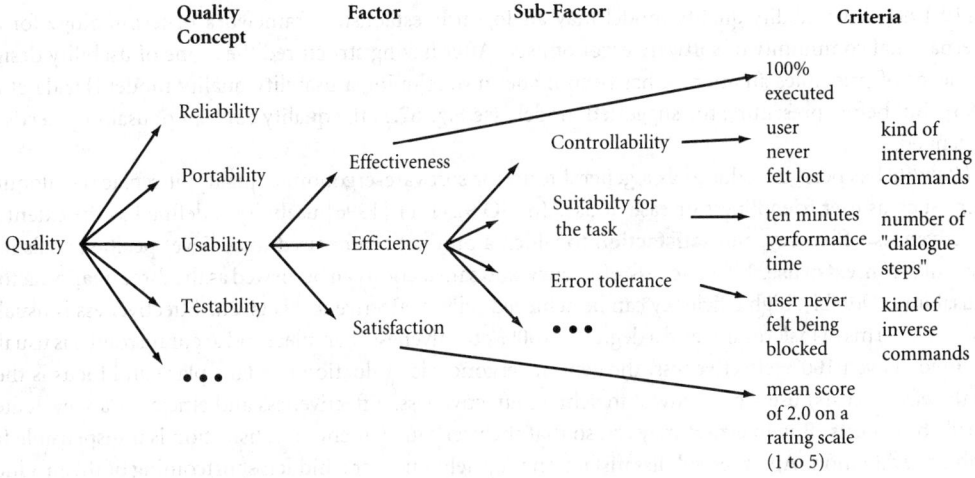

FIGURE 67.2 Usability quality model. (*Source:* Dzida, W. 1995. Standards for user-interfaces. *Comput. Stand. & Interfaces* 17:89–97. With permission.)

- Accessibility
- Reusability
- Portability
- Integrability
- Maintainability
- Suitability for configuration
- Suitability for customization

Characteristics of the organizational interface are highly application dependent and are therefore not an issue of software-ergonomic standardization; this interface will not be further discussed here. However, to be complete, as far as the design of tasks determines this interface (see Fig. 67.1, left interface), principles of task design [ISO 9241-2 1992] can help specify its ergonomic quality. Regarding the organizational interface as a medium between computers of an organization (see Fig. 67.1, right part), ergonomic principles of groupware design should be considered, which include suitability for cooperation, responsiveness, negotiability, and security [Wulf 1993].

The principles help the designer to structure the tradition of thought in ergonomics in abstract concepts, thereby developing a general understanding of user requirements. A specific requirement can be interpreted in terms of a principle, thereby guiding the designer and the user in achieving a common understanding. Principles may also help clarify tradeoffs and priorities between design decisions. Principles can also guide the reader of a standard to apply a specific requirement according to the implied design rationale. Requirements describing specific product attributes (e.g., highlighting of a selected menu option) should be realized in a way that the principle (e.g., of discriminability) is satisfied. A designer who simply obeys a standard requirement may fail to grasp the gist of it and may therefore develop an inappropriate solution. A design problem should not be solved exclusivly by means of a schema or checklist or style guide; rather, the solution should be based on a design rationale [Carroll 1990], which can be a principle derived from a theory of human task performance or human perception.

Usability Quality Model

Last but not least, principles of design contribute to the development of a quality model. A number of software engineering quality models have been discussed; see for instance Boehm et al. [1976] and McCall et al. [1977]. Results of this discussion have been crystalized in an international standard [ISO

12119 1994]. A usability quality model may analogously establish a framework of terminology for an international community of software ergomists. After having structured the scope of usability design by means of principles an attempt has been made in developing a usability quality model [Dzida et al. 1993]. But before presenting the suggested model (see Fig. 67.2) the quality concept of usability needs to be defined.

Usability has been introduced as a general term for software-ergonomic quality; it replaces colloquial terms such as user friendliness or ease of use. In ISO 9241-11 [1996] usability is defined as the extent of **effectiveness, efficiency**, and **satisfaction** to which a product can be used to achieve specified goals in a particular **context of use**. Effectiveness, efficiency, and satisfaction can be viewed as the three quality factors of usability. Notably, high efficiency can be achieved only if effectiveness is given. Effectiveness is usually defined in terms of **task** results, and a degree of 100% effectiveness (complete and accurate result) is usually required. Given 100% effectiveness, the genuine ergonomic evaluation can take place and focus is then on the effort (costs) users must invest to achieve effectiveness. Effectiveness and efficiency are evaluated mostly by experts. But an expert may err, so that the user's judgement on satisfaction is indispensible for usability evaluation. An observed dissatisfaction may help uncover a hidden shortcoming of the product; for measuring subjective usability see Kirakowski and Corbett [1993] and Kirakowski [1995].

The quality model (Fig. 67.2) involves the concept of usability as one concept among many which determine the overall quality of software. The software-engineering concepts (e.g., reliability and testability) determine the degree of effectiveness. The model describes a stepwise operationalization of efficiency, which ends up with a specific measure of usability, called a **criterion**.

Usability Criterion

A usability criterion is defined as a required level of measure, the achievement of which can be verified. To verify this level, the concept of usability is decomposed into its constituent factors: effectiveness, efficiency, and satisfaction. As previously mentioned effectiveness is assumed to be satisfied by the presence of all of the other quality concepts: reliability, portability, etc. Hence, a criterion of 100% effectiveness is postulated as a an inevitable basis for an ergonomic specification of efficiency criteria. Usually, these criteria are derived from subfactors (also referred to as principles) of efficiency. Two types of criteria should be specified during usability requirements analysis:

- Effects of **task performance** observed at the user interface or human outcomes resulting from task performance
- Product attributes

As an example of a required effect of task performance consider the echoing of the selection of a menu option to the user. Note that, although echoing is an effect of task performance, it is not the intended (complete and accurate) final task result.

As an example for a required human outcome, consider the ability of a user to discriminate between active and nonactive menu options. Note here that discrimination is the required outcome.

As an example for a required product attribute, consider the use of brightness coding (highlighting) for menus to facilitate discrimination between active and nonactive menu options. Note that the highlighted menu option is an attribute.

One may argue that echoing a user input is just as much of a product attribute as highlighting a menu option. Indeed, an effect of task performance can be a user-interface attribute. The difference, however, can be seen in the relation of an attribute to a user's activity (performance). When an attribute appears on the display as a consequence of such an activity, it is taken as an effect in the interaction. Effects are, for instance, echo, prompt, error or help message, alert, selection and execution in one dialogue step, and positioning of a cursor. If an attribute is used to design for such an effect, it is taken as a product attribute. Highlighting, for instance, can be used as an attribute to design the system's echo (an effect) to the selection of a menu option (a user performance). Although the distinction between effects and attributes may appear a bit artificial, it will become more useful when we deal with the evaluation of attributes which are task related and those which are neutral (see section on conformity). Effects are always task related. An analogous distinction can be made between human outcomes and attributes. Discrimination is an outcome which can be enabled by an attribute such as highlighting.

These examples illustrate how different kinds of usability criteria may complement each other in a design decision. Hence, when defining the user's list of requirements it may be unsatisfactory to exclusively use product attributes as criteria, since the designer may regard a required attribute as out of date or inconsistent with other design decisions. If the user simply requests a specific attribute, the designer may not know why. The definition of criteria, therefore, should consider the effects of task performance at the user interface and human levels. It should be up to the designer to select an adequate product attribute that fits well with the performance criterion. This is the way standard requirements are intended to be interpreted by the designer.

The purpose of design can also be concisely expressed in terms of usability criteria. A fully detailed design rationale rests on an analysis of the context of use [ISO 9241-11 1992] and the standard task requirements [ISO 9241-2 1992]. A user interface may provide numerous high-tech attributes but the work performed at this interface may nevertheless be of low ergonomic quality. This may be caused by the fact that the tasks are designed without regard for basic ergonomic task requirements. The poor design of a task, of course, accounts for the poor ergonomic quality of the product's context of use but does not bring the usability of a product into discredit. But a really human-centered approach considers not only the design of user-interface attributes, but also the human conditions of work and organization. It is worth mentioning that the introduction of information technology can have effects on the content of jobs and individual interdependencies in an organization. These changes should be taken as an opportunity for redesigning tasks and developing organizations according to ergonomic task requirements (for a model of quality improvement see Fig. 67.4). ISO 9241-2 [1992] not only requires task design to facilitate tasks but also recommends that task design provides an appropriate degree of autonomy to the user in deciding on priority, pace, and procedure. The features of such a task can be mirrored by user-interface attributes of controllability (see dialogue principles).

Structure and Content of the Standards

The quality model may clarify the structure of software-ergonomic standards shown in Fig. 67.3. Similar to this model the standards are hierarchical, beginning with usability as the most general quality concept (Pt. 11), continuing with principles (Pt. 10 and 12), and ending with a series of standards (Pts. 13–17) devoted to specific forms of dialogue. With the quality model in mind, the reader can rapidly find an appropriate standard requirement at the intended level of abstraction. If a specific standard does not provide the guiding information to make a design proposal, the reader should ask for information in one of the two principal standards (dialogue or infomation presentation). The principles encourage freedom of design but remind the designer to meet design principles as closely as possible. In Fig. 67.3 the principal standards, along with the specific standards on dialogue techniques, form the core of usability requirements. Their interpretation is based on task requirements (Pt. 2), which can be derived from a context of use analysis (Pt. 11).

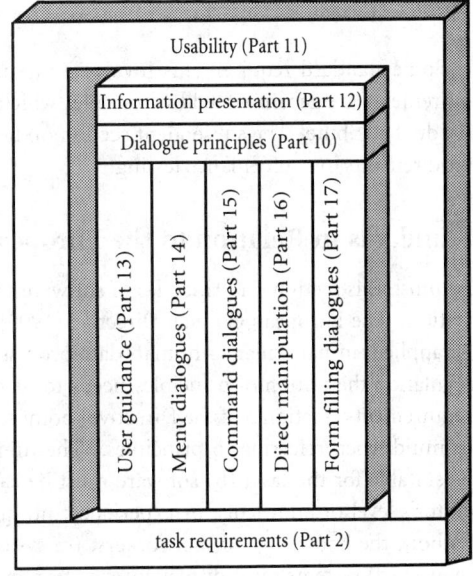

FIGURE 67.3 A structure of the software-ergonomic parts of ISO 9241. (*Source:* Dzida, W. 1995. Standards for user-interfaces. *Comput. Stand. & Interfaces* 17:89–97. With permission.)

Table 67.1 provides a list of all standards to appear unter the title ISO 9241 [1996]: Ergonomic requirements for office work with visual display terminals (VDTs). The specific software-ergonomic parts of this standard are Pts. 10–17.

The differentiation between types of usability criteria helps identify what type of information is provided by a specific standard requirement. Having the types of criteria in mind the reader will immediately know how the requirement can be interpreted and which information needs to be acquired for checking its applicability. A standard contains:

- A required effect of a user's task performance
- A required human outcome
- A product attribute

TABLE 67.1 Parts of ISO 9241, Titles and Status (January 1966)

Part	Title	Status[a]
1	General Introduction	IS
2	Guidance on task requirements	IS
3	Visual display requirements	IS
4	Keyboard requirements	DIS
5	Workstation layout and postural requirements	DIS
6	Environmental requirements	CD
7	Display requirements with reflections	CD
8	Requirements for displayed colors	DIS
9	Requirements for non-keyboard input devices	CD
10	Dialogue principles	IS
11	Guidance on usability	DIS
12	Presentation of information	CD
13	User guidance	CD
14	Menu dialogues	DIS
15	Command dialogues	CD
16	Direct manipulation dialogues	CD
17	From filling dialogues	CD

[a]CD: Committee Draft, DIS: Draft International Standard, IS: Standard. (*Source:* Dzida, W. 1995. Standards for user-interfaces. *Comput. Stand. & Interfaces* 17:89–97. With permission.)

Here are some examples from different standards:

- *Effect of task performance:* "The user actions required to move the cursor from one entry field to the next should be minimized" [ISO 9241-17 1996, par. 6.1.1].
- *Human outcome:* "In order to enable direct manipulation of objects, the system should have areas which can be easily recognized and discriminated by the user ..." [ISO 9241-16 1996, par. 6.2.6].
- *Product attribute:* "If command input is typed, command words should generally not exceed 7 characters" [ISO 9241-15 1996, par. 6.1.4].

Some standard requirements involve a mixture of these three types of information. Nearly all requirements are accompanied by examples which illustrate the requirement in terms of an implemented product attribute. The subsequent section on best practices explains how to deal with types of standard requirements in conformance testing.

Standards in Relation to the European Council Directive

To finish this section on principles of software-ergonomic quality we consider an issue of internationalization. The European Council Directive [1990] requires: "the principles of software ergonomics must be applied, in particular to human data processing" [ECD 1990, p. 18]. For an interpretation of this regulation the notion of principles needs to be clarified. Further requirements in the list of minimum requirements [section 3 of the Directive] point to the background information the authors probably had in mind when referring to principles. The minimum requirements are as follows "(a) software must be suitable for the task; (b) software must be easy to use and, where appropriate, adaptable to the operator's level of knowledge or experience; no quantitative or qualitative checking facility may be used without the knowledge of the workers; (c) systems must provide feedback to workers on their performance; (d) systems must display information in a format and at a pace which are adapted to operators; ..."[ECD 1990, p. 18].

The following is an interpretation of the requirements in terms of principles: Point (a) corresponds to the dialogue principle suitability for the task; although the system's functionality is also addressed. Point (b) is a threefold requirement; the required adaptability corresponds to the dialogue principle suitability for individualization; it also addresses the system's customization. Point (c) corresponds to the dialogue principle self-descriptiveness. Point (d) corresponds to the dialogue principle controllability; although other principles of information presentation are also addressed.

To sum up, when adhering to the principles of software ergonomics the practitioner is well advised to adopt the ISO principles of dialogue [ISO 9241-10 1996] as well as the principles of information presentation [ISO 9241-12 1996].

An additional issue is the legal obligation of the European Council Directive previously listed. All member states of the European Union must transpose the Council Directive into national law. Certainly, national legislation will at least indirectly refer to these international standards, because the European minimum requirements correspond to the minimum ISO requirements. Although an ISO standard originally has the status of recommendation, all standards will be conceived as obligatory requirements within the European software market [Stewart 1992].

Concerning the software-producing companies outside of Europe, there is a question of how they will cope with the specific European standards if they want to deliver products to this market. Software producers will establish European service companies to intensify the contact with the customers as regards requirements analyses, system adaptations, and conformance testing at the users' workplaces [Keil and Carmel 1995]. The next section on best practices is an attempt to give detailed answers to these issues.

67.3 Best Practices

Software-ergonomic standards will be of no value if designers and evaluators do not know how to apply them. Members of the standardization committee repeatedly asked designers and evaluators to judge the applicability of standards. Complaints about the difficulty of interpreting the standards were raised. The main difficulties were in testing products for compliance with the standards. Another concern was raised about the occasions upon which standards must be applied. This section provides help to read and interpret requirements, suggests occasions for conformance testing during the design-use cycle, describes the provisions to be made for testing, and explains how tests can be conducted.

Standards as Guidelines

Software-ergonomic standards are mostly formulated as guidelines rather than precise specifications. The problem with guidelines is that they may become either too detailed and voluminous or too brief (and thus overly generic). The authors of the most comprehensive collection of guidelines predict that designers may be disappointed if they look to guidelines for specific rules but find only general advice instead [Mosier and Smith 1986]. Although the standard guidelines represent the present state of an ergonomically accepted technology, the product attributes involved in the standard lack exact quantitative values. This lack of exactness may make it difficult for the designer to obtain a compliant proposal for a specific design decision, and it may also make it difficult for the evaluator to check a product feature for compliance with a guideline.

Nevertheless, this characteristic of the guidelines need not be regarded as a drawback. The very opposite may be true, in fact, because the guidelines are not at all unclearly stated for a reader who acquires the user requirements as a prerequisite for interpreting the guidelines. This is similar to interpreting another type of standards, the national or international laws, which also require the reader to apply them to a specific state of affairs and its context. To judge conformity, the reader of an ergonomic standard should not expect that a paragraph of a standard can easily be compared with a product attribute. The subsequent text outlines why software-ergonomic standards are like guidelines, what the advantages of guidelines are, and how criteria can be determined so as to enable the designer or the evaluator to apply a standard.

Software-ergonomic standards are formulated as guidelines for several reasons:

- Freedom of design is warranted, i.e., the standard does not specify any particular product attribute.
- The requirements do not imply any specific implementation.

- The requirements do not imply any specific user or user target group.
- They do not presuppose any specific task or organizational setting.

Certainly, the standardization committee should not specify requirements for specific target groups, tasks, or contexts of use. This would produce a tremendous proliferation of standards that would be practically unmanageable. Software-producing companies would protest against standardized attributes, since it is the style of such attributes that should establish the appearance of a product as being unique to a specific company.

Context of Use Analysis

The advantages of the guidelinelike character of standards are achieved at the expense of readers who want to apply them for design or evaluation. These readers first have to define user requirements to provide the basis for an interpretation of a guideline. ISO 9241-11 [1996] serves as a standard that guides the reader in analyzing the context of use of the product to specify the user requirements. Additional guidance for usability context analysis can be retrieved from Bowden and Thomas [1995] and Bevan and Macleod [1994]. Essentially, this analysis provides a specification of user characteristics, their goals and tasks, the equipment (hardware, software, and materials), and the physical and social environments in which the product is or will be used. The specification of users' goals can be converted into a list of user requirements. Other information about the context of use provides the background to explain the ergonomic rationale for the design requirements. In practice, the list of user requirements can become a voluminous document. To give it some structure the designer or the evaluator may ask that the requirements be assigned different weights, so as to resolve tradeoffs or clarify other considerations. Of course, the designer attempts to satisfy all of the requirements and will therefore use the complete list of requirements during the design cycle. The evaluator, however, may want to restrict the scope of evaluation in response to a specific request in order to limit the costs of evaluation. The test of conformity with a standard is a special case of restricted evaluation, because conformance tests consider only the minimum level of usability defined by the standard.

The quality (or usability) of a product cannot be measured in absolute terms. These are always relative to the particular requirements of the context of use. Usually, a product is expected to provide a level of usability higher than the standard requires. The usability of a product is thus relative to this level of user requirements, which may be a matter of negotiation between producer and customer.

The Design-Use Cycle

The analysis of the context of use will normally take place prior to the start of a design cycle. In practice, the starting point is not always clearly defined, except a new product is planned and some software may already be in use. Either a new release of this software is in preparation or the customer is going to choose another product that replaces the current one. Hence, besides the design cycle, an application phase needs to be considered, in order to keep in touch with the reality of software use. A model of the design-use cycle (Fig. 67.4) is introduced to illustrate that software use is regarded as part of an evolutionary process aimed at improving the product by taking into account customer experience. The motivation for this model rests on the fact that many requirements emerge only after the users have gathered some experience with a product. An advantage of the model is that negotiating partners (i.e., producer and customer) can recognize that they have shared responsibilities (see the highlighted areas in Fig. 67.4) during the design-use cycle, which the developer should consider for quality management and the customer should be aware before signing the contract.

The customer's responsibility (in his role as an employer) is addressed in the European Council Directive [ECD 1990]. Article 3 requires an analysis of users' workstations in order to preventively evaluate conditions which may cause users to complain about physical problems, health problems, or mental stress. The test of a product for compliance with international standards could be used as a preventive measure as required by the Directive. The customer will require such a test to be performed in advance by the developer.

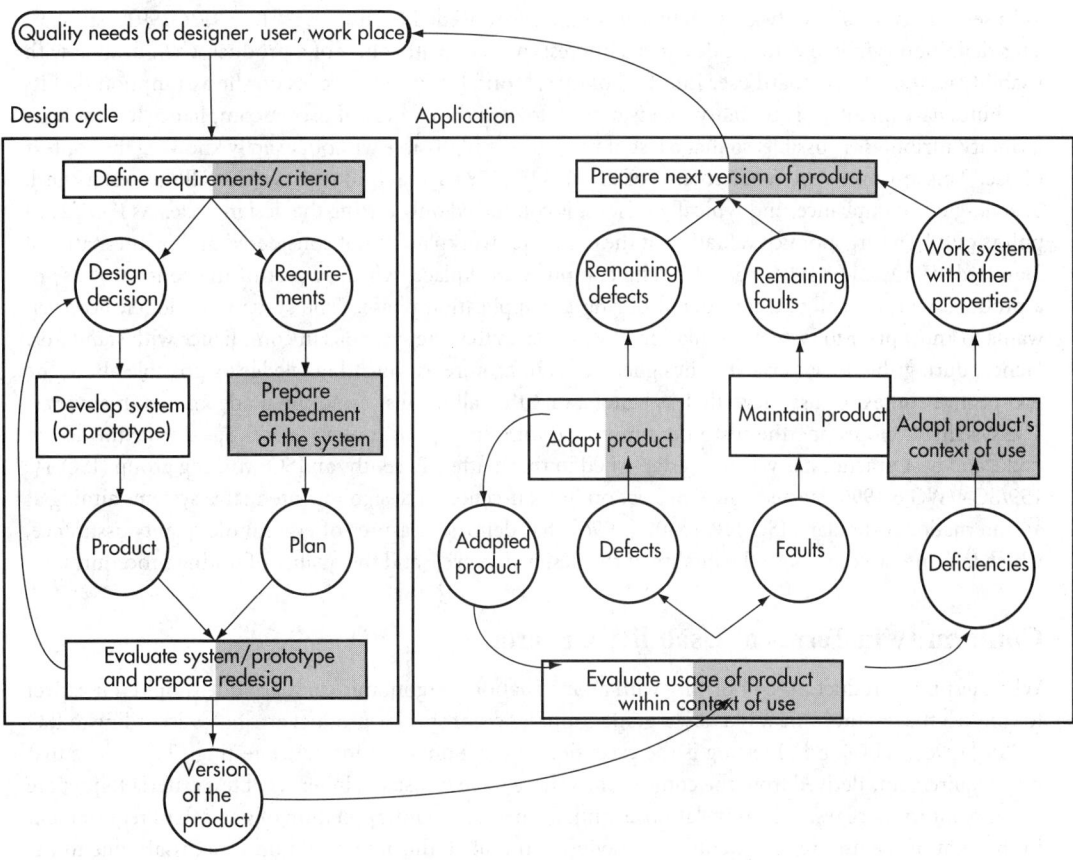

FIGURE 67.4   Model of the design-use cycle.

However, as the model (Fig. 67.4) points out, subsequent usage of a product can also uncover **defects** and **faults**, which would soon initiate an adaptation or a redesign of the product.

In order to manage these quality problems, "effective communications should be maintained to encourage users to discuss their concerns and to ensure timely and effective organization responses" [ISO 9241-2 1992, par. 5]. An organization may respond either by adapting the product or by adapting the product's context of use, so as to embed the system more properly. After the product has been adapted to the user needs, the adaptation should be tested for conformity with the standards, taking into account that the customer will be reponsible for making that test.

After organizational deficiencies have been remedied, new user requirements will evolve, which may give rise for a redesign of the software. From experience we know that organizational conditions of work do change during the useful life of a product. To meet the needs of organizational development the model of the design-use cycle takes account of quality improvements to be evolutionarily achieved (for the software-engineering origin of the model see Floyd et al. [1989]). From the evolutionary character of this process, it is clear that a product's conformity with standards is not achieved once and forever but requires periodical retests during redesign and application. Finally, conformance testing should be introduced as an inevitable measure of quality assurance according to ISO 9001 [1987] and ISO 9000-3 [1991]; for a first interpretation of ISO 9000 [1987] standards as regards usability see Dzida et al. [1993].

Conformity in Terms of Product Attributes

Our model of the design-use cycle can help clarify a confusion caused by two kinds of usability definitions; see the previous definition of ISO 9241-11 [1996] and also ISO 9126 [1991]. ISO 9126 defines usability as "a set of attributes of software which bear on the effort needed for use . . . ." [ISO 9126 1991, par. 4.3]. This definition is in line with the designer's interest to stick to attributes of a product when dealing with usability issues. The standard user interface of OSF/Motif, for instance, relies on the assumption that its attributes have an intrinsic usability that is neutral toward the context of use. A compliance test for such usability attributes is possible, so that a test of the product is possible without exactly knowing the context of use. This test meets the requirements of ISO 12119 [1994]. Practitioners used to call this test a check of style-guide compliance, and typically this test is conducted only during the design cycle. As ISO 12119 points out: "The ergonomic evaluation at the computer workplace is not considered in this International Standard" [ISO 12119 1994, par. 4]. The computer workplace, which is part of the context of use of a product, will normally be considered during the application phase. The system developer, however, wants to know prior to system installation at workplaces that the system is in compliance with standards. Hence, during the design cycle the designer needs to acquire as much knowledge as possible about the prospective context of use, nevertheless being aware that all details cannot be made known in advance. The system developer and the customer must cope with the gaps in context knowledge during the design cycle, and best practice has yet to be established in this matter. Recently, an ISO working group (ISO TC 159/SC 4/WG 6 1996) started a new project on human-centered design for interactive systems aiming at an international standard ISO/NP 13407 [1996] that defines measures of ergonomic quality assurance, which take into account both the quality of the design-use cycle and the quality of testing procedures.

Conformity in Terms of Usability Criteria

When testing a product for conformity with an international ergonomic standard one should not expect to conduct the test the same way as style-guide compliance for specific product attributes according to ISO 12119 [1996]. This kind of testing is the exception in ergonomic conformance testing. The rule is that user requirements derived from the context of use have to be translated in terms of test criteria to prepare for the conformance test. The translation essentially involves an interpretation of a standard requirement in the light of the user requirements. As previously outlined, this interpretation is necessary due to the fact that standard guidelines are formulated neutrally toward users, tasks, and environmental conditions. In principle, there are three possibilities to determine the criteria:

- The standard requirements genuinely represent the criterion (which is the exception). This holds for a required product attribute; for example: Help explanations should vary in type and length [ISO 9241-10 1996]. Type and length are attributes which can be inspected regardless of task and users; i.e., a context of use analysis can be skipped.
- The standard requirement has to be specified with regard to the task at hand. Let us assume that the user wants to avoid the recurrent input of the same data, which instead should easily be available on the display after the first input. The standard [ISO 9241-10 1996] contains a corresponding requirement concerning default values, but this needs to be interpreted in the light of the real task, which may be, for instance, the task of a CAD engineer who works on modeling the geometry of a steel girder and is repeatedly concerned with the values of its flange. The CAD system conforms to the standard if it presents the flange values as defaults.
- The standard requirement has to be specified with regard to task and user needs; this holds for a required human outcome resulting from task performance, for example: "Explanations should assist the user in gaining a general understanding of the dialogue system . . . ." [ISO 9241-10 1996, par. 3.2]; understanding can be assessed only in view of the user and the task. For a user who interacts with an integrated office system the general level of understanding should be much higher than for a user who simply applies a form-filling dialogue. After having determined the required level of understanding, the criterion is specified and the test for conformity with the standard is well prepared.

The software-ergonomic standards, especially Parts 13–17, introduce the concept of the **conditional requirement**, which is a special form of the criterion-oriented approach. A conditional requirement is a sentence formulated in terms of an if-then rule, thereby structuring the sentence into two components, a conditional part and a subsequent guideline part. For example, ISO 9241-14 [1996] contains the following conditional requirement for menu options: "If options can be arranged into conventional or natural groups known to users, options should be organized into levels and menus consistent with that order" [ISO 9241-14 1996, par. 5.1.1]. The if-clause refers to the condition of applying the guideline. Obviously, the task at hand has to be analyzed and the users have to be interviewed in order to determine the criterion, which is a specific interpretation of the guideline.

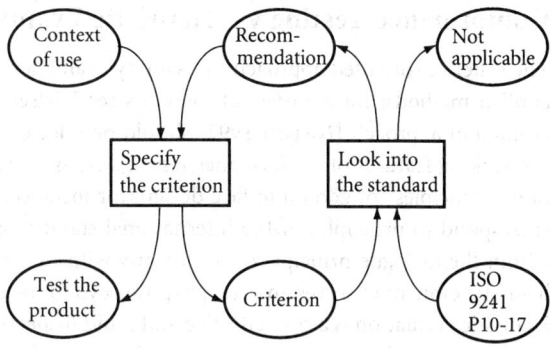

FIGURE 67.5 Conversion of a standard requirement in terms of a criterion.

The criterion-oriented approach on conformance testing is organized as a process mainly determined by two activities: specify the criterion and look into the standards. (See Fig. 67.5.) First, the criterion is determined as regards the requirements of the context of use. If applicable, the criterion is then specified in view of a standard recommendation (set minimum level). Finally, the criterion is compared with relevant product attributes (test the product).

As previously mentioned, the standards require the minimum level of quality. It may occur that the criterion derived from the user requirements includes a higher level of quality than the standard requires. Conformity with the standard is achieved if the usability attribute of the product equals the minimum level of quality.

Some parts of ISO 9241 [1996] do not explicitly phrase requirements or recommendations. For instance, Part 10 (dialogue principles) provides applications of principles which are formulated like guidelines with the exception that the if-clause is not repeatedly formulated to avoid redundancy. For example, the principle of suitability for the task implies that "Help information should be task dependent" [ISO 9241-10 1996, par. 3.1]. Although no conditional clause is explicitly included, it is evident that this guideline can be applied only under the condition that the task has been investigated, so as to specify the criterion prior to the conformance test. When adopting the criterion-oriented approach, conformance testing is possible for all guidelinelike standards [Dzida 1995] regardless of what turn of phrase is used in the standard (conditional requirement or application of a principle).

Figure 67.5 summarizes the steps in interpreting each guideline of a standard. As a final result, a checklist of usability criteria can be achieved, which represents the scope of the conformance tests to be done. The major advantage of this checklist is that it can be made part of the test report, thereby explicitly outlining the scope of conformity having been tested. The reproducibility of the test results is thus warranted. Documenting the conformity of a product in undifferentiated terms can be avoided. If a user requirement changes during the application phase of a product the evaluator can easily pick up the corresponding test criteria formerly used to decide whether a subsequent conformance test is necessary in view of the changed requirement.

The checklist containing usability criteria differs from the typical type of checklist frequently applied in usability evaluation. Typically, checklists contain a mixture of required product attributes and arbitrarily stated performance requirements. It seems to be quite convenient to adopt such a checklist in a variety of investigations regardless of the product under study or the context of use of the product. Opposite this type of checklist, the criterion-oriented checklist is validated for a specific context of use and is legitimized as each item in the list has a defined linkage to a specific task or user requirement. Comparing with this feature the major drawback of a typical checklist containing solely product attributes can be seen in the unspecified linkage of the attributes to user requirements.

Conformance Testing vs. Heuristic Evaluation

The criterion-oriented approach to usability evaluation should also be compared with other frequently applied methods; for a survey of methods see Nielsen and Mack [1994]. In particular, the heuristic evaluation approach [Nielsen 1992] should be selected for a comparison to contrast this method with the striking features of conformance testing. Nielsen classifies usability problems and allocates them to nine "principles" of good interface design, for instance, consistency, feedback, shortcuts. All principles correspond to principles of the international standards; for instance, consistency issues are subsumed within the dialogue principle of conformity with user expectations [ISO 9241-10 1996]. Nevertheless, heuristic evaluation serves another purpose beyond the test for conformity with international standards. Heuristic evaluation is a cost-effective and rapid inspection method like usability walkthroughs, with the main difference being that it is less formal than conformance testing. Inspection of the product can be taken as a *debugging* of usability defects, i.e., identification and diagnosis of the most striking and serious usability problems. Therefore, heuristic evaluation is particularly suited to the inspection of prototypes during the design cycle. Just as conformance testing mostly includes an inspection of the product, so does heuristic evaluation. The major difference lies in the question of what is inspected. A conformance test is an inspection of a required product attribute, i.e., a relevant attribute that fits well with a required effect of task performance or a required human outcome. A heuristic evaluation inspects a usability problem. This approach requires the expert to have the nine principles in mind as a vague quality definition to be matched with a number of product attributes being arbitrarily selected. The product is not necessarily inspected for a selected set of relevant tasks (i.e., a specific context of use is not addressed), although this could be managed by use of scenarios [Carroll 1995]. Hence, the output of heuristic evaluation is not a list of defects that violate specific criteria but that indicate more or less serious usability problems. A shortcoming of heuristic evaluation is that the identified usability problems cannot be fixed for a constructive solution, since it will then be necessary to know a clearly defined user requirement. The advantage of heuristic evaluation, however, is its effectiveness to uncover the major problems in a user interface. Thus, heuristic evaluation contributes to achieve the minimum level of quality, just as conformance tests do.

67.4 Research Issues and Summary

Many standard requirements need further development through empirical investigations. Research issues can easily be found in the standards by looking at the published edition of a Draft International Standard (DIS), which usually contains an annex providing references to source documents for each requirement. Three types of sources are distinguished there: research studies, guidelines, and experts' consensus. For example, the following requirement was selected from guidelines (see Galitz [1985]) to be put into the menu standard [ISO 9241-14 1996, par. 5.3.6]: "If the frequency of option use is known . . . and option groups are small (eight or less), the most frequently used options should be placed first." This requirement may be reasonable for a pull-down menu, but should not be generalized for pop-up menus without further empirical evidence. It may turn out that in a pop-up menu the most frequently used option is more adequately placed at the end of the menu panel, in order to meet Fitts' law (i.e., to minimize the time necessary for moving the mouse to a frequently selected target option). Readers of the ergonomic standards should be made aware of the rich potential for research issues that is raised by the need for empirical evidence to support standard requirements.

Usually, standards will be applied by experts in usability design and evaluation. A significant aspect is that expert judgement is largely based on accumulated practical experience rather than solid empirical data. Although for each requirement the empirical evidence is checked during the development of the standards, the underlying source document will usually be neglected by the practitioner. For many empirical studies, the related experimental results could not always be generalized that far as suggested by the standard requirement. Even if a requirement is based on solid data, a specific application may lead to a divergent interpretation because of the specific circumstances given in the context of use [Landauer and Galotti 1984]. The practitioner will not be concerned with such issues. For a professional researcher, however, the notion that design decisions may be based on judgement rather that on experimental data is repugnant [Smith 1986].

At the designer's workbench design decisions cannot be postponed until supportive data have been obtained from empirical research. Therefore, many design decisions could ex post be investigated in experiments, the results of which then may influence the revision of design and standards. Unfortunately, professional researchers are not yet sufficiently integrated in this process. Also, they may be reluctant to take part in design decisions, thereby accompanying the designer when obtaining from this experience the impetus for further experimental studies. Even if a laboratory researcher is willing to take part in this process, there may be good reasons not to be committed too much because of the limitations of laboratory research to study usability [Eason 1984].

Thus, we may question whether solid empirical foundations for standard requirements can ever be achieved. This is why a design decision, as well as the underlying guideline, will rarely be absolutely true or false but will be more or less appropriate relative to the requirements derived from the context of use of a product.

The series of usability standards [ISO 9241 1996, Pts. 10–17] provides a set of requirements for the minimum ergonomic quality of software products. Although the evidence for some of the requirements may be questioned and may need to be revised in a later review, the worth of the standards should not be underestimated. The set of requirements is based on a balance of different interests among partners in an international market and on a state-of-the-art knowledge in usability research as well. Although the standards are as yet merely conceived as recommendations (except for the European Union) they should be respected throughout the world as a baseline of usability, beyond which the software companies will have sufficient space for developing competitive high-quality products. There may be regions in the world where ergonomic aspects of products and quality of work do not yet play an important role. With the advent of international standards, a worldwide harmonization of work conditions at computer work places may become the most significant effect of this work in the long run.

Defining Terms

Compliance test: An operation that compares relevant attributes of a product with applicable standard requirements for determining the achievement of the level of quality required.

Context of use: The users, goals, tasks, equipment (hardware, software, and materials), and the physical and social environments in which a product is used [ISO 9241-11 1996].

Criterion: A required level of quality measure against which attributes of an object (e.g., a product) or of a process (e.g., the design cycle) are judged to evaluate the level of quality achieved.

Defect: An unintended attribute that impairs the efficient usage of a product.

Design-use cycle: The course of developmental changes through which a product passes from its conception, during its usage, up to the redesign or the termination of its use.

Dialogue: A process in the course of which the user, to perform a given task, inputs data in one or more dialogue steps and receives for each step feedback with regard to the processing of the data concerned.

Direct manipulation: A dialogue technique by which the user directly acts on objects on the screen, e.g., by pointing at them, moving them, and/or changing their physical characteristics (or values) via the use of an input device [ISO 9241-16 1996].

Effectiveness: The accuaracy and completeness with which users achieve specified goals [ISO 9241-11 1996].

Efficiency: The resources expended in relation to the accuracy and completeness with which users achieve goals [ISO 9241-11 1996].

Evaluation: All activities designated to assess the quality of an object (e.g., a product) or a process (e.g., the design cycle) in relation to a criterion, which defines a required level of a quality measure.

Fault: A missing attribute that impairs the effective usage of a product.

Quality: The totality of features and characteristics of a product that bear on its ability to satisfy stated or implied needs [ISO 8402 1994].

Satisfaction: The comfort and acceptability of use [ISO 9241-11 1996].

Task: A specification including the intended result of an activity to be performed on an object (material) with a specified means (method).

Task performance: An activity carried out at a user interface and aimed at completing a task.

Usability: The extent to which a product can be used by specified users to achieve specified goals with effectiveness, efficiency, and satisfaction in a specified context of use [ISO 9241-11 1996].

User interface: An interface that enables information to be passed between a human user and hardware or software components of a computer system [ISO 12119 1994].

References

Berlage, T. 1995. OSF/Motif as a user interface standard. *Comput. Stand. & Interfaces* 17:99–106.

Bevan, N. and Macleod, M. 1994. Usability measurement in context. *Behaviour Inf. Tech.* 13:132–145.

Boehm, B. W., Brown, J. R., and Lipow, M. 1976. Quantitative evaluation of software quality, pp. 529–605. In *Proc. IEEE 2nd Int. Conf. Software Eng.* ACM Press, New York.

Bowden, R. and Thomas, C. 1995. *Usability Context Analysis Guide* ver. 4. National Physical Laboratory, U.K.

Carroll, J. M. 1990. Infinite detail and emulation in an ontologically minimized HCI. In CHI'90 Conf. Proc. *Empowering People.* J. C. Chew and J. Whiteside, eds., pp. 321–328. ACM Press, New York.

Carroll, J. M., ed. 1995. *Scenario-Based Design.* John Wiley and Sons, New York.

CEC 1990. *Social Europe, 2/90*, ed. F. Gutman, Commission of the European Communities. Directorate-General V. Rue de la Loi 200, B-1049 Brussels.

Coutaz, J. 1987. PAC, an object oriented model for dialog design. In INTERACT'87 Conf. Proc. *Human–Computer Interaction.* H. J. Bullinger and B. Shackel, eds., pp. 431–436. Elsevier, Amsterdam.

Dzida, W. 1987. On tools and interfaces. In *Psychological Issues of Human Computer Interaction in the Work Place.* M. Frese, E. Ulich, and W. Dzida, eds., pp. 339–355. North-Holland, Amsterdam.

Dzida, W. 1988. Modellierung und Bewertung von Benutzerschnittstellen. *Software Kurier* 1:13–28.

Dzida, W. 1989. The development of ergonomic standards. *SIGCHI Bull.* 20(3):35–43.

Dzida, W. 1995. Standards for user-interfaces. *Comput. Stand. & Interfaces* 17:89–97.

Dzida, W., Herda, S., and Itzfeldt, W.-D. 1978. User perceived quality of interactive systems. *IEEE Trans. Software Eng.* SE4(4):270–276.

Dzida, W., Wiethoff, M., and Arnold A. G. 1993. ERGO guide—the quality assurance guide to ergonomic software. *Delft University* (The Netherlands) *and GMD* (Germany) *Internal Rep.* GMD, D-53754 Sankt Augustin, Germany.

Eason, K. D. 1984. Toward the experimental study of usability. *Behaviour Inf. Tech.* 3:133–143.

ECD 1990. Council Directive on the minimum safety and health requirements for work with display screen equipment; fifth individual Directive within the meaning of Article 16(1) of Directive 87/391/EEC. No. 90/270/EEC May 29. *Off. J. Eur. Communit.* L156:14–18.

Fähnrich, K.-P. and Ziegler, J. 1985. Workstations using direct manipulation as interaction mode—aspects of design, application and evaluation. In *INTERACT'84 Human–Computer Interaction.* B. Shackel, ed., pp. 693–698. Elsevier, Amsterdam.

Floyd, C., Reisin, F.-M., and Schmidt, G. 1989. STEPS to software development with users. In Proc. ESEC '89, *2nd Eur. Software Eng. Conf.* C. Ghezzi, and J. A. McDermid, eds. Lecture Notes in Computer Science, 387, pp. 48–64. Springer, Heidelberg.

Galitz, W. O. 1985. *Handbook of Screen Format Design.* QED Information Sciences, Wellesley, MA.

Goldberg, A. 1990. Information models, views, and controllers. *Dr. Dobb's J.* 7:54–61, 106–107.

Hartson, H. R. 1989. User-interface management control and communication. *IEEE Software* 1:62–70.

Hartson, H. R. and Hix, D. 1989. Human–computer interface development: concepts and systems for its management. *ACM Comput. Surv.* 21(3):5–92.

IBM 1992. Object-oriented interface design: IBM common user access guidelines. Que Corporation, Carmel, IN.

IFIP 1981. Report of the 1st meeting of the European User Environment Subgroup of the International Federation for Information Processing [IFIP] WG 6.5. GMD, D-53754 Sankt Augustin, Germany, July.

ISO 8402 1994. Quality—Vocabulary.

ISO 9000 1987. Quality management and quality assurance standards—Guidelines for selection and use.

ISO 9000-3 1991. Quality management and quality assurance standards—Part 3: Guidelines for the application of ISO 9001 to the development, supply and maintenance of software.

ISO 9001 1987. Quality systems—Model for quality assurance in design/development, production, installation and servicing.

ISO 9126 1991. Information technology—Software product evaluation—Quality characteristics and guidelines for their use.

ISO 12119 1994. Information technology—Software product evaluation—Quality requirements and testing. 1st ed.

ISO 9241-2 1992. Ergonomic requirements for office work with display terminals (VDTs): guidance on task requirements.

ISO 9241-10 1996. Ergonomic requirements for office work with display terminals (VDTs): Dialogue principles.

ISO CD 9241-11 1996. Ergonomic requirements for office work with display terminals (VDTs): guidance on usability specification and measures.

ISO CD 9241-12 1996. Ergonomic requirements for office work with display terminals (VDTs): presentation of information.

ISO CD 9241-13 1996. Ergonomic requirements for office work with display terminals (VDTs): user guidance.

ISO DIS 9241-14 1996. Ergonomic requirements for office work with display terminals (VDTs): menu dialogues.

ISO CD 9241-15 1996. Ergonomic requirements for office work with display terminals (VDTs): command dialogues.

ISO CD 9241-16 1996. Ergonomic requirements for office work with display terminals (VDTs): direct manipulation dialogues.

ISO CD 9241-17 1996. Ergonomic requirements for office work with display terminals (VDTs): form filling dialogues.

ISO TC 159/SC 4/WG 5 and WG 6. 1996. Ergonomics of human system interaction. N. Butz, Sec. DIN, D-10772 Berlin, Germany.

ISO/NP 13407 1996. New ISO project of ISO/TC 159/SC 4 on: human-centered design for interactive systems.

Keil, M. and Carmel, E. 1995. Customer–developer links in software development. *Commun. ACM* 38:33–44.

Kirakowski, J. 1995. The software usability measurement inventory, background and usage. In *Usability Evaluation in Industry*, P. Jordan, B. Thomas, and R. Weerdmeester. eds., Taylor & Francis.

Kirakowski, J. and Corbett, M. 1993. SUMI: the software usability measurement inventory. *Brit J. Educ. Tech.* 24(3):210–212.

Landauer, T. K. and Galotti, K. M. 1984. What makes a difference when? Comments on Grudin and Barnard. *Human Factors* 26:423–429.

McCall, J. A., Richards, P. K., and Walters, G. F. 1977. Factors in software quality, Vols. I, II, III. *U.S. Rome Air Development Center Rep.* NTIS AD/A-049014, 015, 055.

Moran, T. P. 1981. The command language grammar: a representation for the user interface of interactive computer systems. *Int. J. Man-Machine Stud.* 15:3–50.

Mosier, J. N. and Smith, S. L. 1986. Application of guidelines for designing user interface software. *Behaviour Inf. Tech.* 5:39–46.

Nielsen, J. 1992. Finding usability problems through heuristic evaluation. In *Proc. CHI '92 Conf.* J. Bennett and G. Lynch, eds., pp. 373–380. ACM Press, New York.

Nielsen, J. and Mack, R. L., eds. 1994. *Usability Inspection Methods.* Wiley, New York.

Norman, D. A. 1988. *The Psychology of Everyday Things.* Basic Books, New York.

Olsen, D. R. 1992. *User Interface Management Systems.* Morgan Kaufmann, New York.

OSF 1994. *OSF/Motif Style Guide.* Open Software Foundation. Prentice–Hall, Englewood Cliffs, NJ.

Pangalos, G. J. 1992. Standardization of the user interface. *Comput. Stand. & Interfaces* 14:223–229.

Pfaff, G. E., ed. 1985. *User Interface Management Systems.* Springer, Berlin.

Potosnak, K. 1988. What's wrong with standard user interfaces? *IEEE Software* 5(5):91–92.

Reisig, W. 1985. *Petri Nets. An Introduction.* Springer, Heidelberg.

Rossi, P. H. and Freeman, H. E. 1989. *Evaluation—A Systematic Approach,* 4th ed. Sage, London.

Samuelson, P. 1995. Software compatibility and the law. *Commun. ACM* 38:15–22.

Smith, S. L. 1986. Standards versus guidelines for designing user interface software. *Behaviour Inf. Tech.* 5(1):47–61.

Spring, M. B., Jamison, W., Fithen, K. T., Thomas, P. M, and Pavol, R. A. 1993. Models for a human–computer interaction. In *Encyclopedia of Microcomputers.* A. Kent and J. G. Williams, eds. Vol. 11, pp. 189–218. Marcel Dekker, New York.

Stewart, T. F. M. 1992. The role of HCI standards in relation to the directive. *Displays* 13:125–133.

Stewart, T. F. M. 1990. SIOIS—standard interfaces or interface standards. In INTERACT'90 *Human–Computer Interaction.* D. Diaper et al., eds., pp. xxix–xxxiv. Elsevier, Amsterdam.

SUN 1990. *OPEN LOOK Graphical User Interface Application Style Guidelines.* SUN Microsystems. Addison–Wesley, Reading, MA.

Wegener, H. 1995. The myth of the separable dialogue: software engineering vs. user models. In *Human–Computer Interaction.* K. Nordby et al., eds., pp. 169–172. Chapman & Hall, London.

Weiss, C. H. 1972. *Evaluation Research.* Prentice–Hall, Englewood Cliffs, NJ.

Wulf, V. 1993. Negotiability: a metafunction to support personalizable groupware. In *Human–Computer Interaction—Software and Hardware Interfaces.* G. Salvendy and M. J. Smith, eds., pp. 985–990. Elsevier, Amsterdam.

Further Information

Requests for information concerning international standards should be addressed to one of the ISO members. The complete list can be obtained from:

> ISO Central Secretariat, 1, rue de Varembé, Case postale 56, CH-1211 Genève 20, Switzerland

Information about the current state of development of ISO standards can be retrieved from Internet: http://www.iso.ch/

Members of ISO currently being active in ISO TC 159/SC 4/WG 5 ("Ergonomics of human–system interaction") are as follows:

> **Canada (SCC):** Standards Council of Canada, 45 O'Connor Street, Suite 1200, Ottawa, Ontario K1P 6N7
>
> **Denmark (DS):** Dansk Standard, Baunegaardsvei 73, DK-2900 Hellerup
>
> **France (AFNOR):** Association française de normalisation, Tour Europe, F-92049 Paris La Défense Cedex
>
> **Germany (DIN):** DIN Deutsches Institut für Normung, Burggrafenstraße 6, D-10787 Berlin
>
> **Italy (UNI):** Ente Nationale Italiano di Unificazione, Via Battistotti Sassi 11/b, I-20133 Milano
>
> **Japan (JISC):** Japanese Industrial Standards Committee, Ministry of International Trade and Industry, 1-3-1, Kasumigaseki, Chiyoda-ku, Tokyo 100
>
> **Netherlands (INNI):** Nederlands Normalisati-Institut, Kalfjeslaan 2, NL-2600 GB Delft
>
> **Sweden (SIS):** SIS—Standardiseringen i Sverige, Tegnérgatan 11, S-103 66 Stockholm

United Kingdom (BSI): British Standards Institution, 389 Chiswick High Road, GB-London W4 4AL

U.S. (ANSI): American National Standards Institute, 11 West 42nd Street, 13th floor, New York, NY 10036

More information on standards in computer science and engineering also can be found in the appendices of this Handbook.

Active national member bodies of ISO organize meetings of working groups which mirror the mentioned ISO working group. The meetings are open to anyone who wants to contribute to the current ISO projects either by proposals or by comments to committee drafts or draft international standards.

Most standards need a period of 5–10 years from the first working document up to the final version of an international standard. During this time, use of standards is to be commented in national as well as international journals. A recognized source is the journal *Computer Standards & Interfaces*, published by Elsevier, Amsterdam.

68

Input Devices and Techniques

Robert J. K. Jacob
Tufts University

68.1 Introduction

All aspects of human–computer interaction (HCI), from the high-level concerns of organizational context and system requirements to the conceptual, semantic, and syntactic levels of user-interface design, are ultimately funneled through physical input and output actions and devices. This chapter considers the input half of this physical level of human–computer interaction, the final means by which the user communicates information to the computer. It is also called the lexical level of the design of an interactive system, in contrast to the successively higher syntactic, semantic, and conceptual levels [Foley et al. 1990].

Computer input once consisted of such actions as setting switches and knobs and plugging and unplugging jumper wires in patch boards. For many years after that, the primary form of computer input was the punched card. Users or, more often, specialist keypunch operators punched the input information as holes in paper cards, which could then be read by computer peripherals. Next came the teletype, a device with a typewriter-like keyboard on which the user could type characters and cause corresponding electrical signals to be transmitted to the computer directly. Terminals, keyboards, and displays, the loose descendants of the teletype, continue to provide the principal form of computer input today.

Given the current state of the art, computer input and output are quite asymmetric. The amount of information or bandwidth that is communicated from computer to user is typically far greater than the bandwidth from user to computer. Graphics, animations, audio, and other media can output large amounts of information rapidly, but there are hardly any means of inputting comparably large amounts of information from the user. This is partly due to human abilities: we can receive visual images with very high bandwidth, but we are not very good at generating them. We can generate higher bandwidth with speech and gesture, but computers are not yet adept at interpreting these. User–computer dialogues are thus typically one sided. New input devices and media can help redress this imbalance by obtaining data from the user conveniently and rapidly, but, relative to output, input has been a neglected field of research, particularly in comparison with the great strides made in computer graphics.

0-8493-2909-4/97/$0.00+$.50
© 1997 by CRC Press, Inc.

68.2 Underlying Principles

The fundamental task of computer input is to move information from the brain of the user into the computer. Progress in this area attempts to increase the useful bandwidth across that interface by seeking faster, more natural, and more convenient means for users to transmit information to computers. On the user's side of the communication channel, input is constrained by the nature of human communication organs and abilities; on the computer side, it is constrained only by the input devices and methods that we can invent. Research in input and output centers around the two ends of this channel: the devices and techniques computers can use for communicating with people, and the perceptual abilities, processes, and organs people can use for communicating with computers. It then attempts to find the common ground through which the two can be related by studying new modes of communication that could be used for human–computer communication and developing devices and techniques to use such modes. Basic research seeks theories and principles that can predict user performance in new situations to guide the search for input media and the design of interfaces. In principle, the development of new input/output devices ought to be motivated or guided by the studies of human perceptual facilities and effectors as well as the needs uncovered in studies of existing interfaces. More often, though, the hardware developments have come first, and then HCI researchers try to find uses for the resulting artifacts.

The challenge in this field is, thus, to design new devices and types of dialogues that better fit and exploit the communication-relevant characteristics of humans. In doing so, two significant goals are bandwidth and naturalness. Increasing bandwidth simply means communicating more information per unit of time and, other things being equal, improves the efficiency of user–computer communication.

Naturalness

In seeking naturalness, we attempt to make the user's input actions as close as possible to the user's thoughts that motivated those actions, that is, to reduce the "Gulf of Execution" described by Hutchins et al. [1986], the gap between the user's intentions and the actions necessary to input them into the computer. The motivation for doing this is that it builds on the equipment and skills humans have acquired through evolution and experience and exploits them for communicating with the computer. Direct manipulation interfaces [Shneiderman 1983] have enjoyed great success, particularly with new users, largely because they draw on analogies to existing human skills (pointing, grabbing, moving objects in space), rather than trained behaviors. Virtual reality interfaces, too, gain their strength by exploiting the user's pre-existing abilities and expectations. Navigating through a conventional computer system requires a set of learned, unnatural commands, such as keywords to be typed in or function keys to be pressed. Navigating through a virtual reality system exploits the user's existing, natural navigational commands, such as positioning the head and eyes, turning the body, or walking toward something of interest. The result is to increase the user-to-computer bandwidth of the interface and to make it more natural, because interacting with it is more like interacting with the rest of the world.

Interaction Tasks, Techniques, and Devices

A designer looks at the **interaction tasks** necessary for a particular application [Foley et al. 1990]. Interaction tasks are low-level primitive inputs required from the user, such as entering a text string or choosing a command. For each such task, the designer chooses an appropriate **interaction device** and **interaction technique.** An interaction technique is a way of using a physical device to perform an interaction task. There may be several different ways of using the same device to perform the same task. For example, one could use a mouse to select a command by using a pop-up menu, a fixed menu (palette or toolbox), multiple clicking, circling the desired command, or even writing the name of the command with the mouse. An interaction technique represents an abstraction of some common class of interactive task, such as choosing one of several objects shown on a display screen. Research in interaction techniques studies these primitive elements of human–computer dialogues, which apply across a wide variety of individual applications. Its goal is to add new, high-bandwidth methods to the available store of interaction techniques or dialogue components. Whereas the interaction techniques are specific artifacts that can be applied directly in practical applications, the most useful of them are general enough to be used in a variety of applications, such as the pop-up menu.

In selecting an interaction device and technique for each task in a human–computer interface, simply making an optimal choice for each task individually may lead to a poor overall design, with too many different or inconsistent types of devices or dialogues. Therefore, it is often desirable to compromise on the individual choices to reach a better overall design, not only to avoid surrounding the user with an array of rarely used devices but also to reduce the time penalty incurred in switching between devices. In some situations, the designer has broad freedom to choose input devices appropriate to the task. For example, the cockpit of a new airplane or a military command and control console or a control station for a surgical teleoperator can be outfitted with whatever devices best facilitate operator performance. In many other situations, the designer of a human–computer interface does not have much control over the input hardware environment; they might be designing a piece of software to be used on a standard workstation with a standard or widely available suite of standard input devices, usually a keyboard and mouse. In this case, the designer decides which tasks should be assigned to the mouse and which to the keyboard (or possibly provides the user with synonyms) and which interaction techniques should be used for each task. Here, too, the time penalty for switching the hand from one device to another is a factor: whereas the keyboard might be the optimal input device for choosing a number between one and five, if this task occurs between two mouse tasks, it may be better to provide a graphical menu to save the switching time between devices. With elegant design, it is sometimes possible to provide both interaction techniques as options to the user without compromising the integrity of either.

In many situations, there are additional constraints on the range of input devices and interaction techniques the designer can choose. For example, an interface for a fighter airplane pilot must take into account the fact that the hands are usually already occupied with the task of operating the plane. Users operating under large gravity forces, under the ocean, on a rolling ship, or wearing a bulky spacesuit may impose additional constraints on the input methods one can choose. Interface design for handicapped users constrains the range of input choices in an analogous fashion. In each case, the best choice may not be the same as what would be chosen in an unconstrained situation.

Fitts' Law

User performance with many types of manual input depends on the speed with which the user can move the hand to a target. **Fitts' law** provides a way to predict this and is a key foundation in input design [Card et al. 1983]. It predicts the time required to move based on the distance to be moved and the size of the destination target. The time is proportional to the logarithm of the distance divided by the target width. This leads to a tradeoff between distance and target width: it takes as much additional time to reach a target that is twice as far away as it does to reach one that is half as large. Different manual input devices give rise to different proportionality constants in the equation. Thus, some give better overall performance and others better performance for either long moves or short moves, but the one-for-one tradeoff between distance and target size remains.

Control–Display Ratio

Another way of characterizing many input devices is by their **control-display (C–D) ratio**. This is the ratio between the movement of the input device and the corresponding movement of the object it controls. For example, if a mouse (the control) must be moved 1 in on the desk in order to move a cursor 2 in on the screen (the display), the device has a 1:2 control–display ratio. A high control–display ratio affords greater precision, whereas a low one allows more rapid operation and takes less desk space. An accelerator can be provided, so that the ratio increases dynamically when the user moves faster. This allows more efficient use of desk space, but it can disturb the otherwise straightforward physical relationship between mouse movement and cursor movement [Jellinek and Card 1990]. Of course, with a **direct input device**, such as a touch screen, the C–D ratio is always unity.

68.3 Best Practices

This section surveys the principal types of interaction devices in use today and emerging. Where possible, it is structured around the *output* mechanisms of the user's body rather than the device technology, since the former are more likely to remain constant over time. The principal means of human output or computer input today is through the user's hands, for example, keyboards, mice, gloves, and 3D trackers; these are discussed first. Other limb movements are then considered, followed by voice, and, finally, eye movements and other physiological measurements that may be used as input in the future.

Hands: Discrete Input

Keyboards, attached to workstations, terminals, or portable computers, are one of the principal input devices in use today. Nearly all of them now use a typewriterlike QWERTY keyboard layout, typically augmented with additional keys for moving the cursor, entering numbers, and special functions. Alternative keyboard layouts, such as the Dvorak layout, claim higher typing speed, but they have not been widely accepted because of the pervasiveness of the QWERTY layout. Another alternative that has been introduced is to retain the same assignment of letters to keys but to change the geometrical arrangement of the keys, in order to reduce strain on the hand and wrist during typing. Such a keyboard is typically divided into two halves, one for each hand, and these are pivoted away from each other, toward the respective hands; they may also be sloped upward in the center, better to fit the natural position of the hands.

Today's standard keyboard is widespread and relatively inexpensive to construct. As a result, it has been difficult to displace as the primary means of computer input. In recent years, the chief force serving to displace it has been the shrinking size of computers, as laptops, notebooks, palmtops, and personal digital assistants are being developed. The typewriter keyboard is becoming the largest component of such pocket-sized devices, the one component standing in the way of reducing its overall size, and this is beginning to provide a new driving force for developing alternatives to the keyboard.

As a computer peripheral, the keyboard simply transmits a signal each time a key is depressed (and, possibly, another signal when it is released). Some keyboards transmit a code for the character itself, that is, a if the *a* key is pressed, and A if it is depressed while holding the *shift* key. Other keyboards transmit unencoded signals, that is, an individual signal for the pressing or releasing of each button on the keyboard: the *a* key would be transmitted as a pressing of the *second* key in the *third* row, for example; and a capital A would be transmitted as a sequence of raw events: the pressing of the *shift* key, pressing of the *a* key, releasing of the *shift* key, and so on. Encoding this sequence into a capital A would then be done in the computer. This approach provides the flexibility to define new encodings, new types of shift keys, and new key chord combinations within the software.

Chord Keyboard

Another type of keyboard is the chord keyboard. This is typically designed for one hand and has five keys, one for each finger, plus sometimes additional ones for the thumb (see Fig. 68.1). Instead of pressing single keys, the user can press any combination of keys as a single chord. With five keys, this allows 31 combinations. The chord keyboard was originally introduced along with the mouse, with the intention that the user would use a mouse in the right (or dominant) hand and the one-hand chord keyboard in the other hand [Engelbart and English 1968]. While the mouse has since won widespread acceptance, the chord keyboard has not. Again, as computers become smaller, the benefit of a keyboard that allows touch typing with only five keys may come to outweigh the additional difficulty of learning the chords.

Function Keys

Hardware similar to that of the alphanumeric keyboard may also be used to provide individual, dedicated keys to invoke specific computer commands. These may be permanently labeled, special-purpose pushbuttons, or they may have labels that can be changed under computer control. An extreme, but effective, use of permanently labeled function keys is found in the cash registers of fast-food restaurants, where a

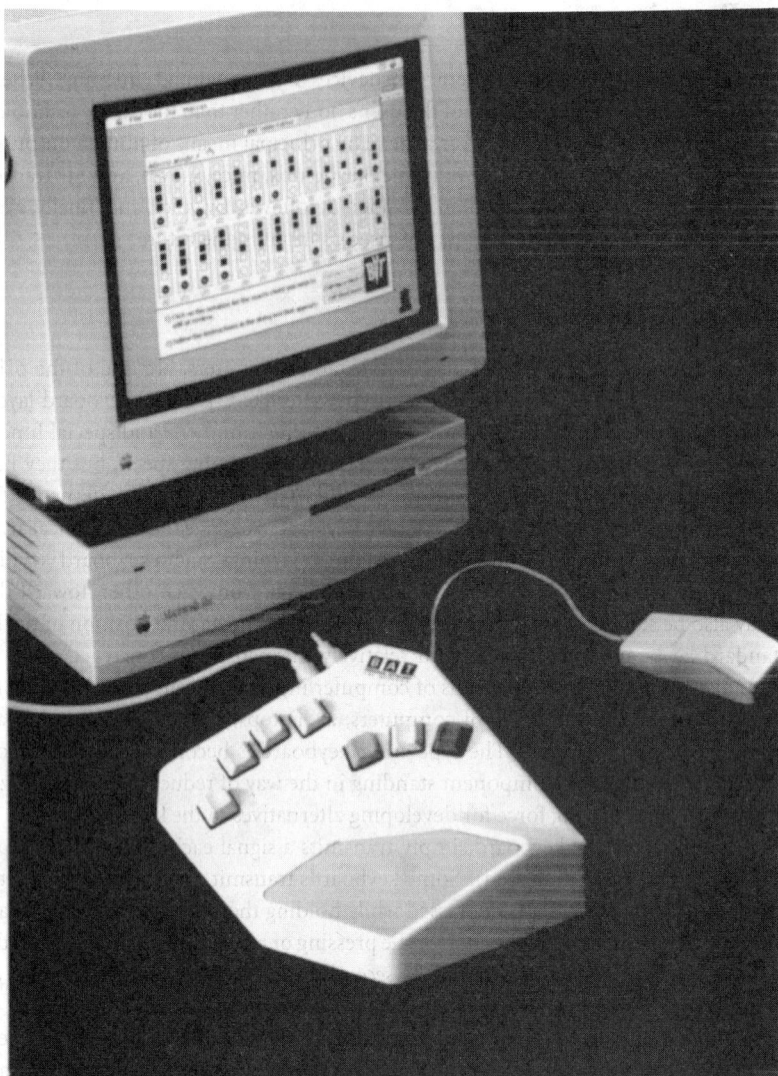

FIGURE 68.1   The Infogrip BAT™ one-handed chord keyboard. This keyboard is used with the left hand. The user places his or her fingers over the four keys on the left and presses one of the three keys on the right with the thumb. By pressing combinations of keys, different numbers, letters, and other symbols can be generated. (*Source:* Photo courtesy of Infogrip, Inc., Ventura, CA.)

large array of special-purpose function keys is provided, one for every possible item that can be purchased. Variable labels for function keys can be provided by placing the keys near the edge of the display and using the adjacent portion of the display to indicate the key labels, or by providing small alphanumeric light emitting diode (LED) or liquid crystal display (LCD) displays above each key. Variable labels can even be provided by a cathode ray tube (CRT) display projected downward onto a half-silvered mirror, so that the labels drawn on the CRT appear to float over the otherwise blank function keys [Knowlton 1977].

Hands: Continuous Input

Whereas keyboards and their variants are the principal means of discrete manual input, a much wider variety of devices is in use for continuous input from the hands. In fact, a number of taxonomies have been proposed for organizing and understanding continuous, manual input devices. The first approaches centered around the idea of logical devices, where devices are grouped by the type of input they provide: for example, locator, string, valuator, choice [Foley et al. 1984]. Another approach organizes continuous input devices by property and the number of dimensions sensed [Buxton 1983]. More recent approaches attempt to incorporate more of the ergonomic differences between seemingly similar devices into the taxonomy, to help guide the selection of the appropriate device for a task [Bleser and Sibert 1990, Mackinlay et al. 1990].

Based on these approaches, devices used for manually operated continuous pointing or locating can be categorized along each of the following dimensions:

- *Type of motion: linear vs. rotary.* For example, a mouse measures linear motion (in two dimensions); a knob, rotary.
- *Absolute or relative measurement.* For example, a mouse measures **relative** motion; a Polhemus magnetic tracker, **absolute**.
- *Physical property sensed: position or force.* A mouse measures position; an isometric joystick, force. For a rotary device, the corresponding properties are angle and torque.
- *Number of dimensions: one, two, or three linear and/or one, two, or three angular.* A mouse measures two linear dimensions, a knob measure one angular dimension, and a Polhemus measures three linear dimensions and three angular.
- *Direct vs. indirect control.* A mouse is **indirect** (you move it on the *table* to point to a spot on the screen); a touch screen is direct (you touch the desired spot on the screen directly).
- *Position vs. rate control.* Moving a mouse changes the *position* of the cursor; moving a rate-control joystick changes the *speed* with which the cursor moves.
- *Integral vs. separable dimensions.* A mouse allows easy, coordinated movement across two dimensions simultaneously (integral), whereas a pair of knobs (as in an Etch-a-Sketch toy) does not (separable) [Jacob et al. 1994].

This covers the range of continuous, manually operated devices. Discrete input devices, such as the keyboard, can be fit into these catagories (and some of the taxonomies discussed also cover discrete devices), but they do not appear in the variety that continuous devices do. Nonmanually operated devices fit less well into the categories; they currently include foot or other body controls, voice input, eye trackers, and a variety of other physiological measuring instruments discussed later.

Given this space of possible continuous manual input devices, we discuss next some of the more common forms of devices currently in use.

One-Dimensional Valuator

A rotary (knob or thumbwheel) or linear (slide) potentiometer may be used for inputting a value along a single axis. Its analog output is converted to a digital computer input each time the computer queries the associated A–D converter. A knob with a digital encoder may also be used. It simply transmits an interrupt signal to the computer each time it is turned a small amount. Unlike an analog potentiometer, such a device typically does not have physical endpoints but turns continuously. The range and meaning of knob movement can be thus arbitrarily modified by the software. In some applications that involve multiple parameters, a single *soft pot* of this type is provided. At any moment, just one of the parameters in the application is chosen to be assigned to this knob and may be adjusted. A dial box is sometimes used in computer graphics; it is an array of several such knobs, often provided with computer-controlled labels. Other input devices may also be used for the task of entering a scalar value, through the use of interaction techniques. For example, a mouse may be used for this job via an on-screen slider; a keyboard may be used by typing in a numeric value.

Two-Dimensional Locator

Today, the mouse is the most widely used device for inputting 2D positions, but it was not the first such device developed. It supplanted devices such as the joystick, trackball, light pen, and arrow keys and, in an early example of the application of HCI research to practice, was demonstrated to give fastest performance and closest approximation to Fitts' law compared to alternative devices at the time [Card et al. 1978]. Despite its popularity, some specific, constrained situations call for alternative devices. For example, the Navy uses trackballs instead of mice on shipboard, because the rolling of the ship makes it difficult to keep a mouse in place. Portable computers use small trackballs, touch-sensitive pads, or tiny joysticks because they are more compact.

The mouse and trackball are relative devices; they report only how far they move, not where they are. They typically generate an interrupt or a piece of serial data each time they move. The data tablet is an absolute locator device that is similar to the mouse in appearance, but the surface on which it operates contains a grid of wires or other sensors that can measure the absolute position of a puck or stylus on the tablet and report it when queried or else in a continuous stream. It is most often seen in graphics and computer-aided design (CAD) applications.

The joystick comes in several varieties. It can be used to control cursor position directly or it can control the rate of speed at which the cursor moves. Since its total range of motion is typically fairly small compared to a display screen, position control is imprecise; however, rate control requires a more complex relationship between the user's action and the result on the display and is therefore more difficult to operate. The joystick can move when it is pushed or, in an isometric joystick, it can remain nearly stationary and simply report the force being applied to it.

Direct input devices obviate the need to relate the position of the device to the position of the cursor on the screen. A touch screen is a device that fits over a CRT or other display and reports the location of finger or stylus touches on its surface. With it, the user can simply point to the desired item on the screen. This requires very little training, but it can be tiring if the user must hold his or her hand up to the screen for a long time. Precision of touchscreens is typically lower than that of other locator devices, though new strategies have been developed to improve the precision attainable with a finger-operated touchscreen [Sears and Shneiderman 1991]. A finger-operated touchscreen can also be used to simulate a keyboard and allows the keys to be relabeled under computer control. However, such a keyboard lacks the tactile feedback of a conventional keyboard, making it slower to operate and a particularly poor choice for eyes-busy applications such as operating a car or airplane. The light pen was a technology used in early graphics systems that also allowed direct pointing on the screen with a light-sensitive stylus.

Modern pen-based systems use technology similar to that of a touchscreen or data tablet, but they typically use the pen as the sole input device. It is used both for location input and for character string input, and, more interestingly, it can also be used for interaction that more closely resembles the way a person would use a regular pen rather than a mouse, such as making circle and arrow gestures to move blocks of text. For entering text, full handwriting recognition is not yet achievable for all users. Block printing of capital letters is possible, but fairly cumbersome. For some users, a compromise works better: using an alphabet of characters specially designed to be easily distinguishable from one another to facilitate computer recognition. Such characters are also typically designed so that each can be drawn with a single stroke without lifting the pen, which makes it easier for the computer to find the boundaries between the letters. It also makes it possible to use a very small input area, in which the input letters are written in succession, on top of one another, for some applications.

Three-Dimensional Locator

The typical 3D locator device functions like the three-dimensional equivalent of a data tablet in that it provides absolute position information along three axes in space, instead of two, either continuously in a stream or each time it is queried. Many such devices also report their orientation, in the form of angles of rotation about the three axes, or yaw, pitch, and roll. The most common such devices (Polhemus and Ascension) use a magnetic signal that is transmitted by a fixed source and received by a sensor held in the user's hand or attached to some object (see Fig. 68.2). Ultrasonic ranging (Logitech) is also used for this purpose. It typically provides less precision, but is more robust in the face of magnetic interference, such as that from a CRT.

While often operated with the hand, the sensor of the 3D tracker is typically a 1-in plastic cube, which can be used in a variety of ways. It can be held in the hand, or attached to a glove, foot, the user's head (as is typically done in virtual reality), or to passive props [Hinckley et al. 1994] or other objects the user will manipulate. A hybrid form of 3D tracker combines a mouse in a single package and allows it to be operated as a mouse while it is located on a table but switches into 3D operation when it is lifted into the air.

Today, all of these 3D devices are still limited compared to a mouse or data tablet—in latency, precision, stability, susceptibility to interference, or number of available samples per second. In addition, they all require that the user hold or attach the small sensor and its trailing wire. Another approach is to use sensors that observe the user, without requiring him or her to hold or wear anything. Camera-based locator devices offer the promise of doing this, but today they are still limited. A single-camera system is limited to its line of sight; more cameras can be added but full coverage of an area may require many cameras and a way to switch among them smoothly. This approach depends on some type of image processing to interpret the picture of the user and extract the desired hand or body position. Small video cameras are beginning to appear as a standard component of graphics workstations; although they are intended for teleconferencing, they will also be useful for this type of 3D input.

Another 3D input device is the Spaceball, which is roughly the 3D analog of an isometric joystick (see Fig. 68.3). It consists of a ball mounted on a fixed platform; the user holds the ball and pushes or twists it

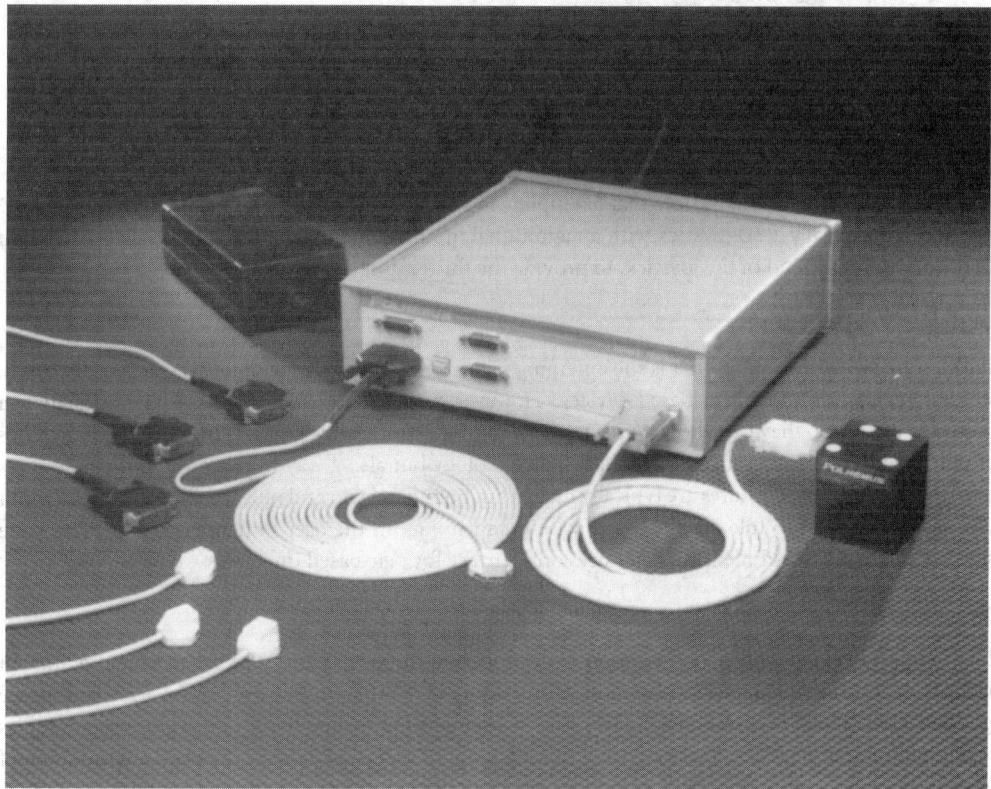

FIGURE 68.2 The Polhemus 3SPACE™ FASTRAK™ 3D magnetic tracker. The device reports the position and orientation in 3D of each of the four sensors (the small white cubes in the foreground), using a magnetic signal sent from the transmitter (the larger black cube on the right). (*Source:* Photo courtesy of Polhemus, Inc., Colchester, VT.)

FIGURE 68.3 The Spaceball Space Controller™ 3D control device. This device operates like a 3D isometric joystick: the user holds the ball and pushes or twists it in the desired direction. (*Source:* Photo courtesy of Spacetec IMC Corporation, Lowell, MA.)

in the desired direction. Finally, note that 3D input can also be achieved with a device referred to as a 3D joystick, which is really a 2D joystick with an additional input device attached to it, typically a knob that can be rotated on the end of the joystick, to provide the third input dimension.

Gesture

Hand gesture is a form of input that is still emerging. The devices used are the same 3D trackers discussed, including magnetic and camera-based devices. However, rather than using them simply to designate a location in 3D space, they can allow a user to make natural, continuous gestures in space. This requires not only a better, nonencumbering 3D tracking technology but also a way to recognize human gestures occurring dynamically. Gestures are typically made with poor precision and repeatability, so a useful input technique would have to tolerate such imprecision and still glean the user's intended action. The same issues arise in using two-dimensional gestures on a surface for pen-based interfaces

Glove

Glove input devices report the configuration of the fingers of the user's hand, also called a hand *posture* in contrast to a gesture, which may involve motion or a sequence of different postures to convey meaning (see Fig. 68.4). The Dataglove uses optical fibers, which attenuate light when bent. Other glove technologies use mechanical sensors. All of these devices typically report a vector containing the bend angle of each of the joints of each finger of the hand. Some also report abduction, the angles formed by the separation of the fingers from each other. Most glove devices combine a 3D tracker, so that they can report the position and orientation of the hand as well as the angle of each finger. From these, it should in principle be possible to derive the exact position in space of each fingertip; however, the accuracy of today's glove device does not always allow this.

FIGURE 68.4 CyberGlove™ 18-sensor instrumented gloves. The gloves report the configuration of the fingers of the user's hand. Note the 3D magnetic sensor incorporated into the wristband of the glove; it reports the position and angle of the hand itself. (*Source:* Photo courtesy of Virtual Technologies, Inc., Palo Alto, CA.)

Two-Handed Input

Aside from touch typing, most of the devices and modes of operation discussed thus far and in use today involve only one hand at a time. People are quite good at manipulating both hands in coordinated or separated tasks, as for example one does in driving a car, piloting an airplane, or performing surgery [Buxton and Myers 1986]. For example, a two-handed approach that simulates the use of a moveable, translucent stencil has been demonstrated to be effective for desktop tasks [Stone et al. 1994].

Other Body Movements

Having considered input from the hand, we consider next other limbs and body movements that can be used as computer input, though, today, they are not nearly as widely used as manual input.

Foot

Simple foot controls are used in automobiles and musical instruments and can readily be used as computer input for discrete or continuous scalar information, using simple input devices. The Mole is a more sophisticated foot-operated input device that provides locator input using a footrest suspended on two sets of pivots [Pearson and Weiser 1986]. Although control is less precise than a manually operated mouse, it leaves the hands free for additional operations.

Head

Head movement can be measured with a 3D tracker and can be used to control cursor position, though this can often require the neck to be held in an awkward fixed position. Another use of head movement is to perform a function more akin to the use of head movement in the natural world—panning and zooming over a display [Hix et al. 1995].

Input for Virtual Reality

Most virtual reality systems rely on the same 3D devices previously discussed, used in combination. They use a 3D magnetic tracker to sense head position and orientation, which then determines the position of the virtual camera, which generates the scene to be displayed in the user's head-mounted display, typically in stereo. The result is the illusion of a realistic, 3D world that surrounds the user wherever he or she looks. The user can reach out into this world and touch the objects in it, using a second 3D tracker attached to the hand (so the computer knows where the user's hand is relative to the displayed world) and, often, a glove (so the computer can detect grasping or other gestures). However, the user will not feel the object when the hand touches it. Mechanisms for providing computer-controlled force and tactile feedback are a topic of current research. An extension of this notion would be to provide virtual tools for input, where the user might first obtain a tool (by reaching for it in the virtual space) and then apply it to a 3D virtual object. A virtual tool can, of course, metamorphose as needed for the job at hand and otherwise improve upon the properties of nonvirtual tools. However, the latency and precision available from today's input devices still fall short of being able to support this smoothly.

Facial Expression

A less obvious form of muscle input is to use the facial expressions of the user. The device for doing this is simply a camera and frame grabber, but image-understanding techniques for interpreting the images into meaningful facial expressions are still emerging. However, much less subtle inputs are also possible. For example, the computer can determine relatively easily from camera input whether the user is still sitting in the chair, facing toward the computer or not, using the telephone, or talking to another person in the room.

Myoelectric Inputs

Beyond physical measurement of limb motions, an emerging technology for muscle input is to measure myoelectric signals from electrodes placed on the user's skin. While currently a research topic, this approach has the potential to provide a more compact, less cumbersome way to measure muscle movements. Such signals can also be detected slightly before the muscle actually begins moving, which can help to reduce overall system latency.

Voice

Another type of input comes from the user's speech. Carrying on a full conversation with a computer as one might do with another person is well beyond the state of the art today and, even if possible, may be a naive goal. Nevertheless, speech can be used as input in several different ways: unrecognized speech, discrete word recognition, and continuous speech recognition.

Unrecognized Speech

Even without understanding the content of the speech, computers can digitize, store, edit, and replay segments of speech in useful ways. Conventional voice mail is an everyday example of this type of function, but far more sophisticated uses of this technology have been developed [Schmandt 1993].

Discrete Word Recognition

Understanding speech as input has been a long-standing area of research. While progress is being made, it is slower than optimists originally predicted, and further work remains in this field. Although the goal of continuous speech recognition remains elusive, unnatural, isolated-word speech recognition can work reasonably well and is appropriate for some tasks. Discrete word recognition requires that the user pause briefly after saying each word. It is a highly unnatural way of speaking, though it can seem appropriate for giving computer commands. Some systems are speaker dependent (they require each particular user to speak the words to be used into the system ahead of time to train the computer), whereas some are speaker independent (they rely on a single set of training data for all users). Performance can also be enhanced by using a restricted grammar. For example, if the first word of each command must be a verb and the second must be a file name, the speech recognizer can use this information to limit the range of possibilities it must examine at each point and thereby provide more accurate results.

Continuous Speech Recognition

One of the most difficult aspects of recognizing continuous speech is simply finding the boundaries between the words. Research continues in the area of continuous speech recognition, with varying degrees of success found in both research and commercial systems. Improved performance can be obtained where the system can be tuned to a particular application domain and input grammar.

Even if the computer could recognize all of the user's words, the problem of understanding natural language is a significant and unsolved one. It can be avoided by using an artificial language of special commands or even a fairly restricted subset of natural language. But, given the current state of the art, the closer the user moves toward full unrestricted natural language, the more difficulties will be encountered.

Multimode Speech Input

Speech is often most useful in conjunction with other input media, providing an additional channel when the user is already occupied. (Driving a car and conducting a conversation is an everyday example.) If the user's hands, feet, and eyes are busy, speech may be the only reasonable choice for some input. However, more interesting cases begin with taking a collection of tasks in a user interface and then allocating them to the range of the user's communication modes. Another use for multiple modes is to combine otherwise ambiguous inputs from several modes (such as pointing and speaking) to yield an unambiguous interpretation of the user's input [Schmandt and Hulteen 1982].

Eye

Although the main role of the eye in most human–computer interaction situations is to receive *output* from the computer, the movements of the user's eye can also be measured and used as input. An eye tracker can measure the visual line of gaze, that is, where the user's eye is pointing in space, and report it to a computer in real time. (See Fig. 68.5). Eye movement-based input, properly used, can provide an unusually fast and natural means of communication, because we move our eyes rapidly and almost unconsciously. However, eye tracking technology today is still only marginally adequate for use in applications; its prime application area is for disabled users, who cannot move their arms or legs.

Using eye movements as input also requires careful design of interaction techniques [Bolt 1981, Jacob 1991]. Eye movements, like other passive inputs to be discussed, are often nonintentional or not conscious, and so they must be interpreted carefully to avoid annoying the user with unwanted responses to his actions, the Midas touch problem. People are not accustomed to operating devices simply by moving their eyes. They expect to be able to look at an item without having the look cause an action to occur. At first it is helpful to be able simply to look at what you want and have it occur without further action; soon, though, it becomes like the Midas touch. Everywhere you look, another command is activated; you cannot look anywhere without issuing a command. Eye movements are an example of the *clutch* problem that

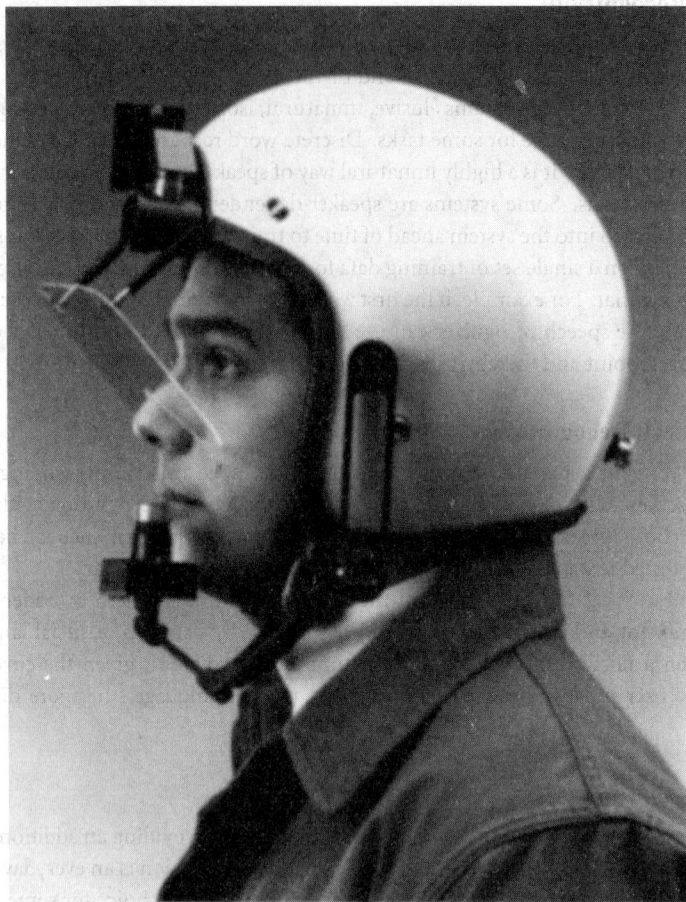

FIGURE 68.5   Applied Science Laboratories helmet-mounted eye tracker. This device measures visual line of gaze or where the user's eye is pointing in space. A tiny camera, located above the user's forehead, views the eye through the half-silvered mirror. A second camera, located near the user's chin, is optionally used to keep a record of what the user saw for later analysis. (*Source:* Photo courtesy of Applied Science Laboratories, Bedford, MA.)

arises in many emerging passive or noncommand forms of input (including speech, gesture, physiological measurement); it requires a way to tell the computer when the device is engaged vs. when the user is using the same communication modality for some other purpose but not talking to the computer.

Passive Measurements

User Behavior

Input may also be obtained from a user without explicit action on the user's part. Behavioral measurements can be made from changes in the user's typing speed, general response speed, manner of moving the cursor, frequency of low-level errors, or other patterns of use. A carefully designed user interface could make intelligent use of such information to modify its dialogue with the user, based on, for example, inferences about the user's alertness or expertise (but note that there is also the potential for abuse of this information). These measures do not require additional input devices, but rather gleaning of additional, typically neglected, information from the existing input stream.

Physiological Measurements

In a similar vein, passive measurements of the user's state may also be made with additional hardware devices. In addition to 3D position tracking and eye tracking, a variety of other physiological characteristics of the user might be measured and the information used to modify the computer's dialogue with its user. Blood pressure, heart rate, respiration rate, eye pupil diameter, and galvanic skin response (the electrical resistance of the skin) are examples of measurements that are relatively easy and comfortable to make, although their accurate instantaneous interpretation within a user–computer dialogue is an open question.

A more difficult measure is an electroencephalogram, although progress has been made in identifying specific evoked potential signals in real time [Wickens et al. 1983]. The most accurate results are currently obtained with a somewhat unwieldy superconducting detector [Lewis et al. 1987], rather than the conventional electrodes, but improvements in this technology can be envisioned.

Direct Connect

Looking well beyond the current state of the art, perhaps the final frontier in user input and output devices, will be to measure and stimulate neurons directly, rather than relying on the body's transducers. This is unrealistic at present, but it may someday be a primary mode of high-performance user–computer interaction. If we view input in HCI as moving information from the brain of the user into the computer, we can see that all current methods require that this be done through the intermediary of some physical action. We strive to reduce the Gulf of Execution, the gap between what the user is thinking and the physical action that must be made to communicate that thought. From this point of view, reducing or eliminating the intermediate physical action ought to improve the effectiveness of the communication. The long-term goal might be to see the computer as a sort of mental prosthesis, where the explicit input and output steps vanish and the communication is direct, from brain to computer.

Other Issues

Relationship to Output

Although this chapter discusses input, it should be clear that many of the newer approaches here are intimately coupled with output. Input devices and their technologies are important but increasingly are meaningful only in context of outputs, especially in more modern, highly interactive forms of interaction. For example, whereas a keyboard makes sense as an isolated input device, a pop-up menu makes sense only when the mouse input and screen output are considered together. In a direct manipulation or graphical interface, the output objects on the display are the principal targets for subsequent input commands, which select and manipulate the displayed objects. Similarly, virtual reality makes sense only when the input from head and hand position sensors controls the moment-to-moment output transmitted to the head-mounted display.

Device Interfaces

A mundane but nagging problem in the area of input is connecting new input devices to a computer. New devices often introduce new, slightly different hardware connections and software protocols for communication. Even superficially similar devices are not yet easily interchangeable and often require essential but fundamentally trivial work to begin using a new device. The communication requirements of many of the input devices discussed here are sufficiently similar and undemanding that a standard physical interface and communication protocol is not a serious technical problem nor would it levy an unreasonable performance penalty. For example, the standard Musical Instrument Digital Interface (MIDI) addresses this problem for both physical connection and simple logical protocol for keyboard-oriented musical instruments, and its dramatic success in expanding the usefulness of electronic musical instruments suggests the benefits.

68.4 Research Issues and Summary: Future Trends

Interaction Style

A new style of interaction that is emerging is noncommand-based interaction [Nielsen 1993]. Whereas other interaction styles await, receive, and respond to explicit inputs from the user, in this approach the computer passively monitors the user and responds as appropriate. Its effect on the field of input is to move from providing objects for the user to *actuate* through specific commands to simply *sensing* the user's body. Jakob Nielsen describes this next generation interaction style:

> The fifth generation user interface paradigm seems to be centered around noncommand-based dialogues. This term is a somewhat negative way of characterizing a new form of interaction but so far, the unifying concept does seem to be exactly the abandonment of the principle underlying all earlier paradigms: that a dialogue has to be controlled by specific and precise commands issued by the user and processed and replied to by the computer. The new interfaces are often not even dialogues in the traditional meaning of the word, even though they obviously can be analyzed as having some dialogue content at some level since they do involve the exchange of information between a user and a computer. The principles shown at CHI'90 which I am summarizing as being noncommand-based interaction are eye tracking interfaces, artificial realities, play-along music accompaniment, and agents. [Nielsen 1990].

This new interaction style will require new devices, interaction techniques, and software approaches to deal with them. Unlike traditional inputs, such as keyboards and mice, the new inputs represent less the intentional actuation of a device or issuance of a command but are more like passive monitoring of the user. This suggests a change from conventional devices to passive equipment that senses the user, such as unobtrusive 3D trackers, hand-measuring devices, remote cameras (plus appropriate pattern recognition), range cameras, eye-movement monitors, and physiological monitors.

Interaction Devices

One clear current need is for 3D tracking with greater accuracy and lower latency than current techniques. A method that freed the user from the wire would also be helpful. Camera-based techniques may solve this problem, or a new technology may be applied to it; both are areas of current research.

Beyond this, we might predict the future of input by looking at some of the characteristics of emerging new computers. The desktop workstation seems to be an artifact of past technology in display devices and in electronic hardware. In the future, it is likely that computers smaller and larger than today's workstation will appear, and the workstation-size machine may disappear. This will be a force driving the design and adoption of future input mechanisms. Small computers are already appearing: laptop and palmtop machines, personal digital assistants, wearable computers, and the like. These are often intended to blend more closely into the user's other daily activities. They will certainly require smaller input devices, and may also require more unobtrusive input mechanisms if they are to be used in settings where the user is simultaneously engaged in other tasks, such as talking to people or repairing a piece of machinery.

At the same time, computers will be getting larger. As display technology improves, as more of the tasks one does become computer based, and as people working in groups use computers for collaborative work, an office-sized computer can be envisioned, with a display that is as large as a desk or wall (and has resolution approaching that of a paper desk). Such a computer leaves considerable freedom for possible input means. If it is a large, fixed installation, then it could accommodate a special-purpose console or cockpit for high-performance interaction. It might also be used in a mode where the large display is fixed, but the user or users move about the room, interacting with each other and with other objects in the room. In that case, while the display may be very large, the input devices would be small and mobile.

Another trend seen in the emergence of virtual reality is that computer input and output is becoming more like interacting with the real world. Instead of inputting strings of characters, users interact with a virtual reality in more natural and expressive ways, moving their heads, hands, or feet. Future input mechanisms might continue this trend toward naturalness and expressivity by allowing users to perform natural gestures or operations and transducing them for computer input. More parts or characteristics of the user's body might be measured for this purpose and then interpreted as input. As a thought experiment along these lines, consider obtaining and interpreting input from the gestures and actions of an orchestra conductor.

Another way to predict the future of computer input devices is to examine the progression that begins with experimental devices used in the laboratory to measure some physical attribute of a person. As such devices become more robust, they may be used as practical medical instruments outside the laboratory. As they become convenient, noninvasive, and inexpensive, they may find use as future computer input devices. The eye tracker is such an example; the physiological monitoring devices discussed may well also turn out to follow this progression.

Finally, in a more practical vein, it is important to remember that there has historically been a long time lag between invention and widespread use of new input or output technologies. Consider the mouse, one of the more successful innovations in input devices, first developed around 1968 [Engelbart and English 1968]. It took approximately 10 years before it was found widely even in very many other research laboratories and perhaps 20 before it was widely used in applications outside the research world. The input mechanisms in use 20 years from now may well be chosen from some of the devices and approaches that today appear to be impractical laboratory curiosities.

Defining Terms

Absolute input device: An input device that reports its actual position, rather than relative movement. A data tablet or Polhemus tracker operates this way (see relative input device).

Control–display ratio: The ratio between the movement a user must make with an input device and the resulting movement obtained on the display. With a large control–display ratio, a large movement is required to effect a small change on the display, affording greater precision. A low ratio allows more rapid operation and takes less desk space.

Direct input device: A device that the user operates directly on the screen or other display to be controlled, such as a touch screen or light pen (see indirect input device).

Fitts' law: A model that predicts time to move the hand or other limb to a target, based on the distance to be moved and the size of the target. The time is proportional to the logarithm of the distance divided by the target width, with constant terms that vary from one device to another.

Indirect input device: A device that the user operates by moving a control that is located away from the screen or other display to be controlled, such as a mouse or trackball (see direct input device).

Interaction device: A hardware computer peripheral through which the user interacts with the computer.

Interaction task: A low-level primitive input to be obtained from the user, such as entering a text string or choosing a command.

Interaction technique: A particular way of using a physical device to perform a generic interaction task. For example, the pop-up menu is an interaction technique for choosing a command or other item from a small set, by means of a mouse and display.

Relative input device: An input device that reports its distance and direction of movement each time it is moved, but cannot report its absolute position. A mouse operates this way (see absolute input device).

References

Bleser, T. W. and Sibert, J. L. 1990. Toto: a tool for selecting interaction techniques. In *Proc. ACM UIST'90 Symp. User Interface Software Tech.*, pp. 135–142. Addison–Wesley/ACM Press, Snowbird, UT.

Bolt, R. A. 1981. Gaze-orchestrated dynamic windows. *Comput. Graphics* 15(3):109–119.

Buxton, W. 1983. Lexical and pragmatic considerations of input structures. *Comput. Graphics* 17(1):31–37.

Buxton, W. and Myers, B. A. 1986. A study in two-handed input. In *Proc. ACM CHI'86 Human Factors Comput. Syst. Conf.*, pp. 321–326.

Card, S. K., English, W. K., and Burr, B. J. 1978. Evaluation of mouse, rate-controlled isometric joystick, step keys, and text keys for text selection on a CRT. *Ergonomics* 21(8):601–613.

Card, S. K., Moran, T. P., and Newell, A. 1983. *The Psychology of Human–Computer Interaction.* Lawrence Erlbaum, Hillsdale, NJ.

Engelbart, D. C. and English, W. K. 1968. A research center for augmenting human intellect, pp. 395–410. In *Proc. 1968 Fall J. Comput. Conf.* AFIPS.

Foley, J. D., van Dam, A., Feiner, S. K., and Hughes, J. F. 1990. *Computer Graphics: Principles and Practice.* Addison–Wesley, Reading, MA.

Foley, J. D., Wallace, V. L. and Chan, P. 1984. The human factors of computer graphics interaction techniques. *IEEE Comput. Graphics Appl.* 4(11):13–48.

Hinckley, K., Pausch, R., Goble, J. C., and Kassell, N. F. 1994. Passive real-world interface props for neurosurgical visualization, pp. 452–458. In *Proc. ACM CHI'94 Human Factors Comput. Syst. Conf.* Addison–Wesley/ACM Press.

Hix, D., Templeman, J. N., and Jacob, R. J. K. 1995. Pre-screen projection: from concept to testing of a new interaction technique, pp. 226–233. In *Proc. ACM CHI'95 Human Factors Comput. Syst. Conf.* Addison–Wesley/ACM Press; http://www.acm.org/sigchi/chi95/Electronic/documnts/papers/dh_bdy.htm [HTML]; http://www.cs.tufts.edu/~jacob/papers/chi95.txt [ASCII].

Hutchins, E. L., Hollan, J. D., and Norman, D. A. 1986. Direct manipulation interfaces. In *User Centered System Design: New Perspectives on human-computer Interaction*, D. A. Norman and S. W. Draper, eds., pp. 87–124. Lawrence Erlbaum, Hillsdale, NJ.

Jacob, R. J. K. 1991. The use of eye movements in human-computer interaction techniques: what you look at is what you get. *ACM Trans. Inf. Syst.* 9(3):152–169.

Jacob, R. J. K., Sibert, L. E., McFarlane, D. C., and Mullen, M. P., Jr. 1994. Integrality and separability of input devices. *ACM Trans. Comput.-Human Interaction* 1(1):3–26; http://www.cs.tufts.edu/~jacob/papers/tochi.txt [ASCII]; http://www.cs.tufts.edu/~jacob/papers/tochi.ps [Postscript].

Jellinek, H. D. and Card, S. K. 1990. Powermice and user performance, pp. 213–220. In *Proc. ACM CHI'90 Human Factors Comput. Sys. Conf.* Addison–Wesley/ACM Press.

Knowlton, K. C. 1977. Computer displays optically superimposed on input devices. *Bell Sys. Tech. J.* 56(3):367–383.

Lewis, G. W., Trejo, L. J., Nunez, P., Weinberg, H., and Naitoh, P. 1987. Evoked neuromagnetic fields: implications for indexing performance. *Biomagnetism 1987, Proc. 6th Int. Conf. Biomagnetism.* Tokyo.

Mackinlay, J. D., Card, S. K., and Robertson, G. G. 1990. A semantic analysis of the design space of input devices. *Human-Comput. Interaction* 5:145–190.

Nielsen, J. 1990. Trip report: CHI'90. *SIGCHI Bull.* 22(2):20–25.

Nielsen, J. 1993. Noncommand user interfaces. *Comm. ACM* 36(4):83–99.

Pearson, G. and Weiser, M. 1986. Of moles and men: the design of foot control for workstations, pp. 333–339. In *Proc. ACM CHI'86 Human Factors Comput. Syst. Conf.*

Schmandt, C. 1993. From desktop audio to mobile access: opportunities for voice in computing. In *Advances in Human-Computer Interaction*, Vol. 4, H. R. Hartson and D. Hix, eds., pp. 251–283. Ablex, Norwood, NJ.

Schmandt, C. and Hulteen, E. A. 1982. The intelligent voice-interactive interface, pp. 363–366. In *Proc. ACM Human Factors Comput. Syst. Conf.*

Sears, A. and Shneiderman, B. 1991. High precision touchscreens: design strategies and comparison with a mouse. *Int. J. Man-Machine Stud.* 43(4):593–613.

Shneiderman, B. 1983. Direct manipulation: a step beyond programming languages. *IEEE Comput.* 16(8):57–69.

Shneiderman, B. 1992. *Designing the User Interface: Strategies for Effective Human-Computer Interaction*, 2nd ed. Addison–Wesley, Reading, MA.

Stone, M. C., Fishkin, K., and Bier, E. A. 1994. The movable filter as a user interface tool, pp. 306–312. In *Proc. ACM CHI'94 Human Factors Comput. Syst. Conf.* Addison–Wesley/ACM Press.

Wickens, C., Kramer, A., Vanasse, L., and Donchin, E. 1983. Performance of concurrent rasks: a psychophysiological analysis of the reciprocity of information-processing resources. *Science* 221:1080–1082.

Further Information

Input is usually seen as part of human–computer interaction, and so information about this area is typically found in general books, journals, or conferences on HCI, rather than in a more specialized venue.

Good introductions to these issues are found in the respective chapters of two standard textbooks in this area, by Shneiderman [1992] and by Foley, van Dam, Feiner, and Hughes [1990].

Research in input devices and techniques is covered in the proceedings of the annual ACM CHI Human Factors in Computing Systems Conference, published by the Association for Computing Machinery (ACM, New York). Other relevant annual conference proceedings include the ACM UIST Symposium on User Interface Software and Technology (ACM), Graphics Interface (Canadian Human–Computer Communications Society, Toronto, Canada), and the Human Factors and Ergonomics Society (HFES, Santa Monica, CA.).

The journal *ACM Transactions on Computer-Human Interaction* (ACM) includes work on input devices and techniques, as do *Human Factors* (published by HFES) and *Human–Computer Interaction* (Lawrence Erlbaum Associates, Inc., Hillsdale, NJ). *ACM Transactions on Graphics* (ACM) publishes a series of articles titled "Interaction Techniques Notebook," which report newly invented interaction techniques.

69

Output Devices and Techniques

Colin Ware
*University of New
Brunswick*

69.1 Introduction

This chapter is devoted to the subject of output devices from a human centered perspective rather than device centered. One reason for taking this approach is that devices are evolving much more rapidly than humans are and only in this way is it possible to say anything that is of more than transitory interest. The primary focus is on device requirements, such as resolution in time and space, based on the properties of human sensory mechanisms. In some instances, this tells us that perception is extremely intolerant of small artifacts, such as the small differences in the colors of large adjacent patches. Equally important, in other instances, perception is tolerant of large artifacts. For example, global shifts in color over an entire display are difficult to detect. Both of these have profound implications for the design of visual displays.

There are many areas in which the distinction between output devices and input devices is becoming blurred. We are used to regarding a screen as a passive output device and a mouse as an input device. However, in the real world many things work in both ways. A sheet of paper or a piece of clay can be used to both record ideas (input) and display them (output). The coupling of input and output is becoming more common in computer interfaces, which makes the separation of the two in this handbook somewhat artificial. Nowhere is this more apparent than with force reflecting devices where, to provide a good sensation of contact, it is recommended that position (input) and force (output) be coupled in a tight loop running at 5000 Hz to simulate a good crisp sensation of contact with a surface.

We begin with some generalizations. Humans get perhaps 70% of all sensory input from vision, with lesser amounts from audition, touch, and the other senses. This is reflected in this chapter by the relative weights given to visual, auditory, and touch output. A visual display can present an excellent picture of state information, localized in space. By contrast, an auditory display presents information that is available

0-8493-2909-4/97/$0.00+$.50
© 1997 by CRC Press, Inc.

regardless of bodily orientation, but since the signal is distributed over time, the message had better be short if it is urgent. Because of these differing qualities, auditory displays are most useful for giving warnings and information about state transitions, whereas visual displays are best for presenting information about the structure of information. Of course, visual displays also allow for state transitions to be perceived through animations and, in general, visual displays will offer the most obvious choice when presenting information.

This chapter is organized according to sensory principles, starting with display issues relating to low-level vision, display resolution, and display uniformity, followed by a discussion of color vision and display issues and guidelines relating to color displays. Next, stereo is discussed and, finally, **virtual reality** (VR) displays. A relatively small amount of space is devoted to auditory and touch displays, respectively, because of their lesser importance in the user interfaces of the present day. In each section the basic sensory principles relating to the display problem are introduced followed by recommendations.

69.2 Visual Displays

We begin with the basic issues relating to display brightness, uniformity, and spatial and temporal resolution. The human eye has an enormous dynamic range. The amount of light reflected from surfaces on a bright day at the beach is about five orders of magnitude higher than the amount available under dim lamp lights. Yet the shapes, layouts, and colors of objects are all perceived in very similar ways under both sets of conditions. Most displays are self-luminous consisting of back lit liquid crystal displays (LCDs) or cathode ray tubes (CRTs). The best of these devices have a dynamic range (the ratio between the maximum and minimum values produced) of a little more than two orders of magnitude, clearly inadequate to reproduce the range of human environment. The reason why these displays work so well is because the eye adapts to the ambient light, and humans are quite insensitive both to the overall level of the illumination and to the overall amount of contrast. Because we must recognize objects in both fog and brilliant sunlight the perceptual system generalizes across these conditions. Thus, we can tolerate extreme variation in the amount of contrast produced by a display.

In practice, when many screen-based devices are viewed under normal room lighting conditions, 15–40% of the light coming from the screen to the user's eye is actually ambient room light reflected by the front surface of the phosphors, or off the screen surface. This means that the effective dynamic range of most devices as they are used in lighted rooms is no better than three or four to one and this is because the devices are not viewed in dark rooms. As a consequence, the best and cheapest ways of getting a better display device in many cases is to shield the display from ambient light. Commercial antiglare screens are neutral density filters, which work because the room light is attenuated twice as it passes through to the screen and is reflected back out again, while the light from the display passes through only once.

Spatial Resolution

Visual acuity is the name given to the ability of the human visual system to resolve fine targets. A standard way of measuring this is to determine how fine a sinusoidal striped pattern can be discriminated from a uniform gray. People with good eyesight are capable of perceiving targets as fine as 50–60 cycles/degree of visual angle when the pattern is of very high contrast.

Figure 69.1 illustrates the spatial sensitivity of the human eye as a function of spatial frequency. Specifically, it illustrates the degree of contrast required for sinusoidal gratings of different spatial

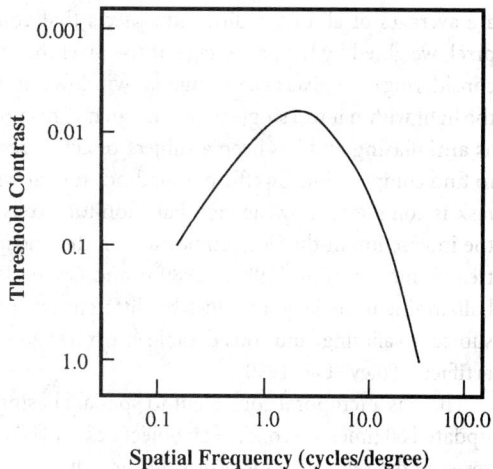

FIGURE 69.1 Spatial contrast sensitivity function of the human visual system. There is a falloff in sensitivity to both detailed patterns (high-spatial frequencies) and to gradually changing gray values (low-spatial frequencies).

frequencies to be perceived. The function can be seen to be an inverted U shape with a peak at about 2 cycles/degree of visual angle. This means that 5-mm stripes at arms, length are optimally visible. The falloff at low spatial frequencies is important since this means that we are insensitive to gradual changes in overall screen luminance. Indeed most CRTs have a brightness falloff toward the edges of as much as 20%, which we barely notice. This display nonuniformity is even more pronounced with rear projection systems due to the construction of screens that project light primarily in a forward direction. This is called the **screen gain**, and a gain of 3.0 means that three times as much light is transmitted in the straight through direction compared to a perfect **Lambertian** diffuser. At other angles, less light is transmitted so that at a 45° off-axis viewing angle only half as much light may be available compared to a perfect diffuser. The result is that for many viewing angles the amount of light across the screen may vary by a factor of two or more even for a fixed viewpoint. Screen gain is also available with front projection with similar nonuniformities as a consequence, although the use of curved screens can compensate to some extent.

The receptors in the human eye have a visual angle of about 0.8′ of arc. High-resolution screens are commonly available that provide 1200 pixels in a 30-cm screen, about 40 pixels/cm. A simple calculation reveals that at about a 50-cm viewing distance, pixels will subtend about 1.5′ of arc, about two times the size of cone receptors in the center of vision. Viewed from 100 cm such a screen has pixels that will be imaged on the retina about the same size as the receptors. This might suggest that we are in reach of the perfect display in terms of spatial resolution. Such a screen would require approximately 80 pixels/cm at normal viewing distances. However, there are a number of **superacuities** where the human visual system is capable of producing performances that imply resolution better than the receptor size. The example of stereo superacuity will be discussed in a later section. The other reason why we may wish to have screens of higher resolution relates to the display artifact known as aliasing.

Aliasing

Sampling theory tell us that a signal can only be reconstructed from its samples if the original signal is sampled at a frequency that is at least two times the highest frequency it contains. It is simplest and fastest to do computer graphics by sampling the scene only once for each pixel. This results in straight lines being drawn, broken up into square pixels, as shown in Fig. 69.2. The effect is known as aliasing or the *jaggies* for static scenes and the *crawlies* for moving scenes. It can be removed by computing pixel color values that are averages of all of the different objects that contribute to the pixel, weighted by the percentage of the pixel they cover. It is like considering each pixel as a rectangular window and averaging all of the light within it to a single uniform patch. This process is known as **antialiasing** and has been a subject of active research in order to find computationally efficient methods for carrying it out. The

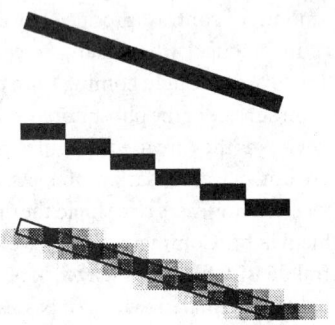

FIGURE 69.2 Aliasing artifacts and antialiasing as a solution.

task is complicated by the fact that monitor pixels are influenced by the distribution of light caused by the interaction of the electron beam with the phosphors on screen in a CRT display. For more complete treatments see Blinn [1989a, 1989b] and Crow [1981]. In the domain of print media, the practice of half-toning to make gray values by different dot patterns having different dot densities and sizes is also subject to aliasing, and considerable effort has gone into the development of algorithms to reduce visible artifacts [Foley et al. 1990].

There is a temporal equivalent to spatial aliasing because of the fact that most screen images are only updated 60 times a second. If an object has moved a considerable distance in that time, then it appears at a series of discrete locations. The solution is to color each pixel according to the percentage contribution of different objects that pass behind it for the duration of the 1/60 s interval. This is often called motion blur or temporal antialiasing. Since this is clearly computationally expensive, stochastic sampling techniques may be used [Cook 1986].

Increasingly, the drawing of static antialiased lines is being supported in hardware by computer graphics systems. However, temporal antialiasing is only done with single frame animation techniques for high-quality computer animations as well as TV commercials and movies.

Temporal Resolution, Refresh, and Update Rates

There is a psychophysical quantity, called the **flicker fusion frequency**, that represents the least rapidly flickering light not perceived as steady. Typically flicker fusion frequency is around 50 Hz for a light that turns completely on and off [Wyszecki and Stiles 1982].

In discussing the performance of monitors it is important to differentiate the **refresh rate** and the **update rate**. The refresh rate is the rate at which a screen is redrawn and it is typically constant (values of 50, 60, 65, or 70 Hz are common). The update rate is the rate at which the image is changed, for example, in an animated scene. In interactive computer animation it is common to have an update rate that is only 10 Hz because of the large amount of material that must be drawn. A rule of thumb states that a 10-Hz update rate is a necessary minimum for smooth animation. However, this clearly depends on factors such as the rate of motion and the size of the objects being animated. If temporal antialiasing is done, the jerky effects from low frame rates will be much reduced.

Color Vision and Color Displays

The single most important fact relating to color displays is that human color vision is trichromatic; our eyes contain three receptors sensitive to different wavelengths. For this reason it is possible to generate a full range of colors using only three sets of lights or printing inks. However, it is much more difficult to exactly specify colors using inks than using lights because, whereas lights can be treated as a simple vector space, inks interact in complex nonlinear ways.

Luminance Specification

Before generalizing to a three-color system it is useful to consider a single-color system for specifying brightness, or **luminance**. This is important not only for simplicity but because the luminance system in human vision gives us most of our information about the shape and layout of objects in space. We can function very well in an achromatic world, as the success of black and white TV attests.

The standard term for specifying how much light is emitted by a self-luminous display is luminance. The luminance standard is part of the international standard for color measurement maintained by the Commission International de L'Eclairage (CIE). The principle on which this standard is based is that of the standard observer who represents the color vision capabilities of all humans, excepting those with specific disabilities such as color blindness. This standard observer is embodied in a set of color matching functions as illustrated in Fig. 69.3. Tables for these functions can be found in Wyszecki and Stiles [1982] and other standard texts. The central function in Fig. 69.3 is the CIE $V(\lambda)$ function, also called $y(\lambda)$, which represents the amount that lights of different wavelengths contribute to the overall sensation of brightness. As can be seen from this curve, short wavelengths (blue) and long wavelengths (red) contribute much less than green wavelengths to the sensation of brightness. To find the luminance of a light source, the spectral distribution of that source is integrated with this luminous efficiency function to get a weighted average that approximates the human sensitivity to that light,

$$L = K_m \int E(\lambda)V(\lambda)\,d\lambda \qquad (69.1)$$

the result is luminance if $K_m = 680$ lm/W and $E(\lambda)$ is measured in watts per unit of solid angle (steradian). For a more complete treatment of this technically complex area see Wyszecki and Stiles [1982].

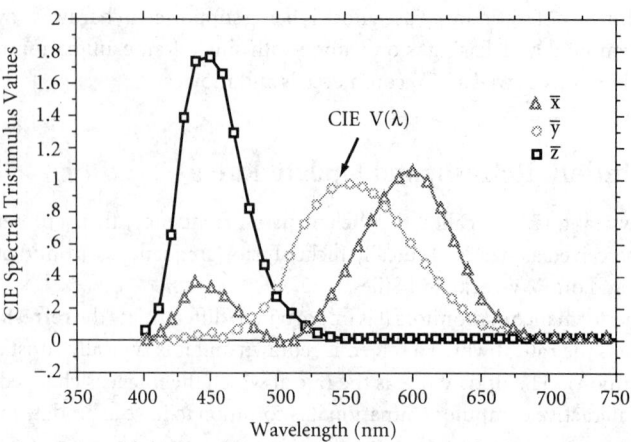

FIGURE 69.3 CIE tristimulus functions. These are used to represent the standard observer in colorimetry. Short wavelengths at the left-hand side appear blue, in the middle they are green, and to the right they are red. We are most sensitive to the green wavelengths around 560 nm.

Color Specification

Full trichromatic color specification is a straightforward extension of luminance specification. To obtain a color specification in CIE tristimulus values we integrate the source spectral energy distribution with all three of the functions shown in Fig. 69.3, x, y, and z. These functions are, in fact, transformations of the shape of the sensitivity functions of three cone receptor types in the human eye. Note that in practice the integrals are done using tables that can be found in Wyszecki and Stiles [1982] and other standard texts. The resulting tristimulus values are denoted by, X, Y, and Z. The Y tristimulus value represents luminance, and L and Y are used interchangeably

$$X = K_m \int E(\lambda)\bar{x}_\lambda \, d\lambda$$

$$Y = K_m \int E(\lambda)\bar{y}_\lambda \, d\lambda \tag{69.2}$$

$$Z = K_m \int E(\lambda)\bar{z}_\lambda \, d\lambda$$

A common way of representing display colors is in terms of CIE chromaticity coordinates on a CIE chromaticity diagram as shown in Fig. 69.4. These coordinates are a given by

$$x = \frac{X}{(X + Y + Z)}$$

$$y = \frac{Y}{(X + Y + Z)} \tag{69.3}$$

A color specification can be given by x, y, Y. A chromaticity diagram is illustrated in Fig. 69.4. The pure spectral hues are given around the boundary of this diagram in nanometers (10^{-9} m).

The gamut of all possible colors is the dark gray region with pure hues at the edge and neutral tones in the center. The triangular region represents the gamut achievable by a particular color monitor, determined by the colors of the phosphors given at the corners of the triangle. Every color within this triangular region is achievable, and every color outside of the triangle is not. This diagram nicely illustrates the tradeoff faced by the designer of color displays. A phosphor that produces a very narrow wavelength band will have

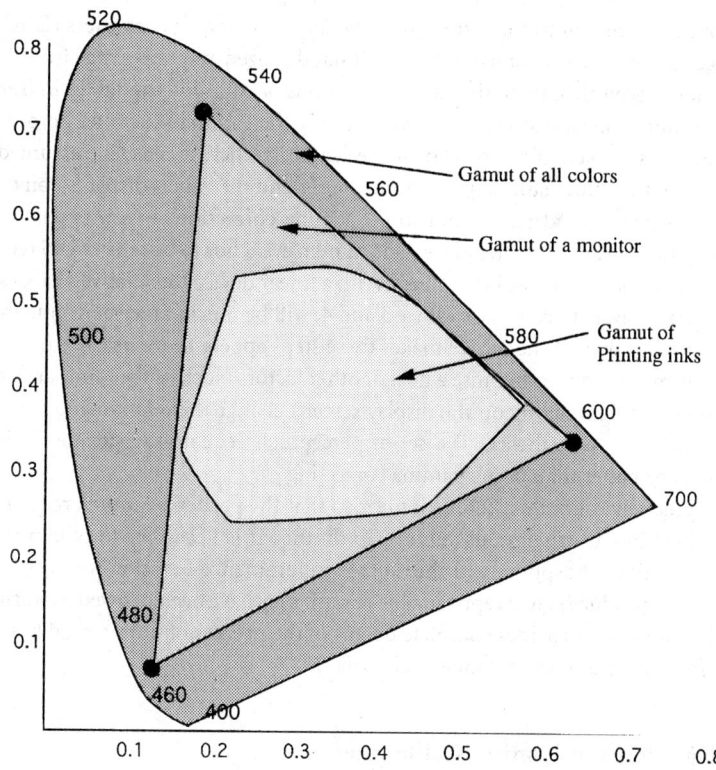

FIGURE 69.4 A CIE chromaticity diagram with a monitor gamut and a printing ink gamut superimposed. It can be seen that the range of available colors with a color printing is smaller than that available with a monitor, and both fall short of providing the full range of color that can be seen.

chromaticity coordinates close to the pure spectral colors, and this will enlarge the triangle. However, this narrowband will also mean that little light is produced. The same principle holds for filter-based displays. Narrowband filters produce less light than broadband filters but allow for the production of more saturated colors. The irregular shape inside the triangle illustrates the gamut of colors obtainable using printing inks. Notice that this set of colors is still smaller, causing difficulties when we wish to try to obtain a hard copy reproduction of the colors on the monitor. This problem is discussed in the section on gamut mapping.

The spacing of colors on the chromaticity diagram is not perceptually uniform; neither is it in X, Y, Z tristimulus coordinates. There are, however, transformations of CIE tristimulus values that produce a space in which equal metric distances are closer to matching equal perceptual differences. Extensive research has been done on the subject of these **uniform color spaces**, which can help in producing perceptually equal color and brightness scales but it should be emphasized that these do not take surrounding color or brightness into account. There are two uniform color space standards, called CIElab and CIEluv, respectively (see Wyszecki and Stiles [1982] for details). The fact that there are two relates to the different requirements of the paint industry and the lighting industry. Work has also been done on using uniform color spaces to produce useful color sequences in map displays [Robertson 1988].

Gamma Correction and Monitor Calibration

Most CRT displays apply a nonlinear mapping between the voltage value input and the luminance output. The function is approximately a gamma function. That is,

$$ScreenLuminance = CV^{\gamma}$$

where V is the voltage value sent to the screen from the digital to analog converters (DACs) that convert frame buffer image values. The value for γ is not at all standardized and may range from 1.5 to 3.0 with a value of 2.5 being fairly typical. One of the earliest functions of color lookup tables in frame buffers was to turn the monitor into a linear output device by inverting the gamma function, a procedure known as **gamma correction**. This is essential for realistic scene synthesis where it is usually assumed that there is a linear relation between the values stored in the frame buffer and the values displayed on the screen.

However, there is often a good reason not to do gamma correction, which is related to the reason monitors were made nonlinear in the first place. The human eye tends to be sensitive to relative changes in light intensity $\delta I / I$ as opposed to absolute values. With a linear device the relative changes in intensity at the low end of the scale, say, between color values 3 and 4, will be much larger perceptually than between 254 and 255. The nonlinearity of the device makes these steps appear more nearly equal. Unfortunately, the situation is made more complex because a host of other factors (such as the luminance of surrounding colors and the amount of light falling on the display screen) change the relationship between luminance and perceived brightness. Nevertheless, if a set of perceptually equal gray steps is desired no gamma correcting is likely to be better than doing gamma correcting.

It is possible to precisely calibrate a color monitor so that the particular inputs required to produce a color may be specified in CIE tristimulus values. To do this, it is necessary to obtain the chromaticity coordinates of the monitor phosphors and the voltage to luminance function for each phosphor. CRT monitors are known to produce stable reproducible color if they have been allowed to warm up for half an hour or more. Cowan [1983] provides complete details of the process of monitor calibration. A method for calibrating print devices is given in Stone et al. [1988].

Chromatic Adaptation and Monitor Calibration

It should be emphasized that, despite the fact that CIE takes the relative sensitivities of the human color receptors into account, it is still a physical standard. It does not take effects such as color contrast and adaptation into account, and these dramatically change the appearance of colors. The process known as chromatic adaptation occurs in the receptors and in the early stages of visual processing. This means that if we are in a room illuminated with a colored light, we gradually become less sensitive to that color (over a period of a few minutes). This can be attributed in part to a shift in the relative sensitivities of the different receptors. For this reason we are mostly unaware that daylight is much bluer than the tungsten light produced from ordinary light bulbs. The practical implication of this is that we can get by with television sets and color printers that are grossly out of calibration. We simply do not notice the overall shift in the colors if they occur over a relatively large area.

Gamut Mapping

At the start of this section, the generalization was made that a three-color output system could produce a complete range of colors. However, this statement is only true if the colors remain within the gamut of the device. Figure 69.4 shows a cross section through the gamut of a CRT device, and it also shows a cross section through a printing ink gamut. It can be seen that the printing ink gamut lies almost entirely within the range of the CRT gamut. Moreover, the ink gamut will vary with the paper it is printed on and the color of the illumination. For this reason there are many screen images that cannot be accurately reproduced on paper.

One solution to this problem would be to simply truncate all colors outside of the printing ink gamut. However, there are much better solutions. Because the eye is relatively insensitive to overall color shifts and overall contrast changes, we can take the gamut from one device and map it into the gamut of the printing inks (or some other device) by compressing and translating it. This is what Stone et al. [1988] call gamut mapping, and the process is designed to preserve the overall color relationships while effectively using the range of the device. However, it should be noted that the original colors will be lost in this process and that after a succession of gamut mappings colors may become quite different from their original values.

Color Information Coding

When considering information display, one of the most important distinctions is between chromatic and luminance information because these types of information are treated quite differently in human perception. Gray scales are not perceived in the same way as rainbow colored scales. A purely chromatic difference is one where two colors of identical luminance, such as red and green, are placed adjacent to one another. Research has shown that we are insensitive to a variety of information if it is presented through purely chromatic changes. This includes shape perception, stereo depth information, shape from shading, and motion. However, chromatic information is of great benefit to us in helping us classify the materials from which objects are made. A number of practical implications arise from the differences in the way luminance and chromatic information is processed in human vision:

- The fact that our spatial sensitivity is lower for chromatic information is important for image compression because it means that less information can be transmitted about hue relative to luminance. Indeed this is built into many image compression standards, including the color television transmission standard [National Television Systems Committee (NTSC)] and the new high-definition TV standard.

- To make text visible it is important to make sure that there is a luminance difference between the color of the text and the color of the background. If the background is likely to be variable it is a good idea to put a contrasting border around the letters. This is commonly done with computer cursors that are created with a white border around a dark center to ensure exactly this kind of contrast.

- When spatial layout is shown either through a stereo display or through motion cues, ensure adequate luminance contrast.

- When fine detail must be shown, for example, with fine lines in a diagram, ensure that there is adequate luminance contrast with the background.

- Chromatic codes are useful for labeling objects belonging to similar classes. Because colors are used for identifying materials in everyday life they are useful in coding information into categories.

- Color (both chromatic and gray scale) can be used as a quantitative code, for example, when a color key is used on a map to represent altitude information. However, there may be problems of distorted readings due to simultaneous contrast effects. Simultaneous contrast is the changes in color appearance of a patch that occur because of the colors of adjacent patches. Certain choices of colors can minimize this [Ware 1988].

A number of empirical studies have shown color coding to be an effective way of identifying information. It is also effective if used in combination with other cues such as shape. For example, targets may be responded to faster if they can be identified by both shape and color differences (see Christ 1975, Stokes et al. [1990], and Silverstein [1987] for useful reviews). Color codes are also useful in the perceptual grouping of objects. Thus, the relationship between a set of different screen objects can be made more apparent by giving them all the same color. However, it is also the case that only a limited number of color codes can be used effectively. The use of more than about 10 will cause the color categories to become blurred. In general, there are complex relationships between the type of symbols displayed, (e.g., point, line, area, or text), the luminance of the display, the luminance and color of the background, and the luminance and color of the symbol [Spiker et al. 1985].

General Issues in Information Coding

The greatest challenge in developing guidelines for information coding is that there are usually effective alternatives, such as color, shape, size, texture, blinking, orientation, and gray value. Although a number of studies do comparisons between one or more coding methods separately, or in combination, there are so many interactions between the task and the complexity of the display that guidelines based on science are not generally practical. However, Tufte provides excellent guidelines for information coding from an aesthetic perspective [Tufte 1983, 1990]. One theoretical concept that has interesting implications is whether or not the coding used can be processed in parallel by the visual system (called **preattentive** discrimination).

Preattentive Codes

The fact that certain coding schemes are processed faster than others is called the popout phenomenon, and this is thought to be due to early preattentive processing by the visual system. Thus, for example, the *shape* of the word "color" is not processed preattentively, and it will be necessary to scan this entire page to determine how many times the word appears. However, if all of the instances of the word "color" were printed in red they would pop out at the viewer (as long as there were not too many other colors on the same page), and if there were less than seven or so instances they would be processed at a single glance. Preattentive processing is done for color, brightness, certain aspects of texture, stereo disparities, and object orientation and size. Codes that are preattentively discriminable are very useful if the rapid search for information is desired [Triesman 1895].

The following visual attributes are known to be preattentive codes and, therefore, useful in differentiating information belonging to different classes:

- Color: use no more than 10 different colors for labeling purposes.
- Orientation: use no more than 10 orientations.
- Blink coding: use no more than 2 blink rates.
- Texture granularity: use no more than 5 grain sizes.
- Stereo depth: the number of depths that can be effectively coded is not known.
- Motion: objects moving out of phase with one another are perceptually grouped.

The number of usable phases is not known.

The Object Display

When the purpose of a display is to allow a user to integrate diverse pieces of information, it may make sense to integrate the information into a single visual object or *glyph* [Wickens 1992]. For example, if the purpose is to represent a pump, the liquid temperature could be shown by changing the color of the pump, the capacity could be shown by the overall size of the pump, and the output pressure might be represented by the changing height of a bar attached to the output pipe. This should allow for a better representation than a set of individual dials showing these attributes separately. However, the disadvantage of the object display is the perceptual distortions that can result from an ill-chosen display mapping. There are unresolved issues such as whether perception of quantity is related to the linear size, area, or volume of a displayed object [Tufte 1990]. For more abstract data representation tasks the choice of whether color, size, orientation, or texture is used to represent a particular data attribute may not be an easy one. Thus, although object displays can often be effective, they must usually be custom designed for each different display problem. In general, this means that the display should somehow match the user's cognitive model of the data [Cole 1986].

In object displays, input and output are becoming integrated. Thus, in the preceding example graphical representation of the pump (and hence an actual pump) could be controlled by using the mouse to directly manipulate the size of the bar representing pump output. For some good examples of the linking of output and input see Ahlberg and Shneiderman [1994].

Stereo Displays

Emerging technologies are making interactive stereoscopic displays more useful. Although stereo displays are part of the virtual reality paradigm discussed later, screen-based stereo displays also can be useful by themselves. There are a number of human factors issues relating to stereo displays that make them difficult to use effectively.

A stereo display uses the ability of the visual system to interpret the differences between images presented to the two eyes as information about the layout of objects in space. Figure 69.5 illustrates the simplest possible stereo display. The eyes are fixated on the vertical line *a*. A second line *b* is closer to *a* in the right eye's image than in the left eye's image. The brain resolves this discrepancy by perceiving the lines as being at different depths.

Retinal disparity is the difference between the angular separation of *a* and *b* in the two eyes (disparity = $\alpha - \beta$).

Vergence is the degree to which the two eyes converge to fixate a target (this is also called phoria).

If the disparity between the two images becomes too great, then *diplopia* occurs. This is the phenomon of seeing a double image. Another way of putting this is that the images are no longer *fused*. However, there is evidence that depth judgments can still be made from a diplopic image, although they will be less accurate. The area

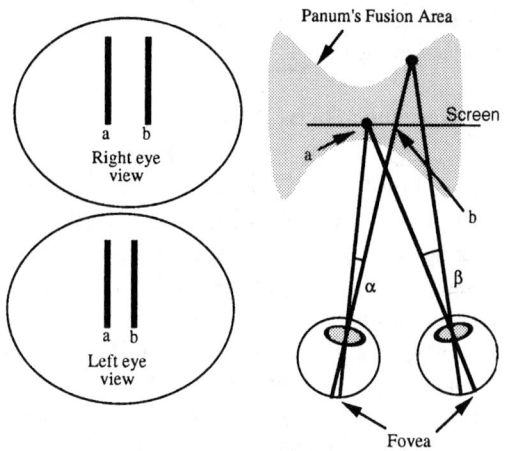

FIGURE 69.5 The geometry of a simple stereo display. Two sets of differently spaced parallel vertical lines are imaged by the two eyes as shown on the left. The plane view on the right shows how this can be resolved into lines at different depths. The two eyes are assumed to be fixated on line *a*. In other words, the center of this line is sharply imaged on the fovea. Panum's fusion area is a term describing the region where images at different depths can be fused into a single image by the brain. The difference between the angles α and β is a measure of stereo disparity.

(or more strictly the volume) within which fusion occurs is called *Panum's fusion area* (see Fig. 69.5). However, the size of Panum's fusion area is highly dependent on a number of visual display parameters such as the exposure duration to the images and the size of the targets. It is also true that depth judgments can be made despite diplopia (in other words, outside of the fusion area), although these are less accurate. For an excellent introductory review of stereo vision from a human factors perspective see Patterson [1992].

The **fovea** is the central region of the retina where the vision is most acute. In the center of the fovea, the maximum disparity before fusion breaks down is only one-tenth of a degree in the worst case, whereas at 6° eccentricity (from the fovea) the limit is one-third of a degree. Consider what these numbers imply for a typical display. Display resolution for conventional flat screen displays is computed by the number of pixels per centimeter, typically about 30 for a high-resolution system. Given a viewing distance of 65 cm and an interpupilary distance of about 6.5 cm we can compute the resolution in depth available in a stereo display. The depth difference corresponding to a 1-pixel horizontal disparity is 10 pixels at this distance. Thus, our typical display can be considered as having 30 pixels/cm in the plane of the screen but only 3 pixels/cm in and out of the screen. The one-tenth of a degree maximum disparity available before diplopia occurs (for points close to fixation) will encompass about 3 pixels, in either direction. Only three depth steps will be discriminable before diplopia occurs, but if we consider objects in front and behind the screen we get a total of six discriminable depth steps for our typical display. At 6° from the fovea this becomes about 20 discriminable depth steps, which would apply to objects at the edge of the screen if we were fixating the center. In other words, depth discrimination is very poor in stereo computer graphics displays and it is typically much worse in **immersion** displays because of the relatively large pixels. Antialiasing techniques can increase the effective resolution, but the 10/1 ratio between horizontal resolution and depth resolution remains in effect at this viewing distance. It should be emphasized that this is the worst case for sharp, well-differentiated objects. The fusion area is larger if the objects are out of focus or if they are moving. Stereopsis is one of the superacuities, disparities smaller than 5 s of arc to be detected [Westheimer 1979] thus the stereo processing mechanism can easily use the information in a screen with 400 pixels/cm at normal viewing distances.

The Vergence-Focus Problem

When we look at objects at different depths two things occur. One is that the vergence of our eyes changes as previously described. The other is that the eyes change in focus to accommodate the difference in depth. The fact that these vergence and focus mechanisms are neurally coupled can be easily demonstrated by covering one eye and looking at a nearby object with the other. The covered eye converges inward even though it cannot see the target. This is focus-driven vergence but it works both ways, vergence can also drive focus. Unfortunately from the point of view of stereo displays, focus is difficult to simulate. We can make images appear out of focus but this does not change the focal length of the lens needed to put things into focus. That is, correct vergence and focus information can be provided only for objects in the plane of the screen. Thus, many VR helmet displays do not use stereo. In general, it is a good idea to keep objects that are subject to interaction near to the plane of the screen in stereo displays. In VR systems this is a distance artificially set by the optics, usually about 2 m in front of the observer. Another way of making stereo displays comfortable is to artificially reduce the disparities from the geometrically correct values. This can be done so that scenes having a lot of depth are given small eye separation values, while objects with little depth have large eye separation values [Ware et al. 1995].

Methods for Achieving Stereo

There are a large number of ways of creating stereo images. Here we shall consider the three most common ways of creating stereo using a monitor display.

Frame Sequential. In the frame sequential method, stereo shutters are used to ensure that the two eyes receive alternate frames of the video image. In one widely used technique, shutters are incorporated in glasses, which synchronize with the monitor via an infrared link. In another technique the polarization of a screen covering the monitor can be changed for alternate frames; in which case the user wears glasses that are polarized differently for each eye. Circular polarization is used since this is not affected by head orientation. Either technique works well if the graphics system can provide at least 50 updates per second to each eye. Systems that generate fewer updates to each eye are not recommended because they cause an irritating flicker. Another problem with this method of stereo presentation is the **ghosting** that occurs primarily because of too slow decay of the green phosphor on the monitor. This means that each eye sees a faint image from the other eye's signal.

Red–Green Anaglyphs. In red–green anaglyphs, the two images are generated by using only the red (for one eye) and green (for the other eye) monitor primaries. The user wears red and green filters over the opposite eyes, which effectively block the undesired images. This is a low-cost solution for experimenting with stereo displays. However, its disadvantages are that only a monochrome image can be displayed and the combination of red and green to the two eyes often produces strange color effects in the combined image.

Mirror Systems. With this method, the screen is divided into two parts, one part for the right eye's image and the other part for the left eye's image. These images are displaced by a system of mirrors (or sometimes prisms) so that the two parts appear superimposed. This is an excellent low-cost solution that creates a high-quality stereo image, with no ghosting but with a sacrifice of the effective display size, since half of the screen must be devoted to each eye. It also provides the most constrained viewing configuration.

69.3 Virtual Reality Displays

The goal of virtual reality displays is to simulate environments so that we can understand and interact with synthetic scenes in ways we have already learned from everyday reality. There has been considerable success

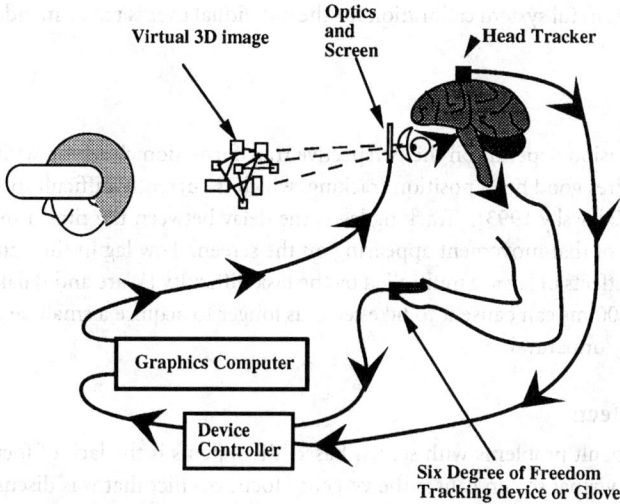

FIGURE 69.6 The two main human–machine control loops in a VR system. In one, a head tracking device is used to update the screen image depending on the user's head virtual viewpoint. In the other, some virtual object is manipulated by the hand.

in generating sounds localized in 3D space. Some success has been achieved in providing visual cues, but touch cues have proved to be the most difficult to display. In this brief analysis the results achieved by these displays are summarized, along with some of the outstanding problems. The term virtual reality is often used to describe highly interactive 3D displays such as those used in 3D video games. However, the more strict definitions relate to a display that has a helmet mounted display coupled with a head tracking system that provides an immersive experience. This is another case where input must be coupled to output in a tight loop because the purpose of the head tracking system is to estimate the user's eye position and to generate a perspective image of a virtual scene from that eye position. Figure 69.6 shows the two main control loops involved in the VR system, one involving a hand tracker and the other involving a head tracker. Ideally, both should be low lag and high precision. Another common element for this kind of display is one or more glove input devices, although the lack of touch sensation has been a major drawback to their utility.

The major advantages of helmet mounted displays are said to be twofold, a sense of presence and ease of learning. The idea is that everyday life skills can be transferred directly without new learning. Unfortunately, although there is considerable skill transfer to tasks performed with helmet mounted displays and data glove inputs, learning is still required for manipulation because of the lack of touch sensation. Some of the most important human factor issues with helmet mounted displays are discussed next.

Resolution

At present a high-resolution screen typically has 1280×1024 pixels. When this is extended over the large field of the helmet-mounted display, the pixels will be at least 20 times larger than those on a conventional monitor at normal viewing distances. To achieve a high-quality wide-angle, high-resolution display, about 5000×5000 pixels would be required for each eye.

Optical Distortions

Because powerful lens systems are required to transform the small screen in the helmet into a wide-angle display, large distortions occur when the user's eye is not placed at the correct optical point relative to the lens system. For more detailed reviews of helmet mounted display technology see Kalawsky [1993] and

Stokes et al. [1990]. Careful system calibration for the individual user is recommended in order to reduce distortions.

Position Tracking

The virtual reality illusion depends on the perspective transformation changing with the positions of the observers. This requires good head position tracking, which is currently difficult to achieve (especially at a reasonable cost) [Kalawsky 1993]. Tracking lag is the delay between the time a movement is made by a user and the result of that movement appearing on the screen. Low lag in the hand tracking system is important since the effects of lag are multiplied by the task difficulty [Ware and Balakrishnan 1994]. This means that a lag of 200 ms can cause it to take seconds longer to acquire a small target, which makes the interaction seem very unnatural.

Depth of Focus Effects

One of the most difficult problems with screen-based VR systems is the lack of focus effects. The fixed focal distance of the virtual screen causes the vergence focus conflict that was discussed earlier. There is no effective technical solution to this problem at present.

Foveal and Peripheral Displays

Resolution is much greater at the fovea in the center of vision than in the periphery. The foveal area subtends about 2° of visual angle, i.e., a disk about 2 cm in diameter held at arm's length. If we move 10° off to the side, the amount of fine detail that can be perceived declines by a factor of 10. It is also the case that color vision mostly takes place in the fovea. However, the periphery of vision is very important to our sense of orientation and to our sense of self-motion. The CAE, Inc. fiber optic helmet mounted display (FOHMD), designed for use in helicopter simulation, has a high resolution area with 1.5′ per pixel positioned by means of an eye tracker so that it is always centered on the fovea [Kalawsky 1993]. This is surrounded by a low-resolution background image 88° horizontally by 66° vertically of 5′ per pixel. The problem with such a display is its inherent complexity since it must incorporate a high-speed eye tracking couple as well as head tracking, together with four image generation systems.

Vection and Motion or Simulator Sickness

Vection is the sensation of self-motion that occurs in many kinds of displays with large visual fields of view. One of the most serious problems with helmet mounted displays is simulator sickness, which may occur in 50% of users or more. It is generally agreed that simulator sickness arises because two different sensory systems are providing conflicting cues. The visual system provides strong cues for the perception of self-motion by the observer through space (vection). The vestibular system of semicircular canals in the inner ear and the otolith organs provide us with information about changes in the motion of the head. In flight simulators, the limited motion of the simulator means that full motion cues cannot be provided. To make the best use of the range of motion available, these simulators typically attempt to match the high-frequency rapid motion changes while drifting back to a neutral position at a rate which is intended to be subthreshold.

In VR systems, if the user only moves within the range of the helmet system natural head motion can be completely accommodated. However, system lag and poor calibration will still cause simulator sickness. Also, most systems have some kind of flying mode to allow for large changes in virtual distance, and these will necessarily result in a vestibular-visual conflict because the flying only results in visual cues to motion without the corresponding vestibular cues.

Simulator sickness is known to get worse with prolonged exposure. It is also known that moving the head in a vehicle that is moving along a convoluted path is an effective way of producing motion sickness (both in real and simulated situations). The weight of a VR helmet can itself be a factor because it alters the cues to head orientation provided by joint and muscle sensors in the body [Lackner and DiZio 1989].

The following guidelines are recommended for the use of immersive helmet mounted systems.

- Users should be introduced to VR systems and simulators with multiple short experiences distributed over a number of days. This kind of scheme will reduce the amount and severity of motion sickness [Biocca 1992].
- The helmet system should be as light as possible, ideally less than a kilogram.
- Users should be instructed to limit head movements, especially when executing maneuvers in virtual vehicles.
- System lag between the tracking device and the graphics update should be less than 50 ms, and the screen update rate should be better than 10 frames/s.
- In addition to the technical difficulties involved in building a high-quality VR display there is a broader range of human factors problems. For instance, people do not like to wear apparatus for extended periods. The apparatus shuts out the rich and varied everyday workspace and most normal interaction with colleagues. There is also the problem that the occupant of VR will also be constrained both by (invisible) objects in the real world and by the tether of wires linking the helmet and other devices to the computing machinery. It is therefore recommended that unless the immersive experience is critical, for example, when guiding remote vehicles or playing video games, VR systems should not be used. Highly interactive screen-based displays may provide a better solution.

Fish Tank Virtual Reality Displays

It is possible to obtain a high quality VR display in a limited space by using a head tracking system coupled with a high-resolution, high-performance workstation. The head tracker is used to estimate the user's eye positions and couple the perspective view to them in such a way that the user has the impression of virtual 3D objects suspended in the vicinity of the monitor screen. Because of its restricted requirements, this kind of display can currently provide much better human factors in many respects than the immersion display [Deering 1992, Arthur et al. 1993]. The angular resolution of the pixels is smaller, and the vergence-focus problems are reduced. However, the sense of immersion is lost, although virtual objects do have a presence in the everyday workspace of the user.

Three-Dimensional Sound Displays

It is possible to synthesize spatially localized sounds with a quality such that spatial localization in the virtual space is almost as good as localization of sounds in the natural environment [Wenzel 1992]. Auditory localization appears to be primarily a two-dimensional phenomenon. That is, observers can localize in horizontal position (azimuth) and elevation angle to some degree of accuracy. Azimuth and elevation accuracies are of the order of 15°. As a practical consequence this means that sound localization is of little use in identifying sources in conventional screen displays. Where localized sounds are really useful is in providing an orienting cue or warning about events occurring behind the user, outside of the field of vision.

There is also a well-known phenomenon called visual capture of sound. Given a sound and an apparent visual source for the sound, for example, a talking face on a cinema screen, the sound is perceived to come from the source despite the fact that the actual source may be off to one side. Thus, visual localization tends to dominate auditory localization when both kinds of cues are present.

69.4 Augmented Reality

Augmented reality displays are displays that add to the surrounding environment rather than blocking it out (as is the case in immersive reality). Thus, for example, the user may wear a semitransparent display that has the effect of projecting labels and diagrams onto objects in the real world. It has been suggested that this may be useful for training people to use complex systems, or for fault diagnosis. For example, when repairing an aircraft engine the names and functions of parts could be made to appear superimposed on the parts seen through the display together with a maintenance record if desired [Feiner et al. 1993]. One of the technical difficulties with this kind of display relates to the fact that the computer must contain a detailed model of the environment, otherwise it is not possible to match the synthetic objects with the real ones. One area where this kind of display has already been used is that of heads-up displays for fighter aircraft. Information about flight paths and various threats in the environment are projected on the screen in front of the pilot as a transparent overlay of the actual environment [Stokes et al. 1990].

69.5 Touch Output Displays

A force display is one in which touch information is synthesized by producing forces on the skin of the operator. However, the touch sensation is extraordinarily complex, involving sensitivity to small shear forces in the skin as well as pressure sensors in the skin and in the joints [MacKenzie 1994]. The present state of the art with such devices is mostly confined to the production of forces based on the position of the manipulator, typically over a small range of motion and with only two or three degrees of freedom. Thus, it is possible to simulate forces on the hand when manipulating some abstract object such as a virtual molecule, but this is far from stimulating the full range of touch sensations. Indeed, there appears to be no physical means by which a complex tactile stimulus can be delivered except in a very localized way. This is probably the biggest obstacle to the full implementation of the virtual reality concept. We can construct a world that we can see and hear but not touch, except in rudimentary ways. The construction of force output devices is extremely technically demanding. They must be stiff in order to be able to create the sensation of solid contact, light so that they have little inertia themselves, and there must be a tight loop between input (position) and output (force). Sigoma [1993] has suggested that having this loop iterate at 5 kHz may be necessary for optimal fine motor control. These high frequencies are necessary to simulate contact with a solid inelastic surface. Nevertheless, it has been shown that force feedback improves performance in certain telerobotic applications when, for example, inserting a peg into a hole [Sheridan 1992]. There is little doubt that the use of force output has potential for applications such as arthroscopic surgery and telerobotics. It may be only a matter of time before we can touch the windows and text on the screen of the display [Akamatsu et al. 1994].

69.6 Auditory Displays

Auditory displays can be divided into two categories: speech and nonspeech. With auditory displays, the problems of display technology are less severe than with visual displays. Inexpensive digital sound output systems exceed the temporal resolution of the human ear. The important issues relate to how sound should be used in a user interface. Of course, speech in everyday life is a tool in dialogue; hearing and speaking are closely coupled. Nevertheless, since contemporary machines have very limited intelligence, this dialogue is very asymmetric.

Nonspeech Audio

The need to resort to auditory output is caused by the problem of visual overload and clutter. Nonspeech sounds may be especially useful for attracting the attention of the user in tasks where the eyes are otherwise occupied, for example, with a direct manipulation interface. Auditory alerting cues have been shown to work well, but only in environments where there is low auditory clutter. However, the number of simple nonspeech alerting signals is especially limiting, and this can easily result in confusion. In an analysis of sound signals in fighter aircraft [Doll and Folds 1985] found that the ground proximity warning and the angle-of-attack warning on an F16 were both an 800-Hz tone. This is especially dangerous since for the former the response would be normally to raise the nose, whereas for the latter the response would normally be to lower the nose.

The problem of lack of variety in simple tones has been addressed by Gaver [1989] with an approach that emphasized the ecological use of sounds. What this means is that sounds are used that have similarity to real-world sounds that have an analogous meaning. For example, copying is accompanied by the sound of pouring water, metaphorically pouring the information into a new container. An empty disc drive sounds like a hollow metal container, and so on.

Speech Output

There are three main uses for speech auditory output delivered either through recorded speech segments of completely synthetic speech. These are warnings, help systems, and data retrieval dialogues, respectively:

- *Warnings.* There has been considerable interest, especially for military applications, in the use of speech in providing warnings to the operators of complex systems. This is an effective use of synthetic speech if the environment is not already acoustically cluttered. Studies relating to such issues as the sex of the speaker have been inconclusive.
- *Help systems.* Synthetic speech is most useful where visual information is not available, for example, in touch-tone phone menu systems. However, it is becoming more and more common in tutorial systems for computer applications to give the application a more personal feel.
- *Data retrieval.* Speech output can be part of a natural language interface for an information providing system. The problems here mostly relate to natural language understanding and are beyond the scope of this chapter.

The rate at which words must be produced to appear natural is quite well defined. Thus, 178 words per minute is found to be intelligible but hurried, whereas a rate of 123 words per minute was found to be distracting and irritatingly slow [Simpson and Marchionda-Frost 1984]. A more natural rate of 156 words per minute is preferred. It is recommended by the U.S. Air Force that synthetic speech be 10 dB above ambient noise levels [Stokes et al. 1990].

69.7 Future Trends

We can expect the high-resolution color raster screen display to continue to be the dominant output device for at least the next decade. From a human factors perspective it matters little if this is a CRT-based device or an LCD device, or some new technology, except where portability is an issue. However, we can expect two current trends to continue. One is the increase in the number of pixels available on the screen up to and beyond the 1080×1920 in the draft high-definition TV standard. We can also expect to see a much greater integration of computing and entertainment systems in the home with digital television. Also activities such as consumer and professional photography are becoming digital, and this will speed the development of personal databases in the home and the workplace.

An increase in the number of both very big and very small screens seems likely. The kind of big screen application we can expect to see more of is the digital white board that allows interactive applications on a wall-sized interface. This allows for group work in ways that are not possible using a small screen [Elrod et al. 1992]. Micromechanical mirror devices can be expected to make large-screen projection systems much cheaper in the near future [Kalawsky 1993]. The other trend is toward the small screen of a personal digital assistant and other computer-based devices. This is part of the ubiquitous computing trend wherein computer power is distributed among numerous special purpose small devices [Weiser 1993].

The extent to which VR displays and stereo displays become more prevalent remains to be seen. Certainly they will become common in the consumer market for games and they will find their way into advanced CAD systems. Higher resolution helmet mounted displays will make a tremendous difference in the usability of these systems, and more applications will be found as a result. However, it is not clear that they will become widespread in more mundane computing environments for the reasons outlined previously. It would be easy to predict the more widespread adoption of stereo displays, which are a mature technology. However, the widespread use of stereo cinematography has been predicted in previous decades without it coming to pass and the next decade may be no exception. The human factors issues of focus and vergence are severe and not likely to be easily solved.

The technology for audio output is already highly evolved. The problem of auditory clutter in the workspace is serious and the solution, to acoustically isolate the users, may also cut them off from important information in the working environment. Nevertheless, there are sufficient obvious advantages for audio output that its use will expand, especially as part of the new developments in multimedia and training systems.

Touch output is an area that is still in the very early stages of evolution and where we can expect to see some significant progress, especially with small-scale low-cost devices. A device called the Phantom^{TM} illustrates the kind of approach that is likely to find many applications although it is not low cost. This device can be used to generate simulated forces in a working volume approximating a cube, 10 in on a side. This will be useful in CAD systems to feel the strain while bending free-form surfaces and other structures, in animation systems to feel the joint constraints while positioning figures, and in telerobotics to control remote devices. There also appears to be considerable potential here for the incorporation of force feedback on the smaller scale of the notebook computer. Providing a limited feel as we click words into place in a word processor would seem to be a natural application.

Defining Terms

Antialiasing: The specification of pixel color values so that they reflect the correct proportions of the colored regions that contribute to that pixel. In temporal antialiasing the amount of time a region of a simulated scene contributes to a pixel is also taken into account.

Augmented reality: The superimposition of artificially generated graphical elements on objects in the environment. Achieved with a see-through head mounted display.

Fish tank virtual reality: A form of virtual reality display that confines the virtual scene to the vicinity of a monitor screen. Typically contrasted with immersion virtual reality.

Flicker fusion frequency: The frequency at which a flickering light is perceived as a steady illumination. Useful in determining the requirements for a visual display.

Fovea: The central part of the retina at which vision is the sharpest. About 2° of visual angle in diameter.

Gamma correction: The correction of nonlinearities of a monitor so that it is possible to specify a color in linear coordinates.

Ghosting: An effect that occurs in some stereo displays due to incomplete separation of the images directed to the two eyes. One eye receives a faint ghost image of the image that is intended for the other eye.

Immersion display: A display in which the user is immersed in an artificial reality. The everyday environment is blocked out.

Lambertian diffuser: A diffuser that spreads incoming light equally in all directions.

Luminance: The standard way of defining an amount of light. This measure takes into account the relative sensitivities of the human eye to light of different wavelengths.

Preattentive processing: Visual stimuli that are processed at an early stage in the visual system in parallel. This processing is done prior to processing by the mechanisms of visual attention.

Refresh rate: The rate at which a computer monitor is redrawn. Sometimes different from the update rate.

Screen gain: A measure of the amount by which a projection video screen reflects light in a preferred direction. The purpose is to give brighter images if viewed from certain positions. There is a corresponding loss in brightness from other viewing positions.

Superacuities: The ability to perceive visual effects with a resolution that is finer than can be predicted from the spacing of receptors in the human eye.

Uniform color space: A transformation of a color specification such that equal metric differences between colors more closely correspond to equal perceptual differences.

Update rate: The rate at which the image on a computer monitor is changed.

Virtual reality: A method of monitoring a user's head position and creating a perceptive view of an artificial world that changes as the user moves, in such a way as to simulate an illusory three-dimensional scene.

References

Ahlberg, C. and Shneiderman, B. 1994. Visual information seeking: Tight coupling of dynamic query filters with starfield displays, pp. 313–317. In *ACM CHI '94 Proc.*

Akamatsu, M., Sato, S., and MacKenzie, I. S. 1994. Multimodal mouse: A mouse-type device with tactile and force display. *Presence* 3(1):73–80.

Arthur, K., Booth, K. S., and Ware, C. 1993. Evaluating human performance for fishtank virtual reality. *ACM Trans. Info. Syst.* 11(3):239–265.

Biocca, F. 1992. Will simulation sickness slow down the diffusion of virtual environment technology. *Presence* 1(3):334–343.

Boynton, R. M. 1979. *Human Color Vision.* Holt, Rinehart and Winston, New York.

Blinn, J. F. 1989a. Return of the jaggy. *IEEE Comput. Graphics Appl.* 9(2):82–89.

Blinn, J. F. 1989b. What we need around here is more aliasing. *IEEE Comput. Graphics Appl.* 9(1):75–79.

Christ, R. E. 1975. Review and analysis of color coding research for visual displays. *Human Factors* 25:71–84.

CIE 1970. CIE document on colorimetry official recommendations. *Commission Internationale de l'Eclairage Publ.* 15.

Cole, W. G. 1986. Medical cognitive graphics, pp. 91–95. In *ACM CHI Proc.*

Cook, R. L. 1986. Stochastic sampling in computer graphics. *ACM Trans. Graphics* 5(1):51–72.

Cowan, W. B. 1983. An inexpensive scheme for calibrations of a color monitor in terms of CIE standard coordinates. *Comput. Graphics* 17(3):315–321.

Crow, F. C. 1981. A comparison of antialiasing techniques. *IEEE Comput. Graphics Appl.* 1(1):40–48.

Deering, M. 1992. High resolution virtual reality. Proc. SIGGRAPH '92. In *Comput. Graphics* 26(2):195–202.

Doll, T. J. and Folds, D. J. 1985. Auditory signals in military aircraft: ergonomic principles versus practice, pp. 111–125. In *Proc. 3rd Symp. Aviation Psych.* Ohio State University, Dept. of Aviation, Columbus.

Elrod, S., Bruce, R., Gold, R., Goldberg, D., Halasz, F., Janssen, W., Lee, D., McCall, K., Pedersen, E., Pier, K., Tang, J., and Wlech, B. 1992. Liveboard: a large interactive display supporting group meetings, presentations and remote collaboration, pp. 599–607. In *ACM CHI '92 Proc.*

Feiner, S. K., MacIntyre, B., and Seligmann, T. 1993. Knowledge based augmented reality. *Commun. ACM* 36(7):52–62.

Foley, J. D., von Dam, A., Feiner, S. K., and Hughes, J. F. 1990. *Computer Graphics: Principles and Practice,* 2nd ed. Addison–Wesley, Reading, MA.

Gaver, W. 1989. The sonic finder: an interface that uses auditory icons. *Human–Comput. Interaction* 4(1):67–94.

Hodges, L. F. 1991. Basic principles of stereographic software development. In *Stereoscopic Displays and Applications II.* SPIE. Vol. 1457.

Kalawsky, R. S. 1993. *The Science of Virtual Reality and Virtual Environments.* Addison–Wesley, Wokingham, England.

Lackner, J. R. and DiZio, P. 1989. Altered sensory-motor control of the head as an etiological factor in space motion sickness. *Perceptual Motor Skills* 68:784–786.

MacKenzie, C. L. 1994. *The Grasping Hand.* Elsevier Science, North-Holland, Amsterdam.

Minsky, M., Ouh-young, M., Steele, J., Brooks, F. P. Jr., and Behensky, M. Feeling and seeing: issues in force display. *Comput. Graphics* 24(2):235–243.

Patterson, R. 1992. Human stereopsis. *Human Factors* 34(2):669–692.

Robertson, P. K. 1988. Perceptual color spaces. Visualizing color gamuts: A user interface for the effective use of perceptual color spaces in data display. *IEEE Comput. Graphics Appl.* 8(5):50–64.

Robinson, R. M., 1989. Eye tracker development on the fiber optic helmet mounted display. In *Helmet-Mounted Display*, pp. 102–109. Vol. 1116. SPIE.

Shenker, M. 1987. Optical design criteria for binocular helmet mounted displays. *SPIE Proc.* Vol. 778.

Sheridan, T. B. 1992. *Telerobotics, Automation, and Human Supervisory Control.* MIT Press, Cambridge, MA.

Sigoma, K. B. 1993. A survey of perceptual feedback issues in dextrous telemanipulation: part I. finger force feedback, pp. 263–270. In *Proc. IEEE Virtual Reality Annu. Int. Symp.* Sept.

Silverstein, D. 1977. Human factors for color display systems. In *Color and the Computer: Concepts, Methods and Research*, pp. 27–61. Academic Press.

Simpson, C. A. and Marchionda-Frost, K. 1984. Synthesized speech rate and pitch effects on intelligibility of warning messages for pilots. *Human Factors* 26:509–517.

Spiker, A., Rogers, S., and Cicinelli, J. 1985. Selecting color codes for a computer-generated topographic map based on perception experiments and functional requirements, pp. 151–158. In *Proc. 3rd Symp. Aviation Psychology*, Ohio State University, Dept. of Aviation, Columbus.

Stokes, A., Wickens, C., and Kite, K. 1990. *Display Technology: Human Factors Concepts.* SAE, Warrendale PA.

Stone, M. C., Cowan, W. B., and Beatty, J. C. 1988. Color gamut mapping and the printing of digital color images. *ACM Trans. Graphics* 7(4):249–292.

Triesman, A. 1985. Preattentive processing in vision. *Comput. Vision, Graphics and Image Process.* 31:156–177.

Tufte, E. R. 1983. *The Visual Display of Quantitative Information.* Graphics Press.

Tufte, E. R. 1990. *Envisioning Information.* Graphics Press.

Valyrus, N. A. 1966. *Stereoscopy.* Focal Press, London.

Ware, C. 1988. Color sequences for univariate maps: theory, experiments and principles. *IEEE Comput. Graphics Appl.* 8(5):41–49.

Ware, C. and Balakrishnan, R. 1994. Object acquisition in VR displays: lag and frame rate. *ACM Trans. Comput. Human Interaction* 1(4):331–357.

Ware, C., Bonner, J., Cater, R., and Knight, W. 1992. Simple animation as a human interrupt. *Int. J. Human–Computer Interaction* 4(4):341–348.

Ware, C. and Franck, G. 1996. Evaluating stereo and motion cues for visualizing information nets in three dimensions. *ACM Trans. Graphics* 15(2):121–139.

Ware, C., Gobrecht, C., and Paton, M. 1995. Algorithm for dynamic disparity adjustment. In IS&T/SPIE Symp. Electron. Imaging, *Science and Technology*, SPIE 2409, pp. 152–159.

Weiser, M. 1993. Some computer science issues in ubiquitous computing. *Commun. ACM* 36(7):75–83.

Wenzel, E. M. 1992. Localization in virtual acoustic displays. *Presence* 1(1):80–107.

Westheimer, G. 1979. Cooperative neural process involved in stereoscopic acuity. *Exp. Brain Res.* 36:585–597.

Wickens, C. D. 1992. *Engineering Psychology and Human Performance*, 2nd ed. Harper Collins.

Wyszecki, G. and Styles, W. S. 1982. *Color Science*, 2nd ed. Wiley, New York.

Interactive Techniques

Jürgen Ziegler
Fraunhofer Institute IAO,
Stuttgart, Germany

70.1 Introduction

The evolution of computer systems and their applications has been tightly linked with, and in many instances driven by, the significant improvements in the techniques through which humans can exchange information with the system. The widespread use of information technology in our professional and private lives would not be conceivable without the advances in human–computer interaction (HCI). The range of user interface techniques available has expanded enormously over the past few decades from physical switches or teletypes to direct manipulation of graphical objects, interactors for virtual three-dimensional (3D) environments, or speech interaction. One can distinguish different generations of computer systems each being connected with a particular user interface paradigm [cf. Gaines and Shaw 1986]. This historical perspective leads from noninteractive batch systems, over line-oriented command language interfaces and full-screen menus and forms, to graphical interfaces and noncommand-based interfaces, which may utilize a wide range of techniques such as voice, gesture, or body motions to allow a more implicit and natural way of interacting with the system [Nielsen 1993, Jacob et al. 1993].

This development can be seen as a continuous broadening of the communication channel between the human user and the system. The increase in bandwidth can be observed on the input side but even more distinctly on the output side of the computer system. In current systems, the availability of two-dimensional (2D) pointing devices still marks the most relevant step toward more flexible input, whereas graphical displays and multimedia have enormously extended the output capabilities of computer systems. Figure 70.1 illustrates the increase in input and output capabilities over different generations of user interfaces. It must be noted, however, that the different technology stages are in reality much less clearly defined and more interwoven than is suggested by distinct generations.

Input and output capabilities cannot be seen in isolation. The next and probably most important, aspect of broadening the communication channel effectively lies in increasing the level of interactivity which determines how tightly the input and output actions are interleaved with each other. Whereas early

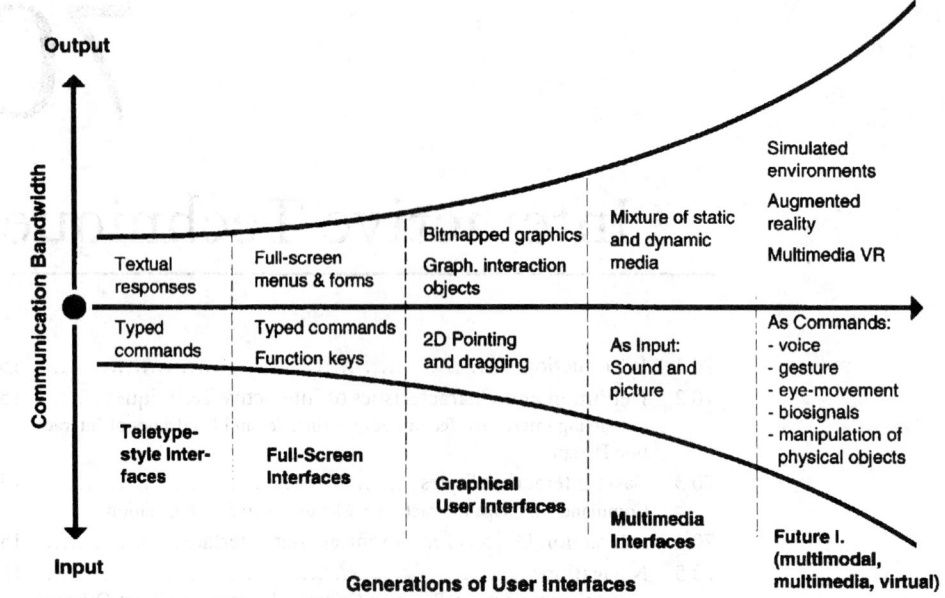

FIGURE 70.1 Different generations of interactive systems with characteristic user interface features, showing the broadening of the communication channel between user and system.

generations of interaction techniques such as command languages or full-screen menu selection followed a conversational paradigm with sequences of clearly separated user inputs and system responses, more recent interaction techniques let the users act in a graphically simulated model world with immediate feedback on their actions, often while they are being performed. High interactivity can be observed, for instance, in the case of drag-and-drop interactions, which can provide complex feedback behavior for guiding the user at a syntactic or even semantic level.

70.2 Definition and Characteristics of Interactive Techniques

The possibilities for interacting with a system are, from a technological point of view, determined by the capabilities of the system's input and output devices and by the interplay between them. Foley et al. [1990] define an interaction technique as a way of using a physical input and output device to perform a generic task in a human–computer dialogue. Generic tasks may correspond to low-level actions such as entering a numeric value or choosing an element from a set of options or to more complex activities, which are composed of elementary steps like setting a number of parameters of an object. A generic task such as selecting an option may be realized through quite different technical means such as a group of physical buttons or a **pop-up menu**. Interaction techniques must be defined, therefore, by the specific type and behavior of the input and output mechanisms used for performing the task. This may result in techniques with complex interactive behavior such as, for example, a scrollable drop-down list where the user can select an entry with the mouse, with cursor keys, or by entering the initial letters of the entry. This growing complexity of modern interaction techniques has led to the need to standardize the look and feel of **interaction objects** or **widgets** in **style guide** documents and to support their implementation through appropriate software components, made available in toolkits such as OSF/Motif, OS/2 Presentation Manager, or Microsoft Windows.

Classifying Interactive Techniques

There have been a variety of approaches to elaborate the properties of different interaction techniques and to categorize them. In one of the earliest books on HCI [Martin 1973] distinguished between user-initiated and system-initiated dialogues. User-initiated techniques such as **command language** input are considered efficient but also error prone and memory burdening. System-initiated dialogues such as **menu** selection or form filling relieve the user's memory and may help to avoid errors but are seen as less efficient. Whereas this basic distinction is still useful, the border is being blurred more and more, especially in modern graphical user interfaces, which offer a rich visual environment while allowing the user to flexibly generate a wide range of different input events. Such hybrid interaction techniques provide good visual guidance but can also be efficient and flexible in use, if designed properly.

Some more recent approaches describe interaction in terms of different underlying paradigms or **metaphors**. Hutchins et al. [1986] propose a distinction between a conversational paradigm and a model world paradigm of interaction. Whereas in the conversational paradigm the interface is seen as a language-based conversation between the user and the system with explicit turn taking, the model world paradigm is based on actions the user can perform in a simulated world (therefore also called manipulation paradigm). Command language interfaces are representative of the conversational paradigm. **Direct manipulation** user interfaces are typical of the manipulation paradigm, embodying both possibilities and constraints for action in a graphically represented model world, which also provides feedback on the effect of an action. Both paradigms have their relative merits and weaknesses. Conversational interfaces may make it difficult, for instance, to maintain a mental model of the system or to refer to entities. Manipulation interfaces can pose problems when dealing with objects currently not visible or when performing repetitive tasks.

Mixed-mode paradigms, therefore, have been proposed to eliminate these shortcomings, such as the collaborative manipulation metaphor [Hutchins 1989], which combines the conversation with an intelligent agent with the manipulation of objects in a model world. This paradigm leads to another important characteristic of interfaces. In conventional interaction styles, the system is mainly seen as a passive instance, which only reacts on the user's input, be that language-based commands or manipulative actions. By incorporating intelligent agents or objects with some simulated behavior, however, the system can also take on an active role. This will allow the user, for instance, to delegate certain tasks to an agent by having a conversation with that agent or to interact with some animated object in a virtual environment which cannot be manipulated in the conventional sense due to certain rules or constraints governing the behavior of that object (e.g., an airplane in a flight simulator). In order to encompass such cases, it seems useful to replace the notion of the manipulation paradigm by a more general actional paradigm where all kinds of user actions can serve for interacting with the system. Figure 70.2 depicts the two dimensions conversation vs. action and activity vs. passivity of the system with example interaction techniques arranged in this space. It must be noted that these dimensions represent continua and not discrete categories. Clicking a

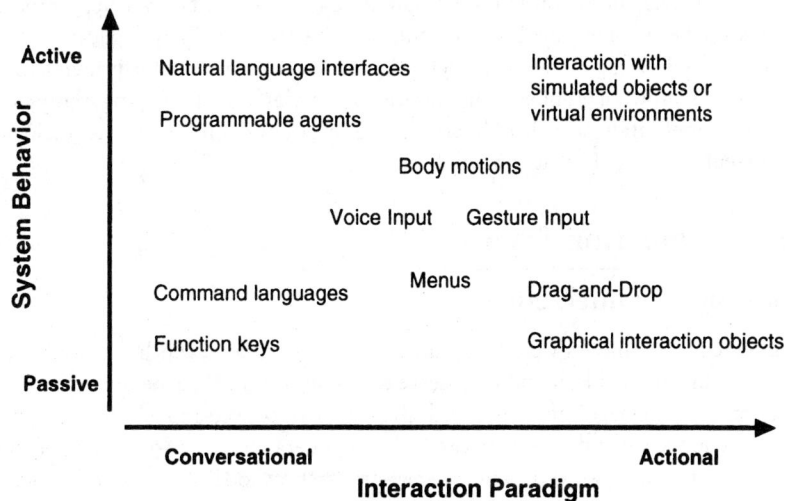

FIGURE 70.2 Dimensions of an interaction space with sample techniques.

push button, for example, in a graphical user interface may have both conversational and actional aspects, whereas a drag-and-drop operation in the same interface is more on the actional end of the continuum.

Principles and Guidelines of Interaction Design

A considerable number of principles and guidelines for designing interaction have been developed, providing the designer with more or less specific advice. Because of the number and diversity of these guidelines only some references to the relevant literature can be given here. General platform-independent guidelines such as the comprehensive texts by Smith and Mosier [1986] or Mayhew [1992] cover a wide range of design issues, addressing areas such as selection and design of interaction techniques, visual design, system messages, or help systems.

Although these general guidelines are not directed at a particular system platform or implementation, various sets of guidelines have been issued by industrial companies or consortia. These so-called style guides describe the design of an interface by means of a particular software toolkit and, therefore, are usually more specific and concrete than platform-independent guidelines. Widely used industrial styleguides are Apple [1992], IBM [1992], Microsoft [1992], and OSF [1992].

The most relevant international standard on HCI is ISO 9241 (for a more detailed description of HCI-related standards and references see Chapter 67 in this Handbook). Currently, there are 19 parts, which are at different stages in the standardization process. Parts 14–19 address different interaction techniques (menu, command, direct manipulation, form filling, question and answer, and natural language dialogues) and provide detailed design guidelines. Part 10 describes general high-level principles for dialogue design, which require an interface to have the following properties:

- Suitability for the task
- Self-descriptiveness
- Controllability
- Conformity with user expectations
- Error tolerance
- Suitability for learning
- Suitability for individualization

Although such general principles contribute to defining commonly agreed goals of user-centered design, they require a great deal of interpretation and are hardly applicable without additional descriptions and examples. Industrial style guides, on the other hand, mainly follow a bottom-up approach providing detailed guidance on the use of the low-level elements of an interface. In many application areas, however, one can see a need to fill the gap between the definition of the elements and the principles for the design of the whole system. A growing number of organizations, therefore, are developing style guides which are more tailored to the application domain of interest, for example, by defining complete dialogue structures for certain recurrent task types in that domain.

70.3 Basic Interaction Styles

Command Language Interaction

The earliest interactive techniques were based on command languages which require the user to enter names from a certain command set and to observe a specific syntax for generating complex command expressions. Command language interaction typically follows the conversational paradigm, although some forms such as function key command entry can be highly interactive. The complexity of command languages ranges from simple keyword or number (e.g., transaction code) entry, over full-fledged artificial languages with complex syntax (such as Unix), to natural languagelike command input with varying degrees of language understanding by the system.

Command languages have received considerable attention in HCI research. Surveys of empirical results can be found in Carroll [1985], Shneiderman [1987], and Barnard and Grudin [1988]. Some of the issues addressed in these studies are not limited to command language interaction, but are also pertinent to design areas such as naming menu entries or finding suitable labels for push buttons.

Command Names

When different people are asked to produce a name for a certain command, there is little agreement in the solutions offered. The name sets generated have little internal structure and are usually not easier to recall than name sets provided by others. Although naming strategies may be used, they are applied in inconsistent ways. The naming behavior of designers in these respects has been found to be very similar to that of users. Although users can cope quite well with arbitrary, even nonsensical names in limited command sets after some training, the application of consistent principles for generating names is important for command memorization and recognition. The following principles are particularly useful for defining well-formed name sets (see, e.g., ISO 9241, part 15):

- Names should be *semantically distinctive* and *unambiguous* in the given system context (e.g., insert and delete are semantically more distinct than add and remove). They should also be *emotionally neutral* (avoid terms such as kill or abort).

- If commands describe a certain combination of superordinate and subordinate concepts, names should be generated by a rule reflecting that *hierarchical relation*. Commands which have semantically opposite effects (such as move forward–move backward) should be described by *congruent or symmetric pairs* of names.

- Abbreviations should be formed following one rule consistently (maybe with at most one secondary rule for few exceptions). The main rules are truncation (e.g., del for delete), dropping vowels (e.g., srt for sort), or using the first letter of each word in a longer phrase (often combined with a hierarchical structure as, e.g., uca for update customer address). Some authors have claimed that truncation is the better strategy for recalling command names, whereas vowel dropping may be more effective for recognizing names.

Command Language Structure

Most studies agree on the relevance of structure in command languages. Green and Payne [1984], for instance, showed that a consistent mapping of the underlying concepts in a text editing task to function and character keys resulted in a better performance than less structured variants. Formal modeling techniques have been developed which can support the structural analysis of command languages. Task-action grammars [Payne and Green 1986] describe the mapping from semantic features to command expressions through attributed generation rules.

Various guidelines have been set up to achieve structured and consistent command languages [e.g., Smith and Mosier [1986], and Shneiderman [1987]. Some of the most relevant recommendations are: the command and its arguments should be consistently ordered; natural languagelike structuring of the command language can be useful especially for nonexperts and where efficiency is not the topmost criterion (this approach is currently also being taken in various scripting and macrolanguages such as HyperTalk™); names for the commands and for the different argument types should be chosen from the same grammatical category, respectively (e.g., the action to be performed should always be represented by verbs); separators between arguments or multiple commands should be consistently used.

Applying Command Interaction

The main disadvantages of command language interaction are high-memory load, typing effort for longer command expressions, and a higher probability of errors. A number of supporting mechanisms have been introduced to alleviate these problems while keeping the benefits such as flexibility or expressive power for expert users. Help panels can act as a memory aid for retrieving command names; command history and editing facilities improve efficiency for repeated or similar activities; command or filename completion saves further keystrokes as soon as the input can be unambiguously identified.

Even though the importance of command language interaction for application users has decreased with the widespread use of menus and graphical interaction styles, they are still relevant for many expert task domains such as controlling an operating system and are often used in conjunction with other techniques, for example, for rapid **navigation** between screen forms in a menu-driven application or for directly entering the name of a text style in a direct manipulation word processor. Commands are often preferred by expert users, particularly for frequently used functions if the number of resulting interaction steps or keystroke actions is appreciably smaller than with other techniques.

Some authors have pointed out that the manipulation style of graphical interfaces lacks some of the advantages of conversational command interaction, such as the possibility to identify unseen objects or support for repetitive actions. This has led to proposals to reintroduce conversational capabilities in manipulation interfaces. From a different perspective, there is an increasing interest in end user programming and task automation, which require either task specifications by demonstration or textual descriptions in the form of scripts or macros. Consequently, one can expect commands to continue to play a role in HCI, particularly as an efficient complement integrated with more visual interaction styles.

Menus

The term menu stands for a wide range of different individual interaction techniques. Menus are used for a large variety of different tasks, have different visual representations and interactive behaviors, and can be operated with basically all input devices available today. Their basic principle is to let the user select one or more choices from a given set of options, which may be represented as text, icons, spoken words, simulated 3D objects, or in any other suitable form.

The development of menus has constituted a significant step toward improving the usability of interactive systems as they decrease memory demands by taking advantage of recognition instead of recall mechanisms and, therefore, are also less error prone than command languages [Shneiderman 1987]. They can also be more efficient if they replace the typing of a longer command by a single selection action. In spite of some studies in which menu systems did not yield better, or even worse performance or acceptance results than, for example, command languages, there is little doubt that the growing complexity of current applications as well as the growing number of applications a single user draws upon would be hard to manage without relieving the user's memory by some form of menu support. Payne [1991] could show that even experienced users have poor free recall for frequently used procedures and rely heavily on visual cues such as those provided by menus.

General Principles of Menu Organization and Selection

The process of searching menu options and making a selection involves different types of cognitive tasks. A first relevant distinction lies in the user's prior knowledge with respect to the target option to be selected. The target may be either explicitly known or only partially specified. In the first case, the user performs to a visual matching process; in the second case, the user must read each option, understand its meaning, and evaluate the item with respect to the task. Menu search based on visual matching is significantly faster than searches which require semantic processing.

Paap and Roske-Hofstrand [1988] further distinguish between identity matching, class inclusion, and equivalence search as basic menu selection tasks. Identity matching corresponds to the visual matching between a known target and the options displayed and can be assumed to occur only with well-known sets of options with conventional names such as lists of countries, etc. Class inclusion tasks require the user to determine whether the target option belongs to a certain, more abstract category, which is displayed (e.g., does the command "show page header" belong to the category View or Format?). Equivalence search occurs if the user has to determine whether the term sought after is equivalent to a certain name displayed on the screen.

A considerable number of empirical studies have addressed the effects of different menu design factors such as organization of items or menu length. Surveys of these results can be found in Paap and Roske-Hofstrand [1988] and Norman [1991]. Although these studies have in some cases led to conflicting results, there are at least some areas of converging findings from which guidelines can be derived such as those in ISO 9241, part 14. Some of the most relevant results and some conclusions for design are presented in the following paragraphs.

Menu Breadth and Depth

The complexity of a menu can be described in terms of the number of items on a single-menu panel (breadth) as well as by the number of levels which must be traversed to select a leaf item (depth). Search time increases as the breadth of a menu grows (as it does with growing depth). However, there are results supporting a linear relationship as well as data suggesting a logarithmic one. Perlman [1984] found a linear relationship between the number of items on a single panel and the time needed to match an option displayed at the beginning of each trial with the corresponding menu entry. Other experimental results, in contrast, suggest a log-linear relation where search time is a function of the log of the number of items. The latter model particularly favors broader menus, i.e., menus with more options at a single level. The difference between the two models might be explained by different underlying search processes: Whereas the linear model works in cases in which the user sequentially reads each single option, the log model can be supported by visual search models in which the display is sampled in a more or less random manner. Again, the type of task the user performs is likely to have an influence on these processes: An information retrieval task with a large number of mainly unknown entries (as, e.g., in an on-line information service) suggests a sequential and more detailed reading of the options, whereas selecting a command from a known set of options is done by visually scanning the display.

The existing studies generally show an advantage of broad menus over deep menus (for the same set of items) although optimal breadth may depend on the search strategy applied (see, e.g., Kiger [1984]). For the design of menu panels in practice, one can suggest to present up to about eight entries if no particular way of organizing the entries, for example, by grouping or alphabetizing them, can be employed [ISO 9241, part 14]. If the entries can be grouped into different categories, then showing as many items on a single panel as is practically feasible while maintaining a clear visual separation of the different groups yields the best results in terms of search time and user preference.

Menu Organization

This design area concerns both the arrangement of items on a single-menu panel and the organization of larger sets of options into hierarchic or other structures. Card [1982] compared alphabetic, categorized (by grouping entries according to their semantic relation), and random ordering of menu items and found alphabetic or categorized organization to be the easiest to search. Perlman [1984] observed that search with an alphabetical order was about twice as fast as for random orders. McDonald et al. [1983] showed advantages for categorical organization, especially when the search targets were only described and not explicitly given by name. The co-occurrence of items in tasks can also be an effective principle of organization, especially if multiple selections are required.

Organized menus generally yield shorter search times except for high levels of practice on the same menu where performance differences consistently disappear. Just because humans can cope with unorganized menus after extensive practice, however, should not be taken as a license to neglect organization in the design of menus. Menu organization can be grounded on content relations of the items as well as on task relations. The right choice will depend on the type of task to be performed and the resulting user goals as well as on a consistent principle of structuring the system as a whole. In an information retrieval task, semantic organization is likely to be more effective, whereas in a system supporting business processes a task-oriented structure clustering the single steps in a larger process may be more appropriate. Careful task analysis is—as in HCI design in general—indispensable. Table 70.1 shows an extended set of principles which can be used for organizing menu structures.

TABLE 70.1   Different Principles for Organizing Menu Structures

Principle of Organization	Description	When to Use?
Semantic categories	Organization according to natural or logical categories (food: milk, bread, meat, etc.; document views: outline, normal, layout)	Most general and widely applicable approach. Suited for grouping information objects (e.g., information retrieval) or functions according to their similarity.
Semantic relations	Items are grouped according to their relation they have with a certain reference object	For navigational menus, e.g., in hypertext structures (goto explanations, references, etc.) or for object-oriented navigation in business applications.
Task structure	All subtasks of a superordinate task are grouped together and possibly displayed in the sequence in which they typically occur	Supports a functional and process-oriented view of the user's tasks. Can serve as checklist of tasks to do. Effective if defined (business) processes are executed.
Frequency of use	Frequently selected items are placed at the top of the list or in higher level menus	For limited number of choices where performance is critical (default selection on most frequently used item).
Conventional order	Order based on convention, often on some underlying concept such as time, size, etc. Examples: months, clothing sizes, etc.	Whenever such an order exists and menu length does not become unwieldy. May be grouped in ordinal ranges.
Alphabetical order	Alphabetized list of items, maybe alphabetic ranges in superordinate menus	For options of same type, value ranges of an attribute (e.g., countries) or if no other way of ordering items is feasible.

Norman [1991] distinguishes menus according to their function, proposing the (semantic) categories pointing (moving to a new node), command control, information display, and data input (data or parameter specification, decision menus). Similar categorizations are used in style guide recommendations for the design of **pull-down menu** bars in graphical user interfaces. The menu bar should consistently show some fixed categories in a certain order (e.g., file, edit, view, etc.). This helps users in orienting themselves quickly in an application's functionality and to form expectations where to find functions of a certain type in new applications.

It has been demonstrated that adding descriptors to menu items, for example, by showing some options of the subsequent menu level can improve user performance under certain conditions. It may be speculated that the type of lookahead mechanism which is provided in pull-down menus through dynamic browsing of the entries of the next level contributes to their usability for similar reasons.

Menu Selection Techniques

Menu items are mainly selected by entering a corresponding identifier (letter or number), by cursor key positioning, or by pointing with a mouse or some other pointing device. Other selection techniques comprise voice input in a speech interface, special keys, or key combinations as shortcuts for pull-down or pop-up menus. If identifiers are entered for item selection, letters are slightly better than numbers provided that items are stable over time and letters are chosen which are compatible with the options [Perlman 1984].

Positioning with a mouse is faster and more accurate than with joystick or cursor keys [Card et al. 1978] whereas both touchscreen input as well as single-letter identifiers were reported to be faster than the mouse in a study by Karat et al. [1986]. Considering the range of selection and manipulation tasks in current user interfaces as well as ergonomic limitations of touchscreens, selecting menus with the mouse is generally the most appropriate mechanism. In pull-down menus, mouse selection should be complemented by keyboard shortcuts (accelerators) for the most frequently used items. By using the mouse for object selection and accelerators for invoking operations on the object, users can effectively use both hands simultaneously and thus achieve higher performance. More recent techniques such as pen input are also efficient means for selecting menu items (for more details see Chapter 81 on input devices in this volume).

Menu Types

The variety of different menu interaction techniques is too large to give a full account of all of them here. Therefore, we will only highlight the major types of menus, keeping in mind that there are many possible variations along the dimensions presented in section 70.2 of this chapter and an enormous number of combinations of output, input, and interaction characteristics. Many interaction objects which are provided by the various toolkits for graphical user interfaces (GUIs) can be considered as menus (such as radio button groups) although they are usually not referred to by that name. We will discuss these GUI elements in a separate section.

For the purpose of classifying the various menu types, a first distinction can be made on the basis of their degree of interactivity. *Modal* menus follow the conversational paradigm introduced earlier, with the system explicitly taking the turn by displaying a menu panel from which the user must choose (or quit) before the interaction can be continued. The historically first—and least interactive—example of modal menus are character-based, *full-screen menus*, which completely replace the previous visual context of the interaction by a set of options. Full-screen menus are still used widely in host-based business applications and should usually offer both selection by identifier entry and selection by cursor keys. Identifier type-ahead and decimal numbering schemes are useful for providing direct access to a target screen. Modal menus may also be realized by overlaying only part of the screen by a menu panel or by a *menu dialogue box* realized as a window in graphical environments. Modal menus require an explicit cancel operation to leave the menu selection context.

In *modeless menus*, options can be selected at any time while working in a specific context (e.g., pull-down menu). Option availability should be controlled dynamically (e.g., by graying out options currently not applicable) depending on the object selected and the state of the current context. Modeless menus increase interactivity mainly by more direct and immediate feedback on the effect of an operation selected. They can be further distinguished with respect to how their visibility is temporally and spatially determined. *Permanent menus* continuously display a full set of options, which can be activated at any time by clicking on them or, e.g., selecting options by function keys. Examples of permanent menus are *menu bars* or *menu palettes* attached to application windows at a fixed position; *floating menus*, which can be placed anywhere on top of application windows; or **task bars**, which provide quick access to frequently used applications or objects. Permanent menus are particularly suitable for speeding up frequently used operations and for providing feedback on parameter settings or modes of operations (such as generator modes in graphics editors) which need to be constantly observed.

Partially permanent menus exhibit either only their top-level entries (e.g., pull-down menus) or show the currently selected option (e.g., drop-down lists). They are usually tied to fixed positions with respect to the screen or to the application window. Further options are displayed on demand by selecting a visible element of the menu structure. Partially permanent menus are appropriate if larger sets of options need to be accessed rapidly. Because of their fixed position, they provide strong visual cues for locating options and support skill development in selecting them.

Temporary menus such as pop-up menus appear only on demand. Their position is usually determined by the location of the pointer at the time they are activated. Option set and availability may depend on the specific object which is represented in the current screen region. Pop-up menus are suitable for smaller sets of options, particularly if the user should keep the object to be manipulated in view. This is often the case for complex displays such as technical diagrams where the cost of visually retrieving the object of interest is high. Although the distance to a pop-up target option is much smaller in comparison to selecting an item from a pull-down menu at the top of the screen or window, selection time in general is not faster with the pop-up menu even on large screens. Apparently, the time gain due to shorter movements is consumed by additional cognitive components needed for orientation and item matching.

Interesting alternatives to the standard menu types have been developed, for example, in the form of *pie menus*, which replace the conventional vertical list of items by arranging them in a circular fashion [Callahan et al. 1988]. *Marking menus*, [Kurtenbach and Buxton 1994] are circular temporary menus operated with a pen. Rapidly drawing a mark in the direction of the target item serves as an accelerator and prevents the menu from popping up.

Direct Manipulation

The concept of direct manipulation (DM) as an interaction style is tightly linked with the development of graphical user interfaces using windows, icons, mouse, and pointing (WIMP) interfaces. It has been pointed out that DM is rather an orienting notion than a clearly circumscribed interaction technique [Hutchins et al. 1986] and spans a large range in a space of possible interaction techniques [Ziegler 1991]. In his original definition of the term, Shneiderman [1983] characterizes DM by the continuous representation of the objects of interest, physical actions instead of textual command input, and rapid, incremental, and reversible operations with immediate feedback.

As opposed to conversational interaction styles, DM relies on visualizing the semantic objects of an application and their relations, thus providing a simulated model world in which the user can act in a quasinatural and spatially oriented manner. Objects and their state are shared between the user and the system through this visual representation and can be referred to by pointing actions. This property, which has been called *interreferential input/output* [Hutchins et al. 1986], facilitates the understanding of the state of the system and of the associated possibilities for action. It can also save input actions such as the need to generate names as input is eliminated. As a drawback, interaction purely based on visual representations prevents the user from addressing objects currently not visible or from specifying action sequences in advance for future execution. These capabilities are often desirable for expert users who know the objects needed for their task and want to manipulate them without additional navigation or who wish to delegate tasks to the system. Mixed-mode interaction styles have been advocated as a way to accommodate these requirements.

In order to create an intuitively comprehensible visual model world of the application domain, metaphors are often employed as a means for drawing on the user's experience and knowledge of the real world. Typical examples of such visual metaphors are desktops for representing file systems or card stacks for embodying structured data records. By suggesting analogies between the interface and some known real-world domain, the user will be supported in understanding the objects of a domain and the relations among them in a more intuitive way and in discovering and exploring the actions which are afforded by the interface. Yet, one should be aware that the analogies drawn between the real world and the metaphorical domain are usually not perfect and can often break down. A typical area where such breakdowns may occur are drag-and-drop operations. The real-world semantics of moving a thing to a different place can take on many meanings in the computer metaphor, depending on the type of the objects involved in the operation, e.g., when copying a file to a volume or creating a new order for a customer in a business application. In these cases, the metaphor provides a general spatial framework for the user's actions rather than a realistic simulation of some noncomputerized domain.

Manipulative Techniques

The mapping from semantic objects to views displayed at the surface of the system mainly determines the range of manipulations which are possible in the model world. If objects and their properties are represented through discrete elements such as text fields or check boxes, pointing can be used for selecting objects or components and for choosing attribute values from a given set. More complex interaction objects may be equipped with handles, which can be displaced in order to change one or more attributes simultaneously and continuously, for example, when reshaping a geometric figure in a graphics editor. It is often this technique of dragging or reshaping objects on the screen, which is called DM in a narrow sense. In general, however, manipulative techniques cover the whole range of spatially oriented operations, which may be distinguished with respect to the function they perform in DM interfaces (all of these manipulations may in principle be performed in 2D as well as in 3D space):

- Pointing takes a position as input for selecting objects or actions.
- Moving changes the position of an object without altering its orientation.
- Orienting changes the angular position of an object with respect to some reference point.
- Drag-and-drop produces a certain geometric relation of two (or more) objects (typically by overlapping or touching each other), which determines a certain operation. The particular path by which this position is reached is usually not relevant for the resulting operation (e.g., a new customer invoice is created by dragging an icon representing a specific customer over a [general] invoice icon).
- Drawing produces a path, which is used either as data or as command input (e.g., for gesture input).
- Shaping changes the form of an object by displacing one or more points, which are significant for its geometrical form.
- Modeling changes the shape of an object by adding or removing picture or volume elements (e.g., when using *virtual clay* for modeling 3D objects).

Whereas this list of manipulative operations is not exhaustive and can be extended, for example, by allowing operations which assemble two different objects into a new one, it indicates the wealth of interactive behaviors which can be achieved on the basis of DM, especially if novel input techniques and operations in 3D are taken into account.

Another important and salient feature of DM is its inherent *object orientation*. Although not necessarily programmed in an object-oriented language, DM interfaces provide the user with an object-oriented view of the application. This is mainly achieved by visually representing the objects of an application and by separating object selection and activation of operations in the user interface. Even in cases where the manipulation of an object is tightly coupled with its selection (as, e.g., in shaping tasks), these two aspects are at least conceptually clearly separated. Treating object selection and operations as two distinct steps helps in minimizing the number of conceptually different operations and in applying them uniformly across a wide range of different object types (*generic* operations). Also, the notion of object classes can be supported by different graphical means of representation such as using the same icon shape for objects of the same conceptual class.

Directness and Affordance

When can we say that a manipulation is also direct? In contrast to the manipulative aspect of an interface, which can be determined through the interactive properties of the system itself, **directness** is mainly seen as a concept which is dependent on both the system and the user's mental representation of the domain and the task. Hutchins et al. [1986] analyze directness in terms of the distance between the user's goals and the concepts and techniques offered by the system (semantic distance) and the relation between the semantic system concepts and their physical appearance (articulatory distance). In addition, they introduce the notion of *direct engagement*, which relates to the user's subjective feeling of being in control or even a part of the model world he or she is interacting with.

Semantic directness is dependent on the distance between the user's mental goals and the semantic objects and operations provided by the system. Semantic directness requires that the system offers the functionality needed through objects and operations which are adequate for the task in terms of level of abstraction, structure, properties, or function performed. A system will be indirect if, for instance, the user has to manipulate several system objects in order to deal with a single coherent entity in the task domain or if several operations have to be performed to achieve a single step in the task. Imagine, for instance, a task involving the creation of business graphics diagrams for which only a graphics editor with basic geometric objects is available. Low semantic directness leads to an increased mental effort for task planning and usually also to lower efficiency due to a higher number of interaction steps.

Articulatory directness concerns the relation between the meaning of an expression and its physical form. In terms of user interface properties, this means that the sensory-motor aspects of the interaction, the form of the input and output elements, and the syntax of performing a coherent task must be compatible with the meaning of the intended action. If, for example, a numeric value can be incremented and decremented by two buttons, one will expect that the increment button sits above the decrement button on the display and not underneath or beside it.

The notion of **affordance** is related to the aspect of directness previously discussed. The affordance of an object depends on the degree to which this object suggests to the user in which way it can be manipulated [cf., Norman 1988]. A button, for example, which appears to be standing out from the screen surface due to its 3D representation, is more likely to suggest pushing it with the mouse than a button represented as a deepening in the screen. Creating affordances by visualizing interaction objects in a suitable manner is therefore an important means for making the user interface more direct and intuitive.

When designing direct manipulation systems, one must consider that directness is depending on the user's knowledge and experience and will therefore change over time in the learning process. What appears direct to the novice may not be appropriate for the more complex action plans of an expert. A certain degree of flexibility and extendability of the system is therefore necessary in order to maintain the user's impression of directness.

Empirical Studies of Direct Manipulation

In a comparison of seven systems, two of which had an iconic, direct manipulation interface, Whiteside et al. [1985] did not find an advantage for the DM style interfaces in terms of performance or user preferences, whereas there were large differerences between the individual instances of the different interaction styles. This study, however, was limited to a short period of time, mainly looking at the initial part of the learning curve. When comparing a DM with a command-driven word processor, Frese et al. [1987] also found no differences in the first of a series of experimental sessions. DM, however, was superior to the command-based system with increasing session number and with more complex tasks, which were performed in later sessions. This result may be due to better retention with a highly visual DM interface and to the typically high consistency of DM systems, which can be shown by the high percentage of procedural rules shared between different functional domains in a DM system.

A number of other studies could also demonstrate the superiority of DM over other interaction styles in such task domains as file manipulation or database handling. In an extensive questionnaire survey, Prümper [1993] had 106 users of graphical interfaces and 244 users of nongraphical interfaces rate their systems with respect to a set of items based on the dialogue principles stated in ISO 9241. GUI users rated their systems better in all of the seven categories, resulting in a positive judgement of system usability, whereas conventional systems overall achieved only neutral ratings and negative responses concerning the possibilities for individualization.

Although these results do not constitute proof of the superiority of DM interfaces in general, it is evident that graphical interfaces based on DM interaction do improve usability. Although a well-designed command language system may still be better than a badly constructed DM interface for a certain task domain, there is little doubt that DM interaction is a pertinent prerequisite for the widespread use of current applications, which are rich in functionality and increasingly integrated.

70.4 Interaction Objects for Graphical User Interfaces

Interaction objects, often also called widgets, represent the basic building blocks of user interfaces and define a specific design for different interaction tasks. Examples of widgets are windows, push buttons, check boxes, icons, or sliders, to name a few. They are usually also provided as software components, facilitating the implementation of the interface. The development of industry standards such as Microsoft Windows, OSF/Motif, or OS/2 Presentation Manager and corresponding style guides defining the behavior and the visual appearance of the interaction objects has contributed significantly to a more consistent design of modern graphical user interfaces. In recent years, these style guides have begun to converge, resulting in rather similar sets of widgets and facilitating the transfer of user knowledge not only between applications, but also between different system platforms.

Current GUI technologies employ some 30 different types of widgets. User interface designers are faced with the problem of selecting interaction objects that are suitable for the task to be accomplished and of using them consistently. Style guides provide useful advice for selecting appropriate interaction objects. In many instances, however, designers will still be left with several options for a certain interaction task. Therefore, a careful analysis of the task characteristics in the given application domain is required. This involves considering a variety of task parameters, such as type and frequency of the task, type of value to be entered, value range and number of possible values, or the potential risk involved with entering a wrong input. Other important factors relate to variance or required flexibility in the task, for instance, whether the user is likely to know the input in advance or will determine it while interacting with the widget.

Several classification schemes have been developed in order to provide a more systematic support in designing or selecting interaction objects. The notion of abstract interaction objects is useful for separating the logical properties of a widget from its concrete appearance and behavior in a particular toolkit. Vanderdonckt and Bodart [1993] describe the following categories and examples of abstract interaction objects:

- Action objects (menu, menu item, cascaded menu, etc.)
- Scrolling objects (scroll bar, scroll arrow, etc.)
- Static objects (label, separator, group box, icon, etc.)
- Control objects (edit field, check box, radio box, list box, push button, etc.)
- Dialogue objects (window, dialogue box, etc.)
- Feedback objects (messsage, progress indicator, cursor, etc.)

In another classification, single action, settings remembering, and drag-and-drop widgets can be distinguished. Single actions may be activated immediately (e.g., by push buttons or menu selection) or be performed continuously while activated (e.g., scrolling a text by pressing a scroll arrow). Settings represent the attribute values of an object and maintain their state until changed by the user or the system. Settings remembering widgets are, for instance, radio or check boxes, drop-down lists, or spin buttons. It must be noted, however, that some widgets are often used for more than one of these functions. A pull-down menu item may activate an action or represent a persistent setting. It is especially important in such cases to make the different types of functions recognizable for the user by structuring the menu in an appropriate manner. With drag-and-drop widgets, the user can move one or more visible objects to another object. In general, the resulting action can depend on the type of the objects involved (moving a new order icon over a customer icon opens an order form with the data of that customer) as well as on the particular geometric relation produced (e.g., when attaching an object to a particular side of another object).

Widget selection rules may be formalized for automatically generating user interfaces from some higher level description of the domain and the tasks. A rule may define, for instance, that for setting a single value of an object attribute with a discrete value range, an abstract interaction object of type *exclusive choice* shall be used. If the number of values to select from is small, this may be realized by a group of radio buttons, otherwise by a list. A further distinction between, say, a drop-down list or a list box must take into account the space available on the screen and, therefore, is dependent on the complete set of items to be presented.

In recent developments of user interface standards, one can observe a trend to provide increasingly complex interaction objects as predefined building blocks. Typical examples of such complex widgets are notebooks with tabs for quickly switching between different pages or tree browsers for interacting with hierarchical structures such as file systems. Rather than just supporting elementary tasks, such widgets provide complete presentation and dialogue structures for performing a coherent set of subtasks. Their interactive behavior can be quite complex and may require substantial initial learning by the user. Standardizing the use of complex widgets, on the other hand, offers the opportunity to make interaction with applications more consistent at a higher level, also supporting navigational aspects rather than just elementary manipulations.

70.5 Navigation

The interaction that the user performs in order to move between different parts of a document, sets of data, or arbitrarily linked pieces of information is usually called navigation. Navigation can be regarded as the user-system interaction "in the large" and distinguished from those parts of the dialogue which actually manipulate data. This distinction is particularly evident in direct manipulation interfaces where only visible objects can be manipulated, requiring navigation steps in order to expose currently hidden objects for subsequent manipulation. As the basic interaction objects become more and more standardized, designing task-oriented, consistent and transparent navigation structures becomes one of the major challenges for the user interface designer. This is especially true for systems typically comprising a large number of different views, such as business information systems or hypertext documents.

In the following, we will use the term *view* for some collection of elements representing an underlying information object (or objects) which can be made visible in a coherent way, e.g., in a window or full-screen display. Navigation may occur within a single view, involving actions such as moving the cursor between different paragraphs of a document or from one data field to another. Navigation between different views involves actions such as moving from a customer data form to a list of invoices sent to that customer or traveling through the different nodes of an on-line help document.

Within-view navigation is dependent on the structure of the underlying object or objects shown in that view, whereas between-view navigation typically involves different underlying objects and is dependent on the relations between those objects. The second type of navigation also applies if a complex object can be shown in several views. In the following, we will focus on aspects of between-view navigation, which determines the overall structure of interactive applications and has a strong impact on the efficiency, transparency, and flexibility of a system.

Navigation Structures

Navigation may be designed according to a number of basic structures which determine the potential dialogue paths through the system (see, e.g., Berk and Devlin [1991]). *Sequences* are mainly used in simple, often system-controlled dialogues, as, e.g., in card-stacklike databases or in guided tours in hypertext systems, and limit the user to basic actions such as moving forward or backward. They are usually not sufficiently flexible for larger applications without additional mechanisms. *Hierarchical* navigation is one of the most commonly used forms and can be found, e.g., in classic treelike menu systems as well as in window hierarchies in modern desktop user interfaces. *Latticelike* navigation provides regularly structured dialogue paths which allow the user to move forward and backward between the different entries of a database as well as left and right in order to see different facets of a single entry (represented, for instance, by a notebooklike interaction object).

The most flexible form of navigation is provided by *network* structures, which represent the main paradigm of hypertext or hypermedia systems [see, e.g., Nielsen 1990]. As arbitrarily linked information networks tend to become confusing even for a smaller number of nodes, various mechanisms for providing additional structure have been explored, such as clustering nodes or distinguishing different types of links. One of the most frequently used approaches is to superimpose a network of free links on a more regular structure such as a hierarchy or a sequential stack. Hierarchies combined with arbitrary links are, for instance, the predominant navigation structure for current on-line help systems. Providing transparent structures is the most important means to prevent the user from getting lost in the dialogue. Navigation aids such as maps or fisheye views may provide additional support but should mainly be considered as supporting measures enhancing the comprehensibility of a well-designed navigation structure.

Another aspect of designing navigation is concerned with different structuring principles, which serve for organizing the system and which can significantly influence the user's understanding. Navigation structures may be based on a functional decomposition of the system, on the objects of an application and their relations, or on arbitrary associations between different pieces of information. In the following, we will discuss some of these principles and their implications for the usability of the system.

Function-Oriented Navigation

In many systems such as conventional database applications, navigation is organized through a hierarchical menu of functions (as, e.g., change address data). The particular data object is either specified along with the selection of the intended operation or through some search mechanism after the operation has been selected. The actual data view (such as the change customer data form) only becomes visible after one or more navigation steps, which are determined by the selection of the intended function, possibly over several levels of hierarchy. We will call this type of navigation **function-oriented**. The resulting dialogue structures are mainly hierarchical, often combined with specific dialogue operations provided for moving around more efficiently in the hierarchy, for example, for going back to the main menu or for directly activating a function in some submenu by using a shortcut.

In function-oriented navigation, the user is guided through the system by concepts which are closely related to specific task goals (change customer data) triggered by well-defined events in the user's work context. On the other hand, function-oriented dialogues may be inflexible when the user's goal changes (e.g., create a new customer entry instead of changing existing data) or is only vaguely defined at the outset (check customer X and decide on appropriate action).

Process-oriented navigation can be seen as an extension of function-oriented dialogues. In this type of navigation, the system provides support for performing a complete set of tasks belonging to a coherent work process (business process). The tasks involved are often dependent on each other. The system can monitor the status of the process and guide the user through the different steps, allowing access only to those functions available in the current state. Although early systems often forced the user to go through a predetermined, rigid sequence of steps, current techniques permit design of such process support in a more flexible manner, for example, by visualizing the tasks to be performed in a kind of graphical task bar through which all tasks with valid preconditions can be activated.

Object-Oriented Navigation

Object-oriented navigation is based on the objects of an application and their relations. The user navigates through different views of an object or between related objects. A typical navigation path may involve selecting a customer icon on the desktop, opening it or performing a search, and choosing a particular customer to be shown in full detail from a window listing all customers or the result of a search. Navigation may continue to a list of orders placed by that customer and on to the items of a specific order. Operations become available in the different views as appropriate.

Views can show a single object instance at different levels of detail, for example, as an icon, as a form showing some of the object's attributes, or as a notebook widget allowing access to the full attribute set. In addition, the user interface must provide views which represent collections of instances of that object class, such as container icons or lists representing all or some instances of the class. Semantic relationships between different object classes are translated into appropriate navigation paths, observing the cardinality of the relations, for example, by introducing list views along the navigation path. The flexibility of the navigation can be greatly enhanced by allowing the user to define additional task-specific views of an object class. At the desktop level, for instance, the user might define icons for frequently dealt with object instances "customer Smith" or subsets filtered according to certain criteria "customers in Washington". Figure 70.3 shows a typical object-oriented navigation structure, using a graphical notation for distinguishing different types of views [Ziegler 1996].

By grounding navigation on the objects of the application domain, transparent dialogue structures can be achieved provided the user is familiar with the semantics of the domain. As objects usually change less frequently over time than the tasks operating on them, object-oriented dialogue structures tend to be more stable. This fact supports consistency within and across applications. There is also a high degree of flexibility, since the user can browse through the different views and look at the relevant data before deciding to perform a particular action. The options for individualization previously mentioned can further enhance flexibility.

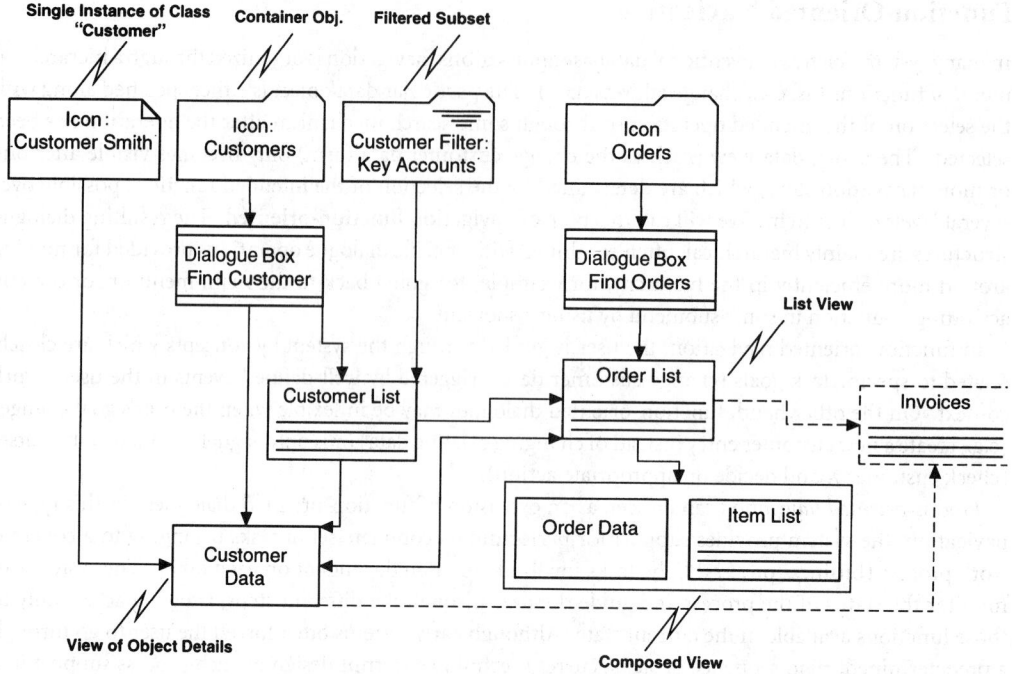

FIGURE 70.3   Example of an object-oriented navigation structure for a typical business application.

Association-Based Navigation in Hypermedia

Hypermedia systems allow navigation between arbitrary units of information (nodes), which can be represented by any combination of media. Nodes may be shown as fixed-sized frames or as scrollable windows. Whereas object-oriented navigation depends on predefined object types of an application domain and the usually invariable relations between those types, navigation in hypermedia structures is typically based on individual pieces of information and the associations between these single pieces.

We will call this type of navigation *association based*. Depending on the particular association by which it is accessed, a single unit of information may take on different roles. A piece of text may serve as an explanation for a statement made in another node and at the same time be part of a hierarchical decomposition of a certain topic. Some systems explicitly use different types of links to structure the navigation space. In issue-based systems, for instance, navigation paths lead from issues to different positions with respect to these issues and from there to arguments for and against those positions.

If not designed properly, complex hypermedia navigation structures incur the danger that users become unsure about where they are, forget their original goals, or completely lose their bearings. Therefore, designers should provide consistent basic navigation mechanisms and regular structures where possible. Regularity can be increased by using, for instance, a hierarchy as the main navigation framework, which can then be extended by links added to the given parent–child associations. Other mechanisms for reducing the complexity of navigation involve, for instance, clustering of sets of related nodes or integrating regular substructures into otherwise arbitrarily linked networks of information. The latter can be observed, for example, in large-scale, distributed hyperstructures such as the World Wide Web where nodes may contain databases with more structured information.

Navigation Support Mechanisms

To facilitate navigation in complex information spaces, a large number of navigation support mechanisms have been proposed. Overview tools provide the user with additional context information showing the current position in relation to some or all other nodes of the structure. Maps use a spatial metaphor in order to provide an overview of the system, sometimes at different levels of detail or combined with a zoom facility. Browsers are useful for exploring standard information structures such as hierarchies or for providing lookahead information about the neighborhood of the current node. Fisheye views are a useful mechanism for visualizing large information structures. They can display the complete context of a piece of information by decreasing the level of detail shown as a function of the distance from the current node.

Guided tours offer one or several predefined paths through an information network to facilitate navigation for the novice or for providing different perspectives on a certain topic. Guide mechanisms should usually allow the user to make excursions from the defined path and to return to the point of departure afterward. Some advanced navigation support mechanisms assign a more active and dynamic role to the system, for example, by means of software agents which collect relevant information or try to recognize the user's plan and provide appropriate navigation links.

70.6 Future Trends in Interactive Techniques

The advances in interactive techniques achieved over the past few years constitute one of the most pertinent factors for the increased usability and widespread use of current information technology. Significant breakthroughs in the application of computers such as the introduction of standard office software in the 1980s or, more recently, the enormous growth of the Internet, are closely linked with interface innovations such as graphical desktop interfaces and easy-to-use web browsers.

The basic interaction techniques, especially for graphical user interfaces, are becoming more and more standardized, leading to a higher consistency of applications both within and across different platforms. Current research and technological developments are indicating some trends which are likely to influence HCI in the future.

There is an increasing interest in analyzing and standardizing more complex and complete interaction tasks by developing user interface design patterns and corresponding software building blocks or frameworks. These patterns address high-level design issues such as navigating consistently in applications.

Mobile systems and the various forms of embedded systems will require new interaction techniques which are tailored to the tasks and constraints of the specific context of use. Examples of such techniques are pen input, handwriting recognition, or interfaces for wearable devices.

The bandwidth of interaction will continue to increase with respect to output as well as input techniques. On the output side, this will lead, for instance, to increased realism in virtual environments. On the input side, recognition of eye movements, gestures, body movements, or biophysical signals are some areas of research for future input techniques. The intention underlying such techniques is to use the full spectrum of human sensorimotor capabilities for interacting with the system. The usability implications of such techniques, though, are just beginning to be explored in HCI research.

Defining Terms

Affordance: Property of an object that makes it obvious how the object is to be used. A push button, for instance, with a three-dimensional appearance which makes it look as if it is protruding from the screen, may indicate push more strongly than a flat button.

Command language: Form of interaction where the user has to type input commands which conform to a particular syntax.

Direct manipulation: An interaction technique in which objects are visually represented on the screen and can be manipulated by the user similarly to the manipulation of a real-world object. Requires the use of a pointing device.

Directness: Degree of correspondence between the user's mental goals and the functions and interaction facilities provided by the system.

Function-oriented interface: A way of structuring the interaction on the basis of the functions of an application. Typically, a function (or entire application) is selected first, and then the data for that function are determined. Function-oriented navigation is based on a functional decomposition of the application, which is often represented as a menu hierarchy.

Interaction object: A component of the interface that is used to represent, contain, and manipulate objects (often also called widgets). Examples are icons, push buttons, check boxes, text fields, etc. Interaction objects are usually provided as interface software components on the different GUI platforms.

Menu: An interaction technique by which the user selects one or more items from a set of options presented by the system in a visual or auditory form.

Metaphor: Originally indicating a rhetoric figure, the concept is used in the design of user interfaces for providing an analogy between some domain familiar to the user and the (typically more abstract) functionality of the system. The best-known example of a visual interface metaphor is the graphical desktop.

Navigation: Aspect of the interaction between user and system by which the user moves around between the different parts of a system. Navigation encompasses both moving between different windows of an application (between-view navigation) as well as moving the focus to one of the elements currently presented (within-view navigation).

Object-oriented interface: A way of structuring the interaction on the basis of the application objects. The object-action paradigm is relevant for the manipulation of objects: The user first selects an object, then the action to be performed on that object. Actions are made available depending on the object the user selects. At the navigation level object-orientation means that the user can navigate through the system on the basis of the application objects and their semantic relationships.

Pop-up menu: A menu that appears when a particular area of the screen is clicked on. It disappears when the user has selected an option.

Pull-down menu: A menu that pulls down like a roller blind from a permanently visible title bar, usually at the top of the display or window.

Style guide: A set of principles and rules for consistently designing user interfaces according to a particular interface style (often for a specific platform). General industry style guides are usually independent of a particular application domain; inhouse or application style guides may provide additional rules for designing particular classes of applications.

Task bar: A (linear) array of buttons which usually floats on top of application windows for providing quick access to frequently needed functions or applications.

Widget: Window gadget, see interaction object.

References

Apple. 1992. *Apple Macintosh Human Interface Guidelines*. Addison–Wesley, Reading, MA.

Barnard, P. J. and Grudin, J. 1988. Command names. In *Handbook of Human–Computer Interaction*. M. Helander, ed., pp. 205–235. Elsevier Science (North-Holland), Amsterdam.

Berk, E. and Devlin, J. 1991. *Hypertext/Hypermedia Handbook*. McGraw–Hill, New York.

Callahan, J., Hopkins, D., Weiser, M., and Shneiderman, B. 1988. An empirical comparison of pie vs. linear menus, pp. 95–100. Proc. of CHI '88, *Conf. on Human Factors in Comput. Syst.* ACM, New York.

Card, S. K. 1982. User perceptual mechanisms in the search of computer command menus, pp. 190–196. *Proc. Human Factors in Comput. Syst.* ACM, New York.

Card, S. K., English, W. K., Burd, B. J. 1978. Evaluation of mouse, rate-controlled isometric joystick, step keys and text keys for text selection on a CRT. *Ergonomics* 21:601–613.

Carroll, J. M. 1985. *What's in a name?* Freeman, New York.

Foley, J. D., van Dam, A., Feiner, S. K., and Hughes, J. F. 1990. *Computer Graphics: Principles and Practices*. Addison–Wesley, Reading, MA.

Frese, M., Schulte-Göcking, H., and Altmann, A. 1987. Lernprozesse in Abhängigkeit von der Trainingsmethode, von Personenmerkmalen und von der Benutzeroberfläche (Direkte Manipulation vs.

konventionelle Interaktion). In W. Schönpflug and M. Wittstock, Hrsg., pp. 377–386. *Software-Ergonomie '87*. Teubner, Stuttgart.

Gaines, B. R. and Shaw, M. L. G. 1986. Foundations of dialog engineering: the development of human–computer interaction. Part. II. *Int. J. Man–Mach. Stud.* 24(2):101–123; (see also Part I 24(1):1–27.

Green, T. R. G. and Payne, S. J. 1984. Organization and learnability in computer languages. *Int. J. Man–Mach. Stud.* 21:7–18.

Hutchins, E. L. 1989. Metaphors for interface design. In *The Structure of Multimodal Dialogue*. M. Taylor, F. Neel, and D. Bouwhuis, eds., pp. 11–28. Elsevier, Amsterdam.

Hutchins, E. L., Hollan, J. D. and Norman, D. A. 1986. Direct manipulation interfaces. In *User Centered System Design*, D. Norman and S. Draper, eds. Laurence Erlbaum, Hillsdale, NJ.

IBM 1992. Object-Oriented Interface Design—IBM Common User Access™ Guidelines. Que Corp., Carnel, IN.

ISO 9241. Ergonomic Requirements for Office Work with Visual Display Terminals. Part 10: Dialogue Principles, Part 14: Menu Dialogues; Part 15: Command Dialogues, International Standards Organisation.

Jacob, R. J. K., Legett, J. J., Myers, B. A., and Pausch, R. 1993. Interaction styles and input/output devices. *Behaviour Inf. Tech.* 12(2):69–79.

Karat, J., Mcdonald, J., and Anderson, M. 1986. A comparison of menu selection techniques: touch panel, mouse and keyboard. *Int. J. Man–Mach. Stud.* 25:73–88.

Kiger, J. I. 1984. The depth/breadth trade-off in the design of menu–driven user interfaces. *Int. J. Man–Mach. Stud.* 20:201–213.

Kurtenbach, G. and Buxton, W. 1994. User learning and performance with marking menus, pp. 258–264. In Proc. CHI '94, *Conf. on Human Factors in Comput. Syst.* ACM, New York.

Martin, J. 1973. *Design of Man–Computer Dialogues*. Prentice–Hall, Englewood Cliffs, NJ.

Mayhew, D. J. 1992. *Principles and Guidelines in Software User Interface Design*. Prentice–Hall, Englewood Cliffs, NJ.

McDonald, J. E., Stone, J. D., and Liebelt, L. S. 1983. Searching for items in menus: the effect of organization and type of target, pp. 834–837. In *Proc. 27th Ann. Meeting Human Factors Soc.*

Microsoft. 1992. The Windows Interface—An Application Design Guide. Microsoft Press, Redmond, WA.

Nielsen, J. 1990. *Hypertext and Hypermedia*. Academic Press, London.

Nielsen, J. 1993. Noncommand user interfaces. *Comm. ACM* 36(4):82–99.

Norman, D. A. 1988. *The Psychology of Everyday Things*. Basic Books, New York.

Norman, K. L. 1991. *The Psychology of Menu Selection: Designing Cognitive Control at the Human/Computer Interface*. Ablex, Norwood, NJ.

OSF 1992. *OSF/Motif ™ Style Guide*, Open Software Foundation Rev. 1.2. Prentice–Hall, Englewood Cliffs, NJ.

Paap, K. and Roske-Hofstrand, R. 1988. Design of menus. In *Handbook of Human–Computer Interaction*. M. Helander, ed., pp. 205–235. Elsevier Science (North-Holland), Amsterdam.

Payne, S. 1991. Display-based action at the user interface. *Int. J. of Man–Mach. Stud.* 35:275–289.

Payne, S. J. and Green, T. R. G. 1986. Task-action grammars: a model for the mental representation of task languages. *Hum.–Comput. Interaction* 2:93–133.

Perlman, G. 1984. Making the right choices with menus, pp. 291–295. In Proc. INTERACT '84, *1st IFIP Conf. Hum.–Comput. Interaction*. London, Sept. 4–7. IFIP, London.

Prümper, J. 1993. Software-evaluation based upon ISO 9241 part 10, pp. 255–265. In Proc. VCHCI '93, *Vienna Conf. Hum.–Comput. Interaction*. Springer–Verlag, Berlin.

Shneiderman, B. 1983. Direct manipulation: a step beyond programming languages. *IEEE Comput.* 16:57–69.

Shneiderman, B. 1987. *Designing the User Interface—Strategies for Effective Human–Computer Interaction*. Addison–Wesley, Reading, MA.

Smith, S. L. and Mosier, J. N. 1986. *Guidelines for Designing User Interface Software*. MITRE, Bedford, MA.

Vanderdonckt, J. and Bodart, F. 1993. Encapsulating knowledge for intelligent automatic interaction objects selection, pp. 424–429. In *Proc. INTERCHI '93*. Amsterdam, April, 24–29. ACM, New York.

Whiteside, J., Jones, S., Levy, P., and Wixon, D. 1985. User performance with command, menu and iconic interfaces, pp. 144–148. In Proc. CHI '85, *Hum. Factors Comput. Syst.*, ACM, New York.

Ziegler, J. 1991. Direct manipulation techniques for the human–computer interface. In *Advances in Computer Graphics VI–Images: Synthesis, Analysis and Interaction*. G. Garcia and I. Herman, eds., pp. 421–448. Springer–Verlag, Berlin.

Ziegler, J. 1996. *Vorgehensweise zum objektorientierten Entwurf graphisch-interaktiver Informationssysteme.* Springer–Verlag, Heidelberg (in German).

Further Information

Interactive techniques have received considerable attention in HCI research and there is a large body of literature relating to the design, implementation, and empirical study of the different techniques. The following publications are useful sources of more detailed information and further references:

Shneiderman [1987] provides a systematic overview of different interaction techniques. Empirical findings as well as design guidelines are presented for each technique.

Helander, M., ed. 1988. *Handbook of Human–Computer Interaction*. North-Holland, Amsterdam. This classic HCI handbook comprises in-depth treatments of various interaction techniques. Good overviews of empirical results, especially for menus and command languages, are presented.

Preece, J., Rogers, Y., Sharp, H., Benyon, D., Holland, S., and Carey, T. 1994. *Human–Computer Interaction*. Addison–Wesley, Wokingham. This textbook provides a comprehensive introduction to HCI, covering different interaction styles and related design issues.

Bellcore. 1995. Design Guide for Multiplatform Graphical User Interfaces. *Bellcore Tech. Rep.* LP-R 13, Dec., can be ordered from. Bellcore Customer Service, 8 Corporate Place, Room 3A-184, Piscataway, NJ 08854-4196. This is a state-of-the-art style guide for the design of graphical user interfaces, covering the relevant interaction objects in the Common User Access™, Motif™, Common Desktop Environment, and Windows 1995 GUI platforms.

71

Multimedia

James L. Alty
Loughborough University,
UK

71.1 Introduction: Media and Multimedia Interfaces

In order to communicate information to other human beings, we need to disturb the environment around us in such a way that those disturbances can be detected by the people with whom we wish to communicate. Furthermore, we need to have previously agreed to the meanings of such disturbances with those with whom we wish to communicate, so that the messages can be understood. In other words, we need to establish a *medium of communication* between ourselves and the target audience. It is in this sense that computer designers talk about *media*.

A medium, therefore, is an agreed mechanism for communication. Some media are very simple in nature, for example, the doorbell on a house. At this simple level there is only one message: someone is at the door. However, one could imagine this medium being developed further in certain circumstances. For example, one ring could mean person A is at the door, two rings would indicate person B, etc. One could even use the bell for Morse code and transmit quite complex messages through it. Why anyone would want to do this is not immediately obvious, but such a system might be useful for someone who was severely disabled and could not easily get to the door.

This simple example illustrates the essential components of a medium: there are the basic tokens or symbols (such as the ringing of the bell), there are agreed structures built from these elements (such as the number of consecutive soundings of the bell, the length of silences, the maximum number of rings in a structure), and there is the assigned meaning to the different structural elements (for example, Morse code). These three components are often termed the symbols (or lexicon), syntax (or structural rules), and semantics (meaning) of the medium. The parallel with language is obvious, and complex media essentially support communication languages. The fourth element of language, pragmatics, is also often present. Pragmatics are concerned with conventions and common usage.

The preceding door bell example also illustrates multiple use of media. The huge leap in communicative power released when the bell is used for Morse code results from developing a mapping between the simple bell and another very powerful medium of communication, our spoken or written language. The system is, in one sense, illustrating the use of multiple media because it can either be used to alert the householder or to transmit a more complex message (or both at the same time).

Multimedia communication is the simultaneous (or sequential) use of more than one medium of communication to transmit information. For human beings, using multiple media is the normal way of communicating. A human being will usually, in parallel, employ spoken language, body language, gesture, and touch to transmit a message to another person. If they are constrained (for example, by having their hands tied) human beings often find it more difficult to communicate. It is interesting to note that when people are observed talking on the telephone, they still use extensive gestures even though no one is observing them.

Just as human beings need media to communicate, so computers (or the designers who write the programs) have to employ media to communicate with their users. This is done through the human–computer interface (HCI). To communicate with users, *output media* are employed. These media need to be comprehended by human beings and often involve words, pictures, or sounds. To interpret what the users are communicating to the computer system, *input media* are used. Such media often require special skills (such as typing skills) to be employed effectively.

The simultaneous use of media to communicate with users on the human–computer interface is termed a *multimedia interface* and can refer to the input media, the output media, or both. Coupling media together can often result in the creation of new media. For example, the coupling of moving video and audio into films or television has created new entertainment media.

Multimedia interfaces are more than a set of interesting new ways of using new technology. Many people believe that such communication is natural and corresponds more closely with how the brain has developed. Marmollin [1992] has described multimedia as exercising "the whole mind." In this viewpoint, the human brain is seen as having evolved in a multisensory environment, where simultaneous input on different channels was essential for survival. Thus, the processing of the human brain has been fine tuned to allow simultaneous sampling and comparison between different channels. When channels agree, a sense of safety and well being is felt. When channels degrade, input from one channel can be used to compensate another. Thus, input channel redundancy (within limits) is thought to be desirable and is a pleasurable experience.

It is important to realize that multimedia design is not just about choosing the obvious medium for a particular communication requirement. Deliberately presenting information in a foreign medium can deliver new and interesting insights into a problem. For example, musical harmony is normally presented through the audio channel, yet new insights into harmonic progressions can be obtained by displaying harmony in a visual manner. A nice example of this is the HarmonySpace application of Holland [1994]. This tool offers both experts and beginners the opportunity of exploring harmony by allowing them to use spatial attributes in their exploration (e.g., nearness, centrality, and shape similarity). Similarly, music can be used to assist in the understanding of computer algorithms or physical processes such as turbulence [Blattner et al. 1992].

One final and important point about multimedia interfaces is their significance for disabled users. The current high emphasis on visual output media can be severely disadvantaging to blind or partially sighted individuals. Designers should exploit the new presentation opportunities offered by the multimedia approach, but they must not forget that their interfaces may be used by someone who may not be able to assimilate all of the channels and they should, therefore, allow sufficient redundancy on channels so that the partial loss of one medium does not fatally affect communication in other media. On the other hand, designers also ought to take advantage of the new aural media by offering specially adapted interfaces for the partially sighted. Some progress has already been made in this area. Edwards [1989] has created a word processor that uses musical tones and synthesized speech. The approach adapts visual interfaces so that blind users can use them. The system provides auditory windows, which signify their position by unique tones when the cursor enters them. Menus are activated from these areas and are spoken. The system is called Soundtrack and can be controlled solely through the audio channel. However, the interface also has a visual manifestation as well and this redundancy can be utilized by a partially sighted person.

71.2 Types of Media

As has already been stated, media can be subdivided into input and output media. These can then be divided according to the sense used to detect them, visual, aural, and haptic (meaning touch) media, which can then be further subdivided again (for example, into language and graphics for visual output media, or sound and music for aural media). Table 71.1 (not intended to be exhaustive) gives some examples of common media.

Currently, haptic media dominate the input media area, whereas visual media dominate the output media field. Aural media are not really fully exploited as yet particularly for input, where voice recognition could offer a flexible and natural interface.

Output Media

Many current output media are reasonably well tuned to human capabilities. They are based on media which have been used between human beings for many years: text, graphics, pictures, video, and sound. Although normal-sized VDU screens do not have quite the same properties as standard-sized office paper, the correspondence is close. Most users, therefore, have little trouble in adjusting to understanding well-designed output visual or aural media. The problems of designing effective output using these media are essentially the same as those in traditional media design (for example, publishing).

One output medium, however, which has not been fully utilized in traditional computer applications, is sound. Although most computers can support quite sophisticated aural output (e.g., music), this is rarely exploited. Gaver [1986] has suggested the use of auditory icons. These are well-known, usually natural, sounds, which have common associations (such as the siren on a police car). Blattner et al. [1989] have further suggested the use of structured *earcons* (based on simple musical motifs). Such motifs (or jingles) are often used in public address systems to precede messages and alert listeners. Alty [1995] has investigated mappings between computer algorithms and music. The internal workings of an algorithm are mapped to musical structures. Thus, in an audiolization of the Bubble Sort Algorithm, moving up the list, swapping elements, and the current state of the list are all mapped into different instruments and rhythmic structures. Disambiguation is further assisted by using stereo output. It appears that people can understand the algorithms from the musical output alone, but more work is needed on which types

TABLE 71.1 Some Common Media

	Aural	Visual	Haptic
Input Media:	Natural sound	Video camera	Keyboard
	Spoken word	TextScan	Mouse
	Synthesized sound	Diagram scan	Tracker ball
		Gesture recognition	Data glove
		Eye tracking	Touch screen
			Foot pedal
			Breathing tube
Output Media:	Natural sound	Written text	Data glove
	Music	Static graphics	Braille pad
	Synthesized sound	Animation	
	Spoken word	Still video	
		Moving video	

of musical mappings are most appropriate. Musical mappings have also been suggested to aid computer program debugging [Francioni et al. 1991].

Another output medium, which has potential but which is underexploited, is 3D vision. The problem is, of course, the present requirement for special glasses. The third dimension has obvious applications in displaying 3D molecules or architectural structures, but it can also be used to improve presentation of other data. Three-dimensional presentation has been used in displaying information in databases. Because in virtual reality systems 3D display is usually essential, it is expected that rapid developments will take place in this area.

Traditional entertainment media usually consist of output media only (films and TV program are examples of these). In computing applications, however, the user is normally able to interact and control the progress of the interaction. This is termed *interactive multimedia*. In the interactive media environments, linking both between different media and within media can be very important, and there has been considerable recent research work on how to link different elements within the various media.

Output text has been transformed through the creation of *hypertext* structures [Nielsen 1990]. Hypertext techniques transform traditional sequential text into a cross-linked structure. Certain words in the text are made *active* in that, when selected, they will transfer the reader to another section of text. In extreme cases the sequential nature of the text is lost, and the text becomes a structure with many paths through it, selectable by the reader. There is disagreement over exactly what constitutes hypertext. Some have argued that any hypertext structure must have the ability of displaying the linking structure explicitly to the user [Halasz 1988]. Others have suggested that all links should be bidirectional. Hypertext linkages across communication networks have now become commonplace, an obvious example being the World Wide Web. Most textual information on the Web is now stored as hypertext (using a language called HTML). Such Web text contains links to other systems on the Web and exceedingly long and complex chains of linkages can be followed.

The term *hypermedia* is often used to describe the creation and support of linkages between different media. Elements of text may link to photographs, movies, or even sound sequences either on local systems or across the communication network. One of the current problems in hypertext and hypermedia structures is navigation. Users can easily become lost in hypermedia space. Nondynamic links can also inhibit exploratory learning.

Input Media

Current input media, in comparison with output media, are cumbersome and unnatural. They require skills (such as keyboard skills) to be used effectively. Furthermore, input media are unusual in the fact that they need to be coupled with some form of output medium to be useful. For example, keyboard input is not effective unless the user receives simultaneous output of what is being typed. In a similar manner, input using a mouse requires visual feedback to be effective. This complicates the analysis of input media.

Recently, there has been more active research on new input media. Developments have been reported on voice recognition (now beginning to reach acceptable levels of performance), gesture and pointing (where the actual visual gestures are tracked by video cameras and interpreted), eye movement (the actual movement of the eye is tracked and can be used as a selection device), lip motion (to assist in speech recognition), facial expression, and handwriting. The research is driven by the current primitive state of input media in contrast to human–human communication.

An interesting feature of many input media is their impreciseness. Voice recognition is often a difficult process because of other extraneous noise; gesture is often vague and ambiguous; lip motion is not read accurately by most human beings. Human beings process such media effectively because they are usually processed in parallel (for example, gesture and lip movement usually accompany speech). The human processing system exploits the redundancy across these channels, comparing input from different channels for confirmation, or seeking support for the interpretation of one channel from another input channel. Some recent experiments have even suggested that human beings combine acoustic and visual information before classifying them separately [Braida 1991]. In other words, separate channel decisions are not made on each input. It is not surprising that much recent input media research work has been concerned with investigating the simultaneous use of a number of input media, and experiments have been carried out to see if recognition can be improved through the use of touch, speech input, and gesture simultaneously.

Experiments on input media have been reported involving the combination of speech recognition with lip reading, gesture with speech, and speech with handwriting [Waibel et al. 1995]. One experiment concerned the simultaneous input of lip reading and voice input. The acoustic input performance was measured in clean and noisy environments. When the acoustic input was clean, word accuracies in excess of 90% were attained. The lip reading performance, on its own, varied between 32 and 47% accuracy, and when used in parallel with the acoustic input had minimal effect on overall accuracy. When the noisy acoustic input was used, however, acoustic recognition on its own fell to around 50% but with lip reading added in parallel, performance improved to over 70%. Thus, adding the lip reading input (which had a relatively poor recognition rate on its own) boosted the recognition rates of acoustic input in noisy environments.

Waibel et al. [1995] have also examined gestural input in some detail. Their input gestures were created by stylus moves on a 2D graphics tablet. They report on the great variability in the way users make gestures. "No matter how many tokens we put in the training database to cover the different gestures used that mean 'delete text' for example, there may always be totally different gestures that are not yet part of the gesture vocabulary" [Waibel et al. 1995]. Bordegoni and Hemmje [1993] have constructed a "dynamic gesture machine" providing graphical feedback in a 3D interface. The system is based on a simple gesture language.

More novel input devices have been reported. Bordegoni and Hemmje [1993] report on the use of a force-input device. This device replaces the mouse in a 3D virtual environment and is based on a SpaceTrackerball, which can not only act as a normal trackerball but can in addition, detect the pressure exerted on the ball for providing 3D movement. A data glove is used, in addition, to detect the position of the hand. Data gloves on their own offer interesting new possibilities particularly in virtual reality applications [Zimmermann et al. 1987].

71.3 Multimedia Hardware Requirements

Multimedia systems inevitably require considerable hardware resources. Such systems need large amounts of memory (both primary and secondary) and any transmission of multimedia data over a network usually requires high bandwidths. It has been estimated that there is a need to be able to handle about 4–5 megabytes of information per second to satisfactorily handle multimedia applications, and this is unlikely to be an overestimate.

The most common multimedia platforms currently in use are based on SUN workstations, personal computers (PCs), or Apple MacIntosh systems. Minimum multimedia enhancements will include a sound board, a super Video Graphics Adapter (VGA) graphics card, and a compact disc-read-only memory (CD-ROM) reader. Originally, there were two defined standards for multimedia computers: multiple processor computer (MPC) level 1 and MPC level 2. MPC level 1 was the first standard, which was superceded by MPC level 2 in 1993. This level-1 standard was not very realistic: a minimum of a 386X processor, 2 megabytes of random-access memory (RAM), a 30 megabyte hard disk, a CD-ROM drive, VGA video (16 colors), an 8-b audio board, and appropriate software. The more realistic level-2 standard specified a 25-MHz 486X processor, 4-megabytes of RAM, 60-megabytes hard disk drive, CD-ROM (300-kilobytes transfer rate), 16-b audio board, video (640 × 480 + 64 K colors), and appropriate software. Recently levels 3 and 4 of the MPC standard have been defined.

Compact-Disc Read-Only Memory Technology

Optical storage media, as typified by the CD-ROM, have provided the much needed increased storage capability required for multimedia applications. The CD-ROM is rather like a traditional long playing record but, in contrast to the traditional record player, the track is read at a constant linear velocity. This means that the rotation speed has to change depending on the position of the track relative to the center. The data is divided into blocks to provide some direct access capability. The approach allows high data volumes to be stored but it reduces direct access possibilities since the data are stored in a long spiral.

The CD-ROM was developed from the audio CD (CD technologies are defined in colored books, the audio technology in the Red book, CD-ROM technology in the Yellow book, and the write-once read-only CD technology, now available on many relatively cheap CD-ROM makers, in the Orange book). These standards define the basic hardware and storage mechanisms. However, there was also a need to standardize the ways in which operating systems access the information. The International Standards Organization (ISO) 9660 standard (originally known as High Sierra) achieves this and allows different operating systems to access the same CD-ROM.

One improvement provided by the ISO standard is the concept of a *session*. A CD-ROM may be built up progressively as sessions are added and when the disk is read, by default, the last session is accessed. This session can access data written in previous sessions or prevent that information from being accessed (for example, an update). This allows the designer to write information to the disk in different sessions (i.e., different times) and to update earlier data.

CD-ROM technology stores data in a manner similar to normal filing systems, using a directory structure to locate the contents. A typical CD-ROM can store up to 650 megabytes of data, which corresponds to 250,000 pages of A4 text, 7,000 full screen images, 72 min of full bandwidth animation or full screen video, or 19 hours of audio. It is important to realize that the high-volume storage capability of CD-ROM disks (>600 megabytes) is achieved at the expense of access times (typically about 170 kilobytes/s though it can be higher on special drives). Thus, the data contained on a CD-ROM are large and need to be properly managed. Although careful placement of files can be used to improve efficiency, most management takes place outside the CD-ROM. The CD-ROM designer should not underestimate the effort involved.

The introduction of low-cost write-once-read-many (WORM) CD-ROM creation systems has recently revolutionized the production process for CD-ROMs. Previously, a designer had to create a binary image of the CD-ROM on magnetic media and deliver this to a CD-ROM manufacturer. From the binary file, a master disk was made from which multiple disks were then pressed. This made low-volume production uneconomical. Now a designer can buy a WORM CD-ROM maker for about the price of a PC, and this has the ability to master individual disks. The availability of this new hardware has expanded the application areas for CD-ROMs. They may now be used as removable media to back up occasional files.

Compact Disc-Interactive (CD-I)

CD-interactive (CD-I), originally developed by Phillips and defined in the Green book, is a special way of storing information on a CD, which provides much more efficient support for interactive multimedia applications. First, the data are stored in a run-time format, which avoids additional processing after retrieval. Second, both digital audio and video (both movie and still) are interleaved in blocks. At compilation time, the audio and video blocks of data are stored in the sequence in which they are expected to be retrieved; thus, the data will usually be under or near the head when required. This avoids excessive head movements, which occur in CD-ROMs, when two different media files need to be accessed at the same time.

Video Storage and Manipulation

Handling video data places very high demands on the multimedia computer system. A typical full video still picture requires about 1 megabyte of memory. If this were to form part of a movie, the system would need to transfer about 30 megabytes/s to give the illusion of motion. Thus, one minute of video could occupy nearly a gigabyte of storage. In addition, huge transfer rates would be required to refresh the memory with new pictures since current hard disks have transfer rates of about 1 megabyte/s. This is clearly not realistic on present hardware, and a key factor in handling these difficulties is video compression.

One of the most common video compression systems used for still images is Joint Photographics Expert Group (JPEG). It can usually achieve compression ratios of about 30:1 without loss. This means that the 30-megabyte/s transfer rate reduces to 1 megabyte/s, just within the capabilities of present hard disk drives, and an image can be stored in about 40 kilobytes. The compression speeds are low.

For moving pictures, the Motion Picture Experts Group (MPEG) compression techniques are used. This achieves compression ratios of about 50:1 without degradation, and the compression and decompression can occur in real time. Using MPEG, a movie can be played from a CD-ROM. Higher degradation is possible with MPEG but results in a loss of image quality.

Other compression techniques exist, such as P∗64, which is used for compressing audio and video in video telephony.

Audio Technology: Digital Audio and the Musical Instrument Digital Interface

Much of the audio activity in multimedia applications is concerned with sound rather than with speech. Most books on multimedia applications hardly mention speech synthesis or voice recognition at all. Voice recognition is not discussed because it is still at an early stage of development and not reliable enough for existing applications. Voice synthesis, on the other hand, works reasonably well and is, perhaps, considered very straightforward. Most personal computers offer standard voice synthesis packages, which work from text input.

Sound is stored in computer systems as digitized audio. Any sounds can be digitized: music, the singing voice, or natural sounds. The input analog signal is sampled at a regular rate and stored. Sampling frequencies vary from 11 to 48 KHz, and the sample size can be 8 or 16 bits. Audio quality requires a sampling rate of 48 KHz, whereas 22 KHz will store speech effectively. Sound editors are available which can work on the digitized waveform. In this way the designer can add reverberation, chorus effects, fading in and out, and alter parameters, such as the volume.

Digital audio, at an appropriate sampling rate (about four times the maximum required audio frequency), stores a nearly faithful representation of the analog waveform. The musical instrument digital interface (MIDI) system, on the other hand, offers quite distinct facilities. In contrast to digital audio this system stores descriptions of a musical score and communicates this information along logical channels to other MIDI devices. Thus, a MIDI keyboard will record note-on and note-off actions, the note value, and the pressure on the key, as well as other control signals, which alter volume, define the timbre (instrument), and define the stereo profile. These signals can then be sent to a MIDI synthesizer, which will reproduce the analog signals necessary to play the score. Each MIDI instrument usually has a MIDI-in, MIDI-thru, an MIDI-out port. A typical MIDI setup is shown in Fig. 71.1.

The keyboard can be used to create basic MIDI instructions (note-on, note-off, etc.), in which case the output is taken from MIDI-out port and processed elsewhere. Alternatively, the internal synthesizer of the keyboard can be used to create sounds from the MIDI signals created by the keyboard, and this results

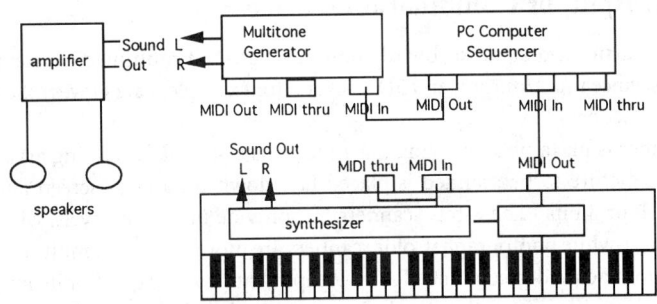

FIGURE 71.1   Typical MIDI connections.

in audio output signals from sound out. The MIDI-in port can also be used by another keyboard, which would send in MIDI signals to be synthesized by the internal synthesizer.

If the MIDI-out port is used from the keyboard, this can either go directly into the MIDI-in port of the multitone generator to produce sounds at the sound out (in this case simply using the multitone generator as a synthesizer instead of the internal synthesizer of the keyboard; connection not shown in the diagram). Or it could be fed into the MIDI-in port of a PC, which is running a MIDI sequencing software package. The resultant MIDI combined output will be stored on the disk of the PC and can then be played at any time sending the output through the MIDI-out port of the PC to the MIDI-in port of the multitone generator, which will simultaneously create the necessary sounds at sound out. Finally, these sounds will be amplified and sent to stereo speakers in the normal way.

The MIDI-thru ports are used to connect to multiple devices. They allow the signals at a MIDI-in port to be presented elsewhere at the same time. For example, the MIDI-thru port on the multitone generator could be connected to the MIDI-in port of the keyboard to give an extra sound (but more likely to a different multitone generator).

It is important to note that MIDI signals are not digitized sounds, they are a sort of musical shorthand. As a result, MIDI files are far more compact than digital audio files (they can be as much as 1000 times more compact). The MIDI system can support up to 16 distinct channels simultaneously. Each channel can be assigned to a different instrument, and the signals can then be sent, in parallel, to a multitimbral tone generator, which recreates the analog sounds. The sounds can be created in a sequencer, that is, a software system which allows a composer to build up the parallel descriptions of each part. Extensive editing facilities are available, and actual musical scores can be produced automatically from the MIDI description. One of the more advanced systems is the CUBASE system [Steinberg 1996].

Using a system such as CUBASE, the composer of the multimedia soundtrack can build up the separate channels and control instrument selection, volume, panning, and special effects. Provided a standard mapping is used, the composer can expect the desired instruments to be selected on playback in another MIDI device. There are a number of standard mappings, one of which is the general instrument sound map (general MIDI level 1), a small portion of which is given in Table 71.2.

TABLE 71.2   Some General MIDI Sound Assignments

Identifier	Sound or Instrument	Identifier	Sound or Instrument
1	Acoustic grand piano	72	Clarinet
14	Xylophone	98	Soundtrack
25	Acoustic electric guitar	125	Telephone ring
41	Violin	128	Gunshot
61	French horn		

Note that not only musical sounds are capable of being reproduced. Other general MIDI sounds include laughing, screaming, heartbeat, door slam, siren, dog, rain, thunder, wind, seashore, bubbles, and many others.

The timing of MIDI compositions can be accurately controlled so that they can be synchronized with other media such as video.

Other Input, Output, or Combination Devices

In addition to the usual input devices (keyboard, mouse, tablet, etc.) multimedia systems may additionally have color scanners, voice input/output, and 3D trackers (for example, data gloves). Some of these devices combine input with feedback.

The flatbed scanner is like a photocopying machine. It is essential for adding artwork to multimedia presentations. The picture to be scanned is placed face down on a glass screen and is then scanned. Scanners can be 1, 4, or 8 bits. The 8-bits scanners can provide up to 256 levels, which will successfully reproduce a black-and-white photograph. Color scanners are more normal in multimedia work. They use three 8-bits values for each red green blue (RGB) component providing a 24-bits image. As well as the color resolution, a scanner has an optical resolution, measured in dots per inch (dots/in). Typical values for optical resolution are 300, 400, or 600 dots/in. Scanners require software to support them, which enables the user to process the image (for example, adjust brightness and contrast).

Three-dimensional trackers can be used to measure the absolute position and orientation of a sensor in space. The tracking is accomplished either by magnetic means or by acoustic means. The yaw, pitch, and roll may also be determined. Such measurements are important if human gesture is to be correctly interpreted. Trackers are often mounted on clothing, for example, data gloves, helmets, or even body suits. The data glove worn on the hand can provide input data on hand position, finger movement, etc.; data gloves are particularly useful in virtual reality environments where they can be used to select virtual objects by touch or grasping. An output feedback system also can be placed in the glove, which provides a sensation of pressure, making the grasping almost lifelike. An interesting example of such an application is the GROPE system [Brooks et al. 1990]. The GROPE system uses a device with force feedback to allow scientists to fit molecules into other molecules. The user can manipulate the computer created objects and actually feel the molecules. Some improvement in performance was noted and users reported that they obtained a better understanding of the physical process of the docking of molecules.

Multimedia Information Delivery: The Multimedia and Hypermedia Experts Group (MHEG) Standard

Standards are currently being developed to allow multimedia information objects to be transmitted and received in real time over wide area networks. The Multimedia and Hypermedia Experts Group (MHEG) standard [Price 1993] is currently a draft ISO standard. The purpose of the MHEG standard is to enable the interchange of structural, spatial, and temporal information related to the composition of multimedia objects. Four modes of interchange have been defined:

- A final storage model during creation and editing of documents
- A format for delivery of final-form digital media
- A format for real-time delivery from servers to clients
- For interapplication exchange of data

The standard describes the generic behavior without specifying the implementation. It uses an object-oriented approach specifying a set of object classes. An element, known as an MHEG engine, codes and decodes objects and delivers them to an application.

71.4 Distinct Application of Multimedia Techniques

Because media are at the heart of all human–computer interaction, an extreme viewpoint would regard all computer interfaces as multimedia interfaces, with single-medium interfaces (such as command languages) being special limiting cases. However, we will restrict the term multimedia interface to apply only to interfaces that employ two or more media either in series or in parallel (or both).

There is a major division in the way multimedia techniques are applied. First, the techniques can be used to front end any computer application using the most appropriate media to transmit the required information to the user. Thus, different media might be used to improve a spreadsheet application, a database front-end retrieval system, an aircraft control panel, or the control desks of a large process control application. On the other hand, multimedia techniques have also created new types of computer applications, particularly in educational and promotional areas. These include information and educational programs. In the front-ending instances, multimedia techniques are enhancing existing interfaces, whereas in the latter fields, applications are being developed which were not previously viable. For example, educational programs analyzing aspects of Picasso's art or Beethoven's music were simply not possible with command-line interfaces. These new applications are quite distinct from more conventional interfaces. They have design problems that are more closely associated with those of movie or television program production.

In both application areas there is a real and serious lack of guidelines as how best to apply multimedia techniques. Just as very fast computers allowed programmers to make mistakes more quickly, so the indiscriminate use of multiple media can confuse users more effectively. The key issues, related to the application of multimedia technology, are therefore more concerned with design than with the technology per se. We are about to enter a decade where the early years will probably be characterized more by bad design than by good design. There is no doubt, however, that good design practice will emerge eventually.

It is not difficult to understand why design is going to be a major issue. The early human–computer interfaces relied almost exclusively on text. Text has been with the human race for a few hundred years. We are all brought up on it, and we have all practiced communicating with it. When new graphics technology allowed programmers to expand their repertoire, the use of diagrams was not a large step either. Human beings were already used to communicating in this way. When color became available, however, the first problems began to occur. Most human beings can appreciate a fine design which uses color in a clever manner. Few would, however, be able to create such a design. Thus, most human beings are not skilled in using color to communicate, and so when programmers tried to add color to their repertoire things went badly wrong. Many gaudy, overcolored interfaces were created before the advice of graphics designers was sought. More recently, the quality of most home video productions shows how important skills will be in using video in interfaces. Such interfaces will require a whole new set of skills, which the average programmer does not currently possess.

The next section examines a case study in multimedia evaluation, and is an example of the type of research work which is needed to progress with the multimedia design problem.

71.5 Case Study: An Investigation into the Effects of Media on User Performance

The application area for this case study is process control. It is thus the first type of application defined in section 71.4. Typically, process control systems are highly dynamic and can be distinguished from traditional computer systems because the application will usually continue whether or not the user intervenes (sometimes with serious consequences). The operator's task in such systems is to ensure that the process continues to operate safely within a predefined economic envelope.

The main task of the interface designer is to present the activities of the dynamic process in such a way that the operators can readily control the process, identify any deviations from the norm in good time, and be provided with guidance as to how to return the system to the normal state. Usually, there are two goals associated with handling problem conditions: first to move the system into a safe (possibly suboptimal) state, and second, to endeavor to return the system to an optimal state.

The process operators multimedia intelligent support environment (PROMISE) project was a large international project, funded by the European Commission, which investigated whether multimedia techniques could assist in the control of such dynamic processes. A multimedia toolset was built, and a series of experiments was carried out to examine the effects of different media on operator performance. Some experiments were performed in the laboratory and others in a live chemical plant. Only the laboratory data will be reported here. Space precludes a complete analysis and so any reader interested in additional information is referred to Alty et al. [1993] for the laboratory work and Alty et al. [1995] for the plant results.

The Laboratory Task

One of the main problems encountered when setting up process control laboratory studies is choosing a task that is reasonably representative of the domain under investigation, that is reasonably straightforward to set up, and that can be learned relatively easily by subjects. Another requirement for our studies was that there must be scope for multimedia exploitation at the interface.

Fortunately, a task already existed that had all of these characteristics, and for which operator behavior had already been extensively studied. This task was Crossman's water bath (Crossman and Cooke 1974), which has been studied on numerous occasions [Sanderson et al. 1989, Moray et al. 1986, Moray and Rotenberg 1989]. The task is deceptively simple. (See Fig. 71.2.)

Crossman's water bath is a simulated hydraulic system. A simple bath contains water, the level of which can be altered by adding water through the in valve (*V in*) or increasing the *outflow* by opening the out valve (*V out*) or both. At any time there will be an outflow from the bath, which could be zero if *V out* is closed. The water therefore has a *Level* in the bath, which will change. The bath is heated (*Heat*). Inside the bath is a further container with a fixed amount of water in it. The temperature of the water in this vessel is continuously measured by *Temp*. Thus, changes to *V in*, *V out*, and *Heat* cause changes to *Temp*, *Level*, and *Outflow*. The subject is given the system in a particular state and asked to stabilize it at a new state (a set of limits within which the variables have to lie after stabilization).

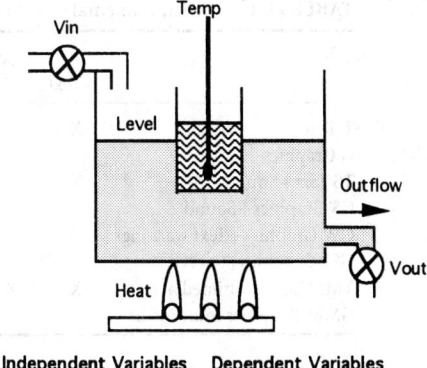

Independent Variables	Dependent Variables
Vin, Vout, Heat	Level, Outflow, Temp

FIGURE 71.2   Crossman's waterbath.

More than 50 subjects were given a brief introduction to the system. The task, and the controls, were explained but without giving any principles of the underlying system. Each session consisted of two halves. In the first-half, the subject had to complete 21 tasks of increasing difficulty (but all the tasks were relatively easy). In the second-half, subjects had to complete a set of more difficult tasks, involving larger moves in the state space and narrower stabilization limits. Tasks were characterized by their *compatibility*, a task descriptor defined by Sanderson et al. [1989]. Tasks had a compatibility from 1 (easy) to 3 (most difficult). In the break between the two halves of the experiment, subjects were tested for their understanding of the state variables. This test was also repeated at the end. All sessions were recorded on video camera, timing statistics were collected by the system, and subjects were asked to verbalize their current beliefs about the system and the reasons for their actions. Sessions could be completely replayed if necessary.

The Media Investigated

A wide range of media were used to create different instances of the controlling interface. The basic media are outlined in Table 71.3, and the actual interfaces are listed in Table 71.4. These eight conditions had five or six subjects in each. Some stylized examples of the interfaces are given in Figs. 71.3(a)–71.3(d).

By comparing results from individual or combinations of elements in Table 71.4, particular variables can be isolated. For example, a comparison of T and G vs. TS and GS can examine the effect of sound.

TABLE 71.3   The Basic Media Used in the Experiment

Medium	Description
Text	Single text values of each variable and the required limits were displayed
Graphics	A graphical representation of the water bath was displayed which reflected the current state. Current values and limits were displayed graphically.
Voice message	A male (or female) voice gave warning messages
Sound	A variable sound of flowing water could be heard which reflected the inflow rate.
Written message	A written message gave warnings.
Scrolling text table	The last 20 values of all variables were displayed in text with the current values at the base. The table continuously scrolled. The limits are shown.
Dynamic graph	A continuously scrolling graph showed the recent history of all of the variables and the current state. Limits were shown as targets.

TABLE 71.4   The Experimental Conditions

	Text	Scroll Table	Sound	Speech Warning	Written Warning	Graphics	History Graph
T Text	X						
G Graphics						X	
TS Text+Sound	X		X				
GS Graphics+Sound			X			X	
GW Graphics+Text warnings					X	X	
GSP Graphics+Speech				X	X	X	
TMM Text multimedia	X	X	X	X	X		
GMM Graphics multimedia			X	X	X	X	X

The main measures taken during the experiment were as follows:

- Time to complete the task
- Total number of user actions
- Number of warning situations entered

In addition, users subjectively rated the interfaces, and debriefing interviews were carried out. A selection of results will be given here, the reader is referred to Alty et al. [1993] for more details.

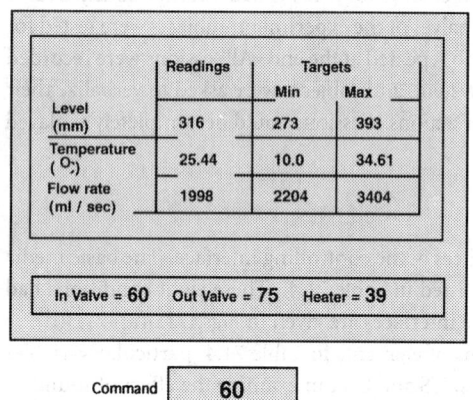

FIGURE 71.3(a)   The text interface.

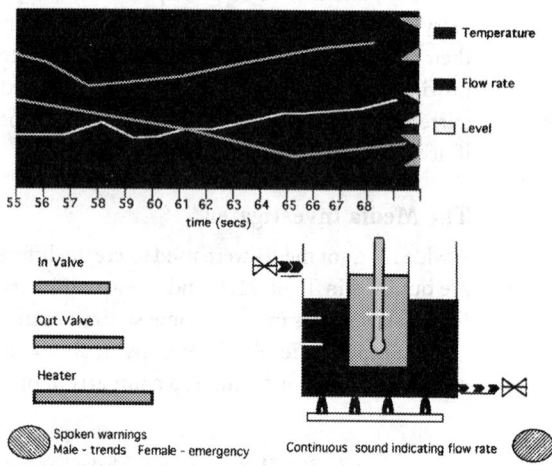

FIGURE 71.3(c)   The graphics multimedia interface.

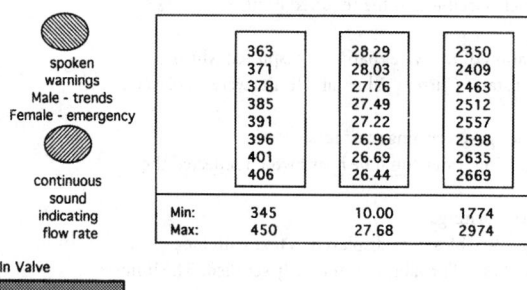

FIGURE 71.3(b)   The text multimedia interface.

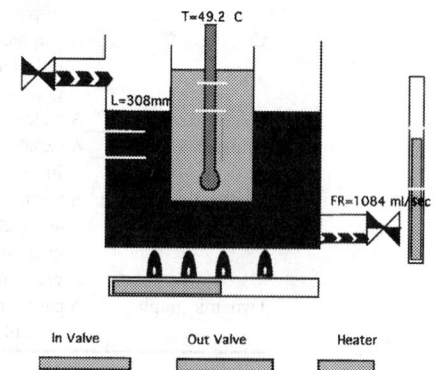

FIGURE 71.3(d)   The graphics interface.

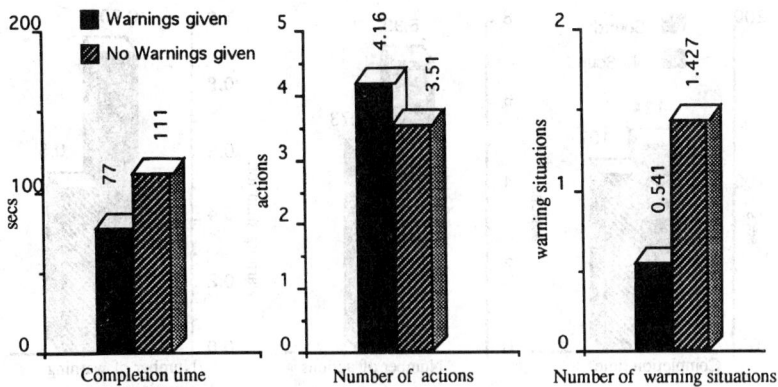

FIGURE 71.4 Effect of warnings.

Effect of Warnings on Performance

If a user strayed outside an acceptable envelope in the state space, warnings (either text or spoken) were given. In some conditions no warning messages were given at all, but the fact that the region had been entered was recorded. Figure 71.4 shows the results for task completion time, number of actions, and number of warning situations encountered.

In the figure it can be seen that warnings improved completion time and caused the number of actions to increase. When warnings were not given, the user spent much more time outside the acceptable envelope. Although the effect on the number of actions was not significant, the other two differences were significant ($p < 0.02$ for completion time and $p < 0.04$ for warning situations). The results are understandable. Receiving a warning enabled a subject to take corrective action immediately. This reduces the completion time but increases the number of actions since warnings may trigger immediate (but possibly inappropriate) responses, which then need further corrective actions. Without warnings, subjects were unaware that they were outside the envelope for considerable periods of time.

Warnings would, therefore, seem to have a beneficial effect. However, an examination of the results of the knowledge questionnaire for both conditions revealed that warnings actually had a detrimental effect on the subject's understanding of the more difficult aspects of the system. Warnings may disrupt thought processes, and users may come to rely on them rather than try to understand the system.

Subjects showed no significant differences in performance as a result of receiving spoken or textual warnings, but they did rate spoken warnings as being the more important. A more detailed analysis of the data revealed that although all subjects rated spoken warnings as important, this was not true for textual warnings. One group had rated them as more important than the other group. The group which rated textual warnings highly also found the tasks difficult. The other group found the tasks much easier. The written warnings need additional processing in comparison with the verbal ones (because a switch has to be made from the visual task in hand) and they can be easily missed. Therefore, we suspect that subjects who found the tasks difficult tended to check the written messages very carefully and, therefore, rated them as more important. Spoken messages, on the other hand, can be processed in parallel with the visual perception of the screen and are rarely missed.

The Effects of Sound

The effects of the presence or absence of sound are shown in Fig. 71.5, where again completion times, number of actions, and the number of warning situations entered are shown. What is immediately striking about these results is an apparent detrimental effect of sound overall (only the effect of warning situations reached significance). A more careful analysis, separating out the performance on the three different levels of task compatibility, however, reveals a much more interesting result. These results are shown in Fig. 71.6.

A clear pattern emerges from these graphs, although the result does not reach significance. Sound appears to have an increasing effect with increasing task complexity and one might speculate why this

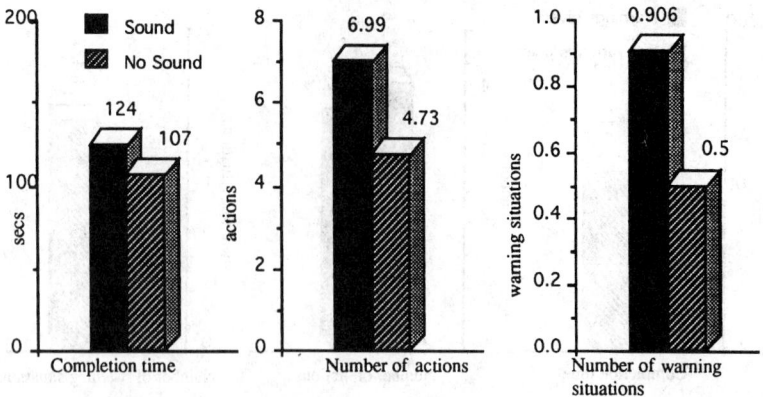

FIGURE 71.5   Effects of sound.

may be so. The moves in state space for category-3 tasks were much greater than the other two. Thus, in general, larger changes occurred in the flow rate of category-3 tasks than for the simpler tasks. These large changes will result in larger sound volume shifts, so that the sound volume may then have been more obvious and usable for category-3 tasks. In the easy tasks, the volume changes were not distinct enough to be useful.

It is obviously important to use media that are able to signal the required differences that the human operator needs for successful control, and in the case of sound this was not happening in the easy tasks. This appears to be a rather obvious result, but an observation of a number of different interfaces has revealed that designers do sometimes use media that cannot assist the operator in disambiguating the output.

Do Multimedia Interfaces Enable Improved Operator Performance?

A basic assumption of multimedia designers is that multimedia interfaces do indeed make a difference to operator performance. But does presentation affect the way information is picked up and consequently reasoned about? Figure 71.7 illustrates the results from the knowledge questionnaires for users who had experienced different multimedia interfaces.

The figure shows the results from the knowledge questionnaire for the six variables, broken down by medium type: text (T), graphics (G), graphics+warnings (GSP+GW), text multimedia (TMM), and graphics multimedia (GMM). A statistical analysis shows that, in the cases of *Outflow* and *V out*, the result is highly significant ($p < 0.01$ in both cases). The other four variables, however, failed to yield a

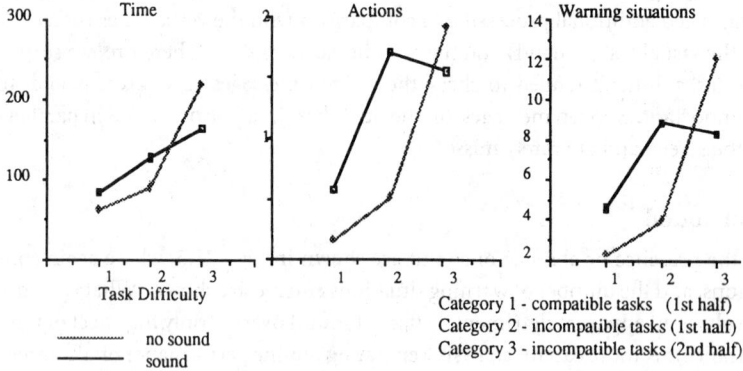

FIGURE 71.6   Task compatibility and sound effects.

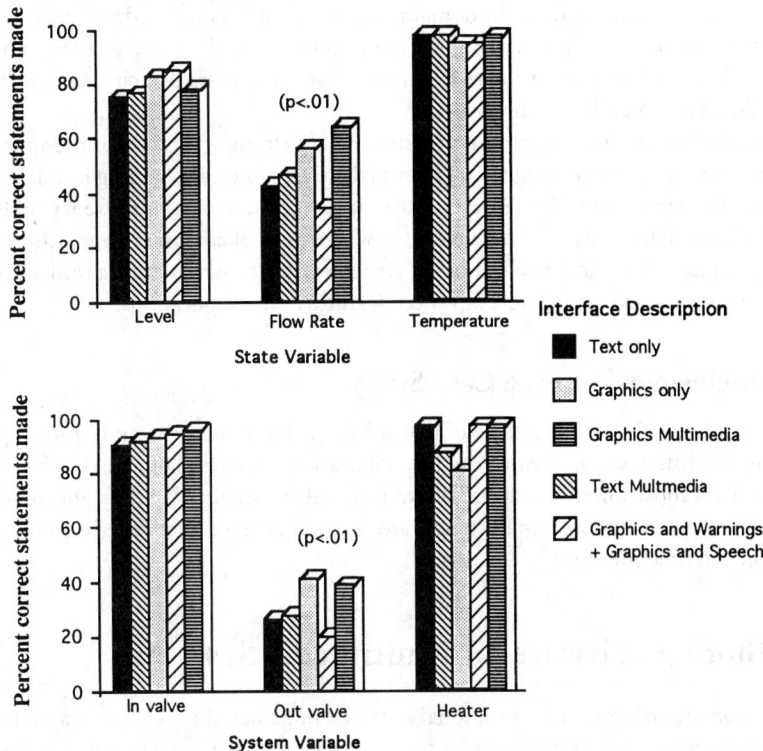

FIGURE 71.7 Overall effect of media on comprehension.

significant result. Thus, the type of medium chosen did affect comprehension in the cases of these two variables (though the analysis does not tell us how).

The figure exhibits some interesting features. Three of the variables have very high levels of comprehension, temperature, in-valve and heater, and a fourth is a reasonably well understood level. The effect of the different media seems to be greatest where understanding is lowest (for example, in the cases of outflow and out valve). How do we explain this effect?

The concepts of heater, temperature, in-valve, and level are easy to understand. The heater is increased, the bath heats up. The in-valve is opened, and the flow into the bath increases. The inflow rate is directly proportional to the setting of the valve. The level variable is a little harder to understand because it depends on the settings of both the in-valve and the out-valve, but users seem to have grasped the basic principles. The outflow and out-valve settings, however, are much harder to understand, and the reasons are in the physics. Although the out-valve may be set at a particular value, this does not result in a fixed outflow value. This is because the actual flow rate will depend on the setting of the out-valve and the level in the bath (i.e., the pressure). Thus setting the out-valve to a particular value does not guarantee a certain outflow rate. Moreover, this outflow rate will change as the bath empties or fills. The level of comprehension is really quite low for both outflow (about 50%) and out-valve (about 30%).

Here is a possible explanation. The concepts of heater, temperature, level, and in-valve are well understood (probably even before the experiment). Thus, it does not matter what medium is used to portray their actions (provided the basic information is exchanged). The users will understand. If the opposite extreme of this situation is considered, where the concepts being portrayed are not understood at all by the users (this condition is not present in this experiment), then one might reasonably postulate that it would not matter what medium was used to show their operation, the users would have great difficulty in understanding what was going on. In other words, the medium would not seriously affect comprehension because there would be no point of contact. The intermediate situation is the interesting one. What if the users understand the basic elements of the domain, but are operating on the edge of their understanding? In the cases of the outflow and the out-valve, the users do understand some aspects but not all (they did not get zero marks in the comprehension test). In such cases, it appears that choosing the right medium did make the difference in understanding.

If this explanation is true, it is an important result and is a strong argument for the appropriate use of multimedia elements in interface design. The intermediate area of understanding just described is a very important one. In process control it represents that area where the operators are trying to understand abnormal conditions in the behavior of the plant, or where users of computer applications are exploring new features. In more traditional applications it represents the situation during initial exposure to the system, or when users are exploring more advanced features of the applications.

Overall Conclusions from the Case Study

This case study is unusual in the literature in that it has tried to examine if multimedia approaches are useful and why. It illustrates how research can be used to tease out aspects of interface design. It provides some evidence that a move toward multimedia interfaces will be beneficial in the right circumstances and provides pointers as to what those circumstances are. It also illustrates that experimentation in this field is quite difficult and time consuming.

71.6 Authoring Software for Multimedia Systems

In the case of promotional software, educational software, or games, the creation of a multimedia application strongly resembles the production of a film or television program, and one talks of multimedia productions. All of the different aspects of the production such as scanned photographs, movie sequences, musical sequences, and text have to be put together into a meaningful whole, all correctly synchronized. For such applications, software to organize and assist in bringing together the contributions from the different media is essential. Such software is often called an authoring tool.

Authoring software enables the designer to create and edit different media, and then put them together in a desired sequence. As with most interface tools, different metaphors are used to ease the design task. One example of a common metaphor used is the time line where all of the events are laid out across the screen in a time sequence. Using a time frame of about 1/30 of a second, elements are positioned on parallel time lines and other events are cued along the time line. Another metaphor is based on the card metaphor (rather like a card address file) where a stack of cards forms the basis of the design. The elements of the multimedia presentation are placed on cards, and these cards are then linked together. A third type of authoring tool is based on an existing toolset such as Visual Basic. Finally other tools are icon based (similar to visual programming).

There are a number of authorware systems on the market. The situation is changing rapidly, and so any detailed comments about tool facilities need to be read with caution. The general principles, however, will apply. Examples of card-based systems include Hypercard and Supercard, tool-based systems include Visual Basic and Toolbook(Windows), time line products include Macromind Director, Action and Animation Works, and an example of an icon-based authoring tool is Authorware. Some tools are available for authoring CD-I applications (MediaMogul).

Which package should be chosen depends on how ambitious the planned multimedia project is. The icon-based tools are the easiest to use, and the time line tools are probably the most complex. The tool-based systems may not provide all of the facilities needed and may require further effort in mastering the tool itself. The card-based tools are relatively simple. There are a huge number of tools on the market. A recent listing of MacIntosh Multimedia products covered over 90 different offerings, though all of these are not authoring tools. Prices for authoring tools vary from a few hundred dollars to over $20,000, though, as one might expect, the buyer often gets what has been paid for.

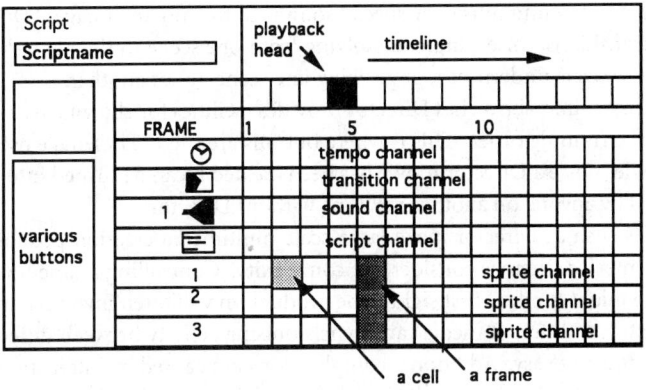

FIGURE 71.8 The director score screen (diagramatic).

An Example Authoring Package: Macromind Director

As an example of a sophisticated authoring tool, Macromedia Director from Macromedia, will be examined [Macromedia 1996]. This is one of the most powerful and complex tools available. As always, there is a tradeoff between the complexity of using the tool and the facilities offered.

In Director, the designer develops what is called the CAST; the CAST contains all of the elements of the production, for example, images, text files, video clips, voice overs, and artwork. These are multimedia elements, and Director has a large number of these predefined, but the designer can design and store new elements in the database (the maximum number has now been raised to 32,000). These new elements can be created using Director tools such as the Paint tool, or they can be imported from Art files, PICT libraries, or QuickTime movie files.

When the CAST has been created, the elements are brought together into the production through what is called the SCORE. The SCORE contains the timing information (as a time line metaphor) and consists of a set of frames, which contain elements from the CAST. Each cell contains information about one CAST member. A frame contains a set of cells, which bring together a number of CAST members for one moment in the movie, everything the viewer sees at a given moment.

A channel is a row of cells, and there can be many channels in the SCORE. Some are special channels such as those for tempo setting, controlling transitions, sound channels (2), palette, and a script channel. A further 48 sprite channels are provided for graphics, animation, movies, and buttons. The scripting channel carries LINGO scripts. A sprite is an image of a cast member. Figure 71.8 illustrates the layout. The playback head shows what part of the script is currently playing.

Animations can be created in real time by moving the element in the scene while recording. Usually the tempo would be slowed down for this. Sprites can be given a layering order so that one will be in the foreground and the others in the background (like overlapping windows). The overlap position can be read in the SCORE. Inanimate objects, which are visible during the animation, have their own time lines, which are filled in alongside the animation.

Director provides a number of techniques for animation. In-between linear will fill in the frames between a start position and an end position of a sprite. In-between special will animate between a start position and an end position along a created curve. It can also accelerate or decelerate effects along the curve. Reverse sequence will play animations backwards. These animation tools are powerful. Film loops can be created and stored as single CAST members. Tools are also provided for animating sets of cast members.

Director provides a large number of tools for using inks, fonts, and colors. Examples include matte, background transparent, reveal and transparent inks, all useful for creating special effects. The palette provides facilities for coloring images. A similar set of tools is available to process sounds. The designer can add voice overs, background music, or special sounds to the moving images. Director provides a set of tools to handle transitions, for example, dissolving from one scene to the next, sliding a scene off the screen revealing a new scene underneath, or growing one scene out of another.

Usually movies are noninteractive, but Director provides facilities for allowing users to interact with the movie and change the running order. This is where buttons are used. These take user input and execute branches in the movie. Once a Director movie has been created it can be turned into a play-only version, which can be viewed by anyone on another machine without Director.

Software packages such as Director make large-scale multimedia creation possible. Underneath the relatively easy user interface there is considerable complexity. Controlling a time line implementation is not as simple as one might think at first sight. The production will often involve a number of processors (probably associated with different media) and synchronising activity between different processors with different clock speeds is not easy. Additional complications are caused by interaction by the user. These may require the freezing of video frames, altering time out delays, and deciding how to terminate, or hold, a sound in midstream.

Director has been used as an example because it currently offers the most complex set of facilities. There are, however, many other excellent tools on the market, for example Authorware Professional or Action. Designers need to balance the facilities offered with cost, ease of use, and their requirements.

71.7 The Future of Multimedia Systems

It is clear that multimedia technology is already available which can greatly augment the choices open to the user-interface designer. These choices will also be available for systems of all sizes as computational speeds increase and hardware costs fall. A corollary of this is the ubiquity of high-bandwidth world wide networks which are capable of transmitting information in a variety of representational forms via multicast or point-to-point connections. For example, the use of the multimedia HTTP network protocol (more commonly known as the World Wide Web) gives an interesting, albeit haphazard, view of multimedia interfaces of the future.

However, it is still the lack of prevalent and wide ranging design criteria which makes multimedia user-interface design an ill-defined and empirically lacking discipline. To counter this, user-centered, rather than technology-centered, research is focusing on the following areas:

- Examining the effect of different media on human cognitive representations, particularly the construction of mental models of represented domains [Faraday 1995, Williams 1996].
- Classifying media in linguistic terms by virtue of their expressiveness [Stenning and Oberlander 1995].
- Matching media to task descriptions [Maybury 1993].

There is no doubt that if user-interface designers are to fully utilize the technology on offer, then suitable design methodologies need to be developed. These must encompass both the cognitive and the goal-oriented aspects of the human–computer system. Without them, multimedia will remain a pragmatic area of HCI application, or worse, will only be fit for use in entertainment systems.

References

Alty, J. L. 1995. Can we use music in computer-human communication?. In *People and Computers X*. M. A. R. Kirby, A. J. Dix, and J. E. Finlay, eds. *Proc. HCI '95*, pp. 409–423. Cambridge Univ. Press, Cambridge.

Alty, J. L., Bergan, M., Craufurd, P., and Dolphin, C. 1993. Experiments using multimedia interfaces in process control: some initial results. *Comput. Graphics* 17(3):205–218.

Alty, J. L., Bergan, J., and Schepens, A. 1995. The design of the PROMISE multimedia system and its use in a chemical plant. In *Multimedia Systems and Applications*. R. A. Earnshaw and J. A. Vince, eds., pp. 53–78. Academic Press. London.

Blattner, M., Greenberg, R., and Kamegai, M. 1992. Listening to turbulence: an example of scientific audiolisation. In *Multimedia Interface Design*, Chap. 6, M. Blattner and R. Dannenberg, eds., pp. 87–102. ACM Press, New York.

Blattner, M., Sumikawa, D., and Greenberg, R. 1989. Earcons and icons: their structure and common design principles. *Hum. Comput. Interaction* 4(1):11–44.

Braida, L. D. 1991. Crossmodal integration in the identification of consonant segments. *J. Exp. Psych.* 43A(3):647–677.

Bordegoni, M. and Hemmje, M. 1993. A dynamic gesture language and graphical feedback for interaction in a 3D user interface. *Proc. EUROGRAPHICS '93*, pp. 1–11.

Brooks, Jnr, F. P., Ouh-Young, M., Batter, J. J., and Kilpatrick, P. 1990. Project GROPE: haptic displays for scientific visualisation. *ACM Computer Graphics* 24(4):177–186.

Burger, J. 1993. *The Multimedia Bible*, pp. 635. Addison–Wesley, Reading, MA.

Crossman, E. R. and Cooke, F. W. 1974. Manual control of slow response systems. In *The Human Operator in Process Control*. E. Edwards and F. Lees, eds., pp. 51–64. Taylor and Francis, London.

Edwards, A. D. N. 1989. Soundtrack: an auditory interface for blind users. *Hum. Comput. Interaction* 4(1):45–66.

Faraday, P. 1995. Evaluating multimedia presentations for comprehension. In *Doctoral Consortium, SIGCHI '95* (Denver, US,: April/May), pp. 49–50. ACM Press, New York.

Francioni, J., Albright, L., and Jackson, J. 1991. Debugging parallel programs using sound. *ACM SIGPLAN Notices* 26(12):68–75.

Gaver, W. W. 1986. Auditory icons: using sound in computer interfaces. *Hum. Comput. Interaction* 2(1):167–177.

Halasz, F. G. 1988. Reflections on notecards: seven issues for the next generation of hypermedia systems. *Commun. ACM* 31(7):836–852.

Holland, S. 1994. Learning about harmony with harmony space: an overview. In *Music Education: An Artificial Intelligence Approach*. Springer–Verlag, London.

Macromedia. 1996. Macromedia Director. Macromedia Inc., San Francisco, CA.

Marmollin, H. 1992. Multimedia from the perspective of psychology. In *Multimedia: Systems Interactions and Applications*. L. Kjelldahl, ed., pp. 39–52. Springer–Verlag, Berlin.

Maybury, M. T. 1993. Planning multimedia explanations using communicative acts. In *Intelligent Multimedia Interface Design*. M. Maybury, ed., pp. 59–74. MIT Press, Cambridge, MA.

Moray, N., Lootsen, P., and Pajak, J. 1986. The acquisition of process control skills. *IEEE Trans. Syst. Man Cybernetics* SMC-16:497–504.

Moray, N. and Rotenberg, I. 1989. Fault management in process control: eye movement and action. *Ergonomics* 32(11):1319–1342.

Nielsen, J. 1990. *Hypertext and hypermedia*, pp. 263. Academic Press, London.

Price, R. 1993. MHEG: an introduction to the future international standard for hypermedia and multimedia object interchange. *Proc. Int. Conf. Multimedia Comput. Syst.* May 1994, Boston, MA.

Sanderson, P. M., Verhage, A. G., and Fuld, R. B. 1989. Statespace and verbal protocol methods for studying the human operator in process control. *Ergonomics* 32(11):1343–1372.

Steinberg, 1996. CUBASE Score for windows. Steinberg soft- and hardware Gmbh CUBASE Score for windows version 3.

Stenning, K. and Oberlander, J. 1995. A cognitive theory of graphical and linguistic reasoning: logic and implementation. *Cognitive Sci.* 19(1):97–140.

Vaughan, T. 1994. *Multimedia: Making it Work*, pp. 560. Osborne, McGraw–Hill.

Waibel, A., Vo, T. M., Duchnowski, P., and Manke, S. 1995. Multimodal interfaces. *Artif. Intell. Rev.*, pp. 1–23.

Williams, D. M. 1996. Multimedia, mental models, and complex domains. In *Doctoral Consortium, SIGCHI '96*. Vancouver, Canada, April, pp. 49–50. ACM Press, New York.

Zimmermann, T. G., Lanier, J., Blanchard, C., Bryson, S., and Harvill, Y. 1987. A hand gesture interface device, pp. 189–192. In *Proc. CHI+GI*.

Further Information

Further general information on multimedia interfaces may be obtained from the following.

Burger, J. 1993. *The Multimedia Bible*, pp. 635. Addison–Wesley.

Nielsen, J. 1990. *Hypertext and Hypermedia*, pp. 263. Academic Press, London.

Vaughan, T. 1994. *Multimedia: Making it Work*, pp. 560. Osborne, McGraw–Hill.

72

Interface Software
Technology

Brad A. Myers*
Carnegie Mellon University

72.1 Introduction

User interface (**UI**) software is often large, complex, and difficult to implement, debug, and modify. One study found that an average of 48% of the code of applications is devoted to the user interface, and that about 50% of the implementation time is devoted to implementing the user interface portion [Myers and Rosson 1992]. As interfaces become easier to use, they become harder to create [Myers 1994]. Today, direct manipulation interfaces [also called graphical user interfaces (GUIs)] are almost universal: one 1993 study found that 97% of all software development on Unix involved a GUI [XBusiness 1994, p. 80]. These interfaces require that the programmer deal with elaborate graphics, multiple ways for giving the same command, multiple asynchronous input devices (usually a keyboard and a locator or pointing device such as a mouse), a *mode free* interface where the user can give any command at virtually any time, and rapid *semantic feedback* where determining the appropriate response to user actions requires specialized information about the objects in the program. Tomorrow's UIs will provide speech and gesture recognition, intelligent agents, and integrated multimedia, and will probably be even more difficult to create. Furthermore, because UI *design* is so difficult, the only reliable way to get good interfaces is to iteratively redesign (and therefore reimplement) the interfaces after user testing, which makes the implementation task even harder.

Fortunately, there has been significant progress in software tools to help with creating UIs, and today, virtually all UI software is created using tools that make the implementation easier. For example, the MacApp system from Apple was reported to reduce development time by a factor of four or five [Wilson

---

*This chapter is revised from an earlier version which appeared as Myers, B. A. 1995. User interface software tools. *ACM Trans. Comput.–Hum. Interaction* 2(1):64–103.

1990]. A study commissioned by NeXT claimed that the average application programmed using the NeXTStep environment wrote 83% fewer lines of code and took one-half the time compared to applications written using less advanced tools, and some applications were completed in one-tenth the time [Booz 1992].

Furthermore, UI tools are a major business. In the Unix market alone, over U.S. $133 million of tools were sold in 1993, which is about 50,000 licenses [XBusiness 1994]. This is a 64% increase over 1992. Forrester Research claims that the total market for UI software tools on all platforms, including vertical tools which include database and UI construction tools, will be 130,000 developers generating U.S. $400 million in revenue. They estimate that this will double each year, growing to 700,000 developers and $1.2 billion by 1996 [DePalma and Woodring 1993].

Mark Hanner from the Meta Group market research firm says that the UI tool market is about to explode. Whereas the first generation of commercial tools was not fully graphical or was not sufficiently powerful, this is no longer true for today's tools. Furthermore, prices for tools have dropped significantly, and fees for run times have been mostly eliminated (so that designers do not have to pay the tool creator for products created using the tools). For the future, there is still a tremendous opportunity for good tools, especially in niche areas such as multimedia, distributed systems, and geographical information systems.

This chapter surveys UI software tools and explains the different types and classifications. However, it is now impossible to discuss *all* UI tools, since there are so many. For example, there are over 100 commercial GUI builders, and many new research tools are reported every year at conferences such as the annual ACM User Interface Software and Technology Symposium (UIST) and the ACM SIGCHI conference. There are also about three doctoral theses on UI tools every year. Therefore, this chapter provides an overview of the most popular approaches, rather than an exhaustive survey.

72.2 Importance of User Interface Tools

There are many advantages to using UI software tools. These can be classified into two main groups. First, the quality of the interfaces might be higher. This is because of the following:

- Designs can be rapidly prototyped and implemented, possibly even before the application code is written.
- It is easier to incorporate changes discovered through user testing.
- More effort can be expended on the tool than may be practical on any single UI since the tool will be used with many different applications.
- Different applications are more likely to have consistent UIs if they are created using the same UI tool.
- It will be easier for a variety of specialists to be involved in designing the UI, rather than having the UI created entirely by programmers. Graphic artists, cognitive psychologists, and human factors specialists may all be involved. In particular, professional UI designers, who may not be programmers, can be in charge of the overall design.
- Undo, Help, and other features are more likely to be available since they might be supported by the tools.

Second, the UI code might be easier and more economical to create and maintain. This is because of the following:

- Interface specifications can be represented, validated, and evaluated more easily.
- There will be less code to write, because much is supplied by the tools.
- There will be better modularization due to the separation of the UI component from the application. This should allow the user interface to change without affecting the application, and a large class of changes to the application (such as changing the internal algorithms) should be possible without affecting the UI.

- The level of programming expertise of the interface designers and implementors can be lower, because the tools hide much of the complexities of the underlying system.
- The reliability of the UI will be higher, since the code for the UI is created automatically from a higher level specification.
- It will be easier to port an application to different hardware and software environments since the device dependencies are isolated in the user interface tool.

72.3 Overview of User Interface Software Tools

Since UI software is so difficult to create, it is not surprising that people have been working for a long time to create tools to help with it. Today, many of these tools and ideas have progressed from research into commercial systems, and their effectiveness has been amply demonstrated. Research systems also continue to evolve quickly, and the models that were popular five years ago have been made obsolete by more effective tools, changes in the computer market (e.g., the demise of OpenLook will take with it a number of tools), and the emergence of new styles of user interfaces such as pen-based computing and multimedia.

Components of User Interface Software

As shown in Fig. 72.1, UI software may be divided into various layers: the **windowing system**, the **toolkit**, and **higher level tools**. Of course, many practical systems span multiple layers.

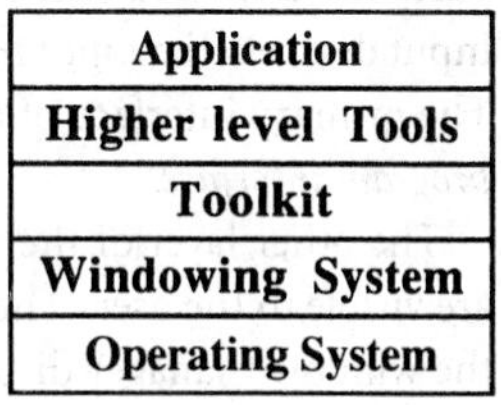

Application
Higher level Tools
Toolkit
Windowing System
Operating System

FIGURE 72.1 The components of user interface software discussed in this chapter.

The windowing system supports the separation of the screen into different (usually rectangular) regions, called *windows*. The X system [Scheifler and Gettys 1986] divides the window functionality into two layers: the window *system*, which is the functional or programming interface, and the **window manager**, which is the user interface. Thus the window system provides procedures that allow the application to draw pictures on the screen and get input from the user, and the window manager allows the end user to move windows around, and is responsible for displaying the title lines, borders, and icons around the windows. However, many people and systems use the name window manager to refer to both layers, since systems such as the Macintosh and Microsoft Windows do not separate them. This chapter will use the X terminology, and use the term windowing system when referring to both layers.

On top of the windowing system is the toolkit, which contains many commonly used **widgets** such as menus, buttons, scroll bars, and text input fields. On top of the toolkit might be higher-level tools, which help the designer use the toolkit widgets. The following sections discuss each of these components in more detail.

Windowing Systems

A windowing system is a software package that helps the user monitor and control different contexts by separating them physically onto different parts of one or more display screens. A survey of various windowing systems was published earlier [Myers 1988b]. Although most of today's systems provide toolkits on top of the windowing systems, as will be explained subsequently, toolkits generally only address the drawing of widgets such as buttons, menus, and scroll bars. Thus, when the programmer wants to draw application-specific parts of the interface and allow the user to manipulate these, the window system interface must be used directly. Therefore, the windowing system's programming interface has significant impact on most user interface programmers.

The first windowing systems were implemented as part of a single program or system. For example, the EMACs text editor [Stallman 1979], and the Smalltalk [Tesler 1981] and DLISP [Teitelman 1979] program-

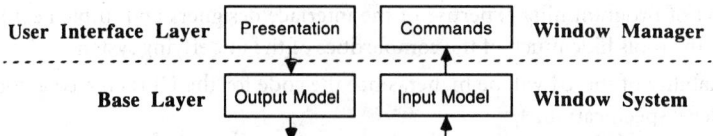

FIGURE 72.2 The windowing system can be divided into two layers, called the base or window system layer, and the user interface or window manager layer. Each of these can be divided into parts that handle output and input.

ming environments had their own windowing systems. Later systems implemented the windowing system as an integral part of the operating system, such as SunView for Suns, and the Macintosh and Microsoft Windows systems. In order to allow different windowing systems to operate on the same operating system, some windowing systems, such as X and Sun's NeWS, operate as separate processes, and use the operating system's interprocess communication mechanism to connect to **applications**.

Structure of Windowing Systems

A windowing system can be logically divided into two layers, each of which has two parts (see Fig. 72.2). The window system, or *base layer*, implements the basic functionality of the windowing system. The two parts of this layer handle the display of graphics in windows (the *output model*) and the access to the various input devices (the *input model*), which usually includes a keyboard and a pointing device such as a mouse. The primary interface of the base layer is procedural, and is called the windowing system's application or *program interface*.

The other layer of the windowing system is the window manager or UI. This includes all aspects that are visible to the user. The two parts of the UI layer are the *presentation*, which consists of the pictures that the window manager displays, and the *commands*, which are how the user manipulates the windows and their contents.

Base Layer

The base layer is the procedural interface to the windowing system. In the 1970s and early 1980s, there were a large number of different windowing systems, each with a different procedural interface (at least one for each hardware platform). People writing software found this to be unacceptable because they wanted to be able to run their software on different platforms, but they would have to rewrite significant amounts of code to convert from one window system to another. The X windowing system [Scheifler and Gettys 1986] was created to solve this problem by providing a hardware-independent interface to windows. X has been quite successful at this, and has driven virtually all other windowing systems out of the workstation hardware market. In the small computer market, the Macintosh runs its own window system or X, and IBM PC-class machines primarily run Microsoft Windows or IBM's Presentation Manager (part of OS/2).

Output Model

The output model is the set of procedures that an application can use to draw pictures on the screen. It is important that all output be directed through the window system so that the graphics primitives can be clipped to the window's borders. For example, if a program draws a line that would extend out of a window's borders, it must be clipped so that the contents of other, independent, windows are not overwritten. Most windowing systems provide special escapes that allow programs to draw directly to the screen, without using the window system's clipping. These operations can be much quicker, but are very dangerous and therefore should seldom be used. Most modern computers provide graphics hardware that is specially optimized to work efficiently with the window system.

In early windowing systems, such as Smalltalk [Tesler 1981] and Sapphire [Myers 1986], the primary output operation was BitBlt (also called RasterOp). These systems primarily supported monochrome screens (each pixel is either black or white). BitBlt takes a rectangle of pixels from one part of the screen and copies it to another part. Various Boolean operations can be specified for combining the pixel values of the source and destination rectangles. For example, the source rectangle can simply replace the destination, or it might be XORed with the destination. BitBlt can be used to draw solid rectangles in either black or white, display text, scroll windows, and perform many other effects [Tesler 1981]. The only additional drawing operation typically supported by these early systems was drawing straight lines.

Later windowing systems, such as the Macintosh and X, added a full set of drawing operations, such as filled and unfilled polygons, text, lines, arcs, etc. These cannot be implemented using the BitBlt operator. With the growing popularity of color screens and nonrectangular primitives (such as rounded rectangles), the use of BitBlt has significantly decreased. It is primarily used now for scrolling and copying off-screen pictures onto the screen (e.g., to implement double buffering).

A few windowing systems allow the full Postscript imaging model to be used to create images on the screen. Postscript provides device-independent coordinate systems and arbitrary rotations and scaling for all objects, including text. Another advantage of using Postscript for the screen is that the same language can be used to print the windows on paper (since many printers accept Postscript). Sun created a version used in the NeWS windowing system, and then Adobe (the creator of Postscript) came out with an official version called Display Postscript, which is used in the NeXT windowing system and is supplied as an extension to the X windowing system by a number of vendors, including DEC and IBM.

All of the standard output models only contain drawing operations for two-dimensional objects. Two extensions to support 3D objects are PEX and OpenGL. PEX is an extension to the X windowing system that incorporates much of the PHIGS graphics standard. OpenGL is based on the GL programming interface that has been used for many years on Silicon Graphics machines. OpenGL provides machine independence for 3D since it is available for various X platforms (SGI, Sun, etc.) and is included as a standard part of new versions of Microsoft Windows.

As shown in Fig. 72.3, the earlier windowing systems assumed that a graphics package would be implemented using the windowing system. For example, the CORE graphics package was implemented on top of the SunView windowing system. All newer systems, including the Macintosh, X, NeWS, NeXT, and Microsoft Windows, have implemented a sophisticated graphics system as *part* of the windowing system.

Input Model

The early graphics standards, such as CORE and PHIGS, provided an input model that does not support the modern, direct manipulation style of interfaces. In those standards, the programmer calls a routine to request the value of a *virtual device* such as a *locator* (pointing device position), *string* (edited text string), *choice* (selection from a menu), or *pick* (selection of a graphical object). The program would then pause waiting for the user to take action. This is clearly at odds with the direct manipulation *mode-free* style, where the user can decide whether to make a menu choice, select an object, or type something.

With the advent of modern windowing systems, a new model was provided: a stream of event records is sent to the window which is currently accepting input. The user can select which window is getting events using various commands, described subsequently. Each event record typically contains the type and value of the event (e.g., which key was pressed), the window to which the event was directed, a timestamp, and the x and y coordinates of the mouse. The windowing system queues keyboard events, mouse button events, and mouse movement events together (along with other special events) and programs must dequeue the events and process them. It is somewhat surprising that, although there has been substantial progress in the output model for windowing systems (from BitBlt to complex 2D primitives to 3D), input is still handled in essentially this same way today as in the original windowing systems, even though there are some well-known unsolved problems with this model:

- There is no provision for special stop-output (control-S) or abort (control-C, command-dot) events, so these will be queued with the other input events.

Sapphire, SunWindows: **Cedar, Macintosh, NeXT:**

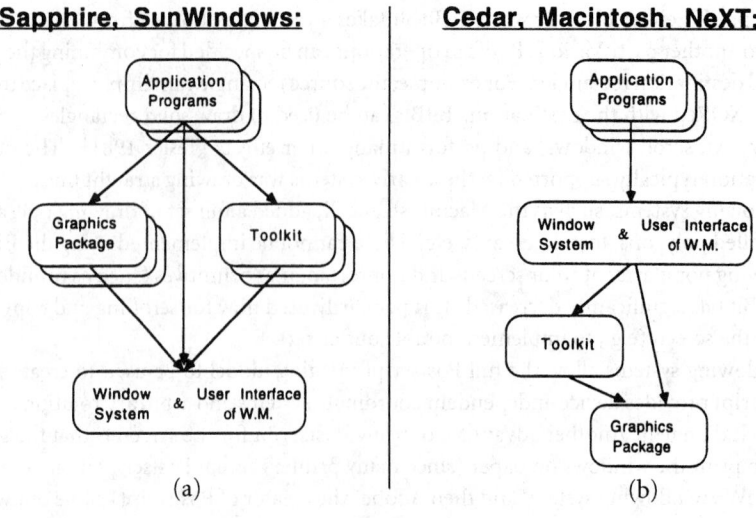

(a)   (b)

NeWS, X:

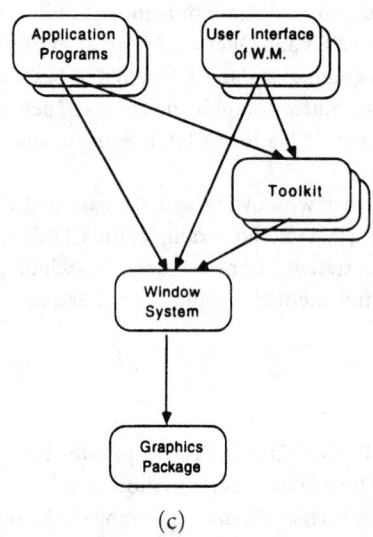

(c)

FIGURE 72.3 Various organizations that have been used by windowing systems. Boxes with extra borders represent systems that can be replaced by users. (a) Early systems tightly coupled the window manager and the window system, and assumed that sophisticated graphics and toolkits would be built on top. (b) The next step in designs was to incorporate into the windowing system the graphics and toolkits, so that the window manager itself could have a more sophisticated look and feel, and so applications would be more consistent. (c) Other systems allow different window managers and different toolkits, while still embedding sophisticated graphics packages.

• The same event mechanism is used to pass special messages from the windowing system to the application. When a window gets larger or becomes uncovered, the application must usually be notified so that it can adjust or redraw the picture in the window. Most window systems communicate this by enqueuing special events into the event stream, which the program must then handle.

- The application must always be willing to accept events in order to process aborts and redrawing requests. If not, then long operations cannot be aborted, and the screen may have blank areas while they are being processed.
- The model is device dependent, since the event record has fixed fields for the expected incoming events. If a 3D pointing device or one with more than the standard number of buttons is used instead of a mouse, then the standard event mechanism cannot handle it.
- Because the events are handled asynchronously, there are many race conditions that can cause programs to get out of synchronization with the window system. For example, in the X windowing system, if you press inside a window and release outside, under certain conditions the program will think that the mouse button is still depressed. Another example is that refresh requests from the windowing system specify a rectangle of the window that needs to be redrawn, but if the program is changing the contents of the window, the wrong area may be redrawn by the time the event is processed. This problem can occur when the window is scrolled.

Although these problems have been known for a long time, there has been little research on new input models (an exception is the Garnet Interactors model [Myers 1990b]).

Communication

In the X windowing system and NeWS, all communication between applications and the window system uses interprocess communication through a network protocol. This means that the application program can be on a different computer from its windows. In all other windowing systems, operations are implemented by directly calling the window manager procedures or through special traps into the operating system. The primary advantage of the X mechanism is that it makes it easier for a person to utilize multiple machines with all their windows appearing on a single machine. Another advantage is that it is easier to provide interfaces for different programming languages: for example the C interface (called xlib) and the Lisp interface (called CLX) send the appropriate messages through the network protocol. The primary disadvantage is efficiency, since each window request will typically be encoded, passed to the transport layer, and then decoded, even when the computation and windows are on the same machine.

User Interface Layer

The UI of the windowing system allows the user to control the windows. In X, the user can easily switch UIs, by killing one window manager and starting another. Popular window managers under X include uwm (which has no title lines and borders), twm, the Motif window manager (mwm), and the OpenLook window manager (olwm). There is a standard protocol through which programs and the base layer communicate to the window manager, so that all programs continue to run without change when the window manager is switched. It is possible, for example, to run applications that use Motif widgets inside the windows controlled by the OpenLook window manager.

A complete discussion of the options for the UIs of window managers was previously published [Myers 1988b]. Also, the video *All the Widgets* [Myers 1990a] has a 30-min segment showing many different forms of window manager UIs.

Some parts of the UI of a windowing system, which is sometimes called its *look and feel*, can apparently be copyrighted and patented. Which parts is a highly complex issue, and the status changes with decisions in various court cases. Good references for more information are the "Legally Speaking" columns of *Communications of the ACM* [Samuelson 1993].

Presentation

The presentation of the windows defines how the screen looks. One very important aspect of the presentation of windows is whether they can *overlap* or not. Overlapping windows, sometimes called *covered* windows, allow one window to be partially or totally on top of another window, as shown in Fig. 72.4. This is also sometimes called the *desktop metaphor*, since windows can cover each other like pieces of paper can

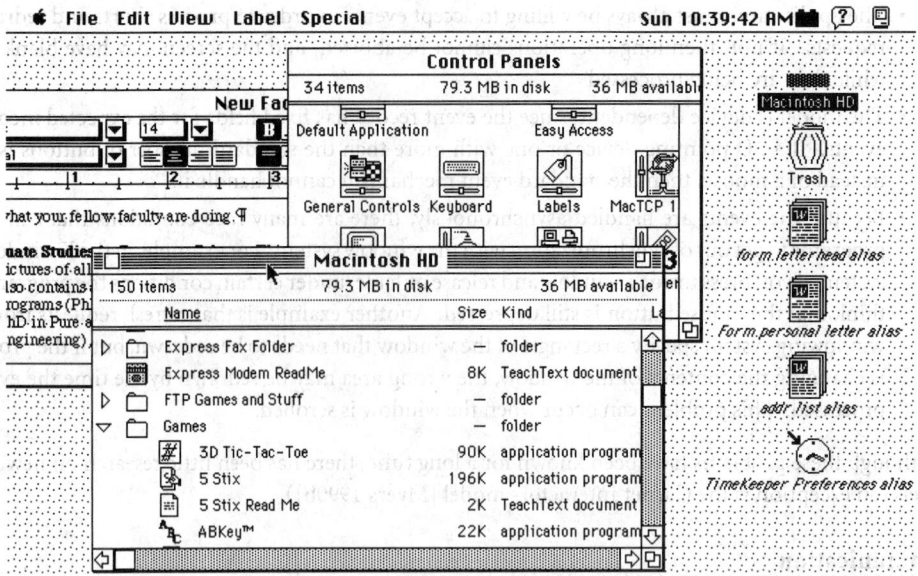

FIGURE 72.4 A screen from the Macintosh showing three windows covering each other, and some icons along the right margin.

cover each other on a desk.[1] The other alternative is called *tiled* windows, which means that windows are not allowed to cover each other. Obviously, a window manager that supports covered windows can also allow them to be side-by-side, but not vice versa. Therefore, a window manager is classified as covered if it allows windows to overlap. The tiled style was popular for awhile, and was used by Cedar [Swinehart et al. 1986], and early versions of the Star [Smith et al. 1982], Andrew [Palay et al. 1988], and Microsoft Windows. A study even suggested that using tiled windows was more efficient for users [Bly and Rosenberg 1986]. However, today tiled windows are rarely seen, because users generally prefer overlapping.

Another important aspect of the presentation of windows is the use of icons (also shown in Fig. 72.4). These are small pictures that represent windows (or sometimes files). They are used because there would otherwise be too many windows to conveniently fit on the screen and manage. Other aspects of the presentation include whether the window has a title line or not, what the background (where there are no windows) looks like, and whether the title and borders have control areas for performing window operations.

Commands

Since computers typically have multiple windows and only one mouse and keyboard, there must be a way for the user to control which window is getting keyboard input. This window is called the input (or keyboard) *focus*. Another term is the *listener* since it is listening to the user's typing. Some systems called the focus the *active window* or *current window*, but these are poor terms since in a multi-processing system, many windows can be actively outputting information at the same time. Window managers provide various ways to specify and show which window is the listener. The most important options are:

- Click-to-type, which means that the user must click the mouse button in a window before typing to it. This is used by the Macintosh.

---

[1]There are usually other aspects to the desktop metaphor, however, such as presenting file operations in a way that mimics office operations, as in the Star office workstation [Smith et al. 1982].

• Move-to-type, which means that the mouse only has to move over a window to allow typing to
it. This is usually faster for the user, but may cause input to go to the wrong window if the user
accidentally knocks the mouse.

Most X window managers (including the Motif and OpenLook window managers) allow the user to choose
which method is desired. However, the choice can have significant impact on the UI of applications. For
example, because the Macintosh requires click-to-type, it can provide a single menu bar at the top, and
the commands can always operate on the focused window. With move-to-type, the user might have to
pass through various windows (thus giving them the focus) on the way to the top of the screen. Therefore,
Motif applications must have a menu bar in each window so the commands will know which window to
operate on.

All covered window systems allow the user to change which window is on top (not covered by other
windows), and usually to send a window to the bottom (covered by all other windows). Other commands
allow windows to be changed in size, moved, created, and destroyed.

Toolkits

A toolkit is a library of widgets that can be called by application programs. A widget is a way of using a
physical input device to input a certain type of value. Typically, widgets in toolkits include menus, buttons,
scroll bars, text type-in fields, etc. Fig. 72.5 shows some examples of widgets. Creating an interface using
a toolkit can only be done by programmers, because toolkits only have a procedural interface.

Using a toolkit has the advantage that the final UI will look and act similarly to other UIs created using
the same toolkit, and each application does not have to rewrite the standard functions, such as menus.

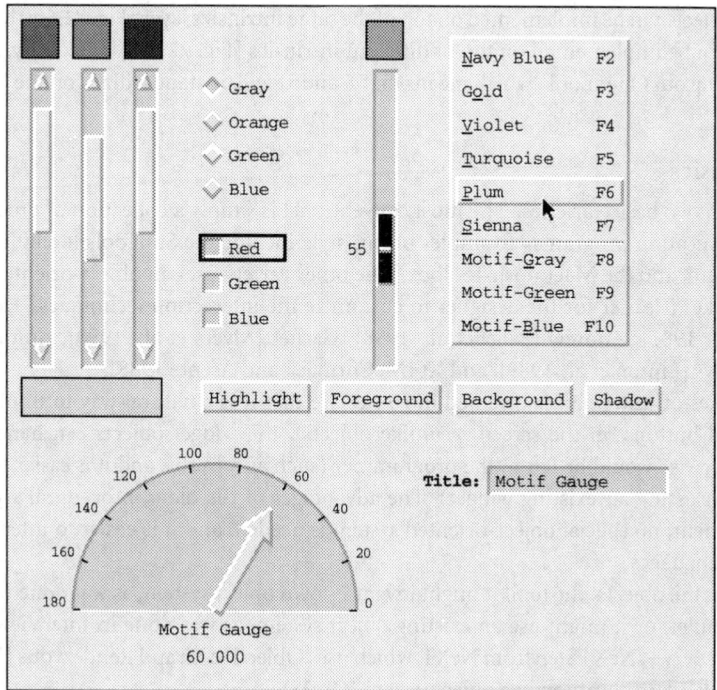

FIGURE 72.5   Some of the widgets with a Motif look-and-feel provided by the
Garnet toolkit.

A problem with toolkits is that the styles of interaction are limited to those provided. For example, it is difficult to create a single slider that contains two indicators, which might be useful to input the upper and lower bounds of a range. In addition, the toolkits themselves are often expensive to create: "The primitives never seem complex in principle, but the programs that implement them are surprisingly intricate" [Cardelli and Pike 1985, p. 199].

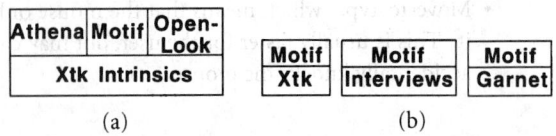

(a)   (b)

FIGURE 72.6   (a) At least three different widget sets that have different looks and feels have been implemented on top of the xt intrinsics. (b) The Motif look-and-feel has been implemented on at least three different intrinsics.

Another problem with toolkits is that they are often difficult to use since they may contain hundreds of procedures, and it is often not clear how to use the procedures to create a desired interface. For example, the documentation for the Macintosh Toolbox now is well over six books, of which about one-third is related to UI programming.

As with the graphics package, the toolkit can be implemented either using or being used by the windowing system (see Fig. 72.3). Early systems provided only minimal widgets (e.g., just a menu), and expected applications to provide others. In the Macintosh, the toolkit is at a low level, and the window manager UI is built using it. The advantage of this is that the window manager can then use the same sophisticated toolkit routines for its UI. When the X system was being developed, the developers could not agree on a single toolkit, so they left the toolkit to be on top of the windowing system. In X, programmers can use a variety of toolkits (for example, the Motif, OpenLook, InterViews [Linton et al. 1989], Garnet [Myers et al. 1990], or tk [Ousterhout 1991] toolkits can be used on top of X), but the window manager must usually implement its UI from scratch.

Because the designers of X could not agree on a single look-and-feel, they created an **intrinsics** layer on which to build different widget sets, which they called xt [McCormack and Asente 1988]. This layer provides the common services, such as techniques for object-oriented programming and layout control. The *widget set* layer is the collection of widgets that is implemented using the intrinsics. Multiple widget sets with different looks and feels can be implemented on top of the same intrinsics layer [Fig. 72.6(a)], or else the same look-and-feel can be implemented on top of different intrinsics [Fig. 72.6(b)]. Recently, Sun announced that it was phasing out OpenLook, which means that X and xt will be standardized on the Motif widget set.

Toolkit Intrinsics

Toolkits come in two basic varieties. The most conventional is simply a collection of procedures that can be called by application programs. Examples of this style include the SunTools toolkit for the SunView windowing system, and the Macintosh Toolbox. The other variety uses an object-oriented programming style, which makes it easier for the designer to customize the interaction techniques. Examples include Smalltalk [Tesler 1981], Andrew [Palay et al. 1988], Garnet [Myers et al. 1990], Amulet [Myers et al. 1995], InterViews [Linton et al. 1989], and Xt [McCormack and Asente 1988].

The advantages of using object-oriented intrinsics are that it is a natural way to think about widgets (the menus and buttons on the screen *seem* like objects), the widget objects can handle some of the chores that otherwise would be left to the programmer (such as refresh), and it is easier to create custom widgets (by subclassing an existing widget). The advantages of the older, procedural style are that it is easier to implement, no special object-oriented system is needed, and it is easier to interface to multiple programming languages.

To implement the objects, the toolkit might invent its own object system, as was done with Xt, Andrew, Garnet, and Amulet, or it might use an existing object system, as was done in InterViews [Linton et al. 1989], which uses C++; NeXTStep from NeXT, which uses Objective-C; and Rendezvous [Hill et al. 1993], which uses the standard common lisp object system (CLOS).

The usual way that object-oriented toolkits interface with application programs is through the use of *call-back procedures*. These are procedures defined by the application programmer that are called when a widget is operated by the end user. For example, the programmer might supply a procedure to be called when the user selects a menu item. Experience has shown that real interfaces often contain hundreds of call-backs, which makes the code harder to modify and maintain [Myers and Rosson 1992]. In addition, different toolkits, even when implemented on the same intrinsics such as Motif and OpenLook, have different call-back protocols. This means that code for one toolkit is difficult to port to a different toolkit. Therefore, research is being directed at reducing the number of call-backs in UI software [Myers 1991b].

Some research toolkits have added novel features to the toolkit intrinsics. For example, Garnet [Myers et al. 1990], Rendezvous [Hill et al. 1993], Bramble [Gleicher 1993], and Amulet [Myers et al. 1995], allow the objects to be connected using *constraints*, which are relationships that are declared once and then maintained automatically by the system. For example, the designer can specify that the color of a rectangle is constrained to be the value of a slider, and then the system will automatically update the color if the user moves the slider.

Widget Set

Typically, the intrinsics layer is look-and-feel independent, which means that the widgets built on top of it can have any desired appearance and behavior. However, a particular widget set must pick a look-and-feel. The video *All the Widgets* shows many examples of widgets that have been designed over the years [Myers 1990a]. For example, it shows 35 different kinds of menus. Like window manager UIs, the widgets' look-and-feel can be copyrighted and patented [Samuelson 1993].

As was previously mentioned, different widget sets (with different looks and feels) can be implemented on top of the same intrinsics. In addition, the same look-and-feel can be implemented on top of different intrinsics. For example, there are Motif look-and-feel widgets on top of the xt, InterViews, and Garnet intrinsics [Fig. 72.6(b)]. Although they all look and operate the same (and so would be indistinguishable to the end user), they are implemented quite differently, and have completely different procedural interfaces for the programmer.

Specialized Toolkits

A number of toolkits have been developed to support specific kinds of applications or specific classes of programmers. For example, the SUIT system [Pausch et al. 1992] (which contains a toolkit and an interface builder), is specifically designed to be easy to learn and is aimed at classroom instruction. Garnet [Myers et al. 1990] provides high-level support for graphical and direct manipulation interfaces, and includes a toolkit, interface builder, and other high-level tools. Amulet [Myers et al. 1995] is designed to let researchers replace or extend the builtin capabilities. Rendezvous [Hill et al. 1993] and Visual Obliq [Bharat and Brown 1994] are designed to make it easier to create applications that support multiple users on multiple machines operating synchronously. Whereas most toolkits provide only 2D interaction techniques, the Brown 3D toolkits [Zeleznik et al. 1991, Stevens et al. 1994] and Silicon Graphics' Inventor toolkit [Strauss and Carey 1992, Wernecke 1994] provide preprogrammed 3D widgets and a framework for creating others. Special support for animations has been added to Artkit, including motion blur, timing, and curved trajectories [Hudson and Stasko 1993].

Tk [Ousterhout 1991] is a popular toolkit for the X window system because it uses an interpretive language called tcl which makes it possible to dynamically change the user interface. Tcl also supports the Unix style of programming where many small programs are glued together.

Virtual Toolkits

Although there are many small differences among the various toolkits, much remains the same. For example, all have some type of menu, button, scroll bar, text input field, etc. Although there are fewer windowing systems and toolkits than there were five years ago, people are still finding it to be a lot of work to convert software from Motif to OpenLook to the Macintosh and to Microsoft Windows. Therefore, a number of systems have been developed that try to hide the differences among the various toolkits, by providing virtual widgets which can be mapped into the widgets of each toolkit. Another name for these tools is *cross-platform development systems*. The programmer writes the code once using the **virtual toolkit** and the code will run without change on different platforms and still look like it was designed with that platform's widgets. For example, the virtual toolkit might provide a single menu routine, which always has the same programmer interface, but connects to a Motif menu, Macintosh menu, or a Windows menu depending on which machine the application is run.

There are two styles of virtual toolkits. In one, the virtual toolkit links to the different *actual* toolkits on the host machine. For example, XVT from XVT Software provides a C or C++ interface that links to the actual Motif, OpenLook, Macintosh, MS-Windows, and OS/2-PM toolkits (and also character terminals) and hides their differences. The second style of virtual toolkit *reimplements* the widgets in each style. For example, Galaxy from Visix Software, Open Interface from NeuronData, and Amulet [Myers et al. 1995] provide libraries of widgets that look like those on the various platforms. The advantage of the first style is that the user interface is more likely to be look-and-feel conformant (since it uses the real widgets). The disadvantage is that the virtual toolkit must still provide an interface to the graphical drawing primitives on the platforms. Furthermore, they tend to only provide functions that appear in all toolkits. Many of the virtual toolkits that take the second approach, for example, Galaxy, provide a sophisticated graphics package and complete sets of widgets on all platforms. However, with the second approach, there must always be a large run-time library, since in addition to widgets that are native to the machine, there is the reimplementation of these same widgets in the virtual toolkit library.

All of the toolkits that work on multiple platforms can be considered virtual toolkits of the second type. For example, SUIT [Pausch et al. 1992] works on X, Macintosh, and Windows; and Garnet [Myers et al. 1990] works on X and the Macintosh. However, these use the same look-and-feel on all platforms (and therefore do not look the same as the other applications on that platform), so they are not classified as virtual toolkits.

Higher Level Tools

Since programming at the toolkit level is quite difficult, there is a tremendous interest in higher level tools that will make the user interface software production process easier. These are discussed next.

Phases

Many higher-level tools have components that operate at different times. The *design time component* helps the UI designer design the UI. For example, this might be a graphical editor, which can lay out the interface, or a compiler to process a UI specification language. The next phase is when the end user is using the program. Here, the *run-time component* of the tool is used. This usually includes a toolkit, but may also include additional software specifically for the tool. Since the run-time component is managing the UI, the term **user interface management system** (**UIMS**) seems appropriate for tools with a significant run-time component.

There may also be an *after-run-time component* that helps with the evaluation and debugging of the UI. Unfortunately, very few UI tools have an after-run-time component. This is partially because tools that have tried, such as MIKE [Olsen and Halversen 1988], discovered that there are very few metrics that can be applied by computers. A new generation of tools trying to evaluate how people will interact with interfaces by automatically creating cognitive models from high-level descriptions of the UI. For example, the USAGE system creates an NGOMSL cognitive model from a high-level UI specification [Byrne et al. 1994] in the UIDE modeling language [Sukaviriya et al. 1993].

Specification Styles

High-level UI tools come in a large variety of forms. One important way that they can be classified is by how the designer specifies what the interface should be. Some tools require the programmer to program in a special-purpose language, some provide an application framework to guide the programming, some automatically generate the interface from a high-level model or specification, and others allow the interface to be designed interactively. Each of these types will be discussed. Of course, some tools use different techniques for specifying different parts of the UI. These are classified by their predominant or most interesting feature.

Language Based

With most of the older UI tools, the designer specifies the UI in a special-purpose language. This language can take many forms, including context-free grammars, state transition diagrams, declarative languages, event languages, etc. The language is usually used to specify the syntax of the UI; i.e., the legal sequences of input and output actions. This is sometimes called the *dialogue*. Green [Green 1986] provides an extensive comparison of grammars, state transition diagrams, and event languages, and Olsen [Olsen 1992] surveys various UIMS techniques.

State Transition Networks. Since many parts of UIs involve handling a sequence of input events, it is natural to think of using a state transition network to code the interface. A transition network consists of a set of states, with arcs out of each state labeled with the input tokens that will cause a transition to the state at the other end of the arc. In addition to input tokens, calls to application procedures and the output to display can also be put on the arcs in some systems. Newman implemented a simple tool using finite state machines in 1968, which handled textual input [Newman 1968]. This was apparently the first UI tool. Many of the assumptions and techniques used in modern systems were present in Newman's tool: different languages for defining the UI and the semantics (the semantic routines were coded in a normal programming language), a table-driven syntax analyzer, and device independence.

State diagram tools are most useful for creating UIs where the UI has a large number of modes (each state is really a mode). For example, state diagrams are useful for describing the operation of low-level widgets (e.g., how a menu or scroll bar works), or the overall global flow of an application (e.g., this command will pop-up a dialogue box, from which you can get to these two dialogue boxes, and then to this other window, etc.). However, most highly interactive systems attempt to be mostly mode free, which means that at each point, the user has a wide variety of choices of what to do. This requires a large number of arcs out of each state, so that state diagram tools have not been successful for these interfaces. In addition, state diagrams cannot handle interfaces where the user can operate on multiple objects at the same time. Another problem is that they can be very confusing for large interfaces, since they get to be a maze of wires and off-page (or off-screen) arcs can be hard to follow.

VAPS, from Virtual Prototypes, is a commercial system that uses the state transition model, and it eliminates the maze-of-wires problem by providing a spreadsheet-like table in which the states, events, and actions are specified. Transition networks have been thoroughly researched, but have not proven particularly successful or useful either as a research or commercial approach.

Context-Free Grammars. Many grammar-based systems are based on parser generators used in compiler development. For example, the designer might specify the UI syntax using some form of BNF. Examples of grammar-based systems are Syngraph [Olsen and Dempsey 1983] and parsers built with YACC and LEX in Unix.

Grammar-based tools, like state diagram tools, are not appropriate for specifying highly interactive interfaces, since they are oriented to batch processing of strings with a complex syntactic structure. These systems are best for textual command languages, and have been mostly abandoned for UIs by researchers and commercial developers.

Event Languages. With event languages, the input tokens are considered to be *events* that are sent to individual event handlers. Each handler will have a condition clause that determines what types of events it will handle, and when it is active. The body of the handler can cause output events, change the internal state of the system (which might enable other event handlers), or call application routines.

Sassafras [Hill 1986] is an event language where the UI is programmed as a set of small event handlers. As described subsequently, the HyperTalk language that is part of HyperCard for the Apple Macintosh can also be considered an event language.

The advantages of event languages are that they can handle multiple input devices active at the same time, and it is straightforward to support nonmodal interfaces, where the user can operate on any widget or object. The main disadvantage is that it can be very difficult to create correct code, since the flow of control is not localized and small changes in one part can affect many different pieces of the program. It is also typically difficult for the designer to understand the code once it reaches a nontrivial size. However, the success of HyperTalk and similar tools shows that this approach is appropriate for small- to medium-size programs.

Declarative Languages.   Another approach is to try to define a language that is declarative (stating what should happen) rather than procedural (how to make it happen). Cousin [Hayes et al. 1985] allows the designer to specify user interfaces in this manner. The user interfaces supported are basically forms, where fields can be text, which is typed by the user, or options selected using menus or buttons. There are also graphic output areas that the application can use in whatever manner desired. The application program is connected to the user interface through *variables*, which can be set and accessed by both. As researchers have extended this idea to support more sophisticated interactions, the specification has grown into full application models, and newer systems are described in the section "Model-Based Automatic Generation."

The layout description languages that come with many toolkits are also a type of declarative language. For example, Motif's user interface language (UIL) allows the layout of widgets to be defined. Since the UIL is interpreted when an application starts, users can (in theory) edit the UIL code to customize the interface. UIL is not a complete language, however, in the sense that the designer must still write C code for many parts of the interface, including any areas containing dynamic graphics and any widgets that change.

The advantage of using declarative languages is that the UI designer does not have to worry about the time sequence of events, and can concentrate on the information that needs to be passed back and forth. The disadvantage is that only certain types of interfaces can be provided this way, and the rest must be programmed by hand in the graphic areas provided to application programs. The kinds of interactions available are preprogrammed and fixed. In particular, these systems provide no support for such things as dragging graphical objects, rubber-band lines, drawing new graphical objects, or even dynamically changing the items in a menu based on the application mode or context. However, these languages are now proving successful as intermediate languages describing the layout of widgets (such as UIL) that are generated by interactive tools.

Constraint Languages.   A number of UI tools allow the programmer to use constraints to define the UI [Borning and Duisberg 1986]. Early constraint systems include Sketchpad [Sutherland 1963], which pioneered the use of graphical constraints in a drawing editor, and Thinglab [Borning 1981], which used constraints for graphical simulation. Subsequently, Thinglab was extended to aid in the generation of UIs [Borning and Duisberg 1986].

The previous discussion of toolkits mentioned the use of constraints as part of the intrinsics of a toolkit. A number of research toolkits now supply constraints as an integral part of the object system (e.g., Garnet and Amulet). In addition, some systems have provided higher level interfaces to constraints, for example C32 [Myers 1991a] allows constraints to be defined using a spreadsheet-like interface.

The advantage of constraints is that they are a natural way to express many kinds of relationships that arise frequently in UIs, for example, that lines should stay attached to boxes, that labels should stay centered within boxes, etc. A disadvantage with constraints is that they require a sophisticated run-time system to solve them efficiently. However, a growing number of research systems are using constraints, and it appears that modern constraint solvers and debugging techniques may solve these problems, so constraints have a great potential to simplify the programming task. As yet, there are no commercial UI tools using general-purpose constraint solvers.

Screen Scrapers.   Some commercial tools are specialized to be *front enders* or *screen scrapers*, which provide a GUI to old programs without changing the existing application code. They do this by providing an in-memory buffer that pretends to be the screen of an old character terminal such as might be attached to an IBM mainframe. When the mainframe application outputs to the buffer, a program the designer writes in a special programming language converts this into an update of a graphical widget. Similarly, when the user operates a widget, the script converts this into the appropriate edits of the character buffer. The leading program of this type is Easel, which also contains an interface builder for laying out the widgets.

Database Interfaces. A very important class of commercial tools supports form-based or GUI-based access to databases. Major database vendors such as Oracle provide tools which allow designers to define the UI for accessing and setting data. Often these tools include interactive form editors (which are essentially interface builders) and special database languages. Fourth-generation languages (4GLs), that support defining the interactive forms for accessing and entering data, also fall into this category.

Visual Programming. Visual programs use graphics and two- (or more) dimensional layout as part of the program specification [Myers 1990c]. Many different approaches to using visual programming to specify UIs have been investigated. Most systems that support state transition networks use a visual representation. Another popular technique is to use dataflow languages. In these, icons represent processing steps, and the data flow along the connecting wires. The UI is usually constructed directly by laying out prebuilt widgets, in the style of interface builders. An example of a visual programming system for creating user interfaces is ProGraph from Pictorius Incorporated. Using a visual language seems to make it easier for novice programmers, but large programs still suffer from the familiar maze-of-wires problem. Other papers (e.g., Myers [1990c]) have analyzed the strengths and weaknesses of visual programming in detail.

Another popular language is Visual Basic from Microsoft. Although this is more of a structure editor for Basic combined with an interface builder and, therefore, does not really count as a visual language, it does make the construction of user interface software easier. Microsoft is pushing Visual Basic as the extension language that people will use to customize and connect all future Windows-based applications.

Summary of Language Approaches. In summary, many different types of languages have been designed for specifying UIs. One problem with all of these is that they can only be used by professional programmers. Some programmers have objected to the requirement for learning a new language for programming just the UI portion [Olsen 1987]. This has been confirmed by market research [XBusiness 1994, p. 29]. Furthermore, it seems more natural to define the graphical part of a UI using a graphical editor. However, it is clear that for the foreseeable future, much of the UI will still need to be created by writing programs, and so it is appropriate to continue investigations into the best language to use for this. Indeed, an entire book is devoted to investigating the languages for programming UIs [Myers 1992b].

Application Frameworks

After the Macintosh Toolbox had been available for a little while, Apple discovered that programmers had a difficult time figuring out how to call the various toolkit functions, and how to ensure that the resulting interface met the Apple guidelines. They therefore created a software system that provides an overall **application framework** to guide programmers. This is called MacApp [Wilson 1990] and uses the object-oriented language Object Pascal. Classes are provided for the important parts of an application, such as the main windows, the commands, etc., and the programmer specializes these classes to provide the application-specific details, such as what is actually drawn in the windows and which commands are provided. MacApp has been very successful at simplifying the writing of Macintosh applications.

Unidraw [Vlissides and Linton 1990] uses a similar approach, but it is more specialized for graphical editors. This means that it can provide even more support. Unidraw uses the C++ object-oriented language and is part of the InterViews system [Linton and Vlissides 1989]. Unidraw has been used to create various drawing and computer-aided design (CAD) programs, and a UI editor [Vlissides and Tang 1991]. The Garnet framework is also aimed at graphical applications, but due to its graphical data model, many of the builtin routines can be used without change (the programmer does not usually need to write methods for subclasses). Even more specialized are various graph programs, such as TGE [Karrer and Scacchi 1990]. These provide a framework in which the designer can create programs that display their data as trees or graphs. The programmer typically specializes the node and arc classes, and specifies some of the commands, but the framework handles layout and the overall control.

An emerging popular approach aims to replace today's large, monolythic applications with smaller components that attach together. For example, you might buy a separate text editor, ruler, paragraph formatter, spell checker, and drawing program, and have them all work together seamlessly. This approach was invented by the Andrew environment [Palay et al. 1988] which provides an object-oriented document model that supports the embedding of different kinds of data inside other documents. These *insets* are unlike data that are cut and pasted in systems like the Macintosh because they bring along the programs that edit them, and therefore can always be edited in place. Furthermore, the container document does not need to know how to display or print the inset data since the original program that created it is always available. The designer creating a new inset writes subclasses that adhere to a standard protocol so the system knows how to pass input events to the appropriate editor. The next generation of operating systems will use this approach extensively: it is the foundation for Microsoft's OLE and Apple's OpenDoc.

All of these frameworks require the designer to write code, typically by creating application-specific subclasses of the standard classes provided as part of the framework.

Another class of systems that might be considered frameworks help create user interfaces that are composed of a series of cards, such as HyperCard from Apple. These systems are discussed subsequently under "Direct Graphical Specification" because their primary interface to the designer is graphical.

Model-Based Automatic Generation

A problem with all of the language-based tools is that the designer must specify a great deal about the placement, format, and design of the UIs. To solve this problem, some tools use *automatic generation* so that the tool makes many of these choices from a much higher level specification. Many of these tools, such as Mickey [Olsen 1989], Jade [Vander Zanden and Myers 1990], and DON [Kim and Foley 1993] have concentrated on creating menus and dialogue boxes. Jade allows the designer to use a graphical editor to edit the generated interface if it is not good enough. DON has the most sophisticated layout mechanisms and takes into account the desired window size, balance, columnness, symmetry, grouping, etc. Creating dialogue boxes automatically has been very thoroughly researched, but there still are no commercial tools that do this.

The User-Interface Design Environment (UIDE) [Sukaviriya et al. 1993] requires that the semantics of the application be defined in a special-purpose language, and therefore might be included with the language-based tools. It is placed here instead because the language is used to describe the functions that the application supports and not the desired interface. UIDE is classified as a model-based approach because the specification serves as a high-level, sophisticated model of the application semantics. In UIDE, the description includes pre- and postconditions of the operations, and the system uses these to reason about the operations and to automatically generate an interface. One interesting feature of UIDE is that the pre- and postconditions are used to automatically generate help. One direction of current research is to make UIDE models easier to create by allowing users to demonstrate some parts of the interface [Frank and Foley 1993].

Another model-based system is HUMANOID [Szekely et al. 1993], which supports the modeling of the presentation, behavior, and dialogue of an interface. The HUMANOID modeling language includes abstraction, composition, recursion, iteration, and conditional constructs to support sophisticated interfaces. The HUMANOID system, which is built on top of the Garnet toolkit [Myers et al. 1990], provides a number of interactive modeling tools to help the designer specify the model. The developers of HUMANOID and UIDE are collaborating on a new combined model called MASTERMIND that integrates their approaches [Neches et al. 1993].

The ITS [Wiecha et al. 1990] system also uses rules to generate an interface. ITS was used to create the visitor information system for the EXPO 1992 World's Fair in Seville, Spain. Unlike the other rule-based systems, the designer using ITS is expected to write many of the rules, rather than just writing a specification that the rules work on. In particular, the design philosophy of ITS is that all design decisions should be codified as rules so that they can be used by subsequent designers, which will hopefully mean that interface designs will become easier and better as more rules are entered. As a result, the designer should never use graphical editing to improve the design, since then the system cannot capture the reason that the generated design was not sufficient.

Although the idea of having the UI generated automatically is appealing, this approach is still at the research level, because the UIs that are generated are generally not good enough. A further problem is that the specification languages can be quite hard to learn and use. Extensive current research is addressing the problems of expanding the range of what can be created automatically (to go beyond dialogue boxes) and to make the model-based approach easier to use.

Direct Graphical Specification

The tools described next all allow the user interface to be defined, at least partially, by placing objects on the screen using a pointing device. This is motivated by the observation that the visual presentation of the user interface is of primary importance in GUIs, and a graphical tool seems to be the most appropriate way to specify the graphical appearance. Another advantage of this technique is that it is usually much easier for the designer to use. Many of these systems can be used by nonprogrammers. Therefore, psychologists, graphic designers, and UI specialists can more easily be involved in the UI design process when these tools are used.

These tools can be distinguished from those that use visual programming since with direct graphical specification, the actual UI (or a part of it) is drawn, rather than being generated indirectly from a visual program. Thus, direct graphical specification tools have been called *direct manipulation programming* since the user is directly manipulating the user interface widgets and other elements.

The tools that support graphical specification can be classified into four categories: **prototyping tools**, those that support a sequence of cards, **interface builders**, and editors for application-specific graphics.

Prototyping Tools. The goal of prototyping tools is to allow the designer to quickly mock up some examples of what the screens in the program will look like. Often, these tools cannot be used to create the real UI of the program; they just show how some aspects will look. This is the chief factor that distinguishes them from other high-level tools. Many parts of the interface may not be operable, and some of the things that look like widgets may just be static pictures. In most prototypers, no real toolkit widgets are used, which means that the designer has to draw simulations that look like the widgets that will appear in the interface. The normal use is that the designer would spend a few days or weeks trying out different designs with the tool, and then completely reimplement the final design in a separate system. Most prototyping tools can be used without programming, so they can, for example, be used by graphic designers.

Note that this use of the term *prototyping* is different from the general phrase *rapid prototyping*, which has become a marketing buzz word. Advertisements for just about all UI tools claim that they support rapid prototyping, by which they mean that the tool helps create the UI software more quickly. The term prototyping is being used in this chapter in a much more specific manner.

Probably the first prototyping tool was Dan Bricklin's Demo program. This is a program for an IBM PC that allows the designer to create sample screens composed of characters and character graphics (where the fixed-size character cells can contain a graphic such as a horizontal, vertical, or diagonal line). The designer can easily create the various screens for the application. It is also relatively easy to specify the actions (mouse or keyboard) that cause transitions from one screen to another. However, it is difficult to define other behaviors. In general, there may be some support for type-in fields and menus in prototyping tools, but there is little ability to process or test the results.

For GUIs, designers often use tools like Macromedia's Director for the Macintosh, which is actually an animation tool. The designer can draw example screens, and then specify that when the mouse is pressed in a particular place an animation should start or a different screen should be displayed. Components of the picture can be reused in different screens, but again the ability to show behavior is limited. HyperCard for the Macintosh is also often used as a prototyping tool.

The primary disadvantage of these prototyping tools is that they cannot create the actual code for the UI. Therefore, the interfaces must be recoded after prototyping. There is also the risk that the programmers who implement the real UI will ignore the prototype. Therefore, a new research tool is trying to provide a quick sketching interface and then convert the sketches into actual widgets [Landay and Myers 1995].

Cards. Many graphical programs are limited to user interfaces that can be presented as a sequence of mostly static pages, sometimes called *frames*, cards, or *forms*. Each page contains a set of widgets, some of which cause transfer to other pages. There is usually a fixed set of widgets to choose from, which have been coded by hand.

An early example of this is Menulay [Buxton et al. 1983], which allows the designer to place text, graphical potentiometers, iconic pictures, and light buttons on the screen and see exactly what the end user will see when the application is run. The designer does not need to be a programmer to use Menulay.

Probably, the most famous example of a card-based system is HyperCard from Apple. There are now many similar programs, such as Tool Book [Asymetrix 1995]. In all of these, the designer can easily create cards containing text fields, buttons, etc., along with various graphic decorations. The buttons can transfer to other cards. These programs provide a scripting language to provide more flexibility for buttons. HyperCard's scripting language is called HyperTalk, and as previously mentioned, is really an event language, since the programmer writes short pieces of code that are executed when input events occur.

Interface Builders. An interface builder allows the designer to create dialogue boxes, menus, and windows that are to be part of a larger UI. These are also called *interface development tools*. Interface builders allow the designer to select from a predefined library of widgets, and place them on the screen using a mouse. Other properties of the widgets can be set using property sheets. Usually, there is also some support for sequencing, such as bringing up subdialogs when a particular button is hit. The Steamer project at BBN demonstrated many of the ideas later incorporated into interface builders and was probably the first object-oriented graphics system [Stevens et al. 1983]. Other examples of research interface builders are DialogEditor [Cardelli 1988], vu [Singh and Green 1988] and Gilt [Myers 1991b]. There are literally hundreds of commercial interface builders. Just two examples are the NeXT Interface Builder and UIM/X from Visual Edge Software, Ltd., for X Windows. Many of the tools previously discussed, such as the virtual toolkits, visual languages, and application frameworks, also contain interface builders.

Interface builders use the actual widgets from a toolkit, so that they can be used to build parts of real applications. Most will generate C code templates that can be compiled along with the application code. Others generate a description of the interface in a language that can be read at run time. For example, UIM/X generates a UIL description. It is usually important that the programmer not edit the output of the tools (such as the generated C code) or else the tool can no longer be used for later modifications.

Although interface builders make laying out the dialogue boxes and menus easier, this is only part of the UI design problem. These tools provide little guidance toward creating good UIs, since they give designers significant freedom. Another problem is that for any kind of program that has a graphics area (such as drawing programs, CAD, visual language editors, etc.), interface builders do not help with the contents of the graphics pane. Also, they cannot handle widgets that change dynamically. For example, if the contents of a menu or the layout of a dialogue box changes based on program state, this must be programmed by writing code. To help with this part of the problem, some interface builders, like UIM/X, provide a C code interpreter.

Data Visualization Tools. An important commercial category of tools is that of dynamic data visualization systems. These tools, which tend to be quite expensive, emphasize the display of dynamically changing data on a computer, and are used as front ends for simulations, process control, system monitoring, network management, and data analysis. The interface to the designer is usually quite similar to an interface builder, with a palette of gauges, graphers, knobs, and switches that can be placed interactively. However, these controls usually are not from a toolkit and are supplied by the tool. Example tools in this category include DataViews from DataViews Corporation, and VAPS from Virtual Prototypes.

Editors for Application-Specific Graphics. When an application has custom graphics, it would be useful if the designer could draw pictures of what the graphics should look like rather than having to write code for this. The problem is that the graphic objects usually need to change at run time, based on the actual data and end user's actions. Therefore, the designer can only draw an *example* of the desired display, which will be modified at run time, and so these tools are called *demonstrational programming* [Myers 1992a]. This distinguishes these programs from the graphical tools of the previous three sections, where the full picture can be specified at design time. As a result of the *generalization* task of converting the example objects into parameterized prototypes that can change at run time, most of these systems are still in the research phase.

Peridot [Myers 1988a] allows new, custom widgets to be created. The primitives that the designer manipulates with the mouse are rectangles, circles, text, and lines. The system generalizes from the designer's actions to create parameterized, object-oriented procedures such as those that might be found in toolkits. Experiments showed that Peridot can be used by nonprogrammers. Lapidary [Myers et al. 1989] extends the ideas of Peridot to allow general application-specific objects to be drawn. For example, the designer can draw the nodes and arcs for a graph program. The DEMO system [Fisher et al. 1992] allows some dynamic, run-time properties of the objects to be demonstrated, such as how objects are created. The Marquise tool [Myers et al. 1993] allows the designer to demonstrate *when* various behaviors should happen, and supports palettes which control the behaviors. Research continues on making these ideas practical.

Specialized Tools

For some application domains, there are customized tools that provide significant high-level support. These tend to be quite expensive, however (i.e., U.S. $20,000–U.S. $50,000). For example, in the aeronautics and real-time control areas, there are a number of high-level tools, such as InterMAPhics from Gallium Software.

72.4 Technology Transfer

UI tools are an area where research has had a tremendous impact on the current practice of software development. Of course, window managers and the resulting GUI style comes from the seminal research at the Stanford Research Institute, Xerox Palo Alto Research Center (PARC), and Massachusetts Institute of Technology (MIT) in the 1970s. Interface builders and card programs such as HyperCard were invented in research laboratories at BBN, the University of Toronto, Xerox PARC, and others. Now, interface builders are at least a U.S. $100 million per year business and are widely used for commercial software development. Event languages, as widely used in HyperTalk and elsewhere, were first investigated in research laboratories. The next generation of environments, such as OLE and OpenDoc, will be based on the component architecture, which was developed in the Andrew environment from Carnegie Mellon University. Thus, whereas some early UIMS approaches such as transition networks and grammars may not have been successful, overall, the UI tool research has changed the way that software is developed.

72.5 Research Issues

Although there are many UI tools, there are many areas in which further research is needed. A report prepared for a Natural Science Foundation (NSF) study discusses future research ideas for UI tools at length [Olsen et al. 1993]. Here, a few of the important ones are summarized.

New Programming Languages

The builtin input/output primitives in today's programming languages support a textual question-and-answer style of UI which is modal and well-known to be poor. Most of today's tools use libraries and interactive programs which are separate from programming languages. However, many of the techniques, such as object-oriented programming, multiple-processing, and constraints, are best provided as *part* of the programming language. Furthermore, an integrated environment, where the graphical parts of an application can be specified graphically and the rest textually, would make the generation of applications much easier. A book discusses how programming languages can be improved to better support user interface software [Myers 1992b].

Increased Depth

Many researchers are trying to create tools that will cover more of the UI, such as application-specific graphics and behaviors. The challenge here is to allow flexibility to application developers while still providing a high level of support. Tools should also be able to support Help, Undo, and Aborting of operations.

Today's UI tools mostly help with the *generation* of the code of the interface, and assume that the fundamental UI design is complete. What are also needed are tools to help with the generation, specification, and analysis of the design of the interface [Landay and Myers 1995]. For example, an important first step in UI design is task analysis, where the designer identifies the particular tasks that the end user will need to perform. Research should be directed at creating tools to support these methods and techniques. These might eventually be integrated with the code generation tools, so that the information generated during early design can be fed into automatic generation tools, possibly to produce an interface directly from the early analyses. The information might also be used to automatically generate documentation and run time help.

Another approach is to allow the designer to specify the design in an appropriate notation, and then provide tools to convert that notation into interfaces. For example, the UAN [Hartson et al. 1990] is a notation for expressing the end user's actions and the system's responses.

Finally, much work is needed in ways for tools to help evaluate interface designs. Initial attempts, such as in MIKE [Olsen and Halversen 1988], have highlighted the need for better models and metrics against which to evaluate the UIs. Research in this area by cognitive psychologists and other user interface researchers (e.g., [Byrne et al. 1994]) is continuing.

Increased Breadth

We can expect the UIs of tomorrow to be different from the conventional window-and-mouse interfaces of today, and tools will have to change to support the new styles. For example, most tools today only deal with 2D objects, but there is already a demand to provide 3D visualizations and animations. New input devices and techniques will probably replace the conventional mouse and menu styles. For example, gesture and handwriting recognition are appearing in mass-market commercial products, such as notepad computers and personal digital assistants such as Apple's Newton (gesture recognition has actually been used since the 1970s in commercial CAD tools). Virtual reality systems, where the computer creates an artificial world and allows the user to explore it, cannot be handled by any of today's tools. In these non-WIMP applications (WIMP stands for windows, icons, menus, and pointing devices), designers will also need better control over the timing of the interface, to support animations and various new media such as video [Nielsen 1993]. Although a few tools are directed at multiple-user applications, there are no direct graphical specification tools, and the current tools are limited by the styles of applications they support.

End User Programming and Customization

One of the most successful computer programs of all time is the spreadsheet. The primary reason for its success is that end users can program (by writing formulas and macros). However, *end user programming* is rare in other applications, and where it exists, usually requires learning conventional programming. For example, AutoCAD provides Lisp for customization. More effective mechanisms for users to customize existing applications and create new ones are needed [Myers 1992b]. However, these should not be built into individual applications as is done today, since this means that the user must learn a different programming technique for each application. Instead, the facilities should be provided at the system level, and therefore should be part of the underlying toolkit. Naturally, since this is aimed at end users, it will not be like programming in C, but rather at some higher level.

The X Business Group predicts that there will be an increased use of tools by end users, rather than professional software developers, which will present enormous opportunities and challenges to tool creators [XBusiness 1994].

Application and User Interface Separation

One of the fundamental goals of user interface tools is to allow better modularization and separation of UI code from application code. However, a recent survey reported that modern toolkits actually make this separation more difficult, due to the large number of call-back procedures required [Myers and Rosson 1992]. Therefore, further research is needed into ways to better modularize the code and how tools can support this.

Tools for the Tools

It is very difficult to create the tools described in this paper. Each one takes an enormous effort. Therefore, work is needed in ways to make the tools themselves easier to create. For example, the Garnet toolkit explored mechanisms specifically designed to make high-level graphical tools easier to create. The Unidraw framework has also proven useful for creating interface builders [Vlissides and Tang 1991]. However, more work is needed.

72.6 Conclusions

The area of UI tools is expanding rapidly. Five years ago, you would have been hard pressed to find any successful commercial higher level tools, but now there are hundreds of different tools, and tools are turning into a billion dollar a year business. Chances are that today, whatever your project is, there is a tool that will help. Tools that are coming out of research laboratories are covering increasingly more of the UI task, are more effective at helping the designer, and are creating better user interfaces. As more companies and researchers are attracted to this area, we can expect the pace of innovation to continue to accelerate. There will be many exciting and useful new tools available in the near future.

Defining Terms

Application or application semantics: The part of the software that is not the user interface.
Application framework: A software architecture, often object oriented, that guides the programmer so that implementing user interface software is easier.
Interface builder: Interactive tool that lays out widgets to create dialogue boxes, menus, and windows that are to be part of a larger user interface. These are also called *interface development tools.*
Intrinsics: The layer of a toolkit on which different widgets are implemented.
Prototyping tool: Allows the designer to quickly mockup some examples of what the screens in the program will look like. Often, these tools cannot be used to create the real user interface of the program; they just show how some aspects will look.
Toolkit: A library of widgets that can be called by application programs.
User interface (UI):   The part of the software that handles the output to the display and the input from the person using the program.
User interface development environments (UIDE):   General term for comprehensive user interface tools.
User interface management system (UIMS):   An older term, not much used now. Sometimes used to cover all user interface tools, but usually limited to tools that handle the sequencing of operations (what happens after each event from the user).
User interface tool: Any software that helps create user interfaces.

Virtual toolkits: Also called *cross-platform development systems*, are programming interfaces to multiple toolkits that allow code to be easily ported to Macintosh, Microsoft Windows, and Unix environments.

Widget: A way of using a physical input device to input a certain type of value. Typically, widgets in toolkits include menus, buttons, scroll bars, text type-in fields, etc.

Window manager: The user interface of the windowing system. Also used to mean the entire windowing system.

Windowing system: Software that separates different processes into different rectangular regions (*windows*) on the screen.

References

Asymetrix. 1995. ToolBook. Asymetrix Corp. 100 100th Ave., NE, Bellevue, WA.

Bharat, K. and Brown, M. H. 1994. Building distributed, multi-user applications by direct manipulation, pp. 71–81. In *Proc. UIST'94, ACM SIGGRAPH Symp. User Interface Software Tech.* Marina del Rey, CA, Nov.

Bly, S. A. and Rosenberg, J. K. 1986. A comparison of tiled and overlapping windows, pp. 101–106. In *Proc. SIGCHI'86 Hum. Factors Comput. Syst.* Boston, MA, April.

Booz 1992. NeXTStep vs. Other Development Environments; Comparative Study. Booz Allen & Hamilton Inc. Rep. available from NeXT Computer, Inc.

Borning, A. 1981. The programming language aspects of Thinglab; a constraint-oriented simulation laboratory. *ACM Trans. Programming Lang. Syst.* 3(4):353–387.

Borning, A. and Duisberg, R. 1986. Constraint-based tools for building user interfaces. *ACM Trans. Graphics* 5(4):345–374.

Buxton, W., Lamb, M. R., Sherman, D., and Smith, K. C. 1983. Towards a comprehensive user interface management system, pp. 35–42. In *Proc. SIGGRAPH'83, Comput. Graphics* 17(3), Detroit, MI, July.

Byrne, M. D., Wood, S. D., Sukaviriya, P., Foley, J. D., and Kieras, D. E. 1994. Automating interface evaluation, pp. 232–237. In *Proc. SIGCHI'94, Hum. Factors Comput. Syst.* Boston, MA, April.

Cardelli, L. 1988. Building user interfaces by direct manipulation, pp. 152–166. In *Proc. UIST'88, ACM SIGGRAPH Symp. User Interface Software Tech.* Banff, Alberta, Canada, Oct.

Cardelli, L. and Pike, R. 1985. Squeak: a language for communicating with mice, pp. 199–204. In *Proc. SIGGRAPH'85, Comput. Graphics.* San Francisco, CA, July.

DePalma, D. A. and Woodring, S. D. 1993. Client/server power tools futures. *Software Strategy Rep.* 4(1):2–13. Forrester Research, Cambridge, MA.

Fisher, G. L., Busse, D. E., and Wolber, D. A. 1992. Adding rule-based reasoning to a demonstrational interface builder, pp. 89–97. In *Proc. UIST'92, ACM SIGGRAPH Symp. User Interface Software Tech.* Monterey, CA, Nov.

Frank, M. R. and Foley, J. D. 1993. Model-based user interface design by example and by interview, pp. 129–137. In *Proc. UIST'93, ACM SIGGRAPH Symp. User Interface Software Tech.* Atlanta, GA, Nov.

Gleicher, M. 1993. A graphics toolkit based on differential constraints, pp. 109–120. In *Proc. UIST'93. ACM SIGGRAPH Symp. User Interface Software Tech.* Atlanta, GA, Nov.

Green, M. 1986. A survey of three dialog models. *ACM Trans. Graphics* 5(3):244–275.

Hartson, H. R., Siochi, A. C., and Hix, D. 1990. The UAN: a user-oriented representation for direct manipulation interface designs. *ACM Trans. Inf. Syst.* 8(3):181–203.

Hayes, P. J., Szekely, P. A., and Lerner, R. A. 1985. Design alternatives for user interface management systems based on experience with COUSIN, pp. 169–175. In *Proc. SIGCHI'85, Hum. Factors Comput. Syst.* San Francisco, CA, April.

Hill, R. D. 1986. Supporting concurrency, communication and synchronization in human-computer interaction—the Sassafras UIMS. *ACM Trans. Graphics* 5(3):179–210.

Hill, R. D., Brinck, T., Patterson, J. F., Rohall, S. L., and Wilner, W. T. 1993. The Rendezvous language and architecture. *Commun. ACM* 36(1):62–67.

Hudson, S. E. and Stasko, J. T. 1993. Animation support in a user interface toolkit: flexible, robust, and reusable abstractions, pp. 57–67. In *Proc. UIST'93, ACM SIGGRAPH Symp. User Interface Software Tech.* Atlanta, GA, Nov.

Karrer, A. and Scacchi, W. 1990. Requirements for an extensible object-oriented tree/graph editor, pp. 84–91. In *Proc. UIST'90, ACM SIGGRAPH Symp. User Interface Software Tech.* Snowbird, UT, Oct.

Kim, W. C. and Foley, J. D. 1993. Providing high-level control and expert assistance in the user interface presentation design, pp. 430–437. In *Proc. INTERCHI'93, Hum. Factors Comput. Syst.* Amsterdam, The Netherlands, April.

Landay, J. and Myers, B. A. 1995. Interactive sketching for the early stages of user interface design, pp. 43–50. In *Proc. SIGCHI'95, Hum. Factors Comput. Syst.* Denver, CO, May.

Linton, M. A., Vlissides, J. M., and Calder, P. R. 1989. Composing user interfaces with InterViews. *IEEE Comput.* 22(2):8–22.

McCormack, J. and Asente, P. 1988. An overview of the X toolkit, pp. 46–55. In *Proc. UIST'88, ACM SIGGRAPH Symp. User Interface Software Tech.* Banff, Alberta, Canada, Oct.

Myers, B. A. 1986. A complete and efficient implementation of covered windows. *IEEE Comput.* 19(9):57–67.

Myers, B. A. 1988a. *Creating User Interfaces by Demonstration.* Academic Press, Boston, MA.

Myers, B. A. 1988b. A taxonomy of user interfaces for window managers. *IEEE Comput. Graphics Appl.* 8(5):65–84.

Myers, B. A. 1990a. All the widgets. *SIGGRAPH Video* Rev. 57.

Myers, B. A. 1990b. A new model for handling input. *ACM Trans. Inf. Syst.* 8(3):289–320.

Myers, B. A. 1990c. Taxonomies of visual programming and program visualization. *J. Visual Lang. Comput.* 1(1):97–123.

Myers, B. A. 1991a. Graphical techniques in a spreadsheet for specifying user interfaces, pp. 243–249. In *Proc. SIGCHI'91, Hum. Factors in Comput. Syst.* New Orleans, LA, April.

Myers, B. A. 1991b. Separating application code from toolkits: eliminating the spaghetti of call-backs, pp. 211–220. In *Proc. UIST'91, ACM SIGGRAPH Symp. User Interface Software Tech.* Hilton Head, SC, Nov.

Myers, B. A. 1992a. Demonstrational interfaces: a step beyond direct manipulation. *IEEE Comput.* 25(8):61–73.

Myers, B. A. ed. 1992b. *Languages for Developing User Interfaces.* Jones and Bartlett, Boston, MA.

Myers, B. A. 1994. Challenges of HCI design and implementation. *ACM Interactions* 1(1):73–83.

Myers, B. A., Giuse, D. A., Dannenberg, R. B., Vander Zanden, B., Kosbie, D. S., Pervin, E., Mickish, A., and Marchal, P. 1990. Garnet: comprehensive support for graphical, highly-interactive user interfaces. *IEEE Comput.* 23(11):71–85.

Myers, B. A., McDaniel, R., Ferrency, A., Mickish, A., Klimovitski, A., and McGovern, A. 1995. The Amulet Reference Manuals. *Computer Science Department, Tech. Rep.* CMU-CS-95-166, Carnegie Mellon University, June. also *Human Computer Interaction Institute* CMU-HCII-95-102. WWW = http://www.cs.cmu.edu/~amulet.

Myers, B. A., McDaniel, R. G., and Kosbie, D.S. 1993. Marquise: creating complete user interfaces by demonstration, pp. 293–300. In *Proc. INTERCHI'93, Hum. Factors Comput. Syst.* Amsterdam, The Netherlands, April.

Myers, B. A. and Rosson, M. B. 1992. Survey on user interface programming, pp. 195–202. In *Proc. SIGCHI'92, Hum. Factors Comput. Syst.* Monterey, CA, May.

Myers, B. A., Vander Zanden, B., and Dannenberg, R. B. 1989. Creating graphical interactive application objects by demonstration, pp. 95–104. In *Proc. UIST'89, ACM SIGGRAPH Symp. User Interface Software Tech.* Williamsburg, VA, Nov.

Neches, R., Foley, J., Szekely, P. S., Sukaviriya, P., Luo, P., Kovacevic, S., and Hudson, S. 1993. Knowledgable development environments using shared design models, pp. 63–70. In *ACM SIGCHI, Proc. 1993 Int. Workshop Intell. User Interfaces.* Orlando, FL, Jan.

Newman, W. M. 1968. A system for interactive graphical programming, pp. 47–54. In *AFIPS Spring J. Comput. Conf.*

Nielsen, J. 1993. Noncommand user interfaces. *Commun. ACM* 36(4):83–99.

Olsen, D. R., Jr. 1987. Larger issues in user interface management. *Comput. Graphics* 21(2):134–137.

Olsen, D. R., Jr. 1989. A programming language basis for user interface management, pp. 171–176. In *Proc. SIGCHI'89, Hum. Factors Comput. Syst.* Austin, TX, April.

Olsen, D. R., Jr. 1992. *User Interface Management Systems: Models and Algorithms.* Morgan Kaufmann, San Mateo, CA.

Olsen, D. R., Jr. and Dempsey, E. P. 1983. Syngraph: a graphical user interface generator, pp. 43–50. In *Proc. SIGGRAPH'83, Comput. Graphics*, Detroit, MI, July.

Olsen, D. R., Jr., Foley, J. D., Hudson, S. E., Miller, J., and Myers, B. 1993. Research directions for user interface software tools. *Behav. Inf. Tech.* 12(2):80–97.

Olsen, D. R., Jr. and Halversen, B. W. 1988. Interface usage measurements in a user interface management system, pp. 102–108. In *Proc. UIST'88, ACM SIGGRAPH Symp. User Interface Software Tech.* Banff, Alberta, Canada, Oct.

Ousterhout, J. K. 1991. An X11 toolkit based on the Tcl language, pp. 105–115. In *Proc. Winter Usenix. Tech. Conf.*

Palay, A. J. et al. 1988. The Andrew toolkit—an overview, pp. 9–21. In *Proc. Winter Usenix Tech. Conf.* Dallas, TX, Feb.

Pausch, R., Conway, M., and DeLine, R. 1992. Lesson learned from SUIT, the simple user interface toolkit. *ACM Trans. Inf. Syst.* 10(4):320–344.

Samuelson, P. 1993. Legally speaking: the ups and downs of look and feel. *Commun. ACM* 36(4):29–35.

Scheifler, R. W. and Gettys, J. 1986. The X window system. *ACM Trans. Graphics* 5(2):79–109.

Singh, G. and Green, M. 1988. Designing the interface designer's interface, pp. 109–116. In *Proc. UIST'88, ACM SIGGRAPH Symp. User Interface Software Tech.* Banff, Alberta, Canada, Oct.

Smith, D. C., Irby, C., Kimball, R., Verplank, B., and Harslem, E. 1982. Designing the Star user interface. *Byte* 7(4):242–282.

Stallman, R. M. 1979. Emacs: The Extensible, Customizable, Self-Documenting Display Editor. *Artificial Intelligence Lab. Tech. Rep.* 519, Massachusetts Institute of Technology, Aug.

Stevens, A., Roberts, B., and Stead, L. 1983. The use of a sophisticated graphics interface in computer-assisted instruction. *IEEE Comput. Graphics Appl.* 3(2):25–31.

Stevens, M. P., Zeleznik, R. C., and Hughes, J. F. 1994. An architecture for an extensible 3D interface toolkit, pp. 59–67. In *Proc. UIST'94, ACM SIGGRAPH Symp. User Interface Software Tech.* Marina del Rey, CA, Nov.

Strauss, P. S. and Carey, R. 1992. An object-oriented 3D graphics toolkit, pp. 341–349. In *Proc. SIG-GRAPH'92, Comput. Graphics.* July.

Sukaviriya, P., Foley, J. D., and Griffith, T. 1993. A second generation user interface design environment: the model and the runtime architecture, pp. 375–382. In *Proc. INTERCHI'93, Hum. Factors Comput. Syst.* Amsterdam, The Netherlands, April.

Sutherland, I. E. 1963. SketchPad: a man-machine graphical communication system, pp. 329–346. In *AFIPS Spring J. Comput. Conf.*

Swinehart, D., Zellweger, P., Beach, R., and Hagmann, R. 1986. A structural view of the Cedar programming environment. *ACM Trans. Programming Lang. Syst.* 8(4):419–490.

Szekely, P., Luo, P., and Neches, R. 1993. Beyond interface builders: model-based interface tools, pp. 383–390. In *Proc. INTERCHI'93, Hum. Factors Comput. Syst.* Amsterdam, The Netherlands, April.

Teitelman, W. 1979. A display oriented programmer's assistant. *Int. J. Man–Mach. Stud.* 11:157–187. Also *Xerox PARC Tech. Rep.* CSL-77-3, Palo Alto, CA, March 8, 1977.

Tesler, L. 1981. The Smalltalk environment. *Byte Mag.* 6(8):90–147.

Vander Zanden, B. and Myers, B. A. 1990. Automatic, look-and-feel independent dialogue creation for graphical user interfaces, pp. 27–34. In *Proc. SIGCHI'90, Hum. Factors Comput. Syst.* Seattle, WA, April.

Vlissides, J. M. and Linton, M. A. 1990. Unidraw: a framework for building domain-specific graphical editors. *ACM Trans. Inf. Syst.* 8(3):204–236.

Vlissides, J. M. and Tang, S. 1991. A Unidraw-based user interface builder, pp. 201–210. In *Proc. UIST'91, ACM SIGGRAPH Symp. User Interface Software Tech.* Hilton Head, SC, Nov.

Wernecke, J. 1994. *The Inventor Mentor.* Addison–Wesley, Reading, MA.

Wiecha, C., Bennett, W., Boies, S., Gould, J., and Greene, S. 1990. ITS: a tool for rapidly developing interactive applications. *ACM Trans. Inf. Syst.* 8(3):204–236.

Wilson, D. 1990. *Programming with MacApp.* Addison–Wesley, Reading, MA.

XBusiness 1994. *Interface Development Technology.* X Business Group, Inc. Fremont, CA.

Zeleznik, R. C. et al. 1991. An object-oriented framework for the integration of interactive animation techniques, pp. 105–122. In *Proc. SIGGRAPH'91, Comput. Graphics.* July.

Further Information

The primary conference for Interface Software Technology is the annual User Interface Software and Technology Symposium (UIST), and the proceedings are available from ACM. Many papers on this topic also appear in the annual ACM SIGCHI conferences.

The World-Wide-Web page http://www.cs.cmu.edu/~bam/toolnames.html contains a comprehensive list of user interface tools, and is frequently updated.

There are not many books specifically about user interface software. Olsen's [1992] *User Interface Management Systems: Models and Algorithms* provides a review of the older UIMS technology, and Myers's [1992b] *Languages for Developing User Interfaces* discusses some research systems and future ideas.

73

The Human Factor in Programming and Software Development

Mary Beth Rosson
Virginia Polytechnic Institute & State University

73.1 Introduction

Behavioral studies of programming were among the earliest in the field of human–computer interaction (HCI) that emerged in the late 1970s. HCI is the study of humans interacting with computing systems, and well before the appearance of modern interactive computer applications, programmers were using text-based command and programming languages to solve complex problems on computers. Cognitive scientists were intrigued by the complex and open-ended nature of software design problems; practitioners were eager to analyze and improve the productivity and reliability of software development. Many of the models and theories of current HCI had their inception in studies of programmers developing, comprehending, or maintaining code.

Research on programmer behavior is often referred to as **software psychology**, a name suggestive of one of the major motivations for research in this area. Cognitive scientists concerned with models of human information processing and with characterizing complex mental representations and strategies recognize software design and development as a rich domain in which to study general problem-solving skills [Newell 1990]. Thus, the mental activities of individual programmers are studied—often in excruciating detail—as they understand programming design problems and develop solutions (the methods and concerns of such analyses are addressed for the more general case of cognitive models in Chapter 63). A related stream of work focuses on the development of expertise in programming and design. The analysis of how programming knowledge and strategies evolve with training or experience serves two goals: a basic research concern with the nature of programming knowledge and a more applied concern with how to best support development of this knowledge through training, documentation, or tools.

0-8493-2909-4/97/$0.00+$.50

While cognitive analyses of programming and software development have established an underlying theoretical framework for work in this area, the interesting questions from an engineering perspective have an applied research agenda. Thus, rather than simply trying to understand the cognition involved in programming and design, researchers have also asked how best to support the cognitive activities in programming. Similarly, researchers have used their understanding of how programming expertise develops to guide the design of training or education.

Applied questions such as these form a subarea of software engineering and face the methodological challenge of conducting empirical work that has meaningful implications for software development activities in the real world. Whereas most analysis of programming behavior has focused on individuals working on small- to medium-sized problems, most programming in the real world takes place in teams coordinating their work on large and complex problems. Recently, however, researchers have begun to tackle the difficult questions of how experts collaborate to solve software design problems and how group and organizational factors facilitate or inhibit this collaboration (see also Chapters 66 and 64).

Recently, attention to programming behavior has expanded from the study of professional or student programmers to include the much more diverse problem of **end user programming** [Nardi 1993]. To some extent this expansion is a natural consequence of the continuing abstraction of programming languages, away from the machine and system software platform in which they operate and toward the problem domains to which they are applied [Fischer 1994]. As languages and tools have begun to refer directly to problem entities and processes, the individuals most familiar with these entities and processes have become more able to create programs that organize them. However, the expansion also reflects the increasing sophistication of the users of computing systems: as professionals become more comfortable with encoding, studying, and archiving their work-related information on computing systems, they expect and demand more power in addressing and manipulating this information.

73.2 Underlying Principles

Prior to addressing how best to support or teach expert programming and design skills, it is useful to overview the research paradigms and theoretical constructs that have been applied to the analysis of programming behavior. The following sections summarize researchers' current understanding of the basic tasks and knowledge comprising programming, as well as some of the characteristics of languages, tools, and social factors that influence how programming activities take place.

Programming Tasks

Fundamentally, computer programming is a design task in which programmers map from a description of a problem in the real world into a concrete set of instructions in some programming language that solves this problem. As in design more generally, the overall design task comprises a number of more specific subtasks. The first of these is *problem understanding*, the analysis of the problem domain, including identification of key problem entities and their interrelationships. To some extent, problem understanding must be a component of any programming situation, in that a programmer must first understand the problem he or she is asked to work on. However, few studies have focused on problem understanding as a task in and of itself, perhaps because of the practical problem of finding programmer participants who possess expertise in specific problem domains [Pennington and Grabowski 1990].

Assuming some understanding of the problem domain, the programmer must next develop a problem *decomposition*, which maps the problem analysis to elements of the software domain. This decomposition will typically occur in an iterative fashion, as the programmer translates problem requirements into subproblems, which are then mapped to modules, submodules, and so on. Once an overall design mapping has been achieved, the process of *coding* can take place; here the programmer implements the software design as an organized body of programming language instructions. After code has been written, the programmer may go through a process of *testing*, in which the implementation is evaluated for completeness, correctness, and efficiency. Errors discovered along the way evoke a *debugging* subtask, during which a programmer must analyze the errors produced and map them back to incorrect design or coding decisions. Finally, once a program has been developed, programmers may be asked to carry out code *maintenance*. Software maintenance can be seen as a *meta* task, in that it may involve reunderstanding of a problem, redesign of a solution, recoding, and so on.

Although researchers have often attempted to analyze one or another of these programming subtasks in detail, it is important to realize that the subtasks will overlap in actual software development situations. In a research setting, for example, a programmer might be given a buggy program to comprehend, making the focus one of code understanding; in the real world, a bug detection process such as this would be interleaved with a corrective process involving changes to the problem decomposition and coding.

At a higher level, programming tasks are also often analyzed as tasks of *composition*, the process of developing a solution to a problem, or *comprehension*, the analysis of a partial or complete solution [Pennington and Grabowski 1990]. Problem understanding, decomposition, and coding are largely composition tasks, whereas testing, debugging, and maintenance emphasize comprehension activities. Of course, even though a particular programming situation may emphasize composition or comprehension, these two general activities again are often interleaved: programmers developing programs comprehend their code as it is developed, and programmers analyzing a program often must reconstruct the composition process that generated it.

Another high-level software design task that cuts across many of the individual programming subtasks is software *reuse*. An experienced programmer will often seek to apply pieces of previous work to the current problem or may take on the subgoal of making aspects of the developing solution reusable for other problems in the future. The reuse can take place at many levels, from that of an abstract design pattern or framework [Deutsch 1989, Gamma et al. 1995] to a concrete code module or fragment [Lange and Moher 1989].

Programming Knowledge

Considerable research attention has been devoted to mapping the knowledge structures possessed by successful programmers. It is generally agreed that for any given problem the relevant knowledge is of at least two sorts: knowledge about the problem domain (e.g., telecommunications, banking) and about the software domain (e.g., decomposition strategies, modularity, data structures, algorithm design). In successful programming, these two sources of knowledge must be integrated: for example, comprehension of a program involves mapping from the text of the program to a programmer's knowledge of programming techniques on the one hand, and understanding of the application domain on the other [Pennington 1987a]. Indeed, the naturalness claims by the advocates of object-oriented programming (OOP) are founded on an assumption that the cognitive distance between problem and software structures is reduced in this programming paradigm [Rosson and Alpert 1990]. Guindon [1990] labels the relevant domain knowledge a **design schema**, an abstract knowledge structure that contains information about a class of design solutions. However, although many researchers have demonstrated the benefit of domain experience, little work has been directed at detailing how it is represented in the minds of experienced programmers.

In contrast, many researchers have discussed the mental representations associated with programming knowledge. A central construct in this has been the **programming plan** [Soloway and Ehrich 1984]: an abstract structure that links the goals of a program to specific computational structures for implementing these goals. For example, most expert programmers are familiar with the notion of a counter or of techniques for supporting input or output in a program. Many empirical studies of programmers have demonstrated plan-based comprehension or generation of program code. Researchers studying the application of plan knowledge also have suggested that experts possess strategies for how to use plans in writing or reading code. More detailed analyses suggest further that programming plans have an articulated structure in which some lines are more salient; this can introduce nonlinearities in the generation or comprehension of plan elements [Rist 1990, Davies 1993]. More recent work in OOP proposes that experts possess knowledge of **design patterns**, a sort of high-level plan that describes communicating objects and classes that are customized to solve a general design problem in a particular context[Gamma et al. 1995]. As yet, however, no empirical work has attempted to determine if experts do in fact possess mental representations corresponding to design patterns.

Soloway and Ehrich also propose that expert knowledge includes rules of programming discourse that describe stylistic guidelines for developing programs independent of languages or applications (e.g., naming variables to reflect their function, avoiding side effects). Finally, at the lowest level is the syntactic knowledge of a particular programming language's conventions and vocabulary [Shneiderman and Mayer 1979].

Schemas, programming plans, and stylistic and syntactic rules encode the declarative knowledge of programmers, the concepts and relationships a programmer will learn and then apply to the solution of new problems. But programmers also possess procedural knowledge comprising strategies for designing, coding, or comprehending a piece of software. At the highest level, programmers are thought to possess knowledge of general design heuristics such as divide-and-conquer, an approach that should be useful in solving complex problems of any sort. More specifically to the software domain, experienced programmers working on routine problems often exhibit a strategy of **systematic expansion**, a breadth-first decomposition in which problems are decomposed into subproblems one level at a time, with each level only a bit more detailed than the previous one [Adelson and Soloway 1985].

The observation of systematic strategies suggests that experts understand the value of abstraction and of developing intermediate levels in their design analysis before jumping to the level of specific program modules or code. At the same time, experts know when to deviate from classical hierarchical decomposition, for example, choosing to analyze a particular subproblem in depth because it is critical to the success of the overall design [Carroll et al. 1979, Jeffries et al. 1981]. Recent work suggests that for complex and novel problems, experts often exhibit **opportunistic planning** [Hayes-Roth and Hayes-Roth 1979], in which subproblems at different levels are analyzed simultaneously with alternation among levels determined by emerging features of the solution [Guindon 1990, Rosson and Carroll in press].

Languages, Notations, and Tools

Programming behavior is an interaction between the problem, the programmer's knowledge and mental capacities, and the languages, tools, or other external representations available. Thus, another thread of research has examined how characteristics of languages or tools influence the programming process.

Perhaps the most heavily studied characteristic of this sort has been the supposed beneficial consequences of structure in a language or environment: for instance, does a structured language that prohibits the use of GO-TO statements [Dijkstra 1968] and supports a block organization produce code that is easier to develop, understand, and maintain? In this case, the relevant research findings have been mixed. Though studies have generally shown a benefit from structured programming techniques, the best control structure for any given situation appears to be highly dependent on the problem, the programming language provided, the programmer's experience, and even the way problem information is received [Curtis 1988, Vessey and Weber 1984].

Empirical studies of programming languages contrasting specific language features (e.g., are logical conditions better than arithmetic ones?) also have not yielded a general analysis of language characteristics. As for studies of control structure, the consequences of various language features depend too much on the details of the problem situation. One study of expert programmers suggests that they have internalized which features are appropriate for different problem situations and may carry out their actual problem solving with reference to a private pseudolanguage containing useful features from a variety of languages [Petre 1990].

A theoretical framework has been offered by Green [1990]. He discusses programming languages as information structures and identifies several cognitive dimensions of programming languages or notations that make them more or less effective for different programming situations. One of these is **viscosity**, or the degree to which a language or notation resists local changes. Highly viscous languages or notations make it difficult for a programmer to experiment with alternatives or to iteratively refine an algorithm, in that a small change may result in a tedious process of adjusting other parts of the program. Another

dimension is **premature commitment,** or the extent to which a decision must be made before its consequences can be seen. Languages that impose a strict ordering on statements—particularly when supported
or enforced by a programming environment—can interfere with a programmer's natural decomposition
of a problem. A third dimension is **role expressiveness,** or the ease with which a programmer can convey
or discover the purpose of a program element. Languages that support extended variable names, for
example, should promote the role expressiveness and consequent readability of program code.

Social and Organizational Factors

Although most behavioral research on programming and software development has studied individual
programmers, programming in the real world typically involves teams of programmers working within
organizations of varying complexity. One obvious consequence of this is that the behavior and results of
the team is a collaborative product of individuals working together. Flor and Hutchins [1991] propose
that team programming activities are best analyzed as **distributed cognition,** wherein the project result is
seen as the result of a complex system of programmers, their internal mental activities, and their shared
externalized task representations. They argue that the analysis of collaborative behavior in a team can
uncover cognitive properties that otherwise are not apparent (e.g., a shared memory of a programming
plan that is available only through shared recall).

One advantage that a team brings to programming problems is that individual team members can bring
divergent perspectives and skills to the problem-solving process. However, the way in which individual
views are expressed and synthesized can vary considerably from team to team and may have significant
effects on team performance. Some teams are highly centralized (i.e., they have a team leader or chief
programmer who has decision-making responsibility); others are decentralized (i.e., leadership responsibilities are distributed among all team members). Researchers [Curtis and Walz 1990, Mantei 1981]
suggest that a decentralized structure should be most effective on small, complex projects that demand
creative solutions and that are not schedule driven, whereas centralized teams are best suited to large but
simple projects run on a tightly controlled schedule. Unfortunately, most projects have a mix of these
characteristics, making it difficult to test these proposals in realistic programming situations.

The characteristics of the organization sponsoring a programming project will also influence the project
results. For example, adherence to a user-centered design process may have a greater positive impact on a
product's ultimate success than any characteristic of the team, the tools recruited by the team, or the skills
of individual team members [Curtis and Walz 1990; see also Chapter 66]. Complex projects that entrain
a complex organization of interdependent programming teams introduce a tremendous communication
overhead that is often not a recognized component of the software development process, but that can be
a significant drain on overall productivity.

73.3 Best Practices

The previous section discussed underlying approaches to describing and predicting the programming and
design behavior of individuals or groups. The current section builds on this descriptive framework to
consider how best to support such activities. It begins by discussing the implications of what is known about
expert behavior in both the initial design development of software and the debugging and maintenance of
existing software. A third section discusses approaches to teaching and learning programming and design
skills. The last section expands the focus to the support of programming teams in the real world.

Software Design and Development

The process of software design is often misconceived as a straightforward activity of problem decomposition. In fact, the process (at least for difficult or novel problems) appears to be much more heterarchical
than this ideal and may involve many false starts and iterations. Thus, the challenge is to build tools or
environments that can facilitate and organize what may often be a rather unstructured process.

Balancing Systematic and Opportunistic Design

The earliest analyses of software design viewed the process as a case of general problem solving characterized by top-down, goal-driven, and hierarchically structured planning and implementation activities [Simon 1973]. From this perspective, programming is seen as a systematic process of stepwise refinement, where a problem is decomposed into successively more detailed problems and subproblems, ultimately resulting in lines of code. Many empirical studies have observed an orderly design process such as this, noting, for example, that the number of intermediate levels and the extent to which a top-down, breadth-first decomposition process is used increases with the experience of the designer (e.g., Adelson and Soloway [1985] and Jeffries et al. [1981]). However, as more studies have been carried out, with a wider range of software professionals working on a wider range of problems and with a wider range of tools and languages, it appears that an important component of expertise is the ability to combine conventional hierarchical decomposition with design episodes that are more heterarchical and opportunistic in nature [Guindon 1990, Visser and Hoc 1990].

Expert designers have been taught systematic decomposition methods and often report their reliance on such approaches when asked about their strategies for solving software design problems [Visser 1987]. At the same time, they are quite willing to depart from these strategies to get their jobs done. Thus, for example, even though the expert behavior described by Jeffries et al. exhibited a high degree of hierarchical decomposition, the experts did at times deviate from a top-down and breadth-first strategy, for example, electing to proceed depth-first for some critical subproblems. Guindon [1990] describes numerous excursions among levels for professional programmers working on an elevator control system. Similar observations are reported by Carroll et al. [1979] in their analysis of a librarian and systems engineer working together to design a library information system and by Visser [1987] studying an expert designing a process-control application.

Cognitive scientists are intrigued by this mix of design strategies because it suggests a highly complex and heuristically guided planning process (see, e.g., Hayes-Roth and Hayes-Roth [1979]). From the perspective of engineering science, however, a more pertinent question concerns the factors associated with systematic vs. more opportunistic problem-solving activities. To the extent that we can identify such factors, we can try to use our understanding to anticipate and facilitate expert design behavior in actual software development situations. Although the empirical work to date is far from conclusive, researchers are beginning to develop a picture of some of the variables likely to influence the systematicity of the design process.

One obvious factor concerns the nature of the problem. Some tasks have an inherently hierarchical structure and thus are more suited to the systematic decomposition approach; the suitability of such an approach can be enhanced by problem specifications that reify its inherent hierarchical structure [Carroll et al. 1980]. However, even more than the inherent structure of the problem, the familiarity of the expert with the various components of the problem seems to influence the degree of nonsystematic design activity.

Paradoxically, from the perspective of predicting and supporting design problem solving, possessing relevant domain experience can both diminish and increase a designer's opportunistic behavior. On the one hand, if a problem is similar to others the designer has solved, the designer will possess schematic knowledge concerning the overall solution structure, the tradeoffs in various approaches, etc. [Guindon 1990]. Such knowledge can be applied to structure the overall solution process, resulting in a relatively systematic and well-ordered decomposition. A designer without such schematic knowledge is much more likely to use an iterative approach in which some subproblems are solved partially in an effort to constrain other subproblems, and in which the developing partial analyses at various levels of abstraction are continually re-examined and refined [Carroll et al. 1979]. On the other hand, a designer with prior experience in a problem domain may also possess specific knowledge at the level of individual plans, pieces of the problem decomposition, or even entire algorithms that may be part of a subproblem's solution. In such a situation, opportunistic episodes occur when a familiar subproblem is recognized and solved immediately in a depth-first fashion simply because the answer is already known [Lange and Moher 1989].

Thus, it appears that prior knowledge can both increase and decrease the extent to which a problem is solved in a systematic top-down, breadth-first fashion. In general, however, research suggests that the degree of heterarchical activity will be greater for problems that are complex and unfamiliar, because these are the ones requiring continual reexamination and refinement of interrelated subproblems. Familiar problems may evoke depth-first solutions in some cases, but the overall structure of the design process is still likely to be well organized.

Features of the external situation may also encourage or discourage systematic vs. opportunistic design behavior. One such feature is the designer's access to reusable software artifacts. Such artifacts can be seen as external representations of design knowledge; the programmer either must already be familiar with the reusable artifact (e.g., as in the case when it was created by the designer or has been used before [Lange and Moher 1989]) or must be able and willing to spend the time learning its capabilities and operation [Fischer 1987]. Like personal domain experience, reusable software can support both systematic and opportunistic activities. Generic, schematic reuseable artifacts, for example, **application frameworks** or general-purpose design patterns [Gamma et al. 1995], are likely to encourage a systematic approach, in that they organize and rationalize an overall solution structure. In contrast, the discovery of relevant widgets or library functions may encourage opportunistic depth-first design behavior.

Characteristics of the language or tools employed in a programming project may also impact programmers' departures from systematic decomposition. Guindon [1990] reports that the use of notes and diagrams was especially important to designers engaged in opportunistic episodes, as it allowed them to keep track of where they were in the overall solution process. An implication is that projects likely to evoke heterarchical design activity (e.g., involving problems that are complex and novel) should be supported by an environment that includes extensive support for creating, referring to, and annotating informal planning documents (see also, Davies [1993]).

In some cases, a programmer may leverage the implicit structuring facilities built into a programming environment to manage a complex iterative and opportunistic design process. For example, many modern operating systems support multitasking, and experts can exploit this by creating and positioning multiple windows to organize their work in a complex design space. In our studies of expert Smalltalk programmers [Rosson and Carroll 1993], we observed the simultaneous use of multiple code browsers (as well as other related interactive tools) to maintain different bits of problem context in a design process that involved considerable switching among interrelated subproblems. Contrast a flexible multitasking environment such as this, where the programmer is in control of structuring the work, to the structured editors for programming languages: such editors may in fact organize and simplify the task of code generation, but perhaps at the cost of forcing premature decisions about control and data structures [Green 1990].

Finally, recall that the social dynamics of a programming team and organization are likely to influence the style of work carried out by team members. Programmers working within a large organization that mandates conventional structured design methods or in teams with a highly centralized structure would normally be expected to employ systematic, top-down and breadth-first design strategies. However, for some problems, in particular those demanding complex and creative solutions, such strategies may be counterproductive. An interesting compromise might be to provide tools or environments that support opportunistic and exploratory design activities on an individual level, while still requiring team members to participate in a well-structured overall process.

The Allure of Object-Oriented Design

Object-oriented design has attracted considerable interest recently from researchers studying programming and design [Détienne and Rist 1995, see Chapters 96 and 108 for general discussions of object-oriented programming and design]. Rosson and Alpert [1990] discuss how several of the fundamental characteristics of OOD map well to our current understanding of expert design behavior, and a number of researchers have begun to assess the putative naturalness of OOD. It is too soon to assess the long-term impact of the paradigm on design problem solving: although some researchers have provided empirical support for benefits anticipated in Rosson and Alpert's analysis, others have documented problems with teaching and applying OOD techniques.

OOD is claimed to reduce the distance between problem and software constructs. In a conventional procedural language such as C, programmers must first develop a problem model and then go through a translation process to map problem requirements onto data structures and procedures. Researchers have shown that the object abstraction, which encapsulates problem information and behaviors relevant to that information, encourages designers to carry out problem understanding and analysis in parallel with development of an object-based software model [Rosson and Gold 1989]. This should facilitate the design process as well as result in designs that are more easily comprehended (i.e., because problem information is more explicitly embedded in the software model).

A comprehensive study by Pennington et al. [1995] provides initial support for a reduced problem solution distance: OOD experts spent little time in explicit problem analysis relative to expert procedural designers but spent considerably more time in defining data and procedural abstractions (objects and methods in OOD terminology). The OOD experts tended to take more time overall in producing their solutions but also produced more complete solutions and thus seemed to be working more efficiently. The other piece of the predicted impact, that an object-oriented design would be more easily comprehended by other experts, is much more difficult to assess empirically, as it is difficult to control for the goodness of a test design across multiple design paradigms (see, e.g., Boehm-Davis et al. [1992]).

Rosson and Alpert also argued that the encapsulation of objects should aid management of a complex solution structure as it evolves: because of the increased modularity, designers are less likely to need information from other subproblems and will be better able to develop an in-depth solution for a given component. At the same time, they should find it easier to shift among subproblems, because the object abstraction should provide a coherent chunk in working memory (or in their external notes or diagrams).

Again, the study of Pennington et al. provides an intriguing glimpse of such effects: OOD experts moved quickly into detailed design (i.e., at the level of individual object and method specification; this is consistent with the observation that these experts analyzed the problem by defining an object space). However, their overall design activity appeared to be relatively top-down, orderly, and balanced, with fewer excursions back up to more abstract concerns. These findings are particularly interesting in light of Green's [1990] suggestion that the encapsulation inherent in object-oriented languages should reduce language viscosity and thus might better support opportunistic design strategies. Given the initial findings of Pennington et al., we might speculate that OOD does support opportunistic behavior of the depth-first sort (e.g., a detailed design for a particularly interesting or important object abstraction) but at the same time may encourage a balanced approach overall (e.g., identifying a relatively complete set of candidate objects early in the process).

A third much discussed feature of OOD is its potential for increasing software reuse (see, e.g., Cox [1986] and Meyer [1988]). From the perspective of design problem solving, the impact of reuse should appear in experts' recruitment (through recall, directed search, or opportunistic recognition) of specific pre-existing abstractions (e.g., existing object classes or frameworks) while solving new problems. Unfortunately, any behavioral assessment of reuse benefit will be entirely confounded with the actual reusability of artifacts developed in object-oriented vs. contrasting paradigms. That is, if it is the case that OOD methods lead to abstractions that are more generally reusable, then we would naturally expect to see experts exploiting such materials. Pennington et al. [1995] did observe that the expert OOD designers reused standard object classes (e.g., lists and sets) in their solutions, whereas the expert procedural designers showed no such tendency; similar findings were reported by Rosson and Gold [1989]. However, it is difficult to assess whether this is due simply to the availability of these reusable artifacts or whether something about the OOD process encourages such reuse. The design problem used by Pennington et al. emphasized data representation, and it may be that other problems with an emphasis on process or control concerns would evoke more reuse among procedural designers.

At the same time that empirical studies are beginning to document some of the possible benefits of OOD on design problem solving, researchers caution against enthusiastic acceptance of this paradigm as the answer to all of our software design problems [Fischer et al. 1995]. Many software development organizations are experiencing great difficulties in teaching OOD methods to designers already skilled in procedural design [Rosson and Carroll 1990], and as yet there is no promising evidence that

OOD methods are more easily learned by novices. There continues to be considerable controversy and discussion concerning the best ways to identify object abstractions and the problems for which object-oriented decompositions are most suited. Finally, even if we assume that OOD will ultimately encourage development of reusable software libraries, there is still significant work to be done on how we can best support designers in the identification and use of such artifacts [Helm and Maarek 1991, Rosson et al. 1991, Rosson and Carroll, in press].

Debugging and Maintaining Software

A central determinant of a programmer's success in correcting or enhancing a piece of software is the programmer's comprehension of the system [Gugerty and Olson 1986, Littman et al. 1986, Pennington and Grabowski 1990]. To diagnose and repair a bug or to introduce a new feature, the programmer must first analyze the existing software model to diagnose how it is faulty or to determine whether and how new material can be incorporated or substituted without compromising the system. Note, however, that debugging or maintenance tasks are highly goal driven: in these tasks, software comprehension is simply a means to an end, in that the goal of debugging or maintenance is not to understand a program in its entirety but rather to manipulate it in some way.

Goal-Directed Software Comprehension

Software comprehension can be defined as the assignment of meaning to the code composing a software system. Of course, it is not sufficient to interpret the meaning of individual statements, or even groups of statements; the programmer also must understand how the specific computational activity will unfold under varying conditions (control flow); and how such activity will manipulate and transform data objects (data flow). In the case of program debugging or modification, the programmer must also understand how the elements of the software map to the requirements of the problem to ensure that removal or revision of a particular software element does not disturb desirable system functionality.

Software comprehension includes both top-down and bottom-up analysis activities. The programmer recruits knowledge of the domain or of programming constructs to develop hypotheses about the meaning of the program; these hypotheses then guide a search for expected program elements [Brooks 1983]. For example, simply reading the name of a program or subroutine (e.g., mail-receipt) may activate relevant schemas or plans that suggest the likely presence of certain types of data or control structures (e.g., a last-in–first-out queue). More detailed hypotheses are generated as comprehension proceeds; for example, the presence of a particular program variable or keyword may lead to a specific hypothesis about associated variables or statements. However, just as for the case of software design, the success of such an organized, top-down process will be a function of the programmers' prior knowledge as well as the typicality of the program under analysis [Widowski and Eyferth 1986]. Thus, at times the comprehension process must include bottom-up analyses, synthesizing individual program statements into larger units, often relying on **mental simulation** to analyze the effect of a group of statements. Of course, this bottom-up analysis often will suggest hypotheses that feed back into the ongoing top-down process.

Of interest here are expert strategies for balancing comprehension effort with code modifications and the resulting implications for how best to support the process. Clearly, the accuracy of a software change and the minimization of its side effects are best assured through careful examination of associated documentation and systematic analysis of the code itself, so that the maintainer can build a complete and coherent mental model of the software [Littman et al. 1986]. However, for large and complex programs, a systematic comprehension strategy is too expensive. Instead, experts adopt a more heuristic approach: they may scan all or parts of the program to look for clues that can focus their analysis process [Jeffries et al. 1981], or they may use external documentation such as a module decomposition chart to develop hypotheses [Koenemann and Robertson 1991]. They then limit their detailed analysis to code perceived as relevant to their modification task [Koenemann and Robertson 1991, Rosson and Carroll, in press]. Only in states of extreme uncertainty are they likely to resort to comprehensive code comprehension as a technique for generating hypotheses [Rosson and Carroll 1993].

The implications of this goal-directed comprehension process are twofold. Programmers working on debugging or maintenance tasks clearly need support for extracting the high-level organization of a complex software system. This may consist of external documentation such as the module charts provided by Koenemann and Robertson. However, given experts' general reluctance to study external documentation, a better approach might be to provide multiple views of the software itself, for example, a structured browsing environment that enables rapid switching between an outline view of the program that lists high-level program information and a detailed (i.e., code-level) view of selected program components. A good example is the standard class hierarchy browser available in most Smalltalk environments that provides separate but coordinated views of class categories, class names organized into a hierarchy, procedure (method) categories, variable and method names, and method code.

At the same time, the details of the code itself should provide clear evidence confirming or disconfirming an expert's current hypothesis. Although it might seem that in-line comments would be an effective means for accomplishing this, most experts reading code spend little time examining comments and focus instead on the code itself, especially on complex code structures [Crosby and Stelovsky 1990]. What appears to be more important is the availability of **beacons**, specific features of the code that are evidence for the presence of particular programming structures [Brooks 1983]. For example, a stereotypical exchange of values in an array suggests the presence of a sort routine. The implications are simple: software developers should be encouraged to incorporate stereotypical structures and to rely on naming conventions as much as possible and should be encouraged to work with role-expressive languages [Green 1990] that allow extended variable naming. Code formatting (e.g., indentation, highlighting of variable names) may also facilitate a goal-directed search process, but only when the formatting conventions are simple, well known, and consistently applied by the developer [Gellenbeck and Cook 1991].

Interactive Debugging

Most behavioral studies of software comprehension, debugging, or maintenance have provided a minimum of environmental support to the participants; these studies were modeled on analyses of text comprehension and thus have focused on programmers' interactions with textual materials. In actual programming situations, however, the programming environment is a critical factor; part of a programmer's expertise is the ability to use the available tools (see Chapter 112 for a general description of programming environments). For example, a study of the debugging anecdotes of professional programmers revealed that a common attribute of serious debugging problems is the inapplicability of available debugging tools (e.g., the debugger slows execution to a point where the bug no longer occurs [Eisenstadt 1993]). Given the practical importance of such tools, it is surprising that few researchers have studied their impact on software debugging or maintenance.

A recent study of software enhancement tasks [Rosson and Carroll 1993, in press] takes a step in this direction. These researchers studied expert Smalltalk programmers working in their natural programming environment, which is populated with sophisticated interactive tools for browsing and manipulating both the code associated with a program and its run-time state and execution behavior. The programmers exhibited behavior generally consistent with programmers asked to analyze or modify textual program listings: they relied on a goal-driven comprehension process, focusing on just those aspects of the software judged to be relevant to their current modification goal. However, they also relied extensively on the tools in the environment to suggest hypotheses and to direct them to other useful information. For example, rather than reading through all of the code relevant to the proper instantiation and initialization of a new window, they analyzed just enough code to take one or a few steps and then tested the incomplete solution to see where the next error would occur, using the resulting debugger messages as a specification for the next piece of code to find and analyse. Rosson and Carroll [1993] termed this highly iterative and incremental analysis process **debugging into existence** and speculated that it reflected a well-developed strategy for minimizing the amount of code comprehension needed to make a program modification.

Of course, such optimizing behavior may not always be desirable. If programmers fail to comprehend large portions of the system being modified, they may introduce unknown side effects. The programmers studied by Rosson and Carroll [1993] were accustomed to the rapid prototyping style of software development, in which the speed of achieving a new level of functionality that can be tested and refined is valued more highly than production of a robust solution. Nonetheless, even these programmers voiced concern about their heavy reliance on tools to refine their analysis and solution of the problem, noting that it was not the accepted approach to such problems (though at the same time they indicated that this was their normal approach). But at least for a certain class of programming situations, i.e., those emphasizing rapid prototyping, this research emphasizes the important role of interactive debugging tools in goal-directed software comprehension.

Coordinating Software and Problem Domains

A component of software comprehension that is especially relevant to software maintenance tasks is the process of mapping between program code and the problem requirements. Pennington [1987a] reports, that high-performing programmers carrying out a modification task were much more likely than low performers to construct coordinated models of the program and of the problem domain; the verbal protocols of the good performers contained evidence of a cross-referencing strategy in which they explicitly connected program elements to features of the problem statement. Good programmers make an effort to understand *why* a particular feature of the solution exists; in Eisenstadt's [1993] study of debugging anecdotes, the most common account of what made a bug *hard* was a large distance between the cause and the effect of the bug.

The implication for software debugging and maintenance is that relevant problem information should be integrated as much as possible with the code statements it engenders. A study by Letovsky, Pinto, Lampert, and Soloway [1987] found that professional code inspectors often test their ability to reconstruct such connections and suggest code changes that would accent problem solution linkages. For example, one suggestion made by inspectors in this study was to replace *tricky* code developed for efficiency reasons with a more conventional solution that was more obviously linked to problem requirements. This raises the interesting possibility that programmers may use standard problem clues that are analogous to the beacons, indicating the presence of particular programming plans.

The integration of problem information with the program code it produces represents a form of design rationale. Techniques for encouraging and supporting the development of design rationale have received considerable attention over the past few years (Moran and Carroll 1996; see also Chapter 66), and although most of this attention has been directed at documenting decisions about functionality or high-level architectural concerns, a few researchers have begun to explore the value of connecting rationale to bits of code using hyperlinks (e.g., documenting the pros and cons of choosing a particular data structure [Rosson and Carroll 1995b]). Of course, such information is simply a modern variant of in-line comments; the difference is that hyperlinks do not obstruct the analysis of the code but rather can be viewed on an as-needed basis. Because of the textual nature of these links, it is also unclear whether programmers would write or read them any more than they write or read other documentation.

Proponents of OOP suggest that its emphasis on problem-oriented rather than system-oriented decomposition might facilitate software comprehension [Rosson and Alpert 1990]. Although this claim may have some face validity, object-oriented systems typically have other characteristics that impede comprehension: the control structure of object-oriented software tends to be highly distributed across a variety of semiautonomous object abstractions which are implemented as part of a complex abstraction hierarchy [Taenzer et al. 1989]. Thus, any advantage accruing from better integration of problem and solution may be masked by complexities stemming from object collaboration and inheritance patterns. An early study contrasting maintenance task performance for object-oriented vs. functional decomposition in fact observed longer maintenance task times for the object-oriented code [Boehm-Davis et al. 1992].

However, the test problems in this program-editing study were quite modest in size (e.g., 100–250 lines) and the researchers commented that their implementation of the object-oriented methodology may not have been ideal. It appears that it is still too soon to conclude whether OOP will facilitate, inhibit, or have no consequence for software comprehension.

Predicting Maintenance Costs

Another strand of work within the general area of software maintenance has focused not on the behavior of programmers comprehending or modifying software but rather on measuring the complexity of the software to be maintained. The goal has been to develop measures, especially metrics computed algorithmically from the code itself, that will predict variations in software comprehension effort and thus be useful in estimating the cost of maintaining the software over its life cycle.

The simplest measure of complexity is a count of the number of lines of code. However, even this simple metric can be difficult to gather reliably, as it requires principled decisions about how to count lines devoted to comments, nonexecutable lines such as headers or data declarations, statements that run over multiple physical lines or that are compound statements such as conditionals or case statements. Furthermore, this simple measure misses complexity issues arising from variations in data structures, both number and variety, choice points, and so on.

Software engineers have proposed a variety of more elaborate metrics that encode information about the elements composing a program. For example, McCabe's [1976] measure of cyclomatic complexity captures the number of distinct control paths through a program. An alternative set of metrics proposed by Halstead [1977] instead quantifies a program's unique elements (i.e., operators and operands). Such metrics appear to do a reasonable job in accounting for complexity differences in small programs but are unable to capture the complexities introduced by module relationships and the overall high-level design of large software systems [Boehm-Davis 1988]. In this respect, one promising direction is the development of metrics that combine low-level and high-level assessments of complexity (see, e.g., Harrison and Cook [1986]).

Teaching and Learning Programming Skills

A long-standing concern of researchers studying the psychology of programming has been how learners acquire the knowledge and skills of programming. To some extent, such research has been carried out to advance theories about the content and nature of expertise in programming; by contrasting novice and expert programming behavior and analyzing how programming behavior changes with increased expertise, researchers can enhance their understanding of asymptotic or expert knowledge [Davies 1993]. Of more interest here is research on the education process itself: how best to teach programming skills, how to facilitate learning subsequent languages after basic skills are established, or, more generally, how to organize a curriculum to educate software professionals.

Learning from Examples

Much programming instruction focuses on conveying the syntactic features of a particular programming language and on specifying and implementing small, well-defined algorithms or data structures [Shaw 1990]. Low-level knowledge such as this is relatively easy to describe and can be exercised and evaluated by relatively small student programs in a relatively small period of time. However, as researchers studying computer science education have pointed out, the much more difficult problem is finding ways to educate novices on *design* skills that will support the analysis, construction, and development of software systems of realistic size [Linn and Clancy 1992a].

One promising approach to teaching design skills relies on student exploration of expert case studies [Linn and Clancy 1992a, 1992b]. This approach is rooted in the observation that successful educators will often model the design process for their students, including, for example, the consideration and decisions concerning alternative designs and perhaps even the intentional commission and recovery from errors or misconceptions [Lalonde 1993]. The goal of the case studies then is to provide such a model in a more systematic fashion and in a way that makes it accessible to a range of students in varied learning settings.

The case studies developed by Linn and Clancy [1992a] consist of a series of documented problem solution examples that cumulate in a way to demonstrate reuse of design approaches. Each case study in the series includes a statement of the problem, a narrative description of an expert's solution written in terms that a student can understand, the actual worked out solution associated with this narrative, study questions that guide students' analysis of the case, and test questions that assess students' final understanding. The case studies illustrate generally useful design heuristics, such as divide-and-conquer, or the use of multiple representations. These researchers have demonstrated that by providing the expert commentary along with the other problem information, students develop a more integrated understanding of programming and learn to attend to aspects of the design process normally ignored or avoided by novices, for example, the consideration of design alternatives and the development of design rationale.

A study by Riecken et al. [1991] offers an interesting perspective on the nature of commentary that might be expected from experts assisting novice programmers. According to these researchers, experts seem to believe that novices need help in understanding the function and behavior of complex programs, whereas novices feel that they should receive more assistance with understanding syntactical structures. This difference in emphasis is consistent with the knowledge level of the two populations: experts have moved beyond syntactic details, whereas novices are often struggling simply to parse the written code. For educators working with experts to develop case studies, the implication is clear: the experts may need explicit encouragement to include scaffolding for the simpler syntactic skills of code reading along with the design-oriented concerns which are salient to them.

Other researchers have taken a more tool-based approach to teaching programming and design skills by example. Carroll and Rosson [1991] discuss the structure and rationale of a View Matcher for teaching Smalltalk programming and design. This tool provides multiple coordinated views of a running application that exemplifies paradigmatic object-oriented design constructs. The multiple views support guided exploration and subsequent refinement of the example; one of the views presents integrative commentary similar to that included in Linn and Clancy's [1992a, 1992b] case studies. More recently, Rosson and Carroll [1996] discuss an extension of their earlier work, the Scenario Browser environment for teaching object-oriented design through **scaffolded examples**. Scaffolded examples are sample problems of realistic size whose complexity is gradually revealed in steps that delineate and reinforce the intrinsic structure of the problem-solving process.

Learning New Languages and Paradigms

Most professional programmers work with more than one programming language over the course of their careers. In general, there is an expectation that as expertise increases, programmers will need less and less support in learning new languages; even at the university level, student programmers are often expected to learn a third or fourth programming language on their own after formal instruction in their first few courses.

In fact, it is relatively easy to demonstrate some degree of savings or positive transfer from the learning of one language to the next. There are many general aspects of programming skill that should facilitate learning a new language: for example, Wu and Anderson [1991] demonstrated that subjects already familiar with Lisp were better able to learn the use of recursion in Prolog, even though the detailed programming plans are quite different (see also Scholtz and Wiedenbeck [1990]). However, the process of transferring one's expertise from one language to another is not always straightforward and can lead to subtle disfluencies in the use of the new language.

As long as programmers are learning a new language that is isomorphic in structure to a language they already know well, the learning will progress smoothly. But if the new language differs in detail, and in particular if it involves the use of different programming plans, expert programmers may need to unlearn the no-longer-applicable aspects of the old language. Scholtz and Wiedenbeck [1992] and Scholtz [1993] demonstrated that even after considerable exposure to a new language, programmers who had prior experience with a different language exhibited programming plans that were appropriate for the prior language but suboptimal for the new language.

This problem is an instance of the general situation that Carroll and Rosson [1987] have termed the **as-similation paradox**. Experienced programmers possess conceptual and strategies knowledge that they will use to incorporate—or assimilate—information about a new language. At the same time that it is aiding them in this fashion, however, the experts' prior programming knowledge may limit or bound their understanding and exploration of the new language. The natural instinct is to try to map from what is known to the new possibilities, but in some instances it is better to simply start over and learn an entirely new method.

It is difficult for designers of documentation and instructional materials to anticipate and address the effects of prior programming knowledge. Instructional designers naturally wish to make the new material seem more familiar and understandable. In a study of a Smalltalk tutorial, Koenemann-Belliveau, Carroll, Rosson and Singley [1994] observed repeated bits of instruction using an analogy to Pascal data types and attributes when introducing the basic concept of an object. On the one hand, these references were easily understood by experienced procedural programmers and thus helped the learning experience to go more smoothly. An instructional designer evaluating the success of the training might well observe this and conclude that the analogy was serving a useful function. On the other hand, the repeated references cumulated across the learning period and produced an inappropriately narrow view of classes and instances. One solution might be to use such analogies initially, but then to carefully expose their limitations later in the instruction.

Evolving the Computer Science Curriculum

Beyond the direct question of how to teach use of a particular programming language or technique, educators have considered how best to teach such skills as part of a general computer science curriculum at the university level. The *ACM/IEEE Computing Curricula '91* report [Tucker et al. 1991] highlights design as one of three fundamental processes in which students should receive training; it also advocates an early emphasis on software engineering methods and concerns. This is in contrast to many traditional computer science programs which have targeted initial instruction more narrowly on low-level programming skills, on programming-in-the-small rather than programming-in-the-large [Shaw 1990].

Many educators have taken these new recommendations as a call for curriculum reform and have begun exploring the impact of new courses, especially at the introductory level. For some, the call for an early emphasis on software engineering concerns has prompted efforts to "turn the curriculum upside-down" [Rosson and Heliotis 1993], incorporating team projects and considerations of social and ethical issues in software development into the first two years of instruction rather than waiting until the third and fourth years [Epstein and Tucker 1993, Tewari and Friedman 1993]. Others advocate a more radical approach, introducing students early on (e.g., in their second semester) to relatively large-scale application-development projects that rely on code libraries and high-level programming environments [Lalonde 1993]. Such an approach allows students to confront from the beginning the problems of understanding problem requirements and of developing and considering design alternatives.

The appearance and general acceptance of the object-oriented paradigm has been a major enabling factor in the evolution toward a broader and more design-centered computer science curriculum [Osborne 1993]. Object-oriented languages typically provide rich libraries of reusable components which can serve both as expert examples of well-conceived abstractions and as the building blocks for larger application-oriented class projects. The object-oriented paradigm also emphasizes the software design concerns of abstraction and reuse, which were among the themes highlighted by *Computing Curricula '91* [Tucker et al. 1991].

Programming in the Real World

The analysis of software development activities in the real world is a challenge. The possible factors affecting the behavior performance are many and complex and often cannot be predicted in advance. As a result, most behavioral studies of programming have avoided these real-world contexts; unfortunately, this means that little data are available concerning the variables influencing individual and team behavior in realistic settings [Curtis and Walz 1990]. However, researchers are beginning to move their focus from relatively artificial laboratory settings into the real world; for example, the recruitment of professional programmers rather than university students as participants in behavioral studies has increased dramatically over the past 10 years [Cook et al. 1993]. A few researchers are also beginning to tackle the problem of untangling the factors influencing the behavior of software development teams.

Software Productivity

One approach to studying the success of software development projects has been to assess the impact of various factors, for example, the use of particular design paradigms, tools, or variations in team composition, on the productivity of the group. This approach depends on reliable measures of software productivity, which are inherently difficult to obtain due to the tradeoffs among many possible project goals [Boehm 1987]. As an obvious example, the speed at which the code is developed trades off against the quality of the code produced. In general, a team will optimize their development process for goals which are emphasized (e.g., speed, reliability, documentation) but at the expense of other goals. Thus, what it means to be productive is necessarily much influenced by the software culture within which the programming team operates.

Software productivity is an assessment of project output as a function of resources input to the project, where output is normally assessed via some sort of code volume metric; for example, one popular measure of productivity is *delivered source instructions per person-month* of effort. Whereas such a measure is clearly simplistic and does not directly address issues of reliability, requirements satisfaction, and so on, it does at least capture at a gross level what a team has produced. The counting of instructions is also a more straightforward measure to calculate than one based on semantic or structural characteristics of the code (e.g., function points). To the extent that the delivery of code is tied to standards for software quality, even a simple measure such as this can be sensitive to quality concerns. Using such measures, researchers have developed models which suggest if project size is held constant, the most important contributing factor may be personnel capability [Boehm 1987].

Exceptional Designers

Curtis et al. [1988] describe a field study of 17 projects in nine companies. These researchers have formulated a layered behavioral model of software development in which issues are identified and analyzed at multiple levels, from psychological factors influencing the individual team member, to group dynamics influencing team and project work, to organizational variables influencing the company and business milieu. One of the variables the researchers analyzed within this framework was the important role played by team members possessing application expertise.

At the level of individual group members, deep application knowledge was observed to be critical to the team's success but not always available. The fortunate teams contained one or more *exceptional designers* or *project gurus*. These individuals were not necessarily high-performing programmers but rather tended to be interdisciplinary and able to integrate overlapping bodies of knowledge to create and maintain an overall vision for the project. The presence of such individuals on a team influenced team dynamics, as they contributed a great amount to group discussions, and quickly formed small but powerful coalitions that helped to control the direction of the team. Beyond controlling team direction, these individuals also played an important role in educating their colleagues; substantial design effort was expended on communicating application expertise and on sharing design visions within and across team boundaries.

Communication and Coordination

Although having smart people, particularly those with deep application knowledge, is critical to the success of software development projects, it is not enough. Even when a team is fortunate enough to contain one or more project gurus, the vision created and maintained by these individuals must still be communicated to others on the team and to other teams in the project. Thus, mechanisms for communication are key determinants of the project's ultimate success.

Curtis et al. [1988] described how communication and coordination processes are crucial in the management of fluctuating and conflicting requirements. Even in cases where the business requirements were stable, specifications sometimes fluctuated because teams applied different interpretations to interdependent requirements and did not sufficiently coordinate their design efforts. At the individual level, instability was sometimes introduced by programmers who went beyond the official requirements to add system features in response to their own programming or design aesthetics. An emphasis on communication and coordination at all phases of the development effort was critical for uncovering and resolving such inconsistencies.

The Curtis et al. [1988] field study also provided several interesting observations concerning effective communication mechanisms. Documentation and formal communication processes seemed to be relatively ineffective as communication media, especially as the size and complexity of a project increased. Instead, project members relied on informal verbal communication and in particular on the development and maintenance of appropriate communication networks. A key ingredient for large projects was the emergence of boundary spanners, project members who could translate the concerns and proposals of one team into language understood by another.

Communication Patterns

Design team meetings are a rich source of information about how team members communicate, and a few researchers have carried out detailed analyses of videotaped design meetings in an effort to better understand and support such communication. For example, Curtis and Walz [1990] describe an analysis in which each utterance was categorized according to its role in design meetings that took place over several months. They found an interesting pattern with respect to utterances representing agreement among participants. In contrast to a simplistic model, which would predict gradually increasing agreement as the team developed and refined a solution, the researchers report an inverted U-shaped curve: agreement gradually increased until the point at which a specification document was released and then began to decrease after that. The researchers acknowledge that many factors may have produced this pattern: for example, it may be necessary to reach consensus to produce the specifications, but then the team can relax and reopen conflicts raised earlier, or it may have been that the initial consensus process addressed a different, perhaps more abstract, level of analysis. However, the work is a first step toward understanding the communication patterns a group might anticipate.

Berlin [1993] examined communication patterns for a specific type of collaboration, that of experts mentoring apprentices about a complex project. She reported an informal conversational style, where the topics and threads are implicitly negotiated by both the expert and the apprentice. The experts volunteer information that they think is key to understanding a system issue but also follow tangents generated by apprentice problem solving or by external artifacts (e.g., details of a piece of code). Thus, the successful mentors were those who possessed not simply expertise but also appropriate collaborative conversational skills. Interestingly, Berlin also observed that one characteristic distinguishing the experts from the apprentices was that they were quicker to seek the expert advice of others when relevant.

Olson and his colleagues [Olson et al. 1992, 1996, Herbsleb et al. 1995] have carried out several analyses of videotaped discussions from design meetings conducted for several projects in different organizations. Like Curtis and Walz, these researchers have defined categories (e.g., issue, alternative, clarification) to use in encoding utterances; they have then examined the frequency of utterance types as well as the transitions from one type of utterance to another. These analyses confirm the importance of communication and coordination; in one study, approximately one-third of discussion time was spent in clarifying the content of questions raised. Another significant portion of meeting time was directed at meeting and project management, again suggesting the importance of better tools for communicating and coordinating the activities of team members. However, these researchers also report considerable structure in the conversations taking place during design meetings: in one analysis over two-thirds of the utterances dealt with discussion of design alternatives and their criteria. The researchers suggest that this relatively structured discussion might be facilitated by a tool that captures and organizes design rationale (see, e.g., Conklin and Begeman [1988]).

73.4 Research Issues and Summary

Behavioral analyses of programming and software development have evolved considerably from the early days when undergraduates were asked to comprehend or generate programs just a few lines in length. Researchers now routinely study experts and are beginning to attend to the behavior of experts working in teams. However, significant research challenges remain, as software technology continues to advance and the nature of programming and programmer populations along with it.

From Plans to Application Frameworks

Most analyses of programming skill consider the development of programming plans to be a fundamental component of expertise. The content of the plans articulated in these analyses has been at a relatively low level, for example, capturing stereotypical mechanisms for initializing variables or for controlling loops or recursions. Such knowledge units are relatively easy to describe and to detect in expert behavior; in particular, they correspond to concrete lines of code. And although even these early studies recognized the importance of domain knowledge [Adelson and Soloway 1985], no serious attempt has yet been made to model such knowledge.

Now, however, the paradigm of software development is maturing. Programmers do still on occasion write code from scratch, and certainly they still must learn basic constructs such as variable initialization and loop control. But with the advent of high-level tools, reusable component libraries, and application builders, the focus of software development is shifting from code generation to code reuse, to programming by construction or composition of parts. Cognitive analyses of programming behavior must similarly raise the level of analysis; the field needs techniques for characterizing and modeling the more abstract and diffuse knowledge structures that support use of these high-level tools.

A promising conceptualization from the OOP community is the notion of *application frameworks*. These are rich collections of interrelated abstractions (classes) that have been designed to work together in stereotypical ways [Deutsch 1989]. Sometimes such frameworks are designed explicitly; more often they emerge from repeated project work within an application domain. An interesting research direction might be to study experts' acquisition and use of such frameworks. The resulting analyses would both suggest ways in which application frameworks might be more usefully designed (i.e., so that they can be better comprehended, extended, and so on) and point to tools or documentation that might be helpful in conveying an application framework and its use to nonexperts. Fischer and his colleagues are already exploring related ideas in their work on domain-oriented design environments [Fischer 1994, Fischer and Lemke 1988].

An Expanding User Population

In response to critics who charged that the behavior of student programmers had little relevance to software development in the real world, researchers have shifted their focus from students to professional programmers. Computer science students are now studied only when the goal is to contrast their relatively undeveloped skills with those of experts, or when the research questions are directed at problems in education. Also in response to critics questioning the validity of results obtained for individual programmers, researchers are expanding their analyses to group behavior, developing techniques for characterizing the distributed and asynchronous problem solving that takes place in software development teams. At the same time, however, an interest in novice programming is re-emerging. But the novices in this case are end users, individuals who have no intention of learning how to program [Nardi 1993]. As a result, an entirely new research area on end user programming has surfaced.

An active research topic in end user programming is the exploration of demonstrational programming systems [Cypher 1993]. Such techniques allow end users to simply show their software what they want to happen as a means of customizing or extending it. The goal is to provide basic programming capabilities without requiring the user to learn a complicated syntax or special terminology. But although programming-by-demonstration systems are rapidly evolving, the research focus has been almost entirely on technology for supporting demonstration capture and interpretation. There is a tremendous opportunity for behavioral researchers to begin characterizing the knowledge and skills acquired by end users interacting with these pseudoprogramming environments.

Coordinating Systematic and Creative Design

Despite the conventional wisdom that robust and reliable systems are best produced by a systematic and hierarchical decomposition, experts often depart from this ideal, especially when presented with a novel or difficult problem. They work at multiple levels simultaneously, building partial solutions to subproblems creatively and opportunistically and then integrating at a later point. The increasing availability of component-based software environments, such as Visual Basic, and sophisticated interactive programming environments, such as Smalltalk, seem likely to encourage such strategies even further. But whereas a highly creative and opportunistic design approach can produce rapid and effective results on an individual or small-group level, these results often must still be integrated and rationalized within a larger project context; we have seen that group members spend a great deal of time communicating and coordinating their work. Thus, a challenge for researchers is to find ways of balancing heterogeneous approaches on the individual level with a more systematic approach at the group or project level.

An obvious approach to such a coordination requirement is documentation: if an individual documents the problem solving experiences that produce a piece of code or a design, group members can understand its rationale and implications and better predict its dependencies or constraints on other aspects of the project. The problem of course is that programmers and designers are notoriously poor at producing and maintaining documentation. One solution might be to employ methods or tools that produce useful documentation implicitly, as a side effect of the design and programming work itself. For example, Rosson and Carroll [1995a, 1995b] describe an application development environment organized around *user scenarios* (see also Chapter 66). In the course of design and implementation, a set of scenarios is successively elaborated from a user task specification into an object-oriented implementation. The result is both a software design and a scenario-based record of how that design was achieved. Although it is too soon to know how useful such implicit documentation will be in integrating specific results within a larger group project, it is clear that scenarios are becoming increasingly popular as a vehicle for sharing design ideas and rationale throughout the software development life cycle [Carroll 1995].

Summary and Conclusions

Software psychology—the behavioral study of programmers and designers—has matured significantly in the past 20 years. Researchers have accepted the challenge of studying real experts working on real programming and design tasks in both individual and group settings. However, many challenges remain. As the software technology available to programmers becomes more powerful and complex, so will the behavior and cognitive processes of the experts who use them; new techniques for modeling and interpreting these activities are required. As the level of abstraction and domain content of programming environments increases, so will the variability of the user population and tasks they attract. Perhaps the greatest challenge, however, is the analysis of collaborative programming and design, as projects struggle to find methods and tools that simultaneously support the skills and strategies of individual programmers and the needs of the group as a whole.

Defining Terms

Application frameworks: A set of classes designed to work together in a stereotypical way in support of a range of related applications. Typically, designers specialize or instantiate the classes of the framework in building new applications.

Assimilation paradox: The observation that users bring prior knowledge to their interactions with systems that at once helps them to make sense of the new information they encounter and places boundaries on their interpretation of that information.

Beacons: Signature lines or expressions that suggest the presence of a programming plan.

Debugging into existence: Creation or modification programming episodes in which the programmer makes progress incrementally, in a trial-and-error fashion guided by feedback from an interactive debugger.

Design patterns: Descriptions of communicating objects and classes that are customized to solve a general design problem in a particular context.

Design schema: Abstract knowledge structures that contain information about a class of design solutions, the elements at each level, and their decomposition into subelements.

Distributed cognition: A complex system of collaborating users, their internal mental activities, and their shared externalized task representations.

End user programming: Interactive systems that support some degree of customization or specialization by the end user without resorting to conventional (i.e., third generation) programming languages.

Mental simulation: Simulating the execution of all or part of a program or algorithm as a means of testing its correctness.

Opportunistic planning: Planning episodes in which subproblems at different levels are analyzed simultaneously, with alternation among levels determined by emerging features of the solution.

Premature commitment: The extent to which a decision about a programming plan or implementation must be made before its consequences can be seen.

Programming plan: An abstract structure that links the goals of a program's specific computational structures for implementing these goals.

Role expressiveness: The ease with which a programmer can convey or discover the purpose of a program element.

Scaffolded examples: Sample problems of realistic size whose complexity is gradually revealed in steps that delineate and reinforce the intrinsic structure of the problem-solving process.

Software psychology: The psychological, especially cognitive, analysis of programming behavior.

Systematic expansion: Breadth-first decomposition in which problems are decomposed into subproblems one level at a time, with each level only a bit more detailed than the previous one.

Viscosity: The degree to which a language or notation resists local changes.

References

Adelson, B. and Soloway, E. 1985. The role of domain experience in software design. *IEEE Trans. Software Eng.* SE-11:233–242.

Altmann, E. M., Larkin, J. H., and John, B. J. 1995. Display navigation by and expert programmer: a preliminary model of memory, pp. 3–10. In CHI'95 Conf., *Proc. Hum. Factors Comput. Syst.* ACM, New York.

Berlin, L. 1993. Beyond program understanding: a look at programming expertise in industry. In *5th Workshop, Empirical Stud. Programming.* C. R. Cook, J. C. Scholtz, and J. C. Spoher, eds., pp. 6–25. Ablex, Norwood, NJ.

Boehm, B. W. 1987. Improving software productivity. *Computer* 20(9):43–57.

Boehm-Davis, D. A. 1988. Software comprehension. In *Handbook of Human–Computer Interaction*, M. Helander, ed., pp. 107–122. North-Holland, New York.

Boehm-Davis, D. A., Holt, R. W., and Schultz, A. C. 1992. The role of program structure in software maintenance. *Int. J. Man–Mach. Stud.* 36(1):21–63.

Brooks, R. 1983. Towards a theory of the comprehension of computer programs. *Int. J. Man–Mach. Stud.* 18:543–554.

Carroll, J. M. 1995. *Scenario-Based Design: Envisioning Work and Technology in System Development.* Wiley, New York.

Carroll, J. M. and Rosson, M. B. 1987. The paradox of the active user. In *Interfacing Thought: Cognitive Aspects of Human–Computer Interaction*, J. M. Carroll, ed., pp. 80–111. MIT Press, Cambridge, MA.

Carroll, J. M. and Rosson, M. B. 1991. Deliberated evolution: stalking the view matcher in design space. *Hum.–Comput. Interaction* 6:281–318.

Carroll, J. M., Thomas, J. C., and Malhotra, A. 1979. Clinical-experimental analysis of design problem solving. *Design Stud.* 1:84–92.

Carroll, J. M., Thomas, J. C., Miller, L. A., and Friedman, H. P. 1980. Aspects of solution structure in design problem solving. *Am. J. Psych.* 93(2):269–284.

Conklin, E. J. and Begeman, M. L. 1988. gIBIS: a hypertext tool for exploratory policy discussion. *ACM Trans. Office Inf. Syst.* 6:303–331.

Cook, C. R., Scholtz, J. C., and Spoher, J. C. 1993. Preface. In *5th Workshop, Empirical Studies of Programming.* Ablex, Norwood, NJ .

Cox, B. J. 1986. *Object Oriented Programming: An Evolutionary Approach.* Addison–Wesley, Reading, MA.

Crosby, M. E. and Stelovsky, J. 1990. How do we read algorithms? A case study. *Computer* 23(1):24–35.

Curtis, B. 1988. Five paradigms in the psychology of programming. In *Handbook of Human–Computer Interaction.* M. Helander, ed., pp. 87–106. North-Holland, New York.

Curtis, B., Krasner, H., and Iscoe, N. 1988. A field study of the software design process for large systems. *Commun. ACM* 31(11):1268–1287.

Curtis, B., Soloway, E. M., Brooks, R. E., Black, J. B., Ehrlich, K., and Ramsey, H. R. 1986. Software psychology: The need for an interdisciplinary program. *Proc. IEEE* 74:1092–1106.

Curtis, B. and Walz, D. 1990. The psychology of programming in the large: team and organizational behavior. In *Psychology of Programming.* J.-M. Hoc, T. R. G. Green, R. Samurçay, and D. J. Gilmore, eds., pp. 253–270. Academic Press, London.

Cypher, A. 1993. *Watch What I Do: Programming by Demonstration.* MIT Press, Cambridge, MA.

Davies, S. P. 1993. Models and theories of programming strategy. *Int. J. Man–Mach. Stud.* 39(2):269–304.

Détienne, F. and Rist, R. 1995. Introduction to this special issue on empirical studies of object-oriented design. *Hum.–Comput. Interaction* 10(2, 3):121–128.

Deutsch, L. P. 1989. Design reuse and frameworks in the Smalltalk-80 system. In *Software Reusability: Applications and Experience.* Vol. 2, T. J. Biggerstaff and A. J. Perlis, eds., pp. 57–72. ACM, New York.

Dijkstra, E. 1968. GO TO considered harmful. *Commun. ACM* 11(3):147–148.

Eisenstadt, M. 1993. Tales of debugging from the front lines. In *5th Workshop, Empirical Studies of Programmers.* C. R. Cook, J. C. Scholtz, and J. C. Spoher, eds., pp. 86–112. Ablex, Norwood, NJ.

Epstein, R. G. and Tucker, A. B. 1993. Introducing object-orientedness into a breadth-first introductory curriculum. *Comput. Sci. Educ.* 4(1):35–44.

Fischer, G. 1987. Cognitive view of reuse and redesign. *IEEE Software* 4(3):60–72.

Fischer, G. 1994. Domain-oriented design environments. In *Automated Software Engineering.* Vol. 1, L. Johnson and A. Finkelstein, eds., pp. 177–203. Kluwer, Boston, MA.

Fischer, G. and Lemke, A. C. 1988. Construction kits and design environments: steps toward human problem-domain communication. *Hum.–Comput. Interaction* 3:179–222.

Fischer, G., Redmiles, D., Williams, L., Puhr, G. I., Aoki, A., and Nakakoji, K. 1995. Beyond object-oriented technology: where current approaches fall short. *Hum.–Comput. Interaction* 10(1):79–119.

Flor, N. V. and Hutchins, E. L. 1991. Analyzing distributed cognition in software teams: a case study of team programming during perfective software maintenance. In *4th Workshop, Empirical Studies of Programmers.* J. Koenemann-Belliveau, T. G. Moher, and S. P. Robertson, eds., pp. 36–64. Ablex, Norwood, NJ.

Gamma, E., Helm, R., Johnson, R., and Vlissides, J. 1995. *Design Patterns: Elements of Reusable Object-Oriented Software.* Addison–Wesley, New York.

Gellenbeck, E. M. and Cook, C. R. 1991. Does signalling help professional programmers read and understand computer programs? In *4th Workshop, Empirical Studies of Programmers.* J. Koenemann-Belliveau, T. G. Moher, and S. P. Robertson, eds., pp. 82–98. Ablex, Norwood, NJ.

Green, T. R. G. 1990. Programming languages as information structures. In *Psychology of Programming,* J.-M. Hoc, T. R. G. Green, R. Samurcay, and D. J. Gilmore, eds., pp. 117–138. Academic Press, London.

Gugerty, L. and Olson, G. M. 1986. Comprehension differences in debugging by skilled and novice programmers. In *Empirical Studies of Programmers*, E. Soloway and S. Iyengar. eds. Ablex, Norwood, NJ.

Guindon, R. 1990. Designing the design process: exploiting opportunistic thoughts. *Hum.–Comput. Interaction* 5:305–344.

Halstead, M. H. 1977. *Elements of Software Science*. North-Holland, New York.

Harrison, W. and Cook, C. 1986. A Micro/Macro Measure of Software Complexity. *Tech. Rep.* TR-86-60-3. Oregon State University, Corvallis.

Hayes-Roth, B. and Hayes-Roth, F. 1979. A cognitive model of planning. *Cognitive Sci.* 3:275–310.

Helm, R. and Maarek, Y. S. 1991. Integrating information retrieval and domain specific approaches for browsing and retrieval in object-oriented class libraries. In *Proc. Object-Oriented Programming, Syst. Appl.*, pp. 47–61. ACM, New York.

Herbsleb, J. D., Klein, H., Olson, G. M., Brunner, H., Olson, J. S., and Harding, J. 1995. Object-oriented analysis and design in software project teams. *Hum.–Comput. Interaction* 10(2, 3):249–292.

Hüni, H., Johnson, R., and Engel, R. 1995. A framework for network protocol software. In *Proc. OOPSLA'95*. ACM, New York.

Hutchins, E. L. in press. *Distributed Cognition*. MIT Press, Cambridge, MA.

Jeffries, R., Turner, A. S., Polson, P., and Atwood, M. E. 1981. The processes involved in designing software. In *Cognitive Skills and Their Acquisition*, J. R. Anderson, ed., pp. 225–283. Erlbaum, Hillsdale, NJ.

Koenemann, J. and Robertson, S. P. 1991. Expert problem solving strategies for program comprehension. In CHI'91 Conf., *Proc. Human Factors Comput. Syst.* S. P. Robertson, J. S. Olson, and G. M. Olson, eds., pp. 125–130. ACM, New York.

Koenemann-Belliveau, J., Carroll, J. M., Rosson, M. B., and Singley, M. K. 1994. Comparative usability evaluation: critical incidents and critical threads, In CHI'94, *Proc. Human Factors Comput. Syst.*, pp. 245–251. ACM, New York.

Lalonde, W. 1993. Making object-oriented concepts play a central role in academic curricula. *Comput. Sci. Educ.* 4(1):13–24.

Lange, B. M. and Moher, T. G. 1989. Some strategies of reuse in an object-oriented programming environment. In CHI '89 Conf., *Proc. Human Factors Comput. Syst.*, pp. 69–74. ACM, New York.

Letorsky, S., Pinto, J., Lampert, R., and Soloway, E. 1987. A cognitive analysis of a code inspection. In *2nd Workshop, Empirical Studies of Programmers*. G. M. Olson, S. Sheppart, and E. Soloway, eds., pp. 231–248. Ablex, Norwood, NJ.

Linn, M. C. and Clancey, M. J. 1992a. Can experts' explanations help students develop program design skills? *Int. J. Man–Mach. Stud.* 36(4):511–552.

Linn, M. C. and Clancey, M. J. 1992b. The case for case studies of programming problems. *Commun. ACM* 35(3):121–132.

Littman, D. C., Pinto, J., Letovsky, S., and Soloway, E. 1986. Mental models and software maintenance. In *Empirical Studies of Programmers*, E. Soloway and S. Iyengar, eds. Ablex, Norwood, NJ.

Mantei, M. 1981. The effect of programming team structures on programming tasks. *Commun. ACM* 21(9):760–768.

McCabe, T. J. 1976. A complexity measure. *IEEE Trans. Software Eng.* SE-2:308–320.

Meyer, B. 1988. *Object-Oriented Software Construction*. Prentice-Hall, New York.

Moran, T. P. and Carroll, J. M. 1996. *Design Rationale: Concepts, Techniques, and Use*. Lawrence Erlbaum Associates, Mahwah, NJ.

Nardi, B. A. 1993. *A Small Matter of Programming*. MIT Press, Cambridge, MA.

Newell, A. 1990. *Unified Theories of Cognition*. Harvard University Press, Cambridge, MA.

Olson, G. M., Olson, J. S., Carter, M. R., and Storrøsen, M. 1992. Small group design meetings: an analysis of collaboration. *Hum.–Comput. Interaction* 7:347–374.

Olson, G. M., Olson, J. S., Storrøsen, M., Carter, M., Herbsleb, J., and Reuter, H. 1996. The structure of activity during design meetings. In *Design Rationale: Concepts, Techniques, and Use*, T. P. Moran and J. M. Carroll, eds., pp. 217–240. Erlbaum, Hillsdale, NJ.

Osborne, M. 1993. Computing curricula 1991 and the case for object-oriented methodology. *Comput. Sci. Educ.* 4(1):25–34.

Pennington, N. 1987a. Comprehension strategies in programming. In *2nd Workshop, Empirical Studies of Programming*. G. M. Olson, S. Sheppard, and E. Soloway, eds., pp. 100–113. Ablex, Norwood, NJ.

Pennington, N. 1987b. Stimulus structures and mental representations in expert comprehension of computer programs. *Cognitive Psychol.* 19:295–341.

Pennington, N. and Grabowski, B. 1990. The tasks of programming. In *Psychology of Programming*, J.-M. Hoc, T. R. G. Green, R. Samurçay, and D. J. Gilmore, eds., pp. 45–62. Academic Press, London.

Pennington, N., Lee, A. Y., and Rehder, B. 1995. Cognitive activities and levels of abstraction in procedural and object-oriented design. *Hum.–Comput. Interaction* 10(2, 3):171–226.

Petre, M. 1990. Expert programmers and programming languages. In *Psychology of Programming*, J.-M. Hoc, T. R. G. Green, R. Samurçay, and D. J. Gilmore, eds., pp. 103–116. Academic Press, London.

Riecken, R. D., Koenemann-Belliveau, J., and Robertson, S. P. 1991. What do expert programmers communicate by means of descriptive commenting? In *4th Workshop, Empirical Studies of Programmers*. J. Koenemann-Belliveau, T. G. Moher, and S. P. Robertson, eds., pp. 177–195. Ablex, Norwood, NJ.

Rist, R. S. 1990. Variability in program design: the interaction of process with knowledge. *Int. J. Man–Mach. Stud.* 33(2):305–322.

Rosson, M. B. and Alpert, S. R. 1990. The cognitive consequences of object-oriented design. *Hum.–Comput. Interaction* 5:345–379.

Rosson, M. B. and Carroll, J. M. 1990. Climbing the Smalltalk mountain. *SIGCHI Bulletin* 21(3):76–79.

Rosson, M. B. and Carroll, J. M. 1993. Active programming strategies in reuse. In Proc. ECOOP '93, *Object-Oriented Programming*, pp. 4–18. Springer–Verlag, Berlin.

Rosson, M. B. and Carroll, J. M. 1995a. Integrating task and software development in object-oriented applications. In CHI '95 Conf., *Proc. Hum. Factors Comput. Syst.* M. B. Rosson and J. Nielsen, eds., pp. 377–384. ACM, New York.

Rosson, M. B. and Carroll, J. M. 1995b. Narrowing the gap between specification and implementation in object-oriented development. In *Scenario-Based Design: Envisioning Work and Technology in System Development*, J. M. Carroll, ed., pp. 247–278. Wiley, New York.

Rosson, M. B. and Carroll, J. M. 1996. Scaffolded examples for learning object-oriented design. *Commun. ACM* 39(4):46–47.

Rosson, M. B. and Carroll, J. M. in press. The reuse of uses in Smalltalk programming. *ACM Trans. Comput.–Hum. Interaction*.

Rosson, M. B., Carroll, J. M., and Bellamy, R. K. E. 1990. Smalltalk scaffolding: a case study in minimalist instruction. In CHI '90 Conf., *Proc. Hum. Factors Comput. Syst.*, pp. 423–429. ACM, New York.

Rosson, M. B., Carroll, J. M., and Sweeney, C. 1991. A View Matcher for reusing Smalltalk classes. In CHI'91, *Proc. Hum. Factors Comput. Syst.*, pp. 277–284. ACM, New York.

Rosson, M. B. and Gold, E. 1989. Problem-solution mapping in object-oriented design. In *Proc. OOPSLA'89*, pp. 7–10. ACM, New York.

Rosson, M. B. and Heliotis, J. E. 1993. The OOPSLA'92 Educators' Symposium (guest editors' introduction). *Comput. Sci. Educ.* 4(1):1–4.

Scholtz, J. 1993. A longitudinal study of transfer between programming languages by experienced programmers. In Proc. HCI'93 Conf., *People and Computers VIII*, pp. 397–410. British Computer Society, Cambridge, UK.

Scholtz, J. and Wiedenbeck, S. 1990. Learning second and subsequent programming languages: a problem of transfer. *Int. J. Hum.–Comput. Interaction* 2:51–71.

Scholtz, J. and Wiedenbeck, S. 1992. The role of planning in learning a new language. *Int. J. Man–Mach. Stud.* 37(2):191–214.

Shaw, M. 1990. Prospects for an engineering discipline of software. *IEEE Software* 7(6):9–12.

Shneiderman, B. and Mayer, R. 1979. Syntactic/semantic interactions in programmer behavior: a model and experimental results. *Int. J. Man–Mach. Stud.* 8(3):219–238.

Simon, H. A. 1973. The structure of ill-structured problems. *Artif. Intell.* 4:181–201.

Soloway, E. and Ehrlich, K. 1984. Empirical studies of programming knowledge. *IEEE Trans. Software Eng.* SE-10:595–609.

Taenzer, D., Ganti, M., and Podar, S. 1989. Problems in object-oriented software reuse. In ECOOP'89, *Proc. European Conf. Object-Oriented Programming*, pp. 25–38. British Computer Society, Cambridge, UK.

Tewari, R. and Friedman, F. L. 1993. A framework for incorporating object-oriented software engineering in the undergraduate curriculum. *Comput. Sci. Educ.* 4(1):45–62.

Tucker, A., Barnes, B., Aiken, R., Barker, K., Bruce, K., Cain, S., Conry, S., Engel, G., Epstein, R., Lidtke, D., Mulder, M., Rogers, J., Spafford, E., and Turner, A. 1991. *Report of the ACM/IEEE CS Joint Curriculum Task Force: Computing Curricula 1991.* ACM and IEEE Press, New York.

Vessey, I. and Weber, R. 1984. Research on structured programming: an empiricist's evaluation. *IEEE Trans. Software Eng.* SE-10(4):397–407.

Visser, W. 1987. Strategies in programming programmable controllers: a field study on a professional programming. In *2nd Workshop, Empirical Studies of Programmers*. G. M. Olson, S. Sheppard, and E. Soloway, eds., pp. 217–230. Ablex, Norwood, NJ.

Visser, W. and Hoc, J.-M. 1990. Expert software design strategies. In *Psychology of Programming*, J.-M. Hoc, T. R. G. Green, R. Samurçay, and D. J. Gilmore, eds., pp. 235–250. Academic Press, London.

Widowski, D. and Eyferth, K. 1986. Comprehending and recalling computer programs of different structural and semantic complexity by experts and novices. In *Human Decision Making and Manual Control*, H.-P. Willumeit, ed. North-Holland Elsevier, Amsterdam.

Wu, Q. and Anderson, J. R. 1991. Knowledge transfer among programming languages. In *Proc. 13th Annu. Conf. Cognitive Sci. Soc.*, pp. 376–381. Erlbaum, Hillsdale, NJ.

Further Information

For a collection of classic studies of programming behavior see:

Curtis, B. 1985. *Tutorial: Human Factors in Software Development.* IEEE Computer Society, Washington, DC.

To sample a variety of empirical studies of programming, see the proceedings of the *Empirical Studies of Programmers*, published by Ablex, Norwood, NJ.

See also:

Curtis, B. 1988. Five paradigms in the psychology of programming. In *Handbook of Human–Computer Interaction*, M. Helander, ed., pp. 87–106. North-Holland, Amsterdam.

Boehm-Davis, D. 1988. Software comprehension. In *Handbook of Human–Computer Interaction*, M. Helander, ed., pp. 107–122. North-Holland, Amsterdam.

Hoc, J. M., Green, T. R. G., Samurçay, R., and Gilmore, D. J. 1990. *Psychology of Programming*. Academic Press, London.

74

On-Line Support Systems: Tutorials, Documentation, and Help

Stuart A. Selber
Texas Tech University

Johndan Johnson-Eilola
Purdue University

Brad Mehlenbacher
*North Carolina State
   University*

74.1 Introduction

On-line support systems help computer users achieve goals and accomplish tasks within the contexts of their **primary work**. Although this definition is extremely broad and includes a wide range of digital forms—from low-end interface elements to high-end hypermedia applications—in this chapter we generally focus on planning, designing, and testing midrange systems: **tutorials, documentation**, and **help**, regardless of their virtual instantiation. We discuss electronic rather than print-based forms because organizations increasingly deliver user support on line for a variety of reasons: to reduce development and production costs, to anticipate distributed computing systems and other environments in which users rarely have easy access to print-based materials, and to benefit from the sophisticated searching and interactive capabilities that on-line environments can provide. In cases where print-based support is still necessary (for example, in packing instructions and in some troubleshooting areas), processes for constructing these documents can be extrapolated from the discussion that follows.

Determining what types of on-line support to provide users is a deceivingly complicated task: they often work with computer technologies in a variety of contexts and with numerous purposes and goals, and single users frequently bridge more than one context, purpose, and goal over time. Consider a technical communication group that purchased a desktop publishing program to help them write, design, and publish documentation. Not surprisingly, they might find useful support in a wide variety of forms. An introductory, on-line tutorial might guide them through basic program functions such as creating files, making templates, and importing graphics. This same tutorial, which might also include an accompanying workbook, could encourage them to examine case examples to learn how discrete portions of the program work together. While using the desktop publishing program itself, the group might employ context-sensitive help to see brief descriptions of available tools. Under certain conditions, they might rely on more extended on-line documentation for feature or process overviews. Still, in other cases, they might link from the on-line documentation to brief tutorials that model particularly complicated procedures.

This scenario illustrates the connections among and differences between the three types of on-line support we discuss in this chapter. Tutorials include the broadest possible topics, with users learning about features and tasks by engaging some combination of explanation, example, and hands-on experimentation. On-line documentation has a narrower pedagogical scope, with users normally consulting reference information for overviews or assistance with task-oriented procedures. On-line help usually has the narrowest focus, with users needing to solve particularly pressing problems as quickly as possible and with minimal interruption.

74.2 Underlying Principles

Developing useful on-line support is a complex, recursive process that should parallel product development in central ways and situate planning, designing, and evaluating as nonlinear, rhetorically based activities. Before beginning this recursive process, at least three foundational areas merit serious attention: the advantages and disadvantages of on-line support, the differences between print-based and on-line support, and the rhetorical frameworks useful for developing on-line support—users, goals, and time/space constraints.

Advantages and Disadvantages of On-Line Support

There are both advantages and disadvantages to creating and delivering support on line. The advantages for designers are increasingly clear. Compared to print-based genres, on-line support is relatively inexpensive to produce, distribute, and update once initial procedures have been stabilized and adopted by development teams. And, because all forms of on-line support require shorter production cycles than their print-based counterparts, designers can work on content issues until just before final products are released (usually one to two weeks). This expanded development period is particularly useful for accurately documenting task-oriented information. The efficiency gains afforded by mature development cycles for on-line support are ultimately noticeable in five key areas: worker productivity, task completion time, the overall time needed for system development, the materials and economic resources needed for production, and maintenance activities such as creating and distributing revisions [Petrauskas 1991].

Users will find that well-designed on-line support can assist their learning goals and task objectives in substantial ways. Obviously, large amounts of reference or database information are more easily and personally used in conjunction with robust software engines that automate or augment search routines. But, beyond the keyword and pattern searching of vast textual and numerical landscapes, the potentially dynamic and interactive nature of on-line support offers many distinct advantages. For example, in on-line tutorials, which are often considered a form of computer-assisted instruction (CAI), computer-assisted learning (CAL), or computer-based training (CBT), users can engage highly interactive, self-paced educational activities at their own convenience and pace, either for immediate performance improvements or for general professional development. In on-line documentation environments, users can usually choose from among many different organizational patterns and navigational systems, as well as customize instructional content by adding personal annotations, trails (via bookmarks), and entirely new texts. And, in on-line help systems, users can stay focused on their primary tasks by employing tightly integrated, context-sensitive assistance at particularly pressing impasses. In addition, all three forms of on-line support can exploit the artificially intelligent operations of computers: in at least basic ways, certain types of on-line support can track user actions and productively respond by adapting what is available or prominent at any one moment.

The potential disadvantages of on-line support, however, are not insignificant. As we will discuss, many of these disadvantages relate to the physical and rhetorical differences between pages and screens and between common ways of instantiating print-based and on-line materials. In addition, there are other potentially problematic issues to consider. Novice users struggling to learn new applications may find little comfort in assistance that is similarly provided on line. Users working with the minimum hardware and/or software requirements needed to run applications may experience performance problems that discourage them from accessing support or from taking advantage of its advanced features. And, poorly designed systems, while conveniently available on line, are often inferior in form and function to well-designed, print-based genres of assistance. Although these disadvantages should not discourage developers from creating on-line support, they should be considered during planning and designing stages.

Differences Between Print-Based and On-Line Support

Important physical and rhetorical differences differentiate print-based and on-line support. Too often, these differences are simply ignored or overlooked, a mistake often resulting in two unfortunate situations: support that fails to exploit a system's unique abilities to store, structure, and retrieve digital information, and/or support that applies those unique abilities in ways that are inappropriate to the needs of users. Designers converting print-based support to on-line support may be tempted to simply dump their source files on line without significantly rethinking core design decisions. Research indicates, however, that such time-saving approaches can be disastrous for end users [Rubens and Krull 1985]. In general, the physical differences between print-based and on-line support relate to resolution, display area, aspect ratio, and presence, whereas the rhetorical differences relate to organizational, navigational, and contextual structures. These differences are summarized in Tables 74.1 and 74.2.

TABLE 74.1 Physical Differences Between Print-Based and On-Line Support

	Pages	Screens
Resolution	70–1200 dots per inch	50–100 dots per inch
Display area	generally larger	generally smaller
Aspect ratio	generally taller than wide	generally wider than tall
Presence	physical static immutable	virtual static dynamic interactive mutable

The *physical* dimensions of on-line support provide a writing and reading space that is qualitatively different than that of pages. Computer screens are much harder to read, for example, because their resolution is typically much lower than professionally printed materials and because the area available for displaying information is often much more limited. Available space for displaying multiple documents simultaneously is also much more limited for screens than printed pages (screen territory vs. desk space). These two constraints, moreover, can limit certain aesthetic dimensions of document and graphic design. In addition, the 4-by-3 aspect ratio of many computer screens—the ratio of their horizontal to vertical dimensions—more closely resembles a movie screen or television monitor than a printed page. This expanded horizon can encourage developers to include more information than they probably should for good readability, and it clearly complicates design principles originally developed for print-based, portrait orientations. The presence of text on screens also differs from that of pages in dramatic ways. Whereas the information on pages is static and immutable, the information on screens can be dynamic and interactive, as well as mutable.

TABLE 74.2 Rhetorical Differences Between Print-Based and On-Line Support

	Pages	Screens
Organizational	linear familiar hierarchical logical/deductive fixed	linear and nonlinear familiar and unfamiliar hierarchical and non-hierarchical logical/deductive associative and dynamic
Navigational	familiar limited static	familiar and unfamiliar robust static and dynamic
Contextual	generally rich	generally poor

Importantly, this malleability provides unique opportunities for customizing support systems and practicing **user-centered design**.

The *rhetorical* dimensions of on-line support—organizational, navigational, and contextual structures—influence the ways in which users understand system logic. Although on-line support is often developed using organizational structures familiar to print-trained readers (linear and hierarchical, for example), it can employ dynamic and associative (weblike) structures that allow users to organize and reorganize information in a wide variety of ways and move between many different levels of instruction. Moreover, on-line support systems can adapt dynamically to user performances, shifting goals, and changing contexts in ways that the technology of print makes impossible. Readers may find that certain navigational structures such as bidirectional links, history trails, and overview maps are similarly more dynamic and at least initially unfamiliar, although many navigational structures such as bookmarks and indices have been appropriated from the realm of print. The idea of both physically and metaphorically navigating through on-line information space, however, departs radically from the user passivity often associated with writing and reading more traditional, printed texts.

Because on-line support is often more dynamic and unfamiliar, its contextual structures can be poor. While using on-line support, individuals lose many traditional user cues—page numbering and tabs, for example—that allow them to work efficiently and effectively with texts, and they can also lose the spatial comfort that holding a printed book can provide. In fact, as Haas [1989] reports, users often have trouble "getting a sense of text," that is, seeing meaning and structure, in extended on-line information spaces. And, she notes that writers commonly struggle with formatting, proofreading, and reorganizing texts that are solely created and examined on-line.

Rhetorical Frameworks for On-Line Support

In developing on-line support, designers must carefully consider the needs and wants of users, including their previous experiences and abilities, their short- and long-term goals, and the environments in which they work. Such a framework for on-line support carefully considers two important areas: the *rhetorical* issues of users, goals, and time/space frames; and the *formal* characteristics of tutorials, documentation, and help.

The main objective of on-line support systems, regardless of their form, is to help *users* achieve *goals* as they negotiate the very real constraints of various *time/space frames*. In fact, these terms—users, goals, and time/space frames—represent key elements in understanding and designing useful systems (see Table 74.3). In working with Table 74.3, remember that many different forms of on-line support may be necessary to adequately help users work across multiple contexts. In addition, each

TABLE 74.3   Rhetorical Frameworks for On-Line Support

	Help	Documentation	Tutorials
Users	expert	intermediate	novice
Goals	narrow/ short term	medium/ short term	broad/ long term
Time/Space Frames	parasitic/ internal	parallel/ internal	encompassing/ external

rhetorical element commonly affects the other two: for example, if a novice user's goal is only to complete a task (and not to understand its broad implications or varied uses), then on-line help may be the best solution, even though this user will not gain a more conceptual understanding. Warning symbols and emergency instructions are two such cases.

Our approach to on-line support asks developers to consider the rhetorical aspects of the left column of Table 74.3—users, goals, and time/space frames—rather than the formal, surface-level characteristics of its top row—help, documentation, and tutorials. Approaches that begin with formal characteristics rather than user contexts are often too brittle to succeed in real-world situations. Users rely heavily on contingent, situated, recursive actions rather than acontextual plans [Beabes and Flanders 1995, Boy 1992, Suchman 1987, Winograd and Flores 1987]. When on-line support development begins with the formal characteristics of different forms of communication rather than the contexts and purposes of use, the resulting systems can appear well designed and executed but generally fail to address the needs and constraints of users; sometimes, the resulting systems force users to remake their work processes and even long-term goals in order to adapt to the forms of on-line instruction [Johnson-Eilola, 1996]. Paradis [1991], for example, analyzes operator's manuals for construction tools to illustrate how decontextualized instructions purchase simplicity and clarity at the cost of human lives. Although there are numerous excellent resources for developing specific forms of support [Carroll 1990], these resources often assume that previous rhetorical analyses have lead developers to a specific form. By carefully analyzing the rhetorical aspects of users, goals, and time/space frames, designers can work to avoid such situations and provide on-line support that is both more appropriate and useful.

Analyzing Users

Of the three types of on-line support—tutorials, documentation, and help—each addresses different kinds of users, including those individuals whose skill sets change from one activity to another. The traditional terms expert, intermediate, and novice can be misleading because there are many different ways to measure someone's knowledge: for example, by specific application, platform, task, and profession. Physicians using a new diagnostic support program may be novices in terms of this specific application, but experts in terms of medical concepts, terminology, and even similar applications. They might need brief, on-screen definitions to remind them of potential drug interactions, but longer term, intensive tutorials (including simulations and self-administered quizzes) to help them restructure patient interviews in ways that the new software program can productively support.

Experts (of whatever type) typically benefit the most from on-line help because they already possess rich structures of understanding; when they ask for assistance, it is often for help in retrieving information from long-term memory. For example, error messages in many Unix versions also function as on-line help by reminding expert users of complicated syntax. Novices, however, usually require more structure and learning aids to understand basic concepts and specific commands and to build mental models on which they can later draw (experts would find such assistance tedious).

Questions to ask when considering user types include the following:

- How much do users know about the software?
- Do users know similar software?
- How much do users know about their goals and tasks?
- How much do they need to know about their goals and tasks?
- Do they need information for short-term (consultative) or long-term (memorized) uses?
- Do they need to learn background information or concepts?
- Are there some areas in which users are experts but others in which they are novices?

Defining and Redefining Goals

Designers should consider the goals of users before developing on-line support systems. In some cases, users in the middle of a complex task may only need minor information to successfully complete their work. For example, an intermediate or expert user of a multimedia environment might, during a large-scale authoring project, need to select a transition effect when splicing two segments of digital video together. A brief, on-line list of available effects and their syntax would constitute appropriate support, providing quick assistance without much cognitive overhead. A new multimedia user, however, might be better served by an interactive tutorial or long-term training session that discusses not only functional commands, but the basics of screen design, graphic arts, music, and learning styles. In the former case, the specific user goal is to code a transition from one video segment to another, a fairly low-level task. In the latter case, the user goal is not merely to get information about transitions, but to learn a broader range of multimedia concepts, of which transitions are only one small part. The information about transition effects referred to by the expert might even be reproduced as part of a lesson for the novice, but with a stronger, pedagogical framework.

When researching the goals of users, these goals must often be divided into subgoals and subtasks appropriate to their current knowledge bases and purposes [Card et al. 1983]. Often, a support suite must help convince users that their immediate goal cannot be accomplished without first undertaking some broader learning. Furthermore, when the goals of users are fairly ambitious but tutorials are required, users may need motivation to keep their attention level high. Such motivation can range from clearly stated objectives that help users track their progress during lessons to institutional rewards for completing a series of lessons.

Questions to ask about user goals include the following:

- What, exactly, do users want to accomplish?
- How might those goals be supported by the software/hardware?
- Are there conceptual issues that users need to learn before using the software/hardware to achieve their goals?
- Should user goals be divided into more manageable subgoals, tasks, or subtasks?
- Will users need feedback (screenshots, audio acknowledgments, and so on) to know that they have achieved their goal?
- Will users need motivation?
- What kinds of user motivation might be appropriate?

Locating Support in Time and Space

Another crucial rhetorical element is the amount of time and screen space available to end users. For example, in word-processing programs, the workspace is normally the screen into which texts are typed and displayed. The workspace is, therefore, the focus of user attention, and the difficulty of refocusing that attention must be considered when determining the type of on-line support to provide. In fact, focus broadens and changes as users move from on-line help (which tightly integrates support and exists only briefly outside the workspace), to on-line documentation (which provides on-screen text paralleling the workspace), to tutorials (which provide a simplified learning workspace, or which construct a learning space separate from the workspace and only infrequently refer to it during exercises or testing). At times, some users may resist redirecting their attention from a workspace to a learning space, even when that action is well worth the effort. As one might expect, this resistance can relate to the knowledge levels and goals of users. For example, individuals who are experts in programming but novices in a specific application may resist attempts to refocus their attention away from the workspace and toward a tutorial during pressing primary work.

Questions to ask about locating support in time and space include the following:

- How much refocusing is required for users to engage the support?
- Do they have both the time and energy to refocus their attention?
- Would users benefit from a simplified version of the workspace for learning?
- Is there sufficient room for on-screen, parallel spaces, or would a workbook or paper-based manual be required?
- Can users be convinced to rethink or broaden their goals to accommodate new foci?
- Are there other, environmental factors that will interfere with some types of support (lighting, noise, technological limitations, and so on)?

74.3 Best Practices

Generally, best practices for developing on-line support can be divided into three broad stages: planning, designing, and evaluating. During planning stages, developers conduct needs assessments and create specifications. These activities provide the necessary foundation for designing on-line support, a process that must address both global issues (such as organizing and maintaining systems) and local issues (such as **chunking** information and providing discourse cues). Evaluating stages include both formative testing— in-process evaluations that may lead to redesign—and summative testing—post-production evaluations that measure end quality [Duffy et al. 1992]. These three stages often occur recursively, as decisions or responses in one area affect the other two.

Planning On-Line Support Systems

Planning is a critical task in the recursive process of developing on-line support systems. Early decisions about form and content will centrally influence subsequent design stages and final products, as will determinations by managers and others about individual and group responsibilities. In fact, although there are many reasons why on-line support systems fail or are only partially successful, in many instances these reasons can be traced to poor or inadequate planning (see Caernarven-Smith [1990] for an annotated bibliography on the benefits of early planning practices). Indications of poor or inadequate planning during development stages include incomplete style and design guidelines, overlapping work duties, missed milestones and deadlines, and significant deviations from standard production practices.

Unforeseeable events will certainly alter schedules and plans in unpredictable ways. It is not uncommon, however, for thoughtful designers to spend 25% of their total project time on planning issues: planning opens lines of communication between managers, designers, and users; coordinates the tasks and activities of project team members; encourages systematic approaches to on-line support; and, over time, saves both human and economic resources.

Assuming that management has already conducted feasibility studies and established the market viability of any new products, at least two essential tasks relate to planning on-line support: conducting needs assessments and developing specifications. A wide range of artifacts often inform these two stages, and are therefore collected before planning tasks begin; among them are: feasibility studies, business plans, product requirements, cost/benefit analyses, mission statements, user surveys, research on competitive products, demographic information, and other market research.

Conducting Needs Assessments

Needs assessments precede specifications and help determine both external (user/customer) and internal (designer/manager) requirements for on-line support systems. For new products, these assessments address broad-based questions, for example: What should our on-line support systems accomplish? What kinds of on-line support systems can we realistically deliver and maintain? What substantial constraints are imposed by the working environments of users? For existing products, the scope of these questions narrows considerably, for instance: What features are the most and least useful? What features require the greatest amount of customer support? What features might be added to or eliminated from the next version? For both external and internal purposes, needs assessments help clarify the main objectives and roles of on-line support.

Like all research projects, needs assessments begin with questions—like the broad and narrow ones just listed—that derive from real organizational, business, and social contexts, both external and internal. Use these questions to help clarify the goals of needs assessments, which should be stated in clear and precise terms.

Next, select empirical methodologies, qualitative and/or quantitative, that can guide your research questions in systematic ways. Useful modes of inquiry include interviews, surveys, user observations, and case studies. Notably, field work in the form of interviews and observations are particularly helpful for understanding the perspectives of users: it occurs within their natural (working) environments [Beabes and Flanders 1995].

Interpret the results of your research carefully and modestly, keeping the results gathered from any single inquiry in perspective. For higher reliability, use a variety of methodologies that allow you to examine problems or issues from many different perspectives: patterns suggested by multiple approaches are generally more reliable. Also, for further confidence in your findings, compare your results with those discussed in any related literature. As with all research projects, fully document and share your results in a completion report. Meeting face-to-face with team members as this report is distributed will allow you to answer questions about your methodologies and findings, as well as limit any possible misreadings of the conclusions.

Task analyses are often an integral part of a thorough needs assessment. They identify the complex range of actions that users must perform, and they further divide these actions into subactions that the planned on-line support systems should discuss. From task analyses, designers gain a clearer sense of one important and constantly evolving triangulation: the information needs of users, their work contexts, and their knowledge bases. By considering the relations between these three areas, designers can better serve as user advocates and representatives of other external interests. Moreover, these activities provide designers with opportunities in which to learn and influence products.

In addition to task analyses, needs assessments also often include media analyses that determine what types of deliverables should be developed in connection with a project: tutorials, documentation, and/or help, paper-based and/or on-line. Also, because they help articulate form and content, needs assessments can suggest the kind of interdisciplinary team that should be formalized. Although internal projects are often handled by programmers and other technical specialists, external projects will require designers who can write for on-line information spaces, edit for both developmental and mechanical concerns, manage on-line projects, test the usability and quality of on-line support, understand user psychology and cognition, design human-computer interfaces, design instructional materials, market products, and so on.

Finally, needs assessments focusing on internal issues can help establish budgets and time lines. Determine what you can afford to design, maintain, and update, and what you can productively accomplish in parallel with product development (see Smith [1994], for approaches to and worksheets for estimating project costs). This information is repeated in specifications for on-line support.

Developing Specifications

Specifications, also commonly called documentation plans or blueprints, outline courses of action for developing on-line support systems, including both product and process descriptions. Product descriptions detail the rhetorical and physical dimensions of deliverables: feature summaries, audience profiles, and system organizations are just some of the areas you should discuss. Process descriptions outline approaches to accomplishing work, including development and production schedules, review protocols, and testing procedures. Although both products and processes may change in a variety of ways, developing specifications is an important planning task: in these documents, you must be explicit about all of the predictable aspects of a project.

There are other important reasons for developing specifications. First, they can help you anticipate problems before they occur and outline workable solutions and contingency plans. They can also help you understand the consequences of any dependencies and risks. Second, specifications can help you articulate the complexity of your task and therefore establish realistic expectations about what can be productively accomplished during development cycles. Third, because specifications include contributions from all team members, they provide early opportunities in which managers, subject-matter experts, and others can provide necessary feedback, express concerns, and raise doubts. In turn, such early participation solidifies both individual and group responsibilities, and it generally increases an organization's commitment to the successful completion of deliverables.

Although there are many different ways to develop specifications, the following 12 elements should always be included:

- Background information
- Audience profiles
- Topic outlines

- Detailed outlines
- Style/design specifications
- Project responsibilities
- Production specifications
- Review protocols
- Development schedules
- Cost estimates
- Dependencies and risks
- Sign-off sheets

Background Information. Include overviews of what you are documenting and what you are developing (their main uses and features), goal and mission statements reflecting managerial, developer, and user concerns, immediate and long-term action plans, scope and purpose statements, instructional goals, the results of your needs assessments and task analyses, common scenarios and contexts in which individuals will use your deliverables, discussions of what your deliverables will not include, associated documents and products, possible authoring tools, and definitions of any specialized language.

Audience Profiles. Move beyond simple demographic information to include user contexts, learning styles, and motivations. Describe the nature of their work, their concerns and biases, their skills both in using computers and in understanding the subject matter you are documenting, what they expect from on-line support systems, their interest levels, their knowledge bases, and any presuppositions they may have about what your support system ought to include.

Topic Outlines. Provide high-level content overviews, including module names, section names, and first-level headings. Include storyboards that map information flow and system logic. Estimate the total number of screens the support will include. Whenever possible, build support systems that allow extension and modification without elaborate revision at the global level.

Detailed Outlines. Expand your topic outline to include any front matter, submodules, subsections, scenarios, examples, index terms, and appendices. Fully describe the contents of all planned information, both textual and graphical. Provide sample sections and example screens.

Style/Design Specifications. Discuss organizational patterns, navigation aids, reference aids, interface designs and grids, linking strategies, aesthetic dimensions, uses of graphics, user cues, emphasis markers, text conventions, style guides, and other standards. Provide thumbnail sketches and sample screens.

Project Responsibilities. List all team members, their titles, and their roles and primary and secondary responsibilities. Identify project leaders for each major phase of development.

Production Specifications. Describe the final physical forms of all deliverables. Consider disks, disk labels, release notes, notes for compact disc read-only memory (CD-ROM) jewel cases, read-me files, packaging materials, quick start guides, workbooks, and marketing literature such as booklets, brochures, catalogs, and data sheets.

Review Protocols. Describe the kinds of reviews planned—**editorial**, **managerial**, and **technical**—and who will conduct these reviews. Describe the kinds of usability tests planned and who will perform these tests. Common methodologies are discussed in the section on evaluating on-line support systems.

Development Schedules. Include schedules and milestones for all project stages (alpha, beta, and final versions), status and review meetings, editing cycles, usability tests, productions tasks, debriefings, and maintenance activities. List all major tasks and their completion dates. Note any critical timing issues and dependencies. Build in slip time and flexibility.

Cost Estimates. Supply project cost estimates. Consider salaries, the amount of work required to produce each screen of information, nondevelopment tasks related to project management, and other resources such as contractors, additional training, computer technologies, and research materials.

Dependencies and Risks. Note all issues that can change schedules and plans and otherwise affect your development processes in unproductive ways. Interview all team members for their concerns. Common dependencies include access to subject-matter experts, accurate and timely reviews, and open lines of communication between team members. Common risks include unstable products, feature changes, and inadequate or unavailable peripheral services for production.

Sign-Off Sheets. Encourage responsibility and accountability by asking team members to approve their tasks during the development of deliverables. Include signature lines, date lines, and spaces for comments and concerns.

When distributing specifications, be sure to clarify the role of your readers in a cover memorandum. Ask them to provide productive criticism at this early stage, and be sure to specify what actions you need of them, such as returning changes and comments by a particular deadline.

Designing On-Line Support Systems

If developers have not planned well, it is likely that they will encounter disastrous results when designing their on-line support systems. Surprisingly, many companies have yet to formalize their specification processes and still approach designs problems in ad hoc fashions. Moreover, design practices, whether they involve building car engines, light switches, door knobs, or on-line support, are very complex activities and often require numerous trial-and-error experiences before central goals are met. That is, designers of on-line support systems must be prepared for ill-defined problem spaces in which meeting the needs of users may be undermined by organizational and other social values, and in which design solutions must be constructed and argued for throughout the development process. Such *wicked* problems—problems having many different possible approaches and solutions—are complicated by a wide range of factors: organizational goals, the task complexity that designers are trying to support, user expectations, hardware and software constraints, the existence of competitive products, and the lack or abundance of available research on particular design features, just to name a few.

The Specification as Springboard

Once a comprehensive specification has been developed, various new constraints and opportunities will immediately demand attention. You may quickly realize that effective design practices require a robust authoring tool, a multidisciplinary design team, constant storyboarding or prototyping (despite careful planning), iterative usability testing, user-centered design perspectives, and direct manipulation features coupled with object-oriented interfaces.

Initially, you must contend with the vast number of available authoring tools that will claim to solve all your complex design problems in elegant ways. Each of these tools, of course, has its own advantages and disadvantages: for example, RoboHelp allows you to link documents in relatively easy ways, HyperCard has a very intuitive scripting language, and WinHelp assumes a good knowledge of Microsoft Word. Ultimately, the features and constraints of your authoring environments will have both productive and unproductive implications for the design of your support systems. For instance, in HyperCard 1.0, it was impossible to create *sticky buttons*, links attached to objects (like words) that stick to those objects in useful ways (even if you scrolled those words off screen). Thus, designers working in this environment either had to accept such tool-based design constraints or find ways of creatively working around them.

During the design process, you will also quickly realize that no single individual has the wide range of skills needed to effectively produce on-line support [Mehlenbacher 1995]. Multimedia authors, for example, must blend many different skills, including those typically associated with instructional design-ers, programmers, architects, cartographers, and rhetoricians [Selber, forthcoming]. Although in every relevant skill area expert advice may not be available or even possible, there may be places in which such advice is absolutely necessary.

Despite the challenges of selecting authoring tools and forming effective teams, most initial design efforts focus on creating a working storyboard. Designers hold meetings and negotiate over intended audiences, the contexts in which a system will be used, the basic functionality of that system, and its ultimate look and feel from a user-interface perspective. Notably, companies such as Wang (in the past) and SAS Institute (currently) often include real users in this initial storyboarding process. Unfortunately, many designers still create product plans without consulting either users or potential competition. And, only very recently has storyboarding necessarily included principles for effective interface design; instead, storyboarding has historically emphasized a system's intended functionality rather than integrating that functionality, interface design, and task-orientation into an overall design process.

Designers should consider the short- and long-term needs of their audiences during all development tasks. Although there is strong industry movement to integrate user-centered design practices and usability testing into the overall design process, research suggests that establishing design quality control remains a difficult practical problem [Benson 1991, Skelton 1992]. Simply defining quality as it relates to planning, designing, and evaluating is a substantial challenge, particularly given the exponential growth and change currently facing developers of on-line support systems.

Finally, many corporations have invested in direct manipulation interfaces and object-oriented programming (OOPs). The *drag and drop* philosophy toward iconic objects has been adopted, for example, by Microsoft and other large vending corporations, and Shneiderman's commitment to direct manipulation has finally been instantiated in contemporary software products [Shneiderman 1987]. Designers should investigate these computing paradigm shifts before developing on-line support and consider which ones might become programming standards for future software releases.

Global Design Issues

It is critical to consider global issues early in the design process. These issues cannot be ignored because it is often difficult, if not impossible, to incorporate them after the development process has begun. Global design issues that merit serious attention include the following:

- Accessibility
- Maintainability
- Support on support
- Organization
- Metaphors and maps
- Navigation
- User control
- Modelessness
- Consistency
- Reversibility and error recovery
- Visual aesthetics
- Feedback structures and context sensitivity

Accessibility. On-line support systems should be easy to access and exit; ideally, users should return to the state in their primary application from which they left when seeking assistance. With the advent of large-screen, high-resolution monitors, this design goal is less difficult to achieve. When appropriate, on-line support can easily be displayed concurrently with primary applications, providing users with the critical ability to keep their original problem contexts in mind while searching for solutions.

Maintainability. Object-oriented approaches to design allow individuals to view codes and scripts as reusable chunks of information that can be copied, extended, and refined, and that can then be shared or protected, hidden or revealed, to multiple designers. This economical perspective allows designers to create high levels of system consistency, integrity, and updateability that are difficult to achieve following more traditional, functional models of programming.

Support on Support.   Although on-line support systems should be easy to use without additional layers of support on support, this goal is frequently difficult to attain. The popular Unix-based Emacs editor, for example, asks users to press ⟨^_⟩ to access help, yet the majority of users do not understand that ⟨^⟩ represents the control key (which may or may not be labeled Control on the keyboard). In addition, users may be dismayed to read four to eight help-on-help panels before accessing actual information for immediate problem solving. Support on support must be as simple and elegant to use as the support system itself. If it is not, a rigorous review of the system is probably needed.

Organization.   Despite the claims of many hypertext and World Wide Web designers, users of support systems often require some form of hierarchical representation and do not feel comfortable browsing or surfing for the information that they need. In short, users of support are goal driven: they want to learn applications; they want to cut-and-paste paragraphs; they want to discover which keystrokes can shorten their procedures. No support user works completely or even primarily in nonlinear ways. Although they may branch from one topic to another for immediate benefits, they normally refuse to browse for extended periods of time to find needed assistance.

Metaphors and Maps.   A thorough task analysis during the planning stage is the best way to perceive how users understand on-line support systems and how those systems will help them accomplish work. Capturing the task expertise of users and their expectations in terms of language, definitions, elaborations, and exceptions to given rules can help designers anticipate mismatches between user expectations and software functionalities. Designers might, for example, begin to question the friendliness of terms such as file, window, and disk, if they closely examined their users' perceptions of those features and their functions [Carroll 1990]. With the Macintosh interface, for instance, we are told that users engage a *virtual* desktop, but few desktops have trashcans on them and most come with telephones and fax machines these days. Generally, well-established metaphors and maps help users identify tools in relation to each other, their current state in a system, and, if designed well, the various paths or selection options available [Winn 1990]. Importantly, interface metaphors and maps are not pedagogically or even politically neutral: they embody particular ways of knowing, working, and learning in on-line information space [Selber 1995, Selfe, C. and Selfe, J. 1994].

Navigation.   It is useful to examine the differences and similarities between terms that describe traditional methods of navigation. Each term—jumping, traversing, searching, browsing, opening, paging, scrolling—reveals various assumptions underlying a designer's perceptions of users, systems, and the user–system interactions involved in navigation. Jumping and traversing, for example, emphasize a system's capabilities or structures. In a hierarchically organized database structure (common to many contemporary on-line information systems), users move down through various levels which are usually more specific categorically. For example, in a complex system about restaurants, the top level might provide users with information about restaurant types. One effective organization, often referred to as functional grouping [Mehlenbacher et al. 1989], would list Chinese, American, Irish, German, Canadian, Japanese, and so on. On selecting American, users would move to a submenu that might list breakfast, lunch, dinner, salads, and soups. One more level into the hierarchy might list specific food types. Another jump down might reveal short descriptions of different dishes within the previously selected categories of ethnicity and meal type.

Similarly, opening, paging and scrolling are generally system-specific terms for describing navigation through on-line information space. Opening derives its roots from the tasks of opening files, subfiles, and more subsubfiles in hierarchical arrangements. Paging suggests that on-line information is sequentially organized in discrete chunks that can be viewed one page at a time. Scrolling connotes a linear, seamless movement up and down (as with ancient scrolls). To clarify assumptions about underlying structure, consider the restaurant system described earlier. Having traversed the hierarchical structure to the list of food descriptions, users might have an opportunity to view a number of relevant meal descriptions. If there are several items, designers will either have to develop another dimension for hierarchical organization (such as spicy and mild) or switch to another organizational model (such as paging). In addition, based on needs assessments, designers must decide if users will need to jump across the hierarchy to see related dishes in other ethnic groups or meal types.

User Control. Users do not appreciate being reminded that they are relatively powerless over the software applications that they use. Hence the common and confident description that users give to reading print-based books rather than browsing on-line: "New technology is like MTV: temporally splintered, short term, and anti-meditative; Generation X, therefore, is superficial." We, as teachers and instructional designers, can appreciate this (arguable) position, but we also recognize the complexity and contradiction that new technologies can introduce into human–computer interactions, modes of community, and communicative expectations. We understand, for example, that historically, Microsoft Disk Operating System (MS-DOS) experts rejected the Macintosh graphical user interface as hiding the complexity of computing from end users, despite the subsequent contemporary efforts by those same individuals to emulate and delineate their recent systems from the same computing philosophy. Obviously, computer users are also consumers, and consumers respond to or reject technologies based on a variety of factors, including their learnability, usability, long-term competitiveness, and institutional guidelines. Lotus 1-2-3's system, in convincing its potential audience that producing graphics was an integral part of selling ideas with data and statistical compilations, buried VisiCalc and its certainty that spreadsheet technology ought to look and feel a particular way and perform x-y-z functions.

Modelessness. Whenever possible, designers should avoid systems or features that make users operate in distinctly new or unfamiliar ways. Heightened expectations about computer interfaces in general encourage users to manipulate objects without necessarily understanding their particular functions. Users expect that they can click on anything and, at the same time, reclick something else to re-establish the state that they only moments ago left. Thus, when users encounter a dialogue box providing three options (OK, save, and cancel), they often expect that selecting the previous window or state will return them to a situation that they remember, understand, and want to work with based on new information (that they have gained from the recently displayed dialog box). Thus, systems like Microsoft Office 6.0 now allow individuals to use the capabilities of many different applications—a spreadsheet, graphics package, and word-processing program—from within the workspace of a single document.

Consistency. On-line support should be consistent from one screen to another, and across the various features that inhabit those screens: for example, text boxes, headings, control buttons, menu-item names, dialogue boxes, and so on. Designers should set parameters early in the design process that establish rules for how all interface features will look and operate as individuals employ support functions. If users are encouraged to develop mental models of how support systems operate, the implications of those models should be consistent across all areas. For example, if pressing the help key displays balloon help in the main workspace, that key should also display balloon help when working in dialogue boxes and in error message windows.

Reversibility and Error Recovery. Users often freely experiment with on-line applications. Although such experimentation has important pedagogical value, it can have terrible implications for support system designers. Therefore, it is critical to build *undo* or backup functions into all systems. Users gain confidence and take learning risks knowing that they can backtrack from any mistakes they have made using unfamiliar applications and support. Such user expectations are not impossible to meet if designers provide ways for users to reverse their procedures and easily recover from errors. These features should be accessible at all levels of support system use.

Visual Aesthetics. One crucial design goal is to produce what is sometimes called subjective user satisfaction. Researchers generally have a difficult time quantifying why some users prefer Windows 95 over the Macintosh operating system (OS) over the Unix Motif interface. Individuals who are comfortable using a program in one platform may not want related on-line support to adopt the visual aesthetics of another. This approach is common on cross-platform applications to both standardize and speed product development, but the negative subjective experience of users may defeat those benefits. Critics of Microsoft Word 6.0 for the Macintosh, for example, frequently complain that its interface mimics that of Windows.

Feedback Structures and Context Sensitivity.   Designers should create system and error messages that are meaningful to people. Such user-centered practices derive from carefully assessing the backgrounds, goals, and work contexts of users to determine the particular support forms and spaces in which assistance will be provided. The ability to move over interface objects and get feedback on their functions, or to click on objects and get useful, related explanations about those objects, is extremely valuable but still relatively rare. Context sensitivity can helpfully shorten feedback loops, allowing users to skip traditional but sometimes distracting methods of finding information about on-screen objects (such as indices and tables of contents).

Local Design Issues

Local design issues, though less critical to the overall success of on-line support systems, are nevertheless important to consider. Research suggests that well-designed systems are sometimes still criticized if users find spelling errors or screens that are cluttered and difficult to read, or if users dislike the ethos or writing style of the on-line support system designers [Wright 1980]. At a minimum, therefore, the local design issues that merit serious attention include the following:

- Levels of explanation
- Chunking
- Discourse cues
- Structured headings
- Bulleting and lists
- Iconic markers
- Typographic legibility
- Negative space

Levels of Explanation.   Different contexts and purposes require different levels of explanation. Highly focused on-line help users typically want very concise, even telegraphic information. Elaborations and explanations are more suited to tutorial users, who must build robust mental models to guide subsequent work. For either type of user, however, an inappropriate level of explanation will certainly impede their performance. In almost every case, users familiar with other forms of technical communication—reports, proposals, and memos, for example—expect simple, concise, active sentence and paragraph structures rather than long, digressing passages (although there are notable exceptions, such as the longer narratives often used in tutorials).

Chunking.   According to Miller [1956], users access information in chunks of seven plus or minus two, and research in document design has extended this finding to the ways users access, process, and retain new information. Therefore, spend adequate design time parsing out information packages that users can consistently access when moving from screen to screen. Chunking can also aid consistency by helping designers separate different kinds of information, in small units, that can be consistently placed across a system. The chunk sizes of information for screens tend to be smaller than those for printed pages because of the numerous physical page/screen issues discussed in the Underlying Principles section. Also, however, chunk sizes vary according to many different rhetorical issues, including user expertise. For example, for an expert Unix user, the command <chmod go+r *.html> may be a single chunk of information. For a novice Unix user, that same command should probably be chunked at the character level, resulting in many more pieces of information.

Discourse Cues. When users of print-based texts read, they rely on certain linear constructs or "grammar[s] of the page" [Selfe 1989, see also Bernhardt 1993]: bookmarks, indices, tables of contents, headers, footers, dog-eared or tabbed pages, visual indications of reading depth, and so on. On-line support systems, however, are foreign ground to print-trained readers; they undermine many well-learned access methods by introducing new concepts and reading actions: for example, keyword searching, artificial intelligence support systems, electronic intelligent support systems (EISSs), and wizard-based learning environments. These systems, while being extensions of hardcopy-based tools in many ways, can complicate the expertise and strategies of print-trained readers.

Designers of on-line support systems, therefore, should provide numerous discourse cues that anchor users in information that they have already processed and that they will process in the future [Charney 1987]. When reading linear texts, it is important for readers to predict their experiences and their texts' realities from one paragraph to the next, from one chapter to the next, or from one frame to the next. That is, even in nonlinear situations, users build connections between what they have just processed and what they are currently processing and, ultimately, what they are expecting to process in the very near future.

Structured Headings. Miller's [1956] argument that chunks of seven items plus or minus two are optimal for processing information is very applicable to this particular design guideline. By definition, headings are concise descriptions of the content that they precede. But, concision in headings is a mixed goal: 2-word headings and 10-word headings are least useful to users. Designers should search for some middle ground between concision and verbosity when labeling the information that follows. Studies also suggest that headings should be consistently placed on all screens, and that they should explicitly spell out what users can expect in the section that follows [Felker 1980, Felker et al. 1981].

Bulleting and Lists. Bulleted and numbered lists provide ways of relating large or complicated bodies of information [Carliner 1987]. These structures distill long or complex concepts that, in traditional sentence form, might appear repetitive or confusing. In addition, in some cases lists can be used as **advance organizers** for sections used in a module or for summaries of information already presented. Bulleted lists should be used for items at the same conceptual level in which ordering is unimportant, whereas numbered lists should be used when ordering is critical (for example, in task-oriented or rank-ordered information).

Iconic Markers. Iconic markers are visual items that highlight important elements (warnings and cautions) or help maintain consistency across different screens in the same class. Such graphical elements help readers skim texts for specific types of information. In this way, iconic markers can help designers create a single system that can assist readers with differing expertise levels, goals, and contexts [Horton 1991, Tufte 1990].

Typographic Legibility. The legibility of on-line type can drastically affect the performance of users. Because readers distinguish letters based on their differences from other letters, serif fonts are generally more readable than san-serif fonts. (Serif fonts, such as Times and Palatino, contain ticks or extensions at the ends of letter strokes, whereas sans-serif fonts, such as Helvetica and Futura, have clean ends.) In some contexts, limitations in screen resolution, screen size, or the characteristics of a particular font contradict this guideline, and so designers must think carefully about type styles. For example, low-resolution screens sometimes make serif fonts illegible. Notably, as reading from high-resolution monitors becomes more like reading from printed pages, guidelines for presenting text on line will better emulate those relating to hardcopy documents [Haas and Hayes 1987]. Moreover, because readers process words, at least in part, by their distinctive shapes, the wider variances of mixed-case words are usually more legible than all-upper-case words.

In addition, for contrast or for short amounts of text, designers can employ less legible type families in on-line environments. Headings in a sans-serif font contrast usefully with body text in a serif font; this contrast helps users distinguish among different classes of on-line text. Finally, the relative degree of openness (or negative space) inside individual letters of fonts can effect legibility. The lowercase letter *a* in Palatino holds more negative space than the same letter in Times, making Palatino more readable in some contexts (such as reading with low-resolution monitors).

Negative Space. Negative space in on-line support systems relates to those screen areas not containing text. Although it might seem wasteful to provide adequate negative space, leaving such space is important for at least two reasons: it helps readers see on-line elements as discrete and separate; and it helps designers limit the amount of information (and number of concepts) they provide on any single screen. Notably, if adequate negative space is provided, only about 1/3 of a printed page will fit on a single screen.

Evaluating On-Line Support Systems

Evaluating usability should be a central component of on-line support system development. Usability refers to the degree of user effectiveness and efficiency in operating on-line support to complete tasks and achieve goals. Although measuring user effectiveness and efficiency is the core goal in usability testing, it is also useful for checking the accuracy, consistency, and completeness of included information. From our perspective, usable systems are accessible, maintainable, visually consistent, comprehensive, accurate, and oriented around the tasks that users must perform [Mehlenbacher 1993].

Historically, the task of evaluating usability has occurred either after on-line support systems were already shipped or in very late stages of product development (beta or postbeta). This kind of **summative evaluation**, aimed at determining the overall quality of systems and their contents, is useful in a number of ways: for determining the ability of users to accomplish their targeted tasks and goals using well-formed products, for methodically collecting user feedback for subsequent version releases, and for comparing the features and designs of finished on-line support systems with those of competing products.

More recent approaches to evaluation, however, expand this rather limited and late-stage approach [Carroll and Rosson 1985]. It is now quite common for developers to intersperse their planning and designing stages with **formative evaluations**—evaluations that contribute in direct ways to the developments and uses of on-line support systems. Viewing evaluation as an integral part of system construction has many important advantages: it enlists real users or user surrogates, thereby adding external (customer-centered) perspectives to otherwise internal (developer-centered) ways of knowing; it helps determine what kinds of major revisions and adjustments might be accomplished prior to significant work investments; and it discourages premature closure on system designs until those designs are clarified and refined in iterative ways by members of a user base.

Evaluating the usability of on-line support systems, either through summative or formative techniques, occurs in both laboratory (controlled) and workplace (natural) contexts, as well as through other research techniques employed at greater distances. Laboratory evaluations allow developers to isolate, observe, and track user task performances in systematic ways. These evaluations primarily provide quantitative feedback measured along predetermined scales, although attitudinal and other types of qualitative information can be collected. Because special equipment is required in these research environments—a physical space, video camera, audio recorder, and/or one-way mirror, for instance—developers with limited usability resources may rely on workplace evaluations and other data collection techniques.

Workplace evaluations take place in the normal environments of users. As such, these evaluations allow developers to examine how well their on-line support systems are employed and understood within the day-to-day contexts of users, their tasks, and their organizations. Notably, workplace evaluations are still informed by systematic approaches to research. In fact, the qualitative feedback gathered in these instances is commonly informed by empirical designs [Bjerknes et al. 1987, Blomberg and Henderson 1990, Piela et al. 1991]. Because workplace evaluations occur at customer sites, they require special access and a high degree of trust between users and developers [Brown and Duguid 1992].

Other research techniques allow developers to supplement usability data collected from laboratory and workplace contexts. Both quantitative and qualitative feedback can be gathered by conducting telephone interviews, administering surveys, compiling literature reviews, and reading related studies. Data collected from all contexts, when examined together, can provide useful directions for both system developments and revisions [MacNealy 1992].

Importantly, the research contexts that developers select will influence their usability results and conclusions in central ways. Experimental studies conducted in controlled laboratories, for example, may yield task completion rates that are either unrealistic or unmeaningful within normal working environments. For this reason, triangulate research contexts whenever possible. Also, the contexts that developers select should support and not drive their research goals, which in turn help them articulate the appropriate forms of evaluation needed for evaluating usability.

Forms of Evaluation

There are many different forms of evaluation available to developers of on-line support systems. We briefly describe nine forms that might be employed during early-, middle-, and late-phase efforts. These forms are summarized in Table 74.4. Although many other techniques exist for testing usability, these represent common, industry-standard approaches. For comprehensive discussions of these and other forms, see Goubil-Gambrell [1992], Nielsen [1993], and Dumas and Redish [1993].

TABLE 74.4 Forms of Evaluation During Development Stages

| | Phase of Development: | | |
Form of Evaluation	Early	Middle	Late
Focus groups	X		
Task analyses	X		
Prototype walkthroughs	X		
Protocol analyses		X	
Performance tests		X	
Contextual inquiries		X	
User-advocate reviews			X
Competitive analyses			X
User surveys			X

Early-Phase Evaluations. Focus groups, task analyses, and prototype walkthroughs are useful forms of evaluation during preplanning, planning, and developmental stages. Focus groups help developers explore the concerns, beliefs, and preferences of real users, as well as understand their perspectives on an organization and its existing products and practices (if any). During focus group sessions, which usually last between 1 and 3 h, 5–10 users respond to a series of preplanned questions formulated by experienced researchers. These sessions are best lead by individuals who have experience with group dynamics and with focusing and refocusing group discussions in both efficient and productive ways.

As discussed in the planning section, task analyses identify the complex range of actions that users must perform, and they further divide and subdivide these actions into subactions that the planned on-line support systems should document and discuss. In these analyses, developers both interview and observe users to understand their typical ways of working and accomplishing tasks. From this information, they extract the working goals of users, the system features and other ancillary information, both print-based and on line, that support these goals, and the full procedures that users follow to accomplish all tasks.

Prototype walkthroughs provide early opportunities in which users can evaluate the planned contents and designs of on-line support systems [Dillard 1992]. As the name of this technique suggests, users explore working models of systems and attempt to accomplish tasks using a limited number of features. From data collected by observing and interviewing users, developers proceed with more stable and aggressive instantiations. Prototype walkthroughs allow designers to collect user feedback prior to investing substantial resources in system development.

Middle-Phase Evaluations. Protocol analyses, performance tests, and contextual inquiries are useful forms of evaluation during pre-beta and beta stages. Protocol analyses ask individuals to verbalize their thoughts while using on-line support systems and/or after they have finished. These thoughts, which are recorded on audiotape and or/videotape, help reveal both confusing and missing information [Lewis 1982]. They also reveal the kinds of expectations that users typically have. Employing this approach, developers target a limited number of representative users and collect richly textured insights into their normally hidden thought processes.

Performance tests measure how well users accomplish a predetermined set of tasks measured along predetermined scales [Liebried and McNair 1992]. Users are tested either in laboratory or workplace contexts, and the data collected from these evaluations is quantitative and/or qualitative. During performance tests, it is important to clearly delineate where user tasks begin and where they end. Common measurements during performance tests include the amount of time taken to complete particular tasks, the number of errors made during task performances, and the most and least frequently used features in a system.

Contextual inquiries examine individual users and how they accomplish tasks in their own working environments [Anderson 1989]. Through observations and interviews, developers construct models of user behavior in real-world settings, determining the ways in which their on-line systems should support those behaviors. Obviously, this technique is also extremely useful during developmental phases.

Late-Phase Evaluations. User-advocate reviews, competitive analyses, and user surveys are useful forms of evaluation during post-beta stages and after on-line support systems have been distributed to customers. As opposed to enlisting real users, user-advocate reviews ask usability specialists to assess the merits and demerits of on-line support systems. Like real users, these specialists can perform a wide range of preplanned tasks identified by developers. Unlike real users, however, they pay particular attention to how well on-line support systems reflect sound human–computer interaction principles.

Competitive analyses examine the strengths, weaknesses, and overall designs of competing products, helping developers differentiate their systems from others in the marketplace. Such analyses often occur in laboratory settings and measure system performances under a wide variety of conditions and with a wide range of hardware and software configurations.

And, user surveys help developers understand customer requirements and systematically collect their feedback. Surveys are generally inexpensive to administer and can provide users with anonymous avenues for providing honest commentary. Importantly, they require researchers who are skilled at developing both valid and useful lines of inquiry, which are both targeted and open ended. User surveys can be administered by mail, electronic mail, over the telephone, or in face-to-face environments [Zimmerman and Muraski 1995].

Procedures for Evaluation

In general, there are eight sequential steps in evaluating on-line support systems:

1. Form an evaluation team
2. Identify evaluation goals
3. Select evaluation methods
4. Develop realistic scenarios
5. Enlist real users
6. Implement the evaluation
7. Analyze the data
8. Distribute the findings

1. Form an evaluation team. Include researchers experienced in conducting usability tests and in employing empirical methods, developers involved in the project, personnel with access to users, product engineers, and marketing specialists.
2. Identify what you hope to accomplish during the evaluation. Clearly state your usability goals in specific terms, paying particular attention to the concerns of all team members, and to what constitutes success and failure during these evaluations.
3. Select appropriate evaluation methods and contexts for accomplishing your objectives. Identify the methods and contexts that are realistically available to you and your evaluation team. Select those which can best support your usability goals.
4. Develop scenarios and real tasks that can help you achieve your evaluation objectives. Make each scenario or task a discrete module that can be easily measured, observed, and/or discussed. Use both direct and indirect approaches. The former asks users to perform specific actions and tasks, whereas the latter provides situations in which users select their actions from within emerging situations. Create a **test checklist** for reference.

5. Enlist real users to participate in the evaluation. Select a range of individuals, including representative users, novices, and experts. The kind and number of users required will depend on the evaluation forms and contexts you select. Be sure to provide incentives for user participation.

6. Implement your evaluation. Explain the process to participants, build a comfort level for them, get signatures on nondisclosure agreements (if required or appropriate), conduct the evaluation, and debrief the participants. Importantly, there are direct connections between your testing techniques and the usefulness of your results. Work carefully and validly.

7. Evaluate the results of your tests. Examine all data—quantitative and qualitative—that you have collected, and consider the effects of form and context on your results. Interpret the results of your research modestly. Be careful not to generalize from what you learn from one or two users working in their own contexts.

8. Formalize and distribute the results of your evaluation. Share your methodologies, results, conclusions, and recommendations with all team members in a completion report.

74.4 Research Issues and Summary

While planning, designing, and evaluating on-line support systems, there are many open research issues that can complicate the development practices we have outlined. From our perspective, the most difficult challenges relate to interpersonal, cultural, and pedagogical areas, both internally (among development team members) and externally (among customers). Specifically, five open research issues that remain under discussed in the literature on on-line support systems include the following: collaboration, diversity in the workplace, translation and internationalization, user motivation and aesthetic response, and designer training.

Collaboration

As we have already mentioned, no one designer is likely to possess all of the diverse skill sets needed to effectively develop on-line support systems. Interdisciplinary collaborative teams, therefore, are commonly assembled in both small and large organizations, and they normally include many different kinds of specialists. Although collaboration has become an integral part of on-line support system development, collaborative activities are not always successful or without their share of complications. As Selber et al. [1996] note, difficulties and failures among and within work groups are numerous and wide ranging for a variety of reasons: group leadership may be ineffective, individual agendas may conflict with group agendas, individuals may be unwilling or unable to discuss differences of opinion, unequal power relations may exist among group members due to various social forces such as organizational rank, and simple logistical issues of coordinating tasks and schedules may exist.

Moreover, the ways in which groups collaborate are expanding in both time and space, and this expansion raises many new questions about the nature of collaboration. Using computer-supported cooperative work tools, designers can now interact on-line in both asynchronous and synchronous modes, and such modes, many claim, have the potential to enrich collaborative activities in both predictable and unpredictable ways (see Greenberg [1991]). Importantly, we would argue that such technological possibilities represent only one factor in fostering productive group work. Changes in organizational structures, group dynamics, and institutional reward systems are also needed for new electronic forms of collaboration to be productively integrated into existing corporate cultures.

Diversity in the Workplace

As workplace diversity increases, on-line support system designers will be working in, and developing products for, groups not adequately addressed by old ways of communicating, most of which were developed for homogeneously segregated audiences. Traditional conventions and procedures for communicating, although often considered normal or standard, commonly represent only one (potentially limiting) approach. This diversity issue is not one of political correctness, but of fundamental concerns for productivity and usability. In some cases, relying on traditional uses of language offends colleagues and users: generic uses of "he," to name but one controversial example, constructs users as explicitly male (see Frank and Treichler [1989] for a comprehensive discussion of language, gender, and professional writing issues). In other cases, communication methods and learning styles that work well for certain groups may actually inhibit the learning of others. The difference in high-context cultures such as the U.S. and low-context cultures such as Japan is one, often-cited example; Japanese readers may find American styles too abrupt, whereas American users may find Japanese styles overly formal (see Hoft [1995] for many additional concerns). General trends of increasing diversity in both the workplace and marketplace will only multiply the need for additional research in this important area.

Translation and Internationalization

The increasing diversity in the workplace already noted relates to translation and internationalization issues in central ways. The growth of global economic systems and marketplaces frequently demands that on-line support system designers recognize that their work will be used by people in other countries and cultures. Although the research and development of translation tools has assisted designers by automating some portions of the translation process, more work is needed before these systems begin to approach the skills of human translators [Spalink 1995]. In addition, although literal translations can sometimes provide partial support for users, broader cultural differences can impede usability to a similar degree. Over the last few decades, researchers have begun to investigate differences in communication styles (direct vs. indirect), learning patterns (visual vs. verbal), and work habits (hierarchical vs. team) evident in different cultures. Although some portions of translation, especially literal translation, can be easily automated or outsourced, cultural differences may reside in the very framework of support systems and their interfaces. Issues of translation and internationalization, therefore, are critical for designers to consider.

User Motivation and Aesthetic Response

Capturing the particular emotive reactions of users to on-line support systems is an important research area that has received limited attention. Less cognitive than affective, subjective user reactions to human–computer interfaces and to the content of on-line support systems can encourage individuals, for example, to embrace one particular application (Windows 3.1) over another (Windows 95), even though the application that they reject may be superior in form and function. Still, with the exception of Shneiderman [1987], who has built a useful but limited likert-scale test aimed at uncovering subjective user reactions to software, few researchers have emphasized the role of emotion and its affect in human–computer interaction. Moreover, it is clear that researchers are not likely to study what their research methods cannot detect, and so various efforts are currently in place to re-examine and re-evaluate the methods employed to gauge user satisfaction. Researchers have begun to contrast and compare more traditional quantitative methods with emerging ethnographic methods [MacNealy 1992, Sullivan and Spilka 1992], and to explore how people work apart from how they work with our tools [Bannon 1995]. As we better understand user preferences, inclinations, and generalizable patterns across cultural, political, and educational levels, user motivation and aesthetic response will gain more attention in on-line support system development.

Developer Training

Because developing on-line support systems is a relatively new type of computer-based work, many degree programs and professional development seminars are rightly tentative in claiming one absolute curriculum that is appropriate for this educational area. In technical communication programs alone, for example, there are different courses of study in English departments, humanities departments, communication and rhetoric departments, and technical journalism departments. This list, of course, does not include those many programs in departments outside the disciplines traditionally interested in language studies: computer science, engineering, human factors, instructional design, psychology, and so on. To complicate curricular matters further, each of these programs privileges particular theoretical perspectives, research methodologies, and ways of knowing that guide how they examine human–computer interaction problems. We would argue, however, that such diversity is a strength rather than weakness given the complexity of most design activities.

Although the orientation of any particular program should reflect the strengths of its faculty and facilities, in general curricula for preparing on-line support system developers should be interdisciplinary and balance production, literacy, and humanistic concerns [Selber 1994]. Even ostensibly simple development tasks, such as selecting background colors or designing icons, require individuals to blend research findings and technical competencies from many different areas: in this case, visual design, psychology, and rhetoric. This example, in addition, highlights the need for programs to balance the kinds of issues they centrally discuss. A focus on production processes would help developers learn the mechanics of selecting background colors and designing icons. Further discussions of literacy, however, might reveal that certain colors connote particular meanings in various cultural contexts or that cultural diversity complicates icon shapes commonly perceived as standard. Even further considerations of humanistic concerns might demonstrate the ways in which icons (and other visual elements) can be unproductively gendered. Without this kind of interdisciplinary work and balanced training, designers will be unprepared to make the kinds of rhetorical judgments we have outlined in this chapter.

Summary

The three development areas we have discussed—planning, designing, and testing—require designers to consider rhetorical as well as technical issues, recursive rather than linear approaches, and qualitative as well as quantitative methods for understanding user behavior. Our increasingly sharper focus on the human dimensions of on-line support highlights the complexity of productively understanding human–computer interactions within the richly textured cultural contexts of users and their work.

Underlying principles encourage designers to consider the advantages and disadvantages of on-line support (paper-based solutions may still be appropriate in certain cases); the physical and rhetorical differences between print-based and on-line support, particularly in terms of organizational, navigational, and contextual structures; and rhetorical frameworks for on-line support, users, their goals, and their time/space frames.

Best practices can be divided into planning, designing, and testing phases, with each of these phases potentially affecting the other two. At a minimum, planning requires designers to conduct needs assessments and develop specifications, two early activities essential to meeting both external (user) and internal (designer/manager) needs. Designing encourages individuals to consider both global and local issues, and to gauge the success of their strategies in terms of user satisfaction. Evaluating includes both formative and summative usability tests, and provides a framework for such testing during early, middle, and late phases of development.

As the technological aspects of on-line support systems and authoring applications continue to evolve, it will be increasingly important for developers to consider the rhetorical dimensions of their work. The best solutions for users will not come solely from understanding technological innovations but from support system environments designed with a wide range of technological, organizational, cultural, and instructional factors in mind.

Defining Terms

Advance organizers: Overview statements, often in bulleted or list form, that provide a summary of what some textual unit (an entire on-line support system, module, or submodule) will contain. They often discuss how that unit is organized and how it might be used given users with differing knowledge bases and goals.

Chunking: A design stage in which developers divide the content of their on-line support systems into discrete, easily identifiable units.

Documentation: A type of on-line support system that provides reference material or discusses extensive procedures. It has a narrower pedagogical scope than tutorials but a broader one than help.

Editorial review: A development stage, usually late in the process, in which on-line support systems are examined for consistency, style, mechanics, and completeness.

Formative evaluations: A type of evaluation occurring while on-line support systems are being developed. In general, they provide feedback that can be used to reconceive or revise systems before they are finished.

Help: A type of on-line support system that provides brief information for solving particularly pressing problems. This kind of support often includes context sensitivity. It has a narrower pedagogical scope than both tutorials and documentation.

Managerial review: A development stage, often occurring many different times during the life of a project, in which on-line support systems are examined for business-related concerns (project costs, customer requests, and so on).

On-line support systems: Digital forms of assistance—namely, tutorials, documentation, and help— that aid computer-based learning and task-oriented activities.

Primary work: Job-related tasks, such as writing a newsletter, managing a budget, or learning a procedure, for which individuals rely on computer technologies for their successful completion.

Summative evaluations: A type of evaluation occurring after on-line support systems are finished or in very late stages of development. In general, they measure the success of final design decisions.

Technical review: A development stage, often occurring many different times during the life of a project, in which on-line support systems are examined for performance issues and by subject-matter experts.

Test checklist: An outline of usability test procedures. It helps remind usability testers of their planned procedures and of the proper order of these procedures.

Tutorials: A type of on-line support system that provides a supportive and safe educational environment for learning processes or products. They usually include interactivity, and the content and pace are controlled, to some degree, by users. Tutorials have a broader pedagogical scope than both documentation and help.

User-centered design: Development practices that privilege rhetorical and humanistic considerations (such as users, their goals, and their social contexts) over technological considerations (such as system capabilities).

References

Anderson, R. 1989. Notes about some experiences with contextual research. *SIGCHI Bull.* 20(4):29–30.

Bannon, L. J. 1995. The politics of design—representing work. *Commun. ACM* 38(9):66–68.

Beabes, M. A. and Flanders, A. 1995. Experiences with using contextual inquiry to design information. *Tech. Commun.* 42(3):409–420.

Benson, T. 1991. Challenging global myths: international quality study. *Ind. Week* 240(19):12–25.

Bernhardt, S. 1993. The shape of text to come: the texture of print on screens. *Coll. Comp. Commun.* 44(2):151–175.

Bjerknes, G., Ehn, P., and Kyng, M. 1987. *Computers and Democracy.* Avebury, London, England.

Blomberg, J. and Henderson, A. 1990. Reflections on participatory design: lessons from the Trillium experience. In *Proc. CHI '90: Human Factors in Computing Systems*, pp. 353–360. ACM, Seattle, WA.

Boy, G. 1992. Computer integrated documentation. In *Sociomedia*, E. Barrett, ed., pp. 507–532. MIT Press, Cambridge, MA.

Brown, J. S. and Duguid, P. 1992. Enacting design for the workplace. In *Usability: Turning Technologies into Tools*, P. S. Adler and T. A. Winograd, eds., pp. 164–197. Oxford University Press, New York.

Caernarven-Smith, P. 1990. Annotated bibliography on costs, productivity, quality, and profitability in technical publishing: 1956–1988. *Tech. Commun.* 37(2):116–121.

Card, S. K., Moran, T. P., and Newell, A. 1983. *The Psychology of Human–Computer Interaction.* Lawrence Erlbaum, Hillsdale, NJ.

Carliner, S. 1987. Lists: the ultimate organizer for engineering writing. In *Writing and Speaking in the Technology Professions: A Practical Guide*, D. Beer, ed., pp. 53–56. IEEE Press, New York.

Carroll, J. 1990. *The Nurnberg Funnel.* MIT Press, Cambridge, MA.

Carroll, J. M. and Rosson, M. B. 1985. Usability specifications as a tool in iterative development. In *Advances in Human–Computer Interaction*, H. R. Hartson, ed., pp. 1–28. Ablex, Norwood, NJ.

Charney, D. 1987. Comprehending non-linear text: the role of discourse cues and reading strategies. In *Hypertext '87 Papers*, pp. 109–120. ACM, Chapel Hill, NC.

Dillard, J. D. 1992. Maximizing documentation usability and product quality through structured rapid prototyping. In *Proc. the 38th Int. Tech. Commun. Conf.*, pp. 119–122. Society for Technical Communication, Arlington, VA.

Duffy, T. M., Palmer, J. E., and Mehlenbacher, B. 1992. *Online Help: Design and Evaluation.* Ablex, Norwood, NJ.

Dumas, J. and Redish, J. A. 1993. *A Practical Guide to Usability Testing.* Ablex, Norwood, NJ.

Felker, D. 1980. *Document design: a review of the relevant research.* American Institutes for Research, Washington, DC.

Felker, D., Pickering, F., Charrow, V., Holland, V., and Redish, J. 1981. *Guidelines for Document Designers.* American Institutes for Research, Washington, DC.

Frank, F. W. and Treichler, P. *Language, Gender, and Professional Writing.* 1989. MLA, New York.

Goubil-Gambrell, P. 1992. A practitioner's guide to research methods. *Tech. Commun.* 39(4):582–591.

Greenberg, S., ed. 1991. *Computer-supported Cooperative Work and Groupware.* Academic Press, New York.

Haas, C. 1989. Seeing it on the screen isn't really seeing it: computer writers' reading problems. In *Critical Perspectives on Computers and Composition Instruction*, G. Hawisher and C. Selfe, eds., pp. 16–29. Teacher's College Press, New York.

Haas, C. and Hayes, J. R. 1987. *Effects of text display variables on reading tasks: computer screen vs. hard copy.* ERIC No. ED 260 387. Communications Design Center, Carnegie Mellon University, Pittsburgh, PA.

Hoft, N. 1995. *International Technical Communication: How to Export Information about High Technology.* Wiley, New York.

Horton, W. 1991. *Illustrating Computer Documentation: The Art of Presenting Information Graphically on Paper and Online.* Wiley, New York.

Johnson-Eilola, J. 1996. *Nostalgic Angels: Rearticulating Hypertext Writing.* Ablex, Norwood, NJ.

Lewis, C. 1982. Using the "thinking-aloud" method in cognitive interface design. In *Res. Rep. 9265-40713*, pp. 1–6. IBM Thomas J. Watson Research Center, Yorktown Heights, NY.

Liebried, K. H. and McNair, C. J. 1992. *Benchmarking: A Tool for Continuous Improvement.* Harper Collins, New York.

MacNealy, M. S. 1992. Research in technical communication: a view of the past and a challenge for the future. *Tech. Commun.* 39(4):533–551.

Mehlenbacher, B. 1993. Software usability: choosing appropriate methods for evaluating on-line systems and documentation. In *SIGDOC 93: 11th Annu. Int. Conf. Proc.*, pp. 209–222. ACM, New York.

Mehlenbacher, B. 1995. Charting the future of technical communication: SIGDOC 94 and the great divide. *J. Comput. Doc.* 19(2):15–21.

Mehlenbacher, B., Duffy, T. M., and Palmer, J. E. 1989. Finding information on a menu: linking menu organization to the user's goals. *Human–Comput. Interaction* 4(3):231–251.

Miller, G. 1956. The magical number seven, plus or minus two: some limits on our capacity for processing information. *Psychol. Rev.* 63:81–87.

Nielsen, J. 1993. *Usability Engineering.* Academic Press, Boston, MA.

Paradis, J. 1991. Text and action: the operator's manual in context and in court. In *Textual Dynamics of the Professions: Historical and Contemporary Studies of Writing in Professional Communities*, C. Bazerman and J. Paradis, eds., pp. 256–278. University of Wisconsin Press, Madison, WI.

Petrauskas, B. 1991. Online reference system design and development. In *Perspectives on Software Documentation: Inquiries and Innovations*, T. Barker, ed., pp. 243–272. Baywood, Amityville, NY.

Piela, P., McKelvey, R., and Mehlenbacher, B. 1991. Integrating usability and participatory design into basic engineering design research. *Eng. Design Res. Center Tech. Rep.* Carnegie Mellon University, Pittsburgh, PA.

Rubens, P. and Krull, R. 1985. Applications of research on document design to on-line displays. *Tech. Commun.* 37(4):29–34.

Selber, S. 1994. Beyond skill building: challenges facing technical communication teachers in the computer age. *Tech. Commun. Q.* 3(4):365–390.

Selber, S. 1995. Metaphorical perspectives on hypertext. *IEEE Trans. Prof. Commun.* 38(2):59–67.

Selber, S. Forthcoming. The politics and practice of media design. In *Foundations for Teaching Technical Communication: Theory, Practice, and Program Design.*, K. Staples and C. Ornatowski, eds. Ablex, Norwood, NJ.

Selber, S., McGavin, D., Klein, W., and Johnson-Eilola, J. 1996. Issues in hypertext-supported collaborative writing. In *Nonacademic Writing: Social Theory and Technology*, A. H. Duin and C. Hansen, eds., pp. 257–280. Lawrence Erlbaum, Hillsdale, NJ.

Selfe, C. 1989. Redefining literacy: the multilayered grammars of computers. In *Critical Perspectives on Computers and Composition Instruction*, G. Hawisher and C. Selfe, eds., pp. 3–15. Teacher's College Press, New York.

Selfe, C. and Selfe, J. 1994. The politics of the interface: power and its exercise in electronic contact zones. *Coll. Comp. Commun.* 45(4):480–504.

Shneiderman, B. 1987. *Designing the User Interface: Strategies for Effective Human–Computer Interaction.* Addison–Wesley, Reading, MA.

Skelton, T. M. 1992. Testing the usability of usability testing. *Tech. Commun.* 39(3):343–359.

Smith, D. 1994. Estimating costs for documentation projects. In *Publications Management: Essays for Professional Communicators*, O. Allen and L. Demming, eds., pp. 143–151. Baywood, Amityville, NY.

Spalink, K. 1995. Document design with translation in mind. *Intercom* 42(7):38–43.

Suchman, L. A. 1987. *Plans and Situated Actions: The Problem of Human–Machine Communication.* Cambridge University Press, New York.

Sullivan, P. and Spilka, R. 1992. Qualitative research in technical communication: issues of value, identity, and use. *Tech. Commun.* 39(4):592–606.

Tufte, E. R. 1990. *Envisioning Information.* Graphics Press, Cheshire, CT.

Winn, W. 1990. Encoding and retrieval of information in maps and diagrams. *IEEE Trans. Prof. Commun.* 33(3):103–107.

Winograd, T. and Flores, F. 1987. *Understanding Computers and Cognition: A New Foundation for Design.* Addison–Wesley, Reading, MA.

Wright, P. 1980. Usability: the criterion for designing written information. In *Processing Visible Language*, Vol. 2, P. A. Koler, M. E. Wrolstad, and H. Bouma, eds., pp. 183–206. Plenum Press, New York.

Zimmerman, D. and Muraski, M. L. 1995. *The Elements of Information Gathering: A Guide for Technical Communicators, Scientists, and Engineers.* Oryx, Phoenix, AZ.

Further Information

In addition to the sources already cited in this chapter, the following books and journals are additional recommended readings. Although the following lists are in no way comprehensive, they represent a starting place for further information about planning, designing, and evaluating effective and usable on-line support systems.

Recommended Books:

Adler, P. S. and Winograd, T. A., eds. 1992. *Usability: Turning Technologies into Tools.* Oxford University Press, New York.

Barker, T., ed. 1991. *Perspectives on Software Documentation: Inquiries and Innovations.* Baywood, Amityville, NY.

Barnum, C. M. and Carliner, S. 1993. *Techniques for Technical Communicators.* Macmillan, New York.

Barrett, E., ed. 1992. *Sociomedia: Multimedia, Hypermedia, and the Social Construction of Knowledge.* MIT Press, Cambridge, MA.

Brooks, T. 1991. *An Introduction to Human–Computer Interaction.* Lawrence Erlbaum, Hillsdale, NJ.

Brockmann, R. 1992. *Writing Better Computer User Documentation: From Paper to Online*, 2nd ed. Wiley, New York.

Doheny-Farina, S., ed. 1988. *Effective Documentation: What We Have Learned from Research.* MIT Press, Cambridge, MA.

Horton, W. 1990. *Designing and Writing Online Documentation: Help Files to Hypertext.* Wiley, New York.

Laurel, B., ed. 1990. *The Art of Human–Computer Interface Design.* Addison–Wesley, Reading, MA.

Norman, D. 1988. *The Design of Everyday Things.* Basic, New York.

Recommended Journals:

ACM Hypertext
Communications of the ACM
Human–Computer Interaction
IEEE Transactions on Professional Communication
International Journal of Human–Computer Interaction
International Journal of Man-Machine Studies
Journal of Computer-Based Instruction
SIGCHI Bulletin
Technical Communication
The Journal of Computer Documentation

VIII

Operating Systems and Networks

Eugene H. Spafford, Section Adviser
Purdue University

O PERATING SYSTEMS FORM THE SOFTWARE INTERFACE between the computer and its applications. This section covers their analysis and design, their performance, and their special challenges in a networked computing environment. Topics include process scheduling, memory management, synchronization, I/O management, scheduling, and Internet protocols. Also important in this area are the complementary concerns for network security and personal privacy.

75

What Is an Operating System?

Raphael Finkel
University of Kentucky

75.1 Introduction

In brief, an operating system is the set of programs that controls a computer. Some operating systems you may have heard of are Unix, Mach, MS-DOS, MS-Windows, Windows/NT, Chicago, OS/2, MacOS, VMS, MVS, and VM. Some of these (Mach and Unix) have been implemented on a wide variety of computers, but most are specific to a particular architecture, such as the Digital Vax, the Intel 286, the Motorola 68000, the IBM 360, and their successors.

Controlling the computer involves software at several levels. We will distinguish kernel services, library services, and application-level services, all of which are part of the operating system. These services can be pictured as in Fig. 75.1. Applications are run by processes, which are linked together with libraries that

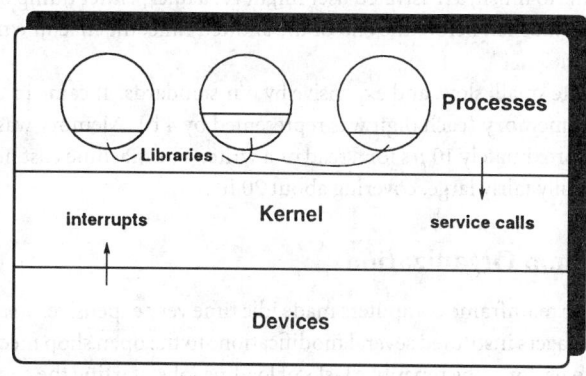

FIGURE 75.1 Operating system services.

perform standard services such as formatting output or presenting information on a display. The kernel supports the processes by providing a path to the peripheral devices. It responds to service calls from the processes and interrupts from the devices.

This chapter discusses how operating systems have evolved, often in response to architectural advances. It then examines the goals and organizing principles of current operating systems. Many books describe operating systems concepts (Deitel [1992], Finkel [1988], Krakowiak [1988], Lane and Mooney [1988], Milenković [1992], Silberschatz and Galvin [1994], Stallings [1992], Tanenbaum [1992]), and specific operating systems (Custer [1993], Kenah and Bate [1984], Kogan and Rawson [1988], Leffler et al. 1989).

75.2 Historical Perspective

Operating systems have undergone enormous change over the years. The changes have been driven primarily by the hardware facilities and their cost and secondarily by the applications that **users** have wanted to run on the computers.

Open Shop Organization

The earliest computers were massive, extremely expensive, and difficult to use. Users would sign up for blocks of time during which they were allowed hands-on exclusive use of the computer. The user would repeatedly load a program into the computer through a device such as a card reader, watch the results, and then decide what to do next.

A typical session on the IBM 1620, a computer in use around 1960, involved several steps in order to compile and execute a program. First, the user would load the first pass of the Fortran compiler. This operation involved clearing main store by typing a cryptic instruction on the console typewriter; putting the compiler, a 10-in stack of punched cards, in the card reader; placing the program to be compiled after the compiler in the card reader; and then pressing the load button on the reader. The output would be a set of punched cards called *intermediate output*. If there were any compilation errors, a light would flash on the console, and error messages would appear on the console typewriter. Assuming everything had gone well so far, the next step would be to load the second pass of the Fortran compiler just like the first pass, putting the intermediate output in the card reader as well. If the second pass succeeded, the output was a second set of punched cards called the *executable deck*. The third step was to shuffle the executable deck slightly, load it along with a massive subroutine library (another 10 in of cards), and observe the program as it ran.

The facilities for observing the results were limited: console lights, output on a typewriter, punched cards, and line-printer output. Frequently, the output was wrong. Debugging often took the form of peeking directly into main store and even patching the executable program by using console switches. If there was not enough time to finish, a frustrated user might get a line-printer dump of main store to puzzle over at leisure. If the user finished before the end of the allotted time, the machine might sit idle until the next reserved block of time.

The IBM 1620 was quite small, slow, and expensive by our standards. It came in three models, ranging from 20 to 60 K digits of **memory** (each digit was represented by 4 b). Memory was built from magnetic cores, which required approximately 10 μs for a read or a write. The machine cost hundreds of thousands of dollars and was physically fairly large, covering about 20 ft^2.

Operator-Driven Shop Organization

The economics of massive mainframe computers made idle time very expensive. In an effort to avoid such idleness, installation managers instituted several modifications to the open shop mechanism just outlined. An **operator** was hired to perform the repetitive tasks of loading jobs, starting the computer, and collecting the output. The operator was often much faster than ordinary users at such chores as mounting cards and magnetic tapes, and so the setup time between job steps was reduced. If the program failed, the operator could have the computer produce a dump. It was no longer feasible for users to inspect main store or patch programs directly. Instead users would submit their runs, and the operator would run them as soon as possible. Each user was charged only for the amount of time the job required.

The operator often reduced setup time by batching similar job steps. For example, the operator could run the first pass of the Fortran compiler for several jobs, save all of the intermediate output, then load the second pass and run it across all of the intermediate output that had been collected. In addition, the operator could run jobs out of order, perhaps charging more for giving some jobs priority over others. Jobs that were known to require a long time could be delayed until night. The operator could always stop a job that was taking too long.

The operator-driven shop organization prevented users from fiddling with console switches to debug and patch their programs. This stage of operating system development introduced the long-lived tradition of the users' room, which had long tables often overflowing with oversized fan-fold paper, and a quietly desperate group of users debugging their programs until late at night.

Off-Line Loading

The next stage of development was to automate the mechanical aspects of the operator's job. First, input to jobs was collected **off line** by a separate computer (sometimes called a *satellite*) whose only task was the transfer from cards to tape. Once the tape was full, the operator mounted it on the main computer. Reading jobs from tape is much faster than reading cards, so less time was occupied with **input/output**. When the computer finished the jobs on one tape, the operator would mount the next one. Similarly, output was generated onto tape, an activity that is much faster than punching cards. This output tape was converted to line-printer listings offline.

A small **resident monitor** program, which remained in main store while jobs were executing, reset the machine after each job was completed and loaded the next one. Conventions were established for control cards to separate jobs and specify their requirements. These conventions were the beginnings of command languages. For example, one convention was to place an asterisk in the first column of control cards, to distinguish them from data cards. The compilation job we just described could be specified in cards that looked like the following:

```
*JOB SMITH           The user's name is Smith.
*    PASS CHESTNUT   Password so others can't use Smith's account
*    OPTION TIME=60  Limit of 60 seconds
*    OPTION DUMP=YES Produce a dump if any step fails.
*STEP FORT1          Run the first pass of the Fortran compiler.
*    OUTPUT TAPE1    Put the intermediate code on tape 1.
*    INPUT FOLLOWS   Input to the compiler comes on the next cards.
     ...             Fortran program
*STEP FORT2          Run the second pass of the Fortran compiler.
*    OUTPUT TAPE2    Put the executable deck on scratch tape 2.
*    INPUT TAPE1     Input comes from scratch tape 1.
*STEP LINK           Link the executable with the Fortran library.
*    INPUT TAPE2     First input is the executable.
*    INPUT TAPELIB   Second input is a tape with the library.
*    OUTPUT TAPE1    Put load image on scratch tape 1.
*STEP TAPE1          Run whatever is on scratch tape 1.
*    OUTPUT TAPEOUT  Put output on the standard output tape.
*    INPUT FOLLOWS   Input to the program comes on the next cards.
     ...             Data
```

The resident monitor had several duties:

- To interpret the command language
- To perform rudimentary accounting
- To provide device-independent input and output by substituting tapes for cards and line printers

This last duty is an early example of information hiding and abstraction: programs would think they were directing output to cards or line printers, but in fact, the output was going elsewhere. Programs would use subroutines provided by the resident monitor for input/output to both logical devices (cards, printers) and physical devices (actual tape drives).

The early operating systems for the IBM 360 series of computer used this style of control. Large IBM 360 installations could cost millions of dollars, so any time that was not spent computing was wasted.

Spooling Systems

Computer architecture advanced throughout the 1960s: Input/output units were designed to run at the same time the computer was computing. They generated an interrupt when they finished reading or writing a record instead of requiring the resident monitor to track their progress. An interrupt causes the computer to save some critical information (such as the current program counter) and to branch to a location specific to the kind of interrupt. Device-service routines, known as **device drivers**, were added to the resident monitor to deal with these interrupts.

Drums, and later, disks were introduced as a secondary storage medium. Now the computer could be computing one job while reading another onto the drum and printing the results of a third from the drum. Unlike a tape, a drum allows programs to be stored anywhere, so there was no need for the computer to execute jobs in the same order in which they were entered. A primitive **scheduler** was added to the resident monitor to sort jobs based on priority and amount of time needed, both specified on control cards. The operator was retained to perform several tasks:

- To mount data tapes needed by jobs (specified on control cards, which caused request messages to appear on the console typewriter)
- To make policy decisions, such as which priority jobs to run and which to hold
- To restart the resident monitor when it failed or was inadvertently destroyed by the running job

This mode of running a computer was known as a **spooling system**, and its resident monitor was the start of modern operating systems. (The word spool originally stood for simultaneous peripheral operations on line, but it is easier to picture a spool of thread, where new jobs are wound on the outside, and old ones are extracted from the inside.) One of the first spooling systems was the Houston automatic spooling program (HASP), an add-on to OS/360 for the IBM 360 computer family.

Batch Multiprogramming

Spooling systems did not make efficient use of all of the hardware's resources. The job that was currently running might not need the entire main store. A job that performed input/output would cause the computer to wait until the input/output was finished. The next software improvement, which occurred in the early 1960s, was the introduction of **multiprogramming**, a scheme in which more than one job is active simultaneously.

Under multiprogramming, while one job is waiting for an input/output operation to complete, another can compute. With luck, no time at all is wasted waiting for input/output. The more jobs run at the same time, the better. However, a **compute-bound** job (one that performs little input/output but much computation) could easily prevent **input/output-bound** jobs (those that perform mostly input/output) from making progress. Competition for the time resource and policies for allocating it are the main theme of Chapter 77.

Multiprogramming also introduces competition for memory. The number of jobs that can be accommodated at one time depends on the size of main store and the hardware available for subdividing that space. In addition, jobs must be secured against inadvertent or malicious interference or inspection by other jobs. It is more critical now that the resident monitor not be destroyed by errant programs, because not one but many jobs suffer if it breaks. In Chapter 80, we will examine policies for memory allocation and how each of them provides security.

The form of multiprogramming we have been describing is often called batch multiprogramming because jobs are grouped by their needs into batches: those that need small memory, those that need customized tape mounts, those that need long execution, and so forth. Each batch might have different priorities and fee structures. Some batches (such as large-memory long-execution jobs) would be scheduled for particular times (such as weekends or late at night). Generally, one job from any batch can run at a time.

Each job is divided into discrete steps. Since job steps are independent, the resident monitor can separate them and apply policy decisions to each step independently. Each step might have its own time, memory, and input/output requirements. In fact, two separate steps of the same job can be performed at the same time if they don't depend on each other. The term **process** was introduced in the late 1960s to mean the entity that performs a single job step. The scheduler creates a new process for each job step. The process will terminate when its step is finished. The operating system (as the resident monitor may now be called) keeps track of each process and its needs. A process may request assistance from the **kernel** by submitting a service call across the **process interface**. Executing programs are no longer allowed to control devices directly; otherwise, they could make conflicting use of devices and prevent the kernel from doing its job. Instead, processes must use service calls to access devices, and the kernel has complete control of the **device interface**.

Allocating resources to processes is not a trivial task. A process might require resources (like tape drives) at various stages in its execution. If a resource is not available, the scheduler might block the process from continuing until later. The scheduler must take care not to block any process forever.

Along with batch multiprogramming came new ideas for structuring the operating system. The kernel of the operating system is composed of routines that manage central store, central processing unit (CPU) time, devices, and other resources. It responds both to requests from processes and to interrupts from devices. In fact, the kernel runs only when it is invoked either from above, by a process, or below, by a device. If no process is ready to run and no device needs attention, the computer sits idle.

Various activities within the kernel share data, but they must not be interrupted when the data are in an inconsistent state. Mechanisms for **concurrency control** were developed to ensure that these activities do not interfere with each other. Chapter 79 introduces the mutual-exclusion and synchronization problems associated with concurrency control and surveys the solutions that have been found for these problems. The MVS operating system for the IBM 360 family was one of the first to use batch multiprogramming.

Interactive Multiprogramming

The next step in the development of operating systems was the introduction of **interactive multiprogramming**. The principal user-oriented input/output device changed in the late 1960s from cards or tape to an interactive terminal. Instead of packaging all of the data that a program might need before it starts running, the interactive user is able to supply input as the program wants it. The data can depend on what the program has produced so far. Among the first terminals were Teletypes, which produced output on paper at perhaps 10 characters/s. Later terminals were called *glass teletypes* because they used a television screen to avoid depending on mechanical parts that made output slow. Like a regular teletype, they could not back up to modify data sitting earlier on the screen. Shortly thereafter, terminals gained cursor addressability, which meant that programs could show entire pages of information and change any character anywhere on a page.

Interactive computing caused a revolution in the way computers were used. Instead of being treated as number crunchers, computers became information manipulators. Interactive text editors allowed users to construct data files on line. These files could represent programs, documents, or data. As terminals improved, so did the text editors, changing from line- or character-oriented interfaces to full-screen interfaces.

Instead of representing a job as a series of steps, interactive multiprogramming (also called **timesharing**) identifies a **session** that lasts from initial connection (*log in*) to the point at which that connection is broken (*log out*). During log in, the user typically gives two forms of identification: a name and a password. The password is not echoed at the terminal, or is at least blackened by overstriking garbage, to avoid disclosing it to onlookers. These data are converted into a **user identifier** that is associated with all of the processes that run on behalf of this user and all of the files they create. This identifier helps the kernel decide whom to bill for services and whether to permit various actions such as modifying files. (We discuss protection in Chapter 90.)

During a session, the user imagines that the resources of the entire computer are devoted to this terminal, even though many sessions may be active simultaneously for many users. Typically, one process is created at log-in time to serve the user. That first process, which is usually a command interpreter, may start others as needed to accomplish individual steps.

Users need to save information from session to session. Magnetic tape is too unwieldy for this purpose. Disk storage became the medium of choice for data storage, for short term (temporary files used to connect steps in a computation), medium term (from session to session), and long term (from year to year). Issues of disk space allocation and backup strategies needed to be addressed to provide this facility.

Interactive computing was sometimes added into an existing batch multiprogramming environment. For example, timesharing option (TSO) was an add-on to the OS/360 operating system. The EXEC-8 operating system for Univac computers included an interactive component, too.

In contrast, many operating systems were designed primarily to support interactive use, with batch facilities added when necessary. TOPS-10 and Tenex (for the Digital PDP-10) and almost all operating systems developed since 1975, including Unix (Digital PDP-11, then Digital VAX, then many platforms), MS-DOS (Intel 8086), OS/2 (Intel 286 family (Kogan and Rawson [1988])), and VMS (Digital VAX (Kenah and Bate [1984])), are designed mainly for interactive use.

Graphical User Interfaces

As computers became less expensive, the time cost of switching from one process to another (which happens frequently in interactive computing) became insignificant. Idle time also became unimportant. Instead, the goal has become to help users get their work done efficiently. This goal led to new software developments, enabled by improved hardware.

Graphics terminals have been advancing since the late 1970s. Personal computers today come with terminals that were quite expensive 5 years ago and practically unheard of 10 years ago. They allow individual control of multicolored pixels; a high-quality monitor can display on the order of a million pixels in a wide range of colors. Pointing devices, particularly the mouse, were developed in the late 1970s. They are linked to the display so that a visible cursor reacts to physical movements of the pointing device. These hardware advances have led to **graphical user interfaces** (**GUIs**).

The earliest GUIs were just rectangular regions of the display that contained, effectively, a cursor-addressable glass teletype. These regions are called windows. The best-known windowing packages are MacOS (Pogue and Schorr [1993]), MS-Windows (for DOS), OS/2 (Kogan and Rawson [1988]), and X Windows (for Unix, VMS, and other operating systems (Reiss and Radin [1992])). Each has developed from simple rectangular models of a terminal to significantly more complex displays.

Programs interact with the hardware by invoking routines in libraries that know how to communicate with the display manager, which itself knows how to place bits on the screen. The early libraries were fairly low level and hard to use; toolkits (in the X Windows environment), especially ones with a fairly small interpreted language (such as Tcl/Tk (Ousterhout [1994])), have eased the task of building good GUI interfaces. The operating systems of these machines often provide for interactive computing but not for multiprogramming. MS-DOS is a good example of this approach. Other operating systems provide multiprogramming as well as interaction and allow the user to start several activities and switch attention to whichever one is currently most interesting.

Distributed Computing

At the same time that displays were improving, networks of computers were being developed. A network requires not only hardware to physically connect machines, but also protocols to use that hardware effectively, operating system support to make those protocols available to processes, and applications that make use of these protocols.

Computers can be connected together by a variety of devices. The spectrum ranges from tight coupling, where several computers share main storage, to very loose coupling, where a number of computers belong to the same international network and can send one another messages.

The ability to send messages between computers has opened new opportunities for operating systems. Individual machines become part of a larger whole, and in some ways, the operating system begins to span networks of machines. Cooperation between machines takes many forms.

- Each machine may offer **network services** to others, such as accepting mail, providing information on who is currently logged in, telling what time it is (quite important in keeping clocks synchronized), allowing users to access machines remotely, and file transfer.
- Machines within the same **site** (typically, those under a single administrative control) may **share file systems** in order to reduce the amount of disk space needed and to allow users to have accounts on multiple machines. Novell nets for MS-DOS and the Sun Network File System for Unix are examples of such arrangements. Shared file systems are the major components of **networked operating systems**.
- Once users have accounts on several machines, they want to associate graphical windows with sessions on different machines. The X Windows package allows processes running on one machine to display on the screen of another.
- Users also want to execute computationally intensive algorithms on many machines in parallel. Packages such as PVM have been built to make it easier to build such applications.
- Standardized ways of presenting data across site boundaries have been developing rapidly. The **file transfer protocol** (**FTP**) service was developed in the early 1970s as a way of transferring files between machines connected on a network. In the early 1990s, the **gopher** service was developed to create a uniform interface for accessing information across the Internet. Information is more general than just files; it can be a request to run a program or to access a database. Each machine can provide a server that responds to connections from any site and communicates a menu of available information. This service was superseded in 1995 by the **World Wide Web**, which supports a GUI to gopher, FTP, and hypertext (documents with links internally and to other documents, often at other sites, and including text, pictures, video, audio, and remote execution of packaged commands).

Of course, all of these forms of cooperation introduce security concerns. Each site has a responsibility to maintain security if for no other reason than to prevent malicious users across the network from using the site as a breeding ground for nasty activity attacking other sites. Security issues are discussed in Chapters 89, 91, and 93; network security in particular is discussed in Chapter 92.

75.3 Goals of an Operating System

During the evolution of operating systems, their purposes have also evolved. At present, operating systems have three major goals:

- Hide details of hardware by creating abstractions
- Manage resources
- Provide a pleasant and effective user interface

We address each of these goals in turn.

Abstracting Reality

We distinguish between the **physical** world of devices, instructions, memory, and time, and the **virtual** world that is the result of abstractions built by the operating system. An **abstraction** is software (often implemented as a subroutine or as a library of subroutines) that hides lower level details and provides a set of higher level functions. Programs that use the abstraction can safely ignore the lower level (physical) details; they need only deal with the higher level (virtual) structures.

Why is abstraction important in operating systems? First, the code needed to control peripheral devices is often not standardized; it can vary from brand to brand, and it certainly varies between, say, disks and tape drives and keyboards. Input/output devices are extremely difficult to program efficiently and correctly. Abstracting devices with a uniform interface make programs easier to write and to modify (for example, to use a different device). Most operating systems provide subroutines called device drivers that perform input/output operations on behalf of programs. The operations are provided at a much higher level than the device itself provides. For example, a program may wish to write a particular block on a disk. Low-level methods involve sending commands directly to the disk to seek to the right block and then undertake memory-to-disk data transfer. When the transfer is complete, the disk interrupts the running program. A low-level program needs to know the format of disk commands, which vary from manufacturer to manufacturer, and must deal with interrupts. In contrast, a program using a high-level routine in the operating system might only need to specify the memory location of the data block and where it belongs on the disk; all of the rest of the machinery is hidden.

Second, the operating system introduces new functions as it abstracts the hardware. In particular, operating systems introduce the **file** abstraction. Programs do not need to deal with disks at all; they can use high-level routines to read and write disk files (instead of disk blocks) without needing to design storage layouts, worry about disk geometry, or allocate free disk blocks.

Third, the operating system transforms the computer hardware into multiple virtual computers, each belonging to a different program. Each program that is running is called a process. Each process views the hardware through the lens of abstraction: memory, time, and other resources are all tailored to the needs of the process. Processes see only as much memory as they need, and that memory does not contain the other processes (or the operating system) at all! They run as if they have all of the CPU cycles on the machine, although other processes and the operating system itself are competing for those cycles. Service calls allow processes to start other processes and to communicate with other processes, either by sending messages or by sharing memory.

Fourth, the operating system can enforce security through abstraction. The operating system must secure both itself and its processes against accidental or malicious interference. Certain instructions of the machine, notably those that halt the machine and those that perform input and output, are moved out of the reach of processes. Memory is partitioned so that processes cannot access each other's memory. Time is partitioned so that even a run-away process will not prevent others from making progress.

For security and reliability, it is wise to structure an operating system so that processes must use the operating system's abstractions instead of dealing with the physical hardware. This restriction can be enforced by the hardware, which provides several **processor states**. Most architectures provide at least two states, called **privileged state** and **nonprivileged state**.

Processes always run in nonprivileged state. Instructions such as those that perform input/output and those that change processor state cause traps when executed in nonprivileged state. Traps are provided to save the current execution context (perhaps on a stack), force the processor to jump to the operating system, and enter privileged state. Once the operating system has finished servicing the trap or interrupt, it returns control to the same process or perhaps to a different one, resetting the computer into nonprivileged state.

The core of the operating system runs in privileged state. All instructions have their usual, physical meanings in this state. The part of the operating system that always runs in privileged state is the kernel of the operating system. It only runs when a process has caused a trap or when a peripheral device has generated an interrupt. Traps do not necessarily represent errors; usually they are **service calls**. Interrupts often indicate that a device has finished servicing a request and is ready for more work. The clock interrupts at a regular rate in order to let the kernel make scheduling decisions.

If the operating system makes use of this dichotomy of states, the abstractions that the operating system provides are presented to processes as service calls, which are like new CPU instructions. Now a program can perform high-level operations with a single instruction. A program that executes a service call generates a trap, which causes a switch to the privileged state of the kernel. (Values on the stack can be used to indicate which service the process is requesting from the kernel.) The advantage of the service-call design over a procedure-call design is that all kernel operations can be protected behind a wall of protection.

Not all operating systems make use of nonprivileged state. MS-DOS, for example, runs all applications in privileged state. Service calls are essentially subroutine calls. Although the operating system provides device and file abstractions, processes may interact directly with disks and other devices. One advantage of this choice is that device drivers can be loaded after the operating system starts; they do not need special privilege. One disadvantage is that viruses can thrive because nothing prevents a program from placing data anywhere it wishes.

Managing Resources

An operating system is not only an abstractor of information, but also an allocator that controls how processes (the active agents) may access resources (passive entities).

A **resource** is a commodity necessary to get work done. The computer's hardware provides a number of low-level resources. Working programs need to reside somewhere in main store (the computer's memory), must execute instructions, and need some way to accept data and present results. These needs are related to the fundamental resources of memory, CPU time, and input/output. The operating system abstracts these resources to allow them to be shared.

In addition to these physical resources, the operating system creates virtual, abstract resources. For example, files are able to store data. They abstract the details of disk storage. **Pseudofiles** (that is, objects that appear to be data files on disk but are in fact stored elsewhere) can also represent devices, processes, communication ports, and even data on other computers. Still higher level resources can be built on top of abstractions. A **database** is a collection of information, stored in one or more files with structure intended for easy access. A **mailbox** is a file with particular semantics.

The resource needs of processes often interfere with each other. Resource managers in the operating system include policies that try to be fair in giving resources to the processes and allow as much computation to proceed as possible. These goals often conflict.

Each resource has its own manager, typically in the kernel. The memory manager allocates regions of main memory for processes. Most modern operating systems use address translation hardware that maps between a process's **virtual addresses** and the underlying **physical addresses**. Only the currently active part of a process's virtual space needs to be physically resident; the rest is kept on backing store (usually a disk) and brought in on demand. The virtual spaces of processes do not usually overlap, although some operating systems also provide **lightweight processes** that share a single virtual space. The memory manager includes policies that determine how much physical memory to grant to each process and which region of physical memory to swap out to make room for other memory that must be swapped in. For more information on virtual memory, see Chapter 80.

The CPU time manager is called the scheduler. Schedulers usually implement a preemptive policy that forces the processes to take turns running. Schedulers categorize processes according to whether they are currently runnable (they may not be if they are waiting for other resources) and their priority.

The file manager mediates process requests such as creating, reading, and writing files. It validates access based on the identity of the user running the process and the permissions associated with the file. The file manager also prevents conflicting accesses to the same file by multiple processes. It translates input/output requests into device accesses, usually to a disk, but often to networks (for remote files) or other devices (for pseudofiles).

The device managers convert standard-format requests into the particular commands appropriate for individual devices, which vary widely among device types and manufacturers. Device managers may also maintain caches of data in memory to reduce the frequency of access to physical devices.

Although we usually treat processes as autonomous agents, it is often helpful to remember that they act on behalf of a higher authority: the human users who are physically interacting with the computer. Each process is usually *owned* by a particular user. Many users may be competing for resources on the same machine. Even a single user can often make effective use of multiple processes.

Each user application is performed by a process. When a user wants to compose a letter, a process runs the program that converts keystrokes into changes in the document. When the user mails that letter electronically, a process runs a program that knows how to send documents to mailboxes.

To service requests effectively, the operating system must satisfy two conflicting goals:

- To let each process have whatever resources it wants
- To be fair in distributing resources among the processes

If the active processes cannot all fit in memory, for example, it is impossible to satisfy the first goal without violating the second. If there is more than one process, it is impossible on a single CPU to give all processes as much time as they want; CPU time must be shared.

To satisfy the computer's owner, the operating system must also satisfy a different set of goals:

- To make sure the resources are used as much as possible
- To complete as much work as possible

These latter goals were once more important than they are now. When computers were all expensive mainframes, it seemed wasteful to let any time pass without a process using it, or to let any memory sit unoccupied by a process, or to let a tape drive sit idle. The measure of success of an operating system was how much work (measured in **jobs**) could be finished and how heavily resources were used. But now, computers have become less expensive; it is no longer critical that computers not sit idle, although efficient use of resources, particularly during peak usage periods, is still important.

User Interface

We have seen how operating systems are creators of abstractions and allocators of resources. Both these aspects are centered on the needs of programmers and the processes that execute programs. But many users are not programmers and are uninterested in the process abstraction and in the interplay between processes and the operating system. They do not care about service calls, interrupts, and devices. Instead, they are interested in what might be termed the *look and feel* of the operating system.

The user interacts with the operating system through the **user interface**. The user interface has seen rapid change over the last 50 years. Some stages along the path are the following:

- The user sits at the computer console interacting with the one process on the machine through switches and a typewriter.
- The user submits a deck of punched cards and receives printed output.
- The user types at a remote teletype interacting with the user's process.
- The user types at a glass teletype (a character display terminal), flipping among activities governed by several processes.
- The user works at a graphics terminal that displays many virtual character terminals at once, each governed by a different process. The graphics terminal includes a pointing device (a mouse or a touch-sensitive screen) that the user can employ to direct attention to a particular virtual terminal.
- The user interacts with application programs through graphical user interfaces that make heavy use of color, multiple fonts, and interactive menus, controlled largely by the pointing device, but also possibly by voice command. The display includes sound, pictures, and video in addition to text.

- The user interacts with application programs running on many different computers networked together, all presenting information on the local display.

The look and feel of the operating system is affected by many components of the user interface. Some of the most important are the command interpreter, the file system, on-line help, and application integration.

Command Interpreter

The user must be able to start, observe, and terminate activities. Generally, we speak of a **command interpreter** that the user deals with. The user prepares **commands** in a syntax acceptable to the command interpreter, which then causes the necessary activity to occur. The syntax accepted by the command interpreter is a major component of the look and feel of a user interface.

In earlier years, the command interpreter was part of the kernel of the operating system. Activities were grouped into jobs, each step of which carried the computation forward. For example, a job might include a compilation step, a linking step, an execution step, and a memory-dump step. A single process would execute the entire job. Failure at one step could cause the process to omit the rest of the steps or to take a different route through the steps. This navigation was under the control of the **job-control language**, which was the language of the command interpreter.

With the advent of timesharing, batch-oriented job-control languages were modified so that the user could interact after each step, seeing the results before deciding what to do next. When operating systems started to let users run multiple interacting processes, the command interpreter was taken out of the kernel and became an ordinary program. Users could then choose among various command interpreters with different characteristics.

More recently, text-oriented job-control languages have been superseded by graphical user interfaces, in which different pictures on the screen represent different activities that may be started, and the user indicates what to do next by clicking the mouse to enter selections. Some of these packages, such as MS-Windows and MacOS, provide a consistent look and feel for all applications. Other packages, such as X Windows (Reiss and Radin [1992]), allow the user to choose among many different *window managers* to affect the look and feel of the applications.

File System

Another component of the user interface is the file structure. Filenaming conventions, such as limits to name lengths, whether names are case sensitive, and the components of multipart names affect the look and feel of the operating system. For example, MS-DOS limits file names to 8 characters and an optional 3-character suffix, and case is not significant. Most versions of Unix do not limit the number or length of components, and case is significant. For more information on DOS and Unix commands and file names, see Appendix E.

File names are understood within some context, which is often a particular account, disk, or the tree of directories. For example, in MS-DOS the context of a file is the disk name (a single letter) followed by a path from the root of that disk's directory tree, with components separated by backslashes, such as `C:\WINDOWS\STARTUP\`. In Unix, the context is specified by a path from the root of the global file system's directory tree, such as `/homes/leslie/current/`.

Most operating systems allow the user to specify a context (by logging in, for example, and then moving to a particular directory). This context is then assumed for all simple file names that are presented by the user (or the user's processes).

On-Line Help

On-line help is available for most operating systems, but is almost never part of the kernel. Therefore, it can be implemented in many different ways and local sites can choose their own strategies.

For example, Unix sites rely heavily on manual pages, which are terse summaries of the commands available and their parameters. Many users find these summaries unpleasant to read. Individual programs may provide specialized on-line help, but they follow no standard format. Often the manual pages refer the user to more complete documentation that may or may not be stored on the computer.

In contrast, MS-Windows has a well-integrated help facility, where each application, once running, is capable of responding to a standard set of help queries in a standard way. Once the user has mastered those queries, on-line help for all programs is available.

Application Integration

Programs need to work together. For example, the linker must be able to understand the format of the files produced by the compiler. If there are several compilers, perhaps for different languages, it is helpful if they all use the same file format so that a single linker can handle them all. In this case, we can say that the different language processors are integrated with respect to their file format. Similarly, if several compilers generate similar debugging information, then a single interactive debugger can be used for programs compiled by all.

Programs that represent their data files in standard (ASCII text) character format generally integrate readily; any text editor can be used to prepare input for a compiler if the former can produce and the latter can accept character format files. The Unix philosophy suggests that programs can generally be integrated in exactly that way. Complex data transformations can be accomplished by connecting several processes in a **pipeline**, each modifying character data in some way and passing it along to the next process.

When applications are not integrated, it is difficult to make them work together. Sometimes, software vendors provide routines that can convert files from the nonstandardized formats used by their competitors to their own formats. For example, Word Perfect can convert files built by Microsoft Word, and vice versa. In addition, vendors try to accommodate the particular data formats of external programs that might produce data that needs to be imported into their programs. For example, a spreadsheet program can usually import data produced by different database programs. Again, these products are not truly integrated; the data conversions are provided as special features in order to enhance their commercial value.

The trend in operating systems is toward increased **integration of applications**. Integration goes beyond filtering the output of one program into the input of another. It incorporates standardized data formats. For example, the executable and linking format (ELF) link-module format is becoming a standard across many different architectures. (ELF was developed by Unix System Laboratories in the late 1980s.) MS-Windows applications generally follow standard formats to allow data and pictures to be transferred interactively between applications. Library routines (called toolkits) for applications written for the X Windows package create applications that share a given look and feel.

75.4 Implementing an Operating System

As mentioned earlier, the core of the operating system is the kernel, a control program that functions in privileged state, reacting to interrupts from external devices and to service requests and traps from processes. Generally, the kernel is a permanent resident of the computer. It creates and terminates processes and responds to their requests for service.

Processes

Each process is represented in the kernel by a collection of data called the **process descriptor**, or a *context block*. A process descriptor includes such information as the following:

- Processor state, stored values of the program counter and registers, is needed to resume execution of the process.
- Scheduling statistics are needed to determine when to resume the process and how much time to let it run.
- Memory allocation, both in main memory and backing store (disk), is needed to accomplish memory management.

- Other resources held, such as tape drives, are needed to manage contention for such resources.
- Open files and pseudofiles (devices, communication ports) are needed to interpret service requests for input and output.
- Accounting statistics are needed to bill users and determine hardware usage levels.
- Privileges are needed to determine if activities such as opening files and executing potentially dangerous service calls should be allowed.
- Abstract process state includes running, ready, waiting for input/output, or some other resource, such as memory.

The process descriptors may be saved in an array, in which case each process can be identified by the index of its descriptor in that array. Other structures are possible, of course, but this concept of **process number** seems to be fairly widespread across operating systems. Some of the information in the process descriptor can be bulky, such as the page tables. Page tables for idle processes may be stored on disk in order to save space in main memory.

Resuming a process, that is, switching control from the kernel back to the process, is a form of **context switching**. It requires that the processor move from privileged to unprivileged state, that the registers and program counter of the process be restored, and that the address-translation hardware be set up to accomplish the correct mappings for this process. Switching back to the kernel is also a context switch; it can happen when the process tries to execute a privileged instruction (including the service-call instruction) or when a device generates an interrupt.

Hardware is often designed to switch context rapidly. For example, the hardware may maintain two sets of registers and address translation data, one for each privilege level. Context switches into the kernel just require moving to the kernel's set of registers. Resuming the same process that was most recently running is also fast. Resuming a different process requires that the kernel load all of the information for the new process into the second set of registers; this activity takes longer. For that reason, a **process switch** is often more expensive than two context switches.

Virtual Machines

Although most operating systems try to present to processes an enhanced and simplified view of the hardware, some take a different tack. They make the process interface look just like the hardware interface, except that the size of memory and the types, numbers, and sizes of input/output devices may be more or less than the physical resources. However, a process is allowed to use all of the machine instructions, even the privileged ones.

Under this organization, the process interface is called a **virtual machine** because it looks just like the underlying machine. The kernel of such an operating system is called a **virtualizing kernel**. Each virtual machine runs its own ordinary operating system.

We will examine virtual operating systems because they elucidate the interplay of traps, context switches, processor states, and the fact that a process at one level is just a data structure at a lower level. Virtualizing kernels were first developed (by IBM in the VM operating system, in the early 1970s) to allow operating system designers to experiment with new versions of an operating system on machines that were too expensive to dedicate to such experimentation. More importantly, virtualizing kernels allow multiple operating systems to run simultaneously on the same machine to satisfy a wide variety of users.

This idea is still valuable. A program that emulates the MS-DOS environment runs as a process under Unix, allowing a Unix user who has MS-DOS programs to run them at the same time as other applications. This emulation is fairly easy, because MS-DOS does not distinguish privileged levels of execution. Mach emulates Unix and can accept modules that emulate other operating systems as well. This emulation is at the library-routine level; service calls are converted to messages directed to a Unix-emulator process that provides all of the services. The NT (Custer [1993]) and OS/2 (Kogan and Rawson [1988]) operating systems for Intel computers also provide for virtual machines running other operating systems.

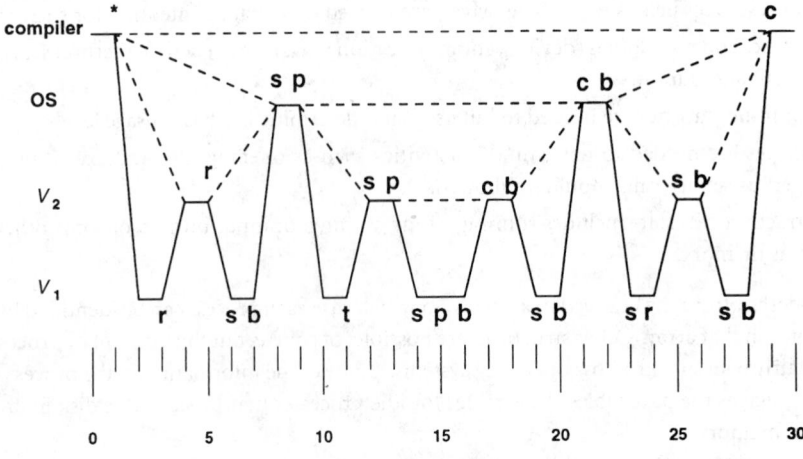

FIGURE 75.2   Emulating a service call.

In a true virtualizing kernel, the hardware executes most instructions (such as arithmetic and data motion) directly. However, privileged instructions, such as the halt instruction, are just too dangerous to let processes use directly. Instead, the virtualizing kernel must run all processes in nonprivileged state to prevent them from accidentally or maliciously interfering with each other and with the kernel itself.

To let each process P imagine it has control of processor states, the kernel keeps track of the virtual processor state of each P, that is, the processor state of the virtual machine that the kernel emulates on behalf of P. This information is stored in P's context block inside the kernel. All privileged instructions executed by P cause traps to the kernel, which then emulates the behavior of the hardware on behalf of P:

- If P is in virtual nonprivileged state, the kernel emulates a trap for P. This emulation puts P in virtual privileged state, although it is still running in physical nonprivileged state. The program counter for P is reset to the proper trap address within P's virtual space.

- If P is in virtual privileged state, the kernel emulates the action of the instruction itself. For example, it terminates P on a halt instruction, and it executes input/output instructions interpretively.

Some dangerous instructions are particularly difficult to emulate. Input/output can be very tricky. Address translation also becomes quite complex. A good test of a virtualizing kernel is to let one of its processes be another virtualizing kernel. For example, consider Fig. 75.2, in which there are two levels of virtualizing kernel, V_1 and V_2, above which sits an ordinary operating system kernel OS, above which a compiler is running.

The compiler executes a single service call (marked ∗) at time 1. As far as the compiler is concerned, OS performs the service and lets the compiler continue (marked c) at time 29. The dashed line at the level of the compiler indicates the compiler's perception that no activity below its level takes place during the interval.

From the point of view of OS, a trap occurs at time 8. This trap appears to come directly from the compiler, as shown by the dashed line connecting the compiler at time 1 and the OS at time 8. OS services the trap (marked s). For simplicity, we assume that it needs to perform only one privileged instruction (marked p) to service the trap, which it executes at time 9. Lower levels of software (which OS cannot distinguish from hardware) emulate this instruction, allowing OS to continue at time 21. It then switches context back to the compiler (marked b) at time 22. The dashed line from OS at time 22 to the compiler at time 29 shows the effect of this context switch.

The situation is more complicated from the point of view of V_2. At time 4, it receives a trap that tells it that its client has executed a privileged instruction while in virtual nonprivileged state. V_2 therefore reflects this trap at time 5 (marked r) back to OS. Later, at time 12, V_2 receives a second trap, this time because its client has executed a privileged instruction in virtual privileged state. V_2 services this trap by emulating the instruction itself at time 13. By time 17, the underlying levels allow it to continue, and at time 18 it switches context back to OS. The last trap occurs at time 25, when its client has attempted to perform a context switch (which is privileged) when in virtual privileged state. V_2 services this trap by changing its client to virtual nonprivileged state and switching back to the client at time 26.

V_1 has the busiest schedule of all. It reflects traps that arrive at times 2, 10, and 23. (The trap at time 23 comes from the context-switch instruction executed by OS.) It also emulates instructions for its client when traps occur at times 5, 14, 19, and 27.

This example demonstrates the principle that each software level is just a data structure as far as its supporting level is concerned. It also shows how a single privileged instruction in the compiler becomes two privileged instructions in OS, which become four in V_2 and eight in V_1. In general, a single privileged instruction at one level might require many instructions at its supporting level to emulate it.

Components of the Kernel

Originally, operating systems were written as a single large program encompassing hundreds of thousands of lines of assembly-language instructions. Two trends have made the job of implementing operating systems less difficult. First, high-level languages have made programming much easier. For instance, over 99% of the Unix operating system is written in C. Complex algorithms can be expressed in a structured, readable fashion, code can be partitioned into modules that interact with each other in a well-defined manner, and compile-time type checking catches most programming errors. Only a few parts of the kernel, such as context switching, need to be written in assembly language.

Second, the discipline of structured programming has suggested a layered approach to designing the kernel. Each layer provides abstractions needed by the layers above it. For example, the kernel can be organized as follows:

- Context- and process-switch services (lowest layer)
- Device drivers
- Resource managers for memory and time
- File system support
- Service call interpreter (highest layer)

For example, the MS-DOS operating system provides three levels: (1) device drivers (the BIOS section of the kernel), (2) a file manager, and (3) an interactive command interpreter (CCP). It supports only one process and provides no security, so there is no need for context-switch services. Because service calls do not need to cross protection boundaries, they are implemented as subroutine calls.

The concept of layering allowed the kernel to be reduced, since much of the work of the operating system need not operate in a protected and hardware-privileged environment. When all of the layers listed are privileged, the organization is called a **macrokernel**. Unix is often implemented in this fashion.

If the kernel only contains code for process creation, interprocess communication, the mechanisms for memory management and scheduling, and the lowest level of device control, the result is a **microkernel**, also called a *communication kernel*. (Mechanisms are distinct from policies, which can be outside the kernel. Policies decide which resources should be allocated in cases of conflict, whereas mechanisms carry out those decisions.) The Mach operating system is a current example of this approach (Rashid [1986]). In this organization, services such as the file system and policy modules for scheduling and memory are relegated to processes. These processes are often referred to as **servers**; the ordinary processes that need those services are called their **clients**. The microkernel itself acts as a client of the policy servers. Servers need to be trusted by their clients, and sometimes they need to execute with some degree of hardware privilege (for example, if they access devices).

The microkernel approach has some distinct advantages:

- It imposes uniformity on the requests that a process might make. Processes need not distinguish between kernel-level and process-level services, since all are provided via messages to servers.
- It allows easier addition of new services, even while the operating system is running, as well as multiple services that cover the same set of needs, so that individual users (and their agent processes) can choose whichever seems best. For example, different file organizations for diskettes are possible; instead of having many file-level modules in the kernel, there can be many file-level servers accessible to processes.
- It allows an operating system to span many machines in a natural way. As long as interprocess communication works across machines, it is generally immaterial to a client where its server is located.
- Services can be provided by teams of servers, any one of which can help any client. This organization relieves the load on popular servers, although it often requires a degree of coordination among the servers on the same team.

Microkernels also have some disadvantages. It is generally slower to build and send a message, accept and decode the reply (taking about 100 μs), than to make a single service call (taking about 1 μs). However, other aspects of service tend to dominate the cost, allowing microkernels to be similar in speed to macrokernels. Keeping track of which server resides on which machine can be complex. This complexity may be reflected in the user interface. The perceived complexity of an operating system has a large effect on its acceptance by the user community.

Recently, people have begun to speak of **nanokernels**, which support only devices and communication ports. They sit at the bottom level of the microkernel, providing services for such other parts of the microkernel as memory management. All of the competing executions supported by the nanokernel are called **threads** to distinguish them from processes. Threads all share kernel memory, and they explicitly yield control in order to let other threads continue. They synchronize with each other by means of primitive locks or more complex semaphores. For more information on processes and threads, see Chapter 76.

Although the trend toward microkernels is unmistakable, macrokernels are likely to remain popular for the forseeable future. Macrokernel versions of Unix are now available for Intel-based personal computers, and the popular MS-DOS and OS/2 (Kogan and Rawson [1988]) operating systems are distinctly macrokernel in design.

75.5 Research Issues and Summary

Operating systems have developed enormously in the last 35 years. Modern operating systems generally have three goals: To hide details of hardware by creating abstractions, to allocate resources to processes, and to provide an effective user interface. Operating systems generally accomplish these goals by running processes in low privilege and providing service calls that invoke the operating system kernel in high-privilege state. The recent trend has been toward increasingly integrated graphical user interfaces that encompass the activities of multiple processes on networks of computers. These increasingly sophisticated application programs are supported by increasingly small operating system kernels.

Current research issues revolve mostly around networked operating systems, including network protocols, distributed shared memory, distributed file systems, mobile computing, and distributed application support. There is also active research in kernel structuring, file systems, and virtual memory.

Defining Terms[1]

Abstraction: Software that hides lower level details and provides a set of higher level functions.

Client: A process that requests services by sending messages to server processes.

---

[1] The defining terms may have more general definitions than given here, and often have other narrow technical definitions. This list indicates how the terms have been used in this chapter.

Command interpreter: A program (usually not in the kernel) that interprets user requests and starts computations to fulfill those requests.

Commands: Instructions in a job-control language.

Compute bound: A process that performs little input/output but needs significant execution time.

Concurrency control: Means to mediate conflicting needs of simultaneously executing threads.

Context switching: The action of directing the hardware to execute in a different context (kernel or process) from the current context.

Database: A collection of files for storing related information.

Device driver: An operating-system module (usually in the kernel) that deals directly with a device.

Device interface: The means by which devices are controlled.

File: A named, long-term repository for data.

FTP: The file transfer protocol service.

Gopher: A network service that connects information providers to their users.

Graphical user interfaces: Interactive programs that make use of a graphic display and a mouse.

Input/output: A resource; ability to interact with peripheral devices.

Input/output bound: A process that spends most of its time waiting for input/output.

Integrated application: An application that agrees on data formats with other applications so they can use each other's outputs.

Interactive multiprogramming: Multiprogramming in which each user deals interactively with the computer.

Job: A set of computational steps packaged to be run as a unit.

Job-control language: A way of specifying the resource requirements of various steps in a job.

Kernel: The privileged core of an operating system, responding to service calls from processes and interrupts from devices.

Lightweight process: A thread.

Macrokernel: A large operating system core that provides a wide range of services.

Mailbox: A file for saving messages between users.

Memory: A resource; ability to store programs and data.

Microkernel: A small privileged operating system core that provides process scheduling, memory management, and communication services.

Multiprogramming: Scheduling several competing processes to run at essentially the same time.

Nanokernel: A very small privileged operating system core that provides simple process scheduling and communication services.

Network services: Services available through the network, such as mail and file transfer.

Networked operating system: An operating system that uses a network for sharing files and other resources.

Nonprivileged state: An execution context that does not allow sensitive hardware instructions to be executed, such as the halt instruction and input/output instructions.

Off line: Handled on a different computer.

Operator: An employee who performs the repetitive tasks of loading and unloading jobs.

Physical: The material on which abstractions are built.

Physical address: A location in physical memory.

Pipeline: A facility that allows one process to send a stream of information to another process.

Privileged state: An execution context that allows all hardware instructions to be executed.

Process: A program being executed; an execution context that is allocated resources such as memory, time, and files.

Process descriptor: A data structure in the kernel that represents a process.

Process interface: The set of service calls available to processes.

Process number: An identifier that represents a process by acting as an index into the array of process descriptors.

Process switch: The action of directing the hardware to run a different process from the one that was previously running.

Processor state: Privileged or nonprivileged state.

Pseudofile: An object that appears to be a file on the disk but is actually stored elsewhere.

Resident monitor: A precursor to kernels; a program that remains in main store during the execution of a job to handle simple requests and to start the next job.

Resource: A commodity necessary to get work done.

Scheduler: An operating system module that manages the time resource.

Server: A process that responds to requests from clients via messages.

Service call: The means by which a process requests service from the kernel, usually implemented by a trap instruction.

Session: The period during which a user interacts with a computer.

Shared file system: Files residing on one computer that can be accessed from other computers.

Site: The set of computers, usually networked, under a single administrative control.

Spooling system: Storing newly arrived jobs on disk until they can be run, and storing output of old jobs on disk until they can be printed.

Thread: An execution context that is independently scheduled, but shares a single address space with other threads.

Time: A resource; ability to execute instructions.

Timesharing: Interactive multiprogramming.

User: A human being physically interacting with a computer.

User identifier: A number or string that is associated with a particular user.

User interface: The facilities provided to let the user interact with the computer.

Virtual: The result of abstraction; opposite of physical.

Virtual address: An address in memory as seen by a process, mapped by hardware to some physical address.

Virtualizing kernel: A kernel that abstracts the hardware to multiple copies that have the same behavior (except for performance) as the underlying hardware.

Virtual machine: An abstraction produced by a virtualizing kernel, similar in every respect but performance to the underlying hardware.

World Wide Web: A network service that allows users to share multimedia information.

References

Custer, H. 1993. *Inside Windows NT*. Microsoft Press.

Deitel, H. M. 1992. *Operating Systems*, 2nd ed. Addison–Wesley, Reading, MA.

Finkel, R. A. 1988. *An Operating Systems Vade Mecum*, 2nd ed. Prentice–Hall, Englewood Cliffs, NJ.

Kenah, L. J. and Bate, S. F. 1984. *Vax/VMS Internals and Data Structures*. Digital Equipment Corporation.

Kogan, M. S. and Rawson, F. L. 1988. The design of operating system/2. *IBM J. Res. Dev.* 27(2):90–104.

Krakowiak, S. 1988. *Principles of Operating Systems*, MIT Press.

Lane, M. G. and Mooney, J. D. 1988. *A Practical Approach to Operating Systems*. Boyd and Fraser.

Leffler, S. J., McKusick, M. K., Karels, M. J., and Quarterman, J. S. 1989. *4.3BSD Unix Operating System*. Addison–Wesley, Reading, MA.

Milenković, M. 1992. *Operating Systems: Concepts and Design*, 2nd ed. McGraw–Hill.

Ousterhout, J. K. 1994. *Tcl and the Tk Toolkit*. Addison–Wesley, Reading, MA.

Pogue, D. and Schorr, J. 1993. *Macworld Macintosh Secrets*. IDG Books Worldwide.

Rashid, R. 1986. Threads of a new system. *Unix Rev.* 4(8):37–49.

Reiss, L. and Radin, J. 1992. *X Window Inside & Out*. McGraw–Hill, New York.

Silberschatz, A. and Galvin, P. B. 1994. *Operating Systems Concepts: Fourth Edition*, 4th ed. Addison–Wesley, Reading, MA.

Stallings, W. 1992. *Operating Systems*. Macmillan, New York.

Tanenbaum, A. S. 1992. *Modern Operating Systems*. Prentice–Hall, Englewood Cliffs, NJ.

76

Thread Management for Shared-Memory Multiprocessors

Thomas E. Anderson
University of California at Berkeley

Brian N. Bershad
University of Washington

Edward D. Lazowska
University of Washington

Henry M. Levy
University of Washington

76.1 Introduction

Disciplined concurrent programming can improve the structure and performance of computer programs on both uniprocessor and multiprocessor systems. As a result, support for *threads*, or lightweight processes, has become a common element of new operating systems and programming languages.

A thread is a sequential stream of instruction execution. A thread differs from the more traditional notion of a heavyweight process in that it separates the notion of execution from the other state needed to run a program (e.g., an address space). A single thread executes a portion of a program, while cooperating with other threads that are concurrently executing the same program. Much of what is normally kept on a per-heavyweight-process basis can be maintained in common for all threads in a single program, yielding dramatic reductions in the overhead and complexity of a concurrent program.

Concurrent programming has a long history. The operation of programs that must handle real-world concurrency (e.g., operating systems, database systems, and network file servers) can be complex and difficult to understand. Dijkstra [1968] and Hoare [1974, 1978] showed that these programs can be simplified when structured as cooperating sequential threads that communicate at discrete points within the program. The basic idea is to represent a single task, such as fetching a particular file block, within a single thread of control, and to rely on the thread management system to multiplex concurrent activities onto the available processor. In this way, the programmer can consider each function being performed by the system separately, and simply rely on automatic scheduling mechanisms to best assign available processing power.

In the uniprocessor world, the principal motivations for concurrent programming have been improved program structure and performance. Multiprocessors offer an opportunity to use concurrency in parallel

0-8493-2909-4/97/$0.00+$.50

programs to improve performance, as well as structure. Moderately increasing a uniprocessor's power can require substantial additional design effort, as well as faster and more expensive hardware components. But, once a mechanism for interprocessor communication has been added to a uniprocessor design, the system's peak processing power can be increased by simply adding more processors. A shared-memory multiprocessor is one such design in which processors are connected by a bus to a common memory.

Multiprocessors lose their advantage if this processing power is not effectively utilized. If there are enough independent sequential jobs to keep all of the processors busy, then the potential of a multiprocessor can be easily realized: each job can be placed on a separate processor. However, if there are fewer jobs than processors, or if the goal is to execute single applications more quickly, then the machine's potential can only be achieved if individual programs can be parallelized in a cost-effective manner. Three factors contribute to the cost of using parallelism in a program:

- **Thread overhead:** The work, in terms of processor cycles, required to create and control a thread must be appreciably less than the work performed by that thread on behalf of the program. Otherwise, it is more efficient to do the work sequentially, rather than use a separate thread on another processor.

- **Communication overhead:** Again in terms of processor cycles, the cost of sharing information between threads must be less than the cost of simply computing the information in the context of each thread.

- **Programming overhead:** A less tangible metric than the previous two, programming overhead reflects the amount of human effort required to construct an efficient parallel program.

High overhead in any of these areas makes it hard to build efficient parallel programs. Costly threads can only be used infrequently. Similarly, if arranging communication between threads is slow, then the application must be structured so that little interthread communication is required. Finally, if managing parallelism is tedious or difficult, then the programmer may find it wise to sacrifice some speedup for a simpler implementation. Few algorithms parallelize well when constrained by high thread, communication, and programming costs, although many can flourish when these costs are low.

Low overhead in these three areas is the responsibility of the thread management system, which bridges the gap between the physical processors (the suppliers of parallelism) and an application (its consumer). In this chapter, we discuss the issues that arise in designing a thread management system to support low-overhead parallel programming for shared-memory multiprocessors. In the next section, we describe the functionality found in thread management systems. Section 76.3 discusses a number of thread design issues. In section 76.4, we survey three systems for shared-memory multiprocessors, Windows NT [Custer 1993], Presto [Bershad et al. 1988], and Multilisp [Halstead 1985], focusing our attention on how they have addressed the issues raised in this chapter.

76.2 Thread Management Concepts

Address Spaces, Threads, and Multiprocessing

An address space is the set of memory locations that can be generated and accessed directly by a program. Address space limitations are enforced in hardware to prevent incorrect or malicious programs in one address space from corrupting data structures in others. Threads provide concurrency within a program, while address spaces provide failure isolation between programs. These are orthogonal concepts, but the interaction between thread management and address space management defines the extent to which data sharing and multiprocessing are supported.

The simplest operating systems, generally those for older style personal computers, support only a single thread and a single-address space per machine. A single-address space is simpler and faster since it allows all data in memory to be accessed uniformly. Separate address spaces are not needed on dedicated systems to protect against malicious users; software errors can crash the system but at least are localized to one user, one machine.

Even single-user systems can have concurrency, however. More sophisticated systems, such as Xerox's Pilot [Redell et al. 1980], provide only one address space per machine, but support multiple threads within that single-address space. Because any thread can access any memory location, Pilot provides a compiler with strong type-checking to decrease the likelihood that one thread will corrupt the data structures of another.

Other operating systems, such as Unix, provide support for multiple-address spaces per machine, but only one thread per address space. The combination of a Unix address space with one thread is called a Unix *process*; a process is used to execute a program. Since each process is restricted from accessing data that belongs to other processes, many different programs can run at the same time on one machine, with errors confined to the address space in which they occur. Processes are able to cooperate by sending messages back and forth via the operating system. Passing data through the operating system is slow, however; only parallel programs that require infrequent communication can be written using threads in disjoint address spaces.

Instead of using messages to share data, processes running on a shared-memory multiprocessor can communicate directly through the shared memory. Some Unix systems allow memory regions to be set up as shared between processes; any data in the shared region can be accessed by more than one process without having to send a message by way of the operating system. The Sequent Symmetry's DYNIX [Sequent 1988] and Encore's UMAX [Encore 1986] are operating systems that provide support for multiprocessing based on shared memory between Unix processes.

More sophisticated operating systems for shared-memory multiprocessors, such as Microsoft's Windows NT and Carnegie Mellon University's Mach operating system [Tevanian et al. 1987] support multiple-address spaces *and* multiple threads within each address space. Threads in the same address space communicate directly with one another using shared memory; threads communicate across address space boundaries using messages. The cost of creating new threads is significantly less than that of creating whole address spaces, since threads in the same address space can share per-program resources. Figure 76.1 illustrates the various ways in which threads and address spaces can be organized by an operating system.

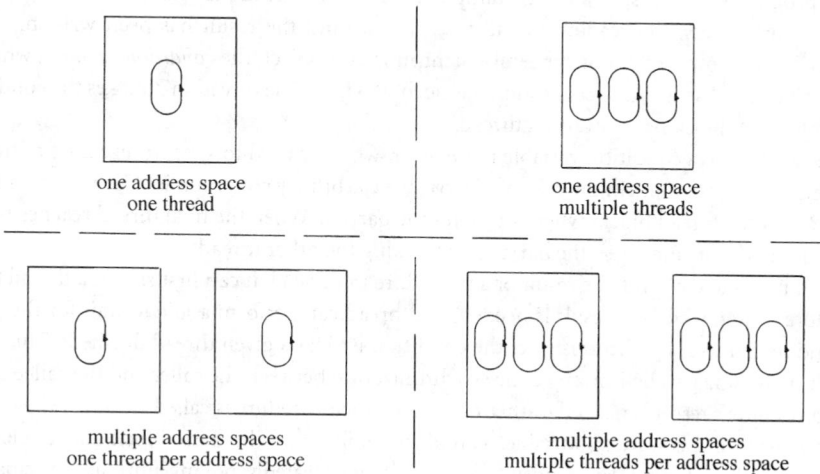

one address space
one thread

one address space
multiple threads

multiple address spaces
one thread per address space

multiple address spaces
multiple threads per address space

FIGURE 76.1 Threads and address spaces. MS-DOS is an example of a one address space, one thread system. A Java run-time engine is an example of one address space with multiple threads. The Unix operating system is an example of multiple address spaces, with one thread per address space. Windows NT is an example of a system that has multiple address spaces and multiple threads per address space.

Basic Thread Functionality

At its most basic level, a thread consists of a program counter (PC), a set of registers, and a stack of procedure activation records containing variables local to each procedure. A thread also needs a control block to hold state information used by the thread management system: a thread can be *running* on a processor, *ready-to-run* but waiting for a processor to become available, *blocked* waiting for some other thread to communicate with it, or *finished*. Threads that are ready-to-run are kept on a *ready-list* until they are picked up by an idle processor for execution. There are four basic thread operations:

- **Spawn:** A thread can create or spawn another thread, providing a procedure and arguments to be run in the context of a new thread. The spawning thread allocates and initializes the new thread's control block and places the thread on the ready-list.
- **Block:** When a thread needs to wait for an event, it may block (saving its PC and registers) and relinquish its processor to run another thread.
- **Unblock:** Eventually, the event for which a blocked thread is waiting occurs. The blocked thread is marked as ready-to-run and placed back on the ready-list.
- **Finish:** When a thread completes (usually by returning from its initial procedure), its control block and stack are deallocated, and its processor becomes available to run another thread.

When threads can communicate with one another through shared memory, *synchronization* is necessary to ensure that threads do not interfere with each other and corrupt common data structures. For example, if two threads each try to add an element to a doubly linked list at the same time, one or the other element may be lost, or the list could be left in an inconsistent state. *Locks* can solve this problem by providing mutually exclusive access to a data structure or region of code. A lock is acquired by a thread before it accesses a shared data structure; if the lock is held by another thread, the requesting thread blocks until the lock is released. (The code that a thread executes while holding a lock is called a *critical section*.) By serializing accesses, the programmer can ensure that threads only see and modify a data structure when it is in a consistent state.

When a program's work is split among multiple threads, one thread may store a result read by another thread. For correctness, the reading thread must block until the result has been written. This data dependency is an example of a more general synchronization object, the *condition variable*, which allows a thread to block until an arbitrary condition has been satisfied. The thread that makes the condition true is responsible for unblocking the waiting thread.

One special form of a condition variable is a *barrier*, which is used to synchronize a set of threads at a specific point in the program. In the case of a barrier, the arbitrary condition is: Have all threads reached the barrier? If not, a thread blocks when it reaches the barrier. When the final thread reaches the barrier, it satisfies the condition and *raises* the barrier, unblocking the other threads.

If a thread needs to compute the result of a procedure in parallel, it can first spawn a thread to execute the procedure. Later, when the result is needed, the thread can perform a *join* to wait for the procedure to finish and return its result. In this case, the condition is: Has a given thread finished? This technique is useful for increasing parallelism, since the synchronization between the caller and the callee takes place when the procedure's result is needed, rather than when the procedure is called.

Locks, barriers, and condition variables can all be built using the basic block and unblock operations. Alternatively, a thread can choose to *spin-wait* by repeatedly polling until an anticipated event occurs, rather than relinquishing the processor to another thread by blocking. Although spin-waiting wastes processor time, it can be an important performance optimization when the expected waiting time is less then the time it takes to block and unblock a thread. For example, spin-waiting is useful for guarding critical sections that contain only a few instructions.

76.3 Issues in Thread Management

This section considers the issues that arise in designing and implementing a thread management system as they relate to the programmer, the operating system, and the performance of parallel programs.

Programmer Issues

Programming Models

The flexibility to adapt to different programming models is an important attribute of thread systems. Parallelism can be expressed in many ways, each requiring a different interface to the thread system and making different demands on the performance of the underlying implementation. At the same time, a thread system that strives for generality in handling multiple models is likely to be well suited to none.

One general principle is that the programmer should choose the most restrictive form of synchronization that provides acceptable performance for the problem at hand. For coordinating access to shared data, messages are a more restrictive, and for many kinds of parallel programs, are a more appropriate form of synchronization than locks and condition variables. Threads share information by explicitly sending and receiving messages to one another, as if they were in separate address spaces, except that the thread system uses shared memory to efficiently implement message passing.

There are some cases where explicit control of concurrency may not be necessary for good parallel performance. For instance, some programs can be structured around a single instruction multiple data (SIMD) model of parallelism. With SIMD, each processor executes the same instruction in lockstep, but on different data locations. Because there is only one program counter, the programmer need not explicitly synchronize the activity of different processors on shared data, thus eliminating a major source of confusion and errors.

Perhaps the simplest programmer interface to the thread system is none at all: the compiler is completely responsible for detecting and exploiting parallelism in the application. The programmer can then write in a sequential language; the compiler will make the transformation into a parallel program. Nevertheless, the compiled program must still use some kind of underlying thread system, even if the programmer does not. Of course, there are many kinds of parallelism that are difficult for a compiler to detect, so automatic transformation has a limited range of use.

Language Support

Threads can be integrated into a programming language; they can exist outside the language as a set of subroutines that explicitly manage parallelism; or they can exist both within and outside the language, with the compiler and programmer managing threads together.

Language support for threads is like language support for object-oriented programming or garbage collection: it can be a mixed blessing. On one hand, the compiler can be made responsible for common bookkeeping operations, reducing programming errors. For example, locks can automatically be acquired and released when passing through critical sections. Further, the types of the arguments passed to a spawned procedure can be checked against the expected types for that procedure. This is difficult to do without compiler support.

On the other hand, language support for threads increases the complexity of the compiler, an important factor if a multiprocessor is to support more than one programming language. Further, the concurrency abstractions provided by a single parallel programming language may not do quite what the programmer wants or needs, making it necessary to express solutions in ways that are unnatural or inefficient.

A reasonable way of getting most of the benefits of language support without many of the disadvantages is to define both a language and a procedural interface to the thread management system. Common operations can be handled transparently by the compiler, but the programmer can directly call the basic thread management routines when the standard language support proves insufficient.

Granularity of Concurrency

The frequency with which a parallel program invokes thread management operations determines its *granularity*. A *fine-grained* parallel program creates a large number of threads, or uses threads that frequently block and unblock, or both. Thread management cost is the major obstacle to fine-grained parallelism. For a parallel program to be efficient, the ratio of thread management overhead to useful computation must be small. If thread management is expensive, then only *coarse-grained* parallelism can be exploited.

More efficient threads allow programs to be finer grained, which benefits both structure and performance. First, a program can be written to match the structure of the problem at hand, rather than the performance characteristics of the hardware on which the problem is being solved. Just as a single-threaded environment on a uniprocessor can prevent the programmer from composing a program to reflect the problem's logical concurrency, a coarse-grained environment can be similarly restrictive. For example, in a parallel discrete-event simulation, physical objects in the simulated system are most naturally represented by threads that simulate physical interactions by sending messages back and forth to one another; this representation is not feasible if thread operations are too expensive.

Performance is the other advantage of fine-grained parallelism. In general, the greater the length of the ready-list, the more likely it is that a parallel program will be able to keep all of the available processors busy. When a thread blocks, its processor can immediately run another thread provided one is on the ready-list. With few threads though, as in a coarse-grained program, processors idle while threads do I/O or synchronize with one another.

The performance of a fine-grained parallel program is less sensitive to changes in the number of processors available to an application. For example, consider one phase of a coarse-grained parallel program that does 50 CPU-min worth of work. If the program creates five threads on a five processor machine, the phase finishes in just 10 min. But, if the program runs with only four processors, then the execution time of the phase *doubles* to 20 min: 10 min with four processors active followed by 10 min with one processor active. (Preemptive scheduling, which could be used to address this problem, has a number of serious drawbacks, which are discussed subsequently.) If the program had originally been written to use 50 threads, rather than 5, then the phase could have finished in only 13 min, a reasonable degradation in performance.

Of course, one could argue that the programmer erred in writing a program that was dependent on having exactly five processors. The program should have been parameterized by the number of processors available when it starts. But, even so, good performance cannot be ensured if that number can vary, as it can on a multiprogrammed multiprocessor. We consider further the issues of multiprogramming in the next section.

Operating System Issues

Multiprogramming

Multiprogramming on a uniprocessor improves system performance by taking advantage of the natural concurrency between computation and I/O. While one program waits for an I/O request, the processor can be running some other program. Because the processor and I/O devices are kept busy simultaneously, more jobs can be completed per unit time than if the system ran only one program at a time.

A multiprogrammed multiprocessor has an analogous advantage. Ideally, periods of low parallelism in one job can be overlapped with periods of high parallelism in another job. Further, multiprogramming allows the power of a multiprocessor to be used by a collection of simultaneously running jobs, none of which by itself has enough parallelism to fully utilize the multiprocessor.

Processor Scheduling

Processor scheduling can be characterized by whether physical processors are assigned directly to threads or are first assigned to jobs and then to threads within those jobs. The first approach, called *one-level*

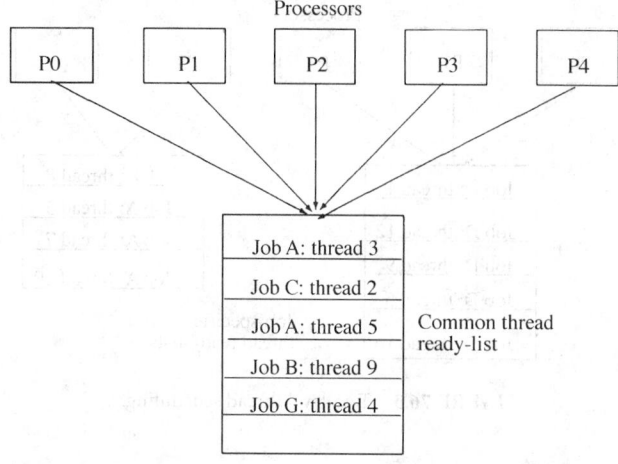

FIGURE 76.2 One-level thread scheduling.

scheduling, makes no distinction between threads in the same job and threads in different jobs. Processors are shared across all runnable threads on the system so that all threads make progress at relatively the same rate. In this case, threads from all jobs are placed on one ready-list that supplies all processors, as shown in Fig. 76.2. Although this scheme makes sense for a uniprocessor operating system, it has some unpleasant performance implications on a multiprocessor.

The most serious problem with one-level scheduling occurs when the number of runnable threads exceeds the number of physical processors, because preemptive scheduling is necessary to allocate processor time to threads in a fair manner. With preemption, a processor can be taken away from one thread and given to another at any time. In a sequential program, preemption has a well-defined effect: the program goes from the running state to the not running state as its one thread is preempted. The effect of preemption on the performance of a sequential program is also well defined: if n CPU-intensive jobs are sharing one processor in a preemptive, round-robin fashion, then each job receives $1/n$th the processor and is slowed down by a factor of n (modulo the preemption and scheduling overhead).

For a parallel program, though, the effects of untimely processor-preemption on performance can be more dramatic. In the previous section, we saw how a coarse-grained program can be slowed down by a factor of two when the number of processors is decreased from five to four. That program exemplified a problem that occurs more generally with preemption and barrier-based synchronization. The program had an implicit barrier, which was the final instruction in the phase. Until all threads reached that instruction, the program could not continue. When one processor was removed, it took twice as long to reach the barrier because not all threads within the job could make progress at an equal rate.

Preemptive multiprocessor scheduling also affects program performance when locks are used, but for a different reason than with barriers. Suppose a thread holding a lock while in a critical section is unexpectedly preempted by the operating system. The lock will remain held until the thread is rescheduled. As threads on other processors try to acquire the lock, they will find it held and be forced to block. It is even possible that, as more threads block waiting for the lock to be freed, the number of that job's runnable threads drops to zero and the application can make no progress until the preempted thread is rescheduled. The overhead of this unnecessary blocking and unblocking slows down the program's execution.

In the previous section, we saw how fine-grained parallelism can improve a program's performance by increasing the chance that a processor will find another runnable thread when its current thread blocks. Unfortunately, a fine-grained parallel program that packs the ready-list interacts badly with the behavior of a one-level scheduler. In particular, when a program's thread blocks in the kernel on an I/O request, the parallelism of the program can only be maintained if the kernel can schedule another of the program's

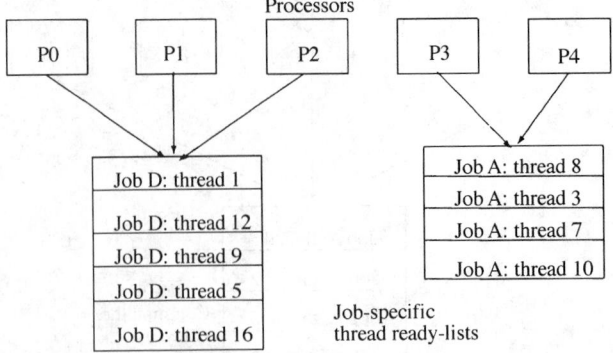

FIGURE 76.3 Two-level thread scheduling.

threads in place of the one that blocked. This benefit, though, comes at the cost of increased preemption activity and diminished overall performance.

The problems of one-level scheduling are addressed by two-level schedulers. With a two-level scheduler, processors are first assigned to a job, and then threads within that job are executed only on the assigned processors. Each job has its own ready-list, which is used only by the job's processors, as shown in Fig. 76.3. Thread preemption may no longer be necessary with a two-level scheduler since a preempted thread will only be replaced by another thread from the same job. Further, for long intervals, a processor runs only threads from the same application, and so the cost of switching between threads is kept low.

In a two-level scheduling system, processors can be allocated to jobs either statically or dynamically. A static two-level scheduler never changes the number of processors given to a job from its initial allocation; if some of those processors are needed by another job, the operating system must preempt all of the job's processors. A dynamic scheduler can adapt the number of processors assigned to each job according to changing conditions.

Dynamic two-level scheduling can give better performance, because it overlaps periods of poor parallelism in one job with periods of high parallelism in another. One difficulty with a dynamic scheduler is that it requires more information from an application describing the current processor requirements. As a result, though, dynamic scheduling can also more easily handle changes in the number of running jobs. For example, when a job finishes, its processors can be reallocated to a running job whose parallelism is increasing. To avoid the problems of one-level scheduling, though, it is crucial that the operating system coordinate with each application when it needs to preempt processors (e.g., to avoid preempting a processor when it would seriously affect performance). A dynamic scheduler always has the option, when it needs processors and no application has any available, of reverting to a static policy.

Kernel- vs User-Level Thread Management

Processor scheduling controls the allocation of processors to jobs. The operating system must be responsible for processor scheduling because processors are a hardware resource and shifting a processor from one job to another involves updating per-processor address space hardware registers. Spawning a thread so that it runs on an already allocated processor, however, does not require modifying privileged state. Thus, thread management and scheduling within a job can be done entirely by the application instead of by the operating system. In this case, thread management operations can be implemented in an application-level library. The library creates virtual processors using the operating system's processor scheduling interface, and schedules the application's threads on top of these virtual processors.

Unlike processor allocation, where a single systemwide scheduling policy can be used, thread scheduling policies benefit from being application specific. Some applications perform well if their threads are scheduled according to some fixed policy, such as first-in–first-out or last-in–first-out, but others need to schedule threads according to fixed, or even dynamically changing priorities. For example, consider a parallel simulation where each simulation object is represented by its own thread. Different objects become sequential bottlenecks at different times in the simulation; the amount of parallelism can be increased by preferentially scheduling these objects' threads.

It is difficult to provide sufficient thread scheduling flexibility with kernel-level threads. While the kernel could define an interface that allows each application to select its thread scheduling policy, it is unlikely that the system designer could foresee all possible application needs.

Thread management involves more than scheduling. A tradeoff exists between user- and kernel-level thread management. A user-level implementation provides more flexibility and better performance; implementing threads in the kernel guarantees a uniformity that eases the integration of threads with system tools.

The downside of having many custom-built thread management systems is that there is no standard thread. By implication, a kernel-level thread management system defines a single, systemwide thread model that is used by all applications. Operating systems that support only one thread model, like those that support only one programming language, can more easily provide sophisticated utilities, such as debuggers and performance monitors. These utilities must rely on the abstraction and often the implementation of the thread model, and a single model makes it easier to provide complete versions of these tools since their cost can be amortized over a large number of applications. Peripheral support for multiple models is possible, but expensive.

A standard thread model also makes it possible for applications to use libraries, or canned software utilities. In the same sense that a standard procedure calling sequence sacrifices speed for the ability to call into separately compiled modules, a standard thread model allows one utility to call into another since they both share the same synchronization and concurrency semantics.

It is important to point out that two-level scheduling does not imply that threads are implemented at the application level; the job-specific ready queues shown in Fig. 76.3 could be maintained either within the operating system or within the application. Also, a user-level thread implementation does not imply two-level scheduling, even though threads *are* being scheduled by the application. This implication only holds in the absence of multiprogramming, or in cases where processors are explicitly allocated to jobs. For example, a user-level thread implementation built on top of Unix processes that share memory suffers from the same problems relating to preemption and I/O as do one-level kernel threads because both are scheduled in a job-independent fashion.

Performance

The performance of thread operations determines the granularity of parallelism that an application can effectively use. If thread operations are expensive, then applications that have inherently fine-grained parallelism must be restructured (if that is even possible) to reduce the frequency of those operations. As the cost of thread operations begins to approach that of a few procedure calls, several issues become performance critical that, for slower operations, would merely be second-order effects.

Simplicity in the thread system's implementation is crucial to performance [Anderson et al. 1989]. There is a performance advantage to building multiple thread systems, each tuned for a single type of application. Even simple features that are needed by only some applications, such as saving and restoring all floating point registers on a context switch, will markedly affect the performance of applications that do not need the functionality. Each context switch takes only tens of instructions; a feature that adds even a few more instructions must have a large compensating advantage to be worthwhile. For example, the ability to preemptively schedule threads within each job makes the thread management system more sluggish at several levels, because preemption must be disabled (and then re-enabled) whenever scheduling decisions are being made. These scheduling decisions are on the critical path of all thread management operations.

Although kernel-level thread management simplifies the generation and maintenance of system tools, it increases the baseline cost of all thread management operations. Just trapping to the operating system can cost as much as the thread operation itself, making a kernel implementation unattractive for high-performance applications. Further, the generality that must be provided by a kernel-level thread scheduler hurts the performance of those applications needing only basic service. Kernel-level threads are less able to cut corners by exploiting application-specific knowledge. With a user-level thread system, the thread management system can be stripped down to provide exactly the functions needed by an application and no more. User-level thread operations also avoid the cost of trapping to the kernel.

Other performance issues have less to do with what a thread system does, than with how it goes about doing it. For example, using a centralized ready-list can limit performance for applications that have extremely fine-grained parallelism. The ready-list is a shared data structure that must be locked to prevent it from being modified by multiple processors simultaneously. Even if the ready-list critical sections consist only of simple enqueue and dequeue operations, they can become a sequential bottleneck, since there is little other work involved in spawning/finishing or blocking/unblocking a thread. An application for which thread overhead is 20% of the total execution time, and half of that overhead is spent accessing the ready-list, then its maximum speedup (the time of the parallel program on P processors divided by the time of the program on one processor) is limited to 10.

The bottleneck at the ready-list can be relieved by giving each processor its own ready-list. In this way, enqueueing and dequeueing of work can occur in parallel, with each processor using a different data structure. When a processor becomes idle, it checks its own list for work, and if that list is empty, it scans other processors' lists so that the workload remains balanced.

Per-processor ready-lists have another nice attribute: threads can be preferentially scheduled on the processor on which they last ran, thereby preserving cache state. Computer systems use caches to take advantage of the principle of *locality*, which says that a thread's memory references are directed to or near locations that have been recently referenced. By keeping references close to the processor in fast cache memory, the average time to access a memory location can be kept low. On a multiprocessor, a thread that has been rescheduled on a different processor will initially find fewer of its references in that processor's cache. For some applications, the cost of fetching these references can exceed the processing time of the thread operation that caused the thread to migrate.

The role of spin-waiting as an optimization technique changes in the presence of high-performance thread operations. If a thread needs to wait for an event, it can block, relinquishing its processor, or spin-wait. A thread must spin-wait for low-level scheduler locks, but in application code a thread should block instead of spin if the event is likely to take longer than the cost of the context switch. Even though context switches can be implemented efficiently, reducing the need to spin-wait, a hidden cost is that context switches also reduce cache locality.

76.4 Three Modern Thread Systems

We now outline three modern thread management systems for multiprocessors: Windows NT, Presto, and Multilisp. The choices made in each system illustrate many of the thread management issues raised in the previous section.

The thread management primitives for each of these systems are shown in Table 76.1. The table is organized to indicate how the primitives in one system relate to those in the others, as well as those provided by the basic thread interface outlined in the Basic Thread Functionality section.

Windows NT is an operating system designed to support Microsoft Windows applications on uniprocessors, shared memory multiprocessors, and distributed systems. Windows NT supports multiple threads within an address space. Its thread management functions are implemented in the Windows NT kernel.

TABLE 76.1 The Basic Operations of Thread Management Systems

Basic	Windows NT	Presto	Multilisp
Spawn	thread_create;thread_resume	Thread::new; Thread::start	(future...)
Block	thread_suspend	Thread::sleep	*Touch unresolved future.*
Unblock	thread_resume	Thread::wakeup	*When future is resolved.*
Finish	thread_terminate	Thread::terminate	*Resolve this future.*

Since NT's underlying thread implementation is shared by all parallel programs, system services such as debuggers and performance monitors can be economically provided.

Windows NT's scheduler uses a priority-based one-level scheduling discipline. Because Windows NT allocates processors to threads in a job-independent fashion, a parallel program running on top of the Windows NT thread primitives (or even a user-level thread management system based on those primitives) can suffer from anomalous performance profiles due to ill-timed preemptive decisions made by the one-level scheduling system.

Presto is a user-level thread management system originally implemented on top of Sequent's DYNIX operating system, but later ported to DEC workstations. DYNIX provides a Presto program with a fixed number of Unix processes that share memory. The Presto run-time system treats these processes as virtual processors and schedules the user's threads among them. Presto's thread interface is nearly identical to Windows NT's.

Presto is distinguished from most other thread systems in that it is structured for flexibility. Presto is easy to adapt to application-specific needs because it presents a uniform object-oriented interface to threads, synchronization, and scheduling. The object-oriented design of Presto encourages multiple implementations of the thread management functions and so offers the flexibility to efficiently accommodate differing parallel programming needs.

Presto has been tuned to perform well on a multiprocessor; it tries to avoid bottlenecks in the thread management functions through the use of per-processor data structures. Presto does not provide true two-level scheduling, even though the thread management functions (e.g., thread scheduling) are implemented in an application library accessible to the user; DYNIX, the base operating system, schedules the underlying virtual processors (Unix processes) any way that it chooses. Although a Presto program can request that its virtual processors not be preempted, the operating system offers no solid guarantee. As a result, kernel preemption threatens the performance of Presto programs in the same was as it does Windows NT programs.

Although Windows NT and Presto are implemented differently, the interfaces to each represent a similar style of parallel programming in which the programmer is responsible for explicitly spawning new threads of execution *and* for synchronizing their access to shared data. This style is not accidental, but reflects the basic function of the underlying hardware: processors communicating through shared memory. One criticism often made of this style is that it forces the programmer to think about coordinating many concurrent activities, which can be a conceptually difficult task.

Multilisp demonstrates how thread support can be integrated into a programming language in order to simplify writing parallel programs. In Multilisp, a multiprocessor extension to LISP, the basic concurrency mechanism is the **future**, which is a reference to a data value that has not yet been computed. The **future** operator can be included in any Multilisp expression to spawn a new thread which computes the value of the expression in parallel. Once the value has been computed, the future *resolves* to that value. In the meantime, any thread that tries to use the future's value in an expression automatically blocks until the future is resolved. The language support provided by Multilisp can be implemented on top of a system like Windows NT or Presto using locks and condition variables.

With Multilisp, the programmer does not need to include any synchronization code beyond the future operator; the Multilisp interpreter keeps track of which futures remain unresolved. By contrast, using the Windows NT or Presto thread primitives, the programmer must add calls to the appropriate synchronization primitives wherever the data is needed. Multilisp, like Presto, uses per-processor ready-lists to reduce contention in scheduling operations.

76.5 Summary

This chapter has examined some of the key issues in thread management for shared-memory multiprocessors.

Shared-memory multiprocessors are now commonplace in both commercial and research computing. These systems can easily be used to increase throughput for multiprogrammed sequential jobs. However, their greatest potential—as yet not fully realized—is for accelerating the execution of single, parallelized programs.

As programmers make use of finer grained parallelism, the design and implementation of the thread management system becomes increasingly crucial. Modern thread management systems must address the programmer interface, the operating system interface, and performance optimizations; language support and scheduling techniques for multiprogrammed multiprocessors are two areas that require further research.

References

Anderson, T. E., Lazowska, E. D., and Levy, H. M. 1989. The performance implications of thread management alternatives for shared memory multiprocessors, pp. 49–60. In *ACM SIGMETRICS Perform. '89 Conf. Meas. Modeling Comput. Syst.* May.

Bershad, B., Lazowska, E., and Levy, H. 1988. PRESTO: a system for object-oriented parallel programming. *Software Prac. Exp.* 18(8):713–732.

Custer, H. 1993. *Inside Windows NT.* Microsoft Press.

Dijkstra, E. W. 1968. Cooperating sequential processes. In *Programming Languages*, pp. 43–112. Academic Press.

Encore. 1986. UMAX 4.2 Programmer's Reference Manual. Encore Computer Corp.

Halstead, R. 1985. Multilisp: A language for concurrent symbolic computation. *ACM Trans. Programming Lang. Syst.* 7(4):501–538.

Hoare, C. A. R. 1974. Monitors: an operating system structuring concept. *Commun. ACM* 17(10):549–557.

Hoare, C. A. R. 1978. Communicating sequential processes. *Commun. ACM* 21(8):666–677.

Redell, D. D., Dalal, Y. K., Horsley, T. R., Lauer, H. C., Lynch, W. C., McJones, P. R., Murray, H. G., and Purcell, S. C. 1980. Pilot: an operating system for a personal computer. *Commun. ACM* 23(2):81–92.

Sequent. 1988. Symmetry Technical Summary. Sequent Computer Systems, Inc.

Tevanian, A., Rashid, R. F., Golub, D. B., Black, D. L., Cooper, E., and Young, M. W. 1987. Mach threads and the Unix kernel: the battle for control, pp. 185–197. In *Proc. USENIX Summer Conf.*

77

Process and Device Scheduling

Robert D. Cupper
Allegheny College

77.1 Introduction[1]

High-level language programmers and computer users, in fact, deal with what is really a virtual computer. That virtual computer they see is facilitated by a software bridge that plays the role of interlocutor between the actual computer hardware and the computer user's environment. This software, described in general in Chapter 75, is the operating system. The computer's operating system (OS) is made up of a group of systems programs that serve two basic ends: (1) to control the allocation and use of the computing system's resources among the various users and tasks and (2) to provide an interface between the computer hardware and the programmer or user that simplifies and makes feasible the creation, coding, debugging, maintenance, and use of applications programs. Thus, the operating system creates and maintains an environment in which users can have programs executed, that is, it provides a structure in which the user can request and monitor execution of his or her programs and can receive the resulting output. To this end, the operating system must make available to the user's program the system resources needed for its execution. These system resources are the processor, primary memory, secondary memory including the file system, and the various devices. Since most modern computing systems are powerful enough to

---

[1]Parts of the Introduction and Background sections of this chapter are reprinted, from this author's contribution to Tucker, A. B., Cupper, R. D., Bradley, W. J., Epstein, R. G., and Kelemen, C. F. 1995. *Fundamentals of Computing II: Abstraction, Data Structures, and Large Software Systems.* McGraw–Hill, New York. With permission.

0-8493-2909-4/97/$0.00+$.50

allow multiple user programs or at least multiple tasks to execute in the same time frame, the operating system must allocate these resources among the potentially competing needs of the multiple tasks in such a way as to ensure that all of the tasks are able to execute to completion. Furthermore, these resources must be allocated so that no one task is unnecessarily or unfairly delayed. This requires that the operating system *schedule* its resources among the various and competing tasks. The detailed characterization of the problem of scheduling computer system resources in a number of settings, the techniques, algorithms, and policies that have been set forth for its solution, and the criteria and method of assessment of the efficacy of these solutions form the subject of this chapter.

The next section establishes the landscape for the discussion with a brief review of methods of delivery of computing services and a look at the essential concept of a **process** —a program in execution—the most basic unit of account in an operating system, and then a brief look at the components of the operating system responsible for the execution of a process. Though this chapter is primarily concerned with the first of the two functions of an operating system, that is, control of the allocation and use of computing system resources, it will become clear that the methods brought to bear on the simultaneous achievement of these two functions cannot treat them as wholly independent.

77.2 Background

Computer service delivery systems may be classified into three groups, which are distinguished by the nature of interaction that takes place between the computer user and his or her program during its processing. These classifications are termed *batch, time-shared,* and *real-time.*

In a batch processing operating system environment, users submit jobs which are collected into a batch and placed on an input queue at the computer where they will be run. In this case, the user has no interaction with the job during its processing, and the computer's response time is the turnaround time, the time from submission of the job until execution is complete and the results are ready for return to the person who submitted the job.

A second mode for delivering computing services is provided by the time-sharing operating system. In this environment, a computer provides computing services to several users concurrently on line. The various users share the central processor, the memory, and other resources of the computer system in a manner facilitated, controlled, and monitored by the operating system. The user, in this environment, has full interaction with the program during its execution, and the computer's response time may be expected to be no more than a few seconds.

The third class, the real-time operating system, is designed to service those applications where response time is of the essence in order to prevent error, misrepresentation, or even disaster. Examples of real-time operating systems are those which handle airlines reservations, machine tool control, and monitoring of nuclear power stations. The systems, in this case, are designed to be interrupted by external signals that require the immediate attention of the computer system.

In fact, many computer operating systems are *hybrids*, providing for more than one of these types of computing service simultaneously. It is especially common to have a background batch system running in conjunction with one of the other two on the same computer system.

Discussion of resource scheduling in this chapter is limited to uniprocessor and **multiprocessor** systems without **network** connections. Resource scheduling in networking and **distributed** computing environments is considered in Chapters 85 and 87. Programs proceed through the computer as processes. Therefore, the various computer system resources are to be allocated to processes. A thorough understanding of that concept is essential in all that follows here. The next section provides the essence; more detail can be found in Chapter 76.

Processes

Most operating systems today are **multiprogramming** systems. Systems such as these, where multiple independent programs are executing, must manage two difficult problems: concurrency and nondeterminacy.

The concurrency problem arises from the coexistence of several active processes in the system during any given interval of time. Nondeterminacy arises from the fact that each process can be interrupted between any two of its steps. The unpredictability of these interruptions, coupled with the randomness that results from processes entering and leaving the system, make it impossible to predict the relative speed of execution of interrelated processes in the system. A mechanism is needed to facilitate thinking about, and ultimately dealing with, the problems associated with concurrency and nondeterminacy. An important part of that mechanism is the conceptual and operational isolation of the fundamental unit of computation that the operating system must manage. This unit is called the *task* or process. Informally, a process or task is a program in execution.

This concept of process facilitates an understanding of the twin problems of concurrency and indeterminacy. Concurrency, as we have seen, occurs whenever there are two or more processes active within the system. Concurrency may be *real*, in the case where there is more than one processor and hence more than one process can execute simultaneously, or *apparent*, whenever there are more processes than processors. In this latter case, it is necessary for the operating system to provide for the switching of processors from one process to another sufficiently rapidly to present the illusion of concurrency to system users. But this is difficult, for whenever a processor is assigned to a new process (called *context switching*), it is necessary to recall where the first process was stopped in order to allow that process, when it gets the processor back, to continue where it left off.

The idea of context switching implies that a particular process can be interrupted. Indeed, a process may be interrupted, as necessary, between individual steps (machine instructions). Such interruptions occur most often when a particular process has used up its quota of processor time or when it has requested and must wait for completion of an input/output (I/O) operation. Nondeterminacy arises from the unpredictable order in which such interruptions can occur.

Since active processes in the system can be interrupted, each process can be in one of three states:

- *Running:* the process is currently executing on a processor.
- *Ready:* the process could use a processor if one were available.
- *Blocked:* the process is waiting for some event, such as I/O completion, to occur.

The relationship between these three states for a particular process is shown in Fig. 77.1.

Here, we see that if a process is currently running and requests I/O, for example, it relinquishes its processor and goes to the blocked state. In order to maintain the illusion of concurrency, each process is assigned a fixed quantum of time, or *time slice*, which is the maximum time a running process can control the processor. If a process is in the running state and does not complete or block before expiration of its time slice, that process is placed in the *ready* state and some other process is granted use of the processor for its quantum of time. A blocked process can move back to the ready state upon completion of the event which blocked it. A process in the ready state becomes running when it is assigned a processor by the system dispatcher.

All of these state changes are interrupt driven. A request for I/O is effected by issuing a supervisor call via an I/O procedure which causes a system interrupt. I/O completion is signaled by an I/O interrupt from a data channel.[2] Time slice exceeded results in an external interrupt from the system's interval timer. And, of course,

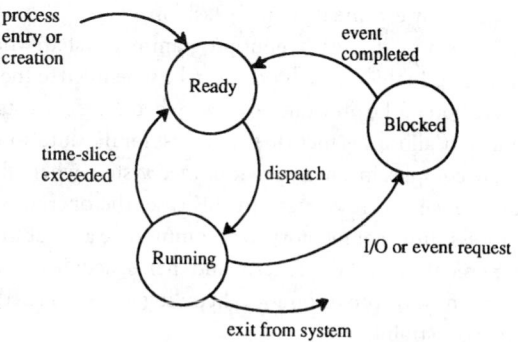

FIGURE 77.1 Process state transitions.

[2] A data channel is a small special purpose computer that executes programs to actually do I/O concurrently with the main processor's program execution.

movement from ready state to running results from the dispatcher giving control of the processor to the most eligible ready process. In each case, when a process gives up the processor, it is necessary to save the particulars of where the process was in its execution when it was interrupted so that it may properly resume later.

Each process within the system is represented by an associated process control block (PCB). The PCB is a data structure containing the essential information about an active process including:

- Process identification (ID)
- Current state of the process
- Register save area
- A pointer to the process's allocated memory area
- Pointers to other allocated resources (disk, printer, etc.)

The latter three contain the information necessary to restart an interrupted process. There is only one set of registers in the system, which is shared by all of the active processes and, therefore, the contents of these registers must be saved just before the context switch. Since memory is space shared, as well as time shared, it is necessary only to save pointers to the locations of the process' memory areas prior to its interruption. Devices vary; some are shareable (for example, disk devices) and so are treated as memory, whereas others (such as the printer) are nonshareable and tied up by a process for as long as it is using them. In either case, it is necessary here to keep track only of the device ID and, perhaps, the current position in a file.

Thus, programs solve problems by being executed; they execute as processes. To this end, the OS must allocate, or schedule, to the process sufficient memory to hold its data and at least the part of the program immediately due for execution, the various devices needed, and a processor. Since there certainly will be multiple processes and possibly even multiple jobs, each made up of processes, it is necessary to have the OS schedule these resources in such a way as to enable all of the jobs to run to completion. The next section deals with scheduling the processor among the processes competing for it to effect the execution of their parent programs.

77.3 Resources and Scheduling

Programs execute as processes using computer system resources including a processor, primary memory, and most likely secondary memory including files and some devices. Thus, in order for the process to execute, it must enter the system and these resources must be allocated to the process. The operating system must schedule the allocation of these resources to a given process so that this process and any others in the system may execute in a timely fashion.

The simplest case is a monoprogrammed system where there is just one program executing in the system at a time. In this case, a process can be scheduled to the system whenever it becomes available following the execution of the previous process. Scheduling consists of determining if sufficient resources are available, and if so, allocating them to the process for the duration of its processing time in the system. The situation is more complex in a multiprogrammed system where there are multiple processes in the system competing for the various resources. In this case, the operating system must schedule and allocate to each active process sufficient memory to accommodate at least the parts of its data and program instructions needed for execution in the near term and then schedule the processor to execute some instructions. In addition, there must be provision for scheduling access to needed files and required devices, all with some sort of time constraint.

Though not all of these resources need to be allocated to a particular process throughout its life in the system, they must be scheduled in such a manner as to be available when needed and in concert, lest that process be stalled for want of one or more resources, tying up other processes waiting for unavailable resources in the interim.

Resource scheduling and allocation is, from the performance point of view, perhaps the most important part of the operating system. Good scheduling must consider the following objectives:

1. Resource allocation that facilitates minimal average turnaround time
2. Resource allocation that facilitates minimal response time
3. Mutual exclusion of processes from nonshareable resources
4. A high level of resource utilization
5. Deadlock prevention, avoidance, or detection

It is clear that these objectives are not necessarily mutually satisfiable. For example, a high level of resource utilization probably will mean a longer average wait for resources, thus lengthening both response and turnaround times. The choice may be, in part at least, a function of the particular service delivery system that the process has entered. A batch system scheduler would favor resource utilization, whereas a time-sharing system would need to be sensitive to response time, and at the extreme, a real-time system would minimize response time at the expense of resource utilization.

An allocation mechanism refers to the implementation of allocations. This includes the data structures used to represent the state of the various resources (shareable or nonshareable, available, busy, or broken), the methods used to assure mutual exclusion in use of nonshareable resources, and the technique for queuing waiting resource requests. The allocation policy refers to the rationale and ramifications of application of the mechanisms. Successful scheduling requires consideration of both.

The practices and policies regarding scheduling each of the resource classes, the processor, primary memory, secondary memory and files, and devices, differ significantly and are next considered in turn. Since the processor is arguably the most important resource—certainly a process could not proceed without it—the discussion turns first to processor scheduling.

77.4 Processor Scheduling

In this section, it is assumed that adequate memory has been allocated to each process and the needed devices are available to allow focus on the problems surrounding the allocation of the processor to the various processes. To distinguish scheduling programs from the input queue for entry into the computing system from the problem of allocating the processor among the active processes already in the system, the term *scheduler* is reserved for the former and *dispatcher* for the latter. The term *active process* refers to a process that has been scheduled, in this sense, into the system from the input queue, that is, has been allocated space in memory and had some of its needed devices allocated.

Development of methods for processor dispatching is motivated by a number of system performance goals, including:

1. Reasonable turnaround and/or response time—here, as previously indicated, the tolerance is governed by the service delivery system (e.g., batch processing vs. time sharing)
2. Predictable performance
3. Good absolute or relative throughput
4. Efficient resource utilization (e.g., low CPU idle time)
5. Proportional resource allocation
6. Reasonable length waiting queues
7. No process must wait forever
8. Satisfaction of real-time constraints

It is evident that these goals sometimes conflict. For example, average response time can be improved, but at the expense of the very longest programs, which are likely to have to wait. Or, minimization of response time could result in poor resource utilization if a long program that has many resources allocated to it is forced to wait for a long while, thus idling the resources allocated to it.

A processor scheduler, CPU scheduler, or dispatcher consists of two parts. The first is a ready queue consisting of the active processes that could immediately use a processor were one available. This queue is made up of all of the processes in the ready state of Fig. 77.1.[3] The other part of the dispatcher is the algorithm used to select the process, from those on the ready queue, to get the processor next. A number of dispatching algorithms have been proposed and tried. These algorithms are classified here into three groups: priority algorithms, rotation algorithms, and multilevel algorithms.

Priority Dispatching Algorithms

These dispatching algorithms may be classified by queue organization, whether they are preemptive or nonpreemptive, and the basis for the priority.

The ready queue may be first-in–first-out (FIFO), priority, or unordered. The queue can be maintained in sorted form which facilitates rapid location of the highest priority process. However, in this case, inserting a new arrival is expensive since, on average, half of the queue will need to be searched to find the correct place for the insertion. Alternatively, new entries can simply and quickly be added to an unsorted ready queue. But, the entire queue must be searched each time the processor is to be allocated to a new process. In fact, a compromise plan might call for a periodic sort, maintaining a short list of new arrivals on the top of the previously sorted queue. In this case, when a new process is to be selected, the priority of the process at the front of the sorted part of the queue is compared with each of the recently arrived unsorted additions, and the processor is assigned to the process of highest priority.

In a nonpreemptive algorithm, the dispatcher schedules the processor to the process at the front of the ready queue and that process executes until it blocks or completes. A preemptive algorithm is the same except that a process, once assigned a processor, will execute until it completes or is blocked, unless a process of higher priority enters the ready queue, in which case, the executing process is interrupted and placed on the ready queue and the now higher priority process is allocated the processor.

First-Come–First-Served (FCFS) Dispatching

When the criteria for priority is arrival time, the dispatching algorithm becomes first-come–first-served (FCFS). In this case, the ready queue is a FIFO queue, processor assignment is made to the process with its PCB at the front of the queue, and new arrivals are simply added to the rear of the queue. The algorithm is easy to understand and implement. FCFS is nonpreemptive and so a process, once assigned a processor, keeps it until it blocks (say, for I/O) or completes. Therefore, the performance of the system in this case is left largely in the hands of fate, that is, how jobs happen to arrive.

Shortest Job First (SJF) Dispatching

The conventional form of the shortest job first (SJF) algorithm is a priority algorithm where the priority is inversely proportional to (user) estimated execution time. The relative accuracy of user estimates is enforced by what is, in effect, a penalty–reward system: too long an estimated execution time puts a job at lower priority than need be, and too short of an estimate is controlled by aborting the job when the estimated time is exceeded, effecting a delay with penalty, by forcing the user to rerun the job. There are preemptive and nonpreemptive forms to the SJF algorithm. In the nonpreemptive form, once a process is allocated a processor, the process runs until completion or block. The preemptive form of the algorithm allows a new arrival to the ready queue with lower estimated running time to preempt a currently executing process with a longer estimated running time.

SJF is clearly advantageous for short jobs and, since a typical execution time distribution is usually weighted toward shorter jobs, especially in an installation providing general computer services, one could argue that a SJF policy would benefit most users. But, as always, there is a tradeoff; in this case, it is that

---

[3] Actually, the ready queue contains representations of the ready processes, that is, the corresponding PCBs (or pointers to them).

long jobs get relatively poor service. This is especially true in the preemptive version where preemption clearly favors short jobs with the result that long jobs can effectively be starved out. Whereas the SJF rule provides the minimum average waiting time, it, like FCFS, appears to apply only to the batch processing system of service delivery.

Priority Dispatching

In this algorithm, process priorities are set based on criteria external to the system, reflecting the importance of the processes. These include such factors as memory size requirements, estimated job execution time, estimated amount of I/O activity, and/or some measure of the importance of the computation as set by the user or the institutional structure. For example, a class of jobs characterized by low execution time estimates combined with minimal resource requirements may be deemed to be of high priority. Similarly, one might argue that particular systems programs, say, device handlers, ought to have high priority. This type of dispatching algorithm can, like those preceding it, be either nonpreemptive or preemptive. In the nonpreemptive form, the highest priority job is assigned to the CPU and run until completion or block. In the preemptive form, the arrival of a higher priority job at the ready queue results in the preemption of the currently running process for the higher priority process. It is clear that priority dispatching serves the highest priority jobs optimally. Thus, to the extent that the externally set priorities reflect actual institutional priorities, priority dispatching is arguably best. But, there are two problems. First, low-priority jobs generally receive poor service. Especially in the preemptive form, low-priority jobs can be indefinitely blocked or starved out from the processor, violating the fairness criterion. Second, to the extent that priorities are set by users, based on the importance or perceived importance of their programs without knowledge of the current system workload mix, the system itself can lose control of its performance parameters. For example, a high-priority process might get good service, but at the cost of other goals such as maximum system throughput or resource utilization.

Dynamic Dispatching Priority Adjustment

Some of the shortcomings of the priority algorithms can be ameliorated by dynamic priority adjustment. In this case, decisions can be based on information about a process that is obtained while the process is in the system. Priorities determined from one of the previously described algorithms can be dynamically adjusted during the life of the process in the system according to a number of criteria such as the number and type of resources currently allocated, accumulated waiting time since the job entered the system, amount of recent processing time, amount of recent I/O activity, total time in the system, etc.

One such plan advanced by Kleinrock [1970] is to allow a process dispatching priority to increase at a rate x while the process is on the ready queue, and rate y while it has a processor assigned to it. Priority then depends on the values of x and y. These could be set externally and differ for different jobs, and/or they could change dynamically over the time a given process is in the system. For example, the starvation problem can be eliminated if x of a high-priority process decreases over time and/or y of a low-priority job increases over time. Use of the SJF rule accompanied by nonlinear functions where x and y decrease over time for a while, then jump to high values, will ensure that no jobs wait very long for service, yet still favors short jobs.

Another dynamic scheme would increase the priority of a process dynamically during periods of high I/O activity, i.e., make priority inversely related to the time interval since the last I/O call. This favoring of I/O-bound jobs is desirable because it compensates for the speed disparity between the more mechanical I/O devices and the higher speed electronic processor. It does this by keeping the devices running, thus minimizing the possibility that the CPU will need to wait for I/O completion. Moreover, the effect on other more compute-bound jobs is minimal because of the limited CPU time required to start I/O operations.

In general, priority algorithms are easy to understand and simple to implement. The main disadvantage is that low-priority jobs tend to get poor service. Moreover, the performance associated with this type of algorithm is not appropriate to some situations. The response times available to processes, especially lower priority programs, would not be acceptable in a time-sharing or real-time environment. Apparently, there is a need to consider another approach to dispatching.

Rotation Algorithms

The essence of the rotation algorithms, as the name implies, is that the CPU is scheduled in rotation so that each job in the ready queue is given some service in order to maintain a reasonable response interval. These algorithms are designed to apply to time-sharing systems.

Simple Round Robin (RR)

In the round robin rotation algorithm, processor time is divided into units called time slices or quanta. The ready queue is treated like a circular queue and each of the processes in the ready queue is given one time slice each rotation. If the process does not complete or block during its quantum, at the end of the quantum, the process is preempted and returned to the end of the ready queue. The idea is to provide response that is reasonably independent of job size and/or priority. In simple RR, all time slices are the same size, say, q. A typical q is 50 ms; the range is 10–100 ms. There is no static or dynamic priority information. Therefore, if there are k processes in the ready queue, each process is scheduled for q out of every kq milliseconds. Thus, users perceive processes running on a processor with $1/k$ of its actual processor speed. Response time is $(k - 1) \times q$ or less.

Clearly, performance is affected by two key parameters, the size of the ready queue k and the quantum size q. Response is inversely related to k, and hence, as the system becomes loaded, response time deteriorates. Whereas k is essentially externally determined, quantum size is a system parameter. If q were infinite, then round robin is FCFS. A large q tends to favor some jobs. If a number of processes in the ready queue are in a blocked state or will block during their quanta, the remaining processes will cycle frequently and the corresponding jobs will run to completion quickly and with excellent response. On the other hand, arrival of new processes, each taking one of the longer quanta, causes the average response to deteriorate substantially. If q is small, these effects are decreased. For example, arrival of a new process to the ready queue will have a much smaller effect on average response time. A small q, in addition to providing a more consistent response, leads to a total waiting time for a job more proportional to the length of the job. At the extreme, a very small q will cause useful execution time of the quantum to be overwhelmed by the context switching delay. This would imply that the size of the quantum ought to be large relative to the time required for a context switch. Turnaround time is also affected by the size of the time slice. Though one might expect that turnaround time would fall as the quantum size increases, this is not always true. Context switches add to turnaround, and, as a very long quantum moves RR toward FCFS, average turnaround can increase substantially. Silberschatz and Galvin [1994] claim that 80% of the CPU executions should be shorter than q.

Round Robin with Priorities

Simple round robin is designed to provide reasonable and fair response time for all active processes in the system. As such, it has no way of recognizing a more important process. Variants of the simple round robin have been proposed to address this problem. *Biased round robin* allows for different length time slices for different processes based on some external priority. A process with a higher priority is assigned a longer q allowing processes to proceed through the system in proportion to their priorities. An alternative, *selfish round robin*, is based on Kleinrock's linearly increasing priority scheme previously discussed (Kleinrock [1970]). Here, x and y are such that $0 \le y < x$. The effect is processes already in execution make entering processes wait in the ready queue until the priority of the new process increases to the value of the priority of the older processes. This will eventually occur since y, the rate of priority increase of an executing process, is less than x, the rate of priority increase of a process in the ready queue.

Cycle-Oriented Round Robin

Simple round robin is subject to considerable performance degradation in terms of response time as more processes enter the system and ultimately the ready queue. This can lead to an increase in response time beyond that which is reasonable and expected for an interactive time-sharing environment. Cycle-oriented round robin was developed to obviate this problem. In this case, the longest tolerable response time, c, is set as the cycle time and becomes the basis for calculation of the quantum length, q. The time slice, $q = c/k$, will guarantee that response time never exceeds the maximum tolerable, c. There are, however, two problems. Process arrivals during the cycle could cause the response time to exceed the acceptable limit. But this is easily resolved by denying new arrivals entry into the ready queue except at the end of a cycle where the time slice size is recalculated. The problem of system overload remains, however, for as k becomes large, q becomes small, and k too large implies that q will be too small, leading to unacceptable overhead from context switching. The solution is to enforce a minimum time slice size.

Multilevel Dispatching

In the Background section, it was pointed out that a single computing system might be designed to provide for two (or more) of the computer service delivery systems, say, time sharing in the foreground and batch in the background. The two delivery systems have different service requirements, and consequently, taken individually, most appropriately would employ different dispatching algorithms. When the two services are available on the same system, it is necessary to account for different scheduling needs and the priority of interactive jobs over background batch jobs. This is accomplished by a multilevel dispatching system where the ready queue is partitioned into two (or more) separate queues, one for the interactive jobs, one for batch jobs, etc. Each process entering the system is assigned to one of the queues based on process type. Then the processes on each of the several queues are scheduled by different algorithms as appropriate, i.e., RR for the interactive job queue and FCFS for the batch job queue. Moreover, a system of preset priorities with preemption is used for scheduling between the multiple queues. For example, there might be three separate queues, one for system processes, one for interactive processes, and one for batch processes. All processes in the system process queue must be completed before any processes from the interactive queue are dispatched to the processor, and similarly, the interactive queue must be empty before any processes from the batch queue are dispatched. A new system process entering the first queue would cause any interactive process or batch process to be preempted, and a new interactive process would preempt a batch process. Alternately, dispatching could occur among the queues on a time slice basis, i.e., so much time for the systems queue, then a quantum of time for the interactive process queue, etc. The time slice given to each queue could, of course, vary, reflecting the priority of processes assigned to that particular queue. For example, after processes in the system queue complete, 75% of the time could be allocated to interactive jobs, and 25% to batch jobs until another system process comes along.

Multilevel Feedback Queue Dispatching

The multilevel queue dispatching described in the last section relies on processes being assigned to one of the queues depending on some external factor, such as process type, and processes once assigned to a particular queue remain there for the duration of their time in the system. Multilevel feedback queue dispatching is similar except that particular processes can move among the separate queues dynamically based on some aspect or aspects of their progress through the system. For example, three queues could be established as before. Here, the queues are designated queue 0, queue 1, and queue 2, with priority highest for the lowest numbered queue, etc. Accordingly, the quantum q_0 would be shortest for the lowest numbered queue, which would be dispatched according to some form of RR algorithm. Queue 1 would also be RR, but with a longer quantum, q_1, and queue 2 FCFS with an even longer quantum, q_2. All new processes enter queue 0; processes that complete or block within the quantum q_0 remain in this highest priority queue 0; those that do not complete or block are moved to queue 1 at q_0 time exceeded, and similarly those in queue 1 that do not complete or block in time q_1 are moved to queue 2. The queues, as before, are scheduled according to a system of preset priorities with preemption. All of the processes in queue 0 must be completed or blocked before any processes from queue 1 are dispatched, and so forth. A new process arriving at queue 1 will preempt a currently executing process from queue 2.

It is evident that the multilevel feedback queue system of dispatching is the most complex. In addition to the three queues and the dispatching algorithms appropriate for each queue, the system requires a description of some algorithm to determine when to move a process to the next lower or next higher priority queue and a method to determine which queue an entering process should initially join.

Process Scheduling in UNIX

The UNIX system uses a system of *decay usage scheduling* that might be characterized in terms of the previously described classes as round robin with multilevel feedback. Basically, each user process begins with a base level priority that puts it into a particular ready queue. Periodically, the priority is recalculated on the basis of recent CPU usage, increasing the priorities of processes inversely with recent CPU usage.

In UNIX, the possible priority levels are divided into two classes (Bach [1986]): kernel priorities and user priorities. Within each class are a number of priority levels as shown in Fig. 77.2. Note that higher priorities are associated with lower priority values.

The basic operation is for the dispatcher to execute a context switch whenever the currently executing process blocks or exits, or when returning to user mode from kernel mode if a process with a priority higher than the process currently allocated the CPU is ready to run.

A process entering the system is assigned the highest user-level priority, called the base-level priority. If there are several processes in the highest user process level, the dispatcher selects the one that has been there longest. The CPU is dispatched to that process. The clock interrupts the executing process, perhaps several times during its quantum. At each clock interrupt, the clock interrupt handler increases an execution time field in the PCB. Approximately once a second (in UNIX system V), the clock interrupt handler applies a decay function to the execution time field in the PCB of each active process in the system. Thus,

$$C(\text{execution time field}) = \text{decay}(C(\text{execution time field}) = C(\text{execution time field})/2.$$

At this same time, the process priority is recalculated based on recent CPU usage,

$$\text{priority} = C(\text{execution time field})/2 + \text{base level priority}.$$

A process that does not block or exit and therefore uses its entire quantum would be assigned a higher number, corresponding to a lower priority. If the process blocks, it is assigned a priority by the system call routine. The priority is not dependent on whether the process is I/O bound or processor bound as in the multilevel feedback queue dispatching previously described, but rather is a function of the particular reason the process blocked. The kernel adjusts the priority of processes returning from kernel mode back

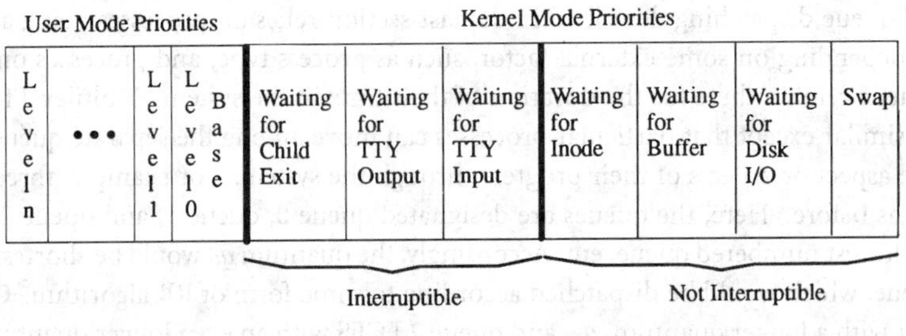

User Mode Priorities				Kernel Mode Priorities						
Level n	•••	Level 1	Level 0 Base	Waiting for Child Exit	Waiting for TTY Output	Waiting for TTY Input	Waiting for Inode	Waiting for Buffer	Waiting for Disk I/O	Swap
				Interruptible			Not Interruptible			

FIGURE 77.2 UNIX process priorities. (*Source:* Adapted from Bach, M. J. 1986. *The Design of the UNIX Operating System*, p. 250. Prentice–Hall, Englewood Cliffs, NJ. With permission.)

to a user-level mode reflecting the fact that it has just had access to kernel resources. Kernel-level priorities can only be obtained via a system call. When the event that caused the block has been completed, the process can resume execution unless preempted by the dispatcher due to the arrival of a process with a higher priority on the ready queue. This could occur upon completion of some event that previously led to block state for some higher priority process or if the periodic priority changes by the clock interrupt handler have made one of the other process's priority higher. In either case, the dispatcher initiates a context switch.

The periodic priority recalculation effectively redistributes the processes among the user-level priority levels. This, along with the policy of dispatching the process longest in the highest priority queue first, assures round robin scheduling for processes in user mode. It should be clear that this scheduler will provide preferential service to highly interactive tasks such as editing because their typically low CPU time to idle time ratio causes their priority level to increase quickly because of their low recent CPU usage.

The UNIX process scheduler includes a feature allowing users to exercise some control over process priority. There is a system call

nice (priority_level)

that permits an additional element in the formula for recalculating priority,

$$\text{priority} = C(\text{execution time field})/2 + \text{base level priority} + \text{nice priority\_level}.$$

This allows the user with a nonurgent process to "nicely" increment the priority calculation resulting in the process moving to a lower level in the user process priority queue. A user cannot use **nice** to raise the priority level of a process or to lower the priority of any other process. Only the superuser can use **nice** to increase a process priority level, and even superuser cannot use **nice** to give another process a lower priority level. But superuser can, of course, kill any process.

Dispatching Algorithms for Real-Time Systems

It is clear that the priority dispatching algorithms are well suited for the batch method of service delivery and the rotation algorithms provide for the performance requirements imposed by multiprogramming time-sharing systems. But what about real-time systems where the constraints on response time are very strict? Real-time systems for critical systems appear as standalone dedicated systems. In this case, when a requesting process arrives, the system must consider the parameters of the request in conjunction with its then-current resource availabilities and accept the request only if it is able to service it within the strict time constraint. However, other real-time processes such as those associated with multimedia, interactive graphics, etc., can be handled by a real-time component of a system also providing time sharing or batch services. But combining real-time applications with others will lead inevitably to a degradation in response and/or turnaround time for the others.

For the combination to be workable, real-time applications must have the highest priority, that priority must not be allowed to deteriorate over time (no aging), and the time required for the dispatcher to interrupt a process and start (or restart) the real-time application must be acceptably small. The dispatch time problem is complicated by the fact that an effectively higher priority process, such as a system call, may be running when the real-time application arrives. One solution to this problem is to interrupt the system call process as soon as possible, that is, when it is not modifying some kernel data structure. This means that even systems applications would contain at least some *interrupt points* where they can be swapped out to make way for a real-time process. Another approach is to use interprocess communication (IPC) primitives (see Chapter 79) to guarantee mutual exclusion for critical kernel data structures, thus allowing systems programs to be interruptible. This latter is the technique used in Solaris 2. According to Silberschatz and Galvin [1994], the Solaris 2 dispatch time with no preemption is around 100 ms, but with preemption and mutual exclusion protection, the dispatch time is closer to 2 ms.

77.5 Memory Scheduling

Memory allocation and scheduling is the function of the memory management subsystem of the operating system. The options include real and virtual memory systems. The mechanisms for scheduling memory include: (1) data structures used to implement free block lists and page and/or segment tables depending on the memory management system structure; (2) cooperation with the processor scheduler to place a process waiting for a memory block, page, or segment in blocked state; and (3) cooperation with the I/O subsystem to queue processes waiting for page or segment transfers to wait for the concomitant I/O service from a disk drive. The details for the memory management subsystem are in Chapter 80 for standalone and networked systems and Chapter 86 for distributed systems.

77.6 Device Scheduling

The devices, sometimes called peripheral devices, that can be attached to a computer system are many and varied. They include terminals, tape drives and printers, plotters, disk drives. These devices differ significantly both in terms of type of operation and speed of operation. In particular, the various devices use different media, encodings, and formats for the data read, written, and stored. On this count, devices can be divided into two groups: block devices that store and transfer information in fixed-sized blocks and character devices that transfer sequences of characters. Disks are block devices; they read, write, and store blocks ranging in size from 128 to 1024 bytes depending on the system (Tanenbaum [1992]). In a block device, each block can be read or written independently of the others. Terminals, tape drives, printers, and network interfaces are character devices. In these cases, data are transferred as a string of characters; there are no block addresses and it is not possible to find a specific character by address on the device.

Devices also differ with respect to the speed of data transmission. A terminal may be able to send perhaps 1000 characters/s as compared with a disk where the transfer rate is closer to a million characters/s. Further, devices differ with respect to the operations they support. A disk drive facilitates arm movement of a head for a seek operation, and a tape drive allows rewind; neither operation is appropriate for the other device.

Devices, in general, consist of two parts: the actual physical device itself and a controller. The actual device is the mechanical part: the turntables and heads for a tape; the disk spindle, platters, heads, and arms for a disk. The controller is the electronic part: a small computer of sorts that contains the circuitry necessary to effect the operation of the one or more similar devices connected to it. Thus, it is the controller that physically initiates the operation of a device, and translates a stream of bits to (input) or from (output) a block of information in the controller's local buffer. In addition, the controller has registers that are used to receive device operation commands and parameters from the operating system and hold status information regarding the device's most recent operation.

Though the buffer in the device's controller effectively separates the slower mechanical operations of a device from the much faster, electronic CPU, data transfer can still extract a delay on system performance. This is true because, for input, eventually the block of data must be transferred from the controller's buffer to main memory for use by the program requesting it. The transfer requires a loop involving the CPU in a character by character transfer. For this reason, many block device controllers contain a direct memory access (DMA) capability. In this instance, the CPU need only send to the controller the main memory start address for the block and the number of characters to send and the controller will effect the transfer asynchronous of CPU operation, until the transfer of the entire block is complete, at which time the controller sends an interrupt to the CPU.

This overview of device operation makes it clear that I/O is one of the more detailed and difficult parts of the programming process. This, plus the obvious need for control over device allocation and utilization, led to the early incorporation of low-level I/O programming as one of the principal functions of the operating system. Thus, it is the responsibility of the I/O subsystem to provide a straightforward interface to user programs, to translate user I/O requests into instructions to the device controller, to handle errors, and to indicate completion of the I/O request to the user program.

In this context, several goals apply to the design of the I/O subsystem. Of course, the I/O subsystem should be efficient since all programs perform at least some I/O, and I/O on the inherently slower devices can often become a bottleneck to system performance. I/O software should provide for device independence in two ways. It should be possible for user programs to be written and translated so as to be independent of a particular device of a given type. That is, it should not be necessary to rewrite or retranslate a program to direct output to one printer used in lieu of another, which is otherwise unavailable. Moreover, it should be possible to have a user program independent even from device type. The program should not have to be rewritten or retranslated to obtain input from a file rather than the keyboard. In the UNIX system, this is effected by treating devices as special files, allowing a uniform naming scheme, a directory path, which allows device independence to include files as well as devices. Similarly, user programs should be free from character code dependence. The user should not need to know or care about the particular codes associated with any one device. In fact, these requirements are equivalent to the goal of uniform treatment of devices. A good user interface should provide for simplicity and therefore minimization of error. The most obvious implication of these goals, especially the last, is that all device specific information, that is, instructions for operation of the device, device character encodings, error handling, etc., should be as closely associated with the specific device or device class as possible. This suggests a layered structure for design of the I/O subsystem.

The structure of the I/O subsystem, then, can be seen as four layers (Lister [1979], Tanenbaum [1992]) proceeding from I/O functions employed in user programs to the device drivers and interrupt handlers that provide low-level program control of the various devices:

1. User program calls to library I/O functions
2. System I/O call interface (device independent)
3. Device drivers (device dependent)
4. Interrupt routines

User programs invoke I/O by means of calls to library functions. These functions, which reside in system libraries and are linked to user programs at execution time and run outside the kernel, provide for two basic services: formatting I/O and setting parameters for the system I/O call interface. An example of a library function that facilitates this is **printf** in UNIX which processes an input format string to format an American Standard Code for Information Interchange (ASCII) string and then invokes the library function **write** to assemble the parameters for the system I/O call interface (Tanenbaum [1992]). These parameters include the name of the logical device or file where I/O is to be done; the type of operation, e.g., read, write, seek, backspace; the amount of data to be transferred (number of characters or blocks); and the source or destination storage location into which (input) or from which (output) a data transfer is to occur.

The system I/O call interface, which is an integral part of the operating system, has three basic functions:

- Link the logical device or file specified in the user level I/O function to an appropriate physical device.
- Perform error checks on the parameters supplied by the user in the I/O function.
- Set up and initiate the request for the physical I/O.

Thus, after ensuring that the I/O parameters in the user request are consistent with the operation requested, e.g., the amount of data in the transfer request is equal to 1 for a character device, or some multiple of the block size for a block device, the I/O call interface assembles the needed parameters into what (Lister [1979]) terms an I/O request block (IORB). The IORB, in addition to the parameters given earlier, will contain pointers to the PCB of the requesting process and an error location for deposit of any error codes should the requested operation turn out to be unsuccessful for any reason.

In order to maintain device independence and to accommodate the unique features and operations associated with the various types of devices, the I/O requests, I/O in progress, and the device characteristics are manifest in a number of data structures, the exact nature of which is, of course, dependent on the system and the computer. These data structures are associated with each device type in so far as possible. A device control block, I/O block, I/O control block (IOCB), unit control block, or channel control block is used to parameterize the characteristics for each device (and/or control unit). Such characteristics include, in addition to the device identification, specific instructions to operate the device, the device status (available, busy, off line), pointers to character translation tables, and a pointer to the PCB of the process which has the device allocated to it. The IOCB also has a pointer to a queue of requests. This request pending queue is a linked list of the IORBs prepared and linked by the system I/O call interface in response to user program I/O requests. There are two possibilities for the structure of the queue of pending requests. It could be organized as a single queue holding IORBs for all devices or as multiple queues, one for each specific device and/or controller type in the system.

The device drivers contain the code for operating the various controllers and devices, utilizing the parameters found in the IOCB. There is a separate device driver for each device type. The device drivers operate in a continuous loop servicing requests from the pending request queue and in turn notifying the user process that requested the I/O when the operation is complete. After selecting a request from the pending request queue, the driver initiates the I/O operation. This is done for some computers by issuing a particular machine instruction such as start I/O (SIO) or, for computers that use memory mapped I/O, writing to registers in the device controller. The device driver then waits for I/O completion.

Upon completion of the physical I/O operation, the device controller generates an interrupt. The corresponding interrupt routine signals the device driver, using semaphores or some other IPC construct. The device driver translates and transmits the data to the destination indicated in the IORB and then signals the requesting user process of I/O completion. The whole process is summarized in Fig. 77.3.

For purposes of scheduling, the various devices can be divided into two groups: shareable and nonshareable. Devices are shareable in the sense that primary memory is shared: more than one process can share space on the device and the several processes' data transmissions to and from the device may be interleaved. In this sense, the disk, like primary memory, is a shareable device. Most other devices are nonshareable. A device is nonshareable because its physical characteristics make it impossible to share. Tape devices are nonshareable because it is impractical to switch tapes between characters. Similarly, line printers are non-shareable because it is not practical to switch the paper between lines. In both cases, attempts to share the devices for output would result in a random intermixing of outputs from the processes writing the output.

Accordingly, techniques for scheduling the two classes of devices differ. Nonshareable devices, by their nature, cannot be dynamically allocated, that is, allocated by requested operation. These devices must be scheduled at a higher level, i.e., when processes enter the system, or at least on a longer term capture-and-release basis. Hence, scheduling in this case is analogous to the high-level scheduling that admits processes to the system from the input queue. Shareable devices, on the other hand, can be scheduled on an operation-by-operation basis. Since requests for these operations appear on an unpredictable and random basis, in exactly the same sense that I/O operations block and unblock process execution, the scheduling of shareable devices is a low-level function, analogous to processor scheduling as previously described.

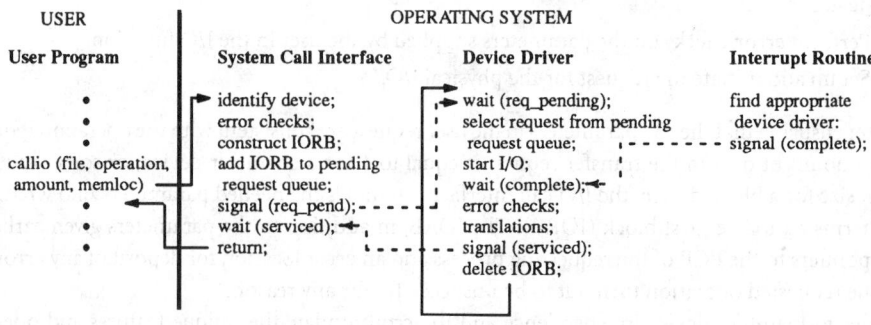

FIGURE 77.3   A sketch of the I/O subsystem. (*Source:* Adapted from Lister, A. M. 1979. *Fundamentals of Operating Systems*, 2nd ed., p. 68. Springer–Verlag, New York. With permission.)

Device scheduling is, at least at first cut, a question of queue organization and management. The problem varies in complexity. The simplest case is a monoprogrammed system where a job is admitted to the system, and needed resources, if available (connected and working), are scheduled to the process until its completion. Of course, even in this case, it should be possible to overlap I/O operations and execution of the main program, requiring interprocess communication in order that the I/O subsystem can notify the main program of completion (or error). Multiprogramming requires that pending request queues be supported both for nonshareable devices that may not be ready (e.g., printer out of paper or tape not mounted on tape device) or otherwise allocated and for shareable devices that may be busy serving another process's request. Thus device scheduling, like process scheduling is, in essence, the question of selecting a request from the queue to be serviced next. The policy decision, as before, becomes choosing between the logical and simple first-come–first-served, or some system based on a priority given to the requesting processes. But there are some additional considerations in the case of shareable devices, such as the disk, to which we now turn.

A shareable device not only has operations from various processes interleaved with one another, but, as a consequence of this interleaving, must have subareas of the medium allocated to these various processes. Therefore, a file name alone, unlike a logical device name in the case of a nonshareable device, is insufficient to specify the particular location of the information on a shareable device. Such data areas are commonly called files and an I/O operation to a device that can accommodate files will need first to be able to find the specific location of the file on the medium. The file maintenance subsystem of the operating system (Chapter 81) keeps a directory of file names and their corresponding locations for this purpose. Therefore, on the occurrence of the first reference to a file operation, the system must reference the directory to obtain the device and the location on that device of the needed file before the actual I/O can be accomplished. This is termed *opening* the file. Since this requires a lookup and usually an additional disk read (to get the directory), it is a time-consuming operation and it would be inefficient to have to do this multiple referencing for every file operation (read or write). Therefore, when a file is opened, a file descriptor is created. The file descriptor contains information to facilitate future references to the same file, including the device identification, the location of the file on that device, a description of the file organization, whether the file is open for read or write, and the position of the last operation. This file descriptor is referenced from the file or device descriptor created by the I/O subsystem when a device or file is allocated to a process, obviating the need to rereference the directory to find the location of the file.

Scheduling Shareable Devices

The most important shareable device in the modern computer system is the disk. Disks provide the majority of the secondary storage used in support of virtual memory and the file system. It is appropriate, therefore, to single out the disk and consider disk scheduling as an example of shareable device scheduling.

A disk is organized as a set of concentric circles called tracks on a magnetic surface. Each track is divided into sector, the physical unit in which data are transferred to or from the disk. A track normally has from 8 to 32 sectors per track (Tanenbaum [1992]). The tracks are read from, or written to, by means of a head attached to a moveable arm allowing placement of the head over any track on the disk. A *floppy disk*, or diskette, consists of one such platter that is removable from the drive unit itself. A so-called *hard disk* consists of a set of such platters stacked vertically around a common spindle. In this case, there are a number of surfaces on both sides of all but the top and bottom platters (which only record information on the inner surfaces). Corresponding to each usable surface, there is a read/write head. These read/write heads are fixed to a comblike set of arms that move the heads together, to position all of the heads over corresponding tracks on all of the disk surfaces. These corresponding tracks taken together are called a *cylinder*. Thus, all of the data stored on a particular cylinder can be referenced without moving the head assembly.

In this environment, information on a disk is referenced by an address consisting of several parts: disk drive number, disk surface, track, and sector. A sector is the smallest physical amount of information that can be read or written in a single disk operation. Sector size is a function of the particular disk drive; sizes vary from 32 to 4096 bytes, but the most common size is 512 bytes (Silberschatz and Galvin [1994]).

To read or write from or to the disk, the heads must be moved to the proper cylinder (track), then wait until the needed sector appears under the head, and then the information can be transferred to or from the control unit buffer. The time for a disk transfer, then, is made up of three components: the time it takes to move the head assembly from its current location to the required cylinder, called *seek* time; the rotational delay needed for the selected sector to come under the read/write heads, called *latency*; and the time for the actual transfer of the data in the sector, called *transfer time*. In order to improve efficiency, data are usually read or written in blocks of one or more sectors. Still, for most disks, the seek time dominates and so one aim of disk scheduling algorithms is to reduce or minimize average disk seek time for each disk operation.

As in the case of other system resource requests, if the needed disk and its controller are available, the disk request is attended to immediately. However, if the disk drive, or its controller, is busy servicing another request, new requests will be queued. Especially in a modern multiprogramming system, since most jobs depend on the disk for program loading and data files, it is likely that the disk request queue will be populated with new requests arriving regularly. When the current request is serviced, the system must choose, from among those on the queue, the next to service. Selection of the next disk request from the queue to service is controlled by a disk scheduling algorithm. Several possibilities exist.

First-Come–First-Served (FCFS)

The most obvious approach is to take disk requests from the disk request queue in the order of their arrival, that is, first-come–first-served. This algorithm is the easiest to understand and implement. Moreover, as in the case of any waiting line, FCFS appears to be naturally fair. Its performance, however, is a function of the likely random cylinder locations of the various requests on the queue. Particularly, when successive requests are for disk blocks whose locations are on cylinders far removed from one another, seek time, which is roughly proportional to the distance that the read/write heads must move, will be significant, leading to a high average seek time and consequently poor performance. Thus, performance is dependent more on the nature and requirements of currently active processes than operating system design. It is possible to improve on the FCFS algorithm.

Closest Cylinder Next (CCN)

The quest for shortest seek time would appear to favor selecting from the disk request queue, the request requiring the minimum head movement, the request at the closest cylinder, i.e., the request requiring the minimum seek time. Closest cylinder next will always provide better performance than FCFS, but like its analog in processor scheduling SJF, CCN, by taking requests found on nearby tracks, including newly arriving requests, tends to provide poor service to requests for disk blocks that happen to start far away from the current read/write head position. Assuming that disk block allocation is evenly distributed across the disk, this is particularly true for requests at the extreme outside and inside cylinders once the heads get positioned near the middle cylinders. Again, following the analogy to SJF, arrival of a string of requests for blocks in nearby cylinders may cause some requests to be starved under the CCN algorithm. CCN gains response time at the expense of fairness, the other extreme from FCFS.

Bidirectional Scan Scheduling (BDS)

The search for requests at relatively nearby tracks without completely abandoning fairness can be accommodated by the bidirectional scan algorithm. Bidirectional scan scheduling works analogously to the scan/seek button on a modern digital audio tuner; that is, the read/write head mechanism begins a scan at one end of the disk (outside or inside) and scans, servicing requests from the queue as the heads come to the corresponding cylinder. The scan continues to the other end of the disk assuring that no requesting process will have to wait too long for service. At the other end of the disk, the head assembly reverses and continues its scan in the opposite direction. This algorithm is sometimes called the elevator algorithm because its operation is nearly analogous to the operation of a building elevator. The analogy is complete except for the fact that for elevator passengers, the elevator's direction at pickup makes a difference depending on whether they are wishing to go up or down, while disk requests are serviced equally well regardless of the direction from which the disk heads arrive. This latter point, however, is significant, and a disadvantage of the BDS scheme. For, if the requests are to disk blocks uniformly placed throughout the disk cylinders, the heads, once they reach one end, are unlikely to find many requests as the first cylinders scanned on the reverse trip are the ones most recently scanned. By the fairness criteria, the requests now most in need of service are at the other end of the disk head travel. This problem is addressed in a variant of BDS.

Circular-Scan Scheduling (CSS)

This algorithm is similar to BDS except that when the scanning heads reach the end of their travel, rather than simply reversing direction, they return to the beginning end of the disk. The effect is a circular scan where the first disk cylinder, in effect, immediately follows the last, and disk requests are provided with a more uniform wait time.

Bounded-Scan Scheduling (BSS)

Both BDS and CSS were characterized as scanning from one end of the disk to the other, that is, from the lowest numbered cylinder to the highest. In fact, both BDS and CSS do too much. They need only scan to the most extreme cylinder represented by the requests in the queue, i.e., scan in one direction as long as there are requests beyond the current location, a bound represented by the block location of the highest or lowest numbered cylinder of a request in the queue. In actuality, both BDS and CSS are implemented in these bounded (in this sense) versions.

Optimal Scheduling

Optimal scheduling requires selecting the next request so that the total seek time is minimal. The problem with optimal scheduling is that the continuing arrival of new disk requests to the queue requires reordering the queue as each request arrives. To the extent that the disk is a heavily used resource with requests arriving continually and often, the computation needed to obtain optimal scheduling is not likely to be worth it. The concept, however, is useful as a reference for comparing the performance of the other algorithms.

Evaluation and Selection of a Disk Scheduling Algorithm

As is the case for the processor scheduling algorithms, performance is related to the request pattern. It is possible to evaluate this performance by estimating or tracing a disk request sequence. To see this, consider a disk with 128 cylinders. For the three patterns of disk requests shown in Fig. 77.4, the total head movements for each of the scheduling algorithms just described are shown in Table 77.1. In each case, assume that the disk head is initially at cylinder 58 and scanning in the direction of increasing cylinder numbers.

Though the FCFS scheme is easy to implement and conveys a sense of fairness, it leads to the poorest performance in each of the three example cases. CCN seems, on cursory examination, to yield the best performance, but as the examples show, this is not always the case. Bounded bidirectional scheduling (BBDS) is as good as or better in two of the three examples. It is clear from the table that performance is a function of the intensity and order of the disk requests, as well as the disk scheduling algorithm. If disk usage is light and the request queue nearly always empty, then choice of a scheduling algorithm is insignificant as they are all effectively the same. Heavy disk usage and short interarrival times of entries on the disk request queue effectively rule out FCFS and make the choice more critical. Since one of the primary uses of the disk is for the file system, disk performance is also influenced by the file space allocation technique. The

Pattern 1	73, 125, 32, 127, 10, 120, 62
Pattern 2	81, 82, 83, 84, 85, 21, 22
Pattern 3	10, 46, 91, 124, 32, 85, 11

FIGURE 77.4 Sample patterns of disk requests.

examples of Table 77.1 suggest that contiguous allocation could result in less head movement and consequently significantly better performance. Pattern 2, with its requests to sequentially numbered cylinders, is associated with better performance with all of the scheduling algorithms considered here. Directories and index blocks, if used, will cause extra disk traffic and, depending on their locations, could affect performance. Since references to files must pass through the directories and index blocks at least part of the time, placement of these near the center tracks of the disk would limit head movement to at most one-half of the total number of cylinders to find the file contents and, consequently, lead to better performance regardless of the scheduling algorithm.

TABLE 77.1 Total Head Movements for Various Scheduling Algorithms

	FCFS	CCN	BDS	CSS	BBDS	BCSS
Pattern 1	540	195	186	223	186	208
Pattern 2	101	91	175	218	91	92
Pattern 3	381	162	186	242	180	216

Scheduling Nonshareable Devices

Nonshareable devices, since they cannot be scheduled dynamically on an operation-by-operation basis, require a different approach to scheduling than the shareable devices. This level of scheduling is particularly important for two reasons: most device types are nonshareable and scheduling of nonshareable resources can lead to **deadlock**, a situation where the system is effectively locked up because none of the active processes can continue until it obtains some resource held exclusively by some other process.

By their very nature, nonshareable devices must be allocated to a single process for the duration of a use session. That is, a tape drive must be allocated to a particular process for the duration of a sequence of input and/or output operations, a printer for the duration of the print output of the process, etc. The duration can be explicitly indicated under program control by *open* and *close* operations, or by default at the entry and exit of a process from the system. Regardless of the origin of the use session, the mechanism is as described previously. The particular device is described by a IOCB, and processes with requests for allocation of the device are queued in the pending requests queue attached to the IOCB. Nonshareability is enforced by an initial value of 1 for the queueing semaphore. Scheduling, then, amounts to waiting for the resource to be freed, then selecting a device request from the pending request queue first-come–first-served, or by some measure of priority. In the latter case, the rationale or criteria for assignment of priorities clearly parallels that for processes in process scheduling. But the mutual exclusion requirement of nonshareability raises again the possibility of deadlock. Therefore, dealing explicitly with this potential must be an integral part of the policy consideration in scheduling of nonshareable devices.

The taxonomy used in this chapter for the classification of computing system resources, processor, primary memory, and devices, with the subclassification of shareable and nonshareable devices, is still not perfect.

Files

A single file may be shared among a number of processes and, in this sense, it too becomes a resource. The problem here is that though a particular file opened in read mode is a shareable resource, that same file, when opened by one process for writing, becomes a nonshareable resource. Thus, allocation of a file to a process's request may proceed only after appropriate read/write mode checks. A file in read mode may be scheduled (i.e., queued) as any other shareable resource, forcing the process to wait (in order) only for availability of the corresponding control unit and device. On the other hand, a process requesting a file already in use in write mode must wait until the process currently writing the file closes the file or exits the system, just as in the case of any other nonshareable resource.

Virtual Devices

Process requests for nonshareable devices can often lead to unavoidable delays, the extent of which is a function of the activity of the process currently holding the needed nonshareable device. In the aggregate, such delays can have a significant adverse effect on overall system performance. There are two possible ways out of this difficulty: increase the number of the offending nonshareable devices, e.g., add more printers; or introduce virtual devices. In this latter case, a process's request for data transfer to a nonshareable device, such as a printer which is allocated elsewhere, is directed instead to some anonymous file on the disk, thus freeing the requesting process from the wait for the otherwise allocated nonshareable device. The file, in this case, then, acts as a virtual printer. Then, a special process, called a spooling daemon or simply spooler, becomes responsible for scheduling and moving the data from the intermediate file to the printer when it becomes available. Of course, operation of the spooler admits yet another opportunity for scheduling—again usually FCFS or priority, depending on the policy adopted.

77.7 Scheduling Policies

In the preceding sections, the various resources of the system were classified, i.e., processors, primary memory, devices, files, and virtual devices, and subclassified into shareable and nonshareable resources. Moreover, the subsystem in which allocation is performed was identified, i.e., process manager, memory manager, I/O subsystem, and file system. In each instance, resource allocation was characterized in terms of some form of a queue and an algorithm for managing the queue. It turns out that algorithms and mechanisms are not enough, in and of themselves, to make optimal scheduling unambiguous. What remains is to establish a policy framework which governs selection of the particular algorithms from the choices, setting and adjusting the priorities where appropriate, and other considerations necessary to keep the overall system running smoothly and serving its customers fairly and in a timely manner. Thus, it is not impossible, or even unlikely that use of the queueing algorithms previously described to allocate resources as they and/or their associated software (e.g., process or memory manager) or hardware (memory, I/O device, or file) become available can lead to a situation where the system is overcommitted to a particular resource (e.g., printer) or resource area (e.g., printers or network connections) leading to diminished throughput for the system or, perhaps more dramatic, to deadlock where the entire system is halted due to a resource allocation state where each of the processes has allocated to it some resource critically needed by another. Perhaps these decisions, the policy decisions, are most difficult, for they are not considered at the algorithmic/queuing level and are less amenable to quantification and algorithmic solution. In this regard, the problem of deadlock appears to be potentially the most debilitating.

Deadlock

Deadlock may occur when system resources are allocated solely on the basis of availability. The simplest example is where process 1 has been allocated nonshareable resource A, say, a tape drive, and process 2 has been allocated nonshareable resource B, say, a printer. Now, if it turns out that process 1 needs resource B (the printer) to proceed and process 2 needs resource A (the tape drive) to proceed and these are the only two processes in the system, each is blocking the other and all useful work in the system stops. This situation is termed deadlock. To be sure, a modern system is likely to have more than two active processes, and therefore the circumstances leading to deadlock are generally more complex; nonetheless, the possibility exists.

Thus, deadlock is a possibility that needs to be considered in resource scheduling and allocation. For this chapter, the concern is with hardware resources such as CPU cycles, primary memory space, and devices such as printers, tape drives, and communications ports, but deadlock can occur in the allocation of logical resources such as files, semaphores, and monitors as well. Coffman et al. [1971], identified four conditions necessary and sufficient for the occurrence of system deadlock:

1. The resources involved are nonshareable.
2. Requesting processes hold already allocated resources while waiting for requested resources.
3. Resources already allocated to a process cannot be preempted.

4. The processes in the system form a circular list where each process in the list is waiting for a resource held by the next process in the list.

There are basically four ways of dealing with the deadlock problem:

1. Ignore deadlock.
2. Detect deadlock and, when it occurs, take steps to recover.
3. Avoid deadlock by cautious resource scheduling.
4. Prevent deadlock by resource scheduling so as to obviate at least one of the four necessary conditions.

Each is considered, in order of decreasing severity, in terms of adverse effects on system performance.

Deadlock Prevention

Since all four of the conditions are necessary for deadlock to occur, it follows that deadlock may be prevented by obviating any one of the conditions. The first condition, i.e., that all resources involved be nonshareable, is difficult to eliminate because some resources, such as the tape drive and printer, are inherently nonshareable. However, this situation can be alleviated in some cases, by spooling requests to a nonshareable device, such as the printer, to a temporary file on the disk for later transfer to the nonshareable device as previously described.

There are two possibilities for elimination of the second condition. First, require that a process request be granted all of the resources it needs at once, prior to execution. The second alternative is to disallow a process from requesting resources whenever it has previously allocated resources. That is, it must finish with those resources previously allocated and relinquish all of them prior to an additional resource request. Both methods assure that a process cannot be in possession of some resources while waiting for others. The first has the disadvantage of causing poor resource utilization by forcing allocation for a period likely to exceed, perhaps considerably, that needed for actual use of the resource. Moreover, depending on the scheduling algorithm, some processes could face starvation from having to wait, perhaps indefinitely, for some resource or resources.

The third condition, nonpreemption, can be alleviated by forcing a process waiting for a resource that cannot immediately be allocated to relinquish all of its currently held resources, so that other processes may use them to finish. Alternatively, the requesting process's request can be satisfied by preempting the requested resource from some currently blocked process, thus effectively eliminating the nonpreemption condition. The problem with these is that some devices are simply not amenable to preemption; a printer, for instance, if preempted will generate a useless page of output interleaved from several processes.

The last condition, the circular list, can be obviated by imposing an ordering on all of the resource types (presumably reflecting the order in which the resource types are likely to be used) and then forcing all processes to request the resources in order. Thus, if a process has resources of type m, it can only request resources of type $n > m$, and if it needs more than one unit of a particular resource type, it must request all of them together. Though this will ensure that the circular list condition is denied, it can hurt resource utilization by requiring that a particular resource be allocated in advance of its logical need.

Deadlock prevention works by forcing rather severe constraints on resource allocation, leading to poor resource utilization and throughput. A less severe approach is to individually consider the implications of each resource request with respect to deadlock.

Deadlock Avoidance

This approach to the deadlock problem employs an algorithm to assess the possibility that deadlock could occur as a result of granting a particular resource request and acting accordingly. This method differs from deadlock prevention, which guarantees that deadlock cannot occur by obviating one of the necessary conditions, and from deadlock detection, in that it anticipates deadlock before it actually occurs. The basic idea is to maintain a status indicator reflecting whether the current situation with respect to resource availability and allocation is safe from deadlock. Then, when a resource request occurs, the system invokes the avoidance algorithm to determine whether granting the request for a set of resources would lead to an unsafe state and if so, denying the request. The most common algorithm, due to Dijkstra [1965], is called the banker's algorithm, so named because the process is analogous to that used by a banker in deciding if a loan can be safely made. The algorithm is shown in Fig. 77.5. Here i and j are the process and resource indices; N and R the number of processes and resources, respectively. Other variables and what they represent include:

`max_need[i, j]`	=	maximum number of resources of type j needed by process i
`total_units[j]`	=	total number of units of resource j available in the system
`allocated[i, j]`	=	number of units of resource j currently allocated to process i
`available_units[i, j]`	=	number of units of resource j available after `allocated[i, j]` units are assigned to process i
`needed[i, j]`	=	remaining need for resource j by process i
`finish[i]`	=	status of process i: 0 if it is not clear that process i can finish, and 1 if process i can finish.

What makes the deadlock avoidance strategy difficult is that granting a resource request that will lead to deadlock may not result in deadlock immediately. Thus, a successful strategy requires some knowledge about possible patterns of future resource needs. In the case of the banker's algorithm, that knowledge is the maximum quantity of each resource type that a particular process will need during its execution. As shown in Fig. 77.5, the algorithm permits requests only when the current request added to the the number of units already allocated is less than that maximum, and then only if granting the request still leaves some path for all of the processes in the system to complete even if every one needs its maximum request.

But this last requirement that each process know its maximum resource needs in advance, an unlikely supposition particularly for interactive jobs, severely limits the applicability of the banker's algorithm. Also, the interactive environment is characterized by a changing number of processes, i.e., N is not set and a varying set of resources R, as units occasionally malfunction and must be taken off line. Further, even if it were to be applicable, Haberman [1969] has shown that execution of the algorithm has complexity proportional to N^2, and since the algorithm is executed each time a resource request occurs, the overhead is significant.

Deadlock Detection

An alternative to the costly prevention and avoidance strategies just outlined is deadlock detection. This approach has two parts:

- An algorithm that tests the system status for deadlock
- A technique to recover from the deadlock

The detection algorithm, which could be invoked in response to each resource request, or if that is too expensive at periodic time intervals, is in many ways similar to that used in avoidance. The basic idea is to check allocations against resource availability for all possible allocation sequences to determine if the system is in a deadlocked state. There is no requirement that the maximum requests a process will need be stated here. The details of the algorithm are shown in Fig. 77.6.

Of course, the deadlock detection algorithm is only half of this strategy. Once a deadlock is detected, there needs to be a way to recover. Several alternatives exist:

```
const int safe = 1;
for (j = 1; j <= R; j++)              /* R resource types */
    available_units[j] = total_units[j];

for (j = 1; j <= R; j++)
    {
    for (i = 1; i <= N; i++)          /* N processes */
        {
        available_units[j] = available_units[j] - allocated[i,j];
           /*allocated[i, j] = number of units of resource j currently
              allocated to process i */
        finish[i] = 0;
           /* initialize - process i may not be able to finish */
        needed[i, j] = max_need[i, j] - allocated[i, j];
           /* needed is remaining need of process i for resource j;
        }
    not_done = 1;
    while (not_done)
        {
        not_done = 0;
        for(i = 1; i <= N; i++)
            if (! finish[i] && needed[i, j] <= available_units[j])
                {
                /* process i can finish */
                finish[i] = 1;
                available_units[j] = available_units[j] + allocated[i, j];
                /* give back process i's resources as done */
                not_done = 1;
                }
        }
    /* Continue loop until a process needed request cannot be met */
    /* Determine if all N processes could be completed. */
    if (allocated_units[j] = total_units[j])
        status = safe;
    else
        status = !safe;
    }
if (status == safe)
    /* allocate the requested resource */
```

FIGURE 77.5 The banker's algorithm for deadlock avoidance. Based on statements of the algorithm in Dijkstra [1965] and Tsichritzis and Bernstein [1974].

```
for (j = 1; j <= R; j++)              /* R resource types */
    available_units[j] = total_units[j];

  for (j = 1; j <= R; j++)
    {
    for (i = 1; i <= N; i++)       /* N processes */
      {
      available_units[j] = available_units[j] - allocated[i,j];
        /*allocated[i, j] = number of units of resource j currently
          allocated to process i */
      finish[i] = 0;
        /* initialize - process i may not be able to finish */
      }
    not_done = 1;
    while (not_done)
      {
      not_done = 0;
      for(i = 1; i <= N; i++)
        if (! finish[i] && request[i, j] <= available_units[j])
          {
          /* process i can finish */
          finish[i] = 1;
          available_units[j] = available_units[j] + allocated[i, j];
          /* give back process i's resources */
          not_done = 1;
          }
      }
  deadlock = 0;
  for (i = 1; i <= N && deadlock == 0; i++)
      if (finish[i] != 1)
          deadlock = 1;
  if (deadlock == 1)
      /* system deadlocked */
```

FIGURE 77.6 Deadlock detection algorithm. Based on statements of the algorithm in Dijkstra [1965] and Tsichritzis and Bernstein [1974].

- Temporarily preempt resources from deadlocked processes. In this case, there must be some criteria for selecting the process and the resource affected. The criteria may include minimization of some cost function, based on parameters such as the number of resources a particular process holds and resource preemptability (e.g., a printer is difficult to temporarily preempt).
- Back off a process to some checkpoint allowing preemption of a needed resource and restarting the process at the checkpoint later. In this case, the simplest way to find a safe checkpoint is to stop the process, return all of its allocated resources, and restart the process from the beginning at a later time.

- Successively kill processes until the system is deadlock free.

These methods are expensive in the sense that the detection algorithm is rerun with each iteration until the system proves to be deadlock free. The detection algorithm, like the avoidance algorithm of Fig. 77.5, has time complexity proportional to N^2. Another potential problem is starvation; care must be taken to ensure that resources are not continually preempted from the same process or that the same process in not repeatedly backed off or killed.

Ignore Deadlock

The last approach, do nothing and hope, reflects the observation that deadlocks and their concomitant effects including possible system crashes do not occur with sufficient frequency to justify the expense in terms of system overhead required to handle deadlock by any of the other three approaches. This is the approach taken in the UNIX operating system.

The Eclectic Approach

In summary, it is clear then that none of the conventional choices for dealing with the potential for deadlock is entirely satisfactory. An alternative strategy is to divide the system resources into classes and to apply the most appropriate technique for dealing with potential deadlocks in each class. Silberschatz and Galvin [1994] suggest four classes:

1. In the case of resources such as the process control blocks used by the system itself, deadlock prevention can be used by forcing resource ordering since there is no contention among pending requests.
2. Primary memory can be shared in the sense that an active process (or part of it) can be swapped out without destroying the process and, therefore, deadlock can be prevented by preemption.
3. For nonshareable resources including devices such as the printer and writeable files, deadlock avoidance can be used as resource requirements become known.
4. Virtual memory space on the disk can be protected from deadlock by avoidance since maximum virtual storage requirements are generally known in advance of execution.

In spite of this, most systems today take the optimistic approach and do nothing to prevent, to avoid, or even to detect deadlock.

77.8 High-Level Scheduling

Process and device scheduling operate in the context of a higher level system scheduling. This higher level scheduling deals with the admission of new processes to the system, setting and changing priorities on the basis of system operation, and generally implementing the policy controls of system operation. In fact, since scheduling system resources is prerequisite to resource allocation, criteria and policy regarding scheduling are intimately related to system resource allocation. To this end, criteria for scheduling decisions match the objectives for the operating system as a whole, that is, to provide good service to processes within a context of operation at a high level of resource utilization. Other factors that cannot be overlooked are the need to assure mutual exclusion for nonshareable resources and prevention or mitigation of starvation and deadlock. A portion of the responsibility for this lays in a system scheduler process. Lister [1979] suggests that the tasks of the scheduler include admission of processes into the system, determination of process priorities, and enforcement of system policies for resource allocation. In the first instance, the scheduler selects batch jobs from an input queue using criteria such as resources required and priority. In a time sharing environment, this task is more difficult since processes enter the system at user login. However, the scheduler can monitor resource utilization and deny access in order to maintain performance criteria for the currently active processes. Second, the scheduler determines the process priorities that govern placement in system queues. Since a primary system scheduling mechanism is manifest in these queues, the scheduler plays a key role here. Finally, the scheduler is responsible for system policies that deal with deadlock and balance, ensuring that no resource class is either over- or underutilized.

The importance of these activities in determining overall system performance suggests that the scheduler itself should be a high-priority process. In particular, the scheduler executes whenever a resource request or release occurs or a process arrives at the system or terminates (Lister [1979]). The lower level schedulers previously discussed are analogously activated by interrupts; both are unpredictable, though interrupts may be expected more frequently than process entry and exit or resource request and release.

The criteria for high-level scheduling decisions are many and varied. System response time is enhanced by minimizing context switches and the associated overhead. Scheduling fewer new processes into the system is an obvious way to achieve this. But such a scheduling policy is likely to result in less than optimal resource utilization as fewer processes (a lower degree of multiprogramming) are together unlikely to use all of the resources declared needed for throughout all of their time in the system. As more processes enter the system, the scheduler has more policy decisions to make when resources are requested and released. The criteria are numerous and their effectiveness is generally affirmed or denied by empirical observation. Several possibilities are summarized in Lister [1979]:

1. Give processes already in possession of a significant quantity of resources priority in resource requests to advance their progress through the system in the hope of reclaiming the allocated resources at the earliest possible time.
2. Give resource rich processes high dispatching priority for the same reason as in 1.
3. Use the working set principle as a criterion for memory allocation.
4. Priorities should reflect process importance; most system processes have higher priorities than user processes; among system processes, the scheduler itself is high priority.
5. Device drivers should have high-priority and drivers for higher speed devices relatively higher priority. This is designed to keep the devices running mitigating the possibility of I/O delays.

In summary, process and device scheduling is a tricky business with goals that are not always mutually consistent. Different possibilities serve different needs. Experimentation and empirical observation are necessary to judge effectiveness in a particular situation.

77.9 Recent Work and Current Research

Though much of the work on process and device scheduling took place during the late 1960s and 1970s with the advent of multiprogramming and time sharing systems and as the subject of operating systems came into its own, technical advances continue to spur new developments. In this section, some of the more significant recent discoveries are summarized and pointers given for further study.

Processor Scheduling

The goals previously listed underlying the development of techniques for processor scheduling are often mutually conflicting. This inevitability has led to the priority and then dynamic dispatching priority adjustment algorithms previously described. Recent work in processor scheduling has continued to focus on refinement of dynamic scheduling schemes.

The problem with priority methods that set priority once at the process's entry into the system is inflexibility, since resource needs do not always match the priorities and separate resources may require resource allocation mechanisms isolated from one another. The dynamic priority adjustment mechanisms described above begin to address these problems. Recent alternatives include fair share schedulers (Henry [1984], Kay and Lauder [1988]), microeconomic schedulers [Waldspurger et al. 1992], lottery schedulers [Waldspurger and Weihl 1994], and time-function scheduling (Fong and Squillante [1995]).

A fair share scheduler is designed to allocate resources to provide each user or group of users with an equitable share of the CPU over some period of time. This is accomplished by the addition of facilities to track the actual amount of processor usage allocated to each process and to dynamically adjust priorities to assure allocation of processor time to each process remains close to established shares.

Henry [1984] describes a fair share scheduler utilized at AT&T Bell Laboratories. This scheme divides users into groups and schedules so that each group is allocated a fair share of CPU time for execution. Within a particular group, processes are scheduled according to the regular dispatcher relative to other processes in the group.

The implementation is a straightforward modification of the standard decay usage scheduling algorithm. In addition to the process's *execution time field*, which is incremented at each clock interrupt, there is another fair share group execution field that is updated at clock interrupt when any process from the group is executing. The fair share group execution field is decayed exactly as the individual process's execution time fields, once a second. Then, process priorities are recalculated using the modified formula

priority = C(execution time field)/2 + base priority + C(fair share group execution field)/2.

The result is that the more CPU time the processes from a particular group have used recently, the higher the priority number, and consequently, the lower the priority for processes in that group. Taken as a whole, ultimately each of the groups, however, gets an equitable share of the CPU time.

Kay and Lauder [1988] describe a fair share scheduler that is designed to work at both the individual user level and the group level. Unlike the AT&T fair share scheduler, the Share scheduler explicitly considers use of resources other than the CPU. The goal is to have a scheduler that is fair to users rather than just processes. Therefore, the approach is designed to schedule resources so that users get a fair share of the machine over time in the sense of balancing an entitlement expressed as a number of shares and recent resource use history. Of course, allocation of shares is an administrative matter, and depends on the organization's entitlement and the user's entitlement within the organization, somehow defined. The implementation is considerably more complex than that of the AT&T algorithm. There are basically three levels of operation. At the first, or user level, a figure representing usage is adjusted by both a decay factor and a measure of total resource consumption. Thus, every $t1$ seconds (several seconds, Kay and Lauder used 4),

$$\text{usage} = \text{usage} \times K1 + \text{charges}$$

where $K1$ is a suitable decay factor, and charges represents the sum of price times quantity for resources used since the last calculation. Subsequently, every $t2$ seconds (e.g., $t2 = 1$ s), process priorities are decayed according to

$$\text{priority} = \text{priority} \times K2 \times (\text{nice} + K2')$$

where $K2$ and $K2'$ are factors reflecting priority decay rate and the effect of the UNIX **nice** command parameter (0 is default). Finally, every $t3$ seconds (1/60–1/100 of a second), a priority recalculation occurs taking into account recent usage and remaining shares

$$\text{priority} = \text{priority} + (\text{usage} \times \text{number of active processes})/(\text{number of shares})^2$$

so that a user's process priority is not only a function of recent usage and entitlement, but also the number of active processes. Thus, a single user's share is effectively spread among the number of currently active processes.

The Share system is also applicable to scheduling in a context where a computer is shared among a number of groups. In this case, shares are allocated to each group in proportion to the division of the machine. This requires more calculations to allocate and reallocate shares within a particular group when a new user logs on subject to keeping the total number of shares allocated to the group constant. The system allows for different decay rates for usage in the various groups. The Share system does provide users with a sense of fairness, but at a cost of additional overhead.

The lottery scheduling scheme of Waldspurger and Weihl [1994] is similar to the Kay and Lauder Share scheduler in the sense that it provides for proportional share resource allocation. In this case, lottery tickets rather than shares, represent the resource entitlements. The scheme is designed to provide active and flexible control of resource allocation in proportion to share allocations.

The scheduler works in the context of a scheduling quantum of 10 ms. Resources are allocated by lottery in the sense that a resource is allocated to the process that holds the winning ticket. The system is fair in that the probability that a particular process will be allocated a particular resource is proportional to the number of tickets held by that process, and it follows that allocation of resources in general will be proportional to the number of tickets held, i.e., to entitlement. This is so because the tickets are fungible in that they represent resource rights independent of the type or mix of resources needed.

Lottery scheduling has the advantage of being relatively straightforward to implement and efficient in that its overhead is comparable to that of a standard time-sharing policy. The scheduling algorithm is, by definition, randomized and, hence, actual allocation at any given moment may not match entitlement shares exactly. But allocations occur every 10 ms and fairness, in the sense of matching entitlements, increases quickly over time as the number of allocations grows. Moreover, the problem of starvation is obviated since any client with tickets will eventually win a lottery and thus be allocated resources.

A more general approach to resource scheduling, called time-function scheduling, has been proposed by Fong and Squillante [1995]. The method uses dynamically changing priorities based on general functions of such factors as the time a given process has waited for a particular resource. The idea is to support a wide and varying range of scheduling objectives by implementing time functions reflecting these objectives. Further, these time functions are effectively variable in response to changes in the system in order to preserve the desired scheduling objectives. Fong and Squillante have found time-function scheduling to provide the same fair share objectives as lottery scheduling with less waiting time variance (resulting from the probabilistic nature of the lottery scheduling).

A number of empirical studies (Ryder [1970], Sherman et al. [1972], Stevens [1968]) have concluded that maximum CPU utilization and system throughput result from a CPU scheduling policy that gives preemptive priority to I/O-bound jobs over compute bound jobs. Kameda [1984] has used a Markovian model of job processing identical to the finite source queueing model to verify that scheduling policies which assure that I/O-bound jobs are given preemptive priority over processor bound jobs provide maximum processor utilization and optimal throughput. Within the confines of the theoretical model and its concomitant assumptions, it is not possible to show that any additional sophistication of the sort previously described can improve upon this result. Kameda concludes that general applicability of the theoretical results is limited by the model and its underlying assumptions.

Other work in processor scheduling includes predictive deadline scheduling and dynamic mixed priority schemes for real-time systems. See Miller [1990, 1992].

Disk Scheduling

Recent work in disk scheduling continues the search for an optimal algorithm through experimental studies and refinement of the well-known algorithms previously described and anticipatory algorithms causing movement of the disk heads in advance of the next disk I/O request. Approaches to the former included extensive simulation studies (Hofri [1980], Teorey and Pinkerton [1972], Teorey [1972]) and analytical modeling (Coffman et al. [1978]).

Continuing in this vein, Geist and Daniel [1987] have proposed a disk scheduling scheme that is a varying mixture of CCN and BDS, designed to obviate both the considerable seek time variances and potential for starvation of CCN and the potential delays for opposite end requests of BDS. The continuum algorithm, $V(R)$, $R \in Q[0, 1]$, takes as its next request the one closest, in a recast sense of closest, to the current position of the head. As with BDS, $V(R)$ keeps track of the current direction of head movement, but in this case, the direction of the just previous seek. The distance to the nearest request in that same direction is the number of cylinders away as before. But the distance to the nearest request in the direction opposite to that of the last seek is R times the total number of cylinders plus the number of cylinders to the request. Note that R is any real number in the interval 0–1. If $R = 0$, $V(R)$ is CCN and if $R = 1$, $V(R)$ is BDS. For values of $0 < R < 1$, $V(R)$ is somewhere between the two.

The implementation requires maintaining two request queues, one for requests in each direction from the current head position. The requests in both queues are kept in order of increasing distance from the current head position. Selection of the next request to process is made by calculating the distance to the first request on each queue, adjusting the distance requiring a change in direction. The request associated with the least adjusted distance is selected, and the current head position and direction are updated. A simulation was used to test different values for R. With respect to improvements in mean waiting time and throughput, the simulation showed $V(0.2)$ to be superior to FCFS and to either pure CCN or BDS.

In an effort to enhance the effectiveness of CCN, Seltzer et al. [1990] and Jacobson and Wilkes [1991] have refined the idea to include rotation time. Thus, shortest access time first (SATF) serves first the request that has the least access time, compiled as the sum of seek and rotation times, from the current head position.

The efficacy of prerequest disk arm movement has been explored by King [1990]. Anticipatory disk arm scheduling policies have been largely overlooked due to the supposition of a continuous substantial request queue length (Denning [1967], Teorey and Pinkerton [1972]). This constant backlog would preclude the existence of time to use to position the disk heads optimally for successive requests. However, Geist and Daniel [1987] and Lynch [1972] reported disk drive utilization rates on the order of 20–35% suggesting that there is time, after all, that might profitably be used for positioning the disk heads in anticipation of the next request. This possibility was investigated by Coffman and Hofri [1978] and Hofri [1980] and more recently by King [1990]. King shows that the potential for performance improvements due to anticipatory scheduling is a function not only of disk idle time, but also the pattern of the locations of the requests over the physical disk space. If the data requests are spaced uniformly over the disk and an FCFS scheduling algorithm is used, prepositioning the disk heads at the middle cylinder results in 25% less head movement on average, which translates to a 13% reduction in mean seek time. In the case of a nonuniform distribution of request data locations on the disk, the optimal anticipatory location is between the most frequently accessed locations, so-called hot-spots, and the midway point; the particular location is a function of the probability of requests at the hot spots. In as much as the locations and probabilities of these hot spots are known (perhaps from system performance data) and the request rate is sufficiently low as to rarely interfere with an anticipatory move, the anticipatory algorithm saves time.

As the frequency of request arrivals increases, so does the possibility that the arriving request will have to wait for completion of an anticipatory seek operation. Ultimately, the demand for disk service could become so high that, in the worst case, disk requests have to wait the entire time needed for anticipatory repositioning. Possible ways out of this dilemma are suggested in King [1990]. The first is to move the heads in short spurts toward the optimal anticipatory location, thus allowing for timely attention to currently arriving disk data requests. The second is to develop technology to facilitate an interruptible disk head seek. The former leads to seek time improvements of 8–10% for an anticipatory version of FCFS.

Simulation was used to assess the effect of anticipatory head motion in the case of a workload representing more realistic data from McNutt [1984]: a workload that reflects the high probability that a given job will generate repeated requests to the same disk location. In this case, especially if the disk request rate is low, few requests will benefit from an anticipatory seek until that particular job completes. This suggests modifying the anticipatory algorithm by including a delay, representing job completion time, between a request and an anticipatory move. This modification, too, yielded a small performance improvement.

Other more complex disk systems with two arms and duplexed disks with separate control and data lines to two disk arms were considered as well. Modifications to the anticipatory algorithm yielded reductions of mean seek distances of 10–29% over more conventional disk scheduling algorithms.

The search for improvements in disk and file system performance has been marked by the development of redundant arrays of inexpensive disks (RAIDs) (Patterson et al. [1988], Patterson [1989]). The idea, as the name implies, is to connect together a number of small disks in order to improve both reliability and performance. Initially, a RAID disk system was made up of a central controller connected to a host computer on the one side and, on the other, to a number of disk controllers, each connected to a series of small disks. Performance is improved by distributing data across a number of disk drives affording the possibility for parallel operation. Reliability is enhanced by striping, that is, storing a stripe of data blocks across a number of drives, thus minimizing the possibility of data loss from drive failure; mirroring, where a complete copy of the data is kept on separate drives; or parity techniques with parity blocks on a dedicated drive or distributed throughout the disk array. Cao et al. [1994] have proposed an architecture where the central controller is replaced by a number of array controller nodes, some of which are dubbed *worker* nodes and connect to local disks, and others *origination* nodes that provide for communication with the host computer. This latter arrangement enhances reliability by obviating the central controller and, thus, the potential that its failure could disable the entire array. The TickerTAIP architecture also has the potential for improved performance by allowing for greater parallelism, e.g., for parity calculation.

The design of the RAID system preserves reliability and minimizes disruption from a failure of any single component, a disk failure, a worker failure, or an originator failure. The latter is assured by an *atomic write* policy, that is, not allowing a write operation to make any changes until there is sufficient redundancy to guarantee successful completion of the write. Ensuring reliability becomes more difficult in the event of simultaneous failure of more than one component, a disaster according to Lampson and Sturgis [1981]. However, following UNIX 4.2 BSD type file systems (McKusick et al. [1994], Cao et al. [1994]) propose controlling the effects of simultaneous component failure by using request sequencing, in essence assuming that no request can begin until all requests on which it depends (e.g., directory or node requests) are complete.

Atomic requests and sequencing enable the TickerTAIP RAID to handle multiple requests requiring queues of requests at worker nodes. This leads to consideration again of request scheduling algorithms, this time, inside the disk array system. Thus, the worker system could use the previously described algorithms such as FCFS, CCN, or SATF. An additional algorithm (Cao et al. [1994]), *batched nearest neighbor* (BNN) is a variant of SATF in that when a worker is available, it takes the entire batch of requests on its queue as a group and applies SATF until the entire batch is served before picking up a new request. BBN provides much of the performance of SATF without the concomitant starvation problem.

Simulations showed that the distributed controller system of Cao et al. [1994] provided greater throughput and lower response times with less CPU power. In addition, the tests showed that when the SATF and BNN algorithms were used, throughput improved substantially over FCFS (more so with SATF) especially when run with real-world-type workloads. Mean response time for the same workload tests were minimum with BNN, probably because of its built-in tendency to mitigate starvation.

Disk scheduling and allocation problems are, of course, closely related to both disk architecture and to the design and implementation of virtual memory and file systems. In particular, see the work of Patterson et al. [1988, 1989], Rosenblum and Ousterhout [1992], and Hartman and Ousterhout [1995], and Chapters 80 and 81 of this Handbook.

Deadlock

Recent work on deadlock appears to be directed primarily at deadlock detection in distributed systems [7, 2]. Refer to Chapter 86, "Distributed File Systems and Distributed Memory," for details.

Defining Terms

Deadlock: "A set of processes is deadlocked if each process in the set is waiting for an event [release] that only another process in the set can cause" (Tanenbaum [1992, p. 242]).

Distributed System: "A *distributed computing system* consists of a number of computers that are connected and managed so that they *automatically* share the job processing load among the constituent computers or separate the job load as appropriate among particularly configured processors. Such a system requires an operating system which, in addition to the typical stand-alone functionality, provides coordination of the operations and information flow among the component computers" (Tucker et al. [1995, p. 403]).

Multiprocessing: "A *multiprocessing* system is a computer hardware configuration that includes more than one independent processing unit" (Tucker et al. [1995, p. 403]).

Multiprogramming: "A *multiprogramming* operating system is an OS which allows more than one active user program (or part of user program) to be stored in main memory simultaneously" (Tucker et al. [1995, p. 403]).

Network: "A *networked computing system* is a collection of physically interconnected computers. The operating system of each of the interconnected computers must contain, in addition to its own stand-alone functionality, provisions for handling communication and transfer of programs and data among the other computers with which it is connected" (Tucker et al. [1995, p. 403]).

Process: "A *process* is a series of operations associated with the execution of a sequence of instructions which effect a particular system or user action" (Tucker et al. [1995, p. 425]).

References

Bach, M. J. 1986. *The Design of the UNIX Operating System*. Prentice–Hall, Englewood Cliffs, NJ.

Badel, D. Z. 1986. The distributed deadlock detection algorithm. *ACM Trans. Comput. Syst.* 4(4):320–337.

Bic, L. and Shaw A. C. 1988. *The Logical Design of Operating Systems*, 2nd ed. Prentice–Hall, Englewood Cliffs, NJ.

Calingaert, P. 1982. *Operating System Elements: A User Perspective*. Prentice–Hall, Englewood Cliffs, NJ.

Cao, P., Lim, S. B., Venkataraman, S., and Wilkes, J. 1994. The TickerTAIP parallel RAID architecture. *ACM Trans. Comput. Syst.* 12(3):236–269.

Chandra, R., Devine, S., Verghese, B., Gupta, A., and Rosenblum, M. 1994. Scheduling and page migration for multiprocessor compute servers. 6th Int. Conf. Architect. Support Programming Lang. and Operating Syst. Proc. *Operating Syst. Rev.* 28(5):12–24.

Chandy, K. M., Haas, L. M., and Misra, J. 1983. Distributed deadlock detection. *ACM Trans. Comput. Syst.* 1(2):144–156.

Coffman, E. G., Jr., Elphick, M., and Shoshani, A. 1971. System deadlocks. *ACM Comput. Surv.* 3(2):67–78.

Coffman, E. G., Jr. and Hofri, M. 1978. A class of FIFO queues arising in computer systems. *Operations Res.* 26(5):864–880.

Coffman, E. and Hofri, M. 1982. On the expected performance of scanning disks. *SIAM J. Comput.* 11:60–70.

Coffman, E. G., Jr. and Kleinrock, L. 1968. Computer scheduling measures and their countermeasures, pp. 11–21. *Proc. AFIPS 32*, SJCC.

Coffman, E. G., Jr. and Denning, P. J. 1973. *Operating Systems Theory*. Prentice–Hall, Englewood Cliffs, NJ.

Daniel, S. and Geist, R. 1983. V-SCAN: an adaptive disk scheduling algorithm, pp. 96–108. *Proc. IEEE Int. Symp. Comput. Syst. Organ.* New Orleans, LA, March.

deJonge, W., Kaashoek, M. F., and Hsieh, W. C. 1993. The logical disk: a new approach to improving file systems. 14th ACM Symp. Operating Syst. Principles Proc. *Operating Syst. Rev.* 27(5):15–28.

Denning, P. J. 1967. Effects of scheduling on file memory operation, pp. 9–21. *Proc. AFIPS Spring J. Comput. Conf. 30*.

Dijkstra, E. W. 1965. Cooperating sequential processes. In *Programming Languages*, F. Gehuys, ed., pp. 43–112. Academic Press, New York.

Fong, L. L. and Squillante, M. S. 1995. Time-function scheduling: a general approach to controllable resource management. 15th ACM Symp. Operating Syst. Principles, Proc. *Operating Syst. Rev.* 29(5):230.

Geist, R. and Daniel, S. 1987. A continuum of disk scheduling algorithms. *ACM Trans. Comput. Syst.* 5(1):77–92.

Habermann, A. N. 1969. Prevention of system deadlocks. *Commun. ACM* 12(7):373–77.

Hartman, J. H. and Ousterhout, J. K. 1993. The zebra striped network file system. 14th ACM Symp. Operating Syst. Principles, Proc. *Operating Syst. Rev.* 27(5):29–43.

Hartman, J. H. and Ousterhout, J. K. 1995. The zebra striped network file system. *ACM Trans. Comput. Syst.* 13(3):274–310.

Hellerstein, J. L. 1993. Achieving service rate objectives with decay usage scheduling. *IEEE Trans. Software Eng.* 19(8):813–825.

Henry, G. J. 1984. The fair share scheduler. *AT&T Bell Lab. Tech. J.* 63(8, P. 2):1845–1857.

Hofri, M. 1980. Disk scheduling: FCFS vs. SSTF revisited. *Commun. ACM* 23(11):645–653.

Hofri, M. 1983. Should the two-headed disk be greedy? Yes, it should. *Inf. Process. Lett.* 16:83–85.

Jacobson, D. M. and Wilkes, J. 1991. Disk scheduling algorithms based on rotational position. Hewlett-Packard Lab. *Tech. Rep.* HPL-CSP-91-7, Palo Alto, CA.

Kameda, H. 1984. Optimality of a central processor scheduling policy. *ACM Trans. Comput. Syst.* 2(1):78–90.

Kay, J. and Lauder, P. 1988. A fair share scheduler. *Commun. ACM* 31(1):44–55.

King, R. 1990. Disk arm movement in anticipation of future requests. *ACM Trans. Comput. Syst.* 8(3):214–229.

Kleinrock, L. 1970. A continuum of time-sharing scheduling algorithms, pp. 453–458. In *Proc. AFIPS Spring J. Comput. Conf. 36*, AFIPS Press, Reston, VA.

Koch, P. D. L. 1987. Disk file allocation based on the buddy system. *ACM Trans. Comput. Syst.* 5(4):352–370.

Lampson, B. W. and Sturgis, H. E. 1981. Atomic transactions. In *Distributed Systems—Architecture and Implementation: An Advanced Course.* Lecture notes in computer science 105, pp. 246–265. Springer–Verlag, New York.

Lane, M. G. and Mooney, J. D. 1989. *A Practical Approach to Operating Systems,* PWS-Kent, Boston, MA.

Lazowska, E. D., Zahorjan, J., Cheriton, D. R., and Zwaenepoel, W. 1986. File access performance of diskless workstations. *ACM Trans. Comput. Syst.* 4(3):238–268.

Lister, A. M. 1979. *Fundamentals of Operating Systems,* 2nd ed. Springer–Verlag, New York.

Lynch, W. C. 1972. Do disk arms move? *Performance Evaluation Rev.* 1:3–16.

Mahlke, S. A., Chen, W. Y., Bringmann, R. A., Hank, R. E., Hwu, W. W., Rau, B. R., and Schlansker, M. S. 1993. Sentinel scheduling: a model for compiler-controlled speculative execution. *ACM Trans. Comput. Syst.* 11(4):376–408.

Mahoney, B. 1994. An "open" oriented file system. *Operating Syst. Rev.* 28(1):48–54.

McCann, C., Viswani, R., and Zahorjan, J. 1993. A dynamic processor allocation policy for multiprogrammed shared-memory multiprocessors. *ACM Trans. Comput. Syst.* 11(2):146–178.

McKusick, M. K., Joy, W. N., Leffler, S. J., and Fabry, R. S. 1984. A fast file system for UNIX. *ACM Trans. Comput. Syst.* 2(3):181–197.

McNutt, B. 1984. A case study of access to VM disk volumes, pp. 175–180. In *CMG Proc.*

Miller, F. W. 1990. Predictive deadline multi-processing. *Operating Syst. Rev.* 24(4):52–62.

Miller, F. W. 1992. The performance of a mixed priority real-time scheduling algorithm. *Operating Syst. Rev.* 26(4):5–13.

Nieh, J. and Lam, M. S. 1995. SMART: a processor scheduler for multimedia applications. 15th ACM Symp. Operating Syst. Principles, Proc. *Operating Syst. Rev.* 29(5):233.

Patterson, D. A., Gibson, G., and Katz, R. H. 1988. A case for redundant arrays of inexpensive disks (RAID). In *Proc. SIGMOD Int. Conf. Manage. Data.* ACM, New York.

Patterson, D. A., Chen, P., Gibson, G., and Katz, R. H. 1989. Introduction to redundant arrays of inexpensive disks (RAID), pp. 112–117. In *Spring COMPCON '89.* IEEE, New York.

Rosenblum, M. and Ousterhout, J. K. 1992. The design and implementation of a log-structured file system. *ACM Trans. Comput. Syst.* 10(1):26–52.

Ruemmler, C. and Wilkes, J. 1993. UNIX disk access patterns, pp. 405–420. In *Proc. Winter USENIX, Conf.,* USENIX Association, Berkeley, CA.

Ryder, K. D. 1970. A heuristic approach to task dispatching. *IBM Syst. J.* 9(3):189–198.

Samadzadeh, M. H. and Koshy, B. S. 1996. A display and analysis tool for process-resource graphs. *Operating Syst. Rev.* 30(1):39–62.

Seltzer, M., Chen, P., and Ousterhout, J. 1990. Disk scheduling revisited, pp. 313–323. In *Proc. Winter USENIX Conf.* USENIX Association, Berkeley, CA.

Sherman, S., Baskett, F., and Browne, J. C. 1972. Trace driven modeling and analysis of CPU scheduling in a multiprogramming system. *Commun. ACM* 15(12):1063–1069.

Silberschatz, A. and Galvin, P. B. 1994. *Operating System Concepts*, 4th ed. Addison–Wesley, Reading, MA.

Stevens, D. F. 1968. On overcoming high priority paralysis in multiprogramming systems: a case history. *Commun. ACM* 11(8):539–541.

Tanenbaum, A. S. 1992. *Modern Operating Systems*. Prentice–Hall, Englewood Cliffs, NJ.

Teorey, T. J. 1972. Properties of disk scheduling policies in multiprogrammed computer systems. In *Proc. AFIPS Fall J. Comp. Conf.*, AFIPS Press, Reston, VA.

Teorey, T. J. and Pinkerton, T. B. 1972. A comparative analysis of disk scheduling policies. *Commun. ACM* 15(3):177–184.

Tsichritzis, D. C. and Bernstein, P. A. 1974. *Operating Systems*, Academic Press, New York.

Tucker, A. B., Cupper, R. D., Bradley, W. J., Epstein, R. G., and Kelemen, C. F. 1995. *Fundamentals of Computing II: Abstraction, Data Structures, and Large Software Systems*, McGraw–Hill, New York.

Waldspurger, C. A., Hogg, T., Huberman, B. A., Kephart, J. O., and Stornetta, W. S. 1992. Spawn: a distributed computational economy, *IEEE Trans. Software Eng.* 18(2):103–117.

Waldspurger, C. A. and Weihl, W. E. 1994. Lottery scheduling: flexible proportional-share resource management. 1st USENIX Symp. Operating Syst. Design Implementation Proc. *Operating Syst. Rev.* (Nov.):1–11.

Further Information

Good introductions to the practical problems in processor and device scheduling are presented in Tanenbaum [1992], Silberschatz and Galvin [1994], and Lister [1979].

Current work is presented at the annual ACM Symposium Operating System Principles, the USENIX Symposium of Operating System Design and Implementation, and the International Conference on Architectural Support for Programming Languages and Operating Systems. Copies of the Proceedings are available from the ACM Special Interest Group on Operating Systems, ACM Headquarters, 1515 Broadway, New York.

The ACM quarterly *Transactions on Computer Systems* reports new developments in computer system scheduling. Also the *Operating System Review*, a publication of the ACM Special Interest Group on Operating Systems, documents current research in the area.

78

Real-Time and Embedded Systems

John A. Stankovic
University of Massachusetts

78.1 Introduction

Real-time systems are defined as those systems in which the correctness of the system depends not only on the logical result of computation, but also on the time in which the results are produced [Stankovic 1988]. Real-time systems span a broad spectrum of complexity from very simple microcontrollers in **embedded systems** (such as a microprocessor controlling an automobile engine) to highly sophisticated, complex, and distributed systems (such as air traffic control for the continental U.S.). Other examples of real-time systems include command and control systems, process control systems, flight control systems, the Space Shuttle avionics system, flexible manufacturing applications, the space station, space-based defense systems, intensive care monitoring, collections of humans/robots coordinating to achieve common objectives (usually in hazardous environments such as undersea exploration or chemical plants), intelligent highway systems, mobile and wireless computing, and multimedia and high-speed communication systems. We are also beginning to see some of these real-time systems adding expert systems [Wright et al. 1986] and other artificial intelligence (AI) technology creating additional requirements and complexities. From this extensive list of applications we can see that real-time and embedded systems technology is a key *enabling* technology for the future in an ever growing domain of important applications.

At least three major trends in the real-time and embedded systems field have had major impacts on its technology. The first is the increased growth and sophistication of embedded systems; the second is the development of more scientific and technological results for **hard real-time systems**; and the third is the advent of distributed multimedia, a **soft real-time system**. In a hard real-time system there is no value to executing tasks after their deadlines have passed. A soft real-time system has tasks that retain some diminished value after their deadlines so these tasks should still be executed, even if they miss their deadlines.

Most embedded systems consist of a small microcontroller and limited software situated within some *product* such as a microwave oven or automobile. Often, the design of embedded systems is severely constrained by power, size, and cost constraints. However, to support increasing sophistication of embedded systems, we now see the common use of powerful microcontrollers and digital signal processor

(DSP) chips, as well as the use of off-the-shelf real-time operating systems and design and debugging tools. Many people involved with embedded systems deal on a daily basis with sensors and data acquisition technology and systems; others construct architectures based on single board computers [many are still 68000 based, but reduced instruction set computer (RISC) processors are beginning to be used more and more] and busses such as the VME bus. Many people are involved with the programming and debugging of embedded systems, largely using the C programming language and cross development and debugging platforms. Embedded systems may or may not have real-time constraints.

In the hard real-time area, many fundamental results have been developed in real-time scheduling, operating systems, architecture and fault tolerance, communication protocols, specification and design tools, formal verification, databases and object-oriented systems. Increased emphasis on all of these areas is expected to continue for the foreseeable future. Many hard real-time systems are embedded systems.

Distributed multimedia has produced a new set of soft real-time requirements and when its potential is fully realized, it will fundamentally change how the world operates. Real-time principles lie at the heart of distributed multimedia, but without the concomitant high-reliability requirements found in safety-critical, hard real-time systems.

78.2 Underlying Principles

Typically, a real-time system consists of a *controlling system* and a *controlled system*. For example, in an automated factory, the controlled system is the factory floor with its robots, assembling stations, and the assembled parts; whereas the controlling system is the computer and human interfaces that manage and coordinate the activities on the factory floor. Thus, the controlled system can be viewed as the *environment* with which the computer interacts.

The controlling system interacts with its environment based on the information available about the environment from various **sensors**. It is imperative that the state of the environment, as perceived by the controlling system, be consistent with the actual state of the environment. Otherwise, the effects of the controlling systems' activities may be disastrous. Hence, periodic monitoring of the environment as well as timely processing of the sensed information is necessary.

Timing correctness requirements in a real-time system arise because of the *physical impact* of the controlling systems' activities upon its environment. For example, if the computer controlling a robot does not command it to stop or turn on time, the robot might collide with another object on the factory floor possibly causing serious damage. In many real-time systems even more severe consequences will result if timing as well as logical correctness properties of the system are not satisfied. For example, consider the effects of nuclear power plants or air traffic control systems failing.

Timing constraints for tasks can be arbitrarily complicated but the most common timing constraints for tasks are either *periodic, aperiodic,* or *sporadic*. A periodic task is one that is activated once every T units of time. The deadline for each activated instance may be less than, equal to, or greater than the period T. An aperiodic task is activated at unpredictable times. A sporadic task is an aperiodic task with an additional constraint that there is a minimum interarrival time between task activations.

Low-level application tasks such as those that process information obtained from sensors, or those that activate elements in the environment (through actuators), typically have stringent timing constraints dictated by the physical characteristics of the environment. A majority of sensory processing is periodic in nature. For example, a radar that tracks flights produces data at a fixed rate. A temperature monitor of a nuclear reactor core should be read periodically to detect any changes promptly. Some of these periodic tasks may exist from the point of system initialization whereas others may come into existence dynamically. The temperature monitor is an instance of a permanent task. An example of a dynamically created task is a (periodic) task that monitors a particular flight; this comes into existence when the aircraft enters an air traffic control region and will cease to exist when the aircraft leaves the region.

More complex types of timing constraints also occur. For example, spray painting a car on a moving conveyor must be started after time t_1 and completed before time t_2. Aperiodic requirements can arise from dynamic events, such as an object falling in front of a moving robot or a human operator pushing a button on a console.

Time related requirements may also be specified in indirect terms. For example, a value may be attached to the completion of each task where the value may increase or decrease with time; or a value may be placed on the *quality* of an answer whereby an inexact but fast answer might be considered more valuable than a slow but accurate answer. In other situations, missing X deadlines might be tolerated, but missing $X + 1$ deadlines cannot be tolerated.

What happens when timing constraints are not met? The answer depends, for the most part, on the type of application. A real-time system that controls a nuclear power plant or one that controls a missile, cannot afford to miss timing constraints of the **critical tasks**. Resources needed for critical tasks in such systems have to be preallocated so that the tasks can execute without delay. In many situations, however, some leeway does exist. For example, even on an automated factory floor, if it is estimated that the correct command to a robot cannot be generated on time, it may be appropriate to command the robot to stop (provided it will not cause other moving objects to collide with it and result in a different type of disaster), or to slow down (thereby dynamically generating more time to produce a correct command). Another example is a periodic task monitoring the position of an aircraft; depending on the aircraft's location and trajectory, missing the processing of one or two radar readings may not cause any problems.

In a real-time system, the characteristics of the various application tasks are usually known a priori and might be scheduled statically or dynamically. Whereas static specification of schedules is typically the case for periodic tasks, the opposite is true for aperiodic tasks. When the periodic temperature monitor of a nuclear reactor senses a problem in the core, it can invoke another (aperiodic) task to activate the appropriate elements of the reactor to correct the problem, for example, to force more coolant into the reactor core. In this case, the deadline for the aperiodic task can be statically determined as a function of the physical characteristics of the reactions within the core. On the other hand, the deadline of a task that controls a robot on a factory floor can be determined dynamically depending on the speed, direction, and weight of the robot. The command to the robot forcing it to turn right, left, or stop should be generated before this deadline.

In a real-time system that is designed in a static manner, the characteristics of the controlled system are assumed to be known a priori and, hence, the nature of activities and the sequence in which these activities take place can be determined off line before the system begins operation. Such systems are quite inflexible even though they may incur lower run-time overheads. In practice, most applications involve a number of components that can be statically specified along with many dynamic components. If handled appropriately, a system with high-resource utilization and low overheads can be produced for such applications.

Although a large proportion of currently implemented real-time systems are static in nature, many next generation systems will have to adopt solutions that are more dynamic and flexible. This is because such systems will be large and complex and they will function in environments that are both uncertain and physically distributed. More importantly, they will have to be maintainable and extensible due to their evolving nature and projected long lifetimes. Because of these characteristics, real-time systems, in general, and systems with the previously described characteristics, in particular, need to be *fast, predictable, reliable,* and *adaptive*.

One long-held misconception about real-time systems is that they only need to be fast to be effective. Basically, being fast is usually a necessary condition, but it is not sufficient. A real-time system has to meet explicit deadlines and being fast on average does not guarantee that a deadline will be met. If a real-time system can be shown to meet its deadlines (using a worst-case rather than an average-case behavior analysis), then we say that it is predictable. Predictability, itself, has many meanings and an entire journal article has been devoted to its meaning [Stankovic and Ramamritham 1990]. For purposes of this chapter it is sufficient to take a simplistic view of predictability. Consider that predictability means that when a task or set of tasks is activated it should be possible to determine their completion time subject to failure assumptions. This must be done taking into account the state of the system (including the state of the operating system and the state of the resources controlled by the operating system) and the tasks' resource needs.

The task of building a real-time system can be very simple or it can be extremely complex. The difficulty depends on the characteristics of the real-time system along five dimensions, which we now discuss.

1. *Granularity of the deadline and laxity of the tasks:* In a real-time system some of the tasks have deadlines and/or periodic timing constraints. If the time between when a task is activated (required to be executed) and when it must complete execution is short, then the deadline is tight (i.e., the granularity of the deadline is small or the deadline is close). This implies that the operating system reaction time has to be short, and the scheduling algorithm to be executed must be fast and very simple. Tight time constraints may also arise when the deadline granularity is large (i.e., from the time of activation), but the amount of computation required is also great. In other words even large granularity deadlines can be tight when the laxity (deadline minus computation time) is small. In many real-time systems tight timing constraints predominate. Consequently, designers focus on developing very fast and simple techniques to react to this type of task activation. In general, the tighter the deadline the more difficult the design task.

2. *Strictness of deadline:* The strictness of the deadline refers to the value of executing a task after its deadline. For a hard real-time task there is no value to executing the task after the deadline has passed. A soft real-time task retains some diminished value after its deadline and so it should still be executed. Very different techniques are usually used for hard and soft real-time tasks. In many cases hard real-time tasks are preallocated and prescheduled resulting in 100% of them making their deadlines. Soft real-time tasks are often scheduled either with nonreal-time scheduling algorithms, with algorithms that explicitly address the timing constraints but aim only at good average-case performance, or with algorithms that combine importance and timing requirements. Hard real-time tasks are more difficult to deal with than soft real-time tasks, and systems which must deal with both types simultaneously are yet even more difficult. Multimedia in a timesharing environment is a soft real-time application, but multimedia in a real-time control environment such as an automated factory must deal with both hard and soft real-time constraints.

3. *Reliability:* Many real-time systems operate under severe reliability requirements. That is, if certain tasks, called critical tasks, miss their deadline then a catastrophe may occur. These tasks are usually guaranteed to make their deadlines by an off-line analysis and by schemes that reserve resources for these tasks even if it means that those resources are idle most of the time. In other words, the requirement for critical tasks should be that all of them always make their deadline (a 100% guarantee), subject to certain failure and workload assumptions. However, too many systems treat all of the tasks that have hard timing constraints as critical tasks (when, in fact, only some of those tasks are truly critical). This can result in erroneous requirements and an overdesigned and inflexible system. It is also common to see hard real-time tasks defined as those with both strict deadlines and critical importance. We prefer to keep a clear separation between these notions because they are not always related. Of course, many other reliability issues must also be resolved, but here we only mention the key issue that deals with timing constraints and reliability.

4. *Size of system and degree of coordination:* Real-time systems vary considerably in size and complexity. In most current real-time systems the entire system is loaded into memory; if there are well-defined phases, each phase is loaded just prior to the beginning of the phase. In many applications, subsystems are highly independent of each other and there is limited cooperation among tasks. The ability to load entire systems into memory and to limit task interactions simplifies many aspects of building and analyzing real-time systems. However, for future large, complex, real-time systems, having completely resident code and highly independent tasks will not always be practical. Moreover, solutions based on virtual memory are not acceptable because of the large degree of unpredictability associated with this technique. Consequently, increased size and coordination raise many new problems that must be addressed and further complicate the notion of predictability. Embedded systems may also have power, physical size, memory, and severe cost constraints adding to the difficulty of their design.

5. *Environment:* The environment in which a real-time system is to operate plays an important role in the design of the system. Many environments are very well defined (a laboratory experiment, an automobile engine, or an assembly line). Designers think of these as deterministic environments (even though they may not be intrinsically deterministic, they are well controlled and assumed to be deterministic). These environments give rise to small, static real-time systems where all deadlines are guaranteed a priori. Even in these simple environments we need to place restrictions on the inputs. For example, a particular assembly line may only be able to cope with five items per minute; given more than that, the system fails. Taking this approach enables an off-line quantitative analysis of the timing properties to be made. Since we know exactly what to expect given the assumptions about the well-defined environment we can usually design and build these systems to be predictable. However, the approaches taken in relatively small, static systems do not scale to other environments which are larger, much more complicated, and less controllable. Consider a next generation real-time system such as a team of cooperating mobile robots on the planet Mars. This system will be large, complex, distributed, adaptive, contain many types of timing constraints, need to operate in a highly nondeterministic environment, and evolve over a long system lifetime. It is not possible to assume that this environment is deterministic or to control it sufficiently well to make it look deterministic. If that were done, the system would be too inflexible and would not be able to react to unexpected events or combinations of events.

78.3 Best Practices

Now that we have presented some of the basic principles of real-time and embedded systems, we can discuss some of the applications in various areas of **real-time computing**, including: real-time scheduling, real-time kernels, real-time architectures and fault tolerance, real-time communications, distributed multimedia, **real-time databases**, real-time formal verification, design and languages, and real-time AI.

Real-Time Scheduling

Real-time scheduling results in recent years have been extensive. Theoretical results have identified worst-case bounds for dynamic on-line algorithms, and complexity results have been produced for various types of assumed task set characteristics. Queueing theoretic analysis has been applied to soft real-time systems covering algorithms based on real-time variations of **first-come–first-serve (FCFS)**, earliest deadline, and least laxity. We have seen the development of scheduling results for imprecise computation (a situation where tasks obtain a greater value the longer they execute, up to some maximum value) [Liu et al. 1991].

More applied scheduling results have also been produced with an extensive set of improvements to the rate monotonic algorithm (this includes the deferrable server and sporadic server algorithms [Sprunt et al. 1989], techniques to address the problem of priority inversion [Sha et al. 1990], and a set of algorithms that perform dynamic on-line planning [Ramamritham et al. 1990]). We have also seen practical application of a priori calculation of static schedules to provide what is called 100% guarantees for critical tasks. Although these a priori analyses are very valuable, system designers should not be lulled into thinking that 100% guarantees mean that no scheduling error can occur. It is important to know that these 100% guarantees are based on many (often unrealistic) assumptions. If the assumptions are a poor match for what can be expected from the environment (more and more likely in a distributed environment), then even with 100% guarantees the system can indeed miss deadlines. Hence, a key issue is to choose an algorithm whose assumptions provide the greatest coverage over what *really* happens in the environment. For all of these scheduling results outlined, the trend has been to deal with increasingly more complicated task set and environment characteristics (e.g., multiprocessing and distributed computing and tasks with precedence constraints).

An exciting trend is the extensive use of schedulability analysis for both static and dynamic real-time systems. For example, the Software Engineering Institute has developed a handbook [Klein et al. 1993] on rate monotonic analysis and gives seminars and tutorials regarding its use in real-time systems.

Real-Time Kernels

One focal point for next generation complex real-time systems is the operating system. The operating system must provide basic support for predictably satisfying real-time constraints, for fault tolerance and distribution, and for integrating time-constrained resource allocations and scheduling across a spectrum of resource types including sensor processing, communications, CPU, memory, and other forms of I/O. Toward this end, at least three major scientific issues need to be addressed:

- The *time dimension* must be elevated to a central principle of the system and should not be simply an afterthought. An especially perplexing aspect of this problem is that most system specification, design, and verification techniques are based on abstraction, which ignores implementation details. This is obviously a good idea; however, in real-time systems, timing constraints are derived from the environment and the implementation. This dilemma is a key scientific issue.

- The basic paradigms found in today's general purpose distributed operating systems must change. Currently, they are based on the notion that application tasks request resources as if they were random processes; operating systems are designed to expect random inputs and to display good average-case behavior. The new paradigm must be based on the delicate balance of *flexibility* and *predictability:* the system must remain flexible enough to allow a highly dynamic and adaptive environment, but at the same time be able to predict and possibly avoid resource conflicts so that timing constraints can be met. This is especially difficult in distributed environments where layers of operating system code and communication protocols interfere with predictability.

- A highly *integrated and time-constrained resource allocation approach* is necessary to adequately address timing constraints, predictability, adaptability, correctness, safety, and fault tolerance. For a task to meet its deadline, resources must be available *in time*, and events must be ordered to meet precedence constraints. Many coordinated actions are necessary for this type of processing to be accomplished on time. The state of the art lacks completely effective solutions to this problem.

For relatively small, less complex, real-time systems, it is often the case that real-time systems are supported by stripped down and optimized versions of timesharing operating systems. To reduce the run-time overheads incurred by the kernel and to make the system *fast*, the kernel underlying the real-time system:

- Has a fast context switch
- Has a small size (with its associated minimal functionality)
- Responds to external interrupts quickly
- Minimizes intervals during which interrupts are disabled
- Provides fixed or variable sized partitions for memory management (i.e., no virtual memory) as well as the ability to lock code and data in memory
- Provides special sequential files that can accumulate data at a fast rate

To deal with timing requirements, the kernel:

- Maintains a real-time clock
- Provides a priority scheduling mechanism
- Provides for special alarms and timeouts
- Permits tasks to invoke primitives to delay by a fixed amount of time and to pause/resume execution

In general, the kernels perform multitasking; intertask communication and synchronization are achieved via standard, well-known primitives such as mailboxes, events, signals, and semaphores. Examples of existing real-time kernels include: QNX, LynxOS, OS-9, VxWorks, and VRTXsa (over 70 commercial real-time kernels exist). Specialized kernels for DSP chips and homegrown kernels for microcontrollers are also still widely found.

Real-time kernels are also being extended to operate in highly cooperative multiprocessor and distributed system environments [Tokuda et al. 1990]. This means that there is an **end-to-end timing requirement** (in the sense that a set of communication tasks must complete before a deadline), i.e., a collection of activities must occur (possibly with complicated precedence constraints) before some deadline. Much research is being done on developing time constrained communication protocols to serve as a platform for supporting this user-level end-to-end timing requirement. However, while the communication protocols are being developed to support host-to-host bounded delivery time, using the current operating system (OS) paradigm of allowing arbitrary waits for resources or events, or treating the operation of a task as a *random process* causes great uncertainty in accomplishing the application level end-to-end requirements. As an example, the Mars project [Kopetz et al. 1989], the Spring project [Stankovic and Ramamritham 1991], and a project at the University of Michigan [Shin 1991] are all attempting to solve this problem. The Mars project uses an a priori analysis and then statically schedules and reserves resources so that distributed execution can be guaranteed to make its deadline. The Spring approach supports dynamic requests for real-time virtual circuits (guaranteed delivery time) and real-time datagrams (best effort delivery) integrated with CPU scheduling so as to guarantee the application level end-to-end timing requirements. The Spring project uses a distributed reflective memory based on a fiber optic ring to achieve lower level predictable communication properties. The Michigan work also supports dynamic real-time virtual circuits and datagrams, but it is based on a general multihop communication subnet.

Research is also being done on developing real-time object-oriented kernels to support the structuring of distributed real-time applications. As far as we know, no commercial products of this type are available. However, due to the major advantages of object orientation, it is likely that many such products will become available in the near future.

The diversity of the applications requiring predictable distributed systems technology will be significant. To handle this diversity, we expect the distributed real-time operating systems must use an *open system* approach, and the applications need to be portable. Regarding the open systems approach, it is important to avoid having to rewrite the operating system for each application area which may have differing timing and fault tolerance requirements. A library of real-time operating system objects might provide the level of functionality, performance, predictability, and portability required. We envision a Smalltalk-like system for hard real time, so that a designer can tailor the OS to the application without having to write everything from scratch. In particular, a library of real-time scheduling algorithms should be available that can be plugged in depending on the run-time task model being used and the load, timing, and fault tolerance requirements of the system.

Regarding the portability of applications, many real-time Unix operating systems are appearing [Furht et al. 1991], and a standard for real-time operating systems, called RT POSIX, is being developed [Gallmeister 1995]. Although such a standard facilitates porting the code, it is still an open issue on how to assess the timing properties of the ported application.

Real-Time Architecture and Fault Tolerance

Real-time systems are usually special purpose. In the past, architectures to support such applications tended to be special purpose too. The current trend is one in which more off-the-shelf components are being used to produce more generic architectures. Although considerable discussion could be given to real-time architectures, we shall consider only briefly how architecture impacts the computation of worst-case execution time and how it supports fault tolerance.

One aspect of architecture for real-time computing is the facility with which the worst-case execution time can be calculated. Worst-case execution times of programs are dependent on the system hardware, the operating system, the compiler used, and the programming language used. Many hardware features that have been introduced to speed up the average-case behavior of programs pose problems when information about worst-case behavior is sought (especially true on many state-of-the-art RISC CPUs). For instance, the ubiquitous caches, pipelining, dynamic random access memories (RAMs), and virtual (secondary) memory, lead to highly nondeterministic hardware behavior. Similarly, compiler optimizations tailored to make better use of these architectural enhancements, as well as more standard techniques such as **constant folding**, which is the replacement of run-time computation by compile-time computation, contribute to poor predictability of code execution times. System interferences due to interrupt handling, shared memory references, and preemptions are additional complications. In summary, any approach to the determination of execution times of real-time programs has many complexities which must be solved.

Many real-time system architectures consist of multiprocessors, networks of uniprocessors, or networks of uni- and multiprocessors. Such architectures have potential for high fault tolerance, but they are also much more difficult to manage in a way such that deadlines are predictably met. Fault tolerance must be designed in at the start, must encompass both hardware and software, and must be integrated with timing constraints. In many situations, the fault tolerant design must be static due to extremely high data rates and severe timing constraints. Ultrareliable systems need to employ proof of correctness techniques to ensure fault tolerance properties [Vytopil 1993]. Primary and backup schedules computed off line are often found in hard real-time systems. We also see new approaches where on-line schedulers predict that timing constraints will be missed, enabling early action on such faults. Dynamic reconfigurability is needed but little progress has been reported in this area. Also, whereas considerable advance has been made in the area of software fault-tolerance, techniques that explicitly take timing into account are lacking.

Since fault tolerance is difficult, the trend is to let experts build the proper underlying support for it. For example, implementing checkpointing, reliable atomic broadcasts, logging, lightweight network protocols, synchronization support for replicas, and recovery techniques, and having these primitives available to applications, then simplifies creating fault tolerant applications. However, many of these techniques have not carefully addressed timing considerations nor the need to be predictable in the presence of failures. Many real-time systems, which require a high degree of fault tolerance, have been designed with significant architectural support, but the design and scheduling to meet deadlines is done statically, with all replicas in lock step. This may be too restrictive for many future applications. What is required is the integration of fault tolerance and real-time scheduling to produce a much more flexible system. For example, the use of the **imprecise computation** model [Liu et al. 1991], or a planning scheduler [Ramamritham et al. 1990] gives rise to a more flexible approach to fault tolerance than static schedules and fixed backup schemes. Adaptive fault tolerance with an explicit interaction with real-time constraints can be found in Bondavali et al. [1993].

Real-Time Communications

Distributed real-time systems require time-constrained message delivery. In many applications the communication protocols and network provide deterministic behavior. An alternative, applicable in other situations, is a best effort approach. Hybrid approaches also exist [Arvind et al. 1991]. Those systems requiring hard guarantees often use time-domain multiple access (TDMA), fiber distributed data interface (FDDI), Institute of Electrical and Electronics Engineers (IEEE) 802.4 token bus or 802.5 token ring [Strosnider and Marchok 1989]. Careful assumptions and analysis accompany the use of these networks to produce a deterministic guarantee. For best effort approaches, variations of carrier sense multiple access/collision detection (CSMA/CD) or window-based schemes can be used [Malcolm and Zhao 1995]. For distributed multimedia [Govindan and Anderson 1991, Clark et al. 1992], timing constraints on the network include end-to-end delays, minimum jitter, and interpacket maximum delays. Other requirements for transmitting audio, video, text, and data traffic include extremely high volume and high speeds. To support these requirements, asynchronous transfer mode (ATM) switches [Newman 1994] have been developed. With the advent of ATM technology we are seeing the projected use of ATMs as the local area network of real-time systems.

Specialized busses are still widely utilized today. The controller area network (CAN) bus for automobiles and the SAFEbus for commercial aircraft [Hoyme and Driscoll 1991] are examples.

Distributed Multimedia

Many real-time control applications such as agile manufacturing and process control operate in highly nondeterministic environments under timing constraints of many types. Significant improvements in these applications can be created by embedding continuous and multimedia support in these applications. For example, in agile manufacturing, remote factories, each consisting of many automated workcells, must coordinate to handle new strategies for incoming product orders, to develop the design of new products, to schedule just-in-time deliveries of manufactured components, to monitor the plant operations, and to collaboratively solve difficult manufacturing floor problems.

To implement solutions cost effectively and to allow multimedia applications direct access to plant operational data, it is envisioned that the same computers would control the time-constrained operations in and across the workcells and support multimedia [Guha et al. 1995]. The backbone network would likely be ATM. Distributed multimedia over ATM networks has enormous potential to provide these applications with teleconferencing for real-time coordination, collaborative design, and a wealth of real-time information such as visual access to remote and local plant operations via cameras, as well as supporting real-time control. These applications would also benefit from an integrated design of a distributed database that contains such information as sensor information and control variables constrained by temporal validity intervals, plant operational data, information on availability of raw materials, inventory of products, customer orders, etc. The confluence of an integrated database, multimedia, and real-time control has great potential for moving application areas such as manufacturing and process control into the next generation.

It is important to point out that although the commercial world is developing multimedia support, most of this work is done in the context of general purpose timesharing systems and **not** integrated with real-time control applications. This difference is significant and will likely require solutions not likely to be developed for commercial multimedia. One reason for this is that the nonreal-time control applications can accept a probability that other applications that are executing can fail (or be late), and so it is likely that simpler solutions and less expensive systems can be built for these applications. This assumption is reasonable for many commercial multimedia-based systems, but not for real-time control applications. On the other hand, real-time control applications have focused on solutions for hard deadline systems, which would be too inefficient if they were used for multimedia. This new, distributed, multimedia, information technology environment will require greater support for soft real-time systems, especially in their continuous and multimedia aspects. Here, new models of guarantees and resource reservations are required.

Distributed multimedia applications require both high-performance operating system features (in order to respond to the high data and processing rates) as well as a **quality-of-service (QOS)** guarantee. A quality-of-service guarantee is a promise to provide a certain level of performance to a multimedia application. Quality of service may be defined in a number of ways even within the same system. Hence, it is important that the OS be flexible enough to accept different types of quality of service requests. Because QOS guarantees are end-to-end, this necessitates an integrated and synergistic approach to the problem, encompassing the scheduling algorithm, the underlying I/O system (e.g., mass storage), the database system, associated operating system components, and the network. The scheduling algorithms must be able to *cosupport* the hard deadline control tasks and the distributed multimedia aspects of the system. Solving the main issues for the I/O system includes interfacing to real-time databases, achieving adequate sustained transfer rates, I/O scheduling, data layout on the disk array, and I/O buffer management. Whereas some aspects of the operating system infrastructure required to meet these needs can be found in today's real-time kernels or as specially developed parts of commercial kernels, OS features typically available today are not sufficient especially in the context of real-time control applications with hard deadlines. Further, much of the current work has concentrated on either single site [Anderson et al. 1992], or network level end-to-end delays; what is required is support for distributed application level to application level processing.

Another issue is call admission in which the system dynamically decides whether to admit a new multimedia session or not. Note that this concept was previously used in hard real-time systems since 1984 [Ramamritham and Stankovic 1984]. In the context of multimedia, call admission requires that we determine that either the session being requested is possible (so we can admit it), or it is not possible even to an acceptable degraded level (so we reject the request, or identify what the degraded service will be). In control applications we may need to override previously admitted sessions or compute tradeoffs to maximize the likelihood of success of the overall mission.

Real-Time Databases

A real-time database is a database system where (at least some) transactions have explicit timing constraints such as deadlines and where data may become invalid with the passage of time. In such a system, transaction processing must satisfy not only the database consistency constraints, but also the timing constraints. Real-time database systems can be found, for instance, in program trading in the stock market, radar tracking systems, battle management systems, and computer integrated manufacturing systems. Some of these systems (such as program trading in the stock market) are soft real-time systems, because missing a deadline is not catastrophic. Usually, research into algorithms and protocols for such systems explicitly addresses deadlines and makes a best effort at meeting deadlines. In soft real-time systems there are no guarantees that specific tasks will make their deadlines.

In real-time databases there is a need for an integrated approach that includes time constrained protocols for concurrency control, conflict resolution, CPU and I/O scheduling, transaction restart and wakeup, deadlock resolution, buffer management, and commit processing. Many protocols based on locking, optimistic, and time-stamped concurrency control have been developed and evaluated in testbed or simulation environments [Abbott and Garcia-Molina 1988]. In most cases the optimistic approaches seem to work best [Huang et al. 1991].

In a typical database system a transaction is a sequence of operations performed on a database. Normally, consistency (serializability), atomicity, and permanence are properties supported by the transaction mechanism. Transaction throughput and response time are the usual metrics. In a soft real-time database, transactions have similar properties, but, in addition, have soft real-time constraints. Metrics include response time and throughput, but also include the percentage of transactions which meet their deadlines, or a weighted value function which reflects the value imparted by a transaction completing on time. On the other hand, in a hard real-time database, not all transactions have serializability, atomicity, and permanence properties. These requirements need to be supported only in certain situations. For example, hard real-time systems are characterized by their close interactions with the environment that they control. This is especially true for subsystems that receive sensory information or that control actuators. Processing involved in these subsystems is such that it is typically not possible to *rollback* a previous interaction with the environment.

Whereas the notion of consistency is relevant here (for example, the interactions of a real-time task with the environment should be consistent with each other), traditional approaches to achieving consistency, involving waits, rollbacks, and aborts are not directly applicable. Instead, compensating transactions may have to be invoked to nullify the effects of previously committed transactions. Also, another transaction property, namely, *permanence*, is of limited applicability in this context. This is because real-time data, such as those arriving from sensors, have limited *lifetimes*: they become obsolete after a certain point in time. Data received from the environment by the lower levels of a real-time system undergo a series of processing steps (e.g., filtering, integration, and correlation). Traditional transaction properties are less relevant at the lowest levels and become more relevant at higher levels in the system. Most hard real-time database systems are main memory databases of small size, with predefined transactions, and handcrafted for efficient performance.

A new trend is the use of active database technology for real-time databases. A **real-time active database (RTADB)** is a system where transactions have timing constraints such as deadlines, where data may become invalid with the passage of time, and where transactions may trigger other transactions. This type of database follows an event–condition–action paradigm subject to timing constraints. RTADBs are in their infancy and no commercial products yet exist, although many products contain *triggers*.

Real-Time Formal Verification, Design, and Languages

Today, when constructing a complex real-time system it is becoming more common to use a formal verification technique [Vytopil 1993, Jahanian and Mok 1986] for certain aspects of the design and to use a commercially available design tool [Kavi 1992]. Many formal verification approaches are available, e.g., those based on Petri nets, temporal logic, timed communicating sequential processes (CSP), probabilistic durational calculus, real-time logic (RTL), or prototype verification system (PVS). Although limitations still exist on the use of these techniques, the value of formal techniques early in the design process has been amply demonstrated. The trend is to develop formalisms that can directly address timing constraints.

Regarding design methods, commercial tools such as STATEMATE [Harel et al. 1990], CARDtools, or Control Shell provide graphical interfaces and many nice database features. Many design methodologies have been extended to deal with real-time systems and recently have included an object-oriented approach [Ellis 1994]. Since tools such as these are so important, continual improvements occur. The future should bring more understandable tools that better and better reflect and support the reliability and timing constraints of real-time systems.

A discussion on real-time languages (those specifically designed for real-time programming and specification) and languages used for real-time systems (assembler, C, ADA, etc.) can be found in Burns and Wellings [1989]. ADA and C are now commonly used for programming many complex real-time systems. Synchronous languages [Halbwachs 1993] are also widely used, mostly in Europe.

Real-Time Artificial Intelligence

Many complex real-time applications now require or will require knowledge-based on-line assistance operating in real time [Paul et al. 1991]. This necessitates a major change to some of the paradigms and implementations previously used by AI researchers. For example, AI systems must be made to run much faster (a necessary but not sufficient condition), allow preemption to reduce latency for responding to new stimuli, attain predictable memory management via incremental garbage collection or by explicit management of memory, include deadlines and other timing constraints in search techniques, develop anytime algorithms (algorithms where a nonoptimal solution is available at any point in time), develop time-driven inferencing, and develop time-driven planning and scheduling. Rules and constraints may also have to be imposed on the design, models, and languages used in order to facilitate predictability, e.g., limit recursion and backtracking to some fixed bound. Coming to grips with what predictability means in such applications is also very important.

In addition to these changes within AI, real-time AI (RTAI) techniques must be interfaced with lower level real-time systems technology to produce a functioning, reliable, and carefully analyzable system. Should the higher level RTAI techniques ignore the system level, treat it as a black box with *general* characteristics, or be developed in an integrated fashion with it so as to best build complex systems? What is the correct interface between these two traditionally separate systems? Integrating RTAI and low-level real-time systems software is quite a challenge because RTAI applications are operating in nondeterministic environments, there is missing or noisy information, some of the control laws are heuristic at best, objectives may change dynamically, partial solutions are sometimes acceptable (so that a tradeoff between the quality of the solution and the time needed to derive it can be made), the amount of processing is significant and highly data dependent, and the execution time of tasks may be difficult to determine. These demanding requirements will drive real-time research for many years to come.

Competing software architectures for real-time AI include production rule architectures, blackboard architectures, and a process trellis architecture. Some real-time AI systems have been built by carefully and severely restricting how production rules and blackboard systems are built and used. Research is ongoing to relax the restrictions so that the power of these architectures can be utilized, but at the same time providing a high degree of predictability. The process trellis architecture, used in the medical domain, is a highly static approach whereas the other two are much more dynamic. The trellis architecture (because it is static) has potential to provide static real-time guarantees for those applications characterized by enough time to completely compute results from a set of inputs before the next set of inputs arrive. This approach is suitable for certain types of real-time AI monitoring systems, but its generality for complex real-time AI systems has not been demonstrated.

In a distributed setting, high-level decision support requires organizing computations with networks of cooperative, semiautonomous agents, each capable of sophisticated problem solving. Theories of communication and organizational structure for groups of cooperative problem solving agents must be developed. These theories must include problem solving under uncertainty and under timing constraints.

78.4 Research Issues and Summary

Many research issues exist in all areas of real-time computing. Many of the problems are of a fundamental nature and others are more applied. It is impossible to list all of the key research issues; instead we identify representative examples of open research problems.

Although many interesting scheduling results have been produced, the state of the art still provides piecemeal solutions. Many realistic issues have not yet been addressed in an integrated and comprehensive manner. The real-time scheduling area still requires analyzable scheduling approaches (it may be a collection of algorithms) that are comprehensive and integrated. For example, the overall approach must be comprehensive enough to handle:

- Preemptable and nonpreemptable tasks
- Periodic and nonperiodic tasks
- Tasks with multiple levels of importance (or a value function)
- Groups of tasks with a single deadline
- End-to-end timing constraints
- Precedence constraints
- Communication requirements
- Resource requirements
- Placement constraints
- Fault tolerance needs
- Tight and loose deadlines
- Normal and overload conditions

The solution must be integrated enough to handle the interfaces between:

- CPU scheduling and resource allocation
- I/O scheduling and CPU scheduling
- CPU scheduling and real-time communication scheduling
- Local and distributed scheduling
- Static scheduling of critical tasks and dynamic scheduling of essential and nonessential tasks

One key issue is the need to provide predictability. Predictability requires bounded operating system primitives, some knowledge of the application, proper scheduling algorithms, and a viewpoint based on a *team* attitude between the operating system and the application. For example, simply having a very primitive kernel that is itself predictable is only the first step. More direct support is needed for developing predictable and fault tolerant real-time applications. One aspect of this support comes in the form of scheduling algorithms. For example, if the operating system is able to perform integrated CPU scheduling and resource allocation in a planning mode so that collections of cooperating tasks can obtain the resources they need at the right time, in order to meet timing constraints, this facilitates the design and analysis of real-time applications. Further, if the operating systems retains information about the importance of a task and what actions to take if the task is assessed as not being able to make its deadline, then a more intelligent decision can be made as to alternative actions, and graceful degradation of the performance of the system can be better supported (rather than a possible catastrophic collapse of the system if no such information is available). Kernels which support retaining and using semantic information about the application are sometimes referred to as *reflective* kernels [Stankovic and Ramamritham 1995].

Basic research is also required in many areas of distributed multimedia, including:

- Specification of quality of service
- Algorithmic and kernel support to actually achieve this guaranteed service and to dynamically negotiate other levels of service, if necessary
- How to perform reservations of sets of resources
- Integrated scheduling across a set of resources (e.g., so that CPU, I/O buffer, disk controller, network bandwidth, and resources at the receiver are reserved together)
- End-to end scheduling
- Atomic guarantees for sets of tasks (This supports the call admission policies needed in multimedia.)

Obviously, achieving complex, real-time systems is nontrivial and will require research breakthroughs in many aspects of system design and implementation. For example, good design methodologies and tools which include programming rules and constraints must be used to guide real-time system developers so that subsequent implementation and *analysis* can be facilitated. This includes proper application decomposition into subsystems and allocation of those subsystems onto distributed architectures. The programming language must provide features tailored to these rules and constraints, must limit its features to enhance predictability, and must provide the ability to specify timing, fault tolerance, and other information for subsequent use at run time. Many language features are continuously being proposed, although few of them are currently used in practice. Execution time of each primitive of the kernel must be bounded and predictable, and the operating system should provide explicit support for all of the requirements including the real-time requirements. New trends in the OS area include the use of microkernels, support for multiprocessors and distributed systems, and real-time thread packages. The architecture and hardware must also be designed to support predictability and facilitate analysis. For example, hardware should be simple enough so that predictable timing information can be obtained. This has implications for how to deal with caching, memory refresh and wait states, pipelining, and some complex instructions, which all contribute to timing analysis difficulties. The resulting system must be scalable to account for the significant computing needs that occur both initially and as the system evolves. An insidious aspect of critical real-time systems, especially with respect to their real-time requirements, is that the weakest link in the entire system can undermine careful design and analysis at other levels. Research is required to address all of these issues in an integrated fashion.

Finally, a number of new trends involve the use of formal verification for real-time systems and the development of real-time databases, real-time object-oriented systems, and real-time artificial intelligence. Since these areas are very new, many open problems exist.

Defining Terms

Constant folding: A compiler optimization technique where run-time computations are replaced by compile-time computations.

Critical real-time task: One in which missing its deadline may cause a catastrophe or total failure of the system.

Embedded system: A system that is a component of a larger system, e.g., an automobile cruise control system or the navigation system of the space shuttle.

End-to-end timing requirement: A single overall timing requirement for a set of tasks that operate with some precedence constraints. This is typical of a distributed set of tasks communicating over a network.

First-come–first-served (FCFS): Scheduling policy that chooses which task to execute based on the earliest time of arrival.

Hard real-time task: One in which there is no value in continuing to execute the task after the deadline has passed.

Imprecise computation: One where if the computation terminates before completion, the intermediate result produced is usable.

Quality-of-service (QOS): Guarantee that is a promise to provide a certain level of performance to a multimedia application.

Real-time active database: One where transactions have timing constraints such as deadlines, where data may become invalid with the passage of time, and where transactions or events may trigger other transactions.

Real-time computing: That computing where the results must be logically correct and produced on time.

Real-time database: One where transactions have timing constraints such as deadlines and where data may become invalid with the passage of time.

Sensor: A device that outputs a signal for the purpose of detecting or measuring a physical property.

Soft real-time task: One that retains some diminished value after its deadline so it should still be executed even if its deadline has passed.

References

Abbott, R. and Garcia-Molina, H. 1988. Scheduling real-time transactions: a performance evaluation, pp. 1–12. *Proc. Very Large Databases Conf.*

Anderson, D., Osawa, O., and Govindan, R. 1992. A file system for continuous media. *ACM Trans. Comput. Syst.* 10(4):311–377.

Arvind, K., Ramamritham, K., and Stankovic, J. 1991. A local area network architecture for communication in distributed real-time systems. *Real-Time Syst.* 3(2).

Bondavali, A., Stankovic, J., and Strigini, L. 1993. Adaptable fault tolerance for real-time systems. *IEEE 3rd Int. Workshop Responsive Comput. Syst.*

Burns, A. and Wellings, A. 1989. *Real-Time Systems and Their Programming Languages.* Addison–Wesley, Reading, MA.

Clark, D., Shenkar, S., and Zhang, L. 1992. Supporting real-time applications in an integrated services packet network: architecture and mechanism. *Proc. ACM SIGCOMM.*

Ellis, J. 1994. *Objectifying Real-Time Systems.* SIGS Books, New York.

Furht, B., Grostick, D., Gluch, D., Rabbat, G., Parker, J., and McRoberts, M. 1991. *Real-Time Unix Systems, Design and Application Guide.* Kluwer Academic, Boston, MA.

Gallmeister, B. 1995. *POSIX.4: Programming for the Real World.* O'Reilly and Associates.

Govindan, R. and Anderson, D. 1991. Scheduling and IPC mechanisms for continuous media. *Proc. Symp. Operating Syst. Principles.* ACM.

Guha, A., Pavan, A., Liu, J., Rastogi, A., and Steeves, T. 1995. Supporting real-time and multimedia applications on the Mercuri testbed. *IEEE J. Selec. Areas Commun.* 13(4):749–763.

Halbwachs, N. 1993. *Synchronous Programming of Reactive Systems.* Kluwer Academic, Boston, MA.

Harel, D., Lachover, H., Naamad, A., Pnueli, A., Politi, M., Sherman, R., Shtull-Trauring, A., and Trakhtenbrot, M. 1990. Statemate: a working environment for the development of complex reactive systems. *IEEE Trans. Software Eng.* 16(4):413–414.

Hoyme, K., Driscoll, K., Herrlin, J., and Radke, K. 1991. ARINC 629 and SAFEbus: data buses for commercial aircraft. *Sci. Honeyweller* 57–70.

Huang, J., Stankovic, J., Ramamritham, K., and Towsley, D. 1991. Experimental evaluation of optimistic concurrency control. *Proc. Very Large Databases Conf.*

Jahanian, F. and Mok, A. 1986. Safety analysis of timing properties in real-time systems. *IEEE Trans. Software Eng.* 12(9):890–904.

Kavi, K. 1992. *Real-Time Systems: Abstractions, Languages, and Design Methodologies.* IEEE Computer Society Press, Los Alamitos, CA.

Klein, M., Ralya, T., Pollak, B., Obenza, R., and Gonzales Harbour, M. 1993. *A Practitioner's Handbook for Real-Time Analysis.* Kluwer Academic, Norwell, MA.

Kopetz, H., Damm, A., Koza, C., and Mulozzani, D. 1989. Distributed fault tolerant real-time systems: the Mars approach. *IEEE Micro.* 9(1):25–40.

Liu, J., Lin, K., Shih, W., Yu, A., Chung, J., and Zhao, W. 1991. Algorithms for scheduling imprecise computations. *IEEE Comput.* 24(5):58–68.

Malcolm, N. and Zhao, W. 1995. Hard real-time communication in multiple access networks. *Real-Time Syst.* 8(1):35–78.

Newman, P. 1994. Traffic management for ATM local area networks. *IEEE Commun. Mag.* 32(8):44–51.

Paul, C., Acharya, A., Black, B., and Strosnider, J. 1991. Reducing problem-solving variance to improve predictability. *Commun. ACM* 34(8).

Ramamritham, K. and Stankovic, J. 1984. Dynamic task scheduling in distributed hard real-time systems. *IEEE Software* 1(3):65–75.

Ramamritham, K., Stankovic, J., and Shiah, P. 1990. Efficient scheduling algorithms for real-time multi-processor systems. *IEEE Trans. Parallel Distributed Comput.* 1(2):184–194.

Sha, L., Rajkumar, R., and Lehoczky, J. 1990. Priority inheritance protocols: an approach to real-time synchronization. *IEEE Trans. Comput.* 39(9):1175–1185.

Shin, K. 1991. HARTS: a distributed real-time architecture. *IEEE Comput.* 24(5).

Sprunt, B., Sha, L., and Lehoczky, J. 1989. Aperiodic task scheduling for hard real-time systems. *Real-Time Syst.* 1:27–60.

Stankovic, J. 1988. Misconceptions about real-time computing: a serious problem for next generation systems. *IEEE Comput.* 21(10).

Stankovic, J. and Ramamritham, K. 1990. What is predictability for real-time systems. *Real-Time Syst. J.* 2:247–254.

Stankovic, J. and Ramamritham, K. 1991. The spring kernel: a new paradigm for real-time systems. *IEEE Software* 8(3):62–72.

Stankovic, J. and Ramamritham, K. 1995. A reflective architecture for real-time operating systems. In *Advances in Real-Time Systems*, pp. 487–507. Prentice–Hall, Englewood Cliffs, NJ.

Strosnider, J. and Marchok, T. 1989. Responsive, deterministic IEEE 802.5 token ring scheduling. *Real-Time Syst.* 1(2):133–158.

Tokuda, H., Nakajima, T., and Rao, P. 1990. Real-time MACH: towards a predictable real-time system. *Proc. USENIX MACH Workshop.*

Vytopil, J. 1993. *Formal Techniques in Real-Time and Fault Tolerant Systems.* Kluwer Academic, Boston, MA.

Wright, M., Green, M., Fiegl, G., and Cross, P. 1986. An expert system for real-time control. *IEEE Software* 3(2):16–24.

Further Information

Tutorial Texts

A good introduction to real-time systems can be found in *Hard Real-Time Systems* by John A. Stankovic and Krithi Ramamritham. More recent research results in the field can be found in a follow-on text entitled *Advances in Real-Time Systems* by the same authors. Yann Hang Lee and C. M. Krishna have edited a tutorial text entitled *Readings in Real-Time Systems* where papers on key real-time concepts are presented. Krishna Kavi has also produced a tutorial text on real-time systems entitled *Real-Time Systems,*

Abstractions, Languages, and Design Methodologies. This latter book approaches real-time systems from the design point of view, whereas the others mentioned take a systems implementation perspective. All of these texts are published by the IEEE Computer Society Press.

Other Information

Proceedings of the Real-Time Systems Symposium are published annually by the IEEE Computer Society. The conference is held each December and research papers are presented.

A new annual IEEE Computer Society Symposium was started in 1995 titled the Real-Time Technology and Applications Symposium. This symposium focuses on the interaction between industry and academia; a proceedings is published.

An archival journal entitled *Real-Time Systems* is published six times a year by Kluwer Academic Publishers. Kluwer also publishes an international book series on real-time computing. There are approximately 20 volumes in this series. For subscription information for the journal or for information on the book series contact: Kluwer Academic Publishers, 101 Philip Drive, Assinippi Park, Norwell, MA 02061.

Embedded Systems Programming is published monthly by Miller Freeman Inc., 600 Harrison St., San Francisco, CA 94107.

Embedded Systems Conference East and Embedded Systems Conference West are each held once per year. These conferences include product exhibitions and technical instruction.

79

Process Synchronization and Interprocess Communication

Craig E. Wills
Worcester Polytechnic Institute

79.1 Introduction

Process **synchronization** (also referred to as process coordination) is a fundamental problem in operating system design and implementation. It is a situation when two or more processes coordinate their activities based on a condition. An example is when one process must wait for another process to place a value in a buffer before the first process can proceed. A specific problem of synchronization is **mutual exclusion**, which requires that two or more concurrent activities do not simultaneously access a shared resource. This resource may be shared data among a set of processes where the instructions that access these shared data form a **critical region** (also referred to as a critical section). A solution to the mutual exclusion problem guarantees that among the set of processes only one process is executing in the critical region at a time.

Processes involved in synchronization are indirectly aware of each other by waiting on a condition that is set by another process. Processes can also communicate directly with each other through **interprocess communication (IPC)**. IPC causes communication to be sent between two (or more) processes. A common form of IPC is message passing.

The origins of process synchronization and IPC are work in concurrent program control by people such as Dijkstra, Hoare, and Brinch Hansen. Dijkstra described and presented a solution to the mutual exclusion problem [Dijkstra 1965] and proposed other fundamental synchronization problems and solutions such as the dining philosophers problem [Dijkstra 1971] and semaphores [Dijkstra 1968]. Brinch Hansen [1972] and Hoare [1972] suggested the concept of a critical region. Brinch Hansen published a classic text with many examples of concurrency in operating systems [Brinch Hansen 1973]. Hoare [1974] provided a complete description of monitors following work by Brinch Hansen [1973].

Modern work in the area is being driven by the development of multithreaded, message-based operating systems executing on multiprocessors. Many of the primitives used for synchronizing processes also work for synchronizing threads. Investigation is being done on mechanisms that work in a multiprocessor

0-8493-2909-4/97/$0.00+$.50

environment and extend to a distributed one. Concurrent programming using synchronization and IPC in a distributed environment, discussed in Chapters 85 and 98, is more complex than the mechanisms described in this chapter because of additional complications. In a distributed environment, there may be a failure by some processes participating in synchronization while others continue, or it is possible for messages to be lost or delayed in an IPC mechanism.

The remainder of this chapter discusses the underlying principles and practices commonly used for synchronization and IPC. The Underlying Principles section identifies fundamental problems needing solutions and issues that arise in considering various solutions. The Best Practices section discusses specific solutions to synchronization and IPC problems and discusses their relative merits in terms of these issues. The chapter concludes with a summary, a glossary of terms that have been defined, references, and sources of further information.

79.2 Underlying Principles

Process synchronization and IPC arose from the need of coordinating concurrent activities in a multiprogrammed operating system. This section defines and illustrates the fundamental synchronization and IPC problems and characterizes the issues on which to compare the solutions.

Synchronization Problems

A fundamental synchronization problem is mutual exclusion, described by Dijkstra [1965]. Lamport [1986] provides a formal treatment of the problem. In this problem, multiple processes wish to coordinate so that only one process is in its critical region of code at any one time. During this critical region, each process accesses a shared resource such as a variable or table in memory. The use of a mutual exclusion primitive for access to a shared resource is illustrated in Fig. 79.1 where two processes have been created that each access a shared global variable through the routine **IncrementX()**.

The routines **BeginRegion()** and **EndRegion()** define a critical region ensuring that **IncrementX()** is not executed simultaneously by both processes. With these routines, the final value of **x** is always two, although the execution order of the two processes is not defined.

To illustrate the need for mutual exclusion, consider the same example without the **BeginRegion()** and **EndRegion()** routines. In this case, the execution order of the statements for each process is time dependent. The final value of **x** may be two if **IncrementX()** is executed to completion for each process, or the value of **x** may be one if the execution of **IncrementX()** is interleaved for the two processes. This example illustrates a **race condition**, where multiple processes access and manipulate the same data with the outcome dependent on the relative timing of these processes. The use of a critical region avoids a race condition.

Many solutions have been proposed for the implementation of these two primitives; the most well known of which are given in the following section. Solutions to the mutual exclusion synchronization problem must meet a number of requirements, which were first set forth in Dijkstra [1965] and are summarized in Stallings [1995]. These requirements are as follows:

- Mutual exclusion must be enforced so that at most one process is in its critical region at any point in time.
- A process must spend a finite amount of time in its critical region.
- The solution must make no assumptions about the relative speeds of the processes or the number of processes.
- A process stopped outside of its critical region must not lead to blocking of other processes.
- A process requesting to enter a critical region held by no other process must be permitted to enter without delay.
- A process requesting to enter a critical region must be granted access within a finite amount of time.

```
int x = 0; /* global shared variable */

main()
{
    CreateProcess(ProcessA);
    CreateProcess(ProcessB);
    /* wait until done */
}

ProcessA()
{
    IncrementX();
    printf("X = %d\n", x);
}

ProcessB()
{
    IncrementX();
    printf("X = %d\n", x);
}

IncrementX()
{
    int Temp;  /* local variable */

    BeginRegion(); /* enter critical region */
    Temp = x;
    Temp = Temp + 1;
    x = Temp;
    EndRegion();   /* exit critical region */
}
```

FIGURE 79.1   Shared variable access handled as a critical region.

Another fundamental synchronization problem is the producer/consumer problem. In this problem, one process produces data to be consumed by another process. Figure 79.2 shows one form of this problem where a *producer* process continually increments a shared global variable and a *consumer* process continually prints out the shared variable. This variable is a fixed-size buffer between the two processes, and hence this specific problem is called the bounded-buffer producer/consumer problem.

The ideal of this example is for the consumer to print each value produced. However, the processes are not synchronized and the output generated is timing dependent. The number 0 is printed 2000 times if the consumer process executes before the producer begins. At the other extreme, the number 2000 is printed 2000 times if the producer process executes before the consumer begins. In general, increasing values of **n** are printed with some values printed many times and others not at all. This example illustrates the need for the producer and consumer to synchronize with each other.

The producer/consumer problem is a specific type of synchronization that is needed between two processes. In general, many types of synchronization between processes can be expressed with the *synchronization graph* in Fig. 79.3. A synchronization graph is a directed graph showing the relative execution

```
int n = 0; /* shared by all processes */

main()
{
    int produce(), consume();

    CreateProcess(produce);
    CreateProcess(consume);
    /* wait until done */
}

produce() /* "produce" values of n */
{
    int i;

    for (i=0; i<2000; i++)
        n++;   /* increment n by 1 */
}

consume() /* "consume" and print values of n */
{
    int i;
    for (i=0; i<2000; i++)
        printf("n is %d\n", n); /* print value of n */
}
```

FIGURE 79.2   Example of producer/consumer synchronization problem.

order for a set of actions (code segments). In the example, actions B and C execute after action A completes, with action D executing after both B and C are complete. Solutions to the synchronization problem need to allow for implementation of this type of problem.

Synchronization Issues

There are many approaches available for solving mutual exclusion and synchronization problems. There are a number of issues concerning the implementation of solutions to these problems, with the primary ones described in the following:

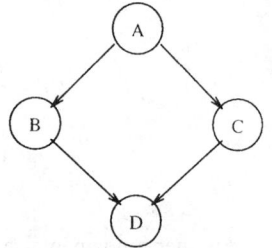

FIGURE 79.3   Synchronization graph for four actions.

- *Processor and store synchronicity:* Solutions to synchronization problems that are processor synchronous require that processors can be made uninterruptible. These solutions work for uniprocessor machines but not for multiprocessors where disabling the interrupts on one processor does not affect another. Solutions to synchronization problems that are store synchronous, which assumes individual references to main memory are atomic, are applicable to both uni- and multiprocessor machines.

- *Busy waiting:* Another issue in synchronization solutions is the consumption of CPU resources while a process is waiting for a condition to occur. Some solutions require **busy waiting**, the continued polling of a condition variable. These solutions are less efficient, particularly on a uniprocessor where busy waiting continues until the time slice of the waiting process is complete.

- *Programmer errors:* Another issue is the potential programming errors inherent in using a particular synchronization approach. Ordering of synchronization primitives is necessary for correct solutions with some primitives. Other primitives for synchronization have been specifically designed to minimize the possibility of such a programmer error.

- *Starvation:* It is possible for synchronization solutions to lead to the condition of **starvation**. Starvation occurs when a process is indefinitely denied access to a resource while other processes are granted access to the resource. Starvation is an issue in mutual exclusion if it is possible for one process to be indefinitely denied access to its critical region while access is granted to other processes.

- *Deadlock:* Another fundamental problem confronted by solutions is **deadlock**. Deadlock is the condition when a set of processes using shared resources or communicating with each other are permanently blocked. Three necessary, but not sufficient, conditions must exist for deadlock to occur among a set of processes sharing resources:
  1. *Mutual exclusion:* A resource may be used by only one process at a time.
  2. *Hold and wait:* Processes holding resources can request new resources.
  3. *No preemption:* A resource given to a process cannot be taken back.
  For deadlock to actually take place, a fourth condition must also exist:
  4. *Circular wait:* A set of processes are in a circular wait; that is, there is a set of processes $\{p_0, p_1, \ldots, p_n\}$ where p_i is waiting for a resource held by p_{i+1}, and p_n is waiting for a resource held by p_0.

When combined with the three necessary conditions, the presence of the circular wait condition indicates that a deadlock has occurred and that no means exists for breaking the deadlock. When designing solutions to synchronization problems such as mutual exclusion, preventing deadlock involves disallowing one or more of the preceding conditions. Specific solutions to the deadlock problem are given in the Best Practices section.

Interprocess Communication Problems

Interprocess communication problems generally involve direct communication between two or more processes, in contrast to synchronization where processes communicate indirectly by waiting on or setting a condition. Thus, communication is indirect between the processes when using synchronization, whereas IPC mechanisms generally communicate by passing messages directly between processes. The primitives for handling messages are

- `send(pid, &message)`
- `pid = receive(&message)`

where `send()` sends the given message to a specific destination process and `receive()` receives the last message sent to it returning the process identifier (id) of the sender.

Message passing is commonly used in operating systems and user applications for communicating between processes. In a message-based operating system where server processes perform many of the functions associated with an operating system, message passing is used to pass requests to the appropriate server and to receive replies, as is shown in the example of Fig. 79.4.

Interprocess Communication Issues

As described in the previous example, IPC mechanisms typically involve the exchange of messages from one process to another. There are a number of issues concerning the implementation of a message passing mechanism, with the primary ones described in the following. Implementations of IPC mechanisms, which address these issues, are described in the Best Practices section.

```
int serverpid;              /* well-known pid of the server */

Server()                    /* server process code */
{
    int pid;                /* process id of the requesting process */
    Message request_msg, reply_msg;

    while (TRUE) {
        pid = receive(&request_msg);        /* receive message */
        /* handle request and build reply message */
        send(pid, &reply_msg);                      /* send the reply */
    }
}

int ServiceRequest(args)        /* service request function with arguments */
{
    Message request_msg, reply_msg;

    /* load args into request_msg */
    send(serverpid, &request_msg);
    (void)receive(&reply_msg);
    return(reply_msg.status);
}
```

FIGURE 79.4   Service request for a message-based operating system.

- *Direct vs. indirect communication:* A key issue in a message passing system is how messages are addressed to their recipient. In a direct message passing scheme, the delivery address is a process id. This approach was illustrated in Fig. 79.4. In an indirect scheme, the address is an intermediate repository that is a drop-off point for the message for later pick up by the recipient.
- *Buffering of data:* Another issue is whether the message passing mechanism allows messages to be buffered if the receiving process is currently not ready to receive a message. If buffering is allowed, then another issue is the size of this buffer.
- *Blocking vs. nonblocking operations:* Related to the issue of buffering is the semantics of the **send()** operation when the message cannot be delivered. The operation can either block until the message can be delivered or the operation can immediately return with an error if blocking would occur. Similarly, the **receive()** operation can be defined to either block and wait for message delivery if no message is available or not block and return an error.
- *Fixed- or variable-sized messages:* Fixed-size messages allow for easier implementation but may require more work for the programmer. In contrast, variable-sized messages are easier to program but require more work to implement.
- *Synchronous vs. asynchronous reception:* Message passing mechanisms allow more process control if messages are received only when the **receive()** operation is invoked. However, some mechanisms allow for communication to be received when it arrives through an asynchronous approach using message handlers.

79.3 Best Practices

This section discusses specific solutions for synchronization and IPC. Each solution contains an example of its use, its relative merits, and problems for which it is useful. More examples of synchronization mechanisms can be found in operating system texts [Silberschatz and Galvin 1994, Stallings 1995, Tanenbaum 1992, Finkel 1988]. Comer [1984] and Tanenbaum [1987] present code for actual operating systems to show how synchronization and IPC mechanisms can be implemented. More concurrent programming examples can be found in Ben-Ari [1982, 1990], Raynal [1986], and Brinch Hansen [1973]. Andrews and Schneider [1983] provides a survey of synchronization and IPC techniques. The chapter concludes with more classic synchronization problems and specific solutions to the deadlock problem.

Synchronization Mechanisms

The mechanisms for synchronization are divided into four types based on their level of implementation and support: software only, hardware support, operating system support, and language support. In addition, hybrid solutions exist that combine more than one of these approaches. Each approach shows a solution to the mutual exclusion problem along with other applicable synchronization problems and discusses the relevant synchronization issues.

Software Solutions

Software-based synchronization solutions require only that multiple processes can access shared global variables. The solutions use these variables to control access to the critical region. Dekker was the first to devise a software solution that correctly handles the mutual exclusion problem among a set of processes. A discussion of this solution is given in Dijkstra [1965]. Peterson [1981] provided a simpler solution of the same problem, which is given in Fig. 79.5 for two processes in terms of the **BeginRegion()** and **EndRegion()** routines.

This solution works correctly on any hardware (uni- or multiprocessor) in which references to main memory are atomic. The main disadvantage is that it requires busy waiting, continued polling of a status variable, before gaining access to the critical region if another process is already in the critical region. The solution can generalize to more processes, but better solutions are available.

Solutions Using Hardware Support

Not all problems of synchronization, particularly ones in the operating system itself, can be handled solely with software-based solutions. As in other aspects of computing, hardware support can not only make the task easier but is also necessary for some levels of synchronization within the operating system.

```
int turn;                         /* whose turn is it? */
int flag[2];                      /* want the mutex? Initially FALSE */

BeginRegion(int pid)              /* pid is 0 or 1 */
{
    int other;                    /* pid of other process */

    other = 1 - pid;             /* the opposite of pid */
    flag[pid] = TRUE;            /* express interest in mutex */
    turn = pid;                   /* set flag */
    while ((turn == pid) && (flag[other] == TRUE))
        ;                         /* busy wait */
}

EndRegion(int pid)
{
    flag[pid] = FALSE;      /* drop interest in mutex */
}
```

FIGURE 79.5 Peterson's solution of mutual exclusion for two processes.

One of the simplest ways to enforce mutual exclusion is to disable hardware interrupts at the start of the critical region, thus ensuring that the process does not give up the CPU (through a context switch) before completing the critical region. When the critical region is done, the process re-enables interrupts. The **BeginRegion()** and **EndRegion()** routines with this approach are shown in Fig. 79.6.

This approach is fast and is used for manipulation of shared operating system data structures on a uniprocessor but in general has many disadvantages for general purpose use:

```
BeginRegion()
{
    DisableInterrupts();
}

EndRegion()
{
    EnableInterrupts();
}
```

FIGURE 79.6 Mutual exclusion by disabling/enabling hardware interrupts.

- Programmers must be careful not to disable interrupts for too long; devices that raise interrupts need to be serviced.
- The programmer must be careful about nesting. Activities that disable interrupts must restore them to their previous settings. In particular, if interrupts are already disabled before entering a critical region, they must remain disabled after leaving the critical region. Code in one critical region may call a routine that executes a different critical region.
- Disabling interrupts prevents all other activities, even though many may never execute the same critical region. Disabling interrupts is like using a sledge hammer; it is a powerful tool but bigger than is needed for most jobs.
- The technique is ineffective on multiprocessor architectures, where disabling interrupts on one processor still allows other processes to run on other processors.

Rather than perform mutual exclusion by controlling interrupts, a common approach is to use special instructions provided by the hardware to implement mutual exclusion. One such instruction is **Test_and_Set**, which is defined by the procedure of Fig. 79.7. It returns the previous value of a

```
int mutex = FALSE;                      /* global variable for mutex */

int Test_and_Set(int *pVar, int value)  /* atomic machine instruction */
{
    int temp;

    temp = *pVar;
    *pVar = value;
    return(temp);
}

BeginRegion()          /* Loop until safe to enter */
{
    while (Test_and_Set(&mutex, TRUE)) ;
        ; /* Loop until return value is FALSE */
}

EndRegion()
{
    mutex = FALSE;
}
```

FIGURE 79.7 Mutual exclusion using the test-and-set machine instruction.

target variable and sets the target to the given value. Most importantly, this instruction is performed in an atomic manner so a context switch cannot occur in the middle of it. In addition, operations performed on two separate processors of a multiprocessor are guaranteed to occur in sequential order because of store synchronicity. Figure 79.7 shows how the **BeginRegion()** and **EndRegion()** primitives are implemented using this instruction. The variable **mutex** is also referred to as a lock variable, and this approach to mutual exclusion is called a **spin lock** because a process spins in an infinite loop waiting for the lock. When the **mutex** variable is set to **FALSE** by a process exiting from its critical region, another process is allowed to gain access to its critical region.

Solutions using other machine instructions are also available. Another common instruction is **EXCH**, which swaps the contents of two memory locations in an atomic fashion. This instruction also can be used to implement mutual exclusion. The advantages of these machine-instruction approaches are their simplicity and the fact they work for any number of processes in either a uni- or a multiprocessor environment. Through the use of more than one mutex variable, multiple critical regions can be easily created.

The primary disadvantage of this approach is its use of busy waiting, thus wasting CPU resources. The use of busy waiting also allows for process starvation if multiple processes are contending for a critical region. Finally, deadlock is possible on a uniprocessor machine if a lower priority process gets interrupted in the middle of its critical region and then a higher priority process tries to gain access to the same critical region. The higher priority process busy waits forever because the lower priority process never runs.

Operating System Support

All of the synchronization approaches shown thus far can be implemented with the bare features of the hardware. The approaches also cause busy waiting if another process already is in its critical region. **Semaphores**, an important synchronization primitive, can be constructed by adding process coordination support to the operating system. Semaphores are data structures consisting of an identifier, a counter, and a queue; where processes waiting on a semaphore are blocked and placed on the queue, processes signaling a semaphore may unblock and remove a process from the queue, and the counter maintains a count of waiting processes.

The concept of a semaphore was first introduced by Dijkstra [1968]. Dijkstra defined two atomic semaphore operations: wait and signal, which he termed the P-operation (for wait; from the Dutch word proberen, to test) and the V-operation (for signal; from the Dutch word verhogen, to increment).

A restricted version of a semaphore, called a *binary semaphore*, limits the value of the counter to 0 and 1. However, the more general case is to use a *counting semaphore*, which has the following properties concerning the counter:

- A nonnegative count always means that the queue is empty.
- A count of negative **n** indicates that the queue contains **n** waiting processes.
- A count of positive **n** indicates that **n** resources are available and **n** requests can be granted without delay.

There are four basic operations defined for creating, deleting, waiting on, and signaling a semaphore.

1. **semid = screate(val)**: Create a semaphore with the given initial value for the counter.
2. **sdelete(semid)**: Delete a semaphore.
3. **swait(semid)**: Wait on a semaphore. Decrement the semaphore counter. If the counter is negative then suspend execution of the process and place it in the semaphore queue.
4. **ssignal(semid)**: Signal a semaphore. Increment the semaphore counter. Make the first process in the semaphore queue ready for execution.

Given these operations, a simple solution to the mutual exclusion problem is shown in Fig. 79.8. Unlike previous solutions, a process waiting for its critical region does not busy wait. While it is waiting, it is in a suspended state, allowing the CPU to perform other activities. Because semaphores are provided by the operating system, they work correctly on either uni- or multiprocessor machines.

```
int semid;

Initialization()
{
    semid = screate(1);    /* initialize the semaphore count to 1 */
}

BeginRegion()
{
    swait(semid);
}

EndRegion()
{
    ssignal(semid);
}
```

FIGURE 79.8   Mutual exclusion using semaphores.

Semaphores also provide a mechanism to solve the bounded-buffer producer/consumer problem, which was introduced in the example of Fig. 79.2. A solution for this problem using semaphores is shown in Fig. 79.9. This solution ensures that the consumer process prints all integer values from 0 to 1999 by alternating between the producer and consumer. The value 0 is printed first because of the way in which the semaphores are initialized.

Although semaphores provide straightforward solutions to both of these problems, semaphores themselves must be implemented in the operating system with lower level primitives such as the **Test_and_Set** machine instruction. Semaphores work well for processes, which are allowed to block, but must not be used in code that cannot block such as interrupt service routines.

Language Constructs

The solutions presented thus far to the mutual exclusion problem each require the programmer to take explicit action to ensure mutual exclusion. To guarantee mutual exclusion some programming languages provide constructs to implicitly guarantee mutual exclusion. One such construct is a **monitor** [Hoare 1974], which permits only one process to be executing in a monitor at a time.

Monitors are a programming language construct, similar to abstract data types in that the programmer defines a set of data types and procedures that can manipulate the data, procedures can be exported to other modules, and the system invokes an initialization routine before execution begins. Monitors differ in that they support guard procedures. When a process invokes a guard procedure, its execution is delayed until no other processes are executing a guard procedure within the monitor. Figure 79.10 revisits the mutual exclusion problem of Fig. 79.1 using a monitor solution written in pidgin C (the C language itself does not support monitors). The definition of a monitor guarantees that the variable **x** is not read and incremented simultaneously.

Monitors are a programming language construct that must be implemented using a lower level facility provided by the hardware or operating system, such as semaphores. Although monitors provide mutual exclusion, they need additional primitives to provide synchronization. To do so, monitors are defined to have condition variables, which are waited on and signaled similar to semaphores. These primitives allow other synchronization problems such as the producer/consumer problem to be implemented with monitors.

Hybrid Solutions

Modern operating systems have migrated from monolithic systems written for a uniprocessor in which critical regions were used to access shared data structures. These critical regions were guarded by setting

```
int n = 0; /* shared by all processes */
int prodid, consid; /* semaphores */

main()
{
    int produce(), consume();

    consid = screate(0);        /* initial count of consumer is 0 */
    prodid = screate(1);        /* initial count of producer is 1 */
    CreateProcess(produce);
    CreateProcess(consume);
    /* wait until done */
}

produce() /* "produce" values of n */
{
    int i;

    for (i=0; i<2000; i++) {
        swait(consid);
        n++;   /* increment n by 1 */
        ssignal(prodid);
    }
}

consume() /* "consume" and print values of n */
{
    int i;
    for (i=0; i<2000; i++) {
        swait(prodid);
        printf("n is %d\n", n); /* print value of n */
        ssignal(consid);
    }
}
```

FIGURE 79.9   Bounded-buffer producer/consumer problem with semaphores.

the interrupt level appropriately. Current operating systems must not only support multiprocessors but also provide real-time capabilities, thus leading to multithreaded designs. In these designs, the use of interrupts to control access to shared data does not work. Rather, these systems have moved to solutions using complex locks, which combine the use of spin locks with the semantics of semaphores.

As one example, the Solaris 2 operating system [Eykholt et al. 1992] uses adaptive mutex locks, a type of complex lock, to protect access to shared data among a set of threads. The adaptive lock starts out executing like a standard spin lock. If the lock is currently free, then the issuing thread immediately obtains the lock. However, if the lock is in use, then the operating system checks the status of the thread holding the lock. If this thread is currently in the run state (as could be the case on a multiprocessor), then the issuing thread continues in a spin lock waiting for what is expected to be a short time for the lock to be released. If the holding thread is not in the run state (as would always be the case on a uniprocessor),

```
MONITOR incr;
EXPORT IncrementX, ReadX;

int x;      /* variable shared by monitor procedures */

GUARD PROCEDURE IncrementX()
{
    int Temp; /* local variable */

    Temp = x;
    Temp = Temp + 1;
    x = temp;
}

GUARD PROCEDURE ReadX()
{
    return(x);
}

Initialization()   /* initialization routine */
{
    x = 0;
}
```

FIGURE 79.10   Mutual exclusion using monitors.

then the issuing thread is suspended until the lock is released. The rationale is to use a spin lock if the wait for the lock is expected to be short and to actually suspend the thread if the wait is expected to be longer.

This hybrid approach tries to minimize overhead and maximize performance. If the size of a critical region is large (hundreds of instructions), then the adaptive mutex lock is less desirable compared to a lock that simply causes a thread to suspend when the lock is not available.

Another type of complex lock used in multithreaded operating systems is a read/write lock. These locks allow either a single writer or multiple readers to simultaneously hold the lock, thus increasing the parallelism when reading of a shared data structure predominates. Writers must wait until all readers have released the lock before obtaining the lock, whereas readers are granted immediate access to the lock in the absence of a writer. To prevent starvation of a writer, all read requests after a write request has been issued are queued until the write request has been satisfied. Starvation of readers in the face of multiple writers is similarly avoided.

Other Solutions

Many other synchronization primitives have been proposed but in general can be expressed in terms of the solutions already given. A sampling of these primitives are critical regions [Hoare 1972, Brinch Hansen 1972], serializers [Atkinson and Hewitt 1979], path expressions [Campbell and Habermann 1974], and event counts and sequencers [Reed and Kanodia 1979].

Interprocess Communication Mechanisms

As with synchronization, a variety of IPC mechanisms are available. The following discusses a number of these mechanisms and how they handle the IPC issues raised in the Underlying Principles section.

Direct Message Passing

The simplest form of message passing is to send messages directly from one process to another. An example of this approach is the low-level message passing facility used in the Xinu operating system [Comer 1984]. The primitive operations used are:

- **`send(pid, msg)`**
- **`msg = receive()`**

where **`send()`** sends a fixed, integer-sized message to a specific process and **`receive()`** returns the last message sent to it. A process can buffer only one message. If **`send()`** detects a message already buffered at the process then it returns immediately with an error, not delivering the message. If no message is buffered then **`send()`** buffers the message and readies the receiving process if it is waiting for a message. The **`receive()`** operation blocks if a message is not available.

Another direct message passing mechanism is implemented in Minix [Tanenbaum 1987]. The primitives for handling messages are:

- **`send(destpid, &message)`**
- **`receive(srcpid, &message)`**

where **`send()`** sends the given message to a specific destination process and **`receive()`** receives a message from a particular process. The source process for **`receive()`** can contain a wildcard value of **ANY**, indicating that messages from any process are accepted. Messages are a fixed size, but there is no buffering. Rather the **`send()`** and **`receive()`** operations *rendezvous* so that both operations block until the receiving process has actually copied the message from the sender. The use of rendezvous explicitly synchronizes the execution of the sending and receiving process.

Mailboxes/Ports

Rather than send directly to process, a more common approach is to define another operating system abstraction called a **mailbox** (also referred to as a port). Mailboxes are buffers that hold messages sent by one process to be received by another process. Thus, there is indirect communication between the two processes. The primitives for handling messages are

- **`send(mailbox, &message)`**
- **`receive(mailbox, &message)`**

where **`send()`** buffers the message in the given mailbox and **`receive()`** removes a message from the mailbox.

As an example, the Unix operating system provides ports to allow for intra- and intermachine communication between processes. Ports often represent well-known services where a server process binds to a port and client processes of the service send requests to the port. The port buffers communication sent to the buffer until it is read by the receiving process. The messages sent to the port can be of variable size.

Message passing can also be implemented by using shared memory and semaphores, illustrating the equivalence of synchronization and IPC primitives. Figure 79.11 shows message passing with shared memory and semaphores to implement a set of mailboxes. The mechanism uses fixed-size messages with each of four mailboxes containing eight message buffers. The **`send()`** operation blocks if there is no buffer space in the mailbox; similarly, the **`receive()`** operation blocks if there is no message available in the mailbox. The mutex semaphore is needed to guarantee that no more than one process tries to send to or receive from a mailbox at the same time. This semaphore would not be needed if only process could send to and receive from a mailbox. The mutex semaphore would also not be needed if a mailbox contained only one buffer slot. This is also an example of a bounded-buffer producer/consumer problem where sending processes produce messages and receiving processes consume them.

```
#define N 8                         /* number of msgs buffered in a mailbox */
#define M 4                         /* number of mailboxes */

Message mailboxes[M][N];            /* shared memory for mailboxes of messages */

int semidMsg[M];                    /* message available */
int semidSlot[M];                   /* slot available */
int semidMutex[M];                  /* controls access to critical region */

Initialization()
{
    int i;

    for (i = 0; I < M; i++) {
        semidMsg[i] = screate(0);   /* no messages are available */
        semidSlot[i] = screate(N);  /* N slots are available for messages */
        semidMutex[i] = screate(1); /* one process can enter region */
        /* initialize indices for inserting/deleting in mailboxes[i] */
    }
}

send(int m, Message *pmsg)          /* send message to mailbox m */
{
    swait(semidSlot[m]);            /* is a slot available */
    swait(semidMutex[m]);           /* enter critical region */
    addmessage(m, pmsg);            /* add msg to circular queue for mailbox m */
    ssignal(semidMutex[m]);         /* exit critical region */
    ssignal(semidMsg[m]);           /* signal message available */
}

receive(int m, Message *pmsg)       /* retrieve message from mailbox m */
{
    swait(semidMsg[m]);             /* is a message available */
    swait(semidMutex[m]);           /* enter critical region */
    removemessage(m, pmsg);         /* remove next msg from queue for mailbox m */
    ssignal(semidMutex[m]);         /* exit critical region */
    ssignal(semidSlot[m]);          /* signal slot available */
}
```

FIGURE 79.11   Message passing with semaphores and shared memory.

Pipes

A special case of IPC is the pipe abstraction available in the Unix operating system. A *pipe* is a unidirectional, stream communication abstraction. One process writes data to the write end of the pipe, and a second process reads data from the read end of the pipe. The pipe itself is a buffer between the two processes that causes the reader to block if no data are available and the writer to block if the buffer is full. As it implements a stream abstraction, there is no notion of fixed-size messages. A pipe is another example of a solution to the bounded-buffer producer/consumer problem where the writing process is a producer and the reading process is a consumer.

```
#define DATA "hello world"
#define BUFFSIZE 1024

int rgfd[2];              /* file descriptors of pipe ends */

main()
{
    char sbBuf[BUFFSIZE];

    pipe(rgfd);           /* create a pipe returning two file desciptors */
    if (fork()) {         /* parent, read data from pipe */
        close(rgfd[1]);   /* close write end */
        read(rgfd[0], sbBuf, BUFFSIZE);
        printf("Pipe contents: %s\n", sbBuf);
        close(rgfd[0]);
    }
    else {                /* child, write data to pipe */
        close(rgfd[0]);   /* close read end */
        write(rgfd[1], DATA, sizeof(DATA));  /* write data to pipe */
        close(rgfd[1]);
        exit(0);
    }
}
```

FIGURE 79.12　　Pipe example in the Unix operating system.

Figure 79.12 shows a simple example of the use of pipes in the Unix operating system. Pipes are typically requested and set up by a Unix command interpreter, but the example shows one process creating another process with **fork()** with a pipe between them. A string of characters is then written to and read from the pipe.

Software Interrupts

Software interrupts are a primitive form of IPC. They are similar to hardware interrupts in that when an interrupt of a process occurs, an interrupt handler routine corresponding to the type of interrupt is invoked. Interrupts are asynchronous so, when an interrupt is received, execution of the process stops and is restarted after the interrupt handling routine has been executed. Software interrupts are sent to a process using the process id of the process. Many interrupts are used for well-known functions such as when the user types the interrupt key, a child process completes, or an alarm scheduled by the process has expired.

Two routines are used to send and handle software interrupts:

- **SendInterrupt(pid, num):** An interrupt of type **num** is sent to process **pid**. In the Unix operating system this routine is **kill()**.
- **HandleInterrupt(num, handler):** This specifies that user supplied **handler** routine should be invoked when interrupt of type **num** occurs. Typical handlers are to ignore the interrupt, terminate the process, or execute a user supplied interrupt handler. In the Unix operating system this routine is **signal()**.

Figure 79.13 shows a sample program for the Unix operating system with software interrupts. It sets up two interrupt handlers for signals 1 and 2 and then goes into an infinite loop where it updates a counter

```
#include <signal.h>

int n;

main(int argc, char **argv)
{
    void InterruptHandler(), InitHandler();

    n = 0;

    signal(SIGINT, InterruptHandler); /* signal 2 */
    signal(SIGHUP, InitHandler);      /* signal 1 */
    while (1) {
        n++;
        sleep(1);                     /* sleep for one second */
    }
}

void InterruptHandler()
{
    printf("The current value of n is %d\n", n);
    exit(0);
}

void InitHandler()
{
    printf("Resetting the value of n to zero\n");
    n = 0;
}

% cc -o signalex signalex.c
% signalex
^C            (interrupt character)
The current value of n is 3
% signalex &
[1] 20822
% kill -1 20822
Resetting the value of n to zero
% kill -2 20822
The current value of n is 19
[1]    Done                   signalex
```

FIGURE 79.13 Software interrupt program and script in the Unix operating system.

and sleeps for 1 s. Figure 79.13 also shows a command line script with this program. Invoking the interrupt character from the command interpreter causes interrupt 2 to be sent to the process. The second invocation of the program causes it to be run in the background with a process id of 20822. The **kill** program is then used to send interrupts to the background process.

Classic Problems

Two classic synchronization problems—the critical region and bounded-buffer producer/consumer—have already been discussed. A slight variation of the producer/consumer problem is to use an unbounded buffer, in which case the producer never blocks because the buffer never fills. Many other classic synchronization problems have been proposed and solved. The following describes two such problems.

Readers/Writers Problem

The readers/writers problem occurs when multiple readers and writers want access to a shared object such as a database. The problem was introduced in Courtois et al. [1971]. In the problem, multiple readers are allowed to access the database simultaneously, but a writer must have exclusive access to the database before performing any updates for consistency. A practical example of this problem is an airline reservation system with many readers and an occasional update of the information.

Figure 79.14 shows a solution to this problem for multiple reader and writer processes with semaphores. The solution allows multiple readers access to the database at a time. A writer process can gain access only after all reader processes have relinquished the database. The solution gives priority to reader processes, who can gain access to the database even if a writer process is already requesting access to the database. Solutions giving priority to the writer processes can also be constructed.

Dining Philosophers Problem

The dining philosophers problem was proposed and solved by Dijkstra [1971]. It consists of five philosophers sitting at a round table. Philosophers each have a bowl of rice in front of them and there is a chopstick in between each bowl (alternately, the problem is described using plates of spaghetti and forks). The problem is illustrated in Fig. 79.15 with the philosophers' bowls labeled A–E and the chopsticks 1–5. These philosophers have two functions in life: think, requiring no interaction with colleagues, and eat, requiring the philosopher to pick up the chopstick on the left and right.

This classic synchronization problem has potential for both deadlock and starvation (literally!). The straightforward solution for a philosopher to eat is to first pick up the left chopstick and then the right chopstick. However, if all philosophers pick up their left chopstick at the same time, they will all deadlock when they go to pick up their right chopstick. A simple modification to this approach is for the philosophers to put down the left chopstick if the right chopstick is not available, wait for some time, and try again. However, there is still a chance that the philosophers will operate in lock step and no philosophers will acquire both chopsticks. This condition is called **livelock** and occurs when attempts by two or more processes (philosophers) to acquire a resource (the left and right chopsticks) run indefinitely without any process succeeding. The dining philosophers problem will be used as a guide in the following discussion on deadlock and starvation.

Deadlock and Starvation

Deadlock occurs when a set of processes using shared resources are permanently blocked trying to gain access to those resources. Classic papers on this topic are Coffman et al. [1971] and Holt [1972]. Zobel and Koch [1988] contains a more up-to-date annotated bibliography on the subject. Deadlock can occur with synchronization such as when two processes each need to gain access to two separate critical regions for execution. If one process gains access to the first critical region and the other process gains access to the second critical region, then these two processes will be in deadlock when they attempt to acquire the other needed critical region.

```
int readercount = 0;          /* number of readers currently reading */
int readermutex;              /* semaphore mutex for reader count */
int dbaccess;                 /* semaphore to control access to database */

main()
{
    readermutex = screate(1);  /* mutex for reader count */
    dbaccess = screate(1);     /* mutex for database */
    CreateProcess(reader);     /* create a reader process */
    CreateProcess(writer);     /* create a writer process */
}

reader()
{
    while (TRUE) {
        swait(readermutex);        /* get access to readercount */
        readercount++;             /* increment count */
        if (readercount == 1)      /* if first reader ... */
            swait(dbaccess);       /* gain access to database */
        ssignal(readermutex);      /* done with count */
        /* read database */
        swait(readermutex);        /* get access to readercount */
        readercount--;             /* decrement count */
        if (readercount == 0)      /* if last reader ... */
            ssignal(dbaccess);     /* relinquish access to database */
        ssignal(readermutex);      /* done with count */
        /* use data read */
    }
}

writer()
{
    while (TRUE) {
        /* generate new data */
        swait(dbaccess);           /* gain access to database */
        /* update database */
        ssignal(dbaccess);         /* relinquish access to database */
    }
}
```

FIGURE 79.14 Solution to the readers/writers problem with semaphores.

There are four principles used for dealing with the issue of deadlock in operating systems. These principles are prevention, detection, avoidance, and recovery and are summarized in Isloor and Marsland [1980]. Solutions to the deadlock problem exemplifying these principles are described in the following. Each of these solutions prevents deadlock by precluding one or more of the four conditions for deadlock given in the Underlying Principles section. The solutions are characterized as being conservative or liberal depending on the degree of concurrency they allow.

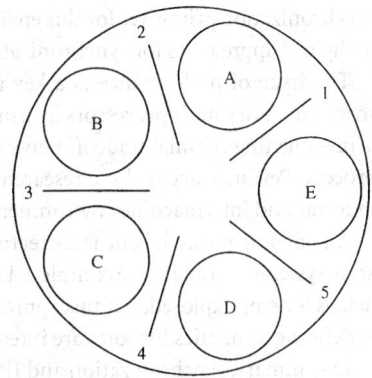

FIGURE 79.15 Illustration of the dining philosophers problem.

The most liberal solution is to allocate requested resources to processes if the resources are available. This approach was given as an initial solution when introducing the dining philosophers problem. As was shown, it can lead to deadlock and requires that a deadlock detection process be periodically invoked. Deadlock detection involves detecting the circular wait condition through any algorithm for detecting cycles in directed graphs. Once deadlock occurs, a deadlock recovery method must be invoked either to take away resources from a victim process or to terminate the process altogether. In the dining philosophers problem, a recovery method would be to take away a chopstick from one of the philosophers.

At the other extreme, the most conservative deadlock prevention solution is *serialization*. This deadlock prevention approach guarantees no deadlock by allowing only one process to acquire resources at a time. In the dining philosophers problem, this approach means that only one philosopher is allowed to eat at any time. The approach prevents any concurrency and can lead to starvation if a philosopher is constantly passed over in obtaining chopsticks.

Another solution that allows more concurrency but still prevents deadlock is *one-shot allocation*. This approach requires that a process obtain all of its resources at once. Using this approach, a dining philosopher must obtain both chopsticks at the same time. It avoids deadlock by preventing the hold and wait condition. This solution prevents deadlock but not starvation. It also requires that a process obtain all of its resources at the same time even if it does not currently need all of them.

A still more liberal solution that prevents deadlock is *hierarchical allocation*. This solution requires that all resources have a number associated with them as in Fig. 79.15. Hierarchical allocation prevents deadlock by requiring that processes can acquire only resources with a higher number than any resource it currently holds. Thus, in Fig. 79.15, each philosopher must first acquire the left chopstick and then the right except for Philosopher E, who must first acquire the right chopstick. This solution is still conservative and requires processes to acquire resources not necessarily in the order they are needed but in the order resources are numbered. It avoids deadlock by preventing the circular wait condition.

As opposed to deadlock prevention, deadlock avoidance policies do not prevent deadlock a priori but monitor the allocation requests as they are made so as not to allow deadlock to occur. A solution using this approach is the *bankers algorithm* (also called the advanced claim algorithm) introduced in Dijkstra [1968]. This solution is the most liberal that avoids deadlock but requires that each process know the maximum number from a class of resources that it may request at any time during execution. The algorithm performs as a banker giving out loans of money in that any resource requests are granted only if they

1. Do not exceed the total number of resources in a class
2. Do not exceed the maximum number of resources for that process
3. Lead to a *safe state* where a sequence of resource deallocations and allocations can allow all processes to complete without deadlock

79.4 Research Issues and Summary

The research issues in synchronization and IPC correspond to the movement toward multithreaded, message-based operating systems running in a multiprocessor environment. The adoption of traditional synchronization primitives for this environment and the exploration of better primitives is leading to work on hybrid approaches for synchronization. This work is also extending to distributed environments.

The issue of performance is a key area as researchers seek to minimize the cost of busy waiting in shared memory multiprocessors by using virtual memory to reduce memory access costs. Other research is investigating optimal tradeoff between the use of busy waiting for a lock vs. suspending the thread or process. Performance is also a research issue for IPC mechanisms in client/server systems as they execute in intra- and intermachine environments.

An ongoing research issue is correctness, particularly as the synchronization problems of modern operating systems become more complex. Using language constructs to better encapsulate the synchronization details is being explored, but this approach can lead to tradeoffs with performance. Other work is looking at defining semantics for software interrupts in a multithreaded design.

In summary, synchronization and IPC are fundamental to multiprogrammed operating system design. Many primitives to solve fundamental problems such as mutual exclusion and producer/consumer exist ranging from software only approaches, to special hardware instructions, to primitives constructed by the operating system and programming languages. Many of the primitives are equivalent in terms of their semantics, and one can be implemented in terms of another. Examples include implementing monitors with semaphores or message passing with shared memory and semaphores. Modern operating systems are using hybrid approaches, which adaptively switch between techniques according to the run-time operating environment.

Defining Terms

Busy waiting: Situation in synchronization when a process continuously polls the status of a condition variable.

Critical region: Set of instructions for a process that access data shared with other processes. Only one process may execute its critical region at a time.

Deadlock: Condition when a set of processes using shared resources or communicating with each other are permanently blocked.

Interprocess communication (IPC): Communication between two (or more) processes directly aware of each other.

Livelock: Condition when attempts by two or more processes to acquire a resource run indefinitely without any process succeeding.

Mailbox: An operating system abstraction containing buffers to hold messages. Messages are sent to and received from the mailbox by processes.

Monitor: Programming language construct providing abstract data types and mutually exclusive access to a set of guard procedures.

Mutual exclusion: A synchronization problem requiring that two or more concurrent activities do not simultaneously access a shared resource.

Race condition: Situation where multiple processes access and manipulate shared data with the outcome dependent on the relative timing of these processes.

Semaphore: Synchronization primitive consisting of an identifier, a counter, and a queue where processes waiting on a semaphore are blocked and placed on the queue; processes signaling a semaphore may unblock and remove a process from the queue, and the counter maintains a count of waiting processes.

Spin lock: Mutual exclusion mechanism where a process spins in an infinite loop waiting for the value of a lock variable to indicate availability.

Starvation: Condition when a process is indefinitely denied access to a resource while other processes are granted access to the resource.

Synchronization: Situation when two or more processes coordinate their activities based upon a condition.

References

Andrews, G. R. and Schneider, F. B. 1983. Concepts and notations for concurrent programming. *Comput. Surv.* 15(1):3–43.

Atkinson, R. and Hewitt, C. 1979. Synchronization and proof techniques for serializers. *IEEE Trans. Software Eng.* 5(1):10–23.

Ben-Ari, M. 1982. *Principles of Concurrent Programming.* Prentice–Hall, Englewood Cliffs, NJ.

Ben-Ari, M. 1990. *Principles of Concurrent and Distributed Programming.* Prentice–Hall, Englewood Cliffs, NJ.

Brinch Hansen, P. 1972. Structured multiprogramming. *Commun. ACM* 15(7):574–578.

Brinch Hansen, P. 1973. *Operating Systems Principles.* Prentice–Hall, Englewood Cliffs, NJ.

Campbell, R. H. and Habermann, A. N. 1974. The specification of process synchronization by path expressions. In *Operating Systems*, E. Gelenbe and C. Kaiser, eds., pp. 89–102. Springer–Verlag, Berlin.

Coffman, E., Elphick, M., and Shoshani, A. 1971. System deadlocks. *ACM Comput. Surv.* 3(2):67–78.

Comer, D. 1984. *Operating System Design, the Xinu Approach.* Prentice–Hall, Englewood Cliffs, NJ.

Courtois, P., Heymans, F., and Parnas, D. L. 1971. Concurrent control with "readers" and "writers." *Commun. ACM* 14(10):667–668.

Dijkstra, E. W. 1965. Solution of a problem in concurrent programming control. *Commun. ACM* 8(9):569.

Dijkstra, E. W. 1968. Co-operating sequential processes. In *Programming Languages*, F. Genuys, ed., pp. 43–112. Academic Press, New York. Reprint of Technical Report EWD-123, Technological University, Eindhoven, The Netherlands (1965).

Dijkstra, E. W. 1971. Hierarchical ordering of sequential processes. *Acta Informatica* 2(1):115–138.

Eykholt, J. R., Kleiman, S. R., Barton, S., Faulkner, S., Shivalingiah, A., Smith, M., Stein, D., Voll, J., Weeks, M., and Williams, D. 1992. Beyond multiprocessing: multithreading the SunOS kernel. In *Proc. Summer USENIX Conf.* USENIX Association, pp. 11–18.

Finkel, R. A. 1988. *An Operating Systems VADE MECUM.* Prentice–Hall, Englewood Cliffs, NJ.

Hoare, C. A. R. 1972. Towards a theory of parallel programming. In *Operating Systems Techniques.* C. A. R. Hoare and R. H. Perrott, eds., pp. 61–71. Academic Press, New York.

Hoare, C. A. R. 1974. Monitors: an operating system structuring concept. *Commun. ACM* 17(10):549–557; Erratum 1975. *Commun. ACM* 18(2):95.

Holt, R. 1972. Some deadlock properties of computer systems. *ACM Comput. Surv.* 4(3):179–196.

Isloor, S. S. and Marsland, T. A. 1980. The deadlock problem: an overview. *IEEE Comput.* 13(9):58–78.

Lamport, L. 1986. The mutual exclusion problem. *J. ACM* 33(2):313–348.

Peterson, G. L. 1981. Myths about the mutual exclusion problem. *Inf. Process. Lett.* 12(3):115–116.

Raynal, M. 1986. *Algorithms for Mutual Exclusion.* Wiley, New York.

Reed, D. P. and Kanodia, R. K. 1979. Synchronization with eventcounts and sequencers. *Commun. ACM* 22(2):81–92.

Silberschatz, A. and Galvin, P. B. 1994. *Operating System Concepts*, 4th ed. Addison–Wesley, Reading, MA.

Stallings, W. 1995. *Operating Systems*, 2nd ed. Prentice–Hall, Englewood Cliffs, NJ.

Tanenbaum, A. 1987. *Operating Systems: Design and Implementation.* Prentice–Hall, Englewood Cliffs, NJ.

Tanenbaum, A. 1992. *Modern Operating Systems.* Prentice–Hall, Englewood Cliffs, NJ.

Zobel, D. and Koch, C. 1988. Resolution techniques and complexity results with deadlocks: a classifying and annotated bibliography. *Operating Syst. Rev.* 22(1):52–72.

Further Information

Many good text books on operating systems, such as those by Silberschatz and Galvin [1994], Stallings [1995], and Tanenbaum [1987] exist, which describe problems, issues, and solutions for synchronization and IPC. The book *Principles of Concurrent and Distributed Programming* by M. Ben-Ari [1990] contains a number of problems and worked out solutions for both concurrent and distributed programming. *Algorithms for Mutual Exclusion* by M. Raynal [1986] presents a comprehensive treatment of solutions for the mutual exclusion problem. "Concepts and Notations for Concurrent Programming" by Andrews and Schneider [1983] provides a survey of processes, synchronization, and interprocess communication.

The Association for Computing Machinery (ACM) Special Interest Group on Operating Systems (SIGOPS) publishes *Operating Systems Review* four times a year. This publication contains work on a variety of operating system topics including synchronization. This group also sponsors the biennial ACM Symposium on Operating Systems Principles that cover the latest developments in the field of operating systems. Its proceedings are published in an issue of *Operating Systems Review*. Another ACM publication, the *ACM Transactions on Computer Systems* is a good source for relevant work.

The USENIX Association sponsors a number of conferences on operating system-related topics. Two general technical conferences are sponsored each year, along with a number of other conferences on specific issues. The group also publishes the journal *Computing Systems*.

80

Virtual Memory

Peter J. Denning
George Mason University

80.1 Introduction

Virtual memory, long a standard feature of nearly every operating system and computer chip, is now invading the Internet through the **World Wide Web (WWW)**. Once the subject of intense controversy, it is now so ordinary that few people think much about it. That this has happened is one of the engineering triumphs of the computer age.

Virtual memory is the simulation of a storage space so large that software programmers and document authors do not need to rewrite their works when the internal structure of a program module, the capacity of a local memory, or the configuration of a network changes. The name, borrowed from optics, recalls the virtual images formed in mirrors and lenses: objects that are not there but behave as if they are. The story of virtual memory, from the Atlas Computer at the University of Manchester in the 1950s to the multicomputers and World Wide Web of the 1990s, is not simply a story of progress in automatic storage allocation; it is a fascinating story of machines helping programmers to protect information, reuse and share objects, and link software components.

80.2 History

From their beginnings in the 1940s, electronic computers had two-level storage systems. The **main memory** was then magnetic cores and is now random access memories **(RAMs)**; the secondary memory was then magnetic drums and is now disks. The processor [central processing unit **(CPU)**] could address only the main memory. A major part of a programmer's job was to devise a good way to divide a program into blocks and to schedule their moves between the levels. The blocks were called *segments* or *pages* and the movement operations *overlays* or *swaps*. This was a complex task even for small programs. The designers of the Atlas Computer at the University of Manchester invented virtual memory in the 1950s to eliminate two looming programming problems: (1) planning and scheduling data transfers between main and **secondary memory** and (2) recompiling programs for each change of size of main memory. They dreamt of automating it all.

0-8493-2909-4/97/$0.00+$.50
© 1997 by CRC Press, Inc.

In 1959, the Atlas computer system was the first working prototype of a virtual memory [Fotheringham 1961, Kilburn et al. 1962). Its designers called it a *one-level storage system*. At the heart of their idea was a radical innovation: a distinction between *address* and *memory location*. It led them to three inventions: (1) They built hardware that automatically translated each address generated by the processor to its current memory location. (2) They devised demand **paging**, an interrupt mechanism triggered by the address translator that moved a missing page of data into the main memory. (3) They built the first replacement algorithm, a procedure to detect and move the least useful pages back to secondary memory.

Despite the success of the Atlas memory system, the literature of 1961 records a spirited debate about the feasibility of automatic storage allocation in general-purpose computers. By that time, Cobol, Algol, Fortran, and Lisp had become the first widely used higher level programming languages. These languages made storage allocation harder because programs were larger, more portable, more modular, and their dynamics more dependent on their input data. Through the 1960s there were dozens of experimental studies that sought to either affirm or deny the hypothesis that operating systems could do a better job at storage allocation than any compiler or programmer [Denning 1970]. The matter was finally laid to rest—in favor of automatic storage allocation—by an extensive study of system performance by an IBM research team [Sayre 1969].

Convinced that virtual memory was the right way to go, the makers of major commercial computers adopted it in the 1960s; these included the IBM 360/67, CDC 7600, Burroughs 5500/6500, RCA Spectra/70, and Multics for the GE 645. By the mid-1970s the IBM 370, DEC VMS, DEC TENEX, and Unix had joined the crowd. These systems all used multiprogramming; their designers turned to virtual memory to solve not only the storage allocation problem but also the more critical memory protection problem. To their designers' dismay, and to the delight of virtual memory's critics, these systems all exhibited **thrashing**, a condition of near-total performance collapse when the multiprogrammed load was too high [Denning 1968]. That triggered a long line of experiments and models seeking to understand thrashing and to design effective load control systems. This was finally accomplished by the late 1970s; near-optimal **throughput** will result when the virtual memory guarantees each active process just enough space to hold its working set [Denning 1980].

Virtual memory attracted hardware designers as well as software designers. In 1965, Maurice Wilkes proposed the *slave memory*, a small high-speed store included in the processor to hold, close by, a small number of most recently used blocks of program code and data. Slave memory used address translation, demand loading, and usage-based replacement. Wilkes said that, by eliminating many data transfers between processor and the main memory, slave memory would allow the system to run within a few percent of the full processor speed at a cost within a few percent of the main memory [Wilkes 1965]. The term *cache memory* replaced *slave memory* in 1968 when IBM introduced cache memory in its 360/85 machine. Cache memory is now a standard principle of computer architecture [Hennessey and Patterson 1990].

If it ended here, this story would already have guaranteed virtual memory a place in history. But the designers of the 1960s were no less inventive than those of the 1950s. Just as the designers of the 1950s sought a solution to the problem of storage allocation, the designers of the 1960s sought solutions to two new kinds of programming problems: (1) shareable, reusable, and recompilable program modules and (2) packages of procedures hiding the internal structure of classes of objects. The first of these led to the **segmented address space**, the second to the architecture that was first called **capability-based addressing** and later **object-oriented programming**.

In 1965 the designers of Multics at the Massachusetts Institute of Technology (MIT) sought systems to support large programs built from separately compiled, shareable modules linked together on demand [Dennis 1965, Organick 1972]. To them, virtual memory as a pure computational storage system was too restrictive; they held that **modular programming** would not become a reality as long as programmers had to bind together manually, by a linking loader or makefile program, the component files of an address space. Their innovation was to add a second dimension of addressing to the virtual address space, enabling it to span segment names as well as within segment addresses. A program could refer to a variable X within a module S by the two-part name (S, X); the symbols S and X were retained by the compiler and converted to the hardware addresses for S and X by the virtual memory on first reference (a *linkage* *fault*). The Multics virtual memory demonstrated sophisticated forms of sharing, reuse, access control, and protection. This innovation, however, did not find its way into general practice; programmers were content with one private, linear address space and a handful of open files. As will be discussed, the World Wide Web [Berners-Lee 1996] is beginning to change this: programs and documents contain *hypertext links:* symbolic pointers to other objects that are not linked until the program references them for the first time.

In 1966 Jack Dennis and Earl Van Horn published a prescient paper that initiated a new line of computer architectures: machines that help programmers create managers of classes of objects. They anticipated what is now called object-oriented programming. They were especially concerned that objects be freely reusable and shareable and, at the same time, be protected from internal access by anyone except their authorized managers. They proposed an extension of virtual memory that would map a process' local name for an object into an internal bit pattern called a *capability;* a capability contained a type indicator, an access code, and a unique name. Their proposal inspired others to build capability machines during the 1970s. These systems included the Plessey 250, IBM System 38, Cambridge CAP, Intel 432, SWARD, and Hydra. In these systems, capabilities were implemented as long addresses (e.g., 64 b), which the hardware protected from alteration. (See Fabry [1974], Myers [1982], and Wilkes and Needham [1979].) The reduced instruction set computer (**RISC**) microprocessor, with its simplified instruction set, rendered capability-managing hardware obsolete by the mid-1980s. But software-managed capabilities, now called **handles**, are indispensable in modern object-oriented programming systems, databases, and distributed operating systems [Chase et al. 1994]. The same conceptual structure has recently reappeared in a proposal to manage objects and intellectual property in the Internet [Kahn and Wilensky 1995]. It is a powerful structure indeed.

You may have wondered why virtual memory, so popular in the operating systems of the 1960s and 1970s, was not present in the personal computer (PC) operating systems of the 1980s. The pundits of the microcomputer revolution proclaimed bravely that personal computers would not succumb to the diseases of the large commercial operating systems; the personal computer would be simple, fast, and cheap. Bill Gates, who said that no user of a personal computer would ever need more than 640K of main memory, brought out the Microsoft Disk Operating System (DOS) in 1982 without most of the common operating system functions, including virtual memory. Over time, however, programmers of personal computers encountered exactly the same programming problems as their predecessors in the 1950s, 1960s, and 1970s. That put pressure on the major PC operating system makers (Apple, Microsoft, and IBM) to add multiprogramming and virtual memory to their operating systems. These makers were able to respond positively because the major chip makers had not lost faith; Intel offered virtual memory and cache in its 80386 chip in 1985; Motorola did likewise in its 68020 chip. Apple offered multiprogramming in its MultiFinder and virtual memory in its System 6 operating system. Microsoft offered multiprogramming in Windows 3.1 and virtual memory in Windows 95. IBM offered multiprogramming and virtual memory in OS/2.

A similar pattern appeared in the early development of distributed-memory multicomputers beginning in the mid-1980s. These machines allowed for a large number of computers, sharing a high-speed interconnection network, to work concurrently on a single problem. Around 1985, Intel and N-Cube introduced the first hypercube machines consisting of 128 component microcomputers. Shortly thereafter, Thinking Machines produced the first commercial supercomputer of this genre, the Connection Machine, with as many as 65,536 component computer chips. These machines soon challenged the traditional supercomputer by offering the same aggregate processing speed at a lower cost [Denning and Tichy 1990]. Their designers initially eschewed virtual memory, believing that address translation and page swapping would seriously detract from the machine's performance. But they quickly encountered new programming problems having to do with synchronizing the **processes** on different computers and exchanging data among them. Without a common address space, their programmers had to pass data in messages. Message operations copy the same data three times: first from the sender's local memory to a local buffer, then across the network to a buffer in the receiver, and then to the receiver's local memory. The designers of these machines began to realize that virtual memory can reduce communication costs by as much as two-thirds because it copies the data once at the time of reference. Tanenbaum [1995] describes a variety of implementations under the topic of distributed shared memory.

The WWW, started in 1991 by Tim Berners-Lee, extends virtual memory to the world. The Web allows an author to embed, anywhere in a document, a **uniform resource locator** (URL), which is an Internet address of a file. The WWW appeals to many people because it replaces the traditional **processor-centered view** of computing with a **data-centered view** that sees computational processes as navigators in an immense space of shared objects. To avoid the problem of URLs becoming invalid when the object's owner moves it to a new machine, Kahn and Wilensky have proposed a two-level mapping scheme that first maps a URL to a handle, locates the machine hosting the object, and then fetches a copy of the object to the local machine [Kahn and Wilensky 1995]. This scheme recalls the Dennis–Van-Horn object-oriented virtual memory of the 1960s but now with worldwide, decentralized mapping systems. With its Java language, Sun Microsystems has extended WWW links to address programs as well as documents; when a Java interpreter encounters the URL of another Java program, it brings a copy of that program to the local machine and executes it [Gilder 1995]. These technologies, now seen as essential for the Internet, vindicate the view of the Multics designers a quarter century ago—that many large-scale computations will consist of many processes roaming a large space of shared objects.

From time to time over the past 40 years, various people have argued that virtual memory is not really necessary because advancing memory technology would soon permit us to have all the random-access main memory we could possibly want. Each new generation of users has discovered that its ambitions for processing, memory, and sharing led it to virtual memory. It is unlikely that today's predictions of the passing of virtual memory will prove to be any more reliable than similar predictions made in 1960, 1965, 1970, 1975, 1980, 1985, 1990, and 1995. Virtual memory accommodates essential patterns in the way people use computers to communicate and share information. It will still be used when we are all gone.

80.3 Structure of Virtual Memory

Figure 80.1 shows a system consisting of a processor, main memory, and secondary memory. Main memory is typically RAM and secondary memory disk. The access time of the RAM is on the order of 0.1–0.01 μs and of the disk 10–100 ms, giving speed ratios from 10^5 to 10^7. In the early computers, the speed ratio was on the order of 10^4. The penalty for referencing an item in secondary memory is even more severe than in the computers of the 1960s. This does not make virtual memory very attractive to those who believe that the primary purpose of virtual memory is to swap pages between main and secondary memory.

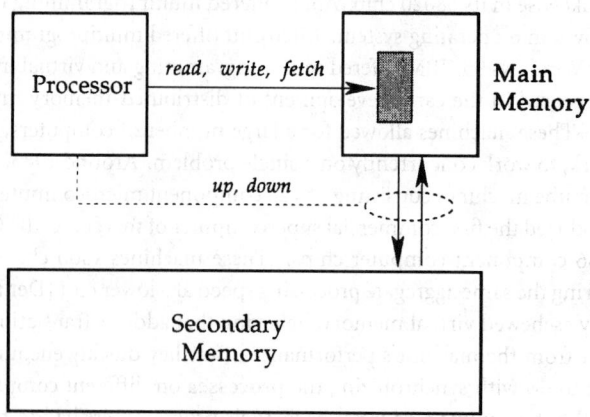

FIGURE 80.1 A processor executes a program from main memory. If the main memory is too small to hold the whole program, portions will be in secondary memory. Without virtual memory, the programmer would have to encode the commands to move blocks of data up and down the memory hierarchy.

Paging

The computer hardware addresses bytes (8-b units of data). Data are usually stored and moved as blocks of contiguous bytes. In the simplest case, called paging, all of the blocks are of the same size, say, 256 bytes. The main memory is divided similarly into blocks of **locations**, called **page frames**. Any page can be loaded in any page frame. We will consider first how a paged virtual memory works and then later examine variations that accommodate segments and other objects of variable sizes.

The set of addresses generatable by the processor is called the *virtual address space*; all programs must be compiled to generate addresses within this space. Similarly, the set of addresses of main memory locations is called the *real memory space*. The virtual memory system defines and maintains a dynamic map f from address to memory space so that the hardware can quickly convert an address x to a memory location $y = f(x)$. The *page table* in which the map is stored has one entry for each page of the address space.

An example will clarify the relationships among these elements. Suppose that the virtual addresses are 32 b, main memory addresses are 24 b, and the page size is $256 = 2^8$ bytes. In any 32-b virtual address, the 8 low-order bits select a byte within a page and the 24 high-order bits select a page. Thus, the address space contains 2^{24} (about 16.8 million) pages and the memory space contains 2^{16} (about 65,000) page frames. The page table f contains 2^{24} entries, each of which is a 16-b word. During address translation, the 24 high-order bits of the virtual address are replaced with the corresponding 16 b from the mapping table. All of this is illustrated in Fig. 80.2.

A master copy of all the pages of the address space resides in the secondary memory; some or all of those pages will be in main memory. As the processor modifies some pages, they will cease to be exact copies of the master; the operating system must write them back before deleting them from main memory. To support this, each page table entry contains a *modified bit* that the mapper turns on automatically during any write to the page.

Sooner or later the processor will generate an address whose page number is not mapped. The mapping unit will detect this and halt, issuing a signal called the *page fault*. In response, the operating system interrupts the running program and invokes a *page fault handler routine* that (1) locates the needed page in the secondary memory, (2) selects a frame of main memory to put that page in, (3) empties that frame,

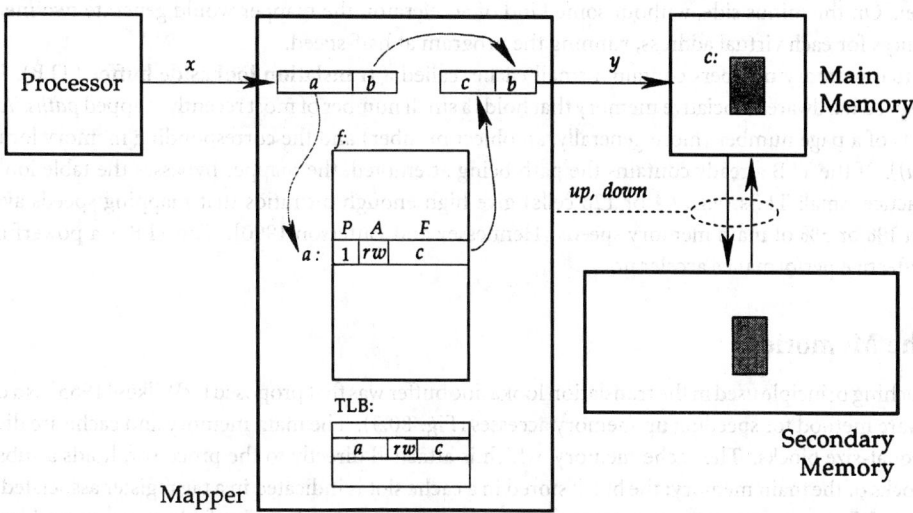

FIGURE 80.2 To convert a virtual address from the process into a real address for the main memory, the mapper refers to a page table f. A presence bit P indicates that the page is present in main memory; if so, the bits in the field F indicate which frame. An access code field A tells whether the page may be read (r) or written (w). A translation lookaside buffer (TLB) accelerates mapping by bypassing the page table on repeat access to pages.

(4) copies the needed page into that frame, and then (5) restarts the interrupted program, allowing it to complete its reference.

The replacement policy (step 2) frees memory by removing pages. The objective is to minimize mistakes, i.e., replacements that are quickly undone when the process recalls the page. This objective is met ideally when the object selected for replacement will not be used again for the longest time among all the loaded objects. To support this, each page table entry contains a *usage bit* that the mapper turns on automatically during any reference to the page. A variety of nonlookahead replacement policies have been studied extensively to see how close they come to this ideal in practice. When the memory space allocated to a process is fixed in size, this usually is least recently used (LRU); when space can vary, it is **working set** (WS) [Denning 1980].

The controller of the channel between main and secondary memory can accept commands of the form *(up, a, b)* and *(down, a, b)*. The up command transfers a page from secondary memory frame *b* into main memory frame *a*. The down command transfers a page from main memory frame *a* to secondary memory frame *b*. The page fault handler routine automatically issues those commands when and as they are needed: in step 3, if the frame has been modified since being loaded and in step 4.

This design makes address translation transparent to the programmer. Since the operating system maintains the contents of the map, it can alter the correspondence between addresses and locations dynamically. A program can now be executed on a wide range of system configurations, from small to large main memories, without recompiling it.

The main memory can also be partitioned among several executing programs. Each one has its own **address map** and can therefore refer only to its own pages. We will say more about multiprogramming later.

Translation Lookaside Buffer

The page tables, which can become quite numerous and large, cannot be stored economically in a local fast memory built in to the mapper. Instead, the mapper contains a pointer to the running process' page table, which is stored in the main memory. On the plus side, this has the additional advantage of simplifying the processor **context-switch** operation because the entire memory state of a process is denoted by one register. On the minus side, without some kind of accelerator, the mapper would generate two memory references for each virtual address, running the program at half-speed.

Virtual memory mappers contain a small cache, called a **translation lookaside buffer** (TLB). It is a high-speed hardware associative memory that holds a small number of most recently mapped *paths*. A path consists of a page number (more generally, an object number) and the corresponding memory location: *(a, f(a))*. If the TLB already contains the path being attempted, the mapper bypasses the table lookups. In practice, small TLBs (e.g., 64 or 128 cells) give high enough hit ratios that mapping speeds average within 1% or 3% of main memory speeds [Hennessey and Patterson 1990]. The TLB is a powerful and cost-effective performance accelerator.

Cache Memories

The caching principle used in the translation lookaside buffer was first proposed by Wilkes [1965] as a direct hardware method for speeding up memory accesses (Fig. 80.3). The main memory and cache are divided into equal-size blocks. The cache memory, which is attached directly to the processor, holds a subset of the blocks of the main memory; the block stored in a cache slot is indicated in a tag register associated with the slot. When the processor generates address *(a, b)*—meaning byte *b* of block *a*—the addressing hardware searches all of the tag registers in parallel for a match on *a*. If there is a match, it addresses byte *b* of that slot. If not, the hardware copies block *a* into a slot, sets the slot's tag to *a*, and then addresses byte *b* of that slot. This is just like paging except that the page table is inverted. (See Hennessey and Patterson [1990].)

Caching is not part of virtual memory but is complementary to it. Caching is simply a method to make a small set of blocks already in main memory immediately accessible to a processor at very high

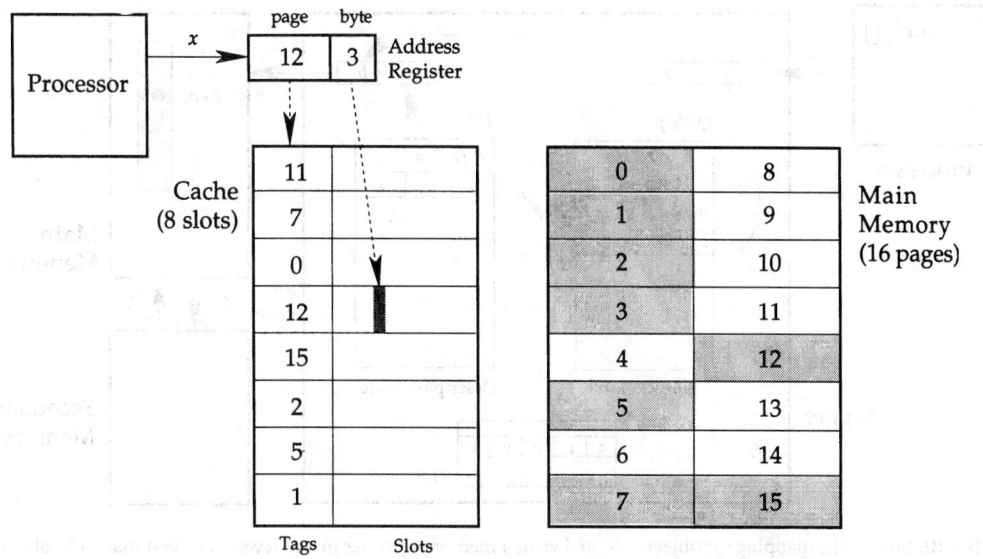

FIGURE 80.3   In a cache memory, the indices of pages (blocks) stored in cache slots are held in tag registers. The address hardware searches the tag registers in parallel for a match on the addressed page, and then uses the remaining address bits to select a byte of that page. The search can be made faster by dividing the tags into 2^m sets, using the m low-order bits of the block number to select the set, and restricting the parallel search to that set. (The figure is drawn for $m = 0$.) This partitions the blocks equally among the sets and thereby limits the number of slots into which a given block may be placed. In the worst case, when 2^m equals the number of cache slots, the set size is 1 and each block can be loaded into one slot only.

speed. It often allows the processor to run at nearly the cache speed while addressing a much larger address space. Most computers today use a three-level memory hierarchy consisting of cache, main memory, and secondary memory.

Object-Oriented Virtual Memory

Many computing environments offer abstractions and functions that require virtual addressing but which are not easily accommodated by paging. These include program objects such as arrays, procedures, structures, processes, message buffers, files, and directories; concurrent processes (threads) with varying **permissions** sharing the same address space; modular programs; and very large address spaces containing many objects shared among many users. The designers of early virtual memories anticipated these uses with segmented and capability-based virtual memories [Dennis 1965, Dennis and Van Horn 1966, Fabry 1974].

The earliest form of object-oriented virtual memory was the segmented address space. It appeared as a collection of named blocks (segments) of various sizes. Each segment was a container for a program object. In the Burroughs B5000 and later series, for example, the Algol compiler created program segments containing procedures and data segments containing array rows [Organick 1973]. The compiler generated virtual addresses of the form *(s, b)*, meaning byte *b* of segment *s*. The size of each segment was explicitly recorded so that the mapper could reject out-of-bounds addresses *b*. A **two-level mapping** scheme converted these virtual addresses into memory addresses. Multics extended segmentation so that the programmer defined the segments. Multics PL/I allowed programmers to refer to operands by symbolic two-part names, and the operating system used a linkage fault to invoke a routine that mapped a symbolic name-pair to its corresponding virtual address *(s, b)* as previously described [Organick 1972].

Today, **object-oriented addressing** has replaced segment addressing. An object can be stored in a memory segment, but the mapping mechanism should be able to verify that an address points to an object

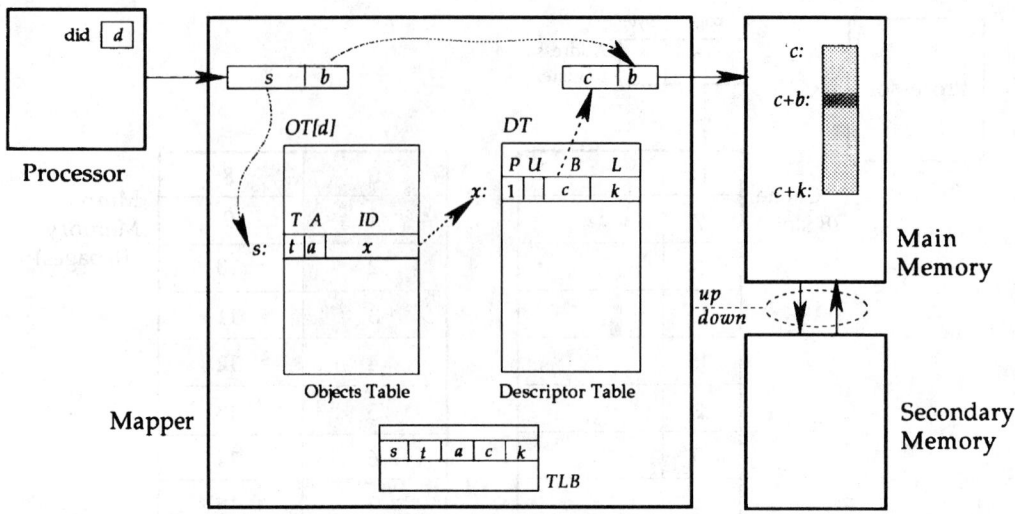

FIGURE 80.4   The mapping for object-oriented virtual memory operates in two levels. The first maps a local object number *s* of type *t* to an object unique identifier *x*, which in turn locates the object's descriptor in a descriptor table. An object's descriptor contains a presence bit *P* and a base-limit pair that designates a memory region of *k* bytes starting at address *c*. The byte number *b* must be less than *k*. A TLB that holds paths *(s, t, a, c, k)* accelerates the mapping. Different processes can have different segment numbers for the same segment: sharing of objects is possible without prior arrangements about the names each process will use internally. Shared objects can be relocated simply by updating the base address *c* in the descriptor.

of the expected type (otherwise producing a protection fault). The two-level mapping mechanism for object-addressing (including segments) is depicted in Fig. 80.4. The mapper follows this routine:

```
processor places (s,b) in address register
if ((t,a,c,k) = LOOKUP(s) undefined)
   then
       (t,a,x) := OT[d,s]
       (P,c,k) := DT[x]
       if (P = 0) then ADDRESS FAULT
       DT[x].U := 1
       LOAD(s,t,a,c,k)
   endif
if (b≥k) then BOUNDS FAULT
if (request not allowed by (t,a)) then PROTECTION FAULT
place c+b in memory address register
```

The operation **LOOKUP(s)** scans all of the TLB cells in parallel and returns the contents of the cell whose key matches **s**. The operation **LOAD** replaces the least recently used cell of TLB with **(s,t,a,c,k)**. The mapper sets the usage bit **U** to 1 whenever the entry is accessed so that the replacement algorithm can detect unused objects.

Object addressing creates a new problem of storage allocation in main memory: finding unused *holes* in which to place object-containing segments loaded from secondary memory. This problem can be alleviated by paging each segment. In Fig. 80.4, this would be implemented by adding a third level of mapping: the base address c points to a page table, and offset b is mapped to a frame in the same way as with pure paging. The combination of segmentation and paging is not very common.

Protection

With virtual memory, the operating system can restrict every process to a domain of least privilege. Only the objects listed in a domain's object table can be accessed by a process in that domain, and only then in accord with the **access codes** stored in the object's handle. In effect, the operating system walls each process off, giving it no chance to read or write the private objects of any other process. This has important benefits for system reliability. Should a process run amok, it can damage only its own objects: a program crash does not imply a system crash. This benefit is so important that many systems use virtual memory even if they allocate enough main memory to hold a process' entire address space.

Multiprogramming

Multiprogramming is a mode of operation in which the main memory is partitioned among the address spaces of different processes. On PCs it allows users to switch among active programs such as word processor, spreadsheet, and print spooler. It also provides a supply of programs ready to be resumed next by the operating system, thus maintaining high CPU efficiency.

Each process must be confined to its assigned region. Virtual memory accomplishes this by allowing a running program to refer only to the objects listed in its object table. Multiprogramming can be done with fixed or variable regions. Fixed regions are easier to implement but variable regions offer better performance. The reason is that the operating system can adjust the size of the region so that the rate of **address faults** stays within acceptable limits. Normally the best performance occurs when the operating system can transfer space from processes with small memory needs to processes with large memory needs. Variable partitions often improve over fixed even when the variation is random [Denning 1980]. System throughput will be near optimal when the virtual memory guarantees each active process just enough space to hold its working set [Denning 1980].

Performance

Although virtual memory simulates a space of objects stored in computational memory, it does not perform as well as if the objects were all stored in a RAM. Virtual memory systems usually detect and retain the most recently used objects in the main memory. Recently used objects thus are likely to be accessible at RAM speed, whereas long-unused objects are likely to be accessible at secondary memory speed. The programmer, therefore, can be confident of highly efficient virtual memory operation whenever programs cluster references to small groups of objects for extended intervals.

The replacement algorithm used by virtual memory to decide which objects to remove from main memory (and implicitly which objects to retain) has the greatest influence on system performance among all of the components of a virtual memory system. This is because most systems move objects into main memory only on demand, and because the mapper is already quite fast.

The main metrics of system performance are throughput and **response time**, and the main metric of memory usage is **space–time**. Throughput is measured as jobs or transactions completed per second, response time as the average number of seconds to complete a job or transaction, and space–time as the total number of byte (or page) seconds accumulated by a job while it holds main memory. If n jobs complete in T seconds, the throughput is $X = n/T$. When the total memory is M bytes, the total space–time available in the system is MT, and therefore the amount per job is $Y = MT/n$. The product of these definitions of X and Y is:

$$M = XY$$

This invariant relation is important because it says that minimizing space–time is the same as maximizing throughput for a given amount of main memory.

When a program has a fixed memory allocation, the minimum space–time occurs when the addressing faults are minimum. The ideal policy—let us call it MIN—replaces the object that will not be used again for the longest time. Unfortunately, such a policy cannot be implemented because the operating system cannot know the future reference pattern of a program. Among the implementable policies, LRU attempts to implement the MIN rule by predicting that time until next reference to an object is same as time since last reference. Although LRU is not as good as MIN, it has been found to be quite robust over a range of programs, typically doing as well or better than other nonlookahead policies, such as first-in-first-out (FIFO). LRU is often used in caches.

If we remove the constraints that memory size is fixed and that replacements occur only at address-fault times, we can do better than MIN. The ideal policy—let us call it variable-space MIN (VMIN)—operates as follows [Prieve and Fabry 1974]. After each object reference, VMIN looks ahead to the moment of the next reference to that object; it compares the space–time that would be accumulated by retaining the object in memory until that moment with the space–time that the addressing fault would generate. If retaining is more expensive, VMIN *immediately* removes the object from main memory. (There is no gain in paying both retention space–time and addressing-fault space–time for that next reference.) Since the retention space–time is proportional to time, we can state VMIN as a simple time-threshold rule: if the forward interval to next reference exceeds a threshold T, immediately replace an object after its current reference; otherwise retain it. The threshold interval T can be adjusted to the relative costs of replacing and retaining.

Although we cannot implement VMIN because we cannot predict future object references, we can implement an approximation that does well. We simply delay VMIN's replace decision until time T has elapsed after the prior reference to an object; equivalently, retain in memory all of the objects that have been referenced in a window of size T looking backwards from the current time. This is the working-set (WS) policy, first defined in 1967 [Denning 1968, 1970, 1980]. The WS and VMIN policies have identical addressing-fault sequences. Empirically, WS has greater space–time than VMIN because it retains objects beyond their last useful references. The best methods for clipping off the resulting spikes in resident set size do not recover more than 5–10% of the overshoot [Denning 1980]. WS is about as good as we can do without lookahead.

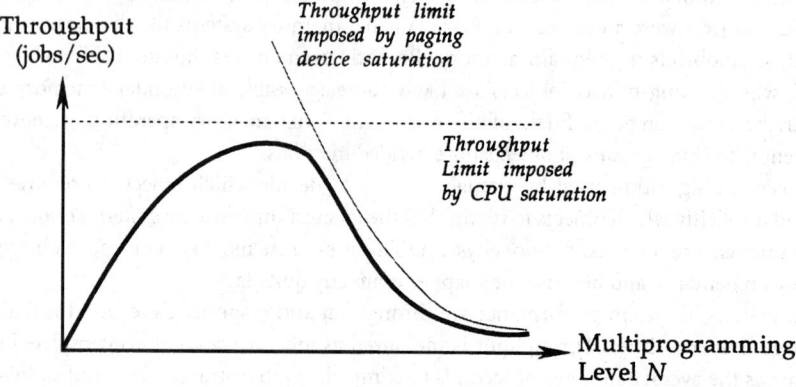

FIGURE 80.5   The system throughput is depicted as a function of the multiprogramming level N, which is the number of active programs among which main memory is partitioned. When N is too large, each program has so little space that it is forced toward a high paging rate. This makes the paging device the bottleneck, which slows down the system, producing thrashing. The ideal load control dynamically adjusts N to be constantly near the peak throughput.

The WS policy works especially well in a multiprogrammed system. Using a common window size T, the system measures every program's working set dynamically. The scheduler admits waiting processes to the active state, one at a time, until the available memory space is filled with working sets. The window size can be adjusted empirically until it maximizes system throughput. System throughput may be improved further, but only by 5–10% at most, by measuring each running process with its own private window size [Denning 1980].

Many early systems using multiprogrammed virtual memory attempted to extend the LRU policy, which works very well in fixed partitions, by lumping all pages in main memory into a global program managed by LRU. This strategy does not have the builtin load control of the working set policy: the scheduler can keep on adding more programs to the active set, reducing the average space available to each program and increasing the paging rate. These policies were therefore subject to thrashing (Fig. 80.5). Thrashing can be avoided by limiting the multiprogramming level either by a fixed limit or by a working-set policy.

80.4 Distributed Shared Memory

Starting in the mid-1980s, Sequent, Intel, Thinking Machines, N-Cube, and then later IBM, Cray, Kendall Square, and a few others introduced commercial multicomputers. These machines allowed for a large number of computers, sharing a high-speed interconnection network, to work concurrently on a single problem. They soon began to challenge the traditional supercomputer by offering the same aggregate processing speed at a lower cost [Denning and Tichy 1990]. But they also introduced a host of new programming problems having to do with synchronizing the processes on the different computers and exchanging data among them. Because these machines offered no common address space among all of the component computers, their programmers had to explicitly pass messages to exchange data. Not only does this complicate the job of programming the machine, it increases message overhead.

The resulting architecture has been referred to not only as virtual memory, but as distributed shared memory [Tannenbaum 1995]. The mapping principles are similar: convert a virtual address to a memory reference on the appropriate computer, using caching and fast interconnect protocols to speed up the mapping. The performance problems are different because the other computers do not form a memory hierarchy, but are instead peers at the same level of memory hierarchy. Typical questions include: (1) Should a page be moved to the computer that most recently referenced it? Or should it be moved only to the computer with the highest density of recent references to it? Or should it be left in the computer that obtained it by page fault? (2) How should a working set be defined when pages are shareable among many processes? How should the system ensure that this working set remains resident even while spread throughout the component memories? (3) How should duplicate copies (replicates) of pages be treated? Questions like these can be answered only by experimenting with the alternatives. They are subjects of considerable attention among designers of these computers and their operating systems.

80.5 The World Wide Web: Virtualizing the Internet

The World Wide Web extends virtual memory to the world. The Web allows an author to embed, anywhere in a document, a uniform resource locator, which is an Internet address of a file. By clicking the mouse on a URL string, the user triggers the operating system to map the URL to the file and then bring a copy of that file from the remote server to the local workstation for viewing. The WWW appeals to many people because it replaces the traditional processor-centered view of computing with a data-centered view that sees computational processes as navigators in an enormous space of shared objects.

A URL is invalidated when the object's owner moves or renames the object. To overcome this problem, Kahn and Wilensky have proposed a scheme that refers to mobile objects by location-independent handles and, with special servers, tracks the correspondence between handles and object locations [Kahn and

Wilensky 1995]. Their method is functionally similar to that described in Fig. 80.4: first it maps a URL to a handle and then it maps the handle to the Internet location of the object. Unlike Fig. 80.4, however, their method does not rely on central databases to store the mapping information.

The WWW is being extended to programs as well as documents. Sun Microsystems has taken the lead with its Java language. The URL of a Java program can be embedded in another program; exercising the link brings the Java program to a local interpreter, which executes it. The Java interpreter is encapsulated so that imported programs cannot access local objects other than those given it as parameters.

80.6 Conclusion

Virtual memory systems are used to meet one or more of these needs:

1. *Automatic storage allocation:* Solving the overlay problem that arises when a program exceeds the size of the computational store available to it. Also includes the problems of relocation and partitioning arising with multiprogramming.

2. *Protection:* Each process is given access to a limited set of objects, its protection domain. The operating system enforces the rights granted in a protection domain by restricting references to the memory regions in which objects are stored and by permitting only the types of reference stated for each object (e.g., read, write, or apply a function). These constraints are easily checked by the hardware in parallel with the main computation. These same principles are being used for efficient implementations of object-oriented programs.

3. *Modular programs:* Programmers should be able to combine separately compiled, reusable, and shareable components into programs without prior arrangements about anything other than interfaces, and without having to link the components manually into an address space.

4. *Object-oriented programs:* Programmers should be able to define managers of classes of objects and be assured that only the manager can access and modify the internal structures of objects [Myers 1982]. Objects should be freely shareable and reusable throughout a distributed system (Chase et al. 1994, Tannenbaum 1995). (This is an extension of the modular programming objective.)

5. *Data-centered programming:* Computations in the World Wide Web tend to consist of many processes navigating through a space of shared, mobile objects. Objects can be bound to a computation only on demand.

6. *Parallel computations on multicomputers:* Scalable algorithms that can be configured at run time for any number of processors are essential to mastery of highly parallel computations on multicomputers. Virtual memory joins the memories of the component machines into a single address space and reduces communication costs by eliminating some of the copying inherent in message passing.

Virtual memory, once the subject of intense controversy, it is now so ordinary that few people think much about it. That this has happened is one of the engineering triumphs of the computer age. Virtual memory accommodates essential patterns in the way people use computers.

Defining Terms

Access control: A means of allowing access to an object based on the type of access sought, the accessor's privileges, and the owner's wishes.

Address map: A table that associates a base address in main memory with an object (or page) number.

Address space: The set of all addresses that a processor can issue in reference to instructions and data.

Addressing fault: An error that halts the mapper when it cannot locate a referenced object in main memory; the fault handler corrects the condition by loading the missing object.

Bounds fault: An error that halts the mapper when it detects that the offset requested into an object exceeds the object's size.

Capability: A systemwide unique identifier for an object; the bits of a capability are protected from alteration.

Context-switch: An operation that switches the CPU from one process to another, by saving all of the CPU registers for the first and replacing them with the CPU registers for the second.

CPU: Central processing unit, or processor.

Data-centered view: A view of computing that emphasizes navigation through a large space of objects.

Handle: A systemwide unique identifier for an object, like a capability without the system guarantee of integrity.

Location space: The set of all hardware addresses of memory locations in RAM.

Main memory: The highest level of the memory hierarchy; all CPU memory references are directed to main memory.

Memory hierarchy: A system of memory devices of different speeds and capacities; allows for trading off between capacity and speed.

Modular programming: Programs are divided into parts that can be shared, reused, and recompiled without affecting other parts of the system as long as the interfaces to modules are unchanged.

Object-oriented addressing: A form of virtual addressing in which object numbers are mapped to memory regions and internal object references are mapped to offsets within an object's memory region.

Object-oriented programming: A form of programming in which data are organized into classes of objects, each with a specific set of functions that can be applied to the objects.

Page frame: A contiguous block of memory locations used to hold a page; a page is a fixed size unit of storage allocation and transfer.

Paging: A method of virtual memory in which address space and location space are paged.

PC: Personal computer.

Permissions: Access rights granted by an object's owner and represented as bits in the object's access code.

Process: An abstraction of the execution of a program, usually represented as the sequence of values of its CPU state as the program traces through its intruction sequence.

Processor-centered view: A view of computing that emphasizes the work of a processor.

Protection fault: An error condition detected by the address mapper when the type of request is not permitted by the object's access code.

Response time: The time from when a command is submitted to a computer until the computer responds with the result.

RAM: Random access memory.

RISC: Reduced instruction set computer (e.g., PowerPC, Sun SPARC, DEC Alpha, MIPS).

Secondary memory: Lower, large capacity level of a memory hierarchy, usually a set of disks.

Segmentation: An approach to virtual memory when the mapped objects were variable-size memory regions rather than fixed-size pages; superseded by object-oriented addressing.

Space–time: The product of the amount of memory and the amount of time used by a process.

Thrashing: A condition of performance collapse in a multiprogramming system when the number of active programs gets too large.

Throughput: The number of jobs (or transactions) per second completed by a computer system.

TLB: Translation lookaside buffer.

Two-level map: A two-tiered mapping table; the upper tier converts local object addresses into system unique handles, and the second tier converts handles to the memory regions containing the objects. Essential for sharing.

URL: Uniform resource locator (in the WWW).

Working-set (WS) policy: A memory allocation strategy that regulates the amount of main memory allocated to a process, so that the process is guaranteed a minimum level of processing efficiency.

World Wide Web (WWW):   A set of servers in the Internet and an access protocol that permits fetching documents by following hypertext links on demand.

References

Berners-Lee, T. 1996. The Web Maestro. *Technology Review*, July.

Chase, J. S., Levy, H. M., Feeley, M. J., and Lazowska, E. D. 1994. Sharing and protection in a single-address-space operating system. *ACM TOCS* 12(4):271–307.

Denning, P. J. 1968. Thrashing: its causes and prevention, pp. 915–922. *Proc. AFIPS FJCC 33*.

Denning, P. J. 1970. Virtual memory. *Comput. Surv.* 2(3):153–189.

Denning, P. J. 1976. Fault tolerant operating systems. *Comput. Surv.* 8(3).

Denning, P. J. 1980. Working sets past and present. *IEEE Trans. on Software Eng.* SE-6(1):64–84.

Denning, P. J. and Tichy, W. F. 1990. Highly parallel computation. *Science* 250:1217–1222.

Dennis, J. B. 1965. Segmentation and the design of multiprogrammed computer systems. *J. ACM* 12(4):589–602.

Dennis, J. B. and Van Horn, E. 1966. Programming semantics for multiprogrammed computations. *ACM Commun.* 9(3):143–155.

Fabry, R. S. 1974. Capability-based addressing. *ACM Commun.* 17(7):403–412.

Fotheringham, J. 1961. Dynamic storage allocation in the Atlas computer, including an automatic use of a backing store. *ACM Commun.* 4(10):435–436.

Gilder, G. 1995. The coming software shift. *Forbes ASAP*, Aug. 5.

Hennessey, J. and Patterson, D. 1990. *Computer Architecture: A Quantitative Approach*. Morgan-Kaufmann.

Kahn, R. and Wilensky, R. 1995. A framework for distributed object services. Technical Note 95-01, Corporation for National Research Initiutives, Reston, VA.

Kilburn, T., Edwards, D. B. G., Lanigan, M. J., and Sumner, F. H. 1962. One-level storage system. *IRE Trans.* EC-11(2):223–235.

Myers, G. J. 1982. *Advances in Computer Architecture*, 2nd ed. Wiley, New York.

Organick, E. I. 1972. *The Multics System: An Examination of Its Structure*. MIT Press, Cambridge, MA.

Organick, E. I. 1973. *Computer System Organization: The B5700/B6700 System*. Academic Press, New York.

Prieve, B. and Fabry, R. 1974. VMIN: an optimal variable space page replacement algorithm. *ACM Commun.* 19(5):295–297.

Sayre, D. 1969. Is automatic folding of programs efficient enough to displace manual? *ACM Commun.* 12(12):656–660.

Tannenbaum, A. S. 1995. *Distributed Operating Systems*. Prentice–Hall, Englewood Cliffs, NJ.

Wilkes, M. V. 1965. Slave memories and dynamic storage allocation. *IEEE Trans.* EC-14:270–271.

Wilkes, M. V. 1975. *Time Sharing Computer Systems*, 3rd ed. Elsevier/North-Holland.

Wilkes, M. V. and Needham, R. 1979. *The Cambridge CAP Computer and Its Operating System*. North-Holland.

81

Secondary Storage and Filesystems

Marshall Kirk
McKusick
Consultant and Writer

81.1 Introduction

The memory on a computer is organized into a hierarchy of storage [Smith 1981]. This storage ranges from small and fast to large and slow. Figure 81.1 shows a typical hierarchy. It is composed of two main parts: the primary store and the secondary store.

The main component of this hierarchy are as follows:

1. The first level of the primary store is the *cache memory*. It is often contained on the same chip as the central processing unit (CPU), or on other nearby chips that can be connected to the CPU with a minimum of delay. Because it must be able to run at close to the speed of the CPU, with access times of as little as a few microseconds, cache memory is typically small, rarely exceeding a few megabytes. The cache is never used for permanent storage; it holds values that are actively being processed by the CPU.

2. The second level of the primary store is the *main memory* on the computer. It currently runs with access times of 60–70 μs that may delay a CPU by 5–100 instruction cycles. The size of main memory ranges from a few megabytes up to several gigabytes. Like the cache, main memory is not used for permanent storage; it holds the active part of running programs. Inactive parts of running programs are swapped out of the main memory to disk when the main memory becomes full. Thus, the size of a program is not constrained to the size of the main memory.

3. The first level of the secondary store is usually built from one or more disk drives. These disk drives are usually connected directly to the computer, though they may be located across a fast

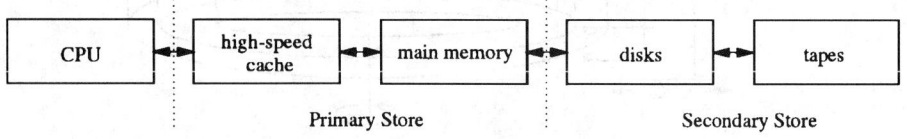

Primary Store   **Secondary Store**

FIGURE 81.1 Computer memory hierarchy.

network on a central storage server. Disks are used for intermediate to long-term data storage. Access time for a fast disk is currently about 10 ms; thus a CPU that needs to access data that is on disk will have to wait thousands of instruction cycles. Modern multitasking operating systems will suspend a program that awaits a disk access and run another program. Some time after the disk access has completed, the program that requested the data will begin to run again.

4. The second level of the secondary store consists of tape drives. The tape drives are used for archival and backup storage. The access time to get to the start of and begin reading a file stored on a robotically managed tape system is several minutes. If a human operator must get involved, the access time takes longer. Historically, actively running programs directly manipulated data on tapes. Today, most applications arrange to have data read from tapes onto disk before beginning to access that data. Tapes are used primarily to archive data that is not currently accessed.

This chapter is concerned with the secondary part of the storage hierarchy; Chapters 18 and 80 discuss primary storage and its management. In particular, disks may be used as a temporary store when the main memory becomes full. This chapter considers disks solely from the perspective of their use as long-term storage media. The first half of this chapter will discuss the hardware used to support secondary storage. The second half of the chapter will discuss filesystem software used to access and manage secondary storage.

81.2 Secondary Storage Devices

Many types of hardware are used to support secondary storage. This section will describe the most commonly used devices—magnetic disks, magneto-optical disks, compact disc–read-only memory (CD-ROM) disks, and various sorts of tape devices.

Magnetic Disks

Disks are the most frequently used form of secondary storage. The most common type of disk is the magnetic disk [White 1980]. Figure 81.2 shows the construction of a typical magnetic disk.

A disk is built from one or more metal platters coated on both sides with an iron oxide compound. The platters are mounted on a spindle connected to a motor that spins them. Data is read and written on the platters by magnetic heads mounted in a head assembly that runs on a track that enables the heads to be

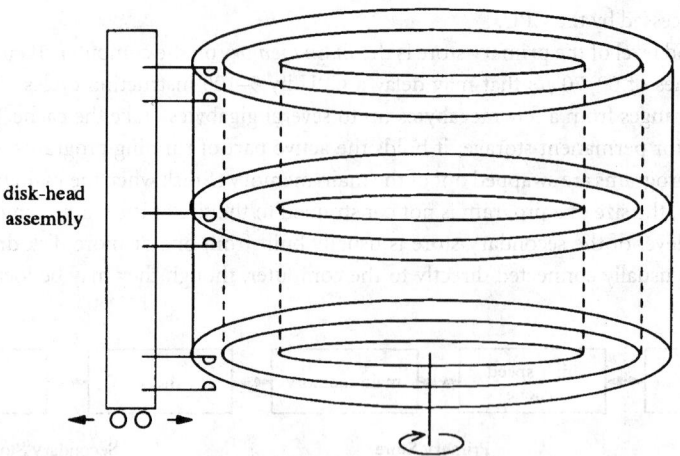

disk-head
assembly

FIGURE 81.2 Construction of a magnetic disk. (*Source:* Adapted from McKusick, M. K., et al. 1989. *The Design and Implementation of the 4.4 BSD UNIX Operating System.* Addison–Wesley, Reading, MA. With permission.)

moved between the inner and outer edges of the platter. The operation of moving the head assembly is referred to as a *seek*. Separate heads scan the top and bottom of each platter. Each platter is divided into a set of concentric circles referred to as *tracks*. Each track is subdivided into fixed sized blocks referred to as *sectors*. The set of all tracks that are accessible from a single seek position are referred to as a *cylinder*.

Moving a head to any particular location on the disk to read or write data is a two-step process. First the head assembly must be moved to the desired cylinder. Once the head assembly is located at the correct cylinder, the head for the requested track is enabled. The second step, called *rotational delay*, is to wait for the desired sector on the disk to rotate under the head. Reading is done by the head by measuring the polarization of the iron oxide. Writing is done by sending a current through the head to generate a magnetic field that polarizes the iron oxide.

The access time for a disk is a function of the seek time plus the rotational delay. If the entire contents of a chunk of data can be stored in the same cylinder as the current head position, then it can be accessed with no seek delay. If the data cannot be placed entirely within the same cylinder, it should be placed on nearby cylinders since seek time is a function of distance. Rotational delay can be minimized by placing data on contiguous sectors within a track. A disk can only read or write data from one head at a time; however, switching between heads is almost instantaneous. If the contents of a chunk of data can be stored contiguously on several tracks within a cylinder, it can be accessed without any rotational delay since the disk can switch between heads at the end of each track.

Early disk technology used platters up to a meter in diameter spinning at a few hundred revolutions per minute. Modern technology has been steadily shrinking the size of the platters so that today they are only a few centimeters in diameter. The smaller platters have allowed the head assemblies to shrink in size and mass. Because the lighter head assemblies can be moved more quickly and have shorter distances to travel, the seek times of modern disks are a hundredth that of early disks. The heads do not touch the disk surface; instead they float on a cushion of air a few micrometers above the surface. The air cushion pushed ahead of the heads creates friction that must be dissipated as heat. Because the amount of friction rises by the cube of the speed of the platter, even a small increase in rotational speed causes the generation of a lot more heat. Over time, the speed of rotation has increased much more slowly than the seek speed. Access times on early disks were dominated by seek times; whereas in modern disks, the rotational delay is the dominating factor.

Data is transferred between the main memory and the disk by a disk controller. Most disk controllers transfer data between the disk and main memory using direct memory access. The CPU issues a command specifying the range of main memory addresses to be written or read and the disk sectors to or from which they should be transferred. The controller is responsible for issuing a seek command to the disk if necessary, waiting for the desired sectors to rotate under the head, transferring the data between the specified main memory locations, and interrupting the CPU to signal that the transfer has been completed.

To improve performance, disk controllers often provide a *track cache*, which holds the contents of the track containing the currently requested sector. For example, consider a file that fits on one track of a disk. Suppose the disk controller has been requested to read the first part of the file, but when the seek completes, the head is sitting over the middle of the file half a rotation away from its beginning. Instead of waiting for the disk to rotate the beginning of the file into position, the controller immediately begins reading the track into its cache. As the beginning of the file passes under the head it is transferred into the main memory as requested and the CPU is notified of its completion. Following the end of the requested data transfer, if there are no further requests awaiting the controller, it reads the remaining quarter of the track into its cache. Since files are often read sequentially, the controller will probably be asked to read the next quarter of the file. Instead of having to wait for the disk to rotate back into position, it can satisfy this request from its cache. Thus, after one disk revolution, the controller can produce any part of the file with no rotational delay. Since the controller can transfer data from its cache to main memory much faster than data is delivered by the disk, the disk cache produces an even greater performance gain than just that measured by eliminating rotational delay.

Using a cache to speed up disk writes is a bit more difficult. If the controller completes a seek half a rotation ahead of the spot that requests the write there is nothing useful that it can do; it must wait for the requested position to rotate into place. Normally the controller waits until the write completes before issuing a completion interrupt to the CPU. Typically, a millisecond or so of CPU processing occurs before it issues the next write request. If the next write request contiguously follows the previous write, then the disk head will have just missed the start of the block and will incur nearly an entire rotation's delay reaching the correct starting point. When writing a large contiguous file the controller will consistently be delayed by a full rotation on each block, which leads to poor throughput. One approach to correcting this problem is to transfer the requested block into the controller cache and then issue the completion interrupt before the block has been placed entirely on the disk. Here, the CPU can prepare and issue the next request for the controller while the previous block is being written. With the new request in hand, the controller can start its transfer to the disk without loss of a revolution.

This approach has a serious failing if the controller does not have a nonvolatile cache. Many applications such as databases depend on the completion interrupt to let them know that critical data such as a transaction log has been stored in a location that will survive a system failure. If the data is only in the volatile controller cache and the system fails before the cache is written to disk, then the database will be unable to recover. Thus, early completion interrupts must be used only if the controller cache uses nonvolatile memory and has software to restart it and write out any incomplete blocks after a system failure.

The capacity of disks has been rising steadily; single disks today hold many gigabytes of data. Unfortunately disks are still able to transfer data from only one location at a time. As the amount of data on the disk grows, this serialized access has become more and more of a bottleneck. To compensate, some systems deliberately use smaller capacity disks to increase the total number of disks on the system, which allows more parallel data access.

Magneto-Optical Disks

Another major type of disk in use today is the magneto-optical disk (MO disk). These disks use polarized plastic and lasers rather than iron oxide and magnetic heads to store and read data. Zero and one bits are recorded on an MO disk with bits of plastic that are aligned with either a left or right polarization. The data are read by shining a laser beam on the plastic and detecting which way the beam is polarized when it reflects back. The direction of polarization of the plastic can be changed only by applying a strong magnetic field to the plastic at a moderately high temperature. A south flux will set the polarization in one direction and a north flux will set it in the other direction. Because of the strength of magnetic flux needed to change the direction of polarization, it is impossible to change the orientation of the magnet fast enough to write an arbitrary set of bits. Thus, writing data on an MO disk is slower because it requires two passes over the track to be written. The first pass is made with a strong magnetic field in one orientation and the laser continuously on at a high enough power to soften the plastic. This pass writes the track to all zeros. The second pass flips the magnet to the opposite polarization and the laser is pulsed for each location that is to be written to a one. Often a third pass will be made to ensure that the data is correctly recorded.

The benefits of the MO disk technology are as follows:

- The read–write mechanism is never close to the media, eliminating the head-crash failures that are common on magnetic disks.
- The media is easily removable much like a CD-ROM, allowing a single drive to archive much data.
- Once written, the media is stable; it will not be erased by stray magnetic fields, be damaged if dropped, or deteriorate over time.

The drawbacks of the MO disk technology are as follows:

- It spins at a tenth the speed of a magnetic disk. Thus, its access time and bandwidth are 10 times slower than a magnetic disk.
- Its capacity is 10 times less than a magnetic disk.
- Its cost per megabyte stored is about double that of a magnetic disk.

As a result, magnetic disks for outnumber the MO disks on computer systems today.

Redundant Array of Inexpensive Disks (RAID)

The biggest problem with magnetic disks is dealing with their inevitable failures. Because the heads fly only micrometers above the disk surface and because of the need to dissipate the heat that this small gap produces, even the best designed disks have a lifetime of only 5–7 years. On a large file server containing 50–100 disks, a disk failure is expected every few weeks. Traditionally, recovery from such failures was handled by use of a tape backup system. Every few weeks a complete copy of each disk would be made on tape. Each day a backup would be made of everything that had changed since the last complete copy had been made. When a disk failed, it would be replaced with a new disk; then the contents of the complete copy, followed by all the daily changes, would be made to complete the recovery process.

There are three problems with this approach:

1. *Capacity limits:* As the capacity of disks has increased, the amount of data that must be backed up on tape has exploded. The large server would have about a terabyte of disk space connected to it. Even with the high-density tapes available today it would take hundreds of tapes to back it up. The server would require several tape drives running continuously just to keep up with the backups.
2. *Data loss:* At best, most systems can only schedule backups once per day. If the disk crashes shortly before it is scheduled for its daily backup, everything that was done that day will be lost. For many businesses, losing even a day's work is unacceptable.
3. *Recovery delay:* Replacing a disk and recovering its contents from backup tapes takes several hours. During the recovery period, the disk is completely unavailable. This recovery delay is often unacceptable for time critical applications.

Two approaches have been taken to avoid these problems. The first of these approaches is a brute force solution called *mirroring*. The number of disks on the system is doubled. Disks are paired off and each pair keeps a copy of its partner. Thus, each time an application does a write, the changed data are written to both disks in the pair. Reads may be done from either disk since they both contain the same data. If one disk in the pair fails, the other disk can continue servicing requests without interruption. When the failed disk is replaced, its initial contents are copied in from its operating partner. Although it may take an hour or two to do the replacement and contents copy, users are unaware of the delay since they are running from the remaining good drive. Additionally, no data is lost since there is no dependence on backup tapes for recovery. Mirroring has traditionally been used for time or business critical data such as that handled by banking and airline reservation systems where the extra cost can be justified.

The second approach to avoiding the tape backup problem is to collect several disks together and use one of them to store a parity of the others. Such an organization is referred to as a redundant array of inexpensive disks (RAID) [Patterson et al. 1987, Chen et al. 1994]. A typical RAID cluster will contain five disks. Four of the disks contain data and the fifth contains a parity of the data on the other four. Each time data is written to any of the other four disks, a new parity must be computed and written to the fifth disk. In practice, the parity is not stored entirely on one disk as it would become the bottleneck when trying to write to one of the other four disks. Instead, each disk in a five-disk RAID cluster would be divided so that 20 percent stores parity and 80 percent stores data that would be covered by parity on the other four disks.

Recovery from disk failure in a RAID cluster is not as instantaneous as it is with mirroring. Access to the RAID cluster must be halted while the broken disk is replaced. The replacement disk is initialized by reading the other four disks in the cluster and computing what value should be put onto the new disk. In data communications, parity can be used to detect errors, but not to correct them. That is because in data communications the receiver does not know which bit is in error. Parity can be used for error correction for a RAID cluster because the cluster knows which disk failed. Thus, it can recompute the correct value for the failed disk using the data on the other drives. Failure recovery on a RAID cluster typically takes 1–2 h. Some RAID implementations will allow applications to continue to have read-only access to the 80 percent of the data that is on the four remaining good disks while the recovery is in progress.

Once the recovery is complete, the cluster returns to the state it was in just before the disk failed. Thus, RAID clusters solve two of the three tape backups problems. They avoid the need for daily backups and they avoid losing data when they fail. Although the data are unavailable during the recovery period, that period is typically about half the time required for a tape recovery. A RAID cluster is considerably cheaper than a mirroring strategy since there is only a 25-percent redundancy of hardware rather than a 100-percent redundancy. Thus, the RAID cluster usage is increasing in nontime critical environments.

Even with mirroring and RAID clusters, the need for tape backups is not completely eliminated. Full backups need to be taken and stored off-site for recovery if a major catastrophe such as a fire destroys the disks in a machine room. Tapes are also needed to recover from user errors where an important file is accidentally deleted. Neither of these problems can be handled by mirroring or RAID clusters.

CD-ROM Disks

The ubiquitous CD-ROM plays a small but important role in computer systems today [Asthana 1995]. Its primary use is as a software distribution medium where it is quickly replacing tapes and floppy disks. It has the benefits of being cheap, reliable, and easily mass produced. It has a much higher capacity than a floppy disk, though lower than that of a tape. It has the random access features of a floppy disk that make it much more convenient than a tape. Because the software is often used directly from the CD-ROM rather than being loaded onto the system disk, large software packages can be used on systems that are otherwise short of disk space. Manufacturers also like CD-ROMs because they tend to reduce software piracy. When diskette-based software is copied onto the system disk before use, the diskette can be passed on to a friend. When the distribution medium is needed to run the software, simultaneous uses of the same software distribution are not possible.

Another niche technology is write-once read-many (WORM) disks. These drives can write to a medium that is initially blank, but only one time. The most common WORM drives write media that can be read in a standard CD-ROM drive. Although the blank medium costs a factor of 10 more than a mass produced CD-ROM, it can be written in half an hour rather than taking several days to produce. The WORM technology is quick and cost effective for distributions of only a few disks and is also useful for producing data archives. The CD-ROM disks are more compact to store than tapes. They are expected to hold data reliably for 50–100 years, compared to tapes which can only hold data reliably for 5–8 years.

Tapes

Tapes remain the most commonly used form of fourth-level storage (see Chapter 69). Early tape technology used 12-in reels of $\frac{1}{2}$-in tape that stored a little over 100 megabytes of data. Current tape technology uses 5×7 cm cartridges with 4-mm tapes that store nearly 10 gigabytes of data. The rule of thumb has been that the largest tapes hold about the same amount of data as the largest disks.

Data transfer to and from tapes tends to be slower than data transfer to and from disks. Random access to tapes is much slower than disks. Even modern 4-mm drives take about 40 s to seek from one end of a tape to the other. The big benefit of tapes is that they are a tenth the cost of disks per megabyte of storage. Also, by installing a robotic tape system with a capacity of several hundred tapes, it is possible to create a file store capable of storing several terabytes of data. Whereas the access time to the data may be a minute or two, it is far cheaper than storing a similar amount of data on disk.

In practice, tape store is generally used as the final repository for data. Recently accessed data is stored on disk where it is more readily available. When the disks become full, the least recently accessed data is copied to tape and deleted from the disk. If it is later needed, it is reloaded onto the disk displacing other less recently accessed data. To maintain reasonable access times, most systems arrange to have enough disk space to keep data disks resident for at least a month.

81.3 Filesystems

Most applications that users run on their computer do not write data directly onto the disk. The operating system provides a filesystem that organizes the data into files. The filesystem is responsible for deciding where the file contents should be placed on the disk. The filesystem provides several important services:

- *Protection:* Most filesystems allow users to control access to their files. At a minimum they can restrict access to themselves, a defined group of other users, or all other users of the system.

- *Organization:* Data in each file can be manipulated independently of data in other files. Data may be added or deleted from one file without affecting the data contained in other files. In particular, the user need not be concerned that one file might run into another file on the disk as they would if they were directly placing the data on the disk themselves.

- *Multiple access:* Modern filesystems allow multiple files to be accessed at the same time and even allow separate programs to consistently access the same file. Consistent access means that if one application writes a file while another is reading the same part of the file, the reader will see the file either before the write has been started, or after it has finished, but not in a partially written state.

- *Space management:* The filesystem tracks the used and free space on the disk. When any file in the filesystem needs to grow, the filesystem finds an appropriate sized piece of free space and allocates it to the file. As long as there is free space left on the disk, the filesystem is prepared to allocate it to any file within the filesystem that wants to grow or to create a new file and allocate the space to that new file.

In addition, a filesystem is expected to optimize the use of disk bandwidth. Thus, it must not only find space to allocate to files, but it must try to find space that is contiguous or at least rotationally close together to minimize the time that it takes to read and write the file. Issues of space management are discussed subsequently.

Directory Structure

Most filesystems allow files to be grouped together into directories. These directories may then be further grouped together into other directories. The files and directories are usually grouped together in a tree hierarchy; Fig. 81.3 shows a typical filesystem hierarchy. The rounded boxes represent directories; the square boxes represent files. The top of the tree is referred to as the **filesystem root**. Files are accessed by giving the set of names of directories from the root of the tree down to the desired file, separated by forward slashes; this name is called the *pathname*. For example, access to the file *mydata* in Fig. 81.3 would use the pathname */users/myhome/mydata* [IEEE 1994].

Many filesystems maintain a current working directory for the user. Instead of having to always specify a file by its complete pathname, the system keeps track of which directory the user is currently referencing,

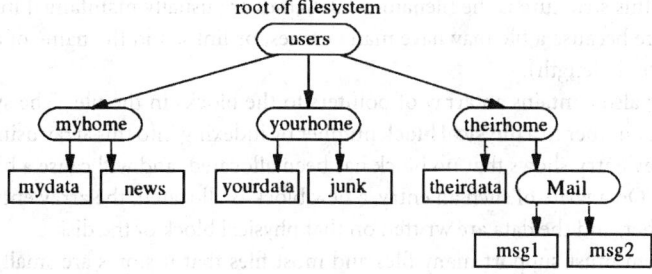

FIGURE 81.3 A set of files and directories. (*Source:* Adapted from McKusick, M. K., et al. 1989. *The Design and Implementation of the 4.4 BSD UNIX Operating System.* Addison–Wesley, Reading, MA. With permission.)

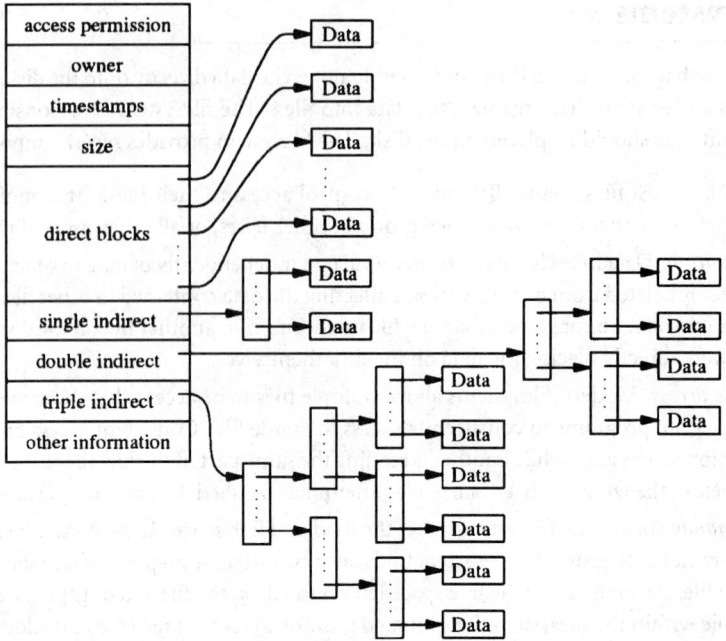

FIGURE 81.4 Extensible data structure used to describe a file. (*Source:* Adapted from McKusick, M. K., et al. 1989. *The Design and Implementation of the 4.4 BSD UNIX Operating System.* Addison–Wesley, Reading, MA. With permission.)

and does all filename translation relative to that directory. The user initially specifies a current directory using a complete pathname. Using the filesystem shown in Fig. 81.3, the user might request that the current directory be set to */users/myhome*. Thereafter, the reference to the file *mydata* can be used without specifying its entire path since it is resident in the current directory.

Describing a File on Disk

To allow both multiple file allocation and random access, most systems uses a data structure similar to that shown in Fig. 81.4 to describe the contents of a file. This structure includes:

- Access permission for the file
- The file's owner
- The time the file was last read and written
- The size of the file in bytes

Notably missing in this structure is the filename. Filenames are usually maintained in directories rather than in this structure because a file may have many names, or links, and the name of a file may be large (often up to 255 bytes in length).

The file structure also contains an array of pointers to the blocks in the file. The system can convert from a **logical block** number to a **physical block** number by indexing into this array using the logical block number. A null array entry shows that no block has been allocated, and will cause a block of zeros to be returned on a read. On a write of such an entry, a new block is allocated, the array entry is updated with the new block number, and the data are written on that physical block of the disk.

Since the filesystem must support many files and most files that it stores are small, the file structure has a small array of pointers for efficient use of space. The first few array entries are allocated in the file structure itself. For typical filesystems, these array entries allow the first 100 kilobytes of data to be located directly using a simple indexed lookup.

For somewhat larger files, Fig. 81.4 shows how the file structure contains a pointer to a single **indirect block** of pointers to data blocks. To find the 100th logical block of a file, the system first fetches the block identified by the indirect pointer, then indexes into it by 100 minus the number of direct pointers, and fetches that data block.

For files that are larger than a few megabytes, this single indirect block is eventually exhausted. These files must resort to using a double indirect block, which is a pointer to a block of pointers to pointers to data blocks. For files of multiple gigabytes, the system uses a triple indirect block, which contains three levels of pointers leading to the data block.

Although indirect blocks appear to increase the number of disk accesses required to reach a block of data, the overhead for this transfer is typically much lower. Most filesystems maintain a memory-based cache of recently read disk blocks. The first time that a block of indirect pointers is needed, it is brought into the filesystem cache. Further accesses to the indirect pointers find the block already resident in memory; thus, they require only a single disk access to reach the data.

The filesystem handles the allocation of new blocks to files as they grow. Simple filesystem implementations, such as those used by early microcomputer systems, allocate files contiguously, one after the next, until the files reach the end of the disk. As files are removed, gaps occur. To reuse this freed space, the system must compact the disk to move all free space to the end. Files can be created only one at a time; to increase the size of a file (other than the last one on the disk), it must be copied to the end and then expanded. For the more complex file structure just described, the locations of the data blocks in each file are given by its block pointers. Although the filesystem may cluster the blocks of a file to improve I/O performance, the file structure can reference blocks scattered anywhere throughout the disk. Thus, multiple files can be written simultaneously, and all disk space can be used without the need for compaction.

Filesystem Input/Output

The filesystem implementation converts the file from the user abstraction as an array of bytes to the structure imposed by the underlying physical medium. Consider a typical medium of a magnetic disk with fixed-sized sectoring. Although the user may wish to write a single byte to a file, the disk supports reading and writing only in multiples of sectors. Here, the system must read the sector containing the byte to be modified, replace the affected byte, and write the sector back to the disk. This operation of converting a random access within an array of bytes to reads and writes of disk sectors is called **block I/O**.

First, the system breaks the user's request into a set of operations to be done on logical blocks of the file. Logical blocks describe block-sized pieces of a file. The system calculates the logical blocks by dividing the array of bytes into filesystem-sized pieces. Thus, if a filesystem's block size is 8192 bytes, logical block 0 would contain bytes 0–8191, logical block 1 would contain bytes 8192–16,383, and so on.

Figure 81.5 shows the flow of information and work required to access the file on the disk. The abstraction shown to the user is an array of bytes. These bytes are collectively described by a file descriptor that refers to some location in the array. The user can request a write operation on the file by presenting the system with a pointer to a buffer, with a request for some number of bytes to be written. Figure 81.5 shows that the requested data need not be aligned with the beginning or end of a disk sector. Further, the size of the request is not constrained to a single disk sector. In the example shown, the user has requested data to be written to parts of logical blocks 1 and 2.

The data in each logical block are stored in a physical block on the disk. A physical block is the location on the disk to which the system maps a logical block. A physical disk block is constructed from one or more contiguous sectors. For a disk with 512-byte sectors, an 8192-byte filesystem block would be built from 16 contiguous sectors. Although the contents of a logical block are contiguous on disk, the logical blocks of the file need not be laid out contiguously.

Returning to our example in Fig. 81.5, we now know which logical blocks are to be updated. Since the disk can transfer data only in multiples of sectors, the filesystem must first arrange to read in the data for any part of the block that is to be left unchanged. The system must arrange an intermediate staging area for the transfer. This staging is done through one or more **system-cache buffers**.

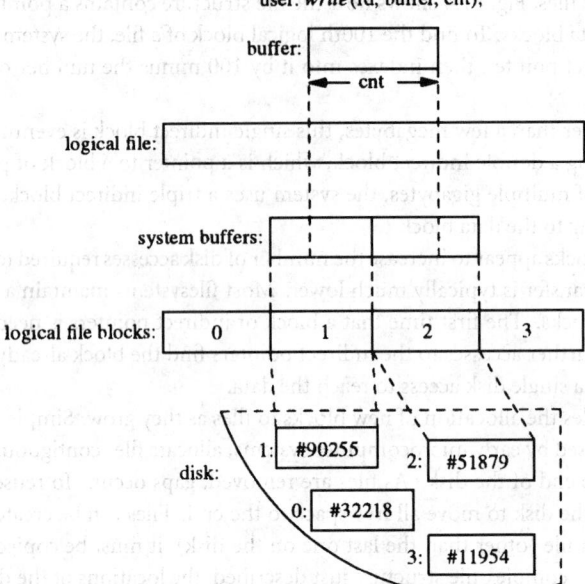

FIGURE 81.5   The block I/O system.   (*Source:* Adapted from McKusick, M. K., et al. 1989. *The Design and Implementation of the 4.4 BSD UNIX Operating System.* Addison–Wesley, Reading, MA. With permission.)

In our example, the user wishes to modify data in logical blocks 1 and 2. The operation iterates over five steps:

1. Allocate a system-cache buffer.
2. Determine the location of the corresponding physical block on the disk.
3. Request the disk controller to read the contents of the physical block into the system-cache buffer and wait for the transfer to complete.
4. Do a memory-to-memory copy from the beginning of the user's I/O buffer to the appropriate portion of the system-cache buffer.
5. Write the block to the disk.

Since the user's request is incomplete, the process is repeated with the next logical block of the file. The system fetches logical block 2 and is able to complete the user's request. Had an entire block been written, the system could have skipped step 3 by simply writing the data to the disk without first reading the old contents. This incremental filling of the write request is transparent to the user's process, because that process is blocked from running during this entire operation. It is also transparent to other processes since the file is locked by this process; any attempted access to the file by any other process will be blocked until this write has completed.

Disk Space Management

The role of the filesystem code is not only to organize the data on a disk but also to minimize the time that it takes to read and write that data. There are two important measurements of access time. The first is the time that it takes to access the data contained within a particular file. The second is the time that it takes to access the data contained within a collection of files that are accessed together.

Examples of collections of files accessed together include files that are collected together in a spool area such as those destined to be sent to a printer or those being batched to be sent as a collection of electronic mail messages. Another example might be a collection of files that makes up the components of a spreadsheet, a document, or a program. Directories provide a strong clue that a set of files will be accessed together. Thus, many filesystems will try to allocate all files contained within a directory in close physical proximity to each other on the disk. If they are accessed together, the disk will not need to make long or multiple seeks to get from one to the next.

Most filesystems put their greatest effort into optimizing the layout of individual files. The default assumption is that the files will be accessed sequentially starting at their beginning. Certain files, such as those that contain a large database, may have randomly scattered accesses. The optimal layout for such files is dependent on their access patterns, which may be known only to the database or application program, or may not be known at all. For such files, the filesystem may allow the database or application to direct the layout of the file. Alternatively, it may simply lay the file out sequentially, and assume that the database or application will attempt to cluster related information within the file to minimize the time it takes to seek between file locations.

For a sequentially accessed file, the usual strategy is to allocate a contiguous piece of space on the disk on which it is stored. When a disk is first put into service, doing sequential allocation for files is easy. The filesystem has a large area of contiguous physical free space, and allocates pieces out of that space for each new file. Unfortunately, this allocation approach quickly uses up all the physically contiguous space. As old files are deleted they return the space that they were using. However, this unused space will be randomly scattered throughout the disk. Eventually the disk will become completely fragmented, with no large contiguous pieces of space remaining, thus making it impossible for the filesystem to allocate new files contiguously.

Filesystems often use defensive algorithms to reduce the rate of fragmentation of the free space on the disk. A key to reducing fragmentation is to observe that most of the files within a filesystem are small. When a new file is created, the filesystem can assume that it will be small. Instead of allocating its initial disk space in a large contiguous area of the disk, the file will be allocated in a smaller fragment that may have been freed when another small file was deallocated. If the file turns out to be small as expected, then it will nicely fit in the small space that it was initially allocated. If it continues to grow, then the filesystem can move it to a fragment left by a somewhat larger file. Only when the file grows large is it finally relocated to the large contiguous space that the filesystem has now been able to hoard. By only allowing large files to be allocated in the large contiguous space, the filesystem is better able to ensure that some large contiguous space will be available when it is needed.

Moving files around on the disk can be a potentially slow and expensive operation if the file must be read and rewritten every time that the filesystem wants to move it. To reduce the relocation cost, the filesystem will attempt to defer writing the file until it has determined its final location on the disk. The deferral is done by holding blocks of the file in system-cache buffers until its size can be determined (see Fig. 81.5). The steps involved in allocating space to a file are as follows:

- When the first block is written, a system-cache buffer is allocated. A single block-sized piece of free disk space is found, the address of that piece of disk space is assigned to the system-cache buffer, but the buffer is not written to the disk.

- If the file continues to grow, a second system-cache buffer is allocated. The filesystem finds a two block-sized piece of free space, frees the single block-sized piece of space originally assigned to the file, and assigns the addresses of the new larger block to the two buffers. As before, neither buffer is written to the disk.

- This process continues until the file has grown to the maximum-sized block allowed (typically the size of a disk track) or the file has ceased to grow at which point the buffers are written to their final destination.

If the system-cache buffers are needed for another purpose, or the application explicitly requests that the file be stored on disk (for example, it is a transaction log that must be in stable storage before the application can proceed) the system-cache buffers always have a location to which they can be written.

The only implication of doing the write early is the loss of performance that comes from writing the data to disk more than once.

This algorithm has the additional benefit that it allows even slowly growing files to be written contiguously. If a file such as one holding accounting records grows at the rate of a few kilobytes per hour, it will be relocated to larger contiguous blocks periodically. Here, the relocation usually involves reading and rewriting the data, since the turnover in system-cache buffers causes them to be flushed before the new disk addresses for the data are assigned. Since the reallocation occurs only a few times per hour, the added I/O overhead does not adversely affect system performance.

To ensure successful layouts, a filesystem must have a quick method for finding contiguous and nearby free blocks on the disk and then allocating them. The most common data structure used for describing the free disk blocks is an array of bits. The blocks on the disk are sequentially numbered; the corresponding bit in the array is set to 1 if the block is being used and 0 if it is free. Block allocation involves setting the bits corresponding to the blocks being allocated; block deallocation involves clearing the bits corresponding to the blocks being freed. Finding other nearby blocks can be done by looking for 0 bits in the array near the location of the most recently allocated block in the file. Clusters of blocks can be identified by looking for strings of 0 bits in the array.

The free disk block array is large for a big disk. Exhaustive searches of the array (for example, when the disk is nearly full and there are few free blocks remaining) would slow filesystem performance unacceptably. Consequently, most filesystems maintain auxiliary data structures that summarize the contents of subranges of this bit array. Such summaries include the number of free blocks and the maximum-sized contiguous piece within each subrange of the bit map. When looking for a block or a cluster of blocks, the filesystem first scans the summary information to find a subrange of the bit map that has the necessary free space. Once it finds the needed space, the filesystem can narrow its search to that subrange of the bit map rather than searching the whole space.

Log-Based Systems

Logging long has been used in database systems to provide recovery after a system failure [Date 1995]. The database periodically does a **checkpoint** of its on-disk data structures to ensure that they are in a consistent state. Following the checkpoint, the database keeps a log of every change that it commits. The log is usually written serially onto a disk that is dedicated to the log. Thus, the log can be written quickly since the disk head is always within no more than a rotational delay from the next location to be written.

If the system fails, the log is used to do a **roll forward** operation on the database to bring it back to consistency. The roll forward works by going through the log and ensuring that all changes that it lists are reflected in the on-disk database state. The time required to recover the database after a system failure is bounded by the frequency of checkpoints. By checkpointing the database and resetting the log every few minutes, the recovery time for a disk failure can be kept to a few minutes or less.

The idea of using a log to recover a database has also been applied to filesystems. Like a database, the filesystem periodically checkpoints its state. It then writes all changes after that point both to its log and to the filesystem itself. When the log becomes full or a time limit is reached, the filesystem checkpoints itself again and resets its log. As with the database, the log is rolled forward after a system failure to ensure that all changes made since the checkpoint are reflected in the filesystem. The log only needs to record modifications to the filesystem. Although it has no effect on the speed with which data can be read from the filesystem, logging does introduce additional overhead when files are written. However, if the log is stored on a separate disk from the rest of the filesystem, writing to the log is seldom the limiting function on the speed with which data are placed into the filesystem.

Although a log provides fast recovery after a system failure, it does not provide the data recovery of mirroring or RAID that can recover from catastrophic disk failure. The reason that logging does not help with catastrophic disk failure is that data may be lost from parts of the disk that the filesystem thinks are stable and hence did not enter in the log.

A more recent filesystem design has taken the use of a log to its logical conclusion. Instead of using a log as an adjunct to an existing filesystem implementation, the entire filesystem is implemented as a log [Rosenblum and Ousterhout 1992]. The log-structured filesystem is being used in commercial products [Wilkes et al. 1995]. The fundamental idea of a log-structured filesystem is to improve filesystem performance by storing all filesystem data in a single, contiguous log. A log-structured filesystem is optimized for writing, and no seek is required between writes, regardless of the file to which the writes belong. It is also optimized for reading files written in their entirety over a brief period of time (as is the norm in workstation environments), because the files are placed contiguously on disk. A log-structured filesystem places files created at the same time together on disk. If an application reads a set of files written at different times, the filesystem potentially will have to seek around on the disk to read them. Log-structured filesystems expect to avoid the need for many seeks when reading files because they assume that all files that are actively being read will reside in the system cache.

The underlying structure of a log-structured filesystem is that of a sequential, append-only log. The disk is statically partitioned into fixed-size contiguous segments that are generally 0.5–1 megabyte. In ideal operation, a log-structured filesystem accumulates **dirty blocks** in memory. When enough blocks have been accumulated to fill a segment, they are written to the disk in a single, contiguous I/O operation. All writes to the disk are appended to the logical end of the log.

Although the log logically grows forever, portions of the log that have already been written must be made available periodically for reuse because the disk is not infinite in length. This process is called *cleaning*, and the utility that performs this reclamation is called the cleaner. The need for cleaning is the reason that the disk is logically divided into segments. Because the disk has reasonably large static areas, it is easy to segregate the portions of the disk that are currently being written from those that are currently being cleaned. The logical order of the log is not fixed, and the log can be viewed as a linked-list of segments, with segments being periodically cleaned, detached from their current position in the log, and reattached after the end of the log.

As a log-structured filesystem writes dirty data blocks to the logical end of the log (that is, into the next available segment), modified blocks will be written to the disk in locations different from those of the original blocks. This behavior is called a **no-overwrite policy**, and it is the responsibility of the cleaner to reclaim space resulting from deleted or rewritten blocks. Generally, the cleaner reclaims space in the filesystem by reading a segment, discarding dead blocks (blocks that belong to deleted files or that have been superseded by rewritten blocks), and rewriting any live blocks to the end of the log.

Cleaning must be done often enough that the filesystem does not fill up; however, the cleaner can have a devastating effect on performance. One study shows that cleaning segments while a log-structured filesystem is active (i.e., writing other segments) can result in a performance degradation of about 35–40 percent for some transaction-processing-oriented applications. This degradation is largely unaffected by how full the filesystem is; it occurs even when the filesystem is half-empty [Seltzer et al. 1995]. Another study shows that typical workstation workloads can permit cleaning during disk idle periods without introducing any user-noticeable latency [Blackwell et al. 1995]. The effect of filesystem cleaning on performance is still hotly debated.

Like a conventional filesystem that has a log associated with it, a log-structured filesystem must periodically do a checkpoint that synchronizes the information on disk so that all disk data structures are completely consistent. The frequency with which checkpoints are done affects the time needed to recover the filesystem after system failure. The more frequently they are done, the shorter the time it takes to recover. Checkpoints must also be taken whenever an application requests that one of its files be moved to stable storage. For example, an editor will usually request that a new version of a file be moved to stable storage before it deletes the old copy of the file. A conventional filesystem must be checkpointed whenever its logging disk becomes full. In the absence of application-requested checkpoints, a log-structured filesystem is only required to checkpoint a segment between the time that it is last written and the time that it is cleaned.

Recovery after a system failure is handled by rolling the filesystem log forward from the last checkpoint. In a conventional filesystem, the changes listed in the log are applied to the filesystem data structures. In a log structured filesystem, the filesystem is the log, so rolling it forward simply means discarding any incomplete operations.

Versioning Systems

Users often want to retain previous versions of files, such as released versions of programs or documents. Usually they use a revision control utility that maintains a database which stores and retrieves the selected versions. Most version control systems only keep a complete copy of the original file. Each successive version is then stored as a set of differences from the previous version. Some version control systems operate by storing each version in a separate file, allowing fast and easy access to older versions (since it is just a matter of finding and reading the desired version). However, this method of version control is wasteful of space, especially if the file being versioned is large and the changes between versions are minimal.

The whole file copy version of revision control can be done by the filesystem itself. Each time a file is opened for writing, the contents of the old file are saved instead of being overwritten. The old file is kept around until the filesystem runs out of space. When additional space is needed for new files, the user must either explicitly delete unneeded earlier versions of files or request the filesystem to reclaim the space from the oldest of the earlier file versions. Log-structured filesystems are particularly good at providing versioning since they never overwrite data in files. Older versions of files can be retrieved by going backward through the log. Space reclamation in a log-structure filesystem is done only by the cleaner process. Thus, an application can control when old versions are reclaimed by controlling when the cleaner process runs. The cleaner process can also be customized to skip over older versions of selected files that the user wishes to retain for a longer period of time.

The user interface for accessing older files may simply provide a list of different versions that are available and allow the user to specify the one that they want. Or it may be more sophisticated, allowing the user travel in time. For example, the user might be able to request all files making up a particular program as they existed at the release date for the program.

Defining Terms

Block I/O: The conversion of application reads and writes of records with arbitrary numbers of bytes into reads and writes that can be done based on the block size and alignment required by the underlying hardware.

Checkpoint: The writing of all modified data associated with a filesystem to stable storage (either non-volatile memory or the disk). A checkpoint ensures that all operations completed before the checkpoint will be recovered following a system failure.

Dirty blocks: In computer systems, modified. A system usually tracks whether or not an object has been modified—is dirty—because it needs to save the object's contents before reusing the space held by the object. For example, in the filesystem, a system-cache buffer is dirty if its contents have been modified. Dirty buffers must be written back to the disk before they are reused.

Filesystem root: The starting point for all absolute pathnames.

Indirect block: A filesystem data structure composed of an array of pointers to disk blocks used to locate the data blocks associated with a file.

Logging: Writing data to a file where existing data are never overwritten; the system thus modifies the file only by appending new data.

Logical block: The sequential fixed-sized pieces of a file. The logical block associated with a given byte offset in a file is calculated by dividing the offset by the filesystem block size. For example, byte 20,000 is located in the third logical block of a file residing on a filesystem with 8-kilobyte blocks.

No-overwrite policy: The filesystem never rewrites existing data in a file. New data is always written into a new location on the disk.

Physical block: The disk sector addresses associated with a logical block of a file. The filesystem finds the contents of a logical block in a file by using the logical block number as an index into an indirect block to find the disk sector address holding the requested data.

Roll forward: Used to recover after a system failure. The operation of rerunning the update operations stored in a log file against a filesystem or database to bring it to a consistent state as of the last update completed in the log.

System-cache buffers: System memory used to hold recently used data. For example, in the filesystem, system-cache buffers are used to hold recently accessed disk blocks.

References

Asthana, P. 1994. The long road to overnight success. *IEEE Spectrum* 31(10):60–66.

Asthana, P. 1995. Superdense optical storage. *IEEE Spectrum* 32(8):25–31.

Blackwell, T., Harris, J., and Seltzer, M. 1995. Heuristic cleaning algorithms in log-structured file systems, pp. 277–288. In *USENIX Assoc. Conf. Proc.* Jan.

Chen, P. M., Lee, E. K., Gibson, G. A., Katz, R. H., and Patterson, D. A. 1994. RAID: high-performance, reliable secondary storage. *ACM Comput. Syst.* 26(2):145–185.

Date, C. J. 1995. *An Introduction to Database Systems*, 6th ed. Addison–Wesley, Reading, MA.

IEEE. 1994. *POSIX: Part 1: System Application Program Interface.* Institute of Electrical and Electronic Engineers, New York.

McKusick, M. K., Karels, M. J., Bostic, K., and Quarterman, J. S. 1996. *The Design and Implementation of the 4.4BSD Operating System.* Addison–Wesley, Reading, MA.

Patterson, D., Garth, G., and Katz, R. 1988. A Case for Redundant Arrays of Inexpensive Disks (RAID). SIGMOD Record 17(3):109–116.

Rosenblum, M. and Ousterhout, J. 1992. The design and implementation of a log-structured file system. *ACM Trans. Comput. Syst.* 10(1):26–52.

Seltzer, M., Smith, K., Balakrishnan, H., Chang, J., McMains, S., and Padmanabhan, V. 1995. File system logging versus clustering: a performance comparison, pp. 249–264. In *USENIX Assoc. Conf. Proc.* Jan.

Silberschatz, A. and Galvin, P. 1994. *Operating System Concepts*, 4th ed. Addison–Wesley, Reading, MA.

Smith, A. J. 1981. Bibliography on file and I/O system optimizations and related topics. *Operating Syst. Rev.* 14(4):39–54.

White, R. M. 1980. Disk storage technology. *Sci. Am.* 243(2):138–148.

Wilkes, J., Golding, R., Staelin, C., and Sullivan, T. 1995. The HP AutoRAID hierarchical storage system. *ACM Operating Syst. Rev.* 29(5):96–108.

Further Information

Chapters 68 and 69 give a good description of the input and output devices used for secondary storage.

A good overview of filesystems can be found in chapter 3 of Silberschatz and Galvin [1994].

Most operating systems today use filesystem designs similar to those found in McKusick et al. [1996]. This chapter summarizes information on file layout on disk described in section 2 of chapter 7; filesystem naming described in section 3 of chapter 7; the traditional disk space management described in section 2 of chapter 8; and a log-structured filesystem described in section 3 of chapter 8 of McKusick et al. [1996].

82

Network Organization
and Topologies

William Stallings
Consultant and Writer

82.1 Transmission Control Protocol/Internet Protocol and Open Systems Interconnection

In this chapter, we examine the communications software needed to interconnect computers, workstations, servers, and other devices across networks. Then we look at some of the networks in contemporary use. When communication is desired among computers from different vendors, the software development effort can be a nightmare. Different vendors use different data formats and data exchange protocols. Even within one vendor's product line, different model computers may communicate in unique ways.

As the use of computer communications and computer networking proliferates, a one at a time special-purpose approach to communications software development is too costly to be acceptable. The only alternative is for computer vendors to adopt and implement a common set of conventions. For this to happen, standards are needed. Such standards would have two benefits:

- Vendors feel encouraged to implement the standards because of an expectation that, because of wide usage of the standards, their products will be more marketable.

- Customers are in a position to require that the standards be implemented by any vendor wishing to propose equipment to them.

It should become clear from the ensuing discussion that no single standard will suffice. Any distributed application, such as electronic mail or client/server interaction, requires a complex set of communications functions for proper operation. Many of these functions, such as reliability mechanisms, are common across many or even all applications. Thus, the communications task is best viewed as consisting of a modular architecture, in which the various elements of the architecture perform the various required functions. Hence, before one can develop standards, there should be a structure, or *protocol architecture,* that defines the communications tasks.

Two protocol architectures have served as the basis for the development of interoperable communications standards: the transmission control protocol/Internet protocol (TCP/IP) protocol suite and the **open systems interconnection (OSI) reference model.** TCP/IP is the most widely used interoperable architecture, especially in the context of **local-area networks** (LANs). In this section, we provide a brief overview of the two architectures.

The Transmission Control Protocol/Internet Protocol Architecture

This architecture is a result of protocol research and development conducted on the experimental packet-switched network, ARPANET, funded by the Defense Advanced Research Projects Agency (DARPA), and is generally referred to as the TCP/IP protocol suite.

The Transmission Control Protocol/Internet Protocol Layers

In general terms, communications can be said to involve three agents: applications, computers, and networks. Examples of applications include file transfer and electronic mail. The applications that we are concerned with here are distributed applications that involve the exchange of data between two computer systems. These applications, and others, execute on computers that can often support multiple simultaneous applications. Computers are connected to networks, and the data to be exchanged are transferred by the network from one computer to another. Thus, the transfer of data from one application to another involves first getting the data to the computer in which the application resides and then getting it to the intended application within the computer.

With these concepts in mind, it appears natural to organize the communication task into four relatively independent layers:

- Network access layer
- Internet layer
- Host-to-host layer
- Process layer

The network access layer is concerned with the exchange of data between an end system (server, workstation, etc.) and the network to which it is attached. The sending computer must provide the network with the address of the destination computer, so that the network may route the data to the appropriate destination. The sending computer may wish to invoke certain services, such as priority, that might be provided by the network. The specific software used at this layer depends on the type of network to be used; different standards have been developed for **circuit switching, packet switching** (e.g., X.25), local-area networks (e.g., Ethernet), and others. Thus, it makes sense to separate those functions having to do with network access into a separate layer. By doing this, the remainder of the communications software, above the network access layer, need not be concerned about the specifics of the network to be used. The same higher layer software should function properly regardless of the particular network to which the computer is attached.

The network access layer is concerned with access to and routing data across a network for two end systems attached to the same network. In those cases where two devices are attached to different networks, procedures are needed to allow data to traverse multiple interconnected networks. This is the function of the Internet layer. The Internet protocol is used at this layer to provide the routing function across multiple networks. This protocol is implemented not only in the end systems but also in routers. A router is a processor that connects two networks and whose primary function is to relay data from one network to the other on its route from the source to the destination end system.

Regardless of the nature of the applications that are exchanging data, there is usually a requirement that data be exchanged reliably. That is, we would like to be assured that all of the data arrive at the destination application and that the data arrive in the order in which they were sent. As we shall see, the mechanisms for providing reliability are essentially independent of the nature of the applications. Thus, it makes sense to collect those mechanisms in a common layer shared by all applications; this is referred to as the host-to-host layer. The transmission control protocol provides this functionality.

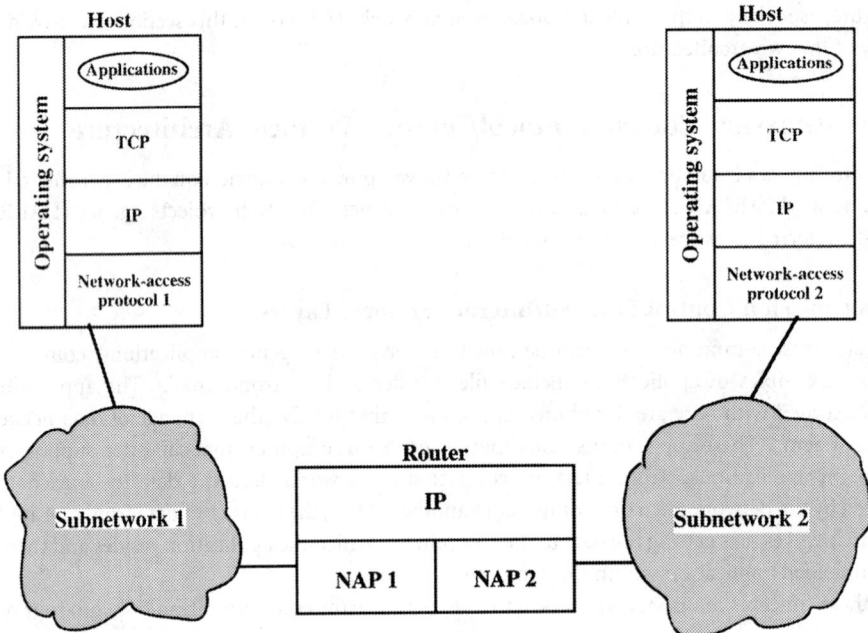

FIGURE 82.1   Communications using the TCP/IP protocol architecture.

Finally, the process layer contains the logic needed to support the various user applications. For each different type of application, such as file transfer, a separate module is needed that is peculiar to that application.

Operation of Transmission Control Protocol and Internet Protocol

Figure 82.1 indicates how these protocols are configured for communications. To make clear that the total communications facility may consist of multiple networks, the constituent networks are usually referred to as *subnetworks*. Some sort of network access protocol, such as the Ethernet logic, is used to connect a computer to a subnetwork. This protocol enables the host to send data across the subnetwork to another host or, in the case of a host on another subnetwork, to a router. IP is implemented in all of the end systems and the routers. It acts as a relay to move a block of data from one host, through one or more routers, to another host. TCP is implemented only in the end systems; it keeps track of the blocks of data to assure that all are delivered reliably to the appropriate application.

For successful communication, every entity in the overall system must have a unique address. Actually, two levels of addressing are needed. Each host on a subnetwork must have a unique global Internet address; this allows the data to be delivered to the proper host. Each process with a host must have an address that is unique within the host; this allows the host-to-host protocol (TCP) to deliver data to the proper process. These latter addresses are known as ports.

Let us trace a simple operation. Suppose that a process, associated with port 1 at host A, wishes to send a message to another process, associated with port 2 at host B. The process at A hands the message down to TCP with instructions to send it to host B, port 2. TCP hands the message down to IP with instructions to send it to host B. Note that IP need not be told the identity of the destination port. All that it needs to know is that the data are intended for host B. Next, IP hands the message down to the network access layer (e.g., Ethernet logic) with instructions to send it to router X (the first hop on the way to B).

To control this operation, control information as well as user data must be transmitted, as suggested in Fig. 82.2. Let us say that the sending process generates a block of data and passes this to TCP. TCP may break this block into smaller pieces to make it more manageable. To each of these pieces, TCP appends

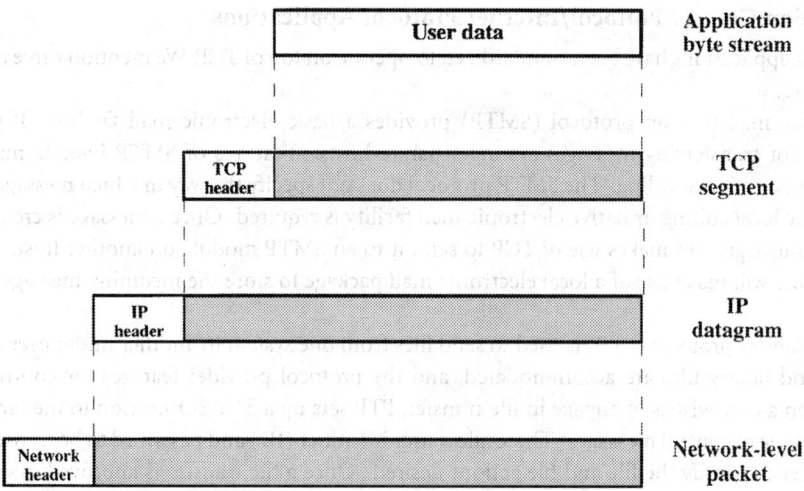

FIGURE 82.2 Protocol data units in the TCP/IP architecture.

control information known as the TCP header, forming a *TCP segment*. The control information is to be used by the peer TCP protocol entity at host *B*. Examples of items that are included in this header include the following:

- *Destination port:* When the TCP entity at *B* receives the segment, it must know to whom the data are to be delivered.
- *Sequence number:* TCP numbers the segments that it sends to a particular destination port sequentially, so that if they arrive out of order, the TCP entity at *B* can reorder them.
- *Checksum:* The sending TCP includes a code that is a function of the contents of the remainder of the segment. The receiving TCP performs the same calculation and compares the result with the incoming code. A discrepancy results if there has been some error in transmission.

Next, TCP hands each segment over to IP, with instructions to transmit it to *B*. These segments must be transmitted across one or more subnetworks and relayed through one or more intermediate routers. This operation, too, requires the use of control information. Thus, IP appends a header of control information to each segment to form an *IP datagram*. An example of an item stored in the IP header is the destination host address (in this example, *B*).

Finally, each IP datagram is presented to the network access layer for transmission across the first subnetwork in its journey to the destination. The network access layer appends its own header, creating a packet, or frame. The packet is transmitted across the subnetwork to router *X*. The packet header contains the information that the subnetwork needs to transfer the data across the subnetwork. Examples of items that may be contained in this header include the following:

- *Destination subnetwork address:* The subnetwork must know to which attached device the packet is to be delivered.
- *Facilities requests:* The network access protocol might request the use of certain subnetwork facilities, such as priority.

At router *X*, the packet header is stripped off and the IP header is examined. On the basis of the destination address information in the IP header, the IP module in the router directs the datagram out across subnetwork 2 to *B*. To do this, the datagram is again augmented with a network access header.

When the data are received at *B*, the reverse process occurs. At each layer, the corresponding header is removed, and the remainder is passed on the next higher layer, until the original user data are delivered to the destination process.

Transmission Control Protocol/Internet Protocol Applications

A number of applications have been standardized to operate on top of TCP. We mention three of the most common here.

The simple mail transfer protocol (SMTP) provides a basic electronic mail facility. It provides a mechanism for transferring messages among separate hosts. Features of SMTP include mailing lists, return receipts, and forwarding. The SMTP protocol does not specify the way in which messages are to be created; some local editing or native electronic mail facility is required. Once a message is created, SMTP accepts the message, and makes use of TCP to send it to an SMTP module on another host. The target SMTP module will make use of a local electronic mail package to store the incoming message in a user's mailbox.

The file transfer protocol (FTP) is used to send files from one system to another under user command. Both text and binary files are accommodated, and the protocol provides features for controlling user access. When a user wishes to engage in file transfer, FTP sets up a TCP connection to the target system for the exchange of control messages. These allow user identifier (ID) and password to be transmitted and allow the user to specify the file and file actions desired. Once a file transfer is approved, a second TCP connection is set up for the data transfer. The file is transferred over the data connection, without the overhead of any headers or control information at the application level. When the transfer is complete, the control connection is used to signal the completion and to accept new file transfer commands.

TELNET provides a remote log-on capability, which enables a user at a terminal or personal computer to log on to a remote computer and function as if directly connected to that computer. The protocol was designed to work with simple scroll-mode terminals. TELNET is actually implemented in two modules: User TELNET interacts with the terminal input/output (I/O) module to communicate with a local terminal. It converts the characteristics of real terminals to the network standard and vice versa. Server TELNET interacts with an application, acting as a surrogate terminal handler so that remote terminals appear as local to the application. Terminal traffic between user and server TELNET is carried on a TCP connection.

The Open Systems Interconnection Model

The open systems interconnection (OSI) reference model was developed by the International Organization for Standardization (ISO) to serve as a framework for the development of communications protocol standards.

Overall Architecture

A widely accepted structuring technique, and the one chosen by ISO, is layering. The communications functions are partitioned into a hierarchical set of layers. Each layer performs a related subset of the functions required to communicate with another system. It relies on the next lower layer to perform more primitive functions and to conceal the details of those functions. It provides services to the next higher layer. Ideally, the layers should be defined so that changes in one layer do not require changes in the other layers. Thus, we have decomposed one problem into a number of more manageable subproblems.

The task of ISO was to define a set of layers and the services performed by each layer. The partitioning should group functions logically and should have enough layers to make each layer manageably small, but it should not have so many layers that the processing overhead imposed by the collection of layers is burdensome.

The resulting OSI architecture has seven layers, which are illustrated in Fig. 82.3. Each computer contains the seven layers. Communication is between applications in the two computers, labeled application X and application Y in the figure. If application X wishes to send a message to application Y, it invokes the application layer (layer 7). Layer 7 establishes a peer relationship with layer 7 of the target computer, using a layer-7 protocol (application protocol). This protocol requires services from layer 6, so the two layer-6 entities use a protocol of their own, and so on down to the physical layer, which actually transmits bits over a transmission medium.

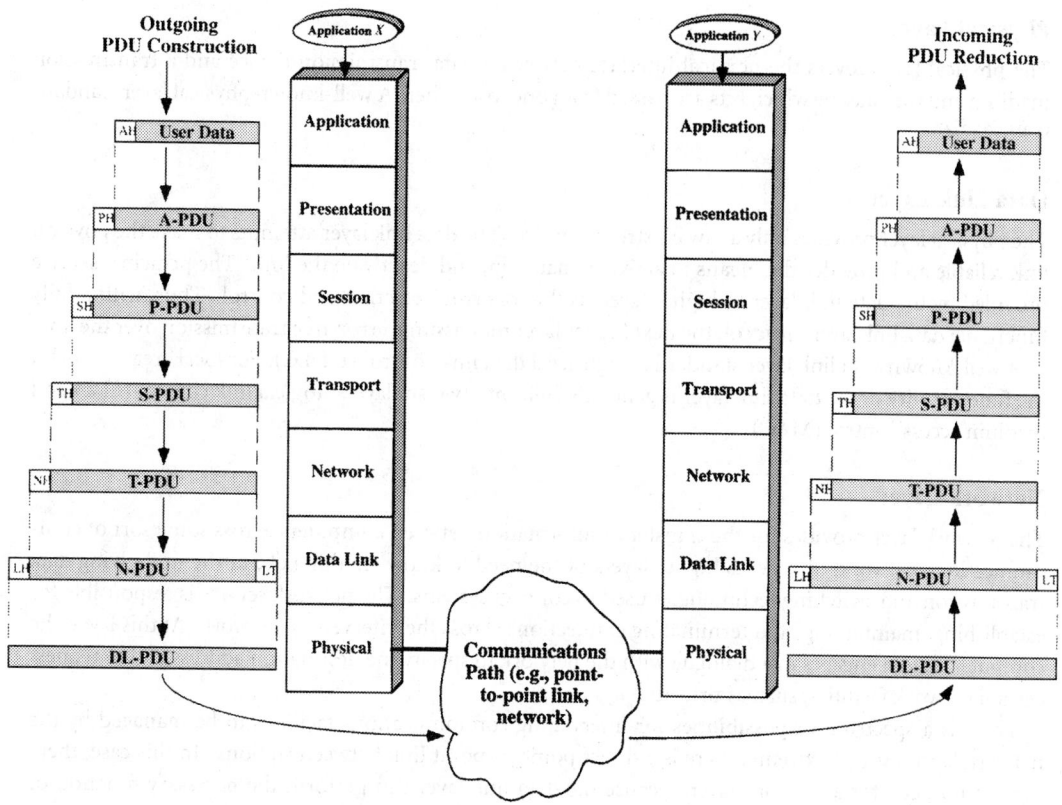

FIGURE 82.3   The OSI environment.

The figure also illustrates the way in which the protocols at each layer are realized. When application *X* has a message to send to application *Y*, it transfers those data to an application layer module. That module appends an application header to the data; the header contains the control information needed by the peer layer on the other side. The original data plus the header, referred to as an application protocol data unit (PDU), are passed as a unit to layer 6. The presentation module treats the whole unit as data and appends its own header. This process continues down through layer 2, which generally adds both a header and a trailer. This layer-2 protocol data unit, usually called a *frame*, is then transmitted by the physical layer onto the transmission medium. When the frame is received by the target computer, the reverse process occurs. As we ascend the layers, each layer strips off the outermost header, acts on the protocol information contained therein, and passes the remainder up to the next layer.

The principal motivation for development of the OSI model was to provide a framework for standardization. Within the model, one or more protocol standards can be developed at each layer. The model defines in general terms the functions to be performed at that layer and facilitates the standards-making process in two ways:

- Because the functions of each layer are well defined, standards can be developed independently and simultaneously for each layer. This speeds up the standards-making process.
- Because the boundaries between layers are well defined, changes in standards in one layer need not affect already existing software in another layer. This makes it easier to introduce new standards.

We now turn to a brief description of each layer and discuss some of the standards that have been developed for each layer.

Physical Layer

The physical layer covers the physical interface between a data transmission device and a transmission medium and the rules by which bits are passed from one to another. A well-known physical layer standard is RS-232-C.

Data Link Layer

The physical layer provides only a raw bit stream service. The data link layer attempts to make the physical link reliable and provides the means to activate, maintain, and deactivate the link. The principal service provided by the data link layer to higher layers is that of error detection and control. Thus, with a fully functional data link layer protocol, the next higher layer may assume error-free transmission over the link.

A well-known data link layer standard is high-level data link control (HDLC). For local area networks, the functionality of the data link layer is generally split into two sublayers: logical link control (LLC) and medium access control (MAC).

Network Layer

The network layer provides for the transfer of information between computers across some sort of communications network. It relieves higher layers of the need to know anything about the underlying data transmission and switching technologies used to connect systems. The network service is responsible for establishing, maintaining, and terminating connections across the intervening network. At this layer, the computer system engages in a dialogue with the network to specify the destination address and to request certain network facilities, such as priority.

There is a spectrum of possibilities for intervening communications facilities to be managed by the network layer. At one extreme, there is a direct point-to-point link between stations. In this case, there may be no need for a network layer because the data link layer can perform the necessary function of managing the link.

Next, the systems could be connected across a single network, such as a circuit-switching or packet-switching network. The lower three layers are concerned with attaching to and communicating with the network; a well-known example is the X.25 standard. The packets that are created by the end system pass through one or more network nodes that act as relays between the two end systems. The network nodes implement layers 1–3 of the architecture. The upper four layers are end-to-end protocols between the attached computers.

At the other extreme, two stations might wish to communicate but are not even connected to the same network. Rather, they are connected to networks that, directly or indirectly, are connected to each other. This case requires the use of some sort of internetworking technique, such as the use of IP.

Transport Layer

The transport layer provides a reliable mechanism for the exchange of data between computers. It ensures that data are delivered error free, in sequence, and with no losses or duplications. The transport layer also may be concerned with optimizing the use of network services and providing a requested quality of service. For example, the session layer may specify acceptable error rates, maximum delay, priority, and security features.

The mechanisms used by the transport protocol to provide reliability are very similar to those used by data link control protocols such as HDLC: the use of sequence numbers, error detecting codes, and retransmission after timeout. The reason for this apparent duplication of effort is that the data link layer deals only with a single, direct link, whereas the transport layer deals with a chain of network nodes and links. Although each link in that chain is reliable because of the use of HDLC, a node along that chain may fail at a critical time. Such a failure will affect data delivery, and it is the transport protocol that addresses this problem.

The size and complexity of a transport protocol depend on how reliable or unreliable the underlying network and network layer services are. Accordingly, ISO had developed a family of five transport protocol standards, each oriented toward a different underlying service.

Session Layer

The session layer provides the mechanism for controlling the dialogue between the two end systems. In many cases, there will be little or no need for session-layer services, but for some applications, such services are used. The key services provided by the session layer include the following:

- *Dialogue discipline:* This can be two-way simultaneous (full duplex) or two-way alternate (half-duplex).
- *Grouping:* The flow of data can be marked to define groups of data. For example, if a retail store is transmitting sales data to a regional office, the data can be marked to indicate the end of the sales data for each department. This would signal the host computer to finalize running totals for that department and start new running counts for the next department.
- *Recovery:* The session layer can provide a checkpointing mechanism, so that if a failure of some sort occurs between checkpoints, the session entity can retransmit all data since the last checkpoint.

ISO has issued a standard for the session layer that includes as options services such as those just described.

Presentation Layer

The presentation layer defines the format of the data to be exchanged between applications and offers application programs a set of data transformation services. For example, data compression or data encryption could occur at this level.

Application Layer

The application layer provides a means for application programs to access the OSI environment. This layer contains management functions and generally useful mechanisms to support distributed applications. In addition, general-purpose applications such as file transfer, electronic mail, and terminal access to remote computers are considered to reside at this layer.

82.2 Network Organization

Traditionally, data networks have been classified as either **wide-area network** (WAN) or local-area network. Although there has been some blurring of this distinction, it is still a useful one. We look first at traditional WANs and then at the more recently introduced higher speed WANs. The discussion then turns to traditional and high-speed LANs.

WANs are used to connect stations over a large area: anything from a metropolitan area to worldwide. LANs are used within a single building or a cluster of buildings. Usually, LANs are owned by the organization that uses them. A WAN may be owned by the organization that uses it (private network) or provided by a third party (public network); in the latter case, the network is shared by a number of organizations.

Traditional Wide-Area Networks

Traditional WANs are switched communications networks, consisting of an interconnected collection of nodes, in which information is transmitted from source station to destination station by being routed through the network of nodes. Figure 82.4 is a simplified illustration of the concept. The nodes are connected by transmission paths.

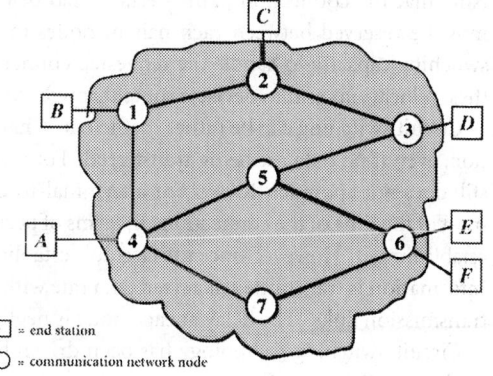

□ = end station
○ = communication network node

FIGURE 82.4   Simple switching network.

Signals entering the network from a station are routed to the destination by being switched from node to node. Two quite different technologies are used in wide-area switched networks: circuit switching and packet switching. These two technologies differ in the way the nodes switch information from one link to another on the way from source to destination.

Circuit Switching

Circuit switching is the dominant technology for both voice and data communications today and will remain so for the foreseeable future. Communication via circuit switching implies that there is a dedicated communication path between two stations. That path is a connected sequence of links between network nodes. On each physical link, a channel is dedicated to the connection. The most common example of circuit switching is the telephone network.

Communication via circuit switching involves three phases, which can be explained with reference to Fig. 82.4. The three phases are as follows:

1. *Circuit establishment:* Before any signals can be transmitted, an end-to-end (station-to-station) circuit must be established. For example, station A sends a request to node 4 requesting a connection to station E. Typically, the link from A to 4 is a dedicated line, so that part of the connection already exists. Node 4 must find the next leg in a route leading to node 6. Based on routing information and measures of availability and perhaps cost, node 4 selects the link to node 5, allocates a free channel [using frequency-division multiplexing (FDM) or time-division multiplexing (TDM)] on that link and sends a message requesting connection to E. So far, a dedicated path has been established from A through 4 to 5. Because a number of stations may attach to 4, it must be able to establish internal paths from multiple stations to multiple nodes. The remainder of the process proceeds similarly. Node 5 dedicates a channel to node 6 and internally ties that channel to the channel from node 4. Node 6 completes the connection to E. In completing the connection, a test is made to determine if E is busy or is prepared to accept the connection.

2. *Information transfer:* Information can now be transmitted from A through the network to E. The transmission may be analog voice, digitized voice, or binary data, depending on the nature of the network. As the carriers evolve to fully integrated digital networks, the use of digital (binary) transmission for both voice and data is becoming the dominant method. The path is: A-4 link, internal switching through 4, 4-5 channel, internal switching through 5, 5-6 channel, internal switching through 6, 6-E link. Generally, the connection is full duplex, and signals may be transmitted in both directions simultaneously.

3. *Circuit disconnect:* After some period of information transfer, the connection is terminated, usually by the action of one of the two stations. Signals must be propagated to nodes, 4, 5, and 6 to deallocate the dedicated resources.

Note that the connection path is established before data transmission begins. Thus, channel capacity must be reserved between each pair of nodes in the path and each node must have available internal switching capacity to handle the requested connection. The switches must have the intelligence to make these allocations and to devise a route through the network.

Circuit switching can be rather inefficient. Channel capacity is dedicated for the duration of a connection, even if no data are being transferred. For a voice connection, utilization may be rather high, but it still does not approach 100%. For a terminal-to-computer connection, the capacity may be idle during most of the time of the connection. In terms of performance, there is a delay prior to signal transfer for call establishment. However, once the circuit is established, the network is effectively transparent to the users. Information is transmitted at a fixed data rate with no delay other than the propagation delay through the transmission links. The delay at each node is negligible.

Circuit-switching technology has been driven by those applications that handle voice traffic. One of the key requirements for voice traffic is that there must be virtually no transmission delay and certainly no variation in delay. A constant signal transmission rate must be maintained, because transmission and reception occur at the same signal rate. These requirements are necessary to allow normal human conversation. Further, the quality of the received signal must be sufficiently high to provide, at a minimum, intelligibility.

Packet Switching

A packet-switching network is a switched communications network that transmits data in short blocks called packets. The network consists of a set of interconnected packet-switching nodes. A device attaches to the network at one of these nodes and presents data for transmission in the form of a stream of packets. Each packet is routed through the network. As each node along the route is encountered, the packet is received, stored briefly, and then transmitted along a link to the next node in the route. Two approaches are used to manage the transfer and routing of these streams of packets: datagram and virtual circuit.

In the datagram approach, each packet is treated independently, with no reference to packets that have gone before. This approach is illustrated in Fig. 82.5(a). Each node chooses the next node on a packet's path, taking into account information received from neighboring nodes on traffic, line failures, and so on. So the packets, each with the same destination address, do not all follow the same route, and they may arrive out of sequence at the exit point. In some networks, the exit node restores the packets to their original order before delivering them to the destination. In other datagram networks, it is up to the destination rather than the exit node to do the reordering. Also, it is possible for a packet to be destroyed in the network. For example, if a packet-switching node crashes momentarily, all of its queued packets

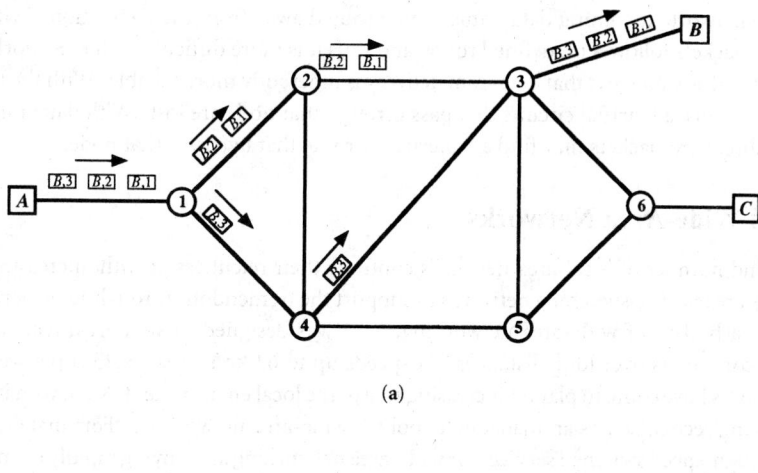

(a)

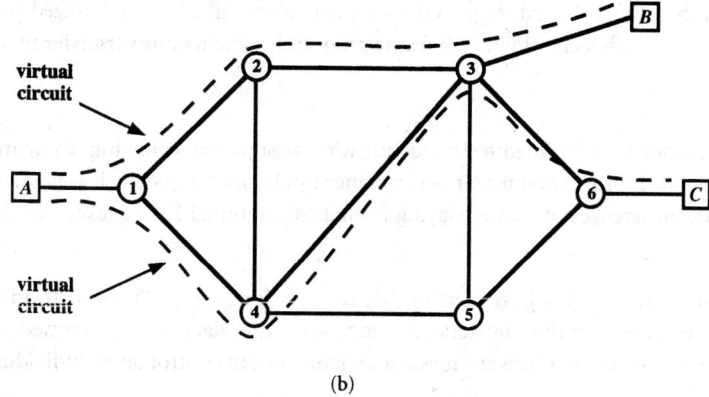

(b)

FIGURE 82.5   Virtual circuit and datagram operation: (a) datagram approach and (b) virtual circuit approach.

may be lost. Again, it is up to either the exit node or the destination to detect the loss of a packet and decide how to recover it. In this technique, each packet, treated independently, is referred to as a datagram.

In the virtual circuit approach, a preplanned route is established before any packets are sent. Once the route is established, all of the packets between a pair of communicating parties follow this same route through the network. This is illustrated in Fig. 82.5(b). Because the route is fixed for the duration of the logical connection, it is somewhat similar to a circuit in a circuit-switching network and is referred to as a virtual circuit. Each packet now contains a virtual circuit identifier as well as data. Each node on the pre-established route knows where to direct such packets; no routing decisions are required. At any time, each station can have more than one virtual circuit to any other station and can have virtual circuits to more than one station.

If two stations wish to exchange data over an extended period of time, there are certain advantages to virtual circuits. First, the network may provide services related to the virtual circuit, including sequencing and error control. *Sequencing* refers to the fact that, because all packets follow the same route, they arrive in the original order. Error control is a service that assures not only that packets arrive in proper sequence but also that all packets arrive correctly. For example, if a packet in a sequence from node 4 to node 6 fails to arrive at node 6, or arrives with an error, node 6 can request a retransmission of that packet from node 4. Another advantage is that packets should transit the network more rapidly with a virtual circuit; it is not necessary to make a routing decision for each packet at each node.

One advantage of the datagram approach is that the call setup phase is avoided. Thus, if a station wishes to send only one or a few packets, datagram delivery will be quicker. Another advantage of the datagram service is that, because it is more primitive, it is more flexible. For example, if congestion develops in one part of the network, incoming datagrams can be routed away from the congestion. With the use of virtual circuits, packets follow a predefined route, and thus it is more difficult for the network to adapt to congestion. A third advantage is that datagram delivery is inherently more reliable. With the use of virtual circuits, if a node fails, all virtual circuits that pass through that node are lost. With datagram delivery, if a node fails, subsequent packets may find an alternative route that bypasses that node.

High-Speed Wide-Area Networks

As the speed and number of local-area networks continue their relentless growth, increasing demand is placed on wide-area packet-switching networks to support the tremendous throughput generated by these LANs. In the early days of wide-area networking, X.25 was designed to support direct connection of terminals and computers over long distances. At speeds up to 64 kb/s or so, X.25 copes well with these demands. As LANs have come to play an increasing role in the local environment, X.25, with its substantial overhead, is being recognized as an inadequate tool for wide-area networking. Fortunately, several new generations of high-speed switched services for wide-area networking are moving rapidly from the research laboratory and the draft standard stage to the commercially available, standardized-product stage. The two most important such technologies are **frame relay** and **asynchronous transfer mode** (ATM).

Frame Relay

Frame relay provides a streamlined technique for wide-area packet switching, compared to X.25 [Black 1994b, Smith 1993]. It provides superior performance by eliminating as much as possible of the overhead of X.25. The key differences of frame relaying from a conventional X.25 packet-switching service are as follows:

- Call control signaling (e.g., requesting that a connection be set up) is carried on a logical connection that is separate from the connections used to carry user data. Thus, intermediate nodes need not maintain state tables or process messages relating to call control on an individual per-connection basis.
- There are only physical and link layers of processing for frame relay, compared to physical, link, and packet layers for X.25. Thus, one entire layer of processing is eliminated with frame relay.

- There is no hop-by-hop flow control and error control. End-to-end flow control and error control is the responsibility of a higher layer, if it is employed at all.

Frame relay takes advantage of the reliability and fidelity of modern digital facilities to provide faster packet switching than X.25. Whereas X.25 typically operates only up to speeds of about 64 kb/s, frame relay is designed to work at access speeds up to 2 Mb/s.

Transmission of data by X.25 packets involves considerable overhead. At each hop through the network, the data link control protocol involves the exchange of a data frame and an acknowledgment frame. Furthermore, at each intermediate node, state tables must be maintained for each virtual circuit to deal with the call management and flow control/error control aspects of the X.25 protocol. In contrast, with frame relay a single user data frame is sent from source to destination, and an acknowledgment, generated at a higher layer, is carried back in a frame.

Let us consider the advantages and disadvantages of this approach. The principal potential disadvantage of frame relaying, compared to X.25, is that we have lost the ability to do link-by-link flow and error control. (Although frame relay does not provide end-to-end flow and error control, this is easily provided at a higher layer.) In X.25, multiple virtual circuits are carried on a single physical link, and link access procedure to frame mode bearer service (LAPB) is available at the link level for providing reliable transmission from the source to the packet-switching network and from the packet-switching network to the destination. In addition, at each hop through the network, the link control protocol can be used for reliability. With the use of frame relaying, this hop-by-hop link control is lost. However, with the increasing reliability of transmission and switching facilities, this is not a major disadvantage.

The advantage of frame relaying is that we have streamlined the communications process. The protocol functionality required at the user–network interface is reduced, as is the internal network processing. As a result, lower delay and higher throughput can be expected. Preliminary results indicate a reduction in frame processing time of an order of magnitude.

The frame relay data transfer protocol consists of the following functions:

- Frame delimiting, alignment, and transparency
- Frame multiplexing/demultiplexing using the address field
- Inspection of the frame to ensure that it consists of an integer number of octets (8-b bytes) prior to zero-bit insertion or following zero-bit extraction
- Inspection of the frame to ensure that it is neither too long nor too short
- Detection of transmission errors
- Congestion control functions

This architecture reduces to the bare minimum the amount of work accomplished by the network. User data are transmitted in frames with virtually no processing by the intermediate network nodes other than to check for errors and to route based on connection number. A frame in error is simply discarded, leaving error recovery to higher layers.

The operation of frame relay for user data transfer is best explained by beginning with the frame format, illustrated in Fig. 82.6. The format is similar to that of other data link control protocols, such as HDLC and LAPB, with one omission: there is no control field. In traditional data link control protocols, the control field is used for the following functions:

- Part of the control field identifies the frame type. In addition to a frame for carrying user data, there are various control frames. These carry no user data but are used for various protocol control functions, such as setting up and tearing down logical connections.
- The control field for user data frames includes send and receive sequence numbers. The send sequence number is used to sequentially number each transmitted frame. The receive sequence number is used to provide a positive or negative acknowledgment to incoming frames. The use of sequence numbers allows the receiver to control the rate of incoming frames (flow control) and to report missing or damaged frames, which can then be retransmitted (error control).

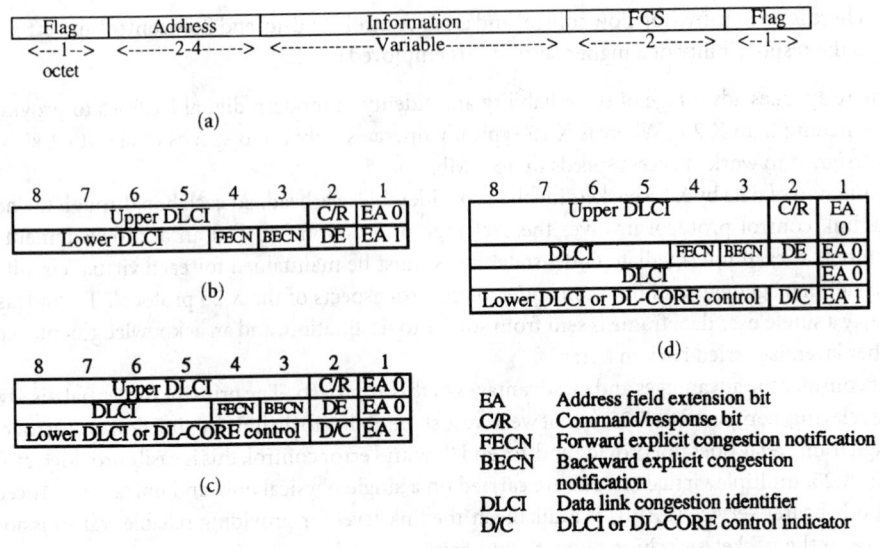

FIGURE 82.6 Frame relay formats: (a) frame format; (b) address field, 2 octets (default); (c) address field, 3 octets; and (d) address field, 4 octets.

The lack of a control field in the frame relay format means that the process of setting up and tearing down connections must be carried out on a separate channel at a higher layer of software. It also means that it is not possible to perform flow control and error control.

The flag and frame check sequence (FCS) fields function as in HDLC and other traditional data link control protocols. The flag field is a unique pattern that delimits the start and end of the frame. The FCS field is used for error detection. On transmission, the FCS checksum is calculated and stored in the FCS field. On reception, the checksum is again calculated and compared to the value stored in the incoming FCS field. If there is a mismatch, then the frame is assumed to be in error and is discarded.

The information field carries higher layer data. The higher layer data may be either user data or call control messages, as explained subsequently.

The address field has a default length of 2 octets and may be extended to 3 or 4 octets. It carries a data link connection identifier (DLCI) of 10, 17, or 24 b. The DLCI serves the same function as the virtual circuit number in X.25: it allows multiple logical frame relay connections to be multiplexed over a single channel. As in X.25, the connection identifier has only local significance: each end of the logical connection assigns its own DLCI from the pool of locally unused numbers, and the network must map from one to the other. The alternative, using the same DLCI on both ends, would require some sort of global management of DLCI values.

The length of the address field, and hence of the DLCI, is determined by the address field extension (EA) bits. The C/R bit is application specific and not used by the standard frame relay protocol. The remaining bits in the address field have to do with congestion control.

Asynchronous Transfer Mode

As the speed and number of local-area networks continue their relentless growth, increasing demand is placed on wide-area packet-switching networks to support the tremendous throughput generated by these LANs. In the early days of wide-area networking, X.25 was designed to support direct connection of terminals and computers over long distances. At speeds up to 64 kb/s or so, X.25 copes well with these demands. As LANs have come to play an increasing role in the local environment, X.25, with its substantial overhead, is being recognized as an inadequate tool for wide-area networking. This has led to increasing interest in frame relay, which is designed to support access speeds up to 2 Mb/s. But, as we look to the not-too-distant future, even the streamlined design of frame relay will falter in the face of a requirement for wide-area access speeds in the tens and hundreds of megabits per second. To accommodate these gargantuan requirements, a new technology is emerging: asynchronous transfer mode (ATM), also known as cell relay [Boudec 1992, Prycker 1993].

Cell relay is similar in concept to frame relay. Both frame relay and cell relay take advantage of the reliability and fidelity of modern digital facilities to provide faster packet switching than X.25. Cell relay is even more streamlined than frame relay in its functionality and can support data rates several orders of magnitude greater than frame relay.

ATM is a packet-oriented transfer mode. Like frame relay and X.25, it allows multiple logical connections to be multiplexed over a single physical interface. The information flow on each logical connection is organized into fixed-size packets, called cells. As with frame relay, there is no link-by-link error control or flow control.

Logical connections in ATM are referred to as virtual channels. A virtual channel is analogous to a virtual circuit in X.25 or a frame-relay logical connection. A virtual channel is set up between two end users through the network and a variable-rate, full-duplex flow of fixed-size cells is exchanged over the connection. Virtual channels are also used for user–network exchange (control signaling) and network–network exchange (network management and routing).

For ATM, a second sublayer of processing has been introduced that deals with the concept of virtual path. A virtual path is a bundle of virtual channels that have the same endpoints. Thus, all of the cells flowing over all of the virtual channels in a single virtual path are switched together.

Several advantages can be listed for the use of virtual paths:

- *Simplified network architecture:* Network transport functions can be separated into those related to an individual logical connection (virtual channel) and those related to a group of logical connections (virtual path).
- *Increased network performance and reliability:* The network deals with fewer, aggregated entities.
- *Reduced processing and short connection setup time:* Much of the work is done when the virtual path is set up. The addition of new virtual channels to an existing virtual path involves minimal processing.
- *Enhanced network services:* The virtual path is internal to the network but is also visible to the end user. Thus, the user may define closed user groups or closed networks of virtual channel bundles.

International Telecommunications Union–Telecommunications Standardization Sector (ITU-T) Recommendation I.150 lists the following as characteristics of virtual channel connections:

- *Quality of service:* A user of a virtual channel is provided with a quality of service specified by parameters such as cell loss ratio (ratio of cells lost to cells transmitted) and cell delay variation.
- *Switched and semipermanent virtual channel connections:* Both switched connections, which require call-control signaling, and dedicated channels can be provided.
- *Cell sequence integrity:* The sequence of transmitted cells within a virtual channel is preserved.
- *Traffic parameter negotiation and usage monitoring:* Traffic parameters can be negotiated between a user and the network for each virtual channel. The input of cells to the virtual channel is monitored by the network to ensure that the negotiated parameters are not violated.

The types of traffic parameters that can be negotiated would include average rate, peak rate, burstiness, and peak duration. The network may need a number of strategies to deal with congestion and to manage existing and requested virtual channels. At the crudest level, the network may simply deny new requests for virtual channels to prevent congestion. Additionally, cells may be discarded if negotiated parameters are violated or if congestion becomes severe. In an extreme situation, existing connections might be terminated.

Recommendation I.150 also lists characteristics of virtual paths. The first four characteristics listed are identical to those for virtual channels. That is, quality of service, switched and semipermanent virtual paths, cell sequence integrity, and traffic parameter negotiation and usage monitoring are all also characteristics

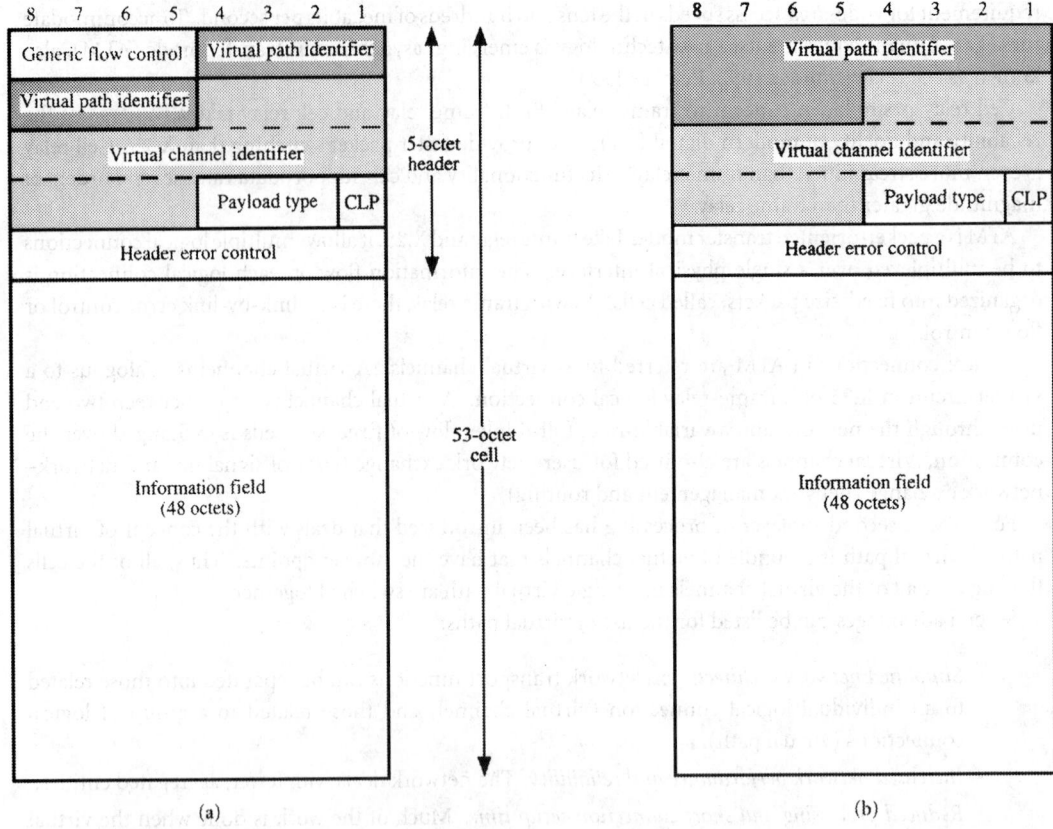

FIGURE 82.7   ATM cell format: (a) user–network interface and (b) network–network interface.

of a virtual path. There are a number reasons for this duplication. First, this provides some flexibility in how the network manages the requirements placed on it. Second, the network must be concerned with the overall requirements for a virtual path and, within a virtual path, may negotiate the establishment of virtual circuits with given characteristics. Finally, once a virtual path is set up, it is possible for the end users to negotiate the creation of new virtual channels. The virtual path characteristics impose a discipline on the choices that the end users may make.

In addition, a fifth characteristic is listed for virtual paths:

- *Virtual channel identifier restriction within a virtual path:* One or more virtual channel identifiers, or numbers, may not be available to the user of the virtual path but may be reserved for network use. Examples would be virtual channels used for network management.

The asynchronous transfer mode makes use of fixed-size cells, consisting of a 5-octet header and a 48-octet information field (Fig. 82.7). There are several advantages to the use of small, fixed-size cells. First, the use of small cells may reduce queuing delay for high-priority cells, because it waits less if it arrives slightly behind a lower priority cell that has gained access to a resource (e.g., the transmitter). Second, it appears that fixed-size cells can be switched more efficiently, which is important for the very high data rates of ATM.

Figure 82.7(a) shows the header format at the user–network interface. Multiple terminals may share a single access link to the network. The generic flow control field is to be used for end-to-end flow control. The details of its application are for further study. The field could be used to assist the customer in controlling the flow of traffic for different qualities of service. One candidate for the use of this field is a multiple-priority level indicator to control the flow of information in a service-dependent manner.

The virtual path identifier and virtual channel identifier fields constitute a routing field for the network. The virtual path identifier indicates a user-to-user or user-to-network virtual path. The virtual channel identifier indicates a user-to-user or user-to-network virtual channel. These identifiers have local significance (as with X.25 and frame relay) and may change as the cell traverses the network.

The payload type field indicates the type of information in the information field. A value of 00 indicates user information; that is, information from the next higher layer. Other values are for further study. Presumably, network management and maintenance values will be assigned. This field allows the insertion of network-management cells onto a user's virtual channel without impacting user's data. Thus, it could provide in-band control information.

The cell loss priority is used to provide guidance to the network in the event of congestion. A value of 0 indicates a cell of relatively higher priority, which should not be discarded unless no other alternative is available. A value of 1 indicates that this cell is subject to discard within the network. The user might employ this field so that extra information may be inserted into the network, with a CLP of 1, and delivered to the destination if the network is not congested. The network sets this field to 1 for any data cell that is in violation of a traffic agreement. In this case, the switch that does the setting realizes that the cell exceeds the agreed traffic parameters but that the switch is capable of handling the cell. At a later point in the network, if congestion is encountered, this cell has been marked for discard in preference to cells that fall within agreed traffic limits.

The header error control (HEC) field is an 8-b error code that can be used to correct single-bit errors in the header and to detect double-bit errors.

Figure 82.7(b) shows the cell header format internal to the network. The generic flow control field, which performs end-to-end functions, is not retained. Instead, the virtual path identifier field is expanded from 8 to 12 b. This allows support for an expanded number of virtual paths internal to the network to include those supporting subscribers and those required for network management.

Traditional Local-Area Networks

The two most widely used traditional LANs are carrier-sense multiple access/collision detection (CSMA/CD) (Ethernet) and token ring.

Carrier-Sense Multiple Access/Collision Detection (Ethernet)

The Ethernet LAN standards was originally designed to work over a bus LAN topology. With the bus topology, all stations attach, through appropriate interfacing hardware, directly to a linear transmission medium, or bus. A transmission from any station propagates the length of the medium in both directions and can be received by all other stations.

Transmission is in the form of frames containing addresses and user data. Each station monitors the medium and copies frames addressed to itself. Because all stations share a common transmission link, only one station can successfully transmit at a time, and some form of medium access control technique is needed to regulate access.

More recently, a star topology has been used. In the star LAN topology, each station attaches to a central node, referred to as the star coupler, via two point-to-point links, one for transmission in each direction. A transmission from any one station enters the central node and is retransmitted on all of the outgoing links. Thus, although the arrangement is physically a star, it is logically a bus: a transmission from any station is received by all other stations, and only one station at a time may successfully transmit. Thus, the medium access control techniques used for the star topology are the same as for bus and tree.

With CSMA/CD, a station wishing to transmit first listens to the medium to determine if another transmission is in progress (carrier sense). If the medium is idle, the station may transmit. It may happen that two or more stations attempt to transmit at about the same time. If this happens, there will be a collision; the data from both transmissions will be garbled and not received successfully. Thus, a procedure is needed that specifies what a station should do if the medium is found busy and what it should do if a collision occurs:

1. If the medium is idle, transmit.
2. If the medium is busy, continue to listen until the channel is idle, then transmit immediately.
3. If a collision is detected during transmission, immediately cease transmitting.
4. After a collision, wait a random amount of time and then attempt to transmit again (repeat from step 1).

Figure 82.8 illustrates the technique. At time t_0, station A begins transmitting a packet addressed to D. At t_1, both B and C are ready to transmit. B senses a transmission and so defers. C, however, is still unaware of A's transmission and begins its own transmission. When A's transmission reaches C, at t_2, C detects the collision and ceases transmission. The effect of the collision propagates back to A, where it is detected some time later, t_3, at which time A ceases transmission.

The Institute of Electrical and Electronics Engineers (IEEE) LAN standards committee has developed a number of versions of the CSMA/CD standard, all under the designation IEEE 802.3. The following options are defined:

- 10-Mb/s bus topology using coaxial cable
- 10-Mb/s star topology using unshielded twisted pair
- 100-Mb/s star topology using unshielded twisted pair
- 100-Mb/s star topology using optical fiber

The last two elements in the list, both known as fast Ethernet, are the newest addition to the IEEE 802.3 standard. Both provide a higher data rate over shorter distances than traditional Ethernet.

Token Ring

The token ring LAN standards operates over a ring topology LAN. In the ring topology, the LAN or metropolitan-area network (MAN) consists of a set of *repeaters* joined by point-to-point links in a closed loop. The repeater is a comparatively simple device, capable of receiving data on one link and transmitting it, bit by bit, on the other link as fast as it is received, with no buffering at the repeater. The links are unidirectional; that is, data are transmitted in one direction only and all oriented in the same way. Thus, data circulate around the ring in one direction (clockwise or counterclockwise).

Each station attaches to the network at a repeater and can transmit data onto the network through the repeater.

As with the bus topology, data are transmitted in frames. As a frame circulates past all of the other stations, the destination station recognizes its address and copies the frame into a local buffer as it goes by. The frame continues to circulate until it returns to the source station, where it is removed.

Because multiple stations share the ring, medium access control is needed to determine at what time each station may insert frames.

The token ring technique is based on the use of a token packet that circulates when all

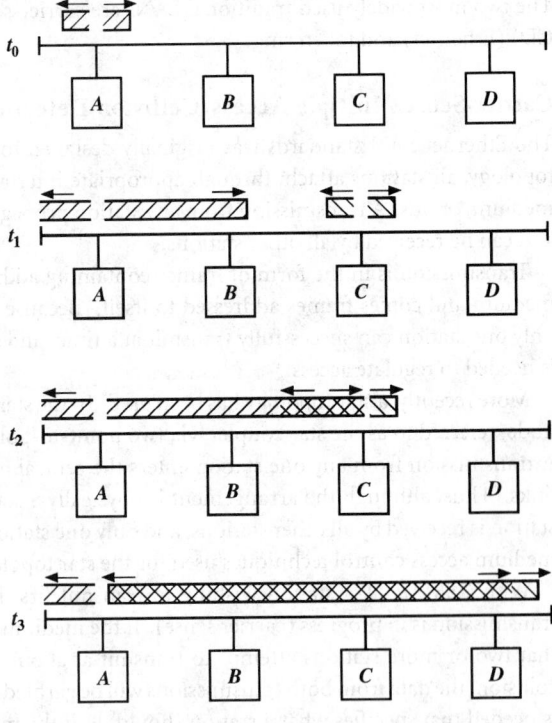

FIGURE 82.8 CSMA/CD operation.

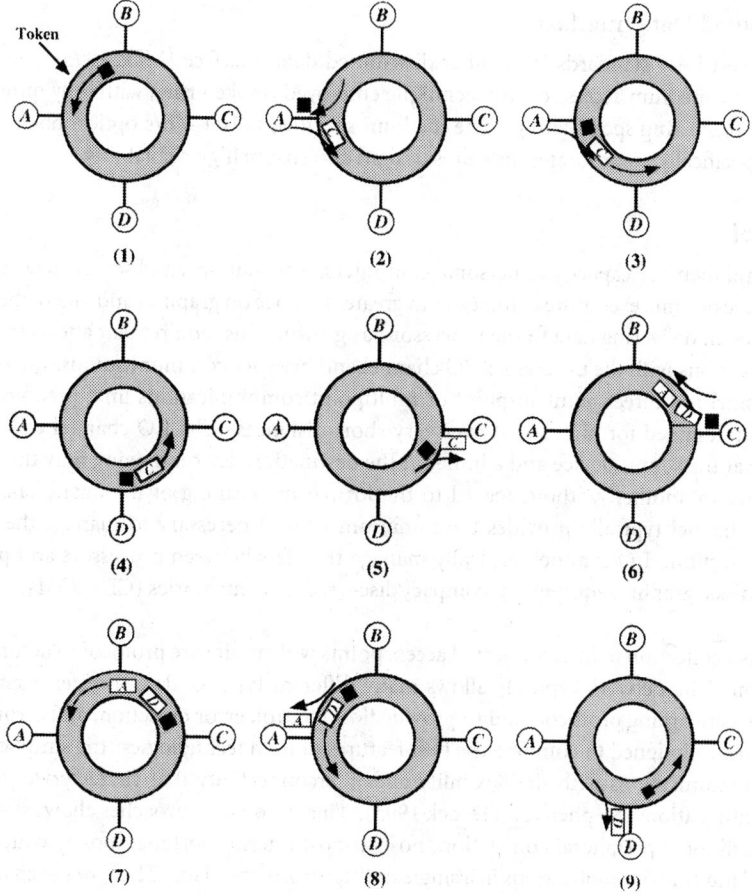

FIGURE 82.9 Token ring operation.

stations are idle. A station wishing to transmit must wait until it detects a token passing by. It then seizes the token by changing 1 b in the token, which transforms it from a token to a start-of-packet sequence for a data packet. The station then appends and transmits the remainder of the fields (e.g., destination address) needed to construct a data packet.

There is now no token on the ring, so other stations wishing to transmit must wait. The packet on the ring will make a round trip and be purged by the transmitting station. The transmitting station will insert a new token on the ring after it has completed transmission of its packet. Once the new token has been inserted on the ring, the next station downstream with data to send will be able to seize the token and transmit. Figure 82.9 illustrates the technique. In the example, *A* sends a packet to *C*, which receives it and then sends its own packets to *A* and *D*.

The IEEE 802.5 subcommittee of IEEE 802 has developed a token ring standard with the following alternative configurations:

- Unshielded twisted pair at 4 Mb/s
- Shielded twisted pair at 4 or 16 Mb/s

High-Speed Local-Area Networks

In recent years, the increasing traffic demands placed on LANs has led to the development of a number of high-speed LAN alternatives. The three most important are fiber distributed data interface (FDDI), Fibre Channel, and ATM LANs.

Fiber Distributed Data Interface

One of the newest LAN standards is the fiber distributed data interface [Mills 1995]. The topology of FDDI is ring. The medium access control technique employed is token ring, with only minor differences from the IEEE token ring specification. The medium specified is 100-Mb/s optical fiber. The medium specification specifically incorporates measures designed to ensure high availability.

Fibre Channel

As the speed and memory capacity of personal computers, workstations, and servers have grown, and as applications have become ever more complex with greater reliance on graphics and video, the requirement for greater speed in delivering data to the processor has grown. This requirement affects two methods of data communications with the processor: I/O channel and network communications.

An I/O channel is a direct point-to-point or multipoint communications link, predominantly hardware based and designed for high speed over very short distances. The I/O channel transfers data between a buffer at the source device and a buffer at the destination device, moving only the user contents from one device to another, without regard to the format or meaning of the data. The logic associated with the channel typically provides the minimum control necessary to manage the transfer plus simple error detection. I/O channels typically manage transfers between processors and peripheral devices, such as disks, graphics equipment, compact disc–read-only memories (CD-ROMs), and video I/O devices.

A network is a collection of interconnected access points with a software protocol structure that enables communication. The network typically allows many different types of data transfer, using software to implement the networking protocols and to provide flow control, error detection, and error recovery.

Fibre Channel is designed to combine the best features of both technologies: the simplicity and speed of channel communications with the flexibility and interconnectivity that characterize protocol-based network communications [Stephens and Dedek 1995]. This fusion of approaches allows system designers to combine traditional peripheral connection, host-to-host internetworking, loosely coupled processor clustering, and multimedia applications in a single multiprotocol interface. The types of channel-oriented facilities incorporated into the Fibre Channel protocol architecture include:

- Data-type qualifiers for routing frame payload into particular interface buffers
- Link-level constructs associated with individual I/O operations
- Protocol interface specifications to allow support of existing I/O channel architectures, such as the small computer system interface (SCSI)

The types of network-oriented facilities incorporated into the Fibre Channel protocol architecture include:

- Full multiplexing of traffic between multiple destinations
- Peer-to-peer connectivity between any pair of ports on a Fiber Channel network
- Capabilities for internetworking to other connection technologies.

Depending on the needs of the application, either channel or networking approaches can be used for any data transfer.

Fibre Channel is based on a simple generic transport mechanism based on point-to-point links and a switching network. This underlying infrastructure supports a simple encoding and framing scheme that in turn supports a variety of channel and network protocols.

The key elements of a Fibre Channel network are the end systems, called *nodes*, and the network itself, which consists of one or more switching elements. The collection of switching elements is referred to as a *fabric*. These elements are interconnected by point-to-point links between ports on the individual nodes and switches. Communication consists of the transmission of frames across the point-to-point links.

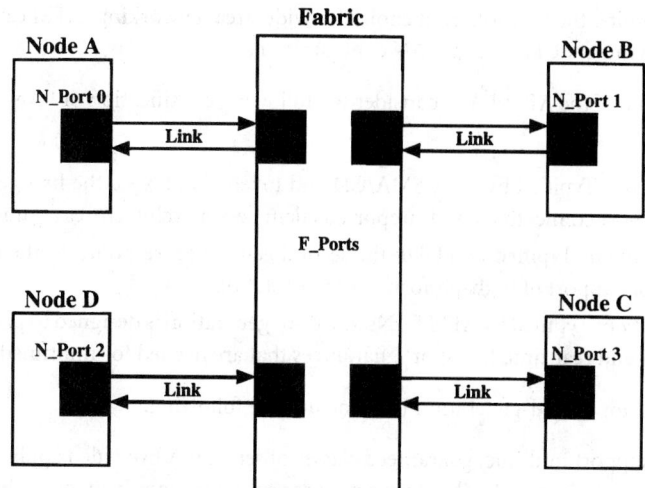

FIGURE 82.10   Fibre Channel port types.

Figure 82.10 illustrates these basic elements. Each node includes one or more ports, called N_Ports, for interconnection. Similarly, each fabric switching element includes one or more ports, called F_ports. Interconnection is by means of bidirectional links between ports. Any node can communicate with any other node connected to the same fabric using the services of the fabric. All routing of frames between N_Ports is done by the fabric. Frames are buffered within the fabric, making it possible for different nodes to connect to the fabric at different data rates.

A fabric can be implemented as a single fabric element, as depicted in Fig. 82.10, or as a more general network of fabric elements. In either case, the fabric is responsible for buffering and routing frames between source and destination nodes.

The Fibre Channel network is quite different from the other LANs that we have examined so far. Fibre Channel is more like a tradional circuit-switching or packet-switching network in contrast to the typical shared-medium LAN. Thus, Fibre Channel need not be concerned with medium access control (MAC) issues. Because it is based on a switching network, the Fibre Channel scales easily in terms of both data rate and distance covered. This approach provides great flexibility. Fibre Channel can readily accommodate new transmission media and data rates by adding new switches and nodes to an existing fabric. Thus, an existing investment is not lost with an upgrade to new technologies and equipment. Further, as we shall see, the layered protocol architecture accommodates existing I/O interface and networking protocols, preserving the pre-existing investment.

Asynchronous Transfer Mode Local-Area Networks

High-speed LANs such as FDDI and Fiber Channel, provide a means for implementing a backbone LAN to tie together numerous small LANs in an office environment. However, there is another solution, known as the ATM LAN, that seems likely to become a major factor in local-area networking [Biagioni et al. 1993, Newman 1994]. The ATM LAN is based on the asynchronous ATM technology used in wide-area networks. The ATM LAN approach has several important strengths, two of which are as follows:

1. The ATM technology provides an open-ended growth path for supporting attached devices. ATM is not constrained to a particular physical medium or data rate. A dedicated data rate between workstations of 155 Mb/s is practical today. As demand increases and prices continue to drop, ATM LANs will be able to support devices at dedicated speeds, which are standardized for ATM, of 622 Mb/s, 2.5 Gb/s, and above.

2. ATM is becoming the technology of choice for wide-area networking. ATM can therefore be used effectively to integrate LAN and WAN configurations.

To understand the role of the ATM LAN, consider the following classification of LANs into three generations:

- *First generation:* Typified by the CSMA/CD and token ring LANs, the first generation provided terminal-to-host connectivity and supported client/server architectures at moderate data rates.
- *Second generation:* Typified by FDDI, the second generation responds to the need for backbone LANs and for support of high-performance workstations.
- *Third generation:* Typified by ATM LANs, the third generation is designed to provide the aggregate throughputs and real-time transport guarantees that are needed for multimedia applications.

Typical requirements for a third-generation LAN include the following:

1. They must support multiple, guaranteed classes of service. A live video application, for example, may require a guaranteed 2-Mb/s connection for acceptable performance, whereas a file transfer program can utilize a *background* class of service.
2. They must provide scalable throughput that is capable of growing both per-host capacity (to enable applications that require large volumes of data in and out of a single host) and aggregate capacity (to enable installations to grow from a few to several hundred high-performance hosts).
3. They must facilitate the interworking between LAN and WAN technology.

ATM is ideally suited to these requirements. Using virtual paths and virtual channels, multiple classes of service are easily accommodated, either in a preconfigured fashion (permanent connections) or on demand (switched connections). ATM is easily scalable by adding more ATM switching nodes and using higher data rates for attached devices. Finally, with the increasing acceptance of cell-based transport for wide-area networking, the use of ATM for a premises network enables seamless integration of LANs and WANs.

The term ATM LAN has been used by vendors and researchers to apply to a variety of configurations. At the very least, ATM LAN implies the use of ATM as a data transport protocol somewhere within the local premises. Among the possible types of ATM LANs are the following:

- *Gateway to ATM WAN:* An ATM switch acts as a router and traffic concentrator for linking a premises network complex to an ATM WAN.
- *Backbone ATM switch:* Either a single ATM switch or a local network of ATM switches interconnect other LANs.
- *Workgroup ATM:* High-performance multimedia workstations and other end systems connect directly to an ATM switch.

These are all pure configurations. In practice, a mixture of two or all three of these types of networks is used to create an ATM LAN.

Figure 82.11 shows an example of a backbone ATM LAN that includes links to the outside world. In this example, the local ATM network consists of four switches interconnected with high-speed point-to-point links running at the standardized ATM rates of 155 and 622 Mb/s. On the premises, there are three other LANs, each of which has a direct connection to one of the ATM switches. The data rate from an ATM switch to an attached LAN conforms to the native data rate of that LAN. For example, the connection to the FDDI network is at 100 Mb/s. Thus, the switch must include some buffering and speed conversion capability to map the data rate from the attached LAN to an ATM data rate. The ATM switch must also perform some sort of protocol conversion from the MAC protocol used on the attached LAN to the ATM cell stream used on the ATM network. A simple approach is for each ATM switch that attaches to a LAN to function as a bridge or router.

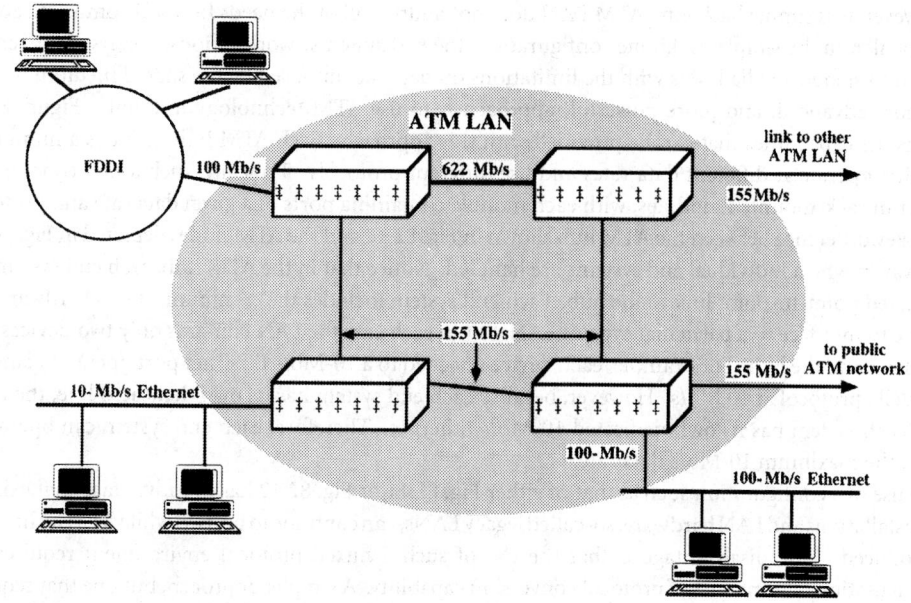

FIGURE 82.11 Example of ATM LAN configuration.

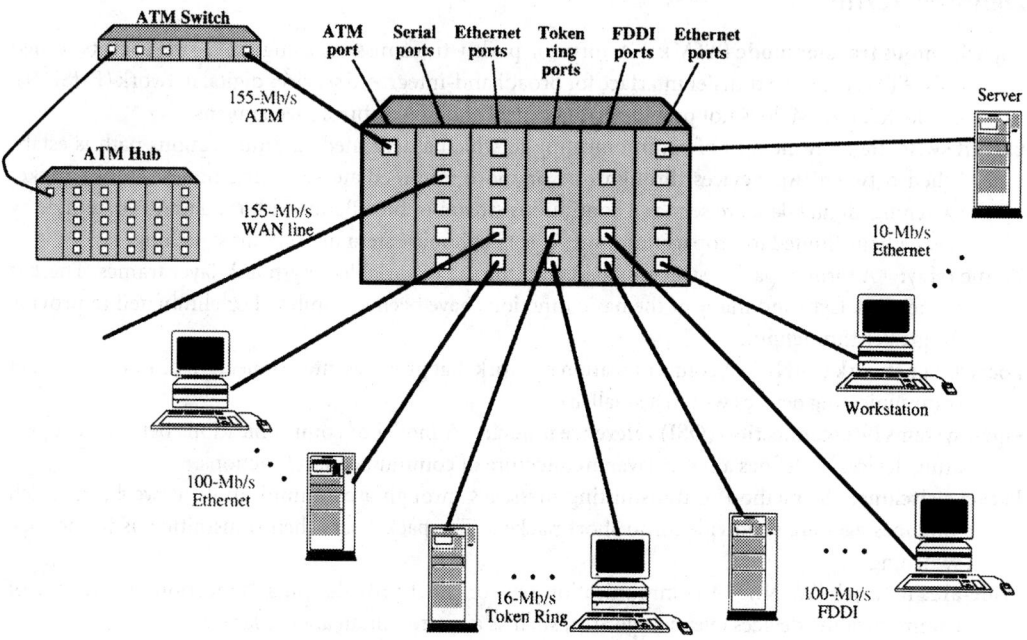

FIGURE 82.12 ATM LAN hub configuration.

An ATM LAN configuration such as that shown in Fig. 82.11 provides a relatively painless method for inserting a high-speed backbone into a local environment. As the on-site demand rises, it is a simple matter to increase the capacity of the backbone by adding more switches, increasing the throughput of each switch, and increasing the data rate of the trunks between switches. With this strategy, the load on individual LANs within the premises can be increased and the number of LANs can grow.

However, this simple backbone ATM LAN does not address all of the needs for local communications. In particular, in the simple backbone configuration, the end systems (workstations, servers, etc.) remain attached to shared-media LANs with the limitations on data rate imposed by the shared medium.

A more advanced, and more powerful, approach is to use ATM technology in a hub. Figure 82.12 suggests the capabilities that can be provided with this approach. Each ATM hub includes a number of ports that operate at different data rates and use different protocols. Typically, such a hub consists of a number of rack-mounted modules, with each module containing ports of a given data rate and protocol.

The key difference between the ATM hub shown in Fig. 82.12 and the ATM nodes depicted in Fig. 82.11 is the way in which individual end systems are handled. Notice that in the ATM hub, each end system has a dedicated point-to-point link to the hub. Each end system includes the communications hardware and software to interface to a particular type of LAN, but in each case the LAN contains only two devices: the end system and the hub! For example, each device attached to a 10-Mb/s Ethernet port operates using the CSMA/CD protocol at 10 Mb/s. However, because each end system has its own dedicated line, the effect is that each system has its own dedicated 10-Mb/s Ethernet. Therefore, each end system can operate at close to the maximum 10-Mb/s data rate.

The use of a configuration such as that of either Fig. 82.11 or Fig. 82.12 has the advantage that existing LAN installations and LAN hardware, so-called legacy LANs, can continue to be used while ATM technology is introduced. The disadvantage is that the use of such a mixed-protocol environment requires the implementation of some sort of protocol conversion capability. A simpler approach, but one that requires that end systems be equipped with ATM capability, is to implement a pure ATM LAN.

Defining Terms

Asynchronous transfer mode (ATM):   A form of packet transmission using fixed-size packets, called cells. ATM is the data transfer interface for broadband-integrated services digital network (B-ISDN). Unlike X.25, ATM does not provide error control and flow control mechanisms.

Circuit switching: A method of communicating in which a dedicated communications path is established between two devices through one or more intermediate switching nodes. Unlike packet switching, digital data are sent as a continuous stream of bits. Bandwidth is guaranteed, and delay is essentially limited to propagation time. The telephone system uses circuit switching.

Frame relay: A form of packet switching based on the use of variable-length link-layer frames. There is no network layer and many of the basic functions have been streamlined or eliminated to provide for greater throughput.

Local area network (LAN):   A communication network that provides interconnection of a variety of data communicating devices within a small area.

Open systems interconnection (OSI) reference model: A model of communications between cooperating devices. It defines a seven-layer architecture of communication functions.

Packet switching: A method of transmitting messages through a communication network, in which long messages are subdivided into short packets. The packets are then transmitted as in message switching.

Wide-area network (WAN):   A communication network that provides interconnection of a variety of communicating devices over a large area, such as a metropolitan area or larger.

References

Biagioni, E., Cooper, E., and Sansom, R. 1993. Designing a practical ATM LAN. *IEEE Network* (March).

Black, U. 1994a. *Emerging Communications Technologies.* Prentice–Hall, Englewood Cliffs, NJ.

Black, U. 1994b. *Frame Relay Networks: Specifications and Implementations.* McGraw–Hill, New York.

Boudec, J. 1992. The asynchronous transfer mode: a tutorial. *Comput. Networks ISDN Syst.* (May).

Comer, D. 1995. *Internetworking with TCP/IP, Volume I: Principles, Protocols, and Architecture.* Prentice–Hall, Englewood Cliffs, NJ.

Halsall, F. 1996. *Data Communications, Computer Networks, and Open Systems.* Addison–Wesley, Reading, MA.

Jain, B. and Agrawala, A. 1993. *Open Systems Interconnection.* McGraw–Hill, New York.

Mills, A. 1995. *Understanding FDDI.* Prentice–Hall, Englewood Cliffs, NJ.

Newman, P. 1994. ATM local area networks. *IEEE Commun. Mag.* (March).

Prycker, M. 1993. *Asynchronous Transfer Mode: Solution for Broadband ISDN.* Ellis Horwood, New York.

Smith, P. 1993. *Frame Relay: Principles and Applications.* Addison–Wesley, Reading, MA.

Stallings, W. 1997a. *Data and Computer Communications,* 5th ed. Prentice–Hall, Englewood Cliffs, NJ.

Stallings, W. 1997b. *Local and Metropolitan Area Networks,* 5th ed. Prentice–Hall, Englewood Cliffs, NJ.

Stephens, G. and Dedek, J. 1995. *Fiber Channel.* Ancot, Menlo Park, CA.

Further Information

For a more detailed discussion of the topics in this chapter, see the following: for TCP/IP, Stallings [1997a] and Comer [1995]; for OSI, Halsall [1996] and Jain and Agrawala [1993]; for traditional and high-speed WANs, Stallings [1997a] and Black [1994a]; for traditional and high-speed LANs, Stallings [1997b].

83

Routing Protocols

83.1 Introduction

Computer networking can be a very confusing field. Terms such as network, subnetwork, domain, **local area network (LAN)**, internetwork, **bridge**, router, and switch are often ill defined. Taking the simplest view, we know that the **data link layer** of a network delivers a packet of information to a neighboring machine, the **network layer** routes through a series of packet switches to deliver a packet from source to destination, and the transport layer recovers from lost, duplicated, and out-of-order packets. But the bridge standards choose to place routing in the data link layer, and the X.25 network layer puts the onus on the network layer to prevent packet loss, duplication, or misordering.

In this chapter we discuss routing protocols, attempting to avoid philosophical questions such as which layer something is, or whether something is an internetwork or a network. For a more complete treatment of these and other questions, see Perlman [1992] and other chapters in this section of this Handbook.

A network consists of several computers interconnected with various types of links. One type is a *point-to-point link*, which connects exactly two machines. It can either be a dedicated link (e.g., a wire connecting the two machines) or a dial-on-demand link, which can be connected when needed (e.g., when there is traffic to send over the link). Another category of link is a **multiaccess link**. Examples are LANs, asynchronous transfer mode (ATM), and X.25. Although X.25 and ATM might be considered networks themselves, they also can be considered as links connecting routers by some protocol running in the routers attached to the ATM or X.25 **cloud**. A multiaccess link presents special challenges to the routing protocol because it has two headers with addressing information. One header (which for simplicity we call the network layer header) gives the addresses of the ultimate source and destination. The other header (which for simplicity we call the data link header) gives the addresses of the transmitter and receiver on that particular link. (See Fig. 83.1, which shows a path incorporating a multiaccess link and a packet with two headers.) Again, the terminology cannot be taken too seriously. If the multiaccess link is something like ATM, some might object to referring to the ATM header as a data link address, since they would claim ATM is a network in its own right, but in the case of a network using ATM as a link, ATM is viewed as a data link.

We start by describing the routing protocols used by so-called bridges, and then describe the routing protocols used by routers. Although the purpose of this chapter is to discuss the algorithms generically, the ones we have chosen are ones that are in widespread use. In many cases we state timer values and field lengths chosen by the implementation, but it should be understood that these are general-purpose

0-8493-2909-4/97/$0.00+$.50
© 1997 by CRC Press, Inc.

LAN

R1   R2

S   D

packet as seen on LAN

D=R2, S=R1	D=D, S=S	data

data link header   network layer header

FIGURE 83.1 A multiaccess link and a packet with two headers.

algorithms. The purpose of this chapter is not as a reference on the details of particular implementations, but to understand the variety of routing algorithms and the tradeoffs between them.

83.2 Bridges

The characteristic of a bridge that differentiates it from a router is that a bridge does its routing in the data link layer, whereas a router does it in the network layer. But that becomes a matter of philosophy and history in which standards body defined a particular algorithm rather than any property of the protocol itself. If the protocol was defined by a data link layer standards body the box implementing it becomes a bridge. If the same protocol were defined by a network layer standards body the box implementing it would become a router. The algorithm itself could in theory be implemented in either layer.

There were routers before there were bridges. What happened was the invention of the so-called local area network, which is a multiaccess link. Unfortunately, the world did not think of a LAN as a multiaccess link, which would be a component in some larger network. Instead, perhaps because the inclusion of the word network into the name local area network, many systems were designed with the assumption that the LAN itself was the entire network, and the systems were designed without a network layer, making these protocols unroutable, at least through the existing network layer protocols.

There are two types of bridging technologies in widespread use. One is known as the transparent bridge. This technology had as a goal complete backward compatibility with existing LAN-only systems. The other technique is known as source route bridging, which can only be considered different from a network layer protocol because the standards committee that adopted it was empowered to define data link protocols, and because the fields necessary for running this protocol were stuck into a header that was defined as a data link header.

Transparent Bridging

The goal of transparent bridging was to invent a box that would interconnect LANs even though the stations were designed with protocols that only worked on a LAN, i.e., they lacked a network layer able to cooperate with routers, devices that were designed for forwarding packets.

The basic idea of a transparent bridge is something that is attached to two or more LANs. On each LAN, the bridge listens promiscuously to all packets, stores them, and forwards each packet onto each other LAN when given permission by the LAN protocol on that LAN. An enhancement is to have the bridge learn, based on addresses in the LAN header, of packets received by the bridge, where the stations reside, so that the bridge does not unnecessarily forward packets. (See Fig. 83.2, where a bridge has learned some of the station addresses.) The bridge learns from the source field in the LAN header, and forwards based on the destination address. For example, in Fig. 83.2, when S transmits a packet with destination

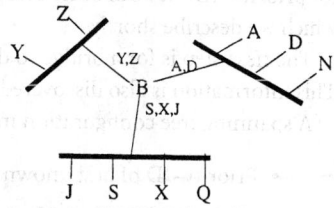

FIGURE 83.2 A bridge learning station addresses.

address D, the bridge learns which interface S resides on, and then looks to see if it has already learned where D resides. If the bridge does not know where D is, then the bridge forwards the packet onto all interfaces (except the one from which the packet was received). If the bridge does know where D is, then the bridge forwards it only onto the interface where D resides, or if the packet arrived from that interface, the bridge discards the packet.

This simple idea only works in a loop-free topology. Loops create problems:

- Packets that will not die: there is no **hop** count in the header, as there would be in a reasonable network layer protocol, to eliminate a packet that is traversing a loop.
- Packets that proliferate uncontrollably: network layer forwarding does not, in general, create duplicate packets, since each router forwards the packet in exactly one direction, and specifies the next recipient. With bridges, a bridge might forward a packet in multiple directions, and many bridges on the LAN might forward a packet. Each time a bridge forwards in multiple directions, or more than one bridge picks up a packet for forwarding, the number of copies of the packet grows.
- If there is more than one path from a bridge to a given station, that bridge cannot learn the location of the station, since packets from that source will arrive on multiple interfaces.

One possibility was to simply declare that bridged topologies must be physically loop free. But that was considered unacceptable for the following reasons:

- The consequences of accidental misconfiguration could be disastrous; any loop in some remote section of the bridged network might spawn such an enormous number of copies of packets that it would bring down the entire bridged network. It would also be very difficult to diagnose and to fix.
- Loops are good, since a loop indicates an alternate path in case of failure.

The solution was the spanning tree algorithm, which is constantly run by bridges to determine a loop-free subset of the current topology. Data packets are only transmitted along the tree found by the spanning tree algorithm. If a bridge or link fails, or if bridges or links start working, the spanning tree algorithm will compute a new tree. The spanning tree algorithm is described in the next section.

Spanning Tree Algorithm

The basic idea behind the spanning tree algorithm is that the bridges agree upon one bridge to be the root of the spanning tree. The tree of shortest paths from that root bridge to each LAN is the calculated spanning tree.

How do the bridges decide on the root? Each bridge comes, at manufacture time, with a globally unique **48-b IEEE 802 address**, usually one per interface. A bridge chooses one of the 48-b addresses that it owns as its identifier (ID). Since each bridge has a unique number, it is a simple matter to choose the one with the smallest number. However, because some network managers like to easily influence which bridge will be chosen, there is a configurable priority value that acts as a more significant field tacked onto the ID. The concatenated number consisting of priority and ID is used in the election, and the bridge with the smallest value is chosen as the root. The way in which the election proceeds is that each bridge assumes itself to be the root unless it hears, through spanning tree configuration messages, of a bridge with a smaller value for priority–ID. News of other bridges is learned through receipt of spanning tree configuration messages, which we describe shortly.

The next step is for a bridge to determine its best path to the root bridge and its own cost to the root. This information is also discovered through receipt of spanning tree configuration messages.

A spanning tree configuration message contains the following, among other information:

- Priority–ID of best known root
- Cost from transmitting bridge to root
- Priority–ID of transmitting bridge

A bridge keeps the best configuration message received on each of its interfaces. The fields in the message are concatenated together, from most significant to least significant, as root's priority–ID to cost to root to priority–ID of transmitting bridge. This concatenated quantity is used to compare messages. The one with the smaller quantity is considered better. In other words, only information about the best known root is relevant. Then information from the bridge closest to that root is considered, and then the priority–ID of the transmitting bridge is used to break ties.

Given a best received message on each interface, B chooses the root as follows:

- Itself, if its own priority–ID beats any of the received value, else
- The smallest received priority–ID value

B chooses its path to the root as follows:

- Itself, if it considers itself to be the root, else
- The minimum cost through each of its interfaces to the best-known root

Each interface has a cost associated with it, either as a default or configured. The bridge adds the interface cost to the cost in the received configuration message to determine its cost through that interface.

B chooses its own cost to the root as follows:

- 0, if it considers itself to be the root, else
- The cost of the minimum cost path chosen in the previous step

B now knows what it would transmit as a configuration message, since it knows the root's priority–ID, its own cost to that root, and its own priority–ID. If B's configuration message is better than any of the received configuration messages on an interface, then B considers itself to be the *designated bridge* on that interface, and transmits configuration messages on that interface. If B is not the designated bridge on an interface, then B will not transmit configuration messages on that interface.

Each bridge determines which of its interfaces are in the spanning tree. The interfaces in the spanning tree are as follows:

- The bridge's path to the root: if more than one interface gives the same minimal cost, then exactly one is chosen. Also, if this bridge is the root, then there is no such interface.
- Any interfaces for which the bridge is designated bridge are in the spanning tree.

If an interface is not in the spanning tree the bridge continues running the spanning tree algorithm, but does not transmit any data messages (messages other than spanning tree protocol messages) to that interface, and ignores any data messages received on that interface.

If the topology is considered a graph with two types of nodes, bridges and LANs, the following is the reasoning behind why this yields a tree:

- The root bridge is the root of the tree.
- The unique parent of a LAN is the designated bridge.
- The unique parent of a bridge is the interface which is the best path from that bridge to the root.

Dealing with Failures

The root bridge periodically transmits configuration messages (with a configurable timer on the order of 1 s). Each bridge transmits a configuration message on each interface for which it is designated, after receiving one on the interface which is that bridge's path to the root. If some time elapses (a configurable value with default on the order of 15 s) in which a bridge does not receive a configuration message on an interface, the configuration message learned on that interface is discarded.

In this way, roughly 15 s after the root or the path to the root has failed, a bridge will discard all information about that root, assume itself to be the root, and the spanning algorithm will compute a new tree.

Eliminating Temporary Loops

In a routing algorithm, the nodes learn information at different times. During the time after a topology change and before all nodes have adapted to the new topology there are temporary loops or temporary partitions (no way to get from some place to some other place). Since temporary loops are so disastrous with bridges (because of the packet proliferation problem), bridges are conservative about bringing an interface into the spanning tree. There is a timer (on the order of 30 s, but configurable). If an interface was not in the spanning tree, but new events convince the bridge the interface should be in the spanning tree, the bridge waits for this timer to expire before forwarding data messages to and from the interface.

Properties of Transparent Bridges

Transparent bridges have some good properties:

- They are plug-and-play, i.e., no configuration is required.
- They fulfill the goal of making no demands on end stations to interact with the bridges in any way.

They have some disadvantages:

- The topology is confined to a spanning tree, which means that some paths are not optimal.
- The spanning tree algorithm is purposely slow about starting to forward on an interface (to prevent temporary loops).

The overhead of the spanning tree algorithm is insignificant. The memory required for a bridge that has k interfaces is about $k * 50$ bytes, regardless of how large the actual network is. The bandwidth consumed per LAN (once the algorithm settles down) is a constant, regardless of the size of the network (since only the designated bridge periodically issues a spanning tree message, on the order of once a second). At worst case, for the few seconds while the algorithm is settling down after a topology change, the bandwidth on a LAN is at most multiplied by the number of bridges on that LAN (since for awhile more than one bridge on that LAN will think it is the designated bridge). The central processing unit (CPU) consumed by a bridge to run the spanning tree algorithm is also a constant, regardless of the size of the network.

Source Route Bridging

Source route bridging was a competing proposal in the IEEE 802 committee. Initially 802.1, the committee standardizing bridges, chose transparent bridges, but the source routing proposal resurfaced in the 802.5 (token ring) committee, as a method of interconnecting token rings.

Source route bridging did not have as a goal the ability to work with existing stations. As such, there is really no technical property of source route bridging that makes it natural for it to appear in the data link layer. It places as much burden on a station as a network layer protocol (e.g., IP, IPX, Appletalk, DECnet). The only reason it is considered a bridging protocol rather than a routing protocol is that it was done within IEEE 802, a committee whose charter was LANs, rather than a network layer committee. The fields to support source route bridging also appear in the LAN header (because it was defined by a data link layer committee), but the actual algorithm could, in theory, have been done in the network layer.

The idea behind source route bridging is that the data link header is expanded to include a route. The stations are responsible for discovering routes and maintaining route caches. Discovery of a route to station D is done by source S launching a special type of packet, an *all-paths explorer* packet, which spawns copies every time there is a choice of path (multiple bridges on a LAN or a bridge with more than two ports). Each copy of the explorer packet keeps a history of the route it has taken. This process, although it might be alarmingly prolific in richly connected topologies, does not spawn infinite copies of the explorer packet for two reasons:

- The maximum length route is 14 hops, and so after 14 hops the packet is no longer forwarded.
- A bridge examines the route before forwarding it onto a LAN, and will not forward onto that LAN if that LAN already appears in the route.

When D receives the (many) copies of the packet, it can choose a path based on criteria such as when it arrived (perhaps indicating the path is faster), or on length of path, or on maximum packet size along the route, which is calculated along with the route.

A route consists of an alternating list of LAN numbers and bridge numbers: 12 b is allocated for the LAN number, and 4 b for the bridge number. The bridge number at 4 b will obviously not distinguish between all of the bridges. Instead, the bridge number only distinguishes bridges that interconnect the same pair of LANs. For example, if the route is LAN A, bridge 3, LAN B, bridge 7, LAN C, and it is received by a bridge on the port that bridge considers to be LAN A, then the bridge looks forward in the route, finds the next LAN number (B), and then looks up the bridge number it has been assigned with respect to the LAN pair (A, B). If it has been assigned 3 for that pair, then it will forward the packet onto the port it has configured to be B.

There are three types of source route bridge packets:

1. Specifically routed: the route is in the header and the packet follows the specified route.
2. All-paths explorer: the packet spawns copies of itself at each route choice, and each copy keeps track of the route it has traversed so far.
3. Single copy broadcast: This acts as an all-paths explorer except that this type of packet is only accepted from ports in the spanning tree and only forwarded to ports in the spanning tree. A single copy will be delivered to the destination, and the accumulated route at the destination will be the path through the spanning tree.

To support the third type of packet, source route bridges run the same spanning tree algorithm as described in the transparent bridge section.

A source route bridge is configured with a 12-b LAN number for each of its ports, along with a 4-b bridge number for each possible pair of ports. In cases where a bridge has too many ports to make it feasible to configure a bridge number for each port pair, many implementations pretend there is an additional LAN inside the bridge, which must be configured with a LAN number, say, n. Paths through the bridge from LAN j to LAN k (where j and k are real LANs) look like they go from j to n to k. Each time a packet goes through such a bridge it uses up another available hop in the route, but the advantage of this scheme is that since no other bridge connects to LAN n, the bridge does not need to be configured with any bridge numbers; it can always use 1.

The algorithm the source route bridge follows for each type of packet is as follows:

- A specifically routed packet is received on the port that the bridge considers LAN j: do a scan through the route to find j. If j is not found, drop the packet. If j is found, scan to the next LAN in the route. If the LAN is k, and this bridge has a port configured as k, then find the bridge number specified between j and k, say, b. If this bridge is configured to be b with respect to the pair (j, k) then forward the packet onto LAN k. Otherwise drop the packet.

- An all-paths explorer packet is received on the port that the bridge considers LAN j: if j is not the last hop in the route, drop the packet. Otherwise, for each other port, forward the packet onto that port unless the destination port's LAN number is already in the accumulated route. If the destination LAN number, say, k, is not in the route, then append (b, k) to the route, where k is the destination LAN number and b is the bridge's number with respect to (j, k). If the route is already full, then drop the packet.

- A single copy broadcast is received on the port that the bridge considers LAN j: if the port from which it was received is not in the spanning tree, drop the packet, otherwise, treat it as an all-paths explorer except do not forward onto ports in the spanning tree.

The standard was written from the point of view of the bridge and did not specify end station operation. For example, there are several strategies end stations might use to maintain their route cache. If S wants to talk to D, and does not have D in its cache, S might send an all-paths explorer. Then D might at that point choose a route from the received explorers, or it might return each one to the source so that the source could make the choice. Or it might choose a route but send an explorer back to the source so that the source could independently make a route choice. Or maybe S, instead of sending an all-paths explorer, might send a single copy explorer, and D might respond with an all-paths explorer.

Properties of Source Route Bridging

Relative to transparent bridges, source route bridges have the following advantages:

- It is possible to get an optimal route from source to destination.
- It is possible to spread traffic load around the network rather than concentrating it into the spanning tree.
- It computes a maximum packet size on the path.
- A bridge that goes down will not disrupt conversations that have computed paths that do not go through that bridge.

Relative to transparent bridges, source route bridges have the following disadvantages:

- In a topology that is not physically a tree, the exponential proliferation of explorer packets is a serious bandwidth drain.
- It requires a lot of configuration.
- It makes end stations more complicated since they have to maintain a route cache.

Since source route bridging is a routing protocol that requires end station cooperation, it must in fairness be compared as well against network layer protocols. Against a network layer protocol such as IP, IPX, DECnet, Appletalk, CLNP, etc., source route bridging has the following advantages:

- It computes the maximum packet size on the path.
- Although it requires significant configuration of bridges, it does not require configuration of endnodes (as in IP, though IPX, DECnet Phase V, and Appletalk also avoid configuration of endnodes).

However, relative to network layer protocols, source route bridging has the following disadvantages:

- The exponential overhead of the all-paths explorer packets.
- The delay before routes are established and data can be exchanged, unless data are carried as an all-paths explorer or single copy broadcast.

83.3 Routers

In this section we discuss network layer protocols and routing algorithms generically. Network layer protocols can be connection oriented or connectionless. A connection-oriented protocol sets up a path, and the routers along the path of a conversation keep state about the information. A connectionless protocol just puts a source and destination address on the packet and launches it. Each packet is self-contained, and is routed independently of other packets from the same conversation. Different packets from the same source to the same destination might take different paths.

Another dimension in which network layer protocols can differ is whether they provide reliable or datagram service. Datagram is best effort service. With a reliable service, the network layer makes sure that every packet is delivered, and refuses to deliver packet n until it manages to deliver $n - 1$.

Examples of datagram connectionless network layers are IPv4, IPv6, IPX, DECnet, CLNP, Appletalk. An example of a datagram connection-oriented network layer is ATM. An example of a reliable connection-oriented network layer is X.25. The last possibility, a reliable connectionless network layer, is not possible, and fortunately there are no examples of standards attempting to accomplish this.

The distinction between connection-oriented and connectionless network layers is blurring. In a connectionless network, routers often keep caches of recently seen addresses, and forward much more efficiently when the destination is in the cache. Usually all but the first packet of a conversation is routed very quickly because the destination is in the router's cache. As such the first packet of the conversation acts as a route setup, and the routers along the path are keeping state, in some sense. Also, there is talk of adding lightweight connections to connectionless network layer protocols for bandwidth reservation or other reasons. In the header would be a field called something like *flow identifier*, which identifies the conversation, and the routers along the path would keep state about the conversation. Another connection-like feature sometimes implemented in routers is header compression, whereby neighbor routers agree on a shorthand for the header of recently seen packets. The first packet of a conversation alerts neighbors to negotiate a shorthand for packets for that conversation.

Whether the network layer is connection oriented or not (even if it is possible to categorize network layers definitively as one or the other) has no relevance to the fact that the network layer needs a routing protocol. Sometimes people think of a connection-oriented network as one in which all of the connections are already established, with a table of input port/connection ID mapping to output port/connection ID, and the only thing the router needs to do is a table lookup of input port/connection ID. If this were the case, then a router in a connection-oriented network would not need a routing protocol, but it is not the case. In order for the mapping table to be created, a route setup packet traverses the path to the destination, and a router has to make the same sort of routing decision as to how to reach the destination as it would on a per-packet basis in a connectionless network. Thus, whether the network is connectionless or not does not affect the type of routing protocol needed. Connectionless network layer protocols differ only in packet format and type of addressing. The type of routing protocol is not affected by the format of data packets and so for the purpose of discussing routing algorithms, it is not necessary for us to pick a specific network layer protocol.

Types of Routing Protocols

One categorization of routing protocols is distance vector vs. link state. Another categorization is intradomain vs. interdomain. We discuss these issues, but first we discuss the basic concepts of addressing and hierarchy in routing and addressing.

Hierarchy

A routing protocol can handle up to some sized network. Beyond that, many factors might make the routing protocol overburdened:

- Memory to hold the routing database
- CPU to compute the routing database
- Bandwidth to transmit the routing information
- The volatility of the information

It takes a routing protocol some amount of time to stabilize to new routes after a topology change. If the topology changes more frequently than the time it takes for the algorithm to settle, things will not work very well.

To support a larger network, portions of the network can be summarized to the outside world. This is similar to breaking the world up into countries, and within a country into states, and within a state into cities. It would be too difficult for the post office to know how to reach every street in the world (assuming that all of the streets had globally unique names), and so within a city, the post office knows all of the streets. But if it does not belong in your city, your post office just forwards it to the appropriate city, unless the letter belongs in a different state (or country). Then the post office just forwards it to the appropriate state (or country). It is routed to the appropriate state, then city, then street, and then finally to the destination.

With a network, assuming addresses are handed out sensibly, a logical circle can be drawn around a portion of the network, and all of the contents of the circle can be summarized with a small amount of information. For example, assuming postal addresses again, all of the U.S. could be summarized as: any addresses that start with the string US. All of Massachusetts, U.S., could be summarized as: any addresses that start with the string US.Mass'. Assuming North America was the logical place to draw a logical circle, then what would be advertised is: any addresses that start with the string US' or Canada' (see Fig. 83.3).

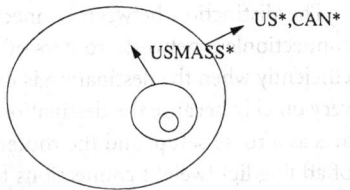

FIGURE 83.3   Network   topology with addresses.

A packet is routed according to the longest matching prefix that has been advertised. For instance, if the packet is addressed to: US.Mass.Acton.Huckleberry Lane.Radia Perlman, and there are prefixes that have been advertised for US*, US.Mass*, then the packet will be routed toward US.Mass*. If the packet originated outside the U.S., then most likely the longest prefix seen by the routing protocol outside the U.S. would be US*'. Once it reached the U.S., the advertisement US.Mass' would be visible.

Hierarchical Addressing

To support hierarchical routing, addresses must be handed out so there is some mechanism for conveniently summarizing addresses in a region. The typical method is to give an organization a block of addresses which all start with the same prefix. The organization might then hand out blocks of the addresses it owns to suborganizations. For example, suppose there are three major backbone Internet providers, and each is given a block of addresses. Say the blocks are xyz*, a*, and b* (there is no reason why the prefix has to be the same length for each provider). The provider that has the block a* has some subscribing regional providers, and gives each of them a block of addresses to give to their customers. Say there are 5 regional providers, and the blocks given out are axyc*, an*, ak*, and adkfjlk*. The regional provider with the block axye* might give out blocks to each subscribing customer network that look like axyc1*, axyc2*, axyc3*, etc. One of those customers with a large network might be careful to assign addresses within the network so that the network can be broken again into pieces that are summarizable. (See Fig. 83.4.)

With this assignment of addresses, provider 1 merely has to advertise: I can reach all addresses of the form xyz*. Typically, the network closer to the backbone advertises * outward, and a network advertises the block summarizing its own addresses toward the backbone. Thus, in Fig. 83.4, R1 would most likely advertise * to R2, and R2 would advertise axye* to R1.

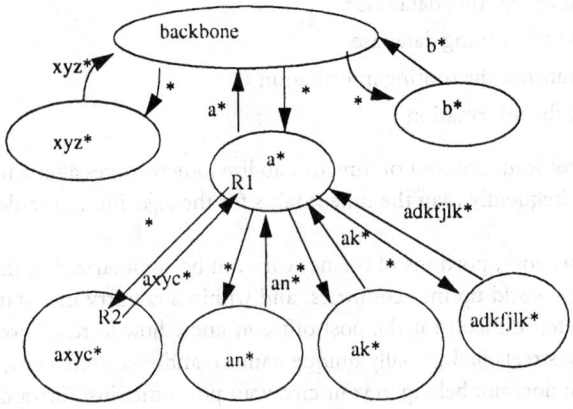

FIGURE 83.4   Block address assignment example.

Domains

What is a domain? It is a portion of a network in which the routing protocol that is running is called an intradomain routing protocol. Between domains one runs an interdomain routing protocol. Well, what is an intradomain protocol? It is something run within a domain. This probably does not help our intuition any.

Originally, the concept of a domain arose around the superstition that routing protocols were so complex that it would be impossible to get the routers from two different organizations to cooperate in a routing protocol, since the routers might have been bought from different vendors, and the metrics assigned to the links might have been assigned according to different strategies. It was thought that a routing protocol could not work under these circumstances. Yet it was important to have reachability between domains.

One possibility was to statically configure reachable addresses from other domains. But a protocol of sorts was devised, known as EGP, which was like a distance vector protocol but without exchanging metrics. It only specified what addresses were reachable, and only worked if the topology of domains was a tree (loop free). EGP placed such severe restraints on the topology, and was itself such an inefficient protocol, that it was clear it needed to be replaced.

In the meantime, intradomain routing protocols were being specified well enough that multivendor operation was considered not only possible but mandatory. There was no particular reason why a different type of protocol needed to run between domains, except for the possible issue of policy-based routing.

The notion of policy-based routing is that it no longer suffices to find a path that physically works, or to find the minimum cost path, but that paths had to obey fairly arbitrary, complex, and eternally changing rules such as a particular country would not want its packets routed via some other particular country. The notion that there should be different types of routing protocols within a domain and between domains makes sense if we agree with all of the following assumptions:

- Within a domain, policy-based routing is not an issue; all paths are legal.
- Between domains, policy-based routing is mandatory, and the world would not be able to live without it.
- Providing for complex policies is such a burden on the routing protocol that a protocol that did policy-based routing would be too cumbersome to be deployed within a domain.

I happen not to agree with any of these assumptions, but these are beliefs, and not something subject to proof either way. Because enough of the world believes these assumptions, different protocols are being devised for interdomain vs. intradomain. At the end of this chapter we discuss some of the interdomain protocols.

Routing Protocols

The purpose of a routing protocol is to compute a *forwarding database*, which consists of a table listing (destination, neighbor) pairs. When a packet needs to be forwarded, the destination address is found in the forwarding table, and the packet is forwarded to the indicated neighbor.

In the case of hierarchical addressing and routing, destinations are not exact addresses, but are rather address prefixes. The longest prefix matching the destination address is selected and routed forward.

The result of the routing computation, the forwarding database, should be the same whether the protocol used is distance vector or link state.

Distance Vector Routing Protocols

One class of routing protocol is known as distance vector. The idea behind this class of algorithm is that each router is responsible for keeping a table (known as a distance vector) of distances from itself to each destination. It computes this table based on receipt of distance vectors from its neighbors. For each destination

D, router R computes its distance to D as follows:

- 0, if R = d
- The configured cost, if D is directly connected to R
- The minimum cost through each of the reported paths through the neighbors.

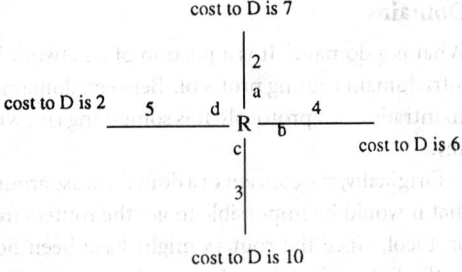

FIGURE 83.5   Example: distance vector protocol.

For example, suppose R has four ports, a, b, c, and d. Suppose also, that the cost of each of the links is, respectively, 2, 4, 3, and 5. On port a, R has received the report that D is reachable at a cost of 7. The other (port, cost) pairs R has heard are (b, 6), (c, 10), and (d, 2). Then the cost to D through port a will be 2 (cost to traverse that link) +7, or 9. The cost through b will be 4 + 6, or 10. The cost through c will be 3 + 10, or 13. The cost through d will be 5 + 2, or 7. So the best path to D is through port d, and R will report that it can reach D at a cost of 7 (see Fig. 83.5).

The spanning tree algorithm is similar to a distance vector protocol in which each bridge is only computing its cost and path to a single destination, the root. But the spanning tree algorithm does not suffer from the count-to-infinity behavior that distance vector protocols are prone to (see next section).

Count-to-Infinity

One of the problems with distance vector protocols is known as the count-to-infinity problem. Imagine a network with 3 nodes, A, B, and C. (See Fig. 83.6.)

Let us discuss everyone's distance to C. C will be 0 from itself, B will be 1 from C, and A will be 2 from C. When C crashes, B unfortunately does not con-

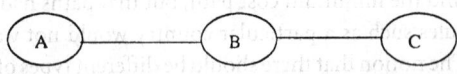

FIGURE 83.6   Example: the count-to-infinity problem.

clude that C is unreachable, but instead goes to its next best path, which is via neighbor A, who claims to be able to reach C at a cost of 2. So now B concludes it is 3 from C, and that it should forward packets for C through A. When B tells A its new distance vector, A does not get too upset. It merely concludes its path (still through B) has gotten a little worse, and now A is 4 from C. A will report this to B, which will update its cost to C as 5, and A and B will continue this until they count to infinity. Infinity in this case, is mercifully not the mathematical definition of infinity, but is instead a parameter (with a definite finite value such as 16). Routers conclude if the cost to something is greater than this parameter, that something must be unreachable.

A common enhancement that makes the behavior a little better is known as *split horizon*. The split horizon rule as usually implemented says that if router R uses neighbor N as its best path to destination D, R should not tell N that R can reach D. This eliminates loops of two routers. For instance, in Fig. 83.6, A would not have told B that A could reach C. Thus, when C crashed, B would conclude B could not reach C at all, and when B reported infinity to A, A would conclude that A could not reach C either, and everything would work as we would hope it would.

Unfortunately, split horizon does not fix loops of 3 or more routers. Referring to Fig. 83.7, and looking at distances to D, when D crashes, C will conclude C cannot reach D. (Because of the split horizon rule, A and B are not reporting to C that they can reach D.)

C will inform A and B that C can no longer reach D. Unfortunately, each of them thinks they have a next-best path through the other. Say A acts first, decides its best path is through B, and that A is now 3 from D. A will report infinity to B (because of split horizon), and report 3 to C.

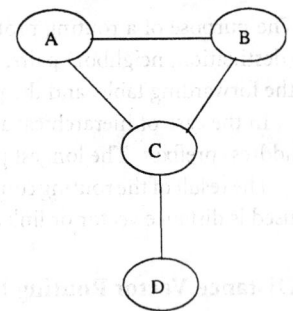

FIGURE 83.7 Example: loops with three or more routers.

B will now (for a moment) think it cannot reach D. It will report infinity to A, but it is too late; A has already reported a finite cost to C. C will now report 4 to B, which will now conclude it is 5 from D.

Even though split horizon does not solve the problem, it is a simple enhancement, never does any harm, does not add overhead, and helps in many cases.

Most of the distance vector protocols in use [router information protocol (RIP) for IP and IPX, and RTMP for Appletalk] are remarkably similar, and we call them the RIP-family of distance vector protocols. These protocols are simple to implement, but are very slow to converge after a topology change. The idea is that routing information is periodically transmitted, quite frequently (on the order of 30 s). Information is discarded if it has not been heard recently (on the order of 2 min). Most implementations only store the best path, and when that path fails they need to wait for their neighbors' periodic transmissions in order to hear about the second best path. Some implementations query their neighbors ("do you know how to reach D?") when the path to D is discarded. In some implementations when R discards its route to D (for instance, it times out), R lets its neighbors know that R has discarded the route, by telling the neighbors R's cost to D is now infinity. In other implementations, after R times the route out, R will merely stop advertising the route, so R's neighbors will need to time out the route (starting from when R timed out the route).

Distance vector protocols need not be periodic. The distance vector protocol in use for DECnet Phases 3 and 4 transmits routing information reliably, and only sends information that has changed. Information on a LAN is sent periodically (rather than collecting acknowledgments from all router neighbors), but the purpose of sending it periodically is solely as an alternative to sending acknowledgments. Distance vector information is not timed out in DECnet, as it is in a RIP-like protocol. Instead there is a separate protocol in which Hello messages are broadcast on the LAN to detect a dead router. If a Hello is not received in time the neighbor router is assumed dead and its distance vector is discarded.

Another variation from the RIP-family of distance vectors is to store the entire received distance vector from each neighbor, rather than only keeping the best report for each destination. Then, when information must be discarded (e.g., due to having that neighbor report infinity for some destination, or due to that neighbor being declared dead) information for finding an alternative path is available immediately.

There are variations proposed to solve the count-to-infinity behavior. One variation has been implemented in border gateway protocol (BGP). Instead of just reporting a cost to destination D, a router reports the entire path from itself to D. This eliminates loops, but has high overhead. Another variation proposed by Garcia-Luna [1989] and implemented in the proprietary protocol EIGRP involves sending a message in the opposite direction of D, when the path to D gets worse, and not switching over to a next best path until acknowledgments are received indicating that the information has been received by the downstream subtree. These variations may improve convergence to be comparable to link state protocols, but they also erode the chief advantage of distance vector protocols, which is their simplicity.

Link State Protocols

The idea behind a link state protocol is that each router R is responsible for the following:

- Identifying its neighbors
- Constructing a special packet known as a *link state packet* (LSP) that identifies R and lists R's neighbors (and the cost to each neighbor)
- Cooperating with all of the routers to reliably broadcast LSPs to all the routers
- Keeping a database of the most recently generated LSP from each other router
- Using the LSP database to calculate routes.

Identifying neighbors and constructing an LSP is straightforward. Calculating routes using the LSP database is also straightforward. Most implementations use a variation of an algorithm attributed to Dijkstra. The tricky part is reliably broadcasting the LSP. The original link state algorithm was implemented in the ARPANET. Its LSP distribution mechanism had the unfortunate property that if LSPs from the same source, but with three different sequence numbers, were injected into the network, these LSPs would turn into a virus. Every time a router processed one of them it would generate more copies, and so the harder the routers worked, the more copies of the LSP would exist in the system. The problem was analyzed and a stable distribution scheme was proposed in Perlman [1983]. The protocol was further refined for the IS-IS routing protocol and copied in OSPF. (See next section.)

One advantage of link state protocols is that they converge quickly. As soon as a router notices one of its links has changed (going up or down), it broadcasts an updated LSP which propagates in a straight line outwards (in contrast with a distance vector protocol where information might sometimes be ping-ponged back and forth before proceeding further, or where propagation of the information is delayed waiting for news from downstream nodes that the current path's demise has been received by all nodes.

Link state protocols have other advantages as well. The LSP database gives complete information, which is useful for managing the network, mapping the network, or constructing custom routes for complex policy reasons [Clark 1989] or for sabotage-proof routing [Perlman 1988].

Reliable Distribution of Link State Packets

Each LSP contains:

- Identity of node that generated the LSP
- A sequence number, large enough to never wrap around except if errors occur (for example, 64 b)
- An age field, estimating time since source generated the LSP
- Other information

Each router keeps a database of the LSP with the largest sequence number seen thus far from each source. The purpose of the age field is to eventually eliminate an LSP from a source which does not exist any more, or which has been down for a very long time. It also serves to get rid of an LSP that is corrupted, or for which the sequence number has reached the largest value.

For each LSP, a router R has a table, for each of R's neighbors, as to whether R and the neighbor N are *in sync* with respect to that LSP. The possibilities are as follows:

- R and N are in sync. R does not need to send anything to N about this LSP.
- R thinks N has not yet seen this LSP. R needs to periodically retransmit this LSP to N until N acknowledges it.
- R thinks N does not know R has the LSP. R needs to send N an acknowledgment (ack) for this LSP.

R goes through the list of LSPs round robin, for each link, and transmits LSPs or acks as indicated. If R sends an ack to N, R changes the state of that LSP for N to be in sync.

The state of an LSP gets set as follows:

- If R receives a new LSP from neighbor N, R overwrites the one in memory (if any) with smaller sequence number, sets send ack for N, and sets send LSP for each of R's other neighbors.
- If R receives an ack for an LSP from neighbor N, R sets the flag for that LSP to be in sync.
- If R receives a duplicate LSP or older LSP from neighbor N, R sets the flag for the LSP in memory (the one with higher sequence number) to send LSP.
- After R transmits an ack for an LSP to N, R changes the state of that LSP to in sync.

If an LSP's age expires, it is important that all of the routers purge the LSP at about the same time. The age is a field that is set to some value by the source, and is counted down. In this way, the source can control how long it will last. If R decides that an LSP's age has expired, R refloods it to R's neighbors (by setting the state to send LSP). If R receives an LSP with the same sequence number as one stored, but the received one has zero age, R sets the LSP's age to 0 and floods it to its neighbors. If R does not have an LSP in memory, and receives one with zero age, R acks it but does not store it or reflood it.

Calculating Routes

Given an LSP database, the most popular method of computing routes is to use some variant of an algorithm attributed to Dijkstra. The algorithm involves having each router compute a tree of shortest paths from itself to each destination. Each node on the tree has a value associated with it which is the cost from the root to that node. The algorithm is as follows:

- Step 0: put yourself, with cost 0, on the tree as root.
- Step 1: examine the LSP of the node X just put on the tree. For each neighbor N listed in X's LSP, add X's cost in the LSP to the cost to X to get some number c. If c is smaller than any path to N found so far, place N tentatively in the tree, with cost c.
- Step 2: find the node tentatively in the tree with smallest associated cost c. Place that node permanently in the tree. Go to step 1.

Interdomain Routing

As stated previously, the only plausible technical difference between a routing protocol designed for inter- vs. intradomain is support of policy-based routing. Policy-based routing is the ability to route according to exotic constraints, such as that one country does not want to see its traffic routed through some other country, some network might be willing to serve as a route-through carrier for only certain classes of traffic, or even that one backbone carrier has special rates at certain times of day and during those times someone would like to see all possible traffic routed via that carrier.

One method of accommodating all of these policies is to have the source compute its own route, perhaps by being given the LSP database. Once the source computes the desired route that meets its constraints, it sets up the path by launching a special path setup packet that specifies the path, travels along the path, and has the router along the path keep track of the route. This is the basic approach taken by IDPR when the route that would be computed by the network would not be appropriate for whatever reason [Steenstrup 1993].

The approach taken by BGP/IDRP [Lougheed 1991] is to use a distance vector protocol in which the entire path is specified (the sequence of domains, not the sequence of routers since the path through a domain is assumed not to be of interest in terms of policy issues). Each router is configured with information that helps it evaluate choices of path to determine how it would like to route to the destination. For instance, if one neighbor says it can reach destination D through path XYZ, and another through path XQV, the router will use its configured information to decide which path it prefers. It chooses one, and that is the path it tells its neighbors. There are also policies that can be configured about which destinations can be reported to which neighbors. The assumption is that if you do not tell a neighbor you can get to a particular destination, that neighbor will not choose to route packets to that destination through you.

A link state approach makes support for many more different types of policies possible than a distance vector approach. With a distance vector approach, either a router chooses one path to the destination and only advertises that (as in BGP/IDRP), or the algorithm has exponential overhead (if every router advertises every possible path to the destination). If the router chooses one path, then it greatly limits the possible paths. For instance, suppose router R can reach destination D through A, B, X or through A, Q, V. Suppose two sources that must get to D through R have different policies. One does not want to route through domain B. The other wants to avoid domain Q. Since R makes the choice, it makes the same choice for all sources. The ability for each source to have a custom route is supposedly one of the primary motivations behind policy-based routing.

Defining Terms

Bridge: A box that forwards information from one link to another but only looks at information in the data link header.

Cloud: An informal representation of a multiaccess link. The purpose of representing it as a cloud is that what goes on inside is irrelevant to what is being discussed. When a system is connected to the cloud it can communicate with any other system attached to the cloud.

Data link layer: The layer that gets information from one machine to a neighbor machine (a machine on the same link).

IEEE 802 address: The 48-b address defined by the IEEE 802 committee as the standard address on 802 LANs.

Hops: The number of times a packet is forwarded by a router.

Local area network (LAN):   A multiaccess link with multicast capability.

MAC address: Synonym for IEEE 802 address.

Medium access control (MAC):   The layer defined by the IEEE 802 committee that deals with the specifics of each type of LAN (for instance, token passing protocols on token passing LANs).

Multiaccess link: A link on which more than two nodes may reside.

Multicast: The ability to transmit a single packet that is received by multiple recipients.

Network layer: The layer that forms a path by concatentation of several links.

References

Clark, D. 1989. Policy routing in internet protocols. RFC 1102, May.

Garcia-Luna-Aceves, J. J. 1989. A unified approach to loop-free routing using distance vectors or link states. *ACM Sigcomm #89 Symp.* Sept.

Lougheed, K., and Rehkter, Y. 1991. A border gateway protocol 3 (BGP-3). RFC 1267, Oct.

Perlman, R. 1983. Fault-tolerant broadcast of routing information. *Comput. Networks* Dec.

Perlman, R. 1988. Network layer protocols with byzantine robustness. *MIT Lab. Computer Science Tech. Rep. #429*, Oct.

Perlman, R. 1992. *Interconnections: Bridges and Routers.* Addison–Wesley, Reading, MA.

Steenstrup, M. 1993. Inter-domain policy routing protocol specification: Version 1. RFC 1479, July.

84

Internetworking

Brian Reid
Digital Equipment
Corporation

Stephen Stuart
Digital Equipment
Corporation

84.1 Introduction

Internetworking is the joining together of two or more computer networks, in order that people on each one can communicate with people on the others. Different networks can potentially be under different management, use different technologies, and have different usage policies. Therefore, the study and practice of internetworking is about policy and its implementation, rather than specific mechanisms or technologies. When two networks are joined and are separately administered, the administration of one network might choose to do something that, if not handled properly, could either disrupt the other network or disrupt communications between them. Practitioners of internetworking develop mechanisms and protocols that can be used to implement policies that are mutually satisfactory to the participants. Policies determine what kind of communication is allowed between networks, what kind of data can be sent from one to the other, and how much trust the management of one network places in the management of the other.

It is inherent in the nature of internetworking that, in general, everybody will want to connect together as one unified internetwork. If there are 100 networks, it is much more useful to have all of them connected into one internetwork than to have 50 of them in one internetwork and 50 in another, with neither internetwork being able to communicate with the other.

Mathematically, internetworking is an equivalence relation. If there are two distinct internetworks, A and B, and if any one of the 50 networks in internetwork A connects to one of the 50 networks in internetwork B, then effectively there is just one overall internetwork: A joined with B, with all 100 networks as members. Internetworking, therefore, has a powerful tendency to form one internetwork to which everything is connected. That singular internetwork is usually referred to as the Internet.

84.2 Internetworking Policy and Mechanism

Within a single administrative domain, e.g., "my network" or "your network," it is realistic to mandate a certain mechanism. For example, a network administrator can require that users of the network use certain protocols, certain communications devices, or certain naming conventions. Between networks, you can negotiate only policies: "if you do X but not Y, then I will permit you to connect." The essence of internetworking thus becomes the formulation and implementation of policies that permit communication across administrative boundaries.

84.3 Communication Across Administrative Boundaries

In order to communicate across an administrative boundary, common principles must be established that define the foundations of communication. While it is possible to build internetworks out of nearly any technology, the phenomenon by which internetworks aggregate into a single Internet has also caused the emergence of a single protocol. If you want to start your own internetwork, you can use any protocol that you and your fellow members agree to. If you want to join the Internet, which is built using transmission control protocol/Internet protocol (TCP/IP) protocols, then you must use TCP/IP protocols. The TCP/IP protocol suite is thus the de facto foundation of communication today, and discussions of internetworking mechanisms and policies are usually phrased in the language of the TCP/IP protocol.

The notion of de facto protocols for any communication makes communication possible without having to negotiate the terms and conditions of communications with each intended communicant. For example, the world's air traffic control is conducted entirely in English, even in countries where English speakers are rare. This eliminates the need to negotiate a language during critical events.

An example of how well-known services and protocols allow communication without a priori administrative negotiation is electronic mail. In order to transmit electronic mail from one host to another, two well-known services (and their corresponding protocols) come into play: Domain Name Service (DNS) and simple mail transfer protocol (SMTP). The specific detailed steps involved in sending mail are well documented in any book on the subject; this is just a summary:

1. Locate the computer to which the mail should be sent. Mail is sent by a *store and forward* mechanism, so the computer to which the mail is sent may not be its ultimate destination, but merely a destination that gets the mail one step topologically closer to delivery to the intended recipient. This involves a lot of DNS inquiry that is discussed in a following section.
2. Make a TCP connection to the destination host, at the *SMTP port* (TCP port 25). This is an example of a *well-known service*. If a host expects to exchange mail using the SMTP protocol, it is obligated to accept connections at the SMTP port. The Internet Assigned Numbers Authority (IANA) publishes the list of registered services and their corresponding port numbers, the most current of which is RFC1700, also known as STD 2).
3. Transfer the mail using the SMTP protocol, as defined in RFC821[1] (and later RFCs, for some extensions to SMTP).

Note that any host is free not to listen on TCP port 25, and should not do so unless it is prepared to receive electronic mail by the SMTP protocol and do something intelligent with it.

This is a simple example of the role of policy and mechanism mentioned in the introduction. There are three mechanisms at work here: Domain Name Service (to find the recipient), TCP (to communicate with the recipient), and SMTP (to know what to say to the recipient). The policy associated with being a member of the Internet is that if you accept communication on TCP port 25, you implement the SMTP mail protocol there. Furthermore, if you accept mail for delivery on that port, you should do something sensible with it. There is no enforcement mechanism that guarantees compliance with this policy, but a computer that implements a nonstandard protocol on TCP port 25 violates the policy agreements that together make

---

[1] *Simple Mail Transfer Protocol*, by Jon Postel; also known as STD 10.

the Internet work. The standards document RFC1880[2] specifies the policies that a well-behaved computer is expected to obey in order to be a member in good standing on the Internet.

84.4 Issues of Security Philosophy, Policy, and Mechanism

Some computers must be kept secure. Some are most useful when they are made completely public. Before implementing any sort of security mechanism, you must have a security policy. It is good, before formulating a security policy, to have a security philosophy.

Security Philosophy

A security philosophy is a realistic consideration of what it is that you are guarding against. Are you trying to keep rogue data out of your computer (for example, if it holds the text of public laws or rules)? Are you trying to keep private data from leaving your computer (for example, if it holds confidential information)? Are you trying to prevent your disk from filling up? Are you trying to prevent certain people from using the computer? Are you trying to keep secret the identities of the people who are using the computer (for example, if it contains information about communicable diseases)? Each one of these security philosophies must have a different policy, though it is possible that all policies will be implemented with a similar mechanism.

Security Policy

With a security philosophy as a guide, the next step is to formulate a security policy. If your goal is to keep out unauthorized users, then your policy could define the process by which a user becomes authorized. If your goal is to prevent unauthorized change to public data, then your policy could define the process by which that data will be changed. Security policies can be classified into two broad policy groups:

- All communication is allowed except that which is specifically restricted.
- All communication is restricted except that which is specifically allowed.

Security Mechanisms

The implementation of a security policy is often embodied in a *firewall*, or *gateway*. The technology on which a firewall is based varies, but falls into two broad categories: packet filters and application relays. Both mechanisms share the characteristic that they exist to demarcate the boundary between the inside and outside of a security domain, and both serve to control the traffic across that boundary.

Packet-Filtering Mechanisms

Packet filters allow TCP and UDP packets to flow between the inside and outside networks. A packet filter is a router that screens each packet to determine whether or not that packet conforms to the security policy. The policy may be as simple as *prevent all communication except electronic mail*, expressed using the packet filter mechanism as:

```
default reject;
between any and any tcp port 25 accept;
```

This prevents all communication except TCP connections whose source or destination port is 25 (the SMTP port). This policy can be gradually expanded to include other types of communication:

```
default reject;
between any and any tcp port 25 accept;
between any and any tcp port 119 accept
```

---

[2] *Internet Official Protocol Standards*, issued by the Internet Architecture Board, edited by Jon Postel; also known as STD 1.

This example prevents all communication except TCP connections with a source or destination port of 25 (as before) or 119, which is used for the transmission of news.

Application Relay Mechanisms

Application relays are often placed on either side (or on both sides) of a packet filter, to further control the flow of information. Relays for electronic mail are a common example. Security policy might dictate that all mail from external sources is required to be relayed through a *bastion host* for delivery to internal destinations. The implementation of this at the packet filter (assuming our bastion host's IP address is 10.0.0.1) is reflected by the following change:

```
default reject;
between host 10.0.0.1 and any tcp port 25 accept;
between any and host 10.0.0.1 tcp port 25 accept;
between any and any tcp port 119 accept;
```

The number of rules for mail has doubled: one rule for the bastion host to deliver mail inbound to internal hosts, and another for internal hosts to relay mail outbound through the bastion host.[3]

Application relays roll up a protocol such as TELNET or file transfer protocol (FTP) to the application layer, allowing access control to be applied or logging information to be recorded. They can be used in conjunction with packet filters, or alone. Single-box firewall solutions are typically based on application relay technology, whereas higher-capacity firewalls tend to cluster application relays around a packet filter (or packet filters).

84.5 Communication Between Mutually Suspicious Partners

It is common for relationships between organizations to include a communications component. Depending on the nature of the relationship—alliance, conflict, vendor/client, etc.—this may be as simple as agreeing to communicate via electronic mail, or else it may involve the installation of dedicated telecommunication circuits that connect internal networks together.

Beyond the usual concerns of privacy, communication with a partner in the absence of trust introduces concerns of unintentionally revealing too much, and of ensuring that communication is genuine (in both directions). Each party to the communication must satisfy itself that the partner at the other end has correctly identified itself, and also must take care that the inbound communication is handled in a way not likely to cause problems should it turn out to be some form of attack.

Encryption and digital signature technology provide mechanisms to ensure that a message is not falsely constructed and its origin is not misrepresented.

84.6 Communication in the Presence of Expected Attacks

Internetworking opens up your network to other networks, and therefore creates the potential for access by unauthorized users, or *attackers*. You probably do not have the luxury of disabling all communication in order to prevent or analyze an attack. Some attacks may have as their goal the disruption of your network; in this case, shutting it down to diagnose or defeat the attack may be the wrong thing to do.

Outsider attacks fall into two broad categories: *penetration attacks* and *denial-of-service attacks*. A penetration attack is one that is trying to access something that normally requires special privileges. A denial-of-service attack is one that is trying to prevent you from being able to use the resources of your own network.

An attack on your network is not quite the same as a stranger wandering the halls of your building, because you can see the stranger and recognize that it is a stranger. An attacker using a computer on your

---

[3]Note that there is additional configuration work required to direct mail through the bastion host.

network may have all of the access privileges that the owner of that computer would have if the owner were sitting there instead. An attacker can dial in over a modem, and then use your network to reach other computers that are not connected to the modem.

A few simple precautions make it much easier to contain or repel an attack. Eliminate the network services that you do not intend to use. Eliminate trust relationships that you do not need. Keep log files of network activity, and review them regularly.

For example, on Unix-based computers, this might include:

- Auditing services started by the daemon *inetd:* fingerd, timed, systatd, ftpd, etc.
- Deciding whether or not the host needs to listen for mail
- Eliminating trust relationships between hosts; disallowing "r" commands and eliminating files such as .rhosts and .netrc

For each service offered, a procedure should be in place to deal with attacks on that service. Taking proactive steps to prepare such procedures in advance of needing them can save valuable time in responding to attacks.

This is a brief summary of the security issues that are intrinsic to internetworking. More in-depth discussions of Internet security can be found in Chapters 90, 91, 92, and 93 of this Handbook. The procedures for dealing with an attack should attempt to strike a balance between preventing damage, continuing to provide services, and finding and eliminating the source of the attack.

84.7 Internetworking Technology

This section describes, and briefly explains, the major technological mechanisms that are used to implement internetworking policies.

The Domain Naming System

Network transport mechanisms deal with *names, addresses,* and *routes.* The name of something is what you call it. Its address is where it is located in your network. The route is how you get to it from where you are. The domain naming system (DNS) is a mechanism for translating names to addresses for various purposes. It is also used for storing information about names other than their address. In the DNS, names are stored in a distributed database, access to which is provided by hosts called *name servers.* A name server is a computer that will look up names for you and tell you what it finds. Some name servers are *authoritative* for some names, which means that they know the truth about those names and do not have to ask any other servers about them. If a server is asked a question about some name and it is not authoritative, it must find a server that is authoritative and relay the question to that server. The information contained in the worldwide DNS is the union of all of the authoritative information in all servers. The complexity of the DNS derives from the fact that it is distributed and decentralized.

Domain names are organized hierarchically. The top level is called the *root,* and it has a few *top-level domains.* The owners of those top-level domains are free to *delegate* those domains to subdomains, and the owners of those subdomains are again free to further delegate to subdomains of their subdomain. There is a set of *root name servers* that are managed cooperatively around the world to contain the same information; these servers will identify the owners of the top-level domains, and identify the servers that are to be asked about them. The root servers are given names in the domain root-servers.net, for example, a.root-servers.net, b.root-servers.net, and so forth. Naturally, if you do not know the address of any root server, you cannot look up the name of anything, including a root server, thus to bootstrap a name server, you must know the address of the root servers as well as their names.

Below the root are various *top-level domains.* Common top-level domains in the U.S. are .com (commercial entities), .edu (educational institutions), .gov (U.S. government), .net, (networks), and .org (other organizations). There are top-level domains for each country as defined by the two-letter ISO country code: .us for the United States, .fr for France, .au for Australia, etc. Each country maintains its own top-level name servers or arranges with somebody to do it for them.

How the Domain Naming System Is Run

The concept of *authority* is central to the DNS. The root name servers have authority over the root domain, whose name is "." Some top-level domains (for example, .com) are handled directly by the root name servers. Other top-level domains (for example, .au) are delegated to other name servers that have the authority for that domain. The determination of who does and does not have authority for a domain is made by the owner of the domain above it.

Delegation of authority under top-level domains is handled by the various registration authorities, typically by or at the direction of a government agency: INRIA in France and the Internet Network Information Center [at the direction of the National Science Foundation (NSF)] in the U.S. The U.S. government delegates the authority of the root servers.

Information about domain name delegations under .com (and some others) can be found by using the *whois* program to query a database maintained on the host *rs.internic.net* (using the well-known WHOIS port and NICNAME protocol described in RFC954[4]). A sample session is included in the following:

```
% whois -h rs.internic.net 'dom digital.com'
   Digital Equipment Corporation (DIGITAL2-DOM)
      250 University Avenue
      Palo Alto, CA 94301-1616

      Domain Name: DIGITAL.COM

      Administrative Contact:
         Reid, Brian K. (BKR) reid@PA.DEC.COM
         (415) 688-1307
      Technical Contact, Zone Contact:
         Digital Equipment Corporation (DEC-NOC) noc@digital.com
         (415) 688-1380 (800) DIGITAL

      Record last updated on 28-Oct-95.
      Record created on 22-Jul-93.

      Domain servers in listed order:

      NS.DEC.COM              204.123.2.42
      CRL.DEC.COM             192.58.206.2
      NS11.DIGITAL.COM        192.208.46.3

   The InterNIC Registration Services Host contains ONLY Internet
   Information (Networks, ASN's, Domains, and POC's).
   Please use the whois server at nic.ddn.mil for MILNET Information.
```

The mechanism whereby authority is delegated is a combination of NS records, which identify name servers, and SOA records, which denote a start-of-authority boundary. In the previous registration example, authority over the digital.com domain is granted to three hosts: ns.dec.com, crl.dec.com, and

---

[4] *NICNAME/WHOIS*, by K. Harrenstein, M. Stahl, and E. Feinler.

ns11.digital.com. It is incumbent upon the maintainers of the domain to ensure that all hosts providing name service for the domain be reasonably consistent.

One of the particularly harmful inconsistencies is known as a *lame delegation*. This occurs when the authority over a domain is extended to a host not prepared to accept it; questions about the name space below that point (hosts or subdomains of the given domain) will fail, causing delays in resolving names, and reflecting poorly on those who run the domain.

The NS records that implement the delegation of authority can be examined using any number of tools; three popular tools are nslookup, dig, and host.

First, a root nameserver (chosen at random) is queried for the NS records for the domain digital.com

```
% host -t ns digital.com. a.root-servers.net
digital.com        NS        CRL.DEC.COM
digital.com        NS        NS11.digital.com
digital.com        NS        NS.DEC.COM
```

The NS records match the hosts listed in the registration. The host ns.dec.com, however, answers differently:

```
% host -t ns digital.com. ns.dec.com
digital.com        NS        crl.dec.com
digital.com        NS        ns1.pa.dec.com
digital.com        NS        ns11.digital.com
digital.com        NS        us1rmc.bb.dec.com
digital.com        NS        ns.dec.com
```

Note that there are two additional name servers not registered with the InterNIC; this is the inverse of the case of lame delegation. One of the extra name servers is, in fact, authoritative over the domain:

```
% host -t soa digital.com.
digital.com        SOA        ns1.pa.dec.com
postmaster.pa.dec.com  (
               1995112900    ;serial (version)
               3600     ;refresh period (1 hour)
               600      ;retry interval (10 minutes)
               259200 ;expire time (3 days)
               7200     ;default ttl (2 hours)
               )
```

Authoritative in this sense means that it is the source of the SOA record, which defines characteristics of the domain that are used by other name servers to determine how long information about the domain can be cached.

How the Domain Name System Is Used

The domain name system was developed to be a distributed database of information about computers on the Internet. Its use has expanded to be somewhat more general than that. By making many different names for the same address, you can provide the illusion that each name has its own computer. This allows the establishment of a presence on the Internet without actually having one. By setting up mail exchanger records for fictitious names, specifying that mail be sent to a gateway machine, it is possible to use Internet naming for mail systems that are not Internet compatible. The ability to process mail for fictitious domains has significantly accelerated the growth of the Internet by making it easy for an organization to join the Internet before it has very many Internet-compatible computers.

Subnetting and Supernetting

The growth of the Internet led to some hard problems in the area of address assignment and routing.

Subnetting (RFC950) was developed to allow large networks (a class A network can contain over 16 million hosts) to be subdivided in a manner that allowed a sensible routing strategy to be employed within a network, by allowing a netmask different from the natural mask associated with a network to be applied to an interface.

TABLE 84.1 CIDR Examples, Showing the Representation of Traditional Classes as CIDRs

CIDR Notation	Network (or Netblock) Represented
16/8	class A network 16.0.0.0
130.180/16	class B network 130.180.0.0
204.123/16	netblock 204.123.0.0–204.123.255.0
204.123.0.128/26	one-quarter of the class C network 204.123.0.0 (the 64 addresses 204.123.0.128–204.123.0.191)
198.55.32/24	class C network 198.55.32.0
198.55.32/21	netblock 198.55.32.0–198.55.39.0
198.55.40/21	netblock 198.55.40.0–198.55.47.0
198.55.32/20	netblock 198.55.32.0–108.55.47.0

Supernetting, later called classless inter-domain routing (CIDR), (RFC1338) was developed to allow smaller networks to be aggregated in a manner that allowed a sensible routing strategy to be developed between networks, and stop (well, slow) routing table growth.

Before CIDR, each individual classful network consumed a routing table entry at the interconnects where large Internet service providers (ISPs) exchange routing information. The exponential rise in the number of organizations with a presence on the Internet was reflected directly in the size of routing tables at the interconnects. In 1992, RFC1338 predicted that the 4775 routes advertised in February of 1992 would grow to over 30,000 within two years.

CIDR provided a methodology for aggregation, allowing the smaller blocks of networks to be combined into larger ones.

Unlike traditional netmask information, CIDR dictates that masks be contiguous, allowing routes to be specified by a base network and mask length. The traditional classes of address can thus be represented as shown in Table 84.1. The last three examples illustrate the aggregation of two blocks into a larger block; two blocks of 8 class C networks are coalesced into a block of 16 class C networks, with a corresponding reduction in the length of the mask.

Changes also had to be made at the routing protocol level, to communicate the mask information necessary to make CIDR work. Internal routing protocols such as (RIP) and exterior routing protocols (BGP) were modified to carry mask information.

Interior Routing Protocols

An interior routing protocol is one that is used inside a network. The study of such protocols is normally considered to be outside the field of internetworking. Examples of interior routing protocols are OSPF, IS-IS, IGRP, and RIP.

Exterior Routing Protocols

Exterior gateway protocols are used to route across administrative boundaries, and, as such, are integral to the study of internetworking. The two most widely used exterior routing protocols are exterior gateway protocol (EGP) and BGP. The border gateway protocol aggregates routes into groups called autonomous systems (ASs) and exchanges routing information as AS paths between ASs rather than between networks themselves.

Like nearly everything else in internetworking, exterior routing is a mixture of policy and mechanism. In the earliest days of the Internet, the policy was that all routers were told all routes. As the Internet evolved, that became impractical, and various forms of *policy-based routing* were explored. Policy-based routing is a technology that allows administrative control over the exchange of routing information between routing devices. When the Internet was under the direct control of the U.S. government, such policy-based routing was implemented using the policy-routing database (PRDB). The PRDB assumed that the Internet had a backbone, or *core*, and that member networks would advertise, onto the core, the routing paths that they offered. The PRDB listed each CIDR that was allowed to be advertised to the core, and what autonomous system was allowed to be the origin or transit AS for that CIDR.

The PRDB was the mechanism for defining routing on the NSFnet backbone. The dismantling of the NSFnet backbone (and with it, the PRDB), and the subsequent definition of network access points (NAPs) and the Routing Arbiter project the technology change that corresponded to the policy change of privatization of the Internet.

Link-Level Interconnect Technologies

The internetworking protocol (IP) is designed to allow communication across administrative boundaries. But, like most other network protocols, it is a layered protocol that can be described with the International Standards Organization (ISO) layering model. Although IP does not exactly match the ISO layering model, the match is good enough, especially at the lower levels. It is the intention of the IP protocol suite that the connection between differently administered networks be made at levels 3 and above (network layer and above). However, it is certainly possible to interconnect networks at level 2 (data link) or level 1 (physical). A level 2 connection would be made by putting both networks on the same Ethernet or FDDI ring or ATM fabric. A level 1 connection would be made by running a wire between the networks being connected. Although such lower level connections can often be made to work adequately from a purely technical standpoint, the administrative, security, and diagnostic problems that arise from lower level interconnects can be insurmountable.

84.8 Internet Service Providers

An Internet service provider (ISP) is a company that provides Internet service to its customers. What it means to provide Internet service varies from provider to provider, but in general the service provided is that the ISP agrees, for a fee, to accept the other end of a data link and to route IP packets across that link on your behalf. The data link can be anything from dialup to gigabits per second; the principle is the same. When you subscribe to an ISP, you (the customer) and they (the provider) must agree on the range of addresses that you will be using, so that they can route traffic that is destined for your address space.

Today there are three different kinds of ISPs. At the top of the hierarchy are the *transit ISPs*. They own large, high-speed data circuits between cities, countries, and continents. They sell long-distance transport in high volumes. In general they restrict their sales to the *midlevel ISPs*. Midlevel ISPs, sometimes called network service providers or NSPs, buy transit from the transit ISPs and sell to corporations and to retail resellers. A corporation with 1000 users could buy Internet service directly from a midlevel ISP. An individual normally would not, but would instead buy service from a *retail ISP*.

A horizontal connection between ISPs of approximately the same rank is called *peering*, and is normally done using the border gateway protocol. A location at which more than two ISPs interconnect, whether or not there is any peering, is usually called a network access point (NAP). Because NAPs have more than one ISP connected to them, they are a good place to put high-volume network servers. The practice of hosting high-volume servers at NAPs is called *collocation*. A company whose business is the operation of high-volume servers is called a *content provider*. An Internet online service is a combination business; it is a retail ISP and a content provider in a single package.

84.9 Summary and Conclusion

The Internet, which has resulted from the internetworking of many thousands of independent computer networks, is an entirely new medium that is rapidly changing the expectations for global communication.

It is not like television, nor radio, nor newspapers, nor magazines, nor anything else. It is an entirely new medium, enabling entirely new forms of communication worldwide.

References

Cheswick, W. and Bellovin, S. 1996. *Firewalls and Internet Security: Repelling the Wily Hacker.* Addison–Wesley, Reading, MA.

Comer, D. E. 1995. *Internetworking with TCP/IP Volume 1: Principles Protocols, and Architecture,* 3rd ed. Prentice–Hall, Englewood Cliffs, NJ.

Comer, D. E. and Stevens, D. L. 1994. *Internetworking with TCP/IP Volume II: Design, Implementation, and Internals,* 2nd ed. Prentice–Hall, Englewood Cliffs, NJ.

Hunt, C. 1992. *TCP/IP Network Administration.* O'Reilly and Associates, Sebastopol, CA.

Liu, C., Peek, J., Jones, R., Buus, B., and Nye, A. 1994. *Managing Internet Information Services.* O'Reilly and Associates, Sebastopol, CA.

85

Overview of Distributed Operating Systems

85.1 Introduction

The advent of distributed systems came hand in hand with that of workstations and personal computers. The presence of many computers interconnected by a network opened up several new possibilities:

1. Every user could have personal dedicated computing cycles, while the network still allowed sharing data or devices through centralized file servers, printers, etc.
2. Reliability could be increased by arranging for computers to take over from each other in the case of crashes.
3. Performance could be increased by allowing software to make use of many processors in parallel.
4. Systems could grow incrementally by adding computers one at a time.

These possibilities triggered research in the new field of *distributed and network operating systems*. The first projects started in the middle 1970s, but the bulk of activity in the area took place in the 1980s. Now, halfway through the 1990s, the activity in distributed systems research seems to be declining.

A distinction between distributed operating systems on the one hand and network operating systems on the other has sometimes been made. A network operating system is essentially a centralized operating system whose components have been distributed over multiple nodes, whereas a distributed system is one in which this distribution, combined with replication, plays a role in achieving fault tolerance as well.

The distribution of components of an operating system over multiple nodes requires splitting up the traditional operating system into its constituent components, leaving only a small amount of machine-dependent and resource-protecting code in a *microkernel* [Accetta et al. 1986, Mullender et al. 1990, Rozier et al. 1988]. Microkernels can thus, to some extent, be viewed as a consequence of introducing network operating system or distributing operating system functionality.

Another distinction is necessary between distributed systems and *parallel* systems. In parallel systems research, the focus is very much on completing computations in minimal time, by exploiting the presence of multiple processing nodes. Distributed systems also exploit parallelism, but there is at least as much concentration on fault tolerance.

Early research in distributed systems focused very much on functionality: better mechanisms for sharing, fault tolerance, communication, and parallelism. Later, attention shifted to keeping the same functionality while improving performance. The declining interest in distributed-systems research today may have a lot to do with the rapid increase in reliability of hardware components during the past 10 years.

85.2 Survey of Distributed Systems Research

In this section we survey the contributions that important projects have made for the distributed systems research community. Only a few of these projects have resulted in complete systems that are in use, but most of them have contributed from their key features to commercial operating systems. It is because of this that we do not survey the important projects one by one, but that, instead, we survey them area by area.

In the following sections we look at some of the important projects in the areas of naming, communication, transactions, and group communication. We do not discuss security in this chapter, even though fault tolerance properly implies security as well. The subject of security is in a separate chapter in this handbook. Research on distributed shared memory has also been left out: distributed shared memory (DSM) makes parallel programs run better, but does not contribute to fault tolerance—it does the opposite, if anything. DSM has been delegated to the parallel processing chapter.

Naming

Amoeba

Amoeba was a research project, initially only of the Vrije Universiteit, later also of CWI, the Centre for Mathematics and Computer Science, both in Amsterdam [Mullender 1985, Mullender et al. 1990, Mullender and Tanenbaum 1986, Tanenbaum et al. 1990]. Together with the V-system[Cheriton 1988], it was one of the first standalone distributed systems.

Amoeba has two levels of naming, which were both innovative. At the system's level, all objects are named using *capabilities*, which are managed in user space. A capability consists of four fields, as illustrated in Fig. 85.1. The *service* field, also known as the service's *port*, identifies the service that manages the object. This field is used by the remote-operations protocol to deliver messages to a server process (see the communication subsection). The *object* field identifies the object to the service, and the *rights* (Rts) field indicates what operations the holder of the capability may carry out on the object. The *check* field prevents forging capabilities; it is calculated by the server, using a *secure hash* of object and rights fields, plus possibly a per-object secret *random number* maintained by the service.

Service ports are 48-b random numbers and, if you know a service's port, you can send messages to it. Services can be made private by keeping their ports secret.

Amoeba uses *request/reply* communication between *clients* and *servers*. A request contains a capability whose port names a service, while its remainder names an object maintained by that service. The Amoeba system finds a server for the service by broadcasting the port and waiting for location information from the servers (clients maintain a server-location cache to save broadcasts). When a server has been found, the request is sent to it; the server processes the request and returns a reply. Replies are addressed to the client's port.

In Amoeba, ports and capabilities are the names for services and objects, respectively. At the operating-system's application programmer's interface (API), they are the only names supported. For operating system and application software development, these fixed-length names are quite convenient.

For human beings, a directory service is available which maps hierarchical path names onto capabilities. A directory entry consists of a name and a *list* of capabilities. Normally, these capabilities all refer to the same object, but carry different rights; they are not all equally powerful.

48 bits	24	8	48
Service	Object	Rts	Check

FIGURE 85.1   Layout of an Amoeba capability.

Depending on the rights in the capability of a directory, a client will be allowed to retrieve subsets of the capabilities in its entries. A powerful capability to a directory allows a client to see most or all of the capabilities, a weak capability allows it to see only small or empty subsets of, presumably, weak capabilities.

DEC Global Name Server (GNS)

The DEC global name server (GNS) [Lampson 1986] is an example of a design that was intended to be scalable to worldwide size and millions of nodes.

The members of the design team (Andrew Birrell, Butler Lampson, Roger Needham, and Michael Schroeder) had been involved with Grapevine, Xerox's name server [Birrell et al. 1982]. One could say that Grapevine succumbed under its own success: its popularity caused it to grow to a size for which it had not been designed, revealing several deficiencies in its scalability [Birrell et al. 1984].

GNS was designed to scale both in size and, as it were, in time. Scaling in size means that the naming database must be able to grow to billions of entries stored at millions of nodes. Scaling in time means that the name space can cope with large structural changes at any level in the naming tree (as, for instance, the unification of East and West Germany and the split up of Czechoslovakia).

The name space of GNS is necessarily hierarchical; no other structure could scale to the desired size. The hierarchy is not necessarily geographically determined; GNS has no problems coping with multi-national organizations. Each directory entry is essentially a list of attribute {*name, value*} pairs. An entry **/nl/utwente/cs/sape** could describe a user and have attributes such as **mailbox** and **certificate**. A user's public-key certificate could then be retrieved under the name **/nl/utwente/cs/sape/certificate**.

Each directory has a unique *directory identifier* (DI). A directory is referred to by its parent via a *directory reference* (DR), which contains a DI. A *full name* in GNS consists of a DI and a *pathname;* the pathname is resolved starting at the directory named by the DI.

The naming database is organized such that every system can retrieve the directories it controls and all parent directories up to the global root by their DIs. However, the flat—and *pure*[1]—name space of DIs alone cannot be used to find directories elsewhere in the naming tree. For this purpose, a DR not only contains the DI of the named directory, but also a list of servers that store copies of the directory. Server names are stored as full names also, and, if one is not careful, looking them up can result in endless lookup loops.

The designers of GNS recognized that availability of the name server is of crucial importance. Directories that are essential for the operation of a system must be available locally. As a consequence, directories near the root of the naming tree will be very highly replicated indeed; the root will be replicated everywhere. With such high degrees of replication, consistent update is not possible.

GNS, therefore, defines a form of loose consistency that may be formulated as follows: "If the supply of updates stops, there will eventually be glorious consistency" [Needham 1993]. This is achieved by making sure that updates have the following properties:

1. Every update eventually reaches every replica.
2. Two updates can be applied in any order and yield identical results (the *set* of updates matters, not the order in which they are made).
3. Updates are *idempotent*: applying an update twice has the same effect as applying it only once.

To achieve property 1, updates are distributed among the copies of a directory by a *sweep algorithm*: A sweep operation visits every directory copy, collects a complete set of updates, and then writes this set back to every copy. Property 3 implies that directory replicas need not keep track of the identity of every update received.

Achieving property 2 is subtle: Normally, you cannot update a directory before it has been created, and so the order of directory-create and directory-update operations does seem to matter. This is solved by

---

[1] Pure names are explained in a subsequent section on naming (section 85.3).

the following ruse: If you update an item and it, or one of the directories on its path, does not exist, then that directory is created and updated, but it is marked *absent.* When a create operation is performed, the item is created if it does not exist, and it is marked *present.* Queries will only find and traverse paths with *present* items.

Furthermore, updates are *time stamped* so that conflicting updates yield the result of the one with the highest time stamp. This also causes the desired idempotency.

DEC GNS was used initially within DEC only, but since it has been adopted by the Open Software Foundation as part of its Distributed Computing Environment, it is becoming more widespread.

Domain Name Server (DNS)

The Domain Name Server [RFC-1035, RFC-1034] is the world's most widely distributed name service. It is used to name and locate hosts, services, and mailboxes on the Internet. It uses a hierarchical name space and pathnames with components separated by dots. Like postal addresses, but unlike most computer-related naming hierarchies, DNS puts the highest level domain at the end to give names, such as **amstel.cs.utwente.nl.** for a machine called Amstel in the Computer-Science Department of the University of Twente in The Netherlands.

Pathnames (at most 255 characters) are composed of a sequence of labels (of at most 63 characters) separated by dots (the character " . "). The root of the name space has an empty name, and so full names end in a dot. Relative names are resolved starting at the current domain. Names that cannot be resolved relative to the current domain are resolved as full names (by appending an imaginary dot).

DNS was designed to scale with the number of hosts and users in the global Internet. Most of the data in the system is expected to change very slowly, but small subsets of the name space may change rapidly, e.g., on the order of seconds to minutes, and DNS should be capable of keeping up with those changes. In doing this, however, DNS depends on the rapidly changing subset being small; if it were large, the update traffic would saturate the Internet.

DNS has a worldwide distributed implementation. Obviously, not all servers can be trusted. Clients can indicate which name servers they trust and name servers in this trusted set are always consulted before any others. Just like GNS, DNS is also based on the assumption that old data are better than no data; consistency guarantees are sacrificed to obtain maximum availability.

Queries are first sent to one of the local name servers. Servers only contain subsets of the name space, and so most names can only be resolved by consulting several servers. Normally, servers do not query other servers on behalf of clients, but they pass the partially resolved query back to the client and let the client resolve the query by iterating over multiple servers. Thus, clients can build up a cache that can tell them which servers serve a particular subdomain of the name space.

Each domain has an administrator who maintains a master naming database. The master database also identifies the master databases of each of the subdomains. An *authoritative* answer to a query is obtained by consulting master databases. Servers not only maintain master databases, but they also cache information from other servers. The system allows administrators to configure their servers to cache certain information from other servers permanently. This mechanism allows names to be resolved even though the domain cannot be reached or the authoritative server is down.

The naming database is maintained by a set of servers. A program which is called a *resolver* queries one or more servers in order to resolve a name. Clients give their queries to a resolver; thus, they do not have to iterate over multiple servers themselves.

Each internal node or leaf node in the name space corresponds to a *resource set*, which is stored by one or more name servers. The name servers make no distinction between internal nodes and leaf nodes (but resolvers do). A resource set contains zero or more *resource records*. A resource record consists of a *type*, a *class*, a *time to live* (TTL), and *type*-dependent (and sometimes also *class*-dependent) *data.*

Types and classes are represented by 16 b. Types and classes are defined per domain (including all subdomains). A new global type or class can only be introduced by the administrator of the root domain. This is the Network Information Center (NIC).

Important types are **A** for host addresses, **CNAME** for naming aliases, **HINFO** for host information [central processing unit (CPU) and operating system], **MX** for mail-handling agents, and **NS** for the authoritative name server for a domain. There are more types. Classes can be used to distinguish between different (sub)networks.

The time to live entry in a resource record tells name servers and resolvers how long it is safe to cache the resource record. When a resource record is cached, the authoritative server is not consulted, and so updates are not seen until the TTL has expired. Administrators must choose the TTL to balance between the two evils of increased name server traffic caused by rapid expiry of caches and increased inconsistency caused by slow expiry.

Queries are for {*name, type*} pairs. To find a mail handler for **sape@cs.utwente.nl**, one posts a query for { **cs.utwente.nl, MX** }. This query will yield a list of mail handlers, for instance:

```
    cs.utwente.nl              preference = 0,
                               mail exchanger = utrhcs.cs.utwente.nl
    cs.utwente.nl              preference = 10,
                               mail exchanger = driene.student.utwente.nl
    utrhcs.cs.utwente.nl       inet address = 130.89.10.247
    driene.student.utwente.nl  inet address = 130.89.220.2
```

In this case, two possible mail handlers are produced, the first one being preferred over the second, and for each one, the **A**-type record is also produced—a useful optimization, since the Internet address will be needed to send the mail message.

The Domain Name Service must be one of the most heavily used distributed applications in the world (along with e-mail, the World Wide Web, and net news) and it works remarkably well. This is quite surprising given the sheer size of the worldwide DNS database today, and the unavoidable variations in professionality of the administrators. The reliability and scalability of DNS show off the skill of its designers.

Plan 9

The group that designed Unix in the late 1960s and early 1970s has built a new operating system named Plan 9 from Bell Laboratories. It is an elegant little system available on CD for academic and noncommercial use [Harcourt 1995]. The Unix philosophy of using the filename space for naming everything[2] has been preserved and developed further which resulted in a very elegant design.

In Plan 9, each server can export a name space. Clients can refer to objects maintained by such a server by name. These name spaces are hierarchical and singly rooted. A process can access the name spaces of several servers by *grafting* them onto its own name space. This grafting is called *mounting*. A *mount table* maintains which server name spaces are mounted where in the naming tree. This resembles the way in which Sun Network File System (NFS) servers can be mounted in Unix.

But where Unix maintains a single mount table per machine, Plan 9 can have a mount table per process. When a process is created, it normally inherits its parent's mount table and then shares it with its parent. However, processes can also inherit a *copy* of the mount table so that mount and unmount operations of the parent are not visible by the child and vice versa. They do this by starting a new *process group*; Plan 9 maintains a mount table per process group. Here, the analogy with Unix file descriptors is enlightening: normally, children inherit the open files from the parent, but, if the parent so chooses, it can modify the set of open files between *fork* and *exec* (using *close, open,* and *dup,* usually). This can be done with mount tables in an analogous way (using *newpgrp,* followed by *mount* and *unmount* operations).

As a result, each process group can create its own private name space. A parent can, for instance, *encapsulate* a child process by mounting an encapsulation server in its root directory before starting it. The encapsulation server can monitor all of the child's file input/output (I/O) requests and name space operations and process (some of) them in the parent's name space. Such encapsulation servers can be very

---

[2]The addition of networking and environment variables to Unix has diluted this philosophy to some extent.

useful for debugging, collecting statistics on application file usage, or for checking out imported software for anomalies such as Trojan horses.

The handle that specifies the server to be mounted is a *connection* in Plan 9 (identified by a file descriptor). Connections to local servers are essentially *pipes*; connections to remote ones are network connections.

There is a standard protocol for accessing the objects maintained by a server. This protocol contains operations normally associated with files: *open, close, read, write, seek,* etc. However, they need not be files. The Domain Name Server in Plan 9, for instance, can present itself as a file system that allows users to open and read a file such as **com/bell-labs/plan 9**. The mouse server implements a file **mouse** which is conventionally mounted as **/dev/mouse** and can be read to give the position of the mouse.

The Plan 9 window system, known as $8^{1}/2$,[3] illustrates the use of the Plan 9 name space very elegantly. The device drivers for screen, keyboard, and mouse are represented by files mounted in **/dev**: **/dev/bitblt** (for bit-blit operations to write to the screen), **/dev/keyboard** and **/dev /mouse.** The window manager, $8^{1}/2$, uses these devices. When $8^{1}/2$ creates a window, it forks the child process that runs in that window, and gives the process a new mount table (by creating a new process group for it); $8^{1}/2$ then mounts itself as a service onto the new process's **/dev** where it implements new versions of **/dev/bitblt**, **/dev/keyboard** and **/dev/mouse**. A process running in a window, thus, does not read from the hardware mouse directly, but from one synthesized by the window manager. The window manager only gives it the mouse clicks and movements related to a particular window. A happy consequence of this organization is that the window manager can run as a window in itself. X-window applications run on Plan 9, through an X server running in an $8^{1}/2$ window.[4]

Plan 9 does not support a global name space in the sense that all objects have the same name everywhere. In fact, it makes explicit use of the fact that different objects have the same name in different places (e.g., **/dev/mouse** in different windows). Rob Pike, a principal designer of Plan 9, claims that, even in global naming schemes, sharing only becomes practical if everybody adheres to certain naming conventions. Naming conventions allow you to find things in the name space by guessing. Plan 9 uses naming conventions explicitly; the user who mounts **/dev/mouse** as **/dev/keyboard** should not be surprised if certain things fail to work properly.

Communication

It is likely that the research into efficient communication for distributed systems in the late 1970s came as a reaction to the cumbersomeness of the standards being developed by the International Organization for Standardization and the Consultative Committee on International Telephony and Telegraphy (CCITT). In any case, the early 1980s saw a race by a number of research groups to develop the fastest protocols for supporting remote procedure call. Early participants in the race were the V systems group at Stanford, led by Cheriton [1988]; and the Amoeba group at the Vrije Universiteit Amsterdam, where van Renesse et al., [1988] made their record attempts for fastest remote procedure call (RPC).

In the mid-1980s, Schroeder and Burrows [1989] thoroughly analyzed the performance of the RPC implementation of DEC Systems Research Center's (SRC's) Firefly multiprocessor. This resulted in a significantly better understanding of the design issues for interprocess communication. Hutchinson and Peterson [1988] at the University of Arizona then designed an extremely flexible framework for building efficient protocol stacks, the *x*-kernel. This is now widely used by researchers in a large number of research systems.

The V System

The V system, during its heyday, ran on SUN workstations connected by an Ethernet. It is believed that the multicast capability of Ethernets inspired a communications infrastructure that made heavy use of it for delivering requests to replicated services.

---

[3] Also after a movie.

[4] It was noticed that X servers were not designed to deal gracefully with resize operations of what they perceive to be their screen.

V uses remote operations for all communication. The common form is a remote operation between two processes, but there is also a form where a request is *multicast* to a set of processes with multiple replies as a result. Requests can thus be addressed to individual processes or to *process groups*.

Processes and process groups in V share a name space; clients sending requests to a process group multicast the request to all its members. The reliability of the multicast is that of the underlying hardware multicast mechanism (i.e., the reliability of Ethernet multicast). A few of the servers, therefore, may not receive the request, but this is viewed as normal and replicated services must take this into account. One, several, or all of the servers can return a response; this is up to the service designer.

The request and reply messages in V consist of a 32-byte *fixed-size message* with an optional *data segment* that can be as large as 16 kilobytes. A minimal message consists of just the fixed-size message. This separation of data into two kinds is useful because it can prevent quite a lot of copying and allows *scatter-gather* operations to collect the data to be sent or to deliver the data to be received. Amoeba does this too.

In the V system, a request is sent and a reply is received with one operation, called *send*. A single system call thus suffices to complete a remote operation, which, in V, is called a *message transaction*. All other interactions with the operation system take place using message transactions.

Amoeba

Amoeba [Mullender 1985, Mullender et al. 1990, Tanenbaum et al. 1990] was designed and implemented as a general-purpose distributed operating system combining fault tolerance to high performance. Its interprocess-communication mechanisms were based on the exclusive use of *remote operations* for all communication: In a remote operation, one process, the client, sends a request to a *service*; exactly one of the service's server processes receives and processes the request and returns a reply.

Requests and replies in Amoeba consist of two parts, a parameter area and a body part, very similar to V. Messages can be of arbitrary length so that a wide range of service types can easily be realized. Also, Amoeba implemented *at-most-once* semantics for its remote operations. Under this regime, requests will normally be executed exactly once. When a failure occurs, however, no reply reaches the client, while the request may not have been carried out at all, or it may have been carried out partially, or even completely. A failure can be caused by a crash of the server process, a crash of the server host or operating system, or a network failure. Sometimes, for instance, when a server cannot be reached, it is possible to provide exact feedback to the client about whether or not its request was acted on, but in most cases a client must find out using indirect means.

In this respect, Amoeba differs from both SUN RPC and V: SUN RPC implements an *at-least-once* strategy in that it retransmits a request until it gets a reply; in the process, it is possible that the request will be executed partially or wholly multiple times. V multicasts a request to multiple servers, with the possibility of multiple, parallel executions.

SUN RPC and V message transactions have the advantage of achieving a higher probability of success, but applications have to use requests with *idempotent* semantics; that is, semantics where executing a request partially or wholly several times, followed by a final, complete execution has the same result as exactly one complete execution. Reading from a file is an example of an idempotent operation: as long as you get the data in the end, semantically, it does not matter if you tried in vain a couple of times.

Amoeba's at-most-once semantics also allows nonidempotent requests to be used (e.g., transfer $10 from my account to that of the Red Cross). The reason for putting so much emphasis on these subtle differences in semantics for different communication mechanisms is that one cannot move an application from a system that implements at-most-once to one that implements at-least-once with impunity. Program correctness can crucially depend on the difference.

Guaranteeing at-most-once behavior during communication with a replicated service requires some care: Suppose a request is sent to one of the servers, but no reply is received. The client may not retransmit to a different server, because the absence of a reply does not indicate that the request was not acted on (the server may have crashed after executing it). At the same time, Amoeba does have to deal with *migrating* server (and client) processes, and so, although retransmissions may not be sent to a different server, it is possible that they are sent to the same server on a different *host*.

Making this work requires individual servers to be named. Therefore, in addition to the *service port* (see preceding section), server processes also have a *unique port*. When a client initiates a remote operation, the system locates a server by broadcasting a *locate message* that contains the server port. Servers respond with a message containing their location and their unique port. After selecting a server (typically by using the first response to arrive), the system commits to that particular server for the duration of the remote operation. If the server migrates, the new location can be found by broadcasting a locate message containing the server's unique port.

Mappings from server port to unique port to current location are, of course, cached. When a host receives a message for a port no longer serviced by it, it returns a not-here reply which invalidates the cache. When a not-here response arrives for the initial transmission of a request, the client may try to locate another server; when it arrives for a retransmission or a control message, the client must try to relocate the server at another address.

Firefly Remote Procedure Call

The Firefly was a shared-memory multiprocessor workstation built at the DEC Systems Research Center in the early 1980s. Its operating system, Taos, implemented a remote-procedure-call mechanism that was used for all communication. When one takes the speed of its processors into account, Firefly RPC may have had the lowest latency implementation of its day. Its performance was the result of carefully crafting the interactions among application threads, kernel threads, and interrupt routines. Schroeder and Burrows [1989] measured and analyzed the performance and their report has provided useful insights into high-performance communication architectures for later generations.

Firefly RPC allows request and reply messages of at most one Ethernet packet in size. The packets must also contain user datagram protocol (UDP) and Internet protocol (IP) headers, so the payload could not exceed 1440 bytes. Remote procedure calls with more data were split up into a sequence of RPCs. To take advantage of the multiprocessor capability of the Firefly, it was common practice to carry out large-data transfers with multiple parallel threads making multiple single-packet RPCs in parallel. The resulting throughput is a significant fraction of the capacity of the Ethernet, but it is achieved by loading five processors at each end almost to capacity. Amoeba, Sprite, and V achieved similar throughput on much less loaded uniprocessors (but with processors that ran three to five times faster).

In their paper, Schroeder and Burrows [1989] measured the performance of a *null RPC* (minimum-size request and reply) and of a *maxresult RPC* (minimum-size request, maximum-size reply). The roundtrip latencies for these were 2.66 ms and 6.35 ms, respectively. With four threads doing maxresult RPCs in parallel, the throughput is 4.65 Mb/s.

A remote procedure call can be separated into four activities, *marshalling, protocol processing, hardware processing*, and *synchronization*.

The code for marshalling parameters into and out of network packets is produced by a *stub compiler* and runs in the application address space. The time needed for marshalling depends on the complexity and size of the parameters to be marshalled. The latency of 6.35 ms for a maxresult RPC was achieved with a (Modula-2) `var array[0..1439] of char` parameter that took 550 μs to marshal; that is, some 10 percent of the RPC latency. For complex parameters of the same size, the marshalling time can be a multiple of this. As more and more 10-Mb/s local area networks (LANs) are now being replaced by much faster networks, the significance of marshalling times will increase. It is, thus, worthwhile to invest in well-designed RPC type systems that use carefully tuned stub compilers.

The protocol processing consists of filling in IP and UDP headers and calculating the UDP checksum. This costs between 50 and 450 μs, depending on packet size. UDP-checksum verification costs the same amount of time.

The hardware processing time can be divided in two parts: the time the driver takes to enqueue the packets and process the interrupt, and the time the hardware itself needs to transmit the packets. The driver time was some 240 μs, and the hardware time 210 μs for a minimum packet and 2880 μs for a maximum packet of 1500 bytes.

Finally, time is needed for synchronization: A user thread must be woken up when its data have arrived and, on the Firefly, an interprocessor interrupt is needed to activate the processor that operates the Ethernet device. The time for this is on the order of 350 μs, where the bulk of the time is used to wake up the receiving thread.

An important thing to notice is that the time the hardware uses to transmit the packets in an RPC call makes up only half of the RPC latency; the other half of the time is spent in software. With faster networks, the software overhead will increase even more. Protocols that spend a large amount of effort to optimize the use of the network hardware are, therefore, in many cases self-defeating. In local-area networks, it pays to use protocols that are as lightweight and simple as possible. The next section describes an excellent project on streamlining protocol stacks.

The *x*-Kernel

The *x*-kernel is a configurable operating system kernel designed specifically to simplify the process of implementing network protocols [Hutchinson et al. 1989]. Its structure allows flexible configuration of protocol stacks, if necessary even at run time, and combines this with excellent performance. This has made it popular in the operating systems research community and, since it became available to researchers, it has been incorporated into several distributed systems.

The *x*-kernel derives its flexibility and performance from several features. The first, and most important, is that there is a uniform interface to all protocol layers. This allows layers to be stacked arbitrarily (although there are, of course, many protocol combinations that make no semantic sense) and it allows one layer in a stack to be replaced by another.

Protocol layers can be *bound late*; that is, a protocol stack can be constructed at run time, when a connection is established. Late binding is exploited through the use of *virtual protocols*. Virtual IP (VIP), for instance, is a protocol layer that provides an IP interface, but uses dynamic binding to other protocol layers to achieve the actual transport. For destinations on the Internet, VIP would use IP itself, but for destinations on the local Ethernet, or dial-up telephone lines, other protocols can be used which provide the best possible performance for the media used.

Another technique exploiting late binding is decomposing protocols into sublayers. A single protocol often combines several functions, for example, *(de)multiplexing, fragmentation,* and *(re)transmission.* Sometimes, higher layers only require a subset of these functions. By decomposing a protocol in separate (dynamically bindable) sublayers, protocol stacks can be composed that have no unnecessary functions or header fields.

Using late binding, a transport protocol can use different lower level protocols, depending on which network is used to reach the destination. An RPC transport protocol, for instance, can use UDP/IP for its data transport when the destination has to be reached over the Internet, but use Ethernet packets directly when the destination is on the same Ethernet. Late binding allows network-dependent optimizations without any loss in flexibility.

Protocol layers in the *x*-kernel have a simple procedural interface. One thread of control can traverse several protocol layers in order to send or deliver packets. This reduces the number of context switches and enhances performance.

Transactions

Locus

The Locus operating system [Walker et al. 1983], developed at the University of California, Los Angeles, (UCLA) in the early 1980s, represents an important advance in distributed systems. It demonstrated that a standard Unix interface can be implemented on a fault-tolerant distributed platform, so that unmodified Unix applications can exploit many of the advantages of an operating system that masks failures.

Two basic mechanisms in Locus are used to assist in failure recovery. One is a replicated storage facility, the other is a nested-transaction mechanism. Replicated storage allows data to survive system and media failures. Transactions allow applications to recover from system crashes.

The system does not replicate computations, so a crash will stop all application processes on a machine. Crash recovery is, therefore, necessary, but the transaction mechanisms can be used to leave the system in a consistent state when the crash occurs.

Replicated files are normally updated consistently, but, when a failure partitions the network, separating replicas, then updates are allowed on the accessible subset of the replicas. This can cause inconsistencies among the replicas. These are detected when the network becomes whole again and then reconciliated.

The automatic reconciliation mechanism tries to repair inconsistencies. When it cannot do so for lack of relevant information, it refers the reconciliation up to a mechanism at a higher level. At higher levels, the reconciliation mechanisms become more specialized and distinguish between directories, mailboxes, database files, and other files. Semantic knowledge of a file type allows more reconciliations to happen. Remaining inconsistencies are referred to the human owner of the file, who is considered to be the ultimate reconciliator.

Locus provides nested transactions [Mueller et al. 1983] for failure atomicity. Nested transactions are transactions within other transactions. The outermost transaction is the *top-level transaction*. When a transaction commits, the resulting state is only visible in the enveloping transaction. When a top-level transaction commits, its effects and the effects of all committed subtransactions become globally visible.

Transactions modify files. Files opened as part of a transaction are locked by that transaction and unlocked when the transaction aborts, or the enveloping top-level transaction either commits or aborts. When a transaction commits, its locks are thus inherited by its supertransaction.

Partitions form a complicating factor in the realization of transactions, because the transaction coordinator may become separated from some of the files that form part of the transaction. Subtransactions, when they are separated from their callers, are simply aborted, and the supertransaction is informed. Transactions are also aborted when they are separated away from a file for which they hold a lock and no other replica of the file is accessible.

If the owner of a transaction is separated from the transaction coordinator, then the owner cannot find out the fate of the transaction until the partition is repaired. Thus, under certain circumstances, applications will be blocked waiting for communication to be restored. This blocking problem occurs in all transaction systems, but in some it is worse than in others.

Locus has been in use as a general-purpose fault-tolerant Unix system for a number of years and it has had a significant influence on many other systems, including Quicksilver.

Quicksilver

The Quicksilver project of the IBM Almaden Research Center [Cabrera and Wyllie 1987, Haskin et al. 1988, Schmuck and Wyllie 1991] has demonstrated that it is possible to build general-purpose support in distributed systems for fault tolerance. The design of Quicksilver was guided by the following principles [Haskin et al. 1988]:

1. Servers and other applications should be resilient to external failures and be able to recover the resources associated with failed components.
2. The operating system should not contain any code to aid the error recovery for particular servers: each server should contain its own recovery code.
3. The point in item 2 notwithstanding, the system should offer a systemwide uniform error recovery architecture to prevent ad hoc proliferation of error recovery mechanisms.
4. There is a mechanism that allows a client to perform a group of logically related activities in interaction with a set of different servers as a single *atomic* operation (all of the activities should succeed, or none).

With respect to fault tolerance, four categories of applications are recognized in Quicksilver: (1) Those that manage volatile internal state that does not have to be recovered in a crash; after a crash, the server is simply started afresh (example: window servers). (2) Servers that manage replicated volatile state; when a single server crashes, it can recover from one of its replicas; when all servers crash, e.g., in a systemwide power failure, they are started afresh (example: the Quicksilver binding agent where servers register themselves so that clients can find them; after a systemwide crash, all servers must reregister). (3) Servers that manage a recoverable state; that is, a state that may not be lost as the result of a crash (example: the file system). (4) Long-running applications that need periodic checkpointing to make their state recoverable (example: simulations).

Quicksilver offers mechanisms for *atomic transactions* to these applications, very similar to the atomic transactions of database systems. Application classes 1 and 2 only use a subset of the mechanisms described subsequently; the others can use the full set.

There are servers that make transaction-based recovery possible:

1. The *transaction manager* is replicated over all nodes and coordinates transaction commit by communicating with other transaction managers.
2. The *log manager* implements a recovery log for the transaction manager's commit log and for servers' recovery data.
3. The *deadlock detector* detects global deadlocks and resolves them by judiciously aborting transactions.

The messages used by clients to communicate with servers carry a *transaction identifier* (*tid*). Servers thus know to which transaction a client request belongs; they can tag the state information they keep with the associated tids. The interprocess communication (IPC) protocols keep track of the servers addressed as part of a particular transaction so that the appropriate transaction managers can be invoked at commit.

The commit protocol messages are used as a mechanism both for transaction synchronization and for failure notification. Before commit, servers maintaining recoverable state make use of the log manager to store the recoverable data.

These three services make up the *recovery manager*: with this recovery manager, Quicksilver concentrates the recovery functions in one place; servers can use them or not use them, according to their needs; applications can choose between transaction-protocol variants, such as one-phase or two-phase, as appropriate to their function.

Servers communicate with their local recovery manager. The recovery managers at different nodes communicate among themselves to achieve atomicity or recovery.

Processes using transactions use the primitives *begin*, *commit*, and *abort* to manage them. Begin allocates a new tid and makes the invoked transaction manager the coordinator for the transaction just begun.

Transactions in Quicksilver typically have an overhead of between 5 and 100 ms above the time required for the operations that were done as part of the transaction.[5] This overhead is a very acceptable price to pay for an excellent, well-structured, fault-tolerant mechanism.

Group Management

The technique of replicating computations over multiple, independently failing processing nodes is not new. It has been in use for a long time in safety-critical real-time applications, such as fly-by-wire aircraft control. In real-time environments, the processors are dedicated to running the application and the replicated computation runs in lock step.

Important techniques for managing fault-tolerance for *non*-real-time applications by replicating computations in more relaxed synchrony were first explored by Birman [1985] in the ISIS system [Birman and Joseph 1987, Joseph and Birman 1986]. The ISIS project has inspired research on the theoretical foundations of replicated computations, causality and virtual synchrony, and models of fault tolerance. This has made it one of the most important projects in distributed systems research.

[5]On an RT-PC.

ISIS

The goal of the ISIS project is to provide a system that automates the "transformation of fault-*intolerant* program specifications into fault-tolerant implementations" [Birman 1985]. This is done by taking a sequential program and replicating its code and data over a number of nodes.

The failure model underlying ISIS is *fail-silent*, that is, processors fail by stopping, not by giving wrong results. The surviving processes find out about the crash using a *failure detector* (which uses *time out* to detect processors that are no longer responding). Network failures are transformed into processor failures by declaring unreachable processors crashed; when the network is repaired, such processors learn about their crash and execute a crash-recovery protocol to synchronize themselves to the rest of the replicated computation again.

Computations manipulate *objects* which are made *resilient* by replicating them over multiple sites. *K*-resiliency means that the replicated object behaves like its nondistributed, sequential counterpart running to completion; that, when *k* or fewer replicas fail, the object continues to accept and process requests and does not block, and that recovering replicas can rejoin the group of replicas; and that, when there are more than *k* failures, the replicas restart when all failures are repaired.

Applications can group operations on objects into (nested) atomic transactions. For this purpose, the system provides operations for starting, committing, and aborting transactions and for locking objects.

Replicated objects coordinate their actions by *broadcasting* the relevant information. The broadcast operations are all *reliable*; that is, if one working replica receives the broadcast, all of them will (see subsection "Group Communication," or Hadzilacos and Toueg [1993]). There are three types of reliable broadcast; they are called *Bcast*, *OBcast*, and *GBcast* and they differ in the *ordering* semantics; that is, in the way delivery of broadcast messages is ordered relative to the delivery of other broadcast messages.

The Bcast primitive achieves a total ordering of broadcast deliveries: if a broadcast message is delivered at one site before another, then it is also delivered before the other at all of the other sites. Such a broadcast operation is known as *atomic reliable broadcast* (see subsection "Group Communication," or Hadzilacos and Toueg [1993]) for details).

The ordering semantics of the Bcast primitive are quite strong: two totally unrelated broadcasts are still forced to be processed in the same order everywhere. Relaxed ordering semantics can be implemented more efficiently. The OBcast primitive is one that does not induce a total order, but instead induces order only on broadcasts that could be related in a cause-and-effect manner. Such broadcast primitives, known as *causal broadcast*, use a logical-time stamp on each message and deliver them in increasing time-stamp order. The logical clock, from which the logical-time stamps are derived, is maintained independently by each process; it is incremented on every broadcast operation and always set to a value that is higher than that in received broadcast messages.[6]

Finally, there is a GBcast primitive which is used to inform the members of a broadcast group (that is, the collection of processes receiving the broadcast messages) of changes in the composition of the group. When a replica joins, it tells the rest of the group with the GBcast operation; when a replica crashes, one of the remaining processes will notice and send a GBcast message on behalf of the crashed process. The GBcast broadcasts are ordered with respect to other broadcast messages: a GBcast message informing of a crash will be delivered after all extant broadcasts (of any kind) from the crashed process have been delivered and a GBcast announcing the joining of the group will be delivered before any messages from the new member. Thus, GBcast messages are totally ordered with respect to group = membership changes; they are also totally ordered with respect to other GBcasts.

ISIS uses OBcast wherever it can, because the extra asynchrony allowed by it causes less waiting of processes for each other. It thus provides for more concurrency. Crashes are rare, so the GBcast operation will only rarely be invoked.

---

[6]Logical clocks are but one way of enforcing causality. Since clock values are rarely the same, most messages will still have a delivery order forced on them even if they are not causally related. A better way to maintain time stamps is the maintenance of *vector clocks* (see Hadzilacos and Toueg [1993] for details).

ISIS applications are built using an object-oriented style of programming. Each object can receive requests from other objects, which are processed and responded to. Each replica of a replicated object will receive all requests (via one of the broadcast primitives) and will also coordinate with the other replicas using broadcast messages.

Knowing which primitive to use in a particular situation is not trivial and ISIS has been criticized for this [Cheriton and Skeen 1993]. There are claims that transactions can be used to manage replicated objects just as well. This may be the case, but the fact remains that ISIS has been more influential in the development of distributed-systems theory and in increasing the understanding of concurrency, fault tolerance, and causality than any other system. The commercial success of ISIS in stock-market applications proves that ISIS certainly is not a toy.

85.3 Best Practice

Most computers are connected to networks now so that all systems are becoming, to a greater or lesser extent, distributed. Most system builders, therefore, need some knowledge of distributed systems in their baggage and distributed-systems research is becoming mixed with other research areas.

A major—probably the major—motivation for distributed systems research used to be the quest for dependable systems, systems that would tolerate failures in order to become more reliable than their parts. This quest has largely succeeded in that we now have a wide range of techniques and algorithms that work.

However, the subsequent integration of such techniques and algorithms in everyday systems has largely failed. We find two important causes for this. One is the reliability of current computer hardware, the other is the difficulty to change systems that have become accepted as standards.

Computer hardware is now very reliable. Disk manufacturers claim mean times between failure of 200,000 h and more so that very few disks ever fail during their operational lifetime. Because of this, in most situations, there is little need for replicated data storage. Highly distributed services, such as electronic mail and the domain name service, have their specialized fault-tolerance mechanisms. In the World Wide Web no fault tolerance exists at the moment, but some replication will likely occur in the next few years. It appears that only a small set of specialized applications and application domains need mechanisms that provide reliability beyond what networked, but nondistributed, systems can give today.

The other reason is that the world is currently burdened with a few operating system standards that cannot easily be extended with fault-tolerance mechanisms without major change. There is such an investment in existing software that any short-term changes are unlikely. In any case, the world's most widely used operating systems have many, more urgent problems to solve before increased fault tolerance will be noticeable.

Failure Models

Distributed systems can grow to a very large size. But large systems tend to be more complicated than small ones and the consequence of a fault is more difficult to control. One of the most desirable properties for a distributed system is that the failure of any component does not cause other components to fail as well.

Fault-tolerant applications must be distributed over multiple processors so that the crash of one or a few processors does not bring down the application as a whole. The state that the application maintains must be distributed as well, with enough added redundancy to recover from failures.

A *failure model* describes what failures are expected to occur. If it is assumed, for instance, that nodes fail by crashing, then recovery is usually simpler than when they can fail by producing erroneous data. In safety-critical applications, such as a fly-by-wire system or a system controlling railway signalling, the failure model typically assumes that a limited number of *arbitrary* failures may occur. Arbitrary failures are also known as *Byzantine* failures—after the problem of the Byzantine generals [Lamport et al. 1982]— or *malicious* failures. For other applications, however, a *fail-stop* model is common: processors fail by stopping. For a more detailed discussion of failure models we refer to Schneider [1993].

Naming

Without a naming mechanism to allow information sharing, a distributed system cannot exist. Names provide a level of indirection in referring to objects, processes, services, and data that is crucial when entities can be relocated or replicated.

Needham [1993] distinguishes between *pure* names and names of the other sort, *impure* names. A pure name only identifies, an impure name also guides: File names (/usr/bin/sort), URLs (http://www.pegasus. esprit.ec.org/sape), and IP addresses (130.89.181.118) are all examples of impure names, because they lead a system toward the location of the named thing. Impure names have the disadvantage that moving an object often requires renaming it.

Pure names do not have this problem. If a name only identifies, there is no relationship between an object's location and its name. Unfortunately, the scale of many systems makes the use of pure names impractical; in the best case, finding the location of an object requires an $O(\sqrt{n})$ search, where n is the number of possible locations [Mullender and Vitányi 1988]. This makes using a pure name from Europe to locate a file in Australia very expensive.

Distributed systems are, for this reason, forced to use impure names. However, some names are impurer than others. While one name space, because of its structure, fixes the location of objects in exactly one location, another name space might allow objects to move within an organization, a local-area network, or a group of nodes.

Name services should be among the most available services in a distributed system. Availability is an even more important requirement for a name service than correctness: When name resolution does not work, not much else can work either, but if a name service occasionally gives wrong or outdated answers, an application might be guided to the wrong place where the error will usually be discovered.

To be very available, a name service usually is highly replicated. Particularly, name services that are accessible worldwide tend to be enormously replicated. The list of top-level domains of the Internet, for example, is replicated at practically every site; that is, hundreds of thousands of times. Under such massive replication, maintaining consistency is impossible.

Most of the data in large name spaces is fairly static, so inconsistencies will be rare. Smaller name spaces, such as a name space for a distributed file system, will have much higher rates of change. Consistency is also more of an issue in such name spaces. Fortunately, the limited size and replication of such name spaces make it easier to maintain consistency.

Schroeder [1993] claims that "an object should have the same name everywhere" so that sharing becomes possible. Pike et al. [1993] have countered that sharing is facilitated primarily by *naming conventions:* Usually, sharing is achieved by knowing where to look for the shared object; that is, you guess most of its name; it is only sometimes that somebody tells you.

The hierarchical name spaces of Digital's GNS, X.500, and DNS are singly rooted; one name has the same meaning everywhere. Plan 9 [Presotto et al. 1991] uses a two-level naming scheme. One name space names *servers*, each server maintains a local name space, and applications construct a name space by *mounting* some of these servers in a private name space. The servers are named in a global name space (which Plan 9 defined somewhat ad hoc), and conventions play an important role in deciding where users are expected to mount certain servers. The result is a name space in which, for the objects that matter, objects have the same name everywhere.

Amoeba [Mullender et al. 1990] was one of the few systems to use pure names. In this system, services were named using 48-b *ports*. Ports were located using broadcast. As a result, Amoeba did not scale to a size beyond a local network composed of bridged Ethernet segments. Within an Amoeba system, however, objects could be arbitrarily relocated, and processes could be migrated without naming inconsistencies.

An excellent example of a very large-scale name service is the design of Digital's Global Name Service [Lampson 1986]. The update operations in GNS were carefully chosen to be commutative and idempotent. Commutativity lets only the set of updates determine the state of the naming database, not the order in which they are applied. The idempotency allows updates to be carried out more than once without effect in the state, so that the distribution of updates does not have to happen too carefully.

Communication

The primary function of the interprocess communication mechanisms in distributed systems is to transport information between processes on the nodes of a distributed system. Processes on the same node can also interact in other ways, such as through shared memory. But even within a node, it is useful to use network communication mechanisms between processes. Doing this achieves three things:

1. The system can be reconfigured in such a way that processes, previously running on one node can run on different nodes.
2. Crashes of one process do not have to bring down another.
3. The message-passing interface between processes provides a convenient place for carrying out sanity checks on the data transmitted.

In effect, the interprocess communication mechanism also functions as a *fire wall* that protects one process from bad influences from another. This fire-wall function is an important one in distributed systems, because it helps to get the property of independent failure, it helps to hide the location of a process or function and, thus, it helps the realization of fault tolerance.

It is almost always assumed that networks can fail. They lose and corrupt packets, and sometimes network links may fail altogether. Corrupted packets are detected by making use of checksums; and they are then treated as lost.

Recovery from lost packets can simply be done by numbering the packets, detecting gaps in the number sequence of received packets, and requesting retransmission. When a connection is set up, packet numbers are usually initialized at zero. It is important that packets from one connection cannot mistakenly be received as part of another. There is this risk especially when a connection is set up to replace one that broke in a host or network crash.

Host and network crashes bring about fundamental uncertainty about the state of the system in the surviving nodes. When receipt of messages is acknowledged by a receiving host which crashes, can the sending host deduce from the receipt of an acknowledgment that its message was acted on? The answer is no, because the recipient may have crashed the microsecond after sending the acknowledgment. Can a host deduce anything from *not* receiving an acknowledgment then? Again, the answer is no; the acknowledgment may have been lost.

When an operation is executed remotely, acknowledgments for the arrival of packets or messages are not useful. The only truly useful information is a message that the operation was completed. This is the *end-to-end argument* of Saltzer et al. [1984] who stated that, in a protocol stack, lower layers can never be used to recover from all of the faults and crashes of higher layers. In other words: the highest level of protocol, which necessarily is in the application domain, must have some error recovery if the application is to be fault tolerant.

The opposite is not true; on the contrary: higher layers of protocol are always used to recover from some of the faults in lower layers: when an IP packet is corrupted, TCP will do the recovery.

The end-to-end argument suggests treating the messages belonging to the execution of a remote operation as belonging together in a single protocol unit, directly used by the application, and this is exactly what most distributed systems do. The protocols that handle such groups are known as *message transaction* [Cheriton 1988, Mullender and Tanenbaum 1986], *remote operations*, or *remote procedure call* [Birrell and Nelson 1984].

The basic idea is that activities in a distributed system are structured around the notion of sending a *request* to carry out an operation to a server and receiving a *response* from that server when the operation is completed. The response serves as the fundamental acknowledgment that the request was carried out. Acknowledgments for message arrival may also be used, but they should be viewed as optimizations only.

Amoeba [Mullender et al. 1990] and the V system [Cheriton 1986] were among the first systems to enter the race for implementing remote operations with very low-response times [Nordmark and Cheriton 1989, van Renesse et al. 1988]. The race was joined by many others, but the most interesting work was published by Schroeder and Burrows [1989], who gave an excellent account for where an implementation

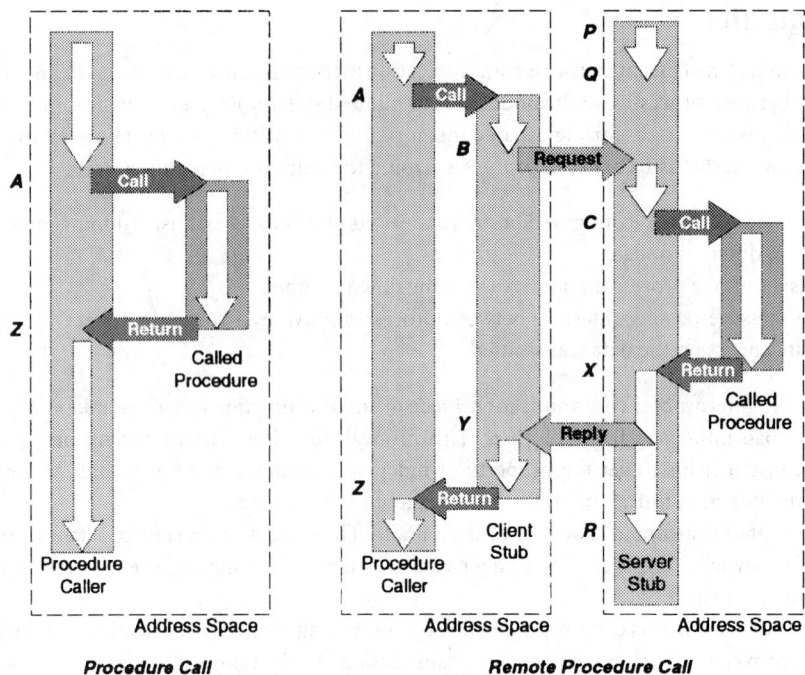

FIGURE 85.2   Structure of remote procedure call.

spends its time and Hutchinson et al. [1989] who built a framework for high-performance protocol stacks that is still being incorporated in many research systems and some commercial ones.

When remote operations are adorned with a mechanism to pass parameters between caller and callee, one has the ingredients for remote procedure call [Birrell and Nelson 1984]. The mechanism is illustrated in Fig. 85.2. On the left is a normal procedure call: The procedure is called at point A by the processor's *call-subroutine* instruction sequence. At point Z, the called procedure returns control to the caller, using the processor's *return-from-subroutine* instruction sequence.

On the right of Fig. 85.2, one can see how remote procedure call works. In the address space of the caller, a *client stub* is linked in, and in the address space of the called procedure, a *server stub* is linked in. On the server side, the server stub begins executing at P; it announces its presence to its clients (presumably via a name server, we shall discuss the mechanisms for this presently) and, at point Q, it asks the system for incoming client messages and blocks.

On the client side, the caller calls the stub at point A, just as it would have called the procedure in the normal procedure-call case. The stub retrieves the parameters from the stack, puts them in a request message, and sends that to the server at point B. This is called *parameter marshalling*. After marshalling, the client stub blocks itself awaiting the reply message.

This unblocks the server stub. The server stub *unmarshals* the parameters—it takes the parameters from the request message, and pushes them onto the stack—and calls the remote procedure at C, using the standard calling sequence. The procedure returns at point X, the server stub retrieves the return parameters, puts them in a reply message, and sends it back to the client at point Y. The server stub does not block at this point. It carries out some clean-up operations, if necessary, and, at point R, it could jump back to point P or Q to await the next client request.

The arrival of the reply message unblocks the client stub, which then retrieves the return parameters for the reply message and it uses them to make a normal return from procedure at point Z.

From the point of view of the calling program and the called procedure, calling a remote procedure appears to be exactly the same as calling a conventional one. This is not the case, however. When remote procedures are used across address space or even machine boundaries, crashes need only affect the caller or the callee. Thus, when remote procedures are used callers must anticipate the possibility that the procedure does not return a value as expected, but that a crash, or communication failure is reported instead.

Another difference between calls within an address space and calls between address spaces is that in the latter case, it is pointless for caller and callee to exchange a pointer to an object in a call. A pointer refers to a different thing in a different address space. Remote procedure call thus imposes restrictions on the kinds of parameters that can be passed that normal procedure-calling sequences do not have.

When remote procedures are used between hosts with different architectures, or between processes written in different programming languages, stubs can be used to convert between the different representations that the parameters may have. This conversion can only be carried out when the parameter types are known.

Given a procedure *signature*,[7] client and server stubs can be built automatically from an interface definition specified in an *interface definition language* (IDL). Interface definition languages are an almost vital tool in building large distributed applications. They specify the interfaces at module boundaries, the fire walls where type checking is so very important.

Examples of IDLs are HP/Apollo's NCS which is now part of the Distributed Computing Environment (DCE) of the Open Software Foundation, SUN RPC [Sun 1985], Mercury [Liskov et al. 1987], Flume [Birrell et al. 1984], Courier [Xerox 1981], and Middl [Roscoe 1994].

In client/server settings, it can be useful to put some of the server functionality in the client stubs. The client stubs then no longer provide a direct mapping of client calls to the stub and remote calls to the server's procedure. When this is the case, we use the term *clerk* or *agent* rather than the word stub.

Clerks can help with the implementation of automatic rebinding, should a server crash [Schroeder 1993]. The clerk provides a good point for this, because semantic knowledge of the service's behavior can be built into it. Alternatively, clerks can try to provide a (degraded) service while a server is down. A name-server stub, for instance, can provide information for its cache which may be obsolete; clients are better off with old data than no data in this case.

Clerks can also help with performance improvements through caching. A client-side cache for a file server can be viewed as a clerk for the file server. Clients make calls to the clerk and the clerk passes some of them on the the server.

Binding

In a distributed system, services do not always have to reside in one location. Reconfiguration can cause services to be moved, or a service can be restarted on a different machine when the original crashes.

Binding and naming are closely related. Binding is the process of mapping the name of a *service* onto a connection with a provider of that service, a *server*.[8] The service can be anything: a mail-delivery service for a particular user, a service that can get a file printed nearby, or another part of a distributed application. A name can vary from something as specific as an IP address plus port number to something as vague as the nearest printer that can render PostScript.

Thus, when a binding is created, a specification of a service is converted into a connection to a particular server. A service can be characterized by two things: one is what it can do for its clients, the service's *function*, the other is how its clients can make it do those things, the service's *interface*.

A service has a *state* and, through the service interface, clients can query or modify that state. The interface describes the syntax and semantics of the interactions between client and service. The semantics describe how the operations that the service can carry out modify the state and and what values will be returned.

---

[7]The signature of a procedure is the procedure's name and type: the number and type of parameters and the type of the return value.

[8]We exercise some care in distinguishing *service* and *server*. Service represents the abstract notion of a set of operations that can be carried out on a set of objects. Server represents a processing entity (a process or processor) that can carry these operations out.

In the binding process, a service is usually sought that has a particular interface and a distinguished state: When delivering mail, we look for a mail service whose state indicates that it works for a particular user; when connecting to a file service, we want one whose state contains the file we aim to read.

A mechanism is needed to *name* the things one binds to, explicitly or implicitly. An example of explicit naming occurs in binding to an NFS file server, the name of the service corresponds directly to the server.[9] Slightly more implicit is the way one names a mail box, sape@cs.utwente.nl, for instance. This name does not directly name a particular server, but a *set* of them. During the binding process, the servers in the set are tried until a binding is established. Bindings can be named even more implicitly: The ANSA Trader [API 1989] allows associative names; for instance, one can name a printer with certain properties: it should be on the fourth floor and it should be capable of printing postscript files.

The binding process consists of three stages. First, a set of servers must be found that implement the service. Then, a connection must be created between client and one of these servers, and finally client and server must initialize their mutual binding state. We shall discuss these in turn.

Finding a server for a service can be done by clients and servers themselves, or with the help of a separate *binding agent*. The V system [Cheriton 1988] and Amoeba [Mullender 1985] are examples of systems where clients locate servers without a separate binding agent. Both systems were designed for use in a local network, and clients found servers by *broadcasting* for the service on the local network. In the V system, service requests were broadcast and, if there were multiple servers for the service, it was up to these servers to decide which one would respond; in some cases all servers would respond. In Amoeba, before a request was sent a *locate* message was broadcast and the first server to respond would be chosen to send the request to.

In far-flung systems, usage of broadcast for the location of a server is not doable. A service with the function of broker is needed to bind clients to servers. The idea is that servers, when they become active, notify the broker service of the service they perform and their location. Naturally, most bindings between clients and servers take place within a node or a local network, but bindings that span the globe occur too. For efficiency, it is common practice to make use of a hierarchy of brokers in such a way that the brokers needed to bring about a binding are as near to client and server as they are to each other.

Binding in large systems cannot be done by making use of a pure[10] name space for the identification of services. Brokerage is best done with a hierarchically organized set of broker services. The Domain Name Service [RFC-1035, RFC-1034] is the almost universally used broker service at the moment.

Binding to the broker presents a bootstrapping problem that is solved by putting the broker at a well-known address, or providing the addresses of a set of brokers in a (well-known) file.

After a server has been found, the second step in the binding process is setting up a connection. This is straightforward and does not need further discussion.

When a connection exists between client and server, negotiation can take place concerning protocol parameters, such as packet sizes, window sizes, network data representations, etc. A step that is becoming increasingly important now that most hosts are connected to the Internet is carrying out an authentication handshake and, if necessary, establishing encryption keys. When all of this is done, a binding exists, and clients can start sending requests to the server.

Client or server crashes result in broken bindings. The role of a server is usually such that, when a client crashes, no attempt is made to create a new binding. The server may, of course, use the report of a broken binding to clean up the connection state. When a server crashes, a client may attempt to bind to another server for the same service in order to be able to continue getting service.

Transactions

Process and node crashes can leave the state of a distributed computation in an inconsistent, unknown state. When a process in a distributed application crashes, the others must find out where that process got to when it crashed in order to do recovery. Consider, as a trivial example, a request made to the bank's computer to transfer a sum of money from one account to another. If that computer fails, it can leave the

---

[9]Provided the server is up; if it is down, binding will fail.

[10]Cf. section 85.3 subsection on naming.

database in at least four states: (1) nothing was done yet, (2) the money was removed from one account but not yet added to the other, (3) the money was added to one account but not yet removed from the other, and (4) the transaction was completed.

Without maintaining extra administration, it is not possible to find out how much of a mess a crashed process leaves behind in a system. This administration could be maintained in an application-dependent manner, but, as it turns out, there are excellent general-purpose mechanisms as well.

In this section and in the next, we discuss these general-purpose mechanisms. In "Group Communication," we show how computations can be replicated and how communication can be structured so that all replicas are guaranteed to receive all relevant information in the correct order.

The mechanisms discussed in this section are based on the notion that distributed applications query and modify a distributed database of some sort and that they structure the update operations in such a manner that, after a crash, a consistent state of the system can always be restored.

The database is organized in such way that updates on it succeed completely or fail completely; that is, if an update fails, it leaves the database in the state it had before the update started. We can call such updates *atomic updates*, because they appear to happen all at once.

Atomicity is also an important structuring mechanism for the management of multiple simultaneous updates. Suppose that, in our bank-account example, two updates on a single bank account happen simultaneously, one depositing and one withdrawing. The updates both proceed by reading the balance, computing the new balance, and writing it back. When the two updates both read before the first writes back, the balance of the account becomes inconsistent, a euphemism for wrong.

The update consists of a group of operations that belong together (a read operation and a write operation, in the example). We call such a group a *transaction*. Database systems usually have operations *transaction-begin* and *transaction-end* to allow applications to indicate the grouping.

By making transactions be—or appear to be—atomic, the effect of simultaneous transactions is to *serialize* the updates; that is, the result would be exactly the same if one transaction finished before the other started. Thus, applications do not have to be aware of concurrent transactions, which makes programming them much simpler.

Transactions have, what is often called, the ACID property: they are *atomic, consistent, isolated*, and *durable*. By consistency, we mean that, provided each transaction by itself maintains consistency, the combination of multiple, concurrent transactions also maintains consistency. Isolation means that transactions do not interfere with each other, they are serialized so that the effect is that of one transaction finishing before the next one starts. Finally, durability means that the updates made by a transaction last: when a transaction finishes successfully, all updates are safely stored on stable media.

So far, all of this is just as relevant to centralized systems as it is to distributed ones. In distributed systems, however, the additional problem is to realize atomic transactions also in the face of failures: host crashes, communication failures, or media failures.

To deal with media failures, data can be replicated. Full replication can be done by *disk mirroring*: storing all data on two, identical, disks. Another popular replication technique is redundant array of inexpensive disks (RAID) [Chen et al. 1988]. Here, a parity disk is added to a small number of disks (say, four) and the blocks on the parity disk consist of the exclusive-or of the corresponding blocks on the data disks. When (a block on) a single disk fails, its contents can be calculated by computing the XOR of the corresponding blocks of the other disks.

Node crashes or communication failures can make it impossible to finish a transaction successfully: a node may store information that is needed to complete the transaction, for instance. Transactions that cannot be completed are *aborted*. Those that can are *committed*.

Atomicity of transactions, like so many other things in computer science, is made possible by introducing a level of indirection: a pointer refers to the data. Updates are made by copying a portion of the databases and modifying it. The updates are then committed by changing the pointer to point to the modified data. The rewrite of a single pointer is an atomic operation (Lampson and Sproull [1979] describe how atomic operations can be implemented on replicated disks); since all data are accessed via the pointer, the whole database undergoes atomic modification as well.

In distributed systems the technique for atomic update is quite similar to the one previously described. The pointers, of course, must be implemented as {server name, data pointer} pairs so that they can refer to remote objects as well as local ones.

A technique known as *two-phase commit* (2PC) works as follows. The database is distributed over a number of servers. A client, wishing to make an update, contacts one of the servers and issues a transaction-begin operation. The client receives a *transaction identifier* and makes read and write requests, labeled with the transaction identifier, to the various database servers. One of the database servers becomes the transaction coordinator; it forms the point where the decision is made to commit or abort the transaction.

All the database servers make their updates as described earlier: both the old state and the new state are stored and the flipping of a pointer will switch between the old and the new contents of the data involved in the transaction. In the distributed case, a *lock* is added. This lock is used to synchronize the commit operations of all of the participants: first all servers lock the data, denying access from outside the transaction, then the pointers are flipped, committing each participant to the transaction's updates, and finally, the participants unlock the data again.

Two-phase locking uses this mechanism as follows. If the client aborts the transaction, the coordinator only has to tell the other servers that the transaction has been aborted. They can then do garbage collection. If the client asks for a commit, the coordinator sends a prepare-to-commit request to all participants. The participants write all buffered data to disk to get ready to flip their pointer, they lock the database, and they send an *okay* reply to the coordinator. Naturally, if a server or a network connection fails, no okay message will be forthcoming.

When the coordinator has received the okay from all participants, it commits the transaction locally (by flipping the pointer) and sends a commit message to all participants. The participants also commit and they can unlock.

This is how transactions are committed using 2PC in normal failure-free circumstances. Let us now consider what havoc failures wreak. Suppose a participant crashes before sending its okay message. In this case, the coordinator will receive one okay too few and refuse to complete the commit. The coordinator will broadcast an abort message to the participants, which leaves them in the state before the transaction started. Now, suppose that the participant crashes after sending the okay. Sending an okay implies that both old and new data are on stable storage, so the transaction can (and probably will) still be committed. The coordinator commits and broadcasts the commit message. When the crashed participant comes back up, it finds its database still locked (the locks are also on stable storage) and it can find out from the coordinator, or another participant, what the outcome of the commit operation was.

Now, suppose the coordinator fails. If it fails before sending the prepare message, the transaction can be aborted without further ado. If it crashes after sending out (some) of the prepare messages, the remaining participants can compare notes. If there is at least one that has not responded okay, then it is safe to abort the transaction. But if all remaining participants have sent their okay, then it is not known whether or not the coordinator had already committed internally. The other participants can only wait until the coordinator comes back up. If the coordinator crashes after sending at least one commit message, then, of course, the other participants can finish the transaction, knowing the coordinator must have already committed internally.

Two-phase commit, thus, can *block* on the failure of the coordinator. The probability that this happens is reasonably small, because the time between receiving the last okay and sending the first commit can be kept quite short.

A technique that further reduces the probability of blocking requires an extra round of communication and is therefore known as *three-phase commit* [Skeen 1982].

Two-phase and three-phase commits are important techniques for making updates in distributed databases atomic. In addition to atomicity mechanisms, however, we also need mechanisms to *serialize* the updates; that is, make sure that the effect of two concurrent updates will be the same as that of either first doing one and then the other, or vice versa.

There are two important approaches to causing serializability, one is based on locking out all access that can cause nonserializable access, the other is to check for serializability at commit time and aborting in case of nonserializability. These approaches are widely know as *pessimistic* and *optimistic*, respectively. Locking out access is pessimistic because it also locks out some accesses which, in retrospect, could be serializable after all; checking at commit time is optimistic because it is used under the assumption that serializability conflicts are rare.

A serializability conflict between two transactions can only occur when they access the same data; otherwise, they are independent and it is obvious that, carrying out the two transactions one after the other or concurrently makes no difference. One transaction only influences another if the former writes data that the latter reads.

This observation suggests a simple test for making the decision whether to allow a transaction to commit in an optimistic setting [Kung and Robinson 1981, Schlageter 1981, Strom and Yemeni 1985]: When a transaction is about to commit, a test is made to see if the intersection of the data read by the current transaction (its *read set*) with the data written by transactions that committed after the start of the current transaction is empty. If it is empty, the current transaction has not read any data that was modified afterwards; therefore, the current transaction would have proceeded no differently if the other transactions had already committed before the current one started and, thus, the current transaction is serializable after the other ones. Transactions under optimistic concurrency control are serialized in commit order.

In pessimistic concurrency control, locking is used to make sure that no transaction proceeds beyond the point where serializability is endangered.

A perfectly safe way of locking is to acquire all of the necessary locks when the transaction begins and to release them when it commits or aborts. This is not possible, however, when it is only during the course of the transaction that the dataset accessed by the transaction becomes known. Transactions thus need to acquire locks dynamically. Locks cannot, however, also be released dynamically, at least not carelessly: Suppose one transaction first modifies datum *a* and then datum *b*, while another transaction does the same in the opposite order. If both transactions acquire and release locks on datum *a* or *b* before acquiring the locks for the other datum, they deadlock.

A technique for dynamic locking that guarantees serializability is *two-phase locking* (2PL), not to be confused with two-phase commit: every transaction consists of two phases, one in which locks are acquired and one in which they are released. No lock may be acquired if even a single one has already been released. As long as it is not known whether more locks are needed, none should be released, so that many transaction systems do not release locks until commit (or abort).

The alternatives of total success or total failure that transactions provide are not always desirable. When a participant in a transaction fails, it is sometimes possible to use an alternative participant without having the whole transaction abort. Mechanisms that support splitting up large transactions into smaller ones are *nested transactions* [Reed 1978]. Subtransactions can commit or abort inside the main transaction. When the main transaction aborts, however, all of its subtransactions are aborted as well. When it successfully commits, however, only the committed subtransactions stay committed.

The ACID properties of transactions make structuring fault-tolerant systems much easier. Transactions have not only been found useful in distributed and centralized databases, but also in distributed file systems and even in operating systems. Examples of systems with atomic properties are Argus from MIT [Liskov and Scheifler 1983], Clouds from Georgia Institute of Technology [McKendry 1984], the Amoeba file system[Mullender and Tanenbaum 1985], Camelot and Avalon from Carnegie-Mellon University (CMU) [Eppinger et al. 1991], and Quicksilver from IBM Almaden [Haskin et al. 1988].

For more information on distributed databases and transactions we refer to books by Bernstein et al. [1987] and by Gray and Reuter [1992]. A compact introduction to distributed transactions can be found in Weihl [1993] and a rigorously formal treatment in Lynch et al. [1993].

Group Communication

In the previous section, we showed how distributed databases can be kept consistent by using transactions that atomically transform the database from one consistent state to another. Many applications can benefit from this way of structuring updates, but not all.

Transactions keep stably stored data consistent, but when applications must manage dynamic data structures consistently and reliably, or deliver exactly the same information to multiple locations, other mechanisms are required.

Group communication forms a class of mechanisms that allows delivering messages to groups of processes or machines. We shall see that group communication allows many more semantic variations than point-to-point communication and that different forms of group communication can be used to solve problems in very different settings.

As a first example, consider the design of a safety-critical control application, such as a fly-by-wire control system. Safety-critical systems must continue to function in the face of processor and communication failures of all kinds, not merely crash failures, but Byzantine failures.

The way in which this is typically done is to run the identical control program on a number of processors. Each of the processors starts out from exactly the same state and is fed with exactly the same information. Consequently, each of the processors will normally produce exactly the same results. These are compared and, if one result differs from the others, the processor with the dissenting result must have failed.

The minimum number of processors for this approach is three—with two processors, when the results differ, one cannot tell which one is wrong—and such an arrangement of three processors is known as *triple-modular redundancy* (TMR). When more processors are used, the configuration is usually labeled *n-modular redundancy* (NMR).

An NMR-based control system will get its inputs from a number of sources (*sensors*) and deliver its outputs to a number of destinations (*actuators*). Each of the sensors will deliver its data to all of the processors, but this, by itself, is not enough to guarantee that the processors will run in lock step. Additionally, all processors must receive the data from the sensors in exactly the same order.

The broadcast system that delivers sensor readings to all processors is known as an atomic broadcast system, because it happens indivisibly: no other broadcast or message delivery can break in and be delivered between the reception of the broadcast by one processor and another.

Another example of an application that uses broadcast is the system that broadcasts the Internet news worldwide. Anyone, anywhere in the world, can send messages and everyone, anywhere, can read them. A news message is labeled with a broad subject classification, the *news group* and with a *subject line* that is supposed to give some clue to its contents. When somebody reacts to a message, they send a *follow-up* message which contains a reference to the original message. Discussions on news net can create long chains of follow-up messages.

When reading the news, it makes sense to read messages in the order in which they follow one another up. It can be confusing to read somebody's reaction to something you have not seen yet. For news delivery, a broadcast system that maintains *causal order* makes much sense. Two messages sent independently can then arrive at different sites in a different order, but a message that causes a follow up must be delivered before the follow up everywhere.

Replicated systems that must withstand crash failures can use broadcast protocols, such as causal broadcast, to organize the communication between replicas and with clients. The participants are then organized as a group of processes, and both communication and membership changes are ordered according to the semantics chosen.

The earliest system that experimented with group communication was ISIS, developed at Cornell under the supervision of Birman [1985]. The idea in ISIS and in follow-up projects worldwide is to create, through reliable, ordered communication an illusion of synchrony: *virtual synchrony*. Examples of other well-known projects that make use of group communication and virtual synchrony are Paralex [Babaoglu et al. 1991], Relacs [Babaoglu et al. 1995], Delta-4 [Veríssimo et al. 1991], and Horus [van Renesse et al. 1995].

Broadcast protocols can be classified according to their ordering semantics, their reliability semantics, and their timing semantics. Ordering is established by making message *reception* and message *delivery* separate operations, so that after a message has been received it is possible to postpone delivery until other messages can be delivered.

Basically, a broadcast protocol is reliable when the following properties are satisfied: (1) If a correct process broadcasts message m, then all correct processes eventually deliver m (*validity*); (2) if a correct process delivers *m*, then all correct processes eventually deliver *m* (*agreement*); (3) for any message *m*, every correct process delivers *m* at most once, and only if some process broadcasts *m* (*integrity*) [Hadzilacos and Toueg 1993].

Reliability, as defined here, is only concerned with *correct* processes; these are the processes that correctly and completely execute the reliable-broadcast protocol at hand. It is thus possible that a process delivers a message *m* and crashes immediately afterwards so that no other process delivers *m*. This process may even act on the information in *m* before crashing. In some cases this is undesirable and, in such cases, reliability can be extended with *uniformity*, where the agreement rule changes into: if a process (correct or not) delivers a message, then all correct processes do so too; and the integrity rule changes into: any message is delivered to a process (correct or faulty) at most once, and only if it was broadcast by a process (correct or faulty).

The ordering semantics can be classified as (1) *no order*; (2) *first-in–first-out* (*FIFO*) *order*: if a process broadcasts *m* before *m'*, then no correct process delivers *m'* before *m*; (3) *causal order*: if the broadcast of *m* causally precedes[11] the broadcast of *m'*, then no correct process delivers *m'* before *m*; (4) *atomic broadcast*: if correct processes *p* and *q* both deliver *m* and *m'*, then *p* delivers *m* before *m'* if and only if *q* delivers *m* before *m'*.

Atomic broadcast does not relate broadcast ordering to delivery order; but it does say that the delivery order must be the same everywhere. Thus, atomicity can be combined with FIFO order, FIFO atomic broadcast, or with causal order, causal atomic broadcast, (causality does imply FIFO, by the way).

Real-time applications require messages being delivered within a bounded time after the broadcast. A protocol that does this is a **timed broadcast protocol**.

Reliability, ordering, and timing requirements can be combined to make, for example, *uniform timed causal atomic broadcast.*

A set of processes in a distributed application can use an appropriate broadcast protocol to maintain a replicated state. When a process crashes, the others must be informed reliably. It is often particularly important that the surviving processes agree on the moment of the crash with respect to the broadcasts made.

The combination of a set of protocols for broadcast (multicast) and a set of protocols to maintain membership state of a group of communicating processes is referred to as group communication. ISIS was the earliest group-communication system and it is still the best known.

Other groups have taken this work further. The Relacs system of Babaoglu et al. [1995] was designed to overcome the problems of scale that were present in ISIS. The Delta-4 project [Powell 1991] has explored group communication in the context of dependable computing.

References

Accetta, M., Baron, R., Bolosky, W., Golub, D., Rashid, R., Tevanian, A., and Young, M. 1986. Mach: a new kernel foundation for UNIX development. *Proc. Summer Usenix Conf.* Atlanta, GA, July.

API. 1989. The *ANSA Reference Manual.* Vol. Release 1.1, Architecture projects management, Poseidon House, Castle Park, Cambridge, UK.

Babaoglu, Ö., Alvisi, L., Amoroso, A., and Davoli, R. 1991. Mapping parallel computations onto distributed systems in paralex. Invited paper, *Proc. IEEE CompEuro '91.* Bologna, Italy, May.

Babaoglu, Ö., Davoli, R., Giachini, L. A., and Baker, M. G. 1995. Relacs: a communication infrastructure for constructing reliable applications in large-scale distributed systems, pp. 612–621. *Proc. 28th Hawaii Int. Conf. Syst. Sci.* II.

Bernstein, P. A., Hadzilacos, V., and Goodman, N. 1987. *Concurrency Control and Recovery in Database Systems.* Addison–Wesley, Reading, MA.

---

[11] An event *e causally precedes* an event *f* if and only if [Hadzilacos and Toueg 1993]: (1) a process executed both *e* and *f*, and in that order; (2) *e* is the broadcast of some message *m*, and *f* is the delivery of *m* at some process; or (3) there is an event *h*, such that *e* precedes *h* and *h* precedes *f*.

Birman, K. P. 1985. Replication and fault tolerance in the ISIS system. *ACM Operating Syst. Rev.* 19(5):79–86. *Proc. 10 Symp. Operating Syst. Principles.* Orcas Island, WA.

Birman, K. P. and Joseph, T. A. 1987. Exploiting virtual synchrony in distributed systems. *ACM Operating Syst. Rev.* 21(5):123–138. *Proc. 11th Symp. Operating Syst. Principles.* Austin, TX.

Birrell, A. D., Lazowska, E. D., and Wobber, E. 1984. Flume—remote procedure call stub generator for Modula-2+. Topaz manpage.

Birrell, A. D., Levin, R., Needham, R. M., and Schroeder, M. D. 1982. Grapevine: an exercise in distributed computing. *Commun. ACM* 25:260–274. [Presented at the 8th ACM Symp. Operating Syst. Principles (1981).]

Birrell, A. D., Levin, R., Needham, R. M., and Schroeder, M. D. 1984. Experience with Grapevine: the growth of a distributed system. *ACM Trans. Comput. Syst.* 2(1):3–23.

Birrell, A. D. and Nelson, B. J. 1984. Implementing remote procedure calls. *ACM Trans. Comput. Syst.* 2:39–59.

Cabrera, L. F. and Wyllie, J. 1987. *QuickSilver Distributed File Services: An Architecture for Horizontal Growth.* Computer Science Department, IBM Almaden Research Center, RJ5578.

Chen, P., Gibson, G., Katz, R. H., Patterson, D. A., and Schulze, M. 1988. Two papers on RAIDs. *Comput. Sci. Div.* EECS, UCB, UCB/CSD 88/479, CA.

Cheriton, D. R. 1986. VMTP: a transport protocol for the next generation of communication systems. *Proc. SIGCOMM '86.* Aug 5–7. ACM.

Cheriton, D. R. 1988. The V distributed system. *Commun. ACM* 31:314–333.

Cheriton, D. R. and Skeen, D. 1993. Understanding the limitations of causally and totally ordered communication. *ACM Operating Syst. Rev.* 27(5): 44–57. *Proc. 14th Symp. Operating Syst. Principles.* Asheville, NC.

Eppinger, J. L., Mummert, L. B., and Spector, A. Z. 1991. *Camelot and Avalon: a Distributed Transaction Facility.* Morgan Kaufmann.

Gray, J. and Reuter, A. 1992. *Transaction Processing: Techniques and Concepts.* Morgan Kaufmann.

Hadzilacos, V. and Toueg, S. 1993. Fault-tolerant broadcasts and related problems. In *Distributed Systems.* S. J. Mullender, ed., 2nd ed., pp. 97–145. ACM Press, New York.

Harcourt. 1995. *Plan 9, Manuals, Documents and CD-ROM.* Harcourt Brace.

Haskin, R., Malachi, Y., Sawdon, W., and Chan, G. 1988. Recovery management in Quicksilver. *ACM Trans. Comput. Syst.* 6(1):82–108.

Hutchinson, N. and Peterson, L. 1988. Design of the x-kernel, pp. 65–75. In *Proc. SIGCOMM '88, Symp. Commun. Architectures and Protocols.* Stanford, CA, Aug.

Hutchinson, N. C., Peterson, L. L., Abbott, M. B., and O'Malley, S. 1989. RPC in the x-kernel: evaluating new design techniques. *ACM Operating Syst. Rev.* 23(5):91–101. *Proc. 12th Symp. Operating Syst. Principles.*

Joseph, T. A. and Birman, K. P. 1986. Low cost management of replicated data in fault-tolerant distributed systems. *ACM Trans. Comput. Syst.* 4(1):54–70.

Kung, H. T. and Robinson, J. T. 1981. On optimistic methods for concurrency control. *ACM Trans. Database Syst.* 6(2):213–226.

Lamport, L., Shostak, R., and Pease, M. 1982. The Byzantine generals problem. *ACM Trans. Programming Lang. Syst.* 4(3):382–401.

Lampson, B. W. 1986. Designing a global name service, pp. 1–10. In *Proc. 5th ACM Annu. Symp. Principles Distributed Comput.* Calgary, Canada, Aug.

Lampson, B. W. and Sproull, R. F. 1979. An open operating system for a single user machine. *ACM Operating Syst. Rev.* 13(5):98–105. *Proc. 7th Symp. Operating Syst. Principles.*

Liskov, B., Bloom, T., Gifford, D., Scheifler, R., and Weihl, W. E. 1987. Communication in the Mercury System. *Programming Methodology Group Memo* 59-1 MIT LCS, Cambridge, MA.

Liskov, B. H., and Scheifler, R. W. 1983. Guardians and actions: linguistic support for robust, distributed programs. *ACM Trans. Programming Lang. Syst.* 5(3):381–404.

Lynch, N. A., Merritt, M., Weihl, W. E., and Fekete, A. 1993. *Atomic Transactions.* Morgan Kaufmann.

McKendry, M. S. 1984. Clouds: a fault-tolerant distributed operating systems. *IEEE Tech. Commun. Distributed Process. Newsletter* 2(6).

Mueller, E. T., Moore, J. D., and Popek, G. J. 1983. A nested transaction mechanism for LOCUS. *ACM Operating Syst. Rev.* 17(5):71–90. *Proc. 9th Symp. Operating Syst. Principles.* Bretton Woods, NH.

Mullender, S. J. 1985. *Principles of Distributed Operating System Design,* Ph.D. Thesis, Vrije Universiteit, Amsterdam, Oct.

Mullender, S. J. and Tanenbaum, A. S. 1985. A distributed file service based on optimistic concurrency control. *ACM Operating Syst. Rev.* 19(5):51–62. *Proc. 10th Symp. Operating Syst. Principles.* Orcas Island, WA.

Mullender, S. J. and Tanenbaum, A. S. 1986. The design of a capability-based distributed operating system. *Comput. J.* 29(4):289–300.

Mullender, S. J., van Rossum, G., Tanenbaum, A. S., van Renesse, R., and van Staveren, J. M. 1990. Amoeba—a distributed operating system for the 1990s. *IEEE Comput.* 23(5).

Mullender, S. J. and Vitányi, P. M. B. 1988. Distributed match-making. *Algorithmica.* 3:367–391.

Needham, R. M. 1993. Names. In *Distributed Systems,* S. J. Mullender, ed., 2nd ed., pp. 315–327. ACM Press, New York.

Nordmark, E. and Cheriton, D. R. 1989. Experiences from VMTP: how to achieve low response time. *Proc. IFIP WG6.1/WG6.4 Int. Workshop Protocols for High-Speed Networks.* H. Rudin and R. Williamson, eds., Zürich, Switzerland, May.

Pike, R., Presotto, D., Thompson, K., Trickey, H., and Winterbottom., P. 1993. The use of name spaces in Plan 9. *ACM Operating Syst. Rev.* 27(2):72–76. *Proc. 5th ACM SIGOPS European Workshop.* Mont Saint-Michel.

Powell, D., ed. 1991. *Delta-4—A Generic Architecture for Dependable Distributed Computing,* ESPRIT Research Rep. Springer Verlag.

Presotto, D., Pike, R., Thompson, K., and Trickey, H. 1991. Plan 9, a distributed system, pp. 43–50. In *Proc. Spring 1991 EurOpen Conf.* Tromsø, Norway.

Reed, D. P. 1978. *Naming and Synchronization in a Decentralized Computer System.* Ph.D. dissertation, MIT. Available as *Tech. Rep.* MIT/LCS/TR-205. Cambridge, MA.

RFC-1034. Domain Names—Concepts and Facilities.

RFC-1035. Domain Names—Implementation and Specification.

Roscoe, T. 1994. Linkage in the Nemesis single address space operating system. *ACM Operating Syst. Rev.* 28(4):48–55.

Rozier, M., Abrossimov, V., Armand, F., Boule, I., Gien, M., Guillemont, M., Hermann, F., Kaiser, C., Langlois, S., Léonard, P., and Neuhauser, W. 1988. CHORUS Distributed Operating Systems. *Chorus Systemes Rep.* CS/TR-88-7.6, Paris.

Saltzer, J. H., Reed, D. P., and Clark, D. D. 1984. End-to-end arguments in system design. *ACM Trans. Comput. Syst.* 2:277–278.

Schlageter, G. 1981. Optimistic methods for concurrency control in distributed database systems. *Proc. VLDB Conf.*

Schmuck, F. and Wyllie, J. 1991. Experience with transactions in QuickSilver. *ACM Operating Syst. Rev.* 25(5). *Proc. 13th Symp. Operating Syst. Principles.* Pacific Grove, CA.

Schneider, F. B. 1993. What good are models and what models are good? In *Distributed Systems.* S. J. Mullender, ed., 2nd ed., pp. 7–26. ACM Press, New York.

Schroeder, M. D. 1993. A state-of-the-art distributed system: computing with BOB. In *Distributed Systems.* S. J. Mullender, ed., 2nd ed., pp. 1–16. ACM Press, New York.

Schroeder, M. D. and Burrows, M. 1989. Performance of Firefly RPC. *ACM Operating Syst. Rev.* 23(5):83–90. *Proc. 12th Symp. Operating Syst. Principles.*

Skeen, D. 1982. *Crash Recovery in a Distributed Database System.* Ph.D. dissertation. University of California, Berkeley.

Strom, R. and Yemeni, S. 1985. Optimistic recovery in distributed systems. *ACM Trans. Comput. Syst.* 3(3):204–226.

Sun. 1985. Remote Procedure Call Protocol Specification. Sun Microsystems, Inc.

Tanenbaum, A. S., van Renesse, R., van Staveren, J. M., Sharp, G. J., Mullender, S. J., Jansen, A. J., and van Rossum, G. 1990. Experiences with the Amoeba distributed operating system. *Commun. ACM* 33(12):46–63.

van Renesse, R., Birman, K. P., Glade, B. B., Guo, K., Hayden, M., Hickey, T., Malki, D., Vaysburd, A., and Vogels, W. 1995. Horus: A Flexible Group Communications System. *Tech. Rep.* TR 95-1500. Cornell University. March.

van Renesse, R., van Staveren, H., and Tanenbaum, A. S. 1988. Performance of the world's fastest distributed operating system. *ACM Operating Sys. Rev.* 22(4):25–34.

Veríssimo, P., Rodrigues, L., and Rufino, J. 1991. The Atomic Multicast protocol (AMp). In *Delta-4—A Generic Architecture for Dependable Distributed Computing*. D. Powell, ed. Springer–Verlag.

Walker, B., Popek, G., English, R., Kline, C., and Thiel, G. 1983. The LOCUS distributed operating system. *ACM Operating Syst. Rev.* 17(5):49–70. *Proc. 9th Symp. Operating Syst. Principles.* Bretton Woods, NH.

Weihl, W. E. 1993. Transaction-processing techniques. S. J. Mullender, ed., 2nd ed., In *Distributed Systems*, pp. 329–352. ACM Press, New York.

Xerox. 1981. Courier: The Remote Procedure Call Protocol, *Xerox Syst. Integration Std.* XSIS-038112, Xerox Corp. Stamford, CT.

86

Distributed File Systems and Distributed Memory

T. W. Doeppner, Jr.
Brown University

86.1 Introduction

The model of a single file system shared by all users of a computer is not only convenient but expected by most computer users. It seems natural to extend this model across multiple computers so that all users on a collection of computers share the same file system, thus forming a *distributed file system*. Similarly, the model of a collection of threads of control sharing the same address space as they cooperate in a computation is attractive for exploiting concurrency. This single-address-space abstraction is certainly the natural model for use on a shared-memory multiprocessor. Its convenience for programming is so compelling that it is used increasingly to take advantage of parallelism on distributed systems, where it is called *distributed memory*.

Primarily because of the ubiquity of Sun's Network File System (NFS), programmers have become accustomed to distributed file systems; they realize how much more convenient such a system is than explicitly copying files across machines. Whereas distributed memory is not so commonplace (most existing implementations are research projects), it shows great promise for taking advantage of the collective power of networked computers for computationally intensive problems.

The traditional means for implementing parallel applications on distributed systems is to use explicit message passing or remote procedure calls. These certainly give programmers full control over the locations of data and processing but also force them to be concerned about such details. The promise of distributed memory is that some of these details can be handled by the underlying implementation. In particular, programmers need not be concerned about the transfer of data among computers. Instead, data appear as needed merely because the program has referenced it. As with traditional virtual memory, exceptionally poor performance can arise, but most programs exhibit reasonable locality of reference and thus work quite well.

In both distributed file and distributed memory systems, the usual implementation model is one of clients obtaining data from servers. Clients typically maintain a cache of data recently obtained from servers: if data have been fetched previously, reads (in file systems) and loads (in memory systems) can often be satisfied directly from the cache without contacting the server. Writes (in file systems) and stores (in memory systems) can be applied to data in the cache and only later made visible to others by updating the server. For file systems, servers are typically distinct from clients and files are permanently assigned to servers, so that clients always contact the same server for the same file. However, an approach pioneered with distributed memory systems [Li and Hudak 1989] and recently adopted for distributed file systems [Anderson et al. 1995] distributes the server role among all of the clients: the home of a shared-memory segment is the client who last modified it.

Our primary concerns are performance, how (and whether) the clients' views of data are kept coherent, and how machine crashes and network outages are handled. Among the performance concerns are minimizing both network traffic and latency of responses to user requests. If the data necessary to handle a user request (such as a read operation on a file or a load from memory) are available locally, latency is minimal, but if not, it can be considerable. However, data can now be transferred over high-speed networks faster than from disks. Thus, it may be quicker to obtain data over the network from the primary storage of a server machine than from a local disk. In either case, since the impact of network traffic on overall performance is more dependent on the number of messages being transmitted than on their size, it is advantageous to transmit data in batches.

In the remainder of this chapter, we first discuss the underlying principles of distributed file and memory systems, including issues of coherency, performance, resilience, naming, replication, disconnected operation, and security. Next is a section on best practices in which we discuss two commercially distributed file systems—Network File System (NFS) and Distributed File System (DFS)—and two research distributed memory systems—IVY and Munin. Finally, we present a summary of the research issues in the field.

86.2 Underlying Principles

What distributed file systems and distributed memory systems have in common is an architectural model in which some number of computers have common access to data. The difference between the two lies primarily in their intended use. Distributed file systems provide data in the form of *files* whose lifetimes are usually far longer than the lifetimes of the processes accessing them. The usual emphasis in a distributed file system is on providing files for the private use of clients and providing shared files for the read-only use of clients. There might be some degree of support for read–write access to shared files, but such use is typically rare. In distributed memory systems the emphasis is on the read–write sharing of data organized into *segments*, and little attention is typically paid to data permanence; the lifetime of data is roughly equal to that of the processes sharing it. In distributed file systems a major concern is resiliency: coping with server and client crashes and network outages. This typically has not been a major concern of distributed memory systems, though there is no reason why it could not be. One usually thinks of the data provided by distributed memory systems as being mapped into the address spaces of the client processes, whereas the files provided by distributed file systems are thought of as being accessed by explicit input/output (I/O) calls (e.g., read and write), but a client operating system could provide a filelike interface to shared memory (though this would be unusual) and files could be mapped into a client's address space (a not uncommon technique).

What is usually desired of distributed file systems and distributed memory systems is that they be **access transparent:** that programs access data on remote computers in the same way as they would locally. Thus, programs use standard read- and write-type file-system calls or use standard load- and store-type operations on memory. This rules out approaches based on explicit file transfer, such as the Internet's file transfer protocol (FTP) and Unix's remote copy (rcp). The underlying implementation is responsible for whatever data movement may be required; the remote data appear to be local to the application program.

Another important property of files is *permanence:* they must be able to continue to exist even when they have no active users. Thus, files are stored on some form of nonvolatile storage, and this ties them to a particular site; files rarely move. Files require administration and maintenance: limitations on the use of available space may have to be enforced. In some circumstances groups of files may have to be moved to different storage devices, perhaps attached to different computers. Files must be backed up, i.e., copied, say to tape, to guard against loss of data because of loss of media or other problems.

It may also be convenient to replicate files and make them available from multiple sources to prevent bottlenecks and protect against loss of access due to server failure. Thus, the location of a file might change over time. An important property of a distributed file system for making such changes of location tolerable is **location transparency:** how one refers to a file, i.e., its name, should not depend on its location.

In the typical distributed memory system, permanence is not an issue. In a number of systems there is no fixed distinction made between clients and servers; instead, a segment has a moveable home, typically on the last computer to have modified it.

In the next few pages we examine the issues that arise in the design of distributed file and memory systems. For both, the idealization against which our designs are compared is the **single-system model:** the behavior observed by parties executing on a distributed system should be identical to the behavior they would observe if all were on a single computer.[1] In practice, some aspects of this ideal are not achievable or not even desirable, but it is our basis for examining the various approaches.

In the next few pages we discuss the major issues in the design of both distributed file systems and distributed memory systems. We start by discussing coherency, first of data, then of file attributes, and then we look at performance. The two concerns are somewhat at odds, and so we examine the interplay. Next we look at resilience, which is also interrelated with the first two concerns. We then look at naming issues, replication, disconnected operation, and finally security issues.

Coherency of Data

A major concern in both distributed file systems and distributed memory systems is coping with concurrent access to files or memory and still providing adequate performance and resilience. Strict adherence to the single-system model can be expensive. However, we can often weaken this model to provide improved performance without making sacrifices in other areas.

One way of achieving the ideal of the single-system model is for a system to be **strictly coherent:** whenever a thread running on a node reads from a file or loads from memory, the value it retrieves is the value produced by the most recent write or store to that location. This, of course, is exactly what happens in the single-system model. What makes strict coherency nontrivial to achieve is, of course, the distributed nature of the underlying architecture: it takes time to make modifications to data visible to all nodes. When a write or store is executed, we say that the value produced becomes *visible* when it can be retrieved by reads or loads from this location by other processors. We distinguish between when an instruction that modifies memory or files is *issued* and when its effect becomes visible to others. Thus, with strict coherency, something must be done to ensure that the effect of a write is visible to the next read to the same location.

Strict coherency, however, turns out to be stronger than necessary. A somewhat weaker requirement still equivalent to the single-system model is **sequential coherency** [Lamport 1979]: the effect of any execution of threads on a collection of nodes is one that could have happened had all been executed on the same processor. The idea here is that a read does not need to return the results of the most recent write if it was possible for that write to have occurred after the read; if it was just an accident of time that the write preceded the read, then there is no reason to require the read to return the write's results. However, if it was no accident that the write preceded the read—if due to synchronization or other mechanisms the write was required to precede the read—then the read must return the write's results.

To see the distinction between strict coherency and sequential coherency, consider the time lines for three nodes in Fig. 86.1. Each

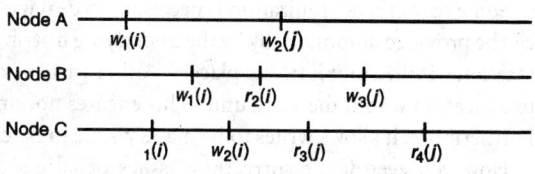

FIGURE 86.1 Concurrent access by three nodes.

[1] The word *computer* can be a bit ambiguous: it can mean, among other things, a standalone system or a single processor from a parallel computer. We often use the word *node* instead, which for our purposes is a system on which the single-system model is easily implemented (for example, a node might be a personal computer or a workstation; a multiprocessor is a node if it is a shared-memory multiprocessor). Thus, a distributed system consists of a number of nodes interconnected by some means for reasonably high-speed communication.

makes the sequence of accesses (reads and writes) indicated by the subscripts; the location accessed is given in parentheses. In a strictly coherent model, $r_1(i)$ of node C must retrieve what was written by $w_1(i)$ of node A, $r_2(i)$ of node B must retrieve what was written by $w_2(i)$ of node C, $r_3(j)$ of node C must retrieve what was written by $w_2(j)$ of node A, and $r_4(j)$ of node C must retrieve what was written by $w_3(j)$ of node B.

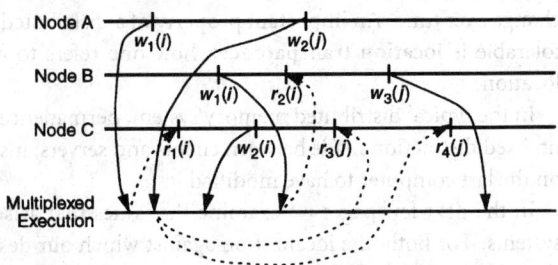

FIGURE 86.2 A possible execution in the sequentially coherent model.

In a sequentially coherent model, however, any interleaving of the accesses of nodes A, B, and C is possible, as long as the relative order of the accesses by each node is preserved. The possible interleaving shown in Fig. 86.2 demonstrates how the accesses might be multiplexed if all were run on a single node. The solid arrows indicate where write accesses would take place, and the dashed arrows indicate which values are retrieved by read accesses. The use of this model could certainly result in a number of possible outcomes of the accesses; if determinism is desired, then explicit synchronization is necessary.

Sequential coherency is certainly a weaker condition than strict coherency: under what circumstances can this weakness be exploited to yield more efficient systems? The fact that loads need not necessarily retrieve the values produced by the most recent stores suggests a technique for improving performance: stores of each processor may be collected into batches and made visible all at once; loads simply retrieve the values produced by the most recent visible stores. We discuss subsequently how we can guarantee that this technique results in sequentially coherent execution.

Unfortunately, there are further complications. Files are typically organized in terms of *blocks*: space may be allocated on disk in blocks and, more importantly for this discussion, files are transferred in block-sized pieces. Similarly, memory is usually organized in terms of *pages*: data are transferred in page-size units. Assume that a file server blindly obeys the requests of its clients to update blocks of a file with blocks supplied by the client (this is, in fact, how most file servers operate). One process might modify one data structure within a file with the knowledge that no other process is accessing that data structure. But some other process might be modifying a different data structure in the same file that happens to share the file-system block occupied by the first data structure. If these processes reside on different nodes, and if both modify their data structures at roughly the same time, a *race* results when each node sends its updated file-system block to the server. The server copies both versions of the file-system block to its (permanent) copy of the file, one at a time, probably in the order received. But since each copy of the block contains only a portion of the total update, the permanent copy of the file on the server will contain the result of one of the two updates, but not both, since the copy of the block it writes second overwrites the changes that came with the first copy. This problem, which occurs in distributed memory systems as well, is known as **false sharing**.

Some sort of synchronization is necessary to deal with the false-sharing problem. This synchronization can be provided automatically by the underlying distributed memory or file system. If a write access to any location within a unit is taking place, no other write or read access is allowed to take place at the same time to a location within the same unit. This ensures not only sequential coherency but also strict coherency. Furthermore, it allows writes to be made visible in batches.

However, even if the correctness issues of false sharing are adequately dealt with, there remains a performance problem: no concurrency of reads and writes to the same unit by different nodes is permitted. In many cases, particularly for file systems, such loss of concurrency is a minor problem since concurrent read/write access is rare. If concurrent access is frequent, however, then this loss of concurrency is serious.

Performance improvements are possible, even with false sharing, if we do not require the distributed file or memory system to take sole responsibility for sequential coherency but require user programs to assist. Thus, we define the notion of **weak coherency**, meaning that sequential coherency can be attained if additional instructions, not needed on a single-processor system, are executed by the program. In sequentially coherent systems we must make certain that, whenever a load takes place, it retrieves a value it could have retrieved on a single-node system. The un-

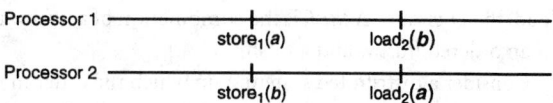

FIGURE 86.3   Potential violation of sequential coherency.

derlying system has no means of determining ahead of time when loads will be taking place and thus must be ready to cope with them at all times. However, a programmer or compiler has knowledge of a program that can be used to advantage. For example, if certain locations are *private*, i.e., are used by only one thread, there is no need to ensure that their values are up to date in all nodes' views. Locations that are shared by multiple threads might be accessed only when they are protected by some sort of synchronization primitive, e.g., a mutex or a semaphore. Since they are not accessed when not so protected and are accessed by only the thread that arranged for the protection when they are protected, we can incorporate into the synchronization primitives code to ensure coherency. In particular, when a thread performs a lock operation to gain mutually exclusive access to a shared data structure, it might issue a *flushr* instruction (defined subsequently) that ensures that subsequent loads can retrieve data recently made visible. As part of an unlock operation, it might issue a *flushw* instruction (also defined subsequently), which ensures that its changes become visible.

How can a distributed file or memory system not be sequentially coherent? Consider the example in Fig. 86.3, which is in terms of a memory system, though it applies equally well to file systems. Here we have two processors, each executing two instructions. Assume that the initial values in locations a and b are both 0. In the sequential coherency model, there are exactly three possible outcomes of the execution of the processors' instructions:

1. Processor 1's load returns 0, processor 2's load returns 1. This happens when processor 1's load occurs before processor 2's store.
2. Processor 1's load returns 1, processor 2's returns 0. This happens when processor 2's load occurs before processor 1's store.
3. Both loads return 1. This happens when neither processor 1's load occurs before processor 2's store nor processor 2's load occurs before processor 1's store.

However, if caching has delayed the effects of stores, has caused loads to retrieve old data, or both, there is a fourth possibility:

4. Both loads return 0. This happens when processor 1's load occurs after processor 2's store and processor 2's load occurs after processor 1's store. Since the effect of the stores is delayed, neither processor's load retrieves the value being set by the other processor's store.

Caching is beneficial for performance, but if it can cause unanticipated results, it must be done with care. In our example we can eliminate the unanticipated results, yet still allow caching, by flushing the cache before each load. We now show that such flushing will always produce sequential coherency.

We first define exactly what we mean by *flush*. We assume that when writes and stores become visible, they do so for all nodes at once; this is true in most, if not all, distributed file and memory systems today and greatly simplifies our exposition. The value produced by a store instruction is not necessarily visible immediately after the instruction is issued. However, once a *flushw* is executed by a processor, the values produced by all stores issued by that processor prior to the issuance of the flushw become visible (if they are not already). A flushw is said to be *ordered* if it causes the values produced by stores to become visible in the same order in which the corresponding stores were issued. When a processor A issues a load instruction to obtain a value from some specific location, the value obtained is the most recent visible value stored there before t time units in the past by some other processor or, if it occurred more recently, the most recent value stored there by processor A. The parameter t is unspecified; it may have an upper bound, but there is a fair amount of uncertainty about its value. However, if a *flushr* is executed immediately prior to the load, then t is zero. A *flush* is the combination of a flushw and a flushr. An *ordered flush* is the combination of an ordered flushw and a flushr.

Consider a distributed system X in which the order in which stored values are made visible is the order in which the stores were issued, and in which each load is preceded by an ordered flush. Is this system sequentially coherent? To show that it is, we must show, for each execution in this system, that there is an equivalent one in a system Y in which all processors execute the same instructions in the same order as they do in X, but in which stores become visible when they are issued.

Our first concern is to define what we mean by an execution in X. Since X is a distributed system, its executions consist of the executions of its component nodes. To simplify the presentation, we assume that each of these nodes consists of a single processor. An execution of a single processor is represented as the sequence of the instructions that were executed. Thus, if α_i is the execution sequence of processor i, then α, an execution of the distributed system, is the collection of all α_i for each processor i.

There is clearly a total order on the instructions in any execution sequence α_i. But because the processors of X execute in parallel, there is no total order on all of the instructions of the components of α; however, a number of partial orders can be defined. In particular, a store instruction supplies the value for a load instruction if the value retrieved from a location by the load was placed there by the store. We say that the store was the *source* of the load.

Our goal is to show that, for any execution α on distributed system X, there is a valid, equivalent execution sequence β on the single-processor system Y consisting of some interleaving of the execution sequences α_i of the component processors of X. Note that the store instructions in α become visible in some particular order. (If the store instructions of two different processors become visible simultaneously, we assume that their effect is as if one became visible before the other; we use this effective order to define our order of becoming visible.) We construct a proposed sequence β of instructions for system Y as follows (again to simplify the presentation, we assume that each of the α_i starts with a store): The first instruction of β is the first visible store of α. Following this in β are all of the initial instructions of each α_i up to, but not including, either the first store or the first load to retrieve a value not produced by the first visible store, whichever comes first. Next in β comes the next visible store from α, followed by all of the subsequent instructions of each α_i starting with the first not previously selected and continuing up to, but not including, either the first store or the first load to retrieve a value not produced by an already selected visible store, whichever comes first. The remaining instructions of β are sequenced accordingly.

We claim that β is not only a valid instruction sequence for the single-processor system Y but also is equivalent to α. To show this, we show that each load in β retrieves the same value as the corresponding load in α.

We first show that if s is the source of l in α (i.e., if store instruction s produced the value fetched by l), then s appears in β before l. This is due to the construction of β: if the source of a load has not been selected for inclusion in β, then the load cannot have been selected either.

Next we show that if a load appears in β, then any conflicting store (i.e., a store that places a value into the location being accessed by the load) that becomes visible after the source of the load became visible must appear after the load in β. To see this, again consider the construction of β. Suppose there is a load l in β, from processor i, for which a conflicting store appears between it and its source store. For this to happen, l could not have been selected for β as part of the instructions selected along with the source store. Thus, during the construction of β, there must have existed some positive number of stores and unsatisfied loads (i.e., loads whose source has not yet been selected for β) in the initial unselected portion of α_i appearing before l.

All of the stores in this portion must have been nonconflicting, because if they had been conflicting, since they occur in α_i and became visible after the source store, they would have overridden the value provided by the source store in α and prevented it from being the source store. If a conflicting store of some other processor had become visible before any of these nonconflicting stores of α_i and had thus been selected in β before l, then it too would have overridden the source store in α. This is because the flush executed before each load guarantees that any store instruction w in α_i appearing before l becomes visible before l is executed, and thus that if a conflicting store of another processor becomes visible before w, it will override l's source before l is executed.

Consider now the loads in the initial unselected portion of α_i. If they accessed the same location as l, they must have the same source store as l, since a different source store would have become visible after the source store of l and thus would have overridden it in α. If any of these load instructions was accessing a different location, both they and l could not have been selected for β until after the sources of these loads became visible. If a conflicting store of l comes before any of these stores and after l's source, then it over-rides this source in α. Thus, there cannot appear between a load in β and its source store a conflicting store.

Since we have shown that the source of a load appears before it in β and that no conflicting stores appear between the source and the load in β, we have what we were after: that loads in β produce the same values as loads in α. Because of this and the fact that the instructions of each α_i appear in the same relative order in β as they do in α_i, each execution in X has an equivalent execution in Y. Thus, X is sequentially coherent.

Coherency of File Attributes

An issue that affects file systems is maintaining file *attributes:* information about files, such as their sizes and the times of the most recent read and write accesses. This information is used often by clients and thus is often cached. Even if the underlying model is strictly coherent, the clients' views of the file system could be at odds with the single-system model if the attributes are not properly maintained.

For example, the following scenario has occurred (and has caused problems) in NFS: file X contains a sequence of records. A process P makes a private copy Y of X and records X's attributes at the time of the copy. P then edits Y, deleting some of the records contained in it (there are no consistency issues, since Y is a private file). P then uses some sort of mechanism to gain mutually exclusive access to X and replaces the contents of X with the contents of Y if X has not been modified (as reflected in the time of last modification stored in its attributes) since the copy Y was produced. If X has been modified (e.g., a new record has been added), then, rather than replacing X, P reproduces the changes it made to Y in the current version of X. Now, suppose a new record is added to X just before P gains mutually exclusive access to X. If P's copy of X's attributes is not appropriately updated, then, regardless of the memory model, P might replace X's contents with Y's rather than merge its edits to Y with the modified X. The result is that the record added to X is lost.

Performance

The primary performance concerns for distributed file and memory systems are minimizing both network traffic and latency of responses to user requests. If the data necessary to handle a user request (such as a read operation on a file or a load from memory) are available locally, latency is minimal, but if not, latency can be considerable. Since the impact of network traffic on overall performance is more dependent on the number of messages being transmitted than on their size, it is advantageous to transmit data in batches.

To reduce latency and network loads, most systems use *caching:* portions of files or segments are automatically maintained on client nodes, either on disk or in primary memory. Assuming that local access is quicker than remote access, this clearly helps reduce average latency, since many reads and writes (and loads and stores) can be handled directly from the cache. This also helps to reduce server and network loads since files and segments tend to be used repeatedly: once a file or segment has been transferred to a node, it is quite likely to be used many times, so that the number of transfers required over the network can be greatly reduced. In fact, many files, though they reside on server nodes, are accessed only on a single client node; thus, there is rarely a need for data transfer between server and client. Furthermore, network bandwidth can often be better utilized with caching, since large amounts of data can be transferred at once. However, data can now be transferred over high-speed networks faster than from disks. Thus, it may be quicker to obtain data over the network from the primary storage of a server machine than from a local disk.

To further improve latency, many systems, particularly file systems, use *prefetching:* a file or portion of a file is fetched before it is needed. Some operating systems provide asynchronous I/O facilities with which one can explicitly request data to be transferred without having to wait for it. One can use multithreaded programming techniques to get the same effect. These approaches require a very knowledgeable programmer and are often difficult to take advantage of. Some operating systems, such as Unix, attempt to predict what data are needed next and fetch it automatically before the program needs it. These techniques are effective only if files are being accessed sequentially (which is the only case for which the operating system can make predictions of data needs), but this happens frequently enough to be quite useful. Prefetching from the disk to primary storage is common in local file systems, and this notion is extended to prefetch from server to client in most distributed file systems.

One uses distributed memory systems to take advantage of parallelism. Thus, the most important measure of the performance of such systems is the speedup obtained when running a program on multiple nodes. This, of course, depends on the program being run, but we can compare actual performance with ideal performance: the time required if the nodes actually shared memory and all memory access times were those of local access.

Resilience

The next major concern is resilience. Failure is easy to cope with in the single-system model: either all components are running or all components are down—a crash crashes everything. Recovering from a crash entails restarting everything; merely resuming operations where things had left off is generally not possible. But this is one aspect of the model that we do not want to mimic in a distributed system: crashes of individual nodes should be handled gracefully by the other nodes. Functionality and data that are available only on the down nodes are lost for the duration of the downtime, but all other functionality and data should continue to be available. Moreover, operations emanating from client nodes that were in progress on a server node when it went down should be able to continue when the server comes back up, if they were not already completed on some other node.

Crash recovery involves first recognizing that there was a crash and then coping with it. Recognizing that there has been a crash is not simple, since a crashed node normally does not broadcast that it is about to crash before doing so. One node might suspect that another is down when nothing is received from the other node, despite expectations, after a reasonable period of time. But this unresponsiveness could be due to a communication failure having nothing to do with the node, or it could arise because the node is merely extremely busy, not down. Thus, the only sure way of determining that a node is down is for it to announce that it has been down as it restarts itself.

Coping with a server crash on a client might be as simple as waiting until the server restarts, then continuing normal operations, as if nothing had happened. Alternatively, the client operation might timeout, passing to the caller the decision on what to do. Yet another approach is to transfer requests from a down server to another server that provides equivalent files (i.e., replicas of the files provided by the down server).

A server's recovery from its own crash depends on how much state information was lost with the crash and hence must be recovered. In the simplest case, servers are *stateless:* there is no state information on the server, thus there is no state information that must be restored, and thus crash recovery involves simply restarting the server.

For servers that maintain state, the state information must be recovered. Such state information could be maintained on nonvolatile storage where it could survive a crash. This is generally not done: doing so would be expensive, and, moreover, losing state information in a crash can be advantageous. Suppose a client crashes while the server is down, thus invalidating its contribution to the server's state. If the server's state were maintained in nonvolatile storage, then, when it comes back to life, it must first determine that the client had crashed and then reset the client's contribution to the state. Since the server's recovery of state information depends on information supplied by clients when the server comes back to life, if the down client provides no information, there is no information that the server must reset.

For a client crash, the general assumption is that the operations in progress on the client cannot be resumed, and thus the client is simply restarted. The onus falls on the server to recognize a client crash and to react by restoring any client-specific state information to some initial value and recovering any resources that may have been allocated to the client. For stateless servers, nothing need be done, but for other types of servers a fair amount of work may be required.

Naming

Our next concern is the naming of files and how it relates to locating files. In the single-system model, we assume that a single tree-structured directory hierarchy (extended to a directed acyclic graph if links are allowed) is used for file naming. Files are identified by their pathnames in the directory hierarchy; this identification also serves as the means for locating files in secondary storage.

The entire name space can be viewed as a disjoint collection of subtrees joined together to form a tree: the root of one subtree is somehow connected to or superimposed on a directory of another. Each server provides some number of these subtrees to its clients. At issue are the mechanisms for joining the subtrees into a single tree and the appearance of the tree to the clients. One approach is for each client to piece together the subtrees itself, independent of the other clients. This approach, which is used by NFS, allows each client node to tailor the combination of subtrees. Thus, though all client nodes might share access to all files, the pathnames of these files might differ across nodes (though, in practice, nodes are typically configured so this is not the case, i.e., any shared file has the same pathname on all nodes).

Another approach is that the connections between subtrees are built into the subtrees themselves. For example, in DFS, subtrees contain links to other subtrees; the subtrees are connected into a single tree whose appearance is identical to all clients.

It is also important that client nodes have private files, containing information to be used only on the node. This could be done by giving each client a subtree and position within the global name space, e.g., /NodeA/..., /NodeB/.... But what is usually done is to provide a private name space so that the same name can refer to different files on different nodes. For example, on Unix systems, whether using NFS or DFS (or both), /etc/passwd refers to the password file on the node on which the pathname is used. This provides what we call **inverse location transparency**: programs referring to node-private files (e.g., /etc/passwd) can use the same pathname regardless of the node on which they are running. The disadvantage is that one cannot easily refer to node-private files on other nodes, but this is rarely important.

Replication

One of the advantages of distribution is that multiple copies of files can be maintained on separate servers, i.e., files can be *replicated*. This can be taken advantage of both for performance and for reliability: the task of providing files to clients can be spread over several servers, thereby reducing the load on each of them. If one server goes down, the others can take over its load.

With location transparency we have the notion that the name of a file or group of files does not tie it to a particular site. Thus, if for administrative or other reasons the files must be moved, they need not be renamed. If the naming technique permits a group of equivalent files to be given a single name, then any of the group can be used to satisfy a read of the files associated with the name. This allows easy **failover** to an alternative server if one holding a copy of the file fails.

Disconnected Operation

Another concern is support for *disconnected operation:* Can a client continue to operate when disconnected from a file server for extended periods of time? For this to take place, those files needed by the client must be somehow cached on the client. In connected operation clients and servers cooperate to maintain the consistency of the various cached copies of files with the copies maintained on servers, but in disconnected operation, such consistency management is not possible. The same sort of consistency management performed in connected operation could be performed for the disconnected case during the possibly brief periods for which a node is connected to the server. The difference, of course, is the time scale involved. In a typical scenario a client node might load its cache from the server, disconnect from the network, and then operate on the cached files for hours or days before reconnecting with the server. Any attempt to provide single-system semantics will fail in this sort of environment.

Security

Our final concern is security. In the single-system model all files and accessors appear to be on the same node and thus access is controlled by a single operating system. There is no opportunity to circumvent security measures other than by successfully masquerading as another user (perhaps by guessing a password). For distributed file systems things are more difficult. File-system data and other information are transmitted across communication networks and are thus subject to being read and modified by malicious third parties. Thus, providing a single-node-system level of security requires measures far beyond those required on such a system.

Providing security across nodes can be expensive. Such expense is justifiable in many, but certainly not all, situations. One approach, as suggested in Hartman and Ousterhout [1995], is to have fairly relaxed security within local clusters of nodes (whose users presumably trust one another) and to apply more stringent security measures to accesses between such clusters.

86.3 Best Practices

In this section we look at the application of the principles discussed in the previous section by examining two commercially available distributed file systems—Sun's Network File System (NFS) and OSF DCE's Distributed File System (DFS)—and two distributed memory systems that are products of university research—IVY from Yale University and Munin from Rice University. NFS, which has been the standard for use in distributed Unix systems for the past decade, is a relatively simple system that has undergone much scrutiny as it has evolved and improved. DFS, which was developed by Transarc and is an outgrowth of their Andrew file system (AFS), is considerably more complex than is NFS and comes closer to achieving our single-system-model ideal in some respects, though not in others. IVY, the seminal research implementation of distributed memory, shows how a virtual memory implementation could be extended to support a strictly coherent distributed memory. Munin takes a somewhat different approach: its model is weakly coherent and sequential coherency is achieved by a variety of techniques, chosen depending on the types of shared data structures and programmer-supplied information on how such data structures are used.

NFS

NFS has four component protocols: a *mount protocol* for making collections of files stored on servers available to client nodes, a *file protocol* for accessing and manipulating files and directories, a *lock* and *status protocol* for locking files over the network and recovering lock state after failures, and an *automount protocol* to support replication and name-space management.

Servers divide their files into disjoint collections called *file systems*, each of which contains a rooted directory tree naming all its members. Servers, as specified in their local /etc/exports files, specify which file systems or subtrees within a file system are available to which clients. By cooperating with a server via the mount protocol, a client can then *mount* a remote file system. This entails superimposing the root of the remote file system on top of a directory in the client's current naming tree. The root of the remote file system (the mounted file system) effectively replaces the mounted-on directory. Thus, the remote file system is attached to the client's naming tree at the mounted-on directory (and the previous contents of that directory are invisible as long as the remote file system remains mounted). The mount protocol additionally provides some minimal security by allowing administrators to specify for each file system on a server which client nodes are allowed to mount it.

Once a client has mounted a remote file system, its processes may access files in this system. This is where the file protocol is used: it provides a remote-procedure-call interface on the server for the client to use. A client process initiates activity on a file by opening it, an action that involves presenting a pathname to its local operating system, which follows the path, into remote file systems as necessary. If the path terminates in a remote file system, a *file handle* is returned to the client operating system identifying the remote file. The server, being stateless, keeps no record of the fact that the file is being used by the client. The file handle is used to identify the file to the server for all operations, including reading and writing.

The original NFS protocol took advantage of a cache of recently accessed file blocks maintained in the client's primary storage, known as the *buffer cache*, which allowed the prefetching of file-system blocks before they are needed (if the file is being accessed sequentially) and the write-behind of modified blocks. The buffer cache has been augmented in the most recent version of the protocol (version 3) with an on-disk cache, used primarily for files that rarely change. The buffer cache suffers from cache-coherency problems. NFS keeps such problems to a normally unnoticeable minimum, but occasionally false sharing and other problems can give the programmer unpleasant surprises.

The NFS approach to cache coherency is for the client to check with the server periodically to determine if items in its cache are stale and, if so, remove them. This is done by maintaining an *attributes cache* on each client that contains, for each file for which the client is caching blocks, the file's attributes, consisting of the file's size and access and modification times. When a file's attributes are fetched, they are tagged with the current time. Blocks of the file in the cache may be used on the client until the attributes expire, usually in a few seconds (the expiration time is a function of how recently the file was last modified; a file modified in the recent past is likely to be modified again soon). When an attempt is made to access blocks in the cache after the attributes have expired, a call is made to the server for the current attributes. If the file has been modified since the previous attributes were fetched, all blocks of the file are removed from the cache (modified blocks are sent back to the server to update the file). Thus, the cache is always close to a consistent state, but its consistency is never guaranteed.

Note that NFS is weakly coherent only if used via its mapped-file interface. On Unix systems, one can map an NFS-provided file into a client's address space. If one is careful to lay out data structures to prevent false-sharing problems, one can use flushr- and flushw-type instructions to achieve program-assisted sequential coherency. Though a flushw-type instruction exists for the file interface (i.e., accessing files via reads and writes), there is no flushr instruction.

An important aspect of performance is the minimization of messages transferred between client and server. NFS is normally layered on top of Sun RPC, which in turn is layered on the user datagram protocol (UDP) of the Internet suite. Sun RPC, particularly when layered on top of UDP, is quite simple: for each request, a response is expected. If the response is not received in a reasonable amount of time, the request is retransmitted, and again and again until a response is received. The usual case, of course, is that the request and response are each transmitted only once, and thus a minimum number of messages is transmitted.

There is a potential problem here. If a response is not received, then the request is retransmitted. If the reason that the response was not received is that the request was lost, then, assuming the retransmitted request arrives at the server, only one instance of the request is executed on the server. But suppose not the first request, but the first response was lost. When the retransmitted request arrives at the server, it is executed, with the result that two instances of the same request are executed on the server. If the request has a cumulative effect, such as transferring $100 from one bank account to another, this double execution would be a serious bug. However, in NFS most requests are designed to be **idempotent**—the effect of executing such a request twice is exactly the same as executing it once. For example, a request to write to a file specifies which file, the location in the file, and the data to be written. Executing the request twice causes the same data to be written to the same place twice—inefficient, but not harmful.

Recent improvements in the performance of transmission control protocol (TCP) implementations have caused some vendors to run NFS on TCP. Although this protocol guarantees reliability, idempotency is still an issue. If a server crashes while performing an operation, the client is uncertain whether the operation took place. Idempotency allows the client simply to repeat the operation once the server is running again, without fear of causing harm.

NFS is the prime example of the advantages of statelessness. Servers running NFS's file protocol need not retain any information about their clients or what their clients are doing: any information about clients maintained on a server is an optimization; it can be regenerated if necessary. Thus, no recovery actions are required of the file protocol on the server after the server has crashed.

As previously mentioned, an issue for the client side is determining *if* the server has crashed. With stateless servers, the client application (as opposed to client operating system) can be oblivious to whether a server has crashed; the operating system simply repeatedly retries file-protocol RPCs to the server until a response is received; the application perceives a slowdown but no loss of function or data. However, via the mount protocol, a client node can specify how a server's crash (or, more accurately, server unresponsiveness) is to be dealt with, by specifying how the remote file system is mounted. One option, known as the *hard mount*, is for the client to retry operations repeatedly, as just described, until they succeed. This has the advantage that, other than a delay in execution, the client process's semantics are unaffected. However, it has the disadvantage that it can cause a client process to block indefinitely, waiting for the result of the operation. Thus, another option, known as the *interruptible hard mount,* allows the remote operation to be aborted while in progress. This has the benefit of releasing a process (and perhaps an entire user session) from the clutches of a server that will be down for an indefinite period but certainly has a dramatic effect on the process's semantics. The final option, the *soft mount,* is that the operation returns an error code indicating that no response was obtained after a reasonable number of retries (note that the operation may or many not have successfully taken place on the server—all the client knows is that there has been no response). This is a useful alternative if the client application takes the trouble to notice such error codes and is prepared to deal with them, but otherwise could cause serious problems. The great advantage of the first two options is that from the client application's point of view, no special treatment is required for server failures—the system is **failure-transparent**.

Coping with a client crash is equally simple since servers are stateless. When a client crashes, whatever was running on it at the time of the crash is lost; when the client restarts, no information is available on what it was doing at the time of the crash. Thus, if the client had state on the server that must be cleaned up after the crash, it cannot direct the cleanup, since it has no knowledge of what this state is. Since servers are stateless, this is not a problem: clients simply restart and servers are oblivious to the fact that anything happened.

However, some sort of state information is required to support file locking: we must keep track of whether files are locked or not. NFS deals with this with a pair of protocols: one for maintaining locks (the *lock protocol*) and one for keeping track of whether clients have restarted (the *status protocol*). A file's server maintains the state information associated with locking. If a client crashes while holding a lock on a file, this lock must be removed. As noted earlier, the server cannot use the lack of communication from the client as an indication that the client has crashed; it requires definite notification, which is supplied when the client restarts (thus, files remain locked while the client is down). Manual intervention may be necessary if the client is down for an extended period of time.

A server crash is more difficult to deal with: the state information concerning locks is kept in volatile storage, which is lost after a crash. If it were kept in nonvolatile storage, then if both a client and the server crashed, there would be difficulties in determining that the client's locks must be removed; the server would realize that certain items are locked but would not realize that not only should they no longer be locked but also that the client is unaware that the items are locked. Thus, when a server crashes, all locks are removed by default. It is then up to the clients, if they remain up, to restore the lock information on the server. When a server restarts after a crash, it informs all its clients via the status protocol that it is now back in operation. It also establishes a grace period during which clients must notify it of locks they had held before the crash, so as to reclaim them. During this period no new lock requests are honored. After the grace period, all state information is recovered (that not recovered, presumably due to down clients, is lost) and the server goes back to normal operation.

The final aspect of NFS has to do with the name space. There are two issues here. It is the responsibility of each node to set up its own name space by mounting into its hierarchy file systems provided by servers. The number of such file systems could be huge, even though, for any one particular node, many are rarely, if ever, accessed. Thus, rather than attempting to use the mount protocol to mount all conceivable file systems when a node is booted, the *automount protocol* is used to mount file systems when needed.

The other issue, also dealt with by the automount protocol, is to support the replication of read-only file systems. If a server providing an important file system, such as the binary images of system commands, crashes, it is important that there be some sort of failover facility so that an alternative server can provide the file system. To accomplish this transparently to the clients, both instances of the file system should appear to have the same name, i.e., the same pathname in a client's directory tree. When a client node mounts a remote file system using the automount protocol, it broadcasts a request asking for providers of the desired file system. There could be many potential providers; the first one whose response is received is chosen. Thus, if one provider is down, another can be used.

The only drawback of this approach occurs when the provider of the important file system crashes while it is mounted on a client's naming tree. There is no support for automatically unmounting the file system and looking for a new provider. Thus, though a new provider is available, it cannot be used. What is done is to unmount file systems automatically if they have not been used after a period of time, typically five minutes. An automount broadcast is then performed on the next access to the file system. This approach works well for file systems that are used sporadically but does not solve the problem for file systems in constant use.

Distributed File System

OSF DCE's DFS goes much farther in some areas than NFS toward achieving the single-system ideal, though not as far in others. Assuming all components (clients, servers, and network) stay up and running, its variance from the ideal is small. DFS achieves this even though it employs large client caches on local disks to hold all files (split into chunks) being used on each client. It maintains consistency via a token-passing algorithm, which produces much state information on servers that must be restored after a server crash and cleared after a client crash.

A DFS installation is an optional part of a DCE *cell,* which is a potentially large (thousand-node) collection of computers sharing a common security database. Each cell supporting DFS has a single DFS name space used by all nodes. As in NFS, the name space consists of a collection of *file sets* (a better term than the ambiguous file system used in NFS and other Unix file systems). The file sets are connected together into a single tree structure by storing mounting information not at each node but in the file sets themselves. This mount information is represented via a form of *symbolic link* that provides the name of the file set mounted at this directory. Unlike NFS, in which the root of the mounted file set replaces the mounted-on directory, here the root of the mounted file set becomes the child of the mounted-on directory. In fact, one mounted-on directory can contain links to any number of file sets.

DFS exploits these links to help provide for the replication of read-only file sets. One can create any number of read-only copies of a file set. Each file set has a short, descriptive name; a read-only copy has the suffix "readonly" at the end of its name. Encoded in the mounting information stored in a file-set link is an indication of whether it is a read–write mount point, so that the read–write version is required, or that it is a read-only mount point, so that any of the read-only replicas may be used.[2]

A special replicated database, the file-set location database (FLDB), provides a mapping from file-set names to the names of the servers that hold the various replicas of the file set. This database is used by client-side DFS code when following a path: when a file-set link is encountered, the client code looks up the file-set name stored in the link in FLDB (these lookups are cached so that name-to-file-set translation is not terribly expensive). Depending on the type of mount point, it selects either the server containing the read–write replica of the file set or any of the servers containing read-only replicas. If the client is using a read-only replica and the server becomes unresponsive (perhaps because it has crashed), the client operating system simply finds from FLDB another server containing a read-only replica and (quietly)

---

[2]There is also a regular mount point that is equivalent to a read–write mount point if it (the mount point) resides in a read–write file set, and to a read-only mount point if it resides in a read-only replica.

switches to it; the client application is oblivious. Thus, when using read-only file sets, failover is automatic. No attempt is made to support read–write replication.

DFS maintains strict coherency. Its clients maintain caches on their local disks that store files in chunks of typically 64 kilobytes each. As in NFS, prefetching and write-behind are used to overlap I/O and computation. Servers maintain the consistency of the chunks from their files that appear on client caches by controlling which clients may read and modify the chunks. This is accomplished via a token-passing algorithm: for a client to perform an operation on some portion of a file in its cache, it must have, from the server, a token that grants it the necessary permission (note that this is not access permission in the sense of whether or not the client is authorized to access the file but merely indicates whether an access can now be done in a strictly coherent fashion).

Various forms of permissions are represented by tokens. There are tokens that allow a client node to read data from a portion of a file and tokens that allow a client node to modify data in a portion of a file. There are tokens for locking a file (or portions of a file), both for shared (read) locks and for exclusive (write) locks. Tokens are required for reading and setting file attributes (such as modification time, access time, and file size). To maintain strict coherency, if any client node is modifying a portion of a file, no other node may be reading or modifying that portion of the file. Of course, any number of nodes may be reading a portion of a file at once, as long as no node is modifying it. This is controlled through the distribution of tokens: to modify a portion of a file in its cache, a client node must obtain a write token from the file's server. To read the data in its cache, the node must have a read token. The server is responsible for making certain that if a write token for a particular portion of a file has been granted, then no read tokens for that portion are outstanding, and so forth. If, for example, a write token is outstanding and some node wishes to read the file, the server contacts the holder of the write token to revoke it and then gives the other node a read token.

Similarly, to modify a file's attributes, a client must have an appropriate token; the server will grant the token only if no tokens are outstanding for reading the attributes. A small problem here is that included with file attributes is the file access time, which should be updated every time the file is read. But doing so requires a *status write* token, which cannot be granted if there are *status read* or status write tokens outstanding. Thus, for single-system semantics, multiple nodes cannot be reading the same portion of a file at once, since doing so would require that they all have status write tokens. Because of this, DFS must deviate from single-system semantics and not maintain the access time of a file exactly as done on a single system. Instead, the update of the access time may be delayed, so that the file may have been accessed some time ago, but the access time does not yet reflect it.

Because DFS must maintain a lot of state information (e.g., what tokens are out), its crash recovery is much more complicated than NFS's. Three independent things can go wrong:

- *A client can crash:* Thus, the server will need to reclaim all of the tokens that were held by the client.
- *A server can crash:* Token information is not held in nonvolatile storage. It must somehow be re-created when the server comes back up.
- *The network can fail:* Though both client and server remain up, neither can communicate with the other.

Suppose that a server is unable to contact a client. If the client has possession of tokens and another client wishes to perform an operation that conflicts with the first client's tokens, then the server would like to revoke the tokens. Since the client is unresponsive, the server must do this unilaterally.

Conversely, suppose that a client is unable to contact a server. As long as the client has what it needs in its cache, then it really has no need to contact the server—it can get along quite well on its own. When the server comes back to life, it recovers its tokens using an approach similar to that used in NFS's lock protocol: the clients notify the servers of the tokens they possess.

Thus, we have two potentially conflicting points of view:

- The client should be able to use its cache even if the server is down or not accessible.
- The server should be able to revoke tokens from a client even if the client is down or not accessible.

If either the server or the client has crashed, then providing the other's point of view is not difficult. But if a network outage occurs and the server and client become separated from each other but both continue to run, then the two points of view conflict with each other.

DFS uses a compromise approach. If the client cannot contact the server, it continues to use its cache until its tokens expire; they are typically good for two hours, though they are normally refreshed every minute or so. However, say the server is actually up and running but is somehow disconnected from the client. If the server has no need to revoke tokens, then it does nothing. But if some other client that is communicating with the server attempts an operation that conflicts with the unresponsive client's tokens, then the server is forced to take action.

If the server has not heard from the client for a few minutes, it can revoke the client's tokens unilaterally. This means that when the client does resume communication with the server, it may discover that not only are some of its tokens no longer good, but some of its modifications to files may be rejected.

To protect client applications from such unexpected bad news, the client-side DFS code causes attempts to modify a file to fail if it has discovered that the server is not responding. A client program can take measures to deal with this problem by repeatedly retrying operations until the server comes back to life. This does not provide the transparency of the NFS hard-mount approach, but it does allow the client to use its cache to satisfy reads while the server is down.

IVY

IVY [Li and Hudak 1989] is one of the earliest distributed memory systems. A research project that included a study of various techniques for handling the visibility of writes while maintaining strict coherency, it helped to establish the viability of the concept of distributed memory by demonstrating impressive performance.

A primary concern was how to maintain strict coherency yet propagate the effects of writes efficiently. All of their solutions involve the notion of a node owning a shared page. Ownership is dynamic: to modify a page, a node must first own it. Other nodes can have copies of a page in their cache; when the owner modifies a page, it *invalidates* the copies in the other nodes' caches by sending them invalidation messages.

To modify a page, a node must first become its owner. This involves first determining the present owner, so that an ownership transfer can be arranged. Thus, some sort of mapping must be established associating page number with owning node. One technique is to have a central server handling the mapping chores: a node attempting to modify a page of which it is not the owner contacts the single mapping node to find the location of the owner, establishes a new mapping giving itself as the owner, and finally contacts the current owner to take over the page. The technique has the advantage of being simple but the disadvantage of creating a performance bottleneck and a single point of failure.

An alternative technique is to distribute the mapping functionality. The simplest implementation of this idea is to spread the mapping chores across all nodes but to have a fixed assignment of page number to node mapping the page to its owner. The developers' experiments showed that this technique was superior to the central mapping technique, but they found it difficult to obtain a fixed assignment that fit all applications well.

The final technique involves a dynamic approach to mapping page number to owner node: each node maintains a table mapping each page to its *probable* owner. Requests for change of ownership are sent to the probable owner who, if not the actual owner, forwards the request to the node it believes to be the owner, and so forth until the actual owner is reached. In the worst case, $N - 1$ messages are required to locate a page's owner, assuming N nodes. The ownership information is updated whenever a node receives an invalidation request, relinquishes ownership, or suffers a write fault, so that the worst-case number of messages for locating a page's owner rarely occurs. Li and Hudak [1989] show that the maximum number of messages required to locate the owner of a single page k times is $O(N + k \log N)$; in practice, a page's owner can be found fairly quickly.

Munin

Munin [Carter et al. 1995], developed at Rice University and the University of Utah, is an example of the use of weak coherency and other techniques to improve the performance of a distributed memory system. Rather than use a single software approach to ensure sequential consistency, multiple approaches are used, depending on the access patterns to the shared data. The loss of concurrency due to false sharing is minimized by using a *write-shared protocol* that merges together modifications made to the same page by separate nodes. In addition, an *update-with-timeout* mechanism removes copies of shared data from caches of nodes that have not used them for a while—this reduces the number of nodes whose caches must be invalidated when there is an update.

Research by the developers suggests that support for five access patterns can significantly improve performance in a distributed memory system:

- *Conventional shared variables* are treated just like shared data in IVY: to modify such a variable, a node must be the sole owner of the page containing it. When modifications occur, invalidation messages are sent to nodes containing copies of the page.

- *Read-only data,* once initialized, are never modified. Thus, no overhead is required to maintain coherency.

- *Migratory data* are used by one node at a time. Thus, once a node that has not been accessing such a data item starts accessing it (via a read or a write), the entire item is transferred to the new node and the original copy in the old node is invalidated.

- *Write-shared data* is a collection of items, each being modified by a single node, that share a page. Without special treatment there would be a false-sharing problem, but, by using the *write-shared protocol,* full concurrency can be achieved: when a node propagates its changes to write-shared data, it transmits merely its changes to the page rather than the entire page. Receivers of these updates can then merge these changes with their own. Conflicting changes can be detected and flagged as run-time errors.

- *Synchronization variables* are specialized implementations of three common synchronization constructs: *locks* (otherwise known as *mutexes*), *barriers,* and *condition variables.* They allow the programmer to take advantage of weak coherency by modifying shared data only when they are protected by the synchronization variables. Sequential coherency is obtained by making modifications visible at appropriate moments (e.g., during an unlock operation).

The net effect of these features of Munin is to decrease significantly the number of messages from those required by strictly coherent distributed memory systems. The developers' measurements show their system to be within 5% of implementations of a number of applications done with explicit message passing and within roughly 30% for others. They argue that further enhancements will improve the latter results substantially.

86.4 Research Issues and Summary

The notion of distributed file systems is a straightforward extension of the notion of local file systems: functionality available on nonnetworked individual computers has been made available to networked computers collectively. But computing has changed a great deal since the days when the only file systems were local file systems—it is no longer enough for data, organized into files, to be made available strictly to a collection of machines statically interconnected with cable. For the notion of a distributed file system to remain relevant, it must be useful in a world in which mobile computing, multimedia, and enterprise computing play key roles.

In mobile computing, client machines are often disconnected from servers. When they are connected, it might be via relatively low-speed links, such as direct and cellular telephone connections. Some commercial products, such as Lotus Notes®, provide some support for data access by mobile users, but they operate in specialized domains (electronic mail, simple databases) and do not provide the general-purpose foundation for constructing applications that is the promise of distributed file systems. Research is under way in adapting distributed file systems to mobile use (e.g., Satyanarayanan et al. [1993] and Huston and Honeyman [1995]). One goal of such research projects is to extend to mobile computing the functionality currently available to clients statically connected to their servers via high-speed networks. The issues being grappled with include providing some degree of coherency (sequential coherency may prove completely infeasible in such environments), loading a client's cache with appropriate files for a lengthy disconnected session, and scheduling communications traffic so that information needed urgently is given top priority. Dealing with coherency may involve resolving the effects of conflicting changes to files by multiple clients. This may be automatic in some cases, but many currently require assistance from the user. Preloading the cache often requires user input. Scheduling communications traffic allows currently needed data to be fetched prior to write-backs of modified data (though this, of course, has an effect on coherency).

Distributed file systems supporting multimedia have stringent performance demands. Not only must data be transferred quickly, but the transmission rate must be reliable so that data arrive before they are needed. A system specialized as a video server developed at Microsoft [Rashid 1994] transmits movies chosen from a library to a large number of clients at an appropriate rate and allows VCR-like functions such as fast forward and reverse. The demands on this system could well prevent its functionality from being subsumed by a general-purpose distributed file system, but some of the techniques used could be of use in general-purpose systems for handling more moderate numbers of clients.

Many of the issues arising in large enterprise systems have to do with scaling, taken here as the number of clients that can use a distributed file system. One point of view is that the collected file systems of a large company (or the entire world, for that matter) should be transparently accessible by anyone from anywhere subject, of course, to security constraints. A relatively small-scale implementation of this view involves users of Transarc's AFS system (the latest incarnation of the forerunner of DFS). Users around the world see one another's files as fitting in a single-directory hierarchy. A program running on a workstation in Providence, RI, can access files in Sydney, Australia, using exactly the same steps as to access files on a local disk. This works because, though the participating sites are widespread, they are relatively few; each site maintains a list of all other sites. If the entire world is to be made available in this fashion, other means will be necessary for sites to locate one another.

With large numbers of clients potentially accessing individual files, improvements in caching and load balancing become essential. Whereas both NFS and DFS support the replication of read-only collections of files, the processes of determining how many replicas to make and where to place them are entirely manual. At the very least, monitoring tools must be available to let an administrator react quickly to changing access patterns, creating and placing replicas, and deleting unneeded ones. Ideally, this sort of load balancing would be automatic.

The promise of distributed memory systems is to provide an easy-to-use programming paradigm for highly parallel systems. For the promise to become reality, the amount of parallelism must approach levels of interest to those with significant computational demands. As the numbers of processors increase, it may become important to integrate some degree of fault tolerance into the model so that individual processor failures do not take down the entire system.

In architecture there are two separate trends, one toward the interconnection of a collection of workstations into a distributed memory system and the other toward a system consisting of a large number of processors (perhaps all in one box) designed from the beginning to be a distributed memory system. In the former, the distributed memory is implemented entirely with software (other than hardware support for virtual memory); the latter typically uses hardware support.

Utilizing a collection of workstations as a parallel computer has been a goal of a number of researchers and many reasonably successful implementations exist (e.g., IVY and Munin). The attraction is that an organization might have a large number of workstations, many of which are unused over long stretches of time (overnight, for example). Among the longstanding issues only partly resolved in current systems are identifying available workstations and coping when workstations become unavailable (i.e., forcing off distributed memory users when the workstation is needed for other work). Adequate resolution of these issues could make this model of computation much more common.

In dedicated distributed memory systems in which processors are physically close to one another, interprocessor communication speeds can be quite high, rivaling processor-memory communication speeds. Such systems are beginning to appear on the market. Among the research issues still requiring attention are means for adequately distributing the parallel components of a computation, balancing the loads of the various processors, and devising and popularizing programming techniques for exploiting large-scale parallelism.

Defining Terms

Access transparency: A system property by which the complications involved in providing access to something (e.g., data) are not apparent to the accessor.

Failover: A system property by which, in the event of a failure of one component, the function provided by that component is taken over by another.

Failure transparency: A system property by which no special actions are required of clients to cope with failures of servers.

False sharing: What happens when two data structures share the same unit of storage, e.g., blocks in file systems in pages in memory systems.

Idempotency: The effect of performing an operation many times in succession (with no conflicting operation appearing between the repetitions) is the same as the effect of performing the operation once.

Inverse location transparency: A system property by which a given name refers to a different item on each node. Thus, a name can be used transparently by a program to refer to information specific to the node on which the program is running.

Location transparency: A system property by which how one refers to something depends on the location of neither the subject nor the object.

Sequential coherency: The property of a distributed file or memory system in which the effect of any execution is an effect that could have happened had all computation taken place on a single processor.

Single-system model: The computational model in which all computation takes place on a single processor.

Strict coherency: The property of a distributed file or memory system in which each read or load retrieves the value produced by the most recent write or store to that location.

Weak coherency: The property of a distributed file or memory system that does not necessarily provide sequential coherency by itself but that can provide sequential coherency if certain additional instructions are executed by any program running on the system.

References

Anderson, T. E., Dahlin, M. D., Neefe, J. M., Patterson, D. A., Roselli, D. S., and Wang, R. Y. 1995. Serverless network file systems, pp. 109–126. In *Proc. 15th ACM Symp. Operating Syst. Principles*. ACM, New York, Dec.

Carter, J. B., Bennett, J. K., and Zwaenepoel, W. 1995. Techniques for reducing consistency-related communication in distributed shared-memory systems. *ACM Trans. Comput. Syst.* 13(3):205–243.

Hartman, J. H. and Ousterhout, J. K. 1995. The zebra striped network file system. *ACM Trans. Comput. Syst.* 13(3):274–310.

Huston, L. B. and Honeyman, P. 1995. Partially connected operation. *USENIX Comput. Syst.* 8(4):365–380.

Lamport, L. 1979. How to make a multiprocessor computer that correctly executes multiprocess programs. *IEEE Trans. Comput.* C-28(9):690–691.

Li, K. and Hudak, P. 1989. Memory coherence in shared memory virtual memory systems. *ACM Trans. Comput. Syst.* 27(4):321–359.

Rashid, R. 1994. Microsoft's tiger media server. *1st Networks Workstations Workshop Rec.* Oct.

Satyanarayanan, M., Kistler, J. J., Mummert, L. B., Ebling, M. R., Kumar, P., and Lu, Q. 1993. Experience with disconnected operation in a mobile environment, pp. 11–28. In *Proc. ACM Symp. Mobile Location-Independent Comput.* USENIX, Berkeley, CA, Aug.

Further Information

Good coverage of a number of the issues covered in this chapter can be found in *Distributed Systems,* edited by Sape Mullender, Addison–Wesley, 1993.

Both ACM's bi-yearly *Symposium on Operating Systems Principles* and the ACM journal *Transactions on Computer Systems* often contain a number of excellent papers on issues related to distributed file and memory systems. Copies of the proceedings and subscription information for the journal can be obtained from ACM, 1515 Broadway, 17th Floor, New York 10036, (212) 869-7440.

The USENIX journal *Computing Systems* also often contains excellent papers on issues related to distributed file and memory systems. Subscription information can be obtained from The MIT Press Journals, 55 Hayward Street, Cambridge, MA 02142, (617) 253-2866, journals-info@mit.edu.

87

Distributed and Multiprocessor Scheduling

Steve J. Chapin
University of Virginia

87.1 Introduction

This chapter discusses central processing unit (CPU) scheduling in parallel and distributed systems. CPU scheduling is part of a broader class of resource allocation problems, and is probably the most carefully studied such problem. The main motivation for multiprocessor scheduling is the desire for increased speed in the execution of a workload. Parts of the workload, called tasks, can be spread across several processors and thus be executed more quickly than on a single processor. In this chapter, we will examine techniques for providing this facility.

The scheduling problem for multiprocessor systems can be generally stated as: "How can we execute a set of tasks T on a set of processors P subject to some set of optimizing criteria C?" The most common goal of scheduling is to minimize the expected run time of a task set. Examples of other scheduling criteria include minimizing the cost, minimizing communication delay, giving priority to certain users' processes, or needs for specialized hardware devices. The scheduling policy for a multiprocessor system usually embodies a mixture of several of these criteria.

Section 87.2 outlines general issues in multiprocessor scheduling and gives background material, including issues specific to either parallel or distributed scheduling. Section 87.3 describes the best practices from prior work in the area, including a broad survey of existing scheduling algorithms and mechanisms. Section 87.4 outlines research issues and gives a summary.

87.2 Issues in Multiprocessor Scheduling

There are several issues that arise when considering scheduling for multiprocessor systems. First, we must distinguish between **policy** and **mechanism**. Mechanism gives us the ability to perform an action; policy

0-8493-2909-4/97/$0.00+$.50
© 1997 by CRC Press, Inc.

decides what we do with the mechanism. Most automobiles have the power to travel at speeds of over 150 km/h (the mechanism), but legal speed limits are usually set well below that (the policy). We will see examples of both scheduling mechanisms and scheduling policies.

Next, we will distinguish between **distributed** and **parallel systems**. Past distinctions have been based on whether an interrupt is required to access some portion of memory; in other words, whether communication between processors is via shared memory (also known as **tightly coupled**) or via message passing (also known as **loosely coupled**). Unfortunately, although this categorization applies well to systems such as shared-memory symmetric multiprocessors (obviously parallel), and networks of workstations (obviously distributed), it breaks down for message passing multiprocessors such as hypercubes. By common understanding, the hypercube is a parallel machine, but by the memory test, it is a distributed system.

The true test of whether a system is parallel or distributed is the support for **autonomy** of the individual nodes. Distributed systems support autonomy, whereas parallel systems do not. A node is autonomous if it is free to behave differently than other nodes within the system.[1] By this test, a hypercube is classified as a parallel machine. There are four components to the autonomy of a multiprocessor system: design autonomy, communication autonomy, execution autonomy, and administrative autonomy.

Design autonomy frees the designers of individual systems from being bound by other architectures; they can design their hardware and software to their own specifications and needs. Design autonomy gives rise to heterogeneous systems, both at the level of the operating system software and at the underlying hardware level. Communication autonomy allows each node to choose what information to send, and when to send it. Execution autonomy permits each processor to decide whether it will honor a request to execute a task. Furthermore, the processor has the right to stop executing a task it had previously accepted. With administrative autonomy, each system sets its own resource allocation policies, independent of the policies of other systems. The local policy decides what resources are to be shared. In effect, execution autonomy allows each processor to have a local scheduling policy; administrative autonomy allows that policy to be different from other processors within the system.

A **task** is the unit of computation in our computing systems, and several tasks working toward a common goal are called a **job**. There are two levels of scheduling in a multiprocessor system: **global scheduling** and **local scheduling** [Casavant and Kuhl 1988]. Global scheduling involves assigning a task to a particular processor within the system. This is also known as mapping, task placement, and matching. Local scheduling determines which of the set of available tasks at a processor runs next on that processor.

Global scheduling takes places before local scheduling, although **task migration**, or dynamic reassignment, can change the global mapping by moving a task to a new processor. To migrate a task, the system freezes the task, saves its state, transfers the saved state to a new processor, and restarts the task. There is substantial overhead involved in migrating a running task.

Given that we have several jobs, each composed of many tasks, competing for CPU service on a fixed set of processors, we have two choices as to how we allocate the tasks to the processors. We can assign several processors to a single job, or we can assign several tasks to a single processor. The former is known as **space sharing**, and the latter is called **time sharing**.

Under space sharing, we usually arrange things so that the job has as many processors as it has tasks. This allows all of the tasks to run to completion, without any tasks from competing jobs being run on the processors assigned to this job. In many ways, space sharing is similar to old-fashioned batch processing, applied to multiprocessor systems. Under time sharing, tasks may be periodically preempted to allow other tasks to run. The tasks may be from the same job or differing jobs. Generally speaking, space sharing is a function of the global scheduling policy, whereas timesharing is a function of local scheduling.

One of the main uses for global scheduling is to perform **load sharing** between processors. Load sharing allows busy processors to offload some of their work to less busy, or even idle, processors. **Load balancing** is a special case of load sharing, in which the goal of the global scheduling algorithm is to keep the load even (or balanced) across all processors. **Sender-initiated** load sharing occurs when busy processors try to find idle processors to offload some work. **Receiver-initiated** load sharing occurs when idle processors seek busy

---

[1] We speak of behavior at the operating system level, not at the application level.

processors. It is now accepted wisdom that load balancing is generally not worth doing, as the small gain in execution time of the tasks is more than offset by the effort expended in maintaining the balanced load.

A global scheduling policy may be thought of as having four distinct parts: the *transfer policy*, the *selection policy*, the *location policy*, and the *information policy*. The transfer policy decides when a node should migrate a task, and the selection policy decides which task to migrate. The location policy determines a partner node for the task migration, and the information policy determines how node state information is disseminated among the processors in the system. For a complete discussion of these components, see Singhal and Shivaratri [1994, ch. 11].

An important feature of the selection policy is whether it restricts the candidate set of tasks to new tasks which have not yet run, or allows the transfer of tasks that have begun execution. *Nonpreemptive* policies only transfer new jobs, whereas *preemptive* policies will transfer running jobs as well. Preemptive policies have a larger set of candidates for transfer, but the overhead of migrating a job that has begun execution is higher than for a new job because of the accumulated state of the running job (such as open file descriptors, allocated memory, etc.).

As the system runs, new tasks arrive while old tasks complete execution (or are served). If the arrival rate is greater than the service rate, then the process waiting queues within the system will grow without bound and the system is said to be unstable. If, however, tasks are serviced as least as fast as they arrive, the queues in the system will have bounded length and the system is said to be **stable**. If the arrival rate is just slightly less than the service rate for a system, it is possible for the additional overhead of load sharing to push the system into instability. A stable scheduling policy does not have this property, and will never make a stable system unstable.

Distributed Scheduling

In most cases, work in distributed scheduling concentrates on global scheduling because of the architecture of the underlying system. Casavant and Kuhl [1988] defines a taxonomy of task placement algorithms for distributed systems, which we have partially reproduced in Fig. 87.1. The two major categories of global algorithms are static and dynamic.

Static algorithms make scheduling decisions based purely on information available at compilation time. For example, the typical input to a static algorithm would include the machine configuration and the number of tasks and estimates of their running time. Dynamic algorithms, on the other hand, take factors into account such as the current load on each processor. Adaptive algorithms are a special subclass of dynamic algorithms, and are important enough that they are often discussed separately. Adaptive algorithms go one step further than dynamic algorithms, in that they may change the policy based on dynamic information. A dynamic load-sharing algorithm might use the current system state information

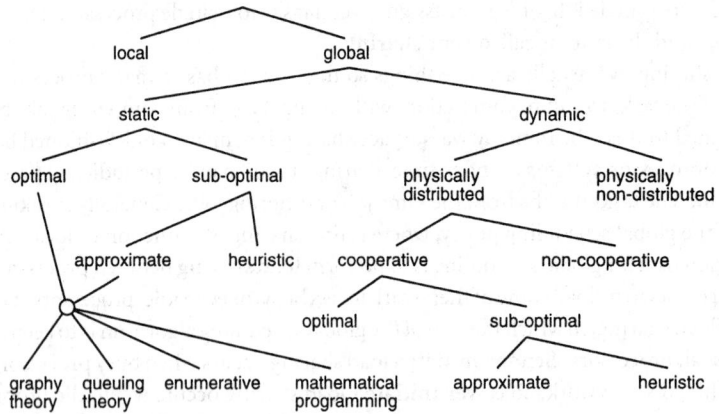

FIGURE 87.1   A taxonomy of distributed scheduling algorithms.

to seek out a lightly loaded host, whereas an adaptive algorithm might switch from sender-initiated to receiver-initiated load sharing if the system load rises above a threshold.

In *physically nondistributed*, or centralized, scheduling policies, a single processor makes all decisions regarding task placement. Under *physically distributed* algorithms, the logical authority for the decision making process is distributed among the processors that constitute the system.

Under *noncooperative* distributed scheduling policies, individual processors make scheduling choices independent of the choices made by other processors. With *cooperative* scheduling, the processors subordinate local autonomy to the achievement of a common goal.

Both static and cooperative distributed scheduling have *optimal* and *suboptimal* branches. Optimal assignments can be reached if complete information describing the system and the task force is available. Suboptimal algorithms are either *approximate* or *heuristic*. Heuristic algorithms use guiding principles, such as assigning tasks with heavy intertask communication to the same processor, or placing large jobs first. Approximate solutions use the same computational methods as optimal solutions, but use solutions that are within an acceptable range, according to an algorithm-dependent metric.

Approximate and optimal algorithms employ techniques based on one of four computational approaches: enumeration of all possible solutions, graph theory, mathematical programming, or queuing theory. In the taxonomy, the subtree appearing below optimal and approximate in the static branch is also present under the optimal and approximate nodes on the dynamic branch; it is elided in Fig. 87.1 to save space.

In future sections, we will examine several scheduling algorithms from the literature in light of this taxonomy.

Scheduling for Shared-Memory Parallel Systems

Researchers working on **shared-memory** parallel systems have concentrated on local scheduling, because of the ability to trivially move processes between processors. There are two main causes of artificial delay that can be introduced by local scheduling in these systems: cache corruption and preemption of processes holding locks:

- *Cache corruption:* As a process runs, the operating system caches several types of information for the process including its working set and recently read file blocks. If this information is not accessed frequently, the operating system will replace it with cache information from other processes.

- *Lock preemption:* Spin locks, a form of busy waiting, are often used in parallel operating systems when contention for a critical section is expected to be low and the critical section is short. The problem of lock preemption occurs when a process that holds the lock is preempted on one processor, while another process waiting to enter the lock is running on a different processor. Until the first process runs again and releases the lock, all of the CPU time used by the second process is wasted.

In an upcoming section, we will examine methods to alleviate or avoid these delays.

87.3 Best Practices

In this section, we will examine the current state of the art in multiprocessor scheduling. We will first consider the techniques used in parallel systems, and then examine scheduling algorithms for message passing systems. Finally, we will study scheduling support mechanisms for distributed systems.

Parallel Scheduling

We will examine three aspects of scheduling for parallel systems: local scheduling for shared-memory systems such as the Sequent Symmetry; static analysis tools that are beneficial for producing global schedules for parallel systems; and dynamic scheduling for **distributed-memory** systems.

Local Scheduling for Parallel Systems

For most shared-memory time-sharing systems, there is no explicit global placement: all processors share the same ready queue, so any task can be run on any processor. In contrast, local scheduling is crucial for these systems, whereas it is nonexistent in space-sharing systems. We will examine several local scheduling techniques for parallel systems. In general, these techniques are attempting to eliminate one of the causes of delay mentioned earlier. All of these techniques are discussed in Singhal and Shivaratri [1994, ch. 17].

Coscheduling, or gang scheduling, schedules the entire pool of subtasks for a single task simultaneously. This can work well with fine-grained applications where communication dominates computation, so that substantial work can be accomplished in a single time slice. Without coscheduling, it is easy to fall into a pattern where subtasks are run on a processor, only to immediately block waiting for communication. In this way, coscheduling combines aspects of both space sharing and time sharing.

Smart scheduling tries to avoid the preemption of a task that holds a lock on a critical section. Under smart scheduling, a process sets a flag when it acquires a lock. If a process has its flag set, it will not be preempted by the operating system. When a process leaves a critical section, it resets its flag.

The Mach operating system uses scheduler hints to inform the system of the expected behavior of a process. Discouragement hints inform the system that the current thread should not run for a while, and hand-off hints are similar to coroutines in that they hand off the processor to a specific thread.

Under affinity-based scheduling, a task is said to have an affinity for the processor on which it last ran. If possible, a task is rescheduled to run on the processor for which it has affinity. This can ameliorate the effects of cache corruption. The disadvantage of this scheme is that it diminishes the chances of successfully doing load sharing because of the desire to retain a job on its current processor. In effect, affinity-based scheduling injects a measure of global scheduling into the local scheduling policy.

Static Analysis

There are several systems that perform static analysis on a task set and generate a static task mapping for a particular architecture. Examples of such systems include Parallax, Hypertool, Prep-P, Oregami, and Pyrros (see Shirazi et al. [1995] for individual papers on these systems). Each of these tools represents the task set as a directed acyclic graph, with the nodes in the graph representing computation steps. Edges in the graph represent data dependencies or communication, where the result of one node is made ready as input for another node. These tools attempt to map the static task graph onto a given machine according to an optimizing criterion (usually, minimal execution time, although other constraints such as minimizing the number of processors used can also be included). The scheduling algorithm then uses some heuristic to generate a near-optimal mapping.

Parallax (Lewis and El-Rewini) is a partitioning and scheduling system that implements seven different heuristic policies. The input to the system is a graph representing the structure of the tasks and a user-selectable representation of the machine architecture. Parallax will then generate schedules based on each of the available heuristics and present the expected results to the user. This system is unique in its ability to permit the user to explore different combinations of scheduling heuristics and machine architecture for a given task set.

Hypertool (Wu and Gajski) takes as input C source code and generates the task graph representing the program. This is distinct from Parallax, where the user must supply the task graph (and may therefore study the behavior of algorithms which have not been explicitly expressed in any particular programming language). Hypertool then schedules the derived task graph on a hypercube.

Sarkar and Hennessy built one of the first tools to extract parallelism from a functional program, partition the individual tasks into jobs, and then place the jobs on a multiprocessor. They developed a new representation for parallel computation called Macro-Dataflow, and applied their work to programs written in SISAL for the VAX.

Prep-P (Berman and Stramm) is a mapping tool that runs in conjunction with the Poker programming environment for the Pringle machine. Prep-P was one of the earliest program mapping tools, and uses a graph description language as input to describe the program structure. Prep-P uses an iterative partitioning algorithm, wherein an initial partitioning is derived and the system repeatedly attempts to improve on the partition by moving a task from one partition into another. Whenever the proposed move results in a lower cost schedule, the move is kept.

Oregami (Lo et al.) is similar to Prep-P in functionality, with the addition of a new model for representing the computation called the temporal communications graph (TCG). The TCG represents each event (computation, message send, or message receipt) as a node within a directed acyclic graph. Thus, it represents a combination of the static task graph with Lamport's process-time graphs.

In many ways, Pyrros (Yang and Gersoulis) represents a merger between several ideas from earlier static scheduling systems. Pyrros uses Sarkar and Hennessy's Macro-Dataflow model, and is targeted for a hypercube architecture. The system takes a Macro-Dataflow graph as input, then performs partitioning and scheduling. It can produce optimal schedules for several restricted classes of algorithms.

Distributed-Memory Systems

In distributed-memory systems, such as hypercube systems, global scheduling is done. Most hypercube systems use space sharing, in that they reserve subcubes of the larger hypercube for use by a single application. Several algorithms have been proposed for this, including that found in Huang et al. [1989]. A typical algorithm maintains a binary tree listing the various sizes of hypercubes available in the system. When a request for an m-dimensional cube is made, the scheduling system searches the tree to see if a hypercube of that exact size is available, and if so, allocates it. If no such hypercube is available, the system splits the smallest hypercube of dimension $>m$ into multiple hypercubes, allocates one, and updates the binary tree to reflect the new set of available hypercubes.

For example, consider a request for 4 processors from a 16-processor hypercube, with all nodes currently free. Four processors comprise a two-dimensional hypercube (a square). The scheduling system would split the 16 processors into two 8-processor cubes, and then split one of the 8-processor cubes into two 4-processor squares. One of the squares would be allocated to the job, leaving one two-dimensional and one three-dimensional hypercube for other jobs.

It is interesting to examine the coexistence of space sharing and time sharing on a single machine. The Intel Paragon is a distributed-memory system that divides its nodes into two partitions: a service partition which runs a general-purpose, full-featured operating system (OSF/1), and a compute partition that runs a special-purpose, highly efficient operating system (SUNMOS). OSF/1 [OSF 1993] implements the full Unix semantics, including time sharing, whereas the Sandia/University of New Mexico operating system (SUNMOS) [Maccabe et al. 1994] provides low-latency communication under a space-sharing paradigm. Users launch their jobs from the service partition, and the scheduling system reserves a portion of the compute partition to run the jobs.

Puma [Wheat et al. 1994] is the successor to SUNMOS, and also combines space sharing and time sharing. Whereas SUNMOS is completely unitasking, Puma will allow multiple tasks from the same job to run on a single node. In this way, Puma implements time sharing over space sharing, and the designers hope to provide improved performance to users.

Distributed Scheduling Algorithms

Many researchers have devised algorithms for task placement in distributed systems. This section categorizes several of these techniques in terms of the taxonomy presented earlier.

Table 87.1 displays information garnered from a survey of existing scheduling algorithms. For each algorithm, an entry indicates whether the method is distributed or centralized, supports heterogeneity, minimizes overhead, or supports scalability. Entries are either Y, N, P, or x, indicating the answer is yes, no, partially, or not applicable, respectively. The remainder of this section contains a brief description of each method, with a discussion of its place in the taxonomy and its salient properties. Interested readers are referred to the cited publications to obtain full details about the algorithms.

Dynamic, Distributed, Cooperative, Suboptimal Algorithms

Blake [1992] describes four suboptimal, heuristic algorithms. Under the first algorithm, nonscheduling (NS), a task is run where it is submitted. The second algorithm is random scheduling (RS), wherein

TABLE 87.1   Summary of Distributed Scheduling Survey

Method	Distributed	Heterogeneous	Overhead	Scalable
Blake [1992] (NS, RS)	Y	N	Y	Y
(ABS, EBS)	Y	N	N	Y
(CBS)	N	N	N	N
Casavant and Kuhl [1984]	Y	N	x	P
Ghafoor and Ahmad [1990]	Y	N	Y	P
Wave scheduling [Van Tilborg and Wittie 1984]	Y	N	x	P
Ni and Abani [1981] (LED)	Y	N	x	N
(SQ)	Y	N	Y	Y
Stankovic and Sidhu [1984]	Y	N	x	P
Stankovic [1985]	Y	N	x	N
Andrews et al. [1982]	Y	x	x	Y
Greedy load sharing [Chowdhury 1990]	Y	N	x	Y
Gao et al. [1984] (BAR)	Y	N	x	N
(BUW)	Y	N	x	N
Stankovic [1984]	Y	N	x	P
Chou and Abraham [1983]	Y	N	x	Y
Bryant and Finkel [1981]	Y	N	x	Y
Casey [1981] (dipstick, bidding)	Y	N	x	N
(adaptive learning)	Y	N	Y	Y
Klappholz and Park [1984]	Y	N	x	Y
Reif and Spirakis [1982]	Y	N	x	N
Ousterhout et al., see Singhal and Shivaratri [1994]	N	N	x	N
Hochbaum and Shmoys [1988]	N	Y	x	x
Hsu et al. [1989]	N	Y	x	x
Stone [1977]	N	Y	x	x
Lo [1988]	N	Y	x	x
Price and Salama [1990]	N	Y	x	x
Ramakrishnan et al. [1991]	N	Y	x	x
Sarkar and Hennessy, in Shirazi et al. [1995]	N	Y	x	x

a processor is selected at random and is forced to run a task. The third algorithm is arrival balanced scheduling (ABS), in which the task is assigned to the processor that will complete it first, as estimated by the scheduling host. The fourth method uses receiver-initiated load balancing, and is called end balanced scheduling (EBS). NS, RS, and ABS use one-time assignment; EBS uses dynamic reassignment.

Casavant and Kuhl [1984] describe a distributed task execution environment for Unix System 7, with the primary goal of load balancing without altering the user interface to the operating system. As such, the system combines mechanism and policy. This system supports execution autonomy, but not communication autonomy or administrative autonomy.

Ghafoor and Ahmad [1990] describe a bidding system that combines mechanism and policy. A module called an information collector/dispatcher runs on each node and monitors the local load and that of the node's neighbors. The system passes a task between nodes until either a node accepts the task or the task reaches its transfer limit, in which case the current node accepts the task. This algorithm assumes homogeneous processors and has limited support for execution autonomy.

Van Tilborg and Wittie [1984] present wave scheduling for hierarchical virtual machines. The task force is recursively subdivided and the processing flows through the virtual machine like a wave, hence the name. Wave scheduling combines a nonextensible mechanism with policy, and assumes the processors are homogeneous.

Ni and Abani [1981] present two dynamic methods for load balancing on systems connected by local area networks: least expected delay and shortest queue. Least expected delay assigns the task to the host with the smallest expected completion time, as estimated from data describing the task and the processors. Shortest queue assigns the task to the host with the fewest number of waiting jobs. These two methods are not scalable because they use information broadcasting to ensure complete information at all nodes. Ni and Abani [1981] also present an optimal stochastic strategy using mathematical programming.

The method described in Stankovic and Sidhu [1984] uses task clusters and distributed groups. Task clusters are sets of tasks with heavy intertask communication that should be on the same host. Distributed groups also have intertask communication, but execute faster when spread across separate hosts. This method is a bidding strategy and uses nonextensible system and task description messages.

Stankovic [1985] lists two scheduling methods. The first is adaptive with dynamic reassignment, and is based on broadcast messages and stochastic learning automata. This method uses a system of rewards and penalties as a feedback mechanism to tune the policy. The second method uses bidding and one-time assignment in a real-time environment.

Andrews et al. [1982] describe a bidding method with dynamic reassignment based on three types of servers: free, preferred, and retentive. Free server allocation will choose any available server from an identical pool. Preferred server allocation asks for a server with a particular characteristic, but will take any server if none is available with the characteristic. Retentive server allocation asks for particular characteristics, and if no matching server is found, a server, busy or free, must fulfill the request.

Chowdhury [1990] describes the greedy load-sharing algorithm. The greedy algorithm uses system load to decide where a job should be placed. This algorithm is noncooperative in the sense that decisions are made for the local good, but it is cooperative because scheduling assignments are always accepted and all systems are working toward a global load-balancing policy.

Gao et al. [1984] describe two load-balancing algorithms using broadcast information. The first algorithm balances arrival rates, with the assumption that all jobs take the same time. The second algorithm balances unfinished work. Stankovic [1984] gives three variants of load-balancing algorithms based on point-to-point communication that compare the local load to the load on remote processors. Chou and Abraham [1983] describe a class of load-redistribution algorithms for processor-failure recovery in distributed systems.

The work presented in Bryant and Finkel [1981] combines load balancing, dynamic reassignment, and probabilistic scheduling to ensure stability under task migration. This method uses neighbor-to-neighbor communication and forced acceptance to load balance between pairs of machines.

Casey [1981] gives an earlier and less complete version of the Casavant and Kuhl taxonomy, with the term *centralised* replacing *nondistributed* and *decentralised* substituting for *distributed*. This paper also lists three methods for load balancing: dipstick, bidding, and adaptive learning, then describes a load-balancing system whereby each processor includes a 2-byte status update with each message sent. The dipstick method is the same as the traditional watermark processing found in many operating systems. The adaptive learning algorithm uses a feedback mechanism based on the run queue length at each processor.

Dynamic Noncooperative Algorithms

Klappholz and Park [1984] describe deliberate random scheduling (DRS) as a probabilistic, one-time assignment method to accomplish load balancing in heavily loaded systems. Under DRS, when a task is spawned, a processor is randomly selected from the set of ready processors, and the task is assigned to the selected processor. DRS dictates a priority scheme for time slicing, and is thus a mixture of local and global scheduling. There is no administrative autonomy or execution autonomy with this system, because DRS is intended for parallel machines.

Reif and Spirakis [1982] present a resource granting system (RGS) based on probabilities and using broadcast communication. This work assumes the existence of either an underlying handshaking mechanism or shared variables to negotiate task placement. The use of broadcast communication to keep all resource providers updated with the status of computations in progress limits the scalability of this algorithm.

Dynamic Nondistributed Algorithms

Ousterhout et al. (see Singhal and Shivaratri [1994]) describe Medusa, a distributed operating system for the Cm* multiprocessor. Medusa uses static assignment and centralized decision making, making it a combined policy and mechanism. It does not support autonomy, nor is the mechanism scalable.

In addition to the four distributed algorithms already mentioned, Blake [1992] describes a fifth method called continual balanced scheduling (CBS), that uses a centralized scheduler. Each time a task arrives, CBS generates a mapping within two time quanta of the optimum and causes tasks to be migrated accordingly. The centralized scheduler limits the scalability of this approach.

Static Algorithms

All of the algorithms in this section are static, and as such, are centralized and without support for autonomy. They are generally intended for distributed-memory parallel machines, in which a single user can obtain control of multiple nodes through space sharing. However, they can be implemented on fully distributed systems.

Hochbaum and Shmoys [1988] describe a polynomial-time, approximate, enumerative scheduling technique for processors with different processing speeds, called the dual-approximation algorithm. The algorithm solves a relaxed form of the bin packing problem to produce a schedule within a parameterized factor ϵ, of optimal. That is, the total run time is bounded by $(1 + \epsilon)$ times the optimal run time.

Hsu et al. [1989] describes an approximation technique, called the critical sink underestimate method. The task force is represented as a directed acyclic graph, with vertices representing tasks and edges representing execution dependencies. If an edge (α, β) appears in the graph, then α must execute before β. A node with no incoming edges is called a *source*, and a node with no outgoing edges is a *sink*. When the last task represented by a sink finishes, the computation is complete; this last task is called the critical sink. The mapping is derived through an enumerative state-space search with pruning, which results in an underestimate of the running time for a partially mapped computation and, hence, the name critical sink underestimate.

Stone [1977] describes a method for optimal assignment on a two-processor system based on a max flow/min cut algorithm for sources and sinks in a weighted directed graph. A maximum flow is one that moves the maximum quantity of goods along the edges from sources to sinks. A minimum cutset for a network is the set of edges with the smallest combined weighting, which, when removed from the graph, disconnects all sources from all sinks. The algorithm relates task assignment to commodity flows in networks, and shows that deriving a max flow/min cut provides an optimal mapping.

Lo [1988] describes a method based on Stone's max flow/min cut algorithm for scheduling in **heterogeneous systems**. This method utilizes a set of heuristics to map from a general system representation to a two-processor system so that Stone's work applies.

Price and Salama [1990] describe three heuristics for assigning precedence-constrained tasks to a network of identical processors. With the first heuristic, the tasks are sorted in increasing order of communication, and then are iteratively assigned so as to minimize total communication time. The second heuristic creates pairs of tasks that communicate, sorts the pairs in decreasing order of communication, then groups the pairs into clusters. The third method, simulated annealing, starts with a mapping and uses probability-based functions to move toward an optimal mapping.

Ramakrishnan et al. [1991] present a refinement of the A* algorithm that can be used either to find optimal mappings or to find approximate mappings. The algorithm uses several heuristics based on the sum of communication costs for a task, the task's estimated mean processing cost, a combination of communication costs and mean processing cost, and the difference between the minimum and maximum processing costs for a task. The algorithm also uses ϵ-relaxation similar to the dual-approximation algorithm of Hochbaum and Shmoys [1988].

Sarkar and Hennessy (in Shirazi et al. [1995]) describe the GR graph representation and static partitioning and scheduling algorithms for single-assignment programs based on the SISAL language. In GR, nodes represent tasks and edges represent communication. The algorithm consists of four steps: cost assignment, graph expansion, internalization, and processor assignment. The cost assignment step estimates the execution cost of nodes within the graph and communication costs of edges. The graph expansion step expands complex nodes, e.g., loops, to ensure that sufficient parallelism exists in the graph to keep all processors busy. The internalization step performs clustering on the tasks, and the processor assignment phase assigns clusters to processors so as to minimize the parallel execution time.

Distributed Scheduling Support Systems

This section describes prior research in scheduling support mechanisms. In this discussion, the terms *local task* and *foreign task* are defined from the point of view of the host executing the task. A local task executes on the host where it originated, without going through the global scheduling system. A foreign task originates at a host different from the one on which it executes. We will examine a commercial load-sharing product, several research prototypes for distributed scheduling, and a distributed operating system.

NetShare

NetShare is a distributed systems construction product of Aggregate Computing, Inc. [Aggregate 1993]. NetShare comprises services that provide resource management and task execution on a heterogeneous local-area network. NetShare has two main components, the resource management subsystem and the task management subsystem.

The resource manager consists of three parts: the resource information server (RIS), the resource agents (RA), and the client side resource library (CSRL). The RIS is a centralized database of information describing resources available within the system, including state information for individual machines. Resource agents run on each machine and advertise their system state to the RIS. Clients use the CSRL to request resource allocation through the RIS. The CSRL is a library of function calls that is linked with individual application programs. There is no scheduling agent external to the applications; they are self-scheduling.

Figure 87.2 shows the interaction between an application, the RIS, and resource agents. In step 1, state information passes from agents to the RIS. In step 2, the application uses the CSRL to query the RIS.

The agent updates consist of the following information:

- The name, architecture, model, and network address of the host
- The name, version, and release of the operating system
- The amount of physical, free virtual, and used virtual memory
- 1-, 5-, and 15-min load averages
- The idle time and CPU usage of the host
- Power rating, based on standard benchmarks
- Number of users
- Two user-definable properties

The two user-definable properties provide a limited extension mechanism for NetShare. Clients query the database through the CSRL, and receive a set of matching records in response. A sample call to the CSRL,

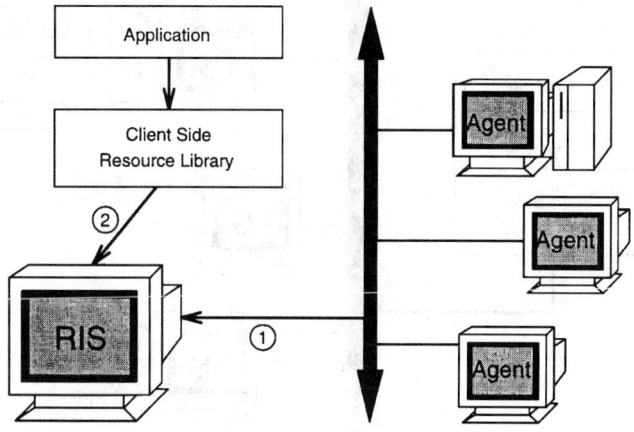

FIGURE 87.2 The NetShare resource management subsystem.

which appears in [Aggregate 1993], is

```
select UNIX_HOST if ((UNIX_HOST:LOAD_5 < 1.0) &&
                     (UNIX_HOST:USERS == 0))
          order by (UNIX_HOST:LOAD_5)
```

This call queries the database for hosts running the Unix operating system, with a 5-min load average less than 1.0, and no active users. The RIS finds the matching set of hosts, and returns the set, sorted by 5-min load average.

The client uses the task management subsystem (TMS) to schedule the individual tasks for execution. The TMS is composed of the task servers (TS) and the client side task library (CSTL). Application programs place individual tasks with calls to the CSTL. Figure 87.3 shows the relationship between the application and the task servers. In the depicted scenario, the client application has selected two servers using the RMS, and has used the TMS to place seven tasks on the servers.

Task servers have limited support for autonomy, in that administrators can set quotas limiting the number of tasks that are either placed by a host (an *export quota*), or that have been accepted from a foreign host (*import quotas*).

Remote Unix, Condor, Butler, and Distributed Batch

Remote Unix and its successor, Condor, were developed at the University of Wisconsin [Litzkow 1987, Bricker et al. 1992]. Butler was developed as part of the Andrew project at Carnegie-Mellon University [Nichols 1987], and distributed batch was developed at the MITRE Corporation [Gantz et al. 1989]. All of these systems attempt to increase utilization and share load across a set of Unix-based workstations, but are less complete systems than NetShare. Therefore, this section groups these systems together and gives a brief description of each.

Remote Unix and Condor use a central resource manager, which gathers information about all of the participating hosts, and a local scheduler per host that controls task execution for that host. Remote Unix has a simple two-level priority scheme for local and foreign tasks, whereas Condor has a policy expression mechanism that provides administrative autonomy. Both Condor and Remote Unix support checkpointing and task migration.

Butler uses a central machine registry and a shared file system to manage a set of homogeneous workstations. All control is centralized. Hosts are dedicated to one task at a time, and are not returned to the free pool until the task completes execution.

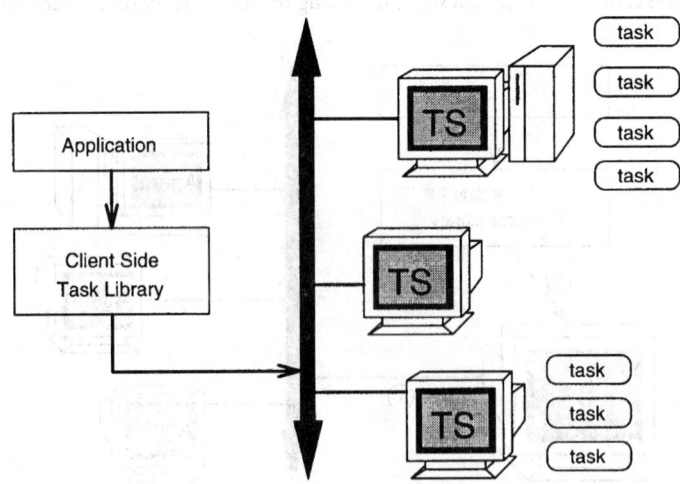

FIGURE 87.3   The NetShare task management subsystem.

Distributed batch runs on a local-area network of 4.2BSD Unix workstations, using centralized storage. Distributed batch contains revocation support in the form of task termination, suspension, and migration. Hosts can be selected based on architecture, operating system version, available memory, local disk configuration, and floating point hardware.

MESSIAHS

Mechanisms effecting scheduling support in autonomous, heterogeneous systems (MESSIAHS) [Chapin and Spafford 1994] are unique among scheduling systems because of their extensive support for autonomy. A MESSIAHS system is structured in a hierarchical fashion, based on administrative domains. This structuring is based on an observation of a social aspect of computing: people are willing to allow outside utilization of their unused resources, as long as they maintain control of the system. This means that the local systems are autonomous, and the local administrators can set their own access policies. An example distributed autonomous system is in Fig. 87.4.

Each node within the MESSIAHS system is a *virtual system*, which represents a subset of the resources of one or more real systems, and has a hierarchical structure modeling the administrative hierarchies of computer systems and institutional organization. Virtual systems can be combined into encapsulating virtual systems. For example, in Fig. 87.4, the university, national laboratory, and industry are each virtual systems, and are collected into a single large distributed (virtual) system. Within the university, national lab, and industry virtual systems are other virtual systems, giving a hierarchical structure. These intermediate groupings may correspond to divisions which contain departments, and the departments may contain research groups, etc. At the lowest level of grouping, each virtual system typically consists of a subset of the capabilities of a single machine.

There is no centralized resource management in MESSIAHS. Instead, each virtual system runs a *scheduling module*, which maintains the system description information for that node. The scheduling module also exchanges service requests with neighboring virtual systems within the hierarchy, and is responsible for starting and stopping jobs.

MESSIAHS provides two interfaces with which administrators can define the scheduling policy for a system: an interpreted language called the MESSIAHS Interface Language (MIL) dedicated to implementing schedulers, and a library of functions for the C programming language. Using these interfaces,

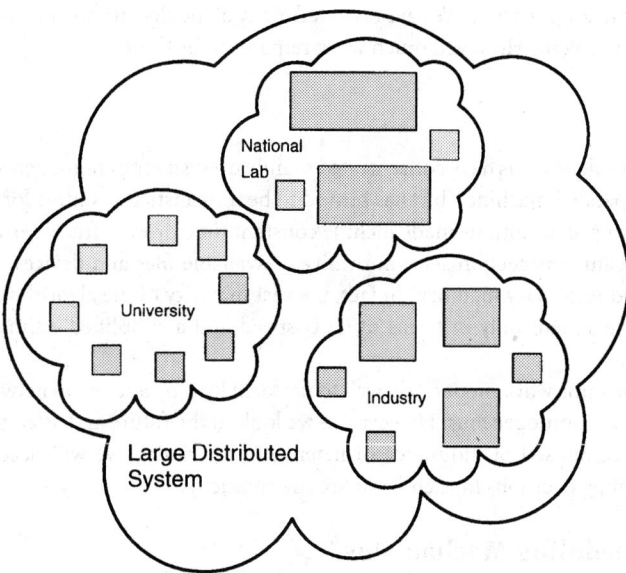

FIGURE 87.4 A sample virtual system.

administrators write small event handlers that are linked into the scheduling system. Thus, MESSIAHS has extensive support for autonomy.

Schizo

The Schizophrenic Workstation system, or Schizo [Swanson et al. 1993], is a distributed operating system personality that is built on top of Mach (see Singhal and Shivaratri [1994], ch. 17). Schizo is intended to use the idle cycles on autonomous workstations which are primarily dedicated to individual users. Like MESSIAHS, Schizo attempts to protect the execution autonomy of the individual nodes.

Schizo has two unique aspects: first, a subset of the nodes in the system, called the *core*, is dedicated to running Schizo and the jobs submitted to the Schizo system. Second, each of the autonomous nodes runs a single server task that allows the autonomous node to switch personalities between a node acting as a dedicated processor for the principal user (or owner) of the machine, or as part of the distributed system. It is from this use of multiple personalities that the authors derived the name of the system.

To preserve the execution autonomy of the participating nodes, Schizo attempts to run its foreign tasks unobtrusively by lowering their priority and by using a special paging algorithm that gives higher priority to the virtual memory pages of local processes. This means that when the owner of a workstation runs a process, that process has higher priority than any Schizo task both for use of the CPU and for use of memory. In addition, Schizo provides a migration facility so that if a node becomes too busy, foreign tasks are moved to other nodes within the system.

The core system periodically polls the noncore nodes to determine their status. When Schizo receives a new task, it attempts to schedule it on a workstation with an acceptable load level that is not currently running any Schizo tasks. If no such workstation can be found, then the task is started on the core system. To ensure that all tasks scheduled by Schizo will run to completion, Schizo makes sure that there are enough resources on the core system to run all of the scheduled tasks in case every noncore node becomes busy. This may mean that Schizo refuses to schedule a task even though it could be run.

The initial evaluation of Schizo indicates that it performs quite well, imposing a negligible performance penalty on local tasks while dramatically increasing the utilization of the individual nodes.

87.4 Research Issues and Summary

The central problem in distributed scheduling is assigning a set of tasks to a set of processors in accordance with one or more optimizing criteria. We have reviewed many of the algorithms and mechanisms developed thus far to solve this problem. However, much work remains to be done.

Algorithms

Until now, scheduling algorithms have concentrated mainly on systems of homogeneous processors. This has worked well for parallel machines, but has proved to be unrealistically simple for distributed systems. Some of the simplifying assumptions made include constant time or even free intertask communication, processors with the same instruction set, uniformity of available files and devices, and the existence of plentiful primary and secondary memory. In fact, the vast majority of the algorithms listed in the survey model the underlying system only in terms of CPU speed and a simplified estimate of interprocessor communication time.

These simple algorithms work moderately well to perform load balancing on networks of workstations that are, in most senses, homogeneous. However, as we look to the future and attempt to build wide-area distributed systems composed of thousands of heterogeneous nodes, we will need policies capable of making good scheduling decisions in such complex environments.

Distributed Scheduling Mechanisms

As we have seen, current scheduling systems do a good job of meeting the technical challenges of supporting the relatively simplistic scheduling policies that have been developed to date. Future work will expand in new directions, especially in the areas of heterogeneity, security, and the social aspects of distributed computing.

Just as scheduling algorithms have not considered heterogeneity, **distributed scheduling mechanisms** have only just begun to support scheduling in heterogeneous systems. Some of the major obstacles to be overcome include differences in the file spaces, speeds of the processors, processor architectures (and possible task migration between them), operating systems and installed software, devices, and memory. Future support mechanisms will have to make this information available to scheduling algorithms to fully utilize a large, heterogeneous distributed system.

The challenges in the preceding list are purely technical. Another set of problems arises from the social aspects of distributed computing. Large-scale systems will be composed of machines from different administrative domains; the social challenge will be to ensure that computations that cross administrative boundaries do not compromise the security or comfort of users inside each domain. To be successful, a distributed scheduling system will have to provide security both for the foreign task and for the local system; neither should be able to inflict harm on the other. In addition, the scheduling system will have to assure users that their local rules for use of their machines will be followed. Otherwise, the computing paradigm will break down, and the large system will disintegrate into several smaller systems under single administrative domains.

Defining Terms

Autonomy: The freedom to be different or behave differently than other nodes within the system.

Centralized mechanisms: Mechanisms in which data are stored at a single node, or which pass all data to a single node for a decision.

Distributed mechanisms: Mechanisms in which decisions are made on the local system, based on data located on that system.

Distributed memory: A system in which processors have different views of memory. Often, each processor has its own memory and cannot directly access another processor's local memory.

Distributed systems: Systems with a high degree of autonomy.

Global scheduling: The assignment of tasks to processors (also called task placement and matching).

Heterogeneous systems: The property of having different underlying machine architecture or systems software.

Job: A group of tasks cooperating to solve a single problem.

Load balancing: A special form of load sharing in which the system attempts to keep all nodes equally busy.

Load sharing: The practice of moving some of the work from busy processors to idle processors. The system does not necessarily attempt to keep the load equal at all processors; instead, it tries to avoid the case where some processors are heavily loaded while others sit idle.

Local scheduling: The decision as to which task, of those assigned to a particular processor, will run next on that processor.

Loosely coupled hardware: A message passing multiprocessor.

Mechanism: The ability to perform an action.

Parallel systems: Systems with a low degree of autonomy.

Policy: A set of rules that decide what action will be performed.

Shared memory: A system in which all processors have the same view of memory. If processors have local memories, then other processors may still access them directly.

Space sharing: A system in which several jobs are each assigned exclusive use of portions of a common resource. For example, if some of the processors in a parallel machine are dedicated to one job, while another set of processors is dedicated to a second job, the jobs are space sharing the CPUs.

Stability: The property of a system that the service rate is greater than or equal to the arrival rate. A stable scheduling algorithm will not make a stable system unstable.

Task: The unit of computation in a distributed system; an instance of a program under execution.

Task migration: The act of moving a task from one node to another within the system.

Tightly coupled hardware: A shared-memory multiprocessor.

Time sharing: A system in which jobs have the illusion of exclusive access to a resource, but in which the resource is actually switched among them.

References

Aggregate. 1993. *Using the NetShare SDK to Build a Distributed Application: A Technical Discussion.* Aggregate Computing, Inc., Minneapolis, MN.

Andrews, G. R., Dobkin, D. P., and Downey, P. J. 1982. Distributed allocation with pools of servers, pp. 73–83. In *Proc. Symp. Principles Distributed Comput.* ACM, Aug.

Blake, B. A. 1992. Assignment of independent tasks to minimize completion time. *Software—Pract. Exp.* 22(9):723–734.

Bond, A. M. and Hine, J. H. 1991. DRUMS: a distributed statistical server for STARS. *Winter USENIX.* Dallas, TX, Jan.

Bricker, A., Litzkow, M., and Livny, M. 1992. Condor Technical Summary. *Department of Computer Science. Tech. Rep.* 1069. University of Wisconsin-Madison, Jan.

Bryant, R. M. and Finkel, R. A. 1981. A stable distributed scheduling algorithm, pp. 314–323. In *Proc. Int. Conf. Distributed Comput. Syst.* IEEE, April.

Casavant, T. L. and Kuhl, J. G. 1984. Design of a loosely-coupled distributed multiprocessing network, pp. 42–45. In *Proc. Int. Conf. Parallel Process.* IEEE, Aug.

Casavant, T. L. and Kuhl, J. G. 1988. A taxonomy of scheduling in general-purpose distributed computing systems. *IEEE Trans. Software Eng.* 14(2):141–154.

Casey, L. M. 1981. Decentralised scheduling. *Aust. Comput. J.* 13(2):58–63.

Chapin, S. J. and Spafford, E. H. 1994. Support for implementing scheduling algorithms using MESSIAHS. *Sci. Programming* 3:325–340.

Chou, T. C. K. and Abraham, J. A. 1983. Load redistribution under failure in distributed systems. *IEEE Trans. Comput.* C-32(9):799–808.

Chowdhury, S. 1990. The greedy load sharing algorithm. *J. Parallel Distributed Comput.* 9:93–99.

Gantz, C. A., Silverman, R. D., and Stuart, S. J. 1989. A distributed batching system for parallel processing. *Software–Pract. Exp.* 19.

Gao, C., Liu, J. W. S., and Railey, M. 1984. Load balancing algorithms in homogeneous distributed systems, pp. 302–306. In *Proc. Int. Conf. Parallel Process.* IEEE, Aug.

Ghafoor, A. and Ahmad., I. 1990. An efficient model of dynamic task scheduling for distributed systems, pp. 442–447. In *Comput. Software App. Conf.* IEEE.

Hochbaum, D. and Shmoys, D. 1988. A polynomial approximation scheme for scheduling on uniform processors: using the dual approximation approach. *SIAM J. Comput.* 17(3):539–551.

Hsu, C. C., Wang, S. D., and Kuo, T. S. 1989. Minimization of task turnaround time for distributed systems. In *Proc. 13th Ann. Int. Comput. Software and Appl. Conf.*

Huang, C. H., Huang, T. L., and Juang, J. Y. 1989. On processor allocation in hypercube systems. In *Proc. 13th Annu. Int. Comput. Software Appl. Conf.*

Klappholz, D. and Park, H. C. 1984. Parallelized process scheduling for a tightly-coupled MIMD machine, pp. 315–321. In *Proc. Int. Conf. Parallel Process.* IEEE, Aug.

Litzkow, M. J. 1987. Remote Unix: Turning idle workstations into cycle servers, pp. 381–384. In *USENIX Summer Conf.* USENIX Association, Berkeley, CA.

Lo, V. M. 1988. Heuristic algorithms for task assignment in distributed systems. *IEEE Trans. Comput.* 37(11):1384–1397.

Maccabe, A. B., McCurley, K. S., Riesen, R., and Wheat, S. R. 1994. SUNMOS for the Intel paragon: a brief user's guide, pp. 245–251. In *Proc. Intel Supercomput. Users' Group*, June. Annual North America Users' Conference.

Ni, L. M. and Abani, K. 1981. Nonpreemptive load balancing in a class of local area networks, pp. 113–118. In *Proc. Comput. Networking Symp.* IEEE, Dec.

Nichols, D. A. 1987. Using idle workstations in a shared computing environment, pp. 5–12. In *Proc. 11th ACM Symp. Operating Syst. Principles*, ACM.

OSF 1993. *Design of the OSF/1 Operating System*. Open Software Foundation. Prentice–Hall, Englewood Cliffs, NJ.

Price, C. C. and Salama, M. A. 1990. Scheduling of precedence-constrained tasks on multiprocessors. *Comput. J.* 33(3):219–229.

Ramakrishnan, S., Cho, I. H., and Dunning, L. 1991. A close look at task assignment in distributed systems, pp. 806–812. In *INFOCOM '91*. Miami, FL, IEEE, April.

Reif, J. and Spirakis, P. 1982. Real time resource allocation in distributed systems, pp. 84–94. In *Proc. Symp. Principles Distributed Comput.* ACM, Aug.

Shirazi, B. A., Hurson, A. R., and Kavi, K. M., eds. 1995. *Scheduling and Load Balancing in Parallel and Distributed Systems*. IEEE Computer Society Press.

Singhal, M. and Shivaratri, N. G. 1994. *Advanced Concepts in Operating Systems*. McGraw–Hill, New York.

Stankovic, J. A. 1984. Simulations of three adaptive, decentralized controlled, job scheduling algorithms. *Comput. Networks* 8(3):199–217.

Stankovic, J. A. 1985. Stability and distributed scheduling algorithms, pp. 47–57. In *Proc. 1985. ACM Comput. Sci. Conf.* ACM, March.

Stankovic, J. A. and Sidhu, I. S. 1984. An adaptive bidding algorithm for processes, clusters and distributed groups, pp. 49–59. In *Proc. Int. Conf. Distributed Comput. Syst.* IEEE, May.

Stone, H. S. 1977. Multiprocessor scheduling with the aid of network flow algorithms. *IEEE Trans. Software Eng.* SE-3(1):85–93.

Stumm, M. 1988. The design and implementation of a decentralized scheduling facility for a workstation cluster, pp. 12–22. In *Proc. 2nd IEEE Conf. Comput. Workstations*. IEEE, March.

Swanson, M., Stoller, L., Critchlow, T., and Kessler, R. 1993. The design of the schizophrenic workstation system, pp. 291–306. In *Proc. Mach III Symp.* USENIX Association.

Theimer, M. M. and Lantz, K. A. 1989. Finding idle machines in a workstation-based distributed system. *IEEE Trans. Software Eng.* 15(11):1444–1458.

Van Tilborg, A. M. and Wittie, L. D. 1984. Wave scheduling—decentralized scheduling of task forces in multicomputers. *IEEE Trans. Comput.* C-33(9):835–844.

Wheat, S. R., Maccabe, A. B., Riesen, R., van Dresser, D. W., and Stallcup, T. M. 1994. PUMA: an operating system for massively parallel systems. *Sci. Programming* 3:275–288.

Zhou, S., Zheng, X., Wang, J., and Delisle, P. 1993. Utopia: a load sharing facility for large, heterogeneous distributed computer systems. *Software—Pract. Exp.* 23(12):1305–1336.

Further Information

Many of the seminal theoretical papers in the area are contained in *Scheduling and Load Balancing in Parallel and Distributed Systems*, edited by Shirazi et al. [1995]. This volume contains many of the papers cited in this chapter, and is an excellent starting point for those interested in further reading in the area.

Advanced Concepts in Operating Systems by Singhal and Shivaratri [1994] contains two chapters discussing scheduling for parallel and distributed systems. These two references contain pointers to much more information than could be presented here.

Descriptions of other distributed scheduling systems may be found in papers describing Stealth [Singhal and Shivaratri 1994, ch. 11], Utopia [Zhou et al. 1993], DRUMS [Bond and Hine 1991], as well as in Theimer and Lantz [1989] and Stumm [1988]. More information about scheduling in distributed operating systems such as Sprite, the V System, Locus, and MOSIX can also be found in Singhal and Shivaratri [1994].

88

Software Support for Heterogeneous Computing

Howard Jay Siegel
Purdue University

Henry G. Dietz
Purdue University

John K. Antonio
Texas Tech University

88.1 Introduction

As a result of advances in high-speed digital communications, researchers have begun to use collections of different high-performance machines in concert to execute computationally intensive application tasks. A single application task often requires a variety of different types of computation (e.g., arithmetic operations on arrays vs. symbolic manipulation). Existing high-performance machines typically achieve only a fraction of their peak performance on certain portions of such application programs, i.e., there is a gap between average sustained performance and the machine's peak performance. One reason for this is that different subtasks of an application can have different computational requirements that are best processed by different types of machine architectures (e.g., see Siegel et al. [1992]). In general, a given machine architecture does not satisfy the computational requirements of all portions of an application equally well. Thus, an important approach to high-performance computing is to construct a **heterogeneous computing (HC)** environment, consisting of a variety of machines interconnected by high-speed links. The focus of this chapter is the software challenges for supporting such an environment.

Figure 88.1 shows a hypothetical example of an application program whose various subtasks are best suited for execution on different machine architectures [Freund and Siegel 1993]. The example is based on a total execution time of 100 units on a baseline serial machine. The first 25 units of time are spent doing computations well suited to execution on a single-instruction-stream/multiple-data-stream (SIMD) machine, the next 40 on a large (e.g., 128 processor) distributed-memory multiple-instruction-stream/multiple-data-stream (MIMD) machine, the next 25 on a vector (pipelined) supercomputer, and the last 10 on a small (e.g., 10 processor) shared-memory MIMD machine. Executing the whole program on an SIMD machine with 16,384 processors may reduce the SIMD portion from 25 units to 0.01 units,

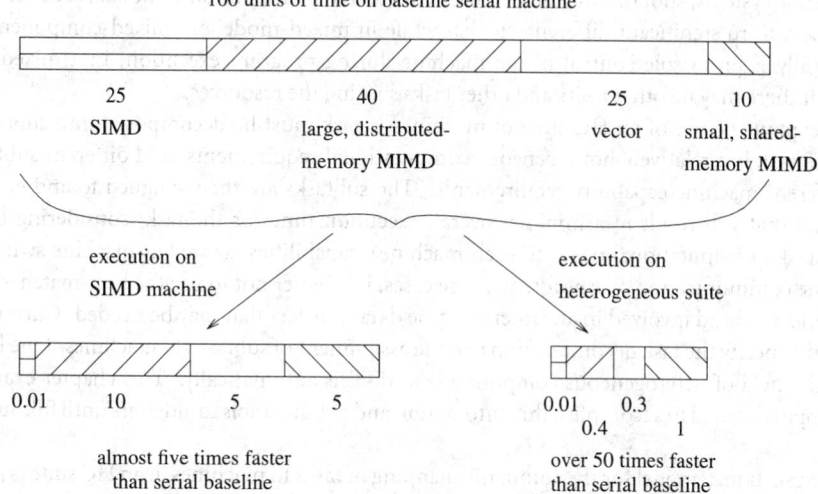

FIGURE 88.1 Hypothetical example of the advantage of using a heterogeneous suite of machines, where the heterogeneous suite time includes intermachine communication overhead. Not drawn to scale.

and also reduce, to varying extents, the other portions of the application execution time. However, the total improvement may only be about a factor of five due to the subtasks that were not as suitable for SIMD mode execution. Alternatively, the use of an HC suite with four different machines, each matched with the computational requirements of the **subtask** for which it is used, can result in an execution over 50 times as fast as the baseline serial machine. However, because multiple machines are being used, any data shared by subtasks executing on different machines must be transferred between the appropriate machines. This intermachine data transfer overhead must be considered as part of the total execution time for the HC suite.

In general, an HC system provides a variety of architectural capabilities, orchestrated to perform an application whose subtasks have diverse execution requirements. There are many types of HC systems. This chapter will concentrate on software support challenges for *mixed-machine* HC systems, where a heterogeneous suite of independent machines of different types are interconnected by a high-speed network [Watson et al. 1994]. Mixed-machine HC has also been referred to as metacomputing [Khokhar et al. 1993].

Other types of HC systems include mixed-mode machines and mixed-component machines. A *mixed-mode* HC system is a single parallel processing machine whose processors are capable of operating in either the synchronous SIMD or asynchronous MIMD mode of parallelism, and can dynamically switch between modes at instruction-level granularity with generally negligible overhead [Siegel et al. 1996b]. (SIMD and MIMD architectures are described in Chapter 20 of this Handbook.) Examples of mixed-mode machines that have been prototyped are PASM, TRAC, OPSILA, Triton, and EXECUBE [Siegel et al. 1996b]. A *mixed-component* HC system is a single machine where each separate component represents one mode of parallelism, and two or more distinct modes are present in the machine [Siegel et al. 1996a]. An example is the SIMD/MIMD image understanding architecture [Weems et al. 1992].

Mixed-machine systems differ from mixed-mode and mixed-component systems in ways that make the effective use of mixed-machine systems more challenging. First, switching execution among machines in a mixed-machine system requires measurable overhead because data shared by subtasks executed on different machines may need to be transferred among machines. Thus, the mixed-machine systems considered in this chapter are assumed to have high-speed connections among machines that make decomposition at the subtask level feasible. However, this intermachine communication overhead must be considered when deciding the best assignment of subtasks to machines in the HC suite. A second difference is that, in mixed-machine systems, subtasks may be executed concurrently on all or on some subset of the machines in the suite. A third significant difference is that while in mixed-mode and mixed-component systems a user typically is given sole control of the machine during program execution, in a mixed-machine environment there may be other users and other tasks sharing the resources.

To fully exploit the use of an HC suite of machines, a task must be decomposed into subtasks, such that each subtask has relatively homogeneous computational requirements, and different subtasks may require different machine capability requirements. The subtasks are then assigned to and executed on the machines that will result in a minimal overall execution time for the task, considering the match of each subtask's computational needs to each machine's capabilities, as well as machine switching and intermachine communication overheads. In some cases, it is better not to use the best-matched machine because of the overhead involved in any intermachine data transfers that may be needed. Currently, users typically must specify the task decomposition and the assignment of subtasks to machines. One long-term pursuit in the field of heterogeneous computing is to do this automatically. This chapter examines the software support needed to accomplish this automation and to build tools to aid users until full automation is possible.

In the next section, a model for the automatic mapping of tasks to machines in an HC suite is presented. This is followed by a discussion of the compiler technology for implementing machine-independent languages suitable for programming HC systems. **Task profiling**, for characterizing applications, and **analytical benchmarking**, for characterizing machines, are then examined. Ways to use these characterizations to match subtasks to machines and establish an execution schedule are then covered. Compiler and operating system technology to support task profiling, analytical benchmarking, and **matching** and **scheduling** is explored. Existing tools to aid HC system users are overviewed. Finally, open research problems in the field are listed.

88.2 Underlying Issues

There are many instances of successful implementations of application tasks across suites of heterogeneous machines. Examples are the simulation of mixing in turbulent convection at the Minnesota Supercomputing Center, interactive rendering of multiple Earth science data sets on the CASA testbed, and using VISTAnet to compute radiation treatment planning for cancer patients [Siegel et al. 1996a]. Typically, however, current users of HC systems must decompose the application task into appropriate subtasks themselves, decide on which machine to execute each subtask, code each subtask specifically for its target machine, and determine the relative execution schedule for the subtasks. The automation of this process is a long-term goal in the field of HC. This automation is important for encouraging and facilitating the use of HC systems, and for improving the actual performance users get from HC systems. Furthermore, research conducted toward this goal should produce tools that will aid the users of HC systems until full automation is possible.

A model for automating the process of task decomposition, matching of subtasks to machines, and scheduling execution is presented in this section. This model will provide an overview of the big picture into which the other sections will fit. A *program* is defined to be an encoding of a given application task. Similarly, a **program segment** is defined to be an encoding of a given subtask.

The first step in using an HC system is to construct the application program. A programming language used in an HC environment must be compilable into efficient code for any machine in the HC suite to allow full flexibility of execution targets. Thus, ideally, this programming language must be machine independent, and supply the compiler with the information it needs to produce efficient code for different target architectures. Furthermore, the program specification should facilitate the decomposition of an application task into subtasks, such that each subtask is homogeneous in terms of its computational requirements. The next section of this chapter addresses the problems involved with the design of such a language and associated compiler. When discussing how to fully automate the process of mapping an application onto an HC suite, the existence of such a language and compiler is assumed.

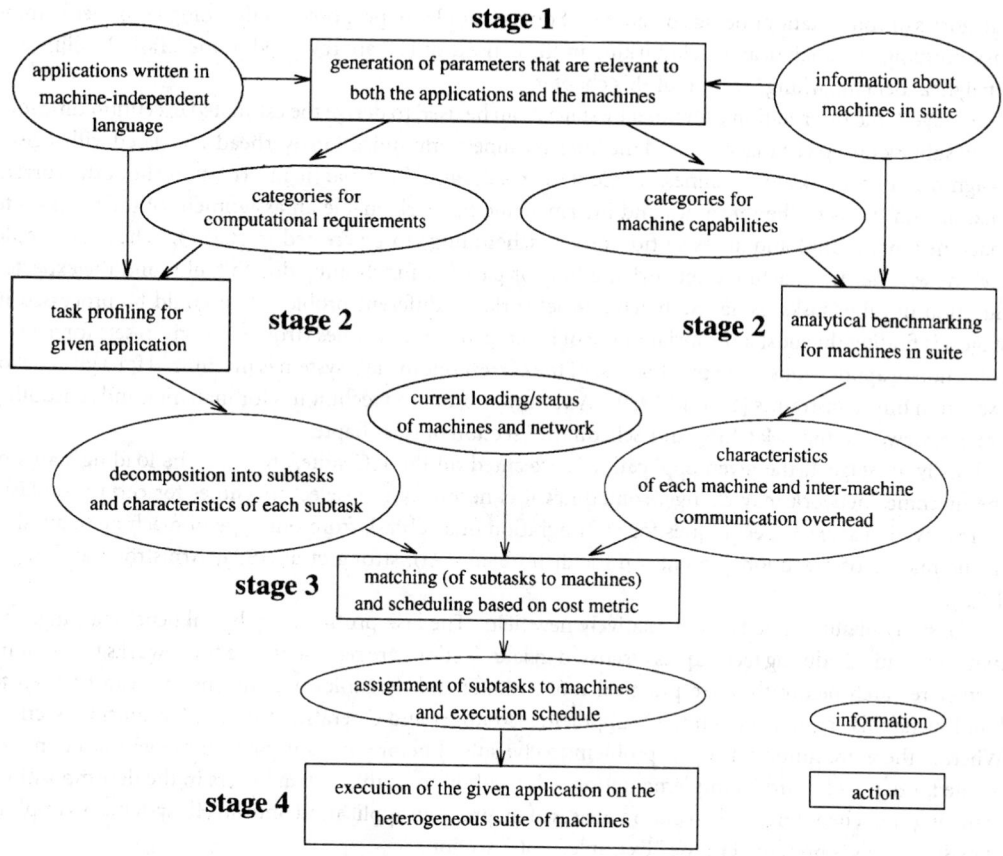

FIGURE 88.2 Model for integrating the software support needed for automating the use of heterogeneous computing systems (based on Siegel et al. [1996a]).

Figure 88.2 overviews the stages of the proposed model for automation, indicating the interactions of the components. The model is from Siegel et al. [1996a], and builds on the one presented in Freund and Siegel [1993]. The model consists of four stages: (1) determination of characterization parameters, (2) task profiling and analytical benchmarking, (3) matching and scheduling, and (4) task execution.

The purpose of stage 1 is to generate a set of parameters that is relevant to both the computational requirements of the applications and the machine capabilities of the HC system. This is done using information about the expected types of application tasks and about the machines in the HC suite. Categories for computational requirements and categories for machine capabilities are derived for each parameter. For example, consider the parameter floating point operations. The computational requirement of the application tasks to be quantified for this parameter is the number of each type of floating point operation needed to perform the calculation. The capabilities of the machines to be quantified for this parameter are the speeds for these different types of floating point operations. Irrelevant parameters are excluded. For example, if the given applications have no floating point operations, then it is not necessary to evaluate the machine's floating point capabilities. The chosen parameters must evolve when new types of applications are added to the set of expected tasks and new types of machines are added to the HC suite.

Stage 2 involves task profiling and analytical benchmarking. Task profiling decomposes the application task into subtasks, each of which is homogeneous with respect to computational requirements. Both the code and the data upon which the specified HC system will operate must be profiled. Analytical benchmarking quantifies how effectively each of the machines in the HC suite performs each of the categories of computations being considered. Some examples of proposed methodologies for performing task profiling and analytical benchmarking in the context of HC are reviewed in the "Task Profiling and Analytical Benchmarking" section of this chapter.

In stage 3, the information generated by stage 2 can be used to derive the estimated execution time for a given subtask on a given machine, and the intermachine communication overhead associated with a given assignment of subtasks to machines. These static results and the dynamic information about the current loading and status of the machines and intermachine network enable an assignment of the subtasks to machines (**matching**) and an execution order (**scheduling**) to be created in stage 3. The status could include whether the machines/network are fully or partially functioning due to faults, and the expected duration of other tasks using the machines/network. A different problem that could be processed in stage 3 is finding the most appropriate suite of heterogeneous machines (from some given set) for a given collection of applications, such that the cost of the corresponding HC system is minimized for a given set of execution time constraints [Freund 1989]. A variety of proposed techniques for matching and scheduling are overviewed in the "Matching and Scheduling" section of this chapter.

Finally, in stage 4, the given application is executed on the HC suite. Because the loading/status of the machines/network may change, sometimes it is necessary to reselect machines for certain subtasks by reactivating stage 3. Techniques for the migration of a subtask from one type of machine to another in the middle of execution present a difficult problem [Armstrong et al. 1994, Armstrong and Siegel 1995].

The study of automatic HC is a relatively new field. The task profiling, analytical benchmarking, and matching and scheduling techniques discussed in later sections are representative frameworks that require further research before they are practical tools. Static and dynamic mechanisms that can be used to build such tools are discussed in the "Supporting Compiler and Operating System Mechanisms" section. Whereas there are numerous open problems to be solved before this automation model can be implemented, progress toward its implementation will result in tools that will aid users in the decomposition, matching and scheduling, and execution phases of mapping an application onto an HC system. A sampling of existing tools is presented in the "Example Tools" section.

88.3 Language Design and Compilation Issues

It may be possible to automatically transform programs written in conventional, sequential, languages into good heterogeneous implementations; however, the technology to do this is currently well beyond the state of the art. Thus, current and foreseeable HC systems rely on a variety of compilation technologies working in concert with languages or libraries specifically designed to simplify the decomposition of a program into relatively heterogeneous parallel segments.

The problem of providing appropriate language and compiler support can be viewed as combining three major subproblems. The first is the development of language design and compiler technology that can work together to generate reasonably efficient code for an HC target. Within the generated code, there are a variety of run-time support library issues unique to HC, and the handling of these issues forms the second subproblem. Finally, the third new challenge involves the fact that the version management and physical distribution of executable code requires methods beyond those needed in a conventional system, blurring the distinction between the *loader* and the compiler.

In any discussion of programming languages or systems, it is useful to distinguish between the **programming model** and the **execution model**. The programming model is simply the model seen by the programmer and expressed in the high-level language code. In contrast, the execution model embodies machine-specific details and is generally expressed either in native assembly language or in language like C or Fortran augmented by calls to machine-dependent parallelism-related functions. Compilation is simply the process of mapping programming model constructs into implementations using the execution model. For an execution model that targets a complex HC system, the primary question is how the programming model can either abstract (i.e., characterize in a general way) or hide this underlying complexity without preventing effective compilation. An overview of a variety of HC programming models appears in Weems et al. [1994].

Perhaps because it can use simpler compiler technology, abstracting heterogeneous properties is currently a far more common approach than hiding these properties from the programmer. In fact, most such systems simply augment a conventional language and compilation system with a set of macros, library functions, or minor language extensions. For example, consider the *parallel virtual machine (PVM)* software system for enabling the development of applications for HC systems (discussed in more detail in the "Example Tools" section) [Beguelin et al. 1993]. Rather than hiding the heterogeneous nature of a target system, the PVM library routines allow the programmer to explicitly test and specify the abstract type of machine upon which each portion of a PVM program will execute. Further, it is the programmer's responsibility to understand and explicitly code for any distinctive properties of machines within the heterogeneous system. Thus, a PVM program to execute on a heterogeneous system composed of a group of workstations and a SIMD supercomputer will almost certainly consist of at least two programs that interact through explicit use of the PVM library: one to be compiled and run on each of the workstations and another for the SIMD supercomputer. Other systems abstracting heterogeneity in this way include SmartNet [Hensgen et al. 1995], p4 [Butler and Lusk 1993], and Linda [Carriero et al. 1992]. These systems are discussed in the "Example Tools" section.

Despite the current emphasis on abstracting, rather than hiding, heterogeneity, it is clear that a programming model that is capable of hiding the heterogeneous nature of the execution model is preferable for most programmers and applications. There are two basic reasons:

- The programmer does not need to understand any characteristics of the target heterogeneous execution model.
- Programs written in this way are potentially mechanically portable.

Although the merit of the first point can be argued on the basis that some machine differences are important to the user (e.g., presence of a video frame buffer on a particular machine), portability is difficult to argue against. Aside from the usual arguments given when discussing portability among individual machines, portability is important here because it is a common property of heterogeneous systems that the precise system configuration immediately available for executing code varies significantly over relatively short time spans. Revisiting the PVM example, where can the SIMD code run if the SIMD supercomputer is down or is busy executing another program?

The primary difficulty in hiding heterogeneity is that it requires significantly stronger compiler technology than is currently available. Thus, a major research focus is both on developing the needed compiler technology and, as a practical necessity, defining language constructs that will make the compiler technology more effective. Language design can facilitate effective compilation in ways described in the next four paragraphs.

Keep the programming model distinct from the execution model. A language that incorporates constructs specific to the architectural features of a particular machine can make program transformation to other targets difficult without simulating these machine features, often greatly increasing execution time. A subtle, but common, example of such machine dependence occurs in languages that assume a particular execution mode. For example, it is remarkably difficult to translate MIMD code into code that will make efficient use of SIMD hardware [Dietz and Krishnamurthy 1993].

Avoid overly specific semantics. The semantics of each construct define constraints on the code that can be generated; thus, constructs should be designed to imply no meaning beyond that which is normally used. For example, the C language specifies that the memory addresses that hold the members of a **struct** are monotonically increasing. Although this fact is rarely used, it is very difficult for a C compiler to prove that this constraint is unnecessary within a particular program; thus, the compiler is prevented from allocating the data in what may be a more efficient layout.

Never block compiler analysis. If the compiler cannot determine the precise meaning of a construct, it cannot transform the construct to better match the target machine. For example, although a programmer might know that two subscripted array references refer to disjoint data items (and hence can be used in parallel), the compiler might not be able to determine this and the programmer might not have a way to inform the compiler.

Allow the programmer to provide *safe* hints. No matter how carefully a compiler is built, the programmer will often know things about his or her program that cannot be derived from the code. Hints about program behavior should be testable; e.g., an assertion that the two subscripted array references above involve only disjoint data items should be accompanied by a method by which the compiler can prove/ensure that this assertion is true. Likewise, performance-oriented hints should make clear what assumptions are being made about the target machine. For example, high-performance Fortran (HPF) [HPFF 1993] data layout directives allow the programmer to suggest an efficient data layout for a particular target machine, but fail to provide any way that the compiler can determine which target machine the programmer had in mind. In fact, the data layout directives of earlier parallel Fortran dialects were more useful than those of HPF in an HC environment because a single program could contain hints for several different target machines and the compiler could easily infer which hints apply to which type of target machine (e.g., hints in Fortran code for Thinking Machines computers are all syntactically comments beginning with **CMF$**, whereas similar hints for MasPar computers all begin with **CMPF**).

Although one generally views a program as being a self-contained entity, there are often many assumptions made about the environment in which the program operates. If the goal is to be the transparent allocation of program code to whatever machine is most suited for its execution, then it is necessary not only that the program directly use only portable constructs, but also that the program's implicit assumptions about the execution environment be supported on all machines. These assumptions about the environment range from the basic functionality of standard library routines to more complex issues involving how files are accessed from any machine within an HC system.

The duplication of standard library routines is generally not a matter of research concern, but can imply significant implementation overhead. For example, many parallel computers come with extensive libraries that have been carefully hand-tuned to compute fast Fourier transforms (FFTs) and similar functions; functionally equivalent routines are often absent from workstation libraries.

Another run-time library issue derives from the fact that different machines often use different representations for the same datum. Several different byte orders are commonly used to represent 32-bit integers, and larger differences exist among the basic formats used to represent floating point numbers. For example, both floating point precision and dynamic range vary, largely because some systems use base 16 vs. base 2 exponents. When must the encoding of data be converted? Further, should it be converted directly from one native format to another or first into a standard interchange format (such as XDR) and then into the target native representation? It has been shown that PVM's use of XDR yields substantial performance penalties [Beguelin et al. 1993].

File (and physical device, e.g., frame buffer) input and output operations are a source of many ponderous problems. Commonly, users of PVM and similar systems assume that the network file system (NFS) will allow user programs to access the same disk files from all machines within an HC system. However, NFS does not guarantee that a file being accessed (especially one being written) by multiple machines will be maintained in a coherent state, nor does NFS provide a user-level means for explicitly flushing buffers to impose a coherent file state. Systems such as Condor [Tannenbaum and Litzcow 1995], which was developed to support migration of a serial task between similar workstations but has recently been extended to work with PVM [Pruyne and Livny 1996], handle the uniform file access problem by creating a daemon process local to the initiating machine that acts as a proxy to perform file operations on behalf of the migrated task. There are a variety of research efforts focusing on the implementation of distributed file systems across heterogeneous clusters using a similar daemon process on each machine.

For much the same reason that uniform file access is complicated by the HC environment, the management of executable object code files is fundamentally different from that in more conventional environments. One question is: when a program (or portion of a program) is to be executed on a particular machine, what is sent there? In homogeneous systems, one commonly sends the object file via NFS access. PVM, and similar systems, offer the option of instead invoking a different object file name on each machine within an HC system; however, the software system does not automatically create/distribute these object files. A more general solution is used in automatic heterogeneous supercomputing (AHS) [Dietz et al. 1993]: when a program is to be executed on a particular machine, a compressed source file is transmitted to the target machine, the source file is compiled there to create a local object file, the object file is executed, and the object file is discarded at the end of execution.

Although the AHS method (discussed in the "Example Tools" section) is viable if compile time is relatively short, each execution of a production program should not require recompilation. For instance, if there are N machines of the same type within the HC system, should there be only one compilation for these N machines, with the resulting object file sent to the machines that need it? Further research needs to be done toward allowing object files to persist so that they will need to be regenerated only when there is no object file available or the object file is found to be out of date. Note that generating and keeping a copy of an object file for every **program segment** on every machine in an HC system is impractical if there are many machines in the system because a typical program execution will use only a few of them. Thus, initiating the execution of a program in an HC system truly involves a combination of normal program loader functions with need-driven redistribution and even recompilation.

In summary, there is a wide range of language and compiler issues for which the ideal solutions are beyond the current state of the art. However, there are alternative or approximate solutions using existing technology to obtain most of the desired characteristics.

88.4 Task Profiling and Analytical Benchmarking

The parallel assessment window system (PAWS) and the distributed heterogeneous supercomputing management system (DHSMS) are briefly overviewed in this section. They represent two example methodologies for performing parts of the task profiling and analytical benchmarking that were defined in the "Underlying Issues" section. In general, existing methodologies make simplifying assumptions and are frameworks rather than fully implemented systems.

The prototype of *PAWS* consists of four tools: the application characterization tool, the architecture characterization tool, the performance assessment tool, and the interactive graphical display tool [Pease et al. 1991]. The *application characterization tool* transforms a given program written in a specific subset of Ada into IF1, an acyclic graphical language that illustrates the program's data dependencies. In IF1, basic operations, such as addition and multiplication, are represented by simple nodes, and complex constructs, such as conditional branches and loops, are represented by compound nodes. By grouping sets of nodes and edges into functions and procedures, the application characterization tool can describe the execution behavior of a given program at various levels. However, this tool does not perform task decomposition based on computational requirements and machine capabilities.

To benchmark machines, the *architecture characterization tool* partitions the architecture into four categories: computation, data movement and communication, input/output (I/O), and control. Each category can be repeatedly partitioned into subsystems until the subsystems in the lowest level can be described by raw timing information. This hierarchical organization of architectural parameters for a specific machine provides a detailed model for determining the operational behavior of each subsystem. The raw timing information of each leaf node of the tree can be user specified or obtained by low-level benchmarking.

The information produced by the architecture characterization tool is used by the *performance assessment tool* to generate timing information for operations on a given machine upon request. Timings for primitive operations are stored within the architecture characterization tool; the performance assessment tool uses these to determine timings for more complicated operations (e.g., complex floating point multiplication). This tool then generates two types of profiles by traversing the task-flow graph produced by the application characterization tool. *Parallelism profiles* represent the applications' theoretical upper bounds of performance (e.g., the maximal number of operations that can be parallelized). *Execution profiles* represent the estimated performance of the applications after they have been partitioned and mapped onto one particular machine (the partitioning and mapping must be specified by the user).

Execution profiles are produced by traversing the application's flow graph and then computing and recording each node's execution time estimate on that machine based on the information provided by the architecture characterization tool. As is the case with most proposed current HC methodologies, information about compound nodes, such as the number of times a loop is to be executed, is assumed to be a derivable constant or estimated by the user and specified as a compiler directive.

Thus, this set of tools performs the functions of task profiling and analytical benchmarking. Furthermore, the execution profiles they produce can be used as part of the matching process of stage 3 of Fig. 88.2. Also, PAWS has an *interactive graphical display tool* that is the menu-driven user interface for accessing the other tools.

The *DHSMS* includes a systematic framework for performing task profiling and analytical benchmarking [Ghafoor and Yang 1993]. DHSMS generates a *universal set of codes (USC)* for task profiling. The USC can also be viewed as a standardized set of benchmarking programs used in analytical benchmarking. Based on the machines in the HC suite, a USC is constructed using a hierarchical structure that is similar to the hardware organizational information maintained by the architectural characterization tool in PAWS. At the highest level of this hierarchical structure, the modes of parallelism for classifying machine architectures are selected. At the second level, finer architectural characteristics, such as the organization of the memory system, can be chosen. This hierarchical structure is organized in such a way that the architectural characteristics at any level are choices for a given category (e.g., type of interconnection network used).

DHSMS assigns a *code type* (i.e., computational characteristics) to each path from the root of the hierarchical structure to a leaf node. Every such path defines a set of architectural features corresponding to the nodes traversed by that path. Mathematically, a USC is defined as a set of code-types $C = \{C_i\}$, where $1 \le i \le K$ and K is the total number of paths from the root to a leaf node. Conceptually, each C_i represents the type of code (i.e., a benchmark program) that is ideally suited for the architectural features indicated by the ith path. The method for generating each benchmark program is still an open problem. Let $v_0(j)$ be the *size of the parallelism* (e.g., maximum possible number of concurrent threads of execution) in the given code block S_j, and let v_i $(1 \le i \le K)$ be a real number between 0 and 1 that indicates how well the code block S_j is matched with the code-type C_i. Then, a *task profiling vector* $\mathbf{V}_j$ for a given code block S_j is defined as $\mathbf{V}_j = [v_0(j), v_1(j), v_2(j), \ldots, v_K(j)]$. The estimation of $\mathbf{V}_j$ is also an open problem in the HC field.

To perform analytical benchmarking, let $b_q(n)$ be the speedup that machine q can achieve compared to a baseline system by executing an optimally matched benchmarking program with the size of parallelism equal to n. Then, analytical benchmarking can be formally defined as a vector $\mathbf{B}(n) = [b_q(n)]$, $q = 1, 2, \ldots, M$, where M is the number of machines available in the HC suite.

I/O benchmarking estimates the I/O overhead of a given architecture as a performance metric that is a function of the amount of data being transmitted through the I/O subsystem. *Network-interface profiles* estimate the overhead of the network due to the protocols for communication and media access. Let $d_q(a_m)$ be the destination-independent expected I/O and network-interface overhead of machine q, when there are a_m units of data transmitted through the mth edge of the data dependence graph of the original program. This can be determined by experimentation and extrapolation. Then, I/O benchmarking and network-interface profiles are defined by the *communication overhead vector* $\mathbf{D}(a_m) = [d_1(a_m), d_2(a_m), \ldots, d_M(a_m)]$. The value of a_m must be derivable from the original application program, be specified by the user as a compiler directive, or be estimated by executing sample data sets on the baseline system.

The information just described is combined into a *task-flow graph (TFG)*, which provides the execution time of each code block S_j on a baseline system and the amount of data transferred between code blocks due to data dependencies. A task profiling vector $\mathbf{V}_j$ is assigned to each code block S_j in the TFG, forming an intermediate *code-flow graph (CFG)*. The number of elements in $\mathbf{V}_j$ and the complexity of task profiling each depend on the number of levels of the hierarchical structure selected by the user. In the final CFG, each code block S_j in the intermediate CFG is associated with an estimated computation time vector $\mathbf{E}_j = [e_1, e_2, \ldots, e_M]$, where e_q is the estimated computation time of code block S_j on machine q and is a function of $\mathbf{V}_j$ and $\mathbf{B}(n)$.

The analytical benchmarking portion of Fig. 88.2 must also produce information about the time required for intermachine data transfers. Let $d^*_{p,q}(a_m)$ be the expected I/O and network-interface overhead when there are a_m units of data transmitted between machine p and machine q [this is a function of $d_p(a_m)$ and $d_q(a_m)$]. Then, in the final CFG, each communication link m between two code blocks in the original TFG is associated with a communication overhead matrix $D^*(a_m) = [d^*_{p,q}(a_m)]$, $1 \le p, q \le M$ (an asterisk is used to distinguish the communication overhead *matrix* D from the communication overhead *vector* D). Any data format conversion overhead also can be added to $d^*_{p,q}(a_m)$. The $M \times M$ matrix $D^*(a_m)$ is assumed to be symmetric along the diagonal. The final CFG can be used in matching and scheduling.

The DHSMS approach is extended in Yang et al. [1993] to include the generation of a representative set of templates (RST) that can characterize the execution behavior of the programs at various levels of detail. Many HC methodologies include a mathematical formulation for task profiling and analytical benchmarking that is similar in concept to that used in DHSMS (e.g., Chen et al. [1993], Freund [1989], Narahari et al. [1994], and Wang et al. [1992]).

In summary, two example frameworks for the task profiling and analytical benchmarking stage were overviewed. Complete practical methods for automatically performing task profiling and analytical benchmarking are still largely open problems. The information that must be generated by this stage is used in the matching and scheduling stage, considered in the next section.

88.5 Matching and Scheduling

For HC systems, in the most general case, matching involves deciding on which machine(s) each subtask should be executed and scheduling involves deciding when to execute a subtask on the machine(s) to which it is mapped [Tan et al. 1995]. Matching and scheduling comprise stage 3 of the model shown in Fig. 88.2.

Matching and scheduling techniques attempt to optimize an objective function subject to a given set of constraints. Minimizing the overall execution time under a cost constraint (based on total dollars available for purchase or usage of machines) or minimizing cost under a performance constraint are two commonly used formulations for HC systems [Chen et al. 1993, Freund 1989, Wang et al. 1992].

Mapping and scheduling research for parallel and distributed computing systems, which is closely related to matching and scheduling problems for HC systems, has focused on how to effectively execute multiple subtasks across a network of sequential processors (e.g., see Atallah et al. [1992]). An extensively studied subclass of problems within the domain of parallel and distributed systems is mapping and scheduling for homogeneous multicomputer (i.e., MIMD) platforms. In this case, the execution time of each subtask is independent of the processor to which it is mapped. For an overview of research in this related area, refer to Norman and Thanisch [1993].

Although some of the existing mapping and scheduling concepts and techniques can be (and have been) applied to matching and scheduling for HC systems, there is a fundamental distinction between mapping and scheduling subtasks for a network of sequential processors (e.g., a network of workstations) and matching and scheduling subtasks for an HC system consisting of various types of parallel computers (e.g., MIMD, SIMD, and vector). In the latter case, a subtask may execute most effectively on a particular type of parallel architecture and matching subtasks to machines of the appropriate type is a more important factor than merely balancing the load among all machines in the suite.

The matching and scheduling problems are typically interdependent. Thus, the solution quality of the best scheduling policy for a given (fixed) matching will generally be inferior to the overall optimal solution in which both scheduling and matching are considered simultaneously. Likewise, solving the matching problem for a given scheduling does not guarantee overall optimality in general.

In the remainder of this section, examples of formulations and techniques for matching and scheduling are overviewed. Throughout this section, only *static* approaches are discussed, in which the matching and scheduling policies are completely determined before execution begins. *Dynamic* approaches, which are beyond the scope of this discussion, have also been proposed, in which decisions are made or modified during the execution of the subtasks. Dynamic approaches are useful in situations where the execution times of the subtasks or the precedence relations among the subtasks are difficult to accurately predict before execution.

In Freund [1989], a mathematical programming formulation for selecting an optimal heterogeneous configuration of machines for a given set of problems under a fixed-cost constraint, known as *optimal selection theory (OST)*, is developed. OST represents one of the first mathematical formulations of the matching problem for HC systems. In the OST framework, program segments are referred to as *code segments*, and are assumed to be totally ordered in time. The OST framework does not explicitly account for times associated with transferring any required data among machines. Thus, the total execution time of the application is equal to the sum of the execution times for all its code segments.

A code segment is defined to be *decomposable* if it can be partitioned into different *code blocks* that can be executed on different machines of the same type concurrently. For simplicity, a linear speedup is assumed when a decomposable code segment is executed on multiple copies of a given machine type. The best matched machine type is assumed for executing each code segment. Also, a sufficient number of machines of each type is assumed to be available.

In the basic mathematical formulation of OST, there is assumed to be $N \geq 1$ code segments and $M \geq 1$ different machine types available for executing these code segments. The number of machines of type j to be used is defined as v_j, the cost of using a machine of type j is c_j, and the estimated time for executing code segment i on machine type j is given by $t_{i,j}$, for all $1 \leq i \leq N, 1 \leq j \leq M$. The formulated optimization problem involves the minimization of the total execution time T, defined subsequently, subject to a given constraint on the total cost of the machines to be used, C. In this formulation, v_j, $1 \leq j \leq M$, are the nonnegative integer-valued optimization variables, and the values of the parameters c_j and $t_{i,j}$ are assumed to be given for all i and j. Thus, the optimization problem associated with OST is stated as follows:

$$\min_{v_1, v_2, \ldots, v_M} \{T\}$$

$$T = \sum_{i=1}^{N} \min_{\substack{1 \leq j \leq M \\ v_j \neq 0}} \left\{ \frac{t_{i,j}}{v_j} \right\}$$

subject to

$$\sum_{j=1}^{M} v_j \geq 1 \quad \text{and} \quad \sum_{j=1}^{M} v_j c_j \leq C$$

The *augmented optimal selection theory (AOST)* [Wang et al. 1992], augments OST by incorporating the performance of code segments for all available machine type choices (not just the best matched machine type) and by assuming a fixed number of machines of each type are available (instead of a sufficient number of machines of each type). The issue of considering all available choices of machines is important in practice because the best matched machine may be unavailable. Also, in practice there may be only a limited number of machines of a given type available for executing a decomposable code segment (i.e., fewer machines than code blocks). A code block to machine assignment technique is proposed to address this situation. The proposed approach assigns code blocks to machines of a given type so as to minimize the time for the last machine to finish executing.

Another extension to this framework, *heterogeneous optimal selection theory (HOST)* [Chen et al. 1993], generalizes AOST by allowing concurrent execution of mutually independent code segments on different types of machines and incorporating the effects of local mapping choices. In general, there can be many possible ways to map a code block onto a given parallel machine. For example, consider the mapping of a code block for the multiplication of two matrices onto a distributed memory parallel machine. The rule used to assign elements of each matrix to the individual processing elements of the machine defines the mapping. In general, the mapping choice used can impact the overall execution time of the code block, e.g., see Siegel et al. [1996c]. For each code block, the HOST framework assumes that the best (minimum execution time) mapping choice is known for each available machine.

OST and the extensions just mentioned represent important frameworks for the matching and scheduling problem. However, one issue not directly accounted for in these frameworks is the possibility for significant communication delay associated with transferring shared data among machines. This issue was addressed in a further refinement to the OST framework known as the *generalized optimal selection theory (GOST)* [Narahari et al. 1994]. In GOST, the most basic code element is a *process*, which is a nondecomposable code segment. It is assumed that an application task consists of several processes modeled by a dependency graph, which is a *directed acyclic graph (DAG)*. Each node of the DAG represents a process and has a number of weights corresponding to the execution times of that process on each machine type for each known mapping onto that machine. Each edge of the DAG represents a dependency between two processes that require communication. Each edge has a number of weights (communication times), one for each reasonable communication path between each possible pair of host machines. An optimal matching and scheduling problem is formulated in which the objective is to assign each process node in the dependency graph to one machine type and to assign a start time so that the completion time of the entire application is minimized. Polynomial-time algorithms are proposed for solving the problem for two special types of DAGs. (This class of problems is known to be NP-complete for general DAGs [Ullman 1975].)

The two types of DAGs considered in [Narahari et al. 1994] are tree DAGs and series-parallel DAGs. A basic property of a *tree DAG* is that there is at most one directed path between every pair of nodes. To define *series-parallel DAGs*, first consider the smallest possible series-parallel DAG, which is two nodes connected with a single directed arc. In general, all series-parallel DAGs can be defined in a recursive manner by replacing an arc of a smaller series-parallel DAG with either a chain of arcs or multiple parallel arcs.

In Tao et al. [1993], three heuristics for matching and scheduling interacting subtasks (modeled by a general DAG) for a given HC suite of machines are presented. Similar to the GOST framework, it is assumed that the expected computation time of each subtask on each machine and the expected amount of communication associated with each arc in the DAG are known. It is also assumed that the bandwidth of each communication link in the HC system and the interface overhead between each pair of machines are known.

The three proposed heuristics are referred to as simulated annealing, tabu search, and stochastic probe. A randomly selected initial matching/scheduling solution is assumed at the beginning of each of these three heuristic searches. The basic actions of the proposed searches are called *moves*. A move involves swapping the current position (i.e., machine location and order of execution on the machine) of two code segments. Only moves that do not violate the precedence relations among the subtasks are allowed. All three algorithms operate by using moves to iteratively search through the space of possible solutions. The three approaches are distinguished by the rules they employ to define the conditions under which a move will be used to generate the next candidate solution in the iterative search for a good solution.

In the simulated annealing approach, moves are allowed that replace the solution of the current iteration with a solution that is of inferior quality (i.e., a solution that would generate a higher overall execution time for the subtasks). This is done in order to avoid entrapment in poor local optima. The likelihood of utilizing such moves in the simulated annealing approach decreases as the number of iterations increases. The tabu search technique compares the solution quality of the current solution with a number of distinct moves from the current solution. However, unlike the simulated annealing approach, the tabu search always selects the best move (i.e., the one that decreases the overall execution time of the subtasks the most) for the next iteration's solution. The stochastic probe technique is a hybrid approach that combines elements of both the greedy search process found in the tabu search and the stochastic nature of the simulated annealing approach. Simulations on randomly generated models were conducted to compare the solution quality and execution times among the three proposed approaches. It is reported that the stochastic probe heuristic yields the best solution quality with the least amount of required computation time.

Although communications/dependencies among subtasks are modeled using DAGs in the approaches developed in Narahari et al. [1994] and Tao et al. [1993], the possibility that a given data set may be available on multiple machines is not considered. This is because the identities of the data sets are not specified in these frameworks. In reality, a particular data set may be required input to several subtasks. For such a case, it is possible for a copy of the data set to exist on multiple machines during the execution of the subtasks. Thus, it is not necessary, in general, to require the data set to be fetched from the machine where it was originally generated (as is done by the two previously described approaches and other similar DAG-like approaches).

In Watson et al. [1994], a framework is developed in which the identity and locations of the data sets are considered. Thus, the time associated with executing a subtask on a particular machine depends on three factors: (1) the execution time of the subtask on the selected machine, (2) which machine(s) contain copies of the data sets the subtask requires as input, and (3) the sizes of the required input data sets. In general, a given machine may contain a data set because it was initially loaded there, it was received from another machine, or it was generated by that machine. There is assumed to be a sequence of subtasks to be executed that are totally ordered in time (as was the assumption for the code segments of the original OST framework). Under this proposed model, the location of the data sets at a given point in time depends on machine choices made for executing the subtasks prior to that point in time. Based on this framework, a matching heuristic for the case of a two-machine HC system is developed.

The matching heuristic of Watson et al. [1994] is conceptualized using a multistage optimization graph in which directed arcs connect each vertex at stage i to two vertices at stage $i + 1$. Each stage of the graph corresponds to one subtask in the sequence of subtasks to be executed. There are two vertices at each stage, which correspond to the two possible machine choices for executing the subtask associated with that stage. The four arcs from stage i to stage $i + 1$ are weighted according to the total time (communication plus computation) associated with executing subtask $i + 1$ on either machine, for each possible machine choice for subtask i.

The objective of the heuristic is to determine the shortest directed path through the multistage optimization graph. This problem is not equivalent to standard shortest path problems (e.g., Moore [1957]), because the weights of the arcs are not constants. In particular, the value of the arc weights connecting stage i to stage $i + 1$ depends on decisions made for all stages up to and including stage i. The heuristic considers the stages in order, and uses a data location table to track the locations of the data sets based on the matching decisions made up to stage i.

The heuristic determines a solution (i.e., a path) for an s-stage multistage graph in $s - 2$ iterations. At each iteration, three successive stages of the graph are reduced to a two-stage representation. For example, during the first iteration a minimum path is found from each vertex in the first stage to each vertex in the third stage (passing through the second stage). The second stage is then removed from the graph, and four new aggregate arcs are used between the first and third stage, representing the minimum cost paths to proceed from the first stage to the third stage for each source/destination pair. The weight of each aggregate arc, which corresponds to the execution time of the underlying path, is updated at each iteration. A data location table is associated with each aggregate arc, and these four tables are also updated at each iteration. These data location tables are used in the next iteration to determine the weights of the arcs that connect the terminal stage of the aggregate arcs to the next stage (based on the input data requirements for the subtask associated with the next stage). This procedure continues in the second and subsequent iterations until only a two-stage graph (with four aggregate arcs) remains after $s - 2$ iterations. The final four aggregate arcs are associated with four distinct paths through the original s-stage graph. The path associated with the aggregate arc having the minimum weight (of the final four) is the solution of the heuristic.

The performance of the heuristic was evaluated using simulated program behaviors. In the simulations, the subtask execution times and the input data set requirements for the subtasks were assigned randomly. The studies indicate that the proposed approach, which has a polynomial time complexity, typically produces assignments with overall execution times that are within 1% of the optimal assignments, which were determined using an exhaustive search that has an exponential time complexity. This research is currently being extended to more than two machines.

Although the field of HC is relatively new, some of the underlying problems in matching and scheduling have been addressed (at least to some extent) in other areas. In the related area of mapping and scheduling for multicomputers, it is noted in Norman and Thanisch [1993] that there is no commonly accepted framework whereby results in the field can be compared. Based on the existing published literature for matching and scheduling, this disturbing trend appears to be occurring in the HC community as well. For example, the heuristics developed in Tao et al. [1993] cannot be directly compared with the heuristic developed in Watson et al. [1994] because two distinct frameworks are used to model the required communications. Furthermore, even though the underlying models for communication used in Narahari et al. [1994] and Tao et al. [1993] are essentially the same, it is difficult to compare the proposed techniques because those developed in Narahari et al. [1994] are applicable for only special types of DAGs. Therefore, the establishment of a common framework within which researchers can compare their findings is an important issue to investigate in the near future.

88.6 Supporting Compiler and Operating System Mechanisms

As outlined in the previous sections, the automatic mapping of programs onto HC environments requires a variety of specialized techniques. However, such an HC software system can only exist if appropriate support software is available. This support generally falls into two categories: compiler technology that facilitates static HC analysis and operating system (OS) technology that provides the dynamic mechanisms required for effective HC execution.

Perhaps the most basic assumptions about compiler technology to enable the automatic use of HC systems are as follows.

- Given a program segment and two potential execution models, it is possible to statically determine which model will result in faster completion.
- Given an arbitrary machine-independent program segment and a target execution model, without user assistance, it is possible for a compiler to generate code implementing the program's functionality in that execution model.

Neither of these basic assumptions is literally valid; however, both are approximately true using the results of recent compiler research.

The static (compile-time) prediction of the execution time for a program on a particular machine is clearly unsolvable in the general case; it is a more complex variation on the classic halting problem. Fortunately, that is not exactly the problem an automatic HC software system needs to solve to determine where a program segment should be executed. The static timing information needed about a pairing of a program segment with a machine is used almost exclusively in comparison to the timing estimates for alternative pairings.

Compiler technology for precisely estimating the execution time of machine code instruction sequences generally involves complex models of pipelined overlap within a processor, as used in Marion [Bradlee et al. 1991]. The *parallel instruction generator generator (PIGG)* described in Cohen [1994], extends this type of pipeline timing model to obtain accurate timing estimates for instruction sequences executing on hardware that contains arrays and hierarchical collections of pipelined structures.

At a coarser granularity (often treating each statement as a single operation), a wide variety of techniques has been developed for approximate timing analysis of larger program segments. The basic approach presented by Sarkar [1989] involves hierarchical management of approximate costs using an intermediate-code graph structure, and this technique has been adapted for a number of parallelizing compilers. An alternative approach, used in AHS, involves translating the program into a generic instruction set, estimating the expected execution count for each type of generic instruction, and then computing the cross product of expected execution counts and the empirically obtained timing for each type of generic instruction on the target machine.

To summarize the comments on compiler timing analysis, techniques are available, but timing variations introduced by cache misses, communication network delays, etc., are not yet effectively modeled. These aspects of prediction are an important area for future research.

When considering code generation for an arbitrary target machine, language design can dramatically affect the degree to which it is practical for a compiler to generate code for an arbitrary execution model. However, there are a variety of specific compiler technologies needed even for the best designed languages.

Compiler generation of code for a wide range of sequential target machines can be accomplished using *template-driven code generator generators*. The front-end of such a compiler first creates a target-independent intermediate representation of the program; typically a variation on abstract syntax trees (ASTs), such as the GNU C Compiler's register transfer language (RTL), is used. Each potential target machine is described by a set of templates that act as a grammar that can be applied to parse the intermediate code generated by the front-end. Each template provides both an intermediate code pattern to match and the information needed to generate the target machine instruction(s) that implement the functionality of that pattern. Thus, modifying a compiler to generate code for a new target machine is simply a matter of creating an appropriate set of templates.

Although template-driven code generation methods can effectively manage the instruction set differences among most sequential machines, adaptation of these methods to manage differences among parallel target machines has proven difficult. One approach is to use an intermediate representation that encapsulates parallelism within each operation in the intermediate form; a good example of this is the vector-oriented intermediate form VCODE [Blelloch and Chatterjee 1990]. VCODE allows conventional template-driven code generation methods to generate code for a wide range of parallel machines, but the resulting code only uses parallelism *within* an individual intermediate vector operation, not across multiple operations. Essentially this same approach is used in many source-to-source translators for parallel languages, where each vector operation is translated into a library call that provides a parallel implementation.

The difficulty in building compilers effectively targeting a wide range of parallel machines lies in the fact that the changes required to restructure parallelism for a different target are often global, rather than localized to a small portion of the intermediate form. It is thus necessary to apply a wide range of sophisticated analysis and transformation techniques to the intermediate form. A good introduction to some of the basic analysis and transformation methods is given by Wolfe [1996], but development of new methods and improvements is at the core of ongoing compiler research. The more insidious problem is that it currently takes a great deal of time and compiler-writing expertise to determine which techniques should be applied in what order for each new parallel target machine. To develop compilers for new parallel machines in a timely manner, a number of researchers are investigating methods to automatically select and order compiler transformations [Cohen 1994].

Although lack of compiler technology of specialized hardware devices may make it impossible to execute certain program segments or program constructs on particular target machines within an HC system, it is useful to note that an automatic HC software system can still function by simply marking such program segments and machine pairings as illegal choices. For example, AHS ensures that a program segment will not be assigned to a machine for which no coding can be generated by using infinity as the expected execution time for that pairing.

Just as the automatic use of HC implies a specific type of compiler support, it also requires a variety of operating system mechanisms. The execution time of a program segment on a particular machine is not just a function of that code and the machine, but also of the load placed on that machine by other processes. The *load average* is a multiplicative factor that indicates how much longer a typical program might take to execute now than if the program was run by itself. Ideally, one might like to automatically update the load average information just before deciding where to run each user program. However, this may be impractical because one needs to obtain the load average for every machine, and a typical HC program run might only use a small fraction of the many machines available. Further research is needed to determine how and when to update tracking of the load average. There is also the complication that, for parallel machines, the concept of load average might not be sufficient.

Once a set of machines within an HC system has been selected for the execution of a program, there is also the need for operating system mechanisms that can initiate and control the execution of the program across these machines. Perhaps the simplest mechanism for such control is the Unix program **rsh**, which initiates a program on a remote machine through a socket connection. Systems like SmartNet and PVM provide a variety of HC system process control functions that essentially duplicate some of the basic operations that Unix supports within a single machine: **ps** (process status), **kill** (terminate a process), etc. There are, however, a variety of complications deriving from the fact that a single HC program can involve many processes on many machines, thus leading to a somewhat cumbersome textual interface. For example, the status of a sequential program running under Unix is easily summarized on a single line, but a similar one-line summary of each program segment executing in an HC environment could easily yield several pages worth of information. The complexity of this interface becomes particularly obvious when one is attempting to debug an HC program using a conventional text-oriented symbolic debugger. Design of better, often graphical, collective debugging tools is a major topic of operating systems research.

Last, but perhaps most important, of the basic operating systems support mechanisms needed is high-performance communication among the machines. All of the operating system features critically rely on intermachine communication, and the minimum usable grain size for an HC program segment is essentially a function of the communication support available.

A wide range of networking hardware may be used in connecting the machines of an HC system: Internet, Ethernet, fiber distributed data interface (FDDI), asynchronous transfer mode (ATM), high-performance parallel interface (HiPPI), or even specialized hardware such as PAPERS [Dietz et al. 1994]. Despite huge differences in hardware structure and performance, the interesting fact is that the type of network hardware seems to have less impact on the HC system network performance than does the software interface.

There are basically two common classes of communication.

- When a large block of data (or a program segment object file) must be transmitted, the primary concern is network bandwidth. In such a case, a high setup overhead is acceptable provided that the data block transmission will occur at near peak bandwidth. Because most conventional networks were designed for block transfers, there is generally little trouble in using the standard network interface mechanisms. For example, PVM's communication model is deliberately based on filling buffers and then sending the entire buffer as a single block; thus, PVM works fairly well for such transfers.

- When short control messages must be sent among the machines of an HC system, the primary concern is minimization of latency. Although the hardware latency of most local area networks is a surprisingly small number of microseconds, using the standard block-oriented software interface often takes on the order of a millisecond [Dietz et al. 1994]. A better software interface is needed. One approach is based on the concept of *active messages* [Von Eicken 1993], in which short messages, each carrying a tuple of function identifier and operands, are sent between machines. Ideally, each transmission is accomplished by the program segment directly polling network hardware device registers; thus, the software overhead of OS calls and socket protocols, and even hardware interrupt latency, is avoided. Direct polling is also used in the PAPERS Library [Dietz et al. 1994]; it requires custom network hardware, but provides both conventional point-to-point communication and a variety of collective operations as single communications.

Of all of the supporting compiler and operating system mechanisms needed, the lack of HC network support in the form of appropriate operating system interfaces and compiler technology for time estimation seems to be the most serious impediment to widespread use of HC. HC networking requirements are different enough from conventional networking needs that sharing the same mechanisms, interfaces, and models works only for the most coarse-grain parallel computations.

88.7 Example Tools

This section overviews examples of software tools that exist or are being developed for HC systems. The tools discussed here aid primarily with stages 3 and 4 of the software support model shown in Fig. 88.2. The functionalities of the tools described in this section tend to evolve and change rapidly; the descriptions here are based on the references given. Some early prototype tools and frameworks have also been developed to aid with stages 1 and 2 of Fig. 88.2. Two examples of such tools are PAWS and DHSMS, which were overviewed in the "Task Profiling and Analytical Benchmarking" section. A survey of related work in distributed queueing and clustering systems, some of which can be applied to HC, is given in Kaplan and Nelson [1993].

Linda was originally implemented for *homogeneous* computing environments such as shared memory parallel computers (e.g., the Sequent Symmetry), distributed memory computers (e.g., the Intel iPSC/2), and local area networks (e.g., a network of workstations). As suggested in Carriero et al. [1992], it is an attractive choice for HC systems as well. In Linda, processes communicate via persistent objects called *tuples*, and not through transient events such as message passing or procedure calls. A process can generate a tuple and place it in a globally shared collection of tuples, the *tuple space*. Tuples can be also removed, read, and evaluated from the tuple space. *Process tuples* incorporate executable code and *data tuples* are passive, ordered collections of data items [Butler et al. 1993]. Although the current version of Linda does not support concurrent interaction among machines in an HC system, the issues that must be resolved to do this are outlined and discussed in Carriero et al. [1992].

Portable programs for parallel processors (p4) is a set of parallel programming tools designed to support portability across a wide range of architectures [Butler and Lusk 1994]. The p4 includes high-level operations that allow certain procedure calls to be replaced with the equivalent p4 calls that are implemented by utilizing system-specific procedures. The long-term goal of this project is to allow a single program to be written for an entire class of systems (e.g., message passing) without requiring the explicit utilization of constructs of the specific system (e.g., Intel Paragon vs. nCUBE 2) in the source code. The p4 function library is linked with the source code to provide functions for message passing, monitors for shared memory, process management, debugging, and language interfacing. Also, p4 can support communication within and among both shared-memory and message-passing machines.

The developers of p4 stress that it is not an abstract tool and that various components of p4 evolved through the development of real applications. As an example, p4 was used in developing a piezoelectric crystal simulation program to coordinate the computations, communications, and I/O among an Intel Touchstone Delta, a Stardent Titan, and a Solbourne workstation. Current and future research directions for p4 include the implementation of Linda with p4 to provide a single high-level programming model [Butler et al. 1993].

The tool *Mentat* has both execution time support facilities and language abstractions that provide a clear separation between the user's application and the target machine [Grimshaw et al. 1994]. This separation is achieved by using an object-oriented language to specify parallelism within the application and compiler technology to handle many of the tedious and time-consuming bookkeeping tasks. Mentat combines a medium-grain dataflow computation model with the object-oriented programming paradigm to produce a system that facilitates hierarchies of parallelism [Grimshaw 1993]. Programs are characterized as directed graphs. The vertices represent computational elements (e.g., class member functions) and the edges model data dependencies between these elements. The idea behind Mentat is to allow the programmer to express the problem in a C++ based language, called *Mentat programming language (MPL)*, which provides many popular features of the C++ language. Mentat uses the dataflow model to exploit the inherent medium-grain parallelism of the program; in addition, the programmer can specify those C++ classes which are themselves of sufficient computational complexity to warrant parallel execution [Grimshaw 1993].

The *run-time system (RTS)* of Mentat, which initially supported execution on homogeneous parallel machines, has been extended to support HC systems. The RTS uses *a virtual macro-dataflow machine* that provides support routines to perform execution time data dependence detection, program graph construction, program graph execution, scheduling, communication, and synchronization. The virtual macro-dataflow machine contains a set of machine-independent components and libraries and a set of machine-dependent components. The virtual macro-dataflow machine can be ported to any supported machine in the HC system by changing only the machine-dependent components, which allows the user to port the application source code to any supported machine without changes.

Experimental studies have been conducted in which matrix multiplication and Gaussian elimination programs were coded in MPL and executed on a network of eight Sun workstations and a 32-node iPSC/2. Although MPL improved the ease of use of the HC system, it was indicated that the performance may not be as good as hand-coded versions that use send and receive protocols. Thus, there is a tradeoff between ease of use and performance. Future work includes the implementation of several optimizations for the MPL compiler.

PVM is a software system that enables an HC system to be used as a coherent, flexible, and concurrent computational resource [Beguelin et al. 1993, Sunderam 1990]. The PVM package includes system-level daemons, called pvmds, which reside on each computer in the HC system, and a library of PVM interface routines.

The pvmds provide services to both local processes and remote processes on other platforms in the HC system. Together, the entire collection of pvmds forms what is called a virtual machine by enabling the HC system to be viewed as a single metacomputer. Three of the major services provided by the pvmds are: process and virtual machine management, communication, and synchronization. Process and virtual machine management involves issues such as computational unit scheduling and placement, configuration and inclusion of remote computers into the host pool, and naming and addressing of computing and resource entities. Communication is accomplished via the use of messages, which are exchanged asynchronously so that a sending process may continue execution without waiting for an acknowledgment from the receiving process. The third major service provided is the synchronization among processes, which can be accomplished by using barriers or by using event rendezvous. The synchronizations may be among multiple processes that are executing on a local machine or may be among processes on different machines.

The second part of the PVM package is a library of interface routines. Applications to be executed on one or more computing platforms in the HC system are able to access these platforms via library calls embedded in imperative procedural languages such as C or FORTRAN. The library routines interact with the pvmd (resident on each machine) to provide services such as communication, synchronization, and process management. The pvmd may provide the requested service alone or in cooperation with other pvmds in the HC system.

X-window analysis and debugging (Xab) is a tool developed for the execution time monitoring of PVM programs [Beguelin et al. 1993]. The Xab tool gives the user direct feedback on what PVM functions the program is executing and how the program is performing in an HC environment. Xab consists of three parts: the Xab library, which contains instrumented PVM routines that are linked to the user's code; a special monitoring process called admon, which receives trace messages from the library routines; and a frontend process, which graphically displays trace events.

Heterogeneous network computing environment (HeNCE) aids users of PVM in decomposing their application into subtasks and deciding how to allocate these subtasks onto the available machines in the HC system [Beguelin et al. 1993]. In HeNCE, the programmer explicitly specifies the parallelism for an application by drawing a directed graph, where nodes represent subtasks written in either FORTRAN or C, and arcs represent dependencies and flow control. There are also four types of control constructs: conditional, looping, fan out, and pipelining.

The user must specify a cost matrix, which represents the cost of executing each subtask on each machine in the HC system. The meaning of the cost parameters is defined by the user (e.g., estimated execution times or utilization costs in terms of dollars). At execution time, HeNCE uses the cost matrix to estimate the most cost effective machine on which to execute each subtask. The current version of HeNCE does not include information to estimate when machines will become available nor the time required for intermachine communications.

Given the graph and the matrix, HeNCE configures a subset of the machines defined in the cost matrix as a virtual machine using PVM constructs. Then HeNCE begins execution of the program. Each node in the graph is realized by a distinct process on some machine. The nodes communicate with each other by sending parameter values needed for execution of a given node, which are specified by the user for each node (subtask). A node obtains parameter values needed to begin execution from predecessors, nodes.

If the immediate predecessors do not have all of the required parameters for a node, earlier predecessors are checked until all required parameters are found. Then the node (subtask) is executed and passes the appropriate parameters onto descendant nodes. HeNCE can trace the execution of the application for display in real time or replay later.

Cluster-M is a recently proposed portable programming tool that is also implementable on an HC suite of computers [Eshaghian and Chen 1995, Eshaghian et al. 1995]. Cluster-M has three main modules: specification, representation, and mapping. In the specification module, machine-independent algorithms are specified based on the Cluster-M portable programming paradigm, and are coded using a portable programming language. Specifications are representable in a form of a multilevel clustered task graph called the *Spec* graph, where the nodes are subtasks and include relative expected execution times, and arcs are weighted by the expected amount of data to be transferred between the subtasks. Similarly, in the representation module, an underlying architecture or an HC suite of computers is represented in a form of a multilevel partitioning of a system graph, called the *Rep* graph, where a node contains the speeds for arithmetic operations for the associated processor, and arcs express the bandwidth for communications between processors. Given an arbitrary Spec graph containing M task modules, and an arbitrary Rep graph of N processors, the mapping module is a portable heuristic tool responsible for a near-optimal mapping of the two graphs in $O(MP)$ time, where $P = \max\{M, N\}$. The mapping module has an interface that can be used with portable network communication tools, such as PVM, for executing portable parallel software across HC suites.

SmartNet is a matcher/scheduler for HC systems that is being designed and developed at NRaD (a Naval laboratory) [Hensgen et al. 1995]. SmartNet's goal is to optimize scheduling criteria, such as minimizing the total time to execute a set of tasks, taking into account the predicted times to compute each task i on each machine j, as well as the latency time for any needed data transfers involved in computing task i on machine j. Its methods are global, general-purpose, scalable, and tunable. SmartNet's algorithms assume that each machine may have previously scheduled tasks either executing or awaiting execution. SmartNet can be configured to achieve a variety of objectives. For example, one objective can be to maximize the overall throughput, rather than to minimize the compute time for any specific task. SmartNet is currently operational and performs the functions previously discussed. It can operate dynamically at execution time, or prior to execution for planning purposes. Various extensions to SmartNet are under development. One is refining SmartNet's ability to use any and all information supplied about affinities for tasks to resources, including a priori, compile-time, run-time, experiential, and user input. Another is adding features so that SmartNet can be used for matching and scheduling communicating subtasks of a given task, such as tracking the location of and use of shared data sets.

Automatic heterogeneous supercomputing (AHS) is a tool that offers some level of support to stages 2, 3, and 4 of the model of Fig. 88.2 (in contrast to the other tools discussed in this section, which support only stages 3 and 4) [Dietz et al. 1993]. AHS uses a method to predict the execution time of subtasks that is a practical form of task profiling and analytical benchmarking. In this approach, data-dependent loop parameters and conditional branch probabilities are approximated by constant values. AHS can use information about the current load on a machine to appropriately weight the expected execution time to account for the load. AHS can estimate the execution time of a specific application program on a group of networked sequential Unix machines (it is not limited to a single machine). The intermachine data transfers are handled by asynchronous communication through a user datagram protocol (UDP) socket. AHS can generate the code for intermachine communication automatically. A proof-of-concept functioning AHS prototype for the MasPar MP-1 and some Unix-based workstations has demonstrated the usefulness of this approach.

88.8 Research Issues and Summary

Before HC can be made available to the average applications programmer in a transparent way, there are numerous challenging open problems that need to be solved in the area of software support. Some progress needs to be made on most of these problems just to facilitate near-optimal practical use of HC systems in a user-specified way. When working on these problems, one needs to consider what information is reasonable to expect from the user and what information can be determined automatically. This section includes a representative list of some of the open problems in software support for HC to convey the types of issues that need to be addressed. These problems are based on Siegel et al. [1996a]; other problems may be found in Khokhar et al. [1993].

To program an HC system, it would be best to have one or more machine-independent programming languages that allow the user to augment the code with compiler directives. The programming language and user-specified directives should be designed to facilitate (1) the compilation of the program into efficient code for any of the machines in the suite, (2) the decomposition of tasks into homogeneous subtasks, (3) the determination of computational requirements of each subtask, and (4) the use of machine-dependent subroutine libraries.

There is also a need for debugging and performance tuning tools that can be used across an HC suite of machines. This involves research in the areas of distributed programming environments and visualization tools.

To implement an automatic HC programming environment, such as the one proposed in the "Underlying Issues" section, a great deal of research for devising practical and theoretically sound methodologies for each component of each stage is required. Furthermore, information about the current loading and status of the machines in the HC suite and the intermachine network ideally should be incorporated into the matching and scheduling decisions. There are many related questions that need to be addressed. What information should be included in the machine/network status (e.g., faulty or not, pending tasks)? How should the current machine/network loading be measured? How can current machine/network loading and status information be incorporated effectively into matching and scheduling decisions? What is the best way to communicate to and structure in machines the loading and status information from other machines and about the network? What is the best way to estimate task/transfer completion time? How often should all this information be updated?

HC-specific operating system support is also needed. This includes techniques for finishing one subtask on a given machine, transferring shared data as needed, and then starting the next subtask on another machine. This occurs at both the local machine level and at the systemwide level.

Methods for dynamic task migration of an executing subtask between different parallel machines due to a machine fault or a need to load rebalance is another area of research where more work is needed. Current research in this area involves how to move an executing subtask between different machines and determining how and when to use dynamic task migration for load rebalancing or fault tolerance.

Research is needed in the area of software support for intermachine data transport, including topics such as the software protocols required, computing the minimum time path between two machines, and devising rerouting schemes in case of faults or heavy loads. Related to this is the data reformatting problem, involving issues such as data type storage formats and sizes, byte ordering within data types, and machines' network-interface buffer sizes.

Finally, there are numerous administrative issues that require software support. These include what to do with priority tasks, what to do with priority users, what to do with interactive tasks, and security.

In summary, even if it is assumed that machines with *peak* high performance can be constructed, achieving *sustained* high performance on real applications is still a very difficult problem. Based on the history of supercomputers, it is reasonable to expect a variety of different architectural designs to be represented in the set of future high-performance machines. Thus, one possible approach to providing sustained high performance on real applications is through the use of HC, where each subtask of an application can be matched to the machine that can execute it most effectively, and multiple machines can be used concurrently to process a single application.

While the uses of existing HC systems demonstrate the significant benefit of HC, the amount of effort currently required to implement an application on an HC system can be substantial. Future research on the open problems just mentioned in the area of software support for HC will improve this situation, will make the use of HC more viable, and will allow HC to realize its inherent potential.

Acknowledgments

The authors thank Arif Ghafoor, Debra Hensgen, Taylor Kidd, Janet M. Siegel, Vaidy Sunderam, Min Tan, and Jaehyung Yang for their assistance. This work was supported in part by Rome Laboratory under grant F30602-96-1-0098 and by NRaD under contract N66001-6142-4235.

Defining Terms

Analytical benchmarking: The quantification of how effectively each machine in an HC suite can perform different categories of computation (i.e., code types). The fact that a given high-performance machine can achieve near-peak performance for only a relatively small set of code types is the underlying motivation for HC. Although some general frameworks and tools for the process of analytical benchmarking have been proposed, more research is needed to achieve complete and effective automation of this process.

Execution model: This view of computation is based on low-level machine-specific constructs (e.g., native assembly language). A related model, called the programming model, attempts to abstract or hide the underlying complexity of the execution model in order to facilitate the development of application code (see programming model definition). For HC, an important challenge is to develop a programming model that can hide as much underlying complexity as possible without preventing effective compilation onto the target execution model.

Heterogeneous computing (HC): The use of a variety of architectural capabilities, orchestrated to perform an application whose subtasks have diverse execution requirements. Three types of HC include mixed-machine, mixed-mode, and mixed-component. Mixed-machine HC is the use of a heterogeneous suite of independent machines of different types that are interconnected by a high-speed network. A mixed-mode HC system is a single parallel processing machine whose processors are capable of operating in either the synchronous SIMD or asynchronous MIMD mode of parallelism, and the processors can dynamically switch between modes. A mixed-component HC system is a single machine where each separate component represents one mode of parallelism, and two or more distinct modes are present in the machine. The focus of this chapter is on software challenges for mixed-machine HC systems.

Matching: The process of assigning subtasks to machines in an HC suite. This process utilizes information generated by task profiling and analytical benchmarking. The matching problem is often formulated as a mathematical optimization problem, where the goal is to minimize a desired objective function (e.g., overall execution time) subject to a set of constraints (e.g., total cost of utilizing or purchasing machines). Matching and scheduling (see scheduling definition) are strongly interrelated. In general, finding an optimal matching and scheduling is an NP-complete problem, and much research is being conducted on heuristic approaches.

Program segment: An encoding of a subtask (see subtask definition).

Programming model: This represents the programmer's view of how computation is performed based on a particular high-level language. In general, the programming model hides machine-specific details of how execution is actually performed on a given platform. Compilation is the process of mapping programming model constructs into implementations using the execution model of a given platform (see execution model definition).

Scheduling: The process of ordering the execution of subtasks assigned to each machine. This process typically assumes that the precedence relations among the subtasks and the execution times of the subtasks are known (or have been estimated). Typically, the objective of the scheduling problem is to minimize the overall execution time of the total task. The solution quality of the scheduling problem depends on how the subtasks are assigned to the machines (see matching definition).

Subtask: A portion of computation associated with a given application task (see program segment definition). For current HC applications, subtasks are typically defined by the programmer. Ideally, for automatic HC, subtasks may be determined through the process of task profiling (see task profiling definition). For this case, the computational requirements of each subtask should be relatively homogeneous.

Task profiling: The process of decomposing an application task into subtasks, each of which is homogeneous with respect to computational requirements, and characterizing those computational requirements. In general, a task and its subtasks consist of both code and data; therefore, both code and data must be considered during the task profiling process. In current uses of HC, tasks are typically decomposed by the application programmer. The complete automation of task profiling is a difficult problem, and represents a long-term goal in the field of HC.

References

Armstrong, J. B. and Siegel, H. J. 1995. Dynamic task migration from SIMD to SPMD virtual machines, pp. 326–333. In *Proc. 1st IEEE Int. Conf. Eng. Complex Comput. Syst.* Nov.

Armstrong, J. B., Siegel, H. J., Cohen, W. E., Tan, M., Dietz, H. G., and Fortes, J. A. B. 1994. Dynamic task migration from SPMD to SIMD virtual machines, pp. 160–169. In *Proc. Int. Conf. Parallel Process.* Vol. II, Aug.

Atallah, M. J., Black, C. L., Marinescu, D. C., Siegel, H. J., and Casavant, T. L. 1992. Models and algorithms for coscheduling compute-intensive tasks on a network of workstations. *J. Parallel Distributed Comput.* 16(4):319–327.

Beguelin, A., Dongarra, J., Geist, A., Manchek, R., and Sunderam, V. 1993. Visualization and debugging in a heterogeneous environment. *IEEE Comput.* 26(6):88–95.

Blelloch, G. E. and Chatterjee, S. 1990. VCODE: a data-parallel intermediate language, pp. 471–480. In *Proc. 3rd Symp. Frontiers Massively Parallel Comput.* Oct.

Bradlee, D. G., Henry, R. R., and Eggers, S. J. 1991. The Marion system for retargetable instruction scheduling, pp. 229–240. In *Proc. ACM SIGPLAN Conf. Programming Lang. Design Implementation.* June.

Butler, R. M., Leveton, A. L., and Lusk, E. L. 1993. p4-Linda: a portable implementation of Linda, pp. 50–58. In *Proc. 2nd Int. Symp. High Performance Distributed Comput.* July.

Butler, R. M. and Lusk, E. L. 1994. Monitors, messages, and clusters: the p4 parallel programming system. *Parallel Comput.* 20:547–564.

Carriero, N., Gelernter, D., and Mattson, T. G. 1992. Linda in heterogeneous computing environments, pp. 43–46. In *Proc. Workshop Heterogeneous Process.* March.

Chen, S., Eshaghian, M. M., Khokhar, A., and Shaaban, M. E. 1993. A selection theory and methodology for heterogeneous supercomputing, pp. 15–22. In *Proc. Workshop Heterogeneous Process.* April.

Cohen, W. E. 1994. *Automatic Construction of Optimizing, Parallelizing, Compilers from Specifications*, Ph.D. dissertation, School of Electrical and Computer Engineering, Purdue University, West Lafayette, IN, Dec.

Dietz, H. G., Cohen, W. E., and Grant, B. K. 1993. Would you run it here … or there? (AHS: automatic heterogeneous supercomputing), pp. 217–221. In *Proc. Int. Conf. Parallel Process.* Vol. II, Aug.

Dietz, H. G., Cohen, W. E., Muhammad, T., and Mattox, T. I. 1994. Compiler techniques for fine-grain execution on workstation clusters using PAPERS. In *Languages and Compilers for Parallel Computing*, K. Pingali, U. Banerjee, D. Gelernter, A. Nicolau, and D. Padua, eds., pp. 31–45. Springer–Verlag, New York.

Dietz, H. G. and Krishnamurthy, G. 1993. Meta-state conversion, pp. 47–56. In *Proc. Int. Conf. Parallel Process.* Vol. II, Aug.

Eshaghian, M. M. and Chen, S. 1995. A fast recursive mapping algorithm. *Concurrency: Pract. Exp.* Spec. Issue Resource Manage. Parallel Distributed Syst. 7(5):391–409.

Eshaghian, M. M., Chen, S., and Wu, Y. 1995. Mapping arbitrary non-uniform task graphs onto arbitrary non-uniform system graphs, pp. 191–195. In *Proc. Int. Conf. Parallel Process.* Vol. II, Aug.

Freund, R. F. 1989. Optimal selection theory for superconcurrency, pp. 699–703. In *Proc. Supercomput. '89*, Nov.

Freund, R. F. and Siegel, H. J. 1993. Heterogeneous processing. *IEEE Comput.* 26(6):13–17.

Ghafoor, A. and Yang, J. 1993. Distributed heterogeneous supercomputing management system. *IEEE Comput.* 26(6):78–86.

Grimshaw, A. S. 1993. Easy-to-use object-oriented parallel processing with Mentat. *IEEE Comput.* 26(5):39–51.

Grimshaw, A. S., Weissman, J. B., West, E. A., and Loyot, E. 1994. Meta systems: an approach combining parallel processing and heterogeneous distributed computing systems. *J. Parallel Distributed Comput.* 21(3):257–270.

Hensgen, D., Moore, L., Kidd, T., Freund, R., Keith, E., Kussow, M., Lima, J., and Campbell, M. 1995. Adding rescheduling to and integrating Condor with SmartNet, pp. 4–12. In *Proc. Heterogeneous Comput. Workshop.* April.

HPFF. 1993. *High Performance Fortran Language Specification, Version 1.0,* High Performance Fortran Forum, Rice University, Houston, TX, May 3.

Kaplan, J. A. and Nelson, M. L. 1993. *A Comparison of Queueing, Cluster and Distributed Computing Systems,* NASA Tech. Mem. 109025. p. 47. Langley Research Center, Hampton, VA, Oct.

Khokhar, A., Prasanna, V. K., Shaaban M., and Wang, C. L. 1993. Heterogeneous computing: challenges and opportunities. *IEEE Comput.* 26(6):18–27.

Moore, E. F. 1957. The shortest paths through a maze, pp. 285–292. In *Proc. Int. Symp. Theory Switching.*

Narahari, B., Youssef, A., and Choi, H. A. 1994. Matching and scheduling in a generalized optimal selection theory, pp. 3–8. In *Proc. Heterogeneous Comput. Workshop.* April.

Norman, M. G. and Thanisch, P. 1993. Models of machines and computation for mapping in multicomputers. *ACM Comput. Surv.* 25(3):263–302.

Pease, D., Ghafoor, A., Ahmad, I., Andrews, D. L., Foudil-Bey, K., Karpinski, T. E., Mikki, M. A., and Zerrouki, M. 1991. PAWS: a performance evaluation tool for parallel computing systems. *IEEE Comput.* 24(1):18–29.

Pruyne, J. and Livny, M. 1996. Interfacing Condor and PVM to harness the cycles of workstation clusters. *J. Future Generations Comput. Syst.* 12(1):67–85.

Sarkar, V. 1989. *Partitioning and Scheduling Parallel Programs for Multiprocessors.* MIT Press, Cambridge, MA.

Siegel, H. J., Antonio, J. K., Metzger, R. C., Tan, M., and Li, Y. A. 1996a. Heterogeneous computing. In *Parallel and Distributed Computing Handbook,* A. Y. Zomaya, ed., pp. 725–761. McGraw–Hill, New York.

Siegel, H. J., Armstrong, J. B., and Watson, D. W. 1992. Mapping computer-vision-related tasks onto reconfigurable parallel processing systems. *IEEE Comput.* 25(2):54–63.

Siegel, H. J., Maheswaran, M., Watson, D. W., Antonio, J. K., and Atallah, M. J. 1996b. Mixed-mode system heterogeneous computing. In *Heterogeneous Computing,* M. M. Eshaghian, ed., pp. 19–65. Artech House, Norwood, MA.

Siegel, H. J., Wang, L., So, J. J., and Maheswaran, M. 1996c. Data parallel algorithms. In *Parallel and Distributed Computing Handbook,* A. Y. Zomaya, ed., pp. 466–499. McGraw–Hill, New York.

Sunderam, V. S. 1990. PVM: a framework for parallel distributed computing. *Concurrency: Pract. Exp.* 2(4):315–339.

Tan, M., Antonio, J. K., Siegel, H. J., and Li, Y. A. 1995. Scheduling and data relocation for sequentially executed subtasks in a heterogeneous computing system, pp. 109–120. In *Proc. Heterogeneous Comput. Workshop.* April.

Tannenbaum, T. and Litzcow, M. 1995. The Condor distributed processing system. *Dr. Dobbs J.* (Feb.): 40–48.

Tao, L., Narahari, B., and Zhao, Y. C. 1993. Heuristics for mapping parallel computations to heterogeneous parallel architectures, pp. 36–41. In *Proc. Workshop Heterogeneous Process.* April.

Ullman, J. 1975. NP-complete scheduling problems. *J. Comput. Syst. Sci.* 10:384–393.

Von Eicken, T. 1993. *Active Messages: An Efficient Communication Architecture for Multiprocessors,* Ph.D. dissertation, University of California, Berkeley, Nov.

Wang, M., Kim, S., Nichols, M. A., Freund, R. F., Siegel, H. J., and Nation, W. G. 1992. Augmenting the optimal selection theory for superconcurrency, pp. 13–22. In *Proc. Workshop Heterogeneous Process.* March.

Watson, D. W., Antonio, J. K., Siegel, H. J., and Atallah, M. J. 1994. Static program decomposition among machines in an SIMD/SPMD heterogeneous environment with non-constant mode switching costs, pp. 58–65. In *Proc. Heterogeneous Comput. Workshop.* April.

Weems, C. C., Riseman, E. M., and Hanson, A. R. 1992. Image understanding architecture: exploiting potential parallelism in machine vision. *IEEE Comput.* 25(2):65–68.

Weems, C. C., Weaver, G. E., and Dropsho, S. G. 1994. Linguistic support for heterogeneous parallel processing: a survey and an approach, pp. 81–88. In *Proc. Heterogeneous Comput. Workshop.* April.

Wolfe, M. 1996. *High Performance Compilers for Parallel Computing,* Addison–Wesley, Redwood City, CA.

Yang, J., Ahmad, I., and Ghafoor, A. 1993. Estimation of execution times on heterogeneous supercomputer architecture, pp. 219–225. In *Proc. Int. Conf. Parallel Process.* Vol. I, Aug.

Further Information

Since 1992, the annual *Proceedings of the Heterogeneous Computing Workshop* has been published by the IEEE Computer Society Press. These proceedings document the latest developments in the field of HC each year. This workshop is the main meeting focused on HC, and is held annually in conjunction with the IEEE International Parallel Processing Symposium.

Two journals have devoted special issues to the field of HC. The June 1993 issue of *IEEE Computer* contains articles related to practical frameworks and tools for HC. The June 1994 issue of *Journal of Parallel and Distributed Computing* contains a broad spectrum of papers on HC, from fundamental concepts, modeling, and techniques, to practical implementations. In addition to special issues, HC and related areas of research are covered periodically in these and other journals and conference proceedings, including: *IEEE Transactions on Computer, IEEE Transactions on Parallel and Distributed Systems, Proceedings of the IEEE International Parallel Processing Symposium,* and *Proceedings of the International Conference on Parallel Processing.*

An overview of the state of the art of HC, including examples of applications that have been implemented on HC systems, is presented in the chapter "Heterogeneous Computing," in the *Parallel and Distributed Computing Handbook,* A. Y. Zomaya, ed., 1996, McGraw–Hill, New York. A recent book devoted to the subject of HC is *Heterogeneous Computing,* M. M. Eshaghian, ed., 1996, Artech House, Norwood, MA. Each chapter covers a single HC topic in considerable detail, and is coauthored by researchers who specialize in that topic.

Security and Privacy Issues in Computer and Communication Systems

89.1 Introduction

This chapter provides an introduction to the concepts of security and privacy in computer-communication systems. Definitions tend to vary widely from one system to another and from one application to another. The definitions used here are intuitively motivated and generally consistent with common usage without trying to be overly precise.

Security is usually considered as encompassing three primary attributes, confidentiality, integrity, and availability, with respect to various information entities such as data, programs, access control parameters, cryptographic keys, and computational resources (including processing and memory). Confidentiality implies that information cannot be read or otherwise acquired except by those to whom such access is authorized. Integrity implies that information cannot be altered except under properly authorized circumstances. Availability implies that resources are available when desired. All three of these attributes must typically be maintained in the presence of malicious users and accidental misuse and ideally also under certain types of system failures.

Secondary attributes include authenticity, nonrepudiability, and accountability, among others. Authenticity of a user, file, or other computational entity implies that the apparent identity of that entity is genuine. Nonrepudiability implies that the authenticity is sufficiently trustworthy that later claims to its falsehood cannot be substantiated. Accountability implies that it is possible to determine what has transpired, in terms of who did what operations on what resources at what time, as desired.

Authorization is the act of granting permission, typically based on authenticated identities and requested needs, with respect to a security policy that determines how authorizations may be granted. A fundamental distinction is made between policy and mechanism: policy implies what must occur or what must not occur, whereas mechanism determines how it is done. One of the fundamental challenges of computer-communication security is establishing the policy unambiguously before carrying out design and implementation and then ensuring that the use of the mechanisms satisfies the required policies.

Security mechanisms can exist at many different hierarchical layers, such as hardware, operating system kernels, network software, database management, and application software. Each such layer may have its own security policy. In this way, it becomes clear (for example) that operating system security depends on hardware mechanisms and on software access-control mechanisms. Furthermore, cryptographic confidentiality and integrity of networked communications typically depend on the security of the underlying storage mechanisms, particularly with respect to protecting keys and exposed unencrypted information. By making careful distinctions among different hierarchical layers of abstraction and different entities in distributed environments, it is possible to associate the proper policies with the appropriate mechanisms and to avoid circular dependencies.

Granularity is an important concept in security. It is possible to specify an access-control policy according to arbitrary granularities, with respect to individual bits, bytes, words, packets, pages, files, hyperfiles, subsystems, systems, collections of related systems, and, indeed, entire networks of systems. Access permissions may themselves be subdivided into individually protecting operations such as read, write, append-only, execute, with the ability to execute without reading or to do a blind append without reading or overwriting. Each functional layer or subsystem may have protections according to its own operations and resources. For example, operations on files, database entities, electronic mail, and network resources such as web pages all tend to be different. Digital commerce adds further security requirements as well. Ideally, each type of object has its own set of protections defined according to the permissible operations on those objects.

Vulnerabilities are rampant in most computer-communication systems. Exploitations of those vulnerabilities are also rampant, including penetrations by outsiders, misuse by insiders, insertion of malicious Trojan horses, personal-computer viruses, financial fraud, and other forms of computer crime, as well as many forms of accidental problems. When exploited, vulnerabilities can lead to a wide range of consequences—for example, resulting from a lack of confidentiality, integrity, availability, authentication, accountability, or other aspect of security, or a lack of reliability. (For numerous examples of vulnerabilities, threats, exploitations, and risks, see Neumann [1995].) Avoiding vulnerabilities is a very difficult matter, because it depends on having sufficiently precise requirements, sufficiently flawless designs, sufficiently correct implementations, sensible system operations, and aware users. All of those require militant adherence to high principles.

Various principles have evolved over the years and are commonly observed, at least in principle! These include the following: (1) separation of concerns according to types of system functions, user roles, and usage modes; (2) minimization of granted privileges; (3) abstraction; (4) encapsulation, as in hiding of implementation detail; (5) avoiding dependence on security by obscurity. An example that encompasses principles 1 and 2 is to avoid using superuser privileges for nonprivileged operations. Overall, good software engineering practice (which includes principles 3 and 4) can contribute considerably to the extent to which a system avoids certain characteristic flaws. Security by obscurity (5) is the generally invalid assumption that your antagonists know less than you do and always will; relying exclusively on that assumption is very dangerous, although security ultimately depends to some extent on staying ahead of your attackers. However, security is greatly increased when subsystem and network operations are well encapsulated, for example, if distributed implementations and remote operations can be hidden from the invoker.

Some pervasive principles have been collected together as the emerging generally accepted system security principles (GSSP) [GSSP 1995], inspired by the National Research Council study, *Computers at Risk* [Clark et al. 1990]. Further effort is being devoted to the establishment of some broad functional principles and detailed principles. A list of pervasive principles is given in Table 89.1

One of the most important problems in ensuring security at every externally visible layer in a hierarchy and in all systems throughout a highly distributed environment is that of ensuring adequate authentication. Identities of all users, subsystems, servers, network nodes, and any other entities that might otherwise be spoofed or subverted must be authenticated with a level of certainty commensurate with the nature of the application, the potential untrustworthiness of the entity, and the risks of compromise. Fixed (reusable) passwords are extremely dangerous, particularly when they routinely traverse unencrypted local or global networks and can be intercepted. One-time tokens of some sort are becoming absolutely essential for authenticating users, systems, and in some cases even subsystems, particularly in distributed systems in which some of the entities are of unknown trustworthiness. Cryptographically based authentication is discussed in Chapter 91.

TABLE 89.1   Generally Accepted System Security Principles (GSSP): Pervasive Principles (March 1995)

 1. *Accountability:* Information security accountability and responsibility should be explicit.
 2. *Awareness:* Principles, standards, conventions, mechanisms (PSCM), and threats should be known to those legitimately needing to know.
 3. *Ethics:* Information distribution and information security administration should respect rights and legitimate interests of others.
 4. *Multidisciplinary:* PSCM should pervasively address technical, administrative, organizational, operational, commercial, educational, and legal concerns.
 5. *Proportionality:* Controls and costs should be commensurate with value and criticality of information, and with probability, frequency, and severity of direct and indirect harm or loss.
 6. *Integration:* PSCM should be coordinated and integrated with each other and with organizational implementation of policies and procedures, creating coherent security.
 7. *Timeliness:* Actions should be timely and coordinated to prevent or respond to security breaches.
 8. *Reassessment:* Security should periodically be reassessed and upgraded accordingly.
 9. *Democracy:* Security should be weighed against relevant rights of users and other affected individuals.
10. *Competency:* Information security professionals should be competent to fulfill their respective tasks.

There are many techniques for enhancing security and reducing the risks associated with compromises of computer security. These techniques necessarily span a wide range, encompassing technological, administrative, and operational measures. Good system design and good software engineering practice can help considerably to increase security but are by themselves inadequate. Considerable burden must also be placed on system administration. In addition, laws and implied threats of legal actions are necessary to discourage misuse and improper user behavior.

Inference and aggregation are problems arising particularly in distributed computer systems and database systems. Aggregation of diverse information items that individually are not sensitive can often lead to highly sensitive conclusions. Inferences can sometimes be drawn from just two pieces of information, even if they are seemingly unrelated. The absence of certain information can also provide information that cannot be gleaned directly from stored data, as can the unusual presence of an encrypted message. Such gleanings are referred to as exploitations of out-of-band information channels. Some channels are called covert channels, because they can leak information that cannot be derived explicitly, for example, as a result of the behavior of exception conditions (covert storage channels) or execution time (covert timing channels). Inferences can often be drawn from bits of gleaned information relating to improperly encapsulated implementations, such as exposed cryptographic keys. Paul Kocher's attack on cryptographic implementations that leak information about keys as a result of the timing behavior of computations is a subtle example of a covert timing channel [Kocher 1995].

Electronic privacy is a socially motivated expectation that computer-communication systems will adequately enforce confidentiality against unauthorized people and that authorized people will behave well enough. Privacy is a meaningless concept in the absence of a clear statement of the expectations that must be satisfied. Enforcement of privacy depends on system security, on adequate laws to discourage misuses that cannot otherwise be prevented, and on a society that is sufficiently orderly to follow the laws. Losses of privacy can have very serious consequences, although those consequences are not the subject of this chapter. See Chapter 93 for more discussion of consequences.

89.2 Conclusions

Attaining adequate security is a challenge in identifying and avoiding potential vulnerabilities and threats (see Neumann [1995]) and in understanding the real risks that those vulnerabilities and threats entail. Security measures should be adopted whenever they protect against significant risks and their overall cost is commensurate with the expected losses. The field of risk management attempts to quantify risks. However, a word of warning is in order: if the techniques used to model risks are themselves flawed, serious danger can result. The risks of risk management may themselves be devastating. See section 7.10 (by Robert Charette) in Neumann [1995].

89.3 Recommendations

Security is a weak-link phenomenon. Weak links can often be exploited by insiders if not by outsiders, with very bad consequences. The challenge of designing and implementing meaningfully secure systems and networks is to minimize the presence of weak links. Considerable experience with past flaws and their exploitations, observance of principles, use of good software engineering methodologies, and extensive peer review are all desirable, but never by themselves sufficient to increase the security of the resulting systems and networks.

References

Amoroso, E. 1994. *Fundamentals of Computer Security Technology.* Prentice–Hall, Englewood Cliffs, NJ.

Cheswick, W. R. and Bellovin, S. M. 1994. *Firewalls and Internet Security.* Addison–Wesley, Reading, MA.

Clark, D. D., Boebert, W. E., Gerhart, S., et al. 1990. *Computers at Risk: Safe Computing in the Information Age.* Computer Science and Technology Board, National Research Council, National Academy Press, Washington, DC.

Dam, K. W., Smith, W. Y., Bollinger, L., et al. 1996. *Cryptography's Role in Securing the Information Society.* Final Report of the National Research Council Crypto Study. National Academy Press, Washington, DC.

Denning, P. J., ed. 1990. *Computers Under Attack: Intruders, Worms, and Viruses.* ACM Press, New York and Addison–Wesley, Reading, MA.

Garfinkel, S. and Spafford, E. 1996. *Practical Unix Security,* 2nd ed. O'Reilly and Associates, Sebastopol, CA.

Gasser, M. 1988. *Building a Secure Computer System.* Van Nostrand Reinhold, New York.

Gasser, M., Goldstein, A., Kaufman, C., and Lampson, B. 1990. The digital distributed system security architecture. *Proc. 12th Nat. Comput. Security Conf.*

GSSP 1995. GSSP: Generally-Accepted System Security Principles, Exposure Draft, 2.0, Nov. 1995, contact wozier@creative.net (Will Ozier).

Hafner, K. and Markoff, J. 1991. *Cyberpunks.* Simon and Schuster, New York.

Hoffman, L. J., ed. 1990. *Rogue Programs: Viruses, Worms, and Trojan Horses.* Van Nostrand Reinhold, New York.

Icove, D., Seger, K., and VonStorch, W. 1995. *Computer Crime.* O'Reilly.

Kocher, P. 1995. Cryptanalysis of Diffie-Hellman, RSA, DSS, and Other Systems Using Timing Attacks. Extended abstract, Dec. 7, 1995.

Landau, S., Kent, S., Brooks, C., Charney, S., Denning, D., Diffie, W., Lauck, A., Miller, D., Neumann, P., and Sobel, D. 1994. *Codes, Keys, and Conflicts: Issues in U.S. Crypto Policy.* ACM Press, New York. Summary available as Crypto Policy Perspectives. *Commun. ACM* 37(8):115–121.

Morris, R. and Thompson, K. 1979. Password security: a case history. *Commun. ACM* 22(11):594–597.

Neumann, P. G. 1995. *Computer-Related Risks.* ACM Press, New York, and Addison–Wesley, Reading, MA.

Neumann, P. G. and Parker, D. B. 1990. A summary of computer misuse techniques. *Proc. 12th Nat. Comput. Security Conf.* Gaithersburg, MD. Oct. 10–13, 1989. National Institute of Standards and Technology.

Russell, D. and Gangemi, G. T. 1991. *Computer Security Basics.* O'Reilly and Associates, Sebastopol, CA.

Stoll, C. 1989. *The Cuckoo's Egg: Tracking a Spy Through the Maze of Computer Espionage.* Doubleday, New York.

Thompson, K. 1984. Reflections on trusting trust. (1983 Turing Award Lecture) *Commun.* ACM 27(8):761–763.

Further Information

In addition to the references, many useful papers on security and privacy issues in computing can be found in the following annual conference proceedings:

Proceedings of the IEEE Security and Privacy Symposia, Oakland, CA, each spring.

Proceedings of the National Information Systems Security Conference, Baltimore, MD, each autumn.

Proceedings of the SEI Conferences on Software Risk, Software Engineering Institute, Carnegie-Mellon University, Pittsburgh, PA.

90

Protection (Security) Models and Policy

Carl E. Landwehr
Naval Research Laboratory

90.1 Introduction

Assets, Vulnerabilities, and Threats

Protection becomes an issue for operating systems and networks when they are used to process information or control systems that represent a significant asset to someone. The value of information may
vary over time, even when the information does not change: tomorrow's weather forecast is more valuable
than yesterday's. The value of the control may also vary: a computer system that controls a nuclear power
plant represents a more significant asset than one that controls a microwave oven. We are naturally more
concerned about protecting more valuable assets. The vulnerabilities of a system are its weak points:
flaws in its design or implementation that could cause it to behave in ways not intended by its builders
or operators. Threats are the agents that, acting against a vulnerability, could cause such misbehavior.
Threats can be external or internal to a system, they may be hostile or benign, and they may be structured
or unstructured. Lightning usually represents an external, benign, unstructured threat. Random operator
errors represent an internal, benign, unstructured threat, and the actions of a subverted operator might
represent an internal, hostile, structured threat. Assets that have significant value are more likely to be
subject to hostile, structured threats.

Implicit and Explicit Security Policies

Secure means, literally, "apart from care." A system is secure if we do not have to worry about it. But
exactly what guarantees do we require of a computer system? When computers began to serve several
users at once, operating systems needed to protect one user's programs and files from the errors of
others. In a military environment, there are requirements for multilevel security that require a system
to separate different levels of classified information from users with different clearances. The security
policy for a system defines what it means for that system to be secure; usually it concerns identification

0-8493-2909-4/97/$0.00+$.50
© 1997 by CRC Press, Inc.

and authentication of users, authorization of actions, and accountability for actions taken. Often, *security policies* are implicit. Most personal computers, as delivered, do not restrict (or even identify) the individuals who use them, and so there is no way to distinguish an authorized access request by a program from an unauthorized one. Nevertheless, users do expect certain properties of their machines, for example, that running a new piece of commerical software should not cause all of their files to be deleted. When security policies are explicit they often specify the behavior desired of the system as a whole, and so the implications for operating system behavior may not be obvious.

Varieties of Models

We use *model* here in the sense of the engineer, rather than the logician. A model of a system represents aspects of it that concern us without representing it in complete detail. An architect's model of a house will accurately reflect the shape and spatial relationships of the parts of the dwelling, but it will not include a functioning furnace or be built of materials whose strength is proportional to those to be used in construction. Security models have been used for two fairly distinct purposes: to represent the security policy a system is intended to enforce and to represent the mechanisms in a computer system that are intended to enforce the policy. Policy models typically represent the flow of authorizations and information; mechanism models represent data structures found in operating systems. Some models combine both purposes.

90.2 Underlying Principles

Concerns about how a system will behave motivate the development of a security policy for a system. A security model represents the security policy, the structure of a system, or both in a form that allows us to reason about the policy and its enforcement.

Security Concerns and Security Policy

Information security concerns conventionally focus on preserving the confidentiality, integrity, and availability of information for those authorized access to it. These concerns are usually considered equivalent to preventing unauthorized disclosure, modification, or withholding (denial of service) of protected information. Parker [1994] suggests that, particularly in a business context, possible loss scenarios should be the basis for information security concerns. He pairs possession with confidentiality, authenticity with integrity, and utility with denial of service as *elements* of information security. For example, if information is encrypted and the key becomes unavailable, its utility may be lost (assuming cryptanalysis is infeasible) even though it remains available. Authenticity, but not necessarily integrity, is compromised if pirated software is altered so it appears to have been produced by someone else. Possession, but not confidentiality, may be lost if a laptop containing the only copy of encrypted data is stolen and held for ransom.

Within the area of confidentiality, security concerns may also depend on the kind or amount of information disclosed to an individual. A firm that provides accounting or legal services to competing clients may need to prevent employees who have seen information concerning one of its clients from viewing similar information from other, competing clients.

This is sometimes known as building a Chinese Wall between sets of employees exposed to different information [Brewer and Nash 1989]. Preventing the undesired aggregation of information can also be a concern. A bank teller may be permitted to have access to information on individual accounts, but not to summaries derived from all accounts, since these might reveal the bank's overall financial status. In a military context, information about the readiness of an individual unit may not be considered highly sensitive, whereas the collective readiness of all units would be.

Security concerns can extend beyond the information in a system to the integrity of the system itself. Errant or malicious programs can overwrite data or programs, reformat disks, or, if the computer is part of a control system, damage property and injure people. Programming errors have, in fact, contributed to fatal accidents [Leveson and Turner 1993]. Concerns such as these typically require attention at the system design level, but their effects can be seen at the level of the operating system.

Security policy is a plan or course of action intended to satisfy a set of security concerns. History teaches that security concerns are unlikely to be satisfied unless they are identified explicitly and a corresponding policy is implemented. A security policy for an organization typically addresses many different security disciplines, including personnel security, administrative security, physical security, and so on. This chapter is concerned with security policies that can be implemented within the scope of a computer system and, more specifically, by an operating system. This focus narrows our concerns significantly.

Access and Authorization Policies

Access and authorization policies specify what behaviors are permitted in terms of the types of access (read, write, execute) one entity (user, process, file) is authorized to have to another. For example, "a user is permitted read, write, or execute access to any file that user creates," or "a process that is permitted to read a file may grant that permission to any other process." Such a policy is called **discretionary** if it leaves the user with the ability to make some choices (i.e., with some discretion) in administering the policy. A mandatory policy, conversely, prohibits the user from making decisions. The use of **mandatory access control** policies is typically associated with military applications, where mandatory security dictates that users not be permitted to read a file classified above their clearance. However, the requirement that every user of a commercial system present a valid password in order to gain access to it is also a form of mandatory security policy. Policies often include both mandatory and discretionary parts: discretionary policy permits the secret-cleared user to decide whether or not to permit other users with secret or higher clearances to read a secret file; mandatory policy prohibits the secret user from placing secret-labeled data where it could be read by users not cleared for it. **Role-based** security policies assign privileges to job functions (roles), rather than individuals; the user inherits the privileges of the job assigned. This approach can simplify the management of privileges, since a new privilege associated with a particular job is automatically inherited by all users who occupy the corresponding role.

Information Flow Policies

Restricting the flow of information is often the purpose of access control policies: a class of users is denied access to some physical object (a file, for example) because information contained in that file is not intended to flow to users in that class. Recognizing this fact, and stating the policy directly in terms of constraints on information flow, can simplify and clarify the resulting security requirements. The classic example is the formulation of military security policy as a lattice of security classes, with the requirement that information can only flow upwards in the lattice. Commercial applications seem more often to involve total isolation of different classes (as in the Chinese Wall policy), but when divisions of two competing firms cooperate on a particular project or proposal, for example, it may be more natural to characterize security policy in terms of permitted and prohibited information flows, rather than accesses. Relationships between component suppliers and system integrators may also lead to security policies formulated most naturally in this way.

Integrity and Availability Policies

Although access control and information flow policies have attracted the most attention and have the best developed models and supporting structures in operating systems and networks, two other kinds of security policies are worth noting here: integrity and availability policies. *Integrity* has many meanings to many groups. In the context of operating systems and networks, protecting integrity usually means protecting data from unauthorized or improper modification. Information flow and access control policies may incorporate integrity concerns. For example, an information flow policy that restricts the flow of information among security classes might be reformulated to restrict the flow of information among integrity classes, where a higher integrity class represents information that is more likely to be accurate or correct, and a lower integrity class represents less reliable information. To preserve the integrity of reliable information, we might prevent low-integrity information from flowing to higher integrity classes.

Similarly, we might protect integrity by controlling read and write accesses to objects with different integrity labels.

Having crucial information available when needed is often more important than preserving its confidentiality or its integrity. Of course, corrupted information may be of little use even if it is available, and relying on compromised information (e.g., a crypto key) can have unpleasant consequences. In the context of operating systems, protecting the availability of information usually means controlling access to resources so that one process cannot indefinitely prevent others from gaining access to it. Systems that serve several users at once usually include precautions against accidental loss of resources but rarely protect against malicious attacks aimed at denial of service. Policies for system availability are rarely distinguished from those for reliability, which typically consider only accidental failures, not malicious behavior.

Security Models

Security policy is at root informal, representing a plan or course of action. A **security model** attempts to capture the security-related behavior and/or structure of a system more formally. It is an abstraction of a system from the standpoint of security and may omit other important aspects that do not affect its security properties. Most security models have focused on confidentiality, rather than integrity or availability.

Policy Models and System Models

A security model may focus either on a system's security policy (e.g., no individual can view a message for which clearance is lacking), leaving open the questions of system design and implementation, or on the system's structure and operations, its mechanisms (e.g., a process running on behalf of a given user can only open a file for reading if there is an entry for that user in the file's access control list with an access mode of *read*). Security policy amounts to the top-level security requirement for a system, so we consider security policy models as models of system requirements. Some models attempt to represent both the requirements and the mechanisms; if such a model is used to guide a development, it will constrain the set of possible implementations more tightly than a model of the requirements only. It is possible to view the distinction between models of requirement and mechanism simply as a difference in the level of system design being modeled. An application-based model may define and refer to entities visible to users (users, messages, files, input and output devices), whereas a mechanism-based model may focus on processes, storage blocks, file locks, and so forth. Difficulties arise when security requirements at the application level are difficult to realize with the mechanisms available at lower levels.

Proving Properties of Models

Whether a particular model focuses on the domain of requirements or the domain of implementations, it should provide a structure with a clear relationship to that domain, so that reasoning about the more simple structure of the model carries over to the relevant real-world domain. A good road map provides a simplified representation of a network of streets and roads that permits a driver to reason correctly about possible paths to a particular destination and, consequently, which way to turn at the next intersection. A successful model of a set of security requirements should permit its user to reason about how a system satisfying those requirements will behave. A security model of the implementation domain should support reasoning about the security-related behavior of the modeled class of mechanisms.

Because security models are artificial—the model consists of abstractions whose properties the modeler is at liberty to define—it is possible to reason about them more rigorously than one can reason about the real world directly. This is a benefit, in that we can state properties we desire of the model (e.g., that its behavior satisfies certain security constraints) and, potentially, determine with certainty whether the property holds for the model or not. But caution is required: a property that we prove holds for the model still may not hold in the real world, if the intended relationships between the model and the real world are violated. If a road map omits, say, a tunnel of which cartographers were unaware, or that was built after the map was printed, drivers may think they cannot cross a river without going over a bridge, when in fact they can. Likewise, proving that a user can never gain access to a file according to a particular security model is of interest only if the system as implemented conforms to that security model.

The **access matrix** model [Lampson 1971, Graham and Denning 1972] is based on abstraction of operating system protection structures. Its simplicity and generality have led to its wide application over many years, and it continues to be a reference point for system developers and users. It includes three primary components: a set of passive **objects** (files, devices, or other entities implemented by the operating system); a set of active **subjects**, which may manipulate the objects; and a set of rules governing the manipulation of objects by subjects, including transformations to the access matrix itself. A subject is a process and a domain (a set of constraints within which the process may access certain objects); every subject is also an object, since it can be read or otherwise manipulated by other subjects. The access matrix is a rectangular array with one row per subject and one column per object. The entry for a particular row and column reflects the permitted modes of access between the corresponding subject and object, typically a subset of {read, write, append, execute}. Although the access matrix is never implemented literally as a matrix (it would be very sparse), it provides a basis for reasoning about accesses within a system. (See Chapter 91, "Authentication, Access Control, and Intrusion Detection.")

Harrison et al. [1976] developed a particular formalization of the access matrix model (the HRU model) and identified the safety property for it: given an initial state for the access matrix and a set of commands, can a particular subject gain a given right a to a particular object? A configuration is safe for a if there is no sequence of transformations that will cause a to be added to an element of the access matrix that does not already contain it. They were able to show that the safety problem for the HRU model is undecidable in general, although it is decidable if the access matrix transformation rules are mono-operational, i.e., if each request to change the access matrix results in only a single change of the form add/remove a row/column or add/remove access right a for a single cell in the matrix. Unfortunately, mono-operational systems exclude most practical security policies, since, for example, they exclude policies that grant the creator of an object particular access rights to that object. The take-grant model [Jones et al. 1976], which represents subjects, objects, and accesses with directed graphs, has a linear time algorithm for safety, but although it has been explored formally in considerable detail [Bishop 1988], it has seen little practical application. Sandhu [1988, 1992] has developed the schematic protection model and typed access matrix models based more closely on the HRU approach, and has been able to narrow the gap between models for which safety can be proven and those reflecting practical access control policies.

Composability of Security Properties

If we define a security property for a system, show that the property holds for a model of that system, and show that the model accurately represents the system, we can be confident that the behavior of that system will enforce that security property. Unfortunately, we cannot be sure that if we connect two such systems together the combination will enforce the same security property that the systems did individually. This is referred to as the composability problem: the combination will only be secure if the security property in question is composable, even if each system individually enforces the specified security property. As computers are increasingly interconnected, security threats and vulnerabilities arising from those interconnections are growing. Finding useful, composable, security properties that would provide a basis for connecting systems without introducing new vulnerabilities is a current research topic.

Enforcement

The concept of a monitor as a system component that controls allocation and use of a particular resource led to the notion of a **reference monitor** as a key component that would control the accesses subjects could gain to objects [Anderson 1972]. When a subject attempts to gain access to an object (e.g., to open a file for reading), the reference monitor checks the access matrix to see whether the requested access is allowed; if not, the request is denied. Three properties are required of a reference monitor: it must be invoked on every access attempt, it must be tamperproof, and it must be small enough to be subject to thorough testing or verification (i.e., it must be correct). A *security kernel*, the hardware and software realization of the reference monitor, is intended to include all of the security enforcement mechanisms in an operating system and nothing else.

General purpose, efficient security kernels proved hard to realize [Landwehr 1983], and partly as a consequence, the notion of a **trusted computing base (TCB)** developed. The TCB, like the security kernel, is intended to include all of the security enforcement mechanisms in the system (thus the TCB boundary is the system's security perimeter) but the TCB is not required to be minimal; it may include nonsecurity-relevant software as well. The Trusted Computer System Evaluation Criteria (TCSEC) [D. of D. 1985] are organized around the TCB concept; they were developed to permit different commercial computer system products to be ranked in terms of how well they realize it. To be ranked at any of the top four of the seven levels established by the TCSEC, the system's documentation must include a security policy model and some evidence that the implementation enforces the constraints of that model.

90.3 Best Practices

A system may in fact be quite secure, but if we cannot convince others of this fact, they will be unlikely to trust valuable assets to be placed under its control. Security models can be used to define what security means for a given system, to guide the system's designers and developers so that an implementation provides the intended security properties, and to convince others that the system is trustworthy. In this section, we first review basic access and authorization models and then security policies and models concerned with the three traditional security properties: confidentiality, integrity, and availability. For each property, we will describe different models developed to permit a determination as to whether a system that corresponds to the model preserves that property.

Access and Authorization Models

The fundamental model for access control is the access matrix model, previously described. The modes of access that are registered in the access matrix at any instant define the authorized accesses between subjects and objects. Because this model corresponds so well both to intuitive notions of controlling the access of people to material objects and to the protection structures implemented by many operating systems, it is widely known and used. We can say that the contents of the access matrix encode the system's security policy and that it is up to the operating system's mechanisms to enforce this policy. Separating policy and mechanism is usually desirable in a system design.

But if we want to reason about whether or not some particular subject can ever read or write some particular object, or whether information can leak across some system boundary, the access matrix model is less than ideal. First, the undecidability of the safety problem limits general results. Second, and closely related, is the problem of Trojan horse programs. Programs, not humans, initiate the operations that alter the access matrix, and a program executed by a given subject can generally exercise all of the rights of that subject, including granting those rights to other subjects. This means that humans who invoke any Trojan horse program are vulnerable to the complete redistribution of their access rights (within the bounds of the authorization scheme) and, unless we are willing to assert that no user will invoke such a program, necessitates a worst-case assumption about propagation of access rights.

To address these two problems, Sandhu [1992] has introduced a version of strong typing to access matrices. Each subject and object is assigned an immutable type at its creation and, with some additional constraints on the structure of commands in the ternary monotonic typed access matrix model, Sandhu shows that its safety is decidable in polynomial time.

A third problem that arises with the access matrix model is the practical difficulty of including all of the relevant objects in the access matrix. In most operating systems, processes, files, and devices may have these kinds of access controls associated with them, but there are often other shared, user-visible data structures and resources (such as file names, file locks, and system clocks) that do not. If the security policy calls for constraining information flow from one set of objects to another in an environment that may include Trojan horses, these uncontrolled resources can act as **covert channels** [Lampson 1973], permitting violations of the security policy even though the controls listed in the access matrix are correctly enforced. The shared resource matrix methodology [Kemmerer, 1983] provides a practical approach to identifying such channels.

Confidentiality Models

Because protecting sensitive information against disclosure is a major concern of military security, and because protecting confidentiality seems to be a more tractable problem than maintaining integrity and availability, confidentiality models dominate the security modeling research literature. The best known and most widely applied of these is the Bell–LaPadula model, which is closely based on the access matrix model, to which it adds the military security lattice, some additional data structures, and a set of definitions and rules [Bell and LaPadula 1975]. Its goal is to establish a formal structure in which even a Trojan horse with access to sensitive data would be unable to leak those data to a user or program that did not already have access to them.

Military Security Structure

Military security classifications are characterized by a linearly ordered set of sensitivity levels (e.g., confidential, secret, top secret) together with a set of compartments. Compartments typically correspond to information on a certain topic or from a certain source that requires special protection. The security label for a particular document or paragraph includes a sensitivity level and a (possibly empty) set of compartments. To be authorized access to a particular document, the user must have a clearance for both the sensitivity level and all of the compartments listed in the label. A sensitivity level and a set of compartments denotes a security level. In general, information is allowed to flow from one security level to another if the sensitivity level of the source is lower than that of the destination (e.g., from confidential to secret) and if the compartment set of the destination level includes all of the compartments of the source level. Recognizing that unclassified, no compartments is the lowest possible security level, from which any information can flow upwards, and top secret, all compartments is the highest level, which can receive information from any other level, we see that the set of security levels together with the flow relation forms a lattice [Denning 1976].

Bell–LaPadula Model

The Bell–LaPadula model associates a security level with each subject and object and identifies a set of access modes that subjects may have to objects, including read, write, and execute. When a subject attempts to read an object, the access matrix is checked, but an additional check is required to determine whether the subject's security level is greater than or equal to that of the object to be read (since it is permissible for a secret level subject to read a confidential level object, for example). Unless the access matrix check [referred to as discretionary access control (DAC)] and the security level check for reading, [referred to as the simple security property of mandatory access control (MAC), or no read-up] both succeed, access is denied. If a subject requests write access to an object, the request is checked against the access matrix for DAC permission, and it is checked to be sure that the security level of the object to be written is equal to the current security level of the subject (referred to as the *-property, or no write-down). This check prevents, for example, a secret level subject from writing secret data into a confidential level object.

The argument that systems conforming to this model preserve security proceeds by identifying a secure state of a system as one in which no users have accesses to objects that would violate the security policy. The so-called basic security theorem (BST) asserts that if a system starts in a secure state and proceeds to new states only through a set of possible transitions, each of which guarantees termination in a secure state if it is initiated from a secure state, then the system will never reach an insecure state. Two noteworthy objections have been raised about this formalization of confidentiality. First, the BST holds regardless of the characterization of a secure state; if we define a secure state to be any state in which all objects could be read or written by all subjects, we can still prove the BST. Consequently, we cannot rely on the fact that the BST has been proven to assure ourselves that we have found an intuitively correct definition of secure state [McLean 1985]. Second, some intuitively insecure systems in fact satisfy the particular definition of security given by Bell and LaPadula [McLean 1990]. It is possible to construct an operation that satisfies the stated security properties but changes the security levels of subjects and objects between states. Thus, while no accesses violate the security policy in either the state before or after the transition, the transition itself has changed the system state in a way that is intuitively insecure, yet not prohibited. These problems, however, have not prevented the model's use as engineering guidance in many system developments.

A major challenge in applying this model rigorously is the covert channel problem mentioned earlier. Although the Bell–LaPadula model is defined only in terms of access controls, its purpose is to enforce an information flow policy: information should not flow from higher security levels to lower ones by any path. Finding all of the potential data structures and hardware devices in a system that could be used to convey information between security domains is a difficult but necessary part of assuring that an implementation based on access control correctly realizes an information flow policy.

Noninterference Model

The noninterference model has also been used in system developments as a basis for showing that a system enforces confidentiality (Goguen and Meseguer [1984], based on earlier work by Feiertag et al. 1977). This model focuses on capturing the confidentiality requirement rather than the mechanisms required of the implementation. By focusing on the properties required to preserve confidentiality at the system specification level, it achieves a cleaner and more intuitive definition of security and permits the implementer more flexibility in realizing the requirements.

This model describes a system in terms of the sequence of system calls (trace) invoked by users at different security levels. System output is defined as a function of the system's input history, the user requesting the output, and the user's clearance. If a user with a given clearance sees the same output at any point in the system's history that would be seen if none of the inputs from users with higher clearances had occurred (i.e., if their inputs were purged from the trace), then the system is secure. That is, the higher level inputs do not interfere in any way with the behavior observed by lower level users; hence the model's name.

Goguen and Meseguer [1984] developed a set of unwinding conditions to help apply this model in a state machine context more commonly used in system specification and development. Proving that the unwinding conditions hold for the state machine's transition functions establishes non-interference. This approach has the benefit that covert storage channels are identified in the normal course of developing the unwinding conditions and proving noninterference; no separate effort is required to identify them as it is in developments based on the Bell–LaPadula model.

Discussion

Applying the access matrix model, Bell–LaPadula model, or noninterference model to an operating system development is likely to lead to a stronger system with fewer security flaws than if security is not given explicit consideration. But there are several issues left open by current models and practices. One we have already touched on: covert channels. Sometimes the issue of covert channels is shrugged off as a consideration of only academic concern. Given certain assumptions about the threats a system is subject to and the requirements it must meet, sometimes this attitude may be appropriate. However, if strict separation is desired between information in different domains, then the concern is not merely covert channels, it is channels of any sort that permit information to cross the boundary between domains. Assuming that some such channels exist, the question becomes: what is their capacity, and what information can be transferred over them? The relevant models of information, and information flow, have their roots in Shannon's information theory [Shannon and Weaver 1949]. Some recent efforts have attempted to model known covert channels in real system designs using information theory [Kang et al. 1995], but this is still a research topic.

All of the models discussed in this section so far are for deterministic systems, where the current state and input uniquely determine the next state. This basis may be appropriate for modeling single computers, but is not fully satisfactory if the security model is to be used to analyze either a network or a system specification, where the current state and input may only determine a set of possible next states (i.e., nondeterministic systems). Substantial research effort has been devoted to developing nondeterministic, including probabilistic, models for properties like noninterference, but none of these has been demonstrated in a practical system development. For a good summary, see McLean [1994].

Note that the security properties intended to be enforced by Bell–LaPadula and non-interference type models are fundamentally policies of strict separation: no information is supposed to flow between incommensurate security domains, and information may flow at most upward from lower security levels to higher ones. When models like these began to be developed in the early to mid-1970s, hardware was expensive, and sharing hardware across security levels seemed essential. With the drastic declines in the cost of hardware in the last decade, separating information (if isolation is what is required) by storing it on physically separate computer systems is a much more realistic approach than it once was. Of course, if absolute separation or simple upward flow does not meet the security requirements of the application, physical separation may not work so well, either. In this case, both walls and gates may be required. The best use of security models such as Bell–LaPadula and noninterference in this case may be to model the walls, and modeling the gates—paths between domains that are sometimes opened intentionally—may require a different approach that can assure that only authorized information passes through. Indeed, the Bell–LaPadula model countenances such flows via the mechanism of a trusted subject that is permitted to violate the *−property but is trusted not to compromise security. Establishing trust in a subject is an activity outside the security model. An early but still relevant approach to organizing a system based on physical separation and encryption rather than trusted software is described by Rushby and Randell [1983]. More recent work to develop a multilevel secure database system based on physical separation and data replication, and to extend the architecture to provide a multilevel secure distributed computing service, is described by Froscher et al. [1994].

Integrity Models

Integrity models reflect the policies and mechanisms intended to assure that data and programs are modified only in an authorized manner. The access matrix model includes controls on writing to subjects and objects, and these can be used for this purpose. For example, only the system administrator may be granted write access to a file used to record the list of authorized users of the system. Of course, if the system administrator ever invokes a Trojan horse program, it could pass this privilege to any other user, because these controls are discretionary.

Lattice-Based Integrity

To limit the ability of Trojan horses to alter data that might be unclassified, yet critical to system operation, Biba [1977] introduced a lattice of integrity levels to the Bell–LaPadula model corresponding to the security lattice. The controls imposed on subjects and objects with respect to their integrity levels are intended to prevent high-integrity data from being modified based on low-integrity data. The integrity level of an object is not necessarily related to its security level. For example, an airline's current flight schedule is readable by all, but should only be written by authorized subjects. Conversely, the actions of any user might append records to the system's audit file, but ability to read the audit file might be limited to system administrators.

In Biba's strict integrity policy a subject can only read up in integrity and write down. The system administrator and the list of authorized users would both be assigned high-integrity levels, so that even if a Trojan horse granted a lower level user access to the critical file, that user could not modify the file. There is still a vulnerability, however, since the Trojan horse could modify the file directly, instead of trying to pass permissions to another user.

Several systems built to meet military security requirements have included both security and integrity levels for subjects and objects, and some practical policies that use integrity levels have been proposed [Lipner 1982, Shirley and Schell 1981, Schell and Denning 1986]. These policies aim primarily at enforcing configuration controls as described in the previous example, rather than assessing whether the output of a particular program or activity is of higher integrity than another. Typical integrity levels for a system might be user, programmer, and administrator (from low to high). Integrity compartments can be used to isolate different functional areas. A properly constructed integrity lattice can be used to model conflict-of-interest classes, as well as enforce a Chinese Wall policy [Sandhu 1993].

Commercial System Integrity

An integrity model based on traditional controls in commercial systems to prevent fraud and assure accuracy in records has also been proposed [Clark and Wilson 1987]. This informal model defines unconstrained data items (UDIs), constrained data items (CDIs), transformation procedures (TPs), integrity verification procedures (IVPs), and Logs and incorporates the notions of separation of duties and certification. UDIs reflect raw, unchecked information that may be entered into the system. CDIs can be altered only by TPs, which may use UDIs and other CDIs as input. A change to a CDI also appends a record to a Log file. IVPs are run periodically against all of the CDIs to assure that they are in a valid state. There is a set of certification assertions (for humans to assure) and enforcement assertions (for machines to assure), including, for example, that the system must maintain the list of relations of the form (*UserID*, TP_i (CDI_a, CDI_b, ...)) relating a user, a TP, and the data objects that TP may reference on behalf of that user. The system must ensure that only executions described in one of these relations are performed. A human must assure that this list of relations is subjected to the separation of duty requirement. The Clark–Wilson model provides highly intuitive structure in which particular applications can be developed, but assuring that an implementation conforms syntactically to the structure of the Clark–Wilson model by itself can do little to assure that the system behaves as intended. In this way, the model is similar to the access matrix model: it provides a structure for expressing a policy; a human must determine whether enforcement of that particular policy will have the desired results.

Availability Models

Assuring availability or providing assured service is often included in lists of security concerns, perhaps because those phrases are brief and positive, but our real concern is to prevent service from being denied by a malicious attack. Work in the fields of dependability and fault-tolerance addresses availability in the positive sense, where system availability may be measured as mean time between failures divided by the sum of mean time to failure and mean time to repair. But these calculations are usually based on failures occurring randomly according to some model and not as the result of malicious actions.

To represent the relevant aspects of preventing denial of service, it is necessary to characterize the agreement between the service provider and the user. Further, the type of resource or service being provided will have to enter into the formulation of the model [Dobson 1992].

In the 1980s, Yu and Gligor developed a specification and verification method for preventing denial of service in the absence of failures and integrity violations [Yu and Gligor 1990]. A user agreement in their model is a set of constraints on the sequences of calls the user may make to the shared service. A user who violates the agreement cannot expect to obtain the intended service, but should not be able to interfere with the service guarantees made to other users who abide by their agreements. The finite waiting time (FWT) policy they propose incorporates the user agreements, a fairness policy (a user will not be blocked forever if that user has many opportunities to make progress), and a simultaneity policy (a user will eventually have all of the opportunities needed to make progress, provided that the user agreements can be satisfied). They apply their temporal-logic-based specification and verification method to a general resource allocator and prove that all users eventually make progress and receive intended service.

Building on this work, Millen [1993] has defined a denial-of-service protection base (DPB) analogous to the concept of a trusted computing base. The TCB is that part of a system that is responsible for the enforcement of confidentiality and integrity properties; the DPB must offer those services whose loss would be viewed as a denial of service. Millen develops this concept as a state machine, drawing on earlier models of resource allocation in operating systems. A DPB is characterized by a resource monitor, a waiting-time policy, and user agreements, and it must satisfy two conditions: each benign process will make progress in accordance with the waiting-time policy (progress), and no non-CPU resource is revoked from a benign process until its time requirement is zero (patience). A resource monitor is built on a set of processes and resource types, each of which has a fixed capacity. Processes may request allocations of particular amounts of particular resources, including the CPU. The state of the resource monitor is given by the current allocation matrix, which determines how many units of each resource type are allocated to each process, the time vector, the space requirement matrix, and the time requirement matrix. The time vector records the real time at which each process is activated or deactivated. The explicit treatment of time in the model permits the definition of waiting-time policies based on maximum waiting time (MWT) and probabilistic waiting times (PWT) in addition to the FWT policy of Yu and Gligor.

Despite this progress in developing denial of service models, it must be acknowledged that these are still research topics. None of these has been applied to production systems as yet, though with increasing concern about denial of service attacks, particularly against networks, these models, or their successors, may soon find application.

90.4 Research Issues and Summary

Models can represent a protection or security policy, the mechanisms that implement a policy, or both. They can provide a basis for reasoning about how a system is intended to behave or how it behaves in fact. Models have been used effectively to guide the design and implementation of systems, to help users understand how a system is intended to work, and to raise the confidence of evaluators and certifiers that a system is safe to operate.

Unless a system is known to contain or control a significant asset, it may not be the target of malicious threats and its vulnerabilities may remain untried. Recognizing security concerns, including loss of possession or confidentiality, integrity or authenticity, and utility or service, is the first step in forming a security policy. As a course of action intended to satisfy a set of security concerns, a security policy is usually stated informally. It may define mandatory controls, discretionary controls, or controls based on users' roles. It may call for two-person control of particular functions or operations in order to secure a system against an attack by a single individual, or for a Chinese Wall to separate employees with knowledge of competing clients' records. An access control policy constrains the objects a user may read, write, or execute, whereas an information flow policy controls whether the data read from one object is permitted to affect data written to another.

Models are typically stated more formally than policies. It is sometimes possible to prove that a formally expressed model has some desired property, such as barring certain kinds of accesses or certain undesired information flows. Protection models represent the protected objects in a system, how users or subjects (their proxies in the computer system) may request access to them, how access decisions are made, and how the rules governing access decisions may be altered. The access matrix model is the primary example of a protection model. Although the **safety property** (determining whether or not a given subject can ever gain a particular access to a given object) is undecidable for this model, its ability to represent a wide variety of security policies and the degree to which its structures match those commonly found in operating systems have kept it in wide use.

Security models have a broader scope than protection models; they may represent the security policy requirements of maintaining confidentiality, integrity, and availability in the face of malicious attacks, possibly including hostile software. The Bell–LaPadula model was developed to capture confidentiality requirements as reflected in the lattice of security levels defined by military classification schemes. Systems built to conform to it implement both mandatory access controls, which normal users cannot modify, and discretionary controls, such as those represented in the access matrix model, which users can change. The mandatory controls are intended to preserve confidentiality of information across security levels even if users execute Trojan horse programs. The noninterference model has similar objectives to the Bell–LaPadula model, but it represents a system in terms of its response to a sequence of calls (a trace). If a user at a given security level sees the same system behavior whether all inputs from higher level users are purged from the system trace or not, then the noninterference property holds and the system is considered secure. Systems have been built to conform to both the noninterference model and the Bell–LaPadula model. Covert channels, mechanisms that conform to the security model but permit information to be transmitted in violation of security policy, are an implementation issue for both models, although a system based on noninterference should not require the separate analysis of storage channels that Bell–LaPadula-based implementations do.

Integrity models have received much less attention than confidentiality models, not because people are less concerned about integrity, but because it has many, often application-specific, interpretations. Biba's integrity model defines a lattice of integrity levels analogous to the lattice of security levels; it controls the modification of data in a way analogous to how the Bell–LaPadula model controls its disclosure. Implementations of this model have been used to enforce configuration control policies. The less formal Clark–Wilson model reflects integrity controls found in commercial systems organized around transactions.

Least developed are availability models, better termed models for denial of service prevention. Both the Yu–Gligor and Millen models require formalizing agreements between the user and the system as to what constitutes delivery of service. Without such agreements the basis for asserting that service has (or has not) been denied is lacking. Neither model, however, has seen significant application.

Confidentiality and integrity controls are typically implemented by some form of reference monitor, which validates each reference by a subject to an object. A trusted computing base comprises all of the security enforcement mechanisms in a computer system, and the system's security perimeter is the boundary of the TCB. Millen's model for preventing denial of service introduces the notion of a denial of service protection base analogous to the TCB.

All of the models discussed here have been developed to reflect the requirements and behavior of a single computer system, yet we live in a world of increasingly interconnected machines. Not surprisingly, many current research issues in this field stem from this difference. These include the development of models that can be applied to nondeterministic specifications, the identification of security properties that are *composable* (i.e., if the property holds for each of two components, then it will also hold for a correctly connected combination of the two), and the development of principles for allocating security requirements among components (decomposition).

Current models also tend to reflect absolute, rather than relative, notions of security. A policy model prescribes that information flow either is or is not permitted between two entities or security levels. In practice, some limited amount of information flow may be tolerable. The tools of information theory are beginning to be applied to quantifying such flows in practical systems.

The representation of security policies oriented toward commercial applications, where integrity and denial of service concerns typically outweigh confidentiality, should lead to improved models in those areas. Assuring that a system actually enforces a specified policy, and that it continues to do so through a series of releases and upgrades, continues to be a difficult problem. Finally, distributed architectures that support freely moving software agents that may ship back not only images and text, but also programs in one form or another to be executed, raise interesting security questions as to how one can be sure that such programs cannot cause damage when executed.

Defining Terms

Access matrix: Matrix with a row for each subject and a column for each subject and each object. Records authorized access modes (such as read, write, execute) subjects may have to objects. Represents the current access authorization policy, but never implemented directly as an array because it would be too sparse.

Covert channel: A channel used to pass information not originally intended for that purpose. Usually applied to channels that permit two cooperating processes to pass information in violation of the security policy, but without violating the system's access controls. For example, if one process suddenly begins to impose heavy performance demands on a system, it may cause responses to other processes' requests to be delayed. This change in delay can be used as a signal.

Discretionary access controls (DAC): Access controls that may be modified by ordinary subjects. The access controls implemented by typical file systems are discretionary in that the owner of the file can grant other users/subjects the right to read, write, or execute the file. A Trojan horse executed by an unwitting subject can alter the discretionary access controls on the associated user's files.

Mandatory access controls (MAC):   Access controls that cannot be modified by ordinary subjects. In military security, mandatory access controls are intended to prevent a user (or a Trojan horse executing on that user's behalf) from moving secret data to an unclassified file, where uncleared users might read it. Special trusted subjects used by system administrators may be permitted to violate these constraints in limited ways.

Object: A passive entity within a system, such as a file, but any entity that can be used to record or transmit data may be considered an object. The mapping of the entities implemented by the system to the subjects and objects of a security model is a crucial step in assuring that the system's behavior will conform to the model.

Reference monitor: A system component that enforces system security policy by checking all accesses initiated by subjects to objects and permitting only those that are consistent with the policy. To be effective, the reference monitor must be tamperproof, enforced on every access, and it must not make mistakes.

Role-based access controls: Access controls associated with a user's job-related responsibilities, rather than with individual identity or security clearance. Like mandatory controls in that a subject may not arbitrarily pass role-based privileges to other users. Provides a convenient way to organize authorizations in relation to job assignments, so that if one individual must assume another's job temporarily or permanently, the individual inherits the role-based authorizations associated with the new assignment.

Safety problem: To determine whether or not a given subject may at some time obtain some specific access authorization to some specified object. For many straightforward security models, the general safety problem is undecidable.

Security model: An abstraction representing security policy, system mechanisms that can enforce it, or both. May be used to inform users how the system behaves, guide the system's design and implementation, and reason about its properties.

Security policy: A plan or course of action intended to satisfy a set of security concerns. In the context of operating systems, a security policy typically seeks to ensure that data are protected against unauthorized disclosure, modification, or withholding. Usually expressed informally, in natural language, they are sometimes formalized. Security policies should be explicit, but often are not.

Subject: An active entity within a system, such as a process or device. Forms a unit for assigning authorizations, usually associated with (acting on behalf of) some user. Subjects are also objects.

Trusted computing base (TCB):   The collection of all mechanisms in a system that enforce security policy. The TCB boundary represents the security perimeter of the system.

References

Anderson, J. P. 1972. Computer Security Technology Planning Study. Vol I, ESD-TR-73-51. ESD/AFSC, (NTIS AD-758 206) Hanscom AFB, Bedford, MA.

Bell, D. E. and LaPadula, L. J. 1975. Secure Computer System: Unified Exposition and MULTICS Interpretation. MTR-2997, MITRE Corp., Bedford, MA.

Biba, K. J. 1977. Integrity Considerations for Secure Computer Systems. MTR-3153 (NTIS AD A039324), MITRE Corp., Bedford, MA.

Bishop, M. 1988. Theft of information in the take-grant protection model, pp. 194–218. In *Comput. Security Found. Workshop.* IEEE Computer Society Press, Los Alamitos, CA.

Brewer, D. F. C. and Nash, M. J. 1989. The Chinese Wall security policy, pp. 215–228. In *Proc. 1989 IEEE Symp. Security Privacy.* IEEE Computer Society Press, Los Alamitos, CA.

Clark, D. D. and Wilson, D. R. 1987. A comparison of commercial and military security policies, pp. 184–194. In *Proc. IEEE Symp. Security Privacy.* IEEE Computer Society Press, Los Alamitos, CA.

Denning, D. E. 1976. A lattice model of secure information flow. *Commun. ACM* 19(5):236–243.

Denning, D. E. 1983. *Cryptography and Data Security.* Addison–Wesley, Reading, MA.

Dobson, J. 1992 . Information and denial of service. In *Database Security V, IFIP Trans A-6*, C. Landwehr and S. Jajodia, eds., pp. 21–46. North–Holland, New York.

D. of D. 1985. Trusted Computer System Evaluation Criteria. Department of Defense, DoD 5200.28-STD, Fort Meade, MD.

Feiertag, R. J., Levitt, K. N., and Robinson, L. 1977. Proving multilevel security of a system design. Proc. 6th ACM Symp. Operating Syst. Principles. *ACM SIGOPS Operating Syst. Rev.* 11(5):57–65.

Froscher, J. N., Kang, M. H., McDermott, J., Costich, O., and Landwehr, C. E. 1994. A practical approach to high assurance multilevel secure computing service, pp. 2–11. In *Proc 10th Annu. Comput. Security Appl. Conf.* IEEE Computer Society Press, Los Alamitos, CA.

Gasser, M. 1988. *Building a Secure Computer System.* Van Nostrand Reinhold, New York.

Goguen, J. and Meseguer, J. 1984. Unwinding and inference control, pp. 75–86. In *Proc. 1984 IEEE Symp. Security Privacy.* IEEE Computer Society Press, Los Alamitos, CA.

Graham, G. S. and Denning, P. J. 1972. Protection—principles and practice, pp. 417–429. In *Proc. AFIPS Spring Jt. Comput. Conf.* AFIPS Press, Arlington, VA.

Harrison, M. A., Ruzzo, W. L., and Ullman, J. D. 1976. Protection in operating systems. *Commun. ACM* 19(8):461–471.

Jones, A. K., Lipton, R. J., and Snyder, L. 1976. A linear time algorithm for deciding security, pp. 337–366. In *Proc. 17th IEEE Symp. Found. Comput.* Houston, TX.

Kang, M. H., Moskowitz, I. S., and Lee, D. C. 1995. A network version of the Pump, pp. 144–154. In *Proc. IEEE Symp. Security Privacy.* IEEE Computer Society Press, Los Alamitos, CA.

Kemmerer, R. A. 1983. Shared resource matrix methodology: an approach to identifying storage and timing channels. *ACM Trans. Comp. Sys.* 1(3):256–277.

Lampson, B. W. 1971. Protection, pp. 437–443. In *5th Princeton Symp. Information Sci Sys*; reprinted *ACM Operating Sys. Rev.* 8(1):18–24.

Lampson, B. W. 1973. A note on the confinement problem. *Commun. ACM* 16(10):613–615.

Landwehr, C. E. 1983. Best available technologies for computer security. *IEEE Comput.* 16(7):86–100.

Leveson, N. and Turner, C. S. 1993. An investigation of the Therac-25 accidents. *IEEE Comput.* 26(7):18–41.

Lipner, S. B. 1982. Nondiscretionary controls for commercial applications, pp. 2–10. In *Proc. IEEE Symp. Security and Privacy.* IEEE Computer Society Press, Los Alamitos, CA.

McLean, J. 1985. A comment on the "basic security theorem" of Bell and LaPadula. *Inf. Proc. Lett.* 20(2): 67–70.

McLean, J. 1990. The specification and modeling of computer security. *IEEE Comput.* 23(1):9–16.

McLean, J. 1994. Security models. In *Encyclopedia of Software Engineering.* J. Marciniak, ed., pp. 1136–1145. Wiley, New York.

Millen, J. K. 1993. A resource allocation model for denial of service protection. *J. Comput. Security* 2(2, 3):89–106.

NRC 1991. Computers at Risk: Safe Computing in the Information Age. National Research Council. National Academy Press, Washington, DC.

Parker, D. B. 1994. Demonstrating the elements of information security with threats, pp. 421–430. In *Proc. 17th Nat. Comput. Security Conf.* NIST, Gaithersburg, MD.

Pfleeger, C. P. 1989. *Security in Computing.* Prentice–Hall, Englewood Cliffs, NJ.

Rushby, J. M. and Randell, B. 1983. A distributed secure system. *IEEE Comput.* 16(7):55–67.

Sandhu, R. S. 1988. The schematic protection model: its definition and analysis for acyclic attenuating schemes. *J. ACM* 35(2):404–432.

Sandhu, R. S. 1992. The typed access matrix model, pp. 122–136. In *Proc. IEEE Symp. on Res. Security Privacy.* IEEE Computer Society Press, Los Alamitos, CA.

Sandhu, R. S. 1993. Lattice-based access control models. *IEEE Comput.* 26(11):9–19.

Schell, R. R. and Denning, D. E. 1986. Integrity in trusted database systems, pp. 30–36. In *Proc. 9th Nat. Comput. Security Conf.* NIST, Gaithersburg, MD.

Shannon, C. E. and Weaver, W. 1949. *The Mathematical Theory of Communication.* University of Illinois Press, Urbana.

Shirley, L. J. and Schell, R. R. 1981. Mechanism sufficiency validation by assignment, pp. 26–32. In *Proc. IEEE Symp. Security Privacy*. IEEE Computer Society Press, Los Alamitos, CA.

Yu, C.-F. and Gligor, V. G. 1990. A specification and verification method for preventing denial of service. *IEEE Trans. Software Eng.* 16(6):581–592.

Further Information

Pfleeger's textbook, *Security in Computing*, provides a comprehensive introduction. Denning's *Cryptography and Data Security* is still a valuable source; the most comprehensive treatment of how to build an operating system to meet security requirements is Gasser's *Building a Secure Operating System*. Those new to the field may also find Computers at Risk, a report available from the National Research Council, a good introduction to computer security issues generally; it includes quite a bit of tutorial material along with its research recommendations.

Much of the key literature in this field is published in the proceedings of annual conferences. The proceedings of the IEEE Symposium on Security and Privacy, the Annual Computer Security Applications Conference, the smaller IEEE Workshop on the Foundations of Computer Security, the European Symposium on Research in Computer Security (ESORICS), and the ACM'S Conference on Computer and Communications Security will reflect the latest research results and applications related to security modeling. Relevant literature can also be found in the *Computer Security Journal*, *IEEE Transactions on Software Engineering*, and *Computers and Security*.

Finally, national and international standards for evaluating the security of computer systems (the U.S. *Trusted Computer System Evaluation Criteria*, the *Canadian Trusted Computer Product Evaluation Criteria*, the *European Information Technology Security Evaluation Criteria*, and the unified *Common Criteria*, now being developed jointly by European, Canadian, and U.S. representatives) contain relevant information for developers of secure systems.

91

Authentication, Access Control, and Intrusion Detection

Ravi S. Sandhu*
George Mason University

Pierangela Samarati
Università degli Studi di Milano

91.1 Introduction

An important requirement of any information management system is to protect information against improper disclosure or modification (known as confidentiality and integrity, respectively). Three mutually supportive technologies are used to achieve this goal. Authentication, access control, and audit together provide the foundation for information and system security as follows. *Authentication* establishes the identity of one party to another. Most commonly, authentication establishes the identity of a user to some part of the system typically by means of a password. More generally, authentication can be computer-to-computer or process-to-process and mutual in both directions. *Access control* determines what one party will allow another to do with respect to resources and objects mediated by the former. Access control usually requires authentication as a prerequisite. The *audit* process gathers data about activity in the system and analyzes it to discover security violations or diagnose their cause. Analysis can occur off line after the fact or it can occur on line more or less in real time. In the latter case, the process is usually called *intrusion detection*. This chapter discusses the scope and characteristics of these security controls.

Figure 91.1 is a logical picture of these security services and their interactions. Access control constrains what a user can do directly as well as what programs executing on behalf of the user are allowed to do. Access control is concerned with limiting the activity of legitimate users who have been successfully authenticated. It is enforced by a reference monitor, which mediates every attempted access by a user (or program executing on behalf of that user) to objects in the system. The reference monitor consults an authorization

*Portions of this paper appeared as Sandhu, R. S. and Samarati, P. 1994. Access control: principles and practice. *IEEE Commun.* 32(9):40–48. © 1994 IEEE. Used with permission.

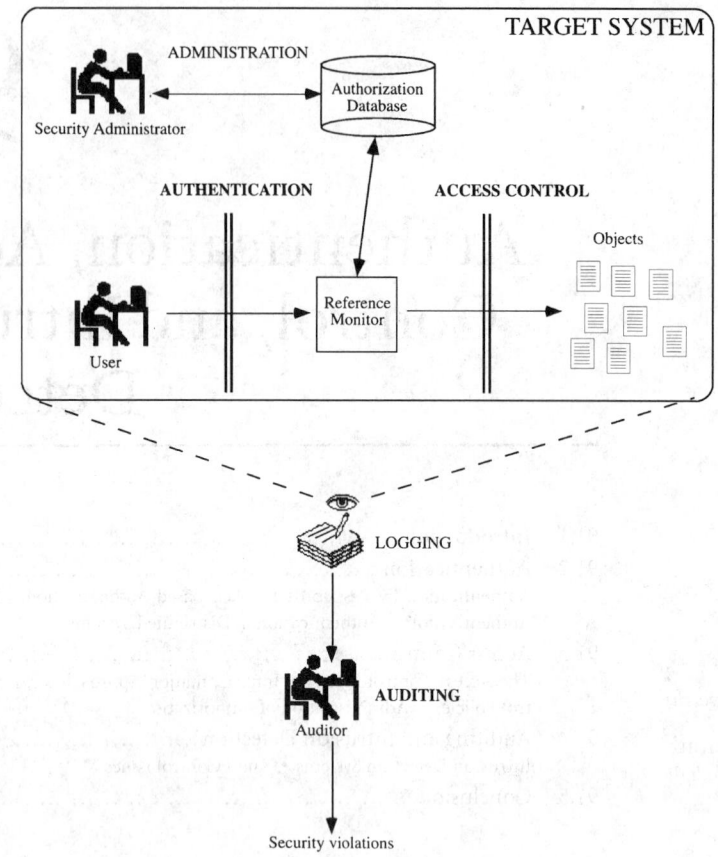

FIGURE 91.1 Access control and other security services.

database to determine if the user attempting to do an operation is actually authorized to perform that operation. Authorizations in this database are administered and maintained by a security administrator. The administrator sets these authorizations on the basis of the security policy of the organization. Users may also be able to modify some portion of the authorization database, for instance, to set permissions for their personal files. Auditing monitors and keeps a record of relevant activity in the system.

Figure 91.1 is a logical picture and should not be interpreted literally. For instance, as we will see later, the authorization database is often stored with the objects being protected by the reference monitor rather than in a physically separate area. The picture is also somewhat idealized in that the separation between authentication, access control, auditing, and administration services may not always be so clear cut. This separation is considered highly desirable but is not always faithfully implemented in every system.

It is important to make a clear distinction between authentication and access control. Correctly establishing the identity of the user is the responsibility of the authentication service. Access control assumes that identity of the user has been successfully verified prior to enforcement of access control via a reference monitor. The effectiveness of the access control rests on a proper user identification and on the correctness of the authorizations governing the reference monitor.

It is also important to understand that access control is not a complete solution for securing a system. It must be coupled with auditing. Audit controls concern a posteriori analysis of all of the requests and activities of users in the system. Auditing requires the recording of all user requests and activities for their later analysis. Audit controls are useful both as a deterrent against misbehavior and as a means to analyze the users' behavior in using the system to find out about possible attempted or actual violations. Auditing can also be useful for determining possible flaws in the security system. Finally, auditing is essential to ensure that authorized users do not misuse their privileges: in other words, to hold users accountable for their actions. Note that effective auditing requires that good authentication be in place; otherwise it is not possible to reliably attribute activity to individual users. Effective auditing also requires good access control; otherwise the audit records can themselves be modified by an attacker.

These three technologies are interrelated and mutually supportive. In the following sections, we discuss, respectively, authentication, access control, and auditing and intrusion detection.

91.2 Authentication

Authentication is in many ways the most primary security service on which other security services depend. Without good authentication, there is little point in focusing attention on strong access control or strong intrusion detection. The reader is surely familiar with the process of signing on to a computer system by providing an identifier and a password. In this most familiar form, authentication establishes the identity of a human user to a computer. In a networked environment, authentication becomes more difficult. An attacker who observes network traffic can replay authentication protocols to masquerade as a legitimate user.

More generally, authentication establishes the identity of one computer to another. Often, authentication is required to be performed in both directions. This is certainly true when two computers are engaged in communication as peers. Even in a client–server situation mutual authentication is useful. Similarly, authentication of a computer to a user is also useful to prevent against spoofing attacks in which one computer masquerades as another (perhaps to capture user identifiers and passwords).

Often we need a combination of user-to-computer and computer-to-computer authentication. Roughly speaking, user-to-computer authentication is required to establish identity of the user to a workstation and computer-to-computer authentication is required for establishing the identity of the workstation acting on behalf of the user to a server on the system (and vice versa). In distributed systems, authentication must be maintained through the life of a conversation between two computers. Authentication needs to be integrated into each packet of data that is communicated. Integrity of the contents of each packet, and perhaps confidentiality of contents, also must be ensured.

Our focus in this chapter is on user-to-computer authentication. User-to-computer authentication can be based on one or more of the following:

- Something the user knows, such as a password
- Something the user possesses, such as a credit-card sized cryptographic token or smart card
- Something the user is, exhibited in a biometric signature such as a fingerprint or voice print

We now discuss these in turn.

Authentication by Passwords

Password-based authentication is the most common technique, but it has significant problems. A well-known vulnerability of passwords is that they can be guessed, especially because users are prone to selecting weak passwords. A password can be snooped by simply observing a user keying it in. Users often need to provide their password when someone else is in a position to observe it as it is keyed in. Such compromise can occur without the user even being aware of it. It is also hard for users to remember too many passwords, especially for services that are rarely used. Nevertheless, because of low cost and low technology requirements, passwords are likely to be around for some time to come.

An intrinsic problem with passwords is that they can be shared, which breaks down accountability in the system. It is all too easy for a user to give their password to another user. Sometimes poor system design actually encourages password sharing because there may be no other convenient means of delegating permissions of one user to another (even though the security policy allows the delegation).

Password management is required to prod users to regularly change their passwords, to select good ones, and to protect them with care. Excessive password management makes adversaries of users and security administrators, which can be counterproductive. Many systems can configure a maximum lifetime for a password. Interestingly, many systems also have a minimum lifetime for a password. This has come about to prevent users from reusing a previous password when prompted to change their password after its maximum life has expired. The system keeps a history of, say, eight most recently used passwords for each user. When asked to change the current password the user can change it eight times to flush the history and then resume reuse of the same password. The response is to disallow frequent changes to a user's password!

Passwords are often used to generate cryptographic keys, which are further used for encryption or other cryptographic transformations. Encrypting data with keys derived from passwords is vulnerable to so-called dictionary attacks. Suppose the attacker has access to known plaintext, that is, the attacker knows the encrypted and plaintext versions of data that were encrypted using a key derived from a user's password. Instead of trying all possible keys to find the right one, the attacker instead tries keys generated from a list of, say, 20,000 likely passwords (known as a dictionary). The former search is usually computationally infeasible, whereas the latter can be accomplished in a matter of hours using commonplace workstations. These attacks have been frequently demonstrated and are a very real threat.

Operating systems typically store a user's password by using it as a key to some cryptographic transformation. Access to the so-called encrypted passwords provides the attacker the necessary known plaintext for a dictionary attack. The Unix system actually makes these encrypted passwords available in a publicly readable file. Recent versions of Unix are increasingly using shadow passwords by which these data are stored in files private to the authentication system. In networked systems, known plaintext is often visible in the network authentications protocols.

Poor passwords can be detected by off-line dictionary attacks conducted by the security administrators. Proactive password checking can be applied when a user changes his or her password. This can be achieved by looking up a large dictionary. Such dictionaries can be very big (tens of megabytes) and may need to be replicated at multiple locations. They can themselves pose a security hazard. Statistical techniques for proactive password checking have been proposed as an alternative [Davies and Ganesan 1993].

Selecting random passwords for users is not user friendly and also poses a password distribution problem. Some systems generate pronounceable passwords for users because these are easier to remember. In principle this is a sound idea but some of the earlier recommended methods for generating pronounceable passwords have been shown to be insecure [Ganesan and Davies 1994]. It is also possible to generate a sequence of one-time passwords that are used one-by-one in sequence without ever being reused. Human beings are not expected to remember these and must instead write them down or store them on laptop hard disks or removable media.

Token-Based Authentication

A token is a credit-card size device that the user carries around. Each token has a unique private cryptographic key stored within it, used to establish the token's identity via a challenge-response handshake. The party establishing the authentication issues a challenge to which a response is computed using the token's private key. The challenge is keyed into the token by the user and the response displayed by the token is again keyed by the user into the workstation to be communicated to the authenticating party. Alternately, the workstation can be equipped with a reader that can directly interact with the token, eliminating the need for the user to key in the challenge and response. Sometimes the challenge is implicitly taken to be the current time, so only the response needs to be returned (this assumes appropriately accurate synchronization of clocks).

The private key should never leave the token. Attempts to break the token open to recover the key should cause the key to be destroyed. Achieving this in the face of a determined adversary is a difficult task. Use of the token itself requires authentication; otherwise the token can be surreptitiously used by an intruder or stolen and used prior to discovery of the theft. User-to-token authentication is usually based on passwords in the form of a personal identification number (PIN).

Token-based authentication is much stronger than password-based authentication and is often called strong as opposed to weak authentication. However, it is the token that is authenticated rather than the user. The token can be shared with other users by providing the PIN, and so it is vulnerable to loss of accountability. Of course, only one user at a time can physically possess the token.

Tokens can use secret key or public key cryptosystems. With secret key systems the computer authenticating the token needs to know the secret key that is embedded in the token. This presents the usual key distribution problem for secret key cryptography. With public key cryptography, a token can be authenticated by a computer that has had no prior contact with the user's token. The public key used to verify the response to a challenge can be obtained with public key certificates. Public key-based tokens have scalability advantages that in the long run should make them the dominant technique for authentication in large systems. However, the computational and bandwidth requirements are generally greater for public vs. secret key systems. Token-based authentication is a technical reality today, but it still lacks major market penetration and does cost money.

Biometric Authentication

Biometric authentication has been used for some time for high-end applications. The biometric signature should be different every time, for example, voice-print check of a different challenge phrase on each occasion. Alternately, the biometric signature should require an active input, for example, dynamics of handwritten signatures. Simply repeating the same phrase every time or using a fixed signature such as a fingerprint is vulnerable to replay attacks. Biometric authentication often requires cumbersome equipment, which is best suited for fixed installations such as entry into a building or room.

Technically the best combination would be user-to-token biometric authentication, followed by mutual cryptographic authentication between the token and system services. This combination may emerge sooner than one might imagine. Deployment of such technology on a large scale is certain to raise social and political debate. Unforgeable biometric authentication could result in significant loss of privacy for individuals. Some of the privacy issues may have technical solutions, whereas others may be inherently impossible.

Authentication in Distributed Systems

In distributed systems, authentication is required repeatedly as the user uses multiple services. Each service needs authentication, and we might want mutual authentication in each case. In practice, this process starts with a user supplying a password to the workstation, which can then act on the user's behalf. This password should never be disclosed in plaintext on the network. Typically, the password is converted to a cryptographic key, which is then used to perform challenge-response authentication with servers in the system. To minimize exposure of the user password, and the long-term key derived from it, the password is converted into a short-term key, which is retained on the workstation, while the long-term user secrets are discarded. In effect these systems use the desktop workstation as a token for authentication with the rest of the network. Trojan horse software in the workstation can, of course, compromise the user's long-term secrets.

The basic principles just outlined have been implemented in actual systems in an amazing variety of ways. Many of the early implementations are susceptible to dictionary attacks. Now that the general nature and ease of a dictionary attack are understood we are seeing systems that avoid these attacks or at least attempt to make them more difficult. For details on actual systems, we refer the reader to Kaufman et al. [1995], Neuman [1994], and Woo and Lam [1992].

91.3 Access Control

In this section we describe access control. We introduce the concept of an access matrix and discuss implementation alternatives. Then we explain discretionary, mandatory, and role-based access control policies. Finally, we discuss issues in administration of authorizations.

The Access Control Matrix

Security practitioners have developed a number of abstractions over the years in dealing with access control. Perhaps the most fundamental of these is the realization that all resources controlled by a computer system can be represented by data stored in objects (e.g., files). Therefore, protection of objects is the crucial requirement, which in turn facilitates protection of other resources controlled via the computer system. (Of course, these resources must also be physically protected so they cannot be manipulated, directly bypassing the access controls of the computer system.)

Activity in the system is initiated by entities known as subjects. Subjects are typically users or programs executing on behalf of users. A user may sign on to the system as different subjects on different occasions, depending on the privileges the user wishes to exercise in a given session. For example, a user working on two different projects may sign on for the purpose of working on one project or the other. We then have two subjects corresponding to this user, depending on the project the user is currently working on.

A subtle point that is often overlooked is that subjects can themselves be objects. A subject can create additional subjects in order to accomplish its task. The children subjects may be executing on various computers in a network. The parent subject will usually be able to suspend or terminate its children as appropriate. The fact that subjects can be objects corresponds to the observation that the initiator of one operation can be the target of another. (In network parlance, subjects are often called initiators, and objects are called targets.)

The subject–object distinction is basic to access control. Subjects initiate actions or operations on objects. These actions are permitted or denied in accord with the authorizations established in the system. Authorization is expressed in terms of access rights or access modes. The meaning of access rights depends on the object in question. For files, the typical access rights are read, write, execute, and own. The meaning of the first three of these is self-evident. Ownership is concerned with controlling who can change the access permissions for the file. An object such as a bank account may have access rights inquiry, credit and debit corresponding to the basic operations that can be performed on an account. These operations would be implemented by application programs, whereas for a file the operations would typically be provided by the operating system.

The access matrix is a conceptual model that specifies the rights that each subject possesses for each object. There is a row in this matrix for each subject and a column for each object. Each cell of the matrix specifies the access authorized for the subject in the row to the object in the column. The task of access control is to ensure that only those operations authorized by the access matrix actually get executed. This is achieved by means of a reference monitor, which is responsible for mediating all attempted operations by subjects on objects. Note that the access matrix model clearly separates the problem of authentication from that of authorization.

An example of an access matrix is shown in Fig. 91.2, where the rights R and W denote read and write, respectively, and the other rights are as previously discussed. The subjects shown here are John, Alice, and Bob. There are four files and two accounts. This matrix specifies that, for example, John is the owner of file 3 and can read and write that file, but John has no access to file 2 or file 4. The precise meaning of ownership varies from one system to another. Usually the owner of a file is authorized to grant other users access to the file as well as revoke access. Because John owns file 1, he can give Alice the R right and Bob the R and W rights, as shown in Fig. 91.2. John can later revoke one or more of these rights at his discretion.

The access rights for the accounts illustrate how access can be controlled in terms of abstract operations implemented by application programs. The inquiry operation is similar to read in that it retrieves information but does not change it. Both the credit and debit operations will involve reading the previous account

	File 1	File 2	File 3	File 4	Account 1	Account 2
John	Own R W		Own R W		Inquiry Credit	
Alice	R	Own R W	W	R	Inquiry Debit	Inquiry Credit
Bob	R W	R		Own R W		Inquiry Debit

FIGURE 91.2   An access matrix.

balance, adjusting it as appropriate, and writing it back. The programs that implement these operations require read and write access to the account data. Users, however, are not allowed to read and write the account object directly. They can manipulate account objects only indirectly via application programs, which implement the debit and credit operations.

Also note that there is no own right for accounts. Objects such as bank accounts do not really have an owner who can determine the access of other subjects to the account. Clearly the user who establishes the account at the bank should not be the one to decide who can access the account. Within the bank different officials can access the account on the basis of their job functions in the organization.

Implementation Approaches

In a large system, the access matrix will be enormous in size, and most of its cells are likely to be empty. Accordingly, the access matrix is very rarely implemented as a matrix. We now discuss some common approaches to implementing the access matrix in practical systems.

Access Control Lists

A popular approach to implementing the access matrix is by means of access control lists (ACLs). Each object is associated with an ACL, indicating for each subject in the system the accesses the subject is authorized to execute on the object. This approach corresponds to storing the matrix by columns. ACLs corresponding to the access matrix of Fig. 91.2 are shown in Fig. 91.3. Essentially, the access matrix column for file 1 is stored in association with File 1, and so on.

By looking at an object's ACL, it is easy to determine which modes of access subjects are currently authorized for that object. In other words, ACLs provide for convenient access review with respect to an object. It is also easy to revoke all access to an object by replacing the existing ACL with an empty one. On the other hand, determining all of the accesses that a subject has is difficult in an ACL-based system. It is necessary to examine the ACL of every object in the system to do access review with respect to a subject. Similarly, if all accesses of a subject need to be revoked all ACLs must be visited one by one. (In practice, revocation of all accesses of a subject is often done by deleting the user account corresponding to that subject. This is acceptable if a user is leaving an organization. However, if a user is reassigned within the organization it would be more convenient to retain the account and change its privileges to reflect the changed assignment of the user.)

Many systems allow group names to occur in ACLs. For example, an entry such as (ISSE, R) can authorize all members of the ISSE group to read a file. Several popular operating systems, such as Unix and VMS, implement an abbreviated form of ACLs in which a small number, often only one or two, of group names can occur in the ACL. Individual subject names are not allowed. With this approach, the ACL has a small fixed size so it can be stored using a few bits associated with the file. At the other extreme, there are a number of access control packages that allow complicated rules in ACLs to limit when and how the access can be invoked. These rules can be applied to individual users or to all users who match a pattern defined in terms of user names or other user attributes.

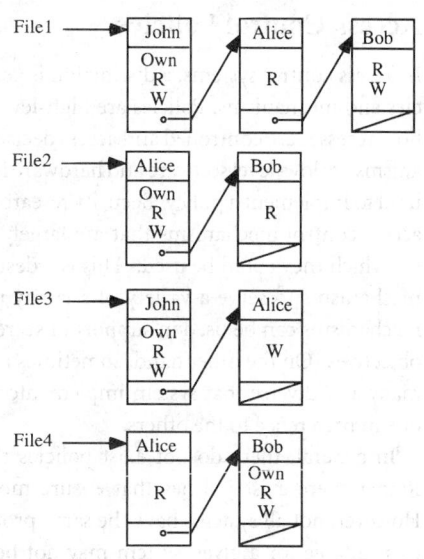

FIGURE 91.3 Access control lists for files in Fig. 91.2.

Capabilities

Capabilities are a dual approach to ACLs. Each subject is associated with a list, called the capability list, indicating for each object in the system the accesses the subject is authorized to execute on the object. This approach corresponds to storing the access matrix by rows. Figure 91.4 shows capability lists for the files in Fig. 91.2. In a capability list approach, it is easy to review all accesses that a subject is authorized to perform by simply examining the subject's capability list. However, determination of all subjects who can access a particular object requires examination of each and every subject's capability list. A number of capability-based computer systems were developed in the 1970s but did not prove to be commercially successful. Modern operating systems typically take the ACL-based approach.

It is possible to combine ACLs and capabili-

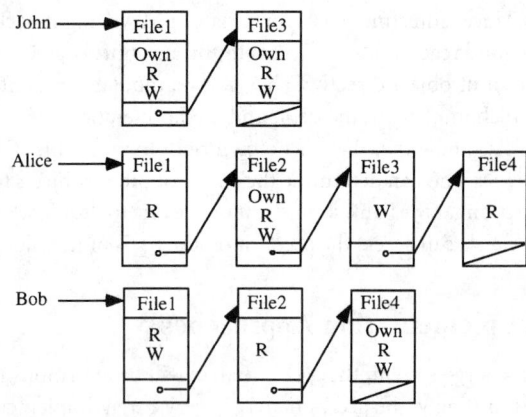

FIGURE 91.4   Capability lists for files in Fig. 91.2.

ties. Possession of a capability is sufficient for a subject to obtain access authorized by that capability. In a distributed system, this approach has the advantage that repeated authentication of the subject is not required. This allows a subject to be authenticated once, obtain its capabilities, and then present these capabilities to obtain services from various servers in the system. Each server may further use ACLs to provide finer grained access control.

Authorization Relations

We have seen that ACL- and capability-based approaches have dual advantages and disadvantages with respect to access review. There are representations of the access matrix that do not favor one aspect of access review over the other. For example, the access matrix can be represented by an authorization relation (or table), as shown in Fig. 91.5. Each row, or tuple, of this table specifies one access right of a subject to an object. Thus, John's accesses to file 1 require three rows. If this table is sorted by subject, we get the effect of capability lists. If it is sorted by object, we get the effect of ACLs. Relational database management systems typically use such a representation.

Access Control Policies

In access control systems, a distinction is generally made between policies and mechanisms. Policies are high-level guidelines that determine how accesses are controlled and access decisions are determined. Mechanisms are low-level software and hardware functions that can be configured to implement a policy. Security researchers have sought to develop access control mechanisms that are largely independent of the policy for which they could be used. This is a desirable goal to allow reuse of mechanisms to serve a variety of security purposes. Often, the same mechanisms can be used in support of secrecy, integrity, or availability objectives. On the other hand, sometimes the policy alternatives are so many and diverse that system implementors feel compelled to choose one in preference to the others.

In general, there do not exist policies that are better than others. Rather there exist policies that ensure more protection than others. However, not all systems have the same protection requirements. Policies suitable for a given system may not be suitable for another. For instance, very strict access control policies, which are crucial to some systems, may be inappropriate for environments where users require

Subject	Access mode	Object
John	Own	File 1
John	R	File 1
John	W	File 1
John	Own	File 3
John	R	File 3
John	W	File 3
Alice	R	File 1
Alice	Own	File 2
Alice	R	File 2
Alice	W	File 2
Alice	W	File 3
Alice	R	File 4
Bob	R	File 1
Bob	W	File 1
Bob	R	File 2
Bob	Own	File 4
Bob	R	File 4
Bob	W	File 4

FIGURE 91.5   Authorization relation for files in Fig. 91.2.

greater flexibility. The choice of access control policy depends on the particular characteristics of the environment to be protected.

We will now discuss three different policies that commonly occur in computer systems as follows:

- Classic discretionary policies
- Classic mandatory policies
- The emerging role-based policies

We have added the qualifier classic to the first two of these to reflect the fact that these have been recognized by security researchers and practitioners for a long time. However, in recent years there is increasing consensus that there are legitimate policies that have aspects of both of these. Role-based policies are an example of this fact.

It should be noted that access control policies are not necessarily exclusive. Different policies can be combined to provide a more suitable protection system. This is indicated in Fig. 91.6. Each of the three inner circles represents a policy that allows a subset of all possible accesses. When the policies are combined, only the intersection of their accesses is allowed. Such a combination of policies is relatively straightforward so long as there are no conflicts where one policy asserts that a particular access *must* be allowed while another one prohibits it. Such conflicts between policies need to be reconciled by negotiations at an appropriate level of management.

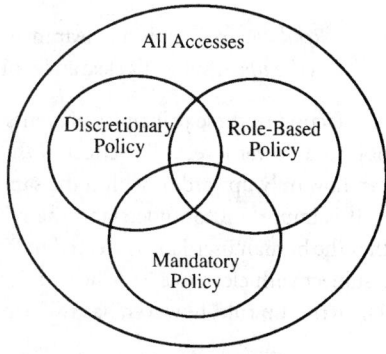

FIGURE 91.6 Multiple access control policies.

Classic Discretionary Policies

Discretionary protection policies govern the access of users to the information on the basis of the user's identity and authorizations (or rules) that specify, for each user (or group of users) and each object in the system, the access modes (e.g., read, write, or execute) the user is allowed on the object. Each request of a user to access an object is checked against the specified authorizations. If there exists an authorization stating that the user can access the object in the specific mode, the access is granted, otherwise it is denied.

The flexibility of discretionary policies makes them suitable for a variety of systems and applications. For these reasons, they have been widely used in a variety of implementations, especially in the commercial and industrial environments.

However, discretionary access control policies have the drawback that they do not provide real assurance on the flow of information in a system. It is easy to bypass the access restrictions stated through the authorizations. For example, a user who is able to read data can pass it to other users not authorized to read it without the cognizance of the owner. The reason is that discretionary policies do not impose any restriction on the usage of information by a user once the user has read it, i.e., dissemination of information is not controlled. By contrast, dissemination of information is controlled in mandatory systems by preventing flow of information from high-level objects to low-level objects.

Discretionary access control policies based on explicitly specified authorization are said to be closed in that the default decision of the reference monitor is denial. Similar policies, called open policies, could also be applied by specifying denials instead of permissions. In this case, for each user and each object of the system, the access modes the user is forbidden on the object are specified. Each access request by a user is checked against the specified (negative) authorizations and granted only if no authorizations denying the access exist. The use of positive and negative authorizations can be combined, allowing the specification of both the accesses to be authorized as well as the accesses to be denied to the users. The interaction of positive and negative authorizations can become extremely complicated [Bertino et al. 1993].

Classic Mandatory Policies

Mandatory policies govern access on the basis of classification of subjects and objects in the system. Each user and each object in the system is assigned a security level. The security level associated with an object reflects the sensitivity of the information contained in the object, i.e., the potential damage that could result from unauthorized disclosure of the information. The security level associated with a user, also called clearance, reflects the user's trustworthiness not to disclose sensitive information to users not cleared to see it. In the simplest case, the security level is an element of a hierarchical ordered set. In the military and civilian government arenas, the hierarchical set generally consists of top secret (TS), secret (S), confidential (C), and unclassified (U), where TS > S > C > U. Each security level is said to dominate itself and all others below it in this hierarchy.

Access to an object by a subject is granted only if some relationship (depending on the type of access) is satisfied between the security levels associated with the two. In particular, the following two principles are required to hold:

> *Read down:* A subject's clearance must dominate the security level of the object being read.
> *Write up:* A subject's clearance must be dominated by the security level of the object being written.

Satisfaction of these principles prevents information in high-level objects (i.e., more sensitive) to flow to objects at lower levels. The effect of these rules is illustrated in Fig. 91.7. In such a system, information can flow only upward or within the same security class.

It is important to understand the relationship between users and subjects in this context. Let us say that the human user Jane is cleared to S and assume she always signs on to the system as an S subject (i.e., a subject with clearance S). Jane's subjects are prevented from reading TS objects by the read-down rule. The write-up rule, however, has two aspects that seem at first sight contrary to expectation:

- First, Jane's S subjects can write a TS object (even though they cannot read it). In particular, they can overwrite existing TS data and therefore destroy it. Because of this integrity concern, many systems for mandatory access control do not allow write up but limit writing to the same level as the subject. At the same time, write up does allow Jane's S subjects to send electronic mail to TS subjects and can have its benefits.

- Second, Jane's S subjects cannot write C or U data. This means, for example, that Jane can never send electronic mail to C or U users. This is contrary to what happens in the paper world, where S users can write memos to C and U users. This seeming contradiction is easily eliminated by allowing Jane to sign to the system as a C or U subject as appropriate. During these sessions, she can send electronic mail to C or U and C subjects.

In other words, a user can sign on to the system as a subject at any level dominated by the user's clearance. Why then bother to impose the write-up rule? The main reason is to prevent malicious software from

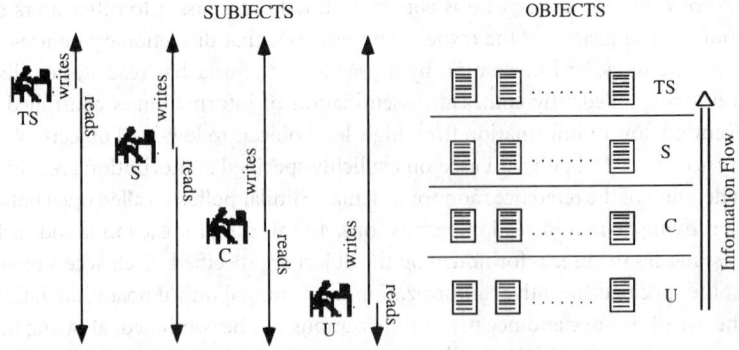

SUBJECTS   OBJECTS

FIGURE 91.7   Controlling information flow for secrecy.

leaking secrets downward from S to U. Users are trusted not to leak such information, but the programs they execute do not merit the same degree of trust. For example, when Jane signs on to the system at U level, her subjects cannot read S objects and thereby cannot leak data from S to U. The write-up rule also prevents users from inadvertently leaking information from high to low.

In addition to hierarchical security levels, categories (e.g., Crypto, NATO, Nuclear) can also be associated with objects and subjects. In this case, the classification labels associated with each subject and each object consist of a pair composed of a security level and a set of categories. The set of categories associated with a user reflect the specific areas in which the user operates. The set of categories associated with an object reflect the area to which information contained in objects are referred. The consideration of categories provides a finer grained security classification. In military parlance, categories enforce restriction on the basis of the need-to-know principle, i.e., a subject should be given only those accesses that are required to carry out the subject's responsibilities.

Mandatory access control can as well be applied for the protection of information integrity. For example, the integrity levels could be crucial (C), important (I), and unknown (U). The integrity level associated with an object reflects the degree of trust that can be placed in the information stored in the object and the potential damage that could result from unauthorized modification of the information. The integrity level associated with a user reflects the user's trustworthiness for inserting, modifying, or deleting data and programs at that level. Principles similar to those stated for secrecy are required to hold, as follows:

Read up: A subject's integrity level must be dominated by the integrity level of the object being read.
Write down: A subject's integrity level must dominate the integrity level of the object being written.

Satisfaction of these principles safeguard integrity by preventing information stored in low objects (and therefore less reliable) to flow to high objects. This is illustrated in Fig. 91.8. Controlling information flow in this manner is but one aspect of achieving integrity. Integrity in general requires additional mechanisms, as discussed in Castano et al. [1994] and Sandhu [1994].

Note that the only difference between Fig. 91.7 and 91.8 is the direction of information flow: bottom to top in the former case and top to bottom in the latter. In other words, both cases are concerned with one-directional information flow. The essence of classical mandatory controls is one-directional information flow in a lattice of security labels. For further discussion on this topic, see Sandhu [1993].

Role-Based Policies

The discretionary and mandatory policies previously discussed have been recognized in official standards, notably, the Orange Book of the U.S. Department of Defense. A good introduction to the Orange Book and its evaluation procedures is given in Chokhani [1992].

There has been a strong feeling among security researchers and practitioners that many practical requirements are not covered by these classic discretionary and mandatory policies. Mandatory policies rise from rigid environments, such as those of the military. Discretionary policies rise from cooperative

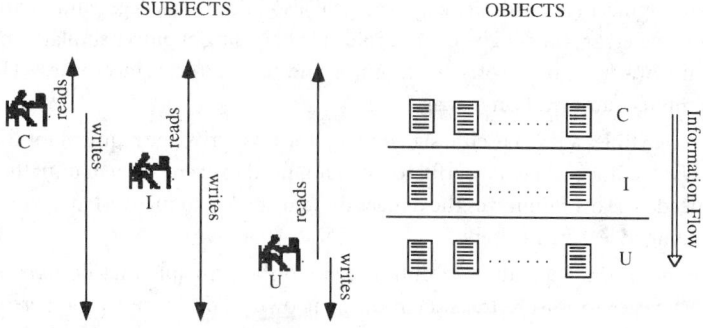

FIGURE 91.8 Controlling information flow for integrity.

yet autonomous requirements, such as those of academic researchers.
Neither requirement satisfies the needs of most commercial enterprises.
Orange Book discretionary policy is too weak for effective control of
information assets, whereas Orange Book mandatory policy is focused on
the U.S. Government policy for confidentiality of classified information.
(In practice the military often finds Orange Book mandatory policies to
be too rigid and subverts them.)

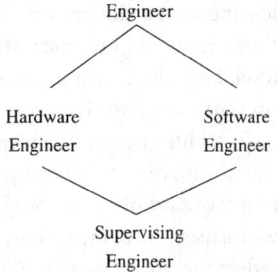

FIGURE 91.9 A role inheritance hierarchy.

Several alternatives to classic discretionary and mandatory policies have
been proposed. These policies allow the specification of authorizations to
be granted to users (or groups) on objects as in the discretionary approach,
together with the possibility of specifying restrictions (as in the mandatory
approach) on the assignment or on the use of such authorizations. One
of the promising avenues that is receiving growing attention is that of role-based access control [Ferraiolo
and Kuhn 1992, Sandhu et al. 1996].

Role-based policies regulate the access of users to the information on the basis of the activities the users
execute in the system. Role-based policies require the identification of roles in the system. A role can
be defined as a set of actions and responsibilities associated with a particular working activity. Then,
instead of specifying all of the accesses each user is allowed to execute, access authorizations on objects are
specified for roles. Users are given authorizations to adopt roles. A recent study by the National Institute
of Standards and Technology (NIST) confirms that roles are a useful approach for many commercial and
government organizations [Ferraiolo and Kuhn 1992].

The user playing a role is allowed to execute all accesses for which the role is authorized. In general, a
user can take on different roles on different occasions. Also, the same role can be played by several users,
perhaps simultaneously. Some proposals for role-based access control allow a user to exercise multiple
roles at the same time. Other proposals limit the user to only one role at a time or recognize that some
roles can be jointly exercised, whereas others must be adopted in exclusion to one another. As yet there are
no standards in this arena, and so it is likely that different approaches will be pursued in different systems.

The role-based approach has several advantages. Some of these are discussed in the following:

- *Authorization management:* Role-based policies benefit from a logical independence in specifying
  user authorizations by breaking this task into two parts, one that assigns users to roles and one
  that assigns access rights for objects to roles. This greatly simplifies security management. For
  instance, suppose a user's, responsibilities change, say, due to a promotion. The user's current
  roles can be taken away and new roles assigned as appropriate for the new responsibilities. If all
  authorization is directly between users and objects, it becomes necessary to revoke all existing
  access rights of the user and assign new ones. This is a cumbersome and time-consuming task.

- *Hierarchical roles:* In many applications, there is a natural hierarchy of roles based on the familiar
  principles of generalization and specialization. An example is shown in Fig. 91.9. Here the roles of
  hardware and software engineer are specializations of the engineer role. A user assigned to the role
  of software engineer (or hardware engineer) will also inherit privileges and permissions assigned
  to the more general role of engineer. The role of supervising engineer similarly inherits privileges
  and permissions from both software-engineer and hardware-engineer roles. Hierarchical roles
  further simplify authorization management.

- *Least privilege:* Roles allow a user to sign on with the least privilege required for the particular task
  at hand. Users authorized to powerful roles do not need to exercise them until those privileges are
  actually needed. This minimizes the danger of damage due to inadvertent errors or by intruders
  masquerading as legitimate users.

- *Separation of duties:* Separation of duties refers to the principle that no users should be given
  enough privileges to misuse the system on their own. For example, the person authorizing a
  paycheck should not also be the one who can prepare them. Separation of duties can be enforced
  either statically (by defining conflicting roles, i.e., roles that cannot be executed by the same user)

or dynamically (by enforcing the control at access time). An example of dynamic separation of duty is the two-person rule. The first user to execute a two-person operation can be any authorized user, whereas the second user can be any authorized user different from the first.

- *Object classes:* Role-based policies provides a classification of users according to the activities they execute. Analogously, a classification should be provided for objects. For example, generally a clerk will need to have access to the bank accounts, and a secretary will have access to the letters and memos (or some subset of them). Objects could be classified according to their type (e.g., letters, manuals) or to their application area (e.g., commercial letters, advertising letters). Access authorizations of roles should then be on the basis of object classes, not specific objects. For example, a secretary can be given the authorization to read and write the entire class of letters instead of being given explicit authorization for each single letter. This approach has the advantage of making authorization administration much easier and better controlled. Moreover, the accesses authorized on each object are automatically determined according to the type of the object without need of specifying authorizations on each object creation.

Administration of Authorization

Administrative policies determine who is authorized to modify the allowed accesses. This is one of the most important, and least understood, aspects of access controls.

In mandatory access control, the allowed accesses are determined entirely on the basis of the security classification of subjects and objects. Security levels are assigned to users by the security administrator. Security levels of objects are determined by the system on the basis of the levels of the users creating them. The security administrator is typically the only one who can change security levels of subjects or objects. The administrative policy is therefore very simple.

Discretionary access control permits a wide range of administrative policies. Some of these are described as follows:

- *Centralized:* A single authorizer (or group) is allowed to grant and revoke authorizations to the users.
- *Hierarchical:* A central authorizer is responsible for assigning administrative responsibilities to other administrators. The administrators can then grant and revoke access authorizations to the users of the system. Hierarchical administration can be applied, for example, according to the organization chart.
- *Cooperative:* Special authorizations on given resources cannot be granted by a single authorizer but needs cooperation of several authorizers.
- *Ownership:* Users are considered owners of the objects they create. The owner can grant and revoke access rights for other users to that object.
- *Decentralized:* In decentralized administration, the owner of an object can also grant other users the privilege of administering authorizations on the object.

Within each of these there are many possible variations.

Role-based access control has a similar wide range of possible administrative policies. In this case, roles can also be used to manage and control the administrative mechanisms.

Delegation of administrative authority is an important area in which existing access control systems are deficient. In large distributed systems, centralized administration of access rights is infeasible. Some existing systems allow administrative authority for a specified subset of the objects to be delegated by the central security administrator to other security administrators. For example, authority to administer objects in a particular region can be granted to the regional security administrator. This allows delegation of administrative authority in a selective piecemeal manner. However, there is a dimension of selectivity that is largely ignored in existing systems. For instance, it may be desirable that the regional security administrator be limited to granting access to these objects only to employees who work in that region. Control over the regional administrators can be centrally administered, but they can have considerable autonomy within their regions. This process of delegation can be repeated within each region to set up subregions, and so on.

91.4 Auditing and Intrusion Detection

Auditing consists of examination of the history of events in a system to determine whether and how security violations have occurred or been attempted. Auditing requires registration or logging of users' requests and activities for later examination. Audit data are recorded in an *audit trail* or *audit log*. The nature and format of these data vary from system to system.

Information that should be recorded for each event includes the subject requesting the access, the object to be accessed, the operation requested, the time of the request, perhaps the location from which the requested originated, the response of the access control system, the amount of resources [central processing unit (CPU) time, input/output (I/O), memory, etc.] used, and whether the operation succeeded or, if not, the reason for the failure, and so on.

In particular, actions requested by privileged users, such as the system and the security administrators, should be logged. First, this serves as a deterrent against misuse of powerful privileges by the administrators as well as a means for detecting operations that must be controlled (the old problem of guarding the guardian). Secondly, it allows control of penetrations in which the attacker gains a privileged status.

Audit data can become voluminous very quickly and searching for security violations in such a mass of data is a difficult task. Of course, audit data cannot reveal all violations because some may not be apparent in even a very careful analysis of audit records. Sophisticated penetrators can spread out their activities over a relatively long period of time, thus making detection more difficult. In some cases, audit analysis is executed only if violations are suspected or their effects are visible because the system shows an anomalous or erroneous behavior, such as continuous insufficient memory, slow processing, or nonaccessibility of certain files. Even in this case, often only a limited amount of audit data, namely, those that may be connected with the suspected violation, are examined. Sometimes the first clue to a security violation is some real-world event which indicates that information has been compromised. That may happen long after the computer penetration occurred. Similarly, security violations may result in Trojan horses or viruses being implanted whose activity may not be triggered until long after the original event.

Intrusion Detection Systems

Recent research has focused on the development of automated tools to help or even to carry out auditing controls. Automated tools can be used to screen and reduce audit data that need to be reviewed by humans. These tools can also organize audit data to produce summaries and measures needed in the analysis. This data reduction process can, for instance, produce short summaries of user behaviors, anomalous events, or security incidents. The auditors can then go over summaries instead of examining each single event recorded. Another class of automated tools is represented by the so-called *intrusion detection systems*. The purpose of these tools is not only to automate audit data acquisition and reduction but also its analysis. Some of the more ambitious efforts attempt to perform intrusion detection in real time.

Intrusion detection systems can be classified as *passive* or *active*. Passive systems, generally operating off line, analyze the audit data and bring possible intrusions or violations to the attention of the auditor, who then takes appropriate actions (see Fig. 91.10). Active systems analyze audit data in real time. Besides bringing violations to the attention of the auditor, these systems may take an immediate protective response on the system (see Fig. 91.11). The protective response can be executed ex post facto, after the violation has occurred, or preemptively, to avoid the violation being perpetrated to completion. This latter possibility depends on the ability of the system to foresee violations. Protective responses include killing the suspected process, disconnecting the user, disabling privileges, or disabling user accounts. The response may be determined in total autonomy by the intrusion detection system or through interactions with the auditors.

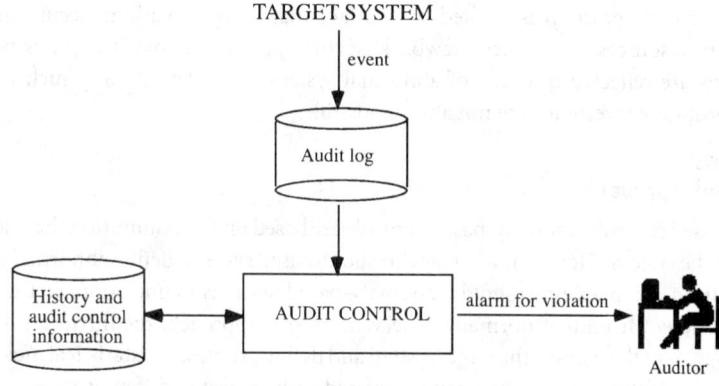

FIGURE 91.10   Passive intrusion detection.

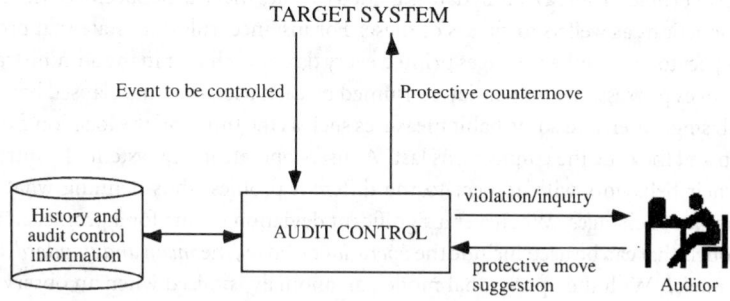

FIGURE 91.11   Active intrusion detection.

Different approaches have been proposed for building intrusion detection systems. No single approach can be considered satisfactory with respect to different kinds of penetrations and violations that can occur. Each approach is appropriate for detecting a specific subset of violations. Moreover, each approach presents some pros and cons determined by the violations that can or cannot be controlled and by the amount and complexity of information necessary for its application. We now discuss the main intrusion detection approaches that have been attempted.

Threshold-Based Approach

The threshold-based approach is based on the assumption that the exploitation of system vulnerabilities involves abnormal use of the system itself. For instance, an attempt to break into a system can require trying several user accounts and passwords. An attempt to discover protected information can imply several, often denied, browsing operations through protected directories. A process infected by a virus can require an abnormal amount of memory or CPU resources.

Threshold-based systems typically control occurrences of specific events over a given period of time with respect to predefined allowable thresholds established by the security officer. For instance, more than three unsuccessful attempts to log in to a given account with the wrong password may indicate an attempt to penetrate that account. Multiple unsuccessful attempts to log in the system, using different accounts, concentrated in a short period of time, may suggest an attempt to break in.

Thresholds can also be established with respect to authorized operations to detect improper use of resources. For instance, a threshold can specify that print requests totaling more than a certain number of pages a day coming from the administrative office is to be considered suspicious. This misuse can be symptomatic of different kinds of violations such as the relatively benign misuse of the resource for personal use or a more serious attempt to print out working data for disclosure to the competition.

The threshold-based approach is limited by the fact that many violations occur without implying overuse of system resources. A further drawback of this approach is that it requires prior knowledge of how violations are reflected in terms of abnormal system use. Determining such connections and establishing appropriate thresholds are not always possible.

Anomaly-Based Approach

Like threshold-based controls, anomaly-based controls are based on the assumption that violations involve abnormal use of the system. However, whereas threshold-based systems define abnormal use with respect to prespecified fixed acceptable thresholds, anomaly-based systems define abnormal use as a use that is significantly different from that normally observed. In this approach, the intrusion detection system observes the behavior of the users in the target system and define profiles, i.e., statistical measures, reflecting the normal behavior of the users. Profiles can be defined with respect to different aspects to be controlled such as the number of events in a user session, the time elapsed between events in a user session, and the amount of resources consumed over a certain period of time or during execution of certain programs.

Construction of profiles from raw audit data is guided by rules that can be specified with respect to single users, objects, or actions as well as to classes of these. For instance, rules can state that profiles should be defined with respect to the number of pages printed every day by each user in the administration office, the number of resources per session and per day consumed by each user, the time elapsed between two log-in sessions for each single user, and some habit measures such as the time and the location from which a user generally logs in and the time the connections last. As users operate in the system, the intrusion detection system learns their behaviors with respect to the different profiles, thus defining what is normal and adapting the profiles to changes. Whenever a significant deviation occurs for a profile, an alarm is raised.

Statistical models that can be used include the *operational model*, the *mean and standard deviation model*, and *time series model*. With the operational model, an anomaly is raised when an observation exceeds a given acceptable threshold. This is similar to the threshold-based approach. With the mean and standard deviation model, an anomaly occurs when the observation falls outside an allowed confidence interval around the mean. For instance, an alarm can be raised if the CPU time consumed during a session for a user falls much below or above the CPU time generally consumed by the same user. With the time series model, an anomaly is raised when an event occurs at a given time at which the probability of its occurring is too low. For instance, a remote night-hour log-in request by a user who has never connected off hours or from outside the building may be considered suspicious.

The main advantage of the anomaly detection approach is that it does not require any a priori knowledge of the target system or of possible flaws from which the system may suffer. However, like the threshold-based approach, it can detect only violations that involve anomalous use. Moreover, some legitimate users may have a very erratic behavior (e.g., logging on and off at different hours or from different locations, varying their activity daily). For such users, no normal behavior can be actually established and misuse by them as well as by masqueraders exploiting their accounts would go undetected. The approach is also vulnerable from insiders who, knowing that behavior profiles are being defined, may either behave in a *bad* way from the beginning or slowly vary their behavior, going from *good* to *bad*, thus convincing the system that the bad behavior is normal.

Rule-Based Approach

In the rule-based approach, rules are used to analyze audit data for suspicious behavior independently from users' behavioral patterns. Rules describe what is suspicious on the basis of known past intrusions or known system vulnerabilities. This approach is generally enforced by means of expert systems encoding knowledge of the security experts about past intrusions in terms of sequences of events or observable properties characterizing violations.

For instance, a rule can specify that a sequence of browsing operations (e.g., **cd**, **ls**, and **more** commands in a Unix environment) coming off hours from a remote location may be symptomatic of an intrusion. Rules can also identify suspicious sequences of actions. For example, that withdrawal of a large amount of money from an account and its deposit back a few days later may be considered suspicious.

The rule-based approach can detect violations that do not necessarily imply abnormal use of resources. Its main limitation is that the expert knowledge encoded in the rules encompasses only known system vulnerabilities and attack scenarios or suspicious events. The system can therefore be penetrated by attackers employing new techniques.

Model-Based Reasoning Approach

The model-based reasoning approach is based on the definition, by the security officers, of models of proscribed intrusion activities [Garvey and Lunt 1991]. Proscribed activities are expressed by means of sequences of user behaviors (single events or observable measures), called scenarios.

Each component of a scenario is therefore a high-level observation on the system and does not necessarily correspond to an audit record (which contains information at a lower level of specification). From these high-level specifications, the intrusion detection system generates, on the basis of specified rules, the corresponding sequences of actions at the level of the audit records. Each audit record produced on the observation of the system is controlled against the specified scenarios to determine if a violation is being carried out. Audit data reduction and analysis can be modeled in such a way that only events relevant to specific scenarios corresponding to intrusions probably being carried out are examined. When the probability of a given scenario being followed passes a specified threshold, an alarm is raised informing the auditor of the suspected violation.

The basis of this approach is essentially the same as the rule-based approach, the main difference being the way in which controls are specified. Whereas in the rule-based approach the security officer must explicitly specify the control rules in terms of the audit data, in the model-based approach the security officer specifies the scenario only in terms of high-level observable properties. This constitutes the main advantage of this approach, which allows the security officer to reason in terms of high-level abstractions rather than audit records. It is the task of the system to translate the scenarios into corresponding rules governing data reduction and analysis.

Like the rule-based approach, this approach can control only violations whose perpetration scenario (i.e., actions necessary to fulfill them) are known. By contrast, violations exploiting unknown vulnerabilities or not yet tried violations cannot be detected.

State Transition-Based Approach

In the state transition-based approach, a violation is modeled as a sequence of actions starting from an initial state to a final compromised state [Ilgun et al. 1995]. A state is a snapshot of the target system representing the values of all volatile, semipermanent, and permanent memory locations. Between the initial and the final states there are a number of intermediate states and corresponding transitions. State transitions correspond to key actions necessary to carry out the violation. Actions do not necessarily correspond to commands issued by users but, instead, refer to how state changes within the system are achieved. A single command may produce multiple actions. Each state is characterized as a set of assertions evaluating whether certain conditions are verified in the system. For instance, assertions can check whether a user is the owner of an object or has some privileges on it, whether the user who caused the last two transitions is the same user, or whether the file to which an action is referred is a particular file. Actions corresponding to state transitions are accesses to files (e.g., read, write, or execute operations), or actions modifying the permissions associated with files (e.g., changes of owners or authorizations), or some of the files' characteristics (e.g., rename operations).

As the users operate in the system, state transitions caused by them are determined. Whenever a state transition causes a final compromised state to be reached, an alarm is raised. The state transition-based approach can also be applied in a real-time active system to prevent users from executing operations that would cause a transition to a compromised state.

The state transition-based approach is based on the same concepts as the rule-based approach and therefore suffers from the same limitations, i.e., only violations whose scenarios are known can be detected. Moreover, it can be used to control only those violations that produce visible changes to the system state. Like the model-based approach, the state transition-based approach provides the advantage of requiring only high-level specifications, leaving the system the task of mapping state transitions into audit records and producing the corresponding control rules. Moreover, because a state transition can be matched by different operations at the audit record level, a single state transition specification can be used to represent different variations of a violation scenario (i.e., involving different operations but causing the same effects on the system).

Other Approaches

Other approaches have been proposed to complement authentication and access control to prevent violations from happening or to detect their occurrence.

One approach consists of preventing, rather than detecting, intrusions. In this class are tester programs that evaluate the system for common weaknesses often exploited by intruders and password checker programs that prevent users from choosing weak or obvious passwords (which may represent an easy target for intruders).

Another approach consists of substituting known bugged commands, generally used as trap doors by intruders, with programs that simulate the commands' execution while sending an alarm to the attention of the auditor. Other trap programs for intruders are represented by fake user accounts with *magic* passwords that raise an alarm when they are used.

Other approaches aim at detecting or preventing execution of Trojan horses and viruses. Solutions adopted for this include integrity checking tools that search for unauthorized changes to files and mechanisms controlling program executions against specifications of allowable program behavior in terms of operations and data flows.

Yet another intrusion detection approach is represented by the so-called keystroke latencies control. The idea behind the approach is that the elapsed time between keystrokes for regularly typed strings is quite consistent for each user. Keystroke latencies control can be used to cope against masqueraders. Moreover, they can also be used for authentication by controlling the time elapsed between the keystrokes when typing the password.

More recent research has interested intrusion detection at the network level [Mukherjee et al. 1994]. Analysis is performed on network traffic instead of on commands (or their corresponding low-level operations) issued on a system. Anomalies can then be determined, for example, on the basis of the probability of the occurrence of the monitored connections being too low or on the basis of the behavior of the connections. In particular, traffic is controlled against profiles of expected traffic specified in terms of expected paths (i.e., connections between systems) and service profiles.

Audit Control Issues

There are several issues that must be considered when employing intrusion detection techniques to identify security violations. These issues arise independently of the specific intrusion detection approach being utilized.

The task of generating audit records can be left to either the target system being monitored or the intrusion detection system. In the former case, the audit information generated by the system may need to be converted to a form understandable by the intrusion detection system. Many operating systems and database systems provide some audit information. However, this information often is not appropriate for security controls because it may contain data not relevant for detecting intrusions and omits details needed for identifying violations. Moreover, the audit mechanism of the target system may itself be vulnerable to a penetrator who might be able to bypass auditing or modify the audit log. Thus, a stronger and more appropriate audit trail might be required for effective intrusion detection.

Another important issue that must be addressed is the retention of audit data. Because the quantity of audit data generated every day can be enormous, policies must be specified that determine when historical data can be discarded.

Audit events can be recorded at different granularity. Events can be recorded at the system command level, at the level of each system call, at the application level, at the network level, or at the level of each keystroke. Auditing at the application and command levels has the advantage of producing high-level traces, which can be more easily correlated, especially by humans (who would get lost in low-level details). However, the actual effect of the execution of a command or application on the system may not be reflected in the audit records and therefore cannot be analyzed. Moreover, auditing at such a high level can be circumvented by users exploiting alias mechanisms or by directly issuing lower level commands. Recording at lower levels overcomes this drawback at the price of maintaining a greater number of audit records (a single user command may correspond to several low-level operations) whose examination by humans (or automated tools) therefore becomes more complicated.

Different approaches can be taken with respect to the time at which the audit data are recorded and, in the case of real-time analysis, evaluated. For instance, the information that a user has requested execution of a process can be passed to the intrusion detection system at the time the execution is required or at the time it is completed. The former approach has the advantage of allowing timely detection and, therefore, a prompt response to stop the violation. The latter approach has the advantage of providing more complete information about the event being monitored (information on resources used or time elapsed can be provided only after the process has completed) and therefore allows more complete analysis.

Audit data recording or analysis can be carried out indiscriminately or selectively, namely, on specific events, such as events concerning specific subjects, objects, or operations, or occurring at a particular time or in a particular situation. For instance, audit analysis can be performed only on operations on objects containing sensitive information, on actions executed off hours (nights and weekends) or from remote locations, on actions denied by the access control mechanisms, or on actions required by mistrusted users.

Different approaches can be taken with respect to the time at which audit control should be performed. Real-time intrusion detection systems enforce control in real time, i.e., analyze each event at the time of its occurrence. Real-time analysis of data brings the great advantage of timely detection of violations. However, because of the great amount of data to analyze and the analysis to be performs, real-time controls are generally performed only on selected data, leaving a more thorough analysis to be performed off line. Approaches that can be taken include the following:

- *Period driven:* Audit control is executed periodically. For example, every night the audit data produced during the working day are examined.
- *Session driven:* Audit control on a user's session is performed when a close session command is issued.
- *Event driven:* Audit control is executed upon occurrence of certain events. For instance, if a user attempts to enter a protected directory, audit over the user's previous and/or subsequent actions is initiated.
- *Request driven:* Audit control is executed upon the explicit request of the security officer.

The intrusion detection system may reside either on the target computer system or on a separate machine. This latter solution is generally preferable because it does not impact the target systems performance and protects audit information and control from attacks perpetrated on the target system. On the other hand, audit data must be communicated to the intrusion detection machine, which itself could be a source of vulnerability.

A major issue in employing an intrusion detection system is privacy. Monitoring user behavior, even if intended for defensive purposes, introduces a sort of Big Brother situation where a centralized monitor is watching everybody's behavior. This may be considered an invasion of individual privacy. It also raises concerns that audited information may be used improperly, for example, as a means for controlling employee performance.

91.5 Conclusion

Authentication, access control, and audit and intrusion detection together provide the foundations for building systems that can store and process information with confidentiality and integrity. Authentication is the primary security service. Access control builds directly on it. By and large, access control assumes authentication has been successfully accomplished. Strong authentication supports good auditing because operations can then be traced to the user who caused them to occur. There is a mutual interdependence between these three technologies, which can be often ignored by security practitioners and researchers. We need a coordinated approach that combines the strong points of each of these technologies rather than treating these as separate independent disciplines.

Acknowledgment

The work of Ravi Sandhu is partly supported by Grant CCR-9503560 from the National Science Foundation and Contract MDA904-94-C-6119 from the National Security Agency at George Mason University.

References

Bertino, E., Samarati, P., and Jajodia, J. 1993. Authorizations in relational database management systems, pp. 130–139. In *1st ACM Conf. Comput. Commun. Security*. Fairfax, VA, Nov.

Castano, S., Fugini, M. G., Martella, G., and Samarati, P. 1994. *Database Security*. Addison–Wesley, Reading, MA.

Chokhani, S. 1992. Trusted products evaluation. *Commun. ACM* 35(7):64–76.

Davies, C. and Ganesan, R. 1993. Bapasswd: a new proactive password checker, pp. 1–15. In *16th NIST-NCSC Nat. Comput. Security Conf.*

Ferraiolo, D. F., Gilbert, D. M., and Lynch, N. 1993. An examination of federal and commercial access control policy needs, pp. 107–116. In *NIST-NCSC Nat. Comput. Security Conf.* Baltimore, MD, Sept.

Ferraiolo, D. and Kuhn, R. 1992. Role-based access controls, pp. 554–563. In *15th NIST-NCSC Nat. Comput. Security Conf.* Baltimore, MD, Oct. 13–16.

Ganesan, R. and Davies, C. 1994. A new attack on random pronouncable password generators, pp. 184–197. In *17th NIST-NCSC Nat. Comput. Security Conf.*

Garvey, T. D. and Lunt, T. 1991. Model-based intrusion detection, pp. 372–385. In *Proc. 14th Nat. Comput. Security Conf.* Washington, DC, Oct.

Ilgun, K., Kemmerer, R. A., and Porras, P. A. 1995. State transition analysis: a rule-based intrusion detection approach. *IEEE Trans. Software Eng.* 21(3):222–232.

Kaufman, C., Perlman, R., and Speciner, M. 1995. *Network Security*. Prentice–Hall, Englewood Cliffs, NJ.

Mukherjee, B., Heberlein, L. T., and Levitt, K. N. 1994. Network intrusion detection. *IEEE Network*, (May/June):26–41.

Neuman, B. C. 1994. Using Kerberos for authentication on computer networks. *IEEE Commun.* 32(9).

Sandhu, R. S. 1993. Lattice-based access control models. *IEEE Comput.* 26(11):9–19.

Sandhu, R. S. 1994. On five definitions of data integrity. In *Database Security VII: Status and Prospects*, T. Keefe and C. E. Landwehr, eds., pp. 257–267. North-Holland.

Sandhu, R. S., Coyne, E. J., Feinstein, H. L., and Youman, C. E. 1996. Role-based access control models. *IEEE Comput.* 29(2):38–47.

Sandhu, R. S. and Samarati, P. 1994. Access control: principles and practice. *IEEE Commun.* 32(9):40–48.

Woo, T. Y. C. and Lam, S. S. 1992. Authentication for distributed systems. *IEEE Comput.* 25(1):39–52.

92

Network and Internet Security

Steven Bellovin
AT&T Bell Laboratories

92.1 Introduction

Why is network security so hard, whereas stand-alone computers remain relatively secure? The problem of network security is hard because of the complex and open nature of the networks themselves.

There are a number of reasons for this. First and foremost, a network is designed to accept requests from outside. It is easier for an isolated computer to protect itself from outsiders because it can demand authentication—a successful log-in—first. By contrast, a networked computer expects to receive unauthenticated requests, if for no other reason than to receive electronic mail. This lack of authentication introduces some additional risk, simply because the receiving machine needs to talk to potentially hostile parties.

Even services that should, in principle, be authenticated often are not. The reasons range from technical difficulty (see the subsequent discussion of routing) to cost to design choices: the architects of that service were either unaware of, or chose to discount, the threats that can arise when a system intended for use in a friendly environment is suddenly exposed to a wide-open network such as the Internet.

More generally, a networked computer offers many different services; a stand-alone computer offers just one: log-in. Whatever the inherent difficulty of implementing any single service, it is obvious that adding more services will increase the threat at least linearly. In reality, the problem is compounded by the fact that different services can interact. For example, an attacker may use a file transfer protocol to upload some malicious software and then trick some other network service into executing it.

Additional problems arise because of the unbounded nature of a network. A typical local area network may be viewed as an implementation of a loosely coupled, distributed operating system. But in single-computer operating systems, the kernel can **trust** its own data. That is, one component can create a control block for another to act on. Similarly, the path to the disk is trustable, in that a read request will retrieve the proper data, and a write request will have been vetted by the operating system.

Those assumptions do not hold on a network. A request to a file server may carry fraudulent user credentials, resulting in access violations. The data returned may have been inserted by an intruder or by an authorized user who is nevertheless trying to gain more privileges. In short, the distributed operating system can not believe anything, even transmissions from the kernel talking to itself.

In principle, many of these problems can be overcome. In practice, the problem seems to be intractable. Networked computers are far more vulnerable than standalone computers.

92.2 General Threats

Network security flaws fall into two main categories. Some services do inadequate authentication of incoming requests. Others try to do the right thing; however, buggy code lets the intruder in. Strong authentication and **cryptography** can do nothing against this second threat; it allows the target computer to establish a well-authenticated, absolutely private connection to a hacker who is capable of doing harm.

Authentication Failures

Some machines grant access based on the network address of the caller. This is acceptable if and only if two conditions are met. First, the trusted network and its attached machines must both be adequately secure, both physically and logically. On a typical local area network (LAN), anyone who controls a machine attached to the LAN can reconfigure it to impersonate any other machine on that cable. Depending on the exact situation, this may or may not be easily detectable. Additionally, it is often possible to turn such machines into eavesdropping stations, capable of listening to all other traffic on the LAN. This specifically includes passwords or even encrypted data if the encryption key is derived from a user-specified password [Gong et al. 1993].

Network-based authentication is also suspect if the network cannot be trusted to tell the truth. However, such a level of trust is not tautological; on typical packet networks, such as the Internet, each transmitting host is responsible for putting its own reply address in each and every packet. Obviously, an attacker's machine can lie—and this often happens.

In many instances, a **topological defense** will suffice. For example, a router at a network border can reject incoming packets that purport to be from the inside network. In the general case, though, this is inadequate; the interconnections of the networks can be too complex to permit delineation of a simple border, or a site may wish to grant privileges—that is, trust—to some machine that really is outside the physical boundaries of the network.

Although **address spoofing** is commonly associated with packet networks, it can happen with circuit networks as well. The difference is in who can lie about addresses; in a circuit net, a misconfigured or malconfigured switch can announce incorrect source addresses. Although not often a threat in simple topologies, in networks where different switches are run by different parties address errors present a real danger. The best-known example is probably the phone system, where many different companies and organizations around the world run different pieces of it. Again, topological defenses sometimes work, but you are still limited by the actual interconnection patterns.

Even if the network address itself can be trusted, there still may be vulnerabilities. Many systems rely not on the network address but on the network *name* of the calling party. Depending on how addresses are mapped to names, an enemy can attack the translation process and thereby spoof the target. See Bellovin [1995] for one such example.

User Authentication

User authentication is generally based on any of three categories of information: something you know, something you have, and something you are. All three have their disadvantages.

The something you know is generally a password or personal identification number (PIN). In today's threat environment, passwords are an obsolete form of authentication. They can be guessed [Klein 1990, Morris and Thompson 1979, Spafford 1989a] picked up by network wiretappers, or simply *social engineered* from users. If possible, avoid using passwords for authentication over a network.

Something you have is a token of some sort, generally cryptographic. These tokens can be used to implement cryptographically strong challenge/response schemes. But users do not like token devices; they are expensive and inconvenient to carry and use. Nevertheless, for many environments they represent the best compromise between security and usability.

Biometrics, or something you are, are useful in high-threat environments. But the necessary hardware is scarce and expensive. Furthermore, biometric authentication systems can be disrupted by biological factors; a user with laryngitis may have trouble with a voice recognition system. Finally, cryptography must be used in conjunction with biometrics across computer networks; otherwise, a recording of an old fingerprint scan may be used to trick the authentication system.

Buggy Code

The Internet has been plagued by buggy network servers. In and of itself, this is not surprising; most large computer programs are buggy. But to the extent that outsiders should be denied access to a system, every network server is a privileged program.

The two most common problems are buffer overflows and shell escapes. In the former case, the attacker sends an input string that overwrites a buffer. In the worst case, the stack can be overwritten as well, letting the attacker inject code. Despite the publicity this failure mode has attracted—the Internet Worm used this technique [Spafford 1989a, 1989b, Eichin and Rochlis 1989, Rochlis and Eichin 1989]—new instances of it are legion. Too many programmers are careless or lazy.

More generally, network programs should check *all* inputs for validity. The second failure mode is simply another example of this: input arguments can contain shell metacharacters, but the strings are passed, unchecked, to the shell in the course of executing some other command. The result is that two commands will be run, the one desired and the one requested by the attacker.

Just as no general solution to the program correctness problem seems feasible, there is no cure for buggy network servers. Nor will the best cryptography in the world help; you end up with a secure, protected communication between a hacker and a program that holds the **back door** wide open.

92.3 Routing

In most modern networks of any significant size, host computers cannot talk directly to all other machines they may wish to contact. Instead, intermediate nodes—switches or routers of some sort—are used to route the data to their ultimate destination. The security and integrity of the network depends very heavily on the security and integrity of this process.

The switches in turn need to know the next hop for any given network address; whereas this can be configured manually on small networks, in general the switches talk to each other by means of **routing protocols**. Collectively, these routing protocols allow the switches to learn the topology of the network. Furthermore, they are dynamic, in the sense that they rapidly and automatically learn of new network nodes, failures of nodes, and the existence of alternative paths to a destination.

Most routing protocols work by having switches talk to their neighbors. Each tells the other of the hosts it can reach, along with associated cost metrics. Furthermore, the information is transitive; a switch will not only announce its directly connected hosts but also destinations of which it has learned by talking to other routers. These latter announcements have their costs adjusted to account for the extra hop.

An enemy who controls the routing protocols is in an ideal position to monitor, intercept, and modify most of the traffic on a network. Suppose, for example, that some enemy node X is announcing a very low-cost route to hosts A and B. Traffic from A to B will flow through X, as will traffic from B to A. Although the diversion will be obvious to anyone who checks the path, such checks are rarely done unless there is some suspicion of trouble.

A more subtle routing issue concerns the return data flow. Such a flow almost always exists, if for no other reason than to provide flow control and error correction feedback. On packet-switched networks, the return path is independent of the forward path and is controlled by the same routing protocols. Machines that rely on network addresses for authentication and authorization are implicitly trusting the integrity of the return path; if this has been subverted, the network addresses cannot be trusted either. For example, in the previous situation, X could easily impersonate B when talking to A or vice versa.

That is somewhat less of a threat on circuit-switched networks, where the call is typically set up in both directions at once. But often, the trust point is simply moved to the switch; a subverted or corrupt switch can still issue false routing advertisements.

Securing routing protocols is hard because of the transitive nature of the announcements. That is, a switch cannot simply secure its link to its neighbors, because it can be deceived by messages really sent by its legitimate and uncorrupted peer. That switch, in turn, might have been deceived by its peers, ad infinitum. It is necessary to have an authenticated chain of responses back to the source to protect routing protocols from this sort of attack.

Another class of defense is topological. If a switch has a priori knowledge that a certain destination is reachable only via a certain wire, routing advertisements that indicate otherwise are patently false. Although not necessarily indicative of malice—link or node failures can cause temporary confusion of the network-wide routing tables—such announcements can and should be dismissed out of hand. The problem, of course, is that adequate topological information is rarely available. On the Internet, most sites are *out there* somewhere; the false hop, if any, is likely located far beyond an individual site's borders. Additionally, the prevalence of redundant links, whether for performance or reliability, means that more than one path may be valid. In general, then, topological defenses are best used at **choke points: fire walls** (section 92.7) and the other end of the link from a site to a network service provider. The latter allows the service provider to be a good network citizen and prevent its customers from claiming routes to other networks.

Some networks permit hosts to override the routing protocols. This process, sometimes called source routing, is often used by network management systems to bypass network outages and as such is seen as very necessary by some network operators.

The danger, though, arises because source-routed packets bypass the implicit authentication provided by use of the return path, as previously outlined. A host that does network address-based authentication can easily be spoofed by such messages. Accordingly, if source routing is to be used, address-based authentication must not be used.

92.4 The Transmission Control Protocol/Internet Protocol (TCP/IP) Protocol Suite

The **transmission control protocol** (**TCP**) suite is the basis for the Internet. Although the general features of the protocols are beyond the scope of this chapter (see Stevens [1995] and Wright and Stevens [1994] for more detail), the security problems of it are less well known.

The most important thing to realize about TCP/IP security is that since IP is a datagram protocol, one cannot trust the source addresses in packets. This threat is not just hypothetical. One of the most famous security incidents—the penetration of Tsutomu Shimomura's machines [Shimomura 1996, Littman 1996]—involved IP address spoofing in conjunction with a TCP sequence number guessing attack.

Sequence Number Attacks

TCP **sequence number attacks** were described in the literature many years before they were actually employed [Morris 1985, Bellovin 1989]. They exploit the predictability of the sequence number field in

TCP in such a way that it is not necessary to see the return data path. To be sure, the intruder cannot get any output from the session, but if you can execute a few commands, it does not matter much if you see their output.

Every byte transmitted in a TCP session has a sequence number; the number for the first byte in a segment is carried in the header. Furthermore, the control bits for opening and closing a connection are included in the sequence number space. All transmitted bytes must be acknowledged explicitly by the recipient; this is done by sending back the sequence number of the next byte expected.

Connection establishment requires three messages. The first, from the client to the server, announces the client's initial sequence number. The second, from the server to the client, acknowledges the first message's sequence number and announces the server's initial sequence number. The third message acknowledges the second.

In theory, it is not possible to send the third message without having seen the second, since it must contain an explicit acknowledgment for a random-seeming number. But if two connections are opened in a short time, many TCP stacks pick the initial sequence number for the second connection by adding some constant to the sequence number used for the first.

The mechanism for the attack is now clear. The attacker first opens a legitimate connection to the target machine and notes its initial sequence number. Next, a spoofed connection is opened by the attacker, using the IP address of some machine trusted by the target. The sequence number learned in the first step is used to send the third message of the TCP open sequence, without ever having seen the second. The attacker can now send arbitrary data to the target; generally, this is a set of commands designed to open up the machine even further.

Connection Hijacking

Although a defense against classic sequence number attacks has now been found [Bellovin 1996], a more serious threat looms on the horizon: **connection hijacking** [Joncheray 1995]. An attacker who observes the current sequence number state of a connection can inject phony packets.

Again, the network in general will believe the source address claimed in the packet. If the sequence number is correct, it will be accepted by the destination machine as coming from the real source. Thus, an eavesdropper can do far worse than simply steal passwords; he or she can take over a session after log in. Even the use of a high-security log-in mechanism, such as a one-time password system [Haller 1994], will not protect against this attack. The only defense is full-scale encryption.

Session hijacking is detectable, since the acknowledgment packet sent by the target cites data the sender never sent. Arguably, this should cause the connection to be reset; instead, the system assumes that sequence numbers have wrapped around and resends its current sequence number and acknowledgment number state.

The **r**-Commands

The so-called **r**-commands— **r**sh and **r**login—use address-based authentication. As such, they are not secure. But too often, the alternative is sending a password in the clear over an insecure net. Neither alternative is attractive; the right choice is cryptography. But that is used all too infrequently.

In many situations, where insiders are considered reasonably trustworthy, use of these commands without cryptography is an acceptable risk. If so, a low-grade fire wall such as a simple **packet filter** *must* be used.

The X Window System

The paradigm for the X window system [Stubblebine and Gligor 1992] is simple: a server runs the physical screen, keyboard, and mouse; applications connect to it and are allocated use of those resources. Put another way, when an application connects to the server, it gains control of the screen, keyboard, and mouse. Whereas this is good when the application is legitimate, it poses a serious security risk if uncontrolled applications can connect. For example, a rogue application can monitor all keystrokes, even those destined for other applications, dump the screen, inject synthetic events, and so on.

There are several modes of access control available. A common default is no restriction; the dangers of this are obvious. A more common option is control by IP address; apart from the usual dangers of this strategy, it allows anyone to gain access on the trusted machine. The so-called **magic cookie** mechanism uses (in effect) a clear-text password; this is vulnerable to anyone monitoring the wire, anyone with privileged access to the client machines, and—often—anyone with network file system access to that machine. Finally, there are some cryptographic options; these, although far better than the other options, are more vulnerable than they might appear at first glance, as any privileged user on the application's machine can steal the secret cryptographic key.

There have been some attempts to improve the security of the X window system [Epstein et al. 1992, Kahn 1995]. The principal risk is the complexity of the protocol: are you sure that all of the holes have been closed? The analysis in Kahn [1995] provides a case in point; the author had to rely on various heuristics to permit operations that seemed dangerous but were sometimes used safely by common applications.

User Datagram Protocol (UDP)

The **user datagram protocol (UDP)** [Postel 1980] poses its own set of risks. Unlike TCP, it is not connection oriented; thus, there is no implied authentication from use of the return path. Source addresses cannot be trusted at all. If an application wishes to rely on address-based authentication, it must do its own checking, and if it is going to go to that much trouble, it may as well use a more secure mechanism.

Remote Procedure Call (RPC), Network Information Service (NIS), and Network File System (NFS)

The most important UDP-based protocol is **remote procedure call (RPC)** [Sun 1988, 1990]. Many other services, such as network information service (NIS) and **network file system (NFS)** [Sun 1989, 1990] are built on top of RPC. Unfortunately, these services inherit all of the weaknesses of UDP and add some of their own. For example, although RPC has an authentication field, in the normal case it simply contains the calling machine's assertion of the user's identity. Worse yet, given the ease of forging UDP packets, the server does not even have any strong knowledge of the actual source machine. Accordingly, no serious action should be taken based on such a packet.

There is a cryptographic authentication option for RPC. Unfortunately, it is poorly integrated and rarely used. In fact, on most systems only NFS can use it. Furthermore, the key exchange mechanism used is cryptographically weak [LaMacchia and Odlyzko 1991].

NIS has its own set of problems; these, however, relate more to the information it serves up. In particular, NIS is often used to distribute password files, which are very sensitive. Password guessing is very easy [Klein 1990, Morris and Thompson 1979, Spafford 1992]; letting a hacker have a password file is tantamount to omitting password protection entirely. Misconfigured or buggy NIS servers will happily distribute such files; consequently, the protocol is very dangerous.

92.5 The World Wide Web

The World Wide Web (WWW) is the fastest-growing protocol on the Internet. Indeed, in the popular press it *is* the Internet. There is no denying the utility of the Web. At the same time, it is a source of great danger. Indeed, the Web is almost unique in that the danger is nearly as great to clients as to servers.

Client Issues

The danger to clients comes from the nature of the information received. In essence, the server tells the client "here is a file, and here is how to display it." The problem is that the instructions may not be benign. For example, some sites supply troff input files; the user is expected to make the appropriate control entries to link that file type to the processor. But troff has shell escapes; formatting an arbitrary file is about as safe as letting unknown persons execute any commands they wish.

The problem of buggy client software should not be ignored either. Several major browsers have had well-publicized bugs, ranging from improper use of cryptography to string buffer overflows. Any of these could result in security violations.

A third major area for concern is **active agents:** pieces of code that are explicitly downloaded to a user's machine and executed. Java [Arnold and Gosling 1996] is the best known, but there are others.

Active agents, by design, are supposed to execute in a restricted environment. Still, they need access to certain resources to do anything useful. It is this conflict, between the restrictions and the resources, that leads to the problems; sometimes the restrictions are not tight enough. And even if they are in terms of the architecture, implementation bugs, inevitable in such complex code, can lead to security holes [Dean and Wallach 1996].

Server Issues

Naturally, servers are vulnerable to security problems as well. Apart from bugs, which are always present, Web servers have a challenging job. Serving up files is the easy part, though this, too, can be tricky; not all files should be given to outsiders.

A bigger problem is the so-called **common gateway interface (CGI) scripts.** CGI scripts are, in essence, programs that process the user's request. Like all programs, CGI scripts can be buggy. In the context of the Web, this can lead to security holes.

A common example is a script to send mail to some destination. The user is given a form to fill out, with boxes for the recipient name and the body of the letter. When the user clicks on a button, the script goes to work, parsing the input and, eventually, executing the system's mailer. But what happens if the user—someone on another site—specifies an odd-ball string for the recipient name? Specifically, what if the recipient string contains assorted special characters, and the shell is used to invoke the mailer?

Administering a WWW site can be a challenge. Modern servers contain all sorts of security-related configuration files. Certain pages are restricted to certain users or users from certain IP addresses. Others must be accessed using particular userids. Some are even protected by their own password files.

Not surprisingly, getting all of that right is tricky. But mistakes here do not always lead to the sort of problem that generates user complaints; hackers rarely object when you let them into your machine.

A final problem concerns the uniform resource locators (URLs) themselves. Web servers are stateless; accordingly, many encode transient state information in URLs that are passed back to the user. But parsing this state can be hard, especially if the user is creating malicious counterfeits.

92.6 Using Cryptography

Cryptography, though not a panacea, is a potent solution to many network security issues. The most obvious use of cryptography is to protect network traffic from eavesdroppers. If two parties share the same secret key, no outsiders can intercept any messages. This can be used to protect passwords, sensitive files being transferred over a network, etc.

Often, though, secrecy is less important than authenticity. Cryptography can help here, too, in two different ways. First, there are cryptographic primitives designed to authenticate messages. Message authentication codes (MACs) are commonly used in electronic funds transfer applications to validate their point of origin.

More subtly, decryption with an invalid key will generally yield garbage. If the message is intended to have any sort of semantic or syntactic content, ordinary input processing will likely reject such messages. Still, care must be taken; noncryptographic **checksums** can easily be confused with a reasonable probability. For example, TCP's checksum is only 16 bits; if that is the sole guarantor of packet sanity, it will fail about once in 2^{16} packets.

Key Distribution Centers

The requirement that every pair of parties share a secret key is in general impractical for all but the smallest network. Instead, most practical systems rely on trusted third parties known as **key distribution centers**

(KDC). Each party shares a long-term secret key with the KDC; to make a secure call, the KDC is asked to (in effect) introduce the two parties, using its knowledge of the shared keys to vouch for the authenticity of the call.

The Kerberos authentication system [Bryant 1988, Kohl and Neuman 1993, Miller et al. 1987, Steiner et al. 1988], designed at Massachusetts Institute of Technology (MIT) as part of Project Athena, is a good example. Although Kerberos is intended for user-to-host authentication, most of the techniques apply to other situations as well.

Each party, known as a *principal*, shares a secret key with the KDC. User keys are derived from a pass phrase; service keys are randomly generated. Before contacting any service, the user requests a **Kerberos ticket-granting ticket (TGT)** from the KDC,

$$K_c[K_{c,\text{tgs}} K_{\text{tgs}}[T_{c,\text{tgs}}]]$$

where $K[X]$ denotes the encryption of X by key K. K_c is the client's key; it is used to encrypt the body of the message. In turn, the body is a ticket-granting ticket, encrypted by a key known only to the server, and an associated session key $K_{c,\text{tgs}}$ to be used along with the TGT. TGTs and their associated session keys normally expire after about 8 hours, and are cached by the client during this time; this avoids the need for constant retyping of the user's password.

The TGT is used to request credentials—tickets—for a service s,

$$s, K_{\text{tgs}}[T_{c,\text{tgs}}], K_{c,\text{tgs}}[A_c]$$

That is, the TGT is sent to the KDC along with an encrypted *authenticator* A_c. The authenticator contains the time of day and the client's IP address; this is used to prevent an enemy from replaying the message.

The KDC replies with

$$K_{c,\text{tgs}}[K_s[T_{c,s}], K_{c,s}]$$

The session key $K_{c,s}$ is a newly chosen random key; $K_s[T_{c,s}]$ is the ticket for user c to access service s. It is encrypted in the key shared by the KDC and s; this assures s of its validity. It contains a lifetime, a session key $K_{c,s}$ that is shared with c, and c's name. A separate copy of $K_{c,s}$ is included in the reply for use by the client. When transmitted by c to s, an authenticator is sent with it, encrypted by $K_{c,s}$; again, this ensures freshness.

Finally, c can ask s to send it a message encrypted in the same session key; this protects the client against someone impersonating the server.

There are several important points to note about the design. First, cryptography is used to create *sealed* packages. Tickets and the like are encrypted along with a checksum; this protects them from tampering. Second, care is taken to avoid repetitive password entry requests; human factors are quite important, as users tend to bypass security measures they find unpleasant. Third, messages must be protected against replay; an attacker who can send the proper message may not need to know what it says. Cut-and-paste attacks are a danger as well, though they are beyond the scope of this chapter.

It is worth noting that the design of cryptographic protocols is a subtle business. The literature is full of attacks that were not discovered until several years after publication of the initial protocol. See, for example, Bellovin and Merritt [1991] and Stubblebine and Gligor [1992] for examples of problems with Kerberos itself.

92.7 Fire Walls

Fire walls [Cheswick and Bellovin 1994] are an increasingly popular defense mechanism. Briefly, a fire wall is an electronic analog of the security guard at the entrance to a large office or factory. Credentials are checked, outsiders are turned away, and incoming packages—electronic mail—is handed over for delivery by internal mechanisms.

The purpose of a fire wall is to protect more vulnerable machines. Just as most people have stronger locks on their front doors than on their bedrooms, there are numerous advantages to putting stronger security on the perimeter. If nothing else, a fire wall can be run by personnel whose job it is to ensure security.

For many sites, though, the real issue is that internal networks *cannot* be run securely. Too many systems rely on insecure network protocols for their normal operation. This is bad, and everyone understands this; too often, though, the choice is between accepting some insecurity or not being able to use the network productively. A fire wall is often a useful compromise; it blocks attacks from a high-threat environment, while letting people use today's technology.

Seen that way, a fire wall works because of what it is not. It is not a general purpose host; consequently, it does not need to run a lot of risky software. Ordinary machines rely on networked file systems, remote log-in commands that rely on address-based authentication, users who surf the Web, etc. A fire wall does none of these things; accordingly, it is not affected by potential security problems with them.

Types of Fire Walls

There are four primary types of fire walls: **packet filters**, dynamic packet filters, **application gateways**, and **circuit relays**. Each has its advantages and disadvantages.

Packet Filters

The cheapest and fastest type of fire wall is the packet filter. Packet filters work by looking at each individual packet, and, based on source address and destination addresses and port numbers, making a pass/drop decision. They are cheap because virtually all modern routers already have the necessary functionality; in effect, you have already paid the price, so you may as well use it. Additionally, given the comparatively slow lines most sites use for external access, packet filtering is fast; a router can filter at speeds higher than, say, a DS1 line (1,500,000 bits/second).

The problem is that decisions made by packet filters are completely context free. Each packet is examined, and its fate decided, without looking at the previous input history. This makes it difficult or impossible to handle certain protocols. For example, file transfer protocol (FTP) [Mills 1985] uses a secondary TCP connection to transfer files; by default, this is an incoming call through the fire wall [Bellovin 1994]. In this situation, the call should be permitted; the client has even sent a message specifying which port to call. But ordinary packet filters cannot cope.

Packet filters must permit not only outgoing packets but also the replies. For TCP, this is not a big problem; the presence of one header bit [the acknowledgment (ACK) bit] denotes a reply packet. In general, packets with this bit set can safely be allowed in, as they represent part of an ongoing conversation. Datagram protocols such as UDP do not have the concept of conversation and hence do not have such a bit, which causes difficulties: when should a UDP packet be allowed in? It is easy to permit incoming queries to known safe servers; it is much harder to identify replies to queries sent from the inside. Ordinary packet filters are not capable of making this distinction. At best, sites can assume that higher numbered ports are used by clients and hence are safe; in general, this is a bad assumption.

Services built on top of Sun's remote procedure call [Sun 1988, 1990] pose a different problem: the port numbers they use are not predictable. Rather, they pick more or less random port numbers and register with a directory server known as the portmapper. Would-be clients first ask portmapper which port number is in use at the moment, and then do the actual call. But since the port numbers are not fixed, it is not possible to configure a packet filter to let in calls to the proper services only.

Dynamic Packet Filters

Dynamic packet filters are designed to answer the shortcomings of ordinary packet filters. They are inherently stateful and retain the context necessary to make intelligent decisions. Most also contain application-specific modules; these do things like parse the FTP command stream so that the data channel can be opened, look inside portmapper messages to decide if a permitted service is being requested, etc. UDP queries are handled by looking for the outbound call and watching for the responses to that port number. Since there is no end-of-conversation flag in UDP, a timeout is needed. This heuristic does not always work well, but, without a lot of application-specific knowledge, it is the only possibility.

Dynamic packet filters promise everything: safety and full transparency. The risk is their complexity; one never knows exactly which packets will be allowed in at a given time.

Application Gateways

Application gateways live at the opposite end of the protocol stack. Each application being relayed requires a specialized program at the fire wall. This program understands the peculiarities of the application, such as data channels for FTP, and does the proper translations as needed.

It is generally acknowledged that application gateways are the safest form of fire wall. Unlike packet filters, they do not pass raw data; rather, individual applications, invoked from the inside, make the necessary calls. The risk of passing an inappropriate packet is thus eliminated.

This safety comes at a price, though. Apart from the need to build new gateway programs, for many protocols a change in user behavior is needed. For example, a user wishing to telnet to the outside generally needs to contact the fire wall explicitly and then redial to the actual destination. For some protocols, though, there is no user visible change; these protocols have their own builtin redirection or proxy mechanisms. Mail and the World Wide Web are two good examples.

Circuit Relays

Circuit relays represent a middle ground between packet filters and application gateways. Because no data are passed directly, they are safer than packet filters. But because they use generic circuit-passing programs, operating at the level of the individual TCP connection, specialized gateway programs are not needed for each new protocol supported.

The best-known circuit relay system is socks [Koblas and Koblas 1992]. In general, applications need minor changes or even just a simple relinking to use the socks package. Unfortunately, that often means it is impossible to deploy it unless a suitable source or object code is available. On some systems, though, dynamically linked run-time libraries can be used to deploy socks.

Circuit relays are also weak if the aim is to regulate outgoing traffic. Since more or less any calls are permissible, users can set up connections to unsafe services. It is even possible to tunnel IP over such circuits, bypassing the fire wall entirely. If these sorts of activities are in the threat model, an application gateway is probably preferable.

Limitations of Fire Walls

As important as they are, fire walls are not a panacea to network security problems. There are some threats that fire walls cannot defend against.

The most obvious of these, of course, is attacks that do not come through the fire wall. There are always other entry points for threats. There might be an unprotected modem pool; there are always insiders, and a substantial portion of computer crime is due to insider activity. At best, internal fire walls can reduce this latter threat.

On a purely technical level, no fire wall can cope with an attack at a higher level of the protocol stack than it operates. Circuit gateways, for example, cannot cope with problems at the simple mail transfer protocol (SMTP) layer [Postel 1982]. Similarly, even an application-level gateway is unlikely to be able to deal with the myriad security threats posed by multimedia mail [Borenstein and Freed 1993]. At best, once such problems are identified a fire wall may provide a place to deploy a fix.

A common question is whether or not fire walls can prevent virus infestations. Although, in principle, a mail or FTP gateway could scan incoming files, in practice it does not work well. There are too many ways to encode files, and too many ways to spread viruses, such as self-extracting executables.

Finally, fire walls cannot protect applications that must be exposed to the outside. Web servers are a canonical example; as previously described, they are inherently insecure, so many people try to protect them with fire walls. That does not work; the biggest security risk is in the service that of necessity must be exposed to the outside world. At best, a fire wall can protect other services on the Web server machine. Often, though, that is like locking up only the bobcats in a zoo full of wild tigers.

92.8 Denial of Service Attacks

Denial of service attacks are generally the moral equivalent of vandalism. Rather than benefitting the perpetrator, the goal is generally to cause pain to the target, often for no better reason than to cause pain.

The simplest form is to flood the target with packets. If the attacker has a faster link, the attacker wins. If this attack is combined with source address spoofing, it is virtually untraceable as well.

Sometimes, denial of service attacks are aimed more specifically. A modest number of TCP open request packets, from a forged IP address, will effectively shut down the port to which they are sent. This technique can be used to close down mail servers, Web servers, etc.

The ability to interrupt communications can also be used for direct security breaches. Some authentication systems rely on primary and backup servers; the two communicate to guard against replay attacks. An enemy who can disrupt this path may be able to replay stolen credentials.

Philosphically, denial of service attacks are possible any time the cost to the enemy to mount the attack is less, relatively speaking, than the cost to the victim to process the input. In general, prevention consists of lowering your costs for processing unauthenticated inputs.

92.9 Conclusions

We have discussed a number of serious threats to networked computers. However, except in unusual circumstances—and they do exist—we do not advocate disconnection. Whereas disconnecting buys you some extra security, it also denies you the advantages of a network connection.

It is also worth noting that complete disconnection is much harder than it would appear. Dial-up access to the Internet is both easy and cheap; a managed connection can be more secure than a total ban that might incite people to evade it. Moreover, from a technical perspective an external network connection is just one threat among many. As with any technology, the challenge is to control the risks while still reaping the benefits.

Defining Terms

Active agents: Programs sent to another computer for execution on behalf of the sending computer.
Address spoofing: Any enemy computer's impersonation of a trusted host's network address.
Application gateway: A relay and filtering program that operates at layer seven of the network stack.
Back door: An unofficial (and generally unwanted) entry point to a service or system.
Checksums: A short function of an input message, designed to detect transmission errors.
Choke point: A single point through which all traffic must pass.
Circuit relay: A relay and filtering program that operates at the transport layer (level four) of the network protocol stack.
Common gateway interface (CGI) scripts: The interface to permit programs to generate output in response to World Wide Web requests.
Connection hijacking: The injection of packets into a legitimate connection that has already been set up and authenticated.
Cryptography: The art and science of secret writing.
Denial of service: An attack whose primary purpose is to prevent legitimate use of the computer or network.

Fire wall: An electronic barrier restricting communications between two parts of a network.

Kerberos Ticket-Granting Ticket (TGT):   The cryptographic credential used to obtain credentials for other services.

Key distribution center (KDC):   A trusted third party in cryptographic protocols that has knowledge of the keys of other parties.

Magic cookie: An opaque quantity, transmitted in the clear and used to authenticate access.

Network file system protocol (NFS):   Originally developed by Sun Microsystems.

Packet filter: A network security device that permits or drops packets based on the network layer addresses and (often) on the port numbers used by the transport layer.

r-Commands: A set of commands (`rsh`, `rlogin`, `rcp`, `rdist`, etc.) that rely on address-based authentication.

Remote procedure call (RPC) protocol: Originally developed by Sun Microsystems.

Routing protocols: The mechanisms by which network switches discover the current topology of the network.

Sequence number attacks: An attack based on predicting and acknowledging the byte sequence numbers used by the target computer without ever having seen them.

Topological defense: A defense based on the physical interconnections of two networks. Security policies can be based on the notions of inside and outside.

Transmission control protocol (TCP):   The basic transport-level protocol of the Internet. It provides for reliable, flow-controlled, error-corrected virtual circuits.

Trust: The willingness to believe messages, especially access control messages, without further authentication.

User datagram protocol (UDP):   A datagram-level transport protocol for the Internet. There are no guarantees concerning order of delivery, dropped or duplicated packets, etc.

References

Arnold, K. and Gosling, J. 1996. *The Java Programming Language.* Addison–Wesley, Reading, MA.

Bellovin, S. M. 1989. Security problems in the TCP/IP protocol suite. *Comput. Commun. Rev.* 19(2):32–48.

Bellovin, S. M. 1994. Fire wall-Friendly FTP. Request for comments (informational) RFC 1579. Internet Engineering Task Force, Feb.

Bellovin, S. M. 1996. Defending against sequence number attacks. RFC 1948. May.

Bellovin, S. M. 1995. Using the domain name system for system break-ins, pp. 199–208. In *Proc. 5th USENIX Unix Security Symp.* Salt Lake City, UT, June.

Bellovin, S. M. and Merritt, M. 1991. Limitations of the Kerberos authentication system, pp. 253–267. In *USENIX Conf. Proc.* Dallas, TX, Winter.

Borenstein, N. and Freed, N. 1993. MIME (Multipurpose Internet Mail Extensions) Part One: Mechanisms for Specifying and Describing the Format of Internet Message Bodies. Request for comments (draft standard) RFC 1521, Internet Engineering Task Force, Sept. (obsoletes RFC 1341; updated by RFC 1590).

Bryant, B. 1988. Designing an authentication system: a dialogue in four scenes. Draft. Feb. 8.

Cheswick, W. R. and Bellovin, S. M. 1994. *Fire walls and Internet Security: Repelling the Wily Hacker.* Addison–Wesley, Reading, MA.

Dean, D. and Wallach, D. 1996. Security flaws in the HotJava web browser. In *Proc. IEEE Symp. Res. Security Privacy.* Oakland, CA, May.

Eichin, M. W. and Rochlis, J. A. 1989. With microscope and tweezers: an analysis of the Internet virus of November 1988, pp. 326–345. In *Proc. IEEE Symp. Res. Security Privacy.* Oakland, CA, May.

Epstein, J., McHugh, J., and Pascale, R. 1992. Evolution of a trusted B3 window system prototype. In *Proc. IEEE Comput. Soc. Symp. Res. Security Privacy.* Oakland, CA, May.

Gong, L., Lomas, M. A., Needham, R. M., and Saltzer, J. H. 1993. Protecting poorly chosen secrets from guessing attacks. *IEEE J. Select. Areas Commun.* 11(5):648–656.

Haller, N. M. 1994. The S/Key one-time password system. In *Proc. Internet Soc. Symp. Network Distributed Syst. Security.* San Diego, CA, Feb. 3.

Joncheray, L. 1995. A simple active attack against TCP. In *Proc. 5th USENIX Unix Security Symp.* Salt Lake City, UT, June.

Kahn, B. L. 1995. Safe use of X window system protocol across a fire wall, pp. 105–116. In *Proc. 5th USENIX Unix Security Symp.* Salt Lake City, UT, June.

Klein, D. V. 1990. Foiling the cracker: a survey of, and improvements to, password security, pp. 5–14. In *Proc. USENIX Unix Security Workshop.* Portland, OR, Aug.

Koblas, D. and Koblas, M. R. 1992. Socks, pp. 77–83. In *Unix Security III Symp.* Baltimore, MD, Sept. 14–17, USENIX.

Kohl, J. and Neuman, B. 1993. The Kerberos Network Authentication Service (V5). Request for comments (proposed standard) RFC 1510, Internet Engineering Task Force, Sept. 1993.

LaMacchia, B. A. and Odlyzko, A. M. 1991. Computation of discrete logarithms in prime fields. *Designs, Codes, Cryptography* 1:46–62.

Littman, J. 1996. *Fugitive Game.* Little, Brown.

Miller, S. P., Neuman, B. C., Schiller, J. I., and Saltzer, J. H. 1987. Kerberos authentication and authorization system. In *Project Athena Technical Plan.* Sec. E.2.1, Massachusetts Institute of Technology, Cambridge, MA, Dec.

Mills, D. 1985. Network Time Protocol NTP. RFC 958, Internet Engineering Task Force, Sept. (obsoleted by RFC1059).

Morris, R. T. 1985. A Weakness in the 4.2BSD Unix TCP/IP Software. AT& T Bell Lab. Computing Science Tech. Rep. 117, Murray Hill, NJ, Feb.

Morris, R. H. and Thompson, K. 1979. Unix password security. *Commun. ACM* 22(11):594.

Postel, J. 1980. User Datagram Protocol. Request for comments (standard) STD 6, RFC 768, Internet Engineering Task Force, Aug.

Postel, J. 1982. Simple Mail Transfer Protocol. Request for comments (standard) STD 10, RFC 821, Internet Engineering Task Force, Aug. (obsoletes RFC 0788).

Rochlis, J. A. and Eichin, M. W. 1989. With microscope and tweezers: the worm from MIT's perspective. *Commun. ACM* 32(6):689–703.

Scheifler, R. W. and Gettys, J. 1992. *X Window System,* 3rd ed. Digital Press, Burlington, MA.

Shimomura, T. 1996. *Takedown.* Hyperion.

Spafford, E. H. 1989a. An analysis of the Internet worm. In *Proc. European Software Eng. Conf.* C. Ghezzi and J. A. McDermid, eds. Lecture notes in computer science, 387, pp. 446–468. Warwick, England, Sept. Springer–Verlag.

Spafford, E. H. 1989b. The Internet worm program: an analysis. *Comput. Commun. Rev.* 19(1):17–57.

Spafford, E. H. 1992. Observations on reusable password choices, pp. 299–312. In *Proc. 3rd USENIX Unix Security Symp.* Baltimore, MD, Sept.

Steiner, J., Neuman, B. C., and Schiller, J. I. 1988. Kerberos: an authentication service for open network systems, pp. 191–202. In *Proc. Winter USENIX Conf.* Dallas, TX.

Stevens, W. R. 1995. *TCP/IP Illustrated,* Vol. 1. Addison–Wesley, Reading, MA.

Stubblebine, S. G. and Gligor, V. D. 1992. On message integrity in cryptographic protocols, pp. 85–104. In *Proc. IEEE Comput. Soc. Symp. Res. Security Privacy.* Oakland, CA, May.

Sun. 1988. RPC: Remote Procedure Call Protocol Specification Version 2. Request for comments (informational) RFC 1057, Internet Engineering Task Force, Sun Microsystems, June (obsoletes RFC 1050).

Sun. 1989. NFS: Network File System Protocol Specification. Request for comments (historical) RFC 1094, Internet Engineering Task Force, Sun Microsystems, March.

Sun. 1990. Network Interfaces Programmer's Guide. SunOS 4.1. Sun Microsystems. Mountain View, CA, March.

Wright, G. R. and Stevens, W. R. 1994. *TCP/IP Illustrated: The Implementation,* Vol. 2. Addison–Wesley, Reading, MA.

93

Malicious Software and Hacking

David Ferbrache
Defence Research Agency

Stuart Mort
Defence Research Agency

93.1 Background

Since the advent of one of the first computer viruses on the IBM personal computer (PC) platform in 1986 the variety and complexity of malicious software has grown to encompass over 5000 viruses on IBM PC, Apple Macintosh, Commodore Amiga, Atari ST, and many other platforms. In addition to viruses a wide range of other disruptions such as Trojan horses, logic bombs, and e-mail bombs have been detected. In each case the software has been crafted with malicious intent ranging from system disruption to demonstration of the intelligence and creativity of the author.

The wide variety of malicious software is complemented by an extensive range of tools and methods designed to support unauthorized access to computer systems, misuse of telecommunications facilities and computer-based fraud. Behind this range of utilities lies a stratified and complex underculture: the computer underground. The underground embraces all age groups, motivations and nationalities, and its activities include software piracy, elite system hacking, pornographic bulletin boards, and virus exchange bulletin boards.

93.2 Culture of the Underground

An attempt to define the computer underground can produce a variety of descriptions from a number of sources. Many consider it a collection of friendless teenagers, who spend their time destroying people's data. To others, it is an elite society of computer gurus, whose expertise is an embarrassment to the legitimate bodies that continually try to extinguish their existence. However, the computer underground is really a collection of computer enthusiasts with as varied a collection of personalities as you would experience in any walk of life.

Not all members of the underground are computer anarchists; many use it as an environment in which to gather information and share ideas. However, many are in the following categories:

0-8493-2909-4/97/$0.00+$.50
© 1997 by CRC Press, Inc.

- *Hackers,* who try and break into computer systems for reasons such as gaining information or destroying data.
- *Malicious software writers,* who create software with a malicious intention. Viruses and Trojan horses are examples.
- *Phreakers,* who hack phones. This is done mainly to gain free phone calls, in order to support other activities such as hacking.

Some have described the inhabitants of the underground as information warriors; this is a too glamorous and inaccurate a term. It is true that many individuals' main cause is the freedom of information. These individuals may gain this information by breaking into a computer system, and extracting the stored information for distribution to any person who wants it. Many try and sell the information; these could be termed information brokers. Virus writers are certainly not information warriors, but may be information destroyers.

Thus, we have the person with the computer, surfing the net. An interesting site is stumbled across, with the electronic equivalent of a barbed-wire fence. Behind this fence there must be something interesting, otherwise, why so much security? The site is probed, in an attempt to challenge the security. Is this just a person's keen interest in the unknown, or is there a deeper malicious intent?

When security is breached, an assessment of the damage must be made. Was the availability of the system damaged? A virus could have destroyed vital files, crucial to the operation of the system. Has the integrity of data been compromised? An employee's salary could have been changed. Confidentiality lost? A company's new idea stolen. The cost of recovering from a security breach can be major: the time spent by an antivirus expert cleaning up machines after an infection, the time lost when employees could not work because their machines were inoperable. The cost mounts up. It is possible for a company dependent on its computer systems to go bankrupt after a security breach. It could also put peoples' lives at risk. The computer underground poses a significant threat to computer systems, of all descriptions, all over the world.

Stereotypes

The underground is a random collection of individuals, communicating over the Internet, bulletin boards, or occasionally face to face. Some individuals amalgamate to form a group. Groups sometimes compete with other groups, to prove they are the best. These competitions usually take the form of who can hack into the most computer systems. T-shirts even get printed to celebrate achievements.

A computer hacking group that did gain considerable recognition was the Legion of Doom (LoD). This group participated in a number of activities, including: obtaining money and property fraudulently from companies by altering computerized information, stealing computer source code from companies and individuals, altering routing information in computerized telephone switches to disrupt telecommunications, and theft or modification of information stored on individuals by credit bureaus. A member of LoD claims that curiosity was the biggest crime they ever committed!

Hacker groups do cause damage and disruption, wasting the resources of system administrators and law enforcement agencies worldwide. It has also been argued that hackers are responsible for closing security loopholes. Many hacker groups such as the Chaos Computer Club state that their members abide by a moral code.

Hacker

Simon Evans is a hypothetical example of a hacker with a broad level of computer expertise. Evans' history is one of persistent attempts to break into networks and systems; attempts which were often successful.

Evans' name first hit the press, when in 1980 a magazine wrote a cover story on an underground group. Evans had met the leader of this group a few years previously. Later in 1980, Evans and this hacker group broke into a computer system at U.S. Leasing. Not content with simple computer system breakins, Evans illegally entered an office of a telecom company in 1981 and stole documents and manuals. Following a tip off, Evans's home was searched and he was arrested, along with his accomplice and the leader of the hacker group. Evans was placed on one year's probation.

During his probation period, Evans managed to gain physical access to some university computers and started using them for hacking purposes. A computer crime unit pursued Evans and he was sentenced to six months in a juvenile prison for breaking probation. In 1984 Evans got a job, working for Great American Merchandising. From this company, he started making unauthorized credit checks, he was reported, and went into hiding.

1985, Evans came out of hiding and enrolled at a computer learning center. He fell for a girl, from whose address he hacked a system at Santa Cruz Operation. The call was traced, and Evans and his girlfriend were arrested. The girlfriend was released and Evans received 3 years probation, during which he married his girlfriend. In 1988, a friend of Evans started talking to the FBI, who subsequently arrested Evans for breaking into Digital Equipment Corporation's systems and stealing software. Evans got a year at a Californian prison, he and his wife then separated.

During Evans's probation in 1992 the FBI started probing again, and Evans went into hiding. In 1994 the Californian Department of Motor Vehicles issued a warrant for Evans' arrest. During the same year, Evans was accused of breaking into a security expert's system in San Diego, and stealing a large amount of information. He left a voice message made by a computer generated voice. Throughout the message he bragged about his expertise, and threatened to kill the security expert with the aid of his friends.

Evans made a mistake, he stored the data he stole from the computer expert on the Well, an on-line conferencing system. This information was spotted, and the security expert was alerted, who then subsequently monitored Evans' activities. Evans was tracked down, arrested and charged with 23 offenses, with a possibility of up to 20 years in prison for each offense.

Although the character described is fictional, the events are based on a real hacker's exploits.

93.3 Techniques and Countermeasures

Malicious Software

Malicious software is specifically written to perform actions that are not desired by the user of a computer. These actions could be passive, displaying a harmless message on the screen, or aggressive, reformatting a hard disk.

The programming abilities required to produce malicious software need not be at genius level. Little experience is required to use the toolkits that are currently available. A number of malicious software authors have taught themselves how to program. Some produce complex programs, which take time to analyze and demonstrate original programming concepts. Much malicious software, however, shows signs of bad programming, and does not execute correctly. Despite the varying quality, malicious software has found its way onto computers worldwide. Malicious software falls into a number of categories.

Trojan Horse. This software pretends to be something it is not. For example, on a disk operating system (DOS) machine, when we type DIR the contents of the current working directory is displayed. If, however, the contents of the directory were deleted, we would be witnessing a Trojan horse in action. It is a program that has the same name as a legitimate piece of software, but when executed, may perform an unexpected malicious act. This malicious act may not occur immediately, but on certain external conditions, for example, the user pressing ctrl-alt-delete (*logic bomb*) or the time being two minutes past midnight (*time bomb*).

Trojan Mule. When a computer is waiting to be logged into, a log-in screen is displayed. A user identification and a password usually needs to be entered in order to gain access to the system. If a piece of software is run that simulates the log-in screen, this would be a Trojan mule. A user would approach the computer, assume the screen was the genuine log-in screen, and enter their user identifier and password.

The Trojan mule would record the data entered and terminate, usually informing the user that the log-in was incorrect. The effect of a Trojan mule is that users' passwords are captured by the person executing the Trojan mule.

Worm. A worm attacks computers that are connected by a network. A worm spreads by attacking a computer, then sending a copy of itself down the network looking for another machine to attack. An important difference exists between a worm and a virus (explained subsequently). A worm makes a copy of itself to spread, which is a standalone entity. A virus makes a copy of itself, but differs in that it needs to attach itself to a program, similar to a parasite attaching to a host. The most infamous example is the Internet worm which attacked computers connected to the Internet on Nov. 2, 1988. It infected over 30% of Internet-connected computers and caused damage estimated at $10–$98 million.

E-Mail Bomb. The e-mail bomb is the electronic equivalent of a letter bomb. When the e-mail is read, an electronic bomb explodes. The result of the explosion may be degradation of system performance due to key system resources being used in the processing of the e-mail message; denial of service because the e-mail program does not filter out certain terminal control codes from e-mail messages, causing the terminal to hang; or something more serious due to the e-mail message containing embedded object code, which in turn contains malicious code (Trojan horse).

Malicious Scripts. These are constructed by the underground to aid an attack on a computer system. The script could take the form of a C program that takes advantage of a known vulnerability in an operating system. It could also be a simplification of a complex command sequence.

Viruses. Viruses have existed for some time and can cause a variety of annoyances to the user. They can produce amusing messages on a user's screen, delete files, and even corrupt the hard disk so that it needs reformatting. Whatever its actions, the virus interferes with the correct operation of the computer without the authorization of the owner.

Many have compared computer viruses to human viruses. Thus the virus writer becomes the equivalent of an enemy waging germ warfare. The most vulnerable computer to virus infection at the moment is the PC running MS-DOS. Viruses do exist that can infect Macintosh, and other types of machines using different operating systems, such as OS/2. Viruses that infect Unix machines are in existence; most are laboratory viruses but there are new reports of one being in the wild, i.e., existing on innocent users machines that have not deliberately installed the virus.

In order to distinguish one virus from another, they are given names by the antivirus industry. Naming conventions vary considerably between antivirus software vendors. A virus author may include a text string in the virus which gives an obvious name, however unprintable. The classic definition of a virus is as follows.

> A virus is a self replicating program that can *infect* other programs, either by modifying them directly or by modifying the environment in which they operate. When an infected file is executed, this will cause virus code within the program to be run.

Boot Sector Viruses

A common form of PC virus is the boot sector virus. When a PC is booted a number of steps are followed. First, the power on self-test (POST) is executed, which tests the integrity of system memory and then initializes the hardware. Information stored in nonvolatile memory is collected, and finally POST sets up the basic input output system (BIOS) address in the interrupt table.

The A: drive is then checked, to see if a disk is present in the drive. This can be seen and heard when the A: drive's motor is started and the light flashes. If a disk is present in the drive, the first sector is read into memory and executed. If no disk is found, then the first sector of the hard disk is read. This sector is known as the master boot sector (MBS). The MBS searches for a pointer to the DOS boot sector (DBS), which is loaded into memory, and control is passed to it.

At this point an opportunity exists for virus infection. A boot sector virus can infect the MBS or the DBS of a hard disk, or the boot sector of the floppy disk. Consider a virus on a floppy first. A floppy with a virus resident on its boot sector is inserted into the A: drive (the original boot sector of the floppy is usually stored elsewhere on the floppy). The machine is booted, and the virus in the boot sector is loaded into memory and executed. The virus searches out the MBS or DBS, depending on the virus' plan, and copies itself to that sector. As with a floppy, the virus usually stores the original MBS or DBS elsewhere on the disk. When the virus has completed execution, it can load the original boot sector and pass control to it, making the actions of the virus invisible to the user. It is important to note that all DOS formatted floppies have a boot sector, even if the floppy is not a system disk.

If the virus infected the MBS of the hard disk (similarly, when the DBS is infected), how does the virus work? The computer is booted from the hard disk, i.e., there's no floppy in the A: drive. The virus code in the MBS is loaded into memory and executed. The virus loads any other sectors that it needs to execute, then loads the original boot sector into memory. The virus is active in memory and can now monitor any floppy disk read/write activity. When an uninfected floppy is detected, it can infect its boot sector. This allows the virus to spread from disk to disk and thus computer to computer.

File Infector Viruses

A file infector virus is basically a program that when executed seeks out another program to infect. When the virus finds a suitable program (the host) it attaches a copy of itself and may alter the host in some way. These alterations ensure that when the host is executed, the attached virus will also be executed. The virus can then seek out another host to infect, and so the process continues.

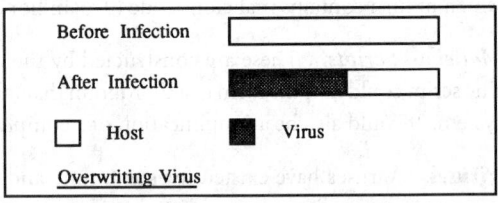

FIGURE 93.1   Overwriting virus.

The virus may attach itself to a host program in a number of ways, the most common types are the following:

>   *Overwriting:* the virus places its code over the host, thus destroying the host (Fig. 93.1). When the virus has finished executing, control is returned to the operating system.
>   *Appending:* the virus places its code at the end of the host (Fig. 93.2). When the host is executed, a jump instruction is usually executed which passes control to the virus. This jump instruction is placed at the start of the host by the virus, the original instructions that were at the start are stored in the body of the virus. During the virus's execution, it replaces the host's original start instructions, and on completion it passes control to these instructions. This process makes the virus invisible to the user until it triggers.
>   *Prepending:* the virus places its code at the start of the host (Fig. 93.3). When the host is executed, the virus is executed first, followed by the host.

Triggers and Payloads

A trigger is the condition that must be met in order for a virus to release its payload, which is the malicious part of the virus. Some viruses simply display a message on the screen, others slow the operation of the

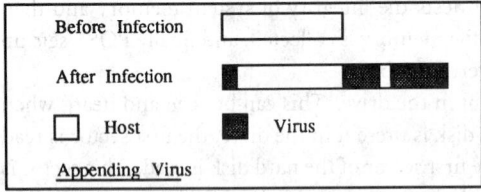

FIGURE 93.2   Appending virus.

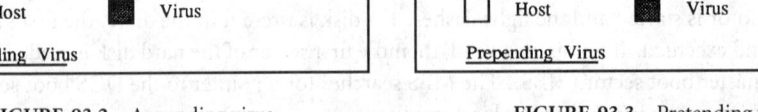

FIGURE 93.3   Pretending virus.

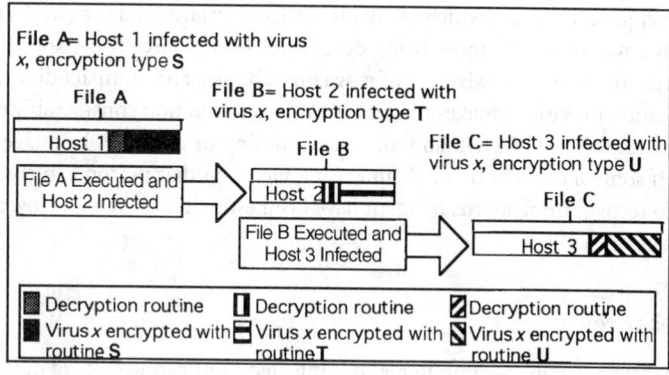

FIGURE 93.4   Polymorphic infection.

computer, the nastier ones delete or corrupt files or reformat the hard disk. The trigger conditions are also only limited by the writer's imagination. It may be that a certain date causes the virus to trigger, a popular day is Friday 13th, or it may be a certain key sequence, such as control-alt-delete.

Virus Techniques

Viruses writers go to great lengths to hide the existence of their viruses. The longer a virus remains hidden, the further its potential spread. Once it is discovered, the virus' trail of infection comes to an end. Common concealment techniques include:

Polymorphism

Polymorphism is a progression from encryption (Fig. 93.4). Virus writers started encrypting their viruses, so that when they were analyzed they appeared to be a collection of random bytes, rather than program instructions. Antivirus software was written that could decrypt and analyze these encrypted viruses. To combat this the writers developed polymorphic viruses.

Polymorphism is the virus' attempt at making itself unrecognizable. It does this by encrypting itself differently every time it infects a new host. The virus can use a different encryption algorithm, as well as a different encryption key when it infects a new host. The virus can now encrypt itself in thousands of different ways.

Stealth

Viruses reveal their existence in a number of ways. An obvious example is an increase in the file size, when an appending or prepending virus infects a host. A file could possibly increase from 1024 bytes long before infection to 1512 bytes after infection. This change could be revealed during a DOS DIR command.

To combat this symptom of the virus' existence, the idea of stealth was created. As was mentioned earlier, the longer a virus remains hidden, the further it spreads. Stealth can be described as a virus' attempt to hide its existence and activities from system services and/or virus detection software.

A virus, for example, to avoid advertising the increase in file size, would intercept the relevant system call and replace it with its own code. This code would take the file size of an infected file, subtract from it the size of the virus, and return the result, the original file size.

Is the Threat of Viruses Real?

Viruses are being written and released every day, in ever increasing numbers. Anyone with access to the Internet can download a virus, even the source code of the virus. These viruses can be run and can spread rapidly between machines. There are widely available electronic magazines such as *40-Hex* that deal with virus writing. They cover new techniques being developed, virus source code, and countermeasures to commercial antivirus software. The existence of magazines, books, and compact disk read-only memory (CD-ROM) information on viruses makes the task of virus construction considerably easier.

If someone has a knowledge of DOS and an understanding of assembly language then that person can write a virus. If someone can boot a PC, and run a file, then that person can create a virus using a toolkit. The costs to recover from a virus incident have been estimated as being as low as $17 and as high as $30,000.

Protection Measures

How can we stop a virus infecting a computer, and if infected, how can we get rid of it before it does any damage? Since prevention is better than cure, a wide range of antivirus software of varying effectiveness is available, commercially and as shareware. When the software has been purchased, follow the instructions. This usually involves checking the machine for viruses first, before installing the software. Antivirus software normally consists of one or more of the following utilities.

Scanner

Every virus (or file for that matter) is constructed from a number of bytes. A unique sequence of these bytes can be selected, which can be used to identify the virus. This sequence is known as the virus' signature. Therefore, any file containing these bytes may be infected with that virus. A scanner simply searches through files looking for this signature.

A scanner is the most common type of antivirus software in use, and is very effective. Unfortunately scanners occasionally produce *false positives*. That is, the antivirus product identifies a file as containing a virus, whereas in reality it is clean. This can occur by a legitimate file containing an identical sequence of bytes to the virus' signature. By contrast, a *false negative* occurs when the antivirus software identifies a file as clean, when in fact it contains a virus.

The introduction of polymorphism techniques complicates the extraction of a signature, and stealth techniques underline the need to operate the scanner in a clean environment. This clean environment is a system booted from the so-called magic object (a write protected clean system diskette). Heuristic scanners have also been developed which analyze executable files to identify segments of code that are typical of a virus, such as code to enable a program to remain resident in memory or intercept interrupt vectors.

Integrity Checkers

Scanners can only identify viruses which have been analyzed and have had a signature extracted. An integrity checker can be used to combat unidentified viruses. This utility calculates a checksum for every file that the user chooses, and stores these checksums in a file. At frequent intervals, the integrity checker is run again on the selected files, and checksums are recalculated. These recalculated values can be compared with the values stored in the file. If any checksums differ then it may be a sign that a virus has infected that file. This may not be the case of course, because some programs legitimately alter files during the course of their execution, and this would result in a different checksum being calculated.

Behavior Blocker

This utility remains in memory while the computer is active. Its task is to alert the user to any suspicious activity. An example would be a program writing to a file. The drawback of this is that user intervention is required to confirm an action to be taken, which can be an annoyance that many prefer to live without.

Fortunately, as viruses increase, so do the number of people taking precautions. With antivirus precautions in place the chance of virus infection can be kept to a minimum.

Virus Construction Kits

These kits allow anyone to create a computer virus. There are a number of types available, offering different functionality. Some use a pull down menu interface (such as the Virus Creation Laboratory), others (such as PS-MPC) use a text configuration file to contain a description of the required virus. Using these tools, anyone can create a variety of viruses in a minimal amount of time.

Hacking

Hacking is the unauthorized access to a computer system. Computer is defined in the broadest sense, and a fine line exists between hacking and telephone phreaking [unauthorized access to telephone switch, private automated branch exchange (PABX) or voice mail]. Routers, bridges, and other network support systems also increasingly use sophisticated computer bases, and are thus open to deliberate attack.

This section provides a hacker's eye view of a target system, indicating the types of probes and data gathering typically undertaken, the forms of penetration attack mounted, and the means of concealing such attacks. An understanding of these techniques is key to the placement of effective countermeasures and auditing mechanisms.

Anatomy of a Hack

An attack can be divided into five broad stages:

1. *Intelligence:* initial information gathering on the target system from bulletin board information swaps, technical journals, and social engineering aimed at extracting key information from current or previous employees. Information collection also includes searching through discarded information (dumpster diving) or physical access to premises.
2. *Reconnaissance:* using a variety of initial probes and tests to check for target accessibility, security measures, and state of maintenance and upgrade.
3. *Penetration:* attacks to exploit known weaknesses or bugs in trusted utilities, the misconfiguration of systems, or the complete absence of security functionality.
4. *Camouflage:* modification of key system audit and accounting information to conceal access to the system, replacement of key system monitoring utilities.
5. *Advance:* subsequent penetration of interconnected systems or networks from the compromised system.

A typical hacking incident will contain all of these key elements. The view seen by a hacker attacking a target computer system is illustrated in Fig. 93.5. There are many access routes which could be used.

Intelligence Gathering

A considerable amount of information is available on most commercial systems from a mix of public and semiopen sources. Examples range from monitoring posts on Usenet news for names, addresses, product information and technical jargon; probing databases held by centers such as the Internet Network Information Center (NIC); to the review of technical journals and professional papers. Information can be exchanged via hacker bulletin boards, shared by drop areas in anonymous FTP servers, or discussed on line in forums such as Internet Relay Chat (IRC).

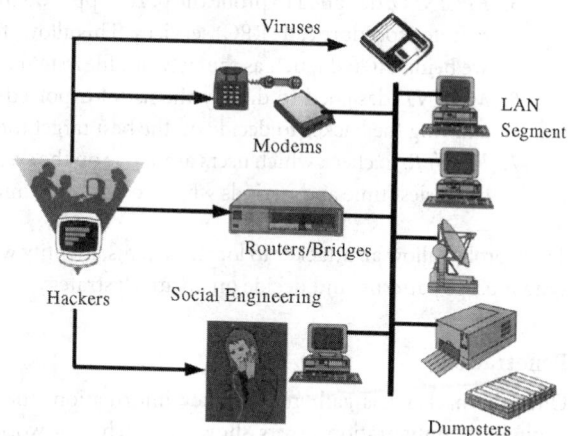

FIGURE 93.5 Possible attacks against IT systems.

Probably the most effective information gathering technique is known as social engineering. This basically consists of masquerade and impersonation to gain information or to trick the target into showing a chink in its security armor. Social engineering ranges from the shared drink in the local bar, to a phone call pretending to be the maintenance engineer, the boss, the security officer, or the baffled secretary who can't operate the system.

Techniques even include a brief spell as a casual employee in the target company. It is often surprising how much temporary staff, even cleaners, tend to be trusted. Physical access attacks include masquerading as legitimate employees (from another branch, perhaps) or maintenance engineers, to covert access using lockpicking techniques and tools (also available from bulletin boards).

Even if physical access to the interior of the building is impossible, access to discarded rubbish may be possible. So-called dumpster diving is a key part of a deliberate attack. Companies often discard key information including old system manuals, printouts with passwords/user codes, organization charts and telephone directories, company newsletters, etc. All this material lends credence to a social engineering attack, and may provide key information on system configuration which helps to identify exploitable vulnerabilities.

Reconnaissance

A wide variety of tools and techniques are available to probe for accessible systems on wide area networks (WAN) such as the Internet. Techniques include the follwing:

1. *Traceroute:* designed to send a series of Internet protocol (IP) packets with increasing time-to-live (TTL) values in order to determine the routing to the target, and to identify intermediate routers and Internet carriers which might be attacked.
2. *DNS dig tools:* designed to query the name-address translation services on the Internet domain name server (DNS) to retrieve a complete listing of all IP addresses within a specified domain. An example might be downloading a list of all MIL domain military systems.
3. *IP scanners:* designed to search a series of IP addresses for active systems. These operate by sending ICMP echo packets to each address in turn (or in random order) and awaiting a reply. These utilities can rapidly locate systems on class B and class C networks (with up to 65,534 and 254 hosts, respectively).
4. *Port scanners:* designed to search a specific system for transmission control protocol (TCP) and user datagram protocol (UDP) ports offering services. The port scanner will attempt to connect to each port in turn, verify whether the connection is accepted, and note which service is being offered. This can also include noting the version string which is sent by utilities such as Telnet (remote log in) and Sendmail [simple mail transfer protocol (SMTP) electronic mail] to check on possible vulnerabilities.
5. *RPCINFO:* designed to probe the portmapper on the remote system, which handles registration of remote procedure call (RPC) services. This allows the hacker to identify which RPC based services are being offered [such as the network filesystem (NFS) or network information system (NIS)].
6. *MOUNT:* designed to display the list of exported filesystems and associated security attributes, allowing the hacker to decide on the best target for NFS attacks.
7. *FINGER:* to check which users are active on the system (ruser is possible substitute), and to decide on busiest time and periods when no system administrator is logged on.

These probes allow an attacker to locate systems, identify which services are offered, gain some idea of the system usage patterns, and decide on an attack strategy.

Penetration

Once the hacker has gathered this key information, then exploitation of security weaknesses begins. If obvious configuration errors show up (such as a world exported NFS filesystem) or access via the anonymous file transfer protocol (FTP) or trivial file transfer protocol (TFTP) to the full filesystem, then

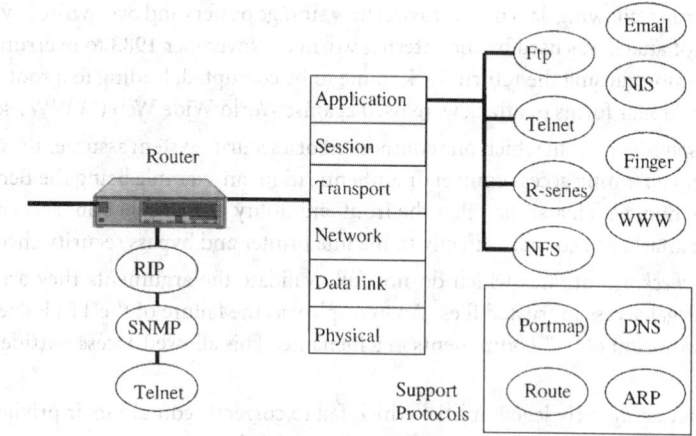

FIGURE 93.6   Network view of system services.

this penetration is rapid. Otherwise the hacker has four courses of action:

1. Try to guess user code and passwords. Common default accounts such as ENGINEER, BIN, SYS, MAINT, GUEST, DIAG, ROOT, FIELD may have weak or default passwords. The hacker may make use of services such as FTP or rexec which do not log failed log-in attempts. Utilities such as *fbomb* use seed lists of common passwords to mount the attack.
2. Try to exploit vulnerabilities in network services such as Sendmail to gain access to the system. Key to the attack is a wide range of attack scripts available on the boards for swapping and trading, together with active participation in full disclosure security discussion lists such as INFOHAX and BUGTRAQ, which openly reveal the details of security holes.
3. Try to exploit weaknesses in the network protocols themselves such as the IP spoofing attack discussed subsequently.
4. Try to break into the network provider's system or network in order to capture user codes and passwords passed in clear across the provider's network. Once captured they can be used to break into the target system; failing which someone on the boards may be able to trade an account for other information.

The proliferation of network services offered by systems, and the increasing intelligence of routers can assist the attacker. Figure 93.6 illustrates the range of services offered to the network by a typical Unix system. While many protocols are supported, each must be adequately secured to prevent system penetration.

Vulnerabilities and Exploitation

The increasing complexity and dynamicity of modern software is one of the key sources of software vulnerabilities. As an example, the Unix operating system now consists of over 2.8 million lines of C code, an estimated 67% of which execute with full privilege. With this size of code base there is a high likelihood of code errors which can open a window for remote exploitation via a network, or local exploitation by a user with an unprivileged account on the local system.

The main source of vulnerabilities are the assumptions made by system programmers about the operating environment of the software, these include:

- *Race conditions:* in which system software competes for access to a shared object. Unix-based operating systems do not support atomic transactions, and as such operations can be interrupted allowing malicious tampering with system resources. Race conditions are responsible for vulnerabilities in utilities such as expreserve, passwd and mail. They have been widely exploited by the group known as the 8-legged groove machine (8lgm).

- *Buffer overruns:* allowing data inputs to overflow storage buffers and over-write key memory areas. This form of attack was used by the Internet worm of November 1988 to overrun a buffer in the fingerd daemon causing the return stack frame to be corrupted, leading to a root privileged shell being run. Similar forms of attack were used against World Wide Web (WWW) servers.
- *Security responsibilities:* in which one component of a security system assumes the other is responsible for implementing access control or authentication, an example being the Berkeley Unix line printer daemon, which assumed that the front end utility (lpr) carried out security checks. This allowed an attacker to connect directly to the line printer and bypass security checks.
- *Argument checking:* utilities which do not fully validate the arguments they are invoked with, allowing illegal access to trusted files. An example was the failure of the TFTP daemon in AIX to check for inclusion of ".." components in a filename. This allowed access outside a secure directory area.
- *Privilege bracketing:* privileged utilities which fail to correctly contain their privilege, and in particular allow users to run commands in a privileged subshell via shell escapes.

Examples of vulnerabilities in the past have included:

- An argument to the log-in command which indicated that authentication had already been carried out by a graphical front end *logintool* and that no further password checks were needed.
- An argument to the log-in command which allowed direct log in with root privileges, or allowed key system files (such as /etc/passwd) to be modified so that they are owned by the unprivileged user logging in.
- A sequence of commands to the sendmail mail program via the SMTP which allowed arbitrary commands to be run with system privileges. This was a new manifestation of an old bug from 1988. The new bug allowed a hacker to cause a mail message to be bounced by the target system and automatically returned to the sender address. This address was a program which would take the mail message as standard input.
- Dynamic library facilities added to newer operating systems allowed an attacker to trick privileged setuid programs into running a Trojanized version of the system library.
- Bugs in the FTP server which allowed a user to begin anonymous (unprivileged) log in, overwrite the buffer containing the user information with the information for a privileged account, and then complete the log-in process. Since the server still believed this was an anonymous log in, no password was requested.

Vulnerabilities often manifest in other forms and on other operating systems. An example is the expreserve bug in the editor recovery software, which was first fixed by Berkeley, then fixed by Sun Microsystems, and finally in a slightly modified form by AT&T.

Penetration scripts circulate widely in the security and hacker community. These scripts effectively deskill hacking, allowing even novices to attack and compromise operating system security. The hacker threat is dynamic and rapidly evolving, new vulnerabilities are discovered, vendors promulgate patches, system administrators upgrade, hackers try again. A single operating system release had 1200 fielded patches, 35 of which were security critical. It is difficult, if not impossible, for system administrators to track patch releases and maintain a secure up-to-date operating system configuration. On a network as large and diverse as the Internet (6.5 million systems, 30 million users in 1995), hackers will always find a vulnerable target which is not correctly configured or upgraded.

Automated Penetration Tools

The task of verifying the security of a system configuration is complex. Security problems originate from insecure manufacturer default configurations; configurations drift as administrators upgrade and maintain the system; vulnerabilities and bugs in software. Tools have been developed to assist in this task, allowing the following:

- Checking of filesystem security settings and configuration files for obvious errors: these include the Computer Oracle & Password System (COPS), Tiger, and Security Profile Inspector (SPI).
- Verifying that operating system utilities are up to date and correctly patched: this is a function included in Tiger and in SPI which use checksum checks on utility files to verify patch installation, and in SATAN and the Internet Security Scanner (ISS) which directly probe for vulnerabilities in network servers.
- Monitoring system configuration to check for malicious alteration of key utilities: this is based on the use of integrity checkers generating checksums of software such as Tripwire.

These tools are extremely powerful, but represent a two-edged sword. Remote probe utilities such as SATAN can be used directly by hackers to scan remote systems for known vulnerabilities. The use of automated attack tools is allowing hackers to screen hundreds of systems within seconds. Initial probes were based on TFTP attack, but now cover a much wider range of protocols including NFS. The NFSBUG security tool allows rapid checking and exploitation of vulnerabilities including mounting of world accessible partitions, forgery of credentials, and exploitation of implementation errors.

Camouflage

After system penetration, hackers will attempt to conceal their presence on the system by altering system log files, audit trails, and key commands. Initial tools such as *cloak* work by altering log files such as *utmp* and *wtmp*, which record system log ins, and altering the system accounting trail.

Second-generation concealment techniques also modified key system utilities such as *ps*, *ls*, *netstat*, and *who* which report information on system state to the administrator. An example is the *rootkit* set of utilities which provides C source code replacements for many key utilities. Once rootkit is installed, the hacker becomes effectively invisible.

Hidden file techniques can also be used to conceal information (using the hidden attribute in DOS, the invisible attribute on the Macintosh, or the " . " prefix on Unix), together with more sophisticated concealment techniques based on steganographic methods such as concealing data in bit-mapped images.

Advance

Once hackers are active on a system there are a number of additional attacks open to them, including the following:

- Exploiting system configuration errors and vulnerabilities to gain additional privilege, such as breaking into the bin, sys, daemon, and root accounts
- Using Trojanized utilities such as telnet to record passwords used to access remote systems
- Implanting trapdoors, which allow easy access to hackers with knowledge of a special password
- Monitoring traffic on the local area network to gather passwords and access credentials passed in clear
- Exhaustive password file dictionary attacks

These attacks aim at breaking into systems on the same network or on networks accessed by users on the local system. Typically, a small network sniffer (such as sunsniff) would be run on the compromised system; this would monitor all TCP-based connections collecting the first 128 bytes of traffic on each virtual circuit in the hope of collecting passwords and user codes. These sniffers log information in hidden files for later recovery.

Vulnerabilities and system configuration errors may also allow the compromise of a privileged account, and subsequent alteration of key utilities such as telnet or FTP may demand passwords from users. These modified utilities can store information for later retrieval. The system log-in utility can be modified to add a simple trapdoor allowing privileged access even if the system administrator has revoked the compromised account originally used for access.

Systems within the same organization often trust each other through the use of explicit trust relationships. A typical example is the r-series protocols from Berkeley which allow a system administrator or user to specify a list of systems which are trusted to log in to corresponding user accounts on the local system without specifying a password. This is very powerful for simplifying administration but very dangerous once a system in the organization is penetrated. The growth of these trust relations over time has been likened to the growth of ivy on old oak trees.

Finally, older systems are susceptible to password cracking style attacks. In these attacks the hacker will take advantage of the one-way encryption algorithm used in Unix to hash plain text passwords to cipher text. This is done by exhaustively encrypting each word in a standard dictionary and then comparing the encrypted version against the stored password cipher text. If the two versions match, then the word can be used as the password for that account. Sophisticated tools are available for both the hacker (to attack accounts) and the system administrator (to proactively check for weak passwords). An example is the crack tool which allows for rule-based transformation of dictionary words to mimic common substitutions, such as replacing I by 1 or O by 0.

Countermeasures

Four major categories of countermeasure are available to counter the hacker threat, they are as follows:

- Firewalls: designed to provide security barriers to prevent unauthorized access to internal systems within an organization from an untrusted network.
- Audit and intrusion detection: designed to provide effective on-line monitoring of systems for unauthorized access.
- Configuration management: to ensure that systems are correctly configured and maintained.
- Community action: to jointly monitor hacker activities and take appropriate action.

Firewall/Gateway Systems

Firewall systems aim to defend internal systems against unauthorized access from an untrusted network. The firewall consists of two main components:

1. Screening or choke router is designed to prevent external access to an internal system other than the bastion host, in addition to restricting incoming protocols to a safer subset such as telnet, FTP and SMTP.
2. Bastion host or firewall authenticates and validates each external access attempt based on a predefined access control list. The bastion host also acts as a proxy by connecting legitimate external users to services on internal systems. Finally, the bastion host will also conceal the structure of the internal network by rewriting outgoing mail addresses and force all outgoing internal connections to be routed through the firewall proxy.

Firewalls are now being evaluated and certified under both the U.S. Trusted Computer Security Evaluation Criteria (TCSEC) and U.K. IT security evaluation criteria (ITSEC). A firewall is an effective barrier defense against external penetration, and if deployed internally within organizations it can also provide protection against insider threats. To be effective a firewall has to be supplemented by strong cryptographic authentication mechanisms based on challenge-response style mechanisms or one-time pads. Without such mechanisms user accounts and passwords will still be carrier unencrypted over Internet provider networks.

Audit and Intrusion Detection Mechanisms

Initial barrier defenses can be supplemented by effective monitoring of networks and systems. Traffic flow analysis can provide an indication of which systems are accessing the organization. Firewall products offer extensive logging of connection attempts, both failed and successful. Intrusion detection systems are being developed which permit user activity profiles to be constructed and allow deviation from established profiles of behavior to be flagged. Examples include the NIDES tool developed at SRI, and the Haystack tool from the U.S. Department of Energy. Key problems with intrusion detection are identifying effective behavior metrics and the processing of heterogeneous audit trails in a wide variety of vendor formats.

Configuration Management

The security of end systems can be improved through strict configuration management, and the use of a variety of proactive configuration checking tools such as COPS, Tiger, and SPI. These tools provide a means for rapidly checking filesystem security settings, verifying patch states, and correcting for the more obvious security blunders. Vendor specific tools such as Asset on the Sun platform are also available. The use of these tools should complement a policy that unnecessary functionality should be disabled on system installation.

Community Action

One of the most effective counters to hacker activity is community action. This takes two main forms: first, collectively reacting to security incidents; second, collectively investigating computer crime. Since the Internet worm of 1988, a series of incident response teams have been set up to deal with network intrusions. An example (and the first of its kind) was the Computer Emergency Response Team (CERT) set up by Carnegie-Mellon University. The CERT provides support to the Internet constituency, providing the following:

- Point of contact for system administrators believing that their system has been compromised
- Means of disseminating security advice and alerts
- Lobbying body for pressuring vendors to fix security problems rapidly
- Central repository of knowledge on the nature of the hacker threat

CERT produces an extensive range of security alerts and advisories giving information on security problems, available fixes, and vendor contact points. The incident response teams work together in a forum known as the Forum for Incident Response and Security Teams (FIRST) to exchange information and techniques. The FIRST group provides a worldwide community of concerned security professionals. FIRST maintains a WWW archive site at www.first.org, which carries advisories and security advice.

The second form of community action is cooperation among law enforcement agencies to investigate computer crime. Most countries in the world now have some form of computer abuse legislation, an example being the Computer Misuse Act in the U.K. The form of legislation varies considerably in its coverage, definition of abuse, extraterritorial extent, and its available penalties. For example, the U.K. act defines three categories of offense:

1. Unauthorized access to computer systems, carrying up to 6 months in prison or a fine
2. Unauthorized access to computer systems with intent to facilitate other criminal acts, such as collecting information from a computer to perpetrate a fraud, carrying up to 5 years in prison
3. Unauthorized modification of data held on a computer system, again with a maximum 5-year penalty

The U.K. act has extraterritorial scope in that a crime is committed if a U.K. system is penetrated or if the penetration attempt originates from the U.K. Clifford Stoll's experiences of attempting to investigate and track a hacker (related in *The Cuckoo's Egg*) clearly indicated the problem of persuading law enforcement agencies in the U.S., Canada, and Europe to work together to track an intruder. In the modern computer world, an attacker can go indirectly through Columbia, China, Nicaragua, and Brazil, making tracking and enforcement a world problem.

Interpol has established a computer crime working group which is attempting to build links between European police forces (including the former Soviet Union), in order to facilitate investigation, share experiences, and unify computer crime legislation.

Phreaking

In contrast to hacking, phreaking is the penetration and misuse of telephone systems. Since the replacement of older crossbar and Strowger switches by digital exchanges, this has become a very nebulous distinction.

Phreaking was born out of the early days of analog switches in which exchanges carried signaling information (such as the number dialed by a subscriber) in-band as part of the voice channel. Signaling information was passed by a series of dual-tone multifrequency (DTMF) codes. The U.S. telephone carriers making use of an international standard Consutative Committee of International Telephony and Telegraphy-5 (CCITT-5), which uses combinations of 700-, 900-, 1100-, 1300-, and 1500-Hz tones to signal line state and number dialed. There was little to prevent a phreaker from generating comparable tones via a PC sound card or a simple oscillator (or box).

In particular, U.S. carriers used a single 2600-Hz tone to signal to trunk equipment that a local call had finished, and that the trunk equipment should be placed in idle mode awaiting the next call. A hacker could inject a 2600-Hz tone, reset the trunk line, and then dial the number required in CCITT-5 format. This technique also allowed the phreaker to evade call billing, and thus proved quite popular with hackers needing to access remote computer systems or bulletin boards.

Colored Boxing

The history of phone phreaking revolved around 2600 Hz. Early stories include a famous phreaker named Captain Crunch, who discovered that a whistle in a breakfast cereal packet generated exactly the correct frequency, to a blind phone phreaker whose perfect pitch allowed him to whistle up the 2600-Hz carrier. A whole spectrum of boxes were built (with plans swapped on the boards) to generate tone sets such as CCITT-5 (a blue box) or the 2200-Hz pulses generated by U.S. coin boxes to signal insertion of coins (a red box) or a combination of both (a silver box). Instructions range from building lineman's handsets (to tap local loops), to ways of avoiding billing by simulating an on-hook condition while making calls, and to ways of disrupting telephone service by current pulses injected on telephone lines.

The move from in-band signaling to common carrier signaling is reducing the risk of boxing attacks. Common carrier signaling carries signaling information for a cluster of calls on a separate digital circuit rather than in-band where it may be susceptible to attack. Older switching equipment in developing countries, and in specialist networks (such as 1-800) are still being targeted.

War Dialers

The worlds of hacking and phone phreaking come together in the war dialer. A war dialer is a device designed to exhaustively scan a range of numbers on a specific telephone exchange looking for modems, fax machines, and interesting line characteristics. War dialers were also capable of randomizing the list of numbers to be searched (to avoid automated detection mechanisms), to automatically log line states, and to automatically capture the log-in screen presented by a remote computer system. While U.S. telecomm charging policies often meant that local calls were free (encouraging a bulletin board culture), to use long distance meant the use of phone phreaking to avoid call billing.

War dialers such as Toneloc and Phonetag offered an effective way of screening over 2000 lines per night. Boards regularly carried the detailed results of these scans for each area code in the U.S.

Modems

A war dial scan was likely to detect many modems with various levels of security, ranging from unprotected to challenge–response authentication. A popular defensive technique was the use of a dial-back facility, in which the user was identified to the modem, which would then ring the user back on a predefined number.

If the modem used the same incoming line to initiate dial-back, a hacker could generate a simulated dial tone to trick the modem into believing that the line had been dropped and that dial-back could begin. The identification of modems also led to a range of other problems:

- Publicly accessible network gateways, which allowed an authorized user to access WAN functionality by dialing in: This led to system intrusion over the Internet which could not be traced back beyond the public dial in.
- Diagnostic modems for computer systems, for PABXs, and for PTT trunk switches: This opened the door to direct attacks on the digital switches with weak password security.

The details of such switches are widely exchanged in the underground (particularly the Unix-based 5-ESS switch from AT&T) with considerable knowledge of the methods for reconfiguration of switches to change quality of service on lines, or to compromise subscriber information. Subscriber information such as the reverse mapping between subscriber number and name/address has been openly sold by phreakers.

Phone Loops

Phone loops refer to a linked pair of subscriber circuits used for testing purposes. Callers dialing both circuits will be automatically interconnected. Numbers for phone loops were widely swapped among phreakers to provide a convenient forum for phreaker/hacker conferences.

PABX Penetration

Private automatic branch exchanges are open to a range of attacks including weak security on diagnostic modem lines and misconfiguration of switches. These systems are now a common target providing a convenient springboard for long-distance attacks. An example might be an attacker who calls into the PABX and then uses private wire circuits belonging to the firm to call out to countries overseas. Direct inward system access (DISA) facilities provide a rich facility set for legitimate company workers outside the office, including call diversion, conferencing, message pickup, etc. If misconfigured these facilities can compromise the security of the company and permit call fraud.

Cellular Phones

Cellular phone technology is still in its infancy in many countries with many analog cellular systems in common usage. Analog cellular phones are vulnerable in a number of areas:

- Call interception: no encryption or scrambling of the call is carried out, calls can therefore be directly monitored by an attacker with a VHF/UHF scanner. U.S. scanners are modified to exclude the cellular phone band, but the techniques for reversing the modification are openly exchanged.
- Signaling interception: signaling information for calls is also carried in clear including the telephone number/electronic serial number (ESN) pair used to authenticate the subscriber. This raises the risk of this information being intercepted and replayed for fraudulent use.
- Reprogramming: commercial phones are controlled by firmware in ROM (or flash ROM) and can be reprogrammed with appropriate interface hardware or access to the manufacturers security code.

Three forms of attack have been described: the simple reprogramming of a cellular phone to an ESN captured on-air or by interrogating a cellular phone; the tumbler, in which a telephone number/ESN pair is randomly generated; and the vampire, in which a modified phone rekeys itself with an ESN intercepted directly over the air.

The use of modified cellular phones provides a linkage between the phreaker community and organized crime. In particular, lucrative businesses have been set up in allowing immigrants to call home at minimal cost on stolen cellular phones or telephone credit cards. Newer digital networks such as GSM are encrypted and not open to the same form of direct interception attack.

Carding

The final category of malicious attack is aimed at the forgery of credit and telephone card information. A key desire is to make free phone calls to support hacking activities. Four techniques have been used:

1. Reading the magnetic stripe on the back of a credit card using a commercial stripe reader. This stripe can then be duplicated and affixed to a blank card or legitimate card.
2. Generating random credit card numbers for a chosen bank with a valid checksum digit. These numbers will pass the simple off-line authentication checks used by vendors for low-value purchases.
3. Using telephone card services to validate a series of randomly generated telephone card numbers generated by modem.
4. Compromising a credit card number in transit over an untrusted network, such as the Internet.

The last category is a growing problem with the increasing range of commercial agencies now attempting to carry out business on the Internet. The introduction of secure electronic funds transfer systems is key to supporting the growth of electronic commerce on the Internet.

93.4 The Future

Computer Security

The level of technical sophistication of computer systems, telecomm switches, router and network infrastructure continues to grow. With increasing complexity comes increasing vulnerability. The focus of hacker attacks has moved with improving security measures, as the attackers seek to find a weak point in system defenses. Common carrier signaling is more secure than in-band, but the digital switches are vulnerable. Firewalls protect end systems but the network infrastructure can be attacked. Security is improving over time, but the level of technical attack sophistication continues to rise.

National Information Infrastructure

The U.S. presidential vision of the information superhighways is leading to growing internetworking and a move toward ubiquitous computing. This move is increasing our use of and dependence on networks. Security will become a key issue on these networks, not just protection against casual penetration but also against deliberate motivated attack by organized crime, terrorists, or anarchists: the beginning of information warfare. The organization of effective coordinated defenses against threats against our infrastructure will be one of the key challenges of the 1990s.

Further Information

Bellovin, S. and Chiswick, B. 1994. *Firewalls and Internet Security.* Addison–Wesley, Reading, MA.
Brunner, J. Shockwave Rider. New York.
Chapman, B. and Zwicky, E. 1995. *Building Internet Firewalls.* O'Reilly & Associates, Sebastopol, CA.
Denning, P. 1990. *Computers Under Attack: Intruders, worms and viruses.* ACM Press, Addison–Wesley, Reading, MA.
Ferbrache, D. 1992. *A Pathology of Computer Viruses.* Springer–Verlag, London, England.

Garfinkel, S. and Spafford, G. 1996. Practical UNIX & Internet Security. O'Reilly & Associates, Sebastopol, CA.

Gibson, W. 1991. *Neuromancer.* Simon & Schuster, New York.

Hoffman, L. 1990. *Rogue Programs: Viruses, Worms and Trojan Horses.* Van Nostrand Reinhold, New York.

Littman, J. 1996. *The Fugitive Game: Online with Kevin Mitnick.* Little, Brown, Boston, MA.

Neumann, P. 1995. *Computer Related Risks.* Addison–Wesley, Reading, MA.

Shimomura, T. and Markoff, J. 1995. *Takedown: The pursuit and capture of Kevin Mitnick, America's most wanted computer outlaw.* Hyperion, New York.

Sterling, B. The Hacker Crackdown: Law and disorder on the electronic frontier. Available via WWW at http://www-swiss.ai.mit.edu/~bal/sterling/.

Stoll, C. 1989. *The Cuckoo's Egg.* Doubleday, Garden City, NY.

Virus Bulletin, Various issues, Virus Bulletin Ltd, England.

Weiner, L. 1995. *Digital Woes: Why we should not depend on software.* Addison–Wesley, Reading, MA.

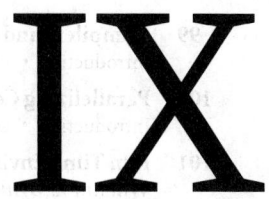

Programming Languages

Kim Bruce, Section Adviser
Williams College

T HE DESIGN SPACE OF PROGRAMMING LANGUAGES is partitioned into paradigms, mechanisms for compiling and run-time management, and language theory. Overall, this section provides a good balance between considerations of language paradigms, implementation issues, and theoretical models. Topics include object-oriented, functional, logic, and imperative programming paradigms, compilers and interpreters, memory management, type systems, and foundational calculi for programming language semantics.

94

Imperative Language Paradigm

Michael J. Jipping
Hope College

Kim Bruce
Williams College

94.1 Introduction

In the 1940s, John von Neumann pioneered the design of basic computer architecture by structuring computers into two major units: a central processing unit (CPU), responsible for computations, and a data storage unit, or memory. This architecture is demand driven, based on a command and instruction-oriented computing model. The basic unit cycle of execution, typically composed of a single instruction, consists of four steps:

1. Obtain the addresses of the result and operands.
2. Obtain the operand data from the operand location(s).
3. Compute the result data from the operand data.
4. Store the result data in the result location.

Note in this sequence how separation of the execution unit from the memory unit has structured the sequence. Data must be located and piped from memory, operated on, and transferred back to memory to be available for the next operation. All operations in a von Neumann machine operate this way, in a stepwise, structured manner. The von Neumann model has been the basis of nearly every computer built since the 1940s.

Imperative programming languages are modeled after the von Neumann model of machine execution and were invented to provide the abstractions of machine components and actions in order to make it easier to program computers. Abstractions such as variables (which model memory cells), assignment statements (which model data transfer), and other language statements are all abstractions of the basic von Neumann approach.

0-8493-2909-4/97/$0.00+$.50

In this chapter we address the fundamental principles underlying imperative programming languages and examine the way the constructs of imperative languages are represented in several languages. We devote special attention to features of more modern imperative programming languages, among them support for abstract data types and newer control constructs such as iterators and exception handling. Examples in this chapter are given in a variety of imperative programming languages, including FORTRAN, Pascal, C, C++, MODULA-2, and Ada 83. In the Best Practices section we explore in more detail the languages FORTRAN IV (chosen for historical reasons), C and C++ (its imperative parts), and Ada 83.

94.2 Data Bindings: Variables, Type, Scope, and Lifetime

In this section we discuss some of the fundamental properties of imperative programming languages. In particular, we address issues related to binding time, the properties of variables, **types**, scope, and lifetime.

Binding Time

We will find it useful to classify many of the differences in programming languages based on the notion of binding time. A **binding** is the association of an attribute to a name. The time at which a binding takes place is an important consideration. There are many times when a binding can occur. Some of these follow:

- Language definition: when the language is designed. An example is the binding of the constant name true to the corresponding Boolean value.
- Language implementation: when a compiler or interpreter is written. An example is the binding of the representation of values of various types.
- Compile time: when a program is being translated into machine language. For example, the type of a variable in a statically typed language is bound at compile time. In statically typed languages, overloaded functions are bound at compile time.
- Load time: when the executable machine language image of the program is loaded into the memory for execution by the execution unit. The location of global variables is bound at load time.
- Procedure or function invocation time: the time a program is being executed. Actual parameters are bound to formal parameters and local variables are bound to locations at procedure invocation time.
- Run time: any time during the execution of a program. A new value can be bound to a variable at run time. In dynamically typed languages, overloaded functions are bound at run time.

As we examine fundamental issues in the definition of imperative programming languages, we will keep in mind the distinctions between languages based on differences in binding time.

Variables

Imperative languages support computation by executing commands whose purpose is to change the underlying state of the computer on which they are executed. The *state* of a computer encompasses the contents of memory and also includes both data which are about to be read from outside of the computer and data which have been output.

Variables are central to the definition of imperative languages as they are objects whose values are dependent on the contents of memory. A variable is characterized by its attributes, which generally include its name, location in memory, value, type, scope, and lifetime.

Depending on context, the meaning of a variable may be considered to be either its value or its location. For instance, in the assignment statement, $x := x + 1$, the meaning of the occurrence of the variable x to the left of the assignment symbol is its location (sometimes called the l-value of x), whereas the meaning of the occurrence on the right side is its value, that is, the value stored at the location corresponding to x (sometimes called the r-value). The location of global variables is bound at load time, whereas the location of local variables and reference parameters is typically bound at procedure entry. The value of the variable can be changed at any point during execution of the program.

Types

Types in programming languages are abstractions which represent sets of values and the operations and relations which are applicable to them. Types can be used to hide the representation of the primitive values of a language, allow type checking at either compile time or run time, help disambiguate overloaded operators, and allow the specification of constraints on the accuracy of computations. Types also can play an important role in compiler optimization.

Types in a programming language include both simple and composite types. The use of *simple types* such as integer, real, Boolean, and character types allows the user to abstract away from the actual computer representation of these values, which may differ from computer to computer. The operations on simple types may or may not be supported directly by the underlying hardware. For instance, many early microprocessors supported only real or floating-point operations in software.

Some languages (e.g., those derived from Pascal) allow the programmer to define their own simple *enumerated* types by simply listing the values of the type. The ordering of elements in this enumeration is significant as these types typically support successor and predecessor functions as well as ordering relations. Later we will discuss mechanisms for supporting **abstract data types**, another way of constructing types which can be used as though they were primitive to a language.

Many languages support the creation of *subrange* types, which allows a programmer to define a new type as a copy of a type with a subset of its values. The new type comes equipped with the same operators as its parent type and is usually compatible with the original type.

Composite or structured data types can be created from simple types using *type constructors*. Typical composite types include arrays, records (or structures), variant records (or unions), sets, subranges, pointer types, and, in a few languages, function or procedure types. For instance, arrays are typically constructed from two types: a subrange type which provides the set of indices of the array, and another type representing the values stored in the array. Not all languages support all these type constructors. For instance, function and procedure types are provided by MODULA-2 but are not available in Ada 83. Many languages support strings as special types of composite types, for instance, as arrays of characters, but they may also be provided as builtin types.

Most imperative languages bind types to variables statically. These bindings are usually specified in declarations, but some languages, such as FORTRAN, allow implicit declaration of variables, with the type binding determined by the name of the identifier (e.g., in FORTRAN if the name starts with I through N then the variable is an integer, otherwise real).

An important issue in type-checking programming languages is type equivalence. When do two terms have equivalent types? The two extremes in the definitions of **type equivalence** are structural and name equivalence:

- Structural equivalence: Two types are said to be *structurally (or domain) equivalent* if they have the same structure. That is, they are built from the same type constructors and builtin types in the same way.
- Name equivalence: Two types are *name equivalent* if they have the same name.

The language C uses structural equivalence, whereas Ada 83 uses name equivalence. There are also a range of possibilities between these two extremes. For instance, Pascal and MODULA-2 use *declaration equivalence*: two types are declaration equivalent if they are name equivalent or they lead back to the same structure declaration by a series of redeclarations.

Inequivalent types may be compatible in certain situations. For instance, two types are assignment compatible if an expression of one type may be assigned to a variable of another. For instance, in Pascal a subrange of integer is assignment compatible with integer, even though the types are not equivalent.

An application of these ideas can be found in the rules for determining whether a particular actual parameter may be used in a procedure call for a particular formal parameter. In Pascal, if the formal parameter is a reference parameter then the actual parameter must be a variable of equivalent type. If the formal parameter is a value parameter then the actual parameter must be assignment compatible.

As mentioned earlier, some languages support the creation of subrange types. The new subrange type is usually assignment compatible with the original type. Because of this compatibility, the new type is called a **subtype** of the parent in Ada. Another mechanism available in Ada, called **derived typing**, defines a new type by constructing an exact copy of a type that already exists. However, the resulting new type is distinct and is not type equivalent or even assignment compatible with the existing type.

The type equivalence rules are the cause of one of the greatest limitations in the use of Pascal. If a formal parameter has an array type, then the actual parameter must have an equivalent type. In particular, the subscript ranges of the two arrays must be identical. Thus, it is impossible to write a procedure in Pascal which can be used to sort different-sized arrays of real numbers. (Actually, the current ANSI standard Pascal provides a special mechanism to allow exceptions to this rule.)

Ada escapes from this problem by designating some properties of types to be static, while others are dynamic. For example, in a type defined to be a subrange of integers, the underlying static type is integer while the subrange bounds are a dynamic property. Only the static properties of types are considered at compile time by the type checker, whereas restrictions due to dynamic properties are checked at run time.

Consider the following Ada declarations as an example of type bindings:

```
    type COINS is (PENNY, NICKEL, DIME, QUARTER);
    subtype SILVER is COINS range (NICKEL..QUARTER);
    type CHANGE is new COINS;

    C1, C2; COINS;
    S: SILVER;
    CH: CHANGE;
```

COINS is an enumerated type, defined by the programmer to allow assignments such as

```
    C1 := DIME;
```

SILVER is a subrange of **COINS**, which includes only the values **NICKEL**, **DIME**, and **QUARTER**. **CHANGE** is a derived type taken from **COINS**.

Because Ada employs name equivalence, only **C1** and **C2** are equivalent, but **S** is assignment compatible with them. If Ada used structural equivalence, then variables **C1**, **C2**, and **CH** would be equivalent.

Scope

The scope of a binding is the area or section of a program in which that particular binding is effective. The method and extent of **scope rules** that define a binding scope will, to a large degree, affect the usefulness and applicability of a language. If, for instance, the rules allow the scope of a binding to be determined by the execution path of a program, the language might be more flexible, yet the code becomes harder to understand.

Scope rules are tied tightly to concepts of binding time. *Static scope rules* determine the scope of a binding at compile time and are based on the lexical structure of the program. *Dynamic scope rules* determine the scope of a binding at run time. Thus, an occurrence of a variable name in a procedure may refer to one variable the first time it is evaluated yet refer to an entirely different variable the next time, depending on the execution path at run time. Most imperative languages use static scope rules.

```
with TEXT_IO; use TEXT_IO;
procedure SCOPED is
    package INT_IO is new INTEGER_IO(integer); use INT_IO;
    I,J: integer;

    procedure P is begin put(J); new_line; end P;

begin
  J := 0;
  I := 10;
  declare      -- Block 1
     J: integer;
  begin
     J := I; -- reference point A
     P;
  end;
  put(J); new_line;
  declare      -- Block 2
     I: Integer
  begin
     I := 5
     J := I + 1; -- reference point B
     P;
  end;
  put(J); new_line;
end;
```

FIGURE 94.1 Scoping rules in Ada.

As an example of scope rules in Ada, consider the code in Fig. 94.1. Static scope rules are determined by the program block structure, which does not change while the program runs. Therefore, the call to procedure **P** prints the variable **J** defined in the outer, main program, no matter where it is called from. Likewise, the assignment in block 1 at reference point A changes **J** from the block and not from the main program. Dynamic scope rules, on the other hand, typically follow dynamic call paths to determine variable bindings. If Ada used dynamic scope rules, the first call to **P** from block 1 would print the value 10 corresponding to the **J** from block 1, whereas the second call to **P** would print the value 3 corresponding to the **J** from the main program.

Execution Units: Expressions, Statements, Blocks, and Programs

An *expression* is a program phrase which returns a value. Expressions are built up from constants and variables using operators. As described earlier, variables may represent two values, depending on context: their location and the value stored at that location. Operators may be builtin, like the arithmetic and comparison operators, or may be user-defined functions.

Reflecting the sequential order of von Neumann computation, an imperative language specifies the order in which operations are evaluated. Typically, evaluation order is determined by precedence rules. A typical precedence rule set for arithmetic expressions might be the following:

1. Subexpressions inside parentheses are evaluated first (according to the precedence rules).
2. Instances of unary negation are evaluated next.
3. Then, multiplication ($*$) and division ($/$) operators are evaluated in left to right order.
4. Finally, addition ($+$) and subtraction ($-$) are evaluated left to right.

Although procedure rules are commonly used by imperative languages, some languages use other conventions to avoid precedence rules. For example, PostScript uses postfix notation for expressions, while LISP uses prefix notation. APL evaluates all expressions from right to left without regard to precedence, using only parentheses to change the evaluation order.

The fundamental unit of execution in an imperative programming language is the *statement*. A statement is an abstraction of machine language instructions, grouped together to form a single logical activity.

The simplest and most fundamental statement in imperative programming languages is the assignment statement. This statement, typically written in the form **x := e** or **x = e** with **x** a variable (or other expression representing a location) and **e** an expression, is usually interpreted by evaluating **e** and copying its value into the location represented by **x**. This is known as the *copy semantics* for assignment.

Less common are languages which use the sharing interpretation of assignment. In these languages, variables generally represent references to objects which contain the actual values. The assignment $x := y$ would then be interpreted as binding the object referred to by y to x rather than its value. Since both variables refer to the same object, they share the same value. If the value of one is changed, the value of the other will also change. This is the *sharing semantics* for assignment.

Declarations and statements may be grouped together to form a *block*. Procedure and function bodies are represented as blocks, whereas **control structures** (discussed subsequently) can also be understood as acting on blocks of statements (generally without declarations). The most general form of a block contains a *declarative* section, which contains the declarations that define the bindings that are effective in the block, and an *executable* section, which contains the statements over which the binding is to hold, i.e., the scope of the declarations.

In so-called block-structured languages (including most languages descended from ALGOL 60, e.g., Pascal, Ada, and C), blocks may be nested. Within any block, therefore, there can be two kinds of bindings in force: *local bindings*, which are specified by the declarative sections associated with the block, and *nonlocal bindings* (also known as *global* bindings), which are bindings defined by declarative sections of blocks within which the specific block is nested.

Consider again the code from Fig. 94.1. The first two assignments of the main program assign **J** from the main program the value 0 and **I** from the main program the value 10. The next assignment assigns the value 10, derived from the global **I**, to the variable **J** from the first inner block. When the definition of the second inner block is encountered, the variable **I** is found in the local scope, while **J** is found in the *outer* scope, that of the main program. The value 6 will be printed for **J** at the end of the main program.

94.3 Control Structures

By adopting the semantics of the basic execution cycle of a von Neumann architecture, an imperative language adopts a strict sequential ordering for its statements. By default, the next statement to execute is the next physical statement in the program. Control structures in imperative languages provide ways to alter this strict sequential ordering. The most common control structures are *conditional structures* and *iterative structures*. *Unconstrained control structures* are also allowed in most languages through the use of goto statements.

Conditional Structures

Conditional control structures (also known as *selection statement*) determine whether or not a block of statements is executed based on the result of one or several tests. These structures fall into one of two classes:

If Statements. All imperative languages include some form of if statement. This control structure provides a text and a single statement or statement block to be executed if the test evaluates to a true value. Optionally, the programmer may provide another block of statements which can be executed only if the test evaluates to false. The following is a simple example from Ada:

```
if (x = 2) then
    y := 3;
else
    y := 6;
end if;
```

The variable **y** is set to either 3 or 6 depending on the value of **x** .

In most languages, if statements can be nested within other control structures, including other if statements. However, nested if statements can result in awkward, deeply nested code. Thus, many languages provide a special construct (e.g., **elsif** in Ada) to represent *else if* constructs without requiring further nesting. The two Ada examples given next are equivalent semantically, though the first, which uses **elsif** , is easier to read than the second, which uses nested conditionals:

```
if (x = 2) then              if (x = 2) then
    y := 3;                      y := 3;
elsif (x = 3) then           else
    y := 15;                     if (x = 3) then
elsif (x = 5) then               y := 15;
    y :=18;                      else
else                             if (x = 5) then
    y := 6;                          y := 18;
end if;                          else
                                     y := 6;
                                 end if;
                               end if;
                             end if;
```

Case Statements. This conditional combines case-by-case expression examination with a restricted multiway conditional. This conditional may be seen to be simply a syntactic convenience, but in many cases its implementation results in a much faster determination at run time of the actual block of code to be executed. Consider the following case statement from Ada:

```
case y is
    when 2 => y := 3;
    when 3 => y := 15;
    when 15 => y := 18;
    when others => y := 6;
end case;
```

An expression (**y** in this case) of an ordinal type occurs after the keyword **case**. Each **when** clause contains a guard, which is a list of one or more constants of the same type as the expression. Most languages require that there be no overlap between these guards. The expression after the keyword **case** is evaluated, and the resulting value is compared to the guards. The block of statements connected with the first matched alternative is executed. If the value does not correspond to any of the guards, the statements in the **others** clause is executed. Note that the semantics of this example is identical to that of the previous example.

The case statement may be implemented in the same way as a multiway if statement, but in most languages it will be implemented via table lookup, resulting in a constant time determination of which block of code is to be executed.

C's switch statement differs from the case previously described in that if the programmer does not explicitly exit at the end of a particular clause of the switch, program execution will continue with the code in the next clause.

Iterative Structures

One of the most powerful features of an imperative language is the specification of *iteration* or statement repetition. *Iterative structures* can be classified as either definite or indefinite, depending on whether the number of iterations to be executed is known before the execution of the iterative command begins:

- Indefinite iteration: The different forms of indefinite iteration control structures differ by where the test for termination is placed and whether the success of the test indicates the continuation or termination of the loop. For instance, in Pascal the **while-do** control structure places the test before the beginning of the loop body (a pretest), and a successful test determines that the execution of the loop shall continue (a continuation test). Pascal's **repeat-until** control structure, on the other hand, supports a posttest, which is a termination test. That is, the test is evaluated at the end of the loop and a success results in termination of the loop.

  Some languages also provide control structures which allow termination anywhere in the loop. The following example is from Ada:

  ```
  loop
     ...
     exit when test;
     ...
  end loop
  ```

  The **exit when test** statement is equivalent to **if test then exit**.

  A few languages also provide a construct to allow the programmer to terminate the execution of the body of the loop and proceed to the next iteration (e.g., C's **continue** statement), whereas some provide a construct to allow the user to exit from many levels of nested loop statements (e.g., Ada's named **exit** statements).

- Definite iteration: The oldest form of iteration construct is the definite or fixed-count iteration form, whose origins date back to FORTRAN. This type of iteration is appropriate for situations where the number of iterations called for is known in advance. A variable, called the *iteration control variable* (ICV), is initialized with a value and then incremented or decremented by regular intervals for each iteration of the loop. A test is performed before each loop body execution to determine if the ICV has gone over a final, boundary value. Ada provides fixed-count iteration as a for loop; an example is shown next.

  ```
  for i in 1..10 loop
      y := y + i;
      z := z * i;
  end loop;
  ```

Here, **i** is initialized to 1, and incremented by 1 for each iteration of the loop, until it exceeds 10. Note that this type of loop is a pretest iterative structure and is essentially syntactic sugar for an equivalent **while** loop.

An ambiguity that may arise with a for loop is what value the iteration control variable has after termination of the loop. Most languages specify that the value is formally undetermined after termination of the loop, though in practice it usually contains either the upper limit of the ICV or the value assigned which first passes the boundary value. Ada eliminates this ambiguity by treating the introduction of the control variable as a variable declaration for a block containing only the for loop.

Some modern programming languages have introduced a more general form of for loop called an *iterator* construct. Iterators allow the programmer to control the scheme for providing the iteration control variable with successive values. The following example is from CLU [Liskov et al. 1977]. We first define the iterator:

```
string_chars = iter (s : string) yields (char);
    index: Int := 1;
    limit: Int := string$size (s);
    while index <= limit do
      yield (string$fetch(s, index));
      index := index + 1;
    end;
  end string_chars;
```

which can be used in a **for** loop as follows:

```
for c: char in string_chars(s) do LoopBody end;
```

When the for loop controlled by an iterator is encountered, control is passed to the iterator, which runs until a **yield** statement is executed. The value associated with the **yield** statement is used as the initial value of the iterator control variable **c**, and the body of the loop is executed. Control is then passed back to the iterator, which resumes execution with the statement following the **yield**. Control is passed to the loop body each time a yield statement is executed and back to the iterator each time the loop body finishes execution. Thus, iterators behave as a restricted form of coroutine, passing control back and forth between the two blocks of code. The loop is terminated when the iterator runs to completion. In the preceding examples this will occur when **index > limit**.

Unconstrained Control Structures: Goto and Exceptions

Unconstrained control structures, generally known as goto constructs, cause control to be passed to the statement labeled by the **identifier** or line number given in the goto statement. Dijkstra [1968] first questioned the use of goto statements in his famous letter, "Goto statement considered harmful," to the editor of the *Communications of ACM*. The controversy over the goto mostly centers on readability of code and handling of the arbitrary transfer of control into and out of otherwise structured sections of program code.

For example, if a goto statement passes control into the middle of a loop block, how is the loop to be initialized, especially if it is a fixed-count loop? Even worse, what happens when a goto statement causes control to enter or exit in the middle of a procedure or function? The problems with readability arise because a program with many goto statements can be very hard to understand if the dynamic (run time) flow of control of the program differs significantly from the static (textual) layout of the program.

Programs with undisciplined use of gotos have earned the name of *spaghetti code* for their similarity in structure to a plate of spaghetti.

Although some argue for the continued importance of goto statements, most languages either greatly restrict their use (e.g., do not allow gotos into other blocks) or eliminate them altogether. In order to handle situations where gotos might be called for, other, more restrictive language constructs have been introduced to make the resulting code more easily readable. These include the **continue** and **exit** statements (particularly labeled **exit** statements) referred to earlier.

Another construct which has been introduced in some languages in order to replace some uses of the goto statement is the *exception*. An exception is a condition or event that requires immediate action on the part of the program. An exception is *raised* or *signaled* implicitly by an event such as arithmetic overflow or an index out of range error, or it can be explicitly raised by the programmer.

The raising of an exception results in a search for an exception *handler*, a block of code defined to handle the exceptional condition and (hopefully) allow normal processing to resume. The search for an appropriate handler generally starts with the routine which is executing when the exception is raised. If no appropriate handler is found there, the search continues with the routine which called the one which contained the exception. The search continues through the chain of routine calls until an appropriate handler is found, or the end of call chain is passed without finding a handler.

If no handler is found the program terminates, but if a handler is found the code associated with the handler is executed. Different languages support different models for resuming execution of the program. The termination model of exception handling results in termination of the routine containing the handler, with execution resuming with the caller of that routine. The continuation model typically resumes execution at the point in the routine containing the handler which occurs immediately after the statement whose execution caused the exception.

The following is an example of the use of exceptions in Ada (which uses the termination model):

```
procedure pop(s: stack) is
   begin
      if empty(s) then raise emptyStack
                  else ...
   end;

procedure balance (parens: string) return boolean is
   pStack: stack
   begin
      ...
      if ... then pop(s) ...
   exception
      when emptyStack => return false
   end
```

Many variations on exceptions are found in existing languages. However, the main characteristics of exception mechanisms are the same. When an exception is raised, execution of a statement is abandoned and control is passed to the nearest handler. (Here "nearest" refers to the dynamic execution path of the program, not the static structure.) After the code associated with the handler is executed, normal execution of the program resumes.

The use of exceptions has been criticized by some as introducing the same problems as goto statements. However, it appears that disciplined use of exceptions for truly exceptional conditions (e.g., error handling) can result in much clearer code than other ways of handling these problems.

We complete our discussion of control structures by noting that, although many control structures exist, only a very few are actually necessary. At the one extreme, simple conditionals and a goto statement are sufficient to replace any control structure. On the other hand, it has been shown [Boehm and Jacopini 1966] that a two-way conditional and a while loop are sufficient to replace any control structure. This result has led some to point out that a language has no need for a goto statement; indeed, there are languages that do not have one.

Procedural Abstraction

Support for abstraction is very useful in programming languages, allowing the programmer to hide details and definitions of objects while focusing on functionality and ease of use. **Procedural abstraction** [Liskov and Guttag 1986] involves separating out the details of an execution unit into a procedure and referencing this abstraction in a program statement or expression. The result is a program that is easier to understand, write, and maintain.

The role of procedural abstraction is best understood by considering the relationships between the four levels of execution units described earlier: expressions, statements, blocks, and programs. A statement can contain several expressions; a block contains several statements; a program may contain several blocks. Following this model, a procedural abstraction replaces one execution unit with another one that is simpler. In practice, it typically replaces a block of statements with a single statement or expression.

The *definition* of a procedure binds the abstraction to a name and to an executable block of statements called the *body*. These bindings are compile-time, declarative bindings. In Ada, such a binding is made by specifying code such as the following:

```
procedure area (height, width: real; result: out real) is
   begin
      result := height * width;
   end;
```

The *invocation* of a procedure creates an activation of that procedure at run time. The *activation record* for a procedure contains data bound to a particular invocation of a procedure. It includes slots for parameters, local variables, other information necessary to access nonlocal variables, and data to enable the return of control to the caller. In languages supporting recursive procedures, more than one activation record can exist at the same time for a given procedure. In those languages, the lifetime of the activation record is the duration of the procedure activation.

Although scoping rules provide access to nonlocal variables, it is generally preferable to access nonlocal information via **parameter** passing. Parameter-passing mechanisms can be classified by the direction in which the information flows: *in parameters*, where the caller passes data to the procedure, but the procedure does not pass data back; *out parameters*, where the procedure returns data values to the caller, but no data are passed in; and *in out parameters*, where data flow in both directions.

Formal parameters are specified in the declaration of a procedure. The *actual parameters* to be used in the procedure activation are specified in the procedural invocation. The procedure passing mechanism creates an association between corresponding formal and actual parameters. The precise information flow which occurs during procedure invocation depends on the parameter passing mechanism.

The association or mapping of formal to actual parameters can be done in one of three ways. The most common method is *positional parameter association*, where the actual parameters in the invocation are matched, one by one in a left-to-right fashion, to the formal parameters in the procedural definition. *Named parameter association* also can be used, where a name accompanies each actual parameter and determines to which formal parameter it is associated. Using this method, any ordering can be used to specify parameter values. Finally, *default parameter association* can be used, where some actual parameter values are given and some are not. In this case, the unmatched formal parameters are simply given a default value, which is generally specified in the formal parameter declaration.

Note that in a procedural invocation, the actual parameter for an in parameter may be any expression of the appropriate type, since data do not flow back, but the actual parameter for either an out or an in out parameter must be a variable, because the data that are returned from a procedural invocation must have somewhere to go.

Parameter passing is usually implemented as being one of copy, reference, and name. There are two copy parameter passing mechanisms. The first, labeled *call-by-value*, copies a value from the actual to the formal parameter before the execution of the procedure's code. This is appropriate for in parameters. A second mode, called *call-by-result*, copies a value from the formal parameter to the actual parameter after the termination of the procedure. This is appropriate for out parameters. It is also possible to combine these two mechanisms, obtaining *call-by-value-result*, providing a mechanism which is appropriate for in out parameters.

The *call-by-reference* passes the address of the actual parameter in place of its value. In this way, the transfer of values is not by copying but occurs by virtue of the formal parameter and the actual parameter referencing the same location in memory. Call-by-reference makes the sharing of values between the formal and actual a two-way, immediate transfer, because the formal parameter becomes an alias for the actual parameter.

Call-by-name was introduced in ALGOL 60 and is the most complex of the parameter passing mechanisms described here. Although it has some theoretical advantages, it is both harder to implement and generally more difficult for programmers to understand. In call-by-name, the actual parameter is re-evaluated every time the formal parameter is referenced. If any of the constituents of the actual parameter expression has changed in value since the last reference to the formal parameter, a different value may be returned at successive accesses of the formal parameter. This mechanism also allows information to flow back to the main program with an assignment to a formal parameter. Although call-by-name is no longer used in most imperative languages, a variant is used in functional languages which employ lazy evaluation (see Chapter 95 of this Handbook).

Several issues crop up when we consider parameters and their use. The first is a problem called *aliasing*, where the same memory location is referenced with two or more names. Consider the following Ada code:

```
procedure MAIN is
   a: integer;
   procedure p(x, y: in out integer) is
      begin
         a := 2;
         x := y + a;
      end;

   begin
      a := 10;
      p(a,a);
      ...
   end;
```

During the call of **p(a,a)** the actual parameter **a** is bound to both of the formal parameters **x** and **y**. Because **x** and **y** are in out parameters, the value for **a** will change after the procedure returns. It is not clear, however, which value **a** will have after the procedure call. If the parameter passing mechanism is call-by-value-result then the semantics of this program depend on the order in which values are copied back to the caller. If they are copied into the parameters from left to right, the value of **a** will be 10 after the call. The results with call-by-reference will be unambiguous (though perhaps surprising to the programmer), with the value of **a** being 4 after the call. In Ada, a parameter specified to be passed as in out may be passed using either call by value-result or call by reference. The preceding code provides an example where, because of aliasing, these parameter passing mechanisms give different answers. Ada terms such programs to be erroneous and considers them not to be legal, even though the compiler may not be able to detect such programs.

Most imperative programming languages support the use of procedures as parameters (Ada is one of the few exceptions). In this case the parameter declaration must include a specification of the number and types of parameters of the procedure parameter. MODULA-2, for example, supports procedure types which may be used to specify procedural parameters. There are few implementation problems in supporting procedure parameters, though the implementation must ensure that nonlocal variables are accessed properly in the procedure passed as a parameter.

There are two kinds of procedural abstractions. One kind, usually known simply as a *procedure*, is an abstraction of a program statement. Its invocation is like a statement, and control passes to the next statement after the invocation. The other type is called a *value returning procedure* or *function*. Functions are abstractions for an operand in an expression. They return a value when invoked, and, upon return, evaluation of the expression containing the call continues.

Many programming languages restrict the values which may be returned by functions. In most cases, this is simply a convenience for the compiler implementor. However, although common in functional languages, most imperative languages which support nested procedures or functions do not allow functions to return other functions or procedures. The reason has to do with the stack-based implementation of block-structured languages. If allowed, it might be possible to return a nested procedure or function which depends on a nonlocal variable which is no longer available when the procedure is actually invoked.

To avoid confusion, most languages allow a name to be bound to only one procedural abstraction within a particular scope. Some languages, however, permit the *overloading* of names. Overloading permits several procedures to have the same name as long as they can be distinguished in some manner. Distinguishing characteristics may include the number and types of parameters or the data type of the return value for a function. In some circumstances, overloading can increase program readability, whereas in others it can make it difficult to understand which operation is actually being invoked.

Program mechanisms to support concurrent execution of program units are discussed in Chapter 98. However, we mention briefly *coroutines* [Marlin 1980], which can be used to support pseudoparallel execution on a single processor. The normal behavior for procedural invocation is to create the procedural instance and its activation record (runtime environment) upon the call and to destroy the instance and the activation record when the procedure returns. With coroutines, procedural instances are first created and then invoked. Return from a coroutine to the calling unit only suspends its execution; it does not destroy the instance. A resume command from the caller results in the coroutine resuming execution at the statement after the last return.

Coroutines provide an environment much like that of parallel programming; each coroutine unit can be viewed as a process running on a single processor machine, with control passing between processes. Despite their interesting nature (and clear advantages in writing operating systems), most programming languages do not support coroutines. MODULA-2 is an example of a language which supports coroutines. As mentioned earlier, iterators can be seen as a restricted case of coroutines.

Data Abstraction

Earlier in this chapter, we introduced the idea of data types as specifying a set of values and operations on them. Here we extend that notion of values and operations to abstract data types and their definitional structures in imperative languages.

The primitive data types of a language are specified by both a set of values and a collection of operations which may be applied to them. Clearly, the set of integers would be useless without the simultaneous provision of operation on those integers. It is characteristic of primitive data types that the programmer is not allowed access to their representations.

Many modern programming languages provide a mechanism for a programmer to specify a new type which behaves as though it were a primitive type. An abstract data type (ADT) is a collection of data objects and operations on those data objects whose representation is hidden in such a way that the new data objects may be manipulated only using the operations provided in the ADT. ADTs abstract away the implementation of a complex data structure in much the same way primitive data types abstract away the details of the underlying hardware implementation.

The *specification* of an ADT presents interface details relevant to the users of the ADT, whereas the *implementation* contains the remaining implementation details that should not be exported to users of the ADT. *Encapsulation* involves the bundling that together of all definitions in the specification of the ADT in one place. Because the specification does not depend on any implementation details, the implementation of the ADT may be included in the same program unit with the specification or it may be contained in a separately compiled unit. This encapsulation of the ADT typically supports information hiding so that the user of the ADT (1) need not know the hidden information in order to use the ADT, and (2) is forbidden from using the hidden information so that the implementation can be changed without impact on correctness to users of the ADT (at least if the specifications of the operations are still satisfied in the new implementation). Of course, one would expect a change in implementation to affect the efficiency of programs using the ADT. A further advantage of information hiding is that, by forbidding direct access to the implementation, it is also possible to protect the integrity of the data structure.

CLU was among the earliest languages providing explicit language support for ADTs through its clusters. Ada also provides facilities to support ADTs via packages. Let us consider an example from MODULA-2, a successor language to Pascal designed by Niklaus Wirth. The stack ADT provides a stack type as well as operations init, push, pop, top, and empty. A MODULA-2 specification for a stack of integers resembles the following:

```
DEFINITION MODULE StackADT;
    TYPE stack;
    PROCEDURE init (VAR s: stack);
    PROCEDURE push (VAR s: stack; elt: INTEGER);
    PROCEDURE pop (VAR s: stack);
    PROCEDURE top (s: stack): INTEGER;
    PROCEDURE empty (s: stack): BOOLEAN
END StackADT.
```

Note the declaration includes the type name and procedural *headers* only. The type **stack** included in the preceding specification is called an *opaque* type in MODULA-2, because users cannot determine the actual implementation of the type from the specification. This ADT specification uses information hiding to get rid of irrelevant detail. Now, this specification can be placed in a separate file and made available to programmers. By including the following declaration:

```
FROM StackADT IMPORT stack, init, push, pop, top, empty;
```

at the beginning of a module, a programmer could use each of these names as though the complete specification was included in the module. Thus, the user can write

```
var s1, s2: Stack;
begin
  push(s1, 15);
  push(s2, 20);
  if not empty(s1) then pop(s1) ...
```

is such a module.

In MODULA-2 the complete definitions of the type **stack** and its associated operations are provided in an implementation module, which typically is stored in a separate file.

```
IMPLEMENTATION MODULE StackADT;
TYPE stack = POINTER TO stackRecord;
     stackRecord = RECORD
                         top: 0..100;
                         values: ARRAY[1..100] of INTEGER;
                         END;
PROCEDURE init(VAR s: stack);
   BEGIN
       ALLOCATE(s, SIZE(stackRecord));
       s^.top :=0
   END
PROCEDURE push (VAR s: stack; elt: INTEGER);
   BEGIN
       s^.top := s^.top + 1;
       s^.value[s^.top] := elt
   END;
PROCEDURE pop(VAR s: stack); ...
PROCEDURE top(s: stack): INTEGER; ...
PROCEDURE empty(s: stack) BOOLEAN; ...
END StackADT.
```

Notice that the type name **stackRecord** is not exported.

The specification module must be compiled before any module that imports the ADT and before its implementation module, but importing modules and the implementation module of the ADT can be compiled in any order. As previously suggested, the implementation is irrelevant to writing and compiling a program using the ADT, though, of course, the implementation must be compiled and present when the final program is linked and loaded in preparation for execution.

There is one important implementation issue which arises with the use the language mechanisms supporting ADTs. When compiling a module which includes variables of an opaque type imported from an ADT (e.g., **stack**), the compiler must determine how much space to reserve for these variables. Either the language must provide a linguistic mechanism to provide the importing module with enough information to compute the size required for values of each type or there must be a default size which is appropriate for every type defined in an ADT. CLU and MODULA-2 use the latter strategy. Types declared as CLU clusters are represented implicitly as pointers, whereas in MODULA-2 opaque types must be represented explicitly using pointer types as in the **Stack ADT** example just given. In either case, the compiler need reserve for a variable of these types only an amount of space sufficient to hold a pointer. The memory needed to hold the actual data pointed to is allocated from the heap at run time. As discussed later, Ada uses a language mechanism to provide size information for each type to importing units.

The definition of ADTs can be *parameterized* in several languages, including CLU, Ada, and C++. Consider the definition of the **stack** ADT. Although the preceding example was specifically given for an integer data type, the implementations of the data type and its operations do not depend essentially on the fact that the stack holds integers. It would be more desirable to provide a parameterized definition of **stack** ADT which can be instantiated to create a stack of any type T.

Allocating space for these parameterized data types raises the same problems as previously discussed for regular ADTs. C++ and Ada resolve these difficulties by requiring parameterized ADTs to be instantiated at compile time, whereas CLU again resolves the difficulty by implementing types as implicit references.

94.4 Best Practices

In this section, we will examine three quite different imperative languages to evaluate how the features of imperative languages have been implemented in each. The example languages are FORTRAN (FORTRAN IV for illustrative purposes), Ada 83, and C++. We chose FORTRAN to give a historical perspective on early imperative languages. Ada 83 is chosen as one of the most important modern imperative languages which supports ADTs. C++ might be considered a controversial choice for the third example language, as it is a hybrid language that supports both ADT-style and object-oriented features. Nevertheless, the more modern feature contained in the C++ language design makes it a better choice than its predecessor, C (though many of the points that will be made about C++ also apply to C). In this discussion we ignore most of the object-oriented features of C++, as they are covered in more detail in Chapter 96.

Data Bindings: Variables, Types, Scope, and Lifetime

Like most imperative languages, all three of our languages use static binding and static scope rules. FORTRAN is unique, however, because it supports the *implicit declaration* of variables. An identifier whose name begins with any of the letters *I* through *L* is implicitly declared to be of type integer, whereas any other identifier is implicitly declared to be of type real. These implicit declarations can be overridden by explicit declaration. Therefore, in the fragment:

```
INTEGER A
I = 0
A = I
B = C + 2.3
```

A and **I** are integer variables, while **B** and **C** are of type real. Most other statically typed languages (including Ada and C++) require *explicit declaration* of each identifier before use. Aside from providing better documentation, these declarations lessen the danger of errors due to misspellings of variable names.

FORTRAN has a relatively rich collection of numerical types, including integer, *double precision* (real), and *complex*. *Logical* (Boolean) is another builtin type, but FORTRAN IV provided no direct support for characters. However, characters could be stored in integer variables. (The *character* data type was added by FORTRAN 77). FORTRAN IV supported arrays of up to three dimensions but did not support records. Strings were represented as arrays of integers. FORTRAN IV did not provide any facilities to define new named types.

Later languages provided much richer facilities for defining data types. Pascal, C, MODULA-2, Ada, and C++ all provided a full range of primitive types as well as constructors for arrays, records (structures in C and C++), variant records (unions in C and C++), and pointers (access types in Ada). All provided facilities for naming new types and for constructing types hierarchically (constructing nested types). Variant records or unions opened up holes in the static type systems of most of these languages, but Ada (and CLU before it) provided restrictions on the access to variants and builtin run time checks in order to prevent type insecurities.

The scope rules for each language, though static in nature, differ significantly. In FORTRAN, the rules are the simplest. The unit of scope for an identifier is either the main program or the procedural unit in which it is declared. Declarations of procedures (subroutines) and functions are straightforward as well in FORTRAN, with no nesting and all parameters passed by reference. Whereas FORTRAN does not support access to nonlocal variables through scoping rules, it allows the programmer to explicitly declare that certain variables are to be more globally available. When two subprograms need to share certain variables, they are listed in a *common* statement which is included in each subprogram. If different combinations of subprograms need to share different collections of variables, several distinct common blocks can be set up, with each subprogram specifying which blocks it wishes access to.

Ada (like Pascal) supports *nested declarations* of procedures and functions. As a result, block structure becomes extremely important to scope determination.

Whereas C and C++ do not provide for nested procedures and functions, they do share with Ada the ability to include declaration statements in local blocks of code. In C++, blocks are syntactically enclosed in bracket {...} symbols, and any declarations that occur between the symbols hold for the duration of the block. Consider the following code:

```
for (i = 0; i<20; i++) {
  int i = 1, j;
  j = 0;
  while (i < 25) {
    j += i*2;
    i ++;
  }
}
```

One might think that when the inner loop is done, the outer loop also will be done, because **i** has the value 26. But since the inner block of statements redeclared **i**, the scope rules state that the new, inner **i** was manipulated, leaving the outer **i** untouched and free to correctly manipulate the for loop.

In FORTRAN IV the lifetime of all variables is the lifetime of the program. As a result, all memory in a program could be statically allocated, including activation records. Because each subprogram has only one activation record, FORTRAN could not support recursion.

Pascal, Ada, C, and C++ all support recursive functions and procedures. In order to support these, implementations generally rely on stack-allocated activation records. Thus, the lifetime of a local variable in a procedure p extends from the call of p to the return to its caller. Each of these languages also supports pointer or access types, generally representing data accessed from the heap (though C and C++ also allow pointers to stack-allocated memory). The lifetime of these variables is generally from the time that the programmer executes a creation instruction until a corresponding destruction statement is executed.

Execution Units

FORTRAN and Ada make a strong distinction between expressions and statements, with expressions simply returning a value, but with statements forming the basic unit for program execution. In C and C++, however, these two units of execution are merged, with statements treated as expressions. The statement **x = 5** assigns the value 5 to the variable x. But, in C++, the = sign is also an operator, and this assignment statement is actually an expression that returns the value being assigned. Thus, the statement **y = x = 5** assigns the value 5 to *both* x and y, because the value 5 is assigned to x and the expression $x = 5$ returns 5, which is assigned to y. Although interesting, it can also be very confusing. Because many expressions will have side effects, the order of evaluation will affect the value returned from an expression. Consider the code

```
if ( (y = ++x) == (x + 6)) { ... }
```

This code actually has two statements embedded in it; first, **++x** increments **x**, then this value is assigned to **y**, then the value assigned is tested against the value of **x + 6**. If the compiler decides to change the order in which the subexpressions are evaluated (a not unheard of occurrence in C++ compilers), it may change whether the guard on the if statement is true or false.

Allowing statements to be part of expressions also means that typographical errors are more likely to give rise to syntactically correct (but logically incorrect) statements. For instance, if one of the = signs in

```
if (x == 6) { ... }
```

is omitted, then it will assign of value 6 to *x* and the conditional will always evaluate to true as all non-0 integers in C and C++ are treated as representing true.

Control Structures

Because FORTRAN was one of the earliest high-level languages, it is not surprising that its control structures are much closer to the underlying machine language instructions. Aside from the do loop (which was similar to the for loop in Pascal and Ada), most other control structures were based on the use of goto statements. Thus, the if statement of FORTRAN IV evaluated an integer expression and, depending on whether the result was negative, zero, or positive, resulted in a jump to one of three statement labels included with the statement. Aside from the usual goto statement, FORTRAN IV also included assigned and computed gotos, which provided some of the flexibility of case statements. FORTRAN 77 and the more recent FORTRAN 90 provide more modern control structures such as the if and while statements of other languages.

The control constructs in Ada are similar to those of Pascal, including if, case, while, and for loops, as well as indefinite loops, which are terminated with exit statements. Several of these were described earlier in the general discussion of control structures.

C and C++ include if statements and a switch construct which is similar to the case statement. The while loop is similar to that in Pascal and Ada, but for the loops in C and C++ are more general than those in most other languages. For loops have the form:

```
for (E₁; E₂; E₃) S;
```

where E_1 is initialization code, E_2 is a test for termination, E_3 contains code to update variables for the next iteration of the loop, and S represents the code to be executed each time through the loop. The test for termination is executed before the update code. Thus, a statement of the form

```
for (i = 1; i < 10, i++) S;
```

will result in S being executed once for each value of **i** from 1 to 10. (The expression $i++$ is an expression which increments the value of **i.**) However, much more flexible statements are also possible

```
for (i = 1; not done and i < 1024; i = 2 * i) S;
```

This statement repeatedly executes **S** while **i** ranges through the powers of 2 from 1 to 1024. If done is ever true, it will terminate early.

Procedural Abstraction

Each of Ada, FORTRAN, C, and C++ provides procedural abstraction. Ada and FORTRAN distinguish between functions and procedures, whereas C and C++ do not since procedures are just functions which return an element of type void. FORTRAN IV also supported single-line statement functions, which could be defined local to a program or subprogram. As noted earlier, Ada, C, and C++ all support recursive functions and procedures, whereas FORTRAN does not.

The languages differ in minor ways in how they return values from functions. FORTRAN, like Pascal, treats the name of the function as a pseudovariable which can be assigned to. An explicit return statement returns control to the calling program unit. When the function returns, the last value stored in the function name is returned as the value of the function. Ada, C, and C++ use return statements of the form return exp to return control to the calling program unit. The value of the expression associated with the return statement is the value returned from the function.

Most programming languages provide system-defined overloaded functions, such as arithmetic operators ($+$, $-$, $*$, etc.) and comparison functions (e.g., $=$, $<$, etc.). Ada and C++ are relatively unusual, though, in allowing user-defined overloading. In both, the compiler must be able to disambiguate at compile time whichever of the versions of the overloaded operator are called for at each of its occurrences. C++ determines which version is called for by looking at the number and types of the actual parameters. Ada goes further and can also use the return type to determine which version works in the particular context in which it is found. Thus, in Ada one may overload the $+$ operator to take two integer parameters and return a user-defined rational type, even though there already exists a builtin version of $+$ which takes two integer parameters and returns an integer. If $+$ occurs in a context in which only an integer result would make sense, the builtin version would be selected. If $+$ occurs in a context in which only a rational value would make sense the user-defined version would be selected. If the system cannot tell which should be used, then an error will occur at compile time.

Unlike FORTRAN, Pascal, and C, both Ada and C++ provide language support for exceptions. Ada and C++ both use the termination model for program resumption after handling the exception.

Data Abstraction and Separate Compilation

FORTRAN provides support for separate compilation of subroutines and functions but provides no type checking across compilation unit boundaries. Thus, the main program may call a function F with two real parameters, but the definition of F in a separately compiled unit may have only one formal parameter, and it might be an integer. Because FORTRAN does not support the definition of new named types, it provides no support for abstract data types.

C and C++ provide slightly better support for separate compilation by allowing the programmer to put external function and procedure declarations in a header file, which may be included into compilation units which use them. The header files are treated as though they were textually part of the compilation unit into which they are included. C provides no support for abstract data types, though C++ does provide strong support through its class facilities. C++ classes give the programmer control over which aspects of a data type the user will be allowed to see and use. Because C++ is described in some depth in Chapter 96, we omit a detailed description here.

Standard Pascal provides no support for separate compilation, though virtually all Pascal implementations provide support for separately compiled units. These units typically include both interface and implementation sections. The interface of a unit may be explicitly imported into another compilation unit with a *uses* statement. This provides for separate, but not independent, compilation without requiring the programmer to create individual header files by hand. These units generally do not provide support for information hiding.

Ada provides both separate compilation and strong support for abstract data types. Like MODULA-2's modules described earlier, Ada packages come in two separately compiled units, the specification and body. Only items listed in the nonprivate part of the package specification are accessible at compile time to units which import the package. The following is an Ada package specification for stacks:

```
package StackADT is
    type stack is private;
    procedure push(s: in out stack; elt: in integer);
    procedure pop(s: in out stack);
```

```
        procedure top(s: in stack) return integer;
        procedure empty(s: in stack) return boolean;
    private
        type stack is record
             top: integer := 0;
             values: array (1..100) of integer
        end record
    end StackADT;
```

The private section of a package specification is necessary to provide a description of private types. This is necessary so that importing programs know how much space to provide for a variable of that type. This is not as clean as the MODULA-2 solution, since any change of representation of the type will require the recompilation of the specification and hence of any program which imports the package. For this particular representation of stack, no initialization routine is necessary because top is initialized to 0 in the declaration of the type.

If the implementation of the private variable is a pointer, then only partial type information need be provided. That is, if we replace the private part of the preceding example by

```
    private
        type stackRecord;
        type stack is access StackRecord;
    end StackADT;
```

then this would provide sufficient information for importing programs to determine the memory needs for a variable of this type (i.e., the amount of space necessary to hold a pointer).

An implementation of the original package specification is given next:

```
    package body StackADT is
       procedure push (s: in out stack; elt: in integer) is
          begin
             s.top := s.top + 1;
             s.value[s.top] := elt
          end push;
       procedure pop(s: in out stack); ...
       function top(s: in stack): integer; ...
       function empty(s: in stack) boolean; ...
    end StackADT;
```

The **StackADT** package specification can be imported into an Ada program unit by including with **StackADT** at the beginning of the unit. Components can be referred to as record components, e.g., **StackADT.stack** and **StackADT.push.** The package name prefix can be omitted if use **StackADT** is also included at the beginning of the unit.

Both Ada and C++ provide mechanisms for supporting parameterized packages (or classes in the case of C++). The C++ template mechanism is quite primitive, with template instantiations being treated as being similar to compile-time macroexpansions. The template is never type checked, only its instantiations. Ada also requires its generic packages to be instantiated at compile time, but the generics are type checked before, rather than after, instantiation. Thus, a generic package can be compiled and later used in another until which does not have access to the implementation.

The following is an example of the header of a generic **BinarySearchTree** package:

```
generic
  type Element is private;
  with function LessThan (x, y: Element) return boolean;
  package BinarySearchTree is
  type BSTree is private;
  ...
  end BinarySearchTree;
```

This can be used in another unit by instantiating it with a type and appropriate function, for example,

```
package PeopleDict is new BinarySearchTree(People, PeopleComp)
```

where **PeopleComp** is a function taking pairs of type **People** and returning a Boolean. **PeopleDict** can then be used like any other package. The ability to require generic package instantiations to include necessary functions and values as well as types ensures that they will not be instantiated with types which do not support the appropriate operations.

94.5 Research Issues and Summary

Research issues in imperative languages in recent years have tended to focus on many of the new constructs presented in this chapter. These include support for exceptions, iterators, abstract data types, and parameterized or generic types. It is fair to say that most current research in programming languages is devoted to implementation and environment issues or to other programming paradigms. There are not many new concepts currently being introduced into imperative programming languages. Many languages which formerly were purely imperative have recently been extended to include object-oriented concepts (e.g., Object Pascal, Objective C, C++, Ada 95). Another series of extensions has provided features for concurrent and distributed programming. Discussions of these two different kinds of extensions can be found in Chapters 96 and 98 of this Handbook.

From our earlier discussion, it is clear that support for abstraction plays an important role in imperative language design and use. Variables abstract away details of memory usage; data types (and in particular abstract data types) abstract from the representation of values to provide support for operations that are independent from the actual implementation; execution units abstract away details of machine instruction execution and expression computation while providing clean interfaces for sharing information between caller and callee.

A second major focus in the development of modern imperative programming languages has been the enrichment of type systems, especially static type systems. Abstract data types can be understood as the enrichment of type systems with so-called existential types, in which the existence of a type is revealed be instantiated with any type (or in Ada and CLU's case any type which comes supplied with the appropriate operations). These more flexible type systems allow for the construction of safe statically typed programming languages which are more expressive than their predecessors. There is hope that we are moving forward to a time when most programmers will see such secure languages as assisting them in their goal of creating correct and efficient software, rather than getting in the way. (See Chapter 104 for a further discussion of type systems.)

We have surveyed the class of programming languages modeled after the sequential organization of the von Neumann architecture. The imperative programming language paradigm is characterized by its sequential, stepwise statement execution.

As discussed in the "Best Practices" section the imperative programming constructs are implemented in a variety of ways in different languages. There are many languages to choose from; choosing the right language for the applications at hand is an important first step to software implementation.

It could be argued that the object-oriented paradigm is simply a minor variation on the imperative paradigm in which remote procedure and function calls replace the more familiar imperative calls. However, the object-oriented paradigm requires an entirely different way of thinking about the organization of a program, with the traditional conception of a program as a series of operations being applied to values replaced in the object-oriented view by an organization of more distributed responsibility. In this view, values (typically referred to as objects) are responsible for knowing how to perform their own operations, and the programmer is responsible for bringing together a group of objects with appropriate capabilities and organizing a program which relies on these distributed capabilities to accomplish a task. Subtyping and inheritance provide important organizing tools and promote code reuse in ways unavailable in traditional imperative languages.

Most programmers today are initially taught to program in imperative languages. Thus, these languages reflect the way that most programmers currently think about algorithm construction and program execution. Whether this will continue in the face of the challenge of the object-oriented paradigm will be interesting to see.

Defining Terms

Abstract data type: A collection of data type and value definitions and operations on those definitions which behaves as a primitive data type. The specifications of these types, values, and operations are generally collected in one place, with the implementations hidden from the user.

Binding: A connection between an abstraction used in the language and a data object as it exists in the computer hardware. The usage, establishment, and number of these bindings characterize the various imperative languages and affect their ease of use and performance.

Control structures: Structures or statements that alter the strict sequential ordering in an imperative program, presenting alternatives to sequential control. Control structures can be conditional, iterative, or unconstrained.

Derived type: A new data type constructed by copying a type that already exists. The resulting new type is distinct and not identified as being copied from the existing type, though operations on the old type are automatically inherited in the new type.

Identifier: The name bound to an abstraction.

Parameters: Data objects passed between the caller and the called procedural abstraction.

Procedural abstraction: Separating out the details of an execution unit in such a way that it may be invoked in a program statement or expression.

Scope rules: Rules in a language that define the area or section of a program in which a particular binding is effective.

Subtype: A new data type defined as a copy of another defined type, typically with a restricted subset of its values. It may generally be used in the same contexts as its parent type.

Type: A collection of values with an associated collection of primitive operations on those values.

Type equivalence: Rules that govern when variables or values from two different data types may be used together.

Variable: An abstraction used in imperative languages for a memory location or cell.

References

Boehm, C. and Jacopini, G. 1966. Flow diagrams, Turing machines, and languages with only two formation rules. *Commun. ACM* 9(5):366–371.

Dijkstra, E. W. 1968. Goto statement considered harmful. *Commun. ACM* 11(3):147–148.

Liskov, B. H. and Guttag, J. V. 1986. *Abstraction and Specification in Program Development.* MIT Press, Cambridge, MA.

Liskov, B., Snyder, A., Atkinson, R., and Schaffert, C. 1977. Abstraction mechanisms in CLU. *IEEE Trans. Software Eng.* SE-5(6):546–558.

Marlin, C. D. 1980. *Coroutines.* Lecture notes in computer science 95. Springer–Verlag, New York.

Further Information

A good examination of imperative languages, as well as other paradigms, can be found in the following texts:

Dershem, H. L. and Jipping, M. J. 1995. *Programming Languages: Structures and Models*, 2nd ed. PWS, Boston, MA.

Louden, K. C. 1993. *Programming Languages: Principles and Practice*. PWS-Kent, Boston, MA.

Pratt, T. W. and Zelkowitz, M. V. 1996. *Programming Languages: Design and Implementation*, 3rd ed. Prentice–Hall, Englewood Cliffs, NJ.

Sebesta, R. 1993. *Concepts of Programming Languages*. 2nd ed. Benjamin-Cummings.

Sethi, R. 1996. *Programming Languages: Concepts and Constructs*, 2nd ed. Addison–Wesley, Reading, MA.

Several journals are devoted to programming languages and language design. *ACM Transactions on Programming Languages and Systems* and *Computer Languages* both feature referred papers on programming languages. *ACM SIGPLAN Notices* is a collection of unreferenced papers from the ACM Special Interest Group on Programming Languages. Proceedings of the ACM conferences, *Principles of Programming Languages* (*POPL*), and *Programming Language Design and Implementation*, provide a good presentation of current research in programming languages.

95

Functional Programming Languages

Benjamin Goldberg
New York University

95.1 Introduction

Functional languages are a class of languages based on the **lambda calculus**, a very simple but powerful model of computation. Proponents claim that the use of a functional language supports faster production of software, shorter programs, and more readable and verifiable code than the use of conventional so-called imperative programming languages. Furthermore, in the research community functional languages have been used as the basis of study on advanced type systems, parallel computing, program optimization, and programming language semantics.

Within the class of functional languages there is substantial variety. In this chapter, we describe three popular languages that are representative of the class: SCHEME, a dialect of LISP; Standard ML; and HASKELL. Although these languages differ in significant ways, they all exhibit the necessary properties in order to be considered functional.

A program written in a functional language consists of function definitions and function applications. As in mathematics, a function is an entity that maps each input to a single output. This is in stark contrast to imperative languages such as C and FORTRAN in which a function is simply a collection of statements

0-8493-2909-4/97/$0.00+$.50
© 1997 by CRC Press, Inc.

which may modify variables, allowing the same input to be mapped to different outputs over the course of the computation.

Consider the factorial function. It is described formally by

$$n! = \begin{cases} 1 & \text{if } n = 0 \\ n(n-1)! & \text{otherwise} \end{cases}$$

In a functional language, in this case Standard ML, the executable definition of factorial is usually written as

```
fun factorial(0) = 1
  | factorial(n) = n * factorial(n-1)
```

and follows directly from the formal definition. In contrast, the factorial function is typically written in an imperative language, in this case American National Standards Institute (ANSI) C, as

```
int factorial(int n)
{ int prod = 1;
   for (int i = 1; i <=n; i++)
     prod = prod * i;
   return(prod);
}
```

where the programmer specifies that a variable holds the running product and is modified in each iteration of the loop. Although a version similar to the first program could be written in most imperative languages, the syntax and traditional programming style of these languages encourage the writing of the second.

Because functional languages, and the programs written in them, are intended to have nice mathematical properties—such as the properties that functions written in functional languages are functions in the true mathematical sense—there is no explicit notion of memory and the modification of memory. As in mathematics, a variable in a functional program simply represents a value, it is not a cell in memory that can be modified. There is no need for an assignment operator, and thus functional languages do not provide one (at least **pure functional languages** do not, see section 95.4). In Standard ML, for example, the operator = does not express assignment but rather an equation. Consider the following statement in C:

```
x = x + 1;
```

It is obvious to a C programmer that this statement specifies the modification of an existing variable **x**. Notice, though, that in mathematics this is an equation that has no solution. There is no **x** that satisfies the equation. The functional programming community has long argued that by departing from a mathematical interpretation of programs, the use of imperative languages leads to complex, poorly understood programs.

By preventing the modification of existing variables, functional languages exhibit what is known as **referential transparency**, which essentially means that in a functional program, equal expressions can be interchanged. Consider, for example, the following expression (in Standard ML):

```
let
   val x = f(a)
in
   ... x + x ...
end
```

This introduces a new variable **x** whose value is the result of the function call **f(a)**, and then evaluates an expression containing the sum **x + x**. Because **x** cannot be modified, a reader of the program can be sure that each occurrence of **x** has the value of **f(a)**, and thus the sum would have the same result as if the programmer had written **f(a) + f(a)**.

In an imperative program, it is difficult for the reader to be sure that the value of **x** had not changed, either by a direct assignment to **x** or indirectly via call to some procedure that modifies the value of **x**. Expressions containing these modifications, either direct or indirect, are known as **side effects** because, aside from returning a value, these expressions have the side effect of changing a variable's value. Functional programmers argue that it is side effects (and the corresponding loss of referential transparency) that lead to incomprehensible large programs.

An additional property that functional languages exhibit is the ability to treat functions as data. That is, functions can be passed as parameters to other functions, returned as the result of function calls, and stored in data structures. Thus, functions are said to be **first-class objects** since their use is no more restricted than other kinds of data. This is attractive for philosophical reasons, since functions are mathematical entities just like integers and Booleans, and for practical reasons, it increases the flexibility of code.

For instance, all functional languages provide a construct for specifying a function value without having to declare the function's name, equivalent to lambda abstractions in the lambda calculus. In Standard ML, such an expression is of the form

```
fn(x) => e
```

and denotes a function whose formal parameter is x and whose body is the expression e. This function value can be used in larger expressions, function calls, etc.

Consider again the factorial function. It might be argued that the formal definition of factorial just given was tailored to suit the recursive nature of the definition of factorial in the functional language, and that a more reasonable and common definition of factorial is

$$n! = \prod_{i=1}^{n} i$$

The product operator $\prod$ is a very useful operator for defining a wide range of functions and has the general form

$$\prod_{i=m}^{n} f(i)$$

for some initial value m, some final value n, and some function f. In a functional language, $\prod$ would be written as a **higher order function**, namely, a function that takes a function as a parameter or returns a function as its result. In particular, $\prod$ takes three parameters, **m**, **n**, and **f** and could be written in Standard ML as

```
fun prod(m,n,f) = if m = n then f(m) else f(m) * prod(m+1,n,f)
```

Thus, factorial can simply be defined as

```
fun fac(n) = prod(1,n, fn i => i)
```

and the exponentiation function computing x^n can be defined as

```
fun power(x,n) = prod(1,n,fn(i) => x)
```

95.2 History of Functional Languages

All functional languages trace their roots to the lambda calculus, developed by the logician Alonzo Church [1941], in the 1930s. This simple model, which describes computation as a series of syntactic conversions between expressions, was developed in order gain a deeper understanding into computation and what it means for functions to be computable, rather than as a programming language (since it obviously predates computers).

The first programming language that at least resembled the lambda calculus was LISP, developed by John McCarthy in the late 1950s [McCarthy et al. 1962]. It differs from the lambda calculus in several important ways: It was dynamically scoped (although McCarthy attributes this to a bug in the initial implementation), and provided an assignment operator. McCarthy states that although the lambda calculus served as an influence on the syntax of LISP, it was not the primary factor in the design of LISP's semantic features [McCarthy 1978]. LISP, however, has had a tremendous influence on modern functional languages. In 1975, Steele and Sussman designed SCHEME [Sussman and Steele 1975], a dialect of LISP that fixed some of the problems of earlier LISPs, such as dynamic scoping, and now its pure subset serves as the most LISP-like of all functional languages.

Another early language to have a great impact on the design of modern functional languages, especially ML and HASKELL, was ISWIM, developed by Landin [1966]. It was an explicit attempt to create a language whose semantics mirrored those of the lambda calculus, provided more convenient syntax and programming features, and was able to be implemented efficiently. Prior to the development of ISWIM, Landin [1964] had developed an abstract machine model, called the SECD machine, which specified how the conversion rules of the lambda calculus could be efficiently executed. Thus, the behavior of ISWIM operators could be described by their effect on the SECD machine.

The visibility of functional languages received a large boost in 1978, when John Backus, the designer of FORTRAN and the recipient of the 1978 A.M. Turing Award (computer science's highest award), chose to describe a new functional language, FP [Backus 1978], in his invited talk upon receiving the award. FP was a language of less expressive power than other functional languages of its time, since it did not provide user-defined higher order functions but rather supplied a fixed number of higher order *combining forms* used to create complex functions out of simple ones, and was heavily influenced by the APL programming language. Despite its limitations, and despite being of little interest today, FP was very influential in attracting researchers to the field of functional programming due Backus's stature, background, and convincing arguments in its favor.

During the 1970s and 1980s, functional languages, both strict and nonstrict, proliferated. Receiving a fair amount of attention and popularity were languages such as ML, SASL, HOPE, Lazy ML, and MIRANDA. Because of this proliferation, there was a movement to create standardized functional languages. The results of these standardization movements were a standardized definition of SCHEME [Rees et al. 1992]; Standard ML [Milner et al. 1990, Milner and Tofte 1991], now the standard strict functional language; and HASKELL [Hudak et al. 1992], now the standard nonstrict functional language. It is these languages that we have chosen to describe in this chapter.

95.3 The Lambda Calculus: Foundation of All Functional Languages

No description of functional languages is complete without the introduction of the lambda calculus, a simple but powerful model of computation. The reader is referred elsewhere in this handbook and to Barendregt [1984] for a detailed description of the lambda calculus. The important points to note about the lambda calculus are the following;

1. All functional languages are simply syntactically sugared versions of the lambda calculus, in some cases typed versions of the lambda calculus. Thus, any property that holds for the lambda calculus, such as its computational power, also holds for functional languages.

2. The lambda calculus is Turing complete: Every computable function can be expressed in the lambda calculus, and thus it is as least as powerful as any other computational model (such as Universal Turing Machines, for example).

3. One of two common evaluation orders in the lambda calculus, applicative order and normal order, have been adopted by almost all functional languages. Those functional languages which use **applicative-order evaluation**, in which the arguments in a function call are evaluated before the body of the function (as is the case with all imperative languages), are called **strict functional languages**. Those functional languages which use **normal-order evaluation**, in which the arguments in a function call are only evaluated if and when needed in the body of the function, are called **nonstrict functional languages**.

4. The first Church–Rosser theorem about the lambda calculus states that no matter which evaluation order is chosen, the result of functional program will be the same as long as the program terminates. Not all evaluation orders are equally likely to terminate, however, and the second Church–Rosser theorem states that the evaluation order that is most likely to lead to termination is normal-order evaluation.

The three languages described here, SCHEME, Standard ML, and HASKELL, are all based on the lambda calculus. They differ primarily in three ways: their syntax, their type systems, and whether they are strict or nonstrict.

95.4 Pure Versus Impure Functional Languages

Of the three functional languages described in this chapter, only one, HASKELL, is purely functional. That is, only HASKELL does not provide any mechanism for performing side effects. Both SCHEME and ML provide mechanisms for performing assignment to variables, although ML's mechanism is far more limited. However, SCHEME and ML deserve to be included in this chapter because good practice dictates that programs written in these languages are generally purely functional and side effects are used only where the programmer considers them absolutely necessary. At the end of the sections on SCHEME and ML, some of their impure features will be described.

A side-effect mechanism that is quite difficult to omit from a language is input/output (I/O). From an external viewpoint, such as the view of the operating system handling I/O requests from a functional program, input and output operations change the state of the input and output buffers (for the terminal, printer, etc.). However, to see why conventional I/O routines, such as read and print, do not support referential transparency within the program, consider

```
let x = read()
in x + x
end
```

where **read()** reads data from the standard input and returns the value read. If referential transparency were preserved, this code could be replaced by

```
read() + read()
```

which is clearly not the case.

SCHEME and ML adopt relatively conventional I/O routines, sacrificing referential transparency in expressions involving I/O. HASKELL, however, uses a more novel approach to support I/O in a referentially transparent manner.

95.5 SCHEME: A Functional Dialect of LISP

SCHEME is a dialect of LISP which differs from the more traditional LISPs (including Common LISP) primarily in that it is *statically scoped.* It is also a smaller, simpler language than most other LISPs. SCHEME, as defined in the IEEE Standard, is not a purely functional language. It supports the **set!** operator, similar to **SETQ** in traditional LISPs, which performs assignment on variables. However, SCHEME programmers tend to write in a functional style, and at least one SCHEME compiler (see Kranz et al. [1986]) performs optimizations specifically targeted at programs written in a functional style.

In this chapter, we will focus on *pure* SCHEME, a subset of SCHEME that is purely functional. Pure SCHEME differs from SCHEME only in that it omits the few side-effect operators that SCHEME provides. By doing so, the mathematical properties of pure SCHEME mirror those of the lambda calculus.

Like all LISPs, SCHEME adopts a prefix notion for all syntactic entities, thus looks strikingly different from conventional languages and other functional languages. The beauty of LISP and SCHEME syntax is that there are very few syntactic rules, thus learning the syntax of the language is trivial. Furthermore, the appearance of SCHEME data structures and SCHEME programs is quite similar, leading to the ability to manipulate programs as data, as is the case with interpreters, compilers, program verifiers, and program transformers.

Also like all LISPs, but unlike the other functional languages described in this chapter, SCHEME has **latent types**, which means that types are associated with values, not variables. Type checking occurs at run time, not compile time (which is why SCHEME is often called a *dynamically typed* language) and a type error is signaled only when a primitive operator (such as **+**, **-**, etc.) has been applied to a value of an inappropriate type. There are no type declarations, and the types of user-defined functions and variables are not specified. Variables can be bound to values of different types over the course of the computation.

SCHEME Data Types

There are two kinds of types in SCHEME (as in LISP), atomic types known as *atoms*, and *pairs*. The atomic types include numbers (floating point numbers and arbitrarily large integers), Booleans (written **#t** and **#f**), character strings, and a type that is peculiar to LISP dialects, namely, symbols. Symbols are objects that have only one property, their name. Two symbols are equivalent if and only if they have the same name. SCHEME symbols are different from those of traditional LISPs, since LISP symbols often have many properties associated with them.

The other kind of type, a pair, is a two-element record. This record is generally referred to as a *cons cell*. Each element can be of any type and is generally implemented as seen in Fig. 95.1(a). The first element is known as the *car* and the second is known as the *cdr*. There is a constant **( )**, called the *empty list.* Any collection of pairs of the form pictured in Fig. 95.1(b) where the cdr of each cons cell is either **( )** or points to another cons cell, is called a *list*. The list is the primary aggregate data structure in SCHEME (and all functional languages). It is a very flexible data structure, since each element of a list can itself be a list. The list pictured in Fig. 95.1(b) would be printed as

```
(3 4 5 6)
```

and the list in Fig. 95.1(c) would be printed as

```
(((1 2) 3) (4 (5 6)))
```

SCHEME Syntax

In SCHEME, every language construct is an expression that returns a value. An expression is either an *atomic expression* or a *combination*. An atomic expression can either be a numeric literal, a string literal, or an identifier (representing both symbols and variables).

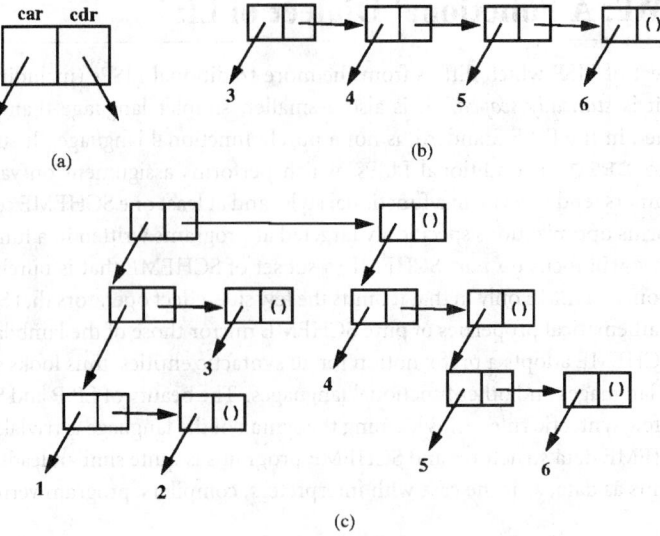

FIGURE 95.1 List structures.

A combination is an expression consisting of an open parenthesis followed by some subexpressions and then a close parenthesis. These represent a range of expressions, including function calls, definitions, conditionals, etc.

A function call is of the form

$$(e_0 \ e_1 \ \ldots \ e_n)$$

where each e_i is an expression. The expression e_0 should evaluate to a function, which is then applied to the values of the expressions $e_1 \ldots e_n$. A call to a function with no arguments is simply written as (e_0). The function call syntax is used for all user-defined functions as well as all predefined functions, including the arithmetic operators.

There are two forms of conditional expressions. The simplest is the **if**, of the form

```
(if e₀ e₁ e₂)
```

If e_0 evaluates to any value other than **#f**, then the value of e_1 is computed and returned as the result of the entire expression. Otherwise, the value of e_2 is computed and returned.

A more general conditional, the **cond** expression, has the form

```
(cond (c₀ e₀)
      (c₁ e₁)
      ...
      (cₙ eₙ))
```

The expressions $c_0, c_1, \ldots, c_n$ are evaluated in order until the first c_i, for some i, evaluates to a value other than **#f**. The value of e_i is then computed and returned. The expression c_n may be replaced by the keyword **else**, in which case the value of e_n is returned if none of $c_0, \ldots, c_{n-1}$ evaluates to a value other than **#f**.

In the construct

```
(quote exp)
```

exp is treated as data—either a symbol, number, or list—instead of an expression to be evaluated. Thus,

```
(quote a)
```

returns the symbol **a**, not the value of the variable **a**. The result of the expression

```
(quote (f a b c))
```

is the list containing the symbols **f**, **a**, **b**, and **c**. It is not a call to function **f** with arguments **a**, **b**, and **c**. Because *quote* is used so extensively in SCHEME and LISP, a syntactic shorthand is provided. The construct **'** *exp* is simply shorthand for **(quote** *exp*)**. Thus, **'a** is equivalent to **(quote a)** and **'(a b c)** is equivalent to **(quote (a b c))**. A nested list is easily specified, for example, **'(a b (c d))**.

SCHEME's **lambda** expression is an expression whose value is a function. It is of the form

```
(lambda (x₁...xₙ) e)
```

and evaluates to a function whose formal parameters are $x_1 \ldots x_n$ and whose body is the expression e. A lambda expression without parameters would be of the form **(lambda () e)**.

A definition of the form

```
(define x e)
```

introduces a new variable x and binds it to the result of evaluating the expression e. The variable x is visible during the evaluation of e, thus allowing for recursive function definitions such as

```
(define fac (lambda (x) (if (= x 0) 1 (* x (fac (- x 1))))))
```

In most implementations, **define** is only allowed at the top level, i.e., not nested inside any other expression. In these cases, the variable introduced is global.

As a syntactic convenience, functions can also be defined using the form

```
(define (f x₁ ... xₙ) e)
```

which is equivalent to **(define** f **(lambda (**x_1 ... x_n**)** e**))**. Thus, the factorial function given previously is generally written

```
(define (fac x) (if (= x 0) 1 (* x (fac (- x 1)))))
```

The **let** construct, of the form

```
(let ((x₁ e₁)
      (x₂ e₂)
      ...
      (xₙ eₙ))
     e)
```

is used to introduce the variables $x_1 \ldots x_n$ and bind them to the values of the expressions $e_1 \ldots e_n$, respectively. The value of e is then computed and returned as the value of the entire **let** expression. The scope of the new variables $x_1 \ldots x_n$ is just the body of e. Thus, these variables cannot be referenced in expressions $e_1 \ldots e_n$. This means that none of $x_1 \ldots x_n$ can be defined recursively.

The **letrec** construct can be used to introduce recursively defined local variables. It has the same form as the **let** construct, except that the keyword **letrec** is used instead of **let**. In this case, the expressions $e_1 \ldots e_n$ are defined in an environment in which each of $x_1 \ldots x_n$ are visible and thus can be referenced. Here is an example of a use of **letrec**,

```
(letrec ((f (lambda (x) (if (= x 0) 1 (g (- x 1)))))
         (g (lambda (y) (if (= y 0) 1 (f (- y 1))))))
   (+ (f 3) (g 5)))
```

where **f** and **g** are mutually recursive functions.

Predefined Functions

SCHEME provides a large number of predefined functions. The usual collection of arithmetic and logical operators, **+**, **-**, **=**, **!=**, **>**, **<**, etc., are provided and can be applied to any numeric values. Examples of their use include **(+ 3 4)**, **(> 6.2 5.1)**, and **(!= 4 5)**.

A function commonly used to create lists is **list**. It takes an arbitrary number of arguments and creates a list containing their values. Thus, for example,

```
(list 'a (+ 2 5) b (list 6 2))
```

would return the list **(a 7 *v* 6 2)**, where *v* is the value of the variable **b**.

The most heavily used list construction function is **cons**. It takes two arguments and, as its name implies, creates a cons cell whose car is the value of the first parameter and whose cdr is the value of the second. For example, here is a function that takes parameters **N** and **M** and constructs the list of integers between **N** and **M**, inclusive.

```
(define (listof N M)
  (cond ((> N M) '())
        (else (cons N (listof (+ N 1) M)))
        ))
```

To access the car and cdr fields of a cons cell, SCHEME provides the functions **car** and **cdr**, respectively. For example, **(car '(3 4 5 6))** returns **3** and **(cdr '(3 4 5 6))** returns the list **(4 5)**. If the first element of a list l_1 is itself a list l_2, then **car** applied to l_1 returns l_2, as one would expect. For example,

```
(car '((1 2) (3 4) 5))
```

returns the list (1 2) and

```
(cdr '((1 2) (3 4) 5))
```

returns ((3 4) 5).

The predicate **null?** is used to test for an empty list. Here **(null? x)** returns **#t** if the value of **x** is the empty list and returns **#f** otherwise. Here is an example of the use of **car**, **cdr**, and **null?**: Given a list of numbers, the function **sumof** returns the sum of the elements of the list.

```
(define (sumof l)
    (cond ((null? l) 0)
          (else (+ (car l) (sumof (cdr l))))
          ))
```

Also, **cons** is useful for constructing lists one element at a time. Another useful predefined function is **append**. It takes as parameters two lists l_1 and l_2 and returns a list containing the elements of l_1 followed by the elements of l_2. For example,

```
(append '(1 2 3 4) '((a b) c d))
```

returns the list **(1 2 3 4 (a b) c d)**. Although **append** is always provided by SCHEME implementations, it is not *primitive* in the sense that it can easily be written in SCHEME.

```
(define (append x y)
   (cond ((null? x) y)
         (else (cons (car x) (append (cdr x) y)))))
```

Another predefined function that can easily be written in SCHEME is **reverse**. This function takes a list *l* and returns a new list with the same elements as *l*, but in reverse order. For example,

```
(reverse '(1 2 (3 4) 5))
```

returns the list **(5 (3 4) 2 1)**. Notice that nested lists, such as the third element of the previous input list, are not recursively reversed. The reverse function can be defined in SCHEME as

```
(define (reverse l)
   (cond ((null? l) '())
         (else (append (reverse (cdr l)) (list (car l))))))
```

Unfortunately, the cost of this function is proportional to the square of the length of the input list. This can be seen by noting that **append** is linear in the size of its argument and is called each time that **reverse** is called recursively. The depth of the recursion in **reverse** is proportional to the length of its argument. A more efficient **reverse**, whose cost is linear in the length of its argument, is

```
(define (reverse l)
   (rev l '()))

(define (rev l accum)
   (cond ((null? l) accum)
         (else (rev (cdr l) (cons (car l) accum)))))
```

One can think of **rev** as successively taking the elements of **l** and putting them at the front of the list **accum**. Thus, when **l** is empty **accum** will contain the elements of **l** in reverse order.

The function **map** is a commonly used predefined function. It takes two parameters, a function f and a list l, and returns a list resulting from applying f to each element of l. For example,

```
(map (lambda (x) (* x 2)) '(3 4 5 6))
```

returns the list **(6 8 10 12)**. It can be written in SCHEME as

```
(define (map f l)
   (cond ((null? l) '())
         (else (cons (f (car l)) (map f (cdr l))))
         ))
```

Impure Features in SCHEME: Assignment and I/O

The most heavily used impure SCHEME construct is **set!**. It is SCHEME's variant of **SETQ** in LISP and is used to modify the value of an existing variable. That is, **(set!** x exp**)** evaluates exp and assigns the result to the variable x. Other side-effect operators include **(set-car!** l exp**)** and **(set-cdr!** l exp**)**, which assign the value of exp to the car and cdr fields of the list l, respectively.

There are a number of I/O routines provided in SCHEME, including those for opening and reading or writing to files. The simplest routines, however, are **(read)** which reads a scheme object (either an atom or a list) from the standard input and returns the object as the result of the call and **(write** exp**)** which writes the value of exp to the standard output. Here **(newline)** starts a new line on the standard output.

95.6 Standard ML: A Strict Polymorphic Functional Language

Standard ML is a popular functional language that uses applicative-order evaluation and has a flexible but static type system. It has a more conventional syntax than SCHEME and provides a pattern-matching facility for programming in an equational style. It also has an exception facility and a sophisticated module system supporting the development of large programs. Several robust implementations of Standard ML implementations exist and are used primarily at universities and research laboratories around the world.

Since the dynamic behavior specified by expressions in ML is based on the lambda calculus, and is thus similar to SCHEME, we will concentrate on ML's syntax and its type system.

Predefined Types in ML

ML provides the usual primitive types, **int**, **real**, **bool**, and **string**. Its aggregate types include lists, tuples, and records. A list is homogeneous, meaning that, unlike SCHEME, all elements of the list must be of the same type. The type for a list of integers is written **int list**, the type for a list of Booleans is written **bool list**, and so on. Literals for lists start and end with square brackets and the elements are separated by commas. Examples of list literals include **[1,2,3]**, **[true,false,true]**, and **[[1,2,3],[4,5,6]]**. The types of these lists are **int list**, **bool list**, and **int list list**, respectively. The literal **[]** denotes the empty list.

A tuple is an ordered collection of elements. Tuples are heterogeneous, their elements can be of different types. A tuple type is written as the element types separated by *****. Thus, **(int * bool * real)** is a tuple type whose first element is an integer, second element is a Boolean, and third element is a real. Tuple literals are written in the same way as list literals, except that parentheses are used instead of square brackets. For example, **(true, 3, [4.2])** denotes a tuple whose type is **bool * int**

* real list. The elements of a tuple are accessed either by position or, more commonly, using patterns as described later in this section.

Records are similar to tuples except that, like in most languages, their elements are named. The type written {**a: int, b:real, c: string**} is a record type with field names **a**, **b**, and **c**, whose types are **int**, **real**, and **string**, respectively.

Being a functional language, ML provides higher order functions. These functions have types like any other object. The type of a function that takes a parameter of type *a* and returns a parameter of type *b* is written *a* -> *b*. Examples of function types are **int -> bool**, **real -> int -> bool**, and **int * real -> bool list**. The **->** is right associative, so the second example is equivalent to **real -> (int -> bool)**. This is a type describing functions that take a **real** as a parameter and return a function taking an **int** as a parameter and returning a **bool**.

Here, **->**, *****, and **list** are known as *type constructors* because they are not types themselves, but rather construct new types (such as **int list** or **bool -> real**) when combined with existing types (such as **int**, **bool**, and **real**).

Expressions in ML

Arithmetic and logical expressions are written using the familiar infix notation. Examples include **a+b**, **4>b**, and **c andalso (d = 5)**. The conditional expression is written

> if *condition* then *exp* else *exp*

and function application is written simply as the juxtaposition of the function and the argument. For example,

> f x

is the application of the function **f** to **x**. Often, ML programmers will put the argument in parentheses as was done in the factorial example in section 95.1. Simply placing parentheses around an expression has no effect on the value of the expression. Function application is left associative, thus

> g 4 5

is equivalent to

> (g 4) 5

List construction and selection are similar to that of SCHEME. The **::** operator is identical to SCHEME's **cons**, so that **x :: xs** returns a list whose first element is **x** and whose subsequent elements are those of the list **xs**. For example, the value of the expression

> 3 :: [4,5,6]

is the list **[3,4,5,6]**.

The **@** operator is identical to SCHEME's **append** function. For example, the value of

> [3,4,5] @ [6,7,8]

is the list **[3,4,5,6,7,8]**.

The ML functions **hd** and **tl** are identical to SCHEME's **car** and **cdr**, respectively. For example, the value of **hd [3,4,5,6]** is **3** and the value of **tl [3,4,5,6]** is **[4,5,6]**.

Function expressions, corresponding to **lambda** expressions in SCHEME, are written in the form

```
fn arg => body
```

Examples are

```
fn x => x + 1
fn a => fn b => a + (b * 2)
```

where **=>** is right associative, and so the second example is equivalent to

```
fn a => (fn b => a + (b * 2)).
```

Declarations in ML

Variables and functions are declared using the **let** construct, much like SCHEME's **let**. It has the form

```
let declaration₁
    declaration₂
    ...
    declarationₙ
in
    exp
end
```

where each *declaration*$_i$ defines a new variable or function, and *exp* is the body of the **let**.

A variable declaration has the form

```
val x = e
```

in which case the expression e is evaluated and the variable x is given the resulting value. A function declaration has the form

$$\text{fun } f x_1 \ldots x_n = e$$

where $x_1 \ldots x_n$ are the formal parameters and e is the body of the function.

Here is an example of a let expression:

```
let val x = 6
    val g = fn z => z + 2
    fun fac n = if n = 0 then 1 else n * fac (n-1)
in
    fac (g x)
end
```

Notice that the variable **g** is bound to a function of type **int -> int**. The use of the keyword **fun** (as in the succeeding line) provides two conveniences: first, the formal parameters appear to the left of the **=**, and second, it supports the definition of recursive functions. In the declaration of **g** using the keyword **val**, **g** cannot appear on the right-hand side of the definition. The keyword **fun** was necessary in the recursive definition of **fac**.

In ML all functions take a single parameter. Thus, the declaration of the function

```
fun f x y = x + y + 2
```

is just shorthand for

```
fun f x = fn y => x + y + 2
```

This function has type **int -> int -> int** and when it is applied to a single argument, it returns a function of type **int -> int**. A function, such as **f**, that can be applied to fewer parameters than appear in the declaration is called a **curried function**, after the logician HASKELL Curry.

Pattern Matching

One of the nicest features of ML is its pattern-matching facility. In function definitions, the formal parameter name can be replaced by a pattern. In the introduction to this chapter, the factorial function was written as

```
fun fac 0 = 1
  |   fac n = n * fac(n-1)
```

in which factorial is defined by two clauses separated by a |. In the first clause, the formal parameter is replaced by the literal **0**. When **fac** is called, if the argument has the value **0**, then the right-hand side of the first clause is evaluated. Otherwise, the formal parameter **n** in the second clause is bound to the value of the argument and the right-hand side of the second clause is evaluated.

Consider a function that computes the sum of the elements of a list.

```
fun sum [] = 0
  |   sum l = hd l + sum (tl l)
```

The literal pattern **[]** in the first clause is used to determine if the argument is the empty list. Instead of using **hd** and **tl** to select the components of **l** in the second clause, **l** could be replaced by a pattern that accomplishes the same thing:

```
fun sum [] = 0
  |   sum (x::xs) = x + sum xs
```

In this case, the pattern **(x::xs)** matches any nonempty list and binds **x** to the head of the list and **xs** to the tail.

A tuple can also be used as a pattern. It was previously mentioned that

```
fun f x y = x + y + 2
```

is a curried function of type **int -> int -> int**, and that it is legal to apply **f** to just one argument. If the programmer knows that **f** will always be called with both arguments, then it is generally more efficient to define **f** as taking a single argument which is a tuple:

```
fun f (x,y) = x + y + 2
```

In this case, **f** has type **int*int->int** and a call to **f** would look like **f(3,4)**. This example also demonstrates how a pattern is used to access the individual elements of a tuple, in this case as **x** and **y**.

Type Definitions

There are several ways to introduce new type names in ML. The simplest way is to create a type synonym, i.e., to define a new name for an existing type. This is accomplished by a declaration of the form

```
type name = type_exp
```

which introduces the new name *name* for the type described by *type_exp*. Some examples are

```
type foo = int * bool * real
type bar = string
type personnel_record = { name: string, salary: int, ss_num:
                          string }
```

No new type is created. Thus, **foo** and **int*bool*real** describe the same type and can be used interchangeably in the program.

New types are created using the **datatype** construct. In its simplest form, a data type declaration specifies all of the elements of the type, much like an enumerated type in PASCAL or ADA.

```
datatype stoplight = Red | Green | Yellow
```

defines a new type **stoplight** whose values are **Red**, **Green**, and **Yellow**.

In the more general form of a data type declaration, the components on the right-hand side can be *value constructors*. Instead of being values themselves, such as **Red** or **Green**, value constructors take parameters and construct values of the new type. Consider,

```
datatype tree = Empty | Leaf of int | Node of tree * tree
```

Here, **Leaf** is a value constructor taking an integer parameter and **Node** is a value constructor taking a tuple of two trees. **Empty**, like **Red**, **Green**, and **Yellow** previously is simply a value constructor that takes no parameters. The declaration of type **tree** says that a value of that type can be the empty tree, a leaf with an integer label, or an interior node with two subtrees.

The expression **(Leaf 5)** constructs a value of type tree which is a leaf node with the label 5. The expression

```
Node (Node (Leaf 5, Node (Leaf 6, Empty)), Leaf 7)
```

constructs the tree shown in Fig. 95.2.

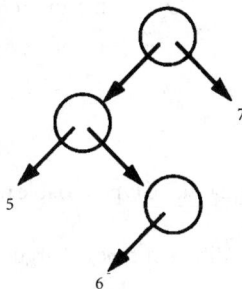

FIGURE 95.2 Tree created by **Node (Node (Leaf 5, Node (Leaf 6, Empty)), Leaf 7)**.

Value constructors can be used in patterns, as in

```
fun drive Red = "stop"
  | drive Green = "go"
  | drive Yellow = "go faster"
```

The type of the function **drive** is **stoplight -> string**. Pattern matching can also be used to select out the parameters associated with value constructors. The fringe function, defined by

```
fun fringe Empty = []
  | fringe (Leaf x) = [x]
  | fringe (Node (left,right)) = fringe(left) @ fringe(right)
```

returns a list of the labels associated with the leaves of a tree. If the tree is empty, then the empty list is returned. If the tree consists of just a leaf, then the variable **x** would be bound to the value of the leaf's label and the list containing **x** would be returned. Otherwise, if the tree consists of a node with left and right subtrees, the variables **left** and **right** are bound to those subtrees and their fringes are computed. The two resulting lists are then appended to form the result. The call

```
fringe (Node (Node (Leaf 5, Node (Leaf 6, Empty)), Leaf 7))
```

would return the list **[5,6,7]**.

Type Variables and Parametric Polymorphism

Consider the **length** function, which computes the length of a list:

```
fun length [] = 0
  | length (x::xs) = 1 + length xs
```

What is it's type? Clearly, it can take a list of any type, since the type of the elements of the list has no effect on its length. Thus, we say that the type of **length** is

$$\forall \alpha. \, \alpha \text{ list} \rightarrow \text{int}$$

which means that for all types α, **length** can take an α list and returns an integer. In ML, this type is written

```
'a -> int
```

where the **'** signifies that **a** is a universally quantified type variable rather than some type **a** previously defined.

Because **length** can be applied to many different types of arguments, as in

```
length [1,2,3] + length [[4,5],[6]] + length [true,false,true]
```

we say that length is **polymorphic** (meaning many shaped). In ML, any object whose type contains a type variable, such as **'a**, is polymorphic. All others are said to be *monomorphic*. This kind of polymorphism is called *parametric polymorphism* because in some theoretical type models, type variables occur as extra formal parameters in a function definition.

ML supplies a **map** function much likes SCHEME's **map**. It can be written as

```
fun map f [] = []
  |  map f (x::xs) = f x :: map f xs
```

The type of **map** is (**'a -> 'b) -> 'a list -> 'b list**, because for any types **'a** and **'b**, **map** takes a function **f** of type **'a -> 'b** and a list of type **'a list** and applies **f** to each element of the list. The result of each application is of type **b**, and since **map** returns the list of the results, the return type of **map** is **'b list**. The result of

```
map (fn n => n+1) [1,2,3] @ map (fn l => length l)
                            [[2.2,3.3],[4.4]]
```

would be **[2,3,4,2,1]**.

In all the ML examples so far, the programmer never specified the types of the functions, variables, or expressions. If desired, one could do so explicitly, as in

```
val a: int list = [1,2,3]
```

and

```
fun f (x:int) (y:real) = (if x = 1 then y + 1.2 else y - 1.7): real
```

In general, however, the ML compiler can infer the types of the functions and variables by the way they are defined and used. This process is called **type inference** and the ML type system, based on work by Hindley and by Milner, ensures that type inference can be safely performed. Furthermore, the type that is inferred for an object is the most general type possible, allowing that object to be used as polymorphically as possible. For example, if type inference had inferred the type of the **length** function to be **int list->int**, then the length function would have been restricted to lists of integers. Instead, type inference infers the more general type **'a list -> int**, allowing **length** to be used on all types of lists.

Type Constructors

Type variables can also be used to parameterize data type declarations. Earlier, we defined a tree type whose leaves were labeled with integers:

```
datatype tree = Empty | Leaf of int | Node of tree * tree
```

Instead, we can write

```
datatype 'a tree = Empty | Leaf of 'a | Node of 'a tree * 'a tree
```

which says that for all types **'a**, an **'a tree** is either empty, a leaf labeled with a value of type **'a**, or a node with two subtrees of type **'a tree**. In this case, **tree** is a type constructor because many different tree types can be constructed by instantiating **'a** with different types. For example, if the programmer writes

```
Node (Leaf 3.2, Empty)
```

the compiler can infer the type of this expression to be **real tree**. Similarly,

```
Node (Leaf[3,4,5], Leaf[4,5,6])
```

describes an **int list tree**. The type variable **'a** can only be instantiated one way within a single type, so that

```
Node (Leaf 4, Leaf true)
```

is illegal.

Interestingly, the type of the expression

```
Empty
```

is **'a tree**, thus **Empty** is polymorphic but is not a function (another example of this is **[]: 'a list**).

Polymorphic functions work well in the presence of type constructors. The **fringe** function seen earlier,

```
fun fringe Empty = []
  |  fringe (Leaf x) = [x]
  |  fringe (Node (left,right)) = fringe(left) @ fringe(right)
```

now has type **'a tree -> 'a list** and can work on any kind of tree.

The ML Module System

In order to support large-scale programs and separate compilation, ML provides a sophisticated module system. As in other languages, a module consists of a *body*, a collection of definitions of types, variables, etc., and an *interface*, specifying which components of the body are visible outside the module. In ML, a module body and a module interface are separate entities. Thus, many different module bodies can share the same interface, and different modules might share a body but have different interfaces.

A module interface, called a *signature* in ML, is described by a signature expression of the form

```
sig  decl₁
     decl₂
     ...
     declₙ
end
```

where each *decl_i* is usually a declaration of the name and type of an object or the name of a type. To give a name to a signature, a declaration of the form

```
signature name = sig_exp
```

is used, where *sig_exp* is a signature expression. For example, the interface for a module implementing a (functional!) stack might be

```
signature STACK = sig
                    type 'a stack
                    val empty: 'a stack
                    val push: ('a * 'a stack) -> 'a stack
                    val pop: 'a stack -> ('a * 'a stack)
                    val isempty: 'a stack -> bool
                    exception stack_underflow
                  end
```

A module body, known as a *structure* in ML, is described by a structure expression of the form

```
struct  def₁
        def₂
        ...
        defₙ
end
```

where each *def_i* is a definition. To give a structure a name, a declaration of the form

```
structure name = struct_exp
```

is used, where *struct_exp* is a structure expression or the name of a previously defined structure.
For example, a structure implementing a stack might look like

```
structure StackImp = struct
                       exception stack_underflow
                       type 'a stack = 'a list
                       val empty = []
                       fun push(x,s) = x::s
                       fun pop [] = raise stack_underflow
```

```
                              |  pop (x::rest) = (x,rest)
                      fun isempty [] = true
                         |  isempty l = false
                end
```

At this point, there is no connection between the signature **STACK** and the structure **StackImp**. Thus, all the components of **StackImp** are visible. To create a stack implementation with the signature **STACK**, we can write:

```
    structure Stack:STACK = StackImp
```

The signature **STACK** can be reused in a different implementation of a stack, as in

```
    structure NewStack : STACK =
       struct
          exception stack_underflow
          datatype 'a stack = empty | non_empty of int * 'a list
          fun push (x,empty) = non_empty (1,[x])
             |  push (x,non_empty(n, s)) = non_empty (n+1, x::s)
          fun pop empty = raise stack_underflow
             |  pop (non_empty(n,x::rest)) = (x, if n=1 then empty else
                                                non_empty(n-1,rest))
          fun isempty empty = true
             |  isempty l = false
       end
```

Once defined, a component x of a module m is referenced by the expression $m.x$. For convenience, the components of a module may be referenced without the module name, if the module is first opened via the command

```
    open name
```

where *name* is the name of the module.

Modules are not values in ML. That is, they cannot be passed to functions, stored in lists, etc. ML does provide something similar to a function from modules to modules. It is called a *functor* and supports code reuse by allowing the definition of a module in terms of other modules. All modules, even those resulting from functor applications, are instantiated at compile time and therefore cannot depend on values computed during execution.

A functor definition has the form

```
    functor  f(s₁ : sig₁, ..., sₙ : sigₙ) : sig_exp = struct_exp
```

where $s_1 \ldots s_n$ are the formal parameter names that will be bound to structures in a functor application and $sig_1 \ldots sig_n$ are the signatures that $s_1 \ldots s_n$ must conform to, respectively. Here, *sig_exp* is the signature of the result of the functor and *struct_exp* is the definition of the resulting structure. For example, suppose one wanted to create an implementation of a queue based on an existing stack implementation, such that the signature of a queue is

```
signature QUEUE =
  sig
      exception queue_underflow
      type 'a queue
      val empty: 'a queue
      val enqueue: ('a * 'a queue) -> 'a queue
      val dequeue: 'a queue -> 'a * 'a queue
      val isempty: 'a queue -> bool
  end
```

Next is a functor definition that takes any structure that conforms to the previous **STACK** signature and creates an implementation of a queue, using the data structures and routines supplied by the stack argument.

```
functor MakeQueue(Stack: STACK): QUEUE =
  struct
      exception queue_underflow
      type 'a queue = 'a Stack.stack * 'a Stack.stack
      val empty = (Stack.empty, Stack.empty)

      fun reverse_stack(from, to) =
        if Stack.isempty from then to                    \
        else let val (x, new_from) = Stack.pop from
          in reverse_stack(new_from, Stack.push(x,to))
          end

      fun enqueue(x,(s1,s2)) = (s1, Stack.push(x,s2))

      fun dequeue (s1,s2) =
        if Stack.isempty s1 then
          if Stack.isempty s2 then raise queue_underflow
          else dequeue (reverse_stack (s2, Stack.empty),
                        Stack.empty)
        else
          let val (x,new_s1) = Stack.pop s1
          in (x, (new_s1,s2))
          end
      fun isempty(s1,s2) = Stack.isempty s1 andalso
                           Stack.isempty s2
  end
```

To create an actual queue module the functor must be invoked, as in

```
structure Queue1 = MakeQueue(Stack)
```

Another implementation of a queue, based on a different stack implementation, is created by

```
structure Queue2 = MakeQueue(NewStack)
```

Functors are commonly used in ML to support separate compilation. They allow a module to be written and compiled despite referring to components of modules that are not yet implemented. Note that, for example, the code for the previous functor **Queue** could have been written, type checked, and compiled before any structure with the signature **STACK** was implemented. Only the signature **STACK** had to exist before compiling **Queue**.

Impurities in ML: References and I/O

As previously mentioned, ML supports conventional I/O, which violates referential transparency. For example,

```
val x = (print "hello"; print "world"; true)
```

where **hello world** would be printed and the value of **x** would be **true**. The semicolon is used to separate statements that are executed sequentially.

The other impure feature of ML is *references*. In conventional languages, these would be considered constant pointers to assignable locations. The expression

```
ref exp
```

allocates a new cell c in memory and places the value of *exp* in c. The address of c is returned as the value of the entire expression. The type of the expression is t **ref**, where t is the type of *exp*. Given, for example, the declaration

```
val x = ref 6
```

the value of **x** is a new cell containing **6**, and the type of **x** is **int ref**.

The value stored in this location may be changed, using the expression

$$exp_1 := exp_2$$

where the value of exp_1 must be a reference of type **ref** t and t is the type of exp_2. Evaluating this expression causes the value of exp_2 to be stored in the location denoted by exp_1. For example,

```
x := 7
```

changes the value referenced by **x** to **7**. The dereference operator is **!**. In the expression **!** *exp*, *exp* must evaluate to a reference and the value of the entire expression is the value contained in the referenced cell. Thus, after the assignment to **x**, the value of **!x** is **7**. Since the value of an expression of type t **ref** is essentially a pointer, references can be used for aliasing. For example, given the code

```
val x = ref 10
val y = x
```

the variable **y**, of type **int ref**, would point to the same location (containing **10**) that **x** does. Thus, the result of the expression

```
(x := !x+1; !y)
```

would be **11**.

95.7 Nonstrict Functional Languages

Before describing HASKELL in detail, it is worthwhile examining the costs and benefits of a nonstrict language, i.e., a language based on normal-order evaluation. Normal-order evaluation specifies that an argument in a function call is evaluated only when the value of the corresponding formal parameter is needed. In most implementations of nonstrict functional languages, any subsequent reference to the formal parameter uses the already computed value of the argument rather than re-evaluating it. This more efficient mechanism for supporting normal-order evaluation is called **lazy evaluation** (it is also sometimes referred to as *call-by-need*). Nonstrict functional languages are often informally referred to as **lazy functional languages**, even though laziness is a property of the implementation rather than the language.

In a nonstrict language, even using lazy evaluation, there is a significant overhead cost to delaying the evaluation of an actual parameter until the corresponding formal parameter is needed. This cost arises due to the fact that an object representing the delayed argument must be constructed when the function is called. This object might be a closure (i.e., a parameterless procedure, generally known as a *thunk*) that will be invoked when the value of the argument is needed, or it might be a graph representation of the delayed expression (this is found in systems that use a technique called graph reduction). In each case, the overhead cost can be substantial.

The benefit of a nonstrict language is that its programs are more likely to terminate: for example, if the evaluation of an argument might never terminate but the argument is not needed by the function. However, since the vast majority of popular programming languages are strict, it might appear that this particular termination issue is unimportant. However, when used properly, nonstrictness frees the programmer from worrying about some control issues, such as interleaving the execution of producer and consumer procedures. The other benefit of using a nonstrict language is that it allows the programmer to create infinite data structures. To illustrate this, consider the following definition:

```
fun numsfrom n = n :: numsfrom (n+1)
```

This function, given an integer argument n, creates the list $[n, n+1, n+2, \ldots]$. In ML, the call

```
numsfrom 1
```

would not terminate until memory was exhausted because all of the (infinite number of) elements of the list would have to be created before the call returned.

In a nonstrict language, the cons function **::** does not evaluate its arguments. Thus, the expression

```
n :: numsfrom (n+1)
```

would create a list whose head is **n** and whose tail is specified by **numsfrom (n+1)** but is left unevaluated. Only when the value of the tail of this list is demanded using the **tl** function is the call **numsfrom (n+1)** actually evaluated. The result of that call, then, is a list whose head is **(n+1)** and whose tail is described by the unevaluated expression **numsfrom (n+2)**. In a nonstrict language, the call

```
numsfrom 1
```

would return almost immediately with a delayed list representing **[1,2,3, ...]**. These infinite, but delayed, lists are generally known as **streams**.

The function

```
fun sumstream s 0 = 0
  |  sumstream s n = hd s + sumstream (tl s) (n - 1)
```

takes a stream *s* and an integer *n* and computes the sum of the first *n* elements of *s*. The result of

```
sumstream (numsfrom 1) 10
```

would compute the sum of the first 10 elements of **(numsfrom 1)**, namely, 55.

From a programmers point of view, the use of infinite data structures provides a nice separation between the production of data (by **numsfrom**, in this case) and the consumption of the data (by **sumstream**). The producer does not need to know how much data the consumer will need, nor does it have to worry about buffering data that is already produced but not consumed. The data is produced only when demanded by the consumer.

A more substantial example is the program that computes the infinite list of primes using the Sieve of Erostosthenes.

```
let fun numsfrom n = n :: numsfrom (n+1)
    fun filter f (x::xs) = if f x then x :: filter f xs
                           else filter f xs
    fun remove_multiples (x :: xs) =
    let fun is_multiple n = (n mod x) <> 0
    in x :: remove_multiples (filter is_mult xs)
    end
in
    remove-multiples (numsfrom 2)
end
```

95.8 HASKELL: A Nonstrict Functional Language

Aside from being a nonstrict functional language, HASKELL features a sophisticated type system that extends the ML-style (i.e., Hindley–Milner) type system to incorporate dynamic overloading.

HASKELL's syntax differs somewhat from that of ML, although the programs have a similar look. A few of the more important syntactic differences are as follows:

- Identifiers representing types and value constructors are capitalized. Identifiers representing type variables and values are not capitalized.
- Function and variable definitions do not begin with a keyword (whereas ML uses **fun** and **val**).

- Type constructors precede their arguments, as in **List Int** (in contrast to **int list** in ML).
- HASKELL uses **:** and **::** in precisely the opposite way from ML. The **:** is the *cons* operator and the **::** is used to associate a type with an expression, as in

```
(4:[5,6]) :: List Int
```

- Indentation can be used to begin and end new blocks. For example, in

```
let f x = let z = x + 3
              in x * y
        y = a * b
    in
        f y
```

the indentation specifies that the names **f** and **y** are defined at the same level.

The HASKELL Class System

HASKELL's most interesting feature, other than its nonstrict semantics, is its *class* system for supporting *dynamic overloading* in a systematic way. The term overloading refers to the ability to give the same name to two distinct entities in a program. Overload resolution is the process by which the use of that name is disambiguated. For example, in ADA the programmer can define two or more different functions with the same name. When encountering that name in a function call, the compiler performs overload resolution to determine which function is being called by examining the number and types of the actual parameters. If the compiler is unable to resolve the overloading, the program is rejected. Because the overload resolution occurs at compile time, this kind of overloading is called *static overloading*.

In almost all languages, there is some form of overloading. For example, most languages use **+** as the name of the addition operators for both integers and reals, even though they are different operators. This is the case in ML, for example. However, the mixing of static overloading and type inference causes a problem in ML. Consider the following definition in ML:

```
fun f x y = x + y
```

The ML compiler rejects this definition because it cannot determine which addition operator, integer or real, is specified by **+**. The parameters **x** and **y** are either integers or reals, but it cannot be determined which. The ML programmer has to provide explicit type information, such asf

```
fun f (x:int) y = x + y
```

In the first (erroneous) definition, the type of **f** is clearly not

$$\forall \alpha. \, \alpha \to \alpha \to \alpha$$

since **+** cannot be applied to every type α. We would like to be able to say that **f** is of type

$$\forall \alpha \text{ for which } + \text{ is defined.} \, \alpha \to \alpha \to \alpha$$

and to allow the programmer to define the meaning of **+** for any type (complex numbers, sets, etc.) desired. Then, when **f** is called, the choice of which **+** to use in the body of **f** depends on the types of the arguments to **f**. Since **f** is polymorphic, albeit in a restricted way, it can be applied to many different types of arguments and thus the choice of **+** has to be made at run time. This kind of overloading is called dynamic overloading and is seen, in a different framework, in object-oriented languages.

HASKELL uses type classes to support dynamic overloading. A type class is a way to specify what operations must be supported by a particular collection of types. For example, the equality class, **Eq**, defined in HASKELL by

```
class Eq a where
    (==) :: a->a->Bool
```

specifies that every type **a** in class **Eq** must provide a definition for the infix equality operator **==** of type **a -> a -> Bool**. One can then write a polymorphic function that uses **==**, for example

```
f :: (Eq a) => a->a->int
f x y = if x == y then 1 else 2
```

The first line gives the type of **f**, which is **a -> a -> int** for any type **a** in class **Eq**.

The notation **(Eq a)** is called a *context* and indicates that **a** is in class **Eq**. Like ML, HASKELL is designed to support type inference. In fact, the first line declaring the type of **f** can be omitted. The HASKELL compiler will infer that the type of the parameters must be in class **Eq**.

User-Defined Types

New types in HASKELL are defined using the **data** construct which is analogous to the **datatype** construct in ML. For example,

```
data IntTree = Empty | Leaf Int | Node IntTree IntTree
```

defines the same integer-labeled tree type seen earlier. HASKELL also provides type constructors, so

```
data Tree a = Empty | Leaf a | Node (Tree a) (Tree a)
```

defines the **Tree** type constructor parameterized by the label type **a**.

Instance Declarations

Once a new type has been created it can be declared to be an *instance* of a type class, in which case the definition of the required operators must be provided. For example, declaring type **IntTree** to be in class **Eq** might look like

```
instance Eq IntTree where
    Empty == Empty                    = True
    Leaf x == Leaf y                  = x == y
    (Node l1 r1) == (Node l2 r2)      = l1 == l2 && r1 = r2
    t1 == t2                          = false
```

The code following the **where** keyword is simply the definition of the **==** operator using pattern matching on the value constructors of **IntTree**. Since **==** is infix, the definition is in infix form.

Once **IntTree** has been declared to be in class **Eq**, an **IntTree** value can be passed to any function expecting a type in class **Eq**.

We would also like to declare that any type constructed from the **Tree** type constructor is in class **Eq**. The declaration

```
instance Eq (Tree a) where
  Empty == Empty                = True
  Leaf x == Leaf y              = x == y
  (Node 11 r1) == (Node 12 r2)  = 11 == 12 && r1 = r2
  t1 == t2                      = false
```

is incorrect because the definition of **==** requires, in the second clause, that the labels **x** and **y** be compared using **==**. Thus, not only must **==** be defined on **(Tree a)** for any type **a**, **==** must also be defined on **a**. Thus, **a** must already be an instance of class **Eq**. The correct instance declaration requires a context as follows:

```
instance (Eq a) => Eq (Tree a) where
  Empty == Empty                = True
  Leaf x == Leaf y              = x == y
  (Node 11 r1) == (Node 12 r2)  = 11 == 12 && r1 = r2
  t1 == t2                      = false
```

This should be read as "For all types **a**, if **a** is in class **Eq** then **(Tree a)** is in class **Eq** with **==** defined as follows ...."

HASKELL also provides a form of inheritance, in which one class can be used to define another class. For example, the class definition

```
class (Eq a) => Ord a where
  (<), (<=), (>=), (>)    :: a->a->Bool
  max, min                :: a->a->a
```

defines the class **Ord** of ordered types in terms of the class **Eq**. In this case, a type **a** can be in class **Ord** if it is in class **Eq** and supports the additional operators previously mentioned. We say that **Eq** is the *superclass* of **Ord** and **Ord** is the *subclass* of **Eq**.

List Comprehensions in HASKELL

Another nice feature of HASKELL is its *list comprehensions*. These are concise expressions for constructing entire lists, resembling set notation in mathematics. For example, the expression

```
[f x | x <- xs]
```

computes the list of all values returned by **(f x)**, for each **x** taken from the list **xs**. Thus

```
[ x * 2 | x <- numsfrom 1]
```

returns the infinite list **[2, 4, 6, ...]**. Naturally, the elements of the list are not computed until needed, just as in the preceding case of the infinite list resulting from **numsfrom 1**.

List comprehensions can also include guards. For example,

```
psort []     = []
psort (x:xs) = psort [y | y <- xs, y < x] ++ [x] ++
                psort [y | y <- xs, y >= x]
```

defines the partition sort (often mislabeled quicksort), where **++** is HASKELL's append operator. The first list comprehension in **psort**

```
[y | y <- xs, y < x]
```

contains the guard **y < x**, so that the only elements **y** taken from **xs** are those that are less than **x**.

Functional I/O in HASKELL

In order for HASKELL's I/O facility to maintain referential transparency, the input to a program is considered a stream, a possibly infinite list. Like other infinite lists, the entire input list is not immediately available at the start of execution, but the elements are supplied over the course of the computation: for example, as the user enters data from the keyboard. Similarly, the output of a program is also a stream. A program, then, can be viewed as a mapping from input streams to output streams. In actuality, HASKELL's I/O system is substantially more complicated in order to support error handling, files, and channels. It makes heavy use of *continuations*, and the reader is referred to a nice introduction to the language in Hudak and Fasel [1992].

95.9 Research Issues in Functional Programming

The functional language research community is very active in a number of areas. Of particular interest is improving the speed of functional language implementations. There are two primary approaches to this: through compiler-based program analysis and optimization techniques and through the parallelization of functional programs. Another area of research is to increase the expressiveness of functional languages, particularly in applications in which side effects are seen as necessary in conventional programs. In this section, we provide a brief description of these research areas and refer the reader to the literature in order to gain a deeper understanding of the issues.

Program Analysis and Optimization

Because of the solid mathematical foundation and well-defined semantics of functional languages, functional programs are particularly good candidates for compile-time analysis and optimization. An analysis technique that has been used in a large number of optimizations is *abstract interpretation* (see Abramsky and Hankin [1987]). It is a technique in which the program is executed at compile time, but rather than operating on the usual kinds of values, it operates on, and returns, only particular pieces of information desired by the compiler. If the desired information is sufficiently restricted, or an approximate answer is acceptable, then the program execution can be shown to terminate and is thus useful at compile time. The example generally used to illustrate abstract interpretation is overly trivial but useful nonetheless: the rule of signs in multiplication. This rule determines the sign of the result of a multiplication, using only the signs of the operands:

```
0 × x = 0 for any x
x × 0 = 0 for any x
```

$$+ \times + \ = \ +$$
$$- \times + \ = \ -$$
$$+ \times - \ = \ -$$

If only the sign of the result of a multiplication is desired, then the rule of signs provides a form of execution that is much less expensive then performing the actual multiplication and then taking the sign of the result.

An early, but still important, application of abstract interpretation arises in *strictness analysis*. Strictness analysis is an analysis used for a nonstrict language in order to transform, where possible, normal-order evaluation into applicative-order evaluation. Essentially, we want to know if a function f will always require the value of its argument, in which case we say that f is *strict*. If we know a function is strict, then we can go ahead and evaluate its argument before calling the function, without changing the result computed by the program.

More formally, we say that f is strict if, even using normal-order evaluation,

$$f(\bot) = \bot$$

where $\bot$ represents a nonterminating computation. Intuitively, this says that if f is applied to an argument whose evaluation would not terminate and yet f does terminate, then it is clear that f does not require the value of that argument and is thus not strict. Abstract interpretation is used to find the strictness property of a function by executing an abstract version f' of f function over the domain of values $\{0, 1\}$ where 0 represents nontermination and 1 represents possible termination. If

$$f'(0) = 0$$

reflects the behavior of f in the previous equation, then we know that f is strict.

Parallel Functional Programming

The attractiveness of functional languages for writing programs for parallel machines arises from the first Church–Rosser theorem. The theorem states that given a function call, the order in which the arguments and the body of the function are evaluated will not effect the final answer, assuming the program terminates. Thus, given the expression

```
(f x) + (g y)
```

the expressions **(f x)** and **(g y)** can be evaluated *in parallel*. In this case, it is clear that both operands to **+** are needed, so that there will be no wasted effort.

There are two major approaches to parallel programming using functional languages (see Kelly [1989] for extended reading in this area). The first involves programming in a standard functional language, such as ML or HASKELL, and using a compiler and run time system that will partition the program into parallel threads and execute them. Because of the Church–Rosser property, it is not difficult for the compiler to determine which expressions can be executed in parallel.

The difficult part is determining the appropriate granularity of the parallel version of the program. Granularity is the measure of the size of the tasks into which the program is decomposed; the finer the granularity, the smaller and more numerous the tasks, and the greater the degree of parallelism. However, there is a cost associated with creating each task, whether due to communication over a network, increased contention for a shared memory, or context switching in the operating system.

The second approach to parallel computing using functional languages involves adding constructs to a functional language for expressing parallelism. These constructs might specify which expressions should be evaluated in parallel (and, thus, which are not worth spawning as their own tasks), which processor an expression should be evaluated on, and, in the case of languages which contain impure features, the creation and use of channels for communication.

Partial Evaluation

Partial evaluation [Bjorner et al. 1988] is the technique where, if part of the input to a program is known at compile time, the program is evaluated as much as possible using the input available. The result is a new version of the program, called the specialized program, that is ready to accept the rest of the input and return the same result as the original program would have on the entire input. The process of specialization creates a more efficient program than the original because the known data have already been integrated into the program, reducing the amount of interpretation that the program has to perform on its input.

State in Functional Programming

An active area of research in functional programming is the development of new techniques for reasoning about state and changes in state in a manner that preserves referential transparency. For example, these techniques, most based on the concept of a monad [Wadler 1992], allow programmers to manipulate arrays in a pure functional language in a style that looks familiar to imperative language programmers [Peyton Jones and Wadler 1993]. Furthermore, this work boosts the efficiency of functional arrays by allowing the arrays to be updated in place, rather than copied each time an update operation is performed (which is normally necessary to preserve referential transparency). This is accomplished by using the type system to encapsulate the array such that it cannot be shared. Thus, since a previous version of the array can never be referenced (because the array is not shared), the new version of the array can be created simply by modifying the previous version.

Defining Terms

Applicative order evaluation: An execution order in which the arguments in a function call are evaluated before the body of the function.

First-class object: An object that can be stored in data structures, passed as arguments, and returned as the result of function calls. In functional languages, functions are first-class objects.

Higher order function: A function that takes another function as a parameter or returns a function as its result.

Lambda calculus: A simple syntactic model of computation equal in power to the Turing Machine.

Latent type system: A type system where types are associated with values, not variables. This usually requires run time type checking which is why latently typed languages such as SCHEME are often referred to as *dynamically typed*.

Lazy evaluation: An evaluation technique for nonstrict functional languages.

Lazy functional language: Informal but common name for a nonstrict functional language.

Nonstrict functional languages: A function language adopting normal-order evaluation.

Normal-order evaluation: An execution order in which the arguments in a functional call are only evaluated if and when needed in the body of the function.

Polymorphism: A property of a languages type system in that an objects type may include type variables which can range over an infinite number of types. Most such polymorphic objects are functions, which can be applied to arguments of many different types.

Pure functional languages: Functional languages that provide absolutely no mechanism for performing side effects, and thus exhibit referential transparency.

Referential transparency: The property of a language that states that equal expressions can be interchanged with each other.

Side effect: A change in the value of a variable as the result of evaluating an expression, e.g., if the expression contains an assignment operation.

Strict functional language: A function language adopting applicative-order evaluation.

Type inference: A process in which the compiler determines the types of objects in a program without the programmer having to declare them explicitly.

References

Abramsky, S. and Hankin, C., eds. 1987. *Abstract Interpretation of Declarative Languages*. Ellis Horwood.

Backus, J. 1978. Can programming be liberated from the von Neumann style? A functional style and its algebra of programs. *Commun. ACM* 21(8):613–641.

Barendregt, H. P. 1984. *The Lambda Calculus: Its Syntax and Semantics*. North-Holland.

Bird, R. and Wadler, P. 1988. *Introduction to Functional Programming*. Prentice–Hall, Englewood Cliffs, NJ.

Bjorner, D., Ershov, A., and Jones, N. 1988. *Partial Evaluation and Mixed Computation*. North-Holland.

Church, A. 1941. *The Calculi of Lambda-Conversion*. Princeton University Press, Princeton, NJ.

Hudak, P. 1989. The conception, evolution, and application of functional programming languages. *ACM Comput. Surv.* 21(3):359–411.

Hudak, P. et al. 1992. Report on the programming language Haskell. *SIGPLAN Notices* 27(5):Section R.

Hudak, P. and Fasel, J. 1992. A gentle introduction to Haskell. *ACM SIGPLAN Notices* 27(5):1–53.

Kelly, P. 1989. *Functional Programming for Loosely-Coupled Multiprocessors*. Pitman.

Kranz, D., Kelsey, R., Rees, J., Hudak, P., Philbin, J., and Adams, N. 1986. Orbit: an optimizing compiler for Scheme, pp. 219–233. In *SIGPLAN '86 Symp. Compiler Construction*. June. *ACM SIGPLAN Notices* 21(7).

Landin, P. 1964. The mechanical evaluation of languages. *Comput. J.* 6(4):308–320.

Landin, P. 1966. The next 700 programming languages. *Commun. ACM* 9(3):157–166.

McCarthy, J. 1978. The history of LISP. In *Proc. ACM SIGPLAN Symp. History Programming Lang.*

McCarthy, J. et al. 1962. *LISP 1.5 Programmers Manual*. MIT Press.

Milner, R. and Tofte, M. 1991. *Commentary on Standard ML*. MIT Press.

Milner, R., Tofte, M., and Harper, R. 1990. *The Definition of Standard ML*. MIT Press.

Paulson, L. 1991. *ML for the Working Programmer*. Cambridge University Press.

Peyton Jones, S. 1987. *The Implementation of Functional Programming Languages*. Prentice–Hall, Englewood Cliffs, NJ.

Peyton Jones, S. and Wadler, P. 1993. Imperative functional programming. In *Proc. 20th ACM Symp. Principles Programming Lang.*

Rees, J., Clinger, W. et al. 1992. Revised Report on the Algorithmic Language Scheme. *Artificial Intelligence Lab. Tech. Rep.* Massachusetts Institute of Technology, Cambridge, Nov.

Sussman, G. and Abelson, H. 1985. *Structure and Interpretation of Computer Programs*. MIT Press.

Sussman, G. and Steele, G. Jr., 1975. Scheme: An Interpreter for Extended Lambda Calculus. *Artificial Intelligence Lab. Tech. Rep. Memo* 349. Massachusetts Institute of Technology, Cambridge.

Ullman, J. 1994. *Elements of ML Programming*. Prentice–Hall, Englewood Cliffs, NJ.

Wadler, P. 1992. The essence of functional programming, pp. 1–14. In *Proc. 19th ACM Symp. Principles Programming Lang.*

Further Information

There are several textbooks that provide an overview of the development and use of functional programming languages. Among these are Bird and Wadler [1988], Paulson [1991], Sussman and Abelson [1985], and Ullman [1994]. The reader is also referred to [Hudak 1989], an excellent survey paper. [Peyton Jones, 1987] provides a description of how functional programming languages are implemented.

There are a number of professional journals that include papers on functional languages. The more eminent of these are *The Journal of Functional Programming, ACM Transactions on Programming Languages and Systems*, and *The Journal of LISP and Symbolic Computation*. Furthermore, recent results in functional programming research can be found in the proceedings of several important annual symposia, including The ACM Symposium on Principles of Programming Languages, The International Conference on Functional Programming, and The ACM Symposium on Programming Language Design and Implementation.

96

The Object-Oriented
Language Paradigm

Raimund Ege
*Florida International
University*

Stuart Hirshfield
Hamilton College

96.1 Introduction

Perhaps because of its relatively recent surge in popularity, **object-oriented programming** (**OOP**) has often been described as a revolutionary new programming paradigm. Such a characterization, though, is only partly accurate.

OOP is, to be sure, a paradigm in the current sense of that term. It embodies a way of organizing and representing knowledge, "a way of viewing the world" [Budd 1991], that encompasses a wide range of programming activities, including program analysis, design, and implementation. The paradigm derives its power from its view of computation as the simulation of real-world entities. That is, according to Dan Ingalls, "Instead of a bit grinding processor... plundering data structures, we have a universe of well-behaved objects that courteously ask each other to carry out their various desires" [Ingalls 1981].

Central to this view of computation is the notion of data abstraction. OOP tools and languages facilitate the description of **objects**: agents that maintain their own internal state (data), perform actions (methods) in their own interest, and interact with other objects (by sending **messages** to one another). Objects can be low-level programming tools, for example, lists, stacks, trees, akin to traditional abstract data types (ADTs). They can also be higher level abstractions that reflect what a program is intended to model: an automated teller machine (ATM), a deck of playing cards, an elevator, or a collection of graphical objects on a screen. Objects in an object-oriented (OO) system resemble the nouns in a textual problem description.

The primary power of OOP derives from the fact that once defined, objects enjoy a type-like status. (As we will explain later, objects are defined via **classes**, which are very much like types). That is, (1) objects can be used without knowing the details of their implementation and can be properly protected from their consumers; (2) objects can be used according to a standard notation, using names, symbols, and operators in conventional ways; and (3) objects can be combined with other objects and types in expressive and efficient ways (composition and hierarchy) to define new, more complex types.

Whereas previous (primarily procedural) languages such as C and PASCAL allowed a programmer to define new types, these were primarily a notational convenience. In such languages, user-defined type names served as shorthands to improve readability and as aids to compilers for recognizing type equivalences. These types, though, did not enjoy the status or flexibility of the builtin types. That is, one could not overload operators to apply to these new types and, more importantly, one could not easily hide the implementation details of a new type from its consumers.

In hindsight, the seeds for the object-oriented paradigm can be found in the languages that developed in the 1970s and 1980s to provide more direct support for building user-defined ADTs. For instance, CLU and ALPHARD were designed in the mid-1970s to support ADTs, including information hiding. CLU, in particular, was influential in the development of both MODULA-2 and ADA-83. These languages, both of which might rightfully be called object-based, allow one to build clearly specified, modular software components which effectively hide their implementation details. ADA, for example, uses the package construct to describe type specifications and subprograms that can belong to a user-defined ADT.

In many ways, the evolution of the OOP paradigm parallels that of the procedural paradigm which preceded it. We saw the procedural paradigm, as manifested in a succession of increasingly high-level programming languages, progress from straight-line code laced with unconditional transfers of control, to block-oriented code that exploited control structures, to finer-grained code built from simple subprograms, to top-down, structured code that relied on parameterized and separately developed libraries of code which distinguished a subprogram's interface from its implementation. OOP allows a programmer to treat data types at a similarly high level. In a sense, OOP is to data abstraction what the procedural paradigm is to algorithm abstraction.

Yet, whereas OOP can clearly be described as a programming paradigm, it is decidedly more evolutionary than revolutionary. Hindsight being what it is, we can now see how the seeds of OOP have been nurtured by developments in many subfields of computer science over the past three decades.

The basic idea of computation as simulation was first popularized in the programming language SIMULA 67, which also introduced the concepts of class, object, and message. Hardware architectures were developed in the 1970s using an object-based approach to describe the components of the machine and their interactions at a higher, more natural level. This development was further motivated by operating systems and applications that depended increasingly on graphical interfaces, which lend themselves directly to object-oriented descriptions. In the entity-relationship database model, in which a collection of information is described in terms of entities, attributes, and relations among entities, we see yet another manifestation of objects (at least the information content aspect of them). Even in knowledge representation schemes that supported work in artificial intelligence (frames, scripts, and semantic networks) we can see clearly this object orientation. Most clearly, the first purely object-oriented language, SMALLTALK, has been around since the early 1970s.

Still, all of the premonitions of OOP did not coalesce into a recognizable influence on programming language and practice until they were motivated to do so by our collective practical experience in software engineering. As we became increasingly adept at exploiting the procedural paradigm and its languages, we recognized more directly their shortcomings. The more we pushed that paradigm, the more difficult it was to support data abstraction to that same degree that algorithm abstraction could be. Further, methods of encapsulation were relatively primitive, thus making if difficult to develop safe, reusable code. Finally, program analysis and design techniques that relied on procedural abstraction as the basis for performing decomposition of complex systems were proving increasingly awkward to apply to domains that were increasingly data, as opposed to algorithm, dependent.

In summary, structured methods flourished in the era of application domains that were either algorithm-centric or information-centric. These same methods prove less useful in an increasingly complex application world, where functionality and information structure have to be considered in context. The concept of an object is a simple and elegant combination of all informational and algorithmic aspects of an application entity.

Today, all of these technologies have matured and all of the motivations have been established to the extent that OOP is a viable, recognizable, and popular approach to software development. Programming languages and commercial development environments abound. SMALLTALK has experienced a commercial rebirth of sorts, and EIFFEL, another pure object-oriented programming language, is gaining popularity. A variety of hybrid OOP languages (like C++, Object PASCAL, and OO COBOL, which are derived from existing higher order languages and extended to incorporate essential OOP features) are dominating the programming language marketplace.

Because OOP is rightly referred to as a paradigm, it has spawned the development of many program analysis and design techniques that support the identification and description of objects in a problem specification. As with all paradigms, OOP languages and techniques are best suited to problems that match the paradigm's view of the world. As the demand for complex, interface-intensive systems—those that can be modeled in real-world terms—increases, so does the use and popularity of OOP.

Finally, if OOP raises the level of abstraction to bridge the gap between programmer and machine, it may be that novice programmers would stand to benefit the most from its use. Indeed, SMALLTALK was developed based on research detailing how young children describe and interact with the world in solving problems. OOP is just now beginning to influence significantly how we teach and learn programming.

There is a growing recognition that object orientation may make learning to program easier for the novice. Universities are teaching object orientation as part of the computer science introduction. However, it is also recognized that experienced programmers and software engineers have to be retrained significantly: not because object orientation is hard, but because of their experience in traditional function-centered (or information-centered) problem solving. Experienced programmers have to, as Bertrand Meyer puts it, re-acquire an object-oriented frame of mind.

Proponents of OOP claim that the paradigm represents the state of the art in terms of bridging the language gap between programmer and machine. It offers the prospect for achieving many of the software quality goals that all programmers aspire to: easily designed, safe, efficient, uniform software.

96.2 Underlying Principles

Much conventional programming wisdom goes by the wayside in OOP. The process-state view of computing (which reflects quite directly both the machine's fetch-execute cycle, and its data processing nature) is the first to go. Thinking about programs as lists of instructions which are executed sequentially and serve to manipulate and change the state of memory is the fundamental strategy of imperative languages. Programming in these languages involves thinking in terms of algorithms, that is, the sequence of actions the computer will perform. OOP languages, on the other hand, permit the programmer to take a broader view and think of a program as a collection of cooperating objects, each of which encapsulates both structure and function.

Each object has three aspects: what it is, what it does, and what it is called. We illustrate these aspects by considering a geometric object, a circle. Most programming languages do not have circles built in as part of the language, and so if we want to write a program that manipulates circles we have to write our own description of circle objects into a program. The descriptive information associated with an object, called its state, includes its properties and the values of these properties. For most objects, the state properties do not change, though the state values may very well be modified during the object's lifetime. For instance, a circle has a radius and a center and at a given time, these might have the values 4.67 and (0, 0). A circle object will always have a radius and a center, though their values will change if the circle's size or location does.

The collection of actions an object may perform on itself is called its behavior. In our example, we may want to use a circle as part of a program that draws on a computer screen, and so we would need to be able to set the center and radius of the object and instruct the circle to draw itself on the screen. Notice that we said a circle would draw itself on the screen. This is an important feature of the object-oriented approach. In a language with an algorithmic approach, one could describe circles by radius and center, as we have, but the description of the actions to perform on a circle would not be as tightly associated with the state data as it is here. That is, the drawing command would not be part of the circle object itself. Encapsulating data and actions together in an object helps us design and understand a program, since, for instance, all circle-related data and actions are collected in one place. If we had to modify a circle object's definition, we would know exactly where the modifications would go, and, furthermore, we would know that we would not have to modify any other part of the program.

Finally, each object has an identity, which serves to distinguish it from all others. In most programming languages, we identify an object by giving it a name, so we might call our circle theCircle.

It is entirely reasonable to assume that a program that would use one circle object might use several. All of these circle objects would be similar, in that they would have the same collection of member data and functions. It would clearly be a wasteful duplication of effort to describe them each separately, since they would differ only in their data values and names. Object-oriented languages allow one to describe a template, if you will, for an entire set of objects. Such a template is called a class.

A class is used to generate a set of objects sharing common properties. Continuing our example, we could describe a class, Circles, by describing the state properties (but generally not the state values) and behaviors of all objects that belong to that class. Then we could describe as many circle objects as we needed merely by declaring that they are instances of the Circles class, that is, they are Circle objects. An object can never exist in isolation; every object must be an instance of some class. This means that if we are going to use an object in a program, we must first provide the class to which the object belongs.

One can think of a class as a means for implementing an abstract data type. That is, to describe a class we identify the properties that any instance of the class (a particular object) must have. These properties consist of state information (referred to as member, or state data) and a collection of behaviors (member functions, or methods) that such objects are capable of performing. Methods typically involve accessing, setting, or otherwise manipulating the object's member data.

Methods are invoked by an object in response to a message, i.e., a request of the object to perform one of its methods. Message passing as a means for invoking methods differs subtly from subprogram invocation in most other programming paradigms. In OOP, there is a formal distinction between the subprogram call (the sending of a message) and the method of invocation (the receiving of the message). It is the responsibility of the receiver of a message to interpret the message, that is, to determine which of its methods to perform in response. So essential is the distinction that OOP has been defined as "programming by sending messages to objects" [Pinson and Weiner 1988].

Taken together, these two mechanisms support the description of full-fledged ADTs in a more general and extensible way than was provided by the first object-based languages. In particular, classes provide an abstraction mechanism which can be used to model a wide range of information structures and information processing agents. Using classes as the primary descriptive vehicle also characterizes an OOP approach to program design that has proven extremely effective for many real-world applications.

Not only are classes a rich descriptive tool, but they are safe, in the software engineering sense of that term. That is, classes effectively encapsulate the abstractions they model by controlling (either implicitly or explicitly, depending on the language being used) access to an object's members. Some of an object's members will be public in nature (visible and accessible to other objects) and some will be private to the object itself (visible and accessible only to other members). Many OOP languages also support a formal distinction between a class's interface and its implementation.

Classes are also an efficient means of representation by virtue of the fact that they can be related to one another both compositionally (an object of one class can serve as a member of an object of another class) and hierarchically (to express is-a relationships). This latter feature is unique to OOP, and it affords a means for describing directly any information that is hierarchical in nature.

In cases where classes are arranged in a hierarchy, a **subclass** is said to **inherit** from its superclasses. That is, instances of the subclass contain (either directly as copies, or indirectly via the superclass) both the data and methods described by the superclass. These same general properties are associated with all subclasses of the particular superclass. Each subclass extends the descriptions of its superclasses by adding specialized member data and methods that differentiate itself from the superclass and from other subclasses.

Going back to our geometric example, we do not have to restrict our objects to circles; we could equally well have objects that are rectangles and triangles, and so on. We could augment our program by including declarations and definitions for the classes Rectangles and Triangles, just as we did when we defined the Circles class. In doing so, we notice that all of these are what we might call geometric objects.

Object-oriented languages allow a class like Rectangles to inherit the member functions and data from a parent class, such as GeometricObjects. For example, since every object in our hierarchy has an extent (a bounding rectangle inside which the object fits as closely as possible), we could declare the data bounds_top, bounds_left, bounds_bottom, bounds_right, and the member function ReportBounds within a superclass named GeometricObjects, and these would apply to all the derived classes Rectangles, Triangles, and Circles. Doing this, we would not need to redeclare these data items in any of these subclasses. They would be inherited from the base class GeometricObjects. Indeed, it is sometimes useful to define a superclass, such as GeometricObjects, knowing full well that we will never be interested in generating direct instances of it. That is, the only instances of GeometricObjects that we will use in our programming are those that are also instances of its subclasses, Circles, Triangles, or Rectangles. In such cases, superclasses are referred to as **abstract classes**.

Inheritance is not only distinctive of OOP languages, but it is also the means by which the paradigm addresses many software engineering concerns. For example, inheritance is an efficient, nonredundant notation for representing information that is hierarchical in nature. It is natural in the sense that it reflects how we humans tend to describe such information (which explains in part why OOP has spawned so many related program design techniques). The efficiency of the notation derives from the ability to reuse code that was used in defining a superclass when defining a subclass. This, in turn, allows the inherited members of classes that are derived from one another to project consistent interfaces to their consumers. That is, member data and functions can be referred to by the same names in both the super- and subclasses.

There are times, though, when a base class action or state might not be appropriate for use by a derived class. All of the classes in our hierarchy would have Draw functions, but drawing a rectangle might involve different strategies than drawing a circle. OOP languages allow a derived class to redefine, or **override**, a member function inherited from a base class. If, for example, we have an object called theShape, and send it a message to draw itself, how is the computer to know which Draw action to use? Whereas procedural languages and object-based ones would resolve this ambiguity statically (that is, based on the declared type of theShape), an OOP language solves this problem at run time, by looking up the class to which theShape belongs, and finding a reference to an appropriate Draw function.

This ability to use the same name for actions on objects of different classes is an example of a kind of **polymorphism** (called inclusion polymorphism) that is common to all OOP languages. In the OOP sense of the term, polymorphism describes the fact that a superclass includes all **instances** of its subclasses. Thus, the notion of a geometric object contains all circles, rectangles, and triangle objects. As a result, an entity can take on different types of information during the course of execution of a program. For example, a geometric object can be a circle, a rectangle, or a triangle at any point during our program.

Other forms of polymorphism which are often supported in OO languages include: overloading (in which two or more functions or operators can share the same name, and so the + operator is allowed to apply to both integer and real numbers), parametric polymorphism (in which a parameter is used to establish choices: for example, generic packages—or C++ class templates—use a parameter to allow another level of abstraction), and coercion polymorphism (C type casting is an example).

These characteristics—objects (as a means for encapsulation of state and behavior), classes (as templates for generating objects), message passing (as the mechanism by which computation takes place), dynamic method invocation (as the means by which methods are bound and interpreted), inheritance (as a technique for modifying classes and creating subclasses), and inclusion polymorphism (which allows expressions of one type to be used in place of expressions of a supertype)—are the essence of OOP. Each contributes significantly to the overall utility of the paradigm, and each allows the paradigm to address one of the many software engineering concerns that motivated it. Whereas different programming languages implement them in various combinations and to varying degrees, any language that implements them all is considered object oriented.

In fact, the meaning of the term object oriented has been recently extended to apply to languages that are not even class based. *Prototype-based* languages are similar to class-based ones, but rather than having classes to provide a template, objects are allowed to inherit instance variables and methods from other objects (usually called prototypes). SELF is an interesting example of this form.

Multimethod systems [exemplified by the common LISP object system (CLOS) and Dylan] choose the appropriate method (dynamically) by looking at the run time types of several arguments, rather than that of a single receiver. As a result, messages are sent in a form that more closely resembles procedure calls: doThis(x,y,z), as opposed to x.doThis(y,z). In this example, doThis may be dispatched on the first and third arguments. That is, rather than going to the object to look up methods, these languages use a dictionary to find the method name and then look for the best fit parameterwise among all methods with the same name. Suffice it to say, much like the paradigm it describes, the term object-oriented programming is a dynamic one, with a variety of still evolving interpretations.

96.3 Best Practices

To illustrate these characteristics of object-oriented programming let us construct a very simple application to deal with queues of packets as they might appear in a network simulation. Packets in our program maintain their name and priority, and allow their observance and comparison. Different kinds of packets are modeled as subclasses, one for packets that carry protocol information (*Ack*) and one for packets that carry data (*Data*). The subclasses specialize how packets are observed. The second set of classes models the *queue* concept. Class *FifoQueue* represents the algorithmic and data abstraction of a standard first-in–first-out queue. Internally it employs a doubly linked list—anchored by a *head* and *tail* member field—to maintain the packets currently queued. The member functions *enter* and *leave* implement the standard protocol of such a queue. The *FifoQueue* class ignores the packet's priority information, but also serves as a superclass for two additional subclasses, *PriQueue* and *QueuePri*, which use the packet's comparison abilities to handle packets of different degrees of importance.

We will develop the example in terms of SMALLTALK, C++, and EIFFEL, three of the major object-oriented programming languages in use today. Our intention here is to provide you with quick overviews of these languages, and to illustrate the different notations and styles for implementing the object-oriented paradigm.

SMALLTALK

SMALLTALK is an integrated programming environment and language that led the OOP evolution, by incorporating the ideas of personal computer, interactive computing, graphical user interface (GUI), and OOP. The underlying design principle was to raise substantially the level of abstraction so that program elements could communicate at a level that was closer to that of human problem solvers.

The language is *purely* object oriented. Every entity in a program is an object, everything from windows and projects to integers. No complex syntax is necessary because everything that can be used within a program is of one type—object. Still, the limited syntax allows for declaring object names and assigning values to them, sending messages, and defining new classes and methods. Control structures are accomplished by message passing, where methods in standard classes define their behavior.

SMALLTALK was originally developed as a research tool, and is thoroughly integrated with its complex and powerful graphical programming environment. It spurred the subsequent development of all current commercial OOP languages, and is enjoying renewed interest, particularly for rapid prototyping and interface intensive applications, as the OOP paradigm comes increasingly into the mainstream of commercial computing. A variety of implementations are available.

Let us develop a SMALLTALK program for our network/packet/queue example. We start by describing the class Packet:

```
Object subclass: #Packet
        instanceVariableNames: 'name priority '
        classVariableNames: ''
        poolDictionaries: ''
        category: 'MyFifoQueue-Examples'
```

A packet is defined as a subclass of *Object*. SMALLTALK enforces a single class hierarchy, i.e., all classes need a superclass. *Object* is the anchor of the existing SMALLTALK class hierarchy. The # character before

Packet is needed as part of SMALLTALK syntax (i.e., it defines a symbol). The Packet class defines two instance variables (member fields): *name* and *priority*, SMALLTALK does not type its member fields; in effect they can refer to an instance of any object within SMALLTALK.

Class variables are fields that do not carry unique values per object, such as **instance variables**, but rather one value per class. For our Packet class we do not need to define any class variables; *pool dictionaries* are another way to refer to variables; and *category* defines that this new class belongs to a set of classes called *MyFifoQueue-Examples*. Categories are used in the SMALLTALK programming environment to quickly find classes.

The packet class needs member functions, i.e., instance methods in SMALLTALK terminology. The first instance method, *list* is defined as follows:

```
!Packet methodsFor: 'printing'!
  list
            Transcript show: name, ' packet: '.
```

The first line constructs the relationship of this method to the packet class (the exclamation marks differentiate it from regular SMALLTALK syntax); the method is part of the *printing* protocol (SMALLTALK categorizes related methods into protocols). The method has name *list*, no arguments are listed. The body of the method consists of a single statement: it instructs to send message *show:* to object *Transcript*. Transcript is a preexisting SMALLTALK object that records text and displays it to the user. In SMALLTALK syntax, a method name that ends with a : (colon) signifies that a parameter is expected: here a string constructed from the *name* of the packet and the text *packet* is sent along with the message as a parameter [the , (comma) is used to concatenate the two parts of the string]. And finally, a SMALLTALK statement is delimited by a . (period).

Next, we define two methods that allow us to compare packets: *less than or equal* written as <=, and *greater than* written as >.

```
!Packet methodsFor: 'access'!
<= aPacket
          ^(priority <= aPacket priority).
> aPacket
          ^(priority > aPacket priority).
priority
               ^priority.
```

Here, <= is defined with a single parameter, i.e., *aPacket*. Again, SMALLTALK does not type method parameters: any object can be sent along as parameter. The body of the first method lists a single *return* statement (^ is used to denote return). Returned is the expression that is listed in parentheses: a comparison of the instance variable *priority* to the priority of the *aPacket* parameter. SMALLTALK hides all instance fields inside an object. Therefore, one packet object has no access to the *priority* of another packet. It is therefore necessary to define an instance method *priority*, which is used by both the <= and > methods. The *priority* method just returns the *priority* instance variable.

And finally, the *initialize* method allows one to set a packet's instance variables. It uses two parameters, one for *name* and one for *priority*. SMALLTALK uses a special style for methods with multiple parameters: they are embedded into the method name: the complete name of this method is *initialize:priority:*, where after each : SMALLTALK expects a parameter:

```
!Packet methodsFor: 'initializing'!
initialize: aName priority: pri
              name := aName.
              priority := pri.
```

Methods can be defined at the instance and at the class level. Instance methods are those that are invoked when an object receives a message. Class methods are typically used to create new instances of a class. For example, a new packet object would be created with

```
Packet new.
```

The *new* method is not defined for class Packet, rather it is inherited from its superclass *Object*. Here *new* creates and returns a new instance of class *Packet*. It can be initialized with our *initialize:priority:* method, as in

```
Packet new initialize: 'one' priority: 10.
```

Since all packets should be initialized when created, we can define a new class method for Packet that enables the creation and initialization of Packet objects in a single method,

```
!Packet class methodsFor: 'initializing'!
new: aName priority: pri
            ^(Packet new initialize: name priority: pri).
```

The name of this method is *new:priority:*, it does not override the *new* method that Packet inherits from its superclass, rather it uses it internally. The body of the *new* class is a single line that proceeds in three steps: (1) it sends the message *new* to the Packet class, which creates a new instance and returns it; (2) the new instance then receives an *initialize:priority:* message to initialize the new packet; and (3) the method then returns the new and initialized object.

Subclass Data is based on the Packet class. It adds two more instance variables, *body* and *length*:

```
Packet subclass: #Data
         instanceVariableNames: 'body length '
         classVariableNames: ''
         poolDictionaries: ''
         category: 'MyFifoQueue-Examples'
```

Since data packets have additional instance variables, it makes sense to define an *initialize* method that is different from the one that is inherited from the packet class. It takes three parameters, one each for the instance variables name, priority and body (the full name of this method is now *initialize:body:priority:*):

```
!Data methodsFor: 'initializing'!
initialize: aName body: aBody priority: pri

        body := aBody.
        length := body size.
        self initialize: aName priority: pri.
```

The *body* instance variable is set from the parameter. The length is calculated by sending a *size* message to the *body* parameter. Note, that while parameters are not typed, there is an assumption that it belongs to a class that defines the *size* method.

The third line in the body shows how one method delegates to another: since *name* and *priority* still need to be set, which we already spelled out in the *initialize:priority:* method in the packet superclass, we send message *initialize: aName priority: pri* to object *self*. *Self* refers to the *Data* object itself that is currently executing the method: it will now execute the method inherited from its superclass. And finally, the *list* method,

```
!Data methodsFor: 'printing'!
list

        super list.
        Transcript show: body.
```

uses the keyword *super*, which is similar to *self*, it allows one to send a message to the current object; however, the corresponding method that is executed is not found in the class of the current object, but rather in its superclass, i.e., the Packet class. In effect, the *list* method defined here, extends the list behavior of the superclass.

The class methods for Data allow for the creation of Data objects with or without an explicit priority:

```
!Data class methodsFor: 'initializing'!
new: aBody
        ^super new initialize: 'Data' body: aBody priority:5.
new: aBody priority: pri
        ^super new initialize: 'Data' body: aBody
        priority: pri.
```

The *super* keyword is needed here. Otherwise, if we would send a new message to the Data class, we would have an infinite loop because we are currently defining the *new* class method for Data. *Super* will ensure that the *new* method defined in the superclass is executed.

Subclass Ack is also based on Packet. It defines no further instance variables, but it redefines the *list* method:

```
Packet subclass: #Ack
        instanceVariableNames: ''
        classVariableNames: ''
        poolDictionaries: ''
        category: 'MyFifoQueue-Examples'

!Ack methodsFor: 'printing'!
list
        Transcript show: 'acknowledged'.
```

The class method for Ack ensures that Ack objects are created with priority 10:

```
!Ack class methodsFor: 'initializing'!
new
        ^(super new initialize: 'Ack' priority: 10).
```

Here, the *new* method overrides the *new* method inherited from its superclass. It also uses the *super* keyword to invoke the *new* method defined by its superclass.

The Packet class hierarchy is now complete. Before we construct our Queue class we need one more helper class: Node, which will be used to implement a doubly linked list. The Node class defines three instance variables: *value* to hold a packet object, *next* to refer to the next node in the list, and *previous* to refer to the previous node in the doubly linked list. The instance methods allow for the manipulation of all instance variables.

```
Object subclass: #Node
        instanceVariableNames: 'value next previous'
        classVariableNames: ''
        poolDictionaries: ''
        category: 'MyFifoQueue-Examples'

!Node methodsFor: 'access'!
next
        ^next.
next: aNode
        next := aNode.
previous
        ^previous.
previous: aNode
        previous := aNode.
value
        ^value.
value: aValue
        value := aValue.
!Node methodsFor: 'initializing'!
initialize: aValue
        value := aValue.
!Node class methodsFor: 'initializing'!
new: aValue
        ^super new initialize: aValue.
```

Now, we can construct the class FifoQueue, which defines two instance variables, *head* to refer to the beginning of the list of nodes, and *tail* to hold on to the end of the list.

```
Object subclass: #FifoQueue
        instanceVariableNames: 'head tail'
        classVariableNames: ''
        poolDictionaries: ''
        category: 'MyFifoQueue-Examples'
```

The first instance method enters a packet object into the queue. It uses a local variable *tmp* for a newly created node object, which is then inserted at the *tail* of the linked list.

```
enter: aPacket
        | tmp |

        tmp := Node new: aPacket.
        tmp previous: tail.
        tail := tmp.
        head notNil
```

```
            ifTrue: [ tmp previous next: tail ]
            ifFalse: [ head := tmp ].
```

Local variables in methods are listed between |. Here *tmp* is created as a new instance of class Node by sending message *new:* to the Node class. The *tmp* receives message *previous:* to attach it to the rest of the linked list content (*tail*). The last statement in the method body is an example of an *if-then-else* construct. First, *head* receives message *notNil*, which returns Boolean objects *true* or *false* depending on whether *head* refers to a Node object or is undefined. The returned Boolean object then receives the *ifTrue:ifFalse:* message. The parameters to *ifTrue:ifFalse:* are delimited by brackets, which denote blocks of SMALLTALK statements. The execution logic is as follows: if *head notNil* returns Boolean object *true*, then the parameter after *ifTrue* is executed; if the returned object is *false*, then the parameter after *ifFalse:* is executed. In effect, we have a regular *if-then-else* construct modeled with object orientation.

The second instance method removes a packet object from the queue. It uses a local variable *it* to store that packet object which is maintained as the value of the *head* node of the list. The instance method finally returns packet *it*, after it has reconnected the list.

```
    leave
                | it |
            it := head value.
            (head = tail)
                    ifTrue: [ head := tail := nil ]
                    ifFalse: [ head next notNil
                            ifTrue: [
                                    head := head next.
                                    head previous: nil.
                            ]
                    ].
            ^it.
```

The third instance method, *list*, allows us to observe the current status of the queue and its contents. It assumes that each object in the list understands a *list* message. In our example, we will store packet objects in the queue, and all packet classes define a *list* method. This illustrates one of the major features of object-oriented programming languages in general, and SMALLTALK in particular: flexibility. At this point we need not worry about which types of packets will actually be stored in a queue: Packet objects, Data objects, or Ack objects are welcome. Moreover, we might even have more packet subclasses in future versions of our software. The *list* method also illustrates a major feature of SMALLTALK, i.e., uncertainty. If we enter objects into the queue that do not understand the list message then this code will fail at the time when the queue is asked to *list*. SMALLTALK represents a very flexible approach to software modeling. As we will see later, other object-oriented programming languages, notably EIFFEL, add considerably more safety and predictability.

```
    list
                | tmp |
            tmp := head.
            [tmp notNil] whileTrue: [
                    tmp value list.
                    tmp := tmp next.
            ].
```

The body of the list method uses a *while* loop: [*tmp notNil*] is a block that is executed for each iteration of the loop; if it results in *true* then the body of the loop (listed after the *whileTrue*: marker) is executed; if it is *false* then the loop terminates.

Now that we have described four classes, it is possible to exercise them. Let us create a queue object and some packets (two Data objects, and two Ack objects), enter them into the queue, and observe the current queue content by sending the appropriate messages:

```
| q w1 c1 w2 c2 |
q := FifoQueue new.
w1 := Data new: 'first packet'.
c1 := Ack new.
w2 := Data new: 'second packet' priority: 6.
c2 := Ack new.
q enter: w1.
q enter: c1.
q enter: w2.
q enter: c2.
q list.
```

The output from *q list* is

```
Data packet: first packet
acknowledged
Data packet: second packet
acknowledged
```

Then, we ask the queue to remove the packets and list them as we go,

```
q leave list.
q leave list.
q leave list.
q leave list.
```

Not surprisingly, the packet objects are stored by the queue object in the order in which they were entered, and then are released in exactly the same order, as we would expect of a first-in–first-out queue abstraction.

To extend our example, we can now extend our queue abstraction by taking the importance (i.e., priority) of packets into account. Class PriQueue is defined as a subclass of Queue, it inherits all instance variables and methods, and specializes the *enter* method:

```
FifoQueue subclass: #PriQueue
        instanceVariableNames: ''
        classVariableNames: ''
        poolDictionaries: ''
        category: 'MyFifoQueue-Examples'.

!PriQueue methodsFor: 'access'!
enter: aPacket

        | tmp p |
```

```
tmp := Node new: aPacket.
tail isNil ifTrue: [
        tail := tmp.
        head := tmp.
] ifFalse: [
        p := tail.
        [ (p notNil) and: (aPacket > (p value)) ] whileTrue: [
        p := p previous
].
p isNil ifTrue: [
        tmp next: head.
        head previous: tmp.
        head := tmp.
] ifFalse: [
        tmp previous: p.
        tmp next: p next.
        (p = tail) ifTrue: [
            tail := tmp.
        ] ifFalse: [
            p next previous: tmp.
        ].
        p next: tmp.
        ]
].
```

The implementation of *enter* for priority queues is significantly more complex. It traverses the existing list for each packet that is to be entered to determine its relative importance. The enter method uses the > instance method defined for packet objects. Again, it is important that all objects stored in a PriQueue understand such a message. The PriQueue class specializes the enter instance method of class Queue. Another way of providing a queue with priority handling would be to specialize the leave instance method rather than enter. Class QueuePri, described next, does that.

```
FifoQueue subclass: #QueuePri
        instanceVariableNames: ''
        classVariableNames: ''
        poolDictionaries: ''
        category: 'MyFifoQueue-Examples'

!QueuePri methodsFor: 'access'!

leave

        | it p |

        it := head.
```

```
head = tail ifTrue: [
        head := tail := nil.
] ifFalse: [
    p := head.
    [p notNil] whileTrue: [
            (p value) > (it value)
                        ifTrue: [ it := p].
        p := p next.
    ].
    it = tail ifTrue: [
            tail := it previous.
            tail next: nil.
    ] ifFalse: [
        it = head ifTrue: [
                head := it next.
                it next previous: nil.
        ] ifFalse: [
                it previous next: it next.
                it next previous: it previous.
        ]
    ]
].
^it value.
```

The strategy for implementing a priority queue is very straightforward. Whenever a packet is to be removed from the queue, we traverse the linked list and determine which of the packets has the highest priority. The good news is that Queue, PriQueue, and QueuePri objects can now be used interchangeably, depending on what kind of queuing strategy is desired. All three classes provide the same protocol, i.e., its objects understand the same set of messages.

In summary, SMALLTALK is actually much more than just a programming language: it is also a very elaborate programming environment that includes a large library of ready to use classes and allows for the interactive and incremental development of SMALLTALK programs. The SMALLTALK language is a truly object-oriented programming language. It hides all informational detail of objects and makes all instance methods freely available. SMALLTALK allows only one form of inheritance, as seen in the example, where a class is defined as a subclass of a single superclass. Other programming languages allow multiple inheritance, where a subclass may have more than one superclass.

SMALLTALK is very flexible, all message requests are resolved when an object receives a message: other programming languages enforce this OOP principle to varying degrees, thus allowing for tradeoffs between flexibility and safety. The creation of objects is defined by programmers. The deletion of objects, however, is left unspecified. SMALLTALK automatically detects if objects are obsolete and reclaims them. This capability of object-oriented run time support systems is called *garbage collection*.

EIFFEL

Our second programming language, EIFFEL, was developed to address software engineering issues in an OOP paradigm. EIFFEL uses a PASCAL style that is both readable and powerful. Implementations exist for a variety of systems.

The classes we have described in our programming example can be directly implemented in EIFFEL. The class Packet, with its instance variables and methods, is described as follows:

```
class Packet
        creation Create
feature {NONE}
        id: STRING;
feature {PACKET}
        priority: INTEGER;
feature
        Create (n: STRING; p: INTEGER) is
        do
                id := n;
                priority := p;
        end; -- Create

        list is
        do
                io.putstring(id);
                io.putstring(" packet: ");
        end;

        infix ">" (other: Packet): BOOLEAN is
        do
                Result := priority > other.priority;
        end;

        infix "<=" (other: Packet): BOOLEAN is
        do
                Result := priority <= other.priority;

        end;
end -- class Packet
```

An EIFFEL class defines *features,* a collective term for *instance variable* and *instance method.* Variables are explicitly typed, e.g., *priority* is of type *INTEGER.* All features can be exported, e.g., *priority* is made visible to class *PACKET*, that is, all *PACKET* objects can see one another's *priority* instance variable value. Exporting to *NONE* has the effect of hiding a feature. If the feature clause does not list a target class, then that feature is implicitly exported to class *ANY*, in effect making it public.

Class *Ack* is defined as a subclass to *Packet.* The *inherit* clause mentions the *rename* and *redefine* subclasses: *rename* allows one to give a feature that is defined in a superclass a different name in a subclass; *redefine* makes explicit that the subclass will provide a specialization of the *list* feature.

```
class Ack
inherit Packet
rename
        Create as Packet_Create
redefine
        list
```

```
           end
        creation Create
        feature
                Create is
                do
                        Packet_Create("Ack", 10);
                end;

                list is
                do
                        io.putstring("acknowledged");
                        io.new_line;
                end;
           end
```

Class Data is just slightly more complicated. It inherits from class Packet, once to redefine the *list* feature, and once to rename the *list* feature into *Packet_list* to be used in the redefined version of *list* in class Data. Note, that SMALLTALK uses the keyword *super*, which is a much cleaner and simpler way to allow a subclass to refer to an instance method defined in the superclass.

```
        class Data
        inherit Packet
                rename Create as Packet_Create
                redefine list
                select list
        end;
                Packet
                    rename Create as Packet_Create,list as Packet_list
        end
        creation Create
        feature
                body: STRING;

                Create (b: STRING; p: INTEGER) is
                do
                        Packet_Create("Data",p);
                        body := b;
                end;

                list is
                do
                        Packet_list;
                        io.putstring(body);
                        io.new_line;
                end;
           end
```

Again, before we can define the FifoQueue class we need the Node class. Class Node is defined here as a simple class with an instance variable *next* of class Packet: this limits our Queues to contain only Packets (of course, EIFFEL also supports parameterized generic classes). Notable here are the extra instance methods *setPrevious* and *setNext* which are necessary to set the next and previous instance variables, despite the fact that these variables are exported. Exporting of variables in EIFFEL only provides read access. As we will see in the section on C++, there exporting a variable, i.e., making it public, allows read and write access. Both the EIFFEL and C++ approaches are in direct contrast to SMALLTALK where all variables are strictly encapsulated inside an object and not accessible at all from the outside.

Also interesting are the types associated with features *next* and *previous*. Both are defined as being of type *Node*. This does not mean that a Node object will contain other Node objects. They will, though, contain the object identifiers of other Node objects. Thinking of object identifiers as pointers to objects yields the conventional linked list metaphor. Moreover, a true object-oriented programming language does not need pointers at all. Since all objects carry their unique identity, that can be used instead. That is why EIFFEL and SMALLTALK do not support pointers !

```
class Node
        creation Create
feature   {ANY}
        value: Packet;
        next: Node;
        previous: Node;

        Create(v: Packet) is
        do
                value := v;
        end;

        setPrevious(n: Node) is
        do
                previous := n;
        end;

        setNext(n: Node) is
        do
                next := n;
        end;
    end
```

Class FifoQueue makes use of these classes to implement our first-in–first-out queue abstraction. The data features *head* and *tail* are hidden, i.e., exported to *NONE*. The function features (or *routines*, in EIFFEL terminology) are exported to class {*ANY*} which represents all classes in EIFFEL, and so are accessible in any client class.

```
class FifoQueue
        feature {NONE}
                head: Node;
                tail: Node;
        feature {ANY}
                enter (p: Packet) is
```

```
                    local
                            tmp: Node;
                    do
                            !!tmp.Create(p);
                            tmp.setPrevious(tail);
                            tail := tmp;
                            if head /= void then
                                    tmp.previous.setNext(tail);
                            else
                                    head := tmp;
                            end;
                    end;

                    leave: Packet is
                    local
                            it: Packet;
                            nil: Node;
                    do
                            it := head.value;
                            if head = tail then
                                    head := nil;
                                    tail := nil;
                            else
                                    if head.next /= void then
                                            head := head.next;
                                            head.setPrevious(nil);
                                    end;
                            end;
                            Result := it;
                    end;

                    list is
                    local
                            tmp: Node;
                    do
                            from
                                    tmp := head
                            until
                                    tmp = void
                            loop
                                    tmp.value.list;
                                    tmp := tmp.next;
                            end;
                    end;

            end
```

As in the SMALLTALK version of the example, we are able to assemble a few objects and send messages. The *!!* notation signifies the creation of a new object. In routine *enter* local object *tmp* is created as an instance of class *Node*. The class of object to be created is known since all variables are strongly typed. Routine *list* uses the *from until* loop construct: before the loop starts, *tmp* is initialized to the value of *head*; the loop will continue to execute its body until *tmp* is undefined, i.e., is equal to *void*.

The next class doubles as the *main* program. It creates a few objects and starts execution. The *!Ack!* notation creates an instance of class *Ack*: here we want to explicitly create an Ack object and store it in a variable *a1* that is of type *Packet*. Of course, this is only legal since *Ack* is a subclass of *Packet*. The rest of the program reflects the same logic as was illustrated in the SMALLTALK example. We enter four objects into the queue, remove them, and observe the queue and its contents.

```
class Main
        creation make
        feature
                make is
                local
                        q: FifoQueue;
                        d1,d2: Packet;
                        a1,a2: Packet;
                do
                        !!q;
                        !Data!d1.Create("first packet", 5);
                        !Data!d2.Create("second packet", 6);
                        !Ack!a1.Create;
                        !Ack!a2.Create;
                        q.enter(d1);
                        q.enter(a1);
                        q.enter(d2);
                        q.enter(a2);
                        q.list;
                        q.leave.list;
                        q.leave.list;
                        q.leave.list;
                        q.leave.list;
                end
```

Classes *PriQueue* and *QueuePri* can be defined as follows. Both classes inherit from *FifoQueue*, one redefining feature *enter* and the other *leave*.

```
class PriQueue
        inherit
                FifoQueue redefine enter
        end
        feature
                enter (p: Packet) is
```

```
                        local
                                tmp,other: Node;
                        do
                                -- code similar in logic to enter method
                                -- of Smalltalk class PriQueue
                        end;
        end

    class QueuePri
            inherit
                    FifoQueue redefine leave
            end
            feature
                    leave: Packet is
                    local
                            it,nil,p: Node;
                    do
                                -- code similar to leave method
                                -- of Smalltalk class QueuePri
                    end;
        end
```

In summary, EIFFEL is a complete and truly object-oriented programming language. It supports a very open and flexible style of encapsulation. All features are private unless explicitly exported, and subclasses have access to inherited features. EIFFEL supports multiple inheritance, where a class can have more than one superclass. EIFFEL also provides garbage collection as a means to reclaim obsolete objects.

EIFFEL supports additional features which are not strictly part of object orientation: it allows generic classes where classes can be parameterized and are instantiated when needed. EIFFEL also supports more formal support for abstract data types: for example it allows one to specify pre- and postconditions to methods, and class invariants, i.e., statements that are true for all instances of a class.

C++

Our third OOP language is C++. C++ is the evolutionary enhancement to the C language, originally developed at Bell Laboratories, to support OOP features. It is most accurately described as a *hybrid* language (i.e., one that extends a traditional procedural language, and remains compatible with that language, to include a variety of OOP features). Basically, this is C with classes, inheritance, and polymorphism.

Typically in C++, a class is separated into a header specification and a body. Next, we list the header for class Packet. The keyword *public* indicates that the instance methods, now called member functions in C++ terminology, are publicly visible. The instance variables (or, as referred to in C++, *member fields*) are *private*. That is, they are not visible outside of this class.

```
    class Packet {
            int priority;
            char *name;
    public:
            Packet(char *n, int p){
```

```
                    name=n;
                    priority=p;
        }
        virtual void list();
        int operator>(Packet &);
        int operator<=(Packet &);
};
```

The class header declares the member fields and member functions along with their accessibility. *Priority* is defined as an integer number (*int*), name is a string (*char* *); the comparisons are defined as operator functions (the & denotes call by reference). The header shows one other important member function, it has the same name as the class and is referred to as the **constructor**. The constructor serves to initialize a new instance of the class when it is created. Here, a new packet needs a name and a priority.

The keyword *virtual* in the declaration of member function *list* enables **dynamic binding** in C++: it is needed to allow inclusion polymorphism to work. Without it, C++ will use static binding as in conventional programming languages.

The class body elaborates the bodies of the member functions:

```
int Packet::operator>(Packet & other) {
        return priority > other.priority;
}

int Packet::operator<=(Packet & other) {
        return priority <= other.priority;
}

void Packet::list() {
        cout << name << " packet: ";
}
```

where *cout* is the predefined output object in C++: it receives message << and prints its parameters.

C++ differs in its support of encapsulation from most other object-oriented programming languages: *private* here does not mean that only a single object has access to its private data, but rather that it is private to the class. All other instances of class Packet have access to each others' *priority* and *name* fields. The member functions that implement the comparison capabilities take advantage of that fact.

Subclass *Ack* is very simple,

```
class Ack: public Packet {
public:
        Ack():Packet("Ack",10){}
        void list() {
                cout << "acknowledged\n";
        }
};
```

Its constructor uses a special syntax to invoke the constructor of its superclass *Packet*. This class header also shows that a member function can (if it is short enough) be fully defined right away.

Similarly, for class *Data*:

```
class Data: public Packet {
        char *body;
        int length;
public:
        Data(char *b, int p = 5):Packet("Data", p){
                length = strlen(b) + 1;
                body = new char [length];
                strcpy(body, b);
        }
        void list();
};

void Data::list() {
        Packet::list();
        cout << body << endl;
}
```

The constructor for class Data is defined with two parameters, one to initialize the *body* field, and one for the *priority*, which will default to 5 if no value is provided in the call. The member function *list* refers to *Packet::list()* which is the *list* function defined in the Packet class: again, SMALLTALK uses a simpler and cleaner approach by using the keyword *super* to refer to a method defined in a superclass. The word *endl* ensures that the output ends with a new line.

The next class *FifoQueue* contains and hides the definition of class *Node* (like EIFFEL, C++ would also support parameterized generic classes). The member fields *head* and *tail* are specified as *protected*. C++ allows one to control explicitly how subclasses have access to inherited members. Protected members are accessible in subclasses, whereas private members are not.

```
class FifoQueue {
protected:
        struct Node {
                Packet & value;
                Node *next, *previous;
                Node(Packet &p):value(p), next(nil),
                        previous(nil){};
        } *head, *tail;
public:
        FifoQueue():head(nil),tail(nil){};
        virtual void enter(Packet &);
        virtual Packet & leave();
        void list();
};
```

Here, the constructor *Node* uses a shortcut to initialize the *value*, *next*, and *previous* member fields; therefore, the body of the constructor can remain empty.

The body of class FifoQueue details the member functions *enter*, *leave*, and *list*:

```
void FifoQueue::enter(Packet &it) {
        Node *tmp = new Node(it);
        tmp->previous = tail;
        tail = tmp;
        if (head)
                tmp->previous->next = tail;
        else
                head = tmp;
}


Packet& FifoQueue::leave() {
        Packet &it = head->value;
        if (head == tail)
                head = tail = nil;
        else if (head->next) {
                head = head->next;
                head->previous = nil;
        }
        return it;
}


void FifoQueue::list() {
        for (Node* tmp = head; tmp; tmp=tmp->next)
                tmp->value.list();
}
```

And finally, we are now able to exercise our objects. This program produces the same output as our previous examples in SMALLTALK and EIFFEL.

```
main(){
        FifoQueue q;
        Data w1("first packet");
        Ack c1, c2;
        Data w2("second packet",6);

        q.enter(w1);
        q.enter(c1);
        q.enter(w2);
        q.enter(c2);

        q.list();

        q.leave().list();
```

```
            q.leave().list();
            q.leave().list();
            q.leave().list();
    }
```

And again, we can use FifoQueue as the base class for subclass PriQueue to refine the *enter* member function,

```
    class PriQueue: public FifoQueue {
    public:
            void enter(Packet &) {

            // logic as before
            }
    };
```

And, of course, a similar class QueuePri is also possible, as follows:

```
    class QueuePri: public FifoQueue {
    public:
            Packet & leave() {

                    // logic as before
            }
    };
```

In summary, C++ (like EIFFEL) provides detailed support for specifying the degree of access to its members. C++ goes even beyond what we illustrated here. It allows one to specify the type of inheritance that is used: public, protected, or private inheritance. All our examples use public inheritance, which propagates the accessibility of members to the subclass. Protected and private inheritance allow one to hide the fact that a class is based on a superclass. C++ supports both single and multiple inheritance. It requires that dynamic binding, i.e., the object-oriented behavior of an object to search for a suitable method for a message at run time, be explicitly requested per member function. C++ uses the keyword *virtual* to request dynamic binding, otherwise it defaults to static binding. C++ leaves memory management to the programmer, as garbage collection is not supported.

96.4 Language Implementation Issues

In evolutionary terms, object-oriented programming languages can be described as either pure or hybrid. The hybrid approach adds object-oriented features on top of a procedural core: C++ and Object-Pascal are representatives of this approach. The pure approach, SMALLTALK and EIFFEL, limits the language features to those that are strictly object oriented. In theoretical terms, the primary distinction between these kinds of languages is seen to lie in their interpretation of the concept of type.

Because of their origins, hybrid languages rely on a traditional approach to typing. That is, the type of something resides with its container. Thus, when we declare an integer variable *i* in C++, for example, as *int i;*, variable *i* is of type integer. In a pure OOP language, such as SMALLTALK or EIFFEL, variables do not exclusively carry the type of what they contain. Rather, the command *Integer new* returns an integer object. That is, the type in this case is seen to reside with the object, and not solely with whatever variable this new object comes to be associated with.

The implications of this to language implementation of this different interpretation of type are significant. For pure OOP languages, the language implementor must simply ensure that every object knows its class (and, of course, somehow its superclasses). This can be implemented either by encoding into classes knowledge of (pointers to) their superclasses, or by allowing classes to directly contain all information that they inherit.

Implementors of hybrid languages must provide support for both notions of type: variables and objects can have type. In C++, object type information is added either via a *vtable* per object (which contains pointers to all functions that apply to the object) or, in newer implementations, by using run time type information (RTTI) extensions.

A second critical distinction between OOP languages is in how they handle the binding of messages to methods. In languages such as SMALLTALK that perform this binding dynamically (at run time), an object receives a message and the search for an appropriate method begins at the class of the object. The search continues through superclasses until the message can be processed. Although this approach affords the programmer tremendous flexibility, it has clear practical downsides.

A more efficient (and increasingly common) approach is to leave the binding choice to the system, i.e., the compiler and linker. That is, the system tries to determine at compile time which method should be invoked for each message sent. In cases where inclusion polymorphism is used (and it may be unclear as to which class to refer to), binding can be performed at run time using techniques varying from simple case statements to a more complex system of virtual method tables. In any case, it is still up to the compiler to detect and to indicate the need for run time binding.

Another important issue to consider in the context of language implementation is the approach one adopts to memory management. OO languages are relatively uniform in their approaches to memory allocation (object creation). Creating objects and all that that entails (determining how much memory to allocate, and the types of member fields, etc.) is performed by the system. Initialization, on the other hand, is left to the programmer. Many languages provide direct support for initializing objects (we have constructors in C++, creation routines in EIFFEL, and initialize methods in SMALLTALK).

There are two common approaches to deleting objects from memory (deallocation). In the programmer-controlled approach, the programmer describes in the class what should happen to objects when they are deleted (destructor). The problem here is that the programmer now needs to know all possible circumstances in which objects of the class will be (ever) used. In practice this has been shown to be a problematic approach. Memory leaks, i.e., memory that is occupied by deleted objects, can occur easily in large bodies of C++ code.

The system-enabled approach to deleting objects uses the concept of garbage collection. All objects that are unknown to other objects are *garbage*. The system needs a way of detecting that fact: reference counters, memory mirroring, mark and sweep, etc., are examples of algorithms that enable it. In principle, the system sweeps all objects constantly to determine which are reclaimable. (Practically speaking, this is quite compute intensive.) To lighten the impact on system performance, garbage collection is typically done either during times of idling or when some threshold of memory usage is reached.

96.5 Research Issues

Object-oriented programming languages have matured dramatically and have gained a significant degree of acceptance over their almost 30 years of history. Perhaps the most dramatic test of the paradigm will be in how it responds to demands imposed on it by the coming advances in computing hardware, architectures, and operating systems. Currently the execution model for the OOP paradigm is strictly sequential and synchronous. Computation starts at one object, which sends a message to the next, and so on. Languages such as Distributed SMALLTALK and JAVA, and architectures such as the common object request broker architecture (CORBA) attempt to go beyond these constraints, pushing OOP into the worlds of distributed and parallel computing.

Distributed SMALLTALK allows objects to exist on more than one site. It also allows messages to be sent across system boundaries. If we are allowed to relax the synchronization constraint, we can conceive of an object sending a message without receiving an immediate response. Then, objects could compute in parallel. Indeed, the paradigm seems naturally to support these notions.

CORBA enables communication among distributed objects, where these objects do not need to be from a single programming language. C++ objects can communicate with SMALLTALK objects. The basic idea is to set up an object request broker that relays messages from a sender to a receiver object. Implementations of distributed SMALLTALK and CORBA are now emerging.

Perhaps the most compelling new entry into the landscape of object-oriented programming languages is JAVA, developed at Sun Microsystems [SUN 1995]. Although its popularity stems from its ability to create highly interactive World Wide Web content, we discuss it here because it is a truly object-oriented programming language.

From an object-oriented perspective, JAVA can be seen as a distillation of many of the good features from SMALLTALK and C++. From C++ it inherits its style of syntax, but with great simplifications: JAVA does not support pointers, no typecasts, only single inheritance, no class templates, no implicit type conversions defined by constructors, and no destructors. Method overloading is supported, but not operator overloading. Emphasis is placed on the readability and understandability of the source code.

From SMALLTALK it inherits its execution model: all objects carry a unique identifier (reference), all methods are dynamically bound (or virtual, in C++ terms), it is compiled into byte codes that are interpreted within the target environment, and all class information is available at run time, which provides additional type-checking capability and robustness. JAVA also provides automatic garbage collection, which simplifies a programmer's task significantly and tends to reduce many errors that are related to memory management. And of course, JAVA has builtin support for networking, such as being aware of Internet protocols (TCP/IP, http, ftp, etc.), and security.

Consider this JAVA version of our Packet class:

```java
class Packet {
        int priority;
        String name;
        Packet(String n, int p) {
                name = n;
                priority = p;
        }
        public void list() {
                System.out.println(name + " packet: ");
        }
        public boolean greater(Packet other) {
                return priority > other.priority;
        }
}
```

The resemblance to C++ is clear: most of the basic syntax, including declarations and control structures, use the C++ style. Missing are pointers and arrays that are based on pointers. JAVA supports actual arrays, but it also features a builtin String class. All object handling is done by reference. Access specifiers (like private or public) are listed per field or function. JAVA comes with a significant set of predefined classes to allow input and output and, of course, to allow one to build *applets* which are JAVA programs that can run within a Web browser that supports a JAVA interpreter.

JAVA insists on a closed-class hierarchy: all classes must have a superclass. If a superclass is not specified in the class declaration then it defaults to class *Object*. In effect, all JAVA objects can be thought of as instances of that class (much as in SMALLTALK). This enables broad run time support, such as automatic garbage collection.

JAVA also supports the notion of *interface*. An interface is simply a specification of the public methods that an object can respond to. The interface does not include any member fields of method bodies. Multiple inheritance is supported for interfaces. Classes can be declared to *implement* interfaces. Interfaces allow the programmer to establish declared relationships between modules, which in turn can change their underlying implementation as the class changes.

In summary, many of the original motivations for the OOP paradigm have been justified by our practical experience with it. OOP has been seen to (1) embody useful analysis and design techniques; (2) revise the traditional software life cycle toward analysis and design, instead of coding, testing, and debugging; (3) encourage software reuse through the development of useful code libraries of related classes; (4) improve the workability of a system so that it is easier to debug, modify, and extend; and (5) appeal to human instincts in problem solving and description, in particular to problems which model real-world phenomena.

Defining Terms

Abstract class: A class that has no direct instances, but is used as a base class from which subclasses are derived. These subclasses will add to its structure and behavior, typically by providing implementations for the methods described in the abstract class.

Class: A description of the data and behavior common to a collection of objects. Objects are instances of classes.

Constructor: An operation associated with a class that creates and/or initializes new instances of the class.

Dynamic binding: Binding performed at run time. In OOP this typically refers to the associating of a particular class with a name, so that the method to be invoked in response to a message can be determined by the class to which it belongs at run time.

Inheritance: A relationship among classes, wherein one class shares the structure or behavior defined in an is-a hierarchy. Subclasses are said to inherit both the data and methods from one or more generalized superclasses; the subclass typically specializes its superclasses by adding to its state data and by redefining its behavior.

Instance: A specific example that conforms to a description of a class. An instance of a class is an object.

Instance variable: The data items that are associated with (and are local to) each instance of a class.

Member function (or **method**): A procedure or function that is defined as part of a class and is invoked in a message passing style. Every instance of a class exhibits the behavior described by a member function/method of the class.

Message: A means for invoking a subprogram or behavior associated with an object.

Object: An object is an instance of a class described by its state, behavior, and identity.

Object-oriented programming: A method of implementation in which a program is described as a sequence of messages to cooperating collections of objects, each of which represents an instance of some class. Classes can be related through inheritance and objects can exhibit polymorphic behavior.

Override: The action that occurs when a method in a subclass with the same name as a method in a superclass takes precedence over the method in the superclass.

Polymorphism (or, **many shapes**): That feature of a variable that can take on values of several different types or a feature of a function that can be executed using arguments of a variety of types.

Subclass (or, **derived class**): A class that inherits variables and methods from another class (called the superclass).

Virtual function (or, **deferred method**): Most generally, a method of a class that must be defined by subclasses to the class. In languages in which dynamic binding is not the default, may also mean that a function is subject to dynamic binding.

References

Agha, G. and Hewitt, C. 1987. Actors: a conceptual foundation for concurrent object-oriented programming. In *Research Directions in Object-Oriented Programming*. B. Schriver and P. Wegner, eds. MIT Press, Cambridge, MA.

Arnold, P., Bodoff, S., Coleman, D., Gilchrist, H., and Hayes, F. 1991. An evaluation of five object-oriented development methods. Hewlett-Packard Laboratories, Bristol, England.

Budd, T. 1991. *An Introduction to Object-Oriented Programming.* Addison–Wesley, Reading, MA.

Cardelli, L. and Wegner, P. 1985. On understanding types, data abstraction, and polymorphism. *ACM Comput. Surv.* 17(4).

Chin, R. and Chanson, S. 1991. Distributed object-based programming systems. *ACM Comput. Surv.* 23(1).

Collins, W. 1992. *Data Structures: An Object-Oriented Approach.* Addison–Wesley, Reading, MA.

Coad, P. and Nocola, J. 1993. *Object-Oriented Programming.* Yourdon Press, Englewood Cliffs, NJ.

Cox, B. 1986. *Object-Oriented Programming: An Evolutionary Approach.* Addison–Wesley, Reading, MA.

Ellis, M. A. and Stroustrup, B. 1984. *The Annotated C++ Reference Manual.* Addison–Wesley, Reading, MA.

Goldberg, A. and Robson, D. 1983. *Smalltalk-80: The Language and Its Implementation.* Addison–Wesley, Reading, MA.

Ingalls, D. 1981. Design principles behind Smalltalk. *Byte* 6(8).

Korson, T. and McGregor, J. 1990. Understanding object-oriented: a unifying paradigm. *Commun. ACM* 33(9).

Lippman, S. 1991. *C++ Primer,* 2nd ed. Addison–Wesley, Reading, MA.

Liskov, B., Snyder, A., Atkinson, R., and Schaffert, C. 1980. Abstraction mechanisms in CLU. In *Programming Logic Design.* A. Wasserman, ed. Computer Society Press, New York.

Madsen, O. 1987. Block structure and object-oriented languages. In *Research Directions in Object-Oriented Programming.* B. Schriver and P. Wegner, eds. MIT Press, Cambridge, MA.

Masini, G., Napoli, A., Colnet, D., Leonard, D., and Tompre, K. 1991. *Object-Oriented Languages.* Academic Press, London, England.

Meyer, B. 1987. Eiffel: programming for reusability and extendability. *SIGPLAN Notices* 22(2).

Meyer, B. 1988. *Object-Oriented Software Construction.* Prentice–Hall, Englewood Cliffs, NJ.

Nygaard, K. and Dahl, O. J. 1981. The Development of the Simula languages. In *History of Programming Languages.* R. Wexelblat, ed. Academic Press, New York.

Peterson, G. 1987. *Tutorial: Object-Oriented Computing,* Vols. 1 and 2. IEEE Computer Society Press.

Pinson, L. and Wiener, R. 1988. *An Introduction to Object-Oriented Programming and Smalltalk.* Addison–Wesley, Reading, MA.

Pinson, L. and Wiener, R. 1990. *Applications of Object-Oriented Programming.* Addison–Wesley, Reading, MA.

Rubin, K. and Goldberg, A. 1992. Object behavior analysis. *Commun. ACM* 35(9).

Schriver, B. and Wegner, P., ed. 1987. *Research Directions in Object-Oriented Programming.* MIT Press, Cambridge, MA.

Shaw, M. 1981. *ALPHARD: Form and Content.* Springer–Verlag, New York.

Snyder, A. 1985. Object-Oriented Programming for Common Lisp. *Rep.* ATC-85-1. Hewlett-Packard. Palo Alto, CA.

Snyder, A. 1993. The essence of objects: concepts and terms. *IEEE Software* 10(1).

Stefik, M. and Bobrow, D. 1986. Object-oriented programming: themes and variations. *Artif. Intell. Mag.* 6(4).

Stroustrup, B. 1986. *The C++ Programming Language.* Addison–Wesley, Reading, MA.

Sun. 1995. The JAVA™ Language: An Overview, Sun Microsystems, http://www.javasoft.com/doc/Overviews/java/.

Wiener, R. 1995. *Software Development Using Eiffel.* Prentice–Hall, Englewood Cliffs, NJ.

Williams, G. 1989. Designing the future: the power of object-oriented programming. *Am. Programmer* 2(7–8).

Wirfs-Brock, R., Wilkerson, B., and Wiener, L. 1990. *Designing Object-Oriented Software.* Prentice–Hall, Englewood Cliffs, NJ.

Further Information

That object-oriented programming has become one of the predominant paradigms is evidenced by the multitude of conferences, journals, text, and World Wide Web sites that are devoted to both general and language-specific topics.

The two most prominent conferences, both of which address a wide range of OOP issues, are the Conference on Object-Oriented Programming Systems, Languages, and Applications (OOPSLA) and the European Conference on Object-Oriented Programming (ECOOP). OOPSLA proceedings are published annually by the ACM as special issues of its *SIGPLAN Notices*. ECOOP proceedings are published in book form by Springer–Verlag.

The Journal of Object-Oriented Programming provides contemporary coverage of OOP languages, applications, and research. There are many other conferences (such as the USENIX C++ Conference), journals (*Eiffel World*), and WWW sites (The Smalltalk Report) devoted to particular languages.

Perhaps the most general of the references listed are Budd [1991], Cox [1986], Peterson [1987], and Stefik and Bobrow [1986].

97

Logic Programming and Constraint Logic Programming

Jacques Cohen
Brandeis University

97.1 Overview

Logic programming (LP) is a language paradigm based on logic. Its constructs are Boolean *implications* (e.g., *q implies p* meaning that *p* is true if *q* is true), compositions using the Boolean operators *and* (called conjunctions) and *or* (called disjunctions). LP can also be viewed as a procedural language in which the procedures are actually Boolean functions, the result of a program always being either true or false. In the case of implications a major restriction applies: when *q* implies *p*, written *p :- q*, then *q* can consist of conjunctions but *p* has to be a singleton, representing the (sole) Boolean function being defined. The Boolean operator *not* is disallowed but there is a similar construct that may be used in certain cases.

The Boolean functions in LP may contain parameters, and the parameter matching mechanism is called **unification**. This type of general pattern matching implies, for example, that a variable representing a formal parameter may be bound to another variable, or even to a complex data structure, representing an

0-8493-2909-4/97/$0.00+$.50
© 1997 by CRC Press, Inc.

actual parameter (and vice versa). When an LP program yields a *yes* answer, the bindings of the variables are displayed indicating that those bindings make the program logically correct and provide a solution to the problem expressed as a logic program.

An important recent extension of LP is **constraint LP** (CLP). In this extension unification can be replaced or complemented by other forms of constraints depending on the domains of the variables involved. For example, in CLP a relationship such as $X > Y$ can be expressed even in the case where X and Y are unbound real variables. As in LP, a CLP program yields answers expressing that the resulting constraints (e.g., $Z < Y + 4$) must be satisfied for the program to be logically correct.

This chapter includes sections describing the main aspects of LP and CLP. It includes examples, historical remarks, theoretical foundations, implementation techniques, metalevel interpretation, and concludes with the most recent proposed extensions to this language paradigm.

97.2 An Introductory Example

Let us consider a simple program written in PROLOG, the main representative among logic programming languages. The program's objective is to check if a date given by the three parameters: *Month* (a string), *Day*, and *Year* (integers) is valid. For example, *date (oct, 15, 1996)* is valid whereas *date (june, 31, 1921)* is not. The program can be expressed as

> date (Month, Day, Year) :- member (Month, [jan, march, may, july, aug, oct, dec]),
>
> > comprised (Day, 1, 31).
>
> date (Month, Day, Year) :- member (Month, [april, june, oct, sept, nov]),
>
> > comprised (Day, 1, 30).
>
> date (feb, Day, Year) :- leap (Year) , comprised (Day, 1, 29).
>
> date (feb, Day, Year) :- comprised (Day , 1, 28).
>
> comprised (Day, Start, End) :- Day >= Start,   Day =< End.
>
> leap (Year) :- (Year / 4) × 4 = Year.

The variables in the preceding program start with a capital letter, e.g., *Month*; in this simple example the constants either start with small case letters (strings) or are integers. The symbol :- can be read as is defined by, and it separates the procedure heading from its body, which consists of calls to other procedures. More precisely, it is convenient to view every procedure as a Boolean function yielding *true* or *false*. The body of such a function is made up of conjunctions of calls to functions. Therefore, a comma separating the calls in the right-hand side correspond to the Boolean operator *and*.

The following remarks also apply to the program. There are multiple definitions of *date*, each of which tests for the corresponding month and the compatible number of days. Therefore, the program is non-deterministic in the sense that several possibilities have to be considered either one at a time, or in parallel. In many cases there might be several solutions to a given program. Also note that the right-hand side of a rule could be empty. In that case the rule indicates that the left-hand side is always true.

Let us for the time being assume that *member (Element, List)* is a builtin function testing if an *Element* is a member of a *List*. The Boolean function *leap* tests if its parameter is a multiple of 4. One could also add a function *year (Year)* which would test if an integer defined by *Year* is positive. Remark that the ordering of the statements in this particular version of the program is important, since a leap year is tested before a nonleap year. Ideally, pure PROLOG programs should be declarative; from a syntactic point of view that means that the order of the rules and the order of the calls in the right-hand sides should not matter. A new version of the program satisfying this criteria is one in which the Boolean function *nonleap (Year) :- (Year / 4) × 4 /= Year* is called in the definition applicable to the month of February, in the case of a nonleap year.

The preceding program is used with queries (equivalent to main programs) to test the validity of given dates. The answers are either *yes* or *no*. Note that in PROLOG one is allowed to have queries containing variables. In the present example, the query *date (X, 31, 1996)* yields as results the successive strings that the variable X is bound to for the months that have 31 days. More specifically, the results are

yes $X = jan$, More?

yes $X = march$, More?

. . .

Yes $X = dec$, More?

No.

Similarly, the query *date (X, 29, 1996)* yields all of the months in a year. Note that the builtin functions such as $=<$ and $>=$ require that their arguments be ground; that is, they cannot contain unbound variables. This is satisfied by the query *date (X, 31, 1996)*. However, a call *date (feb, Y, 1996)* entails a problem which provides an insight on the desirability of extending PROLOG to handle constraints such as it is done in PROLOG IV. The last query fails in standard PROLOG because the variable Y remains unbound. The right answer as provided by a constraint LP processor consists of the two constraints:

Yes $Y >= 1, Y =< 29$, More?

No.

Similarly, the correct answer to the query *date (feb, Y, Z)* submitted to a CLP processor should yield two pairs of constraints corresponding to leap and nonleap years represented by Z. These results are obtained by keeping lists of satisfiable constraints and triggering a failure when the constraints become unsatisfiable. This failure may entail exploring the remaining nondeterministic possibilities.

It is now appropriate to present the recursive Boolean function *member*. It has two parameters: an *Element* and a *List*; the function succeeds if the *Element* appears in the *List*. Lists are represented using the constructor *cons* as in functional languages. This means that a list *[a, b]* can be represented by

$$cons \ (a, \ cons \ (b, \ nil))$$

Note that the *cons* in the preceding representation is simply any user-selected identifier describing a data structure (or record) consisting of two fields: (1) a string and (2) a pointer to another such record or to *nil*. In the case of lists there are special builtin features such as *[a, b]* simplifying their description. The program for *member* consists of two rules:

member (X, cons (X, T)) .

member (X, cons (Y, T)) :- X /= Y, member (X, T).

The first rule states that if the head of a list contains the element X, then *member* succeeds. The second rule uses recursion to inspect the remaining elements of the list. Note that the query *member (X, cons (a, cons(b, nil))*, equivalent to *member (X, [a, b])*, provides two solutions, namely, $X = a$ or $X = b$. It should be obvious that *member (Y, [a, a])* also yields two (identical) solutions.

Finally, notice that *member* can search for more complex data structures. For example: *member (member (a, Y) , Z)* succeeds if the list Z contains an element such as *member (a, U)*; if so, the variable Y is bound to the variable U. The reader should consider the embedded term *member* as the identifier of a record or data structure having two fields. In contrast, the first identifier *member* represents a Boolean function having two parameters. This example illustrates that, in PROLOG, program and data have the same form.

An inquisitive reader will remark that *member* can also be used to place elements in a list if they are not already present in that list. A clue in understanding this property is that the query *member (a, Z)* will bind

Z to a record *cons(a, W)* in which *W* is a new unbound variable created by the PROLOG processor when applying the first rule defining the Boolean function *member.*

97.3 Features of Logic Programming Languages

Summarized next are some of the features whose combination render PROLOG unique among languages:

1. Procedures may contain parameters that are both input and output.
2. Procedures may return results containing unbound variables.
3. **Backtracking** is built in, therefore allowing the determination of multiple solutions to a given problem.
4. General pattern matching capabilities operate in conjunction with a goal-seeking search mechanism.
5. Program and data are presented in similar forms.

The preceding listing of the features of PROLOG does not fully convey the subjective advantages of the language. There are at least three such advantages:

1. Having its foundations in logic, PROLOG encourages the programmer to describe problems in a logical manner that facilitates checking for correctness and, consequently, reduces the debugging effort.
2. The algorithms needed to interpret PROLOG programs are particularly amenable to parallel processing.
3. The conciseness of PROLOG programs, with the resulting decrease in development time, makes it an ideal language for prototyping.

Another important characteristic of PROLOG that deserves extension, and is now being extended, is the ability to postpone variable bindings as much as is deemed necessary (lazy evaluation). Failure and backtracking are triggered only when the interpreter is confronted with a logically unsatisfiable set of constraints. In this respect, PROLOG's notion of variables approaches that used in mathematics.

The price to be paid for the advantages offered by the language amounts to the increasing demands for larger memories and faster central processing units (CPUs). The history of programming language evolution has demonstrated that, with the consistent trend toward less expensive and faster computers with larger memories, this price becomes not only acceptable but also advantageous since the savings achieved by program conciseness and by a reduced programming effort largely compensate for the space and execution time overheads. Furthermore, the quest for increased efficiency of PROLOG programs encourages new and important research in the areas of optimization and parallelism.

97.4 Historical Remarks

The birth of logic programming can be viewed as the confluence of two different research endeavors: one in artificial or natural language processing, and the other in automatic theorem proving. These endeavors contributed to the genesis of the PROLOG language, the principal representative of LP.

Alain Colmerauer, assisted by Philippe Roussel, is credited as the originator of PROLOG, a language that was first developed in the early 1970s and continues to be substantially extended. Colmerauer's contributions stemmed from his interest in language processing using theorem proving techniques. Robert Kowalski was also a major contributor to the development of LP. Kowalski had an interest in logic and theorem proving [Cohen 1988, Bergin 1996]. In their collaboration, Kowalski and Colmerauer became interested in problem solving and automated reasoning using resolution theorem proving.

Kowalski's main research was based on the work of Alan Robinson [1965]. Robinson had the foresight to distinguish the importance of two components in automatic theorem proving: a single inference rule called **resolution**, and the testing for equality of trees called *unification*.

Theorems to be proved using Robinson's approach are placed in a special form consisting of conjunctions of **clauses**. Clauses are disjunctions of positive or negated literals; in the case of Boolean algebra (propositional calculus) the literals correspond to variables. In the more general case of the predicate calculus, literals correspond to a potentially infinite number of Boolean variables, one for each combination of values that the literal has as parameters. A **Horn clause** is one containing (at most) one positive literal; all of the others (if any) are negated.

According to the informal description in the introductory example, a Horn clause corresponds to the definition and body of a Boolean function. The positive literal is the left-hand side of a rule (called *Head*); the negative literals appear in the right-hand side of the rule (called *Body*). A non-Horn clause is one in which (logical) negations can appear qualifying a call in the *Body*. An example of a Horn clause is

date (feb, Day, Year) :- leap (Year), comprised (Day, 1, 29).

since it is equivalent to

leap (Year) and *comprised (Day, 1, 29)* implies *date (feb, Day, Year)*

or

date (feb, Day, Year) or not *(leap (Year))* or not *(comprised (Day, 1, 29))*

If one wished to express the clause

date (feb, Day, Year) :- not leap (Year), comprised (Day, 1, 28).

in which the *not* is the logical Boolean operation, the preceding example would be logically equivalent to

date (feb, Day, Year) and *leap (Year)* or *comprised (Day, 1, 28).,*

which is not a Horn clause since it has two (positive) elements in the head (and there is no procedural equivalent to it).

Kowalski concentrated his research on reducing the search space in resolution-based theorem proving. With this purpose, he developed with Kuehner a variant of the linear resolution algorithm called SL resolution (for linear resolution with selection function [Kowalski and Kuehner 1970]). Kowalski's view is that, from the automatic theorem-proving perspective, this work paved the way for the development of PROLOG. Having this more efficient (but still general) predicate calculus theorem prover available to them, the Marseilles and Edinburgh groups started using it to experiment with problem-solving tasks. To further increase the efficiency of their prover the Marseilles group resorted to daring simplifications that would be inadmissible to logicians. These audacious attempts turned out to open new vistas for the future of LP.

Several formulations for solving a given problem were attempted. Almost invariably, the formulations that happened to be written in Horn clause form turned out to be much more natural than those that used non-Horn clauses. A case in which the Horn clause formulation was particularly effective occurred in parsing strings defined by grammar rules. Recall that context-free grammars have a single nonterminal being defined in the left-hand side of each of its rules, therefore establishing a strong similarity with Horn clauses.

The SL inference mechanism applicable to a slight variant of Horn clauses led to the present PROLOG inference mechanism: **selective linear definite (SLD)** clause resolution. The word definite refers to Horn clauses with exactly one positive literal, whereas general Horn clauses may contain entirely negative clauses. (Since the term Horn clause is more widely used than definite clause the former is often used to denote the latter.)

As a final historical note one should mention that LP gained a renewed impetus by its adoption as the language paradigm for the Japanese Fifth Generation Program. PROLOG and, in particular, its CLP extensions now count on a significant number of loyal and enthusiastic users.

97.5 Resolution and Unification

Resolution and unification appear in different guises in various algorithms used in computer science. This section first describes these two components separately and then their combination as it is used in LP. In doing so it is useful to consider first the case of the propositional calculus (Boolean algebra) in which unification is immaterial. It is well known that there exist algorithms that can always decide if a system of Boolean formulas is satisfiable or not, albeit with exponential complexity.

In terms of the informal example considered in the introduction, one can view resolution as a (non-deterministic) call of a user-defined Boolean function. Unification is the general operation of matching the formal and actual parameters of a call. Consequently, unification does not occur in the case of parameterless Boolean functions.

The **predicate calculus** includes the quantifiers $\forall$ and $\exists$; it may be viewed as a general case of the propositional calculus for which each predicate variable (a literal) can represent a potentially infinite number of Boolean variables. Unification is only used in this latter context. Theorem proving algorithms for the predicate calculus are not guaranteed to provide a yes-or-no answer since they may not terminate.

Resolution

In the propositional calculus a simple form of resolution is expressed by the inference rule

$$\text{if } a \rightarrow b \text{ and } b \rightarrow c \text{ then } a \rightarrow c, \text{ or}$$
$$(\neg a \vee b) \wedge (\neg b \vee c) \rightarrow (\neg a \vee c)$$

Recall that *a implies b* is equivalent to *not a or b*. The final disjunction $\neg a \vee c$ is called a *resolvant*. In particular, resolving $a \wedge \neg a$ implies the empty clause, i.e., falsity.

To better understand the meaning of the empty clause consider the implication $a \rightarrow a$, which is equivalent to *(not a) or a*. This expression is always true; therefore, its negation *not ((not a) or a)* equivalent to *(a and (not a))* is always false. If a Boolean expression is always *true*, its negation is always *false*. Resolution theorem proving consists of showing that if the expression is always true its negation results in contradictions of the type *(a and (not a)* which is always false. The empty clause is simply the resolvant of *(a and (not a)*.

Observe the similarity between resolution and the elimination of variables in algebra, e.g.,

$$a + b = 3 \text{ and } -b + c = 5 \text{ imply } a + c = 8$$

Another intriguing example occurs in matching a procedure definition with its call. Consider, for example,

$$\text{procedure } b; a$$

$$\ldots$$

$$\text{call } b; c$$

in which a is the body of b, and c is the code to be executed after the call of b. If one views the definition of b and its call as complementary, a (pseudo)resolution yields: $a; c$ in which concatenation is noncommutative and the resolution corresponds to replacing a procedure call by its body. Actually, the last example provides an intuitive procedural view of resolution as used in PROLOG.

In the case of pure PROLOG programs, only Horn clauses are allowed. For example, if a, b, c, d, and f are Boolean variables (literals), then

$$b \wedge c \wedge d \to a \quad \text{and} \quad f$$

are readily transformed in Horn clauses since they correspond, respectively, to

$$a \vee \neg b \vee \neg c \vee \neg d \quad \text{and} \quad f$$

where the clause f is called unit clause or a fact. The preceding example is written in PROLOG as

$$a :\text{-} b, c, d. \quad \text{and} \quad f.$$

where the symbols :- and " , " correspond to the logical connectors *only if* and *and*. They are read as: a is true only if b and c and d are true. The above conjunction also requires that f be true; equivalently $b \wedge c \wedge d \to a$ and f are true.

The resolution mechanism applicable to Horn clauses takes as input a conjunction of Horn clauses $H = h_1 \wedge h_2 \wedge \cdots \wedge h_n$, and a query Q, which is the negation of a theorem to be proved. Q consists of the negation of a conjunction of positive literals or, equivalently, a disjunction of negated literals. Therefore, a query is itself in a Horn clause form in which the head is empty.

A theorem is proved by contradiction, namely, the goal is to prove that $H \wedge Q$ is inconsistent, implying that the result of successive resolutions involving the negated literals of Q inevitably—in the case of the propositional calculus—leads to falsity, i.e., the empty clause. In other words, if H implies the nonnegated Q is always true, then H *and* the negated Q is always false.

Consider for example the query *date (oct, 15, 1996)* in our introductory example. Its negation is *not date (oct, 15, 1996)*. This resolves with the first rule yielding the bindings *Month = oct, Day = 15, Year = 1996*. The resolvant is the disjunction *not member (oct, [jan, march, may, july, aug, oct., dec])* or *not comprised (15, 1, 31)*.

Although not elaborated here, the reader can easily find out that the successive resolutions using the definition of *member* will fail since *oct* is a member of the list of months containing 31 days. Similarly, the day *15* is comprised between *1* and *31*. Therefore, the empty (falsity) clause will be reached for both disjuncts of the resolvant.

In what follows the resolution inference mechanism is applied to Horn clauses representing a PROLOG program. One concrete syntax for PROLOG rules is given by:

$$
\begin{array}{ll}
<rule> & ::= <clause> . \,|<unit\ clause> . \\
<clause> & ::= <head> :\text{-} <tail> \\
<head> & ::= <literal> \\
<tail> & ::= <literal> \{, <literal>\} \\
<unit\ clause> & ::= <literal>
\end{array}
$$

where the braces { } denote any number of repetitions (including none) of the sequence enclosed by the brackets <>. First consider the simplest case, where a *literal* is a single letter. For example, consider the following PROLOG program in which rules are numbered for future reference:

1. $a :\text{-} b, c, d.$
2. $a :\text{-} e, f.$
3. $b :\text{-} f.$
4. $e.$
5. $f.$
6. $a :\text{-} f.$

In the first rule, a is the $<head>$, and b, c, d is the $<tail>$, also called the *body*. The fourth and fifth rules are unit clauses, i.e., the body is empty. A query Q is syntactically equivalent to a $<tail>$. For example,

a, e. is a query and it corresponds to the Horn clause $\neg a \vee \neg e$. The result of querying the program is one (or multiple) *yes* or a single *no* answer indicating the success or failure of the query. In this particular example the query a, e. yields two *yes* answers. Note that a is defined by three rules; the second and the last yield the two solutions; the first fails since c is undefined. The successful sequence of the list of goals is

solution 1: $a, e \Rightarrow e, f, e \Rightarrow f, e \Rightarrow e \Rightarrow nil$;

solution 2: $a, e \Rightarrow f, e \Rightarrow e \Rightarrow nil$.

Let us follow in detail the development of the second solution. The negated query *not a or not e* resolves with the sixth rule *a or not f* yielding the resolvant *not f or not e*. Now the last expression is resolved with the f in the fifth rule yielding *not e* as the resolvant. Finally, *not e* is resolved with the e in the fourth rule yielding the empty clause, which implies the falsity of the negation of the query, using the program rules.

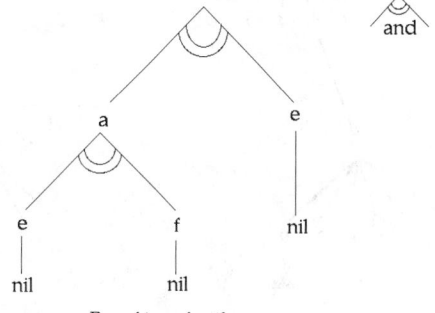

Proof tree for the query a,e

The entire search space is shown in Fig. 97.2 in the form of a tree. In nondeterministic algorithms, that tree is called the tree of choices. Its leaves are nodes representing failures or successes. The internal nodes are labeled with the list of goals that remain to be satisfied. Note that, if the tree of choices is finite, the order of the goals in the list of goals is irrelevant insofar as the presence and number of solutions are concerned. Figure 97.1 shows a proof tree for the first solution of the example. The proof tree is an *and* tree depicting how the proof has been achieved.

FIGURE 97.1   Proof tree.

There are three ways of interpreting the semantics of PROLOG rules and queries. The first is based on logic, in this particular case on Boolean algebra. The literals are Boolean variables, and the rules express formulas. The PROLOG program is viewed as the conjunction of formulas it defines. The query Q succeeds if it can be implied from the program. In a second (called procedural) interpretation of a PROLOG rule, it is assumed that a *<literal>* is a goal to be satisfied. For example, the first rule states that

goal a can be satisfied if goals b, c, and d can be satisfied.

The unit clause states that the defined goal can be satisfied. The program defines a conjunction of goals to be satisfied. The query succeeds if the goals can be satisfied using the rules of the program. Finally, a third interpretation is based on the similarity between PROLOG rules and context-free grammar rules. A PROLOG program is associated with a context-free grammar in which a *<literal>* is a nonterminal and a *<rule>* corresponds to a grammar rule in which the *<head>* rewrites into the *<tail>*; a unit clause is viewed as a grammar rule in which a nonterminal rewrites into the empty symbol ε. Under this interpretation, a query succeeds if it can be rewritten into the empty string.

Although the preceding three interpretations are all helpful in explaining the semantics of this simplified version of PROLOG, the logic interpretation is the most widely used among theoreticians, and the procedural by language designers and implementors.

The algorithms that test if a query Q can be derived from a Horn clause program P can be classified in various manners. An analogy with parsing algorithms is relevant: P corresponds to a grammar G and Q corresponds to the string (of nonterminals) to be parsed, i.e., the sequence of nonterminals that may be rewritten into the empty string using G. *Backward-chaining* theorem provers correspond to top-down parsers and are by far the preferred approach presently used in logic programming. *Forward-chaining* provers correspond to bottom-up parsers. Hybrid algorithms have also been proposed.

In a top-down algorithm the list of goals in a query is examined from left to right and the corresponding (recursive) procedures are successively called, the equivalent of resolution, until the list of goals becomes empty. Note that the algorithm is essentially nondeterministic because there are usually several choices for

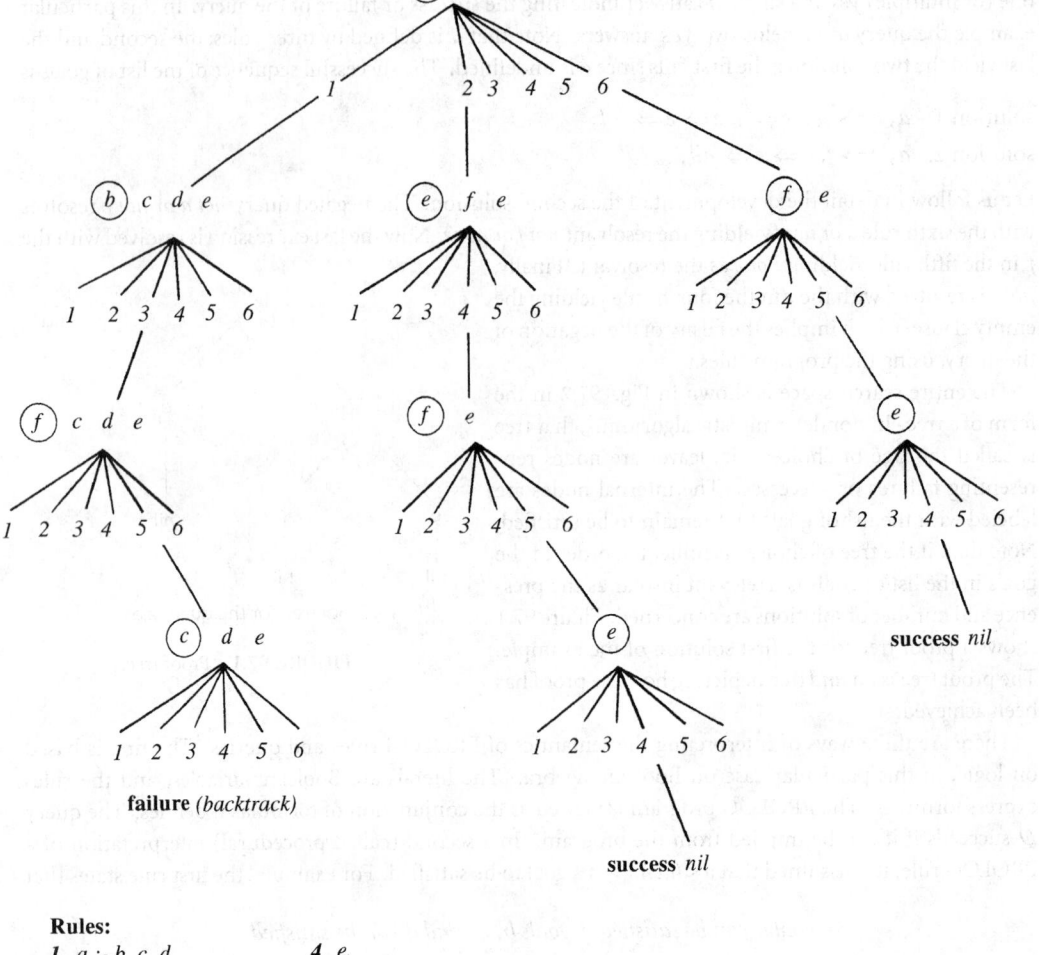

Rules:
1. *a :- b, c, d.*   *4. e.*
2. *a :- e, f.*   *5. f.*
3. *b :- f.*   *6. a :- f.*

Query: *a, e.*   ○ *denotes the head of the list of goals.*

FIGURE 97.2   Tree of choices.

the goals (see rules 1 and 2). Another nondeterministic choice occurs when selecting the (next) element of the list of goals to be considered after processing a goal.

Notice that if one had a program consisting of the sole rule *a :- a.* and the query *a*, the top-down approach would not terminate. The program states that either *a* or ¬*a* are true. Since the query does not specify *a*, it could be either true or false. A semantically correct interpreter should provide the following constraints as answers *a = true* or *a = false*. PROLOG programmers have learned to live with these unpalatable characteristics of top-down provers.

Contrary to what usually happens in parsing, bottom-up provers can be very inefficient unless the algorithm contains selectivity features that prevent the inspection of dead-end paths. In the case of database applications the bottom-up approach using magic sets has yielded interesting results [Minker 1987].

Also note the correspondence between nondeterministic grammars and nondeterministic PROLOG programs. In many applications the programs can be made deterministic and therefore more efficient.

However, ambiguous grammars which correspond to programs having multiple solutions are also useful and therefore nondeterminism has its advantages.

Unification

The operation of unification is akin to that of pattern matching. Robinson's unification algorithm brings rigor to the notion of pattern matching and it has a deep meaning: it replaces a set of fairly intricate predicate calculus axioms specifying the equality of trees by an efficient algorithm, which is easily implementable in a computer. As mentioned earlier, unification can be viewed as a very general matching of actual and formal parameters.

From the logic point of view, unification is used in theorem proving to equate terms which usually result from the elimination of existential quantifiers when placing a predicate calculus formula in clausal form. For example:

$$\forall X \exists Y \text{ such that } p(X, Y) \text{ is always true}$$

is replaced by

$$\forall X \, p(X, g(X))$$

where $g(x)$ is the Skolem function for y (sometimes referred to as an uninterpreted function symbol). The role of unification is to test the equality of literals containing Skolem functions and in so doing, bind values to variables.

Consider, for example, the statement "for all positive integer variables X there is always a variable Y representing the successor of X." The predicate p expressing this statement is $p(X, s(X)) :- integer(X)$ where $s(X)$ is the Skolem function representing Y, the successor of X. This representation is commonly used to specify positive integers from a theoretical perspective. [It is called a Peano representation of integers, e.g., $s(s(0))$ represents the integer 2.]

To show the effect of unification, the definition of a literal in the previous subsection on resolution is now generalized to encompass labeled tree structures.

\<literal\>	::=	*\<composite\>*
\<composite\>	::=	*\<functor\> (\<term\> {,\<term\>})	\<functor\>*
\<functor\>	::=	*\<lower case identifiers\>*
\<term\>	::=	*\<constant\>	\<variable\>	\<composite\>*
\<constant\>	::=	*\<integers and lower case identifiers\>*
\<variable\>	::=	*\<identifiers starting with an upper case letter or _ \>*

Examples of terms are *constant, Var,* 319, *line (point* $(X, 3)$, *point* $(4, 3))$. It is usual to refer to a single rule in a PROLOG program as a clause. A PROLOG procedure (or predicate) is defined as the set of rules whose head has the same *\<functor\>* and arity. Unification tests whether two terms $T1$ and $T2$ can be matched by binding some of the variables in $T1$ and $T2$. The simplest algorithm uses the rules summarized in Table 97.1 to match the terms. If both terms are composite and have the same *\<functor\>*, it recurses on their components. This algorithm only binds variables if it is *absolutely necessary*, so there may be variables that remain unbound. This property is referred to as the most general unifier (mgu).

One can write a recursive function *unify* which, given two terms, tests for the result of unification using the contents of Table 97.1. The unification of the two terms $f(X, g(Y), T)$ and $f(f(a, b), g(g(a, c), Z))$ succeeds with the following bindings $X = f(a, b)$, $Y = g(a, c)$, and $T = Z$. Note that, if one had $g(Y, c)$ instead of $g(a, c)$ in the second term, the binding would have been $Y = g(Y, c)$.

TABLE 97.1

Terms 1↓, 2→	\<constant\> C2	\<variable\> X2	\<composite\> T2
\<constant\> C1	*succeed if* C1 = C2	*succeed with* X2 := C1	*fail*
\<variable\> X1	*succeed with* X1 := C2	*succeed with* X1 := X2	*succeed with* X1 := T2
\<composite\> T1	*fail*	*succeed with* X2 := T1	*succeed if* (1) *T1 and T2 have the same functor and arity* (2) *the matching of corresponding children succeeds*

PROLOG interpreters would usually carry out this circular binding, but would soon get into difficulties since most implementations of the described unification algorithm cannot handle circular structures (called **infinite trees**). Manipulating these structures (e.g., printing, copying) would result in an infinite loop unless the so-called **occur check** were incorporated to test for circularity. This is an expensive test: unification is linear with the size of the terms unified, and incorporation of the occur check renders unification quadratic.

There are versions of the unification algorithm that use a representation called *solved form*. Essentially new variables are created whenever necessary to define terms, e.g., the system:

$$X1 = f(X2, g(X3, X2)) \qquad X5 = g(X4, X2)$$

A solved form is immediately known to be satisfiable since $X2$, $X3$, and $X4$ can be bound to any terms formed by using the function symbols f and g and again replacing (ad infinitum) their variable arguments by the functions f and g. This is called an element of the **Herbrand universe** for the given set of terms. The solved form version of the unification algorithm is presented by Lassez in Minker [1987].

To further clarify the notion of solved forms, consider the equation in the domain of reals specified by the single constraint $X + Y = 5$. This is equivalent to $X = 5 - Y$, which is in solved form and satisfiable for any real value of Y. Basically a constraint in solved form contains definitions in terms of free variables, i.e., those that are not constrained. This form is very useful when unification is extended to be applicable to domains other than trees. For example, linear equations may be expressed in solved form.

Combining Resolution and Unification

Consider now general clauses in which the literals contain arguments which are represented by terms. The result of resolving

$$p(\ldots) \quad \lor \quad r(\ldots)$$

with

$$\neg p(\ldots) \quad \lor \quad s(\ldots)$$

is

$$r(\ldots) \quad \lor \quad s(\ldots)$$

if and only if the arguments of p and $\neg p$ are unifiable and the resulting variables are substituted in the corresponding arguments of r and s.

Using programming language terminology, the combination of resolution and unification corresponds to a Boolean function call in which a variable in the head of a rule is matched (i.e., unified) with its corresponding actual parameter; once this is done the value to which the variable is bound replaces the instances of that variable appearing in the body. Note that a variable can be bound to another variable or to a complex term. Remark also that when unification fails, another head of a rule has to be tried and so on. For example, consider the query $p(a)$ and the program:

$$p(b).$$
$$p(X):-q(X).$$
$$q(a).$$

The first unification of the query $p(a)$ with $p(b)$ fails. The second attempt binds X to a starting a search for $q(a)$ which succeeds. Had the query been $p(Y)$, the two solutions would be $Y = b$ or $Y = a$. The first solution is a consequence of the unit clause $p(b)$, the second follows from binding Y to a (new) variable X in the second clause. A search is then made for $q(X)$, which succeeds with the last clause, therefore binding both Y and X to a.

From a logical point of view, when the result of a resolution is the empty clause (corresponding to the head of a rule with an empty body) and no more literals remain to be examined, the list of bindings is presented as the answer, i.e., the bindings (**constraints**) that result from proving that the query is deductible from the program.

97.6 Procedural Interpretation: Examples

It is worthwhile to first understand the importance of the procedure *append.* as used in symbolic functional languages. In PROLOG the word procedure is a commonly used misnomer since it always corresponds to a Boolean function, as mentioned in the introductory example; *append* is often used to concatenate two lists. Since one has to avoid destructive assignments which have very complex semantic meaning, *append* resorts to copying the first list to be appended and making its last element point to the second list. This preserves the original first list. It should be kept in mind that the avoidance of a destructive assignment forces one to produce results which are *constructed* from a given data using a LISP-like *cons*. Notice that LP also avoids destructive assignment by having variables bound only once during unification.

This section first demonstrates the transformation of a functional specification of *append* into its PROLOG counterpart corresponding to the previous clauses. Examples of inventive usage of *append* are then presented to further illustrate the procedural interpretation of logic programs. Consider the LISP-like function *append* that concatenates two lists, *L1* and *L2*:

function *append (L1, L2 : pLIST): pLIST*;
    if *L1 = nil* then *append := L2*
            else *append := cons (head (L1), append (tail (L1), L2))*;

In the preceding display *L1* and *L2* are pointers to lists; a list is a record containing the two fields *head* and *tail*, which themselves contain pointers to other lists or to atoms. The tail must point to a list or to the special atom *nil*. The constructor *cons (H, T)* creates the list whose *head* and *tail* are, respectively, *H* and *T*.

The function *append* can be rewritten into a procedure having an explicit third parameter *L3* that will contain the desired result. The local variable *T* is used to store intermediate results.

procedure *append (L1, L2: pLIST:* var *L3: pLIST)*;
  begin local *T: pLIST*;
  if *L1 = nil* then *L3 := L2*
        else begin *append (tail (L1), L2, T); L3 := cons (head(L1), T)* end
  end

The former procedure can be transformed into a Boolean function that, in addition to building *L3*, checks if *append* produces the correct result.

function *append* (*L1*, *L2*: *pLIST*: var *L3*: *pLIST*): Boolean;
  begin local *H1*, *T1*, *T*: *pLIST*;
      if *L1* = *nil*   then begin *L3* := *L2*; *append* := true end
              else if {There exists an *H1* and a *T1* such that
                    H1 = *head* (*L1*) and *T1* = *tail* (*L1*)}
                  then begin *append* := *append* (*T1*, *L2*, *T*); *L3* := *cons* (*H1*, *T*) end
                  else *append* := *false*
  end

The Boolean in the conditional surrounded by braces has been presented informally, but it could actually have been programmed in detail. Note that the assignments in the preceding program are executed at most once for each recursive call. The function returns *false* if *L1* is not a list (e.g., if *L1* = *cons* (*a*, *b*) for some atom *b* ≠ *nil*).

One can now transform the last function into a PROLOG-like counterpart in which rules of assignments and conditionals are subsumed by unification and specified by the equality sign. The statement *E1* = *E2* succeeds if *E1* and *E2* can be matched. In addition, some of the variables in *E1* and *E2* may be bound if necessary.

append (*L1*, *L2*, *L3*) is true if *L1* = *nil* and *L3* = *L2*
                          otherwise
append (*L1*, *L2*, *L3*) is true
              if *L1* = *cons* (*H1*, *T1*) and *append* (*T1*, *L2*, *T*) and *L3* = *cons* (*H1*, *T*)
                  otherwise *append* is false.

The reader can now compare the preceding results with the previous description of the subset of PROLOG. This comparison yields the PROLOG program

      append (*L1*, *L2*, *L3*) :- *L1* = *nil*, *L3* = *L2*.

      append (*L1*, *L2*, *L3*) :- *L1* = *cons* (*H1*, *T1*), *append* (*T1*, *L2*, *T*), *L3* = *cons* (*H1*, *T*).

This version is particularly significative since it clearly separates resolution from unification. In this case, the equality sign is the operator that commands a unification between its left and right operands. This could also be done using the unit clause *unify* (*X*, *X*) and substituting *L1* = *nil* by *unify* (*L1*, *nil*), and so on. Replacing *L1* and *L3* with their respective values in the right-hand side of a clause one obtains:

          append (*nil*, *L2*, *L2*).

          append (*cons* (*H1*, *T1*), *L2*, *cons* (*H1*, *T*)) :- *append* (*T1*, *L2*, *T*).

The explicit calls to *unify* have now been replaced by implicit calls that will be triggered by the PROLOG interpreter when it tries to match a goal with the head of a rule. Notice that a <*literal*> now becomes a function name followed by a list of parameters each of which is syntactically similar to a <*literal*>. In the predicate calculus the preceding program corresponds to:

      ∀ *L2* *append* (*nil*, *L2*, *L2*).

      ∀ *H1*, *T1*, *L2*, *T* *append* (*cons* (*H1*, *T1*), *L2*, *cons* (*H1*, *T*)) ∨¬ *append* (*T1*, *L2*, *T*)

A word of caution about types is in order. The preceding version of *append*, would produce a perhaps unwanted result if the list *L2* is a term, say, $g(a)$. For example *append (nil, g (a), Z)* would yield as result $Z = g(a)$. To ensure that this result would not be considered valid one would have to make explicit the calls to functions which test if the two lists to be appended are indeed lists, namely,

$$is\_a\_list\ (nil).$$

$$is\_a\_list\ (cons\ (H,\ T)) :- is\_a\_list\ (T).$$

The Edinburgh PROLOG representation of *cons* (H, T) is $[H \mid T]$, and *nil* is $[\ ]$. The Marseilles counterparts are *H.T* and *nil*. In the Edinburgh dialect, *append* is presented by

$$append\ ([\ ],\ L2,\ L2\ ).$$

$$append\ ([H1 \mid T1\ ],\ L2,\ [H1 \mid T\ ]) :- append\ (T1,\ L2,\ T\ ).$$

The query applicable to the program that uses the term *cons* is *append* (*cons* (*a*, *cons* (*b*, *nil*)), *cons* (*c*, *nil*), *Z*) and it yields *Z* = *cons* (*a*, *cons* (*b*, *cons* (*c*, *nil*))). In Edinburgh PROLOG, the preceding query is stated as *append* ([*a*, *b*], [*c*], *Z*), and the result becomes *Z* = [*a*, *b*, *c*].

A remarkable difference between the original PASCAL-like version and the PROLOG version of *append* is the ability of the latter to determine (unknown) lists that, when appended, yield a given list as a result. For example, the query *append*(*X*, *Y*, [*a*]) yields two solutions: *X* = [ ] *Y* = [*a*] and *X* = [*a*] *Y* = [ ].

The preceding capability is due to the generality of the search and pattern matching mechanism of PROLOG. An often asked question is: Is the generality useful? The answer is definitely yes! A few examples will provide supporting evidence. The first is a procedure for determining a list *LLL*, which is the concatenation of *L* with *L* the result itself being again concatenated with *L*.

$$triple\ (L,\ LLL\ ) :- append\ (L,\ LL,\ LLL\ ),\ append\ (L,\ L,\ LL\ ).$$

Note that the first *append* is executed even though *LL* has not yet been bound. This amounts to copying the list *L* and having the variable *LL* as its last element. After the second append is finished, *LL* is bound, and the list *LLL* becomes fully known. This property of postponing binding times can be very useful. For example, a dictionary may contain entries whose values are unknown. Identical entries will have values that are bound among themselves. When a value is actually determined, all communally unbound variables are bound to that value.

Another interesting example is *sublist* (*X*, *Y*), which is true when *X* is a sublist of *Y*. Let *U* and *W* be the lists at the left and right of the sublist *X*. Then the program becomes

$$sublist\ (X,\ Y) :- append\ (Z,\ W,\ Y),\ append\ (U,\ X,\ Z).$$

where the variables represent the sublists indicated as follows:

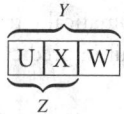

An additional role for *append* is to simulate the behavior of *member* as used in the introductory example. One could state that *member* (*X*, *L*) :- *append* (*Start*, [*X*|*End*], *L*). in which *Start* and *End* are variables representing lists which, when surrounding *X*, produce *L* . This approach is expensive since it may use a quadratic number of *cons*es.

A final example is a bubblesort program. The specification of two adjacent elements *A* and *B* in a list *L* is done by a call:

$$append\ (\_,\ [A,\ B \mid \_],\ L\ )$$

The underscores stand for different variables whose names are irrelevant to the computation, and the notation $[A, B \mid C]$ stands for $cons(A, cons(B, C))$. The rules to bubblesort then become:

$bsort(L, S) :\!\!- append\ (U, [A, B \mid X], L), B < A, append\ (U, [B, A \mid X], M), bsort(M, S).$

$bsort(L, L).$

The first *append* generates all pairs of adjacent elements in L. The literal $B < A$ is a builtin predicate that tests whether B is lexicographically smaller than A. (Note that both A and B must be bound, otherwise a failure occurs! There will be more discussion of this limitation later.) The second *append* reconstructs the modified list, which becomes the argument in a recursive call to *bsort*. If the first clause is no longer applicable, then all pairs of adjacent elements are in order, and the second clause then provides the desired result. This version of bubblesort is space and time inefficient since U, the initial segment of the list, is copied twice at each level of recursion. However, the brevity of the program is indicative of the savings that can be accrued in programming and debugging. It should be now clear to the reader that the translation of a functional program into its PROLOG counterpart is easily done but the reverse translation becomes harder since the functional program has to simulate nondeterminism and backtracking.

A unique property of some PROLOG programs is their ability to perform inverse computations. For example, if $p(X, Y)$ defines a procedure p taking an input X and producing an output Y, it can (in certain cases) determine which input X produces Y. Therefore, if p is a differentiation program it can also perform integration. This is easier said than done, since the use of impure features and simplifications may result in operations that are not correctly backtrackable.

The inverse computation of parsing is string generation, and parsers are now available to perform both operations. A compiler carefully written in PROLOG can also be useful in decompiling. The difficulties encountered in doing the inverse operations are frequently due to the use of impure features.

97.7 Impure Features

In the previous sections, the so-called pure features of PROLOG were described, namely, those that conform with the logic interpretation. A few impure features have been added to the language to make its use more practical. (This situation parallels the introduction of *setq, rplaca,* and other impure LISP functions.) However, a word of warning is in order. Some of these impure features vary from one implementation to another (O'Keefe [1990] is an excellent reference on extralogical features).

The most prominent of these impure features is the **cut**, represented by an exclamation point. Its purpose is to let the programmer change the search control mechanism embodied by the procedure solve in Fig. 97.3. Reconsider the example in section 97.5. Assuming that $a :\!\!- b, !, c, d.$, then in the forward mode, $b, c,$ and d would, as before, be placed in the list of goals and matched with heads of clauses in the database. If, however, a goal following the cut fails (e.g., $c,$ or d), then no further matching of clauses defining a or b would take place. The cut is often used to increase the efficiency of programs and to prevent the consideration of alternate solutions. PROLOG purists avoid the use of cuts.

Another useful predicate is *fail,* which automatically triggers a failure. To implement it, one simply forbids its presence in the database. Other builtin predicates that need to be introduced are the input–output commands *read* and *write.*

Once *cut* and *fail* are available, negation by failure is accomplished by the clauses:

$not\ (X) :\!\!- X, !, fail.$

$not\ (X).$

which fails if X succeeds and vice versa. It is important to note that this artifice does not always follow the rules of true negation in logic. This version of negation illustrates that a term in the head of a rule can appear in the body as a call to a function; therefore, in PROLOG program and data have the same form.

The builtin predicates *assert* and *retract* are used to add or remove a clause from the database; they are often used to simulate assignments. For example, the unit clause *value* (*Variable, Value*). can be asserted using bound parameters, e.g., *value* (*Z, 37*); it can be subsequently retracted to change the actual *Value*, and then reasserted with a new value. Another use of *assert* and *retract* is associated with the builtin function *setof* that collects the multiple answers of a program and places them in a list.

The assignment is introduced using a builtin binary infix operator such as *is*. For example, Y is $X + 1$ is only valid if X has been bound to a number in which case the right-hand side is evaluated and unified with Y; otherwise the *is* fails. To have a fully backtrackable addition, one would have to use CLP. Note that $I = I + 1$ is invalid in CLP as it should be. The equality $Z = X + Y$ in CLP is of course valid with some or all variables unbound.

97.8 Constraint Logic Programming

Major extensions of the unification component of PROLOG became very significant and resulted in a new area of LP called constraint logic programming or CLP. It may well have had its roots in Colmerauer's approach in generalizing the unification algorithm by making it capable of determining the satisfiability of equalities and disequalities of infinite trees (these are actually graphs containing special loops.) However, the notion of backpropagation (Sussman and Steele) dates back to the 1970s. In the case of PROLOG the backtracking mode is triggered only in the case of unsatisfiability of a given unification. A logical extension of this approach is to make similar backtracking decisions for other (builtin) predicates, say, inequalities (i.e., $\leq, \geq, \ldots$) in the domain of rationals.

The first CLP example presented here is the classic program for computing Fibonacci series. Before presenting the program, it is helpful to consider the program's PROLOG counterpart (the annotation *is* corresponds to an assignment and it has been discussed in the section on Impure Features):

$$fib\,(0,1).$$

$$fib\,(1,1).$$

$$fib\,(N,R)\;:\!\text{-}\;N1 \text{ is } N{-}1.$$

$$fib\,(N1,R1\,),$$

$$N2 \text{ is } N{-}2,$$

$$fib\,(N2,R2\,),$$

$$R \text{ is } R1{+}R2.$$

The *is* predicate prevents the program from being invertible: the query *fib* (10, X) succeeds in producing $X = 89$ as a result, but the query *fib* (Y, 89) yields an error, since N is unbound and the assignment to $N1$ is not performed. Note that if we had placed the predicate $N1 \geq 2$ prior to the first recursive call, the query *fib* (Y, 89) would also lead to an error, since the value of N is unbound and the test of inequality cannot be accomplished by the PROLOG interpreter.

The modified CLP version of the program illustrates the invertibility capabilities of CLP interpreters ($N \geq 2$ is a constraint):

$$fib\,(0,1).$$

$$fib\,(1,1).$$

$$fib\,(N,R1{+}R2\,):\!\text{-}\;N \geq 2,$$

$$fib\,(N{-}1,R1\,),$$

$$fib\,(N{-}2,R2\,).$$

The query ?- *fib* (10, *Fib*) yields *Fib* = 89, and the query ?- *fib* (*N*, 89) yields $N = 10$. The latter result is accomplished by solving systems of linear equations and inequations which are generated when an explicit or implicit constraint has to be satisfied. In this example, the matching of actual and formal parameters results in equations. Let us perform an initial determination of those equations. The query ?− *fib* (N, 89) matches only the third clause. New variables $R1$ and $R2$ are created as well as the constraint $R1 + R2 = 89$. The further constraint $N >= 2$ is added to the list of satisfiable constraints which are now $R1 + R2 = 89$ and $N >= 2$.

The recursive calls of *fib* generate further constraints that are added to the previous ones. These are $N1 = N − 1, N2 = N − 2$, and so forth. Recall that each recursive call is equivalent to a call by value in which new variables are created and new constraints are added (see the section on Implementation).

Therefore, unification is replaced by testing the satisfiability of systems of equations and inequations. A nontrivial implementation problem is how to determine if the constraints are satisfiable, only resorting to expensive general methods such as Gaussian elimination and the simplex method as a last resource.

The second example presented is a sorting program. For the purposes of this presentation, it is unnecessary to provide the code of this procedure [Sterling and Shapiro 1994]; *qsort* (L, S) sorts an input list L by constructing the sorted list S. When L is a list of variables, a PROLOG interpreter would fail since unbound variables cannot be compared using the relational operator $\leq$. In CLP, the query: ?− *qsort* ([$X1$, $X2, X3$], S). yields as result $S = [X1, X2, X3], X1 \leq X2, X2 \leq X3$. When requested to provide all solutions, the interpreter will generate all of the permutations of L as well as the applicable constraints.

Jaffar and Lassez [1987] proved that the theoretical foundations of LP languages (see section 97.10) remain valid for CLP languages. Several CLP languages are now widely used among LP practioners. Among them one should mention PROLOG III and IV designed by the Marseilles group, CLP(R) designed in Australia and at IBM, CHIP created by members of the European Reasearch Community, and CLP (BNR) designed in Canada at Bell Northern Research.

The subsequent summary describes the main CLP languages and their domains (see section 97.16):

PROLOG IV: trees, reals, intervals, linear constraints, rationals, finite domains (including Booleans), strings
CLP(R): trees, linear constraints, floating point arithmetic
CHIP: trees, linear constraints, floating point arithmetic, finite domains
CLP(BNR): trees, intervals

The languages considering intervals (defined by their lower and upper bounds) deal with numeric nonlinear constraints; symbolic linear constraints are handled by the first three languages. (Additional information about interval constraints is provided in section 97.13.)

97.9 Applications

The main areas in which LP and CLP have proved successful are summed up in the following:

Symbolic manipulation: Although LISP and PROLOG are currently the main languages in this area, it is probable that a CLP language may replace PROLOG in the next few years. There is a close relationship between the aims of CLP and symbolic languages such as MAPLE, MATHEMATICA, and MACSYMA.

Numerical analysis and operations research: The proposed CLP languages allow their users to generate and refine hundreds of equations and inequations having special characteristics (e.g., the generation of linear equations approximating Laplace's differential equations). The possibility of expressing inequations in a computer language has attracted the interest of specialists in operations research. Difficult problems in scheduling have been solved using CLP in finite domains.

Combinatorics: Nondeterministic languages such as PROLOG have been successful in the solution of combinatorial problems. The availability of constraints extends the scope of problems that can be expressed by CLP programs.

Artificial intelligence applications: Boolean constraints have been utilized in the design of expert systems. Constraints have also been used in natural language processing and in parsing. The increased potential for invertibility makes CLP languages unique in programming certain applications. For example, the inverse operation of parsing is string generation.

Deductive databases: These applications have attracted a considerable number of researchers and developers who are now extending the database (DB) domains to include constraints. The language DATALOG is the main representative and its programs contain only variables or constants (no composite terms are allowed).

Engineering applications: The ease with which CLP can be used for generating and refining large numbers of equations and inequations makes it useful in the solution of engineering problems. Ohm's and Kirchhoff's laws can readily be used to generate equations describing the behavior of electrical circuits.

97.10 Theoretical Foundations

This section provides a summary of the fundamental results applicable to logic programs [Apt 1990, Lloyd 1987]. It will be shown later that these results remain applicable to a wide class of constraint logic programs.

The semantics of LP and CLP are usually defined using either logic or sets. In the logic approach one establishes that given both (1) a Horn clause program P, and (2) a query Q, Q can be shown to be a consequence of P. In other words $\neg P \vee Q$ is always true, or equivalently, using contradiction, $P \wedge \neg Q$ is always false. This relates the logical and operational meaning of programs, i.e., that Q is a consequence of P can be proved by a resolution-based **breadth-first** theorem prover. (This is because a **depth-first** prover could loop in trying to determine a first solution, being therefore incapable of finding other solutions that may well exist.)

The logic-based semantics is accomplished in two steps. The first considers that a program yields a *yes* answer. In that case the results, i.e., the bindings of variables to terms as the result of successive unifications are the constraints that P and Q must satisfy so that Q becomes deducible from P. This is accomplished by Horn clause resolutions that render $P \wedge \neg Q$ unsatisfiable.

The second step of the proof is concerned with logic programs that yield a *no* answer and, therefore, do not specify bindings or constraints. In that case it becomes important to make a stronger statement about the meaning of a clause.

Recall that in the case of *yes* answers, a Horn clause specifies that the *Head* is a consequence of the *Body*, or equivalently that the *Head* is true if the *Body* is true. In the case of a *no* answer the so-called Clark completion becomes applicable. That means that the *Head* is true *if and only if* the *Body* is true. Then the semantics of programs yielding a *no* answer amounts to proving that *not Q* is a consequence of the completion of P. This amounts to considering the implication in *Body implies Head* in every clause of P as being replaced by *Body equivalent to Head*.

It should be remarked that the preceding results are only applicable to queries which do not contain logic negation. For example, the query $\neg\, date(X, Y, Z)$ in the introductory example is invalid, since a negative query is not in Horn clause form. Therefore, only positive queries are allowed and the behavior of the prover satisfies the so-called closed word assumption: only positive queries deducible from the program provide *yes* answers. Recent developments in the so-called nonmonotonic logic extend programs to handle negative queries.

The second approach in defining the semantics of LP and CLP uses sets. Consider the set $S0$ of all unit clauses in a program P. This set involves assigning any variables in these unit clauses to elements of the Herbrand universe. Consider then the clauses whose bodies contain elements of that initial set $S0$. Obviously the *Head* of those clauses is now deducible from the program and the new set $S1$ is constructed by taking the union of $S0$ with the heads of clauses that have been found to be true.

This process continues by computing the sets $S2, S3$ and so forth. Notice that $S(i)$ always contains $S(i - 1)$. Eventually these sets will not change since all the logical information about a finite program P

is contained in them. This is called a least fixed point. Then Q is a consequence of P if and only if each conjunct in Q is in the least fixed point of P.

Consider, for example, the PROLOG program for adding two positive numbers specified by a successor function $s(X)$ denoting the successor of X

$$add\ (0, X, X).$$

$$add\ (s(X), Y, s(Z)) :- add\ (X, Y, Z).$$

First notice the similarity of the preceding example with *append*. Adding 0 to a number X yields X (first rule). Adding the successor of X to a number Y amounts to adding one to the result Z obtained by adding X to Y.

In this simple example the Herbrand universe consists of $0, s(0), s(s(0))$, and so on, namely, the positive natural numbers including the constant 0. The so-called Herbrand base considers all of the literals (in this case *add*) for which a binding of a variable to elements of the Herbrand universe satisfy the program rules. Using the set approach $S0$ consists of all natural numbers since any number can be added to zero. The fixed point corresponds to the infinite set of bindings of X, Y, and Z to elements of the Herbrand universe, which satisfy both rules.

The meaning of a program P and query Q yielding a *no* answer can also be specified using set theory. In that case one starts with the set corresponding to the Herbrand universe $H0$. Then this set is reduced to a smaller set by using once the rules in P. Call this new set $H1$. By applying again the rules in P one obtains $H2$, and so on. In this case there is not necessarily a fixed point Hn. The property pertaining to programs yielding *no* answers then consists of the statement: *not Q* is a logical consequence of the completion of Q if and only if some conjunct of Q is not a member of Hn.

Besides programs yielding *yes* or *no* answers, there are those which loop. The halting problem tells us that we cannot hope to detect all of the programs which will eventually loop. Let us consider, as an example, the program P_1,

$$p(a).$$

$$p(b) :- p(b).$$

As expected, the queries: $Q_1 : p(a)$ and $Q_2 : p(c)$ yield, respectively, *yes* and *no* since $p(a)$ is a consequence of P_1, and $\neg p(c)$ is a consequence of the completion of p_1, i.e.:

$$p(X) \equiv (X = a) \lor (X = b \land p(b))$$

But the interpreter will loop for the query $Q_3 : p(b)$, or when all solutions of $Q_4 : p(X)$ are requested.

As mentioned earlier, the preceding theoretical results can be extended to CLP languages. Jaffar and Lassez [1987] established two conditions that a CLP extension to PROLOG must satisfy so that the semantic meaning (using logic or sets) is still applicable. The first is that the replacement of unification by an algorithm which tests the satisfiability of constraints should always yield a *yes* or *no* answer. This property is called *satisfaction-completeness* and it is obviously satisfied by the unification algorithm in the domain of trees. Similarly, the property applies to systems of linear equations in the domain of rationals and even to polynomial equations but with a significantly larger computational cost.

The second of Jaffar and Lassez's condition is called *solution-compactness*. It basically states that elements of a domain (say, irrational numbers) can be defined by a potentially infinite number of more stringent constraints which bound their actual values by increasingly finer approximations. For example, the real numbers satisfy this requirement. For more detail on Jaffar and Lassez's theory see Jaffar and Maher [1994] and Cohen [1990]. Existing CLP languages satisfy the two conditions established by Jaffar and Lassez.

The beauty of the Jaffar and Lassez metatheory is that they have established conditions under which the basic theorems of logic programming remain valid, provided that the set of proposed axioms specifying constraints satisfy the described properties.

A convenient (although incomplete) taxonomy for CLP languages is to classify them according to their domains or combinations thereof. One can have CLP (rationals), or CLP (Booleans, reals). PROLOG can be described as CLP (trees) and CLP (R) as CLP (reals, trees). A complete specification of CLP language would also have to include the predicates and operations allowed in establishing valid constraints.

From the language-design perspective, the designer would have to demonstrate the correctness of an efficient algorithm implementing the test for constraint satisfiability. This is equivalent to proving the satisfiability of the constraints specified by the axioms.

97.11 Metalevel Interpretation

Metalevel interpretation allows the description of interpreters for the languages (such as LISP or PROLOG) using the languages themselves. In PROLOG, the metalevel interpreter for pure programs consists of a few lines of code. The procedure solve has as a parameter a list of PROLOG goals to be processed. The interpreter assumes that the program rules are stored as unit clauses:

$$clause \ (Head, Body).$$

each corresponding to a rule: *Head :- Body.*, where *Head* is a literal and *Body* is a list of literals. Unit clauses are stored as : *clause* (*Head*, [ ]). The interpreter is:

solve ([ ]).
solve ([*Goal* | *Restgoal*]) :- *solve* (*Goal*), *solve* (*Restgoal*).
solve (*Goal*) :- *clause* (*Goal*, *Body*), *solve* (*Body*).

The first rule states that an empty list of goals is logically correct. (In that case the interpreter should print the latest bindings of the variables.) The second rule states that when processing (i.e., *solv*ing) a list of goals, the *head* and then the *tail* of the list should be processed. The third rule specifies that when a single *Goal* is to be processed one has to lookup the database containing the clauses, and process the *Body* of the applicable clause. In the preceding interpreter metainterpreter unification is implicit. One could write metainterpreters in which the builtin unification is replaced by an explicit sequence of PROLOG constructs using the impure features.

A very useful extension often incorporated into interpreters is the notion of coroutining, or lazy evaluation. The builtin procedure *freeze* (*X*, *P*) tests whether the variable *X* has been bound. If so, *P* is executed; otherwise the pair (*X*, *P*) is placed in a freezer. As soon as *X* becomes bound, *P* is placed at the head of the list of goals for immediate execution.

The procedure *freeze* can be easily implemented by expressing it as a variant of *solve* also written in PROLOG. Although this metalevel programming will of course considerably slow down the execution, this capability can and has been used for fast prototyping extensions to the language [Sterling and Shapiro 1994, Cohen 1990].

Another important application of metalevel programming is partial evaluation. Its objective is to transform a given program (a set of procedures) into an optimized version in which one of the procedures has one or more parameters that are bound to a known value. An example of partial evaluation is the automatic translation of a simple (inefficient) pattern matching algorithm which tests if a given pattern appears in a text. When the pattern is known, a partial evaluator applied to the simple matching algorithm produces the equivalent of the more sophisticated Knuth–Morris–Pratt pattern matching algorithm.

In a metalevel interpreter for a CLP language, a rule is represented by: *clause* (*Head*, *Body*,*Constraints*). corresponding to a rule: *Head :- Body* {*Constraints*}.

The modified procedure solve contains three parameters: (1) the list of goals to be processed, (2) the current set of constraints, and (3) the new set of constraints obtained by updating the previous set. The metalevel interpreter for CLP, written in PROLOG becomes:

$$solve\,([\;\;], C, C).$$

$$solve\,([Goal \mid Restgoal], Previous\_C, New\_C) :-$$

$$solve\,(Goal, Previous\_C, Temp\_C),$$

$$solve\,(Restgoal, Temp\_C, New\_C).$$

$$solve\,(Goal, Previous\_C, New\_C) :-$$

$$clause\,(Goal, Body, Current\_C),$$

$$merge\_constraints\,(Previous\_C, Current\_C, Temp\_C),$$

$$solve\,(Body, Temp\_C, New\_C).$$

The heart of the interpreter is the procedure *merge_constraints*, which merges two sets of constraints: the previous constraints, *Previous_C*, and the constraints introduced by the current clause, *Current_C*. If there is no solution to this new set of constraints, the procedure fails; otherwise, it simplifies the resulting constraints, and it binds any variables which have been constrained to take a unique value. For example, the constraints $X \leq 0 \wedge X \geq 0$ simplify to the constraint $X = 0$, which implies that X can now be bound to 0.

The design considerations which influence the implementation of this procedure will be discussed in section 97.12. Note that the controversial unit logical inference steps per second (LIPS), often used to estimate the speed of PROLOG processors, loses its significance in the case of a constraint language. The number of LIPS is established by counting how many times per second the procedure *clause* is activated; in the case of CLP, this time to process *clause* and *merge_constraints* may vary significantly depending on the constraints being processed.

97.12 Implementation

It is worthwhile to present the basic LP implementation features by describing an interpreter for the simplified PROLOG of section 97.5 written in a Pascal or C-like language. The reader should note the similarities between the metalevel interpreter *solve* of the previous section and the one about to be described.

The rules will be stored sequentially in a database implemented as a one-dimensional array *Rule*[1 . . n] and containing pointers to a special type of linear list. Such a list is a record with two fields, the first storing a letter, and the second being either *nil* or a pointer to a linear list. Let the (pointer) function *cons* be the constructor of a list element, and assume that its fields are accessible via the (pointer) functions *head* and *tail*. The first rule is stored in the database by

$$Rule\,[1] := cons\,('a', cons\,('b', cons\,('c', cons\,('d', nil)))).$$

The fifth rule defining a unit clause is stored as: $Rule\,[5] := cons\,('e', nil)$. Similar assignments are used to store the remaining rules.

The procedure *solve* that has as a parameter a pointer to a linear list is capable of determining whether or not a query is successful. The query itself is the list with which *solve* is first called. The procedure uses two auxiliary procedures *match* and *append*; *match*(A, B) simply tests if the alphanumeric A equals the alphanumeric B; *append* $(Ll, L2)$ produces the list representing the concatenation of Ll with $L2$ (this is equivalent to the familiar *append* function in LISP: it basically copies Ll and makes its last element point to $L2$).

The procedure *solve*, written in a PASCAL-like language, appears in Fig. 97.3. Recall that the variable n represents the number of rules stored in the array *Rule*. The procedure performs a depth-first search of the problem space where the local variable is used for continuing the search in case of a failure. The head of the list of goals L is matched with the head of each rule. If a match is found, the procedure is called recursively with a new list of goals formed by adding (through a call of *append*) the elements of the tail of the matching rule to the goals that remain to be satisfied. When the list of goals is *nil*, all goals have been satisfied, and a success message is issued. If the attempts to match fail, the search is continued in the previous recursion level until the zeroth level is reached in which case no more solutions are possible. For example, the query $a, e.$ is expressed by: *solve (cons ('a', cons ('e', nil)))* and yields the two solutions presented in section 97.5 on resolution of Horn clauses.

```
procedure solve (L : pLIST);
  begin local i: integer;
    if L ≠ nil
    then
      for i := 1 to n do
        if match (head (Rule [i]), head (L)) then
          solve (append (tail (Rule [i]), tail (L)));
    else write('yes')
  end;
```

FIGURE 97.3   An initial version of the interpreter.

Note that, if the tree of choices is finite, the order of the goals in the list of goals is irrelevant insofar as the presence and number of solutions are concerned. Thus, the order of the parameters of *append* in Fig. 97.3 could be switched, and the two existing solutions would still be found. Note that if the last rule were replaced by $a :- f, a.$, the tree of choices would be infinite and solutions similar to the first solution would be found repeatedly. The procedure *solve* in Fig. 97.3 can handle these situations by generating an infinite sequence of solutions. However, had the preceding rule appeared as the first one, the procedure *solve* would also loop, but without yielding any solutions. This last example shows how important the ordering of the rules is to the outcome of a query. This explains Kowalski's dictum program = logic + control, in which control stands for the ordering and (impure) control features such as the cut [Kowalski 1979].

It is not difficult to write a recursive function *unify*, which, given two terms, tests for the result of unification using the contents of Table 97.1 (section 97.5). For this purpose, one has to select a suitable data structure. In a sophisticated version, terms are represented by variable-size records containing pointers to other records, to constants, or to variables. Remark that the extensive updating of linked data structures inevitably leads to unreferenced structures that can be recovered by a garbage collection. It is frequently used in most PROLOG and CLP processors.

A simpler data structure uses linked lists and the so-called Cambridge Polish notation. For example, the term $f(X, g(Y, c))$ is represented by $(f (var x)(g(var y)(const c)))$, which can be constructed with *conses*.

As mentioned in 97.5, if the result of unification results in the binding: $Y := g(Y, c)$ then (most) PROLOG interpreters would soon get into difficulties since most implementations of the described unification algorithm cannot handle circular structures. Manipulating these structures (e.g., printing, copying) would result in an infinite loop unless the so-called occur check were incorporated to test for circularity.

The additional machinery needed to incorporate unification into the procedure *solve* of Fig. 97.3 is described in Cohen [1985]. An important remark is in order: when introducing unification it is necessary to *copy* the clauses in the program and introduce new variables (which correspond to parameters that should be called by value). The frequent copying and updating of lists makes almost mandatory the use of garbage collection which is often incorporated to LP processors.

Warren Abstract Machine (WAM)

D. H. D. Warren, a pioneer in the compilation of PROLOG programs, has proposed in 1983 a set of primitive instructions that can be generated by a PROLOG compiler, usually written using PROLOG. (Warren's approach parallels that of P-code used in early Pascal compilers.) The **Warren abstract machine** (WAM) primitives can be efficiently interpreted using specific machines.

The main data structures used by the WAM are (1) the recursion stack, (2) the heap, and (3) the trail. The heap is used for storing terms and trail for backtracking purposes. The WAM uses the copying approach mentioned in the beginning of this section. A local garbage collector takes advantage of the cut by freeing space in the trail.

The WAM has been used extensively by various groups developing PROLOG compilers. Its primitives are of great efficiency in implementing features such as tail-recursion elimination, indexing of the head of the clause to be considered when processing a goal, the cut, and other extralogical features of PROLOG. A useful reference in describing the WAM is the one by Ait-Kaci [1991]. A recent reference on implementation is by Van Roy [1994].

Parallelism

Whereas for most languages it is fairly difficult to write programs that automatically take advantage of operations and instructions that can be executed in parallel, PROLOG offers an abundance of opportunities for parallelization. There are at least three possibilities for performing PROLOG operations in parallel:

1. *Unification:* Since this is one of the most frequent operations in running PROLOG programs, it would seem worthwhile to search for efficient parallel unification algorithms. Some work has already been done in this area [Jaffar et al. 1992]. However, the results have not been encouraging.
2. *And-parallelism:* This consists of simultaneously executing each procedure in the tail of a clause. For example, in $a(X, Y, U) :- b(X, Z), c(X, Y), d(T, U)$. an attempt is made to continue the execution in parallel for the clauses defining b, c, and d. The first two share the common variable X; therefore, if unification fails in one but not in the other, or if the unification yields different bindings, then some of the labor done in parallel is lost. However, the last clause in the tail can be executed independently since it does not share variables with the other two.
3. *Or-parallelism:* When a given predicate is defined by several rules, it is possible to attempt to apply the rules simultaneously. This is the most common type of parallelism used in PROLOG processors.

Kergommeaux and Codognet [1994] is a recommended survey of parallelism in PROLOG.

Design and Implementation Issues in Constraint Logic Programming

There is an important implementation consideration that appears to be fulfilled in both CLP(R) and PROLOG IV: the efficiency of processing PROLOG programs (without constraints) should approach that of current PROLOG interpreters, i.e., the overhead for *recognizing* more general constraints should be small.

There are three factors that should be considered when selecting algorithms for testing the satisfiability of systems of constraints used in conjunction with CLP processors. They are (1) incrementality, (2) simplification, and (3) canonical forms. The first is a desirable property which allows an increase in efficiency of multiple tests of satisfiability (by avoiding recomputations). This can be explained in terms of the metalevel interpreter for CLP languages described in section 97.11: if the current system of constraints S is known to be satisfiable, the test of satisfiability should be incremental, minimizing the computational effort required to check if the formula remains satisfiable or not. Classical PROLOG interpreters have this property, since previously performed unifications are not recomputed at each inference step. There are modifications of Gaussian methods for solving linear equations which also satisfy this property. This is accomplished by introducing temporary variables and replacing the original system of equations by an equivalent solved form (see section on Unification): *variable = linear terms involving only the temporary variables*.

The simplex method can also be modified to satisfy incrementality. Similarly, the SL resolution method for testing the satisfiability of Boolean equations, and the Gröbner method for testing the satisfiability of polynomial equations, have this property.

In nearly all of the domains considered in CLP it may be possible to replace a set of constraints by a simpler set. This simplification can be time consuming, but is sometimes necessary. The implementor of CLP languages may have to make a difficult choice as to what level of simplification should occur at each step verifying constraint satisfaction. It may turn out that a system of constraints eventually becomes unsatisfiable, and all of the work done in simplification is lost. When a final result has to be output, it becomes essential to simplify it and present it to the reader in the clearest, most readable form.

An important function of simplification is to detect the assignment of a variable to a single value (e.g., from $X \geq 1$ *and* $X \leq 1$ *one infers* $X = 1$). This property is essential when implementing a modified simplex method which detects when a variable is assigned to a single value. Note that this detection is necessary when using lazy evaluation.

The incremental algorithms for testing the satisfiability of linear equations and inequations, as well as that used in the Gröbner method for polynomial equations, are capable of discarding redundant equations; therefore, they perform some simplifications [Sato and Aiba 1993].

The canonical (solved) forms referred to earlier in this section can be viewed as (internal) representations of the constraints which facilitate both the tests of satisfiability and the ensuing simplifications. For example, in the case of the Gröbner method for solving polynomial equations, the input polynomials are internally represented in a normal form, such that variables are lexicographically ordered and the terms of the polynomials are ordered according to their degrees. This ordering is essential in performing the required computations. Also note that if two seemingly different constraints have the same canonical form, only one of them needs to be considered. Therefore, the choice of appropriate canonical forms deserves an important consideration in the implementation of CLP languages [Jaffar and Maher 1994].

Optimization Using Abstract Interpretation

Abstract interpretation is an enticing area of computer science initially developed by Cousot and Cousot [1992]; it consists of considering a subdomain of the variables of a program (usually a Boolean variable, e.g., one representing the evenness or oddness of the final result). Program operations and constructs are performed using only the desired subdomain. Cousot proved that if certain conditions are applicable to the subdomains and the operations acting on their variables, the execution is guaranteed to terminate. Dataflow analyses, partial evaluation, detection of safe parallelism, etc., can be viewed as instances of abstract interpretation. The research group at University of Louvain, Belgium has been active in exploring the capabilities of abstract interpretation in LP and CLP.

97.13 Research Issues

It is worthwhile to classify the numerous extensions of PROLOG into three main categories, namely, those related to (1) resolution beyond Horn clauses, (2) unification, and (3) others, e.g., concurrency. Major extensions of the unification became very significant and resulted in a new area of LP called constraint logic programming or CLP that was dealt with in section 97.8; nevertheless, the more recent addition to CLP dealing with the domain of intervals is discussed in this section.

Resolution Beyond Horn Clauses

Several researchers have suggested extensions for dealing with more general clauses and for developing semantics for negation that are more general than that of negation by failure (see section 97.7). Experience has shown that the most general extension, that is, to the general predicate calculus, poses difficult combinatorial search problems. Nevertheless, substantial progress has been made in extending LP beyond pure Horn clauses. Two such extensions deserve mention: stratified programs and generalized predicate calculus formulas in the *body* part of a clause.

Stratified programs are variants of Horn clause programs that are particularly applicable in deductive databases; true negation may appear in the body of clauses, provided that it satisfies certain conditions. These stratified programs have a clean semantics based on logic and avoid the undesirable features of negation by failure. (See Minker [1987].)

To briefly describe the second extension it is worthwhile to recall that the procedural interpretation of resolution applied to Horn clauses is based on the substitution model: a procedure call consists of replacing the call by the body of the procedure in which the formal parameters are substituted by the actual parameters via unification. The generalization proposed by Ueda and others can use the substitution model to deal with the clauses of the type: *head* :- a general formula in the predicate calculus containing quantifiers and negation.

Concurrent Logic Programming and Constraint Logic Programming

A significant extension of LP has been pursued by several groups. A premise of their effort can be stated as: a programming language worth its salt should be expressive enough to allow its users to write complex but efficient operating systems software (as is the case of the C Language). With that goal in mind they incorporated into LP the concepts of **don't care nondeterminism** as advocated by Dijkstra. The resulting languages are called concurrent LP languages. The variants proposed by these groups were implemented and refined; they have now converged to a common model which is a specialized version of the original designs.

Most of these concurrent languages use special punctuation marks "?" and "|". The question mark is a shorthand notation for *freezes*. For example, the literal $p(X?, Y)$ can be viewed as a form of *freeze* $(X, p(X, Y))$. The vertical bar is called *commit* and usually appears once in the tail of clauses defining a given procedure. Consider, for example,

$$a :- b, c \mid d, e.$$

$$a :- p \mid q.$$

The literals b, c, and d, e are executed using *and* parallelism. However, the computation using *or* parallelism for the two clauses defining a continues only with the clause that first reaches the *commit* sign. For example, if the computation of b, c proceeds faster than p, then the second clause is abandoned, and execution continues with d, e only (see Saraswat 1993).

Interval Constraints

The domain of interval arithmetic has become a very fruitful area of research in CLP. Older has been a pioneer in this area [Older and Vellino 1993]. This domain specifies reals as being defined between lower and upper bounds which can be large integers or rational numbers. The theory of solving most nonlinear and trigonometric equations using intervals guarantees that *if* there is a solution, that solution must lie within the computed intervals. Furthermore, it is also guaranteed that *no* solution exists outside the computed interval or unions of intervals.

The computations involve the operation of *narrowing* that consists of finding new bounds for a quantity denoting the result of an operation (say, $+$, $*$, sin, etc.) involving operands which are also defined by their lower and upper bounds. The narrowing operation also involves intersecting intervals obtained by various computations defining the same variable. The intersection may well fail (e.g., the equality operation applying to operands whose intervals are disjoint). The narrowing is guaranteed to either converge or fail. This, however, may not be sufficient to find possible solutions of interest. One can nevertheless split a given interval into two or more unions of intervals and proceed to find a more precise solution, if one exists. This is akin to enumeration of results in CLP.

The process of splitting is a **don't know nondeterministic** choice, an existing component of LP. The failure of the narrowing operation is analogous to that encountered in CLP when a constraint is unsatisfiable and backtracking occurs. Therefore, there is a natural interaction between LP and the domain of intervals.

Interval arithmetic is known to yield valuable results in computing the satisfiability of nonlinear constraints or in the case of finite domains. Its use in linear constraints is an active area of research since results indicate a poor convergence of narrowing. In the case of polynomials constraints interval arithmetic may well be a strong competitor to Gröebner base techniques.

Constraint Logic Programming Language Design

A current challenge in the design and implementation of CLP is to blend computations in different domains in a harmonious and sound manner. For example, the reals can be represented by intervals whose bounds are floating-point numbers (these have to be carefully implemented to retain soundness due to rounding operations). Actually, floating-point numbers are nothing more than (approximate) very large integers or fractions. This set is, of course, a superset of finite domains, which in turn is a superset of Booleans. Problems in CLP language design that still remain to be solved is how to handle the interaction of these different domains and subdomains. This situation is further complicated by efficiency considerations. Linear inequations, equations, and disequations can be efficiently solved using rational arithmetic but research remains to be done in adapting simplex like methods to deal with interval arithmetic.

97.14 Conclusion

As in most sciences, there has always been a valuable symbiosis among the theoretical and experimental practitioners of computer science including, of course, those working in logic programming. Three examples come to mind: the elimination of the occur test in unifications, the cut, and the *not* operator as defined in PROLOG. These features were created by practical programmers and are here to stay. They provide a vast amount of food for thought for theoreticians. As mentioned earlier, the elimination of the occur test was instrumental in the development of algorithms for unification of **infinite trees**. Although the concept of the cut has resisted repeated attempts for a clean semantic definition, its use is unavoidable in increasing the efficiency of programs. Finally, PROLOG's *not* operator has played a key role in extending logic programs beyond Horn clauses.

CLP is one of the most promising and stimulating new areas in computer science. It amalgamates the knowledge and experience gained in areas as varied as numerical analysis, operations research, artificial languages, symbolic processing, artificial intelligence, logic, and mathematics.

During the past 20 years LP has followed a creative and productive course. It is not unusual for a fundamental scientific endeavor to branch out into many interesting subfields. An interesting aspect of these developments is that LP's original body of knowledge actually branched into subareas, which joined previously existing research areas. For example, CLP is being merged with the area of constraint satisfaction problems (CSP); LP researchers are interested in modal, temporal, intuitionistic, and linear logic; relational database research now includes constraints; operations research and CLP have found previously unexplored similarities and so on.

The several subfields of LP now include research on CLP in various domains, typing, nonmonotonic reasoning, inductive LP, semantics, concurrency, nonstandard logic, abstract interpretation, partial evaluation, blending with functional and with object-oriented language paradigms. It will not be surprising if each of these subfields will become fairly independent from their LP roots and the various specialized groups will organize autonomous journals and conferences. The available literature on LP is abundant and it is likely to be followed by a plentiful number of publications in its autonomous subfields.

Defining Terms

Backtracking: A manner to handle (*don't know*) nondeterministic situations by considering *one* choice at a time and storing information which is necessary to restore a given state of the computation. PROLOG interpreters often use backtracking to implement nondeterministic situations.

Breadth first: A method for traversing trees in which all of the children of a node are considered simultaneously. OR-Parallel PROLOG interpreters use breadth-first traversal.

Clause: A general normal form for expressing predicate calculus formulas. It is a disjunction of literals $(P_1 \lor P_2 \lor \cdots)$ whose arguments are terms. The terms are usually introduced by eliminating existential quantifiers.

Constraint logic programming languages: PROLOG-like languages in which unification is replaced or complemented by constraint solving in various domains.

Constraints: Special predicates whose satisfiability can be established for various domains. Unification can be viewed as equality constraints in the domain of trees.

Cut: An annotation used in PROLOG programs to bypass certain nondeterministic computations.

Depth first: A method for traversing trees in which the leftmost branches are considered first. Most sequential PROLOG interpreters use depth-first traversal.

Don't care nondeterminism: The arbitrary choice of one among multiple possible continuations for a computation.

Don't know nondeterminism: Situations in which there are equally valid choices in pursuing a computation.

Herbrand universe: The set of all terms that can be constructed by combining the terms and constants which appear in a logic formula.

Horn clause: A clause containing (at most) one positive literal. The term *definite clause* is used to denote a clause with exactly one positive literal. PROLOG programs can be viewed as a set of definite clauses in which the positive literal is the head of the rule and the negative literals constitute the body or tail of the rule.

Infinite trees: Trees that can be unified by special unification algorithms which bypass the occur-check. These trees constitute a new domain, different from that of usual PROLOG trees.

Metalevel interpreter: An interpreter written in L for the language L.

Occur-check: A test performed during unification to ensure that a given variable is not defined in terms of itself [e.g., $X = f(X)$ is detected by an occur-check, and unification fails].

Predicate calculus: A calculus for expressing logic statements. Its formulas involve:

- *atoms:* $P(T_1, T_2, \ldots)$ where P is a predicate symbol and T_1 are terms
- *Boolean connectives:* conjunction ($\wedge$), disjunction ($\vee$), implication ($\rightarrow$), and negation ($\neg$).
- *literals:* atoms or their negations
- *quantifiers:* for all ($\forall$), there exists ($\exists$)
- *terms* (also called *trees*): constructed from constants, variables, and function symbols.

Resolution: A single inference step used to prove the validity or predicate calculus formulas expressed as clauses. In its simplest version: $P \vee Q$ and $\neg P \vee R$ imply $Q \vee R$, which is called the resolvant.

SLD resolution: Selective linear resolution for definite clauses inference step used in proving the validity of Horn clauses.

Unification: Matching of terms used in a resolution step. It basically consists of testing the satisfiability or the equality of trees whose leaves may contain variables. Unification can also be viewed as a general parameter matching mechanism.

Warren abstract machine (WAM):   An intermediate (low-level) language that is often used as an object language for compiling PROLOG programs. Its objective is to allow the compilation of efficient PROLOG code.

References

Ait-Kaci, H. 1991. *The WAM: A (Real) Tutorial*. MIT Press.

Apt, K. R. 1990. Logic programming. In *Handbook of Theoretical Computer Science*, J. van Leewun, ed., pp. 493–574. North-Holland, Amsterdam.

Bergin, T. J. 1996. History of programming languages HOPL 2. Addison–Wesley, Reading, MA.

Borning, A. 1981. The programming language aspects of Thing-Lab, a constraint-oriented simulation laboratory. *ACM TOPLAS* 3(4):252–387.

Clocksin, W. F. and Mellish, 1984. *Programming in PROLOG*, 2nd ed. Springer–Verlag, New York.

Cohen, J. 1985. Describing PROLOG by its interpretation and compilation. *Commun. ACM* 28(12):1311–1324.

Cohen, J. 1988. A view of the origins and development of PROLOG. *Commun. ACM* 31(1):26–36.

Cohen, J. 1990. Constraint logic programming languages. *Commun. ACM* (July):52–68.

Cohen, J. and Hickey, T. J. 1987. Parsing and compiling using PROLOG. *ACM Trans. Programming Lang. Syst.* 9(2):125–163.

Colmerauer, A. 1990. An introduction to PROLOG III. *Commun. ACM* 33(7).

Cousot, P. and Cousot, R. 1992. Abstract interpretation and applications to logic programs. Journal of Logic Programming 13(2/3):103–179.

Dincbas, M., Van Hentenryck, P., Simonis, H., Aggoun, A., Graf, T., and Berthier, F. 1988. The constraint logic programming language CHIP, pp. 693–702. In FGCS'88, *Proc. Int. Conf. Fifth Generation Comput. Syst.* Vol. 1. Tokyo, Japan, Dec.

Jaffar, J. and Lassez, J.-L. 1987. Constraint logic programming, pp. 111–119. In *Proc. 14th ACM Symp. Principles Programming Lang.*, Munich.

Jaffar, J. and Maher, M. 1994. Constraint logic programming, a survey. *J. Logic Programming* 503–581.

Jaffar, J., Michaylov, S., and Yap, R. H. C. 1992. The CLP language and system. *ACM Trans. Programming Lang. Syst.* 14(3):339–395.

Kergommeaux, J. C. and Codognet, P. 1994. Parallel LP systems. *Comput. Surv.* 26(3).

Kowalski, R. A. 1979. Algorithm = logic + control. *Commun. ACM* 22(7):424–436.

Kowalski, R. and Kuehner, D. 1970. Resolution with selection function. *Artif. Intell.* 3(3):227–260.

Lloyd, J. W. 1987. *Foundations of Logic Programming.* Springer–Verlag.

Minker, J., ed. 1987. *Foundations of Deductive Databases and Logic Programming.* Morgan Kaufmann.

O'Keefe, R. A. 1990. *The Craft of PROLOG.* MIT Press.

Older, W. and Vellino, A. 1993. Constraint arithmetic on real intervals. In *Constraint Logic Programming: Selected Research.* F. Benhamou and A. Colmerauer, eds., MIT Press.

Robinson, J. A. 1965. A machine-oriented logic based on the resolution principle. *J. ACM* 12(1):23–41.

Saraswat, V. A. 1993. *Concurrent Constraint Programming Languages.* MIT Press.

Sato, S. and Aiba, A. 1993. An application of CAL to robotics. In *Constraint Logic Programming: Selected Research.* F. Benhamou and A. Colmerauer, eds., pp. 161–174. MIT Press.

Shapiro, E. 1989. The family of concurrent LP languages. *Comput. Surv.* 21(3):413–510.

Sterling, L. and Shapiro, E. 1994. *The Art of PROLOG.* MIT Press.

Van Hentenryck, P. 1989. *Constraint Satisfaction in Logic Programming.* Logic programming series, MIT Press, Cambridge, MA.

Van Roy, P. 1994. The wonder years of sequential PROLOG implementation, 1983–1993. *J. Logic Programming* 19(20):385–441.

Warren, D. H. D. 1983. An Abstract PROLOG Instruction Set. *Tech. Note* 309, SRI International, Menlo Park, CA.

Further Information

There are several journals specializing in LP and CLP. Among them we mention: *Journal of Logic Programming*, North-Holland, *New Generation Computing*, Springer–Verlag, and *Constraint*, Kluwer.

Most of the conference proceedings have been published by MIT Press. Recent proceedings on constraints have been published in the Lecture Notes in Computer Science (LNCS) series published by Springer–Verlag. A newsletter is also available (Logic Programming Newsletter, alp@doc.ic.ac.uk)

Among the references provided the following relate to CLP languages: PROLOG III [Colmerauer 1990], CLP(R) [Jaffar et al. 1992], CHIP [Dincbas et al. 1988], CAL [Sato and Aiba 1993], finite domains [Van Hentenryck 1989], and Intervals [Older and Vellino 1993]. The recommended textbooks include Clocksin and Mellish [1984] and Sterling and Shapiro [1994]. The theoretical aspects of LP are well covered in Apt [1990] and Lloyd [1987] and implementation in Ait-Kaci [1991], Kergommeaux and Codognet [1994], Van Roy [1994], and Warren [1983].

98

Concurrent/Distributed Computing Paradigm

Andrew P. Bernat
*The University of Texas at
El Paso*

98.1 Introduction

Concurrent computing is the use of multiple, simultaneously executing processes or tasks to compute an answer or solve a problem. The original motivation for the development of concurrent computing techniques was to allow time-sharing multiple users or jobs on a single computer. Modern workstations use this approach in a substantial manner. Another advantage of concurrent computing, and the reason for much of the current attention to the subject, is that solving a problem using two computers can be up to twice as fast as using just one. Similarly, there is a powerful economic argument for using multiple inexpensive computers to solve a problem which normally requires an expensive supercomputer. Subsequently, we will examine these ideas.

But there is an additional powerful argument for concurrent computing: the world is inherently concurrent. Just as each of us engages in a large number of concurrent tasks (hearing while seeing while reading, etc.), operating systems need to handle many simultaneously executing tasks, robots need to engage in a multiplicity of actions, database systems must simultaneously handle large numbers of users accessing and updating information, etc. Often breaking a problem into concurrent tasks provides a simpler, more straightforward solution.

As an example, consider Conway's problem: input is in the form of 80-character records (card images in the original problem, which gives an idea of how long it has been around), output is to be in the form of 120-character records; each pair of dollar signs $$ is to be replaced by a single dollar sign $ and a space " " is to be added at the end of each input record. In principle, a sequential solution may be developed, but the complications introduced require complex and nonobvious buffer manipulations. Moreover, a concurrent solution consisting of three processes is both simpler and more elegant. The three processes are infinite loops with the following actions:

1. Process1 reads 80-character records into an 81-character buffer, places a space character in location 81, and then outputs single characters from the buffer sequentially.

0-8493-2909-4/97/$0.00+$.50
© 1997 by CRC Press, Inc.

2. Process2 reads single characters and copies them to output, but uses a simple state machine to substitute a single $ for two consecutive $$.
3. Process3 reads single characters, saves them in a buffer and outputs 120-character records.

To develop an implementable solution, we need to decide how the independently executing processes will communicate. A simple, widely used approach is to add two buffers: Buffer1 will store output characters from Process1 to be input by Process2; Buffer2 will store output characters from Process2 to be input by Process3. For simplicity, assume that Buffer1 and Buffer2 each holds a single character. Thus:

1. Process1 reads 80-character records into an 81-character internal buffer, places a space character in location 81, and sequentially places in Buffer1 single characters from the internal buffer.
2. Process2 reads single characters from Buffer1 and places them into Buffer2, but uses a simple state machine to substitute a single $ for two consecutive $$.
3. Process3 reads single characters from Buffer2, saves them in an internal 120-character buffer and outputs 120-character records.

This solution demonstrates the essence of the concurrent paradigm; individual sequential processes cooperate to solve a problem. Cooperation requires that the processes:

1. share information and resources and
2. not interfere during access to shared information or resources.

In the Conway solution, information is readily shared via the buffers. The chief problem is to ensure that access to the two buffers does not conflict, e.g., Process2 should not attempt to retrieve a character from Buffer1 before it has been placed there by Process1 (which would lead to garbage characters) and Process1 should not attempt to place a character into Buffer1 before the previous character has been retrieved by Process2 (which would lead to lost characters).

A simpler example of interference is the simple program (where the statements within the **cobegin— coend** pair are to be executed simultaneously):

```
x := 0
cobegin
  x := x + 1
  x := x + 2
coend
```

Here we ask: what is the value of **x** at the end of execution? Because each assignment statement is actually a sequence of machine-level instructions, various interleavings of these instructions will result in different final values for **x** (1, 2, or 3). Clearly this is unacceptable!

In each of these examples, it is clear that there are **critical regions** in which two (or more) processes have sections of code which may not run concurrently with another process; we must have **mutual exclusion** between the critical regions. In the Conway example, critical regions are:

- Process1 placing a value into Buffer1
- Process2 retrieving a value from Buffer1
- Process2 placing a value into Buffer2
- Process3 retrieving a value from Buffer2

In the simple preceding example the two parallel assignment statements are each critical regions. The essence of avoiding interference is to discover the critical regions and to isolate them. This isolation takes the form of an *entry protocol* before entering the critical region and an *exit protocol* to announce that the critical region has been completed (here the # introduces a comment and the . . . represents the appropriate program code):

```
# entry protocol
...
# critical region code
...
# exit protocol
...
```

This is the basic model used by the *busy-wait* and **semaphore** approaches (discussed subsequently). It is a low-level model in the sense that careful attention must be paid to the placement of the entry and exit protocols in order to ensure that critical regions are properly protected.

Other implementation approaches to concurrency solve the critical region problem by prohibiting any direct interference between concurrent processes. This is done by not allowing any sharing of variables. The **monitor** approach is to put all shared variables or other resources under the control of a single monitor module, which is accessed by only a single process at a time. The **message passing** approach is to share information only through messages passed from process to process. Both of these approaches are also discussed in this chapter.

As well as avoiding interference in variable access, we must also avoid interference in the sharing of resources, e.g., keyboard input for multiple processes. Also, we must ensure that any physical actions of concurrent processes, such as movement of robotic arms, are appropriately synchronized.

Thus, in order to develop concurrent solutions, we require notations to:

1. Specify which portions of our processes may run concurrently
2. Specify which information and resources are to be shared
3. Prevent interference by concurrent processes by ensuring mutual exclusion
4. Synchronize concurrent processes at appropriate points

Further, any proposed solution to a concurrent problem must have certain properties (e.g., see Ben-Ari [1990]):

1. Safety: the property must always be true; examples are (a) noninterference; (b) no **deadlock**, which occurs when no process may continue because all processes are waiting upon conditions which can never occur; and (c) partial correctness—whenever the program terminates, it has the correct answer.
2. Liveness: the property must eventually be true; examples are (a) program terminates (if it also has the correct answer, this is total correctness); (b) no **race**, which occurs when processes continue execution, but no progress is made toward problem solution; and (c) **fairness**, each process should have an opportunity to execute (affected by implementation and scheduler).

The proof that our solutions satisfy these properties is vastly complicated by the concurrent execution: particular orderings of execution may exhibit interference or deadlock while others proceed nicely to termination. Returning to Conway's problem, suppose that Process1 and Process2 are evenly matched. Then each character would be placed by Process1 into Buffer1 and retrieved by Process2 before Process1 is ready to output another character. We test our program and verify that it exhibits the desired correctness properties, lack of deadlock, etc. But if, due to a variation in internal calculations or processor type, Process1 runs faster, then characters will be overwritten and lost; on the other hand, if Process2 runs faster, characters will be repeated. The fact that we tested our program under one particular set of circumstances (even for all possible input) is irrelevant to this issue. Thus, debugging is not satisfactory because of the exponential explosion in the number of possible interleavings that can occur. The only fully satisfactory approach is to use formal methods (techniques which are still predominantly in their development stage), touched on later in this chapter.

This chapter will focus on the software architectures used for concurrency, using a set of archetypical problems and their solutions for illustration. These problems are chosen because of the frequency with which they arise in computing; careful study of actual problems frequently leads to the realization that a seemingly complicated problem is, at heart, one of these archetypes. First we briefly explore hardware architectures and their impact upon software.

98.2 Hardware Architectures

Hardware can influence the synchronization and communication mechanisms primarily through efficiency considerations. **Multiprogramming** is the interleaving of execution of multiple processes on a single computer; time-shared operating systems are examples as are modern UNIX workstations. Although such an approach does not provide the execution speedup discussed in the introduction, it does provide the possibility of elegance and simplicity in problem solution, which is the second argument for the concurrent paradigm.

By employing multiple computers, we have either **multiprocessing**, when the computers share a common memory, or **distributed processing**, when the computers are connected via a network. This chapter focuses on multiprogramming and multiprocessing systems with a short introduction to the additional problems of distributed systems. In addition (but outside the scope of this chapter), a wide variety of hybrid hardware/software approaches exist.

The choice of hardware architecture most directly affects the communication mechanism used: with shared memory multiprocessing, global variables may be used to share information; with distributed memory, communication occurs via messages passed from process to process. When a single computer is used via multiprocessing, either approach may be used efficiently.

98.3 Software Architectures

In order to specify an architecture for implementing concurrency, we must provide the syntax and semantics to:

1. Specify which information and resources are to be shared
2. Specify which portions of our processes may run concurrently
3. Prevent interference by concurrent processes by ensuring mutual exclusion
4. Synchronize concurrent processes at appropriate points

The first feature requires no special notation (shared variables are simply global), and the third and fourth are usually merged into one. A large number of software mechanisms have been proposed to support these features; in this chapter we will explore the most widely used among them:

1. Busy-wait: implementable on virtually any processor without operating system support; this is concurrency without abstractions.
2. Semaphores: historically, this is the oldest satisfactory mechanism.
3. Monitors: these are modules which encapsulate concurrent access to shared data.
4. Message passing: this is a higher level abstraction widely used in distributed systems.

The references at the end of the chapter provide pointers to a number of other mechanisms, such as UNIX fork-join, conditional critical regions, and so on.

Busy-Wait: Concurrency Without Abstractions

We will use a very simple example consisting of two concurrent processes, each with a single critical region. The only assumption that is made is that each memory access is atomic, that is, it proceeds without interruption. Our task is to ensure mutual exclusion; the purpose of the exercise is to demonstrate the care with which a solution must be crafted to ensure the safety and liveness properties discussed previously (following Ben-Ari [1990]).

Our first approach is to ensure that the processes simply take turns in their critical regions:

```
global var turn := 1

process p1
  while true do ->
   # non-critical region
    ...
    # entry protocol
    while turn = 2 do ->
       <nothing>                        # wait for turn
    # critical region
    ...
    # exit protocol
      turn := 2
    # rest of computation
    ...
  end p1

process p2
  while true do ->
   # non-critical region
    ...
    # entry protocol
    while turn = 1 do ->
       <nothing>                        # wait for turn
    # critical region
    ...
    # exit protocol
      turn := 1
    # rest of computation
    ...
  end p2
```

which meets the desired properties but has a fundamental flaw: processes must take turns in entering their critical regions. If **p1** is ready and needs to execute its critical region at a higher rate than **p2**, this cannot take place. The processes are an example of *coroutines*, historically one of the first approaches to concurrency.

If we modify the solution to allow each process to proceed into its critical region if the other process is not in its critical region, and to then notify the other process, we obtain (where **ci** is used to signify that **pi** is in its critical region):

```
    global var c1 := false, c2 := false

    process p1
      while true do ->
        # non-critical region

          ...
        # entry protocol
          while c2 do ->
            <nothing>                    # wait for turn
          c1 := true                     # p1 in critical region
        # critical region

          ...
        # exit protocol
          c1 := false                    # p1 out of critical region
        # non-critical region
    end p1

    process p2
      while true do ->
        # non-critical region

          ...
        # entry protocol
          while c1 do ->
            <nothing>                    # wait for turn
          c2 := true                     # p2 in critical region
        # critical region

          ...
        # exit protocol
          c2 := false                    # p2 out of critical region
        # non-critical region
    end p2
```

but now we have the possibility of violating the mutual exclusion requirement of the critical regions (suppose both **c1** and **c2** are false; **p1** checks via the loop and decides that it may enter; before it sets **c1** to true, **p2** checks via its loop and decides that it may enter).

We may propose to eliminate this disastrous possibility by declaring entry into the critical region before checking:

```
    global var c1 := false, c2 := false

    process p1
      while true do ->
        # non-critical region

          ...
        # entry protocol
          c1 := true                  # signal intent to enter
```

```
          while c2 do ->
            <nothing>              # wait for turn
        # critical region
          ...
        # exit protocol
          c1 := false             # p1 out of critical region
        # non-critical region
    end p1

process p2
  while true do ->
    # non-critical region
      ...
    # entry protocol
      c2 := true              # signal intent to enter
      while c1 do ->
        <nothing>             # wait for turn
    # critical region
      ...
    # exit protocol
      c2 := false             # p1 out of critical region
    # non-critical region
  end p2
```

but now we have raised the possibility of race (when **p1** sets **c1** to true *and* **p2** sets **c2** to true).

A possible solution to this difficulty is to move the announcement statement into the loop, together with a random delay:

```
global var c1 := false, c2 := false

process p1
  while true do ->
    # non-critical region
      ...
    # entry protocol
      c1 := true              # signal intent to enter
      while c2 do ->
        c1 := false           # give up intent if p2 already
                              # in critical region
        <delay>
        c1 := true            # try again
    # critical region
      ...
    # exit protocol
      c1 := false             # p1 out of critical region
    # non-critical region
      ...
```

```
      end p1
      process p2
        while true do ->
          # non-critical region
          ...
          # entry protocol
            c2 := true                # signal intent to enter
            while c1 do ->
              c2 := false             # give up intent if p1 already
                                      # in critical region
              <delay>
              c2 := true              # try again
          # critical region
            ...
          # exit protocol
            c2 := false                      # p2 out of critical region
          # non-critical region
            ...
      end p2
```

But this is not a satisfactory solution, because the solution exhibits race in the (unlikely) case that the two loops proceed in perfect synchronization.

A valid solution may be developed by returning to the concept of taking turns, which ensures mutual exclusion while not requiring alternating turns (thus allowing true concurrency):

```
      global var c1 := false, c2 := false, turn := 1
      process p1
        while true do ->
          # non-critical region
          ...
          # entry protocol
            c1 := true                # signal intent to enter
            turn := 2                 # give p2 priority
            while c2 and turn = 2 do ->
              <nothing>               # wait if p2 in critical region
          # critical region
            ...
          # exit protocol
            c1 := false               # p1 out of critical region
          # non-critical region
            ...
      end p1
      process p2
        while true do ->
          # non-critical region
          ...
```

```
    # entry protocol
      c2 := true                        # signal intent to enter
      turn := 1                         # give p1 priority
      while c1 and turn = 1 do ->
        <nothing>                       # wait if p2 in critical region
    # critical region
      ...
    # exit protocol
      c2 := false                       # p1 out of critical region
    # non-critical region
      ...
  end p2
```

This solution is due to Peterson [1983]; the first valid solution was presented by Dekker.

The importance of the Busy-Wait approach is threefold:

1. It provides a nice introduction to the problems inherent in designing concurrent solutions.
2. It is executable on virtually every machine architecture without further software support and is thus suitable for microcontrollers, etc.
3. Variants are frequently used in hardware implementations.

However, this approach also suffers from two difficulties:

1. It is very inefficient: machine cycles are used in executing empty loops.
2. Programming at such a low-level is highly prone to error.

Semaphores

Dijkstra [1968] presented the first abstract mechanism for synchronization in concurrent programs. The *semaphore*, so named in direct relation to the semaphores used on railroad lines to control traffic over a single track, is a nonnegative integer-valued abstract data type with two operations:

$$P(s): \text{ delay until } s > 0, \quad \text{then } s := s - 1$$
$$V(s): s := s + 1$$

When a process delays on a semaphore, it will be awakened only upon another process executing a **V** operation on that semaphore; thus it uses no machine cycles to check if it can proceed. If more than one process is delaying upon a semaphore, only one (the choice is implementation dependent) can be woken upon a **V** operation at an instant in time.

Additionally, the value of **s** may be set at instance creation via the semaphore declaration; if set to 0, then some process must execute the **V(s)** operation before any processes first executing the **P(s)** operation may continue. With this abstract data type, we thus have a mechanism which handles both interference and synchronization.

Additional notes:

1. These are the *only* two synchronization operations defined; in particular, the value of **s** is not determinable.
2. Implementation of these operations must be either in the hardware or in the (noninterruptible) system kernel.

3. By sleeping while waiting for a semaphore (the delay in **P(s)**), a process does not waste machine cycles by repeatedly checking.

4. The operation names (**P** and **V**) come from the Dutch, passeren (to pass) and vrygeven (to release); sometimes **signal** and **wait** are used in place of **P** and **V**, respectively.

5. Each of the **P** and **V** operations proceeds atomically; that is, it may not be interrupted by another process.

The use of the semaphore in concurrent programming relates directly to the railroad analogy. Each critical section looks like:

```
global var s : semaphore := 1
# entry protocol
  P(s)
# critical region
  ...
# exit protocol
  V(s)
```

The initialization of **s** to 1 ensures that the first process executing **P(s)** will continue. (Deadlock would arise if **s** were initialized to 0.) Only the first process to reach its **P(s)** statement will be allowed to proceed, as subsequent processes will find **s = 0** and will delay. When the first process finishes its critical region, it will execute **V(s)** which sets **s** to 1. One of the waiting processes will be woken up, find **s > 0**, decrement **s** and proceed. Note the importance that these operations are atomic, thus ensuring that two processes cannot wake up and each find **s > 0**.

Semaphores and Producer–Consumer

The producer–consumer problem arises whenever one process is creating values to be used by another process. Examples are Conway's problem, buffers of various kinds, etc. Here we first look at the multi-element buffer version of this problem and then add multiple producers and consumers to this version as a refinement.

```
# define the buffer
  const N := ...        # size
  var buf[N] : int      # buffer
      front := 1        # pointers
      rear := 1
  semaphore empty := N        # counts the number of empty slots
                                in the buffer
            full := 0         # counts the number of items
                                in the buffer

process producer
var x : int
while true do ->
  # produce x
      ...
  P(empty)                    # delay for space in the buffer
  buf[rear] := x              # place value in the buffer
```

```
    V(full)                          # signal that the buffer is non-empty
    rear := rear mod N + 1           # update buffer pointer
  end producer

  process consumer
  var x : int
  while true do ->
    P(full)                          # delay for a value to be in the buffer
    x := buf[front]                  # obtain value
    V(empty)                         # signal that the buffer is not full
    front := front mod N + 1         # update buffer pointer
    # consume x
        ...
  end consumer
```

The buffer processing is conventional; only the actual buffer access must be placed into a critical region because there is no possibility of interference between the assignments to **rear** and **front**. Note also the use of two semaphores, **empty** to signal that the producer may proceed because there is at least one empty slot in the buffer and **full** to signal that the consumer may proceed because there is at least one item in the buffer. Although it is possible to solve this problem with one semaphore, less concurrency would result. Note that the **empty** semaphore is initialized to **N**, the size of the buffer. The producer process can run up to **N** steps ahead of the consumer process.

To allow multiple producers and/or consumers, we must protect the actual buffer operations with additional semaphores to prevent, e.g., two producers from accessing **rear** simultaneously with read and assignment operations. These semaphores, **mutexR** and **mutexF**, guarantee mutual exclusion of access to the **rear** and **front** pointers, respectively. It is not sufficient to use **empty** here, because up to **N** producers will be able to continue through the **P(empty)** statement.

```
    # define the buffer as previously
    semaphore empty := N, full := 0
    semaphore mutexR := 1    # mutual exclusion on rear pointer
              mutexF := 1    # mutual exclusion on front pointer
    process pi               # one for each producer
    var x : int
    while true do ->
      # produce x
        ...
      P(empty)               # delay until space in buffer
      P(mutexR)              # delay until rear pointer is not in use
      # place value in the buffer and modify pointer
        buf[rear] := x; rear := rear mod N + 1
      V(mutexR)              # release rear pointer
      V(full)                # signal that the buffer is non-empty
    end pi

    process ci               # one for each consumer
```

```
    var x : int
    while true do ->
      P(full)                  # delay until value in the buffer
      P(mutexF)                # delay until front pointer is not in use
      # access the value in the buffer and modify pointer
      x := buf[front]; front := front mod N + 1
      P(mutexF)                # release front pointer
      V(empty)                 # signal that there is space in the buffer
      # consume x
        ...
end ci
```

Semaphores and Readers–Writers

The readers–writers model captures the fundamental action of a database:

- No exclusion between readers
- Exclusion between readers and writer
- Exclusion between writers

That is, the software must guarantee only one update of a database record at a time, and no reading of that record while it is being updated.

The simplest semaphore solution is to wait only for the first reader; subsequent readers need not check because no writer could be writing if there is already a reader reading (here **nr** and **nw** are the numbers of active readers and writers, respectively):

```
    ...
    nr := nr + 1
    if nr = 1 -> P(rw)   # if no one is presently reading,
                         # then ensure no one is writing
                         # before proceeding
    # access database
      ...
    nr := nr - 1
    if nr = 0 -> V(rw)   # if no more are reading, possibly wake up
                         # writer, or prepare for next reader
    ...
    P(rw)                # delay until no readers or writers
    # access database
      ...
    V(rw)                # wake up delayed reader or write, or prepare
                         # for next reader or writer
```

This solution gives readers preference over writers: new readers will continually freeze out waiting writers. Extending this solution to other kinds of preferences, such as writer preference or first-come–first-served preference is cumbersome.

A more general approach, easily extended, is known as passing the baton, because control is explicitly handed from process to process. Although a careful explanation of the approach is not given here, the concept is easily summarized. A process must check to ensure that it may legally proceed before doing so; if not the process waits upon a semaphore assigned to it. For example, a writer process would check to see if no readers or writers were executing on the database before it proceeds; if they are it would sleep waiting upon the semaphore assigned to it. When a process is finished accessing the database, it checks the conditions and wakes up (via signaling on the appropriate semaphore) one of the processes waiting upon the condition. This last operation essentially passes the baton from one process to another. The key is that a check is made first, to ensure that it is legal for the other process to wake up. The strength of the passing the baton approach comes when we wish to use its flexibility to develop more general solutions. Details may be found in Andrews [1991].

Semaphores and Dining Philosophers

This charmingly named problem is an example of multiple processes competing for scarce resources. There are five philosophers whose occupation is to philosophize, but who must occasionally eat from a communal pot of spaghetti. A circular table is set with five chairs and one fork between each place (a total of five forks), so that each philosopher has another philosopher on the left and on the right. Each philosopher requires two forks for eating in order to twirl the spaghetti.

An implementation using semaphores (to prevent interference in the use of the forks) is:

```
semaphore fork[0..4] := 1
Philosopheri                    # i runs from 1 to 5
while true do ->
# think
P(fork[i mod 5])
P(fork[(i + 1) mod 5)]
# eat
V(fork[i mod 5])
V(fork[(i + 1) mod 5]
```

A possible scenario is:

1. All philosophers become hungry and sit.
2. Each philosopher reaches for the fork on the left (**P(fork[i mod 5])**).
3. Each philosopher reaches for the fork on the right (**P(fork[(i + 1) mod 5])**).
4. Each philosopher discovers that the fork is missing (having been picked up by the philosopher on the right) and delays.

The result is deadlock. It is not sufficient to simply state the unlikelihood of the scenario; it is a possible scenario and any solution must satisfy the safety properties which prevent deadlock.

There are a number of strategies to eliminate deadlock:

1. Each philosopher, upon discovering a missing fork, returns any fork already picked up and tries again after a random time period. This approach of course decreases, but does not eliminate, the possibility of deadlock. It also introduces the possibility of race if the philosophers proceed in lockstep; it also wastes machine cycles.
2. One philosopher reaches first for the fork on the right. This breaks the symmetry which is a necessary condition for deadlock to occur.
3. A waiter is posted to ensure that at most four philosophers sit at the table; this is the traditional choice used in operating systems, where all system resources must be requested at program startup.
4. A philosopher does not pick up a fork unless assured that both forks are available.

If we choose to add a waiter, a semaphore solution allows at most four philosophers at the table via initializing the value of **waiter** to 4 (thus allowing at most four **P(waiter)** operations before the fifth **P(waiter)** will delay until a **V(waiter)** signifying that a philosopher has left the table):

```
semaphore fork[0..4] := 1
          waiter := 4
Philosopheri                  # i runs from 1 to 5
while true do ->
  # think
  P(waiter)
  P(fork[i mod 5])
  P(fork[(i + 1) mod 5)]
  # eat
  V(fork[i mod 5])
  V(fork[(i + 1) mod 5]
  V(waiter)
```

Difficulties with Semaphores in Software Design

Although the use of semaphores does provide a complete solution to the interference problem, the correctness of the solution is directly dependent on the correct usage of the semaphore operations, operations which are fairly low level and unstructured. Semaphores and shared variables are global to all processes and, like any global data structure, their correct usage requires considerable discipline by the programmer. Additionally, if a large system is to be built, any one implementor is likely responsible for only a portion of the semaphore usage so that correct pairing of **P**s and **V**s may be difficult. Despite this difficulty, semaphores are a widely used construct for concurrency.

Monitors

A more structured approach is to encapsulate the shared data/resources and their operations into a single module called a monitor. A monitor can contain nonexternally accessible data and procedures, which handle the state of resources. External access is strictly controlled through procedure calls to the monitor; mutual exclusion is ensured because procedure execution within the monitor is not concurrent.

Monitors have the traditional advantages of abstract data types, but they must also deal with two issues rising from their use by concurrently executing processes: avoiding interference and providing synchronization. This section illustrates some sample applications of monitors and how they internally handle concurrency.

Returning to the producer–consumer problem, we implement a monitor for handling shared access to the buffer. The monitor requires a synchronization mechanism to ensure that the producer cannot overfill and that the consumer cannot retrieve from the empty buffer. Monitors implement **condition variables** whose values will be queues of processes delayed upon the corresponding condition. Two standard operations defined on conditional variable **cv** are:

> **wait(cv):** Causes the executing process to delay and to be placed at the end of **cv**'s queue; in order to allow eventual awakening of the process, the process must relinquish exclusive access to the monitor when it executes a **wait**.
>
> **signal(cv):** Causes the process at the head of **cv**'s queue to be awakened; if the queue is empty there is no effect.

Although these operations mirror those of semaphores there is a key difference: the **signal** operation has no memory.

Monitors and Producer–Consumer

The buffer monitor can be defined as follows:

```
monitor Buffer
   # define the buffer
      const N := ..                  # size of the buffer
      var buf[N] : int               # buffer
          front := 1                 # buffer pointers
          rear := 1
   # define the condition variables
      var not_full,                  # signaled when count < N
          not_empty : cv             # signaled when count > 0
   procedure deposit(data : int)
      if count = N                   # check for space
         then wait(not_full)         # delay if no space
      buf[rear] := data
      rear := (rear mod N) + 1
      N := N + 1
         signal(not_empty)           # signal non-empty
   end
   procedure fetch(var data : int)
      if count = 0                    # check for not empty
         then wait(not_empty)        # delay if empty
      data := buf[front]
      front := (front mod N) + 1
      N := N - 1
         signal(not_full)            # signal not full
   end Buffer
```

Using this monitor, the producer and consumer tasks can be redone as follows:

```
process Producer
var x : int
while true do ->
   # produce x
       ...
    deposit(x)
end Producer
process Consumer
var x : int
while true do ->
    fetch(x)
   # consume x
       ...
end Consumer
```

It is clear that programming (outside of the monitor) can now be done at a more abstract level, which will lead to more reliable software.

Difficulties with Monitors

However, there are difficulties with monitors as well. Consider the case where we have two consumers, **C1** and **C2**. If the buffer is empty when **C1** executes fetch, then **C1** will delay on **not-empty**. If the producer then executes **deposit** (note that **deposit** and **fetch** cannot run concurrently), it will eventually **signal(not-empty)**, which will wake **C1** up. But if **C2** executes **fetch** before **C1** continues execution and its call to **fetch** proceeds, then **C1** will access an empty buffer. Hence, the **signal** operation must be considered to be a *hint* that proceeding with execution is possible, but not that it is correct. Two approaches are used to solve this problem:

1. Replace the check on the condition variable with a check inside a loop to ensure that the condition is true before execution proceeds. For example:

```
procedure deposit(data : int)
    while count = N do ->      # check for space
        then wait(not_full)    # delay if no space
    buf[rear] := data
    rear := (rear mod N) + 1
    N := N + 1
    signal(not_empty)          # signal non-empty
end
```

2. Give the highest priority to awakening processes so that intervening access to the monitor is not possible; this also requires that the **signal** operation be the last operation executed in any procedure in which it occurs (to ensure that two processes will not be executing within the monitor).

Monitors form the basis for concurrent programming in a number of systems and provide an efficient, high-level mechanism. They have the further advantage, as do other abstract data types or objects, of allowing for local modification and tuning without affecting the remainder of the system.

Message Passing

Consider a hardware architecture with multiple independent computers. Creating a semaphore to be efficiently accessed by processes running in separate computers is a difficult problem. We need a new abstraction for the distributed memory case: *message passing*, in which a sending process outputs a message to a **channel** and a receiving process inputs the message from this same channel. There are a large number of variations of this basic concept depending on the semantics of the operations and the channels. The basic primitives are: (1) channel declaration, (2) **send** <channel> <message>, and (3) **receive** <channel> <variable>.

If both sending and receiving processes block upon reaching their corresponding message passing operation, we have **synchronous** communication; if the sending process may send a message and continue without waiting for receipt, the system is **asynchronous**. Analogies are telephone communication and use of the postal system. The synchronous approach allows for ready synchronization of processes (at the instant of message passing we know where both are in their execution). This was the approach chosen by Hoare [1985] for his communicating sequential processes model and its subsequent implementation in the *Occam* language [Jones and Goldsmith 1988]. If we desire asynchronicity, we can add intermediate buffer processes to the synchronous approach. An advantage of synchronous message passing is that it often simplifies analysis of an algorithm because it is known where the sending and receiving processes are in their execution at the moment the message is passed.

Further variations arise depending on whether channels are one-process-to-one-process, one-to-many, statically instantiated at load time or dynamically created during execution, bidirectional or unidirectional, whether the receiving process must be named by the sending process, etc. However, the basic concept is the same in all cases; only ease of use and efficiency of implementation vary.

Further variations include **remote procedure call** (**RPC**), which is the core of many distributed systems, and **rendezvous**, the approach used in the language Ada. We will further explore these approaches after looking more closely at simple message passing.

Note that, in the message passing approach, there are no shared variables and so interference is not an issue. The critical section issue does not arise because there is no way for concurrent processes to interfere with each other. This is one of the major motivating factors for the use of message passing software architectures.

Message Passing and Producer–Consumer

If our message passing system is asynchronous, we can rely on the system itself to buffer values:

```
channel P2C

process Producer
int x
while true do ->
   # produce x
   send P2C x
end Producer

process Consumer
int x
while true do ->
   receive P2C x
   # consume x
end Consumer
```

Here, the **Producer** can send a message over channel **P2C** and continue producing and sending (up to channel capacity at which point the system will block it) while the **Consumer** will block at the **receive** statement if no messages are available.

If our system is synchronous, then we create a separate buffer process:

```
channel P2B, B2C

process Buffer
# create the buffer
   const N := ..
   var buffer[N] : int
      front := 1
      rear := 1
      count := 0              # number of items in the buffer
   while true do ->
```

```
      if
          # there is room and the producer is sending
            count < n and receive P2B buffer[rear] ->
              count++; rear := rear mod n + 1
        else
            # there are items and the consumer is receiving
              count > 0 and send B2C buffer[front] ->
                count--; front := front mod n + 1
      end Buffer

      process Producer
      var x : int
      while true do ->
        # produce x
          ...
        send P2B x
      end Producer

      process Consumer
      var x : int
      while true do ->
        receive B2C x
        # consume x
          ...
      end Consumer
```

Here the **if** statement is nondeterministic; that is, any **true** clause may be selected. The Boolean conditions in the clauses are called guards. The clauses are:

- If there is room and the producer wishes to send a character
- If there are items to retrieve and the consumer wishes to receive a character

For implementation efficiency reasons, actual programming languages do not allow guards for both input and output statements, and so we must modify, e.g., the buffer and consumer processes to eliminate the output guard:

```
    channel P2B, B2C, C2B

    process Buffer
    # define the buffer
      var buffer[n] : int
      var front := 1
          rear := 1
          count := 0
    while true do - >
      if
        # there is room and the producer is sending
          count < n and receive P2B buffer[rear] ->
```

```
                   count++; rear := rear mod n + 1
         else
           # there are items and the consumer is requesting
             count > 0 and receive C2b buffer[front] ->
                 send B2C buffer[front]
                 count--; front := front mod n + 1
     end Buffer

     process Producer
     var x : int
     while true do ->
       # produce x

             ...
       send P2B x
     end Producer

     process Consumer
     var int : x
     while true do ->
       send C2B NIL               # announce ready for input
       receive B2C x
       # consume x

             ...
     end Consumer
```

Here, the consumer process first announces its intention to receive a value from the buffer process (**send C2B NIL** ; the **NIL** signifying that no message need be actually exchanged) and then actually receives the value (**receive B2C x**).

This program is an example of **client–server** programming; the consumer process is a client of the buffer process; that is, it requests service from the buffer which provides it. Client-server programming is widely used to provide services across a network and is based on the message passing paradigm.

Message Passing and Readers–Writers

The message passing approach to readers-writers is straightforward: do not accept a message from a reader or writer if a writer is accessing; do not accept a message from a writer if a reader is accessing. The solution is simple if we adopt synchronous message passing and the notion of the database as a server:

```
channel Rrequests, Rreceives, Wsends

Reader
  send Rrequests <request message>
  receive Rreceives <data>

Writer
  send Wsends <write message>

Server
```

```
    if
      # there are no writers, accept reader requests
      nw = 0 ->
          receive Rrequests <request message>
          # access the database
          ...
          send Rreceives <data>
      # there are no readers or writers, accept writer
        requests
        nr = 0 and nw = 0 ->
          receive Wsends <write message>
          # modify the database
          ...
```

Message Passing and Dining Philosophers

A message passing version of the dining philosophers problem uses the client–server model and creates a waiter process as a server. Philosophers (client) each request the waiter that they be allowed to sit, obtain both forks and begin eating. The waiter only accepts a philosopher if these actions are possible:

```
    channels getForks, releaseForks

    process Waiter
    var eating[5] := ([5]false)
    while true do ->
      if
        #  philosopher i wishes to eat and philosophers to the left and
           right are not eating
           receive getForks i and
           not (eating[(i + 3) mod 5 + 1] or eating[i mod 5 + 1]) ->
              eating[i] := true
        #  always accept a philosopher who stops eating
           receive releaseForks i ->
              eating[i] := false
    end Waiter

    process Philosopheri          # i from 1 to 5
    while true do ->
      send getForks i
      # eat
      send releaseForks i
      # philosophize
    end Philosopher
```

Message Passing and Semaphore Simulation

Of course, message passing can simulate a semaphore (and vice versa if need be):

```
channels P, V, initSemaphore

process Semaphore
var s : int
receive initSemaphore i
s := i
while true do ->
   if
      # semaphore is non-zero accept P operation
         s > 0 and receive P NIL->
         s--
      # always accept V operation
         receive V NIL ->
         s++
end Semaphore
```

The Remote Procedure Call and Rendezvous Abstractions

The remote procedure call abstraction is widely used to provide client–server services over a distributed network. Revisiting the previous client–server examples, it is clear that the client executes a **send-receive** pair while the server executes a **receive-send** pair. By using the standard procedure model to capture the server's actions, a call statement to capture the client's actions and parameters to capture the messages being sent, we have

```
Client
   ...
   call Server(args)
   ...
Server(formal args)
   ...
   return
```

which mirrors traditional procedure calls. The difference is that the **Server** procedure may be on a machine remote to the **Client** process. Indeed, the server is implemented as a process which is always delayed until a **Client** executes a call. If multiple **Clients** execute calls to a **Server** concurrently, the **Server** must be re-entrant or must provide protection for shared information. The RPC approach forms the basis for distributed systems programs on a wide variety of platforms; its relationship to monitors should be clear.

The calling process and procedure are not truly concurrent in the sense used throughout this chapter in that the calling process delays once the call is made, the procedure does not execute until called, the procedure delays when the return is executed, and the calling process resumes execution only upon the return from the procedure. The model is similar to that of synchronous message passing if the execution of the procedure is viewed as a component of the message passing process (essentially the procedure creates the return message).

We may increase the power of this approach if we modify the procedure into a process and have both processes executing concurrently. When a call is made, execution of the calling process delays while execution of the called process continues until it is ready to accept the call (via a special statement). The called process continues execution, performing actions or calculating values for the return message. The return message is sent back to the caller, the called process continues executing and the calling process resumes execution once the message is received. Because there is an extended time period during which the two processes are synchronized (from called accept through called return), this model of concurrency is termed rendezvous. It is the basis for the model of concurrency used in the Ada language. The Ada model is not symmetric: the calling process must know the name of the process it is calling, but the called process need not know its caller. Accept statements may have guards, as previously discussed for message passing, in order to control acceptance of calls. The complexity of these guards, and their priority, must be carefully followed during program implementation.

There are several advantages of this approach, all based on the possibility of the called routine using multiple accept statements:

1. The called routine may provide different responses to the calling process at different stages of its execution.
2. The called routine may respond differently to different calling processes.
3. The called routine chooses when it will receive a call.
4. Different accept statements may be used to provide different services in a clear fashion (rather than through parameter values).

Difficulties with Message Passing

Message passing systems are frequently inefficient during execution unless the algorithm is carefully developed because messages take time to propagate, and this time is essentially overhead. For example, a single-element buffer version of Conway's problem would spend significantly more time exchanging messages than any other operation.

98.4 Distributed Systems

Besides the difficulties inherent in developing and understanding concurrent solutions, distributed systems contain the fundamental problem of identifying global state. For example, how do we determine if a program has terminated? In the sequential case, this is obvious: we execute the **exit** or **end** statement. In the concurrent case, we must ensure that all processes are ready to terminate. In the multiprocessing case we may do this by checking the ready queue: if it is empty then there are no processes waiting to run, which ensures that no process will ever be added to the ready queue (if no process can run, then there can be no changes to create another ready process). But if we are in a distributed system, there is no single ready queue to examine. If a process is in the suspended queue on its processor, it may be made ready by a message from a process on a different processor.

Similarly, we may still require mutual exclusion on a system resource: how do we ensure access across processors? The solution is to develop a method of determining global state; see, e.g., Ben-Ari [1990].

Whereas a true distributed paradigm has not yet emerged in the programming paradigms domain, it will most likely evolve in the area of operating systems; for more information on distribution in this particular area, readers are encouraged to look at Chapter 85 in this Handbook.

98.5 Formal Approaches

We argued previously that the traditional debugging approach to software verification is of even less value in concurrent programming than in sequential programming due to the need to take into account the enormous number of possible interactions between concurrent processes. Obviously traditional testing only demonstrates the presence of "good" execution histories and is not a mechanism to verify any solution, sequential or concurrent. The use of a trace routine to generate execution histories is a standard sequential technique which becomes infeasible: Consider for example, that n processes each executing m atomic actions would generate $(n * m)!/(m!)^n$ histories. For three processes, each executing only two actions, this is a total of 90 possible histories!

The alternative is to use a formal, mathematically rigorous method to develop your solution and/or to verify a complete solution. Two approaches have been applied to verifying concurrent software:

1. Axiomatic or assertional
2. Process algebraic

The axiomatic approach develops assertions in the predicate logic which characterize the states a computation may be in. The actions of a program are viewed as predicate transformers which move the computation from one state to another. The beginning state is specified by the precondition of the computation, and the final state must be characterized by the postcondition. This approach has been exploited for some time in the sequential paradigm; see Andrews [1991] or Bernstein and Lewis [1993] for comprehensive introductions to the field in the context of concurrency.

The process algebraic approach was pioneered by Hoare [1985], who also pioneered the coarse-grained model of concurrency. The concept is that the interactions between a system and its environment (which is all that is ultimately observable) may be modeled via a mathematical abstraction called a process (this is the abstraction of the computing process as previously used). Processes may be combined via algebraic laws to form systems. Communication between processes is an example of this interaction. By building up a system through these mathematical laws and then transforming the abstract mathematics into an implementable language, one arrives at a correct solution. The language Occam was designed to match the algebraic laws devised by Hoare; transformations exist between these laws and Occam programming constructs (but the transformation is not perfect due to practicalities of implementation) [Hinchey and Jarvis 1995]. A number of subsequent efforts developed process algebras with varying properties [Milner 1989].

Although both approaches are in active use, they are not typically applied in the concurrent paradigm with any greater frequency than they are in the sequential paradigm and they remain primarily research tools. The fundamental difficulty is that theoreticians search for the *fundamental particles* of computing in order to develop mathematical laws enabling formal reasoning. Practical languages are inherently extremely complex mixtures of these fundamental particles and laws in order to have sufficient power to solve real-world problems. Theoretical tools do not yet scale to these large, complex problems.

98.6 Existing Languages with Concurrency Features

A large number of languages have been developed to use the concurrency paradigm; most have remained in the laboratory environment. If the underlying operating system provides the requisite support, then semaphores may be implemented in any language via system calls. Higher level concurrency control structures require modification of the underlying sequential language; for example, Concurrent Pascal [Brinch Hansen 1975] uses monitors, whereas Concurrent C [Gehani and Roome 1986] is based upon the rendezvous. By beginning with a widely used sequential programming language, a designer has a large community from which to draw users to the new language. Languages Ada (concurrency based on the rendezvous) and SR (which includes structures for all of the approaches discussed in this chapter and is therefore particularly useful for exploring concurrent programming [Andrews and Olsson 1993]; also see Hartley [1995], for extensive examples) are examples of sequential languages with concurrent structures included from the initial stages of development.

Object-oriented languages have similarly had concurrency features added. For example, SMALLTALK has the Process and Semaphore classes to provide for the dynamic creation of independent processes and their interaction using the semaphore approach [Goldberg and Robson 1989].

Languages based upon an inherently concurrent model include LINDA (more a language-independent philosophy than a language) [Ahuja et al. 1986] and Occam (synchronous message passing) [Jones and Goldsmith 1988].

98.7 Research Issues and Summary

Whereas it is clear that concurrency is a necessary technique for the solution of many problems, it is also clear that progress must be made in order to ensure its effective application. That this is still a research issue is clear whenever an operating system crashes due to system processes which interfere or we discover someone else in our airplane seat due to concurrent access to the airline's database. This required progress falls into three categories:

1. Theoretical advances must be made to develop formal techniques which scale up to real-world applications. For example, process interference checkers exist, but operate essentially by checking all possible interactions between processes to check for deadlock, etc. This approach rapidly develops combinatorial explosion.
2. Design tools must provide development support for concurrent solutions, for example, debuggers which capture the concurrent computation without overwhelming the user with information.
3. Languages are required with powerful structures to support the correct application of concurrency. For example, the development of concurrent object-oriented languages appears straightforward: simply allow each object to run concurrently since each object is logically autonomous. Regarding the last requirement, however, there are a number of issues which need resolving:
   1. Not all objects need run concurrently as the majority of computation will still be sequential (thereby incurring no scheduler overhead).
   2. If we consider multiple concurrent objects attempting to communicate with the same object: (a) acceptance of a message must delay all other messages in order to correctly preserve the internal state of the object, (b) ordering of message acceptance must be synchronized to ensure computations are correct, and (c) acceptance of messages must occur only at appropriate points in the object's execution.
   3. Inheritance through the class hierarchy creates problems because it will mix this synchronization with object behavior.

The single outstanding problem with concurrency is the development of correct solutions (as it is in all software systems): the state of development of both formal methods and software engineering tools lags behind the sequential world in this regard.

Defining Terms

Asynchronous message passing: The message sending process allows messages to be buffered and the sending process may continue after the send is initiated; the receiving process will block if the message queue is empty.

Channel: The data structure, which may be realized in hardware, over which processes send messages.

Client–server: The software architecture in which clients are able to request services of processes executing on remote machines.

Condition variables: A variable used within a monitor to delay an executing process.

Critical regions: A section of code which must appear to be executed indivisibly.

Deadlock: The state in which processes are waiting for events which can never occur, i.e., the processes cannot progress.

Distributed processing: The use of multiple processors which communicate via a network.

Fairness: Processes will eventually be able to progress, i.e., enter their critical regions.

Message passing: A technique for providing mutual exclusion, communication, and synchronization between concurrent processes via sending messages between processes.

Monitor: An encapsulation of a resource and the operations on that resource which serves to ensure mutual exclusion.

Multiprocessing: The use of multiple processors that share a common memory.

Multiprogramming: Simulating concurrency on a single processor by interleaving instruction execution from multiple processes; time sharing or time slicing.

Mutual exclusion: The property ensuring that a critical region is executed indivisibly.

Race: The state in which two or more processes are continuously competing for resources and neither progresses.

Remote procedure call: The message sending architecture in which processes request services of procedures on remote machines.

Rendezvous: The message passing construct used in the Ada language.

Semaphore: A nonnegative integer-valued variable on which two operations are defined; **P** and **V** to signal intent to enter and exit, respectively, a critical region.

Synchronous message passing: The message sending process requires both sender and receiver to synchronize at the moment of message transmission.

References

Ahuja, S., Carriero, N., and Gelernter, D. 1986. Linda and Friends. *Computer* 19(8):26–34.

Andrews, G. R. 1991. *Concurrent Programming: Principles and Practice*. Benjamin Cummings, New York.

Andrews, G. R. and Olsson, R. A. 1993. *The SR Programming Language*. Benjamin Cummings, New York.

Andrews, G. R. and Schneider, F. B. 1983. Concepts and notations for concurrent programming. *Comput. Surv.* 15(1):3–43; reprinted in Gehani, N. and McGettrick, A. D., eds. 1988. *Concurrent Programming*. Addison–Wesley, New York.

Ben-Ari, M. 1982. *Principles of Concurrent Programming*. Prentice–Hall, London.

Ben-Ari, M. 1990. *Principles of Concurrent and Distributed Programming*. Prentice–Hall, London.

Bernstein, A. J. and Lewis, P. M. 1993. *Concurrency in Programming and Database Systems*. Jones and Bartlett, Boston, MA.

Brinch Hansen, P. 1975. The Programming Language Concurrent Pascal. *IEEE Trans. Software Eng.* 1(2):199–207; reprinted in Gehani, N. and McGettrick, A. D., eds. 1988. *Concurrent Programming*. Addison–Wesley, New York.

Dijkstra, E. W. 1968. The structure of the T. H. E. multiprogramming system. *Commun. ACM* 11:341–346.

Filman, R. E. and Friedman, D. P. 1984. *Coordinated Computing*. McGraw–Hill, New York.

Gehani, N. and McGettrick, A. D., eds. 1988. *Concurrent Programming*. Addison–Wesley, New York.

Gehani, N. H. and Roome, W. D. 1986. Concurrent C. *Software: Pract. and Exp.* 16(9):821–844; reprinted in Gehani, N. and McGettrick, A. D., eds. 1988. *Concurrent Programming*. Addison–Wesley, New York.

Goldberg, A. and Robson, D. 1989. *Smalltalk-80 The Language*. Addison–Wesley, New York.

Hartley, S. J. 1995. *Operating Systems Programming*. Oxford, New York.

Hinchey, M. G. and Jarvis, S. A. 1995. *The CSP Reference Book*. McGraw–Hill, New York.

Hoare, C. A. R. 1985. *Communicating Sequential Processes*. Prentice–Hall, London.

Jones, G. and Goldsmith, M. 1988. *Programming Occam 2*. Prentice–Hall, New York.

Lester, B. P. 1993. *The Art of Parallel Programming*. Prentice–Hall, Englewood Cliffs, NJ.

Milner, R. 1989. *Communication and Concurrency*. Addison–Wesley, New York.

Peterson, G. L. 1983. A new solution to Lamport's concurrent programming problem using small shared variables. *ACM Trans. Programming Lang. Syst.* 5(1):56–55.

Further Information

Further information may be gleaned from a number of sources; particularly recommended are: Andrews and Schneider [1983] for an introduction with a formal axiomatic flair; Andrews [1991] for a comprehensive view of the field, again with an axiomatic flair, targeted to a sophisticated audience, but including a fascinating bibliography with historical notes and extensive problem sets; Ben-Ari [1982] for a nice introduction including problem sets; Ben-Ari [1990] which adds Ada code examples, correctness arguments, and distributed computing; the process algebra approach is developed in Hoare [1985] and Milner [1989]; Filman and Friedman [1984] emphasize the various models of concurrent computation; Lester [1993] provides a comprehensive introduction including efficiency considerations, but without correctness arguments; Bernstein and Lewis [1993] use the axiomatic approach to develop concurrent solutions to a variety of problems with an emphasis on databases; and Gehani and McGettrick [1988] reprint a number of the classic papers in the field.

The journal *Concurrency: Practice and Experience* focuses on practical experience with concurrent machines and concurrent solutions to problems; concurrency is also frequently dealt with in a large number of society journals.

In addition, there are a large number of resources available via the Web which may be discovered through the use of the various search techniques.

99

Compilers and Interpreters

Kenneth C. Louden
San Jose State University

99.1 Introduction

Compilers and interpreters are language translators that have many of the same functions in common, in that both must read and analyze source code. A compiler, however, produces a program equivalent to the source program in a target language, usually object or assembly code but also sometimes C, whereas an interpreter directly executes the source program. Any programming language may be either compiled or interpreted, but languages with significant static properties (e.g., FORTRAN, ADA, C++) are almost always compiled, whereas languages which are more dynamic in nature (e.g., LISP, SMALLTALK) are more likely to be interpreted. Languages that differ substantially from the standard von Neumann model of most architectures (e.g., PROLOG) may also be interpreted rather than compiled. A performance penalty is incurred by interpretation over compilation, and so in cases where speed is critical compilation is to be preferred. By mixing compilation and interpretation, this performance penalty can be reduced, usually to well within an order of magnitude. The advantage to interpretation is that the compilation step is avoided (useful during program development), and an interpreter offers greater control over the execution environment (useful for complex run time environments) and greater flexibility in adapting to different architectures.

The first translators were developed in the 1950s. Prior to the development of high-level languages, a compiler was essentially what is known as a linker today: it *compiled* a collection of machine-language routines from a library to form a single program. A team at IBM under the direction of John Backus is generally credited with developing the first commercial compiler for a high-level language during the period 1954–57 [Backus et al. 1957]. The language translated by this first compiler was FORTRAN, which was designed simultaneously with the compiler (and is also credited with being the first high-level language). Modern translation techniques were first used in Algol60 compilers a few years later (see, e.g., Randell and Russell [1964]), when the relationship of language translation to the theory of finite automata

0-8493-2909-4/97/$0.00+$.50

and context-free grammars became better understood. The study of these subjects was further stimulated by this relationship, and by the early 1970s most of the standard techniques in use today were known. Since then general improvements in translators have come in the following areas: (1) the automation of a significant portion of the construction of a translator; (2) improvements in the speed of the target code, due to the increased application of code improving (or *optimizing*) algorithms; and (3) a greater ability for compilers to be relatively easily **retargeted**, or rewritten for a new target machine language, due in part to automation and in part to a better understanding of the required compiler structure. Improvements have also come in the implementation of special language features, such as exception handling, generics (parametric polymorphism), object-oriented features such as dynamic binding, and parallelization, due to increasing understanding of these mechanisms. One area in which significant theoretical advances have taken place in the past 20 years is in the translation of functional languages, with new algorithms for type checking, type classes, and interpretation by tree reduction (see, e.g., Peyton Jones [1987]). So far, however, these techniques have remained outside of the mainstream of languages and translators.

An important aspect of the technology of translators is its strong interaction with language design. For example, the introduction of block structure in the ALGOL language family gave impetus to the development of stack-based translation algorithms. In turn, the stack-based algorithms influenced the development of language semantics specifically designed to take advantage of the algorithms, such as the lexical scope rule and the principle of syntax-directed translation, in which the semantics of a program (i.e., its execution behavior) is directly reflected in its syntax, or structure, so that translation can be guided by this structure. (This rule could equally well have been formulated as *semantics-based syntax*.) Because of its early appearance, FORTRAN was not designed with these principles in mind and remains a somewhat more difficult language to translate with the standard techniques than later languages of the ALGOL family.

Another aspect to the development of language translators has been the tendency of the computing community to constantly rediscover or recreate translation techniques (other than the basic, well-known ones as outlined in many texts). Primarily, this is due to the lack of detailed documentation of specific translators in the computer science literature, and this in turn is largely due to the commercial and/or proprietary nature of most translators. Two translators that historically were reasonably well documented, and which exerted a corresponding influence on subsequent translator construction, were the portable C compiler [Johnson 1978] and the PASCAL P-compiler [Wirth 1971, Nori et al. 1981]. Lately, the availability of public-domain software from the Internet has greatly improved the opportunity to study existing translators. Particularly well documented and of high quality are the compilers distributed as part of the Gnu software project of the Free Software Foundation [Stallman 1994].

Common to almost all modern compilers is a conceptual structure, in which the tasks performed are divided into **phases**, or logically complete processing steps. The standard phase structure is shown in outline in Fig. 99.1.

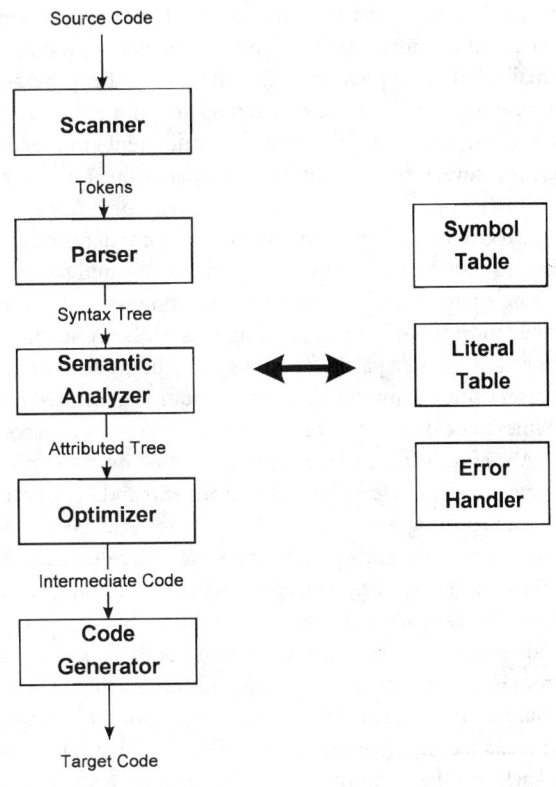

FIGURE 99.1 The phases of a compiler.

The first phase, called the scanner, or lexical analyzer, is the only phase directly involved in reading the source program. It converts the characters of the source program into tokens, or sequences of characters that represent the basic units of program structure. Typical tokens are keywords such as *while*, numeric literals such as 3.14159, and identifiers.

The second phase in a compiler is the parser, or syntactic analyzer, which collects sequences of tokens into complete units such as expressions, statements, and declarations. The output of the parser (either explicitly or implicitly) is a tree or other equivalent data structure representing the structure of the programming unit just recognized. This tree is called the syntax tree.

The third phase in a compiler is the semantic analyzer, which computes attributes or properties of each programming construct as well as its effect on the attributes of other constructs. Typical attributes include data types of expressions and identifiers, memory sizes, and actual or potential memory locations. The semantic analyzer also determines if the construct makes sense according to these attributes, i.e., it performs consistency checks, such as type checking or range checking. The output of the semantic analyzer can be represented by an attributed tree representing the original tree modified by the computed attributes.

The fourth phase in a compiler is called the optimizer in Fig. 99.1. This phase usually generates some form of linear code, called intermediate code, from the representation passed to it by the semantic analyzer. It also applies some forms of code improving algorithms, either before or after generating intermediate code (or both). This phase of compilation is the most variable in different compilers and may even be absent altogether.

The fifth and final compiler phase shown in Fig. 99.1 is the code generator. This phase generates the final target code and may also perform some additional improvements to the code.

All of the phases of a compiler interact with various tables and handlers within the compiler. Figure 99.1 shows three important examples of these other components, the symbol table, the literal table, and the error handler, and indicates their interaction with the phases by a large double-headed arrow. The symbol table maintains the names defined by the program under translation, as well as possible predefined names in the language, and associates the names with their attributes, which may include scope, memory location, data type, and memory size. The symbol table may be monolithic or may be separated into a tree or graph of smaller tables, representing different scopes in the program. A similar table is needed for literal values that appear in the program, such as strings and numeric literals. The third component shown in Fig. 99.1 is the error handler, whose job is to generate different kinds of error messages but, more importantly, to provide error recovery, so that translation may continue (at least as far as is practical) in the presence of errors.

It must be emphasized that the compiler phases are logical units only and may not correspond to any actual grouping of operations within the compiler itself or to any temporal sequencing of these operations. Indeed, it is customary for scanning, parsing, and some semantic analysis to be completely integrated in a single pass over the source code. A compiler may even be one-pass in that all phases including code generation are performed simultaneously (assuming that the language itself permits it). More likely is that there are separate passes for parsing (including scanning), optimization, and code generation (with later passes using the intermediate representation generated by the first pass). If the language does not require names to be declared before use, then a pass is also neccesary to resolve name references.

A useful division of the tasks performed by a compiler is into an analysis part and a synthesis part, sometimes also referred to as the **front end** and **back end**. The analysis part is concerned with analyzing the source program, whereas the synthesis part is concerned with generating the target program. The analysis part depends primarily on the source language, whereas the synthesis part depends primarily on the target language or target machine. Scanning, parsing, and semantic analysis are part of the front end, whereas code generation is part of the back end. Optimization and the generation of intermediate code usually require information about both the source and target and are more difficult to divide into a front end component and a back end component. The more successfully this is done, the easier it is to retarget the compiler. In the best case, a group of compilers are able to share front ends and back ends interchangeably. A popular and effective design for an interpreter consists of a compiler front end and a back end that is an interpreter for the intermediate code produced by the front end. This results in a reasonably efficient interpreter that is also easily retargetable.

99.2 Underlying Principles

Algorithms used in translators are based heavily on computation theory and, to a lesser extent, on formal semantics. Scanners are direct implementations of finite automata that solve string recognition problems through nonrecursive pattern matching. Parsers depend on the theory of context-free grammars and pushdown automata, which solve recursive recognition problems through stack-based pattern matching. Semantic analysis depends on solving sets of tree equations called **attribute grammars**. Code generation and interpretation also can be seen as applications of attribute grammars. It is possible to use even more formal semantic specifications, particularly denotational specifications, of the source and target languages to construct semantic analyzers and code generators (see, e.g., Polak [1981], and Lee [1989]). The advantage to doing so is that the compiler can be proven be correct (i.e., the semantics of the source and target programs are guaranteed to be the same). However, these techniques have not become popular and we do not discuss them further. In the remainder of this section, we will discuss each of the areas mentioned in a little greater detail.

Finite Automata

A finite automaton is an abstract computational machine consisting of a finite number of states and transitions between states based on input symbols. The machine runs by beginning in the starting state, and consuming input symbols while entering new states via corresponding transitions, until either an error state or an accepting state is reached, at which point it may declare success or failure or possibly continue executing. Each state represents stored knowledge about the computation up to that point. A finite automaton can handle arbitrary repetition by a fixed finite set of input symbols, but it cannot handle recursive processes, because that would involve an unpredictable number of states.

Finite automata are the basis for the recognition of tokens within a scanner. Based on the well-known correspondence from computation theory between finite automata and regular expressions, tokens usually are given initially by regular expressions, which are specifications for the string patterns, or **lexemes**, that a token represents. The mathematical theory of regular expressions limits itself to the consideration of three matching operations: the choice between two alternatives, indicated by the vertical bar | (similar to the logical OR operation); the concatenation or sequencing of two strings (with no operator symbol); and the repetition of a pattern, indicated by a postfix asterisk $*$ (sometimes called the closure or Kleene closure operation). Parentheses also are used to group subexpressions together. As an example of the use of regular expressions to represent tokens, the following regular expression for a token represents simple unsigned numbers consisting of a sequence of one or more decimal digits:

$$(0|1|2|3|4|5|6|7|8|9) \quad (0|1|2|3|4|5|6|7|8|9)^*$$

Such a regular expression can be converted into a finite automaton by one of several standard algorithms. The basic method is to use Thompson's construction [Aho et al. 1986, p. 122] to derive a nondeterministic automaton (i.e., one with an unpredictable next state) from the regular expression and then to use the subset construction [Aho et al. 1986, p. 118] to derive an equivalent deterministic automaton from the nondeterministic one. Other algorithms exist that perform this conversion in one step as well as construct an automaton with a minimal number of states. Whereas these algorithms can sometimes be useful for the design of scanners, their primary use is in the construction of scanner generators such as Lex (discussed later in this chapter).

Context-Free Grammars

The theory of context-free grammars extends the ideas of finite automata and regular expressions to recursive situations. A context-free grammar is a collection of named recursive rules of the form $A \rightarrow \alpha_1 | \cdots | \alpha_n$, where A is the name of the rule and the α are strings of tokens and names (including possibly A

itself) representing the different possible choices for the structure of *A*. The names are called **nonterminals** and the tokens are called **terminals**, for technical reasons explained shortly. Grammar rules are also sometimes referred to in the theory as **productions**. Each such rule represents the fact that a structure represented by *A* may become any one of the structures represented by an *α*. The absence of context for these potential replacements of *A* is what makes these grammars context free. It is possible to define grammars that are more general than context-free grammars, and these can be arranged according to increasing generality in what is known as the Chomsky hierarchy. However, context-free grammars are the most useful within the computational constraints of language translation.

Every nonterminal in a context-free grammar defines a set of token strings: the strings of tokens that can be legally parsed by the grammar rule for that nonterminal (and associated rules). The most general structure that is defined by the grammar is usually singled out as the structure representing the entire language, and its associated nonterminal is called the start symbol. A legal parse is represented by a sequence of replacements of nonterminals by choices of right-hand sides of their associated grammar rules. This sequence of replacements is called a derivation. In a derivation, every nonterminal must eventually be eliminated by replacing it with another string, whereas a terminal, once it appears, is never replaced; this may be seen as a justification for the names terminal and nonterminal. In general, there are many possible derivations for the same string, which can be constructed by varying the order in which the replacements are made. Two kinds of derivations that are important for parsing use fixed orders for the replacements. In the first, called a leftmost derivation, the leftmost nonterminal in the current string is always the one to be replaced at the next step. Correspondingly, a rightmost derivation is one in which the rightmost nonterminal is always the one to be replaced at the next step.

Although derivations are useful for expressing exactly which steps are taken in a parse, they do not express the structure of the parsed string very well. A more useful representation of this structure is the parse tree, which represents each terminal and nonterminal by labeled nodes and each replacement step in a derivation by the construction of a set of children of the node label being replaced. In a parse tree, each leaf node is labeled by a token, and each interior node is labeled by a nonterminal. Even parse trees have more detail than is actually needed to determine the meaning of a structure, and a condensed form of a parse tree called a syntax tree is the most useful data structure for translation. In a syntax tree, nodes may have a more diverse structure than in a parse tree: they may have more than one label, and both terminals and nonterminals may be used as labels of interior nodes.

As a brief example of the grammar concepts we have summarized, consider the grammar

$$exp \rightarrow exp + num \mid num$$

which has one nonterminal *exp*, two terminals + and *num*, and a single production with two choices. Here *num* represents a number token with lexemes such as 42 or 7.) An example of a legal string for this grammar is $3 + 4 + 5$. A leftmost (and rightmost) derivation is

$$exp \Rightarrow exp + num \Rightarrow exp + num + num \Rightarrow num + num + num$$

A parse tree for the same string is given in Fig. 99.2. A syntax tree is given in Fig. 99.3.

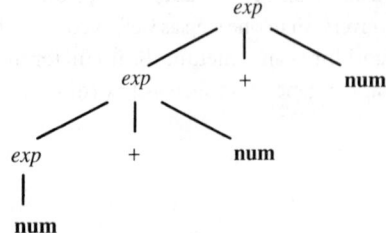

FIGURE 99.2 A parse tree for the string $3 + 4 + 5$.

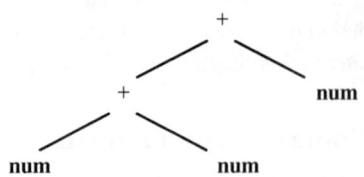

FIGURE 99.3 A syntax tree for the string $3 + 4 + 5$.

Parsing Algorithms

Algorithms designed to match an input string of tokens based on a grammar and, either implicitly or explicitly, to construct a parse or syntax tree are called parsing algorithms. Parsing algorithms come in two general varieties, **top-down** and **bottom-up**. Top-down algorithms construct a parse tree from the root to the leaves by guessing which structures are about to appear based on the next part of the input string and the structure expected to be seen. Bottom-up algorithms construct a parse tree from the leaves to the root by consuming the input and forming a set of subtrees until the next structural element can be guessed from the structures seen so far and the next part of the input string. Because of the recursive nature of context-free grammars, both kinds of algorithms must use a stack, either explicitly or implicitly, to hold partial results. When it is constructed explicitly, this stack is referred to as the parsing stack, and it possibly will contain terminals, nonterminals, or other symbols representing the state of the parser. Because of the nature of their operation, top-down parsers trace out the steps of a leftmost derivation, and bottom-up parsers trace out in reverse the steps of a rightmost derivation.

There are algorithms of both varieties that will parse any context-free grammar. Early's algorithm is the most well-known bottom-up algorithm [Early 1970]. A general top-down algorithm may be found in Graham et al. [1980]. General algorithms may run in significantly slower than linear time (Early's algorithm requires cubic time in general), so they rarely are used in practice. Standard algorithms for top-down parsing are the **LL(k)** algorithms (parsing the input from left to right, giving a leftmost derivation, using k symbols of lookahead), and the **LR(k)** algorithms for bottom-up parsing (parsing the input from left to right, giving a rightmost derivation, using k symbols of lookahead). Both kinds of algorithms require that a grammar satisfy extra conditions to be parsable. The LL(k) algorithms in particular are quite restrictive, although easy to use. The LR(k) algorithms, although less restrictive, are more complex. One top-down parsing method that is more flexible than the LL(k) methods is called **recursive-descent**. A bottom-up algorithm that is simpler than the LR(k) algorithms is called **lookahead LR(1) (LALR(1))** [DeRemer 1971, DeRemer and Pennello 1982] and is normally restricted to one symbol of lookahead. Since these algorithms have proven themselves to be the most effective and easiest to use in practice, we discuss them in a little more detail.

In the method of recursive-descent parsing, the grammar rules are viewed as prescriptions for the code of a set of mutually recursive procedures, one for each nonterminal. Recursive-descent, although suffering from some of the same problems as LL(k) parsing, is more flexible and can use simple ad hoc techniques to solve many of the problems of LL(k) parsing [Wirth 1976]. For instance, simple left recursion, which cannot be handled directly by an LL(k) parser, can be handled in recursive-descent by noting that a left recursive rule $A \rightarrow A\alpha \mid \beta$ is equivalent to a parsing procedure that first recognizes β and then a sequence of zero or more α (since the grammar rule generates strings of the form $\beta\alpha\alpha\ldots$). Thus, a recursive-descent procedure for the grammar $exp \rightarrow exp + num \mid num$ can be written using a while loop, as follows:

```
void exp(void)
{ match(NUM);
  while (nextToken == PLUS)
  { match(PLUS);
    match(NUM);
  }
}
```

An LALR(1) parser uses an explicit parsing stack instead of recursion. The state of a parse can be partially expressed by a finite automaton whose states consist of sets of so-called **items**, each item consisting of a production choice, a distinguished position indicated by a period, repesenting the point of progress in recognizing the rule, and an associated set of lookahead tokens legal at that point in the parse. [In the

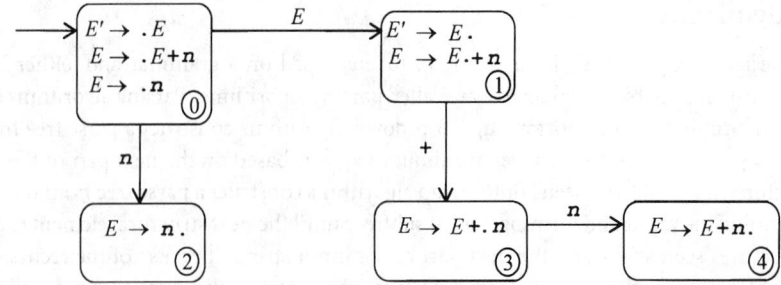

FIGURE 99.4 A DFA of sets of LR(0) items.

following discussion we will use the so-called LR(0) items that lack a lookahead component; although LALR(1) items are more complex, the basics of the LALR(1) algorithm can be understood using these simpler items.]

Consider, for instance, the grammar $exp \rightarrow exp + \mathbf{num} \mid \mathbf{num}$, which we will write for convenience in the form $E \rightarrow E + n \mid n$. There are six LR(0) items: $E \rightarrow .E + n$, $E \rightarrow E. + n$, $E \rightarrow E + .n$, $E \rightarrow E + n.$, $E \rightarrow .n$, and $E \rightarrow n.$. The start of a parse is indicated by beginning in a state represented by a new rule representing a start symbol that cannot appear elsewhere, which we will write as $E' \rightarrow E$ in the example. The corresponding initial item is $E' \rightarrow .E$. Since this rule represents the fact that we may be about to recognize an E, we must also include the items $E \rightarrow .E + n$ and $E \rightarrow .n$ in this state. Transitions to new states are then given by moving the period past the symbols that follow it. For instance, there is a transition on the symbol E from the start state to the state containing the item $E \rightarrow E. + n$, and a transition on the symbol n from the start state to the state containing the item $E \rightarrow n.$. The complete finite automaton of items is given in Fig. 99.4.

This finite automaton has no accepting states and is used only to keep track of the state of the parser. It is used in conjunction with a parser stack that holds the state numbers that the parser has passed through while parsing the input. It is used as follows. First, the initial state is pushed onto the stack. Then, the next input token is consulted. If there is a transition on this token from the current state (on top of the stack), then the token is removed from the input and the new state is pushed onto the stack; this is called a shift operation. If, however, there is an item in the current state of the form $A \rightarrow \alpha.$ (a so-called final item), then this indicates that the string α has already been recognized, and it can be replaced by A; this is called a reduce operation. A reduce operation is performed as follows. The states on the stack corresponding to α are popped from the stack (one state for each symbol in α). The state remaining at the top of the stack must have a transition on A, which is then taken, and the new state is pushed onto the stack.

As an example of this process, consider the automaton of Fig. 99.4, and suppose that the input string is $n + n$. We depict the initial state of the parse as follows:

Parsing Stack	Input
$0	$n + n$

The $ in this representation is used to indicate both the bottom of the stack and the end of the input. The first step in the parse is a shift on n from state 0 to state 2. Then a reduction by $E \rightarrow n$ takes place, and the parser moves to state 1. At that point, the $+$ and n are shifted, and the parser is in state 4. Then a reduction by $E \rightarrow E + n$ is made, popping states 4, 3, and 1, revealing again state 0. Again, the E transition is followed into state 1. At this point the end of the input is encountered, and a reduction by $E' \rightarrow E$ is made, which corresponds to accepting the input. The complete set of actions of the parser is given in Table 99.1,

TABLE 99.1 The Actions of an LALR(1) Parser

Parsing Stack	Input	Action
$0	$n + n$	shift 2
$02	$+n$	reduce $E \rightarrow n$
$01	$+n$	shift 3
$013	n	shift 4
$0134	$	reduce $E \rightarrow E + n$
$01	$	accept

in which shift actions also include the new state number. Table 99.2 shows the LALR(1) parsing table for this simple grammar, which is used by the parser to select the actions indicated in Table 99.1. This table is two dimensional and is indexed by state and lookahead token. Each table entry contains an action; shift entries are indicated by an s and the new state number; reduce entries are indicated by an r and the rule to be reduced; empty entries are errors. Whereas this table is closely related to the automaton of Fig. 99.4,

TABLE 99.2 An LALR(1) Parsing Table

State	Input			Goto
	n	$+$	$\$$	E
0	s2			1
1		s3	accept	
2		$r(E \to n)$	$r(E \to n)$	
3	s4			
4		$r(E \to E + n)$	$r(E \to E + n)$	

the exact entries can be inferred only from a lookahead component, which we did not compute here.

An additional area in Table 99.2 is called the goto area. In this area are the transitions on nonterminals that are performed during reductions. These are essentially the same as shift operations, except that no input is consumed. A parser also may choose to condense this table by the use of *default* entries. For example, since state 2 has only reduce entries by the same rule, this could be made the default, in which case even on input token *n* the parser will perform the given reduction. This has the effect of postponing the declaration of error, but it cannot result in an incorrect parse. A similar default can be used for state 4.

One final bottom-up parsing method that deserves mention is operator-precedence parsing. It is a method that predates the LALR(1) method, but it can be used effectively on expression grammars involving infix operators; see Aho et al. [1986, pp. 203–215] for a description.

Attribute Grammars

Whereas context-free grammars are generally accepted as the standard way to describe the syntax of a programming language, there is no equivalently accepted method for describing semantics. Formal methods for describing semantics, such as operational semantics or denotational semantics, have not met with universal acceptance, and though translators can be derived from such specifications, this is rarely done. Instead, various ad hoc mechanisms are used in translators to perform semantic analysis and code generation or interpretation.

One method that has proven useful for the translator writer is to use a so-called attribute grammar to describe the semantics of a programming language [Knuth 1968]. An attribute grammar associates to each grammar rule a set of equations describing the computational relationships among a set of attributes attached to the symbols in the rule. These attributes can be anything from the data type of a variable to the value of an expression; even the target code generated by a compiler can be represented as a string attribute in an attribute grammar. Most often, attributes are used to represent the static rather than the dynamic properties of programs, and they are viewed as being fixed values attached to the nodes of a syntax tree. Indeed, attribute values are usually written using a dot notation similar to that of record fields, so that $X.a$ means the value of attribute a of symbol X. Attributes may in fact be implemented as fields in syntax tree nodes, or they may be stored in the symbol table or other data structures elsewhere in the translator.

Given a set of attributes $a_1, \ldots, a_k$ and a grammar rule choice $X_0 \to X_1 X_2 \cdots X_n$, the jth attribute at the ith symbol is given in an attribute grammar by an equation of the general form

$$X_i.a_j = f_{ij}(X_0.a_1, \ldots, X_0.a_k, X_1.a_1, \ldots, X_1.a_k, \ldots, X_n.a_1, \ldots, X_n.a_k) \tag{99.1}$$

where f_{ij} is a mathematical function. An attribute grammar is thus written in purely functional style without side effects.

As an example, consider the grammar *exp* → *exp* + **num** | **num**, which we may assume expresses the summation of a series of numbers. An attribute grammar for the numeric value of an expression defined by this grammar is given in Table 99.3. Note that the two instances of the nonterminal *exp* in the first grammar rule must be distinguished by subscripting and that the terminal **num** is assumed to have its numeric value (called *lexval* in Table 99.3) precomputed, possibly by the scanner.

Of particular importance to the translator writer are the kinds of dependencies that the attribute equations create among the attributes of different symbols in a parse tree, since these dependencies determine when and how, or even if, the attributes can be computed during translation. A primary requirement is that the attribute grammar not have any circular dependencies.

TABLE 99.3 An Attribute Grammar for a Simple Expression Grammar

Grammar Rule	Attribute Equations
$exp_1 \rightarrow exp_2 + num$	$exp_1.val = exp_2.val + num.lexval$
$exp \rightarrow num$	$exp.val = num.lexval$

In practical situations, this requirement is virtually guaranteed unless an error has been made. Attributes whose dependencies flow from right to left in the grammar rules [i.e., those whose Eqs. 99.1] all have only the symbol X_0 on the left are called **synthesized attributes**, whereas any other attributes are called **inherited**. Synthesized attributes can be computed bottom-up during a parse or by postorder traversal of the syntax tree, whereas inherited attributes require a more complex computation scheme. Indeed, as the name implies, inherited attributes are often passed down the syntax tree from parent to child, or from sibling to sibling, and so can be computed by some form of modified preorder traversal of the syntax tree.

A great deal of effort can be expended to ensure that all attribute values are computable during the parsing phase, to avoid having to construct the entire syntax tree and to avoid having to make additional passes over the input. The requirements that this places on the attribute grammar vary, depending on the parsing method employed. First, since virtually all parsers read the input from left to right, the attributes must be computable from left to right, and this means that all inherited attributes must depend only on the attribute values of their left siblings. In terms of the Eqs. 99.1, this means that each equation for an inherited attribute ($X_i.a_j$ is on the left and $i > 0$) must have the form

$$X_i.a_j = f_{ij}(X_0.a_1, \ldots, X_0.a_k, X_1.a_1, \ldots, X_1.a_k, \ldots, X_{i-1}.a_1, \ldots, X_{i-1}.a_k) \qquad (99.2)$$

An attribute grammar in which all equations for inherited attributes are of this form is called **L-attributed**. A further requirement for attribute evaluation during parsing is a form of strong noncircularity, in which an order for attribute evaluation can be fixed in advance without incurring any cycles (naturally occurring attribute grammars satisfy this noncircularity requirement, too).

The particulars of the parsing algorithm can also have a significant effect on which attributes are computable during parsing. Recursive-descent parsers are the most flexible; in the recursive routines, inherited attributes can be implemented as passed parameters, whereas synthesized attributes become returned values [Katayama 1984]. Bottom-up parsers most naturally compute synthesized attributes. They do this by maintaining a stack of attribute values in parallel with the parsing stack. New synthesized attributes are computed on this stack at each reduction step. It is also possible to evaluate certain inherited attributes during a bottom-up parse, but this often requires that the grammar be rewritten, so that the attribute equations can be converted to a manageable form. Indeed, it is theoretically possible to rewrite a grammar so that all attributes become synthesized [Knuth 1968]. However, the grammar thus produced bears little resemblance to the original. Thus, in difficult situations, it may be preferable to delay an attribute computation until after the parse and avoid rewriting the grammar into an unrecognizable form.

99.3 Best Practices

The theory of scanning, parsing, and attribute analysis implies that, in principle, a compiler or interpreter can be generated automatically from descriptions of the source language, attributes, and target machine. Whereas a number of compilers have been generated in this way [Farrow 1984], it is not common. This is partially because the necessary tools (particularly for optimization and code generation) are complex and have not become standard, and partially because of the need for efficiency, both in translation and in terms of the generated code, which is more difficult to achieve using general tools. At this writing, it is common to automate only the construction of the scanner and the parser, although at least partial automation of code generation has become more common with the increasing importance of retargetability.

Whereas a large number of scanner and parser generators have been written, only a few have gained broad acceptance. We describe here two sets of tools that have been used in many compilers: the Unix tools Lex and yet another compiler compiler (Yacc) (and their public domain versions Flex and Bison), and the Purdue Compiler Construction Tool Set (PCCTS). Yacc produces an LALR(1), whereas PCCTS produces a recursive descent parser. It should be noted that many commercial compilers and interpreters have been written by hand without the use of any tools at all; in such cases, the parsers are usually written using recursive-descent.

Specifying Syntax Using Regular Expressions and Grammars

In order to automate the task of generating a scanner and a parser, the first step is to specify the tokens using regular expressions and the syntax using context-free grammar rules. Although it is essential to understand the mathematical theory of both of these mechanisms, the theory ignores extensions and features of practical importance. We mention a few of these here.

The theory of regular expression relies on only three operations: concatenation, choice, and repetition. Whereas these are enough to match any string recognizable by a finite automaton, most pattern matching systems extend this set of operators in many ways. As a typical example, consider the regular expression

```
[0-9]+(\.[0-9]+)?
```

which is written using standard conventions for the Unix tools Lex and Grep. This expression specifies a pattern for a simple floating-point constant without exponential part: the expression [0–9] refers to a choice from the range of characters from 0 to 9 (i.e., a digit), the + indicates a repetition of one or more ($a+$ is equivalent to aa^*), the backslash in front of the period *escapes* the metacharacter meaning of the period (otherwise it would match any character), and the question mark indicates an optional component.

Even with these extensions, it is sometimes difficult to write regular expressions for certain patterns, even when such patterns do exist, and one may choose to apply an ad hoc recognition process instead of using a regular expression. A notorious case in point is that of C comments, which can be loosely described as /* (not */) */. The trouble is expressing not */ as a regular expression. More generally, the nonexistence of sequences of more than one character in a string is a difficult property to express as a regular expression. Fortunately, these situations do not occur very often in real languages.

It is also necessary to be aware that the theory of regular expressions can be easily extended to cover simple nonregular situations. For instance, nested comments require recursion and so cannot be directly expressed as a regular expression. Nevertheless, adding a simple counter variable to a scanner permits it to recognize potentially nested comments.

Similar considerations arise in defining syntax using context-free grammars. Consider the following grammar that specifies a simple four-function floating-point calculator program with a single memory:

```
session → asgn nl session | nl
asgn → = expr | expr
expr → expr addop term | term
addop → + | -
term → term mulop factor | factor
mulop → * | /
factor → - factor | ( expr )| NUMBER | m
```

A **session** consists of a sequence of assignments **(asgn)** followed by newlines (indicated in the grammar by **nl**), or just a newline (this is used to end the session). The arithmetic operations have their usual meanings. The optional = sign at the beginning of an **asgn** indicates assignment of the value of the following expression to the memory. The single letter **m** in a **factor** fetches the value of this memory. The token **NUMBER** is the only token with more than one possible lexeme, and it is assumed to be given by the regular expression given earlier in this section.

A sample session with an interpreter for this grammar might look as follows:

```
= 3.14 - 2
        1.14
= m*3.14 + 3
        6.5796
  m*3.14 - 4
            16.6599
```

This represents a computation of the value of the polynomial $x^3 - 2x^2 + 3x - 4$ at the point $x = 3.14$ using Horner's rule $(x^3 - 2x^2 + 3x - 4 = ((x - 2)x + 3)x - 4)$.

This grammar is written in a style called **Backus normal form** or **Backus–Naur form (BNF)**. In it the only metasymbols are the arrow and the vertical bar (sometimes ::= or : or = is used instead of the arrow, and sometimes nonterminals are distinguished more directly from terminals by surrounding them with angle brackets $< \cdots >$). Repetition is expressed by recursion, and optional features are expressed by writing separate choices. This grammar also expresses the associativity and precedences of the operators directly in the grammar: a left recursive grammar rule such as **expr → expr addop term** indicates left associativity, and the different precedence levels **expr**, **term**, and **factor** cause the operators at each level to be given precedences in ascending order (lowest precedence first). When recursion is used only to express repetition, it can be written on the right or the left: the rule for a **session** is written right recursively for more or less arbitrary reasons.

An alternative to this grammar would be to condense it to the following grammar (still in BNF):

```
session → asgn nl session | nl
asgn → = expr | expr
expr → expr op expr
            | - expr
            | ( expr )
            | NUMBER
            | m
op → + | - | * | /
```

This grammar is ambiguous, however, in that the order in which the operations are applied is not specified, and a legal string may have many different parse trees. This grammar can still be used as the basis for the calculator, as long as separate **disambiguating rules** are stated giving the associativities and precedences of the operators. The advantage to this is that the grammar itself becomes shorter and easier to understand.

A different variety of context-free grammar is obtained by adding metasymbols representing optional and repeated constructs. One standard version of this is called **extended BNF (EBNF)**, which uses square brackets $[\cdots]$ to surround optional constructs and braces $\{\cdots\}$ to surround repeated constructs (one could just as well use parentheses, ?, and ∗ as in regular expressions). The calculator grammar in EBNF becomes

```
session→ { asgn nl } nl
asgn → [=] expr
expr → term { addop term }
addop → + | -
term → factor { mulop factor }
mulop → *| /
factor → - factor | ( expr ) | NUMBER | m
```

In such a grammar, the associativity of the operators is now suppressed in favor of showing the repetition directly. A standard convention is to assume left associativity of operators in any rule using the braces (since a session has no operators, the associativity of its rule is of no consequence).

Lex/Flex and Yacc/Bison

Lex [Lesk 1975] and Yacc [Johnson 1975] are scanner and parser generators that are a part of most Unix distributions. Both have public domain versions Flex (Fast Lex [Paxson 1990] based on ideas of Jacobson [1987]) and Bison that run under a variety of operating systems. Each of these programs reads a definition file and produces as output a C souce code file containing a scanning/parsing procedure. The definition file for each has the same basic format,

```
{definitions}
%%
{rules}
%%
{auxiliary routines}
```

We discuss the contents of the definition files first for Lex and then for Yacc, using our running example of a simple calculator program whose tokens and grammar were described previously.

The Lex definition file for the calculator scanner is given in Table 99.4. In this example, the definitions section contains a **# include** directive inside the brackets **%{** and **%}** , and the definitions of the three tokens **digit**, **number**, and **whitespace**, using regular expressions written with previously described metasymbols (**whitespace**, for example, is defined to be a sequence of one or more blanks or tabs). All code inside the special brackets is inserted directly at the beginning of the C output file, thus allowing the user to provide declarations/definitions that may be used by the rest of the C code. In this case, the only insertion is to indicate the inclusion of the file **y.tab.h**. This file can be generated by Yacc, and it contains the definitions of tokens and other globals that permit communication between the Lex-generated scanner and the Yacc-generated parser. In our example (and for one particular version of Yacc), this file is as follows:

```
/* file y.tab.h */
extern double yylval;
extern int yylineno;
# define    NUMBER 258
# define    UNARY 259
```

The subsequent Lex code uses only the definition of **NUMBER** and **yylval** from this file.

TABLE 99.4 A Lex Definition File

```
/************************************************************
CALC.L
************************************************************/
%{
#include "y.tab.h"
%}
digit           [0-9]
number          {digit}+(\.{digit}+)?
whitespace      [\t]+
%%
{whitespace}    { /* skip */ }
{number}        {sscanf(yytext,"% lf",& yylval);
                return NUMBER; }
\n              {return '\n'; }
.               {return yytext[0]; }
```

The rules section specifies the actions that the Lex scanner is to take when each token is recognized. These actions are placed inside a C block and are inserted directly at the appropriate places in the scanner. In Table 99.4 the specified actions are as follows. Whitespace is skipped (empty action). A number causes the scanner to compute the floating point value from the **yytext** string and place it in **yylval** and then return the **NUMBER** token. (The **yytext** string contains the lexeme matched from the input.) The newline character is singled out for special handling, since it is not placed in the **yytext** string; it is returned directly. Finally, the period indicates a default action (it matches any character), and this causes the character value itself to be returned (**yytext[0]**).

This concludes the description of the calculator Lex file in Table 99.4. Note that this file contains no auxiliary routines section, and the **%%** symbol separating this section from the previous rules section is also omitted.

We turn now to a description of the Yacc definition file, which is in Table 99.5. The definitions section contains two lines of C code to be inserted in the output file. The first defines **YYSTYPE** to be **double**; this is the type used to define the Yacc value stack, which needs to be **double**, since expressions compute floating point values. The second line defines a static variable **mem**, which is to be used as the actual memory location for the single calculator memory. The definitions section also contains the token definitions, indicated by the **%token** directive. In this example, only the **NUMBER** token need be defined; other tokens are single character and may be referred to directly. Finally, the definitions section contains a description of the associativity and precedence of the

TABLE 99.5 A Yacc definition file

```
/*************************************************************
CALC.Y
*************************************************************/
%{
#define YYSTYPE double
static double mem;
%}
%token NUMBER
%left '+','-'
%left '*','/'
%left UNARY
%%
session : /* empty */
        | session '\n' { exit(0); }
        | session asgn '\n' {printf("\t%g\n",$2);}
        | session error '\n' {yyerrok;}
        ;
asgn    : '=' expr { mem = $2; $$ = $2;}
        | expr { $$ = $1; }
        ;
expr    : expr '+' expr { $$ = $1 + $3; }
        | expr '-' expr { $$ = $1 - $3; }
        | expr '*' expr { $$ = $1 * $ 3; }
        | expr '/' expr { $$ = $1 / $3; }
        | '-' expr %prec UNARY { $$ = - $2; }
        | '(' expr ')' { $$ = $2; }
        | NUMBER { $$ = $1; }
        | 'm' { $$ = mem; }
        ;
%%
int main()
{yyparse(); return 0;}

void yyerror( char* t)
{fprintf(stderr,"% s\n",t);}
```

arithmetic operators; these are necessary disambiguating rules, since the rules section uses the ambiguous form of the calculator grammar. The order in which the operators are listed determines their precedence (with lowest precedence listed first). The **%left** directive indicates that all operators are left associative. Finally, the **UNARY** token is implicitly defined by the last definition as having the highest precedence. It will be used to give unary minus a higher precedence than any of the binary operators.

The rules section of the Yacc specification contains the grammar in a modified BNF format, with actions contained in braces. Since Yacc is a bottom-up parser generator, it is easiest to use to compute synthesized attributes. In the calculator grammar, the value attribute is synthesized, and it is this attribute that we use Yacc's value stack to compute. The stored memory value, which has an inherited component, is handled directly by using the defined **mem** variable. The action code refers to the attribute values on the value stack by using symbols beginning with **$**. The symbol **$$** refers to the (synthesized) value to be computed for the nonterminal defined by the rule. Each of the symbols **$1**, **$2**, etc., refers to the attribute value computed for each symbol on the right-hand side of the grammar rule. Thus, **$$ = $1 + $3** in the rule **exp : exp + exp** indicates that the value of the first and third symbols (the right-hand expressions) are to be added to get the value of the result expression. This convention allows Yacc rules to be written in a style very close to the synthesized rules of an attribute grammar.

A few changes have been made to the grammar to make it more usable with Yacc. One is that the operators are written directly into the expressions instead of being listed separately; this eliminates the need for a separate character attribute for an operator rule. Two additional changes have been made to the grammar rule for a **session**. First, the rule is written in left recursive instead of right-recursive form. A right-recursive rule causes the parsing stack to grow without limit; this means that a very long session might cause a stack overflow (a similar situation is caused by tail-recursive procedure calls). Thus, all right-recursive rules that do not reflect an associativity requirement should be rewritten in left recursive form. The remaining change is the addition of the rule

```
session : session error '\n' {yyerrok;}
```

to the choices for a session. This is an error handling production. It matches any line that is not a legal assignment to the internally defined Yacc nonterminal **error**, and the action **yyerrok** resets the Yacc parser to accept more input. Yacc also automatically calls a **yyerror** procedure that can print an error message or perform some other action.

Finally, the auxiliary routines section of Table 99.5 contains a **main** procedure, which just calls the parsing procedure **yyparse** which is generated by Yacc. (The scanning procedure generated by Lex is called **yylex** and is automatically called at the appropriate times by **yyparse**.) The **yyerror** procedure is also defined in this section; it simply prints an error message (supplied by Yacc).

Assuming that the Lex definition for the calculator is in the file **calc.l**, and the Yacc definition is in the file **calc.y**, then a running calculator program can be built in Unix with the commands

```
lex -I calc.l
yacc -d calc.y
cc y.tab.c lex.yy.c -ll -ly
```

The file **y.tab.c** is the output file produced by Yacc, and the file **lex.yy.c** is the output file produced by Lex. The option **-I** causes lex to produce an interactive scanner (i.e., one with lazy lookahead), the option **-d** causes Yacc to produce the file **y.tab.h** automatically, and the options **-ll** and **-ly** cause the C compiler to consult the Lex and Yacc libraries when linking (if necessary).

One additional Yacc feature is the verbose option **-v**. If we give the command

```
yacc -v calc.y
```

then a file **y.output** is produced that contains a description of the parsing table used by the Yacc-generated parser, similar to Table 99.2. This can be useful in tracking down exactly the behavior of the parser.

Purdue Compiler Construction Tool Set

PCCTS [Parr et al. 1992] is a set of tools for automatically generating scanners and recursive-descent parsers. Although these parsers suffer from some of the same restrictions as other top-down parsers, PCCTS offers some advantages over Lex and Yacc. First, the parser and scanner generators are more fully integrated, so that only one definition file needs to be created, and this file contains both the token and grammar definitions. Second, there are tools for applying automatic disambiguating rules using either syntactic predicates (checked automatically by the parser), semantic predicates (with predicates supplied by the user), or by increasing the number of lookahead symbols (PCCTS parsers are not restricted to single-symbol lookaheads). Finally, there are additional tools within PCCTS for the automatic generation of both syntax trees and target code.

As with Yacc and Lex, we give a description of a PCCTS definition file using the calculator grammar as an example. The only file needed for PCCTS is shown in Table 99.6. The general form for this file (somewhat simplified) is

```
{header}
{actions or token definitions}
{rules or token definitions}
{actions or token definitions}
```

The header contains C definitions that will go both into the scanner code and the parser code. Actions are C code for auxiliary procedures and data. Token definitions contain regular expressions for tokens, together with actions to be taken on recognition. Rules are grammar rules in a modified EBNF form that also contain actions for execution during the recognition process. Actions must be contained inside the bracketing metasymbols $<< \cdots >>$. Table 99.6 contains a header, an action with the definitions of the mem variable and the main procedure, two token definitions, and five rules. There are no actions or token definitions after the rules.

The header section contains a typedef of **Attrib**, which is the internal name for the returned value of the recursive-descent procedures (this corresponds directly to the Yacc value stack type **YYSTYPE**). The header also contains a definition of the internal macro **zzcr_attr**, which is used whenever a token string is to be converted to a value of type Attrib. There is only one such token—**NUMBER**—and **zzcr_attr** is identified with a call to the C string scan function sscanf that converts a string to a double. The main program is defined in the first action section using the macro

```
ANTLR(session(),stdin);
```

Another tool for language recognition (ANTLR) refers to the parser generation utility of PCCTS. The parameter **session( )** indicates that the start symbol of the grammar is **session**, and so a call to the corresponding procedure begins the parse. The second parameter indicates the input file from which the program is to be taken, in this case **stdin**.

The token definitions consist of the symbol **#token**, followed by the name of the token, a regular expression in quotes defining the token (using conventions similar to Lex and other regular expression processors), and an optional action section. In Table 99.6 there are two tokens defined, **WhiteSpace** and **NUMBER**; the action **zzskip( )** for **WhiteSpace** causes this input to be discarded, and the token **NUMBER** has no action (recall that **zzcr_attr** already tells the scanner how to convert a **NUMBER** to a **double**).

Finally, the rules are given in a modified EBNF form, where $(\cdots)^*$ indicates a repetition of 0 or more times and $\{\cdots\}$ indicates an optional part. Again, as in the previous Yacc solution, we have rewritten the **session** rule so that it is the EBNF equivalent of a left-recursive rule, which saves space on the call stack. We have also included the operator tokens directly in the rules for expressions, saving some steps. Note that PCCTS also allows tokens to be written directly into the rules using double quotes (the \ is used

TABLE 99.6 A PCCTS Definition File

```
/************************************************************
CALC.G
***********************************************************/
#header <<
typedef double Attrib;
#define zzcr_attr(a,tok,t)      sscanf(t,"%lf",a);
>>

<<
static double mem;

int main()
{ ANTLR(session(),stdin);
return 0;
}
>>

#token WhiteSpace "[\t\ ]*" << zzskip(); >>
#token NUMBER "[0-9]+{.[0-9]+}"

session  : (asgn "n" << printf("\t %g\n",$1); >>)*
            "\n" << exit(0); >>
          ;
         : "\=" expr << mem = $2; $asgn = $2; >>
         | expr        << $asgn = $1; >>
          ;
expr     : term << $expr = $1; >>
            ("\+" term << $expr += $2; >>
            | "\-" term << $expr -= $2; >>
            )*
          ;
term     : factor << $term = $1; >>
            ("\*" factor << $term *= $2; >>
            | "\/" factor << $term /= $2; >>
            )*
          ;
factor   : "\ -" factor << $factor = - $2; >>
            | NUMBER << $factor = $1; >>
            | "\(" expr "\)" << $factor = $2; >>
            | "\m" << $factor = mem; >>
          ;
```

to avoid any metasymbol interpretation). Note how the actions for expressions and terms are embedded within the rule to achieve the desired result.

Assuming that the PCCTS definition file is called **calc.g**, a running calculator program can be built with the Unix commands:

```
antlr calc.g
dlg -i parser.dlg scan.c
cc calc.c scan.c err.c
```

The first line calls the main PCCTS parser generator tool, which is called ANTLR. ANTLR generates several files, the most important of which are **calc.c**, containing the parser coded in C, and **parser.dlg**, which is a scanner description specifically intended for input into the PCCTS scanner generator DLG (Deterministic finite automaton-based lexical analyzer generator). DLG is then called in the next line, producing the scanner in the output file **scan.c** (the **-i** option indicates the input will be interactive).

Finally, the C compiler is called on **calc.c**, **scan.c**, and a third file **err.c**, which was also produced by ANTLR.

Dealing with Ambiguity

Although the expectation is that a language grammar should be unambiguous, in that each legal input string will correspond to exactly one syntax tree, there are many practical situations in which it is difficult or impossible to write a grammar unambiguously. In such cases, disambiguating rules must be stated and built into the parser in some way. We have already seen how Yacc can accommodate precedence and associativity disambiguating rules for operators. We give three additional examples of the use of disambiguating rules.

The dangling else ambiguity is typical in languages such as C and PASCAL that allow both simple and compound statements in control structures. In C, the dangling else ambiguity appears in statements like

```
if(e) if(f) {/*then-part*/} else {/*else-part*/}
```

The question is, should the else-part be executed when **f** is false or when **e** is false? The standard answer is that it should be executed when **f** is false (and when **e** is true). This is called the most closely nested disambiguating rule for the if statement, since it implies that the else-part structure is to be associated in the parse tree with the closest previous if statement that does not have an else-part. This also corresponds to always preferring the second choice in the BNF rule

```
if_statement / if ( expression ) statement
             | if ( expression ) statement else statement
```

This disambiguating rule could actually be incorporated directly into the BNF, but it is slightly complicated (and not very illuminating), and the rule as stated is easily implemented in a parser. It is worth noting that languages in which compound statements and simple statements are clearly distinguished (ADA is an example) do not have the dangling else ambiguity.

A second kind of ambiguity is represented by the availability in C++ of function-style casts, where one can write **int(x)** to mean the same thing as the C cast **(int)x**. Now consider the code fragment

```
typedef int (*F)();
void **x;
void p(void)
{ F (*x) ( );
  ...
}
```

Is the first code line inside **p** a cast of global ***x** to type **F**, after which this function value is called, or is it a declaration of a local variable **x** of type "pointer to function returning a value of type **F**?" The C++ standard says it is the latter: any statement that may be interpreted as a declaration *is* a declaration. This means that all grammar rules for declarations are to be preferred to all other grammar rules. Note that this is an ambiguity *between* nonterminals, rather than an ambiguity within a single nonterminal, as with the dangling else ambiguity. It is also impossible to remove this ambiguity directly within the grammar, since the order of rules in a BNF description is immaterial to the language defined.

The final example of an ambiguity that we present here is also one that arises with casts but this time in C. Consider the C statement

```
(t)-x
```

If **t** is a type name declared in a **typedef**, then this is a cast of the value **-x** to type **t**. On the other hand, if **t** is a variable, then this is the value obtained by subtracting **x** from **t**. In order to parse this correctly, it is necessary to consult the symbol table to see if **t** is defined as a type. Thus, the symbol table (or at least the typedef part of it) must be built as parsing proceeds. Unlike the previous examples of ambiguity, this is a context ambiguity that cannot be resolved by purely syntactic means.

In an LR parser, these ambiguities, or parsing conflicts, can be of two different kinds. The first, more common, situation occurs when both a shift and a reduction by a grammar rule choice are called for on the same input token; this is a **shift-reduce conflict**. The other case is when reductions by two different rules are called for on the same input token; this is a **reduce-reduce conflict**. Shift-reduce conflicts are usually resolved by selecting the shift over the reduce. This ensures that the longest possible string will be matched by each rule. The typical example is the dangling else ambiguity, where preferring the shift corresponds exactly to the most closely nested disambiguating rule. Reduce-reduce conflicts, on the other hand, have no natural disambiguating rule. Parsers generally adopt an ad hoc rule, in which the reduction by the lowest numbered rule is preferred (where the rules are numbered in the order they are considered by the parser). The C++ ambiguity between cast expressions and declarations (also described previously) can be resolved in this way in favor of the declarations (as the C++ definition requires) by listing the declaration rules before the expression rules. Yacc adopts both of these disambiguating rules as stated. Unfortunately, the third ambiguity mentioned previously, a cast vs. an arithmetic expression in C, cannot be resolved by either of these means. Typically, this is handled by having the scanner consult the symbol table when recognizing an identifier and returning a different token for a type name than for a variable.

An alternative to these disambiguating rules is to build predicate testing directly into the parser generator in order to give the programmer control over the disambiguating mechanism. This is the case for PCCTS [Parr and Quong 1994], where the ANTLR parser generator allows both syntactic and semantic predicates as disambiguating rules. For instance, the C++ ambiguity between declarations can be resolved by adding a syntactic predicate in the ANTLR definition file as follows:

```
stat: (declaration)? declaration <<···>>
| expression <<···>>
;
```

Here the parentheses and question mark indicate that the parser should try to match a **declaration**, and if that fails it should go on to try to match an **expression**.

The third ambiguity can be solved similarly but with a semantic predicate supplied by the user:

```
var : <<isvar(LATEXT(1))>>? ID <<···>>
   ;
typename
   : <<istype(LATEXT(1))>>? ID <<···>>
   ;
```

Again, the question mark indicates a predicate, but this time the predicate is user-supplied, as indicated by enclosing it in brackets [**LATEXT(1)** is the lexeme of the next token in the input, made available by the scanner].

Attribute Analysis

Yacc and ANTLR restrict the kinds of attributes that can be reasonably computed during a parse, because of the requirements of their parsing algorithms. In cases of difficult attribute computations, these limitations are overcome by either providing ad hoc solutions using external data structures or by constructing an intermediate representation such as a syntax tree during the parse and then writing specialized procedures that perform semantic analysis by traversing the intermediate form in one or more passes. At the time of this writing, tools for automating the computation of general attributes have not been widely used, partially due to the many different varieties of intermediate representations and partially due to the variety and complexity of the semantic attributes of different languages. Some notable systems that do permit the automation of this step include LINGUIST [Farrow 1984], GAG [Kastens et al. 1982] and ELI [Gray et al. 1992]. Code generation is a special case of this problem, and special methods for automating the code generation step have been developed, which we describe next.

Intermediate Representations and Code Generation

The abstract syntax tree is a convenient way to represent the source program within a translator, particularly for semantic atribute analysis. However, it is less suited for code generation. Although target code can be generated as a form of attribute analysis, either during parsing or by traversal of the syntax tree, the quality of this code is usually poor, and further processing must be done before an acceptable level of target code is produced. Typically, this is achieved by generating some form of intermediate code that is closer to the code of the target machine but still abstract enough that the compiler can perform code improving transformations on it. Many choices have been used for this intermediate code. Some of the more common ones are sequences of expression trees, sequences of postfix expressions, an abstract linearized form of syntax tree called three-address code, and actual code for a hypothetical target machine, such as P-code for the stack-based P-machine of many PASCAL-related compilers [Nori et al. 1981]. An example of three-address code corresponding to the source code line `= 3.1 + m/2` for the caluclator language described previously is

```
t1 := m / 2
t2 := 3.1 + t1
m := t2
```

Here the identifiers `t1` and `t2` are temporaries introduced by the compiler that can be thought of as pseudoregisters and later assigned either to actual registers or to temporary locations in memory.

The equivalent (annotated) P-code for this same source code is as follows:

```
ldo r,m    ; load real value onto stack from static location m
ldc r,2.0  ; load constant real value 2.0 onto stack
dvr        ; divide two reals on top of stack, push result
ldc r,3.1  ; load constant real value 3.1 onto stack
adr        ; add two reals on top of stack, push result
sro r,m    ; store real from stack to static location m, pop stack
```

The use of the code of a precisely defined, simple machine such as the P-machine as the intermediate target language has several benefits. First, the compiler can be run in interpreter mode, in which the P-code functions as the target code, and a P-machine simulator is used to execute the P-code on the actual target machine. The target code is then kept as P-code, which is generally much more compact than actual executable machine code. Also, in order to retarget the compiler to a new machine, it is necessary only to rewrite the P-machine simulator. Second, in order to obtain a native-code compiler from the P-code compiler, one need only write a translator from P-code to the native code of the target machine. This is usually done by writing individual procedures for each P-code instruction that perform something like a macroexpansion of each P-code instruction into target code. By itself, this process will produce very poor code, but it can be combined with a static simulation of the P-machine itself, whose stack can be used to discover opportunities for optimization. An improvement on P-code that adds significant attributes for tracking address calculations and optimizing transformations is U-code [Perkins and Sites 1979].

An intermediate language that is compact, capable of efficient interpretation, and easily retargeted to different architectures is of special interest for heterogeneous networks, where the platform on which the intermediate code is to be executed or compiled to native code may not be known in advance. If the intermediate code is given a numeric machine code-like encoding (sometimes called bytecodes), and techniques to improve interpretation performance are used (such as on-the-fly compilation or threaded code interepretation [Bell 1973]), then good perfomance on a variety of machines across a network can be achieved. This is, for example, the basis for JAVA implementations [Gosling 1995].

A similar process of macroexpansion works for three-address code and its variants, although in this case the underlying abstract machine and its state is not made explicit (as it is for P-code), and the accumulated information during code generation must be stored as attributes in data structures associated with the intermediate code. Typically, these include address descriptors, which record the locations at which named quantities (such as variables) can be found, and register descriptors, which record information about the values that can be statically predicted to be in registers at particular points in the target code. Address descriptors can be kept in the symbol table, whereas register descriptors can be maintained in an array indexed by the register numbers.

Unfortunately, both of these code generation schemes spread the information about the target machine throughout the code generator. This means that retargeting the compiler is more costly than if the target machine information were collected compactly in one place. Several approaches have been taken toward doing this.

The first method for improving the retargetability of the code generator involves writing a syntactic description of the macroexpansion process for the intermediate code together with code-emitting actions, much as a Yacc or PCCTS description associates semantic actions with syntactic rules [Glanville and Graham 1978]. This method permits the automatic generation of the code generator from this description, using a tool similar to a parser generator. This method has been expanded in [Ganapathi and Fischer 1985] to include semantic predicates. As an example, consider generating code for the three-address instruction

```
t1 := m / 2
```

A rule that would handle this instruction for the VAX might appear as

```
reg(n)→ adr(a) / const(i) dead?(n) float?(a)
#emit("divd3 #i.,a,rn")
```

This rule establishes that emitting the VAX instruction **divd3 #i.,a,rn** allows the expression consisting of an address **a** divided by a constant **i** to be reduced to a register **n**, provided **n** is dead (i.e., contains a value that is no longer needed), and provided that **a** is the address of a real-valued quantity. Matching this rule to the previous three-address instruction causes **t1** to be identified with **reg(n)**, **m** to be identified with **adr(a)** (thus computing **a** as the address of **m**), and identifying **2** with **const(i)** (thus setting **i = 2**). The resulting line of VAX code generated is

```
divd3 #2.,m,r1
```

In the absence of predicates, such a set of rules describing the target machine could be written as a Yacc input. The difference between this kind of grammar and a grammar describing the syntax of the source language is that the target machine grammar usually comprises thousands rather than hundreds of rules, and the grammar is highly ambiguous, since there are usually many ways to generate target code to achieve a specific effect. Parser generators such as Yacc are inefficient and cumbersome when dealing with such grammars, and a code generator generator specifically tuned to this situation is more appropriately used [Henry 1984].

The second variety of retargeting mechanism we discuss is that used in the portable C compiler [Johnson 1978]. In this mechanism, a special template language is used to describe the process of matching intermediate code to target code. The code generator then consults a table of these templates during code generation in an attempt to find the optimal match.

The third variety of retargeting mechanism is similar to the second, except that both the intermediate code and the target code templates are written in the template language, and the matching routine runs statically, usually during the installation of the compiler, and selects a match between certain sequences of intermediate code and target code that is then fixed once and for all. This method was first described in [Cattell 1980, Leverett et al. 1980] and was part of the Production-Quality Compiler Compiler (PQCC) Project [Stallman 1994]. A related method has been used successfully in the Gnu C++ compiler. In this compiler, the template language is called the register transfer language (RTL). This language is written in Lisp-like prefix form. An example of an instruction template is as follows:

```
(define_insn "divdf3"
  [(set (match_operand:DF 0 "register_operand" "=f")
        (over:DF (match_operand:DF 1 "register_operand" "f")
                 (match_operand:DF 2 "register_operand" "f")))]
  "! TARGET_SOFT_FLOAT"
  "divd3 %1,%2,%0"
  [(set_attr "type" "arith")]
  )
```

Instruction templates are introduced using the **define_insn** operator. This operator is followed by a symbolic name for the instruction template, a pattern with predicates (such as **register_operand**), constraints (such as **f**, indicating the register may be used for floating-point values), and extra conditions (such as **!TARGET_SOFT_FLOAT**, indicating that there must be a hardware floating-point arithmetic unit available in the processor). Following this there is a pattern for the instruction to be generated, with numbers referring to the operands of the pattern (in this example, it is the VAX instruction **divd3** **%1,%2,%0**). Finally, there is a specification of attributes for the template.

Such instruction templates are used to generate actual target code in RTL format directly from C code during a parse. Optimizing steps are then applied directly to the RTL intermediate code, and then templates are again used to generate assembly output. Specialized RTL attribute descriptions are also used to guide the code generation process.

Code Optimization

A code generator that naively expands intermediate code into target code will produce code that is extremely inefficient, both in terms of execution speed and target code size. A production-quality compiler must include processing steps that improves its ability to generate good target code, so that it more nearly resembles the code that would be produced by an assembly language programmer. Such processing steps are usually referred to as optimization, though they almost never produce truly optimal code. Indeed, the production of mathematically optimal code, except in the very simplest cases, is know to be computationally intractable (NP-complete), so to attempt this would result in unacceptably slow compilation speed.

Optimization steps can be built into almost all of the phases of a compiler, from parsing to final target code generation. If an optimization pass is performed separately, it usually occurs after intermediate code generation but before target code generation. Such optimizations are generally *source-level* in that they do not depend heavily on the details of the target machine. Some optimizations, however, are *target-level* and require that the details of the target code be known. Such optimizations are referred to as peephole optimizations, since they were originally (and are sometimes still) performed by examining small sequences of target code and replacing them by more efficient code. This term can be misleading, however, since modern compilers sometimes perform considerably more sophisticated analysis of the target code than the name might imply (an example is the Gnu C++ compiler). One complex aspect of the scheduling of optimizations during compilation is that some optimizations may uncover opportunities for other optimizations and vice versa (this is sometimes called the phase problem). This leads some compilers to repeat certain optimizations in an attempt to catch such cascaded situations.

Optimizations are classified according to the region of the program about which information is gathered in order to perform the code improvement. Local optimizations consider only straight line segments of code, i.e., those not involving jumps or calls. Global optimizations consider the code of a single procedure. Interprocedural optimizations consider the entire program or compilation unit. Most compilers perform some kinds of local optimizations. A more heavily optimizing compiler will perform global optimizations, typically using an information collection method called dataflow analysis, which passes information around a flow graph representing the flow of control through a procedure. It is a rare compiler that performs interprocedural optimization and for a very good reason: such optimizations are largely ineffective unless they are postponed to link time, since many procedures will not be available in a single compilation unit. This means that the linker must be closely coupled with the compiler, in particular the system linker cannot be used. This raises the level of complexity of the compilation environment considerably.

In the remainder of this brief discussion about optimizations, we list the principal sources of code improvement over a naive code generation strategy.

Register Allocation. This is the most important and pervasive issue in the generation of quality code. Keeping temporaries in registers is indispensable, especially for reduced instruction set computer (RISC) architectures. In order to extend register allocation to include local variables, parameters, and global variables, a compiler may permanently allocate certain registers to heavily used quantities. An alternative is to build an *interference graph* and assign registers by graph coloring, permitting noninterfering variables to share the same register. Good global register allocation also requires some form of dataflow analysis in order to identify values that have no further uses in subsequent code and which may be safely overwritten if they are in registers.

Common Subexpression Elimination. This refers to the identification of expression values that are recomputed one or more times in a program and the avoidance of such recomputation by suitable storing and reuse of the computed value. While one might naively think that a good programmer should not write code that contains common subexpressions, in fact, most common subexpressions are due to address computations of array and record references that cannot be expressed adequately in source code and which therefore require elimination by an optimizer. Common subexpressions are relatively easy to identify in straight line code; the principal difficulty in the general case is to determine what subexpressions may have changed between two computations of the same expression. Thus, a compiler may choose to perform only local common subexpression elimination.

Copy Propagation. This optimization tracks code regions in which two or more variables have the same value. After an assignment $x = y$ (in C syntax), x and y have the same value in subsequent expressions until one of them acquires a new value; in this region, any use of x can be replaced by a use of y, which can lead to better code if, for example, y is already in a register. As with common subexpressions, copy statements may exist in the intermediate code even when they are not written by the programmer. A special case of copy propagation is constant propagation, where after an assignment such as $x = 2$, uses of x can be replaced by the constant 2. As with common subexpressions, copy propagation may be done as either a local or a global optimization.

Reduction in Strength. This refers to the replacement of arithmetic expressions by equivalent expressions that execute faster. A typical example is the replacement of multiplication or division by a power of 2 by a shift operation. Another example occurs in loop optimization, where a linear combination calculation is replaced by a simple addition (see Loop Optimization).

Jump Optimization. Opportunities for improving jump code occur when a sequence of jumps can be replaced by a single jump or when a jump can be eliminated by rearranging the code. For instance, the code sequence

```
        goto lab1
        ...
labl:   if x = y goto lab2
lab3:   ...
```

can be replaced by the more efficient

```
    if x = y goto lab2
    goto lab3
    ...
lab3: ...
```

Algebraic Laws.   An optimizer can look for special cases in expressions such as x^*1 and $x + 0$ (replacing these by x). Sometimes it is also worthwhile to replace a computation $x + y$ or x^*y by its commutative equivalent $y + x$ or y^*x. More difficult is to discover opportunities to use distributive laws, such as replacing $x^*y + z^*y$ by the more efficient $(x + z)^*y$.

Loop Optimization.   Loops are traditionally an area where a great deal of attention has been paid to making code as efficient as possible, since programs tend to spend a lot of time in loops. In programs with goto statements it is a nontrivial operation to even discover loops, and this usually must be done by building the flow graph. Fortunately, in modern languages it is reasonable to depend on syntax (such as keywords while and do) to locate loops. Typical loop optimizations include identifying invariant computations inside loops (i.e, those that always yield the same value, regardless of the loop iteration) and moving their computation to just before the loop entry. Another optimization seeks to discover so-called induction variables, which are linear combinations $a^*i + b$, where i is the loop control variable. Since such an expression is incremented by a fixed amount with each iteration, its computation can be reduced in strength to a simple addition.

Constant Folding.   This optimization seeks to replace constant expression (e.g, 3+5) by their resulting values (e.g., 8). Such an optimization could, in theory, replace all computations whose results are predictable at compile time by their results alone. In practice, this would involve repeated applications of constant propagation and folding, and so this optimal result is rarely achieved.

Dead Code Elimination.   This optimization seeks to skip code generation for those statements that are either never reached during execution or whose actions have no effect on the results of the program. The first case happens when compile-time constants are set to select certain actions over others (for example, to suppress the collection of run time statistics). The second occurs if common subexpression elimination or copy propagation makes a computation or assignment unnecessary.

Error Recovery

An important practical problem in the design of a translator is its response to errors in the input program. Although many translators have been constructed so that they stop at the first error encountered, it is generally considered better to attempt to discover as many errors as possible in the input before halting. Thus, the response to errors includes error recovery, so that translation may continue, as well as the generation of informative error messages. A further step in an error handler may also be error repair, where the translator attempts to correct at least some of the errors. Since it is difficult to infer from an erroneous program what the actual intended program was, error repair is usually not attempted, except in special cases, such as missing punctuation.

Errors may be classified into lexical, syntactic, semantic, and run time errors. An interpreter must have a reasonable response to run time errors that involves at least generating an error message and exiting gracefully. A compiler need only recover from static errors, although it may need to generate code to report run time errors. Semantic errors are relatively easy to recover from, since the syntax tree can still be used to guide the remainder of translation. Lexical errors can be simply passed to the parser by the scanner as error tokens. Thus, the principal difficulty in building an error handler occurs in dealing with parsing errors, where the structure of the input is disturbed, and there may be no obvious way to restart the parser.

Several criteria apply to any method for recovery from parse errors. First, it is useful to try to detect errors as early as possible, since continuing to process the input can make it more difficult to determine where the error occurred and to generate an appropriate error message. Second, the translator should attempt to discard as little input as possible when recovering from an error, since this can lead to other errors as well as mislead the programmer. Third, the translator should avoid generating large numbers of error messages caused by a single error. In particular, a translator must never get into an infinite loop when recovering from an error, and this usually requires that at least some input be discarded during error recovery.

A standard technique for error recovery is called panic mode, in which tokens are simply discarded, and the parsing stack is accordingly adjusted until the parse can resume. This method can be made sophisticated enough so that it becomes better than its name implies, and it can be used either as the standard error recovery technique, or as a fall-back technique for a more complex method. The important feature of panic mode is the proper computation of a synchronizing set of tokens, which are used to determine when the parser should stop discarding the input and attempt to resume the parse. Panic mode in recursive-descent parsing is discussed in Wirth [1976, sec. 5.9]. Yacc uses a slightly different method, in which an error pseudotoken is made available with which important error recovery locations can be marked in so-called error productions in the input grammar. For example, in the Yacc definition file of Table 99.5, the error production

```
session : session error '\n' {yyerrok;}
```

was included in the specification. Yacc's behavior on encountering an error is as follows. First, it sets a flag to enter an error phase and then begins popping the parsing stack until a state is found where the the error pseudotoken can be successfully shifted. Then input tokens are discarded until three consecutive tokens can be shifted successfully, whence the error phase is canceled and normal parsing is resumed. The error phase can also be canceled manually by calling the Yacc procedure **yyerrok**. For example, the previous error production for a caculator session causes Yacc to discard all tokens until a newline is reached, and then to resume the parse. The result is that any incorrect line of input is deleted. (Yacc also automatically calls **yyerror** whenever the error phase is entered, which can perform other actions and print an error message.)

Other error recovery and repair mechanisms are discussed in Dion [1982], Graham et al. [1979], and Roehrich [1980].

99.4 Research Issues and Summary

Language translators can be decomposed into the phases of scanning (lexical analysis), parsing (syntactic analysis), semantic analysis, optimization, and code generation. A scanner breaks the input program into tokens by using the theory of regular expressions and finite automata. A parser constructs, implicitly or explicitly, a representation for the syntactic structure of the program by using the theory of context-free grammars. The construction of both a scanner and a parser can be easily automated with tools such as Lex and Yacc or PCCTS. Yacc constructs an LALR(1) bottom-up parser, whereas PCCTS constructs a recursive-descent top-down parser. Bottom-up parsers are generally too complex to construct by hand, but recursive-descent can be used to hand-construct a parser. Finite automation implementations of scanners are also relatively easy to construct by hand.

The semantic analysis and code generation steps of a compiler can be modeled theoretically by an attribute grammar, which expresses in equational form the relationships among the various attributes of language entities. In fact, attribute grammars can be used as a basis for automating the construction of an entire compiler, but this has not become common, possibly because of the complexity of representing the entire semantics of a language as an attribute grammar and possibly because of the difficulty of producing optimized target code. It may be that other semantic definition mechanisms, such as denotational semantics, will result in better automation techniques, but this remains for future study. Current methods typically construct hand-generated semantic analyzers that operate during parsing using auxiliary data structures such as the symbol table or analyzers that perform recursive traversals of a syntax tree.

Some success has been achieved in automating the code generation step, with easy retargeting as the primary goal. These methods include the syntax-based approach of Glanville and Graham [1978] and the semantic approach of Ganapathi and Fischer [1985]. An alternative is the use of a symbolic machine description language to describe the target machine, which is then used by the code generator to produce target code. Effective use of this method has been made in the widely retargeted Gnu C++ compiler.

An important aspect of the automation and retargetability of a compiler is the choice of an appropriate intermediate code representation for the source code. The best choice appears to be a symbolic code for a hypothetical abstract machine. One may then choose to either interpret the intermediate code with a simulator or perform code generation based on a static simulation of this machine. The primary requirements of such an intermediate code are flexibility, security, and the availability of enough information to provide good optimization over a wide variety of target architectures. A significant challenge for future translator technology is to develop a standard intermediate code that can be generated by many different language front ends and that can also be efficiently and safely interpreted and compiled on many different architectures under many different operating systems.

Defining Terms

Attribute grammar: A set of equations associated to the grammar rules of a context-free grammar that define a collection of attributes associated to the terminals and nonterminals of the grammar. Attribute equations are written in purely functional form (i.e., without side effects) and may be solved for the actual attribute values by different kinds of traversals of the parse or syntax tree. Alternatively, attribute values may be computed by replacing the attribute equations with equivalent side-effect-generating code using separate data structures such as the symbol table.

Back end: The part of a compiler that depends only on the target language and is independent of the source language. The back end receives the intermediate code produced by the front end and translates it into the target language.

Backus Normal Form or Backus-Naur Form (BNF): A notation for context-free grammar rules first used in the Algol60 report to describe syntax. It comprises only two metasymbols, usually written $\rightarrow$ or $::=$ and |.

Bottom-up: A parsing algorithm that constructs the parse tree from the leaves to the root. Bottom-up algorithms include LR and LALR parsers, such as those produced by Yacc.

Disambiguating rule: A rule stated separately from the rules of a context-free grammar that specifies the correct choice of syntax tree structure when more than one structure is possible.

Extended BNF (EBNF): Adds bracketing metasymbols [...] and { ...} to BNF to indicate optional and repeated structures, respectively. (These can also be written as (...)? and (...)* to remain consistent with standard regular expression notation.)

Front end: The part of a compiler that depends only on the source language and is independent of the target language. The front end translates and analyzes the source program.

Inherited attribute: An attribute whose value depends on attribute values at syntax tree nodes that are not descendants. A nonsynthesized attribute.

Item: A grammar rule choice with a distinguished position, indicating that a parse has reached that position in attempting to recognize the rule. Sets of items are used by a bottom-up parser to record a state reached during a parse. Items may have 0 or more tokens of lookahead attached to them. LR(0) items contain no lookahead, while LR(1) items contain one token of lookahead.

L-attributed grammar: An attribute grammar whose attributes may be computed by a left to right traversal of the source program. An attribute grammar must be L-attributed for the attributes to be computable during a parse that processes the input from left to right (as most parsers do). Synthesized attributes are always L-attributed.

Lexeme: The actual character string read from the input when recognizing a token. The lexeme of an identifier token is the identifier name.

LL(k):   A top-down parsing algorithm that processes the input from left to right, producing a leftmost derivation using k tokens of lookahead. The term can also be applied to a language that can be unambiguously parsed using this algorithm.

Lookahead LR(1) [LALR(1)]:   The algorithm invented by DeRemer [1971]. The algorithm used in many bottom-up parser generators, including Yacc. A language is also called LALR(1) if it can be parsed unambiguously by the LALR(1) algorithm.

LR(k):   A bottom-up parsing algorithm that processes the input from left to right, producing a rightmost derivation (in reverse) using k tokens of lookahead. The term can also be applied to a language that can be unambiguously parsed using this algorithm.

Nonterminal: A name for a structure defined by a context-free grammar rule. Interior nodes of parse trees are labeled by nonterminals.

Phase: A logical unit of a compiler. Typical phases include scanning, parsing, semantic analysis, and code generation. Phases are to be distinguished from passes, which comprise a complete sequential processing of the input program. Phases may or may not corresponded to physical code units within the compiler.

Production: Another term for a context-free grammar rule or grammar rule choice.

Recursive-descent: A top-down parsing algorithm that translates context-free grammar rules into a set of mutually recursive procedures, with each procedure corresponding to a nonterminal. Recursive-descent parsing is usually the method of choice when writing a parser by hand.

Reduce-reduce conflict: In bottom-up parsers, a property of a state in which a parser has a choice of two productions which can be used to reduce the parsing stack, and both are legal for the amount of lookahead allowed. Reduce-reduce conflicts have no natural disambiguating rule.

Retargeting: The process of changing a compiler to produce target code (assembly or machine code) for a different machine. This may involve rewriting the compiler back end or creating a machine definition file for the new machine.

Shift-reduce conflict: In bottom-up parsers, a property of a state in which a parser has a choice of either reducing the parsing stack using a production or of shifting a token from the input, and both are legal for the amount of lookahead allowed. A natural disambiguating rule is to prefer the shift, thus allowing the parser to match the longest possible input string at each point.

Synthesized attribute: An attribute whose value depends only on the attribute values of descendants in the parse or syntax tree. Synthesized attributes are the easiest to compute during a parse, requiring no special data structures or techniques. The syntax tree itself is the most important example of a synthesized attribute.

Terminal: Another term for a token in a context-free grammar. Leaf nodes of parse trees are labeled by terminals.

Top-down: A parsing algorithm that constructs the parse tree from the root to the leaves. Top-down algorithms include LL parsers and recursive-descent parsers, such as those produced by PCCTS.

References

Aho, A. V., Sethi, R., and Ullman, J. D. 1986. *Compilers: Principles, Techniques, and Tools*. Addison–Wesley, Menlo Park, CA.

Backus, J. et al. 1957. The FORTRAN automatic coding system, pp. 188–198. In *Western Jt. Comput. Conf.* reprinted in 1967. Rosen, S., *Programming Systems and Languages*, pp. 29–47. McGraw–Hill, New York.

Bell, J. R. 1973. Threaded code. *Commun. ACM* 16(6):370–372.

Cattell, R. G. G. 1980. Automatic derivation of code generators from machine descriptions. *ACM Trans. Programming Lang. Syst.* 2(2):173–190.

Davidson, J. W and Fraser, C. W. 1984. Code selection through object code optimization *ACM Trans. Programming Lang. Syst.* 6(4):505–526.

DeRemer, F. L. 1971 Simple LR(k) grammars. *Commun. ACM* 14(7):453–460.

DeRemer, F. L. and Pennello, T. 1982. Efficient computation of LALR(1) lookahead sets. *ACM Trans. Programming Lang. Syst.* 4(4):615–645.

Dion, B. A. 1982. *Locally Least-Cost Error Correctors for Context-Free and Context-sensitive Parser.* University of Michigan Research Press.

Early, J. 1970. An efficient context-free parsing algorithm. *Commun. ACM* 13(2):94–102.

Farrow, R. 1984. Generating a production compiler from an attribute grammar. *IEEE Software* 1(10):77–93.

Ganapathi, M. J. and Fischer, C. N. 1985. Affix grammar driven code generation. *ACM Trans. Programming Lang. Syst.* 7(4):560–599.

Glanville, R. S. and Graham, S. L. 1978. A new method for compiler code generation. In *5th Annu. ACM Symp. Principles Programming Lang.*

Gosling, J. 1995. Java intermediate bytecodes. *ACM SIGPLAN Notices* 30(3):111–118.

Graham, S. L., Haley, C. B., and Joy, W. N. 1979. Practical LR error recover. *SIGPLAN Notices.* 14(8):168–175.

Graham, S. L., Harrison, M. A., and Ruzzo, W. L. 1980. An improved context-free recognizer. *ACM Trans. Programming Lang. Syst.* 2(3):415–462.

Gray, R. W., Heuring, V. P., Levi, S. P., Sloane, A. M., and Waite, W. M. 1992. Eli: a complete, flexible compiler construction system. *Commun. ACM* 35(2):121–131.

Henry, R. R. 1984. *Graham–Glanville Code Generators.* Ph.D. thesis, University of California, Berkeley.

Jacobson, V. 1987. Tuning Unix Lex, or it's not true what they say about Lex. *Proc. Winter Usenix Conf.*

Johnson, S. C. 1975. Yacc—Yet Another Compiler-Compiler. *CS Tech. Rep. 32*, Bell Labs., Murray Hill, NJ.

Johnson, S. C. 1978. A portable compiler: theory and practice, pp. 97–104. In *5th Annu. ACM Symp. Principles Programming Lang.*, ACM Press, New York.

Kastens, U., Hutt, B., and Zimmermann, E. 1982. *GAG: A Practical Compiler Generator.* Lecture notes in computer science 141, Springer–Verlag, New York.

Katayama, T. 1984. Translation of attribute grammars into procedures. *ACM Trans. Programming Lang. Syst.* 6(3):345–369.

Knuth, D. E. 1968. Semantics of context-free languages. *Math. Systems Theory* 2(2):127–145. 1971. Errata 5(1):95–96.

Lee, P. 1989. *Realistic Compiler Generation.* MIT Press, Cambridge, MA.

Lesk, M. 1975. Lex—A Lexical Analyzer Generator. *CS Tech. Rep. 39*, Bell Labs., Murray Hill, NJ.

Leverett, B. W., Cattell, R. G. G., Hobbs, S. O., Newcomer, J. M., Reiner, A. H., Schatz, B. R., and Wulf, W. A. 1980. An overview of the production-quality compiler-compiler project. *IEEE Computer* 13(8):38–40.

Nori, K. V., Ammann, U., Jensen, K., Nägeli, H. H., and Jacobi, Ch. 1981. Pascal P implementation notes. In *Pascal—The Language and Its Implementation*, Barron, D. W., ed. Wiley, Chichester.

Parr, T. J., Dietz, H. G., and Cohen, W. E. 1992. PCCTS reference manual *ACM SIGPLAN Notices* 27(2):88–165.

Parr, T. J. and Quong, R. W. 1994. Adding semantic and syntactic predicates to LL(k): pred-LL(k). *Int. Conf. Compiler Construction*, pp. 263–277. Lecture Notes in Computer Science 786, Springer–Verlag, New York.

Paxson, V. 1990. Flex users manual. (Part of the Gnu ftp distribution).

Perkins, D. R. and Sites, R. L. 1979. Machine independent Pascal code optimization. *ACM SIGPLAN Notices* 14(8):201–207.

Peyton Jones, S. L. 1987. *The Implementation of Functional Programming Languages.* Prentice–Hall, Englewood Cliffs, NJ.

Polak, W. 1981. *Compiler Specification and Verification.* Lecture notes in computer science 124, Springer–Verlag, New York.

Randell, B. and Russell, L. J. 1964. *Algol60 Implementation.* Academic Press, New York.

Roehrich, J. 1980. Methods for the automatic construction of error-correcting parsers. *Acta Informatica* 13(2):115–139.

Stallman, R. 1994. Using and porting Gnu CC. Free Software Foundation Gnu ftp distribution (prep.ai.mit.edu).

Wirth, N. 1971. The design of a Pascal compiler. *Software—Pract. Exp.* 1(4):309–333.

Wirth, N. 1976. *Algorithms + Data Structures = Programs*. Prentice–Hall, Englewood Cliffs, NJ.

Further Information

The standard comprehensive text in compiler design is *Compilers: Principles, Techniques, and Tools* by Alfred V. Aho, Ravi Sethi, and Jeffrey D. Ullman (Addison–Wesley, 1986). Another extremely useful reference, with more detail about implementation techniques, is *Crafting a Compiler with C* by Charles N. Fischer and Richard J. LeBlanc, Jr. (Benjamin-Cummings, 1991). A book that presents a full C compiler in complete detail is *Compiler Design in C* by Allen I. Holub (Prentice–Hall, 1990).

Aside from these and many other texts, the best place to locate the latest information on compiler design is the comp.compilers newgroup on the Internet.

Research papers on language translation can be found in publications by the IEEE and the ACM, particularly the conference proceedings published as part of the *ACM SIGPLAN Notices*, the ACM Annual POPL Conference proceedings, and the *AMC TOPLAS Journal*. For information, contact the ACM at its web site http://www.acm.org or by e-mail at acmhelp@acm.org, and the IEEE at its web site http://www.ieee.org.

General information on interpreters, insofar as they differ from compilers, is more difficult to find in one place. A scheme-based introduction can be found in *Structure and Interpretation of Computer Programs* by H. Abelson and G. J. Sussman with J. Sussman (McGraw–Hill, 1985). Advanced techniques both for the compilation and interpretation of functional languages can be found in *The Implementation of Functional Programming Languages* by Simon L. Peyton Jones (Prentice–Hall, 1987) and *Implementing Functional Languages* by Simon L. Peyton Jones and David Lester (Prentice–Hall, 1992). The latter text concentrates more on implementation issues, whereas the former gives a more theoretical description.

A brief but useful introduction to the use of Lex and Yacc can be found in *The Unix Programming Environment* by Brian W. Kernighan and Rob Pike (Prentice–Hall, 1984). A more detailed study is contained in *Introduction to Compiler Construction with Unix* by A. T. Schreiner and H. G. Friedman, Jr. (Prentice–Hall, 1985). Information about PCCTS can be found in the comp.compilers.tools.pccts newsgroup.

100

Parallelizing Compilers

Michael Wolfe
The Portland Group

100.1 Introduction

The term *parallelizing compiler* typically means a compiler that finds parallelism in a sequential program and generates appropriate code for a parallel computer. More recent parallelizing compilers accept explicitly parallel language constructs, such as array assignments or parallel loops. The parallelization task can be split into two subproblems: identifying potential parallelism, which may be done automatically or by a programmer, and mapping the parallelism onto the target machine. Even if all of the parallelism is identified by the programmer, the mapping process requires compiler analysis and program transformation to generate efficient code.

The original justification for parallelizing compilers was portability. Commercial supercomputers in the 1960s and early 1970s were mostly pipelined sequential processors; many of the applications at that time had been modified and updated over a period of many years. It was not cost-effective to rewrite a large application in a parallel language for the very few parallel computers that became available from time to time. The few parallel languages that were introduced were mostly machine specific, exposing characteristics of the target machine such as the number of processors or pipeline architecture.

A more recent argument for parallelizing compilers is that most parallel languages are poorly defined extensions to sequential languages, and offer no protection against nondeterminism or timing errors. Sequential languages have well-defined semantics and are deterministic; a parallelizing compiler allows parallel execution and preserves the meaning of the program.

The argument against parallelizing compilers is that efficient parallel applications require parallel algorithms. Even the most aggressive parallelizing compiler cannot get efficient parallel performance from an inherently sequential algorithm; instead, the whole algorithm must be replaced, something that cannot be done by a meaning-preserving compiler.

There is truth in all of these arguments. Portability and compatibility are important in the real computing world, and applications are more likely to be migrated piecemeal to a parallel environment rather than all at once with a complete rewrite. Sequential algorithms must be rewritten to get good parallel performance, but that still leaves open the question of how to express the parallel algorithm. Finally, as previously mentioned, even programs written in parallel languages need compiler analysis and transformation to get good performance.

Most of the work in parallelizing compilers has been aimed at Fortran programs, in large part because Fortran programmers are most likely to have compute-bound problems. The same techniques can be and have been applied to other imperative languages, such as C, Pascal, and Ada. Recent work on identifying and using implicit parallelism in single-assignment or applicative languages is also based on the techniques shown here.

100.2 Underlying Principles

A sequential program defines a sequence of operations and the order in which they must be followed. A parallelizing compiler will reorder these operations to get some of them to execute in parallel. In order to have the most flexibility in this reordering, the compiler must find and represent the basic dependence relations between the operations in the program. In a parallel language, the compiler may have an easier time finding the basic dependence relations, but the first step is still the same. The goal is to find the smallest set of dependence relations that preserves the meaning of the program. This means that the compiler must preserve the observed behavior of the program, assuming the program is correct and terminates.

Given the dependence relations and the target machine, the compiler must map the available parallelism onto the target. This can involve a major restructuring of the program, such as reordering statements, changing the order of loops, and so on. A small set of basic reordering transformations is used, driven by different goals depending on the target machine.

Finally, the compiler can generate the target code. This may include the tasks of choosing a scheduling policy, inserting interprocessor synchronization, selecting data memory layouts, and more. After this, the code that will be run on each processor can be optimized as would any uniprocessor program.

Dependence Analysis

The first job of the parallelizing compiler is to find and represent the basic ordering relations that might prevent two operations from executing in parallel. This is called *dependence analysis*; typically a compiler will represent this information in a directed **dependence graph**. The source of a dependence edge in the graph is an operation that must execute before the target of the edge.

As an example, we present the dependence graph for a sequential program fragment in Fig. 100.1. The graph shows the dependence relations at the granularity of statements. The graph includes an edge from statement S_1 to S_2 because S_1 defines a value for **c** and S_2 uses that value; this is called a **flow dependence** or *true dependence* relation because a data value flows from S_1 to S_2. There is also a dependence edge from S_2 to S_3 because S_2 uses an old value of **f** and S_3 defines a new value for **f**; this is called an **antidependence** relation, because it flows in the opposite direction (from the use to the definition) relative to flow dependence. There can also be an **output dependence** relation, where one statement defines a value for a variable and another statement redefines the variable, so that the definitions must occur in order. Note there is no dependence due to **d** in our example; even though S_2 defines **d** and S_3 uses **d**, the statements define and use different elements of the array, so there is no dependence between them.

If we interpret the same example as a declarative program, the dependence graph will change. In a declarative model, the order of the statements in the program does not matter; instead, the program must be executed so that the value in each statement is computed before its result is used. This program would have only flow dependence relations; there would be a dependence edge from S_1 to S_2 due to **c**, as before, and another flow dependence edge from S_3 to S_2 due to **f**, since S_3 defines the value for **f** and S_2 uses that value.

$S_1:$ c = 1
$S_2:$ d[0] = c + f
$S_3:$ f = d[1] + d[2]

FIGURE 100.1 Simple program fragment and its dependence graph.

TABLE 100.1   First Three Iterations of Sample Loop

Statement	i = 1	i = 2	i = 3
S_1:	c[1] = a[1] + b[1]	c[2] = a[2] + b[2]	c[3] = a[3] + b[3]
S_2:	d[1] = d[0] + c[1]	d[2] = d[1] + c[2]	d[3] = d[2] + c[3]
S_3:	f[1] = d[1] + c[2]	f[2] = d[2] + c[3]	f[3] = d[3] + c[4]

To handle loops or recursive procedures, a full dependence graph would have unbounded size. Compilers cannot afford to build such a graph, so each node in the dependence graph typically represents all potential executions of a single operation. Thus, the dependence graph can include cycles.

```
for i = 1 to n do
S₁:   c[i] = a[i] + b[i]
S₂:   d[i] = d[i-1] + c[i]
S₃:   f[i] = d[i] + c[i+1]
  endfor
```

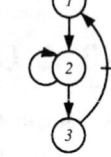

FIGURE 100.2   Sample loop and its dependence graph.

For instance, the first three iterations of the loop in Fig. 100.2 are expanded in Table 100.1. We can see that there is a flow dependence from S_1 to S_2 because S_2 uses the value of **c[i]** that was computed in S_1 in the same iteration of the loop. Similarly, there is a flow dependence from S_2 to S_3 due to a definition and use of **d[i]**. There is also a flow dependence from S_2 to itself because S_2 defines a value of **d[i]** that is used in the same statement on the next iteration of the loop. If the value of **n** is 1, the loop only iterates once and there is no next iteration of the loop; however, to be conservative the compiler assumes that the loop may iterate an unbounded number of times. Finally, there is an antidependence relation from S_3 to S_1 because S_3 uses a value of **c[i+1]** that will be redefined in the next iteration of the loop by S_1. The dependence graph is shown on the right in Fig. 100.2.

Often, dependence analysis is limited to operations within a single execution of one procedure, avoiding the problems of recursion; sometimes the analysis is further limited to a single iteration of a loop, in which case the resulting graph is acyclic. In the general case for loops, a compiler annotates the edges of the graph with information to represent the relative ordering of the operations with respect to the loop iterations.

For scalars, **data dependence** relations can be determined by standard compiler analysis. For arrays, compilers inspect the subscript expressions and loop limits, casting the dependence problem into a mathematical framework. In Fig. 100.2, to find the dependence relation between **c[i]** and **c[i+1]**, the compiler must determine whether there are integers i^d and i^u (representing the value of the loop index **i** at the definition and use) that satisfy the equation $i^d = i^u + 1$. In addition, the solution must lie within the loop limits, $1 \leq i^d, i^u \leq n$. This equation and the loop limit inequalities comprise the *dependence system*.

Most current compilers represent the loop iteration number of the source and target of a dependence relation using a **dependence distance**. The distance d is the integer that satisfies $i^d + d_f = i^u$ (for a flow dependence relation) or $i^u + d_a = i^d$ (for an antidependence relation). In our example, the solutions are $d_f = -1$ and $d_a = +1$; for sequential loops, the dependence distance must not be negative, and so this system must arise from an antidependence relation with distance one.

In this example, finding a solution is trivial. In general, there can be several equations (one for each array dimension) and many unknowns (depending on the loop nesting). For nested loops, the dependence distance will be represented as a vector **d**, with one entry for each loop. The limits of the inner loop can depend on the outer loop, as shown in Fig. 100.3. When the inner loop limits are a simple function of the outer loop index variable, the loop is called **triangular**. In this loop, variable **x** has flow dependence with distance $(1, 0)$, meaning the element $x[i, j]$ assigned at iteration (i, j) will be used at iteration $(i + 1, j)$.

When the dependence distance is zero, we say the dependence is **loop independent**. In a

```
for i = 1 to n do
   for j = 1 to i-1 do
S₁ :   x[i,j] = x[i-1,j] + a[i,j] * y[j]
    endfor
  endfor
```

FIGURE 100.3   Sample doubly nested loop.

sequential program, the **direction of the dependence** is determined by the execution order of the references; the later reference is always dependent on the earlier reference. Thus, in a sequential program, the dependence distance d must be lexicographically positive; that is, the first nonzero element of d must be greater than zero. We say that a **loop carries a dependence** relation if the first nonzero element of the dependence distance appears in the position for that loop.

If the dependence distance is constant, the distance vector can be saved by the compiler as an integer vector. If the distance varies, there are several useful representations, such as the sign but not the magnitude of each element of the distance vector, the minimum and maximum value of each element, or even representing the dependence distance information as a reduced set of inequalities. Different representations vary in precision and cost.

Program Restructuring

Once the dependence information is collected, a parallelizing compiler will try to optimize the performance of the program. Typically, this means identifying parallelism, but there are other aspects to performance, for example, task scheduling policies, memory hierarchy performance, and interprocessor synchronization. Most compilers apply a loop restructuring phase at this point; reordering the loops has the potential of exposing loop-level parallelism or changing its characteristics, such as task granularity.

The most basic loop restructuring optimization is simple parallel loop identification. Executing a loop in parallel typically means that each iteration is treated as a separate task, and the iterations are farmed out or scheduled on different processors. If the loop does not carry any dependence relations, then executing the loop in parallel can be done without adding synchronization among the processors executing different iterations [Allen and Kennedy 1987]. Often a compiler will try to find parallel outer loops, with each iteration turning into a parallel task; this has the advantage of amortizing the overhead of scheduling the task over a larger chunk of parallel work. In Fig. 100.3, the inner loop carries no dependence relations, and so can be marked as a parallel loop. If there are dependence relations carried by the loop, the iterations can be scheduled to multiple processors by adding appropriate interprocessor synchronization or by restructuring the loop. For instance, a global summation can be computed by having each processor compute a local sum, accumulating the global sum at the end of the loop; such a transformation may allow the bulk of the work to proceed without synchronization, but it changes the order of accumulation, which can make a subtle difference in the way floating-point roundoff error is accumulated.

Another important loop restructuring transformations is **loop interchanging**. Simple interchanging of two nested loops is legal if the following three conditions hold:

1. The loops are **tightly nested** (the outer loop contains the inner loop and no other code).
2. The inner loop limits are invariant or triangular in the outer loop.
3. No dependence relation carried by the outer loop has a negative distance in the inner loop.

The dependence relations are modified when the loops are interchanged; in particular, the dependence distance vector elements for the corresponding loops must also be interchanged. Loop interchanging can have several important benefits:

- It may move a parallel loop to the outermost position; this can reduce the cost of scheduling the parallel loop iterations across processors.
- It can move a loop with positive distance vector elements to the outermost position, where it will carry those dependence relations, perhaps allowing the inner loop or loops to execute in parallel.
- It changes the order of memory accesses.

Interchanging the two loops in Fig. 100.3 gives the result in Fig. 100.4. The transformed loop has the same dependence relations, but with dependence distance $(0, 1)$, meaning the outer loop is now a candidate for parallel execution.

Loop interchanging was first explored to enhance parallelism, but the memory access order can be as important to delivered performance. Cache memories are typically loaded in blocks of 4–16 words at a time, to take advantage of spatial and sequential locality. Interchanging loops can improve the **cache hit** ratio by making the inner loop run down the data array elements in the order they are stored. The inner loop in Fig. 100.4 runs down a column of the array, whereas in Fig. 100.3 it runs across a row. For most languages, arrays are stored by rows, which mean Fig. 100.3 will make better use of the cache

```
for j = 1 to n do
  for i = 1 to j-1 do
S₁ : x[i,j] = x[i-1,j] + a[i,j] * y[j]
  endfor
endfor
```

FIGURE 100.4 Sample doubly nested loop after interchanging.

memory by making the inner loop be the *stride-1* loop, accessing memory consecutive memory locations; Fortran stores arrays the other way. In a parallel loop, two iterations may be data independent but may share cache lines; in most parallel machines, as long as the shared cache lines are read-only, there is no performance penalty. However, in common multiprocessor cache implementations, when two processors write to the same cache line, even if the writes are to different words, there is significant communication between the processors to keep the cache lines coherent, degrading overall performance. Interchanging loops to parallelize a nonstride-1 loop can improve performance.

Code Generation

Identifying potential and useful parallelism is an important job for a parallelizing compiler, but the final step is generating efficient executable code; the type of code depends on the parallelism model. For instance, a **cache-coherent** multiprocessor with only a few processors may use a master–worker type of operation. One task acts as the master, executing the sequential code up to the next parallel operation; at that point, the master activates the worker tasks, which work in parallel until the parallel operation is complete. The workers then go back to sleep until again awakened by the master. With only a few parallel tasks, the cost of activating the parallel workers is small. In the case of parallel loops, the iterations must be assigned to the workers. In a *self-scheduled* loop, each worker enters a critical section to get the next available iteration; a critical section is a piece of code which only a single processor can execute at a time. Although self-scheduling can give better load balance among the processors, the cost of synchronizing to enter the critical section can be significant with more than a few processors. In a *prescheduled* loop, the division is done when the workers are activated, allowing the workers to proceed without synchronizing in a scheduling critical section. Variations of self-scheduling assign *chunks* of consecutive iterations to each processor to reduce the overhead of the scheduling critical section.

Synchronization in multiprocessors is often implemented using shared memory; processors can pass values to another in a shared variable. For instance, a shared variable can serve as a critical section lock. Each processor attempting to enter the critical section inspects the lock; if it is zero the lock is free, and the processor can safely enter the critical section by setting the lock to a nonzero value. If the lock is nonzero some other processor is in the critical section, and this processor must wait. To prevent two processors from simultaneously inspecting the lock with a zero value and then both entering the critical section, special swap or test-and-set instructions are used. Also, multiprocessors are designed with hardware coherence mechanisms to ensure that multiple processors writing to the same memory location or to two locations in the same cache line will do so in a well-defined manner.

Massively parallel machines are currently designed with physically distributed memory; for each processor, access to local memory is much faster than access to remote memory. For such machines, a common approach is to generate code so that each processor redundantly executes the scalar code up to the next parallel block. This eliminates the need for a master process to communicate with many workers. Most such schemes use a prescheduling method; in fact, many schemes distribute the data among the processors, with the hope that most of the work on each processor will use only local data, and hence will not need any synchronization between processors. Certain data-parallel programming styles fit this model quite well. The High-Performance Fortran language [Koelbel et al. 1994] was designed with this execution model in mind, to encourage an appropriate programming style.

Massively parallel machines may allow remote memory access, though scalable cache coherence schemes are still a subject of research. Such machines can still use memory locations to synchronize, though it may require more interprocessor network traffic, since updates will not be broadcast automatically. Some machines require software to initiate every message on the interprocessor network. In such cases, the compiler can generate code to simulate remote memory access; however, a remote memory access requires a roundtrip message: one for the request (address) and one for the response (data). An alternative is to have each processor send the required data from its memory to other processors that need it; this requires only a single trip through the network for each processor, but the owner of the data must know what other processors need it.

100.3 Best Practices

The most important characteristic of a parallelizing compiler is that it must preserve the observed behavior of the program; in the best case, this means the answer computed by the program running in parallel should be exactly the same as the answer computed by the same program running sequentially. Parallelizing compilers are designed to preserve this

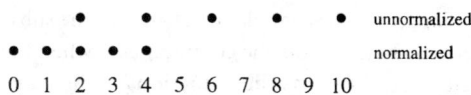

FIGURE 100.5 Normalized and unnormalized iteration spaces for sample loop.

behavior exactly, except for roundoff error differences, when allowed by the programmer. In this section we focus on the mature technology developed for identifying and utilizing parallelism in loops.

Iteration Spaces

An iteration space comprises a set of integer points corresponding to the iterations executed by a nested loop. A single-nested loop has a one-dimensional iteration space. The coordinates assigned to the iterations may be *normalized*; a normalized loop counts up by one, and the lower and upper limits are appropriately adjusted. An unnormalized loop has coordinates equal to the values of the loop index variable for each iteration. The normalized and unnormalized iteration space for the following single loop are shown in Fig. 100.5:

```
for i = 2 to 10 by 2 do
```

For the most part we will restrict ourselves to normalized iteration spaces; a simple transformation converts an unnormalized loop to a normalized one. We represent the loop limits by inequalities: $L \leq i \leq U$; in this case, $0 \leq i \leq 4$.

Doubly nested loops have two-dimensional iteration spaces. In most cases, the limits of all loops are a linear combination of loop invariants and outer loop index variables. When the limits do not fit this model, subsequent parallelization analysis may treat the limit as unbounded, losing precision and restricting some optimizations. Some loops have special iteration space shapes; a doubly nested loop with invariant limits for all loops generates a rectangular iteration space. The loop in Fig. 100.3 generates a triangular iteration space, as shown in Fig. 100.6.

Data Dependence Analysis

Data dependence analysis finds the basic restrictions that prevent parallel execution. For scalars, data dependence analysis can be done with standard compiler analysis. Even in loops, scalar dependence is fast and precise. The example in Fig. 100.7 shows a flow dependence relation from S_1 to S_3 due

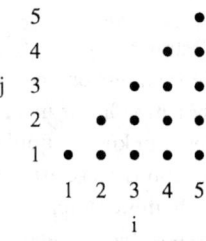

FIGURE 100.6 Triangular iteration space.

to scalar **x**, and a flow dependence from S_3 to S_2 due to scalar **y**. The dependence for **x** is loop independent, since **x** is used in the same iteration as it is assigned, while the dependence for **y** is carried by the loop, since the use occurs on the next iteration. In fact, since the use occurs on the very next iteration, the compiler can determine that the dependence distance is exactly one. This particular loop is a relatively common idiom, using a scalar to save the old value of an array element to be used in the next iteration. These are not the only dependence relations, however; since **x** and **y** are reassigned on each iteration, there are loop carried output dependence relations from S_1 and S_3 to themselves, and so on.

For arrays, the compiler must analyze the subscripts. Current methods handle the common case when the subscripts can be expressed as **affine functions** of the loop index variables and loop invariants. The compiler sets up a system of linear equations to determine if there are integer values of the loop index variables that allow dependence, and adds inequalities to determine if the solution lies within the loop limits. In Fig. 100.8, there are two references to the array **a**. To solve for flow dependence, the dependence system solves for four unknowns, **i** and **j** at the definition **a[i,j]** and at the use **a[j-i,i]**, in the equations:

```
for i = 1 to n do
S₁   :   x = a[i]
S₂   :   a[i] = (y + a[i+1])/2
S₃   :   y = x
     endfor
```

FIGURE 100.7 Scalar dependence example.

```
for i = 1 to n do
    for j = i to n do
S₁   :       a[i,j] = a[j-i,i]
        endfor
    endfor
```

FIGURE 100.8 Sample loop.

$$i^d = j^u - i^u$$

$$j^d = i^u$$

The complete dependence system is shown in matrix form in Fig. 100.9, after adding the loop limit constraints. Note that the loop invariant variable **n** is treated as another unknown in the dependence system. In this system, we can eliminate the equalities by substituting for i^d and j^d in the inequalities. This reduces the system of inequalities to

$$
\begin{pmatrix}
1 & -1 & 0 \\
-1 & 1 & -1 \\
-2 & 1 & 0 \\
1 & 0 & -1 \\
-1 & 0 & 0 \\
1 & 0 & -1 \\
1 & -1 & 0 \\
0 & 1 & -1
\end{pmatrix}
\begin{pmatrix}
i^u \\
j^u \\
n
\end{pmatrix}
\leq
\begin{pmatrix}
-1 \\
0 \\
0 \\
0 \\
-1 \\
0 \\
0 \\
0
\end{pmatrix}
$$

Testing for the existence of a solution in the space defined by the inequalities can be done using several procedures; one common one is Fourier–Motzkin elimination [Williams 1983]. This scheme takes a system of linear inequalities with k unknowns, and projects the system onto a reduced system with one fewer unknown, until there are no unknowns left or an inconsistency is found. Other methods for finding solutions also exist; see Maydan et al. [1991] and Pugh [1992].

In this example, the equalities were eliminated by simple variable substitution; in general, this may not be possible. An algorithm similar to Gaussian elimination, called the generalized greatest common divisor (**GCD**) algorithm [Knuth 1981, Banerjee 1988], uses the equalities to eliminate or replace some unknowns to get a simpler system.

Loop Restructuring

A catalog of restructuring techniques is available for optimizing the performance of loops. In Fig. 100.3, there is a flow dependence carried by the outer loop with distance vector $(1, 0)$. The compiler can *parallelize* the inner loop (mark it for parallel execution), since it does not carry any dependence relations. In the common fork-join style of parallel execution, the program will spawn or awaken the parallel tasks every time the inner loop is reached, and will wait for all of the parallel tasks to complete at the end of the inner loop.

$$\begin{pmatrix} 1 & 0 & 1 & -1 & 0 \\ 0 & 1 & -1 & 0 & 0 \end{pmatrix} \begin{pmatrix} i^d \\ j^d \\ i^u \\ j^u \\ n \end{pmatrix} = \begin{pmatrix} 0 \\ 0 \end{pmatrix}$$

When the inner loop is executed in parallel, the granularity of the tasks is quite small; the cost of spawning or awakening the worker tasks can be quite significant compared to the amount of work done. An alternative is to optimize the program for outer loop parallelism. In this case, the compiler can *interchange* the two loops, as shown in Fig. 100.4, making the outer loop parallel.

$$\begin{pmatrix} -1 & 0 & 0 & 0 & 0 \\ 1 & 0 & 0 & 0 & -1 \\ 1 & -1 & 0 & 0 & 0 \\ 0 & 1 & 0 & 0 & -1 \\ 0 & 0 & -1 & 0 & 0 \\ 0 & 0 & 1 & 0 & -1 \\ 0 & 0 & 1 & -1 & 0 \\ 0 & 0 & 0 & 1 & -1 \end{pmatrix} \begin{pmatrix} i^d \\ j^d \\ i^u \\ j^u \\ n \end{pmatrix} \le \begin{pmatrix} -1 \\ 0 \\ 0 \\ 0 \\ -1 \\ 0 \\ 0 \\ 0 \end{pmatrix}$$

In this case, the choice whether or not to interchange loops was driven by the granularity of parallelism. There

FIGURE 100.9 Dependence system.

are other performance characteristics to consider also; a compiler may choose to parallelize the loop with more iterations, if the loop limits are known, or based on memory reference patterns.

There are simple cases where neither loop can execute in parallel, as shown in Fig. 100.10. There are dependence relations carried by both the inner and outer loop, as shown in Table 100.2.

A compiler can find parallelism by using a generalized **unimodular transformation**; it tries to find a 2×2 **unimodular matrix** T such that $T d$ is lexicographically positive for all dependence distance vectors d, and with some loop that carries no dependence relations. It is easy to find such a matrix so that the outer loop carries all of the dependence relations. Two such matrices are

$$\begin{pmatrix} 1 & 1 \\ 0 & 1 \end{pmatrix} \quad \text{and} \quad \begin{pmatrix} 1 & 1 \\ 1 & 0 \end{pmatrix}.$$

Actually there are many such matrices; the compiler must find one quickly that satisfies as many goals as possible, such as placing the parallel loop at the right-nest level, and considering loop limits, memory references, and other performance characteristics.

The code generation procedure transform the limits for the original loop into limits for the restructured loop. The original limits can be represented by the inequalities $Ai \le c$:

$$\begin{pmatrix} -1 & 0 \\ 1 & 0 \\ 0 & -1 \\ 0 & 1 \end{pmatrix} \begin{pmatrix} i_1 \\ i_2 \end{pmatrix} \le \begin{pmatrix} -2 \\ 99 \\ -2 \\ 99 \end{pmatrix}$$

The new loop indices are $k = T i$ for some transformation matrix T. The limits for the new indices are

```
for i = 2 to 99 do
  for j = 2 to 99 do
    a[i,j] = 0.25 * (a[i-1,j] + a[i+1,j] + a[i,j-1] + a[i,j+1])
  endfor
endfor
```

FIGURE 100.10 Neither loop can execute in parallel.

TABLE 100.2   Data Dependence Relations for Nested Loop Example

Type	From	To	Distance	Carried by
Flow dependence	a[i, j]	a[i - 1, j]	(1,0)	i
Flow dependence	a[i, j]	a[i, j + 1]	(0,1)	j
Antidependence	a[i + 1, j]	a[i, j]	(1,0)	i
Antidependence	a[i, j + 1]	a[i, j]	(0,1)	j

```
for k = 4 to 198 do
  for l = max(2,k-99) to min(99,k-2) do
    a[l,k-l] = 0.25 * (a[l-1,k-l] + a[l+1,k-l] + a[l,k-l-1] + a[l,k-l+1])
  endfor
endfor
```

FIGURE 100.11   Generated code after a linear transformation.

found by solving $AT^{-1}k \leq c$ for limits on each loop. Taking the second transformation matrix from the preceding equation,

$$T^{-1} = \begin{pmatrix} 0 & 1 \\ 1 & -1 \end{pmatrix}.$$

The modified limits are

$$\begin{pmatrix} 0 & -1 \\ 0 & 1 \\ -1 & 1 \\ 1 & -1 \end{pmatrix} \begin{pmatrix} k_1 \\ k_2 \end{pmatrix} \leq \begin{pmatrix} -2 \\ 99 \\ -2 \\ 99 \end{pmatrix}$$

Using the Fourier–Motzkin method to eliminate k_2 finds limits for the outer loop of $4 \leq k_1 \leq 198$; the limits for the inner loop are $\max(2, k_1 - 99) \leq k_2 \leq \min(99, k_1 - 2)$. Within the body of the loop, the original indices i must also be replaced by the modified indices $T^{-1}k$. The generated code for the loop is shown in Fig. 100.11, where the inner loop can be executed in parallel.

Another important restructuring transformation is *tiling* loops to break the iteration space into blocks. One use for tiling is to increase the granularity of a parallel task, by scheduling a tile or chunk of iterations to a processor, rather than scheduling one iteration at a time. Another use is to increase data locality for cache memories, even on a uniprocessor. For instance, a simple matrix multiplication loop is shown in Fig. 100.12. The loops can be reordered to take advantage of parallelism or memory layout, but with any ordering the inner two loops will fetch every element of one of the arrays **a**, **b**, or **c**. When the arrays are larger than the cache memory, this means fetching that array from memory to cache once each time around the outer loop, making memory latency and bandwidth critical performance bottlenecks. The algorithm can be *tiled* by breaking the loop structure into outer loops that step between tiles and inner loops that step between iterations, as shown in Fig. 100.13. Tiling is legal whenever the loops can be interchanged; in fact, tiling may be legal even when the loops cannot be interchanged, if the tile size is smaller than the dependence distance of the outer loop, but that is relatively rare.

Two other important restructuring transformations are loop fission, splitting a loop into two loops, and loop fusion, combining two loops into a single loop. In these and

```
for i = 1 to n do
  for j = 1 to n do
    for k = 1 to n do
      c[i,j] = c[i,j] + a[i,k] * b[k,j]
    endfor
  endfor
endfor
```

FIGURE 100.12   Matrix multiplication loop.

```
for it = 0 to n-1 by tilesize do
   for jt = 0 to n-1 by tilesize do
      for kt = 0 to n-1 by tilesize do
         for i = it+1 to min(n,it+tilesize) do
            for j = jt+1 to min(n,jt+tilesize) do
               for k = kt+1 to min(n,kt+tilesize) do
                  c[i,j] = c[i,j] + a[i,k] * b[k,j]
               endfor
            endfor
         endfor
      endfor
   endfor
endfor
```

FIGURE 100.13   Tiled matrix multiplication loop.

other cases, the transformation is legal as long as it performs the same computations and preserves the dependence relations. Some transformations need more precise dependence information than even a dependence distance vector provides.

Interprocedural Analysis

Some modern compilers analyze programs across procedure boundaries. Such analyses are useful for scalar compilers, such as to determine actual object types of formal arguments in object-oriented languages, or to optimize type inferencing in implicitly typed functional languages. For parallelizing compilers, several important interprocedural analysis methods can be important.

A simple example is interprocedural constant propagation; constant propagation is used to replace variable accesses with constant values, if the compiler can determine that the variable always has constant value at some point. In Fig. 100.14, a compiler can determine that the variable **n** can be replaced by the value 5 in the assignment to **j**; however, **n** is incremented in the loop, so it cannot be replaced in the assignment to **k**. Given the value of **n**, the value of **j** can be computed as 6; since **j** is not modified in the loop, the reference to **j** in the **k** assignment can be replaced by the value 6.

```
n = 5
j = n + 1
loop
   n = n + 1
   k = j + n
endloop
```

FIGURE 100.14 The constant 5 can be propagated to all uses of **n**.

This analysis can be extended across procedure boundaries by constructing an interprocedural call graph, with one node for each procedure and a directed edge between nodes for each procedure call. At each procedure boundary, some actual arguments are bound to formal arguments. The actual arguments may be other variables or expressions; in addition, there may be global variables which are assigned constant values in a calling routine and used in the called routine. In Fig. 100.15, the global variable **n** is assigned a constant value in the main program; this is subsequently bound in procedure **proc** to the formal argument **y**. The literal value 15 is bound to format argument **x**, which is subsequently bound to formal arguments **z** and **y**. To determine whether **x** always has constant value requires inspecting all call sites to **proc**. This information is then used to study whether **y** and **z** have constant values. Even this simple analysis can become almost arbitrarily complex; for instance, procedures can be passed as formal arguments, meaning that constructing the interprocedural call graph requires interprocedural analysis just to determine what routines can call what other routines [Hall and Kennedy 1992]. Mutually recursive routines introduce cycles in the call graph, meaning that a depth-first analysis is insufficient.

The determination of whether an actual argument is constant may depend on both intraprocedural and interprocedural analyses, making it difficult to separate the method into local and global parts. Many heuristics are used to be successful in the common cases without sacrificing compiler speed [Metzger and Stroud 1993].

Another important analysis is determining whether two names might be **aliases** of the same variable or dynamically allocated memory location, when reference arguments or pointers are passed to a procedure. For programs targeted to massively parallel machines, interprocedural analysis can be used to determine not only the values of variables when a procedure is called, but their memory layouts as well. When there are two or more conflicting values or layouts that would otherwise inhibit optimization, the compiler may choose to *clone* the procedure, making two or more copies tailored to the different call sites or situations.

```
global n
n = 5
proc( 15 )
    :
procedure proc( x )
proc2( n, x )
proc2( x, 15 )
    :
procedure proc2( y, z )
    :
```

FIGURE 100.15 Interprocedural constant propagation example.

Multiprocessor Code Generation

The final stage of a parallelizing compiler is generating efficient parallel code. For a shared memory multiprocessor, a common model is to have a single master task execute the sequential parts of the program, and to fork or activate worker tasks when a parallel region is reached. An implementation issue is how to manage the worker tasks; this depends on the efficiency of creating tasks or processes, which often involves an expensive operating system call.

Another issue is scheduling the parallel work onto the worker tasks. As mentioned before, prescheduling means letting the master divide the work into P pieces for the P processors, and activating each of the P workers with its piece. If there is no synchronization between the pieces, each processor can proceed at full speed. In a multiprogrammed system, the program is not guaranteed to have control of all of the processors; one of the worker tasks may be swapped out in favor of some other program. In that case, a self-scheduling method may be used, where the master divides the parallel work into a large number of small pieces, perhaps each consisting of one iteration. Each worker is activated with a pointer to a global scheduler; when it wants another piece of work, the worker executes the scheduler in a critical region.

If there are data dependence relations between the parallel loop iterations, three properties must be satisfied. First, the iterations must be scheduled in a manner that allows parallel work to proceed; for instance, if there is a dependence with distance one, scheduling a chunk of 10 consecutive iterations on one processor would prevent the next processor from proceeding until the first reached its 10th iteration. Second, the processors must actively synchronize with each other; typically this is done by setting and testing memory locations. Third, since the memory is used to pass values between processors, the memory system must satisfy certain consistency principles. For instance, the following loop may be executed in parallel by letting each processor execute one iteration in a self-scheduling manner:

```
for i = 2 to n do
    a[i] = b[i] + c[i]
    d[i] = (a[i] + a[i-1]) * 0.5
endfor
```

Note the dependence relations from the first assignment to the second, with dependence distances (0) and (1). The generated code for a single iteration in a worker process might look like the following:

```
fetch c[i]
fetch b[i]
add
store a[i]
set sync[i]
wait until sync[i-1]
fetch a[i-1]
add
mul
store d[i]
```

This code does not consider the scheduling issues nor the special case of the first iteration which does not have to wait for a predecessor. Here **sync** is a synchronization array, with one element per iteration, which is initialized to zero and set to one when that element of **a[i]** is stored. The **wait until** construct is a *busy-wait* loop, consuming processor cycles until the synchronization flag is set. The memory system must ensure that the value of **sync[i]** is not changed until the value of **a[i]** is stored. In a monolithic memory system, this is not a problem; however, multilevel cache memories and write buffers can cause problems. Very aggressive processors and caches can reorder memory operations, which would allow this program to fail. In such systems, special memory barrier instructions are typically used to cause the processor to stall until all previous fetch or store operations have completed.

Even when the memory system properly implements synchronization operations, the synchronization busy-wait loop itself needs to be optimized. In the preceding example, only one processor is setting each synchronization element and only one processor is testing each element. A busy-wait loop may be implemented using a test-and-set or swap instruction on the synchronization word; to the memory subsystem, these instructions look like a read followed by a write operation. If there are two or more processors waiting at the same synchronization point, the multiple writes by different processors can cause severe thrashing of the cache line among the processors. Not only does this slow down the synchronization, but it consumes memory interconnection bandwidth, affecting the progress of other tasks in the system. To alleviate this problem, the busy-wait loop is usually implemented by a loop that only reads the synchronization word, waiting for the task holding the lock to clear it. This loop will continually fetch a cached copy of the synchronization word, so there is no traffic on the memory bus until the producer clears the synchronization word. At that point, each processor waiting on the word will see the update, and can use a test-and-set or swap (or other means) to resolve which one will get access to the critical section.

Scalable Multicomputer Code Generation

Small-scale systems can provide efficient cache coherence hardware, meaning that the software can assume that all processors have uniform access to the shared memory. Scalable parallel systems typically comprise a large number of nodes, each of which has one or more processors with some memory; the nodes are connected by a high-bandwidth network. Such machines are sometimes called nonuniform memory access (NUMA) machines. They can be broadly cast into two categories:

- Private address space is where each processor has its own hardware memory address space and cannot directly access the memory attached to a different processor; this is often called a *message-passing* machine, since tasks on different processors must communicate by sending software messages.
- Shared address space is where the memory attached to each processor can be directly addressed by other processors.
  The second category is further divided into subcategories, depending on how the shared memory is cached and how coherence is maintained between cached copies:

- Noncache systems have no coherence problems.
- Local-only cache systems do not cache remote memory lines; again, there is no coherence problem, since the only copies are on a single node.
- Noncoherent systems allow remote memory caching, but do not support hardware coherence of the cached copies; such systems must use software to enforce coherence between cached values.
- Scalable coherent cache systems are very much a current research topic.

All scalable systems communicate among the nodes via messages; the differences lie largely in how much of the message creation and handling is visible to the software (and the compiler). In a private address space machine, all messages are created by software.

Scalable coherent cache systems use hardware caches which are filled on a **cache miss**, as with a multiprocessor. However, a multiprocessor uses a single system bus. Cache coherence messages, such as one cache requesting all other caches to flush some locations from their memories, occur one at a time and are visible to all caches, because all caches are attached to the bus. A scalable system uses a more complex communication network; coherence messages may occur in parallel, and should be sent only to caches that are relevant to the message. This requires the caches to keep track of which other caches have copies of the same values. In a scalable coherent cache system, most messages are created by the cache coherence mechanism.

The task for the compiler in such machines is to reduce the number of such messages and to address message latency. One way to reduce the number of messages is to reorganize the program to take advantage of locality. For caching shared address space systems, the number of messages is related to the number of cache misses; if the program is reorganized to have better locality, and thus fewer cache misses, there will be fewer messages. For private address space systems, increased locality means that the data transmitted with a single message will be used many times before it needs to be discarded to make room for other data. Various compiler optimizations have been developed that improve locality; loop tiling is one that we have already seen and which is often used for these systems.

The message count can also be reduced by making most memory references from the local memory. In a private address space system, programs make copies of global data in the local memory, and then refer to the local copies thereafter. As long as there is enough room in the local memory and coherence issues do not arise, the message count can be kept quite low. Even in a shared address space system, programs can ensure that references will be local by making copies of global data in local memory; such a scheme is an implicit admission that the hardware caching mechanism provides insufficient performance. The compiler can transform a program to make local copies of remote data when it determines that such a copy would be used several times before it is updated. Such a transformation is easy to envision in a nested loop environment; interprocedural analysis and optimization can be used for larger program scopes.

An alternate scheme to making copies of global data is to partition the global data among the processor local memories, then to schedule the parallel tasks that use and update the partitions on the appropriate processors. If this can be done, all or most of the data references for each processor will be to the local memory. An aggressive scheme might reorganize the data periodically to take advantage of different reference patterns in various parts of the program. Current research is addressing automatic data partitioning to reduce message traffic. The High-Performance Fortran language accepts user directives to partition the data; a compiler then schedules parallel tasks onto the processor that owns the data used by that task.

Regardless of the scheme used, some messages will remain; the latency of these messages can be tolerated in two ways. If the message can be scheduled before the data is needed, the latency of the message can be hidden behind other computation. In a cached shared address space system, message scheduling reduces to inserting prefetch operations, with the hope that the fetched value will not be evicted from the cache before it is used. Message scheduling is similar to instruction scheduling, except the latencies are larger and usually uncertain, and the objects are often larger than single register values. Message scheduling must also have a method to manage the data buffers on both ends of the message; the message must not be sent before the data is valid, and must not be received into a buffer until all of the old uses of the buffer are complete.

A second method to tolerate message latencies is to keep a pool of waiting tasks on each processor; when a condition is reached where a task needs to wait for a message, the processor can switch to a different task, thus keeping the processor busy while the message is in transit. This is called *multithreading*, where each task is a lightweight process called a *thread*. For shared address space machines where messages are handled in hardware, hardware support is needed to support multithreading. For private address space machines, multithreading can be managed by compiler-generated code. In either case, automatic multithreading is a current research area.

```
a(2:n) = (a(1:n-1) + a(3:n+1))/2

forall( i = 2:n )
   a(i) = (a(i-1) + a(i+1))/2
endforall
```

FIGURE 100.16 Fortran 90 array assignment and equivalent **forall** construct.

Parallel Languages

The techniques developed for parallelizing compilers are also used when handling parallel languages. The Fortran 90 language added parallel array assignments which need parallelization analysis whether compiled for a uniprocessor or parallel machine. For instance, the sequential loop in Fig. 100.7 can be written in one Fortran 90 array assignment statement, as shown in Fig. 100.16. High-Performance Fortran and the Fortran 95 update include a **forall** construct, which allows array assignments to be written using syntax similar to a Fortran **do** loop; this is also shown in the figure. The semantics of array assignments dictate that the right-hand side values are fetched and the expression computed before any left-hand side elements are updated. In this case, it means there is an antidependence from both right-hand side references in the **forall** to the left-hand side. The dependence relations with dependence distances are shown in Table 100.3.

TABLE 100.3 Dependence Relations in Parallel Loop

Type	From	To	Distance
anti	a[i - 1]	a[i]	(-1)
anti	a[i + 1]	a[i]	(1)

```
forall( i = 2:n )
   T[i] = (a[i-1] + a[i+1])/2
endforall
forall( i = 2:n )
   a[i] = T[i]
endforall
```

FIGURE 100.17 Breaking the **forall** up to eliminate dependence problem.

An antidependence with distance of (-1) means that reference **a[i-1]** fetches a value in iteration number i that reference **a[i]** will update in iteration number $(i - 1)$. A sequential loop could not generate a negative dependence distance. If all of the dependence relations had **lexicographically positive** distance vectors, the **forall** could be executed as a sequential loop. Sometimes a program transformation, such as a **linear transformation** of the index set, can be used to transform the **forall** into an equivalent one with lexicographically positive distances. Other transformations and optimizations to generate parallel target code are exactly the same as for parallelizing a sequential program.

In Fig. 100.16, reversing the loop to make one dependence positive will make the other negative; to preserve the dependence relation, the compiler can break the construct into two parts, inserting synchronization between the parts if necessary for parallel execution, as shown in Fig. 100.17. When generating sequential code, another option is for the compiler to recognize that the dependence distance is a small constant and to generate the code in Fig. 100.7 automatically.

Another example of the utility of the analysis here was demonstrated to optimize the functional language Haskell [Anderson and Hudak 1990]. The following mathematical specification of an $n \times n$ matrix a:

$$a_{i,j} = \begin{cases} 1 & \text{if } i = 1, 1 \le j \le n \\ i & \text{if } j = n, 2 \le i \le n \\ a_{i-1,j} + a_{i-1,j+1} + a_{i,j+1} & \text{if } 2 \le i \le n, 1 \le j < n \end{cases}$$

is written in Haskell as shown in Fig. 100.18. Compiling this into correct, efficient code requires checking

```
letrec* a = array((1,1),(n,n))
    ( [ (1,j) := 1 | j <- [1..n] ] ++
      [ (i,n) := i | i <- [2..n] ] ++
      [ (i,j) := a!(i-1,j) + a!(i-1,j+1) + a!(i,j+1)
        | i <- [2..n], j <- [1..n-1] ] )
```

FIGURE 100.18   Haskell array comprehension for a wavefront recurrence.

```
for j = 1 to n do a[1,j] = 1
for i = 2 to n do a[i,n] = i
for i = 1 to n do
    for j = 1 to n-1 do
        a[i,j] = a[i-1,j] + a[i-1,j+1] + a[i,j+1]
```

FIGURE 100.19   Correct efficient sequential code for Haskell example.

that the same array element is not assigned more than once, and determining an order of evaluation for the array. Dependence analysis between the targets of the assignments can be used to find whether an element might be multiply assigned. To determine the evaluation order, the compiler finds which parts of the assignment depend on which others, and looks at dependence distances to determine whether the corresponding loops can be run forward or backward. In this example, the generated code should perform the first two parts of the assignment, in any order, and the final part with the **i** loop running forward and the **j** loop running backward, as shown in Fig. 100.19. The same techniques can also be used to find parallel execution in such programs.

100.4 Research Issues

Parallelizing compilers have been in commercial use since the 1970s; some of the technologies are quite mature. However, there remain many open issues, especially as computers become more widespread and are used for so many different tasks.

Compilers for massively parallel machines are currently available, but a great deal of work still remains in this area. Much of the current work focuses on reducing the frequency and cost of interprocessor communication, by allocating data to eliminate communication or by scheduling communication to hide the latency. Multithreading on each processor uses slack parallelism to hide latency of messages between processors.

Classical data dependence analysis has proven adequate for many parallelization tasks, but it does not allow a compiler to determine when an array can be replaced by a scalar, or when different processors can write to separate copies of an array. More precise array data flow analysis techniques have been developed, but there is not much experience to determine how much benefit there is for the extra cost.

Most parallelizing compiler work has focused on loops and array data structures. Other data structures, such as trees, linked lists, heaps, graphs, and so on, are implemented using dynamic allocation, and are often traversed using recursive procedures. For a parallelizing compiler to be successful on such a program, it must understand the characteristics of the data structure and the traversal.

Programmers frustrated with the weaknesses of parallelizing compilers often propose to solve the problem entirely by introducing a parallel language, or parallel extensions to a sequential language. Intelligent language design is a task not undertaken lightly; the interactions of parallelism with a sequential language can introduce a host of hidden problems. Most such designs are small additions, and assume a nonoptimizing compiler. Actually, a parallel language resolves only one aspect of the parallelizing compiler problem; for best performance, the compiler must still map the program onto the target machine, using dependence analysis and restructuring. A great deal of such work was done for the applicative language SISAL [Cann

1992], with performance results approaching and even surpassing that of mature parallelizing Fortran compilers.

In a perfect world, parallelizing compilers would be as transparent as optimizing compilers are today. The basic problem is that common languages are explicitly sequential, so parallel execution involves a change of paradigm. Higher level languages, more closely aligned with the application rather than with the machine, need not be sequential. For instance, computing the values in a spreadsheet need not be defined strictly sequentially; in fact, updating a spreadsheet in parallel, as long as the dependence relations are satisfied, would be transparent. When the tools used in parallelizing compilers are made available to applications, the world will truly move to parallelism.

100.5 Summary

Originally, parallelizing compilers were designed to discover parallelism in sequential programs. They were most successful when used on truly parallel programs that were written in a sequential language. Even with languages that allow explicit parallelism, parallelizing compilers can restructure the program and map it to the target machine to optimize many aspects of performance.

The three phases in a parallelizing compiler are analysis, restructuring, and code generation. The analysis phase finds the constraints that prevent parallel execution; this is most mature for programs that use arrays and loops, such as linear algebra style programs. The restructuring phase reorders operations to optimize various aspects of performance, such as introducing or eliminating parallelism, reducing synchronization, changing the memory access pattern, and so on. Both these phases benefit from casting the problems into a linear algebra or integer programming framework. Finally, parallel code is generated and optimized so as to improve the overall performance.

Current parallelizing compilers for Fortran and C are reasonably mature for certain programming styles; in fact, feedback from the compiler encourages an appropriate style, further improving the average performance. Future work in parallelizing compilers will address new parallel architectures, dynamic data structures, and higher level languages.

Defining Terms

Affine function: A function of several variables which can be defined as $f(x_1, x_2, x_3, \ldots, x_n) = c_0 + c_1 x_1 + c_2 x_2 + \cdots + c_n x_n$, where $c_0, c_1, \ldots, c_n$ are known constants.

Alias: A relation between two names used in a program that may refer to the same memory location.

Antidependence: A data dependence relation between two statements or operations where the first uses a data element and the second subsequently overwrites it with a new value.

Cache coherence: The scheme used to ensure that copies of the same memory address in different caches have the same value.

Cache hit: A memory fetch in which the value is found in the cache memory.

Cache miss: A memory fetch in which the value is not found in the cache, requiring a fetch from the main memory system.

Data dependence: A relation between two statements or operations where one operation must precede the other because the first produces or uses data that the second uses or overwrites.

Data dependence graph: A graphical representation of the data dependence relations in a program, procedure, or body of code.

Dependence direction: An annotation on data dependence relations, useful when the dependence distance is not constant, defined as the vector of signs of the difference of the **iteration vectors**.

Dependence distance: An annotation on data dependence relations to tell the relative iterations for the statements or operations involved, defined as the difference of the **iteration vectors**.

Flow dependence: A data dependence relation between two statements of operations where the first writes to a data element and the second subsequently reads that value.

Forall: A parallel loop implemented by executing each statement for all values of the index variable before executing the next statement.

GCD: Greatest common divisor.

Iteration vector: An integer vector, with one entry for each loop, that identifies the iteration.

Lexicographically positive: A vector $(x_1, x_2, \ldots, x_n)$ which is not all zeros and in which the first nonzero element is positive.

Linear transformation: A loop restructuring transformation implemented as a linear transformation of the iteration space.

Loop carried dependence: A data dependence relation with a nonzero entry in the distance vector for some loop.

Loop independent dependence: A data dependence relation with all zero entries in the distance vector.

Loop interchanging: A loop restructuring transformation that switches two nested loops.

Output dependence: A data dependence relation between two statements or operations where the first writes to a data element and the second subsequently overwrites it with a new value.

Tightly nested loop: Two (or more) nested loops where the outer loop contains no other code except the inner loop.

Triangular loop: A nested loop where the inner loop limits depend on the outer loop index.

Unimodular matrix: A square matrix with all integer entries, and with determinant equal to ± 1.

Unimodular transformation: A linear transformation using a unimodular transformation matrix.

References

Allen, J. R. and Kennedy, K. 1987. Automatic translation of Fortran programs to vector form. *Trans. Programming Lang. Syst.* 9(4):491–542.

Anderson, S. and Hudak, P. 1990. Compilation of Haskell array comprehensions for scientific computing, pp. 137–149. In *ACM SIGPLAN '90 Conf. Prog. Lang. Design Implementation.* ACM Press, New York.

Bacon, D. F., Graham, S. L., and Sharp, O. J. 1994. Compiler transformations for high-performance computing. *ACM Comput. Surv.* 26(4):345–420.

Banerjee, U. 1988. *Dependence Analysis for Supercomputing.* Kluwer Academic, Norwell, MA.

Banerjee, U. 1993. *Loop Transformations for Restructuring Compilers: The Foundations.* Kluwer Academic, Norwell, MA.

Banerjee, U. 1994. *Loop Parallelization.* Kluwer Academic, Norwell, MA.

Banerjee, U., Eigenmann, R., Nicolau, A., and Padua, D. A. 1993. Automatic program parallelization. *Proc. IEEE* 81(2):211–243.

Cann, D. 1992. Retire Fortran? A debate rekindled. *Commun. ACM* 35(8):81–89.

Hall, M. and Kennedy, K. 1992. Efficient call graph analysis. *Lett. Programming Lang. Syst.* 1(3):227–242.

Knuth, D. 1981. *Seminumerical Algorithms,* 2nd ed. Addison–Wesley, Reading, MA.

Koelbel, C. H., Loveman, D. B., Schreiber, R. S., Steele, G. L., Jr., and Zosel, M. E. 1994. *The High Performance Fortran Handbook.* MIT Press, Cambridge, MA.

Maydan, D. E., Hennessy, J. L., and Lam, M. S. 1991. Efficient and exact data dependence analysis, pp. 1–14. In *ACM SIGPLAN '91 Conf. on Prog. Lang. Design Implementation.* ACM Press, New York.

Metzger, R. and Stroud, S. 1993. Interprocedural constant propagation: an empirical study. *Lett. Programming Lang. Syst.* 3:213–232.

Pugh, W. 1992. A practical algorithm for exact array dependence analysis. *Commun. ACM* 35(8):102–114.

Williams, H. P. 1983. A characterization of all feasible solutions to an integer program. *Discrete Appl. Math.* 5:147–155.

Wolfe, M. 1996. *High Performance Compilers for Parallel Computing.* Addison–Wesley, Reading, MA.

Zima, H. and Chapman, B. 1991. *Supercompilers for Parallel and Vector Computers.* ACM Press, New York.

Further Information

Several recent texts are available covering parallelizing compilers; the book by Michael Wolfe [1996] presents the material from this chapter in much more detail. Hans Zima and Barbara Chapman [1991] summarize the state of the art as of the early 1990s. Utpal Banerjee [1988, 1993, 1994] has written three monographs covering various subareas of parallelizing compilers, focusing on reducing the problem to the mathematical domain.

Two other articles summarizing the techniques used by modern parallelizing compilers are Bacon et al. [1994] and Banerjee et al. [1993]. The monthly journal *IEEE Transactions on Parallel and Distributed Systems* reports recent advances in parallel computing, including languages and compilers.

Several annual conferences and workshops serve as important resources to those interested in the field. The International Conference on Parallel Processing is held in August in the Midwest of the U.S.; recent proceedings are available from CRC Press. The Supercomputing'xx conference is held in the U.S. in November or December annually; proceedings are available from IEEE Computer Society Press. The Languages and Compilers for Parallel Computing workshop is held in August every year, and often contains much more recent advances than formal conferences; proceedings are published by Springer–Verlag in the Lecture Notes in Computer Science monograph series. The International Conference on Supercomputing is held in June or July annually, and moves between Europe, North America, and the South Pacific; it covers applications, systems software, and architectures, and proceedings are available from ACM Press.

101

Run Time Environments and Memory Management

101.1 Introduction

Many computer users never perceive the existence of the run time environment of a programming language, yet its correct operation is essential to the execution of any computer program in the language. A clear knowledge of the run time environment of programming languages is invaluable when writing programs having hard real-time constraints, working with multithreaded operating systems, or writing object-oriented software.

The principal goal of this chapter is to give the reader an overview of the run time environment of programming languages with respect to the calling of procedures and functions and the passing of parameters and return values. After a review of the general principles involved, some examples are presented for the common architectures in use today.

101.2 Principles of Run Time Environments

The purpose of this section is to provide a general introduction to the principles used in designing a run time environment for programming languages with a particular emphasis on procedures and parameters. The sections which follow introduce the basic terminology and major ideas in the design and implementation of such environments. Texts which cover this material in more detail include Sebesta [1996], Louden [1993], Pratt and Zelkowitz [1996], and MacLennan [1987]. Examples are taken primarily from the language C [Kernighan and Ritchie 1988].

Procedures and Functions

In this section we are concerned with the characteristics of a procedure, subprogram, or function. For our purposes such an entity has the following characteristics: (1) it has a single entry point, (2) execution of

0-8493-2909-4/97/$0.00+$.50
© 1997 by CRC Press, Inc.

```
 1 /* bcopy -- copy source block to destination block */
 2 void bcopy (char *s, char *d, int n)
 3    /* s -- pointer to source block      */
 4    /* d -- pointer to destination block */
 5    /* n -- number of bytes to copy      */
 6 {
 7    for (; n > 0; n--)
 8         *(d++) = *(s++);
 9 }
10
11 char hello[] = { "Hello, world!" };
12          /*         1234567890123   */
13 main()
14 {
15    char newhello[50];
16    bcopy(hello,newhello,14);
17    printf("%s\n",newhello);
18 }
```

FIGURE 101.1 The **bcopy** function in American National Standards Institute (ANSI) C.

the calling unit is suspended, and (3) in the absence of a nonlocal goto, control is returned to the caller upon normal completion of the procedure. In the case of a function, a value is returned to the caller.

Figure 101.1 shows a C program that calls a local **bcopy** function to copy a string. Routine **bcopy** is a subroutine or procedure.

Line 16 is said to contain a *call* of the function. When this line is executed, execution of the calling routine, **main** in this case, is explicitly suspended. Program control is transferred to the *body* of routine **bcopy**, in this case everything between the opening brace on line 6 and the closing brace on line 9. Line 2 is called the *header* of the routine; it gives the name of the function, the return type (in this case, the **void** type indicates that the function returns no value), and the names and types of the arguments of the routine.

The variables **s**, **d**, and **n** in line 2 are said to be the *formal arguments* or *formal parameters* of the routine. The variable **hello** in the call on line 16 is said to be the *actual parameter* of the call corresponding to the formal parameter **s**. Similarly, the variable **newhello** on line 16 is the actual parameter of the call corresponding to the formal parameter **d**, and the integer constant **14** is the actual parameter of the call corresponding to the formal parameter **n**. In this example, even though there are only 13 characters in the string to be copied, the constant **14** must be used in the call to be sure to copy the null termination byte with which the C language indicates the end of a string.

In the case of this example when line 16 is executed, control is passed from the calling routine, **main** in this case, to the called routine, **bcopy**, starting at the opening brace on line 6 and continuing until the closing brace is executed (an implicit return). Although this example does not contain one, the called routine can also return to its caller through an explicit **return** statement. After program control has entered line 6 and before the implicit return on line 9, **bcopy** is said to be *active*.

Design Issues

In designing a run time environment for a given programming language, a number of design issues come in to play. The first of these is whether allocation of variables local to a procedure is static or dynamic.

If the language permits recursive calls of a procedure, then each activation of the procedure will have its own set of local variables. In this case, the local variables are allocated dynamically on the run time stack. Referencing of the local environment is done relative to the stack pointer. On the other hand, if variables local to a procedure are required to remember their values from one activation to the next, then the local variables must be allocated statically.

Static allocation generally permits faster referencing of the local environment. However, dynamic allocation makes better use of memory. Languages such as FORTRAN 77 [Zirkel 1994] and COBOL [Welburn 1995] use static allocation by default. Languages such as C/C++ use dynamic allocation by default but permit the programmer to mark any specific local variable as static. In the example shown in Fig. 101.1, the variable **hello** is allocated statically, whereas the variable **newhello** is allocated dynamically.

A second significant design issue is the method or methods by which actual parameters or arguments are passed from the caller to the called routine. These methods are treated in detail in the next section.

A third significant design issue concerns the degree to which it is possible within a procedure to reference nonlocal variables, i.e., variables which are not either passed as parameters or are not declared within the procedure. In the C program given in Fig. 101.1, the variable **hello** is global to the procedure **bcopy**. Because **hello** is global to the entire compilation unit, the variable **hello** could have been directly referenced within **bcopy**. Such nonlocal access is usually avoided, because it complicates the referencing environment, both for the compiler and for those who must read the source code.

A fourth significant design issue concerns *scoping*, the method used to determine how a reference to a nonlocal name within a procedure is resolved. Programming languages use two types of scoping: *static* and *dynamic*.

Most compiled languages that allow nonlocal references, including Ada, C, and C++, use static scoping. With static scoping, reference to a nonlocal name within a procedure is resolved at compile time by referring to the static parent of the procedure: the statically enclosing block or procedure. This process is repeated as needed until the nonlocal name is found.

Within static scoping, there is the further issue of whether static nesting of procedures is allowed. C and C++ do not allow static nesting, but Ada [Ada 1991, Barnes 1995], Pascal [Jensen et al. 1991], and Modula [Carmony 1990] do. To implement static nesting of procedures, as in Ada and Pascal, it is necessary to add another pointer to the activation record, the *static link*. The run time system sets the static link of a procedure to point to the activation record of the statically enclosing procedure. In this way, references to nonlocal variables in the procedure will be resolved in a way that is consistent with the static structure of the program. The dynamic link is not adequate to find the activation record of the statically enclosing block when a procedure is called by a program unit that is not its static parent. In C and C++, where static nesting of procedures is not allowed, no static link is needed, because a nonlocal name referenced within a procedure must be global and can be easily located with no additional information.

The LISP [Steele 1990] and APL [Gilman 1992] programming languages use dynamic scoping, in which reference to a nonlocal name within a procedure is resolved at run time by referring to the activation record of the *dynamic parent* of the procedure, the procedure or block that called the procedure with the nonlocal reference. With dynamic scoping, the program unit whose activation record is consulted to resolve a nonlocal reference cannot be determined statically at compile time, but must determined at run time through the configuration of the stack of *activation records*.

For a more thorough discussion of static and dynamic scoping and the static link, the interested reader should consult Sebesta [1996], Pratt and Zelkowitz [1996], or MacLennan [1987].

Parameter Passing

Most programming languages have based their parameter (or argument) passing convention on how parameter passing is actually implemented. In contrast, Ada bases its parameter passing convention on the semantics of the use of the parameter and leaves it to the compiler implementor to pick an appropriate implementation model.

Ada distinguishes three distinct semantic models for parameter passing: (1) *in*, (2) *out*, and (3) *in–out*. A formal parameter specified as in means that a value will be supplied to the called routine by the caller. No attempt is made by the called routine to change the value of the parameter. A formal parameter tagged as out means that the called routine will set the value of the parameter before any attempt to reference the value of the parameter. Thus, the calling routine must supply a location in which to store the out parameter, but need not initialize that location. A formal parameter tagged as in–out means that the called routine expects the parameter to be initialized and expects to return a value to the calling routine. Thus, the calling routine must supply an initialized location.

Most older languages defined their parameter passing conventions in terms of how they are implemented. The major ones in use today include: (1) *call-by-value*, (2) *call-by-result*, (3) *call-by-value-result*, and (4) *call-by-address* (or *call-by-reference* or *call-by-location*).

In call-by-value, the calling routine supplies a value to the called routine. The formal parameter is treated as a local variable that is initialized at the time of the call to the value supplied by the caller. Thereafter, there is no further link between the actual and formal parameter. The called routine is free to change the value of the formal parameter; this has no effect whatever on the actual parameter either during the activation of the called routine or at return time. Call-by-value can be thought of as a copy-in process.

In contrast, call-by-result can be thought of as a copy-out process. Except for this respect, it is very similar to call-by-value. Again, the formal parameter is treated as an uninitialized local variable. It is the responsibility of the called routine to initialize a result parameter before referencing it. When control is about to be returned to the caller, the final value of a result parameter is copied to the corresponding actual parameter. Except for this final copy, there is no other correspondence between the actual and formal parameters during the duration of the call.

A value-result parameter is one which combines the two previous methods. The formal parameter is treated during the activation of the called routine as a purely local variable, which is initialized at the time of the call to the corresponding actual parameter, and the final value of the formal parameter is copied back to the actual parameter at the time of the return. Thus, call-by-value-result is a copy-in, copy-out process.

In call-by-address (or reference) the calling routine provides the address of each actual argument. In effect, the formal parameter is effectively a constant pointer to the actual argument. References to the formal parameter are compiled to do a dereferencing so that the location of the actual argument is used in all cases. It is impossible to change the value of the pointer, only what the pointer points to. Call-by-address is used by C++ [Stroustrup 1995] for reference parameters, Pascal [Jensen et al. 1991] for **var** parameters, and FORTRAN 77 [Zirkel 1994] for all parameters.

Activation Records

Each time a routine is called, an activation record is created. Although this record may not physically all be stored in the same place, it is still useful to think of it as a single, logical record. Minimally, this record must contain: (1) the return address of the calling routine, (2) space for the formal parameters, and (3) space for the local variables of the called routine. An example of an activation record for the function **bcopy** of Fig. 101.1 is given in Fig. 101.2, which will be discussed in detail in the next section.

If the activation record is allocated statically, then variables local to the called routine can remember their values from one activation to the next. This was a feature of early versions of FORTRAN. The disadvantage of static allocation is that such routines cannot be called recursively.

Most modern languages allocate the activation record on the run time stack, thus allowing recursive routines. C/C++ allow the programmer to specify that some local variables should be allocated statically; by default local variables are allocated dynamically.

At the time of the call, the calling routine must: (1) save its own execution status, including any registers it wants saved; (2) carry out the parameter passing process; (3) pass the return address; and (4) finally transfer control to the called routine. In many modern architectures, steps 3 and 4 are combined into a single machine instruction.

At the time of return, first, the called routine must make available the final values of any result or value-result arguments. Second, if the routine is a function, then the computed value of the function must be made available. Next, the execution status of the caller must be restored. And finally, control must be transferred back to the caller.

Activation Stack

Most modern architectures provide some direct support for procedure calls and parameter passing. A typical *activation stack* using the program of Fig. 101.1 is given in Fig. 101.2.

The figure assumes that execution is on line 7 of the activation of the **bcopy** routine (see Fig. 101.1). In many implementations, a frame pointer register points to the base of the activation record on the run time stack. The stack pointer, of course, points to the logical top of the activation record. Addressing of locals and formal arguments would normally be done relative to the frame pointer, which is constant while the routine is active. Of course, the stack pointer will increase and decrease as the values of temporary variables and expressions are pushed and popped off the run time stack. Thus, using the stack pointer to address locals and temporaries, although possible, is considerably more complicated. Figure 101.2 shows activation records and stack pointer and frame pointer registers for the program listing of Fig. 101.1 at the point just before the **bcopy** routine begins execution.

The dynamic link in the activation record contains the address of the activation record of the routine's caller. In Fig. 101.2, the dynamic link of the **bcopy** routine is shown as an arrow pointing back to the frame pointer location of the activation record of the main program.

In this example, the stack location holding the value **14** (the actual parameter corresponding to the formal parameter **n**) is used as a local variable of the **bcopy** routine. This value will be decremented to **0** as the routine copies the source string to the destination location.

The space for the **newhello** variable is shown on the stack, because this variable is allocated dynamically in the **main** program. The **hello** variable is declared statically in line 11 of Fig. 101.1, so

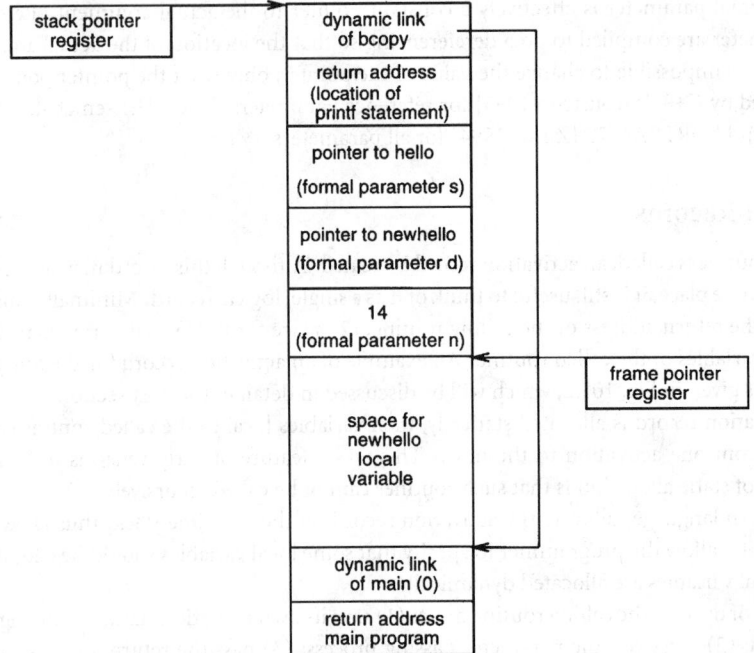

FIGURE 101.2 Activation stack for calling the **bcopy** function.

that stack space for it is not shown in Fig. 101.2. Typically, space for static variables is not allocated on the run time stack, but, instead, in a part of the program that will persist across the activations of a subroutine.

Memory Management

In addition to the use of a run time stack, most modern languages allow dynamic allocation from a form of memory known as the *heap*. In the language C, for example, **malloc** is used to allocate memory from the heap, assigning the address of the memory allocated to a pointer variable, whereas **free** is used to return such memory to the heap. In the language C++, the function **new** is used for allocations and **delete** for deallocations.

Because memory is not usually returned in the reverse order of that in which it was allocated as it is in the stack discipline, some strategy needs to be imposed on managing the heap. Over time, the heap consists of blocks of memory, some of which are still allocated and some of which are free (available to be allocated). By far, the two most common strategies for managing the available storage list are (1) *first fit* and (2) *best fit*.

In the first-fit strategy, the list of available storage is kept by increasing address. This greatly simplifies the task of collapsing adjacent blocks of storage, since the address and size of each block of available storage is known. When a request for a block of storage is made, the available storage list is searched until the first block larger than or equal to the requested size is found. The amount requested is allocated and any excess storage is returned to the heap. The disadvantage to first-fit is that a later request for a large block of storage may fail because large blocks of storage were broken up to satisfy small requests.

In the best-fit strategy, the list of available storage blocks is kept in increasing order of size. When a request for storage is made, the smallest block that will satisfy the request is used. As before, if the block is larger than the requested size the remaining storage is returned to the list. When a block is returned to the heap, it is placed on the list in order of its size. In best fit it is difficult to collapse adjacent blocks of storage without searching the entire available storage list.

In languages without explicit pointers such as LISP [Steele 1990] and SMALLTALK [Goldberg 1985], almost every program action involves allocation or deallocation from the heap. In such languages explicit heap management is deemed too error prone and too tedious. Such languages provide for automatic heap management. Two schemes have been widely used [Sankaran 1994]: *reference counting* and *automatic garbage collection*.

In the reference counting scheme, each heap cell stores a count of the number of pointer references to the cell. This count is incremented and decremented as references to the cell are added and deleted. If the count becomes zero, the cell can be deallocated or returned to the heap, since there are no references to it. Return of a cell to the heap may result in further deallocations.

Reference counting has a number of problems relative to automatic heap management. First, the space consumed by the count is significant. Second, in a language such as LISP a significant amount of time can be consumed. Third, circular lists pose problems for reference counting schemes. In a circular list, every reference count is positive even though all of the list items are no longer in use.

The term garbage refers to an area in the heap that is no longer referenced but cannot be accessed by the program. Garbage collection is the process of reclaiming this space for reuse. Unlike reference counting, garbage collection is a *lazy* method, in that no attempt is made to reclaim garbage as it is generated. When space in the heap goes below a certain threshold, garbage collection is invoked. The most popular method uses the mark-sweep algorithm. First, all cells are marked as garbage. In the mark phase, cells that are currently in use are marked. Finally, all garbage cells are returned in a sweep through the heap.

The problem with garbage collection is that the scheme often fails when it is most needed. When the heap is low on available storage, the system can spend most of its time garbage collecting. One alternative in systems with a large virtual memory is to ignore reclaiming memory, but this sometimes leads to poor paging performance.

A variant that has proved effective in recent years is *generational collection*. This scheme is based on the assumption that most garbage is generated by recent allocations. Cells that are in use longer than a fixed threshold are ignored in further attempts at garbage collection. In this way most of the effort is spent searching only a fraction of memory for new allocations, which significantly reduces the time spent garbage collecting. For further information, the reader should consult Sankaran [Sankaran 1994].

Object-Oriented Paradigm

A recent development in the area of programming languages, the object-oriented programming paradigm, provides the programmer with the facility of data encapsulation and protection, type inheritance, and function polymorphism. Recent languages that support the object-oriented paradigm are SMALLTALK [Goldberg 1985], Ada 95 [Barnes 1995], and C++ [Stroustrup 1995]. The aspects of data encapsulation and type inheritance are provided at compile time by the language compiler and are discussed elsewhere in this volume.

Polymorphism, the ability of a function to handle objects of differing types, is a facility that must be implemented in the run time system. Of the object-oriented languages currently in use, C++ is probably the most widely used.

Figure 101.3 shows a C++ program with function polymorphism. Polymorphism in C++ is expressed through *virtual functions*. In Fig. 101.3, three different object classes are defined. The **Vehicle** class is the *base* class from which the **Auto** and **Bicycle** classes inherit their base properties. The **show** member function of the **Vehicle** class is designated as a **virtual** function, which is a signal to the C++ compiler that the true identity of the function will be deferred to run time, when the type of the object being displayed is known.

The base **Vehicle** class can be used to derive many different classes. Each class derived from the **Vehicle** base class would probably have a **show** function of its own that can best display an object of the derived class. The compiler and run time system determine dynamically at run time the correct **show** function to call in a given context. In lines 35 and 36, an **Auto** object and a **Bicycle** object are declared dynamically with the **new** operation. The **pv** pointer can point to any object of the **Vehicle** base class. In line 39, when **pv** points to an object of the **Auto** class, the run time system calls the **show** member function of the **Auto** class. In line 41, when **pv** points to an object of the **Bicycle** class, the run time system calls the **show** member function of the **Bicycle** class. This selection is performed automatically by the compiler and run time system and does not require explicit action from the programmer.

Virtual functions in C++ are implemented in a number of ways. One of the simplest implementations simply uses indirection. In each object of the class, space is left for a pointer to a table of pointers to the virtual functions of the class, called the *virtual function table* (VFT). When an object of the class is created, the compiler creates the VFT for an object of the class and inserts the address of the VFT in the object. The code for executing a virtual function contains instructions to reference the address of the function through the table of virtual functions.

This process will be clarified by considering again the example of the C++ program of Fig. 101.3. Figure 101.4 shows an **Automobile** and a **Bicycle** object as the compiler would create them in memory. When a call to the virtual function **show** occurs, the pointer to the virtual function table is used to retrieve the address of the function to use in the call. This process is much more efficient than the implementation of virtual functions in the original implementation of SMALLTALK, in which a virtual function call was implemented by a time-consuming search at run time through the program's symbol table.

```
 1  #include <ostream.h>
 2
 3  class Vehicle {
 4  public:
 5      int weight;
 6      virtual void show() {
 7          cout << "Vehicle weight:" << weight << endl;
 8      }
 9      Vehicle( ) { weight = 1000; }
10      Vehicle(int w) {weight = w; }
11  };
12
13  class Auto : public Vehicle {
14  public:
15      int number_of_doors;
16      void show() {
17          cout << "Auto weight:      " << weight << endl;
18          cout << "Number of doors:  " << number_of_doors << endl;
19      }
20      Auto(int nd) : Vehicle(3000) { number_of_doors = nd; }
21  };
22
23  class Bicycle : public Vehicle {
24  public:
25      int number_of_gears;
26      void show() {
27          cout << "Bicycle weight:   " << weight << endl;
28          cout << "Number of gears:  " << number of gears << endl;
29      }
30      Bicycle(int ng = 5) : Vehicle(30) { number_of_gears = ng; }
31  };
32
33  main()
34  {
35      Vehicle *pa = new Auto(4);
36      Vehicle *pb = new Bicycle(10);
37      Vehicle *pv;
38      pv = pa;
39      pv->show();
40      pv = pb;
41      pv->show();
42  }
```

OUTPUT

```
Auto weight:      3000
Number of doors: 4
Bicycle weight:   30
Number of gears:  10
```

FIGURE 101.3   C++ function polymorphism.

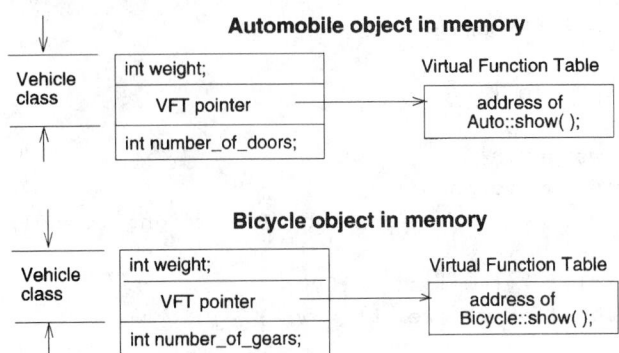

FIGURE 101.4   C++ implementation of virtual functions.

101.3 A Survey of Common Architectures

We cover the subject of run time environments through a description of the support that several current common architectures provide. We will discuss the architectures of the Intel 80x86, the
Motorola 68000, the Sun SPARC, and the IBM RS/6000 (quite similar to the PowerPC [Motorola 1993]).
The reader is expected to be familiar with these architectures, since the coverage here is necessarily
brief.

The Intel 80x86 Architecture

The Intel 80x86 architecture [Mazidi 1995] is probably one of the most widely used computer architectures in the world. Intel developed the 8086 in the mid-1970s as a compatible successor to the 8080 and
8085 *central processing units* (CPUs), which were very popular. They wanted to provide 16-bit processing capability, along with the ability to address more than 64 kilobytes of memory (1 kilobyte is 1024
bytes), both limitations of the 8080/8085 design. The popularity of the Intel 80x86 CPU family was
greatly enhanced when IBM choose the architecture for its personal computer, introduced in 1981. Since
its introduction, the Intel 80x86 architecture has evolved through several different processors, with the
later chips offering additional performance over their predecessors while still supporting the original 8086
instruction set.

Intel 8086 Register Set

Figure 101.5 shows the available registers on the original Intel 8086 CPU. The arithmetic and logic operations take place in the 16-bit **AX**, **BX**, **CX**, and **DX** registers. Each of these registers is subdivided into
two 8-bit registers holding the least and most significant bytes of the 16-bit register.

The 80x86 family of processors uses a *segmented memory model*, in which memory is viewed as being
composed of blocks of cells, called *segments*, of variable size up to 64 kilobytes. In the segmented memory
model, the address of a memory location consists of two parts that are added together: the address of the
segment in which the memory location lies and the *offset* (or distance) from the beginning of that memory
segment to the given memory location.

Four types of registers are used for address calculation: segment registers and *index*, *base*, or *pointer*
registers. A segment register is used to hold the address of the beginning of a memory segment, and
an index, base, or pointer register is used to hold an offset into that segment. The index registers are
(**SI** and **DI**), the base registers are (**BX** and **BP**), and the pointer registers are **SP**, **BP**, and **IP**. The
segment registers are **CS**, **DS**, **SS**, and **ES**. Note that by its name, **BP** register is used as both a base
and a pointer register.

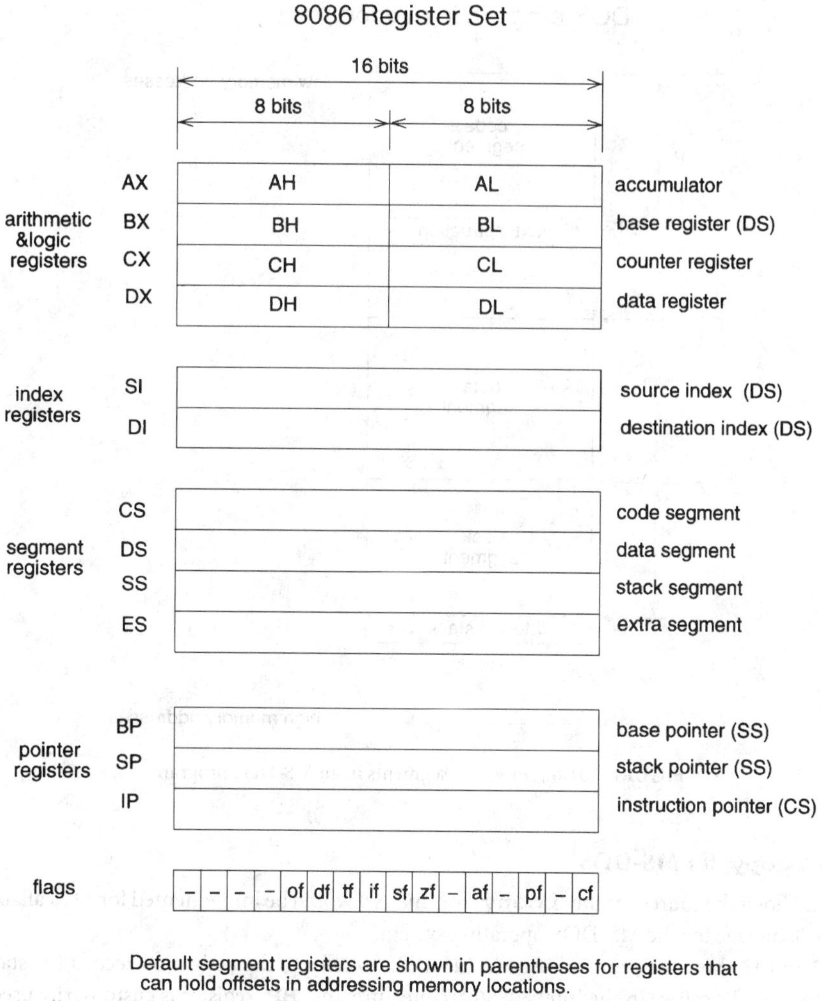

8086 Register Set

Default segment registers are shown in parentheses for registers that
can hold offsets in addressing memory locations.

FIGURE 101.5　Intel 8086 register set.

Intel 80x86 Architecture and the MS-DOS Run Time Environment

With the Intel 80x86 architecture, a program running under Microsoft disk operating system (MS-DOS)
is expected to have three segments:

- A *code segment* consisting of executable code
- A *data segment* holding the data used by the program
- A *stack segment* used by the program for procedure calls and temporary memory allocation

When an MS-DOS program is running, its segments will be arranged as shown in Fig. 101.6. The
CS register locates the beginning of the code segment, and the **IP** register holds the offset of the next
instruction of the code segment to be executed. The **DS** register locates the beginning of the data
segment. The offset of a data item can be expressed directly in an instruction or held in one of the
index, base, or pointer registers. The **SS** locates the beginning of the program's stack segment and the
SP register holds the offset of the top of the stack from the beginning of the stack segment. The Unix
operating system uses a similar scheme.

DOS program in memory

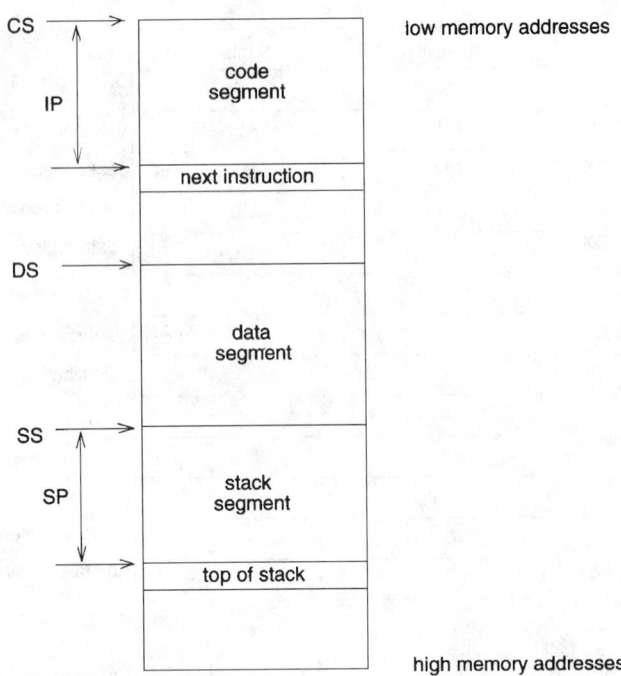

FIGURE 101.6 Program segments in an MS-DOS program.

Function bcopy for MS-DOS

Figure 101.7 shows the source for the **bcopy** routine as it would be implemented for a C-callable routine in assembly language for the MS-DOS operating system.

Lines 12 and 13 show the protocol used in MS-DOS to refer to the activation record (or stack frame) of the routine and its caller. In the Intel 80x86 architecture, the **BP** register is customarily used to point to the activation record, or stack frame, of the currently executing routine. Lines 14, 15, and 16 save the values in the registers at the time of the call on the run time stack. Lines 17, 19, and 20 access the actual parameters of the call. Lines 21 and 23 actually copy the bytes from the source location to the destination. Lines 25, 26, and 27 restore the values in the registers that existed at the time of the call. Line 28 restores the caller's frame pointer, and line 29 returns control to the caller.

The Motorola 68000 Architecture

The Motorola 68000 [Clements 1994] was designed and built in the mid-1970s in an era when it was generally felt that, to achieve superior performance, a processor should provide powerful instructions that could use a variety of addressing modes. Such a processor is usually referred to as a complex instruction set computer (CISC).

The 68000 machine instructions can have zero, one, or two operands. Any of the 12 addressing modes that are meaningful can be applied independently to the source and destination operands of the two-operand instructions.

The 68000 does not use the segmented memory model employed in the Intel 80x86 architecture. Eight of the 68000's registers are *address registers* dedicated to holding memory addresses. One address register holds an entire address, unlike the combination of the segment and offset registers used in the Intel 80x86 architecture.

bcopy-c.ASM

```
 1                    ;   A byte copy routine _bcopy callalbe from MS-DOS C
 2  0000             .MODEL large
 3  0000             .CODE
 4                    ;
 5  0000             PROC _bcopy FAR ; C prefixes every proc name with '_'
 6                   PUBLIC _bcopy
 7                    ;   implements the C function: void bcopy(
 8                    ;       far *SrcBlock;        far ptr to source   [BP+ 6]
 9                    ;       far *DestBlock;       far ptr to dest     [BP+10]
10                    ;       int numbytes);        # of bytes to move  [BP+14]
11                    ;   parameters are pushed from right to left in C
12  0000 55           PUSH   BP          ; save caller's frame pointer
13  0001 8B EC        MOV    BP, SP      ; set our frame pointer
14  0003 1E           PUSH   DS          ; save caller's DS
15  0004 56           PUSH   SI          ; C compiler uses SI, DI for
16  0005 57           PUSH   DI          ;     .. register variables, too
17  0006 8B 4E 0E     MOV    CX,[BP+14]  ; get number of bytes
18  0009 E3 09        JCXZ   byebye      ; leave if number of bytes = 0
19  000B C4 7E 0A     LES    DI,[BP+10]  ; ES:DI points to dest block
20  000E C5 76 06     LDS    SI,[BP+6]   ; DS:SI points to source block
21  0011 FC           CLD                ; df = 0, so SI & DI increment!
22                    ;   NOTE  that we aren't checking for wraparound in SI & DI!!
23  0012 F3> A4       REP    MOVSB       ; do the move
24  0014      byebye:
25  0014 5F           POP    DI          ; restore registers
26  0015 5E           POP    SI
27  0016 1F           POP    DS
28  0017 5D           POP    BP          ; restore caller's frame pointer
29  0018 CB           RET                ; caller removes the parameters
30  0019             ENDP _bcopy
```

FIGURE 101.7 The **bcopy** function in Intel 8086 assembly language.

Apple Computer used the Motorola 68000 family for its Macintosh line of computers until 1994, when it switched to the PowerPC processor [Motorola 1993] to obtain additional performance. The PowerPC 601 CPU is identical to the RS/6000 CPU that we will discuss in a later section. Sun Microcomputer also used the Motorola 68000 family on its Sun3 workstations until these were supplanted by the Sun4s using the SPARC reduced instruction set computer (RISC) architecture.

The MC68000 CPU

The register set for the 68000 CPU is shown in Fig. 101.8. The Motorola 68000 family, like the 80x86 family, has evolved into more powerful processors, with each processor using the same register set. The additional performance of the higher numbered processors in the family comes from wider internal data paths, pipelined instruction execution, and instruction caching.

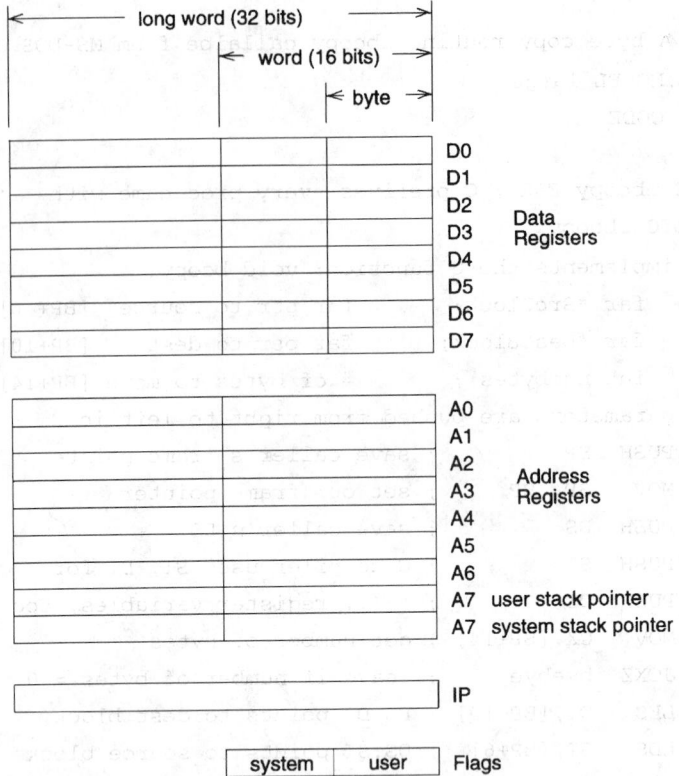

FIGURE 101.8 Motorola 68000 family register set.

The 68000 family CPU has 18 32-bit registers, plus a 16-bit status register. There are 8 data registers for all arithmetic and logic calculations and 9 address registers for addressing memory. The **A7** address register is used for the stack pointer.

Actually, there are two different **A7** registers on the 68000, one for general use (user **A7**) and one for system use (system **A7**). Most programming involves the user version of **A7**. The system version of **A7** (and the system part of the flags register) can be referenced only in supervisor mode. Thus, there are only 8 address registers available for normal user programming. All arithmetic or logic calculations must be performed in the 68000 data registers; address registers cannot be used.

Apple Macintosh Stack Frames

Figure 101.9 contains the **bcopy** procedure recast in 68000 assembly language, callable from C. This example will illustrate the Apple Macintosh stack frame protocol and several features of the 68000 instruction set.

The logic of the procedure is straightforward. Single bytes are copied until the number of bytes to be copied is a multiple of 4, and then long words are copied until all bytes have been copied. The Macintosh subroutine linkage is similar to the Sun3 subroutine linkage in terms of the direction of stack growth and the frame pointer register usage. The stack grows from higher memory locations to lower memory locations, so the predecrement mode is used for the **PUSH** and the postincrement mode is used for the **POP**. The Macintosh operating system uses the **A6** register for the frame pointer, just as the **BP** register is used in the 80x86 architecture.

The **LINK** and **UNLK** instructions were included in the Motorola 68000 instruction set specifically to manage run time activation records. The **LINK** instruction pushes the **A6** register of the caller, sets the value of **A6** to point to the caller's **A6** just pushed, then adds 0 to the stack pointer, **A7**. In

```
 1    ;******************************************************
 2    ;   implements the Macintosh C routine
 3    ;void bcopy(
 4    ;      char* SrcBlock,     pointer to source block
 5    ;      char* DestBlock     pointer to destination block
 6    ;      int NumBytes);      number of bytes to copy
 7    ;******************************************************
 8    ; Alignment restrictions are lenient on the 68000, so a
 9    ; simple algorithm can be used:
10    ;   copy single bytes until the number of bytes left to be
11    ;   copied is a multiple of 4, then copy longs until done.
12    ;******************************************************
13    _bcopy:                     ; set A6 frame ptr
14        LINK      A6, #-0       ; no local variables
15        MOVEM.L   A2-A3/D2-D3,-(A7); push 4 registers
16        MOVE.L    16(A6),A2     ; move source addr to A2
17        MOVE.L    12(A6),A3     ; move dest addr to A3
18        MOVE.L    8(A6),D2      ; move # bytes to copy to D2
19                                ; on 68000, MOVE sets the flags
20        BLE       bcexit        ; leave if nothing to do
21        MOVE.L    D2,D3         ; copy bytes until # bytes to be
22        AND.L     #3,D3         ;  .. copied is a multiple of 4
23        BEQ       copylongs
24    copybytes:
25        MOVE.B    (A2)+,(A3)+   ; copy a byte
26        SUBQ.L    #1,D2         ; decrement D2,
27        SUBQ.L    #1,D3         ;  .. and D3
28        BNE       copybytes
29        TST.L     D2            ; could be 0 left to copy here
30        BEQ       bcexit
31    copylongs:
32        MOVE.L    (A2)+,(A3)+   ; copy a long
33        SUBQ.L    #4,D2         ; keep byte count
34        BNE       copylongs
35    bcexit:
36        MOVEM.L   (A7)+,A2-A3/D2-D3 ; pop the registers
37        UNLK      A6            ; restore caller's frame ptr
38        RTS                     ; return to caller
```

FIGURE 101.9 The **bcopy** function on the Apple Macintosh MC68000.

the Apple Macintosh and Sun3 operating systems, the **A6** register is selected for use as the frame pointer register.

The **bcopy** procedure of Fig. 101.9 uses no local variables, so that the number subtracted from the stack pointer is **0** in this example. If local variables were used, the number of bytes of local variables would be substituted for the **0** to be subtracted from **A7**.

The general form of the Motorola 68000 version of the **bcopy** routine is much like that of the Intel 80x86 version. First, the caller's frame pointer is saved and the current procedure's frame pointer is established (line 14). Then the values in the caller's registers are saved (line 15). The actual parameters of the call are fetched in lines 16, 17, and 18.

The code to do the actual copying occupies lines 21–34. The number of instructions used on the 68000 is large compared to the number of instructions on the Intel 80x86, but this difference is misleading, because the algorithm used here is more complicated than the previous algorithm, which was simply a byte copy.

Line 36 restores the caller's registers, and line 37 restores the caller's frame pointer. The **RTS** in line 38 returns control to the caller.

The Sun SPARC Architecture

The SPARC architecture [Paul 1994] used by Sun Microsystems as the CPU in its Sun4 series of workstations is based directly on the Berkeley RISC-II processor, developed by David Patterson and his students.

In the early 1980s, Patterson came to feel that, rather than developing processors with increasingly complex instruction sets, better performance could be obtained through implementing a reduced instruction set computer, with simpler, more regular machine instructions executing at a higher clock rate.

The SPARC CPU uses the reduced instruction set, the LOAD/STORE architecture, pipelined instruction execution, instruction and data caching, the delayed branch, and a large register set with register windows, all present in Berkeley RISC-II design. Sun Microsystems has added several features to the SPARC architecture not present in the Berkeley RISC-II processor.

SPARC Register Set

The SPARC register set can contain from 48 to 528 32-bit registers. Each register window consists of 24 registers (8 input registers, 8 local registers, and 8 output registers), plus 8 global registers common to all procedures.

One of the innovations in the Berkeley RISC processors was the use of overlapping *register windows* to use in passing parameters in a subroutine call. A sketch of the SPARC register set is shown in Fig. 101.10. Each procedure is allocated a group of 32 registers, separated into four groups: *input* registers, *output* registers, *local* registers, and *global* registers. The CPU manipulates the register windows so that the input registers of the called procedure are identical to the output registers of the calling procedure. In this way, the run time stack is not needed to transmit actual parameters from the caller to the called procedure. This reduces the amount of time required for a procedure call.

The register windows are numbered from the number of windows (**NWINDOWS**) minus 1 down to 0. The SPARC CPU has a processor state register (**PSR**) that contains 4 bits of arithmetic logic unit flags, called the *integer condition codes field*, and a 5-bit field called the *Current Window Pointer* (**CWP**) that holds the number of the current register window being used. Since the output registers of one procedure overlap the input registers of the procedures that it calls, the number of 48 registers allows for two nonoverlapping sets of 16 registers, plus one overlapping set of 8 registers, plus the 8 global registers; that is, a SPARC CPU with 48 registers can implement 2 register windows. A SPARC CPU with 528 registers can implement 32 register windows.

The current window pointer is used by the SPARC CPU to point to the currently active register window. Its value is modified by the programmer through the **save** and **restore** instructions that are discussed subsequently.

The SPARC input registers for a given procedure are denoted by **%i0–%i7**, the local registers of the procedure are denoted by **%l0–%l7**, the output registers are denoted by **%o0–%o7**, and the global

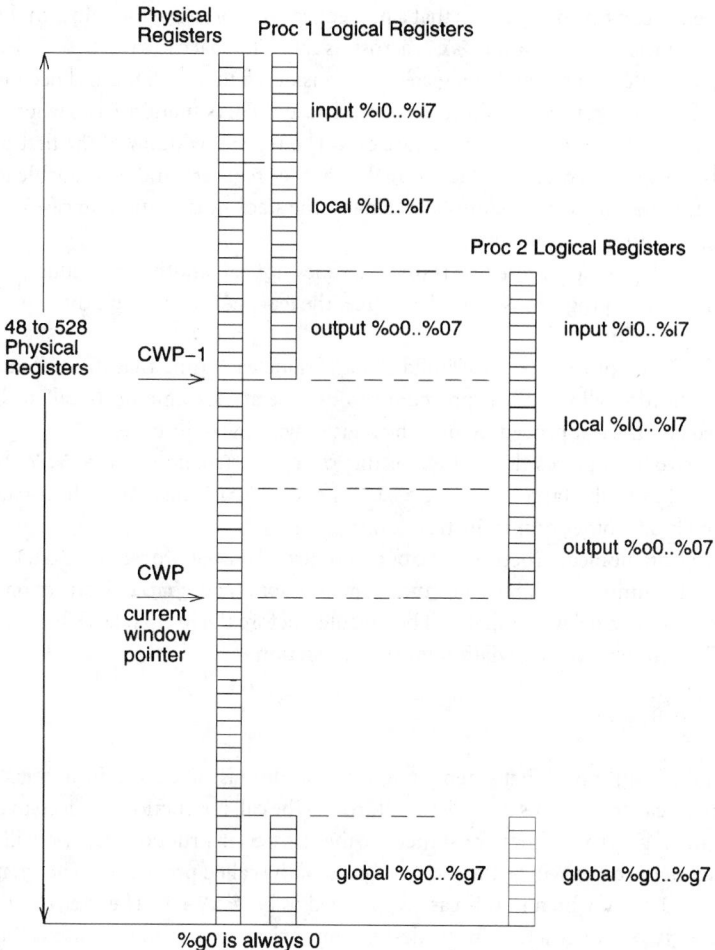

FIGURE 101.10   SPARC register windows.

registers that all procedures can access are denoted by **%g0–%g7**. Several registers are dedicated to special uses. Register **%o6** is the stack pointer register and is usually denoted by **%sp**. The **%i6** register is the procedure's frame pointer register and is denoted by **%fp**. The **%fp** register is actually the caller's stack pointer register because of the register window overlap between the caller's and callee's register windows (see Fig. 101.10).

Register **%o7** is the *link register*, from which the return address of a procedure call can be calculated. The **call** instruction places the address of the **call** instruction in **%o7** before jumping to the target of the call. A procedure would place its first six parameters of a procedure call in registers **%o0–%o5** (in left-to-right order, Pascal or C). Parameters after the sixth must be passed on the stack.

SPARC Procedure Call Protocol

The SPARC procedure call protocol depends on several factors: the semantics of the call/return instructions, the manipulation of the register windows through the **save** and **restore** instructions, and the layout of the SPARC stack frame.

In view of the usage of registers for passing procedure parameters, one might wonder why the SPARC architecture uses a run time stack. The reason is, of course, that there are several situations when the values of some or all of the registers in the procedure's window must be saved and restored. First, the

SPARC procedure call convention specifies that a procedure must preserve the values in the input registers **%i0–%i7**, and the local registers **%l0–%l7**, across its call. The space to save these registers is allotted in the stack frame. The values in the global registers are considered to be volatile and need not be preserved across the call. Second, a *register window overflow* could occur. This situation arises when all of the register windows have been used and there is a procedure call. The register window of the first procedure call in the chain of calls must be saved on the stack to make another register window available for the call. This can happen in a deep nesting of procedure calls. Finally, the stack must be used in case of a procedure call that has more than six input parameters.

In SPARC terminology, a *leaf procedure* is one that does not call another procedure. When a chain of procedure calls made in a program is viewed as a tree, the leaves of the tree are those procedures that do not call any other procedure.

The normal (nonleaf) procedure must build a stack frame each time that it is called, but the SPARC architectural specification allows a leaf procedure to use the stack frame of its caller. No new register window is created and the leaf procedure uses the register window of its caller.

The **call** instruction places the address of the **call** instruction in the **%o7** register, the link register, then branches to the target of the **call**. The **call** is thus a branch instruction and has a branch delay slot like any other branch instruction.

Both the return from nonleaf procedure **ret** and return from leaf procedure **retl** instructions are synthesized from the jump on link register **jmpl** instruction. The **jmpl** instruction is also a branch instruction, and so it has a delay slot that will be executed before the jump takes effect.

The **ret** instruction is an abbreviation for the instruction

```
jmpl    %i7+8,%g0
```

This is based on the assumption that when a nonleaf procedure is called, a shift of register windows has occurred. When the caller performs the call, the address of the call instruction is placed in the caller's **%o7** register. When the register windows are shifted by the **save** instruction that we will discuss shortly, the **%o7** register of the caller will be the **%i7** register of the called procedure. The **jmpl** target is the source operand on the left, which in this case is specified to be **%i7+8**. The **jmpl** instruction places the address of the **jmpl** instruction in the destination register, which in this case is the **%g0** register. Since the **%g0** is permanently zero, its use here indicates that the address of the **jmpl** instruction should not be saved.

The **retl** is for returning from a leaf procedure. It is an abbreviation for the instruction

```
jmpl    %o7+8,%g0
```

This is consistent with the view that the no shift of register windows has occurred with a leaf procedure. The leaf procedure is using the register window of its caller, and so the address of the **call** instruction is expected to be in the **%o7** register where the **call** instruction placed it.

When a nonleaf procedure performs a **call**, the nonleaf procedure does not need to know whether it is calling a leaf procedure or a nonleaf procedure. The problem is resolved by the called procedure's choice between the **ret** and **retl** instructions.

SPARC `save` and `restore` Instructions

The **save** instruction is used as the first instruction of a nonleaf procedure to build its stack frame. As mentioned previously, the stack frame will be used to save registers that must be preserved across the procedure call, in the case of register window overflow or in case the procedure calls a procedure having more than six input parameters.

The SPARC stack frame is shown in Fig. 101.11. The smallest allowable stack frame that a nonleaf procedure can allocate is 92 bytes. This includes 64 bytes to store the procedure's input and local registers, plus a 6-word (24-byte) storage area into which any callee of the procedure can store arguments, and a

1-word (4-byte) location to contain a pointer to an *aggregate return value*. This pointer would be used by a function whose return value was a **struct** or **RECORD** that did not fit in 32 bits.

The **save** instruction for a nonleaf procedure using a minimal stack frame would be

```
save   %sp,-92,%sp
```

The **save** instruction subtracts one from **CWP**, the current register window pointer. If the result of this subtraction is a valid register window number, then the new value is stored into **CWP**; otherwise, a *window overflow* trap is generated that will make a register window available by storing the registers of the earliest procedure in the calling chain.

The **restore** instruction is the reverse of the **save** instruction. The **restore** instruction adds one to **CWP**. If the resulting sum is not a valid register window pointer, then a *window underflow* trap is generated that will terminate execution of the current program.

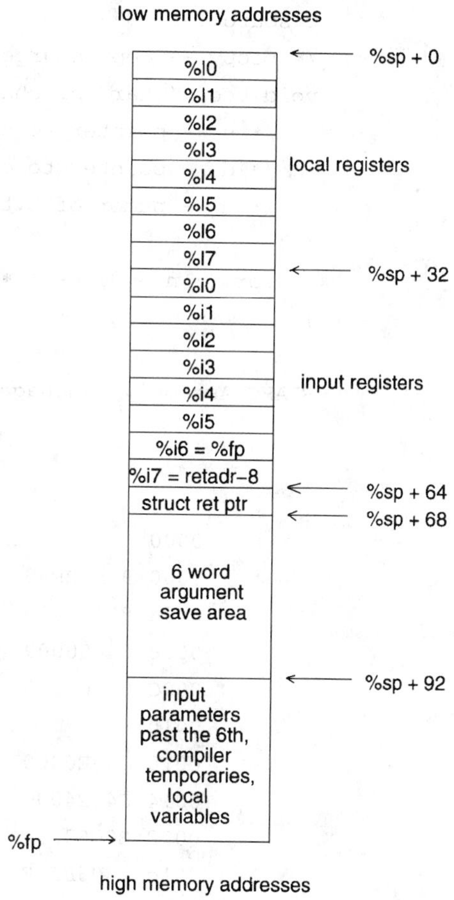

Function **bcopy** as a SPARC Nonleaf Procedure

The source code and the assembly language code for **bcopy** generated by the GNU C compiler is shown in Fig. 101.12. Notice that because of the **save**, **restore**, and **ret** instructions, the procedure is cast as a nonleaf procedure, even though it does not call any other procedure. This is a conservative deci-

FIGURE 101.11   SPARC stack frame.

sion, apparently based on the belief that every procedure should have a stack frame. This sort of conservative design is perhaps one reason why the code that the GNU C generates is so robust.

The parameters of the call are in the input registers now: **%i0** is the **s** parameter, **%i1** is the **d** parameter, and **%i2** is the **n** parameter of the call. The GNU compiler was not able to fill the branch delay slot of the **ble** branch at location **000C** with a useful instruction, so it inserted a no-operation **nop** instruction there. Notice that the **add** instruction at location **0028** that accomplishes the **d++** of the C program is executed in the delay slot of the **bg** instruction at location **0024**. This ensures that the **d** formal parameter has the correct value if the loop is executed again. Notice also that the **restore** instruction at location **0030** is located in the delay slot of the **ret** instruction, so that the **restore** is executed to restore the register window of the caller before the **ret** returns control to the caller.

The IBM RS/6000 Architecture

The America Project that created the RS/6000 design [Motorola 1993] began in 1985 at the IBM T. J. Watson Research Center with most of the IBM 801 team, including John Cocke. The project was named by John Cocke for the America's Cup yachting competition. The goal of the project was to extend the experience gained in building the 801 to produce a processor with improved performance. They eventually settled on an architecture that could consistently execute more than one instruction per clock cycle. John Cocke and Tilak Agerwala coined the phrase *superscalar organization* for such an architecture. Early in

```
C source:
/* bcopy -- copy source block to destination block */
void bcopy(char *s, char *d, int n)
    /* s -- pointer to source block      */
    /* d -- pointer to destination block */
    /* n -- numbr of bytes to copy       */
{
    for (; n > 0; n--) *(d++) = *(s++);
}


SPARC Assembly language:
                         .text
                         .align 4
                         .global_bcopy
        0000            _bcopy:
        0000 9DE3BF90     save    %sp,-112, %sp
        0004 80A6A000     cmp     %i2,0
        0008 04800009     ble     leave
        000C 01000000     nop
        0010            forloop:
        0010 C40E0000     ldub    [%i0],%g2
        0014 C42E4000     stb     %g2,[%i1]
        0018 B0062001     add     %i0,1,%i0
        001c B406BFFF     add     %i2,-1,%i2
        0020 80A6A000     cmp     %i2,0
        0024 14BFFFFB     bg      forloop
        0028 B2066001     add     %i1,1,%i1
        002C            leave:
        002C 81C7E008     ret
        0030 81E80000     restore
```

FIGURE 101.12 Function **bcopy** SPARC assembly language, as generated by the GNU C compiler.

the project, it became clear that the design was not only feasible but offered the possibility of exceptional performance. Rather than develop a prototype in the Research Division, the team approached the IBM development laboratories. In 1986, a group at IBM's Austin, Texas, laboratories began the implementation of a processor based on the ideas of the America Project. The IBM RISC System/6000 was the eventual outcome of that effort. The RS/6000 CPU is essentially identical to the PowerPC 601 CPU.

The RS/6000 achieves its superscalar performance with three concurrently executing processors: the *branch processing unit* (BPU), the *fixed-point processing unit* (FXU), and the *floating-point processing unit* (FPU). The BPU reads instructions from the instruction cache, performs condition register logic and branches, and dispatches any other instructions to the other two processing units. The FXU performs all integer arithmetic, and the FPU performs all floating-point arithmetic. The FPU can perform a floating-point addition and multiplication simultaneously. With a specially configured instruction stream, the RS/6000 CPU can execute five instructions per clock cycle. With the typical mix of instructions generated by a compiler, the instruction execution rate is lower but still superscalar.

Unlike the SPARC architecture, the RS/6000 CPU does not use a delayed branch. It depends instead on its superscalar instruction execution and its multiple-field condition register to minimize branch delays (discussed subsequently).

The RS/6000 distinguishes between leaf and nonleaf procedures, just as the SPARC architecture does. On the RS/6000, a leaf procedure can use the caller's stack frame and stack pointer register (**r1**). A nonleaf procedure is expected to obey the following protocol for building a stack frame:

1. Save the caller's link register in the caller's stack frame at **8(r1)**.
2. Save the condition register at **4(r1)** if condition register fields **CR2**, **CR3**, or **CR4** are modified during the call.
3. Push all nonvolatile floating point registers (**FP 13–FP 31**) used during the call.
4. Push all nonvolatile general purpose registers (**r12–r31**) used during the call.
5. Save the dynamic link (the caller's stack pointer register **r1**; IBM calls it the *back-chain pointer*) at **0(r1)**, and decrement the stack pointer **r1** by the size of the stack frame.

This procedure call protocol results in a stack frame as shown in Fig. 101.13.

The **bcopy** example that we will see next is cast as a nonleaf procedure that builds its stack frame as described previously.

Function **bcopy** as an RS/6000 Nonleaf Procedure

The RS/6000 assembly language code generated by the GNU C compiler for **bcopy** is shown, along with the original C code, in Fig. 101.14. This time, the GNU compiler was forced to compile **bcopy** as a nonleaf procedure by inserting a **printf** call in the C source that has subsequently been removed.

The first job of a nonleaf procedure is to save the caller's link register in the caller's stack frame. The **mflr** instruction at **003C** transfers the value from the link register into **r0**. The **st r0,8(r1)** instruction at **0004** stores the link register value where it is supposed to be, 8 bytes into the stack frame of the caller. The **stu** instruction at location **0008** stores the current value of **r1** (the caller's stack pointer) at the location **-56(r1)** and stores this address (**-56(r1)**) into **r1**. Thus, this instruction reserves the space for the callee's stack frame and sets the dynamic link in one operation. This procedure has no local variables and does not use any of the nonvolatile general purpose or floating-point registers, and so only the obligatory space required for a minimal stack frame is needed.

The actual copying begins at location **000C** where the **d** parameter pointing to the location of the destination block is moved from **r4**, where it was placed by the caller, to **r9**. The **ai. r4,r5,0** moves the contents of **r5** (the **n** parameter, number of bytes to be moved) into **r4**, and sets the flags of the **cr0** field of the condition register, based on the final contents of **r4**. The postpended period (.) indicates that this instruction sets the **cr0** field with its result. The condition register on the RS/6000 has eight separately usable fields of 4 b each. Different fields of the condition registers can be used by instructions to make it easier for the RS/6000 compilers to rearrange instructions to minimize the effect of branches.

If the number of bytes to be moved is nonpositive, the copy loop is skipped by the **ble** instruction, otherwise the **L..5** loop is entered. The first statement of the loop at location **0018** loads the byte pointed to by **0(r3)** (**r3** holds the input parameter **s**, the pointer to the source block) into the least significant byte of **r0** and zeroes out the upper 3 bytes of the register. The **stb r0,0(r9)** stores the least significant byte of **r0** at the location pointed to by **r9** (the **d** pointer to the destination block). The two **ai** instructions at locations **0020** and **0024** move the **s** and **d** pointers forward. The **ai.** instruction at location **0028** decrements the byte count variable **n** and sets the **cr0** field of the condition register. The **bgt** branch at location **0030** continues the loop as long as the decrement at location **0028** leaves the **r4** register greater than zero.

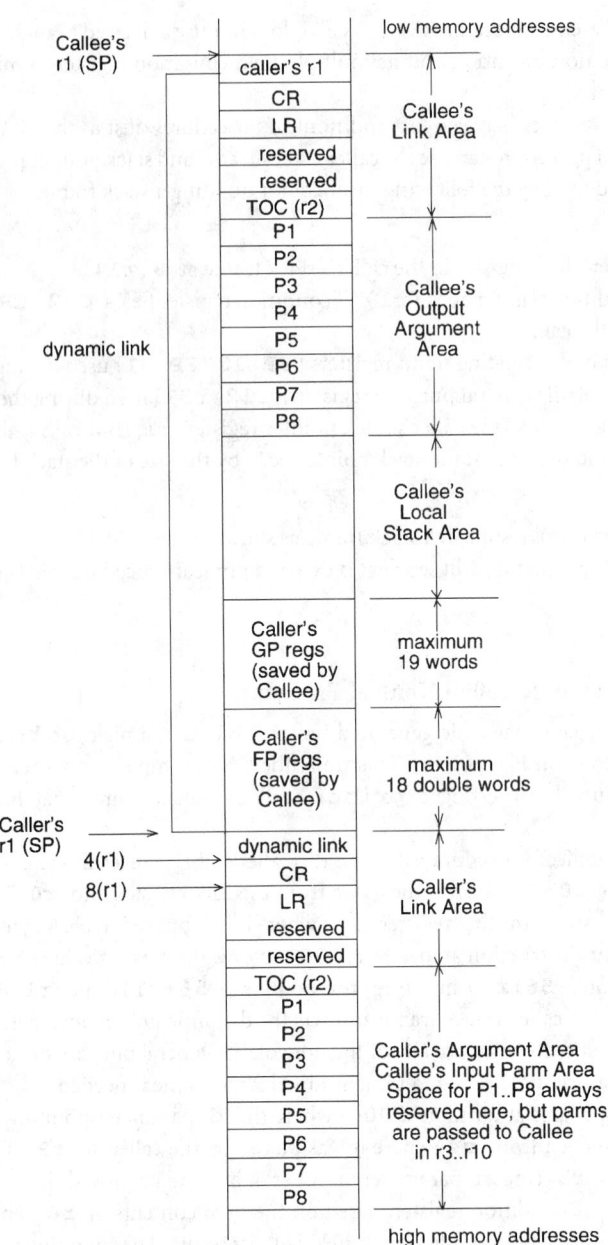

FIGURE 101.13   RS/6000 system stack frame, nonleaf procedure.

The instructions at locations **0034–0038** dismantle **bcopy**'s stack frame and return to the caller. The **ai r1,r1,56** releases the space from the stack frame of **bcopy** and leaves the **r1** stack pointer register pointing to the stack frame of the caller. The **l r0,8(r1)** instruction at location **0038** reloads from **r0** the caller's link **register** value, where it was stored at the beginning of the call. Then the **move to link register** (**mtlr**) **r0** instruction restores the caller's link register from **r0**. This value is the return address for the **bcopy** call, because the **bcopy** routine saved and restored the link register as though **bcopy** was a nonleaf routine. Finally, the **br** branches on the restored link register value to return to the caller.

```
C source:
/* bcopy -- copy source block to destination block */
void bcopy(char *s, char *d, int n)
    /* s -- pointer to source block      */
    /* d -- pointer to destination block */
    /* n -- number of bytes to copy      */
{
    for (; n > 0; n--) *(d++) = *(s++);
}
```

RS/6000 Assembly Language:

```
0000 .bcopy:
0000    mflr r0          ; move link reg into r0
0004    st   r0,8(r1)    ; store LR in caller's stack frame
0008    stu  r1,-56(r1)  ; allocate 56 bytes for my frame
000C    ai   r9,r4,0     ; move parm d into r9
0010    ai.  r4,r5,0     ; move parm n into r4 & set cr0
0014    ble  cr0,L..3    ; bc 4,1,L..3, leave if n <= 0
0018 L..5:
0018    lbz  r0,0(r3)    ; get byte pointed to by r3 aka s
001C    stb  r0,0(r9)    ; store byte @ loc pointed by r9 (d)
0020    ai   r3,r3,1     ; s++
0024    ai   r9,r9,1     ; d++
0028    ai.  r4,r4,-1    ; n-- & set cr0
002C    bgt  cr0,L..5    ; bc 12, 1, L..5, keep looping if n > 0
0034 L..3:
0034    ai   r1,r1,56    ; restore caller's sp, rlse stack frame
0038    l    r0,8(r1)    ; move caller's link register to r0
003C    mtlr r0          ; restore caller's link register LR
0040    br               ; branch on LR to return to caller
```

FIGURE 101.14 Function **bcopy** as a nonleaf procedure on the RS/6000.

101.4 Summary

In this chapter the principles of run time environments have been discussed. The applications of these principles to the major architectures have been shown. An understanding of these principles and their application to a specific architecture is important in developing debuggers, working with multiple threads, and developing other tools which must interact with the run time environment.

One of the continuing areas of research in run time environments is the development of compilers and related tools for both massively parallel and distributed computer systems. The goal is to develop languages and compilers so that programming systems can be developed and debugged on traditional single-CPU workstations and then recompiled and run on either a massively parallel computer system or a distributed computer system. This approach is particularly attractive for the so-called grand challenge problems of science. Wolfe [1996] is one text in this area.

Meek [1995] discusses the work of the International Standard Organization (ISO) in attempting to standardize the notion of a procedure call and parameter passing even in a distributed environment.

Defining Terms

Activation record: A record containing all of the information associated with an activation or call of a procedure or function. This information includes: the return address of the caller, the procedure's parameters and local variables, and the frame pointer of the caller.

Activation stack: A stack of activation records, one for each active procedure call.

Actual parameter or argument: A parameter that appears in a call of the procedure or function.

Call-by-address parameter: A method of parameter transmission whereby the address of the actual parameter is copied to the formal parameter at the time of the call. The formal parameter is effectively a constant pointer variable. Any reference to the formal parameter is treated as a reference to the actual parameter. Also known as call-by-reference or call-by-location.

Call-by-result parameter: A method of parameter transmission whereby the value of the formal parameter is copied back to the actual parameter at the time of the return. Prior to executing the return statement there is no correspondence between the actual and formal parameters.

Call-by-value parameter: A method of parameter transmission whereby the value of the corresponding actual parameter is copied to the formal parameter at the time of the call. Thereafter, there is no correspondence between the actual and formal parameters. In particular, changes to the formal parameter have no effect on the actual parameter.

Call-by-value-result parameter: A method of parameter transmission which combines call-by-value and call-by-result.

Formal parameter or argument: A parameter name that appears in the declaration or header of a procedure or function.

Frame pointer: A register that normally points to the base or beginning of the activation record on the top of the activation stack.

Heap: The portion of memory assigned to an executing program to use for dynamic memory allocation.

Leaf procedure: A procedure that does not call another procedure.

Register window: A collection of registers assigned to an executing process in the SPARC architecture.

Stack pointer: A register that points to the top of the activation stack.

References

Ada 1991. *The Annotated Ada Reference Manual*, 2nd ed. Grebyn, Vienna, VA.

Barnes, J. G. P. 1995. *Programming in Ada 95*. Addison–Wesley, Reading, MA.

Carmony, L. A. 1990. *Modula II*. Computer Science Press, New York.

Clements, A. 1994. *68000 Family Assembly Language*. PWS, Boston, MA.

Gilman, L. 1992. *APL, an Interactive Approach*. Krieger, Malabar, FL.

Goldberg, A. 1985. *Smalltalk-80 : the Language and its Implementation*. Addison–Wesley, Reading, MA.

Jensen, K., Wirth, N., Mickel, A. B., and Miner, J. F. 1991. *Pascal User Manual and Report: ISO Pascal Standard*, 4th ed. Springer–Verlag, New York.

Kernighan, B. W. and Ritchie, D. M. 1988. *The C Programming Language*, 2nd ed. Prentice–Hall, Englewood Cliffs, NJ.

Louden, K. C. 1993. *Programming Languages: Principles and Practice*. PWS-Kent, Boston, MA.

MacLennan, B. J. 1987. *Principles of Programming Languages*, 2nd ed. Holt, Rinehart and Winston, New York.

Mazidi, M. A. 1995. *The 80x86 IBM PC and Compatible Computers: Assembly Languages, Design and Interfacing*. Prentice–Hall, Englewood Cliffs, NJ.

Meek, B. L. 1995. What is a procedure call? *SIGPLAN Notices* 30(9):33–40.

Motorola. 1993. *PowerPC 601: RISC Microprocessor's User's Manual*. Motorola, Phoenix, AZ.

Paul, R. P. 1994. *SPARC Architecture Assembly Language Programming and C*. Prentice–Hall, Englewood Cliffs, NJ.

Pratt, T. W. and Zelkowitz, M. V. 1996. *Programming Languages Design and Implementation*, 3rd ed. Prentice–Hall, Englewood Cliffs, NJ.

Sankaran, N. 1994. Bibliography on garbage collection and related topics. *SIGPLAN Notices* 29(9):149–158.

Sebesta, R. W. 1996. *Concepts of Programming Languages*, 3rd ed. Addison–Wesley, Reading, MA.

Steele, G. L. 1990. *Common LISP: the Language*. Digital Press, Bedford, MA.

Stroustrup, B. 1995. *The C++ Programming Language*, 2nd ed., reprinted with corrections. Addison–Wesley, Reading, MA.

Welburn, T. 1995. *Structured COBOL: Fundamentals and Style*. McGraw–Hill, New York.

Wolfe, M. 1996. *High Performance Compilers for Parallel Computing*. Addison–Wesley, Reading, MA.

Zirkel, G. 1994. *Understanding FORTRAN 77 and 90*. PWS, Boston, MA.

Further Information

The *ACM Transactions on Programming Languages and Systems* (TOPLAS) is probably the leading theoretical journal in the area of run time environments and memory management. However, this area produces fewer theoretical papers than many other areas in programming languages and systems.

A more applied journal is *Software Practice & Experience*, which is aimed at the practitioner rather than the theoretician. The scope of this journal (as stated on the inner cover) is "practical experience with new and established software for both systems and applications." Thus, although the scope is broader, techniques of various architectural features are illustrated. One example might be the simulation of the improvement gained in execution speed by adding register windows to a given architecture.

The journal *Computer Languages* sits squarely between the previous two, in that it contains both theoretical and applied papers, but its scope is more narrowly focused.

Finally, the *Journal of Systems and Software* includes both research papers as well as reports on the state of the art and practical experience. Topic areas include programming methodology and related hardware-software issues, including programming environments.

102

Foundational Calculi for Programming Languages

Benjamin C. Pierce
Indiana University

102.1 Introduction

In the mid-1960s, Landin observed that a complex programming language can be understood in terms of a tiny *core language* capturing the essential mechanisms of some programming style together with a collection of convenient *derived forms* whose behavior is understood by translating them into the core [Tennent 1981]. Landin's core language was the **lambda-calculus**, a formal system in which all computation is reduced to the basic operations of function definition and application. Since the 1960s, the lambda-calculus has seen widespread use in the specification of programming language features, language design and implementation, and the study of type systems. Its importance arises from the fact that it can be viewed simultaneously as a simple programming language in which computations can be described and as a mathematical object about which rigorous statements can be proved.

The lambda-calculus has a strong claim to be a *canonical* model of purely functional computation (the programming paradigm where the only observable properties of an expression are its behavior when applied to arguments). Not only does it capture the ideas of function definition and application in a clear, intuitive way, but all other known models of functional computation—Turing machines, general recursive functions, control structures such as *while* and *goto*, etc.—can be shown to describe exactly the same class of functions. (For a survey of these results, see Davis [1982].) For concurrent and distributed systems, no such canonical model has yet emerged. Instead, many different *process calculi* are being studied, each embodying some particular set of primitives for concurrent computation.

This chapter sketches the definitions and some basic properties of the lambda-calculus and a representative process calculus called the **pi-calculus**.

102.2 Lambda-Calculus

Procedural abstraction is a key feature of most programming languages. Instead of writing the same calculation over and over, we write a procedure or function that performs the calculation abstractly, in terms of one or more named parameters, which we instantiate as needed, providing values for the parameters in each case. For example, we might rewrite an expression such as $(5 \cdot 4 \cdot 3 \cdot 2 \cdot 1) + (7 \cdot 6 \cdot 5 \cdot 4 \cdot 3 \cdot 2 \cdot 1) - (3 \cdot 2 \cdot 1)$ as *Factorial*(5) + *Factorial*(7) − *Factorial*(3), where

$$Factorial(n) \quad = \quad if\ n = 0\ then\ 1\ else\ n \cdot Factorial(n-1)$$

For each nonnegative number n, instantiating the function *Factorial* with the argument n yields a number, the factorial of n, as result. Writing $\lambda n.$ as a shorthand for "the function that, for each n, yields...," we can restate the definition of *Factorial* as

$$Factorial \quad = \quad \lambda n.\ if\ n = 0\ then\ 1\ else\ n \cdot Factorial(n-1)$$

The expression *Factorial*(0) is now read as: "the function '$\lambda n.\ if\ n = 0\ then\ 1\ else\ etc.$' applied to the argument 0," i.e., "the value that results when the bound variable n in the function body '$if\ n = 0\ then\ 1\ else$ $etc.$' is replaced by 0," i.e., "$if\ 0 = 0\ then\ 1\ else\ etc.$," i.e., 1.

In the 1930s, Church invented a mathematical system called the lambda-calculus (or λ-calculus) that embodies this kind of function definition and application in a pure form. In the lambda-calculus *everything* is a function: the arguments accepted by functions are themselves functions and the result returned by a function is another function.

Syntax and Operational Semantics

The syntax of the lambda-calculus comprises the three forms of expression at the top of Fig. 102.1. A variable x by itself is a lambda-expression; the application of a lambda-expression M to another lambda-expression N, written $M\ N$, is a lambda-expression; and the abstraction of a variable x from a lambda-expression M, written $\lambda x.\ M$, is a lambda-expression. L, M, and N are used throughout this chapter to stand for arbitrary lambda-expressions. To avoid writing too many parentheses, application is taken to be

Syntax:

$$
\begin{array}{llll}
L, M, N & ::= & x & \text{variable} \\
& & MN & \text{application} \\
& & \lambda x.\,M & \text{abstraction}
\end{array}
$$

Free variables:

$$
\begin{array}{rcl}
FV(x) & = & \{x\} \\
FV(MN) & = & FV(M) \cup FV(N) \\
FV(\lambda x.\,M) & = & FV(M) - \{x\}
\end{array}
$$

Substitution:

$$
\begin{array}{rcll}
[N/x]x & = & N & \\
[N/x]z & = & z & \text{if } z \neq x \\
[N/x](LM) & = & ([N/x]L)([N/x]M) & \\
[N/x](\lambda z.\,M) & = & \lambda z.\,([N/x]M) & \text{if } z \neq x \text{ and } z \notin FV(N)
\end{array}
$$

Renaming of bound variables:

$$\lambda x.\,M \quad = \quad \lambda y.\,([y/x]M) \quad \text{if } y \notin FV(M)$$

Operational Semantics:

$$(\lambda x.\,M)N \to [N/x]M \quad \text{function application (beta-reduction)}$$

FIGURE 102.1 Syntax and operational semantics of the lambda-calculus.

left-associative, so that $L\, M\, N$ is the same as $(L\, M)\, N$, and the bodies of abstractions extend as far to the right as possible, so that $\lambda x.\, \lambda y.\, x\, y\, x$ is the same as $\lambda x.\, (\lambda y.\, ((x\, y)\, x))$. The variable x is said to be *bound* in the body M of $\lambda x.\, M$.

In its pure form, the lambda-calculus has no builtin constants or operators: no numbers, arithmetic operations, records, loops, sequencing, input/output (I/O), etc. The sole means by which expressions compute is the application of functions to arguments, which is captured formally by the rule at the bottom of Fig. 102.1, traditionally called *beta-reduction*. This rule says that an expression can be reduced by replacing some subexpression of the form $(\lambda x.\, M)N$, called a **redex**, by the result of substituting the argument N for the bound variable x in the body M. For example, $(\lambda x.\, x\, y)\, (u\, v)$ reduces to $u\, v\, y$, while $(\lambda x.\, \lambda y.\, x)\, z\, w$ reduces (by the underlined redex) to $(\lambda y.\, z)\, w$, which further reduces to z. We write $(\lambda x.\, \lambda y.\, x)\, z\, w \rightarrow^* z$ to show that the first expression reduces to the second by some sequence of steps of reduction.

The notion of reduction gives rise to a natural definition of what it means for two expressions to be "the same modulo reduction." M and N are *beta-convertible*, written $M =_\beta N$, if they are identical, or if one reduces to the other, or if they are each convertible to some third expression L. (Formally, this is summarized by saying that the beta-conversion relation is the reflexive, symmetric, transitive closure of beta-reduction.) For example, $(\lambda x.\, x)\, z$ and $(\lambda x.\, \lambda y.\, x)\, z\, w$ are convertible because they both reduce to z.

To make the notions of beta-reduction and conversion completely precise, there is a little technical work to be done. In particular, we must define the *substitution* notation $[N/x]M$. The details, which can be skipped on a first reading, occupy the rest of Fig. 102.1 and of this subsection.

First, we say what it means for variable x to be *free* in an expression M: namely, that x appears at some position in M where it is not bound by an enclosing lambda-abstraction on x. (For example, x is free in $x\, y$ and $\lambda y.\, x\, y$ but not in $\lambda x.\, x$ or $\lambda z.\, \lambda x.\, \lambda y.\, x\, y\, z$.)

An important syntactic convention concerns the inessentiality of bound names. Intuitively, it is clear that the expressions $\lambda x.\, x$ and $\lambda y.\, y$ describe exactly the same function: the function that, given any argument N, returns N. The fact that we use x in one case and y in the other to stand for the argument in the body is of no consequence. This principle is captured by the rule of renaming of bound variables (often called *alpha-conversion*) on the second line from the bottom of Fig. 102.1, which states that we may freely replace the bound variable x by another variable y in a lambda-abstraction $\lambda x.\, M$ as long as y is not among the free variables of M. (The side condition is needed because, for example, we do not want to consider $\lambda x.\, y$ and $\lambda y.\, y$ to be the same function.)

Now we define substitution. Substituting an expression N for a variable x in an expression consisting only of x itself yields N. Substituting N for x in an expression consisting only of a different variable z yields z. To substitute N for x in an application $M\, L$, we substitute in M and L separately. To substitute N for x in a lambda-abstraction $\lambda z.\, M$, we substitute in the body M; however, we do this only when z is not the same as x and z is not one of the free variables of N. The first part of this side condition ensures that we do not allow nonsensical reductions such as $(\lambda x.\, \lambda x.\, x)\, y \rightarrow \lambda x.\, y$, where x is replaced by y even though it is actually bound by an inner abstraction, not the one being reduced. The second prevents a similar kind of mistake where free variables of N are "captured" by abstractions inside a term being substituted into, resulting in reductions such as $(\lambda x.\, \lambda y.\, x)\, y \rightarrow \lambda y.\, y$.

Thus, strictly speaking, the substitution $[N/x]M$ is undefined for some values of N, x, and M. But it is always possible to change the names of bound variables in N and/or M so that the substitution makes sense. For example, we can rewrite $[y/x](\lambda x.\, x)$ as $[y/x](\lambda z.\, z)$, which, by the definition of substitution, equals $\lambda z.\, z$; after renaming the bound variable again, this is the same as $\lambda x.\, x$. It is common practice to elide the renaming steps and say that $[y/x](\lambda x.\, x) = \lambda x.\, x$.

Examples

The lambda-calculus is much more powerful than its tiny definition might suggest. For example, there is no builtin provision for multiargument functions, but it is easy to achieve the same effect using *higher order functions* that yield functions as results. Suppose that M is an expression involving two free variables *x* and *y* and we want to write a function *F* that, for each pair (N, L) of arguments, yields the result of substituting *N* for *x* and *L* for *y* in *M*. Instead of writing $F = \lambda(x, y). M$, as we might in a higher level programming language, we write $F = \lambda x. \lambda y. M$; that is, *F* is a function that, given a value *N* for *x*, yields a function that, given a value *L* for *y*, yields the desired result. We then apply *F* to its arguments one at a time, writing *F N L*, which reduces to $(\lambda y. [N/x]M) L$ and then to $[L/y][N/x]M$. This transformation of multiargument functions into higher order functions is often called Currying after its popularizer, Curry. (It was actually invented by Schönfinkel, but the term Schönfinkeling has not caught on.)

Another common language feature that can easily be encoded in the lambda-calculus is Boolean values and conditionals. Define the lambda-expressions *True* and *False* as follows:

$$True = \lambda t. \lambda f. t$$

$$False = \lambda t. \lambda f. f$$

Both of these expressions are **combinators**; that is, neither contains any free variables. This means that they are inert with respect to substitution: $[N/x] True = True$ no matter what *N* and *x* are. The only way to *interact* with combinators is by applying them to other expressions. For example, we can use application to define a combinator *If* with the property that *If L M N* reduces to *M* when $L = True$ and reduces to *N* when $L = False$:

$$If = \lambda l. \lambda m. \lambda n. l \, m \, n$$

The *If* combinator does not actually do much: *If L M N* means just *L M N*. In effect, the Boolean value *L* itself is the conditional: it takes two arguments and chooses the first (if it is *True*) or the second (if it is *False*). For example, the expression *If True M N* reduces as follows:

If True M N	=	$(\underline{\lambda l. \lambda m. \lambda n. l \, m \, n) \, True} \, M \, N$	by definition
	$\rightarrow$	$(\underline{\lambda m. \lambda n. True \, m \, n) \, M} \, N$	reducing the underlined redex
	$\rightarrow$	$(\underline{\lambda n. True \, M \, n) \, N}$	reducing the underlined redex
	$\rightarrow$	*True M N*	reducing the underlined redex
	=	$(\underline{\lambda t. \lambda f. t) \, M} \, N$	by definition
	$\rightarrow$	$(\underline{\lambda f. M) \, N}$	reducing the underlined redex
	$\rightarrow$	*M*	reducing the underlined redex

(Strictly speaking, this calculation is valid only if the variables *n*, *t*, and *f* are not free in *M* or *N*; otherwise, some extra renaming steps are required.)

We can also write Boolean operators like logical-and as functions:

$$And = \lambda b. \lambda c. b \, c \, False$$

That is, *And* is a function that given two Boolean arguments *b* and *c* returns either *c* (if *b* is *True*) or *False* (if *b* is *False*); thus, *And b c* yields *True* if and only if both *b* and *c* are *True*.

Using Booleans, we can encode pairs of values as lambda-expressions. Define:

$$Pair = \lambda f. \lambda s. \lambda b. b \, f \, s$$

$$1st = \lambda p. p \, True$$

$$2nd = \lambda p. p \, False$$

That is, *Pair M N* is a function that when applied to a Boolean *b* applies *b* to *M* and *N*. By the definition of Booleans, this application yields *M* if *b* is *True* and *N* if *b* is *False*, so that the first and second projection functions *1st* and *2nd* can be implemented simply by supplying the appropriate Boolean. To check that *1st* (*Pair M N*) $=_\beta M$, calculate as follows:

$$
\begin{aligned}
1st\,(Pair\ M\ N) \quad &= \quad 1st\,((\lambda f.\,\lambda s.\,\lambda b.\,b\ f\ s)\ \underline{M}\ N) \qquad && \text{by definition} \\
&\rightarrow \quad 1st\,((\lambda s.\,\lambda b.\,b\ M\ s)\ \underline{N}) \qquad && \text{reducing the underlined redex} \\
&\rightarrow \quad 1st\,(\lambda b.\,b\ M\ N) \qquad && \text{reducing the underlined redex} \\
&= \quad \underline{(\lambda p.\,p\ True)\ (\lambda b.\,b\ M\ N)} \qquad && \text{by definition} \\
&\rightarrow \quad \underline{(\lambda b.\,b\ M\ N)\ True} \qquad && \text{reducing the underlined redex} \\
&\rightarrow \quad True\ M\ N \qquad && \text{reducing the underlined redex} \\
&\rightarrow^* \quad M \qquad && \text{as before}
\end{aligned}
$$

The encoding of numbers as lambda-expressions is only slightly more intricate. Define the Church numerals C_0, C_1, C_2, etc., as follows:

$$
\begin{aligned}
C_0 \quad &= \quad \lambda z.\,\lambda s.\,z \\
C_1 \quad &= \quad \lambda z.\,\lambda s.\,s\ z \\
C_2 \quad &= \quad \lambda z.\,\lambda s.\,s\ (s\ z) \\
&\ \ \vdots \\
C_n \quad &= \quad \lambda z.\,\lambda s.\,\underbrace{s\ (s\ (\ldots (s\ z))\ldots)}_{n\ \text{times}}
\end{aligned}
$$

That is, each number n is represented by a combinator C_n that takes two arguments, z and s (zero and successor), and applies n copies of s to z. As with Booleans and pairs, this encoding makes numbers into active entities: the number n is represented by a function that does something n times, a kind of active unary numeral.

We can define some common arithmetic operations on Church numerals as follows:

$$
\begin{aligned}
Plus \quad &= \quad \lambda m.\,\lambda n.\,\lambda z.\,\lambda s.\,m\,(n\ z\ s)\ s \\
Times \quad &= \quad \lambda m.\,\lambda n.\,m\ C_0\,(Plus\ n)
\end{aligned}
$$

Here, *Plus* is a combinator that takes two Church numerals, m and n, as arguments and yields another Church numeral, i.e., a function that accepts arguments z and s, applies s iterated n times to z (by passing s and z as arguments to n), and then applies s iterated m more times to the result. It is an instructive exercise to check, for example, that *Plus* $C_2\ C_1 =_\beta C_3$. The definition of *Times* uses another trick: since *Plus* takes its arguments one at a time, applying it to just one argument n yields the function that adds n to whatever argument it is given. Passing this function as the second argument to m and zero as the first argument means: apply the function that adds n to its argument, iterated m times, to zero, i.e., add together m copies of n.

To test whether a Church numeral is zero, we must give it a pair of arguments Z and S such that applying S to Z one or more times yields *False*, whereas not applying it at all yields *True*. Clearly, we can take Z to be just *True*. As for S, we use a function that throws away its argument and always returns *False*:

$$
IsZero \quad = \quad \lambda m.\,m\ True\,(\lambda x.\ False)
$$

Surprisingly, it is quite a bit more difficult to subtract using Church numerals. It can be done using the following rather impenetrable *predecessor function*, which, given C_0 as argument, returns C_0 and, given

C_{i+1}, returns C_i

$$Pred \quad = \quad \lambda m.\ 1st\ (m\ Z\ S)$$

where:

$$Z \quad = \quad Pair\ C_0\ C_0$$

$$S \quad = \quad \lambda p.\ Pair\ (2nd\ p)\ (Plus\ (2nd\ p)\ C_1)$$

This definition works by using m as a function to apply m copies of the function S to the starting value Z. Each copy of S takes a pair of numerals ($Pair\ C_i\ C_j$) as its argument and yields ($Pair\ C_j\ C_{j+1}$) as its result. Thus, applying S m times to ($Pair\ C_0\ C_0$) yields ($Pair\ C_0\ C_0$) if $m = 0$ and ($Pair\ C_{m-1}\ C_m$) otherwise. In both cases, the predecessor of m is found in the first component.

Other common data types such as lists, trees, arrays, and variant records can be encoded using similar techniques. Of course, in most programming languages based on the lambda-calculus, such basic data types are added as primitive constants, rather than being encoded.

A lambda-expression containing no redexes is said to be in **normal form**. The lambda-expressions we have seen so far have all shared the property that, independent of the order in which redexes are chosen for reduction, eventually all of the redexes are used up and a normal form is reached. But not all lambda-expressions have this property. For example, the divergent combinator

$$\Omega \quad = \quad (\lambda x.\ x\ x)\ (\lambda x.\ x\ x)$$

can never be reduced to a normal form. It contains just one redex, and reducing this redex yields exactly Ω again!

A similar but more useful example is the so-called Y combinator, which can be used to define recursive functions such as *Factorial*:

$$Y \quad = \quad \lambda f.\ (\lambda x.\ f\ (x\ x))\ (\lambda x.\ f\ (x\ x))$$

The crucial property of Y is that $Y\ F =_\beta F\ (Y\ F)$ for any F, as can be seen by the following calculation:

$$
\begin{aligned}
Y\ F \quad &= \quad \underline{(\lambda f.\ (\lambda x.\ f\ (x\ x))\ (\lambda x.\ f\ (x\ x)))\ F} \\
&\rightarrow \quad \underline{(\lambda x.\ F\ (x\ x))\ (\lambda x.\ F\ (x\ x))} \\
&\rightarrow \quad F\ ((\lambda x.\ F\ (x\ x))\ (\lambda x.\ F\ (x\ x))) \\
&\leftarrow \quad F\ (\underline{(\lambda f.\ (\lambda x.\ f\ (x\ x))\ (\lambda x.\ f\ (x\ x)))\ F}) \\
&= \quad F\ (Y\ F)
\end{aligned}
$$

[Note that it is not quite the case that $Y\ F \rightarrow^* F\ (Y\ F)$: the last arrow goes in the wrong direction. There is a slightly more complicated variant that does have this property.]

Now, suppose we want to write a recursive function definition of the form $F = \langle body\ containing\ F \rangle$, that is, we want to write a definition where the expression on the right-hand side of the equals uses the very function that we are defining, as in the definition of *Factorial*. The intention is that the recursive definition should be unrolled at the point where it occurs; for example, the definition of *Factorial* would intuitively be written

$$
\begin{aligned}
&if\ n = 0\ then\ 1 \\
&else\ n \cdot (if\ n - 1 = 0\ then\ 1 \\
&\qquad else\ n \cdot (if\ n - 2 = 0\ then\ 1 \\
&\qquad\qquad else\ (n - 2) \cdot \ldots))
\end{aligned}
$$

This effect can be achieved by defining $G = \lambda f. \langle body\ containing\ f \rangle$ and $F = Y\ G$, since then

$$
\begin{aligned}
F \quad &= \quad Y\ G \\
&=_\beta \quad G\ (Y\ G) \\
&=_\beta \quad \langle body\ containing\ (Y\ G) \rangle \\
&=_\beta \quad \langle body\ containing\ \langle body\ containing\ (Y\ G) \rangle \rangle \\
& \qquad etc.
\end{aligned}
$$

For example, if we define the factorial function by

$$
\begin{aligned}
Fact \quad &= \quad \lambda fact.\ \lambda n.\ If\ (IsZero\ n)\ C_1\ (Times\ n\ (fact\ (Pred\ n))) \\
Factorial \quad &= \quad Y\ Fact
\end{aligned}
$$

then applying *Factorial* to the Church numeral for 2 leads to the following calculation:

$$
\begin{aligned}
Factorial\ C_2 \quad &= \quad Y\ Fact\ C_2 \\
&=_\beta \quad Fact\ (Y\ Fact)\ C_2 \\
&=_\beta \quad (\lambda fact.\ \lambda n.\ If\ (IsZero\ n)\ C_1\ (Times\ n\ (fact\ (Pred\ n))))\ (Y\ Fact)\ C_2 \\
&=_\beta \quad (\lambda n.\ If\ (IsZero\ n)\ C_1\ (Times\ n\ (Y\ Fact\ (Pred\ n))))\ C_2 \\
&=_\beta \quad If\ (IsZero\ C_2)\ C_1\ (Times\ C_2\ (Y\ Fact\ (Pred\ C_2))) \\
&=_\beta \quad If\ False\ C_1\ (Times\ C_2\ (Y\ Fact\ C_1)) \\
&=_\beta \quad Times\ C_2\ (Y\ Fact\ C_1) \\
&= \quad Times\ C_2\ (Factorial\ C_1)
\end{aligned}
$$

That is, $Factorial\ (2) = 2 \cdot Factorial\ (1)$. This calculation justifies our informal assertion that *Factorial* really implements the factorial function: it can easily be proved that $Factorial\ (C_n) =_\beta C_{!n}$ for each nonnegative number n, where $!n$ is the real factorial of n.

Here are some other well-known combinators:

$$
\begin{aligned}
I \quad &= \quad \lambda x.\ x \\
K \quad &= \quad \lambda x.\ \lambda y.\ x \\
S \quad &= \quad \lambda x.\ \lambda y.\ \lambda z.\ (x\ z)\ (y\ z)
\end{aligned}
$$

I is called the identity combinator because $I\ M =_\beta M$ for any M. K is a combinator that takes two arguments, throws away the second, and returns the first. (We called the same expression *True* before, but we are not thinking of it here as representing a Boolean value.) S *distributes* its third argument to both its first and its second arguments. An interesting property of these three is that between them they contain all of the power of lambda-abstraction, in the sense that for any combinator M there is a combinator N such that $N =_\beta M$ and N can be written using just S, K, I, and application, with no variables or abstractions except those occurring in the definitions of S, K, and I. For example, $False =_\beta K\ I$.

Properties of Reduction

The importance of lambda-calculus in computer science comes from the fact that it is simultaneously powerful enough to form a realistic core for many programming languages and simple enough that its properties can be studied mathematically. Indeed, the study of the lambda-calculus now constitutes a branch of mathematics in its own right. We review a few classical results.

A lambda-expression containing more than one redex can be reduced in more than one way, leading, in general, to different results. For example, $K\ I\ (K\ True\ False)$ reduces in one step to either I or $K\ I\ True$. Here, we can further reduce the second result, yielding I again; but we might worry that there could be some M such that M could be reduced to either N_1 or N_2 in such a way that N_1 and N_2 could never be brought back together by reducing them further. This situation would render the notation $N_1 =_\beta N_2$ nonsensical, since it would be hard to argue that N_1 and N_2 are equal in any behavioral sense. Fortunately, the following theorem, sometimes called the fundamental syntactic property of the lambda-calculus, guarantees that this cannot actually happen, i.e., that reduction in the lambda-calculus is **confluent**.

Theorem 102.1 (Church–Rosser). *If M reduces to two different expressions, N_1 and N_2, then these further reduce to some common expression L. (In symbols: if $M \to^* N_1$ and $M \to^* N_2$, then there is some L such that $N_1 \to^* L$ and $N_2 \to^* L$.)*

Corollary 102.1. *If two terms are beta-convertible, then they both reduce to some common term, i.e., if $M =_\beta N$ then there is some L such that that $M \to^* L$ and $N \to^* L$.*

The latter form of the Church–Rosser property immediately implies the uniqueness of normal forms: if $N_1 =_\beta N_2$ and neither N_1 nor N_2 contains any redexes, then N_1 and N_2 must be identical (modulo names of bound variables). This justifies regarding the normal form of a term (when it has one) as its *meaning* or *value*. For example, we can say that the value of the expression $Plus\ C_3\ C_2$ is C_5.

We have seen that there are some terms, such as Ω, that do not have normal forms. There are also some, such as $K\ I\ \Omega$, that do have a normal form (I) but that can also be reduced indefinitely: $K\ I\ \Omega \to K\ I\ \Omega \to K\ I\ \Omega \to$ etc. We call terms such as Ω nonnormalizable, terms such as $K\ I\ \Omega$ normalizable, and terms like as $Plus\ C_1\ (Plus\ C_2\ C_3)$, for which *every* sequence of reductions ends in a normal form, **strongly normalizing**.

A reduction strategy is a rule specifying which redexes should be reduced first. There are several common reduction strategies (a classic comparison is Plotkin [1975]). The **normal-order** reduction strategy, sometimes called **call-by-name** reduction, always reduces the redex whose λ appears the farthest to the left. The lazy strategy also reduces the leftmost redex but only if that redex is not itself in the body of some abstraction—that is, lazy reduction stops when the expression reaches a **weak head normal form** with no top-level redexes. Similarly, **applicative-order** (or **call-by-value**) reduction always chooses the leftmost redex $(\lambda x.\ M)\ N$, where N is in a particular *evaluated form*: either a variable, or a lambda-abstraction, or a variable applied to some expression in evaluated form (in particular, not a redex). These strategies lead to different choices of which redex to reduce first in the following expression:

$$\lambda v.\ \underline{(\lambda z.\ z)\ ((\lambda w.\ w)\ (x\ (\lambda y.\ y)))} \qquad \text{normal order}$$

$$\lambda v.\ (\lambda z.\ z)\ (\underline{(\lambda w.\ w)\ (x\ (\lambda y.\ y))}) \qquad \text{applicative order}$$

$$\lambda v.\ (\lambda z.\ z)\ ((\lambda w.\ w)\ (x\ (\lambda y.\ y))) \qquad \text{lazy (no reductions allowed)}$$

Applicative-order reduction exactly matches the ordering of function calls in call-by-value programming languages such as SCHEME and ML, whereas lazy reduction approximates the ordering in languages such as HASKELL and ALGOL-60 [cf. Chapter 95 in this Handbook]. Normal-order and lazy reduction are safer strategies than applicative-order, since they never fail to terminate except on nonnormalizable expressions. For example, the expression $K\ I\ \Omega$ leads to the normal form I under normal-order reduction, whereas applicative order leads to an infinite sequence of reductions of Ω.

Theorem 102.2 (Normalization). *A sequence of normal-order reductions beginning from a normalizable term M always terminates in a normal form after a finite number of steps. Similarly, lazy reduction always finds a weak head normal form for any term that has one.*

Operational and Denotational Equivalences

When should we say that two lambda-expressions *behave the same*? Two essentially different sorts of answer can be given.

On the one hand, we can take an operational view, concentrating on how expressions behave under reduction. We have already seen one notion of equivalence that arises in this way, where two expressions are judged equivalent if they are beta-convertible. But we might reasonably wish to extend this notion a little, since there are pairs of expressions such as $(\lambda x. x\, x)(\lambda x. x\, x)$ and $(\lambda x. x\, x\, x)(\lambda x. x\, x)$ that are not beta-convertible, but that nevertheless have arguably the same external behavior (i.e., none at all; just an infinite sequence of internal steps). On the other hand, simply saying "two terms are equivalent if they are convertible or if neither has a normal form," goes too far in the other direction, since there are some terms without normal forms that do behave differently. For example, the expressions $\lambda x. x\, True\, \Omega$ and $\lambda x. x\, False\, \Omega$ are both nonnormalizable, but applying the first to K yields an expression with normal form *True*, whereas applying the second to K yields an expression with normal form *False*. In other words, $\lambda x. x\, True\, \Omega$ and $\lambda x. x\, False\, \Omega$ have the same behavior in isolation (both diverge), but they do not have the same behavior in all contexts.

The idea that equivalent expressions should have the same behavior in all contexts can be formalized using the following notion of **contextual equivalence**, first studied by Morris. First, we introduce the notion of a context: a lambda-expression with a *hole*, written [], into which another expression can be placed. For example, suppose $C[\,]$ is the context $\lambda x. [\,]\, x$. Filling the hole in $C[\,]$ with the expression $x\, y$, written $C[x\, y]$, yields $\lambda x. x\, y\, x$. (Notice how, unlike substitution, the variable x in the expression $x\, y$ is captured by the binder λx.) Two expressions M and N are said to be equivalent when, for any context $C[\,]$, the expression $C[M]$ is normalizable if and only if $C[N]$ is normalizable.

Note that this definition is stated in terms of normalizability and does not explicitly demand that the normal forms of M and N be the same. This notion of *observation* might initially seem too weak. For example, the context $C[\,] = [\,]\, K$ fails to distinguish the expressions $\lambda x. x\, True\, \Omega$ and $\lambda x. x\, False\, \Omega$, since $C[\lambda x. x\, True\, \Omega] \to True$ and $C[\lambda x. x\, False\, \Omega] \to False$ both reduce to normal forms. The discriminating power of contextual equivalence comes from the quantification over all contexts. Here, the more complex context $C[\,] = [\,]\, K\, \Omega\, I$ does distinguish the two processes in question, since $C[\lambda x. x\, True\, \Omega] \to^* K\, True\, \Omega\, \Omega\, I \to^* True\, \Omega\, I \to^* \Omega$ has no normal form, whereas $C[\lambda x. x\, False\, \Omega] \to^* K\, False\, \Omega\, \Omega\, I \to^* False\, \Omega\, I \to^* I$ has the normal form I.

An equivalent way of formulating the intuition behind contextual equivalence is via the notion of **applicative bisimulation**. The idea here is to give a method of testing whether two expressions have observably *different* behavior and regard two expressions as equivalent if they cannot be shown to be different by making any finite sequence of tests. Formally, two combinators M and N are said to be bisimilar only if (1) either both are nonnormalizable or both are normalizable, and (2) for each combinator L, the applications $M\, L$ and $N\, L$ are bisimilar. Two arbitrary expressions M and N, possibly with free variables, are bisimilar if, no matter what combinators are substituted for their free variables, the resulting combinators are bisimilar in the previous sense.

The form of the definition of applicative bisimulation, which takes all expressions to be bisimilar except those that fail one of the two previous conditions, leads to a powerful **coinductive** reasoning technique: to show that two expressions are bisimilar, it suffices to show that no contradiction results from the assumption that they are. For example, to show that $\Omega = (\lambda x. x\, x)(\lambda x. x\, x)$ is bisimilar to $\Omega_1 = (\lambda x. x\, x\, x)(\lambda x. x\, x)$, we reason as follows: If they are not bisimilar, then there must be some sequence of terms $L_1, L_2, \ldots, L_k$ such that $\Omega\, L_1\, L_2 \ldots L_k$ is normalizable and $\Omega_1\, L_1\, L_2 \ldots L_k$ is not, or vice versa. But this cannot be, since both Ω and Ω_1 can reduce only to nonnormalizable terms, no matter what they are applied to. Thus, Ω and Ω_1 are bisimilar.

As this example illustrates, applicative bisimulation is typically much easier to use than contextual equivalence. In order to show directly that Ω and Ω_1 are contextually equivalent, we would have to show that they have the same behavior when placed in an *arbitrary* context; applicative bisimulation allows us to consider just contexts of the form $[\,]\, L_1\, L_2 \ldots$.

Many variants of the definitions of contextual equivalence and applicative bisimulation have been studied. For example, "normalizable" can be replaced by "reaches a normal form under a normal-order reduction strategy" or "...under a lazy strategy," etc.

A somewhat different, denotational perspective on expression equivalence is obtained by returning to the original intuition that lambda-expressions were intended to represent functions, and say that M and N are the same if they represent the same function, i.e., if they map the same inputs to the same outputs. To make this precise, we need to choose some semantic domain D (i.e., some set of functions) and define a denotation function $[\![\,]\!]$ that maps each expression M into an element $[\![M]\!]$ of D. (See Chapter 104 in this Handbook for more discussion of domains and denotation functions.) There are some serious technical problems with defining D and $[\![\,]\!]$, stemming from the fact that, since lambda-expressions take lambda-expressions as arguments and return lambda-expressions as results, each element of D is actually a function from D to D; that is, D must be a larger set than the set of functions from D to D. Trying to construct such a D naively leads to a mathematical paradox, where we end up with a D that is strictly larger than itself. Fortunately, Scott and Plotkin realized in 1969 that, by considering only some of the functions from D to D, the paradox can be avoided. Indeed, the same basic insight can be used to construct many different semantic domains with different properties.

Research Areas

The study of denotational semantic models of the lambda-calculus and related systems has led to a rich research literature, surveyed in textbooks by Barendregt [1984], Gunter [1992], Schmidt [1986], and Winskel [1993] and a shorter article by Gunter and Scott [1990]. One issue that has received considerable attention is the problem of finding *fully abstract* models, in which each lambda-expression is mapped to an element of the model (a mathematical function of some carefully chosen sort) in such a way that two lambda-expressions have equivalent operational behavior if and only if they denote the same element of the model. Operational notions of program equivalence such as applicative bisimulation have begun to receive serious attention only relatively recently (e.g. Gordon [1994]).

Implementation techniques for programming languages based on the lambda-calculus have a long history, from Landin's original SECD machine to more modern proposals such as the G-machine [Peyton Jones and Lester 1992]. A related theoretical development is work on *optimal* reduction strategies, which try to choose redexes so as to reach a normal form as quickly as possible. The lambda-calculus forms a common basis for work on optimization techniques such as partial evaluation [Jones et al. 1993]; related notations are being used as intermediate languages in optimizing compilers for high-level languages such as C.

In impure functional languages such as SCHEME and ML, mutable variables are added to the lambda-calculus, retaining the higher order flavor of the pure calculus while giving up the simple intuition that expressions represent mathematical functions (cf. Chapter 95 in this Handbook). For imperative computation, where evaluation of an expression can also have side effects on mutable variables, it is not yet clear how to reason about equivalence of expressions, since there is no obvious choice for the definition of what is observable. For example, mutable variables may be local to a particular function, and side effects on these are not directly observable in the same way as side effects on global variables, though they can be observed indirectly by observing the input–output behavior of the function.

One of the most active areas of lambda-calculus research is the study of typed lambda-calculi, in which functions are classified according to the types of arguments they can correctly accept and the types of results they can return. Such calculi typically have quite different properties from the untyped calculus presented here. For example, it is typically the case that every term is strongly normalizing (combinators like Ω and Y cannot then be defined). Also, semantic models of typed calculi are often more straightforward to construct. See Chapter 103 in this Handbook.

102.3 Pi-Calculus

The lambda-calculus holds an enviable position: it is recognized as embodying, in miniature, all of the essential features of functional computation. Moreover, other foundations for functional computation, such as Turing machines, have exactly the same expressive power. The inevitability of the lambda-calculus arises from the fact that the only way to observe a functional computation is to watch which output values it yields when presented with different input values.

Unfortunately, the world of concurrent computation is not so orderly. Different notions of what can be observed may be appropriate in different circumstances, giving rise to different definitions of when two concurrent systems have the same behavior; for example, we may wish to observe or ignore the degree of inherent parallelism of a system, the circumstances under which it can deadlock, the distribution of its processes among physical processors, or its resilience to various kinds of failures. Moreover, concurrent systems can be described in terms of many different constructs for creating processes (fork/wait, cobegin/coend, futures, data parallelism, etc.), exchanging information between them (shared memory, rendezvous, message passing, dataflow, etc.), and managing their use of shared resources (semaphores, monitors, transactions, etc.).

This variability has given rise to a large class of formal systems called *process calculi* (sometimes *process algebras*), each embodying the essence of a particular concurrent or distributed programming paradigm. We focus here on one typical process calculus, the pi-calculus (or π-calculus) of Milner et al. [1992] and Milner [1991]. References to some other popular process calculi can be found at the end of the section.

In the pure lambda-calculus, everything is a function; numbers, for example, are encoded as special functions that can be interrogated (by applying them) to find out which number they represent. Analogously, in the pi-calculus, every expression denotes a *process*: a free-standing computational activity, running in parallel with other processes and possibly containing many independent subprocesses. Two processes can interact by exchanging a message on a channel. Indeed, communication along channels is the sole means of computation, just as function application is in the lambda-calculus. The only thing that can be observed about a process's behavior is its ability to send and receive messages.

Syntax and Operational Semantics

The simplest pi-calculus expression is **0**, which denotes a process with no behavior at all. More interestingly, if P is some process expression, then the expression $x(y).P$ denotes a process that waits to read a value y from the channel x and then, having received it, behaves like P. Similarly, $\overline{x}y.P$ denotes a process that first waits to send the value y along the channel x and then, after y has been accepted by some input process, behaves like P.

$P \mid Q$ denotes a process composed of two subprocesses, P and Q, running in parallel. That is, $P \mid Q$ can exhibit all of the observable behaviors (sequences of messages sent and received) of both P and Q, interleaved in any order. Moreover, if P can send a message on some channel x and Q can receive on x, then $P \mid Q$ can perform an internal communication in which the message is exchanged between P and Q.

Placing the restriction operator (νx) before a process expression P ensures that x is a fresh channel in P, i.e., that messages sent and received by P on x will never be mixed with messages sent or received on any other channel created elsewhere, even another channel that happens to be named x. That is, the alphabetic names of channels are unimportant, just as the names of bound variables are unimportant in the lambda-calculus: the process $(\nu x)\overline{y}x.\mathbf{0}$ is completely equivalent to $(\nu z)\overline{y}z.\mathbf{0}$; both introduce a fresh channel, different from all other channels, and send it on y.

Finally, the *replicated process* $!P$ stands for an infinite number of copies of P, all running in parallel. Typically, only a few copies will actually be doing anything at a given moment; $!P$ really should be thought of as a simple notational device for describing processes with infinitely *long* sequences of behaviors. For example, $!(t(w)\overline{x}y.\overline{t}v.\mathbf{0})$ denotes a process that, after it is triggered by someone sending it a message on t, sends a message on x and then triggers another copy of itself by sending another message on t; i.e., it responds to a message on t by sending an infinite stream of messages on x.

When an input or output prefix is followed by **0**, we normally drop the **0**, writing $\overline{x}y$ instead of $\overline{x}y.\mathbf{0}$. Also, to avoid writing too many parentheses, we give input and output prefixes the strongest precedence, replication the next strongest, parallel composition the next, and restriction the weakest, so that $(\nu x)!x(y).a(b) \mid \overline{z}w$ means $(\nu x)((!(x(y).a(b).\mathbf{0})) \mid \overline{z}w.\mathbf{0})$.

Syntax:

$$
\begin{array}{llll}
P, Q, R & ::= & \mathbf{0} & \text{inert process} \\
& & x(y).\, P & \text{input prefix} \\
& & \overline{x}y.\, P & \text{output prefix} \\
& & P \mid Q & \text{parallel composition} \\
& & (\nu x) P & \text{restriction} \\
& & !P & \text{replication}
\end{array}
$$

Renaming of bound variables:

$$
\begin{array}{rcll}
x(y).\, P & = & x(z).\,([z/y]P) & \text{if } z \notin FV(P) \\
(\nu y) P & = & (\nu z)([z/y]P) & \text{if } z \notin FV(P)
\end{array}
$$

Structural Congruence:

$$
\begin{array}{rcll}
P \mid Q & \equiv & Q \mid P & \text{commutativity of parallel composition} \\
(P \mid Q) \mid R & \equiv & P \mid (Q \mid R) & \text{associativity of parallel composition} \\
((\nu x) P) \mid Q & \equiv & (\nu x)(P \mid Q) \quad \text{if } x \notin FV(Q) & \text{scope extrusion} \\
!P & \equiv & P \mid !P & \text{replication}
\end{array}
$$

Operational Semantics:

$$
\begin{array}{rcll}
\overline{x}y.\, P \mid x(z).\, Q & \rightarrow & P \mid [y/z]Q & \text{communication} \\
P \mid R & \rightarrow & Q \mid R & \text{if } P \rightarrow Q \quad\text{reduction under } \mid \\
(\nu x) P & \rightarrow & (\nu x) Q & \text{if } P \rightarrow Q \quad\text{reduction under } \nu \\
P & \rightarrow & Q & \text{if } P \equiv P' \rightarrow Q' \equiv Q \quad\text{structural congruence}
\end{array}
$$

FIGURE 102.2 Syntax and operational semantics of the pi-calculus.

The definitions of free variables and substitution in the pi-calculus are similar to the corresponding definitions in the lambda-calculus. The expressions $x(y).\, P$ and $(\nu x) P$ bind the variable x in the body P. As before, we silently rename bound variables whenever necessary. Substitution is simpler here, because we substitute only variables for variables; we do not substitute processes into other processes.

The operational semantics of pi-calculus expressions is defined as a *reduction* relation, as in the lambda-calculus. We say that P reduces to Q, written $P \rightarrow Q$, if P contains two parallel subprocesses that can communicate on some channel x to become the corresponding subprocesses of Q. This is formalized by the group of rules at the bottom of Fig. 102.2. The first rule, corresponding to the beta-reduction rule in lambda-calculus, defines a primitive step of communication: a redex consisting of an output process $\overline{x}y.\, P$ in parallel with an input process $x(z).\, Q$ reduces to P in parallel with Q, where the data value y is substituted for the bound variable z in Q. For example, $\overline{x}y.\, \overline{y}z \mid x(w).\, w(v)$ reduces to $\overline{y}z \mid y(v)$, which further reduces to $\mathbf{0} \mid \mathbf{0}$, which cannot reduce further. This example underscores the fact that the data value that is passed from sender to receiver during a communication is itself a channel and may later be used by the receiver for communication.

The next two rules specify that if a communication can occur between two subprocesses of a process P then the same communication can still occur when P is placed in parallel with another process Q and, similarly, if a communication can occur within P then the same communication can occur within $(\nu x) P$. Note that we do not allow reductions to occur inside the body of a process that is prefixed by an input or output.

The final rule in Fig. 102.2 allows the two "halves" of a redex to be mixed together with other parallel components of a larger system. For example, in the expression $x(y).\, P \mid \overline{x}y.\, Q \mid \overline{x}zR$, there are two possible communications on x (between the first and second components and between the first and third), but neither has exactly the form of the left-hand side of the communication rule: the second redex has an extra subprocess in the middle and both redexes are in the wrong order, with the input subprocess first. The structural congruence relation $\equiv$ formalizes the intuition that this does not matter, that $x(y).\, P \mid \overline{x}y.\, Q \mid \overline{x}z.\, R$ is just a another way of writing $\overline{x}y.\, Q \mid x(y).\, P \mid \overline{x}z.\, R$ or $\overline{x}z.\, R \mid x(y).\, P \mid \overline{x}y.\, Q$, both of which literally contain redexes. The four rules defining $\equiv$ may be applied any number of times within a process expression. The structural congruence rule for reduction has the effect that P can reduce to Q

whenever P can be rearranged so that it literally contains a redex by which it can reduce to Q. In addition to rearranging the order of parallel compositions—the task of the first two of the structural congruence rules—there are rules for rearranging ν and $!$. The first says that the scope of a ν binding may be enlarged to enable reduction, as in $((\nu z)\overline{x}z. P) \mid x(y). Q \equiv (\nu z)\overline{x}z. P \mid x(y). Q \rightarrow (\nu z)(P \mid [z/y]Q)$. The second formalizes the intuition that $!P$ behaves just the same as an arbitrary number of parallel copies of P: we can move new copies of P out from under the $!$ at will, making them available to participate in communications.

Because of the structural congruence rule, each process expression beginning with an input or an output may, in general, be part of several redexes at once. For example, the expression $(\nu x)\overline{x}y \mid \overline{x}z \mid x(w). \overline{w}v$ contains the redexes $(\nu x)\underline{\overline{x}y} \mid \overline{x}z \mid \underline{x(w). \overline{w}v}$ and $(\nu x)\overline{x}y \mid \underline{\overline{x}z} \mid \underline{x(w). \overline{w}v}$, both of which include $x(w). \overline{w}v$. Reducing either one of these has the effect of destroying the other. Moreover, the resulting processes are irretrievably different: one reduction yields $(\nu x)\overline{x}z \mid \overline{y}v$, whose only further behavior is sending a message on y, whereas the other yields $(\nu x)\overline{x}y \mid \overline{z}v$, which can send a message only on z. This nonconfluence of reduction in the pi-calculus is crucial, since it models the fact that real concurrent programs may often yield different results depending on the order in which various internal events occur. Such timing dependencies may be undesirable (they are often called *race conditions*), but we need a framework in which they *can* occur in order for the assertion that they do *not* occur in a particular process to have any force! A related point is that we are not interested in a conversion relation on processes, as we were in the lambda-calculus: the fact that P and Q can both reduce to R is not sufficient reason to claim that P and Q behave the same, since P and Q also may be able to reduce to other, completely dissimilar, processes.

Examples

As with the lambda-calculus, the simplicity of the pi-calculus is deceptive: the primitive mechanisms of restriction, communication, parallel composition, and replication can be used to model a great variety of programming structures.

One very easy encoding trick allows several values to be sent and received in each message on a channel. Write $\overline{x}\langle y_1, \ldots, y_n \rangle$ for the simultaneous output of the tuple y_1-y_n on x, and $\overline{x}(z_1, \ldots, z_n)$ for the corresponding input process, which accepts a tuple of values and binds them to the variables z_1-z_n. The case where $n = 0$ corresponds to sending a dataless signal on x. These *polyadic* communication prefixes can be encoded in the basic pi-calculus as follows:

$$\overline{x}\langle y_1, \ldots, y_n \rangle. P \quad = \quad (\nu p)\overline{x}p. \overline{p}y_1. \ldots \overline{p}y_n. P \quad \text{choosing } p \notin FV(P)$$

$$x(z_1, \ldots, z_n). Q \quad = \quad x(p).p(z_1). \ldots p(z_n). Q \quad \text{choosing } p \notin FV(Q)$$

That is, to send y_1-y_n on x, we make up a private channel p, send p on x, and then send y_1-y_n, one after the other, on p. Conversely, to read a tuple from x, we read p from x and then read the values from p. By using a fresh p each time we send a tuple on x, we avoid any possible confusion between different processes sending tuples on x at the same time.

The encodings of common data structures in the lambda-calculus all have their counterparts in the pi-calculus. For example, the Boolean values *True* and *False* are encoded by processes that repeatedly accept (over some channel b) a pair of channels t and f and respond by sending a message on either t or f:

$$True (b) \quad = \quad !b(t, f). \overline{t}\langle \rangle$$

$$False (b) \quad = \quad !b(t, f). \overline{f}\langle \rangle$$

If *Test* (b) is defined as $(\nu t)(\nu f)\overline{b}\langle t, f \rangle(t(). P \mid f().Q)$, then the composite process *True* $(b) \mid$ *Test* (b) reduces to *True* $(b) \mid P \mid (\nu t)(\nu f)(0 \mid f().Q)$, whereas *False* $(b) \mid$ *Test* (b) reduces to *False* $(b) \mid Q \mid (\nu t)(\nu f)$ $(0 \mid t(). P)$, where the third subprocess in each case is unable to participate in any further actions. (For notational convenience, some variants of the pi-calculus include extra structural congruence rules like $P \mid 0 \equiv P$ to *garbage collect* such deadlocked subprocesses.) The channel b can be thought of as the *location* of the value *True* (*b*), since *b* is the only means by which other processes can refer to it or interact with it, or as a *reference* to (the process encoding) the value *True*.

The pi-calculus is a quintessentially imperative (i.e., nonfunctional) language, in the sense that we do not expect to get the same result each time if we query a process repeatedly over a channel. A typical example is a *reference cell object*, which maintains a single piece of state, updating it in response to messages sent over a channel w and reporting its current value in response to messages sent over a channel r.

$$
\begin{aligned}
Ref(r, w, i) &= (\nu l)\,\bar{l}\langle i \rangle \mid ReadServer(l, r) \mid WriteServer(l, w) \\
ReadServer(l, r) &= \,!r(c).\,l(v).\,(\bar{c}\langle v \rangle \mid \bar{l}\langle v \rangle) \\
WriteServer(l, w) &= \,!w(c, v').\,l(v).\,(\bar{c}\langle\rangle \mid \bar{l}\langle v' \rangle)
\end{aligned}
$$

The process $Ref(r, w, i)$ comprises three parts: a process waiting to send the initial value i on an internal channel l, which is private to the reference cell; a read server that accepts messages on the channel r and, in response to each, sends back the current value on a result channel c that is included in the message; and a write server that accepts messages containing new values for the reference cell and acknowledges their acceptance on a completion channel c that is included in the message. Each time a read or write request is received, a process is created that reads the current value of the reference cell from the internal channel l, responds to the request by sending something back on the result or completion channel c, and then restores either the existing value or the specified new value of the cell by sending it on l. At any given moment, there will be just one process ready to send on l; this invariant ensures that, for example, the client process

$$
(\nu c)\overline{w}\langle v, c \rangle.c().\,(\nu d)\overline{r}\langle d \rangle.\,d(e).\,Q
$$

will always receive v on d in response to its request message on r, assuming that no other client processes are simultaneously interacting with the same reference cell (note that it waits for the write-completion signal to arrive on c before sending the read request).

As a final illustration of the power of the pi-calculus, here is an encoding of the lambda-calculus itself, with a normal-order reduction strategy. Given a lambda-expression M and a channel p, define the process expression $[\![M]\!](p)$ as follows:

$$
\begin{aligned}
[\![\lambda x.\,M]\!](p) &= p(x, q).\,[\![M]\!](q) \\
[\![x]\!](p) &= \bar{x}\langle p \rangle \\
[\![M\,N]\!](p) &= (\nu q)[\![M]\!](q) \mid ((\nu y)\overline{q}\langle y, p \rangle \mid {!}y(r).\,[\![N]\!](r))
\end{aligned}
$$

$[\![M]\!](p)$, pronounced "the process representing M with argument port p," is an expression that rests dormant until it receives on p a trigger x for its argument and a new argument port q. It then evolves to a new process with argument port q. For example, the lambda-expression $(\lambda x.\,x)z$ is translated as follows:

$$
\begin{aligned}
[\![(\lambda x.\,x)z]\!](p) &= (\nu q)[\![\lambda x.\,x]\!](q) \mid ((\nu y)\overline{q}\langle y, p \rangle \mid {!}y(r).\,[\![z]\!](r)) \\
&= (\nu q)(q(x, q')\bar{x}\langle q' \rangle) \mid ((\nu y)\overline{q}\langle y, p \rangle \mid {!}y(r).\,\bar{z}\langle r \rangle) \\
&\rightarrow (\nu q)(\nu y)\overline{y}\langle p \rangle \mid {!}y(r).\,\bar{z}\langle r \rangle \\
&\rightarrow (\nu q)(\nu y)({!}y(r).\,\bar{z}\langle r \rangle) \mid \bar{z}\langle p \rangle \\
&\text{``=''} \;\; \bar{z}\langle p \rangle \\
&= [\![z]\!](p)
\end{aligned}
$$

Operational Equivalence

The "=" in the penultimate line in the previous calculation deserves some discussion. Informally, it is clear that the process $(\nu q)(\nu y)(!y(r).\,\overline{z}\langle r\rangle)\mid \overline{z}\langle p\rangle$ can communicate only on z, since the replicated input on y appears immediately underneath the binder (νy), which guarantees that there can never be any sender on y. How do we express rigorously such assertions that one process expression has the same behavior as another?

As with the lambda-calculus, there are two basic approaches: either we can try to identify some class of "real processes" and say that two process expressions are equivalent if they denote the same real process, or we can focus on the reduction relation of the process calculus itself and say that two process expressions are equivalent if they have the same reduction behavior in all contexts. Work is proceeding on both fronts, but the situation is much more complex than in the simple world of functional computation (cf. Milner [1990], Baeten and Weijland [1990], Hennessy [1988], and Hoare [1985] for discussion and references). For denotational approaches, the problem is finding a pragmatically satisfying definition of real processes. For operational approaches, the (related) problem is that there are many possible definitions of *behavior*, yielding numerous, subtly different, equivalences [van Glabbeek 1993]. So far, operational techniques have proved most successful, but both remain active research areas.

As in lambda-calculus, the most intuitive way of defining operational equivalence is via some notion of *contextual equivalence*. A *process context* is a process expression with a hole into which another process can be placed. We say that P and Q are equivalent when $C[P]$ and $C[Q]$ have the same observable behavior for each process context $C[\,]$.

In the lambda-calculus, we saw that there were several variations on the definition of contextual equivalence, depending on precisely what *observations* we allow of a lambda-term M in some testing context $C[\,]$. For example, we might choose to observe whether $C[M]$ is normalizable, whether it is normalizable using a normal-order reduction strategy, whether it is strongly normalizing, etc., each choice giving rise to a different notion of equivalence of lambda-terms. In the pi-calculus, there are even more choices. If $C[\,]$ is a testing context for a process P, we might choose to observe, for example:

1. Whether $C[P]$ can eventually perform an input or output action.
2. What *sequences* of input and output actions $C[P]$ can perform.
3. How early or late $C[P]$ becomes committed to producing certain sequences of inputs and outputs (so that, for example, we distinguish the case where $C[P]$ can send on a and then chooses whether to send on b or on c from the case where $C[P]$ chooses first whether to send on a and then b or on a and then c).
4. The circumstances under which $C[P]$ can become deadlocked.

Note that "does $C[P]$ have a normal form?" is *not* a useful kind of observation in the pi-calculus, since it leads us to regard all processes with infinite behaviors as behaviorally identical. In the lambda-calculus, this identification makes sense, since any lambda-expression whose reduction does not terminate can be viewed as an infinite loop. But in the pi-calculus, a process may go on computing forever but still interact usefully with its environment. Real parallel and distributed systems are full of server processes with this property.

As in the lambda-calculus, contextual equivalence between two processes can be difficult to establish, because it demands that they must yield the same observable behavior when placed in an arbitrary testing context $C[\,]$. Fortunately, some useful variants of contextual equivalence can be reformulated in terms of direct conditions on the processes themselves, with no quantification over contexts. For example, the following definition coincides with an observational congruence where the allowed observations are those previously listed second and third.

We say that two process expressions P and Q are **bisimilar** if every action of one can be matched by a corresponding action of the other to reach a bisimilar state. More precisely, (1) if P can output the value a on the channel x and become P', the Q must also be able to output a on x and reach Q' such that P' is bisimilar to Q'; (2) if P can input a value a from the channel x and become P', then Q also must be able to input a from x and reach Q' such that P' is bisimilar to Q'; (3) if P can perform some internal communication to become P', then Q must be able to perform an internal communication to reach a state Q' such that P' and Q' are bisimilar; and three similar clauses where the roles of P and Q are reversed. We are being informal here about what it means that a process can input or output a on x and become another process; this can be formulated precisely using *labeled transition systems*.

Like applicative bisimulation for lambda-terms, the definition of bisimulation for process expressions is given in a coinductive style: two processes are bisimilar if we cannot show that they are not, or, equivalently, if assuming that they are leads to no contradictions. This style of definition gives rise to a powerful technique for proving the bisimilarity of processes, reminiscent of familiar inductive proof techniques used to prove properties of recursive functions in functional programming languages. We illustrate this technique with a simple example.

Suppose we want to show that the process $(\nu q)(\nu y)(!y(r).\,R) \mid S$ is bisimilar to S, as long as R and S do not have q and y among their free variables. Begin by assuming that this is true. To check that there are no contradictions, we choose an arbitrary R and S and check that the three conditions in the definition of bisimulation are satisfied (in both directions). For the first condition, suppose $(\nu q)(\nu y)(!y(r).\,R) \mid S$ can make an output a on some channel x and become P'. Since the subexpression $!y(r).\,R$ begins with an input, it is clear that an output on x must come from S, so that $P' = (\nu q)(\nu y)(!y(r).\,R) \mid S'$ for some S', and that S in isolation can make the same output and become S'. Since $(\nu q)(\nu y)(!y(r).\,R) \mid S'$ and S' are bisimilar (by our original assumption), we have failed to find a contradiction. Conversely, suppose S can make an output a on some channel x to become S'. Then $(\nu q)(\nu y)(!y(r).\,R) \mid S$ can make the same output and become $(\nu q)(\nu y)(!y(r).\,R) \mid S'$, where $(\nu q)(\nu y)(!y(r).\,R) \mid S'$ and S' are again bisimilar by assumption. The other two conditions in the definition of bisimilarity are checked analogously. We have therefore shown the absence of contradictions for an arbitrary R and S, and our original assumption is justified. This example allows us to give a precise meaning to the "=" at the end of the previous section: taking R to be $\overline{z}\langle r \rangle$ and S to be $\overline{z}\langle p \rangle$, we have shown that $(\nu q)(\nu y)((!y(r).\,\overline{z}\langle r \rangle) \mid \overline{z}\langle p \rangle)$ is bisimilar to $\overline{z}\langle p \rangle$.

Many variants of bisimulation have been proposed. One of the most common is so-called weak bisimulation, which relaxes the demand that the processes simulate each other's behavior in lock step and instead regards arbitrarily many steps of internal communication as equivalent to a single step. This equivalence is strictly coarser than the one previously defined (called strong bisimulation when it is necessary to distinguish the two), in the sense that whenever P and Q are strongly bisimilar, they are also weakly bisimilar. In practice, weak bisimulation often is more useful, since we typically want to regard two processes as having the same observable behavior even if one consumes more processor cycles than the other.

Research Areas

The pi-calculus is just one of a large family of process calculi, differing in many details but sharing the basic orientation: focusing on interaction via communication rather than shared variables, on describing concurrent systems using a small set of primitive operators, and on deriving useful algebraic laws for manipulating expressions written using these operators. The first historically, and among the most thoroughly studied, are Milner's *calculus of communicating systems* (CCS) [Milner 1980, 1989], and Hoare's *communicating sequential processes* (CSP) [Hoare 1985]. CCS is the direct predecessor of the pi-calculus: it can be described informally as the *static* fragment of the pi-calculus where the messages exchanged during communication do not contain any data (i.e., every output is of the form $\overline{x}\langle\rangle.\,P$). CSP embodies similar ideas in the form of both a theory and a programming language (Occam). Other members of the family include the variant of CCS described in Hennessy [1988], and Bergstra, Klop, and Baeten's systems, collectively called ACP [Baeten and Weijland 1990]. The rapidly growing number of process calculi has led to interest in taxonomic frameworks such as generalized structured operational semantics [Groote and Vaandrager 1992] and Milner's recent *action structures*, in which many different process calculi can be embedded and their properties compared.

Process calculi are widely used for specification and verification of concurrent systems, especially of communication protocols, both manually and with support from tools such as the Edinburgh Concurrency Workbench [Cleaveland et al. 1993].

Numerous programming languages have combined the concurrency primitives of process calculi with more conventional features for sequential programming; well-known examples include AMBER, CONCURRENT ML, and FACILE. The bare pi-calculus used is the basis of the PICT language. Pi-calculus has also proved useful in the study of concurrent object-oriented languages.

Defining Terms

Applicative-order (call-by-value) reduction: A reduction strategy in which the argument must be reduced to evaluated form before a function application can be reduced.

Bisimulation: An alternative to contextual equivalence, replacing the quantification over all contexts with the more tractable requirement that equivalent expressions should be able to match each others' behavior step-by-step.

Coinduction: A proof technique associated with bisimulation. To prove that two expressions are equivalent, we assume that they are equivalent and show that no contradiction results.

Combinator: A lambda-expression with no free variables.

Confluence: The property of the lambda-calculus that states that the order in which redexes are chosen for reduction does not affect the final normal form that is reached.

Contextual equivalence: Formalizes the intuition that two expressions can be considered the same if they have the same behavior in all contexts.

Lambda-calculus: A core language of functional computation, defined in Fig. 102.1, in which everything is a function and all computation proceeds by function application.

Normal form: A lambda-expression containing no redexes.

Normal-order (call-by-name) reduction: A reduction strategy in which the leftmost redex is always reduced first.

Pi-calculus: A core calculus of message-based concurrency, defined in Fig. 102.2, in which everything is a process and all computation proceeds by communication on channels.

Redex: A subexpression in a form that is ready to be evaluated by a step of reduction.

Strongly normalizing: A lambda-expression for which every sequence of reductions terminates in a normal form.

Weak head normal form: A lambda-expression in which all redexes are inside the bodies of lambda-abstractions.

References

Baeten, J. C. M. and Weijland, W. P. 1990. *Process Algebra*. Cambridge tracts in theoretical computer science 18. Cambridge University Press, Cambridge, England.

Barendregt, H. P. 1984. *The Lambda Calculus*, rev. ed. North–Holland.

Barendregt, H. P. 1990. Functional programming and lambda calculus. In *Handbook of Theoretical Computer Science*. J. van Leeuwen, ed. Vol. B. ch. 7, pp. 321–364. Elsevier/MIT Press.

Cleaveland, R., Parrow, J., and Steffen, B. 1993. The concurrency workbench: a semantics-based tool for the verification of concurrent systems. *ACM Trans. Programming Lang. Syst.* 15(1):36–72.

Davis, M. 1982. *Computability and Unsolvability*. Dover.

Gordon, A. D. 1994. *Functional Programming and Input/Output*. Cambridge University Press, Cambridge, England.

Groote, J. F. and Vaandrager, F. W. 1992. Structured operational semantics and bisimulation as a congruence. *Inf. Comput.* 100:202–260.

Gunter, C. A. 1992. *Semantics of Programming Languages: Structures and Techniqes*. MIT Press, Cambridge, MA.

Gunter, C. A. and Scott, D. S. 1990. Semantic domains. In *Handbook of Theoretical Computer Science*. J. van Leeuwen, ed., Vol. B. 12, pp. 633–674. Elsevier/MIT Press.

Hennessey, M. 1988. *Algebraic Theory of Processes*. MIT Press, Cambridge, MA.

Hindley, J. R. and Seldin, J. P. 1986. *Introduction to Combinators and λ-Calculus*. London Mathematical Society Student Texts, Vol. 1. Cambridge University Press, Cambridge, England.

Hoare, C. A. R. 1985. *Communicating Sequential Processes*. Prentice–Hall, Englewood Cliffs, NJ.

Jones, N. D., Gomard, C. K., and Sestoft, P. 1993. *Partial Evaluation and Automatic Program Generation*. Prentice–Hall International.

Milner, R. 1980. *A Calculus of Communicating Systems*. Lecture notes in computer science, 92. Springer–Verlag.

Milner, R. 1989. *Communication and Concurrency*. Prentice–Hall, Englewood Cliffs, NJ.

Milner, R. 1990. Operational and algebraic semantics of concurrent process. In *Handbook of Theoretical Computer Science*. J. van Jeeuwen, ed., Vol. B, ch. 19, pp. 1201–1242. Elsevier/MIT Press.

Milner, R. 1991. The Polyadic π-Calculus: A Tutorial. *Lab. for Foundations of Computer Science*. Tech. Rep. ECS-LFCS-91-180, Department of Computer Science, University of Edinburgh, UK, Oct. (*Proc. Int. Summer School Logic Algebra Specification*, Marktoberdorf, Aug. Reprinted 1993. *Logic and Algebra of Specification*, F. L. Bauer, W. Brauer, and H. Schwichtenberg, eds. Springer–Verlag).

Milner, R., Parrow, J., and Walker, D. 1992. A calculus of mobile proceses (Parts I and II). *Inf. Comput.* 100:1–77.

Peyton Jones, S. L. and Lester, D. R. 1992. *Implementing Functional Languages*. Prentice–Hall, Englewood Cliffs, NJ.

Plotkin, G. 1975. Call-by-name, call-by-value, and the λ-calculus. *Theor. Comput. Sci.* 1:125–159.

Schmidt, D. A. 1986. *Denotational Semantics: A Methodology for Language Development*. Allyn and Bacon.

Tennent, R. D. 1981. *Principles of Programming Languages*. Prentice–Hall, Englewood Cliffs, NJ.

van Glabbeek, R. J. 1993. The linear time-branching time spectrum II (the semantics of sequential systems with silent moves), pp. 66–81. In *Proc. CONCUR '93*.

van Leeuwen, J., ed. 1990. *Handbook of Theoretical Computer Science*, Vol. B. Elsevier/MIT Press.

Winskel, G. 1993. *The Formal Semantics of Programming Languages: An Introduction*. MIT Press, Cambridge, MA.

Further Information

The standard text for the lambda-caclulus is Barendregt [1984]. Hindley and Seldin [1986] is less comprehensive but somewhat more accessible. Barendregt's article in the *Handbook of Theoretical Computer Science* [1990] is a compact survey. Material on lambda-calculus can also be found in many textbooks on functional programming languages, e.g., Peyton Jones and Lester [1992] and programming language semantics, e.g., Schmidt [1986], Gunter [1992], and Winskel [1993].

The best introduction to the pi-calculus itself is Milner's [1991] tutorial. A deeper and more accessible introduction to many of the same issues can be found in his book on CCS [Milner 1989]. Books by Baeten and Weijland [1990], Hoare [1985], and Hennessy [1988] address other process calculi. Some semantic issues are summarized in Milner [1990].

New work on lambda-calculus and process calculi appears in many journals covering theoretical aspects of computer science, including *Information and Computation, Theoretical Computer Science, Mathematical Structures in Computer Science, Journal of Functional Programming,* and *Transactions on Programming Languages and Systems,* as well as in journals on mathematics and logic. Relevant conferences include Principles of Programming Languages (POPL), Functional Programming and Computer Architecture (FPCA), LISP and Functional Programming (LFP), Logic in Computer Science (LICS), Theoretical Aspects of Computer Science (TACS), the International Conference on the Theory and Practice of Software Development (TAPSOFT), and the International Conference on Concurrency Theory (CONCUR).

103

Type Systems

Luca Cardelli
*Digital Equipment
   Corporation Systems
   Research Center*

103.1 Introduction

The fundamental purpose of a **type system** is to prevent the occurrence of *execution errors* during the running of a program. This informal statement motivates the study of type systems, but requires clarification. Its accuracy depends, first of all, on the rather subtle issue of what constitutes an execution error, which we will discuss in detail. Even when that is settled, the absence of execution errors is a nontrivial property. When such a property holds for all of the program runs that can be expressed within a programming language, we say that the language is **type sound**. It turns out that a fair amount of careful analysis is required to avoid false and embarrassing claims of type soundness for programming languages. As a consequence, the classification, description, and study of type systems has emerged as a formal discipline.

The formalization of type systems requires the development of precise notations and definitions, and the detailed proof of formal properties that give confidence in the appropriateness of the definitions. Sometimes the discipline becomes rather abstract. One should always remember, though, that the basic motivation is pragmatic: the abstractions have arisen out of necessity and can usually be related directly to concrete intuitions. Moreover, formal techniques need not be applied in full in order to be useful and influential. A knowledge of the main principles of type systems can help in avoiding obvious and not so obvious pitfalls, and can inspire regularity and orthogonality in language design.

When properly developed, type systems provide conceptual tools with which to judge the adequacy of important aspects of language definitions. Informal language descriptions often fail to specify the type structure of a language in sufficient detail to allow unambiguous implementation. It often happens that different compilers for the same language implement slightly different type systems. Moreover, many language definitions have been found to be type unsound, allowing a program to crash even though it is judged acceptable by a **typechecker**. Ideally, formal type systems should be part of the definition of all typed

0-8493-2909-4/97/$0.00+$.50

programming languages. This way, typechecking algorithms could be measured unambiguously against precise specifications and, if at all possible and feasible, whole languages could be shown to be type sound.

In this introductory section we present an informal nomenclature for typing, execution errors, and related concepts. We discuss the expected properties and benefits of type systems, and we review how type systems can be formalized. The terminology used in the introduction is not completely standard; this is due to the inherent inconsistency of standard terminology arising from various sources. In general, we avoid the words *type* and *typing* when referring to run time concepts; for example we replace dynamic typing with dynamic checking and avoid common but ambiguous terms such as strong typing. The terminology is summarized in the Defining Terms section.

In section 103.2, we explain the notation commonly used for describing type systems. We review **judgments**, which are formal assertions about the typing of programs, **type rules**, which are implications between judgments, and **derivations**, which are deductions based on type rules. In section 103.3, we review a broad spectrum of simple types, the analog of which can be found in common languages, and we detail their type rules. In section 103.4, we present the type rules for a simple but complete imperative language. In section 103.5, we discuss the type rules for some advanced type constructions: *polymorphism* and *data abstraction*. In section 103.6, we explain how type systems can be extended with a notion of *subtyping*. Section 103.7 is a brief commentary on some important topics that we have glossed over. In section 103.8, we discuss the *type inference* problem, and we present type inference algorithms for the main type systems that we have considered. Finally, section 103.9 is a summary of achievements and future directions.

Execution Errors

The most obvious symptom of an execution error is the occurrence of an unexpected software fault, such as an illegal instruction fault or an illegal memory reference fault.

There are, however, more subtle kinds of execution errors that result in data corruption without any immediate symptoms. Moreover, there are software faults, such as divide by zero and dereferencing *nil*, that are not normally prevented by type systems. Finally, there are languages lacking type systems where, nonetheless, software faults do not occur. Therefore we need to define our terminology carefully, beginning with what is a type.

Typed and Untyped Languages

A program variable can assume a range of values during the execution of a program. An upper bound of such a range is called a **type** of the variable. For example, a variable x of type *Boolean* is supposed to assume only Boolean values during every run of a program. If x has type *Boolean*, then the Boolean expression *not* (x) has a sensible meaning in every run of the program. Languages where variables can be given (nontrivial) types are called **typed languages**.

Languages that do not restrict the range of variables are called **untyped languages:** they do not have types or, equivalently, have a single universal type that contains all values. In these languages, operations may be applied to inappropriate arguments: the result may be a fixed arbitrary value, a fault, an exception, or an unspecified effect. The pure λ-calculus (see Chapter 102) is an extreme case of an untyped language where no fault ever occurs: the only operation is function application and, since all values are functions that operation never fails.

A type system is that component of a typed language that keeps track of the types of variables and, in general, of the types of all expressions in a program. Type systems are used to determine whether programs are **well behaved** (as discussed subsequently). Only program sources that comply with a type system should be considered real programs of a typed language; the other sources should be discarded before they are run.

A language is typed by virtue of the existence of a type system for it, whether or not types actually appear in the syntax of programs. Typed languages are **explicitly typed** if types are part of the syntax, and **implicitly typed** otherwise. No mainstream language is purely implicitly typed, but languages such as ML and Haskell support writing large program fragments where type information is omitted; the type systems of those languages automatically assign types to such program fragments.

Execution Errors and Safety

It is useful to distinguish between two kinds of execution errors: the ones that cause the computation to stop immediately, and the ones that go unnoticed (for a while) and later cause arbitrary behavior. The former are called **trapped errors**, whereas the latter are **untrapped errors**.

An example of an untrapped error is improperly accessing a legal address, for example, accessing data past the end of an array in absence of run time bounds checks. Another untrapped error that may go unnoticed for an arbitrary length of time is jumping to the wrong address: memory there may or may not represent an instruction stream. Examples of trapped errors are division by zero and accessing an illegal address: the computation stops immediately (on many computer architectures).

A program fragment is *safe* if it does not cause untrapped errors to occur. Languages where all program fragments are safe are called **safe languages**. Therefore, safe languages rule out the most insidious form of execution errors: the ones that may go unnoticed. Untyped languages may enforce **safety** by performing run time checks. Typed languages may enforce safety by statically rejecting all programs that are potentially unsafe. Typed languages may also use a mixture of run time and **static checks**.

Although safety is a crucial property of programs, it is rare for a typed language to be concerned exclusively with the elimination of untrapped errors. Typed languages usually aim to rule out also large classes of trapped errors, along with the untrapped ones. We discuss these issues next.

Execution Errors and Well-Behaved Programs

For any given language, we may designate a subset of the possible execution errors as **forbidden errors**. The forbidden errors should include all of the untrapped errors, plus a subset of the trapped errors. A program fragment is said to have **good behavior**, or equivalently to be well behaved, if it does not cause any forbidden error to occur. (The contrary is to have *bad behavior*, or equivalently to be *ill behaved*.) In particular, a well-behaved fragment is safe. A language where all of the (legal) programs have good behavior is called **strongly checked**.

Thus, with respect to a given type system, the following holds for a strongly checked language:

- No untrapped errors occur (safety guarantee).
- None of the trapped errors designated as forbidden errors occur.
- Other trapped errors may occur; it is the programmer's responsibility to avoid them.

Typed languages can enforce good behavior (including safety) by performing static (i.e., compile time) checks to prevent unsafe and ill-behaved programs from ever running. These languages are **statically checked**; the checking process is called **typechecking**, and the algorithm that performs this checking is called the typechecker. A program that passes the typechecker is said to be **well typed**; otherwise, it is **ill typed**, which may mean that it is actually ill behaved, or simply that it could not be guaranteed to be well behaved. Examples of statically checked languages are ML and Pascal (with the caveat that Pascal has some unsafe features).

Untyped languages can enforce good behavior (including safety) in a different way, by performing sufficiently detailed run time checks to rule out all forbidden errors. (For example, they may check all array bounds, and all division operations, generating recoverable exceptions when forbidden errors would happen.) The checking process in these languages is called **dynamic checking**; LISP is an example of such a **dynamically checked language**. These languages are strongly checked even though they have neither static checking, nor a type system.

Even statically checked languages usually need to perform tests at run time to achieve safety. For example, array bounds must in general be tested dynamically. The fact that a language is statically checked does not necessarily mean that execution can proceed entirely blindly.

Several languages take advantage of their static type structures to perform sophisticated dynamic tests. For example Simula67's INSPECT and Modula-3's TYPECASE constructs discriminate on the run time type of an object. These languages are still (slightly improperly) considered statically checked, partially because the dynamic type tests are defined on the basis of the static type system. That is, the dynamic tests for type equality are compatible with the algorithm that the typechecker uses to determine type equality at compile time.

Lack of Safety

By our definitions, a well-behaved program is safe. Safety is a more primitive and perhaps more important property than good behavior. The primary goal of a type system is to ensure language safety by ruling out *all* untrapped errors in all program runs. However, most type systems are designed to ensure the more general good-behavior property, and implicitly safety. Thus, the declared goal of a type system is usually to ensure good behavior of all programs, by distinguishing between well-typed and ill-typed programs.

In reality, certain statically checked languages do not ensure safety. That is, their set of forbidden errors does not include all untrapped errors. These languages can be euphemistically called **weakly checked** (or *weakly typed*, in the literature) meaning that some unsafe operations are detected statically and some are not detected. Languages in this class vary widely in the extent of their weakness. For example, Pascal is unsafe only when untagged variant types and function parameters are used, whereas C has many unsafe and widely used features, such as pointer arithmetic and casting. It is interesting to notice that the first five of the 10 commandments for C programmers [Spencer, H.] are directed at compensating for the weak-checking aspects of C. Some of the problems caused by weak checking in C have been alleviated in C++, and even more have been addressed in Java, confirming a trend away from weak checking. Modula-3 supports unsafe features, but only in modules that are explicitly marked as unsafe, and prevents safe modules from importing unsafe interfaces.

Most untyped languages are, by necessity, completely safe (e.g., LISP). Otherwise, programming would be too frustrating in the absence of both compile time and run time checks to protect against corruption. Assembly languages belong to the unpleasant category of untyped unsafe languages. (See Table 103.1.)

Should Languages Be Safe?

Lack of safety in a language design is motivated by performance considerations (when not introduced by mistake). The run time checks needed to achieve safety are sometimes considered too expensive. Safety has a cost even in languages that do extensive static analysis: tests such as array bounds checks cannot be, in general, completely eliminated at compile time.

Safety, however, is cost effective according to different measures. Safety produces fail-stop behavior in case of execution errors, reducing debugging time. Moreover, safety guarantees the integrity of run time structures, and therefore enables garbage collection. In turn, garbage collection considerably reduces code size and code development time, at the price of some performance.

Thus, the choice between a safe and unsafe language may be ultimately related to a tradeoff between development time and execution time. Although undeniable, the advantages of safety have not yet caused a wide-spread adoption of safe languages. Instead of regarding lack of safety as bad, many developers consider almost safety as almost good, and live with the consequences.

Should Languages Be Typed?

The issue of whether programming languages should have types (even with weak checking) is still subject to some debate. There is little doubt, though, that production code written in untyped languages can be maintained only with great difficulty. From the point of view of maintainability, even weakly

TABLE 103.1 Safety

	Typed	Untyped
Safe	ML	LISP
Unsafe	C	Assembler

checked unsafe languages are superior to safe but untyped languages (e.g., C vs. LISP). Here are the arguments that have been put forward in favor of typed languages, from an engineering point of view:

- *Economy of execution.* Type information was first introduced in programming to improve code generation and run time efficiency for numerical computations, for example, in FORTRAN. In ML, accurate type information eliminates the need for *nil*-checking on pointer dereferencing. In general, accurate type information at compile time leads to the application of the appropriate operations at run time without the need of expensive tests.

- *Economy of small-scale development.* When a type system is well designed, typechecking can capture a large fraction of routine programming errors, eliminating lengthy debugging sessions. The errors that do occur are easier to debug, simply because large classes of other errors have been ruled out. Moreover, experienced programmers adopt a coding style that causes some logical errors to show up as typechecking errors: they use the typechecker as a development tool. (For example, by changing the name of a field when its invariants change even though its type remains the same, so as to get error reports on all its old uses.)

- *Economy of compilation.* Type information can be organized into *interfaces* for program modules, for example as in Modula-2 and Ada. Modules can then be compiled independently of each other, with each module depending only on the interfaces of the others. Compilation of large systems is made more efficient because, at least when interfaces are stable, changes to a module do not cause other modules to be recompiled.

- *Economy of large-scale development.* Interfaces and modules have methodological advantages for code development. Large teams of programmers can negotiate the interfaces to be implemented, and then proceed separately to implement the corresponding pieces of code. Dependencies between pieces of code are minimized, and code can be locally rearranged without fear of global effects. (These benefits can be achieved also by informal interface specifications, but in practice typechecking helps enormously in verifying adherence to the specifications.)

- *Economy of language features.* Type constructions are naturally composed in orthogonal ways. For example, in Pascal an array of arrays models two-dimensional arrays; in ML, a procedure with a single argument that is a tuple of n parameters models a procedure of n arguments. Thus, type systems promote orthogonality of language features, question the utility of artificial restrictions, and thus tend to reduce the complexity of programming languages.

Expected Properties of Type Systems

In the rest of this chapter we proceed under the assumption that languages should be both safe and typed, and therefore that type systems should be employed. In the study of type systems, we do not distinguish between trapped and untrapped errors, nor between safety and good behavior: we concentrate on good behavior, and we take safety as an implied property.

Types, as normally intended in programming languages, have pragmatic characteristics that distinguish them from other kinds of program annotations. In general, annotations about the behavior of programs can range from informal comments to formal specifications subject to theorem proving. Types sit in the middle of this spectrum: they are more precise than program comments, and more easily mechanizable than formal specifications. Here are the basic properties expected of any type system:

- Type systems should be *decidably verifiable*: there should be an algorithm (called a typechecking algorithm) that can ensure that a program is well behaved. The purpose of a type system is not simply to state programmer intentions, but to actively capture execution errors before they happen. (Arbitrary formal specifications do not have these properties.)

- Type systems should be *transparent*: a programmer should be able to predict easily whether a program will typecheck. If it fails to typecheck, the reason for the failure should be self-evident. (Automatic theorem proving does not have these properties.)

- Type systems should be *enforceable*: type declarations should be statically checked as much as possible, and otherwise dynamically checked. The consistency between type declarations and their associated programs should be routinely verified. (Program comments and conventions do not have these properties.)

How Type Systems Are Formalized

As we have discussed, type systems are used to define the notion of well typing, which is itself a static approximation of good behavior (including safety). Safety facilitates debugging because of fail-stop behavior, and enables garbage collection by protecting run time structures. Well typing further facilitates program development by trapping execution errors before run time.

But how can we guarantee that well-typed programs are really well behaved? That is, how can we be sure that the type rules of a language do not accidentally allow ill-behaved programs to slip through?

Formal type systems are the mathematical characterizations of the informal type systems that are described in programming language manuals. Once a type system is formalized, we can attempt to prove a *type soundness* theorem stating that *well-typed programs are well behaved*. If such a soundness theorem holds, we say that the type system is sound. (Good behavior of all programs of a typed language and soundness of its type system mean the same thing.)

In order to formalize a type system and prove a soundness theorem we must in essence formalize the whole language in question, as we now sketch.

The first step in formalizing a programming language is to describe its syntax. For most languages of interest this reduces to describing the syntax of types and *terms*. Types express static knowledge about programs, whereas terms (statements, expressions, and other program fragments) express the algorithmic behavior.

The next step is to define the *scoping* rules of the language, which unambiguously associate occurrences of identifiers to their binding locations (the locations where the identifiers are declared). The scoping needed for typed languages is invariably *static*, in the sense that the binding locations of identifiers must be determined before run time. Binding locations can often be determined purely from the syntax of a language, without any further analysis; static scoping is then called *lexical scoping*. The lack of static scoping is called *dynamic scoping*.

Scoping can be formally specified by defining the set of *free variables* of a program fragment (which involves specifying how variables are bound by declarations). The associated notion of *substitution* of types or terms for free variables can then be defined (see Chapter 102).

When this much is settled one can proceed to define the type rules of the language, which describe a relation *has-type* of the form $M : A$ between terms M and types A. Some languages also require a relation *subtype-of* of the form $A <: B$ between types, and often a relation *equal-type* of the form $A = B$ of type equivalence. The collection of type rules of a language forms its type system. A language that has a type system is called a typed language.

The type rules cannot be formalized without first introducing another fundamental ingredient that is not reflected in the syntax of the language: *static typing environments*. These are used to record the types of free variables during the processing of program fragments; they correspond closely to the symbol table of a compiler during the typechecking phase. The type rules are always formulated with respect to a static environment for the fragment being typechecked. For example, the has-type relation $M : A$ is associated with a static typing environment Γ that contains information about the free variables of M and A. The relation is written in full as $\Gamma \vdash M : A$, meaning that M has type A in environment Γ.

The final step in formalizing a language is to define its semantics as a relation *has-value* between terms and a collection of *results*. The form of this relation depends strongly on the style of semantics that is adopted. In any case, the semantics and the type system of a language are interconnected: the types of a term and of its result should be the same (or appropriately related); this is the essence of the soundness theorem.

The fundamental notions of type system are applicable to virtually all computing paradigms (functional, imperative, concurrent, etc.). Individual type rules can often be adopted unchanged for different paradigms. For example, the basic type rules for functions are the same whether the semantics is call-by-name or call-by-value or, orthogonally, functional or imperative.

In this chapter we discuss type systems independently of semantics. It should be understood, though, that ultimately a type system must be related to a semantics, and that soundness should hold for that semantics. Suffice it to say that the techniques of structural operational semantics (see Chapters 102 and 104) deal uniformly with a large collection of programming paradigms, and fit very well with the treatment found in this chapter.

Type Equivalence

As mentioned above, most nontrivial type systems require the definition of a relation *equal type* of type equivalence. This is an important issue when defining a programming language: when are separately written type expressions equivalent? Consider, for example, two distinct type names that have been associated with similar types:

$$type\ X\ =\ Bool$$

$$type\ Y\ =\ Bool$$

If the type names X and Y match by virtue of being associated with similar types, we have *structural equivalence*. If they fail to match by virtue of being distinct type names (without looking at the associated types), we have *by-name equivalence*.

In practice, a mixture of structural and by-name equivalence is used in most compilers, but the precise mixture is rarely prescribed in the corresponding language definition. In contrast, pure structural equivalence can be easily and precisely defined by means of type rules. Moreover, structural equivalence has unique advantages when typed data has to be stored or transmitted over a network (as in Modula-3). By-name equivalence cannot deal easily with interacting program sources that have been developed and compiled separately in time or space.

We assume structural equivalence in what follows (although this issue does not arise often). If by-name equivalence is desired for a language, one should attempt to write the appropriate type rules: the arbitrary nature of by-name equivalence then becomes apparent. Moreover, satisfactory emulation of by-name equivalence can be obtained within structural equivalence, as demonstrated by the Modula-3 *branding* mechanism.

103.2 The Language of Type Systems

A type system specifies the type rules of a programming language independently of particular typechecking algorithms. This is analogous to describing the syntax of a programming language by a formal grammar, independently of particular parsing algorithms.

It is both convenient and useful to decouple type systems from typechecking algorithms: type systems belong to language definitions, while algorithms belong to compilers. It is easier to explain the typing aspects of a language by a type system, rather than by the algorithm used by a given compiler. Moreover, different compilers may use different typechecking algorithms for the same type system.

As a minor problem, it is technically possible to define type systems that admit only unfeasible typechecking algorithms, or no algorithms at all. The usual intent, however, is to allow for efficient typechecking algorithms.

Judgments

Type systems are described by a particular formalism, which we now introduce. The description of a type system starts with the description of a collection of formal utterances called judgments. A typical judgment has the form:

$$\Gamma \vdash \mathfrak{J} \quad \text{where } \mathfrak{J} \text{ is an assertion; the free variables of } \mathfrak{J} \text{ are declared in } \Gamma.$$

We say that Γ *entails* $\mathfrak{J}$. Here Γ is a static typing environment; for example, an ordered list of distinct variables and their types, of the form $\emptyset, x_1 : A_1, \ldots, x_n : A_n$. The empty environment is denoted by $\emptyset$, and the collection of variables $x_1 \cdots x_n$ declared in Γ is indicated by $dom(\Gamma)$. The form of the *assertion* $\mathfrak{J}$ varies from judgment to judgment, but all the free variables of $\mathfrak{J}$ must be declared in Γ.

The most important judgment, for our present purposes, is the *typing judgment*, which asserts that a term M has a type A with respect to a static typing environment for the free variables of M. It has the form:

$$\Gamma \vdash M : A \qquad M \text{ has type } A \text{ in } \Gamma$$

Examples.

$$\emptyset \vdash true : Bool \qquad\qquad true \text{ has type } Bool$$

$$\emptyset, x : Nat \vdash x + 1 : Nat \qquad x + 1 \text{ has type } Nat, \text{ provided that } x \text{ has type } Nat$$

Other judgment forms are often necessary; a common one asserts simply that an environment is **well formed:**

$$\Gamma \vdash \diamond \qquad \Gamma \text{ is well-formed (i.e., it has been properly constructed)}$$

Any given judgment can be regarded as *valid* (e.g., $\Gamma \vdash true : Bool$) or *invalid* (e.g., $\Gamma \vdash true : Nat$). Validity formalizes the notion of well-typed programs. The distinction between valid and invalid judgments could be expressed in a number of ways; however, a highly stylized way of presenting the set of valid judgments has emerged. This presentation style, based on type rules, facilitates stating and proving technical lemmas and theorems about type systems. Moreover, type rules are highly modular: rules for different constructs can be written separately (in contrast to a monolithic typechecking algorithm). Therefore, type rules are comparatively easy to read and understand.

Type Rules

Type rules assert the validity of certain judgments on the basis of other judgments that are already known to be valid. The process gets off the ground by some intrinsically valid judgment (usually: $\emptyset \vdash \diamond$, stating that the empty environment is well formed).

General form of a type rule:

$$\text{(Rule name) (Annotations)}$$

$$\frac{\Gamma_1 \vdash \mathfrak{J}_1 \ \ldots \ \Gamma_n \vdash \mathfrak{J}_n \ \text{(Annotations)}}{\Gamma \vdash \mathfrak{J}}$$

Each type rule is written as a number of *premise* judgments $\Gamma_i \vdash \mathfrak{J}_i$ above a horizontal line, with a single *conclusion* judgment $\Gamma \vdash \mathfrak{J}$ below the line. When all of the premises are satisfied, the conclusion must hold; the number of premises may be zero. Each rule has a name. [By convention, the first word of the name is determined by the conclusion judgment; for example, rule names of the form "(Val ...)" are for rules whose conclusion is a value typing judgment.] When needed, conditions restricting the applicability of a rule, as well as abbreviations used within the rule, are annotated next to the rule name or the premises.

For example, the first of the following two rules states that any numeral is an expression of type *Nat*, in any well-formed environment Γ. The second rule states that two expressions M and N denoting natural numbers can be combined into a larger expression $M + N$, which also denotes a natural number.

Moreover, the environment Γ for M and N, which declares the types of any free variable of M and N, carries over to $M + N$.

$$\begin{array}{cc}
(\text{Val } n)\ (n = 0, 1, \ldots) & (\text{Val } +) \\[4pt]
\dfrac{\Gamma \vdash \diamond}{\Gamma \vdash n : Nat} & \dfrac{\Gamma \vdash M : Nat \qquad \Gamma \vdash N : Nat}{\Gamma \vdash M + N : Nat}
\end{array}$$

A fundamental rule states that the empty environment is well formed, with no assumptions:

$$(\text{Env } \emptyset)$$

$$\overline{\emptyset \vdash \diamond}$$

A collection of type rules is called a (*formal*) *type system*. Technically, type systems fit into the general framework of *formal proof systems*: collections of rules used to carry out step-by-step deductions. The deductions carried out in type systems concern the typing of programs.

Type Derivations

A derivation in a given type system is a tree of judgments with leaves at the top and a root at the bottom, where each judgment is obtained from the ones immediately above it by some rule of the system. A fundamental requirement on type systems is that it must be possible to check whether or not a derivation is properly constructed.

A **valid judgment** is one that can be obtained as the root of a derivation in a given type system. That is, a valid judgment is one that can be obtained by correctly applying the type rules. For example, using the three rules given previously we can build the following derivation, which establishes that $\emptyset \vdash 1 + 2 : Nat$ is a valid judgment. The rule applied at each step is displayed to the right of each conclusion:

$$\begin{array}{llll}
\dfrac{}{\emptyset \vdash \diamond} & \text{by } (\text{Env } \emptyset) & \dfrac{}{\emptyset \vdash \diamond} & \text{by } (\text{Env } \emptyset) \\[8pt]
\emptyset \vdash 1 : Nat & \text{by } (\text{Val } n) & \emptyset \vdash 2 : Nat & \text{by } (\text{Val } n) \\[6pt]
\multicolumn{2}{c}{\emptyset \vdash 1 + 2 : Nat} & & \text{by } (\text{Val } +)
\end{array}$$

Well Typing and Type Inference

In a given type system, a term M is well-typed for an environment Γ, if there is a type A such that $\Gamma \vdash M : A$ is a valid judgment; that is, if the term M can be given some type.

The discovery of a derivation (and hence of a type) for a term is called the **type inference** problem. In the simple type system consisting of the rules (Env $\emptyset$), (Val n), and (Val $+$), a type can be *inferred* for the term $1 + 2$ in the empty environment. This type is *Nat*, by the preceding derivation.

Suppose we now add a type rule with premise $\Gamma \vdash \diamond$ and conclusion $\Gamma \vdash true : Bool$. In the resulting type system we cannot infer any type for the term $1 + true$, because there is no rule for summing a natural number with a Boolean. Because of the absence of any derivations for $1 + true$, we say that $1 + true$ is *not typeable*, or that it is ill-typed, or that it has a **typing error**.

We could further add a type rule with premises $\Gamma \vdash M : Nat$ and $\Gamma \vdash N : Bool$, and with conclusion $\Gamma \vdash M + N : Nat$ (e.g., with the intent of interpreting *true* as 1). In such a type system, a type could be inferred for the term $1 + true$, which would now be well typed.

Thus, the type inference problem for a given term is very sensitive to the type system in question. An algorithm for type inference may be very easy, very hard, or impossible to find, depending on the type system. If found, the best algorithm may be very efficient, or hopelessly slow. Although type systems are expressed and often designed in the abstract, their practical utility depends on the availability of good type inference algorithms.

The type inference problem for explicitly typed procedural languages such as Pascal is fairly easily solved; we treat it in section 103.8. The type inference problem for implicitly typed languages such as ML is much more subtle, and we do not treat it here. The basic algorithm is well understood (several descriptions of it appear in the literature) and is widely used. However, the versions of the algorithm that are used in practice are complex and are still being investigated.

The type inference problem becomes particularly hard in the presence of **polymorphism** (discussed in section 103.5). The type inference problems for the explicitly typed polymorphic features of Ada, CLU, and Standard ML are treatable in practice. However, these problems are typically solved by algorithms, without first describing the associated type systems. The purest and most general type system for polymorphism is embodied by a λ-calculus discussed in section 103.5. The type inference algorithm for this polymorphic λ-calculus is fairly easy, and we present it in section 103.8. The simplicity of the solution, however, depends on impractically verbose typing annotations. To make this general polymorphism practical, some type information has to be omitted. Such type inference problems are still an area of active research.

Type Soundness

We have now established all of the general notions concerning type systems, and we can begin examining particular type systems. Starting in section 103.3, we review some very powerful but rather theoretical type systems. The idea is that by first understanding these few systems, it becomes easier to write the type rules for the varied and complex features that one may encounter in programming languages.

When immersing ourselves in type rules, we should keep in mind that a sensible type system is more than just an arbitrary collection of rules. Well typing is meant to correspond to a semantic notion of good program behavior. It is customary to check the internal consistency of a type system by proving a type soundness theorem. This is where type systems meet semantics. In the notation of Chapter 104, for denotational semantics we expect that if $\emptyset \vdash M : A$ is valid, then $[\![M]\!] \in [\![A]\!]$ holds (the value of M belongs to the set of values denoted by the type A), and for operational semantics, we expect that if $\emptyset \vdash M : A$ and M reduces to M', then $\emptyset \vdash M' : A$. In both cases the type soundness theorem asserts that well-typed programs compute without execution errors. See Gunter [1992] and Wright and Felleisen [1994] for surveys of techniques, as well as state-of-the-art soundness proofs.

103.3 First-Order Type Systems

The type systems found in most common procedural languages are called **first order**. In type-theoretical jargon this means that they lack type parameterization and type abstraction, which are **second-order** features. First-order type systems include (rather confusingly) higher order functions. Pascal and Algol68 have rich first-order type systems, whereas FORTRAN and Algol60 have very poor ones.

A minimal first-order type system can be given for the untyped λ-calculus described in Chapter 102. There, the untyped λ-abstraction $\lambda x.M$ represents a function of parameter x and result M. Typing for this calculus requires only function types and some base types; we will see later how to add other common type structures.

The first-order typed λ-calculus is called system F_1. The main change from the untyped λ-calculus is the addition of type annotations for λ-abstractions, using the syntax $\lambda x : A.M$, where x is the function parameter, A is its type, and M is the body of the function. (In a typed programming language we would likely include the type of the result, but this is not necessary here.) The step from $\lambda x.M$ to $\lambda x : A.M$ is typical of any progression from an untyped to a typed language: bound variables acquire type annotations.

Since F_1 is based mainly on function values, the most interesting types are function types: $A \to B$ is the type of functions with arguments of type A and results of type B. To get started, though, we also need some basic types over which to build function types. We indicate by *Basic* a collection of such types, and by $K \in Basic$ any such type. At this point basic types are purely a technical necessity, but shortly we will consider interesting basic types such as *Bool* and *Nat*.

The syntax of F_1 is given in Table 103.2. It is important to comment briefly on the role of syntax in typed languages. In the case of the untyped λ-calculus, the context-free syntax describes exactly the legal programs. This is not the case in typed calculi, since good behavior is not (usually) a context-free property. The task of describing the legal programs is taken over by the type system. For example, $\lambda x : K.x(y)$ respects the syntax of F_1 given in Table 103.2, but is not a program of F_1 because it is not well typed, since K is not a function type. The context-free syntax is still needed, but only in order to define the notions of free and bound variables; that is, to define the scoping rules of the language. Based on the scoping rules, terms that differ only in their bound variables, such as $\lambda x : K.x$ and $\lambda y : K.y$, are considered *syntactically identical*. This convenient identification is implicitly assumed in the type rules (one may have to rename bound variables in order to apply certain type rules).

The definition of free variables for F_1 is the same as for the untyped λ-calculus from Chapter 102, simply ignoring the typing annotations.

We need only three simple judgments for F_1; they are shown in Table 103.3. The judgment $\Gamma \vdash A$ is in a sense redundant, since all syntactically correct types A are automatically well formed in any environment Γ. In second-order systems, however, the well formedness of types is not captured by grammar alone, and the judgment $\Gamma \vdash A$ becomes essential. It is convenient to adopt this judgment now, so that later extensions are easier.

Validity for these judgments is defined by the rules in Table 103.4. The rule (Env $\emptyset$) is the only one that does not require assumptions (i.e., it is the only *axiom*). It states that the empty environment is a valid environment. The rule (Env x) is used to extend an environment Γ to a longer environment $\Gamma, x : A$, provided that A is a valid type in Γ. Note that the assumption $\Gamma \vdash A$ implies, inductively, that Γ is valid. That is, in the process of deriving $\Gamma \vdash A$ we must have derived $\Gamma \vdash \diamond$. Another requirement of this rule is that the variable x must not be defined in Γ. We are careful to keep variables distinct in environments, so that when $\Gamma, x : A \vdash M : B$ has been derived, as in the assumption of (Val Fun), we know that x cannot occur in $dom(\Gamma)$.

The rules (Type Const) and (Type Arrow) construct types. The rule (Val x) extracts an assumption from an environment: we use the notation $\Gamma', x : A, \Gamma''$, rather informally, to indicate that $x : A$ occurs somewhere in the environment. The rule (Val Fun) gives the type $A \to B$ to a function, provided that the function body receives the type B under the assumption that the formal parameter has type A. Note how the environment changes length in this rule. The rule (Val Appl) applies a function to an argument: the same type A must appear twice when verifying the premises.

Table 103.5 shows a rather large derivation where all of the rules of F_1 are used.

TABLE 103.2 Syntax of F_1

$A, B ::=$		Types
K	$K \in Basic$	basic types
$A \to B$		function types
$M, N ::=$		Terms
x		variable
$\lambda x : A.M$		function
$M\ N$		application

TABLE 103.3 Judgments for F_1

$\Gamma \vdash \diamond$	Γ is a well-formed environment
$\Gamma \vdash A$	A is a well-formed type in Γ
$\Gamma \vdash M : A$	M is a well-formed term of type A in Γ

TABLE 103.4 Type Rules for F_1

(Env $\emptyset$)

$$\overline{\emptyset \vdash \diamond}$$

(Env x)

$$\frac{\Gamma \vdash A \qquad x \notin dom(\Gamma)}{\Gamma, x : A \vdash \diamond}$$

(Type Const)

$$\frac{\Gamma \vdash \diamond \qquad K \in Basic}{\Gamma \vdash K}$$

(Type Arrow)

$$\frac{\Gamma \vdash A \to B}{\Gamma \vdash A \to B}$$

(Val x)

$$\frac{\Gamma', x : A, \Gamma'' \vdash \diamond}{\Gamma', x : A, \Gamma'' \vdash x : A}$$

(Val Fun)

$$\frac{\Gamma, x : A \vdash M : B}{\Gamma \vdash \lambda x : A.M : A \to B}$$

(Val Appl)

$$\frac{\Gamma \vdash M : A \to B \qquad \Gamma \vdash N : A}{\Gamma \vdash M\ N : B}$$

TABLE 103.5 A Derivation in F_1

$\emptyset \vdash \diamond$	by (Env $\emptyset$)		$\emptyset \vdash \diamond$	by (Env $\emptyset$)	$\emptyset \vdash \diamond$	by (Env $\emptyset$)	$\emptyset \vdash \diamond$ by (Env $\emptyset$)
$\emptyset \vdash K$	by (Type Const)		$\emptyset \vdash K$	by (Type Const)	$\emptyset \vdash K$	by (Type Const)	$\emptyset \vdash K$ by (Type Const)

$$\emptyset \vdash \diamond \quad \text{by (Env } \emptyset\text{)}$$

Derivation:

- $\emptyset \vdash \diamond$ — by (Env $\emptyset$)
- $\emptyset \vdash K$ — by (Type Const)
- $\emptyset \vdash K \to K$ — by (Type Arrow)
- $\emptyset, y : K \to K \vdash \diamond$ — by (Env x)
- $\emptyset, y : K \to K \vdash K$ — by (Type Const)
- $\emptyset, y : K \to K, z : K \vdash \diamond$ — by (Env x)
- $\emptyset, y : K \to K, z : K \vdash y : K \to K$ — by (Val)

- $\emptyset \vdash \diamond$ — by (Env $\emptyset$)
- $\emptyset \vdash K$ — by (Type Const)
- $\emptyset \vdash K \to K$ — by (Type Arrow)
- $\emptyset, y : K \to K \vdash \diamond$ — by (Env x)
- $\emptyset, y : K \to K \vdash K$ — by (Type Const)
- $\emptyset, y : K \to K, z : K \vdash \diamond$ — by (Env x)
- $\emptyset, y : K \to K, z : K \vdash z : K$ — by (Val x)

$$\emptyset, y : K \to K, z : K \vdash y(z) : K \quad \text{by (Val Appl)}$$

$$\emptyset, y : K \to K \vdash \lambda z : K . y(z) : K \to K \quad \text{by (Val Fun)}$$

Now that we have examined the basic structure of a simple first-order type system, we can begin enriching it to bring it closer to the type structure of actual programming languages. We are going to add a set of rules for each new type construction, following a fairly regular pattern. We begin with some basic data types: the type *Unit*, whose only value is the constant *unit*; the type *Bool*, whose values are *true* and *false*; and the type *Nat*, whose values are the natural numbers.

The *Unit* type is often used as a filler for uninteresting arguments and results; it is called *Void* or *Null* in some languages. There are no operations on *Unit*, so we need only a rule stating that *Unit* is a legal type, and one stating that *unit* is a legal value of type *Unit* (Table 103.6).

We have a similar pattern of rules for *Bool*, but Booleans also have a useful operation, the conditional, that has its own typing rule (Table 103.7). In the rule (Val Cond) the two branches of the conditional must have the same type A, because either may produce the result.

The rule (Val Cond) illustrates a subtle issue about the amount of type information needed for type-checking. When encountering a conditional expression, a typechecker has to infer separately the types of N_1 and N_2, and then find a single type A that is compatible with both. In some type systems it might not be easy or possible to determine this single type from the types of N_1 and N_2. To account for this potential typechecking difficulty, we use a subscripted type to express additional type information: if_A is a hint to the typechecker that the result type should be A, and that types inferred for N_1 and N_2 should be separately compared with the given A. In general, we use subscripted types to indicate information that may be useful or necessary for typechecking, depending on the whole type system under consideration. It is often the task of a typechecker to synthesize this additional information. When it is possible to do so, subscripts may be omitted. (Most common languages do not require the annotation if_A.)

The type of natural numbers, *Nat* (Table 103.8), has 0 and *succ* (successor) as generators. Alternatively, as we did earlier, a single rule could state that all numeric constants have type *Nat*. Computations on *Nat* are made possible by the *pred* (predecessor) and *isZero* (test for zero) primitives; other sets of primitives can be chosen.

Now that we have a collection of basic types, we can begin looking at structured types, starting with *product types* (Table 103.9). A product type $A_1 \times A_2$ is the type of pairs of values with first component of type A_1 and second component of type A_2. These components can be extracted with the projections *first* and *second*, respectively. Instead of (or in addition to) the projections, one can use a *with* statement that decomposes a pair M and binds its components to two separate variables x_1 and x_2 in the scope N. The *with* notation is related to pattern matching in ML, but also to Pascal's *with*; the connection with the latter will become clearer when we consider record types.

Product types can be easily generalized to *tuple types* $A_1 \times \cdots \times A_n$, with corresponding generalized projections and generalized *with*.

TABLE 103.6 *Unit* Type

(Type Unit)	(Val Unit)
$\Gamma \vdash \diamond$	$\Gamma \vdash \diamond$
$\Gamma \vdash Unit$	$\Gamma \vdash unit : Unit$

TABLE 103.7 *Bool* Type

(Type Bool)	(Val True)	(Val False)
$\Gamma \vdash \diamond$	$\Gamma \vdash \diamond$	$\Gamma \vdash \diamond$
$\Gamma \vdash Bool$	$\Gamma \vdash true : Bool$	$\Gamma \vdash false : Bool$

(Val Cond)

$$\frac{\Gamma \vdash M : Bool \quad \Gamma \vdash N_1 : A \quad \Gamma \vdash N_2 : A}{\Gamma \vdash (if_A\ M\ then\ N_1\ else\ N_2) : A}$$

TABLE 103.8 *Nat* Type

(Type Nat)	(Val Zero)	(Val Succ)
$\Gamma \vdash \diamond$	$\Gamma \vdash \diamond$	$\Gamma \vdash M : Nat$
$\Gamma \vdash Nat$	$\Gamma \vdash 0 : Nat$	$\Gamma \vdash succ\ M : Nat$

(Val Pred)	(Val IsZero)
$\Gamma \vdash M : Nat$	$\Gamma \vdash M : Nat$
$\Gamma \vdash pred\ M : Nat$	$\Gamma \vdash isZero\ M : Bool$

Union types (Table 103.10) are often overlooked, but are just as important as product types for expressiveness. An element of a union type $A_1 + A_2$ can be thought of as an element of A_1 tagged with a *left* token (created by *inLeft*), or an element of A_2 tagged with a *right* token (created by *inRight*). The tags can be tested by *isLeft* and *isRight*, and the corresponding value extracted with *asLeft* and *asRight*. If *asLeft* is mistakenly applied to a right-tagged value, a trapped error or exception is produced; this trapped error is not considered a forbidden error. Note that it is safe

TABLE 103.9 Product Types

(Type Product)	(Val Pair)
$\dfrac{\Gamma \vdash A_1 \qquad \Gamma \vdash A_2}{\Gamma \vdash A_1 \times A_2}$	$\dfrac{\Gamma \vdash M_1 : A_1 \qquad \Gamma \vdash M_2 : A_2}{\Gamma \vdash \langle M_1, M_2 \rangle : A_1 \times A_2}$
(Val First)	(Val Second)
$\dfrac{\Gamma \vdash M : A_1 \times A_2}{\Gamma \vdash \mathit{first}\, M : A_1}$	$\dfrac{\Gamma \vdash M : A_1 \times A_2}{\Gamma \vdash \mathit{second}\, M : A_2}$

(Val With)
$$\frac{\Gamma \vdash M : A_1 \times A_2 \qquad \Gamma, x_1 : A_1, x_2 : A_2 \vdash N : B}{\Gamma \vdash (\mathit{with}\, (x_1 : A_1,\, x_2 : A_2) := M \mathit{\ do\ } N) : B}$$

to assume that any result of *asLeft* has type A_1, because either the argument is left tagged, in which case the result is indeed of type A_1, or it is right tagged, in which case there is no result. Subscripts are used to disambiguate some of the rules, as we discussed in the case of the conditional.

The rule (Val Case) describes an elegant construct that can replace *isLeft*, *isRight*, *asLeft*, *asRight*, and the related trapped errors. (It also eliminates any dependence of union operations on the *Bool* type). The *case* construct executes one of two branches depending on the tag of M, with the untagged contents of M bound to x_1 or x_2 in the scope of N_1 or N_2, respectively. A vertical bar separates the branches.

In terms of expressiveness (if not of implementation) note that the type *Bool* can be defined as *Unit + Unit*, in which case the *case* construct reduces to the conditional. The type *Int* can be defined as *Nat + Nat*, with one copy of *Nat* for the nonnegative integers and the other for the negative ones. We can define a prototypical trapped error as $\mathit{error}_A = \mathit{asRight}\,(\mathit{inLeft}_A(\mathit{unit})) : A$. Thus, we can build an error expression for each type.

Product types and union types can be iterated to produce tuple types and multiple unions. However, these derived types are rather inconvenient, and are rarely seen in languages. Instead, *labeled* products and unions are used: they go under the names of *record types* and *variant types*, respectively.

A record type is the familiar named collection of types, with a value-level operation for extracting components by name. The rules in Table 103.11 assume the syntactic identification of record types and records up to reordering of their labeled components; this is analogous to the syntactic identification of functions up to renaming of bound variables.

The *with* statement of product types is generalized to record types in (Val Record With). The components of the record M labeled $l_1, \ldots, l_n$ are bound to the variables $x_1, \ldots, x_n$ in the scope of N. Pascal has a similar construct, also called *with*, but where the binding variables are left implicit. (This has the rather unfortunate consequence of making scoping depend on typechecking, and of causing hard-to-trace bugs due to hidden variable clashes.)

Product types $A_1 \times A_2$ can be defined as *Record (first : A_1, second : A_2)*.

Variant types (Table 103.12) are named disjoint unions of types; they are syntactically identified up to reordering of components. The *is l* construct generalizes *isLeft* and *isRight*, and the *as l* construct generalizes

TABLE 103.10 Union Types

(Type Union)	(Val inLeft)	(Val inRight)
$\dfrac{\Gamma \vdash A_1 \qquad \Gamma \vdash A_2}{\Gamma \vdash A_1 + A_2}$	$\dfrac{\Gamma \vdash M_1 : A_1 \qquad \Gamma \vdash A_2}{\Gamma \vdash \mathit{inLeft}_{A_2}\, M_1 : A_1 + A_2}$	$\dfrac{\Gamma \vdash A_1 \qquad \Gamma \vdash M_2 : A_2}{\Gamma \vdash \mathit{inRight}_{A_1}\, M_2 : A_1 + A_2}$
(Val isLeft)	(Val isRight)	
$\dfrac{\Gamma \vdash M : A_1 + A_2}{\Gamma \vdash \mathit{isLeft}\, M : Bool}$	$\dfrac{\Gamma \vdash M : A_1 + A_2}{\Gamma \vdash \mathit{isRight}\, M : Bool}$	
(Val asLeft)	(Val asRight)	
$\dfrac{\Gamma \vdash M : A_1 + A_2}{\Gamma \vdash \mathit{asLeft}\, M : A_1}$	$\dfrac{\Gamma \vdash M : A_1 + A_2}{\Gamma \vdash \mathit{asRight}\, M : A_2}$	

(Val Case)
$$\frac{\Gamma \vdash M : A_1 + A_2 \qquad \Gamma, x_1 : A_1 \vdash N_1 : B \qquad \Gamma, x_2 : A_2 \vdash N_2 : B}{\Gamma \vdash (\mathit{case}_B\, M \mathit{\ of\ } x_1 : A_1 \mathit{\ then\ } N_1 \mid x_2 : A_2 \mathit{\ then\ } N_2) : B}$$

TABLE 103.11 Record Types

(Type Record) (l_i distinct)	(Val Record) (l_i distinct)

$$\frac{\Gamma \vdash A_1 \cdots \Gamma \vdash A_n}{\Gamma \vdash Record(l_1 : A_1, \ldots, l_n : A_n)} \qquad \frac{\Gamma \vdash M_1 : A_1 \cdots \Gamma \vdash M_n : A_n}{\Gamma \vdash record(l_1 = M_1, \ldots, l_n = M_n) : Record(l_1 : A_1, \ldots, l_n : A_n)}$$

(Val Record Select)

$$\frac{\Gamma \vdash M : Record(l_1 : A_1, \ldots, l_n : A_n) \qquad j \in 1..n}{\Gamma \vdash M.l_j : A_j}$$

(Val Record With)

$$\frac{\Gamma \vdash M : Record(l_1 : A_1, \ldots, l_n : A_n) \qquad \Gamma, x_1 : A_1, \ldots, x_n : A_n \vdash N : B}{\Gamma \vdash (with(l_1 = x_1 : A_1, \ldots, l_n = x_n : A_n) := M \, do \, N) : B}$$

TABLE 103.12 Variant Types

(Type Variant) (l_i distinct)	(Val Variant) (l_i distinct)

$$\frac{\Gamma \vdash A_1 \cdots \Gamma \vdash A_n}{\Gamma \vdash Variant(l_1 : A_1, \ldots, l_n : A_n)} \qquad \frac{\Gamma \vdash A_1 \cdots \Gamma \vdash A_n \quad \Gamma \vdash M_j : A_j \quad j \in 1..n}{\Gamma \vdash variant_{(l_1 : A_1, \ldots, l_n : A_n)}(l_j = M_j) : Variant(l_1 : A_1, \ldots, l_n : A_n)}$$

(Val Variant Is)

$$\frac{\Gamma \vdash M : Variant(l_1 : A_1, \ldots, l_n : A_n) \qquad j \in 1..n}{\Gamma \vdash M \, is \, l_j : Bool}$$

(Val Variant As)

$$\frac{\Gamma \vdash M : Variant(l_1 : A_1, \ldots, l_n : A_n) \qquad j \in 1..n}{\Gamma \vdash M \, as \, l_j : A_j}$$

(Val Variant Case)

$$\frac{\Gamma \vdash M : Variant(l_1 : A_1, \ldots, l_n : A_n) \quad \Gamma, x_1 : A_1 \vdash N_1 : B \cdots \Gamma, x_n : A_n \vdash N_n : B}{\Gamma \vdash (case_B \, M \, of \, l_1 = x_1 : A_1 \, then \, N_1 \mid \cdots \mid l_n = x_n : A_n \, then \, N_n) : B}$$

asLeft and *asRight*. As with unions, these constructs may be replaced by a *case* statement, which has now multiple branches.

Union types $A_1 + A_2$ can be defined as *Variant*(*left*: A_1, *right*: A_2). Enumeration types, such as {*red, green, blue*}, can be defined as *Variant*(*red*: *Unit*, *green*: *Unit*, *blue*: *Unit*).

Reference types (Table 103.13) can be used as the fundamental type of mutable locations in imperative languages. An element of *Ref*(*A*) is a mutable cell containing an element of type *A*. A new cell can be allocated by (Val Ref), updated by (Val Assign), and explicitly dereferenced by (Val Deref). Since the main purpose of an assignment is to perform a side effect, its resulting value is chosen to be *unit*. Common mutable types can be derived from *Ref*. Mutable record types, for example, can be modeled as record types containing *Ref* types.

More interestingly, arrays and array operations can be modeled as in Table 103.14, where *Array*(*A*) is the type of arrays of elements of type *A* of some length. (The code uses some arithmetic primitives and local *let* declarations.) The code in Table 103.14 is of course an inefficient implementation of arrays, but it illustrates a point: the type rules for more complex constructions can be derived from the type rule for simpler constructions. The typing rules for array operations shown in Table 103.15 can be easily derived from Table 103.14, according to the rules for products, functions, and refs.

In most programming language, types can be defined recursively. Recursive types are important, since they make all of the other type constructions more useful. They are often introduced implicitly, or without precise explanation, and their characteristics are rather subtle. Hence, their formalization deserves particular care.

The treatment of recursive types requires a rather fundamental addition to F_1: environments are extended to include type variables X. These type variables are used in recursive types of the form $\mu X . A$ (Table 103.16), which intuitively denote solutions to recursive equations of the form $X = A$ where X may occur in A. The operations *unfold* and *fold* are explicit coercions that map between a recursive

TABLE 103.13 Reference Types

(Type Ref)	(Val Ref)
$\dfrac{\Gamma \vdash A}{\Gamma \vdash Ref \, A}$	$\dfrac{\Gamma \vdash M : A}{\Gamma \vdash ref \, M : Ref \, A}$
(Val Deref)	(Val Assign)
$\dfrac{\Gamma \vdash M : Ref \, A}{\Gamma \vdash deref \, M : A}$	$\dfrac{\Gamma \vdash M : Ref \, A \quad \Gamma \vdash N : A}{\Gamma \vdash M := N : Unit}$

TABLE 103.14 An Implementation of Arrays

$Array(A)$	Array type
$\quad Nat \times (Nat \to Ref(A))$	a bound plus a map from indices less than the bound to refs
$array_A(N, M)$	Array constructor (for N refs initialized to M)
$\quad let\ cell_0 : Ref(A) = ref(M)\ and\ \dots$	
$\quad and\ cell_{N-1} : Ref(A) = ref(M)$	
$\quad in\langle N, \lambda x : Nat.if\ x = 0\ then\ cell_0\ else\ if\dots$	
$\qquad else\ if\ x = N - 1\ then\ cell_{N-1}\ else\ error_{Ref(A)}\rangle$	
$bound(M)$	Array bound
$\quad first\ M$	
$M[N]_A$	Array indexing
$\quad if\ N < first\ M$	
$\quad then\ deref((second\ M)(N))$	
$\quad else\ error_A$	
$M[N] := P$	Array update
$\quad if\ N < first\ M$	
$\quad then\ ((second\ M)(N)) := P$	
$\quad else\ error_{Unit}$	

TABLE 103.15 Array Types (Derived Rule)

(Type Array)

$$\frac{\Gamma \vdash A}{\Gamma \vdash Array(A)}$$

(Val Array)

$$\frac{\Gamma \vdash N : Nat \quad \Gamma \vdash M : A}{\Gamma \vdash array(N, M) : Array(A)}$$

(Val Array Bound)

$$\frac{\Gamma \vdash M : Array(A)}{\Gamma \vdash bound\ M : Nat}$$

(Val Array Index)

$$\frac{\Gamma \vdash N : Nat \quad \Gamma \vdash M : Array(A)}{\Gamma \vdash M[N] : A}$$

(Val Array Update)

$$\frac{\Gamma \vdash N : Nat \quad \Gamma \vdash M : Array(A) \quad \Gamma \vdash P : A}{\Gamma \vdash M[N] := P : Unit}$$

type $\mu X.A$ and its unfolding $[\mu X.A/X]A$ (where $[B/X]A$ is the substitution of B for all free occurrences of X in A), and vice versa. These coercions do not have any run time effect (in the sense that $unfold(fold(M)) = M$ and $fold(unfold(M')) = M'$). They are usually omitted from the syntax of practical programming languages, but their existence makes formal treatment easier.

A standard application of recursive types is in defining types of lists and trees, in conjunction with products and union types. The type $List_A$ of lists of elements of type A is defined in Table 103.17, together with the list constructors *nil* and *cons*, and the list analyzer *listCase*.

Recursive types can be used together with record and variant types, to define complex tree structures such as abstract syntax trees. The *case* and *with* statements can then be used to analyze these trees conveniently.

When used in conjunction with function types, recursive types are surprisingly expressive. Via clever encodings, one can show that recursion at the value level is already implicit in recursive types: there is no need to introduce recursion as a separate construct. Moreover, in the presence of recursive types, untyped programming can be carried out within typed languages. More precisely, Table 103.18 shows how to define, for any type A, a divergent element $\perp_A$ of that type, and a fixpoint operator $\mathbf{Y}_A$ for that type. Table 103.19 shows how to encode the untyped λ-calculus within typed calculi. (These encodings are for call-by-name; they take slightly different forms in call-by-value.)

Type equivalence becomes particularly interesting in the presence of recursive types. We

TABLE 103.16 Recursive Types

(Env. X)

$$\frac{\Gamma \vdash \diamond \quad X \notin dom(\Gamma)}{\Gamma, X \vdash \diamond}$$

(Type Rec)

$$\frac{\Gamma, X \vdash A}{\Gamma \vdash \mu X.A}$$

(Val Fold)

$$\frac{\Gamma \vdash M : [\mu X.A/X]A}{\Gamma \vdash fold_{\mu X.A} M : \mu X.A}$$

(Val Unfold)

$$\frac{\Gamma \vdash M : \mu X.A}{\Gamma \vdash unfold_{\mu X.A} M : [\mu X.A/X]A}$$

TABLE 103.17   List Types

$List_A$ $\mu X.Unit + (A \times X)$

$nil_A : List_A$ $fold(inLeft\, unit)$

$cons_A : A \to List_A \to List_A$ $\lambda hd : A.\lambda tl : List_A.fold(inRight\langle hd, tl\rangle)$

$listCase_{A,B} : List_A \to B \to (A \times List_A \to B) \to B$
     $\lambda l : List_A.\lambda n : B.\lambda c : A \times List_A \to B.$
        $case\,(unfold\,l)\,of\,unit :\ Unit\,then\,n \mid p : A \times List_A\,then\,c\,p$

TABLE 103.18   Encoding of Divergence and Recursion via Recursive Types

---

$\perp_A : A$ $(\lambda x : B.\,(unfold_B\, x)\, x)\,(fold_B(\lambda x : B.\,(unfold_B\, x)\, x))$

$Y_A : (A \to A) \to A$ $\lambda f : A \to A.\,(\lambda x : B.\, f\,((unfold_B\, x)\, x))\,(fold_B(\lambda x : B.\, f\,((unfold_B\, x)\, x)))$

where $B \equiv \mu X.X \to A$, for an arbitrary A

TABLE 103.19   Encoding the Untyped λ-Calculus via Recursive Types

---

V $\mu X.X \to X$ the type of untyped λ-terms

$\langle\!\langle x \rangle\!\rangle$ x translation $\langle\!\langle - \rangle\!\rangle$ from untyped λ-terms to V elements

$\langle\!\langle \lambda x.M \rangle\!\rangle$ $fold_V(\lambda x : V.\langle\!\langle M \rangle\!\rangle)$

$\langle\!\langle M N \rangle\!\rangle$ $(unfold_V \langle\!\langle M \rangle\!\rangle \langle\!\langle N \rangle\!\rangle)$

---

have sidestepped several problems here by not dealing with type definitions, by requiring explicit *fold–unfold* coercions between a recursive type and its unfolding, and by not assuming any identifications between recursive types except for renaming of bound variables. In the current formulation we do not need to define a formal judgment for type equivalence: two recursive types are equivalent simply if they are structurally identical (up to renaming of bound variables). This simplified approach can be extended to include type definitions and type equivalence up to unfolding of recursive types [Amadio and Cardelli 1993].

103.4 First-Order Type Systems for Imperative Languages

Imperative languages have a slightly different style of type systems, mostly because they distinguish commands, which do not produce values, from expressions, which do produce values. (It is quite possible to reduce commands to expressions by giving them type *Unit*, but we prefer to remain faithful to the natural distinction.)

As an example of a type system for an imperative language, we consider the untyped imperative language of Chapter 104 extended with variable declarations. This language permits us to study type rules for declarations, which we have not considered so far. The treatment of procedures and data types is very rudimentary in this language, but the rules for functions and data described in section 103.3 can be easily adapted.

The meaning of the features of the imperative language should be self-evident. If not, the reader is advised to skip section 103.4, and come back to it after reading a portion of Chapter 104.

The judgments for our imperative language are listed in Table 103.21. The judgments $\Gamma \vdash C$ and $\Gamma \vdash E : A$ correspond to the single judgment $\Gamma \vdash M : A$ of F_1, since we now have a distinction between commands C and expressions E. The judgment $\Gamma \vdash D \therefore S$ assigns a *signature* S to a *declaration* D; a signature is essentially the type of a declaration. In this simple language a signature consists of a single component, for example, $x : Nat$, and a matching declaration could be var $x : Nat = 3$. In general, signatures would consist of lists of such components, and would look very similar or identical to environments Γ.

Table 103.22 lists the type rules for the imperative language.

The rules (Env . . .), (Type . . .), and (Expr . . .) are straightforward variations on the rules we have seen for F_1. The rules (Decl . . .) handle the typing of declarations. The rules (Comm . . .) handle commands; notice how (Comm Block) converts a signature to a piece of an environment when checking the body of a block.

TABLE 103.20 Syntax of the Imperative Language

$A ::=$		Types
	Bool	Boolean type
	Nat	natural numbers type
	Proc	procedure type (no arguments, no result)
$D ::=$		Declarations
	proc $I = C$	procedure declaration
	var $I : A = E$	variable declaration
$C ::=$		Commands
	$I := E$	assignment
	$C_1; C_2$	sequential composition
	begin D in C end	block
	call I	procedure call
	while E do C end	while loop
$E ::=$		Expressions
	I	identifier
	N	numeral
	$E_1 + E_2$	sum of two numbers
	E_1 not= E_2	inequality of two numbers

TABLE 103.21 Judgments for the Imperative Language

$\Gamma \vdash \diamond$	Γ is a well-formed environment
$\Gamma \vdash A$	A is a well-formed type in Γ
$\Gamma \vdash C$	C is a well-formed command in Γ
$\Gamma \vdash E : A$	E is a well-formed expression of type A in Γ
$\Gamma \vdash D \therefore S$	D is a well-formed declaration of signature S in Γ

TABLE 103.22 Type Rules for Imperative Language

(Env $\emptyset$)   (Env I)

$$\dfrac{}{\emptyset \vdash \diamond} \qquad \dfrac{\Gamma \vdash A \quad I \notin dom(\Gamma)}{\Gamma, I : A \vdash \diamond}$$

(Type Bool)   (Type Nat)   (Type Proc)

$$\dfrac{\Gamma \vdash \diamond}{\Gamma \vdash Bool} \qquad \dfrac{\Gamma \vdash \diamond}{\Gamma \vdash Nat} \qquad \dfrac{\Gamma \vdash \diamond}{\Gamma \vdash Proc}$$

(Decl Proc)   (Decl Var)

$$\dfrac{\Gamma \vdash C}{\Gamma \vdash (\text{proc } I = C) \therefore (I : Proc)} \qquad \dfrac{\Gamma \vdash E : A \quad A \in \{Bool, Nat\}}{\Gamma \vdash (\text{var } I : A = E) \therefore (I : A)}$$

(Comm Assign)   (Comm Sequence)

$$\dfrac{\Gamma \vdash I : A \quad \Gamma \vdash E : A}{\Gamma \vdash I := E} \qquad \dfrac{\Gamma \vdash C_1 \quad \Gamma \vdash C_2}{\Gamma \vdash C_1; C_2}$$

(Comm Block)   (Comm Call)   (Comm While)

$$\dfrac{\Gamma \vdash D \therefore (I : A) \quad \Gamma, I : A \vdash C}{\Gamma \vdash \text{begin } D \text{ in } C \text{ end}} \qquad \dfrac{\Gamma \vdash I : Proc}{\Gamma \vdash \text{call } I} \qquad \dfrac{\Gamma \vdash E : Bool \quad \Gamma \vdash C}{\Gamma \vdash \text{while } E \text{ do } C \text{ end}}$$

(Expr Identifier)   (Expr Numeral)

$$\dfrac{\Gamma_1, I : A, \Gamma_2 \vdash \diamond}{\Gamma_1, I : A, \Gamma_2 \vdash I : A} \qquad \dfrac{\Gamma \vdash \diamond}{\Gamma \vdash N : Nat}$$

(Expr Plus)   (Expr NotEq)

$$\dfrac{\Gamma \vdash E_1 : Nat \quad \Gamma \vdash E_2 : Nat}{\Gamma \vdash E_1 + E_2 : Nat} \qquad \dfrac{\Gamma \vdash E_1 : Nat \quad \Gamma \vdash E_2 : Nat}{\Gamma \vdash E_1 \text{ not= } E_2 : Bool}$$

103.5 Second-Order Type Systems

Many modern languages include constructs for type parameters, type abstraction, or both. Type parameters can be found in the module system of several languages, where a generic module or interface is parameterized by a type to be supplied later. Polymorphic languages such as ML use type parameters more pervasively, at the function level. Type abstraction can similarly be found in conjunction with modules, where it appears as opaque types in interfaces. Languages such as CLU use type abstraction at the data level, to obtain abstract data types. These advanced features can be modeled by so-called second-order type systems.

TABLE 103.23 Syntax of F_2

$A, B ::=$	Types
X	type variable
$A \to B$	function type
$\forall X.A$	universally quantified type
$M, N ::=$	Terms
x	variable
$\lambda x : A.M$	function
$M\,N$	application
$\lambda X.M$	polymorphic abstraction
$M\,A$	type instantiation

Second-order type systems extend first-order type systems with the notion of *type parameters*. A new kind of term, written $\lambda X.M$, indicates a program M that is parameterized with respect to a type variable X that stands for an arbitrary type. For example, the

TABLE 103.24 Judgments for F_2

$\Gamma \vdash \diamond$	Γ is a well-formed environment
$\Gamma \vdash A$	A is a well-formed type in Γ
$\Gamma \vdash M : A$	M is a well-formed term of type A in Γ

identity function for a fixed type A, written $\lambda x : A.x$, can be turned into a parametric identity function by abstracting over A and writing $id \quad \lambda X.\lambda x : X.x$. One can then instantiate such a parametric function to any given type A by a *type instantiation id A*, which produces back $\lambda x : A.x$.

Corresponding to the new terms $\lambda X.M$ we need new *universally quantified* types. The type of a term such as $\lambda X.M$ is written $\forall X.A$, meaning that *forall X*, the body M has type A (here M and A may contain occurrences of X). For example, the type of the parametric identity is $id : \forall X.X \to X$.

The pure second-order system F_2 (Table 103.23) is based exclusively on type variables, function types, and quantified types. Note that we are dropping the basic types K, since we can now use type variables as the basic case. It turns out that virtually any basic type of interest can be encoded within F_2 [Böhm and Berarducci 1985]. Similarly, product types, sum types, existential types, and some recursive types, can be encoded within F_2: polymorphism has an amazing expressive power. Thus, there is little need, technically, to deal with these type constructions directly.

Free variables for F_2 types and terms can be defined in the usual fashion; suffice it to say that $\forall X.A$ binds X in A and $\lambda X.M$ binds X in M. An interesting aspect of F_2 is the substitution of a type for a type variable that is carried out in the type rule for type instantiation, (Val Appl2).

The judgments for F_2 (Table 103.24) are the same ones as for F_1, but the environments are richer. With respect to F_1, the new rules (Table 103.25), are: (Env X), which adds a type variable to the environment;

TABLE 103.25 Type Rules for F_2

(Env $\emptyset$)
$$\frac{}{\emptyset \vdash \diamond}$$

(Env x)
$$\frac{\Gamma \vdash A \qquad x \notin dom(\Gamma)}{\Gamma, x : A \vdash \diamond}$$

(Env X)
$$\frac{\Gamma \vdash \diamond \qquad X \notin dom(\Gamma)}{\Gamma, X \vdash \diamond}$$

(Type X)
$$\frac{\Gamma', X, \Gamma'' \vdash \diamond}{\Gamma', X, \Gamma'' \vdash X}$$

(Type Arrow)
$$\frac{\Gamma \vdash A \quad \Gamma \vdash B}{\Gamma \vdash A \to B}$$

(Type Forall)
$$\frac{\Gamma, X \vdash A}{\Gamma \vdash \forall X.A}$$

(Val x)
$$\frac{\Gamma', x : A, \Gamma'' \vdash \diamond}{\Gamma', x : A, \Gamma'' \vdash x : A}$$

(Val Fun)
$$\frac{\Gamma, x : A \vdash M : B}{\Gamma \vdash \lambda x : A.M : A \to B}$$

(Val Appl)
$$\frac{\Gamma \vdash M : A \to B \quad \Gamma \vdash N : A}{\Gamma \vdash M\,N : B}$$

(Val Fun2)
$$\frac{\Gamma, X \vdash M : A}{\Gamma \vdash \lambda X.M : \forall X.A}$$

(Val Appl2)
$$\frac{\Gamma \vdash M : \forall X.A \quad \Gamma \vdash B}{\Gamma \vdash M\,B : [B/X]A}$$

(Type Forall), which constructs a quantified type $\forall X.A$ from a type variable X and a type A where X may occur; (Val Fun2), which builds a polymorphic abstraction; and (Val Appl2), which instantiates a polymorphic abstraction to a given type, where $[B/X]A$ is the substitution of B for all the free occurrences of X in A. For example, if id has type $\forall X.X \to X$ and A is a type, then by (Val Appl2) we have that $id\,A$ has type $[A/X](X \to X) \equiv A \to$

TABLE 103.26 Existential Types

(Type Exists)	(Val Pack)
$\Gamma, X \vdash A$	$\Gamma \vdash [B/X]M : [B/X]A$
$\overline{\Gamma \vdash \exists X.A}$	$\overline{\Gamma \vdash (pack_{\exists X.A}\, X = B\; with\, M) : \exists X.A}$
(Val Open)	
$\Gamma \vdash M : \exists X.A \quad \Gamma, X, x : A \vdash N : B \quad \Gamma \vdash B$	
$\overline{\Gamma \vdash (open_B\, M\; as\; X, x : A\; in\; N) : B}$	

A. As a simple but instructive exercise, the reader may want to build the derivation for $id(\forall X.X \to X)(id)$.

As extensions of F_2 we could adopt all the first-order constructions that we already discussed for F_1. A more interesting extension to consider is *existentially quantified* types, also known as type abstractions: see Table 103.26.

To illustrate the use of existentials, we consider an **abstract type** for Booleans. As we said earlier, Booleans can be represented as the type $Unit + Unit$. We can now show how to hide this representation detail from a client who does not care how Booleans are implemented, but who wants to make use of *true*, *false* and *cond* (conditional). We first define an interface for such a client to use,

      BoolInterface $\exists Bool.\; Record(true: Bool,\, false: Bool,\, cond: \forall Y.\; Bool \to Y \to Y \to Y)$

This interface declares that there exists a type *Bool* (without revealing its identity) that supports the operations *true*, *false* and *cond* of appropriate types. The conditional is parameterized with respect to its result type Y, which may vary depending of the context of usage.

Next we define a particular implementation of this interface; one that represents *Bool* as $Unit + Unit$, and that implements the conditional via a case statement. The Boolean representation type and the related Boolean operations are packaged together by the *pack* construct.

        boolModule : *BoolInterface*

        $pack_{BoolInterface}\; Bool = Unit + Unit$

        with record(

                $true = inLeft(unit),$

                $false = inRight(unit),$

                $cond = \lambda Y.\lambda x : Bool.\; \lambda y_1 : Y.\; \lambda y_2 : Y.$

                    $case_Y\; x\; of\, x_1 : Unit\; then\; y_1 \mid x_2 : Unit\; then\; y_2)$

Finally, a client could make use of this module by opening it, and thus getting access to an abstract name *Bool* for the Boolean type, and a name *boolOp* for the record of Boolean operations. These names are used in the next example for a simple computation that returns a natural number. (The computation following *in* is, essentially, *if boolOp.true then* 1 *else* 0.)

      $open_{Nat}\; boolModule\; as\; Bool,\, boolOp : Record(true : Bool,\, false : Bool,\, cond : \forall Y.Bool \to Y \to Y \to Y)$

        $in\; boolOp.cond(Nat)(boolOp.true)(1)(0)$

The reader should verify that these examples typecheck according to the rules previously given. Note the critical third assumption of (Val Open) that forbids writing, for example, *boolOp.true* as the body of *open* in the preceding example. Because of that assumption, the abstract name of the representation type (*Bool*) cannot escape the scope of *open*, and therefore values having the representation type cannot escape either. A restriction of this kind is necessary, otherwise the representation type might become known to clients.

103.6 Subtyping

Typed object-oriented languages have particularly interesting and complex type systems. There is little consensus about what characterizes these languages, but at least one feature is almost universally present: **subtyping**. Subtyping captures the intuitive notion of inclusion between types, where types are seen as collections of values. An element of a type can be considered also as an element of any of its supertypes, thus allowing a value (object) to be used flexibly in many different typed contexts.

When considering a subtyping relation, such as the one found in object-oriented programming languages, it is customary to add a new judgment $\Gamma \vdash A <: B$ stating that A is a subtype of B. The intuition is that any element of A is an element of B or, more appropriately, any program of type A is also a program of type B.

One of the simplest type systems with subtyping is an extension of F_1 called $F_{1<:}$. The syntax of F_1 is unchanged, except for the addition of a type *Top* that is a supertype of all types. The existing type rules are also unchanged. The subtyping judgment is independently axiomatized, and a single type rule, called **subsumption**, is added to connect the typing judgment to the subtyping judgment.

The subsumption rule states that if a term has type A, and A is a subtype of B, then the term also has type B. That is, subtyping behaves very much like set inclusion, when type membership is seen as set membership.

The subtyping relation in Table 103.28 is defined as a reflexive and transitive relation with a maximal element called *Top*, which is therefore interpreted as the type of all well-typed terms.

The subtype relation for function types says that $A \rightarrow B$ is a subtype of $A' \rightarrow B'$ if A' is a subtype of A, and B is a subtype of B'. Note that the inclusion is inverted (**contravariant**) for function arguments, while it goes in the same direction (**covariant**) for function results. Simple-minded reasoning reveals that this is the only sensible rule. A function M of type $A \rightarrow B$ accepts elements of type A; obviously it also accepts elements of any subtype A' of A. The same function M returns elements of type B; obviously it returns elements that belong to any supertype B' of B. Therefore, any function M of type $A \rightarrow B$, by virtue of accepting arguments of type A' and returning results of type B', has also type $A' \rightarrow B'$. The latter is compatible with saying that $A \rightarrow B$ is a subtype of $A' \rightarrow B'$.

In general, we say that a type variable occurs contravariantly within another type of F_1, if it always occurs on the left of an odd number of arrows (double contravariance equals covariance). For example, $X \rightarrow Unit$ and $(Unit \rightarrow X) \rightarrow Unit$ are contravariant in X, whereas $Unit \rightarrow X$ and $(X \rightarrow Unit) \rightarrow X$ are covariant in X.

Ad hoc subtyping rules can be added on basic types, such as $Nat <: Int$ [Mitchell 1984].

All of the structured types we considered as extensions of F_1 admit simple subtyping rules; therefore, these structured types can be added to $F_{1<:}$ as well (Table 103.29). Typically, we need to add a single subtyping rule for each type constructor, taking care that the subtyping rule is sound in conjunction with subsumption. The subtyping rules for products and unions work componentwise. The subtyping rules for records and variants operate also lengthwise: a longer record type is a subtype of a shorter record type (additional fields can be forgotten by subtyping), whereas a shorter variant type is a subtype of a longer

TABLE 103.27 Judgments for Type Systems with Subtyping

$\Gamma \vdash \diamond$	Γ is a well-formed environment
$\Gamma \vdash A$	A is a well-formed type in Γ
$\Gamma \vdash A <: B$	A is a subtype of B in Γ
$\Gamma \vdash M : A$	M is a well-formed term of type A in Γ

TABLE 103.28 Additional Rules for $F_{1<:}$

(Sub Refl)
$$\frac{\Gamma \vdash A}{\Gamma \vdash A <: A}$$

(Sub Trans)
$$\frac{\Gamma \vdash A <: B \quad \Gamma \vdash B <: C}{\Gamma \vdash A <: C}$$

(Val Subsumption)
$$\frac{\Gamma \vdash a : A \quad \Gamma \vdash A <: B}{\Gamma \vdash a : B}$$

(Type Top)
$$\frac{\Gamma \vdash \diamond}{\Gamma \vdash Top}$$

(Sub Top)
$$\frac{\Gamma \vdash A}{\Gamma \vdash A <: Top}$$

(Sub Arrow)
$$\frac{\Gamma \vdash A' <: A \quad \Gamma \vdash B <: B'}{\Gamma \vdash A \rightarrow B <: A' \rightarrow B'}$$

TABLE 103.29   Additional Rules for Extensions of $F_{1<:}$

(Sub Product)	(Sub Union)
$\dfrac{\Gamma \vdash A_1 <: B_1 \quad \Gamma \vdash A_2 <: B_2}{\Gamma \vdash A_1 \times A_2 <: B_1 \times B_2}$	$\dfrac{\Gamma \vdash A_1 <: B_1 \quad \Gamma \vdash A_2 <: B_2}{\Gamma \vdash A_1 + A_2 <: B_1 + B_2}$

(Sub Record) (l_i distinct)

$$\frac{\Gamma \vdash A_1 <: B_1 \cdots \Gamma \vdash A_n <: B_n \quad \Gamma \vdash A_{n+1} \cdots \Gamma \vdash A_{n+m}}{\Gamma \vdash Record(l_1 : A_1, \ldots, l_{n+m} : A_{n+m}) <: Record(l_1 : B_1, \ldots, l_n : B_n)}$$

(Sub Variant) (l_i distinct)

$$\frac{\Gamma \vdash A_1 <: B_1 \cdots \Gamma \vdash A_n <: B_n \quad \Gamma \vdash B_{n+1} \cdots \Gamma \vdash B_{n+m}}{\Gamma \vdash Variant(l_1 : A_1, \ldots, l_n : A_n) <: Variant(l_1 : B_1, \ldots, l_{n+m} : B_{n+m})}$$

TABLE 103.30   Environments with Bounded Variables

(Env $X <:$)	(Type $X <:$)	(Sub $X <:$)
$\dfrac{\Gamma \vdash A \quad X \notin dom(\Gamma)}{\Gamma, X <: A \vdash \diamond}$	$\dfrac{\Gamma', X <: A, \Gamma'' \vdash \diamond}{\Gamma', X <: A, \Gamma'' \vdash X}$	$\dfrac{\Gamma', X <: A, \Gamma'' \vdash \diamond}{\Gamma', X <: A, \Gamma'' \vdash X <: A}$

TABLE 103.31   Subtyping Recursive Types

(Type Rec)	(Sub Rec)
$\dfrac{\Gamma, X <: Top \vdash A}{\Gamma \vdash \mu X.A}$	$\dfrac{\Gamma \vdash \mu X.A \quad \Gamma \vdash \mu Y.B \quad \Gamma, Y <: Top, X <: Y \vdash A <: B}{\Gamma \vdash \mu X.A <: \mu Y.B}$

variant type (additional cases can be introduced by subtyping). For example,

$$WorkingAge \quad Variant(student: Unit, adult: Unit)$$

$$Age \quad Variant(child: Unit, student: Unit, adult: Unit, senior: Unit)$$

$$Worker \quad Record(name: String, age: WorkingAge, profession: String)$$

$$Person \quad Record(name: String, age: Age)$$

Then,

$$WorkingAge <: Age$$

$$Worker <: Person$$

Reference types do not have any subtyping rule: $Ref(A) <: Ref(B)$ holds only if $A = B$ (in which case $Ref(A) <: Ref(B)$ follows from reflexivity). This strict rule is necessary because references can be both read and written, and hence behave both covariantly and contravariantly. For the same reason, array types have no additional subtyping rules.

As was the case for F_1, a change to the structure of environments is necessary when considering recursive types. This time, we must add *bounded variables* to environments (Table 103.30). Variables bound by *Top* correspond to our old unconstrained variables. The soundness of the subtyping rule (Sub Rec) for recursive types (Table 103.31) is not obvious, but the intuition is fairly straightforward. To check whether $\mu X.A <: \mu Y.B$ we assume $X <: Y$ and we check $A <: B$; the assumption helps us

TABLE 103.32   Syntax of $F_{2<:}$

$A, B ::=$	Types
X	type variable
Top	the biggest type
$A \rightarrow B$	function type
$\forall X <: A.B$	bounded universally quantified type
$M, N ::=$	Terms
x	variable
$\lambda x : A.M$	function
$M N$	application
$\lambda X <: A.M$	bounded polymorphic abstraction
$M A$	type instantiation

TABLE 103.33   Rules for Bounded Universal Quantifiers

(Type Forall<:)	(Sub Forall<:)

$$\frac{\Gamma, X <: A \vdash B}{\Gamma \vdash \forall X <: A.B} \qquad \frac{\Gamma \vdash A' <: A \quad \Gamma, X <: A' \vdash B <: B'}{\Gamma \vdash (\forall X <: A.B) <: (\forall X <: A'.B')}$$

(Val Fun2<:)	(Val Appl2<:)

$$\frac{\Gamma, X <: A \vdash M : B}{\Gamma \vdash \lambda X <: A.M : \forall X <: A.B} \qquad \frac{\Gamma \vdash M : \forall X <: A.B \quad \Gamma \vdash A' <: A}{\Gamma \vdash M\,A' : [A'/X]B}$$

TABLE 103.34   Rules for Bounded Existential Quantifiers (Derivable)

(Type Exists<:)	(Sub Exists<:)

$$\frac{\Gamma, X <: A \vdash B}{\Gamma \vdash \exists X <: A.B} \qquad \frac{\Gamma \vdash A <: A' \quad \Gamma, X <: A \vdash B <: B'}{\Gamma \vdash (\exists X <: A.B) <: (\exists X <: A'.B')}$$

(Val Pack<:)	(Val Open<:)

$$\frac{\Gamma \vdash C <: A \quad \Gamma \vdash [C/X]M : [C/X]B}{\Gamma \vdash (pack_{\exists X <: A.B}\, X <: A = C\ with\ M) : \exists X <: A.B} \qquad \frac{\Gamma \vdash M : \exists X <: A.B \quad \Gamma \vdash D \quad \Gamma, X <: A, x : B \vdash N : D}{\Gamma \vdash (open_D\, M\ as\ X <: A, x : B\ in\ N) : D}$$

when finding matching occurrences of X and Y in A and B, as long as they are in covariant contexts. A simpler rule asserts that $\mu X.A <: \mu X.B$ whenever $A <: B$ for any X, but this rule is unsound when X occurs in contravariant contexts (e.g., immediately on the left of an arrow).

The bounded variables in environments are also the basis for the extension of F_2 with subtyping, which gives a system called $F_{2<:}$. In this system the term $\lambda X <: A.M$ indicates a program M parameterized with respect to a type variable X that stands for an arbitrary subtype of A. This is a generalization of F_2, since the F_2 term $\lambda X.M$ can be represented as $\lambda X <: Top.M$. Corresponding to the terms $\lambda X <: A.M$, we have bounded type quantifiers of the form $\forall X <: A.B$.

Scoping for $F_{2<:}$ types and terms is defined similarly to F_2, except that $\forall X <: A.B$ binds X in B but not in A, and $\lambda X <: A.M$ binds X in M but not in A.

The type rules for $F_{2<:}$ consist of most of the type rules for $F_{1<:}$ (namely, (Env $\emptyset$), (Env x), (Type Top), (Type Arrow), (Sub Refl), (Sub Trans), (Sub Top), (Sub Arrow), (Val Subsumption), (Val x), (Val Fun), and (Val Appl)), plus the rules for bounded variables (namely, (Env X<:), (Type X<:), and (Sub X<:)), and the ones listed in Table 103.33 for bounded polymorphism.

As for F_2, we do not need to add other type constructions to $F_{2<:}$, since all of the common ones can be expressed within it (except for recursion). Moreover, it turns out that the encodings used for F_2 satisfy the expected subtyping rules. For example, it is possible to encode bounded existential types so that the rules described in Table 103.34 are satisfied. The type $\exists X <: A.B$ represents a *partially abstract type*, whose representation type X is not completely known, but is known to be a subtype of A. This kind of partial abstraction occurs in some languages based on subtyping (e.g., in Modula-3).

Some nontrivial work is needed to obtain encodings of record and variant types in $F_{2<:}$ that satisfy the expected subtyping rules, but even those can be found [Cardelli and Wegner 1985].

103.7 Equivalence

For simplicity, we have avoided describing certain judgments that are necessary when type systems become complex and when one wishes to capture the semantics of programs in addition to their typing. We briefly discuss some of these judgments.

A *type equivalence* judgment, of the form $\Gamma \vdash A = B$, can be used when type equivalence is non-trivial and requires precise description. For example, some type systems identify a recursive type and its unfolding, in which case we would have $\Gamma \vdash \mu X.A = [\mu X.A/X]A$ whenever $\Gamma \vdash \mu X.A$. As another example, type systems with type operators $\lambda X.A$ (functions from types to types) have a reduction rule for operator application of the form $\Gamma \vdash (\lambda X.A) B = [A/X] B$. The type equivalence judgment is usually employed in a *retyping rule* stating that if $\Gamma \vdash M : A$ and $\Gamma \vdash A = B$ then $\Gamma \vdash M : B$.

A *term equivalence* judgment determines which programs are equivalent with respect to a common type. It has the form $\Gamma \vdash M = N : A$. For example, with appropriate rules we could determine that $\Gamma \vdash 2 + 1 = 3 : Int$. The term equivalence judgment can be used to give a typed semantics to programs: if N is an irreducible expression, then we can consider N as the resulting value of the program M.

103.8 Type Inference

Type inference is the problem of finding a type for a term within a given type system, if any type exists. In the type systems we have considered earlier, programs have abundant type annotations. Thus, the type inference problem often amounts to little more than checking the mutual consistency of the annotations. The problem is not always trivial but, as in the case of F_1, simple typechecking algorithms may exist.

A harder problem, called *typability* or **type reconstruction**, consists in starting with an untyped program M, and finding an environment Γ, a type-annotated version M' of M, and a type A such that A is a type for M' with respect to Γ. (A type-annotated program M' is simply one that stripped of all type annotations reduces back to M.) The type reconstruction problem for the untyped λ-calculus is solvable within F_1 by the Hindley–Milner algorithm used in ML [Milner 1978]; in addition, that algorithm has the property of producing a unique representation of all possible F_1 typings of a λ-term. The type reconstruction problem for the untyped λ-calculus, however, is not solvable within F_2 [Wells 1994]. Type reconstruction within systems with subtyping is still largely an open problem, although special solutions are beginning to emerge [Aiken and Wimmers 1993, Eifrig et al. 1995, Gunter and Mitchell 1994, Palsberg 1994].

We concentrate here on the type inference algorithms for some representative systems: F_1, F_2, and $F_{2<:}$. The first two systems have the unique type property: if a term has a type it has only one type. In $F_{2<:}$ there are no unique types, simply because the subsumption rule assigns all of the supertypes of a type to any term that has that type. However, a minimum type property holds: if a term has a collection of types, that collection has a least element in the subtype order [Curien and Ghelli 1992]. The minimum type property holds for many common extensions of $F_{2<:}$ and of $F_{1<:}$ but may fail in the presence of ad-hoc subtypings on basic types.

The Type Inference Problem

In a given type system, given an environment Γ and a term M is there a type A such that $\Gamma \vdash M : A$ is valid? The following are examples:

- In F_1, given $M \equiv \lambda x : K.x$ and any well-formed Γ we have that $\Gamma \vdash M : K \rightarrow K$.
- In F_1, given $M \equiv \lambda x : K.y(x)$ and $\Gamma \equiv \Gamma', y : K \rightarrow K$ we have that $\Gamma \vdash M : K \rightarrow K$.
- In F_1, there is no typing for $\lambda x : B.x(x)$, for any type B.
- However, in $F_{1<:}$ there is the typing $\Gamma \vdash \lambda x : Top \rightarrow B.x(x) : (Top \rightarrow B) \rightarrow B$, for any type B, since x can also be given type Top.
- Moreover, in F_1 with recursive types, there is the typing $\Gamma \vdash \lambda x : B.(unfold_B x)(x) : B \rightarrow B$, for $B \equiv \mu X.X \rightarrow X$, since $unfold_B x$ has type $B \rightarrow B$.
- Finally, in F_2 there is the typing $\Gamma \vdash \lambda x : B.x(B)(x) : B \rightarrow B$, for $B \equiv \forall X.X \rightarrow X$, since $x(B)$ has type $B \rightarrow B$.

(An alternative formulation of the type inference problem requires Γ to be found, instead of given. However, in programming practice one is interested only in type inference for programs embedded in a complete programming context, where Γ is therefore given.)

We begin with the type inference algorithm for pure F_1, given in Table 103.35. The algorithm can be extended in straightforward ways to all of the first-order type structures studied earlier. This is the basis of the typechecking algorithms used in Pascal and all similar procedural languages.

The main routine $Type(\Gamma, M)$, takes an environment Γ and a term M and produces the unique type of M, if any. The instruction *fail* causes a global failure of the algorithm: it indicates a typing error. In this algorithm, as in the ones that follow, we assume that the initial environment parameter Γ is well formed so as to rule out the possibility of feeding invalid environments to internal calls. (For example, we may start with the empty environment when checking a full program.) In any case, it is easy to write a subroutine that checks the well formedness of an environment, from the code we provide. The case for $\lambda x : A.M$ should have a restriction requiring that $x \notin dom(\Gamma)$, since x is used to extend Γ. However, this restriction can be easily sidestepped by renaming, e.g., by making all binders unique before running the algorithm. We omit this kind of restriction from Tables 103.35–103.37.

As an example, let us consider the type inference problem for term $\lambda z : K.y(z)$ in the environment $\emptyset, y : K \to K$, for which we gave a full F_1 derivation in section 103.3. The algorithm proceeds as follows:

$$Type((\emptyset, y : K \to K), \lambda z : K.y(z))$$

$$= K \to Type((\emptyset, y : K \to K, z : K), y(z))$$

$$= K \to (if\ Type((\emptyset, y : K \to K, z : K), y) \equiv Type((\emptyset, y : K \to K, z : K), z)$$

$$\to B\ for\ some\ B\ then\ B\ else\ fail)$$

$$= K \to (if\ K \to K \equiv K \to B\ for\ some\ B\ then\ B\ else\ fail) \qquad (taking\ B \equiv K)$$

$$= K \to K$$

The type inference algorithm for F_2 (Table 103.36) is not much harder than the one for F_1, but it requires a subroutine $Good(\Gamma, A)$ to verify that the types encountered in the source program are well formed. This check is necessary because types in F_2 contain type variables that might be unbound. A substitution subroutine must also be used in the type instantiation case, $M\,A$.

The type inference algorithm for $F_{2<:}$, given in Table 103.37, is more subtle. The subroutine *Subtype* (Γ, A, B) attempts to decide whether A is a subtype of B in Γ, and is at first sight straightforward. It has been shown, though, that *Subtype* is only a semialgorithm: it may diverge on certain pairs A, B that are

---

TABLE 103.35 Type Inference Algorithm for F_1

$Type(\Gamma, x)$
  $if x : A \in \Gamma for\ some\ A\ then\ A\ else\ fail$

$Type(\Gamma, \lambda x : A.M)$
  $A \to Type((\Gamma, x : A), M)$

$Type(\Gamma, M\ N)$
  $if\ Type(\Gamma, M) \equiv Type(\Gamma, N) \to B\ for\ some\ B\ then\ B\ else\ fail$

TABLE 103.36 Type Inference Algorithm for F_2

$Good(\Gamma, X) \qquad X \in dom(\Gamma)$

$Good(\Gamma, A \to B) \qquad Good(\Gamma, A)\ and\ Good(\Gamma, B)$

$Good(\Gamma, \forall X.A) \qquad Good((\Gamma, X), A)$

$Type(\Gamma, x)$
  $if x : A \in \Gamma\ for\ some\ A\ then\ A\ else\ fail$

$Type(\Gamma, \lambda x : A.M)$
  $if\ Good(\Gamma, A)\ then\ A \to Type((\Gamma, x : A), M)\ else\ fail$

$Type(\Gamma, M\ N)$
  $if\ Type(\Gamma, M) \equiv Type(\Gamma, N) \to B\ for\ some\ B\ then\ B\ else\ fail$

$Type(\Gamma, \lambda X.M)$
  $\forall X.Type((\Gamma, X), M)$

$Type(\Gamma, M\ A)$
  $if\ Type(\Gamma, M) \equiv \forall X.B\ for\ some\ X, B\ and\ Good(\Gamma, A)\ then\ [A/X]B\ else\ fail$

TABLE 103.37   Type Inference Algorithm for $F_{2<:}$

$Good(\Gamma, X)$ $X \in dom(\Gamma)$	
$Good(\Gamma, Top)$ $true$	
$Good(\Gamma, A \rightarrow B)$ $Good(\Gamma, A)$ and $Good(\Gamma, B)$	
$Good(\Gamma, \forall X <: A.B)$ $Good(\Gamma, A)$ and $Good((\Gamma, X <: A), B)$	

$Subtype(\Gamma, A, Top)$ $true$

$Subtype(\Gamma, X, X)$ $true$

$Subtype(\Gamma, X, A)$ for $A \neq X, Top$
  if $X <: B \in \Gamma$ for some B then $Subtype(\Gamma, B, A)$ else false

$Subtype(\Gamma, A \rightarrow B, A' \rightarrow B')$
  $Subtype(\Gamma, A', A)$ and $Subtype(\Gamma, B, B')$

$Subtype(\Gamma, \forall X <: A.B, \forall X' <: A'.B')$
  $Subtype(\Gamma, A', A)$ and $Subtype((\Gamma, X' <: A'), [X'/X]B, B')$

$Subtype(\Gamma, A, B)$ false   otherwise

$Expose(\Gamma, X)$ if $X <: A \in \Gamma$ for some A then $Expose(\Gamma, A)$ else fail

$Expose(\Gamma, A)$ A otherwise

$Type(\Gamma, x)$
  if $x : A \in \Gamma$ for some A then A else fail

$Type(\Gamma, \lambda x : A.M)$
  if $Good(\Gamma, A)$ then $A \rightarrow Type((\Gamma, x : A), M)$ else fail

$Type(\Gamma, M N)$
  if $Expose(\Gamma, Type(\Gamma, M)) \equiv A \rightarrow B$ for some A, B
   and $Subtype(\Gamma, Type(\Gamma, N), A)$ then B else fail

$Type(\Gamma, \lambda X <: A.M)$
  if $Good(\Gamma, A)$ then $\forall X <: A.Type((\Gamma, X <: A), M)$ else fail

$Type(\Gamma, M A)$
  if $Expose(\Gamma, Type(\Gamma, M)) \equiv \forall X <: A'.B$ for some X, A', B
   and $Good(\Gamma, A)$ and $Subtype(\Gamma, A, A')$ then $[A/X]B$ else fail

not in subtype relation. That is, the typechecker for $F_{2<:}$ may diverge on ill-typed programs, although it will still converge and produce a minimum type for well-typed programs. More generally, there is no decision procedure for subtyping: the type system for $F_{2<:}$ is undecidable [Pierce 1992]. Several attempts have been made to cut $F_{2<:}$ down to a decidable subset; the simplest solution at the moment consists in requiring equal quantifiers bounds in (Sub Forall<:). In any case, the bad pairs A, B are extremely unlikely to arise in practice. The algorithm is still sound in the usual sense: if it finds a type, the program will not go wrong. The only troublesome case is in the subtyping of quantifiers; the restriction of the algorithm to $F_{1<:}$ is decidable and produces minimum types.

$F_{2<:}$ provides an interesting example of the anomalies one may encounter in type inference. The type inference algorithm given in Table 103.37 is theoretically undecidable but is practically applicable. It is convergent and efficient on virtually all programs one may encounter; it diverges only on some ill-typed programs, which should be rejected anyway. Therefore, $F_{2<:}$ sits close to the boundary between acceptable and unacceptable type systems, according to the criteria enunciated in the introduction.

103.9 Summary and Research Issues

What We Learned

Natural questions for a beginner programmer are: What is an error? What is type safety? What is type soundness? (perhaps phrased, respectively, as: Which errors will the computer tell me about? Why did my program crash? Why does the computer refuse to run my program?). The answers, even informal ones, are surprisingly intricate. We have paid particular attention to the distinction between type safety and type soundness, and we have reviewed the varieties of static checking, dynamic checking, and absence of checking for program errors in various kinds of languages.

The most important lesson to remember from this chapter is the general framework for formalizing type systems. Understanding type systems, in general terms, is as fundamental as understanding BNF (Backus–Naur Form): it is hard to discuss the typing of programs without the precise language of type systems, just as it is hard to discuss the syntax of programs without the precise language of BNF. In both cases, the existence of a formalism has clear benefits for language design, compiler construction, language learning, and program understanding. We described the formalism of type systems, and how it captures the notions of type soundness and type errors.

Armed with formal type systems, we embarked on the description of an extensive list of program constructions and of their type rules. Many of these constructions are slightly abstracted versions of familiar features, whereas others apply only to obscure corners of common languages. In both cases, our collection of typing constructions is meant as a key for interpreting the typing features of programming languages. Such an interpretation may be nontrivial, particularly because most language definitions do not come with a type system, but we hope to have provided sufficient background for independent study. Some of the advanced type constructions will appear, we expect, more fully, cleanly, and explicitly in future languages.

In the latter part of the chapter, we reviewed some fundamental type inference algorithms: for simple languages, for polymorphic languages, and for languages with subtyping. These algorithms are very simple and general, but are mostly of an illustrative nature. For a host of pragmatic reasons, type inference for real languages becomes much more complex. It is interesting, though, to be able to describe concisely the core of the type inference problem and some of its solutions.

Future Directions

The formalization of type systems for programming languages, as described in this chapter, evolved as an application of type theory. Type theory is a branch of formal logic. It aims to replace predicate logics and set theory (which are untyped) with typed logics, as a foundation for mathematics.

One of the motivations for these logical type theories, and one of their more exciting applications, is in the mechanization of mathematics via proof checkers and theorem provers. Typing is useful in theorem provers for exactly the same reasons it is useful in programming. The mechanization of proofs reveals striking similarities between proofs and programs: the structuring problems found in proof construction are analogous to the ones found in program construction. Many of the arguments that demonstrate the need for typed programming languages also demonstrate the need for typed logics.

Comparisons between the type structures developed in type theory and in programming are, thus, very instructive. Function types, product types, (disjoint) union types, and quantified types occur in both disciplines, with similar intents. This is in contrast, for example, to structures used in set theory, such as unrestricted unions and intersections of sets, and the encoding of functions as sets of pairs, that have no correspondence in the type systems of common programming languages.

Beyond the simplest correspondences between type theory and programming, it turns out that the structures developed in type theory are far more expressive than the ones commonly used in programming. Therefore, type theory provides a rich environment for future progress in programming languages.

Conversely, the size of systems that programmers build is vastly greater than the size of proofs that mathematicians usually handle. The management of large programs, and in particular the type structures needed to manage large programs, is relevant to the management of mechanical proofs. Certain type theories developed in programming, for example, for object-orientation and for modularization, go beyond the normal practices found in mathematics, and should have something to contribute to the mechanization of proofs.

Therefore, the cross fertilization between logic and programming will continue, within the common area of type theory. At the moment, some advanced constructions used in programming escape proper type-theoretical formalization. This could be happening either because the programming constructions are ill conceived, or because our type theories are not yet sufficiently expressive: only the future will tell. Examples of active research areas are the typing of advanced object-orientation and modularization constructs and the typing of concurrency and distribution.

Defining Terms

Abstract type: A data type whose nature is kept hidden, in such a way that only a predetermined collection of operations can operate on it.

Contravariant: A type that varies in the inverse direction from one of its parts with respect to subtyping. The main example is the contravariance of function types in their domain. For example, assume $A <: B$ and vary X from A to B in $X \to C$; we obtain $A \to C :> B \to C$. Thus $X \to C$ varies in the inverse direction of X.

Covariant: A type that varies in the same direction as one of its parts with respect to subtyping. For example, assume $A <: B$ and vary X from A to B in $D \to X$; we obtain $D \to A <: D \to B$. Thus, $D \to X$ varies in the same direction as X.

Derivation: A tree of judgments obtained by applying the rules of a type system.

Dynamic checking: A collection of run time tests aimed at detecting and preventing forbidden errors.

Dynamically checked language: A language where good behavior is enforced during execution.

Explicitly typed language: A typed language where types are part of the syntax.

First-order type system: One that does not include quantification over type variables.

Forbidden error: The occurrence of one of a predetermined class of execution errors; typically the improper application of an operation to a value, such as *not* (3).

Good behavior: Same as being well behaved.

Ill typed: A program fragment that does not comply with the rules of a given type system.

Implicitly typed language: A typed language where types are not part of the syntax.

Judgment: A formal assertion relating entities such as terms, types, and environments. Type systems prescribe how to produce valid judgments from other valid judgments.

Polymorphism: The ability of a program fragment to have multiple types (opposite of monomorphism).

Safe language: A language where no untrapped errors can occur.

Second-order type system: One that includes quantification over type variables, either universal or existential.

Static checking: A collection of compile-time tests, mostly consisting of typechecking.

Statically checked language: A language where good behavior is determined before execution.

Strongly checked language: A language where no forbidden errors can occur at run time (depending on the definition of forbidden error).

Subsumption: A fundamental rule of subtyping, asserting that if a term has a type A, which is a subtype of a type B, then the term also has type B.

Subtyping: A reflexive and transitive binary relation over types that satisfies subsumption; it asserts the inclusion of collections of values.

Trapped error: An execution error that immediately results in a fault.

Type: A collection of values. An estimate of the collection of values that a program fragment can assume during program execution.

Type inference: The process of finding a type for a program within a given type system.

Type reconstruction: The process of finding a type for a program where type information has been omitted, within a given type system.

Type rule: A component of a type system. A rule stating the conditions under which a particular program construct will not cause forbidden errors.

Type safety: The property stating that programs do not cause untrapped errors.

Type soundness: The property stating that programs do not cause forbidden errors.

Type system: A collection of type rules for a typed programming language. Same as static type system.

Typechecker: The part of a compiler or interpreter that performs typechecking.

Typechecking: The process of checking a program before execution to establish its compliance with a given type system and therefore to prevent the occurrence of forbidden errors.

Typed language: A language with an associated (static) type system, whether or not types are part of the syntax.

Typing error: An error reported by a typechecker to warn against possible execution errors.

Untrapped error: An execution error that does not immediately result in a fault.

Untyped language: A language that does not have a (static) type system, or whose type system has a single type that contains all values.

Valid judgment: A judgment obtained from a derivation in a given type system.

Weakly checked language: A language that is statically checked but provides no clear guarantee of absence of execution errors.

Well behaved: A program fragment that will not produce forbidden errors at run time.

Well formed: Properly constructed according to formal rules.

Well-typed program: A program (fragment) that complies with the rules of a given type system.

References

Aiken, A. and Wimmers, E. L. 1993. Type inclusion constraints and type inference, pp. 31–41. In *Proc. ACM Conf. Functional Programming Comput. Architecture.*

Amadio, R. M. and Cardelli, L. 1993. Subtyping recursive types. *ACM Trans. Programming Lang. Syst.* 15(4):575–631.

Birtwistle, G. M., Dahl, O.-J., Myhrhaug, B., and Nygaard, K. 1979. Simula Begin. Studentlitteratur.

Böhm, C. and Berarducci, A. 1985. Automatic synthesis of typed λ-programs on term algebras. *Theor. Comput. Sci.* 39:135–154.

Cardelli, L. 1987. Basic polymorphic typechecking. *Sci. Comput. Programming* 8(2).

Cardelli, L. 1994. Extensible records in a pure calculus of subtyping. In *Theoretical Aspects of Object-Oriented Programming*, C. A. Gunter and J. C. Mitchell, eds., pp. 373–425. MIT Press, Cambridge, MA.

Cardelli, L. and Wegner, P. 1985. On understanding types, data abstraction and polymorphism. *ACM Comput. Surv.* 17(4):471–522.

Curien, P.-L. and Ghelli, G. 1992. Coherence of subsumption, minimum typing and type-checking in $F_\leq$. *Math. Struct. Comput. Sci.* 2(1):55–91.

Dahl, O.-J., Dijkstra, E. W., and Hoare, C. A. R. 1972. *Structured Programming.* Academic Press.

Eifrig, J., Smith, S., and Trifonov, V. 1995. Sound polymorphic type inference for objects, pp. 169–184. In *Proc. OOPSLA'95.*

Gunter, C. A. 1992. *Semantics of Programming Languages: Structures and Techniques.* MIT Press, Cambridge, MA.

Girard, J.-Y., Lafont, Y., and Taylor, P. 1989. *Proofs and Types.* Cambridge University Press, Cambridge, England.

Gunter, C. A. and Mitchell, J. C., eds. 1994. *Theoretical Aspects of Object-Oriented Programming.* MIT Press, Cambridge, MA.

Huet, G., ed. 1990. *Logical Foundations of Functional Programming.* Addison–Wesley, Reading, MA.

Jensen, K. 1978. *Pascal User Manual and Report*, 2nd ed. Springer–Verlag, New York.

Liskov, B. H. 1981. *CLU Reference Manual.* Lecture notes in computer science 114. Springer–Verlag, New York.

Milner, R. 1978. A theory of type polymorphism in programming. *J. Comput. Syst. Sci.* 17:348–375.

Milner, R., Tofte, M., and Harper, R. 1989. *The Definition of Standard ML.* MIT Press, Cambridge, MA.

Mitchell, J. C. 1984. Coercion and type inference, pp. 175–185. In *Proc. 11th Annu. ACM Symp. Principles Programming Lang.*

Mitchell, J. C. 1990. Type systems for programming languages. In *Handbook of Theoretical Computer Science*, J. van Leeuwen, ed., pp. 365–458. North-Holland, Amsterdam.

Mitchell, J. C. 1996. *Foundations for Programming Languages.* MIT Press, Cambridge, MA.

Mitchell, J. C. and Plotkin, G. D. 1985. Abstract types have existential type. In *Proc. 12th Annu. ACM Symp. Principles Programming Lang.*

Nordström, B., Petersson, K., and Smith, J. M. 1990. *Programming in Martin-Löf's Type Theory.* Oxford Science.

Palsberg, J. 1994. Efficient inference for object types, pp. 186–195. In *Proc. 9th Annu. IEEE Symp. Logic Comput. Sci.* (To appear in *Inf. Comput.*)

Pierce, B. C. 1992. Bounded quantification is undecidable. In *Proc. 19th Annu. ACM Symp. Principles Programming Lang.*

Reynolds, J. C. 1974. Towards a theory of type structure. In *Proc. Colloquium sur la programmation.* Lecture notes in computer science 19, pp. 408–423. Springer–Verlag, New York.

Reynolds, J. C. 1983. Types, abstraction, and parametric polymorphism. In *Information Processing,* R. E. A. Mason, ed., pp. 513–523. North-Holland, Amsterdam.

Schmidt, D. A. 1994. *The Structure of Typed Programming Languages.* MIT Press, Cambridge, MA.

Spencer, H. The ten commandments for C programmers. annotated ed. (available on the World Wide Web).

Tofte, M. 1990. Type inference for polymorphic references. *Inf. Comput.* 89:1–34.

Wells, J. B. 1994. Typability and type checking in the second-order λ-calculus are equivalent and undecidable, pp. 176–185. In *Proc. 9th Annu. IEEE Symp. Logic Comput. Sci.*

Wijngaarden, V., ed. 1976. *Revised Report on the Algorithmic Language Algol68.*

Wright, A. K. and Felleisen, M. 1994. A syntactic approach to type soundness. *Inf. Comput.* 115(1):38–94.

Further Information

For a complete background on type systems one should read (1) some material on type theory, which is usually rather hard, (2) some material connecting type theory to computing, and (3) some material about programming languages with advanced type systems.

The book edited by Huet [1990] covers a variety of topics in type theory, including several tutorial articles. The book edited by Gunter and Mitchell [1994] contains a collection of papers on object-oriented type theory. The book by Nordström et al. [1990] is a recent summary of Martin-Löf's work. Martin-Löf proposed type theory as a general logic that is firmly grounded in computation. He introduced the systematic notation for judgments and type rules used in this chapter. Girard et al. [1989] and Reynolds [1974] developed the polymorphic λ-calculus (F_2), which inspired much of the work covered in this chapter.

A modern exposition of technical issues that arise from the study of type systems can be found in Gunter [1992], in Mitchell's [1990] article in the *Handbook of Theoretical Computer Science,* in Mitchell [1996], and in the paper by Wright and Felleisen [1994].

Closer to programming languages, rich type systems were pioneered in the period between the development of Algol and the establishment of structured programming [Dahl et al. 1972], and were developed into a new generation of richly typed languages, including Pascal [Jansen 1978], Algol68 [Wijngaarden 1976], Simula [Birtwistle et al. 1979], CLU [Liskov 1981], and ML [Milner et al. 1989]. Reynolds gave type theoretical explanations for polymorphism and data abstraction [Reynolds 1974, Reynolds 1983]. (On that topic, see also Cardelli and Wegner [1985] and Mitchell and Plotkin [1985].) The book by Schmidt [1994] covers several issues discussed in this chapter, and provides more details on common language constructions.

Milner's paper on type inference for ML [Milner 1978] brought the study of type systems and type inference to a new level. It includes an algorithm for polymorphic type inference, and the first proof of type soundness for a (simplified) programming language, based on a denotational technique. A more accessible exposition of the algorithm described in that paper can be found in Cardelli [1987]. Proofs of type soundness are now often based on operational techniques [Tofte 1990, Wright and Felleisen 1994]. Currently, Standard ML is the only widely used programming language with a formally specified type system [Milner et al. 1989].

104

Programming Language Semantics

David A. Schmidt

Kansas State University

104.1 Introduction

A programming language possesses two fundamental features: syntax and semantics. Syntax refers to the appearance of the well-formed programs of the language, and semantics refers to the meanings of these programs. A language's syntax can be formalized by a grammar or syntax chart; such a formalization is found in the back of almost every language manual. A language's semantics should be formalized as well, so that it can appear in the language manual, too. This is the topic of this chapter.

It is traditional for computer scientists to calculate the semantics of a program by using a test-case input and tracing the program's execution with a state table and flow chart. This is one form of semantics, called *operational semantics*, but there are other forms of semantics that are not tied to test cases and traces; we will study several such approaches.

Before we begin, we might ask: What do we gain by formalizing the semantics of a programming language? Before we answer, we might consider the related question: What was gained when language syntax was formalized? The formalization of syntax, via Backus–Naur Form (BNF) rules, produced these benefits:

- The syntax definition standardizes the official syntax of the language. This is crucial to users, who require a guide to writing syntactically correct programs, and to implementors, who must write a correct parser for the language's compiler.

- The syntax definition permits a formal analysis of its properties, such as whether the definition is $LL(k)$, $LR(k)$, or ambiguous.

- The syntax definition can be used as input to a compiler front-end generating tool, such as Yet Another Compiler Compiler (YACC). In this way, the syntax definition is also the implementation of the front end of the language's compiler.

0-8493-2909-4/97/$0.00+$.50
© 1997 by CRC Press, Inc.

There are similar benefits to providing a formal semantics definition of a programming language:

- The semantics definition standardizes the official semantics of the language. This is crucial to users, who require a guide to understanding the programs that they write, and to implementors, who must write a correct code generator for the language's compiler.
- The semantics definition permits a formal analysis of its properties, such as whether the definition is strongly typed, block structured, or single threaded.
- The semantics definition can be used as input to a compiler back-end generating tool, such as Semantics Implementation System (SIS) or MESS [Lee 1989, Mosses 1976]. In this way, the semantics definition is also the implementation of the back end of the language's compiler.

Programming language syntax was studied intensively in the 1960s and 1970s, and presently programming language semantics is undergoing similar intensive study. Unlike the acceptance of BNF as a standard definition method for syntax, it appears unlikely that a single definition method will take hold for semantics—semantics is harder to formalize than syntax, and it has a wider variety of applications.
Semantics definition methods fall roughly into three groups:

- **Operational:** The meaning of a well-formed program is the trace of computation steps that results from processing the program's input. Operational semantics is also called *intensional* semantics, because the sequence of internal computation steps (the intension) is most important. For example, two differently coded programs that both compute factorial have different operational semantics.
- **Denotational:** The meaning of a well-formed program is a mathematical function from input data to output data. The steps taken to calculate the output are unimportant; it is the relation of input to output that matters. Denotational semantics is also called *extensional* semantics, because only the extension—the visible relation between input and output—matters. Thus, two differently coded versions of factorial have nonetheless the same denotational semantics.
- **Axiomatic:** A meaning of a well-formed program is a logical proposition (a specification) that states some property about the input and output. For example, the proposition $\forall x.\, x \geq \mathbf{0} \supset \exists y.\, y = x!$ is an axiomatic semantics of a factorial program.

104.2 A Survey of Semantics Methods

We survey the three semantic methods by applying each of them in turn to the world's oldest and simplest programming language, arithmetic. The syntax of our arithmetic language is:

$$E ::= N \mid E_1 + E_2$$

where N stands for the set of numerals $\{0, 1, 2, \ldots\}$. Although this language has no notion of input data and output data, it does contain the notion of computation, so it will be a useful example for our initial case studies.

Operational Semantics

There are several versions of operational semantics for arithmetic. The one that you learned as a child is called a *term rewriting system*. A term rewriting system uses rewriting rule schemes to generate computation steps. There is just one rewriting rule scheme for arithmetic:

$$N_1 + N_2 \Rightarrow N' \quad \text{where N' is the sum of the numerals } N_1 \text{ and } N_2$$

This rule scheme states that the addition of two numerals is a computation step. One use of the scheme would be to rewrite $1 + 2$ to 3, that is, $1 + 2 \Rightarrow 3$. An operational semantics of a program is the sequence of computation steps generated by the rewriting rule schemes. For example, an operational semantics of the program $(1 + 2) + (4 + 5)$ goes as follows:

$$(1 + 2) + (4 + 5) \Rightarrow 3 + (4 + 5) \Rightarrow 3 + 9 \Rightarrow 12$$

The semantics shows the three computation steps that led to the answer 12. An intermediate expression such as $3 + (4 + 5)$ is a *state*, and so this operational semantics is a trace of the states of the computation.

Perhaps you noticed that another legal semantics for the example is $(1 + 2) + (4 + 5) \Rightarrow (1 + 2) + 9 \Rightarrow 3 + 9 \Rightarrow 12$. The outcome is the same in both cases, but sometimes an operational semantics must be forced to be *deterministic*, that is, a program has exactly one operational semantics.

A **structural operational semantics** is a term rewriting system plus a set of inference rules that state precisely the context in which a computation step can be undertaken.[1] Say that we desire left-to-right computation of arithmetic expressions. This is encoded as follows:

$$N_1 + N_2 \Rightarrow N' \quad \text{where } N' \text{ is the sum of } N_1 \text{ and } N_2$$

$$\frac{E_1 \Rightarrow E_1'}{E_1 + E_2 \Rightarrow E_1' + E_2} \qquad \frac{E_2 \Rightarrow E_2'}{N + E_2 \Rightarrow N + E_2'}$$

The first rule is as before; the second rule states, if the left operand of an addition expression can be rewritten, then the addition expression should be revised to show this. The third rule is the crucial one: if the right operand of an addition expression can be rewritten *and* the left operand is a numeral (that is, it is completely evaluated), then the addition expression should be revised to show this. Working together, the three rules force left-to-right evaluation of expressions.

Now, each computation step must be deduced by these rules. For our example, $(1 + 2) + (4 + 5)$, we must deduce this initial computation step:

$$\frac{1 + 2 \Rightarrow 3}{(1 + 2) + (4 + 5) \Rightarrow 3 + (4 + 5)}$$

Thus, the first step is $(1 + 2) + (4 + 5) \Rightarrow 3 + (4 + 5)$; note that we *cannot* deduce that $(1 + 2) + (4 + 5) \Rightarrow (1 + 2) + 9$. (Try.) The next computation step is justified by this deduction:

$$\frac{4 + 5 \Rightarrow 9}{3 + (4 + 5) \Rightarrow 3 + 9}$$

The last deduction is simply $3 + 9 \Rightarrow 12$, and we are finished. The example shows why the semantics is structural: a computation step, such as an addition, which affects a small part of the overall program, is explicitly embedded into the structure of the overall program.

Operational semantics can also be used to represent internal data structures, such as instruction counters, storage vectors, and stacks. For example, say that our semantics of arithmetic must show that a stack is used to hold intermediate results. Thus, we use a state of the form $\langle s, c \rangle$, where s is the stack and c is the arithmetic expression to be executed. A stack containing n items is written $v_1 :: v_2 :: \ldots :: v_n :: nil$, where v_1 is the topmost item and *nil* marks the bottom of the stack. The c component will be written as a stack as well. The initial state for an arithmetic expression p is written $\langle nil, p :: nil \rangle$, and computation proceeds until the state appears as $\langle v :: nil, nil \rangle$; we say that the result is v.

---

[1] A structural operational semantics is sometimes called a *small-step semantics*, because each computation step is a small step toward the final answer.

The semantics uses three rewriting rules:

$$\langle s,\ N :: c \rangle \Rightarrow \langle N :: s,\ c \rangle$$

$$\langle s,\ E_1 + E_2 :: c \rangle \Rightarrow \langle s,\ E_1 :: E_2 :: add :: c \rangle$$

$$\langle N_2 :: N_1 :: s,\ add :: c \rangle \Rightarrow \langle N' :: s,\ c \rangle \quad \text{where } N' \text{ is the sum of } N_1 \text{ and } N_2$$

The first rule says that a numeral is evaluated by pushing it on the top of the stack. The second rule states that the addition of two expressions is decomposed into first evaluating the two expressions and then adding them. The third rule removes the top two items from the stack and adds them. Here is the previous example, repeated:

$$\langle nil,\ (1 + 2) + (4 + 5) :: nil \rangle$$

$$\Rightarrow \langle nil,\ 1 + 2 :: 4 + 5 :: add :: nil \rangle$$

$$\Rightarrow \langle nil,\ 1 :: 2 :: add :: 4 + 5 :: add :: nil \rangle$$

$$\Rightarrow \langle 1 :: nil,\ 2 :: add :: 4 + 5 :: add :: nil \rangle$$

$$\Rightarrow \langle 2 :: 1 :: nil,\ add :: 4 + 5 :: add :: nil \rangle$$

$$\Rightarrow \langle 3 :: nil,\ 4 + 5 :: add :: nil \rangle \Rightarrow \ldots \Rightarrow \langle 12 :: nil,\ nil \rangle$$

This form of operational semantics is sometimes called a *state transition semantics*, because each rewriting rule operates upon the entire state. With a state transition semantics, there is of course no need for structural operational semantics rules.

The three example semantics just shown are typical of operational semantics. When one wishes to prove properties of an operational semantics definition, the standard proof technique is *induction on the length of the computation*. That is, to prove that a property P holds for an operational semantics, one must show that P holds for all possible computation sequences that can be generated from the rewriting rules. For an arbitrary computation sequence, it suffices to show that P holds no matter how long the computation runs. Therefore, one shows (1) P holds after zero computation steps, that is, at the outset, and (2) if P holds after n computation steps, it holds after $n + 1$ steps. See Nielson and Nielson [1992] for examples.

Denotational Semantics

A drawback of operational semantics is the emphasis it places on state sequences. For the arithmetic language, we were distracted by questions regarding order of evaluation of subphrases, even though this issue is not central to arithmetic. Further, a key aspect of arithmetic, the property that the meaning of an expression is built from the meanings of its subexpressions, was obscured by the operational semantics.

Denotational semantics handles these issues by emphasizing that a program has an underlying mathematical meaning that is independent of whatever computation strategy is taken to uncover it. In the case of arithmetic, an expression such as $(1 + 2) + (4 + 5)$ has the meaning 12, and we need not worry about the internal computation steps that were taken to discover this.

The assignment of meaning to programs is performed in a *compositional* manner: the meaning of a phrase is built from the meanings of its subphrases. We can see this in the denotational semantics of the arithmetic language: first, we note that meanings of arithmetic expressions are natural numbers, $Nat = \{0, 1, 2, \ldots\}$, and we note that there is a binary function, $plus : Nat \times Nat \rightarrow Nat$, which maps a pair of natural numbers to their sum.

The denotational semantics definition of arithmetic is simple and elegant:

$$\mathcal{E} : Expression \to Nat$$

$$\mathcal{E}[\![N]\!] = N$$

$$\mathcal{E}[\![E_1 + E_2]\!] = plus\,(\mathcal{E}[\![E_1]\!], \mathcal{E}[\![E_2]\!])$$

The first line states merely that $\mathcal{E}$ is the name of the function that maps arithmetic expressions to their meanings. Since there are just two BNF constructions for expressions, $\mathcal{E}$ is completely defined by the two equational clauses. The interesting clause is the one for $E_1 + E_2$; it says that the meanings of E_1 and E_2 are combined compositionally by *plus*. Here is the denotational semantics of our example program:

$$\mathcal{E}[\![(1 + 2) + (4 + 5)]\!] = plus\,(\mathcal{E}[\![1 + 2]\!], \mathcal{E}[\![4 + 5]\!])$$

$$= plus\,(plus\,(\mathcal{E}[\![1]\!], \mathcal{E}[\![2]\!]), plus\,(\mathcal{E}[\![4]\!], \mathcal{E}[\![5]\!]))$$

$$= plus\,(3, 9) = 12$$

One might read the preceding example as follows: the meaning of $(1 + 2) + (4 + 5)$ equals the meanings of $1 + 2$ and $4 + 5$ added together. Since the meaning of $1 + 2$ is 3, and the meaning of $4 + 5$ is 9, the meaning of the overall expression is 12. This reading says nothing about order of evaluation or run time data structures—it emphasizes underlying mathematical meaning.

Here is an alternative way of understanding the semantics; write a set of simultaneous equations based on the denotational definition

$$\mathcal{E}[\![(1 + 2) + (4 + 5)]\!] = plus\,(\mathcal{E}[\![1 + 2]\!], \mathcal{E}[\![4 + 5]\!])$$

$$\mathcal{E}[\![1 + 2]\!] = plus\,(\mathcal{E}[\![1]\!], \mathcal{E}[\![2]\!])$$

$$\mathcal{E}[\![4 + 5]\!] = plus\,(\mathcal{E}[\![4]\!], \mathcal{E}[\![5]\!])$$

$$\mathcal{E}[\![1]\!] = 1 \qquad \mathcal{E}[\![2]\!] = 2$$

$$\mathcal{E}[\![4]\!] = 4 \qquad \mathcal{E}[\![5]\!] = 5$$

Now, solve the equation set to discover that $\mathcal{E}[\![(1 + 2) + (4 + 5)]\!]$ is 12.

Since denotational semantics states the meaning of a phrase in terms of the meanings of its subphrases, its associated proof technique is structural induction. That is, to prove that a property P holds for all programs in the language, one must show that the meaning of each construction in the language has property P. Therefore, one must show that each equational clause in the semantic definition produces a meaning with property P. In the case that a clause refers to subphrases (e.g., $\mathcal{E}[\![E_1 + E_2]\!]$), one may assume that the meanings of the subphrases have property P. Again, see Nielson and Nielson [1992] for examples.

Natural Semantics

Recently, a semantics method has been proposed that is halfway between operational semantics and denotational semantics; it is called **natural semantics**. Like structural operational semantics, natural semantics shows the context in which a computation step occurs, and, like denotational semantics, natural semantics emphasizes that the computation of a phrase is built from the computations of its subphrases.

A natural semantics is a set of inference rules, and a complete computation in natural semantics is a single, large derivation. The natural semantics rules for the arithmetic language are

$$N \Rightarrow N$$

$$\frac{E_1 \Rightarrow n_1 \quad E_2 \Rightarrow n_2}{E_1 + E_2 \Rightarrow m} \quad \text{where } m \text{ is the sum of } n_1 \text{ and } n_2$$

Read a configuration of the form $E \Rightarrow n$ as "E evaluates to n." The rules resemble a denotational semantics written in inference rule form; this is no accident: natural semantics can be viewed as a denotational semantics variant where the internal calculations of meaning are made explicit. These internal calculations are seen in the natural semantics of our example expression

$$\frac{\dfrac{1 \Rightarrow 1 \quad 2 \Rightarrow 2}{(1+2) \Rightarrow 3} \quad \dfrac{4 \Rightarrow 4 \quad 5 \Rightarrow 5}{(4+5) \Rightarrow 9}}{(1+2) + (4+5) \Rightarrow 12}$$

Unlike denotational semantics, natural semantics does not claim that the meaning of a program is necessarily mathematical. And unlike structural operational semantics, where a configuration $e \Rightarrow e'$ says that e transits to an intermediate state e', in natural semantics $e \Rightarrow v$ asserts that the final answer for e is v. For this reason, a natural semantics is sometimes called a *big-step semantics*. An interesting drawback of natural semantics is that semantics derivations can be drawn only for terminating programs.

The usual proof technique for proving properties of a natural semantics definition is induction on the height of the derivation trees that are generated from the semantics. Once again, see Nielson and Nielson [1992].

Axiomatic Semantics

An axiomatic semantics produces properties of programs rather than meanings. The derivation of these properties is done by an inference rule set that looks somewhat like a natural semantics.

As an example, say that we wish to calculate even–odd properties of programs in arithmetic and our set of properties is simply $\{is\_even, is\_odd\}$. We can define an axiomatic semantics to do this

$$N : is\_even \text{ if } N \bmod 2 = 0 \qquad N : is\_odd \text{ if } N \bmod 2 = 1$$

$$\frac{E_1 : p_1 \quad E_2 : p_2}{E_1 + E_2 : p_3} \quad \text{where } p_3 = \begin{cases} is\_even & \text{if } p_1 = p_2 \\ is\_odd & \text{otherwise} \end{cases}$$

The derivation of the even–odd property of our example program is

$$\frac{\dfrac{1 : is\_odd \quad 2 : is\_even}{1+2 : is\_odd} \quad \dfrac{4 : is\_even \quad 5 : is\_odd}{4+5 : is\_odd}}{(1+2) + (4+5) : is\_even}$$

In the usual case, the properties to be proved of programs are expressed in the language of predicate logic; see the subsection on axiomatic semantics of the language. Also, axiomatic semantics has strong ties to the *abstract interpretation* of denotational and natural semantics definitions [Abramsky and Hankin 1987, Cousot and Cousot 1977].

104.3 Semantics of Programming Languages

The semantics methods shine when they are applied to a realistic programming language: the primary features of the programming language are proclaimed loudly, and subtle features receive proper mention. Ambiguities and anomalies stand out like the proverbial sore thumb. In this section, we give the semantics of a block-structured imperative language. Emphasis will be placed on the denotational semantics method, but excerpts from the other semantics formalisms will be provided for comparison.

P ∈ Program
D ∈ Declaration
C ∈ Command
E ∈ Expression
I ∈ Identifier = upper-case alphabetic strings
N ∈ Numeral = {0, 1, 2, ...}

P ::= C.
D ::= **proc** I = C
C ::= I := E | C_1 ; C_2 | **begin** D **in** C **end** | **call** I | **while** E **do** C **od**
E ::= N | E_1 + E_2 | E_1 **not=** E_2 | I

FIGURE 104.1   Language syntax rules.

Language Syntax and Informal Semantics

The syntax of the programming language is presented in Fig. 104.1. As stated in the figure, there are four levels of syntax constructions in the language, and the topmost level, Program, is the primary one.[2] Basically, the language is a while-loop language with local, nonrecursive procedure definitions. For simplicity, variables are predeclared and there are just three of them— **X**, **Y**, and **Z**. A program, C., operates as follows: an input number is read and assigned to **X**'s location. Then the body, C, of the program is evaluated, and, on completion, the storage vector holds the results. For example, this program computes n^2 for a positive input n; the result is found in **Z**'s location:

```
begin proc INCR = Z:= Z+X; Y:= Y+1
   in Y:= 0; Z:= 0; while Y not=X do call INCR od end.
```

It is possible to write nonsense programs in the language; an example is **A:=0; call B**. Such programs have no meaning, and we will not attempt to give semantics to them. Nonsense programs are trapped by a typechecker, and an elegant way of defining a typechecker is by a set of typing rules for the programming language; see Chapter 103 for details.

Domains for Denotational Semantics

To give a denotational semantics to the sample language, we must state the sets of meanings, called *domains*, that we use. Our imperative, block-structured language has two primary domains: (1) the domain of storage vectors, called *Store*, and (2) the domain of symbol tables, called *Environment*. There are also secondary domains of Booleans and natural numbers. The primary domains and their operations are displayed in Fig. 104.2.

The domains and operations deserve study. First, the *Store* domain states that a storage vector is a triple. (Recall that programs have exactly three variables.) The operation *lookup* extracts a value from the store, e.g., $lookup(2, \langle 1, 3, 5 \rangle) = 3$, and *update* updates the store, e.g., $update(2, 6, \langle 1, 3, 5 \rangle) = \langle 1, 6, 5 \rangle$. Operation *init_store* creates a starting store. We examine *check* momentarily.

The environment domain states that a symbol table is a list of identifier-value pairs. For example, if variable **X** is the name of location 1, and **P** is the name of a procedure that is a no-op, then the environment that holds this information would appear (**X**, 1) :: (**P**, *id*) :: *nil*, where *id*(*s*) = *s*. (Procedures will be discussed momentarily.) Operation *find* locates the binding for an identifier in the environment, e.g., *find*(**X**, (**X**, 1) :: (**P**, *id*) :: *nil*) = 1, and *bind* adds a new binding, e.g., *bind*(**Y**, 2, (**X**, 1) :: (**P**, *id*) :: *nil*) = (**Y**, 2) :: (**X**, 1) :: (**P**, *id*) :: *nil*. Operation *init_env* creates an environment to start the program.

---

[2]The Identifier and Numeral sets are collections of words—terminal symbols—and not phrase-level *syntax constructions* in the sense of this chapter.

$$Store = \{\langle n_1, n_2, n_3 \rangle \mid n_i \in Nat, i \in 1..3\}$$

$$lookup : \{1, 2, 3\} \times Store \rightarrow Nat$$
$$lookup(i, \langle n_1, n_2, n_3 \rangle) = n_i$$
$$update : \{1, 2, 3\} \times Nat \times Store \rightarrow Store$$
$$update(1, n, \langle n_1, n_2, n_3 \rangle) = \langle n, n_2, n_3 \rangle$$
$$update(2, n, \langle n_1, n_2, n_3 \rangle) = \langle n_1, n, n_3 \rangle$$
$$update(3, n, \langle n_1, n_2, n_3 \rangle) = \langle n_1, n_2, n \rangle$$
$$init\_store : Nat \rightarrow Store$$
$$init\_store(n) = \langle n, 0, 0 \rangle$$
$$check : (Store \rightarrow Store_\perp) \times Store_\perp \rightarrow Store_\perp \text{ where } Store_\perp = Store \cup \{\perp\}$$
$$check(c, a) = \text{if } (a = \perp) \text{ then } \perp \text{ else } c(a)$$

$$Environment = (Identifier \times Denotable)^*$$
$$\text{where } A^* \text{ is a list of } A\text{-elements}, a_1 :: a_2 :: \ldots :: a_n :: nil, n \geq 0$$
$$\text{and } Denotable = \{1, 2, 3\} \cup (Store \rightarrow Store_\perp)$$

$$find : Identifier \times Environment \rightarrow Denotable$$
$$find(i, nil) = 0 ,$$
$$find(i, (i', d) :: rest) = \text{if } (i = i') \text{ then } d \text{ else } find(i, rest)$$
$$bind : Identifier \times Denotable \times Environment \rightarrow Environment$$
$$bind(i, d, e) = (i, d) :: e$$
$$init\_env : Environment$$
$$init\_env = (\mathbf{X}, 1) :: (\mathbf{Y}, 2) :: (\mathbf{Z}, 3) :: nil$$

FIGURE 104.2   Semantic domains.

In the next section, we will see that the job of a command, e.g., an assignment, is to update the store. That is, the meaning of a command is a function that maps the current store to the updated one. (That is why a no-op command is the identity function, $id(s) = s$, where $s \in Store$.) But sometimes commands loop, and no updated store appears. We use the symbol $\perp$, read bottom, to stand for a looping store, and we use $Store_\perp$ to stand for the set of possible outputs of commands. Therefore, the meaning of a command is a function of the form $Store \rightarrow Store_\perp$.

It is impossible to recover from looping, so that if there is a command sequence $C_1; C_2$, and C_1 is looping, then C_2 cannot proceed. The *check* operation is used in the next subsection to watch for this situation.

Finally, here are two commonly used notations. First, functions such as $id(s) = s$ are often reformatted to read $id = \lambda s. s$; in general, for $f(a) = e$, we write $f = \lambda a. e$, that is, we write the argument to the function to the right of the equals sign. This is called *lambda notation* and stems from the *lambda calculus*, an elegant formal system for functions. (See Chapter 102 of this Handbook.) The notation $f = \lambda a. e$ emphasizes that (1) the function $\lambda a. e$ is a value in its own right, and (2) the function's name is f.

Second, it is common to revise a function that takes multiple arguments, for example, $f(a, b) = e$, so that it takes the arguments one at a time: $f = \lambda a. \lambda b. e$. Therefore, if the arity of f was $A \times B \rightarrow C$, its new arity is $A \rightarrow (B \rightarrow C)$. This reformatting trick is called *Currying*, after Haskell Curry, one of the developers of the lambda calculus.

Denotational Semantics of Programs

Figure 104.3 gives the denotational semantics of the programming language. Since the syntax of the language has four levels, the semantics is organized into four levels of meaning. For each level, we define a *valuation function*, which produces the meanings of constructions at that level. For example, at the expression level, the constructions are mapped to their meanings by $\mathcal{E}$.

What is the meaning of the expression, say, **X+5**? This would be $\mathcal{E}[\mathbf{X+5}]$, and the meaning depends on which location is named by **X** and what number is stored in that location. Therefore, the meaning is dependent on the current value of the environment and the current value of the store. Thus, if the current

$$\mathcal{P} : Program \rightarrow Nat \rightarrow Nat_\perp$$
$$\mathcal{P}[\![\mathrm{C.}]\!] = \lambda n.\, \mathcal{C}[\![\mathrm{C}]\!]\, init\_env\, (init\_store\, n)$$
$$\mathcal{D} : Declaration \rightarrow Environment \rightarrow Environment$$
$$\mathcal{D}[\![\textbf{proc}\ \mathrm{I} = \mathrm{C}]\!] = \lambda e.\, bind(\mathrm{I}, \mathcal{C}[\![\mathrm{C}]\!]e, e)$$
$$\mathcal{C} : Command \rightarrow Environment \rightarrow Store \rightarrow Store_\perp$$
$$\mathcal{C}[\![\mathrm{I} := \mathrm{E}]\!] = \lambda e.\, \lambda s.\, update\, (find\, (\mathrm{I}, e), \mathcal{E}[\![\mathrm{E}]\!]e\, s, s)$$
$$\mathcal{C}[\![\mathrm{C_1}; \mathrm{C_2}]\!] = \lambda e.\, \lambda s.\, check\, (\mathcal{C}[\![\mathrm{C_2}]\!]\, e, \mathcal{C}[\![\mathrm{C_1}]\!]e\, s)$$
$$\mathcal{C}[\![\textbf{begin}\ \mathrm{D}\ \textbf{in}\ \mathrm{C}\ \textbf{end}]\!] = \lambda e.\, \lambda s.\, \mathcal{C}[\![\mathrm{C}]\!](\mathcal{D}[\![\mathrm{D}]\!]e)s$$
$$\mathcal{C}[\![\textbf{call}\ \mathrm{I}]\!] = \lambda e.\, find(\mathrm{I}, e)$$
$$\mathcal{C}[\![\textbf{while}\ \mathrm{E}\ \textbf{do}\ \mathrm{C}\ \textbf{od}]\!] = \lambda e.\, \bigcup_{i \geq 0} w_i$$
$$w_0 = \lambda s.\, \perp$$
where
$$w_{i+1} = \lambda s.\, \text{if } \mathcal{E}[\![\mathrm{E}]\!]e\, s \text{ then } check(w_i, \mathcal{C}[\![\mathrm{C}]\!]e\, s) \text{ else } s$$
$$\mathcal{E} : Expression \rightarrow Environment \rightarrow Store \rightarrow (Nat \cup Bool)$$
$$\mathcal{E}[\![\mathrm{N}]\!] = \lambda e.\, \lambda s.\, N$$
$$\mathcal{E}[\![\mathrm{E_1} + \mathrm{E_2}]\!] = \lambda e.\, \lambda s.\, plus(\mathcal{E}[\![\mathrm{E_1}]\!]e\, s, \mathcal{E}[\![\mathrm{E_2}]\!]e\, s)$$
$$\mathcal{E}[\![\mathrm{E_1}\ \textbf{not=}\ \mathrm{E_2}]\!] = \lambda e.\, \lambda s.\, notequals(\mathcal{E}[\![\mathrm{E_1}]\!]e\, s, \mathcal{E}[\![\mathrm{E_2}]\!]e\, s)$$
$$\mathcal{E}[\![\mathrm{I}]\!] = \lambda e.\, \lambda s.\, lookup(find(\mathrm{I}, e), s)$$

FIGURE 104.3 Denotational semantics.

environment is $e_0 = (\mathbf{P}, \lambda s.\, s) :: (\mathbf{X}, 1) :: (\mathbf{Y}, 2) :: (\mathbf{Z}, 3) :: nil$ and the current store is $s_0 = \langle 2, 0, 0 \rangle$, then the meaning of **X+5** is 7,

$$\mathcal{E}[\![\mathbf{X+5}]\!]e_0\, s_0 = plus(\mathcal{E}[\![\mathbf{X}]\!]e_0\, s_0, \mathcal{E}[\![\mathbf{5}]\!]e_0\, s_0)$$
$$= plus(lookup(find(\mathbf{X}, e_0), s_0), 5)$$
$$= plus(lookup(1, s_0), 5) = plus(2, 5) = 7$$

As this simple derivation shows, data structures such as the symbol table and storage vector are modeled by the environment and store arguments. This pattern is used throughout the semantics definition.

As noted in the previous section, a command updates the store. Precisely stated, the valuation function for commands is $\mathcal{C} : Command \rightarrow Environment \rightarrow Store \rightarrow Store_\perp$. For example, for e_0 and s_0 given previously, we see that

$$\mathcal{C}[\![\mathbf{Z:=X+5}]\!]e_0\, s_0 = update(find(\mathbf{Z}, e_0), \mathcal{E}[\![\mathbf{X+5}]\!]e_0\, s_0, s_0) = update(3, 7, s_0) = \langle 2, 0, 7 \rangle$$

But a crucial point about the meaning of the assignment is that it is a function upon stores. That is, if we are uncertain of the current value of store, but we know that the environment for the assignment is e_0, then we can conclude

$$\mathcal{C}[\![\mathbf{Z:=X+5}]\!]e_0 = \lambda s.\, update(3, plus(lookup(1, s), 5), s)$$

That is, the assignment with environment e_0 is a function that updates a store at location 3.

Next, consider this example of a command sequence:

$$\mathcal{C}[\![\mathbf{Z:=X+5;\ call\ P}]\!]e_0\, s_0 = check(\mathcal{C}[\![\mathbf{call\ P}]\!]e_0, \mathcal{C}[\![\mathbf{Z:=X+5}]\!]e_0\, s_0)$$
$$= check(find(\mathbf{P}, e_0), \langle 2, 0, 7 \rangle) = check(\lambda s.\, s, \langle 2, 0, 7 \rangle)$$
$$= (\lambda s.\, s)\langle 2, 0, 7 \rangle = \langle 2, 0, 7 \rangle$$

As noted in the earlier section, the *check* operation verifies that the first command in the sequence produces a proper output store; if so, the store is handed to the second command in the sequence. Also, we see that the meaning of **call P** is the store updating function bound to **P** in the environment.

Procedures are placed in the environment by declarations, as we see in this example: let e_1 denote $(\mathbf{X}, 1) :: (\mathbf{Y}, 2) :: (\mathbf{Z}, 3) :: nil$,

$$\mathcal{C}[\![\mathbf{begin\ proc\ P\ =\ Y:=Y\ in\ Z:=X+5;\ call\ P\ end}]\!]e_1\ s_0$$

$$= \mathcal{C}[\![\mathbf{Z:=X+5;\ call\ P}]\!](\mathcal{D}[\![\mathbf{proc\ P\ =\ Y:=Y}]\!]e_1)s_0$$

$$= \mathcal{C}[\![\mathbf{Z:=X+5;\ call\ P}]\!](bind(\mathbf{P}, \mathcal{C}[\![\mathbf{Y:=Y}]\!]e_1, e_1))s_0$$

$$= \mathcal{C}[\![\mathbf{Z:=X+5;\ call\ P}]\!](bind(\mathbf{P}, \lambda s.\ update(2, lookup(2, s), s), e_1))s_0$$

$$= \mathcal{C}[\![\mathbf{Z:=X+5;\ call\ P}]\!]((\mathbf{P}, id) :: e_1)s_0$$

$$\text{where } id = \lambda s.\ update(2, lookup(2, s), s) = \lambda s.\ s \qquad\qquad (*)$$

$$= \mathcal{C}[\![\mathbf{Z:=X+5;\ call\ P}]\!]e_0\ s_0 = \langle 2, 0, 7 \rangle$$

The equality marked by $(*)$ is significant; we can assert that the function $\lambda s.\ update(2, lookup(2, s), s)$ is identical to $\lambda s.\ s$ by appealing to the *extensionality* law of mathematics: if two functions map identical arguments to identical answers, then the functions are themselves identical. The extensionality law can be used here because in denotational semantics the meanings of program phrases are mathematical— functions. In contrast, the extensionality law cannot be used in operational semantics calculations.

Finally, we can combine our series of little examples into the semantics of a complete program,

$$\mathcal{P}[\![\mathbf{begin\ proc\ P\ =\ Y:=Y\ in\ Z:=X+5;\ call\ P\ end.}]\!]2$$

$$= \mathcal{C}[\![\mathbf{begin\ proc\ P\ =\ Y:=Y\ in\ Z:=X+5;\ call\ P\ end}]\!]init\_env\,(init\_store\,2)$$

$$= \mathcal{C}[\![\mathbf{begin\ proc\ P\ =\ Y:=Y\ in\ Z:=X+5;\ call\ P\ end}]\!]e_1\ s_0$$

$$= \langle 2, 0, 7 \rangle$$

Semantics of the While-Loop

The most difficult clause in the semantics definition is the one for the while-loop. Here is some intuition: to produce an output store, the loop **while** E **do** C **od** must terminate after some finite number of iterations. To measure this behavior, let **while**$_i$ E **do** C **od** be a loop that can iterate at most i times— if the loop runs more than i iterations, it becomes exhausted, and its output is $\bot$. For example, for input store $\langle 4, 0, 0 \rangle$, the loop **while**$_k$ **Y not=X do Y:=Y+1 od** can produce the output store $\langle 4, 4, 0 \rangle$ only when k is greater than 4. (Otherwise, the output is $\bot$.)

It is easy to conclude that the family, **while**$_i$ E **do** C **od**, for $i \geq 0$, can be written equivalently as

while$_0$ E **do** C **od** $=$ *"exhausted"* (that is, its meaning is $\lambda s.\ \bot$)

while$_{i+1}$ E **do** C **od** $=$ **if** E **then** C; **while**$_i$ E **do** C **od else skip fi**

When we refer back to Fig. 104.3, we draw these conclusions:

$$\mathcal{C}[\![\mathbf{while}_0\ E\ \mathbf{do}\ C\ \mathbf{od}]\!]e = w_0$$

$$\mathcal{C}[\![\mathbf{while}_{i+1}\ E\ \mathbf{do}\ C\ \mathbf{od}]\!]e = w_{i+1}$$

Since the behavior of a while-loop must be the union of the behaviors of the **while**$_i$-loops, we conclude that $\mathcal{C}[\![\mathbf{while}\ E\ \mathbf{do}\ C\ \mathbf{od}]\!]e = \bigcup_{i \geq 0} w_i$. The semantic union operation is well defined because each w_i is a function from the set $Store \to Store_\bot$, and a function can be represented as a set of argument-answer pairs. (This is called the *graph of the function*.) Thus, $\bigcup_{i \geq 0} w_i$ is the union of the graphs of the w_i functions.[3]

The definition of $\mathcal{C}[\textbf{while E do C od}]$ is succinct, but it is awkward to use in practice. An intuitive way of defining the semantics is:

$$\mathcal{C}[\textbf{while E do C od}]e = w$$

where $w = \lambda s.$ if $\mathcal{E}[E]e$ s then $check(w, \mathcal{C}[C]e$ $s)$ else s

The problem here is that the definition of w is circular, and circular definitions can be malformed. Fortunately, this definition of w can be claimed to denote the function $\bigcup_{i \geq 0} w_i$ because the following equality holds:

$$\bigcup_{i \geq 0} w_i = \lambda s. \text{ if } \mathcal{E}[E]e \ s \text{ then } check\left(\bigcup_{i \geq 0} w_i, \mathcal{C}[C]e \ s\right) \text{ else } s$$

Thus, $\bigcup_{i \geq 0} w_i$ is a solution—a *fixed point*—of the circular definition, and in fact it is the smallest function that makes the equality hold. Therefore, it is the *least fixed point*.

Typically, the denotational semantics of the while-loop is presented by the circular definition, and the claim is then made that the circular definition stands for the least fixed point. This is called **fixed-point semantics**. We have omitted many technical details regarding fixed-point semantics; these are available in several texts [Gunter 1992, Schmidt 1986, Stoy 1977, Winskel 1993].

Action Semantics

One disadvantage of denotational semantics is its dependence on functions to describe all forms of computation. As a result, the denotational semantics of a large language is often too dense to read and too low level to modify. **Action semantics** is an easy-to-read denotational semantics variant that rectifies these problems by using a family of standard operators to describe standard forms of computation in standard languages [Mosses 1992].

In action semantics, the standard domains are called *facets* and are predefined for expressions (the *functional facet*), for declarations (the *declarative facet*), and for commands (the *imperative facet*). Each facet includes a set of standard operators for consuming values of the facet and producing new ones. The operators are connected together by combinators (*pipes*), and the resulting action semantics definition resembles a dataflow program. For example, the semantics of assignment reads as follows:

$$\text{execute}[I := E] = (\text{find } I \text{ and evaluate}[E]) \text{ then update}$$

One can naively read the semantics as an English sentence, but each word is an operator or a combinator: execute is $\mathcal{C}$, evaluate is $\mathcal{E}$, find is a declarative facet operator, update is an imperative facet operator, and and and then are combinators. The equation accepts as its inputs a declarative facet argument (that is, an environment) and an imperative facet argument (that is, a store) and pipes them to the operators. Thus, find consumes its declarative argument and produces a functional-facet answer, and, independently, evaluate[E] consumes declarative and imperative arguments and produces a functional answer. The and combinator pairs these, and the then combinator transmits the pair to the update operator, which uses the pair and the imperative-facet argument to generate a new imperative result.

The important aspects of an action semantics definition are (1) standard arguments, such as environments and stores, are implicit; (2) standard operators are used for standard computation steps (e.g., find

---

[3] Several important technical details have been glossed over. First, pairs of the form $(s, \bot)$ are ignored when the union of the graphs is performed. Second, for all $i \geq 0$, the graph of w_i is a subset of the graph of w_{i+1}; this ensures the union of the graphs is a function.

$$e \vdash \mathbf{proc}\ I = C \Rightarrow bind(I, (e, C), e) \qquad \frac{e \vdash D \Rightarrow e' \quad e', s \vdash C \Rightarrow s'}{e, s \vdash \mathbf{begin}\ D\ \mathbf{in}\ C\ \mathbf{end} \Rightarrow s'}$$

$$\frac{l = find(I, e) \quad e, s \vdash E \Rightarrow n}{e, s \vdash I := E \Rightarrow update(l, n, s)} \qquad \frac{e, s \vdash C_1 \Rightarrow s' \quad e, s' \vdash C_2 \Rightarrow s''}{e, s \vdash C_1 ; C_2 \Rightarrow s''}$$

$$\frac{(e', C') = find(I, e) \quad e', s \vdash C' \Rightarrow s'}{e, s \vdash \mathbf{call}\ I \Rightarrow s'} \qquad \frac{e, s \vdash E \Rightarrow false}{e, s \vdash \mathbf{while}\ E\ \mathbf{do}\ C\ \mathbf{od} \Rightarrow s}$$

$$\frac{e, s \vdash E \Rightarrow true \quad e, s \vdash C \Rightarrow s' \quad e, s' \vdash \mathbf{while}\ E\ \mathbf{do}\ C\ \mathbf{od} \Rightarrow s''}{e, s \vdash \mathbf{while}\ E\ \mathbf{do}\ C\ \mathbf{od} \Rightarrow s''}$$

FIGURE 104.4　Natural semantics.

$$\text{let } e_0 = (\mathbf{X}, 1) :: (\mathbf{Y}, 2) :: (\mathbf{Z}, 3) :: nil$$
$$s_0 = \langle 2, 0, 0 \rangle, \quad s_1 = \langle 2, 1, 0 \rangle$$
$$E_0 = \mathbf{Y\ not{=}1}, \quad C_0 = \mathbf{Y{:}{=}Y{+}1}$$
$$C_{00} = \mathbf{while}\ E_0\ \mathbf{do}\ C_0\ \mathbf{od}$$

$$\frac{e_0, s_0 \vdash E_0 \Rightarrow true \quad \dfrac{2 = find(\mathbf{Y}, e_0) \quad e_0, s_0 \vdash \mathbf{Y{+}1} \Rightarrow 1}{e_0, s_0 \vdash C_0 \Rightarrow s_1} \quad \dfrac{e_0, s_1 \vdash E_0 \Rightarrow false}{e_0, s_1 \vdash C_{00} \Rightarrow s_1}}{e_0, s_0 \vdash C_{00} \Rightarrow s_1}$$

FIGURE 104.5　Natural semantics derivation.

and update); and (3) combinators connect operators together seamlessly and pass values implicitly. Lack of space prevents a closer examination of action semantics, but see Watt [1991] for an introduction.

The Natural Semantics of the Language

We can compare the denotational semantics of the imperative language with a natural semantics formulation. The semantics of several constructions appear in Fig. 104.4.

A command configuration has the form $e, s \vdash C \Rightarrow s'$, where e and s are the inputs to command C and s' is the output. To understand the inference rules, read them bottom up. For example, the rule for I := E says, given the inputs e and s, one must first find the location l, bound to I, and then calculate the output n, for E. Finally, l, and n are used to update s, producing the output.

The rules are denotational-like, but differences arise in several key constructions. First, the semantics of a procedure declaration binds I not to a function but to an environment-command pair called a *closure*. When procedure I is called, the closure is disassembled, and its text and environment are executed. Since a natural semantics does not use function arguments, it is called a *first-order semantics*. (Denotational semantics is sometimes called a *higher order semantics*.)

Second, the while-loop rules are circular. The second rule states, in order to derive a while-loop computation that terminates in s'', one must derive (1) the test, E is true, (2) the body C, outputs s', and (3) using e and s', one can derive a terminating while-loop computation that outputs s''. The rule makes one feel that the while-loop is running backward from its termination to its starting point, but a complete derivation, such as the one shown in Fig. 104.5, shows that the iterations of the loop can be read from the root to the leaves of the derivation tree.

One important aspect of the natural semantics definition is that derivations can be drawn only for terminating computations. A nonterminating computation is equated with no computation at all.

The Operational Semantics of the Language

A fragment of the structural operational semantics of the imperative language is presented in Fig. 104.6.

For expressions, a computation step takes the form $e \vdash \langle E, s \rangle \Rightarrow E'$, where e is the environment, E is the expression that is evaluated, s is the current store, and E' is E rewritten. In the case of a command

$$e \vdash \langle n_1 + n_2, s \rangle \Rightarrow n_3 \text{ where } n_3 \text{ is the sum of } n_1 \text{ and } n_2$$

$$\frac{e \vdash \langle E, s \rangle \Rightarrow E'}{e \vdash \langle I := E, s \rangle \Rightarrow \langle I := E', s \rangle}$$

$$e \vdash \langle I := n, s \rangle \Rightarrow \mathit{update}(l, n, s) \text{ where } \mathit{find}(I, e) = l$$

$$\frac{e \vdash \langle C_1, s \rangle \Rightarrow \langle C_1', s' \rangle}{e \vdash \langle C_1; C_2, s \rangle \Rightarrow \langle C_1'; C_2, s' \rangle} \qquad \frac{e \vdash \langle C_1, s \rangle \Rightarrow s'}{e \vdash \langle C_1; C_2, s \rangle \Rightarrow \langle C_2, s' \rangle}$$

$$e \vdash \langle \textbf{while } E \textbf{ do } C \textbf{ od}, s \rangle \Rightarrow \langle \textbf{if } E \textbf{ then } C; \textbf{ while } E \textbf{ do } C \textbf{ od else skip fi}, s \rangle$$

$$e \vdash \langle \textbf{call } I, s \rangle \Rightarrow \langle \textbf{use } e' \textbf{ in } C', s \rangle \text{ where } \mathit{find}(I, e) = (e', C')$$

$$\frac{e \vdash \langle C, s \rangle \Rightarrow \langle C', s' \rangle}{e \vdash \langle \textbf{use } e' \textbf{ in } C, s \rangle \Rightarrow \langle \textbf{use } e' \textbf{ in } C', s' \rangle} \qquad \frac{e' \vdash \langle C, s \rangle \Rightarrow s'}{e \vdash \langle \textbf{use } e' \textbf{ in } C, s \rangle \Rightarrow s'}$$

$$e \vdash \textbf{proc } I = C \Rightarrow \mathit{bind}(I, (e, C), e)$$

$$\frac{e \vdash D \Rightarrow e'}{e \vdash \langle \textbf{begin } D \textbf{ in } C \textbf{ end}, s \rangle \Rightarrow \langle \textbf{use } e' \textbf{ in } C, s \rangle}$$

FIGURE 104.6   Structural operational semantics.

C, a step appears $e \vdash \langle C, s \rangle \Rightarrow \langle C', s' \rangle$, because computation on C might also update the store. If the computation step on C uses up the command, the step appears $e \vdash \langle C, s \rangle \Rightarrow s'$.

The rules in the figure are more tedious than those for a natural semantics, because the individual computation steps must be defined, and the order in which the steps are undertaken also must be defined. This complicates the rules for command composition, for example. On the other hand, the rewriting rule for the while-loop merely decodes the loop as a conditional command.

The rules for procedure call are awkward; as with the natural semantics, a procedure I is represented as a closure of the form (e', C'). Since C′ must execute with environment e', which is different from the environment that exists where procedure I is called, the rewriting step for **call** I must retain *two* environments; a new construct, **use** e' **in** C′, remembers that C′ must use e' (and not e). A similar trick is used in **begin** D **in** C **end**.

Unlike a natural semantics definition, a computation can be written for a nonterminating program; the computation is a state sequence of countably infinite length.

An Axiomatic Semantics of the Language

An axiomatic semantics uses properties of stores, rather than stores themselves. For example, we might write the predicate **X** $= 3 \wedge$ **Y** > 0 to assert that the current value of the store contains 3 in **X**'s location and a positive number in **Y**'s location. We write a configuration $\{P\}C\{Q\}$, to assert that if predicate P holds true prior to evaluation of command C, then predicate Q holds upon termination of C (if C does indeed terminate). For example, we can write $\{$**X** $= 3 \wedge$ **Y** $> 0\}$ **Y:=X+Y** $\{$**X** $= 3 \wedge$ **Y** $> 3\}$, and indeed this holds true.

There are three ways of stating the semantics of a command in an axiomatic semantics:

- *Relational semantics:* The meaning of C is the set of P, Q pairs for which $\{P\}C\{Q\}$ holds.
- *Postcondition semantics:* The meaning of C is a function from an input predicate to an output predicate. We write $slp(P, C) = Q$; this means that $\{P\}C\{Q\}$ holds, and for all Q' such that $\{P\}C\{Q'\}$ holds, it is the case that Q implies Q'. This is also called *strongest liberal postcondition semantics*. When termination is demanded also of C, the name becomes **strongest postcondition semantics**.
- *Precondition semantics:* The meaning of C is a function from an output predicate to an input predicate. We write $wlp(C, Q) = P$; this means that $\{P\}C\{Q\}$ holds, and for all P' such that $\{P'\}C\{Q\}$

$$\{[E/I]\,P\}\,I := E\{P\}$$

$$\frac{P \supset P' \quad \{P'\}C\{Q'\} \quad Q' \supset Q}{\{P\}C\{Q\}} \qquad \frac{\{P\}C_1\{Q\} \quad \{Q\}C_2\{R\}}{\{P\}C_1;\,C_2\{R\}}$$

$$\frac{\{P \wedge E\}C_1\{Q\} \quad \{P \wedge \neg E\}C_2\{Q\}}{\{P\}\ \textbf{if}\ E\ \textbf{then}\ C_1\ \textbf{else}\ C_2\ \textbf{fi}\ \{Q\}} \qquad \frac{\{P \wedge E\}C\{P\}}{\{P\}\ \textbf{while}\ E\ \textbf{do}\ C\ \textbf{od}\ \{P \wedge \neg E\}}$$

FIGURE 104.7 Axiomatic semantics.

holds, it is the case that P' implies P. This is also called *weakest liberal precondition semantics*. When termination is demanded also of C, the name becomes **weakest precondition semantics**.

It is traditional to study relational semantics first, so we focus on it here.

If the intended behavior of a program C is written as a pair of predicates P, Q, a relational semantics can be used to verify that $\{P\}C\{Q\}$ holds. For example, we might wish to show that an integer division subroutine **DIV** that takes inputs **NUM** and **DEN** and produces outputs **QUO** and **REM** has this behavior:

$$\{\neg(\textbf{DEN} = 0)\}\ \textbf{DIV}\ \{\textbf{QUO} \times \textbf{DEN} + \textbf{REM} = \textbf{NUM}\}$$

A proof of this claim is a derivation built with the rules in Fig. 104.7.

Figure 104.7 displays the rules for the primary command constructions. The rule for I := E states that a property P about I will hold upon completion of the assignment if $[E/I]\,P$ (that is, P restated in terms of E) holds beforehand. $[E/I]\,P$ stands for the substitution of phrase E for all free occurrences of I in P. For example, $\{\textbf{X} = 3 \wedge \textbf{X+Y} > 3\}\ \textbf{Y:=X+Y}\ \{\textbf{X} = 3 \wedge \textbf{Y} > 3\}$ holds because $[\textbf{X+Y}/\textbf{Y}](\textbf{X} = 3 \wedge \textbf{Y} > 3)$ is $\textbf{X} = 3 \wedge \textbf{X+Y} > 3$.

The second rule lets us weaken a result. For example, since $(\textbf{X} = 3 \wedge \textbf{Y} > 0) \supset (\textbf{X} = 3 \wedge \textbf{X} + \textbf{Y} > 3)$ holds, we deduce that $\{\textbf{X} = 3 \wedge \textbf{Y} > 0\}\ \textbf{Y:=X+Y}\ \{\textbf{X} = 3 \wedge \textbf{Y} > 3\}$ holds.

The properties of command composition are defined in the expected way, by the third rule. The fourth rule, for the if-command, makes a property Q hold upon termination if Q holds regardless of which arm of the conditional is evaluated. Note that each arm of the conditional uses information about the result of the conditional's test.

The most fascinating rule is the last one, for the while-loop. If we can show that a property P is preserved by the body of the loop, then we can assert that no matter how long the loop iterates, P must hold upon termination. P is called the **loop invariant**. The rule is an encoding of a mathematical induction proof: to show that P holds upon completion of the loop, we must prove (1) the basis case: P holds upon loop entry (that is, after zero iterations), and (2) the induction case: if P holds after i iterations, then P holds after $i + 1$ iterations as well. Therefore, if the loop terminates after some number k of iterations, the induction proof ensures that P holds.

Here is an example that shows the rules in action. We wish to verify that

$$\{\textbf{X} = \textbf{Y} \wedge \textbf{Z} = 0\}\ \textbf{while Y not=0 do Y:=Y-1; Z:=Z+1 od}\ \{\textbf{X} = \textbf{Z}\}$$

holds true. The key to the proof is determining a loop invariant; here, a useful invariant is $\textbf{X} = \textbf{Y} + \textbf{Z}$, because $\textbf{X} = \textbf{Y} + \textbf{Z} \wedge \neg(\textbf{Y not=0})$ implies $\textbf{X} = \textbf{Z}$. This leaves us $\{\textbf{X} = \textbf{Y} + \textbf{Z} \wedge$

let P_0 be $X = Y + Z$

P_1 be $X = Y + (Z + 1)$, $\quad P_2$ be $X = (Y - 1) + (Z + 1)$

$E_0 = \textbf{Y not=0}, \quad C_0 = \textbf{Y:=Y-1;Z:=Z+1}$

$$\frac{(P_0 \wedge E_0) \supset P_2 \quad \dfrac{\{P_2\}\textbf{Y:=Y-1}\ \{P_1\} \quad \{P_1\}\textbf{Z:=Z+1}\ \{P_0\}}{\{P_2\}C_0\{P_0\}} \quad P_0 \supset P_0}{\dfrac{\{P_0 \wedge E_0\}C_0\{P_0\}}{\{P_0\}\ \textbf{while}\ E_0\ \textbf{do}\ C_0\ \textbf{od}\ \{P_0 \wedge \neg E_0\}}}$$

FIGURE 104.8 Axiomatic semantics derivation.

Y not=0 } **Y:=Y-1; Z:=Z+1** {**X** = **Y** + **Z**} to prove. We work backward: the rule for assignment gives us: {**X** = **Y** + (**Z** + 1)} **Z:=Z+1** {**X** = **Y** + **Z**}, and we can also deduce that {**X** = (**Y** − 1) + (**Z** + 1)} **Y:=Y-1** {**X** = **Y** + (**Z** + 1)} holds. Since **X** = **Y** + **Z** ∧ **Y not=0** implies **X** = (**Y** − 1) + (**Z** + 1), we can assemble a complete derivation; it is given in Fig. 104.8.

104.4 Applications of Semantics

Increasingly, language designers are using semantics definitions to formalize their creations. An early example was the formalization of a large subset of Ada in denotational semantics [Donzeau-Gouge 1980]. The semantics definition was then prototyped using Mosses's SIS compiler generating system [Mosses 1976]. Scheme is another widely used language which has been given a standardized denotational semantics [Rees and Clinger 1986]. Another notable example is the formalization of the complete Standard ML language in structural operational semantics [Milner et al. 1990].

Perhaps the most significant application of semantics definitions has been to rapid prototyping—the synthesis of an implementation for a newly defined language. Some prototyping systems are SIS [Mosses 1976], PSI [Nielson and Nielson 1988], MESS [Lee 1989], Actress [Brown et al. 1992], and Typol [Despeyroux 1984]. The first two process denotational semantics, the second two process action semantics, and the last handles natural semantics. SIS and Typol are interpreter generators, that is, they interpret a source program with the semantics definition, and PSI, MESS, and Actress are compiler generators, that is, compilers for the source language are synthesized.

A major success of formal semantics is the analysis and synthesis of dataflow analysis and type-inference algorithms from semantics definitions. This subject area, called *abstract interpretation* [Abramsky and Hankin 1987, Cousot and Cousot 1977, Muchnick and Jones 1981], supplies precise techniques for analyzing semantics definitions, extracting properties from the definitions, applying the properties to dataflow and type inference, and proving the soundness of the code-improvement transformations that result. Abstract interpretation provides the theory that allows a compiler writer to prove the correctness of compilers.

Finally, axiomatic semantics is a long-standing fundamental technique for validating the correctness of computer code. Recent emphasis on large-scale and safety-critical systems has again placed the spotlight on this technique. Current research on data-type theory (see Chapter 103 of this Handbook) suggests that a marriage between the techniques of data-type checking and axiomatic semantics is not far in the future.

104.5 Research Issues in Semantics

The techniques in this chapter have proved highly successful for defining, improving, and implementing traditional, sequential programming languages. But new language paradigms present new challenges to the semantics methods.

In the functional programming paradigm, a higher order functional language can use functions as arguments to other functions. This makes the language's domains more complex than those in Fig. 104.2. Denotational semantics can be used to understand these complexities; an applied mathematics called *domain theory* [Gunter 1992, Schmidt 1986] is used to formalize the domains with algebraic equations. For example, the *Value* domain for a higher order, Scheme-like language takes the form

$$Value = Nat + (Value \rightarrow Value)$$

That is, legal values are numbers or functions on values. Of course, Cantor's theorem makes it impossible to find a set that satisfies this equation, but domain theory uses the concept of continuity from topology to restrict the size of *Value* so that a solution can be found as that in the subsection on the semantics of the while-loop, namely

$$Value = \lim_{i \geq 0} V_i, \quad \text{where } \begin{aligned} V_0 &= \{\bot\} \\ V_{i+1} &= Nat \uplus (V_i \to^{ctn} V_i) \end{aligned}$$

where $V_i \to^{ctn} V_i$ denotes the topologically continuous functions on V_i.

Challenging issues also arise in the object-oriented programming paradigm. Not only can objects be parameters (messages) to other objects' procedures (methods), but coercion laws based on *inheritance hierarchies* allow controlled mismatches between actual and formal parameters. Just as an integer actual parameter might be coerced to a floating-point formal parameter, we might coerce an object that contains methods for addition, subtraction, and multiplication to be an actual parameter to a method that expects a formal-parameter object with just addition and subtraction methods. Carelessly defined coercions lead to unsound programs, and so denotational and natural semantics have been used to formalize domain hierarchies and safe coercions for the inheritance hierarchies [Gunter and Mitchell 1994].

Yet another challenging topic is parallelism and communication as it arises in the distributed programming paradigm. Here, multiple processes run in parallel and synchronize through communication. Structural operational semantics has been adapted to formalize systems of processes and to study the varieties of communication the processes might undertake. Indeed, new notational systems have been developed specifically for this subject area. (See Chapter 102 of this Handbook.)

Finally, a long-standing research topic is the relationship between the different forms of semantic definitions. If one has, say, both a denotational semantics and an axiomatic semantics for a programming language, in what sense do the semantics agree? Agreement is crucial, since a programmer might use the axiomatic semantics to reason about the properties of programs, whereas a compiler writer might use the denotational semantics to implement the language. In mathematical logic, one uses the concepts of *soundness* and *completeness* to relate a logic's proof system to its interpretation, and in semantics there are similar notions of *soundness* and *adequacy* to relate one semantics to another [Gunter 1992, Ong 1995].

A standard example is proving the soundness of a structural operational semantics to a denotational semantics: for program P and input v, $(P, v) \Rightarrow v'$ in the operational semantics implies $\mathcal{P}[\![P]\!](v) = v'$ in the denotational semantics. Adequacy is a form of inverse: if $\mathcal{P}[\![P]\!](v) = v'$, and v' is a primitive value (e.g., an integer or Boolean), then $(P, v) \Rightarrow v'$. There is a stronger form of adequacy, called *full abstraction* [Stoughton 1988], which has proved difficult to achieve for realistic languages, although recent progress has been made [Abramsky et al. 1994].

Acknowledgment

Brian Howard and Anindya Banerjee provided helpful criticism.

Defining Terms

Action semantics: A variation of denotational semantics where low-level details are hidden by use of modularized sets of operators and combinators.

Axiomatic semantics: The meaning of a program as a property or specification in logic.

Denotational semantics: The meaning of a program as a compositional definition of a mathematical function from the program's input data to its output data.

Fixed-point semantics: A denotational semantics where the meaning of a repetitive structure, such as a loop or recursive procedure, is expressed as the smallest mathematical function that satisfies a recursively defined equation.

Loop invariant: In axiomatic semantics, a logical property of a while-loop that holds true no matter how many iterations the loop executes.

Natural semantics: A hybrid of operational and denotational semantics that shows computation steps performed in a compositional manner. Also known as a big-step semantics.

Operational semantics: The meaning of a program as calculation of a trace of its computation steps on input data.

Strongest postcondition semantics: A variant of axiomatic semantics where a program and an input property are mapped to the strongest proposition that holds true of the program's output.

Structural operational semantics: A variant of operational semantics where computation steps are performed only within prespecified contexts. Also known as a small-step semantics.

Weakest precondition semantics: A variant of axiomatic semantics where a program and an output property are mapped to the weakest proposition that is necessary of the program's input to make the output property hold true.

References

Abramsky, S. and Hankin, C., eds. 1987. *Abstract Interpretation of Declarative Languages.* Ellis Horwood, Chichester, England.

Abramsky, S., Jagadeesan, R., and Malacaria, P. 1994. Full abstraction for PCF. In *TACS94, Proc. Theor. Aspects Comput. Software.* Lecture notes in computer Science 789, pp. 1–15. Springer.

Apt, K. 1981. Ten years of Hoare's logic: a survey–part 1. *ACM Trans. Programming Lang. Syst.* 3:431–484.

Brown, D. F., Moura, H., and Watt, D. A. 1992. ACTRESS: an action semantics directed compiler generator. In *CC'92, Proc. 4th Int. Conf. Compiler Construction,* Paderborn. Lecture notes in computer science 641, pp. 95–109. Springer–Verlag, New York.

Cousot, P. and Cousot, R. 1977. Abstract interpretation: a unified lattice model for static analysis of programs, pp. 238–252. In *Proc. 4th ACM Symp. Principles Programming Lang.* ACM Press.

Despeyroux, T. 1984. Executable specification of static semantics. In *Semantics of Data Types,* G. Kahn, D. B. MacQueen, and G. Plotkin, eds. Lecture notes in computer science 173, pp. 215–234. Springer–Verlag, New York.

Dijkstra, E. W. 1976. *A Discipline of Programming.* Prentice–Hall, Englewood Cliffs, NJ.

Donzeau-Gouge, V. 1980. On the formal description of Ada. In *Semantics-Directed Compiler Generation.* N. D. Jones, ed. Lecture notes in computer science 94, Springer–Verlag, New York.

Dromey, G. 1989. *Program Derivation.* Addison–Wesley, Sydney.

Gries, D. 1981. *The Science of Programming.* Springer.

Gunter, C. 1992. *Foundations of Programming Languages.* MIT Press, Cambridge, MA.

Gunter, C. and Mitchell, J. 1994 *Theoretical Aspects of Object-Oriented Programming.* MIT Press, Cambridge, MA.

Hennessy, M. 1991. *The Semantics of Programming Languages: An Elementary Introduction Using Structured Operational Semantics.* Wiley, New York.

Hoare, C. A. R. 1969. An axiomatic basis for computer programming. *Commun. ACM* 12:576–580.

Hoare, C. A. R. and Wirth, N. 1973. An axiomatic definition of the programming language Pascal. *Acta Informatica* 2:335–355.

Jones, C. B. 1980. *Software Development: A Rigorous Approach.* Prentice–Hall, Englewood Cliffs, NJ.

Kahn, G. 1987. Natural semantics. In *Proc. STACS '87.* Lecture notes in computer science 247, pp. 22–39. Springer, Berlin.

Lee, P. 1989. *Realistic Compiler Generation.* MIT Press, Cambridge, MA.

Milner, R., Tofte, M., and Harper, R. 1990. *The Definition of Standard ML.* MIT Press, Cambridge, MA.

Morgan, C. 1994. *Programming from Specifications,* 2nd ed. Prentice–Hall. Englewood Cliffs, NJ.

Mosses, P. D. 1976. Compiler generation using denotational semantics. In *Mathematical Foundations of Computer Science.* A. Mazurkiewicz, ed. Lecture notes in computer science 45, pp. 436–441. Springer, Berlin.

Mosses, P. D. 1990. Denotational semantics. In *Handbook of Theoretical Computer Science,* J. van Leeuwen, ed. Vol. B, ch. 11, pp. 575–632. Elsevier.

Mosses, P. D. 1992. *Action Semantics.* Cambridge University Press, Cambridge, England.

Muchnick, S. and Jones, N. D., eds. 1981. *Program Flow Analysis: Theory and Applications.* Prentice–Hall, Englewood Cliffs, NJ.

Nielson, F. and Nielson, H. R. 1988. Two-level semantics and code generation. *Theor. Comput. Sci.* 56(1):59–133.

Nielson, H. R. and Nielson, F. 1992. *Semantics with Applications, a Formal Introduction.* Wiley Professional Computing. Wiley, New York.

Ong, C. H.-L. 1995. Correspondence between operational and denotational semantics. In *Handbook of Computer Science.* S. Abramsky, D. Gabbay, and T. Maibam, eds. Vol. 4. Oxford University Press, Rio de Janeiro, Brazil.

Plotkin, G. D. 1981. A Structural Approach to Operational Semantics. *Tech. Rep.* FN-19, DAIMI, Aarhus, Denmark, Sept.

Rees, J. and Clinger, W. 1986. Revised 3 report on the algorithmic language Scheme. *SIGPLAN Notices* 21:37–79.

Schmidt, D. A. 1986. *Denotational Semantics: A Methodology for Language Development.* Allyn and Bacon.

Schmidt, D. A. 1994. *The Structure of Typed Programming Languages.* MIT Press, Cambridge, MA.

Slonneger, K. and Kurtz, B. 1995. *Formal Syntax and Semantics of Programming Languages: A Laboratory-Based Approach.* Addison–Wesley, Reading, MA.

Stoughton, A. 1988. *Fully Abstract Models of Programming Languages.* Research notes in theoretical computer science. Pitman/Wiley.

Stoy, J. E. 1977. *Denotational Semantics.* MIT Press, Cambridge, MA.

Tennent, R. D. 1991. *Semantics of Programming Languages.* Prentice–Hall International, Englewood Cliffs, NJ.

Watt, D. A. 1991. *Programming Language Syntax and Semantics.* Prentice–Hall International, Englewood Cliffs, NJ.

Winskel, G. 1993. *Formal Semantics of Programming Languages.* MIT Press, Cambridge, MA.

Further Information

The best starting point for further reading is the comparative semantics text of H. R. Nielson and F. Nielson [1992], which thoroughly develops the topics in this chapter. See also the texts by Slonneger and Kurtz [1995] and Watt [1991].

Operational semantics has a long history, but good, modern introductions are Hennessey's [1991] text and Plotkin's report on structural operational semantics [1981]. The principles of natural semantics are documented by Kahn [1987].

Mosses's [1990] paper is a useful introduction to denotational semantics; textbook-length treatments include those by Schmidt [1986], Stoy [1977], Tennent [1991], and Winskel [1993]. Gunter's [1992] text uses denotational-semantics-based mathematics to compare several of the semantics approaches, and Schmidt's [1994] text shows the recent influences of data-type theory on denotational semantics. Action semantics is surveyed by Watt [1991] and defined by Mosses [1992].

Of the many textbooks on axiomatic semantics, one might start with books by Dromey [1989] or Gries [1981]; both emphasize precondition semantics, which is most effective at deriving correct code. Apt's [1981] paper is an excellent description of the formal properties of relational semantics, and Dijkstra's [1976] text is the standard reference on precondition semantics. Hoare's [1969, 1973] landmark papers on relational semantics are worth reading as well. Many texts have been written on the application of axiomatic semantics to systems development; two samples are by Jones [1980] and Morgan [1994].

X

Software Engineering

Steven A. Demurjian, Sr., Section Adviser
The University of Connecticut

THIS SECTION EXAMINES formal specification, design, verification and testing, project management, and other aspects of the software life cycle. Topics include the waterfall and spiral models, software qualities (maintainability, portability, and reuse), formal models and the specification process, the traditional and object-oriented design processes, verification and validation, risk and reliability issues, project scheduling and evaluation, software tools, and interoperability.

Techniques • Comparison of Module-Testing Techniques • Cleanroom • Testing Objects • When Testing Stops

105

Software Process Models

Ian Sommerville
Lancaster University

105.1 Introduction

The development of anything but trivial software systems is a structured activity. Various steps are involved where the software is designed, programmed, and validated. This sequence of activities and their inputs and outputs make up the **software process**. Every organization has its own specific software process, but these individual approaches usually follow some, more abstract, generic process model. These generic **software process models** are the subject of this chapter.

A generic software process model is an abstract representation of the activities and deliverables in the software process. Depending on the level of detail, the model may also show the roles responsible for these activities, the tools used to carry out these activities, communications of different types between activities, and roles and exceptions which must be handled as part of the process. However, in the examples here, the software process is considered at a fairly abstract level and only process activities and their inputs and outputs are discussed.

Software processes are immensely complex. The activities involved in these processes are intellectually demanding and may require significant creativity on the part of the process participants. These processes have a number of attributes as shown in Table 105.1.

If we wish to compare or reason about software processes using these attributes, we must be able to make some assessment of them. Because processes are so complex, a software process model is essential for this purpose. If we wish to improve the software process in some way in order to deliver software more quickly, produce software at lower cost, or deliver software with fewer defects, a defined process model is necessary as a starting point for the improvement process.

Similarly, if we need to communicate and exchange information about software processes, we need a process model. A detailed software process model is an important source of organizational knowledge and is used to communicate organizational standards, procedures, and practices to new engineers and managers.

One of the most important functions of a software process model is to facilitate process management. Process management involves scheduling the process activities, estimating the resources required for each

TABLE 105.1 Software Process Attributes

Process Attribute	Description
Understandability	Is the process clearly and explicitly defined and is the process definition understandable?
Visibility	Does each activity in the process have a well-defined endpoint and results so that progress is clearly visible?
Supportability	To what extent can the process activities be supported by CASE tools?
Acceptability	Do the engineers who are responsible for the software development find the process acceptable and a realistic match to their everyday activities?
Reliability	Is the process designed so that process errors are avoided or trapped before they lead to errors in the software being developed?
Robustness	Do unexpected problems cause process delays or can the process cope with these?
Maintainability	Can the process evolve to meet changing organizational requirements or process improvements for lower costs, higher quality, or faster delivery?
Rapidity	Are there inherent delays in the process which affect the overall development time from system specification to product delivery?

activity, assigning people to carry out the activities, and ensuring that appropriate quality procedures are applied to both the process and the developed product. The process should be designed so that managers can check progress against the plan and accurately judge how resources have been deployed in the process.

Consequently, managers express their plan in terms of their model of the development process. If there is no explicit model, the manager must make assumptions about the process and make estimates on the basis of these assumptions. However, the software developers may actually use a completely different process, so the plan is therefore a misleading guide for project management. An explicit process model, agreed upon by managers and software developers, is an invaluable tool for project communication.

There are, currently, no accepted standards for describing process models. The vast majority of models are expressed informally, using diagrams and descriptive text. Notations used in software design such as data-flow diagrams and entity-relation diagrams may be used to show the flow of work and the relationships between activities, and deliverables. Petri nets have been used to show the timing dependencies between activities, and various more-specialized process description notations have been proposed, although they are not widely used. The best known of these is that suggested by Christie [Christie 1994]. Armenise et al. [Armenise et al. 1992] summarize other process modelling notations.

In the remainder of this chapter, several types of generic process model are discussed. These include the "classical" waterfall model and its derivative, the V-model, models centered on software prototyping, and models which have been developed to accommodate change and uncertainty in a structured way. In the final sections, an assessment is made of the domains where each process model is most applicable, and issues such as process improvement models are briefly discussed.

105.2 Specification-Driven Models

In the 1950s and 1960s, software development was largely an informal process which was undertaken as part of other engineering activities. Rather than being developed according to some formal process standard, software was simply programmed using an informal description of what was required. However, as software systems grew in size and complexity, this informal approach became increasingly inadequate. There were a number of large software project failures where software failed to deliver the required functionality and performance, where the software was unreliable and expensive to maintain, and where the project was completed (or more often, abandoned) years behind schedule.

These failures led a number of organizations to adopt a more structured software development process which was based on the process used for the development of other engineered systems. Thus, there were clearly defined specification, design, and development activities with a "signing-off" process between these activities. When documents were "signed-off," they were deemed to be complete and the next stage of the process could then begin.

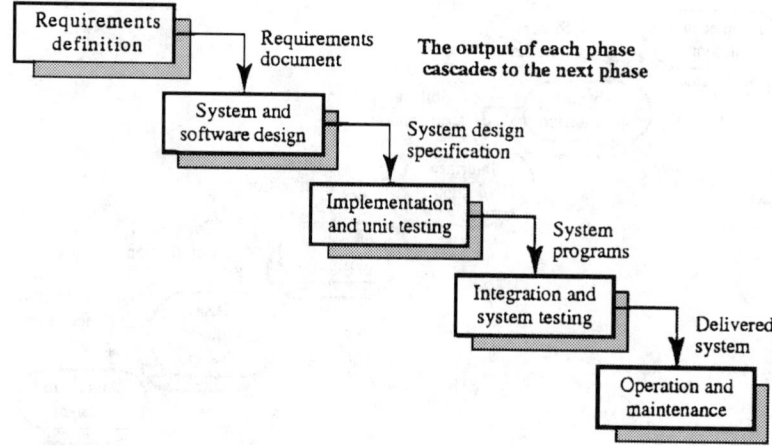

FIGURE 105.1 Waterfall model of the software process.

The first public discussion of this software "life-cycle" was due to Royce [Royce 1970], who described experience with this approach in defense projects. The initial life-cycle model is now termed the **waterfall model**. This model consists of a set of stages, starting with system specification, with results cascading from one stage to another. Figure 105.1 illustrates this waterfall model.

The waterfall model includes the following phases:

1. *Requirements definition.* The services that the system must provide and its operational constraints are defined.
2. *System and software design.* The overall structure of the system is designed and software subsystems are identified. Depending on the organization, the design may be fairly abstract or developed in detail. Structured design methods may be used to develop the software design.
3. *Implementation and unit testing.* The modules making up the system are individually developed in some programming language and tested.
4. *Integration and system testing.* The system modules are integrated into a complete system and this is tested as a whole.
5. *Operation and maintenance.* The software is delivered to the customer and put into use. During its lifetime, it is modified to meet changing requirements and to repair errors discovered in use.

Of course, this is a very abstract view of the software process. Each of these phases must be decomposed into a set of more detailed activities which, in themselves, may be major tasks involving several person-years of work. Further decomposition is necessary to establish the activities undertaken by each engineer. The essence of "waterfall" decomposition is that further cascades of activities should be identified. For example, Fig. 105.2 illustrates the decomposition of the system and software design phase shown in Fig. 105.1. Here, design is considered to involve six separate activities, each of which adds detail to the design. The input to each activity is the document produced by the previous design phase and its output is a document for the next phase. Naturally, each of these design activities might be further decomposed into another cascade, but this would usually only be necessary for large software systems.

The waterfall model of the software process has been widely accepted and, in one form or another, has been incorporated in many process standards such as the US Defense standard, MIL-STD-2167A. In these standards, of course, the model is expressed in detail with the deliverables from each phase carefully defined.

There are numerous variants of the waterfall model which propose different breakdowns of the basic activities of specification, design, implementation, and testing. Perhaps the best known of these is the so-called **V-model** of development (the basis of a German government standard) which explicitly links stages in the development process with validation activities. This is illustrated in Fig. 105.3.

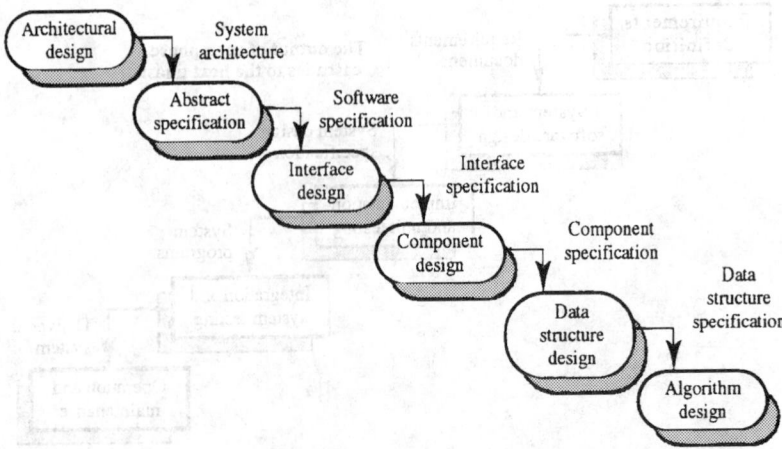

FIGURE 105.2 Waterfall model of the design process.

The V-model maintains the development phases of the waterfall model but links specific validation activities and validation plans with stages in the specification and design process. Therefore, from Fig. 105.3, we can see that the acceptance test plan is generated from the system requirements document and an abstract system specification. This is then used to drive the acceptance testing activity after the software has been delivered to the customer. Similarly, the system integration test plan is associated with the system design activity and the subsystem integration test plan with detailed design.

The V-model illustrates one of the omissions of the simple waterfall model. This model suggests that there is a single output from a phase which is the single input to the next phase. Rather, activities have multiple inputs and outputs which include both product and process information. In the V-model, some of this process information is specified. As part of each activity, a plan for the validation of the results of that activity should be produced. This plan should define the validation process and should set out tests which may be applied to the system to check that the implementation conforms to the system specification.

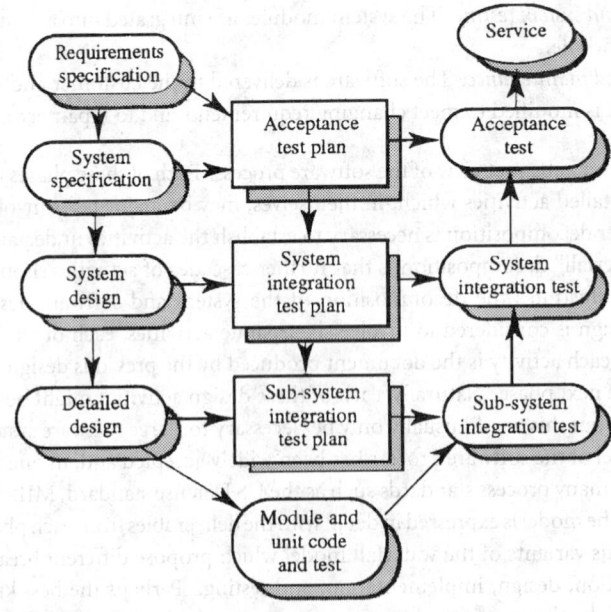

FIGURE 105.3 V-model of development.

The stages shown in the simple waterfall model closely correspond to the stages in the engineering process of developing some product to be manufactured. There are, however, critical differences between software and manufactured products which mean that the waterfall approach is an over-simplified process representation for software development. These differences are:

1. Manufactured products usually deal either with tangible materials (such as machined components, say) or with things which obey physical laws (such as electricity). It is possible to reason about and visualize their operation. By contrast, software deals with information, which is intangible and almost infinitely flexible. People find it very difficult to anticipate what information they need to help them with their work. They cannot specify their requirements with a reasonable level of confidence.
2. There are generic designs for a most manufactured products (e.g., pumps, amplifiers, wings, etc.). Both the specification and the system design can often be expressed as differences from these basic designs. Those involved in specifying and designing the system have a solid and agreed basis for the design. By contrast, there are few "standard" software system designs which are understood by all software engineers. This means that it is usually much more difficult for those involved in the process to communicate about the software which is being designed.
3. There is no real equivalent in software development to the manufacturing stage of product development. Manufacturing involves expensive outlays in raw materials and tooling. Once a product design has been released for manufacture, making changes to the design is extremely expensive and change is usually contemplated only if there are serious design errors which make the product unfit for its original purpose. By contrast, programming is really a more detailed design phase and changes to the software can be accommodated at any stage in the process.

As a consequence of these differences, there tends to be much more design and specification uncertainty in software systems and changes to the system may be required at any stage in the process.

The difficulties in specifying software with any degree of confidence and the inherent flexibility of software led to a modified form of the waterfall model where there is iteration between the stages in the model. As problems are discovered, they are fed back to earlier process stages for correction and revised documents are issued. New customer requirements also result in change and the production of a revised requirements document. This waterfall model with iteration is illustrated in Fig. 105.4.

In principle, the introduction of iteration ought to address the feedback problems of the waterfall model. As problems are discovered in later phases, earlier phases are re-entered and the documents which define the system are modified. Later phases then continue with these modified documents.

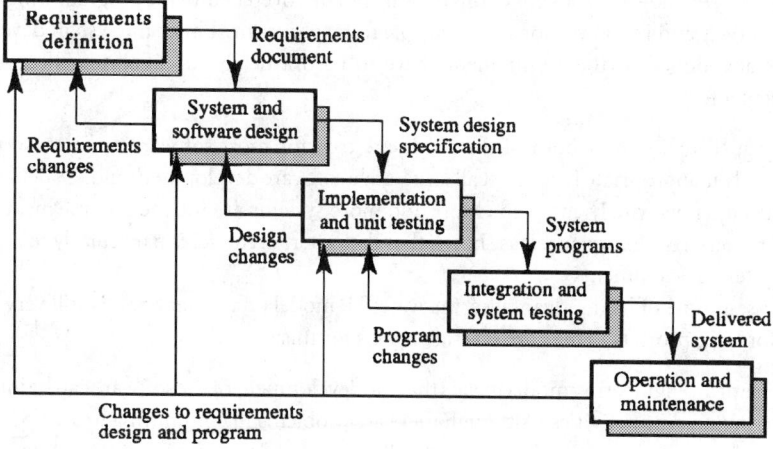

FIGURE 105.4 Iteration in the waterfall model.

The introduction of iteration into the waterfall model is intended to address some of the problems of requirements change but brings with it its own problems:

1. *When should iteration stop?* In principle, the introduction of iteration means that the development process can continue indefinitely. In practice, this is obviously impossible, so there have to be a planned number of iterations in the process. It is very difficult to assess the rate of change of the system, so managers simply plan for some arbitrary number of iterations (normally one or two) during the lifetime of the project. Any further iterations are likely to cause the project to go over budget or to fall behind schedule.

2. *How can change costs be controlled?* There is a high cost involved in analyzing the impact of system changes and reworking the system to incorporate these changes. As changes are, by their very nature, unpredictable, it is very difficult for managers to budget for the costs of iteration.

3. *How can parallel development of the system be supported?* The sequential nature of the waterfall model suggests that, during iteration, activity within a phase should stop until problems have been resolved and revised documents from earlier phases have been produced. In practice, this is completely unrealistic and, normally, development continues while the problems are being resolved. This often results in a great deal of rework, as changes to the system mean that much of this work has to be thrown away.

In small projects, these problems are often addressed informally. Rather than a formal process iteration, those working on the project work together to resolve the difficulties and to make changes to the software. The real problems arise in large projects, particularly those where different parts of the software are developed by different organizations. In these cases, a formal iteration and change mechanism must be introduced. There is much less scope for informal communication between specifiers, designers, and programmers. Iteration in these projects is therefore very expensive and often results in cost and schedule overruns. Customers, therefore, try to avoid process iterations wherever possible.

While this is understandable, there are some circumstances which can arise where this approach can lead to dramatic project failure:

1. Where the system requirements are poorly understood. This may be because the software system is an innovative one and end users have no intuition of how it will be incorporated into their work; it may be because the software is required to integrate hardware which is incompletely specified; it may be because the software is a product in a market which is changing rapidly. Whatever the reason, if requirements are poorly understood, this means that software change is inevitable. The waterfall model (in any form) does not cope well with frequent change.

2. Where the original system requirements are unrealistic and cannot be met, given the available development budget. Problems of this type are not discovered until the programming of the system is underway and initial versions are available for experiment. Either the system development has to be abandoned or the requirements have to be drastically modified to reflect what may be implemented.

It can be argued that one or both of the above are true for most software projects, so the waterfall model is probably inappropriate for almost all large-scale software development and the deficiencies of the waterfall model are now widely accepted. More and more systems are interactive systems, with complex graphical user interfaces. Experience has shown that the waterfall model is particularly inappropriate for this type of system development.

Nevertheless, in spite of its disadvantages, the waterfall model or a variant of it is still very widely used, particularly for large projects. There are several reasons for this:

1. The use of an engineering model means that the development of the software can be integrated with other engineering activities. Although there are problems in committing to a set of requirements or a design, this is sometimes necessary to allow parallel development of the subsystems in a large system.

2. The model is document-based with one or more documents being produced at each stage in the model. This makes the project visible to management and makes it possible to assess progress against budget and schedule estimates.

3. The model supports contractor/subcontractor relationships, which are normal in large projects. As the output of a stage is documented, the contract for developing the next stage may be let to some subcontractor.

The waterfall model of software development is part of many software development standards and is compatible with the process used to develop hardware systems. It allows for process management and it is familiar to engineers from all disciplines. In spite of its deficiencies, it is therefore likely to remain in use for large systems engineering projects for the foreseeable future.

A further reason for the continued use of the waterfall model is the development of "offshore" software engineering [Dedene and De Vreese 1995]. A system is specified in one country but is designed and implemented in some other country with lower labor costs. The document-based nature of the waterfall model and the separation of the development phases means that it may be applied to support this form of development.

The weakness of the waterfall model is its inability to cope with change, so it is most appropriate for systems whose requirements are well understood and which can be specified in detail with some degree of confidence. Generally, it may be successfully applied to small and medium-sized systems which automate well-understood business processes and which are re-implementations of existing systems or prototypes.

105.3 Evolutionary Development Models

The fundamental difficulties with an approach to development based on clearly defined phases such as specification, design, etc. are that stakeholders in a system find it difficult to articulate their requirements in advance and that these requirements change during the development process. The waterfall model forces premature commitment to a set of system requirements. This means that requirements change and rework is almost inevitable. Furthermore, the system which is finally developed may not meet the real needs of the customers buying that system.

To address this problem, an evolutionary approach to development may be adopted. A rudimentary system is initially produced and evolves, according to the customer's needs, to the final required system. An evolutionary approach to development does not require detailed prespecification of the system requirements, nor does it usually involve the (expensive) production of detailed documentation. These stages in this evolutionary approach are:

1. Formulate an outline of the system requirements. This need be neither complete nor consistent but should give developers some guidance as to what the system should do.
2. Develop a system, as rapidly as possible, based on this outline specification.
3. Evaluate this system with users and modify the system until the system functionality meets the users' needs. This involves modifying the initial functionality of the system and adding new functionality as required.

There are several advantages to this evolutionary approach to system development. Users do not have to develop a detailed requirements specification but need only have some general ideas on the software support which they need. A version of the software is delivered quickly so customers can gain business value from it even when it is incomplete. They also find it much easier to refine their requirements when they have a system for experiment. Ideas can be tried out and refined and the interactions between the different parts of the software can be visualized. If the software proves to be unsuitable, it can be discarded at a relatively early stage in the process.

For interactive systems, evolutionary development is particularly important. It is impossible to develop graphical user interfaces according to a waterfall model because of the very high specification uncertainty.

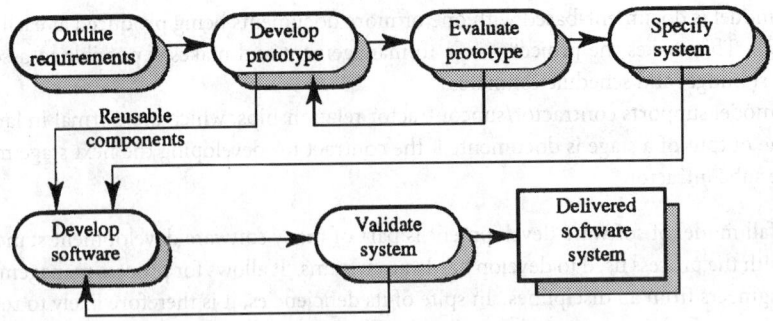

FIGURE 105.5 Throw-away prototyping.

People cannot predict, in advance, whether or not the look and feel of the interface will be acceptable and whether users will be able to access system functionality in an effective way. User interface specification and design is so dependent on the application domain and the particular system users that it must be carried out using some experimental prototype system.

Within this general evolutionary framework, there are two generic models which are commonly used. These are:

1. *Throw-away prototyping.* The objective of this evolutionary development process is to understand the customer's real software requirements. A prototype system is developed, then discarded once this understanding has been achieved.
2. *Evolutionary prototyping.* The objective of this process is to develop a software system for delivery to the customer. The prototype system which is developed becomes the production system used by the customer.

These two approaches to development are illustrated in Figs. 105.5 and 105.6. Although they appear to be similar, the fact that these approaches have different ultimate goals means that they must be approached in quite different ways.

The throw-away prototyping process has the goal of developing a better understanding of the customer's requirements so that a better system specification can be produced. This specification may then be implemented using some other development approach where the software is designed, implemented, and validated according to a waterfall-like process model. As the prototype is not intended for regular use, its development cost may be reduced by leaving out facilities such as error messages and on-line help. Nonfunctional requirements such as performance and reliability may be relaxed.

By contrast, the evolutionary prototyping approach has the goal of producing a final system for delivery to the customer. This system must include all required functionality and must meet the customer's nonfunctional requirements for reliability, availability, usability, etc. These must always be borne in mind when prototyping the system's functionality. As a consequence, the development of this type of prototype is approached quite differently from the development of a throw-away system:

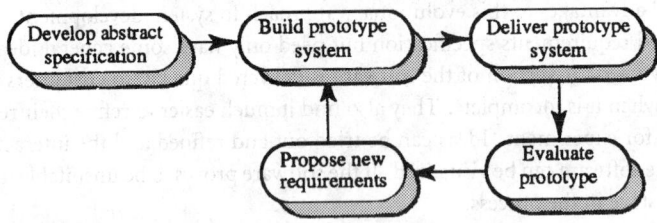

FIGURE 105.6 Evolutionary prototyping.

1. In a throw-away system, the primary objective is to understand the customer's real requirements. Therefore, prototyping should start with those requirements which are *poorly understood* so that customers have as much time as possible for requirements refinement. Well-understood requirements need not be incorporated into the prototype, so it never becomes a fully functional system.
2. In evolutionary prototyping, the primary objective is to deliver useful functionality to the customer as soon as possible. Therefore, prototyping should start with those customer requirements which are *best understood*. Requirements which are less well understood should be added incrementally as customers become familiar with the system.

An evolutionary prototyping approach is now widely used for small and medium size system development, particularly for business systems. In this application domain, prototyping systems such as rapid application development toolkits [Guerrieri 1994] and 4GLs [Wojtkowski and Wojtkowski 1994] have been developed to support the process. These systems make use of the fact that many operations are similar in that they involve retrieving information from a database and formatting that on-screen or in reports. These development systems are therefore centered on a database and include a high-level database query language, tools for screen design, and report generators.

As I have already suggested, evolutionary development is also the normal development approach for interactive systems where the user interface design is a critical part of the system. In this approach, the user interface is designed using an evolutionary approach and the functionality of the system is initially simulated. As the user interface design is refined, more and more functionality is added to the system until a final system is produced. To support this process of development, very high-level languages and environments such as Smalltalk, Lisp or, increasingly, Visual Basic may be used. Alternatively, graphical user interface builders [Colebourne et al. 1993] which allow users to layout interface components and associate functionality with them may be used to support the process.

While an evolutionary development process is more likely to lead to software which is better suited to the *end user's* requirements, it has a number of problems in its own right:

1. It has an end-user focus so critical organizational requirements (such as the need for interoperability, say) may not be given sufficient priority.
2. The constant change to software degrades its structure so that the end result is often difficult and expensive to change. Consequently, the software is expensive to maintain and may have to be completely rewritten after a relatively short lifetime. Organizations are now encountering significant maintenance problems with systems written a few years ago in 4GLs which are no longer supported or which are not available on modern PCs.
3. The process does not have a high visibility and it is difficult for managers to assess how well development is proceeding. As a result, many organizations are reluctant to use an evolutionary approach for large systems where management is the principal problem.
4. It is very difficult to apply this approach to large systems which require a significant infrastructure (e.g., a database and communications) before any system functionality can be provided. If any hardware interfacing is involved, this may have to be specified early to allow the hardware to be designed and manufactured.

Prototyping with user involvement is essential for the development of interactive user interfaces so, for interactive systems in general, evolutionary prototyping is the best generic process to adopt. For small and medium-sized business systems, evolutionary development using 4GLs means that systems can be delivered more quickly and development costs are significantly reduced compared with development in a conventional programming language using the waterfall model. It must be accepted, however, that the lifetime of systems developed using this approach is, inevitably, relatively short.

In applications which have a long lifetime or which have very high performance or reliability requirements, throw-away prototyping may be used for part or all of the system development process. The prototype may be developed using some rapid development method and the final system re-implemented using a programming language which allows better-structured, more-maintainable, and more-efficient systems to be built. For large systems, it is unusual to develop a complete system prototype. Rather, specific parts of the system with high specification uncertainty (such as the user interface) are prototyped before the final system specification is produced.

105.4 Iterative Models

The need for model iteration was discussed earlier in this chapter, where the principal problem with waterfall-type models was the lack of support which they provide for process iteration. Models based on evolutionary development do support iteration but lack visibility and may be difficult to manage. Iterative models are designed to address the need to plan for and accommodate change yet still provide a structured and manageable approach to development.

In this section, I discuss two iterative process models. These are:

1. The incremental development model, where the software is broken down into a set of separately developed increments.
2. The spiral model, where different parts of the system are built in different ways depending on an identification of the risks involved.

These are complementary rather than opposing models, so that a spiral model could be used to develop system increments. I am not aware of any published work which describes such experience, but there is clearly scope for some integration of these approaches.

Incremental Development

The waterfall model of development requires commitment to be made early in the development process. The customer for the system must commit to a set of requirements before design begins, and the designer must commit to particular design strategies before implementation. During the development process, changes in the requirements are normal and require rework of the requirements, design, and implementation.

The incremental approach to development (Fig. 105.7) was suggested by Mills [Mills et al. 1980] as a means of reducing rework in the development process and giving customers some opportunities to delay decisions on their detailed requirements until they had some experience with the system. It is an intrinsic part of the Cleanroom process discussed later in this chapter.

In an incremental development process, customers identify, in outline, the services to be provided by the system and they prioritize these services. That is, they identify which of the services are most important and which are least important to them. A number of delivery increments are then identified where each increment provides some subset of the total system functionality. Naturally, the allocation of services to increments depends on the service priority. The highest-priority services are delivered first to the customer.

After the required set of services has been identified, an overall system architectural design is produced where the set of services is partitioned and assigned to different parts of the system. The subsystems making up the system and their relationships are identified. System services may make use of common facilities such as database facilities, network support, etc., so these are also identified at this stage.

Once the system increments have been identified, the requirements for the services to be delivered in the first increment are defined in detail and that increment is developed using the most appropriate development process. During that development, further requirements analysis for later increments can take place, but no requirements changes for the increment which is currently being developed are accepted. Once an increment is completed and delivered, customers can put it into service. This means that they take early delivery of part of the system functionality. They can experiment with the system, which helps them clarify their requirements for later increments and for later versions of the current increment. As new increments are completed, they are integrated with existing increments so that the system functionality improves with each delivered increment. The common services may be implemented early in the process or may be implemented incrementally as functionality is required by an increment.

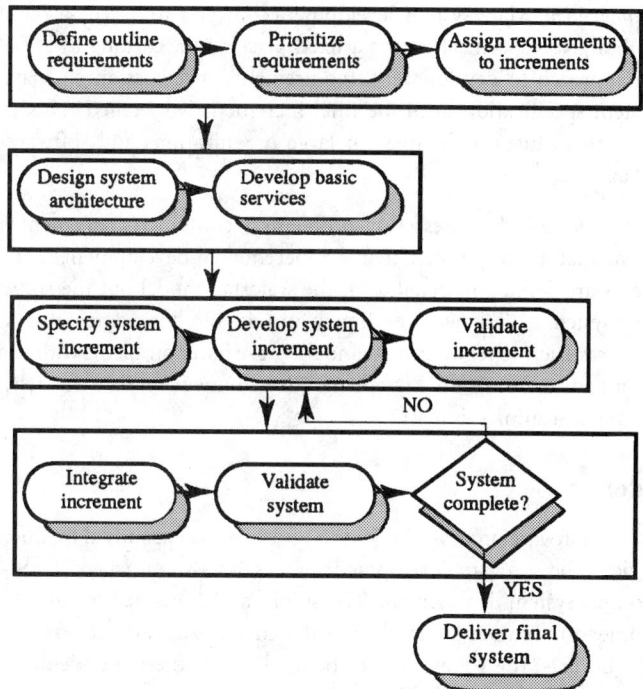

FIGURE 105.7 An incremental development process.

There is no need to use the same process for the development of each increment. Where the services in an increment have a well-defined specification, a waterfall model of development may be used for that increment. Where the specification is unclear, an evolutionary development model may be used.

This incremental development process has a number of advantages:

1. Customers do not have to wait until the entire system is delivered until they can gain value from it. The first increment satisfies their most critical requirements, so the software can be immediately put into use and contribute to the work of the customer buying that software.
2. Customers can use the early increments as a form of prototype and thus gain experience which informs the requirements for later system increments.
3. There is a lower risk of overall project failure. Although problems may be encountered in some increments, it is likely that some will be successfully delivered to the customer.
4. As the highest priority services are delivered first and later increments are integrated with them, it is inevitable that the most important system services receive the most testing. This means that customers are less likely to encounter software failures in the most important parts of the system.

However, this approach to software development does have some problems. These include:

1. *Increment identification.* Increments should be relatively small (no more than 20,000 lines of code) and each increment should deliver some system functionality. It is sometimes difficult to map the customer's requirements neatly onto increments. Requirements are not independent and, in order to satisfy one requirement, several others may also have to be implemented.
2. *Infrastructure provision.* Most systems require an infrastructure which is used by different parts of the system. As requirements are not defined in detail until an increment is to be implemented, it is difficult to identify the detailed functionality required by increments which must be provided in this infrastructure.

3. *Contract management.* Many system development contracts are based on a contractor's developing a system with a given specification for a fixed price and according to a fixed schedule. In the incremental approach to development, requirements specification is delayed so there is not a compete system specification until the final increment is specified. This requires a new form of contract, which causes difficulties for large organizations and software customers such as government agencies.

A variant of this approach which addresses the problems of contract management and common service provision is an incremental delivery rather than an incremental development model. In this case, the customer requirements are defined in detail as in the waterfall model but the software development is structured so that the system is designed, developed, and delivered incrementally. This means that it is easier to identify the infrastructure requirements and there are fewer problems with contract management. However, an important advantage of the incremental development model—namely, the ability to delay detailed requirements specification—is lost.

The Spiral Model

The **spiral model** of the software process, shown in Fig. 105.8, was proposed by Boehm in 1988 [Boehm 1988]. The model views the software development process as a spiral where development spirals from initial conception to final system deployment. The spiral model was developed for use in large defense contractors where some form of waterfall model was the normal software process used and where process standards such as MIL-STD-2167A may have to be applied. It therefore identifies the broad process activities of requirements specification, design, implementation, etc.

The developers of the spiral model recognized the inadequacies of the waterfall approach and designed the model to include activities to resolve design and specification uncertainties. It also allows for regular reviews of progress against well-defined objectives and hence does not suffer from some of the management problems of prototyping approaches.

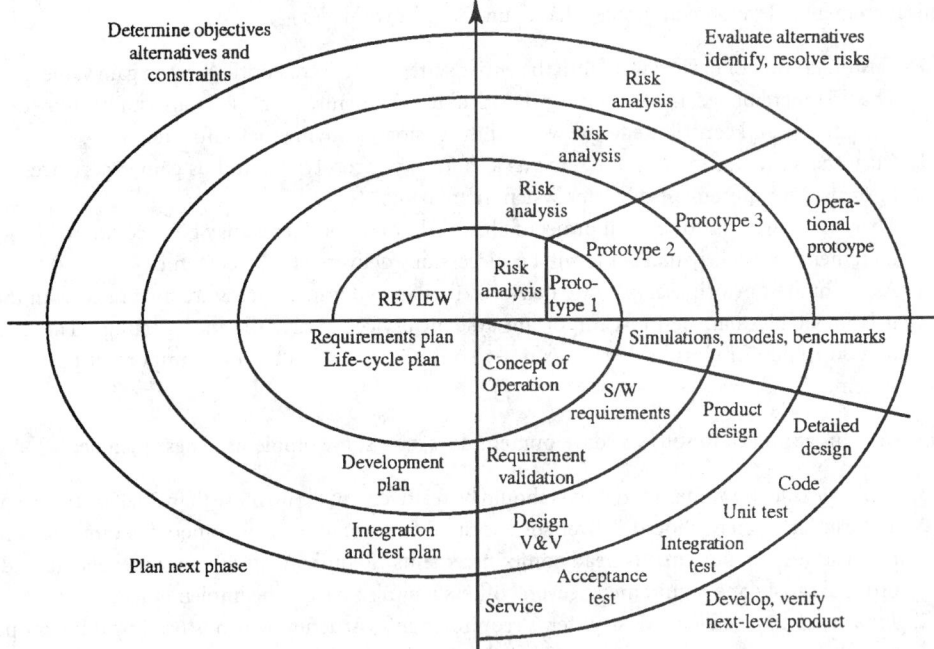

FIGURE 105.8   Boehm's spiral model of the software process. (*Source:* Boehm, B. W. 1988. A spool model of software development and enhancement. *IEEE Computer* 21(5). © IEEE, 1988. With permission.)

The spiral model is a risk-driven model where project risks which are applicable to that phase are identified and resolved before progress is made to the next stage. Each phase of the process corresponds to a round of a spiral. Within that round, there are four stages, shown in separate quadrants in Fig. 105.8:

1. An objective-setting phase, where the objectives of the round and the development constraints are established.
2. A risk-analysis phase, where the project risks are assessed against these objectives and where, if necessary, risk-resolution activities such as prototyping are carried out.
3. A development phase, which may encompass design, programming, and validation activities. Within this phase, a waterfall development model may be adopted.
4. A planning phase, where plans for the next iteration of the spiral are drawn up.

The most important contribution of the spiral model is its explicit recognition of risk in the software process. In the risk-analysis phase, the critical risks which affect that phase of development are identified and work is done to resolve these risks. Risks are considered to be a lack of information, so risk resolution involves information gathering and analysis.

For example, consider a situation where a system is intended to be a "walk up and use" system, where members of the general public use the system to retrieve information of some kind. The obvious risk here is that the user interface is not appropriate for casual use, so that people without computer experience cannot actually retrieve the information that they require. The risk-resolution activities here would involve the creation of mock-up systems and the development of prototypes which are extensively tested before the final development of the system takes place.

The spiral model is an adaptable model which can encompass other process models. For example, for systems where there is a low specification risk, the model can be considered as a waterfall model with no need for prototyping and risk-analysis activities. Where there is a high specification risk, for example, in the development of user interfaces for interactive systems, the model can be considered as an evolutionary development model with the prototyping activities dominating and relatively little development in the third quadrant.

The spiral model was designed in a large defense contracting organization so, as you would expect, it explicitly addresses the needs of project management. The first phase, which sets objectives, identifies alternatives, etc., identifies a baseline for the management process. Progress can be assessed against this. The second phase of risk analysis identifies risks and provides information to managers about these risks. The development phase may produce project documents, and the final phase is explicitly concerned with the management planning of future phases.

There are two main problems with the spiral model:

1. *The difficulties of risk analysis.* This is a very difficult and specialized process and there are relatively few people with this type of experience.
2. *Contractual problems.* This model is explicitly designed to avoid premature decision making, but the nature of system engineering contracts may make this essential. The model can really be used only where there is trust between client and contractor.

The spiral model has been enormously influential in making the notion of risk explicit to the software engineering community. There is now a widespread recognition of the need for software project risk identification and analysis. While this explicit risk analysis is important for all projects, the spiral model itself, with its emphasis on management, is most suited as an alternative to the waterfall model in large system engineering projects.

105.5 Formal Transformation

There are some classes of system, particularly safety-critical systems, where it is very important that the system conforms to its specification. Classically, this conformance is demonstrated by testing the software using test data which is comparable to that which the software must process when it is in use. However, testing can only demonstrate the presence of software errors and cannot prove their absence. Therefore, when it is essential that the software must conform to its specification, it has been argued that the conventional waterfall model, with its emphasis on system validation, is inappropriate.

Rather, an alternative model may be used which does not include an explicit testing phase to discover errors in the implemented system. In this model, a formal mathematical specification is produced and this is systematically transformed into a system implementation. The specification may be mathematically analyzed to discover inconsistencies, and these are removed at this stage. The specification transformations are correctness-preserving, so that the developed software is an exact implementation of the specification.

As current methods of formal specification and analysis do not allow the compilation of a mathematical specification into an efficient implementation, the transformation of a specification into an implementation is a multistage process. Detail is added to the specification at each stage, and the transformations which are carried out may be either automated or manual with some automated assistance.

This formal transformation process is not widely used, although some organizations which develop safety-critical systems (such as railway signalling systems) are now starting to introduce it for part of their software development. However, a variant of this approach is an important part of the Cleanroom process discussed in the next section.

105.6 The Cleanroom Process

The **Cleanroom process** was developed at IBM's Federal Systems Division with the objective of dramatically reducing the number of faults in the software delivered to customers. It combined aspects of the incremental development and the formal transformation model which have already been covered in this chapter. The Cleanroom process takes its name from the cleanroom used in semiconductor fabrication, where the objective is to provide an environment where defects are not introduced into the semiconductor wafers which are being fabricated.

The Cleanroom process was developed from work in the 1970s on structured programming [Linger et al. 1979] and is described in a number of papers by Mills and Linger [Mills et al. 1987, Mills 1988, Cobb and Mills 1990, Linger 1994]. The model is illustrated in Fig. 105.9.

The essential characteristics of the Cleanroom approach to software development are:

1. *Formal software development.* The software is mathematically specified and a development process is used where mathematical arguments are used to demonstrate that the developed software conforms to its specification. This is a weaker approach to the formal transformation model discussed above in that there is no systematic, correctness-preserving specification transformation process. However, the approaches are conceptually similar in that neither model includes a defect-

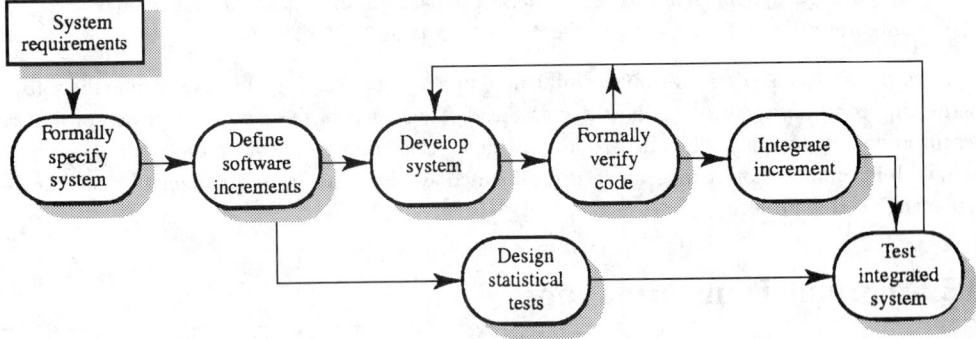

FIGURE 105.9   The Cleanroom process.

testing activity but both rely on mathematical arguments to demonstrate that a program meets its specification.

2. *Incremental development.* An essential characteristic of the Cleanroom process is the incremental development and integration of software. If the software was not structured into increments, it is unlikely that a manageable formal specification and associated mathematical correctness arguments could be produced.

3. *Statistical testing.* The testing process has the goal of validating the reliability of the software rather than defect discovery. Test data are based on an **operational profile**, which is a set of test data which reflects the frequency of actual inputs that the system must process. The number of failures detected processing these inputs reflects the software's reliability. Reliability is predicted using **reliability growth models** [Littlewood 1990].

Reports of the Cleanroom process by its developers suggest that it is very successful in producing software which has a low number of defects. Independent assessment [Selby et al. 1987] confirmed that the Cleanroom process resulted in software which had fewer defects than an approach based on defect testing.

However, the process relies on staff who have the training and ability to work with mathematical specifications and mathematical correctness arguments. This restricts its applicability to organizations which are willing to accept the relatively high training costs involved in introducing this process. It is also unclear how the process may be applied to the development of user interface software, which is an increasingly significant component of most software systems. Formal specification techniques have not been developed for specifying system interaction.

105.7 Process Model Applicability

The most appropriate software process model depends on the organization developing the software, the type of software to be developed, and the capabilities of the staff involved. There is no "ideal" model, and it makes little sense to try and fit all development in an organization to a single approach. The most appropriate process model should be chosen depending on the type of project, the application domain, and the skills and experience of the staff available.

Table 105.2 summarizes the applicability of the different models discussed here.

TABLE 105.2 Process Model Applicability

Process Model	Applicability
Waterfall model	Development of systems whose specification is well understood. This approach may also be used in systems where development is subcontracted, although it is not always technically appropriate in these cases. The waterfall approach may be required by some government organizations.
Throw-away prototyping	Parts of large systems, such as the user interface or expert system components, whose specification cannot be drawn up in advance. Software products where a prototype is developed for test marketing.
Evolutionary development	Interactive systems with a relatively short lifetime. Small to medium-sized business systems based around a database.
Incremental development	The development of large systems whose functionality can be readily partitioned and systems which have well-understood (e.g., a standard DBMS) infrastructure requirements. This approach is particularly appropriate for internal use in an organization; contractual problems are not then an issue.
Spiral model	Software which is part of a large systems engineering project and so involves development by a number of interdisciplinary teams. Again, contractual problems are avoided if the model is used within an organization.
Formal transformations	The development of relatively small safety-critical software systems or systems with very high reliability requirements.
Cleanroom process	Large systems whose functionality can be partitioned and which have very high reliability requirements. Unsuitable for the development of interactive components of these systems.

Large systems normally include subsystems of different types. Rather than impose a single process model for the whole system, each subsystem should be developed according to the most appropriate model. Well-understood parts of the system may be developed using some form of the waterfall model, and those subsystems whose requirements are difficult to predict may be developed using an evolutionary approach.

An issue which has not really been addressed is the relationship between these models and object-oriented development, where there is a blurred boundary between analysis, design, and implementation activities and where there is a significant potential for object reuse. Clearly, incremental development models are suited to this approach, but it is less clear how other process models relate to object-oriented development. This issue is most significant for waterfall models, which are embedded in many standards (such as the US MIL-STD-2167A) and which propose a development process with clear separations between phases. These cannot simply be discarded, and work is required to investigate how to integrate them with object-oriented development.

105.8 Research Issues and Summary

This chapter has discussed a number of generic models of the software process and has suggested where these models are most likely to be appropriate. These have included specification-driven models such as the waterfall model and the V-model, evolutionary and throw-away prototyping, the incremental development and spiral models, and models based on formal software specification.

The three most important factors affecting the choice of process model are:

1. *The project size.* Large projects are dominated by management concerns, so a process model which takes these into account must be used.
2. *The specification uncertainty.* If there is a lot of specification uncertainty, a process based on some form of prototyping must be used for development. In general, the interactive parts of systems always fall into this category.
3. *The nature of the development contract.* The procurer of the software may demand that a particular approach to development is used.

It is now generally recognized that there is no "ideal" process model, so research in this area is not much concerned with the development of new generic models. Rather, current research in the software process is mostly oriented around two related themes:

1. *Process support technology.* What software tools and techniques may be deployed to support the software process and to reduce overall process costs and development schedules?
2. *Process improvement.* How can existing software processes be improved to achieve the same objectives and to develop software with fewer defects?

These themes are closely related as, obviously, one approach to process improvement is process automation where technology is used to take over potentially slow and error-prone human activities.

The notion of **process support technology** is a development of the work in the 1980s on software engineering environments [Taylor et al. 1988, Bott 1989, Thomas 1989, Brown et al. 1992]. These environments were collections of CASE tools to support specific process activities with some degree of integration between these tools. However, while individual CASE tools have been successful in providing support for specific process activities, integrated environments are not widely used. They are not seen as cost-effective by most software development organizations.

There are a number of reasons for this, not least the very rapid change in hardware technology, which has meant that more and more systems are interactive systems for personal computers. These make use of built-in libraries and are often concerned with integrating a number of existing software packages rather than in developing a complete application from scratch. Integrated environments are, in essence, designed to support a waterfall model of development which is not really applicable to this class of system.

It has also been argued that another reason for the lack of use of these large-scale software engineering environments is the fact that they do not provide facilities for process definition and support. Users of these environments should be able to define a detailed model of the software process to be used in a project, the development standards and tools to be used, and the people responsible for each task. The environment should automatically schedule tasks and distribute information, as required, to the engineers involved.

Over the past few years, there has been a great deal of research into this notion of process modelling and associated support technology [Curtis et al. 1992, Krasner et al. 1992, Huff 1995]. There have been a number of experimental environments developed [Finkelstein et al. 1994], and tools such as Process Weaver [Fernström 1993] may be used to provide some measure of process automation. However, at the time of writing (1995), this technology is immature and is not widely used. It is debatable whether it will ever become mainstream software technology, as it does not appear to be particularly suited to the development of small and medium-sized application systems. However, for large software and systems engineering projects which require complex configuration management and where tens and perhaps hundreds of developers must be coordinated, process automation technology may have a role to play.

As discussed above, the notion of process improvement, particularly process improvement for defect reduction, is one which has been widely accepted in a number of industries. As failures in software systems are a result of human design errors rather than material failure (say), there is particular scope for improving products by modifying the software process so that product defects are avoided.

In this area, the most influential work has been done by the Software Engineering Institute at Carnegie Mellon University, which published the **capability maturity model** (CMM) for software process improvement [Humphrey 1988, Paulk et al. 1993, Paulk et al. 1995]. This model identifies and classifies key process activities for large-scale projects and suggests that the capability of an organization is a reflection of the number of these processes which are incorporated in the organizational software development process. Other approaches to maturity assessment, such as the Bootstrap approach [Haase et al. 1994] have also been developed.

The capability maturity model is applicable to the improvement of large-scale processes but less appropriate for small organizations concerned with smaller project development. To address this, Humphrey [Humphrey 1995] has proposed an approach to developing a personal software process and process improvement strategies.

The importance of the software process and software process models is now generally recognized. Evolving software processes to meet new demands for rapid delivery of high-quality interactive software is perhaps the major challenge which we face in the future.

Defining Terms

Capability maturity model: A process improvement model which defines a number of levels of process maturity in terms of the key processes undertaken at these levels.

Cleanroom process: An approach to software development based on fault avoidance. The software is split into increments and each increment is formally specified, verified, and statistically tested.

Operational profile: A set of test data whose frequency reflects the actual usage of the system.

Process support technology: Any CASE tools or methods used to support software process activities.

Reliability growth model: A mathematical model which is used to predict when a given level of software reliability is likely to be reached.

Software process: The activities and their inputs and outputs which are involved in developing a software system from initial conception through to final delivery to a customer.

Software process model: An abstract model of the software process which identifies the principal activities and their deliverables. It may also include information about the tools and development environment used and about the roles of the people responsible for particular activities.

Spiral model: An incremental process model based on cycles where each cycle includes objective setting, risk analysis, development, and planning of the next cycle.

V-model: A derivative of the waterfall model where specification, design, and development activities are explicitly linked to validation activities through deliverables.

Waterfall model: A software process model based on engineering models where the system is specified, designed, implemented, and tested in separate phases.

References

Armenise, P., Bandinelli, S., Ghezzi, C., and Mortenzi, A. 1992. Software process representation languages: survey and assessment. In *Proc. 4th Int. Conf. Software Engineering Knowledge Engineering*. Capri, Italy.

Boehm, B. W. 1988. A spiral model of software development and enhancement. *IEEE Comput.* 21(5):61–72.

Bott, M. F. 1989. *The ECLIPSE Integrated Project Support Environment*. Peter Perigrinus, Stevenage, UK.

Brown, A. W., Earl, A. N., and McDermid, J. A. 1992. *Software Engineering Environments*. McGraw–Hill, London.

Christie, A. 1994. *A Practical Guide to the Technology and Adoption of Software Process Automation*. Software Engineering Institute. Carnegie–Mellon University, Pittsburgh, PA.

Cobb, R. H. and Mills, H. D. 1990. Engineering software under statistical quality control. *IEEE Software* 7(6):44–54.

Colebourne, A., Sawyer, P., and Sommerville, I. 1993. MOG user interface builder: a mechanism for integrating application and user interface. *Interacting Computers* 5(3):315–332.

Curtis, B., Kellner, M. I., and Over, J. 1992. Process modeling. Commun. ACM 35(9):75–90.

Dedene, G. and De Vreese, J.-P. 1995. Realities of off-shore engineering. *IEEE Software* 12(1):35–45.

Fernström, C. 1993. Process Weaver: adding process support to Unix. In *Proc. 2nd Int. Conf. Software Process*, Berlin.

Finkelstein, A., Kramer, J., and Nuseibeh, B., eds. 1994. *Software Process Modelling and Technology*. Wiley, New York.

Guerrieri, E. 1994. Case study: Digital's application generator. *IEEE Software* 11(5):95–96.

Haase, V., Messnarz, R., Koch, G., Kugler, H. J., and Decrinis, P. 1994. Bootstrap: fine tuning process assessment. *IEEE Software* 11(4):25–35.

Huff, K. E. 1995. Software process modeling. In *Trends in Software Process*, A. Fuggetta and A. Wolf, eds., pp. 1–24. Wiley, New York.

Humphrey, W. S. 1988. Characterizing the software process. *IEEE Software* 5(2):73–79.

Humphrey, W. S. 1995. *A Discipline for Software Engineering*. Addison–Wesley, Reading, MA.

Krasner, H., Terrel, J., Linehan, A., Arnold, P., and Ett, W. 1992. Lessons from a learned software process modeling system. *Commun. ACM* 35(9):91–100.

Linger, R. C. 1994. Cleanroom process model. *IEEE Software* 11(2):50–58.

Linger, R. C., Mills, H. D., and Witt, B. I. 1979. *Structured Programming—Theory and Practice*. Addison–Wesley, Reading, MA.

Littlewood, B. 1990. Software reliability growth models. In *Software Reliability Handbook*. P. Rook, ed., pp. 401–412. Elsevier, Amsterdam.

Mills, H. D. 1988. Stepwise refinement and verification in box-structured systems. *IEEE Comput.* 21 (6):23–37.

Mills, H. D., Dyer, M., and Linger, R. 1987. Cleanroom software engineering. *IEEE Software* 4(5):19–25.

Mills, H. D., O'Neill, D., Linger, R. C., Dyer, M., and Quinnan, R. E. 1980. The management of software engineering. *IBM Sys. J.* 24(2):414–477.

Paulk, M. C., Curtis, B., Chrissis, M. B., and Weber, C. V. 1993. Capability maturity model, version 1.1. *IEEE Software* 10(4):18–27.

Paulk, M. C., Weber, C. V., Curtis, B., and Chrissis, M. B. 1995. *The Capability Maturity Model: Guidelines for Improving Software Process*. Addison–Wesley, Reading, MA.

Royce, W. W. 1970. Managing the development of large software systems: concepts and techniques, pp. 1–9. In *Proc. IEEE WESTCON*, Los Angeles, CA.

Selby, R. W., Basili, V. R., and Baker, F. T. 1987. Cleanroom software development: an empirical evaluation. *IEEE Trans. Software Eng.* SE-13(9):1027–1037.

Taylor, R. N., Selby, R. W., Young, M., Belz, F. C., Clarke, L. A., Wileden, J. C., Osterweil, L., and Wolf, A. L. 1988. Foundations for the Arcadia environment architecture. *SIGSOFT Software Engineering Notes* 13(5):1–13.

Thomas, I. 1989. PCTE interfaces: supporting tools in software engineering environments. *IEEE Software* 6(6):15–23.

Wojtkowski, W. G. and Wojtkowski, W. 1994. *4GL Tools and Methods.* Boyd and Fraser, Boston, MA.

Further Information

Process models and the software process in general are covered in most software engineering textbooks [Pressman 1992, Sommerville 1996]. Research in software process issues is covered in recent books such as those by Finkelstein et al. [Finkelstein et al. 1994] and by Fuggetta and Wolf [Fuggetta and Wolf 1996]. There is a series of international and European workshops on the software process and software process technology and, more recently, an international software process conference has been established. Proceedings of the European workshops have been published by Springer; proceedings of the international workshops and process conference are available from the IEEE Computer Society.

The SEI approach to process improvement is well documented [Paulk et al. 1993], and reports of practical experience of the applicability of these models are available directly from the SEI (accessible through the World Wide Web at http://www.sei.cmu.edu/FrontDoor.html). Smaller-scale process improvement is covered in Humphrey's book on personal software processes [Humphrey 1995]. Process measurement and process improvement are discussed in the ami handbook (Addison–Wesley 1995) and in a series of papers by Basili [Basili and Rombach 1988, Basili and Green 1993].

Basili, V. and Green, S. 1993. Software process improvement at the SEL. *IEEE Software* 11(4):58–66.

Basili, V. R. and Rombach, H. D. 1988. The TAME project: towards improvement-oriented software environments. *IEEE Trans. Software Eng.* 14(6):758–773.

Fuggetta, A. and Wolf, A., eds. 1996. *Trends in Software: The Software Process.* Wiley, Chichester, UK.

Pressman, R. S. 1996. *Software Engineering—A Practitioners Approach.* 4th ed. McGraw–Hill, New York.

Sommerville, I. 1996. *Software Engineering.* 5th ed. Addison–Wesley, Wokingham, UK.

106

Software Qualities and Principles

Carlo Ghezzi*
Politecnico di Milano

Mehdi Jazayeri
Technische Universität Wien

Dino Mandrioli
Politecnico di Milano

The goal of any engineering activity is to build something—a product. The civil engineer builds a bridge, the aerospace engineer builds an airplane, and the electrical engineer builds a circuit. The product of software engineering is a *software system* or *application*. It is not as tangible as the other products, but it is a product nonetheless, and it serves a function.

In some ways software products are similar to other engineering products, and in some ways they are very different. The characteristic that perhaps distinguishes software from other engineering products is that software is *malleable*. We can modify the product itself—as opposed to its design—rather easily. This makes software quite different from other products such as cars or ovens.

The **malleability** of software is often misused. Although it is certainly possible to satisfy a new requirement by modifying a bridge or an airplane—for example, to make the bridge support more traffic or the airplane to carry more cargo—such a modification is not taken lightly and certainly is not attempted without first making a design change and verifying the impact of the change extensively. Software engineers, on the other hand, are often asked to perform such modifications on software. Because of its malleability, we seem to think that changing software is easy. In practice, seemingly easy changes have significant and unexpected ramifications.

We may be able to change the code easily with a text editor, but meeting the need for which the change was intended is not necessarily done so easily. Indeed, we must treat software like other engineering products in this regard: a change in software must be viewed as a change in the design rather than in the code, which is just an instance of the product. We can indeed exploit the malleability property, but we need to do it with discipline.

Another characteristic of software is that its creation is human intensive: it requires mostly engineering rather than manufacturing. In most other engineering disciplines, the manufacturing process is considered

*From Ghezzi, C., Jazayeri, M., and Mandrioli, D. 1991. *Fundamentals of Software Engineering*. Prentice-Hall, NJ. Used with permission.

carefully because it determines the final cost of the product. Also, the process has to be managed closely to ensure that defects are not introduced. The same considerations apply to computer hardware products. For software, on the other hand, "manufacturing" is a trivial process of duplication. The software production process deals with design and implementation rather than manufacturing. This process has to meet certain criteria to ensure the production of high-quality software.

Any product is expected to fulfill some need and meet some acceptance standards that set forth the qualities it must have. A bridge performs the function of connecting two disconnected points; one of the qualities it is expected to have is that it will not collapse in a storm or under the weight of a convoy of trucks. In traditional engineering disciplines, the engineer has tools for describing the qualities of the product distinctly from the design of the product. In software engineering, the distinction is not yet so clear. The qualities of the software product are often intermixed in specifications with the design of the product.

To achieve the desired qualities, the construction of any nontrivial product must follow sound *design principles*. Sound design principles apply to any engineering discipline, including software engineering. In applying such principles to software engineering, they must be customized to deal with the peculiar characteristics of software.

Software engineering principles deal with both the *process* of software engineering and the final *product*. The right process will help produce the right product, but the desired product will also affect the choice of which process to use. A traditional problem in software engineering has been the emphasis on either the process or the product to the exclusion of the other. Both are important.

In this chapter, we present general principles that apply throughout the process of software construction and management. Principles, however, are not sufficient. In fact, they are general and abstract statements describing desirable properties of **software processes** and **products**. To apply principles, the software engineer uses appropriate *methods* and specific *techniques* that help incorporate the desired properties into processes and products.

Sometimes, methods and techniques are packaged together to form a **methodology**. The purpose of a methodology is to promote a certain approach to solving a problem by preselecting the methods and techniques to be used. Some important software design methods will be the issue of specific chapters of this Handbook. Many methodologies are supported by software tools. Software tools will be the subject of Chapter 112.

In this chapter, we first examine the qualities that are pertinent to software products and software production processes (sections 106.1 through 106.3) and means to assess them (section 106.4). Next, we present software engineering principles that can be applied to achieve these qualities (section 106.5). We will also refer to other chapters of this Handbook where specific design techniques and methods are described. **Quality assurance** is the process of verifying whether a software product meets the required qualities. We will briefly address the issue of quality assurance (section 106.4) and refer to other chapters of the Handbook that deal specifically with this topic.

Several considerations affect the choice of qualities and principles that must be emphasized. We consider the case where the software is to be developed not just for personal use but for users with little or even no knowledge of computers and software. We also assume that the application is sufficiently large and complex that special effort is required to decompose it into manageable parts. Thus, there is a need for a software development approach that helps to overcome its complexity. Clearly, if the software does not have high quality requirements and its complexity is easily manageable, the need for software engineering principles and techniques diminishes greatly. In general, the choice of principles and techniques is determined by the software quality goals. Software for critical applications, where the effects of errors are serious, even disastrous, impose stricter reliability requirements than noncritical applications.

106.1 Classification of Software Qualities

There are many desirable software qualities. Some of these apply both to the product and to the process used to produce the product. The user wants the software product to be reliable, efficient, and easy to use. The producer of the software wants it to be verifiable, maintainable, portable, and extensible.

The manager of the software project wants the process of software development to be productive and predictable.

In this section, we consider two different classifications of software-related qualities: product versus process and internal versus external.

Product and Process Qualities

We use a *process* to produce the software product. We can also attribute some qualities to the process, although process qualities often are closely related to product qualities. For example, if the process requires careful planning of system test data before any design and development of the system starts, product reliability will increase. Some qualities, such as efficiency, apply both to the product and to the process.

It is interesting to examine the word *product* here. It usually refers to what is delivered to the customer. Even though this is an acceptable definition from the customer's perspective, it is not adequate for the developer who requires a general definition of a software product that encompasses not only the executable code and the user manual that are delivered to the customer but also the requirements, design, source code, test data, etc. With such a definition, all of the artifacts that are produced during the process constitute parts of the product. In fact, it is possible to deliver different subsets of the same product to different customers.

For example, a computer manufacturer might sell to a process control company the object code to be installed in the specialized hardware for an embedded application. It might sell the object code and the user's manual to software dealers. It might even sell the design and the source code to software vendors who modify them to build other products. In this case, the developers of the original system see one product, the salespersons in the same company see a set of related products, and the end user and the software vendor see still other, different products.

Process quality has received increasing attention over time as its impact on product quality has been recognized and observed. This has even led to classifying *software development organizations* on the basis of an evaluation of their processes. For instance, the Software Engineering Institute (SEI) has defined five levels of "maturity" that may be used to characterize the software process quality of an organization. The maturity level of an organization is intended to reflect the ability of the organization to predict their costs and schedules. The levels are called initial, repeatable, defined, managed, and optimizing. The initial level characterizes an organization that has no statistical control over their processes and no predictability. At the second level, the organization has a stable, repeatable process, supported by rigorous project management controls. At the third level, the organization has defined a base process for consistent project management of different projects. At the fourth level of maturity, the process is systematically monitored and its performance is measured with appropriate metrics. At the optimizing level of maturity, the process measurements are used to continuously improve the process. It turns out that most organizations fall into maturity levels 2 and 3. The software maturity models of SEI are being used by organizations to document their "competence" and by customers to qualify their vendors. Process maturity characterization is now being extended to the work of individual software engineers. While the particular way of measuring process maturity and the appropriate metrics to be used are sometimes controversial, the impact of process quality on product quality is generally accepted as substantial.

External Versus Internal Qualities

We can divide software qualities into *external* and *internal* qualities. The external qualities are visible to the users of the system; the internal qualities are those that concern the developers of the system. In general, users of the software care only about the external qualities, but it is the internal qualities—which deal largely with the structure of the software—that help developers achieve the external qualities. For example, the internal quality of verifiability is necessary for achieving the external quality of reliability. In many cases, however, the qualities are related closely and the distinction between internal and external may depend on the user and the delivered product. For instance, a well-documented design is usually an internal quality; in some cases, however, users want the delivery of design documentation as an essential part of the product (e.g., for military products); in this case, this quality becomes external.

106.2 Representative Software Qualities

In this section, we present the most important qualities of software products and processes. Where appropriate, we analyze a quality with respect to the classifications discussed above. Several "software quality models" are proposed in the literature. Most of them share the essential aspects, but differ in emphasis and organization. Quoting the International Standards Organization [ISO 1991], "The maturity of the models, terms, and definitions does not yet allow them to be included in a standard."

Correctness, Reliability, and Robustness

The terms **correctness**, **reliability**, and **robustness** are often used interchangeably to characterize a quality of software that implies that the application performs its functions as expected. At other times, the terms are used with different meanings by different people. Lack of a standard terminology is unfortunate, because these terms deal with important issues. We try to clarify these topics below.

Correctness

A program is *functionally correct* if it behaves according to the specification of the functions it should provide (its *functional requirements*). It is common simply to use the term "correct" rather than "functionally correct"; similarly, in this context, the term "specifications" implies "functional requirements specifications." We will follow this convention when the context is clear.

The definition of correctness assumes that a specification of the system is available and that it is possible to determine unambiguously whether or not a program meets the specification. There are even existing systems for which no specification exists at all. If a specification does exist, it is usually written in an informal style using natural language. Such a specification is likely to contain many ambiguities. Regardless of difficulties with current specifications, however, the definition of correctness is useful. Clearly, correctness is a desirable property for software systems.

Correctness is a mathematical property that establishes the equivalence between the software and its specification. If the specification of the functional requirements is given in a fully rigorous form, then correctness can be assessed systematically and precisely. Correctness can be evaluated through a variety of methods, some stressing an experimental approach (e.g., testing), others stressing an analytic approach (e.g., formal correctness verification). Correctness can be enhanced by using appropriate tools such as high-level languages, particularly those supporting extensive static analysis. Likewise, it can be improved by using standard algorithms or libraries of standard modules rather than inventing new ones.

Reliability

Informally, software is reliable if the user can depend on it.[1] The specialized literature on software reliability defines reliability in terms of statistical behavior—the probability that the software will operate as expected over a specified time interval. For the purpose of this chapter, however, the informal definition is sufficient.

Correctness is an absolute quality: any deviation from the requirements makes the system incorrect, regardless of how minor or serious the consequence of the deviation. The notion of reliability is, on the other hand, relative: if the consequence of a software error is not serious, the incorrect software may still be reliable.

Engineering products are *expected* to be reliable. Unreliable products, in general, disappear quickly from the marketplace. Unfortunately, software products have not achieved this enviable status yet. Software products are commonly released along with a list of "known bugs." Users of software take it for granted that "Release 1" of a product is "buggy." This is a striking symptom of the immaturity of the software engineering discipline.

In classic engineering disciplines, a product is not released if it has "bugs." You do not expect to take delivery of an automobile along with a list of shortcomings or a bridge with a warning not to use the railing.

---

[1] "Dependable" is a term used as a synonym for "reliable."

Design errors are extremely rare and worthy of news headlines. A bridge that collapses may even result in court proceedings against the designers.

On the contrary, software design errors are generally treated as unavoidable. Rather than being surprised at the occurrence of software errors, we *expect* them. Instead of a guarantee of reliability, software buyers receive a disclaimer that the software manufacturer is not responsible for any damages due to product errors. Software engineering can truly be called an engineering discipline only when we can achieve software reliability comparable to the reliability of other products.

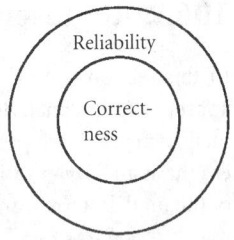

FIGURE 106.1 Relationship between correctness and reliability in the ideal case.

Figure 106.1 illustrates the relationship between reliability and correctness, under the assumption that the functional requirements specification indeed captures all the desirable properties of the application and that no undesirable properties are erroneously specified in it. The figure shows that the set of all reliable programs includes the set of correct programs but not vice versa. In practice, there is an additional problem. In fact, the specification is at best a *model* of what the user wants, but the model may or may not be an accurate statement of the user's needs and actual requirements. All the software can do is meet the specified requirements of the model—it cannot ensure the accuracy of the model.

Thus, Fig. 106.1 represents an idealized situation where the requirements are themselves assumed to be correct, i.e., they are a faithful representation of what the implementation must ensure in order to satisfy the needs of the expected users. Often, however, there are insurmountable obstacles to achieving this goal. The result is that we sometimes have correct applications that are designed for "incorrect" requirements. In this case, correctness of the software is not sufficient to guarantee that the software behaves "as expected by the user." This situation is discussed in the next subsection.

Robustness

A program is robust if it behaves "reasonably" even in circumstances that were not anticipated in the requirements specification—for example, when it encounters incorrect input data or some hardware malfunction (say, a disk crash). A program that assumes perfect input and generates an unrecoverable run-time error if the user types an incorrect command would not be robust. It might be correct, though, if the requirements specification does not state what the action should be in response to incorrect commands. Obviously, robustness is a difficult-to-define quality; after all, if we could state precisely what we should do to make an application robust, we would be able to specify its "reasonable" behavior completely and robustness would become equivalent to correctness (or reliability, in the sense of Fig. 106.1).

Again, an analogy with bridges is instructive. Two bridges connecting two banks of the same river are both "correct" if they each satisfy the stated requirements. If, however, during an unexpected, unprecedented, torrential rain, one collapses and the other one does not, we can call the latter more robust than the former. Notice that the lesson learned from the collapse of the bridge will probably lead to more complete requirements for future bridges, establishing "resistance to torrential rains" as a correctness requirement. In other words, as the phenomenon under study becomes better known, we will approach the ideal case shown in Fig. 106.1, where specifications capture exactly the expected requirements.

The amount of code devoted to robustness depends on the application area. For example, a system written to be used by novice computer users must be more prepared to deal with ill-formatted input than an embedded system that receives its input from a sensor—although, if the embedded system is controlling the space shuttle or life-critical devices, then extra robustness is advisable.

In conclusion, we can see that robustness and correctness are strongly related without a sharp dividing line. If we put a requirement in the specification, its accomplishment becomes an issue of correctness; if we leave it out of the specification, it may become an issue of robustness. The border line between the two qualities is the specification of the system. Finally, reliability comes in because not all incorrect behaviors signify equally serious problems; some incorrect behaviors may be tolerable.

Correctness, robustness, and reliability also apply to the software production process. A process is robust, for example, if it can accommodate unanticipated changes in the environment, such as a new release of the operating system or the sudden transfer of half the employees to another location. A process is reliable if it consistently leads to the production of high-quality products. In many engineering disciplines, considerable research is devoted to the discovery of reliable processes.

Performance

Any engineering product is expected to meet a certain level of **performance**. Unlike the situation in other disciplines, in software engineering we often equate performance with efficiency, but they are not the same. Efficiency is an internal quality and refers to how economically the software utilizes the resources of the computer. Performance, on the other hand, is an external quality based on user requirements. For example, a telephone switch may be required to be able to process 10,000 calls per hour. Efficiency affects, and often determines, the performance of a system.

Performance is important because it affects the usability of the system. If a software system is too slow, it reduces the productivity of the users, possibly to the point of not meeting their needs. If a software system uses too much disk space, it may be too expensive to run. If a software system uses too much memory, it may affect the other applications that are run on the same system because of system thrashing.

Efficiency issues are dependent on current technology. Our view of what is "too expensive" is constantly changing as advances in technology extend the limits. Computers today cost orders of magnitude less than they did only a few years ago, yet they provide orders of magnitude more power.

Performance determines the scalability of a software system. An algorithm that is quadratic may work on small inputs but not work at all for larger inputs. For example, a compiler that uses a register allocation algorithm whose running time is the square of the number of program variables will run slower and slower as the length of the program being compiled increases.

There are several ways to *evaluate* the performance of a system. One method is to analyze the complexity of algorithms used in the software. An extensive theory exists for characterizing the average or worst-case behavior of algorithms, in terms of significant resource requirements such as time and space, or—less traditionally—in terms of number of message exchanges in the case of distributed systems.

Analysis of the complexity of algorithms provides only average or worst-case information, rather than specific information, about a particular implementation. For more specific information, we can use techniques of performance evaluation. The three basic approaches to evaluating the performance of a system are *measurement, analysis*, and *simulation.* We can measure the actual performance of a system by means of hardware and software monitors that collect data while the system is running and allow us to discover bottlenecks in the system. In this case, it is crucial to select input data that lead to representative executions of the system. The second approach is to build a model of the product and analyze it. The third approach is to build a model that simulates the product. Analytic models—often based on queuing theory—are usually easier to build but are less accurate, whereas simulation models are more costly to build but are more accurate. We can sometimes combine the two techniques. At the start of a large project, an analytic model can provide a general understanding of the performance-critical areas of the product, pointing out areas where more thorough study is required; then we can build simulation models of these particular areas.

The notion of performance also applies to a process, in which case we call it **productivity**. Productivity is important enough to be treated as an independent quality and is discussed as such later under timeliness.

Usability

A software system is usable—or user friendly—if its human users find it easy to use. This definition reflects the subjective nature of **usability**. Properties that make an application user-friendly to novice users are different from those desired by expert users. For example, a novice user may appreciate verbose messages, whereas an experienced user will ignore them. Similarly, a nonprogrammer may appreciate the use of menus, whereas a programmer may be more comfortable with typing a textual command.

The user interface is an important component of usability. A software system that presents the novice user with a window interface and a mouse is friendlier than one that requires the user to remember and use a set of one-letter commands. On the other hand, an experienced user might prefer a set of commands that minimize the number of keystrokes rather than a fancy window interface through which he has to navigate to get to the command that he knew all along he wanted to execute.

There is more to usability, however, than the user interface. For example, an embedded software system does not have a human user interface. Instead, it interacts with hardware and perhaps other software systems. In this case, the usability is reflected in the ease with which the system can be configured and adapted to the hardware environment.

In general, the usability of a system depends on the consistency of its user and operator interfaces. Clearly, however, the other qualities mentioned above—such as correctness and performance—also affect usability. A software system that produces wrong answers is not usable, regardless of how fancy its user interface is. Also, a software system that produces answers more slowly than the user requires is not usable even if the answers are displayed in color.

Usability is also discussed under the subject "human factors." Human factors or human engineering plays a major role in many engineering disciplines. For example, automobile manufacturers devote significant effort to deciding the position of the various control knobs on the dashboard. Television manufacturers and microwave oven makers also try to make their products easy to use. User-interface decisions in these classical engineering fields are made, not randomly by engineers, but only after extensive study of user needs and attitudes by specialists in fields such as industrial design or psychology.

Interestingly, ease of use in many of these engineering disciplines is achieved through standardization of the human interface. Once a user knows how to use one television set, he or she can operate almost any other television set. The significant current research and development activity in the area of standard user interfaces for software systems should lead to guidelines for developing more user-friendly systems in the future.

Verifiability

A software system is **verifiable** if its properties can be verified easily. For example, the correctness or the performance of a software system is a property we would be interested in verifying. Verification can be performed either by—possibly formal—analysis methods or through testing. A common technique for improving verifiability is the use of "software monitors," that is, code inserted in the software to monitor various qualities such as performance or correctness. Modular design, disciplined coding practices, and the use of an appropriate programming language all contribute to verifiability.

Verifiability is usually an internal quality, although it sometimes becomes an external quality also. For example, in many security-critical applications, the customer requires the verifiability of certain properties. The highest level of the security standard for a "Trusted Computer System" requires the verifiability of the operating system kernel. Chapter 110 of this book discusses testing and Chapter 109 is devoted to verification.

Maintainability

The term "software maintenance" is commonly used to refer to the modifications that are made to a software system after its initial release. **Maintenance** used to be viewed as merely "bug fixing," and it was distressing to discover that so much effort was being spent on fixing defects. Studies have shown, however, that the majority of time devoted to maintenance is in fact spent on enhancing the product with features that were not in the original specifications or were stated incorrectly.

"Maintenance" is indeed not the proper word to use with software. First, as it is used today, the term covers a wide range of activities, all having to do with modifying an existing piece of software in order to make an improvement. A term that perhaps captures the essence of this process better is "software evolution." Second, in other engineering products, such as computer hardware or automobiles or washing machines, "maintenance" refers to the upkeep of the product in response to the gradual deterioration of parts due to extended use of the product. For example, automobile transmissions are oiled and air filters are dusted and periodically changed. To use the word "maintenance" with software gives the wrong connotation because software does not wear out. Unfortunately, however, the term is used so widely that we will continue using it.

There is evidence that maintenance costs consume over 60% of the total resources of software development organizations. To analyze the factors that affect such costs, it is customary to divide software maintenance into three categories: *corrective, adaptive,* and *perfective* maintenance.

Corrective maintenance has to do with the removal of residual errors present in the product when it is delivered as well as errors introduced into the software during its maintenance. Corrective maintenance accounts for about 20% of maintenance costs.

Adaptive and perfective maintenance are the real sources of change in software; they motivate the introduction of **evolvability** as a fundamental software quality and *anticipation of change* as a general principle that should guide the software engineer. Both of these concepts are discussed below. Adaptive maintenance accounts for nearly another 20% of maintenance costs, whereas over 50% is absorbed by perfective maintenance.

Adaptive maintenance, which accounts for another 20% of maintenance costs, involves adjusting the application to changes in the environment, e.g., a new release of the hardware or the operating system or a new database system. In other words, in adaptive maintenance the need for software changes cannot be attributed to a feature in the software itself, such as the presence of residual errors or the lack of a functionality required by the user. Rather, the software must change because its environment changes.

Finally, perfective maintenance, which accounts for more than 50% of maintenance costs, involves changing the software to improve some of its qualities. Here, changes are due to the need to modify the functions offered by the application, add new functions, improve the performance of the application, make it easier to use, etc. The requests to perform perfective maintenance may come directly from the software engineer, in order to improve the status of the product on the market, or they may come from the customer, to meet some new requirements.

We view maintainability as two separate qualities: **repairability** and evolvability. Software is repairable if it allows the fixing of defects; it is evolvable if it allows changes that satisfy new requirements. The distinction between repairability and evolvability is not always clear. For example, if the requirements specification is vague, it may not be clear whether we are fixing a defect or satisfying a new requirement. In general, however, the distinction between the two qualities is useful.

Repairability

A software system is repairable if it allows the correction of its defects with a limited amount of work. In many engineering products, repairability is a major design goal. For example, automobile engines are built with the parts that are most likely to fail as the most accessible. In computer hardware engineering, there is a subspecialty called repairability, availability, and serviceability (RAS).

In other engineering fields, as the cost of a product decreases and the product assumes the status of a commodity, the need for repairability decreases: it is cheaper to replace the whole thing, or at least major parts of it, than to repair it. For example, in early television sets, you could replace a single vacuum tube. Today, a whole board has to be replaced.

A common technique for achieving repairability in such products is to use standard parts that can be replaced easily. But software parts do not deteriorate. Thus, while the use of standard parts can reduce the cost of software *production*, the concept of replaceable parts does not seem to apply to software repairability. Software is also different in this regard because the cost of software is determined, not by tangible parts, but by human design activity.

Repairability is also affected by the *number* of parts in a product. For example, it is harder to repair a defect in a monolithic automobile body than if the body were made of several regularly shaped parts. In the latter case, we could replace a single part more easily than the whole body. Of course, if the body consisted of too many parts, it would require too many connections among the parts, leading to the probability that the connections themselves might need repair.

An analogous situation applies to software: a software product that consists of well-designed modules is much easier to analyze and repair than a monolithic one. Merely increasing the number of modules, however, does not make a more repairable product. We have to choose the right module structure with the right module interfaces to reduce the need for module interconnections. The right modularization promotes repairability by allowing errors to be confined to few modules, making it easier to locate and remove them.

Repairability can be improved through the use of proper tools. For example, using a high-level language rather than an assembly language leads to better repairability. Also, tools such as debuggers can help in isolating and repairing errors. A product's repairability affects its reliability. On the other hand, the need for repairability decreases as reliability increases.

Evolvability

Like other engineering products, software products are modified over time to provide new functions or to change existing functions. Indeed, the fact that software is so malleable makes modifications extremely easy to apply to an implementation. There is, however, a major difference between software modification and modification of other engineering products. In the case of other engineering products, modifications start at the design level and proceed to implementation of the product. For example, if one decides to add a second story to a house, first one must do a feasibility study to check whether this can be done safely. Then one is required to do a design, based on the original design of the house. Then the design must be approved after assessing that it does not violate the existing regulations. And, finally, the construction of the new part may be commissioned.

In the case of software, unfortunately, people seldom proceed in such an organized fashion. Although the change might be a radical change in the application, too often the implementation is started without doing any feasibility study, let alone a change in the original design. Still worse, after the change is accomplished, the modification is not even documented *a posteriori*; i.e., the specifications are not updated to reflect the change. This makes future changes increasingly difficult to apply.

On the other hand, successful software products are quite long lived. Their first release is the beginning of a long lifetime, and each successive release is the next step in the evolution of the system. If the software is designed with care, and if each modification is thought out carefully, then it can evolve gracefully.

As the cost of software production and the complexity of applications grow, the importance of software evolvability increases. One reason for this is the need to leverage the investment made in the software as the hardware technology advances. Some of the earliest large systems developed in the 1960s are today taking advantage of new hardware, device, and network technologies. For example, the American Airlines SABRE reservation system, initially developed in the middle 1960s, is still evolving with new functionality. This is an amazing feat, considering the increasing performance demands on the system.

Most software systems start out being evolvable, but after years of evolution they reach a state where any major modification runs the risk of "breaking" existing features. In fact, evolvability is achieved by modularization, and successive changes tend to reduce the modularity of the original system. This is even worse if modifications are applied without careful study of the original design and without precise description of changes in both the design and the requirements specification.

Indeed, studies of large software systems show that evolvability decreases with each release of a software product. Each release complicates the structure of the software so that future modifications become more difficult. To overcome this problem, the initial design of the product, as well as any succeeding changes, must be done with evolvability in mind. Evolvability is one of the most important software qualities, and the principles we present in the next part of this chapter will help achieve it.

Evolvability is both a product- and process-related quality. In terms of the latter, the process must be able to accommodate new management and organizational techniques, changes in engineering education, etc.

Reusability

Sometimes it is possible to build software modules or components that may be used in more than one product. Scientific libraries are the best-known examples of **reusable** components. Several large FORTRAN libraries have existed for many years. Users can buy these and use them to build their own products, without having to reinvent or recode well-known algorithms. Indeed, several companies are devoted to producing just such libraries. Another successful example of reusable packages is windowing systems such as X Windows or Motif, for the development of user interfaces.

Reusability is difficult to achieve *a posteriori*. Reusable components must be designed with reusability as a primary design goal. Reusable components are abstractions of useful concepts, are general, and have clear and usable interfaces. A promising technique is the use of object-oriented design, which can unify the qualities of evolvability and reusability.

Languages that support "generic" modules such as Ada and C++ support higher levels of reusability in software components. Most C++ libraries commonly consist of "template" modules. Such language features allow the building of components that are reusable in many different contexts.

Reusability is also a desirable quality at the requirements level. When a new application is conceived, we may try to identify parts that are similar to parts used in a previous application. Thus, we may reuse parts of the previous requirements specification instead of developing an entirely new one.

As discussed above, further levels of reuse may occur when the application is designed or even at the code level. In the latter case, we might be provided with software components that are reused from a previous application. Some software experts claim that in the future new applications will be produced by assembling together a set of ready-made, off-the-shelf components. Software companies will invest in the development of their own catalogues of reusable components so that the knowledge acquired in developing applications will not disappear as people leave but will progressively accumulate in the catalogues. Other companies will invest their efforts in the production of generalized reusable components to be put on the marketplace for use by other software producers.

Reusability applies to the software process as well. Indeed, the various software methodologies can be viewed as attempts to reuse the same process for building different products. The various life cycle models are also attempts at reusing higher-level processes. Another example of reusability in a process is the "replay" approach to software maintenance. In this approach, the entire process is repeated when making a modification. That is, first the requirements are modified, and then the subsequent steps are followed as in the initial product development.

Reusability is a key factor that characterizes the maturity of an industrial field. We see high degrees of reusability in such mature areas as the automobile industry and consumer electronics. For example, in the automobile industry, the engine is often reused from model to model. Moreover, a car is constructed by assembling many components that are highly standardized and used across many models produced by the same industry. Finally, the manufacturing process is often reused. The low degree of reusability in software is an indication of the immaturity of the field.

Portability

Software is portable if it is able to run in different environments. The term "environment" may refer to a hardware platform or a software environment such as a particular operating system. With the proliferation of different processors and operating systems, **portability** has become an important issue for software engineers.

Even within one processor family, portability can be important because of the variations in memory capacity and additional instructions. One way to achieve portability within one machine architecture is to have the software system assume a minimum configuration as far as memory capacity is concerned and use a subset of the machine facilities that are guaranteed to be available on all models of the architecture (such as machine instructions and operating system facilities). But this penalizes the larger models because, presumably, the system can perform better on these models if it does not make such restrictive assumptions. Accordingly, we need to use techniques that allow the software to determine the capabilities of the hardware and to adapt to them. One good example of this approach is the way that UNIX allows programs to interact with many different terminals without explicit assumptions in the programs about the terminals they are using. The X Windows system extends this capability to allow applications to run on any bit-mapped display.

More generally, portability refers to the ability to run a system on different hardware platforms. As the ratio of money spent on software versus hardware increases, portability gains more importance.

Some software systems are inherently machine specific. For example, an operating system is written to control a specific computer, and a compiler produces code for a specific machine. Even in these cases, however, it is possible to achieve some level of portability. Again, UNIX is an example of an operating system that has been ported to many different hardware systems. Of course, the porting effort requires months of work. Still, we can call the software portable because writing the system from scratch for the new environment would require much more effort than porting it.

For many applications, it is important to be portable across operating systems. The operating system provides portability across hardware platforms. Proper modularization is an effective technique for achieving portability.

Interoperability

Interoperability refers to the ability of a system to coexist and cooperate with other systems—for example, a word processor's ability to incorporate a chart produced by a graphing package, or the graphics package's ability to graph the data produced by a spreadsheet, or the spreadsheet's ability to process an image scanned by a scanner.

Rare in software products, interoperability abounds in other engineering products. For example, stereo systems from various manufacturers work together and can be connected to television sets and recorders. In fact, stereo systems produced decades ago accommodate new technologies such as compact disc players, whereas virtually every operating system has to be modified—sometimes significantly—before it can work with the new optical disks.

Once again, the UNIX environment, with its standard interfaces, offers a limited example of interoperability within a single environment: UNIX encourages applications to have a simple, standard interface, which allows the output of one application to be used as the input to another.

The UNIX example also illustrates the limitations of interoperability in current systems: the UNIX standard interface is a primitive, character-oriented one. It is not easy for one application to use structured data—say, a spreadsheet or an image—produced by another application. Also, the UNIX system itself cannot operate in conjunction with other operating systems.

The Object Linking and Embedding (OLE) library offers a practical and effective approach for applications to communicate with each other in the Windows environment. They may also exchange structured and formatted data. The approach depends on standard interfaces and messages that all participating applications understand and process.

A concept related to interoperability is that of an *open system*. An open system is an extensible collection of independently written applications that cooperate to function as an integrated system. An open system allows the addition of new functionality by independent organizations, after the system is delivered. This can be achieved, for example, by releasing the system together with a specification of its "open" interfaces. Any applications developer can then take advantage of these interfaces. Some of the interfaces may be used for communication between different applications or systems. Open systems allow different applications, written by different organizations, to interoperate.

An open system is analogous to a growing organization that evolves over time, adapting to changes in the environment. The importance of interoperability has sparked a growing interest in open systems, prompting standardization efforts in the interfaces for many functions such as database access.

Productivity

Productivity is a quality of the software production process; it measures the efficiency of the process and, as we said before, is the performance quality applied to the process. An efficient process results in faster delivery of the product.

Individual engineers produce software at a certain rate, although there are great variations among individuals of different ability. When individuals are part of a team, the productivity of the team is some function of the productivity of the individuals. Often, the combined productivity is much less than the sum of the parts. Software processes attempt to capitalize on the individual productivity of team members and combine them with the least overhead.

Productivity offers many trade-offs in the choice of a process. For example, a process that requires specialization of individual team members may lead to productivity in producing a certain product but not in producing a variety of products. Software reuse is a technique that leads to the overall productivity of an organization that is involved in developing many products, but a group that produces modules to be used in their as well as in other groups' products will show lower productivity than if they were not concerned with the other groups' products.

Although software productivity is of great interest because of the increasing cost of software, it is difficult to measure. Clearly, we need a metric for measuring productivity—or any other quality—if we are to have any hope of comparing different processes in terms of productivity. Early metrics such as the number of lines of code produced have many shortcomings.

As in other engineering disciplines, efficiency of the process is affected strongly by automation. Modern software engineering tools and environments lead to increases in productivity.

Timeliness

Timeliness is a process-related quality that refers to the ability to deliver a product on time. Historically, timeliness has been lacking in software production processes, leading to the "software crisis," which in turn led to the need for—and birth of—software engineering itself. Even now, many current processes fail to result in a timely product.

The following (real) example is typical of (*circa* 1990) industry practice. The first release of an Ada compiler was promised by a computer manufacturer for a certain date. When the date arrived, the customers who had ordered the product received, instead of the product, a letter stating that because the product still contained many defects, the manufacturer had decided that it would be better to delay delivery rather than deliver a product containing defects. The product was promised for three months later.

After four months, the product arrived, along with a letter stating that many, but not all, of the defects had been corrected. But this time, the manufacturer had decided that it was better to let customers receive the Ada compiler, even though it contained several serious defects, so that the customers could start their own product development using Ada. The value of early delivery at this time had outweighed the cost of delivering a defective product, in the opinion of the manufacturer. In the end, what was delivered was late *and* defective.

Timeliness by itself is not a useful quality, although late delivery often precludes some market opportunities. On-time delivery of a product that is lacking in other qualities, such as reliability or performance, is pointless. Timeliness requires careful scheduling, accurate work estimation, and clearly specified and verifiable milestones. All other engineering disciplines use standard project management techniques to achieve timeliness. There are even many computer-supported project management tools.

Standard project management techniques are difficult to apply in software engineering because of the difficulty in measuring the amount of work required for producing a given piece of software, the difficulty in measuring the productivity of engineers—or even having a dependable metric for productivity—and the use of imprecise and unverifiable milestones. One technique for achieving timeliness is through *incremental delivery* of the product. This technique is illustrated in the following—more successful—example of the delivery of an Ada compiler by a different (real) company from the one described before. This company delivered, very early on, a compiler that supported a very small subset of the Ada language—basically, a subset that was equivalent to Pascal with "packages." The compiler did not support any of the novel features of the language, such as tasking and exception handling. The result was the early delivery of a reliable product. As a consequence, the users started experimenting with the new language and the company took more time to understand the subtleties of the new features of Ada. Over several releases, which took a period of two years, a full Ada compiler was delivered.

Incremental delivery allows the product to become available earlier; and the use of the product helps in refining the requirements incrementally.

Visibility

A software development process is **visible** if all of its steps and its current status are documented clearly. The idea is that the steps and the status of the project are available and easily accessible for external examination.

In many software projects, most engineers and even managers are unaware of the exact status of the project. Some may be designing, others coding, and still others testing, all at the same time. This, by itself, is not bad. Yet, if an engineer starts to redesign a major part of the code just before the software is supposed to be delivered for integration testing, the risk of serious problems and delays will be high.

Visibility is not only an internal quality; it is also external. During the course of a long project, there are many requests for information about the status of the project. Sometimes these require formal presentations on the status, and at other times the requests are informal. Sometimes the requests come from the organization's management for future planning, and at other times they come from the outside, perhaps from the customer. If the software development process has low visibility, either these status reports will not be accurate or they will require a lot of effort to prepare each time.

One of the difficulties of managing large projects is dealing with personnel turnover. In many software projects, critical information about the software requirements and design has the form of "folklore," known only to people who have been with the project either from the beginning or for a sufficiently long time. In such situations, recovering from the loss of a key engineer or adding new engineers to the project is very difficult. In fact, adding new engineers will often reduce the productivity of the whole project while the "folklore" is being transferred slowly from the existing engineers to the new engineers.

Visibility of the process requires not only that all process steps be documented but also that the current status of the intermediate products, such as requirements specifications and design specifications, be maintained accurately; that is, visibility of the product is required also. Intuitively, a product is visible if it is clearly structured as a collection of modules, with clearly understandable functions and easily accessible documentation.

106.3 Quality Requirements in Different Application Areas

The qualities we have described above are generic in the sense that they apply to any software system. But software systems are built to automate a particular application, and these application areas impose more specific requirements on the software. We briefly mention quality requirements in three important areas: information systems, real-time systems, and distributed systems.

Information systems are data oriented and can be characterized on the basis of the way they treat data. Some of the peculiar qualities that characterize information systems are the following:

- *Data integrity.* Under what circumstances will the data be corrupted when the system malfunctions?
- *Data security.* To what extent does the system protect the data from unauthorized access?
- *Data availability.* Under what conditions will the data become unavailable and for how long?

The primary characteristic of real-time systems is that they must respond to events within a predefined and strict time period. For example, in a factory-monitoring system, the software needs to respond to a sudden increase in temperature by immediately setting certain switches or sounding an alarm.

Thus, in addition to the generic software qualities, real-time systems are characterized by how well they satisfy the response-time requirements. Whereas in other systems response time is a matter of performance, in real-time systems response time is one of the correctness criteria. Furthermore, real-time systems are usually used for critical operations (such as patient monitoring, defense systems, and process control) and have stringent reliability requirements.

In the case of highly critical systems, the term *safety* is often used to denote the absence of undesirable behaviors that can cause system hazards. Safety deals with requirements other than the primary mission of a system and requires that the system execute without causing unacceptable risk. Unlike functional requirements, which describe the intended correct behavior in terms of input-output relationships, safety requirements describe what should never happen while the system is executing. In some sense, they are negative requirements: they specify the states the system must never enter.

Advances in processor and network technology have made it possible to build so-called *distributed systems*, which consist of independent computers connected by a communication network. The high-bandwidth, low-error-rate network makes it possible to write distributed software systems, that is, systems with components that run on different computers. Although the generic software qualities apply to distributed software, there are also some new requirements. For example, the software development environment must support users compiling and linking programs on different computers.

Among the qualities of distributed systems are (1) the amount of distribution supported—for example, are the data distributed, or the processing, or both? (2) whether the system can tolerate the partitioning of the network—for example, when the network link makes it impossible for two subsets of the computers to communicate; and (3) whether the system tolerates the failure of individual computers.

One interesting aspect of distributed systems is that they offer new opportunities for achieving some of the qualities discussed. For example, by replicating the same data on more than one computer, we can increase system reliability. Or by distributing the data on more than one computer, we can increase the performance and the reliability of the system.

106.4 Quality Assurance

Once we have decided on the qualities that are the goals of software engineering, we need principles and techniques to help us achieve them. We also need to be able to *verify* whether a given quality has been achieved. Quality assurance is the part of the software production process that deals with ensuring and verifying that desired qualities are achieved.

If a quality is important, it must be *measured* to determine how well it is being achieved. This, in turn, requires that we define each quality precisely so that it is clear what we should be measuring. Without measurements, any claims of improvement are without basis. But without defining a quality precisely, there is no hope that we can measure it precisely—let alone quantitatively.

The established engineering disciplines have standard techniques for measuring quality. For example, the reliability of an amplifier can be measured by determining the range within which it operates. The reliability of a bridge can be measured by the amount of load it can withstand. Indeed, these tolerance levels are released with the product as part of the product specification.

Although some software qualities, such as performance, are measured relatively easily, universally accepted metrics for most qualities do not exist. For example, whether a given system will evolve more easily than another is usually determined subjectively. Nevertheless, metrics are needed, and indeed much research work is currently under way for defining objective metrics.

The current state of practice of quality assurance is a collection of verification and monitoring methods. Some of them are based on objective evaluations such as testing, and others are based on informal procedures such as walkthroughs; a few more ambitious methods aim at exploiting mathematical analysis during software verification. Chapters 108, 109, and 110 relate, to different extents and from different points of view, to the quality assurance problem.

106.5 Software Engineering Principles in Support of Software Quality

In sections 106.2 and 106.3, we have discussed a number of important software qualities. How can we achieve these qualities? In this section, we discuss seven general principles that help in achieving software quality. These principles may be applied throughout the software development process and are not limited to a particular phase of the process. The principles deal with the following: rigor and formality, separation of concerns, modularity, abstraction, anticipation of change, generality, and incrementality. The list, by its very nature, cannot be exhaustive, but it does cover the important areas of software engineering. The principles are of course strongly related and together form a set of guidelines for the engineer to follow.

Rigor and Formality

Software development is a creative activity. There is an inherent tendency in any creative process to be neither precise nor accurate, but rather to follow the inspiration of the moment in an unstructured manner. *Rigor*, on the other hand, is a necessary complement to creativity in every engineering activity: it is only through a rigorous approach that we can produce more reliable products, control their costs, and increase our confidence in their reliability. Rigor does not need to constrain creativity. Rather, it enhances creativity by improving the engineer's confidence in creative results, once they are critically analyzed in the light of a rigorous assessment.

Paradoxically, rigor is an intuitive quality that cannot be defined in a rigorous way. Also, various degrees of rigor can be achieved. The highest degree is what we call *formality*. Thus, formality is a stronger requirement than rigor: it requires the software process to be driven and evaluated by mathematical laws. Formality implies rigor, but the converse is not true: one can be rigorous even in an informal setting.

In every engineering field, the design process proceeds as a sequence of well-defined, precisely stated, and supposedly sound steps. In each step, the engineer follows some method or applies some technique. The methods and techniques applied may be based on some combination of theoretical results derived by some formal modeling of reality, empirical adjustments that take care of phenomena not dealt with by the model, and rules of thumb that depend on past experience. The blend of these factors results in a rigorous and systematic approach—the methodology—that can be easily explained and applied time and again.

There is no need to be always formal during design, but the engineer must know how and when to be formal, should the need arise. For example, the engineer can rely on past experience and rules of thumb to design a short bridge, to be used temporarily to connect the two banks of a creek. She would instead use a mathematical model to verify whether the design is safe if the bridge were a long one that is supposed to stand permanently. She would use a more sophisticated mathematical model if the bridge were exceptionally long, or if it were built in a seismic area. In this case, the mathematical model would consider factors that could be ignored in the previous case.

Another—perhaps striking—example of the interplay between rigor and formality may be observed in mathematics. Textbooks on the calculus of functions are rigorous but seldom formal: theorems are proved carefully, as sequences of intermediate deductions that lead to the final statement. Each deductive step relies on an intuitive justification that should convince the reader of its validity. Rarely is the derivation of a proof stated formally, in terms of mathematical logic. The mathematician is thus satisfied with a rigorous description of the derivation of a proof, without formalizing it completely. In critical cases, however, where the validity of some deduction is unclear, the mathematician may try to formalize the informal reasoning to assess its validity.

These examples show that the engineer (and the mathematician) must be able to understand the level of rigor and formality that should be achieved, depending on the conceptual difficulty of the task and its criticality. The level may even vary for different parts of the same system. For example, critical parts may require a formal description of their intended functions and a formal approach to their assessment. Well-understood and standard parts would require simpler approaches.

The same happens in the case of software engineering. Consider, for instance, the case of software specifications. The description of what a program does may be given in a rigorous way by using natural language; it can also be given formally by providing a formal description in logic. A further advantage of formality, besides supporting rigor, is that formality may be the basis of mechanization of the process. For example, one may hope to use the formal description of the program to create the program (if the program does not yet exist) or to show that the program corresponds to the formal description (if the program and its formal specification exist) or to generate test data.

Traditionally, there is only one phase of software development where a formal approach is used: programming. In fact, programs are formal objects: they are written in a language whose syntax and semantics are fully defined. Programs are formal descriptions that may be automatically manipulated by compilers: they are checked for formal correctness, transformed into an equivalent form in another language (assembly or machine language), "pretty-printed" to improve their appearance, etc. These mechanical operations, which are made possible by the use of formality in programming, can effectively improve the reliability and verifiability of software products.

Rigor and formality are not restricted to programming: they should be applied throughout the software process. Chapter 107 shows these concepts in action in the case of software specification. Chapter 109 does the same for software verification.

So far, our discussion has emphasized the influence of rigor and formality on the reliability and verifiability of software products. Rigor and formality also have beneficial effects on maintainability, reusability, portability, understandability, and interoperability. For example, a rigorous, or even formal, software documentation can improve all of these qualities over informal documentation, which is often ambiguous, inconsistent, and incomplete.

Rigor and formality also apply to software processes. Rigorous documentation of a software process helps in reusing the process in other similar projects. On the basis of this documentation, managers may foresee the steps through which the new project will evolve, assign appropriate resources as needed, etc. Similarly, rigorous documentation of the software process may help maintain an existing product. If the various steps through which the project evolved are documented, one can modify an existing product starting from the appropriate intermediate level of its derivation, not the final code. Finally, if the software process is specified rigorously, managers may monitor it accurately to assess its timeliness and improve productivity.

Separation of Concerns

Separation of concerns allows us to deal with different individual aspects of a problem, so that we can concentrate on each separately. Separation of concerns is a commonsense practice that we try to follow in our everyday life to deal with the difficulties we encounter. The principle should be applied also in software development to master its inherent complexity.

There are many decisions that must be made in the development of a software product. Some of them concern features of the product as such: functions to offer, expected reliability, space and time efficiency, relationship with the environment (special hardware or software resources required), user interfaces, etc. Others concern the development process: development environment, team organization and structure, scheduling, control procedures, design strategies, error recovery mechanisms, etc. Still others concern economic and financial matters. These different decisions may be unrelated to one another. In such a case, it is obvious that they should be treated separately.

Often, however, many decisions are strongly related and interdependent. For instance, a design decision (e.g., caching some data from disk to memory) may depend on the size of the memory of the selected target machine (and hence, the cost of the machine), and this, in turn, may affect the policy for error recovery. When many different design decisions are strongly interconnected, it is difficult to take all the issues into account at the same time; sometimes, they may even require different people. The only way to master the complexity of the project is to separate the different concerns.

There are various ways in which concerns may be separated. First of all, one can separate them in *time*. As an everyday life example, consider a university professor who decides to deal with teaching activities by concentrating classes, seminars, office hours, and department meetings from 9 a.m. to 2 p.m. Monday through Thursday and leaving the rest of the time to research, except for Friday, which is devoted to consulting. Such temporal separation of concerns allows for precise planning of activities and eliminates overhead that would arise through switching from one activity to another in an unconstrained way. Separation of concerns in terms of time is the underlying motivation of the software life cycle, a rational model of the sequence of activities that should be followed in software production (see Chapter 105).

Another type of separation of concerns is in terms of *qualities* that should be treated separately. For example, in the case of software, we might wish to deal separately with the efficiency and the correctness of a given program. One might decide first to design software in such a careful and structured way that its correctness is expected to be guaranteed *a priori* and then to restructure the program partially to improve its efficiency. Similarly, in the verification phase, one might first check the functional correctness of the program and then its performance. Both activities can be done rigorously, applying systematic procedures, or even formally, e.g., using formal correctness proofs and complexity analysis.

Another important type of separation of concerns allows different *views* of the software to be analyzed separately. For example, when we analyze the requirements of an application, it may be helpful to concentrate separately on the data that flow from one activity to another in the system and the flow of control that governs the way different activities are synchronized. Both views help us understand the system we are working on better, although neither one gives a complete view of it.

Still another type of separation of concerns allows us to deal with *parts* of the same system separately; here separation is in terms of size. This is a fundamental concept that we need to exploit to dominate the complexity of software production. It is such an important issue that we prefer to detail it below as a separate point under modularity.

Separation of concerns must be done carefully because it contains an inherent danger: by separating two or more issues, we might miss some global optimization that would be possible by tackling them together. Often, however, our ability to make "optimized" decisions is limited. By combining concerns, we are likely to be overwhelmed by complexity. The complexity of the global problem can be overcome better by concentrating on different aspects separately, even at the expense of missing some potential optimizations.

If two aspects of a problem are intrinsically intertwined (i.e., the problem is not immediately decomposable into separate issues), it is often possible to make some overall design decisions first that effectively separate the concern on the different issues. For example, consider a system where on-line transactions access a database concurrently. In a first implementation of the system, each transaction is supported by the underlying machine by locking the entire database at the start of the transaction and unlocking it at the end. Suppose now that a preliminary performance analysis shows that some transaction, say t_i (which might print some complex report extracting many data from the database), takes longer than we can afford to lock out other transactions. Thus, the problem is to revise the implementation in order to improve its performance yet maintain the overall correctness of the system. Clearly, the two issues— functional correctness and performance—are strongly related. Thus, a first design decision must concern both of them: t_i is no longer implemented as an atomic transaction but is split into several subtransactions $t_{i1}, t_{i2}, \ldots, t_{in}$, each being atomic. The new implementation may affect the correctness of the system, because of the interleaving that may occur between execution of any two subtransactions. Now, however, the two concerns of checking the functional correctness and analyzing performance have been separated, and two independent analyses can be made, maybe even by two different designers with different expertise.

As a final remark, notice that separation of concerns may result in separation of responsibilities in dealing with separate issues. Thus, the principle is the basis for dividing the work on a complex problem into specific work assignments, possibly for different people with different skills. For example, by separating managerial and technical issues in the software process, we allow two types of people to cooperate in a software project. As another example, having separated requirements analysis and specification from other activities in a software life cycle, we may hire specialized analysts with expertise in the application domain, instead of relying on internal resources. The analyst, in turn, may concentrate separately on functional and nonfunctional system requirements.

Modularity

A complex system may be divided into simpler pieces called *modules*. A system that is composed of modules is called *modular*. The main benefit of modularity is that it allows the principle of separation of concerns to be applied in two phases: when dealing with the details of each module in isolation (and ignoring details of other modules) and when dealing with the overall characteristics of all modules and their relationships in order to integrate them into a coherent system. If the two phases are executed in the order mentioned, then we say that the system is designed *bottom up*; the converse denotes *top-down* design.

Modularity is an important property of most engineering processes and products. For example, in the automobile industry, the construction of cars proceeds by assembling building blocks that are designed and built separately. Furthermore, the same parts are often used in successive models, perhaps after minor changes. Most industrial processes are essentially modular, made out of work packages that are combined in simple ways (in sequence or overlapping) to achieve the desired result.

Modularity, however, not only is a desirable design principle but permeates the whole of software production. In particular, there are three goals that modularity tries to achieve in practice: capability of decomposing a complex system, of composing it from existing modules, and of understanding the system in pieces.

The *decomposability* of a system is based on dividing the original problem top down into subproblems and then applying the decomposition to each subproblem recursively. This procedure reflects the well-known Latin motto *divide et impera* (divide and conquer), which describes the philosophy followed by the ancient Romans to dominate other nations: divide and isolate them first and conquer them individually.

The *composability* of a system is based on starting bottom up from elementary components and combining them successively to form the finished system. As an example, a system for office automation may be designed by assembling existing hardware components such as personal workstations, a network, and peripherals; systems software such as the operating system; and productivity tools such as document processors, databases, and spreadsheets. A car is another obvious example of a system that is built by assembling components. Consider first the main subsystems into which a car may be decomposed: the body, the electrical system, the power system, the transmission system, etc. Each of them, in turn, is made out of standard parts; for example, the battery, fuses, cables, etc., form the electrical system. System repair is performed by replacing defective components.

Ideally, in software production we would like to be able to assemble new applications by taking modules from a library and combining them to form the required product. Such modules should be designed with the express goal of being reusable. By using reusable components, we may speed up both the initial system construction and its fine-tuning. For example, it would be possible to replace a component by another that performs the same function but differs in computational resource requirements.

The ability to understand each part of a system separately enhances system modifiability. If the entire system can be understood only in its entirety, modifications are likely to be difficult to apply and the result unreliable. When the need for repair arises, proper modularity helps confine the search for the source of malfunction to single components. Thus, modularity contributes to all qualities such as modifiability and reliability.

To achieve modular composability, decomposability, and understanding, modules must have *high cohesion* and *low coupling*. A module has high cohesion if all of its elements are related strongly. Elements of a module (e.g., statements, procedures, and declarations) are grouped together in the same module for a logical reason, not just by chance; they cooperate to achieve a common goal, which is the function of the module.

In contrast to cohesion, which is an internal property of a module, coupling characterizes a module's relationship to other modules. Coupling is a measure of the amount of mutual dependence among modules (e.g., module A calls a routine provided by module B or accesses a variable declared by module B). If two modules depend on each other heavily, they have high coupling. Ideally, we would like modules in a system to exhibit low coupling, because if two modules are highly coupled, it will be difficult to analyze, understand, modify, test, or reuse them separately. Figure 106.2 provides a graphical view of cohesion and coupling.

A good example of a system that has high cohesion and low coupling is the electric subsystem of a house. Because it is made out of a set of appliances with clearly definable functions and interconnected by simple wires, the system has low coupling. Because each appliance's internal components are there exactly to provide the service the appliance is supposed to provide, the system has high cohesion.

Modular structures with high cohesion and low coupling allow us to see modules as black boxes when the overall structure of a system is described and then deal with each module separately when the module's functionality is described or analyzed. This is just another example of the principle of separation of concerns.

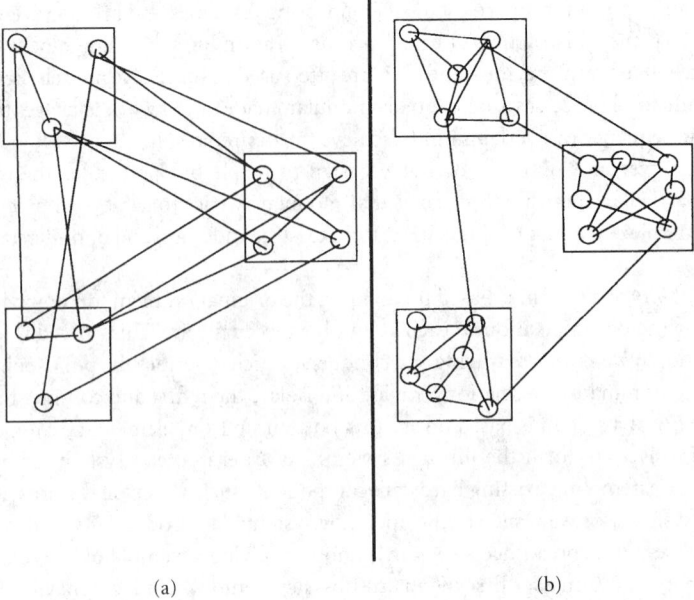

(a)   (b)

FIGURE 106.2 Graphical description of cohesion and coupling. (a) A structure with low cohesion and high coupling. (b) A structure with high cohesion and low coupling. In a software system, nodes may represent data or statements. Arcs represent relations among entities such as data sharing or procedure calls.

Abstraction

Abstraction is a technique for understanding a phenomenon by concentrating on its important attributes and ignoring its other attributes. An abstraction is a model which contains only the relevant aspects of the phenomenon under study. Thus, abstraction is a special case of separation of concerns in which we separate the concern of the important aspects from the concern of the unimportant details.

What we abstract away and consider as unimportant detail depends on the purpose of the abstraction. For example, consider a quartz watch. A useful abstraction for the owner is a description of the effects of pushing its various buttons, which allow the watch to enter various functioning modes and react differently to sequences of commands. A useful abstraction for a watch repairer is a box that can be opened in order to replace the battery. Still other abstractions of the device are useful for understanding the watch and for more serious repair activities or watch design. Thus, there may be many different abstractions of the same reality, each providing a *view* of the reality and serving some specific purpose.

Abstraction is a powerful technique practiced by engineers of all fields for mastering complexity. For example, the representation of an electrical circuit in terms of resistors, capacitors, etc., each characterized by some model in terms of equations, is an idealized abstraction of a device. On the one hand, the equations are a simplified model that approximates the behavior of the real components; on the other hand, the model we build often ignores details such as the fact that there are no "pure" connectors between components and that these connectors should also be modeled in terms of resistors, capacitors, etc. Both facts can be ignored by the designer because the effects they describe are negligible in terms of the results that we wish to observe.

This example illustrates an important general idea: the models we build of phenomena—such as the equations for describing devices—are an abstraction from reality, ignoring certain facts and concentrating on others that are deemed relevant. The same holds for the models built and analyzed by software engineers. For example, when the requirements for a new application are analyzed and specified, software engineers build a model of the proposed application. This model may be expressed in various forms, depending on the required degree of rigor and formality. No matter what language we use for expressing requirements—be it natural language or the formal language of mathematical formulas— what we provide is a model that abstracts away from a number of details that we decide can be ignored safely.

Abstraction permeates the whole of programming. The programming languages that we use are abstractions built on top of the hardware: they provide us with useful and powerful constructs so that we can write (most) programs ignoring such details as the number of bits that are used to represent numbers or the addressing mechanism. This helps us concentrate on the problem to solve rather than the way to instruct the machine on how to solve it. The programs we write are themselves abstractions. For example, a computerized payroll procedure is an abstraction of the manual procedure it replaces: it provides the essence of the manual procedure, not its exact details.

Abstraction is an important principle that applies to both software products and processes. For example, the comments in the header of a procedure are an abstraction that describes the effect of the procedure. When the documentation of the program is analyzed, such comments are supposed to provide all the information that is needed to understand the other parts of the program that use the procedure.

As an example of the use of abstraction in software processes, consider the problem of estimating the cost of a new application. One possible way of doing cost estimation consists of identifying some key factors of the new system and extrapolating from the cost profiles of previous similar systems. The key factors used to perform the analysis are abstractions of the system.

Abstraction is the essence of design and programming and it affects all software qualities. For example, proper abstraction leads to modularity, which aids in achieving maintainability, and reusability.

Anticipation of Change

Software products are subject to frequent changes. We stated earlier that changes are required both to repair software defects and to support new requirements. We therefore identified maintainability as a major software quality.

Achieving software maintainability requires special effort to anticipate how and where the changes are likely to occur. When likely changes are identified, special care must be taken to proceed in a way that will make future changes easy to apply. Software must be designed in such a way that changes that we anticipate in the requirements, or modifications that are planned as part of the design strategy, may be incorporated in the application smoothly and safely. Basically, design choices that are likely to change should be isolated in specific portions of the software so that future changes will be confined to small portions of the software.

Anticipation of change is perhaps the one principle that distinguishes software the most from other types of industrial productions. In many cases, a software application is developed while its requirements are not entirely understood. Often, the application is evolved on the basis of new requirements identified from user feedback. Sometimes, applications are introduced in an environment such as an organizational structure. The environment is affected by the introduction of the application, and this generates new requirements that were not present initially. Anticipation of changes is a principle that we can use to achieve evolvability.

Reusability is another software quality that is strongly affected by anticipation of change. As we saw, a component is reusable if it is directly usable to produce a new product. More realistically, it might undergo slight changes before it can be reused. As such, reusability may be viewed as low-grain evolvability, i.e., evolvability at the component level. If we can anticipate the different contexts in which a software component might be embedded, we may then design the component in a way that context-dependent changes may be accommodated easily. This can be done through the use of generic components. In our discussion of anticipation of change, we focused attention more on software products than on software processes. Anticipation of change, however, also affects the management of the software process. For example, managers can anticipate the effects of personnel turnover. Also, when designing the life cycle of an application, it is important to take the impact of maintenance activities into account. Depending on the anticipated changes, managers must estimate costs and design the organizational structure that will support the evolution of the software.

Generality

The principle of generality may be stated as follows:

> In solving a problem, look for a more general problem that may be hidden behind the problem at hand. It may happen that the generalized problem is not more complex—it may even be simpler—than the original problem. The more general solution has more potential for being reused. The solution may even be already provided by some off-the-shelf package. Also, it may happen that by generalizing a problem you end up designing a module that is invoked at more than one point of the application, rather than having several specialized solutions.

On the other hand, a generalized solution may be more costly, in terms of speed of execution, memory requirements, or development time, than the specialized solution that is tailored to the original problem. Thus, it is necessary to evaluate the trade-offs of generality with respect to cost and efficiency, in order to decide whether it is worthwhile to solve the generalized problem instead of the original problem.

For example, suppose that you want to design an application to handle a small library of cooking recipes. Suppose the recipes have a header—containing information such as a name, a list of ingredients, and cooking information—and a textual part describing how to apply the recipes. Apart from storing recipes in the library, it must be possible to do a sophisticated search for recipes based on available ingredients, maximum calories, etc. Rather than designing a new set of facilities, these searches can be viewed as a special case of a more general set of text-processing facilities, such as those provided by the AWK language under UNIX. Before starting with the design of the specialized set of routines, the designer should consider whether a generalized text processing tool would be more useful. The generalized tool is undoubtedly more reliable than the specialized program to be designed, and it would probably accommodate changes in the requirements or even new requirements. On the negative side, however, there may be a cost of acquisition, and possibly overhead, in the use of the generalized tool.

Generality is a fundamental principle if our goal as software engineers is to develop tools or packages for the market. Such tools as spreadsheets, databases, and word processors are successful because they are general enough to cover the practical needs of most people when they wish to handle their personal business with a computer. Instead of a customized solution for each personal business, it is more economical to use an existing product that fits the need.

Such general-purpose, off-the-shelf products represent a rather general trend in software. Increasingly, general packages that provide standard solutions to common problems are available in many application areas. If the problem at hand may be restated as an instance of a problem solved by a general package, it may be convenient to adopt the package instead of implementing a specialized solution.

This general trend is identical to what happens in other branches of industry. For example, in the early days of automobile technology, it was possible to customize cars according to the specific requirements of the customer. As the field became more industrialized, customers could choose only from a catalogue of models—which correspond to prepackaged solutions—provided by each manufacturer. Nowadays, it is not possible to ask for a personal car design, unless one is ready to pay an enormous amount of money.

At the coding and design level, new programming languages, such as Ada 95 and C++, support the development of generic components in terms of "templates." These facilities allow the principle of generality to be applied in a very concrete fashion. For example, by abstracting the important and common aspects of different data structures, it is possible to write a generic algorithm to work with all those data structures. For example, generic search and sorting routines may be written that apply to sequential data structures such as arrays, lists, and files at no loss of efficiency compared to specialized algorithms. Such application of generality reduces the amount of code that needs to be written and thus increases productivity. It also affords us more time to devote to the quality of the code that we do have to write.

Incrementality

Incremental development proceeds by producing progressively larger increments of the desired product. At each stage, the completed increments form a subset of the final product. Each increment provides additional functionality and brings the currently available subset closer to the desired one. Thus, the desired goal is reached not all in one step but by successively closer approximations to it.

Incrementality applies to many engineering activities. When applied to software, it means that the desired application is produced as an outcome of an evolutionary process.

One way of applying the incrementality principle consists of identifying useful *early subsets* of an application that may be developed and delivered to customers, in order to get *early feedback*. This allows the application to evolve in a controlled manner in cases where the initial requirements are not stable or fully understood. The motivation for incrementality is that in most practical cases there is no way of getting all the requirements right before an application is developed. Rather, requirements emerge as the application—or parts of it—is available for realistic experimentation. Consequently, the sooner we can receive feedback from the customer concerning the usefulness of the application, the easier it is to incorporate the required changes into the product. Thus, incrementality is intertwined with anticipation of change and is one of the cornerstones upon which evolvability may be based.

Incrementality applies to many of the software qualities discussed in the previous sections. We may build increasingly more functional subsets of the target system by progressively adding functions to a kernel of functions, making incrementally more usable, if incomplete, systems. For example, in some business automation systems, some functions could be automated while others are still done manually.

We can also add performance in an incremental fashion. That is, the initial version of the application might emphasize user interfaces and reliability more than performance, and successive releases would then improve space and time efficiency.

When an application is developed incrementally, intermediate stages may constitute *prototypes* of the end product; that is, they are just an approximation of it. Rapid **prototyping** is a way of progressively developing an application hand in hand with the understanding of its requirements. Obviously, a software life cycle based on prototyping is rather different from the typical waterfall model described in Chapter 105, where we first do a complete requirements analysis and specification and then start developing the application. It is based on a more flexible and iterative development model. This difference affects not only the technical aspects of a project but also its organizational and managerial issues.

As we mentioned in connection with anticipation of change, evolutionary software development requires special care in the management of documents, programs, test data, etc., developed for the various versions of software. Each meaningful incremental step must be recorded, documentation must be easily retrieved, changes must be applied in a controlled way, and so on. If this is not done carefully, an intended evolutionary development may quickly turn into undisciplined software development, and all the potential advantages of evolvability would be lost.

106.6 Summarizing Remarks

In this chapter, we have discussed important software engineering qualities and principles. Qualities are the ultimate goal we want to achieve; principles provide the bases for any concrete means to reach the goal.

Software engineering principles, as stated here, might seem too abstract. We emphasized, however, the role of general principles without presenting specific methods, techniques, and tools. As technology evolves, software engineering tools will evolve. Methods and techniques will evolve too—although less rapidly than tools—as our knowledge of software increases. In this picture, principles will be more stable; they constitute the foundation upon which all the rest may be built. Also, they are the keys that must be used by the reader to interpret the concepts discussed in the rest of this handbook. For instance, Chapter 108 presents Object-Oriented design, a currently popular methodology which stresses the qualities of reusability and evolvability as well as modularity, anticipation of changes, generality, and incrementality.

Defining Terms

Cohesion: Property of a modular software system which measures the logical coherence of a module.

Correctness: Software is (functionally) correct if it behaves according to the specification of the functions it should provide.

Coupling: Property of a modular software system which measures the amount of mutual dependence among modules.

Evolvability: Ease of software evolution.

Incremental development: A software process which proceeds by producing progressively larger increments of the desired product. At each stage, the completed increments form a subset of the final product. Each increment provides additional functionality and brings the currently available subset closer to the desired one.

Interoperability: Ability of a software system to coexist and cooperate with other systems.

Maintainability: Ease of maintaining software. It can be further decomposed into evolvability and repairability.

Malleability: Ease of modification of a software product. The product can be modified without modifying its design.

Methodology: A combination of methods and techniques promoting a disciplined approach to software development.

Performance: In software engineering, performance is a synonym for efficiency. It refers to how economically the software utilizes the resources of the computer.

Portability: Software is portable if it is able to run on different machines.

Productivity: Efficiency of the software process.

Prototyping: An incremental development process where intermediate stages constitute executable *prototypes* of the end product; that is, they are just an approximation of it.

Quality assurance: The process of verifying whether a software product meets the required qualities.

Reliability: Software is reliable if the user can depend on it. The mathematical theory of software reliability provides a definition based on statistics and probability principles.

Repairability: A software system is repairable if it allows the correction of its defects with a limited amount of work.

Reusability: Ease of reusing software components in more than one product. Reusability can also refer to other artifacts (such as requirements, design, etc.).

Robustness: Software is robust if it behaves "reasonably" even in circumstances that were not anticipated in the requirements specification.

Software process: Activities through which a software product is developed and maintained.

Software product: All of the artifacts produced by a software process. This definition encompasses not only the executable code and user manuals that are delivered to the customer but also requirements and design documents, source code, test data, etc.

Timeliness: A process-related quality meaning the ability to deliver a product on time.

Usability: A software system is usable—or user friendly—if its human users find it easy to use. This definition reflects the subjective nature of usability.

Verifiability: A software system is verifiable if its properties can be verified easily.

Visibility: A process-related quality meaning that all steps and the current process status are documented clearly.

References

Boehm, B. W., Brown, J. R., Kaspar, H., Lipow, M., MacLeod, G., and Merritt, M. J. 1978. *Characteristics of Software Quality*, Vol. 1 TRW Series on Software Technology. North-Holland, Amsterdam.

Fenton, N. E. 1991. *Software Metrics—A Rigorous Approach*. Chapman & Hall, London.

Garg, P. and Jazayeri, M. 1996. *Process-Centered Software Engineering Environments*. IEEE Computer Society Press.

Ghezzi, C., Jazayeri, M., and Mandrioli, D. 1991. *Fundamentals of Software Engineering.* Prentice–Hall, Englewood Cliffs, NJ.

Humphrey, W. S. 1989. *Managing the Software Process.* Addison–Wesley, Reading, MA.

Humphrey, W. S. 1995. *A Discipline for Software Engineering.* Addison–Wesley, Reading, MA.

ISO/IEC 9126. 1991. Information Technology-Software Product Evaluation. Quality characteristics and guidelines for their use, pp. 12–15.

Jazayeri, M. 1995. Component programming—a fresh look at software components. In *Proc. 5th European Software Engineering Conf.* Lecture Notes in Computer Science 989, pp. 457–478. Springer–Verlag.

Neumann, P. G. 1995. *Computer-Related Risks.* Addison–Wesley, Reading, MA.

Parnas, D. L. 1978. Some software engineering principles. In *Structured Analysis and Design.* State-of-the-art report, INFOTECH Int., pp. 237–247.

Further Information

This chapter is adapted from Chapters 2 and 3 of Ghezzi et al. [1991], which is a general textbook on software engineering.

A classification of software qualities is presented and discussed in detail by Boehm et al. [1978]. The international standard [ISO 9126] also provides a list and a discussion of major software qualities. The book by Fenton [1991] is a complete treatment of quality metrics. Neumann [1995] illustrates the consequences of lack of quality in software. These real-life situations should concern all software professionals.

Two books by Humphrey [1989, 1995] define, respectively, organizational process models and a personal software process. Humphrey [1989] defined the software maturity model. These two books are devoted to improving software quality through better processes and the assessment of the process. Garg and Jazayeri [1996] is a collection of articles on software environments that integrate and automate the software process.

Parnas's work on design methods is the major source of insight into the concepts of separation of concerns, modularity, abstraction, and anticipation of change. In particular, Parnas [1978] illustrates important software engineering principles.

Jazayeri [1995] discusses and gives concrete examples of the power of the principle of generality in design and programming.

107

Formal Models and the Specification Process

Jonathan P. Bowen
The University of Reading

Michael G. Hinchey
*New Jersey Institute of
Technology and
University of Limerick*

107.1 Introduction

Computers do not make mistakes, or so we are told. However, computer software is written by, and hardware systems are designed and assembled by, humans, who certainly *do* make mistakes [Hinchey 1993].

Errors in a computer system may be as a result of misunderstood or contradictory requirements, unfamiliarity with the problem, or simply human error during design or coding of the system. Alarmingly, the costs of maintaining software—the costs of rectifying errors and adapting the system to meet changing requirements or changes in the environment of the system—greatly exceed the original implementation costs.

As computer systems are being used increasingly in safety-critical applications—that is, systems where a failure could result in the loss of human life, mass destruction of property, or significant financial loss—both the media (e.g., [Gibbs 1994]) and various regulatory bodies (see [Bowen and Hinchey 1994, Bowen and Stavridou 1993]) have turned their attention to formal methods and their role in the specification and design phases of system development [Faser et al. 1994].

107.2 Underlying Principles

There can be some confusion over what is meant by a "specification" and a "model." David Parnas differentiates between specification and descriptions or models as follows [Parnas 1995]:

- A **description** is a statement of some of the actual attributes of a product, or a set of products.
- A **specification** is a statement of properties required of a product, or a set of products.

0-8493-2909-4/97/$0.00+$.50
© 1997 by CRC Press, Inc.

- A **model** is a product, neither a description nor a specification. Often it is a product that has some, but not all, of the properties of some "real product."

Others use the terms *specification* and *model* more loosely; a model may sometimes be used as a specification. The process of developing a specification into a final product is one in which a model may be used along the way or even as a starting point.

Formal Methods

Over the last 50 years, computer systems have increased rapidly in terms of both size and complexity. As a result, it is both naive and dangerous to expect a development team to undertake a project without stating clearly and precisely what is required of the system. This is done as part of the **requirements specification** phase of the software life cycle, the aim of which is to describe *what* the system is to do, rather than *how* it will do it.

The use of natural language for the specification of system requirements tends to result in ambiguity and requirements that may be mutually exclusive. Formal methods have evolved as an attempt to overcome such problems, by employing discrete mathematics to describe the function and architecture of a hardware or software system, and various forms of **logic** to reason about requirements, their interactions, and validity.

The term **formal methods** is itself misleading; it originates from formal logic but is now used in computing to refer to a plethora of mathematically based activities [Dean and Hinchey 1995]. For our purposes, a formal method consists of notations and tools with a mathematical basis that are used to unambiguously specify the requirements of a computer system and that support the **proof** of properties of this specification and proofs of correctness of an eventual implementation with respect to the specification.

Indeed, it is true to say that so-called "formal methods" are not so much methods as formal systems. Although the popular formal methods provide a **formal notation**, or formal specification language, they do not adequately incorporate many of the methodological aspects of traditional development methods.

Even the term **formal specification** is open to misinterpretation by different groups of people. Two alternative definitions for "formal specification" are given in a glossary issued by the IEEE [IEEE 1991]:

1. A specification written and approved in accordance with established standards.
2. A specification written in a formal notation, often for use in proof of correctness.

In this chapter we adopt the latter definition, which is the meaning assumed by most formal methods users.

The notation employed as part of a formal method is "formal" in that it has a mathematical semantics so that it can be used to express specifications in a clear and unambiguous manner and allow us to abstract from actual implementations to consider only the salient issues. This is something that many programmers find difficult to do because they are used to thinking about implementation issues in a very concrete manner.

While programming languages are formal languages, they are generally not used in formal specification, as most languages do not have a full formal semantics, and they force us to address implementation issues before we have a clear description of what we want the system to do. Instead, we use the language of mathematics, which is universally understood, well established in notation, and most importantly, enables the generalization of a problem so that it can apply to an unlimited number of different cases [Dijkstra 1981]. Here we have the key to the success of formal specification—one must *abstract* away from the details of the implementation and consider only the essential relationships of the data, and we can model even the most complex systems using simple mathematical objects: e.g., **sets**, **relations**, functions, etc.

At the specification phase, the emphasis is on clarity and precision, rather than efficiency. Eventually, however, one must consider how a system can be implemented in a programming language that, in general, will not support abstract mathematical objects (functional programming languages are an exception) and will be efficient enough to meet agreed requirements and concrete enough to run on the available hardware configuration.

As in structured design methods, the formal specification must be translated to a design—a clear plan for implementation of the system specification—and eventually into its equivalent in a programming language. This approach is known as **refinement.**

The process of **data refinement** involves the transition from abstract data types such as sets, sequences, and mappings to more concrete data types such as arrays, pointers, and record structures, and the subsequent verification that the concrete representation can adequately capture all of the data in the formal specification. Then, in a process known as **operation refinement**, each **operation** must be translated so that it operates on the concrete data types. In addition, a number of **proof obligations** must be satisfied, demonstrating that each concrete operation is indeed a "refinement" of the abstract operation—that is, performing at least the same functions as the abstract equivalent, but more concretely, more efficiently, involving less nondeterminism, etc.

Many specification languages have relatively simple underlying mathematical concepts involved. For example, the Z (pronounced "zed") notation [Spivey 1992] is based on (typed) set theory and first-order predicate logic, both of which *could* be taught at school level. The problem is that many software developers do not currently have the necessary education and training to understand these basic principles, although this is gradually changing as suitable courses are integrated into university curricula [Garlan 1995].

It is important for students who intend to become software developers to learn how to abstract away from implementation detail when producing a system specification. Many find this process of **abstraction** a difficult skill to master. It can be useful for reverse engineering as part of the software maintenance process, to produce a specification of an existing system that requires restructuring [Bowen et al. 1993]. Equally important is the skill of refining an abstract specification towards a concrete implementation, in the form of a program, for example [Morgan 1994], for development purposes.

The process of refinement is often carried out informally because of the potentially high cost of fully formal refinement. Given an implementation, it is theoretically possible, although often intractable, to *verify* that it is correct with respect to a specification, if both are mathematically defined. More usefully, it is possible to *validate* a formal specification by formulating required or expected properties and formally proving, or at least informally demonstrating, that these hold. This can reveal omissions or unexpected consequences of a specification. **Verification** and **validation** are complementary techniques, both of which can expose errors.

107.3 Best Practices

Most engineering disciplines accept mathematics as the underpinning foundations, to allow calculation of design parameters before the implementation of a product [Hoare 1996(b)]. Software engineers have been somewhat slow to accept such principles in practice, despite the very mathematical nature of all software. This is partly because it *is* possible to produce remarkably reliable systems without using formal methods [Hoare 1996(a)].

However, there are now well-documented examples of cases in which a formal approach has been taken to develop significant systems in a beneficial manner, examples which are easily accessible by professionals (e.g., see [Hinchey and Bowen 1995, Hall 1996]). Formal methods, including formal specification and modeling, should be considered as one of the possible techniques to improve software quality, where it can be shown to do this cost-effectively.

In fact, just using formal specification within the software development process has been shown to have benefits in reducing the overall development cost [Bowen and Stavridou 1993]. Costs tend to be increased early in the life cycle, but reduced later on at the programming, testing, and maintenance stages, where correction of errors is far more expensive. A widely publicized example is the IBM CICS (Customer Information Control System) project, where Z was used to specify a portion of this large transaction processing system with an estimated 9% reduction in development costs [Houston and King 1991]. There were approximately half the usual number of errors discovered in the software, leading to increased software quality.

Writing a good specification is something that comes only with practice, despite the existence of guidelines [Gravell 1991]. However there are some good reasons why a mathematical approach may be beneficial in producing a specification:

Precision: Natural language and diagrams can be very ambiguous. A mathematical notation allows the specifier to be very exact about what is specified. It also allows the reader of a specification to identify properties, problems, etc., which may not be obvious otherwise.

Conciseness: A formal specification, although precise, is also very concise compared with an equivalent high-level language program, which is often the first formalization of a system produced if formal methods are not used. Such a specification can be an order of magnitude smaller than the program that implements it, and hence is that much easier to comprehend.

Abstraction: It is all too easy to become bogged down in detail when producing a specification, making it very confusing and obscure to the reader. A formal notation allows the writer to concentrate on the essential features of a system, ignoring those that are implementation details. However, this is perhaps one of the most difficult skills in producing a specification.

Reasoning: Once a formal specification is available, mathematical reasoning is possible to aid in its validation. This is also useful for discussion implications of features, especially within a team of designers.

A design team that understands a particular formal specification notation can benefit from the above improvements in the specification process. It should be noted that much of the benefit of a formal specification derives from the process of producing the specification, as well as the existence of the formal specification after this [Hall 1990].

Specification Languages

The choice of specification language is likely to be influenced by many factors: previous experience, availability of tools, standards imposed by various regulatory bodies, and the particular aspects that must be addressed by the system in question. Another consideration is the degree to which a specification language is executable. This is the subject of some dispute, and the reader is directed elsewhere for a discussion of this topic [Hayes and Jones 1989, Fuchs 1992].

Indeed, the development of any complex system is likely to require the use of multiple notations at different stages in the process and to describe different aspects of a system at various levels of abstraction [Hoare 1987]. As a result, over the last 25 years, the vast majority of the mainstream formal methods have been extended and re-interpreted to address issues of concurrency [Hoare 1985, Milner 1989], real-time behavior [Joseph 1996], and object orientation [Lano and Haughton 1993, Stepney et al. 1992].

There is always, necessarily, a certain degree of trade-off between the expressiveness of a specification language and the levels of abstraction that it supports [Wing 1990]. While certain languages may have wider "vocabularies" and constructs to support the particular situations with which we wish to deal, they are likely to force us towards particular implementations; while they will shorten a specification, they will make it less abstract and more difficult for reasoning [Bowen and Hinchey, 1995(a)].

Formal specification languages can be divided into essentially three classes:

Model-oriented approaches as exemplified by Z [Spivey 1992], VDM [Jones 1991], RAISE [RAISE Language Group 1992], and B [Abrial 1996]. These approaches involve the derivation of an explicit model of the system's desired behavior in terms of abstract mathematical objects.

Property-oriented approaches using **axiomatic semantics** (such as Larch [Guttag et al. 1985]), which use first-order predicate logic to express **preconditions** and **postconditions** of operations over abstract data types, and **algebraic semantics** (such as OBJ [Goguen and Winkler 1988]), which are based on multisorted algebras and relate properties of the system in question to equations over the entities of the system [Hinchey 1993].

Process algebras such as CSP [Hoare 1985] and CCS [Milner 1989], which have evolved to meet the needs of concurrent, distributed, and real-time systems, and which describe the behavior of such systems by describing their algebras of communicating processes.

Unfortunately, it is not always possible to classify a formal specification language in just one of the categories above. LOTOS (Language Of Temporal Ordering Specifications) [Turner 1993], for example, is a combination of ACT ONE and CCS; while it can be classified as an algebraic approach, it exhibits many properties of a process algebra too. Similarly, the RAISE development method is based on extending a model-based specification language (specifically, VDM-SL) with concurrent and temporal aspects.

More visual formalisms, such as Statecharts [Harel 1987], are available and are appealing for industrial use. However, the reasoning aspects and the exact semantics are less well defined. Some specification languages, such as SDL (Specification and Design Language) [Turner 1993], provide particularly good commercial tool support, which is very important for industrial use.

As well as the basic mathematics, a specification language should also include facilities for structuring large specifications. Mathematics alone is all very well in the small, but if a specification is a thousand pages long (and formal specifications of this length exist), there must be aids to organize the inevitable complexity. Z provides the **schema notation** for this purpose, which packages up the mathematics so that it can be reused subsequently in the specification. A number of schema operators, many matching logical connectives, allow recombination in a flexible manner.

A formal specification should also include an informal explanation to put the mathematical description into context and help the reader understand the mathematics. Ideally, the natural language description should be understandable on its own, although the formal text is the final arbiter as to the meaning of the specification. As a rough guide, the formal and informal descriptions should normally be of approximately the same length. The use of mathematical terms should be minimized, unless explanations are being included for didactic purposes.

Specialist and combined languages may be needed for some systems. For example, **hybrid systems** extend the concept of **real-time systems**. In the latter, time must be included, possibly as a continuous variable. In hybrid systems, the number of continuous variables may be increased. This is useful in control systems where a digital computer is responding to real-world analog signals.

Modeling Systems

As previously discussed, the difference between specification and modeling is open to some debate. Different specification languages emphasize and allow modeling to different extents. Algebraic specification eschews the modeling approach, but other specification languages such as Z and VDM actively encourage it.

Some styles of modeling have been formulated for specific purposes. For example, Petri nets may be applied in the modeling of concurrent systems using a specific diagrammatic notation that is quite easily formalizable. The approach is appealing, but the complexity can become overwhelming. Features such as deadlock are detectable, but full analysis can be intractable in practice, since the problem of scaling is not well addressed.

Mathematical modeling allows reasoning about (some parts of) a system of interest [Good and Young 1991]. Here, aspects of the system are defined mathematically, allowing the behavior of the system to be predicted. If the prediction is correct this reinforces confidence in the model. This approach is familiar to many scientists and engineers.

Executable models allow rapid prototyping of systems [Fuchs 1992]. A very high-level programming language such as a functional program or a logic program (which have mathematical foundations) may be used to check the behavior of the system (e.g., see [He 1995]). Rapid prototyping can be useful in demonstrating a system to a customer before the expensive business of building the actual system is undertaken. Again, scientists and engineers are used to carrying out experiments by using such models.

A branch of formal methods known as **model checking** allows systems to be tested exhaustively [McMillan 1993]. Most computer-based systems are far too complicated to test completely because the number of ways the system could be used is far too large. However, a number of techniques, **Binary Decision Diagrams** (BDDs) for example, now allow relatively efficient checking of significant systems, especially for hardware. An extension of this technique, known as **symbolic model checking**, allows even more generality to be introduced.

Mechanical tools exist to handle BDDs and other model-checking approaches efficiently. A tool based on CSP [Hoare 1985], known as FDR (Failure Divergence Refinement), from Formal Systems (Europe) Ltd., allows model checking to be applied to concurrent systems that can be specified in CSP.

Conclusion

Driving forces for best practice include standards, education, training, tools, available staff, certification, accreditation, legal issues, etc. [Bowen and Stavridou 1993]. A full discussion of these is out of the scope of this chapter. Aspects of best practice for specification and modeling depend significantly on the selected specification notation. One of the more popular formal specification notations used in industry is Z. To illustrate the way in which this notation is typically used, a case study using Z is presented in the next section. This demonstrates both some of the underlying principles, and best practice, when employing Z for specification and modeling.

107.4 A Case Study

The Z notation [Spivey 1992] is one of the most widely used formal specification languages and is normally used in a modeling style. An abstract state is first formulated, and then operations on that state are specified. In this section we present a case study using Z to illustrate this style of specification. The example does not exhaustively present the features of Z, but gives a flavor of the style of presentation of a typical Z specification, with extra informal explanation on the notation and conventions where required. Z constructs are introduced as the example is presented and a glossary of Z notation is provided at the end of the chapter for the convenience of the reader. A basic understanding of set theory and logic will help in understanding the specification.

Window management systems are now used extensively for user interfaces to computer systems. The specification given here is of a (fictitious) window system. For more realistic examples of some implemented systems presented in Z, see [Bowen 1996], especially section VI, which includes some actual windows systems, such as the X window system.

Basic Types

Z is a typed language, which allows a certain amount of consistency checking by a mechanical type-checker. However, the only predefined type is the set of integers, denoted $\mathbb{Z}$. Further types must be defined for a particular specification. These **basic types** (also known as **given sets**) may be introduced as follows:

$$[Position, Value]$$

This provides a set of pixel (picture element) positions (e.g., coordinates on a screen), together with possible pixel values (e.g., colors). Note that we are no more specific than this in the specification presented here. It is important not to introduce irrelevant implementation details into a specification, since this restricts the eventual implementor of the system and clutters the specification with information that is not required at a high level of abstraction.

Abbreviation Definitions

It is often useful to include definitions in a specification for commonly used concepts. This helps to reduce the size of the specification and introduces important concepts to the reader in one place, allowing them to be used later within the specification. Pixel maps, relating pixels to their associated values, are an integral part of most window systems. In fact, each pixel has at most one value (assuming it is defined), so we can model a pixel map as a **partial function** from pixel positions to their values:

$$Pixmap == Position \nrightarrow Value$$

Generic Definitions

Z has its own library of "tool-kit" operators, formally defined in terms of more basic mathematical concepts, as presented in [Spivey 1992]. Sometimes it is helpful to extend this library with further **generic definitions** which may be used to define a family of generic constants, applicable to a variety of basic types. Such definitions may be useful for other specifications as well as the one being constructed, allowing reuse of specification components.

For example, a **sequence** of pixel maps may be overlaid in the order given by the sequence to produce a new pixel map. An operator to do this could equally well apply to other partial functions as well as pixel maps, so we can define it generically, using a "distributed overriding" operator:

$$
\begin{array}{l}
\boxed{\begin{array}{l}
=[P, V] \\
\hline
\oplus/ : \mathrm{seq}\ (P \nrightarrow V) \to (P \nrightarrow V) \\
\hline
\oplus/\langle\rangle = \varnothing \\
\forall p : P \nrightarrow V \bullet \oplus/\langle p\rangle = p \\
\forall s, t : \mathrm{seq}\ (P \nrightarrow V) \bullet \oplus/(s \frown t) = (\oplus/s) \oplus (\oplus/t)
\end{array}}
\end{array}
$$

Here, the base cases for the empty sequence $\langle\rangle$ and a singleton sequence $\langle p\rangle$ are considered, followed by the more general case of two arbitrary sequences concatenated together $s \frown t$. Distributed overriding is particularly useful for the specification presented here in specifying the view on a screen of a display, given a sequence of possibly overlapping pixel maps.

Z tool-kit operators normally have a number of laws associated with them, which are helpful in reasoning about specifications. For example, the following law applies for the distributed overriding operator:

$$
p_1, p_2 : Pixmap \vdash \oplus/\langle p_1, p_2\rangle = p_1 \oplus p_2
$$

Such laws must be proved from the original definition. E.g., in this case:

$$
\begin{array}{ll}
\oplus/\langle p_1, p_2\rangle & \\
= \oplus/(\langle p_1\rangle \frown \langle p_2\rangle) & \text{[property of } \frown\text{]} \\
= (\oplus/\langle p_1\rangle) \oplus (\oplus/\langle p_2\rangle) & \text{[by the general case definition]} \\
= p_1 \oplus p_2 & \text{[by the second base case definition, substituting twice]}
\end{array}
$$

Other distributed overriding laws may be found on page 172 of [Bowen 1996].

If the windows in a sequence overlap, it is useful to be able to move selected windows so that their contents may be viewed (or hidden). This is analogous to stuffing a pile of sheets of paper (windows) on a desk (screen). Note that the sheets of paper may be of different sizes and in different positions on the desk.

For example, the following function may be used to move a selected window number in the sequence (if it exists) to the top of the pile (i.e., the end of the sequence). This can also be defined generically:

$$
\begin{array}{l}
\boxed{\begin{array}{l}
=[W] \\
\hline
top: \mathbb{N} \to \mathrm{seq}\ W \to \mathrm{seq}\ W \\
\hline
\forall n : \mathbb{N}; s : \mathrm{seq}\ W \bullet \\
\quad top\ n\ s = \textbf{if}\ n \in \mathrm{dom}\ s\ \textbf{then}\ squash(\{n\} \nmid s) \frown \langle s(n)\rangle\ \textbf{else}\ s
\end{array}}
\end{array}
$$

If the window number n is in the sequence of windows s then it is removed from the sequence (by eliminating that element and squashing the resulting function back into a sequence). This element is then concatenated to the end of the sequence. If the window number is not valid, the sequence of windows is unaffected. The exact technical details require some knowledge of Z, but the above example illustrates the fact that important concepts can be captured formally using relatively short definitions.

In this simple example, we shall ignore the complication of window identifiers. We simply use the position of the window in the sequence to identify it, assuming that the user of the system keeps track of which window is which.

Abstract System State

The window display may be modeled as a sequence of windows against a background "window" which is the size of the display screen itself. The order of the sequence defines which windows are on top in the case of overlapping windows, in ascending order. Only parts of windows that are contained within the background area are displayed.

```
┌─ SYS ────────────────────────────────────────────────────────────
│  windows : seq Pixmap
│  screen, background : Pixmap
│ ─────────────────────────────────────────────────────────────────
│  screen = background ⊕ (dom background ◁ ⊕/windows)
└───────────────────────────────────────────────────────────────────
```

In the specification of the abstract state above, the components *windows* (a sequence of pixel maps), *screen* (as displayed to the user), and *background* (the display if no windows are present), are packaged together in a **schema** box called *SYS*. The **declarations** with their associated type information are above the line and **predicates** defining constraints between these components are (optionally) included below the line. Here, what appears on the display screen is defined in terms of the background pixel map overridden by the sequence of windows in the system, constrained to the background area as defined by its domain of pixel positions.

The screen area is the same as the background area. This can be formalized as follows:

$$SYS \vdash \text{dom } screen = \text{dom } background$$

It is useful to prove such properties correct, either informally, even just mentally, or formally, in order to validate that the specification behaves as expected. Discovering that an expected property does not hold may expose an error in a specification, perhaps in the form of an extra constraint that is required but has been omitted.

Note that the user can see only the display screen. We can specify this view formally by hiding (existentially quantifying) other components in the *SYS* schema to produce a new *View* schema, defined horizontally:

$$View \mathrel{\widehat{=}} SYS \setminus (windows, background)$$

Initially there are no windows in the system:

$$InitSYS \mathrel{\widehat{=}} [SYS' \mid windows' = \langle\rangle]$$

It is important to define the **initial state** and also to ensure that it exists. (Otherwise the system cannot start to operate.) The state at initialization normally consists of the abstract state for the system with some extra constraints. By including *SYS'* in the definition above, all the components in *SYS* are defined, with the extra decoration ′ added to each component name. The prime (or dash) ′ is used by convention in Z to indicate the state *after* an operation. Here we are interested in the state after initialization. Again, we can formulate a property to check that our intuition about the specification is correct:

$$InitSYS \vdash screen' = background'$$

I.e., the screen display at initialization should consist of just the background.

Next we can define general properties about the **change of state** for operations on the system. By convention in Z, Δ ("delta") is used to indicate a change of state where both an unprimed **before state** *SYS* (for example) and a primed **after state** (e.g., *SYS'*) are defined:

$$\Delta\, SYS \mathrel{\widehat{=}} [SYS;\, SYS' \mid background' = background]$$

Here we have added the extra constraint that the background never changes for any operation. This means that we do not have to consider and define this for each individual operation that uses ΔSYS

subsequently, since the predicate *background'* = *background* is automatically included (conjoined) with any other predicates that are defined.

Note that if the before and after states are not related in an operation scheme, then the after state can take on any value. This is the opposite of most programming languages, where unreferenced variables retain their values by default. However this style is useful in specifications since it allows **nondeterminism** to be included easily, where more than one outcome of an operation is allowed. Eventually, of course, the implementor will have to choose a particular outcome, but if the choice is not important at the specification level then leaving the options open gives the implementor a greater choice, possibly allowing different optimization strategies in different implementations of the same specification.

Some operations may leave the state of the system unchanged, for example during a status operation or if an error in the input is detected. Here the Ξ ("xi") convention is used in Z:

$$\Xi\,SYS \mathrel{\widehat{=}} [\,\Delta SYS \mid \theta SYS' = \theta SYS\,]$$

The predicate $\theta SYS' = \theta SYS$ is a shorthand way of ensuring that the tuple formed from all the *SYS* components is the same as that formed from all the primed *SYS'* components.

Operations

In order to use the system, we must have the ability to create windows. These are created on top of all the existing windows. We specify that they must fit within the display background.

```
_ AddWindow0 _____
ΔSYS

window? : Pixmap
_____
dom window? ⊆ dom background
windows' = windows ⌢ (window?)
```

In the definition above, the ΔSYS component automatically includes all the *SYS* and *SYS'* unprimed before and primed after state components, together with the constraint that the background remains unchanged. The *window?* component is an input (as indicated by the Z convention of the added "?"). A **precondition** for the operation is that the window area must be contained within (i.e., be a subset of) the background area. If this is so, then the sequence of windows after the operation has the required window concatenated to it. This means that the window appears on top of any other windows already in the system. Note that by default, predicates on separate lines in a schema are conjoined by using "∧."

The ability to update windows is very useful. This may involve changing the size of the window or its contents or moving it about the screen. Again, the updated window must still fit within the display area.

```
_ Update Window0 _____
ΔSYS

which? : ℕ
window? : Pixmap
_____
which? ∈ dom windows
dom window? ⊆ dom background
windows' = windows ⊕ {which? ↦ window?}
```

Here two inputs are provided, both which window number is to be updated and the new value for the window. The window to be updated must already be in the system and, as for the *AddWindow0* schema, the new window must be within the background for the update to be successful. The sequence of windows is overridden with a new entry for the selected window.

It is desirable to be able to uncover a window which may be partially or even totally obscured by other windows. This can be done by moving the window to the end of the sequence of displayed windows:

```
┌─ ExposeWindow0 ─────────────────────────────────────
│ ΔSYS
│ which? : ℕ
├─────────────────────────────────────────────────────
│ which? ∈ dom windows
│ windows' = top which? windows
└─────────────────────────────────────────────────────
```

Sometimes it is useful to simply rotate the order of the displayed windows, one at a time, moving the bottommost window to the top.

```
┌─ RotateWindows0 ─────────────────────────────────────
│ ΔSYS
├─────────────────────────────────────────────────────
│ windows ≠ ⟨⟩
│ windows' = top 1 windows
└─────────────────────────────────────────────────────
```

Note that sequences are numbered from one updates in Z.

We also wish to be able to delete windows. For instance, we could delete the topmost window (the last window in the sequence):

```
┌─ RemoveTop0 ─────────────────────────────────────────
│ ΔSYS
├─────────────────────────────────────────────────────
│ windows ≠ ⟨⟩
│ windows' = front windows
└─────────────────────────────────────────────────────
```

Alternatively, we may wish to specify which window is to be removed:

```
┌─ RemoveWindow0 ──────────────────────────────────────
│ ΔSYS
│ which? : ℕ
├─────────────────────────────────────────────────────
│ which? ∈ dom windows
│ windows' = squash({which?} ◁ windows)
└─────────────────────────────────────────────────────
```

The above schema definitions give a flavor of the way operations are typically presented in a Z specification. They are intended to illustrate that a number of different operations on a system may be specified succinctly by using Z, providing a suitable abstract state has been formulated.

Error Conditions

The operations covered so far detail what should happen in the event of no errors. Normally operations can also handle error conditions in some controlled manner. It is useful to report the status of an operation. For example, the following reports could be issued:

$Report$ ::= "OK"
         | "Not a window"
         | "No windows"
         | "Invalid window"

Here a **free type** definition defines *Report* to be a set with four possible unique values.

It is helpful to report the fact that the operation was successful if this is the case:

```
┌─ Success ─────────────────────────────────────────────────────
│ rep! : Report
├───────────────────────────────────────────────────────────────
│ rep! = "OK"
└───────────────────────────────────────────────────────────────
```

By convention in Z, "!" indicates an output from an operation.

If errors do occur, then these need to be reported. For example, an invalid window may be specified:

```
┌─ NotAWindow ──────────────────────────────────────────────────
│ Ξ SYS
│ which? : ℕ
│ rep! : Report
├───────────────────────────────────────────────────────────────
│ which? ∉ dom windows
│ rep! = "Not a window"
└───────────────────────────────────────────────────────────────
```

In this case, no change of state occurs, as specified by ΞSYS above. As a precondition, a check is made on whether the window number supplied as an input is not a valid existing window in the system, and if this is so an appropriate report is issued as an output.

It is possible that there are no windows displayed when one is required:

```
┌─ NoWindows ───────────────────────────────────────────────────
│ Ξ SYS
│ rep! : Report
├───────────────────────────────────────────────────────────────
│ windows = ⟨⟩
│ rep! = "No windows"
└───────────────────────────────────────────────────────────────
```

A specified window may not be within the background area:

```
┌─ BadWindow ───────────────────────────────────────────────────
│ Ξ SYS
│ window? : Pixmap
│ rep! : Report
├───────────────────────────────────────────────────────────────
│ ¬ (dom window? ⊆ dom background)
│ rep! = "Invalid window"
└───────────────────────────────────────────────────────────────
```

We may include these errors with the previously defined operations which ignored error conditions, to produce **total** operations:

$$AddWindow\,1 \,\hat{=}\, (AddWindow\,0 \land Success) \lor BadWindow$$

$$Update\,Window\,1 \,\hat{=}\, (Update\,Window\,0 \land Success) \lor BadWindow \lor NotAWindow$$

$$Expose\,Window\,1 \,\hat{=}\, (Expose\,Window\,0 \land Success) \lor NotAWindow$$

$$Rotate\,Windows\,1 \,\hat{=}\, (Rotate\,Windows\,0 \land Success) \lor NoWindows$$

$$RemoveTop\,1 \,\hat{=}\, (RemoveTop\,0 \land Success) \lor NoWindows$$

$$Remove\,Window\,1 \,\hat{=}\, (Remove\,Window\,0 \land Success) \lor NotAWindow$$

Here the schema operators of conjunction ($\wedge$) and disjunction ($\vee$) are used to combine schemas. For both operators, components are merged. If components have the same name, then they must be type-compatible or the specification becomes meaningless. Using schema conjunction, predicates in each schema are logically conjoined. Similarly, if schema disjunction is used, then the predicates in the two schemas are combined using logically disjunction.

The operations are total in that their preconditions are true. This can be checked by calculation, which is a useful way of ensuring that all error conditions have been handled. This is something that is very easily overlooked if only informal specification using natural language and/or diagrams is used.

Status Operations

The contents of an existing window may be of interest:

```
┌─ GetWindow0 ──────────────────────────────────
│ Ξ SYS
│ which? : ℕ
│ window! : Pixmap
├──────────────────────────────────────────────
│ which? ∈ dom windows
│ window! = windows which?
└──────────────────────────────────────────────
```

By using ΞSYS, the state of the system does not change during this operation. Status operations normally have one or more outputs returning some aspect of the state of the system. Here a particular window is returned.

We can make this operation total as well:

$$GetWindow \mathrel{\widehat{=}} (GetWindow0 \wedge Success) \vee NotAWindow$$

Conclusion

Given the abstract state, initial state, and operation schemas defined in this section, the operation of the system consists of starting in the initial state, followed by an arbitrary sequence of the specified operations on the state, as allowed by the preconditions of the operations. If the preconditions of all the operations are true, then any order of operations is allowed.

This section has presented the use of the Z notation [Spivey 1992] in a modeling style, as it is widely used for specifying systems. It should be remembered that Z is a general-purpose specification language and can be used in other styles if desired. However, the use of an abstract state and operations on that state has been found to be a style that is easy to understand (once the notation and conventions have been learned), and this is the approach that is often adopted in practice.

For those wishing to learn Z, there are many textbooks available (e.g., see [Potter et al. 1996, Woodcock and Davies 1996]). There are also books presenting further case studies in Z [Bowen 1996, Hayes 1992]. An international standard for Z is in preparation under ISO/IEC JTC1/SC22 [Brien and Nicholls 1992], but a widely accepted de facto standard for Z, with a matching type-checker called fuzz by the same author, is already available [Spivey 1992]. For information on the practical use of Z, including more advanced structuring techniques such as **promotion**, written by industrial authors, see [Barden et al. 1994].

Good books on the similar VDM approach include [Jones 1991], and [Jones and Shaw 1990] for a selection of case studies.

107.5 Research Issues and Summary

Claims that formal methods can guarantee correct hardware and software, eliminate the need for testing, etc., have led some to believe that formal methods are something almost magical [Hall 1990]. More significantly, beliefs that formal methods are difficult to use, delay the development process, and raise development costs [Bowen and Hinchey 1995(b)] have led many to believe that formal methods offer few advantages over traditional development methods. Formal methods are not a panacea; they are just one of a range of techniques that, when correctly applied, have proven themselves to result in systems of the highest integrity [Bowen and Hinchey 1995(a)].

Method integration is one approach that may aid in the acceptance of formal methods and may help in the technology transfer from academic theory to industrial practice. This has the advantage of providing multiple views of a system, incorporating a graphical representation, that is likely to be more acceptable to nonspecialists, while retaining the ability to propose and prove properties of the systems, and to demonstrate that requirements are contradictory (or otherwise) before implementation.

Hitherto, the uptake of formal methods has been hindered, at least in part, by a lack of tools. A large number of the highly successful projects discussed in [Hinchey and Bowen 1995] required significant investment in tool support. Just as the advent of compiler technology was necessary for the uptake of high-level programming languages, and CASE (Computer Aided Software Engineering) technology provided the impetus for the emergence of structural design methodologies in the 1970s, a significant investment in formal methods tools is required for formal methods to be practical at the level of industrial application [Hinchey and Bowen 1995, Craigen et al. 1995]. In the future, we envisage greater emphasis on IFDSEs (Integrated Formal Development Support Environments) that will support formal specification and development, based on an emerging range of high-quality standalone tools.

Current trends in standardization and the number of standards in the safety-critical domain that are under discussion augur many more future standards in this area. Many of these standards are mentioning formal methods, and others are likely to emphasize formal methods in the future [Bowen and Hinchey 1994]. It should be pointed out that most standards are recommending formal methods rather than mandating them. We hope that this will, however, eventually lead to improvements in safety levels and help protect the lives and resources that are reliant on complex computer systems.

Defining Terms

Formal methods: Techniques, notations, and tools with a mathematical basis, used for specification and reasoning in software or hardware system development.

Formal notation: A language with a mathematical semantics, used for formal specification, reasoning, and proof.

Logic: A scheme for reasoning, proof, inference, etc. Two common schemes are **propositional logic** and **predicate logic**, which is propositional logic generalized with quantifiers. Other logics, such a modal logics, including **temporal logics** which handle time (e.g., TLA, ITL, and more recently Duration Calculus), are also available. Schemes may use **first-order logic** or **higher-order logic**. In the former, functions are not allowed on predicates, simplifying matters somewhat, but in the latter they are, providing greater power. Logics include a calculus which allows reasoning in the logic.

Operation: The performance of some desired action. This may involve the change of state of a system, together with inputs to the operation and outputs resulting from the operation. To specify such an operation, the **before state** (and inputs) and the **after state** (and outputs) must be related with constraining predicates.

Precondition: The predicate which must hold before an operation for it to be successful. Compare **postcondition**, which is the predicate which must hold after an operation.

Predicate: A constraint between a number of variables which produces a truth value (e.g., *true* of *false*).

Proof: A series of mathematical steps forming an argument of the correctness of a mathematical statement or theorem. For example, the **validation** of a desirable property for a formal specification could be undertaken by proving it correct. Proof may also be used to perform a formal **verification** that an implementation meets a specification. A less formal style of reasoning is **rigorous argument**, where a proof outline is sketched informally, which may be done if the effort of undertaking a fully formal proof is not considered cost-effective.

Refinement: The stepwise transformation of a specification towards an implementation (e.g., as a program). Compare **abstraction**, where unnecessary implementation detail is ignored in a specification.

Relation: A connection or mapping between elements in a number of sets. Often two sets (a **domain** and a **range**) are related in a **binary relation**. A special case of a relation is a **function** where individual elements in the domain can only be mapped to at most one element in the range of the function. Functions may be further categorized. For example, a **partial function** may not map all possible elements that could be in the domain of the function, whereas a **total function** maps all such elements.

Set: A collection of distinct objects or **elements**, which are also known as **members** of the set. In a typed language, types may consist of maximal sets, as in the Z notation.

Specification: A description of *what* a system is intended to do, as opposed to *how* it does it. A specification may be *formal* (mathematical) or *informal* (natural language, diagrams, etc.). Compare an **implementation** of a specification, such as a program, which actually performs and executes the actions required by a specification.

State: A representation of the possible values which a system may have. In an abstract specification, this may be modeled as a number of sets. By contrast, in a concrete program implementation, the state typically consists of a number of data structures, such as arrays, files, etc. When modeling sequential systems, each operation may include a **before state** and an **after state** which are related by some constraining predicates. The system will also have an **initial state**, normally with some additional constraints, from which the system starts at initialization.

A glossary of the Z mathematical and schema notation is included for the reader's convenience at the end of the chapter. For more information on the Z notation, see the *Z Reference Manual* [Spivey 1992].

Glossary of Z Notation

Names

a, b	Identifiers
d, e	Declarations (e.g., $a : A;\ b, \ldots : B \ldots$)
f, g	Functions
m, n	Numbers
p, q	Predicates
s, t	Sequences
x, y	Expressions
A, B	Sets
C, D	Bags
Q, R	Relations
S, T	Schemas
X	Schema text (e.g., $d, d \mid p$ or S)

Definitions

$a == x$	Abbreviated definition
$a ::= b \mid \ldots$	Free type definition (or $a ::= b \langle\langle x \rangle\rangle \mid \ldots$)
$[a]$	Introduction of a given set (or $[a, \ldots]$)
$a\_$	Prefix operator
$\_a$	Postfix operator
$\_a\_$	Infix operator

Logic

true	Logical true constant
false	Logical false constant

$\neg p$	Logical negation
$p \wedge q$	Logical conjunction
$p \vee q$	Logical disjunction
$p \Rightarrow q$	Logical implication ($\neg p \vee q$)
$p \Leftrightarrow q$	Logical equivalence ($p \Rightarrow q \wedge q \Rightarrow p$)
$\forall X \bullet q$	Universal quantification
$\exists X \bullet q$	Existential quantification
$\exists_1 X \bullet q$	Unique existential quantification
let $a == x; \ldots \bullet p$	Local definition

Sets and Expressions

$x = y$	Equality of expressions
$x \neq y$	Inequality ($\neg(x = y)$)
$x \in A$	Set membership
$x \notin A$	Nonmembership ($\neg(x \in A)$)
$\emptyset$	Empty set
$A \subseteq B$	Set inclusion
$A \subset B$	Strict set inclusion ($A \subseteq B \wedge A \neq B$)
$\{x, y, \ldots\}$	Set of elements
$\{X \bullet x\}$	Set comprehension
$\lambda X \bullet x$	Lambda-expression—function
$\mu X \bullet x$	Mu-expression—unique value
let $a == x; \ldots \bullet y$	Local definition
if p **then** x **else** y	Conditional expression
$(x, y, \ldots)$	Ordered tuple
$A \times B \times \ldots$	Cartesian product
$\mathbb{P}A$	Power set (set of subsets)
$\mathbb{P}_1 A$	Nonempty power set
$\mathbb{F}A$	Set of finite subsets
$\mathbb{F}_1 A$	Nonempty set of finite subsets
$A \cap B$	Set intersection
$A \cup B$	Set union
$A \setminus B$	Set difference
$\bigcup A$	Generalized union of a set of sets
$\bigcap A$	Generalized intersection of a set of sets
first x	First element of an ordered pair
second x	Second element of an ordered pair
$\#A$	Size of a finite set

Relations

$A \leftrightarrow B$	Relation ($\mathbb{P}(A \times B)$)
$a \mapsto b$	Maplet ((a, b))
$\mathrm{dom}\,R$	Domain of a relation
$\mathrm{ran}\,R$	Range of a relation
$\mathrm{id}\,A$	Identity relation
$Q \,;\, R$	Forward relational composition
$Q \circ R$	Backward relational compositon ($R \,;\, Q$)
$A \triangleleft R$	Domain restriction
$A \ntriangleleft R$	Domain anti-restriction

$A \vartriangleright R$	Range restriction
$A \blacktriangleright R$	Range anti-restriction
$R(\!\!(A)\!\!)$	Relational image
iter n R	Relation composed *n* times
R^n	Same as *iter n R*
$R^\sim$	Inverse of relation (R^{-1})
R^*	Reflexive-transitive closure
R^+	Irreflexive-transitive closure
$Q \oplus R$	Relational overriding (($\mathrm{dom}\,R \vartriangleleft Q) \cup R$)
$a \underline{R} b$	Infix relation

Functions

$A \rightarrowtail B$	Partial functions
$A \rightarrow B$	Total functions
$A \rightarrowtail\!\!\!\rightarrow B$	Partial injections
$A \rightarrowtail B$	Total injections
$A \twoheadrightarrow B$	Partial surjections
$A \twoheadrightarrow B$	Total surjections
$A \rightarrowtail\!\!\!\twoheadrightarrow B$	Bijective functions
$A \nrightarrow B$	Finite partial functions
$A \nrightarrow\!\!\!\rightarrow B$	Finite partial injections
$f\,x$	Function application (or $f(x)$)

Numbers

$\mathbb{Z}$	Set of integers
$\mathbb{N}$	Set of natural numbers $\{0, 1, 2, \ldots\}$
$\mathbb{N}_1$	Set of nonzero natural numbers ($\mathbb{N}\backslash\{0\}$)
$m + n$	Addition
$m - n$	Subtraction
$m * n$	Multiplication
m div n	Division
m mod n	Modulo arithmetic
$m \leq n$	Less than or equal
$m < n$	Less than
$m \geq n$	Greater than or equal
$m > n$	Greater than
succ n	Successor function $\{0 \mapsto 1,\ 1 \mapsto 2, \ldots\}$
$m \mathrel{..} n$	Number range
min A	Minimum of a set of numbers
max A	Maximum of a set of numbers

Sequences

seq A	Set of finite sequences
$\mathrm{seq}_1 A$	Set of nonempty finite sequences
iseq A	Set of finite injective sequences
$\langle\rangle$	Empty sequence
$\langle x, y, \ldots\rangle$	Sequence $\{1 \mapsto x,\ 2 \mapsto y, \ldots\}$
$s \frown t$	Sequence concatenation
$\frown/s$	Distributed sequence concatenation

head s	First element of sequence ($s(1)$)
tail s	All but the head element of a sequence
last s	Last element of sequence ($s(\#s)$)
front s	All but the last element of a sequence
rev s	Reverse a sequence
squash f	Compact a function to a sequence
$A \restriction s$	Sequence extraction ($squash\,(A \lhd s)$)
$s \restriction A$	Sequence filtering ($squash\,(s \rhd A)$)
s prefix *t*	Sequence prefix relation ($s \frown v = t$)
s suffix *t*	Sequence suffix relation ($u \frown s = t$)
s in *t*	Sequence segment relation ($u \frown s \frown v = t$)
disjoint *A*	Disjointness of an indexed family of sets
A partition *B*	Partition an indexed family of sets

Bags

bag *A*	Set of bags or multisets ($A \nrightarrow \mathbb{N}_1$)
$[\![\,]\!]$	Empty bag
$[\![\,x, y, \ldots\,]\!]$	Bag $\{x \mapsto 1,\ y \mapsto 1, \ldots\}$
count C x	Multiplicity of an element in a bag
$C \sharp x$	Same as *count C x*
$n \otimes C$	Bag scaling of multiplicity
$x \in C$	Bag membership
$C \sqsubseteq D$	Subbag relation
$C \uplus D$	Bag union
$C \cup D$	Bag difference
items s	Bag of elements in a sequence

Schema notation

Vertical Schema

$\begin{array}{|l}\hline S \\ \hline d \\ \hline p \\ \hline\end{array}$ — New lines denote ";" and "$\wedge$." The schema name and predicate part are optional. The schema may subsequently be referenced by name in the document.

Axiomatic Definition

$\begin{array}{|l} d \\ \hline p \end{array}$ — The definitions may be nonunique. The predicate part is optional. The definitions apply globally in the document.

Generic Definition

$\begin{array}{|l} [a, \ldots] = \\ d \\ \hline p \\ \hline\end{array}$ — The generic parameters are optional. The definitions must be unique. The definitions apply globally in the document.

$S \mathrel{\widehat{=}} [X]$	Horizontal schema
$[T; \ldots \mid \ldots]$	Schema inclusion
$z.a$	Component selection (given $z : S$)

θS	Tuple of components
$\neg S$	Schema negation
pre S	Schema precondition
$S \wedge T$	Schema conjunction
$S \vee T$	Schema disjunction
$S \Rightarrow T$	Schema implication
$S \Leftrightarrow T$	Schema equivalence
$S \backslash (a, \ldots)$	Hiding of component(s)
$S \restriction T$	Projection of components
$S \mathbin{;} T$	Schema composition (S then T)
$S \gg T$	Schema piping (S outputs to T inputs)
$S[a/b, \ldots]$	Schema component renaming (b becomes a, etc.)
$\forall X \bullet S$	Schema universal quantification
$\exists X \bullet S$	Schema existential quantification
$\exists_1 X \bullet S$	Schema unique existential quantification

Conventions

$a?$	Input to an operation
$a!$	Output from an operation
a	State component before an operation
a'	State component after an operation
S	State schema before an operation
S'	State schema after an operation
ΔS	Change of state (normally $S \wedge S'$)
ΞS	No change of state (normally $[S \wedge S' \mid \theta S = \theta S']$)
$d \vdash p$	Theorem

References

Abrial, J.-R. 1996. *The B-Book*. Cambridge University Press, Cambridge, U.K.

Barden, R., Stepney, S., and Cooper, D. 1994. *Z in Practice*. BCS Practitioner Series. Prentice–Hall, Hemel Hempstead, U.K.

Bowen, J. P. 1996. *Formal Specification and Documentation using Z: A Case Study Approach*. International Thomson Computer Press, London.

Bowen, J. P., Breuer, P. T., and Lano, K. C. 1993. Formal specifications in software maintenance: from code to Z++ and back again. *Inf. Software Tech.* 35(11/12):679–690.

Bowen, J. P. and Hinchey, M. G. 1994. Formal methods and safety-critical standards. *IEEE Comput.* 27(8):68–71.

Bowen, J. P. and Hinchey, M. G. 1995(a). Ten Commandments of Formal Methods. *IEEE Comput.* 28(4):56–63.

Bowen, J. P. and Hinchey, M. G. 1995(b). Seven more myths of formal methods. *IEEE Software* 12(4):34–41.

Bowen, J. P. and Hinchey, M. G., eds. 1995(c). *ZUM '95: The Z Formal Specification Notation*. Lecture Notes in Computer Science 967. Springer–Verlag.

Bowen, J. P. and Stavridou, V. 1993. Safety-critical systems, formal methods and standards. *IEE/BCS Software Eng. J.* 8(4):189–209.

Brien, S. M. and Nicholls, J. E. 1992. *Z Base Standard*. Technical Monograph PRG-107, Oxford University Computing Laboratory, Oxford, U.K.

Craigen, D., Gerhart, S. L., and Ralston, T. J. 1995. Formal methods reality check: Industrial usage. *IEEE Trans. Software Eng.* 21(2):90–98.

Dean, C. N. and Hinchey, M. G. 1995. Introducing formal methods through rôle playing. *ACM SIGCSE Bull.* 27(1):302–306.

Dijkstra, E. W. 1981. Why correctness must be a mathematical concern. In *The Correctness Problem in Computer Science*, R. S. Boyer and J. S. Moore, eds., pp. 1–6. Academic Press, London.

Faser, M. D., Kumar, K., and Vaishnavi, V. K. 1994. Strategies for incorporating formal specifications in software development. *Commun. ACM* 37(10):74–86.

Fuchs, N. E. 1992. Specifications are (preferably) executable. *IEE/BCS Software Eng. J.* 7(5):323–334.

Garlan, D. 1995. Making formal methods effective for professional software engineers. *Inf. Software Tech.* 37(5/6):261–268.

Gaudel, M.-C. and Woodcock, J., eds. 1996. *FME '96: Industrial Benefit and Advances in Formal Methods*. Lecture Notes in Computer Science 1051. Springer–Verlag.

Gibbs, W. W. 1994. Software's chronic crisis. *Sci. Am.* 271(3):86–95.

Goguen, J. and Winkler, T. 1988. *Introducing OBJ3*. SRI International, Menlo Park, CA. Tech. Rep. SRI-CSL-88-9.

Good, D. I. and Young, W. D. 1991. Mathematical methods for digital system development. In *VDM '91: Formal Software Development Methods*, Vol. 2, S. Prehn and W. J. Toetenel, eds. Lecture Notes in Computer Science 552, pp. 406–430. Springer–Verlag.

Gravell, A. 1991. What is a good specification? In *Z User Workshop, Oxford, 1990*, J. E. Nicholls, ed. Workshops in Computing, pp. 137–150. Springer–Verlag, London.

Guttag, J. V., Horning, J. J., and Wing, J. M. 1985. *Larch in Five Easy Pieces*. DEC Systems Research Center, Palo Alto, CA. SRC Report 5.

Hall, J. A. 1990. Seven myths of formal methods. *IEEE Software* 7(5):11–19.

Hall, J. A. 1996. Using formal methods to develop an ATC information system. *IEEE Software* 13(2):66–76.

Harel, D. 1987. Statecharts: A visual formalism for complex systems. *Sci. Comput. Program.* 8:231–274.

Hayes, I. J., ed. 1992. *Specification Case Studies*, 2nd ed. Prentice Hall International Series in Computer Science. Hemel Hempstead, U.K.

Hayes, I. J. and Jones, C. B. 1989. Specifications are not (necessarily) executable. *IEE/BCS Software Eng. J.* 4(6):330–338.

He, J. 1995. *Provably Correct Systems: Modeling of Communication Languages and Design of Optimized Compilers*. McGraw–Hill International Series in Software Engineering, London.

Hinchey, M. G. 1993. Formal methods for system specification: An ounce of prevention is worth a pound of cure. *IEEE Potentials* 12(3):50–52.

Hinchey, M. G. and Bowen, J. P., eds. 1995. *Applications of Formal Methods*. Prentice Hall International Series in Computer Science. Hemel Hempstead, U.K.

Hoare, C. A. R. 1985. *Communicating Sequential Processes*. Prentice Hall International Series in Computer Science. Hemel Hempstead, U.K.

Hoare, C. A. R. 1987. An overview of some formal methods for program design. *IEEE Comput.* 20(9):85–91.

Hoare, C. A. R. 1996(a). How did software get so reliable without proof? In *FME '96: Industrial Benefit and Advances in Formal Methods*. M.-C. Gaudel and J. Woodcock, eds., pp. 1–17. Lecture Notes in Computer Science 1051. Springer–Verlag.

Hoare, C. A. R. 1996(b). The logic of engineering design. *Microprocessing Microprogramming* 41(8/9):525–539.

Houston, I. S. C. and King, S. 1991. CICS project report: Experiences and results from the use of Z in IBM. In *VDM '91: Formal Software Development Methods*, Vol. 1, S. Prehn and W. Toetenel, eds. Lecture Notes in Computer Science 551, pp. 588–596. Springer–Verlag.

IEEE. 1991. IEEE standard glossary of software engineering terminology. In *IEEE Software Engineering Standards Collection*. Elsevier Applied Science, Amsterdam.

Jones, C. B. 1991. *Software Development Using VDM*, 2nd ed. Prentice Hall International Series in Computer Science. Hemel Hempstead, U.K.

Jones, C. B. and Shaw, R., eds. 1990. *Case Studies in Systematic Software Development*. Prentice Hall International Series in Computer Science. Hemel Hempstead, U.K.

Joseph, M., ed. 1996. *Real-Time Systems: Specification, Verification and Analysis*. Prentice Hall International Series in Computer Science. Hemel Hempstead, U.K.

Lano, K. C. and Haughton, H., eds. 1993. *Object-Oriented Specification Case Studies.* Prentice Hall Object-Oriented Series, Hemel Hempstead, U.K.

McMillan, K. L. 1993. *Symbolic Model Checking.* Kluwer Academic Press, Boston.

Milner, R. 1989. *Communication and Concurrency.* Prentice Hall International Series in Computer Science. Hemel Hempstead, U.K.

Morgan, C. 1994. *Programming from Specifications,* 2nd ed. Prentice Hall International Series in Computer Science. Hemel Hempstead, U.K.

Parnas, D. 1995. Using mathematical models in the inspection of critical software. In *Applications of Formal Methods.* M. G. Hinchey and J. P. Bowen, eds., pp. 17–31. Prentice Hall International Series in Computer Science. Hemel Hempstead, U.K.

Potter, B., Sinclair, J., and Till, D. 1996. *Introduction to Formal Specification and Z,* 2nd ed. Prentice Hall International Series in Computer Science. Hemel Hempstead, U.K.

RAISE Language group. 1992. *The RAISE Specification Language.* BCS Practitioner Series. Prentice–Hall, Hemel Hempstead, U.K.

Saiedian, H., ed. 1996. An invitation to formal methods. *IEEE Comput.* 29(4):16–30.

Spivey, J. M. 1992. *The Z Notation: A Reference Manual,* 2nd ed. Prentice Hall International Series in Computer Science, Hemel Hempstead, U.K.

Stepney, S., Barden, R., and Cooper, D., eds. 1992. *Object Orientation in Z.* Workshops in Computing. Springer–Verlag, London.

Turner, K. J., ed. 1993. *Using Formal Description Techniques: An Introduction to Estelle, LOTOS and SDL.* Wiley, Chichester, U.K.

Wing, J. M. 1990. A specifier's introduction to formal methods. *IEEE Comput.* 23(9):8–24.

Woodcock, J. and Davies, J. 1996. *Using Z: Specification, Refinement, and Proof.* Prentice Hall International Series in Computer Science. Hemel Hempstead, U.K.

Further Information

A number of organizations have been established to meet the needs of formal methods practitioners:

- Formal Methods Europe (FME) also organizes a regular conference (e.g., [Gaudel and Woodcock 1996], formerly the VDM symposia) and other activities for users of various formal methods, and publishes a regular newsletter.

- The British Computer Society Special-Interest Group on Formal Aspects of Computing Science (BCS FACS) organizes workshops and meetings on various aspects of formal methods, as well as a series of refinement workshops.

- The Z User Group (ZUG) organizes a regular international conference, historically known as the Z User Meeting (ZUM, e.g., [Bowen and Hinchey 1995(c)]), attracting users of the Z notation from all over the world.

There are now a number of journals devoted specifically to formal methods. These include *Formal Methods in System Design and Formal Aspects of Computing,* published by Springer–Verlag in association with BCS-FACS, and the *FACS Europe* newsletter run jointly by Formal Methods Europe and BCS-FACS. Other European-based journals, such as *The Computer Journal,* the *Software Engineering Journal,* and *Information and Software Technology,* regularly publish articles on, or closely related to, formal methods, and they have run special issues on the subject.

While there are no U.S.-based journals that deal specifically with formal methods, they regularly are featured in periodicals such as *IEEE Computer* (e.g., [Bowen and Hinchey 1994, Bowen and Hinchey 1995(a), Hoare 1987, Saiedian 1996, Wing 1990]), *IEEE Software* (e.g., [Bowen and Hinchey 1995(b), Hall 1990, Hall 1996]), and *Communications of the ACM* (e.g., [Faser et al. 1994]), as well as in journals such as *IEEE Transactions on Software Engineering, ACM Transactions on Software Engineering and Methodology,* and the *Journal of the ACM.*

In addition to the conferences mentioned earlier, the IFIP (International Federation of Information Processing) FORTE international conference concentrates on **Formal Description Techniques** (FDTs). A number of workshops on formal methods (and in particular dealing with industrial usage) have been established, most notably WIFT, the Workshop on Industrial-strength Formal Techniques. Some more wide-ranging conferences give particular attention to formal methods; primary among these are the ICSE (International Conference on Software Engineering) and ICECCS (International Conference on Engineering of Complex Computer Systems) series of conferences. Other specialist conferences in the safety-critical sector, such as COMPASS, SAFECOMP, and SSS (the Safety-critical Systems Symposium) also regularly cover formal methods.

A number of electronic forums are available as on-line newsgroups:

`comp.specification.misc`	Formal specification
`comp.specification.larch`	Larch
`comp.specification.z`	Z notation

In addition, the following electronic mailing lists are available:

`formal-methods-request@cs.uidaho.edu`	Formal methods
`obj-forum-request@comlab.ox.ac.uk`	OBJ
`vdm-forum-request@mailbase.ac.uk`	VDM
`zforum-request@comlab.ox.ac.uk`	Z (gatewayed to `comp.specification.z`)
`zugeis-request@comlab.ox.ac.uk`	Z educational issues

For up-to-date on-line information on formal methods in general, readers are directed to the following World Wide Web URL (Uniform Resource Locator):

`http://www.comlab.ox.ac.uk/archive/formal-methods.html`

108

Software Design

Steven A. Demurjian, Sr.
The University of Connecticut

Software design techniques span a wide spectrum, and they have incrementally evolved as the discipline has matured over the years. In the early 1960s, flowcharts were the most heavily used design technique for programming, and they subsequently evolved through the 1960s and into the mid-1970s into approaches such as data-flow and entity-relationship diagrams. At this same time, parallel efforts began on approaches for design using modules [Parnas 1972] and **abstract data types** (**ADTs**) [Liskov and Zilles 1975, Liskov et al. 1977]. Module concepts were further explored in the late 1970s [Wirth 1977], taking us into the early 1980s, where these design concepts were supported in programming languages such as Smalltalk-80 [Goldberg 1989], Ada [Barnes 1991], and Modula-2 [Wirth 1985]. Starting in the mid-1980s and continuing into the 1990s, there has been the emergence of the object-oriented approach for software design and development, which are being supported today in many different object-oriented programming languages (e.g., Ada 95 [Barnes 1996, Department of Defense 1995], Modula-3 [Harbison 1992], C++ [Stroustrup 1986], Eiffel [Meyer 1992], Smalltalk [Goldberg 1989], Object Pascal [Tesler 1985], Java [Deitel and Deitel 1997], etc.) and database systems (e.g., Ontos [Ontologic 1991], Gemstone [Brett et al. 1989], O2 [Deux et al. 1991], Orion [Kim 1990], ObjectStore [Lamb et al. 1991], etc.). While it would be impossible to review this entire history of software design in a single chapter, we will introduce and trace the important concepts and techniques.

Software design is not an isolated activity, and in the opinion of some, it is one of the most important aspects of the overall design, development, and maintenance process. In an oft-cited article, Fredrick Brooks presents the notion that there is no *silver bullet* or panacea to solve all of the problems related to software design and development [Brooks 1987]. In the article, Brooks establishes a common motivation that crosses application domains, as indicated by the following quotes:

> The hardest single part of building a software system is deciding precisely what to build....
> Therefore, the most important function that the software builder performs for the client is the

iterative extraction and refinement of the product requirements. . . . I would go a step further and assert that it is really impossible for a client, even working with a software engineer, to specify completely, precisely, and correctly the exact requirements of a modern software product before trying some versions of the product. [Brooks 1987, p. 17].

Brooks believes that the focus must be on the design process. Specifically, there is a vacuum or lack of what Brooks calls "great designers," the one or two individuals or software engineers who are head-and-shoulders above the other team members and consequently drive the successful completion of a software system. These great designers must be identified, recognized, and rewarded for their expertise. The key is that we cannot separate the individual (software engineer) from the techniques and processes. Software design approaches are irrelevant without knowledgeable individuals who can utilize and exploit the techniques to their fullest extent. Our discussion of the various approaches and techniques will also include, where appropriate, indications of their strengths and weaknesses.

This chapter contains five sections. To serve as a basis for discussion, section 108.1 introduces the High-Tech Supermarket System, **HTSS**, which is used as an explanation vehicle for the different design approaches presented in this chapter. In section 108.2, we review traditional approaches for software design, including: **top-down**, **bottom-up**, **data-flow diagrams (DFDs)**, **entity-relationship (ER) diagrams**, and **finite-state machines (FSMs)**; highlighting their strengths and weaknesses. The techniques reviewed in section 108.2 are often intended for conceptual software design, used as software engineers first attempt to understand the system components, structure, and interactions. Section 108.3 examines techniques for encapsulation and hiding via **modules** and **ADTs**. These techniques often follow the use of DFDs, ERs, and FSMs, since they allow detailed system structure and interactions to be defined. Section 108.4 expands on encapsulation and hiding concepts with a discussion of **object-oriented design** techniques. A detailed discussion of both basic and advanced concepts coupled with illustrative design examples is provided. Section 108.5 reviews mathematical and analytical design techniques, specifically, **queueing network models**, **time-complexity analysis**, and **simulation models**. These techniques have been a part of computer science and engineering since its earliest days and have been utilized for a wide range of computing-related analysis. They are relevant and important for software design, since they offer the ability to both predict and estimate performance, which is a key concern for software engineers.

108.1 The High-Tech Supermarket System

The High-Tech Supermarket System, **HTSS**, is a modern supermarket that uses the newest and most up-to-date computing technology to support inventory control and to assist customers in their shopping. The purpose of **HTSS** is to utilize computing technology in a positive way to enhance and facilitate the shopping experience for customers by integrating inventory control with the following:

1. the cashier's functions for checking out customers to automatically update inventory when an item is sold,
2. a user-friendly grocery item locator that indicates textually and graphically where items are in the store and if the item is out of stock, and
3. a fast-track deli-orderer (deli orders are entered electronically), with the shoppers allowed to pick up the order weighed and packaged without waiting.

The inventory control aspect of the proposed system would maintain all inventory for the store and alert the appropriate store personnel whenever the amount of an item drops to its reorder limit. The system should also have extensive query capabilities that allow store personnel to investigate the status of the inventory and to track sales for the store over various time periods and other restrictions. Finally, note that **HTSS** and its functional components are based on an actual store that opened in Connecticut, in the spring of 1993. Thus, the concepts that are presented have their basis in an actual "real-world" application.

To support the functional and operational requirements of **HTSS**, from an end-user perspective, there must be a set of user-system interfaces. Possible interfaces include the following:

- Cash register/universal product code (UPC) scanner: Used to process an order, which includes: recording individual items, totaling them, deducting coupons, and taking payment. As each item is scanned, it must be deducted from the inventory, so that values are always consistent and up-to-date.
- Displays for inventory querying: To access and manage the inventory, a separate display is needed. Through this display, orders and updates can be made. Only authorized individuals will be allowed to enter new orders or update the inventory when a shipment arrives.
- Shopper interface for locator: Used by customers to locate where (aisle, shelf) a particular item is displayed in a store.
- Shopper interface for orderer: Through this interface, customers can place orders for the deli (e.g., meats, cheeses, salads, etc.). These orders are then filled and the customer picks up the order at some later time.
- Deli interface for orderer: This interface is needed by store employees that work in the deli department to scan and fill customer orders.

We have chosen this set on the basis of both their differences (they all have unique requirements for their operation) and similarities (they all share common requirements regarding response-time, throughput, and user-friendliness). Response-time and throughput are important for the first two interfaces, since there are likely to be multiple cash registers that must work in parallel with many inventory displays. User-friendliness is also important, for new employees using cash registers, and especially for customers using the different shopper interfaces. Clearly, **HTSS** as an application contains multiple types of data that must interact; performance constraints on throughput and number of concurrent users; persistence for multiple databases, and a wide variety of users with different capabilities and access requirements.

108.2 Traditional Approaches to Design

Traditional design approaches (e.g., top-down, bottom-up, DFDs, etc.) focus on developing a functional characterization of an application. Historically, there are close ties between these approaches and imperative or procedural programming languages such as FORTRAN, Pascal, and C. The reason is that there is a direct correspondence between the design for an application using one of the approaches and its realization as a working piece of software or program, at both a conceptual level and from the perspective of the coding and organizational techniques that are utilized to develop software using an imperative language. This section is a case study of a number of different traditional design approaches—namely, top-down, bottom-up, DFDs, and DRDs, and FSMs. Each approach can be used in many different ways, and is well suited to solving certain kinds of problems, or rather, to develop the solution to a problem from a specific perspective. Each approach can also be used to conceptualize a design at various levels of granularity.

Top-Down and Bottom-Up Design

Regardless of which traditional approach is chosen, the common theme of a functional characterization of the system or application persists. In the case of **HTSS**, a functional list of a subset of the basic system components would include the following:

1. Check-out customers: The actions that must be taken by a cashier to process a customer's order.

2. Locate items: The actions that are taken whenever locating specific items in the store is necessary. This could be initiated as a result of a customer using the item locator/user interface or a bagger needing to check a price or get an item for a customer.
3. Order deli meats, cheeses, and salads: These are the actions that are necessary to support the deli orderer, and to allow the order to be transferred to deli employees for processing.
4. Update and query inventory: These actions are needed by management and stock-room personnel to track and maintain the status of the inventory.

These four major components are determinable based on a *top-down* design examination of the problem at hand. Once these components have been identified, they can each be expressed as a set of detailed tasks via the process of *stepwise refinement*. For **HTSS**, the first component can be refined as the following tasks:

1a. Scan UPCs for all items
1b. Modify inventory
1c. Maintain running total
1d. Subtotal/coupon adjustment
1e. Final total and take payment
1f. Etc. etc. etc.

Tasks 1a, 1b, and 1c are interleaved to process all items for an order, followed by tasks 1d and 1e. Each of these tasks can in turn be refined and expanded by an iterative and incremental process that can evolve the design towards an implementation. For example, as part of task 1e, if a noncash option is chosen, it may be necessary to verify the credit card, automated teller machine (ATM) card, or checking account status. This top-down process proceeds from the general to the specific to arrive at a solution.

The complement of top-down is the *bottom-up* approach which, while still functionally oriented, is driven strongly by information and its usage. For example, given the four components previously reviewed (check-out customers; locate items; order deli meats, cheeses, and salads; and update and query inventory), and the general description of **HTSS** in section 108.1, it is likely that a data structure can be defined that maintains grocery items. In **HTSS**, each **Item** should have a UPC for unique identification, a **Name**, various **Costs** (e.g., **Wholesale** and **Retail**), a **Size** or **Weight**, the **Amount** on shelves or in the stockroom, and so on. Given this information, the major components of **HTSS** can be examined to identify their access requirements. For example:

- UPC scanner: Must scan the UPC on an **Item**, verify it against the database for the inventory, and then return all appropriate information on an **Item** to be used in checking-out a customer's order.
- Locator: Once an **Item** has been selected by a customer, the shelf **Amounts** can be accessed to display quantity and location.
- Inventory control: All of the responsibilities associated with managing the inventory, which include creating new entries for **Items**, updating existing entries, deleting **Items**, querying for both scanner and locator, and so on.

From these requirements, the commonalities can be identified and synthesized in a bottom-up process, to arrive at a set of functions that can support access to **Items**. For example, **Get_UPC_Code()** and **Get_Shelf_Amount()** are two such functions. These low-level functions are used as building blocks to develop higher-level procedures and functions, which can then support the components of **HTSS**.

Whether top-down or bottom-up is utilized for design and implementation, there are still a number of important considerations that are not addressed by either approach. First, as new refinements are made (top-down) or higher-level tasks are determined (bottom-up), there is no way to identify when we are done or whether the design matches the specification. Second, both approaches seem counterproductive with respect to user-interface design, since they are prone to separate system functions from user interface needs. This often leads to user interfaces that are evolved rather than formally planned. Top-down and bottom-up design are both suited to smaller, well-defined problems. Top-down and bottom-up design as principles are very important in many other design approaches. For example, they are both critical when developing solutions using an ADT or module approach, as we will discuss in section 108.3. In addition, in object-oriented approaches (see section 108.4), top-down design for specialization and bottom-up design for generalization are two critical design concepts used in the construction of inheritance hierarchies.

Data-Flow Diagrams

Another approach to design is data-flow diagrams (DFDs), which are used to describe system operations by means of a high-level characterization of information input/output and the identification of major functional actions and informational flow. In the former, the emphasis is on what information must be input, stored, and displayed, so that it can be effectively used. In the latter, the focus is on how the information is used, by displays, individuals, other systems functions, and so on. DFDs as a design technique are very versatile. To represent high-level system behavior, a macroscopic view of an application, DFDs can characterize major system components, as shown for **HTSS** in Fig. 108.1.[1] The major components or *functions* of **HTSS** are represented in a DFD using circles. Actions for *input* by a user or system are found in the rectangular boxes. *Databases*, or repositories of information, are indicated by the parallel lines (open boxes) which enclose a phrase. *Displayed* or output information is identified by the rectangular box with the upper right corner squiggled. The arrows indicate data or information flow, with labels provided to indicate what is flowing between the various portions of a DFD.

The four functions that are represented in Fig. 108.1 correspond to the four of the five user-system interfaces presented for **HTSS** in section 108.1. The **Process Order** function represents all of the actions taken by the **Cashier** to total a customer's order. This includes getting the **Items** from a database that the customer is buying and verifying payment information by means of a **Credit & Check** database. Other functions on the diagram represent inventory controller actions, and requests by shoppers to either locate **Items** in the store or to **Order Deli Items**. Clearly, the DFD as presented is a gross-level description of the major actions for **HTSS**.

DFDs can also be utilized to expand or explore a certain function of a system in greater detail. For **HTSS**, Fig. 108.2 contains a DFD that might represent the **Process Order** function from Fig. 108.1. There are three tasks to process an order, represented by five separate functions. First, each item must be scanned (one function), recorded on the receipt (second function), and updated in the inventory (third function). Once all items have been processed, the second task is to subtotal and subtract

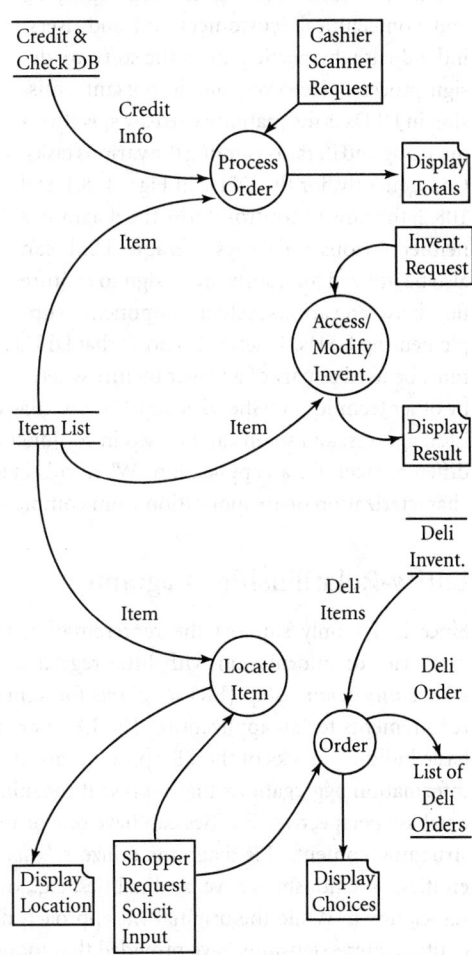

FIGURE 108.1 A macro-level DFD for **HTSS**.

---

[1] We have utilized concepts and notation for DFDs from [Ghezzi et al. 1991].

all appropriate coupons (fourth function). The third and final task completes the order with payment by the customer and verification of valid credit by the cashier (fifth function). To support the five functions, databases are accessed, output is displayed, and flow occurs between them. Overall, the actions in a high-level DFD as given in Fig. 108.1 can be decomposed into greater detail as shown in Fig. 108.2, as the software engineer works in an iterative and incremental process to solve the problem.

DFDs are still very popular today, since they are very easy to use, learn, and understand, even for individuals that do not have a computer science and engineering background. Thus, DFDs are a critical communication medium between the technical (designers and engineers) and nontechnical (customers and end-users) individuals who participate in the software design process. However, one important omission in DFDs is the inability to easily specify sequencing and iteration among the various tasks. Consequently, for the DFDs in Figs. 108.1 and 108.2, the flow of control across the diagram is neither obvious nor always inferable. FSMs can also be utilized for a software design to capture flow between various system components, sup-

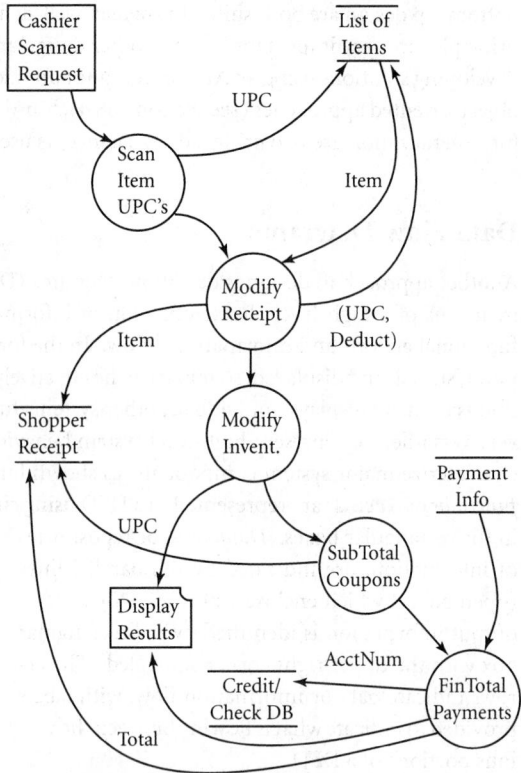

FIGURE 108.2   A micro-level DFD for **HTSS**.

plementing DFDs. Ghezzi also notes that DFDs can only be considered as "... semi-formal notation," and must be used as part of a bigger picture where system structure not represented by DFDs can be captured by other techniques [Ghezzi et al. 1991, pp. 165–166]. I strongly agree with Ghezzi and believe that DFDs, ER diagrams, and so on can be used in conjunction with other software design techniques to describe the different facets for an application. When collected, these various representations correspond to an overall characterization of an application from complementary and supplementary perspectives.

Entity-Relationship Diagrams

Since DFDs only support the representation of information at a coarse granularity level (i.e., large categories of information with little regard to its makeup and content), they can be complemented with *entity-relationship* (ER) *diagrams* for supporting a detailed conceptual modeling of the database requirements for an application. The ER approach was originally proposed by Chen [Chen 1976]. The basic building blocks of the ER approach are entities and relationships. *Entities* are used to model static information aggregations that represent meaningful information components of an application from a database perspective. Entities can have one or more *attributes* associated with them to characterize their structural content. ER diagrams utilize *relationships* to model static information associations between entities. Relationships have cardinalities, e.g., one-to-one, one-to-many, many-to-many, that define the associations. While the original ER approach did not contain the ability to specify inheritance among entities, later extensions have provided that modeling choice. **Inheritance** between entities is available in modern versions of ER diagrams to capture the commonalities that exist from a data/attribute perspective among different entities.

Figure 108.3 is an ER diagram for **HTSS**. In the figure, there are entities for **Item**, **DeliItem**, **CustomerOrder**, **DailySales**, **CreditInfo**, **CreditCard**, **CheckInfo**, and

DebitCard which are shown in rectangular boxes. The attributes for each entity are enclosed in ovals and connected to each entity via lines. For example, the **Item** entity has attributes for **UPC**, **Name**, **W(holesale)Cost**, **R(etail)Cost**, and so on. Relationships are enclosed using diamonds, and include **Order**, **DeliOrder**, and **Sales** in the figure. Order is a one-to-(one-to-many) relationship between **CustomerOrder** and **Item** (one-to-many), and **CustomerOrder** and **DeliOrder** (one-to-one) signifying that one customer order has many **Items** and one **DeliOrder**. A **DeliOrder** is in turn composed of many **DeliItems**. Numbers $(1, n, m)$ are used on the diagram to indicate these cardinalities. Finally, inheritance is used to abstract out commonalities across multiple entities. For example, when paying for an order by a noncash method, the account number, status, and account balance are all common and placed in one entity, **CreditInfo**. The information in this entity can then be inherited by other entities, in this case, **CreditCard**, **CheckInfo**, and **DebitCard**. These other entities in turn have their own unique attributes. Inheritance is represented by lines labeled with **ISA** in Fig. 108.3.

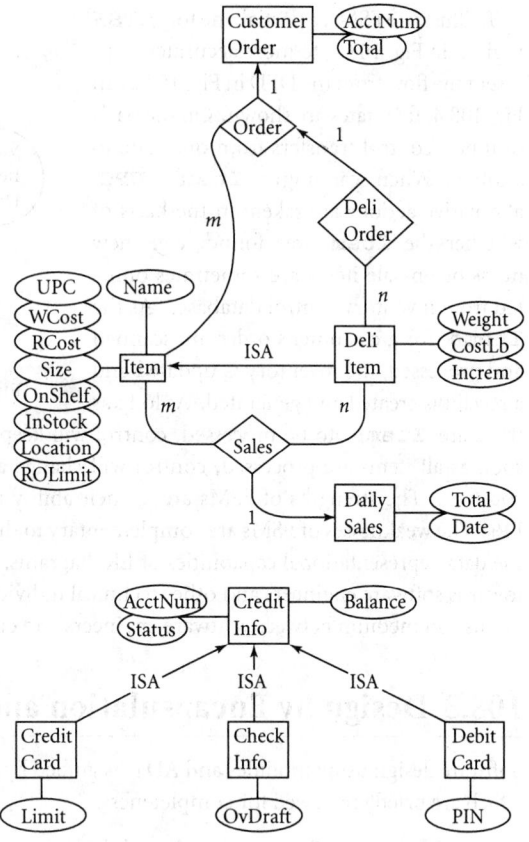

FIGURE 108.3 An ER diagram for **HTSS**.

As a design technique, ER diagrams have many advantages. First, they are an excellent technique for conceptual database design that are easily utilized to represent information, as shown for **HTSS** in Fig. 108.3. As with DFDs, both technical and nontechnical individuals utilize ER diagrams as a means to communicate and exchange ideas on the software design. Second, both the information and its interdependencies can be identified and modeled. Third, by supporting inheritance, generalization is promoted to reduce information redundancies. Despite these advantages, there are also many drawbacks. First, by focusing on information and ignoring functional requirements and usage, it is very possible that one can arrive at an ER diagram that does not meet the needs of the application. Second, ER diagrams lack the ability of DFDs to represent interactions with other system components, e.g., user interfaces, systems functions, etc. This is critical for applications such as **HTSS**, where all of the various and diverse system components are interdependent and must work together.

Finite-State Machines

When developing a software design for an application such as **HTSS**, we have seen that DFDs can capture the flow of information between different system functions, while ER diagrams can represent the database structure and dependencies. One problem with DFDs and ER diagrams is that neither is well suited to the representation of the control aspects of a system. To address this problem, design techniques such as finite state machines (FSMs) can be utilized. Specifically, an FSM can be utilized to capture control, by means of a diagram with states and labeled arcs between them. Each *state* represents a functional point of the design, with each *arc* labeled with different values (phrases) that cause state changes or *transitions* between states.

To illustrate FSMs, a partial one for **HTSS** is given in Fig. 108.4, to more accurately represent the flow from the DFD in Fig. 108.2. In Fig. 108.4, five states are shown. On the basis of input, control transfers from one state to another. When scanning an **Item**'s **UPC**, alternative actions are taken on the basis of whether the **Item** was found, e.g., new items or on-sale items are sometimes omitted from inventory control databases. As the **Items** for a customer's order are scanned and processed, the inventory is updated and a receipt is created and generated. As long as

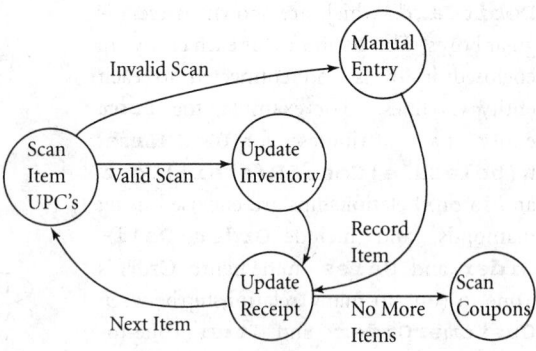

FIGURE 108.4 A FSM for totaling a customer's order.

there are **Items** to be processed, control will keep looping through the left portion of the FSM. As soon as all items are processed, control will change and go on to the next step to scan and deduct all coupons. The strengths of FSMs are in their ability to capture detailed flow to supplement DFDs and ERs. The weaknesses of FSMs are complementary to the advantages of earlier techniques. First, FSMs lack the data-representational capabilities of ER diagrams. Second, they tend to be more detailed and geared towards software engineers and other technical individuals, unlike DFDs and ERs, which are an excellent discussion medium between software engineers and customers.

108.3 Design by Encapsulation and Hiding

Software design using modules and ADTs is guided by a number of classic software engineering concepts, which are briefly reviewed for completeness.

- **Separation of concerns** and **modularity:** Any domain or application can be divided and decomposed into major building blocks and components (separation of concerns). This decomposition allows the application requirements to be further defined and refined, while partitioning these requirements into a set of interacting components (modularity). Changes to the application are (it is hoped) localized. In addition, team-oriented design and development can proceed with different team members concentrating on particular components.

- **Abstraction** and **representation independence:** Through abstraction, the details of an application's components can be hidden, providing a broad perspective on the design. This in turn allows changes to be made to the internal structure and function of each component, achieving representation independence, since the external view of a module/ADT is not impacted.

- **Incrementality/anticipation of change:** The design process at all times is iterative or incremental. This is true whether a given set of modules/ADTs represents an initial or final design for an application. There is an expectation that components will be changed, added, refined, etc., as needed to support evolving requirements.

- **Cohesion/coupling:** An application is cohesive if each component does a single well-defined task. Cohesion has a long history in computing. In the "early days," the rule of thumb was that each procedure or function should be limited to one output page (approximately 60 lines). If so, then the resulting system was deemed to be cohesive. *Coupling* is used to signify the interdependencies of components. Coupling is often considered the complement of cohesion, and an application that minimizes coupling has components that require little or no interaction. When the application also exhibits high cohesion, the end result is a well-defined system with well-understood interactions between its components.

These terms and concepts will occur repeatedly throughout the remainder of this section as modules and ADTs are presented and discussed.

Modules

Module concepts for design were first proposed by Parnas [Parnas 1972] in the early 1970s, and were realized in languages such as Modula-2 and Ada. As a concept, the motivation of modules can be tracked to the examination of existing programs that seemed to share similar solution approaches despite their different domains. Specifically, a given data type (e.g., record, structure, etc.) in a program always seemed to have a set of dedicated procedures and functions that represented its capabilities. Informally, this situation was often organized in separate files during implementation in a programming language such as C. Modules formalized this ad-hoc process by recognizing that most programs are partitionable into discrete program units, with well-defined interactions. Individually, each program unit **encapsulated** the required functionality for a given task, and provided an *interface* for a user, while **hiding** its *implementation*. Representation independence is achieved, since changes can be made to the implementation that have no impact on the interface and its users. Collectively, the ability to encapsulate functionality and to support future changes was the driving force behind introducing modules as an improved design approach.

In many ways, design using modules can mirror the top-down approach of stepwise refinement. The process begins with the identification of major system tasks. These system tasks might correspond to information or operation. For **HTSS**, as shown in Fig. 108.5, modules for system tasks such as **Cashier**, **Scanner** (used by **Cashier**, **Locator**, **Orderer**), and so on, are needed, as well

```
MODULE Cashier;                          MODULE Scanner;
   IMPORT Get_Item(UPC),                    ...
          Get_Credit_Info(AcctNum),       END Scanner;
          Get_Deli_Item(UPC), ...;
   EXPORT Display_SubTot(),               MODULE Locator;
          Display_Total(),                  ...
          Generate_Receipt(), ...;        END Locator;

   {Code for internal data and for       MODULE Orderer;
    procedures and functions}              ...
   END Cashier;                           END Orderer;

MODULE Items;                             MODULE DeliItems;
   IMPORT {various I/O routines};           ...
   EXPORT Get_Item(UPC),                  END DeliItems;
          New_Item(UPC, Name, ...),
          Modify_Item(UPC, Integer),      MODULE CreditCheck;
          Delete_Item(UPC), ...;            ...
   anItem = RECORD                        END CreditCheck;
       UPC : INTEGER;
       NAME: STRING;
       SIZE: REAL;
       ...
   END;

   {Code for Procedures/Functions}
   END Items;
```

FIGURE 108.5 Sample modules for **HTSS**.

as modules for information such as **Items** and **DeliItems** (which are used by many other modules). Modules can also be utilized in a bottom-up direction, by first identifying the lower-level shared modules, and then incrementally combining modules to represent higher-level system functions. A top-down, bottom-up, or even mixed approach to software design using modules is often dictated by the preferences and experiences of software engineers.

In some ways, modules appear to encompass both DFD and ER approaches. Each task in a system will be realized by a module that contains a dedicated set of information and a set of procedures and/or functions that operate on the information to accomplish its task. A subset of a module's information, procedures, and/or functions is identified or tagged for *export* to other modules. Exported portions of a module represent its interface to the "world." Finally, since a module might interact with other modules, a module may *import* information and functionality for either standard actions (e.g., say, for input/output [I/O] of data, printing, strings, etc.), or from other modules of the application. Notice that portions that are imported by one module must have been exported by other modules. Concepts for defining a module and importing (exporting) from (to) other modules are also shown in Fig. 108.5.

When using modules, designers are advised to strive for low coupling and high cohesion. Low coupling implies that the interdependencies of modules with respect to exchanging information is minimal. High cohesion refers to the ability of a module to characterize a single well-defined task. Thus, through modules, controlled sharing is promoted; portions that are not exported from a module are hidden from other modules, and representation independence is facilitated. As we will see shortly, there are strong parallels between modules and ADTs. Through either a top-down or bottom-up approach, modules provide a technique that stresses the breakdown of a software design into logical discrete components. As a technique, modules are intended for software designers and engineers rather than customers and end-users.

Abstract Data Types

ADTs were first proposed by Liskov for the CLU programming language [Liskov and Zilles 1975, Liskov et al. 1977]. An ADT is characterized by a set of operations (procedures and functions) that is often referred to as the **public interface** which represents the behavior of a data type. The **private implementation** of the data type is hidden from the programmer or software engineer who wishes to use the ADT. System-available ADTs have been extensively utilized in programming languages for many years. For example, in Pascal, when using the integer data type, the software engineer is able to utilize all of the appropriate operations against integers (e.g., **+**, **-**, *****, **div**, **mod**, etc.), without needing to know the implementation of integers in the underlying machine-dependent architecture. Designer-defined ADTs are more readily available today in languages such as Ada, C++, and Eiffel.

ADTs are a design technique that allows software engineers to define their own data types. For example, the classic ADT is a **stack** that contains operations for **push**, **pop**, **top**, **initialize**, **is_empty**, and so on. These operations serve as the *public interface* for the software engineer, and they typically include the type of the stack (e.g., integer, real, etc.) and the parameters and return types of the public operations. However, the *private implementation* of **stack**, which includes the implementation of all operations and the data representation (e.g., array or list, etc.), is hidden from the user. Through this division of behavior, ADTs achieve representation independence by means of hidden private implementation, and abstraction by means of visible public interface. Moreover, this allows implementation changes to be made (e.g., say, from an array to a list) as long as these changes are transparent to the public interface (i.e., the operations and their names, parameters, and return types cannot change).

ADTs promote the design and development of applications from the perspective of information and its usage. From an information perspective, there are ties between ADTs and the ER approach. From a usage perspective, the functional characteristics can be explored in either a top-down or a bottom-up direction. However, ADTs take a combined view that focuses on information and its manipulation, which can yield a different design solution than an approach that considers each facet individually.

In the ADT design process, there are a number of considerations that must be addressed, as outlined below:

- Identify the major information units: Determines the ADTs that are needed for an application or system.
- Describe the purpose of each unit: Indicates the overall responsibility for each ADT in the application.
- Define manipulation techniques for each unit: For ADTs, this corresponds to the operations or methods that must be characterized, including the parameters, return type, etc.
- Encapsulate and hide: Representation of each unit and its manipulation are both encapsulated and hidden within the ADT.

In addition to describing the design for individual ADTs, we must also indicate the iterative or cyclical process that can be utilized to arrive at a design solution. We have taken a bottom-up approach to ADT design for **HTSS**, as shown in Figs. 108.6 and 108.7. In Fig. 108.6, the lowest level of ADTs is shown, with an emphasis on the information components for an application. Thus, we have ADTs for **Item**, **DeliItem**,

```
ADT Item;
    PRIVATE DATA: SET OF Item(s), Each Item Contains:
                    UPC, Name, WCost, RCost, OnShelf, InStock,
                    Location, ROLimit;
                  PTR TO Current_Item;
    PUBLIC OPS: Create_New_Item(UPC, ...) : RETURN Status;
                Get_Item_NameCost(UPC) : RETURN (STRING, REAL);
                Modify_Inventory(UPC, Delta) : RETURN Status ;
                Get_InStock_Amt(UPC) : RETURN INTEGER;
                Get_OnShelf_Amt(UPC) : RETURN INTEGER;
                Check_If_On_Shelf(UPC): RETURN BOOLEAN;
                Time_To_Reorder(UPC): RETURN BOOLEAN;
                Get_Item_Profit(UPC): RETURN REAL;
                Get_Item_Location(UPC): RETURN Location;
                ...
END Item;


ADT DeliItem;                              ADT CustomerInfo;
    PRIVATE DATA: SET OF (Item, Weight,        ...
                    CostLb,, Increm);      END CustomerInfo;
    ...
END DeliItem;                              ADT Shelf_Info;

                                              ...
ADT Receipt;                               END Shelf_Info;
    PRIVATE DATA: SET OF Items;
                SET OF Coupons; {An ADT}   ADT Sales_Info;
                SubTotal, Total, PayType;      ...
    ...                                    END Sales_Info;
END Receipt;
```

FIGURE 108.6 Low-level ADTs for **HTSS**.

```
ADT Process_Order; {Middle-Level ADT}
     PRIVATE DATA: {Local variables to process an order.}
     PUBLIC OPS : {What do you think are appropriate?}

     {This ADT uses the ADT/PUBLIC OPS from Item, Deli_Item, Receipt,
      Coupons, and Customer_Info to process and total an Order. each
      Receipt must be cataloged and stored when an Order has been
      completed.}
     ...
END Process_Order;

ADT Sales_Info; {Middle-Level ADT}
     PRIVATE DATA: {Local variables to collate sales information.}
     PUBLIC OPS : {What do you think are appropriate?}

     {This ADT uses the ADT/PUBLIC OPS from Receipt so that
      the sales information for the store can be maintained.}
     ...
END Sales_Info;

ADT Cashier; {High-Level ADT}
     PRIVATE DATA: {Local variables used by a cashier.}
     PUBLIC OPS : {What do you think are appropriate?}

     {This ADT uses the ADT/PUBLIC OPS from the middle-level ADTs
      (Process_Order, Sales_Info, etc.), and from the low-level ADTs.}
     ...
END Cashier;
```

FIGURE 108.7   Middle- and high-level ADTs for **HTSS**.

Receipt, and so on. Given this lowest level, other ADTs can be designed that combine and utilize multiple low-level ADTs, as shown in Fig. 108.7. In this case, the **Process_Order** ADT uses multiple ADTs, while the **Sales_Info** ADT only uses **Receipt**. Current and lower levels are incrementally combined to increase ADT functionality. Eventually, an ADT that describes the uppermost level of system behavior will be specified. Note that a top-down approach to ADTs is also reasonable and feasible.

Clearly, the advantages of ADTs are similar to those of modules. However, since the techniques are somewhat ad hoc, and decisions made at higher levels (for the bottom-up approach) are impacted by lower levels, i.e., if ADTs at the lowest level are wrong, those errors are carried through all subsequent levels. In addition, the lack of inheritance for ADTs will likely result in design redundancies, even though there is **software reuse**.

108.4 Object-Oriented Design

The transition to an object-oriented design from a specification is not a trivial activity. While there have been attempts to map traditional techniques (e.g., DFDs, ER diagrams, modules, etc.) to an object-oriented design, these techniques are ad hoc at best, and they strongly depend on the expertise of the involved software engineers. For example, an ER diagram identifies objects from an informational perspective and the DFD defines functional objects. However, in doing so, an ER diagram will likely not consider *how* the information will be used, while a DFD often ignores *what* detailed information is required. Thus, there is a high likelihood of conflict when attempting to match these two diverse perspectives into a unified object-oriented design.

Instead, most object-oriented design techniques embody their own independent approaches to facilitate the design process, utilizing a unit of encapsulation/abstraction for object-oriented design called an **object type (OT)** or **class**, which has a strong parallel to an ADT. The distinguishing factor between ADTs and the object-oriented approach is the *inheritance* concept which we have previously discussed in our examination of ER diagrams. Inheritance allows controlled sharing between OTs/classes, permitting the passing of data and/or operations from the supertype (superclass) to the subtype (subclass). Regardless of the specific technique, object-oriented design is promoted for a number of important reasons:

- Stresses modularity: Achieved by the OT/class concept and encapsulation.
- Increases productivity: While this is difficult to prove, it is a long-standing claim of object-oriented design.
- Controls information consistency: As for ADTs, this is attained, since hiding allows the access to the private implementation to be managed.
- Promotes software reuse: Software engineers can reuse existing OTs/classes (similar to reusing ADTs) for solving other problems (i.e., the stack is the classic example). In addition, through inheritance, software engineers can define new OTs/classes that acquire the characteristics of existing OTs/classes without violating the hidden implementation.
- Facilitates **software evolution**: The abstraction, encapsulation, hiding, and inheritance of object-oriented design allows minor changes (to hidden or private implementations) to be transparently made, while major increases in functionality can be realized by extending the existing OT/**class library** through inheritance.

Testing cuts across all of these claims: done on an OT-by-OT (class-by-class) basis (modularity); once tested, a OT/class can be used and reused (productivity); testing of changes that occur is limited to the private implementation as long as the public interface has not been changed (evolution). The end result of using object-oriented design is supposed to be a clearer and easier conceptualization of the intended application. Further, the increased emphasis on design (in both time and effort) is offset by a reduction in implementation effort. However, these last two statements are difficult to verify in practice, since as a discipline, the object-oriented paradigm itself is still evolving.

Since object-oriented design is more complicated than the techniques reviewed in previous sections, a more in-depth discussion is provided. We begin by reviewing the key object-oriented design concepts, to establish the needed terminology for the remainder of the presentation. Next, a treatment of the consideration for choosing object types is detailed, since this is one of the key issues in object-oriented design. This is followed by a review of inheritance, focusing on its motivation, usage, and costs. Then, there is an in-depth examination of design considerations and flaws related to both determining object types and utilizing inheritance. All of this material leads to a sample object-oriented design technique, **responsibility-driven design** by means of CRC (class, responsibilities, collaborators) cards using **HTSS**. The section finishes by briefly reviewing other object-oriented design techniques.

Object-Oriented Concepts and Terms

While there is no agreement as yet regarding terminology for object-oriented design, there are a number of core concepts that can be described:

- Object type (OT)/class: Used to model the features (information) and behavior (methods) for an application. Note that public interface and private implementation are as discussed in section 108.3.
- **Information:** The private data of an OT/class. Information represents the different internal data components that define the OT/class and characterize all of its instances.
- **Method:** Contains the definition of the actions required for a particular operation against the private data of an OT/class.
- Encapsulation: The coupling of information and methods within an OT/class.
- Hiding: Controlling access to the information and methods of an OT/class.
- Inheritance: The controlled sharing of information/methods between related OTs/classes of an application, which extends the concepts discussed for ER diagrams. In inheritance relationships, the parent is referred to as the supertype/superclass, while the child is referred to as the subtype/subclass.
- **Inheritance hierarchy:** All inheritance relationships between OTs/classes that share a common parent (or grandparent) form a hierarchy with an identifiable root (ancestor). Inheritance hierarchies are simply the trees that organize the sharing between all related OTs/classes.
- **Instance:** An occurrence of an OT/class, or the actual information/data.
- **Message:** An action (method call) that is initiated by an instance on itself or other instances.
- OT/class library: All OTs/classes and inheritance hierarchies for an application form a common library for use by tools and end-users. Moreover, Chapter 52 contains a detailed discussion of object-oriented databases.

Note that while it is impossible to clearly identify the roots of inheritance in the database area, the classic article on aggregation and generalization as database abstractions by Smith and Smith [Smith and Smith 1977] is a possibility.

Advanced concepts that are important to fully appreciate the potential and power of object-oriented design include **generics** and **dispatching**. A *generic* is a type-parameterizable OT/class. For example, instead of having a stack that is bound to a specific data type (say, **integer**), the stack can require that the data type be provided as part of its initialization. Thus, the creation of a stack [e.g., **Stack(Real)**, **Stack(Char)**, etc.] binds the stack's methods to the appropriate types. *Dispatching* is the run-time or dynamic choice of the method to be called on the basis of type of the calling instance. As a concept, the effective use of dispatching is tightly bound with inheritance, and it offers many benefits: versatility in the design and use of inheritance hierarchies; promotion of reuse and evolution of code, allowing hierarchies to be defined and evolved over time as needs and requirements change; and development of code that is highly generic and easier to debug (and hence reuse/evolve).

Choosing Object Types

The first and most frequent question that is asked by newcomers to object-oriented design is a variant of *How are object types chosen?* Typical (lazy) answers to this question echo software engineering mantra (e.g., make your choices so that the end result is encapsulated with high cohesion and low coupling) or rely on that old-time favorite "As you gain more experience with objects, your ability to identify them will also improve." However, neither answer is really satisfactory. A "better" answer should relate the following:

Choosing object types/classes is not a first step in the design and development process, but rather must follow in a logical fashion from earlier efforts. In practice, the choice must be guided by a specification for an application that contains the intent and requirements. The specification will make use of other software engineering techniques (e.g., DFDs, ER diagrams, etc.), in an effort to define scope and breadth of function, user interfaces, required user/system interactions, and so on. As the specification gains in content and complexity, the relevant object types begin to define themselves as a natural side-effect. One hopes that this leads to an object-oriented design. This in turn will be explored, refined, and evolved into a detailed design, which can then be changed to an implementation.

The moral is that it is unrealistic to 'jump' to an object-oriented design from only a basic understanding of an application. It would be just as unreasonable to make such a jump using any of the other design techniques that are given in this chapter.

Instead, one must acquire an understanding of what is appropriate to put into an object type or class. Three possibilities are illustrated below:

```
                Private Data      Public Interface

Employee        Name              Create_Employee()
OT/Class        Address           Give_Raise(Amount)
                SSN               Change_Address(New_Addr)
                ...               ...


ATM_Log         Acct_Name         Check_Database(Name)
OT/Class        PIN_Number        Verify_PIN(PIN)
                ...               Log_on_Actions(Name, PIN)
                                  Reject()
                                  ...


ATM_User        Action            Log_on_Steps()
OT/Class        Balance           Acct_WithD(Amt)
                WD_Amt            Check_Balance(Number)
                ...               Deposit_Check()
                                  ...
```

The first OT/class has been designed from an information perspective, focusing on the idea that to track **Employees**, standard data and operations are needed. The second OT/class, **ATM_Log**, embodies the functions that take place to log on an individual to an ATM session. Note that even with a functional view, information is needed to capture user input for verifying status. The third OT/class, **ATM_User**, represents a user interface by capturing the different interactions that are supported.[2]

During this design process, there are a number of possible design flaws. First, a software engineer places too much functionality in one OT/class. In this situation, the OT/class is often split into two or more OTs/classes, but it is also possible that the functionality as it is split is absorbed into other, existing, OTs/classes. The latter leads to a second design flaw: an OT/class lacks functionality. In this situation, two or more OTs/classes are often merged to result in an OT/class that exhibits a more cohesive behavior. Again, we emphasize that this has not been an exhaustive treatment of this subject. Rather, it is intended to provide an initial look at the different considerations that have an impact on the choice of object types during design.

Inheritance: Motivation, Usage, and Costs

One of the key aspects that distinguishes object-oriented design from its ADT ancestor is the concept of *inheritance*. To successfully utilize inheritance, an iterative process to identify commonalities (called

---

[2] Any discussion on object-oriented concepts would not be complete without an ATM example! Mercifully, this is the only one in this chapter!

generalization) and distinguish differences (called specialization) is undertaken. For example, in **HTSS**, a first approximation of the different OTs/classes that are needed, concentrating only on information in each OT/class would be as follows:

```
    SnackItem       LiquorItem      MeatItem        ... Other Items ...
      Name            Name            Name

      UPC             UPC             UPC

      ShelfLife       SpecialTaxes    ExpireDate
```

From the previous example, it is clear that two data components (**Name**, **UPC**) are common to all **Item**-related OTs/classes and can be abstracted out (generalized) into the **Item** OT/class as given below:

```
    Item        SnackItem:Item      LiquorItem:Item       MeatItem:Item ...
      Name          ShelfLife           SpecialTaxes          ExpireDate

      UPC
```

The original OTs/classes (e.g., **MeatItem:Item**) inherit private data (e.g., **Name** and **UPC**) from their ancestors (e.g., **Item**).

In general, inheritance-related decisions are often based on the overlaps that exist between information and/or operations across multiple OTs/classes. The end result for an application should be a set of one or more inheritance hierarchies that are extensible, evolvable, and reusable. As previously discussed, specialization is used to define inheritance associations by pushing down differences to lower levels of the hierarchy. In this way, the behavior of the supertype is refined and focused to represent the shared characteristics required by all descendants. For example, in the OT/class hierarchy from **HTSS** below:

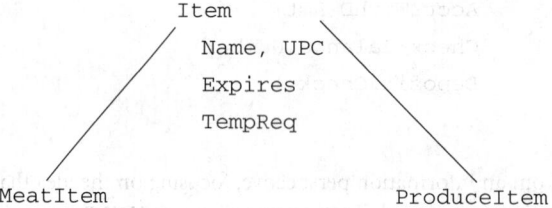

```
                    Item
                    Name, UPC
                    Expires
                    TempReq


    MeatItem                        ProduceItem
```

the **Expires** private data of **Item** should be moved to **MeatItem**, which requires explicit expiration dates, while **TempReq** should be moved to **ProduceItem** to track produce that requires refrigerated versus room-temperature storage. Conversely, generalization operates bottom-up by examining a set of OTs/classes in an attempt to identify commonalities, which are then pushed into a supertype/superclass. For example, in **HTSS**, the OTs/classes:

```
    MeatItem            DeliItem            LiquorItem
      Name, UPC           Name, UPC           Name, UPC

      Expires             Expires             SpecTaxes
```

could be revised into the hierarchy:

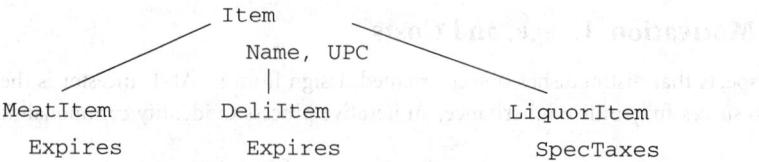

```
                    Item
                    Name, UPC
                      |
    MeatItem        DeliItem            LiquorItem
      Expires         Expires             SpecTaxes
```

which defines a new **Item** supertype/superclass. Note that it might also be necessary to define a new level that is a parent of **MeatItem** and **DeliItem**, and a child of **Item**, e.g., a **PerishItem** for perishable foods. Note also that in these and other examples, the names of variables have been conveniently the same, which, in practice, may not be the case.

Another type of inheritance is based on the concept that OTs/classes with significantly different and unrelated abstract views might require access to the same underlying implementation. For example, in **HTSS**, an **ItemDB** is required to track all items that are sold. There are multiple abstract views for **ItemDB** that can be represented in an inheritance hierarchy:

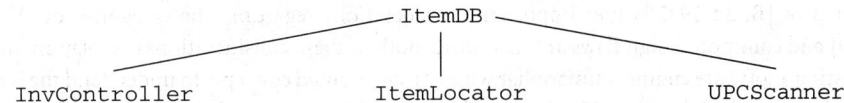

In this example, the interfaces for managing the inventory, locating items, and scanning UPCs all need access to the same information, but they clearly have different intent and usage within **HTSS**. Another type of inheritance is based on a combination of two or more OTs/classes, allowing software engineers to view the design from different perspectives. This can be achieved via multiple inheritance, where **DeliItem** would inherit from both **ProduceItem** and **MeatItem**.

In summary, inheritance can be defined as a relationship where the subtype acquires information and/or operations from a supertype. Inheritance is a transitive operation, allowing behavior to be passed from parent to child to grandchild, etc. Since the subtype is more refined than the supertype, it has been *specialized*. Alternatively, since the supertype contains common characteristics of all of its subtypes, it has been *generalized*. In practice, inheritance is utilized when two or more OTs/classes are functionally and/or informationally related. That is, generalization and specialization decisions can be based on information (as in the previous example), function, or a combination of the two.

The benefits of inheritance are strongly tied to the claims of object-oriented design, and while Budd characterizes these benefits at primarily the implementation level, they clearly also apply to the design level. In the following list from Budd [Budd 1997, pp. 143–144], the parenthetical remarks represent the benefits of inheritance during design:

- Software (design) reusability: When a set of one or more OTs/classes is reused from an earlier effort, there is a high degree of assurance that compiled and tested code/behavior has been provided. (Similar reuse during design accrues corresponding benefits regarding the completeness of a design component in addition to downstream benefits for the implementation.)

- Code (design) sharing: When two or more subtypes (subclasses) of the same supertype (superclasses) can share code as the result of a specialization or generalization, redundancy can be reduced, i.e., the "same" code is implemented/tested only in a single location. (It can be strongly argued that the sharing of code must be identified during the object-oriented design process. Otherwise, the resulting design had many flaws that were not found until the implementation process began.)

- Software components: Promotes the concept of reusable components and the software factory. (Software factory ideas should not be limited to code, since a "component" may represent a portion of a design or an implementation.)

Clearly, design level benefits for inheritance are instrumental in achieving the reuse and evolution claims previously mentioned.

However, these and other benefits noted by Budd do have a cost [Budd 1997, pp. 145–146], which is felt during implementation and runtime for the application. Inheritance associations complicate program data structures, requiring additional compile time (e.g., for overloaded names, multiple inheritance, etc.) while deferring some decisions to runtime (e.g., dispatching). As the complexity of applications increases, the OT/class library size increases, both in numbers and in depth of inheritance hierarchies. This again has an impact on the compile and runtime environments, i.e., there is a heavy cost for dynamic linking. At a more practical level, it is often the case that new and even experienced software engineers define too many operations that have too little functionality in an OT/class. This affects the activation records in the runtime environment, especially when the operations call other operations, thereby requiring nested activation records. In fact, it is often the case that the activation record requires significantly more memory during runtime than the code of a poorly defined "small" operation.

Categories of OTs/Classes and Design Flaws

In Chapter 3 of [Budd 1997] other important considerations regarding the categories of OTs/classes [pp. 48–49] and common design flaws are examined. Both of these considerations can play an important role in assisting a software engineer unfamiliar with object-oriented concepts to understand the issues that impact on designing and defining OTs/classes. There are different categories on how an OT/class is used and/or what an OT/class does in an application. Budd has defined four major categories of OTs/classes:

1. Data managers: This category maintains data or state information to capture and contain the functionality for an application.
2. Data sinks/data sources: This category either produces or processes data, and it may be generated on demand to meet a short-term need.
3. View/observer: This category can serve as an interface for an end-user or might be utilized to collect the public interfaces across multiple OTs/classes into a single view.
4. Facilitator/helper: This category has no functionality when viewed independently, but exists solely to support other OTs/classes. For example, typical facilitators include a string OT/class library, I/O handler OTs/classes, etc.

Note that OTs/classes spanning two or more categories should be decomposed, since they likely contain excessive functionality, i.e., they are noncohesive. While it is important to specify each OT/class and to understand to which category a given OT/class belongs, it is more critical to know the context of the OT/class within the overall application. Why the OT/class has been specified, the role it will play in the application, and the other OTs/classes that it interacts with should all be clearly understood when examining the OT/class.

   Like any other approach, object-oriented design is not intended to allow a software engineer to arrive at the "completed" design in one step from the specification. Rather, object-oriented design and development encourages and promotes an incremental and iterative process, allowing an application to evolve from its specification into an object-oriented design. Thus, when defining OTs/classes in the aforementioned categories, there are a number of common design flaws that might occur after the first few iterations of the design process, including the following:

1. OTs/classes that directly modify the private data of other OTs/classes: This is a **major** error in the use of object-oriented design techniques. To rectify this situation, the public interface must be upgraded to include operations that encapsulate the modification of private data.
2. Too much functionality in one OT/class: In this situation, the obvious solution is to split the OT/class into two or more OT/classes that exhibit higher cohesion. The result should also be lowly coupled.
3. An OT/class that lacks functionality: The reverse of the previous case requires the merging of two or more OTs/classes. Any time OTs/classes are merged, the impact on existing OTs/classes and inheritance hierarchies must be carefully examined.
4. OTs/classes that have unused functionality: In this situation, the key is to understand the reason for the unused functions. Have they been duplicated elsewhere? Were they needed in an earlier prototype? The answers to these questions will dictate the choices made in this case.
5. OTs/classes that duplicate functionality: As in the previous case, it must be understood why the duplication has occurred. Was it due to a problem in the specification? Did two software engineers unintentionally work on defining the same OT/class?

It is expected that the initial attempts at an object-oriented design for an application will not be perfect. The key issue is to learn to recognize imperfections so that they can be corrected in subsequent iterations. The list of common design flaws is not comprehensive; rather, it defines the most commonly occurring errors so that they can be easily eliminated by a novice and avoided as a software engineer gains experience.

The CRC Approach to Object-Oriented Design

Responsibility-driven design [Wirfs-Brock and Wilkerson 1989, Wirfs-Brock et al. 1990] is the the basis of CRC cards [Beck and Cunningham 1989] and is geared towards exploring a design by focusing on responsibilities of different OTs/classes. The key questions are:

> For what actions is this object type responsible?
> What information does this object type share? [Wirfs-Brock and Wilkerson 1989, p. 73]

To support responsibility-driven design, three perspective client models are proposed. *External clients* refer to those classes that a message calls (invoke a method) on a particular class. *Subclass clients* models the inheritance relationship that exists between some classes. *Self-client* indicates that the behavior of a given class is defined by its methods (and perhaps methods inherited from a subclass client). Overall, by focusing on actions and behavior, this approach evokes the spirit of ADTs, while still providing enough capabilities to be useful in real-world situations.

This section presents object-oriented design via the CRC approach that emphasizes responsibility-driven design [Beck and Cunningham 1989, Budd 1997, Wirfs-Brock and Wilkerson 1989, Wirfs-Brock et al. 1990]. The process that is presented focuses on encapsulation and hiding to identify classes and their interdependencies. As an object-oriented design technique, the CRC approach [Beck and Cunningham 1991, Budd 1997] is very simple and straightforward. The premise is to conceptualize the application by identifying the classes (C, a major modeling chunk), its responsibilities (R, what a class does), and its collaborators (C, the other classes that it interacts with). Two sample CRC classes for **HTSS** are shown in Fig. 108.8.

Notice that the *responsibilities* that are defined in Fig. 108.8 are characterized with short phrases (typically three to seven words) that represent the functional capabilities of each class. These responsibilities are meant to be terse, since all information for a class is intended to fit on an index card (or equivalent). Each responsibility will map to one or more methods when the object-oriented design (realized by means of CRC cards) is changed to a detailed design/implementation. Each class also has a list of *collaborators*, which indicate the other classes that interact with the given class. The back of each card is used to identify the private data, thereby placing the information out of sight (hiding at its best!), as has also been shown in the example. The data for **Item** is representative of common information on all grocery items, while the data for **ItemDB** captures the required structures for maintaining a collection of all grocery items in a supermarket's inventory.

Throughout the process of object-oriented design with CRC cards, it is very typical to require a number of iterations, where cards may be split, merged, or refined, as dictated by responsibilities and/or informational needs and requirements. For instance, in the above example, only the **Item** class has been given. In a later iteration, **DeliItem**, **MeatItem**, and other subclasses of **Item** will be defined, which will necessitate changes, e.g., adding or modifying collaborators, moving responsibilities of **Item** to one of its descendants, etc.

Another conceptualization of grocery items in **HTSS** is also possible, to arrive at an inheritance hierarchy that is different from the one given earlier. In this version of the hierarchy, shown in Fig. 108.9, every **Item** in HTSS is specialized as either nonperishable (**NonPItem**) or perishable (**PerishItem**). Each of these two items has descendants to further specialize categories of items. For **PerishItem**, three likely candidates are: **DeliItem**, **MeatItem**, and **ProduceItem**. For **NonPItem**, **PaperItem**, **BeverageItem**, **LiquorItem**, and **SnackItem** are possible. The CRC cards for

```
┌─────────────────────────────────────────────┬──────────────┐
│ Item: A Product Sold in a Supermarket.       │              │
├─────────────────────────────────────────────┤              │
│ Create a New Item                            │              │
│ Return an Item's UPC                         │              │
│ Return an Item's Name                        │   ItemDB     │
│ Return the Quantity of an Item               │              │
│ Return an Item's Reorder Amount              │              │
│ Check the Reorder Status of an Item          │              │
│ Display all Info. on an Item                 │              │
│ Display an Item's Name and Price             │              │
│ Display an Item's Inventory Data             │              │
│ Update an Item's Price                       │              │
└─────────────────────────────────────────────┴──────────────┘
```

Back of Item CRC Card

```
┌─────────────────────────────────────────────────────────────┐
│ Item has:                                                     │
│ UPC, Integer     RetailCost, Real    OnShelf, Integer         │
│ Name, String     InStock, Integer    ROLimit, Integer         │
└─────────────────────────────────────────────────────────────┘
```

```
┌─────────────────────────────────────────────┬──────────────┐
│ ItemDB: All Item's in a Supermarket.         │              │
├─────────────────────────────────────────────┤              │
│ Insert a New Item                            │              │
│ Delete an Existing Item                      │              │
│ Find/Display Item Based on UPC               │  InvControl  │
│ Find/Display Item Based on Name              │  Item        │
│ Find/Display Item Based on Quantity          │              │
│ Find/Display Item Based on Reorder           │              │
│ Print All Items                              │              │
│ Find an Item in Database by UPC              │              │
│ Find an Item in Database by Name             │              │
│ Find the First Item                          │              │
│ Find the Next Item                           │              │
└─────────────────────────────────────────────┴──────────────┘
```

Back of ItemDB CRC Card

```
┌─────────────────────────────────────────────────────────────┐
│ ItemDB has:                                                   │
│ AllItems, Array[Max_Items] of Item                            │
│ Num_Items, Integer                    Curr_Item, Integer      │
└─────────────────────────────────────────────────────────────┘
```

FIGURE 108.8 CRC cards for **Item** and **ItemDB**.

Item and **ItemDB** are as previously given. Only a selected set of CRC cards for the **PerishItem** subhierarchy has been shown in Fig. 108.9.

To provide additional perspective on the usage and utility of CRC cards, Fig. 108.10 contains three additional CRC cards related to **HTSS**. The first is for a UPC **Scanner** to read bar codes when totaling grocery items. The second is related to the **Scanner** to represent the responsibilities and interactions

PerishItem: An Item with a Limited Shelf Life. Superclass: Item	Item ItemDB
Create a New PerishItem Destroy an Existing PerishItem Display a PerishItem's Inventory Data Display a PerishItem's Days Data Update a PerishItem's Days or Environ	
Environ: Enum (RoomTemp, Refrig, Frozen) Days: Subrange 1..21	

DeliItem: Prepared Food Sold at Deli Counter. Superclass: PerishItem	Item ItemDB
Create a New DeliItem Destroy an Existing DeliItem Display a DeliItem's Inventory Data Update a DeliItem's Weight or CostPerLb	
Weight, CostPerLb: Real	

MeatItem: Sold in the Meat Department. Superclass: PerishItem	Item ItemDB
Create a New MeatItem Destroy an Existing MeatItem Display a MeatItem's Inventory Data Display a MeatItem's Type or Cut Update a MeatItem's Weight or CostPerLb Update a MeatItem's Type or Cut	
MeatType: Enum (Beef, Poultry, Lamb, Pork, Fish) MeatCut: Enum (Prime, Regular) Weight, CostPerLb: Real	

ProduceItem: Fruits/Vegetables in Produce Dept. Superclass: PerishItem	Item ItemDB
Create a New ProduceItem Destroy an Existing ProduceItem Display a ProduceItem's Inventory Data Update a ProduceItem's Care Status or SellAmt	
Care: Enum (Cover, H20Spray) SellAmt: Enum (One, Four, Six, Twelve)	

FIGURE 108.9 CRC cards for **PerishItem** and its subclasses.

Scanner: Reads Bar Code and Generates. UPC Equivalent	ItemDB Cashier
Scan Bar Code for UPC	
Check Validity of UPC	
Return Invalid UPC Signal	
Return the Scanned UPC of an Item	
Current UPC	

Cashier: Process a Customer's Order by Ringing Items, Handling Coupons, and Taking Payment	ItemDB Scanner Receipt Credit/Check DB
Ring an Item for Order	
Display Receipt Entry for Item	
Generate Receipt Entry for Item	
Process Coupon for Order	
Display Receipt Entry for Coupon	
Generate Receipt Entry for Coupon	
Calculate Subtotal, Total, Payment, Change	
Display Subtotal, Total, Payment, Change	
Generate Subtotal, Total, Payment, Change for Receipt	
Receipt Entry: UPC, Name, etc. Coupon Value, Payment Type, etc.	

Receipt: Used to Record Receipt Entries While an Order is Being Processed	Cashier
Create a Standard Receipt Entry	
Create a Price/Lb Receipt Entry	
Return Subtotal of Items for Order	
Decrement Subtotal Using Coupon Value	
Return Amount Due for Order	
Create Payment/Change Back Entry	
List of Receipt Entries Subtotal, Total, Payment, etc.	

FIGURE 108.10 CRC cards for a **Scanner**, **Cashier**, and **Receipt**.

of a **Cashier** in this process. The third contains responsibilities to capture and maintain a **Receipt** for the customer's purchases.

Other Object-Oriented Design Techniques

In addition to responsibility-driven design and CRC cards, there are many other popular object-oriented design techniques. Some of the more classic techniques include the Booch method [Booch 1991], the object modeling technique (OMT) [Rumbaugh et al. 1991], the client/server contract approach [Meyer 1988], and a comprehensive object-oriented software engineering approach with use cases [Jacobson et al. 1992]. Another effort has sought to combine these initial techniques into what is termed a Fusion method

[Coleman 1994]. Other newer techniques include the Demeter method for adaptive object-oriented software [Lieberherr 1996], a unification of the Booch method and OMT [Booch and Rumbaugh 1995], and the design pattern approach for categorizing reusable object interactions for multiple application domains [Gamma et al. 1995].

108.5 Mathematical and Analytical Design

From a more quantitative perspective, there is also a wide range of techniques available to mathematically analyze the different components of a software design. These design analysis techniques play a critical role in ensuring that the requirements identified in the specification are attainable in the application design. For example, queueing network models are utilized to estimate performance of systems on the basis of various load and resource access patterns. Also, individual algorithms can be analyzed in detail to arrive at an understanding of their algorithmic complexity with respect to time (absolute) or as compared to other algorithms (relative). Evaluating alternative design solutions to problems is an important activity that occurs during software design. In the remainder of this section, queueing network models, time-complexity analysis, and simulation models are reviewed.

Queueing Network Models

Throughout the history of computer science and engineering, queueing networks have been extensively utilized to model computer systems [Kleinrock 1975, Kleinrock 1976]. Often, after a specification has been written and either prior or concurrent with the software design process, queueing network models can be utilized as an important tool to estimate and predict performance, including potential bottlenecks. In a basic *queueing network model*, jobs arrive and are queued for processing. Processing commences when all required resources are available. During the processing, it may be necessary to wait for a resource (e.g., I/O device), or a job may be returned to the queue if it has exceeded its allowable processing time slice. The former characterizes the *I/O-bound jobs*, the latter, the *computation-bound jobs*. To represent these alternative possibilities, probabilities can be assigned to jobs that correspond to their waiting times, say, *probability r* for the jobs' *I/O waiting time* and *probability* $(1 - r)$ for the jobs' *CPU waiting time*. For design analysis and performance estimation during software design, both open and closed queueing network models can be utilized. In a *closed* model, jobs neither enter nor depart from the network. On the other hand, an *open* queueing network is characterized by one or more jobs coming into the network and by one or more jobs departing from the same network.

The behavior of jobs within the network is characterized by the *probability of transitions* of jobs between servers. When modeling systems, two of the key variables that must be specified are the *scheduling discipline* (such as first-come-first-serve, FCFS) and the *size of the queue*. Typically, we may use first-come-first-serve scheduling and unlimited-capacity queues. For an open queueing network model, a characterization of *job-arrival processes* (such as a Poisson arrival process) must be defined. For a closed queueing network model, the *number of jobs* in the network must be given.

Queueing network models are especially useful during the software design of **HTSS**, since there are so many obvious places where jobs queue up for service. For example, there are multiple cash registers, with multiple customers queued to have their orders processed and totaled. As each order is processed by all cashiers in a concurrent fashion, there are concerns about the performance of simultaneous database access when **Items** are scanned. The overall throughput of the system is critical, to ensure that customers are processed in a timely fashion when the system is under maximum loading. Another aspect of **HTSS** where performance is critical is when deli orders are queued by customers to be filled by deli workers. The number of deli workers (servers), the average number of orders in the queue, and the time needed to fill an average deli order can be used in conjunction to estimate the time delay needed before a customer can proceed to the deli counter to pick up the filled order. The results of queueing network models are an important input to the software design process and can definitely influence and guide software design decisions.

Time-Complexity Analysis

The analysis of the complexity of an algorithm can be used to evaluate the performance of the algorithm [Horowitz and Sahni 1978], and more importantly, from a software design perspective, to compare performance of multiple algorithms. Algorithm analysis can occur during software design even before code has been written. In one approach, equations can be developed which represent the time spent in carrying out an algorithm. In this case, we may be given a description of the algorithm in pseudocode, in software architecture, or in hardware organization, and from this description, we develop *time-complexity equations* which represent the time spent by the algorithm. There are three different types of time-complexity equations that can be defined: the *best-case time* represents the time to execute an algorithm under ideal conditions, the *average-case time* represents the time to execute an algorithm under typical or average conditions, and the *worst-case time* represents the time to execute an algorithm under the exhaustive or worst conditions. During software design, the type of application may dictate the degree of freedom in an algorithm's performance. For example, in a medical, life-critical application, the worst-case time for an algorithm might be the guiding factor, to ensure that lives are not lost under any conditions. In an application such as **HTSS**, the average-case time may be sufficient.

Time-complexity equations may be developed for two different purposes during software design. On the one hand, time-complexity equations may be used for the *case study* of an algorithm, e.g., the best-case, average-case, and worst-case times for the algorithm. On the other hand, time-complexity equations may be used to provide a *relative analysis* of the time spent in different designs of the same algorithm. For example, suppose we wished to compare two different storage/search designs for the database of **Items** in **HTSS**. In one approach, suppose that a sequential data structure (array) is utilized, with **Items** stored in sorted order by **UPC**. Further suppose that for locating an **Item** the best available search method requires an $O(\log n)$ algorithm, where n is the number of **Items**. On the other hand, suppose a heavily indexed data structure is utilized, which keeps indices on data and uses the indices to optimize the storage for fast retrieval. Suppose that the best retrieval method in this case requires an $O(\log_i n)$ algorithm, where i represents the efficiency of the indexing technique. For values of i that exceed 2, the second approach is superior. The tradeoffs between algorithm complexity (both time and space) can be weighed and evaluated by a software engineer to determine the algorithm that best matches the constraints of the problem.

Simulation Models

There are two major types of simulation languages, *continuous simulation languages* and *discrete simulation languages* [Hoover and Perry 1989, Sammet 1969]. In continuous simulation languages, the variables of the simulation change in a uniform fashion for a given uniform increment of time. That is, the variables of the simulation are continuous functions over time. In discrete simulation languages, there is a nonuniform change for the simulation variables over uniform increments of time. That is, the variables of the simulation can be specified as step-functions, where the distance between the steps is discrete. Continuous simulation languages are the oldest, and they have mostly been used for the simulation of analog computers [Sammet 1969].

The GPSS simulation language is characterized as a *block-diagram* or *flowchart-oriented* simulation language [Gordon 1975]. In block-diagram simulation languages, the system to be simulated is decomposed into a fixed number of blocks. Then, to define the simulation structure, the transition or flow between the blocks is specified. The simulation execution involves moving objects between blocks as governed by the transition requirements and restrictions until a termination condition is satisfied. The SIMULA simulation language is characterized as a *process-oriented* simulation language [Hoover and Perry 1989]. In process-oriented simulation languages, the system to be simulated is represented by a fixed number of processes which operate in parallel. The modeling of each process is characterized as a sequence of events. The simulation execution involves moving objects through the process organization of the system. As before, execution terminates when a criterion is met.

Like queueing network models, simulation models can then be utilized to predict and estimate performance under varying system load conditions. Simulation techniques offer a software engineer a more fine-grained estimate of load, since a software design can be decomposed into a number of components that can be analyzed both individually and collectively. Results of simulation models are utilized in a similar way to queueing network models, allowing a software engineer to understand the software design in greater detail. Such an understanding will likely lead to informed and justified decisions during the software design process.

Defining Terms

Abstract data type (ADT): A software design approach that decomposes a problem into components by identifying the public interface and private implementation. An ADT allows software engineers to define their own data types to support the informational and behavioral needs of their applications.

Abstraction: Please see opening paragraph of section 108.3 for a precise definition.

Anticipation of change: Please see opening paragraph of section 108.3 for a precise definition.

Bottom-up design: A software design approach that decomposes a problem into components in a process that proceeds from the low-level specific components to general high-level components.

Class: The major unit of abstraction in the object-oriented design approach, used to model the features (information) and behavior (methods) for an application. Each class is partitionable into a public interface and private implementation. Similar in concept to ADTs, classes can utilize inheritance to expand their capabilities. See also *object type* (OT).

Class library: The collection of all classes and their inheritance hierarchies for an application form a common library for use by software engineers, classes, tools, and end-users.

Cohesion: Please see opening paragraph of section 108.3 for a precise definition.

Coupling: Please see opening paragraph of section 108.3 for a precise definition.

CRC design: An object-oriented design approach based on responsibility-driven design. In CRC design, for each class that is identified, the responsibilities are delineated, and the other classes that the class collaborates with are identified.

Data-flow diagram (DFD): A software design approach that emphasizes the decomposition of a problem on the basis of input information, functional actions, information flow, and output results.

Dispatching: The run-time determination of the method to be called, based on the type of the invoking instance. Dispatching allows code to behave differently at different times, based on who is calling particular methods. Dispatching is tightly bound with inheritance and promotes reuse and evolution of code, allowing inheritance hierarchies to be defined and evolved over time as needs and requirements change.

Encapsulation: The software engineering concept that is utilized by ADTs, modules, and OTs/classes to group together information and methods/operations within a single component. Once these are encapsulated, access to the component can be controlled.

Entity-relationship (ER) diagram: A software design approach that identifies the database components of an application using entities to model information aggregations, attributes on entities to characterize the structural content of information, and relationships between entities to model information associations.

Finite-state machine (FSM): A software design approach that is used to precisely capture control flow using a combination of states, which are the functional components of a design, and arcs, which are labeled with the actions that cause state changes or transitions.

Generic: A type-parameterizable piece of a software or design that will be utilized by multiple components. For example, a linked-list data structure (ADT, module, or OT/class) can be developed generically, independent of the type of element in the list. When the list is declared and then used, it is customized on the basis of type to meet the application's specific needs.

Hiding: Related to encapsulation, hiding involves the private implementation of an ADT, module, or OT/class, which is inaccessible to all other components of an application.

Incrementality: Please see opening paragraph of section 108.3 for a precise definition.

Information: The private data of an OT/class, hidden from view. Information represents the different internal data components that define the OT/class and characterize all of its instances. Information is used by the designer of an OT/class to maintain internal state and is unavailable for direct use by any other components (OTs/classes) of an application.

Inheritance: A modeling technique used in ER diagrams and object-oriented design to represent commonalities that exist between various components in an application. In ER diagrams these components focus on shared information. In object-oriented design, these components may be determined by a combination of shared information and/or behavior (function).

Inheritance hierarchy: All inheritance relationships between OTs/classes that share a common parent (or grandparent) form a hierarchy with an identifiable root (ancestor). Inheritance hierarchies are simply the trees that organize the sharing between all related OTs/classes. In any inheritance relationship between two OTs/classes, the parent is referred to as the supertype/superclass, with the child referred to as the subtype/subclass.

Instance: An occurrence of an OT/class, or the actual information/data. Throughout run-time, instances of OTs/classes are created, exist, and destroyed.

Message: An action (method call) that is initiated by an instance on itself or other instances. While method refers to the compile-time interpretation of an operation on an OT/class, a message is the corresponding run-time concept.

Method: Contains the definition of the actions required for a particular operation against the private data of an OT/class. Methods can be in either the public interface or the private implementation of an OT/class.

Modularity: Please see opening paragraph of section 108.3 for a precise definition.

Module: A software design approach that functionally partitions the components of an application into design/program units, which, like ADTs, have a public interface and private implementation. All services that are available from a module are exported to other modules, while all services that are needed by a module must be imported from other modules.

Object type (OT):   The major unit of abstraction in the object-oriented design approach, used to model the features (information) and behavior (methods) for an application. Each OT is partitionable into a public interface and private implementation. Similar in concept to ADTs, OTs can utilize inheritance to expand their capabilities. See also *class*.

OT library: The collection of all OTs and their inheritance hierarchies for an application form a common library for use by software engineers, OTs, tools, and end-users.

Private implementation: That portion of an ADT, module, or object-oriented type/class that is hidden from the other portions of an application. Critical for achieving representation independence.

Public interface: That portion of an ADT, module, or object-oriented type/class that contains the permissible operations (methods, functions) that are available for use by other portions of an application. Critical for achieving abstraction.

Queueing network models: A mathematically based software analysis/design technique for estimating and predicting performance for computing systems. Queueing models allow software engineers to identify bottlenecks and determine components that are I/O or computation bound, at the earliest stages of the software design process.

Representation independence: Please see opening paragraph of section 108.3 for a precise definition.

Responsibility-driven design: An object-oriented design approach that emphasizes the determination of OT/class behavior based on actions and behavior. In responsibility-driven design the key emphasis is to define what an OT/class does, from the perspective of actions and information that is shared.

Separation of concerns: Please see opening paragraph of section 108.3 for a precise definition.

Simulation models: A software analysis/design technique for predicting and estimating performance under varying system load conditions.

Software evolution: The process that describes the ability to change an application over time as new requirements are identified, when major upgrades occur, or when significant flaws are corrected. As a concept, software evolution is promoted heavily for object-oriented design.

Software reuse: The process that describes the ability to reuse existing software in new applications. When software is reused in its entirety without changes, a gain in productivity is attained. Critical for object-oriented design.

Time-complexity analysis: A software analysis/design technique where the performance of individual algorithms can be precisely determined from a timing perspective. The best-case, average-case, and worst-case times of different algorithms can be compared against one another to assist a software engineer in making the correct choice of an algorithm for an application.

Top-down design: A software design approach that decomposes a problem into components in a process that proceeds from the high-level general components to specific low-level components.

References

Barnes, J. G. P. 1991. *Programming in Ada plus Language Reference Manual*, 3rd ed. Addison–Wesley, Reading, MA.

Barnes, J. G. P. 1996. *Programming in Ada 95*. Addison–Wesley, Reading, MA.

Beck, K. and Cunningham, W. 1989. A laboratory for teaching object-oriented thinking. *Proc. 1989 OOPSLA Conf.*, Oct.

Booch, G. 1991. *Object-Oriented Design With Applications*. Benjamin/Cummings, Redwood City, CA.

Booch, G. and Rumbaugh, J. 1995. *Unified Method for Object-Oriented Development*. Rational Software Corporation, Santa Clara, CA, Tech. Rep.

Bretl, R. et al. 1989. The GemStone Data Management System. In *Object-Oriented Concepts, Databases and Applications*, W. Kim and F. Lochovsky, eds., pp. 283–308. ACM Press, Addison–Wesley, Reading, MA.

Brooks, F. 1987. No silver bullet—essence and accidents of software engineering. *IEEE Comput.* 20(4): 10–19.

Budd, T. 1997. *An Introduction to Object-Oriented Programming*, 2nd ed., Addison–Wesley, Reading, MA.

Chen, P. 1976. The entity-relationship model—toward a unified view of data. *ACM Trans. Database Syst.*, 1(1):9–36.

Coleman, D. 1994. *Object-Oriented Development—The Fusion Method*. Prentice–Hall, Englewood Cliffs, NY.

Deitel, H. and Deitel, P. 1997. *Java: How to Program, 1/e*. Prentice–Hall, Englewood Cliffs, NJ.

Department of Defense, 1995. *Ada 95 Reference Manual*, International Standard, ANSI/ISO/IEC-8652:1995, Jan.

Deux, O. et al. 1991. The O2 system. *Commun. ACM* 34(10):34–48.

Gamma, E. et al. 1995. *Design Patterns: Elements of Reusable Object-Oriented Software*. Addison–Wesley, Reading, MA.

Ghezzi, C., Jazayeri, M., and Mandrioli, D. 1991. *Fundamentals of Software Engineering*. Prentice–Hall, Englewood Cliffs, NJ.

Goldberg, A. 1989. *Smalltalk-80: The Language*. Addison–Wesley, Reading, MA.

Gordon, G. 1975. *The Application of GPSS V to Discrete System Simulation*. Prentice–Hall, Englewood Cliffs, NJ.

Harbison, S. 1992. *Modula-3*. Prentice–Hall, Englewood Cliffs, NJ.

Hoover, S. and Perry, R. 1989. *Simulation: A Problem Solving Approach*. Addison–Wesley, Reading, MA.

Horowitz, E. and Sahni, S. 1978. *Fundamentals of Computer Algorithms*. Computer Science Press, Rockville, MD.

Jacobson, I. et al. 1992. *Object-Oriented Software Engineering: A Use Case Driven Approach*. Addison–Wesley, Reading, MA.

Kim, W. 1990. Object-oriented databases: Definition and research directions. *IEEE Trans. Knowledge Data Eng.* 2(3):327–341.

Kleinrock, L. 1975. *Queueing Systems I*. Wiley, New York.

Kleinrock, L. 1976. *Queueing Systems II*. Wiley, New York.

Lamb, C. et al. 1991. The ObjectStore database system. *Commun. ACM.* 34(10):50–63.

Lieberherr, K. 1996. *Adaptive Object-Oriented Software: The Demeter Method with Propagation Patterns.* PWS Publishing, Boston, MA.

Liskov, B. and Zilles, S. 1975. Specification techniques for data abstraction. *IEEE Trans. Software Eng.* SE-1:7–19.

Liskov, B. et al. 1977. Abstraction mechanisms in CLU. *Commun. ACM* 20(8):564–576.

Meyer, B. 1988. *Object-Oriented Software Construction.* Prentice–Hall, Englewood Cliffs, NJ.

Meyer, B. 1992. *Eiffel: The Language.* Prentice–Hall, Englewood Cliffs, NJ.

Ontologic. 1991. ONTOS object database documentation, Release 2.1. Ontologic, Burlington, MA.

Parnas, D. 1972. A technique for software module specification with examples. *Comm. ACM*, 15(5):330–336.

Pressman, R. 1997. *Software Engineering: A Practitioner's Approach*, 4th ed. McGraw–Hill, New York.

Rumbaugh, J. et al. 1991. *Object-Oriented Modeling and Design.* Prentice–Hall, Englewood Cliffs, NJ.

Sammett, J. R. 1969. *Programming Languages: History and Fundamentals.* Prentice–Hall, Englewood Cliffs, NJ.

Schach, S. 1996. *Classical and Object-Oriented Software Engineering*, 3rd ed. Irwin, Chicago, IL.

Sethi, R. 1996. *Programming Languages: Concepts and Constructs*, 2nd ed. Addison–Wesley, Reading, MA.

Smith, J. and Smith, D. 1977. Database abstractions: Aggregation and generalization. *ACM Trans. Database Syst.* 2(2):105–133.

Sommerville, I. 1996. *Software Engineering*, 5th ed. Addison–Wesley, Reading, MA.

Stroustrup, B. 1986. *The C++ Programming Language.* Addison–Wesley, Reading, MA.

Tesler, L. 1985. *Object Pascal Report.* Apple Computer, Santa Clara, CA.

Wirfs-Brock, R. and Wilkerson, B. 1989. Object-oriented design: a responsibility-driven approach. *Proc. 1989 OOPSLA Conf.*, Oct.

Wirfs-Brock, R., Wilkerson, B., and Weiener, R. 1990. *Designing Object-Oriented Software.* Prentice–Hall, Englewood Cliffs, NJ.

Wirth, N. 1977. Modula: a language for modular multiprogramming. *Software—Practice and Experience* 7:3–35.

Wirth, N. 1985. *Programming in Modula-2*, 3rd ed. Springer–Verlag.

Further Information

The interested reader is referred to software engineering and programming language textbooks for a more in-depth coverage of these and other design techniques. A sampling of representative textbooks includes [Ghezzi et al. 1991, Pressman 1997, Schach 1996, Sethi 1996, Sommerville 1996]. In addition, the two main computing organizations, the Association for Computing Machinery (ACM) and the Institute of Electrical and Electronics Engineers (IEEE) Computer Society, both have publications that are targeted to software engineering and design, discussed along with URLs given below.

Electronically, there are a variety of resources available on different World Wide Web (WWW) sites. R. S. Pressman & Associates, Inc., maintains the WWW site:

> http://www.rspa.com/spi/index.html

Topics of interest for this chapter at this site include: *Software Design Concepts and Methodologies* (section 108.2), *Software Engineering for Real-Time Systems* (section 108.5), and *Object-Oriented Concepts* and *Object-Oriented Analysis and Design* (section 108.4). Upon choosing these topics, the reader is directed to both literature citations and electronic references (both newsgroups and WWW sites).

The ACM maintains the WWW sites:

> http://www.acm.org
> http://www.acm.org/pubs
> http://www.acm.org/sigs

The first site listed is the home page for the ACM. At the pubs site, topics for the magazine *Communications of ACM* and journal *Transactions on Software Engineering and Methodology* are available. At the special interest groups site sigs, the *SIGSOFT* topic is selectable for software engineering.

The IEEE Computer Society also maintains three WWW sites of interest for this chapter:

http://ada.computer.org:80/cshome.htm
http://ada.computer.org:80/pubs/pubs.htm
http://ada.computer.org:80/tab/tab.htm

The first site listed is the home page for the IEEE Computer Society. At the pubs/pubs.htm site, topics for the magazines *Computer* and *IEEE Software* are available. This site also has topics for the journals *Transactions on Software Engineering* and *Transactions on Knowledge & Data Engineering*. The site tab/tab.htm maintains the technical committees supported by the IEEE Computer Society, including the *Software Engineering* topic. Finally, note that all ACM and IEEE Computer Society publications are also available at most major college and university libraries.

109

Verification and Validation

John D. Gannon
University of Maryland

109.1 Introduction

Verification and *validation* are terms that are sometimes used interchangeably. In [Ghezzi et al. 1991], *verification* is used to describe "all activities that are undertaken to ascertain that the software meets its objectives," and *validation* is not used at all. In [Rushby 1993], specification validation is a two-component process of seeking assurance that a specification means something (i.e., is consistent), and that it means what is intended. We use *verification* to describe the process of demonstrating that a description of a software system guarantees particular properties. General properties may be derived from the form of the description (e.g., that functions are total, axioms are consistent, or variables are initialized before they are referenced), and specific properties may be derived from the problem domain. The latter case involves the comparison of two objects, a detailed description of a software system, and a more abstract description of its intended properties.

In section 109.2 we briefly describe validation and verification approaches. Sections 109.3 and 109.4 deal with the verification of general and specific properties of specifications and programs, respectively. We pay particular attention to opportunities for automating verification activities. We conclude with a short discussion of the current verification practices.

109.2 Approaches to Verification

A variety of analysis activities may be used to verify software artifacts. In software inspections, teams of software developers manually examine artifacts for defects. If a requirements document or a design is written in a formal language, it may be possible to use it as a prototype for the system by simulating the description for some test cases. General properties of software artifacts may be verified automatically by static analysis of the artifact. State-exploration or theorem-proving techniques can be used to prove specific properties of system descriptions.

0-8493-2909-4/97/$0.00+$.50
© 1997 by CRC Press, Inc.

Inspections have proven to be an effective method for detecting software defects because they subject a software artifact to the scrutiny of several people, some of whom did not participate in the artifact's design. Early requirements inspections catch errors before they propagate into designs and implementations, making them less costly to repair. Fagan [1976] describes a six-stage inspection process:

- A determination is made that a software artifact is ready for an inspection, and an inspection team is assembled.
- The artifact's author provides reviewers an overview of the artifact.
- The team members individually study the artifact and record potential defects.
- A fixed-length inspection meeting is held. A moderator controls the discussion. The designer presents and explains his work. Participants identify errors (but not solutions), which are recorded by a secretary.
- The author fixes any errors.
- The moderator checks the new version of the artifact and determines if another inspection is necessary.

Successful inspections depend on the experience levels of the participants and the quality of the artifacts. Requirements inspections should include participants who are future users of the system who will help the software developers judge if the system will function as intended. Requirements notations for embedded systems (e.g., the Software Cost Reduction (SCR) notation [Heitmeyer et al. 1995, Heniger 1980], the Requirements State Machine Language [Leveson et al. 1994], and Statecharts [Harel et al. 1990]) describe systems as sets of concurrently executing state machines responding to events in their environments. Finite state machines have precise meanings, but they are also easy to understand because they can be described in tabular or graphic formats.

In order to maximize the benefits of inspections, participants may be given lists of questions about the artifact that they must answer to ensure that they are sufficiently prepared for an inspection. For code inspections, participants may receive checklists of potential errors that they are to check are not present in the implementation.

Simulation of a software artifact helps software developers determine if the system behaves as expected by producing results like those which will be produced by the eventual implementation of the system. Such operational descriptions of systems give recipes for achieving desired results rather than just describing properties of final results. Simulations of state-machine descriptions of systems are easy to perform; however, simulating more detailed descriptions may require developers to sacrifice abstraction in favor of executability. Being able to reverse a simulation may permit analysts to determine how potentially **hazardous states** may be reached [Ratan et al. 1996].

If state machines manipulate few variables with simple data types (i.e., types with finite numbers of values), properties such as deadlock freedom or mutual exclusion can be verified by using state-space enumeration and exploration techniques. More detailed system descriptions with richer data types correspond to infinite-state machines. While more specific properties can be stated and verified for infinite-state machines, analysis techniques must either investigate approximations of the state space by folding states together [Young and Taylor 1989] or reason with compact descriptions of the entire state space (i.e., assertions).

109.3 Verifying Specifications of Systems

With appropriate abstraction, synchronous communicating processes in distributed systems can be described by sets of state-transition diagrams.

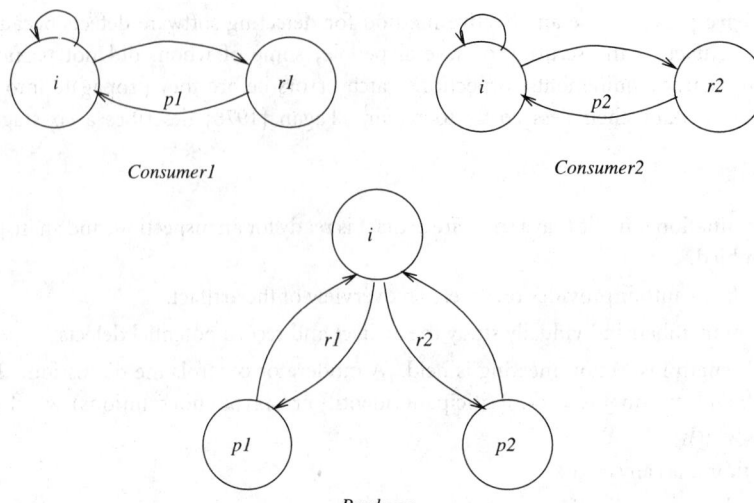

FIGURE 109.1   A simple system with two consumers and one producer.

For example, Fig. 109.1 describes a producer/consumer system with one producer and two consumer processes. *Consumer1* decides either to remain in an idle state (labeled *i*) or to move to a state in which it requests output from the producer (labeled *r1*). When the producer process grants the consumer's request (by entering its state labeled *p1*), *Consumer1* returns to its idle state. *Consumer2* behaves in a similar manner. *Producer* starts in an idle state (also labeled *i*), and moves into a production state (labeled *p1* or *p2*) in response to one of the consumer processes moving into its request state. After satisfying the request, *Producer* returns to its idle state.

General Properties

We can derive general properties for the transitions in our model (e.g., that they are deterministic or total). In Fig. 109.1, unlabeled arcs are considered to be labeled "true." That is, these transitions may always be taken. *Consumer1* has nondeterministic transitions because there are two arcs with the same transition conditions (i.e., true) leaving its idle state and ending in different states. We might always want to ensure that some transition is always enabled from each state. To check this property we compute the logical operation for all the conditions on transitions leaving a state, and we check that its result is identical to true. *Consumer1*'s idle state satisfies this property, but its request state does not, because the only transition leaving this state occurs when *p1* is true (i.e., when the *Producer* is in its state labeled *p1*). Although these are very simple properties, they are valuable checks, particularly for large systems because they do not require construction of the system's state space.

Specific Properties

Reachability Analysis

Reachability analysis is performed to determine if potentially hazardous states (e.g., those representing deadlock or mutual-exclusion failures) are reachable. To perform this analysis, a reachability graph representing the global behavior of the system is constructed and exhaustively examined to determine if a hazardous state is reachable.

Figure 109.2 represents the reachability graph for our producer/consumer system. Each state is labeled with three properties corresponding to properties of *Consumer1*, *Consumer2*, and *Producer*. In the

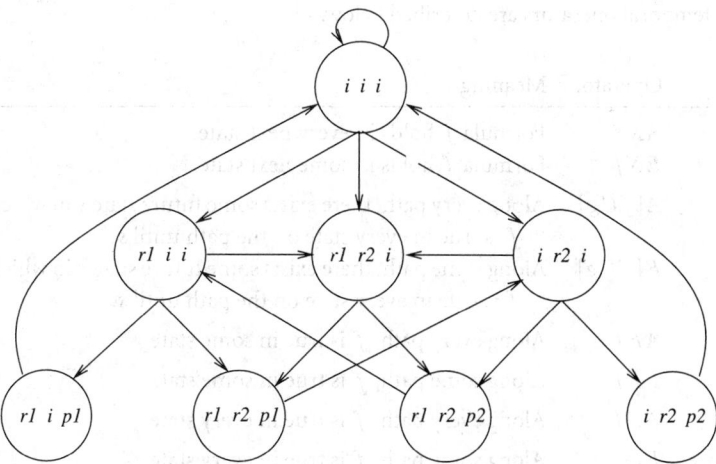

FIGURE 109.2 Reachability graph.

reachability graph's initial state, labeled (i, i, i), each of the processes is in its idle state. Three transitions leave this state, corresponding to either or both *Consumer1* or *Consumer2* issuing requests for output. There are no transitions from the initial state to other states labeled either $(i, i, p1)$ or $(i, i, p2)$ because the producer process waits for either of the consumers to issue a request before moving to a state in which it produces output. Thus we could verify that objects are produced only in response to requests by determining that no reachable states are labeled either $(i, i, p1)$ or $(i, i, p2)$.

Model Checking

Although it is a useful verification technique, reachability analysis can be used only to verify properties specified as propositional logic formulas quantified over all the states in the graph. We might also like to assert properties about sequences of events, e.g., "if the consumer requests an output, the producer always supplies one." Pneuli [1981] showed how **temporal logic** can be used to state such properties and to reason about concurrent systems.

A temporal logic is a propositional logic with additional temporal operators to express concepts such as "always," "eventually," and "until" to assert that formulas are true in all or some future states. Two major types of temporal logic are used in specifications: linear time logic and branching time logic. In linear time logic, states have unique pasts and futures. To prove that a property is invariantly true, the property must be proved over all possible execution paths of the system. In branching time temporal logics, states have unique pasts but many possible futures. Thus, assertions may be made about properties holding on some future executions or on all future executions. The latter assertions are invariants.

Computational tree logic (CTL) is a propositional branching time logic, whose operators permit explicit quantification over all possible futures [Clarke et al. 1986]. The syntax for CTL formulas is summarized below:

1. Every atomic proposition is a CTL formula.
2. If f and g are CTL formulas, then so are: $\sim f$, $f \wedge g$, $f \vee g$, $f \rightarrow g$, AXf, EXf, $A[fUg]$, $E[fUg]$, AFf, EFf, AGf, and EGf.

Note that temporal operators occur only in pairs in which a quantifier A (always) or E (exists) is followed by F (future), G (global), U (until), or X (next). The logical operators have their usual meanings. The meanings of the temporal operators are described below.

Concept	Operator	Meaning
Next	AXf	Formula f holds in every next state.
	EXf	Formula f holds in some next state.
Until	$A[fUg]$	Along every path, there exists some future state s in which g is true, and f is true in every state on the path until s.
	$E[fUg]$	Along some path, there exists some future state s in which g is true, and f is true in every state on the path until s.
Eventually	AFf	Along every path, f is true in some state.
Possibly	EFf	Along some path, f is true in some state.
Invariance	AGf	Along every path, f is true in every state.
Possible invariance	EGf	Along some path, f is true in every state.

The specification "if the consumer requests an output, the producer always supplies one" can be written as the CTL formula: $AG((r1 \rightarrow AF(p1)) \wedge (r2 \rightarrow AF(p2)))$. That is, it is invariantly true that if *Consumer1* makes a request (represented by a state in which *r1* is true), eventually (i.e., along every path starting at such states) the producer supplies an output for *Consumer1* (a state is encountered in which *p1* is true), and similarly for *Consumer2*.

If formula f is true in state s of model M, we write $M, s \models f$. A formula f is true for the model, if it is true in the model's start state, i.e., $M, s_0 \models f$. When we are concerned with a single model, we abbreviate $M, s \models f$ as $s \models f$.

Introduced by Clarke and Emerson [Clarke et al. 1986] and by Quielle and Sifakis [1981], model checking determines the value of a formula f for a particular model by building a reachability graph and computing the set of states in which the formula is true, i.e., $\{s \mid s \models f\}$. For example, formula $AF(f)$ represents the set of states from which a state satisfying f can be reached in *some* number of state transitions along *all* paths from the state.

$$f \vee AXf \vee AX(AXf) \vee \ldots$$

where f is the set of states in which f is true, i.e., the set of states from which an f-state can be reached in zero-state transitions. $AX(f)$ is the set of states all of whose transitions reach an f-state. $AX(AXf)$ is the set of states from which any two state transitions reach an f-state, etc.

This set of states can be computed using the following **least fixpoint** computation:

```
Y = { };
Y' = {s | s ⊨ f};
while ( Y ≠ Y' ) do {
   Y = Y';
   Y' = Y' ∪ {s | all successors of s are in Y};
}
```

By way of example, consider computing the set of states in which $AF(p1)$ for the model in Fig. 109.2. The first iteration computes the set of states in which the formula *p1* holds. This set, which is shaded in Fig. 109.3, corresponds to computing $p1 \vee AX\ false$. During the second iteration, the predecessors of the states already in the set are examined. The state labeled $(r1, i, i)$ is added to the set because the formula is true in all of its successors. At this point, we have computed the set of states satisfying $p1 \vee AX(p1)$. The

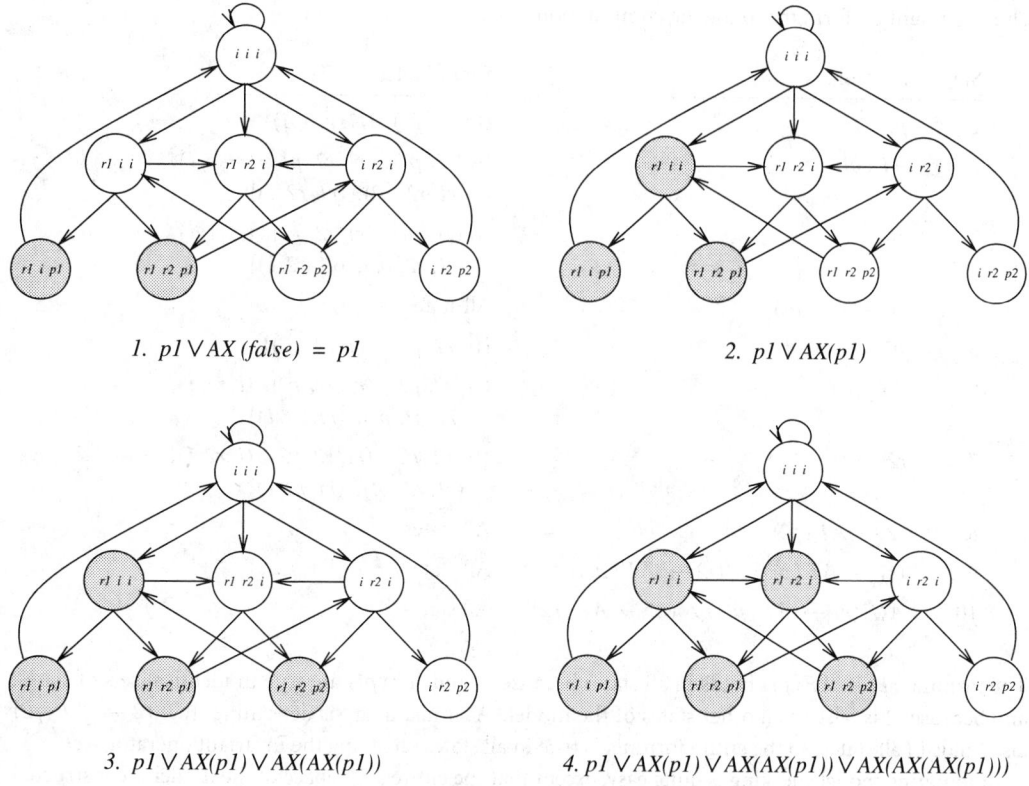

1. $p1 \lor AX \, (false) = p1$

2. $p1 \lor AX(p1)$

3. $p1 \lor AX(p1) \lor AX(AX(p1))$

4. $p1 \lor AX(p1) \lor AX(AX(p1)) \lor AX(AX(AX(p1)))$

FIGURE 109.3 Model checking $s_0 \models AF(p1)$.

remaining iterations are summarized in the following table.

Iteration	State Set	Remarks
1	$\{(r1, i, p1), (r1, r2, p1)\}$	The only states in which $p1$ is true.
2	$\{(r1, i, p1), (r1, r2, p1), (r1, i, i)\}$	$AX(p1)$ is true for the state labeled $(r1, i, i)$ since $p1$ is true in all its successors.
3	$\{(r1, i, p1), (r1, r2, p1), (r1, i, i), (r1, r2, p2)\}$	$AX(AX(p1))$ is true for the state labeled $(r1, i, i)$ since $AX(p1)$ is true in all its successors.
4	$\{(r1, i, p1), (r1, r2, p1), (r1, i, i), (r1, r2, p2), (r1, r2, i)\}$	For the state labeled $(r1, r2, i)$, $AX(AX(AX(p1)))$ is true for one of its successors and $AX(p1)$ is true for the other.
5	$\{(r1, i, p1), (r1, r2, p1), (r1, i, i), (r1, r2, p2), (r1, r2, i)\}$	Each of the possible new predecessors, the states labeled state (i, i, i) and $(i, r2, i)$, have infinitely long paths before reaching a state in which $p1$ is true.

Since no new states are added during the fifth iteration, we reach a fixpoint and the algorithm terminates. Unfortunately, the set of states computed does not contain the start state, s_0 (the state labeled (i, i, i)), so $s_0 \not\models AF(p1)$. However, the model checker gives a specific counter-example (i.e., the loop (i, i, i), $(i, r2, i)$, $(i, r2, p2)$, ...) showing why the formula is not satisfied.

To check $AG((r1 \rightarrow AF(p1)) \land (r2 \rightarrow AF(p2)))$, we calculate the sets of states for which the innermost, simplest formulas hold and work our way outward, calculating sets of states for more complex formulas.

Thus we might perform the following computations:

Step	Formula	Set of States
1	*p1*	$\{(r1, i, p1), (r1, r2, p1)\}$
2	$AF(p1)$	$\{(r1, i, p1), (r1, r2, p1), (r1, i, i),$ $(r1, r2, p2), (r1, r2, i)\}$
3	*r1*	$\{(r1, i, p1), (r1, r2, p1), (r1, i, i),$ $(r1, r2, p2), (r1, r2, i)\}$
4	$r1 \rightarrow AF(p1)$	All states
5	*p2*	$\{(i, r2, p2), (r1, r2, p2)\}$
6	$AF(p2)$	$\{(i, r2, p2), (r1, r2, p2), (i, r2, i),$ $(r1, r2, p1), (r1, r2, i)\}$
7	*r2*	$\{(i, r2, p2), (r1, r2, p2), (i, r2, i),$ $(r1, r2, p1), (r1, r2, i)\}$
8	$r2 \rightarrow AF(p2)$	All states
9	$(r1 \rightarrow AF(p1)) \wedge (r2 \rightarrow AF(p2))$	All states
10	$AG((r1 \rightarrow AF(p1)) \wedge (r2 \rightarrow AF(p2)))$	All states

The formula $r1 \rightarrow AF(p1)$ holds in all states because $r1$ and $AF(p1)$ are true in identical sets of states, and because $r1$ is false in all other states of the model. A similar analysis determines that $r2 \rightarrow AF(p2)$ also holds in all states, so the entire formula is true in all states satisfying the invariant operator AG.

Automating model checking is quite easy, except that the entire state space of the model is constructed before the fixpoint algorithms can be applied. However, model checking can also be done symbolically by manipulating quantified Boolean formulas without constructing a model's state space [McMillan 1993]. To perform symbolic model checking, sets of states and transition relations are represented by formulas, and set operations are defined in terms of formula manipulations. A CTL formula f is evaluated for a model by deriving a propositional logic expression that describes the set of states satisfying f for the model and verifying that the interpretation of the model's initial state satisfies the expression.

109.4 Verifying Programs

General Properties

Probably the best-known property of programs which is verified is that a program is type safe. In statically typed languages, a data type is associated with a variable in a declaration. During its lifetime, the variable may be assigned only values of the same type. The context of each appearance of a variable in a statement implies a type, which can be checked against its declared type. Violations are reported as syntax errors.

Other important properties depend on data and control flow, e.g., each variable is assigned a value before the value is used in an expression. Such properties are verified by static checkers, which analyze a program's syntax tree or control flow graph; no test data are used during these checks. Static checkers fold different states together to make analysis tractable. For example, to check uninitialized variables, we may care only if a variable has been assigned a value or not. Different integer values are all folded to a single "defined" value to reduce the size of the state space. In the following program fragment, **x** will always have a value when control reaches the final write statement, since either **i < j** and **x** is assigned the value 1 or **i >= j** and **x** is assigned the value 2. However, a static checker keeping track of whether or not **x** had been assigned a value on every potential path through the program would conclude that the write statement might be executed with an undefined value.

Statement	Defined Values
`read(i);`	{i}
`read(j);`	{i, j}
`if (i < j)`	
`x = 1;`	{i, j, x}
`fi;`	{i, j} ∩ {i, j, x} = {i, j}
`if (i >= j)`	
`x = 2;`	{i, j, x}
`fi;`	{i, j} ∩ {i, j, x} = {i, j}
`write(x)`	{i, j}

In each if statement, **x** is defined on only one path through the statement. Thus, the static checker intersects the sets of variables defined on each of its paths to determine the set of variables which are certain to have values. This approximation of the state space is called conservative or pessimistically inaccurate because it preserves states which potentially contain errors. In this case, it preserved a state in which x is undefined which can be reached only on an infeasible path (i.e., when $i \geq j \land i < j$). Programmers using static checkers must examine error messages to determine if an anomaly exists before trying to repair their programs.

A syntax-directed definition can be used to describe the analysis needed to verify this property. A syntax-directed definition extends a context-free grammar by associating attributes with grammar symbols. The value of an attribute in a syntax tree is defined by rules associated with each production used at the particular node of the tree. Attributes can be values of any type. In this example, we use two sets of identifiers: In (representing the set of identifiers defined on all paths leading to the current statement) and Out (representing the set of identifiers defined on all paths after executing the current statement). The syntax-directed definition for the simple language is shown below.

Production	Attributes
`P ::= SL`	`SL.In = ∅`
`SL₁ ::= S ';' SL₂`	`S.In = SL₁.In`
	`SL₂.In = S.Out`
	`SL₁.Out = SL₂.Out`
`SL ::= ε`	`SL.Out = SL.In`
`S ::= id '=' Exp`	`S.Out = {id} ∪ S.In`
`S ::= 'if' Exp 'then' SL 'fi'`	`S.In = S.In`
	`S.Out = S.In`
`S ::= 'if' Exp 'then' SL₁ 'else' SL₂ 'fi'`	`SL₁.In = S.In`
	`SL₂.In = S.In`
	`S.Out = SL₁.Out ∩ SL₂.Out`
`S ::= 'while' Exp 'do' SL 'od'`	`SL.In = S.In`
	`S.Out = S.In`
`S ::= 'read' '(''id'')'`	`S.Out = {id} ∪ S.In`
`S ::= 'write' '(''id'')'`	`S.Out = S.In`

The production **P ::= SL** initializes **SL's In** attribute to the empty set. When identical nonterminals appear in a production, the instances are numbered so their attributes can be distinguished. For example, the production **SL ::= S ';' SL** is rewritten as **SL₁ ::= S ';' SL₂**. The set of defined variables available to **S** is the same as that available to **SL₁**. However, the defined variables for **SL₂** are those from **S.Out**. Each **S**-production copies its incoming set of definitions (**In**) to its **Out** attribute and adds any new definitions made in the statement. Thus, for example, the statement **read(i)** adds the variable **i** to the empty set of definitions that reaches it. The set of outgoing definitions from the **if** statement

```
if i < j then x = 1 fi;
```

is {**i,j**}, since the statement contains an execution path on which x is not defined.

Figure 109.4 shows the attributes evaluated on the syntax tree corresponding to our sample program. Since the write statement's **In** attribute contains only {**i,j**}, a static checker would say that definition-before-use property was violated for **x**.

Abstract interpretation [Cousot and Cousot 1976] is a method for computing approximate semantics of programs in order to provide safe answers to questions about their run-time behaviors. In an abstract interpretation of a program, "abstract" values are associated with program variables instead of the actual execution values, and a programming language's operators are redefined to manipulate the abstract values. An abstract interpretation of a program computes a fixpoint approximation of the abstract program state at different points in the program, and properties of the program are verified with respect to these states.

Consider the following example [Cousot and Cousot 1977], which searches a list for a particular value. One property we would like to ensure is that the value of **p** is never **NULL** when it is dereferenced on

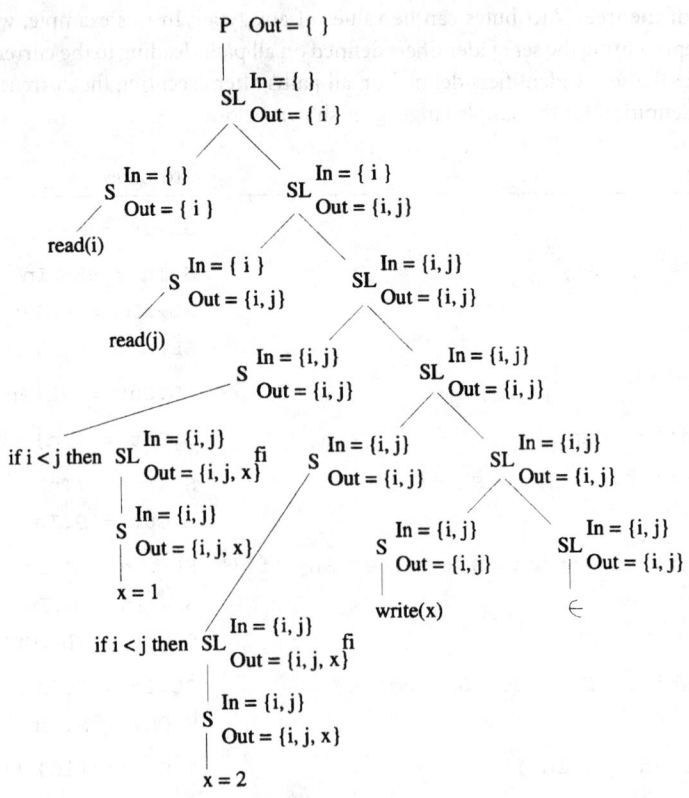

FIGURE 109.4 Attributes evaluated on a syntax tree.

lines 3 or 4.

```
1. p = L; b = TRUE;
2. while (p ≠ NULL & b)
3.     if (p → v == n) then b = FALSE;
4.     else p = p → next;
```

The pointer variable **p** may have one of four possible values: undefined ($\perp$), **NULL**, **nonNULL**, or **NULL** or **nonNULL** ($\top$). These values form the complete lattice in Fig. 109.5. Initially, a pointer variable has the value $\perp$. Pointer variables may be assigned on of the following values: **NULL**, **nonNULL** (i.e., the result produced by a new operation), or the value of another pointer variable. Dereferencing a pointer whose value is **nonNULL** yields the value $\top$, and dereferencing a pointer with any other value yields $\perp$.

Each node in a program's control flow graph defines an output state in terms of its input state. Assignment nodes' output states are identical to their input states except for the value of the variable on the left side of the assignment operation. The output state for a join node is the union of the respective values in its input states. The output state corresponding to the true outcome of a decision node labeled **p! = NULL** is calculated by creating a new state which is identical to the input state except that **p** has value **nonNULL** and intersecting the new state with the input state. The output state corresponding to the false outcome of the decision node is calculated in a similar manner except that **p** has value **NULL** in the newly created state. These computations are depicted in Fig. 109.6, and the definitions for $\cup$ and $\cap$ are given in the following tables. The **X**s in the table for $\cap$ represent error entries corresponding to infeasible paths.

	$x \cup y$			
	$\perp$	NULL	nonNULL	$\top$
$\perp$	$\perp$	NULL	nonNULL	$\top$
NULL	NULL	NULL	$\top$	$\top$
nonNULL	nonNULL	$\top$	nonNULL	$\top$
$\top$	$\top$	$\top$	$\top$	$\top$

	$x \cap y$			
	$\perp$	NULL	nonNULL	$\top$
$\perp$	$\perp$	$\perp$	$\perp$	$\perp$
NULL	$\perp$	NULL	X	NULL
nonNULL	$\perp$	X	nonNULL	nonNULL
$\top$	$\perp$	NULL	nonNULL	$\top$

Since pointer variables have a finite number of abstract values, system states do not have an infinite increasing chain of values.

Figure 109.7 shows the control flow graph for the sample program with edges labeled with computed state values for the pointer variables **p** and **L**. All edges are initially labeled { **p** = $\perp$, **L** = $\perp$} except for the first arc, where we assume **L** has value $\top$. The first time the arc leaving the while statement's join node is reached, the state value is the set labeled "1" in Fig. 109.7. This set of values results from the union of the output state of the assignment statement **p** = **L** and the default program state corresponding to the output state of the if statement's

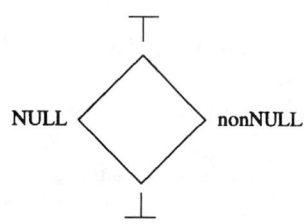

FIGURE 109.5 Lattice of pointer values.

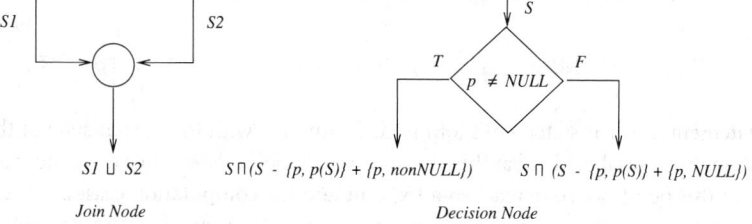

FIGURE 109.6 Computing state values for join and decision nodes.

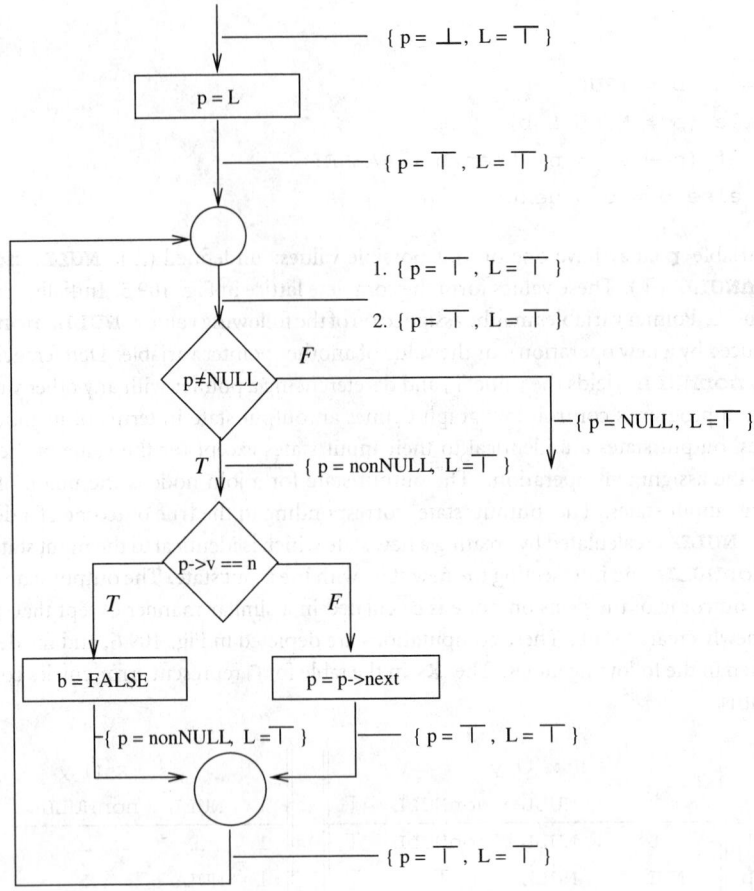

FIGURE 109.7   Calculated state values for sample program.

join node.

```
while join   {p = T, L = T} ∩ {p = ⊥, L = ⊥} = {p = T, L = T}
```

Interpreting the while statement's decision node creates two states:

```
Test succeeds {p = T, L = T} ∩ {p = nonNULL, L = T} = {p = nonNULL, L = T}
Test fails    {p = T, L = T} ∩ {p = NULL, L = T} = {p = NULL, L = T}
```

Since **p** has value **nonNULL** before being dereferenced in the expression **p → next**, the dereference operation yields the value ⊤ which is bound to **p** in the assignment statement. At the if statement's join node, the union of two input states creates the output value.

```
if join   {p = nonNULL, L = T} ∩ {p = T, L = T} = {p = T, L = T}
```

The output statement of the if statement's join node is unioned with the output state of the assignment **p = L**, yielding the state labeled "2" at the while statement's join node. Since this state is identical to the previous state at this point, we have reached a fixpoint and the computation ceases. We can now check that both pointer dereferences (i.e., **p → v** and **p → next**) occurs in input state where **p** has the value **nonNULL**, so we know no **NULL**-value pointers are dereferenced.

Static checkers have difficulty dealing with pointer variables. General properties involving pointers are more easily verified at run time. Two such properties are freedom from memory access errors or **memory leaks**. A memory access error occurs when a reference containing a valid address for a block of storage which has already been freed is used to read from or write to the address. A memory leak occurs when storage is allocated but not freed before the last reference to it is lost. The following program illustrates these problems.

```
void leak(){
    node* q = new node; // the storage referenced by q is a
                        memory leak
}
void remove (node* p){
    delete p; //p is a reference to freed memory
}
void main(){
    node* p = new node;
    p→data = 3;
    remove(p);
    p→data = 4;
    leak();
}
```

In procedure remove, the storage referenced by **p** is freed, but **p** retains its value. When remove returns, **p** is dereferenced to store a value in one of the freed memory locations. When procedure leak is called, **q** is assigned the address of newly allocated storage. When leak returns, **q**'s storage is reclaimed but the storage it referenced remains allocated. Over time, the build-up of state caused by memory leaks can lead to program crashes.

Tools (e.g., Purify [Hasting and Joyce 1992]) redefine memory management routines to record information needed to check these properties. A table corresponding with a single bit for each byte of memory can be used to determine if a memory location is allocated or not. Accesses to unallocated memory are reported. Standard mark and sweep algorithms for **garbage collection** are modified to detect memory leaks. The mark phase recursively follows pointers from the stack and marks referenced heap locations. This phase is conservative because pointers cannot be distinguished from other data, and an integer with a value that appears to be a valid address will cause freed data to be erroneously marked. The sweep phase steps through the heap and reports blocks that are no longer referenced. By labeling each block with the return address of the the functions on the call stack, useful diagnostics about offending statments can be printed.

Specific Properties

Floyd [Floyd 1967] introduced assertional reasoning for sequential programs represented as flowcharts. Hoare formulated this as a logic for program test. A Hoare-triple has the form $\{P\}S\{Q\}$, where P and Q are assertions about program states, and S is a statement. This expression is interpreted as "If P, called the *precondition*, is true before executing S and S terminates normally, then Q, called the *postcondition*, will be true." This concept is called "partial" correctness because S's **termination** is not guaranteed.

Axioms and Rules of Inference

The assignment axiom schema:

$$\text{Assignment:} \quad \{P_y^x\}x = y\{P\}$$

defines the effect of the assignment statement on postcondition P. That is, if we want P to be true after executing the assignment statement $x = y$, then P_y^x, P with all free (i.e., unquantified) occurrences of x replaced by y must be true before executing the assignment. The assignment axiom is a schema that must be instantiated for individual assignment statements. For example, $\{y - 1 \geq 0\}\ x = y - 1\ \{x \geq 0\}$. The assignment axiom allows us to calculate an assertion which describes the set of input states for which the assignment statement will produce the desired result if it terminates. Thus, if we want $x \geq 0$ to hold in all program states after $x = y - 1$ terminates, $y \geq 1$ must hold in all states before this statement executes.

Results for verifying two statements are composed using the following rule of inference:

$$\text{Composition:} \quad \frac{\{P\}S_1\{Q\},\ \{Q\}S_2\{R\}}{\{P\}S_1;\ S_2\{R\}}$$

If the formula above the line (the antecedent) is true, then we may conclude that the statement below the line (the consequent) is true. The rule of composition allows us to combine the results of executing two statements to conclude that if P is true and the execution of S_1 followed by S_2 terminates normally, then R will be true. To reach this conclusion, the antecedent requires us to show that the postcondition of S_1 is the same as the precondition of S_2.

The postcondition of one statement is rarely identical to the precondition of another state, so we have rules of consequence to weaken the postcondition or strengthen the precondition of a statement. The rules of consequence build on predicate logic rules of inference.

$$\text{Rules of consequence:} \quad \frac{\{P\}S\{R\},\ \ R \to Q}{\{P\}S\{Q\}} \quad \frac{P \to R,\ \ \{R\}S\{Q\}}{\{P\}S\{Q\}}$$

Each programming language statement has a separate rule of inference.

$$\text{If statement}_1: \quad \frac{\{P \wedge B\}\ S\ \{Q\},\ \ P \wedge \sim B \to Q}{\{P\}\text{if } B \text{ then } S\{Q\}}$$

$$\text{If statement}_2: \quad \frac{\{P \wedge B\}\ S_1\ \{Q\},\ \ \{P \wedge \sim B\}\ S_2\ \{Q\}}{\{P\}\text{if } B \text{ then } S_1 \text{ else } S_2\ \{Q\}}$$

$$\text{While statement:} \quad \frac{\{P \wedge B\}\ S\ \{P\}}{\{P\}\text{while } B \text{ do } S\ \{P \wedge \sim B\}}$$

A rule of inference captures the statement's semantics. For example, to conclude the consequent of If statement$_2$, we have to show that for each execution path through the statement that if we start execution with assertion P being true and execution of the path terminates normally, then assertion Q will be true. On one path, we assume P and B are true before S_1 is executed, and on the other path we assume P and $\sim B$ are true before S_2 is executed. Each of these paths appears in the left half of Fig. 109.8. Defining a rule for the while statement is more difficult because the number of paths to verify is potentially infinite. The while statement rule of inference resorts to induction to solve this problem. We assume a property P (called the invariant) is true when the while statement begins execution and show that it is still true when S terminates normally. We conclude that P is true after zero or more executions of S, and that B must be false when the statement terminates.

Verifying a Small Program

Consider the following example of a Hoare-style proof of partial correctness of a program which uses repeated subtractions to compute the remainder ($\mathbf{r}$) and quotient ($\mathbf{q}$) obtained be dividing the integer

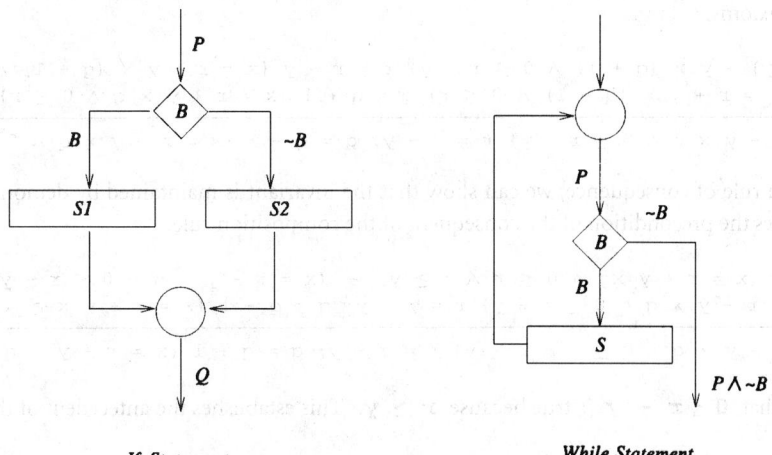

FIGURE 109.8 Flow of control for if and while statements.

x by the integer **y**:

```
{x ≥ 0 ∧ y > 0}
q = 0;
r = x;
while y ≤ r do {
   r = r - y;
   q = q + 1;
   }
{x = r + y* q ∧ 0 ≤ r ∧ r < y}
```

The postcondition characterizes the desired relationship between values in order for **r** and **q** to represent the remainder and quotient, respectively.

For the postcondition to be true, it must be so as a result of application of the while rule of inference. To apply the rule, we must identify the loop invariant (P) in the rule's antecedent. One way to do this is to "remove" $\sim B$ from the postcondition and check if the remainder of the postcondition is invariant. Since B is **y ≤ r**, $\sim B$ is **y > r** or **r < y**, and P is **x = r + y × q ∧ 0 ≤ r**. Hence the inference rule which could be applied would be:

$$\frac{\{x = r + y \times q \land 0 \le r \land r \ge y\} \quad r = r - y;\ q = q + 1 \quad \{x = r + y \times q \land 0 \le r\}}{\{x = r + y \times q \land 0 \le r\} \quad \text{while } r \ge y \text{ do } \{r = r - y\ ;\ q = q + 1\} \quad \{x = r + y \times q \land 0 \le r \land r < y\}}$$

To establish the antecedent to the while rule, we use the assignment axiom for each assignment statement to calculate the property which must be true before each is executed.

```
{x = r + y × (q + 1) ∧ 0 ≤ r} q = q + 1 {x = r + y × q ∧ 0 ≤ r}
{x = (r - y) + y × (q + 1) ∧ 0 ≤ r - y} r = r - y {x = r + y × (q + 1) ∧ 0 ≤ r}
```

The precondition of the first assignment statement can be simplified from **(r - y) + y× (q + 1)** to **r - y × q**. The rule of inference for composition can be applied to compose the results of the assignment axioms:

$$\{x = (r - y) + y \times (q + 1) \land 0 \le r - y\} \ r = r - y \ \{x = r + y \times (q + 1) \land 0 \le r\},$$
$$\{x = r + y \times (q + 1) \land 0 \le r\} \ q = q + 1 \ \{x = r + y \times q \land 0 \le r\}$$

$$\{x = r - y \times q \land 0 \le r - y\} \ r = r - y; \ q = q + 1 \ \{x = r + y \times q \land 0 \le r\}$$

Now, using a rule of consequence, we can show that the invariant is maintained by demonstrating that $P \land B$ implies the precondition of the consequent of the composition rule.

$$(x = r + y \times q \land 0 \le r \land r \ge y) \ \rightarrow \ (x = r - y \times q \land 0 \le r - y),$$
$$\{x = r - y \times q \land 0 \le r - y\} \ r = r - y; \ q = q + 1 \ \{x = r + y \times q \land 0 \le r\}$$

$$\{x = r + y \times q \land 0 \le r \land r \ge y\} \ r = r - y; \ q = q + 1 \ \{x = r + y \times q \land 0 \le r\}$$

We observe that $0 \le r - y$ is true because $r \ge y$. This establishes the antecedent of the while rule of inference.

Now we must determine whether or not the initialization steps in the program make the precondition of the while statement's consequent true. We use the assignment axiom twice to calculate the the property which must be true before each is executed.

$$\{x = x + y \times q \land 0 \le x\} \ r = x \ \{x = r + y \times q \land 0 \le r\}$$
$$\{x = x + y \times 0 \land 0 \le x\} \ q = 0 \ \{x = x + y \times q \land 0 \le x\}$$

After simplifying $x = x + y \times 0$ to true, we use the rule of inference for composition first to compose the results of the two assignment axioms:

$$\{0 \le x\} \ q = 0 \ \{x = x + y \times q \land 0 \le x\},$$
$$\{x = x + y \times q \land 0 \le x\} \ r = x \ \{x = r + y \times q \land 0 \le r\}$$

$$\{0 \le x\} \ q = 0; \ r = x \ \{x = r + y \times q \land 0 \le r\}$$

and then to compose this result with that of the while rule of inference:

$$\{0 \le x\} \ q = 0; \ r = x \ \{x = r + y \times q \land 0 \le r\},$$
$$\{x = r + y \times q \land 0 \le r\} \ \text{while } r \ge y \text{ do } \{ r = r - y; \ q = q + 1 \} \ \{x = r + y \times q \land 0 \le r \land r < y\}$$

$$\{0 \le x\} \ q = 0; \ r = x; \ \text{while } r \ge y \text{ do } \{ r = r - y; \ q = q + 1 \} \ \{x = r + y \times q \land 0 \le r \land r < y\}$$

We use a rule of consequence to show that the program's precondition is stronger than the property we have calculated and must be true before executing the program in order to make the program's postcondition true.

$$(x \ge 0 \land y > 0) \ \rightarrow \ 0 \le x,$$
$$\{0 \le x\}$$
$$q = 0; \ r = x; \ \text{while } r \ge y \text{ do } \{r = r - y; \ q = q + 1\}$$
$$\{x = r + y \times q \land 0 \le r \land r < y\}$$

$$\{x \ge 0 \land y > 0\}$$
$$q = 0; \ r = x; \ \text{while } r \ge y \text{ do } \{r = r - y; \ q = q + 1\}$$
$$\{x = r + y \times q \land 0 \le r \land r < y\}$$

Program Termination

The stated precondition for the program $(x \ge 0 \land y > 0)$ is more restrictive (i.e., describes a smaller set of program states) than the precondition $(x \ge 0)$ we calculated was necessary for the program to execute and produce a set of states satisfying its postcondition. The difference between these assertions highlights the difference between partial and total program correctness. For states satisfying the calculated precondition but not the original precondition (i.e., those in which $\mathbf{x} \geq 0 \land \mathbf{y} \leq 0$), the program would produce the desired result if it halted, but it does not. When values of $\mathbf{y}$ which are less than or equal to 0 are subtracted from $\mathbf{r}$, the difference between $\mathbf{r}$ and $\mathbf{y}$ does not decrease, so the while statement fails to terminate.

To demonstrate that the while statement "while B do S" terminates, we show that B must eventually evaluate to false. To do this, we derive an expression from B whose value is bounded below by 0, and we show that on each path through S the value of the expression decreases. Since it has a lower bound, the expression cannot decrease infinitely, so the while statement must terminate. In our example program, we want to show that $\mathbf{r} \geq \mathbf{y}$ cannot remain true indefinitely. We can form a termination test expression by subtracting $\mathbf{y}$ from both sides of the while statement predicate to obtain $\mathbf{r} - \mathbf{y} \geq 0$. There is only a single path in S on which $\mathbf{r}$ is decremented by $\mathbf{y}$. As long as $\mathbf{y}$ is positive, $\mathbf{r} - \mathbf{y}$ will decrease and the while statement will terminate. Thus, we add the assertion $\mathbf{y} > 0$ to the calculated assertion $\mathbf{x} \geq 0$ to guarantee total correctness.

Advanced Language Features

Arrays. Proofs of programs manipulating scalar variables are relatively straightforward. However, to prove realistic programs, axioms and rules of inference must be devised for all language features. In this section, we discuss arrays and procedure calls, two features that complicate verifications.

Using the axiom of assignment to reason about assignments to variables which are array elements can lead to unsound reasoning. The following code fragment assigns the value 4 to `a[i]` and `a[j]` because the first assignment statement ensures that `i` and `j` have identical values.

```
i = j; a[i] = 3; a[j] = 4;
```

However, using the axioms and rules of inference introduced thus far, we can prove that no matter in what state the program begins execution (i.e., the precondition is true) this code fragment finishes execution with the postcondition `a[i] < a[j]`.

```
{true} i = j; {3 < 4} a[i] = 3; {a[i] < 4} a[j] = 4; {a[i] < a[j]}
```

To avoid this problem, we need to consider an array as a function which maps its indices to values, and an assignment statement as an operation which assigns a new function to the array. For example, `a[i] = 3` assigns a new function to the array `a` which is identical to the old function except that it maps `i` to `3`. That is,

$$\alpha(\mathtt{a},\mathtt{i},\mathtt{x})[\mathtt{j}] = \begin{cases} \mathtt{x} & \text{when } \mathtt{i} = \mathtt{j} \\ \mathtt{a[j]} & \text{when } \mathtt{i} \neq \mathtt{j} \end{cases}$$

Using this definition, we can work out the value of subscripted array expressions, e.g., $\alpha(\alpha(\mathtt{a},\mathtt{3},\mathtt{x}), \mathtt{4},\mathtt{y})[\mathtt{3}] = \alpha(\mathtt{a},\mathtt{3},\mathtt{x})[\mathtt{3}] = \mathtt{x}$. Our new assignment axiom schema is

$$\text{Array assignment:} \quad \left\{ P^a_{\alpha(a,i,x)} \right\} a[i] = x \{P\}$$

With this new axiom and the previous rules of inference, we can reason safely about programs which alter arrays.

```
{α(a,j,4)[i] < α(a,j,4)[j]} a[j] = 4; {a[i] < a[j]}
```

We can simplify α**(a,j,4)[j]** to **4** using the definition of α and continue our verification.

$\{\alpha(\alpha(\alpha(a,i,3),j,4)[i] < 4\}$ a[i] = 3; $\{\alpha(\alpha(a,j,4)[i] < 4\}$

$\{\alpha(\alpha(a,j,3),j,4)[j] < 4\}$ i = j; $\{\alpha(\alpha(\alpha(a,i,3),j,4)[i] < 4\}$

Simplifying α**(α(a,j,3),j,4)[j]** yields the value 4. Thus there are no states in which the program begins execution (i.e., the precondition $4 < 4$ is false) for which this code fragment finishes execution with the postcondition **a[i] < a[j]** .

Procedure Invocations. In verifications involving procedures, our goal is to verify a procedure's body once, and then use this result at each point at which the procedure is invoked. We have two new rules of inference for procedures: one rule handles the substitution of actual parameters for formal parameters, and the other rule relates the procedure's precondition and postcondition to the assertion which must be true after the procedure's invocation [Hoare 1971]. If all our parameters are passed by reference, we can use the following simplified rule of substitution:

$$\text{Substitution:} \quad \frac{\{R\}p(f)\{S\}}{\{R_{k'\ a}^{k\ x}\}p(a)\{S_{k'\ a}^{k\ x}\}}$$

where f and a are the lists of formal and actual parameters, respectively. The procedure's body may not reference nonlocal variables, and each variable in a must be unique. Symbols which are free in R and S but do not appear in the actual parameter list (i.e., k) are renamed. The rule's antecedent requires verification of the procedure's body once using the names of formal parameters.

A procedure's postcondition is rarely identical to the assertion which must be true after the call, since the procedure may be called from many different locations. Thus we need a rule similar to the rule of consequence to adapt the results of the procedure body to the different assertions needed to hold after invocations.

$$\text{Adaptation:} \quad \frac{\{R\}p(a)\{S\}}{\{\exists k\ (R\ \wedge\ \forall a(S \to T))\}p(a)\{T\}}$$

In this rule, the names of actual parameters have a different meaning in R than they do in S and T. The name of a parameter in R represents a value before the call, but the same name in S or T represents a value after the call. These values may be different because parameters are transmitted by reference and may be changed by the procedure's body. Names of actual parameters are free variables in R and universally quantified variables in S and T. Thus even if name appears in R and S or T, its meaning is different. Initial values of variables often appear in a procedure's precondition or postcondition, but not in a or T. These names are existentially quantified because some such value must exist.

By way of example, assume we have verified the body of a procedure **swap(x, y)** whose precondition is **{x = x' $\wedge$ y = y'}** and whose postcondition is **{x = y' $\wedge$ y = x'}**, and we want to verify the following code fragment.

```
{true} a = 1; b = 2; swap(a, b); {a = 2 ∧ b = 1}
```

Having verified the body of swap, we can use the rule of substitution to replace swap's formal parameters with the actual parameters of the call.

$$\frac{\{x = x'\ \wedge\ y = y'\}\ \text{swap(x, y)}\ \{x = y'\ \wedge\ y = x'\}}{\{a = x'\ \wedge\ b = y'\}\ \text{swap(a, b)}\ \{a = y'\ \wedge\ b = x'\}}$$

Substitution's consequent is the antecedent of the rule of adaptation. When we apply adaptation we need to existentially quantify **x'** and **y'** (the initial values of **a** and **b** in the precondition), and universally quantify **a** and **b** (the values of the parameters after the call).

$$\frac{\{a = x' \wedge b = y'\}\ swap(a, b)\ \{a = y' \wedge b = x'\}}{\{\exists,\ x',\ y'\ (a = x' \wedge b = y' \wedge (\forall\ a,\ b,\ (a = y' \wedge b = x') \rightarrow}$$
$$(a = 2 \wedge b = 1))\}$$
$$swap(a, b)\ \{a = 2 \wedge b = 1\}$$

We can pick values for **x′** and **y′** (e.g., **x′ = 1** and **y′ = 2**) to simplify the precondition of the adaptation rule's consequent.

$$(a = 1 \wedge b = 2 \wedge (\forall\ a,b,\ (a = 2 \wedge b = 1) \rightarrow (a = 2 \wedge b = 1)))$$

Clearly, this precondition is established by the sequence of assignment statements.

User-Defined Data Types

Modern programming languages provide special constructs such as classes to implement user-defined data types. These constructs are specifically designed to hide the representation of a value of the type from users who manipulate values of the type only through operations provided by the special constructs. Hoare [Hoare 1972] divided the verification of such programs into two parts.

1. Each operation's preconditions and postconditions are specified using values and operations from well-defined mathematical domains (e.g., sets or lists), and user-level code is verified with these assertions.
2. A representation mapping is defined to relate implementation-level values (e.g., arrays or linked representations) to user-level values. User-level variables in preconditions and postconditions are replaced by the corresponding mapped implementation-level variables, and the implementations of the operations are verified using the techniques described in the previous section.

Guttag et al. [1985] replaced model-oriented, user-level specifications with property-oriented specifications. Property-oriented specifications describe aspects of values in terms of properties they possess. In this approach, called algebraic specification, properties of operations of user-defined types are defined in terms of how they interact with each other.

Algebraic Specifications. Algebraic specifications have syntactic and semantic parts. The syntactic description, often referred to as the type's signature, describes the domains and ranges of the type's operations. For example, some operations on objects of type "stack of integer" are listed below.

```
estack   → Stack
push     Stack × integer → Stack
pop      Stack → Stack
top      Stack → natural
empty    Stack → Boolean
depth    Stack → natural
=        Stack × Stack → Boolean
```

Axioms describe the meanings of operators in terms of how they interact with one another. Axioms appear as equations; each left side contains a composition of operators manipulating implicitly universally quantified variables, and each right side contains a description of how the composition behaves in terms of the type's operators and simple "if-then-else" expressions. The axioms for the operations of type Stack appear below.

```
1.   pop(estack)           =  estack
2.   pop(push(S, X))       =  S
3.   top(estack)           =  0
4.   top(push(S, X))       =  X
5.   empty(estack)         =  true
6.   empty(push(S, X))     =  false
7.   depth(estack)         =  0
8.   depth(push(S, X))     =  depth(S) + 1
9.   T = estack            =  depth(T) = 0
10.  T = push(S, X)       *=  top(T) = X ∧ pop(T) = S
```

where **S** and **T** are objects of types **Stack**, and **X** is an integer. Axiom 2 describes the value computed by pushing an arbitrary value on **Stack S** followed by popping the resulting **Stack** object as being equal to the original value of **S**. Pushing a value on a **Stack** object increases the depth of the object by one according to Axiom 8. Axiom 10 asserts that **Stacks T** and **push(S, X)** are equal if their respective top values (**top(T)** and **X**) and remaining values (**pop(T)** and **S**) are equal.

We can use equational reasoning, replacing a term with an equal term, to validate that the axioms behave as intended. For example, we could check that popping a nonempty **Stack** object decreases its depth by picking a particular **Stack** object (e.g., **push(estack, X)**) and reasoning equationally as follows:

Term	Axiom
(∼(push(estack, X) = estack)→(depth(pop(push(estack, X)))<depth(push(estack, X))))	
(∼(depth(push(estack, X)) = 0)→(depth(pop(push(estack, X)))<depth(push(estack, X))))	9
(∼(depth(estack) + 1 = 0)→(depth(pop(push(estack, X)))<depth(push(estack, X))))	8
(∼(0 + 1 = 0)→(depth(pop(push(estack, X)))<depth(push(estack, X))))	7
true→(depth(pop(push(estack, X)))<depth(push(estack, X))))	1≠0
true→(depth(estack)<depth(push(estack, X))))	2
true→(0<depth(push(estack, X))))	7
true→(0<depth(estack)+1)	8
true→(0<0 + 1)	0<1
true→true	

Axioms are inconsistent when an operation is overspecified. This occurs when two rules can be used to rewrite the same combination of arguments to different values. For example, if we added the following axiom:

```
top(pop(push(S, X))) = X
```

to our previous axioms we would be able to rewrite the term **top(pop(push(estack, 5)))** to two different values.

```
top(pop(push(estack, 5)))  ⇒  5                         New axiom
top(pop(push(estack, 5)))  ⇒  top(estack)  ⇒  0         Axiom 2 then 3
```

Overspecification can be detected by a superposition algorithm [Knuth and Bendix 1970] which uses unification to detect overlapping axioms which produce different results.

Axioms are incomplete when an operation is underspecified (i.e., when no rule can be used to rewrite some combination of arguments). The specification of a type is *sufficiently complete*, if it assigns a value to each term of the type [Guttag and Horning 1978]. All **Stack** values can be built by a finite number of compositions of push operations on **estack** values, since any stack either is empty or is obtainable by pushing some element on some other stack. Operations **estack** and **push** are called *constructors*, and the remaining operations are called *defined* operations.

An algorithm exists for detecting underspecified operations [Huet and Hullot 1982]. The variables on the left side of each axiom must be unique, and a recursive test ensuring that all permutations of constructors may appear in the operation's argument positions must succeed. This test would succeed for Stack's pop operation for any of the following left sides of axioms:

Left sides	Reason for success
`{pop(estack), pop(push(S, X))}`	`All constructors`
`{pop(estack), pop(S)}`	`S` represents `estack` and `push(S', X)`
`{pop(S), pop(push(S, X))}`	`S` represents `estack` and `push(S', X)`
`{pop(S)}`	`S` represents `estack` and `push(S', X)`

Of course, it is much easier to write the right sides of axioms for some of these sets of left sides than for others.

The algorithm of Huet and Hullot works as follows. A set of n-tuples is formed from the n arguments which appear in each of the operation's axioms. The set of arguments in the tuple's first positions is constructed. The test fails if the set does contain either a variable or an instance of each constructor. The n-tuple set is divided into subsets on the basis of which constructor appeared in the first position. Each of these sets is augmented by the set n-tuples with variables in the first position, since the variable could represent an instance of the particular constructor. Each tuple in the set for a constructor with p arguments is transformed as follows. If the first element of the tuple is the constructor, it is replaced by its p arguments forming a new $n - 1 + p$-tuple. If the first element of the tuple is a variable, it is replaced by p fresh variables forming a new $n - 1 + p$-tuple. The test is repeated on each new tuple subset, and succeeds for tuples of length zero.

Figure 109.9 shows the results of applying the test to the **Stack** axioms which define **pop**. The initial set of tuples is **{<estack>, <push(S, X)>}**. The set of arguments in the first positions of these tuples contains an instance of each constructor, so we continue by dividing the set and constructing new tuples. The set of new tuples formed from the tuples with **estack** in the first position is **{< >}**, while the set formed from tuples with **push** in the first position is **{<S, X>}**. The test succeeds for the zero-length tuple. For **{<S, X>}**, the set of arguments in the first position contains a variable, so we form the new set of tuples **{<X>}**. Repeating the step once more leads to a set of zero-length tuples. If one of **push**'s arguments had not been a variable (e.g., **push(estack, X)**), the algorithm would fail because we could not determine a meaning for some terms (e.g., **pop(push(push(estack), 1), 2))**).

To validate sets of axioms, we change each equation of the form $t_1 = t_2$ into a rewrite rule of the form $t_1 \Rightarrow t_2$. A rewrite rule allows the replacement of an instance of t_1 with the corresponding instance of t_2, but it forbids replacement in the opposite direction. Orienting equations transforms an algebraic

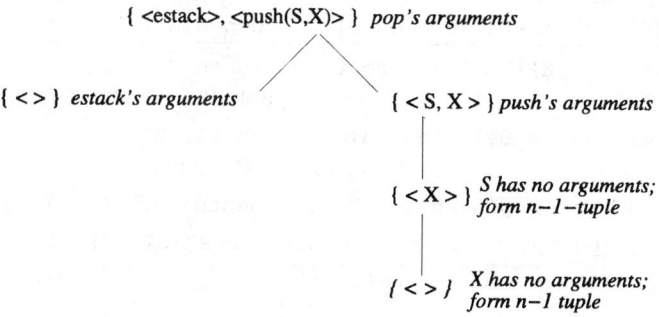

FIGURE 109.9 Verifying that **pop**'s axioms are not underspecified.

specification into a term rewriting system [Dershowitz and Jouannaud 1990] that supports automated verification and validation.

Two crucial properties must hold when equations are oriented. Two terms provably equal by equational reasoning should have a common third term to which both can be rewritten. This property is referred to as **confluence** or *Church–Rosser*. Also, there should be a finite number of rewriting steps that can be applied to a term. This property is referred to as *termination* or *Noetherianity*. To ensure the confluence of a constructor-based specification it is sufficient to avoid overspecification. To ensure sufficient completeness it is necessary, but not sufficient, to avoid underspecification. Although underspecification and overspecification can be checked, the termination of a rewrite system is undecidable [Dershowitz 1987].

Term rewriting allows us to verify that a property is true for all values of a type rather than just testing that the property holds for particular values. However, when we try to verify that popping a nonempty **Stack** object decreases its depth for an arbitrary **Stack** value **S0**, we quickly reach a point where no more rewriting can be performed.

```
(~(S0 = estack)  →  (depth(pop(S0)) < depth(S0)))        = true
(~(depth(S0) = 0))  →  (depth(pop(S0)) < depth(S0)))     = true
```

Equations that cannot be proved by just rewriting may be proved by structural induction [Burstall 1969] or data type induction [Guttag et al. 1978]. Using such techniques, inductive variables are replaced by terms derived from their type's constructors and inductive hypotheses are constructed. If F is a formula to be proved and v is the inductive variable, then for every constructor $c(s_1, \ldots, s_n)$ we prove $F[c(v_1, \ldots, v_n)/v]$, where each v_i is a distinct **Skolem constant**. If $s_i = s$, then $F[v_i/v]$ is an inductive hypothesis for the proof. Our sample proof proceeds by induction on **S0** with two cases: one with **S0** replaced by **estack** with no new inductive hypothesis since **estack** has no arguments, and another with **S0** replaced by **push(S1, X)** in which we assume the original formula with **S0** replaced by **S1** as the inductive hypothesis.

```
imply(negate(depth(S0) = 0))→(depth(pop(S0)) < depth(S0)))            =   true
```

Case S0 = estack

```
(~((depth(estack) = 0))→(depth(pop(estack)) < depth(estack)))         =   true
(~(0 = 0)  →  (depth(pop(estack)) < depth(estack)))                   =   true
(~(true)  →  (depth(pop(estack)) < depth(estack)))                    =   true
(false  →  (depth(pop(estack)) < depth(estack)))                      =   true
(false  →  (depth(estack) < depth(estack)))                          =   true
(false  →  0 < 0)                                                     =   true
(false  →  false)                                                    =   true
true                                                                  =   true
```

Case S0 = push(S1, X)

Inductive hypothesis:
```
   (~(depth(S1) = 0)→(depth(pop(S1)) < depth(S1)))                    =   true
```

```
(~((depth(push(S1, X)) = 0))→(depth(pop(push(S1, X)))
                        < depth(push(S1, X))))                        =   true
(~((succ(depth(S1)) = 0))→(depth(pop(push(S1, X)))
                        < depth(push(S1, X))))                        =   true
(~(false)  →  (depth(pop(push(S1, X))) < depth(push(S1, X))))         =   true
(true  →  (depth(pop(push(S1, X))) < depth(push(S1, X))))            =   true
(true  →  (depth(S1) < succ(depth(S1))))                             =   true
(true  →  true)                                                     =   true
true                                                                  =   true
```

Verifying User-Level Programs. Having defined type **Stack**, we can use the operations of type **Stack** in the preconditions and postconditions of the procedures of the implementation. In the example below, identifier **s** represents each operation's implicit first formal parameter of type **Stack**. The specification of **Push** states that if the value of **s** before invocation is the term **s'** and **Push** terminates normally, then the value of **s** after **Push** will be equal to the term **push(s', x)**

```
class Stack {
private:
  int* v;
  int top;
public:
  void Push(int x) {
    /* pre: s = s'; post: s = push(s', x) */
    ...
  void Pop( );
    /* pre: ~empty(s') ∧ s = s'; post: s = pop(s') */
    ...
};
```

Using the rules of inference for procedure call, we can verify the following code fragment using the stated preconditions and postconditions.

```
{s = A} s.Push(x); s.Pop(); {s = A}
```

First we use the rule of adaptation, to relate **pop**'s precondition and postcondition to the program's postcondition.

$$\frac{\{\sim empty(s') \wedge s = s'\} \; pop(s) \; \{s = pop(s')\}}{\{\exists \, s' \; (\sim empty(s') \wedge s = s' \wedge (\forall \, s, \; s = pop(s') \rightarrow s = A))\} \; pop(s) \; \{s = A\}}$$

Picking **s' = push(A,x)** permits us to begin simplifying the precondition of the adaptation rule's consequent.

Term	Axiom
(~empty(push(A, x)) ∧ s = push(A, x) ∧ (∀ s, s = pop(push(A, x)) → s = A)) ⇒	2
(~empty(push(A, x)) ∧ s = push(A, x) ∧ (∀ s, s = A → s = A)) ⇒	6
(~false ∧ s = push(A, x)) ⇒ s = push(A, x)	

Since **s = push(A, x) → (∃ s' (~empty(s') ∧ s = s' ∧ (∀s, s = pop(s') → s = A)))** we use a rule of consequence to conclude:

$$\frac{\{s = push(A, x) \rightarrow (\exists \, s' \; (\sim empty(s') \wedge s = s' \wedge (\forall s, \; s = pop(s')))), \quad (\exists \, s' \; (\sim empty(s') \wedge s = s' \wedge (\forall s, \; s = pop(s') \rightarrow s = A)))\} \; pop(s) \; \{s = A\}}{\{s = push(A, x)\} \; pop(s) \; \{s = A\}}$$

Applying the rule of adaptation to the invocation of the push operation results in the following rule of inference.

$$\frac{\{s = s'\} \; push(s, x) \; \{s = push(s', x)\}}{\{\exists \, s' \; (s = s' \wedge (\forall \, s, \; s = push(s', x) \rightarrow (s = push(A, x)))\} \; push(s, x) \; \{s = push(A, x)\}}$$

Picking **s′ = A** permits us to simplify the previous precondition to **S = A**. Using rules of consequence and composition, we conclude:

$$\frac{\{s = A\} \text{ push}(s, x) \{s = \text{push}(A, x)\}, \{s = \text{push}(A, x)\} \text{ pop}(s) \{s = A\}}{\{s = A\} \text{ push}(s, x); \text{ pop}(s) \{s = A\}}$$

Verifying Implementation-Level Programs.   In the second part of verifying implementations of user-defined data types, implementations manipulating concrete objects must satisfy preconditions and post-conditions containing terms defined by the axioms. The implementation of the **Stack** operation **Push** is shown below:

```
void Push(int x) {
    /* pre: s = s′; post: s = push(s′,x) */
    s.top = s.top + 1;
    s.v[s.top] = x;
}
```

Hoare [1972] introduced representation mappings to map implementation-level objects to their corresponding user-level objects. In the implementation of type **Stack**, an instance **s** of type **Stack** consisted of an array of integers **s.v** and an integer **s.top** indicating the topmost value. To verify the correctness of an implementation of type **Stack** we define a representation mapping $\mathcal{A}$ which maps an array and an integer to its corresponding user-level value.

1. $\mathcal{A}$(s.v, 0)   =   estack
2. $\mathcal{A}$(s.v, s.top + 1)   =   push($\mathcal{A}$(s.v, s.top), s.v[s.top + 1])

We replace instances of the user-level value **s** with corresponding instances of mapped implementation-level values. The proof obligation for **Push** is

$\{\mathcal{A}$(s.v, s.top) = s′$\}$
    s.top = s.top + 1; s.v[s.top] = x;
    $\{\mathcal{A}$(s.v, s.top) = push(s′, x)$\}$

After using axioms of assignment (for both scalar and array values) and composition, the final step in the verification is an application of a rule of consequence.

$$\frac{\mathcal{A}(s.v, s.top) = s′ \rightarrow (\mathcal{A}(\alpha(s.v, s.top + 1, x), s.top + 1) = \text{push}(s′, x)), \{\mathcal{A}(\alpha(s.v, s.top + 1, x), s.top + 1) = \text{push}(s′, x)\} \ s.top = s.top + 1; s.v[s.top] = x; \{\mathcal{A}(s.v, s.top) = \text{push}(s′, x)\}}{\{\mathcal{A}(s.v, s.top) = s′\} \ s.top = s.top + 1; s.v[s.top] = x; \{\mathcal{A}(s.v, s.top) = \text{push}(s′, x)\}}$$

To show that the antecedent is true, we need to axiomatize the array assignment and subscript operations:

1. newarray[J] = 0
2. α(A,I,X)[J] = (if I = J then X else A[J])

We continue using term rewriting:

Term	Axiom
$\mathcal{A}$(α(s.v, s.top + 1, x), s.top + 1) $\Rightarrow$	Map 2
push($\mathcal{A}$(α(s.v, s.top + 1, x), s.top),	
α(s.v, s.top + 1,x)[s.top + 1]) $\Rightarrow$	Array 2
push($\mathcal{A}$(α(s.v, s.top + 1, x), s.top),	
(if s.top + 1 = s.top + 1 then x else s.v[s.top + 1])) $\Rightarrow$	x = x
push($\mathcal{A}$(α(s.v, s.top + 1, x), s.top), x)	

At this point, we need to reduce $\mathcal{A}\,(\alpha\,(\texttt{s.v, s.top + 1, x}), \texttt{ s.top})$ to $\mathcal{A}\,(\texttt{s.v, s.top})$ to achieve equality. Since the representation mapping only maps values $\texttt{s.v[i]}$ for values of i in the range $1 \leq i \leq \texttt{s.top}$, we can reach this conclusion by proving the following theorem.

Term

```
(i < s.top + 1 → (α(s.v, s.top + 1,x)[i] = s.v[i]))
(i < s.top + 1 → (if s.top + 1 = i then x else s.v[i]) = s.v[i])
(i < s.top + 1 → (if s.top + 1 = i then x = s.v[i] else s.v[i] = s.v[i]))
(i < s.top + 1 → (if s.top + 1 = i then x = s.v[i] else true))
(i < s.top + 1 ∧ s.top + 1 = i → x = s.v[i]) ∧
                          (i < s.top + 1 ∧ ~(s.top + 1 = i) → true)
(false → x = s.v[i]) ∧ (i < s.top + 1 ∧ ~(s.top + 1 = i) → true)
true ∧ (i < s.top + 1 ∧ ~(s.top + 1 = i) → true)
true ∧ true
```

109.5 Current Status

General properties of programs, particularly that variables are initialized before they are used and that no invalid memory references or memory leaks occur, are easy to check automatically. However, verification of the most problem-specific properties is still carried out manually by inspections. As a result, unqualified guarantees about properties cannot be made because the software artifacts inspected may contain deficiencies or the inspectors may not have been thorough enough to find obscure failures.

Verifying specific properties of programs has proved too difficult a task for the average software developer. Program proofs are often more detailed than the programs being verified. When proofs are done manually, they are subject to the same human fallibilities that plague inspections. Automated support for theorem proving includes verification condition generators which apply rules of inference to produce the set of theorems which need to be proved manually by rules of consequence, proof checkers which check that steps in a proof are justified by lemmas from an existing library, and deductive systems which search for proofs by means of simplifications (like term rewriting) and heuristics for generating inductive proofs of necessary lemmas. Theorem provers eliminate errors of omission, but they require proofs of many low-level, uninteresting lemmas. However, skilled users have used theorem provers to verify complex problems such as a Byzantine fault-tolerant algorithm for synchronizing clocks in replicated computers [Rushby and Von Henke 1993]. Such proofs have generally been carried out only for safety-critical applications because even automated proofs require highly skilled experts who know how to use a theorem prover and understand the application domain.

Model checking has proven successful in the design of hardware; it has been used to find bugs in pipelined microprocessors [Burch and Dill 1994] and cache coherence protocols [Clarke et al. 1986]. More recently it has been used to analyze software artifacts, e.g., software architecture designs [Allen and Garlan 1994] and editors [Jackson and Damon 1996], and distributed file system cache coherence protocols [Wing and Vaziri-Farahani 1995]. Software developers may be more likely to understand a proof technique like model checking, which is based on search and which produces counter-examples when proofs fail, than a technique based inductive theorem proving. Model checking an abstraction of a system rather than the system itself raises the level at which we apply formal verification. The key to success in these endeavors is creating an appropriate abstraction of a system so that results obtained from analyzing the abstraction also apply to the system.

Defining Terms

Abstract interpretation: The abstract interpretation of a program is a symbolic interpretation of the program using abstract values rather than actual values. Abstract values represent sets or ranges of actual values, and operators are redefined to manipulate abstract values. At program decision points, separate copies of the program state are created, corresponding to the true and false outcomes of the predicate, and computation continues along each path with the different states. At program join points, values of the states computed along different computation paths are unioned to produce a single state. The union operation for joins must reduce the number of distinct values in the program state in order to compute a fixpoint approximation of the state.

Confluence: A term-rewriting system is confluent if a term t can be rewritten into different forms t_1 and t_2, then we can prove by rewriting that $t_1 = t_2$. For example, the term `top(push(pop(push(estack, 1)), 2)` can be rewritten as either `2` using stack axiom 4 or as `top(push(estack, 2))` using stack axiom 2. We can show these two terms are equal by applying stack axiom 2 to the later term.

Garbage collection: Programming languages that do not require users to explicitly free unneeded storage reclaim this storage through a process called garbage collection. Storage is identified as unneeded (i.e., garbage) when there are no pointers to it. Determining if pointers reference a storage location can be accomplished by reference counting (i.e., explicitly keeping track of the number of pointers to the location) or tracing (i.e., starting with variables on the stack, mark heap locations they reference, and continue marking other heap locations reachable from marked heap locations). The collector either copies reachable values to new storage locations to create a large unmarked region or consolidates adjacent unmarked regions and creates a list of free storage.

Hazardous state: A hazardous state is one in which the occurrence of certain events would lead to a mishap.

Least fixpoint: A fixpoint of a function $f : T \to T$ is an element $t \in T$ such that $f(t) = t$. A least fixpoint of f is the least element of the set of all fixpoints of f.

Memory leak: A memory leak occurs when storage is allocated but not freed before the last reference to it is lost. Memory leaks may cause long-running programs to crash when storage allocation requests fail because of insufficient memory. The storage allocation request triggering the program crash may not be part of a memory leak, so identifying the cause of a memory leak is difficult.

Partial correctness: A program which meets its specification for all specified input values for which it terminates is called partially correct. A program which is partially correct and which terminates for all its inputs is said to be totally correct.

Reachability analysis: In reachability analysis, a graph representing the states and state-transitions of a system is constructed and exhaustively searched to determine if states with particular properties are reachable.

Skolem constant: An automated theorem prover simplifies a formula by replacing a universally quantified variable with a symbolic constant (called a Skolem constant) which represents an arbitrary value of the same type as the variable. For example, if x is a natural number, to prove $(\forall x, x > 0)$, picking a particular natural number to substitute for x might make the formula either true or false. Instead, we pick an arbitrary constant c and try to prove the quantifier-free formula $c > 0$.

Software inspections: Software artifacts are usually verified by people using informal analysis techniques called *inspections*. In inspections, teams of software developers either follow sequences of state changes or possible execution paths resulting from a particular series of events or inputs, or, using a checklist of potential errors, determine if similar errors are present in the artifact.

Temporal logic: A temporal logic is a propositional logic with additional temporal operators to express concepts such as a formula will always be true in the future, or a formula will eventually be true in the future. The value of a temporal logic formula is defined with respect to a finite-state model. If formula f is true in state s of model M, we write $M, s \models f$. A formula f is true for the model if it is true in the model's start state. Temporal logic allows reasoning about state changes rather than just the function computed by the program.

Termination: To demonstrate that the while statement "while B do S" terminates, we show that B must eventually evaluate to false. To do this, we derive an expression from B whose value is bounded below by 0, and we show that on each path through S the value of the expression decreases.

References

Allen, R. and Garlan, D. 1994. Formalizing architectural connection, pp. 71–80. In *Proc. 16th Int. Conf. Software Eng.*

Burch, J. R. and Dill, D. L. 1994. Automatic verification of pipelined microprocessor control. In *Lecture Notes in Computer Science* 818, D. Dill, ed., pp. 68–80. Springer–Verlag.

Burstall, R. 1969. Proving properties of programs by structural induction. *Comput. J.* 12(1):41–48.

Clarke, E., Emerson, E., and Sistla, A. 1986. Automatic verification of finite state concurrent systems using temporal logic specifications. *ACM Trans. Program. Lang. Syst.* 8(2):244–263.

Clarke, E. M., Grumberg, O., Hiraishi, H., Jha, S., Long, D. E., McMillan, K. L., and Ness, L. A. 1990. Verification of the futurebus+ cache coherence protocol, pp. 15–30. In *Proc. 11th Int. Symp. Comput. Hardware Description Lang. Appl.* L. Claesen, ed., North-Holland, Amsterdam.

Cousot, P. and Cousot, R. 1976. Static determination of dynamic properties of programs. In *Proc. "Colloque sur la Programmation."*

Cousot, P. and Cousot, R. 1977. Static determination of dynamic properties of generalized type unions. *SIGPLAN Notices* 12(3):77–94.

Dershowitz, N. 1987. Termination of rewriting. *J. Symb. Comput.* 3:69–116.

Dershowitz, N. and Jouannaud, J. 1990. Rewrite systems. In *Handbook of Theoretical Computer Science B: Formal Methods and Semantics*, J. van Leeuwen, ed. Ch. 6, pp. 243–320. North Holland, Amsterdam.

Fagan, M. E. 1976. Design and code inspections to reduce errors in program development. *IBM Syst. J.* 15(3):182–211.

Floyd, R. W. 1967. Assigning meaning to programs. *Symp. Appl. Math.* 19:19–32.

Ghezzi, C., Jazayeri, M., and Mandrioli, D. 1991. *Fundamentals of Software Engineering.* Prentice–Hall, Englewood Cliffs, NJ.

Guttag, J. V. and Horning, J. J. 1978. The algebraic specification of abstract data types. *Acta Informatica* 10:27–52.

Guttag, J. V., Horning, J. J. and Wing, J. M. 1985. The Larch family of specification languages. *IEEE Software* 2(5):24–36.

Guttag, J. V., Horowitz, E., and Musser, D. 1978. Abstract data types and software validation. *Commun. ACM* 21:1048–1064.

Harel, D., Lachover, H., Naamad, A., Pnueli, A., Politi, M., Sherman, R., Shtull-Trauring, A., and Trakhtenbrot, M. 1990. Statemate: a working environment for the development of complex reactive systems. *IEEE Trans. Software Eng.* 16(4):403–414.

Hastings, R. and Joyce, R. 1992. Purify: fast detection of memory leaks and access errors. In *Proc. Winter 1992 USENIX Conf.*, pp. 125–136.

Heitmeyer, C., Labaw, B., and Kiskis, D. 1995. Consistency checking of scr-style requirements specifications. In *Proc. RE' 95 Int. Symp. Req. Eng.*

Heninger, K. 1980. Specifying software requirements for complex systems: new techniques and their applications. *IEEE Trans. Software Eng.* SE-6(1):2–12.

Hoare, C. A. R. 1971. Procedures and parameters, an axiomatic approach. In *Symposium on the Semantics of Algorithmic Languages*, E. Engler, ed., pp. 102–116. Springer–Verlag.

Hoare, C. A. R. Proof of correctness of data representations. *Acta Inf.* 1(4):271–281.

Huet, G. and Hullot, J.-M. 1982. Proofs by induction in equational theories with constructors. *JCSS* 25(1):239–266.

Jackson, D. and Damon, C. A. 1996. Elements of style: analyzing a software design feature with a counterexample detector, pp. 239–249. In *Proc. 1996 Int. Symp. Software Test. and Anal. (ISSTA)*.

Knuth, D. E. and Bendix, P. B. 1970. *Simple Word Problems in Universal Algebras*, pp. 263–297. Pergamon, Oxford, U.K.

Leveson, N. G., Heimdahl, M. P. E., Hildreth, H., and Reese, J. D. 1994. Requirements specification for process-control systems. *IEEE Trans. Software Eng.* 20(9):684–706.

McMillan, K. L. 1993. *Symbolic Model Checking*. Klewer Academic Publishers, Boston, MA.

Pneuli, A. 1981. A temporal logic of concurrent programs. *Theor. Comput. Sci.* 13:45–60.

Quielle, J. P. and Sifakis, J. 1981. Specification and verification of concurrent systems in cesar. In *Proc. 5th Int. Symp. Program.*

Ratan, V., Partridge, K., Reese, J., and Leveson, N. 1996. Safety analysis tools for requirements specifications. In *Proc. 11th Conf. Comput. Assurance*.

Rushby, J. M. and von Henke, F. 1993. Formal verification of algorithms for critical systems. *IEEE Trans. Software Eng.* 19(1):13–23.

Rushby, J. 1993. *Formal Methods and the Certification of Critical Systems*. SRI International, Palo Alto, CA. Tech Rep.

Wing, J. and Vaziri-Farahani, M. 1995. Model checking software systems: a case study, pp. 128–139. In *Proc. 3rd Symp. Found. Software Eng.*

Young, M. and Taylor, R. N. 1989. Rethinking the taxonomy of fault detection techniques, pp. 53–62. In *Proc. 11th Int. Conf. Software Eng.*

Further Information

The monthly journals *IEEE Transactions on Software Engineering* and *ACM Transactions on Software Engineering and Methodology* contain articles on software verification. Papers on this topic are frequently presented at the International Conference on Software Engineering, the ACM's Foundations of Software Engineering, the ACM's International Conference on Software Analysis and Testing, and the IEEE Conference on COMPuter ASSurance (COMPASS).

Two recent books on the analysis of software systems are

Leveson, N. G. 1995. *Safeware: System Safety and Computers*. Addison–Wesley, Reading, MA.

Rushby, J. 1995. *Formal Methods and the Certification of Critical Systems*. Cambridge University Press, Cambridge, U.K.

Readers interested in automated analysis tools should investigate both automated theorem provers and model checkers. Representative theorem provers include the following:

PVS (Prototype Verification System) is a theorem prover based on classical typed higher-order logic developed at the SRI International Computer Science Laboratory.

EVES unites the Verdi specification language based on set theory and an automated deduction system, called NEVER. This system is available from Mark Saaltink and Dan Craigen of ORA, Ottawa, Ontario, Canada.

LP, the Larch Prover, is an interactive theorem proving system for multisorted first-order logic. It was developed by Stephen Garland and John Guttag at the MIT Laboratory for Computer Science, Cambridge, MA.

Interesting model checkers include:

HyTech (The Cornell HYbrid TECHnology Tool) computes the condition under which a linear hybrid system satisfies a temporal requirement. Hybrid systems are specified as collections of automata with discrete and continuous components.

The SMV (Symbolic Model Verifier) model checker verifies formulas written in a propositional branching-time temporal logic. It is available from Ed Clarke at Carnegie Mellon University, Pittsburgh, PA.

Murphi is a symbolic model checker developed by David Dill at Stanford University, Stanford, CA.

110

Testing: Principles and Practice

Stephen R. Schach*
Vanderbilt University

110.1 Introduction

Software organizations all too frequently have a separate testing phase, after integration and before maintenance. Nothing could be more dangerous from the viewpoint of trying to achieve high-quality software. On the contrary, testing is an integral component of the software process, and an activity that must be carried out throughout the life cycle. During the requirements phase, the requirements must be checked; during the specification phase the specifications must be checked. The design phase requires careful checking at every stage. During the coding phase each module must be tested, and the product as a whole needs testing at the integration phase. After passing the acceptance test, the product is installed and goes into operational mode, and maintenance begins. And hand in hand with maintenance goes repeated testing of modified versions of the product.

*This chapter is based on material taken from Stephen R. Schach, *Classical and Object-Oriented Software Engineering*, 3rd ed. Richard D. Irwin, Chicago. © 1996, pp. 109–133 and 405–420.

There are two types of testing: **execution-based testing** and **nonexecution-based testing**. For example, it is impossible to execute a written specification document; the only alternatives are to review it as carefully as possible or to subject it to some form of analysis. However, once there is executable code, it becomes possible to run test cases, that is, to perform execution-based testing. Nevertheless, the existence of code does not preclude nonexecution-based testing; carefully reviewing code will uncover at least as many faults as running test cases. In this chapter principles and practice of both execution-based and nonexecution-based testing are described.

110.2 Nonexecution-Based Testing

It is not a good idea for the person responsible for drawing up a document to be the only one responsible for reviewing it. Almost everyone has blind spots that allow faults to creep into the document, and those same blind spots prevent the faults from being detected on review. Thus, the review task must be assigned to someone other than the original author of the document. In addition, having only one reviewer may not be adequate; we have all had the experience of reading through a document many times, but failing to detect a blatant spelling error that a second reader picks up almost immediately. This is one of the principles underlying review techniques such as inspections in which a document (such as a specification document or design document) is carefully checked by a team of software professionals with a broad range of skills. The advantage of a review by a team of experts is that the different skills of the participants increase the chances of finding a fault. In addition, a team of skilled individuals working together often generate a synergistic effect.

Inspections

Inspections were first proposed by Fagan for testing designs and code [Fagan 1976]. An inspection has five formal steps. First, an *overview* of the document is given by one of the individuals responsible for producing that document. At the end of the overview session, the document is distributed to the participants. In the second, *preparation*, the participants try to understand the document in detail. Lists of fault types found in recent inspections, with the fault types ranked by frequency, are excellent aids. These lists help team members to concentrate on those areas where most faults have occurred. The third step is the *inspection*. To begin, one participant walks through the document with the inspection team, ensuring that every item is covered, and every branch is taken at least once. Then fault finding commences, with the participants presenting their checklists. The purpose is to find and document the faults, not to correct them. Within one day the leader of the inspection team (the *moderator*) must produce a written report of the inspection, to ensure meticulous follow-through. The fourth stage is the *rework*, in which the individual responsible for that document resolves all faults and problems noted in the written report. The final stage is the *follow-up*. The moderator must ensure that every single issue raised has been satisfactorily resolved, either by fixing the document or by clarifying items that were incorrectly flagged as faults. All fixes must be checked to ensure that no new faults have been introduced [Fagan 1986]. If more than 5% of the material inspected has been reworked, then the team reconvenes for a 100% reinspection.

The inspection should be conducted by a team of four. For example, in the case of a design inspection, the team consists of a moderator, designer, implementer, and tester. The moderator is both manager and leader of the inspection team. There must be a representative of the team responsible for the current phase, as well as a representative of the team responsible for the next phase. The designer is a member of the team that produced the design, whereas the implementer is responsible, either individually or as part of a team, for translating the design into code. Fagan suggests that the tester be any programmer responsible for setting up test cases; it is of course preferable that the tester be a member of the software quality assurance (SQA) group. The IEEE standard recommends a team of between three and six participants [IEEE 1028 1988]. Special roles are played by the *moderator*, the *reader* who leads the team through the design, and the *recorder* who is responsible for producing a written report of the detected faults.

An essential component of an inspection is the checklist of potential faults. For example, the checklist for a design inspection should include items such as: Is each item of the specification document adequately and correctly addressed? For each interface, do the actual and formal arguments correspond? Have error-handling mechanisms been adequately identified? Is the design compatible with the hardware resources, or does it require more hardware than is actually available? Is the design compatible with the software resources—for example, does the operating system stipulated in the specification document have the functionality required by the design?

An important component of the inspection procedure is the record of fault statistics. Faults must be recorded by severity (major or minor—an example of a major fault is one that causes premature termination or damages a database) and by fault type. In the case of a design inspection, typical fault types include interface faults and logic faults. This information can be used in a number of useful ways. First, the number of faults in a given product can be compared with averages of faults detected at the same stage of development in comparable products, giving management an early warning that something is amiss and allowing timely corrective action to be taken. Second, if inspecting the design of two or three modules results in the discovery of a disproportionate number of faults of a particular type, management can begin checking the other modules and take corrective action. Third, if the inspection of the design of a particular module reveals far more faults than were found in any other module in the product, then there is usually a strong case for re-designing that module from scratch. Finally, information regarding the number and types of faults detected at a design inspection will aid the team performing the code inspection of the same module at a later stage.

Fagan's first experiment was performed on a systems product [Fagan 1976]. One hundred person-hours were devoted to inspections, at a rate of two 2-hour inspections per day by a four-person team. Of all the faults that were found during the development of the product, 67% were located by inspections before module testing was started. Furthermore, during the first 7 months of operation, 38% fewer faults were detected in the inspected product than in a comparable product reviewed using informal inspections.

Fagan conducted another experiment, on an applications product [Fagan 1976], and found that 82% of all detected faults were discovered during design and code inspections. A useful side effect of the inspections was that programmer productivity rose because less time had to be spent on module testing. Using an automated estimating model, Fagan determined that as a result of the inspection process, the savings on programmer resources was 25% despite the fact that time had to be devoted to the inspections. In a different experiment Jones found that over 70% of detected faults were discovered by conducting design and code inspections [Jones 1978].

More recent studies produced equally impressive results. In a 6000-line business data processing application, 93% of all detected faults were found during inspections [Fagan 1986]. As reported in [Ackerman et al. 1989], the use of inspections rather than testing during the development of an operating system decreased the cost of detecting a fault by 85%. At the Jet Propulsion Laboratory (JPL), on average each 2-hour inspection has exposed 4 major faults and 14 minor faults [Bush 1990]. Translated into dollar terms, this means a saving of approximately $25,000 *per inspection*. Another JPL study [Kelly et al. 1992] has shown that the number of faults detected decreases exponentially by phase. In other words, with the aid of inspections, faults can be detected early in the software process, thereby saving both time and money.

Inspections are productive only if adequately managed, using appropriate metrics to measure the process. For reasons of space, material on managing inspections and metrics for inspections is omitted here; the reader should consult [Schach 1996] for details.

We turn now to execution-based testing, that is, running test cases.

110.3 Execution-Based Testing

It has been claimed that testing is a demonstration that faults are not present. (A **fault** is the IEEE standard terminology for what is popularly called a **bug**, whereas a **failure** is the incorrect behavior of the product as a consequence of the fault [IEEE 610.12 1990]. Finally, an **error** is the mistake made by the programmer.)

Despite the fact that some organizations spend up to 50% of their software budget on testing, delivered *tested* software is notoriously unreliable.

The reason for this contradiction is simple. As Dijkstra put it, "Program testing can be a very effective way to show the presence of bugs, but it is hopelessly inadequate for showing their absence" [Dijkstra 1972]. What Dijkstra is saying is that if a product is executed with test data and the output is wrong, then the product definitely contains a fault. But if the output is correct, then there still may be a fault in the product; all that particular test has shown is that the product runs correctly on that particular set of test data.

What Should Be Tested?

To be able to describe what properties should be tested, it is first necessary to give a precise description of execution-based testing. According to Goodenough, execution-based testing is a process of inferring certain behavioral properties of a product on the basis, in part, of the results of executing the product in a known environment with selected inputs [Goodenough 1979].

This definition has three troubling implications. First, the definition states that testing is an inferential process. The tester takes the product, runs it with known input data, and examines the output. The tester has to infer what, if anything, is wrong with the product. From this viewpoint, testing is comparable to trying to find the proverbial black cat in a dark room, but without knowing whether or not there is a cat in the room in the first place. The tester has few clues to help find any faults, perhaps 10 or 20 sets of inputs and corresponding outputs, possibly a user fault report, and thousands of lines of code. From this the tester has to deduce if there is a fault, and if so, what it is.

A second problem with the definition arises from the phrase "in a known environment." We can never really know our environment, either the hardware or the software. We can never be certain that the operating system is functioning correctly or that the run-time routines are correct. There may be an intermittent hardware fault in the main memory of the computer. So what is observed as the behavior of the product may in fact be a correct product interacting with a faulty compiler or faulty hardware or some other faulty component of the environment.

The third worrisome part of the definition is the phrase "with selected inputs." In the case of a real-time system, there is frequently no control over the inputs to the system. Consider avionics software. The flight control system has two types of inputs. The first type of input is what the pilot wants the aircraft to do. Thus, if the pilot pulls back on the joystick to climb or opens the throttle to increase the speed of the aircraft, these mechanical motions are transformed into digital signals that are sent to the flight control computer. The second type of input is the current physical state of the aircraft, such as its altitude, speed, and the elevations of the wing flaps. The flight control software uses the values of such quantities to compute what signals should be sent to the components of the aircraft such as the wing flaps and the engines in order to implement the pilot's directives. Whereas the pilot's inputs can easily be set to any desired values simply by setting the aircraft's controls appropriately, the inputs corresponding to the current physical state of the aircraft cannot be so easily manipulated. In fact, there is no way that one can force the aircraft to provide "selected inputs." How then can such a real-time system be tested? The answer is to use a simulator.

A simulator is a working model of the environment in which the product, in this case the flight control software, executes. The flight control software can be tested by causing the simulator to send selected inputs to the flight control software. The simulator has controls that allow the operator to set an input variable to any selected value. Thus, if the purpose of the test is to determine how the flight control software performs if one engine catches fire, then the controls of the simulator are set so that the inputs sent to the flight control software are indistinguishable from what the inputs would be if an engine of the actual aircraft were on fire. The output is analyzed by examining the output signals sent from the flight control software to the simulator. But a simulator can at best be a good approximation of a faithful model of some aspect of the system; it can never be the system itself. Using a simulator means that whereas there is indeed a "known environment," there is little likelihood that this known environment is in every way identical to the actual environment in which the product will be installed.

The preceding definition of testing speaks of "behavioral properties." What behavioral properties must be tested? An obvious answer is: Test whether the product functions correctly. But as will be shown, correctness alone is not sufficient. Before discussing correctness, four other behavioral properties will be considered, namely, utility, reliability, robustness, and performance [Goodenough 1979].

Utility

Utility is the extent to which a user's needs are met when a correct product is used under conditions permitted by its specifications. In other words, a product that is functioning correctly is now subjected to inputs that are valid in terms of the specifications. The user may test, for example, how easy the product is to use, whether the product performs useful functions, and whether the product is cost-effective compared with competing products. Irrespective of whether the product is correct or not, these are vital issues that have to be tested. If the product is not cost-effective, then there is no point in buying it. And unless the product is easy to use, it will not be used at all or it will be used incorrectly. Thus, when considering buying an existing product (including shrink-wrapped software), the utility of the product should be tested first, and if the product fails on that score, then further testing should stop.

Reliability

Another aspect of a product that must be tested is its reliability. **Reliability** is a measure of the frequency and criticality of product failure; recall that a failure is an unacceptable effect or behavior, under permissible operating conditions, that occurs as a consequence of a fault. In other words, it is necessary to know how often the product fails (**mean time between failures**) and how bad the effects of that failure can be. When a product fails, an important issue is how long it takes, on average, to repair it (**mean time to repair**). But often more important is how long it takes to repair the *results* of the failure. This last point is frequently overlooked. Suppose that the software running on a communications front end fails, on average, only once every 6 months, but when it fails it completely wipes out a database. At best the database can be reinitialized to its status when the last checkpoint dump was taken, and the audit trail can then be used to put the database into a state that is virtually up to date. But if this recovery process takes the better part of 2 days, during which time the database and communications front end are inoperative, then the reliability of the product is low, notwithstanding the fact that the mean time between failures is 6 months.

Robustness

Another aspect of every product that requires testing is its **robustness**. Although it is difficult to come up with a precise definition, robustness is essentially a function of a number of factors such as the range of operating conditions, the possibility of unacceptable results with valid input, and the acceptability of effects when the product is given invalid input. A product with a wide range of permissible operating conditions is more robust than a product that is more restrictive. A product should not yield unacceptable results when the input satisfies its specifications; for example, giving a valid command should not have disastrous consequences. A robust product should not crash when the product is *not* used under permissible operating conditions. To test for this aspect of robustness, test data that do not satisfy the input specifications are deliberately input, and the tester determines how badly the product reacts. For example, when the product solicits a name, the tester may reply with a stream of unacceptable characters such as `control-A escape-% ?$#@`. If the computer responds with a message such as `Incorrect data-Try again`, or better, informs the user as to why the data do not conform to what was expected, then it is more robust than a product that crashes whenever the data deviate even slightly from what is required.

Performance

Performance is another aspect of the product that must be tested. For example, it is essential to know the extent to which the product meets its constraints with regard to response time or space requirements. For the on-board computer in a handheld antiaircraft missile, the space constraints of the system may allow only 128 kilobytes (Kb) of main memory to be available for the software. No matter how excellent the software may be, if it needs 256 Kb of main memory then it cannot be used.

Real-time software is characterized by hard time constraints, that is, time constraints of such a nature that if a constraint is not met, information is lost. For example, a nuclear reactor control system may have to sample the temperature of the core and process the data every tenth of a second. If the system is not fast enough to be able to handle interrupts from the temperature sensor every tenth of a second, then data will be lost, and there is no way of ever recovering the data; the next time that the system receives temperature data it will be the current temperature, not the reading that was missed. If the reactor is on the point of a meltdown, then it is critical that all relevant information be both received and processed as laid down in the specifications. With all real-time systems, the performance must meet every time constraint listed in the specifications.

110.4 Correctness

Finally, a definition of **correctness** can be given. A product is correct if it satisfies its output specifications, independent of its use of computing resources, when operated under permitted conditions [Goodenough 1979]. In other words, if input that satisfies the input specifications is provided and the product is given all the resources it needs, then the product is correct if the output satisfies the output specifications.

This definition of correctness, like the definition of testing itself, has worrisome implications. Suppose that a product has been successfully tested against a broad variety of test data. Does this mean that the product is acceptable? Unfortunately, it does not. If a product is correct, all that it means is that it satisfies its specifications. But what if the specifications themselves are incorrect? To illustrate this difficulty, consider the specification shown in Fig. 110.1. The specification states that the input to the sort is an array **p** of **n** integers, whereas the output is another array **q** which is sorted in nondecreasing order. Superficially, the specification seems to be perfectly correct. But consider function **trick_sort** shown in Fig. 110.2. In that function, all **n** elements of array **q** are set to **0**. The function satisfies the specification of Fig. 110.1, and is therefore correct!

What has happened? Unfortunately, the specification of Fig. 110.1 is wrong. What has been omitted is a statement that the elements of **q**, the output array, are a permutation (rearrangement) of the elements of the input array **p**. An intrinsic aspect of sorting is that it is a rearrangement process. And the function of Fig. 110.2 capitalizes on this specification fault. In other words, function **trick_sort** is correct, but the specification of Fig. 110.1 is wrong.

From the preceding example, it is clear that the consequences of specification faults are nontrivial. After all, the correctness of a product is meaningless if its specifications are incorrect. In other words, there is more to testing than just showing that the product is correct.

Testing versus Correctness Proofs

With all the difficulties associated with execution-based testing, computer scientists have tried to come up with other ways of ensuring that the product does what it is supposed to do. One such nonexecution-based alternative that has received considerable attention for more than 25 years is correctness proving.

A correctness proof is a mathematical technique for showing that a product is correct, that is, that it satisfies its specifications. The technique is sometimes termed **verification**. However, the term *verification*

```
Input specification:    p : array of n integers, n > 0.

Output specification:   q : array of n integers such that
                            q[0] ≤ q[1] ≤ ... ≤ q[n - 1]
```

FIGURE 110.1   Incorrect specification for a sort.

```
void trick_sort (int p[], int q[])
{
    for (int i = 0; i < n; i++)
        q[i] = 0;
}
```

FIGURE 110.2 Function **trick_sort** which satisfies specification of Fig. 110.1.

is often used to denote all nonexecution-based techniques, and not only correctness proving. For clarity, this mathematical procedure will be termed **correctness proving**, to remind the reader that what is involved is a mathematical proof process.

An important aspect of correctness proofs is that they should be done in conjunction with design and coding. As Dijkstra put it, "The programmer should let the program proof and program grow hand in hand" [Dijkstra 1972]. For example, when a loop is incorporated into the design a loop invariant is put forward, and as the design is refined, so is the invariant. Developing a product in this way gives the programmer confidence that the product is correct and tends to reduce the number of faults. Quoting Dijkstra again, "The only effective way to raise the confidence level of a program significantly is to give a convincing proof of its correctness" [Dijkstra 1972]. But even if a product is proved to be correct, it must be thoroughly tested as well. To illustrate the necessity for testing in conjunction with correctness proving, consider the following.

In 1969, Naur published a paper on a technique for constructing a product and proving it correct [Naur 1969]. The technique was illustrated by what Naur termed a "line-editing problem," but today would be considered a text-processing problem. It may be stated as follows:

Given a text consisting of words separated by **blank** or **newline** (new line) characters, convert it to line-by-line form in accordance with the following rules:

1. line breaks must be made only where the given text contains a **blank** or **newline**,
2. each line is filled as far as possible, and
3. no line will contain more than **maxpos** characters.

Naur constructed a procedure using his technique, and informally proved its correctness. The procedure consisted of approximately 25 lines of ALGOL code. The paper was then reviewed by Leavenworth in *Computing Reviews* [Leavenworth 1970]. The reviewer pointed out that in the output of Naur's procedure, the first word of the first line is preceded by a blank unless the first word is exactly **maxpos** characters long. Although this may seem a trivial fault, it is a fault that would surely have been detected had the procedure been tested, that is, executed with test data, rather than only proved correct. But worse was to come. London detected three additional faults in Naur's procedure [London 1971]. One is that the procedure did not terminate unless a word longer than **maxpos** characters was encountered. Again, this fault was likely to have been detected if the procedure had been tested. London then presented a corrected version of the procedure and proved formally that the resulting procedure was correct; recall that Naur had used only informal proof techniques.

The next episode in this saga is that Goodenough and Gerhart found three faults that London had not detected, despite his formal proof [Goodenough and Gerhart 1975]. These included the fact that the last word will not be output unless it is followed by a **blank** or **newline**. Yet again, reasonable choice of test data would have detected this fault without much difficulty. In fact, of the total of seven faults collectively detected by Leavenworth, London, and Goodenough and Gerhart, four could have been detected simply by running the procedure on test data, such as the illustrations given in Naur's original paper. The lesson from this saga is clear. Even if a product has been proved to be correct, it must still be tested thoroughly. Furthermore, the case study in this section showed that it is a difficult and error-prone process, even for a 25-line procedure. Is correctness proving an interesting research idea, or is it a powerful software engineering technique whose time has come? This is addressed in the next section.

Correctness Proofs and Software Engineering

A number of software engineering practitioners have put forward reasons why correctness proving should not be viewed as a standard software engineering technique. First, it is claimed that software engineers do not have adequate mathematical training. Second, it is suggested that proving is too expensive to be practical. Third, it is said that proving is too hard. Each of these reasons will be shown to be an oversimplification.

Proofs are usually expressed in first- or second-order predicate calculus, or the equivalent. Not only does this make the proof process simpler for a person, but it allows correctness proving to be done by a computer. To complicate matters further, predicate calculus is now somewhat outdated. To prove the correctness of concurrent software products, techniques using temporal or other modal logics are required [Manna and Pnueli 1992]. There is no doubt that correctness proving requires training in mathematical logic. Fortunately, most computer science majors today either take courses in the requisite material or have the background to learn correctness-proving techniques on the job. Thus colleges are now turning out computer science graduates with sufficient mathematical skills for correctness proving. The claim that practicing software engineers do not have the necessary mathematical training may have been true in the past. However, this claim no longer holds because thousands of computer science majors enter the field each year.

The claim that proving is too expensive for use in software development is also false. On the contrary, the economic viability of correctness proving can be determined on a project-by-project basis using cost–benefit analysis. For example, consider the software for the NASA space station. Human lives are at stake, and if something goes wrong a space shuttle rescue mission may not arrive in time. The cost of proving life-critical space station software correct is large. But the potential cost of a software fault that might be overlooked if correctness proving is not performed is even larger.

The third claim is that correctness proving is too hard. Despite this claim, many nontrivial products have been successfully proved to be correct, including operating system kernels, compilers, and communications systems [Berry and Wing 1985]. Furthermore, many tools such as theorem provers exist to assist in correctness proving. A theorem prover takes as input a product, its input and output specifications, and loop invariants. The theorem prover then attempts to prove mathematically that the product, when given input data satisfying the input specifications, will produce output data satisfying the output specifications.

At the same time, there are some difficulties with correctness proving. For example, how can we be sure that a theorem prover is correct? If the theorem prover prints out **This product is correct**, can we believe it? To take an extreme case, consider the so-called theorem prover shown in Fig. 110.3. No matter what code is submitted to this theorem prover it will print out **This product is correct**. In other words, what reliability can be placed on the output of a theorem prover? One suggestion is to submit a theorem prover to itself and see whether it is correct. Apart from the philosophical implications, a simple way of seeing that this will not work is to consider what would happen if the theorem prover of Fig. 110.3 were submitted to itself for proving. It will, as always, print out **This product is correct**, thereby "proving" its own correctness.

A further difficulty is finding the input and output specifications, and especially the loop invariants, or their equivalents in other logics such as modal logic. Suppose that a product is correct. Unless a suitable invariant for each loop can be found, there is no way of proving the product correct. There are computer-aided software engineering (CASE) tools that can assist in this task. But even with state-of-the-art tools, a software engineer may simply not be able to come up with a correctness proof. One solution to this problem is to develop the product and proof in parallel, as advocated in the previous section. When a loop is designed, an invariant for that loop is specified at the same time. With this approach, it is somewhat easier to prove the code modules correct.

Worse than not being able to find loop invariants, what if the specifications themselves are incorrect? An example of this is function **trick_sort** (Fig. 110.2). A good theorem prover, when given the incorrect specifications of Fig. 110.1, will undoubtedly declare that the function shown in Fig. 110.2 is correct.

```
void theorem_prover (void)
{
    printf ("This product is correct");
}
```

FIGURE 110.3   "Theorem prover."

Manna and Waldinger have stated that "We can never be sure that the specifications are correct" and "We can never be certain that a verification system is correct" [Manna and Waldinger 1978]. These statements from two of the leading experts in the field encapsulate the various points made previously.

Does all this mean that there is no place for correctness proofs in software engineering? Quite the contrary. Proving products correct is an important, and sometimes vital, software engineering tool. Proofs are appropriate where human lives are at stake or where otherwise indicated by cost–benefit analysis: If the cost of proving software correct is less than the probable cost if the product fails, then the product should be proved. However, as the text-processing case study shows, proving alone is not enough. Instead, correctness proving should be viewed as an important component of the set of techniques that must be utilized together to check that a product is correct. Because the aim of software engineering is the production of quality software, correctness proving is indeed an important software engineering technique.

Even when a full formal proof is not justified, the quality of software can be markedly improved through the use of informal proofs. A second way of improving software quality is to insert assertions into the code. Then, if at execution time an assertion does not hold, the product will be halted and the software team can investigate whether the assertion that terminated execution is incorrect, or whether there is indeed a fault in the code that was detected by triggering the assertion. Languages such as Eiffel [Meyer 1992] support assertions directly by means of an **assert** statement. Suppose that an informal proof requires that the value of variable **xxx** be positive at a particular point in the code. Even though the design team may be convinced that there is no way for **xxx** to be negative, for additional reliability they may specify that the statement **assert(xxx > 0)** appear at that point in the code. If **xxx** is less than or equal to zero, execution will terminate, and the situation can then be investigated by the software team. (Unfortunately, **Assert** in C++ is a debugging statement, similar to **assert** in C; it is not part of the language itself.)

Once the users are confident that the product is working correctly, they have the option of switching off assertion checking. This will speed up execution, but if there is a fault that would have been detected by an assertion, it may not be found if assertion checking is switched off. Thus there is a trade-off between run-time efficiency and continuing assertion checking even after the product is in operational mode.

Who Should Perform Execution-Based Testing?

A fundamental issue in execution-based testing is which members of the software development team should be responsible for carrying it out. Suppose a programmer is asked to test a module that he or she has written. Testing has been described by Myers as the process of executing a product with the intention of finding faults [Myers 1979]. Testing is thus a destructive process. On the other hand, the programmer who is doing the testing will ordinarily not wish to destroy his or her work. If the fundamental attitude of the programmer toward the code is the usual protective one, then the chances of that programmer using test data that will highlight faults is considerably lower than if the major motivation were truly destructive.

A successful test is one that finds faults. This, too, poses a difficulty. It means that if the module passes the test, then the test has failed. Conversely, if the module does not perform according to specifications, then the test succeeds. When a programmer is asked to test a module he or she has written, the programmer is being asked to execute the module in such a way that a failure (incorrect behavior) ensues. This goes against the creative instincts of programmers.

An inescapable conclusion of this is that programmers should not test their own modules. After a programmer has been *con*structive and has built a module, testing that module requires the programmer to perform a *de*structive act and to attempt to destroy the very thing that he or she has created. A second reason why execution-based testing should be done by someone else is that the programmer may have misunderstood some aspect of the design or specifications. If testing is done by someone else, such faults may be discovered. Nevertheless, debugging (finding the cause of the failure and correcting the fault) is best done by the original programmer, who is most familiar with the code.

The statement that a programmer should not test his or her own code must not be taken too far. Consider the programming process. The programmer begins by reading the detailed design of the module that may be in the form of a flowchart or, more likely, pseudocode. But whatever technique is used, the programmer must certainly desk-check the module before entering it into the computer. That is, the programmer must try out the flowchart or pseudocode with various test cases, tracing through the detailed design to check that each test case is correctly executed. Only when the programmer is satisfied that the detailed design is correct should the text editor be invoked and the module coded.

Once the module is in machine-readable form, it undergoes a series of tests. First, the programmer attempts to compile the module. When this has been successfully achieved, the following step is to link and load it. Then the programmer attempts to execute the module. If the module executes, then test data are used to determine that the module works successfully, probably the same test data that were used to desk-check the detailed design. Next, if the module executes correctly when correct test data are used, then the programmer tries out incorrect data to test the robustness of the module. When the programmer is satisfied that the module is operating correctly, systematic testing commences. It is this systematic testing that should not be performed by the programmer.

If the programmer should not perform this systematic testing, who should? Independent testing must be performed by the software quality assurance (SQA) group. The key word here is "independent." Only if the SQA group is truly independent of the development team can its members fulfill their mission of ensuring that the product indeed satisfies its specifications, without pressure from software development managers. SQA personnel must report to their own managers and thus protect their independence.

How is systematic testing performed? An essential part of a test case is a statement of the expected output before the test is executed. It is a complete waste of time for the tester to sit at a terminal, execute the module, enter haphazard test data, and then peer at the screen and say, "I guess that looks right." Equally futile is for the tester to plan test cases with great care and then execute each test case in turn, look at the output, and say, "Yes, that certainly looks right." It is far too easy to be fooled by plausible results. If programmers are allowed to test their own code, then there is always the danger that a programmer will see what he or she wants to see. The same danger can occur even when the testing is done by someone else. The solution is for management to insist that before a test is performed, both the test data and the expected results of that test be recorded. After the test has been performed, the actual results obtained should be recorded and compared with the expected results.

Even in small organizations and with small products, it is important that this recording be done in machine-readable form, because test cases should never be thrown away. The reason for this is maintenance. While the product is being maintained, **regression testing** must be performed. Stored test cases that the product has previously executed correctly must be rerun to ensure that the modifications made to add new functionality to the product have not destroyed the product's existing functionality.

110.5 Module Test Case Selection

As pointed out in the previous section, modules undergo two types of testing, namely, informal testing performed by the programmer while developing the module and methodical testing carried out by the SQA group after the programmer is satisfied that the module appears to function correctly. It is this methodical testing that is now discussed. There are in turn two basic types of methodical testing, namely, nonexecution-based testing, in which the module is reviewed by a team, and execution-based testing, in which the module is run against test cases. Techniques for selecting these test cases are now described.

The worst way to test a module is to use haphazard test data. The tester sits in front of the keyboard, and whenever the module requests input, the tester responds with arbitrary data. As will be shown, there is never time to test more than the tiniest fraction of all possible test cases, which can easily number many more than 10^{100}. The few test cases that can be run, perhaps of the order of 1000, are too valuable to waste on haphazard data. Worse, there is a tendency when the machine solicits input to respond more than once with the same data, thus wasting even more test cases. It is clear that test cases must be constructed systematically.

Testing to Specifications versus Testing to Code

There are two basic ways of systematically constructing test data to test a module. The first is to *test to specifications*. This technique is also called **black-box**, data-driven, functional, or input/output-driven testing. In this approach, the code itself is ignored; the only information used in drawing up test cases is the specification document. The other extreme is to *test to code* and to ignore the specification document when selecting test cases. Other names for this technique are **glass-box**, white-box, logic-driven, or path-oriented testing. First, the feasibility of each approach will be considered separately.

Feasibility of Testing to Specifications

Consider the following example. Suppose that the specifications for a certain data processing product state that five types of commission and seven types of discount must be incorporated. Testing every possible combination of just commission and discount requires 35 test cases. It is no use saying that commission and discount are computed in two entirely separate modules, and hence may be tested independently—in black-box testing the product is treated as a black box, and its internal structure is therefore completely irrelevant.

In this example, there are only two factors, namely, commission and discount, taking on 5 and 7 different values, respectively. In any realistic product there will be hundreds, if not thousands, of different factors. Even if there are only 20 factors, each taking on only 4 different values, then a total of 4^{20} or 1.1×10^{12} different test cases have to be examined.

To see the implications of over a trillion test cases, consider how long it would take to test them all. If a team of programmers could be found who could generate, run, and examine test cases at an average rate of one every 30 seconds, then it would take more than a million years to test the product exhaustively.

Thus, exhaustive testing to specifications is impossible in practice because of the combinatorial explosion. There are simply too many test cases to consider. Testing to code is therefore now examined.

Feasibility of Testing to Code

The most common form of testing to code requires that each path through the module must be executed at least once. To see the infeasibility of this, consider the flowchart depicted in Fig. 110.4. Despite the fact that the flowchart appears to be almost trivial, there are over 10^{12} different paths through the flowchart. There are 5 possible paths through the diamond in the center, and the total number of possible paths through the flowchart is therefore $5^1 + 5^2 + \cdots + 5^{18}$ or 4.77×10^{12}. If there can be this many paths through a simple flowchart containing a single loop, it is not difficult to imagine the total number of different paths in a module of reasonable size and complexity, let alone in a large module with many loops. In short, the huge number of possible paths renders exhaustive testing to code as infeasible as exhaustive testing to specifications.

There are additional reasons why testing to code is problematic. Testing to code requires the tester to exercise every path. It is possible to exercise every path without detecting every fault in the product, that is, testing to code is not reliable. To see this, consider the code fragment shown in Fig. 110.5 [Myers 1976]. The fragment was written to test the equality of three integers, **x**, **y**, and **z**, using the false assumption that if the average of three numbers is equal to the first number then the three numbers are equal. Two

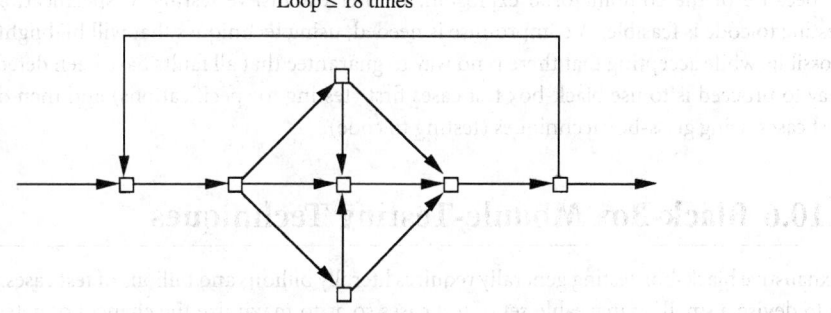

Loop ≤ 18 times

FIGURE 110.4   Flowchart with over 10^{12} possible paths.

```
if ((x + y + z) / 3 == x)
    printf ("x, y, z are equal in value");
else
    printf ("x, y, z are unequal");

Test case 1:     x = 1, y = 2, z = 3
Test case 2:     x = y = z = 2
```

FIGURE 110.5 Incorrect code fragment for determining if three integers are equal, together with two test cases.

```
if (d == 0)                                  x = n / d;
    zero_division_routine;
else
    x = n / d;
        (a)                                          (b)
```

FIGURE 110.6 Two code fragments for computing a quotient.

test cases are also shown in Fig. 110.5. In the first test case the value of the average of the three numbers is 6 / 3 or 2, which is not equal to 1. The product therefore correctly informs the tester that **x**, **y**, and **z** are unequal. The integers **x**, **y**, and **z** are all equal to 2 in the second test case, so the product computes their average as 2, which is equal to the value of **x**, and the product correctly concludes that the three numbers are equal. Thus both paths through the product have been exercised without the fault being detected. Of course, the fault would come to light if test data such as **x = 2, y = 1, z = 3** were used.

A third difficulty with path testing is that a path can be tested only if it is present. Consider the code fragment shown in Fig. 110.6(a). It is clear that there are two paths to be tested, corresponding to the cases **d = 0** and **d ≠ 0**. Next, consider the single statement of Fig. 110.6(b). Now there is only one path, and this path can be tested without the fault being detected. In fact, if a programmer omits checking whether **d = 0** in his or her code, it is likely that the programmer is unaware of the potential danger, and that the case **d = 0** will not be included in the programmer's test data. This problem is an additional argument for having an independent SQA group whose job includes detecting faults of this type.

These examples show conclusively that the criterion "exercise all paths in the product" is not *reliable*, as there exist products for which some data exercising a given path will detect a fault and different data exercising the same path will not. However, path-oriented testing is *valid*, because it does not inherently preclude selecting test data that might reveal the fault.

Because of the combinatorial explosion, neither exhaustive testing to specifications nor exhaustive testing to code is feasible. A compromise is needed, using techniques that will highlight as many faults as possible, while accepting that there is no way to guarantee that all faults have been detected. A reasonable way to proceed is to use black-box test cases first (testing to specifications) and then develop additional test cases using glass-box techniques (testing to code).

110.6 Black-Box Module-Testing Techniques

Exhaustive black-box testing generally requires literally billions and billions of test cases. The art of testing is to devise a small, manageable set of test cases so as to maximize the chances of detecting a fault while minimizing the chances of wasting a test case by having the same fault detected by more than one test case.

Every test case must be chosen to detect a previously un-detected fault. One such black-box technique is equivalence testing combined with boundary value analysis.

Equivalence Testing and Boundary Value Analysis

Suppose the specifications for a database product state that the product must be able to handle any number of records from 1 through 16,383 ($2^{14} - 1$). If the product can handle 34 records and 14,870 records, then the chances are good that it will work fine for, say, 8252 records. In fact, the chances of detecting a fault, if present, are likely to be equally good if any test case from 1 through 16,383 records is selected. Conversely, if the product works correctly for any one test case in the range from 1 through 16,383, then it will probably work for any other test case in the range. The range from 1 through 16,383 constitutes an **equivalence class**, that is, a set of test cases such that any one member of the class is as good a test case as any other member of the class. To be more precise, the specified range of numbers of records that the product must be able to handle defines three equivalence classes as shown in Fig. 110.7. Testing the database product by using the technique of equivalence classes then requires that one test case from each equivalence class be selected. The test case from equivalence class 2 should be handled correctly, whereas error messages should be printed for the test cases from class 1 and class 3.

A successful test case is one that detects a previously undetected fault. To maximize the chances of finding such a fault, a high-payoff technique is **boundary value analysis**. Experience has shown that when a test case that is at or near one side of the boundary of an equivalence class is selected, the probability of detecting a fault increases. Thus, when testing the database product, seven test cases should be selected, as shown in Fig. 110.8.

The preceding example applies to the input specifications. An equally powerful technique is to examine the output specifications. For example, in 1995 the minimum Social Security (OASDI) deduction from any one paycheck permitted by the tax code was $0.00, and the maximum was $3794.40, the latter corresponding to gross earnings of $61,200. Thus, when testing a payroll product, the test cases for the Social Security deduction from paychecks should include input data that are expected to result in deductions of exactly $0.00 and $3794.40. In addition, test data should be set up that might result in deductions of less than $0.00 or more than $3794.40.

In general, for each range (R_1, R_2) listed in either the input or the output specifications, five test cases should be selected, corresponding to values less than R_1, equal to R_1, greater than R_1 but less than R_2, equal to R_2, and greater than R_2. Where it is specified that an item has to be a member of a certain set (for example, "the input must be a letter"), two equivalence classes must be tested, namely, a member of the specified set and a nonmember of the set. Where the specifications lay down a precise value (for example, "the response must be followed by a # sign"), then there are again two equivalence classes, namely, the specified value and anything else.

The use of equivalence classes, together with boundary value analysis, to test both the input specifications and the output specifications is a valuable technique for generating a relatively small set of test data with the potential of uncovering a number of faults that might well remain hidden if less powerful techniques for test data selection were used.

Equivalence class 1: Less than 1 record.
Equivalence class 2: From 1 through 16,383 records.
Equivalence class 3: More than 16, 383 records.

FIGURE 110.7 Equivalence classes for testing database product.

Test case 1:	0 records	Members of equivalence class 1 and adjacent to boundary value
Test case 2:	1 record	Boundary value
Test case 3:	2 records	Adjacent to boundary value
Test case 4:	723 records	Member of equivalence class 2
Test case 5:	16,382 records	Adjacent to boundary value
Test case 6:	16,383 records	Boundary value
Test case 7:	16,384 records	Member of equivalence class 3 and adjacent to boundary value

FIGURE 110.8 Test cases for database product.

110.7 Glass-Box Module-Testing Techniques

In glass-box techniques, test cases are selected by examining the code, rather than the specifications. There are a number of different forms of glass-box testing, including statement, branch, and path coverage.

Structural Testing: Statement, Branch, and Path Coverage

The simplest form of glass-box testing is **statement coverage**, that is, running a series of test cases during which every statement is executed at least once. To keep track of which statements are still to be executed, a CASE tool is required which keeps a record of how many times each statement has been executed over the series of tests. A weakness of this approach is that there is no guarantee that all outcomes of branches are properly tested. To see this, consider the code fragment of Fig. 110.9. The programmer has made a mistake; the compound conditional **s > 1 && t == 0** should have read **s > 1 || t == 0**. The test data shown in the figure allow the statement **x = 9** to be executed without the fault being highlighted.

An improvement over statement coverage is **branch coverage**, that is, running a series of tests to ensure that all branches are tested at least once. Again, a CASE tool is usually needed to help the tester keep track of which branches have or have not been tested. Techniques such as statement or branch coverage are termed "structural tests."

The most powerful form of structural testing is **path coverage**, that is, testing all paths. As shown previously, in a product with loops the number of paths can be very large indeed. As a result, researchers have been investigating ways of reducing the number of paths to be examined while still being able to uncover more faults than would be possible using branch coverage. One way of reducing the number of paths to test is **all-definition-use-path coverage** [Rapps and Weyuker 1985]. In this technique, each occurrence of a variable **qqq**, say, in the source code is labeled either as a *definition* of the variable such as **qqq = 1** or **read (qqq)**, or as a *use* of the variable such as **y = qqq + 3** or **if (qqq < 9) error_b**. All paths between the definition of a variable and the use of that definition are now identified, nowadays by means of an automatic tool. Finally, a test case is set up for each such path. All-definition-use-path coverage is an excellent test technique in that large numbers of faults are frequently detected by relatively few test cases. However, all-definition-use-path coverage does have the disadvantage that the upper bound on the number of paths is 2^d, where d is the number of decision statements (branches) in the product. Examples can be constructed exhibiting the upper bound. However, it has been shown that for real products, as opposed to artificial examples, this upper bound is not reached, and the actual number of paths is proportional to d [Weyuker 1988a]. In other words, the number of test cases needed for all-definition-use-path coverage is generally much smaller than the theoretical upper bound. Thus all-definition-use-path coverage is a practical test case selection technique.

When using structural testing, the situation can arise in which the tester simply cannot come up with a test case that will exercise a specific statement, branch, or path. What may have happened is that there is an infeasible path ("dead code") in the module, that is, a path that cannot possibly be executed for any input data. A tester using statement coverage will soon realize that the item cannot be reached, and the fault would be found.

```
if (s > 1 && t == 0)
    x = 9;

Test case:        s = 2, t = 0.
```

FIGURE 110.9 Code fragment with test data.

Complexity Metrics

The quality assurance viewpoint provides another approach to glass-box testing. Suppose that a manager is told that module **m_1** is more complex than module **m_2**. Irrespective of the precise way in which the term *complex* is defined, the manager will intuitively believe that **m_1** is likely to have more faults than **m_2**. Following this idea, computer scientists have come up with a number of metrics of software complexity as an aid in determining which modules are most likely to have faults. If the complexity of a module is found to be unreasonably high, a manager may direct that the module be redesigned and reimplemented on the grounds that it will probably be cheaper and faster to start from scratch than to attempt to debug a fault-prone module.

A simple metric for predicting numbers of faults is lines of code. The underlying assumption is that there is a constant probability p that a line of code contains a fault. Thus, if a tester believes that, on average, a line of code has a 2% chance of containing a fault, and the module under test is 100 lines long, then this implies that the module is expected to contain two faults, whereas a module that is twice as long is likely to have four faults.

Attempts have been made to find more sophisticated predictors of faults based on measures of product complexity. A typical contender is McCabe's measure of **cyclomatic complexity**, namely, the number of binary decisions (predicates) plus 1 [McCabe 1976]. The cyclomatic complexity is essentially the number of branches in the module. Accordingly, cyclomatic complexity can be used as a metric for the number of test cases needed for branch coverage of a module. This is the basis for so-called "structured testing" [McCabe 1983].

McCabe's metric M can be computed almost as easily as lines of code. In some cases it has been shown to be a good metric for predicting faults; the higher the value of M, the greater the chance that a module contains a fault. For example, Walsh analyzed 276 procedures in the Aegis system, a shipboard combat system [Walsh 1979]. Walsh measured the cyclomatic complexity M and found that 23% of the procedures with M greater than or equal to 10 had 53% of the faults detected. In addition, the procedures with M greater than or equal to 10 had 21% more faults per line of code than the procedures with smaller M values. However, the validity of McCabe's metric has been seriously questioned both on theoretical grounds and on the basis of the many different experiments cited in [Shepperd and Ince 1994].

Musa, Iannino, and Okumoto have analyzed the data available on fault densities [Musa et al. 1987]. They conclude that most complexity metrics show a high correlation with the number of lines of code, or more precisely, the number of deliverable, executable source instructions. In other words, when researchers measure what they believe to be the complexity of a module or product, the result they obtain may largely be a reflection of the number of lines of code, a measure that correlates strongly with the number of faults. In addition, complexity metrics provide little improvement over lines of code for predicting fault rates. Other problems with cyclomatic complexity are discussed in [Weyuker 1988b, Shepperd and Ince 1994].

Code Inspections

At the beginning of this chapter a strong case was made for the use of design inspections. The same arguments hold equally well for code inspections. In brief, the fault-detecting power of this nonexecution-based technique leads to rapid and thorough fault detection. The additional time required for code inspections is more than repaid by increased productivity due to the presence of fewer faults at the integration-testing phase. Furthermore, code inspections have led to a reduction of up to 95% in corrective maintenance costs [Crossman 1982]. Further arguments in favor of inspections are given in the next section.

110.8 Comparison of Module-Testing Techniques

A number of studies have compared strategies for module testing. A major experiment was conducted by Basili and Selby [Basili and Selby 1987]. The techniques compared were black-box testing, glass-box testing, and one-person code reading. The subjects were 32 professional programmers and 42 advanced students. Different results were obtained from the two groups of participants. The professional programmers detected more faults with code reading than with the other two techniques, and the fault detection rate was faster. Two groups of advanced students participated. In one group there was no significant difference among the three techniques; in the other, code reading and black-box testing were equally good, and both outperformed glass-box testing. However, the rate at which students detected faults was the same for all techniques. Overall, code reading led to the detection of more interface faults than did the other two techniques, while black-box testing was most successful at finding control faults. The main conclusion that can be drawn from this experiment is that code inspection is at least as successful at detecting faults as glass-box and black-box testing.

110.9 Cleanroom

The Cleanroom software development technique [Linger 1994] incorporates a number of different techniques, including an incremental life-cycle model, formal techniques for specification and design, and nonexecution-based module-testing techniques such as code reading and code inspections. A critical aspect of the technique is that a module is not compiled until it has passed an inspection. That is, a module should be compiled only after nonexecution-based testing has been successfully accomplished.

The technique has had a number of successes. For example, a prototype automated documentation system was developed for the U.S. Naval Underwater Systems Center using Cleanroom [Trammel et al. 1992]. Altogether 18 faults were detected while the design underwent "functional verification," a review process in which correctness-proving techniques are employed. Informal proofs were used as much as possible; full mathematical proofs were developed only when participants were unsure of the correctness of the portion of the design being inspected. Another 19 faults were detected during walkthroughs of the 1820 lines of FoxBASE code; when the code was then compiled, there were no compilation errors. Furthermore, there were no failures at execution time. This is an additional indication of the power of nonexecution-based testing techniques.

This is certainly an impressive result. However, results that apply to small-scale software products cannot necessarily be scaled up to large-scale software. In the case of Cleanroom, however, results for larger products are also impressive. The relevant metric is the *testing fault rate*, that is, the total number of faults detected per KLOC (thousand lines of code). This is a relatively common metric in the software industry. However, there is a critical difference in the way this metric is computed when Cleanroom is used as opposed to traditional development techniques.

As previously pointed out, when traditional development techniques are used, a module is tested informally by its programmer while it is being developed and thereafter it is tested methodically by the SQA group. Faults detected by the programmer while developing the code are not recorded. However, from the time the module leaves the private workspace of the programmer and is handed over to the SQA group for execution-based and nonexecution-based testing, a tally is kept of the number of faults detected. In contrast, when Cleanroom is used, "testing faults" are counted from the time of compilation. Fault counting then continues through execution-based testing. In other words, when traditional development techniques are used, faults detected informally by the programmer do not count towards the testing fault rate. When Cleanroom is used, faults detected during the inspections and other nonexecution-based testing procedures that precede compilation are certainly recorded, but they do not count towards the testing fault rate.

A report on 17 Cleanroom products appears in [Linger 1994]. For example, Cleanroom was used to develop the 350,000-line Ericsson Telecom OS32 operating system. The product was developed in 18 months by a team of 70. The testing fault rate was only 1.0 fault per KLOC. Another product was the prototype automated documentation system described above; the testing fault rate was 0.0 for the 1820-line program. The 17 products together total nearly 1 million lines of code. The weighted average testing fault rate was 2.3 faults per KLOC, which Linger describes as "a remarkable quality achievement." That praise is certainly not an exaggeration.

110.10 Testing Objects

One of the many reasons put forward for using the object-oriented paradigm is that it reduces the need for testing. Reuse via inheritance is a major strength of the paradigm; once a class has been tested, the argument goes, there is no need to retest it. Furthermore, new methods defined within a subclass of such a tested class have to be tested, but inherited methods need no further testing.

In fact, both these claims are only partially true. In addition, the testing of objects poses certain problems that are specific to object orientation. These issues are discussed in detail in [Schach 1996]; only a brief overview is presented in this section.

Information hiding and the fact that many methods consist of relatively few lines of code can have a significant impact on testing. First consider a product developed using the structured paradigm. Nowadays, such a product generally consists of modules of roughly 50 executable instructions. The interface between a module and the rest of the product is the argument list. There are arguments of two kinds, namely, input arguments that are supplied to the module when it is called, and output arguments that are returned by the module when it returns control to the calling module. Testing a module consists of supplying values to the input arguments, invoking the module, and then comparing the values of the output arguments to the predicted results of the test.

In contrast, a "typical" object will contain perhaps 30 methods, many of which are relatively small, frequently just two or three executable statements [Wilde, Matthews, and Huitt 1993]. Many methods do not return a value to the caller, but rather they change the state of the object. That is, these methods modify attributes (state variables) of the object. The difficulty here is that, to test that the change of state has in fact been correctly performed, it is necessary to send additional messages to the object. For example, consider a bank account object. The effect of method **deposit** is to increase the value of state variable **account_balance**. However, as a consequence of information hiding, the only way to test whether a particular deposit action has been correctly executed is to invoke method **determine_balance** both before and after invoking method **deposit** and see how the bank balance changes.

The situation is worse if the object does not include methods that can be invoked to determine the values of all the state variables. One alternative is to include additional methods for this purpose, and then use conditional compilation to ensure that they are not available other than for testing purposes. The test plan should stipulate that the value of every state variable should be accessible during testing. To satisfy this requirement, additional methods that return the values of the state variables may have to be added to the relevant classes during the design phase. As a result, it will be possible to test the effect of invoking a specific method of an object by querying the value of the applicable state variable.

Surprisingly enough, under certain circumstances inherited methods may still have to be tested. That is, even if a method has been adequately tested, the same method may still require retesting. To make matters even more complex, there are theoretical reasons why it needs to be retested with different test cases [Perry and Kaiser 1990].

It must immediately be pointed out that these complications are not a reason to abandon the object-oriented paradigm. First, these problems arise only through the interaction of methods. Second, it is possible to determine when method retesting is needed [Harrold et al. 1992]. Thus, suppose an instantiation of a class has been thoroughly tested. Any new or redefined methods of a subclass then need to be tested, together with methods that are flagged for retesting because of their interaction with other methods. In short, then, the claim that use of the object-oriented paradigm reduces the need for testing is largely true.

110.11 When Testing Stops

After a product has been successfully maintained for many years it may eventually lose its usefulness and be superseded by a totally different product, in much the same way that electronic valves were replaced by transistors. Alternatively, a product may still be useful, but the cost of porting it to new hardware or of running it under a new operating system may be larger than the cost of constructing a new product, using the old one as a prototype. Thus, finally, the software product is decommissioned and removed from service. Only at that point, when the software has been irrevocably discarded, is it time to stop testing.

Defining Terms

All-definition-use-path coverage: Each occurrence of a variable in the source code is labeled either as a definition of the variable or as a use of the variable. All paths between the definition of a variable and the use of that definition are now identified. A test case is selected and executed for each such path.

Black-box testing: Executing test cases that have been selected solely on the basis of the specification document.

Boundary value analysis: Executing test cases that are at or near the boundary of an equivalence class.

Branch coverage: Selecting and executing a series of test cases to ensure that each branch in the code is tested at least once.

Bug: Inappropriate terminology for a "fault" (see below).

Correctness: A software product is correct if it satisfies its specifications. That is, if input that satisfies the input specifications is provided, then the product is correct if the output satisfies the output specifications.

Correctness proving: A mathematical technique for demonstrating that a product is correct.

Cyclomatic complexity: Essentially, the number of branches in a module. More precisely, it is the number of binary decisions (predicates) plus 1.

Equivalence class: A set of test cases such that any one member of the set is as good a test case as any other member of the set.

Error: A mental mistake made by a programmer. The presence of a fault in a software product is a consequence of an error.

Execution-based testing: Testing by running test cases, that is, by executing code.

Failure: Incorrect behavior of a software product as a consequence of the presence of a fault.

Fault: A mistake in a software product as a consequence of an error made by a programmer.

Glass-box testing: Executing test cases that have been selected solely on the basis of the code.

Inspection: Testing by carefully reviewing a software product.

Mean time between failures: A measure of how often, on average, a product fails.

Mean time to repair: A measure of how long it takes, on average, to repair the effects of a failure.

Nonexecution-based testing: Performing reviews, inspections, analysis, or other forms of testing that do not involve the execution of code.

Path coverage: Selecting and executing a series of test cases to ensure that each path through the code is tested at least once.

Performance: A measure of the extent to which a product meets the constraints stated in the specification document.

Regression testing: Retesting during maintenance to ensure that a change made to the product has not compromised the functionality of the rest of the product.

Reliability: A measure of the frequency and criticality of product failure.

Robustness: A measure of factors such as the range of operating conditions, the possibility of unacceptable results with valid input, and the acceptability of effects when the product is given invalid input.

Statement coverage: Selecting and executing a series of test cases to ensure that every statement in the code is tested at least once.

Utility: The extent to which a user's needs are met when a correct product is used under conditions permitted by its specifications.

Verification: A synonym for correctness proving. Also, informally, all nonexecution-based techniques.

References

Ackerman, A. F., Buchwald, L. S., and Lewski, F. H. 1989. Software inspections: an effective verification process. *IEEE Software* 12(3):31–36.

Basili, V. R. and Selby, R. W. 1987. Comparing the effectiveness of software testing strategies. *IEEE Trans. Software Eng.* SE-13(12):1278–1296.

Berry, D. M. and Wing, J. M. 1985. Specifying and prototyping: some thoughts on why they are successful. In *Formal Methods and Software Development, Proc. Int. Joint Conf. Theory Practice Software Devel.* Vol. 2, pp. 117–128. Springer–Verlag.

Bush, M. 1990. Improving software quality: the use of formal inspections at the Jet Propulsion Laboratory, pp. 196–199. In *Proc. 12th Int. Conf. Software Eng.*, Nice, France.

Crossman, T. D. 1982. Inspection teams, are they worth it? In *Proc. 2nd Nat. Symp. EDP Quality Assurance*, Chicago.

Dijkstra, E. W. 1972. The humble programmer. *Commun. ACM* 15(10):859–866.

Fagan, M. E. 1976. Design and code inspections to reduce errors in program development. *IBM Syst. J.* 15(3):182–211.

Fagan, M. E. 1986. Advances in software inspections. *IEEE Trans. Software Eng.* SE-12(7):744–751.

Goodenough, J. B. and Gerhart, S. L. 1975. Toward a theory of test data selection, pp. 493–510. In *Proc. 3rd Int. Conf. Reliable Software*, Los Angeles, Also published in *IEEE Trans. Software Eng.* SE-1(6):156–173. Revised version: Goodenough, J. B. and Gerhart, S. L. 1977. Toward a theory of test data selection: data selection criteria. In *Current Trends in Programming Methodology*. Vol. 2, R. T. Yeh ed., pp. 44–79. Prentice–Hall, Englewood Cliffs, NJ.

Goodenough, J. B. 1979. A survey of program testing issues. In *Research Directions in Software Technology*, P. Wegner, ed., pp. 316–340. MIT Press, Cambridge, MA.

Harrold, M. J., McGregor, J. D., and Fitzpatrick, K. J. 1992. Incremental testing of object-oriented class structures, pp. 68–80. In *Proc. 14th Int. Conf. Software Eng.*, Melbourne, Australia,

IEEE 610.12. 1990. A glossary of software engineering terminology. *IEEE Computer Society Press*, Los Angeles.

IEEE 1028. 1988. Standard for software reviews and audits. *IEEE Computer Society Press*, Los Angeles.

Jones, T. C. 1978. Measuring programming quality and productivity. *IBM Syst. J.* 17(1):39–63.

Kelly, J. C., Sherif, J. S., and Hops, J. 1992. An analysis of defect densities found during software inspections. *J. Syst. Software* 17(1):111–117.

Leavenworth, B. 1970. Review 19420. *Comp. Rev.* 11(7):396–397.

Linger, R. C. 1994. Cleanroom process model. *IEEE Software* 11(3):50–58.

London, R. L. 1971. Software reliability through proving programs correct. *Proc. IEEE Int. Symp. Fault-Tolerant Comput.*

Manna, Z. and Pnueli, A. 1992. *The Temporal Logic of Reactive and Concurrent Systems*. Springer–Verlag.

Manna, Z. and Waldinger, R. 1978. The logic of computer programming. *IEEE Trans. Software Eng.* SE-4(3):199–229.

McCabe, T. J. 1976. A complexity measure. *IEEE Trans. Software Eng.* SE-2(4):308–320.

McCabe, T. J. 1983. *Structural Testing*. IEEE Computer Society Press, Los Angeles.

Meyer, B. 1992. *Eiffel: The Language*. Prentice–Hall, Englewood Cliffs, NJ.

Musa, J. D., Iannino, A., and Okumoto, K. 1987. *Software Reliability: Measurement, Prediction, Application*. McGraw–Hill, New York.

Myers, G. J. 1976. *Software Reliability: Principles and Practices*. Wiley–Interscience, New York.

Myers, G. J. 1979. *The Art of Software Testing*. John Wiley and Sons, New York.

Naur, P. 1969. Programming by action clusters. *BIT* 9(3):250–258.

Perry, D. E. and Kaiser, G. E. 1990. Adequate testing and object-oriented programming. *J. Object-Oriented Program* 2(1):13–19.

Rapps, S. and Weyuker, E. J. 1985. Selecting software test data using data flow information. *IEEE Trans. Software Eng.* SE-11(2):367–375.

Schach, S. R. 1996. *Classical and Object-Oriented Software Engineering*, 3rd ed. Richard D. Irwin, Chicago.

Shepperd, M. and Ince, D. C. 1994. A critique of three metrics. *J. Syst. Software* 26(9):197–210.

Trammel, C. J., Binder, L. H., and Snyder, C. E. 1992. The automated production control documentation system: A case study in cleanroom software engineering. *ACM Trans. Software Eng. Method* 1(1):81–94.

Walsh, T. J. 1979. A software reliability study using a complexity measure, pp. 761–768. In *Proc. AFIPS Nat. Comp. Conf.*, New York.

Weyuker, E. J. 1988a. An empirical study of the complexity of data flow testing, pp. 188–195. In *Proc. 2nd Workshop Software Testing, Verification, Analysis*. Banff, Canada.

Weyuker, E. J. 1988b. Evaluating software complexity measures. *IEEE Trans. Software Eng.* 14(9):1357–1365.

Wilde, N., Matthews, P., and Huitt, R. 1993. Maintaining object-oriented software. *IEEE Software.* 10(1):75–80.

Further Information

As previously stated, the material of this chapter is taken from S. R. Schach, 1996, *Classical and Object-Oriented Software Engineering*, 3rd ed., Richard D. Irwin. The book contains over 600 references, as well as detailed information regarding conferences, journals, and electronic (Internet and World Wide Web) sources. For reasons of space, only a brief overview of that information can be given here.

The classic work on execution-based testing is *The Art of Software Testing* by G. J. Myers, a work that has had a significant impact on the field of testing. *Software Testing Techniques* by B. Beizer is a compendium on testing, a true handbook on the subject. A similar work is *The Complete Guide to Software Testing* by W. Hetzel.

There are journals dedicated exclusively to testing, such as *Journal of Software Testing and Verification Research*. There are also journals of a more general nature, such as *Communications of the ACM* and *IEEE Software*, in which significant articles on testing are published.

The proceedings of many conferences contain important articles on testing topics. Some cover the entire software engineering field, such as the International Conference on Software Engineering (ICSE). Others are more specialized, such as the Workshop on Testing, Analysis, and Verification (TAV). The proceedings of the Conference on Object-Oriented Programming Systems, Languages, and Applications (OOPSLA) and the European Conference on Object-Oriented Programming (ECOOP) contain articles on testing objects.

The Internet is another valuable source of information on testing. The major relevant Usenet news group is **comp.software.testing**. New newsgroups are continually being approved, so the list of available groups should be scanned for newsgroups that have been formed since the time of writing.

In addition, many organizations maintain information online that can be accessed by mechanisms such as anonymous ftp, Gopher, or the World Wide Web. With regard to the latter, the World Wide Web Virtual Library for Software Engineering can be found at http://ricis.cl.uh.edu/virt-lib/soft-eng.html. An especially valuable source is the Software Engineering Institute at Carnegie-Mellon University. Documents can be obtained by anonymous ftp from ftp.sei.cmu.edu.

111

Development Strategies and Project Management

Roger S. Pressman*
R.S. Pressman & Associates,
Inc.

Successful planning, control, and tracking of a software project is accomplished when a project manager defines an effective development strategy. Once the strategy has been established, software project management commences. The intent of this chapter is to: (1) describe the generic development strategies that are available to software project teams, and (2) present an overview of the tasks required to perform good software project management.

111.1 Development Strategies

A development strategy for **software engineering** integrates a **process model** and the technical methods and tools that populate the model. A process model for software engineering is chosen on the basis of the nature of the project and application, the methods and tools to be used, and the controls and deliverables that are required. Four classes of process models have been widely discussed (and debated). A brief overview of each is presented in the sections that follow.

All software development can be characterized as a problem-solving loop (Fig. 111.1) in which four distinct stages are encountered: status quo, problem definition, technical development, and solution integration [Raccoon 1995]. Status quo represents state of the project; problem definition identifies the specific problem to be solved; technical development solves the problem through the application of some technology; and solution integration delivers the results (e.g., documents, programs, data, new business function, new product) to those who requested the solution in the first place.

*This chapter has been adapted from selected excerpts of R. S. Pressman, *Software Engineering: A Practitioner's Approach*, 4th ed. McGraw–Hill, New York, 1997.

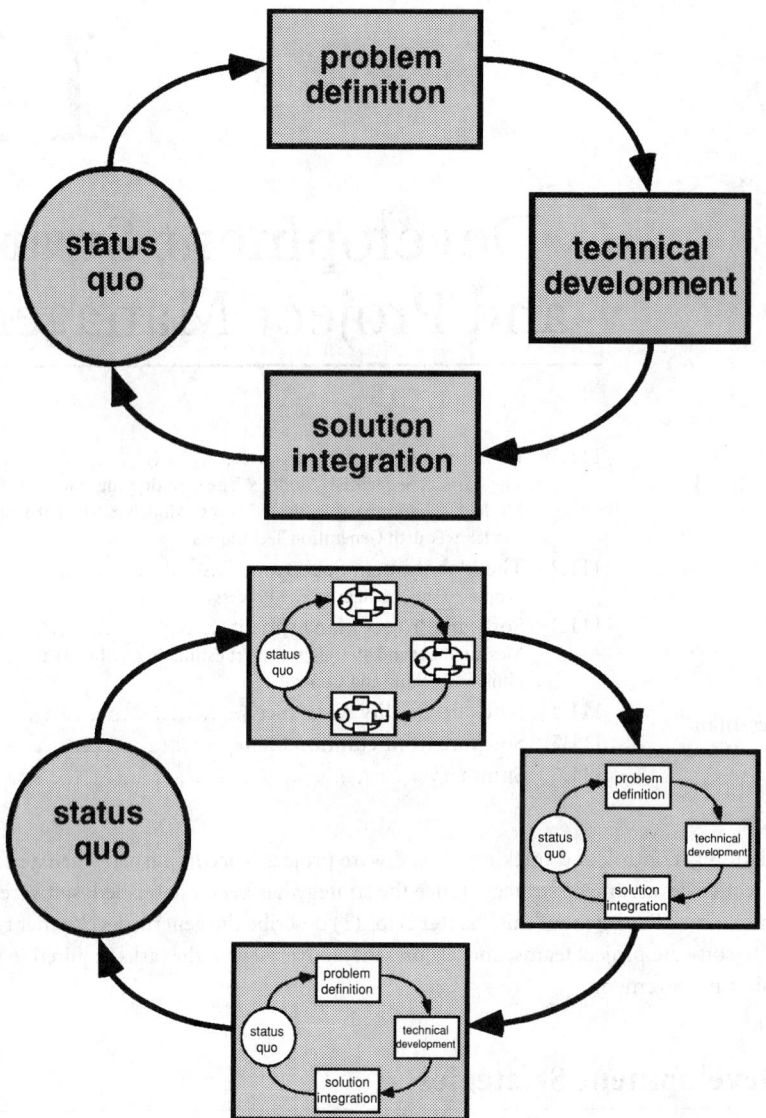

FIGURE 111.1   The phases of a problem-solving loop [Raccoon 1995].

The problem-solving loop described above applies to software engineering work at many different levels of resolution. It can be used at the macro level when the entire application is considered, at a middle level when program components are being engineered, and even at the line of code level.

In the sections that follow, a variety of different process models for software engineering are discussed. Each represents an attempt to bring order to an inherently chaotic activity. It is important to remember that each of the models has been characterized in a way that (one hopes) assists in the control and coordination of a real software project. Each represents a different development strategy, and yet, at their core, all of the models exhibit characteristics of the problem-solving loop described above.

The Linear Sequential Model

Figure 111.2 illustrates the **linear sequential model** for software engineering. Sometimes called the "**classic life cycle**" or the "waterfall model," the linear sequential model demands a systematic, sequential approach to software development that begins at the system level and progresses through analysis, design, coding, **testing**, and **maintenance**. Modeled after the conventional engineering cycle, the linear sequential model encompasses the following activities: system/information engineering and modeling, software **requirements analysis**, design, code generation, and testing and maintenance/**reengineering**.

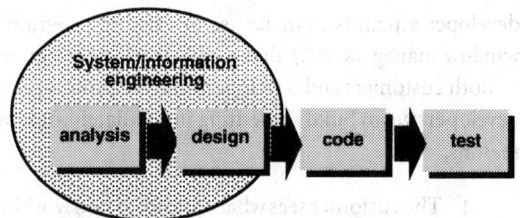

FIGURE 111.2   The linear sequential model.

The linear sequential model is the oldest and the most widely used development strategy. However, criticism of the paradigm has caused even active supporters to question its efficacy [Hanna 1995]. Among the problems that are sometimes encountered when the linear sequential model is applied are the following:

1. Real projects rarely follow the sequential flow that the model proposes. Although the linear model can accommodate iteration, it does so indirectly. As a result, changes can cause confusion as the project team proceeds.
2. It is often difficult for the customer to state all requirements explicitly. The linear sequential model requires this and has difficulty accommodating the natural uncertainty that exists at the beginning of many projects.
3. The customer must have patience. A working version of the program(s) will not be available until late in the project time-span. A major blunder, if undetected until the working program is reviewed, can be disastrous.

In an interesting analysis of actual projects [Bradac et al. 1994], Bradac found that the linear nature of the classic life cycle leads to "blocking states" in which some project team members must wait for other members of the team to complete dependent tasks. In fact, the time spent waiting can exceed the time spent on productive work! The blocking state tends to be more prevalent at the beginning and end of a linear sequential process.

Each of these problems is real. However, the linear development strategy has a definite and important place in software engineering work. It provides a template into which methods for analysis, design, coding, testing, and maintenance can be placed.

The Prototyping Model

Often, a customer defines a set of general objectives for software but does not identify detailed input, processing, or output requirements. In other cases, the developer may be unsure of the efficiency of an algorithm, the adaptability of an operating system, or the form that human–machine interaction should take. In these, and many other situations, a **prototyping** paradigm may offer the best approach.

The prototyping strategy (Fig. 111.3) begins with gathering requirements. Developer and customer meet and define the overall objectives for the software, identify whatever requirements are known, and outline areas where further definition is mandatory. A "quick design" then occurs. The quick design focuses on a representation of those aspects of the software that will be visible to the customer/user (e.g., input approaches and output formats). The quick design leads to construction of a prototype. The prototype is evaluated by the customer/user and is used to refine requirements for the software to be developed. Iteration occurs as the prototype is tuned to satisfy the needs of the customer while enabling the developer to better understand what needs to be done.

Ideally, the prototype serves as a mechanism for identifying software requirements. If a working prototype is built, the

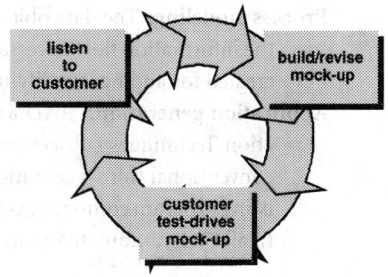

FIGURE 111.3 The prototyping paradigm.

developer attempts to make use of existing program fragments or applies tools (e.g., report generators, window managers, etc.) that enable working programs to be generated quickly.

Both customers and developers like the prototyping paradigm. Users get a feel for the actual system and developers get to build something immediately. Yet, prototyping can also be problematic for the following reasons:

1. The customer sees what appears to be a working version of the software, unaware that the prototype is held together "with chewing gum and baling wire," unaware that in the rush to get it working we have not considered overall software quality or long-term maintainability. When informed that the product must be rebuilt so that high levels of quality can be maintained, the customer cries foul and demands that "a few fixes" be applied to make the prototype a working product. Too often, software development management relents.
2. The developer often makes implementation compromises in order to get a prototype working quickly. An inappropriate operating system or programming language may be used simply because it is available and known; an inefficient algorithm may be implemented simply to demonstrate capability. After a time, the developer may become familiar with these choices and forget all the reasons why they were inappropriate. The less-than-ideal choice has now become an integral part of the system.

Although problems can occur, prototyping can be an effective paradigm for software engineering. The key is to define the rules of the game at the beginning; that is, the customer and developer must both agree that the prototype is built to serve as a mechanism for defining requirements. It is then discarded (at least in part) and the actual software is engineered with an eye toward quality and maintainability.

The RAD Model

Rapid Application Development (RAD) is a linear sequential software development process model that emphasizes an extremely short development cycle. The RAD model is a "high speed" adaptation of the linear sequential model in which rapid development is achieved by using a component-based construction approach. If requirements are well understood and project scope is constrained,[1] the RAD process enables a development team to create a "fully functional system" within very short time periods (e.g., 60–90 days) [Martin 1991]. Used primarily for information systems applications, the RAD approach encompasses the following phases [Kerr and Hunter 1994]:

Business modeling. The information flow among business functions is modeled in a way that answers the following questions: What information drives the business process? What information is generated? Who generates it? Where does the information go? Who processes it?

Data modeling. The information flow defined as part of the business modeling phase is refined into a set of data objects that are needed to support the business. The characteristics (called attributes) of each object are identified and the relationships between these objects are defined.

Process modeling. The data objects defined in the data modeling phase are transformed to achieve the information flow necessary to implement a business function. Processing descriptions are created for adding, modifying, deleting, or retrieving a data object.

Application generation. RAD assumes the use of fourth-generation techniques (Fourth Generation Techniques subsection of the following section). Rather than creating software using conventional third-generation programming languages, the RAD process works to reuse existing program components (when possible) or create reusable components (when necessary). In all cases, automated tools are used to facilitate construction of the software.

---

[1]These conditions are by no means guaranteed. In fact, many software projects have poorly defined requirements at the start. In such cases, prototyping or evolutionary approaches are much better process options.

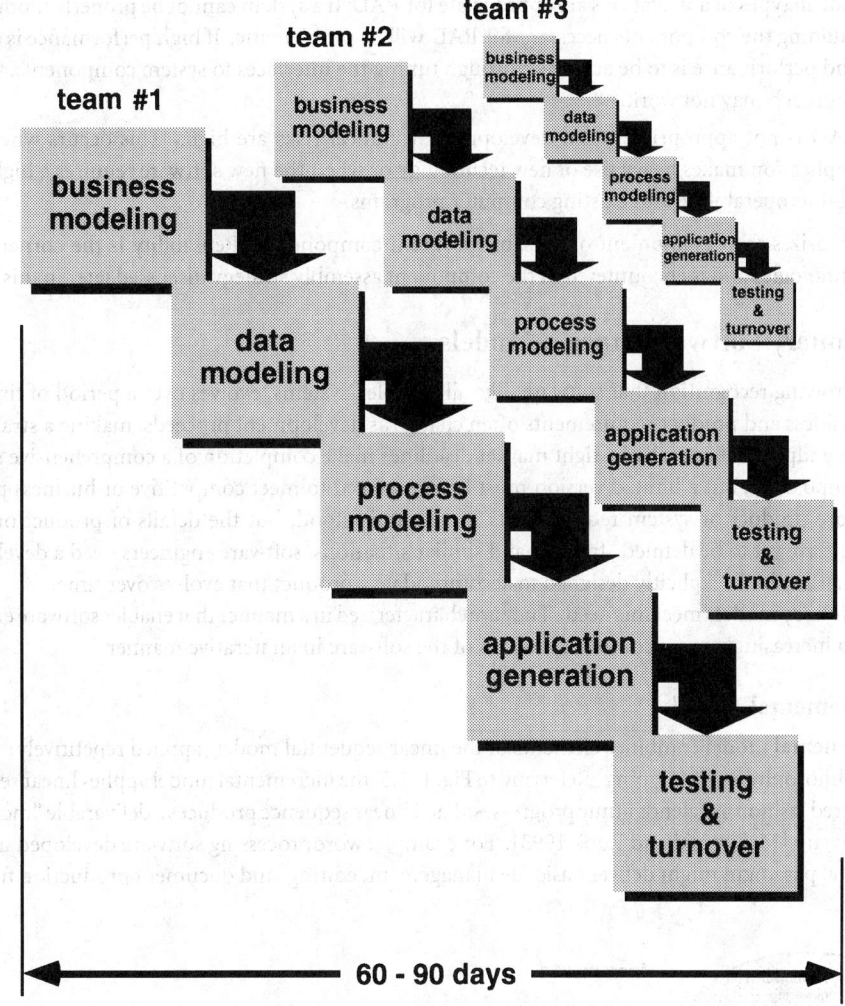

team #1 team #2 team #3

business modeling

data modeling

process modeling

application generation

testing & turnover

60 - 90 days

FIGURE 111.4   The RAD model.

Testing and turnover. Since the RAD process emphasizes reuse, many of the program components have already been tested. This reduces overall testing time. However, new components must be tested and all interfaces must be fully exercised.

The RAD process model is illustrated in Fig. 111.4. Obviously, the time constraints imposed on a RAD project demand "scalable scope" [Kerr and Hunter 1994]. If a business application can be modularized in a way that enables each major function to be completed in less than three months (using the approach described above), it is a candidate for RAD. Each major function can be addressed by a separate RAD team and then integrated to form a whole.

Like all process models, the RAD approach has drawbacks [Butler 1994]:

- For large, but scalable, projects, RAD requires sufficient human resources to create the right number of RAD teams.
- RAD requires developers and customers who are committed to the rapid-fire activities necessary to get a system complete in a much-abbreviated time frame. If commitment is lacking from either constituency, RAD projects will fail.

- Not all types of applications are appropriate for RAD. If a system cannot be properly modularized, building the components necessary for RAD will be problematic. If high performance is an issue, and performance is to be achieved through tuning the interfaces to system components, the RAD approach may not work.

- RAD is not appropriate when development technical risks are high. This occurs when a new application makes heavy use of new technology or when the new software requires a high degree of interoperability with existing computer programs.

RAD emphasizes the development of reusable program components. Reusability is the cornerstone of object technologies and is encountered in the component assembly strategy discussed later in this chapter.

Evolutionary Software Process Models

There is growing recognition that software, like all complex systems, evolves over a period of time [Gilb 1988]. Business and product requirements often change as development proceeds, making a straight line path to an endproduct unrealistic; tight market deadlines make completion of a comprehensive software product impossible, but a limited version must be introduced to meet competitive or business pressure; a set of core product or system requirements is well understood, but the details of product or system extensions have yet to be defined. In these and similar situations, software engineers need a development strategy that has been explicitly designed to accommodate a product that evolves over time.

Evolutionary models meet this need. They are characterized in a manner that enables software engineers to develop increasingly more complete versions of the software in an iterative manner.

The Incremental Model

The **incremental model** combines elements of the linear sequential model (applied repetitively) with the iterative philosophy of prototyping. Referring to Fig. 111.5, the incremental model applies linear sequences in a staggered fashion as calendar time progresses. Each linear sequence produces a deliverable "increment" of the software [McDermid and Rook 1993]. For example, wordprocessing software developed using the incremental paradigm might deliver basic file management, editing, and document production functions

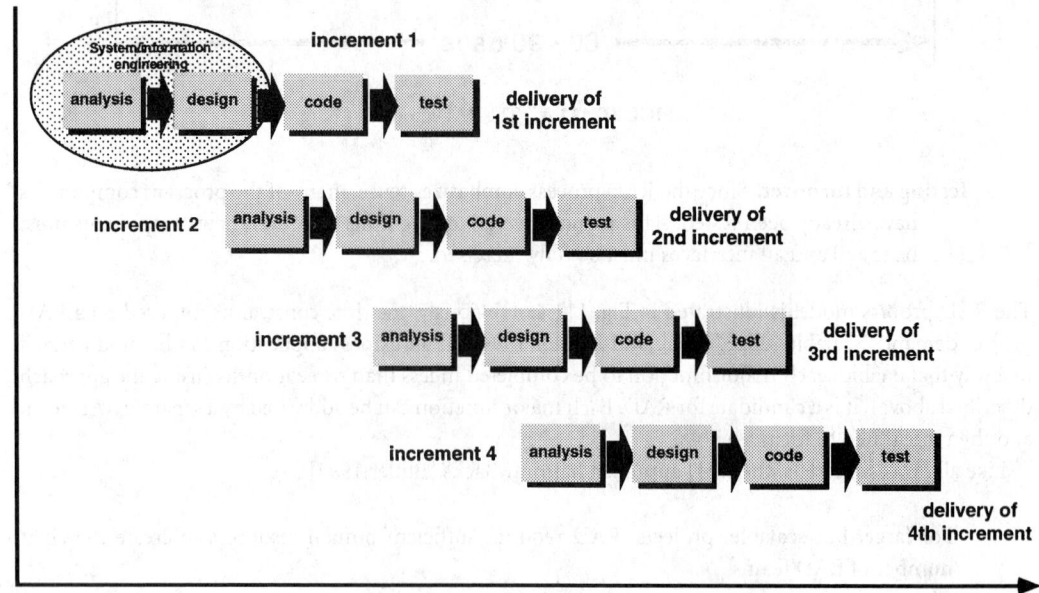

FIGURE 111.5   The incremental model.

in the first increment; more sophisticated editing and document production capabilities in the second increment; spelling and grammar checking in the third increment; and advanced page layout capability in the fourth increment. It should be noted that the process flow for any increment can incorporate the prototyping paradigm.

When an incremental model is used, the first increment is often a *core product*; that is, basic requirements are addressed, but many supplementary features (some known, others unknown) remain undelivered. The core product is used by the customer (or undergoes detailed review). As a result of use and/or evaluation, a plan is developed for the next increment. The plan addresses the modification of the core product to better meet the needs of the customer and the delivery of additional features and functionality. This process is repeated following the delivery of each increment, until the complete product is produced.

Incremental development is particularly useful when staffing is unavailable for a complete implementation by the business deadline that has been established for the project. Early increments can be implemented with fewer people. If the core product is well received, then additional staff (if required) can be added to implement the next increment. In addition, increments can be planned to manage technical risks.

The Spiral Model

The **spiral model**, originally proposed by Boehm [Boehm 1988], is an evolutionary software process model that couples the iterative nature of prototyping with the controlled and systematic aspects of the linear sequential model. It provides the potential for rapid development of incremental versions of the software. Using the spiral model, software is developed in a series of incremental releases. During early iterations, the incremental release might be a paper model or prototype. During later iterations, increasingly more complete versions of the engineered system are produced.

The spiral model is divided into a number of *framework activities*, also called *task regions*. Typically, there are between three and six task regions. Figure 111.6 depicts a spiral model that contains six task regions:

- **Customer communication.** Tasks required to establish effective communication between developer and customer.
- **Planning.** Tasks required to define resources, timelines, and other project-related information.
- **Risk analysis.** Tasks required to assess both technical and management risks.
- **Engineering.** Tasks required to build one or more representations of the application.
- **Construction and release.** Tasks required to construct, test, install, and provide user support (e.g., documentation and training).
- **Customer evaluation.** Tasks required to obtain customer feedback based on evaluation of the software representations created during the engineering stage and implemented during the installation stage.

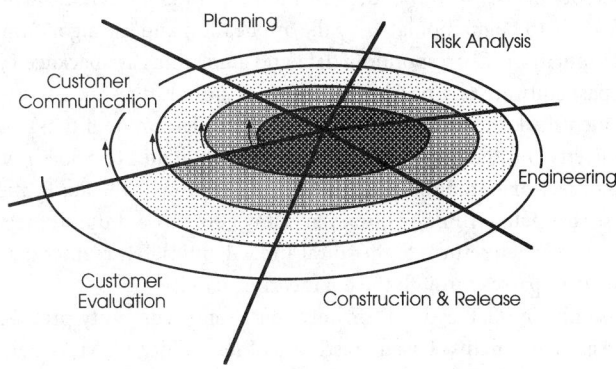

FIGURE 111.6 A typical spiral model.

Each of the regions is populated by a series of work tasks that are adapted to the characteristics of the project to be undertaken. For small projects, the number of work tasks and their formality is low. For larger, more critical projects, each task region contains more work tasks that are defined to achieve a higher level of formality. In all cases, the umbrella activities (e.g., software configuration management and software quality assurance) are performed.

As this evolutionary process begins, the software engineering team moves around the spiral in a clockwise direction, beginning at the center. The first circuit around the spiral might result in development of a product specification; subsequent passes around the spiral might be used to develop a prototype and then progressively more sophisticated versions of the software. Each pass through the planning region results in adjustments to the project plan. Cost and schedule are adjusted on the basis of feedback derived from customer evaluation. In addition, the project manager adjusts the planned number of iterations required to complete the software.

The spiral model is a realistic approach to the development of large-scale systems and software. Because software evolves as the process progresses, the developer and customer better understand and react to risks at each evolutionary level. The spiral model uses prototyping as a risk reduction mechanism but, more importantly, enables the developer to apply the prototyping approach at any stage in the evolution of the product. It maintains the systematic stepwise approach suggested by the classic life cycle but incorporates it into an iterative framework that more realistically reflects the real word. The spiral model demands a direct consideration of technical risks at all stages of the project and, if properly applied, should reduce risks before they become problematic.

But like other paradigms, the spiral model is not a panacea. It may be difficult to convince customers (particularly in contract situations) that the evolutionary approach is controllable. It demands considerable risk assessment expertise, and it relies on this expertise for success. If a major risk is not uncovered and managed, problems will undoubtedly occur. Finally, the model itself is relatively new and has not been used as widely as the linear sequential or prototyping paradigms. It will take a number of years before efficacy of this important new paradigm can be determined with absolute certainty.

The Component Assembly Model

Object technologies provide the technical framework for a component-based process model for software engineering. The object-oriented paradigm emphasizes the creation of classes that encapsulate both data and the algorithms that are used to manipulate the data. If properly designed and implemented, object-oriented classes are reusable across different applications and computer-based system architectures.

The **component assembly model** (Fig. 111.7) incorporates many of the characteristics of the spiral model. It is evolutionary in nature [Nierstrasz 1992], demanding an iterative approach to the creation of software. However, the component assembly model composes applications from prepackaged software components (called "classes" in Fig. 111.7).

The engineering activity begins with the identification of candidate classes. This is accomplished by examining the data that are to be manipulated by the application and the algorithms that will be applied to accomplish the manipulation. Corresponding data and algorithms are packaged into a class.

Classes created in past software engineering projects are stored in a *class library* or repository. Once candidate classes are identified, the class library is searched to determine if these classes already exist. If they do, they are extracted from the library and reused. If a candidate class does not reside in the library, it is engineered using **object-oriented** methods. The first iteration of the application to be built is then composed, using classes extracted from the library and any new classes built to meet the unique needs of the application. Process flow then returns to the spiral and will ultimately reenter the component assembly iteration during subsequent passes through the engineering activity.

The component assembly model leads to software reuse, and reusability provides software engineers with a number of measurable benefits. Based on studies of reusability, QSM Associates reports [Yourdon 1994] component assembly leads to a 70% reduction in development cycle time, an 84% reduction in project cost, and a productivity index of 26.2, compared with an industry norm of 16.9. Although these

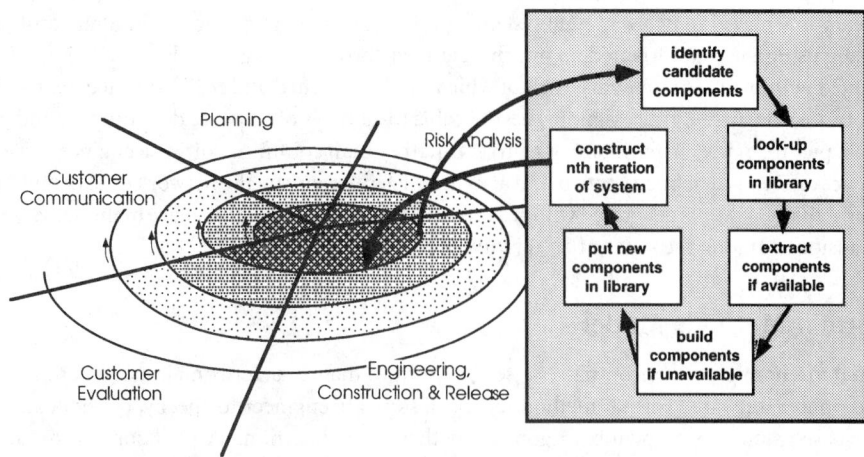

FIGURE 111.7   The component assembly model.

results are a function of the robustness of the component library, there is little question that the component assembly model provides significant advantages for software engineers.

The Concurrent Development Model

The *concurrent development model,* sometimes called *concurrent engineering,* has been described in the following manner by Davis and Sitaram [Davis and Sitaram 1994]:

> Project managers who track project status in terms of the major phases [of the classic life cycle] have no idea of the status of their projects. These are examples of trying to track extremely complex sets of activities using overly simple models. Note that although ... [a large] project is in the coding phase, there are personnel on the project involved in activities typically associated with many phases of development simultaneously. For example, ... personnel are writing requirements, designing, coding, testing, and integration testing [all at the same time]. Software engineering process models by Humphrey and Kellner [Humphrey and Kellner 1989] have shown the concurrency that exists for activities occurring during any one phase. Kellner's more recent work [Kellner 1991] uses statecharts [a notation that represents the states of a process] to represent the concurrent relationship existent among activities associated with a specific event (e.g., a requirements change during late development), but fails to capture the richness of concurrency that exists across all software development and management activities in project ... Most software development process models are driven by time; the later it is, the later in the development process you are. [A concurrent process model] is driven by user needs, management decisions, and review results.

The concurrent process model can be represented schematically as a series of major technical activities, tasks, and their associated states. For example, the *engineering* activity defined for the spiral model is accomplished by invoking the following tasks: prototyping and/or analysis modeling, requirements specification, and design.[2]

The concurrent process model is often used as the paradigm for development of client–server[3] applications. A client–server system is composed of a set of functional components. When applied to client–server, the concurrent process model defines activities in two dimensions [Sheleg 1994]: a system dimension and a component dimension. System level issues are addressed using three activities: **design,** *assembly,* and *use.* The component dimension is addressed with two activities: *design* and *realization.*

---

[2] It should be noted that analysis and design are complex tasks that require substantial discussion.

[3] In client–server applications, software functionality is divided between clients (normally PCs) and a server (a more powerful computer) that typically maintains a centralized database.

Concurrency is achieved in two ways: (1) system and component activities occur simultaneously and can be modeling using the state-oriented approach described above; (2) a typical client–server application is implemented with many components, each of which can be designed and realized concurrently.

In reality, the concurrent process model is applicable to all types of software development and provides an accurate picture of the current state of a project. Rather than confining software engineering activities to a sequence of events, it defines a network of activities. Each activity on the network exists simultaneously with other activities. Events generated within a given activity or at some other place in the activity network trigger transitions among the states of an activity.

The Formal Methods Model

The **formal methods** model encompasses a set of activities that leads to formal mathematical specification of computer software. Formal methods enable a software engineer to specify, develop, and verify a computer-based system by applying a rigorous, mathematical notation. A variation on this approach, called *cleanroom engineering* [Mills et al. 1987, Dyer 1992], is currently applied by some software development organizations.

When formal methods (Chapter 107) are used during development, they provide a mechanism for eliminating many of the problems that are difficult to overcome by using other software engineering paradigms. Ambiguity, incompleteness, and inconsistency can be discovered and corrected more easily— not through *ad hoc* review, but through the application of mathematical analysis. When formal methods are used during design, they serve as a basis for program verification and therefore enable the software engineer to discover and correct errors that might otherwise go undetected.

Fourth Generation Techniques

The term **fourth generation techniques** (4GT) encompasses a broad array of software tools that have one thing in common: each enables the software engineer to specify some characteristic of software at a high level. The tool then automatically generates source code based on the developer's specification. There is little debate that the higher the level at which software can be specified to a machine, the faster a program can be built. The 4GT paradigm for software engineering focuses on the ability to specify software using specialized language forms or a graphic notation that describes the problem to be solved in terms that the customer can understand.

Currently, a software development environment that supports the 4GT paradigm includes some or all of the following tools: nonprocedural languages for database query, report generation, data manipulation, screen interaction and definition, and code generation; high-level graphics capability; and spreadsheet capability. Initially, many of the tools noted above were available only for very specific application domains, but today 4GT environments have been extended to address most software application categories.

Like other paradigms, 4GT begins with a requirements gathering step. Ideally, the customer would describe requirements and these would be directly translated into an operational prototype. But this is unworkable. The customer may be unsure of what is required, may be ambiguous in specifying facts that are known, and may be unable or unwilling to specify information in a manner that a 4GT tool can consume. For this reason, the customer–developer dialog described for other process models remains an essential part of the 4GT approach.

For small applications, it may be possible to move directly from the requirements gathering step to implementation using a nonprocedural *fourth generation language* (4GL). However, for larger efforts, it is necessary to develop a design strategy for the system, even if a 4GL is to be used. The use of 4GT without design (for large projects) will cause the same difficulties (poor quality, poor maintainability, poor customer acceptance) that we have encountered when developing software by using conventional approaches.

Implementation using a 4GL enables the software developer to represent desired results in a manner that results in automatic generation of code to generate those results. Obviously, a data structure with relevant information must exist and be readily accessible by the 4GL.

To transform a 4GT implementation into a product, the developer must conduct thorough testing, develop meaningful documentation, and perform all other solution integration activities that are also required in other software engineering paradigms. In addition, the 4GT-developed software must be built in a manner that enables maintenance to be performed expeditiously.

Like all software engineering paradigms, the 4GT model has advantages and disadvantages. Proponents claim dramatic reduction in software development time and greatly improved productivity for people who build software. Opponents claim that current 4GT tools are not all that much easier to use than programming languages, that the resultant source code produced by such tools is "inefficient," and that the maintainability of large software systems developed using 4GT is open to question.

There is some merit in the claims of both sides, and it is possible to summarize the current state of 4GT approaches:

1. The use of 4GT has broadened considerably over the past decade and is now a viable approach for many different application areas. Coupled with *computer-aided software engineering* (CASE) tools and code generators, 4GT offers a credible solution to many software problems.
2. Data collected from companies that are using 4GT indicate that time required to produce software is greatly reduced for small and intermediate applications and that the amount of design and analysis for small applications is also reduced.
3. However, the use of 4GT for large software development efforts demands as much or more analysis, design, and testing (software engineering activities) to achieve substantial time saving that can be achieved through the elimination of coding.

To summarize, 4GT have already become an important part of software development. When coupled with component assembly approaches, the 4GT paradigm may become the dominant software development strategy as the 21st century dawns.

111.2 The Management Spectrum

Effective software project management focuses on the three P's: *people, problem,* and *process.* The order is not arbitrary. The manager who forgets that software engineering work is an intensely human endeavor will never have success in project management. A manager who fails to encourage comprehensive customer communication early in the evolution of a project risks building an elegant solution for the wrong problem. Finally, the manager who pays little attention to the process runs the risk of inserting competent technical methods and tools into a vacuum.

People

The cultivation of motivated, highly skilled software people has been discussed since the 1960s (e.g., [Cougar and Zawacki 1980, DeMarco and Lister 1987, Weinberg 1988]). The Software Engineering Institute has sponsored a *people management maturity model* "to enhance the readiness of software organizations to undertake increasingly complex applications by helping to attract, grow, motivate, deploy, and retain the talent needed to improve their software development capability" [Curtis 1989].

The people management maturity model defines the following key practice areas for software people: recruiting, selection, performance management, training, compensation, career development, organization, and team and culture development. Organizations that achieve high levels of maturity in the people management area have a higher likelihood of implementing effective software engineering practices.

The Problem

Before a project can be planned, objectives and scope should be established, alternative solutions should be considered, and technical and management constraints should be identified. Without this information, it is impossible to develop reasonable estimates of the cost, a realistic breakdown of project tasks, or a manageable project schedule that provides a meaningful indication of progress.

The software developer and customer must meet to define project objectives and scope. In many cases, this activity occurs as part of a structured customer communication process such as *joint application design* (JAD) [Wood and Silver 1994]. JAD is an activity that occurs in five phases: project definition, research, preparation, the JAD meeting, and document preparation. The intent of each phase is to develop information that helps better define the problem to be solved or the product to be built.

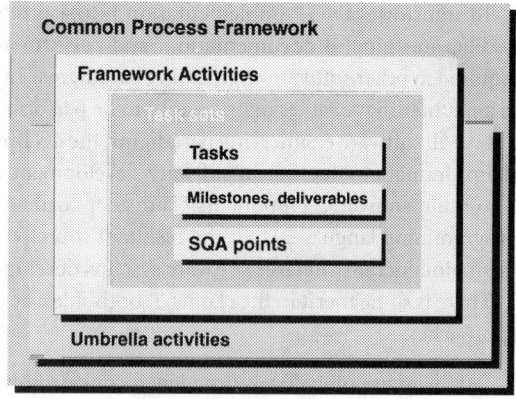

FIGURE 111.8   A common process framework.

The Process

A software process can be characterized as shown in Fig. 111.8. A small number of *framework activities* are applicable to all software projects, regardless of their size or complexity. A number of *task sets*—tasks, milestones, deliverables, and quality assurance points—enable the framework activities to be adapted to the characteristics of the software project and the requirements of the project team. Finally, umbrella activities—such as software quality assurance, software configuration management, and measurement—overlay the process model. Umbrella activities are independent of any one framework activity and occur throughout the process.

In recent years, there has been a significant emphasis on process "maturity." [Paulk et al. 1993] The Software Engineering Institute (SEI) has developed a comprehensive assessment model that is predicated on a set of software engineering capabilities that should be present as organizations reach different levels of process maturity. To determine an organization's current state of process maturity, the SEI uses an assessment questionnaire and a five-point grading scheme. The grading scheme determines compliance with a **capability maturity model** [Paulk et al. 1993] that defines key activities required at different levels of process maturity. The SEI approach provides a measure of the global effectiveness of a company's software engineering practices and establishes five process maturity levels that are defined in the following manner:

> **Level 1, Initial:** The software process is characterized as *ad hoc* and occasionally even chaotic. Few processes are defined, and success depends on individual effort.
>
> **Level 2, Repeatable:** Basic project management processes are established to track cost, schedule, and functionality. The necessary process discipline is in place to repeat earlier successes on projects with similar applications.
>
> **Level 3, Defined:** The software process for both management and engineering activities is documented, standardized, and integrated into an organization-wide software process. All projects use a documented and approved version of the organization's process for developing and maintaining software. This level includes all characteristics defined for level 2.
>
> **Level 4, Managed:** Detailed measures of the software process and product quality are collected. Both the software process and products are quantitatively understood and controlled by using detailed measures. This level includes all characteristics defined for level 3.
>
> **Level 5, Optimizing:** Continuous process improvement is enabled by quantitative feedback from the process and from testing innovative ideas and technologies. This level includes all characteristics defined for level 4.

The five levels defined by the SEI are derived as a consequence of evaluating responses to the SEI assessment questionnaire that is based on the capability maturity model. The results of the questionnaire are distilled to a single numerical grade that provides an indication of an organization's process maturity.

The SEI has associated *key process areas* (KPAs) with each of the maturity levels. The KPAs describe those software engineering functions (e.g., software project planning, requirements management) that must be present to satisfy good practice at a particular level. Each KPA is described by identifying the following characteristics:

- *Goals:* The overall objectives that the KPA must achieve.
- *Commitments:* Requirements (imposed on the organization) that must be met to achieve the goals and provide proof of intent to comply with the goals.
- *Abilities:* Those things that must be in place (organizationally and technically) that will enable the organization to meet the commitments.
- *Activities:* The specific tasks that are required to achieve the KPA function.
- *Methods for monitoring implementation:* The manner in which the activities are monitored as they are put into place.
- *Methods for verifying implementation:* The manner in which proper practice for the KPA can be verified.

Eighteen KPAs (each defined using the structure noted above) are defined across the maturity model and are mapped into different levels of process maturity.

Each of the KPAs is defined by a set of *key practices* that contribute to satisfying its goals. The key practices are policies, procedures, and activities that must occur before a key process area has been fully instituted. The SEI defines *key indicators* as "those key practices or components of key practices that offer the greatest insight into whether the goals of a key process area have been achieved." Assessment questions are designed to probe for the existence (or lack thereof) of a key indicator.

111.3 Software Project Management

Software project management encompasses the following activities: measurement, project estimating, risk analysis, scheduling, tracking, and control. A comprehensive discussion of these topics is beyond the scope of this chapter, but a brief overview of each topic will enable the reader to understand the breadth of management activities required for a mature software engineering organization.

Measurement and Metrics

To be most effective, software metrics should be collected for both the process and the product. Process-oriented metrics [Hetzel 1993, Jones 1991] can be collected during the process and after it has been completed. Process metrics collected during the process focus on the efficacy of quality assurance activities, change management, and project management. Process metrics collected after a project has been completed examine the efficacy of various software engineering activities and productivity. Process measures are normalized using either lines of code or function points [Dreger 1989], so that data collected from many different projects can be compared and analyzed in a consistent manner. Product metrics measure technical characteristics of the software that provide an indication of software quality [Fenton 1991, Zuse 1990, Lorenz and Kidd 1994]. Measures can be applied to models created during analysis and design activities, during code generation, and during testing. The mechanics of measurement and the specific measures to be collected are beyond the scope of this chapter.

Project Estimating

Scheduling and budgets are often dictated by business issues. The role of estimating within the software process often serves as a "sanity check" on the predefined deadlines and budgets that have been established by management. (Ideally, the software engineering organization should be intimately involved in establishing deadlines and budgets, but this is not a perfect or fair world.)

All software project **estimation** techniques require that the project have a bounded scope, and all rely on a high-level functional decomposition of the project and an assessment of project difficulty and complexity. There are three broad classes of estimation techniques [Pressman 1993] for software projects:

Effort estimation techniques. The project manager creates a matrix in which the left-hand column contains a list of major system functions derived using functional decomposition applied to project scope. The top row contains a list of major software engineering tasks derived from the common process framework. The manager (with the assistance of technical staff) estimates the effort required to accomplish each task for each function.

Size-oriented estimation. A list of major system functions derived using functional decomposition applied to project scope. The "size" of each function is estimated by using either lines of code (LOC) or function points (FP). Average productivity data (e.g., function points per person month) for similar functions or projects are used to generate an estimate of effort required for each function.

Empirical models. Using the results of a large population of past projects, an empirical model that relates product size (in LOC or FP) to effort is developed, using a statistical technique such as regression analysis. The product size for the work to be done is estimated and the empirical model is used to generate projected effort.

In addition to the above techniques, a software project manager can develop estimates by analogy; that is, by examining similar past projects and projecting effort and duration recorded for these projects to the current situation.

Risk Analysis

Almost five centuries have passed since Machiavelli said: "I think it may be true that fortune is the ruler of half our actions, but that she allows the other half to be governed by us . . . [fortune] is like an impetuous river . . . but men can make provision against it by dykes and banks." Fortune (we call it risk) is in the back of every software project manager's mind, and that is often where it stays. And as a result, risk is never adequately addressed. When bad things happen, the manager and the project team are unprepared.

In order to "make provision against it," a software project team must conduct risk analysis explicitly. **Risk analysis** [Charette 1990, Jones 1994] is actually a series of steps that enable the software team to perform risk identification, risk assessment, risk prioritization, and risk management. The goals of these activities are: (1) to identify those risks that have a high likelihood of occurrence; (2) to assess the consequence (impact) of each risk should it occur; and (3) to develop a plan for mitigating the risks when possible, monitoring factors that may indicate their arrival, and developing a set of contingency plans should they occur.

Risk identification is a systematic attempt to specify threats to the project plan (estimates, schedule, resource loading, etc.). By identifying known and predictable risks, the project manager takes a first step toward avoiding them when possible and controlling them when necessary.

There are two distinct types of risks for each of the categories that have been presented: generic risks and product-specific risks. Generic risks are a potential threat to every software project. Product-specific risks can be identified only by those with a clear understanding of the technology, the people, and the environment that are specific to the project at hand. To identify product-specific risks, the project plan and the software statement of scope are examined and an answer to the following question is developed: "What special characteristics of this product may threaten our project plan?"

Both generic and product-specific risks should be identified systematically. Gilb [Gilb 1988] drives this point home when he states: "If you don't actively attack the risks, they will actively attack you."

Risk projection, also called *risk estimation,* attempts to rate each risk in two ways—the *likelihood* or probability that the risk is real and the *consequences* of the problems associated with the risk, should it occur. The project planner, along with other managers and technical staff, performs four risk projection activities [Babich 1986]: (1) establish a scale that reflects the perceived likelihood of a risk; (2) delineate the consequences of the risk; (3) estimate the impact of the risk on the project and the product; and (4) note the overall accuracy of the risk projection so that there will be no misunderstandings.

All of the risk analysis activities presented to this point have a single goal—to assist the project team in developing a strategy for dealing with risk. An effective strategy must consider three issues:

- risk avoidance
- risk monitoring, and
- risk management and contingency planning.

The manner in which each of these issues is to be addressed is documented in a plan for *risk mitigation, monitoring, and management.*

Scheduling

Fred Brooks, the well-known author of The Mythical Man-Month [Brooks 1975], was once asked how software projects fall behind schedule. His response was as simple as it was profound: "One day at a time."

The reality of a technical project (whether it involves building a hydroelectric plant or developing an operating system) is that hundreds of small tasks must occur to accomplish a larger goal. Some of these tasks lie outside the mainstream and may be completed without worry about impact on project completion date. Other tasks lie on the "critical path."[4] If these "critical" tasks fall behind schedule, the completion date of the entire project is put into jeopardy.

The objective of the project manager is to define all project tasks, identify the ones that are critical, and then track their progress to ensure that delay is recognized "one day at a time." To accomplish this, the manager must have a schedule that has been defined at a degree of resolution that enables the manager to monitor progress and control the project.

Software project **scheduling** is an activity that distributes estimated effort across the planned project duration by allocating the effort to specific software engineering tasks. It is important to note, however, that the schedule evolves over time. During early stages of project planning, a *macroscopic schedule* is developed. This type of schedule identifies all major software engineering activities and the product functions to which they are applied. As the project gets under way, each entry on the macroscopic schedule is refined into a *detailed schedule.* Here, specific software tasks (required to accomplish an activity) are identified and scheduled.

Scheduling for software development projects can be viewed from two rather different perspectives. In the first, an end-date for release of a computer-based system has already (and irrevocably) been established. The software organization is constrained to distribute effort within the prescribed time frame. The second view of software scheduling assumes that rough chronological bounds have been discussed but that the end-date is set by the software engineering organization. Effort is distributed to make best use of resources and an end-date is defined after careful analysis of the software. Unfortunately, the first situation is encountered far more frequently than the second.

As in all other areas of software engineering, a number of basic principles guide software project scheduling:

Compartmentalization: The project must be compartmentalized into a number of manageable activities and tasks. To accomplish compartmentalization, both the product and the process are decomposed (Chapter 3).

Interdependency: The interdependencies of each compartmentalized activity or task must be determined. Some tasks must occur in sequence, whereas others can occur in parallel. Some activities cannot commence until the work product produced by another is available. Other activities can occur independently.

Time allocation: Each task to be scheduled must be allocated some number of work units (e.g., person-days of effort). In addition, each task must be assigned a start date and a completion

---

[4]The critical path is the sequence of project tasks that must be closely monitored by the project manager.

date that is a function of the interdependencies and whether work will be conducted on a full-time or part-time basis.

Effort validation. Every project has a defined number of staff members. As time allocation occurs, the project manager must ensure that no more than the allocated number of people have been allocated at any given time. For example, consider a project that has three assigned staff members (e.g., 3 person-days are available per day of assigned effort[5]). On a given day, seven concurrent tasks must be accomplished. Each task requires 0.50 person-day of effort. More effort has been allocated than there are people to do the work.

Defined responsibilities: Every task that is scheduled should be assigned to a specific team member.

Defined outcomes: Every task that is scheduled should have a defined outcome. For software projects, the outcome is normally a work product (e.g., the design of a module) or a part of a work product. Work products are often combined in *deliverables*.

Defined milestones: Every task or group of tasks should be associated with a project milestone. A milestone is accomplished when one or more work products have been reviewed for quality and have been approved.

Each of the above principles is applied as the project schedule evolves.

Tracking and Control

Project tracking and control is most effective when it becomes an integral part of software engineering work. A well-defined development strategy should provide a set of milestones that can be used for project tracking. Control focuses on two major issues: quality and change.

To control quality, a software project team must establish effective techniques for software quality assurance, and, to control change, the team should establish a software configuration management framework.

111.4 Software Quality Assurance

In his landmark book on quality, Crosby [Crosby 1979] states:

> The problem of quality management is not what people don't know about it. The problem is what they think they do know . . .
>
> In this regard, quality has much in common with sex. Everybody is for it. (Under certain conditions, of course.) Everyone feels they understand it. (Even though they wouldn't want to explain it.) Everyone thinks execution is only a matter of following natural inclinations. (After all, we do get along somehow.) And, of course, most people feel that problems in these areas are caused by other people. (If only *they* would take the time to do things right.)

There have been many definitions of software **quality** proposed in the literature. For our purposes, software quality is defined as: *Conformance to explicitly stated functional and performance requirements, explicitly documented development standards, and implicit characteristics that are expected of all professionally developed software.*

There is little question that the above definition could be modified or extended. If fact, a definitive definition of software quality could be debated endlessly. But the definition stated above does serve to emphasize three important points:

1. Software requirements are the foundation from which *quality* is assessed. Lack of conformance to requirements is lack of quality.
2. A mature software process model defines a set of development criteria that guide the manner in which software is engineered. If the criteria are not followed, lack of quality will almost surely result.

---

[5]In reality, less than 3 person-days are available because of unrelated meetings, sickness, vacation, and a variety of other reasons. For our purposes, however, we assume 100% availability.

3. There is a set of *implicit requirements* that often goes unmentioned (e.g., the desire for good maintainability). If software conforms to its explicit requirements, but fails to meet implicit requirements, software quality is suspect.

Software quality is designed into a product or system. It is not imposed after the fact. For this reason, **software quality assurance** (SQA) actually begins with the set of *technical methods and tools* that help the analyst to achieve a high quality specification and the designer to develop a high quality design.

Once a specification (or prototype) and design have been created, each must be assessed for quality. The central activity that accomplishes quality assessment is the **formal technical review** (FTR). The FTR—conducted as a *walkthrough* or an *inspection* [Freedman and Weinberg 1990]—is a stylized meeting conducted by technical staff with the sole purpose of uncovering quality problems. In many situations, formal technical reviews have been found to be as effective as testing in uncovering errors in software [Gilb and Graham 1993].

Software testing combines a multistep strategy with a series of test case design methods that help ensure effective error detection. Many software developers use software testing as a quality assurance "safety net." That is, developers assume that thorough testing will uncover most errors, thereby mitigating the need for other SQA activities. Unfortunately, testing, even when performed well, is not as effective as we might like for all classes of errors. A much better strategy is to find and correct errors (using FTRs) before getting to testing.

The degree to which formal *standards and procedures* are applied to the software engineering process varies from company to company. In many cases, standards are dictated by customers or regulatory mandate. In other situations standards are self-imposed. An assessment of compliance to standards may be conducted by software developers as part of a formal technical review, or, in situations where independent verification of compliance is required, the SQA group may conduct its own *audit*.

A major threat to software quality comes from a seemingly benign source: *changes*. Every change to software has the potential for introducing error or creating side effects that propagate errors. The **change control** process contributes directly to software quality by formalizing requests for change, evaluating the nature of change, and controlling the impact of change. Change control is applied during software development and, later, during the software maintenance phase.

Measurement is an activity that is integral to any engineering discipline. An important object of SQA is to track software quality and assess the impact of methodological and procedural changes on improved software quality. To accomplish this, **software metrics** must be collected.

Record keeping and recording for SQA provide procedures for the collection and dissemination of SQA information. The results of reviews, audits, change control, testing, and other SQA activities must become part of the historical record for a project and should be disseminated to development staff on a need-to-know basis. For example, the results of each formal technical review for a procedural design are recorded and can be placed in a "folder" that contains all technical and SQA information about a module.

111.5 Software Configuration Management

Change is inevitable when computer software is built. And change increases the level of confusion among software engineers who are working on a project. Confusion arises when changes are *not* analyzed before they are made, recorded before they are implemented, reported to those who should be aware that they have occurred, or controlled in a manner that will improve quality and reduce error. Babich [Babich 1986] discusses this when he states:

> The art of coordinating software development to minimize . . . confusion is called *configuration management*. Configuration management is the art of identifying, organizing, and controlling modifications to the software being built by a programming team. The goal is to maximize productivity by minimizing mistakes.

Software configuration management (SCM) is an umbrella activity that is applied throughout the software engineering process. Because change can occur at any time, SCM activities are developed to (1) identify change, (2) control change, (3) ensure that change is being properly implemented, and (4) report change to others who may have an interest.

A primary goal of software engineering is to improve the ease with which changes can be accommodated and reduce the amount of effort expended when changes must be made.

111.6 Summary

The role of a software project manager is to understand the scope of the problem to be solved and, knowing this, to select an appropriate development strategy for the problem. Once a strategy is selected, software project management activities are conducted. Project management encompasses the measurement of the process and the product, estimation, risk analysis, scheduling, and tracking. To control the project, software quality assurance and software configuration management also must be conducted.

Defining Terms

Capability maturity model: Defines key activities required at different levels of software process maturity.

Change control: An umbrella process that enables a project team to accept, evaluate, and act on changes in a systematic manner.

Classic life cycle: A linear, sequential approach to process modeling.

Common process framework: A process model that encompasses a limited set of problem-solving activities populated by tasks, milestones, SQA points, and deliverables.

Component assembly model: A process model that encourages construction of software from reusable software components.

Design: An activity that translates the requirements model into a more detailed model that is the guide to implementation of the software.

Errors: A lack of conformance found before software is delivered to the customer.

Estimation: A project planning activity that attempts to project effort and cost for a software project.

Evolutionary model: A process model that is designed with the recognition that software evolves through a number of iterations.

Formal methods: A mathematical approach to the specification and validation of computer-based systems.

Formal technical review: A structured meeting conducted by software engineers and others with the intent of uncovering errors in some deliverable or work product.

Fourth generation techniques: Encompasses a broad array of software tools that enables the software engineer to specify some characteristic of software at a high level of abstraction.

Incremental model: A process model that results in delivery of versions of an application that provide increasingly greater functionality.

Linear sequential model: A process model that defines a set of linear activities for developing computer software.

Maintenance: The activities associated with changes to software after it has been delivered to end-users.

Measurement: Collecting quantitative data about the software or the software engineering process.

Object-oriented: An approach to software development that makes use of a classification approach and packages data and processing together.

Process model: A model that outlines the major activities and work flow for software development and acts as a framework for project management.

Prototyping: The creation of a mock-up of an application with the intent of helping a customer to better identify requirements.

Quality: The degree to which a product conforms to both explicit and implicit requirements.

RAD: A linear sequential software development process model that emphasizes an extremely short development cycle.

Reengineering: A series of activities that transform old systems (with poor maintainability) into software that exhibits high quality.

Requirements analysis: A modeling activity whose objective is to understand what the customer really wants.

Risk analysis: The set of activities that identify and evaluate a potential problem or occurrence that may put a project in jeopardy.

Scheduling: The activity that lays out a timeline for work to be conducted on a project.

Software engineering: A discipline that encompasses process, methods, and tools.

Software metrics: Quantitative measures of the process or the product.

Software quality assurance (SQA): A series of activities that assist an organization in producing high-quality software.

Spiral model: An evolutionary software engineering paradigm.

Testing: A set of activities that attempt to find errors.

Work breakdown structure (WBS): The set of work tasks required to build the software; defined as part of the process model.

References

Babich, W. 1986. *Software Configuration Management.* Addison–Wesley, Reading, MA.

Boehm, B. 1988. A spiral model for software development and enhancement. *Computer* 21(5):61–72.

Bradac, M., Perry, D., and Votta, L. 1994. Prototyping a process monitoring experiment. *IEEE Trans. Software Eng.* 20(10):774–784.

Brooks, M. 1975. *The Mythical Man-Month.* Addison–Wesley, Reading, MA.

Butler, J. 1994. Rapid application development in action. *Managing System Development, Applied Computer Research* 14(5):6–8.

Charette, R. 1990. *Application Strategies for Risk Analysis.* McGraw–Hill, New York.

Cougar, J. and Zawacki, R. 1980. *Managing and Motivating Computer Personnel.* Wiley, New York.

Crosby, P. 1979. *Quality is Free.* McGraw–Hill, New York.

Curtis, B. 1989. People management maturity model. *Proc Intl. Conf. Software Eng.*, Pittsburgh.

Davis, A. and Sitaram, P. 1994. A concurrent process model for software development. *Software Eng. Notes* 19(2):38–51.

DeMarco, T. and Lister, T. 1987. *Peopleware.* Dorset House.

Dreger, J. B. 1989. *Function Point Analysis.* Prentice–Hall, Englewood Cliffs, NJ.

Dyer, M. 1992. *The Cleanroom Approach to Quality Software Development.* Wiley, New York.

Fenton, N. E. 1991. *Software Metrics.* Chapman & Hall, New York.

Freedman, D. and Weinberg, G. 1990. *The Handbook of Walkthroughs, Inspections and Technical Reviews.* Dorset House.

Gilb, T. 1988. *Principles of Software Engineering Management.* Addison–Wesley, Reading, MA.

Gilb, T. and Graham, D. 1993. *Software Inspection.* Addison–Wesley, Reading, MA.

Hanna, M. 1995. Farewell to waterfalls, pp. 38–46. *Software Magazine.* May.

Hetzel, B. 1993. *Making Software Measurement Work.* QED Publishing.

Humphrey, W. and Kellner, M. 1989. Software process modeling: principles of entity process models, pp. 331–342. In *Proc. 11th Intl. Conf. Software Eng.* IEEE Computer Society Press.

Jones, C. 1991. *Applied Software Measurement.* McGraw–Hill, New York.

Jones, C. 1994. *Assessment and Control of Software Risks.* Yourdon Press.

Kellner, M. 1991. Software process modeling support for management planning and control, pp. 8–28. In *Proc. 1st Intl. Conf. Software Process.* IEEE Computer Society Press.

Kerr, J. and Hunter, R. 1994. *Inside RAD.* McGraw–Hill, New York.

Lorenz, M. and Kidd, J. 1994. *Object-Oriented Software Metrics.* Prentice–Hall, Englewood Cliffs, NJ.

Martin, J. 1991. *Rapid Application Development.* Prentice–Hall, Englewood Cliffs, NJ.

McDermid, J. and Rook, P. 1993. Software development process models, pp. 15/26–15/28. In *Software Engineer's Reference Book.* CRC Press, Boca Raton, FL.

Mills, H. D., Dyer, M., and Linger, R. 1987. Cleanroom software engineering. *IEEE Software* X(Y):19–25.

Nierstrasz. 1992. Component-oriented software development. *Commun ACM* 35(9):160–165.

Paulk, M. et al. 1993. *Capability Maturity Model for Software*. Software Engineering Institute, Carnegie Mellon University, Pittsburgh, PA.

Pressman, R. S. 1993. *A Manager's Guide to Software Engineering*. McGraw–Hill, New York.

Raccoon, L. B. S. 1995. The chaos model and the chaos life cycle. *ACM Software Eng. Notes* 20(1):55–66.

Sheleg, W. 1994. Concurrent engineering: a new paradigm for C/S development. *App. Dev. Trends* 1(6):28–33.

Weinberg, G. 1988. *Understanding the Professional Programmer*. Dorset House.

Wood, J. and Silver, D. 1994. *Joint Application Design*, 2nd ed. Wiley, New York.

Yourdon, E. 1994. Software reuse. *App. Dev. Strategies* VI(12):1–16.

Zuse, H. 1990. *Software Complexity*. deGruyer, Berlin.

Further Information

The current state of the art in software engineering can best be determined from monthly publications such as *IEEE Software, Computer,* and the *IEEE Transactions on Software Engineering.* Industry periodicals such as *Application Development Trends* and *Software Development* often contain articles on software engineering topics. The discipline is "summarized" every year in the *Proceedings of the International Conference on Software Engineering,* sponsored by the IEEE and ACM, and is discussed in depth in journals such as *ACM Transactions on Software Engineering and Methodology, ACM Software Engineering Notes,* and *Annals of Software Engineering.*

Many software engineering books have been published in recent years. Some present an overview of the entire process, whereas others delve into a few important topics to the exclusion of others. Three anthologies that cover a wide range of software engineering topics are:

Keyes, J., ed. 1993. *Software Engineering Productivity Handbook*. McGraw–Hil, New York.

McDermid, J., ed. 1993. *Software Engineer's Reference Book*. CRC Press, Boca Raton, FL.

Marchiniak, J. J., ed. 1994. *Encyclopedia of Software Engineering*. Wiley, New York.

An excellent three-volume series written by Weinberg (1992, 1993, 1994. *Quality Software Management.* Dorset House) introduces basis systems thinking and management concepts, explains how to use measurements effectively, and addresses "congruent action," the ability to establish "fit" between the manager's needs, the needs of technical staff, and the needs of the business. It will provide both new and experienced managers with useful information. Fred Brooks (1995. *The Mythical Man-Month,* Anniversary Edition, Addison–Wesley, Reading, MA) has updated his classic book to provide new insight into software project and management issues. S. Purba (1995. *How to Manage a Successful Software Project.* Wiley, New York) presents a number of case studies that indicate why some projects succeed and others fail. E. Bennatan (1995. *Software Project Management in a Client/Server Environment.* Wiley, New York) discussed special management issues associated with the development of client/server systems.

R. House (1988. *The Human Side of Project Management.* Addison–Wesley, Reading, MA) and P. Crosby (1989. *Running Things: The Art of Making Things Happen.* McGraw–Hill, New York) provide practical advice for managers who must deal with human as well as technical problems. Books by T. DeMarco and T. Lister [1987] and G. Weinberg [1988] provide useful insight into software people and the way in which they should be managed.

Pragmatic guidance on project management is presented by F. O'Connell (1994. *How To Run Successful Projects.* Prentice–Hall, Englewood Cliffs, NJ). Still another take on project management in the software world is provided by L. Constantine (1995. *Constantine on Peopleware.* Prentice–Hall, Eaglewood Cliffs, NJ).

A wide variety of information sources on software engineering and the software process is available on the internet. An up-to-date list of World Wide Web references that are relevant to the software process can be found at http://www.rspa.com.

112

Software Tools and Environments

Steven P. Reiss
Brown University

112.1 Introduction

Any system that assists the programmer with some aspect of programming can be considered a programming tool, and a system that assists in some phase of the software development process is called a software tool. A programming environment is a suite of programming tools designed to simplify programming and thereby enhance programmer productivity. A software engineering environment extends this suite to include software tools and the whole software development process.

This article starts with a brief history of the evolution of software tools and an overview that characterizes the different types of tools. It then discusses in more detail the tools available both for programming and for software engineering. We then consider programming environments and how they integrate the various tools and conclude with a look at future trends.

A Brief History

Software tools have been in use almost as long as programs have been written. The first tools, compilers that let programmers use higher-level languages, were soon followed by loaders, which allowed multiple compilations to be combined into a single executable. The advent of time sharing, interactive computing, and more powerful machines in the 1960s and early 1970s led to tools that took advantage of these capabilities. Compilers were sometimes replaced with interpreters, allowing very fast interaction between programmer and program. Text editors were used for program editing in place of punched cards or paper tape. Interactive languages such as Lisp introduced new programming-tool concepts such as interactive editing and debugging as well as a database of information about the program that could be queried by the user. Batch compilers produced a wealth of cross-reference information for program debugging as well as detailed program traces and tracebacks. General-purpose interactive debuggers were introduced, first for assembly-language and then for compiled source language programs.

In the mid-1970s, the UNIX operating system provided an open framework for combining a large collection of programming tools by using a simple character-oriented file system and gave programmers the ability to chain commands. Many now-standard programming tools were introduced in UNIX, including system-building tools such as *make*, version-management tools such as *sccs* and *rcs*, preprocessors such as *lex* and *yacc*, and performance analyzers such as *prof*.

The early 1980s saw the introduction of workstations as a programming aid. The Xerox Palo Alto Research Center in the 1970s had developed a proprietary workstation along with a set of specialized programming environments, one for Lisp [Teitelman 1974], one for Smalltalk [Goldberg and Robson 1983], and one for Mesa, a Pascal-like language [Mitchell et al. 1979]. Later efforts by a wide variety of researchers concentrated on providing environments with rapid turnaround through incremental compilation, generating environments for a number of languages, and providing new tools that exploited the graphics capabilities of the workstations. All these environments also attempted to provide a high degree of integration among the tools, generally by viewing the whole collection of tools as a single system with common data structures.

The late 1980s continued the push toward environments integrating software tools, many of them designed to support large-scale software engineering. Efforts in Europe led to the PCTE standards [Boudier et al. 1989]; efforts in the United States led to proprietary repositories or databases for software engineering from the larger computer companies. Whereas these efforts concentrated on sharing data among tools, other efforts demonstrated that an open environment could be built by having the tools communicate with messages. This technology was simple to use and yielded a high degree of tool integration. Messaging has become the standard for most current programming environments.

Programming environments in the 1990s have had tools that attempt to make better use of the available hardware and to address the problems of large-scale software engineering. As the amount of computing power available to the individual programmer has jumped, tools that would once have been prohibitively expensive, such as those that monitor a program as it runs, have become practical. Advances in graphics hardware and software have led to a range of display-based tools both as front ends to existing tools and as new applications. Finally, as software projects continue to grow in size, new tools have been developed to manage and simplify them.

Software Tools and Productivity

The primary function of software tools and environments has always been to make the programmer more productive. One approach to enhancing productivity is to simplify or automate the programmer's task. For example, compilers raise the level of design and implementation to let the programmer concentrate on higher-level issues, and not the fine points of register usage and instruction timing. Loaders let one build a large system in pieces so that a single change does not require a massive recompilation. System-building tools allow the programmer to specify once the set of commands needed to bring the system up to date; the tool then determines automatically what needs to be recompiled and the system can be rebuilt with a single command. Testing systems automate the running of a large set of test cases and highlight aberrant results. Language-intelligent editors provide the programmer with immediate feedback on syntactic errors and have templates that let a program be built with fewer keystrokes.

Other tools enhance programmer productivity by providing information about the program. Interactive **debuggers** allow programmers to understand a program's behavior and thus find bugs. Cross-reference utilities and their corresponding graphical interfaces, such as class hierarchy browsers or call-graph visualizers, provide information about program structure. Run-time performance monitors show where the application is spending its time.

Software tools also enhance programmer productivity indirectly by facilitating the development of new programming languages, architectures, and methodologies. Compilers and preprocessors such as *lex* and *yacc* are obvious examples of this trend. Other examples are not as obvious; yet it is difficult, for example, to envision working with a language as complex as C++ without a symbolic debugger. Designing today's large software systems on paper without design tools and checkers would be almost impossible.

Characterizing Software Tools and Environments

Software tools can be characterized by the phase of software development and the particular problems they address. Software systems are developed in stages: user requirements, system specifications, top-level and detailed design, coding and testing, and finally maintenance of the system. Different software tools are applicable to different stages.

Most current tools, especially those developed early on, are aimed at programming rather than software engineering. This means that they address programmer productivity issues arising primarily during the coding and testing phases of software development. Within the domain of programming, there are tools geared to the actual writing of programs (editors and preprocessors), tools that automate the translation from source language to machine code (compilers, system builders, and linkers), tools that support debugging (debuggers, memory analyzers, and performance analyzers), and tools that support understanding (**cross-referencers** and program visualizers).

Other tools address programming issues less directly. System-building tools, for example, allow the programmer to define how the system should be put together and automate its construction. **Version-management tools** in effect allow several programmers to work on a system while preserving earlier versions of the same system.

As the importance of the noncoding phases of software engineering has become more apparent and as these phases have become better understood, tools have been developed to simplify and automate the corresponding tasks. **Computer-aided software engineering (CASE) tools** are generally geared toward automating the first few stages of system development, including gathering user requirements, providing usable and consistent specifications, and supporting program and system design. Testing tools are designed to develop test cases automatically, evaluate their effectiveness, and automate the running of a large collection of test cases whenever the system changes. Version-management tools allow simultaneous maintenance and development of multiple instances of a single system.

Software environments are characterized by the kinds of tools they contain and thus by what aspects of software development they address. Programming environments typically integrate a collection of programming tools and are directed mainly toward the coding and maintenance phases, whereas CASE environments typically contain design and specification tools and are directed at the earlier phases (requirements, specifications, and design) of system development.

Software environments can also be distinguished by how their tools interact. Early environments provided either a high degree of integration by including all the tools within the same address space or less integration by using the file system. The key to determining the level and type of integration among the tools in an environment is to look at how the various tools share data. In a single-system environment, the data are in the address space and all tools can access all other tools' data. In a file-based environment such as UNIX, the tools share data only through files in the file system. These early alternatives represent opposite ends of an integration spectrum. Newer environments attempt to bridge this gap, either by sharing all information using a common database or by sharing selective information through messages.

In the next sections of this chapter, we consider the various programming and software tools in turn, discussing the function and evolution of each. We consider first tools directed at programming and then those directed at other aspects of software engineering. Later we discuss integration strategies and their use and effect within programming and software environments.

112.2 Programming Tools

Tools that are used primarily for writing and implementing a program are called programming tools. These tools manage the creation of new code, the edit–compile–debug cycle typically used during development, and the analysis of existing code.

Program Editors

Program editors are tools with which the programmer can create or edit source files. The earliest editors used for programming were simple text editors that knew nothing about the text they were editing and thus gave the programmer little direct assistance. Editors in single-system environments were given additional functionality; in particular, the various dialects of Lisp starting with Interlisp [Teitelman 1974] included **structure editors**, editors that work in terms of the underlying syntactic constructs of the language. Structure editors allow the user to easily choose and replace whole syntactic units and to add syntactic constructs quickly by inserting automatically all the extra keywords and punctuation required.

However, experience with structure editors, which became popular in the early 1980s [Donzeau-Gouge et al. 1984, Fischer et al. 1984, Notkin et al. 1985, Reiss 1985, Teitelbaum and Reps 1981], was generally negative. Programmers think in terms of semantic chunks rather than strict syntactic units, and many low-level constructs such as expressions are easier to enter by typing. Moreover, programmers read their programs with the syntactic sugar and are slowed down by having to write something different from what they read. Some editors that support both typing and structure editing, though, are still in use [Borras et al. 1989].

An alternative to structure editing is the addition of language knowledge to standard text editors. **Language-knowledgable editors** are a compromise between straight text editors and structure editors. They provide such capabilities as automatic indentation, parenthesis checking, and even simple cross-referencing. The primary such editor today is *emacs* [Gosling 1982], which provides automatic indentation for a variety of languages and uses the UNIX tags facility to provide syntactic cross-referencing across files in a directory. More recent versions of the editor color code the different parts of a program (comments, declarations, statements) to enhance understandability and readability.

Program editors also vary in their adaptability to other tools. *emacs*, for example, has been extended to be a front end to a variety of different programming tools for system building, debugging, and other applications. Editors in today's integrated programming environments allow the user not only to invoke other tools but also to issue commands to running tools from within the editor (for instance, setting a breakpoint in the debugger by clicking on a line in the editor).

A simple example from the FIELD environment of such an editor is shown in Fig. 112.1. FIELD, the first programming environment to feature control integration [Reiss 1994], provides visual front ends for a wide variety of programming tools, including all the standard UNIX tools and others for program visualization. The various tools communicate by using messages and a central message server. The editor tool, shown here, features a language-knowledgable text editor on the right side of its main window and an annotation window on the left. Annotations are used to connect the editor with the other tools; in Fig. 112.1, the stop sign indicates a breakpoint in the debugger, the eyeglasses and arrow indicate the debugger focus and the current line, and the × icon indicates a cross-reference.

Compilers

Compilers convert the source code created by an editor into a machine-executable form, whether actual machine code, relative object code, or an internal form that can be interpreted. Compilers differ in the amount of optimization they perform and the level at which they work.

Compiler optimization has improved dramatically over the past few decades. Originally compilers generated code that was far less efficient than handwritten code. Advances in compiler technology and the increasing sophistication of machines have made today's compilers able to generate code superior to what all but the best programmers would write. These advances have allowed high-level languages to be used for a wider variety of applications and encouraged the development of more powerful languages. Optimizing compilers also have made possible the widespread use of reduced instruction set computer (RISC) architectures, which have severe timing constraints and require very efficient register usage to achieve their full potential. Today's optimizing compilers can address these issues.

Most compilers today work at the file level, meaning that they compile one file at a time. Because files can be large and compilers slow, recompilation after a minor change can be expensive. The workstation

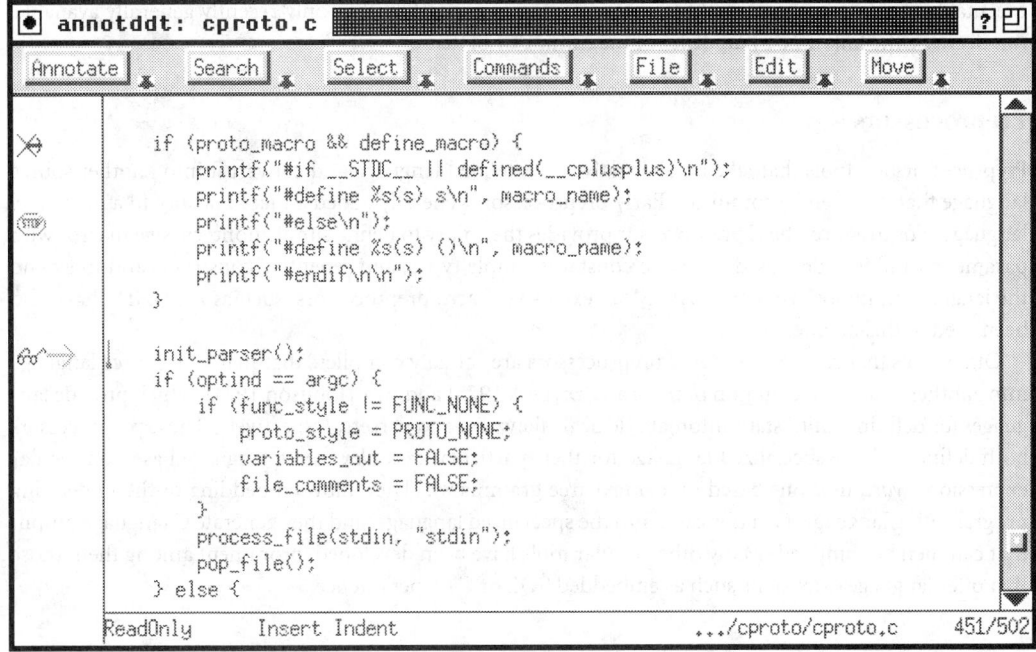

```
if (proto_macro && define_macro) {
    printf("#if __STDC__ || defined(__cplusplus)\n");
    printf("#define %s(s) s\n", macro_name);
    printf("#else\n");
    printf("#define %s(s) ()\n", macro_name);
    printf("#endif\n\n");
}

init_parser();
if (optind == argc) {
    if (func_style != FUNC_NONE) {
        proto_style = PROTO_NONE;
        variables_out = FALSE;
        file_comments = FALSE;
    }
    process_file(stdin, "stdin");
    pop_file();
} else {
```

FIGURE 112.1 FIELD program editor. The area on the left is an annotation window showing annotations tying the editor to the other tools. The arrow, for example, indicates the current debugger line of focus, while the stop sign indicates a breakpoint. (See color version of this figure in the color section of the Handbook.)

programming environments developed in the early 1980s attempted to avoid this implicit delay by introducing incremental compilation, a compilation method that also attempted to make compiler-based environments competitive with interpreter-based environments for handling simple program modifications. Incremental compilation turned out to require a significant amount of memory and time to maintain the necessary structures, such as the parse tree or the symbol table; the resultant overhead and the fact that an incremental compiler was useful only in its own closed environment have strongly discouraged the use of such tools. Moreover, incremental compilers cannot do any real optimization. They are thus found today only in limited applications, typically working at the function level rather than on lower-level constructs in interpreted or partially interpreted environments.

Linking Loaders

Linking loaders combine compiled object files with system libraries to produce an executable file. They typically work in two passes. The first pass determines what object files and library components are required by the resultant executable. For each file or component, the loader assigns run-time addresses and allocates appropriate space in the resultant file. The second pass creates the executable file by copying the object files and relocating them, modifying all absolute addresses to point to the actual address in the resultant file.

Linking loaders have changed little since their introduction around 1960. The major change has been the introduction of incremental linking techniques. Link time, the time spent in constructing the actual executable once the compiler has finished, has become a significant fraction of the edit–compile–debug cycle as the size of executables has grown. Most of this time is spent doing file input and output, because the linker must typically read and write several megabytes of file in order to link even a moderately sized system. Because most changes are minor ones, involving only a small portion of the overall system, an intelligent linker can quickly check to see what has changed and update only the affected portions of the executable. Such incremental linkers, introduced in the late 1980s but only recently generally available, can decrease the link time by an order of magnitude or better.

Preprocessors

Preprocessors are tools that take an alternative or extended language and convert it into another source language that can then be compiled. Early preprocessors generally added a macro facility to a high-level language. For instance, the C preprocessor provides the ability to define simple nonrecursive macros with parameters and is widely used to define constants simple types, and repetitive constructs and to extend the language in minor syntactic ways. More extensive macro preprocessors, such as *m4* UNIX, have also been used in this manner.

Other tools that can be considered preprocessors are actually compilers that map one source language into another. The most common of these are *lex* [Lesk 1975] and *yacc* [Johnson 1974], which provide languages for defining finite-state automata (lexical analyzers) and parsers, respectively. These preprocessors both define a highly specialized language for their particular task (*lex* uses a language based on regular expressions, *yacc*, uses one based on context-free grammars). They allow embedding of the underlying programming language, C in this case, into the specialized language, and they generate C language output that can then be compiled. Many other similar tools have been developed, prominent among them those that offer language extensions such as embedded SQL or C++ persistence.

Cross-Referencers

Cross-referencers relate the use of a name to its definition and are useful in helping programmers to understand and maintain a given system. For example, they can identify the set of call sites for a particular function or the locations in the system at which a given variable is used.

The earliest and still most widely used cross-reference tools are simple text-searching tools, such as *grep* in UNIX, that look for patterns in the source file. These tools are generally accurate if the names concerned are rare or unique across the system and if the program is consistent in its naming and coding styles. More sophisticated tools have been developed recently to obtain more accurate system information; these tools are based on earlier work such as Masterscope in Interlisp and research on databases for storing program information [Linton 1984].

A cross-referencing tool has two phases: generating the information and using it. Cross-reference information is generated either by an independent tool or by a compiler. Using the compiler is generally simpler and faster, because it has already computed the relationships and has accurate information. This is usually not done, however, because there are no standards for producing or storing such information and because, without great care, the amount of information generated can be quite large. The alternative is a separate tool that scans one or more source files to obtain the necessary information. A variety of such tools, which are generally designed for speed rather than accuracy, have been developed. The earliest and simplest is the UNIX tags facility, which uses simple pattern matching to find names and to differentiate functions, types, and variables. More sophisticated systems such as SNiFF [TakeFive Software 1993] use approximate parsing techniques to generate more accurate information. Still others, such as SGI's program database, use complete parsing techniques.

Cross-reference information can either be used by an existing tool, generally an editor, or be provided to the user in a separate tool. *emacs* and other UNIX editors let one use the cross-reference information embodied in the tags facility to jump from a reference to the corresponding definition, moving between files if appropriate. The FIELD editor shown in Fig. 112.1 uses a relational database of cross-reference information gathered from both compilers and separate scanners and allows arbitrary queries that let the user go to referenced lines or display cross-reference information relevant to the current selection.

Cross-reference information can be directly provided to the user either textually or by means of appropriate graphics. Textual displays can be offered in a command line interface, such as in the AT&T's C Information Abstractor [Grass and Chen 1990], or by a separate tool such as FIELD's *xref* or Sun's source browser. Figure 112.2 shows the output of *xref* on a simple query. Common graphical cross-reference tools

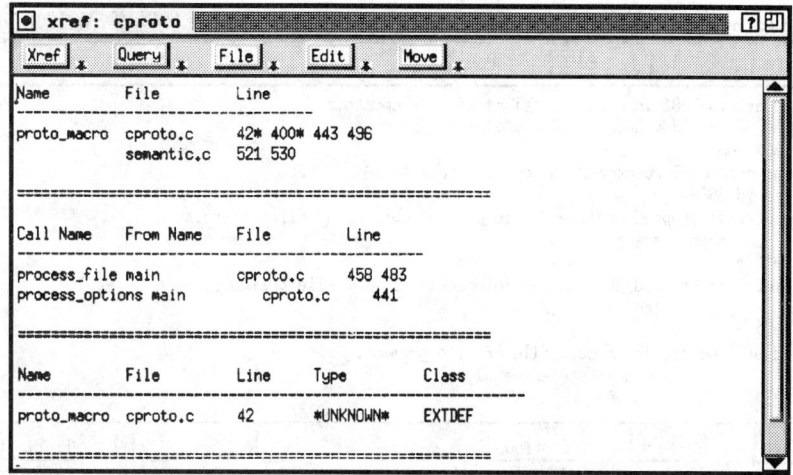

FIGURE 112.2 The results of three queries in FIELD. The first asks for all references to the name **proto_macro** ; the second looks for all calls to functions that have the word **process** in their name; and the third looks for the definition of **proto_macro**.

display standard structures such as the call graph or the class hierarchy for an object-oriented program. As an example, Fig. 112.3 shows a call graph displayed by FIELD's flowview tool with information from the underlying cross-reference database.

Source-Level Debuggers

Debuggers are tools that allow users to control program execution, typically by setting breakpoints and exploring the values of program variables. Source-level debuggers have existed since the 1960s and have

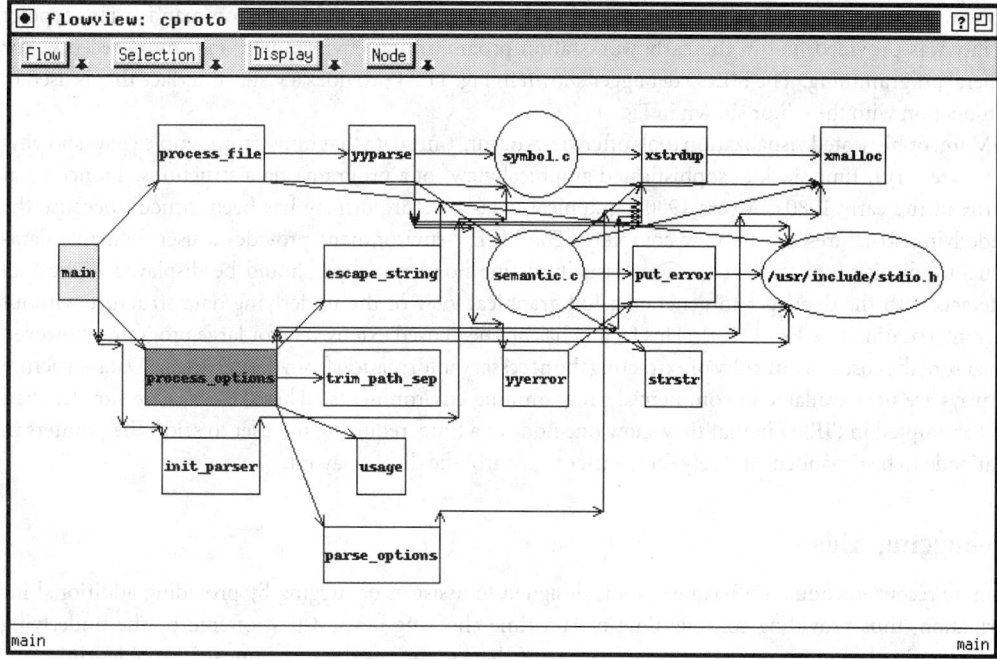

FIGURE 112.3 Call graph display from FIELD based on information in the underlying database. The highlighting reflects the node currently executing the node that is selected. Rectangles indicate functions, ovals represent files. (See color version of this figure in the color section of the Handbook.)

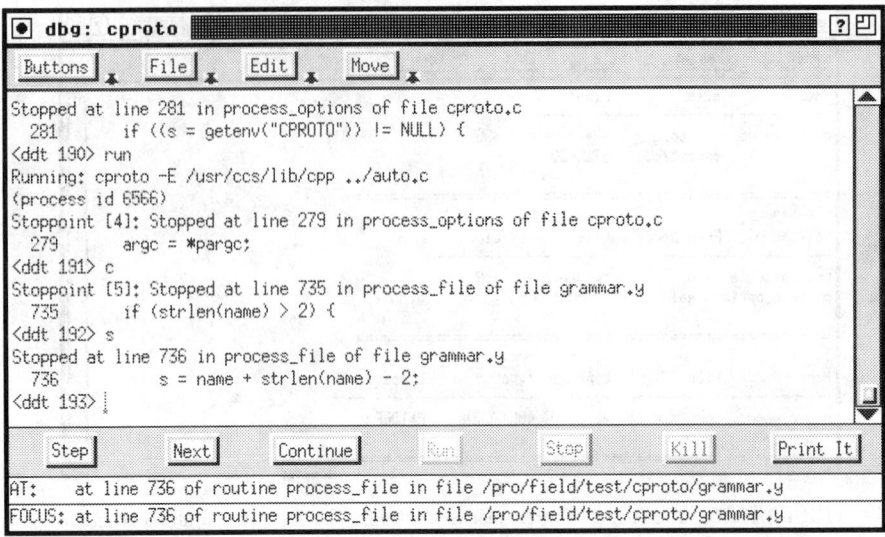

FIGURE 112.4   FIELD debugging tool shows visual interface with user-definable buttons as well as a textual transcript of debugging actions.   (See color version of this figure in the color section of the Handbook.)

not changed significantly since then, although they have become more sophisticated. Those changes that have occurred include the ability to handle optimized (or partially optimized) code, additional commands, the ability to deal with higher-level languages, better command languages, the ability to debug multilanguage programs, and visual interfaces.

The trend recently in debugger technology has been to provide better visualization tools as part of the debugger. The simplest visualization tools provide a view of the program source that is being debugged along with menu or panel buttons for executing the more common debugger commands. This type of facility was provided first in the early workstation programming environment and, more recently, for general programming. The FIELD debugger shown in Fig. 112.4 provides a visual interface that is used in conjunction with the editor shown in Fig. 112.1.

More sophisticated visualization tools offer views of run-time data that range from simple views showing the current run-time stack to sophisticated graphical views of a program's data structures. Pioneered at Xerox in the early 1980s [Myers 1980], graphical data-structure display has been difficult because the underlying structures can be very complex. The FIELD environment provides a user-definable data-structure display tool that let the user visually define how each type should be displayed. The tool interacts with the debugger to generate a full graphical view of the underlying data structure without user intervention (see Fig. 112.5). This facility has not been used extensively for large programs, however, because of the cost and difficulty of extracting the necessary information from the debugger. Data-structure displays are now available in commercial programming environments. These displays are simpler than that attempted in FIELD in that they show one node at a time, requiring the user to select the pointers in that node to be expanded, and rely on the user to control the display layout.

Debugging Aids

A more recent introduction has been tools designed to assist in debugging by providing additional information, thus providing compile-time or run-time checking beyond that offered by the underlying programming language, and to provide a variety of checks (such as detecting unused variables or inconsistent parameters to a function) that are neither made by the compiler nor required by the language but would typically indicate an error in a reasonable program. The advent of better languages (such as ANSI

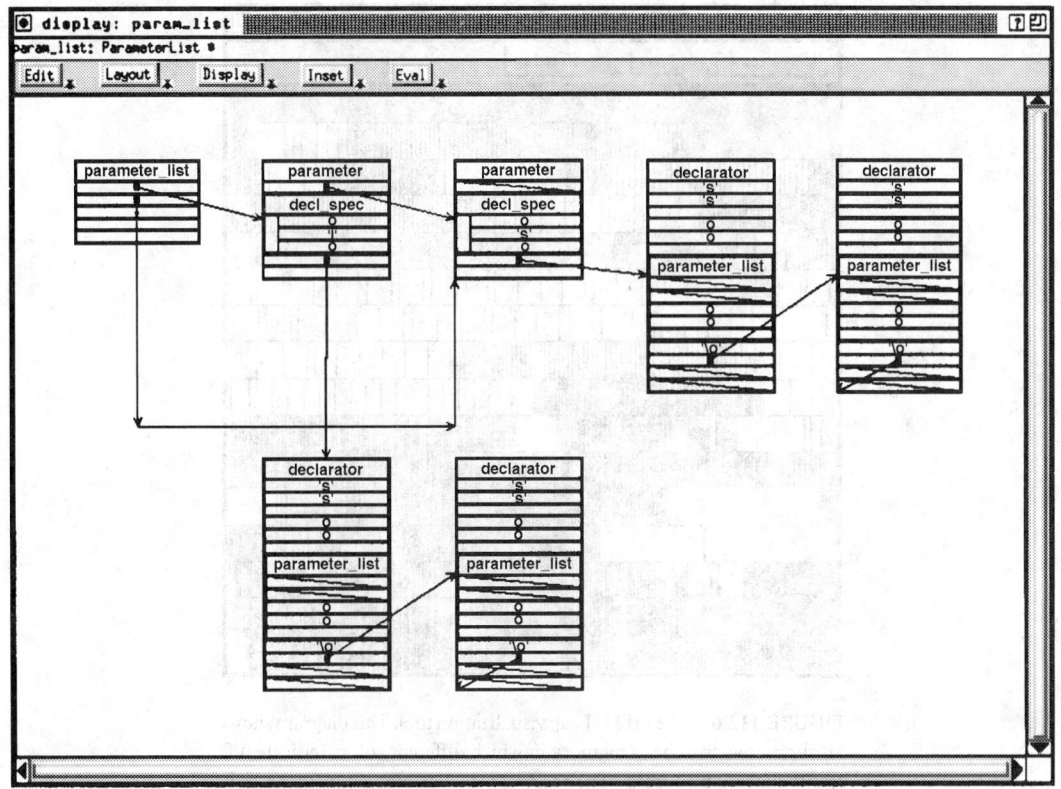

FIGURE 112.5 Data-structure display in FIELD.

C or C++ in place of C) has limited the applicability of the original lint tool, but similar tools that make additional stylistic or semantic checks are still available.

Tools that check information at run time have concentrated on detecting problems in the use of heap memory, because this is where many hard-to-detect problems reside in modern applications. The FIELD memory analyzer tool (see Fig. 112.6) provides a dynamic and user-selectable view of how the program used heap memory and has been used to detect memory leaks as well as anomalous memory allocation behavior. A more sophisticated approach is exemplified by the Purify system [Hastings and Joyce 1992], which checks each memory access to ensure that it is accessing allocated and initialized memory and accumulates information about allocated storage to detect potential memory leaks. This approach has been replicated and expanded in other tools.

Performance Analysis Tools

Run-time information gathering is also useful in understanding a program's performance. Tools that monitor a program while it is running and analyze where it spent its time have proven invaluable both in improving program performance by the identification of bottlenecks and in understanding program behavior. An early such tool is the UNIX *prof* facility. Implemented as a compiler option in most UNIX compilers, *prof* works by sampling the program counter about 100 times a second and incrementing a counter each time a routine is entered. The accumulated statistics are then related back to the source to show approximately how much time was spent in each routine.

More sophisticated program analysis is possible by looking not only at routines but at what routines called those routines. The additional information is used in *gprof* to indicate how much execution time is spent not only in a particular routine but also in that routine and any routine it calls [Graham et al. 1982]. The

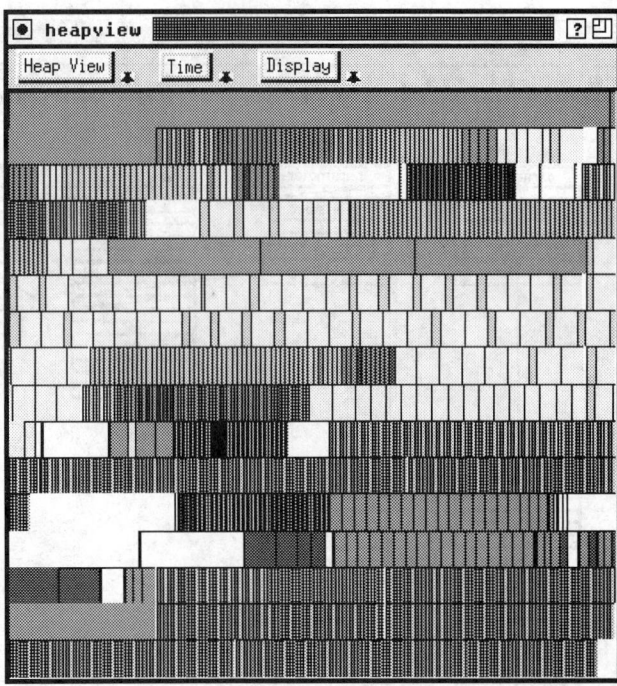

FIGURE 112.6   The FIELD heap visualization tool. This diagram shows an abstract view of heap memory in which different colors indicate different allocation sources. The region at the bottom is a memory leak. (See color version of this figure in the color section of the Handbook.)

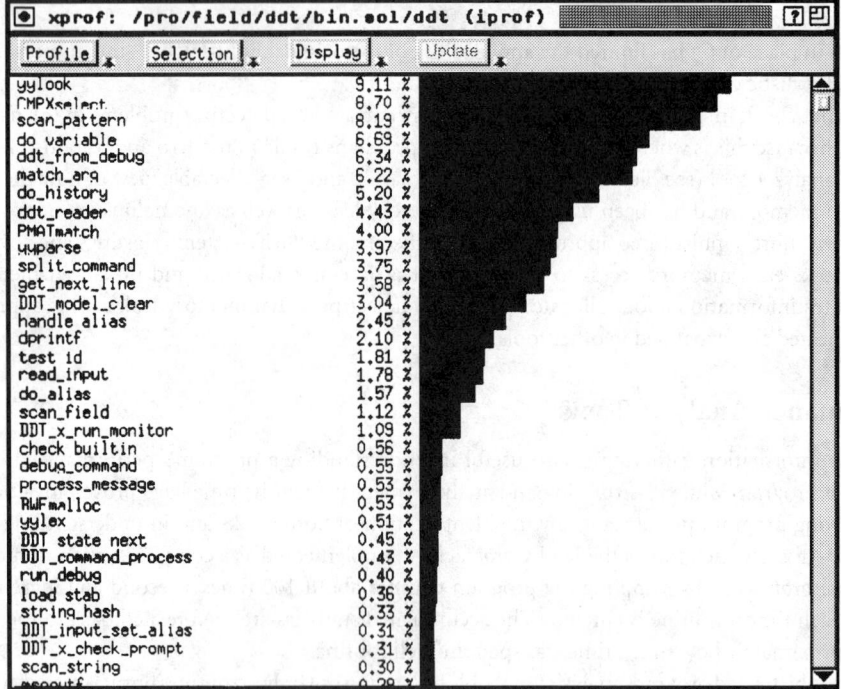

FIGURE 112.7   Performance display showing the amount of time spent in the various routines of a program.

result is a better performance breakdown that permits the faster identification of bottlenecks and a better understanding of program behavior. Further sophistication is provided by the Pixie program analyzer [MIPS Computer Systems 1988], which modifies the executable to count the number of times each basic block (sequence of instructions without a branch) is executed. It thus determines how many times each machine instruction in a program is executed and provides a line-by-line breakdown of where the program spends its time. Modern performance analysis tools combine these techniques with graphical displays to let the programmer easily understand the large amount of resultant data. An example of the graphical output provided by the FIELD profiling interface (using data generated by *gprof*) is shown in Fig. 112.7.

112.3 Software Engineering Tools

A second set of tools addressing software engineering issues is concerned with maintaining the overall system rather than its programming. Existing tools in this area include tools for building the system, tools for managing multiple versions of the system, and tools that directly address different phases of software engineering such as requirements, specifications, design, and testing.

System Building

System-building tools, exemplified by the UNIX tool *make* [Feldman 1979], allow the user to define a system model detailing the relationships among the various source files, the system itself, and the various intermediate products such as object files or preprocessor output files. This model contains information both about dependencies, i.e., which files depend on which other files, and on how to construct the intermediate and final products from their constituents. The construction rules describe the commands or tools to be run and the options to be used in running them. The system model also describes auxiliary commands such as printing the source files or removing the intermediate products.

System-building tools typically use this system model to minimize the commands to be executed after the user has made a set of changes: they determine what the user has changed in the source files and recompile only those files affected. Most of these tools look only at which files have been changed and base their recompilation decisions on that. Some more sophisticated tools, such as SGI's *SmartMake*, actually look at what has been changed in a file to determine if recompilation is in fact required.

The various system-building tools that have been proposed differ primarily in how the system model is specified and what it contains. UNIX *make* uses a declarative system model in which the dependencies are listed either explicitly or by name patterns and in which the rules for building the various products are listed with the dependencies. Macros are provided to simplify the specifications. Different versions of *make* extend this to allow conditional rules as well as macro-based functions. These versions generally consider only what the user actually specifies and do not keep track of the system commands and options used to generate the products, other implicit dependencies that could be important in regenerating a system.

System models based on functional programming, procedural programming, and object-oriented programming have all been used for system-building tools at one point or other, though none has been as successful as *make*'s declarative model. Similarly, several tools have been built, either separately or on top of *make*, that give the user a graphical view of the dependencies. The FIELD tool (Fig. 112.8.) shows only a visualization of the build dependencies; more sophisticated visual tools such as that provided with DEC's FUSE environment let the programmer graphically edit the input files to *make* [Digital Equipment Corporation 1991].

A more significant difference lies in the dependencies created by the tool. System-building tools such as Apollo's DSEE [Leblang and Chase 1984] and its successors construct internal dependencies between the product files and source files on the basis of the version of the tools used and the options used in invoking them. This ensures that object files generated from the same source files at different times are exactly the same. Such systems often maintain a data store of intermediate files so that compilations can be avoided if the work has already been done [Mahler and Lampen 1989].

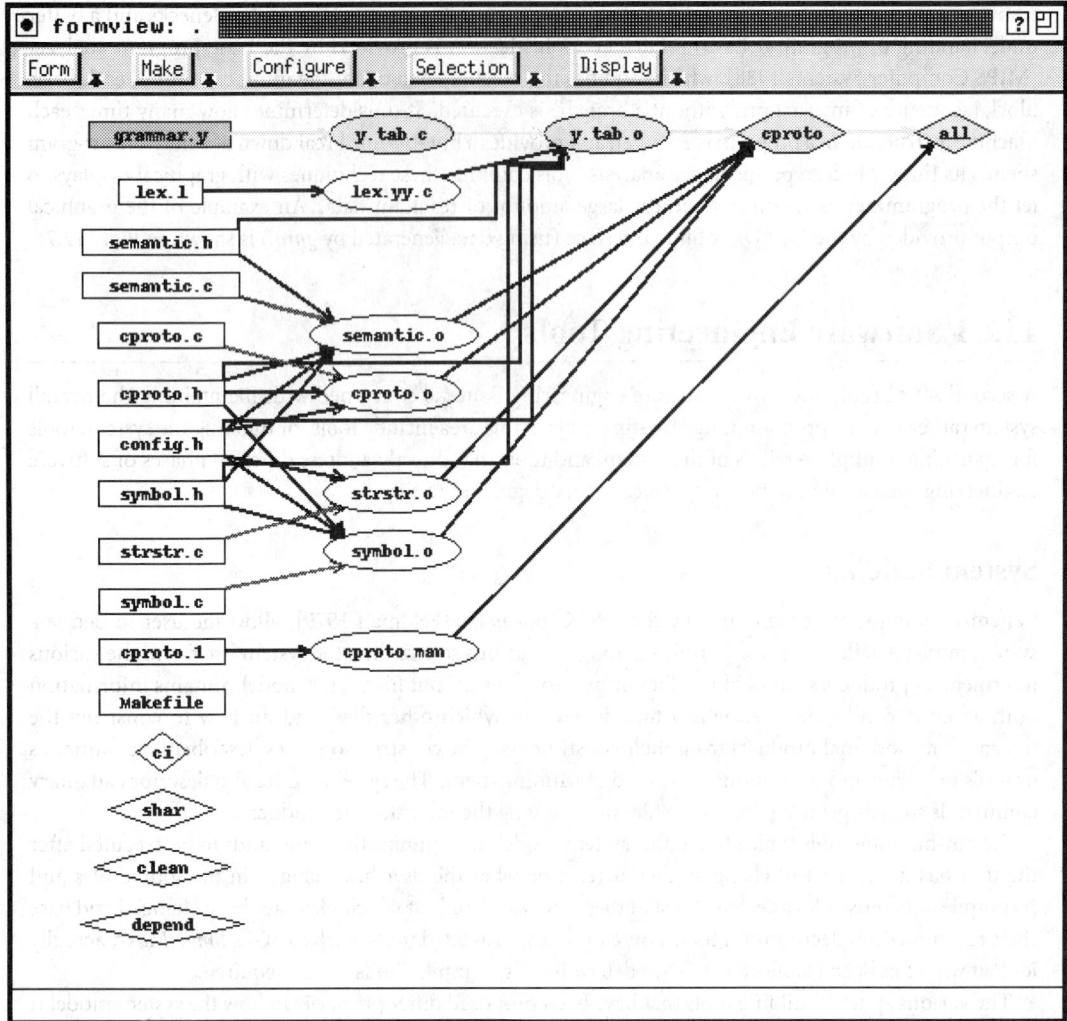

FIGURE 112.8 Graphical dependency visualization in the FIELD interface to the UNIX tool *make*. The color highlighting shows which items have been modified by the programmer and which need to be updated. Elliptical nodes represent intermediate objects, hexagons are systems, and diamonds are virtual *make* targets. The thick arrows represent user-defined dependencies, and the thin ones were induced by *make*. (See color version of this figure in the color section of the Handbook.)

Version Management

Version-management tools manage maintaining multiple versions of source files and access control to these files. Multiple versions of files are useful in large systems where different users might be running different versions. A maintainer of the system then needs access to the source files for each of the running versions in order to handle problems. Multiple versions are also useful during software development in allowing one or more new system releases to be worked on simultaneously, and they are useful in general in preserving the history of software development and thus allow a programmer to retract major changes to a system or to look at the code in a previous state. Access control to source files is convenient for managing multiple programmers working on the same project: each version of each source file is checked out, i.e., currently owned, by one programmer and becomes unwritable by the other programmers.

Version managers differ in several dimensions: whether they work at the file or directory level, the underlying model for access control, and the working model they provide. Early version managers developed for UNIX such as *sccs* [Anonymous 1981] or *rcs* [Tichy 1982] worked at the file level: programmers created versions of individual files and access control was done at the individual file level. More recent systems such as Atria's *ClearCase* work in terms of directories (logical sets of files). Although this does not give as fine a level of control, it provides a more logical and more consistent programmer interface, because a directory is typically a logical portion of a system that should be versioned together.

Similarly, there are two common access control models. Early systems used a transactional model whereby the programmer checked out or locked the various entities in order to work on them and, when done, would check them back into the system as a new version. This provides a firm handle on who is doing what and ensures there are no conflicts among programmers but is at times overly restrictive in preventing multiple programmers from making necessary fixes to the same file. An alternative, more cooperative, model is used in more recent systems by which multiple programmers are allowed to check out the same file at the same time. The first programmer to check in the file can do so. When the other programmers attempt to check the file in, the version manager compares the new file to the one previously checked in to determine if any changes conflict. If not, it merges the two sets of changes to produce a new version of the file; if so, they are resolved by the programmer.

Version-management systems also differ in the interfaces they provide the programmer. Whereas most version managers simply utilize the file system as it is, systems starting with DSEE actually modify the underlying file system to provide a unified view. Here users specify which version of a file they want to use and when they access that file (in its normal location) are given that version. Thus, multiple users can operate in what appears to be the same directory hierarchy on different versions of a file without interfering.

CASE Tools

Although programming tools have been available for a long time, only in the past 10 years or so have tools been available to help directly with other phases of software engineering. Many of these newer tools have attempted to automate the early phases, including the gathering and analysis of user requirements information, the definition of system specifications, and the creation of actual system designs. These tools are commonly called computer-aided software engineering (CASE) tools.

The tools generally available for requirements analysis and specifications are text-based tools that provide additional organizational and referencing capabilities and offer their users the ability to generate all the appropriate paperwork. The tools for dealing with the design aspects of software have generally been graphical tools based on one or more underlying graphical design notations.

A variety of different design notations have been developed for defining the structure and behavior of software systems, including Petri nets, SADT diagrams [Ross 1985], Statecharts [Harel 1984], and object-oriented design using OMT [Rumbaugh et al. 1991] and similar notations. Tools have been created to let the user construct and annotate diagrams of each of these types. Some tool suites, such as that provided by CADRE, permit multiple design notations for different aspects of the software to be combined into a single system model. The more advanced CASE tools allow the user to simulate the underlying models to get some idea of system behavior.

The traditional weakness of these CASE tools is that, although useful for the initial design, they are not particularly helpful during system development. Most of them are capable of generating a code framework based on the design that then serves as a starting point for the actual coding. However, as the programmer modifies the code, the design is not changed, because the system is unable to generate additional code. Moreover, many programmers have found that the cost of maintaining and synchronizing two representations (code and design) is not worth the benefits received.

Code-Generation Tools

Many applications today are written in well-understood domains. For example, many business applications can be viewed as front ends that allow the user to define a query, access a database to get the results of that query, and then present these results in various forms. One trend in programming is to provide software tools designed specifically to handle such applications and thus to greatly simplify their programming. These tools take a high-level description of the specific task as a definition relative to the underlying domain. This description can be specified either textually in a high-level programming language or visually in what is effectively a visual programming language. From this description, the tools generate a complete application. The high-level, domain-specific programming languages used here are often called **fourth-generation languages** (**4GLs**).

The most common applications of these languages are defining user interfaces, defining applications that involve an interface to an underlying database system, and writing applications that tie together a variety of tools. User-interface tools include standalone systems such as the Motif-based *UIMX* and its successors and the visual interfaces in Microsoft's *Visual Basic* and *Visual* C++. A variety of PC-based systems, such as those based on *Lotus Notes*, are available for visually defining applications that access a database system. SGI's *explorer* environment provides a visual language for combining a variety of tools for scientific visualization.

Although code generation tools are not common, they represent one of the most important advances in programming. Once a domain is mature enough to be analyzed and an appropriate 4GL is developed for it, applications involving that domain become significantly simpler to write. Indeed, many such applications can be written by the end users rather than by a staff of programmers.

Testing Tools

Another phase in the software development lifecycle that has been addressed by specific software tools is testing. There are two basic types of tools in this area: those that handle individual test cases and those that manage regression testing. The first set of tools takes a set of test data and analyzes how well it tests the underlying source code. The tools do this by checking what statements have been executed as well as what conditional branches and conditions have been exercised and then report to the user, generally textually, the degree of test coverage. The more sophisticated versions of these tools also do test case generation, analyzing source code or specifications and generating a suite of test cases.

Another set of **testing tools** handles the bookkeeping involved in regression testing. Most testing of large software systems is done by building up a large suite of test cases. Each time a bug or problem is found, an appropriate test case is added to the suite. Whenever a new version of the system is ready, it is tested against each of the test cases in the suite. This is called regression testing, because it is done to ensure that the system does not regress to a state where some test that previously worked no longer does. Regression-testing tools accumulate the suite of test cases and automate their running, checking the output of each test case against the expected output and reporting back which test cases failed. The more sophisticated of these tools handle graphical user interfaces, allowing the testing of programs with visual interfaces.

112.4 Integrating Tools Into an Environment

There are three basic methods of packaging tools to provide an integrated environment. The two simpler schemes are to build the environment either as a single system, as with the various Lisp environments, or as a set of independent tools, as with UNIX. The third method is to have a set of related tools and a way for those tools to communicate. The communication or integration mechanism here is the means for information sharing and coordination among the tools.

Single-system environments, which date back to the 1960s, have typically been developed to support a single programming language by providing a set of integrated facilities. For example, Lisp environments, from the early versions of Interlisp [Teitelman 1974] through those developed for Common Lisp, provide editors, compilers, and a host of debugging facilities as well as other programming tools. Environments for procedural languages developed with the early time-sharing systems, for example, with the various BASIC environments; interactive Fortran environments such as Quiktran also provided a unified set of tools.

Such environments were ushered into the modern age in the 1980s with systems like *Gandalf* from Carnegie Mellon University [Notkin et al. 1985], *PECAN* from Brown University [Reiss 1985], the *Cornell Program Synthesizer* [Teitelbaum and Reps 1981], and *Mentor* from INRIA [Donzeau-Gouge et al. 1984]. Here compiler technology was used to offer syntax-directed editing and incremental compilation for immediate programmer feedback. These systems also introduced workstation-based programming tools such as graphical program views. Today's successors to these environments are the programming systems available on personal computers, such as Symantec's *Think* C++, Borland's *Turbo* C++, and Microsoft's *Visual* C++.

Single-system programming environments can easily offer a high degree of integration, because the tools share the same data structures. At the same time, however, they have several disadvantages, primarily that they are closed systems. It is difficult to add new tools or capabilities to a single-system environment, especially tools designed outside the environment. Even in relatively extensible environments such as the various Lisp systems, incorporating a tool written for one environment into another is quite difficult.

The original alternative to a single-system programming environment was an independent set of tools that share information using files. Early time-sharing environments such as Multics or the Dartmouth Time-Sharing System had separate editors, compilers, and debuggers. The acme of such federated environments is UNIX. Born at Bell Laboratories as a programmer's environment, UNIX has slowly evolved a large set of powerful tools that cover many aspects of the programming process. It has become a mainstay in university and industrial research environments and is a fertile ground for developing new programming tools.

Whereas UNIX has demonstrated that it is relatively easy to add new tools into a federated environment, this approach of using independent tools has two primary disadvantages. The first is the risk of poor performance; because each tool is independent and compartmentalized, there is considerable duplication of effort and excess file input and output. Another disadvantage is that such environments do not give the programmer a consistent, integrated framework. Each tool typically offers its own interface and its own command language, and in addition there is little if any communication among the tools, forcing the programmer to be the integration mechanism. For example, it is the programmer who must correlate line numbers in compiler error messages with the location in the source program.

The third class of programming environment is an integrated programming environment consisting of a set of tools and an integration mechanism tying the tools together. This type of environment offers many of the advantages of both the single-system and the loose-collections-of-tools environments. It is an open environment in that new tools can be developed independently and incorporated later through the integration mechanism. Moreover, by providing a powerful integration mechanism, this approach offers a high degree of coupling among the tools and thus appears to the programmer as a single environment.

Three approaches have been used to achieve the integration of single-system environments with the openness and flexibility of federated environments. These involve different ways for the tools to share information and interfaces.

Data Integration

All environment-integration mechanisms are based on sharing information among the tools. In single-system environments, the sharing is done by allowing the tools to access common data-structures. In federated environments, the sharing is done at the file level, so that, for example, the compiler writes into the object file the information the debugger needs for relating the executable code to the original source.

Using a database system for information sharing is a natural extension of this approach. A program database extends the low-level data-structure sharing used by single-system environments by letting independent tools access a specific set of common data structures in a controlled way. Here the shared data structures are managed by a database system that provides consistency and integrity checking as well as controlled access.

There are two approaches to implementing a programming environment with a program database. In the first, all the tools use the database directly. That is, the tools are designed with the database in mind and use representations that either are stored in or can easily be derived from the database. This has the advantage of efficiency and consistency and is the approach being used to develop Ada programming support environments [Munck et al. 1989], in which an attributed abstract syntax representation is stored in a common database. Here the compiler, debugger, loader, and other tools all access the program as an abstract syntax tree by going through the common database system. The disadvantages of this approach are that existing tools must be rewritten to use the database and that the database representation must be determined before the tools are implemented, so that adding tools not initially anticipated can cause problems.

The second approach to using a program database is to treat it as a "software backplane." Here the tools can use whatever representation is most appropriate: preexisting tools can use their current representations, and new tools can be written to use whatever representation is most efficient. The database system stores a single extensible representation of the data that it maps to the forms needed for each particular application when that application is run. This approach has the advantage of allowing the use of existing tools and of making it easier to write or incorporate new tools in the future. It has the disadvantage that the mappings from the database representation to the application representation can be complex and are not necessarily one-to-one.

The use of a program database in general, however, has disadvantages. The additional system needed to maintain the database complicates the programming environment. Database systems are large, complex programs, and a program database that deals with multiple clients and maintains consistent information is no exception. This strategy also requires that the program representation be well understood before most of the tools are written, because adding new tools that do not fit well with the original definition can be difficult. Finally, program database schemas are generally designed with a particular language in mind, and it is difficult to adapt them to a different language or to accommodate multiple languages simultaneously.

Common-Front-End Tools

A second approach to integration involves the use of a common front end. The most prevalent tool in software development is the text editor, used for creating programs, documentation, system models, etc. It is relatively easy to embed in an editor commands that invoke other tools and use their output. Thus, *emacs* allows the user to invoke the system-build package *make* [Feldman 1979], puts the output of the resultant compilation in a buffer, and lets the user use that buffer to go to the source lines where errors occurred. It also allows the user to invoke the debugger and other tools. This approach gives some sense of integration by providing a common interface for a variety of tools, and it also permits new tools with textual interfaces to be integrated quite simply. It does not, however, provide all the benefits of integration, in that tools really do not share information and the programmer is still aware of the different tools and their uses.

Control Integration

Control integration offers a compromise between integration based on a program database that is too expensive and cumbersome in practice and mechanisms that provide only a common front end and fail to offer the degree of integration programmers desire. It uses a relatively simple mechanism based on broadcast messaging to augment a federated environment and provide a high degree of integration. This mechanism is based on the realization that most tools are compartmentalized and that integration requirements are generally limited to information requests and timely information sharing among the tools.

Control integration is achieved by providing message passing among the various tools. Each tool must be adapted to both send and receive messages. This can be done by modifying the tool or by providing a wrapper around it. Each tool must offer whatever functionality is required of it by the other tools through messages; messaging is centralized and independent of the tools. The environment features a central message server that connects to all the tools. Each tool, when it starts up, registers with the message server by sending it a set of patterns that describe the messages it is interested in. While tools are running, they send messages to the message server. The server in turn compares these messages to all the registered patterns and forwards each message to the tools that have expressed interest in it.

Two types of messages are generally provided for. Command messages are used when one tool wants another to execute a command—for example, when the editor wants the debugger to set a breakpoint. These messages are generally sent synchronously and require a reply from the recipient. Other messages are informational—for example, whenever program execution stops in the debugger, the debugger sends out an informational message indicating the file and line number to all interested tools. These messages are generally sent asynchronously.

Control integration provides many of the benefits we are looking for. The resultant environment is still a set of basically independent tools, yielding a degree of openness unavailable in environments using data integration. Control integration is also relatively inexpensive; both the amount of code needed to support messaging and the number of modifications needed to existing tools are relatively small.

There are, however, several potential disadvantages to control integration. It does not provide an information repository between tools but only connects the tools when they are running. This results in duplication of effort (for example, both the compiler and the cross-reference tool must parse source files). Second, the environment does not speed up the edit–compile–debug cycle of programming: doing this requires additional tools such as an incremental loader. Finally, control integration does not guarantee consistency among the tools. This has to be provided by the tools themselves by the way they use the messaging mechanism.

Actual environments combine these different integration approaches in various ways. The PCTE standard [Boudier et al. 1989], for example, uses control integration and a common set of front-end utilities on top of a data integration basis. FIELD [Reiss 1994] and related environments primarily use control integration but provide specialized repositories to hold long-term data such as cross-reference or configuration-management information.

112.5 Future Tools and Environments

Because software development is difficult and time consuming, researchers continue to develop and extend software tools and environments. What we can expect in the future is better versions of existing tools: faster compilers, incremental loaders, better and more diverse debugging aids, improved editors, etc. We will also see better environments that combine a high degree of tool integration with openness to new tools and languages. In addition, current research is aimed at extending software tool support to such new domains as process tools, groupware, visualization, and program analysis.

Process Tools

One emerging area is tools that help manage the overall process of software development. These **process tools** allow programmers or managers to set guidelines for how software should be developed and then provide tool support both to help programmers follow these guidelines and to enforce the guidelines where necessary. In addition to subsuming system-building tools, process tools can be used to ensure that programmers are informed of changes made by other programmers, that code is tested before it is checked in, and that the proper versions of code are used.

Two primary methods have been proposed for defining the guidelines and how they should be followed. One approach, taken by the Arcadia project in APPL/A, is procedural [Taylor et al. 1989]: the manager writes a program that defines how programmers should interact with the system and with one another. The alternative approach, taken in the Marvel system [Kaiser et al. 1988], is to define the guidelines as a set of rules or triggers. Whenever the system receives notice of an event, it checks its rule base to see if any of the rules should be triggered and takes the appropriate action. Similarly, if the user asks to take an action, the system can check what conditions need to be met before that action can be taken and can initiate them automatically.

Groupware Tools for Programming

Large-scale software engineering is generally done by teams of programmers working together on a single project. Most of the programming tools discussed above are designed with a single user in mind. The primary exception is tools for version management, which facilitate a limited amount of interaction among programmers by controlling access to shared source files and ensuring their consistency. New **groupware tools** allow more advanced coordination among programmers and support group design methodologies as well as group debugging and editing [Ben-Shaul et al. 1992, Kaplan et al. 1992]. Groupware is also present in tools for the early stages of software engineering. The win–win approach to requirements provides an interface and framework whereby programmers, users, and clients can negotiate the requirements or specifications of a system to be developed [Boehm et al. 1995].

Visualization Tools

Current visualization tools, notably class hierarchy displays, call-graph or module displays, and limited data-structure visualization, are not widely used yet because they are display intensive; generating and maintaining the necessary data about an evolving system is time consuming, and they do not address the primary concerns of the programmer—answering specific questions about the software quickly and accurately. Future visualization tools will attempt to avoid these problems. Better graphics capabilities and computational power should make the displays faster and easier to manipulate. Three-dimensional displays will convey more information. Compilers such as Sun's are generating the information needed for visualization cheaply as part of the compilation process. More importantly, visualization efforts are being tied to program databases to let the programmer generate specific queries about the system and visualize the result.

Program Analysis Tools

Another trend that should become visible in future software tools is the increasing use of semantic information for programmer feedback. Program analysis tools undertake a detailed semantic analysis of a system that can then be used for a variety of applications. It can be used to identify potential programming problems (variables that are set but never used, variables that are used but never set, and potential deadlock situations in parallel code), to compare code to formal specifications, identifying places where the two disagree, to assist programmers in merging different versions of a file in which multiple programmers have edited the same function, and to provide automated assistance in reorganizing or reengineering existing code. They can also be used to identify those portions of a program's execution that were relevant to a run-time problem and to identify portions of a program that can run in parallel.

Defining Terms

CASE (computer-aided software engineering) tools: Software tools that address the precoding phases of software engineering, notably requirements, specifications, and design. The more popular such tools are graphical editors for defining system operation in terms of data and control flow and object relationship diagrams.

Control integration: A means of connecting software tools by using message passing. Tools communicate with each other by sending messages through a common message server. Typically a form of broadcast messaging is used where tools tell the server what messages they are interested in and the message server, when it receives a message, rebroadcasts it to the interested tools.

Cross-referencer: A tool that gathers and makes available to other tools and possibly the user information about names in an application. Typically, this information identifies both where names are defined and where they are used. More sophisticated cross-referencers accurately associate references to a name with the proper definition.

Data integration: A means of connecting software tools by using a common database. All intermediate results and tool output are stored in the database for use by other tools. The result is a very tight integration scheme but one that is expensive to implement.

Debugger: A tool that allows the programmer to control and examine the execution of a system.

Emacs: An advanced, customizable, and extensible text editor that is generally the editor of choice on UNIX systems and is available on a wide variety of platforms. Its extensibility has been used to make it language-knowledgeable and to integrate it with other programming tools.

4GL (fourth-generation language): A very high-level, domain-specific language that allows a user to build an application in a well-understood domain quickly and easily with a minimum of programming.

Groupware tools: Tools that allow multiple programmers or designers to work cooperatively. These tools are specifically designed to support geographically distributed cooperation for software development.

Language-knowledgeable editor: A text editor that has been specialized for editing programs in one or more source languages. The knowledge of the language is typically reflected in indentation and in simple syntactic checks such as parenthesis balancing. (See also structure editor.)

lex: A program that takes a description of a set of regular expressions defining different tokens and produces a lexical analyzer for these tokens that can then be integrated into the application.

Linking loader: A programming tool that takes relocatable object files generated by a compiler and integrates them into an executable file. Loaders generally combine single object files representing user compilations with libraries of object files typically representing system libraries.

make: A tool that allows the programmer to define a declarative model of the relationships between source files and generated files and that uses this model to build the system by recompiling only those files that have changed.

Performance-analysis tools: Tools that look at the run-time behavior of a system and determine where the system is spending its time. Such tools are used to identify run-time bottlenecks and to determine what sections of a system should be modified to increase performance.

Process tools: Tools that assist in managing the software development process, for example, controlling how programmers interact or what testing must be done before a file can be incorporated into a system.

Structure editor: A text or graphical editor that allows a source program to be edited in terms of its syntactic constructs. Such editors typically allow the user to select and operate on whole syntactic units and to apply transformations to the underlying parse tree rather than the program text.

Testing tools: Tools that address the testing phase of software engineering, for example tools for generating test cases, tools for managing a suite of test cases, or tools for analyzing the result of testing.

Version-management tools: Tools that allow multiple instances of a system to exist simultaneously, typically by maintaining multiple versions of each source file in the system. These allow the programmer to reconstruct a previous version of the system.

yacc: A program that takes a description of a restricted context-free grammar defining a language and generates a parser for that language that can be integrated into an application.

References

Anonymous. 1981. Source code control system user's guide. In *UNIX Programmer's Manual.*

Ben-Shaul, I. J., Kaiser, G. E., and Heineman, G. T. 1992. An architecture for multi-user software development environments. *Software Eng. Notes* 17(5):149–158.

Boehm, B., Bose, P., Horowitz, E., and Lee, M. J. 1995. Software requirements negotiation and renegotiation aids: a theory-w based spiral approach, pp. 243–254. In *Proc. 17th Intl. Conf. Software Eng.*

Borras, P., Clement, D., Despeyroux, Th., Incerpi, J., Kahn, G., Lang, B., and Pascual, V. 1989. CENTAUR: the system. *SIGPLAN Notices* 24(2):14–24.

Boudier, G., Gallo, F., Minot, R. and Thomas, I. 1989. An overview of PCTE and PCTE+. *SIGPLAN Notices* 24(2):248–257.

Digital Equipment Corporation. 1991. *DEC FUSE for ULTRIX*. Digital Equipment.

Donzeau-Gouge, V., Heut, G., Kahn, G., and Lang, B. 1984. Programming environments based on structured editors: the MENTOR Experience. In *Interactive Programming Environments*, D. R. Barstow, H. E. Shrobe, and E. Sandewall, eds. McGraw–Hill, New York.

Feldman, S. I. 1979. MAKE: a program for maintaining computer programs. *Software Practice Experience* 9(4):255–265.

Fischer, C. N., Pal, A., and Stock, D. L. 1984. The POE language-based editor project. *SIGPLAN Notices* 19(5).

Goldberg, A. and Robson, D. 1983. *Smalltalk-80: the Language and its Implementation*. Addison–Wesley, Reading, MA.

Gosling, J. 1982. *Unix Emacs*. Carnegie-Mellon Computer Science Dept., Pittsburgh, PA.

Graham, S. L., Kessler, P. B., and McKusick, M. K. 1982. gprof: a call graph execution profiler. *SIGPLAN Notices* 17(6):120–126.

Grass, J. E. and Chen, Y.-F. 1990. The C++ information abstractor, pp. 265–275. In *Proc. 2nd USENIX C++ Conf.*

Harel, D. 1984. Statecharts: a visual approach to complex systems. Dept. of Applied Mathematics, Weizmann Institute of Science Rehovot, Israel.

Hastings, R. and Joyce, B. 1992. Purify: fast detection of memory leaks and access errors. In *Proc. Winter Usenix Conf.*

Johnson, S. C. 1974. YACC—yet another compiler compiler. CSTR 32, Bell Laboratories, Murray Hill, NJ.

Kaiser, G. E., Feiler, P. H., and Popovich, S. S. 1988. Intelligent assistance for software development and maintenance. *IEEE Software* 5(3):40–45.

Kaplan, S. M., Tolone, W. J., Carroll, A. M., Bogia, D. P., and Bignoli, C. 1992. Supporting collaborative software development with conversation builder. *Software Eng. Notes* 17(5):11–20.

Leblang, D. B. and Chase, R. P., Jr. 1984. Computer-aided software engineering in a distributed workstation environment. *SIGPLAN Notices* 19(5):104–112.

Lesk, M. E. 1975. LEX—a lexical analyzer generator. CSTR 39. Bell Laboratories, Murray Hill, NJ.

Linton, M. A. 1984. Implementing relational views of programs. *SIGPLAN Notices* 19(5):132–140.

Mahler, A. and Lampen, A. 1989. An integrated toolset for engineering software configurations. *SIGPLAN Notices* 24(2).

MIPS Computer Systems. 1988. *RISCompiler Languages Programmer's Guide*. MIPS Computer Systems.

Mitchell, J. G., Maybury, W., and Sweet, R. 1979. Mesa language manual. *Xerox CSL-79-3*.

Munck, R., Oberndorf, P., Ploedereder, E., and Thall, R. 1989. An overview of DOD_STD_1838A (proposed), the common APSE interface set, Revision A. *SIGPLAN Notices* 24(2):235–247.

Myers, B. A. 1980. Displaying data structures for interactive debugging. *Xerox CSL-80-7*.

Notkin, D., Ellison, R. J., Kaiser, G. E., Kant, E., Habermann, A. N., Ambriola, V., and Montanegero, C. 1985. Special issue on the GANDALF project. *J. Systems Software* 5(2).

Reiss, S. P. 1985. PECAN: program development systems that support multiple views. *IEEE Trans. Soft. Eng.* SE-11:276–284.

Reiss, S. P. 1994. *FIELD: A Friendly Integrated Environment for Learning and Development*. Kluwer Academic Publishers, Boston, MA.

Ross, D. T. 1985. Applications and extensions of SADT. *IEEE Comput.* 18(4):25–35.

Rumbaugh, J., Blaha, M., Premerlani, W., Eddy, F., and Lorensen, W. 1991. *Object-Oriented Modeling and Design*. Prentice–Hall, Englewood Cliffs, NJ.

TakeFive Software. 1993. *SNiFF+ Version 1.0 Reference Guide.* TakeFive Software.

Taylor, R. N., Belz, F. C., Clarke, L. A., Osterweil, L., Selby, R. W., Wileden, J. C., Wolf, A. L., and Young, M. 1989. Foundations for the Arcadia environment architecture. *SIGPLAN Notices* 24(2):1–13.

Teitelbaum, T. and Reps, T. 1981. The Cornell program synthesizer: a syntax-directed programming environment. *Commun. ACM* 24(9):563–573.

Teitelman, W. 1974. *Interlisp Reference Manual,* Xerox.

Tichy, W. 1982. Design, implementation and evaluation of a revision control system. In *Proc. 6th Intl. Conf. Software Eng.*

113

Interactive Software Technology

Peter Wegner
Brown University

The evolution from mainframe to personal computer and network technology is characterized by a paradigm shift from algorithms to interactive models of computation. Sections 113.1–113.3 develop a conceptual framework for interactive computation, showing that models of interaction have observably richer behavior than Turing machines and more completely express the behavior of embedded systems like airline reservation and banking systems. Sections 113.4–113.9 examine specific software architectures for programming in the large, including object-based design, multiple interface models, and technologies for interoperability, design patterns, coordination, and agent-oriented programming. Sections 113.10 and 113.11 present case studies for virtual reality and data information systems, demonstrating the naturalness of interactive models as a basis for computer graphics applications and more generally for empirical computer science.

113.1 Evolution of Computing Paradigms

The problem of driving from Providence to Boston is usually solved by combining *algorithmic* knowledge (a high-level mental map) with *interactive* visual feedback of road signs and road topography. In principle, interactive feedback could be replaced by a noninteractive algorithmic specification so that (assuming no other traffic) a blindfolded person could drive entirely by rules for accelerating, braking, and turning the steering wheel in an entirely algorithmic mode. However, the complexity of such an algorithmic specification is enormous and would, in any case, not handle traffic and other interactively variable factors.

0-8493-2909-4/97/$0.00+$.50
© 1997 by CRC Press, Inc.

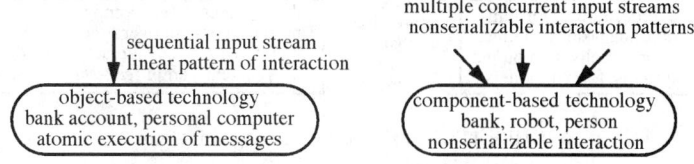

FIGURE 113.1   Interactive inputs for objects and components.

Algorithmic computation is very weak in modeling interactive behavior like driving: it corresponds to blindfolded or autistic behavior in humans. Interactive descriptions are both simpler and more natural than entirely algorithmic specifications even when algorithmic specifications exist, and they can express inherently interactive behavior not expressible by any algorithmic specification. Though algorithms are surprisingly versatile, interaction provides an extra dimension that reduces the complexity of problem solving and expands the power of models of computation so they can express and solve a larger class of problems.

Computing has evolved from machine-language programming through procedure-oriented and structured programming to object-based and component-based programming (Fig. 113.1):

1950s: Machine-language programming: assemblers, hardware-defined action sequences
1960s: Procedure-oriented programming: compilers, programmer-defined action sequences
1970s: Structured programming: software engineering, algorithm architecture
1980s: Object-based programming: personal computers, sequential interaction architecture
1990s: Component-based software technology: concurrent interaction architecture, coordination.

Whereas the transition from machine to procedure-oriented programming simply involves a change in the granularity of actions, the shift from procedure-oriented to object-based systems is more fundamental, involving a shift from algorithmic to interactive computing [Wegner 1995b]. The shift from sequentially interacting objects to concurrent interactions of software components is a further fundamental paradigm shift:

machine language → procedure-oriented language
   quantitative change of scale in the granularity of actions
procedure-oriented → object-based
   qualitative change of expressiveness from algorithms to interactive systems
object-based → component-based
   qualitative change of expressiveness from sequential to concurrent (nonserializable) interaction.

Models of interaction provide a unifying framework for design and analysis of software engineering, artificial intelligence (AI), and database applications. Objects have richer observable behavior than procedures, whereas components with multiple concurrent input streams in turn have richer observable behavior than sequential objects.

Models of interaction provide a unifying conceptual framework for the description and analysis of object- and component-based architectures. They are a tool for exploring software-design models such as OMT, multiple-interface models such as Component Object Model (COM)/OLE, models of interoperability such as Common Object Request Broker Architecture (CORBA), design patterns, coordination languages such as Linda and Gamma, and AI models of planning and control. A case study of interaction architecture for virtual reality is presented, and a final section considers an interaction architecture for a broad class of interactive systems that includes data acquisition systems such as NASA's earth observation system (EOS) and digital library systems.

Objects and algorithms both determine a *contract* between providers and clients of a resource, but objects provide fundamentally richer services to clients that cannot be expressed by algorithms. Algorithms are like sales contracts, guaranteeing an output for every input, while objects are like marriage contracts, describing ongoing contracts for services over time (Fig. 113.2). An object's contract with its clients specifies

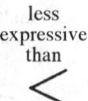

	less expressive than	
function, procedure, algorithm time-independent sales contract given an input, deliver an output closed systems, formal models	$<$	object, agent, actual computer marriage contract over time interactive service, interaction history open systems, empirical computer science

FIGURE 113.2   Sales versus marriage contracts.

its behavior for all contingencies of interaction (in sickness and in health) over the lifetime of the object (till death us do part) [Wegner 1996]. The folk wisdom that marriage contracts cannot be reduced to sales contracts is computationally expressed by interaction not being reducible to algorithms.

Object-based programming has become a dominant technology, but its foundations are shaky: everyone talks about it but no one knows what it is. "Knowing what it is" has proved elusive because of the implicit belief that "what it is" must be defined in terms of algorithms. Irreducibility has the liberating effect of allowing "what it is" to be defined in terms of interactive models rather than algorithms. Component-based software technology is even less mature than object-based technology: it is the technology underlying interoperability, coordination models, pattern theory, and the World Wide Web. Knowing what it is in turn requires liberation from sequential object-based models.

At a more fundamental level, the distinction between algorithms and interaction corresponds to that between closed and open systems and to that between rationalism and empiricism. Interactive models of computation provide a precise characterization of open systems and empirical computer science [Wegner 1995a]. Irreducibility of interactive to algorithmic models implies that empirical computer science is fundamentally richer than models of algorithms and complexity theory, but the cost of greater richness is nonformalizability and incompleteness (in the sense of Godel).

113.2 Models of Interaction

To provide a formal framework for interactive models we extend Turing machines, which model algorithmic computation, to interaction machines, which model objects and software components. Turing machines have a tape that initially contains a finite sequence of input symbols and a state transition mechanism that can read a symbol from the tape, perform a state transition, write a symbol on the tape, and reposition the reading head. Interaction machines extend the Turing machine model by adding input and output actions (read and write statements). Whereas Turing machines require all inputs to appear on the tape prior to the computation and shut out the world during the process of computation, interaction machines allow inputs to be dynamically generated and require inputs to be represented by a potentially infinite stream, since any finite stream can be dynamically extended (see Fig. 113.3).

Interaction machines are intuitively more expressive than Turing machines because they model the passage of time during the process of problem solving, whereas algorithms and Turing machines model only time-independent transformations. This informal difference is formally captured by the difference between finite input tapes and unbounded input streams. Interaction machines correspond to Turing machines with infinite tapes, which are known to be more powerful than Turing machines. Whereas the

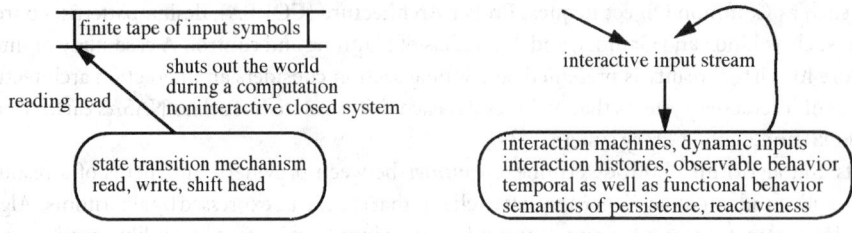

FIGURE 113.3   Turing and interaction machines.

inputs of a Turing machine are countable, the number of potential streams of an interaction machine is nonenumerable.

Interactive adversaries can always extend interactive streams by adding an input. This informal difference between noninteractive and interactive computations is modeled formally by infinite streams as opposed to finite tapes. Though interactive computations are finite in practice, interaction machines and computations with an unbounded number of interaction steps cannot be modeled by the set of all finite sequences (which is enumerable) but must include infinite limit points, because adversaries have the last word and can always extend any finite sequence. The power of adversaries may be used to show how noncomputable functions postulated by the diagonalization argument can be computed by generating sequences that differ from any enumerable sequence in their diagonal position. Environments that generate external input streams not under the control of the interaction machine are modeled mathematically by infinite sets of sequences that include limit points and have a nonenumerable cardinality.

Nonenumerability versus countability expresses the essence of the difference between Turing and interaction machines. "Real" numbers and "real" time are dual nonenumerable abstractions of reality. Real numbers were viewed in the 19th century as models of the infinite divisibility of continuous mathematical and physical space. Interaction machines turn this model inside-out, representing time in the "real" world by infinite input streams. The set of all infinite digit streams is in one-to-one correspondence with both the nonenumerable real numbers and the input streams of an interaction machine.

Though nonenumerability provides a mathematical mechanism for proving the greater power of interactive computing, it does not completely capture the semantics of interaction. Interactive processes are nonalgorithmic even if they have finite time horizons because their inputs are uncontrollable and nondeterministic. The fact that the next step of an interactive process is externally determined causes even finite interactive processes to be nonalgorithmic: interactive processes have lower complexity because they can simply access external data (driving home from work is a simple interactive process but a horrendously complex algorithmic process). The dynamic acquisition of interactive knowledge is a key to

1. informally better performance of practical interactive tasks
2. formally lower complexity of finite tasks in reducing exponential search to polynomial time
3. formally greater expressiveness in modeling tasks with an unbounded number of interactive steps.

Greater expressiveness is the limiting case of lower complexity for finite tasks: nonenumerability is the limiting case of uniform reduction of exponential to polynomial time complexity.

Interaction machines are incomplete in the sense of Godel: their nonenumerable number of true statements cannot be enumerated by a set of theorems. Incompleteness has strong practical consequences: it implies that projects like the fifth-generation computing project that aim to reduce computation to logic fail because their goals cannot in principle be realized. Even a hundredfold increase in effort and a 10-year extension would not have allowed the fifth-generation project to succeed. The view that logic programming is theoretically too weak to model interactive systems, presented by the author at the closing session of the fifth-generation computing project in Tokyo in 1992 [Wegner 1993], has practical technological implications.

Interaction machines provide a precise characterization of empirical computer science, while irreducibility and incompleteness indicate that empirical computer science differs fundamentally from algorithms and relates the "real" world to the "real" numbers. Turing machines have been the dominant computational abstraction for the first 50 years of computer science, but their role in the next 50 years may be less central because they are not powerful enough to capture the behavior over time of objects, software systems, and distributed computation.

Church's thesis that the intuitive notion of computing corresponds to formal computing by Turing machines is seen to be invalid or at least inapplicable, since interaction machines capture persistent, time-dependent behavior of actual computers and software applications more accurately than Turing machines. Interaction machines precisely capture the "empirical computer science" paradigm: interaction expresses observability and models in the natural sciences. The Chomsky hierarchy of machines can be extended beyond Turing machines to synchronous, asynchronous, and nonserializable interaction machines whose expressiveness corresponds respectively to Newtonian, relativistic, and chaos models of physics [Wegner 1996].

113.3 Software Engineering, Artificial Intelligence, and Open Systems

Interactive models provide a common conceptual framework for software engineering and AI. The evolution in AI from logic and search to agent-oriented models is not merely a tactical change but is a strategic paradigm shift from algorithms to more expressive interactive models that fundamentally increases expressive power. The reasoning/interaction dichotomy is precisely that between good old-fashioned AI (GOFAI) and "modern" agent-oriented AI. This paradigm shift is evident not only in research [Agre and Rosenschein 1995] but also in textbooks that systematically reformulate AI in terms of intelligent agents [Russell and Norvig 1994].

Programming in the small (PIS) is algorithmic, whereas programming in the large (PIL) deals not with large programs but with interactive systems. An algorithmic program with a million arithmetic operations is not PIL, whereas medium-size embedded software systems are. PIL is not simply scaled-up PIS; it has qualitatively different program structures and models of computation. The irreducibility of interaction to algorithms implies inexpressibility of PIL by PIS. Scaling up shifts attention from inner activities within components to interaction among components. PIL was observed to differ from PIS as early as the 1960s, but the difference was viewed as a quantitative change of scale. Expressing the difference as a qualitative change of expressiveness explains the observed inability to scale up from algorithms to software systems.

The irreducibility of interactive systems to algorithms confirms Brooks' persuasive argument [Brooks 1995] that there is no silver bullet for simply specifying complex systems. If silver bullets are interpreted as algorithmic or formal specifications, the nonexistence of silver bullets can actually be proved. Systems that persist in time have an interactive essence that cannot be expressed by silver bullets fashioned from algorithmic or formal models. This negative "impossibility result" has a positive liberating effect on explanatory models. Giving up the goal of complete behavior specification requires a psychological adjustment but makes partial system specification respectable. Though a complete elephant cannot be specified, its parts and its forms of behavior (its trunk or mode of eating peanuts) are specifiable. Complete specification must be replaced by the more modest goal of partial specification by interfaces, views, and modes of use.

The idea that physical objects are not completely describable or knowable but that they may have describable parts or views is a basic tenet of the scientific method. Plato's cave metaphor asserts that we are like people in a cave who see only shadows of the real world (incomplete knowledge) on the walls of our cave but not reality itself. Though Plato's denial of the validity of complete knowledge is correct, his inference that partial empirical knowledge is worthless and that only mathematical knowledge is "real" is erroneous. Science is based on acceptance that useful models of the real world can be constructed from shadows on the walls of our cave (images on our retina). Interactive objects can be perceived only by partial interface behaviors, just as real-world objects can be perceived only by observed behaviors.

Software architecture is yet another area whose elusive nature can be explained by the irreducibility of interaction to algorithms. Its definition as "the structure of the components of a program/system, their interrelationships, and principles and guidelines governing their design and evolution over time" [Garlan 1995] indicates a clear focus on interactive interrelations among components as opposed to algorithmic control structures. The analysis of specific architectures such as client/server structures, UNIX pipes, blackboards, and distributed network topologies makes use of algorithmic tools in defining protocols, but software architecture is an art rather than a formal discipline and the complete set of interactive system behaviors determined by software architectures cannot be algorithmically specified. The study of component-based architectures does not fit into the framework of algorithm-based models of computation.

A computing system is said to be open if its computations depend on external information and is said to be closed otherwise. Turing machines with initial tapes and algorithms with initial inputs are *closed*: their

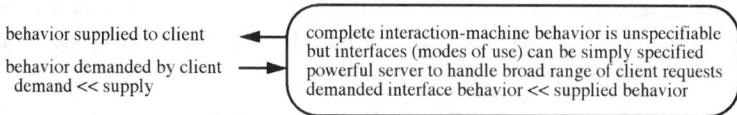

behavior supplied to client

behavior demanded by client

demand << supply

complete interaction-machine behavior is unspecifiable but interfaces (modes of use) can be simply specified powerful server to handle broad range of client requests demanded interface behavior << supplied behavior

FIGURE 113.4   Supplied server behavior versus demanded client behavior.

actions do not depend on external interaction. In contrast, interaction machines are *open*: their actions are influenced by the external world. The distinction between closed and open systems is precisely that between algorithmic and interactive computing. Algorithms are closed because they shut out the world during computation, whereas interactive computing is open in allowing external agents to observe and influence computation.

The complete behavior of servers (or agents) is complex and necessarily open, since it must handle all possible clients, whereas the individual interface demands of clients are often (though not necessarily) simple and closed. Supplied behavior is generally much richer than demanded behavior (see Fig. 113.4).

Useful categories of interactive systems can be classified by the kinds of constraints we impose upon them to make them tractable. Perhaps the most common constraint is that of precluding interaction during a computation, which causes the system to become closed. For example, Turing machines and algorithms are closed systems because their rules of engagement require all inputs to be supplied on an input tape before the beginning of the computation. Any open system can be closed by constraining its rules of engagement to be independent of external effects. Interaction machines require their rules of engagement to permit inputs during the process of computation and are therefore inherently open.

The state transition mechanism of a Turing machine, considered as an isolated system, is open, since inputs occur during execution. Turing machines insulate their state transition mechanism from dynamic interaction by providing a finite tape prior to the start of the computation. Figure 113.5 illustrates an alternative way of closing open systems by an environment that is itself an object. The two objects O1 and O2 are open as isolated systems but become a closed system when composed so that each object talks only to the other. Each object, O1 and O2, interacts as an open system with an arbitrary collection of clients, while the composite system causes each object to constrain the behavior of the other so it interacts with only a specific client.

The set of all possible behaviors of the object O1, viewed as an isolated system, is nonalgorithmic because its input streams can be infinite sequences not generable by any algorithm (not recursively enumerable). When input sequences of O1 are constrained to be those produced by an object O2 in response to a message stream from O1, the behavior of O1 can be tractably described. The object O2 plays the role of a tape, albeit an intelligent tape, in constraining the interaction machine O1 to closed-system behavior. Nonalgorithmicity is not preserved under system composition, and conversely algorithmicity is not preserved under system decomposition.

Though the relation between O1 and O2 appears symmetrical from outside the system, it becomes asymmetrical for observers residing in one of the objects or when the two objects play different roles. For example, O1 may be a server and O2 a client, or O1 an agent and O2 an environment that O1 is exploring or a client on whose behalf the agent is acting. Two-component systems in which each acts as a constraint on the other arise in control theory: one component is the system being controlled and the other is a

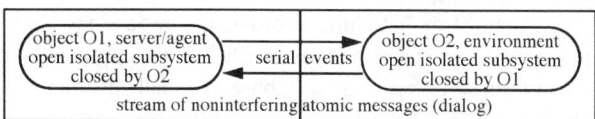

object O1, server/agent open isolated subsystem closed by O2

serial events

object O2, environment open isolated subsystem closed by O1

stream of noninterfering atomic messages (dialog)

FIGURE 113.5   Composition of open subsystems to form a closed system.

controller that may exercise strict control (as in a prison) or loose control (as in a free society). In this context, the idea that an isolated system has richer uncontrolled than controlled behavior is very natural.

113.4 Object-Oriented Design: Sequential Interaction

Object models have many "small" objects with sequential interface and interaction protocols, whereas applications such as airline reservation systems have "heavy" components with multiple interfaces and concurrent interaction protocols. In this section we examine the primitive entities and observable dynamic behavior of object models. The gap between static structure and dynamic behavior is greater for objects than for algorithms, with two distinct levels of execution dynamics: the external dynamics of operation execution is entirely separate from that of inner rule-based algorithm execution. The object modeling technique OMT [Rumbaugh et al. 1990] has an *object model* for describing static object structure, a *dynamic model* that describes interaction histories, and a *functional model* that describes transformation behavior of operations:

> **Object model:** Describes relations among interactive components (nouns)
> > static description of objects, operations, and relations among objects by object diagrams.
> **Dynamic model:** Describes interaction histories of the object model (inter-object dynamics)
> > dynamic inter-object interaction histories (event sequences) touching multiple objects.
> **Functional model:** Describes behavior of specific functions (intra-object dynamics)
> > intra-object transformation behavior at the level of algorithms.

These three levels provide a robust modeling framework for a variety of models of object-oriented design, as pointed out by Jacobson [Jacobson 1991]. They reflect the fact that nouns provide a more direct and more expressive model of the real world than verbs and the further fact that nouns are modeled computationally by the patterns of actions (verbs) that they support. Nouns whose behavior is expressed by patterns of observable actions are naturally modeled by an object model expressing relations among nouns, a dynamic model expressing patterns of interaction, and a functional model specifying individual actions. Externally determined patterns of interaction are not constrained by requirements of algorithm specification.

Dynamic models assume that interaction histories are traces (sequences) of time-independent events. They constrain interactions to a disciplined sequential form but exclude time-dependent and nonserializable interaction histories. Interaction histories can be viewed as test cases analogous to instruction execution histories of algorithm computations. Just as no amount of testing can prove correctness of algorithms, interaction histories can show only the existence of desirable behaviors and cannot prove correctness. The incompleteness of interactive systems implies that proving correctness is not merely hard but impossible. We must be satisfied by showing the existence of desirable behaviors through test cases (sequential interaction histories in the case of OMT) and cannot hope to prove the nonexistence of incorrect behaviors [Wegner 1995a].

Though there is a superficial resemblance among object, dynamic, and functional models and corresponding levels of modeling for algorithms (see Fig. 113.6), object models are very different from flow diagrams and object interaction histories are very different from algorithm execution histories. Flow diagrams have actions as nodes and control paths as edges, whereas object diagrams have objects as nodes and access paths as edges. Algorithm execution histories specify time-independent instruction sequences, whereas system interaction histories specify time-dependent events in real or artificial worlds. Though instruction and operation sequences are both linearly ordered, they have entirely different observable behavior. Instruction sequences define transformations specifiable by computable functions, and operation sequences define processes in time with a richer set of observationally distinct behaviors, as described in section 113.2.

Three-level object design models clearly indicate the role of algorithms as low-level transformation specifications of primitive elements of interaction patterns. Object models focus primarily on patterns of

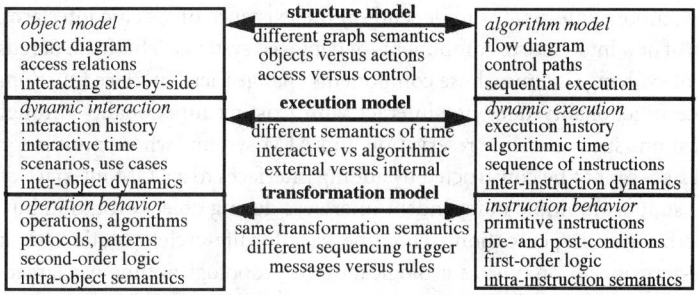

FIGURE 113.6   Three-level model for objects and procedures.

interaction: the functionality of primitive elements is a secondary though still important concern. Restricted sequential patterns of dynamic models are often specifiable by algorithms. But unbounded sequential streams generated by external processes need not be recursively enumerable or deterministic. Because general patterns of interaction are indescribable, the literature on design patterns [Gamma et al. 1994] abandons algorithmic description in favor of informal verbal description of problem and solution structure to describe reusable primitive patterns of object and component interaction. Design patterns are behaviorally simply interaction patterns, but interaction patterns are too low-level to capture user-level regularities (they are the machine language of design patterns). Higher-level patterns are therefore used to describe design patterns, such as subproblem descriptions, trade-offs among design features, and processes of design.

113.5 Multiple-Interface Model: Concurrent Interaction

Objects have fixed sequentially accessible interfaces, whereas components of real-world application systems have multiple interfaces that can be dynamically added and removed and concurrently executed. Such components have richer observable behavior than objects with atomic sequentially executable operations. We first consider multiple independent interfaces that increase the accessible resources and/or the views of a component without introducing concurrency and can be modeled by extending sequential object-based systems, and then we explore components with multiple concurrently accessible interdependent interfaces that require fundamentally new models.

Microsoft's Component Object Model [Brockschmidt 1995] is an architecture for multiple independent interfaces. It extends the client–server model so that components can evolve by dynamically adding new interfaces. Components are specified by interface directories rather than sets of operations of a single interface. Every component must possess an interface I-unknown for managing its interface directory, with operations for adding and deleting interfaces and a "queryinterface" operation that checks the interface directory for the existence of interfaces (see Fig. 113.7).

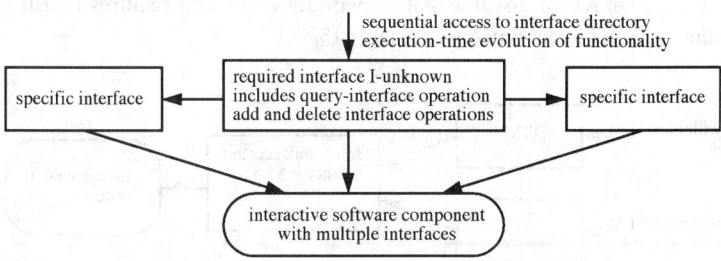

FIGURE 113.7   Multiple interface model for COM/OLE software components.

COM allows execution-time actions to depend on the existence of specific interfaces and permits the dynamic addition of new interfaces to components of deployed systems. This form of reusability simplifies the maintenance of evolving systems whose components change their interface functionality but assumes that new interface functionality does not interact with existing functionality through a shared state. Interactive applications such as airline reservation and ATM systems whose multiple interfaces interact through a shared state cannot be constructed by adding interfaces to a COM interface directory.

COM's objects support multiple independent interfaces during object evolution but not multiple interdependent interfaces during execution. The time scale for life-cycle evolution has larger granularity than that for real-time execution, just as geological or anthropological time has larger granularity than quantum-theoretic time. Models of persistence must handle time at both the macro and the micro levels: there is a deep analogy between interference phenomena in physical models of quantum theory and chaos and computational interference due to nonserializability. Nonserializable interaction among interfaces requires qualitatively different models from COM's creation and deletion of multiple independent interfaces.

Airline reservation systems are prototypical of applications with multiple views sharing a common database. Multiple concurrent interactions with the database have unpredictable functionality, since the state seen by one interface can be unpredictably changed through another interface. Multiple interfaces are an interactive analog of multiple threads: concurrent interactive access parallels concurrent execution. Airline reservation systems have many interfaces that interact through shared data structures:

> Interfaces (modes of use) of airline reservation systems:
> travel agents, making reservations on behalf of clients
> passengers, making direct reservations
> airline desk employees, making inquiries on behalf of clients
> flight attendants, aiding passengers during the flight itself
> accountants, auditing and checking financial transactions
> systems builders, developing and modifying the system.

Travel agent interfaces permit a client to reserve a seat under normal conditions, but this response may be subverted by interaction through other interfaces. The response to a given travel agent request depends on other travel agents, direct reservations by travelers, flight cancelations that cause wholesale transfer from one flight to another, and exogenous events like snowstorms or hurricanes. The response to a message depends on nonlocal events in the external (real) world: the computational analog of action at a distance.

Interdependent interfaces that share a state are a fundamental cause of unpredictable behavior in both physical and computational systems. Unpredictability arises even when access to the shared state is sequential because an interface cannot assume that the state remains undisturbed for successive accesses to the state. However, the problem becomes more acute when access is concurrent because the state seen by an interface may change unpredictably during execution of an operation. We examine this situation for joint bank accounts and show that unpredictability due to conflict among concurrent input streams is the discrete analog of chaos in continuous physical systems.

Suppose that a joint bank account contains $1.5 million and that two clients simultaneously try to withdraw $1 million at different ATMs. Assume that the withdrawal process requires adjusting an investment portfolio and is therefore not instantaneous (see Fig. 113.8).

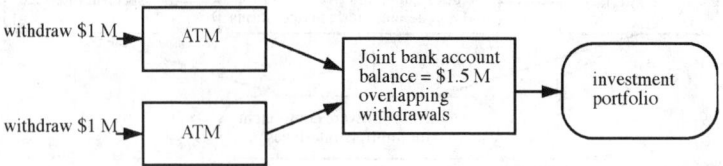

FIGURE 113.8   ATM system as an asynchronous interaction machine.

A transaction system could in principle handle this situation by satisfying only one client and aborting the transaction of the other. But concurrent interactive systems cannot be presumed to be transactionally well behaved: we must model and manage breakdown of transactional behavior, just as psychologists must model and manage ner-

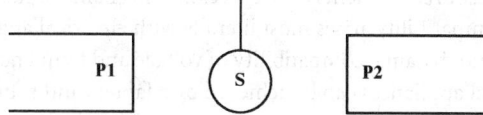

FIGURE 113.9 Chaotic two-dimensional motion of steel pendulum in a magnetic field.

vous breakdowns in people, and physical systems must cope with chaos. Transactions are a computational mechanism for handling system overload caused by multiple simultaneous demands on a resource. The effect of concurrent potentially conflicting operations op1 and op2 of an object can, in the absence of transaction atomicity, be arbitrary and chaotic. Behavior becomes *nonserializable* in that it does not correspond to any sequential execution of the operations op1 and op2. Nonserializability of concurrently executed operations of an object's interface specializes nonserializability of database transactions by considering only atomicity of interface operations but is essentially the same problem.

Though nonserializable behavior is considered undesirable in many contexts, it is more expressive (observably richer) than serializable behavior and can be harnessed for useful purposes. Aborting a transaction or replacing a time-consuming optimal algorithm by an approximate just-in-time algorithm is a technique for managing nonserializable behavior. Aborted transactions and just-in-time algorithms replace ideally desired functionality by less desirable functionality that can be serializably realized.

Transaction management systems that guarantee atomicity are safe but often too conservative. Permitting some degree of controllable nonserializability, just like permitting some unsound behavior in tasks like error checking, can be worthwhile. For example, optimistic concurrency control systems lower their guard on the assumption that no conflicts occur and pay for this by requiring drastic and time-consuming actions when this assumption is violated. High-achieving people, like efficient computers, operate close to the margins of nonserializability and pay the price of greater stress and a higher incidence of nervous breakdowns than persons opting for comfort at the expense of achievement.

Nonatomicity of operations causes unobservable and uncontrollable temporal sensitivity in accessing a shared resource (the object's state) that is analogous to chaos in physics: sensitivity among competing clients to shared computing resources corresponds to sensitivity among competing forces on a shared physical object. The pattern of competing access to shared resources in the bank account example arises in the well-known demonstration of physical chaotic behavior of Fig. 113.9, which illustrates a two-dimensional pendulum (steel ball) in the presence of two magnets. The magnets P1 and P2 correspond to the operations op1 and op2, and the steel ball S corresponds to the state. The magnets exert two streams of impulses on the steel ball just as operations exert streams of impulses on the object's state. Chaos arises when the steel ball passes through regions where the force fields are almost equal in which the motion of the steel ball is extremely sensitive to minute perturbations of the force field. This example illustrates the deep correspondence between chaos and nonserializability and more generally the fact that interaction machines can model phenomena that arise in physical systems.

Chaotic behavior in physics is modeled by nonlinear differential equations. The differential equations for a pendulum have a linear first-order behavior but nonlinear second-order terms when the first-order terms cancel each other. Thus, pendulum behavior is linear when in the force field of one of the magnets but nonlinear in regions where the force fields cancel each other out. Interaction machines are a discrete analog of dynamical systems in physics. Their chaotic behavior is a discrete analog of turbulence in differential equations: both are a consequence of extreme sensitivity to initial conditions.

113.6 Interoperability: Mediated Interaction

Interoperability is the ability of two or more software components to cooperate despite differences in language, interface, and execution platform. It is a scalable form of reusability, extending the reuse of server resources to clients whose accessing mechanisms may be plug-incompatible with sockets of the server. Plug compatibility arises most literally with electrical appliances that require both static compatibility of shape and dynamic compatibility of voltage and frequency. If there is no direct match, interoperability of electrical appliances can be achieved by adapters and transformers.

interoperability = scalable reusability = plug and socket compatibility

As with electrical appliances, incompatibility of software plugs and sockets can be mediated by adapters. Software adapters are part of the interaction architecture rather than the functionality of a software system. Interoperability architectures such as the Common Object Request Broker Architecture (CORBA) are effectively elaborate adapters that provide both static and dynamic compatibility among heterogeneous components. To provide a framework for describing particular architectures such as CORBA, we briefly discuss methods of interface specification and interface bridging.

Interfaces are traditionally specified by types. Static compatibility in strongly typed languages can be determined by type checking. Type compatibility guarantees that the plug of the client fits the socket of the server but not that the service provided is that desired by the caller. Minor type differences between data representations, like that between integers and floating-point numbers, can be handled by coercion. Stronger type differences can be handled by polymorphism, which allows an interface to accept inputs of a variety of types. Heterogeneous components have even stronger differences that cannot be handled by polymorphism. Some of the most powerful systems of interoperation, like UNIX pipes and http World-Wide-Web protocols, are typeless. Interface definition languages with static type-compatibility protocols promote safety and efficiency at a considerable cost in design and implementation complexity. Much of the research in this area is designed to realize the safety of typed systems with acceptable complexity.

Clients and servers from different software platforms or programming languages can talk to each other through *mediators (adapters)* that convert data formats. Mediation in information systems as an organizing principle for interoperation of heterogeneous components is reviewed in Wiederhold [1995]. Though mediators can handle a wide range of differences in data formats and recognized differences of representation, like that between Cartesian and polar coordinates, the general problem of reusing procedure functionality is unsolvable, since functional equivalence is undecidable.

The two major mechanisms for interoperation are interface standardization and interface bridging:

Interface standardization: map client and server interfaces to a common representation
Interface bridging: two-way map between client and server.

Interface standardization is more scalable because m clients and n servers require only $m + n$ maps to a standard interface, compared with $m \times n$ maps for interface bridging. However, interface bridging is more flexible, since it can be tailored to the needs of specific clients and servers. Interface standardization makes explicit the common properties of interfaces, thereby reducing the mapping task, and it separates communication models of clients from those of servers. But predefined standard interfaces preclude supporting new language features not considered at the time of standardization (for example, transactions). Standardized interface systems are closed, while interface bridging systems are open. Architectures for standardized interfaces may be considered a special case of interface-bridging architectures in which the bridge from clients to servers is replaced by two half-bridges from clients to the standard interface and from the standard interface to the servers. We examine CORBA, whose architecture is based on interface standardization and then consider briefly architectures based on interface bridging, like megaprogramming languages.

CORBA [Object Management Group 1995] realizes interface standardization by an object request broker (ORB) that serves as an adapter/transformer. The ORB handles communication among application objects and provides object services (system objects) and common facilities (library objects). Standard interfaces are specified in an interface definition language (IDL) and stored in an interface repository. Interfaces of application objects specified in a variety of object-based languages are mapped to IDL by

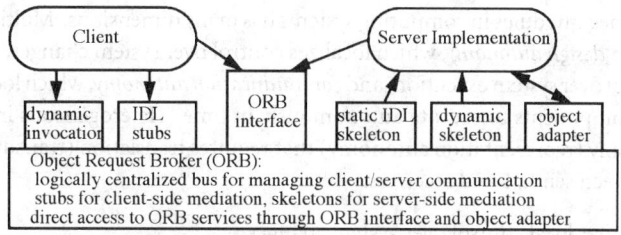

FIGURE 113.10 CORBA interoperability architecture.

a language mapping that automatically maps language-dependent interfaces into language-independent IDL interfaces.

The ORB accepts requests from client objects through an IDL stub or a dynamic invocation for the services of server objects, transmits them for execution to the server, and returns the result to the client. It also accepts calls to system objects from both client and server objects and object-specific calls through an object adapter, as shown in Fig. 113.10.

CORBA provides mediation services (half-bridges) from clients to the ORB and from the ORB to servers. Clients may invoke a service through a static stub or through a dynamic invocation created from the IDL specification at run time. The ORB validates client requests against the IDL interface and dispatches them to the server, where arguments are unpacked (unmarshalled), methods are executed, and results are returned. Server-side software includes object adapters that bind object interfaces and manage object references, and a server skeleton that uses the output of object adapters to map operators to the methods that implement them.

Microsoft's Component Object Model (COM/OLE) [Brockenschmidt 1995] realizes interoperability through a binary (machine-language) standard that specifies multiple interfaces by a pointer to a function table and objects by a principal interface I-unknown for accessing a directory of interfaces through a "queryinterface" function. COM/OLE objects have multiple interfaces that share a common data structure. The multiple-interface model for objects is more scalable than the hierarchical inheritance model of object-oriented programming, providing greater flexibility for both interoperation and extension of object functionality. Class-based inheritance is less scalable than multiple interfaces as an organizing principle for interfaces, both because complex objects are naturally modeled by multiple interfaces and because hierarchies are too restrictive a structuring principle. Use-case models [Jacobson 1991], as well as COM/OLE, elevate collections of interfaces that share a state into a primary structuring principle of software design.

COM/CORBA interoperability, which aims at compatibility between Microsoft and CORBA-compliant components, is being pursued by a broad industry consortium under the auspices of the Object-Management Group (OMG). Proposals for COM/CORBA interoperability aim to realize interoperation by interface bridging. Clients send a request via a surrogate object to a state that marshals the arguments and sends them across a bridge to a server skeleton that unmarshals the arguments and manages the implementation of operations by methods of the target object. The combination of interface standardization within CORBA and interface bridging to COM provides a balance between closed efficiency and open extensibility expressed by the metaphor of closed islands connected by bridges.

The interface-bridging approach to interoperation is illustrated by megaprogramming languages [Wiederhold et al. 1992]. Megamodules capture the functionality of services provided by large organizations such as banks, airline reservation systems, and city transportation systems. They are internally homogeneous, independently maintained software systems that embody the "software community" paradigm, encapsulating not only data but also knowledge, programming traditions, goals and values, and their own terminology, concepts, and interpretation paradigm (called an ontology). Megamodules control their own evolution and can add interface functionality as well as inner functionality. For example, a megamodule for a city transportation system can add new trucking companies, bus companies, and airlines to its interface. Such interface functionality can be modeled for example by COM's add-interface function.

The autonomy of megamodules in computing systems has many dimensions. Multidatabase technology distinguishes between *design autonomy*, which localizes control over system changes, *execution autonomy*, which localizes control over system execution, and *communication autonomy*, which localizes access to data. Asynchrony allows components to execute autonomously in time. Heterogeneous interfaces are still another form of autonomy (representation autonomy) that requires translation (transduction) mechanisms to be interposed between senders and receivers:

Design autonomy: local control over system changes
Execution autonomy: local control over system execution
Communication autonomy: localizes access to data
Representation autonomy: local control over interface representation.

Autonomy is natural in a world where programmers develop heterogeneous modules (megamodules) primarily for local use, ignoring the fact that the application might at some future time interact with remote clients. For example, the San Diego transportation authority might develop a model of transportation without considering that it might at some later time require an interface to airline schedules or transportation systems in Chicago. When an interface is provided, it is unlikely to be compatible with clients using different software platforms and object models. Megaprogramming languages regulate interaction among autonomous modules (megamodules) and where necessary translate messages from the output language of a sender to the input language of a receiver. Megamodules are created and maintained by a software community with a uniform terminology (ontology) that provides a conceptual framework and language for problem solving in an application domain that may differ from the ontology of software communities with whom the megamodule needs to communicate. The problem of communicating among megamodules is similar to that of communicating among countries that speak different languages.

The basic unit of interoperation in the client–server paradigm is the procedure. But procedure-level interoperation is not a sufficient condition, though it is a necessary one, for interoperation of software components [Nierstrasz and Tsichritzis 1996]. Software components may require larger-granularity units of interoperation, since the correspondence between client and server operations may not be one to one. Moreover, interoperation may require preservation of temporal as well as functional properties (order constraints on operations, coordination of inputs from multiple input streams). Such protocol constraints cannot be captured by functional correspondences of individual operations.

113.7 Design Patterns: Specification of Interaction

Patterns are a general tool for describing regularities in any domain of discourse. In the context of interactive modeling, patterns are useful in the description of observable behavior of software components, static structure of software systems, and processes of product development and design. Design patterns are reusable regularities of products or processes of design. The product of design is a component, called a framework, whose observable behavior can in principle be specified as a pattern of interaction. But patterns of interaction are too low-level for designers concerned with regularities that help them create a design from a high-level specification. Design patterns specify high-level regularities at the level of the problem rather than machine-language interaction patterns. Developing a high-level language for patterns of design is harder than developing high-level algorithmic languages. Design patterns are specified in [Gamma et al. 1994] by the problem, the solution, and design trade-offs associated with the problem:

pattern = (name, problem, solution structure, consequences, and trade-offs)

This format for specifying design patterns is independent of both domain and granularity and can be specialized when applied to object-oriented design. Object-oriented patterns are described by their *scope*, which specifies whether the pattern applies to *classes* or *objects*, and by their primary *purpose*, which may be to create a class or object (*creational*), to compose a structure out of components (*structural*), or to specify the interaction of a group of components (*behavioral*).

The model-view-controller (MVC) paradigm has been widely used to illustrate the role of patterns in design. It specifies design by three interdependent perspectives: a *model* that represents the application object, multiple *views* that capture syntactic interfaces, and a *controller* that defines the mode of execution. The model and controller correspond to the object and dynamic models of OMT, whereas MVC views are omitted in OMT but correspond to use cases in Jacobson's object design model. Though the MVC paradigm is a design pattern, its granularity is too great for inclusion by [Gamma et al. 1994] in their design catalog. MVC is described in terms of three component patterns:

> Observer: a pattern to describe one-to-many sharing of a resource (object) by multiple interfaces
> Composite: a pattern to describe composite nested structures
> Strategy: a pattern for run-time selection among multiple implementations of the same algorithm.

Observer describes the many-to-one relation between a model and its views in an abstract reusable way, composite allows nested many-to-one structures, and strategy handles multiple algorithms for realizing controllers associated with views. These three patterns represent independent primitive components that may individually be used in many other contexts that work nicely together in realizing a general form of the MVC paradigm. Each pattern would be very difficult to define by composition of lower-level algorithmic or object-based primitives. Patterns introduce reusable primitive units of behavior that are not easily reducible to or expressible in terms of programming language primitives.

Patterns should be specified so it is easy to determine when they are applicable and should have well-defined dimensions of variability. For example, sort procedures may be viewed as patterns with well-defined applicability and parameters for varying the data and size of the set of elements to be sorted. Design patterns have more complex criteria for applicability that include a description of the class of problems the pattern is designed to solve and a description of the results and trade-offs of applying the pattern. The problem specification and results for sorting can be formally specified, whereas the problem class and effects of interactive patterns cannot generally be formalized and require a sometimes complex qualitative description. Though there are loose analogies in scaling up from algorithmic to interactive patterns, the details of pattern specification are entirely different.

Patterns facilitate codifying of design experience that has in other design domains such as architecture been perceived to be elusive and unformalizable. A systematic method of describing, cataloguing, and using design patterns provides clues to the process of design for both software systems and other artifacts. Frameworks provide an alternative approach that was quite successful in specifying reusable designs at a fairly high level of granularity and was a primary catalyst in developing the pattern concept. In [Gamma et al. 1994], frameworks are viewed as executable implementations of particular application program interfaces (APIs) that may contain implementations of several patterns. They are less flexible in both their granularity and their level of abstraction than implementation-independent design patterns.

Object models use empirical observation of objects of the application domain to model objects of the design [Rumbaugh et al. 1990]. However, designs generally include a richer set of objects than those in the modeled domain and require communication and composition protocols among objects to be modeled. Object-oriented design provides little help with structuring collections of objects into cohesive and reusable subsystems. Patterns for object design are closely tied to objects, classes, and object composition mechanisms. Object interfaces are specified by the set of signatures of an object's operations and determine an object's behavior and its type. Classes specify both the interface and the implementation of objects of a class so that a given interface type can be implemented by several classes. Abstract classes that defer implementation of all their operations to subclasses could in principle be used to realize implementation-independent interface inheritance but would require the client to do the work of implementing all operations. Implementation-independent access to the functionality of an interface is better realized by object composition than by class composition. The design patterns described in [Gamma et al. 1994] include patterns for object composition.

113.8 Coordination: Management of Interaction

Coordination is concerned with the management of interactions among components of a composite system. The entities being coordinated may be user-level services like airline reservations, system-level services like those of an operating system, or algorithm-level services like instructions of a program. Control structures of algorithms are a highly constrained (degenerate) form of coordination. Coordination of interactions among objects and components of a software system is subject to looser constraints and therefore gives rise to richer architectures. Models of coordination are concerned primarily with coordination among components: the term *coordination* has a specialized meaning in this context.

Coordination was popularized by the language Linda [Carriero and Gelernter 1992], which distinguishes between *coordination primitives* of specifying interactions among processes and *computation primitives* for executing algorithms within processes. The study of coordination independently of computation is becoming an important area of research that has spawned several conferences and workshops over the past few years [Ciancarini et al. 1995, Ciancarini and Hankin 1996]. Recent work on coordination models and their applications may be found in [Andreoli et al. 1996].

Linda processes interact through a shared data space called the "tuple space." Processes cannot interact directly; they post output to the tuple space, which may then be used as input by other processes (see Fig. 113.11). Because the tuple space is like a blackboard on which public notices may be posted, this model is referred to as the blackboard model. Linda has four coordination primitives, three of which simply output and input data into the tuple space. Linda may also output live (executing) processes into the tuple space using the "eval" command:

> out(t): output the passive data tuple t to the tuple space
>
> in(t): input a tuple with the value t from the tuple space, removing it from the tuple space. If no tuple with value t exists, the executing process blocks until a tuple with this value becomes available.
>
> rd(t): copy a tuple with the value t from the tuple space, leaving t in the tuple space for others to use. If no tuple with value t exists, the executing process blocks until a tuple with this value becomes available.
>
> eval(t): output a live tuple t and execute it; when execution is completed the tuple becomes passive and can be accessed by in or rd commands of processes that use the result.

Linda's coordination primitives constitute a very simple coordination language that can be combined with a variety of different computation languages like C or Fortran. Inner program structure and outer communication structure can be expressed by distinct and independently specifiable paradigms of coordination. Linda captures a particular (blackboard) coordination architecture that generalizes the producer–consumer paradigm of resource management to multiple processes sharing a pool of common resources. The formal properties and semantics of Linda's coordination primitives are examined in Ciancarini et al. [1995].

The blackboard paradigm is a particular object-based architecture: objects and processes effectively have an internal blackboard whose protocols of inner sharing are like those of blackboards, though they are not a part of the public system semantics. We can think of objects as mechanisms for coordinating

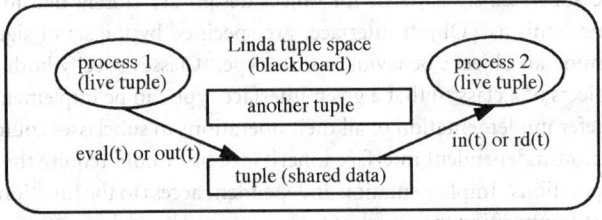

FIGURE 113.11 Coordination in Linda (the blackboard model).

collections of tightly coupled operations by sharing a local state. Sequential object-based languages have a restricted coordination mechanism strictly weaker than that of concurrent languages. Whether or not access to internal blackboards is controlled by a transaction manager is not observable or controllable by the system. Concurrent coordination semantics that guarantees atomicity of communication can be achieved only by sacrificing efficiency. In general, atomicity cannot be guaranteed so that the behavior of concurrently accessible processes with nonserializable inner coordination architectures must be assumed chaotic in the absence of explicit guarantees to the contrary.

The coordination language Gamma [Banatre and Le Metayer 1996] uses multisets (sets with multiple instances of elements) as the basic data structure. Multisets make minimal assumptions about relations among elements compared with arrays, lists, or other common data structures, allowing data structures to be defined without artificial sequentiality. For example, the Gamma reduction rule:

$$max: x, y \rightarrow y \Leftarrow x < y$$

replaces ordered pairs (x, y) by y for any $x < y$ until only the singleton maximum element remains.

This Gamma specification can be implemented either sequentially or concurrently. A sequential implementation is accidentally rather than necessarily sequential, since the implementation mapping loses information about control and does not reflect the specification semantics. In contrast, implementation mappings for procedure-oriented implementations preserve control structure of the source program.

Chemical reactions are an appropriate computation metaphor for abstract computation on multisets: reactions among molecules in a chemical solution are determined solely by chemical affinity independently of any externally imposed order or control structures. The very loose (nonexistent) coordination structure of multisets allows the chemical reaction metaphor to model both the coordination of loosely coupled software components and tightly coupled instructions of an algorithm. Coordination of actions within algorithms can be expressed by adding specialized control structures to multisets.

The elements of multisets may be of any kind whatsoever, including multisets. Multisets with reduction and interaction rules are a model for components, and multisets with multisets as elements can model components. "Second-order Gamma" models multisets with multisets as components and can in principle express the coordination of components that are modeled by multisets within a larger multiset.

Coordination of software components can be viewed as "second-order" computation, whereas execution of instructions is "first-order" computation. Coordination of components deals with the management of behavior of components composed of first-order actions, just as second-order logic is concerned with the behavior of functions and predicates of a first-order language. Second-order logics capture both the mathematics and the intuitive semantics of models of coordination and interaction. The set of all second-order entities (functions, predicates) over an enumerable set such as the integers is nonenumerable, just as are the inputs of an interaction machine. Second-order logics pay the price of not being sound and complete; this follows from the fact that the set of functions and predicates which they manage is nonenumerable.

Because coordination behaviors have more degrees of freedom than algorithm behaviors, coordination models are more varied than algorithm models. Petri nets are a "pure" model of coordination whose transitions model the process of consuming and creating resources independently of the purposes for which the resources are used. A Petri net is a graph structure with two kinds of nodes:

  places that hold tokens representing resources
  transitions that fire by consuming a token from each input place and sending a token to output places.

Tokens are represented by multisets and coordination is realized by rewriting rules for multisets that can easily be modeled in the language Gamma. Petri nets model resource consumption of tasks that require multiple resources to execute, expressing distributed coordination that depends on the availability of resources at distributed locations (places). They abstract away from specific data structures by permitting only a single data structure (tokens) and abstract away from algorithms by modeling resource consumption and creation independently of the task for which resources are used.

Petri net tokens are nonreusable resources better modeled by linear logic than by traditional logics. Linear logic treats rules of inference as resources that are consumed when they are applied. It provides a framework for coordination of nonreusable resources that admits both sequential and parallel coordination, illustrated by *interaction abstract machines* in Andreoli et al. [1993].

Coordination of heterogeneous distributed components can be realized by megaprogramming languages [Wiederhold et al. 1992] that coordinate execution among megamodules by "megaprograms" whose statements specify sequences of coordination actions interspersed with transformation of message data from the format of sending megamodules to that of receivers. Megaprograms for simple coordination tasks can be very simple. However, coordination of distributed concurrently executing tasks with atomicity and real-time constraints can be arbitrarily complex. Megaprograms are examples of "middleware" interspersed between components to realize coordination.

component (resource, megamodule) ↔ mediator (middleware, megaprogram) ↔ component

Coordination elevates middleware to a first-class status, corresponding to the role of management in large organizations. Middleware mediates among software components by transforming data and coordinating actions: middleware for mediating and coordinating software components is examined from several viewpoints in the June 1995 *Computing Surveys* [Wegner 1995]. Garlan defines software architecture in terms of the middleware for coordination and interaction among software components; Nierstrasz and Meijler explore the technology of software component composition; Wiederhold proposes mediators as a uniform mechanism for component coordination; Manola and Heiler examine issues of interoperability; and Sutherland examines middleware for business objects. These diverse papers exhibit a remarkable similarity in their overall goals in exploring middleware mechanisms for component coordination, while demonstrating the complexity of the problem and the diversity of approaches currently being considered.

Explicit coordination becomes more important as systems become large, just as explicit management structures are more important for large than for small companies. However, experience shows that simple typeless coordination systems such as UNIX pipes or http are often more effective than more elaborate strongly typed interface definition languages. Static type-compatibility requirements that promote safety and efficiency appear to have unacceptably high implementation cost in today's coordination technology.

Coordination among heterogeneous components by OMG's CORBA and Microsoft's COM solves the problem of reusability of resources specified in one environment by components in another environment. Most work on interoperability assumes a client–server, object-based model of communication [Nierstrasz and Tsichritzis 1996]. Coordination among heterogeneous distributed components may utilize protocols developed for client–server compatibility in more general contexts of coordination.

Coordination is concerned both with rules for scheduling and firing actions and with communicating and transforming data among components. Executing actions and communicating data, which require very different models (Petri nets and CORBA), need to be integrated:

coordination → firing rules + data exchange

Designers of coordination languages must address the following issues [Hankin 1994].

1. What are the entities being coordinated?
   Procedures, objects, processes, components of a specific type, subsystems.
2. What are the media (architectures) for coordination?
   Blackboard model, client–server, Petri net, Pi calculus, CORBA, COM, UNIX pipes, middleware.
3. What are the protocols and rules of coordination?
   Multiset rewriting rules, message send and receive, ORBs, megaprogramming, HTML.

In [Hankin 1994] it was further suggested that coordination languages be classified according to the dimensions of scalability, encapsulation, decentralization, dynamicity, open-endedness, generativeness, and semantic richness. These dimensions were used to classify and justify design decisions of existing languages. Though these dimensions are useful, the design space for coordination is not well understood and more work is needed to characterize and explore it.

113.9 Agents: Dynamical Systems, Control Theory, and Planning

Agents are proactive components that act on behalf of someone or something else to realize specified goals. Dynamical systems are an interactive abstraction whose continuous form has been widely studied in applied mathematics and whose discrete form models agents. Planning involves choosing a strategy for acting (interacting) that enables an agent to realize goals: plans can be viewed as strategies for controlling agents and therefore as an application of control theory. In exploring trade-offs between interactive and algorithmic models for controlling interactive behavior of agents and dynamical systems, we find that algorithmic, off-line planners have great complexity, whereas on-line, interactive planners sacrifice algorithmicity to realize more effective plans with lower complexity and greater flexibility.

AI focused primarily on logic and algorithms in the 1960s and 1970s but is becoming increasingly interactive. The conflict between logic-based and agent-oriented AI is documented in [Graubard 1988], which traces the demise of early research on neural nets and perceptrons; the increasing dominance of logic-based models spearheaded by research groups at MIT, Carnegie Mellon, and Stanford; and the reemergence of distributed AI in the 1980s. During 1985–1995 AI has increasingly focused on interactive models. Textbooks like [Russell and Norvig 1994], which systematically reworks all subareas of AI in terms of models of intelligent agents, suggest that the agent-oriented perspective is a foundation rather than merely an advanced topic of AI. The 1995 special issue of the journal *Artificial Intelligence* on interaction and agency [Agre and Rosenschein 1995] has over 700 pages of articles exploring the emerging interaction paradigm.

Agents function interactively in an environment that may contain other agents, performing tasks on behalf of other agents or autonomously on their own behalf. They perceive their environment through sensors and act upon it through effectors. Human agents have eyes and ears for sensors and hands, legs, and mouth for effectors, whereas robots have cameras and infrared range finders for sensors and motor-driven arms and wheels for effectors. Whereas software components are typically passively responding servers, agents realize goals by proactive interaction with the environment. Agents may be classified by the degree to which they passively react to or actively control interaction with their environment. Active agents view their environment as a passive domain to be observed or explained, whereas passive server agents have an active environment that directs and controls their actions.

Agents balance built-in cleverness with interactive adaptability in realizing computational goals. The degree to which agents use inner versus interactive cleverness is another dimension of classification: agents range from algorithms that rely entirely on inner cleverness to interactive identity machines that rely entirely on interaction. Agents with inner cleverness are more self-reliant, but interactive utilization of external cleverness, though dependent on external help, is potentially more powerful.

Dynamical systems are agents with a state and a dynamical law that governs how the state changes over time. The intensive study of continuous dynamical systems like Newtonian models of the solar system has over the last 300 years provided many insights applicable to the study of discrete agents in AI and computer science. Continuous differential equations with initial conditions are a discrete analog of algorithms with initial inputs: both determine closed dynamical systems. Physics and computing have primarily considered closed dynamical systems whose initial conditions are given prior to the beginning of the computation. Open dynamical systems that receive unpredictable interactive stimuli do not have a large body of mathematics from which analogous discrete models may be developed.

Dynamical systems with specified patterns of interaction over time can be reduced to closed noninteractive systems by modeling the interaction pattern as an agent that acts as an environment. A controller for a chemical plant or nuclear reactor that regulates a dynamical system to control its behavior may be viewed as a two-agent system: the system being controlled and the system that controls it may be viewed as coupled interacting agents that together form a closed system. Unpredictability (uncontrollability) of the input is an essential ingredient of openness.

There is a close analogy between initial inputs of algorithms and boundary conditions of differential equations. One-point boundary conditions correspond to arguments of a function or the tape of a Turing machine, whereas distributed boundary conditions correspond to closed systems like multitape Turing machines whose inputs are supplied prior to the beginning of the computation.

The classical Newtonian model, which assumes that the universe runs like a clock from initial conditions, is a continuous analog of an "algorithmic" dynamical system. Once the initial condition is given, Newtonian systems become nonempirical because no further observation is needed to predict their behavior. However, closed deterministic models of the universe have interactive open models as subsystems, both because isolated subsystems cannot know about or control interaction with the rest of the universe and because even predictable interactions may simply be too complex to be describable. Thus, interactive subsystems are necessary in describing even closed, deterministic, nonempirical worlds. Control theory studies two-agent systems where one agent, called the controller, controls the behavior of agents (processes) being controlled. Dean and Wellman [1991] apply control theory to planning, where the controller/planner formulates plans to control the behavior of agents to realize specified goals. The controller/planner and agent being controlled are independently designed, loosely coordinated open systems. Planners must maintain a model of agents whose behavior they are controlling, tracking behavior either by monitoring through interactive sensors or prediction by algorithms. Planners that interactively monitor behavior are closed-loop, on-line systems, whereas predictive planners are open-loop, off-line, closed systems. Closed-loop systems are in our sense open and open-loop systems are closed. Figure 113.12 indicates the relation between a controller/planner and the agent (embedded system) that it controls.

Control theory is concerned with both controllability and observability of processes, where a process (dynamical system) is said to be controllable with respect to a set of goal states if it can reach and remain in a goal state and is said to be observable if its state can be identified by observing its outputs. Controllability and observability are dual concepts: constraints on inputs needed to guarantee reaching a goal state are dual to constraints on outputs needed to guarantee observability of the inner state. Conversely, uncontrollability of interactive inputs to a process from the environment is dual to uncontrollability of inner actions of a process by the environment.

Duality of controllability and observability in dynamical systems reflects the inherent duality between agents being controlled and the agents that control them. In a closed-loop environment, each determines an environment for the other: the agent provides an environment for the controller and the controller provides an environment for the agent. Control is realized by messages from the controller to the agent, and observation is realized by messages from the agent to the controller.

Agents controlled by plans can be classified by whether they have a closed algorithmic model or an open interactive model. Open interactive planners are more adaptable: they can react to unpredictable exogenous events like breakdowns, strikes, and earthquakes. Open interactive planners are more expressive and efficient than closed algorithmic models, but closed models are preferred in many contexts because one can reason about their properties. For example, exogenous events are often approximated by probability distributions to transform inherently open systems with unpredictable events into closed systems with

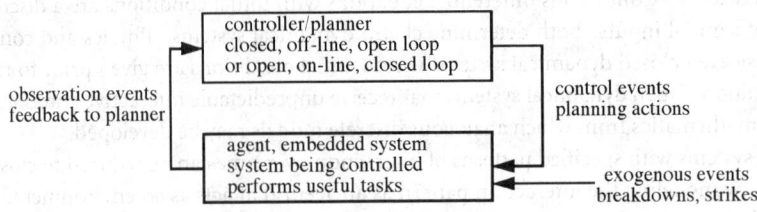

FIGURE 113.12   Controller/planner for an embedded environment.

predictable statistical properties. Agent behavior is often approximated by probability distributions to permit planners to construct off-line plans that will on the average perform well.

Characterization of components of a system by whether they are open or closed tells us something fundamental about their qualitative behavior and expressive power. For example, the design of off-line planners involves single-minded algorithmic considerations, whereas on-line planners trade off reasoning power for greater efficiency, flexibility, and expressive power. The domain of agents, control theory, and planning clearly illustrates the trade-offs between algorithmic tractability and interactive openness that occur in many other AI and software engineering domains.

113.10 Virtual Reality (VR): A Case Study in Interactive Architecture

Paradigms and models of computation can be classified by the relative importance and role of algorithms and interaction in the overall problem-solving task. Autistic systems with no connection to the outside world are completely noninteractive, mathematical and algorithmic models permit just a single interaction, and interactive models support patterns of interaction over time:

> Autistic system: no interaction
> Predicate (pure logic programming): stimulus → binary response
> Function: stimulus → range of response
> Turing machine: initial input tape to final output tape
> Object: input stream to output stream
> Graphical user interface (GUI): input stream to stream of two-dimensional images
> Robots, software components: multiple inputs to multiple outputs
> Virtual reality: multisensory real-time inputs to multisensory real-time outputs.

Predicates, functions, and Turing machines differ in their interactive and algorithmic mechanisms but have the same expressive power. Objects, GUIs, robots, and VR have greater expressive power than Turing machines and progressively richer interaction mechanisms. GUI, robot, and VR paradigms differ in the degree to which the computing agent, its environment, or both are active or passive agents:

> GUI paradigm: the computing agent is passive and the client in the environment is active.
> Robot paradigm: the computing agent (robot) is active and the environment is generally passive.
> VR paradigm: the virtual reality agent and the human client are both active.

VR systems must coordinate a variety of modes of interaction to create the illusion of immersion in a fictitious or remote environment. The illusion is created by coordinated, real-time, multimodal stimulation by a variety of sensory mechanisms:

> visual sensors for visible and electromagnetic stimuli
> auditory sensors of sound waves
> olfactory sensors of chemicals in the atmosphere
> gustatory sensing of chemicals that stimulate the tongue
> haptic sensors of touch including:
> > tactile sensors of temperature, texture, pressure
> > kinesthetic sensors of force by muscles, joints, tendons
> > spatial sensors of limbo/torso positions and angles
> > motion sensors of linear and angular acceleration (inner ear).

To provide a feel for architectures to handle multimodal interactive effects, we briefly examine the structure of VR computers (VRCs). A VRC has two clients (a human user and a model of reality) and two

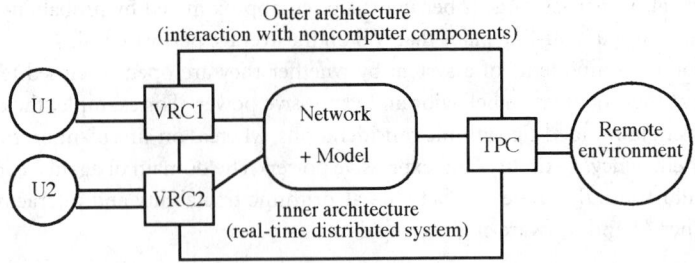

FIGURE 113.13 Interactive architecture for multiagent VR.

primary components:

> User interface manager: transmits multisensory view of model to the user
>> handles changes in perspective due to motion or attitude changes of the user.
> Model manager: manages real-time model updating and environment interaction
>> handles changes of model due to change of the virtual environment.

Changes in perspective are largely model-independent and can be handled independently of the model itself. The model may be stored locally within the VRC, nonlocally within a remote computer, or in a distributed form sharable by multiple agents and/or multiple remote virtual reality sites. The complexity of the model management may range from simple storage of a static local model to time-critical updating of shared distributed information. For slowly changing models, supplying incremental change information may be the most efficient way of updating the model.

A VRC can be conveniently described by loosely coupled user-interface and model management components. Multiagent systems must provide a common distributed model of the environment (for example, a surgical operating room) for a group of users. The task of maintaining such a model is a hard distributed networking problem with stringent real-time constraints that must handle multiple interfaces sharing a common state. However, it is entirely separate from the problem of creating the illusion of immersive interaction for individuals through a VR interface. These two tasks are concerned with different aspects of interaction and together provide a rich environment for case studies in interactive behavior.

Figure 113.13 shows two users U1 and U2, associated VRCs VRC1 and VRC2, and a remote telepresence computer (TPC) interacting with users through a distributed network. VRCs have a VR user interface that handles realistic multisensory immersion and a network interface that handles communication with other users and with the remote environment. The inner architecture of network and model management coordinates interaction among the VRCs and TPC through interfaces that share a common state, while the outer architecture handles high-bandwidth interactions with users to create the effect of immersion and with the remote environment to handle its representation and updating.

113.11 Data Information Systems: A Case Study in Empirical Computer Science

The scientific method operates by data acquisition, discovery of interesting regularities in the data, and the development of models for understanding, predicting, and changing the environment (see Fig. 113.14). Large-scale information systems that follow this paradigm include NASA's earth observation system distributed information system (EOSDIS) and digital library systems that combine data acquisition and user services. Architectures of distributed (or database) information systems (DISs) include a data repository for the mass storage of raw data. The analysis and recognition of data regularities is handled by a feature editor that catalogs the raw data and accumulates feature descriptions as well as summary data in a metadata repository. The process of discovery of significant features of observed data is called data mining.

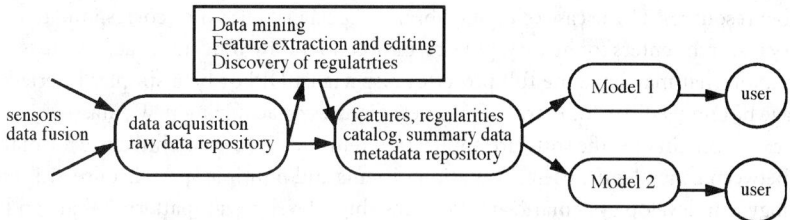

FIGURE 113.14 Data information system architecture.

The metadata are interpreted by multiple models that focus on views of the data for a variety of end-users. Geographers and ecologists focus on different models of raw data of the EOS, whereas anthropologists and historians focus on different aspects of a library corpus. Though end-user views of data are largely independent, cooperative analysis by coordinated loosely coupled specialists can be critical in realizing higher-level goals. Since specialist interfaces are determined by preestablished traditions of their user communities, an interoperation layer is needed to coordinate communication.

DISs must handle three high-level tasks (data acquisition, cataloguing and feature extraction, and multiple user models), each of which has its own research community. Data acquisition requires familiarity with sensor hardware and has historically been handled by hardware engineers. Cataloguing can benefit from techniques of information retrieval, whereas feature extraction (data mining) has been studied using tools of spectral analysis.

User interface modeling is the most software-intensive and also the most neglected of the three DIS tasks. Coordinated access to a data repository by multiple users through a variety of different models or views is a major unsolved problem in both database technology and software engineering. Problems of interoperation have been recognized as a major unsolved problem for which techniques of CORBA, COM, and megaprogramming provide only partial solutions. Standardization on interface definition languages and data exchange formats is needed to make the interoperation problem more tractable. A high-level view of the back end of a DIS with multiple specialist views mediated by interoperation is given in Fig. 113.15.

The support of multiple specialist views of a general-purpose database requires both deep system expertise and domain-specific knowledge in each of the areas being supported. The interoperation layer must solve the very complicated problem of interoperation among components with different domain-specific interface domains. To manage multiple users, the IML must handle transaction management among interoperating system components. The complete design of systems for handling shared access to a database by specialist users is a complex open problem clearly beyond the scope of this paper. The ubiquity of this "scientific method" paradigm is illustrated by the fact that this architecture is common to both NASA's EOS and digital library systems.

NASA's EOS [Short et al. 1995] calls its specialist DISs regional data centers (RDCs). RDCs have local storage, may support local data acquisition and feature editing, and may have multiple users, so that they have the same structure as the principal DIS, though on a smaller scale. The principal DIS provides support for a broad user community, whereas RDCs provide support for specialized user communities that in the case of earth observation may correspond to geographical regions or resource management centers for

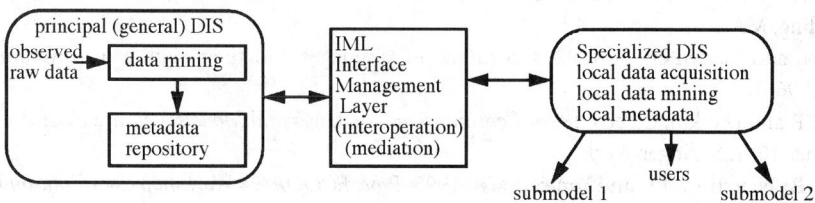

FIGURE 113.15 Principal DIS with multiple specialist DISs.

water or timber resources. In the case of digital libraries, specialist DISs may correspond to universities or to disciplinary research centers for history, physics, or oceanography. End-users access the system through a user workstation that may have the full power of a specialist DIS or be a simpler interface with just a query language but no facilities for extraction of new features or acquiring new data.

Currently each domain-specific software application reinvents its own architecture for managing communication between general-purpose information systems and multiple specialist users. Interactive software technology can develop systematic architectures (high-level design patterns) that can be reused by applications as varied as EOSs and digital libraries to support communities of specialist users in a variety of application domains.

113.12 Conclusion

Interaction machines provide a unifying model for software engineering, AI, and database technology. The distinction between object-based and component-based technology provides a framework for modeling embedded systems with multiple interfaces that share a common state. We have examined a variety of interactive technologies such as OMT, COM/OLE, CORBA, design patterns, planning and control, and case studies of VR, EOSs, and digital libraries, demonstrating that interactive models do indeed provide new perspectives and insights. However, we merely scratch the surface of a very large topic, providing the outlines of a new approach to modeling based on a notion of interactive computation richer than the algorithmic notion of computation of Church and Turing.

The paradigm shift from algorithms in the 1960s and 1970s to interaction in the 1990s is refocusing attention from inner processes of execution to interactive components. Though interactive processes are inherently less amenable to formal analysis than algorithms, heuristic methods of interoperation will play an increasingly important role in the software technology of the 21st century.

References

Agha, G., Wegner, P., and Yonezawa, A., eds. 1993. *Research Directions in Concurrent Object-Oriented Programming*. MIT Press, Cambridge, MA.

Agre, P. and Rosenschein, S., guest eds. 1995. *Computational Research on Interaction and Agency*. Special double issue on artificial intelligence, 72.

Andreoli, J., Ciancarini, P., and Pareschi, R. 1993. Interaction abstract machines. In *Research Directions in Concurrent Object-Oriented Programming*. G. Agha, P. Wegner, and A. Yonezawa, eds., pp. 257–280. MIT Press, Cambridge, MA.

Andreoli, J., Hankin, C., and Le Metayer, D., eds. 1996. *Coordination Programming: Mechanisms, Models, and Semantics*. Imperial College Press, London.

Banatre, J. and Le Metayer, D. 1996. Gamma and the chemical reaction model, ten years after. In *Coordination Programming: Mechanisms, Models, and Semantics*. J. Andreoli, C. Hankin, and D. Le Metayer, eds. Imperial College Press, London.

Blum, M. 1994. *Result Checking*, MIT Distinguished Lecture. MIT Press, Cambridge, MA.

Brockschmidt, K. 1995. *Inside OLE 2*, 2nd ed. Microsoft Press, Redmond, WA.

Brooks, F. 1995. *The Mythical Man-Month: Essays on Software Engineering*, 2nd ed. Addison–Wesley, Reading, MA.

Carriero, N. and Gelernter, D. 1992. Coordination languages and their significance. *Commun. ACM* 35(2):96–107.

Ciancarini, P. and Hankin, C., eds. 1996. *Coordination Languages and Models*. Lecture Notes in Computer Science 1061. Springer–Verlag.

Ciancarini, P. , Nierstrasz, O., and Yonezawa, A. 1995. *Proc. ECOOP '94 Workshop Coordination Languages*. Lecture Notes in Computer Science 924. Springer–Verlag. (articles by D. Gelernter and P. Ciancarini.)

Dean, W. and Wellman, M. 1991. *Planning and Control*. Morgan Kaufman, San Francisco.

Gamma, E., Helm, R., Johnson, R., and Vlissides, J. 1994. *Design Patterns: Elements of Reusable Object-Oriented Software*. Addison–Wesley, Reading, MA.

Garlan, D. 1995. Research directions in software architecture. *Comput. Surveys* 27(2):257–261.

Graubard, S., ed. 1988. *The Artificial Intelligence Debate*. MIT Press, Cambridge, MA.

Jacobson, I. 1991. *Object-Oriented Software Engineering*. Addison–Wesley/ACM Press, Reading, MA.

Milner, R. 1993. Elements of interaction. *Commun. ACM* 36(1):78–89.

Nierstrasz, O. and Tsichritzis, D., eds. 1996. *Object-Oriented Software Composition*. Prentice–Hall, Englewood Cliffs, NJ. (Chapter 3 by D. Konstantas.)

Object Management Group. 1995. *CORBA: Architecture and Specification*, Revision 2.0. Object Management Group, Boston, MA.

Hankin, C., ed. 1994. *Proc. 1st Annu. Workshop Coordination*. Imperial College, Dept. of Computer Science, London. Dec.

Rumbaugh, J., Blaha, M., Premerlani, W., Eddy, F., and Lorensen, W. 1990. *Object-Oriented Modeling and Design*. Prentice–Hall, Englewood Cliffs, NJ.

Russell, S. and Norvig, P. 1994. *Artificial Intelligence: A Modern Approach*. Addison–Wesley, Reading, MA.

Short, N. et al. 1995. Mission to Planet Earth: AI views the world, *IEEE Expert* 10(6):24–34.

Wegner, P., ed. 1995. Research directions in software engineering. *Comput. Surveys* 27(2).

Wegner, P. 1993. Tradeoffs between reasoning and modeling. In *Research Directions in Concurrent Object-Oriented Programming*. G. Agha, P. Wegner, and A. Yonezawa, eds. MIT Press, Cambridge, MA.

Wegner, P. 1995a. Interaction as a basis for empirical computer science. *Comput. Surveys* 27(1):45–48.

Wegner, P. 1995b. Interactive foundations of object-based programming. *IEEE Comput.* 28(10):70–72.

Wegner, P. 1996. Interactive foundations of computing. Report, CS-96-26. Computer Science Department, Brown University, Providence, RI.

Wiederhold, G. 1995. Mediation in information systems. *Comput. Surveys* 27(2):265–267.

Wiederhold, G., Wegner, P., and Ceri, S. 1992. Towards megaprogramming. *Commun. ACM* 35(11):89–99.

Yellin, D. and Strom, R. 1996. *Protocol Specifications and Component adapters*. IBM Report, Yorktown Heights, NY.

Appendixes

Appendix A: Professional Societies in Computing

ACM

"ACM (founded 1947) is an international scientific and educational organization dedicated to advancing the art, science, engineering, and application of information technology, serving both professional and public interests by fostering the open interchange of information and by promoting the highest professional and ethical standards."

ACM has nearly 100,000 members, and its activities include journals and monographs, special interest groups (SIGs), conferences, professional and student chapters, electronic publications, and various outreach projects. More complete information on ACM can be obtained by visiting its Web page, from which the preceding quotation was taken: http://www.acm.org.

The Computing Research Association (CRA)

"The Computing Research Association (CRA) is an association of more than 150 North American academic departments of computer science and computer engineering, industrial laboratories engaging in basic computing research and affiliated professional societies. CRA's mission is to represent and inform the computing research community and to support and promote its interests. CRA seeks to strengthen research and education in the computing fields, expand opportunities for women and minorities and improve public and policy maker understanding of the importance of computing and computing research

0-8493-2909-4/97/$0.00+$.50
© 1997 by CRC Press, Inc.

in our society." More information about the CRA can be obtained by visiting its Web page, from which the preceding quotation was taken: http://www.cra.org.

The Institute of Electrical and Electronics Engineers (IEEE) Computer Society

"Founded in 1946, the IEEE Computer Society has over 100,000 members worldwide. The Computer Society is dedicated to advancing the theory, practice, and application of computers and information processing technology. Through its conferences, tutorials, publications, branch chapters, and standards working groups, the Society serves the interests of a wide range of computer professionals." More information about the IEEE Computer Society can be obtained by visiting its Web page, from which the preceding quotation was taken: http://www.computer.org.

The British Computer Society (BCS)

"The BCS is the Chartered body for Information Technology professionals. Formed in 1957, it has nearly 34,000 members and in May 1990 became a Chartered Engineering Institution. It was also a founding member of the Council for European Professional Informatics Societies (CEPIS). The Society is concerned with the development of computing and its effective application. Under its Royal Charter granted in 1984, it also has responsibilities for education and training, for public awareness, and above all for standards, quality and professionalism." More information about the BCS can be found by visiting its Web page, from which the preceding quotation was taken: http://www.bcs.org.uk.

Computer Professionals for Social Responsibility (CPSR)

"CPSR is a public-interest alliance of computer scientists and others interested in the impact of computer technology on society. As technical experts, CPSR members provide the public and policymakers with realistic assessments of the power, promise, and limitations of computer technology. As concerned citizens, we direct public attention to critical choices concerning the applications of computing and how those choices affect society." More information about CPSR can be found by visiting its Web page, from which the preceding quotation was taken: http://www.cpsr.org/dox/home.html.

The American Association for Artificial Intelligence (AAAI)

"The American Association for Artificial Intelligence is a nonprofit scientific society devoted to the promotion and advancement of artificial intelligence—what constitutes intelligent thought and behavior and how it can be exhibited in computers." More information about AAAI can be found by visiting its Web page, from which the preceding quotation was taken: http://www.aaai.org.

Special Interest Group on Computer Graphics (SIGGRAPH)

"SIGGRAPH is the ACM Special Interest Group on Computer Graphics. Our scope is to promote among our members the acquisition and exchange of information and opinion on the theory, design, implementation, and application of computer-generated graphics and interactive techniques to facilitate communication and understanding." More information about SIGGRAPH can be found by visiting its Web page, from which the preceding quotation was taken: http://www.siggraph.org.

The Society for Industrial and Applied Mathematics (SIAM)

"The goals of SIAM are to advance the application of mathematics to science and industry, promote mathematical research that could lead to effective new methods and techniques for science and industry, and provide media for the exchange of information and ideas among mathematicians, engineers, and scientists. Applied mathematics, in partnership with computing, has become essential in solving many real-world problems. Its methodologies are needed, for example, in modeling physical, chemical, and biomedical phenomena; in designing engineering parts, structures, and systems to optimize performance; in planning and managing financial and marketing strategies; and in understanding and optimizing manufacturing processes." More information about SIAM can be found by visiting its Web page, from which the preceding quotation was taken: http://www.siam.org.

Appendix B:
The ACM Code of Ethics and Professional Conduct*

Preamble

Commitment to ethical professional conduct is expected of every member (voting members, associate members, and student members) of the Association for Computing Machinery (ACM).

This Code, consisting of 24 imperatives formulated as statements of personal responsibility, identifies the elements of such a commitment. It contains many, but not all, issues professionals are likely to face. Section 1 outlines fundamental ethical considerations, whereas section 2 addresses additional, more specific considerations of professional conduct. Statements in section 3 pertain more specifically to individuals who have leadership roles, whether in the workplace or in a volunteer capacity such as with organizations like ACM. Principles involving compliance with this Code are given in section 4.

The Code shall be supplemented by a set of guidelines, which provide explanations to assist members in dealing with the various issues contained in the Code. It is expected that the guidelines will be changed more frequently than the Code.

The Code and its supplemented guidelines are primarily intended to serve as a basis for ethical decision making in the conduct of professional work. Secondarily, they may serve as a basis for judging the merit of a formal complaint pertaining to violation of professional ethical standards.

It should be noted that although computing is not mentioned in the imperatives of section 1.0, the Code is concerned with how these fundamental imperatives apply to one's conduct as a computing professional. These imperatives are expressed in a general form to emphasize that ethical principles that apply to computer ethics are derived from more general ethical principles.

It is understood that some words and phrases in a Code of ethics are subject to varying interpretations, and that any ethical principle may conflict with other ethical principles in specific situations. Questions

*Adopted by ACM Council Oct. 16, 1992; Copyright 1993 by ACM, all rights reserved; reprinted with permission from ACM; originally published in 1993 *Communications of the ACM* 36(2) and also available on the ACM Web site http://www.acm.org.

related to ethical conflicts can best be answered by thoughtful consideration of fundamental principles, rather than reliance on detailed regulations.

1. General Moral Imperatives: *As an ACM member I will ...*

1.1 Contribute to Society and Human Well-Being

This principle concerning the quality of life of all people affirms an obligation to protect fundamental human rights and to respect the diversity of all cultures. An essential aim of computing professionals is to minimize negative consequences of computing systems, including threats to health and safety. When designing or implementing systems, computing professionals must attempt to ensure that the products of their efforts will be used in socially responsible ways, will meet social needs, and will avoid harmful effects to health and welfare.

In addition to a safe social environment, human well-being includes a safe natural environment. Therefore, computing professionals who design and develop systems must be alert to, and make others aware of, any potential damage to the local or global environment.

1.2 Avoid Harm to Others

Harm means injury or negative consequences, such as undesirable loss of information, loss of property, property damage, or unwanted environmental impacts. This principle prohibits use of computing technology in ways that result in harm to any of the following: users, the general public, employees, employers. Harmful actions include intentional destruction or modification of files and programs leading to serious loss of resources or unnecessary expenditure of human resources such as the time and effort required to purge systems of computer viruses.

Well-intended actions, including those that accomplish assigned duties, may lead to harm unexpectedly. In such an event the responsible person or persons are obligated to undo or mitigate the negative consequences as much as possible. One way to avoid unintentional harm is to carefully consider potential impacts on all those affected by decisions made during design and implementation.

To minimize the possibility of indirectly harming others, computing professionals must minimize malfunctions by following generally accepted standards for system design and testing. Furthermore, it is often necessary to assess the social consequences of systems to project the likelihood of any serious harm to others. If system features are misrepresented to users, coworkers, or supervisors, the individual computing professional is responsible for any resulting injury.

In the work environment the computing professional has the additional obligation to report any signs of system dangers that might result in serious personal or social damage. If one's superiors do not act to curtail or mitigate such dangers, it may be necessary to blow the whistle to help correct the problem or reduce the risk. However, capricious or misguided reporting of violations can, itself, be harmful. Before reporting violations, all relevant aspects of the incident must be thoroughly assessed. In particular, the assessment of risk and responsibility must be credible. It is suggested that advice be sought from other computing professionals. See principle 2.5 regarding thorough evaluations.

1.3 Be Honest and Trustworthy

Honesty is an essential component of trust. Without trust an organization cannot function effectively. The honest computing professional will not make deliberately false or deceptive claims about a system or system design, but will instead provide full disclosure of all pertinent system limitations and problems.

A computer professional has a duty to be honest about his or her own qualifications, and about any circumstances that might lead to conflicts of interest.

Membership in volunteer organizations such as ACM may at times place individuals in situations where their statements or actions could be interpreted as carrying the weight of a larger group of professionals.

An ACM member will exercise care to not misrepresent ACM or positions and policies of ACM or any ACM units.

1.4 Be Fair and Take Action Not to Discriminate

The values of equality, tolerance, respect for others, and the principles of equal justice govern this imperative. Discrimination on the basis of race, sex, religion, age, disability, national origin, or other such factors is an explicit violation of ACM policy and will not be tolerated.

Inequities between different groups of people may result from the use or misuse of information and technology. In a fair society, all individuals would have equal opportunity to participate in, or benefit from, the use of computer resources regardless of race, sex, religion, age, disability, national origin, or other such similar factors. However, these ideals do not justify unauthorized use of computer resources nor do they provide an adequate basis for violation of any other ethical imperatives of this Code.

1.5 Honor Property Rights Including Copyrights and Patents

Violation of copyrights, patents, trade secrets, and the terms of license agreements is prohibited by law in most circumstances. Even when software is not so protected, such violations are contrary to professional behavior. Copies of software should be made only with proper authorization. Unauthorized duplication of materials must not be condoned.

1.6 Give Proper Credit for Intellectual Property

Computing professionals are obligated to protect the integrity of intellectual property. Specifically, one must not take credit for other's ideas or work, even in cases where the work has not been explicitly protected by copyright, patent, etc.

1.7 Respect the Privacy of Others

Computing and communication technology enables the collection and exchange of personal information on a scale unprecedented in the history of civilization. Thus there is increased potential for violating the privacy of individuals and groups. It is the responsibility of professionals to maintain the privacy and integrity of data describing individuals. This includes taking precautions to ensure the accuracy of data, as well as protecting it from unauthorized access or accidental disclosure to inappropriate individuals. Furthermore, procedures must be established to allow individuals to review their records and correct inaccuracies.

This imperative implies that only the necessary amount of personal information be collected in a system, that retention and disposal periods for that information be clearly defined and enforced, and that personal information gathered for a specific purpose not be used for other purposes without consent of the individual(s). These principles apply to electronic communications, including electronic mail, and prohibit procedures that capture or monitor electronic user data, including messages, without the permission of users or bona fide authorization related to system operation and maintenance. User data observed during the normal duties of system operation and maintenance must be treated with strictest confidentiality, except in cases where it is evidence for the violation of law, organizational regulations, or this Code. In these cases, the nature or contents of that information must be disclosed only to proper authorities.

1.8 Honor Confidentiality

The principle of honesty extends to issues of confidentiality of information whenever one has made an explicit promise to honor confidentiality or, implicitly, when private information not directly related to the performance of one's duties becomes available. The ethical concern is to respect all obligations of confidentiality to employers, clients, and users unless discharged from such obligations by requirements of the law or other principles of this Code.

2. More Specific Professional Responsibilities: *As an ACM computing professional I will ...*

2.1 Strive to Achieve the Highest Quality, Effectiveness, and Dignity in Both the Process and Products of Professional Work

Excellence is perhaps the most important obligation of a professional. The computing professional must strive to achieve quality and to be cognizant of the serious negative consequences that may result from poor quality in a system.

2.2 Acquire and Maintain Professional Competence

Excellence depends on individuals who take responsibility for acquiring and maintaining professional competence. A professional must participate in setting standards for appropriate levels of competence and strive to achieve those standards. Upgrading technical knowledge and competence can be achieved in several ways: doing independent study; attending seminars, conferences, or courses; and being involved in professional organizations.

2.3 Know and Respect Existing Laws Pertaining to Professional Work

ACM members must obey existing local, state, province, national, and international laws unless there is a compelling ethical basis not to do so. Policies and procedures of the organizations in which one participates must also be obeyed. But compliance must be balanced with the recognition that sometimes existing laws and rules may be immoral or inappropriate and, therefore, must be challenged. Violation of a law or regulation may be ethical when that law or rule has inadequate moral basis or when it conflicts with another law judged to be more important. If one decides to violate a law or rule because it is viewed as unethical, or for any other reason, one must fully accept responsibility for one's actions and for the consequences.

2.4 Accept and Provide Appropriate Professional Review

Quality professional work, especially in the computing profession, depends on professional reviewing and critiquing. Whenever appropriate, individual members should seek and utilize peer review as well as provide critical review of the work of others.

2.5 Give Comprehensive and Thorough Evaluations of Computer Systems and Their Impacts, Including Analysis of Possible Risks

Computer professionals must strive to be perceptive, thorough, and objective when evaluating, recommending, and presenting system descriptions and alternatives. Computer professionals are in a position of special trust and therefore have a special responsibility to provide objective, credible evaluations to employers, clients, users, and the public. When providing evaluations the professional must also identify any relevant conflicts of interest, as stated in imperative 1.3.

As noted in the discussion of principle 1.2 on avoiding harm, any signs of danger from systems must be reported to those who have opportunity and/or responsibility to resolve them. See the guidelines for imperative 1.2 for more details concerning harm, including the reporting of professional violations.

2.6 Honor Contracts, Agreements, and Assigned Responsibilities

Honoring one's commitments is a matter of integrity and honesty. For the computer professional this includes ensuring that system elements perform as intended. Also, when one contracts for work with another party, one has an obligation to keep that party properly informed about progress toward completing that work.

A computing professional has a responsibility to request a change in any assignment that he or she feels cannot be completed as defined. Only after serious consideration and with full disclosure of risks and concerns to the employer or client should one accept the assignment. The major underlying principle here is the obligation to accept personal accountability for professional work. On some occasions other ethical principles may take greater priority.

A judgment that a specific assignment should not be performed may not be accepted. Having clearly identified one's concerns and reasons for that judgment, but failing to procure a change in that assignment, one may yet be obligated, by contract or by law, to proceed as directed. The computing professional's ethical judgment should be the final guide in deciding whether or not to proceed. Regardless of the decision, one must accept the responsibility for the consequences.

However, performing assignments against one's own judgment does not relieve the professional of responsibility for any negative consequences.

2.7 Improve Public Understanding of Computing and Its Consequences

Computing professionals have a responsibility to share technical knowledge with the public by encouraging understanding of computing, including the impacts of computer systems and their limitations. This imperative implies an obligation to counter any false views related to computing.

2.8 Access Computing and Communication Resources Only When Authorized to Do So

Theft or destruction of tangible and electronic property is prohibited by imperative 1.2: "Avoid harm to others." Trespassing and unauthorized use of a computer or communication system is addressed by this imperative. Trespassing includes accessing communication networks and computer systems, or accounts and/or files associated with those systems, without explicit authorization to do so. Individuals and organizations have the right to restrict access to their systems so long as they do not violate the discrimination principle (see 1.4). No one should enter or use another's computer system, software, or datafiles without permission. One must always have appropriate approval before using system resources, including .rm57 communication ports, filespace, other system peripherals, and computer time.

3. Organizational Leadership Imperatives: *As an ACM member and an organizational leader, I will …*

Background Note

This section draws extensively from the draft IFIP Code of Ethics, especially its sections on organizational ethics and international concerns. The ethical obligations of organizations tend to be neglected in most codes of professional conduct, perhaps because these codes are written from the perspective of the individual member. This dilemma is addressed by stating these imperatives from the perspective of the organizational leader. In this context *leader* is viewed as any organizational member who has leadership or educational responsibilities. These imperatives generally may apply to organizations as well as their leaders. In this context *organizations* are corporations, government agencies, and other employers, as well as volunteer professional organizations.

3.1 Articulate Social Responsibilities of Members of an Organizational Unit and Encourage Full Acceptance of Those Responsibilities

Because organizations of all kinds have impacts on the public, they must accept responsibilities to society. Organizational procedures and attitudes oriented toward quality and the welfare of society will reduce harm to members of the public, thereby serving public interest and fulfilling social responsibility. Therefore, organizational leaders must encourage full participation in meeting social responsibilities as well as quality performance.

3.2 Manage Personnel and Resources to Design and Build Information Systems That Enhance the Quality of Working Life

Organizational leaders are responsible for ensuring that computer systems enhance, not degrade, the quality of working life. When implementing a computer system, organizations must consider the personal and professional development, physical safety, and human dignity of all workers. Appropriate human–computer ergonomic standards should be considered in system design and in the workplace.

3.3 Acknowledge and Support Proper and Authorized Uses of an Organization's Computing and Communication Resources

Because computer systems can become tools to harm as well as to benefit an organization, the leadership has the responsibility to clearly define appropriate and inappropriate uses of organizational computing resources. Whereas the number and scope of such rules should be minimal, they should be fully enforced when established.

3.4 Ensure That Users and Those Who Will Be Affected by a System Have Their Needs Clearly Articulated During the Assessment and Design of Requirements; Later the System Must Be Validated to Meet Requirements

Current system users, potential users, and other persons whose lives may be affected by a system must have their needs assessed and incorporated in the statement of requirements. System validation should ensure compliance with those requirements.

3.5 Articulate and Support Policies That Protect the Dignity of Users and Others Affected by a Computing System

Designing or implementing systems that deliberately or inadvertently demean individuals or groups is ethically unacceptable. Computer professionals who are in decision-making positions should verify that systems are designed and implemented to protect personal privacy and enhance personal dignity.

3.6 Create Opportunities for Members of the Organization to Learn the Principles and Limitations of Computer Systems

This complements the imperative on public understanding (2.7). Educational opportunities are essential to facilitate optimal participation of all organizational members. Opportunities must be available to all members to help them improve their knowledge and skills in computing, including courses that familiarize them with the consequences and limitations of particular types of systems. In particular, professionals must be made aware of the dangers of building systems around oversimplified models, the improbability of anticipating and designing for every possible operating condition, and other issues related to the complexity of this profession.

4. Compliance with the Code: *As an ACM member I will ...*

4.1 Uphold and Promote the Principles of This Code

The future of the computing profession depends on both technical and ethical excellence. Not only is it important for ACM computing professionals to adhere to the principles expressed in this Code, each member should encourage and support adherence by other members.

4.2 Treat Violations of This Code As Inconsistent with Membership in the ACM

Adherence of professionals to a Code of ethics is largely a voluntary matter. However, if a member does not follow this Code by engaging in gross misconduct, membership in ACM may be terminated.

Appendix C: Standards-Making Bodies and Standards

International and national standards play an important role in computer science and engineering. Standards help unify the definition and implementation of complex systems, especially in the areas of architecture, human–computer interaction, operating systems and networks, programming languages, and software engineering.

Principal roles in standardization for computer science and engineering are played by the International Standards Organization (ISO), the American National Standards Institute (ANSI), and the Institute of Electrical and Electronics Engineers (IEEE). These organizations are briefly described in the following sections, with pointers to their Web pages provided for further information.

The International Organization for Standardization (ISO)

"The International Organization for Standardization (ISO) is a worldwide federation of national standards bodies from some 100 countries, one from each country. ISO is a non-governmental organization established in 1947. The mission of ISO is to promote the development of standardization and related activities in the world with a view of facilitating the international exchange of goods and services, and to developing cooperation in the spheres of intellectual, scientific, technological and economic activity. ISO's work results in international agreements which are published as International Standards."

There are over 100 member bodies of the ISO, each body representing a particular country's own national standards. Some of the countries and their respective member bodies (in parentheses) in ISO are as follows.

Country	Member Body	Country	Member Body
Australia	(SAA)	Ireland	(NSAI)
Brazil	(ABNT)	Israel	(SII)
Canada	(SCC)	Italy	(UNI)
China	(CSBTS)	Japan	(JISC)
Czech Republic	(COSMT)	Netherlands	(INNI)
Denmark	(DS)	Sweden	(SIS)
Egypt	(EOS)	Switzerland	(SNV)
Estonia	(EVS)	USA	(ANSI)
France	(AFNOR)	Ukraine	(DSTU)
Germany	(DIN)	United Kingdom	(BSI)
India	(BIS)		

0-8493-2909-4/97/$0.00+$.50
© 1997 by CRC Press, Inc.

More information about the ISO can be can be obtained by visiting its Web page: http://www.iso.ch.

The American National Standards Institute (ANSI)

"ANSI is a privately funded federation of leaders representing both the private and public sectors. Our job is to coordinate the U.S. voluntary consensus standards system . . . a system frequently called the most effective and efficient in the world. The strength and effectiveness of the U.S. voluntary standards system rests with the diversity, talent and participation of the ANSI Federation membership. As international standardization continues to affect every business sector, it also impacts the overall quality of life of every American. Consequently, there is a greater need for more representation and participation in the ANSI Federation from every segment of the U.S.

"The ANSI Federation, organized in 1918, is made up of both manufacturing and service businesses, professional societies and trade associations, standards developers, academia, government agencies, and consumer and labor interests, all working together to develop voluntary national consensus standards. Our strength is in the diversity and expertise of our large membership, which now totals over 1700. Operating under its mandate, we have served the U.S. while the strength of our program base and the level of our membership continue to grow."

ANSI standards in computer science exist in the areas of architecture, graphics, and programming languages. The standards committee X3 is accredited by ANSI to maintain standards in information technology, including these areas of computer science. More information about specific ANSI standards in these and other areas can be obtained by visiting the ANSI and X3 Web pages: http://www.ansi.org/docs/home.html and http://www.x3.org.

IEEE Standards

The IEEE also develops standards for certain areas of computer science and engineering, especially the areas of architecture, networks, and software engineering. For more information, visit the IEEE Web page: http://stdsbbs.ieee.org/index.html.

Appendix D:
Hardware and
Networking Standards

The American Standard Code for Information Interchange (ASCII)

The American Standard Code for Information Interchange (ASCII) is a standard representation scheme for text-based information storage and network transfer. The ASCII standard was established in 1968, and the current version of the standard is ANSI X3.110-1983.

Ethernet and Asynchronous Transfer Mode (ATM)

Ethernet is a popular local-area network (LAN) technology that allows computers to exchange information at speeds of 10 and 100 million bits per second (Mb/s). As of 1994, estimates suggest that over 40 million Ethernet connections were in use worldwide. Most current computers come equipped with 10-Mb/s Ethernet connections. The current Ethernet standard is IEEE 802.3 and was first published in 1985. Asynchronous transfer mode (ATM) is a broadband network technology that supports data, voice, and video transmissions at rates up to 150 Mb/s.

For more information about Ethernet and ATM, readers are encouraged to consult Chapter 82 of this Handbook or either of the following Web pages: http://www.host.ots.utexas.edu/ethernet/10quickref/ch1qr_1.html or http://www.yahoo.com/Computers_and_Internet/Computers_and_Networking/ATM.

Floating-Point Arithmetic

Computer implementations of floating-point numbers and arithmetic generally follow the IEEE floating-point standards ANSI/IEEE 754-1985 (R1991) and ANSI/IEEE 854-1988 (R1994). The 754 standard has been adopted by nearly every computer manufacturer since about 1980. It uses a 32- and 64-b binary word as the basis for representing a floating-point number. The 854 standard restates this representation in a radix-independent style.

For more information about floating-point arithmetic, readers should consult Chapter 19 of this Handbook.

Appendix E:
Common System
Command Languages*

Paul W. Ross
Millersville University

Introduction

Since the original IBM Personal Computer® was introduced in 1981, MS-DOS® (PC-DOS® on IBM systems) has become the dominant operating system for computers that use the Intel 8088, 8086, 80286, 80386, 80486, and Pentium® chips. More microcomputers use DOS than any other operating system, and there are thousands of application packages that run under it.

Another common operating system is Unix, or one of its variants. The first version of Unix was created on a Digital Equipment Corporation PDP/7, which was a small machine. The system was developed and refined by Ken Thompson, Dennis Ritchie, and Rudd Cadaday and the first real Unix system for a DEC PDP 11 was completed in 1971. Other modifications occurred over time, with the eventual release of version 6 to the public.

This appendix summarizes the main features of the MS-DOS and Unix command languages. Readers who seek additional details are referred to the readings listed at the end of this appendix.

*Reprinted from *The Handbook of Software for Engineers and Scientists*, Chs. 1, 5, and 6, pp. 3–13, 77–95, and 96–102. 1996. CRC Press, Boca Raton, FL and IEEE Press.

MS-DOS for IBM PCs

Starting MS-DOS is simple. Just turn on the computer. When the computer boot process is complete, you will see a symbol such as:

C>

The letter on the screen followed by the > or greater than symbol is the MS-DOS prompt. It indicates which disk drive is currently engaged. In a floppy-disk system, the drive is designated with the letter A and the second floppy disk drive is the B drive. Hard drives are designated beginning with the letter C and go on through the alphabet. For example, if you have a second hard disk drive, it will be designated as the D drive. Network drives fall after the local hard drives, and CD ROM drives are generally last.

If you want to instruct the computer to read information from another drive, you can switch to another disk drive by typing in B: and pressing the Enter key. Be sure to include the colon. This command shifts the computer from whatever the current drive may be over to drive B.

Types of MS-DOS Commands

MS-DOS responds to a great number of commands for setting computer system parameters, manipulating data files, and executing programs. Depending on the implementation, the commands can be classified as internal and external commands.

The internal commands are those that are built into the operating system itself. They are the commands that are most frequently used and that need to be executed rapidly. Whenever the operating system prompt is visible, these internal commands are available. Under certain circumstances, the commands may be available from within an applications program if it has a MS-DOS Window, as is typical in many word processor and spreadsheet applications.

Less frequently used commands, generally classified as external commands, are known as MS-DOS utility programs. They are not part of the command processor but separately executable programs. They include such things as copying entire disks or sorting data files. This section will discuss the most commonly used MS-DOS commands. There are other less frequently used commands that are described in the MS-DOS reference manual that comes with your computer or through the HELP command. What we describe and discuss here will be enough for the vast majority of situations you will come across.

Files and Disk Management

The file structure of MS-DOS can be viewed as a hierarchy, or tree. It starts at the C> prompt. Each directory can contain files, or directories, or both.

Since a hard disk can hold a large number of files, it is important to have some way of organizing them. This is done with directories and subdirectories to produce a treelike structure where similar files can be grouped together. An example would be to place all of the files associated with a given application package in their own subdirectory.

Subdirectories are created with the MKDIR or MD command. The command format is as follows:

MKDIR [d:]subdirectory

If the command is entered without specifying a disk drive, the default drive is used. Beginning the subdirectory with a backslash (d:\) tells MS-DOS to create the subdirectory starting at the root or main directory. The absence of a leading backslash indicates that you want to create the subdirectory within the current directory. Each subdirectory may contain many subdirectories of its own, in a treelike structure.

Examples of subdirectories are as follows:

MKDIR C:\LEVEL1

This command creates the subdirectory LEVEL1 on drive C starting at the root directory.

MKDIR LEVELN

This command creates subdirectory LEVELN on the default disk in the current directory.

MKDIR C:\LEVEL1\LEVEL2

This command creates subdirectory LEVEL2 on drive C within directory LEVEL1. The subdirectory LEVEL1 must exist before this command is given. Note that successive levels of subdirectories are separated by backslashes.

CHDIR Command

If we want to move to a specific directory, we use the CHDIR (change directory) command. CHDIR, or CD for short, changes the current directory of the specified or default disk drive. The command format is as follows:

CHDIR [[d:]subdirectory]

Removing a Subdirectory

The RMDIR (remove directory), or RD command, removes an empty subdirectory from a disk when it is no longer needed. First, go into the subdirectory and remove all files with the DEL command. The command format for removing the empty directory is then as follows:

RMDIR [d:]subdirectory

Before specifying the subdirectory to be removed, you must go back to at least one subdirectory before the one that you want to remove. For example:

RMDIR \LEVEL1\LEVEL2

This removes the subdirectory LEVEL2 from the LEVEL1 directory on the current disk. Another example is:

RMDIR SUBDIR

This command removes the subdirectory SUBDIR from the current directory.

Files and File Names

A complete file name contains four elements:

1. *The name of the disk drive:* This is A or B on a floppy disk system and is C or higher in a hard drive system. It can be omitted unless you want to refer to a file stored on a disk other than the currently logged disk drive.
2. *The subdirectory name:* This optional feature is a name given to a group of several similarly related files.

3. *The name of the file:* There can be no spaces in a file name. TRAKTWO is fine, but TRAK TWO is not. File names cannot exceed eight characters.
4. *The file extension:* This is an optional extension of the file name. For example, THISFILE.CUR and THISFILE.YTD might contain the current and year-to-date figures, respectively, on a given subject. The extension consists of a period followed by 1–3 characters. You have the option of including an extension or not on the files you create. Most applications packages will provide an extension automatically for their purposes.

Directory Command

To determine what information is stored on your disk, use the DIRectory command. Every disk contains a directory of all of the files stored on it. By typing the MS-DOS command DIR after the prompt and pressing Enter, you will get a list of all of the files stored on the disk in the current drive.

If you wish to see the directory for a different disk, or for a portion (subdirectory) of the files on a disk, then a more specific DIR command is needed. The complete command format for DIR is as follows:

DIR [d:][subdirectory][filename][.ext]

where brackets [" and "] designate optional parts, and the rest of the command format is defined as follows:

- The [d:] is the disk drive; if none is specified, then the computer will assume the current drive.
- The [subdirectory] is the optional name of the subdirectory you want to view.
- The [filename] is the name of the file. If no filename is specified, then all of the files in the specified drive will be displayed.
- The [.ext] is the extension to the filename. Files that are built by application packages often already have extensions. With the files you create, you have the option of adding an extension if you wish.

For example, if we had a subdirectory known as LETTERS on disk drive C, the DIR command to show us all the files on that subdirectory would look like:

DIR C:\LETTERS**

Note the use of the backslash (\) on either side of the subdirectory name. This is functionally equivalent to the / used in Unix (discussed subsequently).

There are two global filename characters, or wildcards. The two wildcard characters are ? and *. The ? symbol can stand for any single character. The * can stand for any string of characters, or a null.

CLS Command

The CLS (clear screen) command clears the display screen. This command also returns the MS-DOS prompt to the top left corner of the screen.

COPY Command

The COPY command is used to make a copy of a file from a source location to a destination or target location. The COPY command works like this:

COPY filename1 filename2

where filename1 is the source and filename2 is the destination. If only one filename is given, it must be the name of a file on another disk or subdirectory with the same name. The general format for the COPY command is as follows:

$$\textbf{COPY} \quad [\text{d:}][\text{subdirectory}]\text{filename}[.\text{ext}] \quad [\text{d:}][\text{subdirectory}]$$

This format copies a file from one drive and/or directory to another disk, or between two directories on the same disk. The first d: in the COPY command indicates the source drive, and the second d: indicates the destination drive.

REN Command

The REN (rename) command changes the name of a file to a new name. The command format for REN is as follows:

$$\textbf{REN} \quad [\text{d:}][\text{subdirectory}]\text{filename}[.\text{ext}] \quad \text{filename}[.\text{ext}]$$

You need to specify the disk drive and subdirectory only for the first file name. If no drive or subdirectory is specified, then MS-DOS will default to the current directory and subdirectory. The second file name will use the drive and subdirectory specified for the first file name.

DEL Command

The DEL command is used when you want to delete an entire file you no longer need. Deleting a file recovers space on the disk for future work. It can also ward off confusion if you have earlier and later versions of the same material. The DEL command permanently deletes one or more files from the disk. The command format is as follows:

$$\textbf{DEL} \quad [\text{d:}][\text{subdirectory}][\text{filename}[.\text{ext}]]$$

If the disk drive is not specified, the default disk drive is used.

TYPE Command

The TYPE command displays the contents of the specified file on the screen. The command format is as follows:

$$\textbf{TYPE} \quad [\text{d:}][\text{subdirectory}]\text{filename}[.\text{ext}]$$

If the drive and subdirectory are not specified, the default drive and subdirectory are assumed. The filename may not contain the global filename character * or ?, because only one file can be displayed at a time.

Formatting Disks

Formatting completely erases any old information from the disk and lays down a skeleton structure to allow the computer to quickly find or store information on the disk. To format a disk, follow these steps:

1. Insert the new unformatted diskette in the A drive.
2. At the MS-DOS prompt type the command:

$$\textbf{FORMAT} \quad [\text{d:}] \quad <\text{Enter}>$$

At this point, the format process takes place. It will take a minute or so. When it is done, you will be told how much space is available on the disk and be asked if you want to format another disk.

DISKCOPY

The COPY command copies single files or groups of files. If you want to copy an entire floppy disk, use the DISKCOPY command. The DISKCOPY command is especially useful for making backup copies of entire floppy disks for archival purposes. The DISKCOPY command copies the contents of the specified source disk drive to the destination drive overwriting *all* of the information previously on the destination diskette. It formats the destination disk at the same time, if it has not already been formatted. The command format is

DISKCOPY [d:] [d:]

The first disk drive specified is the source disk, and the second drive is the destination.

CHKDSK and SCANDISK Commands

Because of minor hardware or software problems, you can start to lose files on a disk. Some of these problems can be located and corrected with the CHKDSK (check disk) or SCANDISK (scan disk) utilities.

The CHKDSK command checks the File Allocation Table on the specified or default disk drive and produces a disk and memory status report. The command format for CHKDSK is as follows:

CHKDSK [d:][filename[.ext]][/F][/V]

If you specify a filename, CHKDSK displays the number of noncontiguous (fragmented) areas occupied by the file(s) in the current directory. The global filename characters * and ? may be used to check selected groups of files. These options may follow the filename.

- /F   This option tells CHKDSK that you want to correct errors found while verifying the disk integrity. This option tells CHKDSK to gather up any lost fragments of files and gives them a name in the form FILEnnnn.CHK, where nnnn is a number. The resulting files can then be deleted so that the space can be restored to the disk for future use.
- /V   This option causes each filename and directory processed to be displayed on the screen, indicating the program's progress.

A more powerful utility provided with recent releases of MS-DOS is the SCANDISK utility. This utility will attempt to repair damaged files, if possible. The syntax is simply

SCANDISK <Enter>

Follow the screen prompts for the various options. The SCANDISK utility takes longer than CHKDSK but will address a wider range of problems than does CHKDSK.

The DEFRAG Utility

As files are alternately created and deleted, files become distributed across the disk instead of being in contiguous locations. As a consequence, it takes longer to load or store files. This problem is known as disk fragmentation. MS-DOS provides a useful utility to reorganize the contents of the disk, placing files in contiguous locations on the disk. The syntax is

DEFRAG <Enter>

Unix

The creators of Unix considered the system's most important role to be that of providing a file system. Virtually everything in Unix is a file. From the user's viewpoint, there are three kinds of files: ordinary disk files, directories, and special files. Ordinary files contain whatever the user chooses to put in them. Most ordinary files are expected to be text files, which are simply strings of characters. Binary files are generally those created by compilers or linkers and contain executable forms of programs.

A directory may be thought of as a table that maps the names of files to their attributes. One of those attributes is, of course, the physical location of the file, but there are other attributes in the table as well. Such additional information includes the type of file (directory, text file, binary file, special file), date of creation, etc. Note that a file in a directory may be another directory, thereby creating a hierarchy of directories and other files. If a directory C is contained in another directory P, then we refer to C as the child of P or a subdirectory of P. We refer to P as the parent of C. Each directory always has at least two entries. The name "." (which does not include the quote marks and is often pronounced dot) refers to the directory itself so that a program may read the current directory by using the name . as an alias to the complete pathname of the directory, as will be defined. The other entry that is in each directory is ".." (which may be used as the name of the parent directory). The topmost directory that has no parent is called the root directory and in that directory . and .. each refer to the same directory, namely, the root.

The same filename may appear in several directories but will always have a unique pathname. The directory hierarchy determines the pathname for a file. The full pathname of a file consists of a sequence of directory names (the root has no name) separated by slashes. Thus, the pathname /tom/dick/harry represents a file named harry contained in the directory dick, which in turn is contained in the directory tom. The directory tom is contained in the root directory. The file harry may be an ordinary file, a directory, or a special file. The pathname for the root directory is simply /.

Getting Started on a Unix System

Logging on and off a Unix system is quite simple. Typically, when assigned an account for a Unix system, a user is given a user identification (userid) and an initial password. Once connected to the system, you simply type your userid at the prompt and then your password when prompted. If such is the case, that would most likely have been made clear when the account was assigned. To log off, the command is *logout* rather than *logoff*. Once logged on, the user is ready to begin issuing commands. For exact details, which are often system specific, consult your systems administrator. Once logged on, the user is ready to begin issuing commands.

Unix Commands: Basic Format

The basic format of Unix commands is straightforward. We shall use the ls command to demonstrate the format of the commands as well as our conventions for describing them.

ls   List contents of directory

All of the following are legitimate ls commands:

```
ls
ls -a
ls -aCl *.doc readme.*
```

The letters following the hyphen are command options. In the preceding example, as in all of our examples here, there are actually many more options than we have indicated. We have included only the most commonly used options and, since not all command options are identical across all Unix systems, only those that are most commonly available.

Wildcards

The asterisks in the last example constitute *wildcards* or **metacharacters**. They are special symbols that can be used to represent a number of files. If you are a MS-DOS user, you are already familiar with wildcards. The wildcards of Unix are quite similar to those of MS-DOS, but those of Unix are more flexible.

There are three types of wildcards that may be used in filenames: the asterisk (∗), which represents any string of characters (including the string of no characters); the question mark (?), which represents any single character; and the character set ([. . .]), which represents any one of the characters included within the square brackets. Within the square brackets, there are further notations that can be used. Hyphens can be used to represent a range of letters. Thus a-g is short for abcdefg. An exclamation mark may be used after the first square bracket to indicate that the character set represents any character except for the characters listed. Thus, [!abc] would represent any character other than a, b, or c. Note that the exclamation point has a different special meaning for the C shell (command processor, like COMMAND.COM in MS-DOS) and so it must be preceded by a backslash (\). In the C shell, we would type [\!abc] to represent any character other than a, b, or c. Note that no implementations of the C **shell** implement this negation feature and so it may not work on your system. If not, you may change to another shell.

Wildcards	
File Name with Wildcards	Will Match
∗	Any filename
a∗	Any filename beginning with a
a∗k	Any filename beginning with a and ending with k
∗a∗	Any filename containing the letter a
a??	Any filename consisting of exactly three letters beginning with a
∗[xyz]	Any filename ending with one of the letters x, y, or z

As a second example, we will consider the *cat* command.

cat [-v] [*file* ...] Concatenates files

Where one of the options is

−v   Lists nonprintable characters.

The *cat* command is also used to view files as explained previously. For example, suppose the file named *start* contains the text

**Four score and seven years ago
our fathers brought forth on this continent**

and the file named *middle* contained the text

a new nation conceived in liberty and dedicated

and the file named *finish* contained the text

to the proposition that all men are created equal.

Then, the command

```
cat start middle finish
```

would cause the following to appear on your screen:

```
Four score and seven years ago
our fathers brought forth on this continent
a new nation conceived in liberty and dedicated
to the proposition that all men are created equal.
```

Redirection and Pipes

If we wished to save the text of the previous example into a separate file we could issue the command

```
cat start middle finish > gettysburg.address
```

That would create a new file (or add to the contents of an existing file) named gettysburg.address that would contain the text as previously displayed. If we wished to view it for confirmation, we might issue the command

```
cat gettysburg.address
```

which would display the file on our screen.

The character > is the **redirection** symbol that directs the system to place the output into a file rather than the standard output (the screen).

Had we put the entire Gettysburg Address into our file instead of just the first sentence, then when the *cat* command displayed the file, most of it would run off the screen before we could view it. To resume this, the command *more* can be used,

```
more gettysburg.address
```

Then, each time the screen fills up, the system generates a pause. Striking the space key will cause the next page of text to be displayed.

Pipes

Commands may be combined in Unix with the **pipe** operator (|), which causes the output of one command to be the input (or first input if there is more than one) of the next command. For example, if we wanted to view the three files *start*, *middle*, and *finish* with the *more* command, we could use three commands

```
more start
more middle
more finish
```

or simply the single command

```
cat start middle finish | more
```

Quotations

Certain characters have special meaning within a shell or command processor; different shells have different sets of special characters. We have already seen that the backslash can be used before a character in order to remove its special meaning. For strings of characters, quotation marks can do something similar. In particular, enclosing a string in single quotes will, with one exception, remove the special meaning from all of the characters in the string. The one exception is the exclamation point (called _bang_) within the C shell. That particular symbol must be preceded by a backslash even when quoted. When using the Bourne or Korn shells, there is no problem with the exclamation point. Enclosing a string within double quotation marks will remove the special function of all symbols except for dollar signs, back quotes, and backslashes (and, for the C shell, exclamation points).

A Basic Minimal Command Set

There are a few key commands that can help you find your way around Unix. One of the most valuable (ultimately) is the **man command**. This will give you the pages in the Unix manual that describe any given command. Recall that our descriptions of commands do not include all of the options and capabilities of commands. The _man_ command will enable you to obtain all of the information you want about a command. For example, if you give the command

```
man man
```

you will see approximately 11 pages of detailed information about the man command. With a little practice, you learn how to scan through the material you do not want or need, but it is useful to have some sort of abbreviated command reference. Many systems will implement a simple help command that summarizes the command set or possibly summarizes the help facilities. We will illustrate methods of combining Unix commands in order to do your own automatic editing a little later.

Copying Files

To copy the file _source_file_ into a (possibly new) file _target_file_ use

cp _source_file target_file_

If _target_file_ already exists, it will be overwritten (but retains its **permissions**). If the _target_file_ does not exist, it will be created.

To copy a list of files into a _target_directory_ use

cp _source_file_ [_file_ ..] _target_directory_

The names in the new directory will be the same as in the old and, as in the previous example, any existing file with the same name will be overwritten.

The following will copy all files in the _source_directory_ into the _target_directory_:

cp [**-R**] _source_directory target_directory_

One of the options to the cp command is

R   Recursively copies subdirectories so that entire directory subtree is copied.

Removing (Deleting) Files

The command

$$\text{rm} \quad [\text{-iR}] \quad \textit{file} \, [\textit{file} ..]$$

removes all of the files listed. If a given file is a directory, then all files in that directory are deleted. The options to the rm command are

i	Interactive mode. Interactively confirms each deletion.
R	Recursively deletes all subdirectories of any directory listed.

Moving or Renaming Files

The command

$$\text{mv} \quad \textit{source\_file} \, [\textit{file} ..] \quad \textit{directory}$$

moves *source_file* and other listed files, if any, into *directory*. This is usually equivalent to copying the designated files and then deleting them.

Creating Directories

To create a new directory named directory-name, which will be a subdirectory of the current directory use

$$\text{mkdir} \quad \textit{directory-name}$$

Moving to a New Directory

The cd commands make directory-name the new current directory from the directory that was accessed when the user logged in. If no directory name is specified, the user's home directory becomes the new current directory.

$$\text{cd} \quad \textit{directory-name}$$

Setting and Changing a File's Permissions

The permissions of the listed files are changed to *mode*.

$$\text{chmod} \quad \textit{mode file} \, [\textit{file}..]$$

If a directory is listed, then all files within that directory have their permissions changed as specified by *mode*. There are two major ways to express the desired mode: symbolic and numeric. For brevity, we will discuss only the symbolic format here. In the symbolic mode format, *mode* c consists of up to three symbols that represent a group of users, an operator, and a permission.

Examples

The command

$$\text{chmod a+rwx myfile}$$

gives full rights for everything to everyone.

Specification of User Group:

u	owner or current user of the file
g	members of same group as the user
w	all other users
a	all users

Operators:

+	Add the permission
–	Remove the permission
=	Set the permission to be . . .

Permissions:

r	read
w	write
x	execute

The command

```
chmod g+rw,o=r mydirectory
```

gives read and write privileges to all members of the owner's group and read-only privileges to all other users. Note that if *mydirectory* is a directory, then the privilege applies to all files in the directory.

Searching for Strings

The three commands grep, fgrep, and egrep are all used to search one or more files for lines matching a pattern. The patterns are given as **regular expressions** (defined subsequently). All three commands do the same thing; the three versions exist only for efficiency. The most general command of the three is egrep and it may be used exclusively. The fastest is fgrep but it can be used only when searching for specific literal strings. If you are searching for a simple pattern, grep can be used and will execute somewhat faster than egrep. Since egrep is the most general, all of our examples will use it.

Regular Expressions

Regular expressions are patterns that describe sets of strings. Regular expressions include ordinary characters and special operators called metacharacters. Ordinary characters are those that are not special operators. The metacharacters may have special meaning depending on their context, or they may behave like ordinary characters. Note that some metacharacters (e.g., * and ?) that are treated as wildcards by the shell do not have the same meaning when used as metacharacters.

Special Operators for Regular Expressions

.	Any single character
*	Zero or more occurrences of preceding regular expression
+	One or more occurrences of preceding regular expression
^	Beginning of line
$	End of line
[. . .]	Any one of the characters in . . . (ranges such as a–z may be used)
[^ . . .]	Any character other than the characters in . . . (ranges such as a–z may be used)
\	Causes the following character to lose its special meaning (if any)
(. . .)	(Valid in egrep only) Used to group regular expressions

Creating and Editing Files

Most user created files are, of course, text files, which can be created and modified by text editors. There are three major editors that are included in Unix environments. There is a line editor called *ed*. Since

Examples of Regular Expressions

Expression	Will match
a.z	Any three-letter string beginning with a and ending with z
a-z	Any single lowercase letter
^a.z	Any three-letter string at the beginning of the line that begins with a and ends with z
[a-zA-Z] *	Any string (including the empty string) consisting entirely of alphabetic characters
^[a-zA-Z]+$	Any line consisting entirely of alphabetic characters (excluding blank) that contains at least one character
^([a-zA-Z] * +) * $ (egrep only)	Any line consisting entirely of alphabetic words separated by blanks (note that the line may not begin with a blank and the last word must be followed by at least one blank)
^([a-zA-Z] * +) * [a-zA-Z]?$ (egrep only)	Any line consisting entirely of alphabetic words separated by blanks (note that the line may not begin with a blank)

line editors are not particularly popular these days, we will omit any discussion of the details. The early full-screen editor, vi, is also falling out of use. One of the most popular editors is the Emacs editor. An excellent book on this is available from O'Reilly Associates (info@ora.com).

Defining Terms

File permissions: Information that is stored with a file that determines what privileges various users have. Privileges are read, write, and execute. Users are divided into the owner, the owner's group, and all other users.

Filter: A command that extracts, inserts, translates, or rearranges the data of its input stream.

Man pages: The description of a command in the on-line manual that is accessible via the *man* command.

Metacharacter: A character that has special meaning to the shell or within a regular expression. Usually represents a set of strings.

Pipe: A mechanism, represented by | that sends the output of one command to another command as input.

Redirection: Changing the standard input (via <) or the standard output (via >) of a command to come from or go to a specified file.

Regular expression: An expression that represents a set of strings. It is used in ed, vi, and grep commands.

Shell: A command processor.

References

Andleigh, P. K. 1990. *UNIX® System Architecture*. PTR Prentice–Hall, Englewood Cliffs, NJ.

Bach, M. J. 1986. *The Design of the UNIX® Operating System*. Prentice–Hall, Englewood Cliffs, NJ.

Bolsky, M. I. and Korn, D. G. 1989. *The Kornshell: Command and Programming Language*. Prentice–Hall, Englewood Cliffs, NJ.

Cameron, D. and Rosenblatt, B. 1996. *Learning Gnu Emacs*, 2nd ed. O'Reilly and Associates, Sebastapol, CA.

Christian, K. and Richter, S. 1994. *The UNIX Operating System*, 3rd ed. Wiley, New York.

Goodheart, B. and Cox, J. 1994. *The Magic Garden Explained: The Internals of UNIX System V Release 4, An Open Systems Design*. Prentice–Hall of Australia, Sydney, Australia.

Hare, C., Dueaney, E., Eckel, G., Lee, S., and Ray, L. 1994. *Inside UNIX®*. Riders, Indianapolis, IN.

Kernighan, B. W. and Pike, R. 1984. *The UNIX® Programming Environment*. Prentice–Hall, Englewood Cliffs, NJ.

Lamb, L. 1990. *Learning the vi Editor*, 5th ed. O'Reilly and Associates, Sebastapol, CA.

Leffler, S. J., McKusick, M. K., Karels, M. J., and Quarterman, J. S. 1989. *The Design and Implementation of the 4.3BSD UNIX Operating System*. Addison–Wesley, Reading, MA.

MS-DOS Reference Manual, Microsoft, Redmond, WA.

Ritchie, D. M. and Thompson, K. 1974. The UNIX Time-Sharing System. *Comm. ACM* 17(7):365–375.

Rosen, K. H., Rosinski, R. R., and Host, D. A. 1994. *Best UNIX Tips Ever*. Osborne McGraw–Hill, Berkeley, CA.

Salus, P. H. 1994. *A Quarter Century of UNIX*. Addison–Wesley, Reading MA.

Sams Development Team. 1994. *UNIX® Unleashed*. Sams, Indianapolis, IN.

Southerton, A. 1993. *Modern UNIX*. Wiley, New York.

Southerton, A. and Perkins, E. C., Jr. 1994. *The UNIX and X Command Compendium*. Wiley, New York.

Topham, D. W. 1992. *Portable UNIX*. Wiley, New York.

Appendix F:
Internet Access
Protocols*

Paul W. Ross
Millersville University

Introduction

The Internet is a super network of universities, research sites, commercial organizations, and government agencies that allows information exchange and sharing of data and resources. The predecessor of the Internet was **ARPANET**, a Department of Defense computer network created in 1969. When the Department of Defense moved its military applications to a new network (MILNET) in the mid-1980s, researchers, and eventually the National Science Foundation, took over the physical network, and the Internet was born. Today, the Internet connects over 10,000 smaller computer networks, over one million individual (host) computers, and over 5 million users. Internet usage is growing at a very rapid rate.

*Reprinted from *The Handbook of Software for Engineers and Scientists.* Chs. 64 and 67, pp. 1432–1455, 1478–1492. 1996. CRC Press, Boca Raton, FL and IEEE Press.

Internet Addresses

Internet addresses fall into two categories : all-numeric **Internet protocol (IP) addresses** (e.g., 192.206. 29.16) or the more common alphanumeric fully qualified domain name or **FQDN** (e.g., cs.millersv.edu). Note that these are not personal **e-mail** addresses; they are the addresses of computers or networks on the Internet. To specify a particular user's Internet address, the individual's userid is prefixed to a domain name (e.g., ross@cs.millersv.edu).

You must specify either an IP address or domain name when using the Internet commands Telnet or file transfer protocol (**FTP**). The domain name has three or four parts, separated by periods. The last part specifies the type of organization that owns the host computer. The most common types are:

edu	College or university
com	Commercial organization
gov	Government agency
mil	Military site
net	Network organization
org	Miscellaneous private organization

The middle portion denotes the particular site (e.g., rice for Rice University). The first part is the actual computer; if the domain name has four parts, the second part usually indicates a local network. For example, cs.millersv.edu is the computer science system (cs) at Millersville University (millersv), which is a university (edu).

To specify the Internet address of an individual user, you give their userid followed by an at sign (@), and then the domain name of the host computer they are on. For example, the postmaster account (postmast) on CS at Millersville University has the Internet address postmast@cs.millersv.edu.

Using Telnet

Telnet is a powerful Internet command that allows remote logon to another computer (remote-host) on the Internet. Telnet is similar to dialing into another computer, except that you are spared the expense and hassle of a modem, communication software, and a long-distance phone call. Note that you must have a valid login id and (usually) a password to use a remote computer. Some computers feature generic userids that allow you to logon (often without supplying a password) and access general information files via a series of menus. Typically, you would initiate a Telnet connections as follows:

Telnet <remote-host domain name>

Using File Transfer Protocol (FTP)

FTP (or anonymous FTP) stands for file transfer protocol. This command is similar to Telnet; you start by logging into a remote system. However, you are then provided with an environment in which you can transfer files between systems, usually from the remote host to your account. To start FTP, enter the FTP command followed by the Internet domain name of the remote host:

FTP <remote host domain name>

For example, to FTP to the University of Illinois, type

FTP vmd.cso.uiuc.edu

You should see a message "Connecting to. . .," and then you will be asked for your login id (identify yourself to the host). For most systems, this is the word anonymous. Type anonymous and press <ENTER>. You will then be asked for a password: this is usually your Internet address; type it in and press <ENTER> again. You should now see a message telling you that you are logged in, along with any general information messages. At this point, you can enter other FTP commands. Some of these commands are as follows:

dir	lists files in current directory on remote host
cd <new dir>	change directory to new dir on remote host
get <filename>	transfer file from remote host to your account
get <filename>	transfer file from remote host and print to screen
put <filename>	transfer file from your account to remote host (only with proper authorization!)
mget <*.ext>	transfer all files of type *.ext from remote host to your account
mput <*.ext>	transfer all files of type *.ext from your account to remote host (if authorized)
quit	exit FTP and logoff remote host

Getting Directory Listings

- dir   Provides a more complete listing of files
- dir *.txt   Will list all files that end in .txt
- ls   Provides a simple listing of files

Maneuvering Through an FTP Site

The directories at an FTP site are organized hierarchically, and maneuvering through them is similar to what you do in Unix or MS-DOS (see Appendix E):

- Look for a file named Index or README at the highest directory level; these files usually provide information as to the type of files located at the FTP site.
- Use the *cd* (change directory) command to go *downward* through directories.
- Use the *cd ..* command to go *upward* through directories.
- Use the *pwd* command to see what directory you are currently viewing.

Using the GET Command

- Use the command *get* to obtain one file:

GET filename-to-get local-filename

- Get filename - (hyphen) obtains a file and lists it to your terminal.
- The name of the filename-to-get should be indicated as it appears at the FTP site, no matter how files are named in your local environment.

- The name of the local-filename should be in a format that your computer can handle:

  On a Unix-based system or DEC VAX: filename.type

  On an IBM system under VM/CMS: filename type A

- The *mget* command may be used to obtain multiple files:

  mget filename1 filename2 filename3

Different File Types: ASCII and Binary

To change between modes:

- Use the command *binary* to set the file type to binary (for transfer of executables or Zipped files).
- Use the command *ASCII* to set the file type to ASCII (text).
- Default file type is ASCII.

How to Identify Files at FTP Sites

File Extension	Operating System	Binary or ASCII?	Program to Translate
com, exe	DOS	Binary	executable file
doc	any	ASCII	text file
gif	any	Binary	GIF viewer program
hqx	Macintosh	ASCII	BinHex 4.0
pit	Macintosh	ASCII	PackIt 3.13
ps	any	ASCII	print to PostScript printer
sit	Macintosh	Binary	Stuffit
tar	Unix	Binary	tar (Unix command)
txt	any	ASCII	text file
uu	Unix	ASCII	uudecode
wp	DOS	Binary	WordPerfect file
z	DOS	Binary	Pack/Unpack
Z		Binary	uncompress
zip	DOS	Binary	PKZIP/PKUNZIP

File Type	Mode to Use for Transfer
Text file	ASCII
Spreadsheet	Binary
Database file	May be Binary or ASCII
Word processor file	May be Binary or ASCII
Program source code	ASCII
Electronic mail messages	ASCII
Backup File	Binary
"Compressed" file	Binary
"Uuencoded" file	ASCII
Executable file	Binary
Postscript file	ASCII

Subscribing to Mailing Lists and Accessing Listservers

What is LISTSERV?

A list server provides a means for groups of computer users with similar interests to communicate among themselves. Subscribing is similar to subscribing to a magazine. Information is automatically delivered to you via electronic mail. You may respond to messages you receive, which creates an interactive environment.

How to Communicate with LISTSERV Servers

Commands are sent to **LISTSERV** hosts via e-mail. LISTSERV hosts usually provide access to a wide range of interest groups. You should send informational commands to LISTSERV@host. By the same token, send mail to be shared with your interest group to Interest-group@host.

Controlling the Mail You Received from a Discussion Group

Use the SET command to control the information you receive:

- SET list-name Mail   (default)

    List mail is distributed to you as mail.
- SET list-name DIGESTS

    Provides you with a digest of mail messages sent to the list. Digests may contain messages from a day, a week, or a month, depending on how the list is handled. Mail messages in the digest are complete.
- SET list-name INDEX

    Provides an index of mail messages sent to the list (by date, time, subject, sender's name, and length of message). You may select and retrieve any of the messages by sending a message to the Listserver.
- SET list-name NOMAIL

    Will suspend the mail you receive from a list; your subscription will be retained. Very useful for suspending your mail from the list when you know you will not have a chance to check your mail (holidays, vacations, etc.).

Obtaining Files from a Listserver

- To obtain a list of files available at a LISTSERV host:

    INDEX filelist

- Files are stored under the LISTSERV host in filelists, which are normally directly associated with the name of discussion groups.
- To obtain a file from a LISTSERV host:

    GET filename filetype filelist

- You obtain the filename and filetype from information received via the INDEX command or information contained in indexes sent to you (if you have indicated SET listname INDEX).
- Indicating a filelist is not necessary, but it saves time. If you do not indicate a filelist, the LISTSERV host will search all of its archived files for the file you have indicated.

Unsubscribing from a Discussion List

- Send the UNSUBSCRIBE command to the LISTSERV:

    UNSUBSCRIBE list-name

- You will receive confirmation of your removal from the discussion group.

Additional Commands to Send to LISTSERV

- LIST   Provides a listing of all discussion groups handled by the particular host to which you send this request.
- REVIEW list-name   Provides information about the mailing list you indicate: who is authorized to review or join the list, whether the list is archived, and a listing of members of the discussion group. (You may not receive a list of all members of a discussion group; members may use the command SET listname CONCEAL to have their name concealed from such a general listing.)

For more information, talk to your system administrator about what news systems you have access to, what restrictions there might be, and so forth. Because of the volume of information that such systems can generate, there may be some limitations at your specific site or from your Internet Service Provider.

Using Search Tools

There are a number of **search tools** available to aid your search for information on the Internet.

Veronica

- Searches **Gopher** sites (over 500) by document *title*.
- Prompts you for keywords.
- Multiple keywords may be used, separated by *and, or, not*.
- Wildcard character may be used, ∗.
- From the list of documents matching your keyword(s), you can retrieve as with any Gopher menu.

Wide Area Information Servers (WAIS)

This is a search tool for databases on the Internet. The following are WAIS steps:

- Select a database to search.
- Form a query by providing keywords.
- Documents are listed, ranked according to number of matches to keyword(s).
- To retrieve a document, select it from the list.

Archie

This is an electronic directory service for FTP sites, which involves two steps:

- Search for information.
- Use FTP to get files of interest.

This service also:

- Scans a database of over 1000 anonymous FTP sites worldwide.
- Provides a list of FTP site addresses and files sorted by host address.

Mosaic and Netscape

Mosaic and Netscape are graphical user interface browsers for the Internet. These systems use what is known as a uniform resource locator (URL). The syntax is typically something like: http://www.amazon .com or http://babbage.millersv.edu.

The symbol http refer to HyperText transfer protocol, which your browser understands. The www means that the link is known to the World Wide Web for accessing hypertext resources on the Internet. These are known as Web sites. The browser can act as a news reader (Netscape) or save and download files (like FTP).

Finding People and Computers

There are two common mechanisms for finding people and computers:

Whois

- Is a directory service to network users.
- Main database is maintained by IRC.
- Will address *known* databases (will not look at all databases at all sites).
- Telnet rs.internic.net.
- Type whois at the prompt.

Netfind

- Known as a white pages tool.
- Provide a name and keywords indicating where a person works.
- Telnet to the location closest to where your target may be (U.S., Canada, etc.)
- Telnet ds.internic.net is an example of one site to use.
- login: netfind.

Ping

The Ping command (like Telnet and FTP) operates on an Internet host computer (as opposed to MAIL and NOTE, which use an individual user's Internet address). The purpose of the Ping command is to determine whether or not a specified host computer is operating and connected to the Internet. A typical use of the Ping command would be: ping acad.fandm.edu. If the remote computer is up and connected, you might see a message similar to the following: Ping #1 response took 0.154 seconds. Successes so far 1. On the other hand, if you see a *time-out* message (e.g. 'Ping #1 timed out'), either the remote (host) computer is not up, or the connection to it has failed. If you see an *unknown host* message, the domain name you specified does not exist or may have a firewall in place to prevent unauthorized access.

Archie

One of the questions that comes up about Internet is: Where do I Telnet/ftp to get a certain type of information? To help address this concern, the Archie database was developed. This resides at several Internet sites, including Rutgers University. To access Archie, select the following under Gopher menus (for more information, see section on Gopher):

Other Information Services

Internet file server (FTP) sites

Search FTP sites (Archie)

You may also access Archie by telnetting to a site offering Archie services, such as Rutgers University:

Telnet archie.rutgers.edu

login with the username "archie"

You will now be in the Archie database. Some useful commands are:

about	What does Archie do?
help	Lists all commands, good for beginners
list	Lists Internet sites in database (over 1000!)
find *word*	Searches for the keyword provided and returns matching filenames with site names (where each file is located)
prog	Searches database for a file
site	Lists files at an archive site
whatis	Searches for a keyword in the database
exit	Quits Archie

Electronic Mail

The mail facility allows you to read, compose, or send a message in the form of a text file to another user on the Internet. To send a message, you must know the Internet address of your intended recipient (e.g., jsmith@um.psu.edu). The command to invoke mail facility varies with the system you are using. Typical mail systems are MAIL on the DEC VAX, and mail, elm, Eudora, or pine on Unix systems.

Gopher

Gopher is a menu-driven information system that is installed at several Internet sites. It is organized hierarchically. As a consequence, a better search engine such as Veronica is usually necessary.

To access Gopher, type the command: gopher. Be patient; the remote system is working when you see messages such as: Connecting . . . , Reading . . . , or System . . . . Once you have successfully accessed Gopher, you will see a menu of information appear.

Veronica

Veronica is an Archie for Gophers, that is, a mechanism for searching (by keywords) titles of Gopher items. The result of a search is a menu that can then be used to select a particular item.

Veronica can be accessed under Gopher. When you choose an area of gopherspace to search, you will be prompted for index keywords for which to search. When matches are found to your search, you will be provided with a menu to access the information.

Usenet News

Usenet News groups are collections of articles (submissions by people reading Usenet news) organized by subject into news groups. They provide a convenient and widely distributed forum for discussion and exchange of ideas. There are over 4500 news groups on a variety of diverse subjects.

Major Categories of Usenet Newsgroups	
comp	Computer science and related topics
news	News network and news software
rec	Hobbies, recreational activities, the arts
sci	Scientific research and applications
talk	Forums for debates on controversial topics
misc	Miscellaneous discussions (may fit in several categories)

Some Particular Newsgroups

alt	Discussions of "alternative ways of looking at things"
bionet	Interest groups for biologists
bit	BITNET listserv discussion groups
biz	Discussions related to business
de	Discussions in German, various topics
ieee	Discussions related to the Institute of Electronic and Electrical Engineers (IEEE)
gnu	Discussions related to the Free Software Foundations (FSF) and the GNU project
k12	Discussions related to teachers and students through high school (K–12)
rec	Recreational interests

Defining Terms

ARPANET: A Department of Defense computer network created in 1969.

Electronic mail: Allows you to compose and send a message in the form of a file to another user on the Internet.

FQDN: Fully qualified domain name (e.g., cs.millersv.edu).

FTP: File transfer protocol.

Gopher: A menu-driven information system that is installed at several Internet sites.

Internet protocol (IP) addresses: Internet address in numeric form (e.g., 192.206.29.16).

LISTSERV: Provides a means for groups of computer users with similar interests to communicate among themselves.

Mosaic and Netscape: Graphical user interface browsers for the Internet.

Search Tools: Veronica, Archie, Gopher, and comparable techniques for searching databases.

Telnet: Allows remote access over the Internet to another computer system.

References

Aboba, B. 1994. *The On-line User's Encyclopedia: Bulletin Boards and Beyond.* Addison–Wesley, Reading, MA.

Braun, E. 1993. *The Complete Internet Director.* Fawcett, New York.

Dern, D. P. 1994. *The Internet Guide for New Users.* McGraw–Hill, New York.

Dougherty, Koman, and Ferguson, 1994. *The MOSAIC Handbook for the X Window System.* (Also versions for Microsoft Windows and the Macintosh.) O'Reilly and Associates, Sebastapol, CA.

Eddings, J. 1994. *How the Internet Works.* Ziff-Davis Press, Emeryville, CA.

Falk, B. 1994. *The Internet Roadmap.* SYBEX, San Francisco, CA.

Fisher, S. 1993. *Riding the Internet Highway.* New Riders, Carmel, IN.

Hahn, H., and Stout, R. 1994. *The Internet Yellow Pages.* Osborne McGraw–Hill, Berkeley, CA.

Kehoe, B. 1994. *Zen and the Art of the Internet: A Beginner's Guide,* 3rd ed. Prentice–Hall, Englewood Cliffs, NJ.

Krol, E. 1994. *The Whole Internet: User's Guide & Catalog,* 2nd ed. O'Reilly and Associates, Sebastapol, CA.

Quarterman, J. S. and Smoot, C.-M. 1994. *The Internet Connection, System Connectivity and Configuration.* Addison–Wesley, Reading, MA.

Smith, R. and Gibbs, M. 1993. *Navigating the Internet.* Sams, Carmel, IN.

Appendix G: Common Programming Languages

This section contains brief descriptions of several major programming languages, with pointers to their standard versions and current Web pages added for further information. Each of these languages is supported by texts and professional references as well as compilers and interpreters. That information is not summarized here; any such summary would be incomplete and soon outdated. Readers who are interested in learning about current texts or implementations for a programming language are encouraged to consult the Web pages and Usenet news groups listed subsequently. Additionally, readers are encouraged to visit the Web pages of the major book publishers as well as their local science and engineering libraries or bookstores.

ADA

ADA was designed during the late 1970s in a collaborative effort sponsored by the U.S. Department of Defense. Its purpose was to provide a common high-level language in which systems programs could be designed and implemented, with special features that support concurrency, data abstraction, and software reuse. ADA was first implemented in the early 1980s and was first standardized in 1983 as a U.S. military standard. Since then, a variety of ADA implementations have emerged and many new features have been added. The current international ADA standard definition is found in ANSI/ISO/IEC 8652-1995.

Features and Applications

EIFFEL programs are written using the philosophy design by contract, which means that each object's state during execution conforms to a predetermined set of preconditions and postconditions. Before a method can be applied to an object, the object must be in a state that satisfies the preconditions. Similarly, after a method has been applied to an object, assurance is guaranteed that the state satisfies the postconditions. Furthermore EIFFEL's type system ensures that most type errors are caught at compile time, and EIFFEL systems provide automatic garbage collection so that programs need not use a destructor to take an object out of use. EIFFEL is implemented on a wide variety of platforms.

Further Information

For more information about EIFFEL, consult the Usenet news group comp.lang.eiffel or the following Web page: http://www.yahoo.com/Computers_and_Internet/Programming Languages/Eiffel.

FORTRAN

Designed by John Backus in 1954, formula translating system (FORTRAN) has become the most widely used scientific and engineering programming language of the past three decades. Its early versions were standardized in 1966 and a more extended version was standardized in 1977. The current FORTRAN standards are defined in ISO/IEC 1539:1991 and ISO/IEC 1539-2:1994 (Part 2: varying length character strings). The major differences between the current standard and its 1977 predecessor are the addition of extensions such as array operators, dynamic memory allocation, derived types and operator overloading, modules, modern control structures, and free format program text.

Features and Applications

FORTRAN is an imperative language, with extensive facilities and libraries to support scientific and engineering applications. Vast amounts of FORTRAN software exist in government and industrial computing laboratories. FORTRAN is implemented efficiently and widely, with compilers available on all contemporary platforms and operating systems. In recent years a significant portion of the FORTRAN community has begun to migrate its applications to other languages, especially C.

Further Information

For more information about FORTRAN, consult the Usenet news group comp.lang.fortran or the following Web page: http://www.yahoo.com/Computers_and_Internet/Programming Languages/Fortran.

JAVA

JAVA is the newest language described in this Appendix. It was designed in the early 1990s by a team at Sun Microsystems headed by James Gosling. Designed to facilitate interactive programming on the Internet and World Wide Web, JAVA was rapidly disseminated among systems programmers at major companies in the technology industry. JAVA programs, or *applets*, can be embedded in other applications, such as Hyper Text markup language (HTML) documents, to provide interactive executable programs for users on the Web.

Features and Applications

According to the description in Sun's white paper, "Java is a simple, object-oriented, distributed, interpreted, robust, secure, architecture neutral, portable, high-performance, multithreaded, and dynamic language." JAVA is based on C++ but excludes much of the baggage that makes C++ cumbersome to use.

Absent from JAVA are pointers; all objects are dynamic, and automatic garbage collection eliminates the need for destructors. Because JAVA is designed for use in networked environments, its designers built facilities for security and robustness into the language.

Further Information

Many computer scientists are optimistic for the future of languages like JAVA, because they promise to support a wholly different and revolutionary programming paradigm from the traditional ones. For more information about JAVA, consult the Usenet news group comp.lang.java or the following Web page: http://www.yahoo.com/Computers_and_Internet/Programming Languages/Java.

LISP

List processor (LISP) was designed by John McCarthy in the late 1950s. LISP has been used predominantly in the artificial intelligence area and developed rapidly throughout the 1960s and 1970s. Two dominant dialects of LISP evolved during that period, MACLISP and INTERLISP. An effort to unify these dialects and develop a single standard resulted in Common LISP, first implemented in the 1980s. Common LISP was finally standardized in 1994 as the standard ANSI X3.226-1994. More recently, object-oriented extensions to LISP have been developed under the rubric Common LISP Object System (CLOS). Thus, one may view CLOS as a hybrid functional/object-oriented programming language. Both Common LISP and CLOS are implemented on a wide range of platforms.

Features and Applications

LISP is a functional programming language, based on the application of functions written in the form of lambda expressions using prefix notation. It is particularly useful in areas of artificial intelligence programming that require the representation of symbolic expressions for mechanical reasoning and knowledge representation. Many illustrations of the functional programming paradigm appear in Chapter 131, and examples of LISP programs appear among the chapters of the Artificial Intelligence section of this Handbook.

Further Information

For more information about LISP and CLOS, consult the Usenet news groups comp.lang.lisp and comp.lang.clos as well as the following Web page: http://www.yahoo.com/Computers_and_Internet/Programming Languages/Lisp.

ML

Meta language (ML) was developed by Robin Milner and others as a functional programming language with imperative features and an unusually advanced concept of type. Its current version was defined in 1983 and is called Standard ML. It is a compiled language whose major applications are in the computer science education and research communities.

Features and Applications

ML has a simple syntax and yet supports data abstraction through its strong static typing system and type inference mechanism, polymorphism, exceptions, and rule-based specifications. ML is widely implemented on the major computing platforms, including Unix, Mac, and PC machines. Standard ML of New Jersey is a free implementation of ML, developed jointly by Princeton University and AT&T.

Further Information

For more information about ML, consult the Usenet news group comp.lang.ml or the following Web page: http://www.yahoo.com/Computers_and_Internet/Programming Languages/ML.

PASCAL

PASCAL was designed by Niklaus Wirth in the early 1970s as a language for teaching principles of computer science and imperative programming. It has been the main language for expressing algorithms in computer science curricula throughout the 1970s and 1980s. However, PASCAL has shown recent signs of obsolescence, because of the rapid rise of the object-oriented programming paradigm and related languages such as C++ and JAVA. PASCAL is widely implemented and is available for most computer platforms. Its current standard version is defined in the document ANSI/ISO/IEC 7185-1990.

Features and Applications

As a language designed for teaching, PASCAL is characterized by a strong type system support for modularity, simple syntax, and robust compile and run time programming environments. Its features have evolved over the past two decades, and nonstandard extensions are available that support a wide range of library functions as well as object-oriented programming. PASCAL was also used as a basis for the design of the language ADA.

Further Information

For more information about PASCAL, consult the Usenet news group comp.lang.pascal.misc or the following Web page: http://www.yahoo.com/Computers_and_Internet/Programming Languages/Pascal.

PERL

PERL is a special-purpose language designed for text processing applications, especially those that require text search, extraction, and text-based reporting. Its syntax is similar to that of C, and it is usually implemented in Unix environments. However, PERL is an interpreted language, designed for rapid prototyping, so that its programs will not run as fast as comparable C programs.

Features and Applications

Optimized for text processing, PERL employs sophisticated pattern matching techniques to speed up text search. It also does not arbitrarily limit the size of a file or the depth of a recursive call, as long as memory is available. PERL has not been standardized by ANSI or ISO at the time of this writing.

Further Information

For more information about PERL, consult the Usenet news group comp.lang.perl.misc or the following Web page: http://www.yahoo.com/Computers_and_Internet/Programming Languages/Perl.

PROLOG

Programming in logic (PROLOG) was developed in the early 1970s by Philippe Roussel. It is primarily an interpreted logic programming language, designed for use in such artificial intelligence applications as problem solving, expert systems, knowledge representation, and natural language processing. PROLOG is implemented on a wide variety of computers, and its general core is defined by the standard ISO/IEC 13211-1:1995.

Features and Applications

The syntax of PROLOG is based on logic expressions, and its semantics is defined using the concepts of *resolution* and *unification*. Chapter 133 in this Handbook provides a tutorial introduction to the logic programming paradigm, with many PROLOG examples provided as illustrations.

Further Information

For more information about PROLOG, consult the Usenet news group comp.lang.prolog or the following Web page: http://www.yahoo.com/Computers_and_Internet/Programming Languages/Prolog.

SCHEME

SCHEME is a dialect of LISP that developed in the 1970s. SCHEME is well suited to educational use, with widespread and inexpensive implementations and a simple syntax and semantics. SCHEME was standardized by ANSI and IEEE in 1991 (ANSI/IEEE 1178-1991).

Features and Applications

SCHEME is distinguished from LISP by its small size, static scoping, and more flexible treatment of functions (i.e., a SCHEME function can be a list element, the value of a variable, the value of an expression, or passed as a parameter).

Further Information

For more information about SCHEME, consult the following Web page: http://www.yahoo.com/Computers_and_Internet/Programming Languages/Scheme.

Tcl/Tk

The Tcl/Tk programming system was developed by John Ousterhout. It has two parts: the programming language Tcl and the toolkit of *widgets* called Tk, which supports the programming of interactive graphical user interfaces (GUIs). A main goal of Tcl/Tk is to support the rapid development and prototyping of such interfaces, so that Tcl programs are usually run in interpretive mode. Tcl/Tk can also be used in coordination with other languages, and it is implemented on a variety of platforms.

Features and Applications

Tcl is an imperative language, with modest support for handling types and data abstractions. It may not be an ideal language for writing large, complex programs; its narrow focus is to facilitate the rapid development of user interfaces. Tk widgets include labels, messages, listboxes, texts, frames, scrollbars, buttons, and other elements that commonly appear in user interfaces. A wide range of applications for languages such as Tcl/Tk are discussed in the Human–Computer Interaction section of this Handbook.

Further Information

For more information about Tcl/Tk, consult the following Web page: http://www.yahoo.com/Computers_and_Internet/Programming Languages/Tcl_Tk.

Appendix H: Common Graphics Languages

GKS and PHIGS

Graphics kernel system (GKS) is a general-purpose 2D graphics system, standardized by ISO and ANSI in the mid-1980s. It emphasizes interactive graphics and device independence. Programmers hierarchical interactive graphics system (PHIGS) is an extension of GKS that includes extensive 3D graphics functions and facilities for creating data structures to model complex objects. PHIGS+ is an extension of PHIGS that fully supports raster systems, and thus accommodates light sources, shading, and reflection in scenes with 3D graphics objects. The functionality of GKS or PHIGS can be accessed through a set of graphics functions called from a program in a language such as ADA, FORTRAN, and C. ISO standards for these language interfaces exist for both GKS (ISO 8651) and PHIGS (ISO 9593).

Further Information

For more information about GKS and PHIGS, consult the following Web page: http://info.mcc.ac.uk/CGU/ITTI/Stds/standards_announce.html.

PostScript

PostScript is both a graphics standard and a programming language for page layout and typesetting text and graphics on laser printers. Because of its wide use, PostScript is implemented on most current printers. Often figures, such as most of the figures in this Handbook, are separately created in PostScript and then embedded in a word processing document at the time it is printed. A wide range of software is available for converting documents into and out of PostScript format. A wide variety of typesetting fonts for text is also available for different PostScript printers and computer platforms.

Further Information

For more information about PostScript, consult the Usenet news group comp.lang.postscript or the following Web page: http://www.yahoo.com/Computers_and_Internet/Programming Languages/PostScript.

0-8493-2909-4/97/$0.00+$.50

X Windows

X Windows is an emerging standard for windowing systems that was developed at MIT. It consists of a library of graphics function calls, called Xlib, written in C, that is freely available. Application programs that require graphics can call functions from this library. The functions in Xlib are simpler than those in GKS or PHIGS. They are also more stylized to the needs of interactive user interface programming, such as creating a window or sampling the mouse pointer. On the other hand, X Windows functions are not as extensive as PHIGS functions in the area of graphics applications.

Further Information

For more information about X Windows, consult the following Web page: http://www.yahoo.com/ Business_and_Economy/Companies/Computers/ Software/Operating_Systems/X.

Appendix I: Common Editing Markup Languages

LaTeX

LaTeX is a markup language and system for document typesetting. It is implemented as a macropackage that extends the TeX system, allowing a wide range of scientific documents to be easily prepared for typesetting. Tex was designed in 1970 by Donald Knuth. Many of the chapters in this Handbook were prepared using LaTeX.

Features and Applications

The LaTeX language can be used to describe the typesetting characteristics (e.g., boldface words, numbered lists) of a document. It is particularly good for describing mathematical formulas, maintaining bibliographies, and managing number streams (such as section numbers, figure numbers, and so on). LaTeX supports the automatic insertion of PostScript figures, and has many other features.

Further Information

For more information about LaTex, consult the Usenet news group comp.text.tex or the following Web page: http://www.ucc.ie/info/TeX/TeXmenu.html.

SGML

Standard generalized markup language (SGML) is a typesetting language for preparing hypertext documents for use on the Web and other interactive electronic media. The language is platform independent, so that SGML documents can be widely interchanged. SGML was standardized by ISO in 1986 as ISO-8879).

Features and Applications

A popular application of SGML is HyperText markup language (HTML), which allows a text to be coded for typesetting (e.g., boldface words or numbered lists) and dynamic linking to other parts of a document of the Web (or another medium).

Further Information

For more information about SGML or HTML, consult the following Web page: http://www.sgmlopen.org.

Indexes

Index of Cited Authors

This index is designed to identify the persons who have made significant contributions to computer science and engineering research and development in the last 50 years. It was derived by collating the names of first and second authors from all cited works that appear throughout the text of the Handbook with the page numbers on which their works are cited. The page numbers in bold indicate opening pages of contributed chapters of this Handbook.

Index of Key Algorithms and Equations

Index of Key Figures
and Tables

Subject Index

Page on which term is defined is indicated in bold.

Page on which term is defined is indicated in bold.

Page on which term is defined is indicated in bold.

Page on which term is defined is indicated in bold.

Page on which term is defined is indicated in bold.

Page on which term is defined is indicated in bold.

Page on which term is defined is indicated in bold.

Page on which term is defined is indicated in bold.

Page on which term is defined is indicated in bold.

Page on which term is defined is indicated in bold.

Page on which term is defined is indicated in bold.

Page on which term is defined is indicated in bold.

Page on which term is defined is indicated in bold.

Page on which term is defined is indicated in bold.

Page on which term is defined is indicated in bold.

Page on which term is defined is indicated in bold.

Page on which term is defined is indicated in bold.

Page on which term is defined is indicated in bold.

Page on which term is defined is indicated in bold.

Page on which term is defined is indicated in bold.

Page on which term is defined is indicated in bold.

Page on which term is defined is indicated in bold.

Page on which term is defined is indicated in bold.

Page on which term is defined is indicated in bold.

Page on which term is defined is indicated in bold.